AF702913

Handbook of

BIOTRANSFORMATIONS OF AROMATIC COMPOUNDS

Handbook of

BIOTRANSFORMATIONS OF AROMATIC COMPOUNDS

B.L. Goodwin

CRC PRESS

Boca Raton London New York Washington, D.C.

Library of Congress Cataloging-in-Publication Data

Handbook of biotransformations of aromatic compounds / Brian L. Goodwin.
p. cm.
Includes bibliographical references and index.
ISBN 0-415-27176-2 (alk. paper)
1. Aromatic compounds—Metabolism. I. Goodwin, Brian L.

QP801.A75H36 2004
572′.5—dc22 2004045488

This book contains information obtained from authentic and highly regarded sources. Reprinted material is quoted with permission, and sources are indicated. A wide variety of references are listed. Reasonable efforts have been made to publish reliable data and information, but the author and the publisher cannot assume responsibility for the validity of all materials or for the consequences of their use.

Neither this book nor any part may be reproduced or transmitted in any form or by any means, electronic or mechanical, including photocopying, microfilming, and recording, or by any information storage or retrieval system, without prior permission in writing from the publisher.

The consent of CRC Press LLC does not extend to copying for general distribution, for promotion, for creating new works, or for resale. Specific permission must be obtained in writing from CRC Press LLC for such copying.

Direct all inquiries to CRC Press LLC, 2000 N.W. Corporate Blvd., Boca Raton, Florida 33431.

Trademark Notice: Product or corporate names may be trademarks or registered trademarks, and are used only for identification and explanation, without intent to infringe.

Visit the CRC Press Web site at www.crcpress.com

© 2005 by CRC Press LLC

No claim to original U.S. Government works
International Standard Book Number 0-415-27176-2
Library of Congress Card Number 2004045488
Printed in the United States of America 1 2 3 4 5 6 7 8 9 0
Printed on acid-free paper

Contents

Author Biography

Brian Goodwin, D.Phil., is retired. He received his M.A. in natural sciences from Cambridge University, England and his D.Phil., from Oxford University, England. He worked as a researcher in the pharmaceutical industry (1958–1961), research worker in the Department of the Regius Professor of Medicine, Oxford University (1961–1967), visiting research worker, Fels Research Institute, Yellow Springs, Ohio (1967–1969), Principal Biochemist, Queen Charlotte's Maternity Hospital, London (1969–1991), Research scientist, Royal Botanic Gardens, Kew, London (1991–1993) and Senior Research Fellow, King's College London (1994–1997). He is also a qualified horticulturist and preacher.

Dr Goodwin's main research interests have been the metabolism of aromatic amino acids and amines and their connection with disease, the metabolic pathways of aromatics and the development of potential drugs.

Dr Goodwin is a Member of the Biochemical Society and a Fellow of the Royal Society of Medicine.

Introduction

Since the earlier work on this subject was published (Goodwin, B.L. [1976] *Handbook of Intermediary Metabolism of Aromatic Compounds*, London: Chapman & Hall), which covered research until about 1972, a great deal of new work has been published, mainly on new compounds, and the need of a new reference work has become apparent, covering research published since then.

Most of the literature cited is found in journals written in English, as most information finds its way into such research journals. However, information quoted in Chemical Abstracts is used when non-English sources are important, for instance Chinese and Japanese literature. In general, reviews and symposia have been excluded as source material; Federation Proceedings is an exception. Since the information in Chemical Abstracts is necessarily incomplete, it is inevitable that much important information is lost when this has been the only available source material.

The aromatic compounds listed in this work are those that contain at least one aromatic C_6 ring, which may be fused with other ring systems. Other aromatic systems (such as pyridine) are excluded; quinoline with benzene fused with pyridine is included. Other ring systems with some aromatic properties with ring systems that do not contain six carbon atoms are also excluded.

It has not been possible to include structural formulae, pharmacological properties, toxicology or medicinal use of the compounds listed in this index. Although structures are not always shown in the original papers, structural formulae of many of the compounds are found in *The Merck Index* Whitehouse Station, N.J.: Merck & Co, Inc, and *Dictionary of Organic Compounds, 6th edition* (1996), London: Chapman & Hall. Almost all of the remainder can be deciphered with the help of Chemical Abstracts.

The data are divided into two sections: Part I Reactions of Individual Compounds, and Part II Enzymes and Reactions, with extensive indexing of Part II. A full bibliography is supplied.

The structure of each compound entry in Part I is:

A[a]

Formed from: B[b]
Formed from C in R[c] W321
Probably formed from: D[d]
→ E[e]
In: S W123
T W456
With H W135[f]
See also: W789[g]
Contrast (in U) W987[h]
→ F[e]
Probably in: V W234[j]
Does not yield G in X W345[k]

[a]The name of the compound listed. Most of these are aromatic compounds, but some are closely related to aromatic compounds that are aromatized by a simple reaction. Acids are named as acids and not as their salts, although the reactions may occur with the ionized forms.

[b]Under the entry for compound B a reaction is shown which forms compound A.

[c]Compound A is formed from compound C in species R; C is usually a simple aliphatic compound, such as acetic acid that is not listed in its own right. The information is quoted from reference W321 (all the W123-type entries illustrate references found in the bibliography).

[d]Under the entry for compound D a reaction is shown which forms compound A, but the evidence is not strong enough to be included under the 'Formed from' heading.

[e]Compound E is formed from compound A.

[f]This reaction has been reported to occur in species S and T in publications W123 and W456 respectively. There is a corresponding entry under compound E of 'Formed from: A'. 'In' indicates that the studies have been carried out in whole organisms, tissue extracts or with purified enzymes obtained from that species. In cases where the source of the enzyme specified has been difficult to establish the alternative form of entry illustrated by 'With H W135' is used, where H is the name of the enzyme. If the identity of the species and enzyme are both unclear, the species name is replaced by 'species unclear'.

[g]Publication W789 reports evidence supporting this reaction, but is not in itself adequate to be quoted as definite evidence. The two main reasons for this are that the identity of the starting compound or the product has not been adequately determined. A second situation is where the reaction A→E is one in a reaction sequence, and either the starting compound studied is not A, or the product detected is not E, but A→E is implicit. This reaction may be one of a network of possible reactions; parallel reactions in such networks are not usually recorded other than under one of the possible reactions. Where structural constraints on compound E require that it originates from compound A, although compound A was not the compound under test, such as benzene→phenyl sulphate where phenol is the necessary intermediate, the published data are counted as definitive evidence and are not included under this heading. This may also apply where the time of appearance and disappearance of compounds in a reaction sequence indicates that A→E is an established reaction.

[h]This heading is rarely used, and is generally reserved for situations where either the reaction is observed in some species but not in others, or when the validity of the reaction has been challenged, such as the biotransformation of saccharin, where the observed reaction in early studies may have been due to the presence of impurities, or where the product has been mis-identified.

[j]This entry is used with the same force as the entries described under the footnote[g], but where there is no literature for a fully authenticated reaction. There is a corresponding entry under compound F of 'Probably formed from: A'.

[k]This is a rarely used entry, where conversion to G in species X has been sought unsuccessfully.

The compound name listed has been chosen primarily for utility; it is more important that the compound can be recognized by the reader than that an accurate systematic name is used. In general, it has been considered that the Chemical Abstracts names are too unwieldy or too unfamiliar for routine use. With lead compounds, a systematic name is also listed (except for those that are too long to be considered useful), as are common alternatives for any compound. In cases where alternatives are relatively simple or are in regular use they are listed in their own right with a reference to the compound name where the data are listed. This arrangement has inevitably led to some inconsistencies of naming. Very occasionally, there is a small discrepancy between the name used at[a] and the product entry when there are some ambiguities, particularly between publications, such as stereochemistry. A similar discrepancy often occurs in the alternative names of compounds listed under trivial names and those used for metabolites; the former are often based on Chemical Abstracts naming which may differ from the names used in the publications reporting the metabolism. Because of these variations in naming in the literature, it is probable that some compounds have been listed under more than one name, although extensive steps have been taken to prevent duplicate entries.

In the previous book it was assumed that conjugation in animals was the final stage of a metabolic sequence prior to excretion, and conjugation reactions were automatically given a definitive reaction step status; since it is no longer certain that this is so, many references to conjugation are now given no more than a 'See also' status.

Some compounds are used routinely in probing the status of P450 systems; when they are used just as probes the literature cited is usually not used unless the data cited in Part 1 are considered to be otherwise inadequate.

Mercapturates are listed as N-Acetyl-S-(L-cysteinyl)-.

'Desmethyl' in a compound name has generally been replaced by 'Demethyl'; 'Desethyl' has been left unaltered. In many instances 'Nor' (when referring to the loss of a methyl group) has been replaced by 'Demethyl'; catecholamines and their analogues such as noradrenaline are exceptions.

'Acetamido' is preferred to 'Acetylamino', and 'Formyl' is used for many aldehydes.

Unless otherwise stated, compound moieties such as 'Phenylpropionic', 'Phenyloctane', 'Phenyloctyl' are straight chain compounds with a terminal phenyl moiety and any side-chain functional group at the other terminal. In general the prefix *n*- is not used for these compounds.

In general, entries such as Hydroxy-demethyl- are preferred to Demethyl-hydroxy; whereas Bisnor-tetrahydro- is used rather than Tetrahydro-bisnor-, particularly with cannabinoids.

Esters may be listed under the name of either the parent acid or parent alcohol.

Plant glucosides are usually β-glucosides; this information is often omitted from the publications cited. Such conjugates are generally named as glucoside, glucuronide etc. without stating the full systematic name.

Certain classes of compounds are named on the basis of alternative numbering systems. As far as possible, this has been rationalized, for instance, cannabinoids, but areas of ambiguity may have led to inadvertent duplication of entries. Stereoisomers are often not listed separately, especially where they are converted into the same product (possibly at different rates). In some cases the research literature cites, for instance, optical isomers that yield a named product, but without sufficient information to identify the stereochemistry. When there is an entry for the product, possibly formed additionally from some other compound, a blurring of the stereochemistry may have been applied to the entries, to avoid misleading the reader. This problem is particularly acute when different systems for describing stereochemistry are used, such as D and L, *d* and *l*, (*R*) and (*S*), + and −, where the information on their interconversion is often not available, with consequent uncertainties.

In these entries the species name is usually either a trivial name like Rat, or the Genus name, without reference to the species name. This is a compromise adopted not only for the sake of consistency and compactness, but also because many studies, especially with microorganisms, either do not identify the microorganisms involved at all, or are identified as a genus with, perhaps a reference number but no species name. In Part 2 the organisms are more fully identified. In work on recombinant organisms, where possible, the organism from which the original genetic material was obtained is quoted.

Little effort has been exerted in rationalizing species and genus names, except where duplication in an entry has become apparent. For instance, Daucus and carrot are used as alternatives, based largely on the preference of the researchers. In general, the name Pseudomonas has not been replaced by Burkholderia, but Corn has been replaced by Zea, since this avoids confusion between corn (Zea) and wheat (Triticum).

Part 1
Reactions of Individual Compounds

A

Abecarnil (isopropyl 6-benzyloxy-4-methoxymethyl-β-carboline-3-carboxylate)
→ 6-Benzyloxy-4-methoxymethyl-β-carboline-3-carboxylic acid
In: Man G2
→ Isopropyl 6-benzyloxy-4-methoxymethyl-β-carboline-3-carboxylic acid
In: Man G2

Abrine
See: N-Methyl-L-tryptophan

Abrusin (8-β-D-glucopyranosyl-5-hydroxy-2-(4-hydroxyphenyl)-6,7-dimethoxy-4*H*-1-benzopyran-4-one)
Formed from: Abrusin-2″-β-D-apioside

Abrusin-2″-β-D-apioside
→ Abrusin
In: Microorganism K574

Acacetin (4′-methylapigenin, 5,7-dihydroxy-4′-methoxyflavone)
Formed from: Apigenin
Kaempferol
Naringenin
→ Acacetin-7-β-D-glucoside
Probably in: Parsley K739
→ Acacetin-β-D-glucuronide
In: Rat G848
→ 5-Hydroxy-4′,7-dimethoxyflavone
Probably in: Citrus A3987

Acacetin-7-β-D-glucoside
Formed from: Acacetin

Acacetin-β-D-glucuronide
Formed from: Acacetin

Acarol (4-bromo-α-(4-bromophenyl)-α-hydroxybenzeneacetic acid)
→ 4,4′-Dibromobenzilic acid
In: Housefly A689
Spider mite A689

Aceanthrylene
→ 1,2-Dihydro-1,2-dihydroxyaceanthrylene
In: Mouse F71

Acebutolol (N-(3-acetyl-4-(2-hydroxy-3-((1-methylethyl)amino)propoxy)phenyl) butanamide)
→ 1-(2-Acetyl-4-aminophenoxy)-2-hydroxy-3-isopropylaminopropane
In: Rat A3921

Aceclofenac ((2-((2,6-dichlorophenyl)amino) phenyl)acetoxyacetic acid)
→ (2-((2,6-Dichloro-4-hydroxyphenyl) amino)phenyl)acetoxyacetic acid
In: Man H739
→ (2-((2,6-Dichlorophenyl)amino)-5-hydroxyphenyl)acetoxyacetic acid
In: Man H739, J47
Monkey J47
Rat J47
→ Diclofenac
In: Man H739, J47
Monkey J47
Rat J47

Acenaphthene
→ 1-Hydroxyacenaphthene
In: Beijerinckia D248
Cunninghamella G725
Microorganism J596
See also H28
→ 6-Hydroxyacenaphthene
In: Cunninghamella G725

Acenaphthene-1,2-diol (*trans* or unspecified)
Probably formed from: Acenaphthylene
1-Hydroxyacenaphthene
→ Acenaphthene-1,2-dione
Probably in: Beijerinckia D248

Cunninghamella G725
→ Acenaphthylene-1,2-diol
Probably in: Beijerinckia D248
→ Naphthalene-1,8-dicarboxylic acid
Probably in: Bacteria J596

Acenaphthene-1,5-diol
Probably formed from: 1-Hydroxyace-naphthene

Acenaphthene-1,2-dione
Probably formed from: Acenaphthene-1,2-diol

Acenaphthenone
Formed from: Fluoranthene
1-Hydroxyacenaphthene

Acenaphthoquinol
Formed from: Acenaphthoquinone

Acenaphthoquinone
→ Acenaphthoquinol
In: Mouse F377
Pig F389
Rabbit H920
Rat H759, J744

Acenaphthylene
→ Acenaphthene-1,2-diol
Probably in: Beijerinckia D248
→ Acenaphthylene-1,2-diol
Probably in: Beijerinckia D248
→ Naphthalene-1,8-dicarboxylic acid
In: Microorganism H28

Acenaphthylene-1,2-diol
Probably formed from: Acenaphthene-1,2-diol
Acenaphthylene

Acenocoumarol (4-nitrowarfarin; both *R*- and *S*-, 3-(β-acetyl-α-(*p*-nitrophenyl)ethyl)-4-hydroxycoumarin)
→ 3-(β-Acetyl-α-(*p*-aminophenyl)ethyl)-4-hydroxycoumarin
In: Man A3159
→ 6-Hydroxyacenocoumarol
In: Man K323
Rat F907
→ 7-Hydroxyacenocoumarol
In: Man K323
Rat F907
→ 8-Hydroxyacenocoumarol
In: Man K323
→ 4-Hydroxy-3-(γ-hydroxy-(*p*-nitrophenyl)butyl)coumarin
In: Man A3159

N-(2-(N-Acetamido)acetoxy)benzylamine
Probably formed from: 2-Acetyl-tetrahydro-3-phenyl-1,2,4-oxadiazin-5-one
→ Benzylamine
Probably in: Rat D762

1-(4-Acetamido-2-acetylphenoxy)-2-hydroxy-3-isopropylaminopropane
Probably formed from: 1-(2-Acetyl-4-aminophenoxy)-2-hydroxy-3-isopropylamino-propane

2-Acetamido-4-aminoanisole
Formed from: 2,4-Diaminoanisole

4-Acetamido-2-aminoanisole
Formed from: 2,4-Diaminoanisole
→ 2,4-Diacetmidoanisole
Probably in: Rat A3704, B645

3′-Acetamido-4-aminoazobenzene
Formed from: 3′-Acetamido-4-methylaminoazobenzene

4′-Acetamido-4-aminoazobenzene
Formed from: 4,4′-Diaminoazobenzene
Probably formed from: 4′-Acetamido-4-methylaminoazobenzene

3-Acetamido-5-aminobenzamide
Probably formed from: 3,5-Diaminobenzamide
→ 3,5-Diacetamidobenzamide
In: Rat B193

4-Acetamido-2-aminobenzoic acid
Probably formed from: 2,4-Diaminobenzoic acid

4′-Acetamido-4-aminobenzophenone
Probably formed from: 4,4′-Diaminobenzophenone
→ 4′-Acetamido-4-amino-3′-hydroxybenzophenone
Probably in: Rat B713

1-Acetamido-2-amino-4-(*p*-fluorobenzyl) aminobenzene
Probably formed from: 1,2-Diamino-4-(*p*-fluorobenzyl)aminobenzene

4′-Acetamido-4-amino-3′-hydroxybenzophenone
Probably formed from: 4′-Acetamido-4-aminobenzophenone

2-Acetamido-4-amino-5-methylphenol
Probably formed from: 2,4-Diamino-5-methylphenol
6-Acetamido-2-amino-3-methylphenol
Probably formed from: 2,6-Diamino-3-methylphenol
2-Acetamido-6-aminotoluene
Formed from: 2,6-Diaminotoluene
→ 2,6-Diacetamidotoluene
In: Rat F56
***o*-Acetamidoaniline**
Formed from: *o*-Nitroacetanilide
o-Phenylenediamine
***m*-Acetamidoaniline**
Formed from: *m*-Nitroacetanilide
Probably formed from: *m*-Phenylenediamine
→ 1,3-Diacetamidobenzene
Probably in: Hamster F764
Rat F764
→ *m*-Phenylenediamine
In: Nocardia G343
***p*-Acetamidoaniline**
Formed from: 4-Acetamidoazobenzene
p-Nitroacetanilide
p-Phenylenediamine
→ N,N-Diacetyl-*p*-phenylenediamine
Probably in: Pseudomonas E393
→ *p*-Phenylenediamine
In: Nocardia G343
***o*-Acetamidoanisole**
See *o*-Methoxyacetanilide
***m*-Acetamidoanisole**
See *m*-Methoxyacetanilide
***p*-Acetamidoanisole**
See *p*-Methoxyacetanilide
4-Acetamidoanthranilic acid
Probably formed from: 5-Acetamido-2-methylaniline
2-Acetamidoanthraquinone
Formed from: 2-Aminoanthraquinone
4-Acetamidoantipyrine (4-acetamido-2,3-dimethyl-1-phenyl-3-pyrazolin-5-one)
Formed from: 4-Aminoantipyrine
4-Acetamidoazobenzene
Formed from: 4-Aminoazobenzene
→ N-Acetyl-*p*-phenylenediamine
Probably in: Pseudomonas E393
→ 4-Aminoazobenzene
In: Pseudomonas E393
→ Aniline
In: Pseudomonas E393
***o*-Acetamidobenzaldehyde**
→ Anthranilaldehyde
In: Ascaris A2036
Moniezia A1755
***m*-Acetamidobenzaldehyde**
Probably formed from: *m*-Acetamidobenzyl alcohol
→ *m*-Aminobenzaldehyde
In: Ascaris A2036
Moniezia A1755
***p*-Acetamidobenzaldehyde**
Formed from: *p*-Acetamidobenzyl alcohol
→ *p*-Aminobenzaldehyde
In: Ascaris A2036
Moniezia A1755
***p*-Acetamidobenzaldehyde thiosemicarbazone** (thiacetazone)
→ *p*-Aminobenzaldehyde thiosemicarbazone
In: Cat G684
Man G684
Mouse G684
Rat G684
***m*-Acetamidobenzamide**
Formed from: *m*-Aminobenzamide
Probably formed from: 3-Acetamido-N-methylbenzamide
→ 5-Acetamido-2-hydroxybenzamide
Probably in: Rat H170
***p*-Acetamidobenzenesulphonamide** (N^4-acetylsulphanilamide)
Formed from: Sulphanilamide
→ N^1,N^4-Diacetylsulphanilamide
Probably in: Bushtail possum C403
Padmelon C403
Tasmanian devil C403
***m*-Acetamidobenzenesulphonic acid**
Formed from: *m*-Aminobenzenesulphonic acid
***p*-Acetamidobenzenesulphonic acid**
Formed from: Sulphanilic acid
***o*-Acetamidobenzoic acid**
See N-Acetylanthranilic acid

_m_-Acetamidobenzoic acid
→ *m*-Aminobenzoic acid
In: Ascaris A2036
Moniezia A1755

_p_-Acetamidobenzoic acid
Formed from: *p*-Acetamidohippuric acid
p-Acetamido-N-(2-diethylaminoethyl) benzamide
Sulphanilic acid
→ *p*-Acetamidobenzoyl-β-D-glucuronic acid
In: Monkey C830
Rabbit H519
See also A3992
→ *p*-Acetamidohippuric acid
In: Ferret A394
Rat B627
Rhesus monkey C830
See also A554, A3992, B36, B201, K377
→ *p*-Aminobenzoic acid
In: Ascaris B2036
Microorganism A261
Moniezia A1755

_p_-Acetamidobenzonitrile
Formed from: *p*-Aminobenzonitrile
→ N-Acetyl-S-(2-(*p*-cyanoanilino)-2-oxoethyl)-L-cysteine
Probably in: Rat D160, F305
→ *p*-Aminobenzonitrile
In: Rat F305
→ *p*-N-Methylsulphinylacetamidobenzonitrile
In: Rat F305
See also D160

_p_-Acetamidobenzoyl-β-D-glucuronic acid
Probably formed from: *p*-Acetamidobenzoic acid

_p_-Acetamidobenzoyl-L-glutamic acid
Formed from: *p*-Aminobenzoyl-L-glutamic acid

1-(_p_-Acetamidobenzoyl)-4-(3,4,5-trimethoxybenzamido)piperidine
Formed from: 1-(*p*-Aminobenzoyl)-4-(3,4,5-trimethoxybenzamido)piperidine

_m_-Acetamidobenzyl alcohol
Probably formed from: *m*-Aminobenzyl alcohol

erythro-4-Acetamidobibenzyl-α,β-diol
Formed from: *trans*-4-Acetamidostilbene
→ 4-Acetamido-α-hydroxybibenzyl
Probably in: Rat E921
→ 4-Acetamido-β-hydroxybibenzyl
Probably in: Rat E921

threo-4-Acetamidobibenzyl-α,β-diol
Formed from: *trans*-4-Acetamidostilbene

2-Acetamidobiphenyl
→ 2-Aminobiphenyl
In: Guinea pig A808
Mouse A808
Rabbit A808
Rat A808

3-Acetamidobiphenyl
→ 3-Acetamido-6-hydroxybiphenyl
In: Rat E169

4-Acetamidobiphenyl
Formed from: N-Hydroxy-4-acetamidobiphenyl
→ 4′-Acetamido-4-hydroxybiphenyl
In: Rat F912
→ 4-Aminobiphenyl
In: Rat F109, F912
See also B178

4-Acetamidobiphenyl nitroxide radical
Probably formed from: N-Hydroxy-4-acetamidobiphenyl

4-Acetamido-2,6-bis(methylthio)phenol
→ 4-Acetamido-2,6-bis(methylthio)phenoxy radical
In: Horseradish J415

4-Acetamido-2,6-bis(methylthio)phenoxy radical
Formed from: 4-Acetamido-2,6-bis(methylthio)phenol

4-Acetamido-4′-bromostilbene
→ N-Hydroxy-4-acetamido-4′-bromostilbene
In: Hamster A2089

4-Acetamido-4′-butoxystilbene
→ N-Hydroxy-4-acetamido-4′-butoxystilbene
In: Hamster A2860
→ 4-Acetamido-4′-(3-hydroxybutoxy) stilbene
In: Hamster A2860

1-(4-Acetamidobutyl)-4-(6-fluorobenzo[_b_]thien-3-yl)piperazine
Probably formed from: 1-(4-((N-Acetylbutyryl)amino)butyl-4-(6-fluorobenzo[*b*]thien-3-yl)piperazine

5-Acetamido-N-butyl-2-hydroxybenzamide
Formed from:
5-Amino-N-butyl-2-(propargyloxy)benzamide

6-Acetamido-2-*sec*-butyl-4-nitrophenol
Formed from: 6-Amino-2-*sec*-butyl-4-nitrophenol

5-Acetamido-3-(4-carboxy-2-methoxyphenylmethyl)-1-methylindole
Probably formed from: 5-Acetamido-3-(2-methoxy-4-((2-methylphenylsulphonylaminocarbonyl)phenylmethyl)-1-methylindole

4-Acetamidocatechol
Formed from: *p*-Acetamidophenol
→ 4-Acetamidocatechol-2-sulphate
In: Man A2671
See also C241
→ *m*-Acetamidophenol
Probably in: Man C241
→ 4-Acetamido-2-methoxyphenol
In: Man A2671
See also C241, F572

4-Acetamidocatechol-2-sulphate
Formed from: 4-Acetamidocatechol

4-Acetamido-2-chlorobenzoic acid
Probably formed from: 4-Amino-2-chlorobenzoic acid

2-Acetamido-4-chloro-5-fluorophenol
Probably formed from: 2-Amino-4-chloro-5-fluorophenol

8β-Acetamido-2-chloro-6-methylergoline
Formed from: Lergotrile

4-Acetamido-3-chloro-α-methylphenylacetic acid
Formed from: 4-Amino-3-chloro-α-methylphenylacetic acid

4-Acetamido-3-chloro-4′-nitrobiphenyl
Probably formed from: 4-Amino-3-chloro-4′-nitrobiphenyl

2-Acetamido-4-chlorophenol
Probably formed from: 2-Amino-4-chlorophenol

2-Acetamido-5-chlorophenol
Formed from: 2-Amino-5-chlorophenol
p-Chloroacetanilide

5-Acetamido-2-chlorophenol
Formed from: *p*-Chloroacetanilide

7-Acetamido-5-(*o*-chlorophenyl)-1,3-dihydro-2*H*-1,4-benzodiazepin-2-one
Formed from: 7-Amino-5-(*o*-chlorophenyl)-1,3-dihydro-2H-1,4-benzodiazepin-2-one

4-Acetamido-4′-chlorostilbene
→ 4′-Chloro-N-hydroxy-4-acetamidostilbene
In: Hamster A2089
→ 4′-Chloro-4-glycollamidostilbene
In: Hamster A2089

α-Acetamidocinnamic acid
→ α-Aminocinnamic acid
In: Brevibacterium E589

4-Acetamido-*o*-cresol
Probably formed from: 4-Amino-*o*-cresol

4-Acetamido-*m*-cresol
Probably formed from: 4-Amino-*m*-cresol
→ 4-Acetamido-*m*-cresyl-β-D-glucuronide
Probably in: Rat B170
→ 4-Acetamido-*m*-cresyl sulphate
Probably in: Rat B170

4-Acetamido-*m*-cresyl-β-D-glucuronide
Probably formed from: 4-Acetamido-*m*-cresol

4-Acetamido-*m*-cresyl sulphate
Probably formed from: 4-Acetamido-*m*-cresol

Acetamidodantroline
Formed from: Aminodantroline

4-Acetamido-2,6-dibromo-2-(S-glutathionyl) phenol
Formed from: 4-Acetamido-2,6-dibromophenol

4-Acetamido-2,6-dibromophenol
→ 4-Acetamido-2,6-dibromo-2-(S-glutathionyl)phenol
In: Rat H716
→ 4-Acetamido-2,6-dibromophenoxy radical
In: Horseradish J415
→ N-Acetyl-3,5-dibromobenzoquinone imine
In: Horseradish H716
Rat H716

4-Acetamido-2,6-di-*tert*-butylphenol
→ 4-Acetamido-2,6-di-*tert*-butylphenoxy radical
In: Horseradish J415

4-Acetamido-2,6-di-*tert*-butylphenoxy radical
Formed from: 4-Acetamido-2,6-di-tert-butylphenol

2-Acetamido-4,6-dichloro-3,5-difluorophenol
Probably formed from: 2-Amino-4,6-dichloro-3,5-difluorophenol

4-Acetamido-2,6-dichloro-2-(S-glutathionyl) phenol
Formed from: 4-Acetamido-2,6-dichlorophenol

4-Acetamido-2,6-dichlorophenol
→ 4-Acetamido-2,6-dichloro-2-(S-glutathionyl)phenol
In: Rat H716
→ 4-Acetamido-2,6-dichlorophenoxy radical
In: Horseradish J415
→ N-Acetyl-3,5-dichlorobenzoquinone imine
In: Horseradish H716
Rat H716

6-Acetamido-2,3-dichlorophenol
Probably formed from: 6-Amino-2,3-dichlorophenol

6-Acetamido-3,4-dichlorophenol
Probably formed from: 6-Amino-3,4-dichlorophenol

4-Acetamido-2,6-dichlorophenoxy radical
Formed from: 4-Acetamido-2,6-dichlorophenol

***p*-Acetamido-N-(2-diethylaminoethyl) benzamide**
Formed from: *p*-Amino-N-(2-diethylamino-ethyl)benzamide
→ *p*-Acetamidobenzoic acid
In: Rat B627
→ *p*-Acetamido-N-(2-diethylaminoethyl) benzamide-N-oxide
In: Rat B627
→ *p*-Acetamido-N-(2-ethylaminoethyl) benzamide
In: Man C293
Rat B627

***p*-Acetamido-N-(2-diethylaminoethyl)benzamide-N-oxide**
Formed from: *p*-Acetamido-N-(2-diethylaminoethyl)benzamide

4-Acetamido-N-(2-diethylaminoethyl)-3-hydroxybenzamide
Formed from: *p*-Amino-N-(2-diethylaminoethyl)benzamide

4-Acetamido-2,6-diethylphenol
→ N-Acetyl-3,5-diethylbenzoquinone imine
In: Horseradish H716
Rat H716
→ 4-Acetamido-2,6-diethylphenoxy radical
In: Horseradish J415

4-Acetamido-2,6-diethylphenoxy radical
Formed from: 4-Acetamido-2,6-diethylphenol

4-Acetamido-2,6-difluorophenol
→ N-Acetyl-3,5-difluorobenzoquinone imine
In: Horseradish H716
Rat H716
→ 4-Acetamido-2,6-difluorophenoxy radical
In: Horseradish J415

4-Acetamido-2,6-difluorophenoxy radical
Formed from: 4-Acetamido-2,6-difluorophenol

7-Acetamido-1,3-dihydro-3-hydroxy-1-methyl-5-phenyl-2*H*-1,4-benzodiazepin-2-one
Probably formed from: 7-Acetamido-1,3-dihydro-1-methyl-5-phenyl-2*H*-1,4-benzodiazepin-2-one
→ 7-Acetamido-1,3-dihydro-3-hydroxy-5-phenyl-2*H*-1,4-benzodiazepin-2-one
Probably in: Rat A352

7-Acetamido-1,3-dihydro-3-hydroxy-5-phenyl-2*H*-1,4-benzodiazepin-2-one
Probably formed from: 7-Acetamido-1,3-dihydro-3-hydroxy-1-methyl-5-phenyl-2*H*-1,4-benzodiazepin-2-one

7-Acetamido-1,3-dihydro-1-methyl-5-phenyl-2*H*-1,4-benzodiazepin-2-one
Formed from: 7-Amino-1,3-dihydro-1-methyl-5-phenyl-2*H*-1,4-benzodiazepin-2-one
→ 7-Acetamido-1,3-dihydro-3-hydroxy-1-methyl-5-phenyl-2*H*-1,4-benzodiazepin-2-one
Probably in: Rat A352
→ 7-Acetamido-1,3-dihydro-5-phenyl-2*H*-1,4-benzodiazepin-2-one
Probably in: Rat A352

7-Acetamido-1,3-dihydro-5-phenyl-2*H*-1,4-benzodiazepin-2-one
Formed from: 7-Amino-1,3-dihydro-5-phenyl-2*H*-1,4-benzodiazepin-2-one

Probably formed from: 7-Acetamido-1,3-dihydro-1-methyl-5-phenyl-2*H*-1,4-benzodiazepin-2-one
→ 7-Amino-1,3-dihydro-5-phenyl-2*H*-1,4-benzodiazepin-2-one
In: Mouse G903
Rat G903

2-Acetamido-7,9-dihydroxyfluorene
Probably formed from: 2-Acetamido-7-hydroxy-9-oxofluorene

3-(5-Acetamido-3,4-dihydroxyphenyl)-2-cyano-N,N-diethylpropenamide
Formed from: Entacapone

4-Acetamido-2,6-diiodophenol
→ N-Acetyl-3,5-diiodobenzoquinone imine
In: Horseradish H716
Rat H716

4-Acetamido-2,6-diiodophenol
→ N-Acetyl-3,5-diiodobenzoquinone imine
In: Horseradish H716
Rat H716

4-Acetamido-2,6-diisopropylphenol
→ N-Acetyl-3,5-diisopropylbenzoquinone imine
In: Horseradish H716
Rat H716

4-Acetamido-2,6-dimethoxyphenol
→ 4-Acetamido-2,6-dimethoxyphenoxy radical
In: Horseradish J415

4-Acetamido-2,6-dimethoxyphenoxy radical
Formed from: 4-Acetamido-2,6-dimethoxyphenol

3′-Acetamido-4-dimethylaminoazobenzene
→ 3′-Acetamido-4-methylaminoazobenzene
In: Rat A776

4′-Acetamido-4-dimethylaminoazobenzene
Formed from: 4-Dimethylamino-4′-(N-methylacetamido)azobenzene
→ 4′-Acetamido-4-methylaminoazobenzene
In: Rat A2833

4-Acetamido-4′-dimethylaminobenzophenone
Probably formed from: 4-Amino-4′-dimethylaminobenzophenone

4-Acetamido-2′,3-dimethylbiphenyl
Formed from: 4-Amino-2′,3-dimethylbiphenyl

2-Acetamido-3,8-dimethylimidazo[4,5-*f*]quinoxaline
Formed from: 2-Amino-3,8-dimethylimidazo[4,5-*f*]quinoxaline

2-Acetamido-N,N-dimethylindoaniline
→ 4′-Dimethylamino-4-hydroxydiphenylamine
In: Rat E214

4-Acetamido-2,6-dimethylphenol
→ N-Acetyl-3,5-dimethyl-*p*-benzoquinone imine
In: Horseradish H716
Rat H716
→ 4-Acetamido-2,6-dimethylphenoxy radical
In: Horseradish J415
→ 2-(S-Glutathionyl)-3,5-dimethylquinol
In: Rat H716

4-Acetamido-2,6-dimethylphenoxy radical
Formed from: 4-Acetamido-2,6-dimethylphenol

4-Acetamido-N-(2,6-dimethylphenyl)benzamide
Formed from: 4-Amino-N-(2,6-dimethylphenyl)benzamide
→ 4-Acetamido-N-(2-hydroxymethyl-6-methylphenyl)benzamide
Probably in: Rat F57

4-Acetamido-2,3-dimethyl-1-phenyl-3-pyrazolin-5-one
See: 4-Acetamidoantipyrine

2-Acetamido-4,6-dinitrobenzaldehyde
Probably formed from: 2,4,6-Trinitrobenzaldehyde

2-Acetamido-4,6-dinitrotoluene
Formed from: 2-Amino-4,6-dinitrotoluene

4-Acetamido-2,6-dinitrotoluene
Formed from: 4-Amino-2,6-dinitrotoluene

2-Acetamido-4,4-diphenyl-5-heptanol
Probably formed from: 2-Amino-4,4-diphenyl-5-heptanol

2-Acetamido-5-ethoxyphenol
Formed from: *p*-Ethoxyacetanilide

5-Acetamido-2-ethoxyphenol
Probably formed from: *p*-Ethoxyacetanilide

4-Acetamido-4′-ethoxystilbene
→ 4-Acetamido-4′-ethoxystilbene-α,β-oxide

In: Hamster A2860
→ 4′-Acetamido-4-hydroxystilbene
In: Hamster A2860
→ N-Hydroxy-4-acetamido-4′-ethoxystilbene
In: Hamster A2860

4-Acetamido-4′-ethoxystilbene-α,β-oxide
Formed from: 4-Acetamido-4′-ethoxystilbene

N-(2-Acetamidoethyl)-4-amino-5-chloro-2-methoxybenzamide
Probably formed from: 4-Amino-N-aminoethyl-5-chloro-2-methoxybenzamide

***p*-Acetamido-N-(2-ethylaminoethyl)benzamide**
Formed from: *p*-Acetamido-N-(2-diethylaminoethyl)benzamide
Probably formed from: *p*-Amino-N-(2-ethylaminoethyl)benzamide

4-(5-(2-Acetamidoethyl)-3-bromo-5-hydroxy-2-oxocyclohexa-1,4-dienoxy)-N-acetyl-3,5-dibromophenethylamine
Formed from: N-Acetyl-3,5-dibromotyramine

2-(2-(N-Acetamido)ethyl)-3′-ethyl-4,4′,5,6′-tetrahydroxybiphenyl
Formed from: N-Acetyl-3,4-dihydroxyphenethylamine
4-Ethylresorcinol

5-(2-Acetamidoethyl)oximino-5-(*p*-trifluoromethylphenyl)valeric acid
Probably formed from: Methyl 5-(2-acetamidoethyl)oximino-5-(*p*-trifluoromethylphenyl)valerate

4-(2-Acetamidoethyl)-*o*-quinone
Formed from: N-Acetyl-3,4-dihydroxyphenethylamine
→ N-Acetyldopamine quinone methide
In: Sarcophaga F386, F793, G184
See also: E975
→ N-Acetylnoradrenaline
In: Manduca E975, G123
Sarcophaga F386, F793, G184

2-Acetamidofluorene
Formed from: 2-Aminofluorene
N-Hydroxy-2-acetamidofluorene
→ 2-Acetamido-1-hydroxyfluorene
In: Man A1869
Monkey A1869
Rabbit B723
Rat A1869, B94, B761
See also E481, F953
→ 2-Acetamido-3-hydroxyfluorene
In: Man A1869
Monkey A1869
Rabbit B723
Rat A1869, B94, B761, D269, E884, G548
See also E481, F953
→ 2-Acetamido-5-hydroxyfluorene
In: Man A1869, C264
Monkey A1869
Rabbit B723, G548
Rat A703, A1869 B94, B761, D269, E884
Trout H643
See also D776, E993
→ 2-Acetamido-6-hydroxyfluorene
In: Rat F341
→ 2-Acetamido-7-hydroxyfluorene
In: Guinea pig A703
Man A1869, C264
Monkey A1869
Mouse A703
Rabbit B723, G548
Rat A703, A1869, B94, B652, B761, D269, E884 G548
Trout H643
See also D776, E481, F953
→ 2-Acetamido-8-hydroxyfluorene
Probably in: Rat D776, E481
→ 2-Acetamido-9-hydroxyfluorene
In: Guinea pig A120
Man C138, C264
Rabbit A120, G548
Rat A120, B94, B288, B761, C138, D269, F341, G548
See also D776, E481, F953
→ 2-Acetyl-2-fluorenyl nitroxide radical
Probably in: Rabbit B184
→ 2-Aminofluorene
In: Guinea pig A808
Man C138, H278
Mouse A808, A3949
Mytilus E713
Rabbit A808, G548
Rat A808, C138, E884, G109, G548, G735, K185
See also A3965

→ 2-Glycollamidofluorene
In: Guinea pig A120
Rabbit A120, B548
Rat G548
See also F638
→ N-Hydroxy-2-acetamidofluorene
In: Guinea pig B179
Hamster C397
Man A1869, C138, C264
Monkey A1869, B179
Mouse A83, C379
Mytilus E713
Rabbit B723, G548
Rat A703, A1869, A2488, A3543, B94, B652, B761, C138, C397, D142, D269, E884, H643
Trout H643
See also E481

3-Acetamidofluorene
Formed from: N-Hydroxy-3-acetamidofluorene
→ 3-Acetamido-2-hydroxyfluorene
In: Guinea pig A2966
Rat A2966
→ 3-Acetamido-7-hydroxyfluorene
In: Guinea pig A2966
Rat A2966
→ 3-Acetamido-9-hydroxyfluorene
In: Guinea pig A2966
Rat A2966

2-Acetamidofluorene nitroxide radical
Probably formed from: N-Hydroxy-2-acetamidofluorene

3-Acetamidofluorene nitroxide radical
Probably formed from: N-Hydroxy-3-acetamidofluorene

2-Acetamidofluoren-1-yl-β-D-glucuronide
Formed from: 2-Acetamido-1-hydroxyfluorene

2-Acetamidofluoren-3-yl-β-D-glucuronide
Formed from: 2-Acetamido-3-hydroxyfluorene

2-Acetamidofluoren-5-yl-β-D-glucuronide
Formed from: 2-Acetamido-5-hydroxyfluorene

2-Acetamidofluoren-7-yl-β-D-glucuronide
Formed from: 2-Acetamido-7-hydroxyfluorene

2-Acetamidofluoren-8-yl-β-D-glucuronide
Formed from: 2-Acetamido-8-hydroxyfluorene

2-Acetamidofluoren-9-yl-β-D-glucuronide
Formed from: 2-Acetamido-9-hydroxyfluorene

6-Acetamido-2-fluoromethyl-3-(*o*-hydroxymethylphenyl)-4(3*H*)-quinazolinone
Probably formed from: N-Acetylafloqualone
6-Amino-2-fluoromethyl-3-(*o*-hydroxymethylphenyl)-4(3*H*)-quinazolinone
→ 2-Fluoromethyl-6-glycollamido-3-(*o*-hydroxymethylphenyl)-4(3*H*)-quinazolinone
Probably in: Monkey C912
Rat C793a

2-Acetamido-5-fluorophenol
Formed from: *p*-Fluoroacetanilide
Probably formed from: 2-Amino-5-fluorophenol
→ 2-Acetamido-5-fluorophenyl-β-D-glucuronide
Probably in: Rat G790, J603

4-Acetamido-2-fluorophenol
Formed from: *m*-Fluoroacetanilide
p-Fluoroacetanilide
Probably formed from: 4-Amino-2-fluorophenol
→ 4-Acetamido-2-fluorophenyl-β-D-glucuronide
Probably in: Rat G790

4-Acetamido-3-fluorophenol
Probably formed from: 4-Amino-3-fluorophenol
→ 4-Acetamido-3-fluorophenyl-β-D-glucuronide
Probably in: Marmoset D952
Rabbit D952
Rat D952, F518, G790
→ 4-Acetamido-3-fluorophenyl sulphate
Probably in: Marmoset D952
Rabbit D952
Rat D952, F518, G790

2-(4-Acetamido-3-fluorophenyl)-5-amino-6,8-difluoro-7-methyl-4*H*-1-benzopyran-1-one
Formed from: 5-Amino-2-(4-amino-3-fluorophenyl)- 6,8-difluoro-7-methyl-4*H*-1-benzopyran-1-one

7-Acetamido-5-(*o*-fluorophenyl)-1,3-dihydro-1-methyl-2*H*-1,4-benzodiazepin-2-one
Formed from: 7-Amino-5-(*o*-fluorophenyl)-1,3-dihydro-1-methyl-2*H*-1,4-benzodiazepin-2-one

2-Acetamido-5-fluorophenyl-β-D-glucuronide
Probably formed from: 2-Acetamido-5-fluorophenol

4-Acetamido-2-fluorophenyl-β-D-glucuronide
Probably formed from: 4-Acetamido-2-fluorophenol

4-Acetamido-3-fluorophenyl-β-D-glucuronide
Probably formed from: 4-Acetamido-3-fluorophenol

4-Acetamido-3-fluorophenyl sulphate
Probably formed from: 4-Acetamido-3-fluorophenol

4-Acetamido-4′-fluorostilbene
→ 4′-Fluoro-4-glycollamidostilbene
In: Hamster A2089
→ N-Hydroxy-4-acetamido-4′-fluorostilbene
In: Hamster A2089

4-Acetamido-3-(S-glutathionyl)phenol
Probably formed from: 4-Amino-3-(S-glutathionyl)phenol

***m*-Acetamidohippuric acid**
Formed from: *m*-Acetamidobenzoic acid

***p*-Acetamidohippuric acid**
Formed from: *p*-Acetamidobenzoic acid
p-Aminohippuric acid

5-Acetamido-2-hydroxybenzamide
Probably formed from: *m*-Acetamidobenzamide

4-Acetamido-2-hydroxybenzoic acid (4-acetamidosalicylic acid)
Formed from: 4-Amino-2-hydroxybenzoic acid
Salicylazosulphapyridine

4-Acetamido-3-hydroxybenzoic acid
Probably formed from: 4-Amino-3-hydroxybenzoic acid
→ 4-Acetamido-3-hydroxybenzoic acid-β-D-glucuronide
Probably in: Man E475

5-Acetamido-2-hydroxybenzoic acid
Formed from: 5-Amino-2-hydroxybenzoic acid

4-Acetamido-3-hydroxybenzoic acid-β-D-glucuronide
Probably formed from: 4-Acetamido-3-hydroxybenzoic acid

4-Acetamido-α-hydroxybibenzyl
Probably formed from: *erythro*-4-Acetamidobibenzyl-α,β-diol

4-Acetamido-β-hydroxybibenzyl
Probably formed from: *erythro*-4-Acetamidobibenzyl-α,β-diol

3-Acetamido-6-hydroxybiphenyl
Formed from: 3-Acetamidobiphenyl

4′-Acetamido-2-hydroxybiphenyl
Probably formed from: 4′-Amino-2-hydroxybiphenyl

4′-Acetamido-3-hydroxybiphenyl
Probably formed from: 4′-Amino-3-hydroxybiphenyl

4′-Acetamido-4-hydroxybiphenyl
Formed from: 4-Acetamidobiphenyl
4′-Amino-4-hydroxybiphenyl

4-Acetamido-4′-(3-hydroxybutoxy)stilbene
Formed from: 4-Acetamido-4′-butoxystilbene

4-(2-Acetamido-1-hydroxyethyl)-*o*-quinone
→ N-Acetylnoradrenaline
In: Sarcophaga F792, F793

2-Acetamido-4-((2-hydroxyethyl)thio)phenol
Formed from: *o*-Acetamidophenol

2-Acetamido-5-((2-hydroxyethyl)thio)phenol
Formed from: *o*-Acetamidophenol

2-Acetamido-6-((2-hydroxyethyl)thio)phenol
Formed from: *o*-Acetamidophenol

2-Acetamido-1-hydroxyfluorene
Formed from: 2-Acetamidofluorene
→ 2-Acetamidofluoren-1-yl-β-D-glucuronide
In: Man H92
Rat D723
→ 2-Amino-1-hydroxyfluorene
In: Rat K185

2-Acetamido-3-hydroxyfluorene
Formed from: 2-Acetamidofluorene

→ 2-Acetamidofluoren-3-yl-β-D-glucuronide
In: Man H92
Rat D723
→ 2-Amino-3-hydroxyfluorene
In: Rat K185

2-Acetamido-5-hydroxyfluorene
Formed from: 2-Acetamidofluorene
→ 2-Acetamidofluoren-5-yl-β-D-glucuronide
In: Man H92
Rat D723
→ 2-Amino-5-hydroxyfluorene
In: Rat K185

2-Acetamido-6-hydroxyfluorene
Formed from: 2-Acetamidofluorene

2-Acetamido-7-hydroxyfluorene
Formed from: 2-Acetamidofluorene
N-Hydroxy-2-acetamidofluorene
→ 2-Acetamidofluoren-7-yl-β-D-glucuronide
In: Man A2086
Monkey A2086
Rat A2086, D723
→ 2-Amino-7-hydroxyfluorene
In: Rat K185

2-Acetamido-8-hydroxyfluorene
Probably formed from: 2-Acetamidofluorene
→ 2-Acetamidofluoren-8-yl-β-D-glucuronide
In: Man H92

2-Acetamido-9-hydroxyfluorene
Formed from: 2-Acetamidofluorene
2-Acetamido-9-oxofluorene
→ 2-Acetamidofluoren-9-yl-β-D-glucuronide
In: Rat D723
→ 2-Acetamido-9-oxofluorene
In: Rat B288
See also A120, F341, G548
→ 2-Amino-9-hydroxyfluorene
In: Rat K185

3-Acetamido-2-hydroxyfluorene
Formed from: 3-Acetamidofluorene

3-Acetamido-7-hydroxyfluorene
Formed from: 3-Acetamidofluorene

3-Acetamido-9-hydroxyfluorene
Formed from: 3-Acetamidofluorene
→ 3-Acetamido-9-oxofluorene
Probably in: Rat A3494
Guinea pig A3494
See also: A2966

4′-Acetamido-3-hydroxy-4-methoxybiphenyl
Probably formed from: 4′-Amino-3-hydroxy-4-methoxybiphenyl

4′-Acetamido-4-hydroxy-3-methoxybiphenyl
Probably formed from: 4′-Amino-4-hydroxy-3-methoxybiphenyl

4′-Acetamido-4-hydroxy-3-methylazobenzene
Probably formed from: 4′-Amino-4-hydroxy-3-methylazobenzene

5-Acetamido-2-hydroxy-N-methylbenzamide
Probably formed from: 3-Acetamido-N-methylbenzamide
→ 5-Acetamido-2-hydroxy-N-methylbenzamide-β-D-glucuronide
Probably in: Rat H170

3-Acetamido-N-hydroxymethylbenzamide
Probably formed from: 3-Acetamido-N-methylbenzamide

5-Acetamido-2-hydroxy-N-methylbenzamide-β-D-glucuronide
Probably formed from: 5-Acetamido-2-hydroxy-N-methylbenzamide

5-Acetamido-1-hydroxymethyl-3-((2-methoxy-4-((2-methylphenyl)sulphonyl)aminocarbonyl)phenylmethyl)indole
Probably formed from: 5-Acetamido-3-((2-methoxy-4-((2-methylphenyl)sulphonyl)aminocarbonyl)phenylmethyl)-1-methylindole
→ 5-Acetamido-3-((2-methoxy-4-((2-methylphenyl)sulphonyl)aminocarbonyl)phenylmethyl)indole
Probably in: Dog J455
Mouse J455
Rat J455

4-Acetamido-N-(2-hydroxymethyl-6-methylphenyl)benzamide
Probably formed from: 4-Acetamido-N-(2,6-dimethylphenyl)benzamide

6-Acetamido-3-(*o*-hydroxymethylphenyl)-2-methylsulphonylmethyl-4(3*H*)-quinazolinone
Probably formed from: N-Acetyl-2-methylsulphonylafloqualone

6-Acetamido-2-hydroxymethyl-3-(*o*-tolyl)-4(3*H*)-quinazolinone
Formed from: N-Acetylafloqualone
Probably formed from: 6-Amino-2-hydroxymethyl-3-(*o*-tolyl)-4(3*H*)-quinazolinone
→ 6-Glycollamido-2-hydroxymethyl-3-(*o*-tolyl)-4(3*H*)-quinazolinone
Probably in: Monkey C793a
Rat C912

2-Acetamido-7-hydroxy-9-oxofluorene
Formed from: 2-Acetamido-9-oxofluorene
→ 2-Acetamido-7,9-dihydroxyfluorene
Probably in: Rat G548

1-(4-Acetamido-3-hydroxyphenoxy)-3-isopropylaminopropan-2-ol (3-hydroxypractolol)
Formed from: Practolol
→ 1-(4-Acetamido-3-hydroxyphenoxy)-3-isopropylaminopropan-2-ol-β-D-glucuronide
Probably in: Rat K122

1-(4-Acetamido-3-hydroxyphenoxy)-3-isopropylaminopropan-2-ol-β-D-glucuronide
Probably formed from: 1-(4-Acetamido-3-hydroxyphenoxy)-3-isopropylaminopropan-2-ol

S-(5-Acetamido-2-hydroxyphenyl)-N-acetyl-L-cysteine
Formed from: S-(5-Acetamido-2-hydroxyphenyl)-L-cysteine
Acetanilide

S-(5-Acetamido-2-hydroxyphenyl)-L-cysteine
Formed from: S-(5-Acetamido-2-hydroxyphenyl)glutathione
→ S-(5-Acetamido-2-hydroxyphenyl)-N-acetyl-L-cysteine
In: Mouse D341
See also A1402, A2671, A3967, B461, C241, D626, D630
→ S-(5-Acetamido-2-hydroxyphenyl)-L-cysteine-β-D-glucuronide
Probably in: Man A1402
→ S-(5-Acetamido-2-hydroxyphenyl) glutathione
In: Mouse D341
→ 5-Acetamido-2-hydroxyphenyl sulphone
Probably in: Hamster A1794
Contrast D626
→ 4-Acetamido-2-thiomethylphenol
Probably in: Hamster D626
Mouse E820
Rat D630

S-(5-Acetamido-2-hydroxyphenyl)-L-cysteine-β-D-glucuronide
Formed from: S-(5-Acetamido-2-hydroxyphenyl)-L-cysteine

S-(5-Acetamido-2-hydroxyphenyl)glutathione
Formed from: *p*-Acetamidophenol
N-Acetyl-*p*-benzoquinone imine
S-(5-Acetamido-2-hydroxyphenyl)-L-cysteine
→ S-(5-Acetamido-2-hydroxyphenyl)-L-cysteine
In: Mouse D341
See also C667, D630, E820
→ 5,5′-Diacetamido-3-(S-glutathionyl)-2,2′-dihydroxybiphenyl
In: Horseradish D988
→ 5,5′-Diacetamido-3,3′-di-(S-glutathionyl)-2,2′-dihydroxybiphenyl
In: Horseradish D988

5-Acetamido-2-hydroxyphenyl sulphone
Probably formed from: S-(5-Acetamido-2-hydroxyphenyl)-L-cysteine

2-(3-Acetamido-2-hydroxy-1-propoxy)-1-(2-propen-1-oxy)benzene
Probably formed from: 2-(3-Amino-2-hydroxy-1-propoxy)-1-(2-propen-1-oxy) benzene

6-(3-Acetamido-2-hydroxypropoxy)-2(1*H*)-quinoline
Probably formed from: 6-(3-Amino-2-hydroxypropoxy)-2(1*H*)-quinoline

2-(3-Acetamido-2-hydroxy-1-propoxy)-1-(2-propen-1-oxy)benzene
Probably formed from: 2-(3-Amino-2-hydroxy-1-propoxy)-1-(2-propen-1-oxy) benzene

6-(3-Acetamido-2-hydroxy-1-propoxy)-2(1*H*)-quinolinone
Probably formed from: 6-(3-Amino-2-hydroxy-1-propoxy)-2(1*H*)-quinolinone

5-Acetamido-2-hydroxyquinoline
Probably formed from: 5-Amino-2-hydroxyquinoline

***trans*-4-Acetamido-3-hydroxystilbene**
Formed from: 4-Acetamidostilbene
→ *trans*-4-Amino-3-hydroxystilbene
Probably in Rat E921

***trans*-4-Acetamido-4′-hydroxystilbene**
Formed from: *trans*-4-Acetamidostilbene
→ *trans*-4-Amino-4′-hydroxystilbene
Probably in Rat E921

7-Acetamido-1-hydroxy-2-(*p*-sulphamoylphenylazo)naphthalene-3,6-disulphonic acid (neoprontosil)
→ Sulphanilamide
In: Rat A114, A3048
See also A3049

5-Acetamido-3-((2-methoxy-4-((2-methylphenyl)sulphonyl)aminocarbonyl) phenyl)methyl)indole
Probably formed from: 5-Acetamido-1-hydroxymethyl-3-((2-methoxy-4-((2-methylphenyl)sulphonylaminocarbonyl)phenyl) methyl)indole

5-Acetamido-3-((2-methoxy-4-((2-methylphenyl)sulphonyl)aminocarbonyl)phenyl)methyl)-1-methylindole
Probably formed from: 5-Amino-3-((2-methoxy-4-((2-methylphenyl)sulphonyl) aminocarbonyl)phenyl)methyl)-1-methylindole
→ 5-Acetamido-3-(4-carboxy-2-methoxyphenylmethyl)-1-methylindole
Probably in: Dog J455
Mouse J455
Rat J455
→ 5-Acetamido-3-((2-methoxy-4-((2-methylphenyl)sulphonyl)aminocarbonyl)phenylmethyl)indole
Probably in: Dog J455
Mouse J455
Rat J455

4-Acetamido-2-methoxyphenol
Formed from: 4-Acetamidocatechol
→ 4-Acetamido-2-methoxyphenyl-β-D-glucuronide
In: Man A2671
See also F572
→ 4-Acetamido-2-methoxyphenyl sulphate
In: Man A2671

4-Acetamido-3-methoxyphenol
Formed from: 4-Acetamidoresorcinol
→ 4-Acetamido-3-methoxyphenyl-β-D-glucuronide
Probably in: Man A1402
→ 4-Acetamido-2-methoxyphenyl sulphate
Probably in: Man A1402

4-Acetamido-2-methoxyphenyl-β-D-glucuronide
Formed from: 4-Acetamido-2-methoxyphenol

4-Acetamido-3-methoxyphenyl-β-D-glucuronide
Formed from: 4-Acetamido-3-methoxyphenol

4-Acetamido-2-methoxyphenyl sulphate
Formed from: 4-Acetamido-2-methoxyphenol

4-Acetamido-3-methoxyphenyl sulphate
Formed from: 4-Acetamido-3-methoxyphenol

4-Acetamido-4′-methoxystilbene
→ N-Hydroxy-4-acetamido-4′-methoxystilbene
In: Hamster A2860

3′-Acetamido-4-methylaminoazobenzene
Probably formed from: 3′-Acetamido-4-dimethylaminoazobenzene
→ 3′-Acetamido-4-aminoazobenzene
Probably in: Rat A776

4′-Acetamido-4-methylaminoazobenzene
Probably formed from: 4′-Acetamido-4-dimethylaminoazobenzene
→ 4′-Acetamido-4-aminoazobenzene
Probably in: Rat A2833

3-Acetamido-4-methylaniline
Formed from: 2,4-Diaminotoluene
→ 2,4-Diacetamidotoluene
In: Hamster A461

5-Acetamido-2-methylaniline
Formed from: 2,4-Diacetamidotoluene
2,4-Diaminotoluene
→ 4-Acetamidoanthranilic acid
Probably in: Rat B423
→ 2,4-Diacetamidotoluene
In: Hamster A461
See also A311, B423, H3, J196

3-Acetamido-N-methylbenzamide
Probably formed from: 3-Amino-N-methylbenzamide

→ *m*-Acetamidobenzamide
Probably in: Rat H170
→ 3-Acetamido-2-hydroxy-N-methylbenzamide
Probably in: Rat H170
→ 3-Acetamido-N-hydroxymethylbenzamide
Probably in: Rat H170

5-Acetamido-2-methylbenzimidazole
Probably formed from: 3,4-Diamino-acetanilide
→ 5-Acetamido-2-methylbenzimidazole-N-oxide
Probably in: Rat J586

5-Acetamido-2-methylbenzimidazole-N-oxide
Probably formed from: 5-Acetamido-2-methylbenzimidazole

4-Acetamido-3-methylbenzoic acid
Probably formed from: 4-Amino-3-methyl-benzoic acid

3-Acetamidomethyl-1,2-diphenyl-2-butyl propionate
Probably formed from: 3-Aminomethyl-1,2-diphenyl-2-butyl propionate

8-Acetamidomethyl-6-methylergoline
Probably formed from: 8β-Aminomethyl-6-methylergoline

4-Acetamido-2-methyl-6-nitrophenol
See: 4-Acetamido-6-nitro-*o*-cresol

6-Acetamido-2-methyl-4-nitrophenol
Formed from: 6-Amino-2-methyl-4-nitrophenol

(*S*)-4-(5-Acetamidomethyl-2-oxooxazolidin-3-yl)-N-carboxymethyl-2-fluoro-N-(2-hydroxyethyl)aniline
Formed from: Linezolid

4-Acetamido-2-methylphenol
Formed from: *m*-Methylacetanilide

4-Acetamido-3-methylphenol
Formed from: N-Acetyl-*o*-toluidine
→ 4-Amino-*m*-cresol
In: Nocardia G343

2-Acetamido-1-methyl-9*H*-pyrido[2,3-*b*]indole
Formed from: 2-Amino-1-methyl-9*H*-pyrido[2,3-*b*]indole

3-Acetamido-1-methyl-5*H*-pyrido[4,3-*b*]indole
Formed from: 3-Amino-1-methyl-5*H*-pyrido[4,3-*b*]indole

4-Acetamido-2-methylsulphinylphenol
Probably formed from: 4-Acetamido-2-methylthiophenol

4-Acetamido-2-methylthiophenol
→ 4-Acetamido-2-methylsulphinylphenol
Probably in: Hamster B639

1-Acetamidonaphthalene
Formed from: α-Naphthylamine

2-Acetamidonaphthalene
Formed from: β-Naphthylamine
→ N-Hydroxy-2-acetamidonaphthalene
In: Dog A880
→ β-Naphthylamine
In: Rat F109

1-Acetamidonaphthalene nitroxide radical
Probably formed from: N-Hydroxy-1-acetamidonaphthalene

2-Acetamidonaphthalene nitroxide radical
Probably formed from: N-Hydroxy-2-acetamidonaphthalene

5-Acetamido-1-naphthol
Formed from: 5-Amino-1-naphthol
→ 5-Acetamido-1-naphthyl-β-D-glucuronide
Probably in: Rat D85
→ 5-Acetamido-1-naphthyl sulphate
Probably in: Rat D85

5-Acetamido-1-naphthyl-β-D-glucuronide
Probably formed from: 5-Acetamido-1-naphthol

1-Acetamido-3-(1-naphthyloxy)-2-propanol
Formed from: 2-Hydroxy-3-(α-naphthyloxy)propylamine

5-Acetamido-1-naphthyl sulphate
Probably formed from: 5-Acetamido-1-naphthol

4-Acetamido-4′-nitroazobenzene
Formed from: 4-Amino-4′-nitroazobenzene

3-Acetamido-5-nitrobenzamide
Formed from: 3-Amino-5-nitrobenzamide

2-Acetamido-4-nitrobenzoic acid
Formed from: 2-Amino-4-nitrobenzoic acid

4-Acetamido-2-nitrobenzoic acid
Formed from: 4-Amino-2-nitrobenzoic acid

4-Acetamido-6-nitro-*o*-cresol (4-acetamido-2-methylphenol)
Probably formed from: 4-Amino-6-nitro-*o*-cresol

6-Acetamido-4-nitro-*o*-cresol
Probably formed from: 6-Amino-4-nitro-*o*-cresol

***d-threo*-2-Acetamido-1-(*p*-nitrophenyl)-1,3-propanediol**
Formed from: D-*threo*-2-Amino-1-(*p*-nitrophenyl)-1,3-propanediol

1-Acetamido-8-nitropyrene
Probably formed from: 1-Amino-8-nitropyrene

2-Acetamido-6-nitro-4-trifluoromethylaniline
Formed from: 3-Nitro-5-trifluoromethyl-*o*-phenylenediamine

2-Acetamido-9-oxofluorene
Formed from: 2-Acetamido-9-hydroxyfluorene
→ 2-Acetamido-9-hydroxyfluorene
In: Rabbit G548
Rat G548
→ 2-Acetamido-7-hydroxy-9-oxofluorene
In: Rabbit G548
Rat G548
→ 2-Aminofluoren-9-one
In: Rabbit G548
→ 2-Glycollamido-9-oxofluorene
In: Rabbit G548
→ N-Hydroxy-2-acetamido-9-oxofluorene
In: Rabbit G548

3-Acetamido-9-oxofluorene
Formed from: 3-Acetamido-9-hydroxyfluorene

Acetamidophen
See: *p*-Acetamidophenol

2-Acetamidophenanthrene nitroxide radical
Probably formed from: N-Hydroxy-2-acetamidophenanthrene

***o*-Acetamidophenol**
Formed from: Acetanilide
o-Aminophenol
→ 2-Acetamido-4-((2-hydroxyethyl)thio)phenol
In: Rat E453
→ 2-Acetamido-5-((2-hydroxyethyl)thio)phenol
In: Rat E453
→ 2-Acetamido-6-((2-hydroxyethyl)thio)phenol
In: Rat E453
→ *o*-Acetamidophenyl sulphate
Probably in: Mouse D532
Rat D532
→ *o*-Aminophenol
In: Ascaris A2036
Microorganism A261
Moniezia A1755
Rat D491
→ 2,5-Dihydroxyacetanilide
In: Mouse D848

***m*-Acetamidophenol** (metacetamol)
Probably formed from: 4-Acetamidocatechol
Acetanilide
m-Aminphenol
→ *m*-Acetamidophenyl-β-D-glucuronide
In: Mouse F572
See also C241, D857
→ *m*-Acetamidophenyl sulphate
In: Man E575
Mouse F572
Rat E575
See also C241, D857
→ *m*-Aminophenol
In: Ascaris A2036
Microorganism A261
Moniezia A1755
Rat D491
→ 2′,3′-Dihydroxyacetanilide
In: Mouse D380
→ 2′,5′-Dihydroxyacetanilide
In: Mouse D380, D814, D848, F572
See also F31
→ 3′,4′-Dihydroxyacetanilide
In: Mouse D380, D814, D848, F572
See also F31
→ S-Glutathionyl-*m*-acetamidophenol
Probably in: Mouse F572
→ N-Hydroxy-*m*-acetamidophenol
In: Mouse F23

***p*-Acetamidophenol** (acetamidophen, paracetamol)
Formed from: *p*-Acetamidophenyl-β-D-glucuronide
Acetanilide
N-Acetyl-*p*-benzoquinone imine

p-Aminophenol
Benorylate
p-(*n*-Butyl)acetanilide
Eterylate
p-Ethoxyacetanilide
p-Fluoroacetanilide
p-Isopropoxyacetanilide
p-Methoxyacetanilide
m-Nitroaniline
Phenazopyridine
Phenetsal
Propachlor
Probably formed from: *m*-Nitroaniline
p-(Thioacetamido)phenol
→ 4-Acetamidocatechol
In: Hamster B53
Man: A1402, A2671
Mouse F572
See also: B897, C241
→ S-(5-Acetamido-2-hydroxyphenyl) glutathione
In: Hamster D850
Horseradish D986
Mouse B823, D341
Rat B287, C667, E223
See also: A262, A263, A1402, A1443, A1507, A1794, A2671, A3232, B287, B348, B542, B712, B823, B969, C202, C241, D175, D626, D630, D822, D938, E696, E820, F572, H553, H801, J558, K60, K85
→ 4-Acetamido-2-methylthiophenol
In: Dog A3739
Hamster B639
Man A3739
Mouse F572
See also C306, C457, F937
→ *p*-Acetamidophenoxy radical
In: Horseradish D350, D490, J415
→ *p*-Acetamidophenyl-β-D-glucuronide
In: Ferret A263
Guinea pig A263, A3132, D938
Hamster A1507, A1794, A3132, D850
Man A31, A587, A1346, A1402, A2221, A2671, A2840 A3132, B685, C241, D458, E412, F482, F779, J228, J239, K33, K85
Mouse A1507, A3132, A3232, B823, D341, D822, E820, F572
Rabbit A263, A2355, E208
Rat A262, A263, A804, A1507, A1780, A1783, A2980, A3132, A3232, A3464, A3967, B287, B461, B969, C1, C202, C667, D62, D175, D425, D812, D822, E696, F338, H553, H801, J550
See also: A3460, B165, C572, D532
Contrast (in rat) B969
→ *p*-Acetamidophenyl sulphate
In: Eubacterium G602
Ferret A263
Guinea pig A263, A3132, D938
Hamster A1507, A1794, A3132, D626, D850
Man A31, A587, A1346, A1402, A2221, A2671, A2840, A3132, A3814, B685, C523, E412, E575, F482, J228, J239, K85
Mouse A1507, A3132, A3232, B823, D341, D822, E820, F572, F937
Rabbit A263, A2355, E208
Rat A262, A263, A1507, A1780, A1783, A3132, A3464, A3967, B287, B461, B969, C1, C202, C667, D62, D175, D425, D460, D812, D822, E575, E696, F338, H553, H801, J550
See also: A3460, B165, C572, D532, E841
→ 4-Acetamidoresorcinol
In: Man A1402
→ Acetanilide
Probably in: Aspergillus E112
→ N-Acetyl-*p*-benzoquinone imine
In: Rat D821
See also A3968
→ *p*-Aminophenol
In: Ambystoma B551
Amphiuma B551
Ascaris A2036
Cynops B551
Hamster B636
Man H465
Microorganism C851
Moniezia A1755
Mouse F572

Nocardia G343
Notophthalamus B551
Penicillium A2341
Rat D81, D491, H194
Salamandra B551
Triturus B551
→ 5,5′-Diacetamido-2,2′-dihydroxybiphenyl
With xanthine oxidase F171
→ 5,5′-Diacetamido-3-(S-glutathionyl)-2,2′-dihydroxybiphenyl
In: Beef E203
Horseradish D986, E203
→ S-(2,5-Dihydroxyphenyl)-L-cysteine
In: Mouse F54, F572
→ *p*-Methoxyacetanilide
In Man A2956, C741
→ Quinol
In: Mouse F572

S-(*p*-Acetamidophenol-3-yl)glutathione
Formed from: *p*-Acetamidophenol

***p*-Acetamidophenoxyacetic acid**
Formed from: Eterylate
N-(4-(2-Hydroxyethoxy)phenyl)acetamide

N-((*p*-Acetamidophenoxy)benzyl)-2,2-dichloroacetanilide
Probably formed from: N-((*p*-Acetamidophenoxy)benzyl)-2,2-dichloro-N-(2-ethoxyethyl)acetamide

N-((*p*-Acetamidophenoxy)benzyl)-2,2-dichloro-N-(2-ethoxyethyl)acetamide
Formed from: 2,2-Dichloro-N-(2-ethoxyethyl)-N-((*p*-glycylaminophenoxy)benzyl)acetamide
→ N-((*p*-Acetamidophenoxy)benzyl)-2,2-dichloroacetamide
Probably in: Rat C803
→ N-((*p*-Acetamidophenoxy)benzyl)-N-(2-ethoxyethyl)glycollamide
Probably in: Rat C803

N-((*p*-Acetamidophenoxy)benzyl)-N-(2-ethoxyethyl)glycollamide
Probably formed from: N-((*p*-Acetamidophenoxy)benzyl)-2,2-dichloro-N-(2-ethoxyethyl)acetamide

2-(4-Acetamidophenoxy)ethyl 2-acetoxybenzoate
See: Eterylate

3-(4-Acetamidophenoxy)-2-hydroxy-N-isopropyl)propylamine
See: Practolol

***p*-Acetamidophenoxyl radical**
Formed from: *p*-Acetamidophenol

4-Acetamidophenyl 2-acetoxybenzoate
See: Benorylate

S-(*m*-Acetamidophenyl)-N-acetyl-L-cysteine
Formed from: Acetanilide

3-(α-(*p*-Acetamidophenyl)-β-acetylethyl)-4-hydroxycoumarin
Formed from: 3-(β-Acetyl-α-(*p*-aminophenyl)ethyl)-4-hydroxycoumarin

4′-Acetamidophenyl 4-aminophenyl sulphone
→ 4′-Acetamidophenyl 4-aminophenyl sulphone-N-β-D-glucuronide
Probably in: Rat A268
→ Bis(*p*-aminophenyl)sulphone
In: Rabbit G966

4′-Acetamidophenyl 4-aminophenyl sulphone-N-β-D-glucuronide
Probably formed from: 4′-Acetamidophenyl 4-aminophenyl sulphone

3-(4-Acetamidophenyl)-3-(2-carboxamidoethyl) tetrahydrofuran-2-one
Formed from: Aminoglutethimide

***d-threo*-1-(*p*-Acetamidophenyl)-2-(dichloroacetamido)-1,3-propanediol**
Probably formed from: D-*threo*-1-(*p*-Aminophenyl)-2-(dichloroacetamido)-1,3-propanediol

4-Acetamidophenyl 2,4-dichlorophenyl ether
Formed from: 4-Aminophenyl 2,4-dichlorophenyl ether

N-(2-Acetamidophenyl)-N′-ethoxycarbonyl-S-methylisothiourea
Formed from: Methyl 1-butylcarbamoylbenzimidazol-2-yl carbamate

α-(*p*-Acetamidophenyl)-α-ethylglutarimide
Formed from: Aminoglutethimide

***m*-Acetamidophenyl-β-D-glucuronide**
Formed from: *m*-Acetamidophenol

***p*-Acetamidophenyl-β-D-glucuronide**
Formed from: *p*-Acetamidophenol
→ *p*-Acetamidophenol

In: Microorganism A2811

N-(2-Acetamidophenyl)-N′-methoxycarbonyl-S-methylisothiourea
Formed from: Methyl 1-(butylcarbamoyl) benzimidazol-2-yl carbamate

***threo*-1-(4-Acetamido)phenyl-2-phenyl-1,2-ethanediol**
Formed from: *cis*-4-Dimethylaminostilbene
trans-4-Dimethylaminostilbene

2-(4-Acetamidophenyl)-1-phenylethanol
Formed from: 4-Dimethylaminobibenzyl

4-Acetamidophenyl-4′-propoxystilbene
→ N-Hydroxy-4-acetamidophenyl-4′ propoxystilbene
In: Hamster A2860
→ 4-Acetamidophenyl-4′-propoxystilbene-α,β-oxide
In: Hamster A2860

4-Acetamidophenyl-4′-propoxystilbene-α,β-oxide
Formed from: 4-Acetamidophenyl-4′ propoxystilbene

5-(*p*-Acetamidophenyl)pyrazin-2(1*H*)-one
→ 5-(*p*-Acetamidophenyl)pyrazin-2(1*H*)-one-β-D-glucuronide
In: Cynomolgus monkey F313
Dog F313
Rat F313
→ 5-(*p*-Aminophenyl)pyrazin-2(1*H*)-one
Probably in: Cynomolgus monkey F313
Dog F313
Rat F313

5-(*p*-Acetamidophenyl)pyrazin-2(1*H*)-one-β-D-glucuronide
Formed from: 5-(*p*-Acetamidophenyl) pyrazin-2(1*H*)-one

***o*-Acetamidophenyl sulphate**
Probably formed from: *o*-Acetamidophenol

***m*-Acetamidophenyl sulphate**
Formed from: *m*-Acetamidophenol

***p*-Acetamidophenyl sulphate**
Formed from: *p*-Acetamidophenol

***p*-Acetamidophenyl 2,4,6-trichlorophenyl ether**
Probably formed from: *p*-Aminophenyl 2,4,6-trichlorophenyl ether

1-Acetamidopyrene
Formed from: 1-Aminopyrene

Acetamidoquinol
See: 2,5-Dihydroxyacetanilide

5-Acetamidoquinoline
Formed from: 5-Aminoquinoline

6-Acetamidoquinoline
Formed from: 6-Aminoquinoline

6-Acetamido-2-quinolone
Probably formed from: 6-Amino-2-quinolone

2-Acetamido-*p*-quinone
Formed from: 2′,5′-Dihydroxyacetanilide
→ 2,5-Dihydroxyacetanilide
In: Mouse D814

4-Acetamido-*o*-quinone
Probably formed from: 3,4-Dihydroxyacetanilide

4-Acetamidoresorcinol
Formed from: *p*-Acetamidophenol
→ 4-Acetamidoresorcinol sulphate
Probably in: Man A1402
→ 4-Acetamido-3-methoxyphenol
Probably in: Man A1402

4-Acetamidoresorcinol sulphate
Probably formed from: 4-Acetamidoresorcinol

4-Acetamidosalicylic acid
See: 4-Acetamido-2-hydroxybenzoic acid

4-Acetamidostilbene
→ *erythro*-4-Acetamidobibenzyl-α,β-diol
In: Rat E921
→ *threo*-4-Acetamidobibenzyl-α,β-diol
In: Rat E921
→ *trans*-4-Acetamido-3-hydroxystilbene
In: Rat E921
→ 4′-Acetamido-4-hydroxystilbene
In: Hamster A2089
Rat E921
→ N-Hydroxy-4-acetamidostilbene
In: Hamster A2089

4-Acetamidostilbene nitroxide radical
Probably formed from: N-Hydroxy-4-acetamidostilbene

4-Acetamido-2-sulphinylmethylphenol
Probably formed from: 4-Acetamido-2-thiomethylphenol

7-Acetamido-2,3,4,5-tetrahydroimidazo[2,1-*b*]quinazolin-2(1*H*)-one
Probably formed from: 7-Amino-2,3,4,5-tetrahydroimidazo[2,1-*b*]quinazolin-2(1*H*)-one

5-Acetamido-2-(4-thiazolyl)benzimidazole
Formed from: Cambendazole
***p*-Acetamidothioanisole**
Formed from: *p*-Acetamidothiophenol
→ *p*-Acetamidothioanisole sulphoxide
In: Horseradish E403
***p*-Acetamidothioanisole sulphoxide**
Formed from: *p*-Acetamidothioanisole
4-Acetamido-2-thiomethylphenol
Probably formed from: S-(5-Acetamido-2-hydroxyphenyl)-L-cysteine
→ 4-Acetamido-2-sulphinylmethylphenol
Probably in: Hamster D626
***p*-Acetamidothiophenol**
→ *p*-Acetamidothioanisole
In: Man C917
S-(6-Acetamido-3-(*o*-tolyl)-4(3*H*)-quinazolinon-2-yl)-N-acetyl-L-cysteine
Formed from: S-(6-Acetamido-3-(*o*-tolyl)-4(3*H*)-quinazolinon-2-yl)-N-acetyl-L-cysteine methyl ester
S-(6-Acetamido-3-(*o*-tolyl)-4(3*H*)-quinazolinon-2-yl)-N-acetyl-L-cysteine methyl ester
→ S-(6-Acetamido-3-(*o*-tolyl)-4(3*H*)-quinazolinon-2-yl)-N-acetyl-L-cysteine
In: Rat D101
5-Acetamido-2-trifluoromethylbenzoic acid
Probably formed from: 3-Methyl-2-trifluoromethylacetanilide
3-Acetamido-2,4,6-triiodophenoxyacetic acid
Probably formed from: 2-(3-Acetamido-2,4,6-triiodophenoxy)ethanol
→ 3-Acetamido-2,4,6-triiodophenoxyacetyl-β-D-glucuronide
In: Rat A2780
3-Acetamido-2,4,6-triiodophenoxyacetyl-β-D-glucuronide
Formed from: 3-Acetamido-2,4,6-triiodophenoxyacetic acid
2-(3-Acetamido-2,4,6-triiodophenoxy)ethanol
Formed from: Iopronic acid
→ 3-Acetamido-2,4,6-triiodophenoxyacetic acid
Probably in: Dog A2788
Man A2871
Rat A2780, A2783, A3381
2-(2-(3-Acetamido-2,4,6-triiodophenoxy) ethoxy)methyl)butyric acid
See: Iopronic acid
Acetanilide
Formed from: Aniline
N-Propylacetanilide
Probably formed from: *p*-Acetamidophenol
→ *o*-Acetamidophenol
In: Aspergillus A1218
See also: A2788, C120
→ *m*-Acetamidophenol
Probably in: Pseudomonas E395
→ *p*-Acetamidophenol
In: Rat A1780, A2106, A2160, A3626, D81
Streptomyces G802
See also: A1125, A2788, A3460, E395, K340
→ S-(*m*-Acetamidophenyl)-N-acetyl-L-cysteine
In: Rat A299
→ S-(5-Acetamido-2-hydroxyphenyl)-N-acetyl-L-cysteine
In: Rat A299
→ Aniline
In: Ascaris A324, A2036
Aspergillus A1218
Gliocladium A1218
Microorganism A261
Moniezia A324, A1755
Nocardia G343
Penicillium A1218
Pseudomonas A347
Rat D81, D491
Rice A3773
Streptomyces A1218
Acetoisovanillone
See: 3′-Hydroxy-4′-methoxyacetophenone
3-Acetonyl-3-hydroxyindole
→ Isatin
In: Rabbit F829
→ Oxindole
In: Rabbit F829
Acetophenone (1-phenylethanone)
Formed from: Acetophenone oxime
Cumene hydroperoxide

(−)-Etomidate
1-Phenylethanol
1-Phenylethylamine
1-Phenylethanol
Probably formed from: 2-Phenyl-2-propanol
→ *cis*- 2′,3′-Dihydro-2′,3′-dihydroxycetophenone
In: Pseudomonas A914
→ *m*-Hydroxyacetophenone
In: Rabbit A287
See also D482
→ *p*-Hydroxyacetophenone
In: Rabbit A287
See also D223, D482
→ 2-Oxo-2-phenylethanol
In: Nitrosomonas H219
Rabbit A287
See also A1745, D223, A365, D482
→ Phenol
In: Pseudomonas E484
→ Phenyl acetate
In: Arthrobacter A2380
Pseudomonas K698
→ 1-Phenylethanol
In: Bacillus G828
Candida G826
Corynebacterium J533, J893
Geotrichum H749
Lactobacillus G148
Micrococcus G828
Rabbit D25, D125
Rat J744
Rhodotorula G826
Saccharomyces G826
Torulopsis G826
Xanthomonas G826

Acetophenone oxime
→ Acetophenone
In: Rabbit E358
→ L-N-Hydroxy-1-phenylethylamine
In: Rat A35, A1452

N-(*o*-Acetophenone)phthalamic acid
Formed from: *o*-(N-Phthalimido) acetophenone

Acetophenone tricarbonylchromium
→ Dihydroacetophenone tricarbonylchromium
Probably in: Bacillus G828
Micrococcus G828

Acetophenon-4′-yl-β-D-glucuronide
Formed from: *p*-Hydroxyacetophenone

Acetophenon-2′-yl sulphate
Probably formed from: *o*-Hydroxyacetophenone

Acetophenon-4′-yl sulphate
Probably formed from: *p*-Hydroxyacetophenone
→ *p*-Hydroxyacetophenone
In: Eubacterium E432

Acetosyringone
See: 4′-Hydroxy-3′,5′-dimethoxyacetophenone

***o*-Acetotoluidide**
See: *o*-Methylacetanilide

***m*-Acetotoluidide**
See: *m*-Methylacetanilide

***p*-Acetotoluidide**
See: *p*-Methylacetanilide

Acetovanillone
See: 4′-Hydroxy-3′-methoxyacetophenone

N-Acetoxy-2-acetamidofluorene
Formed from: N-Hydroxy-2-acetamidofluorene

4-Acetoxy-3-acetoxymethylbenzaldehyde
→ (*R*)-4-Acetoxy-3-acetoxymethylmandelonitrile
In: Almond J302
Apple J302

(*R*)-4-Acetoxy-3-acetoxymethylmandelonitrile
Formed from: 4-Acetoxy-3-acetoxymethylbenzaldehyde

N-Acetoxy-7-acetyl-2-aminofluorene
Formed from: N-Hydroxy-2-acetamido-7-acetylfluorene

N-Acetoxy-2-amino-7-bromofluorene
Formed from: N-Hydroxy-2-acetamido-7-bromofluorene

N-Acetoxy-4-amino-2′,3-dimethylbiphenyl
Formed from: 4-Hydroxylamino-2′,3-dimethylbiphenyl

N-Acetoxy-2-amino-7-ethoxyfluorene
Formed from: N-Hydroxy-2-acetamido-7-ethoxyfluorene

N-Acetoxy-2-amino-7-ethylfluorene
Formed from: N-Hydroxy-2-acetamido-7-ethylfluorene

N-Acetoxy-2-aminofluorene
Formed from: N-Hydroxy-2-acetamidofluorene
→ 2-Aminofluorene
In: Hamster G744

***p*-Acetoxybenzaldehyde**
→ *p*-Acetoxymandelonitrile (both isomers)
In: Almond J194
Sorghum J194, J302

***o*-Acetoxybenzoic acid**
See: Acetylsalicylic acid

***m*-Acetoxybenzoic acid**
→ *m*-Hydroxybenzoic acid
In: Acinetobacter A962

***p*-Acetoxybenzoic acid**
→ *p*-Hydroxybenzoic acid
In: Acinetobacter A962
Rat F753

1-Acetoxy-3,4-benzpyrene
→ 1-Hydroxy-3,4-benzpyrene
Probably in: Rat H290

6-Acetoxy-3,4-benzpyrene
Probably formed from: 6-Hydroxy-3,4-benzpyrene

3-Acetoxy-N-carboxymethyl-2,3-dihydro-2-(*p*-hydroxyphenyl)-1,5-benzothiazepin-4(5*H*)-one
Probably formed from: 3-Acetoxy-N-formylmethyl-2,3-dihydro-2-(*p*-hydroxyphenyl)-1,5-benzothiazepin-4(5*H*)-one

3-Acetoxy-N-carboxymethyl-2,3-dihydro-2-(*p*-methoxyphenyl)-1,5-benzothiazepin-4(5*H*)-one
Probably formed from: 3-Acetoxy-N-formylmethyl-2,3-dihydro-2-(*p*-methoxyphenyl)-1,5-benzothiazepin-4(5*H*)-one

(+)-(2*S*,3*S*)-3-Acetoxy-8-chloro-3,4-dihydro-2-(4-hydroxy-3-methoxyphenyl)-5-(2-(N-methylamino)ethyl)-4-oxo-1,5(2*H*)-benzothiazepine
→ (+)-(2*S*,3*S*)-8-Chloro-3,4-dihydro-3-hydroxy-2-(4-hydroxy-3-methoxyphenyl)-5-(2-(N-methylamino)ethyl)-4-oxo-1,5(2*H*)-benzothiazepine
Probably in: Rat H76

(+)-(2*S*,3*S*)-3-Acetoxy-8-chloro-3,4-dihydro-4-(4-hydroxy-3-methoxyphenyl)-5-(2-(N-methylamino)ethyl)-4-oxo-1,5(2*H*)-benzothiazepine
Probably formed from: N-Norclentiazem

(+)-(2*S*,3*S*)-3-Acetoxy-8-chloro-3,4-dihydro-2-(4-methoxyphenyl)-4-oxo-1,5-benzothiazepin-5(2*H*)-acetic acid
Formed from: Clentiazem

(+)-(2*S*,3*S*)-3-Acetoxy-8-chloro-3,4-dihydro-2-(4-methoxyphenyl)-4-oxo-1,5-benzothiazepin-5(2*H*)-acetonitrile
Formed from: Clentiazem
→ (+)-(2*S*,3*S*)- 8-Chloro-3,4-dihydro-3-hydroxy-2-(4-methoxyphenyl)-4-oxo-1,5-benzothiazepin-5(2*H*)-acetonitrile
Probably in: Rat H76

3′-Acetoxy-7-chloro-1,3-dihydro-5-phenyl-2*H*-1,4-benzodiazepin-2-one
→ 7-Chloro-1,3-dihydro-3-hydroxy-5-phenyl-2*H*-1,4-benzodiazepin-2-one
In: Rat A3036

2-Acetoxy-N-(*p*-chlorophenyl)-3,5-diiodobenzamide
→ N-(*p*-Chlorophenyl)-2-hydroxy-3,5-diiodobenzamide
In: Ascaris A3715
Moniezia A3715
Mouse A3715
Sheep A3715

2-Acetoxy-N-(*p*-chlorophenyl)-5-iodobenzamide
→ N-(*p*-Chlorophenyl)-2-hydroxy-5-iodobenzamide
In: Ascaris A3715
Moniezia A3715
Mouse A3715
Sheep A3715

2-(4-Acetoxy-2,3-dichlorobenzoyl)-5-hydroxythiophene
Formed from: 2-(4-Acetoxy-2,3-dichlorobenzoyl)thiophene

7-Acetoxy-7,8-dihydro-3,4,12-trimethoxy-10,11-methylenedioxy-5*H*-isoindolo[1,2-*b*][3]benzazepin-5-one
→ 7,8-Dihydro-7-hydroxy-3,4,12-trimethoxy-10,11-methylenedioxy-5*H*-isoindolo[1,2-*b*][3]benzazepin-5-one
In: Arthrobacter F1

2-Acetoxy-3,5-diiodobenzoic acid
→ 2-Hydroxy-3,5-diiodobenzoic acid
In: Ascaris A3715
Moniezia A3715
Mouse A3715
Sheep A3715

1-Acetoxy-1-(3,4-dimethoxyphenyl)propan-2-ol
Formed from: 1-(3,4-Dimethoxyphenyl) propane-1,2-diol

2-Acetoxy-1-(3,4-dimethoxyphenyl)propan-1-ol
Formed from: 1-(3,4-Dimethoxyphenyl) propane-1,2-diol

L-5-Acetoxy-2-dimethylamino-4,4-diphenylheptane
See: *l*-α-Acetyl-2-dimethylamino-4,4-diphenyl-5-heptanol

***d-cis*-3-Acetoxy-5-(2-dimethylaminoethyl)-2,3-dihydro-2-(*p*-methoxyphenyl)-1,5-benzothiazepin-4(5*H*)-one**
→ D-*cis*-5-(2-Dimethylaminoethyl)-2,3-dihydro-3-hydroxy-2-(*p*-methoxyphenyl)-1,5-benzothiazepin-4(5*H*)-one
In: Rat A391

3-Acetoxy-N-formylmethyl-2,3-dihydro-2-(*p*-hydroxyphenyl)-1,5-benzothiazepin-4(5*H*)-one
Probably formed from: 3-Acetoxy-N-formylmethyl-2,3-dihydro-2-(*p*-methoxyphenyl)-1,5-benzothiazepin-4(5*H*)-one
→ 3-Acetoxy-N-carboxymethyl-2,3-dihydro-2-(*p*-hydroxyphenyl)-1,5-benzothiazepin-4(5*H*)-one
Probably in: Rat F835

3-Acetoxy-N-formylmethyl-2,3-dihydro-2-(*p*-methoxyphenyl)-1,5-benzothiazepin-4(5*H*)-one
Formed from: Diltiazem
→ 3-Acetoxy-N-carboxymethyl-2,3-dihydro-2-(*p*-methoxyphenyl)-1,5-benzothiazepin-4(5*H*)-one
Probably in: Rat F835
→ 3-Acetoxy-N-formylmethyl-2,3-dihydro-2-(*p*-hydroxyphenyl)-1,5-benzothiazepin-4(5*H*)-one
Probably in: Rat F835
→ N-Formylmethyl-2,3-dihydro-3-hydroxy-2-(*p*-methoxyphenyl)-1,5-benzothiazepin-4(5*H*)-one
Probably in: Rat F835

8-Acetoxy-1-hydroxyanthraquinone
Probably formed from: 8-Acetoxy-1-hydroxy-9-anthrone

8-Acetoxy-1-hydroxy-9-anthrone
Formed from: 1,8-Diacetoxy-9-anthrone
→ 8-Acetoxy-1-hydroxy-9-anthraquinone
Probably in: Man B150
→ 1,8-Dihydroxy-9-anthrone
Probably in: Man B150

N-Acetoxy-2-hydroxypropiophenone
Formed from: *m*-Acetoxybenzaldehyde

N-Acetoxy-2-hydroxyprotriptyline
Probably formed from: 2-Hydroxyprotriptyline

2-Acetoxy-5-iodobenzoic acid
→ 2-Hydroxy-5-iodobenzoic acid
In: Ascaris A3715
Moniezia A3715
Mouse A3715
Sheep A3715

4-(N-Acetoxylamino)biphenyl
Formed from: 4-Hydroxylaminobiphenyl

2-Acetoxylaminofluorene
Formed from: 2-Hydroxylaminofluorene

5-Acetoxylamino-2-methyl-7-methylsulphonylquinoxaline
Probably formed from: 5-Hydroxylamino-2-methyl-7-methylsulphonylquinoxaline

***p*-Acetoxymandelonitrile**
Formed from: *p*-Acetoxybenzaldehyde

8-Acetoxymethylene-6-methyl-9-ergoline
→ Ergotoxine
In: Claviceps A887

2-Acetoxy-N-(3-(*m*-(1-piperidinylmethyl)phenoxy)propyl)acetamide
→ Roxatidine
In: Man D915

1-Acetoxy-2-(2-propen-1-oxy)benzene
Probably formed from: 2-(2-Propen-1-oxy)phenol

N-Acetoxyprotriptyline
Formed from: Protriptyline

***p*-Acetoxysulphinylanisole**
Formed from: *p*-Acetoxythioanisole

7-Acetoxy-Δ^1-tetrahydrocannabinol
Probably formed from: Δ^1-Tetrahydro-7-hydroxycannabinol

***p*-Acetoxythioanisole**
→ *p*-Acetoxysulphinylanisole
In: Horseradish E403

N-Acetyl-S-(N-acetyl-2-afloqualonyl)-L-cysteine
Probably formed from: S-(2-Afloqualonyl)-L-cysteine

N-Acetylafloqualone
Formed from: Afloqualone
→ 6-Acetamido-2-fluoromethyl-3-(*o*-hydroxymethylphenyl)-4(3*H*)-quinazolinone
Probably in: Man C260
Monkey C260
Rat C260
→ 6-Acetamido-2-hydroxymethyl-3-(*o*-tolyl)-4(3*H*)-quinazolinone
Probably in: Monkey C912
→ N-Glycollylafloqualone
Probably in: Monkey C912

N-Acetyl-S-(2-afloqualonyl)-L-cysteine
Formed from: N-Acetyl-S-(2-afloqualonyl)-L-cysteine methyl ester

N-Acetyl-S-(2-afloqualonyl)-L-cysteine methyl ester
→ N-Acetyl-S-(2-afloqualonyl)-L-cysteine
In: Rat C804

5-(5-(3-(N-Acetylalanyl))-5-hydroxy-3-iodocyclohexa-1,4-dienyl lactone)-3-iodotyrosine
Formed from: N-Acetyl-3,5-diiodotyrosine

***l*-α-Acetyl-2-amino-4,4-diphenyl-5-heptanol**
Formed from: *l*-α-Acetyl-2-methylamino-4,4-diphenyl-5-heptanol
→ *l*-α-Acetyl-2-amino-4-(*p*-hydroxyphenyl)-4-phenyl-5-heptanol
Probably in: Monkey A2645, A2979

***l*-α-Acetyl-2-amino-4-(*p*-hydroxyphenyl)-4-phenyl-5-heptanol**
Probably formed from: *l*-α-Acetyl-2-amino-4,4-diphenyl-5-heptanol

1-(2-Acetyl-4-aminophenoxy)-2-hydroxy-3-isopropylaminopropane
Formed from: Acebutolol
→ 1-(4-Acetamido-2-acetylphenoxy)-2-hydroxy-3-isopropylaminopropane
Probably in: Rat A3921

3-(β-Acetyl-α-(*p*-aminophenyl)ethyl)-4-hydroxycoumarin
Formed from: Acenocoumarol
→ 3-(α-(*p*-Acetamidophenyl)-β-acetylethyl)-4-hydroxycoumarin
In: Man A3159

N[5]-Acetylamonafide
Formed from: Amonafide
→ N[5]-Acetyldemethylamonafide
Probably in: Man E887

N[5]-Acetylamonafide-N-oxide
Probably formed from: Amonafide-N-oxide

N-Acetylamphetamine
Formed from: D-Amphetamine

N-Acetylanthranilic acid (*o*-acetamidobenzoic acid)
Formed from: Anthranilic acid
(1*H*)-3-Hydroxy-4-oxoquinaldine
Probably formed from: N-Acetylisatinic acid
→ N-Acetyl-3-hydroxyanthranilic acid
Probably in: Arthrobacter G274
→ Anthranilic acid
In: Ascaris A2036
Moniezia A1755
See also: G274

N-Acetyl-*p*-arsanilic acid (arsacetin)
Formed from: *p*-Arsanilic acid

N-Acetylbatracyclin
Formed from: Batracyclin

N-Acetylbenzidine
Formed from: Benzidine
4,4′-Diacetylbenzidine
→ N-Acetylbenzidine-N′-β-D-glucuronide
In: Man H56

See also C590, H295
→ N-Acetyl-N′-(glutathion-S-yl)benzidine sulphoxide
In: Horseradish K412
→ N-Acetyl-3-hydroxybenzidine
In: Mouse D872
Rat D872
→ N-Acetyl-3′-hydroxybenzidine
In: Mouse D872
Rat D872
→ N-Acetyl-N-hydroxybenzidine
In: Mouse D872
Rat D872
→ N-Acetyl-N′-hydroxybenzidine
In: Mouse D872, J144
Rat D872
→ Benzidine
In: Man H56
Mouse D872
→ 4,4′-Diacetylbenzidine
In: Guinea pig A721
Hamster B129
Man D229, H56, H600
Mouse B129, D872
Rat B129, C590, D229, D872, H600
Salmonella D903

N-Acetylbenzidine-N′-β-D-glucuronide
Formed from: N-Acetylbenzidine

2-Acetylbenzimidazole
Formed from: Thiabendazole

N-Acetyl-2-(3-benzofuryl)ethylamine
Formed from: 2-(3-Benzofuryl)ethylamine

***p*-Acetylbenzoic acid**
Formed from: *p*-Cyanoacetophenone

N-Acetyl-*p*-benzoquinone imine
Formed from: *p*-Acetamidophenol
→ *p*-Acetamidophenol
In: Rat D82, E214
→ S-(*p*-Acetamidophenol-3-yl)glutathione
In: Mouse D82
→ N-Acetyl-*p*-benzosemiquinone imine radical
In: Rat D93
See also D291
→ *p*-Aminophenoxyl radical
In: Rat D798

N-Acetyl-*p*-benzosemiquinone imine radical
Formed from: N-Acetyl-*p*-benzoquinone imine

N-Acetyl-S-(2-benzothiazolyl)-L-cysteine
Formed from: S-Aminocyclohexyl-2-benzothiazolyl sulphenamide
S-Morpholino-2-benzothiazolyl sulphenamide

5-Acetyl-3-(2-(benzoxazol-2-yl)ethyl)-6-methylpyridin-2(1*H*)-one
Probably formed from: 3-(2-(Benzoxazol-2-yl)ethyl)-5-ethyl-6-methylpyridin-2(1*H*)-one
→ 5-Acetyl-3-(2-(4-hydroxybenzoxazol-2-yl)ethyl)-6-methylpyridin-2(1*H*)-one
Probably in: Rat G505
→ 5-Acetyl-3-(2-(6-hydroxybenzoxazol-2-yl)ethyl)-6-methylpyridin-2(1*H*)-one
Probably in: Rat G505
→ 5-Acetyl-3-(2-(7-hydroxybenzoxazol-2-yl)ethyl)-6-methylpyridin-2(1*H*)-one
Probably in: Rat G505

N-Acetyl-S-(2-benzoylthiocarbamoyl)-L-cysteine
Formed from: S-(2-Benzoylthiocarbamoyl)-L-cysteine

O-Acetyl-N-benzoyl-L-tyrosine
Probably formed from: N-benzoyl-L-tyrosine

N-Acetylbenzylamine
Probably formed from: Benzylamine
→ Benzylamine
In: Pseudomonas K790

N-Acetyl-S-benzyl-L-cysteine
Formed from: Benzaldehyde
Benzyl alcohol
S-Benzyl-L-cysteine
Toluene
→ S-Benzyl-L-cysteine
In: Rat C29

N-Acetyl-S-benzyl-L-cysteine sulphoxide
Probably formed from: N-Acetyl-S-benzyl-L-cysteine

N-Acetyl-O^6-benzylguanosine
→ Benzoic acid
In: Mouse G947

4-Acetyl-3-benzyloxy-5-methylbenzoic acid
Formed from: 5-Acetyl-4-benzyloxy-5-methyl-2-pyrone

5-Acetyl-4-benzyloxy-5-methyl-2-pyrone
→ 4-Acetyl-3-benzyloxy-5-methylbenzoic acid
In: Macrophoma E766
N-Acetylbisdemethylchlorpheniramine
Probably formed from: Bisdemethylchlorpheniramine
N-Acetyl-N,N-bisdemethyldoxepin
Probably formed from: Bisdemethyldoxepin
N-Acetylbisdemethyldoxylamine
Probably formed from: Bisdemethyldoxylamine
4-Acetyl-1-(4,4-bis(*p*-fluorophenyl)butyl) piperazine
Formed from: 1-(4,4-bis(*p*-fluorophenyl) butyl)piperazine
N-Acetylbisnorzimetidine
Probably formed from: Bisnorzimetidine
→ N-Acetylbisnorzimetidine-N-oxide
Probably in: Dog B791
Rat B791
N-Acetylbisnorzimetidine-N-oxide
Probably formed from: N-Acetylbisnorzimetidine
9-Acetylboldine
Formed from: Diacetylboldine
N-Acetyl-S-(*o*-bromobenzyl)-L-cysteine
Formed from: *o*-Bromobenzaldehyde
2-Acetyl-6-bromochromone
→ 2-(1-Hydroxyethyl)-6-bromochromone
In: Lactobacillus K425
Saccharomyces K425
Yamadazyma K425
Zygosaccharomyces K425
N-Acetyl-2-bromo-3-(S-(l-cysteinyl))quinol
Probably formed from: 2-Bromo-3-(S-(l-cysteinyl))quinol
N-Acetyl-S-(3-bromo-1,2-dihydro-2-hydroxyphenyl)-L-cysteine
Probably formed from: Bromobenzene-2,3-oxide
N-Acetyl-S-(1-bromo-2-hydroxy-2-phenylethyl)-L-cysteine
Formed from: β-Bromostyrene
N-Acetyl-S-(*p*-bromophenyl)-L-cysteine
Formed from: Bromobenzene
→ S-(*p*-Bromophenyl)-L-cysteine
In: Rat C29
1-(4-((N-Acetyl-N-butyryl)amino)butyl)-4-(6-fluorobenzo[*b*]thien-3-yl)piperazine
Formed from: 3-(4-(4-(6-Fluorobenzo[*b*] thien-3-yl)piperazinyl)butyl)-2,5,5-trimethyl-4-thiazolidinone
→ 1-(4-Acetamidobutyl)-4-(6-fluorobenzo[*b*]thien-3-yl)piperazine
Probably in: Rat J53
→ 1-(4-Butyramidobutyl)-4-(6-fluorobenzo[*b*]thien-3-yl)piperazine
Probably in: Rat J53
N-Acetylcalycotomine
Formed from: N,O-Diacetylcalycotomine
N-Acetylcarbazole
→ Carbazole
In: Pseudomonas J180
3-Acetyl-5-carboxybiphenyl
Formed from: 5-Acetyl-6-phenyl-2-pyrone
6-Acetyl-4-carboxycoumarin
Formed from: 6-Acetyl-4-formylcoumarin
1-Acetylchloramphenicol
Formed from: Chloramphenicol
→ 1,3-Diacetylchloramphenicol
Probably in: Duck G945
3-Acetylchloramphenicol
Formed from: Chloramphenicol
N-Acetyl-*p*-chloroamphetamine
Formed from: *p*-Chloroamphetamine
N-Acetyl-2-(5-chloro-3-benzothienyl)ethylamine
Formed from: 2-(5-Chloro-3-benzothienyl) ethylamine
N-Acetyl-S-(*o*-chlorobenzyl)-L-cysteine
Formed from: *o*-Chlorobenzaldehyde
o-Chlorobenzyl alcohol
o-Chlorobenzyl sulphate
Probably formed from: S-(*o*-Chlorobenzyl)-L-cysteine
N-Acetyl-S-(*m*-chlorobenzyl)-L-cysteine
Formed from: *m*-Chlorobenzaldehyde
Probably formed from: *m*-Chlorobenzyl alcohol
N-Acetyl-S-(*p*-chlorobenzyl)-L-cysteine
Formed from: *p*-Chlorobenzyl sulphate
Probably formed from: *p*-Chlorobenzyl alcohol
4′-Acetyl-4-chlorobiphenyl
Formed from: 4-Chloro-4′-isopropylbiphenyl

→ 4′-Acetyl-4-chloro-3-hydroxybiphenyl
Probably in: Rat A2531

N-Acetyl-2-chloro-6-(S-(L-cysteinyl))-5-hydroxybenzamide
Formed from: 2,6-Dichlorobenzamide
→ 2-Chloro-5-hydroxy-6-methylthiobenzamide
Probably in: Rat F307

4′-Acetyl-4-chloro-3-hydroxybiphenyl
Probably formed from: 4′-Acetyl-4-chlorobiphenyl

4-Acetyl-3-chloro-5-methylbenzoic acid
Formed from: 5-Acetyl-4-chloro-6-methyl-2-pyrone

5-Acetyl-4-chloro-6-methyl-2-pyrone
→ 4-Acetyl-3-chloro-5-methylbenzoic acid
In: Macrophoma E766

N-Acetyl-S-(5-chloro-2-nitrophenyl)-L-cysteine
Probably formed from: S-(5-chloro-2-nitrophenyl)-L-cysteine

N-Acetyl-S-(*p*-chlorophenyl)-L-cysteine
Formed from: Chlorobenzene

N-Acetyl-*m*-chlorophenyl-β-D-glucosaminide
→ *m*-Chlorophenol
In: Beef A1940

N-Acetyl-*p*-chlorophenyl-β-D-glucosaminide
→ *p*-Chlorophenol
In: Beef A1940

2-Acetylchromone
→ 2-(1-Hydroxyethyl)chromone
In: Aspergillus K425
Beauvaria K425
Cunninghamella K425
Geotrichum K425
Lactobacillus K425
Saccharomyces K425
Yamadazyma K425
Zygosaccharomyces K425

N-Acetylchryscandin
Formed from: Chryscandin

(—)-Acetylcorynoloxine
Formed from: (—)-Corynoline

3-Acetylcoumarin
→ 3-(1-Hydroxyethyl)coumarin
In: Rat H416
→ 4-(*o*-Hydroxyphenyl)-2-butanol
In: Rat H416
→ Tetrahydro-2-hydroxy-2-methylbenzopyran
In: Rat H416
Saccharomyces H416

N-Acetyl-S-(3-coumarinyl)-L-cysteine
Formed from: Coumarin

N-Acetyl-S-(4-cyanoanilinomethyl)-L-cysteine
Formed from: 4-Cyano-N,N-dimethylaniline

N-Acetyl-S-(2-(*p*-cyanoanilino)-2-oxoethyl)-L-cysteine
Probably formed from: *p*-Acetamidobenzonitrile

N-Acetyl-S-(1-cyano-2-phenylethyl)-L-cysteine
Formed from: Cinnamonitrile

2-(N-Acetyl-S-(L-cysteinyl)-4-aminophenol
Probably formed from: 4-Amino-2-(S-glutathionyl)phenol

N-Acetyl-S-(L-cysteinyl)-3-bromobenzene
Probably formed from: 4-Bromo-2-(S-glutathionyl)-1,2-dihydro-1-hydroxybenzene

N-Acetyl-S-(L-cysteinyl)-4-bromobenzene
Probably formed from: 5-Bromo-2-(S-glutathionyl)-1,2-dihydro-1-hydroxybenzene

2-(N-Acetyl-S-(L-cysteinyl))-4-bromo-1,2-dihydro-1-hydroxybenzene
Probably formed from: 4-Bromo-2-(S-glutathionyl)-1,2-dihydro-1-hydroxybenzene

2-(N-Acetyl-S-(L-cysteinyl)-5-bromo-1,2-dihydro-1-hydroxybenzene
Probably formed from: 5-Bromo-2-(S-glutathionyl)-1,2-dihydro-1-hydroxybenzene

6-(N-Acetyl-S-(L-cysteinyl))-2-bromoquinol
Formed from: 2-Bromo-6-(S-(L-cysteinyl)) quinol
→ 2-Bromo-6-(S-(L-cysteinyl)) quinol
In: Guinea pig H467
Hamster H467
Mouse H467
Rat H467

10-(N-Acetyl-S-(L-cysteinyl))celiptium
Probably formed from: 10-(S-(L-Cysteinyl)) celiptium

2-(N-Acetyl-S-(L-cysteinyl))-6-chloro-3-hydroxybenzonitrile
Probably formed from: 6-Chloro-2-(S-(L-cysteinyl))-3-hydroxybenzonitrile

2-(N-Acetyl-S-(L-cysteinyl))-1,4-dichlorobenzene
Formed from: *p*-Dichlorobenzene

2-(N-Acetyl-S-(L-cysteinyl))-1,4-dichloro-2,3-dihydro-3-hydroxybenzene
Formed from: *p*-Dichlorobenzene

***trans*-2-(N-Acetyl-S-(L-cysteinyl))-3,5-dichloro-1,2-dihydro-1-hydroxybenzene**
Probably formed from: *trans*-3,5-Dichloro-2-(S-(L-cysteinyl))-1,2-dihydro-1-hydroxybenzene

3-(N-Acetyl-S-(L-cysteinyl))-1,2-dichloro-3,4-dihydro-4-hydroxybenzene
Probably formed from: 1,2-Dichloro-3-(S-glutathionyl)-3,4-dihydro-4-hydroxybenzene

4-(N-Acetyl-S-(L-cysteinyl))-1,2-dichloro-3,4-dihydro-3-hydroxybenzene
Probably formed from: 1,2-Dichloro-4-(S-glutathionyl)-3,4-dihydro-3-hydroxybenzene

5-(N-Acetyl-S-(L-cysteinyl))-1,2-dichloro-4,5-dihydro-4-hydroxybenzene
Probably formed from: 1,2-Dichloro-5-(S-glutathionyl)-4,5-dihydro-4-hydroxybenzene

3-((3-(N-Acetyl-S-(L-cysteinyl))-3,4-dihydro-4-hydroxy-5-trifluoromethyl-2-benzothiazolyl)methyl)-3,4-dihydro-4-oxo-1-phthalazineacetic acid
Probably formed from: 3-((3-(L-Cystein-S-yl))-3,4-dihydro-4-hydroxy-5-trifluoromethyl-2-benzothiazolyl)methyl)-3,4-dihydro-4-oxo-1-phthalazineacetic acid
→ 3-((3-(N-Acetyl-S-(L-cysteinyl))-4-hydroxy-5-trifluoromethyl-2-benzothiazolyl)methyl)-3,4-dihydro-4-oxo-1-phthalazineacetic acid
Probably in: Rat J457

3-((4-(N-Acetyl-S-(L-cysteinyl))-3,4-dihydro-3-hydroxy-5-trifluoromethyl-2-benzothiazolyl)methyl)-3,4-dihydro-4-oxo-1-phthalazineacetic acid
Probably formed from: 3-((4-(S-(L-Cysteinyl))-3,4-dihydro-3-hydroxy-5-trifluoromethyl-2-benzothiazolyl)methyl)-3,4-dihydro-4-oxo-1-phthalazineacetic acid

3′-(N-Acetyl-S-(L-cysteinyl))-2′,3′-dihydro-1′-oxosafrole
Probably formed from: 3′-(S-Glutathionyl)-2′,3′-dihydro-1′-oxosafrole

1-(N-Acetyl-S-(L-cysteinyl))-1,7-dihydroxy-2-hydroxymethyl-5-(3-hydroxypropyl-2,4,6-trimethylindane
Formed from: Illudin S

2-(N-Acetyl-S-(L-cysteinyl))-N-(2-ethyl-6-methylphenyl)-N-(2-methoxy-1-methylethyl) acetamide
Formed from: Metolachlor

3-((3-(N-Acetyl-S-(L-cysteinyl))-4-hydroxy-5-trifluoromethyl-2-benzothiazolyl)methyl)-3,4-dihydro-4-oxo-1-phthalazineacetic acid
Probably formed from: 3-((3-(N-Acetyl-L-cystein-S-yl))-3,4-dihydro-4-hydroxy-5-trifluoromethyl-2-benzothiazolyl)methyl)-3,4-dihydro-4-oxo-1-phthalazineacetic acid

2-(N-Acetyl-S-(L-cysteinyl))-N-isopropyl-N-phenylacetamide
Formed from: 2-(S-(l-Cysteinyl))-N-isopropyl-N-phenylacetamide
→ 2-(N-Acetyl-S-(L-cysteinyl))-N-isopropyl-N-phenylacetamide sulphoxide
In: Rat C24
See also: C402, C407, C421, E790, F505

2-(N-Acetyl-S-(L-cysteinyl))-N-isopropyl-N-phenylacetamide sulphoxide
Formed from: 2-(N-Acetyl-S-(L-cysteinyl))-N-isopropyl-N-phenylacetamide

(4-(2-(N-Acetyl-S-(L-cysteinyl))methylenebutyryl)-2,3-dichlorophenoxy)acetic acid
Probably formed from: (2,3-Dichloro-4-(2-(S-glutathionyl)methylenebutyryl)phenoxy) acetic acid

3-((N-Acetyl-S-(L-cysteinyl))methyl)indole
Formed from: Skatole

4-(N-Acetyl-S-(L-cysteinyl))-4-phenylbutan-2-one
Probably formed from: 4-(S-Glutathionyl)-4-phenylbutan-2-one

7-(N-Acetyl-S-(L-cysteinyl))sirdalud
Formed from: Sirdalud

(N-Acetyl-S-(L-cysteinyl))-2,3,5,6-tetrachlorobenzene
Probably formed from: S-(2,3,5,6-tetrachlorophenyl)cysteinylglycine
→ (N-Acetyl-S-(L-cysteinyl))-2,3,5,6-tetrachlorobenzene sulphoxide
Probably in: Rat H512

(N-Acetyl-S-(L-cysteinyl))-2,3,5,6-tetrachlorobenzene sulphone
Probably formed from: (N-Acetyl-S-(L-cysteinyl))-2,3,5,6-tetrachlorobenzene sulphoxide

(N-Acetyl-S-(L-cysteinyl))-2,3,5,6-tetrachlorobenzene sulphoxide
Probably formed from: (N-Acetyl-S-(L-cysteinyl))-2,3,5,6-tetrachlorobenzene
→ (N-Acetyl-S-(L-cysteinyl))-2,3,5,6-tetrachlorobenzene sulphone
Probably in: Rat H512

***p*-(N-Acetyl-S-(L-cysteinyl))toluene**
Formed from: Toluene

4-(N-Acetyl-S-(L-cysteinyl))-2,4′,5-trichlorobiphenyl
→ 2,4′,5-Trichloro-4-methylsulphonylbiphenyl
In: Rat C666

N-Acetyl-α,β-dehydrotryptophan
Formed from: N-Acetyl-L-tryptophan

N-Acetyl-(*Z*)-α,β-dehydrotryptophanamide
Formed from: N-Acetyl-L-tryptophanamide

N-Acetyl-(*Z*)-α,β-dehydrotryptophanyl-L-phenylalanine
Formed from: N-Acetyl-L-tryptophanyl-L-phenylalanine

N^5-Acetyldemethylamonafide
Probably formed from: N^5-Acetylamonafide

N-Acetyldemethyldextromethorphan
Probably formed from: N-Demethyldextromethorphan

O-Acetyl-O-demethyldextromethorphan
Probably formed from: O-Demethyldextromethorphan

N-Acetyl-N-demethyldoxepin
Probably formed from: N-Demethyldoxepin

N-Acetyldemethyldoxylamine
Probably formed from: Demethyldoxylamine

N-Acetyl-O-demethylpuromycin
Formed from: O-Demethylpuromycin
→ N-Acetylpuromycin
In: Streptomyces D879

N-Acetyl-2,6-diamino-3-(phenylazo)pyridine
Formed from: Phenazopyridine

N-Acetyl-3,5-dibromobenzoquinone imine
Formed from: 4-Acetamido-2,6-dibromophenol

N-Acetyl-3,5-dibromotyramine
→ 4-(5-(2-Acetamidoethyl-3-bromo-5-hydroxy-2-oxocyclohexa-1,4-dienoxy)-N-acetyl-3,5-dibromophenethylamine
In: Horseradish J372

N-Acetyl-3,5-dibromotyrosine
→ 3,4′-Bis(3-(N-acetylalanyl))3′,5,5′-tribromo-4-hydroxydiphenyl ether
In: Horseradish J372

N-Acetyl-S-(3,5-di-*tert*-butyl-4-hydroxybenzyl)-L-cysteine
Probably formed from: 2,6-Di-*tert*-butyl-4-methylphenol

N-Acetyl-3,3′-dichlorobenzidine
Probably formed from: 3,3′-Dichlorobenzidine
→ N,N′-Diacetyl-3,3′-dichlorobenzidine
Probably in: Rat C541

N-Acetyl-3,5-dichlorobenzoquinone imine
Formed from: 4-Acetamido-2,6-dichlorophenol

N-Acetyl-S-(2,4-dichlorophenyl)-L-cysteine
Probably formed from: S-(2,4-Dichlorophenyl)-L-cysteine

N-Acetyl-S-(3,5-dichlorophenyl)-L-cysteine
Probably formed from: S-(3,5-Dichlorophenyl)-L-cysteine

rel-1-Acetyl-4-(4-((2*R*,4*S*)-2-(2,4-dichlorophenyl)-2-(1*H*-imidazol-1-ylmethyl)-1,3-dioxolan-4-yl)methoxy)phenyl)piperazine
See: Ketoconazole

N-Acetyl-3,5-dichlorotyrosine
→ 3,4′-Bis(3-(N-acetylalanyl))-3′,5,5′-trichloro-4-hydroxydiphenyl ether
In: Horseradish J372

N-Acetyl-3,5-dichlorotyrosine methyl ester
→ 3,4′-Bis(3-(N-acetylalanyl))-3′,5,5′-trichloro-4-hydroxydiphenyl ether dimethyl ester
In: Horseradish J372

N-Acetyl-α,β-didehydro-L-tryptophanamide
Formed from: N-Acetyl-L-tryptophanamide
N-Acetyl-3,5-diethylbenzoquinone imine
Formed from: 4-Acetamido-2,6-diethylphenol
N-Acetyl-3,5-difluorobenzoquinone imine
Formed from: 4-Acetamido-2,6-difluorophenol
N-Acetyl-13(*S*)-dihydrodaunorubicin
Formed from: Daunorubicin
***cis*-N-Acetyl-2,3-dihydro-2,3-dihydroxy-L-phenylalanine**
Formed from: N-Acetyl-L-phenylalanine
N-Acetyl-S-(1,2-dihydro-1-hydroxynaphth-2-yl)-L-cysteine
Formed from: *trans*-S-(1,2-Dihydro-1-hydroxynaphth-2-yl)glutathione
Probably formed from: S-(1,2-Dihydro-1-hydroxynaphth-2-yl)-L-cysteine
→ α-Naphthol
In: Rat D694
→ β-Naphthol
In: Rat D694
N-Acetyl-S-(1,2-dihydro-2-hydroxynaphth-1-yl)-L-cysteine
Formed from: S-(1,2-Dihydro-2-hydroxy-1-naphthyl)glutathione
N-Acetyl-S-(9,10-dihydro-9-hydroxyphenanthr-10-yl)-L-cysteine
Probably formed from: S-(9,10-Dihydro-9-hydroxyphenanthr-10-yl)-L-cysteine
N-Acetyl-2,3-dihydro-1*H*-indole
See: N-Acetylindoline
3-Acetyl-4-(3,4-dihydroxy-5-nitrophenyl)butan-2-one-3-sulphonic acid
Probably formed from: 3-(3,4-Dihydroxy-5-nitrobenzylidene)-2,4-pentanedione
2-Acetyl-1,8-dihydroxy-3-(2-oxopropyl) naphthalene
→ Islandicin
In: Penicillium C86
→ Skyrin
In: Penicillium C86
N-Acetyl-3,4-dihydroxyphenethylamine
Formed from: N-Acetyltyramine
3,4-Dihydroxyphenethylamine
→ 2-(2-(N-Acetamido)ethyl)-3′-ethyl-4,4′,5,6′-tetrahydroxybiphenyl
In: Locust G346
→ 4-(2-Acetamidoethyl)-*o*-quinone
In: Drosophila G620
Sarcophaga F90, F386, F411, G184
See also A2556, B897, E407, E975, F964, G123
→ N-Acetyldopamine quinone methide
In: Manduca F68
Periplaneta F68
Sarcophaga F68
→ N-Acetyl-3-hydroxy-4-methoxyphenethylamine
With catechol O-methyltransferase A597
→ N-Acetyl-4-hydroxyphenethylamin-3-yl phosphate
In: Periplaneta A1239
→ N-Acetyl-4-hydroxyphenethylamin-3-yl sulphate
In: Periplaneta A1239
See also E492, E493, E621, F751
→ N-Acetyl-3-methoxytyramine
With catechol O-methyltransferase A597
→ N-Acetylnoradrenaline
In: Drosophila G620
Manduca D797, E975, F68,
Periplaneta F68
Sarcophaga C692, F68, F411
→ 3,4-Dihydroxyphenethylamine
In: Pseudomonas K790
N-Acetyl-3,4-dihydroxy-L-phenylalanine
→ δ-(3-(N-Acetyl)alanyl)-α-hydroxy-cis, cis-muconic semialdehyde
In: Pseudomonas A1090
N-Acetyl-S-(2,5-dihydroxyphenyl)-L-cysteine
Formed from: Bromobenzene
o-Bromophenol
m-Bromophenol
p-Bromophenol
Bromoquinol
S-(2,5-Dihydroxyphenyl)-L-cysteine
→ S-(2,5-Dihydroxyphenyl)-L-cysteine
In: Mouse J321
Rat J321
N-Acetyl-5,6-dihydroxytryptamine
Formed from: 5,6-Dihydroxytryptamine

N-Acetyl-3,5-diiodobenzoquinone methide
Formed from: 4-Acetamido-2,6-diiodophenol
N-Acetyl-3,5-diiodotyrosine
→ 5-(5-(3-(N-acetylalanyl))-5-hydroxy-3-iodocyclohexa-1,4-dienyl lactone)-3-iodo-tyrosine
In: Horseradish J372
→ 3,3′-Bis(3-(N-acetylalanyl))-6,6′-dihydroxy-5,5′-diiodobiphenyl
In: Horseradish J372
N-Acetyl-3,5-diisopropylbenzoquinone imine
Formed from: 4-Acetamido-2,6-diisopropylphenol
N-Acetyl-3,3′-dimethoxybenzidine
Probably formed from: 3,3′-Dimethoxybenzidine
→ N,N-Diacetyl-3,3′-dimethoxybenzidine
Probably in: Rat C541, C623
O-Acetyl-3,4-dimethoxymandelonitrile
Formed from: 3,4-Dimethoxymandelonitrile
N-Acetyl-2,5-dimethoxy-4-propoxyamphetamine
Formed from: 2,5-Dimethoxy-4-propoxyamphetamine
***l*-α-Acetyl-2-dimethylamino-4,4-diphenyl-5-heptanol** (*l*-α-acetylmethadol, *l*-5-acetoxy-2-dimethylamino-4,4-diphenylheptane)
→ *l*-α-Acetyl-2-methylamino-4,4-diphenyl-5-heptanol
In: Man A1491, A3171, H859
Monkey A2665, A2979, A3171
Rat A2088, A2421, A2900, A3171, A3180, A3181, A3900
→ 2-Dimethylamino-4,4-diphenyl-5-heptanol
In: Dog A3171
Monkey A2979
Rat A2088, A3180, A3900
→ 1,5-Dimethyl-3,3-diphenyl-2-ethylidenepyrrolidone
In: Rat A2900
N-Acetyl-2,6-dimethyl-*p*-benzoquinone imine
Formed from: 4′-Acetamido-2′,6′-dimethylphenol
→ N-Acetyl-2,6-dimethyl-*p*-benzosemiquinone imine
In: Rat D93
N-Acetyl-3,4-dimethyl-*p*-benzoquinone imine
Formed from: 4-Hydroxy-3,5-dimethylacetanilide
N-Acetyl-3,5-dimethyl-*p*-benzoquinone imine
Formed from: 4-Acetamido-2,6-dimethylphenol
→ N-Acetyl-3,5-dimethyl-*p*-benzosemiquinone imine
In: Rat D93
→ 4-Amino-2,6-dimethylphenoxyl radical
In: Rat D798
→ 3,5-Dimethyl-*p*-benzoquinone imine
In: Rat F40
N-Acetyl-2,6-dimethyl-*p*-benzosemiquinone imine
Formed from: N-Acetyl-2,6-dimethyl-p-benzoquinone imine
N-Acetyl-3,5-dimethyl-*p*-benzosemiquinone imine
Formed from: N-Acetyl-3,5-dimethyl-*p*-benzoquinone imine
N-Acetyl-1,3-dimethylcarbazole
→ 1,3-Dimethylcarbazole
In: Pseudomonas J180
N-Acetyl-S-(2,4-dinitrophenyl)-L-cysteine
Probably formed from: S-(2,4-Dinitrophenyl)-L-cysteine
N-Acetyl-2,4-dinitrophenyl-β-D-glucosaminide
→ 2,4-Dinitrophenol
In Beef A1940
N-Acetyl-3,3-diphenylpropylamine
Formed from: 3,3-Diphenylpropylamine
N-Acetyldopamine quinone methide
Formed from: 4-(2-Acetamidoethyl)-*o*-quinone
N-Acetyl-3,4-dihydroxyphenethylamine
→ α,β-Dehydro-3,4-dihydroxyphenethylamine
In: Sarcophaga F386
See also F90, F411, G184
O-Acetylephedrine
→ Ephedrine
In: Mouse A3540
N-Acetyl-S-(1-(3-ethenylphenyl)-2-hydroxyethyl)-L-cysteine
Formed from: 1,3-Divinylbenzene

N-Acetyl-S-(1-(4-ethenylphenyl)-2-hydroxyethyl)-L-cysteine
Formed from: 1,4-Divinylbenzene
N-Acetyl-S-(2-(2-ethenylphenyl)-2-hydroxyethyl)-L-cysteine
Formed from: 1,2-Divinylbenzene
N-Acetyl-S-(2-(3-ethenylphenyl)-2-hydroxyethyl)-L-cysteine
Formed from: 1,3-Divinylbenzene
N-Acetyl-S-(2-(4-ethenylphenyl)-2-hydroxyethyl)-L-cysteine
Formed from: 1,4-Divinylbenzene
8-Acetylethidium
Formed from: Ethidium
→ 3,8-Diacetylethidium
Probably in: Rat J573
N-Acetyl-*p*-ethoxyamphetamine
Formed from: *p*-Ethoxyamphetamine
4-Acetyl-3-ethoxy-5-methylbenzoic acid
Formed from: 5-Acetyl-4-ethoxy-6-methyl-2-pyrone
5-Acetyl-4-ethoxy-6-methyl-2-pyrone
→ 4-Acetyl-3-ethoxy-5-methylbenzoic acid
In: Macrophora E766
N-Acetyl-*p*-ethoxyphenyl nitroxide radical
Probably formed from: *p*-Ethoxyacetanilide
N-Acetyl-2-(5-ethyl-3-benzothienyl)ethylamine
Formed from: 2-(5-Ethyl-3-benzothienyl) ethylamine
N-Acetyl-2-(7-ethyl-1,2,3,4-tetrahydronaphth-1-yl)ethylamine
Formed from: 2-(7-Ethyl-1,2,3,4-tetrahydro-naphth-1-yl)ethylamine
N-Acetyl-2-fluorenyl nitroxide radical
Probably formed from: 2-Acetamidoflurorene
N-Acetyl-S-(*o*-fluorobenzyl)-L-cysteine
Formed from: *o*-Flurorobenzaldehyde
N-Acetyl-6-fluorotryptamine
Formed from: 6-Fluorotryptamine
6-Acetyl-4-formylcoumarin
→ 6-Acetyl-4-carboxycoumarin
In: Man J24
N$^{\gamma}$-Acetyl-N^{2}-formyl-5-methoxykynurenamine
Formed from: Melatonin
→ N$^{\gamma}$-Acetyl-5-methoxykynurenamine
In: Rabbit A1212

N′-Acetyl-N-formyl-5-methyl-L-kynurenine
Formed from: N-Acetyl-5-methoxy-L-tryptophan
N-Acetyl-S-(1-(4-formylphenyl)-2-hydroxyethyl)-L-cysteine
Formed from: 1,4-Divinylbenzene
6-O-Acetylgalactosyl(β1-4)xylyl(β)-4-methylumbelliferone
→ Galactosyl(β1-3)-6-O-acetylgalactosyl (β1-4)xylyl(β)-4-methylumbelliferone
In: Escherichia J399
N-(β-Acetylglucosaminido)delaviridine
Formed from: Delaviridine
N-Acetyl-N′-(glutathion-S-yl)benzidine sulphoxide
Formed from: N-Acetylbenzidine
1-Acetyl-1,2,3,3a,8,8a-hexahydro-8a-hydroxy-5-methoxypyrrolo[2,3-*b*]indole
Formed from: Melatonin
N-Acetylhistidyl-alanyl-alanyl-alanyl-*p*-nitroanilide
→ *l*-Alanyl-*p*-nitroanilide
In: Rat C693
2-S-(l-(N-Acetylhomocysteinyl))-N-isopropyl-N-phenylacetamide
Formed from: 2-S-(*l*-Homocysteinyl)-N-isopropyl-N-phenylacetamide
4-(2-Acetylhydrazino)phthalazinone
Formed from: Hydralazine
N-Acetyl-3-hydroxyanthranilic acid
Probably formed from: N-Acetylanthranilic acid
N-Acetyl-3-hydroxybenzidine
Formed from: N-Acetylbenzidine
Probably formed from: 4,4′-Diacetyl-3-hydroxybenzidine
N-Acetyl-3′-hydroxybenzidine
Formed from: N-Acetylbenzidine
N-Acetyl-N-hydroxybenzidine
Formed from: N-Acetylbenzidine
→ 4,4′-Diacetyl-N-hydroxybenzidine
In: Mouse D872
Rat D872
N-Acetyl-N′-hydroxybenzidine
Formed from: N-Acetylbenzidine
N-Acetyl-N′-hydroxybenzidine-N-β-D-glucuronide
→ N-Acetyl-N′-hydroxybenzidine
In: Escherichia J766

Man J766

N-Acetyl-N′-hydroxybenzidine-O-β-D-glucuronide
→ 4,4′-Diacetyl-N′-hydroxybenzidine
In: Mouse D872
Rat D872

5-Acetyl-3-(2-(4-hydroxybenzoxazol-2-yl)ethyl)-6-methylpyridin-2(1*H*)-one
Probably formed from: 5-Acetyl-3-(2-(benzoxazol-2-yl)ethyl)-6-methylpyridin-2(1*H*)-one
→ 5-Acetyl-3-(2-(4-hydroxybenzoxazol-2-yl)ethyl)-6-methylpyridin-2(1*H*)-one O-sulphate
Probably in: Rat G505

5-Acetyl-3-(2-(6-hydroxybenzoxazol-2-yl)ethyl)-6-methylpyridin-2(1*H*)-one
Probably formed from: 5-Acetyl-3-(2-(benzoxazol-2-yl)ethyl)-6-methylpyridin-2(1*H*)-one
→ 5-Acetyl-3-(2-(6-hydroxybenzoxazol-2-yl)ethyl)-6-methylpyridin-2(1*H*)-one O-sulphate
Probably in: Rat G505

5-Acetyl-3-(2-(7-hydroxybenzoxazol-2-yl)ethyl)-6-methylpyridin-2(1*H*)-one
Probably formed from: 5-Acetyl-3-(2-(benzoxazol-2-yl)ethyl)-6-methylpyridin-2(1*H*)-one
→ 5-Acetyl-3-(2-(7-hydroxybenzoxazol-2-yl)ethyl)-6-methylpyridin-2(1*H*)-one O-sulphate
Probably in: Rat G505

5-Acetyl-3-(2-(4-hydroxybenzoxazol-2-yl)ethyl)-6-methylpyridin-2(1*H*)-one O-sulphate
Probably formed from: 5-Acetyl-3-(2-(4-hydroxybenzoxazol-2-yl)ethyl)-6-methylpyridin-2(1*H*)-one

5-Acetyl-3-(2-(6-hydroxybenzoxazol-2-yl)ethyl)-6-methylpyridin-2(1*H*)-one O-sulphate
Probably formed from: 5-Acetyl-3-(2-(6-hydroxybenzoxazol-2-yl)ethyl)-6-methylpyridin-2(1*H*)-one

5-Acetyl-3-(2-(7-hydroxybenzoxazol-2-yl)ethyl)-6-methylpyridin-2(1*H*)-one O-sulphate
Probably formed from: 5-Acetyl-3-(2-(7-hydroxybenzoxazol-2-yl)ethyl)-6-methylpyridin-2(1*H*)-one

N-Acetyl-7-hydroxydemethylchlorpromazine
Probably formed from: Demethyl-7-hydroxychlorpromazine

6-Acetyl-7-hydroxy-2-hydroxymethyl-2-methyl-2*H*-benzo[*b*]pyran
Probably formed from: 6-Acetyl-2-hydroxymethyl-7-methoxy-2-methyl-2*H*-benzo[*b*] pyran

N-Acetyl-4(*S*)-hydroxy-4-(*p*-hydroxyphenyl)-L-threonine
Formed from: Caspofungin
→ 4(*S*)-Hydroxy-4-(*p*-hydroxyphenyl)-L-threonine
Probably in: Man K321

N-Acetyl-S-(1-hydroxyindan-2-yl)-L-cysteine
Probably formed from: S-(1-Hydroxyindan-2-yl)glutathione

N-Acetyl-3-hydroxyindoline
Formed from: N-Acetylindoline

2-Acetyl-7-(2-hydroxy-3-isopropylamino) propoxybenzofuran
See: Befunolol

N-Acetyl-3-hydroxy-3′-methoxybenzidine
Probably formed from: 3-Hydroxy-3′-methoxybenzidine

N-Acetyl-3-hydroxy-4-methoxyphenethylamine
Formed from: N-Acetyl-3,4-dihydroxyphenethylamine

N-Acetyl-3-hydroxy-4-methoxy-L-phenylalanine
Formed from: 3-Hydroxy-4-methoxy-L-phenylalanine
→ 3-Hydroxy-4-methoxy-L-phenylalanine
Probably in: Rat A801

N-Acetyl-S-(2-hydroxy-1-(*p*-methoxyphenyl)-1-propyl)-L-cysteine
Probably formed from: Anethole oxide

N^4-Acetyl-N^1-(2-hydroxy-6-methoxy-4-pyrimidinyl)sulphanilamide
Probably formed from: N^1-(2-Hydroxy-6-methoxy-4-pyrimidinyl)sulphanilamide

N^4-Acetyl-N^1-(6-hydroxy-2-methoxy-4-pyrimidinyl)sulphanilamide
Probably formed from: N^1-(6-Hydroxy-2-methoxy-4-pyrimidinyl)sulphanilamide

N-Acetyl-S-(1-(4-ethenylphenyl)-2-hydroxyethyl)-L-cysteine
Formed from: 1,4-Divinylbenzene

N-Acetyl-S-(2-(2-ethenylphenyl)-2-hydroxyethyl)-L-cysteine
Formed from: 1,2-Divinylbenzene

N-Acetyl-S-(2-(3-ethenylphenyl)-2-hydroxyethyl)-L-cysteine
Formed from: 1,3-Divinylbenzene

N-Acetyl-S-(2-(4-ethenylphenyl)-2-hydroxyethyl)-L-cysteine
Formed from: 1,4-Divinylbenzene

8-Acetylethidium
Formed from: Ethidium
→ 3,8-Diacetylethidium
Probably in: Rat J573

N-Acetyl-*p*-ethoxyamphetamine
Formed from: *p*-Ethoxyamphetamine

4-Acetyl-3-ethoxy-5-methylbenzoic acid
Formed from: 5-Acetyl-4-ethoxy-6-methyl-2-pyrone

5-Acetyl-4-ethoxy-6-methyl-2-pyrone
→ 4-Acetyl-3-ethoxy-5-methylbenzoic acid
In: Macrophora E766

N-Acetyl-*p*-ethoxyphenyl nitroxide radical
Probably formed from: *p*-Ethoxyacetanilide

N-Acetyl-2-(5-ethyl-3-benzothienyl)ethylamine
Formed from: 2-(5-Ethyl-3-benzothienyl) ethylamine

N-Acetyl-2-(7-ethyl-1,2,3,4-tetrahydronaphth-1-yl)ethylamine
Formed from: 2-(7-Ethyl-1,2,3,4-tetrahydro-naphth-1-yl)ethylamine

N-Acetyl-2-fluorenyl nitroxide radical
Probably formed from: 2-Acetamidoflurorene

N-Acetyl-S-(*o*-fluorobenzyl)-L-cysteine
Formed from: *o*-Flurorobenzaldehyde

N-Acetyl-6-fluorotryptamine
Formed from: 6-Fluorotryptamine

6-Acetyl-4-formylcoumarin
→ 6-Acetyl-4-carboxycoumarin
In: Man J24

N$^{\gamma}$-Acetyl-N^{2}-formyl-5-methoxykynurenamine
Formed from: Melatonin
→ N$^{\gamma}$-Acetyl-5-methoxykynurenamine
In: Rabbit A1212

N′-Acetyl-N-formyl-5-methyl-L-kynurenine
Formed from: N-Acetyl-5-methoxy-L-tryptophan

N-Acetyl-S-(1-(4-formylphenyl)-2-hydroxyethyl)-L-cysteine
Formed from: 1,4-Divinylbenzene

6-O-Acetylgalactosyl(β1-4)xylyl(β)-4-methylumbelliferone
→ Galactosyl(β1-3)-6-O-acetylgalactosyl (β1-4)xylyl(β)-4-methylumbelliferone
In: Escherichia J399

N-(β-Acetylglucosaminido)delaviridine
Formed from: Delaviridine

N-Acetyl-N′-(glutathion-S-yl)benzidine sulphoxide
Formed from: N-Acetylbenzidine

1-Acetyl-1,2,3,3a,8,8a-hexahydro-8a-hydroxy-5-methoxypyrrolo[2,3-*b*]indole
Formed from: Melatonin

N-Acetylhistidyl-alanyl-alanyl-alanyl-*p*-nitroanilide
→ *l*-Alanyl-*p*-nitroanilide
In: Rat C693

2-S-(l-(N-Acetylhomocysteinyl))-N-isopropyl-N-phenylacetamide
Formed from: 2-S-(*l*-Homocysteinyl)-N-isopropyl-N-phenylacetamide

4-(2-Acetylhydrazino)phthalazinone
Formed from: Hydralazine

N-Acetyl-3-hydroxyanthranilic acid
Probably formed from: N-Acetylanthranilic acid

N-Acetyl-3-hydroxybenzidine
Formed from: N-Acetylbenzidine
Probably formed from: 4,4′-Diacetyl-3-hydroxybenzidine

N-Acetyl-3′-hydroxybenzidine
Formed from: N-Acetylbenzidine

N-Acetyl-N-hydroxybenzidine
Formed from: N-Acetylbenzidine
→ 4,4′-Diacetyl-N-hydroxybenzidine
In: Mouse D872
Rat D872

N-Acetyl-N′-hydroxybenzidine
Formed from: N-Acetylbenzidine

N-Acetyl-N′-hydroxybenzidine-N-β-D-glucuronide
→ N-Acetyl-N′-hydroxybenzidine
In: Escherichia J766

Man J766

N-Acetyl-N′-hydroxybenzidine-O-β-D-glucuronide
→ 4,4′-Diacetyl-N′-hydroxybenzidine
In: Mouse D872
Rat D872

5-Acetyl-3-(2-(4-hydroxybenzoxazol-2-yl)ethyl)-6-methylpyridin-2(1*H*)-one
Probably formed from: 5-Acetyl-3-(2-(benzoxazol-2-yl)ethyl)-6-methylpyridin-2(1*H*)-one
→ 5-Acetyl-3-(2-(4-hydroxybenzoxazol-2-yl)ethyl)-6-methylpyridin-2(1*H*)-one O-sulphate
Probably in: Rat G505

5-Acetyl-3-(2-(6-hydroxybenzoxazol-2-yl)ethyl)-6-methylpyridin-2(1*H*)-one
Probably formed from: 5-Acetyl-3-(2-(benzoxazol-2-yl)ethyl)-6-methylpyridin-2(1*H*)-one
→ 5-Acetyl-3-(2-(6-hydroxybenzoxazol-2-yl)ethyl)-6-methylpyridin-2(1*H*)-one O-sulphate
Probably in: Rat G505

5-Acetyl-3-(2-(7-hydroxybenzoxazol-2-yl)ethyl)-6-methylpyridin-2(1*H*)-one
Probably formed from: 5-Acetyl-3-(2-(benzoxazol-2-yl)ethyl)-6-methylpyridin-2(1*H*)-one
→ 5-Acetyl-3-(2-(7-hydroxybenzoxazol-2-yl)ethyl)-6-methylpyridin-2(1*H*)-one O-sulphate
Probably in: Rat G505

5-Acetyl-3-(2-(4-hydroxybenzoxazol-2-yl)ethyl)-6-methylpyridin-2(1*H*)-one O-sulphate
Probably formed from: 5-Acetyl-3-(2-(4-hydroxybenzoxazol-2-yl)ethyl)-6-methylpyridin-2(1*H*)-one

5-Acetyl-3-(2-(6-hydroxybenzoxazol-2-yl)ethyl)-6-methylpyridin-2(1*H*)-one O-sulphate
Probably formed from: 5-Acetyl-3-(2-(6-hydroxybenzoxazol-2-yl)ethyl)-6-methylpyridin-2(1*H*)-one

5-Acetyl-3-(2-(7-hydroxybenzoxazol-2-yl)ethyl)-6-methylpyridin-2(1*H*)-one O-sulphate
Probably formed from: 5-Acetyl-3-(2-(7-hydroxybenzoxazol-2-yl)ethyl)-6-methylpyridin-2(1*H*)-one

N-Acetyl-7-hydroxydemethylchlorpromazine
Probably formed from: Demethyl-7-hydroxychlorpromazine

6-Acetyl-7-hydroxy-2-hydroxymethyl-2-methyl-2*H*-benzo[*b*]pyran
Probably formed from: 6-Acetyl-2-hydroxymethyl-7-methoxy-2-methyl-2*H*-benzo[*b*]pyran

N-Acetyl-4(*S*)-hydroxy-4-(*p*-hydroxyphenyl)-L-threonine
Formed from: Caspofungin
→ 4(*S*)-Hydroxy-4-(*p*-hydroxyphenyl)-L-threonine
Probably in: Man K321

N-Acetyl-S-(1-hydroxyindan-2-yl)-L-cysteine
Probably formed from: S-(1-Hydroxyindan-2-yl)glutathione

N-Acetyl-3-hydroxyindoline
Formed from: N-Acetylindoline

2-Acetyl-7-(2-hydroxy-3-isopropylamino)propoxybenzofuran
See: Befunolol

N-Acetyl-3-hydroxy-3′-methoxybenzidine
Probably formed from: 3-Hydroxy-3′-methoxybenzidine

N-Acetyl-3-hydroxy-4-methoxyphenethylamine
Formed from: N-Acetyl-3,4-dihydroxyphenethylamine

N-Acetyl-3-hydroxy-4-methoxy-L-phenylalanine
Formed from: 3-Hydroxy-4-methoxy-L-phenylalanine
→ 3-Hydroxy-4-methoxy-L-phenylalanine
Probably in: Rat A801

N-Acetyl-S-(2-hydroxy-1-(*p*-methoxyphenyl)-1-propyl)-L-cysteine
Probably formed from: Anethole oxide

N⁴-Acetyl-N¹-(2-hydroxy-6-methoxy-4-pyrimidinyl)sulphanilamide
Probably formed from: N¹-(2-Hydroxy-6-methoxy-4-pyrimidinyl)sulphanilamide

N⁴-Acetyl-N¹-(6-hydroxy-2-methoxy-4-pyrimidinyl)sulphanilamide
Probably formed from: N¹-(6-Hydroxy-2-methoxy-4-pyrimidinyl)sulphanilamide

N-Acetyl-5-hydroxy-6-methoxytryptamine
→ 5-Hydroxy-6-methoxytryptamine
In: Pseudomonas K790

N-Acetyl-S-(9-hydroxy-2-methylellipticinium-10-yl)-L-cysteine
Probably formed from: S-(9-Hydroxy-2-methylellipticinium-10-yl)-L-cysteine

N^4-Acetyl-N^1-(5-hydroxymethyl-3-isoxazolyl) sulphanilamide
Probably formed from: N^1-(5-hydroxymethyl-3-isoxazolyl)sulphanilamide

6-Acetyl-2-hydroxymethyl-7-methoxy-2-methyl-2*H*-benzo[*b*]pyran
Formed from: Encecalin
→ 6-Acetyl-7-hydroxy-2-hydroxymethyl-2-methyl-2*H*-benzo[*b*]pyran
Probably in: Melanopus E491

6-Acetyl-2-hydroxymethyl-2-methyl-2*H*-benzo[*b*]pyran
Formed from: 7-Demethylencecalin

N-(3-Acetyl-4-(2-hydroxy-3-((1-methylethyl) amino)propoxy)butanamide
See: Acebutolol

3′-(3-Acetyl-4-hydroxy-1-methyl-2-oxocyclohexa-3,5-dienyl)-2′,4′,6′-trihydroxy-5′-methylacetophenone
Formed from: Usnic acid

N-Acetyl-S-(2-hydroxy-1-(*o*-methylphenyl)ethyl)-L-cysteine
Formed from: *o*-Methylstyrene
o-Methylstyrene oxide

N-Acetyl-S-(2-hydroxy-1-(*m*-methylphenyl)ethyl)-L-cysteine
Formed from: *m*-Methylstyrene
m-Methylstyrene oxide

N-Acetyl-S-(2-hydroxy-1-(*p*-methylphenyl)ethyl)-L-cysteine
Formed from: *p*-Methylstyrene
p-Methylstyrene oxide

N-Acetyl-S-(2-hydroxy-2-(*o*-methylphenyl)ethyl)-L-cysteine
Formed from: *o*-Methylstyrene
o-Methylstyrene oxide

N-Acetyl-S-(2-hydroxy-2-(*m*-methylphenyl)ethyl)-L-cysteine
Formed from: *m*-Methylstyrene
m-Methylstyrene oxide

N-Acetyl-S-(2-hydroxy-2-(*p*-methylphenyl)ethyl)-L-cysteine
Formed from: *p*-Methylstyrene
p-Methylstyrene oxide

N-Acetyl-S-(2-hydroxy-3-(*p*-nitrophenoxy) propyl)-L-cysteine
Probably formed from: S-(2-Hydroxy-3-(*p*-nitrophenoxy)propyl)-L-cysteine

N-Acetyl-4-hydroxyphenethylamin-3-yl phosphate
Formed from: N-Acetyl-3,4-dihydroxyphenethylamine

N-Acetyl-4-hydroxyphenethylamin-3-yl sulphate
Formed from: N-Acetyl-3,4-dihydroxyphenethylamine
4-Hydroxyphenethylamin-3-yl sulphate

N-Acetyl-S-(2-hydroxy-1-phenethyl)-L-cysteine
Formed from: Cinnamonitrile
S-(2-Hydroxy-1-phenethyl)-L-cysteine
S-(2-Hydroxy-1-phenethyl)cysteinylglycine
Probably formed from: β-Bromostyrene
(*S*)-S-(2-Hydroxy-1-phenethyl)glutathione

N-Acetyl-S-(2-hydroxy-2-phenethyl)-L-cysteine
Probably formed from: S-(2-Hydroxy-2-phenethyl)-L-cysteine
→ N-Acetyl-S-phenacyl-L-cysteine
Probably in: Man A3059

N-Acetyl-S-(2-hydroxy-3-phenoxypropyl)-L-cysteine
Formed from: 1,2-Epoxy-3-phenoxy-propane

N-Acetyl-*p*-hydroxyphenyl-β-D-glucosaminide
→ Quinol
In: Beef A1940

N-Acetyl-1-hydroxyphenyl-3-methylaminomethyl-2-phenyl-2-butyl propionate
Probably formed from: 1-Hydroxyphenyl-3-methylaminomethyl-2-phenyl-2-butyl propionate

N-Acetyl-*p*-hydroxyphenyl nitroxide radical
Formed from: *p*-Acetamidophenol

2-Acetyl-7-hydroxypromazine
Formed from: 2-Acetylpromazine

4-((3-(4-Acetyl-3-hydroxy-2-propylphenoxy) propyl)sulphonyl)-γ-hydroxybenzenebutyric acid
Formed from: 4-((3-(4-Acetyl-3-hydroxy-2-propylphenoxy)propyl)sulphonyl)-γ-oxobenzenebutyric acid

4-((3-(4-Acetyl-3-hydroxy-2-propylphenoxy) propyl)sulphonyl)-γ-oxobenzenebutyric acid
→ 4-((3-(4-Acetyl-3-hydroxy-2-propylphenoxy)propyl)sulphonyl)-γ-hydroxybenzenebutyric acid
In: Dog E948
Man E948
Rat E948
→ 4-((3-(4-Acetyl-3-hydroxy-2-propylphenoxy)propyl)sulphonyl) phenylacetic acid
In: Dog E948
Rat E948

4-((3-(4-Acetyl-3-hydroxy-2-propylphenoxy) propyl)sulphonyl)phenylacetic acid
Formed from: 4-((3-(4-Acetyl-3-hydroxy-2-propylphenoxy)propyl)sulphonyl)-γ-oxobenzenebutyric acid

N^4-Acetyl-4-hydroxy-2-sulphanilamidopyrimidine
Probably formed from: 4-Hydroxy-2-sulphanilamidopyrimidine

N-Acetyl-5-hydroxytryptamine
Formed from: 5-Hydroxytryptamine
Melatonin
→ N-Acetyltryptamin-5-yl-β-D-glucuronide
In: Man E441
→ N-Acetyltryptamin-5-yl sulphate
In: Man E441
Rat B941
→ 5-Hydroxytryptamine
In: Pseudomonas K790
See also: G393
→ Melatonin
In Beef C552
Chicken C552, H84
Drosophila E569
Man A585
Rabbit H84
Rat H791, H845

N-Acetyl-6-hydroxytryptamine
→ Pigment
In: Aspergillus C800

N-Acetyl-β-hydroxy-L-tryptophan
Probably formed from: N-Acetyl-L-tryptophan

N-Acetyl-β-hydroxy-L-tryptophanamide
Probably formed from: N-Acetyl-L-tryptophanamide
→ N-Acetyl-β-oxo-L-tryptophanamide
In: Pseudomonas A3617

N-Acetyliminostilbene
→ N-Acetyliminostilbene-10,11-oxide
In: Rat E140

N-Acetyliminostilbene-10,11-oxide
Formed from: N-Acetyliminostilbene

N-Acetylindoline (N-acetyl-2,3-dihydro-1*H*-indole)
→ N-Acetyl-3-hydroxyindoline
In: Cunninghamella D810

N-Acetyliomeglamic acid
Formed from: Iomeglamic acid

N-Acetylisatinic acid
Probably formed from: 1*H*-4-Oxoquinaldine
→ N-Acetylanthranilic acid
Probably in: Arthrobacter G274

O-Acetylmandelamide
Formed from: O-Acetylmandelonitrile
→ O-Acetylmandelic acid
Probably in: Pseudomonas G724

O-Acetylmandelic acid
Formed from: (*S*)-Mandelic acid
Probably formed from: O-Acetylmandelamide

O-Acetylmandelonitrile
→ O-Acetylmandelamide
In: Pseudomonas G724
→ Mandelonitrile
Probably in: Pseudomonas G724

***o*-Acetylmercaptobenzoic acid**
→ *o*-Mercaptobenzoic acid
In: Guinea pig H214

N-Acetylmescaline
Formed from: Mescaline
See: *l*-α-Acetyl-2-dimethylamino-4,4-diphenyl-5-heptanol

l-α-Acetylmethadol
See: l-α-Acetyl-2-dimethylamino-4,4-diphenyl-5-heptanol

N-(N-Acetyl-L-methionyl)-O,O-bis (ethoxycarbonyl)dopamine
→ N-(N-Acetyl-L-methionyl)-3,4-dihydroxyphenethylamine
In: Dog G52
Man F539
See also E952

N-(N-Acetyl-L-methionyl)-3,4-dihydroxyphenethylamine
Formed from: N-(N-Acetyl-L-methionyl)-O,O-bis(ethoxycarbonyl)dopamine
→ 3,4-Dihydroxyphenethylamine
Probably in: Dog G52
Man F539

N-Acetyl-*o*-methoxyamphetamine
Probably formed from: *o*-Methoxy-amphetamine

N-Acetyl-*p*-methoxyamphetamine
Formed from: *p*-Methoxyamphetamine

4′-Acetyl-3-methoxybenzidine
Formed from: 3-Methoxybenzidine
→ 4,4′-Diacetyl-3-methoxybenzidine
Probably in: Rat A1579

N-Acetyl-S-(5-methoxybenzimidazol-2-yl)-L-cysteine
Formed from: Omeprazole

N-Acetyl-2-(5-methoxybenzofuryl)ethylamine
Formed from: 2-(5-Methoxybenzofuryl) ethylamine

N$^{\gamma}$-Acetyl-5-methoxykynurenamine
Formed from: N$^{\gamma}$-Acetyl-N^{2}-formyl-5-methoxykynurenamine

O-Acetyl-*m*-methoxymandelonitrile
Formed from: *m*-Methoxymandelonitrile

4-Acetyl-3-methoxy-5-methylbenzoic acid
Formed from: 5-Acetyl-4-methoxy-6-methyl-2-pyrone

5-Acetyl-4-methoxy-6-methyl-2-pyrone
→ 4-Acetyl-3-methoxy-5-methylbenzoic acid
In: Macrophoma D928, E766

2-Acetyl-6-methoxynaphthalene
→ 2-Hydroxy-2-(6-methoxy-2-naphthyl) propionitrile
In: Almond K6

N-Acetyl-2-methoxy-4-(2′-nitrovinyl)phenyl-D-galactosaminide
→ 2-Methoxy-4-(2′-nitrovinyl)phenol
In: Man C7

1-(3-(4-Acetyl-2-methoxyphenoxy)propyl)-4-(4-fluoro-2-hydroxybenzoyl)piperidine
Formed from: Iloperidone
→ 4-(4-Fluoro-2-hydroxybenzoyl)-1-(3-(4-(1-hydroxyethyl)-2-methoxyphenoxy) propyl)piperidine
Probably in: Dog H447
Man H447
Rat H447

N-Acetyl-*p*-methoxyphenyl-β-D-glucosaminide
→ *p*-Methoxyphenol
In: Beef A1940

N-Acetyl-5-methoxytryptamine
See: Melatonin

O-Acetyl-5-methoxytryptophol
→ 5-Methoxytryptophol
Probably in: Rat C659

N-Acetyl-3-methoxytyramine
Formed from: 3-Methoxytyramine
Probably formed from: N-Acetyl-3,4-dihydroxyphenethylamine

N-Acetyl-3-methoxy-L-tyrosine
Formed from: 3-Methoxy-L-tyrosine

***l*-α-Acetyl-2-methylamino-4,4-diphenyl-5-heptanol**
Formed from:
l-α-Acetyl-2-dimethylamino-4,4-diphenyl-5-heptanol
→ *l*-α-Acetyl-2-amino-4,4-diphenyl-5-heptanol
In: Man A1491, H859
Monkey A2979
Rat A2088, A2900, A3180, A3181, A3900

4′-(N-Acetyl-N-methyl)amino-4-hydroxyazobenzene
Probably formed from: 4-Hydroxy-4′-methylaminoazobenzene

N-Acetyl-3-methylaminomethyl-1,2-diphenyl-2-butyl propionate
Formed from: 3-Methylaminomethyl-1,2-diphenyl-2-butyl propionate

N-Acetyl-S-(*o*-methylbenzyl)-L-cysteine
Formed from: *o*-Methylbenzyl alcohol
o-Xylene

O-Acetyl-N-methylcalycotomine
→ N-Methylcalycotomine
In: Arthrobacter F1

N-Acetyl-3-methylcarbazole
→ 3-Methylcarbazole
In: Pseudomonas J180

N-Acetyl-S-(N-(1-methyl-3,3-diphenylpropylcarbamoyl)-L-cysteine
Probably formed from: S-(N-(1-Methyl-3,3-diphenylpropylcarbamoyl)glutathione

N-Acetyl methylene bis(*o*-chloroaniline)
Formed from: Methylene bis(*o*-chloroaniline)

α-(2-(Acetyl(1-methylethyl)amino)ethyl)-α-(2-chlorophenyl)-1-piperidinebutanamide
See: Bidisomide

N-Acetyl-3′-methyl-4-methylaminoazobenzene
Probably formed from: 3′-Methyl-4-methylaminoazobenzene

N-Acetyl-3-methyl-4-nitrophenyl-β-D-glucosaminide
→ 3-Methyl-4-nitrophenyl
In: Beef A1940

N-Acetyl-*m*-methylphenyl-β-D-glucosaminide
→ *m*-Cresol
In: Beef A1940

N-Acetyl-*p*-methylphenyl-β-D-glucosaminide
→ *p*-Cresol
In: Beef A1940

N-Acetyl-2-methylsulphinylafloqualone
Probably formed from: N-Acetyl-2-methylthioafloqualone
→ N-Acetyl-2-methylsulphonylafloqualone
Probably in: Dog C912
Monkey C912
Rat C793a

N-Acetyl-2-methylsulphonylafloqualone
Probably formed from: N-Acetyl-2-methylsulphinylafloqualone
→ 6-Acetamido-3-(*o*-hydroxymethylphenyl)-2-methylsulphonylmethyl-4(3*H*)-quinazolinone
Probably in: Rat C793a
→ 6-Glycollamido-2-methylsulphonylmethyl-3-(*o*-tolyl)-4(3*H*)-quinazolinone
Probably in: Rat C793a

N-Acetyl-2-methylthioafloqualone
Probably formed from: 2-Methylthioafloqualone
→ N-Acetyl-2-methylsulphinylafloqualone
Probably in: Dog C912
Monkey C912
Rat C793a

N-Acetyl-2-(7-methylthionaphth-1-yl)ethylamine
Formed from: 2-(7-Methylthionaphth-1-yl)ethylamine

N-Acetyl-α-methyltrifluoromethylphenethylamine
Probably formed from: α-Methyltrifluoromethylphenethylamine

N-Acetyl-N-methyltryptamine
Formed from: N-Methyltryptamine

N-Acetyl-α-methyltryptamine
Formed from: α-Methyltryptamine

N-Acetyl-5-methyl-L-tryptophan
→ N′-Acetyl-N-formyl-5-methyl-L-kynurenine
In: Wheat E24

6-Acetylmorphine
Formed from: Heroin
→ 6-Acetylmorphine-3-β-D-glucuronide
In: Man A2907
→ 6,6′-Diacetylpseudomorphine
In: Horseradish A1357
→ Morphine
In: Man H890
See also A1347, A2807

6-Acetylmorphine-3-β-D-glucuronide
Formed from: 6-Acetylmorphine

N-Acetyl-S-(10α-morphinyl)-L-cysteine
Probably formed from: S-(10α-Morphinyl)glutathione

2-Acetylnaphthalene (methyl β-naphthyl ketone)
→ 1-(2-Naphthyl)ethanol
In: Lactobacillus G148

N-Acetyl-S-(1-naphthyl)-L-cysteine
Probably formed from: S-(1-Naphthyl)-L-cysteine

N-Acetyl-S-(*o*-nitrobenzyl)-L-cysteine
Formed from: *o*-Nitrotoluene

N-Acetyl-S-(*p*-nitrobenzyl)-L-cysteine
→ S-(*p*-Nitrobenzyl)-L-cysteine

In: Rat C29

3-Acetyl-2-nitrooestrone
→ 3-Hydroxy-2-nitrooestrone
With glutamate dehydrogenase A2004
→ 2-Nitrooestrone
With glutamate dehydrogenase A2004

N-Acetyl-S-(*p*-nitrophenyl)-L-cysteine
Probably formed from: S-(*p*-Nitrophenyl) glutathione

N-Acetyl-2-nitro-*p*-phenylenediamine
Formed from: 2-Nitro-*p*-phenylenediamine

3-(β-Acetyl-α-(*p*-nitrophenyl)ethyl)-4-hydroxycoumarin
See: Acecoumarol

N-Acetyl-*p*-nitrophenyl-β-D-galactosaminide
→ *p*-Nitrophenol
In: Sheep B500

N-Acetyl-*m*-nitrophenyl-β-D-glucosaminide
→ *m*-Nitrophenol
In: Beef A1940

N-Acetyl-*p*-nitrophenyl-β-D-glucosaminide
→ *p*-Nitrophenol
In: Beef A1940
Man A642, A789, C7
Sheep B500

N-Acetylnoradrenaline
Formed from: 4-(2-Acetamidoethyl)-*o*-quinone
N-Acetyl-3,4-dihydroxyphenethylamine
D-Noradrenaline
→ N-Acetylnoradrenalone
In: Sarcophaga F793

N-Acetylnoradrenalone
Formed from: 4-(2-Acetamido-1-hydroxy-ethyl)-*o*-quinone
N-Acetylnoradrenaline

N-Acetyloctopamine
Formed from: Octopamine

N-Acetyl-S-(2-oxo-2-(*p*-cyanoanilino)ethyl)-L-cysteine
Probably formed from: *p*-Acetamidobenzo-nitrile

N-Acetyl-β-oxo-L-tryptophanamide
Formed from: N-Acetyl-β-hydroxy-L-tryptophanamide

N-Acetyl-S-(pentachlorophenyl)-L-cysteine
Formed from: Hexachlorobenzene
Probably formed from: S-(Pentachlorophenyl)-L-cysteine
→ Pentachlorobenzene
In: Rat C881
→ Pentachlorothiophenol
In: Rat C881
See also C787

N-Acetyl-S-(pentachlorophenyl)-L-cysteine sulphoxide
→ Pentachlorobenzene
In: Rat C881
→ Pentachlorothiophenol
Probably in: Rat C881

N-Acetyl-S-phenacyl-L-cysteine
Probably formed from: N-Acetyl-S-(2-hydroxy-2-phenylethyl)-L-cysteine

N-Acetylphenethylamine
Formed from: Phenethylamine
→ Phenethylamine
In: Pseudomonas K790

N-Acetyl-β-phenethylhydrazine
Formed from: β-Phenethylhydrazine

α-N-Acetyl-δ-N-(*m*-phenoxybenzoyl)-L-ornithine
Formed from: *m*-Phenoxybenzoic acid

1-(*p*-Acetylphenoxy)-2,3,6,7-tetrahydroxy-3,7-dimethyloctane
Probably formed from: 1-(*p*-Acetylphenoxy)-2,6,7-trihydroxy-3,7-dimethyloctane
→ 1-(*p*-Carboxyphenoxy)-2,3,6,7-tetrahydroxy-3,7-dimethyloctane
Probably in Rat A813

1-(*p*-Acetylphenoxy)-2,6,7-trihydroxy-3,7-dimethyloctane
Probably formed from: *trans*-6,7-Epoxy-1-(4′-ethylphenoxy)-3,7-dimethyl-2-octene
→ 1-(*p*-Acetylphenoxy)-2,3,6,7-tetrahydroxy-3,7-dimethyloctane
Probably in Rat A813

1-(*p*-Acetylphenoxy)-3,6,7-trihydroxy-3,7-dimethyloctane
Probably formed from: *trans*-6,7-Epoxy-1-(4′-ethylphenoxy)-3,7-dimethyl-2-octene

N-Acetyl-L-phenylalanine
Formed from: l-Phenylalanine
→ *cis*-N-Acetyl-2,3-dihydro-2,3-dihydroxy-L-phenylalanine
In: Bacterium B871
→ L-Phenylalanine
In: Rat A3520

N-Acetyl-L-phenylalanyl-L-tyrosine
→ N-Acetyl-L-phenylalanyl-L-tyrosine-O-sulphate
In: Rat E197

N-Acetyl-L-phenylalanyl-L-tyrosine-O-sulphate
Formed from: N-Acetyl-L-phenylalanyl-L-tyrosine

1-((N-Acetyl-N-phenylamino)oxy)-1-deoxy-β-D-glucopyranoside
Formed from: 2,3,4,6-Tetra-O-acetyl-1-((N-acetyl-N-phenylamino)oxy)-1-deoxy-β-D-glucopyranoside

N-Acetyl-4-phenylbutylamine
→ 4-Phenylbutylamine
In: Pseudomonas K790

N-Acetyl-S-phenyl-L-cysteine (phenylmercapturic acid)
Probably formed from: S-Phenylglutathione

N-Acetyl-N′-phenyl-2,6-diaminobenzoic acid
Formed from: N-Benzoyl-N′-phenyl-2,6-diaminobenzoic acid

1-(*p*-Acetylphenyl)-3,3-diethyltriazene
→ 1-(*p*-Acetylphenyl)-3-ethyltriazene
In: Mouse D641
→ *p*-Aminoacetophenone
In: Mouse D641

1-(*p*-Acetylphenyl)-3,3-dimethyltriazene
→ 1-(*p*-Acetylphenyl)-3-methyltriazene
In: Mouse C793b, D641
→ *p*-Aminoacetophenone
In: Mouse C793b, D641
→ 1-(*p*-(1-Hydroxyethyl)phenyl)-3,3-dimethyltriazene
In: Mouse C793b

(*S*)-N-Acetyl-1-phenylethylamine
→ (*S*)-1-Phenylethylamine
In: Rhodococcus J326

N-Acetyl-S-(1-phenylethyl)-L-cysteine
Formed from: (1-Bromoethyl)benzene
(2-Bromoethyl)benzene

N-Acetyl-S-(2-phenylethyl)-L-cysteine
Formed from: (2-Bromoethyl)benzene

1-(*p*-Acetylphenyl)-3-ethyltriazene
Formed from: 1-(*p*-Acetylphenyl)-3,3-diethyltriazene

N-Acetyl-L-phenylglycine
→ L-Phenylglycine
In: Fungus K349
Micrococcus B479, B597

1-(*p*-Acetylphenyl)-3-methyltriazene
Formed from: 1-(*p*-Acetylphenyl)-3,3-dimethyltriazene

N-Acetyl-S-((3-phenyl-1,2,4-oxadiazol-5-yl)methyl)-L-cysteine
Probably formed from: S-((3-Phenyl-1,2,4-oxadiazol-5-yl)methyl)glutathione

N-Acetyl-3-phenylpropylamine
→ 3-Phenylpropylamine
In: Pseudomonas K790

5-Acetyl-6-phenyl-2-pyrone
→ 3-Acetyl-5-carboxybiphenyl
In: Macrophoma E766

N-Acetylphosphanilic acid
Formed from: Phosphanilic acid

7-(4-Acetyl-1-piperazinyl)-1-ethyl-6-fluoro-1,4-dihydro-4-oxo-3-quinolinecarboxylic acid
Formed from: 1-Ethyl-6-fluoro-1,4-dihydro-4-oxo-7-(1-piperazinyl)-3-quinolinecarboxylic acid

N-Acetylprizidilol
Formed from: Prizidilol

N-Acetylprocainamide
Formed from: Procainamide

N-Acetylproflavine
Formed from: Proflavine
→ N-Acetylproflavine-N′-β-D-glucuronide
Probably in: Channel catfish J407
Man J142

N-Acetylproflavine-N′-β-D-glucuronide
Probably formed from: N-Acetylproflavine

2-Acetylpromazine
→ 2-Acetyl-7-hydroxypromazine
In: Horse B609
→ 2-(1-Hydroxyethyl)promazine
In: Horse B609

O-Acetylpropranolol
→ Propranolol

In: Man F685
Mouse J222
N-Acetylprotriptyline
Formed from: Protriptyline
N-Acetylpuromycin
Formed from: N-Acetyl-O-demethyl-puromycin
Puromycin
2-(5-Acetylpyridin-2-yl)benzimidazole
Probably formed from: 2-(5-(1-Hydroxyethyl)pyridin-2-yl)benzimidazole
Acetylsalicylic acid (aspirin, 2-acetoxybenzoic acid)
Formed from: Benorylate
Nalbuphine acetylsalicylate
→ Salicylic acid
In: Acinetobacter A962
Ascaris A324, A3040, A3715
Cat C779
Guinea pig H214
Man A3635, C374, C705, C779, D668, E819, F216, F558, G203
Marmoset B201
Moniezia A324, A3040, A3715
Mouse A3715, C779, H921, K148
Pharbitis J917
Pig E259
Rat B201, C237, C779, F588, F753, H921, J845
Sheep A3715
See also A1590, A3602
O-Acetylshikonin
Probably formed from: Shikonin
N^4-Acetylsulphadiazine
Formed from: Sulphadiazine
N^4-Acetylsulphadimethoxine
Formed from: Sulphadimethoxine
N^4-Acetylsulphadoxine
Formed from: Sulphadoxine
N^4-Acetylsulphamerazine
Formed from: Sulphamerazine
N^4-Acetylsulphamethazine
Formed from: Sulphamethazine
N^4-Acetylsulphamethomidine
Formed from: Sulphamethomidine
N^4-Acetylsulphamethoxazole
Formed from: Sulphamethoxazole
N^4-Acetylsulphamonomethoxine
Formed from: Sulphamonomethoxine
→ N^4-Acetylsulphamonomethoxine-N^1-β-D-glucuronide
Probably in: Man F593
N^1-Acetylsulphanilamide
Formed from: Sulphanilamide
N^4-Acetylsulphanilamide
See: *p*-Acetamidobenzenesulphonamide
N^4-Acetylsulphamonomethoxine-N^1-β-D-glucuronide
Probably formed from: N^4-Acetylsulphamonomethoxine
N^4-Acetylsulphapyridine
Formed from: Sulphapyridine
N^4-Acetylsulphatroxazole
Formed from: Sulphatroxazole
N-Acetylsulphisomidine
Formed from: Sulphisomidine
N^4-Acetylsulphisoxazole
Formed from: Sulphisoxazole
N-Acetyl-S-(tetrachloromethylsulphinylphenyl)-L-cysteine
Probably formed from: S-(Tetrachloromethyl-sulphinylphenyl)-L-cysteine
→ N-Acetyl-S-(tetrachloromethylsulphonyl-phenyl)-L-cysteine
Probably in: Rat G688
N-Acetyl-S-(tetrachloromethylsulphonylphenyl)-L-cysteine
Probably formed from: N-Acetyl-S-(tetrachloromethylsulphinylphenyl)-L-cysteine
N-Acetyl-S-(tetrachloromethylthiophenyl)-L-cysteine
Formed from: Pentachlorothioanisole
N-Acetyl-S-(2,3,4,5-tetrachlorophenyl)-L-cysteine
Formed from: 1,2,3,4-Tetrachlorobenzene
N-Acetyl-S-(2,3,5,6-tetrachlorophenyl)-L-cysteine
→ 1,2,4,5-Tetrachlorobenzene
In: Rat C881
→ 2,3,5,6-Tetrachlorothiophenol
In: Rat C881
2-Acetyl-tetrahydro-3-(*p*-hydroxyphenyl)-1,2,4-oxadiazin-5-one
Formed from: 2-Acetyl-tetrahydro-3-phenyl-1,2,4-oxadiazin-5-one

1-Acetyl-1,2,3,4-tetrahydro-6-hydroxyquinoline
Formed from: Quinfamide

N-Acetyl-2-(1,2,3,4-tetrahydro-7-methoxynaphth-1-yl)ethylamine
Formed from: 2-(1,2,3,4-Tetrahydro-7-methoxynaphth-1-yl)ethylamine

N-Acetyl-1,2,3,4-tetrahydro-4-oxoquinoline
Formed from: N-Acetyl-1,2,3,4-tetrahydroquinoline

2-Acetyl-tetrahydro-3-phenyl-1,2,4-oxadiazin-5-one
→ N-(2-(N-Acetamido)acetoxy)benzylamine
In: Rat D762
→ 2-Acetyl-tetrahydro-3-(*p*-hydroxyphenyl)-1,2,4-oxadiazin-5-one
In: Rat D762
→ α-(N-Formamido)-N-glyoxalylbenzamide
In: Rat D762

(8*S*,10*S*)-8-Acetyl-7,8,9,10-tetrahydro-1,6,8,10,11-pentahydroxy-5,12-naphthacenedione
See: Carminomycinone

N-Acetyl-1,2,3,4-tetrahydroquinoline
→ N-Acetyl-1,2,3,4-tetrahydro-4-oxoquinoline
In: Cunninghamella D810

6-Acetyl-3-(1*H*-tetrazol-5-yl)chromone
Probably formed from: 6-(1-Hydroxyethyl)-3-(1*H*-tetrazol-5-yl)chromone

N-Acetyl-*p*-thioanisidine
Formed from: *p*-Thioanisidine

N-Acetyl-*p*-thioanisidine sulphone
Formed from: *p*-Thioanisidine sulphone

N-Acetyl-*p*-thioanisidine sulphoxide
Formed from: *p*-Thioanisidine sulphoxide

2-Acetylthiomethyl-4-(4-methylphenyl)-4-oxobutanoic acid
→ 4-(4-Methylphenyl)-4-oxo-2-thiomethylbutanoic acid
In: Rat J168

1-((*S*)-3-Acetylthio-2-methylpropanoyl)-L-prolyl-L-phenylalanine
→ 1-((*S*)-3-Mercapto-2-methylpropanoyl)-L-prolyl-L-phenylalanine
In: Rat D937

N-Acetyl-*o*-tolidine
Formed from: *o*-Tolidine
→ 4-Acetamido-3-methylphenol
In: Fusarium F814
→ N,N′-Diacetyl-*o*-tolidine
Probably in: Rat C471, C541

N-Acetyl-*p*-toluenesulphonamide
Formed from: *p*-Toluenesulphonamide

N-Acetyl-β-(*o*-tolyloxy)alanine
Formed from: Mephenesin

N-Acetyl-S-(2,3,5-trichlorophenyl)-L-cysteine
Formed from: 1,2,4-Trichlorobenzene

N-Acetyl-S-(2,4,5-trichlorophenyl)-L-cysteine
Formed from: 1,2,4-Trichlorobenzene

N-Acetyltrovafloxacin
Formed from: Trovafloxacin
→ N-Acetyltrovafloxacin-β-D-glucuronide
Probably in: Man J140

N-Acetyltrovafloxacin-β-D-glucuronide
Probably formed from: N-Acetyltrovafloxacin

N-Acetyltryptamine
Formed from: Methyl tryptamine dithiocarbamate
Tryptamine

N-Acetyltryptamin-5-yl-β-D-glucuronide
Formed from: N-Acetyl-5-hydroxytryptamine

N-Acetyltryptamin-5-yl sulphate
Formed from: N-Acetyl-5-hydroxytryptamine

N-Acetyl-L-tryptophan
Formed from: L-Tryptophan
→ N-Acetyl-α,β-dehydrotryptophan
In: Chromobacterium H248
→ N-Acetyl-β-hydroxy-L-tryptophan
Probably in: Pseudomonas A3007
→ N-Formylacetylkynurenine
In: Wheat C559, E24
→ Indole-3-glyoxal
In: Pseudomonas A3009

N-Acetyl-L-tryptophanamide
→ N-Acetyl-α,β-dehydrotryptophanamide
In: Chromobacterium H248
Pseudomonas A3087, A3617
→ N-Acetyl-β-hydroxy-L-tryptophanamide
In: Pseudomonas A3617
See also: A3007
→ Indole-3-glyoxal
In: Pseudomonas A3009
→ 5-(3-Indolyl)-2-methyl-2-oxazoline-4-carboxamide

In: Pseudomonas C511
→ L-Tryptophan
In: Rat A3520
N-Acetyl-L-tryptophanyl-L-phenylalanine
→ N-Acetyl-(*Z*)-α,β-dehydrotryptophanyl-L-phenylalanine
In: Chromobacterium J462
O-Acetyltryptophol
Formed from: Tryptophol
N-Acetyltyramine
Formed from: Tyramine
→ N-Acetyl-3,4-dihydroxyphenethylamine
In Agaricus J460
Schistocerca B826
→ N-Acetyltyramine-4-β-D-glucuronide
Probably in: Chicken A411
Guinea pig A411
Rabbit A411
→ Tyramine
In: Aspergillus A3888
Candida A3822
Citrobacter A3888
Enterobacteria A3889
Epidermophyton A3888
Escherichia A3888
Fungus A3889
Klebsiella A3888
Microsporum A3888
Morganella A3888
Mucor A3888
Mycobacteria A3889
Penicillium A3888
Proteus A3888
Pseudomonas A3888
Staphylococcus A3888
Streptococcus A3888
Trichophyton A3888
N-Acetyltyramine-4-β-D-glucuronide
Formed from: N-Acetyltyramine
N-Acetyl-L-tyrosine
Probably formed from: L-Tyrosine
→ Di-(N-acetyl-L-tyrosine)
In: Horseradish J327
→ Tetra-(N-acetyl-L-tyrosine)
In: Horseradish J327
→ Tri-(N-acetyl-L-tyrosine)
In: Horseradish J327
→ L-Tyrosine
In: Rat A3520, C29
N-Acetyl-L-tyrosyl-alanyl-alanyl-alanyl-*p*-nitroanilide
→ Alanyl-*p*-nitroanilide
In: Rat C693
N-Acetylviloxazine
Formed from: Viloxazine
N-Acetylzonisamide
Formed from: Zonisamide
Acid Orange 6 (4-((2,4-dihydroxyphenyl)azo)benzenesulphonic acid)
→ 2,4-Dihydroxyaniline
In: Cautobacter K314
→ Sulphanilic acid
In: Cautobacter K314
Acid Orange 7 (4-((2-hydroxynaphthalenyl)azo)benzenesulphonic acid, orange II)
→ 1-Amino-2-naphthol
In: Cautobacter K314
Proteus A317
Shigella G767
See also: F152
→ Sulphanilic acid
In: Cautobacter K314
Proteus A317
Shigella G767
See also: F152
Acid Orange 12 (Orange RN, 6-hydroxy-5-phenylazo-2-naphthalenesulphonic acid)
→ 1-Amino-2-hydroxynaphthalene-6-sulphonic acid
In: Cautobacter K314
Pig A2821
→ Aniline
In: Cautobacter K314
Pig A2821
→ 2-Hydroxy-1-(*p*-hydroxyphenylazo)naphthalene-6-sulphonic acid
In: Pig A2821
Acid Red 88 (4-((2-hydroxy-1-naphthalenyl)azo)-1-naphthalenesulphonic acid)
→ 4-Aminonaphthalene-1-sulphonic acid
In: Cautobacter K314
→ 1-Amino-2-naphthol
In: Cautobacter K314

Acid Red 114 (8-((3,3′-dimethyl-4′-(((4-methylphenyl)sulphonyl)oxy)phenyl)azo)(((1,1′-biphenyl)-4-yl)azo)-7-hydroxy-1,3-naphthalenedisulphonic acid)
→ *o*-Tolidine
In: Dog B739
Rat B739

Acid Red 151 (4-((4(2-hydroxy-1-naphthalenyl)phenyl)azobenzenesulphonic acid)
→ *p*-Phenylenediamine
In: Cautobacter K314
→ Sulphanilic acid
In: Cautobacter K314

Acid Yellow 36 (metanil yellow, 3-((4-(phenylamino)phenyl)azo)benzenesulphonic acid)
→ 3-Aminobenzenesulphonic acid
In: Rat C87, F379
See also C116
→ 4-Aminodiphenylamine
In: Rat C87, F379
See also C116

Acitretin (etretin, (2*E*, 4*E*, 6*E*, 8*E*)-9-(4-methoxy-2,3,6-trimethylphenyl)-3,7-dimethoxy-2,4,6,8-nonatetraenoic acid)
Formed from: Etretinate
Isoacitretin
→ Acitretin-β-D-glucuronide
In: Rat G495
→ Acitretinoyl CoA
In: Man K525
→ O-Demethylacitretin
Probably in: Man F118
→ Etretinate
In: Man G376, K525
Rat G376
→ 9-(3-Hydroxymethyl-4-methoxy-2,6-dimethylphenyl)-3,7-dimethylnona-2,4,6,8-tetraenoic acid
Probably in: Man F118
→ Isoacitretin
In: Man G869
Rat G856, H174
See also F58, F118, G858
→ *trans*, *trans*-8-(4-Methoxy-2,3,6-trimethylphenyl)-2,6-dimethylocta-5,7-dienoic acid
In: Man G869
Rat G495
See also: F118

Acitretin-β-D-glucuronide
Formed from: Acitrexin

Acitretinoyl CoA
Formed from: Acitretin
→ Butyl acitretinate
In: Man K525
→ Etretinate
In: Man K525
→ Hexyl acitretinate
In: Man K525
→ Methyl acitretinate
In: Man K525
→ Propyl acitretinate
In: Man K525

Aclacinomycin A
Formed from: Aclacinomycin B
Aklavinone
→ Aklavinone
In: Man C594
→ Bisanhydroaklavinic acid
In: Man C594
→ Bisanhydroaklavinone
In: Guinea pig B497
Hamster B497
Man C594
Monkey B497
Mouse B497
Rabbit B497
Rat B497
→ N-Demethylaclacinomycin
In: Guinea pig B497
Hamster B497
Monkey B497
Mouse B497
Rabbit B497
Rat B497
→ 7-Deoxyaklavinone
In: Man C594
Rat B154, B229, B497
Vibrio C509
→ 7,7′-Dideoxy-7,7′-biaklavinone
In: Guinea pig B497
Hamster B497

Man C594
Monkey B497
Mouse B497
Rabbit B497
Rat B497
Vibrio C509
→ Dihydroaclacinomycin A (2 sugar moiety-bound isomers)
In: Guinea pig B497
Hamster B497
Man C594
Monkey B497
Mouse B497
Rabbit B497
Rat B497

Aclacinomycin B
Formed from: Aklavinone
→ Aclacinomycin A
In: Streptomyces D17

Acridine
→ Acridone
In: Rat D227
→ *trans*-1,2-Dihydro-1,2-dihydroxyacridine
In: Cunninghamella H268
→ 3,4-Dihydro-3,4-dihydroxyacridine
Probably in Rat D227
→ 2-Hydroxyacridine
In: Cunninghamella H268

4′-(9-Acridinylamino)methanesulphon-*m*-anisidide
See: Amsacrine

4′-(9-Acridinylimino)-3′-methoxy-2′,5′-cyclohexadien-1′-one
Probably formed from: $N^{4'}$-(9-Acridinyl)-$N^{1'}$-methanesulphonyl-3′-methoxy-2′,5′-cyclohexadien-1′,4′-diimine

$N^{4'}$-(9-Acridinyl)-$N^{1'}$-methanesulphonyl-3′-methoxy-2′,5′-cyclohexadien-1′,4′-diimine
Formed from: Amsacrine
→ 4′-(9-Acridinylimino)-3′-methoxy-2′,5′-cyclohexadien-1′-one
Probably in: Rat D127
→ Amsacrine
In: Hamster F129

Acridone (9-10H)-acridinone
Formed from: Acridine

Acrivastine ((2*E*)-3-(6-((1*E*)-1-(4-methylphenyl)-3-(1-pyrrolidinyl)-1-propenyl)-2-pyrrolidinyl)-2-propenoic acid)
→ 3-(6-(3-(N-(3-Carboxypropyl)amino)-1-(*p*-methylphenyl)-1-propenyl)-2-pyridinyl)-2-propenoic acid
In: Dog G382
→ Dihydroacrivastine
In: Dog G382

Acronine (acrocynine, 3,12-dihydro-6-methoxy-3,3,12-trimethyl-7*H*-pyrano[2,3-*c*]acridan-7-one)
→ N-Demethylacronine
In: Guinea pig A1646
Mouse A1646
Rabbit A1646
→ O-Demethylacronine
In: Guinea pig A1646
Mouse A1646
Rabbit A1646
→ 9-Hydroxyacronine
In: Aspergillus A1218
Cunninghamella A1218
Dog A1646
Man A1646
Mouse A1646
Pig A1646
Rabbit A1646
Streptomyces A1218
→ 11-Hydroxyacronine
In: Aspergillus A1218
Cunninghamella A1218
Dog A1646
Man A1646
Mouse A1646
Pig A1646
Rabbit A1646
Streptomyces A1218
→ 3-Hydroxymethyl-3-demethylacronine
In: Mouse A1646
Rabbit A1646

Actinocin (2-amino-4,6-dimethyl-3-oxo-3*H*-phenoxazine)
Formed from: 3-Hydroxy-4-methylanthranilic acid

Actinomycin (general heading)
Formed from: 3-Hydroxy-4-methylanthranilic acid

(+)-N-(1-((Adamantan-1-yl)methyl)-2,3,4,5-tetrahydro-2,4-dioxo-5-phenyl-1,5-benzodiazepin-3-yl)-N′-(*p*-hydroxyphenyl)urea
Formed from: (+)-N-(1-((Adamantan-1-yl)methyl)-2,3,4,5-tetrahydro-2,4-dioxo-5-phenyl-1,5-benzodiazepin-3-yl)-N′-phenylurea

(+)-N-(1-((Adamantan-1-yl)methyl)-2,3,4,5-tetrahydro-2,4-dioxo-5-phenyl-1,5-benzodiazepin-3-yl)-N′-phenylurea
→ (+)-N-(1-((Adamantan-1-yl)methyl)-2,3,4,5-tetrahydro-2,4-dioxo-5-phenyl-1,5-benzodiazepin-3-yl)-N′-(*p*-hydroxyphenyl) urea
In: Dog H926, J414
Man H926
Rat H926, J414
→ (+)-N-(2,3,4,5-Tetrahydro-1-((3-hydroxyadamantan-1-yl)methyl)-2,4-dioxo-5-phenyl-1,5-benzodiazepin-3-yl)-N′-phenylurea
In: Dog H926, J414
Man H926
Rat H926, J414
→ (+)-N-(2,3,4,5-Tetrahydro-1-((4-hydroxyadamantan-1-yl)methyl)-2,4-dioxo-5-phenyl-1,5-benzodiazepin-3-yl)-N′-phenylurea
In: Dog H926
Man H926
Rat H926

N-Adenylylanthranilic acid
Formed from: Anthranilic acid
→ Cyclo-(anthraniloyl-phenylalanyl)
Probably in Penicillium K906

N-Adenylylphenylalanine
Formed from: L-Phenylalanine
→ Cyclo-(anthraniloyl-phenylalanyl)
Probably in Penicillium K906

Adenosylcobalamin
Formed from: Vitamin B12

2-(2′-6N-Adenyl)acetamidoestradiol 3-methyl ether
Formed from: 2-Bromoacetamidoestradiol 3-methyl ether

Adiantifoline ((6a*S*)-9-(4,5-dimethoxy-2-(((1*S*)-1,2,3,4-tetrahydro-6,7-dimethoxy-2-methyl-1-isoquinolinyl)methyl)phenoxy)-5,6,6a,7-tetrahydro-1,2,3,10-tetramethoxy-6-methoxy-6-methyl-4*H*-dibenzo[*de*,*g*] quinoline)
Formed from: Reticuline
Thalicarpine

Adinazolam (8-chloro-N,N-dimethyl-6-phenyl-4*H*-[1,2,4]triazolo[4,3-*a*][1,4]benzodiazepine)
→ Demethyladinazolam
In: Man E489

Adiphenine (α-diphenylacetic acid 2-(diethylamino)ethyl ester)
→ Diphenylacetic acid
In: Mouse B401
Rat B401, C313

D-(−)-Adrenaline (epinephrine)
Formed from: Adrenaline-3-sulphate
Dipivaloyl adrenaline
d-Metanephrine
d-Noradrenaline
p-Sympatol
→ D-Adrenaline quinone
Probably in: Beef A624
Corcyra A2556
Man A63
Mushroom B897
→ Adrenaline-O-sulphate (position unspecified)
In: Dog A3444
Eubacterium E432, G602
Guinea pig A3444
Man C322, C523, D583, D607, D736, D977, D979
Monkey A3444
Rabbit A3444
Rat A3444, B941, C185, C441, E962
Contrast (in beef) D607
→ 3,4-Dihydroxymandelaldehyde
In: Horse A254
Hymenolepis A3815
Man A724
Micrococcus K420
Rat A724, A1053
Salmo A1137
Scyliorhinus A1137

See also: B384, C549, F2, F212, H115
→ 1-(3,4-Dihydroxyphenyl)-2-methyliminoethanol
In: Arthrobacter B360
→ Metanephrine
In: Bufo C549
Cod F212
Dog A849
Horse A254
Man A630, A724, A3831
Monkey H783
Pig B813, C346, H783
Rabbit A2497
Rat A486, A724, A1054, D704, H115, K683
Salmo A1137
Scyliorhinus A1137
Streptomyces K652
Trichostrongylus F2
→ N-Methyladrenaline
In: Beef A1463
Dog C657
Man A2823
Monkey A2823
Rat A80, A2823
→ Paranephrine
In: Rat A486

L-Adrenaline
Formed from: *m*-Sympatol
→ L-Metanephrine
In: Rat A486
→ L-Paranephrine
In: Rat A486

D-Adrenaline quinone
Formed from: D-Adrenaline

Adrenaline-3-sulphate
This and/or Adrenaline-4-sulphate formed from D-Adrenaline
→ D-Adrenaline
In: Man E177

Adriamycin (10-((3-amino-2,3,6-trideoxy-α-L-lyxo-hexopyranosyl)oxy)-7,8,9,10-tetrahydro-6,8,11-trihydroxy-8-(hydroxyacetyl)-1-methoxy-5,12-naphthacenedione)
Formed from: Adriamycin octanoylhydrazone
Daunorubicin
Daunorubicinol
4′-O-Tetrahydropyranyladriamycin
Formed from: Propionic acid in Streptomyces A3476
→ Adriamycin aglycone
In: Man A2230, A3035
→ Adriamycinol
In: Man A37, A1989, A2230, A3035, B25
Rabbit A50, A139, B169
Rat C36
→ Adriamycin-5,12-quinol
In: Beef G401
→ 8-(1,2-Dihydroxyethyl)-7,8,9,10-tetrahydro-6,8,11-trihydroxy-1-methoxy-5,12-naphthacenedione
In: Rat A520
→ 7,8,9,10-Tetrahydro-6,8,11-trihydroxy-8-hydroxyacetyl-1-methoxy-5,12-naphthacenedione
In: Cow B371
Man A1989
Rat A520
Streptomyces A2755, A3537

Adriamycin aglycone (adriamycinone)
Formed from: Adriamycin
Adriamycin octanoylhydrazone
→ Dihydroadriamycinone
In: Streptomyces B571, B600

Adriamycin octanoylhydrazone
→ Adriamycin
In: Mouse B631
Rabbit B631
→ Adriamycin aglycone
In: Mouse B631

Adriamycinol (10(3-amino-2,3,6-trideoxy-α-L-lyxo-hexopyranosyl)oxy)-8-((1*S*)-1,2-dihydroxyethyl)-7,8,9,10-tetrahydro-6,8,11-trihydroxy-1-methoxy-5,12-naphthacenedione)
Formed from: Adriamycin
→ Deoxyadriamycinol aglycone
Probably in: Man A1989
→ 8-(1,2-Dihydroxyethyl)-7,8,9,10-tetrahydro-6,8,11-trihydroxy-1-methoxy-5,12-naphthacenedione
Probably in: Man A1989

Adriamycinol aglycone
See: 8-(1,2-Dihydroxyethyl)-7,8,9,10-tetrahydro-6,8,11-trihydroxy-1-methoxy-5,12-naphthacenedione
Adriamycinone
See: Adriamycin aglycone
Adriamycin-5,12-quinol
Formed from: Adriamycin
Aflatoxicol (aflatoxin R_0, (1*S*,6a*R*,9a*S*)-2,3,6a,9a-tetrahydro-4-methoxycyclopenta[*c*]furo[3′,2′:4,5]furo[2,3-*h*][1]benzopyran-11(1*H*)-one)
Formed from: Aflatoxin B_1
→ Aflatoxin B_1
In: Duck A582
Orange navelworm K690
Rat D322
Rainbow trout A3881
→ Aflatoxicol M_1
In: Dog A2098
Orange navelworm K690
Aflatoxicol M_1 ((1*R*,6a*R*,9a*R*)-2,3,6a,9a-tetrahydro-1,9a-dihydroxy-4-methoxycyclopenta[*c*]furo[3′,2′:4,5]furo[2,3-*h*][1]benzopyran-11(1*H*)-one)
Formed from: Aflatoxicol
Aflatoxin B_1
→ Aflatoxicol M_1-β-D-glucuronide
Probably in: Rainbow trout D226
Aflatoxicol M_1-β-D-glucuronide
Probably formed from: Aflatoxicol M_1
Aflatoxin B_1 (6a*R*,9a*S*)-2,3,6a,9a-tetrahydro-4-methoxycyclopenta[*c*]furo[3′,2′:4,5]furo[2,3-*h*][1]benzopyran-1,11(1*H*)-dione)
Formed from: Aflatoxicol
Aflatoxicol
Aflatoxin B_2
Aflatoxin G_1
Aflatoxin Q_1
O-Benzylsterigmatocystin
O-Ethylsterigmatocystin
O-Methylsterigmatocystin
O-Propylsterigmatocystin
Sterigmatocystin
Versicolorin A
Formed from: Acetate in Aspergillus A1826
Norsolorinic acid in Aspergillus A1826
→ Aflatoxicol
In: Chicken K690
Drosophila D464
Duck A582
Goat D974
Mouse K690
Navel orangeworm K690
Petroselinum G283
Rainbow trout A3881
Streptococcus D517
→ Aflatoxicol M_1
In: Trout D226
→ Aflatoxin B_{2a}
In: Drosophila D464
Monkey D802
Mouse D802
Navel orangeworm K690
Rat D802
Streptococcus D517
→ Aflatoxin B_1-2,3-oxide
Probably in: Hamster A1028
Rainbow trout D74
Rat A334, A1028
→ Aflatoxin B_1-8,9-oxide
In: Man F879, J87
Woodchuck J477
See also: D42, D420, D972
→ Aflatoxin M_1
In: Beef A1318, A2537, A2600, J676
Buffalo A2600
Chicken A929, A1523
Drosophila D464
Goat A2600, D974
Guinea pig A954
Man A1089
Monkey A983, A1523, D802
Mouse D802, E593
Navel orangeworm K690
Rainbow trout A3881
Rat A1112, A1203, A1295, A1523, A2673, A3015, A3832, C152, D42, D74, D790, D802, D803
Sheep A2600
See also: D745
→ Aflatoxin P_1
Man A1089, A3027

Monkey A983, A3027, D802
Mouse A2673, A3027, D802, E593
Rat A2673, D790, D802, D803
Contrast (in chicken, monkey and rat) A1523
→ Aflatoxin Q_1
In: Beef A2537
Chicken A1523
Goat D974
Man A1316, A3027, D420, E494
Monkey A1294, A1316, A1523, A3027, A3832, D802, E414
Mouse D802
Rat A1295, A1523, A2673, A3015, A3027, A3832, C152, D42, D74, D802, D803
→ Aflatoxin Γ
In: Aspergillus A1566
→ Aflatoxin Δ
In: Aspergillus A1566
→ 1,2-Dihydro-1,2-dihydroxyaflatoxin B_1
In: Man D420
See also J676
→ 2-(S-Glutathionyl)-2,3-dihydro-3-hydroxyaflatoxin B_1
Probably in: Rat A3733
→ 8-(S-Glutathionyl)-8,9-dihydro-9-hydroxyaflatoxin B_1
In: Mouse E593
Rat C852
See also C810, J676
→ 2-(N^7-Guanyl)-2,3-dihydro-3-hydroxyaflatoxin B_1
In: Rat D790
→ 2-Hydroxyaflatoxin B_1
In: Rat A482
→ 3-Hydroxyaflatoxin B_1
In: Monkey A1581
→ 4-Hydroxyaflatoxin B_1
In: Rat A482

Aflatoxin B_2 (6a*R*,9a*S*)-2,3,6a,8,9,9a-hexahydro-4-methoxycyclopenta[*c*]furo[3′,2′:4,5]furo[2,3-*h*][1]benzopyran-1,11(1*H*)-dione)
Formed from: Dihydro-O-methylsterigmatocystin
Dihydrosterigmatocystin
Sterigmatocystin
Versicolorin B
Probably formed from: 1,2-Dihydro-1-hydroxy-O-methylsterigmatocystin
→ Aflatoxin B_1
In: Duck A3584
→ Aflatoxin M_2
Probably in: Aspergillus D745
→ 2-Hydroxyaflatoxin B_2
In: Rat A482
→ 4-Hydroxyaflatoxin B_2
In: Rat A482

Aflatoxin B_{2a} (6a*S*,8*R*,9a*S*)-2,3,6a,8,8,9a-hexahydro-8-hydroxy-4-methoxycyclopenta[*c*]furo[3′,2′:4,5]furo[2,3-*h*][1]benzopyran-1,11(1*H*)-dione)
Formed from: Aflatoxin B_1

Aflatoxin G_1 (7a*R*,10a*S*)-3,4,7a,10a-tetrahydro-5-methoxy-1*H*,12*H*-furo[3′,2′:4,5]furo[2,3-*h*]pyrano[3,4-*c*][1]benzopyran-1,12-dione)
Formed from: O-Methylsterigmatocystin
Sterigmatocystin
→ Aflatoxin B_1
In: Aspergillus E618
→ Aflatoxin G_2
In: Aspergillus E618
→ 2-Hydroxyaflatoxin G_1
In: Rat A482
→ 4-Hydroxyaflatoxin G_1
In: Rat A482

Aflatoxin G_2 (7a*R*,10a*S*)-3,4,7a,9,10.10a-hexahydro-5-methoxy-1*H*,12*H*-furo[3′,2′:4,5]furo[2,3-*h*]pyrano[3,4-*c*][1]benzopyran-1,12-dione)
Formed from: Aflatoxin G_1
1,2-Dihydro-1-hydroxy-O-methylsterigmatocystin
Dihydro-O-methylsterigmatocystin
Dihydrosterigmatocystin

Aflatoxin M_1 (6a*R*,9a*R*)-2,3,6a,9a-tetrahydro-4-methoxycyclopenta[*c*]furo[3′,2′:4,5]furo[2,3-*h*][1]benzopyran-1,11-dione)
Formed from: Aflatoxin B_1
→ Aflatoxin M_1-β-D-glucuronide
In: Chicken A929
→ 4-O-Demethylaflatoxin M_1
Probably in: Mouse E593

Aflatoxin M_2 (2,3,6a,8,9,9a-hexahydro-4-methoxycyclopenta[*c*]furo[3′,2′:4,5]furo[2,3-*h*][1]benzopyran-1,11-dione)

Aflatoxin P_1 (6a*R*,9a*S*)-2,3,6a,9a-tetrahydro-4-hydroxycyclopenta[*c*]furo[3′,2′:4,5]furo[2,3-*h*][1]benzopyran-1,11-dione)
Formed from: Aflatoxin B_1
→ Aflatoxin P_1-β-D-glucuronide
Probably in: Rat E593

Aflatoxin Q_1 (3*S*,6a*R*,9a*S*)-2,3,6a,9a-tetrahydro-3-hydroxy-4-methoxycyclopenta[*c*]furo[3′,2′:4,5]furo[2,3-*h*][1]benzopyran-1,11-dione)
Formed from: Aflatoxin B_1
→ Aflatoxin B_1
In: Rat A3832

Aflatoxin R_0
See: Aflatoxicol

Aflatoxin Γ
Formed from: Aflatoxin B_1

Aflatoxin Δ
Formed from: Aflatoxin B_1

Aflatoxin B_1 dialcohol
Formed from: 8,9-Dihydro-8,9-dihydroxyaflatoxin B_1

Aflatoxin M_1-β-D-glucuronide
Formed from: Aflatoxin M_1

Aflatoxin P_1-β-D-glucuronide
Probably formed from: Aflatoxin P_1

Aflatoxin B_1-2,3-oxide
Probably formed from: Aflatoxin B_1

Aflatoxin B_1-8,9-oxide
Formed from: Aflatoxin B_1
→ 8,9-Dihydro-8,9-dihydroxyaflatoxin B_1
In: Rat H945
See also D42, D420, D972
→ 8-(S-Glutathionyl)-8,9-dihydro-9-hydroxyaflatoxin B_1
In: Rat D722

Afloqualone (6-amino-2-fluoromethyl-3-(2-methylphenyl)-4(3*H*)-quinazolinone)
→ N-Acetylafloqualone
In: Man C260, C624
Monkey C260, C912
Rat C260, C793a
→ S-(2-Afloqualonyl)-L-cysteine
In: Rat C804
See also C793a
→ 6-Amino-2-fluoromethyl-3-(*o*-hydroxymethylphenyl)-4(3*H*)-quinazolinone
In: Dog C912
Man C260, C624
Monkey C260
Rat C260
See also C793a
→ 6-Amino-2-hydroxymethyl-3-(*o*-tolyl)-4(3*H*)-quinazolinone
In: Dog C912
See also C793a
→ N-Glycollylafloqualone
In: Man C624
Rat C793a
→ 3′-Hydroxyafloqualone
In: Dog C912
→ 4′-Hydroxyafloqualone
In: Dog C912

S-(2-Afloqualonyl)-L-cysteine
Formed from: N-Acetyl-S-(2-afloqualonyl)-L-cysteine methyl ester
Afloqualone
→ N-Acetyl-S-(N-acetyl-2-afloqualonyl)-L-cysteine
Probably in: Dog C912
Monkey C912
→ 2-Methylthioafloqualone
Probably in: Rat C793a, C804

Afrormosin
See: 7-Hydroxy-4′,6-dimethoxyisoflavone

Agroclavine (8,9-didehydro-6,8-dimethylergoline)
Formed from: 4-Dimethylallyl-N-methyl-L-tryptophan
Probably formed from: Chanoclavine I
→ Elymoclavine
In: Claviceps B572, B809, C796
See also A752
→ Ergosine
In: Claviceps B572
→ Isosetoclavine
In: Claviceps A1553
See also H750
→ Setoclavine
In: Claviceps A1553

See also H750
(17*R*,21a)-Ajmalan-17,21-diol
See: Ajmaline
Ajmalicine (16,17-didehydro-19-methyloxayohimban-1-carboxylic acid methyl ester)
Formed from: Cathenamine
4,21-Dehydrogeissoschizine
Geissoschizine
Tryptamine
Ajmaline ((17*R*, 21a)-ajmalan-17,21-diol)
Formed from: Strictosidine
Aklanonic acid (9,10-dihydro-4,5-dihydroxy-3-(1-hydroxy-3-oxo-1-pentenyl)-9,10-dioxo-2-anthraceneacetic acid)
Probably formed from: Aklanonic acid methyl ester
Formed from glucose in Streptomyces D742
→ Aklanonic acid methyl ester
In: Streptomyces F811
→ 7-Deoxy-ε-rhodomycinone
In: Streptomyces D210
→ ε-Rhodomycinone
In: Streptomyces D210, D742
→ Pigment IP/II
In: Streptomyces D210
Aklanonic acid methyl ester
Formed from: Aklanonic acid
→ Aklanonic acid
Probably in: Streptomyces D210
→ Aklaviketone
In: Streptomyces F811
Aklaviketone (1*R*,2*R*)-2-ethyl-1,2,3,4,6,11-hexahydro-2,5,7-trihydroxy-4,6,11-trioxo-1-naphthacenecarboxylic acid methyl ester)
Formed from: Aklanonic acid methyl ester
→ Aklavinone
In: Streptomyces F811
→ Cinerubin A
In: Streptomyces F259, F357
→ Maggiemycin
Probably in: Streptomyces F811
→ ε-Rhodomycinone
In: Streptomyces F259, F357
Aklavinone (2-ethyl-1,2,3,4,6,11-hexahydro-2,4,5,7-tetrahydroxy-6,11-dioxo-1-naphthacenecarboxylic acid methyl ester)
Formed from: Aclacinomycin A
Aklaviketone
→ Aclacinomycin A
In: Streptomyces D170
See also: B571
→ Aclacinomycin B
In: Streptomyces D170
→ Aklavinone-7-rhodosaminide
Probably in: Man C594
→ Bisanhydroaklavinone
In: Streptomyces B443
→ 10-Decarbomethoxyaklavinone
In: Streptomyces B571
→ 7-Deoxyaklavinone
In: Streptomyces B443, D170
→ ε-Rhodomycinone
In: Streptomyces B443, B571, B598, F811
Aklavinone glycoside
Formed from: Aklavinone
Aklavinone-7-rhodosaminide
Probably formed from: Aklavinone
Akuammicine ((19*E*)-2,16,19,20-tetradehydrocuran-17-oic acid methyl ester)
Formed from: L-Tryptophan
Alachlor (N-chloroacetyl-2,6-diethyl-N-methoxymethylaniline)
→ N-Chloroacetyl-2,6-diethylaniline
In: Chaetomium A1560
Cunninghamella G755
Mouse F61
Monkey F61
Rat F61
See also F528
→ 1-Chloroacetyl-7-ethyl-2,3-dihydroindole
In: Chaetomium A1560
→ N-Chloroacetyl-6-ethyl-2-(1-hydroxyethyl)-N-methoxymethylaniline
In: Mouse F61
Monkey F61
Rat F61
→ 2,6-Diethyl-N-methoxymethylaniline
In: Chaetomium A1560

Rat D943

Alangiside (11*S*,12*R*,12a*S*,13a*R*)-12-ethenyl-11-(β-D-glucopyranosyloxy)-5,11,12,12a,-13,13a-hexahydro-2-hydroxy-3-methoxy-6*H*,8*H*-benzo[*a*]pyrano[3,4-*g*] quinolizin-8-one)
Formed from: Deacetylipecoside

L-Alanyl-L-alanyl-β-naphthylamide
→ β-Naphthylamine
In: Rabbit A1589
Rat A523

4-(N-(β-Alanyl)aminoethyl)-*o*-quinone
Probably formed from: N-(β-Alanyl)-3,4-dihydroxyphenethylamine
→ N-(β-Alanyl)dopamine quinone methide
In: Sarcophaga F875, F793
→ N-(β-Alanyl)noradrenaline
In: Sarcophaga F792, F793, F875

4-(2-(N-Alanyl)amino-1-hydroxyethyl)-*o*-quinone
→ N-Alanylnoradrenalone
In: Sarcophaga F792, F793

4′-(d-Alanyl)aminophenyl 4-aminophenyl sulphone
→ Bis(*p*-aminophenyl)sulphone
In: Rabbit G966

4′-(l-Alanyl)aminophenyl 4-aminophenyl sulphone
→ Bis(*p*-aminophenyl)sulphone
In: Rabbit G966

N-(l-Alanyl)-3,4-dihydroxyphenethylamine
→ 3,4-Dihydroxyphenethylamine
Probably in: Dog A1887
Guinea pig A1887
Monkey A1887
Mouse A1887
Rabbit A1887
Rat A1887

N-(β-Alanyl)-3,4-dihydroxyphenethylamine
→ 4-(N-(β-Alanyl)aminoethyl)-*o*-quinone
Probably in: Drosophila G620
→ N-(β-Alanyl)noradrenaline
In: Sarcophaga F875

N-(β-Alanyl)dopamine quinone methide
Formed from: 4-(N-(β-Alanyl)-2-aminoethyl)-*o*-quinone

L-Alanyl-N-formyl-L-kynurenine
Formed from: L-Alanyl-L-tryptophan

5-(3-(l-Alanyl)-3-(S-glutathionyl)-*o*-quinone
Probably formed from: 5-(S-Glutathionyl)-3,4-dihydroxy-L-phenylalanine

L-Alanyl-L-leucyl-β-naphthylamide
→ β-Naphthylamine
In: Rabbit A1589

L-Alanyl-β-naphthylamide
→ β-Naphthylamine
In: Agave A2745
Bacteria C146
Beef A526
Flavobacterium C73
Glycine A1596
Man A1397, A3067
Pig A3067
Pseudomonas G643
Rabbit A1589, B334
Rat A523
Streptococcus C277

L-Alanyl-*p*-nitroanilide
Formed from: N-Acetyl-L-histidyl-L-alanyl-L-alanyl-L-alanyl-*p*-nitroanilide
N-Acetyl-L-tyrosyl-L-alanyl-L-alanyl-L-alanyl-*p*-nitroanilide
3-Carboxypropionyl-L-alanyl-L-alanyl-L-alanyl-*p*-nitroanilide
Succinyl-β-alanyl-L-alanyl-L-alanyl-*p*-nitroanilide
Succinyl-L-alanyl-glycyl-L-alanyl-*p*-nitroanilide
Succinyl-L-alanyl-L-leucyl-L-alanyl-*p*-nitroanilide
Succinyl-L-asparaginyl-L-alanyl-L-alanyl-L-alanyl-*p*-nitroanilide
Succinyl-glycyl-L-alanyl-L-alanyl-*p*-nitroanilide
Succinyl-L-leucyl-L-alanyl-L-alanyl-*p*-nitroanilide
→ *p*-Nitroaniline
In: Bacillus A1006
Bacteria C146
Nocardia G343
Rat C173, C844
Streptococcus C277

N-(β-Alanyl)noradrenaline
Formed from: 4-((N-β-Alanyl)aminoethyl)-*o*-quinone

N-(β-Alanyl)-3,4-dihydroxyphenethylamine
→ N-(β-Alanyl)noradrenalone
In: Sarcophaga F793

N-Alanylnoradrenalone
Formed from: 4-(2-(N-Alanyl)amino-1-hydroxyethyl)-*o*-quinone

N-(β-Alanyl)noradrenalone
Formed from: N-(β-Alanyl)noradrenaline

L-Alanyl-L-phenylalanine
→ L-Phenylalanine
In: Brevibacterium E587

4-(3-(D-Alanyl))-*o*-quinone
Formed from: 3,4-Dihydroxy-D-phenylalanine

4-(3-(L-Alanyl))-*o*-quinone
Formed from: 3,4-Dihydroxy-L-phenylalanine
Probably formed from: 3-Aminotyrosine
3,4-Dihydroxy-D-phenylalanine
→ 5,6-Dihydroxyindole
In: Bombyx H429

4-(3-Alanyl)-*o*-quinone methyl ester
Formed from: 3,4-Dihydroxy-L-phenylalanine methyl ester

L-Alanyl-L-tryptophan
→ L-Alanyl-N-formyl-L-kynurenine
In: Wheat E24
→ L-Tryptophan
In: Brevibacterium E587

β-Alanyltyramine
Formed from: Tyramine

L-Alanyl-L-tyrosine
→ L-Tyrosine
In: Brevibacterium E587

Albendazole ((5-(prolylthio)-1*H*-benzimidazol-2-yl)carbamic acid methyl ester)
Formed from: Albendazole sulphoxide
Netobimin
→ Albendazole sulphoxide
In: Ascaris A3849
Beef C20, G550, H499, H525, K26
Dog F905
Goat H499
Man F905
Microorganism G77, H169
Moniezia A3849
Mouse A3849, C20
Pig E843
Rat C20, F905, H143, J588, K749
Sheep C20, G7, G163, G550, H499, J827
→ Methyl (5-(methylsulphinyl)-1*H*-benzimidazol-2-yl)carbamate
In: Beef C20
Mouse C20
Rat C20
Sheep C20

Albendazole sulphone
Formed from: Albendazole sulphoxide
→ 2-Amino-5-(propylsulphonyl)-1*H*-benzimidazole
Probably in: Beef C20
Mouse C20
Rat C20
Sheep C20
→ Methyl 5-((2-hydroxypropyl)sulphonyl)-1*H*-benzimidazol-2-yl)carbamate
Probably in: Beef C20
Mouse C20
Rat C20
Sheep C20
→ Methyl 5-((3-hydroxypropyl)sulphonyl)-1*H*-benzimidazol-2-yl)carbamate
Probably in: Beef C20
Mouse C20
Rat C20
Sheep C20
→ Methyl (6-hydroxy-5-(propylsulphonyl)-1*H*-benzimidazol-2-yl)carbamate
Probably in: Beef C20
Mouse C20
Rat C20
Sheep C20

Albendazole sulphoxide
Formed from: Albendazole
→ Albendazole
In: Ascaris A3849
Microorganism G77
Moniezia A3849
Mouse A3849
→ Albendazole sulphone
In: Ascaris A3849
Dog F905
Man F905
Microorganism G77

Moniezia A3849
Mouse A3849
Rat F905
Sheep G7
See also C20, H525
→ 2-Amino-5-(propylsulphinyl)-1*H*-benzimidazole
Probably in: Beef C20
Mouse C20
Rat C20
Sheep C20

Albuterol
See N-(*tert*-Butyl)-4-hydroxy-3-hydroxymethylphenylethanolamine

Alclofenac (4-allyloxy-3-chlorophenylacetic acid, 3-chloro-4-(2-propenyloxy)benzeneacetic acid)
→ Alclofenac oxide
In: Man A3905
→ 3-Chloro-4-hydroxyphenylacetic acid
In: Mouse A820
Pig A820
Rat A820

Alclofenac oxide
Formed from: Alclofenac

Alfentanyl (N-(1-(2-(4-ethyl-4,5-dihydro-5-oxo-1*H*-tetrazol-1-yl)ethyl)-4-(methoxymethyl)-4-piperidinyl)-N-phenylpropanamide)
→ Demethylalfentanyl
In: Dog E838, E895
Rat E838
→ N-(4-(Methoxymethyl)-4-piperidinyl)-N-phenylpropanamide
In: Dog E838, E895
Man H176
Rat E838, E895
→ *p*-Propionamidophenol
In: Dog E895
Man H176
Rat E895
→ Propionanilide
In: Man H176

Alinidine (N-(2,6-dichlorophenyl)-4,5-dihydro-N-(2-propenyl)-1*H*-imidazol-2-amine)
→ Clonidine
In: Man C454

Alizarin (1,2-dihydroxyanthraquinone)
→ Alizarin-β-D-glucuronide
In: Man G524, G968, J416, J783
→ Alizarin primveroside
In: Morinda J636
→ 2-β-D-Glucopyranosyloxy-1-hydroxyanthraquinone
In: Streptomyces A1497, B462, B463
See also E681, K889
→ 2-Hydroxyanthraquinon-1-yl sulphate
Probably in: Beef A1211

Alizarin-2-β-D-glucoside
See: 2-β-D-Glucopyranosyloxy-1-hydroxyanthraquinone

Alizarin-β-D-glucuronide
Formed from: Alizarin

Alizarin primeveroside
Formed from: Alizarin

Allocryptopine (5,7,8,15-tetrahydro-3,4-dimethoxy-6-methyl[1,3]benzodioxolo[5,6-*e*][2]benzazecin-14(6*H*)-one)
Formed from: Berberine
N-Methylcanadine
Tetrahydroberberine
cis-7,8,13,13a-Tetrahydro-N-methylberbinium
→ Chelerythrine
In: Macleaya C696

Alloimperatorin (9-hydroxy-4-(3-methyl-2-butenyl)-7*H*-furo[3,2-*g*][1]benzopyran-7-one)
Formed from: Mevalonic acid in Thamnosma A1013

Alloimperatorin methyl ether (9-methoxy-4-(3-methyl-2-butenyl-7*H*-furo[3,2-*g*][1]benzopyran-7-one)
Formed from: L-Phenylalanine
Formed from: Acetic acid, Glycine in Thamnosma A1012
→ Alloimperatorin methyl ether epoxide
Probably in Thamnosma A1010

Alloimperatorin methyl ether epoxide
Probably formed from: Alloimperatorin methyl ether

1-(4-Allylamino-6-(2,3-dihydroxypropyl)amino-1,3,5-triazin-2-yl)-4-(bis(*p*-fluorophenyl)methyl)piperazine
Formed from: Almitrine

→ 1-(4,6-Bis((2,3-dihydroxypropyl)amino)-1,3,5-triazin-2-yl)-4-(bis(*p*-fluorophenyl)methyl)piperazine
Probably in: Man G187

N-Allylamphetamine
Formed from: N-Allyl-N-methylamphetamine
N,N-Diallylamphetamine

4-Allylanisole
Formed from: 4-Allylphenol

Allylbenzene
→ Allylbenzene-2,3-oxide
In: Rat B528
→ Cinnamyl alcohol
In: Rat A2998
→ (1-Hydroxyallyl)benzene
In: Rat A2998

Allylbenzene-2,3-oxide
Formed from: Allylbenzene
→ 2-(S-Glutathionyl)-3-phenyl-1-propanol
In: Guinea pig G963
Mouse G963
Rat G963
→ 3-Phenylpropane-1,2-diol
In: Guinea pig G86, G963
Mouse G86, G963
Rabbit G963
Rat G963
See also: B528

4-Allylcatechol
Formed from: Safrole
Probably formed from: Eugenol
→ 4-Allylcatechol-2′,3′-oxide
In: Rat B528
→ 4-Allyl-2-methoxyphenol
In: Rat C564
→ Eugenol
In: Rat C564
See also: A3567
→ 4-Propylcatechol
Probably in: Rat A3567

4-Allylcatechol-2′,3′-oxide
Formed from: 4-Allylcatechol
→ 4-(2,3-Dihydroxypropyl)catechol
Probably in: Rat B528

N-Allyl-7,8-dihydro-14-hydroxymorphinone
See: Naloxone

(−)-N-Allyl-3,6β-dihydroxymorphinan
Formed from: Levallorphan

5-Allyl-5-(3,4-dihydroxyphenyl)barbituric acid
Probably formed from: *p*-Hydroxyalphenal

5-Allyl-2,2′-dihydroxy-5′-(1-propen-1(*E*)-yl) biphenyl
Probably formed from: Dihydromagnolol
→ 5-Allyl-2,2′-dihydroxy-5′-propylbiphenyl
Probably in: Rat D625

5-Allyl-2,2′-dihydroxy-5′-propylbiphenyl
Probably formed from: 5-Allyl-2,2′-dihydroxy-5′-(1-propen-1(*E*)-yl)biphenyl

6-Allyl-8β-(((3-dimethylaminopropyl)amino) carbonyl)ergoline
Formed from: Cabergoline

6-Allylergoline-8β-carboxylic acid
Formed from: Cabergoline
→ Ergoline-8β-carboxylic acid
Probably in: Man G832
Monkey G832
Rat G832

1-((6-Allylergolin-8β-yl)carbonyl)-1-(3-(dimethylamino)propyl)-3-ethylurea
See: Cabergoline

1-((6-Allylergolin-8β-yl)carbonyl)-3-ethylurea
Formed from: Cabergoline

(−)-N-Allyl-3-hydroxymorphinan
See: Levallorphan

1-(2-Allyl-4-hydroxyphenoxy)propane-2,3-diol
Probably formed from: 1-(2-Allylphenoxy)propane-2,3-diol

5-Allyl-3-methoxycatechol
Formed from: Myristicin

1-Allyl-3-methoxy-4,5-methylenedioxybenzene
See: Myristicin

4-Allyl-2-methoxyphenol
See: Eugenol

5-Allyl-2-methoxyphenol
Probably formed from: 4-Allylcatechol

N-Allyl-N-methylamphetamine
→ N-Allylamphetamine
In: Man K37
→ N-Methylamphetamine
In: Man K37

4-Allyl-1,2-methylenedioxybenzene
See: Safrole

N-Allylnormorphine
See: Nalorphine
3-Allyloxy-1,2-benzothiazole 1,1-dioxide
See: Oryzemate
4-Allyloxy-3-chlorophenylacetic acid
See: Alclofenac
4-Allylphenol (chavicol)
Formed from: L-Phenylalanine
Probably formed from: Estragole
→ 4-Allylanisole
In: Ocimum K692
Pimpinella J188
→ 4-Allylphenol-2,3-oxide
Probably in: Rat B528
→ *p*-Coumaryl alcohol
In: Penicillium H389, J16, K761
→ *p*-Coumaryl alcohol
In: Penicillium H389, J16, K761
4-Allylphenol-2,3-oxide
Formed from: 4-Allylphenol
→ *p*-(2,3-Dihydroxypropyl)phenol
Probably in: Rat B528
1-(2-Allylphenoxy)-3-aminopropan-2-ol
Formed from: Alprenolol
→ 1-(2-Allylphenoxy)lactic acid
Probably in: Dog A3505
Guinea pig A3505
Rat A3504, A3505
→ 1-(2-Allylphenoxy)propan-2,3-diol
Probably in: Dog A3505
Guinea pig A3505
Rat A3504, A3505
1-(2-Allylphenoxy)-3-isopropylaminopropan-2-ol
See: Alprenolol
1-(2-Allylphenoxy)lactic acid
Probably formed from: 1-(2-Allylphenoxy)-3-aminopropan-2-ol
1-(2-Allylphenoxy)propan-2,3-diol
Probably formed from: 1-(2-Allylphenoxy)-3-aminopropan-2-ol
→ 1-(2-Allyl-4-hydroxyphenoxy)propan-2,3-diol
Probably in: Rat A3504
5-Allyl-5-phenylbarbituric acid
See: Alphenal
5-Allyl-5′-(1-propen-(*E*)-yl)2,2′-dihydroxybiphenyl
Probably formed from: Dihydromagnolol

Allyl *o*-sulphamoylbenzoate
Formed from: Oryzemate
4-Allylveratrole
Formed from: Eugenol
Almitrine (6-(4-(bis(4-fluorophenyl)methyl) piperazinyl)-N,N′-di-(2-propenyl)-1,3,5-triazine-2,4-diamine)
→ 1-(4-Allylamino-6-(2,3-dihydroxypropyl) amino-1,3,5-triazin-2-yl)-4-(bis(*p*-fluorophenyl)methyl)piperazine
In: Man G187
→ Deallylalmitrine
In: Man G187
Aloe-emodin (1,8-dihydroxy-3-hydroxymethylanthraquinone)
Formed from: Aloin
Chrysophanol
Aloe-emodin bianthrone (4,4′,5,5′-tetrahydroxy-2,2′-bis(hydroxymethyl)-(9,9′-bianthracene)-10,10′(9*H*,9′*H*)-dione
Formed from: Barbaloin
Aloenin (6-(4-hydroxy-2-β-D-glucosyloxy-6-methylphenyl)-4-methoxy-2-pyrone)
→ 6-(2,4-Dihydroxy-6-methylphenyl)-4-methoxy-2-pyrone
In: Rat C105
→ 7-Hydroxy-2,5-dimethylchromone
In: Rat C105
Aloesin
See: Aloeresin B
Aloeresin A (7-hydroxy-8-(2-O-((2*E*)-3-(4-hydroxyphenyl)-1-oxo-2-propenyl-β-D-glucopyranosyl)-5-methyl-2-(2-oxopropyl-4*H*-1-benzopyran-4-one)
→ Aloeresin B
In: Microorganism G226
→ Cinnamic acid
In: Microorganism G226
Aloeresin B (aloesin, 8-β-D-glucopyranosyl-7-hydroxy-5-methyl-2-(2-oxopropyl)-4*H*-1-benzopyran-4-one)
Formed from: Aloeresin A
→ Aloesone
In: Microorganism G226
→ 7-Hydroxy-2-(2-hydroxypropyl)-5-methylchromone
In: Microorganism G226

Aloesaponarin 1 (3,8-dihydroxy-1-methyl-9,10-anthracenedione)
Formed from: Aloesaponol 1

Aloesaponol 1 (5,6,7,8-tetrahydro-3,6,9-trihydroxy-1-methyl-8-oxo-2-anthracenecarboxylic acid methyl ester)
Formed from: Acetate in Aloe A3140
→ Aloesaponarin 1
In: Aloe A3140
→ Laccaic acid D methyl ester
Probably in: Aloe A3140

Aloesone (7-hydroxy-5-methyl-2-(2-oxopropyl)-4*H*-1-benzopyran-4-one)
Formed from: Aloeresin B

Aloin (probably a mixture, mainly barbaloin)
→ Aloe-emodin
In: Man J215

Alphenal (5-allyl-5-phenylbarbituric acid)
→ Alphenal-2′,3′-oxide
In: Rat A2423
→ *p*-Hydroxyalphenal
In: Rat A2423

Alphenal-2′,3′-oxide
Formed from: Alphenal
→ 2′,3′-Dihydro-2′,3′-dihydroxyalphenal
Probably in: Rat A2423

Alpidem (6-chloro-2-(4-chlorophenyl)-N,N-dipropylimidazo[1,2-*a*]pyridine-3-acetamide)
→ 6-Chloro-2-(*p*-chlorophenyl)-N-(2-hydroxypropyl)-N-propylimidazo[1,2-*a*]pyridine-3-acetamide
In: Rat G55
→ Depropylalpidem
In: Rat G55

Alpinigenine (4b*R*,11b*S*,13*R*)-4b,5,6,7,11b,13-hexahydro-1,2,9,10-tetramethoxy-5-methyl[2]benzopyrano[3,4-*a*][3]benzazepin-13-ol)
Formed from: Muramine

Alprazolam (8-chloro-1-methyl-6-phenyl-4*H*-[1,2,4]triazolo[4,3-*a*][1,4]benzodiazepine)
→ 4-Hydroxyalprazolam
In: Man G883, J746
Monkey G883
Mouse G883
Rat G883
→ α-Hydroxyalprazolam
In: Man G883, J746
Monkey G883
Mouse G883
Rat G883

Alprenolol (1-(2-allylphenoxy)-3-isopropylamino-propan-2-ol)
→ 1-(2-Allylphenoxy)-3-aminopropan-2-ol
In: Dog A3505
Guinea pig A3505
Rat A3504, A3505
→ Alprenolol dihydrodiol
In: Dog A3505
Guinea pig A3505
Rat A3504, A3505
→ *p*-Hydroxyalprenolol
In: Dog A1259
Guinea pig A3505
Man A1259
Rat A1259, A3504, A3505

Alprenolol dihydrodiol
Formed from: Alprenolol

Alternariol (3,4′,5-trihydroxy-6′-methyldibenzo-α-pyrone)
Formed from: Alternariol methyl ether
Formed from: Acetic acid in Alternaria C448
→ 3,4′-Dihydroxy-5-methoxy-6′-methyldibenzo-α-pyrone
In: Alternaria E301, G473

Alternariol methyl ether
→ Alternariol
In: Pig E602, E747
→ Alternariol methyl ether-β-D-glucuronide
In: Pig E602, E747

Alternariol methyl ether-β-D-glucuronide
Formed from: Alternariol methyl ether

Alterporriol A (5*S*,6*R*,7*S*,8*R*)-5,6,7,8-tetrahydro-4,4′,5,6,6′,7,8-heptahydroxy-2,2′-dimethoxy-7,7′-dimethyl-(1,1′-bianthracene)-9,9′,10,10′-tetrone)
Formed from: Altersolanol A
Macrosporin

Altersolanol A (1*R*,2*S*,3*R*,4*S*)-1,2,3,4-tetrahydro-1,2,3,4,5-pentahydroxy-7-methoxy-2-methyl-9,10-anthracenedione)
→ Alterporriol A
In: Alternaria G466

Amaranth
See: 2-Hydroxy-1-(4-sulpho-1-naphthylazo)naphthalene-3,6-disulphonic acid

Amaranthin
Formed from: Betanin
(2*S*)-2,3-Dihydro-5,6-dihydroxyindole-2-carboxylic acid-5-β-D-glucuronide
3,4-Dihydroxy-L-phenylalanine
→ Celosianin II
In: Chenopodium G255
→ O-Demethylcelosianin II
In: Chenopodium G255
→ 3‴-Demethoxycelosianin II
In: Chenopodium G255
→ 5‴-Methoxycelosianin II
In: Chenopodium G255

Ambroxol (2-amino-3,5-dibromo-N-(4-hydroxycyclohexyl)benzylamine)
→ 3,5-Dibromoanthranilic acid
In: Man J794
→ 6,8-Dibromo-1,2,3,4-tetrahydro-3-(*trans*-4-hydroxycyclohexyl)quinazoline
In: Man J794

Amethocaine
See: Tetracaine

Amethopterin
See: Methotrexate

Amiben
See: 3-Amino-2,5-dichlorobenzoic acid

N-Amidino-2-(2,6-dichlorophenyl)acetamide
See: Guanfacine

Amidino-3-(2,6-dimethylphenyl)urea
Formed from: Lidamidine

Amidino-N-α-(2-naphthylsulphonylglycyl)phenylalanine piperide
Formed from: N-α-(2-Naphthylsulphonylglycyl)phenylalanine piperide

***p*-Amidinophenol**
Formed from: Pentamidine

1-(*p*-Amidinophenoxy)-5-(*p*-hydroxyamidino)phenoxy)pentane
Formed from: Pentamidine
→ Bis(*p*-(hydroxyamidino)phenoxy)pentane
Probably in: Rat G682

5-(*p*-Amidinophenoxy)pentanoic acid
Probably formed from: 5-(*p*-Amidinophenoxy)pentanol

5-(*p*-Amidinophenoxy)pentanol
Formed from: Pentamidine
→ 5-(*p*-Amidinophenoxy)pentanoic acid
Probably in: Rat G682

1-Amido-N-(*o*-carboxybenzenesulphonamido)glutamic acid
Probably formed from: 5-Amido-γ-(2,3-dihydro-1,1-dioxido-3-oxo-1,2-benzisothiazol-2-yl)glutaric acid

5-Amido-α-(2,3-dihydro-1,1-dioxido-3-oxo-1,2-benzisothiazol-2-yl)glutaric acid
Probably formed from: 3-(2,3-Dihydro-1,1-dioxido-3-oxo-1,2-benzisothiazol-2-yl)-2,6-dioxopiperidine

5-Amido-γ-(2,3-dihydro-1,1-dioxido-3-oxo-1,2-benzisothiazol-2-yl)glutaric acid
Probably formed from: 3-(2,3-Dihydro-1,1-dioxido-3-oxo-1,2-benzisothiazol-2-yl)-2,6-dioxopiperidine
→ 1-Amido-N-(*o*-carboxybenzenesulphonamido)glutamic acid
Probably in: Rat C388
→ γ-(2,3-Dihydro-1,1-dioxido-3-oxo-1,2-benzisothiazol-2-yl)glutaric acid
Probably in: Rat C388

α-Amido-3-phenoxybenzyl 2-(*p*-chlorophenyl)isovalerate
Formed from: Fenvalerate

Amiflamine (4-dimethylamino-2,α-dimethylphenethylamine)
→ Demethylamiflamine
In: Man D287

Amifloxacin (6-fluoro-1,4-dihydro-1-(methylamino)-7-(4-methyl-1-piperazinyl)-4-oxo-3-quinolinecarboxylic acid)
→ Amifloxacin-N-oxide
In: Monkey D679
Rat D679

→ 6-Fluoro-1,4-dihydro-1-methylamino-4-oxo-7-(1-piperazinyl)-3-quinolinecarboxylic acid
In: Monkey D679

Amifloxacin-N-oxide
Formed from: Amifloxacin

Amineptine (7-(10,11-dihydro-5*H*-dibenzo[*a*,*d*]cyclohepten-5-yl)amino)heptenoic acid)
→ 5-(5*H*-dibenzo[*a,d*]cycloheptan-5-yl)aminoheptanoic acid
In: Dog G202
Man G202
Rat G202

5-((2-Aminoacetamido)methyl)-1-(4-chloro-2-(*o*-chlorobenzoyl)phenyl)-N,N-dimethyl-1*H*-1,2,4-triazole-3-carboxamide (rilmazafone)
→ (8-Chloro-2-(*o*-chlorophenyl)-N,N-dimethyl-4*H*-1,2,4-triazolo[1,5-*a*][1,4]benzodiazepine-2-carboxamide
In: Dog E871a
Monkey E292, E850
→ 8-Chloro-6-(*o*-chlorophenyl)-N,N-dimethyl-4*H*-1,2,4-triazolo[1,5-*a*][1,4]benzodiazepine-2-carboxamide
In: Monkey E850

***o*-Aminoacetophenone**
Formed from: *o*-Formamidoacetophenone

***p*-Aminoacetophenone**
Formed from: 1-(*p*-Acetylphenyl)-3,3-diethyltriazene
1-(*p*-Acetylphenyl)-3,3-dimethyltriazene
p-Aminobenzoic acid
p-Nitroacetophenone
→ *p*-Methylaminoacetophenone
In: Rabbit C333
→ Levorin
In: Streptomyces C350

4-((Aminoacetyl)amino)-N-(2,6-dimethylphenyl)benzamide
Formed from: 4-(((Ethylamino)acetyl)amino)-N-(2,6-dimethylphenyl)benzamide

5-Aminoacetyl-8-chloro-5,10-dihydro-11*H*-dibenzo[*d,c*][1,4]diazepin-11-one
Formed from: 8-Chloro-5,10-dihydro-5-((2-hydroxyethyl)aminoacetyl)-11*H*-dibenzo[*d,c*][1,4]diazepin-11-one

4-Amino-N-aminoethyl-5-chloro-2-methoxybenzamide
Formed from: 4-Amino-5-chloro-N-ethylaminoethyl-2-methoxybenzamide
→ N-(2-Acetamidoethyl)-4-amino-5-chloro-2-methoxybenzamide
Probably in: Dog A3346
Man A3346
Rat A3346
→ 4-Amino-5-chloro-N-ethenyl-2-methoxybenzamide
Probably in: Dog A3346
Man A3346
Rat A3346

5-Amino-2-(4-amino-3-fluorophenyl)-6,8-difluoro-7-methyl-4*H*-1-benzopyran-1-one
→ 2-(4-Acetamido-3-fluorophenyl)-5-amino-6,8-difluoro-7-methyl-4*H*-1-benzopyran-1-one
In: Rat K471

5-Amino-3-((4′-((7-amino-1-hydroxy-3-sulpho-2-naphthalenyl)azo)(1,1′-biphenyl)-4-yl)azo)-4-hydroxy-2,7-naphthalenedisulphonic acid
See: Direct Blue 2

2-Amino-5-(*p*-anisidino)benzoquinone di-*p*-methoxyphenylimine
Formed from: *p*-Methoxyaniline

2-Aminoanthracene
→ 2-Anthracenyl nitroxide radical
In: Rabbit B184

1-Aminoanthraquinone
Formed from: 1-Methylaminoanthraquinone
→ 1-Aminoanthraquinone-β-D-glucuronide
Probably in: Sheep C808
→ 1-Amino-2-hydroxyanthraquinone
Probably in: Sheep C808
→ 1-Amino-4-hydroxyanthraquinone
Probably in: Sheep C808

2-Aminoanthraquinone
→ 2-Acetamidoanthraquinone
In: Rat A3818
→ 2-Formamidoanthraquinone
In: Rat A3818

1-Aminoanthraquinone-β-D-glucuronide
Probably formed from: 1-Aminoanthraquinone

4-Aminoantipyrine (4-amino-2,3-dimethyl-1-phenyl-3-pyrazolin-5-one)
Formed from: L-Leucyl-4-aminoantipyrine
Probably formed from: 4-Formamidoantipyrine
4-Methylaminoantipyrine
→ 4-Acetamidoantipyrine
In: Guinea pig A1941
Man A1265, A1939, A1941
Rabbit A1941
Rat A1941, E488
See also A910, A911, A1110, A1124, A1140, A1141, A1161, A3163, B10, F467, H167
→ 4-Hydroxyantipyrine
Probably in: Man A910, A911
Rat A1141
→ Rubrazonic acid
In: Rat E488
Does not yield 4-formamidoantipyrine in guinea pig, man, rabbit and rat A1941

4-Amino-5-(1-aziridinyl)-2-nitrobenzamide
Formed from: 5-(1-Aziridinyl)-2,4-dinitrobenzamide

***p*-Aminoazobenzenearsonic acid**
See: *p*-Arsanilic acid

***o*-Aminobenzaldehyde**
See: Anthranilaldehyde

***o*-Aminobenzamide**
See: Anthranilamide

4-Aminoazobenzene
Formed from: 4-Acetamidoazobenzene
4-Methylaminoazobenzene
→ 4-Acetamidoazobenzene
In: Hamster F522, G125
Mammal C812
→ 4′-Amino-4-hydroxyazobenzene
Probably in: Rat A1979, A1981, A3898
→ Aniline
In: Bacillus A3342, J366
Pseudomonas E395
Rat G133, G380, G511
→ 4-Azobenzenyl nitroxide radical
In: Rabbit B184
→ *p*-Phenylenediamine
In: Bacillus A3342, J366
Pseudomonas E395
Rat G133, G380, G511

***o*-Aminobenzaldehyde** (anthranilaldehyde)
Formed from: *o*-Acetamidobenzaldehyde
o-Nitrobenzaldehyde

***m*-Aminobenzaldehyde**
Formed from: *m*-Acetamidobenzaldehyde
m-Aminobenzyl alcohol
m-Nitrobenzaldehyde

***p*-Aminobenzaldehyde**
Formed from: *p*-Acetamidobenzaldehyde
p-Nitrobenzaldehyde

***p*-Aminobenzaldehyde thiosemicarbazone**
Formed from: *p*-Acetamidobenzaldehyde thiosemicarbazone

***p*-Aminobenzamide**
Formed from: *p*-Aminobenzonitrile
1-(*p*-Carboxamidophenyl)-3,3-dimethyltriazene
p-Nitrobenzoic acid
→ *p*-Aminobenzoic acid
Probably in: Rhodococcus H952

***p*-Aminobenzenesulphonamide**
See: Sulphanilamide

***o*-Aminobenzenesulphonic acid**
See: Orthanilic acid

***m*-Aminobenzenesulphonic acid** (metanilic acid)
Formed from: Acid Yellow 36
m-Nitrobenzenesulphonic acid
→ *m*-Acetmidobenzenesulphonic acid
In: Rabbit E399
→ *m*-Aminophenol
In: Pseudomonas E408
→ 2,3-Dihydroxybenzenesulphonic acid
Probably in: Alcaligenes K293

***p*-Aminobenzenesulphonic acid**
See: Sulphanilic acid

2-Aminobenzimidazole
Formed from: Benzimidazole 2-methylcarbamate
2-(3-Ethoxycarbonyl-2-ureido)aniline
2-(3-Ethoxycarbonyl-2-ureido)-1-(2-ureido)benzene
2-(3-Methoxycarbonyl-2-ureido)aniline
2-(3-Methoxycarbonyl-2-ureido)-1-(2-ureido)benzene
Methyl benzimidazol-2-ylcarbamate
→ 2-Amino-5-hydroxybenzimidazole

Probably in: Mouse A260
→ Benzimidazole
In: Melon A3065
See also A1478, A3064

2-Aminobenzimidazol-5-yl-β-D-glucuronide
Probably formed from: 2-Amino-5-hydroxybenzimidazole

(2-Amino-1*H*-benzimidazol-5-yl)phenylmethanone
Formed from: Mebendazole

2-Aminobenzimidazol-5-yl sulphate
Probably formed from: 2-Amino-5-hydroxybenzimidazole

***o*-Aminobenzoic acid**
See Anthranilic acid

***m*-Aminobenzoic acid**
Formed from: *m*-Acetamidobenzoic acid
m-Aminobenzonitrile
o-Aminophenol
Ethyl *m*-aminobenzoate
m-Nitrobenzoic acid
Probably formed from: *m*-Aminobenzyl alcohol
3-Carboxy-4′-dimethylaminoazobenzene
→ *m*-Acetamidobenzoic acid
In: Rabbit E399, H519
See also A3783, A3835, A3963, D383
→ *m*-Aminobenzoyl-β-D-glucuronide
In: Rabbit H519
→ *m*-Aminobenzyl alcohol
In: Coriolus G821
Desulfovibrio K615
→ 5-Amino-1,2-dihydro-1,2-dihydroxybenzoic acid
Probably in: Pseudomonas B465
→ 5-Amino-2-hydroxybenzoic acid
In: Pseudomonas H299

***p*-Aminobenzoic acid** (PABA)
Formed from: *p*-Acetamidobenzoic acid
p-Aminobenzoyl-L-glutamic acid
p-Amino-N-(2-diethylaminoethyl) benzamide
N-Benzoyl-L-tyrosyl-*p*-aminobenzoic acid
2-(4-Carboxyphenylazo)-4,5-dimethylphenol
1-(*p*-Carboxyphenylazo)-4-naphthol
Chorismic acid
2-Diethylaminoethyl *p*-aminobenzoate
Ethyl *p*-aminobenzoate
10-Formylfolic acid
10-Formylfolic tetraglutamate
Isochorismic acid
p-Methylaminobenzoic acid
p-Methyl red
p-Nitrobenzoic acid
Pteroic acid
Probably formed from: *p*-Aminobenzamide
Aniline
→ *p*-Acetamidobenzoic acid
In: Balsam pear J322
Escherichia J520
Fruit bat A3867
Hamster F114, F115
Man A3253, A3310, B163, F174, H204, J505, J507
Marmoset B201, B802
Monkey A1997
Mouse F174, G841, J405
Pig A1997
Rabbit A43, A554, A869, A1988, A2875, A3253, B110, C78, E399, H519, K377
Rana H529
Rat A1997, A3620, A3646, A3647, B36, B74, B110, B201, H152, J634
Sheep A2788, J348
Shrimp K485
Tree shrew A1997
See also A3992, D383
→ *p*-Aminoacetophenone
In: Streptomyces C350
→ *p*-Aminobenzoyl CoA
In: Beef H740
Desulphobacterium G155
→ *p*-Aminobenzoyl-β-D-glucuronic acid
In: Marmoset B201
Rabbit H519
Rat A1846, A3620, B36, B201, B740
See also: A3992
→ 4-Amino-2-decaprenylbenzoic acid
Probably in: Rat A147
→ 4-Amino-1,2-dihydro-1,2-dihydroxybenzoic acid
In: Pseudomonas B465
→ *p*-Aminohippuric acid

In: Marmoset B201
Rabbit A554, K377
Rat A1846, A3620, A3646, A3647, B36, B201 F693
→ 4-Amino-3-hydroxybenzoic acid
Probably in: Pseudomonas A2554
→ 4-Amino-2-nonaprenylbenzoic acid
Probably in: Rat A147
→ 4-Amino-2-octaprenylbenzoic acid
Probably in: Rat A147
→ *p*-Aminophenol
In: Agaricus D770, E162
→ 4-Amino-2-hydroxybenzoic acid
In: Rat A3646, A3647
→ *p*-(D-Glucosylamino)benzoic acid
In: Aerobacter A1055
→ *p*-(D-Glucuronamino)benzoic acid
Probably in Rat B740
→ *p*-Hydroxybenzoic acid
Probably in: Rat A3620

o-Aminobenzonitrile
Formed from: Benzimidazole
o-Nitrobenzonitrile
→ Anthranilamide
In: Rhodococcus H952
→ Anthranilic acid
In: Rhodococcus H952

m-Aminobenzonitrile
Formed from: *m*-Nitrobenzonitrile
→ *m*-Aminobenzoic acid
In: Rhodococcus H952

p-Aminobenzonitrile
Formed from: *p*-Acetamidobenzonitrile
p-Cyanophenyl azide
p-Benzylaminobenzonitrile
p-Nitrobenzonitrile
Probably formed from: *p*-Methylaminobenzonitrile
→ *p*-Acetamidobenzonitrile
Probably in: Rabbit E399
See also: D160
→ *p*-Aminobenzamide
In: Rhodococcus H952
→ 4-Amino-3-hydroxybenzonitrile
Probably in: Rat D160, F305
→ *p*-Hydroxylaminobenzonitrile
In: Rabbit F839
→ *p*-Methylaminobenzonitrile
Probably in: Rat F305

p-Aminobenzonitril-3-yl sulphate
Probably formed from: 4-Amino-3-hydroxybenzonitrile

3-Aminobenzophenone
Formed from: 3-Nitrobenzophenone
→ *m*-Aminophenyl phenylcarbinol
Probably in: Saccharomyces F201

4-Aminobenzophenone
Formed from: 4-Nitrobenzophenone
→ *p*-Aminophenyl phenylcarbinol
Probably in: Saccharomyces F201

6-Aminobenzothiazole
Formed from: 6-(*p*-Dimethylaminophenyl) azobenzothiazole

3-Amino-1,2,4-benzotriazine
Formed from: 3-Amino-1,2,4-benzotriazine-1-oxide

3-Amino-1,2,4-benzotriazine-1,4-dioxide
See: Tirapazamine

3-Amino-1,2,4-benzotriazine-1-oxide
Formed from: Tirapazamine
→ 3-Amino-1,2,4-benzotriazine
In: Mouse F747, H32
See also: E972

1-Amino-1*H*-benzotriazole
Formed from: 1-Benzylamino-1*H*-benzotriazole
→ 1*H*-Benzotriazole
In: Guinea pig H836

p-Aminobenzoyl CoA
Formed from: *p*-Aminobenzoic acid
→ Benzoyl CoA
In: Desulphobacterium G155

4-Amino-3-benzoyl-2,5-dihydro-1,2-dimethyl-5-oxopyrrole
Formed from: Premazepam
→ 4-Amino-3-benzoyl-2,5-dihydro-2-hydroxymethyl-1-methyl-5-oxopyrrole
Probably in: Rat C834
→ 4-Amino-3-benzoyl-2,5-dihydro-2-methyl-5-oxopyrrole
Probably in: Dog C834

4-Amino-3-benzoyl-2,5-dihydro-2-hydroxymethyl-1-methyl-5-oxopyrrole
Probably formed from: 4-Amino-3-benzoyl-2,5-dihydro-1,2-dimethyl-5-oxopyrrole

4-Amino-3-benzoyl-2,5-dihydro-2-hydroxy-2-methyl-5-oxopyrrole
Probably formed from: 4-Amino-3-benzoyl-2,5-dihydro-2-methyl-5-oxopyrrole

4-Amino-3-benzoyl-2,5-dihydro-2-methyl-5-oxopyrrole
Probably formed from: 4-Amino-3-benzoyl-2,5-dihydro-1,2-dimethyl-5-oxopyrrole
→ 4-Amino-3-benzoyl-2,5-dihydro-2-hydroxy-2-methyl-5-oxopyrrole
Probably in: Dog C834

***m*-Aminobenzoyl-β-D-glucuronic acid**
Formed from: *m*-Aminobenzoic acid

***p*-Aminobenzoyl-L-glutamic acid**
Formed from: Folic acid
Probably formed from: Methotrexate
→ *p*-Acetamidobenzoyl-L-glutamic acid
In: Mouse J405
→ *p*-Aminobenzoic acid
In Pseudomonas A1061

2-Amino-5(6)-benzoylimidazole
Probably formed from: 2-Amino-5(6)-(α-hydroxybenzyl)imidazole

2-(*p*-Aminobenzoyloxy)benzoic acid
→ Salicylic acid
In: Man A291

3-(*o*-Aminobenzoyl)propionic acid
Formed from: Ethyl indole-3-propionylglycinate

1-(*p*-Aminobenzoyl)-4-(3,4,5-trimethoxybenzamido)piperidine
→ 1-(*p*-Acetamidobenzoyl)-4-(3,4,5-trimethoxybenzamido)piperidine
In: Rat A1146

1-Amino-3,4-benzpyrene
Formed from: 1-Nitro-3,4-benzpyrene

6-Amino-3,4-benzpyrene
Formed from: 6-Nitro-3,4-benzpyrene

2-Amino-1-(N-benzylacetamido)imidazole
Formed from: Benzimidazole

***o*-Aminobenzyl alcohol**
Formed from: *o*-Nitrobenzyl alcohol
Probably formed from: Quinifuryl
o-Toluidine
→ *o*-Acetamidobenzyl alcohol
Probably in: Rat B170
→ Anthranilic acid
Probably in: Rat B170

***m*-Aminobenzyl alcohol**
Formed from: *m*-Aminobenzoic acid
m-Nitrobenzyl alcohol
Probably formed from: *m*-Toluidine
→ *m*-Acetamidobenzyl alcohol
Probably in: Rat A3783, A3963
→ *m*-Aminobenzaldehyde
In: Arion G804
Helix G804
Limax G804
→ *m*-Aminobenzoic acid
Probably in: Rat A3783, A3963, B27

***p*-Aminobenzyl alcohol**
Formed from: *p*-Nitrobenzyl alcohol
→ *p*-Aminobenzaldehyde
In: Arion G804
Helix G804
Limax G804
Mycobacterium A150

2-Amino-6-benzyl-3-ethoxycarbonyl-4,5,6,7-tetrahydrothieno[2,3-*c*]dihydropyridinium
Formed from: 2-Amino-6-benzyl-3-ethoxycarbonyl-4,5,6,7-tetrahydrothieno[2,3-*c*] pyridine

2-Amino-6-benzyl-3-ethoxycarbonyl-4,5,6,7-tetrahydrothieno[2,3-*c*]pyridine
→ 2-Amino-6-benzyl-3-ethoxycarbonyl-4,5,6,7-tetrahydrothieno[2,3-*c*] dihydropyridinium
In: Rat A899

4-Amino-N-benzyl-N-(*p*-hydroxyphenyl)-1-methylpiperidine
Formed from: Bamipine
→ 4-Amino-N-benzyl-N-(*p*-hydroxyphenyl)-1-methylpiperidine-β-D-glucuronide
In: Rat E310
→ 4-Amino-N-benzyl-N-(*p*-hydroxyphenyl)piperidine
Probably in: Rat E310

4-Amino-N-benzyl-N-(*p*-hydroxyphenyl)-1-methylpiperidine-β-D-glucuronide
Formed from: 4-Amino-N-benzyl-N-(*p*-hydroxyphenyl)-1-methylpiperidine
4-Amino-N-benzyl-N-(*p*-hydroxyphenyl)piperidine
Probably formed from: 4-Amino-N-benzyl-N-(*p*-hydroxyphenyl)-1-methylpiperidine
4-Amino-N-(1-benzyl-4-piperidinyl)-5-chloro-2-methoxybenzamide
See: Clebopride
4-Amino-N-(1-benzyl-4-piperidinyl-N-oxide)-5-chloro-2-methoxybenzamide
Formed from: Clebopride
N-(*p*-Aminobenzyl)-*p*-toluidine
Formed from: *p*-Toluidine
2-Aminobiphenyl
Formed from: 2-Acetamidobiphenyl
→ 2-Aminobiphenyl-N-glucuronide
In: Man H541, J416
→ 2-Amino-3-hydroxybiphenyl
In: Dog G389
Guinea pig E608, G389
Hamster E608, G389
Mouse E608, G389
Rat E608, G389
→ 2-Amino-5-hydroxybiphenyl
In: Dog G389
Guinea pig E608, G389
Hamster E608
Mouse E608, G389
Rat E608, G389
3-Aminobiphenyl
→ 3-Aminobiphenyl-N-glucuronide
In: Man H541
→ 3-Amino-2-hydroxybiphenyl
In: Rat E169
→ 3-Amino-4-hydroxybiphenyl
In: Rat E169
→ 3-Amino-6-hydroxybiphenyl
In: Rat E169
4-Aminobiphenyl
Formed from: 4-Acetamidobiphenyl
4-Formamidobiphenyl
4-Nitrobiphenyl
→ 4-Aminobiphenyl-N-glucuronide
In: Man H541, J416
Rat D288, F309
→ 4-Acetamidobiphenyl
In: Guinea pig F109, J408
Hamster F114, F115, F563, G757
Man F172, J408
Mouse J408
Rabbit F109, J408
Rat F109, G1, J408
→ 4′-Amino-2-hydroxybiphenyl
Hamster B172
Mouse B172
Rat B172, J408
→ 4-Amino-3-hydroxybiphenyl
In: Guinea pig J408
Hamster B172, J408
Mouse B172, J408
Rabbit J408
Rat B172, J408
→ 4′-Amino-3-hydroxybiphenyl
Probably in: Rat G1
→ 4′-Amino-4-hydroxybiphenyl
In: Guinea pig B172, J408
Hamster B172, J408
Mouse B172, J408
Rabbit J408
Rat B172, G1, J408
→ 4,4′-Azobis(biphenyl)
In: Man J95
→ N-(4-Biphenylyl)oxamic acid
In: Rat J408
→ 4-Formamidobiphenyl
In: Guinea pig F109
Rabbit F109
Rat F109
→ 4-Hydroxylaminobiphenyl
In: Beef C727
Dog A880, C727
Guinea pig B172
Hamster B172
Monkey A880
Mouse B172
Rat B172
See also A1972
→ 4-Methylaminobiphenyl
In: Rabbit F687
→ 4-Nitrosobiphenyl
In: Ram C360

2-Aminobiphenyl-N-glucuronide
Formed from: 2-Aminobiphenyl
3-Aminobiphenyl-N-glucuronide
Formed from: 3-Aminobiphenyl
4-Aminobiphenyl-N-glucuronide
Formed from: 4-Aminobiphenyl
2-Aminobiphenyl-5-yl-β-D-glucuronide
Formed from: 2-Amino-5-hydroxybiphenyl
4-Aminobiphenyl-4′-yl nitroxide radical
Formed from: Benzidine
2-Aminobiphenyl-3-yl sulphate
Formed from: 2-Amino-3-hydroxybiphenyl
2-Aminobiphenyl-5-yl sulphate
Formed from: 2-Amino-5-hydroxybiphenyl
4-Amino-N,N-bis(2-chloroethyl)-1-methoxy-2-methylnaphthalene
See: Mitoclomine
2-Amino-1,1-bis(*p*-chlorophenyl)propane
→ 1,1-Bis(*p*-chlorophenyl)propan-2-one
Probably in: Mouse A2190
Insect A2190
→ 1,1-Bis(*p*-chlorophenyl)prop-1-ene
Probably in: Mouse A2190
Insect A2190
7-Amino-3,6-bis(2,4-dihydroxyphenyl)-2-phenoxazone
Formed from: Resorcinol
6-Amino-11,13-bis(β-D-glucopyranosyl)-12,13-dihydro-1-hydroxy-5*H*-indolo[2,3-*a*]pyrrolo[3,4-*c*]carbazole-5,7(6*H*)-dione
Probably formed from: 6-Amino-13-(β-D-glucopyranosyl)-12,13-dihydro-1,11-dihydroxy-5*H*-indolo[2,3-*a*]pyrrolo[3,4-*c*]carbazole-5,7(6*H*)-dione
4-Amino-2,5-bis(S-glutathionyl)phenol
Probably formed from: 4-Amino-2-(S-glutathionyl)phenol
→ 4-Amino-2,3,5-tris(S-glutathionyl)phenol
Probably in: Rat G670
4-Amino-3′-bromoazobenzene
Formed from: 3′-Bromo-4-methylaminoazobenzene
→ 4-Amino-3′-bromo-4′-hydroxyazobenzene
Probably in: Rat A776
4-Amino-4′-bromoazobenzene
Formed from: 4′-Bromo-4-methylaminoazobenzene
2-Amino-3-(*p*-bromobenzoyl)benzeneacetic acid
See: Bromfenac
2-(2-Amino-5-bromobenzoyl)pyridine
Formed from: Bromazepam
→ 2-(2-Amino-5-bromo-3-hydroxybenzoyl)pyridine
In: Guinea pig A2834
Rabbit A2834
Rat A2834
See also A716, A924, A1121, A1439, A2213
→ 2-Amino-5-bromo-α-pyridylbenzyl alcohol
In: Guinea pig A2834
Rabbit A2834
Rat A2834
N-(4-Amino-3-bromobenzyl)-2-bromo-4-methylaniline
Formed from: 2-Bromo-4-methylaniline
***trans*-2-Amino-5-bromo-6-(3,4-dihydro-3,4-dihydroxyphenyl)-4-pyridinone**
Formed from: Bropirimine
4-Amino-5-bromo-N-(2-(diethylamino)ethyl)-2-methoxybenzamide
See: Bromopride
4′-Amino-3-bromo-4-hydroxyazobenzene
Probably formed from: 4-Amino-3′-bromoazobenzene
2-(2-Amino-5-bromo-3-hydroxybenzoyl)pyridine
Formed from: 2-(2-Amino-5-bromobenzoyl)pyridine
→ 2-(2-Amino-5-bromo-3-hydroxybenzoyl)pyridine-β-D-glucuronide
In: Dog A716
Rabbit A716
Rat A716
2-(2-Amino-5-bromo-3-hydroxybenzoyl)pyridine-β-D-glucuronide
Formed from: 2-(2-Amino-5-bromo-3-hydroxybenzoyl)pyridine
4-Amino-2-bromophenol
Formed from: *m*-Bromoaniline
4-Amino-3-bromophenol
Formed from: *o*-Bromoaniline
2-Amino-5-bromo-6-phenyl-4-pyrimidinone
See: Bropirimine

2-Amino-5-bromo-α-pyridylbenzyl alcohol
Formed from: 2-(2-Amino-5-bromobenzoyl)pyridine

2-Amino-5-bromo-3-sulphatoxybenzonitrile
Formed from: Diethyl 4-((4-bromo-2-cyanophenyl)carbamoyl) benzylphosphonate

2-Amino-3-bromo-5-trifluoromethylphenol
Formed from: 2-Bromo-4-trifluoromethylaniline
→ 2-Amino-3-bromo-5-trifluoromethylphenyl-β-D-glucuronide
In: Rat J301
→ 2-Amino-3-bromo-5-trifluoromethylphenyl sulphate
In: Rat J301

2-Amino-3-bromo-5-trifluoromethylphenyl-β-D-glucuronide
Formed from: 2-Amino-3-bromo-5-trifluoromethylphenol

2-Amino-3-bromo-5-trifluoromethylphenyl sulphate
Formed from: 2-Amino-3-bromo-5-trifluoromethylphenol

4-Amino-2-(4-butanoylhexahydro-1*H*-1,4-diazepin-1-yl)-6,7-dimethoxyquinazoline
→ 4-Amino-2-(4-butanoylhexahydro-1*H*-1,4-diazepin-1-yl)-6,7-dimethoxyquinazoline-β-D-glucuronide
In: Rabbit C373
→ 4-Amino-2-(4-butanoylhexahydro-1*H*-1,4-diazepin-1-yl)-7-hydroxy-6-methoxyquinazoline
In: Rabbit C373

4-Amino-2-(4-butanoylhexahydro-1*H*-1,4-diazepin-1-yl)-6,7-dimethoxyquinazoline-β-D-glucuronide
Formed from: 4-Amino-2-(4-butanoylhexahydro-1*H*-1,4-diazepin-1-yl)-6,7-dimethoxyquinazoline

4-Amino-2-(4-butanoylhexahydro-1*H*-1,4-diazepin-1-yl)-7-hydroxy-6-methoxyquinazoline
Formed from: 4-Amino-2-(4-butanoylhexahydro-1*H*-1,4-diazepin-1-yl)-6,7-dimethoxyquinazoline
→ 4-Amino-2-(4-butanoylhexahydro-1*H*-1,4-diazepin-1-yl)-7-hydroxy-6-methoxyquinazoline-N-β-D-glucuronide
In: Rat C373
→ 4-Amino-2-(4-butanoylhexahydro-1H-1,4-diazepin-1-yl)-7-hydroxy-6-methoxyquinazoline-O-β-D-glucuronide
In: Rat C373

4-Amino-2-(4-butanoylhexahydro-1*H*-1,4-diazepin-1-yl)-7-hydroxy-6-methoxyquinazoline-N-β-D-glucuronide
Probably formed from: 4-Amino-2-(4-butanoylhexahydro-1*H*-1,4-diazepin-1-yl)-7-hydroxy-6-methoxyquinazoline

4-Amino-2-(4-butanoylhexahydro-1*H*-1,4-diazepin-1-yl)-7-hydroxy-6-methoxyquinazoline-O-β-D-glucuronide
Formed from: 4-Amino-2-(4-butanoylhexahydro-1*H*-1,4-diazepin-1-yl)-7-hydroxy-6-methoxyquinazoline

4-Amino-3-butoxybenzoic acid
Formed from: Benoxinate
→ 4-Amino-3-butoxybenzoic acid-β-D-glucuronide
Probably in: Man E475
→ 4-Amino-3-butoxybenzoylglycine
In: Man E475
→ 4-Amino-3-hydroxybenzoic acid
Probably in: Man E475

4-Amino-3-butoxybenzoic acid-β-D-glucuronide
Probably formed from: 4-Amino-3-butoxybenzoic acid

4-Amino-3-butoxybenzoylglycine
Formed from: 4-Amino-3-butoxybenzoic acid

6-Amino-2-*sec*-butyl-4-nitrophenol
Formed from: 2-*sec*-Butyl-4,6-dinitrophenol
→ 6-Acetamido-2-*sec*-butyl-4-nitrophenol
In: Azotobacter A3824

5-Amino-N-butyl-2-(propargyloxy)benzamide
→ 5-Acetamido-N-butyl-2-hydroxybenzamide
In: Rat A1139

4-(4-Aminobutyl)-*o*-quinone
Probably formed from: 4-(3,4-dihydroxyphenyl)butylamine

4-Amino-2-(4-butyrylhexahydro-1*H*-1,4-diazepin-1-yl)-6-hydroxy-7-methoxyquinazoline
Formed from: Bunazosin
→ 4-Amino-2-(4-butyrylhexahydro-1*H*-1,4-diazepin-1-yl)-6-hydroxy-7-methoxyquinazoline-β-D-glucuronide
Probably in: Man G881

4-Amino-2-(4-butyrylhexahydro-1*H*-1,4-diazepin-1-yl)-7-hydroxy-6-methoxyquinazoline
Formed from: Bunazosin

4-Amino-2-(4-butyrylhexahydro-1*H*-1,4-diazepin-1-yl)-7-hydroxy-6-methoxyquinazoline-β-D-glucuronide
Probably in: Man G881

4-Amino-2-(4-butyrylhexahydro-1*H*-1,4-diazepin-1-yl)-6-hydroxy-7-methoxyquinazoline-β-D-glucuronide
Probably formed from: 4-Amino-2-(4-butyryl-hexahydro-1*H*-1,4-diazepin-1-yl)-6-hydroxy-7-methoxyquinazoline

4-Amino-2-(4-butyrylhexahydro-1*H*-1,4-diazepin-1-yl)-7-hydroxy-6-methoxyquinazoline-β-D-glucuronide
Probably formed from: 4-Amino-2-(4-butyryl-hexahydro-1*H*-1,4-diazepin-1-yl)-7-hydroxy-6-methoxyquinazoline

Aminocarb (4-Dimethylamino-*m*-tolyl N-methylcarbamate)
→ Methylamino-*m*-tolyl N-methylcarbamate
In: Crayfish B232

2-Amino-α-carboline (2-amino-1*H*-pyrido[2,3-*b*]indole)
→ 2-Amino-3- hydroxy-α-carboline
In: Man H697b
Mouse H697b
Rat H697b
→ 2-Amino-6-hydroxy-α-carboline
In: Man H697b
Mouse H697b
Rat H697b
→ 2-Hydroxylamino-α-carboline
Probably in: Man H697b
Mouse H697b
Rat H697b

4′-Amino-3-carboxyazobenzene
Formed from: 3-Carboxy-4′-methylaminoazobenzene
Probably formed from: 4′-Amino-3-formylazobenzene

4′-Amino-4-carboxyazobenzene
Formed from: 4-Carboxy-4′-methylaminoazobenzene

2-Amino-1-(4-carboxy-2,5-dimethoxyphenyl)propane
Probably formed from: 2-Amino-1-(4-hydroxymethyl-2,5-dimethoxyphenyl)propane

2-Amino-2′-carboxydiphenyl ether
Probably formed from: 2-Amino-2′-hydroxymethyldiphenyl ether
→ 10,11-Dihydro-11(10*H*)-oxodibenz[*b,f*]-1,4-oxazepine
Probably in: Rat C495

6-Amino-5-((1-carboxyethyl)oxy)-1,3-cyclohexadiene-1-carboxylic acid
→ Anthranilic acid
In: Escherichia E400

6-Amino-N-(1-(1*H*)-(1-carboxy-1-heptyl))imidazol-4-yl)phthalamidic acid
→ 3-Aminophthalic acid
In: Xanthobacter J253

4-Amino-2-(4-(2-carboxy-2-hydroxybutyryl)piperazin-1-yl-6,7-dimethoxyquinazoline
Formed from: Terazocin

4-Amino-3-carboxy-5-hydroxy-1-(*p*-sulphophenyl)pyrazole (aminopyrazolone)
Formed from: 3-Carboxy-5-hydroxy-1-(*p*-sulphophenyl)-4-(*p*-sulphophenylazo)pyrazole
→ Sulphanilic acid
In: Rat A2528

1-(4-Amino-5-carboxymethyl-2-chlorophenethyl)-4-(1,2-benzisothiazol-3-yl)piperazine sulphoxide
Probably formed from: Ziprasidone sulphoxide

2-Amino-3-carboxymethyl-3,4-dihydro-6-piperidinoquinazoline
Formed from: 2,3,4,5-Tetrahydro-7-piperidinoimidazo[2,1-*b*]quinazolin-2(1*H*)-one

4-Amino-N-(6-carboxy-4-methyl-2-pyrimidinyl)benzenesulphonamide
Probably formed from: 4-Amino-N-(6-hydroxymethyl-4-methyl-2-pyrimidinyl) benzenesulphonamide

4-Amino-N-(5-carboxypentyl)-3,5-dichlorophenylethanolamine
Formed from: Picumeterol
→ 4-Amino-N-(3-carboxypropyl)-3,5-dichlorophenylethanolamine
Probably in: Dog H323
Rat H323

2-Amino-3-carboxyphenyl sulphate
Formed from: 3-Hydroxyanthranilic acid

4-Amino-N-(3-carboxypropyl)-3,5-dichlorophenylethanolamine
Probably formed from: 4-Amino-N-(5-carboxypentyl)-3,5-dichlorophenylethanolamine

4-Amino-N-(4-carboxy-2-pyrimidinyl) benzenesulphonamide
Probably formed from: 4-Amino-N-(4-hydroxymethyl-2-pyrimidinyl)benzenesulphonamide

3-Aminocatechol
Formed from: *o*-Aminophenol

4-Aminocatechol
Formed from: 4-Amino-2-hydroxybenzoic acid
5-Amino-2-hydroxybenzoic acid
m-Aminophenol
p-Aminophenol

4-Amino-3′-chloroazobenzene
Formed from: 3′-Chloro-4-methylaminoazobenzene
→ 4-Amino-3′-chloro-4-hydroxyazobenzene
Probably in: Rat A776

4-Amino-4′-chloroazobenzene
Formed from: 4′-Chloro-4-methylaminoazobenzene

4-Amino-2-chlorobenzaldehyde
Probably formed from: 4-Amino-2-chlorobenzyl alcohol

2-Amino-4-chloro-1,5-benzenedisulphonamide
Formed from: Hydrochlorthiazide

2-Amino-5-chlorobenzhydrol
Formed from: Inabenfide

2-Amino-5-chlorobenzoic acid
Formed from: 5-Chloroindole-3-acetic acid
→ 2-Amino-5-chlorophenol
In: Agaricus E162
→ Anthranilic acid
In: Desulfomonile G152

4-Amino-2-chlorobenzoic acid
Formed from: Chloroprocaine
Probably formed from: 3-Chloro-4-methylaniline
→ 4-Acetamido-2-chlorobenzoic acid
In: Bluegill sunfish J230
→ 4-Amino-2-chlorophenol
In: Agaricus E162

4-Amino-2-chlorobenzoic acid 2-(diethylamino)ethyl ester
See: Chloroprocaine

2-Amino-5-chlorobenzophenone
Formed from: Chlordiazepoxide
7-Chloro-1,3-dihydro-3-hydroxy-5-phenyl-2*H*-1,4-benzodiazepin-2-one
Probably formed from: 2-Amino-5-nitrobenzophenone
→ 5-Chloro-2-methylaminobenzophenone
Probably in: Dog A3843
Man A3843
Rat A3843

2-Amino-5-chlorobenzophenone amidinohydrazide
Formed from: 2-Amino-5-chlorobenzophenone N-hydroxyamidinohydrazide
→ 2-Amino-5-chlorobenzophenone N-hydroxyamidinohydrazide
In: Rabbit G939
Rat G939
→ 2-Amino-5-chloro-4′-hydroxybenzophenone amidinohydrazide
Probably in: Rabbit G939
Rat G939

2-Amino-5-chlorobenzophenone N-hydroxyamidinohydrazide
Formed from: 2-Amino-5-chlorobenzophenone amidinohydrazide
→ 2-Amino-5-chlorobenzophenone amidinohydrazide

In: Rabbit G939
Rat G939

2-Amino-5-chlorobenzoxazole (zoxazolamine)
→ 2-Amino-5-chloro-6-hydroxybenzoxazole
In: Rat A2160, C258, D39

4-Amino-2-chlorobenzyl alcohol
Formed from: 2-Chloro-4-methylaniline
→ 4-Amino-2-chlorobenzaldehyde
Probably in: Rat G580

N-(4-Amino-2-chlorobenzyl)-2-chloro-4-methylaniline
Formed from: 2-Chloro-4-methylaniline

3-Amino-4-(*p*-chlorobenzylideneamino)-5-(*p*-chlorophenyl)-4*H*-1,2,4-triazene
Formed from: Robenidine

N-(2-Amino-6-chlorobenzyl)-N-methylglycinemorpholide
Formed from: Fominoben

4-Amino-5-chloro-N-diethylaminoethyl-2-hydroxybenzamide
Formed from: Metoclobamide

4-Amino-5-chloro-N-diethylaminoethyl-2-methoxybenzamide
See: Metoclopramide

4-Amino-5-chloro-N-((2-diethylamino)ethyl)-2-((methylsulphinyl)ethoxy)benzamide
Formed from: 4-Amino-5-chloro-N-((2-diethylamino)ethyl)-2-((methylthio)ethoxy) benzamide
→ 4-Amino-5-chloro-N-((2-diethylamino)ethyl)-2-((methylthio) ethoxy)benzamide
In: Rat G899

4-Amino-5-chloro-N-((2-diethylamino)ethyl)-2-((methylthio)ethoxy)benzamide
Formed from: 4-Amino-5-chloro-N-((2-diethylamino)ethyl)-2-((methylsulphinyl) ethoxy)benzamide
→ 4-Amino-5-chloro-N-((2-diethylamino)ethyl)-2-((methylsulphinyl)ethoxy)benzamide
In: Rat G899

4-Amino-5-chloro-N-ethenyl-2-methoxybenzamide
Probably formed from: 4-Amino-N-aminoethyl-5-chloro-2-methoxybenzamide

4-Amino-5-chloro-2-ethoxy-N-((4-((4-fluorophenyl)methyl)-2-morpholinyl)methyl)benzamide
See: Mosapride

4-Amino-5-chloro-2-ethoxy-3-hydroxy-N-((1,4-oxazin-2-yl)methyl)benzamide
Probably formed from: 4-Amino-5-chloro-2-ethoxy-N-((1,4-oxazin-2-yl)methyl) benzamide

4-Amino-5-chloro-2-ethoxy-3-hydroxy-N-((5-oxo-1,4-oxazin-2-yl)methyl)benzamide
Probably formed from: 4-Amino-5-chloro-2-ethoxy-N-((5-oxo-1,4-oxazin-2-yl)methyl) benzamide

4-Amino-5-chloro-2-ethoxy-N-((1,4-oxazin-2-yl)methyl)benzamide
Formed from: Mosapride
→ 4-Amino-5-chloro-2-ethoxy-3-hydroxy-N-((1,4-oxazin-2-yl)methyl) benzamide
Probably in: Rat G984
→ 4-Amino-5-chloro-2-ethoxy-N-((5-oxo-1,4-oxazin-2-yl)methyl) benzamide
Probably in: Rat G984

4-Amino-5-chloro-2-ethoxy-N-((5-oxo-1,4-oxazin-2-yl)methyl)benzamide
Probably formed from: 4-Amino-5-chloro-2-ethoxy-N-((1,4-oxazin-2-yl)methyl) benzamide
→ 4-Amino-5-chloro-2-ethoxy-3-hydroxy-N-((5-oxo-1,4-oxazin-2-yl)methyl)benzamide
Probably in: Rat G984

4-Amino-5-chloro-N-ethylaminoethyl-2-hydroxybenzamide
Formed from: 4-Amino-5-chloro-N-ethylaminoethyl-2-methoxybenzamide

4-Amino-5-chloro-2-ethylaminoethyl-2-methoxybenzamide
Formed from: Metoclopramide
→ 4-Amino-N-aminoethyl-5-chloro-2-methoxybenzamide
In: Dog A3346
Rabbit A1799
See also: A2400

→ 4-Amino-5-chloro-N-ethylaminoethyl-2-hydroxybenzamide
In: Rabbit A1799
→ 4-Amino-5-chloro-N-(2′-hydroxyethyl)-2-methoxybenzamide
In: Rabbit A1799

2-Amino-N-(2-chloroethyl)-6-nitro-N-propyl-4-trifluoromethylaniline
Formed from: Fluchloralin

2-Amino-5-chloro-2′-fluorobenzophenone
Formed from: Ethyl loflazepate

6-Amino-9-(2-chloro-6-fluorobenzyl)purine
→ 6-Amino-9-(2-chloro-6-fluorobenzyl) purine-N-oxide
In: Chicken A2733
→ 2-Chloro-6-fluorobenzyl alcohol
In: Mouse A2735
Rat A2735

6-Amino-9-(2-chloro-6-fluorobenzyl)purine-N-oxide
Formed from: 6-Amino-9-(2-chloro-6-fluorobenzyl)purine

4-Amino-5-chloro-N-(1-(3-(3-fluoro-4-hydroxyphenoxy)propyl)-3-methoxy-4-piperidinyl)-2-methoxybenzamide
Formed from: Cisapride

***cis*-4-Amino-5-chloro-N-(1-(3-(4-fluoro-2-hydroxyphenoxy)propyl)-3-methoxy-4-piperidinyl)-2-methoxybenzamide**
Formed from: Cisapride

2-Amino-4-chloro-5-fluorophenol
Formed from: 3-Chloro-4-fluoroaniline
→ 2-Acetamido-4-chloro-5-fluorophenol
Probably in: Dog A3942
Rat A3942
→ 2-Amino-4-chloro-4-fluorophenyl sulphate
In: Dog A3942
Rat A3942

5-Amino-2-chloro-4-fluorophenol
Formed from: N-(4-Chloro-2-fluoro-5-((1-methyl-2-propynyl)oxy)phenyl)-3,4,5,6-tetrahydrophthalimide
→ 5-Amino-2-chloro-4-fluorophenyl-β-D-glucuronide
Probably in: Rat G568, G704
→ 5-Amino-2-chloro-4-fluorophenyl sulphate
Probably in: Rat G568, G704

***cis*-4-Amino-5-chloro-N-(1-(3-(4-fluorophenoxy)propyl)-3-hydroxy-4-piperidinyl)-2-methoxybenzamide**
Formed from: Cisapride

5-Amino-2-chloro-4-fluorophenyl-β-D-glucuronide
Probably formed from: 5-Amino-2-chloro-4-fluorophenol

2-Amino-4-chloro-5-fluorophenyl sulphate
Probably formed from: 2-Amino-4-chloro-5-fluorophenol

5-Amino-2-chloro-4-fluorophenyl sulphate
Probably formed from: 5-Amino-2-chloro-4-fluorophenol

4α-Amino-8-chloro-1,2,3,4,4a,10a-hexahydro-10*H*-benzopyrano[3,2-*c*]pyridin-10-ylacetic acid lactam
Formed from: 4α-Amino-8-chloro-1,2,3,4,4a,10a-hexahydro-2-methyl-10*H*-benzopyrano[3,2-*c*]pyridin-10-ylacetic acid lactam

4α-Amino-8-chloro-1,2,3,4,4a,10a-hexahydro-2-methyl-10*H*-benzopyrano[3,2-*c*]pyridin-10-ylacetic acid lactam
→ 4α-Amino-8-chloro-1,2,3,4,4a,10a-hexahydro-10*H*-benzopyrano[3,2-*c*] pyridin-10-ylacetic acid lactam
In: Dog B629
Rat B629

4′-Amino-3-chloro-4-hydroxyazobenzene
Probably formed from: 4-Amino-3′-chloroazobenzene

2-Amino-5-chloro-4′-hydroxybenzophenone amidinohydrazone
Probably formed from: 2-Amino-5-chlorobenzophenone amidinohydrazone

2-Amino-5-chloro-6-hydroxybenzoxazole
Formed from: 2-Amino-5-chlorobenzoxazole

4-Amino-5-chloro-N-(1-(α-hydroxy)benzyl-4-piperidinyl)-2-methoxybenzamide
Formed from: Clebopride

4-Amino-5-chloro-N-(2′-hydroxyethyl)-2-methoxybenzamide
Formed from: 4-Amino-5-chloro-N-ethylaminoethyl-2-methoxybenzamide
Metoclopramide

2-Amino-7-chloro-3-hydroxy-3*H*-phenoxazine
Probably formed from: 2-Amino-7-chloro-3*H*-phenoxazin-3-one

4-Amino-5-chloro-N-(4-(N-hydroxypiperidyl))-2-methoxybenzamide
Formed from: 4-Amino-5-chloro-2-methoxy-N-(4-piperidyl)benzamide

***cis*-4-Amino-5-chloro-N-(1-(3-hydroxypropyl)-3-methoxy-4-piperidinyl)-2-methoxybenzamide**
Formed from: Cisapride

N-(*p*-((2-Amino-5-chloro-4-hydroxyquinazolin-6-yl)aminomethyl)benzoyl)diglutamylglutamic acid
Probably formed from: N-(*p*-((2-Amino-5-chloro-4-hydroxyquinazolin-6-yl)aminomethyl)benzoyl)23glutamylglutamic acid
→ N-(*p*-((2-Amino-5-chloro-4-hydroxyquinazolin-6-yl)aminomethyl)benzoyl) triglutamylglutamic acid
Probably in: Pig F760

N-(*p*-((2-Amino-5-chloro-4-hydroxyquinazolin-6-yl)aminomethyl)benzoyl)glutamic acid
→ N-(*p*-((2-Amino-5-chloro-4-hydroxyquinazolin-6-yl)aminomethyl) benzoyl)glutamylglutamic acid
In: Pig F760

N-(*p*-((2-Amino-5-chloro-4-hydroxyquinazolin-6-yl)aminomethyl)benzoyl)glutamylglutamic acid
Probably formed from: N-(*p*-((2-Amino-5-chloro-4-hydroxyquinazolin-6-yl)aminomethyl)benzoyl)glutamic acid
→ N-(*p*-((2-Amino-5-chloro-4-hydroxyquinazolin-6-yl)aminomethyl) benzoyl)diglutamylglutamic acid
Probably in: Pig F760

N-(*p*-((2-Amino-5-chloro-4-hydroxyquinazolin-6-yl)aminomethyl)benzoyl)tetraglutamyl glutamic acid
Probably formed from: N-(*p*-((2-Amino-5-chloro-4-hydroxyquinazolin-6-yl)aminomethyl)benzoyl)triglutamylglutamic acid

N-(*p*-((2-Amino-5-chloro-4-hydroxyquinazolin-6-yl)aminomethyl)benzoyl)triglutamylglutamic acid
Probably formed from: N-(*p*-((2-Amino-5-chloro-4-hydroxyquinazolin-6-yl)aminomethyl)benzoyl)diglutamylglutamic acid
→ N-(*p*-((2-Amino-5-chloro-4-hydroxyquinazolin-6-yl)aminomethyl)benzoyl) tetraglutamylglutamic acid
Probably in: Pig F760

N-(*p*-((2-Amino-5-chloro-4-hydroxyquinazolin-6-yl)methylamino)benzoyl)diglutamylglutamic acid
Probably formed from: N-(*p*-((2-Amino-5-chloro-4-hydroxyquinazolin-6-yl)methylamino)benzoyl)glutamic acid
→ N-(*p*-((2-Amino-5-chloro-4-hydroxyquinazolin-6-yl)methylamino) benzoyl)triglutamylglutamic acid
Probably in: Pig F760

N-(*p*-((2-Amino-5-chloro-4-hydroxyquinazolin-6-yl)methylamino)benzoyl)glutamic acid
→ N-(*p*-((2-Amino-5-chloro-4-hydroxyquinazolin-6-yl)methylamino)benzoyl)diglutamylglutamic acid
Probably in: Pig F760

N-(*p*-((2-Amino-5-chloro-4-hydroxyquinazolin-6-yl)methylamino)benzoyl)tetraglutamylglutamic acid
Probably formed from: N-(*p*-((2-Amino-5-chloro-4-hydroxyquinazolin-6-yl)methylamino)benzoyl)triglutamylglutamic acid

N-(*p*-((2-Amino-5-chloro-4-hydroxyquinazolin-6-yl)methylamino)benzoyl)triglutamylglutamic acid
Probably formed from: N-(*p*-((2-Amino-5-chloro-4-hydroxyquinazolin-6-yl) methylamino)benzoyl)diglutamylglutamic acid
→ N-(*p*-((2-Amino-5-chloro-4-hydroxyquinazolin-6-yl)methylamino) benzoyl)tetraglutamylglutamic acid
Probably in: Pig F760

4-Amino-5-chloro-2-methoxybenzamide
Formed from: Metoclopramide

4-Amino-5-chloro-2-methoxybenzoic acid
Formed from: Clebopride

2-((4-Amino-5-chloro-2-methoxybenzoyl)amino) acetic acid
Formed from: Metoclopramide

4-Amino-5-chloro-2-methoxy-N-(3-methoxy-2-oxo-4-piperidinyl)benzamide
Probably formed from: 4-Amino-5-chloro-2-methoxy-N-(3-methoxy-4-piperidinyl) benzamide

4-Amino-5-chloro-2-methoxy-N-(5-methoxy-2-oxo-4-piperidinyl)benzamide
Probably formed from: 4-Amino-5-chloro-2-methoxy-N-(3-methoxy-4-piperidinyl) benzamide

4-Amino-5-chloro-2-methoxy-N-(3-methoxy-4-piperidinyl)benzamide
Formed from: Cisapride
→ 4-Amino-5-chloro-2-methoxy-N-(3-methoxy-2-oxo-4-piperidinyl)benzamide
Probably in: Rat F926
→ 4-Amino-5-chloro-2-methoxy-N-(5-methoxy-2-oxo-4-piperidinyl)benzamide
Probably in: Rat F926

4-Amino-5-chloro-2-methoxy-N-(2-oxo-4-piperidinyl)benzamide
Probably formed from: 4-Amino-5-chloro-2-methoxy-N-(2-oxo-4-piperidinyl)benzamide

4-Amino-5-chloro-2-methoxy-N-4-piperidinyl) benzamide (debenzylclebopride)
Formed from: Clebopride
→ 4-Amino-5-chloro-N-(4-(N-hydroxypiperidinyl))-2-methoxybenzamide
In: Rabbit B180
→ 4-Amino-5-chloro-2-methoxy-N-(2-oxo-4-piperidinyl)benzamide
Probably in: Rabbit B66
→ 4-Amino-5-chloro-2-methoxy-N-(4′-(Δ^1-piperidinyl-N-oxide))benzamide
In: Rabbit B180

4-Amino-5-chloro-2-methoxy-N-(4′-(Δ^1-piperidinyl-N-oxide))benzamide
Formed from: 4-Amino-5-chloro-2-methoxy-N-(4-piperidinyl)benzamide

4-Amino-3-chloro-α-methylphenylacetic acid
Formed from: Pirprofen
→ 4-Acetamido-3-chloro-α-methylphenylacetic acid
In: Man C41
Mouse C41
Rat C41

4-Amino-3-chloro-4′-nitrobiphenyl
Formed from: 4,4′-Dinitrobiphenyl
→ 4-Acetamido-3-chloro-4′-nitrobiphenyl
Probably in: Rat A721

2-Amino-4-chlorophenol
Formed from: *m*-Chloroaniline
→ 2-Acetamido-4-chlorophenol
Probably in: Pseudomonas K2
→ 2-Amino-4-chloromuconic semialdehyde
In: Pseudomonas J255, J329

2-Amino-5-chlorophenol
Formed from: 2-Amino-5-chlorobenzoic acid
p-Chloroaniline
1-Chloro-4-nitroaniline
p-Chloronitrobenzene
Probably formed from: *p*-Chlorophenylhydroxylamine
→ 2-Acetamido-5-chlorophenol
In: Rat G742
See also B799, C879, F729, G519, K2
→ 2-Amino-5-chlorophenyl sulphate
In: Monkey F729
Mouse F729
Rat F729
→ 5-Chloropicolinic acid
In: Comamonas K80

4-Amino-2-chlorophenol
Formed from: 4-Amino-2-chlorobenzoic acid
m-Chloroaniline
Probably formed from: *p*-Chloroaniline

4-Amino-3-chlorophenol
Formed from: *o*-Chloroaniline
→ 4-Amino-3-chlorophenyl-β-D-glucuronide
In: Rat J447
→ 4-Amino-3-chlorophenyl sulphate
In: Rat J447

5-Amino-2-chlorophenol
Probably formed from: *p*-Chloroaniline

2-Amino-7-chloro-3*H*-phenoxazin-3-one
Formed from: *p*-Chloroaniline
→ 2-Amino-7-chloro-3-hydroxy-3*H*-phenoxazine
Probably in: Bacillus A3080

2-Amino-N-(*o*-chlorophenyl)benzamide
Formed from: Mecloqualone

N-(4-Amino-2-chlorophenyl)benzamide
Formed from: N-(2-chloro-4-nitrophenyl) benzamide

2-Amino-7-chloro-5-phenyl-3*H*-1,4-benzodiazepine-4-oxide
Formed from: Chlordiazepoxide

γ-Amino-β-(*p*-chlorophenyl)butyric acid (baclofen)
→ β-(*p*-Chlorophenyl)-γ-hydroxybutyric acid
In: Rat A3273

N-(4-Amino-2-chlorophenyl)-5-chloro-2-hydroxybenzamide
Formed from: 5-Chloro-2-(2-chloro-4-nitrophenyl)-2-hydroxybenzamide

2-Amino-2-(*o*-chlorophenyl)cyclohexanone
See: N-Demethylketamine

2-Amino-2-(*o*-chlorophenyl)cyclohex-3-en-1-one
Probably formed from: N-Demethylketamine

2-Amino-2-(*o*-chlorophenyl)cyclohex-5-en-1-one
Probably formed from: N-Demethylketamine

7-Amino-5-(*o*-chlorophenyl)-1,3-dihydro-2*H*-1,4-benzodiazepin-2-one
Formed from: Clonazepam
→ 7-Acetamido-5-(*o*-chlorophenyl)-1,3-dihydro-2*H*-1,4-benzodiazepin-2-one
In: Man B945

8-Amino-6-(*o*-chlorophenyl)-1,2-dihydro-2-(N-methylpiperazinylmethyl)-1*H*,4*H*-imidazo [1,2-5][1,4]benzodiazepin-1-one
Formed from: Loprazolam

4-Amino-1-(*p*-chlorophenyl)-2,3-dimethylbutan-2-ol
Formed from: 1-(*p*-chlorophenyl)-2,3-dimethyl-4-methylaminobutan-2-ol

4-Amino-3-chlorophenyl-β-D-glucuronide
Formed from: 4-Amino-3-chlorophenol

5-Amino-4-chloro-2-phenyl-3(2*H*)-pyridazinone
See: Chloridazon

N-(4-Amino-2-chlorophenyl)salicylamide
Formed from: N-(2-chloro-4-nitrophenyl) salicylamide

2-Amino-5-chlorophenyl sulphate
Formed from: 2-Amino-5-chlorophenol

4-Amino-3-chlorophenyl sulphate
Formed from: 4-Amino-3-chlorophenol

4-Amino-7-chloroquinoline
Formed from: Chloroquine

2-Amino-3-chloro-5-trifluoromethylphenol
Formed from: 2-chloro-4-trifluoromethyl-aniline
→ 2-Amino-3-chloro-5-trifluoromethyl-phenyl-β-D-glucuronide
In: Rat J551
→ 2-Amino-3-chloro-5-trifluoromethylphenyl sulphate
In: Rat J551

2-Amino-3-chloro-5-trifluoromethylphenyl-β-D-glucuronide
Formed from: 2-Amino-3-chloro-5-trifluoromethylphenol

2-Amino-3-chloro-5-trifluoromethylphenyl sulphate
Formed from: 2-Amino-3-chloro-5-trifluoromethylphenol

6-Aminochrysene
Formed from: 6-Nitrochrysene
→ 6-Amino-1,2-dihydro-1,2-dihydroxychrysene
Probably in: Man F167
Mouse E658
Contrast: A1779

α-Aminocinnamic acid
Formed from: α-Acetamidocinnamic acid
→ L-Phenylalanine
With L-phenylalanine dehydrogenase E587

2-Amino-*m*-cresol (2-amino-3-methylphenol)
Formed from: 4-Methyl-3-nitrophenol
o-Toluidine
→ 2-Amino-3-methylmuconic semialdehyde
In: Pseudomomas J255, J329

2-Amino-*p*-cresol (2-amino-4-methylphenol)
→ 2-Amino-4-methylmuconic semialdehyde
In: Escherichia K583
Pseudomomas J329

3-Amino-*p*-cresol (3-amino-4-methylphenol)
Formed from: 4-Methyl-3-nitrophenol

4-Amino-*o*-cresol (4-amino-2-methylphenol)
Formed from: *m*-Toluidine
Probably formed from: 4′-Dimethylamino-4-hydroxy-3-methylazobenzene
4-Dimethylamino-3′-methylazobenzene
→ 4-Acetamido-*o*-cresol
Probably in: Rat A3783, B27

4-Amino-*m*-cresol (4-amino-3-methylphenol)
Formed from: 4-Acetamido-3-methylphenol
o-Toluidine
→ 4-Acetamido-*m*-cresol
Probably in: Rat B170
→ 4-Amino-*m*-cresyl-β-D-glucuronide
Probably in: Rat B170

→ 4-Amino-*m*-cresyl sulphate
Probably in: Rat B170

5-Amino-*o*-cresol (5-amino-2-methylphenol)
Formed from: 2-Methyl-5-nitrophenol

6-Amino-*m*-cresol (2-amino-5-methylphenol)
→ 2-Amino-5-methylmuconic semialdehyde
In: Pseudomomas J255, J329
→ 2-Amino-3,4,4a,10a-tetrahydro-4a,7-dimethylphenoxazin-3-one
In: Streptomyces F209

4-Amino-*m*-cresyl-β-D-glucuronide
Probably formed from: 4-Amino-*m*-cresol

4-Amino-*m*-cresyl sulphate
Probably formed from: 4-Amino-*m*-cresol

2-Amino-5-cyanophenol
Probably formed from: 5-Cyano-2-methylaminophenol
→ 2-Amino-5-cyanophenyl sulphate
Probably in: Mouse D367

2-Amino-5-cyanophenyl sulphate
Probably formed from: 2-Amino-5-cyanophenol

S-Aminocyclohexyl-2-benzothiazolyl sulphenamide
→ N-Acetyl-S-(2-benzothiazolyl)-L-cysteine
In: Rat H767
→ 2-Mercaptobenzothiazole
In: Rat H767

2-Amino-5-cyclopropylcarbinolimidazole
Probably formed from: 2-Amino-5-cyclopropylcarbonylimidazole

2-Amino-5-cyclopropylcarbonylimidazole
Formed from: Ciclobendazole
→ 2-Amino-5-cyclopropylcarbinolimidazole
Probably in: Rat A3457

***cis*-5-Amino-1-cyclopropyl-7-(3,5-dimethyl-1-piperazinyl)-6,8-difluoro-1,4-dihydro-4-oxo-3-quinolinecarboxylic acid**
See: Sparfloxacin

Aminodantroline
Formed from: Dantroline
→ Acetamidodantroline
In: Hamster C538
Mouse C538
Rat C538

4-Amino-2-decaprenylbenzoic acid
Probably formed from: *p*-Aminobenzoic acid

6β-Amino-6-deoxydihydromorphine
Formed from: Azidomorphine

4-Amino-3-((4′-((2,4-diamino-5-methylphenyl)azo)(1,1′-biphenyl)-4-yl)azo)-5-hydroxy-6-(phenylazo)-2,7-naphthalenedisulphonic acid
See: Direct Black 4

4-Amino-3-((4-((2,4′-diaminophenyl)azo)(1,1′-biphenyl)-4-yl)azo)-5-hydroxy-6-(phenylazo)-2,7-naphthalenedisulphonic acid
See: Direct Black 38

2-Amino-3,5-dibromo-N-cyclohexyl-N-methylbenzylamine
See: Bromhexine

2-Amino-3,5-dibromo-α-(((1,1-dimethylethyl) amino)methyl)benzenemethanol
See: Ambroxol

2-Amino-3,5-dibromo-N-(4-hydroxycyclohexyl)benzylamine
See: Bromhexine

2-Amino-3,5-di-*tert*-butylphenol
→ 3,5-Di-*tert*-butyl-*o*-quinone
In: Streptomyces F209

4-Amino-2,6-dichlorobenzaldehyde
Probably formed from: 4-Amino-2,6-dichlorobenzyl alcohol

α-Amino-2,6-dichlorobenzaldoxime
→ 2,6-Dichlorobenzonitrile
In: Microorganism A197

3-Amino-2,5-dichlorobenzoic acid (amiben)
→ 3-Amino-2,5-dichlorobenzoic acid-N-glucoside
Probably in: Foxtail A3508

4-Amino-3,5-dichlorobenzoic acid
Probably formed from: 4-Amino-3,5-dichlorophenylglyoxylic acid
→ 4-Amino-3,5-dichlorobenzoyl-β-D-glucuronide acid
In: Dog D203
→ 4-Amino-3,5-dichlorohippuric acid
In: Beef J591
Rat D203, H930, J440

3-Amino-2,5-dichlorobenzoic acid-N-glucoside
Formed from: 3-Amino-2,5-dichlorobenzoic acid

2-Amino-2′,5-dichlorobenzophenone
Probably formed from: 2′,5-Dichloro-2-(2-hydroxyethylamino)acetamidobenzophenone
→ 2-Amino-2′,5-dichloro-3-hydroxybenzophenone
Probably in: Mouse A828
Rat A828

4-Amino-3,5-dichlorobenzoyl-β-D-glucuronic acid
Formed from: 4-Amino-3,5-dichlorobenzoic acid

4-Amino-2,6-dichlorobenzyl alcohol
Formed from: 2,6-Dichloro-4-methylaniline
→ 4-Amino-2,6-dichlorobenzaldehyde
Probably in: Rat G580

N-(4-Amino-2,6-dichlorobenzyl)-2,6-dichloro-4-methylaniline
Formed from: 2,6-Dichloro-4-methylaniline

N-(4-Amino-2,6-dichlorobenzylidene)-2,6-dichloro-4-methylaniline
Formed from: 2,6-Dichloro-4-methylaniline

2-Amino-4,6-dichloro-3,5-difluorophenol
Formed from: 3,5-dichloro-2,4-difluoroaniline
→ 2-Acetamido-4,6-dichloro-3,5-difluorophenol
Probably in: Rat J260
→ 2-Amino-4,6-dichloro-3,5-difluorophenyl sulphate
In: Rat J260

2-Amino-4,6-dichloro-3,5-difluorophenyl sulphate
Formed from: 2-Amino-4,6-dichloro-3,5-difluorophenol

4-Amino-2,6-dichloro-3-fluorophenol
Formed from: 3,5-dichloro-2,4-difluoroaniline
→ 4-Amino-2,6-dichloro-3,5-difluorophenyl-β-D-glucuronide
In: Rat J260
→ 4-Amino-2,6-dichloro-3,5-difluorophenyl sulphate
In: Rat J260

4-Amino-2,6-dichloro-3,5-difluorophenyl-β-D-glucuronide
Formed from: 4-Amino-2,6-dichloro-3-fluorophenol

4-Amino-2,6-dichloro-3,5-difluorophenyl sulphate
Formed from: 4-Amino-2,6-dichloro-3-fluorophenol

4-Amino-3,5-dichloro-α-(((1,1-dimethylethyl)amino)methyl)benzenemethanol
See: Clenbuterol

4-Amino-3,5-dichlorohippuric acid
Formed from: 4-Amino-3,5-dichlorobenzoic acid

2-Amino-2′,5-dichloro-3-hydroxybenzophenone
Probably formed from: 2-Amino-2′,5-dichlorobenzophenone
→ 2-Amino-2′,5-dichloro-3-hydroxybenzophenone-β-D-glucuronide
Probably in: Mouse A828
Rat A828

2-Amino-2′,5-dichloro-3-hydroxybenzophenone-β-D-glucuronide
Probably formed from: 2-Amino-2′,5-dichloro-3-hydroxybenzophenone

4-Amino-3,5-dichloromandelic acid
Formed from: Clenbuterol
→ 4-Amino-3,5-dichlorophenylglyoxylic acid
Probably in: Beef J591
Dog D203
Rat J440, J591

4-Amino-3,3′-dichloro-4′-nitrosodiphenylmethane
Probably formed from: N-Hydroxy-4,4′-methylenebis(*o*-chloroaniline)

4-Amino-2,6-dichlorophenol
→ 2,6-Dichloroquinol
In: Flavobacterium G641

6-Amino-3,4-dichlorophenol
Formed from: 3,4-Dichloroaniline
→ 6-Acetamido-3,4-dichlorophenol
Probably in: Rabbit C879
Rat C879
→ 6-Amino-3,4-dichlorophenyl sulphate
Probably in: Monkey J156
Mouse J156
Rat J156

6-Amino-2,3-dichlorophenol
Formed from:3,4-Dichloroaniline
→ 6-Acetamido-2,3-dichlorophenol
Probably in: Rabbit C879
Rat C879

→ 6-Amino-2,3-dichlorophenyl sulphate
Probably in: Monkey J156
Mouse J156
Rat J156

4-Amino-3,5-dichlorophenylglyoxylic acid
Probably formed from: 4-Amino-3,5-dichloromandelic acid
→ 4-Amino-3,5-dichlorobenzoic acid
Probably in: Beef J591
Dog D203
Rat D203, J440

6-Amino-2,3-dichlorophenyl sulphate
Probably formed from: 6-Amino-2,3-dichlorophenol

6-Amino-3,4-dichlorophenyl sulphate
Probably formed from: 6-Amino-3,4-dichlorophenol

4′-Amino-2′,5-dichlorosalicylanilide
Formed from: 2′,5-Dichloro-4′-nitrosalicylanilide

***p*-Amino-N-(2-diethylaminoethyl)benzamide**
(procainamide)
→ *p*-Acetamido-N-(2-diethylaminoethyl) benzamide
In: Man A450, A1264, A1518, A2218, A2403, A2455, A2562, A3310, A3857, K438
Monkey A450
Rat A2510, A3857, B277, B627, C839, D823, F234, J634
→ 4-Acetamido-N-(2-diethylaminoethyl)-3-hydroxybenzamide
In: Rat B627
→ *p*-Aminobenzoic acid
In: Man A3310
See also: A450, B627
→ *p*-Amino-N-(2-diethylaminoethyl)-N-hydroxybenzamide
In: Man C653, E879, F549, G97, K438
Rat C653, E897
→ *p*-Amino-N-(2-ethylaminoethyl)benzamide
In: Man B412, K438
See also C293
→ N-(2-Diethylaminoethyl)-*p*-nitrobenzamide
In: Man K438

***p*-Amino-N-(2-diethylaminoethyl)-N-hydroxybenzamide**
Formed from: *p*-Amino-N-(2-diethylaminoethyl)benzamide

4-Amino-3,5-diethylphenol
Probably formed from: 2,6-diethylaniline
→ 3,5-Diethylbenzoquinone imine
Probably in: Mouse F528
Rat F528

2-Aminodifluorenylamine
Formed from: 2-Aminofluorene

4-Amino-2,3-difluorophenol
Formed from: 2,3-Difluoroaniline

4-Amino-2,5-difluorophenol
Formed from: 2,5-Difluoroaniline

4-Amino-2,6-difluorophenol
Formed from: 3,5-Difluoroaniline

4-Amino-3,5-difluorophenol
Formed from: 2,6-Difluoroaniline

4-Amino-1,2-dihydro-1,2-dihydroxybenzoic acid
Formed from: *p*-Aminobenzoic acid

5-Amino-1,2-dihydro-1,2-dihydroxybenzoic acid
Probably formed from: *m*-Aminobenzoic acid

6-Amino-1,2-dihydro-1,2-dihydroxybenzoic acid
Probably formed from: Anthranilic acid

6-Amino-1,2-dihydro-1,2-dihydroxychrysene
Probably formed from: 6-Aminochrysene

1-Amino-4,5-dihydro-4,5-dihydroxypyrene
Formed from: *trans*-4,5-Dihydro-4,5-dihydroxy-1-nitropyrene

7-Amino-1,3-dihydro-3-hydroxy-5-phenyl-2*H*-1,4-benzodiazepin-2-one
Probably formed from: 7-Amino-1,3-dihydro-5-phenyl-2*H*-1,4-benzodiazepin-2-one

7-Amino-2,3-dihydro-8-hydroxy-5-phenyl-2*H*-1,4-benzodiazepin-2-one
Probably formed from: 7-Amino-1,3-dihydro-5-phenyl-2*H*-1,4-benzodiazepin-2-one

7-Amino-1,3-dihydro-1-methyl-5-phenyl-2*H*-1,4-benzodiazepin-2-one
Formed from: Nimetazepam
→ 7-Acetamido-1,3-dihydro-1-methyl-5-phenyl-2*H*-1,4-benzodiazepin-2-one
In: Rat A352

N-(4-((2-Amino-4,7-dihydro-4-oxo-3*H*-pyrrolo[2,3-*d*]pyrimidin-5-yl)ethyl)benzoyl)-L-glutamic acid
→ N-(4-((2-Amino-4,5,6,7-tetrahydro-4,6-dioxo-3*H*-pyrrolo[2,3-*d*]pyrimidin-5-yl)ethyl)benzoyl)-L-glutamic acid
In: Mouse J157

4-(((2-Amino-1,4-dihydro-4-oxo-6-pteridinyl)methyl)amino)benzoic acid
See: Pteroic acid

7-Amino-1,3-dihydro-5-phenyl-2*H*-1,4-benzodiazepin-2-one
Formed from: 7-Acetamido-1,3-dihydro-5-phenyl-2*H*-1,4-benzodiazepin-2-one
1,3-Dihydro-7-nitro-5-phenyl-2*H*-1,4-benzodiazepin-2-one
→ 7-Acetamido-1,3-dihydro-5-phenyl-2*H*-1,4-benzodiazepin-2-one
In: Man B104
Mouse G903
Rat A297, A352, D113, G903
See also: A2466
→ 7-Amino-1,3-dihydro-3-hydroxy-5-phenyl-2*H*-1,4-benzodiazepin-2-one
Probably in: Dog A1763
→ 7-Amino-1,3-dihydro-8-hydroxy-5-phenyl-2*H*-1,4-benzodiazepin-2-one
Probably in: Rabbit A3876

2-Amino-3,4-dihydroxybenzoic acid
See: 3,4-Dihydroxyanthranilic acid

2-Amino-3,6-dihydroxybenzoic acid
See: 3,6-Dihydroxyanthranilic acid

2′-Amino-2,3-dihydroxybiphenyl
Probably formed from: Carbazole
→ 6-(*o*-Aminophenyl)-2-hydroxy-6-oxohexa-2,4-dienoic acid
Probably in: Pseudomonas H21, H765
→ Catechol
Probably in: Pseudomonas H21, H765

4′-Amino-3,4-dihydroxybiphenyl
Probably formed from: 4′-Amino-4-hydroxybiphenyl
→ 4′-Amino-3-hydroxy-4-methoxybiphenyl
Probably in: Rat G1
→ 4′-Amino-4-hydroxy-3-methoxybiphenyl
Probably in: Rat G1

7-Amino-3,6-dihydroxy-4,5-diethoxyphthalide
Probably formed from: 7-Amino-3-hydroxy-4,5,6-triethoxyphthalide
→ 7-Amino-3,6-dihydroxy-4,5-diethoxyphthalide-β-D-glucuronide
Probably in: Man E105
→ 7-Amino-3,6-dihydroxy-4,5-diethoxyphthalide-O-sulphate
Probably in: Man E105

7-Amino-3,6-dihydroxy-4,5-diethoxyphthalide-β-D-glucuronide
Probably formed from: 7-Amino-3,6-dihydroxy-4,5-diethoxyphthalide

7-Amino-3,6-dihydroxy-4,5-diethoxyphthalide-O-sulphate
Probably formed from: 7-Amino-3,6-dihydroxy-4,5-diethoxyphthalide

2-Amino-5,6-dihydroxy-N,N-dipropyltetralin
→ 2-Amino-6-hydroxy-5-methoxy-N,N-dipropyltetralin
In: Rat C895

8-(4-Amino-2,3-dihydroxy-1-methylbutyl)amino-4-hydroxymethyl-2,6-dimethoxy-5-(3-trifluoromethylphenoxy)quinoline-N-oxide
Probably formed from: 8-(4-Amino-3-hydroxy-1-methylbutyl)amino-2,6-dimethoxy-4-methyl-5-(3-trifluoromethylphenoxy)quinoline
→ 8-(4-Amino-1-methyl-2,3-dioxobutyl)amino-2-hydroxy-4-hydroxymethyl-6-methoxy-5-(3-trifluoromethylphenoxy)quinoline-N-oxide
Probably in: Rat H135

8-(4-Amino-2,3-dihydroxy-1-methylbutyl)amino-2,5,6-trihydroxy-4-hydroxymethylquinoline
Probably formed from: 8-(4-Amino-3-hydroxy-1-methylbutyl)amino-2,5,6-rihydroxy-4-methylquinoline

8-Amino-2,6-dihydroxy-4-methyl-5-(3-trifluoromethylphenoxy)quinoline
Probably formed from: 8-Amino-2,6-dimethoxy-4-methyl-5-(3-trifluoromethylphenoxy)quinoline

1-(4-Amino-6-(2,3-dihydroxypropyl)amino-1,3,5-triazin-2-yl)-4-bis(*p*-fluorophenyl)methyl)piperazine
Probably formed from: Deallylalmitrine

8-Amino-5,6-dihydroxyquinoline
Probably formed from: 8-Amino-5-hydroxy-6-methoxyquinoline
4-Amino-2,6-diiodophenol
Formed from: 2,6-Diiodo-4-nitrophenol
N-(3-Amino-2,6-diiodophenyl)-N-methylglutaramidic acid
Formed from: Iomeglamic acid
→ Methyl N-(3-amino-2,6-diiodophenyl)-N-methylglutaramidate
Probably in: Man A683
N-(3-Amino-4,5-diiodophenyl)-N-methylglutaramidic acid
Formed from: Iomeglamic acid
N-(3-Amino-4,6-diiodophenyl)-N-methylglutaramidic acid
Formed from: Iomeglamic acid
→ Methyl N-(3-amino-4,6-diiodophenyl)-N-methylglutaramidate
Probably in: Man A683
2-Amino-1-(2,5-dimethoxy-4-methylphenyl)propane
See: 2,5-Dimethoxy-4-methylamphetamine
8-Amino-2,6-dimethoxy-4-methyl-5-(3-trifluoromethylphenoxy)quinoline
Formed from: 8-(4-Amino-1-methylbutyl)amino-2,6-dimethoxy-4-methyl-5-(3-trifluoromethylphenoxy)quinoline
→ 8-Amino-2,6-dihydroxy-4-methyl-5-(3-trifluoromethylphenoxy)quinoline
Probably in: Rat H135
→ 2,6-Dihydroxy-8-hydroxylamino-4-hydroxymethyl-5-(3-trifluoromethylphenoxy)quinoline
Probably in: Rat H135
→ 6-Hydroxy-2-methoxy-4-methyl-8-nitroso-5-(3-trifluoromethylphenoxy)quinoline
Probably in: Rat H135
5-Amino-2-(3,4-dimethoxyphenyl)-2-isopropylvaleronitrile
Probably formed from: 2-(3,4-Dimethoxyphenyl)-2-isopropyl-5-methylaminovaleronitrile
4-Amino-6,7-dimethoxy-2-(1-piperazinyl)quinazoline
Formed from: Prazocin
Terazocin
4-Amino-N-(5,6-dimethoxy-4-pyrimidinyl)benzenesulphonamide
See: Sulphadimidine
1-(4-Amino-6,7-dimethoxy-2-quinazolinyl)hexahydro-4-(1-oxobutyl)-1*H*-1,4-diazepine
See: Bunazosin
4-Amino-6,7-dimethoxy-2-quinazolinyl)-4-(tetrahydro-2-furanyl)carbonyl)piperazine
See: Terazocin
4-Amino-6,7-dimethoxyquinoline (amiquinisin)
→ 4-Amino-3-hydroxy-6,7-dimethoxyquinoline
In: Man B539
Rat B539
2-Amino-4′-dimethylaminoazobenzene
→ N,N-Dimethyl-*p*-phenylenediamine
In: Rat G380
→ *o*-Phenylenediamine
In: Rat G380
3-Amino-4′-dimethylaminoazobenzene
→ N,N-Dimethyl-*p*-phenylenediamine
In: Rat G380
→ *m*-Phenylenediamine
In: Rat G380
4-Amino-4′-dimethylaminoazobenzene
→ N,N-Dimethyl-*p*-phenylenediamine
In: Rat G133, G380
→ *p*-Phenylenediamine
In: Rat G133, G380
2-Amino-5-dimethylaminobenzoic acid
Formed from: 2-Carboxy-4-dimethylaminoazobenzene
4-Amino-4′-dimethylaminobenzophenone
Probably formed from: 4′-Dimethylamino-4-methylaminobenzophenone
→ 4-Acetamido-4′-dimethylaminobenzophenone
Probably in: Rat B713
→ 4-Amino-4′-methylaminobenzophenone
Probably in: Rat B713
→ 4-Amino-4′-dimethylamino-3-hydroxybenzophenone
Probably in: Rat B713
5-Amino-2-(2-(dimethylamino)ethyl)-1*H*-benz[*d*,*e*]isoquinoline
See: Amonafide

4-Amino-4′-dimethylamino-3-hydroxybenzophenone
Probably formed from: 4-Amino-4′-dimethylaminobenzophenone

4-Amino-2′,3-dimethylbiphenyl
→ 4-Acetamido-2′,3-dimethylbiphenyl
In: Hamster G757

2-Amino-3,8-dimethylimidazo[4,5-*f*]quinoxaline
→ 2-Acetamido-3,8-dimethylimidazo[4,5-*f*]quinoxaline
In: Rat G414
→ 2-Amino-3,8-dimethylimidazo[4,5-*f*]quinoxaline-2-N-β-D-glucuronide
In: Man J369
Rat F195, G414
→ 2-Amino-3,8-dimethylimidazo[4,5-*f*]quinoxaline-2-N-sulphate
In: Man J369
Rat F195, F232, G414
→ 2-Amino-5-hydroxy-3,8-dimethylimidazo[4,5-*f*]quinoxaline
In: Rat G414
See also: F195, F232
→ 2-Amino-8-hydroxymethyl-3-methylimidazo[4,5-*f*]quinoxaline
In: Rat G414
→ 2-Hydroxylamino-3,8-dimethylimidazo[4,5-*f*]quinoxaline
In: Rat G414

2-Amino-3,8-dimethylimidazo[4,5-*f*]quinoxaline-2-N-β-D-glucuronide
Formed from: 2-Amino-3,8-dimethylimidazo[4,5-*f*]quinoxaline

2-Amino-3,8-dimethylimidazo[4,5-f]quinoxaline-2-N-sulphate
Formed from: 2-Amino-3,8-dimethylimidazo[4,5-*f*]quinoxaline

4-Amino-N-(4,5-dimethyl-3-isoxazyl)benzenesulphonamide
See: Sulphatroxazole

2-Amino-4,6-dimethyl-3-oxo-3*H*-phenoxazine
See: Actinocin

2-Amino-4,5-dimethylphenol
→ 2-Amino-4,5-dimethylmuconic semialdehyde
In: Pseudomonas J329

4-Amino-3,5-dimethylphenol
→ 4-Amino-3,5-dimethylphenyl sulphate
In: Man J764

2-Amino-4,5-dimethylphenol
Formed from: 2-(*p*-Carboxyphenylazo)-4,5-dimethylphenol

2-Amino-1-(2,6-dimethylphenoxy)propane
→ 2,6-Dimethylphenol
In: Rabbit E510

4-Amino-2,6-dimethylphenoxyl radical
Formed from: N-Acetyl-3,5-dimethyl-*p*-benzoquinone radical

4-Amino-N-(2,6-dimethylphenyl)benzamide
Formed from: 4-(((Diethylamino)acetyl)amino)-N-(2,6-dimethylphenyl)benzamide
→ 4-Acetamido-N-(2,6-dimethylphenyl)benzamide
In: Mouse E951
Rat F57

2-Amino-N-(2,6-dimethylphenyl)butanamide
Probably formed from: N-(2,6-Dimethylphenyl)-2-ethylaminobutanamide

4-Amino-2,3-dimethyl-1-phenyl-3-pyrazolin-5-one
See: 4-Aminoantipyrine

4-Amino-3,5-dimethylphenyl sulphate
Formed from: 4-Amino-3,5-dimethylphenol

4-Amino-N-(2,6-dimethyl-4-pyrimidinyl)benzenesulphonamide
See: Sulphisomidine

5-Amino-2,4-dinitrobenzamide
Formed from: 5-(1-Aziridinyl)-2,4-dinitrobenzamide

4-Amino-2,6-dinitrobenzoic acid
Probably formed from: 2,4,6-Trinitrobenzoic acid

2-Amino-4,6-dinitrophenol
Formed from: Picric acid

2-Amino-4,6-dinitrotoluene
Formed from: 2,4,6-Trinitrotoluene
→ 2,6-Diamino-4-nitrotoluene
In: Man E349
See also: G719

4-Amino-2,6-dinitrotoluene
→ 4-Amino-2-hydroxylamino-6-nitrotoluene
Probably in: Pseudomonas H957
→ 2,4-Diamino-6-nitrotoluene

Probably in: Escherichia A2624
Man E349
Methanococcus H215
Pseudomonas A2624, J181
Rat D614, E349
Veillonella A2624

4-Aminodiphenylamine
Formed from: Acid yellow 36

1-Amino-4,4-diphenyl-2-butanol
Formed from: Drobuline

2-Amino-4,4-diphenyl-5-heptanol
Formed from: 2-Methylamino-4,4-diphenyl-5-heptanol
→ 2-Acetamido-4,4-diphenyl-5-heptanol
Probably in: Man A1123
Rat A1123, A2384

2-Amino-4,5-diphenyloxazole
Formed from: 2-(2-Hydroxyethyl)amino-4,5-diphenyloxazole
→ 2-Amino-hydroxy-4,5-diphenyloxazole
Probably in: Man F30

2-(2-Aminoethoxy)-3-chlorobiphenyl (2-chloro-6-phenylphenoxyethylamine)
→ 2-(2-Aminoethoxy)-3-chlorobiphenyl-β-D-glucuronide
In: Rat A1890
→ 3-Chloro-2-formylmethoxybiphenyl
In: Rat A1890
→ 3-Chloro-2-hydroxybiphenyl
In: Rat A1890

2-(2-Aminoethoxy)-3-chlorobiphenyl-β-D-glucuronide
Formed from: 2-(2-Aminoethoxy)-3-chlorobiphenyl

2-(2-Aminoethoxy)-3,5-dichlorobiphenyl (2,4-dichloro-6-phenylphenoxyethylamine)
→ 2-(2-Aminoethoxy)-3,5-dichlorobiphenyl-β-D-glucuronide
In: Rat A1890
→ 3,5-Dichloro-2-formylmethoxybiphenyl
In: Rat A1890
→ 3,5-Dichloro-2-hydroxybiphenyl
In: Rat A1890

2-(2-Aminoethoxy)-3,5-dichlorobiphenyl-β-D-glucuronide
Formed from: 2-(2-Aminoethoxy)-3,5-dichlorobiphenyl

1-(2-(2-Aminoethoxy)-5-hydroxyphenyl)butanol
Probably formed from: 1-(*o*-(2-Aminoethoxy)phenyl)butanol
→ 1-(2-(2-Aminoethoxy)-5-hydroxyphenyl)butanol-β-D-glucuronide
Probably in: Rat A3324

1-(2-(2-Aminoethoxy)-5-hydroxyphenyl)butanol-β-D-glucuronide
Probably formed from: 1-(2-(2-Aminoethoxy)-5-hydroxyphenyl)butanol

2-((2-Aminoethoxy)methyl)-5-carboxy-4-(*o*-chlorophenyl)-3-ethoxycarbonyl-6-methylpyridine
Probably formed from: 2-((2-Aminoethoxy)methyl)-4-(*o*-chlorophenyl)-3-ethoxycarbonyl-5-methoxycarbonyl-6-methylpyridine
→ 2-((2-Aminoethoxy)methyl)-3,5-dicarboxy-4-(*o*-chlorophenyl)- 6-methylpyridine
Probably in: Dog F41
Rat F41

2-((2-Aminoethoxy)methyl)-3-carboxy-4-(*o*-chlorophenyl)-5-methoxycarbonyl-6-methylpyridine
Probably formed from: 2-((2-Aminoethoxy)methyl)-4-(*o*-chlorophenyl)-3-ethoxycarbonyl-5-methoxycarbonyl-6-methylpyridine
→ 2-((2-Aminoethoxy)methyl)-4-(*o*-chlorophenyl)-3-ethoxycarbonyl-6-methylpyridine
Probably in: Dog E836
Rat E836

2-((2-Aminoethoxy)methyl)-4-(*o*-chlorophenyl)-1,4-dihydro-3-(2-hydroxyethoxy)carbonyl-5-methoxycarbonyl-6-methylpyridine
Formed from: Amlodipine

2-((2-Aminoethoxy)methyl)-4-(*o*-chlorophenyl)-3-ethoxycarbonyl-1,4-dihydro-5-methoxycarbonyl-6-methylpyridine
Formed from: Amlodipine

2-((2-Aminoethoxy)methyl)-4-(*o*-chlorophenyl)-3-ethoxycarbonyl-5-methoxycarbonyl-6-methylpyridine
Formed from: Amlodipine

→ 2-((2-Aminoethoxy)methyl)-5-carboxy-4-(*o*-chlorophenyl)-3-ethoxycarbonyl-6-methylpyridine
Probably in: Man E848, F41
Rat E836, F41
→ 3-Carboxy-2-((carboxymethoxy)methyl-4-(*o*-chlorophenyl)-5-methoxycarbonyl-6-methylpyridine
Probably in: Dog F78
Man E848
Rat F78
→ 4-(*o*-Chlorophenyl)-3-ethoxycarbonyl-2-hydroxymethyl-5-methoxycarbonyl-6-methylpyridine
Probably in: Man E848, F41, F78
Rat F41

2-((2-Aminoethoxy)methyl)-4-(*o*-chlorophenyl)-3-ethoxycarbonyl-6-methylpyridine
Probably formed from: 2-((2-Aminoethoxy)methyl)-3-carboxy-4-(*o*-chlorophenyl)-5-methoxycarbonyl-6-methylpyridine

2-((2-Aminoethoxy)methyl)-3,5-dicarboxy-4-(*o*-chlorophenyl)-6-hydroxymethylpyridine
Probably formed from: 2-((2-Aminoethoxy)methyl)-3,5-dicarboxy-4-(*o*-chlorophenyl)-6-methylpyridine

2-((2-Aminoethoxy)methyl)-3,5-dicarboxy-4-(*o*-chlorophenyl)-6-methylpyridine
Probably formed from: 2-((2-Aminoethoxy) methyl)-5-carboxy-4-(*o*-chlorophenyl)-3-ethoxycarbonyl-6-methylpyridine
→ 2-((2-Aminoethoxy)methyl)-3,5-dicarboxy-4-(*o*-chlorophenyl)-6-hydroxymethylpyridine
Probably in: Dog F41

1-(*o*-(2-Aminoethoxy)phenyl)butanol
Probably formed from: *o*-(2-Isopropyl-aminoethoxy)butyrophenone
→ 1-(2-(2-Aminoethoxy)-5-hydroxyphenyl) butanol
Probably in: Rat A3324
→ 1-(*o*-(2-Aminoethoxy)phenyl)butanol-β-D-glucuronide
Probably in: Rat A3324

1-(o-(2-Aminoethoxy)phenyl)butanol-β-D-glucuronide
Formed from: Panomifene
Probably formed from: 1-(*o*-(2-Aminoethoxy)phenyl)butanol
1-(4-(2-Aminoethoxy)phenyl)-3,3,3-trifluoro-1,2-diphenylpropene

N-(2-Aminoethyl)acridine-4-carboxamide
Formed from: N-(2-(Methylamino)ethyl) acridine-4-carboxamide

7-(2-Aminoethylamino)-3-carboxy-6-fluoro-N-(*p*-fluorophenyl)quinolone
Formed from: Difloxacin

7-(2-Aminoethylamino)-1-cyclopropyl-6-fluoro-1,4-dihydro-4-oxoquinoline-3-carboxylic acid
Probably formed from: 3′-Oxociprofloxacin

***p*-Amino-N-(2-ethylaminoethyl)benzamide**
Formed from: *p*-Amino-N-(2-diethylaminoethyl)benzamide

7-(2-Aminoethylamino)-1-ethyl-6-fluoro-1,4-dihydro-4-oxoquinoline-3-carboxylic acid
Probably formed from: 1-Ethyl-6-fluoro-1,4-dihydro-4-oxo-7-(3-oxopiperazinyl) quinoline-3-carboxylic acid
→ 7-Amino-1-ethyl-6-fluoro-1,4-dihydro-4-oxoquinoline-3-carboxylic acid
Probably in: Dog C104
Monkey C104
Rat C104

***p*-(2-Aminoethyl)benzaldehyde**
Probably formed from: *p*-(2-Aminoethyl) benzylamine

α-(1-Aminoethyl)benzenemethanol
See: Norephedrine

(α*S*)-α-((1*S*)1-Aminoethyl)benzenemethanol
See: D-Norpseudoephedrine

Aminoethyl benzhydryl ether
Formed from: Benzhydryl methylaminoethyl ether

***p*-(2-Aminoethyl)benzylamine**
→ *p*-(2-Aminoethyl)benzaldehyde
Probably in: Pig A3693

N-(2-Aminoethyl)-2-carboxamido-1,4-benzodioxan
Formed from: Idazoxan
→ 1,4-Benzodioxan-2-carboxylic acid

Probably in: Rat F284

(+)-(2*S*, 3*S*)-5-(2-Aminoethyl)-8-chloro-3,4-dihydro-3-hydroxy-2-(4-hydroxy-3-methoxyphenyl)-4-oxo-1,5(2*H*)-benzothiazepine
Probably formed from: (+)-(2*S*, 3*S*)-8-Chloro-3,4-dihydro-3-hydroxy-2-(4-hydroxy-3-methoxyphenyl)-5-(2-(N-methylamino)ethyl)-4-oxo-1,5(2*H*)-benzothiazepine

1-(2-Aminoethyl)-7-chloro-5-(2-fluorophenyl)-1,3-dihydro-2*H*-1,4-benzodiazepin-2-one
Formed from: 7-Chloro-1-(2-aminoethyl)- 5-(2-fluorophenyl)-1,3-dihydro-2*H*-1,4-benzodiazepin-2-one

2-(2-Aminoethyl)-3-(*p*-chlorophenyl)-3-hydroxyphthalimidine
Formed from: Mazindol
→ 2-Carboxymethyl-3-(*p*-chlorophenyl)-3-hydroxyphthalimidine
Probably in: Dog A3673
Man A3673

10β-Aminoethyl-10,11-dihydro-5-methyl-11-oxo-5*H*-dibenzo[*b,e*][1,4]diazepine
Formed from: 10,11-Dihydro-5-methyl-10β-methylaminoethyl-11-oxo-5*H*-dibenzo[*b,e*][1,4]diazepine
→ 10β-Aminoethyl-10,11-dihydro-11-oxo-5*H*-dibenzo[*b,e*][1,4]diazepine
Probably in: Man A712

10β-Aminoethyl-10,11-dihydro-11-oxo-5*H*-dibenzo[*b,e*][1,4]diazepine
Probably formed from: 10β-Aminoethyl-10,11-dihydro-5-methyl-11-oxo-5*H*-dibenzo[*b,e*][1,4]diazepine

2-Aminoethyl 5-(5,5-dimethyl-2-oxo-1,3,2-dioxaphosphorinan-2-yl)-1,4-dihydro-2,6-dimethyl-4-(*m*-nitrophenyl)-3-pyridinecarboxylate
Probably formed from: 2-(Phenylamino)ethyl 5-(5,5-dimethyl-2-oxo-1,3,2-dioxaphosphorinan-2-yl)-1,4-dihydro-2,6-dimethyl-4-(*m*-nitrophenyl)-3-pyridinecarboxylate
→ 2-Aminoethyl 5-(5,5-dimethyl-2-oxo-1,3,2-dioxaphosphorinan-2-yl)-2,6-dimethyl-4-(*m*-nitrophenyl)-3-pyridinecarboxylate
Probably in: Man H625
→ Carboxymethyl 5-(5,5-dimethyl-2-oxo-1,3,2-dioxaphosphorinan-2-yl)-1,4-dihydro-2,6-dimethyl-4-(*m*-nitrophenyl)-3-pyridinecarboxylate
Probably in: Man H523
→ 2-Hydroxyethyl 5-(5,5-dimethyl-2-oxo-1,3,2-dioxaphosphorinan-2-yl)-1,4-dihydro-2,6-dimethyl-4-(*m*-nitrophenyl)-3-pyridinecarboxylate
Probably in: Man H523

2-Aminoethyl 5-(5,5-dimethyl-2-oxo-1,3,2-dioxaphosphorinan-2-yl)-2,6-dimethyl-4-(*m*-nitrophenyl)-3-pyridinecarboxylate
Probably formed from: 2-Aminoethyl 5-(5,5-dimethyl-2-oxo-1,3,2-dioxaphosphorinan-2-yl)-1,4-dihydro-2,6-dimethyl-4-(*m*-nitrophenyl)-3-pyridinecarboxylate

4-(2-Aminoethyl)-1-ethyl-3,3-diphenyl-2-pyrrolidinone
Probably formed from: 3-Oxodoxapram

7-Amino-1-ethyl-6-fluoro-1,4-dihydro-4-oxoquinoline-3-carboxylic acid
Probably formed from: 7-(2-Aminoethylamino)-1-ethyl-6-fluoro-1,4-dihydro-4-oxoquinoline-3-carboxylic acid

3-Amino-5-ethyl-8-β-D-glucuronosylamino-6-phenylphenanthradinium
Formed from: Ethidium

8-Amino-5-ethyl-3-β-D-glucuronosylamino-6-phenylphenanthradinium
Formed from: Ethidium

6-Amino-1-ethyl-2-methyl-7-nitro-5-trifluoromethylbenzimidazole
Formed from: Dinitramine
→ 6-Amino-2-methyl-7-nitro-5-trifluoromethylbenzimidazole
Probably in: Aspergillus A1002

7-Amino-2-ethyl-5-methylsulphonyl-1-propylbenzimidazole
Formed from: Nitralin

5-(2-Aminoethyl)oximino-5-(*p*-trifluoromethylphenyl)valeric acid
Probably formed from: 5-Hydroxy-*p*-trifluoromethylvalerophenone (*E*)-O-(2-aminoethyl) oxime

Methyl 5-(2-aminoethyl)oximino-5-(*p*-trifluoromethylphenyl)valerate
→ 5-(Carboxymethyl)oximino-5-(*p*-trifluoromethylphenyl)valeric acid
Probably in: Man D24
→ 5-(2-Hydroxyethyl)oximino-5-(*p*-trifluoromethylphenyl)valeric acid
Probably in: Man D24
→ 5-Oximino-5-(*p*-trifluoromethylphenyl valeric acid
Probably in: Man D24
→ 5-Oxo-5-(*p*-trifluoromethylphenyl)valeric acid
Probably in: Man D24

6-Amino-N-(1-ethylpropyl-3,4-dimethyl-2-nitro-N-nitrosoaniline
Formed from: N-(1-Ethylpropyl-3,4-dimethyl-2,6-dinitro-N-nitrosoaniline

4-(2-Aminoethyl)-*o*-quinone
Formed from: 3,4-Dihydroxyphenethylamine

3-Amino-α-ethyl-2,4,6-triiodobenzenepropanoic acid
See: Iopanoic acid

1-Aminofluoranthene
Formed from: 1-Nitrofluoranthene

2-Aminofluoranthene
Formed from: 2-Nitrofluoranthene

3-Aminofluoranthene
Formed from: 3-Nitrofluoranthene
Probably formed from: 3-Hydroxylaminofluoranthene

7-Aminofluoranthene
Formed from: 7-Nitrofluoranthene

8-Aminofluoranthene
Formed from: 8-Nitrofluoranthene

2-Aminofluorene (2-fluorenamine)
Formed from: 2-Acetamidofluorene
N-Acetoxy-2-aminofluorene
2-Benzamidofluorene
2-Butyramidofluorene
2-Formamidofluorene
2-Nitrofluorene
2-Succinamidofluorene
Probably formed from: 2-Hydroxylamino-fluorene
→ 2-Acetamidofluorene
In: Aeromonas J677
Balsam pear J322
Enterobacter J816
Enterobius K174
Escherichia J520
Guinea pig F109
Hamster F114, F115, F563, G757
Hen F694
Lactobacillus K208
Man A3253, F172, H278
Mouse D340, G841, J405
Mytilus E713
Pseudomonas J652
Rabbit A3253, C78, F109
Rana H529
Rat F109
Shrimp K485
See also: E481, F953
→ 2-Aminodifluorenamine
In: Horseradish D448
With prostaglandin H synthase D448
→ 2-Aminofluorene-β-D-glucuronide
In: Man J416
→ 2-Amino-5-hydroxyfluorene
In: Dog D738
Man D738
Rat D738
→ 2-Amino-7-hydroxyfluorene
In: Dog D738
Man D738
Rat D738
→ Azofluorene
In: Horseradish C754, D448
Ram C754
With prostaglandin H synthase D448
→ 2-Fluorenyl nitroxide radical
In: Rabbit B184
→ 2-Formamidofluorene
In: Guinea pig F109
Oyster F890
Rabbit F109
Rat F109
→ 2-Hydroxylaminofluorene
In: Dog D738
Guinea pig B179
Hamster C397
Man D738

Monkey B179
Mouse C397
Rat C397, D738
See also: C754, D776
→ 2-Nitrosofluorene
In: Horseradish C754
Oyster F890
Ram C360
See also: D448

2-Aminofluorene-β-D-glucuronide
Formed from: 2-Aminofluorene

2-Aminofluoren-9-one
Formed from: 2-Acetamido-9-oxofluorene

4-Amino-4′-fluoroazobenzene
Formed from: 4′-Fluoro-4-methylaminoazobenzene

2-Amino-4-fluorobenzoic acid
See: 4-Fluoroanthranilic acid

2-Amino-5-fluorobenzoic acid
See: 5-Fluoroanthranilic acid

N-(4-Amino-3-fluorobenzyl)-2-fluoro-4-methylaniline
Formed from: 2-Fluoro-4-methylaniline

2-Amino-4-(1-(2-fluorobiphenyl-4-yl)ethyl)thiazole
Formed from: 4-(1-(2-Fluorobiphenyl-4-yl)ethyl)-2-methylaminothiazole

4′-Amino-3-fluoro-4-hydroxyazobenzene
Probably formed from: 3-Fluoro-4-hydroxy-4′-methylaminoazobenzene

2-Amino-4-(1-(2-fluoro-4′-hydroxybiphenyl-4-yl)ethyl)thiazole
Probably formed from: 4-(1-(2-Fluoro-4′-hydroxybiphenyl-4-yl)ethyl)-2-methylaminothiazole

6-Amino-2-fluoromethyl-3-(*o*-hydroxymethylphenyl)-4(3*H*)-quinazolinone
Formed from: Afloqualone
→ 6-Acetamido-2-fluoromethyl-3-(*o*-hydroxymethylphenyl)-4(3*H*)-quinazolinone
Probably in: Man C624
Rat C793a

6-Amino-2-(fluoromethyl)-3-(2-methylphenyl)-4(3*H*)-quinazolinone
See: Afloqualone

2-Amino-3-fluorophenol
Formed from: *o*-Fluoroaniline
→ 2-Amino-3-fluorophenyl sulphate
Probably in: Rat G790

2-Amino-4-fluorophenol
Formed from: *m*-Fluoroaniline
→ 2-Amino-4-fluorophenyl sulphate
Probably in: Rat G790

2-Amino-5-fluorophenol
Formed from: *p*-Fluoroaniline
→ 2-Acetamido-5-fluorophenol
Probably in: Rat G790
→ 2-Amino-5-fluorophenyl-β-D-glucuronide
Probably in: Rat G790, J558, J603
→ 2-Amino-5-fluorophenyl sulphate
Probably in: Rat G790, J558, J603

4-Amino-2-fluorophenol
Formed from: *m*-Fluoroaniline
→ 4-Acetamido-2-fluorophenol
Probably in: Rat G790
→ 4-Amino-2-fluorophenyl-β-D-glucuronide
Probably in: Rat G790
→ 4-Amino-2-fluorophenyl sulphate
Probably in: Rat G790

4-Amino-3-fluorophenol
Formed from: *o*-Fluoroaniline
→ 4-Acetamido-3-fluorophenol
Probably in: Marmoset D952
Rabbit D952
Rat D952, F518, G790
→ 4-Amino-2-fluorophenyl-β-D-glucuronide
Probably in: Marmoset D952
Rabbit D952
Rat D952, F518, G790
→ 4-Amino-2-fluorophenyl sulphate
Probably in: Marmoset D952
Rabbit D952
Rat D952, F518, G790

4-Amino-5-(*o*-fluorophenyl)-1,3-dihydro-1-methyl-2*H*-1,4-benzodiazepin-2-one
Formed from: Flunitrazepam
→ 4-Acetamido-5-(*o*-fluorophenyl)-1,3-dihydro-1-methyl-2*H*-1,4-benzodiazepin-2-one
In: Man A3604

2-Amino-5-fluorophenyl-β-D-glucuronide
Probably formed from: 2-Amino-5-fluorophenol

4-Amino-2-fluorophenyl-β-D-glucuronide
Probably formed from: 4-Amino-2-fluorophenol

4-Amino-3-fluorophenyl-β-D-glucuronide
Probably formed from: 4-Amino-3-fluorophenol

2-Amino-3-fluorophenyl sulphate
Probably formed from: 2-Amino-3-fluorophenol

2-Amino-4-fluorophenyl sulphate
Probably formed from: 2-Amino-4-fluorophenol

2-Amino-5-fluorophenyl sulphate
Probably formed from: 2-Amino-5-fluorophenol

4-Amino-2-fluorophenyl sulphate
Probably formed from: 4-Amino-2-fluorophenol

4-Amino-3-fluorophenyl sulphate
Probably formed from: 4-Amino-3-fluorophenol

L-2-Amino-3-(*p*-fluorophenyl)valeric acid
Formed from: 3-(*p*-Fluorophenyl)-2-oxovaleric acid

9-Amino-8-fluoro-1,2,3,4-tetrahydro-1-hydroxy-2,4-methanoacridine
Formed from: 9-Amino-8-fluoro-1,2,3,4-tetrahydro-2,4-methanoacridine
→ 9-Amino-8-fluoro-1,2,3,4-tetrahydro-1-hydroxy-2,4-methanoacridine-1-β-D-glucuronide
In: Rat G833, G873
→ 9-Amino-8-fluoro-1,2,3,4-tetrahydro-1-oxo-2,4-methanoacridine
Probably in: Rat G873

9-Amino-8-fluoro-1,2,3,4-tetrahydro-1-hydroxy-2,4-methanoacridine-1-β-D-glucuronide
Formed from: 9-Amino-8-fluoro-1,2,3,4-tetrahydro-1-hydroxy-2,4-methanoacridine

9-Amino-8-fluoro-1,2,3,4-tetrahydro-2,4-methanoacridine
→ 9-Amino-8-fluoro-1,2,3,4-tetrahydro-1-hydroxy-2,4-methanoacridine
In: Rat G833, G873
→ 9-Amino-8-fluoro-1,2,3,4-tetrahydro-2,4-methanoacridine-N-β-D-glucuronide
In: Rat G833, G873

9-Amino-8-fluoro-1,2,3,4-tetrahydro-2,4-methanoacridine-N-β-D-glucuronide
Formed from: 9-Amino-8-fluoro-1,2,3,4-tetrahydro-2,4-methanoacridine

9-Amino-8-fluoro-1,2,3,4-tetrahydro-1-oxo-2,4-methanoacridine
Probably formed from: 9-Amino-8-fluoro-1,2,3,4-tetrahydro-1-hydroxy-2,4-methanoacridine

4-Amino-4′-formamidodiphenyl sulphone
Formed from: 4,4′-Diformamidodiphenyl sulphone
→ Bis(*p*-aminophenyl) sulphone
In: Dog A521
Man A1722
Monkey A521

α-Amino-2-(formylamino)-γ-oxobenzenebutanoic acid
See: N-Formyl-D-kynurenine
N-Formyl-L-kynurenine

4′-Amino-3-formylazobenzene
Probably formed from: 4-Amino-3′-hydroxymethylazobenzene
→ 4′-Amino-3-carboxyazobenzene
Probably in: Rat A3835

4-Amino-2-(4-(2-furoyl)piperazin-1-yl)-6,7-dimethoxyquinazoline
See: Prazocin

β-(5-Amino-2-furyl)-*p*-carboxystyrene
Formed from: *p*-Carboxy-β-(5-nitro-2-furyl)styrene

6-Amino-13-(β-D-glucopyanosyl)-12,13-dihydro-1,11-dihydroxy-5*H*-indolo[2,3-*a*]pyrrolo[3,4-*c*]carbazole-5,7(6*H*)-dione
Formed from: 6-Formamido-13-(β-D-glucopyanosyl)-12,13-dihydro-1,11-dihydroxy-5*H*-indolo[2,3-*a*]pyrrolo[3,4-*c*]carbazole-5,7(6*H*)-dione
→ 6-Amino-11,13-bis(β-D-glucopyanosyl)-12,13-dihydro-1-hydroxy-5*H*-indolo[2,3-*a*]pyrrolo[3,4-*c*]carbazole-5,7(6*H*)-dione
Probably in: Rat J600

2-Amino-10-glucuronophenothiazine
Probably formed from: 2-Aminophenothiazine

4-Amino-2-(S-glutathionyl)phenol
Formed from: *p*-Aminophenol

→ 2-(N-Acetyl-L-cystein-S-yl)-4-aminophenol
Probably in: Rat K60
→ 4-Amino-2,5-bis(S-glutathionyl)phenol
Probably in: Rat G670

4-Amino-3-(S-glutathionyl)phenol
Formed from: *p*-Aminophenol
→ 4-Acetamido-3-(S-glutathionyl)phenol
Probably in: Rat G670

Aminoglutethimide (α-(*p*-aminophenyl)-α-ethylglutarimide)
→ 3-(4-Acetamidophenyl)-3-(2-carboxamidoethyl)tetrahydrofuran-2-one
In: Guinea pig E834
Man E834
Rabbit E834
Rat E834
→ α-(*p*-Acetamidophenyl)-α-ethylglutarimide
In: Guinea pig C807, E834
Man A666, E834
Rabbit E834
Rat C10, C807, E834
→ 2-(4-Amino-3-hydroxyphenyl)-2-ethylglutarimide
In: Rat C10
→ α-(*p*-Aminophenyl)-α-ethyl-β-hydroxyglutarimide
In: Guinea pig E834
Rabbit E834
→ α-(*p*-Aminophenyl)-α-ethyl-γ-hydroxyglutarimide
In: Guinea pig E834
Rabbit E834
→ 2-(*p*-Aminophenyl)glutarimide
In: Rat C10
→ 2-(*p*-Aminophenyl)-2-(1-hydroxyethyl) glutarimide
In: Guinea pig E834
Rabbit E834
Rat C10
→ α-Ethyl-α-(*p*-hydroxylaminophenyl) glutarimide
In: Guinea pig C807
Man D46
Rat C807
→ α-Ethyl-α-(*p*-nitrophenyl)glutarimide
In: Guinea pig C807
Rat C10, C807

4-Amino-2-(hexahydro-4-(3-hydroxybutyryl)-1*H*-1,4-diazepin-1-yl)-6,7-dimethoxyquinazoline
Formed from: Bunazocin
→ 4-Amino-2-(hexahydro-4-(3-hydroxybutyryl)-1*H*-1,4-diazepin-1-yl)-6,7-dimethoxyquinazoline-β-D-glucuronide
Probably in: Man G881

4-Amino-2-(hexahydro-4-(3-hydroxybutyryl)-1*H*-1,4-diazepin-1-yl)-6,7-dimethoxyquinazoline-β-D-glucuronide
Probably formed from: 4-Amino-2-(hexahydro-4-(3-hydroxybutyryl)-1*H*-1,4-diazepin-1-yl)-6,7-dimethoxyquinazoline

8-(6-Aminohexylamino)-6-methoxy-4-methylquinoline
Formed from: *p*-Aminobenzoic acid
p-Nitrohippuric acid
Probably formed from: 8-(6-Ethylaminohexylamino)-6-methoxy-4-methylquinoline
p-Aminohippuric acid
→ *p*-Acetamidohippuric acid
In: Mouse A3964

4-Amino-1-hydroxyanthracene
See: 1-(*p*-Sulphophenylazo)-4-anthranol

1-Amino-2-hydroxyanthraquinone
Probably formed from: 1-Aminoanthraquinone
→ 1-Amino-2-hydroxyanthraquinone-β-D-glucuronide
Probably in: Sheep C808

1-Amino-4-hydroxyanthraquinone
Probably formed from: 1-Aminoanthraquinone
→ 1-Amino-4-hydroxyanthraquinone-β-D-glucuronide
Probably in: Sheep C808

1-Amino-2-hydroxyanthraquinone-β-D-glucuronide
Probably formed from: 1-Amino-2-hydroxyanthraquinone

1-Amino-4-hydroxyanthraquinone-β-D-glucuronide
Probably formed from: 1-Amino-4-hydroxyanthraquinone

4′-Amino-4-hydroxyazobenzene
Probably formed from: 4-Aminoazobenzene

4-Hydroxy-4′-methylaminoazobenzene
→ *p*-Aminophenol
In: Rat G133, G511
→ *p*-Phenylenediamine
In: Rat G133, G511

β-Amino-4-hydroxybenzenepropanoic acid
See: β-Tyrosine

2-Amino-5-hydroxybenzimidazole
Probably formed from: 2-Aminobenzimidazole
→ 2-Aminobenzimidazol-5-yl-β-D-glucuronide
Probably in: Mouse A260
→ 2-Aminobenzimidazol-5-yl sulphate
Probably in: Mouse A260

2-Amino-3-hydroxybenzoic acid
See: 3-Hydroxyanthranilic acid

2-Amino-5-hydroxybenzoic acid
See: 5-Hydroxyanthranilic acid

3-Amino-4-hydroxybenzoic acid
→ Aminoquinol
In: Candida J461

3-Amino-5-hydroxybenzoic acid
Formed from: 3-Dehydro-3-deoxy-5-aminoshikimic acid
→ Rifamycin B
In: Nocardia F99, F612
→ Rifamycin O
In: Nocardia F99
→ Rifamycin S
In: Nocardia F99

4-Amino-2-hydroxybenzoic acid (*p*-aminosalicylic acid)
Formed from: *p*-Aminobenzoic acid
→ 4-Acetamido-2-hydroxybenzoic acid
In: Hamster F115
Man A663, A1518
Rabbit A1037, C78
→ 4-Aminocatechol
In: Agaricus D770, E162
Trichosporon D232

4-Amino-3-hydroxybenzoic acid
Probably formed from: *p*-Aminobenzoic acid
4-Amino-3-butoxybenzoic acid
→ 4-Acetamido-3-hydroxybenzoic acid
Probably in: Man E475
→ 4-Amino-3-hydroxybenzoic acid-β-D-glucuronide
Probably in: Man E475

5-Amino-2-hydroxybenzoic acid (5-aminosalicylic acid)
Formed from: *m*-Aminobenzoic acid
Azodisalicylic acid
Mordant Orange I
Mordant Yellow 10
Salicylazosulphapyridine
Sulphasalazine
→ 5-Acetamido-2-hydroxybenzoic acid
In: Man F414, F924
Microorganism C128, E648, E807, F509, F472
Mouse J405
Rat C425, J576
See also: D570
→ *cis*-4-Amino-6-carboxy-2-oxohexa-3,5-dienoic acid
In: Pseudomonas G480
→ 4-Aminocatechol
In: Trichosporon D232
→ 4-Amino-2-oxo-*cis, cis*-hepta-3,5-diendioic acid
Probably in: Pseudomonas H299
→ 5-Aminosalicylic acid-N-glucoside
In: Man F456, F924
Rat F456
→ 5-Formamido-2-hydroxybenzoic acid
In: Rat F924
→ Gentisic acid
In: Man E215
→ Salicylic acid
In: Man E215

4-Amino-3-hydroxybenzoic acid-β-D-glucuronide
Probably formed from: 4-Amino-3-hydroxybenzoic acid

4-Amino-3-hydroxybenzonitrile
Probably formed from: *p*-Aminobenzonitrile
→ 4-Aminobenzonitril-3-yl sulphate
Probably in: Rat D160, F305

2-Amino-6-hydroxybenzothiazole
Formed from: Riluzole

1-Amino-3-hydroxy-1*H*-benzotriazole
Probably formed from: 1-Benzylamino-1*H*-benzotriazole

1-Amino-4-hydroxy-1*H*-benzotriazole
Probably formed from: 1-Benzylamino-1*H*-benzotriazole

2-Amino-5(6)-(α-hydroxybenzyl)imidazole
Formed from: Mebendazole
→ 2-Amino-5(6)-benzoylimidazole
Probably in: Eel J134
Rat C369

2-Amino-3-hydroxybiphenyl
Formed from: 2-Aminobiphenyl
→ 2-Aminobiphenyl-3-yl sulphate
In: Guinea pig E608
Hamster E608
Mouse E608
Rat E608

2-Amino-5-hydroxybiphenyl
Formed from: 2-Aminobiphenyl
→ 2-Aminobiphenyl-5-yl-β-D-glucuronide
In: Guinea pig E608
Hamster E608
Mouse E608
Rat E608
→ 2-Aminobiphenyl-5-yl sulphate
In: Guinea pig E608
Hamster E608
Mouse E608
Rat E608

3-Amino-2-hydroxybiphenyl
Formed from: 3-Aminobiphenyl

3-Amino-4-hydroxybiphenyl
Formed from: 3-Aminobiphenyl

3-Amino-6-hydroxybiphenyl
Formed from: 3-Aminobiphenyl

4′-Amino-2-hydroxybiphenyl
Formed from: 4-Aminobiphenyl
→ 4′-Acetamido-2-hydroxybiphenyl
Probably in: Rat G1, J408

4-Amino-3-hydroxybiphenyl
Formed from: 4-Aminobiphenyl

4′-Amino-3-hydroxybiphenyl
Probably formed from: 4-Aminobiphenyl
→ 4′-Acetamido-3-hydroxybiphenyl
Probably in: Rat G1

4′-Amino-4-hydroxybiphenyl
Formed from: 4-Aminobiphenyl
→ 4′-Acetamido-4-hydroxybiphenyl
Probably in: Guinea pig J408
Hamster J408
Mouse J408
Rabbit J408
Rat G1, J408
→ 4′-Amino-3,4-dihydroxybiphenyl
Probably in: Rat G1

5-(2-Amino-1-hydroxybutyl)-8-hydroxycarbostyril
Formed from: Procaterol

2-Amino-3-hydroxy-α-carboline
Formed from: 2-Amino-α-carboline

2-Amino-6-hydroxy-α-carboline
Formed from: 2-Amino-α-carboline

4-Amino-3-hydroxy-6,7-dimethoxyquinoline
Formed from: 4-Amino-6,7-dimethoxyquinoline

4-Amino-4′-hydroxy-3,3′-dimethylbiphenyl
Probably formed from: 4-Amino-3,3′-dimethylbiphenyl

2-Amino-5-hydroxy-3,8-dimethylimidazo[4,5-*f*]quinoxaline
Formed from: 2-Amino-3,8-dimethylimidazo[4,5-*f*]quinoxaline
→ 2-Amino-5-hydroxy-3,8-dimethylimidazo[4,5-*f*]quinoxaline-β-D-glucuronide
Probably in: Rat F195
→ 2-Amino-5-hydroxy-3,8-dimethylimidazo[4,5-*f*]quinoxaline-O-sulphate
Probably in: Rat F195, G414
→ 2-Amino-5-hydroxy-8-hydroxymethyl-3-methylimidazo[4,5-*f*]quinoxaline
Probably in: Rat F195, G414

2-Amino-5-hydroxy-3,8-dimethylimidazo[4,5-*f*]quinoxaline-β-D-glucuronide
Probably formed from: 2-Amino-5-hydroxy-3,8-dimethylimidazo[4,5-*f*]quinoxaline

2-Amino-5-hydroxy-3,8-dimethylimidazo[4,5-*f*]quinoxaline-O-sulphate
Probably formed from: 2-Amino-5-hydroxy-3,8-dimethylimidazo[4,5-*f*]quinoxaline

2-Amino-N-(3-hydroxy-2,6-dimethylphenyl) butanamide
Probably formed from: 2-Ethylamino-N-(3-hydroxy-2,6-dimethylphenyl)butanamide
N-(3-Hydroxy-2,6-dimethylphenyl)-2-propylaminobutanamide

2-Amino-N-(4-hydroxy-2,6-dimethylphenyl) butanamide
Probably formed from: 2-Ethylamino-N-(4-hydroxy-2,6-dimethylphenyl)butanamide
N-(4-Hydroxy-2,6-dimethylphenyl)-2-propylaminobutanamide

2-Amino-hydroxy-4,5-diphenyloxazole
Probably formed from: 2-Amino-4,5-diphenyloxazole

2-Amino-1-hydroxyfluorene
Formed from: 2-Acetamido-1-hydroxyfluorene
Probably formed from: 2-Hydroxylaminofluorene

2-Amino-3-hydroxyfluorene
Formed from: 2-Acetamido-3-hydroxyfluorene
Probably formed from: 2-Hydroxylaminofluorene

2-Amino-5-hydroxyfluorene
Formed from: 2-Acetamido-5-hydroxyfluorene
2-Aminofluorene

2-Amino-7-hydroxyfluorene
Formed from: 2-Acetamido-7-hydroxyfluorene
2-Aminofluorene

2-Amino-9-hydroxyfluorene
Formed from: 2-Acetamido-9-hydroxyfluorene

2-Amino-4-hydroxy-4-(4-hydroxy-3-methoxyphenyl)butyric acid
Formed from: Vanillin

4′-Amino-4-hydroxy-3-hydroxymethylazobenzene
Probably formed from: 4′-Amino-4-hydroxy-3-methylazobenzene

11-Amino-10-hydroxy-1-(2-(1-hydroxy-1-methylethyl)-4-thiazolyl)-2-methyl-3,6-dioxo-11-phenyl-8-phenylmethyl-2,4,7-triazadodecane
Probably formed from: [5S-(5R*, 8R*, 10R*, 11R*)]-10-Hydroxy-1-(2-(1-hydroxy-1-methylethyl)-4-thiazolyl)-2-methyl-3,6-dioxo-8,11-bis(phenylmethyl)-2,4,7,12-tetraazatridecan-13-oic acid 5-thiazolylmethyl ester

2-Amino-5-hydroxy-8-hydroxymethyl-3-methylimidazo[4,5-*f*]quinoxaline
Probably formed from: 2-Amino-5-hydroxy-3,8-dimethylimidazo[4,5-*f*]quinoxaline
→ 2-Amino-5-hydroxy-8-hydroxymethyl-3-methylimidazo[4,5-*f*]quinoxaline-β-D-glucuronide
Probably in: Rat F195

2-Amino-5-hydroxy-8-hydroxymethyl-3-methylimidazo[4,5-*f*]quinoxaline-β-D-glucuronide
Probably formed from: 2-Amino-5-hydroxy-8-hydroxymethyl-3-methylimidazo[4,5-*f*]quinoxaline

4-Amino-5-hydroxy-3-((4′-((4-hydroxyphenyl)azo)(1,1′-biphenyl)-4-yl)azo)-6-(phenylazo)-2,7-naphthalenedisulphonic acid
See: Direct Green 1

***cis*-2-Amino-1-hydroxyindane**
Formed from: 2-Amino-1-indanone
→ *trans*-2-Amino-1-hydroxyindane
In: Rabbit A449
→ *cis*-1,2-dihydroxyindane
In: Rabbit A449
→ *trans*-1,2-dihydroxyindane
In: Rabbit A449

***trans*-2-Amino-1-hydroxyindane**
Formed from: *cis*-2-Amino-1-hydroxyindane
2-Amino-1-indanone
→ *cis*-1,2-dihydroxyindane
In: Rabbit A449
→ *trans*-1,2-dihydroxyindane
In: Rabbit A449

2-Amino-5-hydroxyindane
Formed from: 2-Aminoindane

5-Amino-4-hydroxy-3-iodobenzamide
Probably formed from: 5-Amino-4-hydroxy-3-iodobenzonitrile

5-Amino-4-hydroxy-3-iodobenzonitrile
Probably formed from: 4-Hydroxy-3-iodo-5-nitrobenzonitrile
→ 5-Amino-4-hydroxy-3-iodobenzamide

Probably in: Rat C297

4-Amino-2-hydroxylamino-6-nitrotoluene
Probably formed from: 4-Amino-2,6-dinitrotoluene

4′-Amino-4-hydroxy-3-methoxyazobenzene
Probably formed from: 4-Amino-3′-methoxyazobenzene

4′-Amino-3-hydroxy-4-methoxybiphenyl
Probably formed from: 4′-Amino-3,4-dihydroxybiphenyl
→ 4′-Acetamido-3-hydroxy-4-methoxybiphenyl
Probably in: Rat G1

4′-Amino-4-hydroxy-3-methoxybiphenyl
Probably formed from: 4′-Amino-3,4-dihydroxybiphenyl
→ 4′-Acetamido-4-hydroxy-3-methoxybiphenyl
Probably in: Rat G1

2-Amino-6-hydroxy-5-methoxy-N,N-dipropyltetralin
Formed from: 2-Amino-5,6-dihydroxy-N,N-dipropyltetralin

8-Amino-5-hydroxy-6-methoxyquinoline
Probably formed from: 5-Hydroxyprimaquine
→ 8-Amino-5,6-dihydroxyquinoline
Probably in: Dog C66

4-Amino-3-hydroxy-4′-methylaminobenzophenone
Probably formed from: 4-Amino-4′-methylaminobenzophenone
→ 4-Acetamido-3-hydroxy-4′-methylaminobenzophenone
Probably in: Rat B713

4-Amino-3′-hydroxymethylazobenzene
Probably formed from: 4-Amino-3′-methylazobenzene
→ 4′-Amino-3-formylazobenzene
Probably in: Rat A3835

4′-Amino-4-hydroxy-3-methylazobenzene
Probably formed from: 4-Amino-3′-methylazobenzene
4-Hydroxy-3-methyl-4′-methylaminoazobenzene
→ 4′-Acetamido-4-hydroxy-3-methylazobenzene
Probably in: Rat A3783, E872

→ 4′-Amino-4-hydroxy-3-hydroxymethylazobenzene
Probably in: Rat B27

8-(4-Amino-3-hydroxy-1-methylbutyl)amino-5,6-dihydroxy-2-methoxy-4-methylquinoline
Probably formed from: 8-(4-Amino-3-hydroxy-1-methylbutyl)amino-2,6-dimethoxy-4-methyl-5-(3-trifluoromethylphenoxy)quinoline
→ 8-(4-Amino-3-hydroxy-1-methylbutyl) amino-2,5,6-trihydroxy-4-methylquinoline
Probably in: Rat H135

8-(4-Amino-3-hydroxy-1-methylbutyl)amino-2,6-dihydroxy-4-methyl-5-(3-trifluoromethylphenoxy)quinoline
Probably formed from: 8-(4-Amino-3-hydroxy-1-methylbutyl)amino-2-hydroxy-6-methoxy-4-methyl-5-(3-trifluoromethylphenoxy)quinoline
→ 8-(2,3-Dioxo-5-methylpyrrolidin-5-yl) amino-2,6-dihydroxy-4-methyl-5-(3-trifluoromethylphenoxy)quinoline
Probably in: Rat H135

8-(4-Amino-3-hydroxy-1-methylbutyl)amino-2,6-dimethoxy-4-methyl-5-(3-trifluoromethylphenoxy)quinoline
Formed from: 8-(4-Amino-1-methylbutyl) amino-2,6-dimethoxy-4-methyl-5-(3-trifluoromethylphenoxy)quinoline
→ 8-(4-Amino-2,3-dihydroxy-1-methylbutyl)amino-4-hydroxymethyl-2,6-dimethoxy-5-(3-trifluoromethylphenox-y)quinoline-N-oxide
Probably in: Rat H135
→ 8-(4-Amino-3-hydroxy-1-methylbutyl) amino-5,6-dihydroxy-2-methoxy-4-methylquinoline
Probably in: Rat H135
→ 8-(4-Amino-3-hydroxy-1-methylbutyl) amino-2-hydroxy-6-methoxy-4-methyl-5-(3-trifluoromethylphenoxy)quinoline
Probably in: Rat H135
→ 8-(4-Amino-1-methyl-3-oxobutyl)amino-6-hydroxy-4-hydroxymethyl-2-methoxy-5-(3-trifluoromethylphenoxy)quinoline
Probably in: Rat H135

→ 8-(4-Amino-1-methyl-2-oxobutyl)amino-6-hydroxy-2-methoxy-4-methyl-5-(3-trifluoromethylphenoxy)quinoline
Probably in: Rat H135

8-(4-Amino-3-hydroxy-1-methylbutyl)amino-6-hydroxyquinoline
Formed from: 8-(4-Amino-1-methylbutyl)amino-6-hydroxyquinoline

8-(4-Amino-3-hydroxy-1-methylbutyl)amino-2-hydroxy-6-methoxy-4-methyl-5-(3-trifluoromethylphenoxy)quinoline
Probably formed from: 8-(4-Amino-3-hydroxy-1-methylbutyl)amino-2,6-dimethoxy-4-methyl-5-(3-trifluoromethylphenoxy)quinoline
→ 8-(4-Amino-3-hydroxy-1-methylbutyl)amino-2,6-dihydroxy-4-methyl-5-(3-trifluoromethylphenoxy)quinoline
Probably in: Rat H135
→ 8-(4-Amino-2-hydroxy-1-methyl-3-oxobutyl)amino-2-hydroxy-4-hydroxymethyl-6-methoxy-5-(3-trifluoromethylphenoxy)quinoline
Probably in: Rat H135
→ 8-(4-Amino-1-methyl-3-oxobutyl)amino-2-hydroxy-6-methoxy-4-methyl-5-(3-trifluoromethylphenoxy)quinoline
Probably in: Rat H135

8-(4-Amino-3-hydroxy-1-methylbutyl)amino-5-hydroxy-6-methoxyquinoline
Probably formed from: 5-hydroxyprimaquine
→ 8-(4-Amino-1-methyl-3-oxobutyl)amino-5-hydroxy-6-methoxyquinoline
Probably in: Rat H136

8-(4-Amino-3-hydroxy-1-methylbutyl)amino-2,5,6-trihydroxy-4-methylquinoline
Probably formed from: 8-(4-Amino-3-hydroxy-1-methylbutyl)amino-2,6-dimethoxy-4-methyl-5-(3-trifluoromethylphenoxy)quinoline
→ 8-(4-Amino-2,3-dihydroxy-1-methylbutyl)amino-2,5,6-trihydroxy-4-hydroxymethylquinoline
Probably in: Rat H135
→ 8-(4-Amino-1-methyl-3-oxobutyl)amino-2,5,6-trihydroxy-4-methylquinoline
Probably in: Rat H135

2-Amino-1-(4-hydroxymethyl-2,5-dimethoxyphenyl)propane
Probably formed from: 2,5-Dimethoxy-4-methylamphetamine
→ 2-Amino-1-(4-carboxy-2,5-dimethoxyphenyl)propane
Probably in: Rabbit A1401

2-Amino-2′-hydroxymethyldiphenyl ether
Formed from: Dibenz[*b,f*]-1,4-oxazepine
→ 2-Amino-2′-carboxydiphenyl ether
Probably in: Rat C495

2-Amino-5-hydroxy-3-methylimidazo[4,5-*f*]quinoline
Formed from: 2-Amino-3-methylimidazo[4,5-*f*]quinoline
→ 2-Amino-5-hydroxy-3-methylimidazo[4,5-*f*]quinoline-5-β-D-glucuronide
In: Cynomolgus monkey G673
→ 2-Amino-5-hydroxy-3-methylimidazo[4,5-*f*]quinoline-5-sulphate
In: Cynomolgus monkey G673

2-Amino-5-hydroxy-3-methylimidazo[4,5-*f*]quinoline-5-β-D-glucuronide
Formed from: 2-Amino-5-hydroxy-3-methylimidazo[4,5-*f*]quinoline

2-Amino-5-hydroxy-3-methylimidazo[4,5-*f*]quinoline-5-sulphate
Formed from: 2-Amino-5-hydroxy-3-methylimidazo[4,5-*f*]quinoline

2-Amino-8-hydroxymethyl-3-methylimidazo[4,5-*f*]quinoxaline
Formed from: 2-Amino-3,8-dimethylimidazo[4,5-*f*]quinoxaline
→ 2-Amino-8-hydroxymethyl-3-methylimidazo[4,5-*f*]quinoxaline-2-sulphate
Probably in: Rat G414

2-Amino-8-hydroxymethyl-3-methylimidazo[4,5-*f*]quinoxaline-2-sulphate
Probably formed from: 2-Amino-8-hydroxymethyl-3-methylimidazo[4,5-*f*]quinoxaline

4-Amino-N-(2-(4-hydroxymethyl-6-methyl)pyrimidinyl)benzenesulphonamide
Formed from: Sulphadimidine

→ 4-Amino-N-(2-(4-hydroxymethyl-6-methyl)pyrimidinyl)benzenesulphonamide-β-D-glucuronide
Probably in: Man F954

4-Amino-N-(6-hydroxymethyl-4-methyl-2-pyrimidinyl)benzenesulphonamide
Formed from: Sulphamerazine
→ 4-Amino-N-(6-carboxy-4-methyl-2-pyrimidinyl)benzenesulphonamide
Probably in: Horse F928

4-Amino-N-(2-(4-hydroxymethyl-6-methyl)pyrimidinyl)benzenesulphonamide-β-D-glucuronide
Probably formed from: 4-Amino-N-(2-(4-hydroxymethyl-6-methyl)pyrimidinyl)benzenesulphonamide

8-(4-Amino-2-hydroxy-1-methyl-3-oxobutyl)amino-4-formyl-2-hydroxy-6-methoxy-5-(3-trifluoromethylphenoxy)quinoline
Probably formed from: 8-(4-Amino-2-hydroxy-1-methyl-3-oxobutyl)amino-2-hydroxy-4-hydroxymethyl-6-methoxy-5-(3-trifluoromethylphenoxy)quinoline

8-(4-Amino-2-hydroxy-1-methyl-3-oxobutyl)amino-2-hydroxy-4-hydroxymethyl-6-methoxy-5-(3-trifluoromethylphenoxy)quinoline
Probably formed from: 8-(4-Amino-3-hydroxy-1-methylbutyl)amino-2-hydroxy -6-methoxy-4-methyl-5-(3-trifluoromethylphenoxy)quinoline
→ 8-(4-Amino-2-hydroxy-1-methyl-3-oxobutyl)amino-4-formyl-2-hydroxy-6-methoxy-5-(3-trifluoromethylphenoxy)quinoline
Probably in: Rat H135
→ 8-(4-Amino-1-methyl-2,3-dioxobutyl)amino-2-hydroxy-4-hydroxymethyl-6-methoxy-5-(3-trifluoromethylphenoxy)quinoline
Probably in: Rat H135

8-(4-Amino-3-hydroxy-1-methyl-2-oxobutyl)amino-6-hydroxy-2-methoxy-4-methyl-5-(3-trifluoromethylphenoxy)quinoline
Probably formed from: 8-(4-Amino-3-hydroxy-1-methylbutyl)amino-2,6-dimethoxy-4-methyl-5-(3-trifluoromethylphenoxy)quinoline

2-Amino-4-hydroxy-1-methyl-6-phenylimidazo[4,5-*b*]pyridine
Formed from: 2-Amino-1-methyl-6-phenylimidazo[4,5-*b*]pyridine

4-Amino-7-hydroxy-10-methylpteroic acid
Formed from: 4-Amino-10-methylpteroic acid

4-Amino-N-(4-hydroxymethyl-2-pyrimidinyl)benzenesulphonamide
Formed from: Sulphamerazine
→ 4-Amino-N-(4-carboxy-2-pyrimidinyl)benzenesulphonamide
Probably in: Horse F928

4-Amino-N-(4-hydroxy-6-methyl-2-pyrimidinyl)benzenesulphonamide
Formed from: Sulphamerazine

N-(*p*-((2-Amino-4-hydroxy-5-methylquinazolin-6-yl)aminomethyl)benzoyl)diglutamylglutamic acid
Probably formed from: N-(*p*-((2-Amino-4-hydroxy-5-methylquinazolin-6-yl)aminomethyl)benzoyl)glutamylglutamic acid
→ N-(*p*-((2-Amino-4-hydroxy-5-methylquinazolin-6-yl)aminomethyl)benzoyl)triglutamylglutamic acid
Probably in: Pig F760

N-(*p*-((2-Amino-4-hydroxy-5-methylquinazolin-6-yl)aminomethyl)benzoyl)glutamic acid
→ N-(*p*-((2-Amino-4-hydroxy-5-methylquinazolin-6-yl)aminomethyl)benzoyl)glutamylglutamic acid
Probably in: Pig F760

N-(*p*-((2-Amino-4-hydroxy-5-methylquinazolin-6-yl)aminomethyl)benzoyl)glutamylglutamic acid
Formed from: N-(*p*-((2-Amino-4-hydroxy-5-methylquinazolin-6-yl)aminomethyl)benzoyl)glutamic acid
→ N-(*p*-((2-Amino-4-hydroxy-5-methylquinazolin-6-yl)aminomethyl)benzoyl)diglutamylglutamic acid
Probably in: Pig F760

N-(*p*-((2-Amino-4-hydroxy-5-methylquinazolin-6-yl)aminomethyl)benzoyl)tetraglutamylglutamic acid
Probably formed from: N-(*p*-((2-Amino-4-hydroxy-5-methylquinazolin-6-yl)aminomethyl)benzoyl)triglutamylglutamic acid

N-(*p*-((2-Amino-4-hydroxy-5-methylquinazolin-6-yl)aminomethyl)benzoyl)triglutamylglutamic acid
Probably formed from: N-(*p*-((2-Amino-4-hydroxy-5-methylquinazolin-6-yl)aminomethyl)benzoyl)diglutamylglutamic acid
→ N-(*p*-((2-Amino-4-hydroxy-5-methylquinazolin-6-yl)aminomethyl)benzoyl)tetraglutamylglutamic acid
Probably in: Pig F760

N-(*p*-((2-Amino-4-hydroxy-5-methylquinazolin-6-yl)-N-methylaminomethyl)benzoyl)diglutamylglutamic acid
Probably formed from: N-(*p*-((2-Amino-4-hydroxy-5-methylquinazolin-6-yl)-N-methylaminomethyl)benzoyl)glutamylglutamic acid
→ N-(*p*-((2-Amino-4-hydroxy-5-methylquinazolin-6-yl)-N-methylaminomethyl)benzoyl)triglutamylglutamic acid
Probably in: Pig F760

N-(*p*-((2-Amino-4-hydroxy-5-methylquinazolin-6-yl)-N-methylaminomethyl)benzoyl)glutamic acid
→ N-(*p*-((2-Amino-4-hydroxy-5-methylquinazolin-6-yl)-N-methylaminomethyl)benzoyl)glutamylglutamic acid
In: Pig F760

N-(*p*-((2-Amino-4-hydroxy-5-methylquinazolin-6-yl)-N-methylaminomethyl)benzoyl)glutamylglutamic acid
Formed from: N-(*p*-((2-Amino-4-hydroxy-5-methylquinazolin-6-yl)-N-methylaminomethyl)benzoyl)glutamic acid
→ N-(*p*-((2-Amino-4-hydroxy-5-methylquinazolin-6-yl)-N-methylaminomethyl)benzoyl)diglutamylglutamic acid
Probably in: Pig F760

N-(*p*-((2-Amino-4-hydroxy-5-methylquinazolin-6-yl)-N-methylaminomethyl)benzoyl)pentaglutamylglutamic acid
Probably formed from: N-(*p*-((2-Amino-4-hydroxy-5-methylquinazolin-6-yl)-N-methylaminomethyl)benzoyl)tetraglutamylglutamic acid

N-(*p*-((2-Amino-4-hydroxy-5-methylquinazolin-6-yl)-N-methylaminomethyl)benzoyl)tetraglutamylglutamic acid
Probably formed from: N-(*p*-((2-Amino-4-hydroxy-5-methylquinazolin-6-yl)-N-methylaminomethyl)benzoyl)triglutamylglutamic acid
→ N-(*p*-((2-Amino-4-hydroxy-5-methylquinazolin-6-yl)-N-methylaminomethyl)benzoyl)pentaglutamylglutamic acid
Probably in: Pig F760

N-(*p*-((2-Amino-4-hydroxy-5-methylquinazolin-6-yl)-N-methylaminomethyl)benzoyl)triglutamylglutamic acid
Probably formed from: N-(*p*-((2-Amino-4-hydroxy-5-methylquinazolin-6-yl)-N-methylaminomethyl)benzoyl)diglutamylglutamic acid
→ N-(*p*-((2-Amino-4-hydroxy-5-methylquinazolin-6-yl)-N-methylaminomethyl)benzoyl)tetraglutamylglutamic acid
Probably in: Pig F760

6-Amino-2-hydroxymethyl-3-(*o*-tolyl)-4(3*H*)-quinazolinone
Formed from: Afloqualone
→ 6-Acetamido-2-hydroxymethyl-3-(*o*-tolyl)-4(3*H*)-quinazolinone
Probably in: Monkey C912
Rat C793a

1-Amino-2-hydroxynaphthalene
See: 1-Amino-2-naphthol

1-Amino-4-hydroxynaphthalene
See: 4-Amino-1-naphthol

2-Amino-1-hydroxynaphthalene
See: 2-Amino-1-naphthol

5-Amino-1-hydroxynaphthalene
See: 5-Amino-1-naphthol

5-Amino-2-hydroxynaphthalene
See: 5-Amino-2-naphthol

7-Amino-2-hydroxynaphthalene
See: 7-Amino-2-naphthol

1-Amino-2-hydroxynaphthalene-3,6-disulphonic acid
Formed from: 2-Hydroxy-1-(4-sulpho-1-naphthylazo)naphthalene-3,6-disulphonic acid

2-Hydroxy-1-(2,4,5-trimethylphenylazo)naphthalene-3,6-disulphonic acid

1-Amino-2-hydroxynaphthalene-6-sulphonic acid
Formed from: Acid Orange 12
2-Hydroxy-1-(*p*-sulphophenylazo) naphthalene-6-sulphonic acid

2-Amino-1-hydroxynaphthalene-4-sulphonic acid
Formed from: 1-Hydroxy-2-(5-sulpho-2,4-xylylazo)naphthalene-4-sulphonic acid

5-Amino-2-hydroxy-3-nitrobenzoic acid
Probably formed from: 2-Hydroxy-3,5-dinitrobenzoic acid

2-Amino-3-hydroxy-5-nitrobenzophenone
Formed from: 1,3-Dihydro-7-nitro-5-phenyl-2H-1,4-benzodiazepin-2-one

2-Amino-4-hydroxy-4-(*o*-nitrophenyl)butyric acid
Formed from: *o*-Nitrobenzaldehyde

2-Amino-4-hydroxy-4-(*p*-nitrophenyl)butyric acid
Formed from: *p*-Nitrobenzaldehyde

2-Amino-4-hydroxy-4-phenylbutyric acid
Formed from: Benzaldehyde

4-Amino-3-hydroxyphenyl 2,4-dichlorophenyl ether
Probably formed from: 4-Amino-3-methoxyphenyl 2,4-dichlorophenyl ether

1-Amino-2-hydroxy-2-phenylethane
See: Phenylethanolamine

2-(4-Amino-3-hydroxyphenyl)-2-ethylglutarimide
Formed from: Aminoglutethimide
→ 2-(4-Amino-3-sulphatoxyphenyl)-2-ethylglutarimide
In: Rat C10

2-Amino-6-(*p*-hydroxyphenyl)-1-methylimidazo[4,5-*b*]pyridine
Formed from: 2-Amino-1-methyl-6-phenylimidazo[4,5-*b*]pyridine
→ 2-Amino-6-(*p*-hydroxyphenyl)-1-methylimidazo[4,5-*b*]pyridine-β-D-glucuronide
In: Guinea pig H345
Hamster H345
Monkey H88
→ 2-Amino-6-(*p*-hydroxyphenyl)-1-methylimidazo[4,5-*b*]pyridine-O-sulphate
In: Guinea pig H345
Hamster H345
Monkey H88
Rat G443, H345, H417

2-Amino-6-(*p*-hydroxyphenyl)-1-methylimidazo[4,5-*b*]pyridine-β-D-glucuronide
Formed from: 2-Amino-6-(*p*-hydroxyphenyl)-1-methylimidazo[4,5-*b*]pyridine

2-Amino-6-(*p*-hydroxyphenyl)-1-methylimidazo[4,5-*b*]pyridine-O-sulphate
Formed from: 2-Amino-6-(*p*-hydroxyphenyl)-1-methylimidazo[4,5-*b*]pyridine

3-Amino-2-hydroxy-3-phenylpropionic acid
See: Phenylisoserine

***p*-Amino-β-hydroxyphenylpyruvic acid**
Formed from: *p*-Aminophenylserine

8-(3-Amino-2-hydroxy)propoxy-3,4-dihydro-3-hydroxy-2*H*-1-benzopyran
Probably formed from: 8-(3-Amino-2-hydroxy)propoxy-3,4-dihydro-3-nitroxy-2*H*-1-benzopyran

8-(3-Amino-2-hydroxy)propoxy-3,4-dihydro-3-nitrooxy-2*H*-1-benzopyran
Formed from: Nipradilol
→ 8-(3-Amino-2-hydroxy)propoxy-3,4-dihydro-3-hydroxy-2*H*-1-benzopyran
Probably in: Rat D882

2-(3-Amino-2-hydroxy-1-propoxy)-1-(2-propen-1-oxy)benzene
Formed from: Oxprenolol
→ 2-(3-Acetamido-2-hydroxy-1-propoxy)-1-(2-propen-1-oxy)benzene
Probably in: man D664
→ 2-(2-Hydroxy-3-methylamino-1-propoxy)-1-(2-propen-1-oxy)benzene
Probably in: Dog A1181

6-(3-Amino-2-hydroxypropoxy)-2(1*H*)-quinolinone
→ 6-(3-Acetamido-2-hydroxypropoxy)-2(1*H*)-quinolinone
Probably in: Dog J154
Rat J154

4-(2-Amino-1-hydroxyprop-1-yl)-*o*-quinone
Probably formed from: α-Methylnoradrenaline

1-Amino-6-hydroxypyrene
Probably formed from: 1-Aminopyrene

1-Amino-8-hydroxypyrene
Probably formed from: 1-Aminopyrene

4-Amino-N-(4-hydroxy-2-pyrimidinyl) benzenesulphonamide
Formed from: Sulphadiazine

N-(*p*-((2-Amino-4-hydroxyquinazolin-6-yl)aminomethyl)benzoyl)diglutamylglutamic acid
Probably formed from: N-(*p*-((2-Amino-4-hydroxyquinazolin-6-yl)aminomethyl) benzoyl)glutamylglutamic acid
→ N-(*p*-((2-Amino-4-hydroxyquinazolin-6-yl)aminomethyl)benzoyl)triglutamylglutamic acid
Probably in: Pig F760

N-(*p*-((2-Amino-4-hydroxyquinazolin-6-yl)aminomethyl)benzoyl)glutamic acid
→ N-(*p*-((2-Amino-4-hydroxyquinazolin-6-yl)aminomethyl)benzoyl)glutamylglutamic acid
In: Pig F760

N-(*p*-((2-Amino-4-hydroxyquinazolin-6-yl) aminomethyl)benzoyl)glutamylglutamic acid
Formed from: N-(*p*-((2-Amino-4-hydroxyquinazolin-6-yl)aminomethyl) benzoyl)glutamic acid
→ N-(*p*-((2-Amino-4-hydroxyquinazolin-6-yl)aminomethyl)benzoyl)diglutamylglutamic acid
Probably in: Pig F760

N-(*p*-((2-Amino-4-hydroxy-6-quinazolinyl) aminomethyl)benzoyl)-L-glutamic acid
See: 5,8-dideazaisofolic acid

N-(*p*-((2-Amino-4-hydroxyquinazolin-6-yl) aminomethyl)benzoyl)pentaglutamylglutamic acid
Probably formed from: N-(*p*-((2-Amino-4-hydroxyquinazolin-6-yl)aminomethyl) benzoyl)tetraglutamylglutamic acid

N-(*p*-((2-Amino-4-hydroxyquinazolin-6-yl) aminomethyl)benzoyl)tetraglutamylglutamic acid
Probably formed from: N-(*p*-((2-Amino-4-hydroxyquinazolin-6-yl)aminomethyl)benzoyl)triglutamylglutamic acid
→ N-(*p*-((2-Amino-4-hydroxyquinazolin-6-yl)aminomethyl)benzoyl)pentaglutamylglutamic acid
Probably in: Pig F760

N-(*p*-((2-Amino-4-hydroxyquinazolin-6-yl) aminomethyl)benzoyl)triglutamylglutamic acid
Probably formed from: N-(*p*-((2-Amino-4-hydroxyquinazolin-6-yl)aminomethyl) benzoyl)diglutamylglutamic acid
→ N-(*p*-((2-Amino-4-hydroxyquinazolin-6-yl)aminomethyl)benzoyl)tetraglutamylglutamic acid
Probably in: Pig F760

N-(*p*-((2-Amino-4-hydroxyquinazolin-6-yl)methoxy)benzoyl)diglutamylglutamic acid
Probably formed from: N-(*p*-((2-Amino-4-hydroxyquinazolin-6-yl)methoxy)benzoyl) glutamic acid
→ N-(*p*-((2-Amino-4-hydroxyquinazolin-6-yl)methoxy)benzoyl)triglutamylglutamic acid
Probably in: Pig F760

N-(*p*-((2-Amino-4-hydroxyquinazolin-6-yl)methoxy)benzoyl)glutamic acid
→ N-(*p*-((2-Amino-4-hydroxyquinazolin-6-yl)methoxy)benzoyl)diglutamylglutamic acid
In: Pig F760

N-(*p*-((2-Amino-4-hydroxyquinazolin-6-yl)methoxy)benzoyl)glutamylglutamic acid
→ N-(*p*-((2-Amino-4-hydroxyquinazolin-6-yl)methoxy)benzoyl)diglutamylglutamic acid
Probably in: Pig F760

N-(*p*-((2-Amino-4-hydroxyquinazolin-6-yl)methoxy)benzoyl)tetraglutamylglutamic acid
Probably formed from: N-(*p*-((2-Amino-4-hydroxyquinazolin-6-yl)methoxy)benzoyl)triglutamic acid

N-(*p*-((2-Amino-4-hydroxyquinazolin-6-yl)methoxy)benzoyl)triglutamylglutamic acid
Probably formed from: N-(*p*-((2-Amino-4-hydroxyquinazolin-6-yl)methoxy)benzoyl) diglutamylglutamic acid
→ N-(*p*-((2-Amino-4-hydroxyquinazolin-6-yl)methoxy)benzoyl)tetraglutamylglutamic acid
Probably in: Pig F760

N-(*p*-((2-Amino-4-hydroxyquinazolin-6-yl) methylamino)benzoyl)diglutamylglutamic acid
Probably formed from: N-(*p*-((2-Amino-4-hydroxyquinazolin-6-yl)methylamino)benzoyl)glutamylglutamic acid
→ N-(*p*-((2-Amino-4-hydroxyquinazolin-6-yl)methylamino)benzoyl)triglutamylglutamic acid
Probably in: Pig F760

N-(*p*-((2-Amino-4-hydroxyquinazolin-6-yl)methylamino)benzoyl)glutamic acid
→ N-(*p*-((2-Amino-4-hydroxyquinazolin-6-yl)methylamino)benzoyl)glutamylglutamic acid
In: Pig F760

N-(*p*-((2-Amino-4-hydroxyquinazolin-6-yl)methylamino)benzoyl)glutamylglutamic acid
Formed from: N-(*p*-((2-Amino-4-hydroxyquinazolin-6-yl)methylamino)benzoyl) glutamic acid
→ N-(*p*-((2-Amino-4-hydroxyquinazolin-6-yl)methylamino)benzoyl)diglutamylglutamic acid
Probably in: Pig F760

N-(*p*-((2-Amino-4-hydroxyquinazolin-6-yl) methylamino)benzoyl)pentaglutamylglutamic acid
Probably formed from: N-(*p*-((2-Amino-4-hydroxyquinazolin-6-yl)methylamino) benzoyl)tetraglutamylglutamic acid

N-(*p*-((2-Amino-4-hydroxyquinazolin-6-yl)methylamino)benzoyl)tetraglutamylglutamic acid
Probably formed from: N-(*p*-((2-Amino-4-hydroxyquinazolin-6-yl)methylamino)benzoyl)triglutamylglutamic acid
→ N-(*p*-((2-Amino-4-hydroxyquinazolin-6-yl)methylamino)benzoyl)pentaglutamylglutamic acid
Probably in: Pig F760

N-(*p*-((2-Amino-4-hydroxyquinazolin-6-yl) methylamino)benzoyl)triglutamylglutamic acid
Probably formed from: N-(*p*-((2-Amino-4-hydroxyquinazolin-6-yl)methylamino) benzoyl)diglutamylglutamic acid
→ N-(*p*-((2-Amino-4-hydroxyquinazolin-6-yl)methylamino)benzoyl)tetraglutamylglutamic acid
Probably in: Pig F760

N-(*p*-((2-Amino-4-hydroxyquinazolin-6-yl) methylaminomethyl)benzoyl)diglutamylglutamic acid
Probably formed from: N-(*p*-((2-Amino-4-hydroxyquinazolin-6-yl)methylaminomethyl)benzoyl)glutamylglutamic acid
→ N-(*p*-((2-Amino-4-hydroxyquinazolin-6-yl)methylaminomethyl)benzoyl)triglutamylglutamic acid
Probably in: Pig F760

N-(*p*-((2-Amino-4-hydroxyquinazolin-6-yl) methylaminomethyl)benzoyl)glutamic acid
→ N-(*p*-((2-Amino-4-hydroxyquinazolin-6-yl)methylaminomethyl)benzoyl)glutamylglutamic acid
In: Pig F760

N-(*p*-((2-Amino-4-hydroxyquinazolin-6-yl) methylaminomethyl)benzoyl)glutamylglutamic acid
Formed from: N-(*p*-((2-Amino-4-hydroxyquinazolin-6-yl)methylaminomethyl)benzoyl) glutamic acid
→ N-(*p*-((2-Amino-4-hydroxyquinazolin-6-yl)methylaminomethyl)benzoyl) diglutamylglutamic acid
Probably in: Pig F760

N-(*p*-((2-Amino-4-hydroxyquinazolin-6-yl) methylaminomethyl)benzoyl)pentaglutamylglutamic acid
Probably formed from: N-(*p*-((2-Amino-4-hydroxyquinazolin-6-yl)methylaminomethyl)benzoyl)tetraglutamylglutamic acid

N-(*p*-((2-Amino-4-hydroxyquinazolin-6-yl) methylaminomethyl)benzoyl)tetraglutamylglutamic acid
Probably formed from: N-(*p*-((2-Amino-4-hydroxyquinazolin-6-yl)methylaminomethyl)benzoyl)triglutamylglutamic acid
→ N-(*p*-((2-Amino-4-hydroxyquinazolin-6-yl)methylaminomethyl)benzoyl)pentaglutamylglutamic acid
Probably in: Pig F760

N-(*p*-((2-Amino-4-hydroxyquinazolin-6-yl)methylaminomethyl)benzoyl)triglutamylglutamic acid
Probably formed from: N-(*p*-((2-Amino-4-hydroxyquinazolin-6-yl)methylaminomethyl)benzoyl)diglutamylglutamic acid
→ N-(*p*-((2-Amino-4-hydroxyquinazolin-6-yl)methylaminomethyl)benzoyl) tetraglutamylglutamic acid
Probably in: Pig F760

N-(*p*-(((2-Amino-4-hydroxyquinazolin-6-yl)methyl)-N,N-dimethylamino)benzoyl) diglutamylglutamic acid
Probably formed from: N-(*p*-(((2-Amino-4-hydroxyquinazolin-6-yl)methyl)-N,N-dimethylamino)benzoyl)glutamylglutamic acid
→ N-(*p*-(((2-Amino-4-hydroxyquinazolin-6-yl)methyl)-N,N-dimethylamino)benzoyl) triglutamylglutamic acid
Probably in: Pig F760

N-(*p*-(((2-Amino-4-hydroxyquinazolin-6-yl)methyl)-N,N-dimethylamino)benzoyl) glutamic acid
→ N-(*p*-(((2-Amino-4-hydroxyquinazolin-6-yl)methyl)-N,N-dimethylamino) benzoyl)glutamylglutamic acid
in: Pig F760

N-(*p*-(((2-Amino-4-hydroxyquinazolin-6-yl)methyl)-N,N-dimethylamino)benzoyl) glutamylglutamic acid
Formed from: N-(*p*-(((2-Amino-4-hydroxyquinazolin-6-yl)methyl)-N,N-dimethylamino)benzoyl)glutamic acid
→ N-(*p*-(((2-Amino-4-hydroxyquinazolin-6-yl)methyl)-N,N-dimethylamino)benzoyl) diglutamylglutamic acid
Probably in: Pig F760

N-(*p*-(((2-Amino-4-hydroxyquinazolin-6-yl)methyl)-N,N-dimethylamino)benzoyl) pentaglutamylglutamic acid
Probably formed from: N-(*p*-(((2-Amino-4-hydroxyquinazolin-6-yl)methyl)-N,N-dimethylamino)benzoyl)tetraglutamylglutamic acid

N-(*p*-(((2-Amino-4-hydroxyquinazolin-6-yl)methyl)-N,N-dimethylamino)benzoyl) tetraglutamylglutamic acid
Probably formed from: N-(*p*-(((2-Amino-4-hydroxyquinazolin-6-yl)methyl)-N,N-dimethylamino)benzoyl)triglutamylglutamic acid
→ N-(*p*-(((2-Amino-4-hydroxyquinazolin-6-yl)methyl)-N,N-dimethylamino)benzoyl)pentaglutamylglutamic acid
Probably in: Pig F760

N-(*p*-(((2-Amino-4-hydroxyquinazolin-6-yl)methyl)-N,N-dimethylamino)benzoyl) triglutamylglutamic acid
Probably formed from: N-(*p*-(((2-Amino-4-hydroxyquinazolin-6-yl)methyl)-N,N-dimethylamino)benzoyl)diglutamylglutamic acid
→ N-(*p*-(((2-Amino-4-hydroxyquinazolin-6-yl)methyl)-N,N-dimethylamino)benzoyl) tetraglutamylglutamic acid
Probably in: Pig F760

N-(*p*-(((2-Amino-4-hydroxyquinazolin-6-yl)methyl)-N-formylamino)benzoyl)diglutamylglutamic acid
Formed from: N-(*p*-(((2-Amino-4-hydroxyquinazolin-6-yl)methyl)-N-formylamino) benzoyl)glutamic acid
→ N-(*p*-(((2-Amino-4-hydroxyquinazolin-6-yl)methyl)-N-formylamino)benzoyl) triglutamylglutamic acid
Probably in: Pig F760

N-(*p*-(((2-Amino-4-hydroxyquinazolin-6-yl)methyl)-N-formylamino)benzoyl)glutamic acid
→ N-(*p*-(((2-Amino-4-hydroxyquinazolin-6-yl)methyl)-N-formylamino)benzoyl) diglutamylglutamic acid
In: Pig F760

N-(*p*-(((2-Amino-4-hydroxyquinazolin-6-yl)methyl)-N-formylamino)benzoyl) tetraglutamylglutamic acid
Probably formed from: N-(*p*-(((2-Amino-4-hydroxyquinazolin-6-yl)methyl)-N-formylamino)benzoyl)triglutamylglutamic acid

N-(*p*-(((2-Amino-4-hydroxyquinazolin-6-yl)methyl)-N-formylamino)benzoyl) triglutamylglutamic acid
Probably formed from: N-(*p*-(((2-Amino-4-hydroxyquinazolin-6-yl)methyl)-N-formylamino)benzoyl)diglutamylglutamic acid
→ N-(*p*-(((2-Amino-4-hydroxyquinazolin-6-yl)methyl)-N-formylamino)benzoyl) tetraglutamylglutamic acid
Probably in: Pig F760

N-(*p*-((2-Amino-4-hydroxyquinazolin-6-yl)methylthio)benzoyl)glutamic acid
→ N-(*p*-((2-Amino-4-hydroxyquinazolin-6-yl)methylthio)benzoyl)triglutamylglutamic acid
In: Pig F760

N-(*p*-((2-Amino-4-hydroxyquinazolin-6-yl)methylthio)benzoyl)pentaglutamylglutamic acid
→ N-(*p*-((2-Amino-4-hydroxyquinazolin-6-yl)methylthio)benzoyl)trtraglutamylglutamic acid
Probably in: Pig F760

N-(*p*-((2-Amino-4-hydroxyquinazolin-6-yl)methylthio)benzoyl)tetraglutamylglutamic acid
Probably formed from: N-(*p*-((2-Amino-4-hydroxyquinazolin-6-yl)methylthio)benzoyl)triglutamylglutamic acid
→ N-(*p*-((2-Amino-4-hydroxyquinazolin-6-yl)methylthio)benzoyl)pentaglutamylglutamic acid
Probably in: Pig F760

N-(*p*-((2-Amino-4-hydroxyquinazolin-6-yl)methylthio)benzoyl)triglutamylglutamic acid
Formed from: N-(*p*-((2-Amino-4-hydroxyquinazolin-6-yl)methylthio) benzoyl)glutamic acid
→ N-(*p*-((2-Amino-4-hydroxyquinazolin-6-yl)methylthio)benzoyl)tetraglutamylglutamic acid
Probably in: Pig F760

5-Amino-2-hydroxyquinoline
Formed from: 5-Aminoquinoline
→ 5-Acetamido-2-hydroxyquinoline
Probably in: Rabbit G994

8-Amino-6-hydroxyquinoline
Probably formed from: 8-(4-Amino-1-methylbutylamino)-6-hydroxyquinoline
→ 5,6-Dihydroxy-8-hydroxylaminoquinoline
Probably in: Rat H136

***trans*-4-Amino-3-hydroxystilbene**
Probably formed from: *trans*-4-Acetamido-3-hydroxystilbene

***trans*-4′-Amino-4-hydroxystilbene**
Probably formed from: *trans*-4′-Acetamido-4-hydroxystilbene

5-((4′-((7-Amino-1-hydroxy-3-sulpho-2-naphthalenyl)azo)(1,1′-biphenyl)-4-yl)azo)-2-hydroxybenzoic acid
See: Direct Brown 2

4-Amino-3-((4′-((1-hydroxy-4-sulpho-2-naphthalenyl)azo)-3,3′-dimethoxy(1,1′-biphenyl)-4-yl) azo)-1-naphthalenesulphonic acid
See: Direct Violet 32

7-Amino-3-hydroxy-4,5,6-triethoxyphthalide
Probably formed from: 7-Amino-4,5,6-triethoxyphthalide
→ 7-Amino-3,6-dihydroxy-4,5-diethoxyphthalide
Probably in: Dog E105
Mouse E105
Rat E105
→ 7-Amino-3-hydroxy-4,5,6-triethoxyphthalide-β-D-glucuronide
Probably in: Dog E105
Mouse E105
Rat E105

7-Amino-3-hydroxy-4,5,6-triethoxyphthalide-β-D-glucuronide
Probably formed from: 7-Amino-3-hydroxy-4,5,6-triethoxyphthalide

2-Aminoimidazo[4,5-*f*]quinoline
Formed from: 2-Amino-3-methylimidazo[4,5-*f*]quinoline
→ 2-Aminoimidazo[4,5-*f*]quinoline-N^2-β-D-glucuronide
Probably in: Cynomolgus monkey G673

2-Aminoimidazo[4,5-*f*]quinoline-N^2-β-D-glucuronide
Probably formed from: 2-Aminoimidazo[4,5-*f*]quinoline

(N(*E*))-(N-(((2-(((2-(Aminoiminomethyl)amino)-4-thiazolyl)methyl)thio)ethyl)amino)methylene)-4-bromobenzenesulphonamide
See: Ebrotidine

N-(Amino(imino)methyl)-N^1-(*m*-chlorophenyl) urea
Probably formed from: N-(*m*-Chlorophenyl)-N^1-(imino(methylamino)methyl)urea

1-Aminoindane
→ Indan-1-one
In Rabbit F295
Rat F295

2-Aminoindane
→ *cis*-2-Amino-1-hydroxyindane
In Rabbit A449
Rat A449
→ *trans*-2-Amino-1-hydroxyindane
In Rabbit A449
Rat A449
→ 2-Amino-5-hydroxyindane
In Rabbit A449
Rat A449
→ 2-Indanol
In Rabbit F295
→ Indan-2-one
In Rabbit F295
Rat F295

2-Amino-1-indanone
→ *cis*-2-Amino-1-hydroxyindanc
In Rabbit A449
→ *trans*-2-Amino-1-hydroxyindane
In Rabbit A449

5-Aminoindazole
Formed from: 5-(*p*-Dimethylaminophenyl) azoindazole

5-Aminoindole
→ 5-Amino-L-tryptophan
Probably in: Carrot B874
Nicotiana B874

α-Amino-1*H*-indole-3-butanoic acid
See: Homotryptophan

4-Amino-3′-iodoazobenzene
Formed from: 3′-Iodo-4-methylaminoazobenzene

4-Amino-4′-iodoazobenzene
Formed from: 4′-Iodo-4-methylaminoazobenzene

4-Amino-2-iodophenol
Formed from: *m*-Iodoaniline

4-Amino-3-iodophenol
Formed from: *o*-Iodoaniline

Aminoisofenphos
See: N-Deisopropylisofenphos

2-Amino-1-(4-isopropylsulphonylphenyl)propanol
Formed from: Suloctidil
→ 1,2-Dihydroxy-1-(4-isopropylsulphonyl-phenyl)propane
Probably in: Man A2910

Aminomethaqualone (3-(4-amino-2-methoxyphenyl)-2-methyl-4(3*H*)-quinazolinone)
Formed from: Nitromethaqualone

4-Amino-3′-methoxyazobenzene
Formed from: 3′-Methoxy-4-methylaminoazobenzene
→ 4′-Amino-4-hydroxy-3-methoxyazobenzene
Probably in: Rat A776

4-Amino-4′-methoxyazobenzene
Formed from: 4′-Methoxy-4-methylaminoa-zobenzene

5-Amino-3-((2-methoxy-4-((2-methylphenyl) sulphonyl)aminocarbonyl)phenylmethyl)-1-methylindole
Formed from: Zafirlukast
→ 5-Acetamido-3-((2-methoxy-4-(2-methyl-phenyl)sulphonylaminocarbonyl) phenylmethyl)-1-methylindole
Probably in: Dog J455
Mouse J455
Rat J455

8-Amino-6-methoxy-4-methylquinoline
Formed from: 8-(6-Diethylaminohexylamino)-6-methoxy-4-methylquinoline
→ 8-Amino-6-methoxy-4-methylquinoline-N-β-D-glucuronide
Probably in: Rat F477

8-Amino-6-methoxy-4-methylquinoline-N-β-D-glucuronide
Probably formed from: 8-Amino-6-methoxy-4-methylquinoline

4-Amino-2-methoxyphenol
Probably formed from: 3,4-Dimethoxyaniline

(*S*)-(+)-2-Amino-4-(*p*-methoxyphenyl)butane
Formed from: 4-(*p*-Methoxyphenyl)-2-butanone

4-Amino-3-methoxyphenyl 2,4-dichlorophenyl ether
Formed from: Chlomethoxynil
→ 4-Amino-3-hydroxyphenyl 2,4-dichlorophenyl ether
Probably in: Rat C358

3-(4-Amino-2-methoxyphenyl)-2-hydroxymethyl-4(3*H*)-quinazolinone
Probably formed from: 2-Hydroxymethyl-3-(2-methoxy-4-nitrophenyl)-4(3*H*)-quinazolinone
→ 3-(4-Amino-2-methoxyphenyl)-2-hydroxymethyl-4(3*H*)-quinazolinone-β-D-glucuronide
Probably in: Rat C437

3-(4-Amino-2-methoxyphenyl)-2-hydroxymethyl-4(3*H*)-quinazolinone-β-D-glucuronide
Probably formed from: 3-(4-Amino-2-methoxyphenyl)-2-hydroxymethyl-4(3*H*)-quinazolinone

3-(4-Amino-2-methoxyphenyl)-2-methyl-4(3*H*)-quinazolinone
See: Aminomethaqualone

3′-(((2*S*)-2-Amino-3-(4-methoxyphenyl)-1-oxopropyl)amino)-3′-deoxy-N,N-dimethyladenosine
See: Puromycin

4-Amino-4′-(N-methylacetamido)azobenzene
Formed from: 4′-(N-Methylacetamido)-4-methylaminoazobenzene

4-Amino-4′-methylaminobenzophenone
Probably formed from: 4-Amino-4′-dimethylaminobenzophenone
→ 4-Amino-3-hydroxy-4′-methylaminobenzophenone
Probably in: Rat B713
→ 4,4′-Diaminobenzophenone
Probably in: Rat B713

4-Amino-3′-methylazobenzene
Formed from: 3′-Methyl-4-methylaminoazobenzene
→ 4-Amino-3′-hydroxymethylazobenzene
Probably in: Rat A3835
→ 4′-Amino-4-hydroxy-3-methylazobenzene
Probably in: Rat E872

4-Amino-4′-methylazobenzene
Formed from: 4′-Methyl-4-methylaminoazobenzene

***p*-Aminomethylbenzaldehyde**
Formed from: *p*-Aminomethylbenzylamine

3-Amino-N-methylbenzamide
Formed from: Ecabapide
→ 3-Acetamido-N-methylbenzamide
Probably in: Rat H170

2-Amino-3-methylbenzoic acid
Probably formed from: 2,6-Dimethylaniline

4-Amino-3-methylbenzoic acid
Formed from: 2,4-Dimethylaniline
→ 4-Amino-3-methylhippuric acid
In: Dog F240
→ 4-Acetamido-3-methylbenzoic acid
Probably in: Dog F240

***p*-Aminomethylbenzylamine**
→ *p*-Aminomethylbenzaldehyde
In: Pig A3693

8-(4-Amino-1-methylbutyl)amino-5,6-dihydroxyquinoline
Probably formed from: 8-(4-Amino-1-methylbutyl)amino-6-hydroxyquinoline
5-Hydroxyprimaquine

8-(4-Amino-1-methylbutyl)amino-2,6-dimethoxy-4-methyl-5-(3-trifluoromethylphenoxy) quinoline
→ 8-Amino-2,6-dimethoxy-4-methyl-5-(3-trifluoromethylphenoxy)quinoline
In: Rat H135
→ 8-(4-Amino-3-hydroxy-1-methylbutyl) amino-2,6-dimethoxy-4-methyl-5-(3-trifluoromethylphenoxy)quinoline
In: Rat H135
→ 8-(4-Amino-1-methylbutyl)amino-2,5,6-trihydroxy-4-methylquinoline
In: Rat H135
→ 8-(4-Carboxy-2-butyl)amino-2,6-dihydroxy-4-methyl-5-(3-trifluoromethylphenoxy)quinoline
In: Rat H135
→ 8-(4-Carboxy-2-butyl)amino-5-hydroxy-2,6-dimethoxy-4-methylquinoline

In: Rat H135

8-(4-Amino-1-methylbutyl)amino-6-hydroxyquinoline
Formed from: Primaquine
→ 8-(4-Amino-3-hydroxy-1-methylbutyl)amino-6-hydroxyquinoline
In: Rat H136
→ 8-Amino-6-hydroxyquinoline
Probably in: Dog C66
→ 8-(4-Amino-1-methylbutyl)amino-5,6-dihydroxyquinoline
In: Mouse C648
→ 8-(4-Amino-1-methylbutyl)amino-6-hydroxyquinoline-N-oxide
In: Rat H136

8-(4-Amino-1-methylbutyl)amino-6-hydroxyquinoline-N-oxide
Formed from: 8-(4-Amino-1-methylbutyl)amino-6-hydroxyquinoline

8-(4-Amino-1-methylbutyl)amino-2,5,6-trihydroxy-4-methylquinoline
Formed from: 8-(4-Amino-1-methylbutyl)amino-2,6-dimethoxy-4-methyl-5-(3-trifluoromethylphenoxy)quinoline

7-Amino-4-methylcoumarin
Formed from: Benzoyl-L-arginyl-4-methylcoumarin-7-amide
N-*tert*-Butoxycarbonyl-L-leucyl-L-seryl-L-threonyl-L-arginyl-4-methylcoumarin-7-amide
N-*tert*-Butoxycarbonyl-L-valyl-L-leucyl-L-lysyl-4-methylcoumarin-7-amide
N-*tert*-Butoxycarbonyl-L-valyl-L-prolyl-L-arginyl-4-methylcoumarin-7-amide
7-Ethylmino-4-methylcoumarin
→ 7-Amino-4-methylcoumarin-β-D-glucuronide
In: Man H541

7-Amino-4-methylcoumarin-β-D-glucuronide
Formed from: 7-Amino-4-methylcoumarin

8β-Aminomethyl-1,6-dimethylergoline
Formed from: Metergoline
→ 8β-Aminomethyl-6-methylergoline
Probably in: Rat C436

2-Aminomethyl-4-(1,1-dimethylethyl)-6-iodophenol
→ 2-Aminomethyl-4-(1,1-dimethylethyl)-6-iodophenyl-β-D-glucuronide
Probably in: Man A3917
→ 2-Aminomethyl-4-(1,1-dimethylethyl)-6-iodophenyl sulphate
In: Dog A3917
Man A3917
Rat A3917

2-Aminomethyl-4-(1,1-dimethylethyl)-6-iodophenyl-β-D-glucuronide
Probably formed from: 2-Aminomethyl-4-(1,1-dimethylethyl)-6-iodophenol

2-Aminomethyl-4-(1,1-dimethylethyl)-6-iodophenyl sulphate
Formed from: 2-Aminomethyl-4-(1,1-dimethylethyl)-6-iodophenol

2-Amino-5-methyl-4,6-dinitrophenol
Probably formed from: 4-Amino-2,6-dinitrotoluene

4-Amino-3-methyl-2,6-dinitrophenol
Probably formed from: 2-Amino-4,6-dinitrotoluene

8-(4-Amino-1-methyl-2,3-dioxobutyl)amino-4-formyl-2,6-dihydroxy-5-(3-trifluoromethylphenoxy)quinoline
Probably formed from: 8-(4-Amino-1-methyl-2,3-dioxobutyl)amino-4-formyl-2-hydroxy-6-methoxy-5-(3-trifluoromethylphenoxy)quinoline

8-(4-Amino-1-methyl-2,3-dioxobutyl)amino-4-formyl-2-hydroxy-6-methoxy-5-(3-trifluoromethylphenoxy)quinoline
Probably formed from: 8-(4-Amino-1-methyl-2,3-dioxobutyl)amino-2-hydroxy-4-hydroxymethyl-6-methoxy-5-(3-trifluoromethylphenoxy)quinoline
→ 8-(4-Amino-1-methyl-2,3-dioxobutyl)amino-4-formyl-2,6-dihydroxy-5-(3-trifluoromethylphenoxy)quinoline
Probably in: Rat H135

8-(4-Amino-1-methyl-2,3-dioxobutyl)amino-2-hydroxy-4-hydroxymethyl-6-methoxy-5-(3-trifluoromethylphenoxy)quinoline
Probably formed from: 8-(4-Amino-1-methyl-2,3-dioxobutyl)amino-2-hydroxy-4-

hydroxymethyl-6-methoxy-5-(3-trifluoromethylphenoxy)quinoline
→ 8-(4-Amino-1-methyl-2,3-dioxobutyl)amino-4-formyl-2-hydroxy-6-methoxy-5-(3-trifluoromethylphenoxy)quinoline
Probably in: Rat H135

8-(4-Amino-1-methyl-2,3-dioxobutyl)amino-2-hydroxy-4-hydroxymethyl-6-methoxy-5-(3-trifluoromethylphenoxy)quinoline-N-oxide
Probably formed from: 8-(4-Amino-2,3-dihydroxy-1-methylbutyl)amino-4-hydroxymethyl-2,6-dimethoxy-5-(3-trifluoromethylphenoxy)quinoline-N-oxide

3-Aminomethyl-1,2-diphenyl-2-butanol
Probably formed from: 3-Aminomethyl-1,2-diphenyl-2-butyl propionate

3-Aminomethyl-1,2-diphenyl-2-butyl propionate
Formed from: 3-Methylaminomethyl-1,2-diphenyl-2-butyl propionate
→ 3-Aminomethyl-1,2-diphenyl-2-butanol
Probably in: Man A1683
→ 3-Acetamidomethyl-1,2-diphenyl-2-butyl propionate
Probably in: Man A1683

2-Amino-N-(1-methyl-1,2-diphenylethyl)acetamide
See: Remacemide

4-Amino-3-methylhippuric acid
Formed from: 4-Amino-3-methylbenzoic acid

4-Aminomethyl-7-hydroxycoumarin
Formed from: 4-Aminomethyl-7-methoxycoumarin

3-Aminomethyl-1-hydroxyphenyl-2-phenyl-2-butyl propionate
Probably formed from: 1-Hydroxyphenyl-3-methylaminomethyl-2-phenyl-2-butyl propionate

2-Amino-3-methylimidazo[4,5-*f*]quinoline
→ 2-Aminoimidazo[4,5-*f*]quinoline
In: Cynomolgus monkey G673
→ 2-Hydroxylamino-3-methylimidazo[4,5-*f*]quinoline
In: Cynomolgus monkey G673
→ 2-Amino-3-methylimidazo[4,5-*f*]quinoline-N^2-β-D-glucuronide
In: Cynomolgus monkey G673
→ 2-Amino-3-methyl-7(6*H*)-oxoimidazo[4,5-*f*]quinoline
In: Cynomolgus monkey G673
→ 2-Amino-5-hydroxy-3-methylimidazo[4,5-*f*]quinoline
In: Cynomolgus monkey G673
→ 2-Amino-3-methylimidazo[4,5-*f*]quinoline-N^2-sulphate
In: Cynomolgus monkey G673

2-Amino-3-methylimidazo[4,5-*f*]quinoline-N^2-β-D-glucuronide
Formed from: 2-Amino-3-methylimidazo[4,5-*f*]quinoline

2-Amino-3-methylimidazo[4,5-*f*]quinoline-N^2-sulphate
Formed from: 2-Amino-3-methylimidazo[4,5-*f*]quinoline

3-Aminomethylindole
Probably formed from: L-Tryptophan

***p*-Aminomethyl-N-isopropylbenzamide**
Probably formed from: Procarbazine

2-Amino-3-methylmercaptobiphenyl
Formed from: N-Hydroxy-2-acetamidobiphenyl

4-Amino-3-methylmercaptobiphenyl
Formed from: N-Hydroxy-4-acetamidobiphenyl

4-Aminomethyl-7-methoxycoumarin
→ 4-Aminomethyl-7-hydroxycoumarin
In: Man K339

8β-Aminomethyl-6-methylergoline
Probably formed from: 8β-Aminomethyl-1,6-dimethylergoline
→ 8β-Acetamidomethyl-6-methylergoline
Probably in: Rat C436

2-Amino-6-methyl-4-nitrophenol
See: 6-Amino-4-nitro-*o*-cresol

4-Amino-6-methyl-2-nitrophenol
See: 4-Amino-6-nitro-*o*-cresol

6-Amino-2-methyl-4-nitrophenol
Formed from: 2-Methyl-4,6-dinitrophenol
→ 6-Acetamido-2-methyl-4-nitrophenol
In: Azotobacter A3824

6-Amino-2-methyl-7-nitro-5-trifluoromethylbenzimidazole
Probably formed from: 6-Amino-1-ethyl-2-methyl-7-nitro-5-trifluoromethylbenzimidazole

8-(4-Amino-1-methyl-3-oxobutyl)amino-2,6-dihydroxy-4-hydroxymethyl-5-(3-trifluoromethylphenoxy)quinoline
Probably formed from: 8-(4-Amino-1-methyl-3-oxobutyl)amino-6-hydroxy-4-hydroxymethyl-2-methoxy-5-(3-trifluoromethylphenoxy)quinoline

8-(4-Amino-1-methyl-3-oxobutyl)amino-6-hydroxy-4-hydroxymethyl-2-methoxy-5-(3-trifluoromethylphenoxy)quinoline
Probably formed from: 8-(4-Amino-3-hydroxy-1-methylbutyl)amino-2,6-dimethoxy-4-methyl-5-(3-trifluoromethylphenoxy)quinoline
→ 8-(4-Amino-1-methyl-3-oxobutyl) amino-2,6-dihydroxy-4-hydroxymethyl-5-(3-trifluoromethylphenoxy)quinoline
Probably in: Rat H135
→ 8-(5-Methyl-2,3-dioxopyrrolidin-5-yl) amino-6-hydroxy-4-hydroxymethyl-2-methoxy-5-(3-trifluoromethylphenoxy) quinoline
Probably in: Rat H135

8-(4-Amino-1-methyl-3-oxobutyl)amino-2-hydroxy-6-methoxy-4-methyl-5-(3-trifluoromethylphenoxy)quinoline
Probably formed from: 8-(4-Amino-3-hydroxy-1-methylbutyl)amino-2-hydroxy-6-methoxy-4-methyl-5-(3-trifluoromethylphenoxy)quinoline

8-(4-Amino-1-methyl-3-oxobutyl)amino-5-hydroxy-6-methoxyquinoline
Probably formed from: 8-(4-Amino-3-hydroxy-1-methylbutyl)amino-5-hydroxy-6-methoxyquinoline
→ 5,6-Dihydroxy-8-(3-hydroxy-5-methyl-2-oxopyrrolidin-1-yl)quinoline
Probably in: Rat H136
→ 5,6-Dihydroxy-8-(5-methyl-2-oxopyrrolidin-1-yl)quinoline-N-oxide
Probably in: Rat H136

8-(4-Amino-1-methyl-3-oxobutyl)amino-2,5,6-trihydroxy-4-methylquinoline
Probably formed from: 8-(4-Amino-3-hydroxy-1-methylbutyl)amino-2,5,6-trihydroxy-4-methylquinoline

2-Amino-3-methyl-7(6*H*)-oxoimidazo[4.5-*f*]quinoline
Formed from: 2-Amino-3-methylimidazo[4.5-*f*]quinoline
→ 2-Amino -7(6*H*)-oxoimidazo[4.5-*f*]quinoline
Probably in: Cynomolgus monkey G673

2-Amino-3-methylphenol
See: 2-Amino-*m*-cresol

2-Amino-4-methylphenol
See: 2-Amino-*p*-cresol

2-Amino-5-methylphenol
See: 6-Amino-*m*-cresol

3-Amino-4-methylphenol
See: 3-Amino-*p*-cresol

4-Amino-2-methylphenol
See: 4-Amino-*o*-cresol

4-Amino-3-methylphenol
See: 4-Amino-*m*-cresol

5-Amino-2-methylphenol
See: 5-Amino-*o*-cresol

2-(4-Amino-3-methylphenyl)benzothiazole
→ 2-(4-Amino-3-methylphenyl)-6-hydroxybenzothiazole
In: Man K600

O-(4-Amino-3-methylphenyl) O,O-dimethylphosphorothioate
Formed from: Fenitrothion
→ O-(4-Amino-3-methylphenyl) O-methylphosphorothioate
Probably in: Coriolus E639

2-(4-Amino-3-methylphenyl)-6-hydroxybenzothiazole
Formed from: 2-(4-Amino-3-methylphenyl)benzothiazole

2-Amino-1-methyl-6-phenylimidazo[4,5-*b*]pyridine
→ 2-Amino-4-hydroxy-1-methyl-6-phenylimidazo[4,5-*b*]pyridine
In: Rat H345
→ 2-Amino-6-(*p*-hydroxyphenyl)-1-methylimidazo[4,5-*b*]pyridine
In: Guinea pig H345
Hamster H345
Man H179
Monkey H88

Mouse H179
Rat G443, H179, H345, H417
→ 2-Amino-1-methyl-6-phenylimidazo[4,5-*b*]pyridine-β-D-glucuronide
In: Guinea pig H345
Hamster H345
Rat H345
→ 2-Hydroxylamino-1-methyl-6-phenylimidazo[4,5-*b*]pyridine
In: Guinea pig H345
Man H179
Monkey H88
Mouse H179
Rat G443, H179, H345, H417

2-Amino-1-methyl-6-phenylimidazo[4,5-*b*]pyridine-β-D-glucuronide
Formed from: 2-Amino-1-methyl-6-phenylimidazo[4,5-*b*]pyridine

O-(4-Amino-3-methylphenyl) O-methylphosphorothioate
Probably formed from: O-(4-Amino-3-methylphenyl) O,O-dimethylphosphorothioate

2-Amino-2-methyl-1-phenylpropane
See: Phentermine

3-Amino-4-methylphenyl-6-phenylpyridine
Formed from: Minaprine

2-Amino-1-methyl-6-(propylsulphonyl)-1*H*-benzimidazole
Probably formed from: 2-Amino-5-(propylsulphonyl)-1*H*-benzimidazole

4-Amino-10-methylpteroic acid
Formed from: Methotrexate
→ 4-Amino-7-hydroxy-10-methylpteroic acid
In: Microorganism A396

2-Amino-1-methyl-9*H*-pyrido[2,3-*b*]indole
→ 2-Acetamido-1-methyl-9*H*-pyrido[2,3-*b*]indole
In: Hamster F563

3-Amino-1-methyl-5*H*-pyrido[4,3-*b*]indole
→ 3-Acetamido-1-methyl-5*H*-pyrido[4,3-*b*]indole
In: Hamster F563
→ 3-Hydroxylamino-1-methyl-5*H*-pyrido[4,3-*b*]indole
In: Mouse B173
Rat B636, C197

4-Aminomethyl-*o*-quinone
Probably formed from: 3,4-Dihydroxybenzylamine

2-Aminomethyl-1,2,3,4-tetrahydro-6-hydroxymethyl-7-nitroquinoline
Formed from: Oxamniquine

2-Aminomethyl-1,2,3,4-tetrahydro-6-methyl-7-nitroquinoline
→ 2-Carboxy-1,2,3,4-tetrahydro-6-methyl-7-nitroquinoline
In: Mouse A3711
Rat A3711

(*R*)-3-Amino-3-methyl-N-(2,3,4,5-tetrahydro-2-oxo-1-((2-(1*H*-tetrazol-5-yl)(1,1′-biphenyl)-4-yl)methyl)-1*H*-1-benzazepin-3-yl) butamide
→ (*R*)-3-Amino-3-methyl-N-(2,3,4,5-tetrahydro-2-oxo-1-((2-(1*H*-tetrazol-5-yl)(1,1′-biphenyl)-4-yl)methyl)-1*H*-1-((7-hydroxybenzazepin-3-yl)butamide
In: Dog H724
Rat H724

(*R*)-3-Amino-3-methyl-N-(2,3,4,5-tetrahydro-2-oxo-1-((2-(1*H*-tetrazol-5-yl)(1,1′-biphenyl)-4-yl)methyl)-1*H*-1-((7-hydroxybenzazepin-3-yl)butamide
Formed from: (*R*)-3-Amino-3-methyl-N-(2,3,4,5-tetrahydro-2-oxo-1-((2-(1*H*-tetrazol-5-yl)(1,1′-biphenyl)-4-yl)methyl)-1*H*-1-benzazepin-3-yl)butamide

2-Amino-4-methyl-5-trifluoromethylphenol
Formed from: 3-methyl-4-trifluoromethylaniline
→ 2-Amino-4-methyl-5-trifluoromethylphenyl-β-D-glucuronide
Probably in: Rat J785
→ 2-Amino-4-methyl-5-trifluoromethylphenyl sulphate
Probably in: Rat J785

6-Amino-2-methyl-3-trifluoromethylphenol
Formed from: 3-methyl-4-trifluoromethylaniline
→ 6-Amino-2-methyl-3-trifluoromethylphenyl-β-D-glucuronide
Probably in: Rat J785
→ 6-Amino-2-methyl-3-trifluoromethylphenyl sulphate

Probably in: Rat J785

2-Amino-4-methyl-5-trifluoromethylphenyl-β-D-glucuronide
Formed from: 2-Amino-4-methyl-5-trifluoromethylphenol

6-Amino-2-methyl-3-trifluoromethylphenyl-β-D-glucuronide
Formed from: 6-Amino-2-methyl-3-trifluoromethylphenol

2-Amino-4-methyl-5-trifluoromethylphenyl sulphate
Formed from: 2-Amino-4-methyl-5-trifluoromethylphenol

6-Amino-2-methyl-3-trifluoromethylphenyl sulphate
Formed from: 6-Amino-2-methyl-3-trifluoromethylphenol

2-Amino-10-(3-morpholinopropionyl) phenothiazine
Formed from: Moricizine
→ 2-Aminophenothiazine
Probably in: Man J85, J747

1-Aminonaphthalene
See: α-Naphthylamine

2-Aminonaphthalene
See: β-Naphthylamine

2-Aminonaphthalene-1-sulphonic acid (Tobias acid)
→ Salicylic acid
In: Pseudomonas E102, F185

4-Aminonaphthalene-1-sulphonic acid
Formed from: Acid Red 88

5-Aminonaphthalene-1-sulphonic acid
→ 5-Amino-1-naphthol
In: Pseudomonas E408

1-Amino-2-naphthol
Formed from: Acid Orange 7
Acid Red 88
Acid Red 151
α-Naphthylamine

1-Amino-4-naphthol
Formed from: 1-(*p*-Carboxyphenylazo)-4-naphthol
1-(*p*-Sulphophenylazo)-4-naphthol

2-Amino-1-naphthol
Formed from: β-Naphthylamine
Probably formed from: 2-Naphthylhydroxylamine
→ 2-Aminonaphth-1-yl sulphate
Probably in: Rat B773

5-Amino-1-naphthol
Formed from: 5-Aminonaphthalene-1-sulphonic acid
→ 5-Acetamido-1-naphthol
In: Rat D85
→ 5-Aminonaphth-1-yl-β-D-glucuronide
In: Rat D85
→ 5-Aminonaphth-1-yl sulphate
In: Rat D85

5-Amino-2-naphthol
Formed from: β-Naphthylamine

7-Amino-2-naphthol
Formed from: β-Naphthylamine

2-Amino-1,4-naphthoquinone
→ 2,4-Diamino-1-naphthol
In: Rat E214

5-Aminonaphth-1-yl-β-D-glucuronide
Formed from: 5-Amino-1-naphthol

2-Aminonaphth-1-yl sulphate
Formed from: 2-Amino-1-naphthol

5-Aminonaphth-1-yl sulphate
Formed from: 5-Amino-1-naphthol

4-Amino-3′-nitroazobenzene
Formed from: 4-Methylamino-3′-nitroazobenzene

4-Amino-4′-nitroazobenzene
→ 4-Acetamido-4′-nitroazobenzene
In: Bacillus E276

3-Amino-5-nitrobenzamide
Formed from: 3,5-Dinitroazobenzene
→ 3-Acetamido-5-nitrobenzamide
In: Rat B193
→ 3-Amino-5-nitrobenzoic acid
Probably in: Rat B193
→ 3,5-Diaminobenzamide
In: Rat B193

2-Amino-4-nitrobenzoic acid
Probably formed from: 2,4-Dinitrobenzoic acid
→ 2-Acetamido-4-nitrobenzoic acid
Probably in: Rat B578, H515
→ 2,4-Diaminobenzoic acid
Probably in: Rat H515

2-Amino-6-nitrobenzoic acid
Probably formed from: 2,6-Dinitrobenzoic acid

3-Amino-5-nitrobenzoic acid
Probably formed from: 3-Amino-5-nitrobenzoic acid
3,5-Dinitrobenzoic acid

4-Amino-2-nitrobenzoic acid
Probably formed from: 2,4-Dinitrobenzoic acid
→ 4-Acetamido-2-nitrobenzoic acid
Probably in: Rat H854

5-Amino-2-nitrobenzoic acid
Formed from: 5-(*o*-Carboxy-N-benzoyl) amino-2-nitrobenzoic acid

2-Amino-5-nitrobenzophenone
Formed from: 1,3-Dihydro-7-nitro-5-phenyl-2*H*-1,4-benzodiazepin-2-one
→ 2-Amino-5-chlorobenzophenone
Probably in: Man A3843
→ 2-Formamido-5-nitrobenzophenone
Probably in: Rabbit A3876

2-Amino-4-nitrobenzyl alcohol
Probably formed from: 2,4-Dinitrobenzyl alcohol

2-Amino-6-nitrobenzyl alcohol
Probably formed from: 2,6-Dinitrobenzyl alcohol

4-Amino-2-nitrobenzyl alcohol
Probably formed from: 2,4-Dinitrobenzyl alcohol

4-Amino-6-nitro-*o*-cresol
Probably formed from: 4,6-Dinitro-*o*-cresol
→ 4-Acetamido-6-nitro-*o*-cresol
Probably in: Rat C470

6-Amino-4-nitro-*o*-cresol (2-amino-6-methyl-4-nitrophenol)
Probably formed from: 4,6-Dinitro-*o*-cresol
→ 6-Acetamido-4-nitro-*o*-cresol
In: Rat C470
→ 2,4-Diamino-6-methylphenol
Probably in: Rat C470

4-Amino-4′-nitrodiphenyl sulphone
Formed from: Bis(*p*-aminophenyl) sulphone

3-Amino-9-nitrofluoranthene
Formed from: 3,9-Dinitrofluoranthene

2-Amino-3-nitrophenol
Formed from: *o*-Nitroanisole
→ 2-Amino-3-nitrophenyl-β-D-glucuronide
In: Tomato F851

2-Amino-4-nitrophenol
Formed from: 2,4-Dinitrophenol

2-Amino-5-nitrophenol
Formed from: *p*-Nitroaniline

4-Amino-2-nitrophenol
Formed from: 2,4-Dinitrophenol

4-Amino-3-nitrophenol
Formed from: *o*-Nitroaniline
→ 4-Amino-3-nitrophenyl-β-D-glucuronide
In: Tomato F851

2-Amino-3-nitrophenyl-β-D-glucuronide
Formed from: 2-Amino-3-nitrophenol

4-Amino-3-nitrophenyl-β-D-glucuronide
Formed from: 4-Amino-3-nitrophenol

D-*threo*-2-Amino-1-(*p*-nitrophenyl)-1,3-propanediol
Formed from: Chloramphenicol
→ D-*threo*-2-Acetamido-1-(*p*-nitrophenyl)-1,3-propanediol
Probably in: Duck G945

1-Amino-8-nitropyrene
Formed from: 1,8-Dinitropyrene
→ 1-Acetamido-8-nitropyrene
Probably in: Rabbit E953
→ 1,8-Diaminopyrene
Probably in: Rabbit E953

4-Amino-4′-nitrosodiphenylmethane
4,4′-Diaminodiphenylmethane

2-Amino-4-nitrotoluene
Formed from: 2,4-Dinitrotoluene
Probably formed from 4-Nitro-2-nitrosotoluene
→ 2,4-Diaminotoluene
Probably in: Desulphovibrio G827
Pseudomonas J196
Rat F324, H854
→ 4,4′-Dinitro-2,2′-azoxytoluene
Probably in: Mucrosporium A2716
→ 2-Methyl-5-nitroacetanilide
In: Hamster A461
Microorganism D319
Pseudomonas J196
Rat B607, F324

See also: A2716, B885
→ 4-Nitro-*o*-quinone
In: Phanerochaete G436

4-Amino-2-nitrotoluene
Formed from: 2,4-Dinitrotoluene
Probably formed from 2-Nitro-4-nitrosotoluene
→ 2,4-Diaminotoluene
Probably in: Desulphovibrio G827
Dog A1703
Microorganism C263, G542
Monkey A1703
Mouse D319
Rabbit A1703
Rat A1703, B885
Veillonella A2624
→ 2,2′-Dinitro-4,4′-azoxytoluene
Probably in: Mucrosporium A2716
→ 4-Methyl-3-nitroacetanilide
In: Pseudomonas J196
Rat B607

N-((2-Amino-3-nitro-5-trifluoromethyl)phenyl)methanesulphinamide
Probably formed from: 3-Nitro-5-trifluoromethyl-*o*-phenylenediamine

4-Amino-2-nonaprenylbenzoic acid
Probably formed from: *p*-Aminobenzoic acid

(*R*)-13β-Amino-5,6,7,8,9,10,11,12-octahydro-5α-methyl-5,11-methanobenzocyclodecen-3-ol
See: Dezocine

2-Amino-1,2,3,4,4a,5,6,10b-octahydro-6-propylbenzo[*f*]quinoline
Formed from: Proterguride

4-Amino-2-octaprenylbenzoic acid
Probably formed from: *p*-Aminobenzoic acid

2-Amino-17β-oestradiol
→ 2-Amino-17β-oestradiol-3-sulphate
In: Beef A1211

4-Amino-17β-oestradiol
→ 4-Amino-17β-oestradiol-3-sulphate
In: Beef A1211

2-Amino-17β-oestradiol-3-sulphate
Formed from: 2-Amino-17β-oestradiol

4-Amino-17β-oestradiol-3-sulphate
Formed from: 4-Amino-17β-oestradiol

2-Amino-α-oxobenzeneacetic acid
See: Isatinic acid

2-Amino-5-oxocyclohex-1-enecarboxyl CoA
Formed from: Anthraniloyl CoA

2-(2-Amino-1-oxoethyl)-3-(*p*-chlorophenyl)-3-hydroxyphthalimidine
Probably formed from: 5-(*p*-Chlorophenyl)-2,5-dihydro-5-hydroxy-3*H*-imidazol[2,1-*a*]isoindol-3-one

2-Amino-7(6*H*)-oxoimidazo[4,5-*f*]quinoline
Probably formed from: 2-Amino-3-methyl-7(6*H*)-oxoimidazo[4,5-*f*]quinoline

9-Amino-3-oxo-3*H*-phenoxazine-1-carboxylic acid
See: Cinnabarinic acid

10-(4-N-(5-Aminopentanoic acid)-1-piperidino)carbonyloxycamptothecin
Formed from: Irinotecan

8-((1-Aminopent-4-yl)amino)-6-methoxyquinoline
See: Primaquine

4-(5-Aminopentyl)-*o*-quinone
Probably formed from: 5-(3,4-Dihydroxyphenyl)pentylamine

***p*-Aminophenethylamine**
→ *p*-Aminophenylacetaldehyde
In: Alcaligenes H367

***o*-Aminophenol**
Formed from: *o*-Acetamidophenol
Anthranilic acid
Aniline
3-Hydroxyanthranilic acid
o-Nitrophenol
Probably formed from: 2-Hydroxyazobenzene
→ *m*-Aminobenzoic acid
In: Microorganism G917
→ 3-Aminocatechol
In: Candida E204
Trichosporon A902
See also: E565
→ α-Aminomuconic semialdehyde
In: Bacterium H915
Escherichia K583
Pseudomonas A3636, J255, J329
See also: J699
→ *o*-Aminophenol
In: Pseudomonas K520
→ 2-Amino-3*H*-phenoxazin-3-one
In: Man D940
Pseudomonas K520

→ *o*-Aminophenyl-β-D-glucuronide
In: Guinea pig A862
Monkey A1997
Mouse C266
Pig A1997
Rabbit A43, A1997
Rat A1997, A2002, A3827, B116, B740
Tree shrew A1997
→ *o*-Aminophenyl sulphate
In: Man E575
Rat E575
→ Aminoquinol
In: Streptomyces F190
→ *o*-(D-Glucuronamino)phenol
Probably in: Rat B740
→ *o*-Nitrophenol
In: Serratia K143

m-Aminophenol
Formed from: *m*-Acetamidophenol
m-Aminobenzenesulphonic acid
Aniline
m-Nitrophenol
→ *m*-Acetamidophenol
Probably in: Sugar beet A633
→ 4-Aminocatechol
In: Candida K653
Trichosporon A902
See also: E204, E565
→ *m*-Aminophenyl sulphate
In: Man E575
Rat E575

p-Aminophenol
Formed from: *p*-Acetamidophenol
p-Aminobenzoic acid
4′-Amino-4-hydroxyazobenzene
Aniline
4′-Dimethylamino-4-hydroxyazobenzene
p-Ethoxyaniline
p-Fluoroaniline
5-(*p*-Hydroxyanilino)-1,2,3,4-thiatriazole
4-Hydroxyazobenzene
p-Hydroxyglycollanilide
4-Hydroxy-4′-methylaminoazobenzene
p-Hydroxyoxanilic acid
p-Methoxyaniline
Probably formed from: 4-Aminophenyl diethyl phosphorothioate
p-Chloroaniline
p-Hydroxymethaphenilene
p-(Hydroxphenyl)benzylamine
Isopropyl *p*-hydroxycarbanilate
p-Methylaminophenol
p-Nitrophenol
p-Propionamidophenol
→ *p*-Acetamidophenol
In: Guinea pig F109
Hamster G757
Rabbit E399, F109
Rat B174, D81, F109, G670, K60
Sheep A2788
See also: A2160, A2811, B165, C530, K520
→ 4-Aminocatechol
In: Candida E204, K653
Trichosporon A902
See also: E565
→ 4-Amino-2-(S-glutathionyl)phenol
In: Rat G670
→ 4-Amino-3-(S-glutathionyl)phenol
In: Horseradish G4
Rat G670
See also: K60
→ *p*-Aminophenoxy radical
In: Horseradish C544
Ram C544
→ *p*-Aminophenyl-β-D-glucuronide
In Pig A3460
Rat A3460, B165, K60
Sheep A3460
See also: D532
→ *p*-Aminophenyl sulphate
In Pig A3460
Rat A3460, B165
Sheep A3460
See also: A508, C530, D532
→ *p*-Formamidophenol
In: Guinea pig F109
Rabbit F109
Rat F109, K60
→ *p*-Nitrophenol
In: Serratia K143
→ Phenol

Probably in: Pseudomonas A3100

2-Aminophenothiazine
Probably formed from: 2-Amino-10-(3-morpholinopropionyl)phenothiazine
→ 2-Amino-10-glucuronophenothiazine
Probably in: Man J747

2-Amino-3*H*-phenoxazin-3-one
Formed from: *o*-Aminophenol

***p*-Aminophenoxy radical**
Formed from: N-Acetyl-*p*-benzoquinone imine
p-Aminophenol

3-Amino-4-phenoxy-5-sulphamoylbenzoic acid
Formed from: Bumetanide

***p*-Aminophenylacetaldehyde**
Formed from: *p*-Aminophenethylamine

2-Aminophenylacetic acid
Formed from: Carbazole

***p*-Amino-L-phenylalanine**
Formed from: Chorismic acid
Probably formed from: Shikimic acid
→ *p*-Aminophenylpyruvic acid
In: Bushbean E500
Flavobacterium C206
→ Aureothin
In: Microorganism A1236
Streptomyces A704

5-(*p*-Aminophenylazo)-3-methylrhodanine
Formed from: 3-Methyl-5-(*p*-nitrophenylazo)rhodanine

N-(*p*-Aminophenyl)benzamide

N-(*p*-Nitrophenyl)benzamide

3-Amino-1-phenylbutane
Formed from: Labetalol

2-Amino-2-phenylbutan-1-ol
Probably formed from: 2-Methylamino-2-phenylbutan-1-ol

L-2-Amino-4-phenylbutyric acid
(homophenylalanine, γ-phenylbutrine)
→ Homophenylalanine aldoxime
In: Brassica K66
→ 2-Oximino-4-phenylbutyric acid
Probably in: Brassica H677
→ 3-Phenylpropionaldoxime
Probably in: Tropaeolum A162

N-(*p*-Aminophenyl)-5-chloro-2-hydroxybenzamide
Formed from: 5-Chloro-2-hydroxy-N-(*p*-nitrophenyl)benzamide

1-((3-(*p*-Aminophenyl)-1-cycloheptylureido)methyl)-3-((1-cycloheptyl-3-(*p*-dimethylaminophenyl)ureido)methyl)benzene
Formed from: 1-((1-Cycloheptyl-3-(*p*-dimethylaminophenyl)ureido)methyl)-3-((1-cycloheptyl-3-(*p*-methylaminophenyl) ureido)methyl)benzene

1-((3-(*p*-Aminophenyl)-1-cycloheptylureido)methyl)-3-((1-cycloheptyl-3-(*p*-methylaminophenyl)ureido)methyl)benzene
Probably formed from: 1,3-Bis((1-cycloheptyl-(3-(*p*-methylaminophenyl) ureido)methyl)benzene
→ 1,3-Bis((3-(*p*-aminophenyl)-1-cycloheptylureido)methyl)benzene
In: Dog H130
Rat G581, H130

D-*threo*-1-(*p*-Aminophenyl)-2-(dichloroacetamido)-1,3-propanediol
Formed from: Chloramphenicol
2-(Dichloroacetamido)-1-(*p*-nitrosophenyl)-1,3-propanediol
→ D-*threo*-1-(*p*-Acetamidophenyl)-2-(dichloroacetamido)-1,3-propanediol
In: Man E898
See also: C618
→ 2-(Dichloroacetamido)-1-(*p*-hydroxylaminophenyl)-1,3-propanediol
In: Rat D811

4-Aminophenyl 2,5-dichloro-4-hydroxyphenyl ether
Probably formed from: 4-Aminophenyl 2,4-dichlorophenyl ether

4-Aminophenyl 2,4-dichlorophenyl ether
Formed from: Nitrofen
→ 4-Acetamidophenyl 2,4-dichlorophenyl ether
In: Rat C543
→ 4-Aminophenyl 2,5-dichloro-4-hydroxyphenyl ether
Probably in: Rat C543

4-Aminophenyl diethyl phosphorothioate
Formed from: Diethyl 4-nitrosophenyl phosphorothioate

→ *p*-Aminophenol
In: Microorganism A2435

4-(3-Aminophenyl)-1,4-dihydro-2,6-dimethylpyridine-3,5-dicarboxylic acid 3-(2-(N-benzyl-N-methylamino))ethyl ester 5-methyl ester
Formed from: 1,4-Dihydro-2,6-dimethyl-4-(3-nitrophenyl)pyridine-3,5-dicarboxylic acid 3-(2-(N-benzyl-N-methylamino))ethyl ester 5-methyl ester
→ 4-(3-Aminophenyl)-1,4-dihydro-2,6-dimethylpyridine-3,5-dicarboxylic acid 3-(2-methylamino)ethyl ester 5-methyl ester
Probably in: Dog A2096
Rat A2096

4-(3-Aminophenyl)-1,4-dihydro-2,6-dimethylpyridine-3,5-dicarboxylic acid 3-methyl ester 5-(2-methylamino)ethyl ester
Formed from: Methyl-4-(*m*-nitrophenyl) pyridine-3,5-dicarboxylic acid 3-methyl ester 5-(2-methylamino) ethyl ester
Probably formed from: 4-(3-Aminophenyl)-1,4-dihydro-2,6-dimethylpyridine-3,5-dicarboxylic acid 3-(2-(N-benzyl-N-methylamino))ethyl ester 5-methyl ester
1,4-Dihydro-2,6-dimethyl-4-(3-nitrophenyl)pyridine-3,5-dicarboxylic acid 3-methyl ester 5-(2-methylamino)ethyl ester
→ 4-(3-Aminophenyl)- 2,6-dimethylpyridine-3,5-dicarboxylic acid 3-methyl ester 5-(2-methylamino)ethyl ester
Probably in: Dog A2096
Rat A2096

4-(3-Aminophenyl)-1,4-dihydro-2,6-dimethylpyridine-3,5-dicarboxylic acid 3-methyl ester 5-(3-piperidinyl) ester
Probably formed from: Debenzylbenidipine
→ 4-(3-Aminophenyl)- 2,6-dimethylpyridine-3,5-dicarboxylic acid 3-methyl ester 5-(3-piperidinyl) ester
Probably in: Rat H927
→ 4-(3-Aminophenyl)-1,4-dihydro-2,6-dimethylpyridine-3,5-dicarboxylic acid 3-(3-piperidinyl) ester
Probably in: Rat H927

4-(3-Aminophenyl)-1,4-dihydro-2,6-dimethylpyridine-3,5-dicarboxylic acid 3-(3-piperidinyl) ester
Probably formed from: 4-(3-Aminophenyl)-1,4-dihydro-2,6-dimethylpyridine-3,5-dicarboxylic acid 3-methyl ester 5-(3-piperidinyl) ester

1-(*p*-Aminophenyl)-1,3-dihydroxyacetone
Formed from: *p*-Aminophenyl)serinol

4-(3-Aminophenyl)-2,6-dimethylpyridine-3,5-dicarboxylic acid 3-ethyl ester 5-methyl ester
Probably formed from: 1,4-Dehydronitrendipine

4-(3-Aminophenyl)-2,6-dimethylpyridine-3,5-dicarboxylic acid 3-methyl ester 5-(2-methylamino)ethyl ester
Probably formed from: 4-(3-Aminophenyl)-1,4-dihydro-2,6-dimethylpyridine-3,5-dicarboxylic acid 3-methyl ester 5-(2-methylamino)ethyl ester

4-(3-Aminophenyl)-2,6-dimethylpyridine-3,5-dicarboxylic acid 3-methyl ester 5-(3-piperidinyl) ester
Probably formed from: 4-(3-Aminophenyl)-1,4-dihydro-2,6-dimethylpyridine-3,5-dicarboxylic acid 3-methyl ester 5-(3-piperidinyl) ester

4-(*o*-Aminophenyl)-2,4-dioxobutyric acid
Formed from: L-Kynurenine

α-(*p*-Aminophenyl)-α-ethylyglutarimide
See: Aminoglutethimide

α-(*p*-Aminophenyl)-α-ethyl-β-hydroxyglutarimide
Formed from: Aminoglutethimide

α-(*p*-Aminophenyl)-α-ethyl-γ-hydroxyglutarimide
Formed from: Aminoglutethimide

***o*-Aminophenyl-β-D-glucuronide**
Formed from: *o*-Aminophenol

***p*-Aminophenyl-β-D-glucuronide**
Formed from: *p*-Aminophenol

2-(*p*-Aminophenyl)glutarimide
Formed from: Aminoglutethimide
→ 2-(*p*-Aminophenyl)-3-hydroxyglutarimide
Probably in: Rat C10
→ 2-(*p*-Aminophenyl)-4-hydroxyglutarimide
Probably in: Rat C10

4-Aminophenyl 4′-glycylaminophenyl sulphone
→ Bis(*p*-Aminophenyl)sulphone

In: Rabbit G966

o-Aminophenylglyoxylic acid
Formed from: Indole-3-acetic acid

o-Aminophenylhydracrylic acid
Formed from: quinoline

4-(m-Aminophenyl)-3-(2-hydroxyethoxy)carbonyl-6-hydroxymethyl-5-isopropoxycarbonyl-2-methylpyridine
Probably formed from: 3-(2-Hydroxyethoxy)carbonyl-6-hydroxymethyl-5-isopropoxycarbonyl-2-methyl-4-(*m*-nitrophenyl)pyridine

4-(m-Aminophenyl)-3-(2-hydroxyethoxy) carbonyl-5-hydroxy-2-isopropoxycarbonyl-2,6-dimethylpyridine
Probably formed from: 4-(*m*-Aminophenyl)-3-(2-hydroxyethoxy)carbonyl-5-isopropoxy-carbonyl-2,6-dimethylpyridine

4-(m-Aminophenyl)-3-(2-hydroxyethoxy) carbonyl-5-isopropoxycarbonyl-2,6-dimethylpyridine
Probably formed from: 3-(2-Hydroxyethoxy)carbonyl-5-isopropoxycarbonyl-2,6-dimethyl-4-(*m*-nitrophenyl)pyridine
→ 4-(*m*-Aminophenyl)-3-(2-hydroxyethoxy)carbonyl-5-(1-hydroxy-2-propoxy)carbonyl-2,6-dimethylpyridine
Probably in: Dog G189
Monkey G189

2-(p-Aminophenyl)-2-(1-hydroxyethyl)glutarimide
Formed from: Aminoglutethimide

2-(p-Aminophenyl)-3-hydroxyglutarimide
Probably formed from: 2-(*p*-Aminophenyl)glutarimide

2-(p-Aminophenyl)-4-hydroxyglutarimide
Probably formed from: 2-(*p*-Aminophenyl)glutarimide

11β-(p-Aminophenyl)-17β-hydroxy-17α-(1-(3-hydroxy)propynyl)oestra-4,9-dien-3-dione
Probably formed from: 17β-Hydroxy-17α-(1-(3-hydroxy)propynyl)-11β-(*p*-methylaminophenyl)oestra-4,9-dien-3-dione

4-Aminophenyl-4′-hydroxylaminophenyl sulphone
Formed from: Bis(*p*-aminophenyl) sulphone

4-(m-Aminophenyl)-2-hydroxymethyl-5-isopropoxycarbonyl-3-(2-methoxyethoxycarbonyl-6-methylpyridine
Probably formed from: 2-Hydroxymethyl-5-isopropoxycarbonyl-3-(2-methoxyethoxy) carbonyl-6-methyl-4-(*m*-nitrophenyl) pyridine

6-(o-Aminophenyl)-2-hydroxy-6-oxohexa-2,4-dienoic acid
Probably formed from: 2′-Amino-2,3-dihydroxybiphenyl
→ Anthranilic acid
Probably in: Pseudomonas H21

4-Aminophenyl 4′-(D-leucyl)aminophenyl sulphone
→ Bis(*p*-aminophenyl) sulphone
In: Rabbit G966

4-Aminophenyl 4′-(L-leucyl)aminophenyl sulphone
→ Bis(*p*-Aminophenyl) sulphone
In: Rabbit G966

4-Aminophenyl 4′-(L-lysyl)aminophenyl sulphone
→ Bis(*p*-Aminophenyl) sulphone
In: Rabbit G966

p-Aminophenyl 2-nitro-4-trifluoromethylphenyl ether
Formed from: *p*-Nitrophenyl 2-nitro-4-trifluoromethylphenyl ether

5-(2-Aminophenyl)-5-oxopentanoic acid
Formed from: Carbazole

6-(2-Aminophenyl)-6-oxohexanoic acid
Formed from: Carbazole

4-Aminophenyl 4′-(D-phenylalanyl)aminophenyl sulphone
→ Bis(*p*-aminophenyl) sulphone
In: Rabbit G966

4-Aminophenyl 4′-(L-phenylalanyl)aminophenyl sulphone
→ Bis(*p*-aminophenyl) sulphone
In: Rabbit G966

m-Aminophenyl phenylcarbinol
Probably formed from: 3-Aminobenzophenone

p-Aminophenyl phenylcarbinol
Probably formed from: 4-Aminobenzophenone

4-Aminophenyl phosphonic acid
See: Phosphanilic acid

2-Amino-1-phenyl-1-propanol
Formed from: Cathinone
Probably formed from: 2-Ethylamino-1-phenyl-1-propanol

2-Amino-1-phenyl-1-propanone
See: Cathinone

3-(2-Aminophenyl)propionic acid
Formed from: Carbazole

5-(*p*-Aminophenyl)pyrazine
Probably formed from: 5-(*p*-Aminophenyl)pyrazine-2(1*H*)-one

5-(*p*-Aminophenyl)pyrazin-2(1*H*)-one
Probably formed from: 5-(*p*-Acetamidophenyl)pyrazin-2(1*H*)-one
→ 5-(*p*-Aminophenyl)pyrazine
Probably in: Cynomolgus monkey F313
Dog F313
Rat F313

***p*-Aminophenylpyruvic acid**
Formed from: *p*-Aminophenylalanine

N-(*p*-Aminophenyl)salicylamide
Formed from: N-(*p*-Nitrophenyl)salicylamide

***erythro*-*p*-Amino-L-phenylserine**
→ *p*-Amino-β-hydroxyphenylpyruvic acid
In: Flavobacterium C206
→ Aureothin
In: Microorganism A1236
Streptomyces A704

***threo*-*p*-Amino-L-phenylserine**
→ Aureothin
In: Microorganism A1236

***p*-Aminophenylserinol**
→ 1-(*p*-Aminophenyl)-1,3-dihydroxyacetone
In: Flavobacterium C206

***o*-Aminophenyl sulphate**
Formed from: *o*-Aminophenol

***m*-Aminophenyl sulphate**
Formed from: *m*-Aminophenol

***p*-Aminophenyl sulphate**
Formed from: *p*-Aminophenol

3-((*p*-Aminophenyl)sulphonamido)-5-hydroxymethyl-4-methylisoxazole
Formed from: Sulphatroxazole

***p*-Aminophenyl 2,4,6-trichlorophenyl ether**
Formed from: *p*-Nitrophenyl 2,4,6-trichlorophenyl ether
→ *p*-Acetamidophenyl 2,4,6-trichlorophenyl ether
Probably in: Carp F703
→ *p*-Formamidophenyl 2,4,6-trichlorophenyl ether
Probably in: Carp F703

L-2-Amino-3-phenylvaleric acid (β-ethyl-L-phenylalanine)
Formed from: 2-Oxo-3-phenylvaleric acid

L-2-Amino-5-phenylvaleric acid
Formed from: 2-Oxo-5-phenylvaleric acid

3-Aminophthalhydrazide
See: Luminol

1-Aminophthalazine
→ 4-Amino-1(2*H*)-phthalazinone
In Beef E222
Rabbit E222

4-Amino-1(2*H*)-phthalazinone
Formed from: 1-Aminophthalazine

3-Aminophthalic acid
Formed from: 6-Amino-N-(1-(1*H*)-(1-carboxy-1-(*n*-heptyl))imidazol-4-yl) phthalamidic acid

5-Amino-*m*-phthalic acid
Formed from: 5-Nitro-*m*-phthalic acid

10-((4-Amino-1-piperidino)carbonyloxy)-7-ethylcamptothecin
Formed from: Irinotecan

(8*S*,10*S*)-8-(3-Aminopropanoyloxyacetyl)-10-((2,6-dideoxy-2-fluoro-α-L-talopyranosyl)oxy)-7,8,9,10-tetrahydro-6,8,11-trihydroxy-1-methoxy-5,12-naphthacenedione
→ (8*S*,10*S*)-10-((2,6-Dideoxy-2-fluoro-α-L-talopyranosyl)oxy)-7,8,9,10-tetrahydro-6,8,11-trihydroxy-8-hydroxyacetyl-1-methoxy-5,12-naphthacenedione
In: Rat J77

2-(3-Aminopropionyl)aniline
See: Kynuramine

2-Amino-2′,6′-propionylxylidide
See: Tocainide

2-Aminopropiophenone
Formed from: 2-(Ethylamino)propiophenone
2-(Methylamino)propiophenone
→ Norephedrine
In: Man A1167
See also: A103

4-(3-Aminopropylamino)-7-chloroquinoline
Probably formed from: 7-Chloro-4-((3-ethylamino)propylamino)quinoline

5-Aminopropyl-5*H*-dibenzo[*a*,*d*]cycloheptene
Formed from: Protriptyline

9-(3-Aminopropyl)-9*H*-fluorene-9-carboxamide
Formed from: Indecainide
→ 6-Hydroxy*spiro*(9*H*-fluorene-3′,9-piperidine)-2′-one
Probably in: Dog D372
Rat D372

2-(2-Aminopropyl)-4-methyl-*p*-quinone
Probably formed from: 3,6-Dihydroxy-4-methylamphetamine

4-(2-Aminopropyl)-*o*-quinone
Probably formed from: 3,4-Dihydroxyamphetamine

4-(3-Aminopropyl)-*o*-quinone
Probably formed from: 3-(3,4-Dihydroxyphenyl)propylamine

2-Amino-5-(propylsulphinyl)-1*H*-benzimidazole
Probably formed from: Albendazole sulphoxide
→ 2-Amino-5-(propylsulphonyl)-1*H*-benzimidazole
Probably in: Beef C20
Mouse C20
Rat C20
Sheep C20

2-Amino-5-(propylsulphonyl)-1*H*-benzimidazole
Probably formed from: Albendazole sulphone
2-Amino-5-(propylsulphinyl)-1*H*-benzimidazole
→ 2-Amino-1-methyl-6-(propylsulphonyl)-1*H*-benzimidazole
Probably in: Beef C20
Mouse C20
Rat C20
Sheep C20

10-(3-Aminopropyl)-2-trifluoromethyl-10*H*-phenothiazine
Formed from: Fluoracizine
Probably formed from: 10-(3-Methylamino-propyl)-2-trifluoromethyl-10*H*-phenothiazine
10-(3-(4-Methylpiperazin-1-yl)propyl)-2-trifluoromethyl-10*H*-phenothiazine
→ 10-(3-Aminopropyl)-2-trifluoromethyl-10*H*-phenothiazine sulphoxide
Probably in: Dog A69
Man A69

10-(3-Aminopropyl)-2-trifluoromethyl-10*H*-phenothiazine sulphoxide
Probably formed from: 10-(3-Aminopropyl)-2-trifluoromethyl-10*H*-phenothiazine

Aminopterin (N-(4-(((2,4-diamino-6-pteridinyl)methyl)amino)benzoyl)-L-glutamic acid)
→ 7-Hydroxyaminopterin
In: Rabbit D644

Aminopyrazolone
See: 4-Amino-3-carboxy-5-hydroxy-1-(*p*-sulphophenyl)pyrazole

1-Aminopyrene
Formed from: 1-Formamidopyrene
1-Nitropyrene
1-Nitrosopyrene
→ 1-Acetamidopyrene
In: Goldfish H877
Guinea pig F109
Microorganism E595, F841
Rabbit F109
Rat D271, E595, F109
See also: D194
→ 1-Amino-6-hydroxypyrene
Probably in: Rat D281, D333
→ 1-Amino-8-hydroxypyrene
Probably in: Rat D281, D333
→ 1-Formamidopyrene
In: Goldfish H877
Guinea pig F109
Microorganism F841
Rabbit F109
Rat F109

4-Aminopyrene
Formed from: 4-Nitropyrene

1-Aminopyrene-4,5-dione
Formed from: 1-Nitropyrene-4,5-dione

1-(3-Amino-2-pyridinyl)-4-((6-hydroxy-5-methoxy-1*H*-indol-2-yl)carbonyl)piperazine
Probably formed from: 1-(3-Amino-2-pyridinyl)-4-((5-methoxy-1*H*-indol-2-yl)carbonyl)piperazine

1-(3-Amino-2-pyridinyl)-4-((5-methoxy-1*H*-indol-2-yl)carbonyl)piperazine
Formed from: Ateviridine
→ 1-(3-Amino-2-pyridinyl)-4-((6-hydroxy-5-methoxy-1*H*-indol-2-yl)carbonyl) piperazine
In: Rat J449
→ 1-(3-Amino-2-pyridinyl)-4-((5-methoxy-1*H*-indol-2-yl)carbonyl)piperazine-N-sulphate
In: Rat J449

1-(3-Amino-2-pyridinyl)-4-((5-methoxy-1*H*-indol-2-yl)carbonyl)piperazine-N-sulphate
Formed from: 1-(3-Amino-2-pyridinyl)-4-((5-methoxy-1*H*-indol-2-yl)carbonyl) piperazine

2-Amino-1*H*-pyrido[2,3-*b*]indole
See: 2-Amino-α-carboline

***m*-Aminopyrimethamine**
Formed from: *m*-Azidopyrimethamine

Aminopyrine (4-dimethylaminoantipyrine, pyramidone)
→ Aminopyrine radical
In: Rat D567
→ 2′,3′-Dihydro-2′,3′-dihydroxyaminopyrine
In: Phenylobacterium B838
→ 4-Dimethylamino-3-hydroxymethyl-2-methyl-1-phenyl-3-pyrazolin-5-one
In: Rat D328
→ 4-Dimethylamino-3-methyl-1-phenyl-3-pyrazolin-5-one
In: Man H167
→ 4-Formamidoantipyrine
In: Guinea pig A1941, A2372
Man A1931, A1941, A2362, B10, H167
Rabbit A1941, A1942, A2372
Rat A1942
Tulipa K539
→ 4-Methylaminoantipyrine
In: Cunninghamella A859
Man A910, A911, B10, H167, J559
Monkey A1997
Pig A1997
Rabbit A1996
Rat A1125, A1997, A3163, C669, D98, E488
Tree shrew A1997
With catalase B378
→ 4-Methylrubazoic acid
In: Man A910, A911

Aminopyrine radical
Formed from: Aminopyrine

Aminoquinol
3-Amino-4-hydroxybenzoic acid
o-Aminophenol
→ 2′,5′-Dihydroxyacetanilide
Probably in: Streptomyces F190

5-Aminoquinoline
→ 5-Acetamidoquinoline
In: Rabbit G994
→ 5-Amino-2-hydroxyquinoline
In: Rabbit G994

6-Aminoquinoline
→ 6-Acetamidoquinoline
In: Rabbit H370

4-Aminoquinoline-N-oxide
4-Hydroxylaminoquinoline-N-oxide

6-Amino-2-quinolone
Formed from: 6-(4-(3,4-Dimethoxybenzoyl)-1-piperazinyl)-3,4-dihydro-2(1*H*)-quinolone
→ 6-Acetamido-2-quinolone
Probably in: Monkey F318
Mouse F318
Rat F318
→ 2,6-Dihydroxyquinoline
Probably in: Man F318
Rat F318

2-Aminoresorcinol
Formed from: 2-Nitroresorcinol

***p*-Aminosalicylic acid**
See: 4-Amino-2-hydroxybenzoic acid

5-Aminosalicylic acid
See: 5-Amino-2-hydroxybenzoic acid

5-Aminosalicylic acid-N-glucoside
Formed from: 5-Amino-2-hydroxybenzoic acid

4-Aminostilbene
→ 4-Stilbenyl nitroxide radical
In: Rabbit B184

4-Aminostilben-4′-yl nitroxide radical
Formed from: 4,4′-Diaminostilbene

***o*-(1-Amino-2-sulphamoylethyl)phenol**
Formed from: Zonisamide

→ *o*-(1-Amino-2-sulphamoylethyl)phenyl sulphate
Probably in: Rat F594

o-(1-Amino-2-sulphamoylethyl)phenyl sulphate
Probably formed from: *o*-(1-Amino-2-sulphamoylethyl)phenol

2-(4-Amino-3-sulphatoxyphenyl)-2-ethylglutarimide
Formed from: 2-(4-Amino-3-hydroxyphenyl)-2-ethylglutarimide

5-((4′-((1-Amino-4-sulpho-2-naphthalenyl)azo)(1,1′-biphenyl)-4-yl)azo)-2-hydroxybenzoic acid
See: Direct Orange 8

5-(Aminosulphonyl)-4-chloro-2-((2-furanylmethyl)amino)benzoic acid
See; Furosemide

4-Amino-2,3,5,6-tetrachlorophenol
Probably formed from: 2,3,5,6-Tetrachloroaniline

S-(Aminotetrachlorophenyl)glutathione
Formed from: Pentachloronitrobenzene

4-Amino-2,3,5,6-tetrafluorophenol
Probably formed from: 2,3,5,6-Tetrafluoroaniline

9-Amino-1,2,3,4-tetrahydroacridine
See: Tacrine

9-Amino-1,2,3,4-tetrahydro-1-acridinol
See: Velnacrine

2-Aminotetrahydro-6,7-dihydroxynaphthalene
Formed from: Diacetyl 2-aminotetrahydro-6,7-dihydroxynaphthalene
Dibenzoyl 2-aminotetrahydro-6,7-dihydroxynaphthalene
Diisobutyryl 2-aminotetrahydro-6,7-dihydroxynaphthalene
Dipivaloyl 2-aminotetrahydro-6,7-dihydroxynaphthalene
→ 2-Aminotetrahydro-6-hydroxy-7-methoxynaphthalene
Probably in: Rat C251

2-Amino-3,4,4a,10a-tetrahydro-4a,7-dimethylphenoxazin-3-one
Formed from: 6-Amino-*m*-cresol

N-(4-((2-Amino-4,5,6,7-tetrahydro-4,6-dioxo-3*H*-pyrrolo[2,3-*d*]pyrimidin-5-yl)ethyl)benzoyl)-L-glutamic acid
Formed from: N-(4-((2-Amino-4,7-dihydro-4-oxo-3*H*-pyrrolo[2,3-*d*]pyrimidin-5-yl)ethyl) benzoyl)-L-glutamic acid

2-Aminotetrahydro-6-hydroxy-7-methoxynaphthalene
Probably formed from: 2-Aminotetrahydro-6,7-methoxynaphthalene

7-Amino-2,3,4,5-tetrahydroimidazo[2,1-*b*] quinazolin-2(1*H*)-one
Probably formed from: 7-(4-Carboxybutyl) amino-2,3,4,5-tetrahydroimidazo[2,1-*b*] quinazolin-2(1*H*)-one
→ 7-Acetamido-2,3,4,5-tetrahydroimidazo[2,1-*b*]quinazolin-2(1*H*)-one
Probably in: Rat H463

8-Amino-1,2,3,4-tetrahydro-2-methyl-4-phenylisoquinoline
See: Nomifensine

1-Aminotetralin
→ α-tetralone
In: Rabbit F295
Rat F295

2-Aminotetralin
→ β-tetralone
In: Rabbit F295
Rat F295

***o*-Aminothioanisole**
Formed from: *o*-Aminothiophenol

***o*-Aminothiophenol**
→ *o*-Aminothioanisole
In: Man D926

N-(Aminothioxomethyl)-4-hydroxy-2-methyl-2*H*-1,2-benzothiazine-3-carboxamide-1,1-dioxide
Probably formed from: 4-Hydroxy-N-((2-hydroxypropyl)aminothioxomethyl)-2-methyl-2*H*-1,2-benzothiazine-3-carboxamide-1,1-dioxide

4-Aminotoluene-2-sulphonic acid
Formed from: 4-Nitrotoluene-2-sulphonic acid

4-Amino-*m*-tolyl N-methylcarbamate
Formed from: 4-Methylamino-*m*-tolyl N-methylcarbamate

2-Amino-4,6,7-trichloro-3*H*-phenoxazin-3-one
Probably formed from: 3,4-Dichloroaniline

7-Amino-4,5,6-triethoxyphthalide
Formed from: Tritoqualine
→ 7-Amino-3-hydroxy-4,5,6-triethoxyphthalide
In: Dog E105
Mouse E105
Rat E105

4-Amino-2-trifluoromethylbenzonitrile
Formed from: Casodex

4-Amino-2-trifluoromethylphenol
Formed from: *p*-Trifluoromethylaniline

4-Amino-3-trifluoromethylphenol
Formed from: 4-Nitro-3-trifluoromethylphenol

2-Amino-1-(*m*-trifluoromethylphenyl)ethanol
Formed from: Tetrahydro-2-(3-trifluoromethylphenyl)-1,4-oxazine

2-Amino-4-trifluoromethylphenyl *p*-nitrophenyl ether
Formed from: *p*-Nitrophenyl 2-nitro-4-trifluoromethylphenyl ether

4-Amino-2,3,5-trifluorophenol
Formed from: 2,3,6-Trifluoroaniline

4-Amino-2,3,5-tris(S-glutathionyl)phenol
Formed from: 4-Amino-2,5-bis(S-glutathionyl)phenol

5-Amino-L-tryptophan
Probably formed from: 5-Aminoindole

3-Amino-L-tyrosine
→ 4-(3-L-Alanyl)-*o*-quinone
Probably in: Man A63

3-Aminoumbelliferone
→ Novobiocin
In: Streptomyces A945

4-Amino-4′-ureidodiphenyl sulphone
→ 4-Nitro-4′-ureidodiphenyl sulphone
In: Chicken A970
Rat A970

4-Amino-*o*-xylene
See: 3,4-Dimethylaniline

4-Amino-3,5-xylyl methylcarbamate
Probably formed from: 4-Methylamino-3,5-xylyl methylcarbamate

Amiodarone ((2-butyl-3-benzofuranyl)(4-(2-(diethylamino)ethoxy)-3,5-diiodophenyl)methanone)
→ Desethylamiodarone
In: Dog F591
Hamster H772
Man C478, D249, G837, K324
Rabbit D825, E865
Rat E865

Amiquinisin
See: 4-Amino-6,7-dimethoxyquinoline

Amitraz (N'-(2,4-dimethylphenyl)-N-(((2,4-dimethylphenyl)imino)methyl)-N-methylmethanimidamine)
→ 2,4-Dimethylaniline
Probably in: Rat D943

Amitriptyline (10,11-dihydro-N,N-dimethyl-5*H*-dibenzo[*a*,*d*]cycloheptene-$\Delta^{5,\gamma}$-propylamine)
Formed from: Amitriptyline-N-oxide
→ Amitriptyline-N-oxide
In: Cunninghamella H485
Dog A1695, A2075
Man A1442, E894, G472
→ Amitriptyline-β-D-glucuronide
In: Man C614, E555, F237, F523, G509, H541, J137, J416, K58
Rat C614
→ 1,2-Dihydro-1,2-dihydroxyamitriptyline
In: Man E894
→ 2-Hydroxyamitriptyline
In: Cunninghamella H485
Man E894
Rat E892, E931
→ 3-Hydroxyamitriptyline
In: Cunninghamella H485
Man E894
Rat E892
→ 10-Hydroxyamitriptyline
In: Cunninghamella H485
Dog A1695, A2075
Man A3913, C26, E894, F523, G472, J70
Rat D739, E892, E931, H887
→ 11-Hydroxyamitriptyline
In: Cunninghamella H485
→ Nortriptyline
In: Cunninghamella D185, H485

Dog A1695
Man A3913, C26, C886, E894, G472, J70, J531, K415
Rat D739, H887, K415
See also: A888, A2075, E931, E892, G188

Amitriptyline-β-D-glucuronide
Formed from: Amitriptyline
→ 10-Hydroxyamitriptyline-N-β-D-glucuronide
In: Man F523

Amitriptyline-N-oxide
Formed from: Amitriptyline
→ Amitriptyline
In: Man G188, G472
Rat C558, E892
→ 10-Hydroxyamitriptyline-N-oxide
In: Man A3768, G472
See also: G188

Amlodipine (2-((2-aminoethoxy)methyl)-4-(*o*-chlorophenyl)-3-ethoxycarbonyl-1,4-dihydro-5-methoxycarbonyl-6-methylpyridine)
→ Demethylamlodipine
In: Rat E836, F41, F78
→ 2-((2-Aminoethoxy)methyl)-4-(*o*-chlorophenyl)-1,4-dihydro-3-(2-hydroxyethoxy)carbonyl-5-methoxycarbonyl-6-methylpyridine
In: Rat E836, F41, F78
→ 2-((2-Aminoethoxy)methyl)-4-(*o*-chlorophenyl)-3-ethoxycarbonyl-1,4-dihydro-5-hydroxycarbonyl-6-methylpyridine
In: Rat F41
→ 2-((2-Aminoethoxy)methyl)-4-(*o*-chlorophenyl)-3-ethoxycarbonyl-5-methoxycarbonyl-6-methylpyridine
In: Dog E836, F41, F78
Man E848, F41, F78
Rat F41, F78
→ 2-((Carboxymethoxy)methyl)-4-(*o*-chlorophenyl)-1,4-dihydro-3-ethoxycarbonyl-5-methoxycarbonyl-6-methylpyridine
In: Dog E836, F78
Rat E836, F78

Amodiaquine (7-chloro-4-(3′-diethylamino-4′-hydroxyanilino)quinoline)
→ Amodiaquine-N-oxide
In: Rat H159
→ Desethylamodiaquine
In: Man D529, E326, H159
Rat H159
See also: A278
→ 5′-(S-Glutathionyl)amodiaquine
In: Rat H159

Amodiaquine-N-oxide
Formed from: Amodiaquine

Amonafide (5-amino-2-(2-(dimethylamino)ethyl)-1*H*-benz[*d*,*e*]isoquinoline)
→ N^5-Acetylamonafide
In: Man E887
→ Amonafide-N-oxide
In: Man E887

Amonafide-N-oxide
Formed from: Amonafide
→ N^5-Acetylamonafide-N-oxide
Probably in: Man E887

Amosulalol (5-(1-hydroxy-2-((2-(*o*-methoxyphenoxy)ethyl)amino)ethyl)-2-methylbenzenesulphonamide)
→ 5-(1-Hydroxy-2-((2-(4-hydroxy-2-methoxyphenoxy)ethyl)amino)ethyl)-2-methylbenzenesulphonamide
In: Dog D1
Monkey D369
Rat D1
→ 5-(1-Hydroxy-2-((2-(5-hydroxy-2-methoxyphenoxy)ethyl)amino)ethyl)-2-methylbenzenesulphonamide
In: Dog D1
Man D369
Monkey D369
Rat D1
→ 5-(1-Hydroxy-2-((2-(*o*-hydroxyphenoxy)ethyl)amino)ethyl)-2-methylbenzenesulphonamide
In: Dog D1
Rat D1, D369
→ 5-(1-Hydroxy-2-((2-(*o*-methoxyphenoxy)ethyl)amino)ethyl)-2-hydroxymethylbenzenesulphonamide
In: Dog D1

Rat D1, D369
→ *o*-Methoxyphenoxyacetic acid
In: Dog D1, D369
Monkey D369
Rat D1

Amoxapine (2-chloro-11-(1-piperazinyl) dibenz[*b,f*][1,4]oxazepine)
→ 7-Hydroxyamoxapine
In: Rat B953

Amoxicillin (6-(((2*R*)-amino-(4-hydroxyphenyl) acetyl)amino)-3,3-dimethyl-7-oxo-4-thia-1-azabicyclo[3.2.0]heptane-2-carboxylic acid)
Formed from: D-*p*-Hydroxyphenylglycine

Amperozide 4-(4,4-bis(*p*-fluorophenyl)butyl)-N-ethyl-1-piperazinecarboxamide)
→ 4,4-(Bis(*p*-fluorophenyl)-butanoic acid
In: Rat H745
→ 4,4-(Bis(*p*-fluorophenyl))-*n*-butanol
In: Rat H745
→ 4-(4,4(Bis(*p*-fluorophenyl))butyl)-1-(N-ethylcarboxamido)-2,5-dihydroxypiperazine
In: Dog H745
Man H745
Rat H745
→ N-(4-(Bis(*p*-fluorophenyl))-4-hydroxybutyl)-N′-(N-ethylcarboxamido)-N-(2-hydroxyethyl)ethylenediamine
In: Rat H745
→ N-(4-(Bis(*p*-fluorophenyl))-4-hydroxybutyl)-N′-(N-ethylcarboxamido)-N′-(2-hydroxyethyl)ethylenediamine
In: Rat H745
→ Desethylamperozide
In: Dog H745
Man H745

(*R*)-(−)-Amphetamine
Probably formed from: Demethyldeprenyl
→ Benzyl methyl ketoxime
Probably in: Rabbit A2901
→ *p*-Hydroxy-(-)-amphetamine
In: Rat A101, A1593
→ α-Methyl-β-phenethylhydroxylamine
In: Guinea pig A388, C5
Rabbit A388, A2034, A2901, C5, D91
Rat C5
→ L-Norephedrine
In: Guinea pig A388
Rabbit A388
→ Phenylacetone
In: Guinea pig C5
Rabbit C5
Rat C5

(*S*)-(+)-Amphetamine (d-α-methylphenethylamine; this isomer is assumed unless the (*R*)-isomer is stated)
Formed from: Amphetaminil
Aponeuron
N-Benzylamphetamine
N-Benzylideneamphetamine
Benzyl methyl ketoxime
N-(*n*-Butyl)amphetamine
Captagon
Clobenorex
N-(2-Cyanoethyl)amphetamine
Fenetylline
N-Hydroxy-N-propylamphetamine
N-Isopropylamphetamine
Mefenorex
Mesocarb
N-Methylamphetamine
α-Methyl-β-phenethylhydroxylamine
α-Methyl-L-phenylalanine
α-Methyl-N-(*n*-propylidene)benzeneethanamine-N-oxide
Morforex
2-Nitro-1-phenylpropane
Prenylamine
N-Propylamphetamine
Probably formed from: N-Hydroxy-N-methylamphetamine
→ N-Acetylamphetamine
In: Mycobacterium B380
See also: A3878, C68
→ Benzyl methyl ketoxime
In: Cunninghamella F345
Rabbit A1166, A1398
See also: A915, A2453, A2719, A2850, A3878, C68
→ *m*-Hydroxyamphetamine
In: Rat A3272
→ *p*-Hydroxy-d-amphetamine
In: Horse A915

Man B147, J544
Marmoset B147
Monkey B147
Mouse A3160
Rat A77, A101, A315, A1593 A2015, A2348, A2495, A2975, A3160, A3289, B7, B147, E76, K20, K673
Vervet B147
→ N-Methylamphetamine
In: Rabbit E6
Rat A80
→ α-Methyl-β-phenethylhydroxylamine
In: Cunninghamella F345
Guinea pig A388
Rabbit A388, A1166, A1692, A2034, A3621
See also: A3878
→ D-Norephedrine
In: Beef A1101
Guinea pig A388
Man A1473
Rabbit A388
Rat A2495
See also: A528, C300, C663
→ Phenylacetone
Guinea pig C5
Rabbit A3621, C5, J98
Rat C5, J98

Amphetaminil (α-((1-methyl-2-phenylethyl) amino)benzeneacetonitrile)
→ Amphetamine
In: Man A609

Amphetamin-4-yl-β-D-glucuronide
Formed from: *p*-Hydroxyamphetamine

Amphetamin-4-yl sulphate
Formed from: *p*-Hydroxyamphetamine

Ampicillin (6-(((2*R*)-aminophenylacetyl)amino)-3,3-dimethyl-7-oxo-4-thia-1-azabicyclo[3.2.0]heptane-2-carboxylic acid)
Formed from: D-Phenylglycine methyl ester
Talampicillin

Ampiroxicam (ethyl 1-((2-methyl-1,1-dioxido-3-(2-pyridinylamino)carbonyl)-2*H*-1,2-benzothiazin-4-yl)oxy)ethylcarbonic acid ester)
→ Piroxicam
In: Man F896

Amsacrine (4′-(9-acridinylamino) methanesulphon-*m*-anisidide)
Formed from: $N^{4'}$-(9-Acridinyl)-$N^{1'}$-methanesulphonyl-3′methoxy-2′,5′-cyclohexadiene-1′,4′-diimine
→ $N^{4'}$-(9-Acridinyl)-$N^{1'}$-methanesulphonyl-3′methoxy-2′,5′-cyclohexadiene-1′,4′-diimine
In: Rat D127
→ 5′-(S-Glutathionyl)amsacrine
In: Mouse G236, F444
Rat G236
→ 6′-(S-Glutathionyl)amsacrine
In: Mouse G236, F444
Rat G236, G987
→ 4-Methyl-5-(N-methyl)carboxamido-9-anilinoacridine
In: Mouse G236
Rat G236

Amygdalin (mandelonitrile-β-gentiobioside)
→ Mandelonitrile
In: Davallia K910
Prunus G648
→ Prunasin
In: Cherry H695, K910
Mucor J528
Penicillium J528

Amyl L-γ-glutamyl dopamide
→ 3,4-Dihydroxyphenethylamine
In: Dog A3945

***p-n*-Amylphenol**
See: *p*-*n*-Pentylphenol

Analog 11
See: 1,1-Dichloro-*cis*-2,3-diphenylcyclopropane

4-Androstene-3,17-dione
→ 3-Hydroxy-17-methoxy-1,3,5(10)-oestratriene
In: Clostridium A625
→ Oestrone
In: Man A2259

Anethole
See: *trans*-*p*-Methoxypropenylbenzene

Anethole oxide
Probably formed from: *p*-Methoxypropenylbenzene

→ N-Acetyl-S-(2-hydroxy-1-(*p*-methoxyphenyl)-1-propyl)-L-cysteine
Probably in: Mouse H725
Rat H725
→ 1,2-Dihydroxy-1-(*p*-methoxyphenyl) propane
Probably in: Mouse H725
Rat H725
→ 1-(*p*-Methoxyphenyl)acetone
Probably in: Mouse H725
Rat H725

Anhydrotetracycline (1-dimethylamino-1,4,4a,5,12,12a-hexahydro-2,4a,6,7-tetrahydroxy-11-methyl-4,5-dioxonaphthacene-3-carboxamide)
→ 12-Dehydrotetracycline
In: Streptomyces K900, K938

3′,4′-Anhydrovinblastine
→ Vinamidine
In: Catharanthus E701
→ Vinblastine
In: Catharanthus B99, C398, E701
→ Hydroxyvinamidine
In: Catharanthus E701

Anhydrovinblastine-N-oxide
Probably formed from: Vinblastine

3′,4′-Anhydrovinblastinium
Formed from: Catharanthine
Vindoline

Anigorufone (2-hydroxy-9-phenyl-1*H*-phenalen-1-one)
Formed from: Cinnamic acid
7-(3,4-Dihydroxyphenyl)-1-phenylhepta-1,3-dien-5-one

Aniline (benzenamine)
Formed from: 4-Acetamidoazobenzene
Acetanilide
Acid Orange 12
p-Aminoazobenzene
Benzanilide
Benzimidazole-2-methylcarbamate
N-Benzoylaniline
N-Benzylaniline
4-Bis(2-bromoethyl)aminoazobenzene
4-Bis(2-bromoethyl)amino-2-methylazobenzene
Butyranilide
3-Carboxanilido-5,6-dihydro-2-methyl-1,4-oxathiin
2-Carboxy-4-dimethylaminoazobenzene
5-Chloro-2-hydroxy-N-phenylbenzamide
2,4-Diaminoazobenzene
N-(2,4-Dichlorobenzyl)aniline
N-(2,6-Dichlorobenzyl)aniline
4-Dimethylaminoazobenzene
4-Dimethylamino-2-methylazobenzene
2,5-Dimethylfuran-3-carboxanilide
3,3-Dimethyl-1-phenyltriazene
Diphenylamine
1,2-Diphenylhydrazine
1,3-Diphenyl-1-triazene
Glutathione sulphinanilide
Glycollanilide
Hydrazobenzene
2-Hydroxyazobenzene
4-Hydroxyazobenzene
Isopropyl carbanilate
Methaphenilene
4-Methylaminoazobenzene
N-Methylaniline
N-Methyl-N-nitrosoaniline
Nitrobenzene
N-(*p*-Nitrobenzoyl)aniline
N-(*p*-Nitrobenzyl)aniline
Phenylhydrazine
Phenylhydroxylamine
N-Phenyllinoleamide
Propionanilide
N-Propylalanylanilide
Salicylanilide
p-Toluoyl chloride phenylhydrazone
N-(2,4,6-Trifluorobenzyl)aniline
N-(2,4,6-Trimethylbenzyl)aniline
Valeranilide
Probably formed from: N-Ethylaniline
Formanilide
o-Hydroxyazobenzene
→ Acetanilide
In: Apocynum A2619
Arthrobacter E276
Bacillus E276
Catharanthus A2619
Conium A2619

Cordyceps E276
Cunninghamella A1218
Hen F694
Nocardia G343
Paracoccus A2824
Pig A3460
Rabbit C78
Rainbow trout J582
Rat A3460, B165, B174, D81, D282
Sheep A2788, A3460, D282
See also: E393, E395
→ *p*-Aminobenzoic acid
Probably in: Desulphobacterium G155
→ *o*-Aminophenol
In: Beef E60
Boophilus A1795
Mouse D532
Nephila A1795
Pig A3460
Rat A3460, B174, D532
Sheep A2788, A3460
Rainbow trout J582, K436
See also: A1129, A1218, A1250, A2019, A2714, A2821, E395, K520
→ *m*-Aminophenol
In: Apocynum A2619
Catharanthus A2619
Conium A2619
See also: A2019, A2714
→ *p*-Aminophenol
In: Beef E60
Boophilus A1795
Cunninghamella A1218
Man A114, A572, B64, B145, D135, D283, D557
Monkey A1997
Mouse D532
Nephila A1795
Pig A1997, A3460
Rabbit A351
Rainbow trout J639
Rat A114, A312, A572, A1881, A1997, A3460, B165, B174, D98, D532, H273
Sheep A2788, A3460, B64
Tree shrew A1997
With catalase B940
With haemoglobin D43
With myoglobin H614
See also: A335, A1125, A1129, A1250, A1592, A2019, A2120, A2420, A2714, A2821, K520, K539
→ Aniline-N-β-D-glucuronide
In: Man H541
Rat D288
See also: A1881
→ Aniline-N-sulphate
In: Guinea pig E456, H455
Man J571
Mouse H455
Rabbit H455, J914
Rat E45, H455
→ Anthranilic acid
In: Rhodococcus D273, D595
→ Azobenzene
In: Geotrichum A690
Horseradish A661
Pseudomonas H12
→ Benzene
In: Desulphovibrio G827
→ *o*-Bromoaniline
In: Pseudomonas H12
→ *p*-Bromoaniline
In: Pseudomonas H12
→ Catechol
In: Comamonas J622
Nocardia A1520, A2663
Pseudomonas E164
See also: K520
→ γ-Glutamylaniline
In: Boophilus A1795
→ N-Methylaniline
In: Rabbit A115, C156, C333, E6
Pig K923
→ Nitrobenzene
In: Nitrosomonas H219
→ *p*-Nitrophenol
In: Rat A650
→ Nitrosobenzene
With chloroperoxidase A3579
→ Phenylhydroxylamine
In: Rabbit A1746, F839
Rainbow trout G32

Rat B174, C489
See also: C797
→ Phenyl nitroxide radical
In: Rabbit B184
→ Pigment
In: Geotrichum A654
→ Quinol
Probably in: Pseudomonas K520

Aniline-N-β-D-glucuronide
Formed from: Aniline

Aniline mustard
See: N,N-(Bis(2-chloroethyl)aniline

Aniline-N-sulphate
Formed from: Aniline

4-(N-Anilino)piperidine
Probably formed from: 4-(N-Propionylanilino)piperidine

Aniracetam (1-(*p*-anisoyl-2-pyrrolidinone)
→ *p*-Anisamidobutyric acid
In: Rat K464
→ Anisic acid
In: Rat K464

Anisaldehyde (*p*-methoxybenzaldehyde)
Formed from: Anisic acid
Anisoin
Anisyl alcohol
4-Hydroxy-4′-methoxybenzoin
p-Methoxyphenylglycol
Probably formed from: *p*-Methoxybenzylamine
p-Methoxycinnamic acid
Pyrilamine
→ Anisic acid
In: Acinetobacter E663
Beef C508
Desulphovibrio F666
Man J24
Rat H761
Rhodopseudomonas G246
Streptomyces E653
See also: 2744, E737, G865
→ Anisoin
In: Pseudomonas K424
→ Anisyl alcohol
In: Dunaliella G459
Euglena G422
Mung bean J910
See also: A2744, A3779
→ 2-Hydroxy-*p*-methoxypropiophenone
In: Pseudomonas K524
→ *p*-Methoxymandelonitrile
In: Almond G325, J194, K6
Loquat K6
Peach K6
Sorghum J194
→ (1*R*,2*S*)-1-(*p*-Methoxyphenyl)propane-1,2-diol
In: Saccharomyces E487

***p*-Anisamidobutyric acid**
Formed from: Aniracetam

Anisic acid (*p*-methoxybenzoic acid)
Formed from: Aniracetam
Anisaldehyde
Anisoin
erythro-2-(*o*-Chlorophenyl)-2-(4-(*p*-methoxybenzyl)-1-piperazinyl)-1-phenylethanol
p-Hydroxybenzoic acid
Isovanillic acid
p-Methoxybenzonitrile
p-Methoxybenzoyl-L-alanine
p-Methoxyhippuric acid
p-Methoxyphenylpropylamine
trans-p-Methoxypropenylbenzene
p-Methoxypropylbenzene
Probably formed from: *p*-Methoxycinnamic acid
3-(*p*-Methoxyphenyl)propionic acid
→ Anisaldehyde
In: Clostridium G416
Coniophora A3779
Neurospora A750
Pleurotus A3779
Polystictus A3779
Schizophyllum A3779
With NAD(P)-dependent benzaldehyde oxidoreductase E248
→ Anisole
In: Sporotrichum B73
→ Anisoyl CoA
In: Beef H740

Pseudomonas G729
→ Anisoyl-β-D-glucose
In: Quercus K737
→ Anisoyl-β-D-glucuronic acid
In: Mouse H725
Rabbit H519
Rat H725, J549
→ Anisyl alcohol
In: Corynespora G821
→ *p*-Hydroxybenzoic acid
In: Acetobacterium F655
Chaetomium K896
Microorganism B321
Nocardia A960, F266
Pseudomonas A563, A903, A2953, E957, K800, K801
Vanilla F824
Wheat A709
Xerocomus K896
See also: A989, B2
→ Isovanillic acid
In: Nocardia A960, F266
→ *p*-Methoxyhippuric acid
In: Man E812
Mouse H725
Rabbit H519
Rat A3927, H725
Contrast (rat) J549

p-Anisidine
See: *p*-Methoxyaniline

Anisoin (4,4′-dimethoxybenzoin, 2-hydroxy-1,2-bis(methoxyphenyl)ethanone)
Formed from: Anisaldehyde
→ Anisaldehyde
In: Psudomonas F221
→ Anisic acid
In: Psudomonas E737

Anisole (methyl phenyl ether)
Formed from: Anisic acid
Phenol
→ *o*-Bromoanisole
In: Corallina E597
→ *p*-Bromoanisole
In: Corallina E597
→ *cis*-2,3-Dihydro-2,3-dihydroxyanisole
In: Pseudomonas J656, K699
See also: H251
→ Guaiacol
In: Aspergillus, A1218, A2849
Cunninghamella A1218, A2849
Curvularia A2849
Gliocladium A1218
Helicostylum A1218, A2849
Mucor A2849
Pseudomonas H251
Rat D122, G500
Rhizopus, A1218, A2849
Sporotrichum A2849
Streptomyces, A1218
Trichothecium A2849
→ 3-Methoxycatechol
In: Pseudomonas F283
→ *m*-Methoxyphenol
In: Rat D122
→ *p*-Methoxyphenol
In: Aspergillus, A1218, A2849
Cunninghamella A1218, A2849
Methylosinus D513
Mucor A2849
Nitrosomonas H631
Penicillium A1218
Rat A2160, D122, G500
Rhizopus, A2849
→ Phenol
In: Aspergillus, A1218
Beijerinckia H251
Catharanthus A2619
Cunninghamella A859, A1218
Helicostylum A1218
Man D135
Methylosinus D513
Microorganism G917
Nitrosomonas H251
Pseudomonas H251
Rat A2160, G500
Rhizopus, A1218
Streptomyces A1218

Anisoyl CoA
Formed from: Anisic acid

Anisoyl-β-D-glucose
Formed from: Anisic acid

Anisoyl-β-D-glucuronic acid
Formed from: Anisic acid
1-(*p*-Anisoyl-2-pyrrolidinone
See: Aniracetam
Anisyl alcohol (*p*-methoxybenzyl alcohol)
Formed from: Anisaldehyde
Anisic acid
1-(3,4-Diethoxyphenyl)-1-hydroxy-2-(*p*-methoxyphenyl)ethane
p-Hydroxybenzyl alcohol
p-Methylanisole
→ Anisaldehyde
In: Acinetobacter E663
Arion G804
Botrytis G613
Helix G804
Limax G804
Phanerochaete D467, F257, H613
Pleurotus F867, G668, K373
Pseudomonas A3837
Rhodopseudomonas D197
See also: E653
Anitrazafen
See: 5,6-Bis-(*p*-methoxyphenyl)-3-methyl-1,2,4-triazine
Anol
See: *p*-(1-Propenyl)phenol
Anthocyanidin (general entry)
→ Anthocyanidin-3-D-glucoside
In: Hippeastrum A2523
Tulipa A2523
Anthocyanidin-3-D-glucoside
Formed from: Anthocyanidin
Anthracene
→ Anthracene-1,2-oxide
In: Rat D792
→ Anthraquinone
In: Cunninghamella K11
Curvularia K11
Phanerochaete G418
Salmonella D805
→ Catechol
In: Aspergillus K190
→ (+)-*cis*-1,2-Dihydro-1,2-dihydroxyanthracene
In: Beijerinckia A2413, A2680
Pseudomonas A2680
→ *trans*-1,2-Dihydro-1,2-dihydroxyanthracene
In: Cunninghamella D107
Rabbit A2413, E576
Rat A2413
See also: D805
→ 2,3-Dihydroxyanthracene
In: Pseudomonas K278
→ 3-Hydroxy-2-naphthoic acid
In: Microorganism G680
Pseudomonas K278
→ 9-Methylanthracene
In: Rat E612
9(10*H*)-Anthracenone
See: Anthrone
Anthracene-1,2-oxide
Formed from: Anthracene
→ *trans*-1,2-Dihydro-1,2-dihydroxyanthracene
In: Rat D261
2-Anthracenyl nitroxide radical
Formed from: 2-Aminoanthracene
Anthraflavic acid
See: Anthraflavin
Anthraflavin (anthraflavic acid, 2,6-dihydroxyanthraquinone)
→ Anthraflavin-β-D-glucuronide
In: Cinchona K889
Man G524, G968, J416, J783
Streptomyces B462
→ 6-β-D-Glucopyranosyloxy-1-hydroxyanthraquinone
In: Streptomyces A1497
See also: B462
Anthraflavin-6-β-D-glucoside
See: 6-β-D-Glucopyranosyloxy-1-hydroxyanthraquinone
Anthraflavin-β-D-glucuronide
Formed from: Anthraflavin
9-Anthraldehyde
See: 9-Formylanthracene
Anthramycin (3-((11*R*,11a*S*)-5,10,11,11a-tetrahydro-9,11-dihydroxy-8-methyl-5-oxo-1*H*-pyrrolo[2,1-*c*][1,4]benzodiazepin-2-yl-2-propenamide)
Formed from: 3,4-Dihydroxy-L-phenylalanine

L-Tryptophan
L-Tyrosine
Anthranil (2,1-benzisoxazole)
Formed from: Indole
Anthranilaldehyde (*o*-aminobenzaldehyde)
Formed from: *o*-Acetamidobenzaldehyde
o-Nitrobenzaldehyde
Probably formed from: *o*-Formamidobenzaldehyde
Anthranilamide (*o*-aminobenzamide)
Formed from: *o*-Aminobenzonitrile
Anthranilic acid
→ 3-Hydroxyanthranilamide
In: Rat A1411, A2891, D149
→ 5-Hydroxyanthranilamide
In: Rat A754, A1411, A2891
Anthranilamid-3-yl sulphate
Formed from: 3-Hydroxyanthranilamide
Anthranilamid-5-yl sulphate
Formed from: 5-Hydroxyanthranilamide
Anthranilhydroxamic acid
→ Anthranilamide
In: Guinea pig C897
Rabbit C897
Anthranilic acid (*o*-Aminobenzoic acid)
Formed from: N-Acetylanthranilic acid
o-Aminobenzonitrile
6-Amino-5-((1-carboxyethyl)oxy)-1,3-cyclohexadiene-1-carboxylic acid
2-Amino-5-chlorobenzoic acid
Aniline
Bentazon
Carbazole
4-Bis(2-bromoethyl)amino-2′-carboxy-2-methylazobenzene
4-Bis(2-bromopropyl)amino-2′-carboxy-2-methylazobenzene
2-(4′-Bis(2″-bromopropyl)aminophenylazo)benzoic acid
2′-Carboxy-4-bis(2-bromopropyl)aminoazobenzene
Chorismic acid
Cinnamyl anthranilate
N-Formylanthranilic acid
Indole
Isatin
L-Kynurenine
β-Methylanthranilic acid
β-Methyl-L-kynurenine
Methyl red
o-Nitrobenzoic acid
Quinoline
Probably formed from: *o*-Aminobenzyl alcohol
o-Aminophenylglyoxylic acid
6-(*o*-Aminophenyl-2-hydroxy-6-oxohexa-2,4-dienoic acid
3-(*o*-Carboxyphenylimino)isoxazolidine
Indole-3-carboxylic acid
→ N-Acetylanthranilic acid
In: Pseudomonas C120
→ N-Adenylylanthranilic acid
In: Penicillium K906
→ Anthranilamide
In: Rat A2891, D296
→ Anthranilic acid-β-D-glucoside
In: Zea A2207
Raphanus B440
→ Anthraniloyl CoA
In: Pseudomonas F65, G342, G642, K742
→ Anthraniloyl-β-D-glucuronic acid
In: Guinea pig A865
Rabbit H519
Rat B740
See also: A3827
→ 6-Amino-1,2-dihydro-1,2-dihydroxybenzoic acid
Probably in: Pseudomonas B465
→ *o*-Aminophenol
In: Agaricus D770, E162
See also: C120
→ Anthraniloyl CoA
In: Pseudomonas F65, G342, G642
→ Anthraniloyl-β-D-glucuronic acid
In: Guinea pig A865
Rabbit H519
Rat B740
See also: A3827
→ Arborine
In: Glycosmis A538
→ Auranthine
In: Penicillium E362
→ Catechol

In: Aspergillus A1299
Micrococcus D56
Pseudomonas C120, G434
See also: F417
→ Cyclopeptine
In: Penicillium G535
→ *o*-(D-Glucuronamino)benzoic acid
Probably in: Rat B740
→ 3-Hydroxyanthranilic acid
In: Rat A3653, F676
Trichosporon C393
→ 5-Hydroxyanthranilic acid
In: Rat A3653, D296
→ N-Malonylanthranilic acid
In: Arachis K727
→ N-Methylanthranilic acid
In: Choisya F872
Ruta F872, K733
→ Phenazine-1-carboxylic acid
Probably in: Pseudomonas A958
→ Pyrocatechuic acid
In: Aspergillus A1299, B327, D231
Trichosporon B790, C393, E270
→ 1,2,3,4-Tetrahydro-2-methyl-4-quinazolinone
In: Rat A1516

Anthranilic acid-β-D-glucoside
Formed from: Anthranilic acid

l-3-Anthraniloylalanine
See: l-Kynurenine

Anthraniloyl CoA
Formed from: Anthranilic acid
→ 2-Amino-5-oxocyclohex-1-enecarboxyl CoA
In: Azoarcus K215
→ Benzoyl CoA
In: Pseudomonas G642
→ Dihydroanthraniloyl CoA
Probably in: Pseudomonas F465
Thauera J175
→ 5-Hydroxyanthraniloyl CoA
In: Pseudomonas H605
See also: F465

Anthraniloyl-β-D-glucuronic acid
Formed from: Anthranilic acid

Anthrapurpurin (1,2,7-trihydroxyanthraquinone)
→ Anthrapurpurin-β-D-glucoside
In: Cinchona K889

Anthrapurpurin-β-D-glucoside
Formed from: Anthrapurpurin

Anthraquinone (9,10-anthracenedione)
Formed from: Anthracene
Anthrone
9-Hydroxymethylanthracene
10-Methylene-9-anthrone
→ Benzoic acid
Probably in: Microorganism D452
→ Phthalic acid
In: Phanerochaete G418

Anthrarufin (1,5-dihydroxyanthraquinone)
→ Anthrarufin-β-D-glucoside
In: Cinchona K889

Anthrarufin-β-D-glucoside
Formed from: Anthrarufin

Anthrone (9(10*H*)-anthracenone)
Formed from: 9,10-Dihydroxyanthracene
→ Anthraquinone
In: Pseudomonas H254

Antibiotic A23187 (5-(methylamino)-2-(((2*R*,3*R*,6*S*,8*S*,9*R*,11*R*)-3,9,11-trimethyl-8-(1*S*)-1-methyl-2-oxo-2-(1*H*-pyrrol-2-yl)ethyl)-1,7-dioxaspiro[5.5]undec-2-yl)methyl)-4-benzoxazolecarboxylic acid)
Formed from glucose in S. chartreusis C347

Antipyrine (phenazone, 2,3-dimethyl-1-phenyl-3-pyrazolin-5-one)
→ 2,3-Dihydro-2,3-dihydroxyantipyrine
In: Microorganism A3169
→ 3,4-Dihydro-3,4-dihydroxyantipyrine
In: Guinea pig A1432
Man A1432
Rat A1432
→ 4-Hydroxyantipyrene
In: Beef K299
Guinea pig A1432
Man A64, A1432, A3796, B366, B394, C217, C236, C370, C415, C418, D632, D719, H700
Rabbit C418
Rat A685, A1432, A1705, A3645, A3850, B366, C236, C415, C418, F455, G546, J769

See also: B305
→ 3-Hydroxymethyl-2-methyl-1-phenyl-3-pyrazolin-5-one
In: Beef K299
Man A3796, B366, B394, C217, C236, C370, C415, D632, D719, H700
Rat A685, A1705, A3645, A3850, B366, C236, C415, F455, G546, J769
See also: A1432, B351
→ 1-(*o*-Hydroxyphenyl)-2,3-dimethyl-3-pyrazolin-5-one
Probably in: Sugar beet B351
→ 1-(*p*-Hydroxyphenyl)-2,3-dimethyl-3-pyrazolin-5-one
In: Man B911, C236
Rat B395, C236
Sugar beet B351
→ 3-Methyl-1-phenyl-3-pyrazolin-5-one
In: Beef K299
Man A761, A2862, A3796, B366, B394, C217, C370, C415, C418, C481, D632, H700
Rabbit C418
Rat A1365, A1705, A3645, A3850, B366, C415, C418, C481, F455, G546, J769

Antipyrylazo III (3,6-bis(2,3-dihydro-1,5-dimethyl-3-oxo-2-phenyl-1*H*-pyrazol-4-yl)azo)-4,5-dihydroxy-2,7-naphthalenedisulphonic acid)
→ Free radical
In: Rat C105

Antrafenine (2-(4-(3-trifluoromethylphenyl) piperazinyl)ethyl 2-(7-trifluoromethyl-4-quinolinylamino)benzoate)
→ 2-(7-Trifluoromethyl-4-quinolinyl) anthranilic acid
In: Dog B473
Man B473
Rat B473
→ 1-(*m*-Trifluoromethylphenyl)piperazine
In: Mouse D551
Rat D308, D551
→ 2-(4-(*m*-Trifluoromethylphenyl) piperazinyl)ethanol
In: Dog B473
Man B473
Rat B473

Aphelandrine ((3*S*,3a*S*,15*S*)-3,3a,6,7,8,9,10,11,12,13,14,15-dodecahydro-3-(4 hydroxyphenyl-4*H*-1,16-etheno-5,15-(propaniminoethano)furo[3,4-*l*][1,5,10] triazacyclohexadecine-4,21-dione)
→ 30-Hydroxyaphelandrine
In: Aphelandra H351

Aphelandrine-30,31-quinone
Formed from: 30-Hydroxyaphelandrine

Apiforol (dihydronaringenin, leucoapigeninidin)
Formed from: Naringenin

Apigenin (4′,5,7-trihydroxyflavone)
Formed from: Apigenin-7-O-(6-O-malonylglucoside)
Chrysin
Naringenin
Probably formed from: Apigenin-7-O-glucoside
Apigenin-7-O-glucoside
Apigenin-4′-sulphate
→ Acacetin
In: Robinia C100
See also: A3987
→ Apigenin-4′-O-glucoside
In: Allium G248
→ Apigenin-7-O-glucoside
In: Chrysanthemum J320
Parsley C721, K739
Petunia K303
Spirodela A1635
See also: A1030, K889
→ Apigenin-β-D-glucuronide
In: Man G968, J416, J783
→ Genkwanin
In: Oryza J190
→ *p*-Hydroxybenzoic acid
In: Rhizobium H280
→ Luteolin
In: Parsley C772
Rat J300
Sinningia E570
Zea D947
See also: A1635
→ Phenylacetic acid
In: Rhizobium H280
→ Phloroglucinol

In: Rhizobium H280
→ Vitexin
Probably in: Spirodela A1030

Apigenin-7-apioglucoside
See: Apiin

Apigenin-4′,7-disulphate
→ Apigenin-4′-sulphate
In: Helix F102
→ Apigenin-7-sulphate
In: Helix F102

Apigenin-4′-O-glucoside
Formed from: Apigenin

Apigenin-7-O-glucoside
Formed from: Apigenin
→ Apigenin
In: Man J821
Pig K309
Secale K512
→ Apigenin-7-O-(6-O-malonylglucoside)
In: Cicer D445
Parsley C721, D684
→ Apiin
In: Parsley K827

Apigenin-β-D-glucuronide
Formed from: Apigenin

Apigenin-7-O-(6-O-malonylglucoside)
Formed from: Apigenin
→ Apigenin
In: Parsley C721

Apigenin-4′-sulphate
Formed from: Apigenin-4′,7-disulphate
→ Apigenin
Probably in: Helix F102

Apigenin-7-sulphate
Formed from: Apigenin-4′,7-disulphate

Apiin (apigenin-7-apioglucoside)
Formed from: Apigenin-7-O-glucoside
→ Malonylapiin
In: Parsley D684

7-O-(β-D-Apiofuranosyl(1→2)-β-D-glucosyl) biochanin A
Formed from: Biochanin A-7-β-D-glucoside

7-O-(β-D-Apiofuranosyl(1→2)-β-D-glucosyl) formononetin
Formed from: Formononetin-7-β-D-glucoside

7-O-(β-D-Apiofuranosyl(1→2)-β-D-glucosyl) luteolin
Formed from: Luteolin-7-β-D-glucoside

7-O-(β-D-Apiofuranosyl(1→2)-β-D-glucosyl) naringenin
Formed from: Prunin

O-(β-D-Apiofuranosyl(1→2)-β-D-glucosyl)-*p*-nitrophenol
Formed from: *p*-Nitrophenyl-β-D-glucoside

O-(β-D-Apiofuranosyl(1→2)-β-D-glucosyl)-*p*-vinylphenol
Formed from: *p*-Vinylphenyl-β-D-glucoside

Apiracohl (2-(1-phthalazinyl)hydrazinecarboxylic acid ethyl ester)
→ Hydralazine
In: Rat A1148
→ 3-Oxotriazolophthalazine-4-β-D-glucuronide
In: Rat A1148

Apoatropine (α-methylenebenzeneacetic acid (3-endo)-8-methyl-8-azabicyclo[3.2.1]oct-3-yl ester)
Formed from: Hyoscyamine
→ 6-β-Hydroxyapoatropine
In: Hyoscyamus E350

Apocodeine ((6a*R*)-5,6,6a,7-tetrahydro-10-methoxy-6-methyl-4*H*-dibenzo[*de*,*g*]quinolin-11-ol)
Formed from: 10,11-Dimethoxyaporphine
→ Apomorphine
In: Rat A1088, A1673

Apogossypol ((+)-3,3′-dimethyl-5,5′-bis(1-methylethyl)-(2,2′-binaphthalene)-1,1′,6,6′,7,7′-hexol)
Formed from: Gossypol

Apohyoscine (α-methylenebenzeneacetic acid (1α,2β,4β,5α,7β)-9-methyl-3-oxa-9-azatricyclo[3.3.1.0^{2,4}]non-7-yl ester)
Formed from: l-Phenylalanine

Apomorphine ((6a*R*)-5,6,6a,7-tetrahydro-6-methyl-4*H*-dibenzo[*de*,*g*]quinolin-10,11-diol)
Formed from: Apocodeine
Diisobutyrylapomorphine
Dipivaloylapomorphine
Dipropionylapomorphine
→ Apomorphine-β-D-glucuronide

In: Rat K595
→ Apomorphine quinone
Probably in: Man A63
Rat A1088
→ O-Methylapomorphine
In: Rat A2597

Apomorphine-β-D-glucuronide
Formed from: Apomorphine

Apomorphine quinone
Probably formed from: Apomorphine

Aponeuron (N-(α-methylphenethyl)-2-phenylglycinonitrile)
→ Amphetamine
In: Man A528, A592

Aponorscopolamine (α-methylenebenzeneacetic acid (1α,2β,4β,5α,7β)-3-oxa-9-azatricyclo[3.3.1.0^{2,4}]non-7-yl ester)
Probably formed from: Aposcopolamine

Aposcopolamine (apohyoscyamine, α-methylenebenzeneacetic acid (1α,2β,4β,5α,7β)-9-methyl-3-oxa-9-azatricyclo[3.3.1.0^{2,4}]non-7-yl ester)
Formed from: L-Phenylalanine
Scopolamine
→ Aponorscopolamine
Probably in: Guinea pig G19
Mouse G19
Rat G19

Apovincaminic acid
Formed from: Ethyl apovincaminate
→ Dihydroxyapovincaminic acid
Probably in: Rat A2810

Aprindine (N-2,3-dihydro-1*H*-inden-2-yl)-N,N′-diethyl-N-phenyl-1,3-propanediamine)
→ Desethylaprindine
In: Dog A36
Man A36
Monkey A36
Pig A36
→ 3-Diethylaminopropylaniline
In: Dog A36
Man A36
Monkey A36
Pig A36
→ N-(3-Diethylaminopropyl)-2-aminoindane
In: Dog A36
Man A36
Monkey A36
Pig A36
→ Hydroxyaprindine
In: Dog A36
Man A36
Monkey A36
Pig A36

Aprophen (α-methyl-α-phenylbenzeneacetic acid 2-(diethylamino)ethyl ester)
→ 2,2-Diphenylpropionic acid
In: Rat G142
→ 2,2-Diphenylpropionic acid 2-(N-ethyl-N-(2-hydroxyethyl)aminoethyl ester
In: Rat G142

Aptocaine (α-methyl-N-(2-methylphenyl)-1-pyrrolidineacetamide)
→ *o*-Toluidine
In: Rabbit A2934

O-α-L-Arabinofuranosyl-(1→3)-O-(2-O-feruloyl-α-L-arabinofuranosyl)-(1→5)-L-arabinofuranose
→ Ferulic acid
In: Aspergillus H381

(α*R*)-α-((6-O-α-L-Arabinopyranosyl-β-D-glucopyranosyl)oxy)benzeneacetonitrile
See: Vicianin

Arborine (1-methyl-2-(phenylmethyl)-4(1*H*)-quinazolinone)
Formed from: Anthranilic acid
L-Phenylalanine
Phenylacetic acid

Arbutin (*p*-hydroxyphenyl-β-D-glucoside)
Formed from: Quinol
→ Quinol
In: Man J364
→ Quinol diglucoside
Probably in: Rauwolfia G726

Arctigenin ((3*R*,4*R*)-4-((3,4-dimethoxyphenyl)methyl)dihydro-3-((4-hydroxy-3-methoxyphenyl)methyl)-2(3*H*)-furanone)
Formed from: Arctiin
Coniferyl alcohol
O-Demethylarctigenin
Matairesinol
→ O-Demethylarctigenin
Probably in: Rat H48

Arctiin ((3*R*,4*R*)-4-((3,4-dimethoxyphenyl)methyl)-3-((4-(β-D-glucopyranosyloxy)-3-methoxyphenyl)methyl)dihydro-2(3*H*)-furanone)
→ Arctigenin
In: Rat H48

L-Arginyl-L-arginyl-β-naphthylamide
→ β-Naphthylamine
In: Man A2149

L-Arginyl-β-naphthylamide
→ β-Naphthylamine
In: Beef A526
Crithidia A1102
Glycine A1596
Man A1219, A1397, A2149
Pseudomonas G643
Rabbit A1589, B334

L-Arginyl-*p*-nitroanilide
→ *p*-Nitroaniline
In: Nocardia G343

Aricept
See: Donepezil

Aricine ((19α,20α)-16,17-didehydro-7-methoxy-19-methyloxayohimban-16-carboxylic acid methyl ester)
Formed from: Strictosidine

Arildone (4-(6-(2-chloro-4-methoxyphenylox-y)hexyl)-3,5-heptanedione)
→ 2-Chloro-4-methoxyphenol
In: Mouse C389

Aristolic acid I
→ 8-Hydroxy-3,4-methylenedioxy-1-phenanthrenecarboxylic acid
Probably in: Rat E784

Aristolochic acid I (8-methoxy-6-nitrophenanthro[3,4-*d*]-1,3-dioxole-5-carboxylic acid)
Formed from: L-Tyrosine
→ Aristolic acid I
In: Rat E784
→ Aristololactam I
In: Rat D311, E784
→ Aristolochic acid Ia
In: Rat E784
→ 8-Hydroxy-3,4-methylenedioxy-1-phenanthrenecarboxylic acid
Probably in: Rat E784

Aristolochic acid Ia (8-hydroxy-6-nitrophenanthro[3,4-*d*]-1,3-dioxole-5-carboxylic acid)
Formed from: Aristolochic acid I

Aristolochic acid II (6-nitrophenanthro[3,4-*d*]-1,3-dioxole-5-carboxylic acid)
→ Aristololactam II
In: Rat D311, E784
→ 3,4-Methylenedioxy-1-phenanthrenecarboxylic acid
In: Rat E784

Aristololactam I (8-methoxybenzo[*f*]-1,3-benzodioxolo[6,5,4-*cd*]indol-5(6*H*)-one)
Formed from: Aristolochic acid I
→ Aristololactam Ia
Probably in: Rat E784

Aristololactam Ia (8-hydroxybenzo[*f*]-1,3-benzodioxolo[6,5,4-*cd*]indol-5(6*H*)-one)
Probably formed from: Aristololactam I
Aristololactam II

Aristololactam II (benzo[*f*]-1,3-benzodioxolo[6,5,4-*cd*]indol-5(6*H*)-one)
Formed from: Aristolochic acid II
→ Aristololactam Ia
Probably in: Rat E784

Armepavine (1,2,3,4-tetrahydro-1-(*p*-hydroxybenzyl)-6,7-dimethoxy-2-methylisoquinoline)
→ *p*-Hydroxybenzyl alcohol
In: Nelumbo D566
→ 1,2,3,4-Tetrahydro-6,7-dimethoxy-2-methylisoquinoline
In: Nelumbo D566

Arogenate
See: Pretyrosine

Aromadendrin (dihydrokaempferol, 3-hydroxynaringenin, katuranin, 3,4′,5,7-tetrahydroxyflavanone)
Formed from: Naringenin
→ Aromadendrin-3-glucoside
In: Hippeastrum E636
→ Aromadendrin-7-glucoside
In: Matthiola A3890
→ Dihydroquercetin
In: Dianthus C392
Haplopappus A200
Matthiola B679

Petroselinum C772
→ Kaempferol
In: Dianthus G761
Petroselinum B957
Petunia D950
Tulipa F162
→ 3,4-*cis*-Leucopelargonidin
In: Cryotomeria E758
Dianthus G482
Larch J328
Matthiola D782
→ Pelargonidin
Probably in: Silene B404

Aromadendrin-3-glucoside
Formed from: Aromadendrin

Aromadendrin-7-glucoside
Formed from: Aromadendrin

Arphamenine A ((αR)-α-((3S)-3-amino-6-(aminoiminomethyl)amino)-2-oxohexyl) benzenepropanoic acid)
Probably formed from: Benzylsuccinic acid

Arsacetin
See: N-Acetyl-*p*-arsanilic acid

***p*-Arsanilic acid** (*p*-aminoazobenzenearsonic acid)
Formed from: 4-Dimethylaminoazobenzene-4′-arsonic acid
→ N-Acetyl-*p*-arsanilic acid
In: Guinea pig K16
Pig F877

Arsenazo III (3,6-bis(2-arsonophenyl)azo)-4,5-dihydroxy-2,7-naphthalenedisulphonic acid)
→ Free radical
In: Rat G105

Arteflene (4-((1*Z*)-2-(2,4-bis(trifluoromethyl) phenyl)ethenyl)-4,8-dimethyl-2,3-dioxabicyclo[3.3.1]nonan-7-one)
→ *cis*-4-(2,4-Bis(trifluoromethyl)phenyl) but-3-en-2-one
In: Rat K76
→ *cis*-2,4-Bis(trifluoromethyl)cinnamic acid
In: Rat K76
→ 8-Hydroxyarteflene
In: Rat K76

β-Artelinic acid (4-((((3*R*,5a*S*,6-*R*,8a*S*,9*R*,10*S*,12*R*,12a*R*)-decahydro-3,6,9-trimethyl-3,10-epoxy-12*H*-pyrano[4,3-*j*]-1,2-benzodioxepin-10-yl)oxy)methyl)benzoic acid)
→ *p*-Hydroxymethylbenzoic acid
Probably in: Man J725

Aspalatone (maltol acetylsalicylate)
→ Maltol salicylate
In: Rat H777

Asparaginyl-β-naphthylamide
→ β-Naphthylamine
In: Beef A526

Aspartame (L-aspartyl-L-phenylalanine methyl ester)
→ L-Phenylalanine
Probably in: Dog A2939
Man A2939, H223
Monkey A2939
Mouse A2939
Rabbit A2939
Rat A2939
Pig H223

L-Aspartyl-β-naphthylamide
→ β-Naphthylamine
In: Bacteria C146
Beef A526
Man A1219
Streptomyces C277

L-Aspartyl-L-phenylalanine
→ L-Phenylalanine
In: Brevibacterium E587

L-Aspartyl-L-phenylalanine methyl ester
See: Aspartame

Asperphenamate (N-benzoyl-L-phenylalanine (2*S*)-2-(benzoylamino)-3-phenylpropyl ester)
Probably formed from: N-Benzoylphenylalanine
N-Benzoylphenylalaninol

Aspirin
See: Acetylsalicylic acid

Aspulvinone C ((5*Z*)-3-(3,4-dihydro-2,2-dimethyl-2*H*-1-benzopyran-6-yl)-5-((3,4-dihydro-7-hydroxy-2,2-dimethyl-2*H*-1-benzopyran-6-yl)methylene)-4-hydroxy-2(5*H*)-furanone)
Formed from: Aspulvinone G

Aspulvinone D ((5*Z*)-3-(3,4-dihydro-7-hydroxy-2,2-dimethyl-2*H*-1-benzopyran-6-yl)-4-

hydroxy-5-((4-hydroxy-3-(3-methyl-2-butenyl)phenyl)methylene)-2(5*H*)-furanone)
Formed from: Aspulvinone G

Aspulvinone E ((5*Z*)-4-hydroxy-3-(4-hydroxyphenyl)-5-((4-hydroxyphenyl)methylene)-2(5*H*)-furanone)
→ Aspulvinone I
In: Aspergillus K829

Aspulvinone F ((5*Z*)-3-(3,4-dihydro-7-hydroxy-2,2-dimethyl-2*H*-1-benzopyran-6-yl)-5-((2,3-dihydro-2-(1-hydroxy-1-methylethyl)-5-benzofuranyl)methylene)-4-hydroxy-2(5*H*)-furanone)
Formed from: Aspulvinone G

Aspulvinone G ((5*Z*)-3-(2,4-dihydroxyphenyl)-4-hydroxy-5-((4-hydroxyphenyl)methylene)-2(5*H*)-furanone)
→ Aspulvinone C
In: Aspergillus K829
→ Aspulvinone D
In: Aspergillus K829
→ Aspulvinone F
In: Aspergillus K829

Aspulvinone H ((5*Z*)-4-hydroxy-3-(4-hydroxy-3-(3-methyl-2-butenyl)phenyl)-5-((4 hydroxy-3-(3-methyl-2-butenyl)phenyl)methylene)-2(5*H*)-furanone)
Formed from: Aspulvinone I

Aspulvinone I ((5*Z*)-4-hydroxy-5-((4-hydroxy-3-(3-methyl-2-butenyl)phenyl)methylene)-3-(4-hydroxyphenyl)-2(5*H*)-furanone)
Formed from: Aspulvinone E
→ Aspulvinone H
Probably in: Aspergillus K829

Astemizole (1-((4-fluorophenyl)methyl)-N-(1-(2-(4-methoxyphenyl)ethyl)-4-piperidinyl)-1*H*-benzimidazole-2-amine)
→ Demethylastemizole
In: Dog E425
Man E425
Rat E425
→ *p*-Fluorobenzoic acid
In: Man E425
→ 5-Hydroxyastemizole
In: Dog E425
Rat E425
→ 6-Hydroxyastemizole
In: Dog E425
Man E425
Rat E425
→ *p*-Methoxyphenylacetic acid
In: Dog E425
Rat E425
→ 2-(1-(2-(*p*-Methoxyphenyl)ethyl)piperidin-4-yl)aminobenzimidazole
In: Dog E425
Rat E425

Astringin (3-((1*E*)-2-(3,4-dihydroxyphenyl)ethenyl)-5-hydroxyphenyl-β-D-glucopyranoside)
→ 3,5-Dihydroxybenzaldehyde
In: Pseudomonas F370
→ Protocatechualdehyde
In: Pseudomonas F370

Asulam (methyl *p*-aminobenzenesulphonylcarbamate)
→ Methyl *p*-acetamidobenzenesulphonylcarbamate
In: Rat C752
→ Sulphanilamide
Probably in: Rat C752

Atenolol (1-(4-carbamoylmethylphenoxy)-3-isopropylaminopropan-2-ol)
→ Hydroxyatenolol
In; Dog A3052
Man A3052
Rat A3052

Ateviridine (1-(3-(ethylamino)-2-pyridinyl)-4-((5-methoxy-1*H*-indol-2-yl)carbonyl)piperazine)
→ 1-(3-Amino-2-pyridinyl)-4-((5-methoxy-1*H*-indol-2-yl)carbonyl)piperazine
In: Rat J449
→ 1-(1-(3-Ethylamino-4-hydroxy-2-pyridinyl)-4-((5-methoxy-1*H*-indol-2-yl)carbonyl)piperazine
Probably in: Rat J449
→ 1-(3-Ethylamino-2-pyridinyl)-4-((5-hydroxy-1*H*-indol-2-yl)carbonyl)piperazine
In: Rat J449
→ 1-(3-Ethylamino-2-pyridinyl)-4-((6-hydroxy-1*H*-indol-2-yl)carbonyl)piperazine
Probably in: Rat J449

→ 1-(3-Ethylamino-2-pyridinyl)-4-((7-hydroxy-1*H*-indol-2-yl)carbonyl) piperazine
Probably in: Rat J449

Atorvastatin ((β*R*,δ*R*)-2-(4-fluorophenyl)-β,δ-dihydroxy-5-(1-methylethyl)-3-phenyl-4-((phenylamino)carbonyl)-1*H*-pyrrole-1-heptanoic acid)
→ 2-(*p*-Fluorophenyl)-β,δ-dihydroxy-4-((*o*-hydroxyphenyl)amino)carbonyl-5-(1-methylethyl)-3-phenyl-1*H*-pyrrole-1-heptanoic acid
In: Dog J736
Man K329
Mouse J433
Rat J736
→ 2-(*p*-Fluorophenyl)-β,δ-dihydroxy-4-(((*p*-hydroxyphenyl)amino)carbonyl-5-(1-methylethyl)-3-phenyl-1*H*-pyrrole-1-heptanoic acid
In: Dog J736
Man K329
Mouse J433
Rat J736
→ 2-(*p*-Fluorophenyl)-β,δ-dihydroxy-5-(1-methylethenyl)-3-phenyl-4-((phenylamino)carbonyl-1*H*-pyrrole-1-heptanoic acid
In: Mouse J433
→ 2-(*p*-Fluorophenyl-5-(1-methylethyl)-3-phenyl-4-((phenylamino)carbonyl)-1*H*-pyrrole-1-pentanoic acid
In: Mouse J433

Atorvastatin lactone
→ *o*-Hydroxyatorvastatin lactone
In: Man K329
→ *p*-Hydroxyatorvastatin lactone
In: Man K329

Atracurium (2,2′-(1,5-pentanediylbis(oxy(3-oxo-3,1-propanediyl)))bis(1-(3,4-dimethoxyphenyl)methyl)-1,2,3,4-tetrahydro-6,7-dimethoxy-2-methylisoquinolinium)
→ N-(2-Carboxyethyl)laudanosinium
In: Cat B976
→ 5-Hydroxypentyl N-(2-carboxyethyl) laudanosinium ester
In: Cat B976

Atropic acid (α-methylenebenzeneacetic acid)
Formed from: 2,2-Diphenylethylene

Atropine
See: Hyoscyamine

Auramycin A
Formed from: Auramycin B
Auramycinone

Auramycin B
Formed from: Auramycinone
→ Auramycin A
In: Streptomyces D17

Auramycinone ((1*R*,2*R*,4*S*)-1,2,3,4,6,11-hexahydro-2,4,5,7-tetrahydroxy-2-methyl-6,11-dioxo-1-naphthacenecarboxylic acid methyl ester)
→ Auramycin A
In: Streptomyces D170
→ Auramycin B
In: Streptomyces D170
→ 7-Deoxyauramycinone
In: Streptomyces D170
→ 11-Hydroxyauramycinone
In: Streptomyces D170
→ 10-Hydroxy-9-methyldaunomycin
In: Streptomyces D170

Auranthine ((−)-6,7,7a,8-tetrahydroquinazolino[3′,2′:1,6]pyrido[2,3-*b*][1,4]benzodiazepin-9,16-dione)
Formed from: Anthranilic acid

Aureosidin (3′,4,4′,6-tetrahydroaurone)
Formed from: 2′,3,4,4′,6′-Pentahydroxychalcone
2′,4,4′,6′-Tetrahydroxychalcone

Aureothamin (4-((2*E*)-3-(4-aminophenyl)-2-methyl-2-propenylidene)-2-methoxy-3,5-dimethyl-6-((2*R*,4*Z*)-tetrahydro-2-furanyl)-4*H*-pyran-4-one)
→ Aureothin
In: Microorganism A1236

Aureothin (aureotine, 2-methoxy-3,5-dimethyl-6-((2*R*,4*Z*)-tetrahydro-4-((2*E*)-2-methyl-3-(4-nitrophenyl)-2-propenylidene)-2-furanyl)-4*H*-pyran-4-one)
Formed from: *p*-Amino-L-phenylalanine
erythro-*p*-Amino-L-phenylserine
threo-*p*-Amino-L-phenylserine

Aureothamin

Aureotine

See: Aureothin

(1*S*,3*S*)-Austrocortilutein ((1*S*,3*S*)-1,2,3,4-tetrahydro-1,3,8-trihydroxy-6-methoxy-3-methyl-9,10-anthracenedione)

Formed from: (*S*)-Torosachrysone gentiobioside

(1*S*,3*S*)-Austrocortirubin ((5*S*,7*S*)-5,6,7,8-tetrahydro-5,7,9,10-tetrahydroxy-2-methoxy-7-methyl-1,4-anthracenedione)

Formed from: (*S*)-Torosachrysone gentiobioside

Autumnaline ((1*S*)-1,2,3,4-tetrahydro-1-(2-(3-hydroxy-4,5-dimethoxyphenyl)ethyl)-6-methoxy-2-methyl-7-isoquinolinol)

→ Colchicine

In: Colchicum J458

Averufin ((2*S*,6*S*)-3,4,5,6-tetrahydro-7,9,11-trihydroxy-2-methyl-2,6-epoxy-2*H*-anthra[2,3-*b*]oxocin-8,13-dione)

→ Versiconal acetate

In: Aspergillus D57

→ Versicolorin B

Probably in: Aspergillus F260

8-Azabicyclo[3.2.1]octan-3-yl 3,5-dichlorobenzoate

Formed from: 8-Methyl-8-azabicyclo[3.2.1] octan-3-yl 3,5-dichlorobenzoate

1-(3-Azabicyclo[3.3.0]oct-3-yl)-3-(4-carboxybenzenesulphonyl)urea

Probably formed from: 1-(3-Azabicyclo[3.3.0]oct-3-yl)-3-(4-hydroxymethylbenzenesulphonyl)urea

1-(3-Azabicyclo[3.3.0]oct-3-yl)-3-(4-hydroxymethylbenzenesulphonyl)urea

→ 1-(3-Azabicyclo[3.3.0]oct-3-yl)-3-(4-carboxybenzenesulphonyl)urea

Probably in: Man D169, H488

Azacyclonol

See: Diphenyl 4-piperidinylmethanol

Azaperol (α-(4-fluorophenyl)-4-(4-(2-pyridinyl)-1-piperazinebutanol)

Formed from: Azaperone

Azaperone (α-(4-fluorophenyl)-4-(4-(2-pyridinyl)-1-piperazinyl)-1-butanone)

→ Azaperol

In: Pig A2468

→ 5′-Hydroxyazaperone

In: Horse H754

Azapetine (6,7-dihydro-6-(2-propenyl)-5*H*-dibenz[*c*,*e*]azepine)

→ Dehydrozapetine

In: Rabbit A3909

1-Azaphenanthrene-9,10-oxide

→ S-(9,10-Dihydro-9-hydroxy-1-azaphenanthr-10-yl)glutathione

In: Rat C472

See also: D235

→ S-(9,10-Dihydro-10-hydroxy-1-azaphenanthr-9-yl)glutathione

In: Rat C472

4-Azaphenanthrene-9,10-oxide

→ S-(9,10-Dihydro-9-hydroxy-4-azaphenanthr-10-yl)glutathione

In: Rat C472

See also: D235

→ S-(9,10-Dihydro-10-hydroxy-4-azaphenanthr-10-yl)glutathione

In: Rat C472

Azapropazone (5-(dimethylamino)-9-methyl-2-propyl-1*H*-pyrazolo[1,2-*a*][1,2,4]benzotriazine-1,3(2*H*)-dione)

→ 6-Hydroxyazapropazone

In: Man A439

→ 8-Hydroxyazapropazone

In: Man D692

Azelastine (4-((4-chlorophenyl)methyl)-2-(hexahydro-N-methyl-1*H*-azepin-4-yl)-1(2*H*)-phthalazinone)

→ 2-(1-Carboxy-5-methylamino-2-pentyl)-4-(*p*-chlorobenzyl)-1(2*H*)-phthalazinone

In: Guinea pig G387

→ 3-(1-Carboxy-5-methylamino-2-pentyl)-4-(*p*-chlorobenzyl)-1(2*H*)-phthalazinone

In: Guinea pig G387

→ 4-(*p*-Chlorobenzyl)-6-hydroxy-2-(N-methylperhydroazepinyl-(4))-1(2*H*)-phthalazinone

In: Guinea pig D765

Rat D765

→ 4-(*p*-Chlorobenzyl)-7-hydroxy-2-(N-methylperhydroazepinyl-(4))-1(2*H*)-phthalazinone

In: Guinea pig D765

Rat D765
→ 4-(*p*-Chlorobenzyl)-2-(N-methyl-7-oxoperhydroazepinyl-(4))-1(2*H*)-phthalazinone
In: Guinea pig D765
Rat D765
→ Demethylazelastine
In: Guinea pig D765, G387
Man J737, J787
Rat D765

α-Azidopenicilloic acid
Formed from: Azidocillin

Azidocillin (2*S*,5*R*,6*R*)-6-(((2*R*)-azidophenylacetyl)amino)-3,3-dimethyl-7-oxo-4-thia-1-azabicyclo[3.2.1]heptane-2-carboxylic acid)
→ α-Azidopenicilloic acid
In: Dog A695
Rat A695

2-Azido-1-hydroxyindane
Formed from: 2-Azidoindane

Azidoindane
→ 2-Azido-1-hydroxyindane
In: Pseudomonas K207

17α-Azidomethyloestradiol
→ 17α-Hydroxymethyloestradiol
In: Streptomyces A2225

Azidomorphine
→ 6β-Amino-6-deoxydihydromorphine
In: Man A3050

***m*-Azidopyrimethamine** (5-(4-chloro-3-azidophenyl)-6-ethyl-2,4-pyrimidinediamine)
→ *m*-Aminopyrimethamine
In: Mouse F319

2,2′-Azinodi-(3-ethylbenzothiazoline-6-sulphonic acid)
→ 2,2′-Azinodi-(3-ethylbenzothiazoline-6-sulphonic acid) cation radical
In: Coriolus K677
→ 3-(3-Ethyl-6-sulphobenzothiazol-2-ylidenehydrazono)-4-oxo-1,5-cyclpoexadiene-1-carboxylic acid
In: Polyporus E134
→ 6-(3-Ethyl-6-sulphobenzothiazol-2-ylidenehydrazono)-3-oxo-1,4-cyclohexadiene-1-carboxylic acid
In: Polyporus E134

2,2′-Azinodi-(3-ethylbenzothiazoline-6-sulphonic acid) cation radical
Formed from: 2,2′-Azinodi-(3-ethylbenzothiazoline-6-sulphonic acid)

Azinphos-methyl (O,O-dimethyl S-((4-oxo-1,2,3-benzotriazin-3(4*H*)-yl)methyl) phosphorodithioate, guthion)
→ Benzazimide
In: Rat B283
→ O-Methyl S-(4-oxo-1,2,3-benzotriazin-3(4*H*)-yl)methyl) phosphorodithioate
In: Housefly A615

5-(Aziridinyl)-2,4-dinitrobenzamide
→ 4-Amino-5-(1-aziridinyl)-2-nitrobenzamide
In: Rat A1684
→ 5-Amino-2,4-dinitrobenzamide
In: Rat A1684
→ 2-(1-Aziridinyl)-5-nitroaniline
In: Rat A2503
→ N-(2-Chloroethyl)-2,4-dinitroaniline
In: Rat A2503
→ 2,4-Dinitroaniline
In: Rat A2503
→ N-(2-Hydroxyethyl)-2,4-dinitroaniline
In: Rat A2503
→ 5-(2-Hydroxyethyl)-2,4-dinitrobenzamide
In: Rat A1684

2-(1-Aziridinyl)-5-nitroacetanilide
Formed from: 2-(1-Aziridinyl)-5-nitroaniline

2-(1-Aziridinyl)-5-nitroaniline
Formed from: 1-(1-Aziridinyl)-2,4-dinitrobenzene
→ 2-(2-Chloroethyl)amino-5-nitroaniline
Probably in: Rat A2503
→ 2-(1-Aziridinyl)-5-nitroacetanilide
In: Rat A2503

Azobenzene (diphenyldiazene)
Formed from: Aniline
Azoxybenzene
→ Aniline
In: Ascaris A304, A324
Microorganism D887
Moniezia A324, A1785
→ Azoxybenzene
In: Guinea pig G258, G259

Hamster G258, G259
Mouse G258, G259
Rabbit G258, G259
Rat G258, G259
See also: H12
→ Hydrazobenzene
In: Rat D887

Azobenzenyl nitroxide radical
Formed from: 4-Aminoazobenzene

***p*-Azobenzonitrile**
Formed from: *p*-Nitrobenzonitrile

Azodisalicylic acid (3,3′-azobis(6-hydroxybenzoic acid))
→ 5-Amino-2-hydroxybenzoic acid
In: Man D570

4,4′-Azobis(biphenyl)
Formed from: 4-Aminobiphenyl

2,2′-Azobisfluorene
See: Azofluorene

3,3′-Azobis(6-hydroxybenzoic acid
See: Azodisalicylic acid

Azodiphenylmethane
Formed from: 4,4′-Diaminodiphenylmethane
→ Azoxydiphenylmethane
Probably in: Rat G777

Azofluorene (2,2′-azobisfluorene)
Formed from: 2-Aminofluorene

Azonafide (2-(2-(dimethylamino)ethyl)-1,2-dihydro-3*H*-dibenz[*de*,*h*]isoquinoline-1,3-dione)
→ Azonafide-N-oxide
In: Rat H935
→ 2-Carboxmethyl-1,2-dihydro-3*H*-dibenz[*de*,*h*]isoquinoline-1,3-dione
In: Rat H935
→ Demethylazonafide
In: Rat H935

Azonafide-N-oxide
Formed from: Azonafide

Azo-procarbazine (N-isopropyl-α-(2-methylazo)-*p*-toluamide)
→ *p*-Formyl-N-isopropylbenzamide
In: Rat C35
→ *p*-Hydroxymethyl-N-isopropylbenzamide
In: Rat C35
→ N-Isopropyl-α-(2-methyl-NNO-azoxy)-*p*-toluamide
In: Rat C35
→ N-Isopropyl-α-(2-methyl-ONN-azoxy)-*p*-toluamide
In: Rat C35

ω,ω′-Azotoluene (bis(phenylmethyl)diazene)
→ ω,ω′ -Azoxytoluene
In: Guinea pig G258, G259
Hamster G258, G259
Mouse G258, G259
Rabbit G258, G259
Rat G258, G259

Azoxybenzene (diphenyldiazene-1-oxide)
Formed from: Azobenzene
→ Azobenzene
In: Saccharomyces H894
→ Nitrobenzene
Probably in: Pseudomonas H12

***p*-Azoxybenzonitrile**
Formed from: *p*-Nitrobenzonitrile

Azoxybischlorobenzene
Formed from: *p*-Chlorophenylhydroxylamine

Azoxydiphenylmethane
Formed from: Azodiphenylmethane

***o,o′*-Azoxytoluene**
Formed from: *o*-Toluidine

ω,ω′-Azozytoluene
Formed from: ω,ω′-Azotoluene

B

Baccatin III ((2a*R*,4*S*,4a*S*,6*R*,9*S*,11*S*,12-*S*,12a*R*,12b*S*)-6,12b-bis(acetyloxy)-12-(benzoyloxy)-1,2a,3,4,4a,6,9,10,11,12,12a,12b-dodecahydro-4,9,11-trihydroxy-4a,8,13,13-tetramethyl-7,11-methano-5*H*-cyclodeca[3,4]benz[1,2-*b*]oxet-5-one)
  Formed from: 10-Deacetylbaccatin III
    Taxol
  → Deacetylbaccatin III
    Probably in: Eucalyptus K597

Baclofen
  See: γ-Amino-β-(*p*-chlorophenyl)butyric acid

Baicalein (5,6,7-trihydroxyflavone)
  Formed from: Baicalin
  → Baicalin
    In: Scutellaria K402
  → Baicalein-β-D-glucoside
    In: Nicotiana K556
  → Baicalein-O-sulphate
    In: Eubacterium H354

Baicalein-β-D-glucoside
  Formed from: Baicalein

Baicalein-7-β-D-glucuronide
  See: Baicalin

Baicalein-O-sulphate
  Formed from: Baicalein

Baicalin (baicalein-7-β-D-glucuronide)
  → Baicalein
    In: Microorganism J525
    Scutellaria H747
  → Baicalin-O-sulphate
    In: Eubacterium H354

Baicalin-O-sulphate
  Formed from: Baicalin

Bakuchiol (4-((1*E*,3*S*)-3-ethenyl-3,7-dimethyl-1,6-octadienyl)phenol)
  Formed from: *p*-Coumaric acid

Bambuterol (dimethylcarbamic acid 5-(2-(1,1-dimethylethyl)amino)-1-hydroxyethyl)-1,3-phenylene ester)
  → N-*tert*-Butyl-3-hydroxy-5-dimethylcarbamoyloxyphenylethanolamine
    In: Dog F128
    Guinea pig F128
    Man F128
    Mouse F128
    Rabbit F128
    Rat F121, F128
  → Demethylbambuterol
    In: Rat F121

Bamethan
  See: N-Butyl-3,4-dihydroxyphenethylamine

Bamipine (1-methyl-N-phenyl-N-(phenylmethyl)-4-piperidinamine)
  → 4-Amino-N-benzyl-N-(*p*-hydroxyphenyl)-1-methylpiperidine
    In: Rat E310

Banamite (benzoyl chloride 2,4,6-trichlorophenylhydrazone)
  → Benzaldehyde 2,4,6-trichlorophenylhydrazone
    In: Spider mite A1407
  → Benzaldoxime
    In: Spider mite A1407
  → Benzoic acid
    In: Spider mite A1407
  → 2,4,6-Trichloroaniline
    In: Spider mite A1407
  → 2,4,6-Trichlorophenylhydrazine
    In: Spider mite A1407

Barbaloin ((10*S*)-10-β-D-glucopyranosyl-1,8-dihydroxy-3-(hydroxymethyl)-9(10*H*)-anthracenone)
  → Aloe-emodin bianthrone

In: Microorganism F44
→ 1,8-Dihydroxy-3-hydroxymethylanthrone
In: Microorganism F44
Eubacterium H68

Barnidipine ((4*S*)-1,4-dihydro-2,6-dimethyl-4-(3-nitrophenyl)-3,5-pyridinedicarboxylic acid methyl (3*S*)-1-(phenylmethyl)-3-pyrrolidinyl ester)
→ (1-Benzyl-3-pyrrolidinyl) methyl 2,6-dimethyl-4-(*m*-nitrophenyl)pyridine-3,5-dicarboxylate
In: Man J84, J257
→ (4*S*)-1,4-Dihydro-2,6-dimethyl-4-(*m*-nitrophenyl)pyridine-3,5-dicarboxylic acid
In: Dog H518
Man J84
Rat H518
→ Methyl 3-carboxy-1,4-dihydro-2,6-dimethyl-4-(*m*-nitrophenyl)pyridine-5-carboxylate
In: Man J84, J257
→ (+)-(3′*S*,4*S*)-Methyl 3-pyrrolidinyl 1,4-dihydro-2,6-dimethyl-4-(*m*-nitrophenyl) pyridine-3,5-dicarboxylate
In: Dog H518
Man J84, J257
Rat H518

Barnon (flamprop-isopropyl, isopropyl N-benzoyl-N-(3-chloro-4-fluorophenyl)-2-aminopropionate)
→ N-Benzoyl-N-(3-chloro-4-fluorophenyl)-2-aminopropionic acid
In: Dog A2035, A3057
Rat A2035, A3057
→ 1-Hydroxyisopropyl N-benzoyl-N-(3-chloro-4-fluorophenyl)-2-aminopropionate
In: Dog A2035, A3057
Rat A2035, A3057
→ Isopropyl N-(3-chloro-4-fluorophenyl)-N-(*m*-hydroxybenzoyl)-2-aminopropionate
In: Dog A2035
Rat A2035
→ Isopropyl N-(3-chloro-4-fluorophenyl)-N-(*p*-hydroxybenzoyl)-2-aminopropionate
In: Dog A2035, A3057
Rat A2035, A3057

Batatasin I (3,4-dimethoxy-6-(2-phenylethyl)-1,2-benzenediol)
Formed from: *p*-Coumaric acid
Resveratrol

Batatasin III (3-(2-(3-hydroxyphenyl)ethyl)-5-methoxyphenol)
Formed from: *m*-Hydroxyphenylpropionic acid
3,3′,5-Trihydroxybibenzyl

Batracyclin (8-aminoisoindolo[1,2-*b*]quinazolin-12(10*H*)-one)
→ N-Acetylbatracyclin
In: Man J759

Baycaron
See: Mefruside

Baygon
See: Isopropoxyphenyl N-methylcarbamate

(*R*,*R*)-Bebeerine (2,3,13a,14,15,16,25,25a-octahydro-18,29-dimethoxy-1,14-dimethyl-13*H*-4,6:21,24-dietheno-8,12-metheno-1*H*-pyrido[3′,2′:14,15][1,11]dioxacycloeicosino[2,3,4-*ij*]isoquinoline-9,19-diol)
Formed from: (*R*)-N-Methylcoclaurine

Beclamide (N-benzyl-3-chloropropionamide)
→ Benzoic acid
In: Man A3507, H776
→ N-Benzyl-3-chloro-2-hydroxypropionamide
In: Man A3507
→ 3-Chloro-N-(*m*-hydroxybenzyl) propionamide
In: Man H776
→ 3-Chloro-N-(*p*-hydroxybenzyl) propionamide
In: Man A3507, H776

Befunolol (2-acetyl-7-(2-hydroxy-3-isopropylamino)propoxybenzofuran)
→ 4-Hydroxybefunolol
In: Guinea pig B675
Hamster B675
Mouse B675
Rabbit B675
Rat B675
→ 2-(1-Hydroxyethyl)-7-(2-hydroxy-3-isopropylaminopropoxy)benzofuran
In: Dog A2203, B374
Guinea pig B374, B675
Hamster B675

Man B374, B675
Monkey B374
Mouse B374, B675
Rabbit B374, B675
Rat B374
In: Erythrocyte A2189

Bemitradine (8-(2-ethoxyethyl)-7-phenyl [1,2,4]triazolo[1,5-*c*]pyrimidin-5-amine)
→ Desethylbemitradine
In: Man F314
→ *p*-Hydroxybemitradine
In: Man F314

Benadryl
See: Diphenhydramine

Benazepril ((3*S*)-3-(((1*S*)-1-(ethoxycarbonyl)-3-phenylpropyl)amino)-2,3,4,5-tetrahydro-2-oxo-1*H*-1-benzazepine-1-acetic acid)
→ Benazeprilat
In: Dog J262
Man F904

Benazeprilat ((3*S*)-3-(((1*S*)-1-carboxy-3-phenylpropyl)amino)-2,3,4,5-tetrahydro-2-oxo-1*H*-1-benzazepine-1-acetic acid)
Formed from: Benazepril

Bencyclane (1-benzyl-1-(3-dimethylaminopropoxy)cycloheptane)
→ 1-Benzyl-1-(3-dimethylaminopropoxy)-4-hydroxycycloheptane
In: Man A3506
Rat A3506

Bendiocarb (2,2-dimethyl-1,3-benzodioxol-4-ol methylcarbamate)
→ 2,2-Dimethylbenzo-1,3-dioxolo-4-ol
In: Corn rootworm E490
Man C176
Rat C176
→ 2,2-Dimethylbenzo-1,3-dioxol-4-ol N-hydroxymethylcarbamate
In: Man C176
Rat C176

Benfluron (5-(2-(dimethylamino)ethoxy)-7*H*-benzo[*c*]fluoren-7-one)
→ Benfluron-N-oxide
In: Rat D204, K387
See also: E312
→ Dihydrobenfluron-N-oxide
Species unclear E312
→ 5-(2-(Dimethylamino)ethoxy)-7-hydroxy-7*H*-benzo[*c*]fluorene
In: Rat D204, K387
→ 9-Hydroxybenfluron
In: Rat K387
→ 5-(2-(Methylamino)ethoxy)-7-hydroxy-7*H*-benzo[*c*]fluoren-7-one
In: Rat K387

Benfluron-N-oxide
Formed from: Benfluron

Benfuracarb (2,3-dihydro-2,2-dimethyl-7-benzofuranyl N-(N-2-(ethoxycarbonyl)ethyl)-N-(isopropylsulphenamoyl)-N-methylcarbamate)
→ Carbofuran
In: Bean D806
Corn D806
Cotton D806
→ 2,3-Dihydro-7-hydroxy-2,2-dimethylbenzofuran
In: Bean D806
Corn D806
Cotton D806
→ 3-Hydroxybenfuracarb
In: Bean D806
Corn D806
Cotton D806

Benidipine ((4R)-1,4-dihydro-2,6-dimethyl-4-(*m*-nitrophenyl)pyridine-3,5-dicarboxylic acid 3-(1-benzyl-3-piperidinyl) ester 5-methyl ester)
→ Debenzylbenidipine
In: Dog F11
Man J105
Rat F11, H927
→ Dehydrobenidipine
In: Man J105
Rat H927
→ 1,4-Dihydro-2,6-dimethyl-4-(*m*-nitrophenyl)pyridine-3,5-dicarboxylic acid 3-methyl ester
In: Dog F11
Man J105
Rat F11, H927

Benodanil
See: 2-Iodobenzanilide

Benomyl
See: 1-Butylcarbamoyl)benzimidazol-2-yl carbamate

Benorylate (4-acetamidophenyl 2-acetoxybenzoate)
→ Acetylsalicylic acid
In: Microorganism A2811
→ *p*-Acetamidophenol
In: Man A743
Microorganism A2811
Rabbit A2355
→ Phenetsal
In: Man F517
→ Salicylic acid
In: Man A743
Rabbit A2355
Rat A2349

(*R*)-Benoxaprofen (2-(4-chlorophenyl)-α-methyl-5-benzoxazoleacetic acid; this isomer is assumed if not stated)
→ (*S*)-Benoxaprofen
In: Man A3919, B769
Rat B769
→ (*R*)-Benoxaprofen-β-D-glucuronic acid
In: Man E886, J613
Monkey E886
Rabbit E886
Rat J377, J788
→ (*R*)-2-(*p*-Chlorophenyl)-α-methyl-5-benzoxazoleacetyltaurine
In: Rat J377

(*S*)-Benoxaprofen
→ (*S*)-Benoxaprofen-β-D-glucuronic acid
In: Man E886
Monkey E886
Rabbit E886
See also: A3037
→ (*S*)-2-(*p*-Chlorophenyl)-α-methyl-5-benzoxazoleacetyltaurine
In: Rat J377

(*R*)-Benoxaprofen-β-D-glucuronide
Formed from: (*R*)-Benoxaprofen

(*S*)-Benoxaprofen-β-D-glucuronide
Formed from: (*S*)-Benoxaprofen

Benoxinate (2-diethylaminoethyl 4-amino-3-butoxybenzoate)
→ 4-Amino-3-butoxybenzoic acid
In: Man E475

Benserazide (1-seryl-2-(2,3,4-trihydroxybenzyl) hydrazine)
→ 2-(2,4-dihydroxy-3-methoxybenzyl)-1-serylhydrazine
Probably in: Pig B117, B813

Bentazon (3-isopropyl-1*H*-2,1,3-benzothiadiazin-4(3*H*)-one-2,2-dioxide)
→ Anthranilic acid
In: Crab A1042

Benz[*j*]aceanthrylene
→ Benz[*j*]aceanthrylen-2-one
In: Man J688
Rat J688
→ *trans*-1,2-Dihydro-1,2-dihydroxybenz[*j*] aceanthrylene
In: Man J688
Mouse E455
Rat J688, J804
→ 9,10-Dihydro-9,10-dihydroxybenz[*j*] aceanthrylene
In: Mouse E455
Rat J688
→ 11,12-Dihydro-11,12-dihydroxybenz[*j*]aceanthrylene
In: Mouse E455
→ 10-Hydrozybenz[*j*]aceanthrylene
In: Mouse E455
Rat J688
→ 8-Hydroxybenz[*j*]aceanthrylen-1-one
In: Rat J688

Benz[*l*]aceanthrylene
→ 1,2-Dihydro-1,2-dihydroxybenz[*l*] aceanthrylene
In: Mouse E455
→ 4,5-Dihydro-4,5-dihydroxybenz[*l*] aceanthrylene
In: Mouse E455
→ 7,8-Dihydro-7,8-dihydroxybenz[*l*] aceanthrylene
In: Mouse E455

Benz[*j*]aceanthrylen-2-one
Formed from: Benz[*j*]aceanthrylene

Benz[*c*]acridine-7-carboxylic acid
Probably formed from: 7-Hydroxymethylbenz[*c*]acridine

Benzalacetone (benzylideneacetone, 1-phenyl-1-buten-3-one)
Formed from: *p*-Coumaroyl CoA
→ 4-(*p*-Hydroxyphenyl)butan-2-one
In: Raspberry J26

Benzaldehyde
Formed from: Benzoic acid
Benzoin
Benzoyladenylate
N-Benzylalanine
Benzyl alcohol
Benzylamine
1-Benzylaminobenzo-1*H*-triazole
N-Benzylaniline
p-Benzylaminobenzonitrile
N-Benzylcarbazole
N-Benzyl-N,α-dimethylphenethylamine
N-Benzyl-N,1-dimethyl-2-propynylamine
N-Benzyl-*p*-chloroalanine
N-Benzyl-*p*-nitroaniline
N-Benzyl-*p*-toluidine
N-Benzyl-2,4,6-tribromoaniline
N-Benzyl-2,4,6-trichloroaniline
N-Benzyl-2,4,6-trifluoroaniline
N-Benzyl-2,4,6-trimethylaniline
N-(2-Hydroxy-2-phenylethyl)-N-methyl-N-nitrosobenzylamine
Mandelonitrile
N-Methylbenzylamine
N-Methyl-N-nitrosobenzylamine
Styrene
Probably formed from: Benzydamine
N-Benzyl-*p*-chloro-N-methylbenzamide
Dibenzylamine
Dibenzylaniline
N,N-Dibenzylbenzamide
Donepezil
(+)-Etomidate
(−)-Etomidate
Pargyline
trans-4-Phenyl-3-buten-2-one
Phenylglyoxylic acid
Tribenzylamine
Tripelennamine
→ N-Acetyl-S-benzyl-L-cysteine
In: Rat E513
→ 2-Amino-4-hydroxy-4-phenylbutyric acid
In: Pseudomonas D449
→ Benzalsuccinic acid
Probably in: Azoarcus K238
→ Benzoic acid
In: Acinetobacter D487, F366
Arion G804
Arthrobacter J33
Euglena G422
Flavobacterium E355
Heliothis F962
Helix G804
Hyphomicrobium C123
Limax G804
Lithospermum G229
Man J24
Manduca F962
Methylosinus F404
Microorganism K642
Mouse G804
Nicotiana J819
Pleurotus H241
Pseudomonas A30, D211, J707, K493
Rabbit B457, E674
Rat B191, B457
Rhodopseudomonas G246
Saccharomyces E336
Serratia G197
Spodoptera F962
Streptomyces C284
See also: A1726, A2744, A3219, A3389, B927, F598
→ Benzoin
In: Pseudomonas K424
→ Benzyl alcohol
In: Cabbage F931
Candida J192
Chlorella K176
Cranberry A3545
Horse A2174
Dunaliella G459, K176
Euglena G422
Isochrysis K176
Lactobacillus G148
Man F87
Microorganism K642

Mung bean J207
Mycobacterium A150
Nannochloris K176
Nitrosomonas H219
Phanerochaete F866
Pleurotus H241
Porphyridium K176
Rabbit B386, B457, K569
Rat B457, J744, H759
Saccharomyces E336, H425
Yeast H98
See also: A2744, D487, E674
→ *l*-Ephedrine
In: Ephedra A168
→ 1-Hydroxy-1-phenyl-2-propanone
In: Candida H771, J192
Saccharomyces C367, E333, E336, J192
Yeast H98
Zygosaccharomyces K611
→ Mandelonitrile ((*R*) and (*S*))
In: Almond G325, J194, J302, K6, K320
Apple J302
Apricot J302
Peach K6
Prunus E353, H781, J302
Sorghum H208, J194
→ 1-Oxo-1-phenyl-2-propanol
In: Pseudomonas K524
See also: G607
→ 2-Oxo-1-phenyl-1-propanol
In: Saccharomyces G437
→ D-Phenylserine
In: Arthrobacter G140
→ *erythro*-1-Phenylpropane-1,2-diol
Probably in: Saccharomyces E487

Benzaldehyde 2,4,6-trichlorophenylhydrazine
Formed from: Banamite

Benzaldoxime
Formed from: Banamite
→ Benzyl-β-D-sulphatoglucose
In: Brassica F657

Benzalphthalide
→ Benzoic acid
In: Pseudomonas H111
→ Phthalic acid
In: Pseudomonas H111

Benzalsuccinic acid
Probably formed from: Benzaldehyde

Benzamide
Formed from: Benzoic acid
Thiobenzamide sulphoxide
→ Benzoic acid
In: Ascaris A2037
Aspergillus A640
Moniezia A2037
Mouse A2037
Nocardia G343
Pseudomonas H435
Rhodococcus J857
Sheep A2037

Benzamidine (benzenecarboximidamide)
Formed from: Benzamidoxime
N-Methylbenzamidine
→ Benzamidoxime
In: Rabbit C573, E238, E830, F230, F768, G798, G990
Rat G990

Benzamidocatechol
→ 4-Benzamido-2-methoxyphenol
With catechol O-methyltransferase A597
→ 5-Benzamido-2-methoxyphenol
With catechol O-methyltransferase A597

2-Benzamidofluorene
→ 2-Aminofluorene
In: Rat A808
→ Benzoic acid
In: Rat A808

4-Benzamido-1-(4-(5-hydroxyindol-3-yl)-4-oxobutyl)piperidine
Formed from: 4-Benzamido-1-(4-(indol-3-yl)-4-oxobutyl)piperidine

4-Benzamido-1-(4-(indol-3-yl)-4-oxobutyl) piperidine
→ 4-Benzamido-1-(4-(5-hydroxyindol-3-yl)-4-oxobutyl)piperidine
In: Monkey B550
Rat B550
→ 4-Benzamidopiperidine
In: Monkey B550
Rat B550

4-Benzamido-2-methoxyphenol
Formed from: 4-Benzamidocatechol

5-Benzamido-2-methoxyphenol
Formed from: 4-Benzamidocatechol
1-(4-Benzamido)-3-methyltriazene
Formed from: 1-(4-Benzamido)-3,3-dimethyltriazene
4-Benzamido-2-oxopiperidine
Formed from: 4-Benzamidopiperidine
2-Benzamidophenol
Formed from: Benzanilide
4-Benzamidophenol
Formed from: Benzanilide
N-(2-Benzamidophenyl)-N′-ethoxycarbonyl-S-methylisothiourea
Formed from: Methyl 1-(butylcarbamoyl) benzimidazol-2-yl carbamate
N-(2-Benzamidophenyl)-N′-methoxycarbonyl-S-methylisothiourea
Formed from: Methyl 1-(butylcarbamoyl) benzimidazol-2-yl carbamate
4-Benzamidopiperidine
Formed from: 4-Benzamido-1-(4-(indol-3-yl)-4-oxobutyl)piperidine
Indoramin
→ 4-Benzamido-2-oxopiperidine
Probably in: Monkey B550
3-(2-(4-Benzamidopiperidino)ethyl)indole
See: Indoramin
Benzamidoxime (N-hydroxybenzamidine)
Formed from: Benzamidine
→ Benzamidine
In: Rabbit E358
Benzanilide (N-phenylbenzamide)
→ Aniline
In: Ascaris A2037
Moniezia A2037
Mouse A2037
Sheep A2037
→ 2-Benzamidophenol
In: Rat B970
→ 4-Benzamidophenol
In: Rat B970
→ Benzoic acid
In: Ascaris A2037
Moniezia A2037
Mouse A2037
Sheep A2037
→ 4-Hydroxybenzanilide
In: Rat B970
→ Salicylanilide
In: Rat B970
Benzanthracene (1,2-benzanthracene, benz[*a*] anthracene)
Formed from: Benzanthracene-5,6-oxide
Benzanthracene-8,9-oxide
7-Methylbenzanthracene
→ Benzanthracene-5,6-oxide
In: Man A976
Rat A70, C283, E364
See also: A13 A306, B376, C267, D108, H285, H779, H793, K482
→ Benzanthracene-8,9-oxide
In: Rat C283
→ *cis*-1,2-Dihydro-1,2-dihydroxybenzanthracene
In: Beijerinckia A2227, D225
→ *trans*-1,2-Dihydro-1,2-dihydroxybenzanthracene
In: Man H285
→ *trans*-3,4-Dihydro-3,4-dihydroxybenzanthracene
In: Cunninghamella H329
Man: H285
Rat K482
→ *cis*-8,9-Dihydro-8,9-dihydroxybenzanthracene
In: Beijerinckia D225
→ *trans*-8,9-Dihydro-8,9-dihydroxybenzanthracene
In: Cunninghamella H329
Hamster A13
Man A13, D108, H285, H793
Mouse D108
Rat A13, A68, A306, C229, D108, K482
Saccharomyces B353
Tilapia H113
See also: B376
→ *cis*-10,11-Dihydro-10,11-dihydroxybenzanthracene
In: Beijerinckia D225
→ *trans*-10,11-Dihydro-10,11-dihydroxybenzanthracene
In: Cunninghamella H329
Man D108, H285, H793
Mouse D108

Mycobacterium H779
Rat D108, K482
→ 1-Hydroxybenzanthracene
In: Dog D108
Man D108
Mouse D108
Rat D108
→ 9-Hydroxybenzanthracene
In: Hamster A13
Man A13
Rat A13
→ 11-Hydroxybenzanthracene
In: Dog D108
Man D108
Mouse D108
Rat D108
→ 7-Methylbenzanthracene
In: Rat F534
See also: F742, G491

Benzanthracene-5,6-oxide
Formed from: Benzanthracene
→ Benzanthracene
In: Rat A350
→ *trans*-5,6-Dihydro-5,6-dihydroxybenzanthracene
In: Man A976, A2486, A3217, B63
Rat A70, A95, A2486, A3031, A3166, B63, B869
Mouse A2486
See also: A13, A68, A306
→ S-(5,6-Dihydro-5,6-dihydroxybenzanthr-5-yl)glutathione
In: Man A976
Cynomolgus monkey A3952
Rat A70, C472
See also: A306, D235, D842
→ S-(5,6-Dihydro-5,6-dihydroxybenzanthr-6-yl)glutathione
In: Rat C472

Benzanthracene-8,9-oxide
Formed from: Benzanthracene
→ Benzanthracene
In: Rat A350
→ *trans*-8,9-Dihydro-8,9-dihydroxybenzanthracene
In: Rat A95

Benzanthracene-3,4-quinone
Formed from: *trans*-3,4-Dihydro-3,4-dihydroxybenzanthracene

Benzanthr-3-yl-β-D-glucuronide
Formed from: 3-Hydroxybenzanthracene

Benzarone (2-ethyl-3-(*p*-hydroxybenzoyl) benzofuran)
Formed from: Benzbromarone
→ 2-Ethyl-5-hydroxy-3-(*p*-hydroxybenzoyl)benzofuran
Probably in: Dog E789
Man E789
Rat E789
→ 3-(*p*-Hydroxybenzoyl)-2-(1-hydroxyethyl)benzofuran
In: Dog E789
Man E789
Rat E789

Benzazimide (1,2,3-benzotriazin-4(1*H*)-one)
Formed from: Azinphos-methyl
Gutoxon

Benzbromarone ((3,5-dibromo-4-hydroxyphenyl) (2-ethyl-3-benzofuranyl)methanone)
→ Benzarone
In: Man C67
Contrast (in man): F495
→ 1′-Hydroxybenzbromarone
In: Man F495, F619, G835
→ 2′-Hydroxybenzbromarone
Probably in: Man G835
→ 6-Hydroxybenzbromarone
In: Man G835
See also: F619

Benzene
Formed from: Aniline
Benzoic acid
1,3-Diphenyl-1-triazene
Ferulic acid
O-Methyl phenylphosphonic acid
Phenylhydrazine
Phenylmercuric acetate
p-Toluic acid
Triphenyllead
Probably formed from: Benzenesulphonic acid
→ Benzene oxide
In: Rat J351

→ *cis*-1,2-Dihydro-1,2-dihydroxybenzene
In: Microorganism J512
Moraxella A638
Pseudomonas A1611, B678, F103
Rhodococcus H873
See also: E274, G653
→ *trans*-1,2-Dihydro-1,2-dihydroxybenzene
In: Rat A2100, H818
See also: E274, F568, G295
→ *trans, trans*-Muconic acid
In: Chimpanzee G639
Cynomolgus monkey G639
Mouse D358
Rat H99
→ Phenol
In: Chimpanzee G639
Cunninghamella A1218
Cynomolgus monkey G639
Gliocladium A1218
Hamster A1872
Man F533, K345, K489
Methylococcus F278
Mouse F533, H453, J386, K345
Nitrosomonas D921, H219, H631
Penicillium A1218
Rabbit A1872, A1996, B90, D605, F148, K345
Rat A569, A1872, A2100, A2160, B120, D695, E220, F553, F568, F686, G295, H453, H818, K345
Streptomyces A1218
See also: D50, F600, J321
→ S-Phenylglutathione
In: Rat F568, G295
See also: A2100
→ Toluene
In: Man F853
See also: A3363, K536
→ N^7-Phenylguanine
In: Rat E751

Benzenecarboximidamide
See: Benzamidine

Benzene-1,2-dicarboxylic acid
See: Phthalic acid

Benzene-1,3-dicarboxylic acid
See: Isophthalic acid

Benzene-1,4-dicarboxylic acid
See: Terephthalic acid

1,4-Benzenedimethanol
Formed from: *p*-Methylbenzyl alcohol

Benzene oxide (7-oxabicyclo[4.1.0]hepta-2,4-diene)
Formed from: Benzene
→ *S*-(1,2-Dihydro-1,2-dihydroxyphenyl) glutathione
In: Sheep A2282

(+)-(*Z*)-7-((1*R*,2*S*,3*S*,4*S*)-3-Benzenesulphonamidobicyclo[2.2.1]hept-2-yl)-5-heptenoic acid
→ (*Z*)-5-((1*R*,2*S*,3*S*,4*S*)-3-Benzenesulphonamidobicyclo[2.2.1]hept-2-yl)-3-pentenoic acid
In: Rat H713

5-((1*R*,2*S*,3*S*,4*S*)-3-Benzenesulphonamidobicyclo[2.2.1]hept-2-yl)pentanoic acid
Probably formed from: (*Z*)-5-((1*R*,2*S*,3*S*,4*S*)-3-Benzenesulphonamidobicyclo[2.2.1]hept-2-yl)-3-pentenoic acid

(*Z*)-5-((1*R*,2*S*,3*S*,4*S*)-3-Benzenesulphonamido bicyclo[2.2.1]hept-2-yl)-3-pentenoic acid
Formed from: (+)-(*Z*)-7-((1*R*,2*S*,3*S*,4*S*)-3-Benzenesulphonamidobicyclo[2.2.1]hept-2-yl)-5-heptenoic acid
→ 5-((1*R*,2*S*,3*S*,4*S*)-3-Benzenesulphonamido bicyclo[2.2.1]hept-2-yl)pentanoic acid
Probably in: Rat H713
→ 3-((1*R*,2*S*,3*S*,4*S*)-3-Benzenesulphonamido bicyclo[2.2.1]hept-2-yl)propenoic acid
Probably in: Rat H713

3-((1*R*,2*S*,3*S*,4*S*)-3-Benzenesulphonamidobicyclo [2.2.1]hept-2-yl)propenoic acid
Probably formed from: (*Z*)-5-((1*R*,2*S*,3*S*,4*S*)-3-Benzenesulphonamidobicyclo[2.2.1]hept-2-yl)-3-pentenoic acid
→ 3-((1*R*,2*S*,3*S*,4*S*)-3-Benzenesulphonamido -6-hydroxybicyclo[2.2.1]hept-2-yl) propenoic acid
Probably in: Rat H713

3-((1*R*,2*S*,3*S*,4*S*)-3-Benzenesulphonamido-6-hydroxybicyclo[2.2.1]hept-2-yl)propenoic acid
Probably formed from: 3-((1*R*,2*S*,3*S*,4*S*)-3-

Benzenesulphonamidobicyclo[2.2.1]hept-2-yl)propenoic acid

Benzenesulphonanilide
→ N-(*p*-Hydroxyphenyl)benzenesulphonanilide
In: Rat A2160

Benzenesulphonic acid
→ Benzene
Probably in: Rhodococcus H618
→ Catechol
In: Alcaligenes H297, K775
Pseudomonas J657
→ 2,3-Dihydroxybenzenesulphonic acid
Probably in: Alcaligenes K293
→ Phenol
In: Pseudomonas E408

Benzenesulphonyl-4-hydroxypiperidine
Formed from: Benzenesulphonylpiperidine

Benzenesulphonylpiperidine
→ Benzenesulphonyl-4-hydroxypiperidine
In: Rat C23

Benzenethiol
See: Thiophenol

Benzhexol (1-cyclohexyl-1-phenyl-3-piperidinopropan-1-ol)
→ 1-(2-Hydroxy)cyclohexyl-1-phenyl-3-piperidinopropan-1-ol
In: Man A3039
→ 1-(3-Hydroxy)cyclohexyl-1-phenyl-3-piperidinopropan-1-ol
In: Man A3039
→ 1-(4-Hydroxy)cyclohexyl-1-phenyl-3-piperidinopropan-1-ol
In: Man A3039

Benzhydrol (diphenylcarbinol)
Formed from: Benzophenone
Bis(*p*-chlorophenyl)carbinol
Diphenylmethane
1,1,1′,1′-Tetraphenyldimethyl ether
→ Benzhydrol-β-D-glucuronide
In: Rat B40
→ Benzophenone
Probably in: Pseudomonas A2541, A2831
→ Mandelic acid
In: Pseudomonas A2831

Benzhydrol-β-D-glucuronide
Formed from: Benzhydrol

Benzhydryl methylaminoethyl ether
Formed from: Diphenhydramine
→ Aminoethyl benzhydryl ether
In: Man A1554

1-(Benzhydryloxyethyl)piperidino-4-acetic acid
Formed from: 1-(Benzhydryloxyethyl) piperidino-4-ethyl acetate

1-(Benzhydryloxyethyl)piperidino-4-ethyl acetate
→ 1-(Benzhydryloxyethyl)piperidino-4-acetic acid
In: Man K182
Rat K82

Benzidine (4,4′-diaminobiphenyl)
Formed from: N-Acetylbenzidine
4,4′-Dinitrobiphenyl
Direct Black 4
Direct Black 38
Direct Blue 2
Direct Brown 2
Direct Green 1
Direct Orange 1
Direct Orange 8
Direct Red 28
Hydrazobenzene
→ N-Acetylbenzidine
In: Guinea pig A721, B129
Hamster B129
Man A3253, D229, H295, H600
Mouse B129, D872
Rabbit A3253, C78
Rat B129, C500, D229, D872, H600
Salmonella D903
→ 4-Aminobiphenyl-4′yl nitroxide radical
In: Rabbit B184
→ Benzidine di-imide
In: Horseradish F959
→ Benzidine-N-β-D-glucuronide
In Dog H38
Man H295, H541
Rat C590
→ 3-(S-Glutathionyl)benzidine
In: Rabbit D963
→ N-Methylbenzidine
In: Rabbit E687

Benzidine di-imide
Formed from: Benzidine

Benzidine-N-β-D-glucuronide
Formed from: Benzidine
Benzil (diphenylethanedione)
Formed from: Ditazole
→ Benzoic acid
In: Rabbit A2363
→ Benzoin
Probably in: Rabbit A2363
→ Hydroxybenzil
Probably in: Man F30
Benzilic acid (α-hydroxy-α-phenylbenzeneacetic acid)
Probably formed from: O-(*n*-Propyl)benzilic acid
Benzimidazole (2-nitro-N-(phenylmethyl)-1*H*-imidazole-1-acetamide)
Formed from: 2-Aminobenzimidazole
→ *o*-Aminobenzontrile
In: Melon A3065
See also: A1478, A9064
→ Benzimidazolylcobamide
In: Clostridium A2272
Eubacterium B467
→ 5-Hydroxybenzimidazole
In: Rat A1994
→ 1-Methylbenzimidazole
With catechol O-methyltransferase A1849
Benzimidazole-2-methylcarbamate
→ 2-Aminobenzimidazole
In: Melon A1478
→ Aniline
In: Melon A1478
2(3*H*)-Benzimidazolinone
Formed from: Droperidol
3-(2-(2-Benzimidazolyl)benzyloxy)ethyl)-6-hydroxyquinazolin-4-one
Probably formed from: 3-(2-(2-Benzimidazolyl)benzyloxy)ethyl)quinazolin-4-one
3-(2-(2-Benzimidazolyl)benzyloxy)ethyl) quinazolin-2,4-dione
Formed from: 3-(2-(2-Benzimidazolyl) benzyloxy)ethyl)quinazolin-4-one
→ 3-(2-Hydroxyethyl)quinazolin-2,4-dione
Probably, in an unspecified species E202
3-(2-(2-Benzimidazolyl)benzyloxy)ethyl) quinazolin-4-one
→ 3-(2-(2-Benzimidazolyl)benzyloxy)ethyl)quinazolin-2,4-dione
Probably, in an unspecified species E202
→ 3-(2-(2-Benzimidazolyl)benzyloxy)ethyl)-6-hydroxyquinazolin-4-one
Probably, in an unspecified species E202
→ 3-(2-Hydroxyethyl)quinazolin-4-one
In an unspecified species E202
→ 11-Oxo-1H-isoindolo[2,1-*a*]benzimidazole
In an unspecified species E202
Benzimidazolylcobamide
Formed from: Benzimidazole
2-(2-((3-(1*H*-Benzimidazol-2-yl)-3-β-D-glucuronyloxypropyl)methylamino)ethyl)-6-fluoro-1,2,3,4-tetrahydro-2-hydroxy-1-isopropylnaphthalene
Probably formed from: 2-(2-((3-(1*H*-Benzimidazol-2-yl)propyl)methylamino)ethyl)-6-fluoro-1,2,3,4-tetrahydro-2-hydroxy-1-isopropylnaphthalene
Benzimidazol-2-ylglyoxal
Formed from: Thiabendazole
2-(2-Benzimidazolyl)-4-hydroxymethyl-2-phenyl-1,3-dioxolone
Probably formed from: Oxapadol
2-(2-((3-(1*H*-Benzimidazol-2-yl)propyl) amino)ethyl)-6-fluoro-1,2,3,4-tetrahydro-2,4-dihydroxy-1-isopropylnaphthalene
Probably formed from: 2-(2-((3-(1*H*-Benzimidazol-2-yl)propyl)methylamino)ethyl)-6-fluoro-1,2,3,4-tetrahydro-2,4-dihydroxy-1-isopropylnaphthalene
2-(2-((3-(1*H*-Benzimidazol-2-yl)propyl)amino) ethyl)-6-fluoro-1,2,3,4-tetrahydro-2-hydroxy-1-isopropylnaphthalene
Probably formed from: N-Demethylmibefradil
→ 6-Fluoro 2-(2-(3-(5-β-D-glucuronyloxy-1*H*-benzimidazol-2-yl)propyl)amino)ethyl-1,2,3,4-tetrahydro-2-hydroxy-1-isopropylnaphthalene
Probably in: Rat G311
2-(2-((3-(1*H*-Benzimidazol-2-yl)propyl) amino)ethyl)-6-fluoro-1,2,3,4-tetrahydro-2-hydroxy-1-isopropyl-4-oxonaphthalene
Probably formed from: 2-(2-((3-(1*H*-

Benzimidazol-2-yl)propyl)amino)ethyl)-6-fluoro-1,2,3,4-tetrahydro-2,4-dihydroxy-1-isopropylnaphthalene
N-Demethylmibefradil

2-(2-((3-(1*H*-Benzimidazol-2-yl)propyl)methyl amino)ethyl)-6-fluoro-1,2,3,4-tetrahydro-2,4-dihydroxy-1-isopropylnaphthalene
Formed from: 2-(2-((3-(1*H*-Benzimidazol-2-yl)propyl)methylamino)ethyl)-6-fluoro-1,2,3,4-tetrahydro-2-hydroxy-1-isopropylnaphthalene
→ 2-(2-((3-(1*H*-Benzimidazol-2-yl)propyl) amino)ethyl)-6-fluoro-1,2,3,4-tetrahydro-2,4-dihydroxy-1-isopropylnaphthalene
Probably in: Rat G311
→ 2-(2-((3-(1*H*-Benzimidazol-2-yl)propyl) amino)ethyl)-6-fluoro-1,2,3,4-tetrahydro-2-hydroxy-1-isopropyl-4-oxonaphthalene
Probably in: Cynomolgus monkey J103
Marmoset J103
Rabbit J103
Rat J103

2-(2-((3-(1*H*-Benzimidazol-2-yl)propyl)methyl amino)ethyl)-6-fluoro-1,2,3,4-tetrahydro-2-hydroxy-1-isopropylnaphthalene
Formed from: Mibefradil
→ 2-(2-((3-(1*H*-Benzimidazol-2-yl) propyl)methylamino)ethyl)-6-fluoro-1,2,3,4-tetrahydro-2,4-dihydroxy-1-isopropylnaphthalene
In: Cynomolgus monkey J103
Man J103
Marmoset J103
Rabbit J103
Rat J103
See also: G311
→ 2-(2-((3-(1*H*-Benzimidazol-2-yl)-3-β-D-glucuronyloxypropyl)methylamino)ethyl)-6-fluoro-1,2,3,4-tetrahydro-2-hydroxy-1-isopropylnaphthalene
Probably in: Rat G311
→ 6-Fluoro-1,2,3,4-tetrahydro-2-hydroxy-2-(2-((3-(4-hydroxy-1*H*-benzimidazol-2-yl)propyl)methylamino)ethyl)-1-isopropylnaphthalene
In: Cynomolgus monkey J103
Man J103
Marmoset J103
Rabbit J103
Rat J103
See also: G311
→ 6-Fluoro-1,2,3,4-tetrahydro-2-hydroxy-2-(2-((3-(5-hydroxy-1*H*-benzimidazol-2-yl)propyl)methylamino)ethyl)-1-isopropylnaphthalene
In: Cynomolgus monkey J103
Man J103
Marmoset J103
Rabbit J103
Rat J103
→ 6-Fluoro-1,2,3,4-tetrahydro-2-hydroxy-1-isopropyl-2-((N-methylamino)ethyl) naphthalene
In: Cynomolgus monkey J103
Man J103
Marmoset J103
Rabbit J103
Rat J103

Benz[*d*]isothiazoline-1,1-dioxide
→ *o*-Carboxybenzenesulphonamide
In: Man A3244
Rat A3244
→ Saccharin
In: Man A3244
Rat A3244
→ *o*-Sulphamoylbenzyl alcohol
In: Man A3244
Rat A3244

1-(1,2-Benzisothiazol-3-yl)piperazine
Formed from: Ziprasidone

1-(1,2-Benzisothiazol-3-yl)piperazine sulphone
Formed from: Tiospirone

5-(2-(4-(1,2-Benzisothiazol-3-yl)-1-piperazin-1-yl)ethyl)-1-chloro-1,3-dihydro-2*H*-indol-2-one
See: Ziprasidone

1,2-Benzisoxazole
→ 2,3-Dihydro-1,2-benzisoxazol-3-one
Probably in: Rabbit F304

2,1-Benzisoxazole
See: Anthranil

Benznidazole (2-nitro-N-(phenylmethyl)-1*H*-imidazole-1-acetamide)
→ 2-Amino-1-(N-benzylacetamido)imidazole
In: Mouse E504

Rat D199

Benzocaine
See: Ethyl *p*-aminobenzoate

Benzocinnoline (benzo[*c*]cinnoline)
Formed from: Benzocinnoline-N-oxide

Benzocinnoline-N-oxide
→ Benzocinnoline
In: Saccharomyces H894

5,6-Benzocoumarin
Formed from: Phenanthrene

7,8-Benzocoumarin
Formed from: Phenanthrene

Benzocyclobutene (bicyclo[4.2.0]octa-1,3,5,7-tetraene; the identity of the compound studied may have been benzocyclobutane)
→ Benzocyclobuten-1-ol
In: Pseudomonas G678

Benzocyclobuten-1-ol
Formed from: Benzocyclobutene
→ Benzocyclobuten-1-one
In: Pseudomonas G678

Benzocyclobuten-1-one
Formed from: Benzocyclobuten-1-ol

1,2-Benzodinitrile
See Phthalonitrile

1,3-Benzodinitrile
See Isophthalonitrile

1,4-Benzodinitrile
See Terephthalonitrile

Benzodioxan-2-carboxylic acid
Probably formed from: N-(2-Aminoethyl)-2-carboxamido-1,4-benzodioxan

***cis*-1-(4-(1,3-Benzodioxolo-5-yl)-4-methoxy-1-cyclohexyl)-4-methoxy-1-cyclohexyl)-hydroxy-4-(*m*-hydroxyphenyl) piperidine**
Probably formed from: *cis*-1-(4-(1,3-Benzodioxolo-5-yl)-4-methoxy-1-cyclohexyl)-4-methoxy-1-cyclohexyl)-4-(*m*-methoxyphenyl)piperidine

***cis*-1-(4-(1,3-Benzodioxolo-5-yl)-4-methoxy-1-cyclohexyl)-4-methoxy-1-cyclohexyl)-4-(*m*-hydroxyphenyl)piperidine**
Probably formed from: *cis*-1-(4-(1,3-Benzodioxolo-5-yl)-4-methoxy-1-cyclohexyl)-4-methoxy-1-cyclohexyl)-4-(*m*-methoxyphenyl)piperidine

***cis*-1-(4-(1,3-Benzodioxolo-5-yl)-4-methoxy-1-cyclohexyl)-4-methoxy-1-cyclohexyl)-4-(*m*-methoxyphenyl)piperidine**
→ *cis*-1-(4-(1,3-Benzodioxolo-5-yl)-4-methoxy-1-cyclohexyl)-4-methoxy-1-cyclohexyl)-4-(*m*-hydroxyphenyl)piperidine
Probably in: Rat H688
→ *cis*-1-(4-(1,3-Benzodioxolo-5-yl)-4-methoxy-1-cyclohexyl)-4-methoxy-1-cyclohexyl)-hydroxy-4-(*m*-hydroxyphenyl)piperidine
Probably in: Rat H688

N-(1(*R*)-1,3-Benzodioxol-5-yl)butyl-3,3-diethyl-2(*S*)-((2-hydroxy-4-methyl-1-piperazinyl) carbonyl)phenoxy-4-oxo-1-azetidinecarboxamide
Formed from: N-(1(*R*)-1,3-Benzodioxol-5-yl)butyl-3,3-diethyl-2(*S*)-((4-methyl-1-piperazinyl)carbonyl)phenoxy-4-oxo-1-azetidinecarboxamide

N-(1(*R*)-1,3-Benzodioxol-5-yl)butyl-3,3-diethyl-2(*S*)-((3-hydroxy-4-methyl-1-piperazinyl) carbonyl)phenoxy-4-oxo-1-azetidinecarboxamide
Formed from: N-(1(*R*)-1,3-Benzodioxol-5-yl)butyl-3,3-diethyl-2(*S*)-((4-methyl-1-piperazinyl)carbonyl)phenoxy-4-oxo-1-azetidinecarboxamide

N-(1(*R*)-1,3-Benzodioxol-5-yl)butyl-3,3-diethyl-2(*S*)-1-(((2-methylamino)ethyl)amino) carbonyl)phenoxy-4-oxo-1-azetidinecarboxamide
Formed from: N-(1(*R*)-1,3-Benzodioxol-5-yl)butyl-3,3-diethyl-2(*S*)-((4-methyl-1-piperazinyl)carbonyl)phenoxy-4-oxo-1-azetidinecarboxamide

N-(1(*R*)-1,3-Benzodioxol-5-yl)butyl-3,3-diethyl-2(*S*)-((4-methyl-1-piperazinyl)carbonyl) phenoxy-4-oxo-1-azetidinecarboxamide
→ N-(1(*R*)-1,3-Benzodioxol-5-yl)butyl-3,3-diethyl-2(*S*)-((2-hydroxy-4-methyl-1-piperazinyl)carbonyl)phenoxy-4-oxo-1-azetidinecarboxamide
In: Rhesus monkey J270
→ N-(1(*R*)-1,3-Benzodioxol-5-yl)butyl-3,3-diethyl-2(*S*)-((3-hydroxy-4-methyl-1-piperazinyl)carbonyl)phenoxy-4-oxo-1-azetidinecarboxamide

In: Rhesus monkey J270
→ N-(1(*R*)-(1,3-Benzodioxol-5-yl)butyl-3,3-diethyl-2(*S*)-1-(((2-methylamino)ethyl) amino)carbonyl)phenoxy-4-oxo-1-azetidinecarboxamide
In: Rhesus monkey J270
→ N-(1(*R*)-1,3-Benzodioxol-5-yl)butyl-3,3-diethyl-2(*S*)-((4-methyl-1-piperazinyl) carbonyl)phenoxy-4-oxo-1-azetidinecarboxamide-N-oxide
In: Rat J268, J270
→ N-(1(*R*)-1,3-Benzodioxol-5-yl) hydroxybutyl-3,3-diethyl-2(*S*)-((4-methyl-1-piperazinyl)carbonyl)phenoxy-4-oxo-1-azetidinecarboxamide
In: Rhesus monkey J270
→ 3,3-Diethyl-N-(1(*R*)-(3,4-dihydroxyphenyl) butyl-2(*S*)-((4-methyl-1-piperazinyl) carbonyl)phenoxy-4-oxo-1-azetidinecarboxamide
In: Man J268
Monkey J268, J270
→ N-(2-Ethylbut-1-en-1-yl)-N′-(1-(3,4-methylenedioxyphenyl)but-1-yl)urea
In: Rhesus monkey J270
→ N-(*p*-Hydroxybenzoyl)-N′-methylpiperazine
In: Rhesus monkey J270

N-(1(*R*)-1,3-Benzodioxol-5-yl)butyl-3,3-diethyl-2(*S*)-((4-methyl-1-piperazinyl)carbonyl) phenoxy-4-oxo-1-azetidinecarboxamide-N-oxide
Probably formed from: N-(1(*R*)-1,3-Benzodioxol-5-yl)butyl-3,3-diethyl-2(*S*)-((4-methyl-1-piperazinyl)carbonyl)phenoxy-4-oxo-1-azetidinecarboxamide

(3*R*,4*R*)-4-(1,3-Benzodioxol-5-ylmethyl)dihydro-3-((3,4,5-trimethoxyphenyl)methyl)-2(3*H*)-furanone
See: Yatein

(6*R*)-6-((1*E*)-2-(1,3-benzodioxol-5-yl)ethenyl)-5,6-dihydro-4-methoxy-2*H*-pyran-2-one
See: Methystictin

N-(1(*R*)-1,3-Benzodioxol-5-yl)hydroxybutyl-3,3-diethyl-2(*S*)-((4-methyl-1-piperazinyl)carbonyl) phenoxy-4-oxo-1-azetidinecarboxamide
Formed from: N-(1(*R*)-1,3-Benzodioxol-5-yl)butyl-3,3-diethyl-2(*S*)-((4-methyl-1-piperazinyl)carbonyl)phenoxy-4-oxo-1-azetidinecarboxamide

4-(2-((2-(1,3-Benzodioxol-5-yl)-1-methylethyl) amino)-1-hydroxyethyl)-1,2-benzenediol
See: Protokylol

(1*S*)-1-(1,3-Benzodioxol-5-yl)-1,2,3,4-tetrahydro-6,7-dimethoxy-2-methylisoquinoline
See: (−)-Cryptostyline I

Benzofenac
See: 4-Benzyloxy-3-chlorophenylacetic acid

7,8-Benzoflavone (2-phenyl-4*H*-naphtho[1,2-*b*]pyran-4-one)
→ 7,8-Dihydro-7,8-dihydroxy-7,8-benzoflavone
Probably in: Man H58
→ 7,8-Benzoflavone-5,6-oxide
Probably in: Man H58
→ 6-Hydroxy-7,8-benzoflavone
Probably in: Man H58
→ 7-Hydroxy-7,8-benzoflavone
Probably in: Man H58

7,8-Benzoflavone-5,6-oxide
Probably formed from: 7,8-Benzoflavone
→ 5,6-Dihydro-5,6-dihydroxy-7,8-benzoflavone
Probably in: Man H58

Benzo[*b*]fluoranthrene
→ 1,2-Dihydro-1,2-dihydroxybenzo[*b*] fluoranthrene
In: Rat C205
→ 10,11-Dihydro-10,11-dihydroxybenzo[*b*]fluoranthrene
In: Rat C205
→ 4-Hydroxybenzo[*b*]fluoranthrene
In: Rat C205
→ 5-Hydroxybenzo[*b*]fluoranthrene
In: Rat C205
→ 6-Hydroxybenzo[*b*]fluoranthrene
In: Rat C205

Benzo[*j*]fluoranthrene
→ 9,10-Dihydro-9,10-dihydroxybenzo[*j*] fluoranthrene
In Salmonella B737

Benzo[*k*]fluoranthrene
→ 8,9-Dihydro-8,9-dihydroxybenzo[*k*] fluoranthrene

In Salmonella B737
→ Hydroxyfluoanthenic acid
In: Microorganism J303

3(2*H*)-Benzofuranone
See: Coumaranone

(*S*-(*R,*S**))-4-((1-(((1-(5-Benzofuranyl)butyl) amino)carbonyl)-3,3-diethyl-4-oxo-2-azetidinyl)oxy)benzeneacetic acid**
→ (*S*-(*R**,*S**))-4-((1-(((1-(5-Benzofuranyl) butyl)amino)carbonyl)-3,3-diethyl-4-oxo-2-azetidinyl)oxy)benzeneacetyl-β-D-glucuronic acid
In: Rat J64
Rhesus monkey J64

(S-(*R,*S**))-4-((1-(((1-(5-Benzofuranyl)butyl) amino)carbonyl)-3,3-diethyl-4-oxo-2-azetidinyl)oxy)benzeneacetyl-β-D-glucuronic acid**
Formed from: (S-(*R**,*S**))-4-((1-(((1-(5-Benzofuranyl)butyl)amino)carbonyl)-3,3-diethyl-4-oxo-2-azetidinyl)oxy)benzeneacetic acid

N-((Benzofuran-5-yl)methylcarbonyl)glycine
Probably formed from: 5-Carboxymethylbenzofuran

N-(1-(2-(5-Benzofurazanyl)ethyl)-3,4-dihydro-4-hydroxy*spiro*((2*H*)-1-benzopyran-2,4′-piperidin)-6-yl)methanesulphonamide
Formed from: N-(1-(2-(5-Benzofurazanyl) ethyl)-3,4-dihydro-4-oxo*spiro*((2*H*)-1-benzopyran-2,4′-piperidin)-6-yl)methanesulphonamide

N-(1′-(2-(5-Benzofurazanyl)ethyl)-3,4-dihydro-4-oxo*spiro*(2*H*-(1-benzopyran-2,4′-piperidin)-6-yl)methanesulphonamide
→ N-(1′-(2-(5-Benzofurazanyl)ethyl)-3,4-dihydro-4-hydroxy*spiro*(2*H*-(1-benzopyran-2,4′-piperidin)-6-yl)methanesulphonamide
In: Dog G577
Rat G577
→ N-(1′-(3,4-Dihydro-4-oxo*spiro*(2*H*-(1-benzopyran-2,4′-piperidin)-6-yl)methanesulphonamide
In: Rat G577

2-(3-Benzofuryl)ethylamine
→ N-Acetyl-2-(3-benzofuryl)ethylamine
In: Man K490

Benzoic acid
Formed from: N-Acetyl-O[6]-benzoylguanosine
Banamate
Beclamide
Benzaldehyde
Benzalphthalide
Benzamide
2-Benzamidofluorene
Benzanilide
Benzil
Benzoin
Benzonitrile
N-Benzoyl-L-alanine
β-Benzoyl-L-alanine
N-Benzoyl-L-aminobutyric acid
N-Benzoylaniline
N-Benzoyl-N-(1-carboxyethyl)-3,4-dichloroaniline ethyl ester
N-Benzoyl-N′-(*p*-chlorobenzyl)hydrazine
Benzoyl CoA
Benzoyl-β-D-glucuronic acid
N-Benzoylindoline
N-Benzoyl-*p*-nitroaniline
3-Benzoyloxy-N-(*tert*-butyl)-2-hydroxypropylamine
3-Benzoyloxy-N-(*tert*-butyl)-2-methoxypropylamine
3-Benzoyloxy-2-(1,1-dimethyl-2-(((morpholino)carbonyl)amino)ethyl)-2-hydroxypropylamine
N-Benzoyl-N′-phenyl-2,6-diaminobenzoic acid
O-Benzoylthiamine
Benzyl chloride
Benzylhydrazine
S-Benzyl-N-malonyl-L-cysteine
N-Benzyl-N′-picolinylpiperazine
2,2-Bis(hydroxymethyl)-3,3-dimethylbutyl benzoate
o-Bromobenzoic acid
m-Bromobenzoic acid
m-Chlorobenzoic acid
5-Chloro-2-hydroxy-6-oxo-6-phenylhexa-2,4-dienoic acid
N-(2-Chloro-4-nitrophenyl)benzamide
N-(*o*-Chlorophenyl)benzamide
p-Chlorophenyl benzoate

1-(*p*-Chlorophenyl)-4-dimethylamino-2,3-dimethylbutan-2-ol
m-Cresol
Cyclohexanecarboxylic acid
Cyclohex-3-enecarboxylic acid
Decylbenzene
Dibenzothiophene
Dibenzoyl 2-amino-tetrahydro-6,7-dihydroxynaphthalene
Dibenzoylepinine
Dibenzyl diselenide
Didehydro-4-methyl-5-phenyl-1,3,4-thiadiazolidine-2-thione
N,N-Diethylbenzamide
2,4-Dihydroxy-6-oxo-6-phenylhexa-2,4-dienoic acid
3,3-Diphenyl-N-(1-phenylethyl) propylamine
Dodecylbenzene
D-Ephedrine
L-Ephedrine
N-Ethylbenzamide
Hexylbenzene
Hippuric acid
N-Hydroxy-2-benzamidofluorene
m-Hydroxybenzoic acid
p-Hydroxybenzoic acid
4-Hydroxybenzoin
2-Hydroxy-6-oxo-6-phenylhexa-2,4-dienoic acid
D-Isopropylnorephedrine
Isoxathion
Methyl benzoate
p-Methylbenzyl benzoate
β-Methylstyrene
1-Methyl-3-thiobenzoylthiourea
N-(*p*-Nitrophenyl)benzamide
Nonylbenzene
D-Norephedrine
Octylbenzene
n-Pentanedecylbenzene
Phenylacetic acid
L-Phenylalanine
Phenyl benzoate
β-Phenylhydracrylic acid
1-Phenylpentan-1-ol
Phenylpyruvic acid
Phthalic acid
Pivhydrazine
D-Pseudoephedrine
L-Pseudoephedrine
Quinol
Saddamine
Shikimic acid
Tersavid
N-(*p*-Toluoyl)indoline
Tribenoside
Tropacocaine

Probably formed from: Anthraquinone
γ-Benzoylbutyric acid
N-Benzoylalanine
N-Benzoyl-L-aminobutyric acid
N^6-Benzyladenosine
Dihydrokawain
γ-Hydroxy-γ-phenylpropyl carbamate
4-Hydroxy-*trans*-stilbene
Kawain
Mebendazole
3-Oxo-3-phenylpropionic acid
Phenol
Quinol
Tribenoside

→ Benzaldehyde
In: Clostridium G416
Cranberry A3545
Nocardia G207, H955
Pleurotus H241
With NAD(P)-dependent benzaldehyde oxidoreductase E248

See also: E743

→ Benzamide
In: Streptomyces E279

→ Benzene
In: Microorganism D737

→ Benzoyladenylate
In: Neurospora A750

→ N-Benzoylaspartic acid
In: Barley A1726

→ Benzoyl CoA
In: Bacillus H893
Beef H740
Desulfitomaculum K210

Hypericum J509
Man K220
Microorganism H500, J213
Pseudomonas F65, G225, G342, G729, G983, K191
Rat A1369
Rhodopseudomonas A1370, E605
→ Benzoyl-β-D-glucuronic acid
In: Bat A884, A1451
Bandicoot D914
Dog A3771
Echidna D914
Ferret A394, A3046, A3771
Hamster A3771
Horse A915, B801
Marmoset B201
Mouse A651
Padmelon D914
Possum D914
Potoroos D914
Rabbit H519
Rat A651, A2104, A3771, C381, D914, F693
Sugar glider D914
Tasmanian devil D914
→ Benzoyl-β-D-glucose
In: Aconitum F445
Cherry C756
Dioscoreophyllum F445
Glycine A3585, A3679
Glycyrrhiza F445
Ipomoea D70
Mung bean A3585
Nicotiana F445
Quercus K737
Raphanus B440, B599
→ Benzoylhydroxylamate
In: Neurospora A750
→ O-Benzoyl-L-malic acid
In: Barley A1726
→ Benzoyltaurine
In: Bluegill E582
Catfish F554
Flounder E900
Minnow E582
Perch E582
Trout E582, E911
→ Benzyl alcohol
In: Desulphovibrio K615
Hormoconis F651
→ Benzyl benzoate
In: Cranberry A3545
→ Catechol
In: Brevibacterium A681
Micrococcus D56
Moraxella A2193
Pseudomonas D472
Streptomyces B763
→ Cyclohexanecarboxylic acid
In: Microorganism A3105
→ Cyclohex-1-enecarboxylic acid
In: Microorganism A3105
→ 2,5-Dichlorobenzoic acid
In: Caldariomyces J373
→ 1,2-Dihydro-1,2-dihydroxybenzoic acid
In: Pseudomonas B465, G434
→ Hippuric acid
In: Bat A3356
Bandicoot D914
Civet A1312
Dog A3771
Echidna D914
Elephant A2387
Ferret A394, A3046, A3771
Genet A1312
Hamster A3771
Horse A915, B801
Lion A1312
Man A126, A1772, A3507, G834
Marmoset B201
Mouse A651
Padmelon D914
Possum D914
Potoroos D914
Rabbit A553, E674, H519
Rat A651, A1772, A1991, A2104, A2869, A3730, A3771, A3833, B56, B201, C308, C381, D914, F693, G299, G545, H148, H722, J908
Sugar glider D914
Tasmanian devil D914
Contrast (fruit bat) A884

→ *m*-Hydroxybenzoic acid
In: Flavobacterium E355
Man E65
Pseudomonas D771
→ *p*-Hydroxybenzoic acid
In: Apocynum A2619
Aspergillus A640, A1606, B927, D542, D593
Byssochlamys C76
Catharanthus A2619
Conium A2619
Cunninghamella A1218
Curvularia A1218
Man E65
Microorganism G167
Millipede A3389
Nocardia F266
Rhodotorula D620
Trichosporon B368
See also A1011, A2401, A3219, B2
→ 2-Hydroxycyclohexanecarboxylic acid
In: Moraxella A2193
→ Methyl benzoate
In: Antirrhinum K650
Candida J721
→ *cis,cis*-Muconic acid
In: Corynebacterium K81
→ Ornithuric acid
In: Chicken A1156
Coturnix A1156
Pheasant A1156
Quail E516
Tuatara B449
→ 3-Oxo-3-phenylpropionic acid
In: Horse C183
See also: B801, B964
→ Salicylic acid
In: Aspergillus A1218
Bacillus A2217
Cunninghamella A1218
Man E65
Nicotiana H681
Oryza H593
Potato J691
See also: F615, J819

Benzoin (2-hydroxy-1,2-diphenylethanone)
Formed from: Benzaldehyde
Benzil
→ Benzaldehyde
In: Pseudomonas F221
→ Benzoic acid
In: Rabbit A2363
→ *meso*-Hydrobenzoin
In: Rabbit A2363
→ *threo*-Hydrobenzoin
In: Rabbit A2363

1,2-Benzoisothiazol-3(2*H*)-one 1,1-dioxide
See: Saccharin

Benzo[*b*]naphtho[1,2-*d*]thiophene
→ Benzo[*b*]naphtho[1,2-*d*]thiophene sulphone
In: Rat E166
→ *trans*-Dihydrodihydroxybenzo[*b*]naphtho[1,2-*d*]thiophene
In: Rat E166

Benzo[*b*]naphtho[2,1-*d*]thiophene
→ Benzo[*b*]naphtho[2,1-*d*]thiophene sulphone
In: Rat E166
→ *cis*-Dihydrodihydroxybenzo[*b*]naphtho[2,1-*d*]thiophene
In: Sphingomonas K97

Benzo[*b*]naphtho[2,3-*d*]thiophene
→ Benzo[*b*]naphtho[2,3-*d*]thiophene-6,11-quinone
In: Rat E166
→ Benzo[*b*]naphtho[2,3-*d*]thiophene sulphone
In: Rat E166

Benzo[*b*]naphtho[2,3-*d*]thiophene-6,11-quinone
Formed from: Benzo[*b*]naphtho[2,3-*d*]thiophene

Benzo[*b*]naphtho[1,2-*d*]thiophene sulphone
Formed from: Benzo[*b*]naphtho[1,2-*d*]thiophene
→ Hydroxybenzo[*b*]naphtho[1,2-*d*]thiophene sulphone
Probably in: Rat E166

Benzo[*b*]naphtho[2,1-*d*]thiophene sulphone
Formed from: Benzo[*b*]naphtho[2,1-*d*]thiophene
Hydroxybenzo[*b*]naphtho[2,1-*d*]thiophene sulphone
Probably in: Rat E166

Benzo[*b*]naphtho[2,3-*d*]thiophene sulphone
Formed from: Benzo[*b*]naphtho[2,3-*d*]thiophene
Benzonitrile
→ Benzoic acid
In: Cryptococcus K694
Fusarium A3364
Klebsiella G430
Nocardia A1367, A3420, D725
Pseudomonas K184
Rhodococcus G172, H952, J857
→ 2,3-Dihydro-2,3-dihydroxybenzonitrile
In: Pseudomonas J656
See also: F283
→ *o*-Hydroxybenzonitrile
In: Rat A2160, D122
→ *m*-Hydroxybenzonitrile
In: Nitrosomonas H631
Rat A2160, D122, H305
→ *p*-Hydroxybenzonitrile
In: Nitrosomonas H631
Rat A2160, D122, H305
Benzo[*c*]phenanthrene
→ Benzo[*c*]phenanthrene-3,4-oxide
In: Rat E381, E459, F748, F936
See also: E483, F945
→ Benzo[*c*]phenanthrene-5,6-oxide
In: Rat E381, E459, F748, F936
Benzo[*c*]phenanthrene-3,4-oxide
Formed from: Benzo[*c*]phenanthrene
→ *trans*-3,4-Dihydro-3,4-dihydroxybenzo[*c*]phenanthrene
Probably in: Rat E483
Benzo[*c*]phenanthrene-5,6-oxide
Formed from: Benzo[*c*]phenanthrene
→ *trans*-5,6-Dihydro-5,6-dihydroxybenzo[*c*]phenanthrene
In: Rat E381
Benzo[*c*]phenanthridine
→ (+)-(3*S*,4*R*)-*cis*-3,4-Dihydro-3,4-dihydroxybenzo[*c*]phenanthridine
In: Sphingomonas K97
→ (+)-(9*S*,10*R*)-*cis*-9,10-Dihydro-9,10-dihydroxybenzo[*c*]phenanthridine
In: Sphingomonas K97
Benzophenone (diphenylmethanone)
Formed from: Benzhydrol
Cinnarizine
Probably formed from: 2-Hydroxy-2,2-diphenylethylmethylnitrosamine
→ Benzhydrol
Probably in: Man A3384
Rat K676
→ *p*-Hydroxybenzophenone
In: Rat B347, K676
Benzophenone diphenylhydrazone
Formed from: Tetrabenazine
2*H*-1-Benzopyran-2-one
See: Coumarin
4*H*-1-Benzopyran-4-one
See: 4-Chromone
2-(5*H*-[1]Benzopyrano[2,3-*b*]pyridin-7-yl)propionic acid
→ 2-(5*H*-[1]Benzopyrano[2,3-*b*]pyridin-7-yl)propionyl 3-acetylglucoside
In: Mouse A2657, A2676
→ 2-(5*H*-[1]Benzopyrano[2,3-*b*]pyridin-7-yl)propionyl-β-D-3-acetylglucuronic acid
In: Rat A2657
→ 2-(5*H*-[1]Benzopyrano[2,3-*b*]pyridin-7-yl)propionyl-β-D-glucose
In: Mouse G127
→ 2-(5*H*-[1]Benzopyrano[2,3-*b*]pyridin-7-yl)propionyl-β-D-glucuronic acid
In: Guinea pig G127
Mouse G127
Rabbit G127
Rat G127
2-(5*H*-[1]Benzopyrano[2,3-*b*]pyridin-7-yl)propionyl 3-acetylglucoside
Formed from: 2-(5*H*-[1]Benzopyrano[2,3-*b*]pyridin-7-yl)propionic acid
2-(5*H*-[1]Benzopyrano[2,3-*b*]pyridin-7-yl)propionyl-β-D-3-acetylglucuronic acid
Formed from: 2-(5*H*-[1]Benzopyrano[2,3-*b*]pyridin-7-yl)propionic acid
2-(5*H*-[1]Benzopyrano[2,3-*b*]pyridin-7-yl)propionyl-β-D-glucose
Formed from: 2-(5*H*-[1]Benzopyrano[2,3-*b*]pyridin-7-yl)propionic acid
2-(5*H*-[1]Benzopyrano[2,3-*b*]pyridin-7-yl)propionyl-β-D-glucuronic acid
Formed from: 2-(5*H*-[1]Benzopyrano[2,3-*b*]pyridin-7-yl)propionic acid

'(–)-*anti*-Benzo[*a*]pyrene-(7*R*,8*S*)-diol-9*S*,10*R*-epoxide glutathione conjugate' (possibly a mixture)
→ '(–)-*anti*-Benzo[*a*]pyrene-(7*R*,8*S*)-diol-9*S*,10*R*-epoxide cysteinylglycine conjugate'
With γ-glutamyl transpeptidase K286

'(–)-*anti*-Benzo[*a*]pyrene-(7*R*,8*S*)-diol-9*S*,10*R*-epoxide cysteinylglycine conjugate'
Formed from: '(–)-*anti*-Benzo[*a*]pyrene-(7*R*,8*S*)-diol-9*S*,10*R*-epoxide glutathione conjugate'

Benzo[*c*]quinoline
See: Phenanthridine

Benzo[*f*]quinoline
→ Benzo[*f*]quinoline-N-oxide
In: Bullhead F275
Rat C811, E613, F275
→ 5,6-Dihydro-5,6-dihydroxybenzo[*f*]quinoline
In: Rat E613
→ 7,8-Dihydro-7,8-dihydroxybenzo[*f*]quinoline
In: Bullhead F275
Rat C811, E613, F275
→ 9,10-Dihydro-9,10-dihydroxybenzo[*f*]quinoline
In: Bullhead F275
Rat C811, E613, F275
→ 7-Hydroxybenzo[*f*]quinoline
In: Bullhead F275
Rat C811, E613, F275
→ 9-Hydroxybenzo[*f*]quinoline
In: Bullhead F275
Rat F275

Benzo[*h*]quinoline
→ 5,6-Dihydro-5,6-dihydroxybenzo[*h*]quinoline
In: Rat C811
→ 7,8-Dihydro-7,8-dihydroxybenzo[*h*]quinoline
In: Rat C811
→ 7-Hydroxybenzo[*h*]quinoline
In: Rat C811

Benzo[*f*]quinoline-N-oxide
Formed from: Benzo[*f*]quinoline

Benzoquinone
See: *p*-Quinone

3-(*o*-Benzoquinon-4-yl)acrylic acid
Formed from: Caffeic acid

Benzothiazole
→ 2-Hydroxybenzothiazole
In: Guinea pig J9
Rat J9
→ 2-Methylmercaptoaniline
In: Guinea pig G15

***S*-(2-Benzothiazolyl)-L-cysteine**
Probably formed from: *S*-(2-Benzothiazolyl)glutathione
→ 2-Mercaptobenzothiazole
In: Eubacterium D356
Rat F597
See also: E853

1-(2-Benzothiazolyl)-1,3-dimethylurea (methabenzthiazuron)
→ 1-(2-Benzothiazolyl)-3-hydroxymethyl-1-methylurea
In: Onion A172
Hypocrea A3197
Triticum A172
→ 1-(6-Hydroxy-2-benzothiazolyl)-1,3-dimethylurea
In: Hypocrea A3197
→ 2-Methylamino-2-benzothiazole
In: Hypocrea A3197

***S*-(2-Benzothiazolyl)glutathione**
Formed from: 2-Methylthiobenzothiazole
→ *S*-(2-Benzothiazolyl)-L-cysteine
Probably in: E853

1-(2-Benzothiazolyl)-3-hydroxymethyl-1-methylurea
Formed from: 1-(2-Benzothiazolyl)-1,3-dimethylurea
→ 1-(2-Benzothiazolyl)-1-methylurea
In: Onion A172
Triticum A172

1-(2-Benzothiazolyl)-1-methylurea
Formed from: 1-(2-Benzothiazolyl)-3-hydroxymethyl-1-methylurea

1-(1,2-Benzothiazol-3-yl)-3-oxopiperazine sulphone
Probably formed from: 1-(1,2-Benzothiazol-3-yl)piperazine sulphone

1-(1,2-Benzothiazol-3-yl)piperazine
Formed from: Ziprasidone

→ 1-(1,2-Benzothiazol-3-yl)piperazine sulphoxide
Probably in: Man J279

1-(1,2-Benzothiazol-3-yl)piperazine sulphone
Probably formed from: 1-(1,2-Benzothiazol-3-yl)piperazine sulphoxide
→ 1-(1,2-Benzothiazol-3-yl)-3-oxopiperazine sulphone
Probably in: Man J279

1-(1,2-Benzothiazol-3-yl)piperazine sulphoxide
Probably formed from: 1-(1,2-Benzothiazol-3-yl)piperazine
→ 1-(1,2-Benzothiazol-3-yl)piperazine sulphone
Probably in: Man J279

N-(1-Benzo[*b*]thien-2-ylethyl)-N-hydroxyurea
See: Zileuton

N-(1-Benzo[*b*]thien-2-ylethyl)urea
Formed from: Zileuton
→ N-(1-Benzo[*b*]thien-2-ylethyl)urea sulphoxide
Probably in: Man H469

N-(1-Benzo[*b*]thien-2-ylethyl)urea sulphoxide
Probably formed from: N-(1-Benzo[*b*]thien-2-ylethyl)urea

Benzo[*b*]thiophene
→ Benzo[*b*]thiophene sulphoxide
In: Rat G284

Benzo[*b*]thiophene-2,3-dione
Probably formed from: 2,3-Dihydroxybenzo[*b*]thiophene

Benzo[*b*]thiophene sulphoxide
Formed from: Benzo[*b*]thiophene

Benzo[*b*]thiophen-3-ylacetic acid
Formed from: 3-(2-Dimethylaminoethyl) benzo[*b*]thiophene
Probably formed from: 3-(Benzo[*b*] thiophen-3-yl)pyruvic acid
→ Benzo[*b*]thiophen-3-ylacetylglycine
In: Rat A2874

Benzo[*b*]thiophen-3-ylacetylglycine
Formed from: Benzo[*b*]thiophen-3-ylacetic acid

3-(Benzo[*b*]thiophen-3-yl)alanine
→ 3-(Benzo[*b*]thiophen-3-yl)pyruvic acid
In: Rat A2874

3-(Benzo[*b*]thiophen-3-yl)lactic acid
Formed from: 3-(Benzo[*b*]thiophen-3-yl)pyruvic acid

3-(Benzo[*b*]thiophen-3-yl)pyruvic acid
Formed from: 3-(Benzo[*b*]thiophen-3-yl)alanine
→ Benzo[*b*]thiophen-3-ylacetic acid
Probably in: Rat A2874
→ 3-(Benzo[*b*]thiophen-3-yl)lactic acid
In: Rat A2874

1,2,3-Benzotriazin-4(1*H*)-one
See: Benzazimide

1*H*-Benzotriazole
Formed from: 1-Amino-1*H*-benzotriazole
1-Ethyl-1*H*-benzotriazole
1-Methyl-1*H*-benzotriazole
→ 4-Hydroxy-1*H*-benzotriazole
In: Rat C240
→ 5-Hydroxy-1*H*-benzotriazole
In: Rat C240

1,3-Benzoxazine-2,4(3*H*)-dione (carsalem)
Formed from: Ethyl 2-carbamoyloxybenzoate

Benzoxazole
→ 2,3-Dihydrobenzoxazol-2-one
Probably in: Rabbit F304

3-(2-(Benzoxazol-2-yl)ethyl)-5-ethenyl-6-methyl-pyridin-2(1*H*)-one
Probably formed from: 3-(2-(Benzoxazol-2-yl)ethyl)-5-(1-hydroxyethyl)-6-methylpyridin-2(1*H*)-one

3-(2-(Benzoxazol-2-yl)ethyl)-5-ethyl-6-hydroxy-methylpyridin-2(1*H*)-one
Formed from: 3-(2-(Benzoxazol-2-yl)ethyl)-5-ethyl-6-methylpyridin-2(1*H*)-one
→ 3-(2-(Benzoxazol-2-yl)ethyl)-5-(1-hydroxyethyl)-6-hydroxymethylpyridin-2(1*H*)-one
In: Man G867
Rat G867

3-(2-(Benzoxazol-2-yl)ethyl)-5-ethyl-6-methylpyridin-2(1*H*)-one
→ 5-Acetyl-3-(2-(benzoxazol-2-yl)ethyl)-6-methylpyridin-2(1*H*)-one
Probably in: Rat G505
→ 3-(2-(Benzoxazol-2-yl)ethyl)-5-ethyl-5,6-dihydro-5,6-dihydroxy-6-methylpyridin-2(1*H*)-one

In: Man G867
Rat G867
→ 3-(2-(Benzoxazol-2-yl)ethyl)-5-ethyl-6-hydroxymethylpyridin-2(1*H*)-one
In: Man G505, G867
Rat G867
→ 3-(2-(Benzoxazol-2-yl)ethyl)-5-(1-hydroxyethyl)-6-methylpyridin-2(1*H*)-one
In: Man G867
Rat G867
→ 5-Ethyl-3-(2-(6-hydroxybenzoxazol-2-yl)ethyl)-6-methylpyridin-2(1*H*)-one
In: Man G505, G867
Rat G867
→ 5-Ethyl-3-(2-(7-hydroxybenzoxazol-2-yl)ethyl)-6-methylpyridin-2(1*H*)-one
In: Man G867
Rat G867

3-(2-(Benzoxazol-2-yl)ethyl)-5-(1-hydroxyethyl)-6-hydroxymethylpyridin-2(1*H*)-one
Probably formed from: 3-(2-(Benzoxazol-2-yl)ethyl)-5-ethyl-6-hydroxymethylpyridin-2(1*H*)-one

3-(2-(Benzoxazol-2-yl)ethyl)-5-(1-hydroxyethyl)-6-methylpyridin-2(1*H*)-one
Formed from: 3-(2-(Benzoxazol-2-yl)ethyl)-5-ethyl-6-methylpyridin-2(1*H*)-one
→ 3-(2-(Benzoxazol-2-yl)ethyl)-5-ethenyl-6-methylpyridin-2(1*H*)-one
Probably in: Man G867
Rat G867

Benzoylacetic acid
See: 3-Oxophenylpropionic acid

Benzoyladenylate
Formed from: Benzoic acid
→ Benzaldehyde
In: Neurospora A750

N-Benzoyl-L-alanine
→ Benzoic acid
In: Corynebacteriun E145

β-Benzoyl-L-alanine
→ Benzoic acid
In: Pseudomonas B248
→ β-Benzoylpyruvic acid
In: Hansenula D978

N-Benzoyl-L-aminobutyric acid
→ Benzoic acid
In: Corynebacteriun E145

N-Benzoylaniline
→ Aniline
In: Rabbit K435
→ Benzoic acid
In: Rabbit K435

N-Benzoyl-*p*-anisidine
Formed from: N-Benzyl-*p*-anisidine

Benzoyl-L-arginamide
→ Benzoyl-L-arginine
In: Oryza D743

Benzoyl-L-arginine
Formed from: Benzoyl-L-arginamide
Benzoyl-L-arginine ethyl ester
Benzoyl-L-arginyl 4-methylcoumarin-7-amide
Benzoyl-L-arginyl-β-naphthylamide
Benzoyl-L-arginyl-*p*-nitroanilide

Benzoyl-L-arginine ethyl ester
→ Benzoyl-L-arginine
In: Oryza D743

Benzoyl-L-arginyl-4-methylcoumarin-7-amide
→ 7-Amino-4-methylcoumarin
In: Oryza D743
→ Benzoyl-L-arginine
In: Oryza D743

Benzoyl-L-arginyl-β-naphthylamide
→ Benzoyl-L-arginine
In: Glycine A1596
Hen G654
Rabbit B787
→ β-Naphthylamine
In: Glycine A1596
Hen G654
Rabbit B334, B524, B787

Benzoyl-L-arginyl-*p*-nitroanilide
→ Benzoyl-L-arginine
In: Bacillus F34
Glycine A1596
Oryza D525, D743
Vigna C171
→ *p*-Nitroaniline
In: Bacillus F34
Glycine A1596
Oryza D525, D743
Vigna C171

N-Benzoylaspartic acid
Formed from: Benzoic acid
N-Benzoyl-4-(α-benzyl)-4-hydroxypiperidine
Formed from: N-Benzoyl-α-benzylpiperidine
N-Benzoyl-α-benzylpiperidine
→ N-Benzoyl-4-(α-benzyl)-4-hydroxypiperidine
In: Rat C23
2-Benzoyl-6-bromochromone
→ 6-Bromo-2-(α-hydroxybenzoyl)chromone
In: Lactobacillus K425
Saccharomyces K425
Yamadazyma K425
Zygosaccharomyces K425
N-Benzoylbutylamine
→ N-Benzoyl-3-hydroxybutylamine
In: Rat C23
N-Benzoyl-3-*sec*-butyl-1-oxa-4-azaspiro[4.4] nonane
→ N-Benzoyl-3-(ω-hydroxy-*sec*-butyl)-1-oxa-4-azaspiro[4.4]nonane
In: Beauveria K678
γ-Benzoylbutyraldehyde
Formed from: 2-Methoxy-6-oxo-6-phenyl-hexa-2,4-dienoic acid
→ γ-Benzoylbutyric acid
Probably in: Pseudomonas E616
γ-Benzoylbutyric acid
Formed from: Biphenyl
Probably formed from: γ-Benzoyl-butyraldehyde
→ Benzoic acid
Probably in: Achromobacter A706
Pseudomonas E41
N-Benzoyl-N-(1-carboxyethyl)-3,4-dichloroaniline
Formed from: N-Benzoyl-N-(1-carboxyethyl)-3,4-dichloroaniline ethyl ester
N-Benzoyl-N-(1-carboxyethyl)-3,4-dichloroaniline ethyl ester (benzoylprop-ethyl, Suffix)
→ Benzoic acid
In: Wheat A1319
→ N-Benzoyl-N-(1-carboxyethyl)-3,4-dichloroaniline
In: Wheat A1319
→ N-Benzoyl-3,4-dichloroaniline
In: Wheat A1319
Benzoyl chloride 2,4,6-trichlorophenylhydrazone
See: Banamite
N-Benzoyl-*p*-chloroaniline
Formed from: N-Benzyl-*p*-chloroaniline
N^2-Benzoyl-N^6-(*p*-chlorobenzoyl)lysine
Formed from: *p*-Chlorobenzoic acid
N^6-Benzoyl-N^2-(*p*-chlorobenzoyl)lysine
Formed from: *p*-Chlorobenzoic acid
N^2-Benzoyl-N^5-(*p*-chlorobenzoyl)ornithine
Formed from: *p*-Chlorobenzoic acid
N^5-Benzoyl-N^2-(*p*-chlorobenzoyl)ornithine
Formed from: *p*-Chlorobenzoic acid
N-Benzoyl-N-(3-chloro-4-fluorophenyl)-2-aminopropionic acid
Formed from: Barnon
→ N-Benzoyl-N-(3-chloro-4-fluorophenyl)-2-aminopropionyl-β-D-glucuronic acid
In: Dog A2035
Rat A2035
N-Benzoyl-N-(3-chloro-4-fluorophenyl)-2-aminopropionyl-β-D-glucuronic acid
Formed from: N-Benzoyl-N-(3-chloro-4-fluorophenyl)-2-aminopropionic acid
2′-Benzoyl-4′-chloroisonicotinanilide
Formed from: Inabenfide
→ 2′-Benzoyl-4′-hydroxyisonicotinanilide
Probably in: Rat E780
2-Benzoylchromone
→ 2-(α-Hydroxybenzoyl)chromone
In: Aspergillus K425
Beauveria K425
Cunninghamella K425
Lactobacillus K425
Rhodotorula K425
Saccharomyces K425
Yamadazyma K425
Zygosaccharomyces K425
Benzoyl CoA
Formed from: *p*-Aminobenzoyl CoA
Anthraniloyl CoA
Benzoic acid
(*R*)-Benzylsuccinyl CoA
Gentisoyl CoA
m-Hydroxybenzoyl CoA
Phenylglyoxylic acid
Salicyloyl CoA

Probably formed from: *p*-Hydroxybenzoyl CoA
→ Benzoic acid
In: Pseudomonas K191
→ 1-Benzoyl-2,3-dipalmitoylglycerol
In: Rat F749
→ 1-Benzoyl-2-hexadecylglycerol
In: Rat F749
→ Cyclohexa-1,5-diene-1-carboxyl CoA
In: Pseudomonas H19
Thauera J527, J812, K566
See also: J175
→ Hippuric acid
In: Rat A1369
→ *m*-Hydroxybenzoyl CoA
In: Pseudomonas H397
→ 3,4,5,6-Tetrahydrobenzoyl CoA
In: Pseudomonas G481, G642
→ 2,4,6-Trihydroxybenzophenone
In: Hypericum J509

N-Benzoylcyclohexylamine
→ N-Benzoyl-3-hydroxycyclohexylamine
In: Rat C23

N-Benzoyl-3,4-dichloroaniline
Formed from: N-Benzoyl-N-(1-carboxyethyl)-3,4-dichloroaniline ethyl ether

N-Benzoyl-O-(2-diethylaminoethyl)-L-tyrosine
Formed from: Tiropramide
→ N-Benzoyl-O-(2-ethylaminoethyl)-L-tyrosine
Probably in: Rat F97
→ N-Benzoyl-L-tyrosine
Probably in: Rat F97

N-Benzoyl-O-(2-diethylaminoethyl)-L-tyrosyl N-(3-hydroxypropyl)propylamide
Formed from: Tiropramide
→ N-Benzoyl-O-(2-diethylaminoethyl)-L-tyrosyl N-propenamide
Probably in: Rat F97

N-Benzoyl-O-(2-diethylaminoethyl)-L-tyrosyl N-propenamide
Probably formed from: N-Benzoyl-O-(2-diethylaminoethyl)-L-tyrosyl N-(3-hydroxypropyl)propylamide

5-Benzoyl-2,3-dihydro-1*H*-pyrrolizine-1-carboxylic acid
See: Ketorolac

1-Benzoyl-2,3-dipalmitoylglycerol
Formed from: Benzoyl CoA

Benzoylecgonine (3-(benzoyloxy)-8-methyl-8-azabicyclo[3.2.1]octane-2-carboxylic acid)
Formed from: Cocaine
Probably formed from: Cocaethylene
→ Benzoylnorecgonine
Probably in: Dog A1793
Rat A1344
→ Cocaethylene
Probably in: Rat J611

N-Benzoyl-O-(2-ethylaminoethyl)-L-tyrosine
Probably formed from: N-Benzoyl-O-(2-diethylaminoethyl)-L-tyrosine

N-Benzoyl-O-(2-ethylaminoethyl)tyrosyl propylamide
Probably formed from: Desethyltiropramide

Benzoylformic acid
See: Phenylglyoxylic acid

Benzoylgentiobioside
Probably formed from: Benzoylglucose

Benzoyl-β-D-glucose
Formed from: Benzoic acid
→ Benzoylgentiobioside
Probably in: Aconitum F445
Coffea F445
Dioscoreophyllum F445
Nicotiana F445
→ O-Benzoylmalic acid
Probably in: Glycine A3679

Benzoyl-β-D-glucuronic acid
Formed from: Benzoic acid
→ Benzoic acid
In: Beef A1029
Escherichia A1029

Benzoylglycine
See: Hippuric acid

3-Benzoyl-4-(N-(N-glycollyl)glycyl)amino-1,2-dimethylpyrrole
Formed from: Premazepam

Benzoylglycyl-L-lysine
→ Hippuric acid
With carboxypeptidase B A243

N-Benzoylheptamethyleneimine
→ N-Benzoyl-5-hydroxyheptamethyleneimine
In: Rat C23

1-Benzoylhexadecylglycerol
Formed from: Benzoyl CoA
N-Benzoylhexamethyleneimine
→ N-Benzoyl-3-hydroxyhexamethyleneimine
In: Rat C23
→ N-Benzoyl-4-hydroxyhexamethyleneimine
In: Rat C23
N-Benzoylhex-4-enylamine
→ N-Benzoyl-3-hydroxyhex-4-enylamine
In: Rat C23
N-Benzoylhydrazine
Formed from: N-Benzoyl-N′-(*p*-chlorobenzyl)hydrazine
N-Benzoyl-3-hydroxybutylamine
Formed from: N-Benzoylbutylamine
N-Benzoyl-3-(ω-hydroxy-*sec*-butyl)-1-oxa-4-azaspiro[4.4]nonane
Formed from: N-Benzoyl-3-*sec*-butyl-1-oxa-4-azaspiro[4.4]nonane
N-Benzoyl-3-hydroxycyclohexylamine
Formed from: N-Benzoylcyclohexylamine
N-Benzoyl-5-hydroxyheptamethyleneimine
Formed from: N-Benzoylheptamethyleneimine
N-Benzoyl-3-hydroxyhexamethyleneimine
Formed from: N-Benzoylhexamethyleneimine
N-Benzoyl-4-hydroxyhexamethyleneimine
Formed from: N-Benzoylhexamethyleneimine
N-Benzoyl-3-hydroxyhex-4-enylamine
Formed from: N-Benzoylhex-4-enylamine
N^2-Benzoyl-N^5-*m*-hydroxybenzoylornithine
Probably formed from: N^2-Benzoyl-N^5-*m*-phenoxybenzoylornithine
N^5-Benzoyl-N^2-*m*-hydroxybenzoylornithine
Probably formed from: N^5-Benzoyl-N^2-*m*-phenoxybenzoylornithine
Benzoyl-4′-hydroxyisonicotinanilide
Probably formed from: 2′-Benzoyl-4′-chloroisonicotinanilide
Benzoylhydroxylamate
Formed from: Benzoic acid
N-Benzoyl-3-hydroxy-N-methylbutylamine
Formed from: N-Benzoyl-N-methylbutylamine
N-Benzoyl-3-hydroxy-3-methylbutylamine
Formed from: N-Benzoyl-3-methylbutylamine
N-Benzoyl-4-hydroxy-4-methylpiperidine
Formed from: N-Benzoyl-4-methylpiperidine
2′-Benzoyl-2-hydroxy-4′-nitroacetanilide
Formed from: 1,3-Dihydro-7-nitro-5-phenyl-2*H*-1,4-benzodiazepin-2-one
N-Benzoyl-4-hydroxypiperidine
Formed from: N-Benzoylpiperidine
N-Benzoyl-3-hydroxypropylamine
Formed from: N-Benzoylpropylamine
N-Benzoyl-3-hydroxypyrrolidine
Formed from: N-Benzoylpyrrolidine
N-Benzoyliminostilbene
→ N-Benzoyliminostilbene-10,11-oxide
In: Rat E140
N-Benzoyliminostilbene-10,11-oxide
Formed from: N-Benzoyliminostilbene
N-Benzoylindoline
→ Benzoic acid
In: Cunninghamella D810
→ Indoline
In: Cunninghamella D810
N-Benzoylmalic acid
Formed from: Benzoic acid
Probably formed from: Benzoylglucose
Benzoyl-3-methoxy-5-methylbenzoic acid
Formed from: 5-Benzoyl-4-methoxy-6-methyl-2-pyrone
5-Benzoyl-4-methoxy-6-methyl-2-pyrone
→ Benzoyl-3-methoxy-5-methylbenzoic acid
In: Macrophoma E766
3-Benzoyl-α-methylbenzeneacetic acid
See: (*R*)-Ketoprofen
N-Benzoyl-3-methylbutylamine
→ N-Benzoyl-3-hydroxy-3-methylbutylamine
In: Rat C23
N-Benzoyl-N-methylbutylamine
→ N-Benzoyl-3-hydroxy-N-methylbutylamine
In: Rat C23
3-Benzoyl-4-(N-(N-methyl)glycyl)amino-1,2-dimethylpyrrole
Formed from: Premazepam
N-Benzoyl-4-methylpiperidine
→ N-Benzoyl-4-hydroxy-4-methylpiperidine
In: Rat C23
3-Benzoylmorphine
→ Morphine
In: Man K517
N-Benzoyl-*p*-nitroaniline
→ Benzoic acid

In: Rabbit K435
→ *p*-Nitroaniline
In: Rabbit K435

3-Benzoyl-1-methylpyridinium
Formed from: 3-Benzoylpyridine
→ 3-(α-Hydroxybenzoyl)-1-methylpyridinium
Probably in: Rat C664

3-Benzoyl-1-methylthiourea
Formed from: 1-Methyl-3-thiobenzoylthiourea
→ 3-Benzoyl-1-methylurea
Probably in: Pythium F154

3-Benzoyl-1-methylurea
Formed from: 3-Benzoyl-1-methylthiourea

6-Benzoyl-2-naphthol
→ 6-Benzoylnaphth-2-yl sulphate
In: Beef A1211

6-Benzoylnaphth-2-yl sulphate
Formed from: 6-Benzoyl-2-naphthol

Benzoylnorecgonine
Formed from: Norcocaine
Probably formed from: Benzoylecgonine

3-Benzoyloxy-N-(*tert*-butyl)-2-hydroxypropylamine
→ Benzoic acid
In: Rabbit H563

3-Benzoyloxy-N-(*tert*-butyl)-2-methoxypropylamine
→ Benzoic acid
In: Rabbit H563

3-Benzoyloxy-N-(1,1-dimethyl-2-(((morpholino)carbonyl)amino)ethyl)-2-hydroxypropylamine
→ Benzoic acid
In: Rabbit H563

N-(2-Benzoyloxyethyl)norfenfluramine
→ Norfenfluramine
In: Guinea pig A1152
Man A427
→ N-(2-Hydroxyethyl)norfenfluramine
In: Man A427

4-Benzoylphenol
Formed from: 4-Benzoylphenyl N-methylcarbamate

N²-Benzoyl-N⁵-*m*-phenoxybenzoylornithine
Formed from: *m*-Phenoxybenzoic acid
→ N²-Benzoyl-N⁵-*m*-hydroxybenzoylornithine
Probably in: Chicken E941

N⁵-Benzoyl-N²-*m*-phenoxybenzoylornithine
Formed from: *m*-Phenoxybenzoic acid
→ N⁵-Benzoyl-N²-*m*-hydroxybenzoylornithine
Probably in: Chicken E941

N-Benzoylphenylalanine
Formed from: L-Phenylalanine
→ Asperphenamate
Probably in: Penicillium E376
→ N-Benzoylphenylalaninol
Probably in: Penicillium B808, E376

N-Benzoyl-L-phenylalanine (2*S*)-2-(benzoylamio)-3-phenylpropyl ester
See: Asperphenamate

N-Benzoylphenylalaninol
Formed from: N-Benzoylphenylalanine
→ Asperphenamate
Probably in: Penicillium E376

N-Benzoyl-N′-phenyl-2,6-diaminobenzoic acid
→ N-Acetyl-N′-phenyl-2,6-diaminobenzoic acid
In: Guinea pig A2401
Rat A2401
→ Benzoic acid
In: Guinea pig A2401
Rat A2401

4-Benzoylphenyl N-methylcarbamate
→ 4-Benzoylphenol
In: Dog A694

2-(3-Benzoylphenyl)propionic acid
→ 2-(3-Benzoylphenyl)propionyl-β-D-glucoside
In: Panax F255
→ 2-(3-Benzoylphenyl)propionylsophoroside
In: Panax F255

2-(3-Benzoylphenyl)propionyl-β-D-glucoside
Formed from: 2-(3-Benzoylphenyl)propionic acid

2-(3-Benzoylphenyl)propionylsophoroside
Formed from: 2-(3-Benzoylphenyl)propionic acid

N-Benzoylpiperidine
→ N-Benzoyl-4-hydroxypiperidine
In: Rat C23

N-Benzoylpropylamine
→ N-Benzoyl-3-hydroxypropylamine

In: Rat C23

4-(2-Benzoylpropyl)-1-(2-ethoxy-2-phenylethyl)piperazine
See: Eprazinone

2-Benzoylpyridine
→ 2-(α-Hydroxybenzoyl)pyridine
In: Rabbit K569

3-Benzoylpyridine
→ 3-Benzoylpyridine-N-oxide
In: Rat C664
→ 3-(α-Hydroxybenzoyl)pyridine
In: Rabbit B431, K569
Rat C664

4-Benzoylpyridine
→ 4-(α-Hydroxybenzoyl)pyridine
In: Hamster F153
Monkey F241
Mouse E821
Rabbit B431, D125, K569

3-Benzoylpyridine-N-oxide
Formed from: 3-Benzoylpyridine
→ 3-(α-Hydroxybenzyl)pyridine-N-oxide
In: Rat C664

Benzoylpseudotropeine
See: Tropacocaine

N-Benzoylpyrrolidine
→ N-Benzoyl-3-hydroxypyrrolidine
In: Rat C23

β-Benzoylpyruvic acid
Formed from: β-Benzoyl-L-alanine

Benzoyltaurine
Formed from: Benzoic acid

9-Benzoyl-1,2,3,4-tetrahydro-6-hydroxycarbazole-3-carboxylic acid
Formed from: Oxarbazole

N-Benzoyl-1,2,3,4-tetrahydro-4-hydroxyisoquinoline
Formed from: N-Benzoyl-1,2,3,4-tetrahydroisoquinoline
→ N-Benzoyl-1,2,3,4-tetrahydro-4-oxoisoquinoline
Probably in: Cunninghamella D810

N-Benzoyl-1,2,3,4-tetrahydro-4-hydroxyquinoline
Formed from: N-Benzoyl-1,2,3,4-tetrahydroquinoline
→ N-Benzoyl-1,2,3,4-tetrahydro-4-oxoquinoline
Probably in: Cunninghamella D810

N-Benzoyl-1,2,3,4-tetrahydroisoquinoline
→ N-Benzoyl-1,2,3,4-tetrahydro-4-hydroxyisoquinoline
In: Cunninghamella D810

9-Benzoyl-1,2,3,4-tetrahydro-6-methoxycarbazole-3-carboxylic acid
See: Oxarbazole

N-Benzoyl-1,2,3,4-tetrahydro-4-oxoisoquinoline
Probably formed from: N-Benzoyl-1,2,3,4-tetrahydro-4-hydroxyisoquinoline

N-Benzoyl-1,2,3,4-tetrahydro-4-oxoquinoline
Probably formed from: N-Benzoyl-1,2,3,4-tetrahydro-4-hydroxyquinoline

N-Benzoyl-1,2,3,4-tetrahydroquinoline
→ N-Benzoyl-1,2,3,4-tetrahydro-4-hydroxyquinoline
In: Cunninghamella D810

O-Benzoylthiamine
→ Benzoic acid
In: Man A775
Rat A775
See also: A1126

N-Benzoyl-*p*-toluidine
Formed from: N-Benzyl-*p*-toluidine

N-Benzoyl-2,4,6-trimethylaniline
Formed from: N-Benzyl-2,4,6-trimethylaniline

Benzoyl trimethyl germane
→ α-Hydroxybenzoyl trimethyl germane
In: Bacillus G770
Candida G770
Corynebacterium G770
Micrococcus G770
Nocardia G770
Pichia G770
Rhodotorula G770
Saccharomyces G770
Torulopsis G770
Xanthomonas G770

Benzoyl trimethyl silane
→ α-Hydroxybenzoyl trimethyl silane
In: Bacillus G770, G828
Candida G770
Corynebacterium G770
Deleya G828
Micrococcus G770, G828
Nocardia G770

Pichia G770
Rhodotorula G770
Saccharomyces G770
Torulopsis G770
Xanthomonas G770

N-Benzoyl-L-tyrosine
Formed from: N-Benzoyl-L-tyrosyl-*p*-aminobenzoic acid
Probably formed from: N-Benzoyl-O-(2-diethylaminoethyl)tyrosine
→ O-Acetyl-N-benzoyl-L-tyrosine
Probably in: Rat F97

N-Benzoyl-L-tyrosyl-*p*-aminobenzoic acid
→ *p*-Aminobenzoic acid
In: Guinea pig A2978
Rat A2978
→ N-Benzoyl-L-tyrosine
In: Guinea pig A2978
Rat A2978

N-Benzoyl-L-tyrosyl dipropylamide
Formed from: Tiropramide
N-Benzoyl-L-tyrosyl propylamide
Probably in: Rat F97

N-Benzoyl-L-tyrosyl propylamide
N-Benzoyl-L-tyrosyl dipropylamide

Benzphetamine
See: N-Benzyl-N,α-dimethylphenethylamine

3,4-Benzpyrene (benzpyrene, benzo[*a*]pyrene)
Formed from: 3,4-Benzpyrene-4,5-oxide
3,4-Benzpyrene-7,8-oxide
6-Nitro-3,4-benzpyrene
→ 3,4-Benzpyrene-4,5-oxide
In: Guinea pig B812
Rabbit B286
Rat A70, A739, C204
See also: A68, A955, A1661, A1714, A2009, A2327, A2379, A2394, A2589, A2814, A3093, A3127, A3164, A3277, A3412, A3422, A3598, A3649, A3681, A3746, A3770, A3782, A3790, A3847, A3937, B122, B134, B376, B387, B509, B559, B561, B656, B657, B670, C139, C309, C321, C390, C427, C683, D51, D465, E226, F150, G445, G972, H10, J94, J311, J503, K562
→ 3,4-Benzpyrene-7,8-oxide
Probably in: Asterias G972
Barnacle B744
Beef B937
Bluegill E610
Bullhead D465, E610
Camel H103
Catfish J311
Cunninghamella B235, B353
Duck K562
Goldfish D465
Guinea pig C683
Hamster A13, A1644, A2009, A3625, A3746, A3770, A3937, B134
Ictalurus C139
Man A13, A1644, A2379, A2406, A3321, A3744, A3847, B310, B350, B559, C138, C321, G360, H10, J94, J835, K162
Mirror carp G147
Mouse A3681, A3746, D51, F150, F376, G445
Mullet C141
Mycobacterium H779
Rabbit A2327, A2394, A3790, B286, E576
Rat A13, A68, A955, A1644, A1661, A1714, A2589, A3093, A3127, A3164, A3234, A3277, A3412, A3598, A3625, A3649, A3681, A3746, A3782, B376, B561, B656, B670, C138, C204, C229, C390, C427, D465, E226
Rhesus monkey A2814
Scup D965
Sea anemone J503
Sole C473
Tilapia H113
Trout E610, J311
→ 3,4-Benzpyrene-9,10-oxide
Probably in: Asterias G972
Barnacle B744
Beef B937
Bluegill E610
Bullhead D465, E610
Catfish J311
Cunninghamella B235, B353
Duck K562
Goldfish D465
Guinea pig C683
Hamster A13, A1644, A2009, A3625, A3746, A3770, A3937, B134
Ictalurus C139

Man A13, A1644, A2379, A2406, A3321, A3744, A3847, B310, B350, B559, C138, C321, G360, H10, J94, J835
Mirror carp G147
Mouse A3681, A3746, D51, F130, F376, G445
Mullet C141
Rabbit A2327, A2394, A2729, A3790, B286, B509, C309
Rat A13, A68, A955, A1644, A1661, A1714, A2589, A3093, A3127, A3164, A3277, A3412, A3598, A3625, A3649, A3681, A3746, A3782, B376, B561, B656, B670, C138, C204, C229, C390, C427, C520, D465, E226, G445
Rhesus monkey A2814
Scup D965
Sea anemone J503
Sole C473
Trout E610, J311

→ 3,4-Benzpyrene-1,6-quinone
In: Bullhead D465
Catfish J311
Cunninghamella B235, B353
Duck K562
Glycine B692, D241
Goldfish D465
Man A2406, B122, H10, J835
Mouse A3681
Parsley B692
Phanerochaete E86
Pycnoporus J907
Rat A955, A2379, A3164, A3277, A3598, A3649, A3681, A3782, C204, D465, E226
Rhesus monkey A2814
Sea anemone J503
Starfish G972
Trout J311
See also: D207, G601

→ 3,4-Benzpyrene-3,6-quinone
In: Bullhead D465
Catfish J311
Cunninghamella B235
Duck K562
Glycine B692, D241
Goldfish D465
Man G360, H10, J835
Mouse G445
Parsley B692
Phanerochaete E86
Pycnoporus J907
Rabbit A2394, C309
Rat A955, A2379, A3164, A3277, A3412, A3649, A3782, B656, C204, D465, E226, G445
Rhesus monkey A2814
Sea anemone J503
Trout J311
See also: A2145, D207

→ 3,4-Benzpyrene-6,12-quinone
In: Duck K562
Glycine B692, D241
Man A2406, H10, J835
Monkey A2814
Parsley B692
Phanerochaete E86
Pycnoporus J907
Rat A2379, A3412, A3782, B656, C204, E226, G101
Sea anemone J503
Starfish G972
See also: D207, G601

→ Chrysene-4,5-dicarboxylic acid
In: Mycobacterium H779

→ *cis*-4,5-Dihydro-4,5-dihydroxy-3,4-benzpyrene
In: Selenastrum D730

→ *cis*-7,8-Dihydro-7,8-dihydroxy-3,4-benzpyrene
In: Selenastrum D730

→ *cis*-9,10-Dihydro-9,10-dihydroxy-3,4-benzpyrene
In: Beijerinckia A2227

→ *cis*-10,11-Dihydro-10,11-dihydroxy-3,4-benzpyrene
In: Selenastrum D732

→ 1-Hydroxy-3,4-benzpyrene
In: Man J94
Rat A458, C204
Sole C473

→ 3-Hydroxy-3,4-benzpyrene
In: Barnacle B744

Beef B937
Bluegill E480
Bullhead D465
Camel H103
Candida B775
Catfish J311
Crow C307
Cunninghamella B235
Duck K562
Egret C307
Goldfish D465
Guinea pig B812, C383
Hamster A13, A1644, A1658, A2009, A3262, A3412, A3625, A3770, A3937, B134, E480
Ictalurus C139
Killifish B259
Kite C307
Man A13, A114, A1644, A1869, A2379, A2406, A3744, B122, B350, C138, E480, G360, H10, J94, J835
Mouse A3681, D51, E480, F150, F376, G445
Mullet C141
Pig A1997
Pigeon C307
Procambarus J492
Rabbit A2327, A2394, A2729, A3790, B286, C309
Rat A13, A68, A114, A458, A955, A1644, A1714, A1869, A1997, A2379, A2454, A2589, A3127, A3164, A3277, A3412, A3625, A3649, A3681, A3782, B477, B561, B656, B657, C138, C204, C226, C520, D444, D465, E226, E480, G445
Rhesus monkey A1869, A1997, A2814
Sea anemone J503
Scup D965
Sole C473, G374
Tilapia H113
Tree shrew A1997
Trout A2145, J311

See also: A2120, A2127, C668

→ 4-Hydroxy-3,4-benzpyrene
In: Rat C204

→ 7-Hydroxy-3,4-benzpyrene
In: Man B122, J835
Mouse G445, F376
Rat A458, A2454, A3127, C204, G445
Scup D965

→ 8-Hydroxy-3,4-benzpyrene
In: Rainbow trout A1424

→ 9-Hydroxy-3,4-benzpyrene
In: Barnacle B744
Camel H103
Candida B775
Cunninghamella B235, B353
Duck K562
Goldfish D465
Glycine E416
Guinea pig C683
Hamster A3262, A3770, A3412, B134
Ictalurus C139
Man A2379, A2406, A3744, B122, B387, G360, J94, J835
Mouse A3681, F150, G445, F376
Mullet C141
Rabbit A2327, A3790, C309
Rat A458, A955, A2454, A3127, A3164, A3412, A3598, A3649, A3681, A3782, B561, B656, B657, B959, C204, D465, E226, G445
Rhesus monkey A2814
Saccharomyces A3326
Scup D965
Sole C473
Tilapia H113

→ 6-Hydroxymethyl-3,4-benzpyrene
In: Rat A142, A815, B586

→ 6-Oxy-3,4-benzpyrene radical
In: Rat A1896, A2298

Benz[*e*]pyrene

→ *trans*-4,5-Dihydro-4,5-dihydroxybenz[*e*]pyrene
In: Hamster C340
Rat B523, C767, D110

→ *trans*-9,10-Dihydro-9,10-dihydroxybenz[*e*]pyrene
In: Rat B523, C767, D110

→ 1-Hydroxybenz[*e*]pyrene
In: Rat C767, D110

→ 3-Hydroxybenz[*e*]pyrene

In: Hamster C340
Rat C767, D110

3,4-Benzpyrene-4,5-oxide
Formed from: 3,4-Benzpyrene
→ 3,4-Benzpyrene
In: Rat A350, A2601, A3527, B558
→ *trans*-4,5-Dihydro-4,5-dihydroxy-3,4-benzpyrene
In: Clam B69
Crab B69
Dogfish B69
Eel B69
Flounder B69
Lobster B69
Man A2486, A3217, B63, B220, C386, C507
Mouse A2486, B648, D609
Rabbit A1974
Rat A739, A1894, A1964, A2486, A2728, A3031, A3166, A3461, A3902, B33, B63, B142, B220, B869, C189, D140, D628, F374
Sheepshead B69
Skate B69
Stingray B69
See also: A68, A532, A1714, A2009, A2589, B812
→ S-(4,5-Dihydro-4-hydroxy-3,4-benzpyren-5-yl)glutathione
In: Man E928
Rat C472, D842
See also: A1964, A1974, A2282, A3440, A3461, A3790, A3902, D235, D628, D912, E434
→ S-(4,5-Dihydro-5-hydroxy-3,4-benzpyren-4-yl)glutathione
In: Man E928
Rat C472, D842

3,4-Benzpyrene-7,8-oxide
Formed from: 3,4-Benzpyrene
→ 3,4-Benzpyrene
In: Rat B558
→ *trans*-7,8-Dihydro-7,8-dihydroxy-3,4-benzpyrene
In: Man A3217, B63, B220
Rat A3031, A3166, B63, B142, B567, B869
→ S-(7,8-Dihydro-7-hydroxy-3,4-benzpyren-8-yl)glutathione
Probably in: at D912

3,4-Benzpyrene-9,10-oxide
Formed from: 3,4-Benzpyrene
→ *trans*-9,10-Dihydro-9,10-dihydroxy-3,4-benzpyrene
In: Man A3217
Rat A3031, A3166, B142
See also: A13, A68, A1644, A1714, A2009, A2589

3,4-Benzpyrene-11,12-oxide
Probably formed from: 3,4-Benzpyrene
→ *trans*-11,12-Dihydro-11,12-dihydroxy-3,4-benzpyrene
In: Man A3217, B63
Rat A3031, B63, B142

3,4-Benzpyrene-1,6-quinone
Formed from: 3,4-Benzpyrene
1-Hydroxy-3,4-benzpyrene
→ 1,6-Dihydroxy-3,4-benzpyrene
Probably in: Rat G601

3,4-Benzpyrene-3,6-quinone
Formed from: 3,4-Benzpyrene
3,6-Dihydroxy-3,4-benzpyrene
→ 3,6-Dihydroxy-3,4-benzpyrene
In: Rat A3637, B674, B691, D317, D713
See also: A3622, G601

3,4-Benzpyrene-4,5-quinone
Probably formed from: 4,5-Dihydro-4,5-dihydroxy-3,4-benzpyrene

3,4-Benzpyrene-6,12-quinone
Formed from: 3,4-Benzpyrene
→ 6,12-Dihydroxy-3,4-benzpyrene
Probably in: Rat G601

3,4-Benzpyren-1-yl-β-D-glucuronide
Formed from: 1-Hydroxy-3,4-benzpyrene

3,4-Benzpyren-2-yl-β-D-glucuronide
Formed from: 2-Hydroxy-3,4-benzpyrene

3,4-Benzpyren-3-yl-β-D-glucuronide
Formed from: 3-Hydroxy-3,4-benzpyrene
→ 3-Hydroxy-3,4-benzpyrene
In: Bacteroides D953
Clostridium D953
Escherichia D953
Peptococcus D953
Rat D673

3,4-Benzpyren-4-yl-β-D-glucuronide
Formed from: 4-Hydroxy-3,4-benzpyrene

3,4-Benzpyren-5-yl-β-D-glucuronide
Formed from: 5-Hydroxy-3,4-benzpyrene

3,4-Benzpyren-6-yl-β-D-glucuronide
Formed from: 6-Hydroxy-3,4-benzpyrene

3,4-Benzpyren-7-yl-β-D-glucuronide
Formed from: 7-Hydroxy-3,4-benzpyrene

3,4-Benzpyren-8-yl-β-D-glucuronide
Formed from: 8-Hydroxy-3,4-benzpyrene

3,4-Benzpyren-9-yl-β-D-glucuronide
Formed from: 9-Hydroxy-3,4-benzpyrene
→ 9-Hydroxy-3,4-benzpyrene
In: Clostridium D953
Escherichia D953

3,4-Benzpyren-10-yl-β-D-glucuronide
Formed from: 10-Hydroxy-3,4-benzpyrene

3,4-Benzpyren-11-yl-β-D-glucuronide
Formed from: 11-Hydroxy-3,4-benzpyrene

3,4-Benzpyren-12-yl-β-D-glucuronide
Formed from: 12-Hydroxy-3,4-benzpyrene

3,4-Benzpyren-1-yl sulphate
Formed from: 1-Hydroxy-3,4-benzpyrene

3,4-Benzpyren-2-yl sulphate
Formed from: 2-Hydroxy-3,4-benzpyrene

3,4-Benzpyren-3-yl sulphate
Formed from: 3-Hydroxy-3,4-benzpyrene
→ 3-Hydroxy-3,4-benzpyrene
In: Limpet A1897

3,4-Benzpyren-4-yl sulphate
Formed from: 4-Hydroxy-3,4-benzpyrene

3,4-Benzpyren-6-yl sulphate
Formed from: 6-Hydroxy-3,4-benzpyrene

3,4-Benzpyren-7-yl sulphate
Formed from: 7-Hydroxy-3,4-benzpyrene
→ 7-Hydroxy-3,4-benzpyrene
In: Clostridium D953

3,4-Benzpyren-8-yl sulphate
Formed from: 8-Hydroxy-3,4-benzpyrene

3,4-Benzpyren-9-yl sulphate
Formed from: 9-Hydroxy-3,4-benzpyrene
→ 9-Hydroxy-3,4-benzpyrene
In: Escherichia D953

3,4-Benzpyren-10-yl sulphate
Formed from: 10-Hydroxy-3,4-benzpyrene

3,4-Benzpyren-12-yl sulphate
Formed from: 12-Hydroxy-3,4-benzpyrene

Benztropine ((3-*endo*)-3-(diphenylmethoxy)-8-methyl-8-azabicyclo[3.2.1]octane)
→ Benztropine-N-oxide
In: Rat H312
→ Demethylbenztropine
In: Rat H312
→ *p*-Hydroxybenztropine
In: Rat H312

Benztropine-N-oxide
Formed from: Benztropine
→ *p*-Hydroxybenztropine-N-oxide
Probably in: Rat H312

Benzydamine (1-benzyl-3-(3-dimethylaminopropoxy)-1*H*-indazole)
Formed from: Benzydamine-N-oxide
→ Benzaldehyde
Probably in: Man D678
Rabbit A826
Rat B417
→ Benzydamine-N-oxide
In: Man D678
Rabbit A826
Rat B417, G989
→ 1-Benzyl-3-(3-dimethylaminopropoxy)-hydroxy-1*H*-indazole
In: Man D678
→ 1-Benzyl-1*H*-indazolone
In: Rabbit A826
Man D678
→ Demethylbenzydamine
In: Man D678
Rabbit A826
Rat B417
→ 3-(3-Dimethylaminopropoxy)-1-(*p*-hydroxybenzyl)-1*H*-indazole
In: Rabbit A826
Rat B417
→ 3-(3-Dimethylaminopropoxy)-1*H*-indazole
In: Man D678

Benzydamine-N-oxide
Formed from: Benzydamine
→ Benzydamine
In: Rat B307

Benzyl acetate
Formed from: Benzyl alcohol
→ Benzyl alcohol
Probably in: Man E398

→ *cis*-2,3-Dihydro-2,3-dihydroxybenzyl acetate
In: Pseudomonas J656
See also: F283

N^6-Benzyladenine
Formed from: N^6-Benzyladenosine

N^6-Benzyladenosine
Formed from: N^6-Benzyladenosine-5′-phosphate
→ Benzoic acid
Probably in: Rat B149
→ N^6-Benzyladenine
In: Rat B149

N^6-Benzyladenosine-5′-phosphate
→ N^6-Benzyladenosine
In: Rat B149

1-Benzylagroclavine
→ 1-Benzylelymoclavine
In: Claviceps D237

Benzyl alcohol
Formed from: Benzaldehyde
Benzoic acid
Benzyl butyl phthalate
S-Benzyl O-ethyl phenylphosphonothioate
Benzyl mandelate
Benzyl nicotinate
m-Chlorobenzyl alcohol
Cyclopropylbenzene
Dibenzyl sulphoxide
N-((3,4-Dihydro-7-(piperazin-1-yl)-1(1*H*)-isoquinolinon-2-yl)acetyl)-3(*S*)-(ethynyl)-β-alanine benzyl ester
Peroxyphenylacetic acid
Toluene
Tripelennamine
Probably formed from: Benzyl acetate
Phenylacetone
→ N-Acetyl-S-benzyl-L-cysteine
In: Rat C98
See also: E309, E674
→ Benzaldehyde
In: Acinetobacter E473, E663, J621
Botrytis G613
Escherichia K491
Horse A2174
Man F87, F598
Methylosinus F404
Mycobacterium A150
Nitrosomonas H219
Phanerochaete F257, H613
Pig F389
Pleurotus F867, G668, H241
Pseudomonas A1366, A3837, J219
Rhodopseudomonas D197
Thauera H610
See also: A1726, D211, E309, G652, H219, J819
→ Benzyl acetate
In: Clarkia J521, K240
Musk melon K561
→ Benzyl benzoate
In: Cranberry A3545
→ Benzyl-β-D-glucoside
In: Cherry C756
Sorghum K313
See also: F786
→ Benzyl sulphate
In: Rat E630

Benzylaldoxime
Formed from: L-Phenylalanine

Benzylamine
Formed from: N-Acetylbenzylamine
Benzyl glucosinolate
Benzylhydrazine
N-Benzylphthalamidic acid
Dibenzylamine
Probably formed from: N-(2-(N-Acetamido)acetoxy)benzylamine
Demethylpargyline
N-Methylbenzylamine
→ N-Acetylbenzylamine
In: Cunninghamella C68
→ Benzaldehyde
In: Beef A574, A967, F395
Hedgehog A434
Man A821, A1901, A2010, A2835, A3957
Pig A72, A1033, A1185, A2587, B206, J833
Pseudomonas C517
Rabbit B457
Rat A1355, A1760, A2835, A3948, B457, F217, F816
Sheep A535

See also: G149
→ N-Methylbenzylamine
In: Rabbit E6

p-Benzylaminobenzonitrile
→ *p*-Aminobenzonitrile
In: Hamster G951
→ Benzaldehyde
In: Hamster G951

1-Benzylamino-1*H*-benzotriazole
→ Benzaldehyde
In: Guinea pig H836
→ 1-Amino-1*H*-benzotriazole
In: Guinea pig H836
→ 1-Amino-3-hydroxy-1*H*-benzotriazole
Probably in: Guinea pig H836
→ 1-Amino-4-hydroxy-1*H*-benzotriazole
In: Guinea pig H836

2-(Benzylamino)ethyl 5-(5,5-dimethyl-2-oxo-1,3,2-dioxaphosphorinan-2-yl)-1,4-dihydro-2,6-dimethyl-4-(*m*-nitrophenyl)-3-pyridinecarboxylate
Formed from: Efonidipine

p-Benzylaminophenol
Formed from: N-Benzylaniline
Probably formed from: N-Benzyl-*p*-hydroxy-N-(3-(2-hydroxy-2-methyl-1-propoxy)-2-(1-pyrrolidino)propyl)aniline

2-Benzylamino-1-propanol
Probably formed from: N-Benzyl-N-phenoxyisopropylamine

6-Benzylaminopurine
→ β-(6-Benzylaminopurin-9-yl)alanine
In: Lupinus C770
Pisum C770
Spinacia C770

β-(6-Benzylaminopurin-9-yl)alanine
Formed from: 6-Benzylaminopurine

2-Benzylaminopyridine
Formed from: Tripelennamine

N-Benzylamphetamine (norbenzphetamine, N-benzyl-α-methylphenethylamine)
Formed from: N-Benzyl-N,α-dimethylphenethylamine
→ Amphetamine
In: Rabbit A2717
See also: C433
→ N-(1-Benzylethyl)-α-phenylnitrone
In: Rabbit A2717
→ N-Benzyl-N-hydroxyamphetamine
In: Rabbit A2717

N-Benzylaniline
Formed from: Dibenzylaniline
→ Aniline
In: Ferret D367
Guinea pig D367, E227
Hamster D367, E227, G951
Mouse D367, E227
Rabbit D367, E227
Rat D367, E227
→ Benzaldehyde
In: Ferret D367
Guinea pig D367
Hamster D367, G951
Mouse D367
Rabbit D367
Rat D367, E227
→ N-Benzoylaniline
In: Hamster D367, G951
Rabbit D367, E227
→ *p*-Benzylaminophenol
In: Ferret D367
Guinea pig D367, E227
Hamster D367, E227
Mouse D367, E227
Rabbit D367, E227
Rat D367, E227
→ α,N-Diphenylnitrone
In: Ferret D367
Guinea pig D367, E227
Hamster D367, E227, G951
Mouse D367, E227
Rabbit D367, E227
Rat D367, E227

N-Benzylbenzamide
Probably formed from: N,N-Dibenzylbenzamide

Benzyl benzoate
Formed from: Benzoic acid
Benzyl alcohol

N-Benzyl-3-*sec*-butyl-1-oxa-4-azaspiro[4.4]nonane
→ N-Benzyl-3-(ω-hydroxy-*sec*-butyl)-1-oxa-4-azaspiro[4.4]nonane
In: Beauveria K678

2(*R*)-Benzyl-5-((2(*S*)-*tert*-butylaminocarbonyl) piperazino)-N-(2,3-dihydroxy-1(*S*)-indanyl)-4(*S*)-hydroxypentanamide
Probably formed from: 3-Hydroxyindinavir

2(*R*)-Benzyl-5-((2(*S*)-*tert*-butylaminocarbonyl) piperazino)-4(*S*)-hydroxy-N-(2(*R*)-hydroxy-1(*S*)-indanyl)pentanamide
Formed from: Indinavir

Benzyl *n*-butyl phthalate
→ Benzyl alcohol
In: Rat K261
→ Phthalic acid monobenzyl ester
In: Rat K261
→ Phthalic acid mono-*n*-butyl ester
In: Rat K261

Benzyl *n*-butyl sulphide
→ Benzyl *n*-butyl sulphoxide
In: Helminthosporium G427
Mortierella G427

Benzyl *tert*-butyl sulphide
→ Benzyl *tert*-butyl sulphoxide
In: Helminthosporium G427
Mortierella G427

Benzyl *n*-butyl sulphone
Formed from: Benzyl *n*-butyl sulphoxide

Benzyl *n*-butyl sulphoxide
Formed from: Benzyl *n*-butyl sulphide
→ Benzyl *n*-butyl sulphone
Probably in: Helminthosporium G427
Mortierella G427

Benzyl *tert*-butyl sulphoxide
Formed from: Benzyl *n*-butyl sulphide

N-Benzylcarbazole
→ Benzaldehyde
In: Hamster H391
→ Carbazole
In: Hamster H391
→ N-Benzylhydroxycarbazole
In: Hamster H391

N-Benzyl-3-carboxamido-4-hydroxyquinolinium
Formed from: N-Benzyl-3-carboxamidoquinolinium

N-Benzyl-3-carboxamidoquinolinium
→ N-Benzyl-3-carboxamido-4-hydroxyquinolinium
In: Rabbit D208

3-Benzyl-4-carboxy-4-hydroxythiazolidin-2-thione
Formed from: S-(N-Benzylthiocarbamoyl)-L-cysteine

N-Benzyl-N-(2-((3-carboxypropyl)amino)-3-(2-methyl-1-propoxy)propyl)aniline
Formed from: Bepridil
Probably formed from: N-Benzyl-N-(3-(2-methyl-1-propoxy)-2-(2-oxo-1-pyrrolidino)propyl)aniline

8-Benzyl-(+)-catechin
→ 8-Benzyl-3′-O-methyl-(+)-catechin
In: Pig D112

Benzyl chloride
→ Benzoic acid
In: Marmoset E58
→ Benzyl-L-cysteine
In: Marmoset E58

N-Benzyl-*p*-chloroaniline
→ Benzaldehyde
In: Ferret D637
Guinea pig D637
Hamster D637, G951
Mouse D637
Rabbit D637
Rat D637
→ N-Benzoyl-*p*-chloroaniline
In: Hamster D637, G951
Rabbit D637
→ N-Benzylidene-*p*-chloroaniline
In: Rabbit H263
→ *p*-Chloroaniline
In: Ferret D637
Guinea pig D637
Hamster D637, G951
Mouse D637
Rabbit D637
Rat D637
→ N-(*p*-Chlorophenyl)-α-phenylnitrone
In: Guinea pig D637
Hamster D637, G951
Mouse D637
Rat D637

N-Benzyl-*p*-chlorobenzamide
Formed from: N-Benzyl-*p*-chloro-N-methylbenzamide

N-Benzyl-3-chloro-2-hydroxypropionamide
Formed from: Beclamide

→ N-Benzyl-3-chloro-2-hydroxypropionamide-β-D-glucuronide
In: Man A3507

N-Benzyl-*p*-chloro-N-methylbenzamide
→ Benzaldehyde
Probably in: Rat G13
→ N-Benzyl-*p*-chlorobenzamide
In: Rat G13
→ *p*-Chloro-N-methylbenzamide
In: Rat G13

2-Benzyl-4-chlorophenol
→ 2-Benzyl-4-chlorophenyl-β-D-glucuronide
In: Rat E131
→ 4-Chloro-2-(*p*-hydroxybenzyl)phenol
Probably in: Rat E131

2-Benzyl-4-chlorophenyl-β-D-glucuronide
Formed from: 2-Benzyl-4-chlorophenol

N-Benzyl-N-(*p*-chlorophenyl)hydroxylamine
Probably formed from: N-(*p*-chlorophenyl)-α-phenylnitrone

N-Benzyl-3-chloropropionamide
See: Beclamide

Benzyl cyanide
See: Phenylacetonitrile

4-Benzyl-1-cyclopropyl-2,3-dihydropyridinium
Probably formed from: 4-Benzyl-1-cyclopropyl-1,2,3,6-tetrahydropyridine

4-Benzyl-1-cyclopropyl-1,2,3,6-tetrahydropyridine
→ 4-Benzyl-1-cyclopropyl-2,3-dihydropyridinium
Probably In: Beef J11

S-Benzyl-L-cysteine
Formed from: N-Acetyl-S-benzyl-L-cysteine
→ N-Acetyl-S-benzyl-L-cysteine
In: Marmoset E58
Rat A3043

S-Benzyl-L-cysteinyl-*p*-nitroanilide
→ *p*-Nitroaniline
In: Escherichia E715

N-Benzyldaunorubicin
Formed from: N,N-Dibenzyldaunorubicin
→ N-Benzyldaunorubicinol
Probably in: Rat C539
→ Daunorubicin
In: Rat C539, D88

N-Benzyldaunorubicinol
Probably formed from: N-Benzyldaunorubicin

4-Benzyldehydrocyclopeptine
→ 4-Benzyl-3′-hydroxydehydrocyclopeptine
In: Penicillium A1633

O⁶-Benzyl-2′deoxyguanosine
→ O⁶-Benzylguanine
In: Rat H322

Benzyldiazonium ion
Formed from: 1-Phenylazo-2-naphthol

Benzyl N,N-di-*sec*-butylsulphinylcarbamate
Formed from: Drepamon
Probably formed from: Benzyl N,N-di-*sec*-butylsulphonylcarbamate

Benzyl N,N-di-*sec*-butylsulphonylcarbamate
→ Benzyl N,N-di-*sec*-butylsulphinylcarbamate
Probably in: Barnyard grass A1827
Rice A1827

S-Benzyl N,N-di-*sec*-butylthiocarbamate
See: Drepamon

2-(3-Benzyl-3,4-dihydro-3,4-dihydroxy-2*H*-1-benzopyran-7-yl)-4-trifluoromethylbenzoic acid
Formed from: 2-(3*S*,4*R*)-3-Benzyl-3,4-dihydro-4-hydroxy-2*H*-1-benzopyran-7-yl)-4-trifluoromethylbenzoic acid

2-((3*S*,4*R*)-3-Benzyl-3,4-dihydro-4-hydroxy-2*H*-1-benzopyran-7-yl)-4-trifluoromethylbenzoic acid
→ 2-(3-Benzyl-3.4-dihydro-3,4-dihydroxy-2*H*-1-benzopyran-7-yl)-4-trifluoromethylbenzoic acid
In: Cynomolgus monkey J755
Rat J755
→ 2-((3*S*,4*R*)-3-Benzyl-3,4-dihydro-4-hydroxy-2*H*-1-benzopyran-7-yl)-4-trifluoromethylbenzoic acid-4-β-D-glucuronide
In: Cynomolgus monkey J755
Rat J755
→ 2-((3*S*,4*R*)-3-Benzyl-3,4-dihydro-4-hydroxy-2*H*-1-benzopyran-7-yl)-4-trifluoromethylbenzoyl-β-D-glucuronic acid
In: Cynomolgus monkey J755
Rat J755
→ 2-((3*S*)-3-Benzyl-3,4-dihydro-4-oxo-2*H*-1-benzopyran-7-yl)-4-trifluoromethylbenzoic acid
In: Cynomolgus monkey J755

Rat J755
→ 2-((3*S*,4*R*)-3,4-Dihydro-4-hydroxy-3-(α-hydroxybenzyl)-2*H*-1-benzopyran-7-yl)-4-trifluoromethylbenzoic acid
In: Cynomolgus monkey J755
Rat J755
→ 2-((3*S*,4*R*)-3,4-Dihydro-4-hydroxy-3-(*p*-hydroxybenzyl)-2*H*-1-benzopyran-7-yl)-4-trifluoromethylbenzoic acid
In: Cynomolgus monkey J755
Rat J755

2-((3*S*,4*R*)-3-Benzyl-3,4-dihydro-4-hydroxy-2*H*-1-benzopyran-7-yl)-4-trifluoromethylbenzoic acid-4-β-D-glucuronide
Formed from: 2-((3*S*,4*R*)-3-Benzyl-3,4-dihydro-4-hydroxy-2*H*-1-benzopyran-7-yl)-4-trifluoromethylbenzoic acid

2-((3*S*,4*R*)-3-Benzyl-3,4-dihydro-4-hydroxy-2*H*-1-benzopyran-7-yl)-4-trifluoromethylbenzoyl-β-D-glucuronic acid
Formed from: 2-((3*S*,4*R*)-3-Benzyl-3,4-dihydro-4-hydroxy-2*H*-1-benzopyran-7-yl)-4-trifluoromethylbenzoic acid

2-Benzyl-3,7-dihydro-6-(*p*-hydroxyphenyl)-8-benzylimidazo[1,2-*a*]pyrazin-3-one
→ 3-Benzyl-5-(*p*-hydroxyphenyl)-2-((*p*-hydroxyphenylacetyl)amino)pyrazine
In: Renilla K813, K822

2-((3*S*)-3-Benzyl-3,4-dihydro-4-oxo-2*H*-1-benzopyran-7-yl)-4-trifluoromethylbenzoic acid
Formed from: 2-((3S,4R)-3-Benzyl-3,4-dihydro-4-hydroxy-2*H*-1-benzopyran-7-yl)-4-trifluoromethylbenzoic acid

N-Benzyl-3,4-dihydroxyphenethylamine
→ N-Benzyl-3-hydroxy-4-methoxyphenethylamine
In: Rat A486
→ N-Benzyl-4-hydroxy-3-methoxyphenethylamine
In: Rat A486

1-Benzyl-1-(3-dimethylaminopropoxy) cycloheptane
See: Bencyclane

1-Benzyl-1-(3-dimethylaminopropoxy)-4-hydroxycycloheptane
Formed from: Bencyclane
→ 1-Benzyl-1-(3-dimethylaminopropoxy)-4-oxocycloheptane
In: Man A3506
Rat A3506

1-Benzyl-1-(3-dimethylaminopropoxy)-1*H*-indazole
See: Benzydamine

1-Benzyl-1-(3-dimethylaminopropoxy)-4-oxocycloheptane
Formed from: 1-Benzyl-1-(3-dimethylaminopropoxy)-4-hydroxycycloheptane

N-Benzyl-N′,N″-dimethylguanidine
See: Bethanidine

Benzyl dimethyl nonylammonium
→ Benzyl methyl nonylamine
In: Rat D638

N-Benzyl-N,α-dimethylphenethylamine (benzphetamine, N-benzyl-N-methylamphetamine)
→ Benzaldehyde
In: Rabbit J98
Rat J98
→ N-Benzylamphetamine
In: Cat C562
Dog C562
Guinea pig C562
Hamster A1658
Macaque A2117
Man C798, D135, E52
Mouse C562
Quail C562
Rabbit A570, A1996, C562
Rat C433, C562, C669, D39, D98,
Trout C562
See also: J98
→ N-Benzyl-N,α-dimethylphenethylamine-N-oxide
In: Mouse D446
Pig D466
→ N-Benzyl-*m*-hydroxy-N-methylamphetamine
In: Man E52
Rat C376
→ N-Benzyl-*p*-hydroxy-N-methylamphetamine
In: Man E52
Rat C376, C433

→ N-Methylamphetamine
In: Man A589, E52, J692
Rat C376, C433
→ Phenylacetone
In: Rabbit J98
Rat J98

N-Benzyl-N,α-dimethylphenethylamine-N-oxide
Formed from: N-Benzyl-N,α-dimethylphenethylamine

4-Benzyl-1-(2,6-dimethylphenoxy)-10-(hexahydro-2,4-dioxopyrimidin-1-yl)-7-(*p*-hydroxybenzyl)-11-methyl-2,9-dioxo-3,8-diazadodecane
Probably formed from: 4,7-Dibenzyl-1-(2,6-dimethylphenoxy)-10-(hexahydro-2,4-dioxopyrimidin-1-yl)-11-methyl-2,9-dioxo-3,8-diazadodecane

7-Benzyl-1-(2,6-dimethylphenoxy)-10-(hexahydro-2,4-dioxopyrimidin-1-yl)-4-(*p*-hydroxybenzyl)-11-methyl-2,9-dioxo-3,8-diazadodecane
Probably formed from: 4,7-Dibenzyl-1-(2,6-dimethylphenoxy)-10-(hexahydro-2,4-dioxopyrimidin-1-yl)-11-methyl-2,9-dioxo-3,8-diazadodecane

4-Benzyl-1-(2,6-dimethylphenoxy)-10-(hexahydro-4-hydroxy-2-oxopyrimidin-1-yl)-7-(*p*-hydroxybenzyl)-11-methyl-2,9-dioxo-3,8-diazadodecane
Probably formed from: 4,7-Dibenzyl-1-(2,6-dimethylphenoxy)-10-(hexahydro-4-hydroxy-2-oxopyrimidin-1-yl)-11-methyl-2,9-dioxo-3,8-diazadodecane

7-Benzyl-1-(2,6-dimethylphenoxy)-10-(hexahydro-4-hydroxy-2-oxopyrimidin-1-yl)-4-(*p*-hydroxybenzyl)-11-methyl-2,9-dioxo-3,8-diazadodecane
Probably formed from: 4,7-Dibenzyl-1-(2,6-dimethylphenoxy)-10-(hexahydro-4-hydroxy-2-oxopyrimidin-1-yl)-11-methyl-2,9-dioxo-3,8-diazadodecane

4-Benzyl-1-(2,6-dimethylphenoxy)-10-(hexahydro-2-oxopyrimidin-1-yl)-7-(*p*-hydroxybenzyl)-11-methyl-2,9-dioxo-3,8-diazadodecane
Probably formed from: 4,7-Dibenzyl-1-(2,6-dimethylphenoxy)-10-(hexahydro-2-oxopyrimidin-1-yl)-11-methyl-2,9-dioxo-3,8-diazadodecane

4-Benzyl-1-(2,6-dimethylphenoxy)-10-(hexahydro-2-oxopyrimidin-1-yl)-7-(α-hydroxybenzyl)-11-methyl-2,9-dioxo-3,8-diazadodecane
Probably formed from: 4,7-Dibenzyl-1-(2,6-dimethylphenoxy)-10-(hexahydro-2-oxopyrimidin-1-yl)-11-methyl-2,9-dioxo-3,8-diazadodecane

7-Benzyl-1-(2,6-dimethylphenoxy)-10-(hexahydro-2-oxopyrimidin-1-yl)-4-(*p*-hydroxybenzyl)-11-methyl-2,9-dioxo-3,8-diazadodecane
Probably formed from: 4,7-Dibenzyl-1-(2,6-dimethylphenoxy)-10-(hexahydro-2-oxopyrimidin-1-yl)-11-methyl-2,9-dioxo-3,8-diazadodecane

1-Benzyl-3-(3-dimethylaminopropoxy)-hydroxy-1*H*-indazole
Formed from: Benzydamine

N-Benzyl-N′-(2,2-dimethylpropionyl)hydrazine
See: Pivhydrazine

N-Benzyl-N,1-dimethyl-2-propynylamine
→ Benzaldehyde
Probably in: Rat E645
→ N-Benzyl-1-methyl-2-propynylamine
In: Rat E645
→ N-Methylbenzylamine
In: Rat E645

N-Benzyl-N′,N′-dimethyl-N-(2-pyridyl)ethylenediamine
See: Tripelennamine

Benzyl dimethyl sulphide
Formed from: Benzyl methyl sulphide

1-Benzylelymoclavine
Formed from: 1-Benzylagroclavine

3-O-Benzyl-1-(4-ethoxy-3-methoxyphenyl)glycerol β-(4-formyl-2-methoxyphenoxy) ether
→ 3-Benzyloxy-1-(4-ethoxy-3-methoxyphenyl)propan-1,2-diol
In: Phanerochaete D18

N-Benzyl-3-ethyl-7-hydroxy-1-oxa-4-azaspiro[4.4]nonane
Formed from: N-Benzyl-3-ethyl-1-oxa-4-azaspiro[4.4]nonane

N-Benzyl-3-ethyl-1-oxa-4-azaspiro[4.4]nonane
→ N-Benzyl-3-ethyl-7-hydroxy-1-oxa-4-azaspiro[4.4]nonane
In: Beauveria K678

N-(1-Benzylethyl)-α-phenylnitrone
Formed from: N-Benzylamphetamine
S-Benzyl O-ethyl phenylphosphonothioate (Inezin)
→ Benzyl alcohol
In: Piricularia A431
→ Dibenzyl disulphide
In: Piricularia A431
→ O-Ethyl phenylphosphonothioate
In: Piricularia A431
→ Toluene-α-sulphonic acid
In: Piricularia A431
Benzyl ethyl sulphide
→ Benzyl ethyl sulphoxide
In: Helminthosporium G427
Mortierella G427
Benzyl ethyl sulphone
Probably formed from: Benzyl ethyl sulphoxide
Benzyl ethyl sulphoxide
Formed from: Benzyl ethyl sulphide
→ Benzyl ethyl sulphone
Probably in: Helminthosporium G427
Mortierella G427
5-Benzyl-3-furoic acid
Formed from: *cis*-Resmethrin
trans-Resmethrin
→ 5-(*p*-Hydroxybenzyl)-3-furoic acid
Probably in: Rat A2044
→ 5-(α-Hydroxybenzyl)-3-furoic acid
Probably in: Rat A2044
5-Benzyl-3-furylcarboxylic acid
Formed from: 5-Benzyl-3-furylmethanol
→ 5-Benzyl-3-furylcarboxylglycine
In: Rat B14
→ 5-Benzyl-3-furylcarboxyl-β-D-glucuronic acid
In: Rat B14
→ 5-(*p*-Hydroxybenzyl)-3-furylcarboxylic acid
Probably in: Rat B14
5-Benzyl-3-furylcarboxyl-β-D-glucuronic acid
Probably formed from: 5-Benzyl-3-furylcarboxylic acid
5-Benzyl-3-furylcarboxylglycine
Probably formed from: 5-Benzyl-3-furylcarboxylic acid

5-Benzyl-3-furylmethanol
Formed from: Kadethrin
cis-Resmethrin
trans-Resmethrin
→ 5-Benzyl-3-furylcarboxylic acid
In: Rat B14
→ 5-(α-Hydroxybenzyl)-3-hydroxymethylfuran
Probably in: Rat B14
5-Benzyl-3-furylmethyl (1*R*,*cis*(*E*))-3-(2′carboxy-4-sulphydrylbut-1-en-1-yl)-2,2-dimethyl-1-cyclopropanecarboxylate
Formed from: Kadethrin
5-Benzyl-3-furylmethyl *cis*-chrysanthemate
See: *cis*-Resmethrin
Benzyl glucosinolate
→ Benzylamine
In: Lepidium A825
→ Benzyl isothiocyanate
In: Lepidium A825
→ Benzyl thiocyanate
In: Lepidium A825, A999, B594
→ Phenylacetonitrile
In: Lepidium A825, A999
Benzyl-β-D-glucoside
Formed from: Benzyl alcohol
Benzyl-β-D-glucuronide
Probably formed from: Benzyl alcohol
Benzyl L-γ-glutamyl dopamide
→ 3,4-Dihydroxyphenethylamine
In: Dog A3945
N-Benzylguanidine
Formed from: N-Benzyl-N′-methylguanidine
O^6-Benzylguanidine
Formed from: O^6-Benzyl-2′-deoxyguanosine
5-Benzylhydantoin (D or L, as indicated)
Formed from: N-Carbamoyl-D-phenylalanine
N-Carbamoyl-L-phenylalanine
→ N-Carbamyl-D-phenylalanine
In: Flavobacterium F938
→ N-Carbamyl-L-phenylalanine
In: Flavobacterium F938
→ D-Phenylalanine
In: Pseudomonas E391, E450
→ L-Phenylalanine
In: Flavobacterium A3000, F940

Benzylhydrazine
Formed from: Isocarboxazide
→ Benzoic acid
In: Rat A2104, A3922
→ Benzylamine
In: Microorganism A3616

N-Benzyl-N-hydroxyamphetamine
Formed from: N-Benzylamphetamine

N-Benzyl-*m*-hydroxyamphetamine
Probably formed from: N-Benzyl-*m*-hydroxy-N-methylamphetamine

N-Benzyl-*p*-hydroxyamphetamine
Probably formed from: N-Benzyl-*p*-hydroxy-N-methylamphetamine
→ N-Benzyl-4-hydroxy-3-methoxyamphetamine
Probably in: Man E52
Rat C376

3-Benzyl-4-hydroxybutanoic acid
Formed from: 3-Methyl-4-phenyl-3-butenoic diethylamide

N-Benzyl-3-(ω-hydroxy-*sec*-butyl)-1-oxa-4-azaspiro[4.4]nonane
Formed from: N-Benzyl-3-*sec*-butyl-1-oxa-4-azaspiro[4.4]nonane

N-Benzylhydroxycarbazole
Formed from: N-Benzylcarbazole

4-Benzyl-3′-hydroxydehydrocyclopeptine
Formed from: 4-Benzyldehydrocyclopeptine

N-Benzyl-*p*-hydroxy-N-(3-(2-hydroxy-2-methyl-1-propoxy)-2-(2-hydroxy-1-pyrrolidino)propyl) aniline
Probably formed from: N-Benzyl-*p*-hydroxy-N-(3-(2-hydroxy-2-methyl-1-propoxy)-2-(1-pyrrolidino)propyl)aniline
→ N-Benzyl-*p*-hydroxy-N-(3-(2-hydroxy-2-methyl-1-propoxy)-2-(2-oxo-1-pyrrolidino)propyl)aniline
Probably in: Man E944
Monkey E944
Rabbit E944

N-Benzyl-*p*-hydroxy-N-(3-(2-hydroxy-2-methyl-1-propoxy)-2-(2-oxo-1-pyrrolidino)propyl)aniline
Probably formed from: N-Benzyl-*p*-hydroxy-N-(3-(2-hydroxy-2-methyl-1-propoxy)-2-(2-hydroxy-1-pyrrolidino)propyl)aniline

N-Benzyl-*p*-hydroxy-N-(3-(2-hydroxy-2-methyl-1-propoxy)-2-(1-pyrrolidino)propyl)aniline
Probably formed from: N-Benzyl-*p*-hydroxy-N-(3-(2-hydroxy-2-methyl-1-propoxy)-2-(1-pyrrolidino)propyl)aniline
→ N-Benzyl-*p*-hydroxy-N-(3-(2-hydroxy-2-methyl-1-propoxy)-2-(2-hydroxy-1-pyrrolidino)propyl)aniline
Probably in: Mouse E944
→ *p*-Benzylaminophenol
Probably in: Man E944
Monkey E944
Mouse E944
Rabbit E944
Rat E944

N-Benzyl-*p*-hydroxy-N-(3-hydroxy-2-(1-pyrrolidino)propyl)aniline
Probably formed from: N-Benzyl-*p*-hydroxy-N-(3-(2-methyl-1-propoxy)-2-(1-pyrrolidino)propyl)aniline

N-Benzyl-7-hydroxy-3-isopropyl-1-oxa-4-azaspiro[4.4]nonane
Formed from: N-Benzyl-3-isopropyl-1-oxa-4-azaspiro[4.4]nonane

N-Benzyl-4-hydroxy-3-methoxyamphetamine
Probably formed from: N-Benzyl-*p*-hydroxyamphetamine

N-Benzyl-3-hydroxy-4-methoxyphenethylamine
Formed from: N-Benzyl-3,4-dihydroxyphenethylamine

N-Benzyl-4-hydroxy-3-methoxyphenethylamine
Formed from: N-Benzyl-3,4-dihydroxyphenethylamine

N-Benzyl-*m*-hydroxy-N-methylamphetamine
Formed from: N-Benzyl-N,α-dimethylphenethylamine
→ N-Benzyl-*m*-hydroxyamphetamine
Probably in: Rat C376

N-Benzyl-*p*-hydroxy-N-methylamphetamine
Formed from: N-Benzyl-N,α-dimethylphenethylamine
→ N-Benzyl-*p*-hydroxyamphetamine
Probably in: Man E52
Rat C376, C433

Benzyl 3-hydroxy-2-methylbutyrate
Formed from: Benzyl 2-methyl-3-oxobutyrate

5-Benzyl-3-hydroxymethylfuran
Formed from: Kadethrin

N-Benzyl-7-hydroxy-2-methyl-1-oxa-4-azaspiro[4.4]nonane
Formed from: N-Benzyl-2-methyl-1-oxa-4-azaspiro[4.4]nonane

N-Benzyl-7-hydroxy-3-methyl-1-oxa-4-azaspiro[4.4]nonane
Formed from: N-Benzyl-3-methyl-1-oxa-4-azaspiro[4.4]nonane

N-Benzyl-*p*-hydroxy-N-(3-(2-methyl-1-propoxy)-2-(2-oxo-1-pyrrolidino)propyl) aniline
Probably formed from: N-Benzyl-N-(3-(2-methyl-1-propoxy)-2-(2-oxo-1-pyrrolidino)propyl)aniline

N-Benzyl-N-(3-(2-hydroxy-2-methyl-1-propoxy)-2-(2-oxo-1-pyrrolidino)propyl) aniline
Probably formed from: N-Benzyl-N-(3-(2-methyl-1-propoxy)-2-(2-oxo-1-pyrrolidino)propyl)aniline

N-Benzyl-N-(3-(2-hydroxy-2-methyl-1-propoxy)-2-(1-pyrrolidino)propyl)aniline
Formed from: Bepridil
→ N-(3-(2-Hydroxy-2-methyl-1-propoxy)-2-(1-pyrrolidino)propyl)aniline
Probably in: Man E944
Monkey E944
Mouse E944
Rabbit E944
Rat E944

N-Benzyl-*p*-hydroxy-N-(3-(2-methyl-1-propoxy)-2-(1-pyrrolidino)propyl)aniline
Formed from: Bepridil
→ N-Benzyl-*p*-hydroxy-N-(3-(2-hydroxy-2-methyl-1-propoxy)-2-(1-pyrrolidino) propyl)aniline
Probably in: Man E944
Monkey E944
Mouse E944
Rabbit E944
Rat E944
→ N-Benzyl-*p*-hydroxy-N-(3-hydroxy-2-(1-pyrrolidino)propyl)aniline
Probably in: Rat E944

N-Benzyl-N-(3-hydroxy-2-(2-oxo-1-pyrrolidino)propyl)aniline
Probably formed from: N-Benzyl-N-(3-(2-methyl-1-propoxy)-2-(2-oxo-1-pyrrolidino)propyl)aniline

4-(2-Benzyl-4-hydroxyphenoxy)butyric acid
Probably formed from: 4-(2-Benzyl-4-hydroxyphenoxy)-N-methylbutylamine

4-(2-Benzyl-4-hydroxyphenoxy)butylamine
Probably formed from: 4-(2-Benzyl-4 -hydroxyphenoxy)-N-methylbutylamine

N-Benzyl-N-(*p*-hydroxyphenoxy)isopropylamine
Probably formed from: N-Benzyl-N-phenoxyisopropylamine

4-(2-Benzyl-4-hydroxyphenoxy)-N-methylbutylamine
Formed from: 4-(*o*-Benzylphenoxy)-N-methylbutylamine
→ 4-(2-Benzyl-4-hydroxyphenoxy)butylamine
Probably in: Dog D989
Mouse D989
Rabbit D989
Rat D989
→ 4-(2-Benzyl-4-hydroxyphenoxy)butyric acid
Probably in: Dog D989
Mouse D989
Rabbit D989
Rat D989

4-(2-Benzyl-5-hydroxyphenoxy)-N-methylbutylamine
Formed from: 4-(*o*-Benzylphenoxy)-N-methylbutylamine

3-Benzyl-5-(*p*-hydroxyphenyl)-2-((*p*-hydroxyphenylacetyl)amino)pyrazine
Formed from: 2-Benzyl-3,7-dihydro-6-(*p*-hydroxyphenyl)-8-benzylimidazo[1,2-*a*] pyrazin-3-one

N-Benzyl-7-hydroxy-3-phenyl-1-oxa-4-azaspiro[4.4]nonane
Formed from: N-Benzyl-3-phenyl-1-oxa-4-azaspiro[4.4]nonane

N-Benzyl-7-hydroxy-3-propyl-1-oxa-4-azaspiro[4.4]nonane
Formed from: N-Benzyl-3-propyl-1-oxa-4-azaspiro[4.4]nonane

N-Benzyl-3-(2-hydroxypropyl)-1-oxa-4-azaspiro[4.4]nonane
Formed from: N-Benzyl-3-propyl-1-oxa-4-azaspiro[4.4]nonane
N-Benzyl-N-(4-hydroxy-2-pyridyl)-N′,N′-dimethylethylenediamine
Formed from: Tripelennamine
N-Benzyl-N-(2-(2-hydroxy-1-pyrrolidino)-3-(2-methyl-1-propoxy)propyl)aniline
Formed from: Bepridil
→ N-Benzyl-N-(3-(2-methyl-1-propoxy)-2-(2-oxo-1-pyrrolidino)propyl)aniline
Probably in: Man E944
Monkey E944
Mouse E944
Rabbit E944
Rat E944
→ N-Benzyl-N-(2-(2-hydroxy-1-pyrrolidino)-3-(2-methyl-1-propoxy)propyl)aniline
Probably in: Monkey E944
1-Benzyl-2-hydroxyquinolinium
Formed from: 1-Benzylquinolinium
1-Benzyl-4-hydroxyquinolinium
Formed from: 1-Benzylquinolinium
Benzylideneacetone
See: Benzalacetone
N-Benzylideneamphetamine
→ Amphetamine
In: Guinea pig C222
N-Benzylidene-*p*-chloroaniline
Formed from: N-Benzyl-*p*-chloroaniline
Benzylidenemalononitrile
→ Benzylmalononitrile
In: Man B260
7-Benzylidenenaltrexone
→ 7-Benzylidenenaltrexone-β-D-glucuronide
In: Rat J97a
7-Benzylidenenaltrexone-β-D-glucuronide
Formed from: 7-Benzylidenenaltrexone
1-Benzyl-1*H*-indazolone
Formed from: Benzydamine
N-Benzyl-3-isopropyl-1-oxa-4-azaspiro[4.4]nonane
→ N-Benzyl-7-hydroxy-3-isopropyl-1-oxa-4-azaspiro[4.4]nonane
In: Beauveria K678
Benzyl isopropyl sulphide
→ Benzyl isopropyl sulphoxide
In: Helminthosporium G427
Mortierella G427
Benzyl isopropyl sulphone
Probably formed from: Benzyl isopropyl sulphoxide
Benzyl isopropyl sulphoxide
Formed from: Benzyl isopropyl sulphide
→ Benzyl isopropyl sulphone
Probably in: Helminthosporium G427
Mortierella G427
Benzyl isothiocyanate
Formed from: Benzyl glucosinolate
→ S-(N-Benzylthiocarbamoyl)glutathione
In: Rat A3133
See also: C372, F292
O-Benzyllactaldehyde
→ 3-Benzyloxy-2-hydroxybutyronitrile
In: Almond K320
O-Benzyllactic acid
See: 2-Benzyloxypropionic acid
2-Benzylmalic acid
Formed from: Phenylpyruvic acid
→ 3-Benzylmalic acid
In: Nasturtium A162
3-Benzylmalic acid
Formed from: 2-Benzylmalic acid
→ 2-Amino-4-phenylbutyric acid
Probably in: Nasturtium A162
→ Benzylsuccinic acid
Probably in: Chromobacterium E20
Benzylmalononitrile
Formed from: Benzylidenemalononitrile
S-Benzyl-L-malonyl-L-cysteine
→ Benzoic acid
In: Rat F908
→ Benzyl methyl sulphoxide
In: Rat F908
Benzyl mandelate
→ Benzyl alcohol
In: Man A3172, A3394
→ Mandelic acid
In: Man A3172, A3394
Benzyl mercaptan
Formed from: Dibenzyl disulphide
→ Benzyl sulphoxide

In: Mouse D446
Pig D446

***l*-N-Benzyl-β-methoxy-3-trifluoromethylphenethylamine**
→ β-Methoxy-3-trifluoromethylphenethylamine
In: Dog A631
Man A631
Monkey A631
Rat A631
→ α-Methoxy-*m*-trifluoromethylphenylacetic acid
In: Dog A631
Monkey A631

2-(N-Benzyl-N-methylamino)ethyl methyl 4-(*m*-aminophenyl)-1,4-dihydro-2,6-dimethylpyridine-3,5-dicarboxylate
Formed from: Nicardipine

2-(N-Benzyl-N-methylamino)ethyl methyl 1,4-dihydro-2,6-dimethyl-4-(*m*-nitrophenyl) pyridine-3,5-dicarboxylate
See: Nicardipine

2-(N-Benzyl-N-methylamino)ethyl methyl 2,6-dimethyl-4-(*m*-nitrophenyl)pyridine-3,5-dicarboxylate
Formed from: Nicardipine

N-Benzyl-N-methylamphetamine
See: N-Benzyl-N,α-dimethylphenethylamine

8-Benzyl-3′-O-methyl-(+)-catechin
Formed from: 8-Benzyl-(+)-catechin

N-Benzyl-N′-methylguanidine
Formed from: Bethanidine
→ N-Benzylguanidine
In: Dog A2001
Rat A2001

Benzyl methyl ketoxime (1-phenyl-2-propanone oxime)
Formed from: D-Amphetamine
α-Methyl-β-phenethylhydroxylamine
Probably formed from: L-Amphetamine
→ Amphetamine
In: Mycobacterium B380
→ α-Hydroxybenzyl methyl ketoxime
In: Rabbit A3315
→ 2-Nitro-1-phenylpropane
In: Rabbit A3315
Rat A1982
→ Phenylacetone
In: Mycobacterium B380
Rabbit A1398, A3315
Rat A1982

Benzyl methyl nonylamine
Formed from: Benzyl dimethyl nonylammonium

N-Benzyl-2-methyl-1-oxa-4-azaspiro[4.4]nonane
→ N-Benzyl-7-hydroxy-2-methyl-1-oxa-4-azaspiro[4.4]nonane
In: Beauveria K678

N-Benzyl-3-methyl-1-oxa-4-azaspiro[4.4]nonane
→ N-Benzyl-7-hydroxy-3-methyl-1-oxa-4-azaspiro[4.4]nonane
In: Beauveria K678

Benzyl 2-methyl-3-oxobutyrate
→ Benzyl 3-hydroxy-2-methylbutyrate
In: Candida K940
Endomycopsis K940
Hansenula K940
Lipomyces K940
Pichia K940
Rhodotorula K940
Saccharomyces K940

N-Benzyl-α-methylphenethylamine
See: N-Benzylamphetamine

α-Benzyl-N-methylphenethylamine
→ α-Benzylphenethylamine
In: Rat C595

N-Benzyl-N-methylpropargylamine
See: Pargyline

N-Benzyl-N-(3-(2-methyl-1-propoxy)-2-(2-oxo-1-pyrrolidino)propyl)aniline
Probably formed from: N-Benzyl-N-(2-(2-hydroxy-1-pyrrolidino)-3-(2-methyl-1-propoxy)propyl)aniline
→ N-Benzyl-N-(2-(3-carboxypropylamino)-3-(2-methyl-1-propoxy)propyl)aniline
Probably in: Man E944
Mouse E944
→ N-Benzyl-*p*-hydroxy-N-(3-(2-methyl-1-propoxy)-2-(2-oxo-1-pyrrolidino)propyl) aniline
Probably in: Man E944
Monkey E944
Mouse E944
Rabbit E944

Rat E944
→ N-Benzyl-N-(3-(2-hydroxy-2-methyl-1-propoxy)-2-(2-oxo-1-pyrrolidino)propyl) aniline
Probably in: Man E944
→ N-(3-Hydroxy-2-(2-oxo-1-pyrrolidino) propyl)aniline
Probably in: Monkey E944
→ N-(3-(2-Methyl-1-propoxy)-2-(2-oxo-1-pyrrolidino)propyl)aniline
Probably in: Monkey E944
Mouse E944

N-Benzyl-1-methyl-2-propynylamine
Formed from: N-Benzyl-N,1-dimethyl-2-propynylamine

3-Benzyl-N-methylpyridinium
Formed from: 3-Benzylpyridine

4-Benzyl-1-methylpyridinium
Probably formed from: 4-Benzyl-1,2,3,6-tetrahydro-1-methylpyridine

Benzyl methyl sulphide
→ Benzyl methyl sulphoxide
In: Helminthosporium G427
Mortierella G427
Mouse D446
Pig D446
→ Benzyl dimethyl sulphide
In: Mouse E677

Benzyl methyl sulphone
Probably formed from: Benzyl methyl sulphoxide

Benzyl methyl sulphoxide
Formed from: S-Benzyl-N-malonyl-L-cysteine
Benzyl methyl sulphide
→ Benzyl methyl sulphone
In: Helminthosporium G427
Mortierella G427
Rat F908

3-O-Benzylmorphine
→ 3-O-Benzylmorphine-6-β-D-glucuronide
In: Man K517

3-O-Benzylmorphine-6-β-D-glucuronide
Formed from: 3-O-Benzylmorphine

Benzyl nicotinate
→ Benzyl alcohol
In: Man J141

N-Benzyl-*p*-nitroaniline
→ Benzaldehyde
In: Hamster G951
→ *p*-Nitroaniline
In: Hamster G951

N-Benzyloxyamphetamine
→ *p*-Hydroxyamphetamine
In: Cunninghamella F893

***m*-Benzyloxybenzaldehyde**
→ *m*-Benzyloxybenzoic acid
In: Rat H761
→ *m*-Benzyloxy-2-hydroxypropiophenone
In: Pseudomonas K524

***p*-Benzyloxybenzaldehyde**
→ *p*-Benzyloxybenzoic acid
In: Rat H761

***m*-Benzyloxybenzoic acid**
Formed from: *m*-Benzyloxybenzaldehyde

***p*-Benzyloxybenzoic acid**
Formed from: *p*-Benzyloxybenzaldehyde

N-Benzyloxycarbonyl-2,3-dehydrotryptophan
Formed from: N-Benzyloxycarbonyl-tryptophan

N-Benzyloxycarbonyltryptophan
→ N-Benzyloxycarbonyl-2,3-dehydrotryptophan
In: Chromobacterium A2770

4-Benzyloxy-3-chlorobenzoic acid
Probably formed from: 4-Benzyloxy-3-chlorophenylglyoxylic acid

4-Benzyloxy-3-chlorobenzoic acid
Formed from: 4-Benzyloxy-3-chlorophenylacetic acid
→ 4-Benzyloxy-3-chlorophenylglyoxylic acid
In: Dog C88
Man C88
Rat C88

4-Benzyloxy-3-chlorophenylacetic acid
(benzofenac)
→ 4-Benzyloxy-3-chloromandelic acid
In: Dog C88
Man C88
Rat C88
→ 4-Benzyloxy-3-chlorophenylacetylglycine
In: Dog C88
Man C88
Rat C88

→ 3-Chloro-4-hydroxyphenylacetic acid
In: Dog C88
Man C88
Rat C88

4-Benzyloxy-3-chlorophenylacetylglycine
Formed from: 4-Benzyloxy-3-chlorophenylacetic acid

4-Benzyloxy-3-chlorophenylglyoxylic acid
Probably formed from: 4-Benzyloxy-3-chloromandelic acid
→ 4-Benzyloxy-3-chlorobenzoic acid
Probably in: Dog C88
Man C88
Rat C88

2-Benzyloxyethanol
Formed from: 3-Benzyloxy-1-(4-ethoxy-3-methoxyphenyl)propan-1,2-diol

3-Benzyloxy-1-(4-ethoxy-3-methoxyphenyl) propan-1,2-diol
Formed from: 3-O-Benzyl-1-(4-ethoxy-3-methoxyphenyl)glycerol β-(4-formyl-2-methoxyphenoxy) ether
→ 2-Benzyloxyethanol
In: Phanerochaete D18
→ 4-Ethoxy-3-methoxybenzyl alcohol
In: Phanerochaete D18

3-Benzyloxy-2-hydroxybutyronitrile
Formed from: O-Benzyllactaldehyde

3-Benzyloxy-2-hydroxy-4-phenylbutyronitrile
Formed from: O-Benzylphenyllactaldehyde

***m*-Benzyloxy-2-hydroxypropiophenone**
Formed from: *m*-Benzyloxybenzaldehyde

4-Benzyloxy-3-methoxybenzaldehyde
→ 4-Benzyloxy-3-methoxybenzyl alcohol
In: Phanerochaete F866

4-Benzyloxy-3-methoxybenzyl alcohol
Formed from: 4-Benzyloxy-3-methoxybenzaldehyde

6-Benzyloxy-4-methoxymethyl-β-carboline-3-carboxylic acid
Formed from: Abecarnil
→ 6-Benzyloxy-4-methoxymethyl-β-carboline-3-carboxyl-β-D-glucuronic acid
In: Man G2

6-Benzyloxy-4-methoxymethyl-β-carboline-3-carboxyl-β-D-glucuronic acid
Formed from: 6-Benzyloxy-4-methoxymethyl-β-carboline-3-carboxylic acid

5-(Benzyloxymethyl)hydantoin
→ N-Carbamyl-O-benzylserine
In: Flavobacterium F938

***p*-Benzyloxyphenol**
→ *p*-Hydroxybenzaldehyde
In: Acinetobacter F51

7-Benzyloxyphenoxazone
See: Resorufin benzyl ester

2-Benzyloxy-1-propanol
Formed from: 2-Benzyloxypropionic acid

2-Benzyloxypropionic acid (O-benzyllactic acid)
→ 2-Benzyloxy-1-propanol
In: Gloeosporium D935
Glomerella D935

7-Benzyloxy-4-trifluoromethylcoumarin
→ 7-Hydroxy-4-trifluoromethylcoumarin
In: Man K350

Benzylpenicillin (penicillin G)
Formed from: Phenylacetyl CoA
S-(Phenylacetyl)glutathione
Phenylacetylglycine
→ Benzylpenicilloic acid
In: Klebsiella G193

Benzylpenicilloic acid
Formed from: Benzylpenicillin

α-Benzylphenethylamine
Formed from: α-Benzyl-N-methylphenethylamine

N-Benzyl-N-phenoxyisopropylamine
Formed from: Phenoxybenzamine
→ 2-Benzylamino-1-propanol
Probably in: Rat A1993
→ N-Benzyl-N-(*p*-hydroxyphenoxy) isopropylamine
Probably in: Dog A2992
Man A1993
Rat A1993
→ Phenoxyisopropylamine
Probably in: Dog A1993

4-(*o*-Benzylphenoxy)-N-methylbutylamine
→ 4-(2-Benzyl-4-hydroxyphenoxy)-N-methylbutylamine
In: Dog D989

Mouse D989
Rabbit D989
Rat D989
→ 4-(2-Benzyl-5-hydroxyphenoxy)-N-methylbutylamine
In: Dog D989
Mouse D989
Rabbit D989
Rat D989
→ 4-(*o*-(*p*-Hydroxybenzyl)phenoxy)-N-methylbutylamine
In: Dog D989
Mouse D989
Rabbit D989
Rat D989

***p*-Benzylphenylhydroxylamine**
Formed from: N-Hydroxy-*p*-benzylacetanilide

O-Benzylphenyllactaldehyde
→ 3-Benzyloxy-2-hydroxy-4-phenylbutyronitrile
In: Almond K320

N-Benzyl-3-phenyl-1-oxa-4-azaspiro[4.4]nonane
→ N-Benzyl-7-hydroxy-3-phenyl-1-oxa-4-azaspiro[4.4]nonane
In: Beauveria K678

Benzyl phenyl sulphide
→ Benzyl phenyl sulphoxide
In: Helminthosporium G427
Mortierella G427

Benzyl phenyl sulphoxide
Formed from: Benzyl phenyl sulphide

Benzylphthalamidic acid
→ Benzylamine
In: Xanthobacter J253
→ Phthalic acid
In: Xanthobacter J253

N-Benzyl-N′-picolinylpiperazine
→ Benzoic acid
In: Dog E549, E550
Man E549
Rat E549, E550
→ N-Benzylpiperazine
In: Dog E549
Man E549
Rat E549

N-Benzylpiperazine
Formed from: N-Benzyl-N′-picolinylpiperazine

***erythro*-2-(4-Benzylpiperidino)-1-(4-hydroxy-3-methoxyphenyl)-1-propanol**
Formed from: *erythro*-2-(4-Benzylpiperidino)-1-(*p*-hydroxyphenyl)-1-propanol
→ *erythro*-2-(4-Benzylpiperidino)-1-(4-hydroxy-3-methoxyphenyl)-1-propanol-β-D-glucuronide
Probably in: Rat A3408
→ *erythro*-2-(4-(*p*-Hydroxybenzyl)piperidino)-1-(4-hydroxy-3-methoxyphenyl)-1-propanol
Probably in: Rat A3408

***erythro*-2-(4-Benzylpiperidino)-1-(4-hydroxy-3-methoxyphenyl)-1-propanol-β-D-glucuronide**
Probably formed from: *erythro*-2-(4-Benzylpiperidino)-1-(4-hydroxy-3-methoxyphenyl)-1-propanol

***erythro*-2-(4-Benzylpiperidino)-1-(*p*-hydroxyphenyl)-1-propanol** (ifenprodil)
→ *erythro*-2-(4-Benzylpiperidino)-1-(4-hydroxy-3-methoxyphenyl)-1-propanol
In: Rabbit A3409
Rat A3408
→ *erythro*-2-(4-Benzylpiperidino)-1-(*p*-hydroxyphenyl)-1-propanol-β-D-glucuronide
In: Rat A3408
→ *erythro*-2-(4-(*p*-Hydroxybenzyl)piperidino)-1-(*p*-hydroxyphenyl)-1-propanol
In: Rabbit A3409
Rat A3408, A3409

***erythro*-2-(4-Benzylpiperidino)-1-(*p*-hydroxyphenyl)-1-propanol-β-D-glucuronide**
Formed from: *erythro*-2-(4-Benzylpiperidino)-1-(*p*-hydroxyphenyl)-1-propanol

N^1-Benzyl-N^2-pivaloylhydrazine
See: Tersavid

Benzyl propyl sulphide
→ Benzyl propyl sulphoxide
In: Helminthosporium G427
Mortierella G427

N-Benzyl-3-propyl-1-oxa-4-azaspiro[4.4]nonane
→ N-Benzyl-7-hydroxy-3-propyl-1-oxa-4-azaspiro[4.4]nonane
In: Beauveria K678
→ N-Benzyl-3-(2-hydroxypropyl)-1-oxa-4-azaspiro[4.4]nonane
In: Beauveria K678

Benzyl propyl sulphoxide
Formed from: Benzyl propyl sulphide

3-Benzylpyridine
→ 3-Benzyl-N-methylpyridinium
In: Rabbit E824

(1-Benzyl-3-pyrrolidinyl) methyl 2,6-dimethyl-4-(*m*-nitrophenyl)pyridine-3,5-dicarboxylate
Formed from: Barnidipine

1-Benzylquinolinium
→ 1-Benzyl-2-hydroxyquinolinium
In: Rabbit D208
→ 1-Benzyl-4-hydroxyquinolinium
In: Rabbit D208

Benzyl selenocyanate
→ Dibenzyl diselenide
In: Rat G282
→ Phenylmethaneselenic acid
In: Rat G282

O-Benzyl-L-serine
Formed from: N-Carbamyl-O-benzyl-L-serine

O-Benzylsterigmatocystin
→ Aflatoxin B_1
In: Aspergillus K249

(*R*)-(+)-Benzylsuccinic acid
Formed from: (*R*)-Benzylsuccinyl CoA
Toluene
Probably formed from: Benzylmalic acid
→ Arphamenine A
Probably in: Chromobacterium E20
→ (*R*)-Benzylsuccinyl CoA
In: Thauera K302

(*R*)-Benzylsuccinyl CoA
Formed from: (*R*)-(+)-Benzylsuccinic acid
→ Benzoyl CoA
In: Thauera K302
→ (*R*)-(+)-Benzylsuccinic acid
In: Thauera K302

Benzyl sulphate
Formed from: Benzyl alcohol

Benzyl-β-D-sulphatoglucoside
Formed from: Benzaldoxime

Benzyl sulphoxide
Formed from: Benzyl mercaptan

4-Benzyl-1,2,3,6-tetrahydro-1-methylpyridine
→ 4-Benzyl-1-methylpyridinium
Probably in: Man F395
Rat F816

7-Benzyl-3-thia-7-azabicyclo[3.3.1]nonane
→ 7-Benzyl-3-thia-7-azabicyclo[3.3.1]nonane sulphoxide
In: Dog H333

7-Benzyl-3-thia-7-azabicyclo[3.3.1]nonane sulphoxide
Formed from: 7-Benzyl-3-thia-7-azabicyclo[3.3.1]nonane

S-(N-Benzylthiocarbamoyl)-L-cysteine
Formed from: S-(N-Benzylthiocarbamoyl)-L-cysteinylglycine
→ N-Acetyl-S-(N-benzylthiocarbamoyl)-L-cysteine
In: Guinea pig C372
Rabbit C372
Rat A3133
See also: F292
→ 3-Benzyl-4-carboxy-4-hydroxythiazolidin-2-thione
In: Guinea pig C372
Rabbit C372

S-(N-Benzylthiocarbamoyl)-L-cysteinylglycine
Formed from: S-(N-Benzylthiocarbamoyl)glutathione
→ S-(N-Benzylthiocarbamoyl)-L-cysteine
In: Rat A3133
See also: C372, F292

S-(N-Benzylthiocarbamoyl)glutathione
Formed from: Benzyl isothiocyanate
→ S-(N-Benzylthiocarbamoyl)-L-cysteinylglycine
In: Rat A3133
See also: C372, F292

Benzyl thiocyanate
Formed from: Benzyl glucosinolate

N-Benzyl-*p*-toluidine
→ Benzaldehyde
In: Ferret D637
Guinea pig D637

Hamster D637, G951
Mouse D637
Rabbit D637
Rat D637
→ N-Benzoyl-*p*-toluidine
In: Ferret D637
Guinea pig D637
Hamster D637, G951
Mouse D637
Rabbit D637
Rat D637
→ N-(*p*-Methylphenyl)-α-phenylnitrone
In: Ferret D637
Hamster D637, G951
→ *p*-Toluidine
In: Ferret D637
Guinea pig D637
Hamster D637, G951
Mouse D637
Rabbit D637
Rat D637

Benzyl *p*-tolyl sulphide
→ Benzyl *p*-tolyl sulphoxide
In: Helminthosporium G427
Mortierella G427

Benzyl *p*-tolyl sulphoxide
Formed from: Benzyl *p*-tolyl sulphide

N-Benzyl-2,4,6-tribromoaniline
→ Benzaldehyde
In: Hamster G951
→ 2,4,6-Tribromoaniline
In: Hamster G951

N-Benzyl-2,4,6-trichloroaniline
→ Benzaldehyde
In: Hamster G951
→ 2,4,6-Trichloroaniline
In: Hamster G951

N-Benzyl-2,4,6-trifluoroaniline
→ Benzaldehyde
In: Hamster G951
→ 2,4,6-Trifluoroaniline
In: Hamster G951

N-Benzyl-2,4,6-trimethylaniline
→ Benzaldehyde
In: Hamster G951
→ N-Benzoyl-2,4,6-trimethylaniline
In: Hamster G951
→ α-Phenyl-N-(2,4,6-trimethylphenyl)nitrone
In: Hamster G951
→ 2,4,6-Trimethylaniline
In: Hamster G951

N³-Benzyluracil
Formed from: N³-Benzyluridine

N³-Benzyluridine
→ N³-Benzyluracil
In: Mouse G114

4-O-Benzylvanillic acid
→ 4-O-Benzylvanillyl alcohol
In: Nocardia K406

4-O-Benzylvanillyl alcohol
Formed from: 4-O-Benzylvanillic acid

α-Benzylveratryl alcohol
→ Veratraldehyde
In: Bjerkandera F610
Phanerochaete F610

Bepridil (β-((2-methylpropoxy)methyl)-N-phenyl-N-(phenylmethyl)-1-pyrrolidineethanamine)
→ N-Benzyl-N-(2-((3-carboxypropyl)amino)-3-(2-methyl-1-propoxy)propyl)aniline
In: Man G62
Monkey G62
Mouse G62
Rabbit G62
Rat G62
→ N-Benzyl-*p*-hydroxy-N-(3-(2-methyl-1-propoxy)-2-(1-pyrrolidino)propyl)aniline
In: Man E944, G62
Monkey E944, G62
Mouse E944, G62
Rabbit E944, G62
Rat E944, G62
→ N-Benzyl-N-(3-(2-hydroxy-2-methyl-1-propoxy)-2-(1-pyrrolidino)propyl)aniline
In: Man E944, G62
Monkey E944, G62
Mouse E944, G62
Rabbit E944, G62
Rat E944, G62
→ N-Benzyl-N-(2-(2-hydroxy-1-pyrrolidino)-3-(2-methyl-1-propoxy)propyl)aniline
In: Man E944
Monkey E944

Mouse E944
Rabbit E944
Rat E944
→ N-(*p*-Hydroxyphenyl)benzylamine
In: Man G62
Monkey G62
Mouse G62
Rabbit G62
Rat G62
→ N-(3-(2-Methyl-1-propoxy)-2-(1-pyrrolidino)propyl)aniline
In: Man E944, G62
Monkey E944, G62
Mouse E944, G62
Rabbit E944, G62
Rat E944, G62

Berbamunine ((1*S*)-1,2,3,4-tetrahydro-1-((4-(2-hydroxy-5-(((1*R*)-1,2,3,4-tetrahydro-7-hydroxy-6-methoxy-2-methyl-1-isoquinolinyl)methyl)phenoxy)phenyl)methyl)-6-methoxy-2-methyl-7-isoquinolinol)
Formed from: Coclaurine
N-Methylcoclaurine

Berberine (5,6-dihydro-9,10-dimethoxybenzo[*g*]-1,3-benzodioxolo[5,6-*a*]quinolizinium)
Formed from: Columbamine
3,4-Dihydroxy-L-phenylalanine
Norcoclaurine
Reticuline
Tetrahydroberberine
L-Tyrosine
Probably formed from: 7,8-Dihydroberberine
→ Allocryptopine
In: Corydalis H585
→ Tetrahydroberberine
In: Corydalis G136, H585
Dicentra J496
→ Thalictricavine
In: Corydalis H585

Bergapten (4-methoxy-7*H*-furo[3,2-*g*][1]benzopyran-7-one)
Probably formed from: Bergaptol
Psoralen
→ Isopimpinellin
In: Heracleum A2582

Bergaptol (4-hydroxy-7*H*-furo[3,2-*g*][1]benzopyran-7-one)
→ Bergapten
In: Parsley D983
Ruta A3572

Bergenin ((2*R*,3*S*,4*S*,4a*R*,10b*S*)-3,4,4a,10b-tetrahydro-3,4,8,10-tetrahydroxy-2-(hydroxymethyl)-9-methoxypyrano[3,2-*c*][2]benzopyran-6(2*H*)-one)
→ 3,5-Dihydroxy-4-methoxybenzoic acid
In: Erwinia A466

Berupipam ((5*S*)-5-(5-bromo-2,3-dihydro-7-benzofuranyl)-8-chloro-2,3,4,5-tetrahydro-3-methyl-1*H*-3-benzazepin-7-ol)
→ Berupipam-β-D-glucuronide
In Dog H198
Man H198
Mouse H198
Rat H198

Berupipam-β-D-glucuronide
Formed from: Berupipam

Besipirdine (N-(*n*-propyl)-N-(4-pyridinyl)-1*H*-indolamine)
→ Depropylbesipirdine
In: Cunninghamella J158
→ 5-Hydroxybesipirdine
In: Cunninghamella J158

Betacyanin (general name)
Formed from: 3,4-Dihydroxy-L-phenylalanine

Betalamic acid ((2*S*,4*E*)-1,2,3,4-tetrahydro-4-(oxoethylidene)-2,6-pyridine(carboxylic acid)
Formed from: 3,4-Dihydroxy-L-phenylalanine

Betanidin ((2*S*)-4-(2-((2*S*)-2-carboxy-2,3-dihydro-5,6-dihydroxy-1*H*-indol-1-yl)ethenyl)-2,3-dihydro-2,6-pyridinedicarboxylic acid)
Formed from: 3,4-Dihydroxy-L-phenylalanine
→ Betanin
In: Dorotheanthus J185
→ Gomphrenin I
In: Dorotheanthus J185

Betanin ((2*S*)-4-(2-((2*S*)-2-carboxy-2,3-dihydro-5-(β-D-glucopyranosyloxy)-6-hydroxy-1*H*-indol-1-yl)ethenyl)-2,3-dihydro-2,6-pyridinedicarboxylic acid)
Formed from: Betanidin
→ Amaranthin
In: Celosia A1325

Bethanidine (N-benzyl-N′,N″-dimethylguanidine)
→ N-Benzyl-N′-methylguanidine
In: Dog A2001
Rat A2001
→ N-(Hydroxybenzyl)-N′,N″-dimethylguanidine
In: Dog A2001
Rat A2001

Bezafibrate (2-(4-(2-(4-chlorobenzamido)ethyl)phenoxy)-2-methylpropionic acid)
→ Hydroxybezafibrate
In: Rat B768

BHT
See: 2,6-Di-*tert*-butyl-4-methylphenol

Biantrazole (losoxantrone, 7-hydroxy-2-(2-((2-hydroxyethyl)amino)ethyl)-5-((2-((2-hydroxyethyl)amino)ethyl)amino)anthra[1,9-*cd*]pyrazol-6-(2*H*)-one)
→ 5-(2-((Carboxymethyl)amino)ethyl)amino)-7-hydroxy-(2-((2-hydroxyethyl)amino) ethyl)anthra[1,9-*cd*]pyrazol-6(2*H*)-one
In: Man G698, H175
→ 2-(2-((Carboxymethyl)amino)ethyl)-7-hydroxy-5-((2-((2-hydroxyethyl)amino) ethyl)amino)anthra[1,9-*cd*]pyrazol-6(2*H*)-one
In: Man G698, H175

Bibenzyl (1,1′-(1,2-ethanediyl)bisbenzene)
Formed from: N-Nitrosodibenzylamine
→ 2,3-Dihydroxybibenzyl
Probably in: Pseudomonas J10
→ 1,2-Diphenylethanol
In: Mortierella G427

2,2′-Bi(dihydrodeoxymorphine)
Formed from: Dihydrodeoxymorphine

Bidisomide (α-(2-(acetyl(1-methylethyl) amino)ethyl)-α-(2-chlorophenyl)-1 piperidine-butanamide)
→ Deisopropylbidisomide
In: Man H321
See also: K115
→ Hydroxybidisomide
In: Man H321

Bifluranol (*erythro*-α′-ethyl-3,3′-difluoro-4,4′-dihydroxy-α-methylbibenzyl)
→ Bifluranol-β-D-glucuronide
In: Dog B921
Ferret B921
Rat B921
→ Bifluranol-O-sulphate
In: Dog B921
Ferret B921
Rat B921

Bifluranol-di-β-D-glucuronide
Formed from: Bifluranol-β-D-glucuronide
Bifluranol-di-O-sulphate
Bifluranol-O-sulphate

Bifluranol-β-D-glucuronide
Formed from: Bifluranol
→ Bifluranol-di-β-D-glucuronide
In: Dog B921
Ferret B921
Rat B921
→ Bifluranol-β-D-glucuronide phosphate
Probably in: Dog B921
Ferret B921
→ Bifluranol-β-D-glucuronide sulphate
Probably in: Dog B921
Ferret B921
Rat B921

Bifluranol-β-D-glucuronide phosphate
Probably formed from: Bifluranol-β-D-glucuronide

Bifluranol-β-D-glucuronide sulphate
Probably formed from: Bifluranol-β-D-glucuronide

Bifluranol-O-sulphate
Formed from: Bifluranol
→ Bifluranol-di-O-sulphate
In: Dog B921
Ferret B921

Bifuralol (2-(2-((1,1-dimethylethyl)amino)-1-hydroxyethyl)-7-ethylbenzofuran)
→ 1′-Hydroxybifuralol
In: Man D604

Binifibrate (3-pyridinecarboxylic acid 2-(2-(4-chlorophenoxy)-2-methyl-1-oxopropoxy)-1,3-propanediyl ester)
→ Clofibric acid
In: Rat G403

2,2′-Binormorphine
Formed from: Normorphine

Biochanin A (5,7-dihydroxy-4′-methoxyisoflavone)
Formed from: Biochanin A-7-glucoside
→ Biochanin A-glucoside (structure unclear)
In: Rabbit A571
→ Biochanin A-7-β-D-glucoside
In: Cicer K738
→ Biochanin A-β-D-glucuronide
In: Rabbit A571
→ Biochanin A sulphate
In: Man J865
→ 5,7-Dihydroxy-4′-methoxyisoflavanone
In: Fusarium C446, C474, D744, E724
→ Genistein
In: Ascochyta D744
Fusarium C474, E537
Man J865
→ Homogentisic acid
In: Armillaria H831
→ 5-Hydroxy-4′,7-dimethoxyisoflavone
In: Alfalfa G322
→ *p*-Hydroxyphenylacetic acid
In: Ascochyta D744
→ Phloroglucinol
In: Armillaria H831
→ Pratensein
In: Ascochyta D744
Chickpea E433, E537, H4, K751, K799
→ 5,6,7-Trihydroxy-4′-methoxyisoflavone
In: Microorganism J627
→ Vestitone
In: Chickpea E433, H4, K751, K799

Biochanin B
See: Formononetin

Biochanin A-glucoside (structure unclear)
Formed from: Biochanin A

Biochanin A-7-β-D-glucoside
Formed from: Biochanin A
Biochanin A-7-glucoside-6″-malonate
→ Biochanin A
In: Ascochyta D744
Fusarium D474
→ 7-O-(β-D-Apiofuranosyl(1→2)-β-D-glucosyl)biochanin A
In: Parsley K827
→ Biochanin A-7-glucoside-6″-malonate
In: Cicer D445

Biochanin A-7-glucoside-6″-malonate
Formed from: Biochanin A-7-glucoside
→ Biochanin A-7-glucoside
In: Fusarium D474

Biochanin A-β-D-glucuronide
Formed from: Biochanin A

Biochanin A sulphate
Formed from: Biochanin A

Biphenyl
Formed from: Dibenzothiophene
→ γ-Benzoylbutyric acid
In: Microorganism A963
Pseudomonas E41
→ *cis*-2,3-Dihydro-2,3-dihydroxybiphenyl
In: Pseudomonas G783, K613
→ *cis*-3,4-Dihydro-2,3-dihydroxybiphenyl
In: Pseudomonas K613
→ 2-Hydroxybiphenyl
In: Absidia A3427, B41
Aspergillus A1218, B549
Asterias A1379
Avocado A11
Buccinium A2913
Burkholderia K422
Cancer A1379
Candida B775
Cirolana A2913
Cunninghamella A3427, B41, B549
Debaryomyces J824
Ferret A2108
Gerydon A1379
Gilbertella A3427, B41
Gliocladium A3427, B41, B549
Guinea pig A2911
Hamster A323, A1371, A1658, A3323
Helicostylum A1218, B549
Homarus A1379
Mouse A2108, A3323, A3849, B615, B617
Mucor D602
Mytilus A1379
Ophiocomina A2913
Pigeon G885
Pig A2650
Pseudomonas K422
Pycnoporus K689

Raja A1379
Rabbit A2147, A2911, A3323,
Rainbow trout J582
Rat A1371, A1753, A2108, A2109, A2127, A2160, A2371, A2660, A2724, A3231, A3323, A3680, A3808, B195, B615, B617, B628, C291, D122, G885
Salvelinus A1379
Streptomyces A1218
Syncephalastrum A3427, B41
Trichosporon K682
Zooplankton A1379
See also: B546
→ 3-Hydroxybiphenyl
In: Burkholderia K422
Candida B775
Cunninghamella A3427, B41
Guinea pig A2911
Hamster A3323
Mouse A3323, B615, B617
Mucor D602
Nocardia B797
Pig A2650
Pseudomonas K422
Rabbit A613, A2911, A3323
Rat A2109, A2160, A2660, A3231, A3680, A3808, B195, B615, B617, B628, D122
Trichosporon K682
→ 4-Hydroxybiphenyl
In: Absidia A3427, B41, B434
Aspergillus A1218, B434, B549, C860
Asterias A1379
Avocado A11
Buccinium A2913
Cancer A1379
Candida B775
Cirolana A2913
Cunninghamella A1218, A3427, B41, B434, B549
Curvularia A1218
Cyanobacterium B383
Debaryomyces J824
Ferret A2108
Gerydon A1379
Gilbertella A3427, B41
Gliocladium B549
Guinea pig A2911
Hamster A323, A1371, A1658, A3323
Helicostylum A1218
Homarus A1379
Monkey A1997
Mouse A2108, A3323, A3849, B615, B617
Mucor D602
Mytilus A1379
Ophiocomina A2913
Pig A1997, A2650
Pigeon G885
Pycnoporus K689
Rabbit A570, A2147, A2911, A3323
Rainbow trout J582
Raja A1379
Rat A1371, A1753, A1997, A2108, A2109, A2127, A2160, A2371, A2660, A2724, A3231, A3264, A3680, A3808, B195, B615, B617, B628, B651, C291, D122, D413, G885
Salvelinus A1379
Streptomyces A1218
Tree shrew A1997
Trichosporon K682
Yeast A9
Zooplankton A1379
→ 3-(*p*-Hydroxyphenyl)muconolactone
In: Aspergillus H23
→ 6-Phenylpicolinic acid
In: Pseudomonas B328

4-Biphenylamine
See: 4-Aminobiphenyl

(1,1′-Biphenyl)-2,2′-dicarboxylic acid
See: Diphenic acid

3,3′-((1,1′-Biphenyl)-4,4′-diylbis(azo))bis(4-amino-1-naphthacenesulphonic acid)
See: Direct Red 28

Biphenyl-2,5-quinone
Formed from: 2,5-Dihydroxybiphenyl

4-Biphenylylacetic acid
Formed from: 4-(4-Biphenylyl)-3-hydroxybutyric acid
4-Ethynylbiphenyl
Fenbufen
Probably formed from: N-(1,1-Bis(hydroxymethyl)ethyl)-4-biphenylylacetamide

→ N-(4-Biphenylylacetyl)taurine
In: Dog A1108
→ 4′-Hydroxybiphenyl-4-ylacetic acid
Probably in: Rabbit A3631
Rat A1108, A3631

N-(4-Biphenylylacetyl)taurine
Formed from:4-Biphenylylacetic acid

3-(4-Biphenylyl)butyronitrile
→ 2-Hydroxy-2-(4′-Hydroxy-4-biphenylyl)propionic acid
In: Rat C747
→ 3-(4′-Hydroxy-4-biphenylyl)butyric acid
In: Rat C747
→ 3-(3′,4′-Dihydroxy-4-biphenylyl)-3-hydroxybutyronitrile
In: Rat C747

(1,1′-Biphenyl)-4,4′-diamine
See: Benzidine

N-(3-(1,1′-Biphenyl)-4-yl)-N-((diphenoxyphosphinyl)methyl-L-alanyl)-β-alanine
→ N-(3-(1,1′-Biphenyl)-4-yl)-N-((*o*-phenylphosphonyl)methyl-L-alanyl)-β-alanine
In: Dog H243
Man H243
Monkey H243
Rat H243

Biphenyl-2-yl-β-D-glucuronide
Formed from: 2-Hydroxybiphenyl

Biphenyl-3-yl-β-D-glucuronide
Formed from: 3-Hydroxybiphenyl

Biphenyl-4-yl-β-D-glucuronide
Formed from: 4-Hydroxybiphenyl

2-(4-Biphenylyl)-1-(S-glutathionyl)ethanol
Probably formed from: *p*-Phenylstyrene oxide

4-(4-Biphenylyl)-3-hydroxybutyric acid
→ 4-Biphenylylacetic acid
In: Guinea pig A2637
Man A2637
Rat A2637

4-(Biphenyl-4-yl)-4-hydroxybutyric acid
Formed from: Fenbufen

2-(Biphenyl-4-yl)lactic acid
Probably formed from: 2-(Biphenyl-4-yl)-2-propanol

4-Biphenylyl N-methylcarbamate
→ 4-Hydroxybiphenyl
In: Man A476

N-(4-Biphenylyl)oxamic acid
Formed from: 4-Aminobiphenyl

4-(Biphenyl-4-yl)-4-oxobutyric acid
See: Fenbufen

N-(3-(1,1′-Biphenyl)-4-yl)-N-((O-phenylphosphonyl)methyl-L-alanyl)-β-alanine
Formed from: N-(3-(1,1′-Biphenyl)-4-yl)-N-((diphenoxyphosphinyl)methyl-L-alanyl)-β-alanine
→ N-(3-(1,1′-Biphenyl)-4-yl)-N-((phosphonyl)methyl-L-alanyl)-β-alanine
In: Dog H243
Man H243
Monkey H243
Rat H243

N-(3-(1,1′-Biphenyl)-4-yl)-N-((phosphonyl)methyl-L-alanyl)-β-alanine
Formed from: N-(3-(1,1′-Biphenyl)-4-yl)-N-((O-phenylphosphonyl)methyl-L-alanyl)-β-alanine

2-(Biphenyl-4-yl)-1-propanol
Formed from: 4-Isopropylbiphenyl
→ 2-(Biphenyl-4-yl)-2-propionic caid
Probably in: Rat A3055

2-(Biphenyl-4-yl)-2-propanol
Formed from: 4-Isopropylbiphenyl
→ 2-(Biphenyl-4-yl)lactic acid
Probably in: Rat A3055
→ 2-(4′-Hydroxybiphenyl-4-yl)-2-propanol
Probably in: Rat A3055

2-(Biphenyl-4-yl)propionic acid
Probably formed from: 2-(Biphenyl-4-yl)-1-propanol
→ 2-(4′-Hydroxybiphenyl-4-yl)propionic acid
Probably in: Rat A3055

Biphenyl-2-yl sulphate
Formed from: 2-Hydroxybiphenyl

Biphenyl-3-yl sulphate
Formed from: 3-Hydroxybiphenyl

Biphenyl-4-yl sulphate
Formed from: 4-Hydroxybiphenyl

Bis(4-acetoxyphenyl)methylenecyclohexane
→ Bis(*p*-Hydroxyphenyl) methylenecyclohexane

Probably in: Sporotrichum A2630

3,3′-Bis(3-(N-acetylalanyl))-6,6′-dihydroxy-5,5′-diiodobiphenyl
Formed from: N-Acetyl-3-5-diiodotyrosine

3,4′-Bis(3-(N-acetylalanyl))-3′,5,5′-tribromo-4-hydroxydiphenyl ether
Formed from: N-Acetyl-3,5-dibromotyrosine

3,4′-Bis(3-(N-acetylalanyl))-3′,5,5′-trichloro-4-hydroxydiphenyl ether
Formed from: N-Acetyl-3,5-dichlorotyrosine

3,4′-Bis(3-(N-acetylalanyl))-3′,5,5′-trichloro-4-hydroxydiphenyl ether dimethyl ester
Formed from: N-Acetyl-3,5-dichlorotyrosine methyl ester

5,5′-Bis(3-L-alanyl)-2,2′-dihydroxybiphenyl
See: L-Dityrosine

3,4′-Bis(3-alanyl)-3′,5,5′-tribromo-4-hydroxydiphenyl ether
Formed from: 3,5-Dibromo-L-tyrosine

3,4′-Bis(3-(N-acetylalanyl))-3′,5,5′-trichloro-4-hydroxydiphenyl ether
Formed from: N-Acetyl-3,5-dichlorotyrosine

1,5-Bis(4-amidinophenoxy)-2-pentane
See: Pentamidine

1,5-Bis(*p*-amidinophenoxy)-2-pentanol
Formed from: Pentamidine

1,5-Bis(*p*-amidinophenoxy)-3-pentanol
Formed from: Pentamidine

4,4′-Bis(4-amino-3-chlorophenyl)-3,3′-dichloroazobenzene
Formed from: 3,3′-Dichlorobenzidine

5,5′-Bis(2-aminoethyl)-2,2′-dihydroxy-3,3′-dimethoxybiphenyl
Formed from: 3-Methoxytyramine

8,8′-Bis(4-amino-1-methylbutyl)amino)-5,5′-biquinoline
Formed from: Primaquine

Bis(2-amino-3-nitro-5-trifluoromethylphenyl) diazene
Probably formed from: 3-Nitro-5-trifluoro-methyl-*o*-phenylenediamine
→ Bis(2-amino-3-nitro-5-trifluoromethylphenyl)diazene-N-oxide
Probably in: Streptomyces B13

Bis(2-amino-3-nitro-5-trifluoromethylphenyl) diazene-N-oxide
Probably formed from: Bis(2-amino-3-nitro-5-trifluoromethylphenyl)diazene

1,3-Bis((3-(*p*-aminophenyl)-1-cycloheptylureido) methyl)benzene
Formed from: 1-((3-(*p*-Aminophenyl)-1-cycloheptylureido)methyl)-3-((1-cycloheptyl-3-(*p*-methylaminophenyl) ureido)methyl)benzene

N,N′-Bis(*p*-aminophenyl)-N,N′-dimethylethylenediamine
→ Free radical
Probably in: Rat A856

Bis(*p*-aminophenyl) sulphone (dapsone)
Formed from: 4′-Acetamidophenyl) 4-aminophenyl sulphone
4′-(D-Alanyl)aminophenyl 4-aminophenyl sulphone
4′-(L-Alanyl)aminophenyl 4-aminophenyl sulphone
4-Amino-4′-formamidodiphenyl sulphone
4-Aminophenyl) 4′-glycylaminophenyl sulphone
4-Aminophenyl) 4′-(D-leucyl)aminophenyl sulphone
4-Aminophenyl) 4′-(L-leucyl)aminophenyl sulphone
4-Aminophenyl) 4′-(L-lysyl)aminophenyl sulphone
4-Aminophenyl) 4′-(D-phenylalanyl)amino-phenyl sulphone
4-Aminophenyl) 4′-(L-phenylalanyl)amino-phenyl sulphone
→ 4′-Acetamidophenyl) 4-aminophenyl sulphone
In: Dog A1537
Goat A3766
Man A1518, G111
Rat A268, A1277
→ 4-Amino-4′-nitrodiphenyl sulphone
In: Rat A970
→ 4-Aminophenyl-4′-hydroxylaminophenyl sulphone
In: Dog A906
Man A906, E685, G111, J19, K59
Rabbit A906

Rat G943
→ Bis(p-aminophenyl) sulphone-N-β-D-glucuronide
In: Man H541
Rat A268
→ Bis(p-aminophenyl) sulphone sulphamate
In: Rat A268

Bis(p-aminophenyl) sulphone-N-β-D-glucuronide
Formed from: Bis(p-Aminophenyl) sulphone

Bis(p-aminophenyl) sulphone sulphamate
Formed from: Bis(p-Aminophenyl) sulphone

Bisanhydro-13-dihydrodaunomycinone
Formed from: Daunorubicin

Bisanhydroaklavinic acid
Formed from: Aclacinomycin A

Bisanhydroaklavinone
Formed from: Aclacinomycin A
Aklavinone

4,4′-Bisazoxybiphenyl
Formed from: N-Hydroxy-4-acetamidobiphenyl

4,4′-Bisazoxychlorobenzene
Formed from: p-Chloro-N-hydroxyacetanilide

2,2′-Bisazoxyfluorene
Formed from: N-Hydroxy-2-acetamidofluorene

4-Bis(2-bromoethyl)aminoazobenzene
→ Aniline
In: Mouse A92
Rat A92
→ N,N-Di-(2-bromoethyl)-p-phenylenediamine
In: Mouse A92
Rat A92

4-Bis(2-bromoethyl)amino-2-methylazobenzene
→ Aniline
In: Mouse A92
Rat A92
→ N,N-Di-(2-bromoethyl)-3-methyl-p-phenylenediamine
In: Mouse A92
Rat A92

2-(4′-Bis(2″-bromopropyl)aminoazo)benzoic acid
→ Anthranilic acid
In: Rat A2351
→ N,N-Di-(2-bromopropyl)-p-phenylenediamine
In: Rat A2351

3,3′-Bis(5-carboxy-2-hydroxy-3-methoxyphenyl)-4,4′-dihydroxy-5,5′-dimethoxybiphenyl
Formed from: 5,5′-Dicarboxy-2,2′-dihydroxy-3,3′-dimethoxybiphenyl

2′,4-Bis(carboxymethoxy)-4′-(3-hydroxymethyl-2-butenyloxy)dihydrochalcone
Probably formed from: 2′,4-Biscarboxymethoxy-4′-(3-methyl-2-butenyloxy)dihydrochalcone
→ 4-(3-Carboxy-2-butenyloxy)-2′,4-bis(carboxymethoxy)dihydrochalcone
Probably in: Rat B372

2′,4-Bis(carboxymethoxy)-4′-(3-methyl-2-butenyloxy)dihydrochalcone
Formed from: 2′-Carboxymethoxy-4,4′-bis(3-methyl-2-butenyloxy)dihydrochalcone
→ 2′-Bis(carboxymethoxy)-4′-(3-hydroxy-methyl-2-butenyloxy)dihydrochalcone
Probably in: Rat B372

1,4-Bis(2-(carboxymethylamino)ethylamino)-5,8-dihydroxyanthraquinone
Formed from: 1-(2-(Carboxymethylamino) ethylamino)-5,8-dihydroxy-4-(2-(2-hydroxyethylamino)ethylamino) anthraquinone

1,4-Bis(carboxymethyl)benzene
Formed from: 1,4-Divinylbenzene
→ p-(Carboxymethyl)phenylacetylglycine
Probably in: Rat F881

2,4-Bis(carboxymethyl)phenol
Formed from: 5-Carboxymethylbenzofuran
→ 3-Carboxymethyl-4-hydroxyphenylacetylglycine
Probably in: Dog H866
Man H866
Mouse H866
Rabbit H866
Rat H866

N^2, N^5-Bis(p-chlorobenzoyl)ornithine
Formed from: p-Chlorobenzoic acid

Bis(p-chlorobenzyl)amine
→ Bis(p-chlorobenzyl)hydroxylamine
In: Rat J523
→ p-Chlorobenzaldehyde
In: Rat J523
→ p-Chlorobenzylamine

In: Rat J523

Bis(*p*-chlorobenzyl)hydroxylamine
Formed from: Bis(*p*-chlorobenzyl)amine
→ Bis(*p*-chlorobenzyl)nitrone
Probably in: Rat J523

1,3-Bis(*p*-chlorobenzylideneamino)guanidine
See: Robenidine

Bis(*p*-chlorobenzyl)nitrone
Probably formed from: Bis(*p*-chlorobenzyl)hydroxylamine

4-(Bis(2-chloroethyl)amino)phenol
Formed from: N,N-(Bis(2-chloroethyl))aniline

***p*-(Bis(2-chloroethyl)amino)phenylacetic acid**
Formed from: Chlorambucil
3,4-Dehydrochlorambucil

4-(Bis(2-chloroethyl)amino)-L-phenylalanine
See: L-Phenylalanine mustard

4-(Bis(2-chloroethyl)amino)phenylbutyric acid
See: Chlorambucil

N,N-(Bis(2-chloroethyl))aniline (aniline mustard)
→ 4-(Bis(2-chloroethyl)amino)phenol
In: Rat A2929
See also: A2928

N,N-Bis(2-chloroethyl)-*p*-hydroxybenzenamine
See: *p*-Hydroxyaniline mustard

Bis(5-chloro-2-hydroxyphenyl)methane
See: Dichlorophen

Bis(5-chloro-2-nitrobenzene)disulphide
Formed from: 5-Chloro-2-nitrothiophenol

2,2-Bis(*p*-chlorophenyl)acetaldehyde
Formed from: 1-Chloro-2,2-bis(*p*-chlorophenyl)ethane
→ 2,2-Bis(*p*-chlorophenyl)acetic acid
Probably in: Mouse C675
→ 2,2-Bis(*p*-chlorophenyl)ethanol
Probably in: Mouse C675

2,2-Bis(*p*-chlorophenyl)acetic acid
Formed from: 2,2-Bis(*p*-chlorophenyl)ethanol
1,1-Dichloro-2,2-bis(*p*-chlorophenyl)ethane
1,1,1-Trichloro-2,2-bis(*p*-chlorophenyl)ethane
Probably formed from: 2,2-Bis(*p*-chlorophenyl)acetaldehyde
3,3-Bis(*p*-chlorophenyl)pyruvic acid
→ Bis(*p*-chlorophenyl)acetyl-L-alanine
In: Hamster A1380, A1742
Mouse A1380, A1742
→ Bis(*p*-chlorophenyl)acetyl-β-D-glucoside
In: Glycine H402
→ Bis(*p*-chlorophenyl)acetyl-β-D-glucuronic acid
In: Hamster A1380, A1742
Mouse A1380, A1742
→ Bis(*p*-chlorophenyl)acetylglycine
In: Hamster A1380
Mouse A1380, A1742
Rat A1742
→ Bis(*p*-chlorophenyl)acetyl-L-serine
In: Hamster A1742
Mouse A1742
→ Bis(*p*-chlorophenyl)acetyltaurine
In: Lobster C38
→ Bis(*p*-chlorophenyl)-2-hydroxyacetic acid
Probably in: Mouse C439
→ Bis(*p*-chlorophenyl)methane
In: Pseudomonas A2541
See also: A531, A3397

Bis(*p*-chlorophenyl)acetyl-L-alanine
Formed from: Bis(*p*-chlorophenyl)acetic acid

Bis(*p*-chlorophenyl)acetyl-β-D-glucoside
Formed from: Bis(*p*-chlorophenyl)acetic acid

Bis(*p*-chlorophenyl)acetyl-β-D-glucuronic acid
Formed from: Bis(*p*-chlorophenyl)acetic acid

Bis(*p*-chlorophenyl)acetylglycine
Formed from: Bis(*p*-chlorophenyl)acetic acid

Bis(*p*-chlorophenyl)acetyl-L-serine
Formed from: Bis(*p*-chlorophenyl)acetic acid

Bis(*p*-chlorophenyl)acetyltaurine
Formed from: Bis(*p*-chlorophenyl)acetic acid

Bis(*p*-chlorophenyl)carbinol (4,4′-dichlorobenzhydrol)
Formed from: Bis(*p*-chlorophenyl)methane
4,4′-Dichlorobenzophenone
→ Benzhydrol
In: Pseudomonas A2541
→ 4,4′-Dichlorobenzophenone
In: Pseudomonas A2541
Trout A3397
See also: A531, A3521

2,2-Bis(*p*-chlorophenyl)ethanol
Formed from: 1,1-Bis(*p*-chlorophenyl)ethylene

Probably formed from: 2,2-Bis(*p*-chlorophenyl)acetaldehyde
→ Bis(*p*-chlorophenyl)acetic acid
In: Rat A928
See also: A3397

1,1-Bis(*p*-chlorophenyl)ethylene
Formed from: 1-Chloro-2,2-bis(*p*-chlorophenyl)ethylene
→ 2,2-Bis(*p*-chlorophenyl)ethanol
In: Rat A928

2,2-Bis(*p*-chlorophenyl)-2-hydroxyacetic acid
Formed from: 2,2-Bis(*p*-chlorophenyl)acetic acid

Bis(*p*-chlorophenyl)methane
Formed from: Bis(*p*-chlorophenyl)acetic acid
→ Bis(*p*-chlorophenyl)acetic acid
In: Pseudomonas A2541
See also: A531
→ Bis(*p*-chlorophenyl)carbinol
In: Pseudomonas A2541
See also: A531
→ *p*-Chlorophenylglycollaldehyde
In: Pseudomonas A2541

1,1-Bis(*p*-chlorophenyl)-2-nitropropane
See: Prolan

1,1-Bis(*p*-chlorophenyl)propan-2-one
Probably formed from: 2-Amino-1,1-bis(*p*-chlorophenyl)propane
→ 3,3-Bis(*p*-chlorophenyl)pyruvic acid
Probably in: Insect A2190
Mouse A2190

1,1-Bis(*p*-chlorophenyl)prop-1-ene
Probably formed from: 2-Amino-1,1-bis(*p*-chlorophenyl)propane

3,3-Bis(*p*-chlorophenyl)pyruvic acid
Probably formed from: 1,1-Bis(*p*-chlorophenyl)propan-2-one
→ Bis(*p*-chlorophenyl)acetic acid
Probably in: Insect A2190
Mouse A2190
→ 4,4′-Dichlorobenzophenone
Probably in: Insect A2190
Mouse A2190

2,3-Bis(CoA-S-yl)-4-dimethylamino-5-(S-glutathionyl)phenol
Probably formed from: *p*-Dimethylaminophenol

N[1],N[10]-Bis(*p*-coumaroyl)spermidine
Probably formed from: *p*-Coumaric acid
→ Lunarine
In: Lunaria J502

Bis(*p*-cyanophenyl) phosphate
→ *p*-Hydroxybenzonitrile
In: Rat A2591

1,3-Bis((1-cycloheptyl-3-(*p*-dimethylaminophenyl) ureido)methyl)benzene
→ 1-((1-Cycloheptyl-3-(*p*-dimethylamino phenyl)ureido)methyl)-3-((1-cycloheptyl-3-(*p*-methylaminophenyl)ureido)methyl) benzene
In: Dog H130
Rat G581, H130

1,3-Bis((1-cycloheptyl-3-(*p*-methylaminophenyl) ureido)methyl)benzene
Formed from: 1-((1-Cycloheptyl-3-(*p*-dimethylaminophenyl)ureido)methyl)-3-((1-cycloheptyl-3-(*p*-methylaminophenyl) ureido)methyl)benzene
→ 1-((3-(*p*-Aminophenyl)-1-cycloheptyl ureido)methyl)-3-((1-cycloheptyl-3-(*p*-methylaminophenyl)ureido)methyl) benzene
Probably in: Rat G581, H130

2,5-Bis(S-cysteinyl)-3,4-dihydroxy-L-phenylalanine
Formed from: 5-(S-Cysteinyl)-3,4-dihydroxy-L-phenylalanine
→ 2,4,5-Tris(S-cysteinyl)-3,4-dihydroxy-L-phenylalanine
In: Horseradish B762

Bisdeallylalmitrine
Formed from: Deallylalmitrine

Bisdebutylbucromarone
Formed from: Debutylbucromarone

Bisdechlorogeodin ((+)-4-hydroxy-6′-methoxy-6-methyl-3,4′-dioxo*spiro*(benzofuran-2(3*H*),1′-[2,5]cyclohexadiene)-2′-carboxylic acid methyl ester)
Formed from: Sulochrin

Bisdeisopropyldisopyramide
Probably formed from: Deisopropyl-disopyramide

Bisdeisopropyltolterodine
Formed from: Deisopropyltolterodine

→ 3-(2-Hydroxy-5-methylphenyl)-3-(*p*-hydroxyphenyl)propylamine
Probably in: Rat J419

Bisdemethylamiflamine
Formed from: Demethylamiflamine

Bisdemethylazonafide
Probably formed from: Demethylazonafide

Bisdemethylbenzydamine
Formed from: Demethylbenzydamine

Bisdemethylbromazepine
Formed from: Demethylbromazepine

Bisdemethylchlorotrianisene
Probably formed from: Demethylchlorotrianisene

Bisdemethylchlorpheniramine
Formed from: Demethylchlorpheniramine
→ N-Acetylbisdemethylchlorpheniramine
Probably in: Man F902
Rat F902

Bisdemethylchlorpromazine
Formed from: Demethylchlorpromazine
Prochlorperazine
Probably formed from: N-(γ-(2-Chloro-10-phenothiazinyl)propyl)ethylenediamine
N-(γ-(2-Chloro-10-phenothiazinyl)propyl)-N′-methylethylenediamine
Perphenazine
→ Bisdemethylchlorpromazine sulphoxide
In: Rat A2881, A3925
See also: A69, A693, A901
→ 7-Hydroxybisdemethylchlorpromazine
Probably in: Man A693
Rabbit A901

Bisdemethylchlorpromazine sulphone
Formed from: Bisdemethylchlorpromazine sulphoxide

Bisdemethylchlorpromazine sulphoxide
Formed from: Bisdemethylchlorpromazine
→ Bisdemethylchlorpromazine sulphone
In: Rat A2881, A3925

Bisdemethylcitalopram
Formed from: Demethylcitalopram
→ Bisdemethylcitalopram-β-D-glucuronide
Probably in: Man J758

Bisdemethylcitalopram-β-D-glucuronide
Probably formed from: Bisdemethylcitalopram

N,O-Bisdemethyldextromethorphan
See: 3-Hydroxymorphinan

N,O-Bisdemethyldiltiazem
Probably formed from: N-Demethyldiltiazem
→ N,O-Bisdemethyldiltiazem-β-D-glucuronide
Probably in: Man D126
→ N,O-Bisdemethyldiltiazem sulphate
Probably in: Man D126

N,O-Bisdemethyldiltiazem-β-D-glucuronide
Probably formed from: N,O-Bisdemethyldiltiazem

N,O-Bisdemethyldiltiazem sulphate
Probably formed from: N,O-Bisdemethyldiltiazem

5,6-Bis-O-demethyldonezepil
Probably formed from: 5-O-Demethyldonezepil
→ 5,6-Bis-O-demethyldonezepil-5-β-D-glucuronide
Probably in: Rat J789
→ 5,6-Bis-O-demethyldonezepil-6-β-D-glucuronide
Probably in: Rat J789

5,6-Bis-O-demethyldonezepil-5-β-D-glucuronide
Probably formed from: 5,6-Bis-O-demethyldonezepil

5,6-Bis-O-demethyldonezepil-6-β-D-glucuronide
Probably formed from: 5,6-Bis-O-demethyldonezepil

Bisdemethyldothiepin
Formed from: Northiaden

Bisdemethyldoxepin
Formed from: Demethyldoxepin
→ N-Acetyl-N,N-bisdemethyldoxepin
Probably in: Cunninghamella J782

Bisdemethyldoxylamine
Formed from: Demethyldoxylamine
→ N-Acetylbisdemethyldoxylamine
Probably in: Man E785
Rat E11
Squirrel monkey E11

Bisdemethylimipramine (DDMI)
Formed from: Demethylimipramine

N,O-Bisdemethyllevomepromazine
Probably formed from: O-Demethyllevomepromazine

C,O-Bisdemethyllevormeloxifene
Probably formed from: Demethyl-levormeloxifene

4,5′-Bisdemethyllevosemotiadil
Probably formed from: 5′-Demethyllevosemotiadil

Bisdemethylloperamide
Formed from: Demethylloperamide
→ Hydroxybisdemethylloperamide
Probably in: Rat B275

Bisdemethylmifepristone
Formed from: Demethylmifepristone

4′,6-Bisdemethylpapaverine
Formed from: 4′-Demethylpapaverine
6-Demethylpapaverine

Bisdemethylpromazine (γ-(phenothiazinyl(10)) propylamine)
Formed from: N-Hydroxybisnorpromazine
Perazine
Probably formed from: Demethylperazine
Demethylpromazine
→ Bisdemethylpromazine sulphoxide
Probably in: Rat A69

Bisdemethylpromazine sulphone
Formed from: Bisdemethylpromazine sulphoxide

Bisdemethylpromazine sulphoxide
Probably formed from: Bisdemethylpromazine
→ Bisdemethylpromazine sulphone
In: Man A2881

N,O-Bisdemethylpyrilamine
Probably formed from: N-Demethylpyrilamine
O-Demethylpyrilamine

4,5′-Bisdemethylsemotiadil
Probably formed from: 5′-Demethylsemotiadil
→ 4,5′-Bisdemethylsemotiadil-β-D-glucuronide
Probably in: Rat H313

4,5′-Bisdemethylsemotiadil-β-D-glucuronide
Probably formed from: 4,5′-Bisdemethylsemotiadil

Bisdemethyltamoxifen
Probably formed from: Demethyltamoxifen

3′,7-Bisdemethyltofizopam
Probably formed from: 7-Demethyltofizopam

4′,7-Bisdemethyltofizopam
Probably formed from: 7-Demethyltofizopam

Bisdemethyltopotecan
Formed from: Bisdemethyltopotecan
→ Bisdemethyltopotecan-β-D-glucuronide
Probably in: J704

Bisdemethyltopotecan-β-D-glucuronide
Probably formed from: Bisdemethyltopotecan

3′,5′-Bisdemethyltrimethoprim
Probably formed from: 3′-Demethyltrimethoprim

N,N-Bisdemethylvenlafaxine
Formed from: N-Demethylvenlafaxine
→ N,N,O-Trisdemethylvenlafaxine
Probably in: Man G564
Rhesus monkey G564

N,O-Bisdemethylvenlafaxine
Probably formed from: N-Demethylvenlafaxine
→ N,O-Bisdemethylvenlafaxine-β-D-glucuronide
Probably in: Man G564
Mouse G564
Rat G564
Rhesus monkey G564
→ N,O-Bisdemethylvenlafaxine-O-sulphate
Probably in: Man G564
Mouse G564
Rat G564
Rhesus monkey G564

N,O-Bisdemethylvenlafaxine-β-D-glucuronide
Probably formed from: N,O-Bisdemethylvenlafaxine

N,O-Bisdemethylvenlafaxine-O-sulphate
Probably formed from: N,O-Bisdemethylvenlafaxine

Bisdesethylamodiaquine
Probably formed from: Desethylamodiaquine

Bisdesethylbromerguride
Formed from: Desethylbromerguride

Bisdesethylentacapone
Formed from: Desethylentacapone

N,N-Bisdesethylethacizin
Probably formed from: N-Desethylethacizin

N-Bisdesethyllisuride
Formed from: Desethyllisuride

Bis(3,5-dichloro-2-hydroxyphenyl) sulphide
Formed from: Bis(3,5-dichloro-2-hydroxyphenyl) sulphoxide

→ Bis(3,5-dichloro-2-hydroxyphenyl) sulphide-β-D-glucuronide
Probably in: Man C151

Bis(3,5-dichloro-2-hydroxyphenyl) sulphide-β-D-glucuronide
Probably formed from: Bis(3,5-dichloro-2-hydroxyphenyl) sulphide

Bis(3,5-dichloro-2-hydroxyphenyl) sulphoxide
→ Bis(3,5-dichloro-2-hydroxyphenyl) sulphide
In: Man C151
→ Bis(3,5-dichloro-2-hydroxyphenyl) sulphoxide-β-D-glucuronide
Probably in: Man C151

Bis(3,5-dichloro-2-hydroxyphenyl) sulphoxide-β-D-glucuronide
Probably formed from: Bis(3,5-dichloro-2-hydroxyphenyl) sulphoxide

N,N′-Bis(3,4-dichlorophenyl)-N-hydroxyhydrazine
Formed from: 3,3′,4,4′-Tetrachloroazobenzene

1,3-Bis(3,4-dichlorophenyl)triazene
Formed from: 3,4-Dichloroaniline

4,4′-Bis(diethylaminoethoxy)hexoestrol
→ 4-Ethylaminoethoxyhexoestrol
In: Man B430

3,6-Bis(2,3-dihydro-1,5-dimethyl-3-oxo-2-phenyl-1*H*-pyrazol-4-yl)azo)-4,5-dihydroxy-2,7-naphthalenedisulphonic acid
See: Antipyrylazo III

2,3-Bis(α,4-dihydroxy-3-methoxybenzyl)succinic dilactone
Formed from: Ferulic acid

3,6-Bis(2,4-dihydroxyphenyl)-7-hydroxy-2-phenoxazone
Formed from: Resorcinol

1,3-Bis(2,4-dihydroxyphenyl)-1,3-propanedione
→ 1-(2,4-Dihydroxyphenyl)-(4-hydroxy-2-methoxyphenyl)-1,3-propanedione
In: Glycyrrhiza K886

(2*R*,2*R*′,3*R*,3′*R*,4*R*)-2,2′-bis(3,4-dihydroxyphenyl)-3,3′,4,4′-tetrahydro-(4,8′-bi-2*H*-1-benzopyran)-3,3′,5,5′,7,7′-hexol
See: Procyanidin B_2

(2*R*,2*R*′,3′*R*,3*S*,4*S*)-2,2′-bis(3,4-dihydroxyphenyl)-3,3′,4,4′-tetrahydro-(4,8′-bi-2*H*-1-benzopyran)-3,3′,5,5′,7,7′-hexol
See: Procyanidin B_4

(2*R*,2*R*′,3*S*,3′*S*,4*S*)-2,2′-bis(3,4-dihydroxyphenyl)-3,3′,4,4′-tetrahydro-(4,8′-bi-2*H*-1-benzopyran)-3,3′,5,5′,7,7′-hexol
See: Procyanidin B_3

2,2-Bis(*p*-(2,3-dihydroxypropoxy)phenyl)propane
Formed from: 2,2-Bis(*p*-(2,3-epoxypropoxy)phenyl)propane
→ 2-(*p*-(2-Carboxymethoxy)phenyl)-2-(*p*-(2,3-dihydroxypropoxy)phenyl)propane
Probably in: Mouse B704

1-(4,6-Bis((2,3-dihydroxypropyl)amino)-1,3,5-triazin-2-yl)-4-(bis(*p*-fluorophenyl)methyl) piperazine
Probably formed from: 1-(4-Allylamino-6-((2,3-dihydroxypropyl)amino)-1,3,5-triazin-2-yl-4-(bis(*p*-fluorophenyl)methyl)piperazine

1,2-Bis(3,4-dimethoxyphenyl)-1,3-propanediol
→ 3,4-Dimethoxyphenylglycol
In: Phanerochaete C401
→ Veratraldehyde
In: Phanerochaete C401

4,4′-Bis(dimethylamino)benzophenone
See: Michler's ketone

1,4-Bis(2-dimethylaminoethyl)amino-5,8-dihydroxyanthraquinone
Formed from: 1-(2-Dimethylaminoethyl) amino-4-(2-(dimethylamino-N-oxide)ethyl) amino-5,8-dihydroxyanthraquinone

1,4-Bis(2-(dimethylamino-N-oxide)ethyl)amino-5,8-dihydroxyanthraquinone
→ 1-(2-Dimethylaminoethyl)amino-4-(2-(dimethylamino-N-oxide)ethyl)amino-5,8-dihydroxyanthraquinone
In: Rat J753

4,4-Bis(5,5-dimethyl-1,3,2-dioxaphosphorinan-2-yl)-*m*-fluorobutyrophenone
→ 4,4-Bis(5,5-dimethyl-1,3,2-dioxaphosphorinan-2-yl)-3-fluoro-hydroxybutyrophenone
In: Rat J285
→ 4,4-Bis(5,5-dimethyl-1,3,2-dioxaphosphorinan-2-yl)-1-(*m*-fluorophenyl)-1-butanol
In: Rat J285
→ 4-(5,5-Dimethyl-1,3,2-dioxaphosphorinan-2-yl)-*m*-fluoro-4-(5-hydroxymethyl-5-methyl-1,3,2-dioxaphosphorinan-2-yl) butyrophenone
In: Rat J285

→ 1-(3-Fluoro-5-(S-glutathionyl)-4,5-dihydro-4-hydroxyphenyl-4,4-bis(5,5-dimethyl-1,3,2-dioxaphosphorinan-2-yl)-1-butanone
Probably in: Rat J285

4,4-Bis(5,5-dimethyl-1,3,2-dioxaphosphorinan-2-yl)-3-fluoro-hydroxybutyrophenone
Formed from: 4,4-Bis(5,5-dimethyl-1,3,2-dioxaphosphorinan-2-yl)-*m*-fluorobutyrophenone

4,4-Bis(5,5-dimethyl-1,3,2-dioxaphosphorinan-2-yl)-1-(*m*-fluorophenyl)-1-butanol
Formed from: 4,4-Bis(5,5-dimethyl-1,3,2-dioxaphosphorinan-2-yl)-*m*-fluorobutyrophenone

5-((3,5-Bis(1,1-dimethylethyl)-4-hydroxyphenyl)methyl)-4-thiazolidinone
See: Tazofelone

2,2-Bis(*p*-(2,3-epoxypropoxy)phenyl)propane
→ 2,2-Bis(*p*-(2,3-dihydroxypropoxy)phenyl)propane
In: Mouse B704
→ 2,2-Bis(*p*-(2,3-epoxypropoxy)phenyl)propane
In: Mouse B704
→ 2-(*p*-(2-Carboxy-2-hydroxyethoxy)phenyl)-2-(*p*-(2-hydroxy-3-sulphonylpropoxy)phenyl)propane
In: Mouse B704
→ 2-(*p*-(2,3-Dihydroxypropoxy)phenyl)-2-(*p*-hydroxyphenyl)propane
In: Mouse B704

1,2-Bis(3-ethoxycarbonyl-2-thioureido)benzene
(thiophanate)
Formed from: Methyl 1-(butylcarbamoyl)benzimidazol-2-yl carbamate
→ 2-(3-Ethoxycarbonyl-2-thioureido)aniline
In: Mouse A260
→ 2-(3-Ethoxycarbonyl-2-thioureido)-1-(3-ethoxycarbonyl-2-ureido)benzene
In: Mouse A260
→ 2-(3-Ethoxycarbonyl-2-thioureido)-1-(2-ureido)benzene
In: Mouse A260

1,2-Bis(3-ethoxycarbonyl-2-ureido)benzene
Formed from: 2-(3-Ethoxycarbonyl-2-thioureido)-1-(3-ethoxycarbonyl-2-ureido)benzene
→ 2-(3-Ethoxycarbonyl-2-ureido)-1-(2-ureido)benzene
In: Mouse A260

Bis(*o*-ethoxyphenyl) phosphate
Probably formed from: Bis(*o*-ethoxyphenyl) 9-((2-phosphonomethoxy)ethyl)adenine

Bis(*o*-ethoxyphenyl) 9-((2-phosphonomethoxy)ethyl)adenine
→ Bis(*o*-ethoxyphenyl) phosphate
Probably in: Rat J136

O,O-Bis(*o*-ethoxyphenyl) O-phenyl phosphate
→ *o*-(α-Hydroxyethyl)phenol
In: Mouse D244
Rat D244

2,2-Bis(*p*-fluorophenyl)acetic acid
Probably formed from: 4,4-Bis(*p*-fluorophenyl)butanoic acid

4,4-Bis(*p*-fluorophenyl)butanoic acid
Formed from: Amperozide
1-(4,4-Bis(*p*-fluorophenyl)butyl)butanoic acid
1-(4,4-bis(*p*-fluorophenyl)butyl)piperazine
Lidoflazine
Penfluridol
→ 2,2-Bis(*p*-fluorophenyl)acetic acid
Probably in: Dog A3915
Man A3915
Rat A3915
→ 4,4-Bis(*p*-fluorophenyl)butanoyl-β-D-glucuronic acid
In: Rabbit H745
→ 4,4-Bis(*p*-fluorophenyl)-4-hydroxybutanoic acid
In: Dog A2699
Rat A2699

4,4-Bis(*p*-fluorophenyl)butanoyl-β-D-glucuronic acid
Formed from: 4,4-Bis(*p*-fluorophenyl)butanoic acid

4,4-Bis(*p*-fluorophenyl)-*n*-butanol
Formed from: Amperozide

1-(4,4-Bis(4-fluorophenyl)butyl)-4-(4-chloro-3-(trifluoromethyl)phenyl)-4-piperidinol
See: Penfluridol

4-(4,4-Bis(4-fluorophenyl)butyl)-N-(2,6-dimethylphenyl)-1-piperazineacetamide
See: Lidoflazine

4-(4,4-Bis(*p*-fluorophenyl)butyl)-1-(N-ethylcarboxamido)-2,5-dihydroxypiperazine
Formed from: Amperozide

4-(4,4-bis(*p*-fluorophenyl)butyl)-N-Ethyl-1-piperazinecarboxamide
See: Amperozide

1-(4,4-Bis(*p*-fluorophenyl)butyl)-4-methylpiperazine
Formed from: 1-(4,4-bis(*p*-fluorophenyl) butyl)piperazine

4-(4,4-Bis(*p*-fluorophenyl)butyl)-1-(2-oxopiperazine)aceto-2′,6′-xylidide
Formed from: Lidoflazine

1-(4,4-Bis(*p*-fluorophenyl)butyl)piperazine
Formed from: Lidoflazine
→ 4-Acetyl-1-(4,4-bis(*p*-fluorophenyl) butyl)piperazine
In: Dog A2699
Rat A2699
→ 4,4-Bis(*p*-fluorophenyl)butanoic acid
In: Dog A2699
Rat A2699
→ 1-(4,4-Bis(*p*-fluorophenyl)butyl)-4-formylpiperazine
In: Dog A2699
Rat A2699
→ 1-(4,4-Bis(*p*-fluorophenyl)butyl)-4-methylpiperazine
In: Dog A2699
Rat A2699

1-(4,4-Bis(*p*-fluorophenyl)butyl)-4-formylpiperazine
Probably formed from: 1-(4,4-Bis(*p*-fluorophenyl)butyl)piperazine

4,4-Bis(*p*-fluorophenyl)-4-hydroxybutanoic acid
Probably formed from: 4,4-Bis(*p*-fluorophenyl)butanoic acid

N-(4-(Bis(*p*-fluorophenyl)-4-hydroxybutyl)-N′-carboxymethyl-N′-(N-ethylcarboxamido) ethylenediamine
Probably formed from: N-(4-(Bis(*p*-fluorophenyl)-4-hydroxybutyl)-N′-(N-ethylcarboxamido)-N′-(2-hydroxyethyl) ethylenediamine

N-(4-(Bis(*p*-fluorophenyl))-4-hydroxybutyl)-N′-(N-ethylcarboxamido)-N-(2-hydroxyethyl) ethylenediamine
Formed from: Amperozide

N-(4-(Bis(*p*-fluorophenyl))-4-hydroxybutyl)-N′-(N-ethylcarboxamido)-N′-(2-hydroxyethyl) ethylenediamine
Formed from: Amperozide
→ N-(4-(Bis(*p*-fluorophenyl)-4-hydroxybutyl)-N′-carboxymethyl-N′-(N-ethylcarboxamide)ethylenediamine
Probably in: Rat H745

Bis(*p*-fluorophenyl)(hydroxymethyl)(methyl) silane
Formed from: Flusilazole
→ Bis(*p*-fluorophenyl) (hydroxymethyl) (methyl) silane-β-D-glucuronide
In: Goat K144

Bis(*p*-fluorophenyl)(hydroxymethyl)(methyl) silane-β-D-glucuronide
Formed from: Bis(*p*-fluorophenyl) (hydroxymethyl) (methyl) silane

Bis(*p*-fluorophenyl)methanol
Formed from: Flunarizine
→ Bis(*p*-fluorophenyl)methyl-β-D-glucuronide
In: Man G310
Rat G310

1-(Bis(*p*-fluorophenyl)methyl)-4-((2,3-dihydroxy-3-phenyl)propyl)piperazine
Probably formed from: 1-(Bis(*p*-fluorophenyl)methyl)-4-((2,3-epoxy-3-phenyl)propyl)piperazine

1-(Bis(*p*-fluorophenyl)methyl)-4-((2,3-epoxy-3-phenyl)propyl)piperazine
Probably formed from: Flunarizine
→ 1-(Bis(*p*-fluorophenyl)methyl)-4-((2,3-dihydroxy-3-phenyl)propyl) piperazine
Probably in: Man G310

Bis(*p*-fluorophenyl)methyl-β-D-glucuronide
Formed from: Bis(*p*-fluorophenyl)methanol

1-(Bis(*p*-fluorophenyl)methyl)-4-(2-hydroxy-3,4-dimethoxybenzyl)piperazine
Formed from: 1-(Bis(*p*-fluorophenyl)methyl)-4-(2,3,4-trimethoxybenzyl)piperazine

1-(Bis(*p*-fluorophenyl)methyl)-4-(3-hydroxy-2,4-dimethoxybenzyl)piperazine
Formed from: 1-(Bis(*p*-fluorophenyl)methyl)-4-(2,3,4-trimethoxybenzyl)piperazine

1-(Bis(*p*-fluorophenyl)methyl)-4-(4-hydroxy-2,3-dimethoxybenzyl)piperazine
Formed from: 1-(Bis(*p*-fluorophenyl)methyl)-4-(2,3,4-trimethoxybenzyl)piperazine

1-(Bis(*p*-fluorophenyl)methyl)piperazine
Formed from: 1-(Bis(*p*-fluorophenyl)methyl)-4-(2,3,4-trimethoxybenzyl)piperazine
Flunarizine

Bis(*p*-fluorophenyl)(methyl)silanol
Formed from: Flusilazole

(Bis(*p*-fluorophenyl)methyl(1*H*-1,2,4-triazol-1-ylmethyl)silane
See: Flusilazole

1-(Bis(*p*-fluorophenyl)methyl)-4-(2,3,4-trimethoxybenzyl)piperazine
→ 1-(Bis(*p*-fluorophenyl)methyl)-4-(2-hydroxy-3,4-dimethoxybenzyl)piperazine
In: Rat G353
→ 1-(Bis(*p*-fluorophenyl)methyl)-4-(3-hydroxy-2,4-dimethoxybenzyl)piperazine
In: Rat G353
→ 1-(Bis(*p*-fluorophenyl)methyl)-4-(4-hydroxy-2,3-dimethoxybenzyl)piperazine
In: Rat G353
→ 1-(Bis(*p*-fluorophenyl)methyl)piperazine
In: Rat G353
→ 4,4′-Difluorobenzophenone
Probably in: Rat G353
→ Trimetazidine
Probably in: Rat G353
→ 2,3,4-Trimethoxybenzaldehyde
Probably in: Rat G353

6,8-Bis(C-glucosyl)-4′,5,7-trihydroxyflavanone
See: Vicenin

8,9-Bis(S-glutathionyl)-7,10-dihydroxy-2-(2-((2-hydroxyethyl)amino)ethyl)-5-((2-(methylamino)ethyl)amino)anthra[1,9-*c,d*]pyrazol-6(2*H*)-one
Probably formed from: 9-(S-Glutathionyl)-7,10-dihydroxy-2-(2-((2-hydroxyethyl)amino)ethyl)-5-((2-(methylamino)ethyl)amino)anthra[1,9-*c,d*]pyrazol-6(2*H*)-one

1,4-Bis(S-glutathionyl)-2-hydroxy-17β-oestradiol
Probably formed from: S-(2-hydroxy-17β-oestradiol-4-yl)glutathione

6,7-Bis(S-glutathionyl)mitroxantrone
Formed from: 6-(S-Glutathionyl)mitroxantrone

Bishomotryptophan
→ β-Hydroxybishomotryptophan
In: Pseudomonas B84

1,5-Bis(*p*-(hydroxyamidino)phenoxy)pentane
Probably formed from: 1-(*p*-Amidinophenoxy)-5-(*p*-(hydroxyamidino)phenoxy)pentane

2-Bis(2-hydroxyethyl)amino-4,5-diphenyloxazole
See: Ditazole

***p*-(N,N-Bis(2-hydroxyethyl)amino)phenylacetic acid**
Probably formed from: 4-(*p*-(N,N-Bis(2-hydroxyethyl)amino)phenyl)butyric acid
→ *p*-((2-Hydroxyethylidene)amino)phenylacetic acid
Probably in: Rat A2041

4-(Bis(2-hydroxyethyl)amino)-L-phenylalanine
Formed from: L-Phenylalanine mustard

4-(*p*-(N,N-Bis(2-hydroxyethyl)amino)phenyl)butyric acid
Formed from: Chlorambucil
→ 4-(*p*-((2-Hydroxyethylidene)amino)phenyl)butyric acid
Probably in: Rat A2041
→ *p*-(N,N-Bis(2-hydroxyethyl)amino)phenylacetic acid
Probably in: Rat A2041

N,N-Bis(2-hydroxyethyl)-3,4,5-trimethoxybenzamide
Probably formed from: Tetrahydro-4-(3,4,5-trimethoxybenzoyl)-1,4-oxazine
→ N-(2-Carboxymethyl)-N-(2-hydroxyethyl)-3,4,5-trimethoxybenzamide
Probably in: Man C304

2,4-Bis(hydroxylamino)-6-nitrotoluene
Formed from: 4-Hydroxylamino-2,6-dinitrotoluene

1,2-Bis(4-hydroxy-3-methoxyphenyl)ethylene
Formed from: *erythro*-1,2-Bis(4-hydroxy-3-methoxyphenyl)-1,3-propanediol

→ Vanillin
In: Pseudomonas F244

(1*E*,6*E*)-1,7-Bis(4-hydroxy-3-methoxyphenyl)-1,6-heptadiene-3,5-dione
See: Curcumin

(*R*-(*R,*R**))-2,3-Bis((4-hydroxy-3-methoxyphenyl)methyl)-1,4-butanediol**
See: (−)-Secoisolariciresinol

***erythro*-1,2-Bis(4-hydroxy-3-methoxyphenyl)-1,3-propanediol**
→ 3-Hydroxy-1,2-bis(4-hydroxy-3-methoxyphenyl)-1-propanone
In: Pseudomonas F244
→ 1,2-Bis(4-hydroxy-3-methoxyphenyl) ethylene
In: Pseudomonas F244

1,3-Bis(hydroxymethyl)benzene
Formed from: *m*-Hydroxymethylbenzaldehyde

1,4-Bis(hydroxymethyl)benzene
Formed from: *p*-Hydroxymethylbenzaldehyde

3-(N,N-Bis(hydroxymethyl))carbamoyl-7-chloro-1,3-dihydro-1-methyl-5-phenyl-2*H*-1,4-benzodiazepin-2-one
Probably formed from: 7-Chloro-1,3-dihydro-3-(N-hydroxymethyl-N-methyl) carbamoyl-1-methyl-5-phenyl-2*H*-1,4-benzodiazepin-2-one
→ 3-(N,N-Bis(hydroxymethyl))carbamoyl-7-chloro-1,3-dihydro-5-phenyl-2*H*-1,4-benzodiazepin-2-one
Probably in: Dog D639
Monkey D639
Mouse D639
Rat D639

3-(N,N-Bis(hydroxymethyl))carbamoyl-7-chloro-1,3-dihydro-5-phenyl-2*H*-1,4-benzodiazepin-2-one
Probably formed from: 3-(N,N-Bis(hydroxymethyl))carbamoyl-7-chloro-1,3-dihydro-1-methyl-5-phenyl-2*H*-1,4-benzodiazepin-2-one

2,2-Bis(hydroxymethyl)-3,3-dimethylbutyl benzoate
Formed from: *tert*-Butylbicycloorthobenzoate
→ Benzoic acid
In: Mouse E803
→ 2-Formyl-2-hydroxymethyl-3,3-dimethylbutyl benzoate
In: Mouse E803

N-(1,1-Bis(hydroxymethyl)ethyl)-4-biphenylylacetamide
Formed from: 4-Methyl-2-(4-phenylbenzyl)-2-oxazoline-4-methanol
→ Formed from: 4-Biphenylylacetic acid
Probably in: Rat A1108

1,8β-Bis(hydroxymethyl)-10α-methoxy-6-methylergoline
Probably formed from: 8β-Hydroxymethyl-10α-methoxy-1,6-dimethylergoline
→ 1-Formyl-8β-hydroxymethyl-10α-methoxy-6-methylergoline
Probably in: Dog E167
Rat E167

1,8-Bis(hydroxymethyl)naphthalene
Formed from: 1-Hydroxymethyl-8-methylnaphthalene
→ Naphthalene-1,8-dicarboxylic acid
Probably in: Pseudomonas A1482

5-(2,5-Bis(hydroxymethyl)phenoxy)-2,2-dimethylpentanoic acid
Probably formed from: 5-(2-Hydroxymethyl-5-methylphenoxy)-2,2-dimethylpentanoic acid
5-(5-Hydroxymethyl-2-methylphenoxy)-2,2-dimethylpentanoic acid
→ 5-(2,5-Bis(hydroxymethyl)phenoxy)-2,2-dimethylpentanoic acid-β-D-glucuronide
Probably in: Hamster J565
Rat J564, J565
→ 5-(5-Carboxy-2-hydroxymethylphenoxy)-2,2-dimethylpentanoic acid
Probably in: Hamster J565
Rat J564, J565

5-(2,5-Bis(hydroxymethyl)phenoxy)-2,2-dimethylpentanoic acid-β-D-glucuronide
Probably formed from: 5-(2,5-Bis(hydroxymethyl)phenoxy)-2,2-dimethylpentanoic acid

2,2-Bis(4-hydroxy-3-methylphenyl)propane
→ 1-Hydroxy-2,2-bis(4-hydroxy-3-methylphenyl)propane
In: Bacillus G623
→ 4′-Hydroxy-3′-methylacetophenone
In: Bacillus G623
→ 4-Hydroxy-3-methylbenzoic acid
In: Bacillus G623

Bis(4-hydroxyphenyl) ether
Formed from: 4-Hydroxyphenyl phenyl ether
→ 4-Hydroxy-3-methoxyphenyl *p*-hydroxyphenyl ether
Probably in: Guinea pig E10
→ Quinol
In: Sphingomonas J882

5,5-Bis(*p*-hydroxyphenyl)hydantoin
Probably formed from: 5-(*p*-hydroxyphenyl)-5-phenylhydantoin
→ 5,5-Bis(*p*-hydroxyphenyl)hydantoin-β-D-glucuronide
In: Man A2110
Rat A2110
→ 5-(3,4-Dihydroxyphenyl)-5-(*p*-hydroxyphenyl)hydantoin
In: Man A2110
Rat A2110

5,5-Bis(*p*-hydroxyphenyl)hydantoin-β-D-glucuronide
Formed from: 5,5-Bis(*p*-hydroxyphenyl) hydantoin

Bis(*p*-hydroxyphenyl)methylenecyclohexanone
Formed from: Bis(4-acetoxyphenyl) methylenecyclohexanone
→ Bis(*p*-hydroxyphenyl)methylenecyclohexanone-β-D-glucuronide
Probably in: Sporotrichum A2630
→ Bis(*p*-hydroxyphenyl)methylenecyclohexanone-β-D-4-O-methylglucoside
Probably in: Sporotrichum A2630

Bis(*p*-hydroxyphenyl)methylenecyclohexanone-β-D-glucuronide
Probably formed from: Bis(*p*-hydroxyphenyl)methylenecyclohexanone

Bis(*p*-hydroxyphenyl)methylenecyclohexanone-β-D-4-O-methylglucoside
Probably formed from: Bis(*p*-hydroxyphenyl)methylenecyclohexanone

3,3-Bis(*p*-hydroxyphenyl)-7-methyl-2-indolinone
Formed from: 3,3-Bis(*p*-hydroxyphenyl)-7-methyl-2-indolinone disulphate

3,3-Bis(*p*-hydroxyphenyl)-7-methyl-2-indolinone disulphate
→ 3,3-Bis(*p*-hydroxyphenyl)-7-methyl-2-indolinone
In: Helix B243
Microorganism B243

5,6-Bis(*p*-hydroxyphenyl)-3-methyl-1,2,4-triazine
Probably formed from: 5-(*p*-hydroxyphenyl)-6(*p*-methoxyphenyl)-3-methyl-1,2,4-triazine
→ 5,6-Bis(*p*-hydroxyphenyl)-3-methyl-1,2,4-triazine-N-oxide
Probably in: Rat B396
→ 3-Hydroxymethyl-5,6-bis(*p*-hydroxyphenyl)-1,2,4-triazine
Probably in: Rat B396

5,6-Bis(*p*-hydroxyphenyl)-3-methyl-1,2,4-triazine-N-oxide
Probably formed from: 5,6-Bis(*p*-hydroxyphenyl)-3-methyl-1,2,4-triazine

2,2-Bis(*p*-hydroxyphenyl)propane (bisphenol A)
Formed from: 2,2-Bis(*p*-hydroxyphenyl) propane-β-D-glucuronide
Tetrabromobisphenol A
→ 2,2-Bis(*p*-hydroxyphenyl)propane-β-D-glucuronide
In: Rat K500, K544, K651
→ 1,2-Bis(*p*-hydroxyphenyl)propanol
In: Bacterium H119
→ *p*-Hydroxyacetophenone
In: Bacillus G623
Microorganism K529
→ 1-Hydroxy-2,2-bis(*p*-hydroxyphenyl) propane
In: Bacillus G623
Bacterium H119
→ *p*-Hydroxybenzoic acid
In: Bacillus G623
Microorganism K529
→ 2-(*p*-Hydroxyphenyl)-2-(*p*-sulphatoxyphenyl)propane
In: Man K344

2,3-Bis(*p*-hydroxyphenyl)propane-1,2-diol
Formed from: 1-Hydroxy-2,2-bis(*p*-hydroxyphenyl)propane

2,2-Bis(*p*-hydroxyphenyl)propane-β-D-glucuronide
Formed from: 2,2-Bis(*p*-hydroxyphenyl) propane
→ 2,2-Bis(*p*-hydroxyphenyl)propane
In: Rat K651

1,2-Bis(*p*-hydroxyphenyl)-2-propanol
Formed from: 2,2-Bis(*p*-hydroxyphenyl) propane

→ 4,4′-Dihydroxy-α-methylstilbene
In: Bacterium H119
→ *p*-Hydroxyacetophenone
In: Bacterium H119
→ *p*-Hydroxybenzaldehyde
In: Bacterium H119

2,2-Bis(*p*-hydroxyphenyl)propionic acid
Formed from: 1-Hydroxy-2,2-bis(*p*-hydroxyphenyl)propane

Bis(*p*-hydroxyphenyl)pyridyl-2-methane
Formed from: Bis(*p*-hydroxyphenyl)pyridyl-2-methane disulphide

Bis(*p*-hydroxyphenyl)pyridyl-2-methane disulphide
→ Bis(*p*-hydroxyphenyl)pyridyl-2-methane
In: Microorganism A2711

3,3′-Bisindolylmethaneaspartic acid
Formed from: Indole-3-acetic acid

2,2-Bis(3-indolyl)indoxyl
Formed from: Indole

1,2-Bis-(5-methoxy-2-benzimidazolyl)-1,2-ethanediol
→ 1-(5-Hydroxy-2-benzimidazolyl)-2-(5-methoxy-2-benzimidazolyl)-1,2-ethanediol
In: Mouse A1415

1,2-Bis-(3-methoxycarbonyl-2-thioureido)benzene
Formed from: Methyl 1-(butylcarbamoyl) benzimidazol-2-yl carbamate
→ 2-(3-Methoxycarbonyl-2-thioureido)aniline
In: Mouse A260
→ 2-(3-Methoxycarbonyl-2-thioureido)-1-(3-methoxycarbonyl-2-ureido)benzene
In: Mouse A260
→ 2-(3-Methoxycarbonyl-2-thioureido)-1-(2-thioureido)benzene
In: Mouse A260

1,2-Bis-(3-methoxycarbonyl-2-ureido)benzene
Formed from: 2-(3-methoxycarbonyl-2-thioureido)-1-(3-methoxycarbonyl-2-ureido)benzene
→ 2-(3-Methoxycarbonyl-2-ureido)benzene
In: Mouse A260
→ 2-(3-Methoxycarbonyl-2-ureido)-1-(2-ureido)aniline
In: Mouse A260

5,6-Bis-(*p*-methoxyphenyl)-3-methyl-1,2,4-triazine (anitrazafen)
→ 5-(*p*-Hydroxyphenyl)-6-(3-methoxyphenyl)-3-methyl-1,2,4-triazine
In: Rat B396
→ 6-(*p*-Hydroxyphenyl)-5-(3-methoxyphenyl)-3-methyl-1,2,4-triazine
In: Rat B396

Bis-(*p*-methoxyphenyl)urea
Formed from: 3-Diethylamino-O-(*p*-methoxyphenylcarbamoyl)propiophenone oxime

4,5:4′,5′-Bis(methylenedioxy)biphenyl-2,2′ quinone
Formed from: 2,2′-Dihydroxy-4,5:4′,5′-bis(methylenedioxy)biphenyl

α-(2-(Bis(1-methylethyl)amino)ethyl)-α-phenyl-2-pyridineacetamide
See: Disopyramide

1,3-Bis(muconolacto-3-yl)benzene
Formed from: *m*-Terphenyl

Bis(*p*-nitrophenyl) phosphate
→ *p*-Nitrophenol
In: Escherichia A1511
Rat A2591

Bisnorargemonine ((5*S*)-5,6,11,12-tetrahydro-3,8-dimethoxy-13-methyldibenzo[*a*,*e*]cycloocten-5,11-imine-2,9-diol)
Formed from: Norcoclaurine

4″,5″-Bisnorcannabidiol-3″,7-dioic acid
Probably formed from: Cannabidiol-5″,7-dioic acid

4″,5″-Bisnorcannabidiol-3″-oic acid
Probably formed from: 5″-Norcannabidiol-4″-oic acid
→ 3″,4″,5″-Tris-2″,6-dihydroxynorcannabidiol
Probably in: Dog G113
Man G113
Rat G113
→ 3″,4″,5″-Tris-2″,7-dihydroxynorcannabidiol
Probably in: Dog G113
Man G113
Rat G113
→ 4″,5″-Bis-6-hydroxynorcannabidiol-3″-oic acid
Probably in: Dog G113
Guinea pig B458
Man G113
Rat G113
→ 4″,5″-Bis-7-hydroxynorcannabidiol-3″-oic acid
Probably in: Dog G113
Guinea pig B458

Man G113
Rat G113
→ 3″,4″,5″-Trisnorcannabidiol-2″-oic acid
Probably in: Dog G113
Man F503, G113
Rat G113

Bisnorchlorphenoxamine
Formed from: Norchlorphenoxamine

N,N-Bisnorclentiazem
Formed from: N-Norclentiazem

N,O-Bisnorclentiazem
Probably formed from: N-Norclentiazem
→ N,N,O-Trisnorclentiazem
Probably in: Rat H76

4″,5″-Bisnor-6,7-dihydroxycannabidiol-3″-oic acid
Probably formed from: 4″,5″-Bisnor-7-hydroxycannabidiol-3″-oic acid
6,7-Dihydroxycannabidiol-5″-oic acid
→ 2″,3″,4″,5″-Tetranor-6,7-dihydroxycannabidiol-1″-oic acid
Probably in: Dog F557

Bisnor-2,10-dihydroxytrimipramine
Probably formed from: Bisnor-2-hydroxytrimipramine

4″,5″-Bisnor-6-hydroxycannabidiol-3″-oic acid
Probably formed from: 4″,5″-Bisnorcannabidiol-3″-oic acid
6β-Hydroxycannabidiol
6-Hydroxycannabidiol-5″-oic acid
→ 4″,5″-Bisnor-6-oxocannabidiol-3″-oic acid
Probably in: Dog F557, G113
Rat G113
→ 2″,3″,4″,5″-Tetranor-6-hydroxycannabidiol-1″-oic acid
Probably in: Dog F557

4″,5″-Bisnor-7-hydroxycannabidiol-3″-oic acid
Probably formed from: 4″,5″-Bisnorcannabidiol-3″-oic acid
7-Hydroxycannabidiol
7-Hydroxycannabidiol-5″-oic acid
→ 4″,5″-Bisnor-3″,7-dicarboxycannabidiol
Probably in: Dog G113
Rat G113
→ 4″,5″-Bisnor-6,7-dihydroxycannabidiol-3″-oic acid
Probably in: Dog G113
Rat G113

Bisnormethotrimeprazine
Formed from: Normethotrimeprazine

Bisnorpromethazine
Formed from: Norpromethazine

Bisnorpromethazine sulphone
Probably formed from: Norpromethazine sulphone

Bisnorprothanon
Formed from: Norprothanon

4″,5″-Bisnor-Δ¹-tetrahydrocannabinol
See: Δ¹-Tetrahydropropylcannabinol

4″,5″-Bisnor-Δ⁶-tetrahydrocannabinol
See: Δ⁶-Tetrahydropropylcannabinol

4″,5″-Bisnor-Δ¹-*trans*-tetrahydrocannabinol-3″,7-dioic acid
Formed from: Δ¹-Tetrahydrocannabinol
Probably formed from: 4″,5″-Bisnor-Δ¹-tetrahydro-7-hydroxycannabinol-3″-oic acid

4″,5″-Bisnor-Δ⁶-tetrahydrocannabinol-3″,7-dioic acid
Probably formed from: 4″,5″-Bisnor-Δ⁶-tetrahydrocannabinol-3″-oic acid

4″,5″-Bisnor-Δ¹-tetrahydrocannabinol-3″-oic acid
Formed from: Δ¹-Tetrahydrocannabinol
Δ¹-Tetrahydro-4″-hydroxycannabinol
Probably formed from: Δ¹-Tetrahydrocannabinol-5″-oic acid
Δ¹-Tetrahydro-4″-hydroxybisnorcannabinol
→ 4″,5″-Bisnor-Δ¹-tetrahydro-6α-hydroxycannabinol-3″-oic acid
Probably in: Mouse D345
→ 4″,5″-Bisnor-Δ¹-tetrahydro-6β-hydroxycannabinol-3″-oic acid
Probably in: Man C294
Mouse D345, F526
Rabbit B89
→ 4″,5″-Bisnor-Δ¹-tetrahydro-7-hydroxycannabinol-3″-oic acid
Probably in: Man C294
Mouse D345, F526
→ 1″,2″-Dehydro-4″,5″-bisnor-Δ¹-tetrahydrocannabinol-3″-oic acid
Probably in: Man D456
→ 2″,3″,4″,5″-Tetranor-Δ¹-tetrahydrocannabinol-1″-oic acid
Probably in: Mouse D345

→ 3″,4″,5″-Trisnor-Δ^1-tetrahydrocannabinol-2″-oic acid
Probably in: Mouse F526

4″,5″-Bisnor-Δ^6-tetrahydrocannabinol-3″-oic acid
Formed from: Δ^6-Tetrahydrocannabinol
Δ^6-Tetrahydrocannabinol-5″-oic acid
→ 4″,5″-Bisnor-Δ^6-tetrahydrocannabinol-3″,7-dioic acid
Probably in: Rat A3676
→ 1″,2″-Dehydro-4″,5″-bisnor-Δ^6-tetrahydrocannabidiol-3″,7-dioic acid
Probably in: Rat A3676
→ 2″,3″,4″,5″-Tetranor-Δ^6-tetrahydrocannabinol-1″-oic acid
Probably in: Rat D345

4″,5″-Bisnor-Δ^1-tetrahydro-6β,7-dihydroxycannabinol-3″-oic acid
Probably formed from: 4″,5″-Bisnor-Δ^1-tetrahydro-6β-hydroxycannabinol-3″-oic acid

4″,5″-Bisnor-Δ^1-tetrahydro-6α-hydroxycannabinol-3″-oic acid
Probably formed from: 4″,5″-Bisnor-Δ^1-tetrahydrocannabinol-3″-oic acid

4″,5″-Bisnor-Δ^1-tetrahydro-6β-hydroxycannabinol-3″-oic acid
Probably formed from: 4″,5″-Bisnor-Δ^1-tetrahydrocannabinol-3″-oic acid
Δ^1-Tetrahydro-6β-cannabinol
→ 4″,5″-Bisnor-Δ^1-tetrahydro-6β,7-dihydroxycannabinol-3″-oic acid
Probably in: Rabbit B89
→ 4″,5″-Bisnor-Δ^1-tetrahydro-6-oxocannabinol-3″-oic acid
Probably in: Mouse D345

4″,5″-Bisnor-Δ^1-tetrahydro-7-hydroxycannabinol-3″-oic acid
Probably formed from: 4″,5″-Bisnor-Δ^1-tetrahydrocannabinol-3″-oic acid
Δ^1-Tetrahydro-7-cannabinol
→ 4″,5″-Bisnor-Δ^1-tetrahydrocannabinol-3″,7-dioic acid
Probably in: Mouse D345

4″,5″-Bisnor-Δ^1-tetrahydro-6-oxocannabinol-3″-oic acid
Probably formed from: 4″,5″-Bisnor-Δ^1-tetrahydro-6β-hydroxycannabinol-3″-oic acid

Bisnortilidine
Formed from: Nortilidine

Bisnortrimipramine
Formed from: Nortrimipramine

Bisnoryangonin (4-hydroxy-6-(4-hydroxystyryl)-2-pyrone)
Formed from: *p*-Coumaroyl CoA
→ Hispidin
In: Polyporus A1332

Bisnorzimetidine
Formed from: Norzimetidine
→ N-Acetylbisnorzimetidine
Probably in: Dog B791
Rat B791

Bisphenol A
See: 2,2-Bis(*p*-hydroxyphenyl)propane

Bisphenol A bis(2,3-dihydroxypropyl) ether
Formed from: Bisphenol A 4-(2,3-dihydroxypropyl) ether 4′-glycidyl ether

Bisphenol A bisglycidyl ether
→ Bisphenol A 4-(2,3-dihydroxypropyl) ether 4′-glycidyl ether
In Man K112, K113
Mouse K112, K113
Rat K112, K113

Bisphenol A 4-(2,3-dihydroxypropyl) ether 4′-glycidyl ether
Formed from: Bisphenol A bisglycidyl ether
→ Bisphenol A bis(2,3-dihydroxypropyl) ether
In Man K112, K113
Mouse K112, K113
Rat K112, K113

1,2′-Bis(N-phenylacetamidoxy)ethoxyethane
→ Glycollanilide
In: Rat C45
→ 2-(N-Phenylacetamidoxy)ethanol
In: Rat C45
→ 2-(N-Phenylacetamidoxy)ethoxyethanol
In: Rat C45

N^2,N^5-Bis(phenylacetyl)ornithine
Probably formed from: Phenylacetic acid

N,N-Bis(phenylcarbamoylmethyl) dimethyl ammonium
→ Carboxymethyl dimethyl phenylcarbamoylmethyl ammonium
Probably in: Rat A269

Bis(phenylmethyl)diazene
See: ω,ω-Azotoluene

Bisphenyl 9-((2-phosphonomethoxy)ethyl)adenine
→ Diphenyl phosphate
Probably in: Rat J136

1,2′-Bis(N-phenyl-3-propionamidoxy)ethoxyethane
→ (3-Hydroxypropion)anilide
In: Rat C45
→ 2-(N-Phenyl-3-propionamidoxy)ethanol
In: Rat C45
→ 2′-(N-Phenyl-3-propionamidoxy)ethoxyethanol
In: Rat C45

Bis(phthalazin-1-yl)amine
Formed from: Hydralazine

Bis(2,3,5,6-tetrachlorophenyl) disulphide
Probably formed from: 2,3,5,6-Tetrachlorothiophenol

4-(*p*-(N,N-Bis(2-(thiocyano)ethyl)amino)phenyl)butyric acid
Formed from: Chlorambucil

3,5-Bis(trifluoromethyl)benzoic acid
→ 3,5-Bis(trifluoromethyl)benzoyl-β-D-glucuronic acid
In: Rat J549

3,5-Bis(trifluoromethyl)benzoyl-β-D-glucuronic acid
Formed from: 3,5-Bis(trifluoromethyl)benzoic acid

***cis*-2,4-Bis(trifluoromethyl)cinnamic acid**
Formed from: Arteflene
→ *cis*-2,4-Bis(trifluoromethyl)cinnamoylglycine
Probably in: Rat K76

***cis*-2,4-Bis(trifluoromethyl)cinnamoylglycine**
Probably formed from: *cis*-2,4-Bis(trifluoromethyl)cinnamic acid

***cis*-2,4-Bis(trifluoromethyl)phenyl)but-3-en-2-one**
Formed from: Arteflene

(2*R*)-(1-(*R*)-3,5-Bis(*p*-trifluoromethyl)phenylethoxy)-3-(*S*)-(*p*-fluorophenyl)-4-(3-(5-oxo-4*H*-1,2,4-triazolo)methyl)morpholine
Formed from: (2*R*)-(1-(*R*)-3,5-Bis(*p*-trifluoromethyl)phenylethoxy)-3-(*S*)-(*p*-fluorophenyl)-4-(3-(1-phosphoryl-5-oxo-4*H*-1,2,4-triazolo)methyl)morpholine

(2*R*)-(1-(*R*)-3,5-Bis(*p*-trifluoromethyl)phenylethoxy)-3-(*S*)-(*p*-fluorophenyl)-4-(3-(1-phosphoryl-5-oxo-4*H*-1,2,4-triazolo)methyl)morpholine
→ (2*R*)-(1-(*R*)-3,5-Bis(*p*-trifluoromethyl)phenylethoxy)-3-(*S*)-(*p*-fluorophenyl)-4-(3-(5-oxo-4*H*-1,2,4-triazolo)methyl)morpholine
In: Dog J778
Man J778
Rat J778

1-(2-(Bis(*p*-trifluoromethylphenyl)methoxy)ethyl)-1,2,5,6-tetrahydropyridine-3-carboxylic acid
→ 1-(2-(Bis(*p*-trifluoromethylphenyl)methoxy)ethyl)pyridiniumcarboxylic acid
In: Man G809

1-(2-(Bis(*p*-trifluoromethylphenyl)methoxy)ethyl)pyridiniumcarboxylic acid
Formed from: 1-(2-(Bis(*p*-trifluoromethyl phenyl)methoxy)ethyl)-1,2,5,6-tetrahydropyridine-3-carboxylic acid

2,8-Bis(trifluoromethyl)quinoline-4-carboxylic acid
Formed from: Mefloquine

1,4-Bis(3-(3,4,5-trimethoxybenzoyl)propyl)piperazine
See: Dilazep

Bithionol (2,2′-thiobis(4,6-dichlorophenol))
→ Bithionol-β-D-glucuronide
In: Man A3319
→ Bithionol sulphoxide
In: Ascaris A3849
Moniezia A3849
Mouse A3849

Bithionol-β-D-glucuronide
Formed from: Bithionol

Bithionol sulphoxide
Formed from: Bithionol

Bitolterol (4-methylbenzoic acid 4-(2-((1,1-dimethylethyl)amino)-1-hydroxyethyl)-1,2-phenylene ester)
→ N-(*tert*-Butyl)noradrenaline
In: Dog A1987
Man A1987
Rat A1987, A3718
→ *p*-Toluic acid
In: Rat A1987

Boldine ((6a*S*)-5,6,6a,7-tetrahydro-1,10-dimethoxy-6-methyl-4*H*-dibenzo[*de*,*g*]quinoline-2,9-diol)
Formed from: (+)-Isoboldine
Norboldine
Reticuline

Bornaprolol (1-(2-bicyclo[2.2.1]hept-2-ylphenoxy)-3-(1-methylethyl)amino)-2-propanol)
→ Bornaprolol-β-D-glucuronide
In: Man F589

Bornaprolol-β-D-glucuronide
Formed from: Bornaprolol

Bosentan (4-(1,1-dimethylethyl)-N-(6-(2-hydroxyethoxy)-5-(2-methoxyphenoxy)-(2,2′-bipyrimidin)-4-yl)benzenesulphonamide)
→ O-Demethylbosentan
In: Man J732
→ N-(6-(2-Hydroxyethoxy)-5-(2-methoxyphenoxy)-(2,2′-bipyrimidin)-4-yl)-4-(2-hydroxymethyl-2-propyl)benzenesulphonamide
In: Man J732

Brassinine ((1*H*-indol-3-ylmethyl)carbamodithioic acid methyl ester)
→ Methyl (3-indolylethyl)dithiocarbamate sulphoxide
In: Phoma G132

Brimodinine (5-bromo-6-(imidazolidinylideneamino)quinoxaline)
→ 5-Bromo-6-guanidinoquinoxaline
In: Dog H743
Man H743
Monkey H743
Rabbit H743
Rat H743
→ 4′,5′-Dehydrobrimodinine
In: Dog H743
Monkey H743
Rabbit H743
Rat H743
→ 2,3-Dihydrobrimodinine
In: Man H743
Monkey H743
Rabbit H743
Rat H743
→ 3,4-Dihydro-3-oxobrimodinine
In: Man H743
Monkey H743
Rabbit H743
Rat H743
→ 4′-Oxobrimodinine
In: Man H743

Brofoxine (6-bromo-1,4-dihydro-4,4-dimethyl-2*H*-3,1-benzoxazin-2-one)
→ 6-bromo-1,4-dihydro-4-hydroxymethyl-4-methyl-2*H*-3,1-benzoxazin-2-one
In: Rat A3595
→ 1,4-Dihydro-6-hydroxy-4,4-dimethyl-2*H*-3,1-benzoxazin-2-one
In: Rat A3595
→ 8-Hydroxybrofoxine
In: Rat A3595

Bromazepam (7-bromo-1,3-dihydro-5-(2′-pyridyl)-2*H*-1,4-benzodiazepin-2-one)
→ 2-(2-Amino-5-bromobenzoyl)pyridine
In: Dog A716, A1121
Man A924, A1121, A1439
Guinea pig A2834
Rabbit A716, A1121, A2213, A2834
Rat A716, A1121, A1788, A2834
→ Bromazepam-3′-mercapturic acid
In: Rat A2590
→ Bromazepam-4-oxide
In: Dog A1121
Man A1121
Mouse A1121
Rat A1121
→ 3-Hydroxybromazepam
In: Dog A1121
Man A1121, A1439, C866
Mouse A1121
Rabbit A2213
Rat A1121, A1788
→ 5′-Hydroxybromazepam
In: Rabbit A2213
→ 9-Hydroxybromazepam
In: Rabbit A2213
→ Methylthiobromazepam
In: Rat A1788, A2590

Bromazepam-4-oxide
Formed from: Bromazepam

Bromazepam-3′-mercapturic acid
Formed from: Bromazepam

Bromazepine (2-((4-bromophenyl)phenylmethoxy)-N,N-dimethylethanamine)
→ Bromazepine-N-oxide
In: Man B556
→ *p*-Bromobenzhydrol
In: Man B556
→ Demethylbromazepine
In: Man B556

Bromazepine-N-oxide
Formed from: Bromazepine

Bromerguride (N′-((8α)-2-bromo-9,10-didehydro-6-methylergolin-8-yl)-N,N-diethylurea)
→ Desethylbromerguride
In: Dog F337
Guinea pig F337
Monkey F337
Rat F337

Bromfenac (2-amino-3-(*p*-bromobenzoyl) benzeneacetic acid)
→ Bromfenac-N-β-D-glucuronide
In: Rat J428

Bromfenac-N-β-D-glucuronide
Formed from: Bromfenac

Bromhexine (2-amino-3,5-dibromo-N-cyclohexylbenzylamine)
→ 3,5-Dibromoanthranilic acid
In: Dog A763
Man A763
Mouse A763
Rabbit A763
Rat A763

2-Bromoacetamidoestrone 3-methyl ether
→ 2-(2′-6N-Adenyl)acetamidoestradiol 3-methyl ether
In: Man C174

***p*-Bromoacetanilide**
Formed from: *p*-Bromoaniline

***o*-Bromoacetophenone**
→ 1-(*o*-Bromophenyl)ethanol
In: Corynebacterium J893

***m*-Bromoacetophenone**
→ 1-(*m*-Bromophenyl)ethanol
In: Corynebacterium J893

***p*-Bromoacetophenone**
→ 1-(*p*-Bromophenyl)ethanol
In: Corynebacterium J893
Lactobacillus G148

***o*-Bromoaniline**
Formed from: Aniline
o-Bromonitrobenzene
→ 4-Amino-3-bromophenol
In: Rat H273

***m*-Bromoaniline**
Formed from: *m*-Bromonitrobenzene
→ 4-Amino-2-bromophenol
In: Rat H273

***p*-Bromoaniline**
Formed from: Aniline
p-Bromonitrobenzene
→ *p*-Bromoacetanilide
In: Rabbit E399
→ *p*-Bromonitrosobenzene
In: Pea C797
With chloroperoxidase A3579
→ *p*-Bromophenylhydroxylamine
In: Rabbit F839

***o*-Bromoanisole**
Formed from: Anisole
o-Bromophenol

***p*-Bromoanisole**
Formed from: Anisole
p-Bromophenol
→ *p*-Bromophenol
In: Methylosinus D513

5-Bromoanthranilic acid
Formed from: 5-Bromoindole

***o*-Bromobenzaldehyde**
Probably formed from: *o*-Bromobenzoic acid
→ N-Acetyl-S-(*o*-bromobenzyl)-L-cysteine
In: Rat C553
→ *o*-Bromobenzoic acid
In: Acinetobacter F366
→ *o*-Bromobenzyl alcohol
In: Horse C508
See also: E743

***m*-Bromobenzaldehyde**
Formed from: *m*-Bromobenzoic acid
→ *m*-Bromobenzoic acid
In: Rat H761
→ *m*-Bromobenzyl alcohol
Probably in: Nocardia E743
→ *m*-Bromomandelonitrile
With (S)-oxynitrilase F680

p-Bromobenzaldehyde
Formed from: *p*-Bromobenzoic acid
→ *p*-Bromobenzoic acid
In: Beef C508
Rat H761
Yeast C508
→ *p*-Bromobenzyl alcohol
In: Horse C508
Yeast C508
See also: E743
→ *p*-Bromomandelonitrile
In: Almond J194
Sorghun J194
→ 4,4′-Dibromobenzoin
In: Pseudomonas K424

p-Bromobenzamidine
→ *p*-Bromobenzamidoxine
In: Rabbit E830, F768

p-Bromobenzamidoxime
Formed from: *p*-Bromobenzamidine

7-Bromobenzanthracene
→ 7-Bromo-3,4-dihydro-3,4-dihydroxybenzanthracene
In: Rat C576
→ 7-Bromo-5,6-dihydro-5,6-dihydroxybenzanthracene
In: Rat C576
→ 7-Bromo-8,9-dihydro-8,9-dihydroxybenzanthracene
In: Rat C576
→ 7-Bromo-10,11-dihydro-10,11-dihydroxybenzanthracene
In: Rat C576

Bromobenzene
→ N-Acetyl-S-(*p*-bromophenyl)-L-cysteine
Probably in: Rat A73
→ N-Acetyl-S-(2,5-dihydroxyphenyl)-L-cysteine
In: Rat G626
→ Bromobenzene-2,3-oxide
Probably in: Rat F917
Guinea pig F917
→ Bromobenzene-3,4-oxide
In: Rat C876
See also: C456, D563, E878, E891, F917, G884
→ *cis*-3-Bromo-1,2-dihydro-1,2-dihydroxybenzene
In: Pseudomonas J656
→ *trans*-3-Bromo-1,2-dihydro-1,2-dihydroxybenzene
In: Guinea pig E878
Rat C831, D402, D563, E878, F525
→ *o*-Bromophenol
In: Guinea pig E878, G884
Man E947
Mouse B212, E947
Rat A2160, B825, C55, C831, C899, D122, D563, E878, G884
→ *m*-Bromophenol
In: Guinea pig E878, G884
Rabbit A1825
Rat A2160, D122, E878, F525, G884
→ *p*-Bromophenol
In: Guinea pig E878, G884
Man E947
Mouse B212, E947
Nitrosomonas H631
Rabbit A1825, C55
Rat A2160, B825, C831, C899, D122, E878, F525, G884
See also: A73
→ 2,5-Dimethoxythioanisole
In: Rat G626

Bromobenzene-2,3-oxide
Probably formed from: Bromobenzene
→ N-Acetyl-S-(3-bromo-1,2-dihydro-2-hydroxyphenyl)-L-cysteine
Probably in: Rat F917

Bromobenzene-3,4-oxide
Formed from: Bromobenzene
→ 4-Bromo-1,2-dihydro-1,2-dihydroxybenzene
Probably in: Rat D563
→ 5-Bromo-2-(S-glutathionyl)-1,2-dihydro-1-hydroxybenzene
Probably in: Rat C876, F917
→ 4-Bromo-2-(S-glutathionyl)-1,2-dihydro-1-hydroxybenzene
Probably in: Guinea pig E891
Rat E891, F917

***p*-Bromobenzenesulphonamide**
Formed from: Ebrotidine

***p*-Bromobenzhydrol**
Formed from: Bromazepine
→ *p*-Bromobenzophenone
Probably in: Man B556

***o*-Bromobenzoic acid**
Formed from: *o*-Bromobenzaldehyde
2-Bromobiphenyl
→ Benzoic acid
In: Microorganism K269
→ *o*-Bromobenzaldehyde
Probably in: Nocardia E743
→ Catechol
In: Pseudomonas F247, G434

***m*-Bromobenzoic acid**
Formed from: *m*-Bromobenzaldehyde
m-Bromobenzonitrile
o-Bromophenol
→ Benzoic acid
In: Desulfomicrobium J491
→ *m*-Bromobenzaldehyde
In: Nocardia G207
See also: E743, K615
→ 5-Bromo-1,2-dihydro-1,2-dihydroxybenzoic acid
Probably in: Pseudomonas B465

***p*-Bromobenzoic acid**
Formed from: *p*-Bromobenzaldehyde
p-Bromobenzonitrile
→ *p*-Bromobenzaldehyde
In: Nocardia G207
See also: E743, K615
→ *p*-Bromobenzoyl CoA
In: Pseudomonas G729, K191
→ *p*-Bromohippuric acid
In: Rat F693
→ *p*-Hydroxybenzoic acid
In: Alcaligenes E335
Arthrobacter D544, E752
Pseudomonas B695, E359

***m*-Bromobenzonitrile**
Formed from: *m*-Bromobenzaldehyde
→ *m*-Bromobenzoic acid
In: Fusarium A3364
Nocardia A3420, D725

***p*-Bromobenzonitrile**
→ *p*-Bromobenzoic acid
In: Fusarium A3364
Nocardia A3420, D725

***p*-Bromobenzophenone**
Probably formed from: *p*-Bromobenzhydrol
→ *p*-Hydroxybenzophenone
Probably in: Man B556

***p*-Bromobenzoyl CoA**
Formed from: *p*-Bromobenzoic acid
→ *p*-Hydroxybenzoyl CoA
In: Acinetobacter G896
Arthrobacter H279
Pseudomonas G920, H645

6-Bromo-3,4-benzpyrene
→ 6-Bromo-4,5-dihydro-4,5-dihydroxy-3,4-benzpyrene
In: Rat C505
→ 6-Bromo-7,8-dihydro-7,8-dihydroxy-3,4-benzpyrene
In: Rat C505

***o*-Bromobenzyl alcohol**
Formed from: *o*-Bromobenzaldehyde

***m*-Bromobenzyl alcohol**
Formed from: *m*-Bromobenzaldehyde

***p*-Bromobenzyl alcohol**
Formed from: *p*-Bromobenzaldehyde

(2-Bromobenzylidene)malononitrile
→ (2-Bromobenzyl)malononitrile
In: Man B260

(3-Bromobenzylidene)malononitrile
→ (3-Bromobenzyl)malononitrile
In: Man B260

(2-Bromobenzyl)malononitrile
Formed from: (2-Bromobenzylidene) malononitrile

(3-Bromobenzyl)malononitrile
Formed from: (3-Bromobenzylidene) malononitrile

2-Bromobiphenyl
→ *o*-Bromobenzoic acid
In: Pseudomonas E41
→ 2-Bromo-3-hydroxybiphenyl
In: Rat B448
→ 2-Bromo-4-hydroxybiphenyl
In: Rat B448
→ 2-Bromo-5-hydroxybiphenyl

In: Rat B448

3-Bromobiphenyl
→ 3-Bromo-4-hydroxybiphenyl
In: Rat B448
→ 3′-Bromo-4-hydroxybiphenyl
In: Rat B448

4-Bromobiphenyl
→ 4′-Bromo-3-hydroxybiphenyl
In: Rat B448
→ 4′-Bromo-4-hydroxybiphenyl
In: Hen A3893
Rat B448, C291
→ 3-(*p*-Bromophenyl)muconolactone
In: Aspergillus H23

2-Bromo-3,5-bis(S-glutathionyl)quinol
Probably formed from: 2-Bromo-3-(S-glutathionyl)quinol

2-Bromo-3,6-bis(S-glutathionyl)quinol
Probably formed from: 2-Bromo-3-(S-glutathionyl)quinol

1-Bromo-2-(2-bromoethyl)benzene
→ 1-Bromo-2-(2-bromoethyl)-3,4-dihydro-3,4-dihydroxybenzene
In: Pseudomonas H780

1-Bromo-2-(2-bromoethyl)-3,4-dihydro-3,4-dihydroxybenzene
Formed from: 1-Bromo-2-(2-bromoethyl) benzene

4-Bromo-α-(4-bromophenyl)-α-hydroxybenzeneacetic acid
See: Acarol

Bromobuterol (4-amino-3,5-dibromo-α-(((1,1-dimethylethyl)amino)methyl)benzenemethanol)
→ 3,5-Dibromo-α-(((1,1-dimethylethyl) amino)methyl)-4-hydroxylaminobenzene-methanol
In: Pig J452

Bromobutide (2-bromo-3,3-dimethyl-N-(1-methyl-1-phenylethyl)butyramide)
→ 2-Bromo-N-(2-hydroxy-1-methyl-1-phenylethyl)-3,3-dimethylbutyramide
In: Rat D215
→ 2-Bromo-4-hydroxy-3,3-dimethyl-N-(1-methyl-1-phenylethyl)butyramide
In: Mouse D193
Rat D193, D215
→ 2-Bromo-N-(1-(*p*-hydroxyphenyl)-1-methylethyl)-3,3-dimethylbutyramide
In: Mouse D193
Rat D193, D215
→ 3,3-Dimethyl-N-(1-methyl-1-phenylethyl)butyramide
In: Mouse D193
Rat D193, D215

2-Bromo-3-carboxy-2-hydroxy-3-methyl-N-(1-methyl-1-phenylethyl)butyramide
Probably formed from: 2-Bromo-4-hydroxy-3,3-dimethyl-N-(1-methyl-1-phenylethyl)-butyramide

6-Bromo-(+)-catechin
→ 6-Bromo-3′-O-methyl-(+)-catechin
In: Pig D112

8-Bromo-(+)-catechin
→ 8-Bromo-3′-O-methyl-(+)-catechin
In: Pig D112

3-Bromocatechol
Formed from: *m*-Bromofluorobenzene
o-Bromophenol
→ 2-Bromo-*cis*,*cis*-muconic acid
In: Pseudomonas A3769

4-Bromocatechol
Formed from: 1-Bromo-3,4-methylenedioxybenzene
4-Bromo-1,2-dihydro-1,2-dihydroxybenzene
m-Bromophenol
p-Bromophenol
Probably formed from: 5-Bromosalicylic acid
→ 4-Bromo-2-methoxyphenol
In: Rat A486
With catechol O-methyltransferase A597
→ 5-Bromo-2-methoxyphenol
In: Rat A486
With catechol O-methyltransferase A597

2-Bromo-4-chloroaniline
Formed from: *p*-Chloroaniline
→ 2,6-Dibromo-4-chloroaniline
Probably in: Caldaryomyces C845

5-Bromo-4-chloro-3-hydroxyindole
Formed from: 5-Bromo-4-chloro-3-indolyl-β-D-galacoside

2-Bromo-4-(2-chloro-4-hydroxy-5-methoxyphenyl)-9-hydroxymethyl-6*H*-thieno[3,2-*f*]-1,2,4-triazolo[4,3-*a*]-1,4-diazepine
Probably formed from: 4′-Hydroxy-5′-methoxybrotizolam

2-Bromo-4-(2-chloro-4-hydroxyphenyl)-9-hydroxymethyl-6*H*-thieno[3,2-*f*]-1,2,4-triazolo[4,3-*a*]-1,4-diazepine
Probably formed from: 2-Bromo-4-(*o*-chlorophenyl)-9-hydroxymethyl-6*H*-thieno[3,2-*f*]-1,2,4-triazolo[4,3-*a*]-1,4-diazepine

5-Bromo-4-chloro-3-indolyl-β-D-galacoside
→ 5-Bromo-4-chloro-3-hydroxyindole
In: Petunia B495

2-Bromo-1-chloro-4-nitrobenzene
→ 2-Bromo-1-(S-glutathionyl)-4-nitrobenzene
In: Rat H516

2-Bromo-4-chlorophenol
→ 2-Bromo-4-chlorophenyl-β-D-glucuronide
In: Rat B962
→ 5-(4-Bromo-2-chlorophenoxy)-3-methoxy-*o*-quinone
In: Rhizoctonia D690
→ 4-(4-Bromo-2-chlorophenoxy)-2,6-dimethoxyphenol
In: Rhizoctonia D690
→ 4-(4-(4-Bromo-2-chlorophenoxy)-2,6-dimethoxyphenoxy)-2,6-dimethoxyphenol
In: Rhizoctonia D690
→ *p*-Chlorophenol
In: Desulfitobacterium K247

4-(4-Bromo-2-chlorophenoxy)-2,6-dimethoxyphenol
Formed from: 4-Bromo-2-chlorophenol
4-Hydroxy-3,5-dimethoxybenzoic acid

4-(4-(4-Bromo-2-chlorophenoxy)-2,6-dimethoxyphenoxy)-2,6-dimethoxyphenol
Formed from: 4-Bromo-2-chlorophenol
4-Hydroxy-3,5-dimethoxybenzoic acid

5-(4-Bromo-2-chlorophenoxy)-3-methoxy-*o*-quinone
Formed from: 2-Bromo-4-chlorophenol
4-Hydroxy-3,5-dimethoxybenzoic acid

2-Bromo-4-chlorophenyl-β-D-glucuronide
Formed from: 2-Bromo-4-chlorophenol

7-Bromo-5-(*o*-chlorophenyl)-1,3-dihydro-2*H*-1,4-benzodiazepin-2-one (phenazepam)
→ 7-Bromo-5-(*o*-chlorophenyl)-1,3-dihydro-3-hydroxy-2*H*-1,4-benzodiazepin-2-one
In: Rat A3698, B106

7-Bromo-5-(*o*-chlorophenyl)-1,3-dihydro-3-hydroxy-2*H*-1,4-benzodiazepin-2-one
Formed from: 7-Bromo-5-(*o*-chlorophenyl)-1,3-dihydro-2*H*-1,4-benzodiazepin-2-one

7-Bromo-5-(*o*-chlorophenyl)-2,3-dihydro-2-hydroxymethyl-1*H*-1,4-benzodiazepine
Probably formed from: 7-Bromo-5-(*o*-chlorophenyl)-2,3-dihydro-2-methoxymethyl-2*H*-1,4-benzodiazepine

7-Bromo-5-(*o*-chlorophenyl)-2,3-dihydro-2-hydroxymethyl-1-methyl-1*H*-1,4-benzodiazepine
Formed from: 7-Bromo-5-(*o*-chlorophenyl)-2,3-dihydro-2-methoxymethyl-1-methyl-1*H*-1,4-benzodiazepine

6-Bromo-4-(*o*-chlorophenyl)-1,2-dihydro-3-hydroxy-1-methylquinazoline
Probably formed from: 6-Bromo-4-(*o*-chlorophenyl)-1,2-dihydro-1-methylquinazoline

7-Bromo-5-(*o*-chlorophenyl)-2,3-dihydro-2-methoxymethyl-1*H*-1,4-benzodiazepine
Formed from: 7-Bromo-5-(*o*-chlorophenyl)-2,3-dihydro-2-methoxymethyl-1-methyl-1*H*-1,4-benzodiazepine
→ 7-Bromo-5-(*o*-chlorophenyl)-2,3-dihydro-2-hydroxymethyl-1*H*-1,4-benzodiazepine
Probably in: Dog D579
Man D579
Rabbit D579

7-Bromo-5-(*o*-chlorophenyl)-2,3-dihydro-2-methoxymethyl-1-methyl-1*H*-1,4-benzodiazepine (metaclazepam)
→ 7-Bromo-5-(*o*-chlorophenyl)-2,3-dihydro-2-hydroxymethyl-1-methyl-1*H*-1,4-benzodiazepine
In: Dog D579
Man D579
Rabbit D579
→ 7-Bromo-5-(*o*-chlorophenyl)-2,3-dihydro-2-methoxymethyl-1*H*-1,4-benzodiazepine
In: Dog D579
Man D579

Rabbit D579
→ 6-Bromo-4-(*o*-chlorophenyl)-1,2-dihydro-1-methylquinazoline
In: Dog D579
Man D579
Rabbit D579
→ 6-Bromo-4-(*o*-chlorophenyl)-2-methylquinazoline
In: Dog D579
Man D579
Rabbit D579

6-Bromo-4-(*o*-chlorophenyl)-1,2-dihydro-1-methylquinazoline
Formed from: 7-Bromo-5-(*o*-chlorophenyl)-2,3-dihydro-2-methoxymethyl-1-methyl-1*H*-1,4-benzodiazepine
→ 6-Bromo-4-(*o*-chlorophenyl)-1,2-dihydro-3-hydroxy-1-methylquinazoline
Probably in: Dog D579
Man D579
Rabbit D579
→ 6-Bromo-4-(*o*-chlorophenyl)-1,2-dihydroquinazoline
Probably in: Dog D579
Man D579
Rabbit D579

6-Bromo-4-(*o*-chlorophenyl)-1,2-dihydroquinazoline
Probably formed from: 6-Bromo-4-(*o*-chlorophenyl)-1,2-dihydro-1-methylquinazoline

2-Bromo-4-(*o*-chlorophenyl)-6-hydroxy-9-hydroxymethyl-6*H*-thieno[3,2-*f*]-1,2,4-triazolo[4,3-*a*]-1,4-diazepine
Probably formed from: 2-Bromo-4-(*o*-chlorophenyl)-9-hydroxymethyl-6*H*-thieno[3,2-*f*]-1,2,4-triazolo[4,3-*a*]-1,4-diazepine
6-Hydroxybrotizolam

2-Bromo-4-(*o*-chlorophenyl)-9-hydroxymethyl-6*H*-thieno[3,2-*f*][1,2,4]triazolo[4,3-*a*][1,4]diazepine (α-hydroxybrotizolam)
Formed from: Brotizolam
→ 2-Bromo-4-(2-chloro-4-hydroxyphenyl)-9-hydroxymethyl-6*H*-thieno[3,2-*f*]-1,2,4-triazolo[4,3-*a*]-1,4-diazepine
Probably in: Dog E91
Man E91
Rat E91
→ 2-Bromo-4-(*o*-chlorophenyl)-6-hydroxy-9-hydroxymethyl-6*H*-thieno[3,2-*f*]-1,2,4-triazolo[4,3-*a*]-1,4-diazepine
Probably in: Dog E91
Man E91
Rat E91
→ 2-Bromo-4-(*o*-chlorophenyl)-6*H*-thieno[3,2-*f*]-1,2,4-triazolo[4,3-*a*]-1,4-diazepine
Probably in: Rat E91

6-Bromo-4-(*o*-chlorophenyl)-2-methylquinazoline
Formed from: 7-Bromo-5-(*o*-chlorophenyl)-2,3-dihydro-2-methoxymethyl-1-methyl-1*H*-1,4-benzodiazepine

2-Bromo-4-(*o*-chlorophenyl)-6*H*-thieno[3,2-*f*]-1,2,4-triazolo[4,3-*a*]-1,4-diazepine
Formed from: 2-Bromo-4-(*o*-chlorophenyl)-9-hydroxymethyl-6*H*-thieno[3,2-*f*][1,2,4]triazolo[4,3-*a*][1,4]diazepine

***m*-Bromocinnamide**
Formed from: Cinromide

***m*-Bromocinnamic acid**
Formed from: Cinromide

***p*-Bromocinnamic acid**
→ *p*-Bromostyrene
In: Microorganism F390

Bromocriptine ((5′α)-2-bromo-12′-hydroxy-2′-(1-methylethyl)-5′-(2-methylpropyl)ergotaman-3′,6′,18′-trione)
→ 2-Bromoisolysergamide
In: Dog C715
Man C715
Monkey C715
Mouse C715
Rat C715
→ 2-Bromolysergamide
In: Dog C715
Man C715
Monkey C715
Mouse C715
Rat C715
→ 8′α-Hydroxybromocriptine
In: Rat C607
→ 8′β-Hydroxybromocriptine
In: Rat C607
→ 8′α-Hydroxyisobromocriptine
In: Rat C607
→ 8′β-Hydroxyisobromocriptine

In: Rat C607
→ 7′,8′-Seco-8′-carboxybromocriptine
In: Rat C607
→ 7′,8′-Seco-8′-carboxyisobromocriptine
In: Rat C607

2-Bromo-3-(S-(L-cysteinylglycyl))quinol
Formed from: 2-Bromo-3-(S-glutathionyl) quinol
→ 2-Bromo-3-(S-(L-cysteinyl))quinol
Probably in: Rat G941
→ 2*H*-8-Bromo-(3-glycinyl)-7-hydroxy-1,4-benzothiazine
Probably in: Rat G941

2-Bromo-3-(S-(L-cysteinyl))quinol
Probably formed from: 2-Bromo-3-(S-(L-cysteinylglycyl))quinol
→ N-Acetyl-2-bromo-3-(S-(L-cysteinyl)) quinol
Probably in: Rat G941

2-Bromo-6-(S-(L-cysteinyl))quinol
Formed from: 6-(N-Acetyl-S-(L-cysteinyl))-2-bromoquinol
→ 6-(N-Acetyl-S-(L-cysteinyl))-2-bromoquinol
In: Guinea pig H467
Hamster H467
Mouse H467
Rat H467

4-Bromo-2,5-dichlorophenol
Formed from: Leptophos

O-(4-Bromo-2,5-dichlorophenyl) O-methyl phenylphosphonate
Formed from: Leptophos

O-(4-Bromo-2,5-dichlorophenyl) O-methyl phenylphosphonothioate
See: Leptophos

N′-((8α)-2-Bromo-9,10-didehydro-6-methylergolin-8-yl)-N,N-diethylurea
See: Bromerguride

1-Bromo-2,3-difluorobenzene
→ 1-Bromo-2,3-difluoro-5,6-dihydro-5,6-dihydroxybenzene
In: Pseudomonas J536

1-Bromo-2,3-difluoro-5,6-dihydro-5,6-dihydroxybenzene
Formed from: 1-Bromo-2,3-difluorobenzene

(5*S*)-5-(5-Bromo-2,3-dihydro-7-benzofuranyl)-8-chloro-2,3,4,5-tetrahydro-3-methyl-1*H*-3-benzazepin-7-ol
See: Berupipam

7-Bromo-3,4-dihydro-3,4-dihydroxybenzanthracene
Formed from: 7-Bromobenzanthracene

7-Bromo-5,6-dihydro-5,6-dihydroxybenzanthracene
Formed from: 7-Bromobenzanthracene

7-Bromo-8,9-dihydro-8,9-dihydroxybenzanthracene
Formed from: 7-Bromobenzanthracene

7-Bromo-10,11-dihydro-10,11-dihydroxybenzanthracene
Formed from: 7-Bromobenzanthracene

***cis*-3-Bromo-1,2-dihydro-1,2-dihydroxybenzene**
Formed from: Bromobenzene

***trans*-3-Bromo-1,2-dihydro-1,2-dihydroxybenzene**
Formed from: Bromobenzene

4-Bromo-1,2-dihydro-1,2-dihydroxybenzene
Probably formed from: Bromobenzene-3,4-oxide
→ 4-Bromocatechol
In: Rat D563, F525

5-Bromo-1,2-dihydro-1,2-dihydroxybenzoic acid
Probably formed from: *m*-Bromobenzoic acid

6-Bromo-4,5-dihydro-4,5-dihydroxy-3,4-benzpyrene
Formed from: 6-Bromo-3,4-benzpyrene

6-Bromo-7,8-dihydro-7,8-dihydroxy-3,4-benzpyrene
Formed from: 6-Bromo-3,4-benzpyrene

4-Bromo-2,3-dihydro-2,3-dihydroxytoluene
→ 3-Bromo-6-methylcatechol
In: Pseudomonas K348

***cis*-7-Bromo-1,2-dihydro-1,2-dihydroxynaphthalene**
Formed from: 2-Bromonaphthalene
→ 7-Bromo-1,2-dihydroxynaphthalene
In: Pseudomonas J370

***cis*-8-Bromo-1,2-dihydro-1,2-dihydroxynaphthalene**
Formed from: 1-Bromonaphthalene
→ 8-Bromo-1,2-dihydroxynaphthalene
In: Pseudomonas J370

1-Bromo-*trans*-4,5-dihydro-4,5-dihydroxypyrene
Formed from: 1-Bromopyrene

1-Bromo-*trans*-9,10-dihydro-9,10-dihydroxypyrene
Formed from: 1-Bromopyrene-9,10-oxide

6-Bromo-1,4-dihydro-4,4-dimethyl-2*H*-3,1-benzoxazin-2-one
See: Brofoxine

6-Bromo-1,4-dihydro-4-hydroxymethyl-4-methyl-2*H*-3,1-benzoxazin-2-one
Formed from: Brofoxine

7-Bromo-1,3-dihydro-5-(2′-pyridyl)-2*H*-1,4-benzodiazepin-2-one
See: Bromazepam

3-Bromo-2,5-dihydroxybenzoic acid
→ 2-Bromo-6-carboxy-4-hydroxy-*cis*,*cis*-muconic semialdehyde
In: Pseudomonas F649

3-Bromo-4,5-dihydroxybenzonitrile
Probably formed from: 3,5-Dibromo-4-hydroxybenzonitrile
→ 3,4,5-Trihydroxybenzonitrile
Probably in: Wheat A746

2-Bromo-3,4′-dihydroxybiphenyl
Probably formed from: 2-Bromo-3-hydroxybiphenyl

2-Bromo-4,4′-dihydroxybiphenyl
Probably formed from: 2′-Bromo-4-hydroxybiphenyl

2-Bromo-4′,5-dihydroxybiphenyl
Probably formed from: 2′-Bromo-4-hydroxybiphenyl

3-Bromo-4,4′-dihydroxybiphenyl
Probably formed from: 3-Bromo-4-hydroxybiphenyl

4′-Bromo-3,4-dihydroxybiphenyl
Probably formed from: 4′-Bromo-4-hydroxybiphenyl

2-Bromo-3,6-dihydroxy-4-methoxybenzaldehyde
Formed from: 5-Bromo-4-hydroxy-3-methoxybenzaldehyde

7-Bromo-1,2-dihydroxynaphthalene
Formed from: *cis*-7-Bromo-1,2-dihydro-1,2-dihydroxynaphthalene

8-Bromo-1,2-dihydroxynaphthalene
Formed from: *cis*-8-Bromo-1,2-dihydro-1,2-dihydroxynaphthalene

3-Bromo-4′,5-dihydroxysalicylanilide
Probably formed from: 3,5-Dibromo-4′-hydroxysalicylanilide

2-Bromo-4,5-dimethoxybenzyl alcohol
Formed from: Veratryl alcohol

***trans*-ω-Bromo-3,4-dimethoxystyrene**
Formed from: 3,4-Dimethoxycinnamic acid

3′-Bromo-4-dimethylaminoazobenzene
→ 3′-Bromo-4-methylaminoazobenzene
In: Rat A776
→ 3-Bromo-4′-dimethylamino-4-hydroxyazobenzene
In: Rat A776

4′-Bromo-4-dimethylaminoazobenzene
→ 4′-Bromo-4-methylaminoazobenzene
In: Rat A1979
→ 4′-Dimethylamino-4-hydroxyazobenzene
In: Rat A1979

3-Bromo-4′-dimethylamino-4-hydroxyazobenzene
Formed from: 3′-Bromo-4-dimethylaminoazobenzene

5-Bromo-2-dimethylaminophenol
Formed from: *p*-Bromo-N,N-dimethylaniline

***p*-Bromo-N,N-dimethylaniline**
Formed from: *p*-Bromo-N,N-dimethylaniline-N-oxide
→ 5-Bromo-2-dimethylaminophenol
Probably with cytochrome c B185
→ *p*-Bromo-N-methylaniline
In: Rabbit F839
Rat D578

***p*-Bromo-N,N-dimethylaniline-N-oxide**
→ *p*-Bromo-N,N-dimethylaniline
With cytochrome c B185
→ *p*-Bromo-N-methylaniline
With cytochrome c B185

2-Bromo-3,3-dimethyl-N-(1-methyl-1-phenylethyl)butyramide
See: Bromobutide

(1-Bromoethyl)benzene
→ N-Acetyl-S-(1-phenylethyl)-L-cysteine
In: Rat A3708

(2-Bromoethyl)benzene
→ N-Acetyl-S-(1-phenylethyl)-L-cysteine
In: Rat A3708
→ N-Acetyl-S-(2-hydroxy-2-phenylethyl)-L-cysteine

In: Rat A3708

1-Bromo-4-ethylbenzene
→ 1-(*p*-Bromophenyl)ethanol
In: Mortierella D969

S-(*p*-(2-Bromoethyl)phenyl)glutathione
Probably formed from: *p*-Nitrophenethyl bromide

4-Bromo-17-ethynyl-2-hydroxyoestradiol
Probably formed from: 4-Bromo-17-ethynyloestradiol

4-Bromo-17-ethynyloestradiol
→ 4-Bromo-17-ethynyl-2-hydroxyoestradiol
Probably in: Hamster D514

2-Bromofluorene
→ 2-Bromofluorenone
In: Pseudomonas H672
→ 4-Bromophthalic acid
In: Pseudomonas H672

2-Bromofluorenone
Formed from: 2-Bromofluorene

***m*-Bromofluorobenzene**
→ 3-Bromocatechol
In: Pseudomonas F76
→ 4-Bromo-2-fluorophenol
In: Rat G953

3-(4-Bromo-2-fluorobenzyl-7-chloro-2,4-dioxo-1,2,3,4-tetrahydroquinazolin-1-yl)acetic acid
See: Zenarestat

4-Bromo-2-fluorophenol
Formed from: *m*-Bromofluorobenzene
→ 4-Bromo-2-fluorophenyl-β-D-glucuronide
In: Rat G953
→ 4-Bromo-2-fluorophenyl sulphate
In: Rat G953

4-Bromo-2-fluorophenyl-β-D-glucuronide
Formed from: 4-Bromo-2-fluorophenol

4-Bromo-2-fluorophenyl sulphate
Formed from: 4-Bromo-2-fluorophenol

3-Bromogentisic acid
→ 6-Bromo-4-hydroxy-2-oxohepta-3,5-dienoic acid
In: Rhodococcus H441

4-Bromo-2-(S-glutathionyl)-1,2-dihydro-1-hydroxybenzene
Probably formed from: Bromobenzene-3,4-oxide
→ N-Acetyl-S-(L-cysteinyl)-3-bromobenzene
Probably in: Guinea pig E891
→ 2-(N-Acetyl-S-(L-cysteinyl))-4-bromo-1,2-dihydro-1-hydroxybenzene
Probably in: Guinea pig E891

5-Bromo-2-(S-glutathionyl)-1,2-dihydro-1-hydroxybenzene
Probably formed from: Bromobenzene-3,4-oxide
→ N-Acetyl-S-(L-cysteinyl)-4-bromobenzene
Probably in: Guinea pig E891
→ 2-(N-Acetyl-S-(L-cysteinyl))-5-bromo-1,2-dihydro-1-hydroxybenzene
Probably in: Guinea pig E891

2-Bromo-1-(S-glutathionyl)-4-nitrobenzene
Formed from: 2-Bromo-1-chloro-4-nitrobenzene

2-Bromo-3-(S-glutathionyl)quinol
Formed from: Bromoquinol
→ 2-Bromo-3,5-bis(S-glutathionyl)quinol
Probably in: Rabbit F165
Rat E889
→ 2-Bromo-3,6-bis(S-glutathionyl)quinol
Probably in: Rabbit F165
Rat E889
→ 2-Bromo-3-(S-(L-cysteinylglycyl))quinol
In: Beef F691
With γ-glutamyltranspeptidase F691
See also: G941

2-Bromo-5-(S-glutathionyl)quinol
Formed from: Bromoquinol

2-Bromo-6-(S-glutathionyl)quinol
Formed from: Bromoquinol

2*H*-8-Bromo-(3-glycinyl)-7-hydroxy-1,4-benzothiazine
Probably formed from: 2-Bromo-3-(S-(L-cysteinylglycyl))quinol

5-Bromo-6-guanidinoquinoxaline
Formed from: Brimodinine

***p*-Bromohippuric acid**
Formed from: *p*-Bromobenzoic acid

3-Bromo-4-hydroxybenzaldehyde
Formed from: *p*-Hydroxybenzaldehyde
→ 2,3-Dibromo-4,5-dihydroxybenzaldehyde
Probably in: Odonthalia A3989
→ 3,5-Dibromo-4-hydroxybenzaldehyde
In: Odonthalia A3989

5-Bromo-2-hydroxybenzoic acid
See: 5-Bromosalicylic acid

3-Bromo-4-hydroxybenzonitrile
Probably formed from: 3,5-Dibromo-4-hydroxybenzonitrile
→ *p*-Hydroxybenzonitrile
Probably in: Wheat A746

***cis*-3-Bromo-3-(*p*-hydroxybenzoyl)acrylic acid**
Formed from: Cytemba

6-Bromo-2-(α-hydroxybenzyl)chromone
Formed from: 2-Benzyl-6-bromochromone

2-Bromo-3-hydroxybiphenyl
Formed from: 2-Bromobiphenyl
→ 2-Bromo-3,4′-dihydroxybiphenyl
Probably in: Rat B448

2′-Bromo-4-hydroxybiphenyl
Probably formed from: 2-Bromobiphenyl
→ 2-Bromo-4,4′-dihydroxybiphenyl
Probably in: Rat B448
→ 2-Bromo-4′,5-dihydroxybiphenyl
Probably in: Rat B448

2-Bromo-5-hydroxybiphenyl
Formed from: 2-Bromobiphenyl

3-Bromo-4-hydroxybiphenyl
Formed from: 3-Bromobiphenyl
→ 3-Bromo-4,4′-dihydroxybiphenyl
Probably in: Rat B448

3′-Bromo-4-hydroxybiphenyl
Formed from: 3-Bromobiphenyl

4′-Bromo-3-hydroxybiphenyl
Formed from: 4-Bromobiphenyl

4′-Bromo-4-hydroxybiphenyl
Formed from: 4-Bromobiphenyl
→ 4′-Bromo-3,4-dihydroxybiphenyl
Probably in: Poultry A3893

4-Bromo-2-hydroxycyclohexa-3,5-dienyl-L-cysteine
→ *m*-Bromophenol
In: Guinea pig G884
Rat G884

2-Bromo-4-hydroxy-3,3-dimethyl-N-(1-methyl-1-phenylethyl)butyramide
Formed from: Bromobutide
→ 2-Bromo-3-carboxy-2-hydroxy-3-methyl-N-(1-methyl-1-phenylethyl)butyramide
Probably in: Mouse D193
Rat D193, D215
→ 2-Bromo-4-hydroxy-3,3-dimethyl-N-(1-methyl-1-phenylethyl)butyramide-β-D-glucuronide
Probably in: Mouse D193
Rat D193

2-Bromo-4-hydroxy-3,3-dimethyl-N-(1-methyl-1-phenylethyl)butyramide-β-D-glucuronide
Probably formed from: 2-Bromo-4-hydroxy-3,3-dimethyl-N-(1-methyl-1-phenylethyl)butyramide

6-Bromo-2-(1-hydroxyethyl)chromone
Formed from: 2-Acetyl-6-bromochromone

2-Bromo-4-hydroxy-N-(1-(4-hydroxy-3-methoxyphenyl)-1-methylethyl)-3,3-dimethylbutyramide
→ 4-Hydroxy-N-(1-(4-hydroxy-3-methoxyphenyl)-1-methylethyl)-3,3-dimethylbutyramide
Probably in: Rat D215

2-Bromo-4-hydroxy-N-(1-(*p*-hydroxyphenyl)-1-methylethyl)-3,3-dimethylbutyramide
Probably formed from: 2-Bromo-N-(1-(*p*-hydroxyphenyl)-1-methylethyl)-3,3-dimethylbutyramide
→ 2-Bromo-4-hydroxy-N-(1-(*p*-hydroxyphenyl)-1-methylethyl)-3,3-dimethylbutyramide-β-D-glucuronide
Probably in: Mouse D193
Rat D193
→ 4-Hydroxy-N-(1-(*p*-hydroxyphenyl)-1-methylethyl)-3,3-dimethylbutyramide
Probably in: Mouse D193
Rat D193, D215

2-Bromo-4-hydroxy-N-(1-(*p*-hydroxyphenyl)-1-methylethyl)-3,3-dimethylbutyramide-β-D-glucuronide
Probably formed from: 2-Bromo-4-hydroxy-N-(1-(*p*-hydroxyphenyl)-1-methylethyl)-3,3-dimethylbutyramide

2-Bromo-4-hydroxy-5-methoxybenzaldehyde
→ 2-Bromo-4-hydroxy-5-methoxybenzoic acid
In: Microorganism E664
→ 2-Bromo-4-hydroxy-5-methoxybenzyl alcohol
In: Microorganism E664

5-Bromo-4-hydroxy-3-methoxybenzaldehyde
→ 5-Bromo-4-hydroxy-3-methoxybenzoic acid
In: Burkholderia K107

→ 2-Bromo-3,6-dihydroxy-4-methoxybenzaldehyde
In: Burkholderia K107

2-Bromo-4-hydroxy-5-methoxybenzoic acid
Formed from: 2-Bromo-4-hydroxy-5-methoxybenzaldehyde

5-Bromo-4-hydroxy-3-methoxybenzoic acid
Formed from: 5-Bromo-4-hydroxy-3-methoxybenzaldehyde

2-Bromo-4-hydroxy-5-methoxybenzyl alcohol
Formed from: 2-Bromo-4-hydroxy-5-methoxybenzaldehyde

2-Bromo-N-(1-(4-hydroxy-3-methoxyphenyl)-1-methylethyl)-3,3-dimethylbutyramide
Probably formed from: 2-Bromo-N-(1-(*p*-hydroxyphenyl)-1-methylethyl)-3,3-dimethylbutyramide
→ 2-Bromo-4-hydroxy-N-(1-(4-hydroxy-3-methoxyphenyl)-1-methylethyl)-3,3-dimethylbutyramide
Probably in: Mouse D193
Rat D215
→ N-(1-(4-Hydroxy-3-methoxyphenyl)-1-methylethyl)-3,3-dimethylbutyramide
Probably in: Rat D215

3-Bromo-4-hydroxy-4′-methylaminoazobenzene
Probably formed from: 3′-Bromo-4-methylaminoazobenzene

2-Bromo-4-hydroxymethylphenol
Formed from: 2-Bromo-4-methylphenol

2-Bromo-N-(2-hydroxy-1-methyl-1-phenylethyl)-3,3-dimethylbutyramide
Formed from: Bromobutide

2-Bromo-N-(1-(*p*-hydroxyphenyl)-1-methylethyl)-3,3-dimethylbutyramide
Formed from: Bromobutide
→ 2-Bromo-4-hydroxy-N-(1-(*p*-hydroxyphenyl)-1-methylethyl)-3,3-dimethylbutyramide
Probably in: Mouse D193
Rat D193, D215
→ 2-Bromo-N-(1-(4-hydroxy-3-methoxyphenyl)-1-methylethyl)-3,3-dimethylbutyramide
Probably in: Rat D215
→ 2-Bromo-N-(1-(*p*-hydroxyphenyl)-1-methylethyl)-3,3-dimethylbutyramide-β-D-glucuronide
Probably in: Mouse D193
Rat D193

2-Bromo-N-(1-(*p*-hydroxyphenyl)-1-methylethyl)-3,3-dimethylbutyramide-β-D-glucuronide
Probably formed from: 2-Bromo-N-(1-(*p*-hydroxyphenyl)-1-methylethyl)-3,3-dimethylbutyramide

5-Bromo-2-hydroxyquinol
Formed from: 4-Bromoresorcinol

3-Bromo-5-hydroxysalicylanilide
Formed from: 3,5-Dibromosalicylanilide

3-(3-Bromo-2-(5-hydroxy-2-(1*H*-tetrazol-5-yl)phenyl)benzofuran-5-ylmethyl)-2-butyl-5-chloro-3*H*-imidazole-4-carboxylic acid
Formed from: 3-(3-Bromo-2-(2-(1*H*-tetrazol-5-yl)phenyl)benzofuran-5-ylmethyl)-2-butyl-5-chloro-3*H*-imidazole-4-carboxylic acid

5-Bromo-6-(imidazolidinylideneamino)quinoxaline
See: Brimodinine

2-Bromoindane
→ *trans*-2-Bromoindanol
Probably in: Pseudomonas K207

***trans*-2-Bromoindanol**
Formed from: Indene
Probably formed from: 2-Bromoindane

2-Bromoindene
→ Indene-1,2-oxide
In: Rat A2886

5-Bromoindole
Formed from: Indole
→ 5-Bromoanthranilic acid
In: Jasminum B70
Tecoma C170

5-Bromoindoxyl
Formed from: 5-Bromoindoxyl acetate

5-Bromoindoxyl acetate
→ 5-Bromoindoxyl
In: Mouse D582

2-Bromoisolysergamide
Formed from: Bromocriptine
→ 2-Bromoisolysergic acid
Probably in: Dog C715
Man C715
Mouse C715

Rat C607, C715
Rhesus monkey C715

2-Bromoisolysergic acid
Probably formed from: 2-Bromoisolysergamide

4-Bromoisoquinoline
Formed from: 4-Bromoisoquinoline-N-oxide

4-Bromoisoquinoline-N-oxide
→ 4-Bromoisoquinoline
In: Saccharomyces H894

2-Bromolysergamide
Formed from: Bromocriptine
→ 2-Bromolysergic acid
Probably in: Dog C715
Man C715
Mouse C715
Rat C607, C715
Rhesus monkey C715

2-Bromolysergic acid
Probably formed from: 2-Bromolysergamide

2-Bromo-D-lysergic diethylamide
→ 2-Bromo-D-lysergic ethylamide
In: Rat A98
→ 2-Bromo-D-norlysergic diethylamide
In: Rat A98

2-Bromo-D-lysergic ethylamide
Formed from: 2-Bromo-D-lysergic diethylamide
→ 2-Bromo-D-norlysergic ethylamide
In: Rat A98

***p*-Bromomandelic acid**
→ *p*-Bromophenylglyoxylic acid
In: Rhodotorula H90

***m*-Bromomandelonitrile**
Formed from: *m*-Bromobenzaldehyde

***p*-Bromomandelonitrile**
Formed from: *p*-Bromobenzaldehyde

***cis*-3-Bromo-3-(*p*-methoxybenzoyl)acrylic acid**
See: Cytemba

5-Bromo-4-methoxycatechol
Formed from: 5-Bromo-4,5-methylenedioxyanisole

4-Bromo-2-methoxy-17β-oestradiol
Formed from: 4-Bromo-17β-oestradiol

4-Bromo-2-methoxyphenol
Formed from: 4-Bromocatechol

5-Bromo-2-methoxyphenol
Formed from: 4-Bromocatechol

3′-Bromo-4-methylaminoazobenzene
Formed from: 3′-Bromo-4-dimethylaminoazobenzene
→ 4-Amino-3′-bromoazobenzene
In: Rat A776
→ 3-Bromo-4-hydroxy-4′-methylaminoazobenzene
Probably in: Rat A776

4′-Bromo-4-methylaminoazobenzene
Formed from: 4′-Bromo-4-dimethylaminoazobenzene
→ 4-Amino-4′-bromoazobenzene
In: Rat A1979

2-Bromo-4-methylaniline
→ N-(4-Amino-3-bromobenzyl)-2-bromo-4-methylaniline
In: Rat G518
→ 2,2′-Dibromo-4,4′-dimethylhydrazobenzene
In: Rat G518

***p*-Bromo-N-methylaniline**
Formed from: *p*-Bromo-N,N-dimethylaniline
p-Bromo-N,N-dimethylaniline-N-oxide

6-Bromo-3′-O-methyl-(+)-catechin
Formed from: 6-Bromo-(+)-catechin

8-Bromo-3′-O-methyl-(+)-catechin
Formed from: 8-Bromo-(+)-catechin

3-Bromo-6-methylcatechol
Formed from: 4-Bromo-2,3-dihydro-2,3-dihydroxytoluene

2-Bromo-4,5-methylenedioxyanisole
→ 5-Bromo-4-methoxycatechol
In: Rat A3947

1-Bromo-3,4-methylenedioxybenzene
→ 4-Bromocatechol
In: Rat A3947

2-Bromo-4-methylphenol
→ 2-Bromo-4-hydroxymethylphenol
In: Pseudomonas D696

***p*-Bromomethylphenylglyoxylic acid**
→ *p*-Methylbenzoylthiamine pyrophosphoric acid
In: Pseudomonas E714

(*E*)-4-(*p*-Bromomethylphenyl)-2-oxo-3-butenoic acid
→ *p*-Methylcinnamic acid
In: Yeast F279

***p*-Bromo-α-methylstyrene oxide**
See: 2-(*p*-Bromophenyl)propylene oxide

4-Bromo-N-(2-methylsulphinylethylamino ethylene)benzenesulphonamide
→ 4-Bromo-N-(2-methylsulphonylethylaminoethylene)benzenesulphonamide
Probably in: Man H448

4-Bromo-N-(2-methylsulphonylethylamino ethylene)benzenesulphonamide
Probably formed from: 4-Bromo-N-(2-methylsulphinylethylaminoethylene)benzenesulphonamide

5-Bromo-α-naphthaldehyde
→ 5-Bromo-α-naphthoic acid
In: Man J24

1-Bromonaphthalene
→ *cis*-8-Bromo-1,2-dihydro-1,2-dihydroxynaphthalene
In: Pseudomonas J370

2-Bromonaphthalene
→ *cis*-7-Bromo-1,2-dihydro-1,2-dihydroxynaphthalene
In: Pseudomonas J370

5-Bromo-α-naphthoic acid
Formed from: 5-Bromo-α-naphthaldehyde

8β-(5-Bromonicotinoylhydroxymethyl)-10α-methoxy-1,6-dimethylergoline
See: Nicergoline

***o*-Bromonitrobenzene**
→ *o*-Bromoaniline
In: Saccharomyces F201

***m*-Bromonitrobenzene**
→ *m*-Bromoaniline
In: Saccharomyces F201

***p*-Bromonitrobenzene**
→ *p*-Bromoaniline
In: Saccharomyces F201

***p*-Bromonitrosobenzene**
Formed from: *p*-Bromoaniline

2-Bromo-D-norlysergic diethylamide
Formed from: 2-Bromo-D-lysergic diethylamide

2-Bromo-D-norlysergic ethylamide
Probably formed from: 2-Bromo-D-lysergic ethylamide

2′-Bromonorreticuline
→ 2′-Bromopapaverine
In: Papaver E777

2-Bromo-17β-oestradiol
→ 2-Hydroxy-17β-oestradiol
Probably in: Hamster D712

4-Bromo-17β-oestradiol
→ 4-Bromo-2-methoxy-17β-oestradiol
Probably in: Hamster D712

2′-Bromopapaverine
Formed from: 2′-Bromonorreticuline

***o*-Bromophenol**
Formed from: Bromobenzene
2,4-Dibromophenol
Phenol
Salicyl alcohol
→ N-Acetyl-S-(2,5-dihydroxyphenyl)-L-cysteine
In: Rat G626
→ *o*-Bromoanisole
In: Phanerochaete G782
→ *m*-Bromobenzoic acid
In: Microorganism G917
→ 3-Bromocatechol
In: Bacillus J340
→ 2,4-Dibromophenol
Probably in: Ulva K153
→ 2,6-Dibromophenol
Probably in: Ulva K153
→ 2,5-Dimethoxythioanisole
In: Rat G626
→ Phenol
In: Microorganism G331, J395

***m*-Bromophenol**
Formed from: Bromobenzene
4-Bromo-2-hydroxycyclohexa-3,5-dienyl-L-cysteine
→ N-Acetyl-S-(2,5-dihydroxyphenyl)-L-cysteine
In: Rat G626
→ 4-Bromocatechol
In: Bacillus J340
Guinea pig E891

Rat E891
→ Bromoquinol
In: Guinea pig E891
Rat E891
→ 2,5-Dimethoxythioanisole
In: Rat G626
→ Phenol
In: Microorganism J395

p-Bromophenol
Formed from: *p*-Bromoanisole
Bromobenzene
2,4-Dibromophenol
p-Hydroxybenzaldehyde
p-Hydroxybenzoic acid
p-Hydroxybenzyl alcohol
p-Hydroxyphenylacetic acid
Phenol
→ N-Acetyl-S-(2,5-dihydroxyphenyl)-L-cysteine
In: Rat G626
→ *p*-Bromoanisole
In: Guinea pig A115
Phanerochaete G782
→ 4-Bromocatechol
In: Guinea pig E891
Protobacteria J466
Rat C837, D391, E891, F525
See also: E878
→ *p*-Bromophenyl-β-D-glucuronide
In: Rat B962
→ 2,4-Dibromophenol
Probably in: Ulva K153
→ 2,5-Dimethoxythioanisole
In: Rat G626
→ Phenol
In: Microorganism J395
→ Quinol
In: Rat J81

4-(α-(p-Bromophenoxy)benzyl)-1-methylpyridinium
Probably formed from: 4-(α-(*p*-Bromophenoxy)benzyl)-1,2,3,6-tetrahydro-1-methylpyridine

4-(α-(p-Bromophenoxy)benzyl)-1,2,3,6-tetrahydro-1-methylpyridine
→ 4-(α-(*p*-Bromophenoxy)benzyl)-1-methylpyridinium
Probably in: Rat F816

N-((p-Bromophenyl)acetyl)-N-hydroxyurea
Formed from: N-((*p*-Bromophenyl)acetyl)urea

N-((p-Bromophenyl)acetyl)-N-methylurea
Formed from: N-((*p*-Bromophenyl)acetyl)urea

N-((p-Bromophenyl)acetyl)urea
→ N-((*p*-Bromophenyl)acetyl)-N-hydroxyurea
In: Rat K39
→ N-((*p*-Bromophenyl)acetyl)-N-methylurea
In: Rat K39
→ 4-(*p*-Bromophenyl)-3-oxapyrrolidine-2,5-dione
In: Rat K39

p-Bromo-L-phenylalanine
→ *p*-Bromophenypyruvic acid
In: Phaseolus E500

S-(p-Bromophenyl)-L-cysteine
Formed from: N-Acetyl-S-(*p*-(bromophenyl)-L-cysteine

(4*S*,3*R*)-*trans*-N-(p-Bromophenyl)-2,3-dehydro-3-hydroxy-4,5-diphenyl-1-pyrazolidinecarboxamide-β-D-glucuronide
Formed from: (4*S*,3*R*)-*trans*-N-(*p*-Bromophenyl-3-oxo-4,5-diphenyl-1-pyrazolidinecarboxamide

(*Z*)-3-(4-Bromophenyl)-N,N-dimethyl-3-(3-pyridyl)allylamine
See: Zimelidine

γ-(4-Bromophenyl)-N,N-dimethyl-2-pyridinepropanamine
See: Brompheniramine

1-(o-Bromophenyl)ethanol
Formed from: *o*-Bromoacetophenone

1-(m-Bromophenyl)ethanol
Formed from: *m*-Bromoacetophenone

1-(p-Bromophenyl)ethanol
Formed from: *p*-Bromoacetophenone
1-Bromo-4-ethylbenzene

2-Bromo-1-phenylethanol
Formed from: Styrene

1-(m-bromophenyl)ethylamine
Formed from: Octyl 1-(*m*-bromophenyl) ethylamine oxalamate

4-(2-(m-Bromophenyl)ethyl)-1-methylpyridinium
Probably formed from: 4-(2-(*m*-Bromophenyl) ethyl)-1,2,3,6-tetrahydro-1-methylpyridine

(2*E*)-3-Bromophenyl-N-ethyl-2-propenamide
See: Cinromide

4-(2-(*m*-Bromophenyl)ethyl)-1,2,3,6-tetrahydro-1-methylpyridine
→ 4-(2-(*m*-Bromophenyl)ethyl)-1-methylpyridinium
Probably in: Rat F816

***p*-Bromophenyl-β-D-glucuronide**
Formed from: *p*-Bromophenol

2-(*m*-Bromophenyl)-1-(S-glutathionyl)ethanol
Probably formed from: *m*-Bromostyrene oxide

2-(*p*-Bromophenyl)-1-(S-glutathionyl)ethanol
Probably formed from: *p*-Bromostyrene oxide

***p*-Bromophenylglyoxylic acid**
Formed from: *p*-Bromomandelic acid

***p*-Bromophenylhydracrylic acid**
Formed from: Cinnamic acid

***p*-Bromophenylhydroxylamine**
Formed from: *p*-Bromoaniline

4-(4-(4-Bromophenyl)-4-hydroxy-1-piperidinyl)-1-(4-fluorophenyl)-1-butanone
See: Bromperidol

4-*m*-Bromophenyl-1-methylpyridinium
Probably formed from: 4-*m*-Bromophenyl-1,2,3,6-tetrahydro-1-methylpyridine

3-(*p*-Bromophenyl)muconolactone
Formed from: 4-Bromobiphenyl

4-(*p*-Bromophenyl)-3-oxapyrrolidine-2,5-dione
Formed from: N-((*p*-Bromophenyl)acetyl)urea

4-Bromophenyl-1,3-oxathiolane
→ 4-Bromophenyl-1,3-oxathiolane sulphoxide
In: Man G579

4-Bromophenyl-1,3-oxathiolane sulphoxide
Formed from: 4-Bromophenyl-1,3-oxathiolane

(4*S*,3*R*)-*trans*-N-(*p*-Bromophenyl)-3-oxo-4,5-diphenyl-1-pyrazolidinecarboxamide
→ (4*S*,3*R*)-*trans*-N-(*p*-Bromophenyl)-2,3-dehydro-3-hydroxy-4,5-diphenyl-1-pyrazolidinecarboxamide-β-D-glucuronide
In: Dog H443
Mouse H443
Rat H443

2-((4-Bromophenyl)phenyl(methoxy)-N,N-dimethylethanamine
See: Bromazepine

(*R*)-2-(*p*-Bromophenyl)propane-1,2-diol
Formed from: 2-(*p*-Bromophenyl)propylene oxide

2-Bromo-1-phenylpropane-1,3-diol
Formed from: *trans*-Cinnamyl alcohol

2-(*p*-Bromophenyl)propylene oxide (*p*-bromo-α-methylstyrene oxide)
→ 2-(*p*-Bromophenyl)propane-1,2-diol ((*R*) assumed)
In: Aspergillus J682
Pig H753

3-(*p*-Bromophenyl)-3-(3-pyridyl)acraldehyde
→ 3-(*p*-Bromophenyl)-3-(3-pyridyl)acrylic acid
In: Dog B791
Rat B791

3-(*p*-Bromophenyl)-3-(3-pyridyl)acrylic acid
Formed from: 3-(*p*-Bromophenyl)-3-(3-pyridyl)acraldehyde
→ 3-(*p*-Bromophenyl)-3-(3-pyridyl)acrylic acid N-oxide
In: Dog B791
Rat B791

3-(*p*-Bromophenyl)-3-(3-pyridyl)acrylic acid N-oxide
Probably formed from: 3-(*p*-Bromophenyl)-3-(3-pyridyl)acrylic acid

***p*-Bromophenylpyruvic acid**
Formed from: *p*-Bromophenyl-L-alanine

4-(*m*-Bromophenyl)-1,2,3,6-tetrahydro-1-methylpyridine
→ 4-(*m*-Bromophenyl)-1-methylpyridinium
Probably in: Man F395

4-Bromophthalic acid
Probably in: 2-Bromofluorene

Bromopride (4-amino-5-bromo-N-(2-(diethylamino)ethyl)-2-methoxybenzamide)
→ Desethylbromopride
In: Dog C250

1-Bromopyrene
→ 1-Bromo-*trans*-4,5-dihydro-4,5-dihydroxypyrene
In: Rat E913
→ 1-Bromo-*trans*-9,10-dihydro-9,10-dihydroxypyrene
In: Rat E913

1-Bromopyrene-9,10-oxide
→ 1-Bromo-*trans*-9,10-dihydro-9,10-dihydroxypyrene
In: Rat E913

Bromoquinol
Formed from: *m*-Bromophenol
→ N-Acetyl-S-(2,5-dihydroxyphenyl)-L-cysteine
In: Rat G626
→ 2-Bromo-3-(S-glutathionyl)quinol
In: Rabbit F165
Rat D834, E889
→ 2-Bromo-5-(S-glutathionyl)quinol
In: Rat E889
→ 2-Bromo-6-(S-glutathionyl)quinol
In: Rat E889
→ 2-Bromoquinol-β-D-glucuronide
In: Rabbit F165
→ 2,5-Dimethoxythioanisole
In: Rat G626

2-Bromoquinol-β-D-glucuronide
Formed from: 2-Bromoquinol

3-Bromoquinoline
Formed from: 3-Bromoquinoline-N-oxide

3-Bromoquinoline-N-oxide
→ 3-Bromoquinoline
In: Saccharomyces H894

4-Bromoresorcinol
→ 5-Bromo-2-hydroxyquinol
In: Pseudomonas A2172, A2391

5-Bromosalicylic acid (5-bromo-2-hydroxybenzoic acid)
→ 4-Bromocatechol
Probably in: Pseudomonas E146

β-Bromostyrene
Formed from: Cinnamic acid
→ N-Acetyl-S-(1-bromo-2-hydroxy-2-phenylethyl)-L-cysteine
In: Marmoset A3481
Rabbit A3481
Rat A3481
→ N-Acetyl-S-(2-hydroxy-1-phenylethyl)-L-cysteine
In: Marmoset A3481
Rabbit A3481
Rat A3481
→ Mandelic acid
In: Marmoset A3481
Rabbit A3481
Rat A3481

***p*-Bromostyrene**
Formed from: *p*-Bromocinnamic acid

***m*-Bromostyrene oxide**
→ 2-(*m*-Bromophenyl)-1-(S-glutathionyl) ethanol
Probably in: Sheep A2282

***p*-Bromostyrene oxide**
→ 2-(*p*-Bromophenyl)-1-(S-glutathionyl) ethanol
Probably in: Sheep A2282

***p*-Bromotsulphinylanisole**
Formed from: *p*-Bromothioanisole
→ *p*-Bromosulphonylanisole
Probably in: Mortierella D759

***p*-Bromosulphonylanisole**
Probably formed from: *p*-Bromosulphinylanisole

Bromosulphophthalein
See: Sulphobromophthalein

3-(3-Bromo-2-(2-(1*H*-tetrazol-5-yl) phenyl)benzofuran-5-ylmethyl)-2-butyl-5-chloro-3*H*-imidazole-4-carboxylic acid
→ 3-(3-Bromo-2-(5-hydroxy-2-(1*H*-tetrazol-5-yl)phenyl)benzofuran-5-ylmethyl)-2-butyl-5-chloro-3*H*-imidazole-4-carboxylic acid
In: Streptomyces H203
→ 3-(3-Bromo-2-(2-(1*H*-tetrazol-5-yl) phenyl)benzofuran-5-ylmethyl)-5-chloro-2-(1-hydroxybutyl)-3*H*-imidazole-4-carboxylic acid
In: Streptomyces H203
→ 3-(3-Bromo-2-(2-(1*H*-tetrazol-5-yl) phenyl)benzofuran-5-ylmethyl)-5-chloro-2-(2-hydroxybutyl)-3*H*-imidazole-4-carboxylic acid
In: Streptomyces H203
→ 3-(3-Bromo-2-(2-(1*H*-tetrazol-5-yl) phenyl)benzofuran-5-ylmethyl)-5-chloro-2-(3-hydroxybutyl)-3*H*-imidazole-4-carboxylic acid
In: Streptomyces H203

3-(3-Bromo-2-(2-(1*H*-tetrazol-5-yl)phenyl)benzofuran-5-ylmethyl)-5-chloro-2-(1-hydroxybutyl)-3*H*-imidazole-4-carboxylic acid
Formed from: 3-(3-Bromo-2-(2-(1*H*-tetrazol-5-yl)phenyl)benzofuran-5-ylmethyl)-2-butyl-5-chloro-3*H*-imidazole-4-carboxylic acid

3-(3-Bromo-2-(2-(1*H*-tetrazol-5-yl)phenyl)benzofuran-5-ylmethyl)-5-chloro-2-(2-hydroxybutyl)-3*H*-imidazole-4-carboxylic acid
Formed from: 3-(3-Bromo-2-(2-(1*H*-tetrazol-5-yl)phenyl)benzofuran-5-ylmethyl)-2-butyl-5-chloro-3*H*-imidazole-4-carboxylic acid

3-(3-Bromo-2-(2-(1*H*-tetrazol-5-yl)phenyl)benzofuran-5-ylmethyl)-5-chloro-2-(3-hydroxybutyl)-3*H*-imidazole-4-carboxylic acid
Formed from: 3-(3-Bromo-2-(2-(1*H*-tetrazol-5-yl)phenyl)benzofuran-5-ylmethyl)-2-butyl-5-chloro-3*H*-imidazole-4-carboxylic acid
→ 3-(3-Bromo-2-(2-(1*H*-tetrazol-5-yl)phenyl)benzofuran-5-ylmethyl)-5-chloro-2-(3-oxobutyl)-3*H*-imidazole-4-carboxylic acid
Probably in: Streptomyces H203

3-(3-Bromo-2-(2-(1*H*-tetrazol-5-yl)phenyl)benzofuran-5-ylmethyl)-5-chloro-2-(3-oxobutyl)-3*H*-imidazole-4-carboxylic acid
Probably formed from: 3-(3-Bromo-2-(2-(1*H*-tetrazol-5-yl)phenyl)benzofuran-5-ylmethyl)-5-chloro-2-(3-hydroxybutyl)-3*H*-imidazole-4-carboxylic acid

***o*-Bromothioanisole**
Formed from: *o*-Bromothiophenol

***p*-Bromothioanisole**
Formed from: *p*-Bromothiophenol
→ *p*-Bromosulphinylanisole
In: Mortierella D759

***m*-Bromothiophenol**
Probably formed from: 1,3-Dibromobenzene

***cis*-2-Bromo-α,α,α-trifluoro-4,5-dihydro-4,5-dihydroxytoluene**
Formed from: *m*-Bromo-α,α,α-trifluorotoluene

***cis*-6-Bromo-α,α,α-trifluoro-2,3-dihydro-2,3-dihydroxytoluene**
Formed from: *m*-Bromo-α,α,α-trifluorotoluene

***m*-Bromo-α,α,α-trifluorotoluene**
→ *cis*-2-Bromo-α,α,α-trifluoro-4,5-dihydro-4,5-dihydroxytoluene
In: Pseudomonas J536
→ *cis*-6-Bromo-α,α,α-trifluoro-2,3-dihydro-2,3-dihydroxytoluene
In: Pseudomonas J536

***o*-Bromothiophenol**
→ *o*-Bromothioanisole
In: Man D926

***p*-Bromothiophenol**
→ *p*-Bromothioanisole
In: Man D926

2-Bromo-4-trifluoromethylaniline
→ 2-Amino-3-bromo-5-trifluoromethylphenol
In: Rat J301
→ 2-Bromo-4-trifluoromethylphenylhydroxylamine
In: Rat J301
→ 2-Bromo-4-trifluoromethylphenylsulphamate
In: Rat J301

2-Bromo-4-trifluoromethylphenylhydroxylamine
Formed from: 2-Bromo-4-trifluoromethylaniline

2-Bromo-4-trifluoromethylphenylsulphamate
Formed from: 2-Bromo-4-trifluoromethylaniline

3-Bromo-L-tyrosine
Probably formed from: L-Tyrosine
→ 3,5-Dibromo-L-tyrosine
Probably in: Aplysina H696

11-Bromovincamine (11-bromo-14,15-dihydro-14-hydroxyeburnamenine-14-carboxylic acid methyl ester)
→ 11-Bromovincaminic acid
In: Dog C513
Man C513

11-Bromovincaminic acid (11-bromo-14,15-dihydro-14-hydroxyeburnamenine-14-carboxylic acid)
Formed from: 11-Bromovincamine

Bromoxynil
See: 3,5-Dibromo-4-hydroxybenzonitrile

Bromoxynil octanoate (3,5-dibromo-4-hydroxybenzonitrile octanoate ester)
→ 3,5-Dibromo-4-hydroxybenzonitrile

In: Wheat A746
Bromperidol (4-(4-(4-bromophenyl)-4-hydroxy-1-piperidinyl)-1-(4-fluorophenyl)-1-butanone)
→ Bromperidol-β-D-glucuronide
In: Man C622
See also: C343
→ *p*-Fluorobenzoylpropionic acid
In: Rat J740
→ *p*-Fluorophenylacetic acid
In: Dog C343
Man C343
Rat C343
Bromperidol-β-D-glucuronide
Formed from: Bromperidol
Brompheniramine (α-(4-bromophenyl)-N,N-dimethyl-2-pyridinepropanamine)
→ Brompheniramine-N-oxide
In: Cunninghamella H456
→ Demethylbrompheniramine
In: Cunninghamella H456
Brompheniramine-N-oxide
Formed from: Brompheniramine
Bropirimine (2-amino-5-bromo-6-phenyl-4-pyrimidinone)
→ *trans*-2-Amino-5-bromo-6-(3,4-dihydro-3,4-dihydroxyphenyl)-4-pyrimidinone
In: Man J450
→ *m*-Hydroxybropirimine
In: Man J450
→ *p*-Hydroxybropirimine
In: Man J450
Brotizolam (2-bromo-4-(2-chlorophenyl)-9-methyl-6*H*-thieno[3,2-*f*][1,2,4]triazolo[4,3-*a*][1,4]diazepine)
→ 2-Bromo-4-(*o*-chlorophenyl)-9-hydroxymethyl-6*H*-thienyo[3,2-*f*]-1,2,4-triazolo[4,3-*a*]-1,4-diazepine
In: Dog E91
Man E91, J259
Monkey E91
Mouse G186
→ 9-Demethylbrotizolam
Probably in: Mouse G186
→ 6-Hydroxybrotizolam
In: Man E91, J259
Monkey E91
Mouse G186
Rat E91
→ *p*-Hydroxybrotizolam
In: Dog E91
Rat E91
Brucine (2,3-dimethylstrychnidin-10-one)
Formed from: Brucine-N-oxide
→ 16-Hydroxybrucine
Probably in: Microorganism G927
Brucine-N-oxide
→ Brucine
In: Microorganism G927
Rat J876
BSP
See: Sulphobromophthalein
Bucindolol (2-(2-hydroxy-3-((2-(1*H*-indol-3-yl)-1,1-dimethylethyl)amino)propoxy) benzonitrile)
→ 5-Hydroxybucindolol
In: Man D634
Bucromarone (2-(4-(3-(dibutylamino)propoxy)-3,5-dimethylbenzoyl)-4*H*-1-benzopyran-4-one)
→ Debutylbucromarone
In: Mouse F715
Rat F715
→ 2-(4-Hydroxy-3,5-dimethylbenzoyl) chromone
In: Mouse F715
Rat F715
Bucloxic acid
See: 4-(3-Chloro-4-cyclohexylphenyl)-4-oxobutyric acid
Budipine (1-(1,1-dimethylethyl)-4,4-diphenylpiperidine)
→ Budipine-N-oxide
In: Rat G415
→ Hydroxybudipine
In: Rat G415
Budipine-N-oxide
Formed from: Budipine
→ Hydroxybudipine-N-oxide
Probably in: Rat G415
Budralazine (1-(2-(1,3-Dimethyl-2-butenylidene)hydrazino)phthalazine)
→ Hydralazine
In: Rat A1816, A3713
→ 3-Methyl-*s*-triazolo[3,4-*a*]phthalazine

In: Rat A1816, A2544, A3713
→ 1(2*H*)-Phthalazinone
In: Rat A1503, A1816
→ *s*-Triazolo[3,4-*a*]phthalazine
In: Rat A1503, A1816, A2544

Bufotenine
See: 5-Hydroxy-N,N-dimethyltryptamine

Bufuralol (α-(*tert*-butylamino)methyl)-7-ethyl-2-benzofuranmethanol)
→ α-Carboxy-7-ethyl-2-benzofuranmethanol
In: Dog A2806
Man A2806
Rat A2806
→ 1′-Hydroxybufuralol
In: Dog A2806
Man A2806, C270, E283, H775, J762
Mouse K161
Rat A2806, D875, K51
Rhesus monkey H775
→ 2′-Hydroxybufuralol
In: Dog A2806
Man A2806, C572
Rat A2806
→ 4-Hydroxybufuralol
In: Dog A2806
Man A2806, E283
Rat A2806
→ 6-Hydroxybufuralol
In: Dog A2806
Man A2806, E283
Rat A2806

Bulbocapnine
Formed from: Reticuline

Bumetanide (3-(*n*-butylamino)-4-phenoxy-5-sulphamoylbenzoic acid)
→ 3-Amino-4-phenoxy-5-sulphamoylbenzoic acid
In: Man A2335
Rat A1992
→ 3-((3-Hydroxy-*n*-butyl)amino)-4-phenoxy-5-sulphamoylbenzoic acid
In: Rat A1992
→ 3-((4-Hydroxy-*n*-butyl)amino)-4-phenoxy-5-sulphamoylbenzoic acid
In: Man A2335
Rat A1992, C611, D507

Bunazosin (1-(4-amino-6,7-dimethoxy-2-quinazolinyl)hexahydro-4-(1-oxobutyl)-1*H*-1,4-diazepine)
→ 4-Amino-2-(4-butyrylhexahydro-1*H*-1,4-diazepin-1-yl)-6-hydroxy-7-methoxyquinazoline
In: Man G881
→ 4-Amino-2-(4-butyrylhexahydro-1*H*-1,4-diazepin-1-yl)-7-hydroxy-6-methoxyquinazoline
In: Man G881
→ 4-Amino-2-(hexahydro-4-(3-hydroxybutyryl-1*H*-1,4-diazepin-1-yl)-6,7-dimethoxyquinazoline
In: Man G881
→ Bunazosin-N-β-D-glucuronide
In: Man G881

Bunazosin-N-β-D-glucuronide
Formed from: Bunazocin

Bunitrolol (2-(3-((1,1-dimethylethyl)amino)-2-hydroxypropoxy)benzonitrile)
→ 3-(*o*-Cyanophenoxy)lactic acid
In: Mouse B278
→ 3-(*o*-Cyanophenoxy)-1,2-propanediol
In: Mouse B278
→ *o*-Hydroxybenzonitrile
In: Rat B278
→ 4-Hydroxybunitrolol
In: Japanese monkey H850
Man H586
Marmoset H850
Mouse B278
Rat B278, G82, G486, H534

Bunolol
See: 5-(3-*tert*-Butylamino-2-hydroxypropoxy)-1,2,3,4-tetrahydro-1-oxonaphthalene

Bupivacaine (1-butylpipecolo-2′,6′-xylidide)
→ 1-Butyl-3′-hydroxypipecolo-2′,6′-xylidide
In: Man J881
Rat A255
→ 1-Butyl-4′-hydroxypipecolo-2′,6′-xylidide
In: Man J881
Monkey A255
→ 2,6-Dimethylaniline
Probably in: Monkey A255
→ Pipecolo-2′,6′-xylidide

In: Man J881, K14
Monkey A255

Bupranolol (1-(2-chloro-5-methylphenoxy)-3-(1,1-dimethylethyl)amino)-2-propanol)
→ Hydroxybupranolol
In: Man H536

Buprenorphine ((αS,5α,7α)-17-(cyclopropylmethyl)-α-(1,1-dimethylethyl)-4,5-epoxy-18,19-dihydro-3-hydroxy-6-methoxy-α-methyl-6,14-ethenomorphinan-7-methanol)
→ Buprenorphine-β-D-glucuronide
In: Man J74, J416, J783
Monkey H861
Rat A2350, B584, G857, H539, J97a
→ Decyclopropylmethylbuprenorphine
In: Man D382, H950, J783
Rat A2350, A2970, B584, D364, G857

Buprenorphine-β-D-glucuronide
Formed from: Buprenorphine

Bupropion (wellbutrin, 2-*tert*-butylamino-*m*-chloropropiophenone)
→ 2-*tert*-Butylamino-1-(*m*-chlorophenyl)-1-propanol (*erythro* and *threo*)
In: Man E229, D908
Mouse E229
Rat E229
→ *m*-Chlorobenzoic acid
In: Man E229
→ *m*-Chloro-2-(2-(hydroxymethyl)-2-propylamino)propiophenone
In: Man E229, D908
Mouse E229
Rat E229
→ Dihydrobupropion (*erythro* and *threo*)
In: Guinea pig E854
Man D888, H640, K132
→ 3-Hydroxybupropion
In: Man H640
→ *p*-Hydroxybupropion (position unclear)
Probably in: Man H640
→ ω-Hydroxybupropion
In: Man D888, H640, K132, K136
See also: E854

Butachlor (N-(butoxymethyl)-2-chloro-N-(2,6-diethylphenyl)acetamide)
→ 4-Hydroxybutachlor
In: Man G886

Butamoxane (2-(butylaminomethyl)-1,4-benzodioxane)
→ 6-Hydroxybutamoxane
In: Rat A1390
→ 7-Hydroxybutamoxane
In: Rat A1390

Butanilicaine (2-(butylamino-N-(2-chloro-6-methylphenyl)acetamide)
→ 2-Chloro-6-methylaniline
In: Man C677
Rat H465

6-(Butan-2-on-3-yl)-2,4-dihydroxy-3,5-dimethylbenzaldehyde
→ Citrinin
In: Penicillium B560, B694

6-Butanoylmorphine
Formed from: 3,6-Dibutanoylmorphine
→ Morphine
In: Rat E232

Butaperazine (2-butyryl-10-(3-(4-methylpiperazin-1-yl)propyl)phenothiazine)
→ Butaperazine sulphoxide
In: Dog A295
Rat A295
→ Hydroxybutaperazine
In: Dog A295
Rat A295
→ 2-(1-Hydroxybutyl-10-(3-(4-methylpiperazin-1-yl)propyl)phenothiazine
In: Dog A295
Rat A295

Butaperazine sulphone
Probably formed from: Butaperazine sulphoxide

Butaperazine sulphoxide
Formed from: Butaperazine
→ Butaperazine sulphone
Probably in: Dog A295
Rat A295

Butazolidine
See: Phenylbutazone

Butein
See: 2′,3,4,4′-Tetrahydroxychalcone

***p*-(But-1-en-1-yl)phenol**
Formed from: *p*-(*n*-Butyl)phenol

But-2-en-1-yl phenyl sulphide
→ But-2-en-1-yl phenyl sulphoxide

In: Mortierella D78

But-2-en-1-yl phenyl sulphoxide
Formed from: But-2-en-1-yl phenyl sulphide

Butorphanol (*l*-N-cyclobutylmethyl-3,14-dihydroxymorphinan)
→ 3,14-Dihydroxy-N-(3-hydroxycyclobutylmethyl)morphinan
In: Man B112
→ 3,14-Dihydroxymorphinan
In: Man B112

***p*-(*n*-Butoxy)acetanilide**
→ *p*-Acetamidophenol
In: Cunninghamella F590
Mouse A3502
→ *p*-(*n*-Butoxy)aniline
In: Mouse C609
→ *p*-(*n*-Butoxy)-N-hydroxyacetanilide
In: Mouse A3502
→ *p*-(2-Hydroxybutoxy)acetanilide
In: Cunninghamella F590

***p*-(*n*-Butoxy)aniline**
Formed from: *p*-(*n*-Butoxy)acetanilide

1-(3-Butoxy-2-carbamoyloxypropyl)-5-ethyl-5-(4-hydroxy-3-methoxyphenyl)-(1*H*,3*H*,5*H*)-pyridine-2,4,6-trione
Probably formed from: *p*-Hydroxyfebarbamate

***trans*-2-(Butoxycarbonylethyl)-3-(*p*-hydroxyphenyl)-N-phenyllactam**
→ *trans*-2-(Carboxyethyl)-3-(*p*-hydroxyphenyl)-N-phenyllactam
In: Microorganism K110

N-*tert*-Butoxycarbonyl-L-leucyl-L-seryl-L-threonyl-L-arginyl-4-methylcoumarin-7-amide
→ 7-Amino-4-methylcoumarin
In: Rice D743

N-*tert*-Butoxycarbonyl-L-valyl-L-leucyl-L-lysyl-4-methylcoumarin-7-amide
→ 7-Amino-4-methylcoumarin
In: Rice D743

N-*tert*-Butoxycarbonyl-L-valyl-L-prolyl-L-arginyl-4-methylcoumarin-7-amide
→ 7-Amino-4-methylcoumarin
In: Rice D743

4-*n*-Butoxycatechol
Formed from: 4-*n*-Butoxyphenol
→ 4-*n*-Butoxy-*o*-quinone
In: Mushroom H907

7-Butoxycoumarin
→ Umbelliferone
In: Rat C160

2-(4-Butoxy-2,3-dichlorobenzoyl)-5-hydroxythiophene
Formed from: 2-(4-Butoxy-2,3-dichlorobenzoyl)thiophene

2-(4-Butoxy-2,3-dichlorobenzoyl)thiophene
→ 2-(4-Butoxy-2,3-dichlorobenzoyl)-5-hydroxythiophene
In: Man K165

2-Butoxy-N-(2-diethylaminoethyl)-4-quinolinecarboxamide
See: Dibucaine

4-(((Butoxy)ethoxy)ethoxy)methyl-5-propylcatechol
Formed from: Piperonyl butoxide

2-Butoxy-N-(2-ethylamino)ethyl-4-quinolinecarboxamide
Formed from: Dibucaine

***p*-(*n*-Butoxy)-N-hydroxyacetanilide**
Formed from: *p*-(*n*-Butoxy)acetanilide
→ *p*-(*n*-Butoxy)phenylhydroxylamine
In: Mouse C609

N-(Butoxymethyl)-2-chloro-N-(2,6-diethylphenyl)acetamide
See: Butachlor

1-Butoxynaphthalene
→ α-Naphthol
In: Rat B582
→ 1-(2-Hydroxybutoxy)naphthalene
In: Rat B582
→ 1-(3-Hydroxybutoxy)naphthalene
In: Rat B582

***p*-*n*-Butoxyphenol**
→ *p*-*n*-Butoxycatechol
In: Mushroom H907

3-Butoxy-1-phenoxy-2-propanol
See: Febuprol

***p*-(*n*-Butoxy)phenylhydroxylamine**
Formed from: *p*-(*n*-Butoxy)-N-hydroxyacetanilide

4-*n*-Butoxy-*o*-quinone
Formed from: 4-*n*-Butoxycatechol

Butter yellow
See: 4-Dimethylaminoazobenzene

Buturon (N′-(4-chlorophenyl)-N-methyl-N-(1-methyl-2-propynyl)urea)
→ *p*-Chloroaniline
In: Rat A3743
→ *p*-Chlorophenylurea
In: Rat A3743

N-(*n*-Butylalanyl)-*o*-toluidine
→ *o*-Toluidine
In: Mouse A830

Butyl *p*-aminobenzoate
→ Butyl *p*-sulphamidobenzoate
In: Eubacterium H439
Haemophilus H439
Klebsiella H439

4-(Butylamino)benzoic acid 2-(dimethylamino)ethyl ester
See: Tetracaine

1-*tert*-Butylamino-3-(5-carboxy-2-chlorophenoxy)-2-propanol
Formed from: 1-*tert*-Butylamino-3-(2-chloro-5-hydroxymethylphenoxy)-2-propanol

1-*tert*-Butylamino-3-(2-chloro-5-hydroxymethylphenoxy)-2-propanol
Formed from: 1-*tert*-Butylamino-3-(2-chloro-5-methylphenoxy)-2-propanol
→ 1-*tert*-Butylamino-3-(5-carboxy-2-chlorophenoxy)-2-propanol
In: Rat A948
→ 1-*tert*-Butylamino-3-(2-chloro-5-hydroxymethylphenoxy)-2-propanol-β-D-glucuronide
Probably in: Rat A948

1-*tert*-Butylamino-3-(2-chloro-5-hydroxymethylphenoxy)-2-propanol-β-D-glucuronide
Probably formed from: 1-*tert*-Butylamino-3-(2-chloro-5-hydroxymethylphenoxy)-2-propanol

2-Butylamino-N-(2-chloro-6-methylphenyl) acetamide
See: Butanilicaine

1-*tert*-Butylamino-3-(2-chloro-5-methylphenoxy)-2-propanol
→ 1-*tert*-Butylamino-3-(2-chloro-5-hydroxymethylphenoxy)-2-propanol
In: Rat A948

2-*tert*-Butylamino-1-(*m*-chlorophenyl)-1-propanol
Formed from: Bupropion
→ 2-*tert*-Butylamino-1-(*m*-chlorophenyl)-1-propyl-β-D-glucuronide
In: Man E229

2-*tert*-Butylamino-1-(*m*-chlorophenyl)-1-propyl-β-D-glucuronide
Formed from: 2-*tert*-Butylamino-1-(*m*-chlorophenyl)-1-propanol

1-*tert*-Butylamino-3-(2,3-dimethylphenoxy)-2-propanol
→ 1-*tert*-Butylamino-3-(2,3-dimethylphenoxy)-2-propyl-β-D-glucuronide
In: Monkey A2369
Rabbit A2369
Rat A2369
→ 1-*tert*-Butylamino-3-(4-hydroxy-2,3-dimethylphenoxy)-2-propanol
In: Monkey A2369
Rabbit A2369
Rat A2250, A2369
→ 1-*tert*-Butylamino-3-(5-hydroxy-2,3-dimethylphenoxy)-2-propanol
Probably in: Rat A2250
→ 1-*tert*-Butylamino-3-(2-hydroxymethyl-3-methylphenoxy)-2-propanol
In: Monkey A2369
Rabbit A2369
Rat A2369
→ 1-*tert*-Butylamino-3-(3-hydroxymethyl-3-methylphenoxy)-2-propanol
In: Monkey A2369
Rabbit A2369
Rat A2369
→ 2,3-Dimethylphenol
In: Monkey A2369
Rabbit A2369
Rat A2369
→ 2,3-Dimethylphenoxypropan-2,3-diol
In: Monkey A2369
Rabbit A2369
Rat A2369

1-*tert*-Butylamino-3-(2,3-dimethylphenoxy)-2-propyl-β-D-glucuronide
Formed from: 1-*tert*-Butylamino-3-(2,3-dimethylphenoxy)-2-propanol

1-*tert*-Butylamino-3-(4-hydroxy-2,3-dimethylphenoxy)-2-propanol
Formed from: 1-*tert*-Butylamino-3-(2,3-dimethylphenoxy)-2-propanol
→ 1-*tert*-Butylamino-3-(4-hydroxy-2,3-dimethylphenoxy)-2-propyl-β-D-glucuronide
In: Monkey A2369
Rabbit A2369
Rat A2369
→ 1-*tert*-Butylamino-3-(4-hydroxy-3-hydroxymethyl-2-methylphenoxy)-2-propanol
In: Monkey A2369
Rabbit A2369
Rat A2369
→ 4-(3-*tert*-Butylamino-2-hydroxypropoxy)-2,3-dimethylphenyl sulphate
In: Monkey A2369
Rabbit A2369
Rat A2369
→ 2,3-Dimethylquinol
Probably in: Monkey A2369
Rabbit A2369
Rat A2369
→ 3-(4-Hydroxy-2,3-dimethylphenoxy)-1-(2-hydroxymethyl-2-propylamino)-2-propanol
Probably in: Monkey A2369
Rabbit A2369
Rat A2369

1-*tert*-Butylamino-3-(5-hydroxy-2,3-dimethylphenoxy)-2-propanol
Probably formed from: 1-*tert*-Butylamino-3-(2,3-dimethylphenoxy)-2-propanol

1-*tert*-Butylamino-3-(4-hydroxy-2,3-dimethylphenoxy)-2-propyl-β-D-glucuronide
Probably formed from: 1-*tert*-Butylamino-3-(4-hydroxy-2,3-dimethylphenoxy)-2-propanol

1-*tert*-Butylamino-3-(4-hydroxy-3-hydroxymethyl-2-methylphenoxy)-2-propanol
Probably formed from: 1-*tert*-Butylamino-3-(4-hydroxy-2,3-dimethylphenoxy)-2-propanol

1-*tert*-Butylamino-3-(2-hydroxymethyl-3-methylphenoxy)-2-propanol
Formed from: 1-*tert*-Butylamino-3-(2,3-dimethylphenoxy)-2-propanol

1-*tert*-Butylamino-3-(3-hydroxymethyl-2-methylphenoxy)-2-propanol
Formed from: 1-*tert*-Butylamino-3-(2,3-dimethylphenoxy)-2-propanol
→ 1-*tert*-Butylamino-3-(3-hydroxymethyl-2-methylphenoxy)-2-propyl-β-D-glucuronide
Probably in: Monkey A2369
Rabbit A2369
Rat A2369
→ 3-Hydroxy-2-methylbenzyl alcohol
Probably in: Monkey A2369
Rabbit A2369
Rat A2369

1-*tert*-Butylamino-3-(3-hydroxymethyl-2-methylphenoxy)-2-propyl-β-D-glucuronide
Probably formed from: 1-*tert*-Butylamino-3-(3-hydroxymethyl-2-methylphenoxy)-2-propanol

8-(3-*tert*-Butylamino-2-hydroxypropoxy)-5-carboxycoumarin
Formed from: 8-(3-*tert*-Butylamino-2-hydroxypropoxy)-5-hydroxymethylcoumarin

5-(3-*tert*-Butylamino-2-hydroxypropoxy)-3,4-dihydrocarbostyril
See: Carteolol

4-(3-*tert*-Butylamino-2-hydroxypropoxy)-2,3-dimethylphenyl sulphate
Formed from:1-*tert*-Butylamino-3-(4-hydroxy-2,3-dimethylphenoxy)-2-propanol

8-(3-*tert*-Butylamino-2-hydroxypropoxy)-5-hydroxymethylcoumarin
Formed from: 8-(3-*tert*-Butylamino-2-hydroxypropoxy)-5-methylcoumarin
→ 8-(3-*tert*-Butylamino-2-hydroxypropoxy)-5-carboxycoumarin
In: Dog A3406
Rat A3406

8-(3-*tert*-Butylamino-2-hydroxypropoxy)-5-methylcoumarin
→ 8-(3-*tert*-Butylamino-2-hydroxypropoxy)-5-hydroxymethylcoumarin
In: Dog A3406
Rat A3406

5-(3-*tert*-Butylamino-2-hydroxypropoxy)-1,2,3,4-tetrahydro-1-hydroxynaphthalene
Formed from: 5-(3-*tert*-Butylamino-2-hydroxypropoxy)-1,2,3,4-tetrahydro-1-

oxonaphthalene
→ 5-(3-*tert*-Butylamino-2-hydroxypropoxy)-1,2,3,4-tetrahydro-1-hydroxynaphthalene-β-D-glucuronide
Probably in: Man A1711, A3522
Rat A3404
→ 5-(3-*tert*-Butylamino-2-hydroxypropoxy)-1,2,3,4-tetrahydro-1-hydroxynaphthalene sulphate
Probably in: Rat A3404
→ 5-(3-*tert*-Butylamino-2-hydroxypropoxy)-1,2,3,4-tetrahydro-1-oxonaphthalene
In: Man A1333
Rat A1333

5-(3-*tert*-Butylamino-2-hydroxypropoxy)-1,2,3,4-tetrahydro-1-hydroxynaphthalene-β-D-glucuronide
Probably formed from: 5-(3-*tert*-Butylamino-2-hydroxypropoxy)-1,2,3,4-tetrahydro-1-hydroxynaphthalene

5-(3-*tert*-Butylamino-2-hydroxypropoxy)-1,2,3,4-tetrahydro-1-hydroxynaphthalene sulphate
Probably formed from: 5-(3-*tert*-Butylamino-2-hydroxypropoxy)-1,2,3,4-tetrahydro-1-hydroxynaphthalene

5-(3-*tert*-Butylamino-2-hydroxypropoxy)-1,2,3,4-tetrahydro-1-oxonaphthalene (bunolol)
Formed from: 5-(3-*tert*-Butylamino-2-hydroxypropoxy)-1,2,3,4-tetrahydro-1-hydroxynaphthalene
→ 5-(3-*tert*-Butylamino-2-hydroxypropoxy)-1,2,3,4-tetrahydro-1-hydroxynaphthalene
In: Man A1333, A1711, A3369, A3522
Rat A1333, A3404
→ 5-(3-*tert*-Butylamino-2-hydroxypropoxy)-1,2,3,4-tetrahydro-1-oxonaphthalene-β-D-glucuronide
In: Man A1711, A3522
Rat A3404
→ 5-(3-*tert*-Butylamino-2-hydroxypropoxy)-1,2,3,4-tetrahydro-1-oxonaphthalene sulphate
In: Man A3522
Rat A3404
→ Hydroxybunolol
In: Dog A3309
See also: A3369, A3404
→ 1,2,3,4-Tetrahydro-5-hydroxy-1-oxonaphthalene
In: Dog A3045
→ 1,2,3,4-Tetrahydro-1-oxonaphth-5-yloxyacetic acid
In: Dog A3309
Rat A3404
→ β-(1,2,3,4-Tetrahydro-1-oxonaphth-5-yloxy)lactic acid
In: Dog A3404
Rat A3404

5-(3-*tert*-Butylamino-2-hydroxypropoxy)-1,2,3,4-tetrahydro-1-oxonaphthalene-β-D-glucuronide
Formed from: 5-(3-*tert*-Butylamino-2-hydroxypropoxy)-1,2,3,4-tetrahydro-1-oxonaphthalene

5-(3-*tert*-Butylamino-2-hydroxypropoxy)-1,2,3,4-tetrahydro-1-oxonaphthalene sulphate
Formed from: 5-(3-*tert*-Butylamino-2-hydroxypropoxy)-1,2,3,4-tetrahydro-1-oxonaphthalene

2-(Butylaminomethyl)-1,4-benzodioxane
See: Butamoxane

α-(*tert*-Butylamino)methyl)-7-ethyl-2-benzofuranmethanol
See: Bufuralol

4-Butylaminomethyl-7-hydroxycoumarin
Formed from: 4-Butylaminomethyl-7-methoxycoumarin

2-Butylaminomethyl-8-methoxy-1,4-benzodioxane
See: 8-Methoxybutamoxane

4-Butylaminomethyl-7-methoxycoumarin
→ 4-Butylaminomethyl-7-hydroxycoumarin
In Man K339

3-(*n*-Butylamino)-4-phenoxy-5-sulphamoylbenzoic acid
See: Bumetanide

N-(*n*-Butyl)amphetamine
→ Amphetamine
In: Mycobacterium B380
→ N-(*n*-Butyl)-*p*-hydroxyamphetamine
In: Rat B7, G402

N-(*n*-Butyl)aniline
Formed from: N-(*n*-Butyl)-N-methylaniline

N-(*sec*-Butyl)aniline
Formed from: N-*sec*-Butyl-N-methylaniline
N,N-Di-*sec*-butylaniline

***p-tert*-Butylanisole**
→ 2-*tert*-Butylquinol
In: Man D22
***p*-Butylbenzaldehyde**
→ *p*-Butylbenzoic acid
In: Beef C508
***n*-Butylbenzene**
→ 1-Phenyl-1-butanol
In: Rat H689
→ 1-Phenyl-3-butanol
In: Rat H689
***sec*-Butylbenzene**
→ 2-Phenyl-2-butanol
Probably in: Rainbow trout F443
→ 3-Phenyl-1-butanol
Probably in: Rainbow trout F443
***tert*-Butyl benzoate-3,4-oxide**
→ *tert*-Butyl 3,4-dihydro-4-(S-glutathionyl)3-hydroxybenzoate
Probably in: Sheep A2282
(2-Butyl-3-benzofuranyl)(4-(2-(diethylamino)ethoxy)-3,5-diiodophenyl)methanone
See: Amiodarone
***p*-Butylbenzoic acid**
Formed from: *p*-Butylbenzaldehyde
***p-tert*-Butylbenzoic acid**
Probably formed from: *p-tert*-Butylbenzyl alcohol
→ *p-tert*-Butylbenzoyl-β-D-glucuronic acid
In: Rat J549
→ *p-tert*-Butylhippuric acid
Probably in: Rat C764
Contrast: in rat J549
→ 2-(*p*-Carboxyphenyl)-2-methylpropan-1-ol
Probably in: Guinea pig C764
Rat C764
1-Butyl-1*H*-benzotriazole
→ 1-(3-Hydroxybutyl-1-yl)-1*H*-benzotriazole
In: Rat C240
2-Butylbenzotriazole
→ 2-Butyl-4,5-dihydro-4,5-dihydroxybenzotriazole
In: Rat C419
→ 2-Butyl-4-hydroxybenzotriazole
In: Rat C419
→ 2-Butyl-5-hydroxybenzotriazole
In: Rat C419
2-*tert*-Butylbenzotriazole
→ 2-*tert*-Butyl-4,5-dihydro-4,5-dihydroxybenzotriazole
In: Rat C419
→ 2-*tert*-Butyl-4-hydroxybenzotriazole
In: Rat C419
→ 2-*tert*-Butyl-5-hydroxybenzotriazole
In: Rat C419
***p-tert*-Butylbenzoyl-β-D-glucuronic acid**
Formed from: *p-tert*-Butylbenzoic acid
***p-tert*-Butylbenzyl alcohol**
Formed from: *p-tert*-Butylbenzene
→ *p-tert*-Butylbenzoic acid
Probably in: Guinea pig C764
Rat C764
***tert*-Butylbicycloorthobenzoate**
→ 2,2-Bis(hydroxymethyl)-3,3-dimethylbutyl benzoate
In: Mouse E803
→ 2-Formyl-2-hydroxymethyl-3,3-dimethylbutyl benzoate
In: Mouse E803
2-*tert*-Butyl-6-(2-*tert*-butyl-4-methoxyphenoxy)-4-methoxyphenol
Formed from: 2-*tert*-Butyl-4-methoxyphenol
Butylcaptax (2-(butylthio)benzothiazole)
→ 2-Hydroxyamylthiobenzothiazole
In: Cotton A723
***p*-(*tert*-Butylcarbonyloxy)benzaldehyde**
→ *p*-(*tert*-Butylcarbonyloxy)mandelonitrile (both isomers)
In: Almond J194, J302
Apple J302
Cherry J302
Sorghum J194
***p*-(*tert*-Butylcarbonyloxy)mandelonitrile**
Formed from: *p*-(*tert*-Butylcarbonyloxy)benzaldehyde
5-(2(*S*)-N-(*tert*-Butyl)carboxamido)piperazinyl)-N-(2(*R*),3-dihydroxy-1(*S*)-indanyl)-4(*S*)-hydroxy-2(*R*)-phenylmethyl)pentanamide
Probably formed from: 5-(2(*S*)-N-(*tert*-Butyl)carboxamido)piperazinyl)-4(*S*)-hydroxy-N-(2(*R*)-hydroxy-1(*S*)-indanyl)-2(*R*)-phenylmethyl)pentanamide

5-((2(*S*)-N-(*tert*-Butyl)carboxamido)piperazinyl)-4(*S*)-hydroxy-N-(2(*R*)-hydroxy-1(*S*)-indanyl)-2(*R*)-phenylmethyl)pentanamide
Formed from: 5-((2(*S*)-N-(*tert*-Butyl) carboxamido)-4-(3-pyridylmethyl) piperazinyl)-4-(*S*)-hydroxy-N-(2(*R*)-hydroxy-1(*S*)-indanyl)-2(*R*)-phenylmethyl) pentanamide
→ 5-((2(*S*)-N-(*tert*-Butyl)carboxamido) piperazinyl)-N-(2(*R*),3-dihydroxy-1(*S*)-indanyl)-4(*S*)-hydroxy-2(*R*)-phenyl-methyl)pentanamide
Probably in: Dog H538
Man H538
Rat H538

5-((2(*S*)-N-(*tert*-Butyl)carboxamido)-4-(3-pyridylmethyl)piperazinyl)-4-(*S*)-hydroxy-N-(2(*R*)-hydroxy-1(*S*)-indanyl)-2(*R*)-phenylmethyl)pentanamide
→ 5-((2(*S*)-N-(*tert*-Butyl)carboxamido)piperazinyl)-4(*S*)-hydroxy-N-(2(*R*)-hydroxy-1(*S*)-indanyl)-2(*R*)-phenylmethyl)pentanamide
In: Man H538
→ 5-((2(*S*)-N-(*tert*-Butyl)carboxamido)-4-(3-pyridyl-N-oxide-methyl)piperazinyl)-N-(2(*R*),3-dihydroxy-1(*S*)-indanyl)-4(*S*)-hydroxy-2(*R*)-phenylmethyl)pentanamide
In: Man H538

5-((2(*S*)-N-(*tert*-Butyl)carboxamido)-4-(3-pyridyl-N-oxide-methyl)piperazinyl)-N-(2(*R*),3-dihydroxy-1(*S*)-indanyl)-4(*S*)-hydroxy-2(*R*)-phenylmethyl) pentanamide
Formed from: 5-((2(*S*)-N-(*tert*-Butyl) carboxamido)-4-(3-pyridylmethyl)piperazinyl)-4(*S*)-hydroxy-N-(2(*R*)-hydroxy-1(*S*)-indanyl)-2(*R*)-phenylmethyl)pentanamide

2-*n*-Butyl-5-carboxy-4-chloro-1-((2′-(1*H*-tetrazol-5-yl)-1,1′-biphenyl-4-yl)-1*H*-imidazole
Formed from: 2-*n*-Butyl-4-chloro-5-formyl-1-((2′-(1*H*-tetrazol-5-yl)-1,1′-biphenyl-4-yl)-1*H*-imidazole

2-*n*-Butyl-5-carboxy-4-chloro-1-((2′-(1*H*-tetrazol-5-yl)-1,1′-biphenyl-4-yl)methyl)-1*H*-imidazole
Formed from: 2-*n*-Butyl-4-chloro-5-hydroxymethyl-1-((2′-(1*H*-tetrazol-5-yl)-1,1′-biphenyl-4-yl)methyl)-1*H*-imidazole
→ 2-*n*-Butyl-5-carboxy-4-chloro-1-((2′-(1*H*-tetrazol-5-yl)-1,1′-biphenyl-4-yl)methyl)-1*H*-imidazole-β-D-glucuronide
Probably in: Man G250
Monkey G250
Rat G250

2-*n*-Butyl-5-carboxy-4-chloro-1-((2′-(1*H*-tetrazol-5-yl)-1,1′-biphenyl-4-yl)methyl)-1*H*-imidazole-β-D-glucuronide
Probably formed from: 2-*n*-Butyl-5-carboxy-4-chloro-1-((2′-(1*H*-tetrazol-5-yl)-1,1′-biphenyl-4-yl)methyl)-1*H*-imidazole

1-Butyl-3-(*p*-carboxyphenylsulphonyl)urea
Probably formed from: 1-Butyl-3-(*p*-hydroxymethylphenylsulphonyl)urea

4-Butyl-4-(β-carboxypropionyloxymethyl)-1,2-diphenyl-3,5-pyrazolidinedione
See: Suxibuzone

3-*sec*-Butylcatechol
Formed from: *o*-*sec*-Butylphenol

4-*sec*-Butylcatechol
Formed from: 4-*sec*-Butyl-2-nitrophenol

3-*tert*-Butylcatechol
Formed from: *o*-*tert*-Butylphenol

4-*tert*-Butylcatechol
Probably formed from: 4-*tert*-Butylphenol
→ 4-*tert*-Butylcatechol-β-D-glucuronide
In: Mouse J799
Rat J799
→ 4-*tert*-Butylcatechol sulphate
In: Mouse J799
Rat J799
→ 4-*tert*-Butyl-2-methoxyphenol
Probably in: Mouse J799
Rat J799
→ 4-*tert*-Butyl-*o*-quinone
In: Mushroom B897

4-*tert*-Butylcatechol-β-D-glucuronide
Formed from: 4-*tert*-Butylcatechol

4-*tert*-Butylcatechol sulphate
Formed from: 4-*tert*-Butylcatechol

N-Butyl-*p*-chlorobenzamide
Formed from: N-Butyl-*p*-chloro-N-methylbenzamide

4-*tert*-Butyl-(3-(2-chloroethyl)ureido)benzene
→ N-(*p*-(2-Hydroxy-1,1-dimethylethyl) phenyl)urea

In: Mouse H936

2-Butyl-4-chloro-5-formyl-1-(2′-(1*H*-tetrazol-5-yl)-1,1′-biphenyl-4-yl)-1*H*-imidazole
→ 2-Butyl-5-carboxy-4-chloro-1-(2′-(1*H*-tetrazol-5-yl)-1,1′-biphenyl-4-yl)-1*H*-imidazole
In: Man H146

N-*tert*-Butyl-2-chloro-4-hydroxy-5-methoxyphenylethanolamine
Formed from: N-*tert*-Butyl-*o*-chlorophenylethanolamine

2-*n*-Butyl-4-chloro-5-hydroxymethyl-1-((2′-(1*H*-tetrazol-5-yl)-1,1′-biphenyl-4-yl)hydroxymethyl)-1*H*-imidazole
Formed from: 2-*n*-Butyl-4-chloro-5-hydroxymethyl-1-((2′-(1*H*-tetrazol-5-yl)-1,1′-biphenyl-4-yl)methyl)-1*H*-imidazole

2-*n*-Butyl-4-chloro-5-hydroxymethyl-1-((2′-(1*H*-tetrazol-5-yl)-1,1′-biphenyl-4-yl)methyl)-1*H*-imidazole (losartan)
→ 2-*n*-Butyl-5-carboxy-4-chloro-1-((2′-(1*H*-tetrazol-5-yl)-1,1′-biphenyl-4-yl) hydroxymethyl)-1*H*-imidazole
In: Man G250, H146, H193
Monkey G250
Rat G250
→ 2-*n*-Butyl-4-chloro-5-hydroxymethyl-1-((2′-(1*H*-tetrazol-5-yl)-1,1′-biphenyl-4-yl)methyl)-1*H*-imidazole-β-D-glucuronide
In: Man G250
Monkey G250
Rat G250
→ 2-*n*-Butyl-4-chloro-5-hydroxymethyl-1-((2′-(1*H*-tetrazol-5-yl)-1,1′-biphenyl-4-yl) hydroxymethyl)-1*H*-imidazole
In: Man G250, H153
Monkey G250
Rat G250
→ 4-Chloro-2-(1-hydroxy-*n*-butyl)-5-hydroxymethyl-1-((2′-(1*H*-tetrazol-5-yl)-1,1′-biphenyl-4-yl)methyl)-1*H*-imidazole
In: Man G250
Monkey G250
Rat G250
→ 4-Chloro-2-(3-hydroxy-*n*-butyl)-5-hydroxymethyl-1-((2′-(1*H*-tetrazol-5-yl)-1,1′-biphenyl-4-yl)methyl)-1*H*-imidazole
In: Man G250
Monkey G250
Rat G250

2-*n*-Butyl-4-chloro-5-hydroxymethyl-3-((2-(1*H*-tetrazol-5-yl)-1,1′-biphenyl-4-yl)methyl)-1*H*-imidazole
→ 2-*n*-Butyl-4-chloro-5-hydroxymethyl-3-((2-(1*H*-tetrazol-5-yl)-1,1′-biphenyl-4-yl)methyl)-1*H*-imidazole-β-D-glucuronide
In: Dog G693
Monkey G47, G693

2-*n*-Butyl-4-chloro-5-hydroxymethyl-1-((2′-(1*H*-tetrazol-5-yl)-1,1′-biphenyl-4-yl)methyl)-1*H*-imidazole-β-D-glucuronide
Formed from: 2-*n*-Butyl-4-chloro-5-hydroxymethyl-1-((2′-(1*H*-tetrazol-5-yl)-1,1′-biphenyl-4-yl)methyl)-1*H*-imidazole

2-*n*-Butyl-4-chloro-5-hydroxymethyl-3-((2′-(1*H*-tetrazol-5-yl)-1,1′-biphenyl-4-yl)methyl)-1*H*-imidazole-β-D-glucuronide
Formed from: 2-*n*-Butyl-4-chloro-5-hydroxymethyl-3-((2′-(1*H*-tetrazol-5-yl)-1,1′-biphenyl-4-yl)methyl)-1*H*-imidazole

N-Butyl-*p*-chloro-N-methylbenzamide
→ *p*-Chloro-N-methylbenzamide
In: Rat G13
→ N-Butyl-*p*-chlorobenzamide
In: Rat G13

***n*-Butyl 2-chloro-5-nitrobenzoate**
→ *n*-Butyl 2-(S-glutathionyl)-5-nitrobenzoate
In: Rat H516

***tert*-Butyl 2-chloro-5-nitrobenzoate**
→ *tert*-Butyl 2-(S-glutathionyl)-5-nitrobenzoate
In: Rat H516

4-*tert*-Butyl-2-chlorophenol
Formed from: Crufomate
→ 4-*tert*-Butyl-2-chlorophenyl-β-D-glucuronide
In: Rat A3338, A3351
Sheep A3001
→ 2-Chloro-4-(2-hydroxy-1,1-dimethylethyl)phenol
In: Rat A3338, A3351

N-*tert*-Butyl-*o*-chlorophenylethanolamine
→ N-*tert*-Butyl-2-chloro-4-hydroxy-5-methoxyphenylethanolamine

In: Rat B776
→ *o*-Chloromandelaldehyde
Probably in: Dog B776
Guinea pig B776
Rabbit B776

4-*tert*-Butyl-2-chlorophenyl-β-D-glucuronide
Formed from: 4-*tert*-Butyl-2-chlorophenol

4-*tert*-Butyl-2-chlorophenyl methyl methylphosphphoramidate
See: Crufomate

4-*tert*-Butyl-2-chlorophenyl methyl phosphate
Formed from: Crufomate

4-*tert*-Butyl-2-chlorophenyl methylphosphphoramidate
Formed from: Crufomate
→ 4-*tert*-Butyl-2-chlorophenyl phosphate
Probably in: Sheep A2995

4-*tert*-Butyl-2-chlorophenyl phosphate
Probably formed from: 4-*tert*-Butyl-2-chlorophenyl methylphosphphoramidate

2-(4-*tert*-Butylcyclohexyl)-3-hydroxy-1,4-naphthoquinone
→ 3-Hydroxy-2-(4-(1-hydroxymethyl isopropyl)cyclohexyl)-1,4-naphthoquinone
In: Man G985

2-Butyl-4,5-dihydro-4,5-dihydroxybenzotriazole
Formed from: 2-Butylbenzotriazole

2-*tert*-Butyl-4,5-dihydro-4,5-dihydroxybenzotriazole
Formed from: 2-*tert*-Butylbenzotriazole

***tert*-Butyl 3,4-dihydro-4-(S-glutathionyl)-3-hydroxybenzoate**
Probably formed from: *tert*-Butyl benzoate-3,4-oxide

7-*tert*-Butyl-2,5-dihydroxy-3,3-dimethylcoumaran
Probably formed from: 7-*tert*-Butyl-5-hydroxy-3,3-dimethylcoumaran

N-Butyl-3,4-dihydroxyphenethylamine (bamethan)
→ N-Butyl-3,4-dihydroxyphenethylamine-β-D-glucuronide
In: Rat D344

N-Butyl-3,4-dihydroxyphenethylamine-β-D-glucuronide
Formed from: N-Butyl-3,4-dihydroxyphenethylamine

N-(*tert*-Butyl)-4-(3,4-dihydroxyphenyl)-4-phenyl-2-butylamine
Probably formed from: N-(*tert*-Butyl)-4-(*p*-hydroxyphenyl)-4-phenyl-2-butylamine
→ N-(*tert*-Butyl)-4-(4-hydroxy-3-methoxyphenyl)-4-phenyl-2-butylamine
Probably in: Dog D784, F677

4-*tert*-Butyl-2,6-dimethyl-3,5-dinitroacetanilide
Formed from: 4-*tert*-Butyl-2,6-dimethyl-3,5-dinitroaniline

4-*tert*-Butyl-2,6-dimethyl-3,5-dinitroaniline
Formed from: Musk xylene
→ 4-*tert*-Butyl-2,6-dimethyl-3,5-dinitroacetanilide
In: Rat K13

***sec*-Butyl-4,6-dinitrophenol** (dinoseb)
→ 6-Amino-2-*sec*-butyl-4-nitrophenol
In: Azotobacter A3824

N-(*tert*-Butyl)-4,4-diphenyl-2-butylamine
See: Terodiline

4′-*tert*-Butyl-4-(4-(diphenylmethoxy) piperidino)butyrophenone
See: Ebastine

4-Butyl-1,2-diphenyl-3,5-pyrazolidinedione
See: Phenylbutazone

17-O-Butyl-7-ethylcamptothecin-21-(2-dimethylamino)ethylamide
→ 7-Ethylcamptothecin-21-(2-dimethylamino)ethylamide
In: Rat H632

17-O-*tert*-Butyl-7-ethylcamptothecin-21-(2-dimethylamino)ethylamide
→ 7-Ethylcamptothecin-21-(2-dimethylamino)ethylamide
In: Rat H632

N-(*tert*-Butyl)-3-(*o*-fluorobenzoyloxy)-2-hydroxypropylamine
→ *o*-Fluorobenzoic acid
In: Rabbit H563

Butyl gallate
→ Butyl gallate-O-sulphate
In: Eubacterium G475

Butyl gallate-O-sulphate
Formed from: Butyl gallate

***n*-Butyl L-γ-glutamyl dopamide**
→ 3,4-Dihydroxyphenethylamine
In: Dog A3945

n-Butyl 2-(S-glutathionyl)-5-nitrobenzoate
Formed from: *n*-Butyl 2-chloro-5-nitrobenzoate

***tert*-Butyl 2-(S-glutathionyl)-5-nitrobenzoate**
Formed from: *tert*-Butyl 2-chloro-5-nitrobenzoate

2-*tert*-Butyl-5-(S-glutathionyl)quinol
Formed from: 2-*tert*-Butylquinol

2-*tert*-Butyl-6-(S-glutathionyl)quinol
Formed from: 2-*tert*-Butylquinol

***p*-*tert*-Butylhippuric acid**
Probably formed from: *p*-*tert*-Butylbenzoic acid

N-(*n*-Butyl)-*p*-hydroxyamphetamine
Formed from: N-(*n*-Butyl)amphetamine

N-(*n*-Butyl)-*p*-methoxyamphetamine
→ N-(*n*-Butyl)-4-hydroxy-3-methoxyamphetamine
Probably in: Rat A1998, B7, G402

2-Butyl-4-hydroxybenzotriazole
Formed from: 2-Butylbenzotriazole

2-Butyl-5-hydroxybenzotriazole
Formed from: 2-Butylbenzotriazole

2-*tert*-Butyl-4-hydroxybenzotriazole
Formed from: 2-*tert*-Butylbenzotriazole

2-*tert*-Butyl-5-hydroxybenzotriazole
Formed from: 2-*tert*-Butylbenzotriazole

N-*tert*-Butyl-3-hydroxy-5-dimethylcarbamoyloxyphenylethanolamine
Formed from: Bambuterol
→ Terbutaline
In: Dog F128
Guinea pig F128
Man F128
Mouse F128
Rabbit F128
Rat F128

7-*tert*-Butyl-5-hydroxy-3,3-dimethylcoumaran
Probably formed from: 2-*tert*-Butyl-6-(2-hydroxy-1,1-dimethylethyl)quinol
→ 7-*tert*-Butyl-3,5-dihydroxy-3,3-dimethylcoumaran
Probably in: Dog A1079
Man A1079
Rat A1079

2-*tert*-Butyl-6-(2-hydroxy-1,1-dimethylethyl) quinol
Formed from: 2,6-Di-*tert*-butylquinol
→ 7-*tert*-Butyl-5-hydroxy-3,3-dimethylcoumaran
Probably in: Dog A1079
Man A1079
Rat A1079

N-*tert*-Butyl-4-hydroxy-4,4-diphenyl-2-butylamine
Formed from: Terodiline

Butyl 9-hydroxyfluorenyl-9-carboxylate
→ 3-Hydroxybutyl 9-hydroxyfluorenyl-9-carboxylate
In: Hordeum A1509
→ 9-Hydroxyfluorenyl-9-carboxylic acid
In: Hordeum A1509

2-*tert*-Butyl-4-hydroxy-6-(2-(hydroxymethyl) isopropyl)-4-methylcyclohexa-2,5-dienone
Probably formed from: 2,6-Di-*tert*-butyl-4-hydroxy-4-methylcyclohexa-2,5-dienone

N-*tert*-Butyl-4-hydroxy-3-hydroxymethylphenyl ethanolamine (albuterol, salbutamol)
→ Salbutamol-β-D-glucuronide
In: Beef D344
Rat J580
→ Salbutamol sulphate
In: Man A2082, C322, C523, G523, J91, J862

N-(*n*-Butyl)-4-hydroxy-3-methoxyamphetamine
Probably formed from: N-(*n*-Butyl)-*p*-hydroxyamphetamine

***n*-Butyl-3-hydroxy-4-methoxyphenylethanolamine**
Formed from: *n*-Butylnoradrenaline

***n*-(2-Butyl)-3-hydroxy-4-methoxyphenylethanolamine**
Formed from: *n*-(2-Butyl)noradrenaline

N-(*tert*-Butyl)-4-(4-hydroxy-3-methoxyphenyl)-4-phenyl-2-butylamine
Probably formed from: N-*tert*-Butyl-3,4-dihydroxyphenyl)-4-phenyl-2-butylamine

***p*-*tert*-Butyl-N-hydroxymethylbenzamide**
Formed from: *p*-*tert*-Butyl-N-methylbenzamide

6-*tert*-Butyl-2-(1-hydroxymethyl)isopropyl-4-methylene-2,5-cyclohexadienone
Formed from: 2,6-Di-*tert*-butyl-4-methylene-2,5-cyclohexadienone

6-*tert*-Butyl-2-(1-hydroxymethyl)isopropyl-4-methylphenol
Formed from: 2,6-Di-*tert*-butyl-4-methylphenol

1-Butyl-3-(*p*-hydroxymethylphenylsulphonyl)urea
Formed from: Tolbutamide
→ 1-Butyl-3-(*p*-carboxyphenylsulphonyl)urea
Probably in: Rabbit C826

N-(*tert*-Butyl)-*o*-hydroxyphenylethanolamine
See: Tulobuterol

N-(*tert*-Butyl)-4-(*p*-hydroxyphenyl)-4-phenyl-2-butylamine
Formed from: Terodiline
→ N-(*tert*-Butyl)-4-(3,4-dihydroxyphenyl)-4-phenyl-2-butylamine
Probably in: Dog D784, F677
Rat D774
→ 1-(*p*-Hydroxyphenyl)-1-phenyl-3-butanone
Probably in: Dog D784

1-Butyl-3′-hydroxypipecolo-2′,6′-xylidide
Formed from: Bupivacaine

1-Butyl-4′-hydroxypipecolo-2′,6′-xylidide
Formed from: Bupivacaine

Butylidenephthalide
→ S-(L-Cysteinyl)butylidenephthalide
In: Mouse K644

N-(*n*-Butyl)-*p*-methoxyamphetamine
→ N-(*n*-Butyl)-*p*-hydroxyamphetamine
In: Man J727

3-(*tert*-Butyl)-5-methoxycatechol
Formed from: 3-(*tert*-Butyl)-5-methoxyphenol

4-(*tert*-Butyl)-5-methoxycatechol
Formed from: 3-(*tert*-Butyl)-4-methoxyphenol
→ 4-*tert*-Butyl-5-methoxy-*o*-quinone
Probably in: Rat F35

2-*tert*-Butyl-4-methoxyphenol
→ 2-*tert*-Butyl-6-(2-*tert*-butyl-4-methoxyphenoxy)-4-methoxyphenol
In: Horseradish G503
→ 3,3′-Di-*tert*-butyl-2,2′-dihydroxy-5,5′-dimethoxybiphenyl
In: Horseradish A3956, G503
Rat A3956, C584, D476
→ 2-*tert*-Butyl-4-methoxycatechol
In Rat D587, G39
→ 2-*tert*-Butyl-4-methoxyphenyl-β-D-glucuronide
In: Man G524
Rat F35
→ 2-*tert*-Butylquinol
In: Rat D883, F35
See also: G39, G503

3-*tert*-Butyl-4-methoxyphenol
→ 3-*tert*-Butyl-4-methoxyphenyl-β-D-glucuronide
In: Man J490
Rat F35
→ 3-*tert*-Butyl-4-methoxyphenyl sulphate
In: Rat F35

4-*tert*-Butyl-2-methoxyphenol
Probably formed from: 4-*tert*-Butylcatechol

4-*tert*-Butyl-3-methoxyphenol
Probably formed from: 4-*tert*-Butylcatechol

2-*tert*-Butyl-4-methoxyphenyl-β-D-glucuronide
Formed from: 2-*tert*-Butyl-4-methoxyphenol

3-*tert*-Butyl-4-methoxyphenyl-β-D-glucuronide
Formed from: 3-*tert*-Butyl-4-methoxyphenol

3-*tert*-Butyl-4-methoxyphenyl sulphate
Probably formed from: 3-*tert*-Butyl-4-methoxyphenol

5-*tert*-Butyl-2-methoxyquinol
Formed from: 5-*tert*-Butyl-2-methoxyquinone

5-*tert*-Butyl-2-methoxyquinone
→ 5-*tert*-Butyl-2-methoxyquinol
In: Sporotrichum B301

4-*tert*-Butyl-5-methoxy-*o*-quinone
Probably formed from: 4-*tert*-Butyl-5-methoxycatechol

2-*tert*-Butyl-6-methylaniline
→ 2-*tert*-Butyl-6-methylnitrosobenzene
In: Rat D943

N-*n*-Butyl-N-methylaniline
→ N-*n*-Butylaniline
In: Rabbit A3484
→ N-*n*-Butyl-N-methylaniline-N-oxide
In: Rabbit A3484
→ N-Methylaniline
In: Rabbit A3484

N-*sec*-Butyl-N-methylaniline
→ N-*sec*-Butylaniline
In: Rabbit A3484
→ N-*sec*-Butyl-N-methylaniline-N-oxide
In: Rabbit A3484

N-*n*-Butyl-N-methylaniline-N-oxide
Formed from:N-*n*-Butyl-N-methylaniline
N-*sec*-Butyl-N-methylaniline-N-oxide
Formed from: N-*sec*-Butyl-N-methylaniline
***p-tert*-Butyl-N-methylbenzamide**
→ *p-tert*-Butyl-N-hydroxymethylbenzamide
In: Mouse C319
2-*tert*-Butyl-6-methylnitrosobenzene
Formed from: 2-*tert*-Butyl-6-methylaniline
2-*tert*-Butyl-5-methylthioquinol
Formed from: 2-*tert*-Butylquinol
2-*tert*-Butyl-5-methylthioquinol
Formed from: 2-*tert*-Butylquinol
2-*tert*-Butyl-6-methylthioquinol
Formed from: 2-*tert*-Butylquinol
4-*sec*-Butyl-2-nitrophenol
→ 4-*sec*-Butylcatechol
In: Pseudomonas E607, H65
Butyl *p*-nitrophenyl ether
→ 3-Hydroxybutyl *p*-nitrophenyl ether
In: Rabbit A3047
→ 4-Hydroxybutyl *p*-nitrophenyl ether
In: Rabbit A3047
→ *p*-Nitrophenol
In: Rat A119
N-Butylnoradrenaline
→ N-Butyl-3-hydroxy-4-methoxyphenylethanolamine
In: Rat A486
→ N-Butylnormetanephrine
In: Rat A486
N-(2-Butyl)noradrenaline
→ N-(2-Butyl)-3-hydroxy-4-methoxyphenylethanolamine
In: Rat A486
→ N-(2-Butyl)normetanephrine
In: Rat A486
N-(*tert*-Butyl)noradrenaline (colterol)
Formed from: Bitolterol
N-(*tert*-Butyl)noradrenaline-β-D-glucuronide
→ N-(*tert*-Butyl)normetanephrine
In: Rat A486
N-(*tert*-Butyl)noradrenaline-β-D-glucuronide
Formed from: N-(*tert*-Butyl)noradrenaline
N-*tert*-Butylnorchlorcyclizine
→ N-(β-Hydroxy-*tert*-butyl)norchlorcyclizine
In: Rat A573
→ Norchlorcyclizine
In: Rat A573
N-Butylnormetanephrine
Formed from: N-Butylnoradrenaline
N-(2-Butyl)normetanephrine
Formed from: N-(2-Butyl)noradrenaline
N-(*tert*-Butyl)normetanephrine
Formed from: N-(*tert*-Butyl)noradrenaline
→ N-(*tert*-Butyl)normetanephrine-β-D-glucuronide
Probably in: Rat A3718
N-(*tert*-Butyl)normetanephrine-β-D-glucuronide
Probably formed from: N-(*tert*-Butyl) normetanephrine
***cis-tert*-Butyloxycarbonyl-2,3-dihydro-2,3-dihydroxy-L-phenylalanine**
Formed from: *tert*-Butyloxycarbonyl-L-phenylalanine
→ *tert*-Butyloxycarbonyl-2,3-dihydroxy-L-phenylalanine
In: Bacterium B871
***tert*-Butyloxycarbonyl-2,3-dihydroxy-L-phenylalanine**
Formed from: *cis-tert*-Butyloxycarbonyl-2,3-dihydro-2,3-dihydroxy-L-phenylalanine
N-*tert*-Butyloxycarbonyl-3,4-dihydroxy-L-phenylalanine
Formed from: N-*tert*-Butyloxycarbonyl-L-tyrosine
N^2-(Butyloxycarbonyl)-N^3-(2-(3-formamidinophenyl)-4-hydroxyisoxazolin-5(*R*)-yl)acetyl)-2,3-diaminopropionic acid
Formed from: Demethylroxifiban
N^2-(Butyloxycarbonyl)-N^3-(2-(3-(4-formamidinophenyl)-5-hydroxyisoxazolin-5(*R*)-yl)acetyl)-2,3-diaminopropionic acid
Formed from: Demethylroxifiban
***tert*-Butyloxycarbonyl-L-phenylalanine**
→ *cis-tert*-Butyloxycarbonyl-2,3-dihydro-2,3-dihydroxy-L-phenylalanine
In: Bacterium B871
N-*tert*-Butyloxycarbonyl-L-tyrosine
→ N-*tert*-Butyloxycarbonyl-3,4-dihydroxy-L-phenylalanine
In: Aspergillus A2771

Butylphen
See: *p-tert*-Butylphenol
***p-n*-Butylphenol**
→ *p*-(But-1-en-1-yl)phenol
In: Penicillium J836
→ *p-n*-Butylphenyl sulphate
In: Man K405
→ 1-(*p*-Hydroxyphenyl)-*n*-butanol
In: Pseudomonas F381
***o-sec*-Butylphenol**
Formed from: *o-sec*-Butylphenyl N-methylcarbamate
→ 3-*sec*-Butylcatechol
In: Pseudomonas E958, J392
***p-sec*-Butylphenol**
→ *p-sec*-Butylphenyl sulphate
In: Man K405
→ 2-(*p*-Hydroxyphenyl)but-1-ene
In: Penicillium J836
→ 2-(*p*-Hydroxyphenyl)but-2-ene
In: Penicillium J836
***o-tert*-Butylphenol**
→ 3-*tert*-Butylcatechol
In: Pseudomonas E958
***m-tert*-Butylphenol**
→ *m-tert*-Butylphenyl sulphate
In: Man E575
Rat E575
***p-tert*-Butylphenol** (butylphen)
→ 4-*tert*-Butylcatechol
Probably in: Mushroom B897
→ *p-tert*-Butylphenyl-β-D-glucuronide
In: Man J490
Rat B644
→ *p-tert*-Butylphenyl sulphate
In: Man E575, G524, K405
Rat B644, E575
***o-sec*-Butylphenyl carbamate**
Formed from: *o-sec*-Butylphenyl N-methylcarbamate
***p-tert*-Butylphenyl diphenyl phosphate**
→ *p*-(2-Carboxy-2-propyl)phenyl diphenyl phosphate
In: Cunninghamella D627
→ Diphenyl phosphate
In: Cunninghamella D627
→ Phenol
In: Cunninghamella D627
***p-tert*-Butylphenyl-β-D-glucuronide**
Formed from: *p-tert*-Butylphenol
1-(4-*tert*-Butylphenyl)-4-(4-(hydroxy(diphenyl)methyl)piperidin-1-yl)-1-butanone
Formed from: Terfenadine
***o-sec*-Butylphenyl N-hydroxymethylcarbamate**
Formed from: *o-sec*-Butylphenyl N-methylcarbamate
***tert*-Butyl phenyl ketone**
→ 2,2-Dimethyl-1-phenylpropanol
In: Bacillus G770
Candida G770
Corynebacterium G770
Micrococcus G770
Nocardia G770
Pichia G770
Rhodotorula G770
Saccharomyces G770
Torulopsis G770
Xanthomonas G770
***o-sec*-Butylphenyl N-methylcarbamate**
→ *o-sec*-Butylphenol
In: Aspergillus A1811, A1813
Cladosporium A1811, A1813
Coniothyrium A1813
Fusarium A1813
Penicillium A1813
→ *o-sec*-Butylphenyl carbamate
In: Aspergillus A1811, A1813
Cladosporium A1811, A1813
Coniothyrium A1813
Fusarium A1813
Penicillium A1813
→ *o-sec*-Butylphenyl N-hydroxymethylcarbamate
In: Aspergillus A1811, A1813
Cladosporium A1811, A1813
Coniothyrium A1813
Fusarium A1813
Penicillium A1813
→ *o*-(1-Hydroxy-1-methylpropyl)phenyl N-methylcarbamate
In: Aspergillus A1811
→ *erythro-o*-(2-Hydroxy-1-methylpropyl) phenyl N-methylcarbamate

In: Aspergillus A1811
See also: A1812, A1813
→ *threo-o*-(2-Hydroxy-1-methylpropyl)phenyl N-methylcarbamate
In: Aspergillus A1811
See also: A1813
→ *o*-(3-Hydroxy-1-methylpropyl)phenyl N-methylcarbamate
In: Aspergillus A1811
See also: A1813, A1815
→ *o*-(1-Hydroxymethylpropyl)phenyl N-methylcarbamate
In: Aspergillus A1811

***p*-*n*-Butylphenyl sulphate**
Formed from: *p*-*n*-Butylphenol

***p*-*sec*-Butylphenyl sulphate**
Formed from: *p*-*sec*-Butylphenol

***m*-*tert*-Butylphenyl sulphate**
Formed from: *m*-*tert*-Butylphenol

***p*-*tert*-Butylphenyl sulphate**
Formed from: *p*-*tert*-Butylphenol

(*S*)-3-*n*-Butylphthalide
Formed from: *o*-*n*-Pentylbenzoic acid

1-Butylpipecolo-2′,6′-xylidide
See: Bupivacaine

2-*tert*-Butylquinol
Formed from: *p*-*tert*-Butylanisole
2-*tert*-Butyl-4-methoxyphenol
→ 2-*tert*-Butyl-5-(S-glutathionyl)quinol
In: Rat G39
→ 2-*tert*-Butyl-6-(S-glutathionyl)quinol
In: Rat G39
→ 2-*tert*-Butyl-5-methylthiophenol
In: Rat G39
→ 2-*tert*-Butyl-6-methylthiophenol
In: Rat G39
→ 2-*tert*-Butylquinol sulphate
Probably in: Rat F35
→ 2-*tert*-Butylquinone
In: Horseradish G503
Rat D883
→ *tert*-Butyl semiquinone radical
In: Rat G666

2-*tert*-Butylquinol sulphate
Formed from: 2-*tert*-Butylquinol
2-*tert*-Butylquinone
Formed from: 2-*tert*-Butylquinol
→ *tert*-Butyl semiquinone radical
In: Rat G666

4-*tert*-Butylquinone
Formed from: 4-*tert*-Butylcatechol

Butyl salicylate
→ Salicylic acid
In: Guinea pig D144

***tert*-Butyl semiquinone radical**
Formed from: 2-*tert*-Butylquinol
2-*tert*-Butylquinone

β-*n*-Butylstyrene oxide
→ (*R*,*R*)-*threo*-1,2-dihydroxy-1-phenylhexane
In: Rabbit J17

Butyl *p*-sulphamidobenzoate
Formed from: Butyl *p*-aminobenzoate

***p*-*tert*-Butylsulphinylanisole**
Formed from: *p*-*tert*-Butylthioanisole

***n*-Butylsulphinylbenzene**
Formed from: *n*-Butylthiobenzene

***p*-*tert*-Butylsulphinylbenzene**
Formed from: *p*-*tert*-Butylthiobenzene

2-(*tert*-Butylsulphonylmethyl)phenylpropionic acid
Formed from: Remikiren

***n*-Butyl-Δ^6-tetrahydrocannabinol**
→ *n*-Butyl-Δ^6-tetrahydro-5α-hydroxycannabinol
In: Mouse F291
→ *n*-Butyl-Δ^6-tetrahydro-5β-hydroxycannabinol
In: Mouse F291
→ *n*-Butyl-Δ^6-tetrahydro-7-hydroxycannabinol
In: Mouse F291

***n*-Butyl-Δ^6-tetrahydrocannabinol-7-oic acid**
Probably formed from: *n*-Butyl-Δ^6-tetrahydro-7-hydroxycannabinol
→ *n*-Butyl-Δ^6-tetrahydro-5α-hydroxycannabinol-7-oic acid
Probably in: Mouse F291
→ *n*-Butyl-Δ^6-tetrahydro-5β-hydroxycannabinol-7-oic acid
Probably in: Mouse F291
→ Δ^6-Tetrahydro-(2-hydroxy-*n*-butyl) cannabinol-7-oic acid
Probably in: Mouse F291

→ Δ^6-Tetrahydro-(3-hydroxy-*n*-butyl) cannabinol-7-oic acid
Probably in: Mouse F291

_n_-Butyl-Δ^6-tetrahydro-5α-hydroxycannabinol
Formed from: *n*-Butyl-Δ^6-tetrahydrocannabinol

_n_-Butyl-Δ^6-tetrahydro-5β-hydroxycannabinol
Formed from: *n*-Butyl-Δ^6-tetrahydrocannabinol

_n_-Butyl-Δ^6-tetrahydro-7-hydroxycannabinol
Formed from: *n*-Butyl-Δ^6-tetrahydrocannabinol

_n_-Butyl-Δ^6-tetrahydro-5α-hydroxycannabinol-7-oic acid
Probably formed from: *n*-Butyl-Δ^6-tetrahydrocannabinol-7-oic acid

_n_-Butyl-Δ^6-tetrahydro-5β-hydroxycannabinol-7-oic acid
Probably formed from: *n*-Butyl-Δ^6-tetrahydrocannabinol-7-oic acid

p-tert-Butylthioanisole
→ *p-tert*-Butylsulphinylanisole
In: Mortierella D759

_n_-Butylthiobenzene
Formed from: *n*-Butylsulphinylbenzene

p-tert-Butylthiobenzene
→ *p-tert*-Butylsulphinylbenzene
In: Mortierella D759

2-(Butylthio)benzothiazole
See: Butylcaptax

4-(Butylthio)catechol
Formed from: *p*-(Butylthio)phenol
→ 4-(Butylthio)-*o*-quinone
In: Mushroom H907

_p_-(Butylthio)phenol
→ 4-(Butylthio)catechol
In: Mushroom H907

4-(Butylthio)-_o_-quinone
Formed from: 4-(Butylthio)catechol

p-tert-Butyltoluene
→ *p-tert*-Butylbenzyl alcohol
In: Guinea pig C764
Rat C764
→ 2-Methyl-2-(*p*-tolyl)propan-1-ol
In: Guinea pig C764
Rat C764

1-Butyl-3-(_p_-tolylsulphonyl)urea
See: Tolbutamide

Butyl 2-((_m_-trifluoromethylphenyl)amino) benzoate
→ Flufenamic acid
In: Rat B499

5-_tert_-Butyl-2,4,6-trinitro-_m_-xylene
See: Musk xylene

1-(4-Butyramidobutyl)-4-(6-fluorobenzo[_b_]thien-3-yl)piperazine
Probably formed from: 1-(4-((N-Acetyl-N-butyryl)amino)butyl)-4-(6-fluorobenzo[*b*] thien-3-yl)piperazine

3-(((2-Butyramido)cyclopentyl)formamidomethyl)-2′-(1_H_-tetrazol-5-yl)biphenyl
Formed from: Irbesartan

2-Butyramidofluorene
→ 2-Aminofluorene
In: Guinea pig A808
Man A808
Mouse A808
Rabbit A808
Rat A808

Butyranilide
→ Aniline
In: Rice A3773

Butyrophenone (1-phenyl-1-butanone)
Probably formed from: 1-Phenyl-1-butanol

3-Butyrylchloramphenicol
Formed from: Chloramphenicol

1-_n_-Butyryl-4-cinnamylpiperazine
→ 1-Butyryl-4-(1,2-dihydro-1,2-dihydroxy-1-phenyl-3-propyl)piperazine
In: Guinea pig A3589
→ 1-*n*-Butyryl-4-(*p*-hydroxycinnamyl) piperazine
In: Rat A787, A1403
See also: A3589
→ Cinnamic acid
Probably in: Rat A1403
→ 4-Cinnamyl-1-(1-hydroxybutyryl) piperazine
In: Guinea pig A3589
→ 4-Cinnamyl-1-(2-hydroxybutyryl) piperazine
In: Guinea pig A3589
→ 1-Cinnamylpiperazine

In: Rat A787

1-Butyryl-4-(1,2-dihydro-1,2-dihydroxy-1-(*p*-hydroxyphenyl)-3-propyl)piperazine
Probably formed from: 1-Butyryl-4-(1,2-dihydro-1,2-dihydroxy-1-phenyl-3-propyl) piperazine

1-Butyryl-4-(1,2-dihydro-1,2-dihydroxy-1-phenyl-3-propyl)piperazine
Formed from: 1-*n*-Butyryl-4-cinnamylpiperazine
→ 1-Butyryl-4-(1,2-dihydro-1,2-dihydroxy-1-(*p*-hydroxyphenyl)-3-propyl)piperazine
Probably in: Guinea pig A3589

1-Butyryl-4-(3,4-dihydroxycinnamyl)piperazine
Probably formed from: 1-Butyryl-4-(*p*-hydroxycinnamyl)piperazine
→ 1-Butyryl-4-(4-hydroxy-3-methoxycinnamyl)piperazine
Probably in: Mouse A3589

1-*n*-Butyryl-4-(*p*-hydroxycinnamyl)piperazine
Formed from: 1-*n*-Butyryl-4-cinnamylpiperazine
→ 1-*n*-Butyryl-4-(3,4-dihydroxycinnamyl) piperazine
In: Mouse A3589
→ 1-(*p*-Hydroxycinnamyl) piperazine
Probably in: Rat A787, A1403

1-*n*-Butyryl-4-(4-hydroxy-3-methoxycinnamyl) piperazine
Probably formed from: 1-*n*-Butyryl-4-(3,4-dihydroxycinnamyl)piperazine

2-Butyryl-10-(3-(4-methylpiperazin-1-yl) propyl)phenothiazine
See: Butaperazine

6-Butyrylmorphine
See: 6-Butanoylmorphine

Butyryl propranolol
→ Propranolol
In: Rat J878

***n*-Butyrylsalicylic acid**
→ Salicylic acid
In: Acinetobacter A962

C

Cabergoline (1-((6-allylergolinyl-8β-yl)carbonyl)-1-(3-(dimethylamino)propyl)-3-ethylurea)
→ 6-Allyl-8β-(((3-dimethylaminopropyl) amino)carbonyl)ergoline
In: Man G832
Monkey G832
Rat G832
→ 1-((6-Allylergolin-8β-yl)carbonyl)-3-ethylurea
In: Man G832
Monkey G832
→ 6-Allylergoline-8β-carboxylic acid
In: Man G832
Monkey G832
Rat G832
→ 1-((3-Dimethylamino)propyl)-1-(ergolin-8β-yl)carbonyl-3-ethylurea
In: Rat G832

Caffealdehyde (3,4-dihydroxycinnamaldehyde)
Formed from: Caffeoyl CoA

Caffeic acid (3,4-dihydroxycinnamic acid)
Formed from: Cichoric acid
Chlorogenic acid
m-Coumaric acid
Ferulic acid
Luteolin
Methyl caffeate
Rosmarinic acid
Tannin
Probably formed from: 3,4-Dihydroxy-L-phenylalanine
→ 3-(*o*-Benzoquinon-4-yl)acrylic acid
In: Mushroom H907
See also: A218, A1332, A1636, A1930, A2433, A3114, A3190, E458, F664, H948
→ Caffeic acid-4-α-D-glucoside
Probably in: Bacillus H234
→ Caffeic acid-4-β-D-glucoside
In: Nicotiana B442
Rose K809
Tomato B846
See also: B819, G897, K556, K889
→ Caffeic acid 4-sulphate
Probably in: Rat J686
→ Caffeoyl CoA
In: Cephalocereus H633
Cucumber A2514
Glycine A2148, A2514
Leek A2514
Marrow A2514
Pea A2514
Pepper A2514
Petunia A2845
Phaseolus A2514
Polyporus A226
Potato A204
Pseudomonas B298
Tomato A3149
Wheat K484
Willow D466
See also: A3147
→ Caffeoyl-β-D-glucose
In: Chenopodium G255
Potato D70
Raphanus B440
→ 3-(2-Carboxyethenyl)-*cis,cis*-muconic acid
In: Hydrogenophaga J218
→ *m*-Coumaric acid
In: Mouse A647
See also: A1035, J686
→ *p*-Coumaric acid
In: Microorganism D737, G432
→ Curcumin
In: Curcuma A1147

→ Dehydrodicaffeic acid dilactone
In: Inonotus A3016, A3019
→ Dihydrocaffeic acid
In: Microorganism D894
Pseudomonas A1448
Rat A557
→ 3,4-Dihydroxystyrene
In: Bacillus J623
Brettanomyces H356
Cladosporiun A2675
Erwinia G797
Fusarium G797
Lactobacillus H897
Klebsiella G797, J853
Microorganism F390
Rat A557
See also: A2690
→ 4-Ethylcatechol
In: Brettanomyces H356
In: Rat A557
→ Ferulic acid
In: Alfalfa B487, G273
Apple D258
Aspen G624
Bamboo A169
Cortaderia A187
Glycine A2436, K728
Japanese black pine A1634, J304
Loblolly pine A1634, J304
Nicotiana A3437, A3757, A3778, B490
Populus A3477
Rat A187, A486, A557
Silene C351
Spinach C94
Spinach beet A2315
Streptomyces K652
Wheat J39, K585
See also: A206, A2869, C100, F182, F367, J190, J686
→ Hypoxoside
In: Hypoxis G408
→ *p*-Hydroxyphenylacetic acid
In: Pseudomonas A1448
→ Isoferulic acid
In: Cortaderia A187
Populus A3477
Rat A486
Tobacco A3757
→ Maclurin
Probably in: Anemarrhena B806
→ Phloretic acid
In: Pseudomonas A1448
→ Quercetin
Probably in: Hydrangea A1927
→ Rosmarinic acid
In: Coleus A3313

***cis*-Caffeic acid**
→ *cis*-Caffeic acid-4-β-D-glucoside
In: Sphagnum J384
→ 6,7-Dihydroxycoumarin
In: Mushroom D598

Caffeic acid-α-D-glucoside
Probably formed from: Caffeic acid

Caffeic acid-4-β-D-glucoside
Formed from: Caffeic acid

***cis*-Caffeic acid-4-β-D-glucoside**
Formed from: *cis*-Caffeic acid

Caffeic acid-4-sulphate
Formed from: Caffeic acid

Caffeoyl CoA
Formed from: Caffeic acid
Chlorogenic acid
p-Coumaroyl CoA
Rosmarinic acid
→ Caffealdehyde
In: Glycine A2576
→ Caffeoylgalactarate
In: Secale K722
→ Caffeoylglucarate
In: Secale K722
→ Caffeoylglucarolactone
In: Secale K722
→ Caffeoyl *p*-hydroxyphenyllactic acid ester
In: Coleus G420, J349
→ Caffeoylshikimic acid
In: Secale K722
→ Chlorogenic acid
In: Nicotiana A3147, A3758
Potato A3145, A3755
Secale K722
Stevia A3758
Tomato A3149, K723

See also: A3475
→ Cyanidin
Probably in: Silene B404
→ Eriodictyol
In: Daucus D698
Glycine E579
Spinach D596, D867
Tulipa A2704, A3792
→ Feruloyl CoA
Daucus F94, K731
Loblolly pine J304
Parsley F160
See also: K625
→ 2′,3,4,4′,6′-Pentahydroxychalcone
In: Buckwheat E180
Rye E662
Tulipa B377
→ Rosmarinic acid
In: Coleus E660, G257, G420, H17
→ 2′,3,4,4′-Tetrahydroxychalcone
In: Glycine F789

Caffeoyl-3′-dephospho CoA
→ Feryloyl-3′-dephospho CoA
In: Daucus K731

Caffeoylgalactarate
Formed from: Caffeoyl CoA

Caffeoylglucarate
Formed from: Caffeoyl CoA

Caffeoylglucarolactone
Formed from: Caffeoyl CoA

Caffeoyl-β-D-glucose
Formed from: Caffeic acid
Probably formed from: *p*-Coumaroyl-D-glucose
→ Chlorogenic acid
In: Sweet potato F143
See also: F145
→ O-Demethylcelosianin II
In: Chenopodium G255

Caffeoyl *p*-hydroxyphenyllactic acid ester
Formed from: Caffeoyl CoA
p-Coumaroyl *p*-hydroxyphenyllactic acid ester
p-Hydroxyphenyllactic acid
→ Rosmarinic acid
In: Coleus G747, J349

Caffeoyl-N-hydroxysuccinamide
Formed from: *p*-Coumaroyl-N-hydroxysuccinamide
→ N-Hydroxy-3-(*o*-quinon-4-yl) acryloylsuccinamide
In: Mushroom H907

Caffeoyl methylamide
Formed from: Dihydrocaffeoyl methylamide

Caffeoylquinic acid
See: Chlorogenic acid

Caffeoylshikimic acid
Formed from: Caffeoyl CoA

2-O-Caffeoyltartaric acid
Formed from: Cichoric acid
Probably formed from: *p*-Coumaroyl tartrate
→ Caffeoyl tartrate quinone
In: Grape F701
See also: E458

Caffeoyl tartrate quinone
Formed from: Caffeoyl tartrate

Calycosin (3′,7-dihydroxy-4′-methoxyisoflavone)
Formed from: Formononetin
→ Maackiain
In: Clover A3474

Camalexine (3-(2-thiazolyl)-1*H*-indole)
→ N-Formyl-4-hydroxy-2-((thiazol-2-yl) carbonyl)aniline
In: Rhizoctonia K392
→ 5-Hydroxycamalexin
In: Rhizoctonia K392
→ N-(2-Hydroxyethyl)indole-3-carboxamide
In: Rhizoctonia K392
→ 3-(Tetrahydrooxazol-2-yl)indole
In: Rhizoctonia K392

Camazepam (7-chloro-3-dimethylcarbamoyloxy-1-methyl-5-phenyl-1,3-dihydro-2*H*-1,4-benzodiazepin-2-one)
→ 7-Chloro-1,3-dihydro-3-(N,N-dimethyl) carbamoyloxy)-5-(*p*-hydroxyphenyl)-1-methyl-2*H*-1,4-benzodiazepin-2-one
In: Rat E57
→ 7-Chloro-1,3-dihydro-3-(N,N-dimethyl) carbamoyl-5-phenyl-2*H*-1,4-benzodiazepin-2-one
In: Dog D639
Monkey D639

Mouse D639
Rat D639, E57
→ 7-Chloro-1,3-dihydro-3-(N-hydroxymethyl-N-methyl)carbamoyl-1-methyl-5-phenyl-2*H*-1,4-benzodiazepin-2-one
In: Rat E57
See also: D639
→ 7-Chloro-1,3-dihydro-3-hydroxy-1-methyl-5-phenyl-2*H*-1,4-benzodiazepin-2-one
In: Dog D639
Monkey D639
Mouse D639
Rat D639, E57
→ 7-Chloro-1,3-dihydro-1-methyl-3-(N-methyl)carbamoyl-5-phenyl-2*H*-1,4-benzodiazepin-2-one
In: Dog D639
Monkey D639
Mouse D639
Rat D639, E57

Cambendazole (isopropyl 2-(4-thiazolyl)-5-benzimidazolecarbamate)
→ 5-Acetamido-2-(4-thiazolyl)benzimidazole
In: Pig A3510
Sheep A3510
→ 4-Glutathionylcambendazole
In: Hamster B49
→ 1-Hydroxyisopropyl 2-(4-thiazolyl)-5-benzimidazolecarbamate
In: Pig A3510
Sheep A3510
→ Isopropyl 2-carboxamido-5-benzimidazolecarbamate
In: Pig A3510
Sheep A3510
→ Isopropyl 2-(1-formamido-2-hydroxyethyl)-5-benzimidazolecarbamate
In: Pig A3510
Sheep A3510
→ Isopropyl 2-formamidomethyl-5-benzimidazolecarbamate
In: Pig A3510
Sheep A3510
→ Isopropyl 2-(1-hydroxyethyl)-5-benzimidazolecarbamate
In: Pig A3510
Sheep A3510
→ Isopropyl 2-(2-hydroxy-2-methylsulphonylethyl)-5-benzimidazolecarbamate
In: Pig A3510
Sheep A3510

Camostat (N,N-dimethylcarbamoylmethyl 4-(4-guanidinobenzoyloxy)phenylacetate)
→ 4-Guanidinobenzoic acid
In: Dog G854
Man C338, G854
Rat G854
→ 4-(4-Guanidinobenzoyloxy)phenylacetic acid
In: Dog G854
Man G854
Rat G854

Camptosar
See: Irinotecan

Camptothecin ((4*S*)-4-ethyl-4-hydroxy-1*H*-pyrano[3′,4′:6,7]indolizino[1,2-*b*]quinoline-3,14(4*H*,12*H*)-dione)
Formed from: Strictosamine
Strictosidine

Canadine
See: Tetrahydroberberine

Candesartan (2-ethoxy-1-((2′-(1*H*-tetrazol-5-yl)(1,1′-biphenyl)-4-yl)methyl)-1*H*-benzimidazole-7-carboxylic acid)
Formed from: Candesartan cilexitil
→ Candesartan-N-β-D-glucuronide
Probably in: Rat J220
→ Candesartan-O-β-D-glucuronide
In: Rat J220

Candesartan cilexitil
→ Candesartan
In: Rat J220

Candesartan-N-β-D-glucuronide
Probably formed from: Candesartan

Candesartan-O-β-D-glucuronide
Formed from: Candesartan

Candoxatril (*cis*-4-(((1-((2*S*)-3-((2,3-dihydro-1*H*-1-inden-5-yl)oxy)-2-(2-methoxyethoxy)methyl)-3-oxopropyl)cyclopentyl)carbonyl)aminocyclohexanecarboxylic acid)
→ 5-Hydroxyindane
Probably in: Mouse J286

Cannabichromene (2-methyl-2-(4′-methyl-3′-pentenyl)-7-pentyl-2*H*-1-benzopyran-5-ol)
→ Cannabichromene-3′,4′-oxide
In: Cat F608
Gerbil F608
Guinea pig F608
Hamster F608
Mouse F608
Rabbit F608
Rat F608
→ 1′-Hydroxycannabichromene
In: Cat F608
Gerbil F608
Guinea pig F608
Hamster F608
Mouse F608, G335
Rabbit F608
Rat F608
→ 1″-Hydroxycannabichromene
In: Cat F608
Gerbil F608
Guinea pig F608
Hamster F608
Mouse F608, G335
Rabbit F608, G335
Rat F608
→ 2′-Hydroxycannabichromene
In: Cat F608
Gerbil F608
Guinea pig F608
Hamster F608
Mouse F608, G335
Rabbit F608
Rat F608
→ 2″-Hydroxycannabichromene
In: Cat F608
Gerbil F608
Guinea pig F608
Hamster F608
Mouse F608, G335
Rabbit F608, G335
Rat F608
→ 3″-Hydroxycannabichromene
In: Cat F608
Gerbil F608
Guinea pig F608
Hamster F608
Mouse F608, G335
Rabbit F608, G335
Rat F608
→ 4″-Hydroxycannabichromene
In: Cat F608
Gerbil F608
Guinea pig F608
Hamster F608
Mouse F608, G335
Rabbit F608, G335
Rat F608
→ 5′-Hydroxycannabichromene
In: Cat F608
Gerbil F608
Guinea pig F608
Hamster F608
Mouse F608, G335
Rabbit F608, G335
Rat F608
→ 5″-Hydroxycannabichromene
In: Cat F608
Gerbil F608
Guinea pig F608
Hamster F608
Mouse F608, G335
Rabbit F608, G335
Rat F608
→ 6′-Hydroxycannabichromene
In: Cat F608
Gerbil F608
Guinea pig F608
Hamster F608
Mouse F608, G335
Rabbit F608, G335
Rat F608

Cannabichromene-3′,4′-oxide
Formed from: Cannabichromene
→ 3′,4′-Dihydro-3′,4′-dihydroxycannabichromene
Probably in: Cat F608
Gerbil F608
Guinea pig F608
Hamster F608
Mouse F608
Rabbit F608

Rat F608

Cannabichromenic acid ((+)-5-hydroxy-2-methyl-2-(4-methyl-3-pentenyl)-7-pentyl-2*H*-1-benzopyran-6-carboxylic acid)
Formed from: Cannabigerolic acid

Cannabidiol (2-((1*R*,6*R*)-3-methyl-6-(1-methylethenyl)-2-cyclohexen-1-yl)-5-pentyl-1,3-benzenediol)
Probably formed from: Cannabinol
→ Cannabidiol-β-D-glucuronide
In: Mouse A1660, A2085, A2989
Rat C805
See also: F503, G460
→ Cannabichromene
In: Mouse A2989
→ Cannabielsoin
In: Guinea pig F43, F194, G426
Mouse F194
Rabbit F194
Rat F194
→ Cannabinol
In: Mouse A2989
→ 8,9-Dihydro-8,9-dihydroxycannabidiol
In: Rabbit F708
→ 3,5-Dihydroxy-4-(4-isopropenyl-1-methylcyclohexen-3-yl)phenylbutyric acid
In: Man F503
→ 6,7,7a,8,9,11a-Hexahydro-1,7,dihydroxy-7,10-dimethyldibenzo[*b*,*d*]oxepin
In: Guinea pig H673
→ 1″-Hydroxycannabidiol
In: Cat F709
Guinea pig B458, F709
Hamster F709
Man F7503
Mouse F709
Rabbit F708
Rat A2611
→ 2″-Hydroxycannabidiol
In: Gerbil F709
Guinea pig F709
Hamster F709
Man F503
Mouse F709
Rabbit F708
Rat A2611
→ 3″-Hydroxycannabidiol
In: Cat F709
Gerbil F709
Guinea pig B458, F709
Hamster F709
Mouse F709
Rabbit F708, F709
Rat A2611, F709
→ 4″-Hydroxycannabidiol
In: Cat F709
Dog F497, G460
Gerbil F709
Guinea pig B458, F709
Hamster F709
Mouse F709
Rabbit F708, F709
Rat A2611, F709
Syncephalastrum A2595
→ 5″-Hydroxycannabidiol
In: Cat F709
Dog F497, G460
Gerbil F709
Guinea pig F709
Hamster F709
Rabbit F708, F709
Rat A2611
See also: A2085, F503, G113
→ 6α-Hydroxycannabidiol
In: Cat F709
Gerbil F709
Guinea pig B458, F709
Hamster F709
Mouse F709
Rabbit F709
Rat A2611, F709
See also: A2085, A2989, F503, F557, F708, G113, G460
→ 6β-Hydroxycannabidiol
In: Gerbil F709
Guinea pig B458, F709
Hamster F709
Mouse F709
Rabbit F709
Rat A2611, F709
See also: A2777, F497
→ 7-Hydroxycannabidiol

In: Cat F709
Dog F557, G113
Gerbil F709
Guinea pig B458, F43, F709
Hamster F709
Man F503, G113
Mouse A1660, A2085, A2989, F709
Rabbit F708, F709
Rat A995, F509, G113
See also: A2777, G608
→ 10-Hydroxycannabidiol
In: Gerbil F709
Guinea pig F709
Hamster F709
Rabbit F708
Rat F709
→ 3″,4″,5″-Trisnor-2″-hydroxycannabidiol
Probably in: Man G460
→ Propylcannabinol
In: Mouse A2989

Cannabidiol dimethyl ether
→ Cannabidiol-1,2-oxide dimethyl ether
In: Guinea pig G426

Cannabidiol-5″,7-dioic acid
Probably formed from: 7-Hydroxycannabidiol-5″-oic acid
5-Hydroxycannabidiol-7-oic acid
→ 4″,5″-Bisnorcannabidiol-3″,7-dioic acid
Probably in: Dog F557
Rat G608

Cannabidiol-β-D-glucuronide
Formed from: Cannabidiol

Cannabidiolic acid (2,4-dihydroxy-3-((1*R*,6*R*)-3-methyl-6-(1-methylethenyl)-2-cyclohexen-1-yl)-6-pentylbenzoic acid)
Formed from: Cannabigerolic acid
Cannabinerolic acid
→ Cannabinolic acid
Probably in: Cannabis A205

Cannabidiol methyl ether
→ Cannabielsoin methyl ether
In: Guinea pig G426

Cannabidiol-7-oic acid
Formed from: 7-Hydroxycannabidiol
→ 8,9-Dihydro-8,9-dihydroxycannabidiol-7-oic acid
Probably in: Man G113
→ 1″-Hydroxycannabidiol-7-oic acid
Probably in: Man F503, G113
→ 2″-Hydroxycannabidiol-7-oic acid
Probably in: Man F503, G113
Mouse A2085
Rat G113
→ 3″-Hydroxycannabidiol-7-oic acid
Probably in: Man F503, G113
Mouse A2085
Rat G113
→ 4″-Hydroxycannabidiol-7-oic acid
Probably in: Guinea pig B458
Man F503, G113
Mouse A2085
→ 5″-Hydroxycannabidiol-7-oic acid
Probably in: Rat G113
→ 10-Hydroxycannabidiol-7-oic acid
Probably in: Rat G113

Cannabidiol-1,2-oxide dimethyl ether
Formed from: Cannabidiol dimethyl ether

Cannabielsoin (cannabielsoin A?)
Formed from: Cannabidiol

Cannabielsoin methyl ether
Formed from: Cannabidiol methyl ether

Cannabigerol (2-((2*E*)-3,7-dimethyl-2,6-octadienyl)-5-pentyl-1,3-benzenediol)
→ Cannabigerol-2′,3′-oxide
In: Cat F704
Gerbil F704
Guinea pig F704
Hamster F704
Mouse F704
Rabbit F704
Rat F704
→ Cannabigerol-6′,7′-oxide
In: Cat F704
Gerbil F704
Guinea pig F704
Hamster F704
Mouse F704
Rabbit F704
Rat F704
→ 1″-Hydroxycannabigerol
In: Cat F704
Gerbil F704

Guinea pig F704
Hamster F704
Mouse F704
Rabbit F704
Rat F704
→ 2″-Hydroxycannabigerol
In: Cat F704
Gerbil F704
Guinea pig F704
Hamster F704
Mouse F704
Rabbit F704
Rat F704
→ 3″-Hydroxycannabigerol
In: Cat F704
Gerbil F704
Guinea pig F704
Hamster F704
Mouse F704
Rabbit F704
Rat F704
→ 4′-Hydroxycannabigerol
In: Cat F704
Gerbil F704
Guinea pig F704
Hamster F704
Mouse F704
Rabbit F704
Rat F704
→ 4″-Hydroxycannabigerol
In: Cat F704
Gerbil F704
Guinea pig F704
Hamster F704
Mouse F704
Rabbit F704
Rat F704
→ 5″-Hydroxycannabigerol
In: Cat F704
Gerbil F704
Guinea pig F704
Hamster F704
Mouse F704
Rabbit F704
Rat F704
→ 8′-Hydroxycannabigerol
In: Cat F704
Gerbil F704
Guinea pig F704
Hamster F704
Mouse F704
Rabbit F704
Rat F704
→ 9′-Hydroxycannabigerol
In: Cat F704
Gerbil F704
Guinea pig F704
Hamster F704
Mouse F704
Rabbit F704
Rat F704

Cannabigerolic acid (2-((2*E*)-3,7-dimethyl-2,6-octadienyl)-2,4-dihydroxy-6-pentylbenzoic acid)
→ Cannabichromenic acid
In: Cannabis A205, J861
→ Cannabidiolic acid
In: Cannabis A205, J861
→ Cannabigerolic acid monomethyl ether
In: Cannabis A205

Cannabigerolic acid monomethyl ether
Formed from: Cannabigerolic acid

Cannabigerol-2′,3′-oxide
Formed from: Cannabigerol
→ 2′,3′-Dihydro-2′,3′-dihydroxycannabigerol
Probably in: Cat F704
Gerbil F704
Guinea pig F704
Hamster F704
Mouse F704
Rabbit F704
Rat F704

Cannabigerol-6′,7′-oxide
Formed from: Cannabigerol
→ 6′,7′-Dihydro-6′,7′-dihydroxycannabigerol
Probably in: Cat F704
Gerbil F704
Guinea pig F704
Hamster F704
Mouse F704
Rabbit F704
Rat F704

Cannabinerolic acid (3-((2*Z*)-3,7-dimethyl-2,6-octadienyl)-2,4-dihydroxy-6-pentylbenzoic acid)
→ Cannabidiolic acid
In: Cannabis J205

Cannabinol (6,6,9-trimethyl-3-pentyl-6*H*-dibenzo[*b*,*d*]pyran-1-ol)
Formed from: Δ^1-Tetrahydrocannabinol
$\Delta^{3,4}$-Tetrahydrocannabinol
Δ^6-Tetrahydrocannabinol
→ Cannabidiol
Probably in: Man A675
→ Cannabinol-β-D-glucuronide
In: Pinellia H72
Rat C805
→ 1″-Hydroxycannabinol
In: Cat G143
Gerbil G143
Guinea pig B458, G143
Hamster G143
Mouse G143
Rabbit G143
Rat A3405, A3436, G143
→ 2″-Hydroxycannabinol
In: Cat G143
Gerbil G143
Guinea pig G143
Hamster G143
Mouse G143
Rabbit G143
Rat A3436, G143
→ 3″-Hydroxycannabinol
In: Cat G143
Gerbil G143
Guinea pig B458, G143
Hamster G143
Mouse G143
Rabbit G143
Rat A3436, G143
→ 4″-Hydroxycannabinol
In: Cat G143
Gerbil G143
Guinea pig B458, G143
Hamster G143
Mouse G143
Rabbit G143
Rat A3405, A3436, A3767, G143
Syncephalastrum A2595
→ 5″-Hydroxycannabinol
In: Cat G143
Gerbil G143
Guinea pig G143
Hamster G143
Mouse G143
Rabbit G143
Rat A3436, A3767, G143
→ 7-Hydroxycannabinol
In: Cat G143
Gerbil G143
Hamster G143
Mouse A3405, G143, H67
Rabbit G143
Rat A949, A3436, A3405, A3767, G143
See also: A2256, A2989, A2994
→ 9′-Hydroxycannabinol
In: Pinellia H72

Cannabinol-β-D-glucuronide
Formed from: Cannabinol

Cannabinolic acid (1-hydroxy-6,6,9-trimethyl-3-pentyl-6*H*-dibenzo[*b*,*d*]pyran-4-carboxylic acid)
Formed from: Cannabidiolic acid

Cannabinol-7-oic acid
Formed from: Δ^1-Cannabinol
7-Oxocannabinol
→ Cannabinol-7-oic acid-β-D-glucuronide
Probably in: Mouse A2989, A3405

Cannabinol-7-oic acid-β-D-glucuronide
Probably formed from: Cannabinol-7-oic acid

Canthin-6-one (6*H*-indolo[3,2,1-*de*][1,5]naphthyridin-6-one)
Formed from: β-Carboline-1-propionic acid
→ Canthin-6-one-3-oxide
In: Ailanthus E304
→ 1-Hydroxycanthin-6-one
In: Ailanthus E304

Canthin-6-one-3-oxide
Formed from: Canthin-6-one

Capsaicin ((6*E*)-N-((4-hydroxy-3-methoxyphenyl)methyl)-8-methoxy-6-nonenamide)
Formed from: Isoeugenol
→ N-(4,5-Dihydroxy-3-methoxybenzyl)-(*E*)-6-nonenamide

In: Rat H410
→ ω-Hydroxycapsaicin
In: Aspergillus H410
Rabbit H410
Rat H410
→ 4-Hydroxy-3-methoxybenzylamine
In: Rat G450

Captagon (fenethylline, 3,7-dihydro-1,3-dimethyl-7-(2-((1-methyl-2-phenylethyl)amino)ethyl)-1*H*-purine-2,6-dione)
→ Amphetamine
In: Man A528
See also: A1043

Carbamazepine (5*H*-dibenz[*b*,*f*]azepine-5-carboxamide)
→ 10-Carbamoyl-9-hydroxymethylacridan
In: Man D550, K102
→ Carbamazepine-N-β-D-glucuronide
In: Man A1967, A3453, J101
Rat A2649, H701
→ Carbamazepine-10,11-oxide
In: Man A542, A1420, A2202, A3122, A3455, C27, J101, K102
Mouse J271
Rabbit J909
Rat A1420, A1797, A2078, A2202, C27, D813, H701, K102
See also: A413, A3453, C3, D550, E318, G379
→ S-(L-Cysteinyl)carbamazepine
Probably in: Mouse J271
→ 2,3-Dihydro-2,3-dihydroxycarbamazepine
In: Man C3
→ Dihydrodihydroxycarbamazepine (3 isomers; structures unclear)
In: Man A986, A1967
Rat A986
→ 3-S-(Glutathionyl)-2,3-dihydro-2-hydroxycarbamazepine
In: Rat H701
→ 2-Hydroxycarbamazepine
In: Man C3, D550, J101
Mouse J271
Rat H701
→ 3-Hydroxycarbamazepine
In: Man A1967, A3453, C3, D550, J101, K102
Mouse J271, K102
Rat C3, H701
→ 10-Hydroxycarbamazepine
In: Man K102
→ Hydroxy-methylsulphinylcarbamazepine
In: Man C3
→ Iminostilbene
In: Man A2202
Rat A890, A2202, A2797
→ Methylsulphonylcarbamazepine
Probably in: Mouse J271

Carbamazepine-N-β-D-glucuronide
Formed from: Carbamazepine

Carbamazepine-10,11-oxide
Formed from: Carbamazepine
→ Carbamazepine-10,11-oxide-N-β-D-glucuronide
In: Rat H701
See also: J271
→ Carbamazepine-10-yl-β-D-glucuronide
Probably in: Man G379
→ *cis*-10,11-Dihydro-10,11-dihydroxycarbamazepine
Probably in: Man K417
→ *trans*-10,11-Dihydro-10,11-dihydroxycarbamazepine
In: Man C27
Rat A2294, A3283, C27
See also: A413, A1420, A2202, A3453, K102
→ Iminostilbene-10,11-oxide
In: Rat A2294, A3283

Carbamazepine-10,11-oxide-N-β-D-glucuronide
Formed from: Carbamazepine-10,11-oxide

3-Carbamoyl-7-chloro-1,3-dihydro-1-methyl-5-phenyl-2*H*-1,4-benzpdiazepin-2-one
Probably formed from: 7-Chloro-1,3-dihydro-3-(N-hydroxymethyl)carbamoyl-1-methyl-5-phenyl-2*H*-1,4-benzpdiazepin-2-one

3-Carbamoyl-7-chloro-1,3-dihydro-5-phenyl-2*H*-1,4-benzodiazepin-2-one
Probably formed from: 7-Chloro-1,3-dihydro-3-(N-hydroxymethyl)carbamoyl-5-phenyl-2*H*-1,4-benzodiazepin-2-one

4-Carbamoyl-7-chloro-1,3,4,5-tetrahydro-1-methyl-5-phenyl-2*H*-1,4-benzodiazepin-2-one
See: Uxepam

4-Carbamoyl-7-chloro-1,3,4,5-tetrahydro-5-phenyl-2*H*-1,4-benzodiazepin-2-one
Formed from: Uxepam

2-Carbamoyl-1-hydroxyindane
Formed from: 2-Carbamoylindane

10-Carbamoyl-9-hydroxymethylacridan
Formed from: Carbamazepine

2-Carbamoylindane
→ 2-Carbamoyl-1-hydroxyindane
In: Pseudomonas K207

1-(4-Carbamoylmethylphenoxy)-3-isopropylaminopropan-2-ol
See: Atenolol

Carbamoylmethyl-4-(3,4,5-trimethoxycinnamoyl)piperazine
Formed from: 1-(N-Isopropylcarbamoyl-methyl)-4-(3,4,5-trimethoxycinnamoyl)piperazine

1-(2-Carbamoyloxy-3-(carboxymethoxy)propyl)-5-ethyl-5-phenyl-(1*H*,3*H*,5*H*)-pyrimidine-2,4,6-trione
Probably formed from: 1-(2-Carbamoyloxy-3-(2-hydroxyethoxy)propyl)-5-ethyl-5-phenyl-(1*H*,3*H*,5*H*)-pyrimidine-2,4,6-trione

1-(2-Carbamoyloxy-3-(2,3-dihydroxybutoxy)propyl)-5-ethyl-5-phenyl-(1*H*,3*H*,5*H*)-pyrimidine-2,4,6-trione
Probably formed from: 1-(2-Carbamoyloxy-3-(3-hydroxybutoxy)propyl)-5-ethyl-5-phenyl-(1*H*,3*H*,5*H*)-pyrimidine-2,4,6-trione
→ 1-(2-Carbamoyloxy-3-(2-hydroxyethoxy)propyl)-5-ethyl-5-phenyl-(1*H*,3*H*,5*H*)-pyrimidine-2,4,6-trione
Probably in: Man D26

1-(2-Carbamoyloxy-3-(2-hydroxybutoxy)propyl)-5-ethyl-5-(*p*-hydroxyphenyl)-(1*H*,3*H*,5*H*)-pyrimidine-2,4,6-trione
Probably formed from: 1-(2-Carbamoyloxy-3-(2-hydroxybutoxy)propyl)-5-ethyl-5-phenyl-(1*H*,3*H*,5*H*)-pyrimidine-2,4,6-trione

1-(2-Carbamoyloxy-3-(3-hydroxybutoxy)propyl)-5-ethyl-5-(*p*-hydroxyphenyl)-(1*H*,3*H*,5*H*)-pyrimidine-2,4,6-trione
Probably formed from: 1-(2-Carbamoyloxy-3-(3-hydroxybutoxy)propyl)-5-ethyl-5-phenyl-(1*H*,3*H*,5*H*)-pyrimidine-2,4,6-trione
p-Hydroxyfebarbamate
→ 1-(2-Carbamoyloxy-3-(3-oxobutoxy)propyl)-5-ethyl-5-(*p*-hydroxyphenyl)-(1*H*,3*H*,5*H*)-pyrimidine-2,4,6-trione
Probably in: Man D26

1-(2-Carbamoyloxy-3-(2-hydroxybutoxy)propyl)-5-ethyl-5-phenyl-(1*H*,3*H*,5*H*)-pyrimidine-2,4,6-trione
→ 1-(2-Carbamoyloxy-3-(2-hydroxybutoxy)propyl)-5-ethyl-5-(*p*-hydroxyphenyl)-(1*H*,3*H*,5*H*)-pyrimidine-2,4,6-trione
Probably in: Man D26

1-(2-Carbamoyloxy-3-(3-hydroxybutoxy)propyl)-5-ethyl-5-phenyl-(1*H*,3*H*,5*H*)-pyrimidine-2,4,6-trione
Formed from: Febarbamate
→ 1-(2-Carbamoyloxy-3-(2,3-dihydroxybutoxy)propyl)-5-ethyl-5-phenyl-(1*H*,3*H*,5*H*)-pyrimidine-2,4,6-trione
Probably in: Man D26
→ 1-(2-Carbamoyloxy-3-(3-hydroxybutoxy)propyl)-5-ethyl-5-(*p*-hydroxyphenyl)-(1*H*,3*H*,5*H*)-pyrimidine-2,4,6-trione
Probably in: Man D26
→ 1-(2-Carbamoyloxy-3-(3-oxobutoxy)propyl)-5-ethyl-5-phenyl-(1*H*,3*H*,5*H*)-pyrimidine-2,4,6-trione
Probably in: Man D26

1-(2-Carbamoyloxy-3-(4-hydroxybutoxy)propyl)-5-ethyl-5-phenyl-(1*H*,3*H*,5*H*)-pyrimidine-2,4,6-trione
Formed from: Febarbamate
→ 3-(5-Ethyl-5-phenyl-(1*H*,3*H*,5*H*)-pyrimidine-2,4,6-trion-1-yl)-2-carbamoyloxypropoxybutyric acid
Probably in: Man D26

1-(2-Carbamoyloxy-3-(2-hydroxyethoxy)propyl)-5-ethyl-5-(*p*-hydroxyphenyl)-(1*H*,3*H*,5*H*)-pyrimidine-2,4,6-trione
Probably formed from: 1-(2-Carbamoyloxy-3-(2-hydroxyethoxy)propyl)-5-ethyl-5-phenyl-(1*H*,3*H*,5*H*)-pyrimidine-2,4,6-trione

1-(2-Carbamoyloxy-3-(2-hydroxyethoxy)propyl)-5-ethyl-5-phenyl-(1*H*,3*H*,5*H*)-pyrimidine-2,4,6-trione
Formed from: Febarbamate
Probably formed from: 1-(2-Carbamoyloxy-3-(2,3-dihydroxybutoxy)propyl)-5-ethyl-5-

phenyl-(1*H*,3*H*,5*H*)-pyrimidine-2,4,6-trione
→ 1-(2-Carbamoyloxy-3-(carboxymethoxy)propyl)-5-ethyl-5-(*p*-hydroxyphenyl)-(1*H*,3*H*,5*H*)-pyrimidine-2,4,6-trione
Probably in: Man D26
→ 1-(2-Carbamoyloxy-3-(2-hydroxyethoxy)propyl)-5-ethyl-5-phenyl-(1*H*,3*H*,5*H*)-pyrimidine-2,4,6-trione
Probably in: Man D26
Rat C368

1-(2-Carbamoyloxy-3-hydroxypropyl)-5-ethyl-5-(4-hydroxy-3-methoxyphenyl)-(1*H*,3*H*,5*H*)-pyrimidine-2,4,6-trione
Probably formed from: 3-(5-Ethyl-5-(*p*-hydroxyphenyl)-(1*H*,3*H*,5*H*)-pyrimidine-2,4,6-trion-1-yl)-2-carbamoyloxypropanol

1-(2-Carbamoyloxy-3-hydroxypropyl)-5-ethyl-5-(*p*-hydroxyphenyl)-(1*H*,3*H*,5*H*)-pyrimidine-2,4,6-trione
Probably formed from: *p*-Hydroxyfebarbamate

1-(2-Carbamoyloxy-3-hydroxypropyl)-5-ethyl-5-phenyl-(1*H*,3*H*,5*H*)-pyrimidine-2,4,6-trione
Formed from: Febarbamate

2-Carbamoyloxymethyl-3-carboxy-4-(2,3-dichlorophenyl)-1,4-dihydro-5-methoxycarbonyl-6-methylpyridine
Formed from: 3-Isopropyl 5-methyl 2-carbamoyloxymethyl-4-(2,3-dichlorophenyl)-1,4-dihydro-6-methylpyridine-3,5-dicarboxylate

2-Carbamoyloxymethyl-5-carboxy-4-(2,3-dichlorophenyl)-1,4-dihydro-3-isopropoxycarbonyl-6-methylpyridine
Formed from: 3-Isopropyl 5-methyl 2-carbamoyloxymethyl-4-(2,3-dichlorophenyl)-1,4-dihydro-6-methylpyridine-3,5-dicarboxylate

2-Carbamoyloxymethyl-5-carboxy-4-(2,3-dichlorophenyl)-6-hydroxymethyl-3-isopropylcarbonylpyridine
Probably formed from: 2-Carbamoyloxymethyl-5-carboxy-4-(2,3-dichlorophenyl)-3-isopropylcarbonyl-6-methylpyridine

2-Carbamoyloxymethyl-5-carboxy-4-(2,3-dichlorophenyl)-3-isopropoxycarbonyl-6-methylpyridine
Formed from: 3-Isopropyl 5-methyl 2-carbamoyloxymethyl-4-(2,3-dichlorophenyl)-6-methylpyridine-3,5-dicarboxylate
→ 2-Carbamoyloxymethyl-5-carboxy-4-(2,3-dichlorophenyl)-6-hydroxymethyl-3-isopropylcarbonylpyridine
Probably in: Dog F493
Rat F493

2-Carbamoyloxymethyl-3-carboxy-4-(2,3-dichlorophenyl)-5-methoxycarbonyl-6-methylpyridine
Probably formed from: 3-Isopropyl 5-methyl 2-carbamoyloxymethyl-4-(2,3-dichlorophenyl)-6-methylpyridine-3,5-dicarboxylate

2-Carbamoyloxymethyl-5-(3,5-dichlorophenyl)sulphinyl-4-(hydroxyisopropyl)-1-(4-pyridyl)methyl-1*H*-imidazole
Probably formed from: 2-Carbamoyloxymethyl-5-(3,5-dichlorophenyl)thio-4-(hydroxyisopropyl)-1-(4-pyridyl)methyl-1*H*-imidazole

2-Carbamoyloxymethyl-5-(3,5-dichlorophenyl)sulphinyl-4-isopropyl-1-(4-pyridyl)methyl-1*H*-imidazole-N-oxide
Probably formed from: 2-Carbamoyloxymethyl-5-(3,5-dichlorophenyl)sulphinyl-4-isopropyl-1-(4-pyridyl)methyl-1*H*-imidazole

2-Carbamoyloxymethyl-5-(3,5-dichlorophenyl)sulphinyl-4-isopropyl-1-(4-pyridyl)methyl-1*H*-imidazole
Formed from: 2-Carbamoyloxymethyl-5-(3,5-dichlorophenyl)thio-4-isopropyl-1-(4-pyridyl)methyl-1*H*-imidazole
→ 2-Carbamoyloxymethyl-5-(3,5-dichlorophenyl)sulphinyl-4-isopropyl-1-(4-pyridyl)methyl-1*H*-imidazole-N-oxide
Probably in: Rat J417
→ 2-Carbamoyloxymethyl-5-(3,5-dichlorophenyl)sulphonyl-4-isopropyl-1-(4-pyridyl)methyl-1*H*-imidazole
Probably in: Rat J417

2-Carbamoyloxymethyl-5-(3,5-dichlorophenyl)sulphonyl-4-isopropyl-1-(4-pyridyl)methyl-1*H*-imidazole
Formed from: 2-Carbamoyloxymethyl-5-(3,5-dichlorophenyl)thio-4-isopropyl-1-(4-pyridyl)methyl-1*H*-imidazole

2-Carbamoyloxymethyl-5-(3,5-dichlorophenyl)thio-4-(hydroxyisopropyl)-1-(4-pyridyl)methyl-1*H*-imidazole
Formed from: 2-Carbamoyloxymethyl-5-(3,5-dichlorophenyl)thio-4-isopropyl-1-(4-pyridyl)methyl-1*H*-imidazole
→ 2-Carbamoyloxymethyl-5-(3,5-dichlorophenyl)sulphinyl-4-(hydroxyisopropyl)-1-(4-pyridyl)methyl-1*H*-imidazole
Probably in: Rat J417

2-Carbamoyloxymethyl-5-(3,5-dichlorophenyl)thio-4-isopropyl-1-(4-pyridyl)methyl-1*H*-imidazole
→ 2-Carbamoyloxymethyl-5-(3,5-dichlorophenyl)sulphinyl-4-isopropyl-1-(4-pyridyl)methyl-1*H*-imidazole
In: Rat J417
→ 2-Carbamoyloxymethyl-5-(3,5-dichlorophenyl)thio-4-(hydroxyisopropyl)-1-(4-pyridyl)methyl-1*H*-imidazole
In: Rat J417
→ 2-Carbamoyloxymethyl-5-(3,5-dichlorophenyl)thio-4-isopropyl-1-(4-pyridyl)methyl-1*H*-imidazole-N-oxide
In: Rat J417

2-Carbamoyloxymethyl-5-(3,5-dichlorophenyl)thio-4-isopropyl-1-(4-pyridyl)methyl-1*H*-imidazole-N-oxide
Formed from: 2-Carbamoyloxymethyl-5-(3,5-dichlorophenyl)thio-4-isopropyl-1-(4-pyridyl)methyl-1*H*-imidazole

1-(2-Carbamoyloxy-3-(3-oxobutoxy)propyl)-5-ethyl-5-(4-hydroxy-3-methoxyphenyl)-(1*H*,3*H*,5*H*)-pyrimidine-2,4,6-trione
Probably formed from: 1-(2-Carbamoyloxy-3-(3-oxobutoxy)propyl)-5-ethyl-5-(*p*-hydroxyphenyl)-(1*H*,3*H*,5*H*)-pyrimidine-2,4,6-trione

1-(2-Carbamoyloxy-3-(3-oxobutoxy)propyl)-5-ethyl-5-(*p*-hydroxyphenyl)-(1*H*,3*H*,5*H*)-pyrimidine-2,4,6-trione
Probably formed from: 1-(2-Carbamoyloxy-3-(3-oxobutoxy)propyl)-5-ethyl-5-phenyl-(1*H*,3*H*,5*H*)-pyrimidine-2,4,6-trione
→ 1-(2-Carbamoyloxy-3-(3-oxobutoxy)propyl)-5-ethyl-5-(4-hydroxy-3-methoxyphenyl)-(1*H*,3*H*,5*H*)-pyrimidine-2,4,6-trione
Probably in: Man D26

1-(2-Carbamoyloxy-3-(3-oxobutoxy)propyl)-5-ethyl-5-phenyl-(1*H*,3*H*,5*H*)-pyrimidine-2,4,6-trione
Probably formed from: 1-(2-Carbamoyloxy-3-(3-hydroxybutoxy)propyl)-5-ethyl-5-phenyl-(1*H*,3*H*,5*H*)-pyrimidine-2,4,6-trione
→ 1-(2-Carbamoyloxy-3-(3-oxobutoxy)propyl)-5-ethyl-5-(*p*-hydroxyphenyl)-(1*H*,3*H*,5*H*)-pyrimidine-2,4,6-trione
Probably in: Man D26

2-Carbamoyloxy-2-phenylpropionic acid
Probably formed from: 2-Phenylpropane-1,3-diol-1-carbamate

1-(4-Carbamoyloxypiperidino)-4-hydroxy-6,7-dimethoxyphthalazine
Formed from: 4-Hydroxycarbazeran

α-Carbamoyl-3-phenoxybenzyl 2-(*p*-chlorophenyl)-3-methylbutyrate
Formed from: Fenvalerate

1-(2-(4-Carbamoylphenoxy)ethylamino)-3-(4-hydroxy-2-methylphenoxy)propan-2-ol
Formed from: Tolamolol
→ 1-(2-(4-Carbamoylphenoxy)ethylamino)-3-(4-hydroxy-2-methylphenoxy)propan-2-ol-β-D-glucuronide
In: Dog A637
Guinea pig A637
Mouse A637
Rabbit A637
Rat A637
→ 1-(2-(4-Carbamoylphenoxy)ethylamino)-3-(4-hydroxy-2-methylphenoxy)propan-2-ol sulphate
In: Mouse A637
Rabbit A637
Rat A637

1-(2-(4-Carbamoylphenoxy)ethylamino)-3-(4-hydroxy-2-methylphenoxy)propan-2-ol-β-D-glucuronide
Probably formed from: 1-(2-(4-Carbamoylphenoxy)ethylamino)-3-(4-hydroxy-2-methylphenoxy)propan-2-ol

1-(2-(4-Carbamoylphenoxy)ethylamino)-3-(4-hydroxy-2-methylphenoxy)propan-2-ol sulphate
Probably formed from: 1-(2-(4-

Carbamoylphenoxy)ethylamino)-3-(4-hydroxy-2-methylphenoxy)propan-2-ol

α-Carbamoyl-3-phenoxybenzyl 2-(*p*-chlorophenyl)-3-methylbutyrate
Formed from: Fenvalerate
→ α-Carboxy-3-phenoxybenzyl 2-(*p*-chlorophenyl)-3-methylbutyrate
Probably in: Cabbage D585

1-Carbamoyl-α-phenyl-2-piperidineacetic acid
Probably formed from: Methyl 1-carbamoyl-α-phenyl-2-piperidineacetate

3-Carbamoyl-2-phenylpropionaldehyde
Formed from: Felbamate

N-Carbamyl-O-benzylserine
Formed from: 5-(Benzyloxymethyl)hydantoin
→ O-Benzyl-L-serine
In: Flavobacterium F938

N-Carbamyl-3,4-dimethoxy-D-phenylalanine
Formed from: 5-(3,4-Dimethoxybenzyl) hydantoin
→ 5-(3,4-Dimethoxybenzyl)hydantoin
In: Flavobacterium F938

N-Carbamyl-3,4-dimethoxy-L-phenylalanine
Formed from: 5-(3,4-Dimethoxybenzyl) hydantoin
→ 5-(3,4-Dimethoxybenzyl)hydantoin
In: Flavobacterium F938
→ 3,4-Dimethoxy-L-phenylalanine
In: Flavobacterium F938

N-Carbamyl-3,4-methylenedioxy-D-phenylalanine
Formed from: 5-(3,4-Methylenedioxybenzyl)hydantoin
→ 5-(3,4-Methylenedioxybenzyl)hydantoin
In: Flavobacterium F938

N-Carbamyl-3,4-methylenedioxy-L-phenylalanine
Formed from: 5-(3,4-Methylenedioxybenzyl) hydantoin
→ 5-(3,4-Methylenedioxybenzyl)hydantoin
In: Flavobacterium F938
→ 3,4-Methylenedioxy-L-phenylalanine
In: Flavobacterium F938

N-Carbamyl-D-phenylalanine
Formed from: D-5-Benzylhydantoin
→ D-5-Benzylhydantoin
In: Flavobacterium F938

N-Carbamyl-L-phenylalanine
Formed from: L-5-Benzylhydantoin
→ L-5-Benzylhydantoin
In: Flavobacterium F938
→ L-Phenylalanine
In: Flavobacterium F938

N-Carbamyl-L-tryptophan
Formed from: L-5-Indolylmethylhydantoin
→ L-5-Indolylmethylhydantoin
In: Flavobacterium F938
→ L-Tryptophan
In: Flavobacterium F938

N-Carbamyl-D-tyrosine
Formed from: 5-(*p*-Hydroxybenzyl)hydantoin
→ 5-(*p*-Hydroxybenzyl)hydantoin
In: Flavobacterium F938

N-Carbamyl-L-tyrosine
Formed from: 5-(*p*-Hydroxybenzyl)hydantoin
→ 5-(*p*-Hydroxybenzyl)hydantoin
In: Flavobacterium F938
→ L-Tyrosine
In: Flavobacterium F938

Carbaryl
See: 1-Naphthyl methylcarbamate

Carbazeran (1-(4-(ethylcarbamoyloxy)piperidino)-6,7-dimethoxy)phthalazine)
→ Demethylcarbazeran
In: Dog D161
See also: G852
→ 1-(4-(Ethylcarbamoyloxy)-hydroxypiperidino)-6,7-dimethoxyphthalazine
In: Man D161
→ 4-Hydroxycarbazeran
In: Baboon D361
Dog D161
Man D361
See also: G852
Contrast (in dog) D361
→ 1-(4-Hydroxypiperidino)-6,7-dimethoxyphthalazine
In: Guinea pig G852

Carbazole
Formed from: N-Acetylcarbazole
N-Benzylcarbazole
1-Carboxycarbazole
3-Carboxycarbazole
N-Methylcarbazole
→ 2′-Amino-2,3-dihydroxybiphenyl

Probably in: Pseudomonas H21
→ 2-Aminophenylacetic acid
In: Ralstonia K442
→ 6-(2-Aminophenyl)-6-oxohexanoic acid
In: Ralstonia K442
→ 5-(2-Aminophenyl)-5-oxopentanoic acid
In: Ralstonia K442
→ 3(2-Aminophenyl)propionic acid
In: Ralstonia K442
→ Anthranilic acid
In: Pseudomonas H94, H765
Ralstonia K442
→ 2-Carboxyindole-3-acetic acid
In: Ralstonia K442
→ 3-Hydroxycarbazole
In: Pseudomonas H83
See also: G90, H672
→ 4-Hydroxyquinoline
In: Ralstonia K442
→ Indole-3-acetic acid
In: Pseudomonas H765
→ Indole-2,3-dicarboxylic acid
In: Ralstonia K442
→ 3-Methyleneindole-2-carboxylic acid
In: Ralstonia K442

1-(Carbazol-(4)-oxy)-3-((β-D-glucuronosyl) ((2-methoxyphenoxy)ethyl)amino)-2-propanol
Formed from: Carvedilol

1-(9*H*-Carbazol-4-yloxy)-3-((2-(2-methoxyphenoxy)ethyl)amino)-2-propanol
See: Carvedilol

Carbendazim
See: Methyl benzimidazol-2-ylcarbamate

Carbenicillin (α-carboxybenzylpenicillin)
Formed from: Carfecillin

4-Carbethoxy-N-methyl-4-phenylpiperidine
See: Meperidine

Carbidopa
See: L-α-Hydrazino-3,4-dihydroxy-α-methylphenylpropionic acid

Carbidopa quinone
Probably formed from: L-α-Hydrazino-3,4-dihydroxy-α-methylphenylpropionic acid

N-Carbobenzoxy-3,4-dihydroxy-L-phenylalanine
Formed from: N-Carbobenzoxy-L-tyrosine

Carbobenzoxy-L-glutamyl-L-tyrosine
→ Carbobenzoxy-L-glutamyl-L-tyrosine-O-sulphate
In: Rat E197

Carbobenzoxy-L-glutamyl-L-tyrosine-O-sulphate
Formed from: Carbobenzoxy-L-glutamyl-L-tyrosine

Carbobenzoxyglycyl-L-tyrosine
→ Carbobenzoxyglycyl-L-tyrosine-O-sulphate
In: Rat E197

Carbobenzoxyglycyl-L-tyrosine-O-sulphate
Formed from: Carbobenzoxyglycyl-L-tyrosine

Carbobenzoxy-L-lysine
Formed from: Carbobenzoxy-L-lysyl-L-alanine
Carbobenzoxy-L-lysyl-L-alanyl-L-alanine
Carbobenzoxy-L-lysyl-L-alanyl-L-alanyl-L-alanine
Carbobenzoxy-L-lysylglycylglycylglycylglycine
Carbobenzoxy-L-lysylglycine
Carbobenzoxy-L-lysyl-L-leucine
Carbobenzoxy-L-lysyl-L-phenylalanine

Carbobenzoxy-L-lysyl-L-alanine
→ Carbobenzoxy-L-lysine
In: Rice D743

Carbobenzoxy-L-lysyl-L-alanyl-L-alanine
→ Carbobenzoxy-L-lysine
In: Rice D743

Carbobenzoxy-L-lysyl-L-alanyl-L-alanyl-L-alanine
→ Carbobenzoxy-L-lysine
In: Rice D743

Carbobenzoxy-L-lysylglycylglycylglycylglycine
→ Carbobenzoxy-L-lysine
In: Rice D743

Carbobenzoxy-L-lysylglycine
→ Carbobenzoxy-L-lysine
In: Rice D743

Carbobenzoxy-L-lysyl-L-leucine
→ Carbobenzoxy-L-lysine
In: Rice D743

Carbobenzoxy-L-lysyl-L-phenylalanine
→ Carbobenzoxy-L-lysine
In: Rice D743
→ L-Phenylalanine
In: Rice D743

N-Carbobenzoxy-L-tyrosine
→ N-Carbobenzoxy-3,4-dihydroxy-L-phenylalanine
In: Aspergillus A2711
S-(α-(Carbethoxy)benzyl) O,O-dimethyl phosphorothioate
Formed from: Phenthioate
S-(α-(Carbethoxy)benzyl) O-methyl phosphorodithioate
Formed from: Phenthioate
→ S-(α-(Carbethoxy)benzyl) O-methyl phosphorothioate
Probably in: Housefly A2661
Mouse A2661
→ S-(α-(Carboxy)benzyl) O-methyl phosphorodithioate
Probably in: Housefly A2661
Mouse A2661
S-(α-(Carbethoxy)benzyl) O-methyl phosphorothioate
Probably formed from: S-(α-(Carbethoxy) benzyl) O-methyl phosphorodithioate
4-Carbethoxy-4-(4-hydroxy-3-methoxyphenyl)-N-methylpiperidine
Probably formed from: 4-Carbethoxy-4-(*p*-hydroxyphenyl)-N-methylpiperidine
4-Carbethoxy-4-(*p*-hydroxyphenyl)-N-methylpiperidine
Probably formed from: Meperidine
→ 4-Carbethoxy-4-(4-hydroxy-3-methoxyphenyl)-N-methylpiperidine
Probably in: Rat A2659
N-Carbethoxyiminostilbene
→ N-Carbethoxyiminostilbene-10,11-oxide
In: Rat E140
N-Carbethoxyiminostilbene-10,11-oxide
Formed from: N-Carbethoxyiminostilbene
Carbofuran (2,3-dihydro-2,2-dimethylbenzofuran-7-yl methylcarbamate)
Formed from: Benfuracarb
2,3-Dihydro-2,2-dimethyl-7-benzofuranyl N-(di-*n*-butylaminosulphenyl) N-methylcarbamate
2,3-Dihydro-2,2-dimethyl-7-benzofuranyl N-morpholinosulphenyl N-methylcarbamate
→ 2,3-Dihydro-7-hydroxy-2,2-dimethyl-7-benzofuran
In: Achromobacter E21
Brussels sprout F433
Cauliflower F433
Nicotiana A402
Sugar beet F433
See also: A1948
→ 2,3-Dihydro-2,2-dimethyl-7-benzofuranyl carbamate
Probably in: Cotton A3871
Zeu A3871
→ 2,3-Dihydro-2,2-dimethyl-7-benzofuranyl N-hydroxymethylcarbamate
In: Trichogaster B731
See also: A1315, A3871, C359
→ 3-Hydroxycarbofuran
In: Aspergillus J641
Earthworm A788
Fusarium J641
Nicotiana A402
Pine A1315
Trichogaster B731
Zea A952, A2579
See also: A1948, A3081, A3871, B960, C359, F433
β-Carboline
See: Norharman
β-Carboline-3-carboxylic acid
Formed from: Propyl β-carboline-3-carboxylate
β-Carboline-1-propionic acid
Formed from: L-Tryptophan
→ Canthin-6-one
In: Ailanthus E304
5-(5-Carbonyloxy-β-D-glucuronido-2-hydroxymethylphenoxy)-2,2-dimethylpentanoic acid
Probably formed from: 5-(5-Carboxy-2-hydroxymethylphenoxy)-2,2-dimethylpentanoic acid
Carbostyril
See: 2-Hydroxyquinoline
Carbosulfan
See: 2,3-Dihydro-2,2-dimethyl-7-benzofuranyl N-(di-*n*-butylaminosulphenyl)-N-methylcarbamate

Carbosulfan sulphone
Formed from: 2,3-Dihydro-2,2-dimethyl-7-benzofuranyl N-(di-*n*-butylaminosulphenyl)-N-methylcarbamate

5-Carboxamido-5*H*-dibenzo[*a*,*d*]cycloheptene
See: Cytenamide

7-Carboxamido-2,3-dihydro-5,6-dihydroxy-1-(4-phenylbut-2-yl)indole
Formed from: Labetalol
→ 7-Carboxamido-2,3-dihydro-5,6-dihydroxy-1-(4-phenylbut-2-yl)indole-5-β-D-glucuronide
Probably in: Monkey G533
→ 7-Carboxamido-2,3-dihydro-5,6-dihydroxy-1-(4-phenylbut-2-yl)indole-6-β-D-glucuronide
Probably in: Monkey G533

7-Carboxamido-2,3-dihydro-5,6-dihydroxy-1-(4-phenylbut-2-yl)indole-5-β-D-glucuronide
Probably formed from: 7-Carboxamido-2,3-dihydro-5,6-dihydroxy-1-(4-phenylbut-2-yl) indole

7-Carboxamido-2,3-dihydro-5,6-dihydroxy-1-(4-phenylbut-2-yl)indole-6-β-D-glucuronide
Probably formed from: 7-Carboxamido-2,3-dihydro-5,6-dihydroxy-1-(4-phenylbut-2-yl) indole

7-Carboxamido-5,6-dihydroxy-1-(4-phenylbut-2-yl)indole
Formed from: Dilevalol
→ 7-Carboxamido-5,6-dihydroxy-1-(4-phenylbut-2-yl)indole-5-β-D-glucuronide
Probably in: Cynomolgus monkey H55
Man H55
→ 7-Carboxamido-5,6-dihydroxy-1-(4-phenylbut-2-yl)indole-6-β-D-glucuronide
Probably in: Cynomolgus monkey H55
Man H55

7-Carboxamido-5,6-dihydroxy-1-(4-phenylbut-2-yl)indole-5-β-D-glucuronide
Probably formed from: 7-Carboxamido-5,6-dihydroxy-1-(4-phenylbut-2-yl)indole

7-Carboxamido-5,6-dihydroxy-1-(4-phenylbut-2-yl)indole-6-β-D-glucuronide
Probably formed from: 7-Carboxamido-5,6-dihydroxy-1-(4-phenylbut-2-yl)indole

7-Carboxamido-5,6-dihydroxy-1-(4-phenylbut-2-yl)indole-5-β-D-glucuronide
Probably formed from: 7-Carboxamido-5,6-dihydroxy-1-(4-phenylbut-2-yl)indole

1-(*p*-Carboxamidophenyl)-3,3-dimethyltriazene
→ *p*-Aminobenzamide
In: Mouse C49
→ *p*-(3,3-Dimethyltriazeno)benzoic acid
In: Mouse C49

***p*-Carboxamidophenyl sulphate**
Formed from: *p*-Hydroxybenzamide

2-Carboxamido-1,2,3,4-tetrahydroisoquinoline
Formed from: N-Hydroxydebrisoquine

3-Carboxanilido-5,6-dihydro-2-methyl-1,4-oxathiin (carboxin, vitavax)
→ Aniline
In: Nocardia A2663
→ 3-Carboxanilido-5,6-dihydro-2-methyl-1,4-oxathiin sulphoxide
In: Arachis D52
Barley A1689
Wheat A1689
→ Malonanilic acid
In: Arachis D52

3-Carboxanilido-5,6-dihydro-2-methyl-1,4-oxathiin sulphone
Formed from: 3-Carboxanilido-5,6-dihydro-2-methyl-1,4-oxathiin sulphoxide

3-Carboxanilido-5,6-dihydro-2-methyl-1,4-oxathiin sulphoxide
Formed from: 3-Carboxanilido-5,6-dihydro-2-methyl-1,4-oxathiin
→ 3-Carboxanilido-5,6-dihydro-2-methyl-1,4-oxathiin sulphone
In: Barley A1689
Wheat A1689
See also: D52

Carboxin
See: 3-Carboxanilido-5,6-dihydro-2-methyl-1,4-oxathiin

9-Carboxyanthracene
Formed from: 9-Formylanthracene

_o_-Carboxybenzalpyruvic acid
Probably formed from: 1-Hydroxy-2-naphthoic acid
→ *o*-Formylbenzoic acid
With aldolase C730
See also: H813, J674

8-Carboxybenzanthracene
Probably formed from: 8-Formylbenzanthracene

_o_-Carboxybenzenesulphonamide
(*o*-sulphamoylbenzoic acid)
Formed from: Benz[*d*]isothiazoline-1,1-dioxide
o-Toluenesulphonamide
Unchanged in man and rat A2039

_p_-Carboxybenzenesulphonamide (*p*-sulphamoylbenzoic acid)
Formed from: 2,3-Dihydro-3-hydroxybenz[*d*]isothiazole-1,1-dioxide
p-Formylbenzenesulphonamide
Mafemide

_p_-Carboxybenzenesulphonic acid
Formed from: *p*-Formylbenzenesulphonic acid
→ Protocatechuic acid
In: Comamonas G244

N-(_o_-Carboxybenzenesulphonyl)glutamine
Probably formed from: 2-Carboxy-N-(2,6-dioxopiperidin-3-yl)benzenesulphonamide

3-(_p_-Carboxybenzenesulphonyl)-1-(2-_endo_-hydroxy-3-_endo_-bornyl)urea
1-((1*R*)-2-*endo*-Hydroxy-3-*endo*-bornyl)-3-(*p*-(α-hydroxy)toluenesulphonyl)urea

5-(_p_-Carboxybenzenesulphonyloxy)-2-hydroxybenzenesulphonic acid
Formed from: 2-Hydroxy-5-(*p*-hydroxymethylbenzenesulphonyloxy) benzenesulphonic acid

3-Carboxy-1,2-bezisoxazole
Formed from: Zonisamide

2-Carboxybenzoic acid
See: Phthalic acid

3-Carboxybenzoic acid
See: Isophthalic acid

4-Carboxybenzoic acid
See: Terephthalic acid

4-Carboxybenzophenone
→ 4-Carboxy-1,2-dihydro-1,2-dihydroxybenzophenone
In: Pseudomonas F697

5-(_o_-Carboxy-N-benzoyl)amino-2-nitrobenzoic acid
→ 5-Amino-2-nitrobenzoic acid
In: Xanthobacter J253
→ Phthalic acid
In: Xanthobacter J253

S-(α-Carboxybenzyl) O,O-dimethyl phosphorodithioate
Formed from: Phenthoate

S-(α-Carboxybenzyl) O-methyl phosphorodithioate
Probably formed from: S-(α-(Carbethoxy) benzyl) O-methyl phosphorodithioate
→ S-(α-Carboxybenzyl) phosphorothioate
Probably in: Housefly A2661
Mouse A2661

α-Carboxybenzylpenicillin
See: Carbenicillin

S-(α-Carboxybenzyl) phosphorothioate
Probably formed from: S-(α-Carboxybenzyl) O-methyl phosphorodithioate

4-Carboxybiphenyl
Formed from: 4-Formylbiphenyl
→ 4-Carboxybiphenyl-β-D-glucuronide
In: Rat C515
→ 4-Carboxy-1,2-dihydro-1,2-dihydroxybiphenyl
In: Pseudomonas F697

2′-Carboxy-4-bis(2-bromoethyl)aminoazobenzene
→ Anthranilic acid
In: Mouse A92
Rat A92
→ N,N-Di(2-bromoethyl)-*p*-phenylenediamine
In: Mouse A92
Rat A92

2′-Carboxy-4-bis(2-bromoethyl)amino-2-methylazobenzene
→ Anthranilic acid
In: Mouse A92
Rat A92
→ N,N-Di-(2-bromoethyl)-3-methyl-*p*-phenylenediamine
In: Mouse A92
Rat A92

2′-Carboxy-4-bis(2-bromopropyl)amino-2-methylazobenzene
→ Anthranilic acid
In: Baboon A92
Mouse A92
Rat A92
→ N,N-Di-(2-bromopropyl)-3-methyl-*p*-phenylenediamine
In: Baboon A92
Mouse A92
Rat A92

4′-(3-Carboxy-2-butenyloxy)-2′,4-bis(carboxymethoxy)dihydrochalcone
Probably formed from: 2′,4-Bis(carboxymethoxy)-4′-(3-hydroxymethyl-2-butenyloxy)chalcone

6-(*cis*-3-Carboxy-2-buten-1-yl)-2′,4′,5,7-tetrahydroxyisoflavone
Probably formed from: 4″-Hydroxyluteone

8-(4-Carboxy-2-butyl)amino-2,6-dihydroxy-4-methyl-5-(3-trifluoromethylphenoxy)quinoline
Formed from: 8-(4-Amino-1-methylbutyl)amino-2,6-dimethoxy-4-methyl-5-(3-trifluoromethylphenoxy)quinoline

8-(2-Carboxy-2-butyl)amino-5-hydroxy-2,6-dimethoxy-4-methylquinoline
Formed from: 8-(4-Amino-1-methylbutyl)amino-2,6-dimethoxy-4-methyl-5-(3-trifluoromethylphenoxy)quinoline
→ 8-(2-Carboxy-2-butyl)amino-2,5,6-trihydroxy-4-methylquinoline
Probably in: Rat H135

7-(4-Carboxy-2-butyl)amino-2,3,4,5-tetrahydroimidazo[2,1-*b*]quinazolin-2(1*H*)-one
Probably formed from: 2,3,4,5-Tetrahydro-7-(2-oxopiperidino)imidazo[2,1-*b*]quinazolin-2(1*H*)-one
→ 7-Amino-2,3,4,5-tetrahydroimidazo[2,1-*b*]quinazolin-2(1*H*)-one
Probably in: Rat H463
→ 7-(2-Carboxyethyl)amino-2,3,4,5-tetrahydroimidazo[2,1-*b*]quinazolin-2(1*H*)-one
Probably in: Rat H463

8-(2-Carboxy-2-butyl)amino-2,5,6-trihydroxy-4-methylquinoline
Probably formed from: 8-(2-Carboxy-2-butyl)amino-5-hydroxy-2,6-dimethoxy-4-methylquinoline

5-(4-Carboxybutyl)-1-(3,4-dichlorobenzyl)biguanide
Formed from: 1-(3,4-Dichlorobenzyl)-5-octylbiguanide

1-(4-Carboxybutyl)-N-ethyl-N-methyl-4-propyl-4-piperidinecarboxamide
Probably formed from: 5′-Hydroxysameride

N-(4-Carboxybutyl)-2-hydroxy-3-(*o*-tolyloxy)propylamine
Formed from: N-(4-(Ethoxycarbonyl)butyl)-2-hydroxy-3-(*o*-tolyloxy)propylamine

3-(3′-Carboxybutyl)-2-methyl-1,4-naphthoquinone
Formed from: Vitamin K_1

6-(3′-Carboxybutyl)-2,3,5-trimethyl-1,4-benzoquinone
Probably formed from: α-Tocopherol

1-Carboxycarbazole
Formed from: 1-Formylcarbazole
→ Carbazole
In: Pseudomonas J180

3-Carboxycarbazole
Formed from: 3-Formylcarbazole
→ Carbazole
In: Pseudomonas J180

4-Carboxycarbostyril
Formed from: 4-Formylcarbostyril

5-Carboxy-3-(5-carboxy-2-hydroxy-3-methoxyphenyl)-*o*-quinone
Probably formed from: 5,5′-Dicarboxy-2,2′-dihydroxy-3,3′-dimethoxybiphenyl

3-Carboxy-5-(carboxymethoxy)carbonyl-2,6-dimethyl-4-(*m*-nitrophenyl)pyridine
Probably formed from: 3-Carboxy-5-(2-hydroxyethoxy)carbonyl-2,6-dimethyl-4-(*m*-nitrophenyl)pyridine

3-Carboxy-2-((carboxymethoxy)methyl)-4-(*o*-chlorophenyl)-1,4-dihydro-5-methoxycarbonyl-6-methylpyridine
Probably formed from: 2-((Carboxymethoxy)methyl)-4-(*o*-chlorophenyl)-3-ethoxycarbonyl-1,4-dihydro-5-methoxycarbonyl-6-methylpyridine

5-Carboxy-2-((carboxymethoxy)methyl)-4-(*o*-chlorophenyl)-3-ethoxycarbonyl-6-hydroxymethylpyridine
Probably formed from: 5-Carboxy-2-((carboxymethoxy)methyl)-4-(*o*-chlorophenyl)-3-ethoxycarbonyl)-6-methylpyridine

5-Carboxy-2-((carboxymethoxy)methyl)-4-(*o*-chlorophenyl)-3-ethoxycarbonyl-6-methylpyridine
Probably formed from: 2-((Carboxymethoxy) methyl)-4-(*o*-chlorophenyl)-3-ethoxycarbonyl-6-methylpyridine
→ 5-Carboxy-2-((carboxymethoxy)methyl)-4-(*o*-chlorophenyl)-3-ethoxycarbonyl-6-hydroxymethylpyridine
Probably in: Dog E836

5-Carboxy-2-((carboxymethoxy)methyl)-4-(*o*-chlorophenyl)-3-ethoxycarbonyl-6-methylpyridine
Probably formed from: 2-((Carboxymethoxy) methyl)-4-(*o*-chlorophenyl)-3-ethoxycarbonyl-5-methoxycarbonyl-6-methylpyridine

3-Carboxy-2-((carboxymethoxy)methyl)-4-(*o*-chlorophenyl)-5-methoxycarbonyl-6-methylpyridine
Probably formed from: 2-((2-Aminoethoxy) methyl)-4-(*o*-chlorophenyl)-3-ethoxycarbonyl-5-methoxycarbonyl-6-methylpyridine

5-Carboxy-7-(1-carboxy-1-methyl)ethyl)-2,3-dihydro-2-hydroxy-3,3-dimethylbenzofuran
Formed from: 2,6-Di-*tert*-butyl-4-methylphenol

3-Carboxy-7-chloro-1,2,4-benzothiadiazine-1,1-dioxide
Formed from: 7-Chloro-3-hydroxymethyl-1,2,4-benzothiadiazine-1,1-dioxide

3-Carboxy-7-(4-(*p*-chlorobenzyl)piperazin-1-yl)propoxy-4-methylcoumarin
Probably formed from: 7-(4-(*p*-Chlorobenzyl) piperazin-1-yl)propoxy-3-hydroxymethyl-4-methylcoumarin

4-Carboxy-4′-chlorobiphenyl
Formed from: 4-Chloro-4′-isopropylbiphenyl
→ 4-Carboxy-4′-chloro-3′-hydroxybiphenyl
Probably in: Rat A2531

4-Carboxy-4′-chloro-3′-hydroxybiphenyl
Probably formed from: 4-Carboxy-4′-chlorobiphenyl

N-(2-Carboxy-3-chlorophenyl)anthranilic acid
Probably formed from: N-(3-Chloro-2-formylphenyl)anthranilic acid

1-Carboxy-2-(2-chloro-2-phenylethenyl)-3,3-dimethylcyclopropane
Formed from: Flumethrin

2-Carboxy-4-(*o*-chlorophenyl)-3-ethoxycarbonyl-5-methoxycarbonyl-6-methylpyridine
Probably formed from: 4-(*o*-Chlorophenyl)-3-ethoxycarbonyl-2-hydroxymethyl-5-methoxycarbonyl-6-methylpyridine

5-Carboxy-4-(*o*-chlorophenyl)-3-ethoxycarbonyl-6-methyl-2-(4-(methylimidazo[4,5-*c*]pyrid-1-yl)phenyl)pyridine
Probably formed from: 1,4-Dehydromodipafant

3-Carboxy-4-(*o*-chlorophenyl)-2-hydroxymethyl-5-methoxycarbonyl-6-methylpyridine
Probably formed from: 4-(*o*-Chlorophenyl)-3-ethoxycarbonyl-2-hydroxymethyl-5-methoxycarbonyl-6-methylpyridine
→ 3,5-Dicarboxy-4-(*o*-chlorophenyl)-2-hydroxymethyl-6-methylpyridine
Probably in: Dog F41

5-Carboxy-2-(3-(4-(*m*-chlorophenyl)piperazin-1-yl)propyl)-2,4-dihydro-4-(2-phenoxyethyl)-3*H*-1,2,4-triazol-3-one
Probably formed from: 2-(3-(4-(*m*-Chlorophenyl)piperazin-1-yl)propyl)-2,4-dihydro-5-(1-hydroxyethyl)-4-(2-phenoxyethyl)-3*H*-1,2,4-triazol-3-one

***o*-Carboxy-*cis*-cinnamic acid**
Formed from: 1,2-Dihydroxynaphthalene

5-Carboxy-2-cyano-6-hydroxymethyl-3-methoxycarbonyl-4-(*m*-nitrophenyl)pyridine
Probably formed from: 5-Carboxy-2-cyano-3-methoxycarbonyl-6-methyl-4-(*m*-nitrophenyl)pyridine

5-Carboxy-2-cyano-3-methoxycarbonyl-6-methyl-4-(*m*-nitrophenyl)pyridine
Probably formed from: Deisopropylnilvadipine
→ 5-Carboxy-2-cyano-6-hydroxymethyl-3-methoxycarbonyl-4-(*m*-nitrophenyl) pyridine
Probably in: Dog E818
Rat E818

4-Carboxy-2-decaprenylphenol
Formed from: *p*-Hydroxybenzoic acid

35-Carboxy-35-demethylmontelukast
Probably formed from: (*R*)-1-((1-(3-(2-(7-Chloro-2-quinolinyl)ethenyl)phenyl)-3-(2-(1,2-dihydroxy-2-propyl)phenyl)propyl)thiomethyl) cyclopropaneacetic acid

5′-Carboxy-5-(5,5′-dicarboxy-2′-hydroxy-3,3′-dimethoxy-2-biphenyloxy)-2,2′-dihydroxy-3,3′-dimethoxybiphenyl
Formed from: 5,5′-Dicarboxy-2,2′-dihydroxy-3,3′-dimethoxybiphenyl

4-Carboxy-2′,5′-dichlorobiphenyl
Formed from: 2,5-Dichloro-4′-isopropylbiphenyl

2-Carboxy-2′,6′-dichloro-4′-hydroxy-3′-hydroxymethyldiphenylamine
Probably formed from: 2-Carboxy-2′,6′-dichloro-3′-hydroxymethyldiphenylamine

2-Carboxy-2′,6′-dichloro-3′-hydroxymethyldiphenylamine
Formed from: Meclofenamic acid
→ 2-Carboxy-2′,6′-dichloro-4′-hydroxy-3′-hydroxymethyldiphenylamine
Probably in: Man A3561
→ 2′,3-Dicarboxy-2,6-dichlorodiphenylamine
Probably in: Man A3561

2-Carboxy-2′,6′-dichloro-4-hydroxy-3′-methyldiphenylamine
Formed from: Meclofenamic acid

2-Carboxy-2′,6′-dichloro-4′-hydroxy-3′-methyldiphenylamine
Formed from: Meclofenamic acid

***o*-((3-Carboxy-2,6-dichloro-4-hydroxyphenyl) amino)benzoic acid**
Probably formed from: *o*-((3-Carboxy-2,6-dichlorophenyl)amino)benzoic acid
→ *o*-((2,6-Dichloro-4-hydroxyphenyl) amino)benzoic acid
Probably in: Man E707

2-Carboxy-2′,6′-dichloro-3′-methyldiphenylamine
See: Meclofenamic acid

4-Carboxy-2,3-dichlorophenoxyacetic acid
Formed from: Tienilic acid

***o*-((3-Carboxy-2,6-dichlorophenyl)amino)benzoic acid**
Probably formed from: *o*-((2,6-Dichloro-3-hydroxymethylphenyl)amino)benzoic acid
→ *o*-((3-Carboxy-2,6-dichloro-4-hydroxyphenyl)amino)benzoic acid
Probably in: Man E707

3-Carboxy-4-(2,3-dichlorophenyl)-1,4-dihydro-2-hydroxymethyl-5-methoxycarbonyl-6-methylpyridine
Formed from: Felodipine

3-Carboxy-4-(2,3-dichlorophenyl)-5-ethoxycarbonyl-1,4-dihydro-2-hydroxymethyl-6-methylpyridine
Formed from: Felodipine

3-Carboxy-4-(2,3-dichlorophenyl)-5-ethoxycarbonyl-2,6-dimethylpyridine
Probably formed from: Didehydrofelodipine
→ 3-Carboxy-4-(2,3-dichlorophenyl)-5-ethoxycarbonyl-2-hydroxymethyl-6-methylpyridine
Probably in: Rat E846
→ 4-(2,3-Dichlorophenyl)-3-ethoxycarbonyl-2,6-dimethylpyridine
Probably in: Dog E862
Man E862
Rat E862

3-Carboxy-4-(2,3-dichlorophenyl)-5-ethoxycarbonyl-2-hydroxymethyl-6-methylpyridine
Probably formed from: 3-Carboxy-4-(2,3-dichlorophenyl)-5-ethoxycarbonyl-2-hydroxymethyl-6-methylpyridine
→ 2,3-Dicarboxy-4-(2,3-dichlorophenyl)-5-ethoxycarbonyl-6-methylpyridine
Probably in: Rat E846

3-Carboxy-4-(2,3-dichlorophenyl)-2-hydroxymethyl-5-methoxycarbonyl-6-methylpyridine
Probably formed from: 3-Carboxy-4-(2,3-dichlorophenyl)-5-methoxycarbonyl-2,6-dimethylpyridine
→ 2,3-Dicarboxy-4-(2,3-dichlorophenyl)-5-methoxycarbonyl-6-methylpyridine
Probably in: Rat E846

3-Carboxy-4-(2,3-dichlorophenyl)-5-methoxycarbonyl-2,6-dimethylpyridine
Probably formed from: Didehydrofelodipine

→ 3-Carboxy-4-(2,3-dichlorophenyl)-2-hydroxymethyl-5-methoxycarbonyl-6-methylpyridine
Probably in: Dog E862
Man E862
Rat E862
→ 4-(2,3-Dichlorophenyl)-3-methoxycarbonyl-2,6-dimethylpyridine
Probably in: Dog E862
Man E862
Rat E862

3-Carboxy-4-(*o*-difluoromethoxyphenyl)-2-(α-hydroxy)methyl-5-methoxycarbonyl-6-methylpyridine
Formed from: Ryosidine

4-Carboxy-1,2-dihydro-1,2-dihydroxybenzophenone
Formed from: 4-Carboxybenzophenone

4-Carboxy-1,2-dihydro-1,2-dihydroxybiphenyl
Formed from: 4-Carboxybiphenyl

7-Carboxy-3,4-dihydro-8-hydroxy-3-methylisocoumarin
Formed from: Ochratoxin B

2-Carboxy-3,4-dihydro-3-(*o*-methylphenyl)-4-oxopyrido[2,3-*d*]pyrimidine
Probably formed from: 3,4-Dihydro-2-hydroxymethyl-3-(*o*-methylphenyl)-4-oxopyrido[2,3-*d*]pyrimidine

3-Carboxy-1,8-dihydroxy-6-methoxyanthraquinone
Formed from: Physcion

4-Carboxy-3,5-dihydroxyphenyl sulphate
Probably formed from: 2,4,6-Trihydroxybenzoic acid

5-Carboxy-2,2′-dihydroxy-3,3′,5′-trimethoxybiphenyl
Formed from: 5,5′-Dicarboxy-2,2′-dihydroxy-3,3′-dimethoxybiphenyl

4-Carboxy-2,5-dimethoxyamphetamine
Formed from: 4-Hydroxymethyl-2,5-dimethoxyamphetamine

6-Carboxy-3,8-dimethoxybenzofuran-1,4-quinone
Formed from: 5,5′-Dicarboxy-2,2′-dihydroxy-3,3′-dimethoxybiphenyl

4-Carboxy-6,7-dimethoxycoumarin
Formed from: 4-Formyl-6,7-dimethoxycoumarin

5-Carboxy-2-(3,4-dimethoxyphenyl)-2,3-dihydro-3-hydroxymethyl-7-methoxybenzofuran
Formed from: 2-(3,4-Dimethoxyphenyl)-5-formyl-2,3-dihydro-3-hydroxymethyl-7-methoxybenzofuran
→ 5-Carboxy-2-(3,4-dimethoxyphenyl)-3-formyl-2,3-dihydro-7-methoxybenzofuran
In: Phanerochaete B796
→ Veratric acid
In: Phanerochaete B796

5-Carboxy-2-(3,4-dimethoxyphenyl)-3-formyl-2,3-dihydro-7-methoxybenzofuran
Formed from: 5-Carboxy-2-(3,4-dimethoxyphenyl)-2,3-dihydro-3-hydroxymethyl-7-methoxybenzofuran

2-Carboxy-4-dimethylaminoazobenzene
→ 2-Amino-5-dimethylaminobenzoic acid
In: Rat B28
→ Aniline
In: Rat B28

3-Carboxy-4′-dimethylaminoazobenzene
Probably formed from: 4′-Dimethylamino-3-formylbenzoic acid
→ 3-Carboxy-4′-methylaminoazobenzene
In: Rat A776
See also: A3783, A3835, E872
→ *m*-Aminobenzoic acid
Probably in: Rat A3835

4-Carboxy-6-dimethylaminocarbostyril
Formed from: 6-Dimethylamino-4-formylcarbostyril

4-Carboxy-6-dimethylaminocoumarin
Formed from: 6-Dimethylamino-4-formylcoumarin

3-(4-Carboxy-1,1-dimethylbutyl)-6,6a,7,8,10,10a-hexahydro-1,9-dihydroxy-6,6-dimethyl-9*H*-dibenzo[*b*,*d*]pyran
Probably formed from: 3-(6-Carboxy-1,1-dimethylhexyl)-6,6a,7,8,10,10a-hexahydro-1,9-dihydroxy-6,6-dimethyl-9*H*-dibenzo[*b*,*d*]pyran
→ 3-(2-Carboxy-1,1-dimethylethyl)-6,6a,7,8,10,10a-hexahydro-1,9-dihydroxy-6,6-dimethyl-9*H*-dibenzo[*b*,*d*]pyran
Probably in: Nocardia B37

3-(4-Carboxy-1,1-dimethylbutyl)-6,6a,7,8,10,10a-hexahydro-1-hydroxy-6,6-dimethyl-9*H*-dibenzo[*b*,*d*]pyran-9-one
Probably formed from: 7′-Oxonabilone
→ 3-(2-Carboxy-1,1-dimethylethyl)-6,6a,7,8,10,10a-hexahydro-1-hydroxy-6,6-dimethyl-9*H*-dibenzo[*b*,*d*]pyran-9-one
Probably in: Nocardia A3010, B37

3-(2-Carboxy-1,1-dimethylethyl)-6,6a,7,8,10,10a-hexahydro-1,9-dihydroxy-6,6-dimethyl-9*H*-dibenzo[*b,d*]pyran
Probably formed from: 3-(4-Carboxy-1,1-dimethylbutyl)-6,6a,7,8,10,10a-hexahydro-1,9-dihydroxy-6,6-dimethyl-9*H*-dibenzo[*b*,*d*]pyran

3-(2-Carboxy-1,1-dimethylethyl)-6,6a,7,8,10,10a-hexahydro-1-hydroxy-6,6-dimethyl-9*H*-dibenzo[*b,d*]pyran-9-one
Probably formed from: 3-(4-Carboxy-1,1-dimethylbutyl)-6,6a,7,8,10,10a-hexahydro-1-hydroxy-6,6-dimethyl-9*H*-dibenzo[*b*,*d*]pyran-9-one

3-(6-Carboxy-1,1-dimethylhexyl)-6,6a,7,8,10,10a-hexahydro-1,9-dihydroxy-6,6-dimethyl-9*H*-dibenzo[*b,d*]pyran
→ 3-(4-Carboxy-1,1-dimethylbutyl)-6,6a,7,8,10,10a-hexahydro-1,9-dihydroxy-6,6-dimethyl-9*H*-dibenzo[*b,d*]pyran
Probably in: Nocardia B37

2-Carboxy-4′-dimethylaminoazobenzene
See: Methyl red

4-Carboxy-4′-dimethylaminoazobenzene
See: *p*-Methyl red

3-(6-Carboxy-1,1-dimethylhexyl)-6,6α,7,8,10,10α-hexahydro-1,9-dihydroxy-6,6-dimethyl-9*H*-dibenzo[*b,d*]pyran
Probably formed from: Dihydro-7′-oxonabilone

5′-Carboxy-4,8-dimethylpsoralen
Formed from: 4,5′,8-Trimethylpsoralen

3-(4-Carboxy-1,3-dioxobutyl)-4-hydroxy-6-phenyl-2-pyrone
Formed from: Norwogonin

2-Carboxy-2-(1,3-dioxolan-2-yl)-6-nitro-4-(2-oxopyrrolidin-1-yl)-2*H*-1-benzopyran
Probably formed from: 2-(1,3-Dioxolan-2-yl)-2-formyl-6-nitro-4-(2-oxopyrrolidin-1-yl)-2*H*-1-benzopyran

2-Carboxy-N-(2,6-dioxopiperidin-3-yl)benzenesulphonamide
Probably formed from: 3-(2,3-Dihydro-1,1-dioxido-3-oxo-1,2-benzisothiazo-2-yl)-2,6-dioxopiperidine
→ N-(*o*-Carboxybenzenesulphonyl)glutamine
Probably in: Rat C388

4-Carboxy-2-diprenylphenol
See: 3-Geranyl-4-hydroxybenzoic acid

6-Carboxy-2-((di-*n*-propyl)amino)-5-hydroxytetralin
Formed from: 2-((Di-*n*-propyl)amino)-5-hydroxy-6-methyltetralin
→ 6-Carboxy-5-hydroxy-2-((*n*-propyl)amino)tetralin
Probably in: Rat D819

(3*R*,4*R*)-3-((1-Carboxyethenyl)oxy)-4-hydroxy-1,5-cyclohexadiene-1-carboxylic acid
See: Chorismic acid

4-(2-(2-Carboxy)ethenyl)phenyl-2,6-dimethyl-3,5-pyridinedicarboxylic acid diethyl ester
Probably formed from: 1,4-Dehydrolacidipine

5-(4-(2-Carboxyethoxy)-3-ethoxyphenyl)-2,4-thiazolidinedione
Probably formed from: 5-(4-(3,4-Dihydroxypentyloxy)-3-ethoxyphenyl)-2,4-thiazolidinedione

4′-((2-Carboxy)ethoxy)-2′-methoxyacetophenone
Formed from: Iloperidone

5-Carboxy-2-(4-ethoxy-3-methoxyphenyl)-2,3-dihydro-3-hydroxymethyl-7-methoxybenzofuran
Formed from: 5-Carboxy-2-(4-ethoxy-3-methoxyphenyl)-3-formyl-2,3-dihydro-7-methoxybenzofuran

5-Carboxy-2-(4-ethoxy-3-methoxyphenyl)-3-formyl-2,3-dihydro-7-methoxybenzofuran
→ 5-Carboxy-2-(4-ethoxy-3-methoxyphenyl)-2,3-dihydro-3-hydroxymethyl-7-methoxybenzofuran
In: Phanerochaete B796
→ 3,5-Dicarboxy-2-(4-ethoxy-3-methoxyphenyl)-2,3-dihydro-7-methoxybenzofuran
In: Phanerochaete B796

→ 4-Ethoxy-3-methoxybenzoic acid
In: Phanerochaete B796

1-(4-Carboxy-2-ethoxyphenoxy)propan-2-ol
Formed from: 3-Ethoxy-4-propoxybenzoic acid

1-(4-Carboxy-2-ethoxyphenoxy)propan-3-ol
Formed from: 3-Ethoxy-4-propoxybenzoic acid

7-(2-Carboxyethyl)amino-2,3,4,5-tetrahydroimidazo[2,1-*b*]quinazolin-2(1*H*)-one
Probably formed from: 7-(4-Carboxybutyl) amino-2,3,4,5-tetrahydroimidazo[2,1-*b*] quinazolin-2(1*H*)-one

α-Carboxy-7-ethyl-2-benzofuranmethanol
Formed from: Bufuralol

1-Carboxy-1-ethyl N-benzoyl-N-(3-chloro-4-fluorophenyl)-2-aminopropionate
Probably formed from: 1-Hydroxyisopropyl N-benzoyl-N-(3-chloro-4-fluorophenyl)-2-aminopropionate

N-(2-Carboxy-1-ethylcarbonyl)-N′-(2,6-dimethylphenyl)urea
Probably formed from: N-(2,6-Dimethylphenyl)-N′-(N-methylsuccinamido)urea

2-Carboxyethyl 3-carboxy-2,6-dimethyl-4-(*m*-nitrophenyl)pyridine-5-carboxylate
Probably formed from: 2-Carboxyethyl 2-hydroxypropyl 2,6-dimethyl-4-(*m*-nitrophenyl)pyridine-3,5-dicarboxylate
→ 2-Carboxyethyl 3-carboxy-2-hydroxymethyl-6-methyl-4-(*m*-nitrophenyl) pyridine-5-carboxylate
Probably in: Dog G593
Rat G593

2-Carboxyethyl 3-carboxy-2-hydroxymethyl-6-methyl-4-(*m*-nitrophenyl)pyridine-5-carboxylate
Probably formed from: 2-Carboxyethyl 3-carboxy-2,6-dimethyl-4-(*m*-nitrophenyl) pyridine-5-carboxylate

1-Carboxyethyl (*S*)-2-carboxymethyl-7-fluoro-3,4-dihydro-3-oxo-2*H*-quinoxaline-1-carboxylate
Probably formed from: 1-Carboxyethyl (*S*)-2-ethyl-7-fluoro-3,4-dihydro-3-oxo-2*H*-quinoxaline-1-carboxylate

4-(1-Carboxyethyl)-4′-chlorobiphenyl
Formed from: 4-Chloro-4′-isopropylbiphenyl

4-(1-Carboxyethyl)-4′-chloro-3′-hydroxybiphenyl
Probably formed from: 4-(1-Carboxyethyl)-4′-chlorobiphenyl

10-Carboxyethyl-2-chloro-7-hydroxyphenothiazine
Probably formed from: 10-Carboxyethyl-2-chlorophenothiazine

N-(2-Carboxyethyl)-3-chloroiminodibenzyl
Formed from: 3-Chlorodemethylimipramine

10-Carboxyethyl-2-chlorophenothiazine
Formed from: Chlorpromazine
Demethylchlorpromazine
→ 10-Carboxyethyl-2-chloro-7-hydroxyphenothiazine
Probably in: Man A693
→ 10-Carboxymethyl-2-chlorophenothiazine
Probably in: Man A693
→ 10-Carboxyethyl-2-chlorophenothiazine sulphoxide
Probably in: Man A693

10-Carboxyethyl-2-chlorophenothiazine sulphoxide
Probably formed from: 10-Carboxyethyl-2-chlorophenothiazine

N-(2-Carboxyethyl)-3-(*o*-chlorophenoxy)-2-hydroxypropylamine
Formed from: 3-(*o*-Chlorophenoxy)-N-(2-(ethoxycarbonyl)ethyl)-2-hydroxypropylamine

6-(2-Carboxyethyl)chromone-2-carboxylic acid
Formed from: 6-*n*-Propylchromone-2-carboxylic acid

5-(2-Carboxyethyl)-5*H*-dibenzo[*a*,*d*]cycloheptene
Formed from: Protiptyline

4-(1-Carboxyethyl)-2′,5′-dichlorobiphenyl
Formed from: 2,5-Dichloro-4′-isopropylbiphenyl

N-(2-Carboxyethyl)-N′-(2,6-dimethylphenyl)urea
Formed from: Recainam

1-Carboxyethyl (*S*)-2-ethyl-7-fluoro-3,4-dihydro-3-oxo-2*H*-quinoxaline-1-carboxylate
Probably formed from: 1-Hydroxy-2-propyl (*S*)-2-ethyl-7-fluoro-3,4-dihydro-3-oxo-2*H*-quinoxaline-1-carboxylate
→ 1-Carboxyethyl (*S*)-2-carboxymethyl-7-fluoro-3,4-dihydro-3-oxo-2*H*-quinoxaline-1-carboxylate
Probably in: Man K29

4-(1-Carboxyethyl)-2-fluorobiphenyl
See: Flurbiprofen

3-(2-Carboxyethyl)-3-(*p*-fluorophenyl)-5-isobenzofurancarbonitrile
Formed from: Citalopram
→ 3-(2-Carboxyethyl)-3-(*p*-fluorophenyl)-5-isobenzofurancarbonitrile-β-D-glucuronide
Probably in: Man J758

3-(2-Carboxyethyl)-3-(*p*-fluorophenyl)-5-isobenzofurancarbonitrile-β-D-glucuronide
Probably formed from: 3-(2-Carboxyethyl)-3-(*p*-fluorophenyl)-5-isobenzofurancarbonitrile

5-(2-Carboxyethyl)-6-hydroxy-4-methoxybenzofuran
Formed from: 5-Methoxypsoralen

N-(2-Carboxyethyl)-2-hydroxy-3-(α-naphthyloxy)propylamine
Formed from: N-((2-Ethoxycarbonyl)ethyl)-2-hydroxy-3-(α-naphthyloxy)propylamine

***trans*-2-(2-Carboxyethyl)-3-(*p*-hydroxyphenyl)-N-phenyllactam**
Formed from: *trans*-2-(Butoxycarbonylethyl)-3-(*p*-hydroxyphenyl)-N-phenyllactam
trans-2-(Ethoxycarbonylethyl)-3-(*p*-hydroxyphenyl)-N-phenyllactam
trans-2-(Heptoxycarbonylethyl)-3-(*p*-hydroxyphenyl)-N-phenyllactam
trans-2-(Hexoxycarbonylethyl)-3-(*p*-hydroxyphenyl)-N-phenyllactam
trans-3-(*p*-Hydroxyphenyl)-2-methoxycarbonylethyl-N-phenyllactam
trans-3-(*p*-Hydroxyphenyl)-2-methoxyethoxycarbonylethyl-N-phenyllactam
trans-3-(*p*-Hydroxyphenyl)-2-pentoxycarbonylethyl-N-phenyllactam
trans-3-(*p*-Hydroxyphenyl)-N-phenyl-2-propoxycarbonylethyl-N-phenyllactam
trans-3-(*p*-Hydroxyphenyl)-N-phenyl-2-(2,2,2-trifluoroethoxycarbonylethyl)-N-phenyllactam

2-(2-Carboxyethyl) 2-hydroxypropyl 2,6-dimethyl-4-(*m*-nitrophenyl)pyridine-3,5-dicarboxylate
Probably formed from: 2-Hydroxypropyl 3-hydroxypropyl 2,6-dimethyl-4-(*m*-nitrophenyl)pyridine-3,5-dicarboxylate
→ 2-Carboxyethyl 3-carboxy-2,6-dimethyl-4-(*m*-nitrophenyl)pyridine-5-carboxylate
Probably in: Dog G593
Rat G593

2-(2-Carboxyethyl)-6-hydroxy-2,5,7,8-tetramethylchroman
Formed from: α-Tocopherol

N-(2-Carboxyethyl)-2-hydroxy-3-(*o*-tolyloxy)propylamine
Formed from: N-(2-Ethoxycarbonyl)-2-hydroxy-3-(*o*-tolyloxy)propylamine

N-(2-Carboxyethyl)iminodibenzyl
Formed from: Demethylimipramine

N-(2-Carboxyethyl)laudanosinium
Formed from: Atracurium

4-(1-Carboxyethyl)-3-methyl-1-phenyl-3-pyrazolin-5-one
Probably formed from: 4-(1-Isopropyl)-3-methyl-1-phenyl-3-pyrazolin-5-one

N-Carboxyethyl-4-methylsulphonyl-2,6-dinitro-N-propylamine
Probably formed from: Nitralin

10-(2-Carboxyethyl)phenothiazine
Formed from: Demethylpromazine
Promazine

N-(1-(*R*)-(Carboxy)ethyl)-L-phenylalanine
Formed from: L-Phenylalanine

N-(1-(*R*)-(Carboxy)ethyl)-L-phenylalaninol
Formed from: L-Phenylalaninol

4-(2-Carboxyethyl)phenyl *trans*-4-aminomethylcyclohexanecarboxylate
→ Phloretic acid
Probably in: Rat A3407

***p*-(N-(2-Carboxyethyl)-N-propylsulphamoyl)benzoic acid**
Formed from: *p*-(N-(3-Hydroxypropyl)-N-propylsulphamoyl)benzoic acid

4-(2-Carboxyethyl)-*o*-quinone
Formed from: Dihydrocaffeic acid
→ Dihydroesculetin quinone
In: Mushroom K140

2-(3-Carboxyethyl)-3-((2-(1*H*-tetrazol-5-yl)(1,1′-biphenyl)-4-yl)methyl)-1,3-diaza*spiro*[4,4]non-1-en-4-one
Formed from: Irbesartan

α-(2-Carboxyethylthio)-*p*-chloro-N-methylbenzylamine-S,S-dioxide
Formed from: Chlormezanone (artifact?)
→ *p*-Chlorobenzaldehyde

Probably in: Man E130
→ *p*-Chloro-N-methylbenzaldimine
Probably in: Man E130

7-Carboxy-4-(*p*-fluorophenyl)-1-isopropyl-2(1*H*)-quinazolinone
Formed from: 4-(*p*-Fluorophenyl)-7-hydroxymethyl-1-isopropyl-2(1*H*)-quinazolinone

5-Carboxy-3-formyl-2-(4-hydroxy-3,5-dimethoxyphenyl)-7-methoxycoumarone
Probably formed from: 5-Carboxy-2-(4-hydroxy-3,5-dimethoxyphenyl)-3-hydroxymethyl-7-methoxycoumarone
→ 5-Carboxy-2-(4-hydroxy-3,5-dimethoxyphenyl)-7-methoxycoumarone
Probably in: Fusarium E174

2-Carboxy-2-(S-glutathionyl)-2-phenylethanol
Probably formed from: 1-Carboxystyrene oxide

4-Carboxy-2-heptaprenylphenol
Formed from: *p*-Hydroxybenzoic acid

6-(7-Carboxyheptyl)-2,3-dimethoxy-5-methylquinol
Probably formed from: 6-(7-Carboxyheptyl)-2,3-dimethoxy-5-methylquinone
→ 6-(7-Carboxyheptyl)-2,3-dimethoxy-5-methylquinol-β-D-glucuronide
In: Dog E119
Rat E119
→ 6-(7-Carboxyheptyl)-2,3-dimethoxy-5-methylquinol sulphate
In: Dog E119
Rat E119

6-(7-Carboxyheptyl)-2,3-dimethoxy-5-methylquinol-β-D-glucuronide
Formed from: 6-(7-Carboxyheptyl)-2,3-dimethoxy-5-methylquinol

6-(7-Carboxyheptyl)-2,3-dimethoxy-5-methylquinol sulphate
Formed from: 6-(7-Carboxyheptyl)-2,3-dimethoxy-5-methylquinol

6-(7-Carboxyheptyl)-2,3-dimethoxy-5-methylquinone
Probably formed from: 6-(9-Carboxynonyl)-2,3-dimethoxy-5-methylquinone
→ 6-(7-Carboxyheptyl)-2,3-dimethoxy-5-methylquinol
Probably in: Dog E119
Rat E119
→ 6-(5-Carboxypentyl)-2,3-dimethoxy-5-methylquinone
Probably in: Dog D764, E119
Rat D764, E119

N-(1-(1*H*)-(1-Carboxy-1-(*n*-heptyl))imidazol-4-yl)-6-fluorophthalamidic acid
→ 3-Fluorophthalic acid
In: Xanthobacter J253, K793

N-(1-(1*H*)-(1-Carboxy-1-(*n*-heptyl))imidazol-4-yl)phthalamidic acid
→ Phthalic acid
In: Xanthobacter J253

4-Carboxy-2-hexaprenyl-6-methoxyphenol
Formed from: *p*-Hydroxybenzoic acid
→ Ubiquinone 30
In: Saccharomyces B870

4-Carboxy-2-hexaprenylphenol
Formed from: *p*-Hydroxybenzoic acid
→ 2-Hexaprenyl-6-methoxyphenol
In: Saccharomyces A3353

2-Carboxy-3-hydroxybenzo[*b*]thiophene
Probably formed from: 2-Formyl-3-hydroxybenzo[*b*]thiophene

3-Carboxy-8-hydroxychromone
Probably formed from: 8-Hydroxy-3-hydroxymethylchromone
→ 8-Hydroxychromone
Probably in: Dog A1761

4-Carboxy-6-hydroxycoumarin
Formed from: 4-Formyl-6-hydroxycoumarin

5-Carboxy-2-(4-hydroxy-3,5-dimethoxyphenyl)-3-hydroxymethyl-7-methoxycoumarin
Formed from: 5-Formyl-2-(4-hydroxy-3,5-dimethoxyphenyl)-3-hydroxymethyl-7-methoxycoumarin
→ 5-Carboxy-2-(4-hydroxy-3,5-dimethoxyphenyl)-3-hydroxymethyl-7-methoxycoumarone
Probably in: Fusarium E174

5-Carboxy-2-(4-hydroxy-3,5-dimethoxyphenyl)-3-hydroxymethyl-7-methoxycoumarone
Probably formed from: 5-Carboxy-2-(4-hydroxy-3,5-dimethoxyphenyl)-3-hydroxymethyl-7-methoxycoumarin
→ 5-Carboxy-3-formyl-2-(4-hydroxy-3,5-dimethoxyphenyl)-7-methoxycoumarone

Probably in: Fusarium E174

5-Carboxy-2-(4-hydroxy-3,5-dimethoxyphenyl)-7-methoxycoumarone
Probably formed from: 5-Carboxy-3-formyl-2-(4-hydroxy-3,5-dimethoxyphenyl)-7-methoxycoumarone

2-Carboxy-4-hydroxy-5,8-dioxoanthraquinone
Probably formed from: 8-Hydroxyrhein

9-Carboxy-10-(2-hydroxyethoxy)anthracene
Formed from: 2-(((10-(2-Hydroxyethoxy)-9-anthracenyl)methyl)amino)-1,3-propanediol

3-Carboxy-5-(2-hydroxyethoxy)carbonyl-2,6-dimethyl-4-(*m*-nitrophenyl)pyridine
Probably formed from: 3-(2-Hydroxyethoxy)carbonyl-5-isopropoxycarbonyl-2,6-dimethyl-4-(*m*-nitrophenyl)pyridine
→ 3-Carboxy-5-(carboxymethoxy) carbonyl-2,6-dimethyl-4-(*m*-nitrophenyl)pyridine
Probably in: Monkey G189

2-(*p*-(2-Carboxy-2-hydroxyethoxy)phenyl)-2-(*p*-(2,3-dihydroxypropoxy)phenyl)propane
Probably formed from: 2-(*p*-(2,3-Dihydroxypropoxy)phenyl)-2-(*p*-hydroxyphenyl)propane

2-(*p*-(2-Carboxy-2-hydroxyethoxy)phenyl)-2-(*p*-(2-hydroxy-3-sulphonylpropoxy)phenyl) propane
Formed from: 2,2-Bis(2,3-epoxypropoxy) phenyl)propane
→ 2-(*p*-(2-Carboxymethoxy)phenyl)-2-(*p*-(2-hydroxy-3-sulphonylpropoxy) phenyl)propane
Probably in: Mouse B704

***o*-(2-Carboxy-2-hydroxyethoxy)-ω-phenylpropiophenone**
Probably formed from: *o*-(2-hydroxy-3-propionaldoxy)-ω-phenylpropiophenone

2-(2-Carboxy-2-hydroxyethoxy)-1-(2-propen-1-oxy)benzene
Formed from: Oxprenolol

3′-(1-Carboxy-1-hydroxyethoxy)-2-trifluoromethylbenzanilide
Probably formed from: 3′-(1-Hydroxy-2-propoxy)-2-trifluoromethylbenzanilide

2-(1-Carboxy-1-hydroxyethyl)-6-(1-hydroxy-1-methylethyl)naphthalene
Probably formed from: 2-(1,2-Dihydroxy-1-methylethyl)-6-(1-hydroxy-1-methylethyl)naphthalene

4-Carboxy-3-hydroxy-1-(3-hydroxy-4-methoxyphenyl)-4,4-dimethyl-1-butene
Probably formed from: 3,5-Dihydroxy-1-(3-hydroxy-4-methoxyphenyl)-4,4-dimethyl-1-pentene

1-Carboxy-N-((1*R*,2*R*)-2-hydroxy-1-(hydroxymethyl)-2-(4-(methylsulphonyl) phenyl)ethyl)formamide
See: Thiamphenicol oxamic acid

3-Carboxy-N-(2-hydroxy-1-(*p*-hydroxyphenyl)-1-methylethyl)butyramide
Probably formed from: 3-Carboxy-N-(2-hydroxy-1-methyl-1-phenylethyl)-3-methylbutyramide
→ 3-Carboxy-N-(2-hydroxy-1-(*p*-hydroxyphenyl)-1-methylethyl) butyramide-β-D-glucuronide
Probably in: Mouse D417
Rat D193

3-Carboxy-N-(2-hydroxy-1-(*p*-hydroxyphenyl)-1-methylethyl)butyramide-β-D-glucuronide
Probably formed from: 3-Carboxy-N-(2-hydroxy-1-(*p*-hydroxyphenyl)-1-methylethyl)butyramide

3-Carboxy-4-hydroxymandelic acid
→ 3-Carboxy-4-hydroxyphenylglyoxylic acid
Probably in: Reseda A1600

3-Carboxy-4-hydroxy-5-methoxybenzoic acid
See: 5-Carboxyvanillic acid

2-(3-Carboxy-6-hydroxy-5-methoxyphenyl)ethanol
Probably formed from: 2-(5-Formyl-2-hydroxy-3-methoxyphenyl)-3-hydroxy-1-(4-hydroxy-3,5-dimethoxyphenyl)-1-propanone

2-(5-Carboxy-2-hydroxy-3-methoxyphenyl)-6-methoxy-*p*-quinone
Probably formed from: 5,5′-Dicarboxy-2,2′-dihydroxy-3,3′-dimethoxybiphenyl
Vanillic acid

6-(α-Carboxy-α-hydroxymethyl)-7-hydroxy-5-methoxycoumarin
Formed from: 5-Methoxypsoralen

6-(5′-Carboxy-3′-hydroxy-3′-methylpentyl)-2,3,5-trimethyl-1,4-benzoquinone
Formed from: α-Tocopherol

5-(5-Carboxy-2-hydroxymethylphenoxy)-2,2-dimethylpentanoic acid
Probably formed from: 5-(2,5-Bis(hydroxymethyl)phenoxy)-2,2-dimethylpentanoic acid
→ 5-(5-Carbonyloxy-β-D-glucuronido-2-hydroxymethylphenoxy)-2,2-dimethylpentanoic acid
Probably in: Hamster J565
Rat J564, J565
→ 5-(5-Carboxy-2-hydroxymethylphenoxy)-2,2-dimethylpentanoyl-β-D-glucuronic acid
Probably in: Rat J564

5-(5-Carboxy-2-hydroxymethylphenoxy)-2,2-dimethylpentanoyl-β-D-glucuronic acid
Probably formed from: 5-(5-Carboxy-2-hydroxymethylphenoxy)-2,2-dimethylpentanoic acid

3-(4-(1-Carboxy-1-hydroxymethyl)phenoxy)-1-isopropylaminopropan-2-ol
Probably formed from: 3-(4-(1,2-Dihydroxyethyl)phenoxy)-1-isopropylaminopropan-2-ol
→ 3-(4-Hydroxymethyl)phenoxy)-1-isopropylaminopropan-2-ol
Probably in: Rat A1750

3-Carboxy-N-(2-hydroxy-1-methyl-1-phenylethyl)-3-methylbutyramide
Probably formed from: 3-Carboxy-3-methyl-N-(1-methyl-1-phenylethyl)butyramide
→ 3-Carboxy-N-(2-hydroxy-1-(*p*-hydroxyphenyl)-1-methylethyl)butyramide
Probably in: Mouse D417

5-Carboxy-2-hydroxy-3-(10′(*Z*)-pentadecenyl)-1,4-benzoquinone
See: Maesanin

3-Carboxy-4-hydroxyphenylacetic acid
See: 5-Carboxymethylsalicylic acid

3-(5-Carboxy-2-hydroxyphenyl)-N,N-diisopropyl-3-phenylpropylamine
Formed from: 3-(2-Hydroxy-5-hydroxymethylphenyl)-N,N-diisopropyl-3-phenylpropylamine
→ 3-(5-Carboxy-2-hydroxyphenyl)-N,N-diisopropyl-3-phenylpropylamine-β-D-glucuronide
Probably in: Dog J419
Mouse J419
→ 3-(5-Carboxy-2-hydroxyphenyl)-N-isopropyl-3-phenylpropylamine
Probably in: Dog J419
Mouse J419

3-(5-Carboxy-2-hydroxyphenyl)-N,N-diisopropyl-3-phenylpropylamine-β-D-glucuronide
Probably formed from: 3-(5-Carboxy-2-hydroxyphenyl)-N,N-diisopropyl-3-phenylpropylamine

3-Carboxy-2-hydroxyphenyl-D-glucoside
Formed from: Pyrocatechuic acid

3-Carboxy-4-hydroxyphenyl-D-glucoside
Formed from: Gentisic acid

4-Carboxy-3-hydroxyphenyl-D-glucoside
Formed from: 2,4-Dihydroxybenzoic acid

3-Carboxy-4-hydroxy-L-phenylglycine
Formed from: 3-Carboxy-4-hydroxyphenylglyoxylic acid

3-Carboxy-4-hydroxyphenylglyoxylic acid
Probably formed from: 3-Carboxy-4-hydroxymandelic acid
→ 3-Carboxy-4-hydroxy-L-phenylglycine
In: Iris A1600
Reseda A1600

N′-Carboxy-*o*-hydroxyphenylhydrazone
Probably formed from: *o*-Hydroxy-N′-methoxycarbonylphenylhydrazone
→ N′-Carboxy-*o*-hydroxyphenylhydrazone-O-sulphate
Probably in: Rat E599

N′-Carboxy-*o*-hydroxyphenylhydrazone-O-sulphate
Probably formed from: N′-Carboxy-*o*-hydroxyphenylhydrazone

3-(5-Carboxy-2-hydroxyphenyl)-3-(*p*-hydroxyphenyl)-N,N-diisopropylpropylamine
Probably formed from: 3-(2-Hydroxy-5-hydroxymethylphenyl)-3-(*p*-hydroxyphenyl)-N,N-diisopropylpropylamine

3-(5-Carboxy-2-hydroxyphenyl)-N-isopropyl-3-phenylpropylamine
Probably formed from: 3-(5-Carboxy-2-hydroxyphenyl)-N,N-diisopropyl-3-phenylpropylamine

→ 3-(5-Carboxy-2-hydroxyphenyl)-N-isopropyl-3-phenylpropylamine-β-D-glucuronide
Probably in: Dog J419
Mouse J419

3-(5-Carboxy-2-hydroxyphenyl)-N-isopropyl-3-phenylpropylamine-β-D-glucuronide
Probably formed from: 3-(5-Carboxy-2-hydroxyphenyl)-N-isopropyl-3-phenylpropylamine

3-Carboxy-4-(*o*-hydroxyphenyl)-5-methoxycarbonyl-2,6-dimethylpyridine
Formed from: Ryosidine

3-Carboxy-N-(1-(*p*-hydroxyphenyl)-1-methylethyl)-3-methylbutyramide
Probably formed from: 3-Carboxy-3-methyl-N-(1-methyl-1-phenylethyl)butyramide

3-Carboxy-4-hydroxyphenylpyruvic acid
→ 3-Carboxy-L-tyrosine
In: Iris A1600
Reseda A1600

3-Carboxy-5-hydroxyphenyl sulphate
Formed from: 3,5-Dihydroxybenzoic acid

6-Carboxy-5-hydroxy-2-((*n*-propyl)amino)tetralin
Probably formed from: 6-Carboxy-2-((di-*n*-propyl)amino)-5-hydroxytetralin

3-Carboxy-1-hydroxy-5*H*-pyrido[3,2-*a*]phenoxazin-5-one
See: Xanthommatin

5-Carboxy-3-hydroxy-*o*-quinone
Formed from: Gallic acid

3-Carboxy-5-hydroxy-1-(*p*-sulphophenyl)-4-(*p*-sulphophenylazo)pyrazole (tartrazine)
→ 4-Amino-3-carboxy-5-hydroxy-1-(*p*-sulphophenyl)pyrazole
In: Proteus A272, A317
Rat A1701, A2528
Shigella G767
See also: F152
→ Sulphanilic acid
In: Proteus A272, A317
Rat A1701, A2528
Shigella G767
See also: F152
→ *p*-Sulphophenylhydrazine
In: Rat A1701

2-Carboxy-4-hydroxy-α-tetralone
→ Catalpalactone
In: Catalpa A2736
→ Catalponol
In: Catalpa A2736
→ 4,9-Dihydroxy-α-lapachone
In: Catalpa A2736

2-Carboxyindole-3-acetic acid
Formed from: Carbazole

7-Carboxylumichrome
Formed from: Riboflavine

8-Carboxylumichrome
Formed from: Riboflavine

3-Carboxymandelic acid
Probably formed from: 3-Carboxyphenylacetic acid
3-Carboxyphenylpyruvic acid
→ 3-Carboxyphenylglyoxylic acid
Probably in: Iris A1600, A3601
Reseda A1600

3-Carboxymethenyl-4-mercapto-N-(α-(methoxycarbonyl)-*o*-chlorobenzyl)piperidine
Formed from: Clopidogrel

***p*-(Carboxymethoxy)acetanilide**
Probably formed from: *p*-(2-Hydroxybutoxy)acetanilide
p-(2-Hydroxyethoxy)acetanilide

2-(((10-Carboxymethoxy-9-anthracenyl)methyl)amino)-1,3-propanediol
Formed from: 2-(((10-(2-Hydroxyethoxy)-9-anthracenyl)methyl)amino)-1,3-propanediol

***p*-Carboxymethoxybenzaldehyde**
→ *p*-Carboxymethoxybenzoic acid
In: Beef C508

***p*-Carboxymethoxybenzoic acid**
Formed from: *p*-Carboxymethoxybenzaldehyde

5-(α-(Carboxymethoxy)benzyl)-1-methyl-1*H*-pyrazole
Formed from: Cizolirtine

2′-Carboxymethoxy-4,4′-bis(3-methyl-2-butenyloxy)dihydrochalcone
Formed from: 2′-Carboxymethoxy-4,4′-bis(3-methyl-2-butenyloxy)chalcone

2′-Carboxymethoxy-4,4′-bis(3-methyl-2-butenyloxy)chalcone
→ 2′,4-Biscarboxymethoxy-4′-(3-methyl-2-butenyloxy)dihydrochalcone
In: Rat B372
→ 2′-Carboxymethoxy-4,4′-bis(3-methyl-2-butenyloxy)dihydrochalcone
In: Rat B372

3-Carboxy-5-methoxycarbonyl-2,6-dimethyl-4-(2-nitrophenyl)pyridine
Probably formed from: 3,5-Di-(methoxycarbonyl)-2,6-dimethyl-4-(2-nitrophenyl)pyridine

4-Carboxy-5-methoxycarbostyril
Formed from: 4-Formyl-5-methoxycarbostyril

4-Carboxy-6-methoxycarbostyril
Formed from: 4-Formyl-6-methoxycarbostyril

2-Carboxymethoxy-3-chlorobiphenyl
Formed from: 3-Chloro-2-formylmethoxybiphenyl

1-Carboxymethoxy-1-(*p*-chlorophenyl)-1-phenylethane
Probably formed from: 1-(*p*-Chlorophenyl)-1-formylmethoxy-1-phenylethane

3-Carboxy-8-methoxychromone
Formed from: 3-Hydroxymethyl-8-methoxychromone
→ 8-Methoxychromone
Probably in: Dog A1761
Rat A2113

4-Carboxy-6-methoxycoumarin
4-Formyl-6-methoxycoumarin

3-((4-Carboxymethoxy)-2,3-dichlorobenzoyl)thiophene
→ 3-((4-Carboxymethoxy)-2,3-dichlorobenzoyl)thiophene sulphoxide
Probably in: Rat G284

3-((4-Carboxymethoxy)-2,3-dichlorobenzoyl)thiophene sulphoxide
Probably formed from: 3-((4-Carboxymethoxy)-2,3-dichlorobenzoyl)thiophene

2-Carboxymethoxy-3,5-dichlorobiphenyl
Formed from: 3,5-Dichloro-2-formylmethoxybiphenyl

5-Carboxymethoxy-3,4-dihydrocarbostyril
Formed from: Carteolol

2-((Carboxymethoxy)methyl)-4-(*o*-chlorophenyl)-3-ethoxycarbonyl-1,4-dihydro-5-methoxycarbonyl-6-methylpyridine
Formed from: Amlodipine
→ 3-Carboxy-2-((carboxymethoxy)methyl)-4-(*o*-chlorophenyl)-1,4-dihydro-5-methoxycarbonyl-6-methylpyridine
Probably in: Dog E836, F41, F78
→ 2-((Carboxymethoxy)methyl)-4-(*o*-chlorophenyl)-3-ethoxycarbonyl-5-methoxycarbonyl-6-methylpyridine
Probably in: Dog E836
Man E836

2-((Carboxymethoxy)methyl)-4-(*o*-chlorophenyl)-3-ethoxycarbonyl-5-methoxycarbonyl-6-methylpyridine
Probably formed from: 2-((Carboxymethoxy)methyl)-4-(*o*-chlorophenyl)-3-ethoxycarbonyl-1,4-dihydro-5-methoxycarbonyl-6-methylpyridine
→ 5-Carboxy-2-((carboxymethoxy)methyl)-4-(*o*-chlorophenyl)-3-ethoxycarbonyl-6-methylpyridine
Probably in: Dog E836

2-(2-(3-Carboxy-4-methoxy-5-methyl)pyridinyl)methylsulphinyl-5-methoxybenzimidazole
Probably formed from: 2-(2-(3-Hydroxymethyl-4-methoxy-5-methyl)pyridinyl)methylsulphinyl-5-methoxybenzimidazole

2-(2-(5-Carboxy-4-methoxy-3-methyl)pyridinyl)methylsulphinyl-5-methoxybenzimidazole
Formed from: 2-(2-(5-Hydroxymethyl-4-methoxy-3-methyl)pyridinyl)methylsulphinyl-5-methoxybenzimidazole

ω-(4-Carboxy-2-methoxyphenoxy)acetovanillone
Probably formed from: 3-(4-Carboxy-2-methoxyphenoxy)-1-vanillyl-1-propanol
→ 4′-Hydroxy-3′-methoxyacetophenone
In: Pseudomonas E832
→ Vanillic acid
In: Pseudomonas E832

2-(4-Carboxy-2-methoxyphenoxy)-6-methoxy-*p*-quinone
Formed from: Vanillic acid

3-(4-Carboxy-2-methoxyphenoxy)-1-vanillyl-1-propanol
→ ω-(4-Carboxy-2-methoxyphenoxy) acetovanillone
Probably in: Pseudomonas E832

2-(*p*-(2-Carboxymethoxy)phenyl)-2-(*p*-(2,3-dihydroxypropoxy)phenyl)propane
Probably formed from: 2,2-Bis(*p*-(2,3-dihydroxypropoxy)phenyl)propane

2-(*p*-(2-Carboxymethoxy)phenyl)-2-(*p*-(2-hydroxy-3-sulphonylpropoxy)phenyl)propane
Probably formed from: 2-(*p*-(2-Carboxy-2-hydroxyethoxy)phenyl)-2-(*p*-(2-hydroxy-3-sulphonylpropoxy)phenyl)propane

3-(4-Carboxy-2-methoxyphenylmethyl)-5-(cyclopentoxycarbonylamino)-1-methylindole
Formed from: Zafirlukast

2(*R*)-((4-Carboxymethoxy)phenyl)methyl-5(*S*)-(1,1-dimethylethoxycarbonylamino)-4(*S*)-hydroxy-N-(2(*R*)-hydroxy-1(*S*)-indanyl)hexamide
Formed from: 5(*S*)-(1,1-Dimethylethoxycarbonylamino)-4(*S*)-hydroxy-N-(2(*R*)-hydroxy-1(*S*)-indanyl)-2(*R*)-(4-(2-(4-morpholinyl) ethoxy)phenyl)methyl-6-phenylhexamide

2-(Carboxymethoxy)-1-(2-propen-1-oxy)benzene
Formed from: Oxprenolol
→ *o*-Hydroxyphenylacetic acid
Probably in: Man D664

3-Carboxy-4′-methylaminoazobenzene
Formed from: 3-Carboxy-4′-dimethylaminoazobenzene
Probably formed from: 3-Formyl-4′-methylaminoazobenzene
→ 4′-Amino-3-carboxyazobenzene
In: Rat A776

4-Carboxy-4′-methylaminoazobenzene
Formed from: *p*-Methyl red
→ 4′-Amino-4-carboxyazobenzene
In: Rat A1979

1-(2-(Carboxymethylamino)ethylamino)-5,8-dihydroxy-4-(2-(2-hydroxyethylamino) ethylamino)anthraquinone
Formed from: Mitoxantrone
→ 1,4-Bis(2-(carboxymethylamino) ethylamino)-5,8-dihydroxyanthraquinone
In: Man E153, G48
Pig G48, G231
Rat G48

5-((2-((Carboxymethyl)amino)ethyl)amino)-7-hydroxy-(2-((2-hydroxyethyl)amino)ethyl) anthra[1,9-*cd*]pyrazol-6(2*H*)-one
Formed from: Biantrazole

2-(2-((Carboxymethyl)amino)ethyl)-5-((2-((carboxymethyl)amino)ethyl)amino)-7-hydroxyanthra[1,9-*cd*]pyrazol-6(2*H*)-one
Probably formed from: 2-(2-((Carboxymethyl) amino)ethyl)-7-hydroxy-5-((((2-hydroxyethyl) amino)ethyl)amino)anthra[1,9-*cd*]pyrazol-6(2*H*)-one

2-(2-((Carboxymethyl)amino)ethyl)-7-hydroxy-5-((((2-hydroxyethyl)amino)ethyl)amino) anthra[1,9-*cd*]pyrazol-6(2*H*)-one
→ 2-(2-((Carboxymethyl)amino)ethyl)-5-((2-((carboxymethyl)amino)ethyl)amino)-7-hydroxyanthra[1,9-*cd*]pyrazol-6(2*H*)-one
Probably in: Man G698

2-(1-Carboxy-5-methylamino-2-pentyl)-4-(*p*-chlorobenzyl)-1(2*H*)-phthalazinone
Formed from: Azelastine

3-(1-Carboxy-5-methylamino-2-pentyl)-4-(*p*-chlorobenzyl)-1(2*H*)-phthalazinone
Formed from: Azelastine

***o*-Carboxy-N-methylbenzenesulphonamide**
Formed from: Isoxicam

5-Carboxymethylbenzofuran
Formed from: Darifenacin
→ N-((Benzofuran-5-yl)methylcarbonyl) glycine
In: Dog H866
Man H866
Rat H866
→ 2,4-Bis(carboxymethyl)phenol
In: Dog H866
Man H866
Mouse H866
Rabbit H866
Rat H866

4-(3-Carboxy-2-methylbenzyl)imidazole
Formed from: 4-(3-Hydroxymethyl-2-methylbenzyl)imidazole

3-Carboxymethyl-5-(2-carboxymethylethenyl)-2-(4-hydroxy-3-methoxyphenyl)-7-methoxybenzofuran
Formed from: Methyl ferulate

6-Carboxymethyl-(+)-catechin
→ 6-Carboxymethyl-3′-O-methyl-(+)-catechin
In: Pig D112

8-Carboxymethyl-(+)-catechin
→ 8-Carboxymethyl-3′-O-methyl-(+)-catechin
In: Pig D112

5-Carboxymethyl-4′-chloro-3-biphenylyl-β-D-glucuronide
Formed from: 4′-Chloro-5-hydroxy-3-biphenylylacetic acid

5-Carboxymethyl-4′-chloro-3-biphenylyl sulphate
Formed from: 4′-Chloro-5-hydroxy-3-biphenylylacetic acid

***p*-(N-(Carboxymethyl)-N-(2-chloroethyl)amino)benzoic acid**
Probably formed from: *p*-(N-(2-Chloroethyl)-N-(2-hydroxyethyl)amino)benzoic acid
→ *p*-(2-Chloroethylamino)benzoic acid
Probably in: Rat A2041

10-Carboxymethyl-2-chlorophenothiazine
Probably formed from: 10-Carboxyethyl-2-chlorophenothiazine
→ 10-Carboxymethyl-2-chlorophenothiazine sulphoxide
Probably in: Man A693

10-Carboxymethyl-2-chlorophenothiazine sulphoxide
Probably formed from: 10-Carboxymethyl-2-chlorophenothiazine

2-Carboxymethyl-3-(*p*-chlorophenyl)-3-hydroxyphthalimidine
Probably formed from: 2-(2-Aminoethyl)-3-(*p*-chlorophenyl)-3-hydroxyphthalimidine

4-O-(Carboxymethyl)-1-deoxy-1,4-dihydro-4-hydroxy-1-oxorifamycin γ-lactone
See: Rifamycin O

3-Carboxymethyl-2′,4′-difluorobiphenyl
Formed from: Flobufen

2-Carboxymethyl-1,2-dihydro-3*H*-dibenz[*de*,*h*]isoquinoline-1,3-dione
Formed from: Azonafide

N-Carboxymethyl-2,3-dihydro-3-hydroxy-2-(*p*-hydroxyphenyl)-1,5-benzothiazepin-4(5*H*)-one
Probably formed from: N-Formylmethyl-2,3-dihydro-3-hydroxy-2-(*p*-hydroxyphenyl)-1,5-benzothiazepin-4(5*H*)-one

N-Carboxymethyl-2,3-dihydro-3-hydroxy-2-(*p*-methoxyphenyl)-1,5-benzothiazepin-4(5*H*)-one
Probably formed from: N-Formylmethyl-2,3-dihydro-3-hydroxy-2-(*p*-methoxyphenyl)-1,5-benzothiazepin-4(5*H*)-one

7-(3-Carboxymethyl-1,3-dihydroxypropyl)phenyl)methoxy-4-phenylnaphtho[2,3-*c*]furan-1(3*H*)-one
Formed from: 7-(3-(Tetrahydro-4-hydroxy-2*H*-pyran-4-yl)phenyl)methoxy-4-phenylnaphtho[2,3-*c*]furan-1(3*H*)-one

6-Carboxymethyl-7,8-dimethoxycoumarin methyl ester
Formed from: 8-Methoxypsoralen

10-Carboxymethyl-7,8-dimethylisoalloxazine
Formed from: Riboflavine

Carboxymethyl 5-(5,5-dimethyl-2-oxo-1,3,2-dioxaphosphorinan-2-yl)-1,4-dihydro-2,6-dimethyl-4-(*m*-nitrophenyl)-3-pyridinecarboxylate
Formed from: Efonidipine
Probably formed from: 2-Aminoethyl 5-(5,5-dimethyl-2-oxo-1,3,2-dioxaphosphorinan-2-yl)-1,4-dihydro-2,6-dimethyl-4-(*m*-nitrophenyl)-3-pyridinecarboxylate
→ Carboxymethyl 5-(5,5-dimethyl-2-oxo-1,3,2-dioxaphosphorinan-2-yl)-2,6-dimethyl-4-(*m*-nitrophenyl)-3-pyridinecarboxylate
Probably in: Man H523
Rat H625

Carboxymethyl 5-(5,5-dimethyl-2-oxo-1,3,2-dioxaphosphorinan-2-yl)-2,6-dimethyl-4-(*m*-nitrophenyl)-3-pyridinecarboxylate
Probably formed from: Carboxymethyl 5-(5,5-dimethyl-2-oxo-1,3,2-dioxaphosphorinan-2-yl)-1,4-dihydro-2,6-dimethyl-4-(*m*-nitrophenyl)-3-pyridinecarboxylate

Carboxymethyl dimethyl phenylcarbamoylmethyl ammonium
Probably formed from: N,N-Bis(phenylcarbamoylmethyl) dimethylammonium

4-(5-Carboxy-5-methyl-1,3,2-dioxaphosphorinan-2-yl)-4-(5,5-dimethyl-1,3,2-dioxaphosphorinan-2-yl)-*m*-fluorobutyrophenone
Probably formed from: *m*-Fluoro-4-(5-hydroxymethyl-5-methyl-1,3,2-dioxaphosphorinan-

2-yl)-4-(5,5-dimethyl-1,3,2-dioxaphosphorinan-2-yl)butyrophenone

N-((1-Carboxy-1-methylethoxy)carbonyl)-α-hydroxy-β-phenyl-β-alanyl)-10-deacetylbaccatin III
Probably formed from: N-((1-Formyl-1-methylethoxy)carbonyl)-α-hydroxy-β-phenyl-β-alanyl)-10-deacetylbaccatin III

3-(4-(1-Carboxy-1-methylethoxy)phenyl)-5-(3-pyridyl)-1,2,4-oxadiazole
Formed from: 3-(4-(1-Ethoxycarbonyl-1-methylethoxy)phenyl)-5-(3-pyridyl)-1,2,4-oxadiazole

4-(1-Carboxy-1-methylethyl)-2-chlorophenyl methyl methylphosphoramidate
Probably formed from: 2-Chloro-4-(2-hydroxy-1,1-dimethylethyl)phenyl methyl methylphosphoramidate
→ 4-(1-Carboxy-1-methylethyl)-2-chlorophenyl methyl methylphosphoramidate-β-D-glucuronide
In: Sheep A3001
→ 4-(1-Carboxy-1-methylethyl)-2-chlorophenyl methyl phosphate
Probably in: Sheep A2995
→ 4-(1-Carboxy-1-methylethyl)-2-chlorophenyl methyl phosphoramidate
Probably in: Sheep A2995
→ 2-(3-Chloro-4-hydroxyphenyl)-2-methylpropionic acid
Probably in: Sheep A2995

4-(1-Carboxy-1-methylethyl)-2-chlorophenyl methyl methylphosphoramidate-β-D-glucuronide
Probably formed from: 4-(1-Carboxy-1-methylethyl)-2-chlorophenyl methyl methylphosphoramidate

4-(1-Carboxy-1-methylethyl)-2-chlorophenyl methyl phosphate
Probably formed from: 4-(1-Carboxy-1-methylethyl)-2-chlorophenyl methyl methylphosphoramidate
→ 4-(1-Carboxy-1-methylethyl)-2-chlorophenyl phosphate
Probably in: Sheep A2995

4-(1-Carboxy-1-methylethyl)-2-chlorophenyl methyl phosphoramidate
Probably formed from: 4-(1-Carboxy-1-methylethyl)-2-chlorophenyl methyl methylphosphoramidate

4-(1-Carboxy-1-methylethyl)-2-chlorophenyl phosphate
Probably formed from: 4-(1-Carboxy-1-methylethyl)-2-chlorophenyl methyl phosphate

***p*-(4-(*p*-((1-Carboxy-1-methyl)ethyl)phenyl-2-hydroxy)butoxy)benzoic acid**
Probably formed from: *p*-(2-Hydroxy-4-(*p*-((1-hydroxymethyl-1-methyl)ethyl)phenyl)butoxy)benzoic acid

1-(4-(1-Carboxy-1-methylethyl)phenyl)-4-(4-(hydroxy(diphenyl)methyl)piperidin-1-yl)-1-butanol (terfenadine acid)
Formed from: 4-(4-(Hydroxy(diphenyl)methyl)piperidin-1-yl)-1-(4-(1-(hydroxymethyl)-1-methylethyl)phenyl)-1-butanol

***o*-(2-Carboxy-1-methylethyl)phenyl N-methylcarbamate**
Probably formed from: *o*-(3-Hydroxy-1-methylpropyl)phenyl N-methylcarbamate

3-(N′-Carboxymethyl)formylureido-4-hydroxy-2-methyl-2*H*-1,2-benzothiazine-1,1-dioxide
Formed from: Sudoxicam
→ 3-Formylureido-4-hydroxy-2-methyl-2*H*-1,2-benzothiazine-1,1-dioxide
In: Dog A2137
Monkey A2137
Rat A2137

5-Carboxymethyl-3-hydroxybenzofuran
Probably formed from: (*S*)-2-(1-(2-(2,3-Dihydro-3-hydroxybenzofuran-5-yl)ethyl)-3-pyrrolidinyl)-2,2-diphenylacetamide

1-(3-Carboxymethyl-4-hydroxy-2,5-dimethoxy-6-(2-(1-pyrrolidinyl)ethoxy)phenyl)-3-methylurea
Formed from: Carocainide

N-(2-Carboxymethyl)-N-(2-hydroxyethyl)-3,4,5-trimethoxybenzamide
Probably formed from: N,N-Bis(2-hydroxyethyl)-3,4,5-trimethoxybenzamide

6-Carboxymethyl-7-hydroxy-5-methoxycoumarin
Formed from: 5-Methoxypsoralen

6-Carboxymethyl-7-hydroxy-8-methoxycoumarin
Formed from: 8-Methoxypsoralen
→ 7,α-Dihydroxy-8-methoxy-2-oxo-2*H*-1-benzopyran-6-acetic acid
Probably in: Dog A3911

3-Carboxymethyl-5-hydroxy-2-methylindole
→ 3-Carboxymethyl-5-oxo-2-methylindole
In: Man J427

3-Carboxymethyl-4-hydroxyphenylacetylglycine
Probably formed from: 2,4-Bis (carboxymethyl)phenol

(*S*)-2-(1-(2-(3-Carboxymethyl-4-hydroxyphenyl) ethyl)-3-pyrrolidinyl)-2,2-diphenylacetamide
Probably formed from: (*S*)-2-(1-(4-Hydroxy-3-(2-hydroxyethyl)phenyl)ethyl)-3-pyrrolidinyl)-2,2-diphenylacetamide

4-Carboxymethyl-2-hydroxyphenyl-β-D-glucuronide
Probably formed from: 3,4-Dihydroxyphenylacetic acid

4-Carboxymethyl-2-hydroxyphenyl sulphate
Probably formed from: 3,4-Dihydroxyphenylacetic acid

2-(4-Carboxymethyl-4-hydroxypiperidin-1-yl)-6-cyano-1,2,3,4-tetrahydronaphthalene
Formed from: (+)-N-(1′-(6-Cyano-1,2,3,4-tetrahydro-2(*R*)-naphthalenyl)-3,4-dihydro-4(*R*)-hydroxy*spiro*(2*H*-1-benzopyran-2,4′-piperidin)-6-yl) methanesulphonamide

(*S*)-4-((3-Carboxymethyl-1*H*-indol-5-yl)methyl)-2-oxazolidine
Probably formed from: (*S*)-4-((3-Formylmethyl-1*H*-indol-5-yl)methyl)-2-oxazolidine

6-Carboxymethyl-3′-O-methyl-(+)-catechin
Formed from: 6-Carboxymethyl-(+)-catechin

8-Carboxymethyl-3′-O-methyl-(+)-catechin
Formed from: 8-Carboxymethyl-(+)-catechin

Carboxymethyl methyl 1,4-dihydro-2,6-dimethyl-4-(*m*-nitrophenyl)pyridine-3,5-dicarboxylate
Probably formed from: Methyl 2-oxoethyl 1,4-dihydro-2,6-dimethyl-4-(*m*-nitrophenyl) pyridine-3,5-dicarboxylate
→ Carboxymethyl methyl 2,6-dimethyl-4-(*m*-nitrophenyl)pyridine-3,5-dicarboxylate
Probably in: Man D889

Carboxymethyl methyl 2,6-dimethyl-4-(*m*-nitrophenyl)pyridine-3,5-dicarboxylate
Probably formed from: Carboxymethyl methyl 1,4-dihydro-2,6-dimethyl-4-(*m*-nitrophenyl)pyridine-3,5-dicarboxylate

2-Carboxymethyl-1-methyl-5-(*p*-methylbenzoyl)pyrrole
See: Tolmetin

3-Carboxy-3-methyl-N-(1-methyl-1-phenylethyl)butyramide
Probably formed from: 4-Hydroxy-3,3-dimethyl-N-(1-methyl-1-phenylethyl) butyramide
→ 3-Carboxy-N-(1-(*p*-hydroxyphenyl)-1-methylethyl)-3-methylbutyramide
Probably in: Rat D215
→ 3-Carboxy-N-(2-hydroxy-1-methyl-1-phenylethyl)-3-methylbutyramide
Probably in: Mouse D417

5-(Carboxymethyl)oximino-5-(*p*-trifluoromethylphenyl)valeric acid
Probably formed from: 5-(2-Aminoethyl) oximino-5-(*p*-trifluoromethylphenyl)valeric acid
5-Hydroxy-*p*-trifluoromethylphenylvalerophenone (*E*)-O-(carboxymethyl) oxime

4-Carboxymethyloxindole
Formed from: Ropinirole

3-Carboxy-2-methyl-4-oxobenzothiazine-1,1-dioxide
Formed from: Piroxicam

3-Carboxymethyl-5-oxo-2-methylindole
Formed from: 3-Carboxymethyl-5-hydroxy-2-methylindole

3-(5′-Carboxy-3′-methyl-2′-pentenyl)-2-methyl-1,4-naphthoquinone
Probably formed from: Vitamin K_1

1-Carboxymethylphenoxathiin-10,10-dioxide
Probably formed from: 1-(1-Hydroxyethyl)phenoxathiin-10,10-dioxide

5-(5-Carboxy-2-methylphenoxy)-2,2-dimethylpentanoic acid
Formed from: 5-(5-Hydroxymethyl-2-methylphenoxy)-2,2-dimethylpentanoic acid
→ 5-(5-Carboxy-2-methylphenoxy)-2,2-dimethylpentanoic acid-β-D-glucuronide

Probably in: Hamster J565
Rat J565

5-(5-Carboxy-2-methylphenoxy)-2,2-dimethylpentanoic acid-β-D-glucuronide
Probably formed from: 5-(5-Carboxy-2-methylphenoxy)-2,2-dimethylpentanoic acid

3-(4-Carboxymethylphenoxy)-1-isopropylaminopropan-2-ol
Probably formed from: 3-(4-(2-Hydroxyethyl)phenoxy)-1-isopropylaminopropan-2-ol

***p*-(Carboxymethyl)phenylacetylglycine**
Probably formed from: 1,4-Bis(carboxymethyl)benzene

4-(2′-(Carboxymethyl)phenylamino)-7-chloroquinoline
Formed from: Glifanan

4-(1-(3-Carboxy-2-methylphenyl)ethyl)-1*H*-imidazole
Probably formed from: 4-(1-(3-Hydroxy-methyl-2-methylphenyl)ethyl)-1*H*-imidazole

1-Carboxymethyl-4-phenylimidazolidin-2-one
Probably formed from: 1-Carboxymethyl-4-phenyl-2-thioimidazolidine
→ 4-Phenylimidazolidin-2-one
Probably in: Cow H733

3-Carboxy-5-methylphenyl N-methylcarbamate
Formed from: 3,5-Dimethylphenyl N-methyl-carbamate
→ 3-Hydroxy-5-methylbenzoic acid
Probably in: Rabbit A1477

3-Carboxy-2-methyl-1-phenyl-3-pyrazolin-5-one
Formed from: 3-Hydroxymethyl-2-methyl-1-phenyl-3-pyrazolin-5-one

***p*-Carboxymethylphenyl sulphate**
Probably formed from: *p*-Hydroxyphenylacetic acid

1-Carboxymethyl-4-phenyl-2-thioimidazolidine
Formed from: Levamisol
→ 1-Carboxymethyl-4-phenylimidazolidin-2-one
Probably in: Cow H733

N-Carboxymethylphthalamidic acid
→ Phthalic acid
In: Xanthobacter J253

2-Carboxy-4-(4-methyl-1-piperazinyl)-10*H*-thieno[2,3-*b*][1,5]benzodiazepine
Probably formed from: 2-Hydroxymethyl-4-(4-methyl-1-piperazinyl)-10*H*-thieno[2,3-*b*][1,5]benzodiazepine
→ 2-Carboxy-4-(4-methyl-1-piperazinyl)-10*H*-thieno[2,3-*b*][1,5]benzodiazepine-N-oxide
Probably in: Dog J150
Man J150
Mouse J150
Rat J150
Rhesus monkey J150
→ 2-Carboxy-4-(1-piperazinyl)-10*H*-thieno[2,3-*b*][1,5]benzodiazepine
Probably in: Man J150
Rhesus monkey J150

2-Carboxy-4-(4-methyl-1-piperazinyl)-10*H*-thieno[2,3-*b*][1,5]benzodiazepine-N-oxide
Probably formed from: 2-Carboxy-4-(4-methyl-1-piperazinyl)-10*H*-thieno[2,3-*b*][1,5]benzodiazepine
→ 2-Carboxy-4-(4-methyl-1-piperazinyl)-10*H*-thieno[2,3-*b*][1,5]benzodiazepine-N-oxide-β-D-glucuronide
Probably in: Rhesus monkey J150

2-Carboxy-4-(4-methyl-1-piperazinyl)-10*H*-thieno[2,3-*b*][1,5]benzodiazepine-N-oxide-β-D-glucuronide
Probably formed from: 2-Carboxy-4-(4-methyl-1-piperazinyl)-10*H*-thieno[2,3-*b*][1,5]benzodiazepine-N-oxide

8-(3-Carboxy-1-methylpropylamino)-6-methoxyquinoline
Formed from: 8-(3-Formyl-1-methylpropylamino)-6-methoxyquinoline

2-(5-Carboxymethylpyridin-2-yl)benzimidazole
Formed from: 2-(5-(1-Hydroxyethyl)pyridin-2-yl)benzimidazole

5-((4-(2-(5-Carboxymethyl)-2-pyridinyl)ethoxy)phenyl)methyl)-2,4-thiazolidinedione
Probably formed from: 5-((4-(2-(5-(2-Hydroxyethyl)-2-pyridinyl)ethoxy)phenyl)methyl)-2,4-thiazolidinedione
→ 5-((4-(2-(5-Carboxy-2-pyridinyl)ethoxy)phenyl)methyl)-2,4-thiazolidinedione

Probably in: Rat G946
→ 5-((4-(2-(5-Carboxymethyltauryl-2-pyridinyl)ethoxy)phenyl)methyl)-2,4-thiazolidinedione
In: Rat G946

3-(5-Carboxy-3-methyl-1*H*-pyrrol-2-ylmethylene)-1,3-dihydroindol-2-one
Probably formed from: 1,3-Dihydro-3-(5-hydroxymethyl-1*H*-pyrrol-2-ylmethylene)indol-2-one
→ 3-(5-Carboxy-3-methyl-1*H*-pyrrol-2-ylmethylene)-1,3-dihydroindol-2-one-β-D-glucuronide
Probably in: Dog K336
Man K336
Monkey K336
Mouse K336
Rat K336

3-(5-Carboxy-3-methyl-1*H*-pyrrol-2-ylmethylene)-1,3-dihydroindol-2-one-β-D-glucuronide
Probably formed from: 3-(5-Carboxy-3-methyl-1*H*-pyrrol-2-ylmethylene)-1,3-dihydroindol-2-one

4-Carboxymethyl-*o*-quinone
Formed from: 3,4-Dihydroxyphenylacetic acid

4-O-(Carboxymethyl)rifamycin
See: Rifamycin B

5-Carboxymethylsalicylic acid (3-carboxy-4-hydroxyphenylacetic acid)
Probably formed from: 6-Carboxymethyl-3-(1*H*-tetrazol-5-yl)chromone

2-Carboxy-5-(S-methylsulphonamido)indole
Formed from: Delaviridine

5-((4-(2-(5-Carboxymethyltauryl-2-pyridinyl)ethoxy)phenyl)methyl)-2,4-thiazolidinedione
Formed from: 5-((4-(2-(5-Carboxymethyl)-2-pyridinyl)ethoxy)phenyl)methyl)-2,4-thiazolidinedione

6-Carboxymethyl-3-(1*H*-tetrazol-5-yl)chromone
Formed from: 6-Ethyl-3-(1*H*-tetrazol-5-yl)chromone
→ 5-Carboxymethylsalicylic acid
Probably in: Guinea pig A3486

2-((Carboxymethyl)thio)-N-(3-(3-(piperidinomethyl)phenoxy)propyl)acetanilide
Formed from: 2-(2-Hydroxyethyl-1-thio)-N-(3-(3-(piperidinomethyl)phenoxy)propyl)acetamide

N-(3-Carboxy-1-methyl-1,2,4-triazol-5-yl)-3-((*m*-piperidin-1-ylmethyl)phenoxy)propylamine
Formed from: Loxtidine

3-Carboxymethyl-1-(3-trifluoromethylphenyl)quinazoline-2,4(1*H*,3*H*)-dione
Formed from: 3-(2-Hydroxyethyl)-1-(3-trifluoromethylphenyl)quinazoline-2,4(1*H*,3*H*)-dione

5-Carboxymethylvanillin (5-formyl-2-hydroxy-3-methoxyphenylacetic acid)
Probably formed from: Dehydrodivanillin
→ 5-Carboxyvanillin
Probably in: Microorganism D540

***p*-Carboxy-β-(5-nitro-2-furyl)styrene**
→ *p*-(5-Amino-2-furyl)-β-carboxystyrene
In: Rabbit G297
→ *p*-Carboxyphenyl)-5-cyano-3-oxo-1,4-pentadiene
In: Rabbit G297

S-(3-Carboxy-4-nitrophenyl)glutathione
Probably formed from: 2,5-Dinitrobenzoic acid

S-(4-Carboxy-2-nitrophenyl)glutathione
Probably formed from: 3,4-Dinitrobenzoic acid

S-(4-Carboxy-3-nitrophenyl)glutathione
Probably formed from: 2,4-Dinitrobenzoic acid

4-Carboxy-2-nonaprenylphenol
Formed from: *p*-Hydroxybenzoic acid

6-(9-Carboxynonyl)-2,3-dimethoxy-5-methylquinol
Probably formed from: 6-(9-Carboxynonyl)-2,3-dimethoxy-5-methylquinone
→ 6-(9-Carboxynonyl)-2,3-dimethoxy-5-methylquinol-β-D-glucuronide
Probably in: Dog E119
Rat E119
→ 6-(9-Carboxynonyl)-2,3-dimethoxy-5-methylquinol sulphate
Probably in: Dog E119
Rat E119

6-(9-Carboxynonyl)-2,3-dimethoxy-5-methylquinol-β-D-glucuronide
Probably formed from: 6-(9-Carboxynonyl)-2,3-dimethoxy-5-methylquinol

6-(9-Carboxynonyl)-2,3-dimethoxy-5-methylquinol sulphate
Probably formed from: 6-(9-Carboxynonyl)-2,3-dimethoxy-5-methylquinol

6-(9-Carboxynonyl)-2,3-dimethoxy-5-methylquinone
Formed from: Idebenone
→ 6-(7-Carboxyheptyl)-2,3-dimethoxy-5-methylquinol
Probably in: Dog D764, E119
Rat D764, E119
→ 6-(9-Carboxynonyl)-2,3-dimethoxy-5-methylquinol
Probably in: Dog E119
Rat E119

5-(2-(9-Carboxynonyl)phenyl)-4,6-dithianonanedioic acid
Probably formed from: 5-(2-(12-Hydroxydodecyl)phenyl)-4,6-dithianonanedioic acid
→ 5-(2-(5-Carboxypentyl)phenyl)-4,6-dithianonanedioic acid
Probably in: Guinea pig E873

4-Carboxy-2-octaprenylphenol (4-hydroxy-3-octaprenylbenzoic acid)
Formed from: *p*-Hydroxybenzoic acid

2-Carboxy-N-(2-oxopiperidin-3-yl)benzenesulphonamide
Formed from: Supidimide

2-Carboxy-4-oxo-2-prenyl-1-tetralone
Formed from: *o*-Succinylbenzoic acid

6-(5-Carboxypentanoxy)-β-naphthaldehyde
→ 6-(5-Carboxypentanoxy)-β-naphthoic acid
In: Man J24

6-(5-Carboxypentanoxy)-β-naphthoic acid
Formed from: 6-(5-Carboxypentanoxy)-β-naphthaldehyde

4-Carboxypentaprenylphenol
Formed from: *p*-Hydroxybenzoic acid

8-(5-Carboxypentylamino)-6-methoxy-4-methylquinoline
Formed from: 8-(6-Diethylaminohexylamino)-6-methoxy-4-methylquinoline

α-(((5-Carboxypentyl)amino)methyl)-4-hydroxy-1,3-benzenedimethanol
Probably formed from: 4-Hydroxy-α-(((6-hydroxyhexyl)amino)methyl)-1,3-benzenedimethanol

1-(5-Carboxypentyl)-5-(3,4-dichlorobenzyl) biguanide
Formed from: 1-(3,4-Dichlorobenzyl)-5-octylbiguanide
→ 1-(3-Carboxypropyl)-5-(3,4-dichlorobenzyl)biguanide
Probably in: Dog J380, K61
Rat K61

6-(Carboxypentyl)-2,4-dihydroxybenzaldehyde
Probably formed from: 6-(Carboxypentyl)-2,4-dihydroxybenzoic acid

6-(Carboxypentyl)-2,4-dihydroxybenzoic acid
Formed from: Zearalenone
→ 6-(Carboxypentyl)-2,4-dihydroxybenzaldehyde
Probably in: Fusarium A1817
→ 3,5-Dihydroxyphenylacetic acid
Probably in: Fusarium A1817
→ 2,3,5-Trihydroxyphenylacetic acid
Probably in: Fusarium A1817

6-(5-Carboxypentyl)-2,3-dimethoxy-5-methylquinol
Probably formed from: 6-(5-Carboxypentyl)-2,3-dimethoxy-5-methylquinone
→ 6-(5-Carboxypentyl)-2,3-dimethoxy-5-methylquinol-β-D-glucuronide
Probably in: Dog E119
Rat E119
→ 6-(5-Carboxypentyl)-2,3-dimethoxy-5-methylquinol sulphate
Probably in: Dog D764, E119
Rat D764, E119

6-(5-Carboxypentyl)-2,3-dimethoxy-5-methylquinol-β-D-glucuronide
Probably formed from: 6-(5-Carboxypentyl)-2,3-dimethoxy-5-methylquinol

6-(5-Carboxypentyl)-2,3-dimethoxy-5-methylquinol sulphate
Probably formed from: 6-(5-Carboxypentyl)-2,3-dimethoxy-5-methylquinol

6-(5-Carboxypentyl)-2,3-dimethoxy-5-methylquinone
Probably formed from: 6-(7-Carboxyheptyl)-2,3-dimethoxy-5-methylquinone
→ 6-(5-Carboxypentyl)-2,3-dimethoxy-5-methylquinol
Probably in: Dog E119
Rat E119
→ 6-(3-Carboxypropyl)-2,3-dimethoxy-5-methylquinone
Probably in: Dog D764, E119
Rat D764, E119

1-(5-Carboxypentyl)-N-ethyl-N-methyl-4-phenyl-4-piperidinearboxamide
Probably formed from: 6′-Hydroxysameridine
→ 1-(3-Carboxypropyl)-N-ethyl-N-methyl-4-phenyl-4-piperidinearboxamide
Probably in: Rat K43

5-(2-(5-Carboxypentyl)phenyl)-4,6-dithianonanedioic acid
Probably formed from: 5-(2-(9-Carboxynonyl)phenyl)-4,6-dithianonanedioic acid
→ 5-(2-(3-Carboxypropyl)phenyl)-4,6-dithianonanedioic acid
Probably in: Guinea pig E873

***o*-Carboxyphenoxyacetic acid**
Formed from: *o*-Formylphenoxyacetic acid
o-Formylphenoxypropionic acid

***m*-Carboxyphenoxyacetic acid**
Formed from: *m*-Formylphenoxyacetic acid

***p*-Carboxyphenoxyacetic acid**
Formed from: *p*-Formylphenoxyacetic acid

α-Carboxy-3-phenoxybenzyl 2-(*p*-chlorophenyl)-3-methylbutyrate
Probably formed from: α-Carbamoyl-3-phenoxybenzyl 2-(*p*-chlorophenyl)-3-methylbutyrate

1-(*p*-Carboxyphenoxy)-10-(*p*-chlorophenoxy) decane
→ 1-(*p*-Carboxyphenoxy)-10-(*p*-chlorophenoxy)decane cholesteryl ester
In: Rat C310

1-(*p*-Carboxyphenoxy)-10-(*p*-chlorophenoxy) decane cholesteryl ester
Formed from: 1-(*p*-Carboxyphenoxy)-10-(*p*-chlorophenoxy)decane

1-(2-(*p*-Carboxyphenoxy)ethylamino)-3-(2-methylphenoxy)propan-2-ol
Formed from: Tolamolol

***o*-Carboxyphenoxypropionic acid**
Formed from: *o*-Formylphenoxypropionic acid

1-(*p*-Carboxyphenoxy)-2,3,6,7-tetrahydroxy-3,7-dimethyloctane
Probably formed from:1-(*p*-Acetylphenoxy)-2,3,6,7-tetrahydroxy-3,7-dimethyloctane
→ 1-(*p*-Carboxyphenoxy)-2,3,6,7-tetrahydroxy-3,7-dimethyloctane-β-D-glucuronide
Probably in: Rat A813

1-(*p*-Carboxyphenoxy)-2,3,6,7-tetrahydroxy-3,7-dimethyloctane-β-D-glucuronide
Probably formed from: 1-(*p*-Carboxyphenoxy)-2,3,6,7-tetrahydroxy-3,7-dimethyloctane

2-Carboxyphenylacetic acid
See: Homophthalic acid

3-Carboxyphenylacetic acid
→ 3-Carboxymandelic acid
Probably in: Iris A3601

3-Carboxy-L-phenylalanine
Formed from: 3-Carboxyphenypyruvic acid
Shikimic acid
→ 3-Carboxymandelic acid
Probably in: Iris A1600, A3601
Reseda A1600
→ *m*-Hydroxymethylphenylalanine
Probably in: Iris A3601

4-(2′-Carboxyphenylamino)-7-chloroquinoline
Formed from: Glifanan

3*H*-2-(4-Carboxyphenylamino)methyl-1,2-dihydro-1-pyrrolizinone
Formed from: 3*H*-1,2-Dihydro-2-(4-methylphenylamino)methyl-1-pyrrolizinone

2-(*p*-Carboxyphenylazo)-4,5-dimethylphenol
→ *p*-Aminobenzoic acid
In: Ascaris A304, A324
Cunninghamella A859
Moniezia A324, A1741, A1785
→ 2-Amino-4,5-dimethylphenol
In: Ascaris A304, A324
Cunninghamella A859
Moniezia A324, A1741, A1785

1-(*p*-Carboxyphenylazo)-4-naphthol
→ *p*-Aminobenzoic acid
In: Pseudomonas D180
→ 1-Amino-4-naphthol
In: Pseudomonas D180

O-(N-(*p*-Carboxyphenyl)carbamoyl) ethyl mycophenolate
→ Mycophenolic acid
In: Dog D518
Guinea pig D518
Mouse D518
Rabbit D518
Rat D518

1-(*p*-Carboxyphenyl)-5-cyano-3-hydroxy-1-pentene
Formed from: 1-(*p*-Carboxyphenyl)-5-cyano-1-pentene

1-(*p*-Carboxyphenyl)-5-cyano-3-oxo-1,4-pentadiene
Formed from: *p*-Carboxy-β-(5-nitro-2-furyl)styrene
→ 1-(*p*-Carboxyphenyl)-5-cyano-3-oxo-1-pentene
Probably in: Rabbit G297

1-(*p*-Carboxyphenyl)-5-cyano-3-oxo-1-pentene
Probably formed from: 1-(*p*-Carboxyphenyl)-5-cyano-3-oxo-1,4-pentadiene
→ 1-(*p*-Carboxyphenyl)-5-cyano-3-hydroxy-1-pentene
In: Rabbit G297
→ 1-(*p*-Carboxyphenyl)-5-cyano-3-pentanone
In: Rabbit G297

1-(*p*-Carboxyphenyl)-5-cyano-3-pentanol
Formed from: 1-(*p*-Carboxyphenyl)-5-cyano-3-pentanone

1-(*p*-Carboxyphenyl)-5-cyano-3-pentanone
Probably formed from: 1-(*p*-Carboxyphenyl)-5-cyano-3-oxo-1-pentene
→ 1-(*p*-Carboxyphenyl)-5-cyano-3-pentanol
In: Rabbit G297

1-(*p*-Carboxyphenyl)-3,3-dimethyltriazene
→ 1-(*p*-Carboxyphenyl)-3-methyltriazene
In: Rat A112

1-(*p*-Carboxyphenyl)-3,3-dimethyltriazene ethyl ester
→ 1-(*p*-Carboxyphenyl)-3-methyltriazene ethyl ester
In: Rat A112

5-(*p*-Carboxyphenyl)-5-ethylbarbituric acid
Formed from: 5-Ethyl-5-(*p*-tolyl)barbituric acid

***o*-Carboxyphenyl-β-D-glucoside**
Formed from: Salicylic acid

***p*-Carboxyphenyl-β-D-glucoside**
Formed from: *p*-Hydroxybenzoic acid
→ *p*-Hydroxybenzoic acid
In: Lithospermum F369

***o*-Carboxyphenyl-β-D-glucuronide**
Formed from: Salicylic acid

***p*-Carboxyphenyl-β-D-glucuronide**
Formed from: *p*-Hydroxybenzoic acid

3-Carboxy-L-phenylglycine
Formed from: 3-Carboxyphenylglyoxylic acid

***m*-Carboxyphenylglycol**
Probably formed from: *m*-Hydroxymethyl-phenylglycol

3-Carboxyphenylglyoxylic acid
Probably formed from: 3-Carboxymandelic acid
→ 3-Carboxy-L-phenylglycine
In: Iris A1600, A3601
Reseda A1600

3-Carboxyphenyl-5-hydroxymethyl-2-oxazolidinone
Formed from: Toloxatone

3-(*o*-Carboxyphenylimino)isoxazolidine
Formed from: 2,3-Dihydro-9*H*-isoxazolo [3,2-*b*]quinazolin-9-one
→ Anthranilic acid
Probably in: Rat A3074

2-((2-Carboxyphenyl)methoxy)-6-hydroxy-4-(1-hydroxy-1-methylethyl)-1-methyl-7-oxabicyclo[2.2.1]heptane
Probably formed from: α,6,8-Trihydroxycin-methylin

2-((2-Carboxyphenyl)methoxy)-4-(1-hydroxy-1-methylethyl)-1-methyl-7-oxabicyclo[2.2.1] heptane
Probably formed from: α,8-Dihydroxycin-methylin

2-((2-Carboxyphenyl)methoxy)-4-(1-(hydroxymethyl)ethyl)-1-methyl-7-oxabicyclo[2.2.1]heptane
Probably formed from: α,9-Dihydroxycinmethylin

2-((2-Carboxyphenyl)methoxy)-6-hydroxy-1-methyl-4-(1-methylethyl)-7-oxabicyclo[2.2.1]heptane
Probably formed from: α,6-Dihydroxycinmethylin

***m*-Carboxyphenyl N-methylcarbamate**
Formed from: *m*-Hydroxymethylphenyl N-methylcarbamate
→ *m*-Hydroxybenzoic acid
Probably in: Housefly A1209
Rabbit A2666
Rat A1209

2-(*p*-Carboxyphenyl)-2-methylpropan-1-ol
Probably formed from: *p-tert*-Butylbenzoic acid
→ 2-(*p*-Carboxyphenyl)-2-methylpropionic acid
Probably in: Rat C764

2-(*p*-Carboxyphenyl)-2-methylpropionic acid
Probably formed from: 2-(*p*-Carboxyphenyl)-2-methylpropan-1-ol

5-(*o*-Carboxyphenyl)-1-methyl-3-(3-pyridyl)-1*H*-1,2,4-triazole
Formed from: 5-(*o*-Hydroxymethylphenyl)-1-methyl-3-(3-pyridyl)-1*H*-1,2,4-triazole

5-(*p*-Carboxyphenyl)-1-methylpyrrole-2-acetic acid
Formed from: 5-(*p*-Hydroxymethylphenyl)-1-methylpyrrole-2-acetic acid

1-(*p*-Carboxyphenyl)-3-methyltriazene
Formed from: 1-(*p*-Carboxyphenyl)-3,3-dimethyltriazene

1-(*p*-Carboxyphenyl)-3-methyltriazene ethyl ester
Formed from: 1-(*p*-Carboxyphenyl)-3,3-dimethyltriazene ethyl ester

4-(*o*-Carboxyphenyl)-4-oxobutyric acid
See: *o*-Succinylbenzoic acid

(*E*)-2-(1-(*p*-Carboxyphenyl)-3-(2-oxopyrrolidin-1-yl)propenyl)pyridine
Probably formed from: (*E*)-2-(1-(*p*-Carboxyphenyl)-3-(1-pyrrolidinyl)propenyl)pyridine

5-(*p*-Carboxyphenyl)-5-phenylhydantoin
Formed from: 5-(*p*-(Hydroxymethyl)phenyl)-5-phenylhydantoin

2-(*p*-Carboxyphenyl)propionic acid
Formed from: Suprofen
Probably formed from: 2-(4-(2-Carboxypropyl)phenyl)propionic acid

N-((1*S*)-1-Carboxy-3-phenylpropyl)-L-alanyl-L-proline
See: Enlaprilat

(4*S*-(4α,7α,(*R),12bβ))-7-(*S*-(1-Carboxy-3-phenylpropyl)amino)-1,2,3,4,6,7,8,12b-octahydro-6-oxopyrido[2,1-*a*][2]benzazepine-4-carboxylic acid**
Formed from: (4*S*-(4α,7α,(*R**),12bβ))-7-(*S*-(1-Ethoxycarbonyl-3-phenylpropyl)amino)-1,2,3,4,6,7,8,12b-octahydro-6-oxopyrido[2,1-*a*][2]benzazepine-4-carboxylic acid

3-(6-(1-(*p*-Carboxyphenyl)-3-(1-pyrrolidinyl)-1-propenyl)-2-pyridinyl)propionic acid
Probably formed from: Dihydroacrivastine

3-Carboxyphenylpyruvic acid
→ 3-Carboxymandelic acid
Probably in: Iris A3601
→ 3-Carboxy-L-phenylalanine
In: Iris A1600
Reseda A1600

(*E*)-2-(1-(*p*-Carboxyphenyl)-3-(1-pyrrolidinyl)propenyl)pyridine
Formed from: Triprolidine
→ (*E*)-2-(1-(*p*-Carboxyphenyl)-3-(2-oxopyrrolidin-1-yl)propenyl)pyridine
Probably in: Dog G512
Mouse G510

4-(*o*-Carboxyphenyl)succinyl CoA
→ 1,4-Dihydroxy-2-naphthoic acid
In: Escherichia D710
Mycobacterium D710

***o*-Carboxyphenyl sulphate**
Formed from: Salicylic acid

***m*-Carboxyphenyl sulphate**
Formed from: *m*-Hydroxybenzoic acid

***p*-Carboxyphenyl sulphate**
Formed from: *p*-Hydroxybenzoic acid

3-(*p*-Carboxyphenyl)sulphonyl-1-(hexahydroazepin-1-yl)urea
Probably formed from: 1-(Hexahydroazepin-1-yl)-3-(*p*-hydroxymethylphenyl)sulphonylurea

N-(*p*-Carboxyphenylsulphonyl)-5-methyl-2-pyrazoline-1-carboxamide
Formed from: N-(*p*-Hydroxymethylphenylsulphonyl)-5-methyl-2-pyrazoline-1-carboxamide

N-(*p*-Carboxyphenylsulphonyl)-N′-2-(1-phenylpropyl)urea
Formed from: N-(*p*-Hydroxymethylphenylsulphonyl)-N′-2-(1-phenylpropyl)urea

4-(5-(*p*-Carboxyphenyl)-3-trifluoromethyl-1*H*-pyrazol-1-yl)benzenesulphonamide
Formed from: 4-(5-((*p*-Hydroxymethyl)phenyl)-3-trifluoromethyl-1*H*-pyrazol-1-yl)benzenesulphonamide
→ 4-(5-(*p*-Carboxyphenyl)-3-trifluoromethyl-1*H*-pyrazol-1-yl)benzenesulphonamide-β-D-glucuronide
Probably in: Man K19
Monkey K123
Mouse K123
Rabbit K123
Rat K30

4-(5-(*p*-Carboxyphenyl)-3-trifluoromethyl-1*H*-pyrazol-1-yl)benzenesulphonamide-β-D-glucuronide
Probably formed from: 4-(5-(*p*-Carboxyphenyl)-3-trifluoromethyl-1*H*-pyrazol-1-yl)benzenesulphonamide

2-Carboxy-4-(1-piperazinyl)-10*H*-thieno[2,3-*b*][1,5]benzodiazepine
Probably formed from: 2-Carboxy-4-(4-methyl-1-piperazinyl)-10*H*-thieno[2,3-*b*][1,5]benzodiazepine

3-(2-Carboxy-1-propenyl)-2,2-dimethylcyclopropanecarboxylic acid (3-(*p*-hydroxyphenoxy)phenyl)methyl ester
Probably formed from: 3-(2-Carboxy-1-propenyl)-2,2-dimethylcyclopropanecarboxylic acid (3-phenoxyphenyl)methyl ester
→ 3-(2-Carboxy-1-propenyl)-2-hydroxymethyl-2-methylcyclopropanecarboxylic acid (3-(*p*-hydroxyphenoxy)phenyl)methyl ester
Probably in: Rat C63

3-(2-Carboxy-1-propenyl)-2,2-dimethylcyclopropanecarboxylic acid (3-phenoxyphenyl)methyl ester
Formed from: 3-(2-Hydroxymethyl-1-propenyl)-2,2-dimethylcyclopropanecarboxylic acid (3-phenoxyphenyl)methyl ester
→ 3-(2-Carboxy-1-propenyl)-2,2-dimethylcyclopropanecarboxylic acid (3-(*p*-hydroxyphenoxy)phenyl)methyl ester
Probably in: Rat C63
→ 3-(2-Carboxy-1-propenyl)-2-hydroxymethyl-2-methylcyclopropanecarboxylic acid (3-phenoxyphenyl)methyl ester
Probably in: Rat C63

3-(2-Carboxy-1-propenyl)-2-hydroxymethyl-2-methylcyclopropanecarboxylic acid (3-(*p*-hydroxyphenoxy)phenyl)methyl ester
Probably formed from: 3-(2-Carboxy-1-propenyl)-2,2-dimethylcyclopropanecarboxylic acid (3-(*p*-hydroxyphenoxy)phenyl)methyl ester

3-(2-Carboxy-1-propenyl)-2-hydroxymethyl-2-methylcyclopropanecarboxylic acid (3-phenoxyphenyl)methyl ester
Probably formed from: 3-(2-Carboxy-1-propenyl)-2,2-dimethylcyclopropanecarboxylic acid (3-phenoxyphenyl)methyl ester

2-(3-Carboxypropionyl)benzoyl CoA
Formed from: *o*-Succinylbenzoic acid

5-(3-Carboxypropionyl)-2-(*p*-chlorophenyl)oxazole-4-acetic acid
Probably formed from: 2-(*p*-Chlorophenyl)-5-(ω-hydroxybutyryl)oxazole-4-acetic acid

3-Carboxy-1-propionyl 1-(2,3-dichloro-4-hydroxyphenyl) O-(2-(diethylamino)ethyl)ketoxime
Probably formed from: 3-Carboxy-1-propionyl 1-(2,3-dichloro-4-methoxyphenyl) O-(2-(diethylamino)ethyl)ketoxime

3-Carboxy-1-propionyl 1-(2,3-dichloro-4-methoxyphenyl) O-(2-(diethylamino)ethyl)ketoxime
Formed from: Diclofurime
→ 3-Carboxy-1-propionyl 1-(2,3-dichloro-4-hydroxyphenyl) O-(2-(diethylamino)ethyl)ketoxime
Probably in: Dog D428
Rat D428

→ 3-Carboxy-1-propionyl 1-(2,3-dichloro-4-methoxyphenyl) O-(2-(ethylamino)ethyl) ketoxime
Probably in: Dog D428
Rat D428

3-Carboxy-1-propionyl 1-(2,3-dichloro-4-methoxyphenyl) O-(2-(ethylamino)ethyl) ketoxime
Probably formed from: 3-Carboxy-1-propionyl 1-(2,3-dichloro-4-methoxyphenyl) O-(2-(diethylamino)ethyl)ketoxime
→ 3-Carboxy-1-propionyl 1-(2,3-dichloro-4-hydroxyphenyl) O-(2-(ethylamino)ethyl) ketoxime
Probably in: Dog D428
Rat D428

3-Carboxypropionyl-L-alanyl-L-alanyl-L-alanyl-*p*-nitroanilide
→ L-Alanyl-*p*-nitroanilide
In: Rat C844

3,4-*trans*-4-(4-(((3-Carboxypropyl)amino) ethoxy)phenyl)-7-methoxy-2,2-dimethyl-3-phenylchromane
Probably formed from: 3,4-*trans*-7-methoxy-2,2-dimethyl-4-(4-((2-oxopyrrolidin-1-yl)ethoxy)phenyl)-3-phenylchromane

3-(6-(3-(N-(3-Carboxypropyl)amino)-1-(*p*-methoxyphenyl)-1-propenyl)-2-pyridinyl)-2-propenoic acid
Formed from: Acrivastine

3-(3-Carboxypropylamino)-4-phenoxy-5-sulphamoylbenzoic acid
Probably formed from: 3-(4-Hydroxy-*n*-butyl) amino-4-phenoxy-5-sulphamoylbenzoic acid

2-(3-Carboxypropylamino)-1-phenylpentane
Probably formed from: N-(1-Phenylpent-2-yl)pyrrolidone

4-(2-Carboxy-2-propyl)-2-chlorophenol
Probably formed from: 2-Chloro-4-(2-hydroxy-1,1-dimethylethyl)phenol
→ 4-(2-Carboxy-2-propyl)-2-chlorophenyl-β-D-glucuronide
Probably in: Rat A3338

4-(2-Carboxy-2-propyl)-2-chlorophenyl-β-D-glucuronide
Probably formed from: 4-(2-Carboxy-2-propyl)-2-chlorophenol

1-(3-Carboxypropyl)-5-(3,4-dichlorobenzyl) biguanide
Probably formed from: 1-(5-Carboxypentyl)-5-(3,4-dichlorobenzyl)biguanide

6-(3-Carboxypropyl)-2,3-dimethoxy-5-methylquinol
Probably formed from: 6-(3-Carboxypropyl)-2,3-dimethoxy-5-methylquinone
→ 6-(3-Carboxypropyl)-2,3-dimethoxy-5-methylquinol-β-D-glucuronide
Probably in: Dog E119
Rat E119
→ 6-(3-Carboxypropyl)-2,3-dimethoxy-5-methylquinol sulphate
Probably in: Dog E119
Rat E119

6-(3-Carboxypropyl)-2,3-dimethoxy-5-methylquinol-β-D-glucuronide
Probably formed from: 6-(3-Carboxypropyl)-2,3-dimethoxy-5-methylquinol

6-(3-Carboxypropyl)-2,3-dimethoxy-5-methylquinol sulphate
Probably formed from: 6-(3-Carboxypropyl)-2,3-dimethoxy-5-methylquinol

6-(3-Carboxypropyl)-2,3-dimethoxy-5-methylquinone
Probably formed from: 6-(5-Carboxypentyl)-2,3-dimethoxy-5-methylquinone
→ 6-(3-Carboxypropyl)-2,3-dimethoxy-5-methylquinol
Probably in: Dog D764, E119
Rat D764, E119
→ 6-(3-Carboxypropyl)-2-hydroxy-3-methoxy-5-methylquinone
Probably in: Dog E119
Rat E119

6-(3-Carboxypropyl)-2-hydroxy-3-methoxy-5-methylquinol
Probably formed from: 6-(3-Carboxypropyl)-2-hydroxy-3-methoxy-5-methylquinone
→ 6-(3-Carboxypropyl)-2-hydroxy-3-methoxy-5-methylquinol-β-D-glucuronide
Probably in: Dog E119
Rat E119

6-(3-Carboxypropyl)-2-hydroxy-3-methoxy-5-methylquinol-β-D-glucuronide
Probably formed from: 6-(3-Carboxypropyl)-

2-hydroxy-3-methoxy-5-methylquinol

1-(3-Carboxypropyl)-N-ethyl-N-methyl-4-phenyl-4-piperidinecarboxamide
Probably formed from: 1-(5-Carboxypentyl)-N-ethyl-N-methyl-4-phenyl-4-piperidinecarboxamide

6-(3-Carboxypropyl)-2-hydroxy-3-methoxy-5-methylquinone
Probably formed from: 6-(3-Carboxypropyl)-2,3-dimethoxy-5-methylquinone
→ 6-(3-Carboxypropyl)-2-hydroxy-3-methoxy-5-methylquinol
Probably in: Dog E119
Rat E119

5-(2-Carboxypropyl)-1-naphthylacetic acid
Formed from: 5-Isobutyl-1-naphthylacetic acid

***p*-(2-Carboxypropyl)phenylacetic acid**
Formed from: 5-Isobutylphenylacetic acid

***p*-(2-Carboxy-2-propyl)phenyl diphenyl phosphate**
Formed from: *p-tert*-Butylphenyl diphenyl phosphate
→ *p*-(2-Carboxy-2-propyl)phenyl *p*-hydroxyphenyl phenyl phosphate
Probably in: Cunninghamella D627
→ *p*-(2-Carboxy-2-propyl)phenyl phenyl phosphate
Probably in: Cunninghamella D627
→ 2-(*p*-Hydroxyphenyl)-2-methylpropionic acid
Probably in: Cunninghamella D627

5-(2-(3-Carboxypropyl)phenyl)-4,6-dithianonanedioic acid
Probably formed from: 5-(2-(5-Carboxypentyl)phenyl)-4,6-dithianonanedioic acid

***p*-(2-Carboxy-2-propyl)phenyl *p*-hydroxyphenyl phenyl phosphate**
Probably formed from: *p*-(2-Carboxy-2-propyl)phenyl diphenyl phosphate

***p*-(2-Carboxy-2-propyl)phenyl phenyl phosphate**
Probably formed from: *p*-(2-Carboxy-2-propyl)phenyl diphenyl phosphate

2-(4-(2-Carboxypropyl)phenylpropionic acid
Formed from: (*R*)-2-(4-Isobutylphenyl) propionic acid
Probably formed from: 2-(4-(2-Hydroxy-2-methylpropyl)phenyl)propionic acid
→ 2-(*p*-Carboxyphenyl)propionic acid
Probably in: Man A3560

5-((4-(2-(5-Carboxy-2-pyridinyl)ethoxy)phenyl)methyl)-2,4-thiazolidinedione
Probably formed from: 5-((4-(2-(5-Carboxymethyl-2-pyridinyl)ethoxy)phenyl)methyl)-2,4-thiazolidinedione
→ 5-((4-(2-(5-Carboxytauryl-2-pyridinyl)ethoxy)phenyl)methyl)-2,4-thiazolidinedione
In: Rat G946
→ 5-((4-(2-(5-Hydroxy-2-pyridinyl)ethoxy)phenyl)methyl)-2,4-thiazolidinedione
Probably in: Rat G946

4-Carboxy-*o*-quinone
Formed from: 3,4-Dichlorobenzoic acid
Protocatechuic acid

Carboxy-*p*-quinone
Formed from: Gentisic acid

1-Carboxystyrene oxide
→ 2-Carboxy-2-(S-glutathionyl)-2-phenylethanol
Probably in: Sheep A2282

5-((4-(2-(5-Carboxytauryl-2-pyridinyl)ethoxy)phenyl)methyl)-2,4-thiazolidinedione
Formed from: 5-((4-(2-(5-Carboxy-2-pyridinyl)ethoxy)phenyl)methyl)-2,4-thiazolidinedione

1-Carboxy-1,2,3,4-tetrahydro-6,7-dihydroxy-1-methylisoquinoline
See: Salsolinol-1-carboxylic acid

2-Carboxy-1,2,3,4-tetrahydro-6-methyl-7-nitroquinoline
Formed from: 2-Aminomethyl-1,2,3,4-tetrahydro-6-methyl-7-nitroquinoline

4-Carboxy-2-tetraprenylphenol
Formed from: *p*-Hydroxybenzoic acid

N-(5-Carboxy-2-thiazolyl)-4-hydroxy-2-methyl-2*H*-1,2-benzothiazine-3-carboxamide-1,1-dioxide
Formed from: 4-Hydroxy-N-(5-hydroxymethyl-2-thiazolyl)-2-methyl-2*H*-1,2-benzothiazine-3-carboxamide-1,1-dioxide

3-Carboxy-*s*-triazolo[3,4-*a*]phthalazine
Formed from: 3-Hydroxymethyl-*s*-triazolo[3,4-*a*]phthalazine

→ 3-Carboxy-*s*-triazolo[3,4-*a*]phthalazine-β-D-glucuronic acid
Probably in: Rat A2816

3-Carboxy-*s*-triazolo[3,4-*a*]phthalazine-β-D-glucuronic acid
Probably formed from: 3-Carboxy-*s*-triazolo[3,4-*a*]phthalazine

3-Carboxy-1,6,8-trihydroxyanthraquinone
Probably formed from: 3-Formyl-1,6,8-trihydroxyanthraquinone

3-Carboxy-2,4,6-trihydroxyphenylacetic acid
Formed from: Phellamurin
→ 3-Carboxy-2,4,6-trihydroxybenzoic acid
Probably in: Aspergillus A3306

4-Carboxy-2-triprenylphenol
Formed from: *p*-Hydroxybenzoic acid

3-Carboxy-L-tyrosine
Formed from: 3-Carboxy-4-hydroxyphenylpyruvic acid
Shikimic acid

5-Carboxyvanillic acid (3-carboxy-4-hydroxy-5-methoxybenzoic acid)
Probably formed from: Dehydrodivanillic acid
→ Vanillic acid
Probably in: Microorganism D540

5-Carboxyvanillin (5-formyl-2-hydroxy-3-methoxybenzoic acid)
Probably formed from: 5-Carboxymethyl-vanillin

5-(2-Carboxyvinyl)-2,3-dihydro-2-(4-hydroxy-3-methoxyphenoyl)-3-hydroxymethyl-7-methoxybenzofuran
Formed from: Dehydrodiconiferyl alcohol
→ 5-(2-(4-Hydroxy-3-methoxyphenyl)vinyl)ferulic acid
Probably in: Pseudomonas F77

1-Carboxyxanthone
Formed from: Xanthone-1-acetic acid

Carebastine (4-(4-(4-(diphenylmethoxy)-1-piperidinyl)-1-oxobutyl)-α,α-dimethylbenzeneacetic acid)
Formed from: 4′-(1-Hydroxymethyl-1-methylethyl)-4-(4-(diphenylmethoxy)piperidino)butyrophenone

Carfecillin (α-phenoxycarbonylbenzylpenicillin)
→ Carbenicillin
In: Man A3396
→ Phenol
In: Dog B531
Man A3396, B531
Rat B531

Carminomycin 1 ((8*S*,10*S*)-8-acetyl-10-((3-amino-2,3,6-trideoxy-α-L-lyxo-hexopyranosyl)oxy)-7,8,9,10-tetrahydro-1,6,8,11-tetrahydroxy-5,12-naphthacenedione)
→ Daunorubicin
In: Streptomyces B600

Carminomycinone ((8*S*,10*S*)-8-acetyl-7,8,9,10-tetrahydro-1,6,8,10,11-pentahydroxy-5,12-naphthacenedione)
Probably formed from: ε-Rhodomycinone
→ Daunorubicin
In: Streptomyces B571

Carocainide (N-(4,7-dimethoxy-6-(2-(1-pyrrolidinyl)ethoxy)-5-benzofuranyl)-N′ methylurea)
→ Carocainide-N-oxide
In: Man D633
→ 1-(2,5-Dimethoxy-3-carboxymethyl-4-hydroxy-6-(2-(1-pyrrolidinyl)ethoxy)phenyl)-3-methylurea
In: Man D633

Carocainide-N-oxide
Formed from: Carocainide

Carpipramine (1′-(3-(10,11-dihydro-5*H*-dibenz[*b*,*f*]azepin-5-yl)propyl)-(1,4-bipiperidine)-4′-carboxamide)
→ 2-Hydroxycarpipramine
In: Man D368
→ 2″-Hydroxycarpipramine
In: Man D368
→ 10-Hydroxycarpipramine
In: Man D368
→ Iminodibenzyl
In: Man D368

Carprofen ((*R*)-and (S)-6-chloro-α-methylcarbazole-2-acetic acid)
→ Carprofen-β-D-glucuronide
In: Dog B728, H938, J406
Horse H407, J406
Man B728, F24, J406
Rat B728, G46, J406
Sheep J406
→ 7-Hydroxycarprofen

In: Dog B728
Man B728
Rat B728
→ 8-Hydroxycarprofen
In: Dog B728
Man B728
Rat B728
→ α-Hydroxycarprofen
In: Dog B728
Rat B728

Carprofen-β-D-glucuronide
Formed from: Carprofen

Carsalem
See: 1,3-Benzoxazine-2,4-(3*H*)-dione

Carteolol (5-(3-*tert*-butylamino-2-hydroxypropoxy)-3,4-dihydrocarbostyril)
→ 5-Carboxymethoxy-3,4-dihydrocarbostyril
In: Dog A3118
→ 3,4-Dihydro-5-hydroxycarbostyril
In: Dog A3118
Rat A3118
→ 8-Hydroxycarteolol
In: Dog A3118
Rat A3118, J296

Carthamin
Formed from: Precarthamin

Carvacrol (5-isopropyl-2-methylphenol)
→ Carvacrol-β-D-glucuronide
In: Man G524, G783
→ Carvacrol sulphate
In: Mouse H959
→ 2-Isopropyl-5-methylquinol
In: Mucor A832

Carvacrol-β-D-glucuronide
Formed from: Carvacrol

Carvacrol sulphate
Formed from: Carvacrol

Carvedilol (1-(9*H*-Carbazol-4-yloxy)-3-(((2-methoxyphenoxy)ethyl)amino)-2-propanol)
→ 1-(Carbazol-(4)-oxy)-3-((β-D-glucuronosyl)((2-methoxyphenoxy)ethyl) amino)-2-propanol
In: Dog J443
Mouse J443
Rat J443
See also: G99
→ Carvedilol-9-β-D-glucuronide
In: Dog J443
Mouse J443
Rat J443
→ Carvedilol-O-β-D-glucuronide
In: Dog G99, J443
Mouse J443
Rat G99, J443
→ Demethylcarvedilol
In: Dog J443
Man H594, J272
Mouse J443
Rat G958
→ N-(2,3-Dihydroxypropyl)-2-(*o*-methoxyphenoxy)ethylamine
In: Dog J443
Mouse J443
Rat J443
→ 1-Hydroxycarvedilol
In: Dog J443
Man J272
Mouse J443
Rat F584, G33, G958
→ 3-Hydroxycarvedilol
In: Dog J443
→ 4′-Hydroxycarvedilol
In: Man H594, J272
Mouse J443
Rat G958
→ 5′-Hydroxycarvedilol
In Man H594, J272
Mouse J443
→ 8-Hydroxycarvedilol
In: Dog J443
Man J272
Mouse J443
Rat F584, G33, G958, J443

Carvedilol-9-β-D-glucuronide
Formed from: Carvedilol

Carvedilol-O-β-D-glucuronide
Formed from: Carvedilol

Casodex (4′-cyano-3-(*p*-fluorophenylsulphonyl)-2-hydroxy-2-methyl-3′-(trifluoromethyl) propionanilide)
→ 4-Amino-2-trifluoromethylbenzonitrile

In: Mouse G592
Rat G592
→ Casodex-β-D-glucuronide
In: Dog G592
Man G592
Rabbit G592
Rat G592, H200
→ 4′-Cyano-3-(4-fluoro-3-hydroxyphenylsulphonyl)-2-hydroxy-2-methyl-3′-trifluomethylpropionanilide
In: Dog G592
Man G592
Rat G592
See also: G708, H200
→ 3-(*p*-Fluorophenylsulphonyl)-2-methyllactic acid
In: Mouse G592
Rat G592

Casodex-β-D-glucuronide
Formed from: Casodex

Caspofungin (1-((4*R*,5*S*)-5-((2-aminoethyl) amino)-N^2-(10,12-dimethyl-1-oxotetradecyl)-4-hydroxy-L-ornithine)-5-((3*R*)-3-hydroxy-L-ornithine)pneumocandin B_o)
→ N-Acetyl-4(*S*)-hydroxyphenyl)-L-threonine
In: Man K321

Catalpalactone (3-(5,6-dihydro-6,6-dimethyl-2-oxo-2-pyran-3-yl)-1(3*H*)-isobenzofuranone)
Formed from: *o*-Succinylbenzoic acid
4-(*o*-Carboxyphenyl)-4-oxobutyric acid

Catalponol ((2*R*,4*S*)-3,4-dihydro-4-hydroxy-2-(3-methyl-2-butenyl)-1(2*H*)-naphthalenone)
Formed from: 2-Carboxy-4-hydroxy-α-tetralone

Catalponone ((2*R*)-2,3-dihydro-2-(3-methyl-2-butenyl)-1,4-naphthalenedione)
Formed from: *o*-Succinylbenzoic acid

(+)-Catechin (catechuic acid, *cis*-3,3′,4′-5,7-pentahydroxyflavan)
Formed from: Cinnamic acid
2,3-*trans*-3,4-*cis*-3,3′,4,4′-5,7-Hexahydroxyflavan
Probably formed from: 3-Palmitoyl-(+)-Catechin
→ (+)-Catechin-α-D-glucoside
In: Bacillus H234
→ (+)-Catechin-7-O-β-D-glucoside
In: Lobelia H686
→ (+)-Catechin-β-D-glucuronide
In: Man C486
Marmoset C249
Rat C249, C311, E245, K645
→ (+)-Catechin quinone
Probably in: Acacia J187
Banana A218
Botrytis A1930
Olive A2360
Potato A3190
Tea A972
Trametes K106
→ (+)-Catechin-4′-sulphate
In: Eubacterium G475
See also C311, C486
→ Dehydrodicatechin A
With peroxidase A896
→ Dehydrodicatechin B_1
With peroxidase A896
→ Dehydrodicatechin B_2
With peroxidase A896
→ Dehydrodicatechin B_3
With peroxidase A896
→ Dehydrodicatechin B_4
With peroxidase A896
→ (−)-Epicatechin
In: Uncaria A1162
→ 1-(3,4-Dihydroxyphenyl)-3-(2,4,6-trihydroxyphenyl)propan-2-ol
In: Microorganism C880
→ 8-Hydroxycatechin
With phenol oxidase A896
→ 4-Hydroxy-3-methoxyphenylacetic acid
In: Rat A1375
→ *m*-Hydroxyphenylhydracrylic acid
In: Monkey A1378
→ *m*-Hydroxyphenylpropionic acid
In: Man C486
Monkey A1378
→ δ-(*m*-Hydroxyphenyl)-γ-valerolactone
In: Monkey A1378
Rat C781
→ Procyanidin polymer
Probably in: Sorghum A3641
→ Protocatechuic acid

In: Bradyrhizobium G292, J484
Chaetomium G814
Rhizobium E742
→ 3,4′,5,7-Tetrahydroxy-3′-methoxyflavan
In: Man C486, C722
Marmoset C249
Pig D112
Rat B534, C249, C311, C781, E245, K645
See also: C312
→ 3,3′,5,7-Tetrahydroxy-4′-methoxyflavan
In: Rat K645
→ 2,4,6-Trihydroxybenzoic acid
In: Bradyrhizobium G292, J484
Chaetomium G814
Rhizobium E742

(+)-Catechin-4′,5-disulphate
Probably formed from: (+)-Catechin-4′-sulphate

(+)-Catechin-α-D-glucoside
Probably formed from: (+)-Catechin

(+)-Catechin-7-O-β-D-glucoside
Formed from: (+)-Catechin

(+)-Catechin-β-D-glucuronide
Formed from: (+)-Catechin

(+)-Catechin quinone
Formed from: (+)-Catechin

(+)-Catechin-4′-sulphate
Formed from: (+)-Catechin
→ (+)-Catechin-4′,5-disulphate
Probably in: Eubacterium G475

Catechol (1,2-dihydroxybenzene, pyrocatechol)
Formed from: Aniline
Anthracene
Anthranilic acid
Benzenesulphonic acid
Benzoic acid
o-Bromobenzoic acid
o-Chlorobenzoic acid
Dehydroshikimic acid
cis-1,2-Dihydro-1,2-dihydroxybenzene
trans-1,2-Dihydro-1,2-dihydroxybenzene
2,3-Dihydro-2,3-dihydroxy-3-nitrobenzene
2,3-Dihydroxybenzoic acid
2,3-Dihydroxyphenylacetic acid
3,4-Dihydroxy-L-phenylalanine
l-Ephedrine
Ferulic acid
o-Fluorobenzoic acid
Guaiacol
o-Iodobenzoic acid
o-Methoxybenzoic acid
1,2-Methylenedioxybenzene
Nitrobenzene
o-Nitrophenol
Phenol
Protocatechuic acid
Pyrocatechuic acid
Salicylaldehyde
Salicylic acid
Shikimic acid
Vanillin
Veratrole
Probably formed from: 2′-Amino-2,3-dihydroxybiphenyl
→ 3,4-Dihydroxy-L-phenylalanine
In: Citrobacter J512
Erwinia A708, B292, D488, E959
Symbiobacterium G603
→ Diphenylenedioxide-2,3-quinone
In: Spinach K880
→ Guaiacol
In: Man G658
Nicotiana A3778, B490
Phanerochaete G782
Rat A486
Spinach beet A2315
Streptomyces K652
With catechol O-methyltransferase A597, C461
→ *m*-Hydroxybenzoic acid
In: Microorganism G917
→ 2-Hydroxymuconic semialdehyde
In: Alcaligenes J653
Amanita G119
Bacillus A1342
Penicillium F639
Pseudomonas A768, D771, G355, G647, H272, J515, J649, J903, K5, K204
Rhodococcus J221
See also: A546, D484, J255, J329

→ *o*-Hydroxyphenyl-α-D-glucoside
Probably in: Bacillus H234
→ *o*-Hydroxyphenyl-β-D-glucoside
In: Datura A1725
Gardenia E338
Manduca G816
Periplaneta G816
→ *o*-Hydroxyphenyl-β-D-glucuronide
In: Chicken C629
Rat B962, K595
→ *o*-Hydroxyphenyl sulphate
In: Man E575
Rat E575
→ *cis*,*cis*-Muconic acid
In: Acinetobacter J390
Alcaligenes A3769, K775
Aspergillus C342
Brevibacterium, A681
Burkholderia K445
Candida E204
Hydrogenophaga J218
Moraxella A638
Penicillium F639
Phanerochaete H284
Pseudomonas A3769, D771, G647, H506, J702
Rhizobium D672, F647
Rhodococcus D247, J703
Streptomyces B877, C510
Trichosporon C84
→ Polycatechol
In: Bjerkandera K565
→ Protocatechuic acid
In: Clostridium H576, J184
→ Pyrogallol
In: Candida E204, K653
Paecilomyces F199
Trichosporon C84
See also: G817
→ *o*-Quinone
In: Amasya H567
Apple J678
Horseradish E603
Pear J678
See also: A63, A218, A972, A1730, A2360, A2433, A3114, A3135, A3974, B897, C519, E369, E407, F151, F274, F664, F944, G130, G620, H570, H948, J313, K96, K677
→ 2,3,4′-Trihydroxy-3′-methoxybiphenyl
With lactoperoxidase J352
→ Xylobiosylcatechol
In: Penicillium J487
→ β-Xylosylcatechol
In: Penicillium J487

Catechol-4-sulphonic acid
See: 3,4-Dihydroxybenzenesulphonic acid

Catenarin (1,4,5,7-tetrahydroxy-2-methylanthraquinone)
→ 5-Hydroxyquestin
In: Aspergillus C334
→ Islandicin
In: Penicillium A1046
→ Rubroskyrin
In: Penicillium A1046
→ Skyrin
In: Penicillium A1046

Catharanthine (3,4-didehydroibogamine-18-carboxylic acid methyl ester)
Formed from: Strictosidine
→ 3′,4′-Anhydrovinblastinium
In: Catharanthus E745
Horseradish E722
→ Vinblastine
In: Catharanthus A3608

Catharine
Formed from: 3′,4′-Dehydrovinblastine

Catharinine
Formed from: Vinblastine
→ 16-Deoxycatharinine-(15*H*)-15,16-oxide
Probably in: Man F231

Cathenamine (20,21-didehydroajamalicine)
Probably formed from: Strictosidine
→ Ajmalicine
In: Catharanthus A3069
→ 19-Epiajmalicine
In: Catharanthus A3069
→ Tetrahydroalstonine
In: Catharanthus A3069

Cathinone ((2*R*)-and (2*S*)-isomers; latter is usually understood, 2-amino-1-phenyl-1-propanone))
→ Norephedrine
In: Man E77

→ D-Pseudonoroephedrine
In: Man E77
→ L-Pseudonoroephedrine
In: Man E77
Cefcanel daloxate
→ Mandelylglycine
In: Man G532
→ N-Mandelyl-2-aminoethanol
In: Man G532
Celecoxib (4-(5-(4-methylphenyl)-3-trifluoromethyl-1*H*-pyrazol-1-yl) benzenesulphonamide)
→ Hydroxycelecoxib
In: Rabbit K123
→ 4-(5-((*p*-Hydroxymethyl)phenyl)-3-trifluoromethyl-1*H*-pyrazol-1-yl) benzenesulphonamide
In: Cynomolgus monkey K123
Dog J781
Man K19
Mouse K123
Rabbit K123
Rat K30
Rhesus monkey K123
Celiptium (N^2-methyl-9-hydroxyellipticinium)
→ Celiptium-9-β-D-glucuronide
In: Man D102
Rat D102
→ 10-(S-(L-Cysteinyl))celiptium
In: Man D102
Rat D102
Celiptium-9-β-D-glucuronide
Formed from: Celiptium
Celosianin II
Formed from: Amaranthin
Feruloyl-β-D-glucose
Centchroman (*rel*-1-(2-(4-((3*R*,4*R*)-3,4-dihydro-7-methoxy-2,2-dimethyl-3-phenyl-2*H*-1-benzopyran-4-yl)phenoxy)ethyl) pyrrolidine)
→ Demethylcentchroman
In: Rat H531
Centpiperalone (2-piperazino-3*H*-4-quinazolone)
→ 6-Hydroxycentpiperalone
In: Monkey A2560
Cephaeline ((1*R*)-1-(((2*S*,3*R*,11b*S*)-3-ethyl-1,3,4,6,7,11b-hexahydro-9,10-dimethyl-2*H*-benzo[*a*]quinolizin-2-yl)methyl)-1,2,3,4-tetrahydro-7-methoxy-6-isoquinolinol)
Formed from: Deacetylisoipecoside
→ Emetine
Probably in: Alangium A3154
Cephaelis A3154, A3861
Cephalexin (6*R*,7*R*)-7-(((2*R*)-aminophenylacetyl) amino)-3-methyl-8-oxo-5-thia-1-azabicyclo[4.2.0]oct-2-ene-2-carboxylic acid)
Formed from: D-Phenylglycine methyl ester
Cephalomannine
Probably formed from: Phenylisoserine
Cerivastatin (*E*)-(+)-(3*R*,5*S*)-7-(4-fluorophenyl)-2,6-diisopropyl-5-(methoxymethyl)pyrid-3-yl)-3,5-dihydroxyhept-6-enoic acid)
→ Demethylcerivastatin
In: Dog J426
Man J135
→ Cerivastatin-β-D-glucuronide (3 isomers)
In: Dog J426
→ Cerivastatin lactone
In: Dog J426
→ (*E*)-(5*S*)-7-(4-(*p*-Fluorophenyl)-2,6-diisopropyl-5-(methoxymethyl)pyrid-3-yl)-5-hydroxyhept-6-enoic acid
In: Mouse J426
→ 5-(4-(*p*-Fluorophenyl)-2,6-diisopropyl-5-(methoxymethyl)pyrid-3-yl)-3-hydroxy-pent-4-enoic acid
In: Mouse J426
Rat J426
→ (*E*)-(+)-(3*R*,5*S*)-7-(4-(*p*-Fluorophenyl)-6-(1-hydroxy-2-propyl)-2-isopropyl-5-(methoxymethyl)pyrid-3-yl)-3,5-dihydroxyhept-6-enoic acid
In: Man J135
Cerivastatin-β-D-glucuronide
Formed from: Cerivastatin
Cerivastatin lactone
Formed from: Cerivastatin
→ Demethylcerivastatin lactone
Probably in: Dog J426
Mouse J426
Rat J426
Cetraxate (4-(((*trans*-4-(aminomethyl)cyclohexyl) carbonyl)oxy)benzenepropanoic acid)
→ Phloretic acid

In: Rabbit D691
Chaetoglobosin A
Formed from: L-Tryptophan
Chaetoglobosin B
Formed from: L-Tryptophan
Chanoclavine I ((2*E*)-2-methyl-3-((4*R*,5*R*)-1,3,4,5-tetrahydro-4-(methylamino)benz[*cd*]indol-5-yl)-2-propen-1-ol, assumed unless otherwise stated)
Probably formed from: 4-Dimethylallyl-L-tryptophan
→ Agroclavine
In: Claviceps A893
See also: A752
→ Chanoclavine I-O-β-D-fructofuranoside
In: Claviceps F512, F516
→ Chanoclavine I-β-galactoside
In: Aspergillus H11
→ Elymoclavine
In: Claviceps A1132, C93
See also: A752
(−)-Chanoclavine II ((2*E*)-2-methyl-3-((4*R*,5*S*)-1,3,4,5-tetrahydro-4-(methylamino)benz[*cd*]indol-5-yl)-2-propen-1-ol)
Probably formed from: 4-Dimethylallyl-L-tryptophan
Chanoclavine I aldehyde ((2*E*)-2-methyl-3-((4*R*,5*R*)-1,3,4,5-tetrahydro-4-(methylamino)benz[*cd*]indol-5-yl)-2-propenal)
→ Elymoclavine
In: Claviceps A1132
Chanoclavine I-O-β-D-fructofuranoside
Probably formed from: Chanoclavine I
Chanoclavine-β-galactoside
Formed from: Chanoclavine
Chavicol
See: 4-Allylphenol
Cheilanthifoline ((6a*S*)-6,6a,11,14-tetrahydro-8-methoxy-12*H*-benzo[*a*]-1,3-benzodioxolo[4,5-*g*]quinolizin-9-ol)
Formed from: Reticuline
Chelerythrine (1,2-dimethoxy-12-methyl[1,3]benzodioxolo[5,6-*c*]phenanthridinium)
Formed from: Allocryptopine
Dihydrochelerythrine
Tetrahydroberberine
cis-7,8,13,13a-Tetrahydro-N-methylberbinium
(+)-Chelidonine (5b*R*,6*S*,12b*S*)-5b,6,7,12b,13,14-hexahydro-13-methyl[1,3]benzodioxolo[5,6-*c*]-1,3-dioxolo[4,5-*i*]phenanthridin-6-ol)
→ Dehydrochelidonine
In: Corydalis H70
→ Dihydrosanguinarine
In: Corydalis H70
Chelirubine (5-methoxy-13-methyl[1,3]benzodioxolo[5,6-*c*]-1,3-dioxolo[4,5-*i*]phenanthridinium)
Formed from: Sanguinarine
Probably formed from: Dihydrochelirubine
→ Macarpine
In: Macleaya C696
Chinoform
See: 5-Chloro-8-hydroxy-7-iodoquinoline
Chlomethoxynil (2,4-dichlorophenyl 3-methoxy-4-nitrophenyl ether)
→ 4-Amino-3-methoxyphenyl 2,4-dichlorophenyl ether
In: Rat C358
→ 2,4-Dichlorophenol
In: Rat C358
→ 2,4-Dichlorophenyl 6-hydroxy-3-methoxy-4-nitrophenyl ether
In: Rat C358
→ 2,4-Dichlorophenyl 3-hydroxy-4-nitrophenyl ether
In: Rat C358
α-Chloracetamidocinnamic acid
→ L-Phenylalanine
In: Bacillus J356
Chlorambucil (4-(*p*-(bis(2-chloroethyl)amino)phenylbutyric acid)
→ *p*-(Bis(2-chloroethyl)amino)phenylacetic acid
In: Rat A1665, B320
→ 4-(*p*-(Bis(2-hydroxyethyl)amino)phenyl)butyric acid
Probably in: Man J545
Rat A2041
→ 4-(*p*-(N,N-Bis(2-(thiocyano)ethyl)amino)phenyl)butyric acid
In: Man J545
→ *p*-(2-Chloroethylamino)phenylacetic acid
Probably in: Rat A2041

→ *p*-(N-(2-Chloroethyl)-N-(2-hydroxyethyl) amino)benzoic acid
Probably in: Rat A2041
→ 3,4-Dehydrochlorambucil
In: Rat A3959
→ 4-(*p*-(N-(2-Hydroxyethyl)-N-(2-(thiocyano)ethyl)amino)phenyl)butyric acid
In: Man J545
→ *p*-Morpholinophenylacetic acid
Probably in: Rat A2041

Chloramphenicol (D-*threo*-2-(dichloroacetamido)-1-(*p*-nitrophenyl)-1,3-propanediol)
→ 1-Acetylchloramphenicol
In: Duck G945
Klebsiella A460
Streptomyces A2476
→ 3-Acetylchloramphenicol
In: Klebsiella A460
Streptomyces A2476
→ D-*threo*-2-Amino-1-(*p*-nitrophenyl)-1,3-propanediol
In: Duck G945
Man E898
Rainbow trout D351
→ D-*threo*-1-(*p*-Aminophenyl)-2-(dichloroacetamido)-1,3-propanediol
In: Man B860, E898
Rat C618
→ 3-Butyrylchloramphenicol
In: Streptomyces A2476
→ Chloramphenicol-β-D-glucuronide
In: Cat C562
Dog C562
Duck G945
Guinea pig C562
Man C654, E898
Mouse C562
Quail C562
Rabbit C562
Rat A65, A2511, A2329, B962, C562, C618, C672, H539
Trout C562, D351
See also: B167
→ Chloramphenicol-2-oxamaldehyde
In: Rat A3906, B601
See also: A2496, B564, C618, E898, G945
→ Chloramphenicol-O-sulphate
In: Duck G945
Goat E939
→ Dechlorochloramphenicol
In: Rat C593
→ 2-(Dichloroacetamido)-3-hydroxy-3-(*p*-nitrophenyl)-1-propanal
In: Rat C31
See also: H609
→ 2-(Dichloroacetamido)-1-(*p*-nitrosophenyl)-1,3-propanediol
In: Man K689
→ 2-Glycollamido-1-(*p*-nitrophenyl)-1,3-propanediol
In: Duck G945
Rainbow trout D351
→ 3-Isobutyrylchloramphenicol
In: Streptomyces A2476
→ 3-Isovalerylchloramphenicol
In: Streptomyces A2476
→ *p*-Nitrobenzaldehyde
Probably in: Man H565
Rat H565
→ *p*-Nitrobenzyl alcohol
In: Rat C31
→ 3-Propionylchloramphenicol
In: Streptomyces A2476

Chloramphenicol-β-D-glucuronide
Formed from: Chloramphenicol

Chloramphenicol-2-oxamaldehyde
Formed from: Chloramphenicol
→ Chloramphenicol-2-oxamic acid
In: Rat B601
See also: A3906, B564, C618, E898, G945
→ 2-Glycollamido-1-(*p*-nitrophenyl)-1,3-propanediol
Probably in: Man E898
Rat C618

Chloramphenicol-2-oxamic acid
Formed from: Chloramphenicol-2-oxamaldehyde
→ Chloramphenicol-2-oxamylethanolamine
Probably in: Duck G945
→ Chloramphenicol-2-oxamylglycine
Probably in: Duck G945

Chloramphenicol-2-oxamylethanolamine
Formed from: Chloramphenicol-2-oxamic acid
Chloramphenicol-2-oxamylglycine
Formed from: Chloramphenicol-2-oxamic acid
Chloramphenicol-O-sulphate
Formed from: Chloramphenicol
Chlorazepate (chlordiazepate)
→ 7-Chloro-1,3-dihydro-5-phenyl-2*H*-1,4-benzodiazepin-2-one
In: Man A3120
→ 6-Chloro-4-phenyl-2(1*H*)-quinazolinone
In: Rat A282
Chlorcyclizine (1-(*p*-chloro-α-phenylbenzyl)-4-methylpiperazine)
→ N-(*p*-Chlorobenzhydryl)ethylenediamine
In: Rat A576
→ Norchlorcyclizine
In: Rat A522, A573
Chlordiazepate
See: Chlorazepate
Chlordiazepoxide (librium, 7-chloro-2-methylamino-5-phenyl-3*H*-1,4-benzodiazepine-4-oxide)
→ 2-Amino-5-chlorobenzophenone
In: Man A765
→ 2-Amino-7-chloro-5-phenyl-3*H*-1,4-benzodiazepine-4-oxide
In: Man A1246, A3318
→ 7-Chloro-1,3-dihydro-5-phenyl-2*H*-1,4-benzodiazepin-2-one-4-oxide
In: Man A1246
Chlordimeform (N′-(4-chloro-*o*-tolyl)-N,N-dimethylformamidine)
→ 4-Chloro-2-methylaniline
In: Rat D443
→ 1-(4-Chloro-*o*-tolyl)-3,3-dimethylurea
In: Mouse A2534
→ Demethylchlordimeform
In: Cabbage looper A1838
Housefly A779
Man 2047
Mouse A2534
Spider mite A1944
Chlorfenvinphos (2-chloro-1-(2,4-dichlorophenyl) ethenyl diethyl phosphate)
→ 2-Chloro-1-(2,4-dichlorophenyl)ethenyl ethyl phosphate
In: Man E653
Rabbit E653
Rat E653
Chloridazon (5-amino-4-chloro-2-phenyl-3(*H*)-pyridazinone)
→ *m*-Hydroxychloridazon
In: Papaver B351
→ *p*-Hydroxychloridazon
In: Papaver B351
Sugar beet B351
Chlormezanone (2-(4-chlorophenyl)tetrahydro-3-methyl-4*H*-1,3-thiazin-4-one-1,1-dioxide)
→ α-(2-Carboxyethylthio)-*p*-chloro-N-methylbenzylamine-S,S-dioxide (a possible artifact) E130
→ *p*-Chlorobenzaldehyde
In: Man J687
→ 2-(N-(*p*-Chlorobenzoyl)-N-methylcarbamoyl)ethylsulphonic caid
In: Mouse A3044
Rat A3044
***o*-Chloroacetanilide**
Formed from: *o*-Chloroaniline
→ *o*-Chloroaniline
In: Nocardia G343
***m*-Chloroacetanilide**
Formed from: *m*-Chloroaniline
***p*-Chloroacetanilide**
Formed from: *p*-Chloroaniline
p-Chloro-N-hydroxyacetanilide
→ 2-Acetamido-5-chlorophenol
In: Hamster A802
Rat G742
See also: B237
→ 5-Acetamido-2-chlorophenol
In: Hamster A802
→ *p*-Chloroaniline
In: Nocardia G343
Rat G742
See also: B178
→ *p*-Chloroglycollanilide
Probably in: Rat F729
→ *p*-Chloro-N-hydroxyacetanilide
In: Hamster A48, A802

o-Chloroacetophenone
→ *o*-Chlorophenyl acetate
In: Alcaligenes G107
→ 1-(*o*-Chlorophenyl)ethanol
In: Corynebacterium J893

m-Chloroacetophenone
Formed from: 1-(*m*-Chlorophenyl)ethanol
→ *m*-Chlorophenyl acetate
In: Alcaligenes G107
→ 1-(*m*-Chlorophenyl)ethanol
In: Corynebacterium J893
Geotrichum H749

p-Chloroacetophenone
Formed from: 1-(*p*-Chlorophenyl)ethanol
Probably formed from: 3,3-Dichloro-2-(*p*-chlorophenyl)propionic acid
→ *p*-Chlorobenzaldehyde
Probably in: Ralstonia K542
→ *p*-Chlorophenyl acetate
In: Alcaligenes G107
→ 1-(*p*-Chlorophenyl)ethanol
In: Corynebacterium J893
Geotrichum H749
Lactobacillus G148
Rabbit D125

N-Chloroacetyl-2,6-diethylaniline
Formed from: Alachlor
→ 1-Chloroacetyl-7-ethyl-2,3-dihydroindole
In: Chaetomium A1560
→ 2,6-Diethylaniline
In: Chaetomium A1560
Mouse F528
Rat F528

N-Chloroacetyl-2,6-diethyl-N-methoxymethylaniline
See: Alachlor

N-Chloroacetyl-2,6-di-(1-hydroxyethyl)-N-methoxymethylaniline
Probably formed from: N-Chloroacetyl-6-ethyl-2-(1-hydroxyethyl)-N-methoxymethylaniline

1-Chloroacetyl-7-ethyl-2,3-dihydroindole
Formed from: Alachlor
N-Chloroacetyl-2,6-diethylaniline

N-Chloroacetyl-6-ethyl-2-(1-hydroxyethyl)aniline
Probably formed from: N-Chloroacetyl-6-ethyl-2-(1-hydroxyethyl)-N-methoxymethylaniline

N-Chloroacetyl-6-ethyl-2-(1-hydroxyethyl)-N-methoxymethylaniline
Formed from: Alachlor
→ N-Chloroacetyl-2,6-di-(1-hydroxyethyl)-N-methoxymethylaniline
Probably in: Monkey F61
Mouse F61
Rat F61
→ N-Chloroacetyl-6-ethyl-2-(1-hydroxyethyl)aniline
Probably in: Monkey F61
Mouse F61
Rat F61
→ N-Chloroacetyl-6-ethyl-N-methoxymethyl-2-vinylaniline
In: Cunninghamella G755

N-Chloroacetyl-6-ethyl-N-methoxymethyl-2-vinylaniline
Formed from: Alachlor
N-Chloroacetyl-6-ethyl-2-(1-hydroxyethyl)-N-methoxymethylaniline

N-Chloroacetyl-L-tryptophan
→ L-Tryptophan
In: Rat C29

p-Chloroamphetamine
→ N-Acetyl-*p*-chloroamphetamine
Cunninghamella F345
→ 3-Chloro-4-hydroxyamphetamine
In: Rat A2186
→ *p*-Chloronorephedrine
In: Rat A3002
→ *p*-Chloro-N-(1-methylacetonylidene) amphetamine
In: Mycobacterium B508
→ 3,4-Dimethoxyamphetamine
In: Rat A3179

o-Chloroaniline
Formed from: *o*-Chloroacetanilide
5-Chloro-N-(*o*-chlorophenyl)-2-hydroxybenzamide
N-(*o*-Chlorophenyl)benzamide
N-(*o*-Chlorophenyl)salicylamide
o-Chloronitrobenzene
o-Chloropropionanilide
o-Chlorophenylhydrazono)-3-methyloxazol-5-one

→ 4-Amino-3-chlorophenol
In: Rat H273, J447
→ *o*-Chloroacetanilide
In: Guppy G853
Paracoccus A2824
→ *o*-Chloroaniline-β-D-glucuronide
In: Rat F530, J447
→ 3-Chlorocatechol
In: Rhodococcus F258, F548
→ 2,2′-Dichloroazobenzene
In: Horseradish A661
→ *o*-Chloronitrobenzene
In: Serratia K143
→ *o*-Chlorophenylsulphamic acid
In: Rat J447

m-Chloroaniline
Formed from: *m*-Chloronitrobenzene
m-Chloropropionanilide
→ 2-Amino-4-chlorophenol
In: Fusarium A3855
See also: K2
→ 4-Amino-2-chlorophenol
In: Rat H273
→ *m*-Chloroacetanilide
In: Guppy G853
Paracoccus A2824
Rat F530
→ *m*-Chloroaniline-β-D-glucuronide
In: Rat F530
→ 4-Chlorocatechol
In: Rhodococcus F258, F548
→ *m*-Chloronitrobenzene
In: Serratia K143
→ 3,3′-Dichloroazobenzene
In: Horseradish A661

p-Chloroaniline
Formed from: Buturon
N-Benzyl-*p*-chloroaniline
p-Chloroacetanilide
p-Chloro-N-(*p*-chlorobenzyl)aniline
p-Chloro-N-methylaniline
p-Chloronitrobenzene
1-(*p*-Chlorophenylazo)piperidine
1-(*p*-Chlorophenyl)-3,3-dimethyltriazene
p-Chlorophenylhydroxylamine
p-Chlorophenyl azide
p-Chloropropionanilide
Diflubenzuron
Monolinuron
Sulofenur
→ 2-Amino-5-chlorophenol
In: Fusarium A3855
Man C879
Monkey F729
Mouse F729
Rabbit C879
Rat F729, G742
See also: A2019, A2713, A2714, G519, G871, F731, K2
→ 4-Amino-2-chlorophenol
Probably in: Rat A2019, A2713, A2714
→ 5-Amino-2-chlorophenol
Probably in: Rat A2019, A2713, A2714
→ 2-Amino-7-chloro-3*H*-phenoxazin-3-one
In: Bacillus A3080
Bacterium A1077
→ *p*-Aminophenol
Probably in: Rat A2713
→ 2-Bromo-4-chloroaniline
In: Caldariomyces C845
→ *p*-Chloroacetanilide
In: Azotobacter B218
Bacillus A3080
Fusarium A700, B237
Guinea pig F109
Guppy G853
Man G519
Monkey F729
Paracoccus A2824
Rabbit E399, F109
Rat F109, F530, G729, G742
Rhodosporidium B799
Streptomyces A2938
→ *p*-Chloroaniline-N-sulphate
In: Guinea pig E456
→ 4-Chlorocatechol
In: Rhodococcus F548
→ *p*-Chloroformanilide
In: Guinea pig F109
Rabbit F109
Rat F109
Streptomyces A2938

→ *p*-Chloro-N-methylaniline
In: Pig K923
Rabbit C333
→ *p*-Chloronitrobenzene
In: Serratia K143
→ *p*-Chloronitrosobenzene
In: Caldariomyces A3579, A3838, C845
Man C879
→ *p*-Chlorooxanilic acid
In: Rat G742
→ *p*-Chlorophenylhydroxylamine
In: Azotobacter B218
Fusarium A700
Pea C797
Rabbit F839
Rat A3703
Red-winged balckbird A3703
Trout G32
→ N-(*p*-Chlorophenyl)succinimide
In: Phanerochaete E317
→ *p*-Chloropropionanilide
In: Azotobacter B218
Bacillus A3080
→ 2,4-Dichloroaniline
In: Caldariomyces C845
Man G847
Rat G742
See also: G519, G871
→ 4,4′-Dichloroazobenzene
In: Fusarium A700
Geotrichum A661
Horseradish A661
→ 4,4′-Dichlorodiazoaminobenzene
In: Paracoccus A2538

o-Chloroaniline-β-D-glucuronide
Formed from: *o*-Chloroaniline

m-Chloroaniline-β-D-glucuronide
Formed from: *m*-Chloroaniline

p-Chloroaniline-N-sulphate
Formed from: *p*-Chloroaniline

3-(p-Chloroanilino)-10-(p-chlorophenyl)-2,10-dihydro-2-isopropyliminophenazine
See: Clofazimine

o-Chloroanisole
→ *o*-Chlorophenol
In: Microorganism K269
Sporomusa J199

m-Chloroanisole
→ *m*-Chlorophenol
In: Sporomusa J199

p-Chloroanisole
→ *p*-Chlorophenol
In: Methylosinus D513
Sporomusa J199

5-Chloroanthranilic acid
Probably formed from: N-Formyl-5-chloroanthranilic acid

5-(4-Chloro-3-azidophenyl)-6-ethyl-2,4-pyrimidinediamine
See: *m*-Azidopyrimethamine

o-Chlorobenzaldehyde
Formed from: *o*-Chlorobenzylamine
o-Chlorostyrene
Probably formed from: *o*-Chlorobenzoic acid
Clobenorex
→ N-Acetyl-S-(*o*-chlorobenzyl)-L-cysteine
In: Rat C553, E296
→ *o*-Chlorobenzyl alcohol
In: Chlorella K176
Dunaliella G459
Euglena G422
Horse C508
Microorganism K269
Saccharomyces H425
See also: E743
→ (1*R*,2*S*)-1-(*o*-Chlorophenyl)propane-1,2-diol
In: Saccharomyces E487

m-Chlorobenzaldehyde
Formed from: *m*-Chlorobenzoic acid
m-Chlorobenzyl alcohol
m-Chlorobenzylamine
m-Chlorostyrene
→ N-Acetyl-S-(*m*-chlorobenzyl)-L-cysteine
In: Rat E296
→ *m*-Chlorobenzoic acid
In: Euglena G422
Rat H761
See also H953
→ *m*-Chlorobenzyl alcohol
In: Chlorella K176
Dunaliella G459, K176

Euglena G422
Isochrysis K176
Nannochloris K176
Porphyridium K176
Saccharomyces H425
See also: E743
→ *m*-Chloro-2-hydroxypropiophenone
In: Pseudomonas K524
→ *m*-Chloromandelonitrile
With (*S*)-oxynitrilase F680

p-Chlorobenzaldehyde
Formed from: Chlormezanone
p-Chlorobenzoic acid
p-Chlorobenzyl alcohol
p-Chloro-N-(*p*-chlorobenzyl)aniline
p-Chlorostyrene
Probably formed from: N-Benzoyl-N′-(*p*-chlorobenzyl)hydrazine
α-(2-Chloroethylthio)-*p*-chloro-N-methylbenzylamine-S,S-dioxide
p-Chloroacetophenone
→ *p*-Chlorobenzoic acid
In: Beef C508
Euglena G422
Rat H761
Yeast C508
See also: E130, J687, K542
→ *p*-Chlorobenzyl alcohol
In: Chlorella K176
Dunaliella G459, K176
Euglena G422
Horse C508
Nannochloris K176
Porphyridium K176
Rabbit B386, D125
Saccharomyces C508, H425
See also: E743
→ *p*-Chloro-2-hydroxypropiophenone
In: Pseudomonas K524
→ *p*-Chloromandelonitrile
With (*S*)-oxynitrilase F680
→ (1*R*,2*S*)-1-(*p*-Chlorophenyl)propane-1,2-diol
In: Saccharomyces E487
→ 4,4′-Dichlorobenzoin
In: Pseudomonas K424

o-Chlorobenzaldoxime
→ *o*-Chlorobenzyl-β-D-sulphatoglucoside
In: Brassica F657

o-Chlorobenzamide
Formed from: *o*-Chlorobenzonitrile
→ *o*-Chlorobenzoic acid
In: Aspergillus A640

m-Chlorobenzamide
→ *m*-Chlorobenzoic acid
In: Aspergillus A640

p-Chlorobenzamide
→ *p*-Chlorobenzoic acid
In: Aspergillus A640

p-Chlorobenzamidine
→ *p*-Chlorobenzamidoxime
In: Rabbit E830, F768

p-Chlorobenzamidoxime
Formed from: *p*-Chlorobenzamidine

2-Chlorobenzanilide
→ 2-Chloro-4′-hydroxybenzanilide
In: Rhizopus A485

Chlorobenzene
Formed from: β-Lindane
γ-Lindane
→ N-Acetyl-S-(*p*-chlorophenyl)-L-cysteine
In: Rat G932
→ Chlorobenzene-2,3-oxide
Probably in: Rat A2267
→ Chlorobenzene-3,4-oxide
Probably in: Rat A2267
→ *cis*-3-Chloro-1,2-dihydro-1,2-dihydroxybenzene
In: Pseudomonas J656, K699
→ *o*-Chlorophenol
In: Aspergillus A1218
Cunninghamella A1218
Gliocladium A1218
Man E947
Mouse E947
Rat A2160, A2320, C899, D900, G932
Rhizopus A1218
Streptomyces A1218
See also: D59, F278, F857, G627, J344
→ *m*-Chlorophenol
In: Nitrosomonas H631

Rat A2160, A2320, D900, G932
→ *p*-Chlorophenol
In: Aspergilus A1218
Cunninghamella A1218
Man E947
Mouse E947
Nitrosomonas H631
Rat A2160, A2320, C899, D900, G932
→ *o*-Chlorothioanisole
In: Rat G932
→ *m*-Chlorothioanisole
In: Rat G932
→ *p*-Chlorothioanisole
In: Rat G932

Chlorobenzene-2,3-oxide
Probably formed from: Chlorobenzene
→ *trans*-3-Chloro-1,2-dihydro-1,2-dihydroxybenzene
In: Rat A2267
→ S-(3-Chloro-1,2-dihydro-2-hydroxyphenyl)glutathione
Probably in: Sheep A2282

Chlorobenzene-3,4-oxide
Probably formed from: Chlorobenzene
→ *trans*-4-Chloro-1,2-dihydro-1,2-dihydroxybenzene
In: Rat A2267
→ S-(4-Chloro-1,2-dihydro-2-hydroxyphenyl)glutathione
Probably in: Sheep A2282

Chlorobenzenesulphonamide
Formed from: Chlorpropamide

***p*-Chlorobenzenesulphonic acid**
Probably formed from: *p*-Chlorophenyl methyl sulphone
→ 3,4-Dihydroxybenzenesulphonic acid
Probably in: Alcaligenes K293

1-(*p*-Chlorobenzenesulphonyl)-3-(2-hydroxypropyl)urea
Formed from: Chlorpropamide

1-(*p*-Chlorobenzenesulphonyl)-3-(3-hydroxypropyl)urea
Formed from: Chlorpropamide

***p*-Chlorobenzenesulphonylurea**
Formed from: Chlorpropamide

***o*-Chlorobenzhydrol**
Probably formed from: *o*-Chlorobenzophenone

***p*-Chlorobenzhydrol**
Probably formed from: *p*-Chlorobenzophenone

N-(*p*-Chlorobenzhydryl)ethylenediamine
Formed from: Chlorcyclizine

***o*-Chlorobenzoic acid**
Formed from: *o*-Chlorobenzamide
o-Chlorobenzonitrile
3-(*o*-Chlorobenzoyloxy)-N-(1,1-dimethyl-2-(((morpholino)carbonyl)amino)ethyl)-2-hydroxypropylamine
o-Chlorobenzyl alcohol
2-Chlorobiphenyl
6-(*o*-Chlorophenyl)-2-hydroxy-6-oxohexa-2,4-dienoic acid
erythro-2-(*o*-Chlorophenyl)-2-(4-(*p*-methoxybenzyl)-1-piperazinyl)-1-phenylethanol
3-(5-(*o*-Chlorophenyl)-4-pyridyl)-1,2,4-oxadiazole
o-Chlorotoluene
2,3-Dichlorobenzoic acid
2,5-Dichlorobenzoic acid
2,6-Dichlorobenzoic acid
2,2′-Dichlorobiphenyl
2,4′-Dichlorobiphenyl
Ticlopidine
Probably formed from: 2,6-Dichlorobenzoic acid
→ Catechol
In: Pseudomonas F247, F401, G434, H212, J173
→ *o*-Chlorobenzaldehyde
Probably in: Mucor E743
Nocardia E743
Rhizopus E743
→ *o*-Chlorobenzoyl CoA
In: Rhodopseudomonas E605
→ *o*-Chlorobenzoyl-β-D-glucuronic acid
In: Rabbit H519
Contrast (in rat) J549
→ *o*-Chlorobenzyl alcohol
In: Coriolus G821
→ 3-Chlorocatechol

In: Pseudomonas G448
→ 6-Chloro-1,2-dihydro-1,2-dihydroxybenzoic caid
Probably in: Pseudomonas B465
→ *o*-Chlorohippuric acid
In: Rabbit H519
Rat A2610, A2644, A3927, C52, E778, J549, K507
→ 2-Chloro-4-hydroxybenzoic acid
In: Aspergillus D593
→ Pyrocatechuic acid
In: Pseudomonas F401

m-Chlorobenzoic acid
Formed from: Bupropion
m-Chlorobenzaldehyde
m-Chlorobenzamide
m-Chlorobenzonitrile
3-Chlorobiphenyl
o-Chlorophenol
6-(*m*-Chlorophenyl)-2-hydroxy-6-oxohexa-2,4-dienoic acid
4-Chlorophthalic acid
3,5-Dichlorobenzoic acid
3,3′-Dichlorobiphenyl
→ Benzoic acid
In: Desulfomicrobium J491
Desulfomonile G152, H508, H642
Microorganism J341
→ *m*-Chlorobenzaldehyde
In: Clostridium G416
Nocardia G207
See also: E743, J491, K615
→ *m*-Chlorobenzoyl CoA
In: Beef H740
→ *m*-Chlorobenzoyl-β-D-glucuronic acid
In: Rabbit H519
→ 3-Chlorocatechol
In: Alcaligenes G922
Microorganism A2237
Pseudomonas C855, F392
See also: B11, G434, H265
→ 4-Chlorocatechol
In: Alcaligenes G922
Pseudomonas A1192, C855
→ 5-Chloro-1,2-dihydro-1,2-dihydroxybenzoic caid
Probably in: Pseudomonas B465, F392
→ *m*-Chlorohippuric acid
In: Man E229
Rabbit H519
→ 3-Chloro-4-hydroxybenzoic acid
In: Aspergillus D593
See also: A640
→ 5-Chlorosalicylic acid
In: Bacillus A1192
→ *m*-Hydroxybenzic acid
In: Pseudomonas A488

p-Chlorobenzoic acid
Formed from: *p*-Chlorobenzaldehyde
p-Chlorobenzamide
p-Chlorobenzonitrile
p-Chlorobenzophenone
p-Chlorobenzotrichloride
p-Chlorobenzoyl-L-alanine
4-Chlorobiphenyl
p-Chlorocinnamic acid
p-Chloro-N,N-diethylbenzamide
p-Chloro-N-ethylbenzamide
1-(*p*-Chlorophenyl)-4-dimethylamino-2,3-dimethylbutan-2-ol
6-(*p*-Chlorophenyl)-2-hydroxy-6-oxohexa-2,4-dienoic acid
p-Chlorotoluene
3,4-Dichlorobenzoic acid
4,4′-Dichlorobenzophenone
4,4′-Dichlorobiphenyl
Lofepramine
Robenidine
1,1,1-Trichloro-2,2-bis(*p*-chlorophenyl)ethane
Zomepirac
Probably formed from: 2-(N-(*p*-Chlorobenzoyl)-N-methylcarbamoyl) ethylsulphonic acid
p-Chlorohippuric acid
→ N^2-Benzoyl-N^6-(*p*-chlorobenzoyl)lysine
In: Chicken A2253
→ N^6-Benzoyl-N^2-(*p*-chlorobenzoyl)lysine
In: Chicken A2253
→ N^2-Benzoyl-N^5-(*p*-chlorobenzoyl)ornithine
In: Chicken A2253
→ N^5-Benzoyl-N^2-(*p*-chlorobenzoyl)ornithine

In: Chicken A2253
→ N^2,N^5-Bis(*p*-chlorobenzoyl)ornithine
In: Chicken A2253
→ *p*-Chlorobenzaldehyde
In: Clostridium G416
Nocardia G207
See also: E743
→ *p*-Chlorobenzoyl CoA
In: Beef H740
Pseudomonas G364, G729, H216, H645
→ *p*-Chlorobenzoyl-β-D-glucuronic acid
In: Mouse A3044
Rabbit H519
Rat A3044
Contrast (rat) J549
→ 4-Chlorocatechol
Probably in: Pseudomonas B11, G434, H265
→ *p*-Chlorohippuric acid
In: Dog A2029
Man A2029, E130
Monkey B107
Mouse A3044
Rabbit H519
Rat A2029, A3044, F693, H475, J549, K507
See also: D496
→ *p*-Hydroxybenzic acid
In: Acinetobacter K141
Alcaligenes E290, E335
Arthrobacter A2790, D544, D621, E752
Bacterium G454
Nocardia B226
Pseudomonas B695, D473, E359, G156, G205, K7

o-Chlorobenzonitrile
→ *o*-Chlorobenzamide
In: Rhodococcus G172, H952
→ *o*-Chlorobenzoic acid
In: Nocardia D725
Rhodococcus G172, H952

m-Chlorobenzonitrile
→ *m*-Chlorobenzoic acid
In: Fusarium A3364
Nocardia A3420, D725
Rhodococcus G172

p-Chlorobenzonitrile
→ *p*-Chlorobenzoic acid
In: Fusarium A3364
Nocardia A3420, D725
Rhodococcus G172, H952

o-Chlorobenzophenone
→ Benzophenone
Probably in: Man A3384
→ *o*-Chlorobenzhydrol
Probably in: Man A3384

p-Chlorobenzophenone
Formed from: Hydroxyzine
→ *p*-Chlorobenzhydrol
Probably in: Rat A2197
→ *p*-Chlorobenzoic acid
In: Pseudomonas E41
→ 4′-Chloro-4-hydroxybenzophenone
Probably in: Rat A2197

2-(5-Chloro-3-benzothienyl)ethylamine
→ N-Acetyl-2-(5-chloro-3-benzothienyl) ethylamine
In: Man K490

p-Chlorobenzotrichloride
→ *p*-Chlorobenzoic acid
In: Rat D496
→ α,α′,4,4′-Tetrachlorostilbene
In: Rat D496

5-Chlorobenzoxazolin-2-one
See: Chlorzoxazone

p-Chlorobenzoyl-L-alanine
→ *p*-Chlorobenzoic acid
In: Corynebacterium E145

o-Chlorobenzoyl CoA
Formed from: *o*-Chlorobenzoic acid

m-Chlorobenzoyl CoA
Formed from: *m*-Chlorobenzoic acid

p-Chlorobenzoyl CoA
Formed from: *p*-Chlorobenzoic acid
2,4-Dichlorobenzoyl CoA
→ *p*-Hydroxybenzoic acid
In: Pseudomonas G364
→ *p*-Hydroxybenzoyl CoA
In: Acinetobacter G896
Arthrobacter H279
Corynebacterium J178
Pseudomonas G920, H216, H645, K191

4-(*p*-Chlorobenzoyl)-6,7-dimethoxyisoquinoline
Formed from: 4-(*p*-Chlorobenzyl)-6,7-dimethoxyisoquinoline

5-(*p*-Chlorobenzoyl)-1,4-dimethyl-1*H*-pyrrole-2-acetic acid
See: Zomepirac

N-(*p*-Chlorobenzoyl)ethylenediamine
Probably formed from: N-(*p*-Chlorobenzoyl)-N′-(2-hydroxyethyl)ethylenediamine

***o*-Chlorobenzoyl-β-D-glucuronic acid**
Formed from: *o*-Chlorobenzoic acid

***m*-Chlorobenzoyl-β-D-glucuronic acid**
Formed from: *m*-Chlorobenzoic acid

***p*-Chlorobenzoyl-β-D-glucuronic acid**
Formed from: *p*-Chlorobenzoic acid

N-(*p*-Chlorobenzoyl)-N′-(2-hydroxyethyl)ethylenediamine
Formed from: Moclobemide
→ N-(*p*-Chlorobenzoyl)ethylenediamine
Probably in: Man G707
Rat F831, G707

1-(*p*-Chlorobenzoyl)-5-hydroxy-2-methylindole-3-acetic acid
Formed from: 1-(*p*-Chlorobenzoyl)-5-methoxy-2-methylindole-3-acetic acid

5-(*p*-Chlorobenzoyl)-4-hydroxymethy-1-methyl-1*H*-pyrrole-2-acetic acid
Formed from: Zomepirac

2-(*p*-Chlorobenzoyl)-5-hydroxythiophene
Formed from: 2-(*p*-Chlorobenzoyl)thiophene

***o*-Chlorobenzoylisonicotinic acid amidoxime**
Formed from: 3-(5-(*o*-Chlorophenyl)-4-pyridyl)-1,2,4-oxadiazole

1-(*p*-Chlorobenzoyl)-5-methoxy-2-methylindole-3-acetic acid (indomethacin)
Formed from: Indomethacin farnesil
Proglumetacin
Tropesin
→ 1-(*p*-Chlorobenzoyl)-5-hydroxy-2-methylindole-3-acetic acid
In: Man J615
See also: C223
→ 5-Methoxy-2-methylindole-3-acetic acid
In: Pig J248
See also: C223

2-(N-(*p*-Chlorobenzoyl)-N-methylcarbamoyl)ethylsulphonic acid
Formed from: Chlormezanone
→ *p*-Chlorobenzoic acid
Probably in: Mouse A3044
Rat A3044
→ *p*-Chloro-N-methylbenzamide
Probably in: Mouse A3044
Rat A3044

***p*-Chlorobenzoylmethyl demethylimipramine**
See: Lofepramine

1-(*p*-Chlorobenzoyl)-2-methyl-5-methoxy-3-indolylacetylhydroxamic acid
See: Oxametacin

3-(*p*-Chlorobenzoyl)-2-methylphenylacetic acid
Formed from: 4-(3-(*p*-Chlorobenzoyl)-2-methylphenyl)butanoic acid

4-(3-(*p*-Chlorobenzoyl)-2-methylphenyl)butanoic caid
→ 3-(*p*-Chlorobenzoyl)-2-methylphenylacetic acid
In: Dog A3060
Mouse A3060
Rat A3060
→ 4-(3-(*p*-Chloro-α-hydroxybenzyl)-2-methylphenyl)butanoic caid
In: Man A3060

2-(4-(*p*-Chlorobenzoyl)phenoxy)-2-methylpropionic acid (fenofibric acid)
Formed from: Fenofibrate
→ 2-(4-(*p*-Chlorobenzoyl)phenoxy)-2-methylpropionyl-β-D-glucuronic acid
In: Dog E920
Guinea pig E920
Man F535
Rat E920
→ 2-(4-(*p*-Chloro-α-hydroxybenzyl)phenoxy)-2-methylpropionic acid
In: Dog E920
Guinea pig E920
Rat E920

2-(4-(*p*-Chlorobenzoyl)phenoxy)-2-methylpropionic aciid 1-methylethyl ester
See: Fenofibrate

2-(4-(*p*-Chlorobenzoyl)phenoxy)-2-methylpropionyl-β-D-glucuronic acid
Formed from: 2-(4-(*p*-Chlorobenzoyl)phenoxy)-2-methylpropionic acid

4-(*p*-Chlorobenzoyl)piperidine
Formed from: 4-(4-(*p*-Chlorobenzoyl)piperidino)-4′-fluorobutyrophenone
→ 4-(*p*-Chlorobenzoyl)piperidone
Probably in: Rat A2126

4-(4-(*p*-Chlorobenzoyl)piperidino)-4′-fluorobutyrophenone
→ 4-(*p*-Chlorobenzoyl)piperidine
In: Rat A2126

4-(*p*-Chlorobenzoyl)piperidone
Probably formed from: 4-(*p*-Chlorobenzoyl)piperidine

2-(*p*-Chlorobenzoyl)thiophene
→ 2-(*p*-Chlorobenzoyl)-5-hydroxythiophene
In: Rat F130

***o*-Chlorobenzyl alcohol**
Formed from: *o*-Chlorobenzaldehyde
o-Chlorobenzoic acid
o-Chlorobenzylidene malononitrile
→ *o*-Chlorobenzoic acid
In: Rat E778
See also: H953
→ S-(*o*-Chlorobenzyl)-L-cysteine
In: Rat E778
See also: C553, E296
→ *o*-Chlorobenzyl-β-D-glucuronide
In: Rat E778
→ *o*-Chlorobenzyl-β-D-xylopyranoside
In: Coriolus G821

***m*-Chlorobenzyl alcohol**
Formed from: *m*-Chlorobenzaldehyde
→ N-Acetyl-S-(*m*-chlorobenzyl)-L-cysteine
Probably in: Rat E296
→ Benzyl alcohol
In: Desulfomicrobium J491
→ *m*-Chlorobenzaldehyde
In: Rhodopseudomonas D197
See also: H953

***p*-Chlorobenzyl alcohol**
Formed from: *p*-Chlorobenzaldehyde
→ N-Acetyl-S-(*p*-chlorobenzyl)-L-cysteine
Probably in: Rat E296
→ *p*-Chlorobenzaldehyde
In: Rhodopseudomonas D197

***o*-Chlorobenzylamine**
→ *o*-Chlorobenzaldehyde
In: Sheep A535

***m*-Chlorobenzylamine**
→ *m*-Chlorobenzaldehyde
In: Sheep A535

***p*-Chlorobenzylamine**
Formed from: Bis(*p*-chlorobenzyl)amine

N-(*p*-Chlorobenzyl)-N-(*p*-chlorophenyl)hydroxylamine
Formed from: *p*-Chloro-N-(*p*-chlorobenzyl)aniline
→ N-(*p*-Chlorobenzyl)-N-(*p*-chlorophenyl)nitrone
Probably in: Rat J523

N-(*p*-Chlorobenzyl)-N-(*p*-chlorophenyl)nitrone
Probably formed from: N-(*p*-Chlorobenzyl)-N-(*p*-chlorophenyl)hydroxylamine

1-(*p*-Chlorobenzyl)-1-cyclopentyl-3-(3,4-dihydroxyphenyl)urea
Probably formed from: Pencycuron

1-(*p*-Chlorobenzyl)-1-cyclopentyl-3-(4-hydroxy-3-methylthiophenyl)urea
Probably formed from: 1-(*p*-Chlorobenzyl)-1-cyclopentyl-3-(*p*-hydroxyphenyl)urea
→ 1-(*p*-Chlorobenzyl)-1-(3-hydroxycyclopentyl)-3-(4-hydroxy-3-methylthiophenyl)urea
Probably in: Rabbit C413

1-(*p*-Chlorobenzyl)-1-cyclopentyl-3-(*p*-hydroxyphenyl)urea
Formed from: Pencycuron
→ 1-(*p*-Chlorobenzyl)-1-cyclopentyl-3-(*p*-hydroxyphenyl)urea-β-D-glucuronide
In: Rabbit C413
→ 1-(*p*-Chlorobenzyl)-1-cyclopentyl-3-(4-hydroxy-3-methylthiophenyl)urea
Probably in: Rabbit C413, C428
→ 1-(*p*-Chlorobenzyl)-1-(3-hydroxycyclopentyl)-3-(*p*-hydroxyphenyl)urea
Probably in: Rabbit C413, C428
→ 1-(*p*-Chlorobenzyl)-3-(*p*-hydroxyphenyl)urea
Probably in: Rabbit C413, C428

1-(*p*-Chlorobenzyl)-1-cyclopentyl-3-(*p*-hydroxyphenyl)urea-β-D-glucuronide
Formed from: 1-(*p*-Chlorobenzyl)-1-cyclopentyl-3-(*p*-hydroxyphenyl)urea

1-(*p*-Chlorobenzyl)-1-cyclopentyl-3-phenylurea
See: Pencycuron

S-(*o*-Chlorobenzyl)-L-cysteine
Formed from: *o*-Chlorobenzyl alcohol
→ N-Acetyl-S-(*o*-chlorobenzyl)-L-cysteine
Probably in: Rat C553, E296, E778

4-(*p*-Chlorobenzyl)-6,7-dihydroxyisoquinoline
Probably formed from: 4-(*p*-Chlorobenzyl)-6-hydroxy-7-methoxyisoquinoline
→ 4-(*p*-Chlorobenzyl)-6-hydroxy-7-methoxyisoquinoline
In: Rat E817
→ 4-(4-Chloro-3-hydroxybenzyl)-6,7-dihydroxyisoquinoline
Probably in: Rat E817

4-(*p*-Chlorobenzyl)-6,7-dimethoxyisoquinoline
→ 4-(*p*-Chlorobenzoyl)-6,7-dimethoxyisoquinoline
In: Rat E817
→ 4-(*p*-Chlorobenzyl)-6,7-dimethoxyisoquinoline-N-oxide
In: Rat E817
→ 4-(*p*-Chlorobenzyl)-6-hydroxy-7-methoxyisoquinoline
In: Man C825
Mouse D426
Rat D426, E817
→ 4-(*p*-Chlorobenzyl)-7-hydroxy-6-methoxyisoquinoline
In: Man C825
Mouse D426
Rat D426, E817
→ 4-(4-Chloro-3-hydroxybenzyl)-6,7-dimethoxyisoquinoline
In: Rat E817
→ Isopapaverine
In: Silene E680

4-(*p*-Chlorobenzyl)-6,7-dimethoxyisoquinoline-N-oxide
Formed from: 4-(*p*-Chlorobenzyl)-6,7-dimethoxyisoquinoline

***o*-Chlorobenzyl-β-D-glucuronide**
Formed from: *o*-Chlorobenzyl alcohol

4-(*p*-Chlorobenzyl)-2-(hexahydro-N-methylazepin-4-yl)-6-hydroxy-1(2*H*)-phthalazinone
Formed from: Azelastine

4-(*p*-Chlorobenzyl)-2-(hexahydro-N-methylazepin-4-yl)-7-hydroxy-1(2*H*)-phthalazinone
Formed from: Azelastine

4-(*p*-Chlorobenzyl)-2-(hexahydro-N-methyl-7-oxoazepin-4-yl)-1(2*H*)-phthalazinone
Formed from: Azelastine

N-(*p*-Chlorobenzyl)hydrazine
Formed from: N-Benzoyl-N′-(*p*-chlorobenzyl)hydrazine

1-(*p*-Chlorobenzyl)-1-(3-hydroxycyclopentyl)-3-phenylurea
Formed from: Pencycuron

1-(*p*-Chlorobenzyl)-1-(3-hydroxycyclopentyl)-3-(4-hydroxy-3-methylthiophenyl)urea
Probably formed from: 1-(*p*-Chlorobenzyl)-1-cyclopentyl-3-(4-hydroxy-3-methylthiophenyl) urea

1-(*p*-Chlorobenzyl)-1-(3-hydroxycyclopentyl)-3-(*p*-hydroxyphenyl)urea
Probably formed from: 1-(*p*-Chlorobenzyl)-1-cyclopentyl-3-(*p*-hydroxyphenyl)urea
→ 1-(*p*-Chlorobenzyl)-1-(3-hydroxycyclopentyl)-3-(*p*-hydroxyphenyl)urea-β-D-glucuronide
Probably in: Rabbit C413

1-(*p*-Chlorobenzyl)-1-(3-hydroxycyclopentyl)-3-(*p*-hydroxyphenyl)urea-β-D-glucuronide
Probably formed from: 1-(*p*-Chlorobenzyl)-1-(3-hydroxycyclopentyl)-3-(*p*-hydroxyphenyl) urea

4-(*p*-Chlorobenzyl)-6-hydroxy-7-methoxyisoquinoline
Formed from: 4-(*p*-Chlorobenzyl)-6,7-dimethoxyisoquinoline
4-(*p*-Chlorobenzyl)-6,7-dihydroxyisoquinoline
→ 4-(*p*-Chlorobenzyl)-6,7-dihydroxyisoquinoline
Probably in: Mouse D426
Rat D426, E817

4-(*p*-Chlorobenzyl)-7-hydroxy-6-methoxyisoquinoline
Formed from: 4-(*p*-Chlorobenzyl)-6,7-dimethoxyisoquinoline

1-(*p*-Chlorobenzyl)-3-(*p*-hydroxyphenyl)urea
Probably formed from: 1-(*p*-Chlorobenzyl)-1-cyclopentyl-3-(*p*-hydroxyphenyl)urea

***o*-Chlorobenzylidenemalononitrile** (CS)
→ *o*-Chlorobenzyl alcohol
In: Rat E778
→ (*o*-Chlorobenzyl)malononitrile
In: Man B260
→ *o*-Chlorophenylacetic acid
In: Rat E778
→ 3-(*o*-Chlorophenyl)-2-cyanopropionic acid
In: Rat E778

***p*-Chlorobenzylidenemalononitrile**
→ (*p*-Chlorobenzyl)malononitrile
In: Man B260

(*o*-Chlorobenzyl)malononitrile
Formed from: *o*-Chlorobenzylidenemalononitrile
→ *o*-Chlorophenylacetic acid
In: Rat E778
→ 3-(*o*-Chlorophenyl)-2-cyanopropionic acid
In: Rat E778
→ 3-(*o*-Chlorophenyl)propionamide
In: Rat E778

(*p*-Chlorobenzyl)malononitrile
Formed from: *p*-Chlorobenzylidenemalononitrile

N-(*o*-Chlorobenzyl)-α-methylphenethylamine
See: Clobenzorex

4-(*p*-Chlorobenzyl)-1-methylpyridinium
Probably formed from: 4-(*p*-Chlorobenzyl)-tetrahydro-1-methylpyridine

***o*-(*m*-Chlorobenzyloxy)acetophenone**
Formed from: Croconazole
→ *o*-(*m*-Chlorobenzyloxy)benzoic acid
Probably in: Rabbit F123
Rat F507
→ 2-(*o*-(*m*-Chlorobenzyloxy)phenyl)-2-oxoethanol
Probably in: Rabbit F123
Rat F507
→ *o*-(3-Chloro-5-hydroxybenzyloxy)acetophenone
Probably in: Rabbit F123
→ *o*-Hydroxyacetophenone
Probably in: Rat F507

***o*-(*m*-Chlorobenzyloxy)benzoic acid**
Probably formed from: *o*-(*m*-Chlorobenzyloxy)acetophenone
→ 2-(*m*-Chlorobenzyloxy)-5-hydroxybenzoic acid
Probably in: Rabbit F123

1-(2-(4-Chlorobenzyloxy)-2-(2,4-dichlorophenyl)ethyl)tetrahydro-2,4-dioxoimidazole
Formed from: Econazole
→ 1-(2-(4-Chlorobenzyloxy)-2-(2,4-dichlorophenyl)-2-hydroxyethyl)tetrahydro-2,4-dioxoimidazole
Probably in: Monkey C780
Rat C780

1-(2-(4-Chlorobenzyloxy)-2-(2,4-dichlorophenyl)ethyl)tetrahydro-2,5-dioxoimidazole
Probably formed from: 1-(2-(4-Chlorobenzyloxy)-2-(2,4-dichlorophenyl)ethyl)tetrahydro-2-hydroxy-5-oxoimidazole

1-(2-(4-Chlorobenzyloxy)-2-(2,4-dichlorophenyl)ethyl)tetrahydro-2-hydroxy-5-oxoimidazole
Formed from: Econazole
→ 1-(2-(4-Chlorobenzyloxy)-2-(2,4-dichlorophenyl)ethyl)tetrahydro-2,5-dioxoimidazole
Probably in: Man B715
Monkey C780
Rat C780
→ 1-(2-(4-Chlorobenzyloxy)-2-(2,4-dichlorophenyl)-2-hydroxyethyl)tetrahydro-2,5-dioxoimidazole
Probably in: Monkey C780
Rat C780

1-(2-(4-Chlorobenzyloxy)-2-(2,4-dichlorophenyl)-2-hydroxyethyl)tetrahydro-2,4-dioxoimidazole
Probably formed from: 1-(2-(4-Chlorobenzyloxy)-2-(2,4-dichlorophenyl)ethyl)tetrahydro-2,4-dioxoimidazole

1-(2-(4-Chlorobenzyloxy)-2-(2,4-dichlorophenyl)-2-hydroxyethyl)tetrahydro-2,5-dioxoimidazole
Probably formed from: 1-(2-(4-Chlorobenzyloxy)-2-(2,4-dichlorophenyl)ethyl)tetrahydro-2-hydroxy-5-oxoimidazole

3-(*o*-Chlorobenzyloxy)-N-(1,1-dimethyl-2-(((morpholino)carbonyl)amino)ethyl)-2-hydroxypropylamine
→ *o*-Chlorobenzoic acid
In: Rabbit H563

2-(*m*-Chlorobenzyloxy)-5-hydroxybenzoic acid
Probably formed from: *o*-(*m*-Chlorobenzyloxy)benzoic acid

1-(1-(2-(*m*-Chlorobenzyloxy)-5-hydroxyphenyl)vinyl-1*H*-imidazole
Formed from: Croconazole

***o*-(*m*-Chlorobenzyloxy)phenylglycol**
Probably formed from: 2-(*o*-(*m*-Chlorobenzyloxy)phenyl)-2-oxoethanol
→ *o*-(3-Chloro-5-hydroxybenzyloxy)phenylglycol
Probably in: Rabbit F123

2-(*o*-(*m*-Chlorobenzyloxy)phenyl)-2-oxoethanol
Probably formed from: *o*-(*m*-Chlorobenzyloxy)acetophenone
→ *o*-(*m*-Chlorobenzyloxy)phenylglycol
Probably in: Rabbit F123
→ 2-(*o*-(3-Chloro-5-hydroxybenzyloxy)phenyl)-2-oxoethanol
Probably in: Rabbit F123

1-(1-(2-(*m*-Chlorobenzyloxy)phenyl)vinyl)-1*H*-imidazole
See: Croconazole

7-(4-(*p*-Chlorobenzyl)piperazin-1-yl)propoxy-3,4-dimethylcoumarin
See: Picumast

7-(4-(*p*-Chlorobenzyl)piperazin-1-yl)propoxy-3-hydroxymethyl-4-methylcoumarin
→ 3-Carboxy-7-(4-(*p*-chlorobenzyl)piperazin-1-yl)propoxy-4-methylcoumarin
Probably in: Man F413

***o*-Chlorobenzyl sulphate**
→ N-Acetyl-S-(*o*-chlorobenzyl)-L-cysteine
In: Rat C553

***p*-Chlorobenzyl sulphate**
→ N-Acetyl-S-(*p*-chlorobenzyl)-L-cysteine
In: Rat C553

***o*-Chlorobenzyl-β-D-sulphatoglucoside**
Formed from: *o*-Chlorobenzaldoxime

4-(*p*-Chlorobenzyl)tetrahydro-1-methylpyridine
→ 4-(*p*-Chlorobenzyl-1-methylpyridinium
Probably in: Rat F816

***o*-Chlorobenzyl-β-D-xylopyranoside**
Formed from: *o*-Chlorobenzyl alcohol

2-Chlorobiphenyl
→ *o*-Chlorobenzoic acid
In: Alcaligenes A1822
Pseudomonas H265
→ *cis*-2′-Chloro-2,3-dihydro-2,3-dihydroxybiphenyl
In: Burkholderia K547
Comamonas K187
Pseudomonas K187
→ 2-Chloro-4-hydroxybiphenyl
In: Rat A3954
See also: B448
→ 2-Chloro-5-hydroxybiphenyl
In: Rat B448
→ 6-(*o*-Chlorophenyl)-2-hydroxy-6-oxohexa-2,4-dienoic acid
Probably in: Acinetobacter B125
Alcaligenes B125

3-Chlorobiphenyl
→ *m*-Chlorobenzoic acid
In: Pseudomonas H265
→ *cis*-3′-Chloro-2,3-dihydro-2,3-dihydroxybiphenyl
In: Burkholderia K547
Comamonas K187
Pseudomonas K187
→ 3′-Chloro-4-hydroxybiphenyl
In: Rat A3954, B448

4-Chlorobiphenyl
→ *p*-Chlorobenzoic acid
In: Achromobacter A706
Alcaligenes A1822
Microorganism A963
Pseudomonas E41, H265
→ *cis*-4′-Chloro-2,3-dihydro-2,3-dihydroxybiphenyl
In: Comamonas K187
Pseudomonas K187
→ 4′-Chloro-2-hydroxybiphenyl
In: Rat H755
→ 4-Chloro-3-hydroxybiphenyl
In: Rat A3410
→ 4′-Chloro-3-hydroxybiphenyl
In: Pigeon G419

Rat A2747, H755
→ 4′-Chloro-4-hydroxybiphenyl
In: Cow A1576, A2046
Goat A1576, A2046
Guinea pig D301
Hen A3893
Lichen A2478
Man E938
Pig A2302
Pigeon G419
Rabbit A2053
Rat A2168, A2747, A3410, A3463, A3954, B448, C291, E938, H755
Rhizopus A959
See also: A701
→ 6-(*p*-Chlorophenyl)-2-hydroxy-6-oxohexa-2,4-dienoic acid
In: Microorganism H816

2′-Chlorobiphenyl-4-acetic acid
Formed from: Itanoxone

2′-Chlorobiphenyl-4-carboxylic acid
Formed from: Itanoxone

4′-Chlorobiphenyl-4-yl-β-D-glucuronide
Formed from: 4′-Chloro-4-hydroxybiphenyl

1-Chloro-2,2-bis(*p*-chlorophenyl)ethane
→ 2,2-Bis(*p*-chlorophenyl)acetaldehyde
In: Mouse C675
→ 1-Chloro-2,2-bis(*p*-chlorophenyl)ethylene
In: Trout A3397

1-Chloro-2,2-bis(*p*-chlorophenyl)ethylene
(DDMU)
Formed from: 1-Chloro-2,2-bis(*p*-chlorophenyl)ethane
1,1-Dichloro-2,2-bis(*p*-chlorophenyl) ethane
→ 1,1-Bis(*p*-chlorophenyl)ethylene
In: Rat A928
Contrast (in trout) A3397

3-Chloro-N,N-bisdemethylimipramine
Formed from: 3-Chloro-N-demethylimipramine

8-Chloro-5-bis(formylmethyl)aminoacetyl)-5,10-dihydro-11*H*-dibenzo[*d*,*c*][1,4]diazepin-11-one
Formed from: 8-Chloro-5,10-dihydro-5-(bis(2-hydroxyethyl)aminoacetyl)-11*H*-dibenzo[*d*,*c*][1,4]diazepin-11-one

1-Chloro-2,2-bis(*p*-hydroxyphenyl)ethylene
Probably formed from: 1,1-Dichloro-2,2-bis(*p*-hydroxyphenyl)ethylene

7-Chloro-3-carbamoyl-1,3-dihydro-1-methyl-5-phenyl-2*H*-1,4-benzodiazepin-2-one
Probably formed from: 7-Chloro-1,3-dihydro-1-methyl-3-(N-methyl)carbamoyl-5-phenyl-2*H*-1,4-benzodiazepin-2-one
→ 7-Chloro-3-carbamoyl-1,3-dihydro-5-phenyl-2*H*-1,4-benzodiazepin-2-one
Probably in: Rat E57

7-Chloro-3-carbamoyl-1,3-dihydro-5-phenyl-2*H*-1,4-benzodiazepin-2-one
Probably formed from: 7-Chloro-3-carbamoyl-1,3-dihydro-1-methyl-5-phenyl-2*H*-1,4-benzodiazepin-2-one

4-Chlorocarbostyril
Formed from: 4-chloroquinoline

3-Chlorocatechol
Formed from: *o*-Chloroaniline
o-Chlorobenzoic acid
m-Chlorobenzoic acid
cis-3-Chloro-1,2-dihydro-1,2-dihydroxybenzene
m-Chlorofluorobenzene
6-Chloro-2-nitrophenol
o-Chlorophenol
m-Chlorophenol
p-Chlorophenol
3-Chlorosalicylic acid
3,5-Dichlorocatechol
→ 3-Chloro-2-hydroxy-*cis*,*cis*-muconic semialdehyde
In: Microorganism A2237
Sphingomonas J654, K221
See also: G355, J648, J903, K445
→ 2-Chloro-*cis*,*cis*-muconic acid
In: Pseudomonas A3769, B754, J702
See also: B11, G922
→ 2-Hydroxy-*cis*,*cis*-muconic acid
In: Pseudomonas J344, J515, K5

4-Chlorocatechol
Formed from: *m*-Chloroaniline
p-Chloroaniline
m-Chlorobenzoic acid
p-Chlorobenzoic acid
5-Chloro-2-(2,4-dichlorophenoxy)phenol

5-Chloro-2-hydroxybenzoic acid
4-Chloro-2-methoxyphenol
4-Chloro-2-nitrophenol
m-Chlorophenol
p-Chlorophenol
4-Chlorosalicylic acid
5-Chlorosalicylic acid
2,4-Dichlorobenzoic acid
2,5-Dichlorobenzoic acid
4,5-Dichlorocatechol
Probably formed from: *p*-Chlorobenzoic acid
2,4-Dichlorobenzoic acid
3,5-Dichlorocatechol
4,5-Dichlorophenol
→ 4-Chloro-2-hydroxy-*cis*,*cis*-muconic semialdehyde
In: Burkholderia J903, K445
Pseudomonas G355, G647
→ 5-Chloro-2-hydroxy-*cis*,*cis*-muconic semialdehyde
In: Pseudomonas G657
See also: H272, J515, J649, J653, K5
→ 3-Chloro-*cis*,*cis*-muconic acid
In: Alcaligenes A3769
Pseudomonas A3769, B754, G647, J702
See also: B11, H357
→ 4-Chloro-2-methoxyphenol
In: Rat A486
With catechol O-methyltransferase A597
→ 5-Chloro-2-methoxyphenol
In: Rat A486
With catechol O-methyltransferase A597
→ 5-Chloropyrogallol
Probably in: Candida E204
Trichosporon A902

4-Chloro-2-(*o*-chlorobenzoyl)-N-methyl-N′-glycylglycinanilide (peptido-aminobenzophenone)
→ 7-Chloro-5-(*o*-chlorophenyl)-1,3-dihydro-1-methyl-2*H*-1,4-benzodiazepin-2-one
In: Dog A3940, B115

2-(3-Chloro-4-(4-chlorobenzoyl)phenyl)-*as*-triazine-3,5(2*H*,4*H*)-dione
→ 2-(3-Chloro-4-(4-chloro-α-hydroxybenzyl) phenyl-*as*-triazine-3,5(2*H*,4*H*)-dione
In: Chicken A1984

***p*-Chloro-N-(*p*-chlorobenzyl)aniline**
→ *p*-Chloroaniline
In: Rat J523
→ *p*-Chlorobenzaldehyde
In: Rat J523
→ N-(*p*-Chlorobenzyl)-N-(*p*-chlorophenyl) hydroxylamine
In: Rat J523

2-(3-Chloro-4-(4-chloro-α-hydroxybenzyl) phenyl-*as*-triazine-3,5(2*H*,4*H*)-dione
Formed from: 2-(3-Chloro-4-(4-chlorobenzoyl)phenyl-*as*-triazine-3,5(2*H*,4*H*)-dione

7-Chloro-5-(2-chloro-4-hydroxyphenyl)-1,3-dihydro-2*H*-1,4-benzodiazepin-2-one
Formed from: 7-Chloro-5-(*o*-chlorophenyl)-1,3-dihydro-2*H*-1,4-benzodiazepin-2-one

8-Chloro-6-(2-chloro-4-hydroxyphenyl)-N,N-dimethyl-4*H*-1,2,4-triazolo[1,5-*a*][1,4] benzodiazepine-2-carboxamide
Probably formed from: 8-Chloro-6-(*o*-chlorophenyl)-N,N-dimethyl-4*H*-1,2,4-triazolo[1,5-*a*][1,4]benzodiazepine-2-carboxamide

8-Chloro-6-(2-chloro-4-hydroxyphenyl)-N-hydroxymethyl-N-methyl-4*H*-1,2,4-triazolo[1,5-*a*][1,4]benzodiazepine-2-carboxamide
Probably formed from: 8-Chloro-6-(*o*-chlorophenyl)-N,N-dimethyl-4*H*-1,2,4-triazolo[1,5-*a*][1,4]benzodiazepine-2-carboxamide
→ 8-Chloro-6-(2-chloro-4-hydroxyphenyl)-N-hydroxymethyl-4*H*-1,2,4-triazolo[1,5-*a*][1,4]benzodiazepine-2-carboxamide
Probably in: Dog E871a

8-Chloro-6-(2-chloro-hydroxyphenyl)-1-hydroxymethyl-4*H*-*s*-triazolo[4,3-*a*] [1,4]benzodiazepine
Probably formed from: 8-Chloro-6-(*o*-chlorophenyl)-1-hydroxymethyl-4*H*-*s*-triazolo[4,3-*a*][1,4]benzodiazepine

8-Chloro-6-(2-chloro-4-hydroxyphenyl)-N-hydroxymethyl-4*H*-1,2,4-triazolo[1,5-*a*][1,4]benzodiazepine-2-carboxamide
Probably formed from: 8-Chloro-6-(2-chloro-4-hydroxyphenyl)-N-hydroxymethyl-N-methyl-4*H*-1,2,4-triazolo[1,5-*a*][1,4] benzodiazepine-2-carboxamide

8-Chloro-6-(2-chloro-3-hydroxyphenyl)-4*H*-imidazo[1,5-*a*][1,4]benzodiazepine-3-carboxamide
Formed from: 8-Chloro-6-(*o*-chlorophenyl)-4*H*-imidazo[1,5-*a*][1,4]benzodiazepine-3-carboxamide

8-Chloro-6-(2-chloro-4-hydroxyphenyl)-4*H*-imidazo[1,5-*a*][1,4]benzodiazepine-3-carboxamide
Formed from: 8-Chloro-6-(*o*-chlorophenyl)-4*H*-imidazo[1,5-*a*][1,4]benzodiazepine-3-carboxamide

8-Chloro-6-(2-chloro-5-hydroxyphenyl)-4*H*-imidazo[1,5-*a*][1,4]benzodiazepine-3-carboxamide
Formed from: 8-Chloro-6-(*o*-chlorophenyl)-4*H*-imidazo[1,5-*a*][1,4]benzodiazepine-3-carboxamide

8-Chloro-6-(2-chloro-hydroxyphenyl)-1-methyl-4*H*-*s*-triazolo[4,3-*a*][1,4]benzodiazepine
Formed from: Triazolam

8-Chloro-6-(2-chloro-4-hydroxyphenyl)-N-methyl-4*H*-1,2,4-triazolo[1,5-*a*][1,4]benzodiazepine-2-carboxamide
Probably formed from: 8-Chloro-6-(*o*-chlorophenyl)-N-methyl-4*H*-1,2,4-triazolo[1,5-*a*][1,4]benzodiazepine-2-carboxamide

8-Chloro-6-(2-chloro-4-hydroxyphenyl)-4*H*-1,2,4-triazolo[1,5-*a*][1,4]benzodiazepine-2-carboxamide
Probably formed from: 8-Chloro-6-(*o*-chlorophenyl)-4*H*-1,2,4-triazolo[1,5-*a*][1,4]benzodiazepine-2-carboxamide

5-Chloro-N-(2-chloro-4-nitrophenyl)-2-hydroxybenzamide
→ N-(4-Amino-2-chlorophenyl)-5-chloro-2-hydroxybenzamide
In: Ascaris A2037
Moniezia A2037
Mouse A2037
Sheep A2037

7-Chloro-5-(*o*-chlorophenyl)-1,3-dihydro-2*H*-1,4-benzodiazepin-2-one (2′-chloronordiazepam)
Formed from: Cloxazolam
Probably formed from: 7-Chloro-5-(*o*-chlorophenyl)-1,3-dihydro-1-methyl-2*H*-1,4-benzodiazepin-2-one
→ 7-Chloro-5-(2-chloro-4-hydroxyphenyl)-1,3-dihydro-2*H*-1,4-benzodiazepin-2-one
Probably in: Mouse A828
Rat A828
→ 7-Chloro-5-(*o*-chlorophenyl)-1,3-dihydro-3-hydroxy-2*H*-1,4-benzodiazepin-2-one
In: Dog A1087, A3864
See also: A828, A3940
→ 7-Chloro-5-(*o*-chlorophenyl)-1,3-dihydro-9-hydroxy-2*H*-1,4-benzodiazepin-2-one
In: Dog A3864

7-Chloro-5-(*o*-chlorophenyl)-1,3-dihydro-3-hydroxy-2*H*-1,4-benzodiazepin-2-one
(lorazepam)
Formed from: 7-Chloro-5-(*o*-chlorophenyl)-1,3-dihydro-2*H*-1,4-benzodiazepin-2-one
Lorazepam-3-acetate
Lormetazepam
→ 7-Chloro-5-(*o*-chlorophenyl)-1,3-dihydro-2*H*-1,4-benzodiazepin-2-on-3-yl-β-D-glucuronide
In: Cat A1882
Man A2679, A2819, A2908
Pig A927
Rat A2414, B962
See also: A1087
→ 6-Chloro-4-(*o*-chlorophenyl)-2(1*H*)-quinazoline-2-carboxylic acid
In: Man A2679
Rat A2414
→ 6-Chloro-4-(*o*-chlorophenyl)-2(1*H*)-quinazolinone
In: Man A2679
Rat A282, A2414
→ Lorazepam dihydrodiol
In: Rat A2414
→ Hydroxylorazepam
In: Man A2679
Rat A2414
→ Hydroxymethyllorazepam
In: Rat A2414

7-Chloro-5-(*o*-chlorophenyl)-1,3-dihydro-2*H*-1,4-benzodiazepin-2-on-3-yl-β-D-glucuronide
Formed from: 7-Chloro-5-(*o*-chlorophenyl)-1,3-dihydro-3-hydroxy-2*H*-1,4-benzodiazepin-2-one

7-Chloro-5-(*o*-chlorophenyl)-1,3-dihydro-9-hydroxy-2*H*-1,4-benzodiazepin-2-one
Formed from: 7-Chloro-5-(*o*-chlorophenyl)-1,3-dihydro-2*H*-1,4-benzodiazepin-2-one

7-Chloro-5-(*o*-chlorophenyl-1,3-dihydro-3-hydroxy-1-methyl-2*H*-1,4-benzodiazepin-2-one
See: Lormetazepam

7-Chloro-5-(*o*-chlorophenyl)-1,3-dihydro-1-methyl-2*H*-1,4-benzodiazepin-2-one
Formed from: 4-Chloro-2-(*o*-chlorobenzoyl)-N-methyl-N′-glycylglycinanilide
→ 7-Chloro-5-(*o*-chlorophenyl)-1,3-dihydro-2*H*-1,4-benzodiazepin-2-one
Probably in: Dog A3940, B115

8-Chloro-6-(*o*-chlorophenyl)-N,N-dimethyl-4*H*-1,2,4-triazolo[1,5-*a*][1,4]benzodiazepine-2-carboxamide
Formed from: 5-((2-Aminoacetamido)methyl)-1-(4-chloro-2-(*o*-chlorobenzoyl)phenyl-N,N-dimethyl-1*H*-1,2,4-triazole-3-carboxamide
→ 8-Chloro-6-(2-chloro-4-hydroxyphenyl)-N,N-dimethyl-4*H*-1,2,4-triazolo[1,5-*a*][1,4]benzodiazepine-2-carboxamide
Probably in: Cynomolgus monkey E850
→ 8-Chloro-6-(2-chloro-4-hydroxyphenyl)-N-hydroxymethyl-N-methyl-4*H*-1,2,4-triazolo[1,5-*a*][1,4]benzodiazepine-2-carboxamide
Probably in: Dog E871a
→ 8-Chloro-6-(*o*-chlorophenyl)-4-hydroxy-N,N-dimethyl-4*H*-1,2,4-triazolo[1,5-*a*][1,4]benzodiazepine-2-carboxamide
Probably in: Cynomolgus monkey E850
→ 8-Chloro-6-(*o*-chlorophenyl)-N-methyl-4*H*-1,2,4-triazolo[1,5-*a*][1,4]benzodiazepine-2-carboxamide
Probably in: Cynomolgus monkey E292, E850
Dog E871a
→ 8-Chloro-6-(*o*-chlorophenyl)-4*H*-1,2,4-triazolo[1,5-*a*][1,4]benzodiazepine-2-carboxylic acid
Probably in: Cynomolgus monkey E292, E850
Dog E871a

5-Chloro-N-(*o*-chlorophenyl)-2-hydroxybenzamide
→ *o*-Chloroaniline
In: Mouse A2037
→ 5-Chlorosalicylic acid
In: Mouse A2037

4-Chloro-α-(4-chlorophenyl)-α-hydroxybenzeneacetic acid 1-methylethyl ester
See: Chloropropylate

8-Chloro-6-(*o*-chlorophenyl)-4-hydroxy-N,N-dimethyl-4*H*-1,2,4-triazolo[1,5-*a*][1,4]benzodiazepine-2-carboxamide
Probably formed from: 8-Chloro-6-(*o*-chlorophenyl)-N,N-dimethyl-4*H*-1,2,4-triazolo[1,5-*a*][1,4]benzodiazepine-2-carboxamide

8-Chloro-6-(*o*-chlorophenyl)-4-hydroxy-1-hydroxymethyl-4*H*-*s*-triazolo[4,3-*a*][1,4]benzodiazepine
Probably formed from: 8-Chloro-6-(*o*-chlorophenyl)-1-hydroxymethyl-4*H*-*s*-triazolo[4,3-*a*][1,4]benzodiazepine

8-Chloro-6-(*o*-chlorophenyl)-4-hydroxy-4*H*-imidazo[1,5-*a*][1,4]benzodiazepine-3-carboxamide
Formed from: 8-Chloro-6-(*o*-chlorophenyl)-4*H*-imidazo[1,5-*a*][1,4]benzodiazepine-3-carboxamide

8-Chloro-6-(*o*-chlorophenyl)-1-hydroxymethyl-4*H*-*s*-triazolo[4,3-*a*][1,4]benzodiazepine
Formed from: Triazolam
→ 8-Chloro-6-(2-chloro-hydroxyphenyl)-1-hydroxymethyl-4*H*-*s*-triazolo[4,3-*a*][1,4]benzodiazepine
Probably in: Rat A3702
→ 8-Chloro-6-(*o*-chlorophenyl)-4-hydroxy-1-hydroxymethyl-4*H*-*s*-triazolo[4,3-*a*][1,4]benzodiazepine
Probably in: Dog A2422, A3702
Rat A3702
→ 8-Chloro-6-(*o*-chlorophenyl)-1-hydroxymethyl-4*H*-*s*-triazolo[4,3-*a*][1,4]benzodiazepine
Probably in: Dog A2422

8-Chloro-6-(*o*-chlorophenyl)-4-hydroxy-N-methyl-4*H*-1,2,4-triazolo[1,5-*a*][1,4]benzodiazepine-2-carboxamide
Probably formed from: 8-Chloro-6-(*o*-chlorophenyl)-4*H*-1,2,4-triazolo[1,5-*a*][1,4]benzodiazepine-2-carboxamide

6-Chloro-2-(*p*-chlorophenyl)-N-(2-hydroxypropyl)imidazo[1,2-*a*]pyridine-3-acetamide
Probably formed from: Depropylalpidem

6-Chloro-2-(*p*-chlorophenyl)-N-(2-hydroxypropyl)-N-propylimidazo[1,2-*a*]pyridine-3-acetamide
Formed from: Alpidem

8-Chloro-6-(*o*-chlorophenyl)-4-hydroxy-4*H*-1,2,4-triazolo[1,5-*a*][1,4]benzodiazepine-2-carboxamide
Probably formed from: 8-Chloro-6-(*o*-chlorophenyl)-4-hydroxy-4*H*-1,2,4-triazolo[1,5-*a*][1,4]benzodiazepine-2-carboxamide
8-Chloro-6-(*o*-chlorophenyl)-N-methyl-4*H*-1,2,4-triazolo[1,5-*a*][1,4]benzodiazepine-2-carboxamide

8-Chloro-6-(*o*-chlorophenyl)-4*H*-imidazo[1,5-*a*][1,4]benzodiazepine-3-carboxamide
→ 8-Chloro-6-(*o*-chlorophenyl)-4-hydroxy-4*H*-imidazo[1,5-*a*][1,4]benzodiazepine-3-carboxamide
In: Dog C617
→ 8-Chloro-6-(2-chloro-3-hydroxyphenyl)-4-hydroxy-4*H*-imidazo[1,5-*a*][1,4]benzodiazepine-3-carboxamide
In: Dog C617
→ 8-Chloro-6-(2-chloro-4-hydroxyphenyl)-4-hydroxy-4*H*-imidazo[1,5-*a*][1,4]benzodiazepine-3-carboxamide
In: Dog C617
→ 8-Chloro-6-(2-chloro-5-hydroxyphenyl)-4-hydroxy-4*H*-imidazo[1,5-*a*][1,4]benzodiazepine-3-carboxamide
In: Dog C617

8-Chloro-6-(*o*-chlorophenyl)-N-methyl-4*H*-1,2,4-triazolo[1,5-*a*][1,4]benzodiazepine-2-carboxamide
Probably formed from: 8-Chloro-6-(*o*-chlorophenyl)-N,N-dimethyl-4*H*-1,2,4-triazolo[1,5-*a*][1,4]benzodiazepine-2-carboxamide
→ 8-Chloro-6-(2-chloro-4-hydroxyphenyl)-N-methyl-4*H*-1,2,4-triazolo[1,5-*a*][1,4]benzodiazepine-2-carboxamide
Probably in: Cynomolgus monkey E850
→ 8-Chloro-6-(*o*-chlorophenyl)-4-hydroxy-N-methyl-4*H*-1,2,4-triazolo[1,5-*a*][1,4]benzodiazepine-2-carboxamide
Probably in: Cynomolgus monkey E850
→ 8-Chloro-6-(*o*-chlorophenyl)-4*H*-1,2,4-triazolo[1,5-*a*][1,4]benzodiazepine-2-carboxamide
Probably in: Cynomolgus monkey E850
Dog E871a

6-Chloro-4-(*o*-chlorophenyl)quinazoline-2-carboxylic acid
Formed from: 7-Chloro-5-(*o*-chlorophenyl)-1,3-dihydro-3-hydroxy-2*H*-1,4-benzodiazepin-2-one

6-Chloro-4-(*o*-chlorophenyl)-1,3-dihydroquinazoline
Formed from: 7-Chloro-5-(*o*-chlorophenyl)-1,3-dihydro-3-hydroxy-2*H*-1,4-benzodiazepin-2-one

8-Chloro-6-(*o*-chlorophenyl)-4*H*-*s*-triazolo[4,3-*a*][1,4]benzodiazepine
Probably formed from: 8-Chloro-6-(*o*-chlorophenyl)-1-hydroxymethyl-4*H*-*s*-triazolo[4,3-*a*][1,4]benzodiazepine

8-Chloro-6-(*o*-chlorophenyl)-4*H*-1,2,4-triazolo[1,5-*a*][1,4]benzodiazepine-2-carboxamide
Probably formed from: 8-Chloro-6-(*o*-chlorophenyl)-N-methyl-4*H*-1,2,4-triazolo[1,5-*a*][1,4]benzodiazepine-2-carboxamide
→ 8-Chloro-6-(2-chloro-4-hydroxyphenyl)-4*H*-1,2,4-triazolo[1,5-*a*][1,4]benzodiazepine-2-carboxamide
Probably in: Cynomolgus monkey E850
→ 8-Chloro-6-(*o*-chlorophenyl)-4-hydroxy-4*H*-1,2,4-triazolo[1,5-*a*][1,4]benzodiazepine-2-carboxamide
Probably in: Cynomolgus monkey E850

8-Chloro-6-(*o*-chlorophenyl)-4*H*-1,2,4-triazolo[1,5-*a*][1,4]benzodiazepine-2-carboxylic acid
Probably formed from: 8-Chloro-6-(*o*-chlorophenyl)-N,N-dimethyl-4*H*-1,2,4-triazolo[1,5-*a*][1,4]benzodiazepine-2-carboxamide

α-Chloro-β-(3-chloro-*o*-tolyl)propionic acid
Formed from: α-Chloro-β-(3-chloro-*o*-tolyl)propionitrile

α-Chloro-β-(3-chloro-*o*-tolyl)propionitrile
→ α-(3-Chloro-*o*-tolyl)acetic acid
In: Bean A1269
→ α-Chloro-β-(3-chloro-*o*-tolyl)propionic acid
In: Bean A1269
→ β-(3-Chloro-*o*-tolyl)α-hydroxypropionitrile
In: Bean A1269

***o*-Chlorocinnamic acid**
→ 3-(*o*-Chlorophenyl)propionic acid
In: Microorganism K269
→ *o*-Chlorostyrene
In: Microorganism F390

***m*-Chlorocinnamic acid**
→ *m*-Chlorostyrene
In: Microorganism F390

***p*-Chlorocinnamic acid**
→ *p*-Chlorobenzoic acid
In: Curvularia G821
→ *p*-Chlorocinnamyl alcohol
In: Coriolus G821
→ *p*-Chlorostyrene
In: Daucus K244
Microorganism F390

***p*-Chlorocinnamyl alcohol**
Formed from: *p*-Chlorocinnamic acid
→ 3-(*p*-Chlorophenyl)propanol
Probably in: Coriolus G821

4-Chloro-*o*-cresol
See: 4-Chloro-2-methylphenol

6-Chloro-5-cyclohexylindane-1-carboxylic acid
→ *cis*-6-Chloro-5-(3′-hydroxycyclohexyl) indane-1-carboxylic acid
In: Mouse A1235
Rat A1220, A1234
→ *cis*-6-Chloro-5-(4′-hydroxycyclohexyl) indane-1-carboxylic acid
In: Mouse A1235
Rat A1220, A1234
→ *trans*-6-Chloro-5-(4′-hydroxycyclohexyl) indane-1-carboxylic acid
In: Rat A1220, A1234

3-Chloro-4-cyclohexylphenylacetic acid
Formed from: 4-(3-Chloro-4-cyclohexylphenyl)-4-oxobutyric acid

3-Chloro-4-cyclohexylphenylglyoxylic acid
Formed from: Fenclorac

4-(3-Chloro-4-cyclohexylphenyl)-4-oxobutyric acid
(bucloxic acid)
→ 3-Chloro-4-cyclohexylphenylacetic
In: Dog A1749
Man A1749
Monkey A1749
→ 4-(3-Chloro-4-(*ax*-4-hydroxycyclohexyl)phenyl)-4-hydroxybutyric acid
In: Man A1749
Monkey A1749
Pig A1749
→ 4-(3-Chloro-4-(*eq*-4-hydroxycyclohexyl)phenyl)-4-hydroxybutyric acid
In: Man A1749
Monkey A1749
Pig A1749
→ 3-Chloro-4-(*eq*-3-hydroxycyclohexyl) phenylacetic acid
In: Dog A1749
Man A1749
Monkey A1749
→ 3-Chloro-4-(*ax*-4-hydroxycyclohexyl) phenylacetic acid
In: Man A1749
Pig A1749
Rat A1749
→ 3-Chloro-4-(*eq*-4-hydroxycyclohexyl) phenylacetic acid
In: Man A1749
Pig A1749
Rat A1749

***p*-Chloro-N-cyclopropylbenzamide**
Formed from: *p*-Chloro-N-cyclopropyl-N-methylbenzamide

6-Chloro-4(*S*)-cyclopropyl-3,4-dihydro-4-((6-hydroxy-2-pyridyl)ethynyl)quinazolin-2(1*H*)-one
Formed from: 6-Chloro-4(*S*)-cyclopropyl-3,4-dihydro-4-((2-pyridyl)ethynyl)quinazolin-2(1*H*)-one

6-Chloro-4(*S*)-cyclopropyl-3,4-dihydro-8-hydroxy-4-(2-pyridyl)ethynyl)quinazolin-2(1*H*)-one
Formed from: 6-Chloro-4(*S*)-cyclopropyl-3,4-dihydro-4-((2-pyridyl)ethynyl)quinazolin-2(1*H*)-one

6-Chloro-4(*S*)-cyclopropyl-3,4-dihydro-4-((2-pyridyl)ethynyl)quinazolin-2(1*H*)-one
→ 6-Chloro-4(*S*)-cyclopropyl-3,4-dihydro-4-((6-hydroxy-2-pyridyl)ethynyl)quinazolin-2(1*H*)-one
In: Rat H201
→ 6-Chloro-4(*S*)-cyclopropyl-3,4-dihydro-8-hydroxy-4-((2-pyridyl)ethynyl)quinazolin-2(1*H*)-one
In: Rat H201

(*S*)-6-Chloro-4-(cyclopropylethynyl)-1,4-dihydro-4-trifluoromethyl-2*H*-3,1-benzoxazin-2-one
See: Efavirenz

(*S*)-6-Chloro-4-(cyclopropylethynyl)-3,4-dihydro-4-trifluoromethyl-2(1*H*)-quinazolinone
→ (*S*)-6-Chloro-4-(cyclopropylethynyl)-3,4-dihydro-8-hydroxy-4-trifluoromethyl-2(1*H*)-quinazolinone
In: Rat K554

(*S*)-6-Chloro-4-(cyclopropylethynyl)-3,4-dihydro-8-hydroxy-4-trifluoromethyl-2(1*H*)-quinazolinone
Formed from: (*S*)-6-Chloro-4-(cyclopropylethynyl)-3,4-dihydro-4-trifluoromethyl-2(1*H*)-quinazolinone
→ (*S*)-6-Chloro-4-(cyclopropylethynyl)-3,4-dihydro-8-hydroxy-4-trifluoromethyl-2(1*H*)-quinazolinone-β-D-glucuronide
In: Rat K554

(*S*)-6-Chloro-4-(cyclopropylethynyl)-3,4-dihydro-8-hydroxy-4-trifluoromethyl-2(1*H*)-quinazolinone-β-D-glucuronide
Formed from: (*S*)-6-Chloro-4-(cyclopropylethynyl)-3,4-dihydro-8-hydroxy-4-trifluoromethyl-2(1*H*)-quinazolinone

***p*-Chloro-N-cyclopropyl-N-methylbenzamide**
→ *p*-Chloro-N-cyclopropylbenzamide
In Rat G13
→ *p*-Chloro-N-methylbenzamide
In Rat G13

7-Chloro-1-cyclopropylmethyl-1,3-dihydro-3-hydroxy-5-phenyl-2*H*-1,4-benzodiazepin-2-one
Formed from: 7-Chloro-1-cyclopropylmethyl-1,3-dihydro-5-phenyl-2*H*-1,4-benzodiazepin-2-one
→ 7-Chloro-1-cyclopropylmethyl-1,3-dihydro-3-hydroxy-5-phenyl-2*H*-1,4-benzodiazepin-2-on-3-yl-β-D-glucuronide
In: Monkey A3748
→ 7-Chloro-1,3-dihydro-3-hydroxy-5-phenyl-2*H*-1,4-benzodiazepin-2-one
Probably in: Monkey A3748

7-Chloro-1-cyclopropylmethyl-1,3-dihydro-5-phenyl-2*H*-1,4-benzodiazepin-2-one
→ 7-Chloro-1-cyclopropylmethyl-1,3-dihydro-3-hydroxy-5-phenyl-2*H*-1,4-benzodiazepin-2-one
In: Man A3775, F765
Monkey A3748
Rat F409
→ 7-Chloro-1,3-dihydro-5-phenyl-2*H*-1,4-benzodiazepin-2-one
In: Man A3541, A3775, B687, F765
Monkey A3748
Rat A831, F409

7-Chloro-1-cyclopropylmethyl-5-(*o*-fluorophenyl)-1,3-dihydro-2*H*-1,4-benzodiazepin-2-one
See: Flutoprazepam

7-Chloro-1-cyclopropylmethyl-1,3-dihydro-3-hydroxy-5-phenyl-2*H*-1,4-benzodiazepin-2-on-3-yl-β-D-glucuronide
Formed from: 7-Chloro-1-cyclopropylmethyl-1,3-dihydro-3-hydroxy-5-phenyl-2*H*-1,4-benzodiazepin-2-one

2-Chloro-6-(S-cysteinyl)benzonitrile
Formed from: 2,6-Dichlorobenzonitrile
→ 2-Chloro-6-sulphydrylbenzonitrile
In: Beef F586

6-Chloro-2-(S-cysteinylglycyl)-3-hydroxybenzonitrile
Probably formed from: 6-Chloro-2-(S-glutathionyl)-3-hydroxybenzonitrile
→ 6-Chloro-2-(S-cysteinyl)-3-hydroxybenzonitrile
Probably in: Chicken F302

6-Chloro-2-(S-cysteinyl)-3-hydroxybenzonitrile
Probably formed from: 6-Chloro-2-(S-cysteinylglycyl)-3-hydroxybenzonitrile
→ 2-(N-Acetyl-S-cysteinyl)-6-chloro-3-hydroxybenzonitrile
Probably in: Chicken F302
→ 6-Chloro-3hydroxy-2-sulphydrylbenzonitrile
Probably in: Chicken F302
Goat F296
Rat F296

4′-Chlorodehydrocyclopeptine
→ 4′-Chloro-3′-hydroxydehydrocyclopeptine
In: Penicillium A1633

3-Chloro-N-demethyl-2-hydroxyimipramine
Formed from: 3-Chloro-2-hydroxyimipramine
Probably formed from: 3-Chloro-N-demethylimipramine

3-Chloro-N-demethyl-7-hydroxyimipramine
Formed from: 3-Chloro-N-demethylimipramine
→ 3-Chloro-N-demethyl-7-hydroxyimipramine-β-D-glucuronide
Probably in: Rat H702
→ 3-Chloro-N-demethyl-7-hydroxyimipramine sulphate
Probably in: Man J153
Rat H702

3-Chloro-N-demethyl-8-hydroxyimipramine
Formed from: 3-Chloro-N-demethylimipramine
→ 3-Chloro-N,N-bisdemethylimipramine
In: Man J200

3-Chloro-N-demethyl-10-hydroxyimipramine
Probably formed from: 3-Chloro-N-demethylimipramine

3-Chloro-N-demethyl-N-hydroxyimipramine
Formed from: 3-Chloro-N-demethylimipramine

3-Chloro-N-demethylimipramine
→ N-(2-Carboxyethyl)-3-chloroiminodibenzyl
In: Rabbit C379
→ 3-Chloro-N,N-bisdemethylimipramine
In: Rat J829
→ 3-Chloro-N-demethyl-2-hydroxyimipramine
In: Man A1962
Rabbit C379
→ 3-Chloro-N-demethyl-8-hydroxyimipramine
In: Man J200
See also: A1962, H656
→ 3-Chloro-N-demethyl-10-hydroxyimipramine
Probably in: Rabbit C379
→ 3-Chloro-N-demethyl-N-hydroxyimipramine
In: Rabbit C379

3-Chloro-N-demethyl-7-hydroxyimipramine-β-D-glucuronide
Probably formed from: 3-Chloro-N-demethyl-7-hydroxyimipramine

3-Chloro-N-demethyl-7-hydroxyimipramine sulphate
Probably formed from: 3-Chloro-N-demethyl-7-hydroxyimipramine

1-Chlorodibenzo-*p*-dioxin
→ 4-Chloro-1,2-dihydro-1,2-dihydroxydibenzo-*p*-dioxin
In: Beijerinckia B358
→ 1-Chloro-hydroxydibenzo-*p*-dioxin
In: Rat A3809

2-Chlorodibenzo-*p*-dioxin
Probably formed from: 2,3,7-Trichlorodibenzo-*p*-dioxin
→ 3-Chloro-1,2-dihydro-1,2-dihydroxydibenzo-*p*-dioxin
Probably in: Beijerinckia B358
→ 2-Chloro-hydroxydibenzo-*p*-dioxin
In: Rat A3809

2-Chlorodibenzofuran
→ 2-Chloro-3-hydroxydibenzofuran
In: Rat G128
→ 2-Chloro-7-hydroxydibenzofuran
In: Rat G128

3-Chlorodibenzofuran
→ 4-Chloro-2,2′,3-trihydroxybiphenyl
In: Pseudomonas G253
→ 4′-Chloro-2,2′,3-trihydroxybiphenyl
In: Pseudomonas G253

4-Chlorodibenzofuran
→ 3′-Chloro-2,2′,3-trihydroxybiphenyl
In: Sphingomonas J382

8-Chlorodibenz[*b*,*f*][1,4]oxazepine-10(11*H*)-carboxylic acid 2-(3-(ethylsulphonyl)-1-oxopropyl)hydrazide
→ 8-Chloro-hydroxydibenz[*b*,*f*][1,4]oxazepine-10(11*H*)-carboxylic acid 2-(3-(ethylsulphonyl)-1-oxopropyl) hydrazide
In: Man G850
Rat G850
→ 1-(5-Chloro-2-((*o*-hydroxymethyl)phenoxy)phenyl)-3-(4-thiahexanamid-S,S-dioxide)urea
In: Man G850
Rat G850

8-Chlorodibenz[*b,f*][1,4]oxazepine-10(11*H*)-carboxylic acid hydrazide
Formed from: 8-Chlorodibenz[*b,f*][1,4] oxazepine-10(11*H*)-carboxylic acid 2-(1-oxo-3-(4-pyridinyl)propyl) hydrazide

8-Chlorodibenz[*b,f*][1,4]oxazepine-10(11*H*)-carboxylic acid 2-(1-oxo-3-(pyridine-N-oxide-4-yl)propyl)hydrazide
Formed from: 8-Chlorodibenz[*b,f*][1,4] oxazepine-10(11*H*)-carboxylic acid 2-(1-oxo-3-(4-pyridinyl)propyl) hydrazide

8-Chlorodibenz[*b,f*][1,4]oxazepine-10(11*H*)-carboxylic acid 2-(1-oxo-3-(pyridine-N-oxide-4-yl)propyl)hydrazide
Formed from: 8-Chlorodibenz[*b,f*] [1,4]oxazepine-10(11*H*)-carboxylic acid 2-(1-oxo-3-(4-pyridinyl)propyl) hydrazide

8-Chlorodibenz[*b,f*][1,4]oxazepine-10(11*H*)-carboxylic acid 2-(1-oxo-3-(4-pyridinyl) propyl)hydrazide
→ 8-Chlorodibenz[*b,f*][1,4]oxazepine-10(11*H*)-carboxylic acid hydrazide
In: Rat G850
→ 8-Chlorodibenz[*b,f*][1,4]oxazepine-10(11*H*)-carboxylic acid 2-(1-oxo-3-(pyridine-N-oxide-4-yl)propyl)hydrazide
In: Man G850
Rat G850
→ 8-Chlorodibenz[*b,f*][1,4]oxazepine-10(11*H*)-carboxylic acid 2-(1-oxo-3-(4-pyridinyl)propyl)hydrazide-β-D-glucuronide
In: Man G850
→ 8-Chloro-hydroxydibenz[*b,f*][1,4] oxazepine-10(11*H*)-carboxylic acid 2-(1-oxo-3-(4-pyridinyl)propyl)hydrazide
In: Man G850
→ 1-(5-Chloro-2-((*o*-hydroxymethyl)phenoxy) phenyl-3-(N-(3-(4-pyridinyl)propionamido) urea
In: Man G850
Rat G850

8-Chlorodibenz[*b,f*][1,4]oxazepine-10(11*H*)-carboxylic acid 2-(1-oxo-3-(4-pyridinyl)propyl)hydrazide-β-D-glucuronide
Formed from: 8-Chlorodibenz[*b,f*][1,4] oxazepine-10(11*H*)-carboxylic acid 2-(1-oxo-3-(4-pyridinyl)propyl)hydrazide

5-Chloro-2-(2,4-dichlorophenoxy)phenol
→ 4-Chlorocatechol
In: Rat A3503
→ 2,4-Dichlorophenol
In: Rat A3503
→ 2′,4,4′-Trichloro-2,3-dihydroxydiphenyl ether
Probably in: Rat A3503
→ 2′,4,4′-Trichloro-2,5-dihydroxydiphenyl ether
Probably in: Rat A3503
→ 2′,4,4′-Trichloro-2,6-dihydroxydiphenyl ether
Probably in: Rat A3503
→ 4,4′,6-Trichloro-2,2′-dihydroxydiphenyl ether
Probably in: Rat A3503
→ 4,4′,6-Trichloro-2,3′-dihydroxydiphenyl ether
Probably in: Rat A3503

2-Chloro-1-(3,4-dichlorophenyl)ethanol
Formed from: 3,4-Dichlorophenacyl chloride

2-Chloro-1-(2,4-dichlorophenyl)ethenyl diethyl phosphate
See: Chlorfenvinphos

2-Chloro-1-(2,4-dichlorophenyl)ethenyl ethyl phosphate
Formed from: Chlorfenvinphos

***p*-Chloro-O-diethylaminoethylbenzaldoxime**
→ *p*-Chloro-O-diethylaminoethylbenzaldoxime-N-oxide
In: Dog A2526

***p*-Chloro-O-diethylaminoethylbenzaldoxime-N-oxide**
Formed from: *p*-Chloro-O-diethylaminoethylbenzaldoxime

7-Chloro-1-(2-diethylaminoethyl)-5-(2-fluorophenyl)-1,3-dihydro-2*H*-1,4-benzodiazepin-2-one (flurazepam)
→ 7-Chloro-1-(2-ethylaminoethyl)-5-(2-fluorophenyl)-1,3-dihydro-2*H*-1,4-benzodiazepin-2-one

In: Cat D325
Rat A873
→ 7-Chloro-1-(2-((N-ethyl-N-(2-hydroxyethyl)) amino)ethyl)-5-(2-fluorophenyl)-1,3-dihydro-2*H*-1,4-benzodiazepin-2-one
In: Cat D325
→ 7-Chloro-5-(2-fluorophenyl)-1-(2-hydroxyethyl)-1,3-dihydro-2*H*-1,4-benzodiazepin-2-one
In: Man A1499
See also: A2317
→ 7-Chloro-5-(*o*-fluorophenyl)-2,3-dihydro-2-oxo-1*H*-1,4-benzodiazepine-1-aldehyde
In: Man C601

7-Chloro-4-(3-diethylamino-4-fluoroanilino) quinoline
→ 7-Chloro-4-(3-ethylamino-4-fluoroanilino)quinoline
In: Rat H159

7-Chloro-4-((3-diethylamino)propylamino) quinoline
See: Chloroquine

***p*-Chloro-N,N-diethylbenzamide**
→ *p*-Chlorobenzoic acid
In: Rat H475

2-Chloro-2,6-diethyl-N-(butoxymethyl)acetanilide
See: Butachlor

8-Chloro-5-(2,3-dihydrobenzofuran-7-yl)-2,3,4,5-tetrahydro-7-hydroxy-3-methyl-1*H*-3-benzazepine
Formed from: (+)-8-Chloro-5-(2,3-dihydrobenzofuran-7-yl)-2,3,4,5-tetrahydro-7-methoxymethoxy-3-methyl-1*H*-3-benzazepine
→ 8-Chloro-5-(2,3-dihydrobenzofuran-7-yl)-2,3,4,5-tetrahydro-7-hydroxy-3-methyl-1*H*-3-benzazepine-β-D-glucuronide
In: Rat G296
→ 8-Chloro-5-(2,3-dihydro-3-hydroxybenzofuran-7-yl)-2,3,4,5-tetrahydro-7-hydroxy-3-methyl-1*H*-3-benzazepine
Probably in: Rat G72

8-Chloro-5-(2,3-dihydrobenzofuran-7-yl)-2,3,4,5-tetrahydro-7-hydroxy-3-methyl-1*H*-3-benzazepine-β-D-glucuronide
Formed from: 8-Chloro-5-(2,3-dihydrobenzofuran-7-yl)-2,3,4,5-tetrahydro-7-hydroxy-3-methyl-1*H*-3-benzazepine

(+)-8-Chloro-5-(2,3-dihydrobenzofuran-7-yl)-2,3,4,5-tetrahydro-7-methoxymethoxy-3-methyl-1*H*-3-benzazepine
→ 8-Chloro-5-(2,3-dihydrobenzofuran-7-yl)-2,3,4,5-tetrahydro-7-hydroxy-3-methyl-1*H*-3-benzazepine
In: Rat G72
→ 8-Chloro-5-(2,3-dihydro-3-hydroxybenzofuran-7-yl)-2,3,4,5-tetrahydro-7-methoxymethoxy-3-methyl-1*H*-3-benzazepine
In: Rat G72

8-Chloro-5,10-dihydro-5-(bis(2-hydroxyethyl) aminoacetyl)-11*H*-dibenzo[*d*,*c*][1,4]diazepin-11-one
→ 8-Chloro-5-(bis(formylmethyl)aminoacetyl)-5,10-dihydro-11*H*-dibenzo[*d*,*c*][1,4] diazepin-11-one
In: Rat F833
→ 8-Chloro-5,10-dihydro-11*H*-dibenzo[*d*,*c*][1,4]diazepin-11-one
In: Rat F833
→ 8-Chloro-5,10-dihydro-5-((2-hydroxyethyl) aminoacetyl)-11*H*-dibenzo[*d*,*c*][1,4] diazepin-11-one
In: Rat F833
→ 8-Chloro-5-glycollyl-5,10-dihydro-11*H*-dibenzo[*d*,*c*][1,4]diazepin-11-one
In: Rat F833

8-Chloro-5,10-dihydro-11*H*-dibenzo[*d*,*c*][1,4] diazepin-11-one
Formed from: 8-Chloro-5,10-dihydro-5-(bis(2-hydroxyethyl)aminoacetyl)-11*H*-dibenzo[*d*,*c*][1,4]diazepin-11-one

1-(8-Chloro-10,11-dihydrodibenzo[*b*,*f*]thiepin-10-yl)-4-methylpiperazine
See: Octoclothepine

***cis*-3-Chloro-1,2-dihydro-1,2-dihydroxybenzene**
Formed from: Chlorobenzene
→ 3-Chlorocatechol
In: Pseudomonas K699

***trans*-3-Chloro-1,2-dihydro-1,2-dihydroxybenzene**
Formed from: Chlorobenzene-2,3-oxide

***trans*-4-Chloro-1,2-dihydro-1,2-dihydroxybenzene**
Formed from: Chlorobenzene-3,4-oxide

5-Chloro-1,2-dihydro-1,2-dihydroxybenzoic acid
Formed from: *m*-Chlorobenzoic acid

6-Chloro-1,2-dihydro-1,2-dihydroxybenzoic acid
Probably formed from: *o*-Chlorobenzoic acid

1-(1-(2-(3-Chloro-4,5-dihydro-4,5-dihydroxybenzyloxy)phenyl)vinyl)imidazole
Formed from: Croconazole

***cis*-2′-Chloro-2,3-dihydro-2,3-dihydroxybiphenyl**
Formed from: 2-Chlorobiphenyl

***cis*-3′-Chloro-2,3-dihydro-2,3-dihydroxybiphenyl**
Formed from: 3-Chlorobiphenyl

***cis*-4′-Chloro-2,3-dihydro-2,3-dihydroxybiphenyl**
Formed from: 4-Chlorobiphenyl

3-Chloro-1,2-dihydro-1,2-dihydroxydibenzo-*p*-dioxin
Probably formed from: 2-Chlorodibenzo-*p*-dioxin

4-Chloro-1,2-dihydro-1,2-dihydroxydibenzo-*p*-dioxin
Probably formed from: 1-Chlorodibenzo-*p*-dioxin

7-Chloro-1,3-dihydro-3-(N,N-di(hydroxymethyl))carbamoyl-N-methyl-5-phenyl-2*H*-1,4-benzodiazepin-2-one
Probably formed from: 7-Chloro-1,3-dihydro-3-(N-hydroxymethyl-N-methyl)carbamoyl-1-methyl-5-phenyl-2*H*-1,4-benzodiazepin-2-one
→ 7-Chloro-1,3-dihydro-3-(N,N-di(hydroxymethyl))carbamoyl-5-phenyl-2*H*-1,4-benzodiazepin-2-one
Probably in: Rat E57

7-Chloro-1,3-dihydro-3-(N,N-di(hydroxymethyl))carbamoyl-5-phenyl-2*H*-1,4-benzodiazepin-2-one
Probably formed from: 7-Chloro-1,3-dihydro-3-(N,N-di(hydroxymethyl))carbamoyl-N-methyl-5-phenyl-2*H*-1,4-benzodiazepin-2-one

***cis*-7-Chloro-1,2-dihydro-1,2-dihydroxynaphthalene**
Formed from: 2-Chloronaphthalene
→ 7-Chloro-1,2-dihydroxynaphthalene
In: Pseudomonas J370

***cis*-8-Chloro-1,2-dihydro-1,2-dihydroxynaphthalene**
Formed from: 1-Chloronaphthalene
→ 8-Chloro-1,2-dihydroxynaphthalene
In: Pseudomonas J370

7-Chloro-5-(3,4-dihydro-3,4-dihydroxyphenyl)-1,3-dihydro-3-hydroxy-2*H*-1,4-benzodiazepin-2-one
Formed from: 7-Chloro-1,3-dihydro-3-hydroxy-5-phenyl-2*H*-1,4-benzodiazepin-2-one

2-Chloro-5,6-dihydro-5,6-dihydroxyquinoline
Formed from: 2-Chloroquinoline

2-Chloro-7,8-dihydro-7,8-dihydroxyquinoline
Formed from: 2-Chloroquinoline

4-Chloro-2,3-dihydro-2,3-dihydroxytoluene
→ 3-Chloro-6-methylcatechol
In: Pseudomonas K348

7-Chloro-1,3-dihydro-3-dimethylcarbamoyl-5-(*p*-hydroxyphenyl)-1-methyl-2*H*-1,4-benzodiazepin-2-one
Formed from: Camazepam

7-Chloro-1,3-dihydro-3-(N,N-dimethyl)carbamoyl-5-phenyl-2*H*-1,4-benzodiazepin-2-one
Formed from: Camazepam
→ 7-Chloro-1,3-dihydro-3-(N-dimethyl)carbamoyl-5-phenyl-2*H*-1,4-benzodiazepin-2-one
In: Dog D639
Monkey D639
Mouse D639
Rat D639

7-Chloro-1,3-dihydro-1,3-dimethyl-5-phenyl-2*H*-1,4-benzodiazepin-2-one (3-methyldiazepam)
→ 6-Chloro-1-methyl-4-phenyl-2(1*H*)-quinazolinone
In: Rat A282
→ 6-Chloro-4-phenyl-2(1*H*)-quinazolinone
In: Rat A282

8-Chloro-5-(2,3-dihydro-3-hydroxybenzofuran-7-yl)-2,3,4,5-tetrahydro-7-hydroxy-1*H*-3-benzazepine
Probably formed from: 8-Chloro-5-(2,3-dihydro-3-hydroxybenzofuran-7-yl)-2,3,4,5-tetrahydro-7-hydroxy-3-methyl-1*H*-3-benzazepine

8-Chloro-5-(2,3-dihydro-3-hydroxybenzofuran-7-yl)-2,3,4,5-tetrahydro-7-hydroxy-3-methyl-1*H*-3-benzazepine
Probably formed from: 8-Chloro-5-(2,3-dihydrobenzofuran-7-yl)-2,3,4,5-tetrahydro-7-hydroxy-3-methyl-1*H*-3-benzazepine

→ 8-Chloro-5-(2,3-dihydro-3-hydroxybenzofuran-7-yl)-2,3,4,5-tetrahydro-7-hydroxy-1*H*-3-benzazepine
Probably in: Rat G72

8-Chloro-5-(2,3-dihydro-3-hydroxybenzofuran-7-yl)-2,3,4,5-tetrahydro-7-methoxymethoxy-1*H*-3-benzazepine
Probably formed from: 8-Chloro-5-(2,3-dihydro-3-hydroxybenzofuran-7-yl)-2,3,4,5-tetrahydro-7-methoxymethoxy-3-methyl-1*H*-3-benzazepine

8-Chloro-5-(2,3-dihydro-3-hydroxybenzofuran-7-yl)-2,3,4,5-tetrahydro-7-methoxymethoxy-3-methyl-1*H*-3-benzazepine
Formed from: (+)-8-Chloro-5-(2,3-dihydrobenzofuran-7-yl)-2,3,4,5-tetrahydro-7-methoxymethoxy-3-methyl-1*H*-3-benzazepine
→ 8-Chloro-5-(2,3-dihydro-3-hydroxybenzofuran-7-yl)-2,3,4,5-tetrahydro-7-methoxymethoxy-1*H*-3-benzazepine
Probably in: Rat G72

7-Chloro-1,3-dihydro-3-hydroxy-1,3-dimethyl-5-phenyl-2*H*-1,4-benzodiazepin-2-one
(3-methyldiazepam)
→ 6-Chloro-1-methyl-4-phenyl-2(1*H*)-quinazolinone
In: Rat A282
→ 6-Chloro-4-phenyl-2(1*H*)-quinazolinone
In: Rat A282

8-Chloro-5,10-dihydro-5-((2-hydroxyethyl)aminoacetyl)-11*H*-dibenzo[*d*,*c*][1,4]diazepin-11-one
Formed from: 8-Chloro-5,10-dihydro-5-(bis(2-hydroxyethyl)aminoacetyl)-11*H*-dibenzo[*d*,*c*][1,4]diazepin-11-one
→ 5-Aminoacetyl-8-chloro-5,10-dihydro-11*H*-dibenzo[*d*,*c*][1,4]diazepin-11-one
In: Rat F833

7-Chloro-1,3-dihydro-3-hydroxy-5-(4-hydroxy-3-methoxyphenyl)-2*H*-1,4-benzodiazepin-2-one
Probably formed from: 7-Chloro-1,3-dihydro-3-hydroxy-5-(*p*-hydroxyphenyl)-2*H*-1,4-benzodiazepin-2-one

(+)-(2*S*,3*S*)-8-Chloro-3,4-dihydro-3-hydroxy-2-(4-hydroxy-3-methoxyphenyl)-5-(2-(N-methylamino)ethyl)-4-oxo-1,5(2*H*)-benzothiazepine
Probably formed from: (+)-(2*S*,3*S*)-3-Acetoxy-8-chloro-3,4-dihydro-2-(4-hydroxy-3-methoxyphenyl)-5-(2-(N-methylamino)ethyl)-4-oxo-1,5(2*H*)-benzothiazepine
→ (+)-(2*S*,3*S*)-5-(2-Aminoethyl)-8-chloro-3,4-dihydro-3-hydroxy-2-(4-hydroxy-3-methoxyphenyl)-4-oxo-1,5(2*H*)-benzothiazepine
Probably in: Rat H76

(+)-(2*S*,3*S*)-8-Chloro-3,4-dihydro-2-(4-hydroxy-3-methoxyphenyl)-4-oxo-1,5-benzothiazepine-5(2*H*)-acetic acid
Probably formed from: Deacetylclentiazem

7-Chloro-4,5-dihydro-3-(5-(1-hydroxy-1-(hydroxymethyl)ethyl)-5-methyl-1,2,4-oxadiazol-3-yl)-6*H*-imidazo[1,5-*a*][1,4]benzodiazepin-6-one
Probably formed from: 7-Chloro-4,5-dihydro-3-(5-(1-hydroxy-1-methylethyl)-5-methyl-1,2,4-oxadiazol-3-yl)-6*H*-imidazo[1,5-*a*][1,4]benzodiazepin-6-one

7-Chloro-1,3-dihydro-3-hydroxy-5-(*m*-hydroxyphenyl)-2*H*-1,4-benzodiazepin-2-one
Formed from: 7-Chloro-1,3-dihydro-3-hydroxy-5-phenyl-2*H*-1,4-benzodiazepin-2-one

7-Chloro-1,3-dihydro-3-hydroxy-5-(*p*-hydroxyphenyl)-2*H*-1,4-benzodiazepin-2-one
Formed from: 7-Chloro-1,3-dihydro-3-hydroxy-5-phenyl-2*H*-1,4-benzodiazepin-2-one
→ 7-Chloro-1,3-dihydro-3-hydroxy-5-(*p*-hydroxyphenyl)-2*H*-1,4-benzodiazepin-2-one-β-D-glucuronide
Probably in: Mouse E909
Rat A492, D431
→ 7-Chloro-1,3-dihydro-3-hydroxy-5-(4-hydroxy-3-methoxyphenyl)-2*H*-1,4-benzodiazepin-2-one
Probably in: Rat F852, H149

7-Chloro-1,3-dihydro-3-hydroxy-5-(*p*-hydroxyphenyl)-2*H*-1,4-benzodiazepin-2-one-β-D-glucuronide
Probably formed from: 7-Chloro-1,3-dihydro-3-hydroxy-5-(*p*-hydroxyphenyl)-2*H*-1,4-benzodiazepin-2-one

7-Chloro-1,3-dihydro-3-hydroxy-5-(*p*-hydroxyphenyl)-1-methyl-2*H*-1,4-benzodiazepin-2-one
(*p*-hydroxytemazepam)
Probably formed from: 7-Chloro-1,3-dihydro-3-hydroxy-1-methyl-5-phenyl-2*H*-1,4-benzodiazepin-2-one

→ *p*-Hydroxytemazepam-β-D-glucuronide
Probably in: Rat D111, D431

(+)-(2*S*,3*S*)-8-Chloro-3,4-dihydro-3-hydroxy-2-(4-methoxyphenyl)-4-oxo-1,5-benzothiazepine-5(2*H*)-acetonitrile
Probably formed from: (+)-(2*S*,3*S*)-3-Acetoxy-8-chloro-3,4-dihydro-3-hydroxy-2-(4-methoxyphenyl)-4-oxo-1,5-benzothiazepine-5(2*H*)-acetonitrile

7-Chloro-1,3-dihydro-3-(N-hydroxymethyl)carbamoyl-1-methyl-5-phenyl-2*H*-1,4-benzodiazepin-2-one
Probably formed from: 7-Chloro-1,3-dihydro-3-(N-hydroxymethyl-N-methyl)carbamoyl-1-methyl-5-phenyl-2*H*-1,4-benzodiazepin-2-one
→ 7-Chloro-1,3-dihydro-3-(N-hydroxymethyl)carbamoyl-5-phenyl-2*H*-1,4-benzodiazepin-2-one
Probably in: Dog D639
Monkey D639
Mouse D639
Rat D639, E57
→ 3-Carbamoyl-7-chloro-1,3-dihydro-1-methyl-5-phenyl-2*H*-1,4-benzodiazepin-2-one
Probably in: Dog D639
Monkey D639
Mouse D639
Rat D639

7-Chloro-1,3-dihydro-3-(N-hydroxymethyl)carbamoyl-5-phenyl-2*H*-1,4-benzodiazepin-2-one
Probably formed from: 7-Chloro-1,3-dihydro-3-(N-hydroxymethyl)carbamoyl-1-methyl-5-phenyl-2*H*-1,4-benzodiazepin-2-one
Probably formed from: 7-Chloro-1,3-dihydro-3-(N-hydroxymethyl-N-methyl)carbamoyl-5-phenyl-2*H*-1,4-benzodiazepin-2-one
→ 3-Carbamoyl-7-chloro-1,3-dihydro-5-phenyl-2*H*-1,4-benzodiazepin-2-one
Probably in: Dog D639
Monkey D639
Mouse D639
Rat D639

N-(5-Chloro-3,4-dihydro-8-hydroxy-3-methylcoumarin-7-oyl)aspartic acid
Probably formed from: 2′,3′-Dihydro-2′,3′-dihydroxyochratoxin A

7-Chloro-4,5-dihydro-3-(5-(1-hydroxy-1-methylethyl)-5-methyl-1,2,4-oxadiazol-3-yl)-6*H*-imidazo[1,5-*a*][1,4]benzodiazepin-6-one
Formed from: 7-Chloro-4,5-dihydro-5-methyl-3-(5-(1-methylethyl)-1,2,4-oxadiazol-3-yl)-6*H*-imidazo[1,5-*a*][1,4]benzodiazepin-6-one
→ 7-Chloro-4,5-dihydro-3-(5-(1-hydroxy-1-(hydroxymethyl)ethyl)-5-methyl-1,2,4-oxadiazol-3-yl)-6*H*-imidazo[1,5-*a*][1,4]benzodiazepin-6-one
Probably in: Dog G959
Monkey G959
Rat G959

7-Chloro-1,3-dihydro-3-(N-hydroxymethyl-N-methyl)carbamoyl-1-methyl-5-phenyl-2*H*-1,4-benzodiazepin-2-one
Formed from: Camazepam
→ 3-(N,N-Bis(hydroxymethyl)carbamoyl-7-chloro-1,3-dihydro-1-methyl-5-phenyl-2*H*-1,4-benzodiazepin-2-one
Probably in: Dog D639
Monkey D639
Mouse D639
Rat D639, E57
→ 7-Chloro-1,3-dihydro-3-(N-hydroxymethyl)carbamoyl-1-methyl-5-phenyl-2*H*-1,4-benzodiazepin-2-one
Probably in: Dog D639
Monkey D639
Mouse D639
Rat D639, E57
→ 7-Chloro-1,3-dihydro-3-(N-hydroxymethyl-N-methyl)carbamoyl-5-phenyl-2*H*-1,4-benzodiazepin-2-one
Probably in: Dog D639
Monkey D639
Mouse D639
Rat D639, E57

7-Chloro-1,3-dihydro-3-(N-hydroxymethyl-N-methyl)carbamoyl-5-phenyl-2*H*-1,4-benzodiazepin-2-one
Probably formed from: 7-Chloro-1,3-dihydro-3-(N-hydroxymethyl-N-methyl)carbamoyl-1-methyl-5-phenyl-2*H*-1,4-benzodiazepin-2-one
→ 7-Chloro-1,3-dihydro-3-(N-hydroxymethyl)carbamoyl-5-phenyl-2*H*-1,4-benzodiazepin-2-one

Probably in: Rat E57

7-Chloro-1,3-dihydro-3-hydroxy-1-methyl-5-phenyl-2*H*-1,4-benzodiazepin-2-one
(N-methyloxazepam, temazepam)
Formed from: 7-Chloro-1,3-dihydro-1-methyl-5-phenyl-2*H*-1,4-benzodiazepin-2-one
→ 7-Chloro-1,3-dihydro-3-hydroxy-5-(*p*-hydroxyphenyl)-1-methyl-2*H*-1,4-benzodiazepin-2-one
Probably in: Rat D11, D431
→ 7-Chloro-1,3-dihydro-3-hydroxy-1-methyl-5-phenyl-2*H*-1,4-benzodiazepin-2-one-β-D-glucuronide
In: Man F700
See also: D111, D431
→ 7-Chloro-1,3-dihydro-3-hydroxy-5-phenyl-2*H*-1,4-benzodiazepin-2-one
In: Guinea pig A471
Man F700
Mouse J28
Rat B857
See also: E57
→ 6-Chloro-1-methyl-4-phenyl-2(1*H*)-quinazolinone
In: Rat A282
→ 6-Chloro-4-phenyl-2(1*H*)-quinazolinone
In: Rat A282

7-Chloro-1,3-dihydro-hydroxy-3-methyl-5-phenyl-2*H*-1,4-benzodiazepin-2-one
Formed from: 7-Chloro-1,3-dihydro-3-methyl-5-phenyl-2*H*-1,4-benzodiazepin-2-one

7-Chloro-1,3-dihydro-3-hydroxy-1-methyl-5-phenyl-2*H*-1,4-benzodiazepin-2-one-β-D-glucuronide
Formed from: 7-Chloro-1,3-dihydro-3-hydroxy-1-methyl-5-phenyl-2*H*-1,4-benzodiazepin-2-one

7-Chloro-1,3-dihydro-3-hydroxy-1-methyl-5-phenyl-2*H*-1,4-benzodiazepin-2-one-β-D-glucuronide
Formed from: 7-Chloro-1,3-dihydro-3-hydroxy-1-methyl-5-phenyl-2*H*-1,4-benzodiazepin-2-one

7-Chloro-1,3-dihydro-3-hydroxy-5-phenyl-2*H*-1,4-benzodiazepin-2-one (oxazepam)
Formed from: 3′-Acetoxy-7-chloro-1,3-dihydro-5-phenyl-2*H*-1,4-benzodiazepin-2-one
7-Chloro-1,3-dihydro-3-hydroxy-1-methyl-5-phenyl-2*H*-1,4-benzodiazepin-2-one
3-Hydroxyhalazepam
Mendon
Oxazepam acetate
Oxazepam *n*-butyrate
Oxazepam diethylacetate
Oxazepam isobutyrate
Oxazepam α-methyl-β-phenylpropionate
Oxazepam β-phenylpropionate
Oxazepam γ-phenylpropionate
Oxazepam propionate
Oxazepam succinate half ester
Oxazepam succinate half ester
Oxazepam succinate methyl ester
Probably formed from: 7-Chloro-1-cyclopropylmethyl-1,3-dihydro-3-hydroxy-5-phenyl-2*H*-1,4-benzodiazepin-2-one
→ 2-Amino-5-chlorobenzophenone
In: Man A492
Pig A492
→ 7-Chloro-5-(3,4-dihydro-3,4-dihydroxyphenyl)-1,3-dihydro-3-hydroxy-2*H*-1,4-benzodiazepin-2-one
In: Rat H149
See also: H773
→ 7-Chloro-1,3-dihydro-3-hydroxy-5-(*m*-hydroxyphenyl)-2*H*-1,4-benzodiazepin-2-one
In: Rat D852, H149
→ 7-Chloro-1,3-dihydro-3-hydroxy-5-(*p*-hydroxyphenyl)-2*H*-1,4-benzodiazepin-2-one
In: Man A492
Mouse E909
Rat A492, D852, H149, H773
Pig A492
See also: A831, D111
→ 7-Chloro-1,3-dihydro-3-hydroxy-5-phenyl-2*H*-1,4-benzodiazepin-2-one-3-β-D-glucuronide
In: Dog C633
Man A492, C633, H933
Monkey C633
Mouse E909
Rabbit D337
Rat A492, B962, C672, D111, D431, H149
Pig A492, A927, A3669

→ 7-Chloro-1,3-dihydro-3-hydroxy-5-phenyl-2*H*-1,4-benzodiazepin-2-one-O-sulphate
In: Rat H149
→ 6-Chloro-4-phenyl-2(1*H*)-quinazolinecarboxylic acid
In: Man H773
Mouse E909, H773
Rat H149, H773
→ 6-Chloro-4-phenyl-2(1*H*)-quinazolinone
In: Man A492
Rat A282, D852
Pig A492
See also: A3036

7-Chloro-1,3-dihydro-9-hydroxy-5-phenyl-2*H*-1,4-benzodiazepin-2-one
Formed from: 7-Chloro-1,3-dihydro-5-phenyl-2*H*-1,4-benzodiazepin-2-one-4-oxide
→ 7-Chloro-1,3-dihydro-9-hydroxy-5-phenyl-2*H*-1,4-benzodiazepin-2-one-4-oxide
Probably in: A579

7-Chloro-1,3-dihydro-5-(4-hydroxyphenyl)-2*H*-1,4-benzodiazepin-2-one
→ 7-Chloro-1,3-dihydro-5-(4-hydroxyphenyl)-2*H*-1,4-benzodiazepin-2-one-β-D-glucuronide
Probably in: Rat D111, D431, H149
→ 7-Chloro-1,3-dihydro-5-(4-hydroxyphenyl)-2*H*-1,4-benzodiazepin-2-one sulphate
Probably in: Monkey A3748
Rat H149

7-Chloro-1,3-dihydro-3-hydroxy-5-phenyl-2*H*-1,4-benzodiazepin-2-one-3-β-D-glucuronide
Formed from: 7-Chloro-1,3-dihydro-3-hydroxy-5-phenyl-2*H*-1,4-benzodiazepin-2-one

7-Chloro-1,3-dihydro-5-(4-hydroxyphenyl)-2*H*-1,4-benzodiazepin-2-one-β-D-glucuronide
Probably formed from: 7-Chloro-1,3-dihydro-5-(4-hydroxyphenyl)-2*H*-1,4-benzodiazepin-2-one

7-Chloro-1,3-dihydro-3-hydroxy-5-phenyl-2*H*-1,4-benzodiazepin-2-one-O-sulphate
Formed from: 7-Chloro-1,3-dihydro-3-hydroxy-5-phenyl-2*H*-1,4-benzodiazepin-2-one

7-Chloro-1,3-dihydro-5-(4-hydroxyphenyl)-2*H*-1,4-benzodiazepin-2-one sulphate
Probably formed from: 7-Chloro-1,3-dihydro-5-(4-hydroxyphenyl)-2*H*-1,4-benzodiazepin-2-one

7-Chloro-1,3-dihydro-5-(4-hydroxyphenyl)-1-methyl-2*H*-1,4-benzodiazepin-2-one
Formed from: 7-Chloro-1,3-dihydro-1-methyl-5-phenyl-2*H*-1,4-benzodiazepin-2-one
→ 7-Chloro-1,3-dihydro-5-(4-hydroxyphenyl)-1-methyl-2*H*-1,4-benzodiazepin-2-one-β-D-glucuronide
Probably in: Rat D111

7-Chloro-1,3-dihydro-3-hydroxy-5-phenyl-1-methyl-2*H*-1,4-benzodiazepin-2-one-β-D-glucuronide
Formed from: 7-Chloro-1,3-dihydro-3-hydroxy-5-phenyl-1-methyl-2*H*-1,4-benzodiazepin-2-one

7-Chloro-1,3-dihydro-5-(4-hydroxyphenyl)-1-methyl-2*H*-1,4-benzodiazepin-2-one-β-D-glucuronide
Formed from: 7-Chloro-1,3-dihydro-5-(4-hydroxyphenyl)-1-methyl-2*H*-1,4-benzodiazepin-2-one

7-Chloro-1,3-dihydro-9-hydroxy-5-phenyl-2*H*-1,4-benzodiazepin-2-one-4-oxide
Probably formed from: 7-Chloro-1,3-dihydro-9-hydroxy-5-phenyl-2*H*-1,4-benzodiazepin-2-one

7-Chloro-1,3-dihydro-5-(4-hydroxyphenyl)-2*H*-1,4-benzodiazepin-2-one sulphate
Probably formed from: 7-Chloro-1,3-dihydro-5-(4-hydroxyphenyl)-2*H*-1,4-benzodiazepin-2-one

S-(3-Chloro-1,2-dihydro-2-hydroxyphenyl) glutathione
Probably formed from: Chlorobenzene-2,3-oxide

S-(4-Chloro-1,2-dihydro-2-hydroxyphenyl) glutathione
Probably formed from: Chlorobenzene-3,4-oxide

8-Chloro-2,4-dihydro-6-(*m*-hydroxyphenyl)-1*H*-*s*-triazol[4,3-*a*][1,4]benzodiazepin-1-one
Probably formed from: 8-Chloro-6-(*m*-hydroxyphenyl)-1*H*-*s*-triazol[4,3-*a*][1,4] benzodiazepine

8-Chloro-2,4-dihydro-6-(*p*-hydroxyphenyl)-1*H*-*s*-triazol[4,3-*a*][1,4]benzodiazepin-1-one
Probably formed from: 8-Chloro-6-(*p*-hydroxyphenyl)-1*H*-*s*-triazol[4,3-*a*][1,4]benzodiazepine
8-Chloro-2,4-dihydro-6-phenyl-1*H*-*s*-triazol[4,3-*a*][1,4]benzodiazepin-1-one
→ 5-Chloro-2-(2,3-dihydro-3-oxo-4*H*-1,2,4-triazol-4-yl)-4′-hydroxybenzophenone
Probably in: Dog A258

5-Chloro-2,3-dihydro-3-(hydroxy-(2-thienyl)methylene)-2-oxo-1*H*-indol-1-ylcarboxamide
See: Tenidab

7-Chloro-1,3-dihydro-3-(N-methyl)carbamoyl)-5-phenyl-2*H*-1,4-benzodiazepin-2-one
Probably formed from: 7-Chloro-1,3-dihydro-3-(N,N-dimethyl)carbamoyl)-5-phenyl-2*H*-1,4-benzodiazepin-2-one
7-Chloro-1,3-dihydro-1-methyl-3-(N-methyl)carbamoyl)-5-phenyl-2*H*-1,4-benzodiazepin-2-one

5-((3-Chloro-6,11-dihydro-6-methyl-5,5-dioxodibenzo[*c*,*f*]-1,2-thiazepin-11-yl) amino)pentanoic acid
Formed from: Tianeptine
→ 3-((3-Chloro-6,11-dihydro-6-methyl-5,5-dioxodibenzo[*c*,*f*]-1,2-thiazepin-11-yl) amino)propionic acid
Probably in: Man F576

3-((3-Chloro-6,11-dihydro-6-methyl-5,5-dioxodibenzo[*c*,*f*]-1,2-thiazepin-11-yl) amino)propionic acid
Probably formed from: 5-((3-Chloro-6,11-dihydro-6-methyl-5,5-dioxodibenzo[*c*,*f*]-1,2-thiazepin-11-yl)amino)pentanoic acid

7-Chloro-1,3-dihydro-1-methyl-3-(N-methyl) carbamoyl)-5-phenyl-2*H*-1,4-benzodiazepin-2-one
Formed from: Camazepam
→ 7-Chloro-3-carbamoyl-1,3-dihydro-1-methyl-5-phenyl-2*H*-1,4-benzodiazepin-2-one
Probably in: Rat E57
→ 7-Chloro-1,3-dihydro-3-(N-hydroxy-methyl)carbamoyl)-1-methyl-5-phenyl-2*H*-1,4-benzodiazepin-2-one
Probably in: Rat E57
→ 7-Chloro-1,3-dihydro-3-(N-methyl) carbamoyl)-5-phenyl-2*H*-1,4-benzodiazepin-2-one
Probably in: Rat E57

7-Chloro-4,5-dihydro-5-methyl-3-(5-(1-methylethyl)-1,2,4-oxadiazol-3-yl)-6*H*-imidazo[1,5-*a*][1,4]benzodiazepin-1-one
→ 7-Chloro-4,5-dihydro-3-(5-(1-hydroxy-1-methylethyl)-5-methyl-1,2,4-oxadiazol-3-yl)-6*H*-imidazo[1,5-*a*][1,4]benzodiazepin-1-one
In: Dog G959
Monkey G959
Rat G959

7-Chloro-2,3-dihydro-1-methyl-5-phenyl-1*H*-1,4-benzodiazepine (medazepam)
→ 7-Chloro-1,3-dihydro-1-methyl-5-phenyl-2*H*-1,4-benzodiazepin-2-one
In: Man A3629
→ 7-Chloro-2,3-dihydro-5-phenyl-1*H*-1,4-benzodiazepine
In: Rat A1876
→ 6-Chloro-1-methyl-4-phenyl-2(1*H*)-quinazolinone
In: Rat A282

7-Chloro-1,3-dihydro-1-methyl-5-phenyl-2*H*-1,4-benzodiazepin-2-one (diazepam)
Formed from: 7-Chloro-2,3-dihydro-1-methyl-5-phenyl-1*H*-1,4-benzodiazepine
l-4,5-Dihydrodiazepam
Uxepam
→ 7-Chloro-1,3-dihydro-3-hydroxy-1-methyl-5-phenyl-2*H*-1,4-benzodiazepin-2-one
In: Beauveria G706
Man A719, A2219, A2709, A2899, J93, J299
Mouse A1115
Rat J28
See also: A1266, A2264, A2858
→ 7-Chloro-1,3-dihydro-5-(4-hydroxyphenyl)-1-methyl-2*H*-1,4-benzodiazepin-2-one
In: Beauveria G706
Rat D111, D431
See also: A2858

→ 7-Chloro-1,3-dihydro-5-phenyl-2*H*-1,4-benzodiazepin-2-one
In: Beauveria G706
Cunninghamella D185
Man A719, A1073, A1305, A2212, A2219, A2264, A2709, A2795, A2899, A3012, A3151, A3370, J93
Monkey A1237
Mouse J28
Rat A1266, A1876, C639, D111, D431
See also: A2858, A3006, A3629, B541, E158
→ 6-Chloro-1-methyl-4-phenyl-2(1*H*)-quinazolinone
In: Rat A282
→ 6-Chloro-4-phenyl-2(1*H*)-quinazolinone
In: Rat A282

7-Chloro-1,3-dihydro-3-methyl-5-phenyl-2*H*-1,4-benzodiazepin-2-one
→ 7-Chloro-1,3-dihydro-hydroxy-3-methyl-5-phenyl-2*H*-1,4-benzodiazepin-2-one
In: Rat A1266
→ 6-Chloro-4-phenyl-2(1*H*)-quinazolinone
In: Rat A282

3-Chloro-6,11-dihydro-6-methyl-5,5,11-trioxodibenzo[*c*,*f*]-1,2-thiazepine
Formed from: Tianeptine
→ 3-Chloro-6,11-dihydro-5,5,11-trioxodibenzo[*c*,*f*]-1,2-thiazepine
Probably in: Man F576

8-Chloro-5,10-dihydro-5-oxazolyl-11*H*-dibenzo[*d*,*c*][1,4]diazepin-11-one
Probably formed from: 8-Chloro-5-glyoxylyl-5,10-dihydro-11*H*-dibenzo[*d*,*c*][1,4]diazepin-11-one

7-Chloro-1,3-dihydro-2-oxo-5-phenyl-1*H*-1,4-benzodiazepine-3-carboxylic acid
See: Mendon

5-Chloro-2-(2,3-dihydro-3-oxo-4*H*-1,2,4-triazol-4-yl)benzophenone
Probably formed from: 5-Chloro-2-(4*H*-1,2,4-triazol-4-yl)benzophenone
→ 5-Chloro-2-(2,3-dihydro-3-oxo-4*H*-1,2,4-triazol-4-yl)-2′-hydroxybenzophenone
Probably in: Dog A258
→ 5-Chloro-2-(2,3-dihydro-3-oxo-4*H*-1,2,4-triazol-4-yl)-4′-hydroxybenzophenone
Probably in: Dog A258
→ 5-Chloro-2-(2,3,4,5-tetrahydro-3,5-dioxo-1*H*-1,2,4-triazol-4-yl)benzophenone
Probably in: Dog A258

5-Chloro-2-(2,3-dihydro-3-oxo-4*H*-1,2,4-triazol-4-yl)-2′-hydroxybenzophenone
Probably formed from: 5-Chloro-2-(2,3-dihydro-3-oxo-4*H*-1,2,4-triazol-4-yl)benzophenone
8-Chloro-6-(*p*-hydroxyphenyl)-4*H*-*s*-triazolo[4,3-*a*][1,4]benzodiazepine

5-Chloro-2-(2,3-dihydro-3-oxo-4*H*-1,2,4-triazol-4-yl)-4′-hydroxybenzophenone
Probably formed from: 8-Chloro-2,4-dihydro-6-(*p*-hydroxyphenyl)-1*H*-*s*-triazolo[4,3-*a*][1,4]benzodiazepin-1-one
5-Chloro-2-(2,3-dihydro-3-oxo-4*H*-1,2,4-triazol-4-yl)benzophenone

7-Chloro-2,3-dihydro-5-phenyl-1*H*-1,4-benzodiazepine
Formed from: 7-Chloro-1-cyclopropylmethyl-2,3-dihydro-5-phenyl-1*H*-1,4-benzodiazepine
7-Chloro-2,3-dihydro-1-methyl-5-phenyl-1*H*-1,4-benzodiazepine
→ 7-Chloro-1,3-dihydro-5-phenyl-2*H*-1,4-benzodiazepin-2-one
In: Rat A1876

7-Chloro-1,3-dihydro-5-phenyl-2*H*-1,4-benzodiazepin-2-one
Formed from: Chlorazepate
7-Chloro-1-cyclopropylmethyl-1,3-dihydro-5-phenyl-2*H*-1,4-benzodiazepine-2-one
7-Chloro-2,3-dihydro-1-methyl-5-phenyl-1*H*-1,4-benzodiazepine
7-Chloro-1,3-dihydro-1-methyl-5-phenyl-2*H*-1,4-benzodiazepin-2-one
7-Chloro-2,3-dihydro-5-phenyl-1*H*-1,4-benzodiazepine
Halazepam
Mendon
Pinazepam
Probably formed from: 7-Chloro-2,3-dihydro-1-methyl-5-phenyl-1*H*-1,4-benzodiazepine
→ 7-Chloro-1,3-dihydro-3-hydroxy-5-phenyl-2*H*-1,4-benzodiazepin-2-one
In: Mouse A1115, A3269, J28

Rat A3269
See also: A719, A831, A1305, A3006, A3012, A3150, A3370, A3775, C639, D431, F765, G706
→ 7-Chloro-1,3-dihydro-5-(4-hydroxyphenyl)-2*H*-1,4-benzodiazepin-2-one
Probably in: Monkey A3748
Rat A831, D111

7-Chloro-1,3-dihydro-5-phenyl-2*H*-1,4-benzodiazepin-2-one-4-oxide (demoxepam)
Formed from: Chlordiazepoxide
→ 7-Chloro-1,3-dihydro-9-hydroxy-5-phenyl-2*H*-1,4-benzodiazepin-2-one
In: Dog A579

8-Chloro-2,4-dihydro-6-phenyl-1*H*-*s*-triazol[4,3-*a*][1,4]benzodiazepin-1-one
Formed from: 8-Chloro-6-phenyl-1*H*-*s*-triazol[4,3-*a*][1,4]benzodiazepine
→ 8-Chloro-2,4-dihydro-6-(*p*-hydroxyphenyl)-1*H*-*s*-triazol[4,3-*a*][1,4]benzodiazepin-1-one
In: Dog A258, A303
Guinea pig A303
Mouse A303
Rabbit A303
Rat A258, A301, A303

6-Chloro-3,4-dihydro-7-sulphamoyl-2*H*-1,2,4-benzothiadiazine-1,1-dioxide
See: Hydrochlorthiazide

3-Chloro-6,11-dihydro-5,5,11-trioxodibenzo[*c*,*f*]-1,2-thiazepine
Probably formed from: 3-Chloro-6,11-dihydro-6-methyl-5,5,11-trioxodibenzo[*c*,*f*]-1,2-thiazepine

4′-Chlorodihydrowarfarin
Formed from: 4′-Chlorowarfarin

5-Chloro-2,4-dihydroxyaniline
Probably formed from: 5-Chloro-6-hydroxybenzoxazolin-2-one

1-Chloro-2,4-dihydroxybenzene
See: 4-Chlororesorcinol

3-Chloro-2,5-dihydroxybenzoic acid
See: 3-Chlorogentisic acid

4-Chloro-2,5-dihydroxybenzoic acid
→ 6-Carboxy-3-chloro-4-hydroxy-*cis*,*cis*-muconic semialdehyde
In: Pseudomonas F649

5-Chloro-2,3-dihydroxybenzoic acid
Probably formed from: 5-Chlorosalicylic acid

5-Chloro-3,4-dihydroxybenzoic acid
Formed from: 5-Chloro-4-hydroxy-3-methoxybenzoic acid

6-Chloro-2,3-dihydroxybenzonitrile
Formed from: 2,6-Dichlorobenzonitrile

2′-Chloro-2,3-dihydroxybiphenyl
Formed from: 2,2′-Dichlorobiphenyl

2′-Chloro-3,4-dihydroxybiphenyl
Probably formed from: 2′-Chloro-4-hydroxybiphenyl

2-Chloro-4,4′-dihydroxybiphenyl
Formed from: 2′-Chloro-4-hydroxybiphenyl

2-Chloro-4′,5-dihydroxybiphenyl
Probably formed from: 2′-Chloro-4-hydroxybiphenyl

3-Chloro-4,4′-dihydroxybiphenyl
Probably formed from: 3′-Chloro-4-hydroxybiphenyl

4′-Chloro-2,3-dihydroxybiphenyl
Formed from: 4′-Chloro-2-hydroxybiphenyl

4′-Chloro-2,3-dihydroxybiphenyl
Formed from: 4′-Chloro-2-hydroxybiphenyl
2,4′-Dichlorobiphenyl

4′-Chloro-2,5-dihydroxybiphenyl
Formed from: 4′-Chloro-2-hydroxybiphenyl
4′-Chloro-3-hydroxybiphenyl

4-Chloro-3,4′-dihydroxybiphenyl
Formed from: 4,4′-Dichloro-3-hydroxybiphenyl
Probably formed from: 4′-Chloro-4-hydroxybiphenyl

4′-Chloro-3,4-dihydroxybiphenyl
Formed from: 4′-Chloro-3-hydroxybiphenyl
4′-Chloro-4-hydroxybiphenyl
→ 4′-Chloro-3-hydroxy-4-methoxybiphenyl
In: Rabbit A2053
See also: A247, A2747
→ 4′-Chloro-4-hydroxy-3-methoxybiphenyl
In: Rabbit A2053
See also: A247, A2747

2-Chloro-3,7-dihydroxydibenzofuran
Probably formed from: 2-Chloro-3-hydroxydibenzofuran

4′-Chloro-3,5-dihydroxy-4-methoxybiphenyl
Probably formed from: 4′-Chloro-3-hydroxy-4-methoxybiphenyl

2-Chloro-1,6-dihydroxy-8-methoxy-3-methyl-9,10-anthracenedione
See: 2-Chloroquestin

5-Chloro-7,8-dihydroxy-2-methylquinoline
→ 5-Chloro-7,8-dihydroxy-2-methylquinoline-7-β-D-4-O-methylglucoside
In: Sporotrichum A2630
→ 5-Chloro-7,8-dihydroxy-2-methylquinoline-8-β-D-4-O-methylglucoside
In: Sporotrichum A2630

5-Chloro-7,8-dihydroxy-2-methylquinoline-7-β-D-4-O-methylglucoside
Formed from: 5-Chloro-7,8-dihydroxy-2-methylquinoline

5-Chloro-7,8-dihydroxy-2-methylquinoline-8-β-D-4-O-methylglucoside
Formed from: 5-Chloro-7,8-dihydroxy-2-methylquinoline

7-Chloro-1,2-dihydroxynaphthalene
Formed from: *cis*-7-Chloro-1,2-dihydro-1,2-dihydroxynaphthalene

8-Chloro-1,2-dihydroxynaphthalene
Formed from: *cis*-8-Chloro-1,2-dihydro-1,2-dihydroxynaphthalene

6-Chloro-2,3-dihydroxy-1,4-naphthoquinone
Probably formed from: 6-Chloro-2,3-dimethoxy-1,4-naphthoquinone

3-Chloro-4,5-dihydroxyphenylacetic acid
Probably formed from: 3-Chloro-4-hydroxyphenylacetic acid

4-Chloro-2,3-dihydroxyphenylacetic acid
Formed from: 4-Chloro-2-hydroxyphenylacetic acid

2-(2-Chloro-3,4-dihydroxyphenyl)-2-(*p*-chlorophenyl)acetic acid
Probably formed from: 2-(2-Chloro-3-hydroxyphenyl)-2-(*p*-chlorophenyl)acetic acid
→ 2-(2-Chloro-4-hydroxy-3-methoxyphenyl)-2-(*p*-chlorophenyl)acetic acid
Probably in: Chicken A2368

1-(3-Chloro-4,6-dihydroxyphenyl)-3-(2-(3,3-dimethyl-1-azetidinyl)ethyl)imidazolin-2-one
Probably formed from: 1-(3-Chloro-4-hydroxyphenyl)-3-(2-(3,3-dimethyl-1-azetidinyl)ethyl)imidazolin-2-one
→ 1-(3-Chloro-4,6-dihydroxyphenyl)-3-(2-(3,3-dimethyl-1-azetidinyl)ethyl)imidazolin-2-one-β-D-glucuronide
Probably in: Dog D384
Rat D384, D645

1-(3-Chloro-4,6-dihydroxyphenyl)-3-(2-(3,3-dimethyl-1-azetidinyl)ethyl)imidazolin-2-one-β-D-glucuronide
Probably formed from: 1-(3-Chloro-4,6-dihydroxyphenyl)-3-(2-(3,3-dimethyl-1-azetidinyl)ethyl)imidazolin-2-one

N-(*p*-((5-Chloro-2,4-dihydroxyquinazolin-6-yl)aminomethyl)benzoyl)diglutamylglutamic acid
Probably formed from: N-(*p*-((5-Chloro-2,4-dihydroxyquinazolin-6-yl)aminomethyl)benzoyl)glutamylglutamic acid
→ N-(*p*-((5-Chloro-2,4-dihydroxyquinazolin-6-yl)aminomethyl)benzoyl)triglutamylglutamic acid
Probably in: Pig F760

N-(*p*-((5-Chloro-2,4-dihydroxyquinazolin-6-yl)aminomethyl)benzoyl)glutamic acid
→ N-(*p*-((5-Chloro-2,4-dihydroxyquinazolin-6-yl)aminomethyl)benzoyl)glutamylglutamic acid
In: Pig F760

N-(*p*-((5-Chloro-2,4-dihydroxyquinazolin-6-yl)aminomethyl)benzoyl)glutamylglutamic acid
Probably formed from: N-(*p*-((5-Chloro-2,4-dihydroxyquinazolin-6-yl)aminomethyl)benzoyl)glutamic acid
→ N-(*p*-((5-Chloro-2,4-dihydroxyquinazolin-6-yl)aminomethyl)benzoyl)diglutamylglutamic acid
Probably in: Pig F760

N-(*p*-((5-Chloro-2,4-dihydroxyquinazolin-6-yl)aminomethyl)benzoyl)triglutamylglutamic acid
Probably formed from: N-(*p*-((5-Chloro-2,4-dihydroxyquinazolin-6-yl)aminomethyl)benzoyl)diglutamylglutamic acid

N-(*p*-((5-Chloro-2,4-dihydroxyquinazolin-6-yl)methylamino)benzoyl)diglutamylglutamic acid
Probably formed from: N-(*p*-((5-Chloro-2,4-dihydroxyquinazolin-6-yl)methylamino)benzoyl)glutamylglutamic acid

→ N-(*p*-((5-Chloro-2,4-dihydroxyquinazolin-6-yl)methylamino)benzoyl)triglutamylglutamic acid
Probably in: Pig F760

N-(*p*-((5-Chloro-2,4-dihydroxyquinazolin-6-yl)methylamino)benzoyl)glutamic acid
→ N-(*p*-((5-Chloro-2,4-dihydroxyquinazolin-6-yl)methylamino)benzoyl) glutamylglutamic acid
In: Pig F760

N-(*p*-((5-Chloro-2,4-dihydroxyquinazolin-6-yl)methylamino)benzoyl)glutamylglutamic acid
Probably formed from: N-(*p*-((5-Chloro-2,4-dihydroxyquinazolin-6-yl)methylamino) benzoyl)glutamic acid
→ N-(*p*-((5-Chloro-2,4-dihydroxyquinazolin-6-yl)methylamino)benzoyl) diglutamylglutamic acid
Probably in: Pig F760

N-(*p*-((5-Chloro-2,4-dihydroxyquinazolin-6-yl)methylamino)benzoyl) tetraglutamylglutamic acid
Probably formed from: N-(*p*-((5-Chloro-2,4-dihydroxyquinazolin-6-yl)methylamino) benzoyl)triglutamylglutamic acid

N-(*p*-((5-Chloro-2,4-dihydroxyquinazolin-6-yl)methylamino)benzoyl)triglutamylglutamic acid
Probably formed from: N-(*p*-((5-Chloro-2,4-dihydroxyquinazolin-6-yl)methylamino) benzoyl)diglutamylglutamic acid
→ N-(*p*-((5-Chloro-2,4-dihydroxyquinazolin-6-yl)methylamino)benzoyl)tetraglutamylglutamic acid
Probably in: Pig F760

3′-Chloro-2′,5-dihydroxy-3,7,8-trimethoxyflavone
See: Chloroflavonin

1-Chloro-3,4-dimethoxybenzene
See: 4-Chloroveratrole

2-Chloro-3,4-dimethoxybenzoic acid
→ 2-Chloro-4-hydroxy-3-methoxybenzoic acid
In: Polyporus A1117

5-Chloro-3,4-dimethoxybenzoic acid
→ 5-Chloro-3-hydroxy-4-methoxybenzoic acid
In: Polyporus A1117

6-Chloro-2,3-dimethoxy-1,4-naphthalenediol diacetate
See: Lonapalene

6-Chloro-2,3-dimethoxy-1,4-naphthoquinone
Formed from: Lonapalene
→ 6-Chloro-2,3-dihydroxy-1,4-naphthoquinone
Probably in: Rat G175

3′-Chloro-4-dimethylaminoazobenzene
→ 3′-Chloro-4-methylaminoazobenzene
In: Rat A776
→ 3-Chloro-4′-dimethylamino-4-hydroxyazobenzene
In: Rat A776

4′-Chloro-4-dimethylaminoazobenzene
→ 4′-Chloro-4-methylaminoazobenzene
In: Rat A1979
→ 4′-Dimethylamino-4-hydroxyazobenzene
In: Rat A1979

2-Chloro-11-(2-dimethylaminoethoxy) dibenzo[*b*,*f*]thiepine
See: Zotepine

2-Chloro-α-(2-(dimethylamino)ethyl)-α-phenylbenzenemethanol
See: Clofedanol

7-Chloro-4-(3′-diethylamino-4′-hydroxyanilino)quinoline
See: Amodiaquine

3-Chloro-4′-dimethylamino-4-hydroxyazobenzene
Formed from: 3′-Chloro-4-dimethylaminoazobenzene

3-Chloro-2-dimethylamino-1,4-naphthoquinone
→ 3-Chloro-2-methylamino-1,4-naphthoquinone
In: Rat D663

2-Chloro-10-(3-dimethylaminopropyl) phenothiazine
See: Chlorpromazine

***p*-Chloro-N,N-dimethylaniline**
→ *p*-Chloro-N-methylaniline
In: Rabbit F839
Rat D578

***p*-Chloro-N,N-dimethylbenzamide**
→ *p*-Chloro-N-methylbenzamide
In: Rat G13, G683

4-Chloro-α,α-dimethylbenzeneethanamine
See: Chlorphentermine

7-Chloro-3-dimethylcarbamoyloxy-1-methyl-5-phenyl-1,3-dihydro-2*H*-1,4-benzodiazepin-2-one
See: Camazepam

***p*-Chloro-α,α-dimethylphenethylamine**
(chlorphentermine)
→ 2-(*p*-Chlorophenyl)-1,1-dimethylethylhydroxylamine
In: Guinea pig A177
Man A177
Monkey A177
Rabbit A177
→ 1-(*p*-Chlorophenyl)-2-hydroxylamino-2-methylpropane
In: Man A1258
Rabbit A21, A1166
→ 1-(*p*-Chlorophenyl)-2-methyl-2-nitropropane
In: Man A177, A1258
Rabbit A21, A177

***p*-Chloro-N,N-dimethylphenoxyethylamine**
→ *p*-Chloro-N-methylphenoxyethylamine
In: Rat A2421

1-Chloro-2,4-dinitrobenzene
→ S-(2,4-Dinitrophenyl)glutathione
In: Ascaris A3038
Guinea pig G352
Man J599
Moniezia A3038
Mytilus G689
Rabbit C878
Rat G352, H516, J583
See also: A2974, A3509, A3953, B69, B702, C685, C901, D303, E434, J102
→ 2-Chloro-5-nitroaniline
In: Mycobacterium J916
→ 4-Chloro-3-nitroaniline
In: Mycobacterium J916

4-Chlorodioxindole
Formed from: 4-Chloroindole-3-acetic acid

5-Chlorodioxindole-3-acetic acid
Formed from: 5-Chloroindole-3-acetic acid

2-(4-((2-Chloro-1,2-diphenylethenyl)phenoxy)-N,N-diethylethanamine
See: Clomiphene

2-(4-((1*E*)-2-Chloro-1,2-diphenylethenyl)phenoxy)-N,N-diethylethanamine
See: Enclomiphene

2-Chloroemodin
→ 2-Chloroquestin
In: Aspergillus C334

1,1′,1″-(1-Chloro-4-ethenyl-2-ylidene)tris(4-methoxybenzene)
See: Chlorotrianisene

2-Chloro-1-(3-ethoxy-4-nitrophenoxy)-4-trifluoromethylbenzene
See: Oxyfluorofen

***p*-(2-Chloroethylamino)benzoic acid**
Probably formed from: *p*-(N-(Carboxymethyl)-N-(2-chloroethyl)amino)benzoic acid

7-Chloro-1-(2-ethylaminoethyl)-5-(2-fluorophenyl)-1,3-dihydro-2*H*-1,4-benzodiazepin-2-one
Formed from: 7-Chloro-1-(2-diethylaminoethyl)-5-(2-fluorophenyl)-1,3-dihydro-2*H*-1,4-benzodiazepin-2-one
→ 1-(2-Aminoethyl)-7-chloro-5-(2-fluorophenyl)-1,3-dihydro-2*H*-1,4-benzodiazepin-2-one
In: Rat A873

7-Chloro-4-(3-ethylamino-4-fluoroanilino)quinoline
Formed from: 7-Chloro-4-(3-diethylamino-4-fluoroanilino)quinoline

2-(2-Chloroethyl)amino-5-nitroacetanilide
Probably formed from: 2-(2-Chloroethyl)amino-5-nitroaniline

2-(2-Chloroethyl)amino-5-nitroaniline
Probably formed from: 2-(1-Aziridinyl)-5-nitroaniline
→ 2-(2-Chloroethyl)amino-5-nitroacetanilide
Probably in: Rat A2503

4-(2-Chloroethyl)aminophenol
Formed from: N-(2-Chloroethyl)aniline

***p*-(2-Chloroethylamino)phenylacetic acid**
Probably formed from: Chlorambucil

7-Chloro-4-(3-ethylamino)propylamino)quinoline
Formed from: Chloroquine
→ 4-(3-Aminopropylamino)-7-chloroquinoline
Probably in: Man C239

p-Chloro-N-ethylaniline
Formed from: p-Chloro-N-ethyl-N-methylaniline

N-(2-Chloroethyl)aniline
→ 4-(2-Chloroethyl)aminophenol
In: Rat A2929
Species unclear A2928

p-Chloro-N-ethylbenzamide
Formed from: p-Chloro-N-ethyl-N-methylbenzamide
→ p-Chlorobenzoic acid
In: Rat H475

1-Chloro-4-ethylbenzene
→ 1-(p-Chlorophenyl)ethanol
In: Mortierella D969
→ 2-(p-Chlorophenyl)ethanol
In: Mortierella D969

6-(2-Chloroethyl)-1,4-dihydroxy-2-hydroxymethyl-2,5,7-trimethylindane
Formed from: Illudin S

N-(2-Chloroethyl)-2,4-dinitroaniline
→ S-(N-(2,4-Dichlorophenyl)-2-aminoethyl)mercapturic acid
Probably in: Rat A2503

N-(2-Chloroethyl)-2,6-dinitro-N-propyl-4-(trifluoromethyl)benzenamine
See: Fluchloralin

1-(2-Chloroethyl)-2-ethyl-7-nitro-5-trifluoromethylbenzimidazole
Formed from: Fluchloralin

p-(N-(2-Chloroethyl)-N-(2-hydroxyethyl)amino)benzoic acid
Probably formed from: Chlorambucil
→ p-(N-Carboxymethyl-N-(2-chloroethyl)amino)benzoic acid
Probably in: Rat A2041

7-Chloro-1-(2-((N-ethyl-N-(2-hydroxyethyl))amino)ethyl)-5-(2-fluorophenyl)-1,3-dihydro-2*H*-1,4-benzodiazepin-2-one
Formed from: 7-Chloro-1-(2-diethylaminoethyl)-5-(2-fluorophenyl)-1,3-dihydro-2*H*-1,4-benzodiazepin-2-one

7-Chloro-4-(3-(ethyl(hydroxyethyl)amino)-4-fluoroanilino)quinoline
→ 7-Chloro-4-(3-(ethyl(hydroxyethyl)amino)-4-fluoroanilino)quinoline-β-D-glucuronide
In: Rat H159

7-Chloro-4-(3-(ethyl(hydroxyethyl)amino)-4-fluoroanilino)quinoline-β-D-glucuronide
Formed from: 7-Chloro-4-(3-(ethyl(hydroxyethyl)amino)-4-fluoroanilino)quinoline

7-Chloro-4-(3-(ethyl(hydroxyethyl)amino)-5-(S-glutathionyl)-4-hydroxyanilino)quinoline
Formed from: 7-Chloro-4-(3-(ethyl(hydroxyethyl)amino)-4-hydroxyanilino)quinoline

7-Chloro-4-(3-(ethyl(hydroxyethyl)amino)-4-hydroxyanilino)quinoline
→ 7-Chloro-4-(3-(ethyl(hydroxyethyl)amino)-5-(S-glutathionyl)-4-hydroxyanilino)quinoline
In: Rat H159
→ 7-Chloro-4-(3-(ethyl(hydroxyethyl)amino)-4-hydroxyanilino)quinoline-β-D-glucuronide
In: Rat H159

7-Chloro-4-(3-(ethyl(hydroxyethyl)amino)-4-hydroxyanilino)quinoline-β-D-glucuronide
Formed from: 7-Chloro-4-(3-(ethyl(hydroxyethyl)amino)-4-hydroxyanilino)quinoline

2-Chloro-N-(2-ethyl-6-hydroxymethylphenyl)-N-(2-hydroxy-1-methylethyl)acetamide
Probably formed from: 2-Chloro-N-(2-ethyl-6-methylphenyl)-N-(2-hydroxy-1-methylethyl)acetamide
→ 2-Chloro-N-(2-ethyl-6-hydroxymethylphenyl)-N-(2-hydroxy-1-methylethyl)acetamide-β-D-glucuronide
In: Bluegill sunfish G779

2-Chloro-N-(2-ethyl-6-hydroxymethylphenyl)-N-(2-hydroxy-1-methylethyl)acetamide-β-D-glucuronide
Probably formed from: 2-Chloro-N-(2-ethyl-6-hydroxymethylphenyl)-N-(2-hydroxy-1-methylethyl)acetamide

2-Chloro-N-(2-ethyl-6-hydroxymethylphenyl)-N-(2-methoxy-1-methylethyl)acetamide
Formed from: Metolachlor
→ 2-Chloro-N-(2-ethyl-6-hydroxymethylphenyl)-N-(2-methoxy-1-methylethyl)acetamide-β-D-glucuronide
In: Bluegill sunfish G779
→ 2-Chloro-N-(2-(1-hydroxy)ethyl-6-hydroxymethylphenyl)-N-(2-methoxy-1-methylethyl)acetamide

Probably in: Rhizoctonia F871
Syncephalastrum F871

2-Chloro-N-(2-ethyl-6-hydroxymethylphenyl)-N-(2-methoxy-1-methylethyl)acetamide-β-D-glucuronide
Probably formed from: 2-Chloro-N-(2-ethyl-6-hydroxymethylphenyl)-N-(2-methoxy-1-methylethyl)acetamide

***p*-Chloro-N-ethyl-N-methylaniline**
→ *p*-Chloro-N-ethylaniline
In: Guinea pig C569
Hamster C569
Mouse C569
Rabbit C569
Rat C569
→ *p*-Chloro-N-ethyl-N-methylaniline-N-oxide
In: Guinea pig C569
Hamster C569
Mouse C569
Rabbit C569
Rat C569
→ *p*-Chloro-N-methylaniline
In: Guinea pig C569
Hamster C569
Mouse C569
Rabbit C569
Rat C569

***p*-Chloro-N-ethyl-N-methylaniline-N-oxide**
Formed from: *p*-Chloro-N-ethyl-N-methylaniline

***p*-Chloro-N-ethyl-N-methylbenzamide**
→ *p*-Chloro-N-ethylbenzamide
In: Rat G13
→ *p*-Chloro-N-methylbenzamide
In: Rat G13

N-(2-Chloroethyl)-N-(1-methyl-2-phenoxyethyl)benzenemethanamine
See: Phenoxybenzamine

2-Chloro-N-(2-ethyl-6-methylphenyl)acetamide
Formed from: Metolachlor

2-Chloro-N-(2-ethyl-6-methylphenyl)-N-(2-hydroxy-1-methylethyl)acetamide
Formed from: Metolachlor
→ 2-Chloro-N-(2-ethyl-6-methylphenyl)-N-(2-hydroxy-1-methylethyl)acetamide-β-D-glucuronide
Probably in: Bluegill sunfish G779
→ 2-Chloro-N-(2-ethyl-6-hydroxymethylphenyl)-N-(2-hydroxy-1-methylethyl)acetamide
Probably in: Bluegill sunfish G779
Cunninghamella H847

2-Chloro-N-(2-ethyl-6-methylphenyl)-N-(2-hydroxy-1-methylethyl)acetamide-β-D-glucuronide
Probably formed from: 2-Chloro-N-(2-ethyl-6-methylphenyl)-N-(2-hydroxy-1-methylethyl) acetamide

2-Chloro-N-(2-ethyl-6-methylphenyl)-N-(2-methoxy-1-methylethyl)acetamide
See: Metolachlor

3-(1-(*p*-(2-Chloroethyl)phenyl)butyl)-4-hydroxycoumarin
See: Clocoumarol

6-(3-(*p*-(2-Chloroethyl)phenyl) 1-hydroxy-2-oxohexylidene)-2,4-cyclohexadien-1-one
Formed from: Clocoumarol

2-(*p*-(2-Chloroethyl)phenyl)-1-hydroxypentyl *o*-hydroxyphenyl ketone
Formed from: Clocoumarol
→ 2-(*p*-(2-Chloroethyl)phenyl)pentyl *o*-hydroxyphenyl ketone
Probably in: Rat A3928

2-(*p*-(2-Chloroethyl)phenyl)pentyl *o*-hydroxyphenyl ketone
Probably formed from: 2-(*p*-(2-Chloroethyl)phenyl)-1-hydroxypentyl *o*-hydroxyphenyl ketone

Chloroflavonin (3′-chloro-2′,5-dihydroxy-3,7,8-trimethoxyflavone)
Formed from: Dechlorochloroflavonin

3-Chloro-4-fluoroaniline
→ 2-Amino-4-chloro-5-fluorophenol
In: Dog A3942
Rat A3942

***m*-Chlorofluorobenzene**
→ 3-Chlorocatechol
In: Pseudomonas F76
→ 4-Chloro-2-fluorophenol
In: Rat G953

3-Chloro-4-fluorobenzoic acid
→ *p*-Fluorobenzoic acid
In: Corynebacterium J178
→ 4-Fluorocatechol
Probably in: Pseudomonas G434

2-Chloro-6-fluorobenzyl alcohol
Formed from: 6-Amino-9-(2-chloro-6-fluorobenzyl)purine

3-Chloro-4-fluorocatechol
Formed from: 3-Chloro-4-fluorophenol

4-Chloro-3-fluorocatechol
Formed from: 4-Chloro-3-fluorophenol

4-Chloro-5-fluorocatechol
Formed from: 3-Chloro-4-fluorophenol
4-Chloro-3-fluorophenol

N-(4-Chloro-2-fluoro-5-hydroxyphenyl)-1-sulpho-1,2-cyclohexanedicarboximide
Probably formed from: N-(4-Chloro-2-fluoro-5-((1-methyl-2-propynyl)oxy)phenyl)-1-sulpho-1,2-cyclohexanedicarboximide

3-Chloro-2-fluoro-4-methoxybenzaldehyde
Probably formed from: 2-Fluoro-4-methoxybenzaldehyde
→ 3,5-Dichloro-2-fluoro-4-methoxybenzaldehyde
Probably in: Bjerkandera H914

3-Chloro-5-fluoro-4-methoxybenzaldehyde
Probably formed from: 3-Fluoro-4-methoxybenzaldehyde

N-(4-Chloro-2-fluoro-5-((1-methyl-2-propynyl)oxy)phenyl)-1-sulpho-1,2-cyclohexanedicarboximide
Formed from: N-(4-Chloro-2-fluoro-5-((1-methyl-2-propynyl)oxy)phenyl)-3,4,5,6-tetrahydrophthalimide
→ N-(4-Chloro-2-fluoro-5-hydroxyphenyl)-1-sulpho-1,2-cyclohexanedicarboximide
Probably in: Rat G568

N-(4-Chloro-2-fluoro-5-((1-methyl-2-propynyl)oxy)phenyl)-3,4,5,6-tetrahydrophthalimide
→ 5-Amino-2-chloro-4-fluorophenol
In: Rat G568, G704
→ N-(4-Chloro-2-fluoro-5-((1-methyl-2-propynyl)oxy)phenyl)-1-sulpho-1,2-cyclohexanedicarboximide
In: Rat G568

3-Chloro-4-fluorophenol
→ 3-Chloro-4-fluorocatechol
In: Trichosporon H420
→ 4-Chloro-5-fluorocatechol
In: Trichosporon H420
→ 3-Chloro-4-fluorophenyl-β-D-glucuronide
In: Rat H468, H522
→ 3-Chloro-4-fluorophenyl sulphate
In: Rat H468, H522

4-Chloro-2-fluorophenol
Formed from: *m*-Chlorofluorobenzene
→ 4-Chloro-2-fluorophenyl-β-D-glucuronide
In: Rat G953
→ 4-Chloro-2-fluorophenyl sulphate
In: Rat G953

4-Chloro-3-fluorophenol
→ 4-Chloro-3-fluorocatechol
In: Trichosporon H420
→ 4-Chloro-5-fluorocatechol
In: Trichosporon H420
→ 4-Chloro-3-fluorophenyl-β-D-glucuronide
In: Rat H468, H522
→ 4-Chloro-3-fluorophenyl sulphate
In: Rat H468, H522

7-Chloro-5-(*o*-fluorophenyl)-1,3-dihydro-2*H*-1,4-benzodiazepin-2-one
Formed from: 7-Chloro-1-cyclopropylmethyl-5-(*o*-fluorophenyl)-1,3-dihydro-2*H*-1,4-benzodiazepin-2-one
Probably formed from: 7-Chloro-5-(*o*-fluorophenyl)-1,3-dihydro-1-(2,2,2-trifluoroethyl)-2*H*-1,4-benzodiazepin-2-one
→ 7-Chloro-5-(*o*-fluorophenyl)-1,3-dihydro-2*H*-1,4-benzodiazepin-2-one-β-D-glucuronide
Probably in: Man D336

7-Chloro-5-(*o*-fluorophenyl)-1,3-dihydro-2*H*-1,4-benzodiazepin-2-one-β-D-glucuronide
Probably formed from: 7-Chloro-5-(*o*-fluorophenyl)-1,3-dihydro-2*H*-1,4-benzodiazepin-2-one

7-Chloro-5-(*o*-fluorophenyl)-1,3-dihydro-3-hydroxy-2*H*-1,4-benzodiazepin-2-one
Probably formed from: 7-Chloro-5-(*o*-fluorophenyl)-1,3-dihydro-3-hydroxy-1-(2,2,2-trifluoroethyl)-2*H*-1,4-benzodiazepin-2-one

7-Chloro-5-(*o*-fluorophenyl)-1,3-dihydro-3-hydroxy-1-(2,2,2-trifluoroethyl)-2*H*-1,4-benzodiazepin-2-one
Probably formed from: 7-Chloro-5-(*o*-fluorophenyl)-1,3-dihydro-1-(2,2,2-trifluoroethyl)-2*H*-1,4-benzodiazepin-2-one

→ 7-Chloro-5-(*o*-fluorophenyl)-1,3-dihydro-3-hydroxy-2*H*-1,4-benzodiazepin-2-one
Probably in: Hamster D392
Man D336
Mouse D843
→ 7-Chloro-5-(*o*-fluorophenyl)-1,3-dihydro-3-hydroxy-1-(2,2,2-trifluoroethyl)-2*H*-1,4-benzodiazepin-2-one-β-D-glucuronide
Probably in: Man D336

7-Chloro-5-(*o*-fluorophenyl)-1,3-dihydro-3-hydroxy-1-(2,2,2-trifluoroethyl)-2*H*-1,4-benzodiazepin-2-one-β-D-glucuronide
Probably formed from: 7-Chloro-5-(*o*-fluorophenyl)-1,3-dihydro-3-hydroxy-1-(2,2,2-trifluoroethyl)-2*H*-1,4-benzodiazepin-2-one

7-Chloro-5-(*o*-fluorophenyl)-2,3-dihydro-2-oxo-1*H*-1,4-benzodiazepine-1-aldehyde
Formed from: 7-Chloro-1-(2-diethylaminoethyl)-5-(*o*-fluorophenyl)-1,3-dihydro-2*H*-1,4-benzodiazepin-2-one

7-Chloro-5-(2-fluorophenyl)-2,3-dihydro-2-oxo-1*H*-1,4-benzodiazepine
See: Decarboxyloflazepate

7-Chloro-5-(2-fluorophenyl)-2,3-dihydro-2-oxo-1*H*-1,4-benzodiazepine-3-carboxylic acid
See: Loflazepate

7-Chloro-5-(*o*-fluorophenyl)-1,3-dihydro-1-(2,2,2-trifluoroethyl)-2*H*-1,4-benzodiazepin-2-one
Formed from: Quazepam
→ 7-Chloro-5-(*o*-fluorophenyl)-1,3-dihydro-3-hydroxy-1-(2,2,2-trifluoroethyl)-2*H*-1,4-benzodiazepin-2-one
Probably in: Hamster D392
Man D336
Mouse D392, D843
→ 7-Chloro-5-(*o*-fluorophenyl)-1,3-dihydro-2*H*-1,4-benzodiazepin-2-one
Probably in: Hamster D392
Man D336, D492
Mouse D392, D843

7-Chloro-5-(*o*-fluorophenyl)-1,3-dihydro-1-(2,2,2-trifluoroethyl)-2*H*-1,4-benzodiazepine-2-thione
See: Quazepam

N-(3-Chloro-4-fluorophenyl)-N-(3,4-dihydroxybenzoyl)-2-aminopropionic acid
Probably formed from: N-(3-Chloro-4-fluorophenyl)-N-(*p*-hydroxybenzoyl)-2-aminopropionic acid

3-Chloro-4-fluorophenyl-β-D-glucuronide
Formed from: 3-Chloro-4-fluorophenol

4-Chloro-2-fluorophenyl-β-D-glucuronide
Formed from: 4-Chloro-2-fluorophenol

4-Chloro-3-fluorophenyl-β-D-glucuronide
Formed from: 4-Chloro-3-fluorophenol

N-(3-Chloro-4-fluorophenyl)-N-(*m*-hydroxybenzoyl)-2-aminopropionic acid
Probably formed from: Isopropyl N-(3-chloro-4-fluorophenyl)-N-(*m*-hydroxybenzoyl)-2-aminopropionate

N-(3-Chloro-4-fluorophenyl)-N-(*p*-hydroxybenzoyl)-2-aminopropionic acid
Probably formed from: Isopropyl N-(3-chloro-4-fluorophenyl)-N-(*p*-hydroxybenzoyl)-2-aminopropionate
→ N-(3-Chloro-4-fluorophenyl)-N-(3,4-dihydroxybenzoyl)-2-aminopropionic acid
Probably in: Dog A2035
Rat A2035

7-Chloro-5-(2-fluorophenyl)-1-(2-hydroxyethyl)-1,3-dihydro-2*H*-1,4-benzodiazepin-2-one
Formed from: 7-Chloro-1-(2-diethylaminoethyl)-5-(2-fluorophenyl)-1,3-dihydro-2*H*-1,4-benzodiazepin-2-one
→ 7-Chloro-5-(2-fluorophenyl)-3-hydroxy-1-(2-hydroxyethyl)-1,3-dihydro-2*H*-1,4-benzodiazepin-2-one
In: Mouse A2317
Rat A2317

7-Chloro-5-(2-fluorophenyl)-3-hydroxy-1-(2-hydroxyethyl)-1,3-dihydro-2*H*-1,4-benzodiazepin-2-one
Probably formed from: 7-Chloro-5-(2-fluorophenyl)-1-(2-hydroxyethyl)-1,3-dihydro-2*H*-1,4-benzodiazepin-2-one

8-Chloro-6-(*o*-fluorophenyl)-4-hydroxy-1-hydroxymethyl-4*H*-imidazo[1,5-*a*][1,4]benzodiazepine
Formed from: 8-Chloro-6-(*o*-fluorophenyl)-1-hydroxymethyl-4*H*-imidazo[1,5-*a*][1,4]benzodiazepine
4-Hydroxymidazolam

8-Chloro-6-(*o*-fluorophenyl)-1-hydroxymethyl-4*H*-imidazo[1,5-*a*][1,4]benzodiazepine (1′-hydroxymidazolam)
Formed from: Midazolam
→ 8-Chloro-6-(*o*-fluorophenyl)-4-hydroxy-1-hydroxymethyl-4*H*-imidazo[1,5-*a*][1,4] benzodiazepine
In: Man J43
Rat J43
→ 1′-Hydroxymidazolam-β-D-glucuronide
Probably in: Man J225

5-Chloro-1-(*p*-fluorophenyl)-3-(piperidin-4-yl) indole
Formed from: Sertindole

6-Chloro-4-(*o*-fluorophenyl)-2(1*H*)-quinazolinone
Formed from: Ethyl loflazepate

3-Chloro-4-fluorophenyl sulphate
Formed from: 3-Chloro-4-fluorophenol

4-Chloro-2-fluorophenyl sulphate
Formed from: 4-Chloro-2-fluorophenol

4-Chloro-3-fluorophenyl sulphate
Formed from: 4-Chloro-3-fluorophenol

***p*-Chloroformanilide**
Formed from: *p*-Chloroaniline
Diflubenzuron

3-Chloro-2-formylmethoxybiphenyl
Formed from: 2-(2-Aminoethoxy)-3-chlorobiphenyl
→ 2-Carboxymethoxy-3-chlorobiphenyl
In: Rat A1890
→ 3-Chloro-2-(2-hydroxyethyoxy)biphenyl
In: Rat A1890

6-Chloro-5-formylmethyloxindole
Formed from: Ziprasidone
→ 6-Chlorooxindol-5-ylacetic acid
Probably in: Man J279, K470

***p*-Chloro-α-formyl-α-phenylbenzyl alcohol**
Probably formed from: *p*-Chloro-α-hydroxymethyl-α-phenylbenzyl alcohol
→ α-(*p*-Chlorophenyl)mandelic acid
Probably in: Rat F448

N-(3-Chloro-2-formylphenyl)anthranilic acid
Probably formed from: N-(3-Chloro-2-hydroxymethylphenyl)anthranilic acid
→ N-(2-Carboxy-3-chlorophenyl)anthranilic acid
Probably in: man J161

Chlorogenic acid (caffeoylquinic acid)
Formed from: Caffeoyl CoA
Caffeoyl-D-glucose
5′-*p*-Coumaroylquinic acid
→ Caffeic acid
In: Aspergillus K909, K911
→ Caffeoyl CoA
In: Potato A3145, A3755
Tomato A3149
→ 2-O-Caffeoylgalactarate
In: Tomato K721
→ 2-O-Caffeoylglucarate
In: Tomato K721
→ Chlorogenic acid quinone
Probably in: Apple A3974
Artichoke F664
Avocado A1636
Banana A218, A3974
Botrytis A1930
Coriolus A3974
Glomerella A3974
Grape E458
Ipomoea J822
Mango A3114
Mushroom A3974
Olive A2360
Pear A1007
Potato A3190
Sorghum A2433
Tramates A3974
Vaccinium H948
→ Chlorogenic acid-O-sulphate
In: Eubacterium G475
→ Feruloylquinic acid
In: Nicotiana B490
Parsley F160
→ Isochlorogenic acid
In: Ipomoea F179, F180

Chlorogenic acid quinone
Formed from: Chlorogenic acid

Chlorogenic acid-O-sulphate
Formed from: Chlorogenic acid

3-Chlorogentisic acid (3-chloro-2,5-dihydroxybenzoic acid)
→ 6-Chloro-4-hydroxy-2-oxohepta-3,5-dienoic acid

In: Rhodococcus H441

6-Chloro-2-(S-glutathionyl)-3-hydroxybenzonitrile
Formed from: 2,6-Dichlorobenzonitrile
→ 6-Chloro-2-(S-cysteinylglycyl)-3-hydroxybenzonitrile
Probably in: Chicken F302

(*S*)-6-Chloro-4-(2-(S-glutathionyl)-4-hydroxy-3-oxo-1-buten-1-yl)-1,4-dihydro-4-trifluoromethyl-2*H*-3,1-benzoxazin-2-one
Formed from: Efavirenz

2-Chloro-6-(S-glutathionyl)quinol
Formed from: 2,6-Dichloroquinol
→ Chloroquinol
In: Phanerochaete K138

4′-Chloro-4-glycollamidostilbene
Formed from: 4-Acetamido-4′-chlorostilbene

***p*-Chloroglycollanilide**
Probably formed from: *p*-Chloroacetanilide
→ *p*-Chlorooxanilic acid
Probably in: Man F731
Monkey F279, F731
Mouse F279, F731
Rat F279, F731

8-Chloro-5-glycollyl-5,10-dihydro-11*H*-dibenzo[*d*,*c*][1,4]diazepin-11-one
Formed from: 8-Chloro-5,10-dihydro-5-(bis(2-hydroxyethyl)aminoacetyl)-11*H*-dibenzo[*d*,*c*][1,4]diazepin-11-one
→ 8-Chloro-5-glyoxylyl-5,10-dihydro-11*H*-dibenzo[*d*,*c*][1,4]diazepin-11-one
Probably in: Rat F833

8-Chloro-5-glyoxylyl-5,10-dihydro-11*H*-dibenzo[*d*,*c*][1,4]diazepin-11-one
Probably formed from: 8-Chloro-5-glycollyl-5,10-dihydro-11*H*-dibenzo[*d*,*c*,][1,4]diazepin-11-one
→ 8-Chloro-5-oxalyl-5,10-dihydro-11*H*-dibenzo[*d*,*c*][1,4]diazepin-11-one
Probably in: Rat F833

5-Chloro-4-guanidino-2,1,3-benzothiadiazole
Formed from: Sirdalud

(6a*S*-*trans*)-11-Chloro-6,6a,7,8,9,13b-hexahydro-12-hydroxy-7-methyl-5*H*-benzo[*d*]naphth[2,1-*b*]azepine
→ (6a*S*-*trans*)-11-Chloro-6,6a,7,8,9,13b-hexahydro-12-hydroxy-7-methyl-5*H*-benzo[*d*]naphth[2,1-*b*]azepine-β-D-glucuronide
In: Man G960
Monkey G671
Rat G671

(6a*S*-*trans*)-11-Chloro-6,6a,7,8,9,13b-hexahydro-12-hydroxy-7-methyl-5*H*-benzo[*d*]naphth[2,1-*b*]azepine-β-D-glucuronide
Formed from: (6a*S*-*trans*)-11-Chloro-6,6a,7,8,9,13b-hexahydro-12-hydroxy-7-methyl-5*H*-benzo[*d*]naphth[2,1-*b*]azepine

***o*-Chlorohippuric acid**
Formed from: *o*-Chlorobenzoic acid

***m*-Chlorohippuric acid**
Formed from: *m*-Chlorobenzoic acid

***p*-Chlorohippuric acid**
Formed from: *p*-Chlorobenzoic acid
→ *p*-Chlorobenzoic acid
In: Corynebacterium E145

***p*-Chlorohydracrylanilide**
Formed from: *p*-Chloropropionanilide

2-Chlorohydroquinone
See: Chloroquinol

4′-Chloro-N-hydroxy-4-acetamidostilbene
Formed from: 4-Acetamido-4′-chlorostilbene

***p*-Chloro-N-hydroxyacetanilide**
Formed from: *p*-Chloroacetanilide
→ 4,4′-Bisazoxychlorobenzene
In: Rabbit B178
Rat B178
→ *p*-Chloroacetanilide
In: Rabbit B178
Rat B178
→ *p*-Chloro-N-hydroxyacetanilide-β-D-glucuronide
In: Rabbit C588
Rat A1840
→ *p*-Chloro-N-hydroxyacetanilide-O-sulphate
In: Rat A1840
→ *p*-Chloronitrosobenzene
In: Rabbit B178
Rat B178, E249

***p*-Chloro-N-hydroxyacetanilide-β-D-glucuronide**
Formed from:*p*-Chloro-N-hydroxyacetanilide

***p*-Chloro-N-hydroxyacetanilide-O-sulphate**
Formed from: *p*-Chloro-N-hydroxyacetanilide

3-Chloro-4-hydroxyamphetamine
Formed from: *p*-Chloroamphetamine
***p*-Chloro-N-hydroxyaniline**
See: *p*-Chlorophenylhydroxylamine
4-Chloro-3-hydroxyanthranilic acid
Formed from: 6-Chlorotryptophan
3-Chloro-4-hydroxybenzaldehyde
→ 3-Chloro-4-hydroxybenzoic acid
In: Microorganism E664
→ 3-Chloro-4-hydroxybenzyl alcohol
In: Microorganism E664
2-Chloro-4′-hydroxybenzanilide
Formed from: 2-Chlorobenzanilide
2-Chloro-4-hydroxybenzoic acid
Formed from: *o*-Chlorobenzoic acid
→ Chloroquinol
In: Candida J461
2-Chloro-5-hydroxybenzoic acid
→ *m*-Hydroxybenzoic acid
In: Microorganism K269
3-Chloro-2-hydroxybenzoic acid
See: 3-Chlorosalicylic acid
3-Chloro-4-hydroxybenzoic acid
Formed from: *m*-Chlorobenzoic acid
3-Chloro-4-hydroxybenzaldehyde
Formed from: 3,5-Dichloro-4-hydroxybenzoic acid
→ *o*-Chlorophenol
In: Microorganism G632
See also: G291
→ Chloroquinol
In: Candida J461
→ *p*-Hydroxybenzoic acid
In: Desulfitobacterium J232
Desulfomicrobium J491
4-Chloro-2-hydroxybenzoic acid
See: 4-Chlorosalicylic acid
5-Chloro-2-hydroxybenzoic acid
See: 5-Chlorosalicylic acid
4′-Chloro-4-hydroxybenzophenone
Probably formed from: *p*-Chlorobenzophenone
5-Chloro-6-hydroxybenzoxazolin-2-one
Formed from: Chlorzoxazone
→ 5-Chloro-2,4-dihydroxyaniline
Probably in: Man C295

3-Chloro-4-hydroxybenzyl alcohol
Formed from: 3-Chloro-4-hydroxybenzaldehyde
4-(4-Chloro-3-hydroxybenzyl)-6,7-dihydroxyisoquinoline
Probably formed from: 4-(*p*-Chlorobenzyl)-6,7-dihydroxyisoquinoline
4′-Chloro-2′-(α-hydroxybenzyl)isonicotinanilide
See: Inabenfide
4-(3-(*p*-Chloro-α-hydroxybenzyl)-2-methylphenyl)butanoic acid
Formed from: 4-(3-(*p*-Chlorobenzoyl)-2-methylphenyl)butanoic acid
***o*-(3-Chloro-5-hydroxybenzyloxy)acetophenone**
Probably formed from: *o*-(*m*-Chlorobenzyloxy)acetophenone
***o*-(3-Chloro-5-hydroxybenzyloxy)phenylglycol**
Probably formed from: *o*-(*m*-Chlorobenzyloxy)phenylglycol
2-(*o*-(3-Chloro-5-hydroxybenzyloxy)phenyl)-2-oxoethanol
Probably formed from: 2-(*o*-(*m*-Chlorobenzyloxy)pheny)-2-oxoethanol
4-Chloro-2-(*p*-hydroxybenzyl)phenol
Probably formed from: 2-Benzyl-4-chlorophenol
2-(4-(*p*-Chloro-α-hydroxybenzyl)phenoxy)-2-methylpropionic acid
Probably formed from: 2-(4-(*p*-Chlorobenzoyl)phenoxy)-2-methylpropionic acid
→ 2-(4-(*p*-Chloro-α-hydroxybenzyl)phenoxy)-2-methylpropionic acid-β-D-glucuronide
Probably in: Dog E920
Guinea pig E920
Rat E920
2-(4-(*p*-Chloro-α-hydroxybenzyl)phenoxy)-2-methylpropionic acid-β-D-glucuronide
Probably formed from: 2-(4-(*p*-Chloro-α-hydroxybenzyl)phenoxy)-2-methylpropionic acid
3-Chloro-N-(*m*-hydroxybenzyl)propionamide
Formed from: Beclamide
→ 3-Chloro-N-(*m*-hydroxybenzyl) propionamide-β-D-glucuronide
In: Man H776
→ 3-Chloro-N-(*m*-hydroxybenzyl) propionamide sulphate

In: Man H776

3-Chloro-N-(*p*-hydroxybenzyl)propionamide
Formed from: Beclamide
→ 3-Chloro-N-(*p*-hydroxybenzyl) propionamide-β-D-glucuronide
In: Man A3507, H776
→ 3-Chloro-N-(*p*-hydroxybenzyl) propionamide sulphate
In: Man A3507, H776
→ *p*-Hydroxybenzoic acid
Probably in: Man A3507

3-Chloro-N-(*m*-hydroxybenzyl)propionamide-β-D-glucuronide
Formed from: 3-Chloro-N-(*m*-hydroxybenzyl)propionamide

3-Chloro-N-(*p*-hydroxybenzyl)propionamide-β-D-glucuronide
Formed from: 3-Chloro-N-(*p*-hydroxybenzyl)propionamide

3-Chloro-N-(*m*-hydroxybenzyl)propionamide sulphate
Formed from: 3-Chloro-N-(*m*-hydroxybenzyl)propionamide

3-Chloro-N-(*p*-hydroxybenzyl)propionamide sulphate
Formed from: 3-Chloro-N-(*p*-hydroxybenzyl)propionamide

2′-Chloro-4-hydroxybiphenyl
Formed from: 2-Chlorobiphenyl
→ 2′-Chloro-3,4-dihydroxybiphenyl
Probably in: Pigeon G885
Rat G885
Sheep G885
→ 2′-Chloro-4,4′-dihydroxybiphenyl
Probably in: Rat B448
→ 2′-Chloro-4′,5-dihydroxybiphenyl
Probably in: Rat B448

2-Chloro-5-hydroxybiphenyl
Formed from: 2-Chlorobiphenyl

3-Chloro-2-hydroxybiphenyl
Formed from: 2-(2-Aminoethoxy)-3-chlorobiphenyl
2-(2-Aminoethoxy)-3,5-dichlorobiphenyl

3′-Chloro-4-hydroxybiphenyl
Formed from: 3-Chlorobiphenyl
→ 3′-Chloro-4,4′-dihydroxybiphenyl
Probably in: Rat B448

4′-Chloro-2-hydroxybiphenyl
Formed from: 4-Chlorobiphenyl
→ 4′-Chloro-2,3-dihydroxybiphenyl
In: Rat H755
→ 4′-Chloro-2,5-dihydroxybiphenyl
In: Rat H755

4-Chloro-3-hydroxybiphenyl
Formed from: 4-Chlorobiphenyl

4′-Chloro-3-hydroxybiphenyl
Formed from: 4-Chlorobiphenyl

4,4′-Dichlorobiphenyl
→ 4′-Chloro-2,5-dihydroxybiphenyl
In: Rat H755
→ 4′-Chloro-3,4-dihydroxybiphenyl
In: Rat H755

4′-Chloro-4-hydroxybiphenyl
Formed from: 4-Chlorobiphenyl
4,4′-Dichlorobiphenyl
→ 4′-Chlorobiphenyl-4-yl-β-D-glucuronide
In: Rat A3463
→ 4′-Chloro-3,4-dihydroxybiphenyl
In: Rabbit A2053
Rat A247, H755
See also: A701, A1576, A2046, A2302, A2747, A3410, A3893, G419
→ 4′-Chloro-3,4′-dihydroxybiphenyl
Probably in: Frog A2530
Rat A2530, A3410
→ 4′-Chloro-4-methoxybiphenyl
Probably in: Lichen A2478

4′-Chloro-5-hydroxy-3-biphenylylacetic acid
Formed from: 4′-Chloro-5-methoxy-3-biphenylylacetic acid
→ 5-Carboxymethyl-4′-chloro-3-biphenylyl-β-D-glucuronide
In: Rabbit A3555
→ 5-Carboxymethyl-4′-chloro-3-biphenylyl sulphate
In: Guinea pig A3555
Monkey A3555
Mouse A3555
Rat A3524
→ 4′-Chloro-5-hydroxy-3-biphenylylacetyl-*β*-D-glucuronic acid
Probably in: Rabbit A3555

4′-Chloro-5-hydroxy-3-biphenylylacetyl-β-D-glucuronic acid
Formed from: 4′-Chloro-5-hydroxy-3-biphenylylacetic acid

4-Chloro-2-(1-hydroxy-*n*-butyl)-5-hydroxymethyl-1-((2′-(1*H*-tetrazol-5-yl)-1,1′-biphenyl-4-yl)methyl)-1*H*-imidazole
Formed from: 2-*n*-Butyl-4-chloro-5-hydroxymethyl-1-((2′-(1*H*-tetrazol-5-yl)-1,1′-biphenyl-4-yl)methyl)-1*H*-imidazole

4-Chloro-2-(3-hydroxy-*n*-butyl)-5-hydroxymethyl-1-((2′-(1*H*-tetrazol-5-yl)-1,1′-biphenyl-4-yl)methyl)-1*H*-imidazole
Formed from: 2-*n*-Butyl-4-chloro-5-hydroxymethyl-1-((2′-(1*H*-tetrazol-5-yl)-1,1′-biphenyl-4-yl)methyl)-1*H*-imidazole

***cis*-6-Chloro-5-(3′-hydroxycyclohexyl)indane-1-carboxylic acid**
Formed from: 6-Chloro-5-(cyclohexyl)indane-1-carboxylic acid

***cis*-6-Chloro-5-(4′-hydroxycyclohexyl)indane-1-carboxylic acid**
Formed from: 6-Chloro-5-(cyclohexyl)indane-1-carboxylic acid

***trans*-6-Chloro-5-(4′-hydroxycyclohexyl)indane-1-carboxylic acid**
Formed from: 6-Chloro-5-(cyclohexyl)indane-1-carboxylic acid
→ 6-Chloro-5-(4′-oxocyclohexyl)indane-1-carboxylic acid
Probably in: Rat A1220, A1234

3-Chloro-4-(*eq*-3-hydroxycyclohexyl)phenylacetic acid
Formed from: 4-(3-Chloro-4-cyclohexylphenyl)-4-oxobutyric acid

3-Chloro-4-(*ax*-4-hydroxycyclohexyl)phenylacetic acid
Formed from: 4-(3-Chloro-4-cyclohexylphenyl)-4-oxobutyric acid

3-Chloro-4-(*eq*-4-hydroxycyclohexyl)phenylacetic acid
Formed from: 4-(3-Chloro-4-cyclohexylphenyl)-4-oxobutyric acid

4-(3-Chloro-4-(*ax*-4-hydroxycyclohexyl)phenyl)-4-hydroxybutyric acid
Formed from: 4-(3-Chloro-4-cyclohexylphenyl)-4-oxobutyric acid

4-(3-Chloro-4-(*eq*-4-hydroxycyclohexyl)phenyl)-4-hydroxybutyric acid
Formed from: 4-(3-Chloro-4-cyclohexylphenyl)-4-oxobutyric acid

4′-Chloro-3′-hydroxydehydrocyclopeptine
Formed from: 4′-Chlorodehydrocyclopeptine

3-Chloro-2-hydroxydemethylimipramine
Formed from: 3-Chloro-2-hydroxyimipramine

1-Chloro-hydroxydibenzo-*p*-dioxin
Formed from: 1-Chlorodibenzo-*p*-dioxin

2-Chloro-hydroxydibenzo-*p*-dioxin
Formed from: 2-Chlorodibenzo-*p*-dioxin

2-Chloro-3-hydroxydibenzofuran
Formed from: 2-Chlorodibenzofuran
→ 2-Chloro-3,7-dihydroxydibenzofuran
Probably in: Rat G128

2-Chloro-7-hydroxydibenzofuran
Formed from: 2-Chlorodibenzofuran

8-Chloro-hydroxydibenz[*b*,*f*][1,4]oxazepine-10(11*H*)-carboxylic acid 2-(3-(ethylsulphonyl)-1-oxopropyl)hydrazide
Formed from: 8-Chlorodibenz[*b*,*f*][1,4]oxazepine-10(11*H*)-carboxylic acid 2-(3-(ethylsulphonyl)-1-oxopropyl)hydrazide

8-Chloro-hydroxydibenz[*b*,*f*][1,4]oxazepine-10(11*H*)-carboxylic acid 2-(1-oxo-3-(4-pyridinyl)propyl)hydrazide
Formed from: 8-Chlorodibenz[*b*,*f*][1,4]oxazepine-10(11*H*)-carboxylic acid 2-(1-oxo-3-(4-pyridinyl)propyl)hydrazide

2-Chloro-4-hydroxy-3,5-dimethoxybenzaldehyde
→ 2-Chloro-4-hydroxy-3,5-dimethoxybenzoic acid
In: Platycephalus H318
→ 2-Chloro-4-hydroxy-3,5-dimethoxybenzyl alcohol
In: Platycephalus H318

2-Chloro-4-hydroxy-3,5-dimethoxybenzoic acid
Formed from: 2-Chloro-4-hydroxy-3,5-dimethoxybenzaldehyde

2-Chloro-4-hydroxy-3,5-dimethoxybenzyl alcohol
Formed from: 2-Chloro-4-hydroxy-3,5-dimethoxybenzaldehyde

2-Chloro-4-(2-hydroxy-1,1-dimethylethyl)phenol
Probably formed from: 2-Chloro-4-(2-hydroxy-1,1-dimethylethyl)phenyl methyl methylphosphoramidate

→ 4-(2-Carboxy-2-propyl)-2-chlorophenol
Probably in: Rat A3338
→ 2-Chloro-4-(2-hydroxy-1,1-dimethylethyl)phenol-2′-β-D-glucuronide
Probably in: Rat A3338
→ 2-Chloro-4-(2-hydroxy-1,1-dimethylethyl)phenyl-β-D-glucuronide
Probably in: Rat A3338, A3351
Sheep A3001

2-Chloro-4-(2-hydroxy-1,1-dimethylethyl)phenol-2′-β-D-glucuronide
Probably formed from: 2-Chloro-4-(2-hydroxy-1,1-dimethylethyl)phenol

2-Chloro-4-(2-hydroxy-1,1-dimethylethyl)phenyl-β-D-glucuronide
Probably formed from: 2-Chloro-4-(2-hydroxy-1,1-dimethylethyl)phenol

2-Chloro-4-(2-hydroxy-1,1-dimethylethyl)phenyl formylphosphoramidate
Formed from: 2-Chloro-4-(2-hydroxy-1,1-dimethylethyl)phenyl methylphosphoramidate

2-Chloro-4-(2-hydroxy-1,1-dimethylethyl)phenyl methyl methylphosphoramidate
Formed from: Crufomate
→ 4-(1-Carboxy-1-methylethyl)-2-chlorophenyl methyl methylphosphoramidate
Probably in: Rat A3338, A3351
Sheep A2995
→ 2-Chloro-4-(2-hydroxy-1,1-dimethylethyl)phenol
Probably in: Sheep A2995
→ 2-Chloro-4-(2-hydroxy-1,1-dimethylethyl)phenyl methyl methylphosphoramidate-β-D-glucuronide
In: Sheep A3001
→ 2-Chloro-4-(2-hydroxy-1,1-dimethylethyl)phenyl methyl phosphate
Probably in: Rat A3338
Sheep A2995
→ 2-Chloro-4-(2-hydroxy-1,1-dimethylethyl)phenyl methyl phosphoramidate
Probably in: Rat A3338
Sheep A2995
→ 2-Chloro-4-(2-hydroxy-1,1-dimethylethyl)phenyl methylphosphoramidate
Probably in: Sheep A2995

2-Chloro-4-(2-hydroxy-1,1-dimethylethyl)phenyl methyl methylphosphoramidate-β-D-glucuronide
Formed from: 2-Chloro-4-(2-hydroxy-1,1-dimethylethyl)phenyl methyl methylphosphoramidate

2-Chloro-4-(2-hydroxy-1,1-dimethylethyl)phenyl methyl phosphate
Probably formed from: 2-Chloro-4-(2-hydroxy-1,1-dimethylethyl)phenyl methyl methylphosphoramidate
→ 2-Chloro-4-(2-hydroxy-1,1-dimethylethyl)phenyl phosphate
Probably in: Sheep A2995

2-Chloro-4-(2-hydroxy-1,1-dimethylethyl)phenyl methyl phosphoramidate
Probably formed from: 2-Chloro-4-(2-hydroxy-1,1-dimethylethyl)phenyl methyl methylphosphoramidate

2-Chloro-4-(2-hydroxy-1,1-dimethylethyl)phenyl methylphosphoramidate
Probably formed from: 2-Chloro-4-(2-hydroxy-1,1-dimethylethyl)phenyl methyl methylphosphoramidate
→ 2-Chloro-4-(2-hydroxy-1,1-dimethylethyl)phenyl formylphosphoramidate
Probably in: Sheep A2995

2-Chloro-4-(2-hydroxy-1,1-dimethylethyl)phenyl phosphate
Probably formed from: 2-Chloro-4-(2-hydroxy-1,1-dimethylethyl)phenyl methyl phosphate

***p*-Chloro-N-hydroxy-α,α-dimethylphenethylamine**
Formed from: 2-(*p*-Chlorophenyl)-1,1-dimethylnitroethane
→ *p*-Chloro-N-hydroxy-α,α-dimethylphenethylamine-β-D-glucuronide
In: Rat A2057

***p*-Chloro-N-hydroxy-α,α-dimethylphenethylamine-β-D-glucuronide**
Formed from: *p*-Chloro-N-hydroxy-α,α-dimethylphenethylamine

2-Chloro-3-hydroxy-1,3-diphenyl-1-propene
Formed from: 1,1-Dichloro-*cis*-2,3-diphenylcyclopropane
→ 2-Chloro-3-oxo-1,3-diphenyl-1-propene
Probably in: Rat F573

→ 2-Chloro-3-hydroxy-3-(*p*-hydroxyphenyl)-1-phenyl-1-propene
Probably in: Rat F573

3-Chloro-2-(2-hydroxyethoxy)biphenyl
Formed from: 3-Chloro-2-formylmethoxybiphenyl

***p*-Chloro-N-(2-(N-(2-hydroxyethyl)amino)ethyl)benzamide**
Formed from: Moclobemide

2-Chloro-N-(2-(1-hydroxyethyl)-6-hydroxymethylphenyl)-N-(2-methoxy-1-methylethyl)acetamide
Probably formed from: 2-Chloro-N-(2-ethyl-6-hydroxymethylphenyl)-N-(2-methoxy-1-methylethyl)acetamide

2-Chloro-N-(2-(1-hydroxyethyl)-6-methylphenyl)-N-(2-methoxy-1-methylethyl)acetamide
Probably formed from: Metolachlor

5-Chloro-2-hydroxyhippuric acid
Formed from: 5-Chlorosalicylic acid

1-(5-Chloro-2-(4-hydroxy-2-hydroxymethylphenoxy)phenyl)-3-(N-(4-thiahexanamid-S,S-dioxide)urea
Probably formed from: 1-(5-Chloro-2-(2-hydroxymethylphenoxy)phenyl)-3-(N-(4-thiahexanamid-S,S-dioxide)urea

N-(3-Chloro-4-hydroxy-2-hydroxymethylphenyl)anthranilic acid
Probably formed from: N-(3-Chloro-4-hydroxy-2-methylphenyl)anthranilic acid
→ N-(3-Chloro-4-hydroxy-2-hydroxymethylphenyl)anthranilic acid-β-D-glucuronide
Probably in: Man H853

N-(3-Chloro-4-hydroxy-2-hydroxymethylphenyl)anthranilic acid-β-D-glucuronide
Probably formed from: N-(3-Chloro-4-hydroxy-2-hydroxymethylphenyl)anthranilic acid

2-Chloro-3-hydroxy-3-(*p*-hydroxyphenyl)-1-phenyl-1-propene
Probably formed from: 2-Chloro-3-hydroxy-1,3-diphenyl-1-propene

3-Chloro-2-hydroxyimipramine
(2-hydroxyclomipramine)
Formed from: 3-Chloroimipramine
→ 3-Chloro-2-hydroxy-N-demethylimipramine
In: Man J200

3-Chloro-8-hydroxyimipramine
Formed from: 3-Chloroimipramine

2-Chloro-1-hydroxyindane
Formed from: 2-Chloroindane

5-Chloro-8-hydroxy-7-iodoquinoline (chinoform, clioquinol, entero-vioform, iodochlorhydroxyquin)
Formed from: 5-Chloro-7-iodoquinolin-8-yl-β-D-glucuronide
→ 5-Chloro-7-iodoquinolin-8-yl-β-D-glucuronide
In: Dog A2318
Guinea pig A2947
Man A2947
Rabbit A2947
Rat A424, A2947, A3654, D11, F239
→ 5-Chloro-7-iodoquinolin-8-yl sulphate
In: Dog A2318
Guinea pig A2947
Rabbit A2947
Rat A424, A2947, A3654, D11, F239
→ 5-Chloro-8-hydroxyquinoline
In: Dog A2318
Rat A3350

3-Chloro-4-hydroxy-4′-isopropylbiphenyl
Formed from: 4-Chloro-4′-isopropylbiphenyl

4-Chloro-3-hydroxy-4′-isopropylbiphenyl
Formed from: 4-Chloro-4′-isopropylbiphenyl

2-Chloro-4-hydroxy-5-methoxybenzaldehyde
→ 2-Chloro-4-hydroxy-5-methoxybenzoic acid
In: Microorganism E664
→ 2-Chloro-4-hydroxy-5-methoxybenzyl alcohol
In: Microorganism E664

5-Chloro-4-hydroxy-3-methoxybenzaldehyde
→ 5-Chloro-4-hydroxy-3-methoxybenzoic acid
In: Microorganism E664
→ 5-Chloro-4-hydroxy-3-methoxybenzyl alcohol
In: Microorganism E664

2-Chloro-4-hydroxy-3-methoxybenzoic acid
Formed from: 2-Chloro-3,4-dimethoxybenzoic acid

2-Chloro-4-hydroxy-5-methoxybenzoic acid
Formed from: 2-Chloro-4-hydroxy-5-methoxybenzaldehyde

5-Chloro-3-hydroxy-4-methoxybenzoic acid
Formed from: 5-Chloro-3,4-dimethoxybenzoic acid

5-Chloro-4-hydroxy-3-methoxybenzoic acid
Formed from: 5-Chloro-4-hydroxy-3-methoxybenzaldehyde
→ 5-Chloro-3,4-dihydroxybenzoic acid
In: Desulfomicrobium J491
Nocardia A960
→ Vanillic acid
In: Desulfomicrobium J491

2-Chloro-4-hydroxy-5-methoxybenzyl alcohol
Formed from: 2-Chloro-4-hydroxy-5-methoxybenzaldehyde

5-Chloro-4-hydroxy-3-methoxybenzyl alcohol
Formed from: 5-Chloro-4-hydroxy-3-methoxybenzaldehyde

4′-Chloro-3-hydroxy-4-methoxybiphenyl
Formed from: 4′-Chloro-3,4-dihydroxybiphenyl
→ 4′-Chloro-3,5-dihydroxy-4-methoxybiphenyl
Probably in: Rat A247

4′-Chloro-4-hydroxy-3-methoxybiphenyl
Formed from: 4′-Chloro-3,4-dihydroxybiphenyl

2-Chloro-7-hydroxy-8-methoxyphenothiazine
Probably formed from: 7-Hydroxy-8-methoxychlorpromazine

2-(2-Chloro-4-hydroxy-3-methoxyphenyl)-2-(*p*-chlorophenyl)acetic acid
Probably formed from: 2-(2-Chloro-3,4-dihydroxyphenyl)-2-(*p*-chlorophenyl)acetic acid

3-(2-Chloro-4-hydroxy-3-methoxyphenyl)-2-methyl-4(3*H*)-quinazolinone
Probably formed from: 3-(2-Chloro-4-hydroxyphenyl)-2-methyl-4(3*H*)-quinazolinone

β-Chloro-3-hydroxy-4-methoxystyrene
Formed from: β-Chloro-*p*-methoxystyrene

3-Chloro-4-hydroxy-4′-methylaminoazobenzene
Probably formed from: 3′-Chloro-4-methylaminoazobenzene

***p*-Chloro-N-hydroxy-N-methylaniline**
Formed from: *p*-Chloro-N-methylaniline

2-(3-Chloro-2-hydroxymethylanilino)nicotinic acid
Formed from: Clonixin

***p*-Chloro-N-hydroxymethylbenzamide**
Formed from: *p*-Chloro-N-methylbenzamide

7-Chloro-3-hydroxymethyl-1,2,4-benzothiadiazine-1,1-dioxide
Formed from: Diazoxide
→ 3-Carboxy-7-chloro-1,2,4-benzothiadiazine-1,1-dioxide
In: Dog A1232
Man A1232
Monkey A1232

4-Chloro-1-(4-hydroxy-3-methylbenzyl)-2-(*p*-hydroxyphenyl)imidazole-3-acetic acid
Probably formed from: 4-Chloro-1-(4-hydroxy-3-methylbenzyl)-2-phenylimidazole-3-acetic acid

4-Chloro-1-(4-hydroxy-3-methylbenzyl)-2-phenylimidazole-3-acetic acid
Formed from: 4-Chloro-1-(4-methoxy-3-methylbenzyl)-2-phenylimidazole-3-acetic acid
→ 4-Chloro-1-(4-hydroxy-3-methylbenzyl)-2-(*p*-hydroxyphenyl)imidazole-3-acetic acid
In: Dog C529
Rat C529

4-Chloro-5-hydroxymethyl-2-butyl-3-((2′-(1*H*-tetrazol-5-yl)(1,1′-biphenyl-4-yl)methyl)-1*H*-imidazole
→ 4-Chloro-5-hydroxymethyl-2-butyl-3-((2′-(1*H*-tetrazol-5-yl)(1,1′-biphenyl-4-yl) methyl)-1*H*-imidazole-N^2-β-D-glucuronide
In: Dog G693
Monkey G693

4-Chloro-5-hydroxymethyl-2-butyl-3-((2′-(1*H*-tetrazol-5-yl)(1,1′-biphenyl-4-yl)methyl)-1*H*-imidazole-N^2-β-D-glucuronide
Formed from: 4-Chloro-5-hydroxymethyl-2-butyl-3-((2′-(1*H*-tetrazol-5-yl)(1,1′-biphenyl-4-yl)methyl)-1*H*-imidazole

4-Chloro-2-hydroxymethylphenoxyacetic acid
Probably formed from: 4-Chloro-2-methylphenoxyacetic acid

1-(5-Chloro-2-((*o*-hydroxymethyl)phenoxy) phenyl)-3-(N-(3-(4-pyridinyl)propionamido) urea
Formed from: 8-Chlorodibenz[*b*,*f*][1,4] oxazepine-10(11*H*)-carboxylic acid 2-(1-oxo-3-(4-pyridinyl)propyl)hydrazide

1-(5-Chloro-2-((*o*-hydroxymethyl)phenoxy) phenyl)-3-(4-thiahexanamid-S,S-dioxide)urea
Formed from: 8-Chlorodibenz[*b*,*f*][1,4] oxazepine-10(11*H*)-carboxylic acid 2-(3-(ethylsulphonyl)-1-oxopropyl)hydrazide
→ 1-(5-Chloro-2-((4-hydroxy-2-hydroxy-methyl)phenoxy)phenyl)-3-(4 thiahexanamid-S,S-dioxide)urea
Probably in: Man G850
Rat G850

N-(3-Chloro-2-hydroxymethylphenyl)anthranilic acid
Formed from: N-(3-Chloro-2-methylphenyl) anthranilic acid
→ N-(3-Chloro-2-formylphenyl)anthranilic acid
Probably in: Man C95, H853
→ N-(3-Chloro-2-hydroxymethylphenyl) anthranilic acid-β-D-glucuronide
Probably in: Man B844, H853
Rabbit B844
Rat B844

N-(3-Chloro-4-hydroxy-2-methylphenyl) anthranilic acid
Formed from: N-(3-Chloro-2-methylphenyl) anthranilic acid
→ N-(3-Chloro-4-hydroxy-2-methylphenyl) anthranilic acid-β-D-glucuronide
In: Man B844, B879, H853
Rabbit B844
→ N-(3-Chloro-2-methyl-4-sulphatoxyphenyl) anthranilic acid
In: Rat B844
→ N-(3-Chloro-4-hydroxy-2-hydroxymethyl-phenyl)anthranilic acid
Probably in: Man H853

N-(3-Chloro-5-hydroxy-2-methylphenyl) anthranilic acid
Formed from: N-(3-Chloro-2-methylphenyl) anthranilic acid
→ N-(3-Chloro-5-hydroxy-2-methylphenyl)-5-hydroxyanthranilic acid
Probably in: Man J161

N-(3-Chloro-2-hydroxymethylphenyl)anthranilic acid
Formed from: N-(3-Chloro-2-methylphenyl) anthranilic acid
→ N-(3-Chloro-2-formylphenyl)anthranilic acid
Probably in: Man J161
→ N-(3-Chloro-2-hydroxymethylphenyl)-5-hydroxyanthranilic acid
Probably in: Man J161

N-(3-Chloro-4-hydroxy-2-methylphenyl) anthranilic acid-β-D-glucuronide
Probably formed from: N-(3-Chloro-4-hydroxy-2-methylphenyl)anthranilic acid

N-(3-Chloro-2-hydroxymethylphenyl)anthranilic acid-β-D-glucuronide
Probably formed from: N-(3-Chloro-2-hydroxymethylphenyl)anthranilic acid

***p*-Chloro-α-hydroxymethyl-α-phenylbenzyl alcohol**
Probably formed from: *p*-Chloro-α-methyl-α-phenylbenzyl alcohol
→ *p*-Chloro-α-formyl-α-phenylbenzyl alcohol
Probably in: Rat F448

1-(3-Chloro-4-hydroxymethylphenyl)-3,3-dimethylurea
Formed from: Chlorotoluron

N-(3-Chloro-2-hydroxymethylphenyl)-5-hydroxyanthranilic acid
Probably formed from: N-(3-Chloro-2-hydroxymethylphenyl)anthranilic acid

1-(3-Chloro-4-hydroxymethylphenyl)-3-methylurea
Probably formed from: 1-(3-Chloro-4-methylphenyl)-3-methylurea

***m*-Chloro-2-(2-(hydroxymethyl)-2-propylamino)propiophenone**
Formed from: Bupropion
→ 1-(*m*-Chlorophenyl)-2-(2-(hydroxymethyl)-2-propylamino)-1-propanol
Probably in: Man E229

2-Chloro-5-hydroxy-6-methylthiobenzamide
Probably formed from: N-Acetyl-2-chloro-6-(S-(L-cysteinyl))-5-hydroxybenzamide

6-Chloro-3-hydroxy-2-methylthiobenzonitrile
Probably formed from: 6-Chloro-3-hydroxy-2-sulphydrylbenzonitrile
→ 6-Chloro-3-hydroxy-2-methylthiobenzonitrile-O-sulphate
Probably in: Beef F586

6-Chloro-3-hydroxy-2-methylthiobenzonitrile-O-sulphate
Probably formed from: 6-Chloro-3-hydroxy-2-methylthiobenzonitrile

4-Chloro-N′-(5-(4-hydroxy-4-methylvaleric acid))-N′-methyl-*m*-benzenedisulphonamide
Formed from: 4-Chloro-N′-methyl-N′-(tetrahydro-2-methyl-5-oxofurfur-2-yl)-*m*-benzenedisulphonamide

2-Chloro-3-hydroxy-1,4-naphthoquinone
Formed from: 2-Chloro-1,4-naphthoquinone

5-Chloro-2-hydroxy-N-(*p*-nitrophenyl)benzamide
→ N-(*p*-Aminophenyl)-5-chloro-2-hydroxybenzamide
In: Ascaris A2037
Moniezia A2037
Mouse A2037
Sheep A2037
→ 5-Chlorosalicylic acid
In: Ascaris A2037
Mouse A2037

5-Chloro-2-hydroxy-6-oxo-6-phenylhexa-2,4-dienoic acid
→ Benzoic acid
In: Burkholderia K522

2-Chloro-7-hydroxyphenothiazine
Probably formed from: 2-Chlorophenothiazine

2-Chloro-N-hydroxyphenothiazine (identity challenged)
Formed from: 2-Chlorophenothiazine

2-Chloro-4-hydroxyphenoxyacetic acid
Formed from: 2,4-Dichlorophenoxyacetic acid

3-(3-Chloro-4-hydroxyphenoxy)pyrrolidine-1-(N-formyl)carboxamide
Probably formed from: 3-(3-Chloro-4-hydroxyphenoxy)pyrrolidine-1-(N-methyl) carboxamide

3-(3-Chloro-4-hydroxyphenoxy)pyrrolidine-1-(N-methyl)carboxamide
Formed from: 3-(3-Chlorophenoxy) pyrrolidine-1-(N-methyl)carboxamide
→ 3-(3-Chloro-4-hydroxyphenoxy)pyrrolidine-1-(N-formyl)carboxamide
Probably in: Dog A2114
Man A2114

3-Chloro-4-hydroxyphenyl acetate
→ *p*-Hydroxyphenyl acetate
In: Desulfitobacterium K247

3-Chloro-4-hydroxyphenylacetic acid
Formed from: Alclofenac
4-Benzyloxy-3-chlorophenylacetic acid
p-Chlorophenylacetic acid
→ 3-Chloro-4,5-dihydroxyphenylacetic acid
Probably in: Escherichia G817
→ *p*-Hydroxyphenylacetic acid
In: Desulfitobacterium J232

4-Chloro-2-hydroxyphenylacetic acid
Probably formed from: *p*-Chlorophenylacetic acid
→ 4-Chloro-2,3-dihydroxyphenylacetic acid
In: Pseudomonas C233

4-Chloro-3-hydroxyphenylacetic acid
Formed from: *p*-Chlorophenylacetic acid

N-((4-Chloro-2-hydroxyphenyl)amino) carbonyl-2,6-difluorobenzamide
Formed from: Diflubenzuron

N-((4-Chloro-3-hydroxyphenyl)amino) carbonyl-2,6-difluorobenzamide
Formed from: Diflubenzuron

5-Chloro-2-hydroxy-N-phenylbenzamide
→ Aniline
In: Ascaris A2037
Mouse A2037
Sheep A2037
→ 5-Chlorosalicylic acid
In: Ascaris A2037
Mouse A2037
Sheep A2037

2-(2-Chloro-3-hydroxyphenyl)-2-(*p*-chlorophenyl)acetic acid
Probably formed from: 2-(*o*-Chlorophenyl)-2-(*p*-chlorophenyl)acetic acid
→ 2-(2-Chloro-3,4-dihydroxyphenyl)-2-(*p*-chlorophenyl)acetic acid
Probably in: Chicken A2368

2-(2-Chloro-4-hydroxyphenyl)-2-(*p*-chlorophenyl) acetic acid
Probably formed from: 2-(*o*-Chlorophenyl)-2-(*p*-chlorophenyl)acetic acid

2-(2-Chloro-5-hydroxyphenyl)-2-(*p*-chlorophenyl) acetic acid
Probably formed from: 2-(*o*-Chlorophenyl)-2-(*p*-chlorophenyl)acetic acid

4-Chloro-3-hydroxyphenyl 4-chlorophenyl sulphone
Formed from: 4,4′-Dichlorophenyl sulphone
→ 4-Chloro-3-hydroxyphenyl 4-chlorophenyl sulphone-β-D-glucuronide
In: Rat H707

4-Chloro-3-hydroxyphenyl 4-chlorophenyl sulphone-β-D-glucuronide
Formed from: 4-Chloro-3-hydroxyphenyl 4-chlorophenyl sulphone

1-(4-Chloro-2-hydroxyphenyl)-3-(4,5-dichloro-2-hydroxyphenyl)urea
Probably formed from: 1-(4-Chlorophenyl)-3-(4,5-dichloro-2-hydroxyphenyl)urea

1-(4-Chloro-3-hydroxyphenyl)-3-(4,5-dichloro-2-hydroxyphenyl)urea
Probably formed from: 1-(4-Chlorophenyl)-3-(4,5-dichloro-2-hydroxyphenyl)urea

1-(4-Chloro-2-hydroxyphenyl)-3-(3,4-dichlorophenyl)urea
Formed from: 1-(*p*-Chlorophenyl)-3-(3,4-dichlorophenyl)urea

1-(4-Chloro-3-hydroxyphenyl)-3-(3,4-dichlorophenyl)urea
Formed from: 1-(*p*-Chlorophenyl)-3-(3,4-dichlorophenyl)urea

N-(3-Chloro-4-hydroxyphenyl)-N′-(4,5-dihydro-1-methyl-4-oxo-1*H*-imidazol-2-yl)urea
Formed from: Fenobam
→ N-(3-Chloro-4-hydroxyphenyl)-N′-(4,5-dihydro-4-oxo-1*H*-imidazol-2-yl)urea
Probably in: Rat H350

N-(3-Chloro-4-hydroxyphenyl)-N′-(4,5-dihydro-4-oxo-1*H*-imidazol-2-yl)urea
Probably formed from: N-(3-Chloro-4-hydroxyphenyl)-N′-(4,5-dihydro-1-methyl-4-oxo-1*H*-imidazol-2-yl)urea

1-(3-Chloro-4-hydroxyphenyl)-3-(2-(3,3-dimethyl-1-azetidinyl)ethyl)imidazolin-2-one
Formed from: Zetidoline
→ 1-(3-Chloro-4,6-dihydroxyphenyl)-3-(2-(3,3-dimethyl-1-azetidinyl)ethyl) imidazolin-2-one
Probably in: Dog D384
Rat D384, D645
→ 1-(3-Chloro-4-hydroxyphenyl)-3-(2-(3,3-dimethyl-1-azetidinyl)ethyl)imidazolin-2-one-β-D-glucuronide
Probably in: Dog D384
Rat D384, D645

1-(5-Chloro-2-hydroxyphenyl)-3-(2-(3,3-dimethyl-1-azetidinyl)ethyl)imidazolin-2-one
Formed from: Zetidoline

1-(3-Chloro-4-hydroxyphenyl)-3-(2-(3,3-dimethyl-1-azetidinyl)ethyl)imidazolin-2-one-β-D-glucuronide
Probably formed from: 1-(3-Chloro-4-hydroxyphenyl)-3-(2-(3,3-dimethyl-1-azetidinyl)ethyl)imidazolin-2-one

1-(4-Chloro-2-hydroxyphenyl)-3,3-dimethyltriazene
Probably formed from: 1-(4-Chlorophenyl)-3,3-dimethyltriazene

3-(4-Chloro-2-hydroxyphenyl)-1,1-dimethylurea
Formed from: 3-(4-Chlorophenyl)-1,1-dimethylurea
→ 3-(4-Chloro-2-hydroxyphenyl)-1,1-dimethylurea-β-D-glucuronide
Probably in: Bean A940
→ 3-(4-Chloro-2-hydroxyphenyl)-1-methylurea
Probably in: Bean A940

3-(4-Chloro-2-hydroxyphenyl)-1,1-dimethylurea-β-D-glucuronide
Probably formed from: 3-(4-Chloro-2-hydroxyphenyl)-1,1-dimethylurea

N-(3-Chloro-4-hydroxyphenyl)-N′-(imino(methylamino)methyl)urea
Probably formed from: N-(*m*-Chlorophenyl)-N′-(imino(methylamino)methyl)urea

4-Chloro-2-(*p*-hydroxyphenyl)-1-(4-methoxy-3-methylbenzyl)imidazole-3-acetic acid
Formed from: 4-Chloro-1-(4-methoxy-3-methylbenzyl)-2-phenylimidazole-3-acetic acid

1-Chloro-2-(*p*-hydroxyphenyl)-2-(*p*-methoxyphenyl)ethylene
Probably formed from: 1,1-Dichloro-2-(*p*-hydroxyphenyl)-2-(*p*-methoxyphenyl)ethane
→ 1-Chloro-2-(*p*-hydroxyphenyl)-2-(*p*-methoxyphenyl)ethylene-β-D-glucuronide
Probably in: Goat C198

1-Chloro-2-(*p*-hydroxyphenyl)-2-(*p*-methoxyphenyl)ethylene-β-D-glucuronide
Probably formed from: 1-Chloro-2-(*p*-hydroxyphenyl)-2-(*p*-methoxyphenyl)ethylene

2-(3-Chloro-4-hydroxyphenyl)-2-methylpropionic acid
Probably formed from: 4-(1-Carboxy-1-methylethyl)-2-chlorophenyl methyl methylphosphoramidate
→ 2-(3-Chloro-4-hydroxyphenyl)-2-methylpropionic acid-β-D-glucuronide
Probably in: Sheep A3001

2-(3-Chloro-4-hydroxyphenyl)-2-methylpropionic acid-β-D-glucuronide
Probably formed from: 2-(3-Chloro-4-hydroxyphenyl)-2-methylpropionic acid

3-(2-Chloro-3-hydroxyphenyl)-2-methyl-4(3*H*)-quinazolinone
Formed from: Mecloqualone

3-(2-Chloro-4-hydroxyphenyl)-2-methyl-4(3*H*)-quinazolinone
Formed from: Mecloqualone
→ 2-(2-Chloro-4-hydroxy-3-methoxyphenyl)-2-methyl-4(3*H*)-quinazolinone
Probably in: Man A1247

4-Chloro-3-hydroxyphenyl methyl sulphone
Probably formed from: *p*-Chlorophenyl methyl sulphone
→ 4-Chloro-3-hydroxyphenyl methyl sulphone-β-D-glucuronide
Probably in: Rat A1946, A2639
→ 4-Chloro-3-hydroxyphenyl methyl sulphone sulphate
Probably in: Rat A1946, A2639

4-Chloro-3-hydroxyphenyl methyl sulphone-β-D-glucuronide
Probably formed from: 4-Chloro-3-hydroxyphenyl methyl sulphone

4-Chloro-3-hydroxyphenyl methyl sulphone sulphate
Probably formed from: 4-Chloro-3-hydroxyphenyl methyl sulphone

3-(4-Chloro-2-hydroxyphenyl)-1-methylurea
Probably formed from: 3-(4-Chloro-2-hydroxyphenyl)-1,1-dimethylurea
→ 3-(4-Chloro-2-hydroxyphenyl)-1-methylurea-β-D-glucuronide
Probably in: Bean A940

3-(4-Chloro-2-hydroxyphenyl)-1-methylurea-β-D-glucuronide
Probably formed from: 3-(4-Chloro-2-hydroxyphenyl)-1-methylurea

1-(3-Chloro-4-hydroxyphenyl)piperazine
Formed from: 1-(*m*-Chlorophenyl)piperazine
→ 1-(3-Chloro-4-hydroxyphenyl)piperazine-β-D-glucuronide
In: Rat G862
→ 1-(3-Chloro-4-hydroxyphenyl)piperazine sulphate
In: Rat G862

1-(3-Chloro-4-hydroxyphenyl)piperazine-β-D-glucuronide
Formed from: 1-(3-Chloro-4-hydroxyphenyl)piperazine

1-(3-Chloro-4-hydroxyphenyl)piperazine sulphate
Formed from: 1-(3-Chloro-4-hydroxyphenyl)piperazine

3-((4-(4-Chloro-3-hydroxyphenyl)piperazin-1-yl)methyl)-1*H*-pyrrolo[2,3-*b*]pyridine sulphate
Probably formed from: 3-((4-(4-Chlorophenyl)piperazin-1-yl)methyl)-1*H*-pyrrolo[2,3-*b*]pyridine

2-(3-(4-(3-Chloro-4-hydroxyphenyl)piperazin-1-yl)propyl)-5-ethyl-2,4-dihydro-4-(2-phenoxyethyl)-3*H*-1,2,4-triazol-3-one
Formed from: Nefazodone

5-Chloro-6-hydroxy-3-phenylthioindole-2-carboxamide
Formed from: 5-Chloro-3-phenylthioindole-2-carboxamide

5-Chloro-3-(*p*-hydroxyphenylthioindole-2-carboxamide
Formed from: 5-Chloro-3-phenylthioindole-2-carboxamide

8-Chloro-4-hydroxy-6-phenyl-4*H*-*s*-triazolo[4,3-*a*][1,4]benzodiazepine
Formed from: 8-Chloro-6-phenyl-4*H*-*s*-triazolo[4,3-*a*][1,4]benzodiazepine
→ 5-Chloro-2-(4*H*-1,2,4-triazol-4-yl) benzophenone
Probably in: Dog A258
Man A258

8-Chloro-6-(*m*-hydroxyphenyl)-4*H*-*s*-triazolo[4,3-*a*][1,4]benzodiazepine
Formed from: 8-Chloro-6-phenyl-4*H*-*s*-triazolo[4,3-*a*][1,4]benzodiazepine
→ 8-Chloro-2,4-dihydro-6-(*m*-hydroxyphenyl)-4*H*-*s*-triazolo[4,3-*a*][1,4]benzodiazepin-1-one
Probably in: Dog A258

8-Chloro-6-(*p*-hydroxyphenyl)-4*H*-*s*-triazolo[4,3-*a*][1,4]benzodiazepine
Formed from: 8-Chloro-6-phenyl-4*H*-*s*-triazolo[4,3-*a*][1,4]benzodiazepine
→ 8-Chloro-2,4-dihydro-6-(*p*-hydroxyphenyl)-4*H*-*s*-triazolo[4,3-*a*][1,4]benzodiazepin-1-one
Probably in: Dog A258
Rat A258
→ 8-Chloro-2-(2,3-dihydro-3-oxo-4*H*-1,2,4-triazol-4-yl)-2′-hydroxybenzophenone
Probably in: Dog A258

8-Chloro-6-(*p*-hydroxyphenyl)-1*H*-*s*-triazolo[4,3-*a*][1,4]benzodiazepin-1-one
Probably formed from: 8-Chloro-6-phenyl-1*H*-*s*-triazolo[4,3-*a*][1,4]benzodiazepin-1-one

1-(4-Chloro-2-hydroxyphenyl)urea
Formed from: 1-(*p*-Chlorophenyl)urea
→ 1-(4-Chloro-2-hydroxyphenyl)urea-β-D-glucuronide
In: Cotton A170

1-(4-Chloro-2-hydroxyphenyl)urea-β-D-glucuronide
Formed from: 1-(4-Chloro-2-hydroxyphenyl)urea

***m*-Chloro-2-hydroxypropiophenone**
Formed from: *m*-Chlorobenzaldehyde

***p*-Chloro-2-hydroxypropiophenone**
Formed from: *p*-Chlorobenzaldehyde

7-Chloro-3-hydroxy-5-(pyrrol-2-yl)-2(3*H*)-oxo-1,4-benzodiazepine
Formed from: 7-Chloro-5-(pyrrol-2-yl)-2(3*H*)-oxo-1,4-benzodiazepine

7-Chloro-5-(α-hydroxypyrrol-2-yl)-2(3*H*)-oxo-1,4-benzodiazepine
Formed from: 7-Chloro-5-(pyrrol-2-yl)-2(3*H*)-oxo-1,4-benzodiazepine

2-Chloro-5-hydroxyquinol
Formed from: 4-Chlororesorcinol
2,5-Dichloroquinol
→ 2-Hydroxy-*p*-quinone
In: Burkholderia J705

6-Chloro-2-hydroxyquinol
→ 2-Chloro-4-hydroxy-*cis*,*cis*-muconic acid
In: Azotobacter H602
Streptomyces H347

5-Chloro-8-hydroxyquinoline
Formed from: 5-Chloro-8-hydroxy-7-iodoquinoline
Probably formed from: 5-Chloro-8-hydroxyquinoline-β-D-glucuronide
→ 5-Chloro-8-hydroxyquinoline-β-D-glucuronide
In: Rat A3654
→ 5-Chloro-8-hydroxyquinoline-O-sulphate
In: Rat A3654

8-Chloro-2-hydroxyquinoline
Formed from: 8-Chloroquinoline

5-Chloro-8-hydroxyquinoline-β-D-glucuronide
Formed from: 5-Chloro-8-hydroxyquinoline
→ 5-Chloro-8-hydroxyquinoline
Probably in: Rat A3654

5-Chloro-8-hydroxyquinoline-O-sulphate
Formed from: 5-Chloro-8-hydroxyquinoline

β-Chloro-*p*-hydroxystyrene
Formed from: β-Chloro-*p*-methoxystyrene

6-Chloro-3-hydroxy-2-sulphydrylbenzonitrile
Probably formed from: 6-Chloro-2-(S-(L-cysteinyl))-3-hydroxybenzonitrile
→ 6-Chloro-3-hydroxy-2-methylthiobenzonitrile
Probably in: Beef F586
Goat F296
Rat F296

5-Chloro-4-(2-imidazolin-2-ylamino)-2,1,3-benzothiadiazole
See: Sirdalud
3-Chloroimipramine (clomipramine)
→ 3-Chloro-N-demethylimipramine
In: Glycine K359
Man A3270, D119, H656
Rat A3178, D60, H691, J829
→ 3-Chloroimipramine-N-β-D-glucuronide
In: Man H162
→ 3-Chloroimipramine-N-oxide
In: Man A1962
→ 3-Chloro-2-hydroxyimipramine
In: Man A1962, J200
→ 3-Chloro-8-hydroxyimipramine
In: Man A1962, J200
3-Chloroimipramine-N-β-D-glucuronide
Formed from: 3-Chloroimipramine
3-Chloroimipramine-N-oxide
Formed from: 3-Chloroimipramine
1-Chloroindane
→ Indene-1,2-oxide
Probably in: Rat A2886
2-Chloroindane
→ 2-Chloro-1-hydroxyindane (*cis* and *trans*)
In: Pseudomonas K207
7-Chloroindole-3-acetamide
Formed from: 7-Chloro-L-tryptophan
→ Indole-3-acetamide
Probably in: Pseudomonas B424
4-Chloroindole-3-acetic acid
→ 4-Chlorodioxindole
In: Bradyrhizobium H679
5-Chloroindole-3-acetic acid
→ 2-Amino-5-chlorobenzoic acid
In: Bradyrhizobium H679
→ 5-Chlorodioxindole-3-acetic acid
In: Bradyrhizobium H679
7-Chloroindole-3-pyruvic acid
Formed from: 7-Chloro-L-tryptophan
5-Chloro-7-iodoquinolin-8-yl-β-D-glucuronide
Formed from: 5-Chloro-8-hydroxy-7-iodoquinoline
→ 5-Chloro-8-hydroxy-7-iodoquinoline
With β-glucuronidase A878
5-Chloro-7-iodoquinolin-8-yl sulphate
Formed from: 5-Chloro-8-hydroxy-7-iodoquinoline
2-Chloro-N-isopropylacetamide
See: Propachlor
***p*-Chloro-N-isopropylbenzamide**
Formed from: *p*-Chloro-N-isopropyl-N-methylbenzamide
4-Chloro-4′-isopropylbiphenyl
→ 4′-Acetyl-4-chlorobiphenyl
In: Rat A2531
→ 4-Carboxy-4′-chlorobiphenyl
In: Cunninghamella A3078
See also A2531
→ 4-(1-Carboxyethyl)-4′-chlorobiphenyl
In: Frog A3078
Goldfish A3078
See also: A2531
→ 3-Chloro-4-hydroxy-4′-isopropylbiphenyl
In: Rat A2531
→ 4-Chloro-3-hydroxy-4′-isopropylbiphenyl
In: Rat A2531
***p*-Chloro-N-isopropyl-N-methylbenzamide**
→ *p*-Chloro-N-isopropylbenzamide
In: Rat G13
→ *p*-Chloro-N-methylbenzamide
In: Rat G13
***p*-Chlorolactanilide**
Formed from: *p*-Chloropropionanilide
***o*-Chloromandelaldehyde**
Probably formed from: N-*tert*-Butyl-*o*-chlorophenylethanolamine
→ *o*-Chloromandelic acid
Probably in: Dog B776
Rabbit B776
→ *o*-Chlorophenylglycol
Probably in: Guinea pig B776
***o*-Chloromandelic acid**
Probably formed from: *o*-Chloromandelaldehyde
***p*-Chloromandelic acid**
→ *p*-Chlorophenylglyoxylic acid
In: Rhodotorula H90
***m*-Chloromandelonitrile**
Formed from: *m*-Chlorobenzaldehyde
***p*-Chloromandelonitrile**
Formed from: *p*-Chlorobenzaldehyde

4′-Chloro-4-methoxybiphenyl
Probably formed from: 4′-Chloro-4-hydroxybiphenyl

4′-Chloro-5-methoxy-3-biphenylylacetic acid
→ 4′-Chloro-5-hydroxy-3-biphenylylacetic acid
In: Dog A3555
Guinea pig A3555
Man A3555
Mouse A3555
Rabbit A3555
Rat A3555
→ 4′-Chloro-5-methoxy-3-biphenylylacetyl-β-D-glucuronic acid
In: Man A3555
Rat A3555
→ 2-(4′-Chloro-5-methoxy-3-biphenylyl)-2-hydroxyacetic acid
In: Dog A3555

4′-Chloro-5-methoxy-3-biphenylylacetyl-β-D-glucuronic acid
Formed from: 4′-Chloro-5-methoxy-3-biphenylylacetic acid

2-(4′-Chloro-5-methoxy-3-biphenylyl)-2-hydroxyacetic acid
Formed from: 4′-Chloro-5-methoxy-3-biphenylylacetic acid

4-Chloro-1-(4-methoxy-3-methylbenzyl)-2-phenylimidazole-3-acetic acid
→ 4-Chloro-1-(4-hydroxy-3-methylbenzyl)-2-phenylimidazole-3-acetic acid
In: Dog C529
Rat C529
→ 4-Chloro-2-(*p*-hydroxyphenyl)-1-(4-methoxy-3-methylbenzyl)imidazole-3-acetic acid
In: Dog C529
Rat C529

4-Chloro-1-methoxynaphthalene
Formed from: 1-Methoxynaphthalene

2-Chloro-4-methoxyphenol
Formed from: Arildone
→ 2-Chloro-4-methoxyphenyl sulphate
In: Mouse C389
→ Chloroquinol
Probably in: Mouse C389

4-Chloro-2-methoxyphenol
Formed from: 4-Chlorocatechol
→ 4-Chlorocatechol
In: Arthrobacter C527
→ 4-Chloroveratrole
In: Arthrobacter C527

5-Chloro-2-methoxyphenol
Formed from: 4-Chlorocatechol

4-(6-(2-Chloro-4-methoxyphenyloxy)hexyl)-3,5-heptanedione
See: Arildone

2-Chloro-4-methoxyphenyl sulphate
Formed from: 2-Chloro-4-methoxyphenol

β-Chloro-*p*-methoxystyrene
→ β-Chloro-*p*-hydroxystyrene
In: Rat A3662
→ β-Chloro-3-hydroxy-4-methoxystyrene
In: Rat A3662

3-Chloro-4-methylacetanilide
Probably formed from: 3-Chloro-4-methylaniline

***p*-Chloro-N-(1-methylacetonyl)amphetamine**
Probably formed from: *p*-Chloro-N-(1-methylacetonylidene)amphetamine

***p*-Chloro-N-(1-methylacetonylidene)amphetamine**
Formed from: *p*-Chloroamphetamine
→ *p*-Chloro-N-(1-methylacetonyl)amphetamine
Probably in: Mycobacterium B508

3′-Chloro-4-methylaminoazobenzene
Formed from: 3′-Chloro-4-dimethylaminoazobenzene
→ 4-Amino-3′-chloroazobenzene
In: Rat A766
→ 3-Chloro-4-hydroxy-4′-methylaminoazobenzene
Probably in: Rat A766

4′-Chloro-4-methylaminoazobenzene
Formed from: 4′-Chloro-4-dimethylaminoazobenzene
→ 4-Amino-4′-chloroazobenzene
In: Rat A1979

5-Chloro-2-methylaminobenzophenone
Probably formed from: 2-Amino-5-chlorobenzophenone

2-Chloro-11-(2-methylaminoethoxy)dibenzo[*b*,*f*]thiepin
→ 2-Chloro-11-(2-methylaminoethoxy)dibenzo[*b*,*f*]thiepin-N-sulphate

In: Rat E51

2-Chloro-11-(2-methylaminoethoxy) dibenzo[*b*,*f*]thiepin-N-sulphate
Formed from: 2-Chloro-11-(2-methylaminoethoxy)dibenzo[*b*,*f*]thiepin

3-Chloro-2-methylamino-1,4-naphthoquinone
Formed from: 3-Chloro-2-dimethylamino-1,4-naphthoquinone

7-Chloro-2-methylamino-5-phenyl-3*H*-1,4-benzodiazepine-4-oxide
See: Chlordiazepoxide

***p*-Chloro-N-methylaniline**
Formed from: *p*-Chloroaniline
p-Chloro-N,N-dimethylaniline
p-Chloro-N-ethyl-N-methylaniline
→ *p*-Chloroaniline
In: Hen B815
Man B815
Rabbit A1916
Rat A3703
→ *p*-Chloro-N-hydroxy-N-methylaniline
In: Rat A3603

2-Chloro-4-methylaniline
→ 4-Amino-2-chlorobenzyl alcohol
In: Rat G580
N-(4-Amino-2-chlorobenzyl)-2-chloro-4-methylaniline
In: Rat G580
→ 2-Chloro-4-methylnitrosobenzene
In: Rat G580
→ 2,2′-Dichloro-4,4′-dimethylazobenzene
In: Rat G518, G580
→ 2,2′-Dichloro-4,4′-dimethylhydrazobenzene
In: Rat G518, G580

2-Chloro-6-methylaniline
Formed from: Butanilicaine
→ 2,2′-Dichloro-6,6′-dimethylazobenzene
In: Horseradish A661

3-Chloro-2-methylaniline
→ 3,3′-Dichloro-2,2′-dimethylazobenzene
In: Geotrichum A661
Horseradish A661

3-Chloro-4-methylaniline
Formed from: Solan
→ 4-Amino-2-chlorobenzoic acid
Probably in: Bluegill sunfish J230
→ 3,3′-Dichloro-4,4′-dimethylazobenzene
In: Geotrichum A661
Horseradish A661
→ 3-Chloro-4-methylacetanilide
In: Bluegill sunfish J230
See also: A3263, B237

4-Chloro-2-methylaniline (4-chloro-*o*-toluidine)
Formed from: Chlordimeform
Probably formed from: 4-Chloro-2-methylformanilide
→ 4-Chloro-2-methylphenylhydroxylamine
In: Rat B82
→ 4,4′-Dichloro-2,2′-dimethylazobenzene
In: Geotrichum A661
Horseradish A661
Rat B82

5-Chloro-2-methylaniline
→ 5,5′-Dichloro-2,2′-dimethylazobenzene
In: Geotrichum A661
Horseradish A661

***p*-Chloromethylbenzaldehyde**
Formed from: *p*-Chloromethylphenylglyoxylic acid

***p*-Chloro-N-methylbenzaldimine**
Probably formed from: α-(2-Carboxyethylthio)-*p*-chloro-N-methylbenzylamine-S,S-dioxide
→ *p*-Chloro-N-methylbenzamide
Probably in: Man E130

***p*-Chloro-N-methylbenzamide**
Formed from: N-Benzyl-*p*-chloro-N-methylbenzamide
N-Butyl-*p*-chloro-N-methylbenzamide
p-Chloro-N-cyclopropyl-N-methylbenzamide
p-Chloro-N,N-dimethylbenzamide
p-Chloro-N-ethyl-N-methylbenzamide
p-Chloro-N-isopropyl-N-methylbenzamide
p-Chloro-N-methyl-N-propylbenzamide
Probably formed from: 2-(N-(*p*-Chlorobenzoyl)-N-methylcarbamoyl)ethylsulphonic acid
p-Chloro-N-methylbenzaldimine
→ *p*-Chloro-N-hydroxymethylbenzamide
In: Mouse C319

3-Chloro-2-methylbenzoic acid
→ 4-Chloro-3-methylcatechol
In: Pseudomonas G657

5-Chloro-2-methylbenzoic acid
→ *o*-Toluic acid
In: Desulfomonile G152

7-Chloro-3-methyl-1,2,4-benzothiadiazine-1,1-dioxide
See: Diazoxide

6-Chloro-α-methylcarbazole-2-acetic acid
See: Carprofen

3-Chloro-4-methylcatechol
Probably formed from: 2-Chloro-3-methylphenol

3-Chloro-6-methylcatechol
Formed from: 4-Chloro-2,3-dihydro-2,3-dihydroxytoluene
p-Chlorotoluene
Probably formed from: 2-Chloro-5-methylphenol

4-Chloro-3-methylcatechol
Formed from: 3-Chloro-2-methylbenzoic acid

4-Chloro-5-methylcatechol
Formed from: 4-Chloro-5-methyl-2-nitrophenol

5-Chloro-3-methylcatechol
Formed from: 4-Chloro-2-methylphenol

5-Chloro-4-methylcatechol
Probably formed from: 4-Chloro-3-methylphenol

***p*-Chloromethylcinnamaldehyde**
Formed from: (*E*)-4-(*p*-chloromethylphenyl)-2-oxo-3-butenoic acid

2-Chloro-6-methylergoline-8β-acetic acid
Formed from: Lergotrile

2-Chloro-6-methylergoline-8β-acetonitrile
See: Lergotrile

4-Chloro-2-methylformanilide
Formed from: Demethylchlordimeform
→ 4-Chloro-2-methylaniline
Probably in: Cabbage looper A1838
Housefly A779
Man A2047
Mouse A2534
Rat A1821
Spider mite A1944
→ N-Formyl-5-chloroanthranilic acid
Probably in: Mouse A2534
Rat A1821

4-Chloro-5-methyl-2-nitrophenol
→ 4-Chloro-5-methylcatechol
In: Pseudomonas E607, H65

2-Chloro-4-methylnitrosobenzene
Formed from: 2-Chloro-4-methylaniline

2-Chloro-3-methylphenol
→ 3-Chloro-4-methylcatechol
Probably in: Pseudomonas A2391

2-Chloro-5-methylphenol
→ 3-Chloro-6-methylcatechol
Probably in: Pseudomonas A2391

4-Chloro-2-methylphenol (4-chloro-*o*-cresol)
Formed from: 2-(4-Chloro-2-methylphenoxy)propionic acid
→ 5-Chloro-3-methylcatechol
In: Acinetobacter C232
Proteobacteria J466

4-Chloro-3-methylphenol
→ 5-Chloro-4-methylcatechol
Probably in: Pseudomonas A2391

4-(3-(2-Chloro-10*H*-phenothiazin-10-yl)propyl)-1-piperazineethanol
See: Perphenazine

4-Chloro-2-methylphenoxyacetic acid
Formed from: 4-Chloro-2-methylphenoxyacetic acid bisethyleneglycol ester
→ 4-Chloro-2-hydroxymethylphenoxyacetic acid
Probably in: Barley G813

4-Chloro-2-methylphenoxyacetic acid bisethyleneglycol ester
→ 4-Chloro-2-methylphenoxyacetic acid
In: Barley G813

1-(2-Chloro-5-methylphenoxy)-3-(1,1-dimethylethyl)amino)-2-propanol
See: Bupranolol

***p*-Chloro-N-methylphenoxyethylamine**
Formed from: *p*-Chloro-N,N-dimethylphenoxyethylamine

2-(4-Chloro-2-methylphenoxy)propionic acid (mecoprop)
→ 4-Chloro-2-methylphenol
In: Alcaligenes J911
Burkholderia J911
Microorganism G183

N-(3-Chloro-2-methylphenyl)anthranilic acid
(tolfenamic acid)
→ N-(3-Chloro-2-hydroxymethylphenyl) anthranilic acid
In: Man B844, C95, H853, J161
Rabbit B844
Rat B844
→ N-(3-Chloro-4-hydroxy-2-methylphenyl) anthranilic acid
In: Man B844, C95, H853, J161
Rabbit B844
Rat B844
→ N-(3-Chloro-5-hydroxy-2-methylphenyl) anthranilic acid
In: Man J161
→ N-(3-Chloro-2-methylphenyl) anthraniloylglycine
In: Rabbit B844
→ N-(3-Chloro-2-methylphenyl)anthraniloyl-β-D-glucuronic acid
In: Man B879, J161
→ N-(3-Chloro-2-methylphenyl)-5-hydroxyanthranilic acid
In: Man J161

N-(3-Chloro-2-methylphenyl)anthraniloyl-β-D-glucuronic acid
Formed from: N-(3-Chloro-2-methylphenyl) anthranilic acid

N-(3-Chloro-2-methylphenyl)anthraniloylglycine
Formed from: N-(3-Chloro-2-methylphenyl)anthranilic acid

7-Chloro-1-methyl-5-phenyl-1*H*-1,5-benzodiazepino-2,4(3*H*,5*H*)-dione
See: Clobazam

***p*-Chloro-α-methyl-α-phenylbenzyl alcohol**
Formed from: Clemastine
→ *p*-Chloro-α-hydroxymethyl-α-phenylbenzyl alcohol
Probably in: Rat F448

1-(3-Chloro-4-methylphenyl)-3,3-dimethylurea
See: Chlorotoluron

***p*-Chloromethylphenylglyoxylic acid**
→ *p*-Chloromethylbenzaldehyde
In: Pseudomonas E714

4-Chloro-2-methylphenylhydroxylamine
Formed from: 4-Chloro-2-methylaniline

N-(3-Chloro-2-methylphenyl)-5-hydroxyanthranilic acid
Formed from: N-(3-Chloro-2-methylphenyl)anthranilic acid

N-(3-Chloro-2-methylphenyl)-N′-(2-methylpropyl)sulphinylurea
Formed from: N-(3-Chloro-2-methylphenyl)-N′-(2-methylpropyl)thiourea

N-(3-Chloro-2-methylphenyl)-N′-(2-methylpropyl)thiourea
→ N-(3-Chloro-2-methylphenyl)-N′-(2-methylpropyl)sulphinylurea
In: Rat J330

1-(3-Chloro-4-methylphenyl)-3-methylurea
Formed from: Chlorotoluron
→ 1-(3-Chloro-4-hydroxymethylphenyl)-3-methylurea
Probably in: Wheat A3869

(*E*)-4-(*p*-Chloromethylphenyl)-2-oxo-3-butenoic acid
→ *p*-Chloromethylcinnamaldehyde
In: Yeast F279
→ *p*-Methylcinnamic acid
In: Yeast F279

6-Chloro-1-methyl-4-phenyl-2(1*H*)-quinazolinone
Formed from: 7-Chloro-1,3-dihydro-1,3-dimethyl-5-phenyl-2*H*-1,4-benzodiazepin-2-one
7-Chloro-1,3-dihydro-3-hydroxy-1,3-dimethyl-5-phenyl-2*H*-1,4-benzodiazepin-2-one
7-Chloro-1,3-dihydro-3-hydroxy-1-methyl-5-phenyl-2*H*-1,4-benzodiazepin-2-one
7-Chloro-2,3-dihydro-1-methyl-5-phenyl-1*H*-1,4-benzodiazepine
7-Chloro-1,3-dihydro-1-methyl-5-phenyl-2*H*-1,4-benzodiazepin-2-one

2-Chloro-10-(3-(4-methyl-1-piperazinyl)propyl)-10*H*-phenothiazine
See: Prochlorperazine

***p*-Chloro-N-methyl-N-propylbenzamide**
→ *p*-Chloro-N-methylbenzamide
In: Rat G13
→ *p*-Chloro-N-propylbenzamide
In: Rat G13

N-(3-Chloro-2-methyl-4-sulphatoxyphenyl) anthranilic acid
Probably formed from: N-(3-Chloro-4-hydroxy-2-methylphenyl)anthranilic acid

4-Chloro-2-methylsulphinylaniline
Probably formed from: 4-Chloro-2-(methylthio)aniline
→ 4-Chloro-2-methylsulphonylaniline
Probably in: Mucor F36

1-Chloro-3-methylsulphinyl-4-nitrobenzene
Probably formed from: 1-Chloro-3-methylthio-4-nitrobenzene
→ 1-Chloro-3-methylsulphonyl-4-nitrobenzene
Probably in: Mucor F36

4-Chloro-2-methylsulphonylaniline
Probably formed from: 4-Chloro-2-methylsulphinylaniline

1-Chloro-3-methylsulphonyl-4-nitrobenzene
Probably formed from: 1-Chloro-3-methylsulphinyl-4-nitrobenzene

4-Chloro-N′-methyl-N′-(tetrahydro-2-methyl-5-oxofurfur-2-yl)-*m*-benzenedisulphonamide
→ 4-Chloro-N′-(5-(4-hydroxy-4-methylvaleric acid))-N′-methyl-*m*-benzenedisulphonamide
In: Rat A741

3-Chloro-2-(methylthio)aniline
Probably formed from: 1-Chloro-2-methylthio-3-nitrobenzene

4-Chloro-2-(methylthio)aniline
Probably formed from: 1-Chloro-3-methylthio-4-nitrobenzene
→ 4-Chloro-2-methylsulphinylaniline
Probably in: Mucor F36

4-Chloro-3-methylthioaniline
Probably formed from: 1-Chloro-2-methylthio-4-nitrobenzene

1-Chloro-2-methylthio-3-nitrobenzene
Formed from: 1,2-Dichloro-3-nitrobenzene
→ 3-Chloro-2-(methylthio)aniline
Probably in: Mucor D463

1-Chloro-2-methylthio-4-nitrobenzene
Formed from: 1,2-Dichloro-4-nitrobenzene
→ 4-Chloro-3-(methylthio)aniline
Probably in: Mucor B955

1-Chloro-3-methylthio-4-nitrobenzene
Formed from: 1,3-Dichloro-4-nitrobenzene
Probably formed from: 5-Chloro-2-nitrothiophenol
→ 4-Chloro-2-(methylthio)aniline
Probably in: Mucor D463, F36
→ 1-Chloro-3-methylsulphinyl-4-nitrobenzene
Probably in: Mucor F36

***m*-Chloro-2-methyl-*p*-valerotoluidide**
See: Solan

Chloromonilicin
Formed from: Moniliphenone

4-Chloro-N-(2-(4-morpholinyl)ethyl)benzamide
See: Moclobemide

1-Chloronaphthalene
→ *cis*-8-Chloro-1,2-dihydro-1,2-dihydroxynaphthalene
In: Pseudomonas J370
→ 4-Chloro-α-naphthol
In: Pig A2266
See also: A1829

2-Chloronaphthalene
→ *cis*-7-Chloro-1,2-dihydro-1,2-dihydroxynaphthalene
In: Pseudomonas J370
→ 3-Chloro-β-naphthol
In: Pig A2266
→ 4-Chlorosalicylic acid
In: Pseudomonas B899, B905
See also: C147

4-Chloro-α-naphthol
Formed from: 1-Chloronaphthalene

3-Chloro-β-naphthol
Formed from: 2-Chloronaphthalene

7-Chloro-β-naphthol
Probably formed from: 2,7-Dichloronaphthalene

2-Chloro-1,4-naphthoquinone
→ 2-Chloro-3-hydroxy-1,4-naphthoquinone
In: Pseudomonas F46, K897

2-(7-Chloro-1,8-naphthyridin-2-yl)-3-((4-hydroxypiperidin-1-yl)carbonylmethyl) isoindolin-1-one
Formed from: Pazinadone

Chloroneb
See: 2,5-Dichloro-1,4-dimethoxybenzene

2-Chloro-4-nitroaniline
Formed from: N-(2-Chloro-4-nitrophenyl)benzamide
N-(2-Chloro-4-nitrophenyl)salicylamide
2-Chloro-5-nitroaniline
Formed from: 1-Chloro-2,4-dinitrobenzene
4-Chloro-3-nitroaniline
Formed from: 1-Chloro-2,4-dinitrobenzene
2-Chloro-5-nitroaniline
2-Chloro-5-nitrobenzaldehyde
→ 2-(S-Glutathionyl)-5-nitrobenzaldehyde
In: Rat H516
4-Chloro-3-nitrobenzaldehyde
→ 4-Chloro-3-nitrobenzyl alcohol
In: Rat H759
***o*-Chloronitrobenzene**
Formed from: *o*-Chloroaniline
→ *o*-Chloroaniline
In: Rat F530
→ S-(*o*-Nitrophenyl)glutathione
In: Rat F530
***m*-Chloronitrobenzene**
Formed from: *m*-Chloroaniline
→ *m*-Chloroaniline
In: Rat F530
See also: K2
***p*-Chloronitrobenzene**
Formed from: *p*-Chloroaniline
p-Chloronitrosobenzene
→ 2-Amino-5-chlorophenol
In: Comamonas K80
→ *p*-Chloroaniline
In: Fusarium A700
Man G519, G847
Rat F530, G871
See also: H2, K2
→ 2-Chloro-5-nitrophenol
In: Man G519, G847
Rat G871
→ *p*-Chloronitrosobenzene
In: Rhodosporidium B799
→ S-(*p*-Nitrophenyl)glutathione
In: Rat F530, H516
2-Chloro-5-nitrobenzonitrile
→ 2-(S-Glutathionyl)-5-nitrobenzonitrile
In: Rat H516
2-Chloro-5-nitrobenzophenone
→ 2-(S-Glutathionyl)-5-nitrobenzophenone
In: Rat H516
4-Chloro-3-nitrobenzyl alcohol
Formed from: 4-Chloro-3-nitrobenzaldehyde
2-Chloro-5-nitrophenol
Formed from: *p*-Chloronitrobenzene
4-Chloro-2-nitrophenol
→ 4-Chlorocatechol
In: Pseudomonas E607, E726, H65
6-Chloro-2-nitrophenol
→ 3-Chlorocatechol
In: Pseudomonas E607, H65
N-(2-Chloro-4-nitrophenyl)benzamide
→ N-(4-Amino-2-chlorophenyl)benzamide
In: Ascaris A2037
Moniezia A2037
Mouse A2037
Sheep A2037
→ Benzoic acid
In: Ascaris A2037
Moniezia A2037
Mouse A2037
Sheep A2037
→ 2-Chloro-4-nitroaniline
In: Ascaris A2037
Moniezia A2037
Mouse A2037
Sheep A2037
S-(5-Chloro-2-nitrophenyl)-L-cysteine
Probably formed from: S-(5-Chloro-2-nitrophenyl)glutathione
→ N-Acetyl-S-(5-chloro-2-nitrophenyl)-L-cysteine
Probably in: Mucor F36
→ 5-Chloro-2-nitrothiophenol
Probably in: Mucor F36
S-(5-Chloro-2-nitrophenyl)glutathione
→ S-(5-Chloro-2-nitrophenyl)-L-cysteine
Probably in: Mucor F36
N-(2-Chloro-4-nitrophenyl)salicylamide
→ N-(4-Amino-2-chlorophenyl)salicylamide
In: Ascaris A2037
Moniezia A2037
Mouse A2037
Sheep A2037

→ 2-Chloro-4-nitroaniline
In: Mouse A2037
→ Salicylic acid
In: Mouse A2037

m-Chloronitrosobenzene
→ N-(*m*-Chlorophenyl)isobutyrohydroxamic acid
In: Rat J429

p-Chloronitrosobenzene
Formed from: *p*-Chloroaniline
p-Chloro-N-hydroxyacetanilide
p-Chloronitrobenzene
p-Chlorophenylhydroxylamine
→ *p*-Chloronitrobenzene
In: Pea C797
See also: A700, C879
→ N-(*p*-Chlorophenyl)glycolhydroxamic acid
In: Rat B130
→ *p*-Chlorophenylhydroxylamine
In: Rhodosporidium B799
Yeast B207
→ N-(*p*-Chlorophenyl)isobutyrohydroxamic acid
In: Rat J429

5-Chloro-2-nitrothiophenol
Probably formed from: S-(5-Chloro-2-nitrophenyl)-L-cysteine
→ Bis(5-chloro-2-nitrobenzene)disulphide
In: Mucor F36
→ 1-Chloro-3-methylthio-4-nitrobenzene
Probably in: Mucor F36

2-Chloro-5-nitro-1-trifluoromethylbenzene
→ 2-(S-Glutathionyl)-5-nitro-1-trifluoromethylbenzene
In: Rat H516

2′-Chloronordiazepam
See: 7-Chloro-5-(*o*-chlorophenyl)-1,3-dihydro-2*H*-1,4-benzodiazepin-2-one

p-Chloronorephedrine
Formed from: *p*-Chloroamphetamine

16α-Chloro-17β-oestradiol
Formed from: 16α-Chlorooestrone

16α-Chlorooestrone
→ 16α-Chloro-17β-oestradiol
In: Rat A1749

p-Chlorooxanilic acid
Formed from: *p*-Chloroaniline
Probably formed from: *p*-Chloroglycollanilide

6-Chlorooxindol-5-ylacetic acid
Probably formed from: 6-Chloro-5-formylmethyloxindole
→ 6-Chlorooxindol-5-ylacetyl-β-D-glucuronic acid
In: Man J279

6-Chlorooxindol-5-ylacetyl-β-D-glucuronic acid
Formed from: 6-Chlorooxindol-5-ylacetic acid

1-(2-(6-Chlorooxindol-5-yl)ethyl)-4-(2-sulphonamido)benzoylpiperazine
Probably formed from: Ziprasidone sulphoxide

1-((5-Chloro-2-oxo-3(2H)-benzothiazolyl)acetyl)piperazine
Formed from: Tiaramide
→ 4-((5-Chloro-2-oxo-3(2*H*)-benzothiazolyl)acetyl)piperazinesulphonic acid
In: Guinea pig H455
Man H455, J571
Monkey H455
Mouse C571, H455
Rabbit H455, J914
Rat C571, E45, E51, H455
See also: C487

4-((5-Chloro-2-oxo-3(2H)-benzothiazolyl)acetyl)-1-piperazineacetic acid
Formed from: Tiaramide
→ 4-((5-Chloro-2-oxo-3(2*H*)-benzothiazolyl)acetyl)-1-piperazineacetic acid-N-oxide
In: Man C301
See also: B981, C298

4-((5-Chloro-2-oxo-3(2H)-benzothiazolyl)acetyl)-1-piperazineacetic acid-N-oxide
Formed from: 4-((5-Chloro-2-oxo-3(2*H*)-benzothiazolyl)acetyl)-1-piperazineacetic acid

4-((5-Chloro-2-oxo-3(2H)-benzothiazolyl)acetyl)-1-piperazineethanol
See: Tiaramide

4-((5-Chloro-2-oxo-3(2H)-benzothiazolyl)acetyl)piperazinesulphonic acid
Formed from: 1-((5-Chloro-2-oxo-3(2*H*)-benzothiazolyl)acetyl)piperazine

6-Chloro-5-(4′-oxocyclohexyl)indane-1-carboxylic acid
Probably formed from: *trans*-6-Chloro-5-(4′-hydroxycyclohexyl)indane-1-carboxylic acid

2-Chloro-3-oxo-1,3-diphenyl-1-propene
Probably formed from: 2-Chloro-3-hydroxy-1,3-diphenyl-1-propene
1,1-Dichloro-*cis*-2,3-diphenylcyclopropane

6-Chloro-4-oxo-10-propyl-4*H*-pyrano[3,2-*g*]quinoline-2,8-dicarboxylic acid
→ 6-(S-Glutathionyl)-4-oxo-10-propyl-4*H*-pyrano[3,2-*g*]quinoline-2,8-dicarboxylic acid
In: Rat D370

***p*-Chlorophenethylamine**
→ *p*-Chlorophenylacetaldehyde
In: Alcaligenes H367
See also: F305

***o*-Chlorophenol**
Formed from: *o*-Chloroanisole
Chlorobenzene
3-Chloro-4-hydroxybenzoic acid
1-(*o*-Chlorophenyl)ethanol
2,3-Dichlorophenol
2,5-Dichlorophenol
2,6-Dichlorophenol
Lindane
2,2′,4-Trichlorodiphenyl ether
→ *m*-Chlorobenzoic acid
In: Microorganism G917
→ 3-Chlorocatechol
In: Bacillus J340
Candida E204, K653
Nocardia A1192
Proteobacteria J466
Pseudomonas E731, G627
Trichosporon A902
See also: E565, F857, J344
→ *o*-Chlorophenyl-β-D-glucuronide
In: Bitterling C406
Bream C406
Carp C406
Goldfish C406
Gudgeon C406
Guppy C406
Minnow C406
Perch C406
Roach C406
Rudd C406
Stickleback C406
Tench C406
→ *o*-Chlorophenyl sulphate
In: Bitterling C406
Bream C406
Carp C406
Goldfish C406
Gudgeon C406
Guppy C406
Man E575
Minnow C406
Perch C406
Rat B61, E575
Roach C406
Rudd C406
Stickleback C406
Tench C406
→ 3-Chloro-L-tyrosine
In: Escherichia B292
→ Phenol
In: Desulfitobacterium K217
Microorganism K584

***m*-Chlorophenol**
Formed from: N-Acetyl-*m*-chlorophenyl-β-D-glucosaminide
m-Chloroanisole
Chlorobenzene
m-Chlorophenoxyacetic acid
1-(*m*-Chlorophenyl)ethanol
2,3-Dichlorophenol
3,5-Dichlorophenol
→ 3-Chlorocatechol
In: Pseudomonas E731
→ 4-Chlorocatechol
In: Candida K653
Nocardia A1192
Rhodococcus F258, F383
Rhodotorula A982
Trichosporon A902
See also: A2391, E204, E565 G817
→ *m*-Chlorophenyl sulphate
In: Man E575
Rat B61, B941, E575

→ 2-Chloro-L-tyrosine
In: Erwinia B292

***p*-Chlorophenol**
Formed from: N-Acetyl-*p*-chlorophenyl-β-D-glucosaminide
2-Bromo-4-chlorohenol
p-Chloroanisole
Chlorobenzene
p-Chlorophenoxyacetic acid
m-(*p*-Chlorophenoxy)benzoic acid
p-Chlorophenyl benzoate
1-(*p*-Chlorophenyl)ethanol
Chlorphenesin carbamate
Clofibride
4,4′-Dichlorodiphenyl ether
2,4-Dichlorophenol
3,4-Dichlorophenol
2,4,4′,5-Tetrachlorodiphenyl ether
→ 3-Chlorocatechol
In: Comamonas J468
→ 4-Chlorocatechol
In: Acinetobacter C232
Candida E204, K653
Penicillium H82
Proteobacteria J466
Pseudomonas E731, G627
Rhodococcus F258, F383
Rhodotorula A982
Trichosporon A902
See also: B897, E565, F895, G932
→ 4-(*p*-Chlorophenoxy)-2,6-dimethoxyphenol
In: Rhizoctonia D690
→ 4-(4-(*p*-Chlorophenoxy)-2,6-dimethoxyphenoxy)-2,6-dimethoxyphenol
In: Rhizoctonia D690
→ 4-(4-(4-(*p*-Chlorophenoxy)-2,6-dimethoxyphenoxy)-2,6-dimethoxyphenoxy)-2,6-dimethoxyphenol
In: Rhizoctonia D690
→ 4-(4-(4-(4-(*p*-Chlorophenoxy)-2,6-dimethoxyphenoxy)-2,6-dimethoxyphenoxy)-2,6-dimethoxyphenoxy)-2,6-dimethoxyphenol
In: Rhizoctonia D690
→ 5-(*p*-Chlorophenoxy)-3-methoxy-*o*-quinone
In: Rhizoctonia D690
→ *p*-Chlorophenyl-β-D-glucuronide
In: Mouse C80
Rat B555, B644
Zebra fish E813
→ *p*-Chlorophenyl sulphate
In: Cat A2686
Rat B61, B644, B941, E575, K735
Zebra fish E813
→ Quinol
In: Rat J81

2-Chlorophenothiazine
Formed from: Chlorpromazine
→ 2-Chloro-7-hydroxyphenothiazine
Probably in: Man A693
→ 2-Chloro-N-hydroxyphenothiazine
In: Guinea pig A322
Rabbit A322
Contrast (mis-identification claim) A2923, A3488
→ 2-Chlorophenothiazine-N-hydroperoxide
In: Guinea pig A322
Rabbit A322
Contrast (mis-identification claim) A2923, A3488
→ 2-Chlorophenothiazine-N-oxide
In: Guinea pig A322
Rabbit A322
→ 2-Chlorophenothiazine sulphoxide
Probably in: Man A693

2-Chlorophenothiazine-N-hydroperoxide (identity challenged)
Formed from: 2-Chlorophenothiazine

2-Chlorophenothiazine-N-oxide
Formed from: 2-Chlorophenothiazine

2-Chlorophenothiazine sulphoxide
Probably formed from: 2-Chlorophenothiazine

N-(γ-(2-Chloro-10-phenothiazinyl)propyl) ethylenediamine
Formed from: Perphenazine
Prochlorperazine
→ Bisdemethylchlorpromazine
Probably in: Rat A69

N-(γ-(2-Chloro-10-phenothiazinyl)propyl)-N′-hydroxymethylethylenediamine
Formed from: Perphenazine

N-(γ-(2-Chloro-10-phenothiazinyl)propyl)-N′-methylethylenediamine
Formed from: Prochlorperazine
→ Bisdemethylchlorpromazine
Probably in: Rat A69

***p*-Chlorophenoxyacetamide**
Formed from: Iproclozide

***m*-Chlorophenoxyacetic acid**
→ *m*-Chlorophenol
In: Microorganism H104

***p*-Chlorophenoxyacetic acid**
Formed from: Chlorphenesin carbamate
2,4-Dichlorophenoxyacetic acid
→ *p*-Chlorophenol
Probably in: Azotobacter F895

1-(*p*-Chlorophenoxyacetyl)-2-isopropylhydrazine
See: Iproclozide

3-(3-Chlorophenoxy)-4-((aminocarbonyl)amino)butanoic acid
Probably formed from: 3-(3-Chlorophenoxy)-4-(((methylamino)carbonyl)amino)butanoic acid

***m*-(*p*-Chlorophenoxy)benzoic acid**
→ *p*-Chlorophenol
In: Pseudomonas F696, G206
→ *m*-Hydroxybenzoic acid
Probably in: Pseudomonas G206

4-(*p*-Chlorophenoxy)-2,6-dimethoxyphenol
Formed from: *p*-Chlorophenol
4-Hydroxy-3,5-dimethoxybenzoic acid

4-(4-(*p*-Chlorophenoxy)-2,6-dimethoxyphenoxy)-2,6-dimethoxyphenol
Formed from: *p*-Chlorophenol
4-Hydroxy-3,5-dimethoxybenzoic acid

4-(4-(4-(*p*-Chlorophenoxy)-2,6-dimethoxyphenoxy)-2,6-dimethoxyphenoxy)-2,6-dimethoxyphenol
Formed from: *p*-Chlorophenol
4-Hydroxy-3,5-dimethoxybenzoic acid

4-(4-(4-(4-(*p*-Chlorophenoxy)-2,6-dimethoxyphenoxy)-2,6-dimethoxyphenoxy)-2,6-dimethoxyphenoxy)-2,6-dimethoxyphenol
Formed from: *p*-Chlorophenol
4-Hydroxy-3,5-dimethoxybenzoic acid

3-(*o*-Chlorophenoxy)-N-(2-(ethoxycarbonyl)ethyl)-2-hydroxypropylamine
→ N-(2-Carboxyethyl)-3-(*o*-chlorophenoxy)-2-hydroxypropylamine
In Rabbit H563

2-(*p*-Chlorophenoxy)isobutyric acid
See: Clofibric acid

***p*-Chlorophenoxylactic acid**
Formed from: Chlorphenesin carbamate

5-(*p*-Chlorophenoxy)-3-methoxy-*o*-quinone
Formed from: *p*-Chlorophenol
4-Hydroxy-3,5-dimethoxybenzoic acid

3-(3-Chlorophenoxy)-4-(((methylamino)carbonyl)amino)butanoic acid
Formed from: 3-(3-Chlorophenoxy)pyrrolidine-1-(N-methyl)carboxamide
→ 3-(3-Chlorophenoxy)-4-((aminocarbonyl)amino)butanoic acid
Probably in: Dog A2114
Man A2114
Rat A2114

S-(*p*-Chlorophenoxy)-2-methylpropanoyl)glutathione
Formed from: Clofibryl-β-D-glucuronide

3-(4-Chlorophenoxy)-1,2-propanediol 1-carbamate
See: Chlorphenesin carbamate

3-(3-Chlorophenoxy)pyrrolidine-1-carboxamide
Formed from: 3-(3-Chlorophenoxy)pyrrolidine-1-(N-methyl)carboxamide
→ Hydroxy-3-(3-chlorophenoxy)pyrrolidine-1-carboxamide
Probably in: Rat A2114

3-(3-Chlorophenoxy)pyrrolidine-1-(N-methyl)carboxamide
→ 3-(3-Chloro-4-hydroxyphenoxy)pyrrolidine-1-(N-methyl)carboxamide
In: Dog A2114
Man A2114
Rat A2114
→ 3-(3-Chlorophenoxy)-4-(((methylamino)carbonyl)amino)butanoic acid
In: Dog A2114
Man A2114
Rat A2114
→ 3-(3-Chlorophenoxy)pyrrolidine-1-carboxamide
In: Dog A2114
Man A2114
Rat A2114

Chlorophentermine
See: *p*-Chloro-α,α-dimethylphenethylamine
***m*-Chlorophenylacetaldehyde**
→ *m*-Chlorophenylacetic acid
In: Rat H761
***p*-Chlorophenylacetaldehyde**
Formed from: *p*-Chlorophenethylamine
→ *p*-Chlorophenylacetic acid
In: Rat H761
***p*-Chlorophenylacetaldoxime**
→ *p*-Chlorophenylacetonitrile
In: Bacillus K411
***o*-Chlorophenyl acetate**
Formed from: *o*-Chloroacetophenone
***m*-Chlorophenyl acetate**
Formed from: *m*-Chloroacetophenone
***p*-Chlorophenyl acetate**
Formed from: *p*-Chloroacetophenone
***o*-Chlorophenylacetic acid**
Formed from: *o*-Chlorobenzylidene-malononitrile
o-Chlorobenzylmalononitrile
***m*-Chlorophenylacetic acid**
Formed from: *m*-Chlorophenylacetaldehyde
***p*-Chlorophenylacetic acid**
Formed from: Bis(*p*-chlorophenyl)methane
p-Chlorophenylacetaldehyde
p-Chlorophenylacetonitrile
p-Chloro-L-phenylalanine
Probably formed from: 3,3-Dichloro-2-(*p*-chlorophenyl)propionic acid
→ 3-Chloro-4-hydroxyphenylacetic acid
In: Pseudomonas B798, C233
→ 4-Chloro-2-hydroxyphenylacetic acid
In: Pseudomonas C233
→ 4-Chloro-3-hydroxyphenylacetic acid
In: Pseudomonas B798, C233
→ *p*-Chlorophenylacetyl-β-D-glucuronic acid
In: Pigeon A395
→ *p*-Chlorophenylacetylglycine
In: Ferret A394, A3046
Hen A395
Pigeon A395
Rat A2144
→ *p*-Chlorophenylacetylornithine
In: Hen A395
→ *p*-Chlorophenylacetyltaurine
In: Ferret A394, A3046
Hen A395
Pigeon A395
→ 3,4-Dihydroxyphenylacetic acid
In: Pseudomonas B798, E163
***o*-Chlorophenylacetone**
→ 1-(*o*-Chlorophenyl)-2-propanol
In: Rabbit C25
***p*-Chlorophenylacetone**
→ 1-(*p*-Chlorophenyl)-2-propanol
In: Rabbit C25
ω-Chlorophenylacetone
→ ω-Chlorophenylacetone cyanhydrin
In: Almond J334
Apple J334
ω-Chlorophenylacetone cyanhydrin
Formed from: ω-Chlorophenylacetone
***p*-Chlorophenylacetonitrile**
Formed from: *p*-Chlorophenylacetaldoxime
→ *p*-Chlorophenylacetic acid
In: Rhodococcus G172
***p*-Chlorophenylacetyl-β-D-glucuronic acid**
Formed from: *p*-Chlorophenylacetic acid
***p*-Chlorophenylacetylglycine**
Formed from: *p*-Chlorophenylacetic acid
***p*-Chlorophenylacetylornithine**
Formed from: *p*-Chlorophenylacetic acid
***p*-Chlorophenylacetyltaurine**
Formed from: *p*-Chlorophenylacetic acid
***o*-Chloro-L-phenylalanine**
→ *o*-Chlorophenylpyruvic acid
In: Bushbean E444, E500, E640
***m*-Chloro-L-phenylalanine**
→ *m*-Chlorophenylpyruvic acid
In: Bushbean E444, E500, E640
***p*-Chloro-L-phenylalanine**
→ *p*-Chlorophenylacetic acid
In: Rat A2144
→ *p*-Chlorophenylpyruvic acid
In: Bushbean E444, E500, E640
N-((*p*-Chlorophenyl)amino)carbonyl-2,6-difluorobenzamide
See: Diflubenzuron
N-((*p*-Chlorophenyl)amino)carbonyl-2,6-difluoro-3-hydroxybenzamide
Formed from: Diflubenzuron

N-(((4-Chlorophenyl)amino)carbonyl)-2,3-dihydro-1*H*-indene-5-sulphonamide
See: Sulofenur

***p*-Chlorophenyl azide**
→ *p*-Chloroaniline
In: Mouse G8

1-(*p*-Chlorophenylazo)-4-hydroxypiperidine
Formed from: 1-(*p*-Chlorophenylazo) piperidine

1-(*p*-Chlorophenylazo)piperidine
→ *p*-Chloroaniline
In: Rat A2922
→ 1-(*p*-Chlorophenylazo)-4-hydroxypiperidine
In: Rat A2922
→ 1-(*p*-Chlorophenylazo)-1,2,3,4-tetrahydropyridine
In: Rat A2922

1-(*p*-Chlorophenylazo)-1,2,3,4-tetrahydropyridine
Formed from: 1-(*p*-Chlorophenylazo) piperidine

N-(*o*-Chlorophenyl)benzamide
→ Benzoic acid
In: Ascaris A2037
Moniezia A2037
Mouse A2037
Sheep A2037
→ *o*-Chloroaniline
In: Ascaris A2037
Moniezia A2037
Mouse A2037
Sheep A2037

***p*-Chlorophenyl benzoate**
→ Benzoic acid
In: Pseudomonas K206
→ *p*-Chlorophenol
In: Pseudomonas K206

1-(*p*-Chloro-α-phenylbenzyl)-4-methylpiperazine
See: Chlorcyclizine

(3-Chlorophenyl)carbamic acid 1-methylethyl ester
See: Chlorpropham

2-(*o*-Chlorophenyl)-2-(*p*-chlorophenyl)acetic acid
Formed from: 1,1-Dichloro-2-(*o*-chlorophenyl)-2-(*p*-chlorophenyl)acetic acid
1,1-Dichloro-2-(*o*-chlorophenyl)-2-(*p*-chlorophenyl)ethane
1,1,1-Trichloro-2-(*o*-chlorophenyl)-2-(*p*-chlorophenyl)acetic acid
→ 2-(*o*-Chlorophenyl)-2-(*p*-chlorophenyl) acetylglycine
In: Rat A991
→ 2-(*o*-Chlorophenyl)-2-(*p*-chlorophenyl) acetyl-L-serine
In: Rat A991
→ 2-(*o*-Chlorophenyl)-2-(*p*-chlorophenyl) glycollic acid
Probably in: Dog E876
→ 2-(*o*-Chloro-3-hydroxyphenyl)-2-(*p*-chlorophenyl)acetic acid
Probably in: Chicken A2368
Rat A991
→ 2-(*o*-Chloro-4-hydroxyphenyl)-2-(*p*-chlorophenyl)acetic acid
Probably in: Chicken A2368
Man A1526
Rat A991
→ 2-(*o*-Chloro-5-hydroxyphenyl)-2-(*p*-chlorophenyl)acetic acid
Probably in: Rat A991
→ 2,4′-Dichlorobenzhydrol
Probably in: Rat A991
→ Methyl 2-(*o*-chlorophenyl)-2-(*p*-chlorophenyl)acetate
Probably in: Chicken A2368

2-(*o*-Chlorophenyl)-2-(*p*-chlorophenyl) acetylglycine
Formed from: 2-(*o*-Chlorophenyl)-2-(*p*-chlorophenyl)acetic acid

2-(*o*-Chlorophenyl)-2-(*p*-chlorophenyl)acetyl-L-serine
Formed from: 2-(*o*-Chlorophenyl)-2-(*p*-chlorophenyl)acetic acid

2-(*o*-Chlorophenyl)-2-(*p*-chlorophenyl)glycollic acid
Probably formed from: 2-(*o*-Chlorophenyl)-2-(*p*-chlorophenyl)acetic acid

3-(*o*-Chlorophenyl)-2-cyanopropionic acid
Formed from: *o*-Chlorobenzylidene-malononitrile
o-Chlorobenzylmalononitrile

(*S*)-(+)-6-(*o*-Chlorophenyl)-3-cyclopropanecarbonyl-2,3,4,5-tetrahydro-8,11-dimethyl-8*H*-

pyrido[4′,3′:4,5]thieno[3,2-*f*][1,2,4]triazolo[4,3-*a*][1,4]diazepine
→ 6-(*o*-Chlorophenyl)-2,3,4,5-tetrahydro-8,11-dimethyl-8*H*-pyrido[4′,3′:4,5]thieno[3,2-*f*][1,2,4]triazolo[4,3-*a*][1,4]diazepine
In: Rhesus monkey J66

***p*-Chlorophenyl-2,3-dehydroisovaleric acid**
Probably formed from: *p*-Chlorophenylisovaleric acid

1-(*p*-Chlorophenyl)-3-(4,5-dichloro-2-hydroxyphenyl)urea
Formed from: 1-(*p*-Chlorophenyl)-3-(4,5-dichlorophenyl)urea
→ 1-(4-Chloro-2-hydroxyphenyl)-3-(4,5-dichloro-2-hydroxyphenyl)urea
Probably in: Rat A2876
→ 1-(4-Chloro-3-hydroxyphenyl)-3-(4,5-dichloro-2-hydroxyphenyl)urea
Probably in: Rat A2876

1-(*p*-Chlorophenyl)-3-(4,5-dichlorophenyl)urea
→ 1-(4-Chloro-2-hydroxyphenyl)-3-(4,5-dichlorophenyl)urea
Probably in: Rat A2876
→ 1-(4-Chloro-3-hydroxyphenyl)-3-(4,5-dichlorophenyl)urea
Probably in: Rat A2876
→ 1-(*p*-Chlorophenyl)-3-(4,5-dichloro-2-hydroxyphenyl)urea
Probably in: Rat A2876

5-(*o*-Chlorophenyl)-1,3-dihydro-2*H*-benzodiazepin-2-one
→ 5-(*o*-Chlorophenyl)-1,3-dihydro-3-hydroxy-2*H*-benzodiazepin-2-one
In: Mouse A3269
Rat A3269

5-(*p*-Chlorophenyl)-2,3-dihydro-2,5-dihydroxy-5*H*-imidazo[2,1-*a*]isoindole
Formed from: Mazindol

5-(*p*-Chlorophenyl)-2,5-dihydro-2,5-dihydroxy-3*H*-imidazo[2,1-*a*]isoindol-3-one
Probably formed from: 5-(*p*-Chlorophenyl)-2,5-dihydro-5-hydroxy-3*H*-imidazo[2,1-*a*]isoindol-3-one

10-(*p*-Chlorophenyl)-2,10-dihydro-10-(*p*-hydroxyanilino)-2-isopropyliminophenazine
Formed from: Clofazimine

5-(*o*-Chlorophenyl)-1,3-dihydro-3-hydroxy-2*H*-benzodiazepin-2-one
Formed from: 5-(*o*-Chlorophenyl)-1,3-dihydro-2*H*-benzodiazepin-2-one

5-(*p*-Chlorophenyl)-2,5-dihydro-5-hydroxy-3*H*-imidazo[2,1-*a*]isoindol-3-one
Formed from: Mazindol
→ 2-Amino-1-oxoethyl)-3-(*p*-chlorophenyl)-3-hydroxyphthalimidine
Probably in: Dog A3673
Man A3673
Rat A3673
→ 5-(*p*-Chlorophenyl)-2,5-dihydro-2,5-dihydroxy-3*H*-imidazo[2,1-*a*]isoindol-3-one
Probably in: Rat A1999

N-(*m*-Chlorophenyl)-N′-(4,5-dihydro-5-hydroxy-1-methyl-4-oxo-1*H*-imidazol-2-yl)urea
Formed from: Fenobam

3-(4′-Chlorophenyl)-3,4-dihydro-6-hydroxy-2-methyl-4-oxopyrido[2,3-*d*]pyrimidine
Formed from: 3-(4′-Chlorophenyl)-3,4-dihydro-2-methyl-4-oxopyrido[2,3-*d*]pyrimidine
→ 3-(4′-Chlorophenyl)-3,4-dihydro-2-methyl-4-oxopyrido[2,3-*d*]pyrimidin-6-yl-β-D-glucuronide
In: Rat A2163

10-(*p*-Chlorophenyl)-2,10-dihydro-2-isopropyliminophenazin-3-yl-β-D-glucuronide
Formed from: Clofazimine

N-(*m*-Chlorophenyl)-N′-(4,5-dihydro-1-methyl-4-oxo-1*H*-imidazol-2-yl)urea
See: Fenobam

3-(4′-Chlorophenyl)-3,4-dihydro-2-methyl-4-oxopyrido[2,3-*d*]pyrimidine
→ 3-(4′-Chlorophenyl)-3,4-dihydro-6-hydroxy-2-methyl-4-oxopyrido[2,3-*d*]pyrimidine
In: Rat A2163, A2235
→ 3-(4′-Chlorophenyl)-3,4-dihydro-2-methyl-4-oxopyrido[2,3-*d*]pyrimidine-8-oxide
In: Rat A2163, A2235
→ 3-(4′-Chlorophenyl)-3,4-dihydro-2-methyl-4-oxopyrrolo[2,3-*d*]pyrimidine
In: Rat A2163, A2235
→ 3-(4′-Chlorophenyl)-3,4,7,8-tetrahydro-2-methyl-4,7-dioxopyrido[2,3-*d*]pyrimidine
In: Rat A2163, A2235

3-(4′-Chlorophenyl)-3,4-dihydro-2-methyl-4-oxopyrido[2,3-*d*]pyrimidine-8-oxide
Formed from: 3-(4′-Chlorophenyl)-3,4-dihydro-2-methyl-4-oxopyrido[2,3-*d*]pyrimidine

3-(4′-Chlorophenyl)-3,4-dihydro-2-methyl-4-oxopyrido[2,3-*d*]pyrimidin-6-yl-β-D-glucuronide
Formed from: 3-(4′-Chlorophenyl)-3,4-dihydro-6-hydroxy-2-methyl-4-oxopyrido[2,3-*d*]pyrimidine

3-(4′-Chlorophenyl)-3,4-dihydro-2-methyl-4-oxopyrrolo[2,3-*d*]pyrimidine
Formed from: 3-(4′-Chlorophenyl)-3,4-dihydro-2-methyl-4-oxopyrido[2,3-*d*] pyrimidine

5-(*o*-Chlorophenyl)-1,3-dihydro-7-nitro-2*H*-1,4-benzodiazepin-2-one
See: Clonazepam

***erythro*-2-(*o*-Chlorophenyl)-2-(4-(3,4-dihydroxybenzyl)-1-piperazinyl)-1-phenylethanol**
Probably formed from: *erythro*-2-(*o*-Chlorophenyl)-2-(4-(*p*-hydroxybenzyl)-1-piperazinyl)-1-phenylethanol
→ *erythro*-2-(*o*-Chlorophenyl)-2-(4-(3-hydroxy-4-methoxybenzyl)-1-piperazinyl)-1-phenylethanol
Probably in: Rat A3927
→ *erythro*-2-(*o*-Chlorophenyl)-2-(4-(4-hydroxy-3-methoxybenzyl)-1-piperazinyl)-1-phenylethanol
Probably in: Rat A3927

***cis*-(−)-2-(2-Chlorophenyl)-5,7-dihydroxy-8-(3-hydroxy-1-methyl-4-piperidinyl)-4*H*-1-benzopyran-1-one**
See: Flavopyridol

2-(*p*-Chlorophenyl)-2,3-dihydroxyisovaleric acid
Probably formed from: 2-(*p*-Chlorophenyl)-3-hydroxyisovaleric acid

2-(*p*-Chlorophenyl)-2,4-dihydroxyisovaleric acid
Probably formed from: 2-(*p*-Chlorophenyl)-4-hydroxyisovaleric acid

N-(*p*-Chlorophenyl)-α,4-dihydroxy-3-methoxy-N-(1-(1-methylethyl)-4-piperidinyl) benzeneacetamide
Probably formed from: N-(*p*-Chlorophenyl)-4-hydroxy-3-methoxy-N-(1-(1-methylethyl)-4-piperidinyl)benzeneacetamide

***p*-Chlorophenyl 3,4-dihydroxyphenyl ether**
Probably formed from: *p*-Chlorophenyl *p*-hydroxyphenyl ether
→ *p*-Chlorophenyl 3-hydroxy-4-methoxyphenyl ether
Probably in: Rat E881
→ *p*-Chlorophenyl 4-hydroxy-3-methoxyphenyl ether
Probably in: Rat E881

1-(*p*-Chlorophenyl)-4-dimethylamino-2,3-dimethylbutan-2-ol
→ Benzoic acid
In: Rabbit A2252
Rat A2252
→ *p*-Chlorobenzoic acid
In: Rabbit A2252
→ 1-(*p*-Chlorophenyl)-2,3-dimethyl-4-methylaminobutan-2-ol
In: Rabbit A2252
Rat A2252

1-(3-Chlorophenyl)-2-((1,1-dimethylethyl)amino)-1-propanone
See: Bupropion

1-(*o*-Chlorophenyl)-3-dimethylamino-1-phenyl-1-propene
Formed from: Clofedanol
→ 3-(*o*-Chlorophenyl)-3-phenylacrolein
Probably in: Man A3384

3-(*p*-Chlorophenyl)-1-dimethylamino-3-(2-pyridyl)propane
See: Chlorpheniramine

1-(3-Chlorophenyl)-3-(2-(3,3-dimethyl-1-azetidinyl)ethyl)imidazolin-2-one
See: Zetidoline

2-(*p*-Chlorophenyl)-1,1-dimethylethylhydroxylamine
Formed from: *p*-Chloro-α,α-dimethylphenethylamine

1-(*p*-Chlorophenyl)-2,3-dimethyl-4-methylaminobutan-2-ol
Formed from: 1-(*p*-Chlorophenyl)-4-dimethylamino-2,3-dimethylbutan-2-ol
→ 4-Amino-1-(*p*-chlorophenyl)-2,3-dimethylbutan-2-ol
In: Rabbit A2252
Rat A2252

2-(*p*-Chlorophenyl)-1,1-dimethylnitroethane (nitrochlorphentermine)
→ *p*-Chloro-N-hydroxy-α,α-dimethylphenethylamine
In: Rat A2057

1-(*p*-Chlorophenyl)-3,3-dimethyltriazene
→ *p*-Chloroaniline
In: Rat A2019, A2713, A2714
→ 1-(4-Chloro-2-hydroxyphenyl)-3,3-dimethyltriazene
Probably in: Rat A2714

3-(*p*-Chlorophenyl)-1,1-dimethylurea (monuron)
→ 3-(4-Chloro-2-hydroxyphenyl)-1,1-dimethylurea
In: Bean A940
→ 3-(*p*-Chlorophenyl)-1-methylurea
In: Mouse C47

2-(*p*-Chlorophenyl)-1,3-dithiolane
→ 2-(*p*-Chlorophenyl)-1,3-dithiolane-1-oxide
In: Rabbit F355

2-(*p*-Chlorophenyl)-1,3-dithiolane-1,3-dioxide
Formed from: 2-(*p*-Chlorophenyl)-1,3-dithiolane-1-oxide

2-(*p*-Chlorophenyl)-1,3-dithiolane-1-oxide
Formed from: 2-(*p*-Chlorophenyl)-1,3-dithiolane
→ 2-(*p*-Chlorophenyl)-1,3-dithiolane-1,3-dioxide
In: Rabbit F355

1-(*o*-Chlorophenyl)ethanol
Formed from: *o*-Chloroacetophenone
→ *o*-Chlorophenol
In: Alcaligenes G107

1-(*m*-Chlorophenyl)ethanol
Formed from: *m*-Chloroacetophenone
→ *m*-Chloroacetophenone
In: Geotrichum H749
→ *m*-Chlorophenol
In: Alcaligenes G107

1-(*p*-Chlorophenyl)ethanol
Formed from: *p*-Chloroacetophenone
1-Chloro-4-ethylbenzene
→ *p*-Chloroacetophenone
In: Geotrichum H749
→ *p*-Chlorophenol
In: Alcaligenes G107

2-(*p*-Chlorophenyl)ethanol
Formed from: 1-Chloro-4-ethylbenzene

2-Chloro-1-phenylethanol
Formed from: Phenacyl chloride

4-(*o*-Chlorophenyl)-3-ethoxycarbonyl-1,4-dihydro-5-(N-(3-hydroxy-2-pyridyl)carbamoyl-6-methyl-2-(4-(methylimidazo[4,5-*c*]pyrid-1-yl)phenyl)pyridine
Probably formed from: Modipafant

4-(*o*-Chlorophenyl)-3-ethoxycarbonyl-1,4-dihydro-5-(N-(3-hydroxy-2-pyridyl)carbamoyl-6-methyl-2-(4-(methylimidazo[4,5-*c*]pyrid-1-yl)phenyl)pyridine-N-oxide
Probably formed from: Modipafant-N-oxide

4-(*o*-Chlorophenyl)-3-ethoxycarbonyl-2-hydroxymethyl-5-methoxycarbonyl-6-methylpyridine
Probably formed from: 2-((2-Aminoethoxy)methyl)-4-(*o*-chlorophenyl)-3-ethoxycarbonyl-5-methoxycarbonyl-6-methylpyridine
→ 2-Carboxy-4-(*o*-chlorophenyl)-3-ethoxycarbonyl-5-methoxycarbonyl-6-methylpyridine
Probably in: Dog F41
Man F41
Rat F41
→ 3-Carboxy-4-(*o*-chlorophenyl)-2-hydroxymethyl-5-methoxycarbonyl-6-methylpyridine
Probably in: Dog F41
Man F41

4-(*o*-Chlorophenyl)-3-ethoxycarbonyl-5-(N-(3-hydroxy-2-pyridyl)carbamoyl-6-methyl-2-(4-(methylimidazo[4,5-*c*]pyrid-1-yl)phenyl)pyridine
Probably formed from: 1,4-Dehydromodipafant

S-(*p*-Chlorophenyl) O-ethyl ethanephosphonodithioate
→ *p*-Chlorothiophenol
In: Rat A1946

4-(2-(*m*-Chlorophenyl)ethyl)-1-methylpyridinium
Probably formed from: 4-(2-(*m*-Chlorophenyl)ethyl)-1,2,3,6-tetrahydro-1-methylpyridine

4-(2-(*m*-Chlorophenyl)ethyl)-1,2,3,6-tetrahydro-1-methylpyridine
→ 4-(2-(*m*-Chlorophenyl)ethyl)-1-methylpyridinium

Probably in: Rat F816

4-(*p*-Chlorophenyl)-1-(4-(*p*-fluorophenyl)-4-oxobutyl)pyridine
Formed from: Haloperidol

1-(*p*-Chlorophenyl)-1-formylmethoxy-1-phenylethane
Formed from: Chlorphenoxamine
→ 1-Carboxymethoxy-1-(*p*-chlorophenyl)-1-phenylethane
Probably in: Man E495
→ 1-(*p*-Chlorophenyl)-1-(2-hydroxyethoxy)-1-phenylethane
Probably in: Man E495

2-(*p*-Chlorophenyl)-5-(2-furyl)oxazole-4-acetic acid
Formed from: Ethyl 2-(*p*-chlorophenyl)-5-(2-furyl)oxazole-4-acetate
→ 2-(*p*-Chlorophenyl)-5-(2-furyl)oxazole-4-acetyl-β-D-glucuronic acid
In: Dog E877
Rabbit E877
Rat E877
→ 2-(*p*-Chlorophenyl)-5-(ω-hydroxybutyryl) oxazole-4-acetic acid
In: Dog E877
Rabbit E877
Rat E877

2-(*p*-Chlorophenyl)-5-(2-furyl)oxazole-4-acetyl-β-D-glucuronic acid
Probably formed from: 2-(*p*-Chlorophenyl)-5-(2-furyl)oxazole-4-acetic acid

2-(*m*-Chlorophenyl)-1-(S-glutathionyl)ethanol
Probably formed from: *m*-Chlorostyrene oxide

***o*-Chlorophenyl-β-D-glucuronide**
Formed from: *o*-Chlorophenol

***p*-Chlorophenyl-β-D-glucuronide**
Formed from: *p*-Chlorophenol

***o*-Chlorophenylglycol**
Probably formed from: *o*-Chloromandelaldehyde

N-(*p*-Chlorophenyl)glycolhydroxamic acid
Formed from: *p*-Chloronitrosobenzene

***p*-Chlorophenylglycollaldehyde**
Formed from: Bis(*p*-chlorophenyl)methane

***p*-Chlorophenylglyoxylic acid**
Formed from: *p*-Chloromandelic acid

2-(2-Chlorophenylhydrazono)acetoacetic acid
Formed from: 4-(2-Chlorophenylhydrazono)-3-methyloxazol-5-one

4-(2-Chlorophenylhydrazono)-3-methyloxazol-5-one (drazoxolon)
→ *o*-Chloroaniline
In: Aerobacter A417
Pseudomonas A417
→ 2-(2-Chlorophenylhydrazono)acetoacetic acid
In: Aerobacter A417
Pseudomonas A417

***erythro*-2-(*o*-Chlorophenyl)-2-(4-(*p*-hydroxybenzyl)-1-piperazinyl)-1-phenylethanol**
Formed from: *erythro*-2-(*o*-Chlorophenyl)-2-(4-(*p*-methoxybenzyl)-1-piperazinyl)-1-phenylethanol
→ *erythro*-2-(*o*-Chlorophenyl)-2-(4-(3,4-dihydroxybenzyl)-1-piperazinyl)-1-phenylethanol
Probably in: Rat A3927
→ *erythro*-2-(*o*-Chlorophenyl)-2-(4-(*p*-hydroxybenzyl)-1-piperazinyl)-1-phenylethanol-β-D-glucuronide
Probably in: Rat A3927

***erythro*-2-(*o*-Chlorophenyl)-2-(4-(*p*-hydroxybenzyl)-1-piperazinyl)-1-phenylethanol-β-D-glucuronide**
Probably formed from: *erythro*-2-(*o*-Chlorophenyl)-2-(4-(*p*-hydroxybenzyl)-1-piperazinyl)-1-phenylethanol

β-(*p*-Chlorophenyl)-γ-hydroxybutyric acid
Formed from: γ-Amino-β-(*p*-chlorophenyl)butyric acid

2-(*p*-Chlorophenyl)-5-(ω-hydroxybutyryl) oxazole-4-acetic acid
Probably formed from: 2-(*p*-Chlorophenyl)-5-(2-furyl)oxazole-4-acetic acid
→ 5-(3-Carboxypropionyl)-2-(*p*-chlorophenyl)oxazole-4-acetic acid
Probably in: Dog E877
Rabbit E877
Rat E877

N-(*p*-Chlorophenyl)-2-hydroxy-3,5-diiodobenzamide
Formed from: 2-Acetoxy-N-(*p*-chlorophenyl)-3,5-diiodobenzamide

N-(*p*-Chlorophenyl)-5-hydroxy-3,4-dimethoxy-N-(1-(1-methylethyl)-4-piperidinyl) benzeneacetamide
Probably formed from: N-(*p*-Chlorophenyl)-4-hydroxy-3-methoxy-N-(1-(1-methylethyl)-4-piperidinyl)benzeneacetamide

1-(*p*-Chlorophenyl)-1-(2-hydroxyethoxy)-1-phenylethane
Probably formed from: 1-(*p*-Chlorophenyl)-1-formylmethoxy-1-phenylethane

1-(*p*-Chlorophenyl)-3-(2-hydroxyethyl) imidazolidin-2-one
Formed from: Zetidoline

5-(*p*-Chlorophenyl)-5-hydroxy-5*H*-imidazo [2,1-*a*]isoindole
Formed from: Mazindol

N-(3-(*m*-Chlorophenyl)-3-hydroxy-3*H*-indol-2-yl)-3-aminopropionic acid
Formed from: Ciclazindole

N-(*p*-Chlorophenyl)-2-hydroxy-5-iodobenzamide
Formed from: 2-Acetoxy-N-(*p*-chlorophenyl)-5-iodobenzamide

2-(*p*-Chlorophenyl)-2-hydroxyisovaleric acid
Probably formed from: 2-(*p*-Chlorophenyl) isovaleric acid

2-(*p*-Chlorophenyl)-3-hydroxyisovaleric acid
Probably formed from: 2-(*p*-Chlorophenyl) isovaleric acid
→ 2-(*p*-Chlorophenyl)-2,3-dihydroxyisovaleric acid
Probably in: Dog D443
Mouse C127
Rat C127
→ 2-(*p*-Chlorophenyl)-3-hydroxyisovaleric acid-β-D-glucuronide
Probably in: Dog D443

2-(*p*-Chlorophenyl)-4-hydroxyisovaleric acid
Probably formed from: 2-(*p*-Chlorophenyl) isovaleric acid
→ 2-(*p*-Chlorophenyl)-2,4-dihydroxyisovaleric acid
Probably in: Rat B233, E286
Quail E286

2-(*p*-Chlorophenyl)-3-hydroxyisovaleric acid-β-D-glucuronide
Probably formed from: 2-(*p*-Chlorophenyl)-3-hydroxyisovaleric acid

***p*-Chlorophenylhydroxylamine** (*p*-chloro-N-hydroxyaniline)
Formed from: *p*-Chloroaniline
p-Chloronitrosobenzene
→ 4,4′-Azoxybischlorobenzene
In: Man F342
→ 2-Amino-5-chlorophenol
In: Rhodosporidium B799
→ *p*-Chloroaniline
In: Man F342
Rhodosporidium B799
→ *p*-Chloronitrosobenzene
In: Fusarium A700
Man F342
Pea C797
With chloroperoxidase A3838

1-(*p*-Chlorophenyl)-2-hydroxylamino-2-methylpropane
Formed from: *p*-Chloro-α,α-dimethylphenethylamine
→ 1-(*p*-Chlorophenyl)-2-hydroxylamino-2-methylpropane-β-D-glucuronide
In: Rabbit A21

1-(*p*-Chlorophenyl)-2-hydroxylamino-2-methylpropane-β-D-glucuronide
Formed from: 1-(*p*-Chlorophenyl)-2-hydroxylamino-2-methylpropane

***erythro*-2-(*o*-Chlorophenyl)-2-(4-(3-hydroxy-4-methoxybenzyl)-1-piperazinyl)-1-phenylethanol**
Probably formed from: *erythro*-2-(*o*-Chlorophenyl)-2-(4-(3,4-dihydroxybenzyl)-1-piperazinyl)-1-phenylethanol
→ *erythro*-2-(*o*-Chlorophenyl)-2-(4-(3-hydroxy-4-methoxybenzyl)-1-piperazinyl)-1-phenylethanol-β-D-glucuronide
Probably in: Rat A3927

***erythro*-2-(*o*-Chlorophenyl)-2-(4-(4-hydroxy-3-methoxybenzyl)-1-piperazinyl)-1-phenylethanol**
Probably formed from: *erythro*-2-(*o*-Chlorophenyl)-2-(4-(3,4-dihydroxybenzyl)-1-piperazinyl)-1-phenylethanol
→ *erythro*-2-(*o*-Chlorophenyl)-2-(4-(4-hydroxy-3-methoxybenzyl)-1-piperazinyl)-1-phenylethanol-β-D-glucuronide
Probably in: Rat A3927

***erythro*-2-(*o*-Chlorophenyl)-2-(4-(3-hydroxy-4-methoxybenzyl)-1-piperazinyl)-1-phenylethanol-β-D-glucuronide**
Probably formed from: *erythro*-2-(*o*-Chlorophenyl)-2-(4-(3-hydroxy-4-methoxybenzyl)-1-piperazinyl)-1-phenylethanol

***erythro*-2-(*o*-Chlorophenyl)-2-(4-(4-hydroxy-3-methoxybenzyl)-1-piperazinyl)-1-phenylethanol-β-D-glucuronide**
Probably formed from: *erythro*-2-(*o*-Chlorophenyl)-2-(4-(4-hydroxy-3-methoxybenzyl)-1-piperazinyl)-1-phenylethanol

N-(*p*-Chlorophenyl)-3-hydroxy-4-methoxy-N-(1-(1-methylethyl)-4-piperidinyl) benzeneacetamide
Probably formed from: N-(*p*-Chlorophenyl)-*p*-hydroxy-N-(1-(1-methylethyl)-4-piperidinyl)benzeneacetamide

N-(*p*-Chlorophenyl)-4-hydroxy-3-methoxy-N-(1-(1-methylethyl)-4-piperidinyl) benzeneacetamide
Probably formed from: N-(*p*-Chlorophenyl)-*p*-hydroxy-N-(1-(1-methylethyl)-4-piperidinyl)benzeneacetamide
→ N-(*p*-Chlorophenyl)-α,4-dihydroxy-3-methoxy-N-(1-(1-methylethyl)-4-piperidinyl)benzeneacetamide
Probably in: Dog D94
→ N-(*p*-Chlorophenyl)-5-hydroxy-3,4-dimethoxy-N-(1-(1-methylethyl)-4-piperidinyl)benzeneacetamide
Probably in: Dog D94
Man D94
Rat D94

***p*-Chlorophenyl 3-hydroxy-4-methoxyphenyl ether**
Probably formed from: *p*-Chlorophenyl 3,4-dihydroxyphenyl ether

***p*-Chlorophenyl 4-hydroxy-3-methoxyphenyl ether**
Probably formed from: *p*-Chlorophenyl 3,4-dihydroxyphenyl ether

N-(*p*-Chlorophenyl)-*p*-hydroxy-N-(1-(1-methylethyl)-4-piperidinyl)benzeneacetamide
Formed from: Lorcainide
→ N-(*p*-Chlorophenyl)-3-hydroxy-4-methoxy-N-(1-(1-methylethyl)-4-piperidinyl) benzeneacetamide
Probably in: Dog D94
→ N-(*p*-Chlorophenyl)-4-hydroxy-3-methoxy-N-(1-(1-methylethyl)-4-piperidinyl) benzeneacetamide
Probably in: Dog D94
Man D94
Rat D94

2-(4-((*p*-Chlorophenyl)hydroxymethyl)phenoxy)-2-methylpropionic acid
Formed from: Fenofibrate

1-(*m*-Chlorophenyl)-2-(2-(hydroxymethyl)-2-propylamino)-1-propanol
Probably formed from: *m*-Chloro-2-(2-(hydroxymethyl)-2-propylamino) propiophenone

1-(1-(*m*-Chlorophenyl)-1-hydroxy-2-methylprop-2-yl)imidazole
Formed from: 1-(1-(*m*-Chlorophenyl)-2-methyl-1-oxoprop-2-yl)imidazole

3-(*p*-Chlorophenyl)-1-hydroxymethylurea
Formed from: 3-(*p*-Chlorophenyl)-1-methylurea
→ *p*-Chlorophenylurea
Probably in: Cotton A170

3-(3-(*m*-Chlorophenyl)-3-hydroxy-2-oxindol-1-yl)propionic acid
Probably formed from: 3-(3-(*m*-Chlorophenyl)-3-hydroxy-2-oxindol-1-yl)propylamine

3-(3-(*m*-Chlorophenyl)-3-hydroxy-2-oxindol-1-yl)propylamine
Formed from: Ciclazindole
→ 3-(3-(*m*-Chlorophenyl)-3-hydroxy-2-oxindol-1-yl)propionic acid
Probably in: Man A3842

6-(*o*-Chlorophenyl)-2-hydroxy-6-oxohexa-2,4-dienoic acid
Probably formed from: 2-Chlorobiphenyl
→ *o*-Chlorobenzoic acid
In: Burkholderia K522

6-(*m*-Chlorophenyl)-2-hydroxy-6-oxohexa-2,4-dienoic acid
Probably formed from: 2-Chlorobiphenyl
→ *m*-Chlorobenzoic acid
In: Burkholderia K522

6-(*p*-Chlorophenyl)-2-hydroxy-6-oxohexa-2,4-dienoic acid
Probably formed from: 4-Chlorobiphenyl
→ *p*-Chlorobenzoic acid

Probably in: Burkholderia
Microorganism H816
→ 6-(*p*-Chlorophenyl)pyridine-2-carboxylic acid
Probably in: Pseudomonas G252

p-Chlorophenyl o-hydroxyphenyl ether
Formed from: *p*-Chlorophenyl phenyl ether

p-Chlorophenyl p-hydroxyphenyl ether
Formed from: *p*-Chlorophenyl phenyl ether
→ *p*-Chlorophenyl 3,4-dihydroxyphenyl ether
Probably in: Rat E881

4-(p-Chlorophenyl)-2-(p-hydroxyphenyl)thiazol-5-ylacetic acid
Formed from: Fentiazac

4-(p-Chlorophenyl)-4-hydroxypiperidine
Formed from: Haloperidol
Loperamide

4-(4-(p-Chlorophenyl)-4-hydroxypiperidino)-4′-fluorobutyrophenone
See: Haloperidol

N-(m-Chlorophenyl)-N′-(imino(methylamino)methyl)urea
Formed from: Fenobam
→ N-(Amino(imino)methyl)methyl)-N′-(*m*-chlorophenyl)urea
Probably in: Rat H350
→ N-(3-Chloro-4-hydroxyphenyl)-N′-(imino(methylamino)methyl)urea
Probably in: Rat H350

N-(p-Chlorophenyl)-N′-(indene-5-sulphonyl)urea
Formed from: Sulofenur

N-(p-Chlorophenyl)-N′-(indene-6-sulphonyl)urea
Formed from: Sulofenur

N-(m-Chlorophenyl)isobutyrohydroxamic acid
Formed from: *m*-Chloronitrosobenzene

N-(p-Chlorophenyl)isobutyrohydroxamic acid
Formed from: *p*-Chloronitrosobenzene

2-(p-Chlorophenyl)isovaleric acid
Formed from: Fenvalerate
→ *p*-Chlorophenyl-2,3-dehydroisovaleric acid
Probably in: Mouse C127
Rat C127
→ 2-(*p*-Chlorophenyl)-2-hydroxyisovaleric acid
Probably in: Quail E286
Rat E286
→ 2-(*p*-Chlorophenyl)-3-hydroxyisovaleric acid
Probably in: Dog D443
Mouse C127
Rat C127
→ 2-(*p*-Chlorophenyl)-4-hydroxyisovaleric acid
Probably in: Quail E286
Rat B233, E286
→ 2-(*p*-Chlorophenyl)isovaleric acid-β-D-glucoside
In: Bean D565
Cabbage D565
Cotton D565
Cucumber D565
Tomato D565
→ 2-(*p*-Chlorophenyl)isovaleric acid-β-D-glucuronide
Probably in: Dog D443
→ Cholesteryl (2*R*)-2-(*p*-chlorophenyl) isovalerate
Probably in: Mouse E823

2-(p-Chlorophenyl)isovaleric acid-β-D-glucoside
Formed from: 2-(*p*-Chlorophenyl)isovaleric acid
→ 2-(*p*-Chlorophenyl)isovaleric acid-β-D-(2-β-D-glucopyranosyl)glucoside
Probably in: Cotton D565
Tomato D565
→ 2-(*p*-Chlorophenyl)isovaleric acid-β-D-(6-β-D-glucopyranosyl)glucoside
Probably in: Cabbage D565
Cotton D565
Cucumber D565
Kidney bean D565
Tomato D565
→ 2-(*p*-Chlorophenyl)isovaleric acid-β-D-(6-malonyl)glucoside
Probably in: Cabbage D565
Cotton D565
Cucumber D565
Kidney bean D565
Tomato D565
→ 2-(*p*-Chlorophenyl)isovaleric acid-β-D-(6-β-D-xylopyranosyl)glucoside
Probably in: Cotton D565

2-(*p*-Chlorophenyl)isovaleric acid-β-D-(2-β-D-glucopyranosyl)glucoside
Probably formed from: 2-(*p*-Chlorophenyl) isovaleric acid-β-D-glucoside

2-(*p*-Chlorophenyl)isovaleric acid-β-D-(6-β-D-glucopyranosyl)glucoside
Probably formed from: 2-(*p*-Chlorophenyl) isovaleric acid-β-D-glucoside

2-(*p*-Chlorophenyl)isovaleric acid-β-D-glucuronide
Probably formed from: 2-(*p*-Chlorophenyl) isovaleric acid

2-(*p*-Chlorophenyl)isovaleric acid-β-D-(6-malonyl)glucoside
Probably formed from: 2-(*p*-Chlorophenyl) isovaleric acid-β-D-glucoside

2-(*p*-Chlorophenyl)isovaleric acid-β-D-(6-β-D-xylopyranosyl)glucoside
Probably formed from: 2-(*p*-Chlorophenyl) isovaleric acid-β-D-glucoside

α-(*p*-Chlorophenyl)mandelic acid
Probably formed from: *p*-Chloro-α-formyl-α-phenylbenzyl alcohol

***erythro*-2-(*o*-Chlorophenyl)-2-(4-(*p*-methoxybenzyl)-1-piperazinyl)-1-phenylethanol**
→ Anisic acid
In: Rat A3927
→ *o*-Chlorobenzoic acid
In: Rat A3927
→ *erythro*-2-(*o*-Chlorophenyl)-2-(4-(*p*-hydroxybenzyl)-1-piperazinyl)-1-phenylethanol
In: Rat A3927
→ *erythro*-2-(*o*-Chlorophenyl)-2-(4-(*p*-methoxybenzyl)-1-piperazinyl)-1-phenylethanol-β-D-glucuronide
In: Rat A3927
→ *erythro*-2-(*o*-Chlorophenyl)-2-(1-piperazinyl)-1-phenylethanol
In: Rat A3927
→ *p*-Methylbenzylpiperazine
In: Rat A3927

***erythro*-2-(*o*-Chlorophenyl)-2-(4-(*p*-methoxybenzyl)-1-piperazinyl)-1-phenylethanol-β-D-glucuronide**
Formed from: *erythro*-2-(*o*-Chlorophenyl)-2-(4-(*p*-methoxybenzyl)-1-piperazinyl)-1-phenylethanol

1-(2-((4-Chlorophenyl)methoxy)-2-(2,4-dichlorophenyl)ethyl)-1*H*-imidazole
See: Econazole

3-(*p*-Chlorophenyl)-1-methoxy-1-methylurea
See: Monolinuron

2-(*o*-Chlorophenyl)-2-methylaminocyclohexanone
See: Ketamine

(*R*)-2-(*p*-Chlorophenyl)-α-methyl-5-benzoxazoleacetyltaurine
Formed from: (*R*)-Benoxaprofen

(*S*)-2-(*p*-Chlorophenyl)-α-methyl-5-benzoxazoleacetyltaurine
Formed from: (*S*)-Benoxaprofen

N-(*p*-Chlorophenyl)-1-(methylethyl)-4-piperidinamine
Formed from: Lorcainide

N-(*p*-Chlorophenyl)-N-(1-(1-methylethyl)-4-piperidinyl)benzeneacetamide
See: Lorcainide

α-(*o*-Chlorophenyl)-α-methylmalonic acid
→ 2-(*o*-Chlorophenyl)propionic acid
In: Alcaligenes G716

α-(*m*-Chlorophenyl)-α-methylmalonic acid
→ 2-(*m*-Chlorophenyl)propionic acid
In: Alcaligenes G716

α-(*p*-Chlorophenyl)-α-methylmalonic acid
→ 2-(*p*-Chlorophenyl)propionic acid
In: Alcaligenes G716

1-(*p*-Chlorophenyl)-2-methyl-2-nitropropane
Formed from: *p*-Chloro-α,α-dimethylphenethylamine

1-(1-(*m*-Chlorophenyl)-2-methyl-1-oxoprop-2-yl)imidazole
→ 1-(1-(*m*-Chlorophenyl)-1-hydroxy-2-methylprop-2-yl)imidazole
In: Man J61

N′-(4-Chlorophenyl)-N-methyl-N-(1-methyl-2-propynyl)urea
See: Buturon

1-(2-Chlorophenyl)-4-(2-(1-methyl-1*H*-pyrazol-4-yl)ethyl)piperazine
See: Empiprazole

2-(4-Chlorophenyl)tetrahydro-3-methyl-4*H*-1,3-thiazin-4-one-1,1-dioxide
See: Chlormezanone

4-(*o*-Chlorophenyl)-1-methylpyridinium
Probably formed from: 4-(*o*-Chlorophenyl)-

1,2,3,6-tetrahydro-1-methylpyridine

4-(*m*-Chlorophenyl)-1-methylpyridinium
Probably formed from: 4-(*m*-Chlorophenyl)-1,2,3,6-tetrahydro-1-methylpyridine

4-(*p*-Chlorophenyl)-1-methylpyridinium
Probably formed from: 4-(*p*-Chlorophenyl)-1,2,3,6-tetrahydro-1-methylpyridine

3-(*p*-Chlorophenyl)-N-methyl-3-(2-pyridyl) propionamide
Probably formed from: Demethylchlorpheniramine

3-(*o*-Chlorophenyl)-2-methyl-4(3*H*)-quinazolinone
See: Mecloqualone

***p*-Chlorophenyl methyl sulphoxide**
Formed from: *p*-Chlorothiophenol
→ *p*-Chlorophenyl methyl sulphone
Probably in: Rat A1946

***p*-Chlorophenyl methyl sulphone**
Probably formed from: *p*-Chlorophenyl methyl sulphoxide
→ 4-Chloro-3-hydroxyphenyl methyl sulphone
Probably in: Rat A1946, A2639
→ *p*-Chlorobenzenesulphonic acid
Probably in: Rat A1946, A2639

3-(*p*-Chlorophenyl)-1-methylurea
Formed from: 3-(*p*-Chlorophenyl)-1,1-dimethylurea
Monolinuron
→ 3-(*p*-Chlorophenyl)-1-hydroxymethylurea
Probably in: Cotton A170
→ 3-(*p*-Chlorophenyl)urea
In: Mouse C47

2-Chlorophenyl 4-nitrophenyl ether
Formed from: Nitrofen

2-(*p*-Chlorophenyl)-1,3-oxathiolane
→ 2-(*p*-Chlorophenyl)-1,3-oxathiolane sulphoxide
In Pig F176
Rat F176

2-(*p*-Chlorophenyl)-1,3-oxathiolane sulphone
Probably formed from: 2-(*p*-Chlorophenyl)-1,3-oxathiolane sulphoxide

2-(*p*-Chlorophenyl)-1,3-oxathiolane sulphoxide
Formed from: 2-(*p*-Chlorophenyl)-1,3-oxathiolane
→ 2-(*p*-Chlorophenyl)-1,3-oxathiolane sulphone
Probably in: Pig F176
Rat F176

2-(5-(*p*-Chlorophenyl)pentyl)oxiranecarboxyl CoA
→ 2-(5-(*p*-Chlorophenyl)pentyl) oxiranecarboxylic acid
In: Rat J261

2-(5-(*p*-Chlorophenyl)pentyl)oxiranecarboxylic acid
Formed from: 2-(5-(*p*-Chlorophenyl) pentyl)oxiranecarboxyl CoA

2-Chloro-6-phenylphenoxyethylamine
See: 2-(2-Aminoethoxy)-3-chlorobiphenyl

3-(*o*-Chlorophenyl)-3-phenylacrolein
Probably formed from: 3-(*o*-Chlorophenyl)-3-dimethylamino-1-phenyl-1-propene

1-(*p*-Chlorophenyl)-1-phenylethanol
Probably formed from: Chlorphenoxamine

***p*-Chlorophenyl phenyl ether**
→ *p*-Chlorophenyl *o*-hydroxyphenyl ether
In: Rat E881
→ *p*-Chlorophenyl *p*-hydroxyphenyl ether
In: Rat E881

2-(1-(4-Chlorophenyl)-1-phenylethoxy)-N,N-dimethylethanamine
See: Chlorphenoxamine

2-(2-(4-((4-Chlorophenyl)phenylmethyl)-1-piperazinyl)ethoxy)ethanol
See: Hydroxyzine

N-(*p*-Chlorophenyl)-α-phenylnitrone
Formed from: N-Benzyl-*p*-chloroaniline
→ N-Benzyl-N-(*p*-chlorophenyl) hydroxylamine
Probably in: Hamster D637
Rabbit D637

3-(4-Chlorophenyl)-1-phenyl-1*H*-pyrazole-4-acetic acid
See: Lonazolac

4-(*p*-Chlorophenyl)-2-phenylthiazol-5-ylacetic acid
See: Fentiazac

N-(*o*-Chlorophenyl)phthalamidic acid
→ Phthalic acid
In: Xanthobacter K793

1-(*o*-Chlorophenyl)piperazine
Formed from: Enpiprazole

1-(*m*-Chlorophenyl)piperazine
Formed from: Etoperidone
Nefazodone

Trazodone
→ 1-(3-Chloro-4-hydroxyphenyl)piperazine
In: Man J888
Rat G862

3-((4-(4-Chlorophenyl)piperazin-1-yl)methyl)-1*H*-pyrrolo[2,3-*b*]pyridine
→ 3-((4-(4-Chloro-3-hydroxyphenyl)piperazin-1-yl)methyl)-1*H*-pyrrolo[2,3-*b*]pyridine sulphate
Probably in: Man K40
Monkey K40
Rat K40

***erythro*-2-(*o*-Chlorophenyl)-2-(1-piperazinyl)-1-phenylethanol**
Formed from: *erythro*-2-(*o*-Chlorophenyl)-2-(4-(*p*-methoxybenzyl)-1-piperazinyl)-1-phenylethanol
→ *erythro*-2-(*o*-Chlorophenyl)-2-(1-piperazinyl)-1-phenylethyl-β-D-glucuronide
In: Rat A3927

***erythro*-2-(*o*-Chlorophenyl)-2-(1-piperazinyl)-1-phenylethyl-β-D-glucuronide**
Formed from: *erythro*-2-(*o*-Chlorophenyl)-2-(1-piperazinyl)-1-phenylethanol

2-(3-(4-(3-Chlorophenyl)-1-piperazinyl)propyl)-4,5-diethyl-2,4-dihydro-3*H*-1,2,4-triazol-3-one
See: Etoperidone

2-(3-(4-(*m*-Chlorophenyl)piperazin-1-yl)propyl)-2,4-dihydro-5-(1-hydroxyethyl)-4-(2-phenoxyethyl)-3*H*-1,2,4-triazol-3-one
Formed from: Nefazodone
→ 5-Carboxy-2-(3-(4-(*m*-chlorophenyl) piperazin-1-yl)propyl)-2,4-dihydro-4-(2-phenoxyethyl)-3*H*-1,2,4-triazol-3-one
Probably in: Man G872

N-(*p*-Chlorophenyl)-N-(4-piperidinyl) benzeneacetamide
Formed from: Lorcainide

(1*R*,2*S*)-1-(*o*-Chlorophenyl)propane-1,2-diol
Formed from: *o*-Chlorobenzaldehyde

(1*R*,2*S*)-1-(*p*-Chlorophenyl)propane-1,2-diol
Formed from: *p*-Chlorobenzaldehyde

2-(*p*-Chlorophenyl)propane-1,2-diol
Formed from: *p*-Chlorophenyl)propylene oxide

1-(*o*-Chlorophenyl)-2-propanol
Formed from: *o*-Chlorophenylacetone

1-(*p*-Chlorophenyl)-2-propanol
Formed from: *p*-Chlorophenylacetone

3-(*p*-Chlorophenyl)propanol
Probably formed from: *p*-Chlorocinnamyl alcohol

2-(*p*-Chlorophenyl)propionamide
Formed from: 2-(*p*-Chlorophenyl) propionitrile
→ 2-(*p*-Chlorophenyl)propionic acid
Probably in: Rhodococcus G172

3-(*o*-Chlorophenyl)propionamide
Formed from: *o*-Chlorobenzylmalononitrile

2-(*o*-Chlorophenyl)propionic acid
Formed from: α-(*o*-Chlorophenyl)-α-methylmalonic acid

2-(*m*-Chlorophenyl)propionic acid
Formed from: α-(*m*-Chlorophenyl)-α-methylmalonic acid

2-(*p*-Chlorophenyl)propionic acid
Formed from: α-(*p*-Chlorophenyl)-α-methylmalonic acid
Probably formed from: 2-(*p*-Chlorophenyl) propionamide

3-(*o*-Chlorophenyl)propionic acid
Formed from: *o*-Chlorocinnamic acid

2-(*p*-Chlorophenyl)propionitrile
→ 2-(*p*-Chlorophenyl)propionamide
In: Rhodococcus G172

2-(*p*-Chlorophenyl)propylene oxide
→ 2-(*p*-Chlorophenyl)propane-1,2-diol
In: Pig H753

6-(*p*-Chlorophenyl)pyridine-2-carboxylic acid
Probably formed from: 6-(*p*-Chlorophenyl)-2-hydroxy-6-oxohexa-2,4-dienoic acid

4-(4-(*p*-Chlorophenyl)pyridin-1-yl))-1-(*p*-fluorophenyl)-1-butanone
Probably formed from: 4-(4-(*p*-Chlorophenyl)-1,2,3,6-tetrahydropyridin-1-yl))-1-(*p*-fluorophenyl)-1-butanone

3-(5-(*o*-Chlorophenyl)-4-pyridyl)-1,2,4-oxadiazole
→ *o*-Chlorobenzoic acid
In: Rat 2610, A2644
→ *o*-Chlorobenzoylisonicotinic acid amidoxime
In: Rat 2610, A2644

3-(*p*-Chlorophenyl)-3-(2-pyridyl)-*n*-propanol
Formed from: Chlorpheniramine

3-(*p*-Chlorophenyl)-3-(2-pyridyl)propionaldehyde
Formed from: Chlorpheniramine

3-(*p*-Chlorophenyl)-3-(2-pyridyl)propionic acid
Formed from: Chlorpheniramine
→ 3-(*p*-Chlorophenyl)-3-(2-pyridyl)propionyl-β-D-glucuronic acid
Probably in: Rat F902

3-(*p*-Chlorophenyl)-3-(2-pyridyl)propionyl-β-D-glucuronic acid
Probably formed from: 3-(*p*-Chlorophenyl)-3-(2-pyridyl)propionic acid

***o*-Chlorophenylpyruvic acid**
Formed from: *o*-Chloro-L-phenylalanine

***m*-Chlorophenylpyruvic acid**
Formed from: *m*-Chloro-L-phenylalanine

***p*-Chlorophenylpyruvic acid**
Formed from: *p*-Chloro-L-phenylalanine

6-Chloro-4-phenyl-2(1*H*)-quinazoline
Formed from: 7-Chloro-1,3-dihydro-3-hydroxy-5-phenyl-2*H*-1,4-benzodiazepin-2-one

3-(*o*-Chlorophenyl)quinazolinone
Formed from: Mecloqualone

6-Chloro-4-phenyl-2(1*H*)-quinazolinone
Formed from: Chlorazepate
7-Chloro-1,3-dihydro-1,3-dimethyl-5-phenyl-2*H*-1,4-benzodiazepin-2-one
7-Chloro-1,3-dihydro-3-hydroxy-1,3-dimethyl-5-phenyl-2*H*-1,4-benzodiazepin-2-one
7-Chloro-1,3-dihydro-3-hydroxy-1-methyl-5-phenyl-2*H*-1,4-benzodiazepin-2-one
7-Chloro-1,3-dihydro-3-hydroxy-5-phenyl-2*H*-1,4-benzodiazepin-2-one
7-Chloro-1,3-dihydro-1-methyl-5-phenyl-2*H*-1,4-benzodiazepin-2-one
7-Chloro-1,3-dihydro-3-methyl-5-phenyl-2*H*-1,4-benzodiazepin-2-one

N-(*o*-Chlorophenyl)salicylamide
→ *o*-Chloroaniline
In: Mouse A2037
Sheep A2037
→ Salicylic acid
In: Mouse A2037
Sheep A2037

N-(*p*-Chlorophenyl)succinimide
Formed from: *p*-Chloroaniline

***o*-Chlorophenylsulphamic acid**
Formed from: *o*-Chloroaniline

***o*-Chlorophenyl sulphate**
Formed from: *o*-Chlorophenol

***m*-Chlorophenyl sulphate**
Formed from: *m*-Chlorophenol

***p*-Chlorophenyl sulphate**
Formed from: *p*-Chlorophenol

5-Chloro-3-phenylsulphinylindole-2-carboxamide
Formed from: 5-Chloro-3-phenylthioindole-2-carboxamide
→ 5-Chloro-3-phenylsulphonylindole-2-carboxamide
In: Rhesus monkey G584

4-((*p*-Chlorophenyl)sulphinyl)phenol
Probably formed from: 4-((*p*-Chlorophenyl)thio)phenol
→ 4-((*p*-Chlorophenyl)sulphonyl)phenol
Probably in: Cotton B591

O-(4-((*p*-Chlorophenyl)sulphinyl)phenyl O-ethyl S-propylphosphorothioate
Formed from: O-(4-((*p*-Chlorophenyl)thio)phenyl) O-ethyl S-propylphosphorothioate
→ O-(4-((*p*-Chlorophenyl)sulphonyl)phenyl) O-ethyl S-propylphosphorothioate
In: Cotton B591

1-((*p*-Chlorophenyl)sulphonyl)-3-(2-hydroxypropyl)urea
Formed from: Chlorpropamide

1-((*p*-Chlorophenyl)sulphonyl)-3-(3-hydroxypropyl)urea
Formed from: Chlorpropamide

5-Chloro-3-phenylsulphonylindole-2-carboxamide
Formed from: 5-Chloro-3-phenylsulphinylindole-2-carboxamide

1-(*p*-Chlorophenyl)sulphonyl-3-propylurea
See: Chlorpropamide

(*p*-Chlorophenyl)sulphonylurea
Formed from: Chlorpropamide

4-((*p*-Chlorophenyl)sulphonyl)phenol
Probably formed from: 4-((*p*-Chlorophenyl)sulphinyl)phenol

O-(4-((*p*-Chlorophenyl)sulphonyl)phenyl) O-ethyl S-propylphosphorothioate
Formed from: O-(4-((*p*-Chlorophenyl)sulphinyl)phenyl) O-ethyl S-propylphosphorothioate

10-(*m*-Chlorophenyl)-6,8,9,10-tetrahydrobenzo[*b*][1,8]naphthyridine-5(7*H*),6-dione
Probably formed from: 10-(*m*-Chlorophenyl)-6,8,9,10-tetrahydro-6-hydroxybenzo[*b*][1,8]naphthyridin-5(7*H*)-one

10-(*m*-Chlorophenyl)-6,8,9,10-tetrahydrobenzo[*b*][1,8]naphthyridin-5(7*H*)-one
→ 10-(*m*-Chlorophenyl)-6,8,9,10-tetrahydro-6-hydroxybenzo[*b*][1,8]naphthyridin-5(7*H*)-one
In: Cynomolgus monkey J685
Dog J685
Man J685
Rat J685
→ 10-(*m*-Chlorophenyl)-6,8,9,10-tetrahydro-7-hydroxybenzo[*b*][1,8]naphthyridin-5(7*H*)-one
In: Cynomolgus monkey J685
Dog J685
Man J685
Rat J685
→ 10-(*m*-Chlorophenyl)-6,8,9,10-tetrahydro-9-hydroxybenzo[*b*][1,8]naphthyridin-5(7*H*)-one
In: Cynomolgus monkey J685
Dog J685
Man J685
Rat J685

6-(*o*-Chlorophenyl)-2,3,4,5-tetrahydro-8,11-dimethyl-8*H*-pyrido[4′,3′:4,5]thieno[3,2-*f*][1,2,4]triazolo[4,3-*a*][1,4]diazepine
Formed from: (*S*)-(+)-6-(*o*-Chlorophenyl)-3-cyclopropanecarbonyl-2,3,4,5-tetrahydro-8,11-dimethyl-8*H*-pyrido[4′,3′:4,5]thieno[3,2-*f*][1,2,4]triazolo[4,3-*a*][1,4]diazepine

10-(*m*-Chlorophenyl)-6,8,9,10-tetrahydro-6-hydroxybenzo[*b*][1,8]naphthyridin-5(7*H*)-one
Formed from: 10-(*m*-Chlorophenyl)-6,8,9,10-tetrahydrobenzo[*b*][1,8]naphthyridin-5(7*H*)-one
→ 10-(*m*-Chlorophenyl)-6,8,9,10-tetrahydro-6-hydroxybenzo[*b*][1,8]naphthyridin-5(7*H*)-one-β-D-glucuronide
In: Cynomolgus monkey J685
Dog J685
Man J685
Rat J685
→ 10-(*m*-Chlorophenyl)-6,8,9,10-tetrahydrobenzo[*b*][1,8]naphthyridine-5(7*H*),6-dione
Probably in: Cynomolgus monkey J685
Dog J685
Man J685
Rat J685

10-(*m*-Chlorophenyl)-6,8,9,10-tetrahydro-7-hydroxybenzo[*b*][1,8]naphthyridin-5(7*H*)-one
Formed from: 10-(*m*-Chlorophenyl)-6,8,9,10-tetrahydrobenzo[*b*][1,8]naphthyridin-5(7*H*)-one

10-(*m*-Chlorophenyl)-6,8,9,10-tetrahydro-9-hydroxybenzo[*b*][1,8]naphthyridin-5(7*H*)-one
Formed from: 10-(*m*-Chlorophenyl)-6,8,9,10-tetrahydrobenzo[*b*][1,8]naphthyridin-5(7*H*)-one

10-(*m*-Chlorophenyl)-6,8,9,10-tetrahydro-6-hydroxybenzo[*b*][1,8]naphthyridin-5(7*H*)-one-β-D-glucuronide
Formed from: 10-(*m*-Chlorophenyl)-6,8,9,10-tetrahydro-6-hydroxybenzo[*b*][1,8]naphthyridin-5(7*H*)-one

3-(4′-Chlorophenyl)-3,4,7,8-tetrahydro-2-methyl-4,7-dioxopyido[2,3-*d*]pyrimidine
Formed from: 3-(4′-Chlorophenyl)-3,4-dihydro-2-methyl-4-oxopyrido[2,3-*d*]pyrimidine

4-(*o*-Chlorophenyl)-1,2,3,6-tetrahydro-1-methylpyridine
→ 4-(*o*-Chlorophenyl)-1-methylpyridinium
Probably in: Beef F395
Man F395

4-(*m*-Chlorophenyl)-1,2,3,6-tetrahydro-1-methylpyridine
→ 4-(*m*-Chlorophenyl)-1-methylpyridinium
Probably in: Beef F395
Man F395

4-(*p*-Chlorophenyl)-1,2,3,6-tetrahydro-1-methylpyridine
→ 4-(*p*-Chlorophenyl)-1-methylpyridinium
Probably in: Beef F395
Man F395

4-Chlorophenyl-1,2,3,6-tetrahydropyridine
Formed from: 4,5-Dehydrohaloperidol-N-oxide

4-(4-(*p*-Chlorophenyl)-1,2,3,6-tetrahydropyridin-1-yl)-1-(*p*-fluorophenyl)-1-butanone
Formed from: Haloperidol

→ 4-(4-(*p*-Chlorophenyl)pyridin-1-yl))-1-(*p*-fluorophenyl)-1-butanone
Probably in: Man K545

3-(*p*-Chlorophenyl)thiazolo[3,2-*a*]benzimidazole-2-acetic acid
Formed from: 3-(*p*-Chlorophenyl) thiazolo[3,2-*a*]benzimidazole-2-acetic acid-*β*-D-glucuronide
3-(*p*-Chlorophenyl)thiazolo[3,2-*a*] benzimidazole-2-acetyl-β-D-glucuronic acid
→ 3-(*p*-Chlorophenyl)thiazolo[3,2-*a*] benzimidazole-2-acetic acid-β-D-glucuronide
In: Monkey C632
→ 3-(*p*-Chlorophenyl)thiazolo[3,2-*a*] benzimidazole-2-acetyl-β-D-glucuronic acid
In: Dog C632
Monkey C632
Rat C632

3-(*p*-Chlorophenyl)thiazolo[3,2-*a*]benzimidazole-2-acetic acid-β-D-glucuronide
Formed from: 3-(*p*-Chlorophenyl) thiazolo[3,2-*a*]benzimidazole-2-acetic acid
→ 3-(*p*-Chlorophenyl)thiazolo[3,2-*a*] benzimidazole-2-acetic acid
With β-glucuronidase C632

3-(*p*-Chlorophenyl)thiazolo[3,2-*a*]benzimidazole-2-acetyl-β-D-glucuronic acid
Formed from: 3-(*p*-Chlorophenyl) thiazolo[3,2-*a*]benzimidazole-2-acetic acid
→ 3-(*p*-Chlorophenyl)thiazolo[3,2-*a*] benzimidazole-2-acetic acid
With β-glucuronidase C632

2-(*p*-Chlorophenyl)thiazol-4-ylacetic acid
See: Fenclozic acid

2-(*p*-Chlorophenyl)thiazol-4-ylacetyl-β-D-glucuronic acid
Formed from: Fenclozic acid

2-(*p*-Chlorophenyl)thiazol-4-ylacetyltaurine
Formed from: Fenclozic acid

***p*-Chlorophenylthioacetic acid**
Formed from: 4-(*p*-Chlorophenylthio)butanol

4-(*p*-Chlorophenylthio)butanol
→ *p*-Chlorophenylthioacetic acid
In: Dog B242
Rat B242

5-Chloro-3-phenylthioindole-2-carboxamide
→ 5-Chloro-3-(*p*-hydroxyphenylthio)indole-2-carboxamide
In: Rhesus monkey G584
→ 5-Chloro-6-hydroxy-3-phenylthioindole-2-carboxamide
In: Rhesus monkey G584
→ 5-Chloro-3-phenylsulphinylindole-2-carboxamide
In: Rhesus monkey G584

S-(((4-Chlorophenyl)thio)methyl) O,O-diethyl phosphorodithioate
See: Trithion

4-((*p*-Chlorophenyl)thio)phenol
Formed from: O-(4-((*p*-Chlorophenyl)thio)phenyl) O-ethyl S-propylphosphorothioate
→ 4-((*p*-Chlorophenyl)sulphinyl)phenol
Probably in: Cotton B591

O-(4-((*p*-Chlorophenyl)thio)phenyl) O-ethyl S-propylphosphorothioate
→ 4-((*p*-Chlorophenyl)thio)phenol
In: Cotton B591
→ O-(4-((*p*-Chlorophenyl)sulphinyl)phenyl) O-ethyl S-propylphosphorothioate
In: Cotton B591

8-Chloro-6-phenyl-4*H*-*s*-triazolo[4,3-*a*][1,4] benzodiazepine
→ 8-Chloro-2,4-dihydro-6-phenyl-4*H*-*s*-triazolo[4,3-*a*][1,4]benzodiazepin-1-one
In: Dog A258, A303
Guinea pig A303
Man A258
Mouse A303
Rabbit A303
Rat A258, A301, A302
→ 8-Chloro-4-hydroxy-6-phenyl-4*H*-*s*-triazolo[4,3-*a*][1,4]benzodiazepine
In: Dog A258, A303
Guinea pig A303
Man A258
Mouse A303
Rabbit A303
Rat A258, A297, A301
→ 8-Chloro-6-(*m*-hydroxyphenyl)-4*H*-*s*-triazolo[4,3-*a*][1,4]benzodiazepine
In: Dog A258

→ 8-Chloro-6-(*p*-hydroxyphenyl)-4*H*-*s*-triazolo[4,3-*a*][1,4]benzodiazepine
In: Dog A258, A303
Guinea pig A303
Man A258
Mouse A303
Rabbit A303
Rat A258, A297, A301, A303
→ 8-Chloro-6-phenyl-1*H*-*s*-triazolo[4,3-*a*][1,4]benzodiazepin-1-one
In: Rat A297

8-Chloro-6-phenyl-1*H*-*s*-triazolo[4,3-*a*][1,4] benzodiazepin-1-one
Formed from: 8-Chloro-6-phenyl-4*H*-*s*-triazolo[4,3-*a*][1,4]benzodiazepine
→ 8-Chloro-6-(*p*-hydroxyphenyl)-1*H*-*s*-triazolo[4,3-*a*][1,4]benzodiazepin-1-one
Probably in: Rat A297

***p*-Chlorophenylurea**
Formed from: Buturon
3-(*p*-Chlorophenyl)-1-hydroxymethylurea
Diflubenzuron
→ 1-(4-Chloro-2-hydroxyphenyl)urea
In: Cotton A170

1-Chlorophthalazine
→ 4-Chloro-1(2*H*)-phthalazinone
In: Beef E222
Rabbit E222

4-Chloro-1(2*H*)-phthalazinone
Formed from: 1-Chlorophthalazine

4-Chlorophthalic acid
→ *m*-Chlorobenzoic acid
In: Bacillus D31
Microorganism D31
→ 4,5-Dihydro-4,5-dihydroxyphthalic acid
Probably in: Pseudomonas K922

4-Chloropinselin (2,8-dihydro-4-chloro-6-methyl-9-oxo-9*H*-xanthene-1-carboxylic acid methyl ester)
Formed from: Moniliphenone

4-Chloro-5-((piperazin-1-yl)ethyl)oxindole
Formed from: Ziprasidone

Chloroprocaine (4-amino-2-chlorobenzoic acid 2-(diethylamino)ethyl ester)
→ 4-Amino-2-chlorobenzoic acid
In: Rabbit G517, H563

3-Chloro-4-(2-propenyloxy)benzeneacetic acid
See: Alclofenac

***o*-Chloropropionanilide**
→ *o*-Chloroaniline
In: Rice A3773
Taraxacum A230

***m*-Chloropropionanilide**
→ *m*-Chloroaniline
In: Rice A3773
Taraxacum A230

***p*-Chloropropionanilide**
→ *p*-Chloroacetanilide
In: Rabbit A377
→ *p*-Chloroaniline
In: Rice A3773
Taraxacum A230
→ *p*-Chlorohydracrylanilide
In: Rabbit A377

Chloropropylate (4-chloro-α-(4-chlorophenyl)-α-hydroxybenzeneacetic acid 1-methylethyl ester)
→ 4,4′-Dichlorobenzilic acid
In: Housefly A689
Spider mite A689

***p*-Chloro-N-propylbenzamide**
Formed from: *p*-Chloro-N-methyl-N-propylbenzamide

N-(3-Chloropropyl)-α-methylphenethylamine
See: Mefenorex

5-Chloropyrogallol (5-chloro-1,2,3-trihydroxybenzene)
Probably formed from: 4-Chlorocatechol

7-Chloro-5-(pyrrol-2-yl)-2(3*H*)-oxo-1,4-benzodiazepine
→ 7-Chloro-5-(α-hydroxypyrrol-2-yl)-2(3*H*)-oxo-1,4-benzodiazepine
In: Dog G515
Man G515
Rat G515
→ 7-Chloro-3-hydroxy-5-(pyrrol-2-yl)-2(3*H*)-oxo-1,4-benzodiazepine
In: Dog G515
Rat G515

2-(3-Chloro-4-(1-pyrrolyl)phenyl)propionic acid
Formed from: Pirprofen
→ 2-(3-Chloro-4-(1-pyrrolyl)phenyl) propionyl-β-D-glucuronic acid

Probably in: Man C41
Mouse C41
Rhesus monkey C41
→ 2-(3-Chloro-4-(1-pyrrolyl)phenyl) propionyltaurine
Probably in: Mouse C41
Rat C41

2-(3-Chloro-4-(1-pyrrolyl)phenyl)propionyl-β-D-glucuronic acid
Formed from: 2-(3-Chloro-4-(1-pyrrolyl) phenyl)propionic acid

2-(3-Chloro-4-(1-pyrrolyl)phenyl)propionyltaurine
Formed from: 2-(3-Chloro-4-(1-pyrrolyl) phenyl)propionic acid

2-Chloroquestin (2-chloro-1,6-dihydroxy-8-methoxy-3-methyl-9,10-anthracenedione)
Formed from: 2-Chloroemodin

Chloroquine (7-Chloro-4-((3-diethylamino) propylamino)quinoline)
→ 4-Amino-7-chloroquinoline
In: Man C239
→ 7-Chloro-4-((3-ethylamino) propylamino)quinoline
In: Dog C806
Man C239
Rabbit G629
Rat K88
→ Chloroquine-N-oxide
In: Man A3727, C239

Chloroquine-di-N-oxide
Formed from: Chloroquine-N-oxide

Chloroquine-N-oxide
Formed from: Chloroquine
→ Chloroquine-di-N-oxide
In: Man A3727, C239

Chloroquinol (2-chlorohydroquinone)
Formed from: 2-Chloro-6-(S-glutathionyl) quinol
2-Chloro-4-hydroxybenzoic acid
3-Chloro-4-hydroxybenzoic acid
2,4-Dichlorophenol
3,4-Dichlorophenol
Probably formed from: 2-Chloro-4-methoxyphenol
→ Chloroquinol sulphate
Probably in: Mouse C389

Chloroquinol sulphate
Probably formed from: Chloroquinol

2-Chloroquinoline
→ 2-Chloro-5,6-dihydro-5,6-dihydroxyquinoline
In: Pseudomonas J537
→ 2-Chloro-7,8-dihydro-7,8-dihydroxyquinoline
In: Pseudomonas J537

4-Chloroquinoline
→ 4-Chlorocarbostyril
In: Agrobacterium G778

8-Chloroquinoline
→ 8-Chloro-2-hydroxyquinoline
In: Rhodococcus K767

2-((7-Chloro-4-quinolinyl)amino)benzoic acid
See: Glafenic acid

2-((7-Chloro-4-quinolinyl)amino)benzoic acid 2,3-dihydroxypropyl ester
See: Glafenine

(*R*)-1-((1-(3-(2-(7-Chloro-2-quinolinyl)ethenyl) phenyl)-3-(2-(1,2-dihydroxy-2-propyl) phenyl)propyl)thiomethyl)cyclopropaneacetic acid (36-hydroxymontelukast)
Formed from: Montelukast
→ 35-Carboxy-35-demethylmontelukast
Probably in: Man H804

(*S*)-(3-(((3-(2-(7-Chloroquinolin-2-yl)-(*E*)-ethenyl)phenyl)-3-dimethylamino-3-oxopropylthio)methyl)thio)propionic acid
See: Verlukast

(*S*)-(3-(((3-(2-(7-Chloroquinolin-2-yl)-(*E*)-ethenyl)phenyl)-3-(N-hydroxymethyl-N-methylamino)-3-oxopropylthio)methyl)thio) propionic acid
Formed from: Verlukast

4-Chlororesorcinol (1-chloro-2,4-dihydroxybenzene)
→ 2-Chloro-5-hydroxyquinol
In: Pseudomonas A2391

5-Chlorosaccharin
Unchanged in rat A3245

3-Chlorosalicylic acid (3-chloro-2-hydroxybenzoic acid)
Formed from: 3′-Chloro-2,2′,3-trihydroxybiphenyl
→ 3-Chlorocatechol

In: Pseudomonas H506
See also: E146

4-Chlorosalicylic acid (4-chloro-2-hydroxybenzoic acid)
Formed from: 2-Chloronaphthalene
4′-Chloro-2,2′,3-trihydroxybiphenyl
→ 7-Carboxy-4-chloro-2-oxohept-3,5-dienoic acid
Probably in: Pseudomonas C147
→ 4-Chlorocatechol
In: Trichosporon D232
See also: E146

5-Chlorosalicylic acid (5-chloro-2-hydroxybenzoic acid)
Formed from: *m*-Chlorobenzoic acid
5-Chloro-N-(2-chlorophenyl)-2-hydroxybenzamide
5-Chloro-2-hydroxy-N-(*p*-nitrophenyl) benzamide
5-Chloro-2-hydroxy-N-phenylbenzamide
Meseclazone
Seclazone
→ 7-Carboxy-4-chloro-2-oxohept-3,5-dienoic acid
Probably in: Bacillus B224
Pseudomonas C147
→ 4-Chlorocatechol
In: Pseudomonas H506
Trichosporon D232
See also: E146
→ 5-Chloro-2,3-dihydroxybenzoic acid
Probably in: Bacillus A1192
→ 5-Chloro-2-hydroxyhippuric acid
In: Man A2880
→ 5-Chlorosalicylic acid-β-D-glucuronide
In: Man A2880
→ Salicylic acid
In: Desulfomonile G152

5-Chlorosalicylic acid-β-D-glucuronide
Formed from: 5-Chlorosalicylic acid

***o*-Chlorostyrene**
Formed from: *o*-Chlorocinnamic acid
→ *o*-Chlorobenzaldehyde
In: Coprinus K431
Man K431
→ *o*-Chlorostyrene oxide
In: Coprinus K431
Man K431

***m*-Chlorostyrene**
Formed from: *m*-Chlorocinnamic acid
→ *m*-Chlorobenzaldehyde
In: Coprinus K431
Man K431
→ *m*-Chlorostyrene oxide
In: Coprinus K431
Man K431

***p*-Chlorostyrene**
Formed from: *p*-Chlorocinnamic acid
→ *p*-Chlorobenzaldehyde
In: Coprinus K431
Man K431
→ *p*-Chlorostyrene oxide
In: Coprinus K431
Man K431

***o*-Chlorostyrene oxide**
Formed from: *o*-Chlorostyrene

***m*-Chlorostyrene oxide**
Formed from: *m*-Chlorostyrene
→ 2-(*m*-Chlorophenyl)-1-(S-glutathionyl)ethanol
Probably in: Sheep A2282

***p*-Chlorostyrene oxide**
Formed from: *p*-Chlorostyrene

N-Chlorosulphamethoxazole
Formed from: Sulphamethoxazole

4-Chloro-5-sulphamoylanthranilic acid
See: Saluamine

4-Chloro-3-sulphamoylbenzamide
Formed from: Tripamide

2-(N-(4-Chloro-3-sulphamoylbenzenesulphonyl)-N-methylaminomethyl)-2-methyltetrahydrofuran
See: Mefruside

4-Chloro-3-sulphamoylbenzoic acid
Formed from: Tripamide

4-Chloro-3-sulphamoylbenzoyl-(N′-acetyl) hydrazide
Formed from: Tripamide

***p*-Chlorosulphinylanisole**
Formed from: *p*-Chlorothioanisole

4-Chloro-7-sulphobenzofuran
→ 4-(S-Glutathionyl)-7-sulphobenzofuran
Probably in: Antechinus J102

2-Chloro-6-sulphydrylbenzonitrile
Probably formed from: 2-Chloro-6-(S-cysteinyl)benzonitrile
→ 2-Chloro-6-sulphydrylbenzonitrile-β-D-glucuronide
Probably in: Beef F586
→ 6-Chloro-3-hydroxy-2-sulphydrylbenzonitrile
Probably in: Beef F586

2-Chloro-6-sulphydrylbenzonitrile-β-D-glucuronide
Probably formed from: 2-Chloro-6-sulphydrylbenzonitrile

6-Chloro-2,3,4,5-tetrahydro-1*H*-3-benzazepine
Formed from: 6-Chloro-2,3,4,5-tetrahydro-3-methyl-1*H*-3-benzazepine
→ 6-Chloro-2,3,4,5-tetrahydro-N-hydroxy-1*H*-3-benzazepine
Probably in: Dog E924

***trans*-5-Chloro-2,3,3a,12b-tetrahydro-1*H*-dibenz[2,3:6,7]oxepino[4,5-*c*]pyrrolidine**
Formed from: *trans*-5-Chloro-2,3,3a,12b-tetrahydro-2-methyl-1*H*-dibenz[2,3:6,7]oxepino[4,5-*c*]pyrrolidine

***trans*-5-Chloro-2,3,3a,12b-tetrahydro-1*H*-dibenz[2,3:6,7]oxepino[4,5-*c*]pyrrolidine-2-carbamate-β-D-glucuronide**
Formed from: *trans*-5-Chloro-2,3,3a,12b-tetrahydro-2-methyl-1*H*-dibenz[2,3:6,7]oxepino[4,5-*c*]pyrrolidine

6-Chloro-2,3,4,5-tetrahydro-7,8-dihydroxy-1-(*p*-hydroxyphenyl)-1*H*-3-benzazepine
See: Fenoldopam

5-Chloro-2-(2,3,4,5-tetrahydro-3,5-dioxo-1*H*-1,2,4-triazol-4-yl)benzophenone
Probably formed from: 5-Chloro-2-(2,3-dihydro-3-oxo-4*H*-1,2,4-triazol-4-yl)benzophenone

6-Chloro-2,3,4,5-tetrahydro-N-hydroxy-1*H*-3-benzazepine
Probably formed from: 6-Chloro-2,3,4,5-tetrahydro-1*H*-3-benzazepine
→ 6-Chloro-2,3,4,5-tetrahydro-N-hydroxy-1*H*-3-benzazepine-β-D-glucuronide
Probably in: Dog E924

6-Chloro-2,3,4,5-tetrahydro-N-hydroxy-1*H*-3-benzazepine-β-D-glucuronide
Probably formed from: 6-Chloro-2,3,4,5-tetrahydro-N-hydroxy-1*H*-3-benzazepine

(*R*)-(+)-7-Chloro-2,3,4,5-tetrahydro-8-hydroxy-3-methyl-1-phenyl-1*H*-3-benzazepine
→ (*R*)-(+)-7-Chloro-2,3,4,5-tetrahydro-8-hydroxy-3-methyl-1-phenyl-1*H*-3-benzazepine-β-D-glucuronide
In: Man G960

(*R*)-(+)-7-Chloro-2,3,4,5-tetrahydro-8-hydroxy-3-methyl-1-phenyl-1*H*-3-benzazepine-β-D-glucuronide
Formed from: (*R*)-(+)-7-Chloro-2,3,4,5-tetrahydro-8-hydroxy-3-methyl-1-phenyl-1*H*-3-benzazepine

8-Chloro-2,3,4,5-tetrahydro-7-hydroxy-3-methyl-5-phenyl-1*H*-3-benzazepine
→ 8-Chloro-2,3,4,5-tetrahydro-7-hydroxy-3-methyl-5-phenyl-1*H*-3-benzazepine-β-D-glucuronide
In: Man H250
Rat G45

8-Chloro-2,3,4,5-tetrahydro-7-hydroxy-3-methyl-5-phenyl-1*H*-3-benzazepine-β-D-glucuronide
Formed from: 8-Chloro-2,3,4,5-tetrahydro-7-hydroxy-3-methyl-5-phenyl-1*H*-3-benzazepine

6-Chloro-2,3,4,5-tetrahydro-3-methyl-1*H*-3-benzazepine
→ 6-Chloro-2,3,4,5-tetrahydro-1*H*-3-benzazepine
In: Dog E859, E924
Rat E859
→ 6-Chloro-2,3,4,5-tetrahydro-3-methyl-1*H*-3-benzazepine-N-oxide
In: Dog E859
Rat E859

6-Chloro-2,3,4,5-tetrahydro-3-methyl-1*H*-3-benzazepine-N-oxide
Formed from: 6-Chloro-2,3,4,5-tetrahydro-3-methyl-1*H*-3-benzazepine

***trans*-5-Chloro-2,3,3a,12b-tetrahydro-2-methyl-1*H*-dibenz[2,3:6,7]oxepino[4,5-*c*]pyrrolidine**
→ *trans*-5-Chloro-2,3,3a,12b-tetrahydro-1*H*-dibenz[2,3:6,7]oxepino[4,5-*c*]pyrrolidine
In: Rat F660
→ *trans*-5-Chloro-2,3,3a,12b-tetrahydro-1*H*-dibenz[2,3:6,7]oxepino[4,5-*c*]pyrrolidine-2-carbamate-β-D-glucuronide
In: Rat F660

→ *trans*-5-Chloro-2,3,3a,12b-tetrahydro-2-methyl-1*H*-dibenz[2,3:6,7]oxepino[4,5-*c*]pyrrolidine-N-oxide
In: Rat F660

trans-5-Chloro-2,3,3a,12b-tetrahydro-2-methyl-1*H*-dibenz[2,3:6,7]oxepino[4,5-*c*]pyrrolidine-N-oxide
Formed from: *trans*-5-Chloro-2,3,3a,12b-tetrahydro-2-methyl-1*H*-dibenz[2,3:6,7]oxepino[4,5-*c*]pyrrolidine

7-Chloro-1,2,3,5-tetrahydro-1-methyl-2-oxo-5-phenyl-4*H*-1,4-benzodiazepine-4-carboxamide
See: Uxepam

7-Chloro-1,3,4,5-tetrahydro-5-phenyl-2*H*-1,4-benzodiazepin-2-one
Formed from: *l*-4,5-Dihydrodiazepam

Chlorothalonil (2,4,5,6-tetrachloroisophthalonitrile)
→ Trichloro-S-glutathionylisophthalonitrile (3 isomers)
Probably in: Ictalurus G220
→ 2,5,6-Trichloro-4-methylthioisophthalonitrile
In: Azomonas G674
See also: J453, K171

***o*-Chlorothioanisole**
Formed from: Chlorobenzene

***m*-Chlorothioanisole**
Formed from: Chlorobenzene
m-Chlorothiophenol

***p*-Chlorothioanisole**
Formed from: Chlorobenzene
p-Chlorothiophenol
→ *p*-Chlorosulphinylanisole
In: Horseradish E403
Mortierella D759

***m*-Chlorothiophenol**
→ *m*-Chlorothioanisole
In: Man D926

***p*-Chlorothiophenol**
Formed from: S-(*p*-Chlorophenyl) O-ethyl ethanephosphonodithioate
Trithion sulphoxide
→ *p*-Chlorosulphinylanisole
In: Rat A1946
See also: A2639
→ *p*-Chlorothioanisole
In: Man D926
Rat C121
Tetrahymena E96
→ *p*-Chlorothiophenyl-β-D-glucuronide
In: Rat A1946, A2639
→ *p*-Chlorothiophenyl sulphate
Probably in: Rat A2639

***p*-Chlorothiophenyl-β-D-glucuronide**
Formed from: *p*-Chlorothiophenol

***p*-Chlorothiophenyl sulphate**
Probably formed from: *p*-Chlorothiophenol

2-Chloroticlopidine
Formed from: Ticlopidine

***o*-Chlorotoluene**
→ *o*-Chlorobenzoic acid
In: Thauera H616
→ *o*-Chlorobenzyl alcohol
In: Burkholderia H953
Pseudomonas H953

***m*-Chlorotoluene**
→ *m*-Chlorobenzyl alcohol
In: Burkholderia H953
Pseudomonas H953

***p*-Chlorotoluene**
→ *p*-Chlorobenzoic acid
In: Thauera H616
→ 3-Chloro-6-methylcatechol
In: Pseudomonas F67

4-Chloro-*o*-toluidine
See: 4-Chloro-2-methylaniline

2-(3-Chloro-*o*-toluidino)nicotinic acid
See: Clonixin

Chlorotoluron (1-(3-chloro-4-methylphenyl)-3,3-dimethylurea)
→ 1-(3-Chloro-4-hydroxymethylphenyl)-3,3-dimethylurea
In: Wheat A3869
→ 1-(3-Chloro-4-methylphenyl)-3-methylurea
In: Wheat A3869

α-(3-Chloro-*o*-tolyl)acetic acid
Formed from: α-Chloro-β-(3-chloro-*o*-tolyl)propionitrile

N′-(4-Chloro-*o*-tolyl)-N,N-dimethylformamidine
See: Chlordimeform

1-(4-Chloro-*o*-tolyl)-3,3-dimethylurea
Formed from: Chlordimeform
→ 1-(4-Chloro-*o*-tolyl)-3-methylurea

Probably in: Mouse A2534
N′-(4-Chloro-*o*-tolyl)formamidine
Formed from: Demethylchlordimeform
β-(3-Chloro-*o*-tolyl)-α-hydroxypropionitrile
Formed from: α-Chloro-β-(3-chloro-*o*-tolyl)propionitrile
N′-(4-Chloro-*o*-tolyl)-N-methylformamidine
See: Demethylchlordimeform
1-(4-Chloro-*o*-tolyl)-3-methylurea
Probably formed from: 1-(4-Chloro-*o*-tolyl)-3,3-dimethylurea
→ 1-(4-Chloro-*o*-tolyl)urea
Probably in: Mouse A2534
1-(4-Chloro-*o*-tolyl)urea
Probably formed from: 1-(4-Chloro-*o*-tolyl)-3-methylurea
Chlorotrianisene (1,1′,1″-(1-chloro-4-ethenyl-2-ylidene)tris(4-methoxybenzene))
→ Demethylchlorotrianisene
In: Rabbit A3466
Rat B647
5-Chloro-2-(4*H*-1,2,4-triazol-4-yl)benzophenone
Probably formed from: 8-Chloro-4-hydroxy-6-phenyl-4*H*-*s*-triazolo[4,3-*a*][1,4] benzodiazepine
→ 5-Chloro-2-(2,3-dihydro-3-oxo-4*H*-1,2,4-triazol-4-yl)benzophenone
Probably in: Dog A258
Man A258
8-Chloro-2-(4*H*-1,2,4-triazol-4-yl)benzophenone
Formed from: Estazolam
2-Chloro-1-(2,4,5-trichlorophenyl)ethanol
Formed from: Gardona
2,4,5-Trichlorophenacyl chloride
(*E*)-2-Chloro-1-(2,4,5-trichlorophenyl)vinyl dimethyl phosphate
Formed from: Gardona
(*Z*)-2-Chloro-1-(2,4,5-trichlorophenyl)vinyl dimethyl phosphate
See: Gardona
2-Chloro-1-(2,4,5-trichlorophenyl)vinyl methyl phosphate (demethyl tetrachlorovinphos)
Formed from: Gardona
→ 2′,4′,5′-Trichloroacetophenone
In: Chicken A3198
See also: A3807
→ 2,4,5-Trichlorophenacyl chloride
In: Chicken A3198
See also: A3807
2-Chloro-4-trifluoromethylaniline
→ 2-Amino-3-chloro-5-trifluoromethylphenol
In: Rat J551
→ 2-Chloro-4-trifluoromethylaniline-β-D-glucuronide
In: Rat J551
→ 2-Chloro-4-trifluoromethylhydroxylamine
In: Rat J551
→ 2-Chloro-4-trifluoromethylsulphamate
In: Rat J551
2-Chloro-4-trifluoromethylaniline-β-D-glucuronide
Formed from: 2-Chloro-4-trifluoromethylaniline
2-Chloro-5-trifluoromethylphenol
Formed from: *p*-Chloro-α,α,α-trifluortoluene
N-(4-(2-Chloro-4-trifluoromethylphenoxy)-2-ethoxyphenyl)acetanilide
Probably formed from: 4-(2-Chloro-4-trifluoromethylphenoxy)-2-ethoxyaniline
→ N-(4-(2-Chloro-4-trifluoromethylphenoxy)-2-hydroxyphenyl)acetanilide
Probably in: Rat A2748
4-(2-Chloro-4-trifluoromethylphenoxy)-2-ethoxyaniline
Formed from: Oxyfluorofen
→ N-(4-(2-Chloro-4-trifluoromethylphenoxy)-2-ethoxyphenyl)acetanilide
Probably in: Rat A2748
N-(4-(2-Chloro-4-trifluoromethylphenoxy)-2-hydroxyphenyl)acetanilide
Probably formed from: N-(4-(2-Chloro-4-trifluoromethylphenoxy)-2-ethoxyphenyl) acetanilide
5-(2-Chloro-4-trifluoromethylphenoxy)-2-nitrophenol
Formed from: Oxyfluorofen
2-Chloro-4-trifluoromethylphenylhydroxylamine
Formed from: 2-Chloro-4-trifluoromethylaniline
→ 2-Chloro-4-trifluoromethylphenylhydroxylamine-β-D-glucuronide
Probably in: Rat J551

2-Chloro-4-trifluoromethylphenylhydroxylamine-β-D-glucuronide
Probably formed from: 2-Chloro-4-trifluoromethylphenylhydroxylamine

4-(4-Chloro-3-trifluoromethylphenyl)-4-hydroxypiperidine
Formed from: Penfluridol

2-Chloro-4-trifluoromethylsulphamate
Formed from: 2-Chloro-4-trifluoromethylaniline

N-(2-Chloro-4-(trifluoromethyl)phenyl)valine cyano(3-phenoxyphenyl)methyl ester
See: Fluvalinate

***p*-Chloro-α,α,α-trifluorotoluene**
→ 2-Chloro-5-trifluoromethylphenol
In: Rat C555

5-Chloro-1,2,3-trihydroxybenzene
See: 5-Chloropyrogallol

3′-Chloro-2,2′,3-trihydroxybiphenyl
Formed from: 4-Chlorodibenzofuran
→ 3-Chlorosalicylic acid
In: Sphingomonas J382

4-Chloro-2,2′,3-trihydroxybiphenyl
Formed from: 3-Chlorodibenzofuran
→ Salicylic acid
In: Pseudomonas G253

4′-Chloro-2,2′,3-trihydroxybiphenyl
Formed from: 3-Chlorodibenzofuran
→ 4-Chlorosalicylic acid
In: Pseudomonas G253

5-Chloro-2,4′,6-trihydroxy-2′,4-dimethoxy-6′-methylbenzophenone
See: Griseophenone B

2-Chloro-4,5,6-tris(methylthio)isophthalonitrile
Probably formed from: 2,5-Dichloro-4,6-bis(methylthio)isophthalonitrile

6-Chlorotryptophan
→ 4-Chloro-3-hydroxyanthranilic acid
In: Guinea pig B475
Rat B475

7-Chloro-L-tryptophan
→ 7-Chloroindole-3-acetamide
In: Pseudomonas B424
→ 7-Chloroindole-3-pyruvic acid
In: Pseudomonas C706

2-Chlorotyramine
Formed from: 2-Chloro-L-tyrosine

3-Chlorotyramine
Formed from: 3-Chloro-L-tyrosine

2-Chloro-L-tyrosine
Formed from: *m*-Chlorophenol
→ 2-Chlorotyramine
In: Micrococcus C62

3-Chloro-L-tyrosine
Formed from: *o*-Chlorophenol
L-Tyrosine
→ 3-Chlorotyramine
In: Micrococcus C62

4-Chloroveratrole (1-chloro-3,4-dimethoxybenzene)
Formed from: 4-Chloro-2-methoxyphenol

4′-Chlorowarfarin
→ 4′-Chlorodihydrowarfarin
In: Rabbit G276

Chlorphenesin carbamate (3-(4-chlorophenoxy)-1,2-propanediol 1-carbamate)
→ Chlorphenesin carbamate-β-D-glucuronide
In: Guinea pig A3347
Rat A2288
→ *p*-Chlorophenol
In: Guinea pig A3288
Rat A3288
→ *p*-Chlorophenoxyacetic acid
In: Guinea pig A3347
→ *p*-Chlorophenoxylactic acid
In: Guinea pig A3288
Rat A3288

Chlorphenesin carbamate-β-D-glucuronide
Formed from: Chlorphenesin carbamate

Chlorpheniramine (3-(*p*-chlorophenyl)-1-dimethylamino-3-(2-pyridyl)propane)
→ 3-(*p*-Chlorophenyl)-3-(2-pyridyl)-*n*-propanol
In: Dog A3931
Man F902
Rat F902
See also: E805
→ 3-(*p*-Chlorophenyl)-3-(2-pyridyl) propionaldehyde
In: Rabbit E649
→ 3-(*p*-Chlorophenyl)-3-(2-pyridyl)propionic acid
In: Dog A3931
Man F902

Rat F902
→ Chlorpheniramine-β-D-glucuronide
In: Man G18
→ Chlorpheniramine-N-oxide
In: Cunninghamella H456
Man F902
Pig G78
Rat F902
→ Demethylchlorpheniramine
In: Cunninghamella H456
Dog A3931
Man F902
Pig G78
Rabbit E649, E805
Rat F902

Chlorpheniramine-β-D-glucuronide
Formed from: Chlorpheniramine

Chlorpheniramine-N-oxide
Formed from: Chlorpheniramine

Chlorphenoxamine (2-(1-(4-chlorophenyl)-1-phenylethoxy)-N,N-dimethylethanamine)
→ Chlorphenoxamine-N-oxide
Probably in: Man E495
→ 1-(*p*-Chlorophenyl)-1-formylmethoxy-1-phenylethane
Probably in: Man E495
→ 1-(*p*-Chlorophenyl)-1-phenylethanol
Probably in: Man E495
→ Norchlorphenoxamine
In: Man E495

Chlorphenoxamine-N-oxide
Probably formed from: Chlorphenoxamine

Chlorphentermine (4-chloro-α,α-dimethylbenzeneethanamine)
→ N-Hydroxychlorphentermine
In: Rabbit A2245
See also: A3022
→ Nitrosochlorphentermine
In: Rabbit A2245

Chlorpromazine (2-chloro-10-(3-dimethylaminopropyl)phenothiazine)
Formed from: Chlorpromazine-N-oxide
→ 1-Carboxyethyl-2-chlorophenothiazine
In: Man A693
Rabbit E822
→ 2-Chlorophenothiazine
In: Man A693
→ Chlorpromazine-β-D-glucuronide
In: Man E932
Rat C614
→ Chlorpromazine-N-hydroperoxide
Probably in: Man A2826
Contrast (wrong identity) A2923, A3488
→ Chlorpromazine-N-oxide
In: Guinea pig A901, A2992, C619
Horse A901
Man A693, A901, A1059, G579
Rabbit A901
Rat B54, B663, C126
Sealion A901
→ Chlorpromazine radical
Species unclear K386
→ Chlorpromazine sulphoxide
In: Beef A3445
Guinea pig A901 A1353, A1754, C619
Horse A901
Man A693, A901, A1059, C619, F137, G579
Rabbit A901
Sealion A901
See also: F317, F480, G27
→ Demethylchlorpromazine
In: Cunninghamella D185
Guinea pig A901, A2992, C619
Horse A901
Man A693, A901, A1059, C619, D119, D860
Rabbit A901, B54, B663
Rat A69, H691, J829
Sealion A901
See also: A1754, A2881, F317, F480, G27
→ 2-Hydroxychlorpromazine
In: Man A693
→ 7-Hydroxychlorpromazine
In: Guinea pig A901
Horse A901
Man A693, A901, A1059, C619, K515
Rabbit A901
Sealion A901
See also: F317, F480, G27
→ 8-Hydroxychlorpromazine
In: Man A693

→ Promazine
In: Man A2723, B753, D860
Monkey A2723

Chlorpromazine-N,S-dioxide
Formed from: Chlorpromazine-N-oxide
Probably formed from: Chlorpromazine sulphoxide
→ Demethlchlorpromazine sulphoxide
In: Dog F480
Man F480

Chlorpromazine-β-D-glucuronide
Formed from: Chlorpromazine

Chlorpromazine-N-oxide
Formed from: Chlorpromazine
→ Chlorpromazine
In: Dog F480
Man F480
Rat F317, G27
→ Chlorpromazine-N,S-dioxide
In: Dog F480
Man F480
Rat G27

Chlorpromazine-7,8-quinone
Formed from: 7,8-Dihydroxychlorpromazine

Chlorpromazine radical
Formed from: Chlorpromazine

Chlorpromazine sulphoxide
Formed from: Chlorpromazine
→ Chlorpromazine-N,S-dioxide
Probably in: Man A693

Chlorpromazin-6-yl-β-D-glucuronide
Formed from: 6-Hydroxychlorpromazine

Chlorpromazin-7-yl-β-D-glucuronide
Formed from: 7-Hydroxychlorpromazine

Chlorpromazin-8-yl-β-D-glucuronide
Formed from: 8-Hydroxychlorpromazine

Chlorpropamide (1-(*p*-chlorophenyl)sulphonyl-3-propylurea)
→ 1-(*p*-Chlorobenzenesulphonamide
In: Man A578, A590
→ 1-(*p*-Chlorophenyl)sulphonyl-3-(2-hydroxypropyl)urea
In: Man A578, A590
Rat A1291
→ 1-(*p*-Chlorophenyl)sulphonyl-3-(3-hydroxypropyl)urea
In: Man A578, A590
Rat A1291
→ *p*-Chlorophenylsulphonylurea
In: Man A578, A590
Rat A1291

Chlorpropham ((3-chlorophenyl)carbamic acid 1-methylethyl ester)
→ 4-Hydroxychlorpropham
In: Rat J659

Chlortetracycline ((4*S*,4a*S*,5a*S*,6*S*,12a*S*)-7-chloro-4-(dimethylamino)-1,4,4a,5a,6,11,12a-octahydro-3,6,10,12,12a-pentahydroxy-6-methyl-1,11-dioxo-2-naphthacenecarboxamide)
→ 4-Epichlortetracycline
In: Hen K669
→ Isochlortetracycline
In: Hen K669

Chlorzoxazone (5-chloro-2(1*H*)-benzoxazolone)
→ 5-Chloro-6-hydroxybenzoxazolin-2-one
In: Rat C295
→ 6-Hydroxybenzoxazolin-2-one
In: Rat C295
→ 6-Hydroxychlorzoxazone
In: Beef H862
Cat H862
Dog H862
Ferret H862
Horse H862
Macaque H862
Man H862, J89, J299, J376
Mouse H862
Pig H862
Rabbit H862
Rat H862

Choisyine (1,2-dihydro-4,10-dimethoxy-α,α-dimethyldifuro[2,3-*b*:3′,2′-*f*]quinoline-2-methanol)
Formed from: Dictamnine
4-Methoxy-3-prenylquinol-2-one

Cholesteryl (2*R*)-2-(*p*-chlorophenyl)isovalerate
Probably formed from: 2-(*p*-chlorophenyl) isovalerate

Chorismic acid ((3*R*,4*R*)-3-((1-carboxyethenyl) oxy)-4-hydroxy-1,5-cyclohexadiene-1-carboxylic acid)
→ *p*-Aminobenzoic acid
In: Streptomyces D734, F111

→ *p*-Amino-L-phenylalanine
In: Streptomyces A2461, B158
→ Anthranilic acid
Probably in: Pseudomonas A958

(+)-3-Chromanol ((+)-dihydro-3-hydroxybenzopyran)
Formed from: 3-Oxochroman
Precocene II

4-Chromanol (dihydro-4-hydroxybenzopyran)
Formed from: 4-Chromanone
→ 4-Chromanone
In: Hanster K443
Rabbit A2032

3-Chromanone
See: 3-Oxochroman

4-Chromanone (dihydrobenzopyran-4-one)
Formed from: 4-Chromanol
→ 4-Chromanol
In: Apergillus G102
→ 4-Chromone
In: Apergillus G102

4-Chromone (1-benzopyran-4-one)
Formed from: 4-Chromanone

Chrysanthemin (cyanidin-3-glucoside)
Formed from: Cyanidin
→ Cyanidin-3-acetylglucoside
In: Zinnia G996
→ Cyanidin-3,5-diglucoside
In: Silene B410
→ Cyanidin-3-rhamnosylglucoside
In: Silene B405, B409
→ Paeonidin-3-glucoside
In: Vitis H886

Chrysazin
See: 1,8-Dihydroxyanthraquinone

Chrysazin diacetate (1,8-dihydroxy-9,10-anthracenedione diacetate)
Formed from: 1,8,9-Trihydroxyanthracene-1,8-diacetate

Chrysazin-8-β-D-glucoside
See: 8-β-D-Glucopyranosyloxy-1-hydroxyanthraquinone

Chryscandin (3-(((2*S*)-2-amino-3-(4-methoxyphenyl)-1-oxopropyl)amino)-1-(6-amino-9*H*-purin-9-yl)-1,3-dideoxy-β-D-ribofuranuronic acid)
→ N-Acetylchryscandin
In: Streptomyces D879

Chrysene
Formed from: 5-Methylchrysene
6-Methylchrysene
Probably formed from: *trans*-4,5-Dihydro-4,5-dihydroxy-3,4-benzpyrene
→ Chrysene-1,2-oxide
In: Rat E522, F933
See also: B285, B376, B772, D36, D588, E275, E571, F449
→ Chrysene-3,4-oxide
In: Rat E522, F933
See also: B285, B376, B772, D36, E275, E571, F449, K97
→ Chrysene-5,6-oxide
In: Rat E522, F933
See also: B285, B376, D36, E275
→ (+)-(3*S*,4*R*)-*cis*-3,4-Dihydro-3,4-dihydroxychrysene
In: Sphingomonas K97
→ 1-Hydroxychrysene
In: Man E571, F449
Hamster E571, F449
→ 2-Hydroxychrysene
In: Man E571, F449
Hamster E571, F449
→ 3-Hydroxychrysene
In: Man E571, F449
Hamster E571, F449
→ 4-Hydroxychrysene
In: Man E571, F449
Hamster E571, F449
→ 6-Methylchrysene
In: Rat G118

Chrysene-4,5-dicarboxylic acid
Formed from: 3,4-benzpyrene

Chrysene-1,2-oxide
Formed from: Chrysene
→ *trans*-1,2-Dihydro-1,2-dihydroxychrysene
Probably in: Mouse B376, D36
Rat B285, B772, E275

Chrysene-3,4-oxide
Formed from: Chrysene
→ *trans*-3,4-Dihydro-3,4-dihydroxychrysene
Probably in: Mouse B376, D36
Rat B285, B772, E275

Chrysene-5,6-oxide
Formed from: Chrysene
→ *trans*-5,6-Dihydro-5,6-dihydroxychrysene
In: Rat E275
See also: B285, B376, D36

2-((6-Chrysenylmethyl)amino)-2-hydroxymethylpropionic acid
Formed from: Crisnatol

2-((6-Chrysenylmethyl)amino)-2-methyl-1,3-propanediol
See: Crisnatol

S-((Chrysen-5-yl)methyl)-L-cysteine
Formed from: 5-Hydroxychrysene sulphate

Chrysin (5,7-dihydroxy-2-phenyl-4*H*-1-benzopyran-4-one)
→ Apigenin
In: Rat J300
→ Chrysin-β-D-glucuronide
In: Man G968, J783, J805
Rat H539, J805
→ Chrysin sulphate
In: Man J805, K124
Rat J805
→ Norwogonin
In: Pseudomonas A1328

Chrysin-β-D-glucuronide
Formed from: Chrysin

Chrysin sulphate
Formed from: Chrysin

Chrysoeriol (4′,5,7-trihydroxy-3′-methoxyflavone)
Formed from: Luteolin
→ Chrysoeriol-4′-glucoside
Probably in: Rose K809
→ Chrysoeriol-7-β-D-glucoside
In: Parsley K739
→ 4′,5-Dihydroxy-3′,7-dimethoxyflavone
Probably in: Calamondin orange A3987

Chrysoeriol-4′-glucoside
Formed from: Chrysoeriol

Chrysoeriol-7-β-D-glucoside
Formed from: Chrysoeriol

Chrysophanol (1,8-dihydroxy-3-methylanthraquinone)
Formed from: Chrysophanol-8-O-β-D-glucopyranoside
Emodin
Emodinanthrone
→ Aloe-emodin
In: Rat J421
→ Islandicin
In: Penicillium A1046

Chrysophanol anthrone (1,8-dihydroxy-3-methyl-9(10*H*)-anthracenone)
Formed from: Emodinanthrone

Chrysophanol-8-O-β-D-glucopyranoside
→ Chrysophanol
In: Microorganism K527

Cibenzoline (2-(2,2-diphenylcyclopropyl)-4,5-dihydro-1*H*-imidazole)
→ 4,5-Dehydrocibenzoline
In: Dog C488, D845
Man D853, K52
Rat C488, K52
→ *p*-Hydroxycibenzoline
In: Dog D845, K52
Man D853
Rat K52

Cichoric acid ((2*R*,3*R*)-O-dicaffeoyltartaric acid)
→ Caffeic acid
Echinacea K633
→ 2-O-Caffeoyltartaric acid
Echinacea K633

Ciclazindole (10-(3-chlorophenyl)-2,3,4,10-tetrahydropyrimido[1,2-*a*]indol-10-ol)
→ N-(3-(*m*-Chlorophenyl)-3-hydroxy-3*H*-indol-2-yl)-3-aminopropionic acid
In: Man A3842
→ 3-(*m*-Chlorophenyl)-3-hydroxy-3*H*-2-oxindol-1-yl)propylamine
In: Man A3842

Cicletanine (3-(4-chlorophenyl)-1,3-dihydro-6-methylfuro[3,4-*c*]pyridin-7-ol; both isomers)
→ Cicletanine-β-D-glucuronide
In: Man K462
Rat H451, K462
→ Cicletanine-O-sulphate
In: Man K462
Rat H451, K462

Cicletanine-β-D-glucuronide
Formed from: Cicletanine

Cicletanine-O-sulphate
Formed from: Cicletanine

Ciclobendazole (methyl 5-cyclopropylcarbonyl-2-benzimidazolecarbamate)
→ 2-Amino-5-cyclopropylcarbonyl-2-benzimidazole
In: Rat A3457
→ 6-Hydroxyciclobendazole
In: Dog A3457
Rat A3457

Cicloprofen
See: *l*-α-Methylfluorene-2-acetic acid

Ciglitazone (5-(4-(1-Methylcyclohexylmethoxy)benzyl)-2,4-thiazolidinedione)
→ 2′-Hydroxyciglitazone
In: Rat C753
→ *cis*-3′-Hydroxyciglitazone
In: Rat C753
→ *trans*-3′-Hydroxyciglitazone
In: Rat C753
→ *cis*-4′-Hydroxyciglitazone
In: Rat C753
→ *trans*-4′-Hydroxyciglitazone
In: Rat C753

Cilostamide (N-cyclohexyl-4-((1,2-dihydro-2-oxo-6-quinolinyl)oxy)-N-methylbutanamide)
→ Demethylcilostamide
In: Rat D165
→ 4-(1,2-Dihydro-2-oxo-6-quinolyloxy)-N-methylbutyramide
In: Rat D165
→ 2,6-Dihydroxyquinoline
In: Rat D165
→ 2′-Hydroxycilostamide
In: Rat D165
→ 3′-Hydroxycilostamide
In: Rat D165
→ 4′-Hydroxycilostamide
In: Rat D165
→ 4-Oxocilostamide
In: Rat D165

Cilostazol (6-(4-(1-cyclohexyl-1*H*-tetrazol-5-yl)butoxy)-3,4-dihydro-2(1*H*)-quinolinone)
→ 3,4-Dehydrocilostazol
In: Man D755
Rat D755
→ 3,4-Dihydro-6-hydroxycarbostyril
In: Dog D755
Man D755
Rat D755
→ 2″-*cis*-Hydroxycilostazol
In: Dog D755
Rat D755
→ 2″-*trans*-Hydroxycilostazol
In: Man D755
→ 3″-*cis*-Hydroxycilostazol
In: Dog D755
Rat D755
→ 3″-*trans*-Hydroxycilostazol
In: Dog D755
Rat D755
→ 4-Hydroxycilostazol
In: Dog D755
Rat D755
→ 4″-*cis*-Hydroxycilostazol
In: Dog D755
Man D755
Rat D755
→ 4″-*trans*-Hydroxycilostazol
In: Dog D755
Rat D755

Cimetropium (9-(cyclopropylmethyl)-7-((2*S*)-3-hydroxy-1-oxo-2-phenylpropoxy)-9-methyl-3-oxa-9-azoniatricyclo[3.3.1.0^{2,4}]nonane)
→ De(hydrohydroxy)cimetropium
In: Guinea pig G234
Hamster G234
Mouse G234
Rat G234
→ De(hydroxymethyl)cimetropium
In: Hamster G234
Rat G234
→ N-Demethylcimetropium
In: Guinea pig G234
Hamster G234
Mouse G234
Rat G234
→ *p*-Hydroxycimetropium
In: Hamster G234
Rat G234

(8α,9*R*)-Cinchonan-6′,9-diol
See: Cupreine

(9*S*)-Cinchonan-6′,9-diol
See: Cupreidine

(8α,9*R*)-Cinchonan-9-ol
See: Cinchonidine
(8α)-Cinchonan-9-one
See: Cinchonidinone
6′-Cinchonidine
See: Cupreine
Cinchonidine ((8α,9*R*)-cinchonan-9-ol)
Formed from: Cinchoninone
→ Cinchonidinone
In: Cinchona E357
→ 2′-Hydroxycinchonidine
In: Guinea pig G508
Rabbit G508
Cinchonidinone ((8α)-cinchonan-9-one)
Formed from: Cinchonidine
Cinchonine ((9*S*)-cinchonan-9-ol)
Formed from: Cinchoninone
→ Cinchoninone
In: Cinchona E357
→ 2′-Hydroxycinchoninone
In: Guinea pig G508
Rabbit G508
Cinchoninone (cinchonan-9-one)
Formed from: Cinchonine
→ Cinchonidine
In: Cinchona E357
→ Cinchonine
In: Cinchona E357
Cinepazide (1-(N-pyrrolidinocarbonylmethyl)-4-(3,4,5-trimethoxycinnamoyl)piperazine)
→ 4-(3-Hydroxy-4,5-dimethoxycinnamoyl)-1-(N-pyrrolidinocarbonylmethyl)piperazine
In: Dog A1789
Man A1789
Rat A1789
→ 4-(4-Hydroxy-3,5-dimethoxycinnamoyl)-1-(N-pyrrolidinocarbonylmethyl)piperazine
In: Dog A1789
Man A1789
Rat A1789
→ 1-(N-(2-Oxopyrrolidino)carbonylmethyl)-4-(3,4,5-trimethoxycinnamoyl)piperazine
In: Dog A1789
Man A1789
Rat A1789
Cinerubin A
Formed from: Aklaviketone
Cinerubin B
ε-Pyrromycinone
→ 7-Deoxycinerubinone A
In: Aeromonas A870
Cinerubin B
Formed from: ε-Pyrromycinone
→ Cinerubin A
In: Streptomyces D17
Cinmethylin (*exo*-1-methyl-4-(1-methylethyl)-2-((2-methylphenyl)methoxy)-7-oxabicyclo[2.2.1]heptane)
→ α-Hydroxycinmethylin
In: Goat F98
Rat D942
→ 6-Hydroxycinmethylin
In: Goat F98
→ 8-Hydroxycinmethylin
In: Rat D942
→ 9-Hydroxycinmethylin
In: Goat F98
Rat D942
→ *o*-Hydroxymethylbenzoic acid
In: Goat F98
Rat D942
→ 2-((3-Hydroxy-2-methylphenyl)methoxy)-1-methyl-4-(1-methylethyl)-7-oxabicyclo[2.2.1]heptane
In: Goat F98
→ 2-((4-Hydroxy-2-methylphenyl)methoxy)-1-methyl-4-(1-methylethyl)-7-oxabicyclo[2.2.1]heptane
In: Goat F98
→ 2-((5-Hydroxy-2-methylphenyl)methoxy)-1-methyl-4-(1-methylethyl)-7-oxabicyclo[2.2.1]heptane
In: Goat F98
Cinnabarinic acid (9-amino-3-oxo-3*H*-phenoxazine-1-carboxylic acid)
Formed from: 3-Hydroxyanthranilic acid
Cinnamaldehyde (3-phenyl-2-propenal)
Formed from: Cinnamic acid
Cinnamoyl CoA
Cinnamyl alcohol
Probably formed from: Cinnarizine
Flunarizine
→ Cinnamic acid
In: Acinetobacter E663

Arion G804
Arthrobacter J33
Euglena G422
Helix G804
Limax G804
Man J24
Mouse F432, G804
Pseudomonas J707
Rhodopseudomonas G246
Serratia G197
→ Cinnamyl alcohol
In: Cucumber A2432
Dunaliella G459
Euglena G422
Mung bean J910
Pyrococcus K222
Wheat J39
See also: B67
→ Colchicine
In: Colchicum D980
→ 2-(S-Glutathionyl)-3-phenylpropionaldehyde
Probably in: Armyworm E389
→ 2-Hydroxy-4-phenylbut-3-enonitrile
In: Almond K6
Loquat K6
Peach K6

Cinnamic acid (3-phenyl-2-propenoic acid)
Formed from: Aloeresin A
Cinnamaldehyde
Ethyl cinnamate
Ferulic acid
N-(2-hydroxyethyl)cinnamide
Naftiline
n-Pentadecylbenzene
L-Phenylalanine
4-Phenylbut-3-enoic acid
4-Phenylbutyric acid
1-Phenylheptane
L-Phenyllactic acid
1-Phenylnonane
1-Phenyltridecane
Probably formed from: 1-Butyryl-4-cinnamylpiperazine
3-Phenylpropionic acid
→ Angiorufone
In: Angiozanthos H157
→ 2-Bromophenylhydracrylic acid
In: Corallina E597
→ β-Bromostyrene
In: Corallina E597
→ Catechin'
In: Chamaecyparis A3223
Salix A3223
→ Cinnamaldehyde
In: Neurospora A750
Nocardia G207
See also: A2517, A2744, B67, K222
→ Cinnamoyl CoA
In: Cephalocereus H633
Clostridium K534
Glycine A2148, A2514
Pea A2514
Petunia A2845
→ Cinnamoyl-β-D-glucose
In: Cestrum A3475, B493
Chenopodium G255
Coleus B493
Pilea B493
Quercus K737
Raphanus B440
Sweet potato D70, F145
→ Cinnamoylglycine
In: Man A1539
Rat A2198
→ Cinnamyl alcohol
In: Corynespora G821
→ *o*-Coumaric acid
In: Melilotus A145
See also: A1466
→ *m*-Coumaric acid
Probably in: Orchis C734
→ *p*-Coumaric acid
In: Aconitum F445
Cestrum A182, A3475
Coffea F445
Coleus G749, J349
Dioscoreophyllum F445
Glycyrrhiza F445
Lunaria J502
Mentha A206
Nicotiana F445

Oryza H593
Pimpinella H667
Potato D964
Rat A2869
Sorghum K916
See also: A1466, A1927, B493, E28
→ Cyanidin
Probably in: Petunia A2739
→ L-Ephedrine
In: Ephedra A168
→ Epicatechin
In: Aesculus A3223
Rubus A3223
→ Eugenol
In: Ocimum A1494
→ Gallic acid
In: Hydrangea A1927
→ Hircinol
In: Orchis C734
→ Hirsutidin
In: Flax A1134
→ Isoorientin
In: Flax A1134
→ Kaempferol
In: Hydrangea A1927
→ Lucenin
In: Flax A1134
→ Malvidin
In: Flax A1134
→ Methyl cinnamate
In: Lentinus A214
→ Orientin
In: Flax A1134
→ Phenylacetic acid
In: Nocardia A1329
See also: B363
→ L-Phenylalanine
In: Endomyces E316
Rhodotorula H555
See also: J12
→ β-Phenylhydracrylic acid
Probably in: Man A126, A1539, D205
Microorganism A3189, B363
Nicotiana J819
Oryza H593
Rat A1403, A2198, A2869
Streptomyces C818
→ 3-Phenylpropionic acid
In: Microorganism J12
Streptomyces C818
See also: A1329, B363, C646
→ Proanthocyanidin A_2
In: Aesculus A3223
→ Procyanidin B_2
In: Aesculus A3223
Rubus A3223
→ Procyanidin B_3
In: Chamaecyparis A3223
Salix A3223
→ Procyanidin B_4
In: Rubus A3223
→ Scopoletin
In: Helianthus J818
→ Styrene
In: Catharanthus K244
Cryptococcus H400
Pichia G594
→ 1,2,3,4-Tetrahydro-1-(4-hydroxy-3-methoxyphenethyl)-2-methylisoquinoline
Probably in: Colchicum A606
→ (*S*)-Tropic acid
Probably in: Datura A2557
→ Tylophorine
In: Tylophora A1734
→ Vicenin
In: Flax A1134

Cinnamide

Formed from: N-(2-Hydroxyethyl)cinnamide

Cinnamonitrile

→ N-Acetyl-S-(1-cyano-2-phenylethyl)-L-cysteine
In: Rat B630
→ N-Acetyl-S-(1-hydroxy-2-phenylethyl)-L-cysteine
In: Rat B630

Cinnamoyl CoA

Formed from: Cinnamic acid
→ Cinnamaldehyde
In: Glycine A2576
See also: D980
→ Cinnamoyltyramine
In: Nicotiana D558

→ Pinosylvine
In: Barlia C734
Dioscorea D136
Epipactis G138
Pine B313, H607
→ 2′,4′,6′-Trihydroxychalcone
In: Cephalocereus H633

Cinnamoyl-β-D-glucose
Formed from: Cinnamic acid
→ Cinnamoylquinic acid
In: Sweet potato F143
→ *p*-Coumaroyl-β-D-glucose
Probably in: Sweet potato F145

Cinnamoylglycine
Formed from: Cinnamic acid

Cinnamoylquinic acid
Formed from: Cinnamoyl-β-D-glucose

Cinnamoyltropine
→ 6β-Hydroxycinnamoyltropine
In: Hyocyamus E350

Cinnamoyltyramine
Formed from: Cinnamoyl CoA
Tyramine

Cinnamyl acetate
Probably formed from: Cinnamyl alcohol
→ Cinnamaldehyde
In: Cinnamomum B67
→ Cinnamyl acetate epoxide
In: Pseudomonas K167

Cinnamyl acetate epoxide
Formed from: Cinnamyl acetate

Cinnamyl alcohol (3-phenyl-2-propen-1-ol)
Formed from: Allylbenzene
Cinnamaldehyde
Cinnamic acid
Cinnamyl anthranilate
Cinnamyl formate
→ 2-Bromo-1-phenylpropane-1,3-diol
In: Corallina E597
→ Cinnamaldehyde
In: Acinetobacter E473, E663, J621
Arion G804
Botrytis G613
Corynebacterium J893
Eucalyptus G660
Glycine A2325
Helix G804
Limax G804
Pine G484
Pleurotus F867, G668
Rhodopseudomonas D197
Trametes K590
Wheat G465
→ Cinnamyl acetate
Probably in: Cinnamomum B67
→ Cinnamyl alcohol epoxide
In: Pseudomonas K167
→ Cinnamyl alcohol-O-sulphate
In: Mouse H959
→ 3-Phenylpropan-1-ol
Probably in: Dunaliella G459
Euglena G422
Pyrococcus K222

Cinnamyl alcohol epoxide
Formed from: Cinnamyl alcohol

Cinnamyl alcohol-O-sulphate
Formed from: Cinnamyl alcohol

Cinnamyl anthranilate
→ Anthranilic acid
In: Mouse D753
→ Cinnamyl alcohol
In: Mouse D753

Cinnamyl formate
→ Cinnamyl alcohol
In: Escherichia B886

4-Cinnamyl-1-(1-hydroxybutyryl)piperazine
Formed from: 1-*n*-Butyryl-4-cinnamylpiperazine
→ 1-(1-Hydroxybutyryl)-4-(*p*-hydroxycinnamyl)piperazine
Probably in: Guinea pig A3589

4-Cinnamyl-1-(2-hydroxybutyryl)piperazine
Formed from: 1-Butyryl-4-cinnamylpiperazine
→ 1-(2-Hydroxybutyryl)-4-(*p*-hydroxycinnamyl)piperazine
Probably in: Guinea pig A3589

***trans*-3-(*trans*-Cinnamyloxy)-4-hydroxycinnamic acid**
Formed from: *p*-Coumaric acid

1-Cinnamylpiperazine
Formed from: 1-*n*-Butyryl-4-cinnamylpiperazine
Probably formed from: Cinnarizine

Flunarizine
Cinnarizine (((*E*)-1-diphenylmethyl)-4-(3-phenyl-2-propyl)piperazine)
→ Benzophenone
In: Rat G720, H35
See also: H852
→ Cinnamaldehyde
Probably in: Man H852
→ 1-Cinnamylpiperazine
Probably in: Man H852
→ 1-(Diphenylmethyl)-4-(3-(*p*-hydroxyphenyl)-2-propenyl)piperazine
In: Rat G720, H35
Man H852
→ 1-(Diphenylmethyl)piperazine
In: Rat G720, H35
See also: H852
→ 1-((*p*-Hydroxyphenyl)phenylmethyl)-4-(3-phenyl-2-propenyl)piperazine
In: Rat G720, H35
Man H852
Cinnoline (1,2-benzodiazine)
→ 4-Hydroxycinnoline
In: Arthrobacter J33
Rabbit A3695
Rat A3695
Cinpropazide (N-isopropyl-2-(4-(3,4,5-trimethoxycinnamoyl)-1-piperazinyl)acetamide)
→ 2-(4-(4-Hydroxy-3,5-dimethoxycinnamoyl)-1-piperazinyl)-N-isopropyl)acetamide
In: Dog A1400
→ 2-(4-(3,4,5-Trimethoxycinnamoyl)-1-piperazinyl)acetamide
In: Dog A1400
Cinromide ((2*E*)-3-bromophenyl-N-ethyl-2-propenamide)
→ *m*-Bromocinnamamide
In: Monkey D925
→ *m*-Bromocinnamic acid
In: Monkey D925
Ciprofibrate
See: 2-(4-(2,2-Dichlorocyclopropyl)phenoxy-2-methylpropanoic acid
Ciprofloxacin (1-cyclopropyl-6-fluoro-1,4-dihydro-4-oxo-7-(1-piperazinyl)-3-quinoline-carboxylic acid)
→ Ciprofloxacin-N-sulphate
In: Man E230
→ N-Formylciprofloxacin
In: Man E230
→ 2′-Oxociprofloxacin
In: Man E230
→ 3′-Oxociprofloxacin
In: Beef F955
Monkey E251
Ciprofloxacin-N-sulphate
Formed from: Ciprofloxacin
Ciramadol ((−)-*cis*-2-(α-dimethylamino-*m*-hydroxybenzyl)cyclohexanol)
→ Ciramadol-β-D-glucuronide (2 isomers)
In: Man D885
Monkey B990
Rat B990
→ Hydroxyciramadol
In: Monkey B990
→ Hydroxymethoxyciramadol
In: Monkey B990
→ Norciramadol
In: Monkey B990
Rat B990
→ *m*-(*cis*-Octahydro-3-methyl-2*H*-1,3-benzazocin-4-yl)phenol
In: Monkey B990
Ciramadol-β-D-glucuronide
Formed from: Ciramadol
Cisapride (*cis*-4-amino-5-chloro-N-(1-(3-(4-fluorophenoxy)propyl)-3-methoxy-4-piperidinyl)-2-methoxybenzamide)
→ 4-Amino-5-chloro-N-(1-(-3-(3-fluoro-4-hydroxyphenoxy)propyl)-3-methoxy-4-piperidinyl)-2-methoxybenzamide
In: Dog E927
Man E927, K479
Rat E926
→ 4-Amino-5-chloro-N-(1-(-3-(4-fluoro-2-hydroxyphenoxy)propyl)-3-methoxy-4-piperidinyl)-2-methoxybenzamide
In: Man K479
→ *cis*-4-Amino-5-chloro-N-(1-(3-(4-fluorophenoxy)propyl)-3-hydroxy-4-piperidinyl)-2-methoxybenzamide
In: Man K479

→ *cis*-4-Amino-5-chloro-N-(1-(3-hydroxypropyl)-3-methoxy-4-piperidinyl)-2-methoxybenzamide
In: Dog E927
Man E927, K479
Rat E926
→ 4-Amino-5-chloro-2-methoxy-N-(3-methoxy-4-piperidinyl)benzamide
In: Dog E927
Man E927, K57, K137
Rat E926
→ Cisapride-N^4-β-D-glucuronide
In: Rat E926
→ Cisapride-N-oxide
In: Man K479
→ *p*-Fluorophenol
In: Rat E926
See also: E927
→ 3-(*p*-Fluorophenoxy)propionaldehyde
Probably in: Man E927, K57
→ 2-Hydroxycisapride
In: Man K57
Rat E926
→ 3-Hydroxycisapride
In: Rat E926
→ 3′-Hydroxycisapride
In: Rat E620, E926

Cisapride-N^4-β-D-glucuronide
Formed from: Cisapride

Cisapride-N-oxide
Formed from: Cisapride

Citalopram (3-(3-dimethylaminopropyl)-3-(4-fluorophenyl)-5-isobenzofurancarbonitrile)
→ 3-(2-Carboxyethyl)-3-(*p*-fluorophenyl)-5-isobenzofurancarbonitrile
In: Man J758
→ Citalopram-β-D-glucuronide
In: Man J758
→ Citalopram-N-oxide
In: Man J758, K301
→ Demethylcitalopram
In: Man J758, K301

Citalopram-β-D-glucuronide
Formed from: Citalopram

Citalopram-N-oxide
Formed from: Citalopram

Citrinin (3*R*,*trans*)-4,6-dihydro-8-hydroxy-3,4,5-trimethyl-6-oxobenzopyran-7-carboxylic acid)
Formed from: 6-(Butan-2-on-3-yl)-2,4-dihydroxy-3,5-dimethylbenzaldehyde
→ Dihydrocitrinone
In: Rat D13

Cizolirtine (N,N-dimethyl-2-((1-methyl-1*H*-pyrazol-5-yl)phenylmethoxy)ethanamine)
→ Cizolirtine-N-oxide
In: Dog J743
Rat J743
→ 5-(α-(2-Dimethylaminoethoxy)benzyl)-1*H*-pyrazole
Probably in: Dog J743
Rat J743
→ 5-(α-(Carboxymethoxy)benzyl)-1-methyl-1*H*-pyrazole
In: Dog J743
Rat J743
→ 5-(α-(2-Methylaminoethoxy)benzyl)-1-methyl-1*H*-pyrazole
In: Dog J743
Rat J743

Cizolirtine-N-oxide
Formed from: Cizolirtine

(+)-Clausenamide ((3*R*,4*S*,5*S*)-*rel*-3-hydroxy-5-((*R*)-hydroxyphenylmethyl)-1-methyl-4-phenyl-2-pyrrolidinone)
→ 4-Hydroxyclausenamide
In: Rat K198
→ 6-Hydroxyclausenamide
In: Rat K198
→ 3-Hydroxy-5-(hydroxy-(*m*-hydroxyphenyl)methyl)-1-methyl-4-phenyl-2-pyrrolidinone
Probably in: Rat K198
→ 3-Hydroxy-4-(*o*-hydroxyphenyl)-5-(hydroxyphenylmethyl)-1-methyl-2-pyrrolidinone
Probably in: Rat K198

Clavicipitic acid
Formed from: 4-Dimethylallyl-L-tryptophan

Cleavamine ((7*R*)-5-ethyl-1,4,7,8,9,10-hexahydro-2*H*-3,7-methanoazacycloundecino[5,4-*b*] indole)
→ Cleavamine epoxide
Probably in: Polyporus C417

Cleavamine epoxide
Probably formed from: Cleavamine
Clebopride (4-amino-N-(1-benzyl-4-piperidinyl)-5-chloro-2-methoxybenzamide)
→ 4-Amino-N-(1-benzyl-4-piperidinyl-N-oxide)-5-chloro-2-methoxybenzamide
In: Rabbit B180
→ 4-Amino-5-chloro-N-(1-(α-hydroxy) benzyl-4-piperidinyl)-2-methoxybenzamide
In: Dog A3904
→ 4-Amino-5-chloro-2-methoxybenzoic acid
In: Dog A3904
Rabbit A3982
→ 4-Amino-5-chloro-2-methoxy-N-(4-piperidinyl)benzamide
In: Dog A3904
Rabbit A3904, A3982
Rat A3904
See also: B66
→ Clebopride-N^4-β-D-glucuronide
In: Rabbit A3904
→ S-(2-Ornithylamino-4-methoxy-5-(1-methyl-4-piperidinylamino) carboxyphenyl)glutathione
In: Rat F683
Clebopride-N^4-β-D-glucuronide
Formed from: Clebopride
Clemastine ((2*R*)-2-((1*R*)-1-(4-chlorophenyl)-1-phenylethoxy)ethyl)-1-methylpyrrolidine)
→ *p*-Chloro-α-methyl-α-phenylbenzyl alcohol
In: Rat F448
→ 3′,4′-Dihydro-3′,4′-dihydroxyclemastine
In: Rat F448
→ 3′-Hydroxyclemastine
In: Rat F448
Clenbuterol (4-amino-3,5-dichloro-α-(((1,1-dimethylethyl)amino)methyl) benzenemethanol)
→ 4-Amino-3,5-dichloromandelic acid
In: Beef J591
Rat J440
See also: C203, H930
→ Clenbuterol-β-D-glucuronide
Probably in: Beef J580
→ Clenbuterol sulphamate
In: Beef J591
Rat J440
→ 3,5-Dichloro-α-(((1,1-dimethylethyl) amino)methyl)-4-hydroxylaminobenzenemethanol
In: Beef J591
Pig J452
Rat H855, H930, J440
→ Hydroxyclenbuterol
In: Beef J591
Rat J440
See also: J580
Clenbuterol-β-D-glucuronide
Probably formed from: Clenbuterol
Clenbuterol sulphamate
Formed from: Clenbuterol
Clentiazem ((2*S*, 3*S*)-3-acetyloxy-8-chloro-5-(2-(dimethylamino)ethyl)-2,3-dihydro-2-(4-methoxyphenyl)-1,5-benzothiazepin-4-(5*H*)-one)
→ (+)-(2*S*, 3*S*)-3-Acetoxy-8-chloro-3,4-dihydro-2-(4-methoxyphenyl)-4-oxo-1,5-benzothiazepin-5(2*H*)-acetic acid
In: Rat H76
→ (+)-(2*S*, 3*S*)-3-Acetoxy-8-chloro-3,4-dihydro-2-(4-methoxyphenyl)-4-oxo-1,5-benzothiazepin-5(2*H*)-acetonitrile
In: Rat H76
→ Deacetylclentiazem
In: Rat H76
→ N-Norclentiazem
In: Rat H76
→ O-Norclentiazem
In: Rat H76
Clioquinol
See: 5-Chloro-8-hydroxy-7-iodoquinoline
Clobazam (7-chloro-1-methyl-5-phenyl-1*H*-1,5-benzodiazepino-2,4(3*H*,5*H*)-dione)
→ Demethylclobazam
In: Dog B22
Man B22
Monkey B22
Rat B22
→ 3′,4′-Dihydro-3′,4′-dihydroxyclobazam
In: Man B22
Monkey B22
Rat B22
→ *p*-Hydroxyclobazam

In: Dog B22
Man B22
Monkey B22
Rat B22

Clobenzorex (N-(*o*-chlorobenzyl)-α-methylphenethylamine)
→ Amphetamine
In: Man K234, K401
Rat A422
→ *o*-Chlorobenzaldehyde
Probably in: Rat A422

Clocoumarol (3-(1-(*p*-(2-chloroethyl)phenyl)butyl)-4-hydroxycoumarin)
→ 6-(3-(*p*-(2-Chloroethyl)phenyl)-1-hydroxy-2-oxohexylidene)-2,4-cyclohexadien-1-one
In: Rat A2896
→ 6-(3-(*p*-(2-Chloroethyl)phenyl)-1-hydroxypentyl *o*-hydroxyphenyl ketone
In: Rat A3928

Clofazimine (3-(*p*-chloroanilino)-10-(*p*-chlorophenyl)-2,10-dihydro-2-isopropyliminophenazine)
→ 10-(*p*-Chlorophenyl)-2,10-dihydro-3-(*p*-hydroxyanilino)-2-isopropyliminophenazine
In: Man C22
→ 10-(*p*-Chlorophenyl)-2,10-dihydro-2-isopropyliminophenazin-3-yl-β-D-glucuronide
In: Man C22
→ Clofazimine hydrate-β-D-glucuronide
In: Man B982

Clofazimine hydrate-β-D-glucuronide
Formed from: Clofazimine

Clofedanol (2-chloro-α-(2-(dimethylamino)ethyl)-α-phenylbenzenemethanol)
→ *o*-Chlorobenzophenone
In: Man A3384
→ 1-(*o*-Chlorophenyl)-1,3-dimethylamino-1-phenyl)-1-propene
In: Man A3384

Clofibrate (ethyl clofibrate)
→ Clofibric acid
In: Dog A1903, C589
Ferret C589
Man C589, H465
Rabbit C589
Rat A1903, A1969, C589, C677

Clofibric acid (clofibrate, 2-(*p*-chlorophenoxy)isobutyric acid)
Formed from: Binifibrate
Clofibrate
Clofibride
Clofibryl CoA
→ Clofibryl-β-D-glucuronic acid
In: Cat B148, C589
Dog A1903, B148, C589
Ferret B148, C589
Man B148, C589, G349
Rabbit A3499, B148, B253, C589
Rat A1903, A1969, B148, B555, C589, E937
See also: C631, D973, J416, J783
→ Clofibryltaurine
In: Cat B148, C589
Dog B148, C589
Ferret B148, C421, C589

Clofibride (4-hydroxy-N,N-dimethylbutyramide 2-(*p*-chlorophenoxy)isobutyrate)
→ *p*-Chlorophenol
In: Rat B555
→ Clofibric acid
In: Rat B555

Clofibryl CoA
→ Clofibric acid
In: Rat J261

Clofibryl-β-D-glucuronic acid
Formed from: Clofibric acid
→ S-(*p*-Chlorophenoxy-2-methylpropanoyl)glutathione
In: Rat H141

Clofibryltaurine
Formed from: Clofibric acid

Clomipramine
See: 3-Chloroimipramine

Clomiphene (2-(4-(2-chloro-1,2-diphenylethenyl)phenoxy)-N,N-diethylethanamine)
→ Clomiphene-N-oxide
In: Rat C656
→ Desethylclomiphene
In: Rat C656
→ 4-Hydroxyclomiphene
In: Rat C656, E476

→ 4′-Hydroxyclomiphene
In: Rat E476
Clomiphene-N-oxide
Formed from: Clomiphene
Clonazepam (5-(*o*-chlorophenyl)-1,3-dihydro-7-nitro-2*H*-1,4-benzodiazepin-2-one)
→ 7-Amino-5-(*o*-chlorophenyl)-1,3-dihydro-2*H*-1,4-benzodiazepin-2-one
In Man A2841, B945
Monkey B5
Rat D152
See also: A838
→ 3-Hydroxyclonazepam
In: Rat D152
Clonidine (N-(2,6-dichlorophenyl)-4,5-dihydro-1*H*-imidazol-2-amine)
Formed from: Alinidine
→ *p*-Hydroxyclonidine
In: Man A2906
Clonixin (2-(3-chloro-*o*-toluidino)nicotinic acid)
→ 2-(3-Chloro-2-hydroxymethylanilino) nicotinic acid
In: Dog A728
Monkey A728
Rat A728
→ 4′-Hydroxyclonixin
In: Dog A728
Monkey A728
Rat A728
→ 5-Hydroxyclonixin
In: Dog A728
Monkey A728
Rat A728
(*R*)-Clopidogrel ((*R*) assumed unless otherwise stated)
Probably formed from: (*S*)-Clopidogrel
→ 3-Carboxymethenyl-4-mercapto-N-(α-(methoxycarbonyl)-*o*-chlorobenzyl) piperidine
In: Man K646
→ (*R*)-Clopidogrel acid
Probably in: Rat K332
(*S*)-Clopidogrel ((α*S*)-α-(2-chlorophenyl)-6,7-dihydrothieno[3,2-*c*]pyridine-5(4*H*)-acetic acid methyl ester)
→ (*S*)-Clopidogrel acid
In: Rat K332
→ (*R*)-Clopidogrel
Probably in: Rat K332
(*R*)-Clopidogrel acid ((α*R*)-α-(2-chlorophenyl)-6,7-dihydrothieno[3,2-*c*]pyridine-5(4*H*)-acetic acid)
Probably formed from: (*R*)-Clopidogrel
(*S*)-Clopidogrel acid
Formed from: (*S*)-Clopidogrel
Cloprostenol ((5*Z*)-rel-7-((1*R*,2*R*,3*R*,5*S*)-2-((1*E*,3*R*)-4-(3-chlorophenoxy)-3-hydroxy-1-butenyl)-3,5-dihydroxycyclopentyl-5-heptenoic acid)
→ Dinorcloprostenol
In: Marmoset A3845
→ Tetranorcloprostenol
In: Rat A3845
Cloxazolam (10-chloro-11b-(*o*-chlorophenyl)-2,3,4,5,6,7,11b-hexahydrobenzo[6,7]-1,4-diazepino[5,4-*b*]oxazol-6-one)
→ 7-Chloro-5-(*o*-chlorophenyl)-1,3-dihydro-2*H*-1,4-benzodiazepin-2-one
In: Mouse A828
Rat A828
→ 2′,5-Dichloro-2-(2-hydroxyethylamino) acetamidobenzophenone
In: Mouse A828
Rat A828
Clozapine (8-chloro-11-(4-methyl-1-piperazinyl)-5*H*-dibenzo[*b*,*e*][1,4]diazepine)
Formed from: Clozapine-N-oxide
→ Clozapine-N-β-D-glucuronide
In: Man H162, H541, J153
→ Clozapine-N-oxide
In: Guinea pig K255
Man H390, J147, J153, K276, K602
Mouse K255
Rat K255, K695
→ 8-Dechloro-8-hydroxyclozapine
In: Man J153
See also: H702, J227
→ 8-Dechloro-8-methylthioclozapine
In: Man J153
See also: J227
→ Demethylclozapine
In: Guinea pig K255
Man H390, J147, J153, K276, K602
Mouse K255

Rat K255
See also: H702, J227
→ 7-Hydroxyclozapine
In: Man H390, J153
Rat H702
→ 7-Methylthioclozapine
In: Man J153
→ 3′-Oxoclozapine
In: Man J153

Clozapine-N-β-D-glucuronide
Formed from: Clozapine

Clozapine-N-oxide
Formed from: Clozapine
→ Clozapine
In: Rat J227

Clozic (2-((4′-chloro[1,1′-biphenyl]-4-yl)methoxy-2-methylpropanoic acid)
→ Clozic-β-D-glucuronide
In: Hamster C708

Clozic-β-D-glucuronide
Formed from: Clozic

2-(CoA-S-yl)-4-dimethylamino-3,5-bis(S-glutathionyl)phenol
Probably formed from: *p*-Dimethylaminophenol

Cocaethylene ((1*R*,2*R*,3*S*,5*S*)-3-(benzoyloxy)-8-methyl-8-azabicyclo[3.2.1]octane-2-carboxylic acid ethyl ester)
Formed from: Cocaine
Probably formed from: Benzoylecgonine
→ Benzoylecgonine
Probably in: Rat J611

Cocaine ((1*R*,2*R*,3*S*,5*S*)-3-(benzoyloxy)-8-methyl-8-azabicyclo[3.2.1]octane-2-carboxylic acid ethyl ester)
Formed from: Cocaine-N-oxide
Probably formed from: Cocaethylene
→ Benzoylecgonine
In: Dog A1793, D478
Guinea pig G561
Man F21, G336, G582
Mouse A2726, G582
Rabbit A2726, D478
Rat A1249, A1344, A2726, A3092
Squirrel monkey K23
See also: A1720, A3721
→ Ethylcocaine
In: Man G336
→ Hydroxycocaine
In: Rat E201
→ Norcocaine
In: Dog A1793
Guinea pig G561
Man A3547, G38
Mouse A2726, C560, E809
Rabbit A2726
Rat A1249, A2726, A3092, E201, H599, H623
See also: A3721

Cocaine-N-oxide
→ Cocaine
In: Rat A3721

Cocculidine ((3β)-1,6-didehydro-3,15-dimethoxyerythrinan)
Formed from: Isococculidine
Norprotosinomenine
→ Isococculidine
In: Erythrina A3588

Coclaurine (1,2,3,4-tetrahydro-7-hydroxy-1-(*p*-hydroxybenzyl)-6-methoxyisoquinoline)
Formed from: Higenamine
N-Methylcoclaurine
Norcoclaurine
→ Berbamunine
In: Berberis G736
→ Crotsparine
In: Croton A185
→ Crotsparinine
In: Croton A185
→ 1-(3,4-Dihydroxybenzyl)-1,2,3,4-tetrahydro-7-hydroxy-6-methoxyisoquinoline
In: Berberis F855
→ Methylcoclaurine
In: Berberis K780, K933
Eschscholtzia K933
Peumus K933
See also: B673, H629
→ Sparsiflorine
In: Croton A185

Codamine ((1*S*)-1-((3,4-dimethoxyphenyl) methyl)-1,2,3,4-tetrahydro-6-methoxy-2-methyl-7-isoquinolinol)
Formed from: Laudanosine

Codeine (5α,6α)-7,8-didehydro-4,5-epoxy-3-methoxy-17-methylmorphinan-6-ol)
Formed from: Codeinone
Morphine
→ Codeine-6-β-D-glucuronide
In: Guinea pig F521
Man H933, J74
Mouse F521
Rabbit F521
Rat F521
→ Codeine-7,8-oxide
In: Rat B345
→ Codeine-N-oxide
In: Guinea pig A3729
Horseradish C414
→ Codeinone
In: Guinea pig D419, G51
Hamster K443
Papaver K756
Pseudomonas F815, H892
Rabbit H920
Rat D419
See also: K311
→ 14β-Hydroxycodeine
In: Pseudomonas K311
→ Morphine
In: Candida G412
Guinea pig D419, F521
Man J244
Mouse F521
Rabbit F521
Rat A1963, C608, D419, F521, F746, H537
See also: A270, C354, G51
→ Norcodeine
In: Candida G412
Cunninghamella D185
Dog B138
Guinea pig B138, F521
Man B138, H726
Mouse F521
Rabbit B138, F521
Rat B138, C608, C669, F521
See also: A270, A696

Codeine-6-β-D-glucuronide
Formed from: Codeine

Codeine-7,8-oxide
Formed from: Codeine

Codeine-N-oxide
Formed from: Codeine

Codeinone (5α)-7,8-didehydro-4,5-epoxy-3-methoxy-17-methylmorphinan-6-one)
Formed from: Codeine
Probably formed from: O-Northebaine-6-ethyl ether
Thebaine
→ Codeine
In: Papaver A3141, B205, K756, K757
Pseudomonas H785
Rat J744
See also: A270, C354, K311
→ Hydrocodone
In: Pseudomonas H892, J635
See also: K311
→ 14β-Hydroxycodeinone
In: Pseudomonas K311

Codorphone ((5α,8β)-17-(cyclopropylmethyl)-4,5-epoxy-8-ethyl-3-methoxymorphinan-6-one)
→ 17-Cyclopropylmethyl-4,5α-epoxy-8*β*-(2-hydroxyethyl)-3-methoxymorphinan-6-one
In: Dog C636
→ Decyclopropylmethylcodorphone
In: Dog C636
Man C636
Rat C636
→ Demethylcodorphone
In: Dog C636
Rat C636
→ Dihydrocodorphone
In: Dog C636
Rat C636

Coenzyme Q_6
See: Ubiquinone 30

Coenzyme Q_8
See: Ubiquinone 40

Coenzyme Q_9
See: Ubiquinone 45

Coenzyme Q_{10}
See: Ubiquinone 50

Colchicine (N-((7*S*)-5,6,7,9-tetrahydro-1,2,3,10-tetramethoxy-9-oxobenzo[*a*]heptalen-7-yl) acetamide)
Formed from: Autumnaline
Cinnamaldehyde
Deacetylcolchicine
3-Phenylpropionaldehyde
Probably formed from: Isoandrocymbine
→ O^2-Demethylcolchicine
In: Mammal A1540
Streptomyces B227
→ O^3-Demethylcolchicine
In: Bacterium J458
Mammal A1540
Streptomyces B227
→ O^{10}-Demethylcolchicine
In: Mouse A1540
Rat A1540

Colterol
See: N-(*tert*-Butyl)noradrenaline

Columbamine (5,6-dihydro-2-hydroxy-3,9,10-trimethoxydibenzo[*a*,*g*]quinolizinium)
Formed from: Nococlaurine
Reticuline
Tetrahydrocolumbamine
→ Berberine
In: Berberis D536
→ Palmatine
In: Berberis E256, E268
→ Tetrahydrocolumbamine
In: Corydalis C136

Columnidin
Formed from: Eriodictyol

Conidendrin (α-conidendrol, tsugaresinol, (3a*R*,4*S*,9a*R*)-3a,4,9,9a, tetrahydro-6-hydroxy-4-(4-hydroxy-3-methoxyphenyl)-7-methoxynaphtho[2,3-*c*] furan-1(3*H*)-one)
→ Protocatechuic acid
In: Acinetobacter G275

α-Conidendrol
See: Conidendrin

Coniferaldehyde (ferulaldehyde, 4-hydroxy-3-methoxycinnamaldehyde)
Formed from: Coniferyl alcohol
Eugenol
Ferulic acid
→ Coniferaldehyde-4-glucoside
In: Rose K809
→ Coniferyl alcohol
In: Aralia G815
Beech E558
Forsythia A848
Glycine A2325, B922
Horse A3152
Populus D54
Spruce B922, G781
Swede A216
See also: A1687, A2744, G821
→ *cis*-Coniferyl alcohol
In: Beech E558
→ Dihydroconiferaldehyde
Probably in: Trametes A2744
→ 3,4-Dihydroxy-5-methoxycinnamaldehyde
In: Liquidambar K217
→ Ferulic acid
In: Pseudomonas J707
→ Feruloyl CoA
In: Glycine B922
Spruce B922

Coniferaldehyde-4-glucoside
Formed from: Coniferaldehyde

Coniferin (4-O-glucosylconiferyl alcohol)
Formed from: Coniferyl alcohol
→ Coniferyl alcohol
In: Picea K833
Pinus H308

Coniferyl alcohol (4-hydroxy-3-methoxycinnamyl alcohol)
Formed from: Coniferaldehyde
Coniferin
β-Coniferyl alcohol
Dehydrodiconiferyl alcohol
Eugenol
Guaiacyglycerol
→ Arctigenin
In: Forsythia F661
→ Coniferaldehyde
In: Acinetobacter E473, E663
Aralia G815
Cinchona E357
Forsythia A848, A3152

Horse A3152
Picea G781
Pleurotus F867
Swede A216
See also: A3255
→ Coniferin
In: Forsythia A3365
In: Rose K809
→ Dehydrodiconiferyl alcohol
In: Abies G745
Larix G745
Picea G745
Pinus G628, G745
→ Dihydroconiferyl alcohol
Probably in: Trametes A2744
→ Epipinoresinol
In: Forsythia F661
→ Hydnocarpin
In: Silybum A3221
→ Isoconiferin
In: Forsythia A3365
→ Phillygenin
In: Forsythia F661
→ Pinoresinol
In: Abies G745
Forsythia H286
Larix G745
Picea G745
Pinus G628, G745
Sesamum J516
→ Silybin
In: Horseradish A3221
Silybum A3221

***cis*-Coniferyl alcohol**
Formed from: Coniferaldehyde

Coptisine (6,7-dihydro-bis[1,3]benzodioxolo [5,6-*a*:4′,5′-*g*]quinolizinium)
→ Tetrahydrocoptisine
In: Corydalis G136

Coreximine ((13a*S*)-5,8,13,13a-tetrahydro-3,10-dimethoxy-6*H*-dibenzo[*a*,*g*]quinolizine-2,11-diol)
Formed from: Reticuline
Probably formed from: 2,3,10,11-Tetrahydroxyberberine

Corilagin (cyclic 3,6-((1*R*)-4,4′,5,5′,6,6′-hexahydroxy(1,1′-biphenyl)-2,2′-dicarboxylate-β-D-glucopyranose)
→ Corilagin-O-sulphate
In: Eubacterium G475

Corilagin-O-sulphate
Formed from: Corilagin

Coripalline
See: 1,2,3,4-Tetrahydro-7-hydroxy-6-methoxy-2-methylisoquinoline

Corydaline (13*S*,13a*R*)-5,8,13,13a,tetrahydro-2,3,9,10-tetramethoxy-13-methyl-6*H*-dibenzo[*a*,*g*]quinolizine)
Formed from: 3,4-Dihydroxy-L-phenylalanine
Mesocorydaline
13-Methylpalmatine
Palmatine
Tetrahydropalmatine
→ Mesocorydaline
In: Corydalis H585
→ 13-Methylpalmatine
In: Corydalis H585

Corydalubine
Formed from: Palmatine

(−)-Corynoline (5b*R*,6*S*,12b*R*)-5b,6,7,12b,13,14-hexahydro-5b,13-dimethyl[1,3]benzodioxolo[5,6-*i*]phenanthridin-6-ol)
→ (−)-Acetylcorynoloxine
In: Corydalis H70

(+)-Corynoline (5b*R*,6*S*,12b*R*)-5b,6,7,12b,13,14-hexahydro-5b,13-dimethyl[1,3]benzodioxolo[5,6-*i*]phenanthridin-6-ol)
→ (+)-Corynoloxine
In: Corydalis H70

(+)-Corynoloxine (4*S*,5a*S*,10*R*,12a*R*)-5a,6,12,12a-tetrahydro-12a,15-dimethoxy-4*H*-bis[1,3]benzodioxolo[4,5*c*:5′,6′-*g*][1]benzopyran-4,12-imine)
Formed from: (+)-Corynoline

Cotarnine (5,6,7,8-tetrahydro-4-methoxy-6-methyl-1,3-dioxolo[4,5-*g*]isoquinolin-5-ol)
Formed from: Narcotine
Noscapine
Tritoqualine

***o*-Coumaraldehyde** (*o*-hydroxycinnamaldehyde)
Formed from: *o*-Coumaric acid
o-Coumaroyl CoA

***m*-Coumaraldehyde** (*m*-hydroxycinnamaldehyde)
Formed from: *m*-Coumaric acid
***p*-Coumaraldehyde** (*p*-hydroxycinnamaldehyde)
Formed from: *p*-Coumaric acid
p-Coumaroyl CoA
p-Coumaryl alcohol
Probably formed from: 3-(*p*-Hydroxyphenyl)propanol
→ *p*-Coumaryl alcohol
In: Forsythia A1062
Glycine A2325, B922
Populus D54
Spruce B922, G781
Wheat H510
See also: A2744
Coumaranone (3(2*H*)-benzofuranone)
→ *o*-Hydroxyphenylacetic acid
In: Man K327
***o*-Coumaric acid**
Formed from: Cinnamic acid
Probably formed from: *o*-Methoxycinnamic acid
→ *o*-Coumaraldehyde
In: Neurospora A750
→ *o*-Coumaroyl CoA
In: Glycine A2148
Tomato A3149
→ *o*-Coumaroyl-D-glucose
In: Sweet potato D70
→ *o*-Coumaryl alcohol
In: Corynespora G821
→ 2,3-Dihydroxycinnamic acid
In: Pseudomonas A934
→ *o*-Hydroxystyrene
In: Microorganism F390
→ Melilotic acid
Probably in: Man G954
Rat E13, G605
***cis-o*-Coumaric acid**
Formed from: Coumarin
***m*-Coumaric acid**
Formed from: Caffeic acid
m-Tyrosine
Probably formed from: Cinnamic acid
→ *m*-Coumaraldehyde
In: Neurospora A750
Nocardia G207
→ *m*-Coumaric acid-α-D-glucoside
Probably in: Bacillus H234
→ *m*-Coumaric acid-O-sulphate
Probably in: Rat J686
→ *m*-Coumaroyl CoA
In: Glycine A2148
Pseudomonas B298
Tomato A3149
→ *m*-Coumaroyl-β-D-glucose
In: Quercus K737
→ *m*-Hydroxyphenylhydracrylic acid
Probably in: Pseudomonas H426
→ *m*-Hydroxyphenylpropionic acid
In: Clostridium A653
Lactobacillus A653
Proteus A653
See also: C734, D136, J686
→ *m*-Hydroxystyrene
In: Microorganism F390
***p*-Coumaric acid** (*trans*-*p*-hydroxycinnamic acid)
Formed from: Caffeic acid
Cinnamic acid
O-(5-O-((*E*)-*p*-Coumaroyl)-α-L-arabinofuranosyl)(1→3)-O-β-D-xylopyranosyl(1→4)-D-xylopyranose
p-Coumaroyl-β-D-glucose
O-(5-O-((*E*)-*p*-Coumaroyl)(O-β-D-xylopyranosyl(1→2))-O-α-L-arabinofuranosyl(1→3))-O-β-D-xylopyranosyl(1→4)-D-xylopyranose
Daidzein
Genistein
p-Hydroxyphenyllactic acid
Methyl *p*-coumarate
Naringenin
L-Tyrosine
Umbelliferone
O-β-D-Xylopyranosyl(1→1)-O-(5-O-((*E*)-*p*-coumaroyl)-α-L-arabinofuranosyl)(1→3))-O-β-D-xylopyranosyl(1→4)-D-xylopyranose
Probably formed from: Phloretic acid
p-(1-Propenyl)phenol
→ Bakuchiol
In: Psoralea D515

See also: D862
→ Batatasin I
Dioscorea D136
→ N^1,N^{10}-Bis(*p*-coumaroyl)spermidine
Probably in: Lunaria J502
→ Caffeic acid
In: Alnus A3343
Bean F45
Mushroom H907
Mucuna E769
Mytilus B314
Parsley A1100
Potato D964
Sorghum A2433
See also: A206, A1333, A1930, A2869, A3136, A3475, E458, F842
→ *trans*-3-(*trans*-Cinnamyloxy)-4-hydroxycinnamic acid
In: Trametes G224
→ *p*-Coumaraldehyde
In: Neurospora A750
See also: A2744
→ *cis-p*-Coumaric acid
Probably in: Sphagnum J179
→ *p*-Coumaric acid-α-D-glucoside
Probably in: Bacillus H234
→ *p*-Coumaric acid-4-β-D-glucoside
In: Cinchona K889
Tomato B846
→ *p*-Coumaroyl CoA
In: Aubergine A2514
Centaurium J306
Cephalocereus H633
Cucumber A2514
Forsythia A1062
Glycine A2148, A2514
Leek A2514
Marrow A2514
Pea A2514
Pepper A2514
Petunia A2845
Phaseolus A2514
Polyporus A226
Poplar C740
Potato A204
Pseudomonas B298
Tomato A3149
Wheat K585
Willow D466
See also: A206, A1332, A1930
→ *p*-Coumaroyl-β-D-glucose
In: Cestrum B493
Chenopodium G255
Nicotiana K576
Quercus K737
Raphanus B440
Sweet potato D70
Tomato B846
See also: E28, F445, G897
→ 5-(*p*-Coumaroyl)quinic acid
In: Periploca A207
See also: A3475
→ *p*-Coumaryl alcohol
Probably in: Alnus A3144
Betula A3144
→ Curcumin
In: Curcuma A1147
→ *p*-Ethylphenol
In: Lactobacillus H52
Pediococcus H52
→ Hydrangea glycoside A
Probably in: Hydrangea A1927
→ *p*-Hydroxybenzaldehyde
In: Lithospermum G229
See also: G432, H725
→ (−)-(*trans-p*-Hydroxycinnamoyl)lupinine
In: Lupinus A1945
→ *p*-Hydroxyphenylhydracrylic acid
Probably in: Acinetobacter C817
Man A126
Polyporus A2573
Rat A1334
Rhizoctonia A1466
Streptomyces C818
Trametes A2744
→ *p*-Hydroxystyrene
In: Bacillus H346, J623
Brettanomyces E263
Catharanthus K244
Cladosporium A2675, G797
Daucus K244
Erwinia G797

Fusarium E115
Klebsiella G797, J853
Lactobacillus H897, K567
Microorganism F390
Nicotiana K244
Polyporus A2690
Pseudomonas H364
Rhizoctonia G797
Saccharomyces G924
→ Mangiferin
In: Anemarrhena A3380, B562
→ Phloretic acid
In: Rat A1334
See also: A2571, A2767, G432, H725
→ *trans*-Sphagnum acid
In: Sphagnum J179
→ Umbelliferone
Probably in: Hydrangea A1927

***cis-p*-Coumaric acid**
Probably formed from: *p*-Coumaric acid
→ *cis-p*-Coumaric acid-4-β-D-glucoside
In: Sphagnum J384
See also: J179
→ *cis-p*-Coumaroyl CoA
In: Glycine A2148
→ *p*-Hydroxystyrene
In: Cladosporium A2675

***m*-Coumaric acid-α-D-glucoside**
Probably formed from: *m*-Coumaric acid

***p*-Coumaric acid-α-D-glucoside**
Probably formed from: *p*-Coumaric acid

***p*-Coumaric acid-4-β-D-glucoside**
Formed from: *p*-Coumaric acid

***cis-p*-Coumaric acid-4-β-D-glucoside**
Formed from: *cis-p*-Coumaric acid

***m*-Coumaric acid-O-sulphate**
Probably formed from: *m*-Coumaric acid

Coumarin (2*H*-1-benzopyran-2-one)
Formed from: Phenanthrene
→ N-Acetyl-S-(3-coumarinyl)-L-cysteine
In: Rat G286
→ *cis-o*-Coumaric acid
In: Aerobacter A653
Rat G605, G954
→ 3,4-Dihydroxycoumarin
In: Pseudomonas A934
Rat G605
→ 3-Hydroxycoumarin
In: Beef H154
Gerbil G73, G604
Hamster G604
Man A5, G73, G604, G954
Mouse J913
Rabbit A5
Rat A5, C863, G73, G604, G605, J913
See also: B468, B850, F914, G956
→ 4-Hydroxycoumarin
In: Guinea pig G956
Rat G605, G954, G956
→ 5-Hydroxycoumarin
In: Gerbil G73, G604
Hamster G604
Man G73, G604
Rat C863, G604, G605
→ 6-Hydroxycoumarin
In: Gerbil G73
Man G73
Rat C863, G605
→ 8-Hydroxycoumarin
In: Gerbil G73
Man G73
Rat C863, G605
→ *o*-Hydroxyphenylacetaldehyde
In: Beef H154
Gerbil G73, G338, G604
Hamster G604, J809
Man G73, G604, G954
Mouse J913, K258
Rat G73, G338, G604, G605, G954, J913
→ *o*-Hydroxyphenylacetic acid (not via 3-hydroxycoumarin)
In: Rat C863, F914
→ *o*-Tyrosol
In: Rat C863
→ Umbelliferone
In: Baboon A5, B850
Beef H154, H544
Cat A5
Catharanthus A2619
Chicken B468
Conium A2619
Cunninghamella A1218

Dog A5
Gerbil G73, G604, H544
Gliocladium A1218
Guinea pig G956
Hamster A5, G604, H546
Helicostylum A1218
Man A5, A3553, A3682, D153, D559, G604, G956, H542, H544, J796, K272, K325
Monkey G956, H544
Mouse A5, G956, H222, J913, K258, K325
Pig A5
Rabbit A5, G335, H544
Rat A114, A335, C863, G73, G605, G954, J796, J913
Streptomyces A1218

Coumarin-4-yl-β-D-glucoside
Formed from: 4-Hydroxycoumarin

Coumarin-4-yl-β-D-glucuronide
Formed from: 4-Hydroxycoumarin

S-(3-(7′-Coumaroxy)-2-hydroxy-1-propyl) glutathione
Formed from: 7-Glycidoxycoumarin

***p*-Coumaroyl agmatine**
Formed from: *p*-Coumaroyl CoA

O-(5-O-((*E*)-*p*-Coumaroyl-α-L-arabinofuranosyl)(1→3)-O-β-D-xylopyranosyl(1→4)-D-xylopyranose
→ *p*-Coumaric acid
In: Cynodon F865
Neocallimastix F865, G327, G759
Penicillium J499

O-(*p*-Coumaroyl)choline
Formed from: *p*-Coumaroyl-D-glucose

***o*-Coumaroyl CoA**
Formed from: *o*-Coumaric acid
→ *o*-Coumaraldehyde
In: Eucalyptus H330

***m*-Coumaroyl CoA**
Formed from: *m*-Coumaric acid
→ 3,3′,5-Trihydroxystilbene
In: Barlia C734

***p*-Coumaroyl CoA**
Formed from: *p*-Coumaric acid
p-Coumaroylquinic acid
p-Coumaroylshikimic acid
→ Bis-noryangonin
In: Petroselinum A149, A2618
→ Caffeoyl CoA
In: Parsley F66
→ *p*-Coumaraldehyde
In: Forsythia A1062
Glycine A2576
Populus D54
Spruce B922
→ *p*-Coumaroylagmatine
In: Barley B811
→ *p*-Coumaroyl-3,4-dihydroxyphenethylamine
In: Nicotiana F72
→ *p*-Coumaroyl 3,4-dihydroxyphenyllactic acid ester
In: Coleus H17, J349
→ *p*-Coumaroylepilupinine
In: Lupinus H43
→ *p*-Coumaroylgalactarate
Secale K722
→ *p*-Coumaroylglucarate
Secale K722
→ *p*-Coumaroylglucarolactone
Secale K722
→ *p*-Coumaroyl *p*-hydroxyphenyllactic acid ester
In: Coleus H17, J349
→ 5′-*p*-Coumaroylquinic acid
In: Nicotiana A3758
Potato A3755
Secale K722
Stevia A3758
Tomato A3149
→ *p*-Coumaroylshikimic acid
In: Agrostemma B202
Betula B202
Palus B202
Catharanthus B202
Capsicum B202
Caragana B202
Catalpa B202
Cichorium B202
Coffea B202
Coleus B202
Datura B202
Drosophyllum B202

Forsythia B202
Galium B202
Geum B202
Juglans B202
Linum B202
Lonicera B202
Lycopersicon B202
Malus B202
Nicotiana B202
Petroselinum B202
Phaseolus B202
Pimpinella B202
Pinus B202
Rauwolfia B202
Rhazya B202
Rheum B202
Rubia B202
Salix B202
Secale K722
Sedum B202
Solanum B202
Stevia B202
Contrast (in Ceratopteris) B202
→ *p*-Coumaroyltyramine
In: Nicotiana D558, F72
→ 6′-Deoxychalcone
In: Glycine E661
→ 5,6-Dihydro-4-hydroxy-6-(4-hydroxystyryl)-2-pyrone
In: Petroselinum A2618
See also: A3362
→ *p*-Hydroxybenzalacetone
In: Haplopappus A3362
Petroselinum A2618
Raspberry J26
Petroselinum A2618
→ *p*-Hydroxybenzaldehyde
In: Pseudomonas K92
→ Naringenin
In: Daucus D698
Glycine E579
Glycyrrhiza E619
Haplopappus A3362
Humulus J887
Petroselinum A149, A656, A2618
Phaseolus C813
Spinach D596, D867
Tulipa A2704, A3792
See also: B404, B957
→ Pinosylvine
In: Humulus J887
→ Resveratrol
In: Barlia C734
Dioscorea D136
Epipactis G138
Picea D213
Pine B313
→ 2′,4,4′,6′-Tetrahydroxychalcone
In: Buckwheat E180
Calamondin F419
Dianthus C392
Parsley B352, K808
Pueraria F37
Rye E662
Tulipa B377
→ 2′,4,4′,-Trihydroxychalcone
In: Glycine F789
Glycyrrhiza E619
Pueraria F37
See also: A2373

cis-p-Coumaroyl CoA
Formed from: *cis-p*-Coumaric acid

***p*-Coumaroyl-3,4-dihydroxyphenethylamine**
Formed from: *p*-Coumaroyl CoA
3,4-Dihydroxyphenethylamine

***p*-Coumaroyl 3,4-dihydroxyphenyllactic acid ester**
Formed from: *p*-Coumaroyl CoA
p-Coumaroyl *p*-hydroxyphenyllactic acid ester
3,4-Dihydroxyphenyllactic acid ester
→ Rosmarinic acid
In: Coleus G747, J349

***p*-Coumaroylepilupinine**
Formed from: *p*-Coumaroyl CoA

***p*-Coumaroylgalactarate**
Formed from: *p*-Coumaroyl CoA

***p*-Coumaroylglucarate**
Formed from: *p*-Coumaroyl CoA

***p*-Coumaroylglucarolactone**
Formed from: *p*-Coumaroyl CoA

o-Coumaroyl-D-glucose
Formed from: *o*-Coumaric acid
m-Coumaroyl-β-D-glucose
Formed from: *m*-Coumaric acid
p-Coumaroyl-D-glucose
Formed from: *p*-Coumaric acid
Probably formed from: Cinnamoyl-D-glucose
→ Caffeoyl-D-glucose
Probably in: Ipomoea F145, F842
→ *p*-Coumaric acid
In: Raphanus K724
→ O-(*p*-Coumaroyl)choline
In: Raphanus K720
→ Coumaroylglucuronosylglucose
In: Chenopodium G255
→ *p*-Coumaroylquinic acid
In: Sweet potato F143
→ 3‴-Demethylcelosianin II
In: Chenopodium G255
→ 1,2-Di-O-(*p*-coumaroyl)glucose
In: Raphanus K724
Coumaroylglucuronosylglucose
Formed from: *p*-Coumaroyl-D-glucose
p-Coumaroyl p-hydroxyphenyllactic acid ester
Formed from: *p*-Coumaroyl CoA
p-Hydroxyphenyllactic acid
→ Caffeoyl *p*-hydroxyphenyllactic acid ester
In: Coleus G747, J349
→ *p*-Coumaroyl 3,4-dihydroxyphenyllactic acid ester
In: Coleus G747, J349
p-Coumaroyl-N-hydroxysuccinamide
→ Caffeoyl-N-hydroxysuccinamide
In: Mushroom H907
5′-p-Coumaroylquinic acid
Formed from: *p*-Coumaric acid
p-Coumaroyl CoA
p-Coumaroyl-β-D-glucose
→ Chlorogenic acid
In: Daucus E536
Periploca A207
See also: A182
→ Coumaroyl CoA
Probably in: Potato A3145

p-Coumaroylshikimic acid
Formed from: *p*-Coumaroyl CoA
→ *p*-Coumaroyl CoA
In: Cichorium B202
p-Coumaroyl tartrate
→ Caffeoyl tartrate
Probably in: Grape E458
p-Coumaroyltyramine
Formed from: *p*-Coumaroyl CoA
Tyramine
O-(5-O-((E)-p-Coumaroyl)(O-β-D-xylopyranosyl(1→2))-O-α-L-arabinofuranosyl(1→3))-O-β-D-xylopyranosyl(1→4)-D-xylopyranose
→ *p*-Coumaric acid
In: Neocallimastix G327
o-Coumaryl alcohol
Formed from: *o*-Coumaric acid
p-Coumaryl alcohol
Formed from: 4-Allylphenol
p-Coumaraldehyde
Probably formed from: *p*-Coumaric acid
→ *p*-Coumaraldehyde
In: Glycine A2325
Spruce G781
→ Rhododendrin
In: Alnus A3144
Betula A3144
Coumestrol (3,9-dihydroxy-6*H*-benzofuro[3,2-*c*][1]benzopyran-6-one)
Formed from: Daidzein
Crassifoline ((1*S*)-1,2,3,4-tetrahydro-1-((3-hydroxy-4-methoxyphenyl)methyl)-7-methoxy-2-methyl-8-isoquinolinol)
→ Tetrahydro-1,9-dihydroxy-2,10-dimethoxy-protoberberine
In: Eschscholtzia H680
Credazine (3-(*o*-methylphenoxy)pyridazine)
→ 3-(*o*-Hydroxynethylphenoxy)pyridazine
In: Rat A3372
→ 3-(4-Hydroxy-2-methylphenoxy)pyridazine
In: Rat A3372
o-Cresol (2-methylphenol)
Formed from: Di-*o*-cresyl phosphate
Ferulic acid
4-Hydroxy-3-methylbenzoic acid
o-Methylanisole

o-Methylphenoxyacetic acid
Toluene
o-Toluic acid
Tri-*o*-cresyl phosphate
→ *o*-Cresyl-β-D-glucuronide
In: Man F903
Rana E841
Zebra fish E813
→ *o*-Cresyl sulphate
In: Beef D607
Man D607, E575, F903, K405
Rana E841
Rat E575
Xenopus E841
Zebra fish E813
→ 4-Hydroxy-3-methylbenzoic acid
In: Microorganism G278
→ *o*-Methylanisole
In: Guinea pig A115
→ 3-Methylcatechol
In: Bacillus A2397
Candida E204
Microorganism C776, E31
Pseudomonas E958, J392
Proteobacteria J466
Trichosporon A902
See also: A768, E565, F163, F857, J344
→ Methylquinol
In: Microoranism C776, E31
Rana E841
Xenopus E841
→ 2-Methylresorcinol
In: Rana E841
Xenopus E841
→ 4-Methylresorcinol
In: Microoranism C776, E31
→ 3-Methyl-L-tyrosine
In: Escherichia B292
→ Poly-*o*-cresol
Bjerkandera K565
→ Salicyl alcohol
Probably in: Rat G129
→ Salicylic acid
In: Rana E841
Xenopus E841
→ *m*-Toluic acid
In: Microoranism G917

***m*-Cresol** (3-methylphenol)
Formed from: N-Acetyl-*m*-methylphenyl-β-D-glucosaminide
m-Cresyl N-methylcarbamate
2-Hydroxy-6-methylbenzoic acid
m-Hydroxyphenylacetic acid
m-Methylanisole
Toluene
→ Benzoic acid
In: Microorganism G981
→ *m*-Cresyl-β-D-glucuronide
In: Bitterling C406
Bream C406
Carp C406
Goldfish C406
Gudgeon C406
Man F903
Minnow C406
Perch C406
Roach C406
Rudd C406
Stickleback C406
Tench C406
→ *m*-Cresyl sulphate
In: Bitterling C406
Bream C406
Carp C406
Goldfish C406
Gudgeon C406
Guppy C406
Man F903, K405
Minnow C406
Mouse H959
Perch C406
Rat B61, B941
Roach C406
Rudd C406
Stickleback C406
Tench C406
See also: A1209
→ *m*-Hydroxybenzyl alcohol
In: Penicillium A1468
See also: A707, A2155, C406, H665
→ *m*-Hydroxybenzylsuccinic acid
In: Desulfobacterium K292

→ 4-Hydroxy-2-methylbenzoic acid
In: Microorganism G981
→ *m*-Methylanisole
In: Guinea pig A115
→ 4-Methylcatechol
In: Bacillus A2397
Candida K653
Trichosporon A902, G196
See also: A768, A943, A2391, E204, E565, G817
→ Methylquinol
In: Penicillium A1468
See also: A2314
→ 2-Methyl-L-tyrosine
In: Escherichia B292

***p*-Cresol** (4-methylphenol)
Formed from: N-Acetyl-*p*-methylphenyl-β-D-glucosaminide
p-Cymene
Ferulic acid
p-Hydroxyphenylacetic acid
p-Methylanisole
m-(*p*-Methylphenoxy)benzoic acid
Piperitinone
Safrole
Toluene
L-Tyrosine
Formed from: 1,8-Cineole in bushtail possum A3934
Probably formed from: *p*-Toluic acid
→ *p*-Cresyl-β-D-glucoside
In: Manduca G897
→ *p*-Cresyl-β-D-glucuronide
In: Man F903
→ *p*-Cresyl sulphate
In: Beef D607
Man D607, D977, D979, E575, F903, K405
Rat B61, B941, E575, K735
See also: K120
→ 2,2′-Dihydroxy-5,5′-dimethylbiphenyl
In: Pisum K404
→ *p*-Hydroxybenzyl alcohol
In: Achromobacter F868
Aspergillus H20
Bacterium F89
Pseudomonas A3276, A3829, A3870, B853, D696, E361, E396, F88, F381, K203, K289, K410, K841
See also: E176, G180, H616, J680, K210
→ *p*-Methylanisole
In: Guinea pig A115
→ 4-Methylcatechol
In: Aspergillus E731
Avocado A3975
Bacillus A2397
Candida E204, K653
Microorganism E175
Mushroom H907
Potato D964
Pseudomonas E731
Trichosporon A902
See also: A7, A768, A943, A1301, A1930, A2433, A3190, B897, E565, F892, G817
→ Pummerer′s ketone
In: Pisum K404

***o*-Cresyl di-(*o*-hydroxymethylphenyl) phosphate**
Formed from: Di-*o*-cresyl *o*-hydroxymethylphenyl phosphate

***o*-Cresol 2,3-dihydroxypropyl ether**
Formed from: *o*-Cresol glycidyl ether

***o*-Cresol glycidyl ether**
→ *o*-Cresol 2,3-dihydroxypropyl ether
In: Man K112, K113
Mouse K112, K113
Rat K112, K113
→ 1-(S-Glutathionyl)-3-(*o*-methylphenoxy)propan-2-ol
In: Man K112
Mouse K112
Rat K112

***p*-Cresyl-β-D-glucoside**
Formed from: *p*-Cresol

***o*-Cresyl-β-D-glucuronide**
Formed from: *o*-Cresol

***m*-Cresyl-β-D-glucuronide**
Formed from: *m*-Cresol

***p*-Cresyl-β-D-glucuronide**
Formed from: *p*-Cresol

***m*-Cresyl N-hydroxymethylcarbamate**
Formed from: *m*-Cresyl N-methylcarbamate

***o*-Cresyl *o*-hydroxymethylphenyl phosphate**
Probably formed from: Di-*o*-cresyl *o*-hydroxymethylphenyl phosphate
→ Saligenin cyclic *o*-cresyl phosphate
Probably in: Cat D375

***m*-Cresyl N-methylcarbamate** (*m*-tolyl N-methylcarbamate, tsumacide)
→ *m*-Cresol
In: Bean A1209
Housefly A1209
Rabbit A2666
Rat A1209
→ *m*-Cresyl N-hydroxymethylcarbamate
In: Housefly A1209
→ 4-Hydroxy-3-methylphenyl N-methylcarbamate
In: Housefly A1209
Rat A1209
→ *m*-Hydroxyphenyl N-methylcarbamate
In: Bean A1209
Housefly A1209
Rat A1209
See also: A2666

***o*-Cresyl sulphate**
Formed from: *o*-Cresol

***m*-Cresyl sulphate**
Formed from: *m*-Cresol

***p*-Cresyl sulphate**
Formed from: *p*-Cresol

Crisnatol (2-((6-chrysenylmethyl)amino)-2-methyl-1,3-propanediol)
→ 2-((6-Chrysenylmethyl)amino)-2-hydroxymethylpropionic acid
In: Rat F741
→ *trans*-1,2-Dihydro-1,2-dihydroxycrisnatol
In: Man G490
Rat F741
→ 1-Hydroxycrisnatol
In: Man G490
Rat F741

Crixivan
See: Indinavir

Croconazole (1-(1-(2-(3-chlorobenzyloxy) phenyl)vinyl)-1*H*-imidazole)
Formed from: Croconazole-β-D-glucuronide
→ *o*-(*m*-Chlorobenzyloxy)acetophenone
In: Rabbit F123
→ 1-(1-(2-(*m*-Chlorobenzyloxy)-5-hydroxyphenyl)vinyl)-1*H*-imidazole
In: Rabbit F123
→ 1-(1-(2-(3-Chloro-4,5-dihydro-4,5-dihydroxybenzyloxy)phenyl)vinyl)-1*H*-imidazole
In: Rabbit F123
→ Croconazole-β-D-glucuronide
In: Rabbit F490
→ 1-(*o*-Hydroxyphenyl)-1-(1*H*-imidazol-1-yl)ethylene
In: Rabbit F123
Rat F507

Croconazole-β-D-glucuronide
Formed from: Croconazole
→ Croconazole
With β-glucuronidase F490

Cromakalim (6-cyano-3,4-dihydro-2,2-dimethyl-*trans*-(4-(2-oxo-1-pyrrolidinyl)-2*H*-benzo[*b*]pyran-3-ol)
→ Cromakalim-β-D-glucuronide
In: Monkey G308
Rat G308
→ 3-Hydroxycromakalim
In: Monkey G308
Rat G308
→ 4-Hydroxycromakalim
In: Monkey G308
Rat G308
→ 6-Cyano-3,4-dihydro-*trans*-4-(γ-hydroxybutyramido)-2,2-dimethyl-2*H*-benzo[*b*] pyran-3-ol
In: Monkey G308
Rat G308

Cromakalim-β-D-glucuronide
Formed from: Cromakalim

Croneton
See: 2-Ethylthiomethylphenyl N-methylcarbamate

Crotsparine ((8′a*S*)-2′,3′,8′,8′a,-tetrahdro-6′-hydroxy-5-methoxyspiro(2,5-cyclohexadiene-1,7′(1′*H*)-cyclopent[*ij*] isoquinolin)-4-one)
Formed from: Coclaurine
→ N-Methylcrotsparine
Probably in: Croton B432

Crotsparinine ((8′a*S*)-2′,3′,8′,8′a,-tetrahdro-6′-hydroxy-5-methoxyspiro(2-cyclohexene-1,7′(1′*H*)-cyclopent[*ij*]isoquinolin)-4-one)
Formed from: Coclaurine

Crufomate (4-*tert*-butyl-2-chlorophenyl methyl methylphosphoramidate)
→ 4-*tert*-Butyl-2-chlorophenol
In: Sheep A2995, A3001
Rat A3338, A3351
→ 4-*tert*-Butyl-2-chlorophenyl methyl phosphate
In: Sheep A2995
Rat A3351
→ 4-*tert*-Butyl-2-chlorophenyl methylphosphoramidate
In: Sheep A2995
Rat A3338
→ 2-Chloro-4-(2-hydroxy-1,1-dimethylethyl)phenyl methyl methylphosphoramidate
In: Sheep A2995, A3001
Rat A3338, A3351

Cryptoechinuline A ((3*Z*)-3-((2-(1,1-dimethyl-2-propenyl)-6-(3-methyl-2-butenyl)-1*H*-indol-3-yl)methylene)-6-methylene-2,5-piperazinedione)
Formed from: L-Tryptophan

(−)-Cryptpstyline I ((1*S*)-1-(1,3-benzodioxol-5-yl)-1,2,3,4-tetrahydro-6,7-dimethoxy-2-methylisoquinoline)
Formed from: Cryptpstyline I immonium salt
3,4-Dihydroxyphenethylamine
3,4-Dimethoxyphenethylamine
Vanillin

Cryptpstyline I immonium salt
→ (−)-Cryptpstyline I
In Cryptpstylis A1515

CS
See: *o*-Chlorobenzylidenemalononitrile

Cumaldehyde (*p*-isopropylbenzaldehyde)
Formed from: Cumyl alcohol
→ cumic acid
In: Mouse F432
See also: A2602, A2692, A2694, A3934, C574, F457
→ Cumyl alcohol
In: Rabbit F457
→ *p*-Isopropylmandelonitrile (both isomers)
In: Almond J194, J302
Apple J302
Cherry J302
Sorghum J194

Cumene
See: Isopropylbenzene

Cumene hydroperoxide (1-methyl-1-phenylethyl hydroperoxide)
→ Acetophenone
In: Mycobacterium J916
→ Cumyl alcohol
In: Mycobacterium J916
Pisum K404
→ Dicumenyl peroxide
In: Pisum K404

Cumic acid (*p*-isopropylbenzoic acid)
Formed from: Cumaldehyde
→ Cumylglycine
In: Rat C574
→ 2,3-Dihydrocumic acid
In: Pseudomonas A2694
→ 2,3-Dihydro-2,3-dihydroxycumic acid
Probably in: Pseudomonas A2602, A2692
→ *p*-(1-Hydroxy-2-propyl)benzoic acid
Probably in: Rabbit F457
Rat C574
→ *p*-(2-Hydroxy-2-propyl)benzoic acid
Probably in: Rabbit F457
Rat C574

Cumyl alcohol (*p*-isopropylbenzyl alcohol)
Formed from: Cumaldehyde
Cumene hydroperoxide
p-Cymene
→ Cumaldehyde
In: Acinetobacter E663
See also: A2602, A2692, A2694, A3934

Cumylglycine
Formed from: Cumic acid

Cupreidine ((9*S*)-cinchonan-6′,9-diol)
→ Cupreidinone
In: Cinchona E357

Cupreidinone
Formed from: Cupreidine

Cupreine (6′-cinchonidine, (8α,9*R*)-cinchonan-6′,9-diol)
→ 6′-Hydroxycinchonidinone
In: Cinchona E357

Curcumin ((1*E*,6*E*)-1,7-bis(4-hydroxy-3-methoxyphenyl)-1,6-heptadiene-3,5-dione)
Formed from: Caffeic acid
p-Coumaric acid
Ferulic acid
L-Phenylalanine
→ Curcumin-β-D-glucuronide
In: Mouse J577
→ Ferulic acid
In: Rat A3492
→ Tetrahydrocurcumin
In: Mouse J577
Rat A3492

Curcumin-β-D-glucuronide
Formed from: Curcumin

Cyanidin (3,3′,4′,5,7-pentahydroxyflavylium)
Probably formed from: 2,3-*trans*-3,4-*cis*-3,3′,4,4′,5,7-Hexahydroxyflavan
→ Chrysanthemin
In: Cabbage A1733
Carrot E280
Matthiola D575
Petunia A3191
Senecio J722
Silene B404, B407
See also: A2739, A3088
→ Cyanidin-3-galactoside
In: Carrot H787
→ Delphinidin
Probably in: Petunia A215, A2739, A2740
→ Paeonidin
In: Petunia C743
Vitis H886
See also: A215, A2739

Cyanidin-3-acetylglucoside
Formed from: Chrysanthemin

Cyanidin-5-caffeoylglucoside-3-glucoside
Formed from: Cyanidin-3,5-diglucoside

Cyanidin-3-(4-caffeoylrhamnosyl(1 → 6)) glucoside
Formed from: Cyanidin-3-rhamnosylglucoside

Cyanidin-3-(4-caffeoylrhamnosyl(1 → 6)) glucoside-5-glucoside
Formed from: Cyanidin-5-glucoside-3-rhamnosylglucoside

Cyanidin-3,5-diglucoside
Formed from: Chrysanthemin
→ Cyanidin-5-caffeoylglucoside-3-glucoside
In: Gentiana K788
→ Cyanidin-5-glucoside-3-rhamnosylglucoside
In: Silene B409

Cyanidin-3-galactoside (idein)
Formed from: Cyanidin

Cyanidin-3-glucoside
See: Chrysanthemin

Cyanidin-5-glucoside-3-rhamnosylglucoside
Formed from: Cyanidin-3,5-diglucoside
Cyanidin-3-rhamnosylglucoside
→ Cyanidin-3-(4-caffeoylrhamnosyl(1 → 6))glucoside-5-glucoside
In: Silene B411

Cyanidin-3-rhamnosylglucoside
Formed from: Chrysanthemin
→ Cyanidin-3-(4-caffeoylrhamnosyl(1 → 6))glucoside
In: Silene B411
→ Cyanidin-5-glucoside-3-rhamnosylglucoside
In: Silene B408, B410

***p*-Cyanoacetophenone**
→ *p*-Acetylbenzoic acid
In: Rhodococcus H952

***o*-Cyanobenzaldehyde**
See: *o*-Cyanobenzaldehyde

***m*-Cyanobenzaldehyde**
See: *m*-Cyanobenzaldehyde

***p*-Cyanobenzaldehyde**
See: *p*-Cyanobenzaldehyde

***p*-Cyanobenzamidine**
→ *p*-Cyanobenzamidoxine
In: Rabbit E830, F768

***p*-Cyanobenzamidoxine**
Formed from: *p*-Cyanobenzamidine

***o*-Cyanobenzoic acid**
Probably formed from: Phthalonitrile

***m*-Cyanobenzoic acid**
Formed from: *m*-Formylbenzonitrile

Isophthalonitrile
→ *m*-Cyanobenzoyl CoA
In: Beef H740
→ Isophthalic acid
In: Rhodococcus J825

p-Cyanobenzoic acid
Formed from: *p*-Formylbenzonitrile
Probably formed from: Terephthalonitrile
→ *p*-Cyanobenzoyl CoA
In: Beef H740
→ *p*-Cyanobenzoyl-β-D-glucuronic acid
In: Rabbit H519
→ *p*-Cyanohippuric acid
In: Rabbit H519

m-Cyanobenzoyl CoA
Formed from: *m*-Cyanobenzoic acid

p-Cyanobenzoyl CoA
Formed from: *p*-Cyanobenzoic acid

p-Cyanobenzoyl-β-D-glucuronic acid
Formed from: *p*-Cyanobenzoic acid

(4-Cyanobenzylidene)malononitrile
→ (4-Cyanobenzyl)malononitrile
In: Man B260

(4-Cyanobenzyl)malononitrile
Formed from: (4-Cyanobenzylidene) malononitrile

(*E*)-2-Cyano-N,N-diethyl-3-(3,4-dihydroxy -5-nitrophenyl)propenamide
See: Entacapone

(*E*)-2-Cyano-N,N-diethyl-3-(4-hydroxy-3-methoxy-5-nitrophenyl)propenamide
Formed from: Entacapone

6-Cyano-3,4-dihydro-*trans*-4-(γ-hydroxybutyramido)-2,2-dimethyl-2*H*-benzo[*b*]pyran-3-ol
Formed from: Cromakalim

(5*R*)-3-(2-((1*S*)-3-Cyano-1,2-dihydroxypropyl)benzothiazol-6-yl)-5-methoxymethyl-2-oxazolidinone
Formed from: (5*R*)-3-(2-((1*S*)-3-Cyano-1-hydroxypropyl)benzothiazol-6-yl)-5-methoxymethyl-2-oxazolidinone

4-Cyano-4-(3,4-dimethoxyphenyl)-N,5-dimethylhexylamine
Formed from: 5-(N-(3,4-Dimethoxyphenethyl)methylamino)-2-(3,4-dimethoxyphenyl)-2-isopropylvaleronitrile
→ 4-Cyano-4-(3,4-dimethoxyphenyl)-N-formyl-5-methylhexylamine
Probably in: Rat E943
→ 4-Cyano-4-(3,4-dimethoxyphenyl)-5-methylhexylamine
Probably in: Rat E943
→ 4-Cyano-4-(4-hydroxy-3-methoxyphenyl)-N,5-dimethylhexylamine
Probably in: Rat E943

4-Cyano-4-(3,4-dimethoxyphenyl)-N-formyl-5-methylhexylamine
Probably formed from: 4-Cyano-4-(3,4-dimethoxyphenyl)-N,5-dimethylhexylamine

4-Cyano-4-(3,4-dimethoxyphenyl)-5-methylhexylamine
Probably formed from: 4-Cyano-4-(3,4-dimethoxyphenyl)-N,5-dimethylhexylamine

5-Cyano-2-dimethylaminophenyl sulphate
Formed from: 5-Cyano-2-dimethylaminophenol

4-Cyano-N,N-dimethylaniline
→ N-Acetyl-S-(4-cyanoanilinomethyl)-L-cysteine
In: Mouse D290, D367
Rat D290
→ 5-Cyano-2-methylaminophenol
In: Mouse D367
→ 4-Cyano-N-methylaniline
In: Rabbit D40

4-Cyano-N,N-dimethylaniline-N-oxide
→ 4-Cyano-N-methylaniline
In: Rabbit D40

3-Cyano-1-(1,4-dioxobut-2-en-2-yl)-6-(6-(3-(3α-hydroxy-6,8-dioxabicyclo[3.2.1]octanyl))pyridin-2-ylmethoxy)naphthalene
Probably formed from: (*1S*, 5*R*)-3-Cyano-1-(3-furyl)-6-(6-(3-(3α-hydroxy-6,8-dioxabicyclo[3.2.1]octanyl))pyridin-2-ylmethoxy) naphthalene
→ 3-Cyano-6-(6-(3-(3α-hydroxy-6,8-dioxabicyclo[3.2.1]octanyl))pyridin-2-ylmethoxy)-1-hydroxy-4-oxobut-2-en-2-yl)naphthalene
Probably in: Rat J29
Rhesus monkey J29

N-(2-Cyanoethyl)amphetamine (fenproporex)
→ Amphetamine

In: Man A408, E143

2-Cyano-N-ethyl-N-(2-hydroxyethyl)-3-(3,4-dihydroxy-5-nitrophenyl)propenamide
Formed from: Entacapone

4′-Cyano-3-(4-fluoro-3-hydroxyphenylsulphonyl)-2-hydroxy-2-methyl-3′-trifluoromethylpropionanilide
Formed from: Casodex
→ 4′-Cyano-3-(4-fluoro-3-hydroxyphenylsulphonyl)-2-hydroxy-2-methyl-3′-trifluoromethylpropionanilide-β-D-glucuronide
In: Dog G592
Man G592
Rat G592

4′-Cyano-3-(4-fluoro-3-hydroxyphenylsulphonyl)-2-hydroxy-2-methyl-3′-trifluoromethylpropionanilide-β-D-glucuronide
Formed from: 4′-Cyano-3-(4-fluoro-3-hydroxyphenylsulphonyl)-2-hydroxy-2-methyl-3′-trifluoromethylpropionanilide

4′-Cyano-3-(*p*-fluorophenyl)-2-hydroxy-2-methyl-3′-(trifluoromethyl)propionanilide
See: Casodex

(1*S*,5*R*)-3-Cyano-1-(3-furyl)-6-(6-(3-(3α-hydroxy-6,8-dioxabicyclo[3.2.1]octanyl))pyridin-2-ylmethoxy)naphthalene
→ 3-Cyano-1-(1,4-dioxobut-2-en-2-yl)-6-(6-(3-(3α-hydroxy-6,8-dioxabicyclo[3.2.1]octanyl))pyridin-2-ylmethoxy)naphthalene
In: Rat J29
Rhesus monkey J29

(2*Z*,4*E*)-4-cyano(β-D-glucopyranosyloxy)methylene)-2-hexenedioic acid
See: Triglochinin

6-Cyano-2-(4-(2-(3-glutathionyl-2-hydroxy-5-methanesulphonamidophenyl)-2-hydroxyethyl)-4-hydroxypiperidin-1-yl)-1,2,3,4-tetrahydronaphthalene
Formed from: (+)-N-(1′-(6-Cyano-1,2,3,4-tetrahydro-2(*R*)-naphthalenyl)-3,4-dihydro-4(*R*)-hydroxy*spiro*(2*H*-1-benzopyran-2,4′-piperidin)-6-yl)methanesulphonamide

6-Cyano-4-(2-(3-glutathionyl-2-hydroxy-5-methanesulphonamidophenyl)-2-hydroxyethyl)-1,2,3,4-tetrahydronaphthalene
Formed from: (+)-N-(1′-(6-Cyano-1,2,3,4-tetrahydro-2(*R*)-naphthalenyl)-3,4-dihydro-4(*R*)-hydroxy*spiro*(2*H*-1-benzopyran-2,4′-piperidin)-6-yl)methanesulphonamide

***p*-Cyanohippuric acid**
Formed from: *p*-Cyanobenzoic acid

3-Cyano-5-hydroxycamalexin
Probably formed from: 5-Hydroxycamalexin

3-Cyano-6-(6-(3-(3α-hydroxy-6,8-dioxabicyclo[3.2.1]octanyl))pyridin-2-ylmethoxy)-1-hydroxy-4-oxobut-2-en-2-yl)naphthalene
Probably formed from: 3-Cyano-1-(1,4-dioxobut-2-en-2-yl)-6-(6-(3-(3α-hydroxy-6,8-dioxabicyclo[3.2.1]octanyl))pyridin-2-ylmethoxy)naphthalene

2-Cyano-1-hydroxyindane
Formed from: 2-Cyanoindane

3-Cyano-5-hydroxyindole
Formed from: 5-Hydroxycamalexine

2-Cyano-5-(1-hydroxyisopropoxycarbonyl)-3-methyoxycarbonyl-6-methyl-4-(*m*-nitrophenyl)pyridine
Probably formed from: Dehydronilvadipine

4-Cyano-4-(4-hydroxy-3-methoxyphenyl)-N,5-dimethylhexylamine
Probably formed from: 4-Cyano-4-(3,4-dihydroxyphenyl)-N,5-dimethylhexylamine

2-Cyano-*p*-hydroxy-3-(N-methylpyrrol-2-yl)-3-oxopropionanilide
Formed from: 2-Cyano-3-(N-methylpyrrol-2-yl)-3-oxopropionanilide

***m*-Cyano-2-hydroxypropiophenone**
Formed from: *m*-Formylbenzonitrile

***p*-Cyano-2-hydroxypropiophenone**
Formed from: *p*-Formylbenzonitrile

(5*R*)-3-(2-((1*S*)-3-Cyano-1-hydroxypropyl)benzothiazol-6-yl)-4-hydroxy-5-hydroxymethyl-2-oxazolidinone
Probably formed from: (5*R*)-3-(2-((1*S*)-3-Cyano-1-hydroxypropyl)benzothiazol-6-yl)-5-hydroxymethyl-2-oxazolidinone

(5*R*)-3-(2-((1*S*)-3-Cyano-1-hydroxypropyl)benzothiazol-6-yl)-4-hydroxy-5-methoxymethyl-2-oxazolidinone
Formed from: (5*R*)-3-(2-((1*S*)-3-Cyano-1-hydroxypropyl)benzothiazol-6-yl)-5-methoxymethyl-2-oxazolidinone

(5*R*)-3-(2-((1*S*)-3-Cyano-1-hydroxypropyl) benzothiazol-6-yl)-5-hydroxymethyl-2-oxazolidinone
Formed from: (5*R*)-3-(2-((1*S*)-3-Cyano-1-hydroxypropyl)benzothiazol-6-yl)-5-methoxymethyl-2-oxazolidinone
→ (5*R*)-3-(2-((1*S*)-3-Cyano-1-hydroxypropyl)benzothiazol-6-yl)-4-hydroxy-5-hydroxymethyl-2-oxazolidinone
Probably in: Rat J284

(5*R*)-3-(2-((1*R*)-3-Cyano-1-hydroxypropyl)benzothiazol-6-yl)-5-methoxymethyl-2-oxazolidinone
Formed from: (5*R*)-3-(2-(3-Cyano-1-oxopropyl)benzothiazol-6-yl)-5-methoxymethyl-2-oxazolidinone
→ (5*R*)-3-(2-(3-Cyano-1-oxopropyl) benzothiazol-6-yl)-5-methoxymethyl-2-oxazolidinone
In: Dog H939
Guinea pig H939
Mouse H939
Pig H939
Rat H939
Rhesus monkey H939
Squirrel monkey H939

(5*R*)-3-(2-((1*S*)-3-Cyano-1-hydroxypropyl) benzothiazol-6-yl)-5-methoxymethyl-2-oxazolidinone
Formed from: (5*R*)-3-(2-(3-Cyano-1-oxopropyl)benzothiazol-6-yl)-5-methoxymethyl-2-oxazolidinone
(5*R*)-3-(2-(3-Cyanopropyl)benzothiazol-6-yl)-5-methoxymethyl-2-oxazolidinone
→ (5*R*)-3-(2-((1*S*)-3-Cyano-1,2-dihydroxypropyl)benzothiazol-6-yl)-5-methoxymethyl-2-oxazolidinone
In: Rat J284
→ (5*R*)-3-(2-((1*S*)-3-Cyano-1-hydroxypropyl)benzothiazol-6-yl)-4-hydroxy-5-methoxymethyl-2-oxazolidinone
In: Rat J284
→ (5*R*)-3-(2-((1*S*)-3-Cyano-1-hydroxypropyl)benzothiazol-6-yl)-5-hydroxymethyl-2-oxazolidinone
In: Rat J284
→ (5*R*)-3-(2-(3-Cyano-1-oxopropyl)benzothiazol-6-yl)-5-methoxymethyl-2-oxazolidinone
In: Pig H939
Rat H939
Rhesus monkey H939

2-Cyanoindane
→ 2-Cyano-1-hydroxyindane
In: Pseudomonas K207

4-Cyano-3-methoxy-5-methylbenzoic acid
Formed from: 5-Cyano-4-methoxy-6-methyl-2-pyrone

5-Cyano-4-methoxy-6-methyl-2-pyrone
→ 4-Cyano-3-methoxy-5-methylbenzoic acid
In: Macrophoma E766

5-Cyano-2-methylaminophenol
Formed from: 4-Cyano-N,N-dimethylaniline
→ 2-Amino-5-cyanophenol
Probably in Mouse D367
→ 5-Cyano-2-methylaminophenyl sulphate
Probably in Mouse D367

5-Cyano-2-methylaminophenyl sulphate
Probably formed from: 5-Cyano-2-methylaminophenol

4-Cyano-N-methylaniline
Formed from: 4-Cyano-N,N-dimethylaniline
4-Cyano-N,N-dimethylaniline-N-oxide

2-(5-Cyano-N-methyl-3,5-dimethyl-3-azapentanamido)-5-nitrobenzophenone
→ Nimetazepam
In: Dog A2102
Mouse A2102
Rat A2016, A2101, A2102

2-Cyano-3-(N-methylpyrrol-2-yl)-3-oxopropionanilide
→ 2-Cyano-*p*-hydroxy-3-(N-methylpyrrol-2-yl)-3-oxopropionanilide
In: Baboon E935
Dog E935
Hamster E935
Monkey E935
Mouse E935
Rat E935

4-Cyanomethyl-*o*-quinone
Formed from: 3,4-Dihydroxybenzyl cyanide

2-Cyano-1-methyl-3-(*p*-(1,4,5,6-tetrahydro-4-methyl-6-oxopyridazin-3-yl)phenyl)guanidine
Unchanged in dog, mouse and rat F561

3-Cyano-2-methyl-3-(3,4,5-trimethoxyphenyl)-6-hexanal
Probably formed from: Norgallopamil
→ 3-Cyano-2-methyl-3-(3,4,5-trimethoxyphenyl)-6-hexanol
Probably in: Man F527
Rat F527
→ 4-Cyano-5-methyl-4-(3,4,5-trimethoxyphenyl)hexanoic acid
Probably in: Man F527
Rat F527

4-Cyano-5-methyl-4-(3,4,5-trimethoxyphenyl) hexanoic acid
Probably formed from: 3-Cyano-2-methyl-3-(3,4,5-trimethoxyphenyl)-6-hexanal

3-Cyano-2-methyl-3-(3,4,5-trimethoxyphenyl)-6-hexanol
Probably formed from: 3-Cyano-2-methyl-3-(3,4,5-trimethoxyphenyl)-6-hexanal

N-(3-Cyano-2-methyl-3-(3,4,5-trimethoxyphenyl)-6-hexyl)methylamine
Formed from: Gallopamil

1-Cyanonaphthalene
See: 1-Naphthonitrile

2-Cyanonaphthalene
See: 2-Naphthonitrile

(5*R*)-3-(2-(3-Cyano-1-oxopropyl)benzothiazol-6-yl)-5-methoxymethyl-2-oxazolidinone
Formed from: (5*R*)-3-(2-((1*R*)-3-Cyano-1-hydroxypropyl)benzothiazol-6-yl)-5-methoxymethyl-2-oxazolidinone
(5*R*)-3-(2-((1*S*)-3-Cyano-1-hydroxypropyl)benzothiazol-6-yl)-5-methoxymethyl-2-oxazolidinone
→ (5*R*)-3-(2-((1*R*)-3-Cyano-1-hydroxypropyl)benzothiazol-6-yl)-5-methoxymethyl-2-oxazolidinone
In: Pig H939
Rat H939
Rhesus monkey H939
→ (5*R*)-3-(2-((1*S*)-3-Cyano-1-hydroxypropyl)benzothiazol-6-yl)-5-methoxymethyl-2-oxazolidinone
In: Dog H939
Guinea pig H939
Pig H939
Mouse H939
Rat H939
Rhesus monkey H939
Squirel monkey H939

N-(3-(3-Cyano-5-oxopyrazolo[1,5-*a*]pyrimidin-7-yl)phenyl)acetamide
Formed from: N-(3-(3-Cyano-5-oxopyrazolo[1,5-*a*]pyrimidin-7-yl)phenyl)-N-ethylacetamide

N-(3-(3-Cyano-5-oxopyrazolo[1,5-*a*]pyrimidin-7-yl)phenyl)-N-ethylacetamide
→ N-(3-(3-Cyano-5-oxopyrazolo[1,5-*a*] pyrimidin-7-yl)phenyl)acetamide
In: Man J299

***o*-Cyanophenol**
See: *o*-Hydroxybenzonitrile

***m*-Cyanophenol**
See: *m*-Hydroxybenzonitrile

***p*-Cyanophenol**
See: *p*-Hydroxybenzonitrile

α-Cyano-3-phenoxybenzyl 2-(*p*-chlorophenyl) isovalerate
See: Fenvalerate

3-(*o*-Cyanophenoxy)lactic acid
Formed from: Bunitrolol

3-(*o*-Cyanophenoxy)-1,2-propanediol
Formed from: Bunitrolol

***o*-Cyanophenylacetic acid**
Formed from: α-Cyano-*o*-tolunitrile

***p*-Cyanophenyl azide**
→ *p*-Aminobenzonitrile
In: Mouse G8

2-Cyano-1-(1-phenylcyclohexyl)piperidine
Formed from: Phencyclidine

4-Cyanophenyl dimethyl phosphate
See: Cyanoxon

O-4-Cyanophenyl O,O-dimethyl phosphorothioate
See: Cyanox

2-(*p*-Cyanophenyl)-1,3-dithiolane
→ 2-(*p*-Cyanophenyl)-1,3-dithiolane-1-oxide
In: Rabbit F355

2-(*p*-Cyanophenyl)-1,3-dithiolane-1-oxide
Formed from: 2-(*p*-Cyanophenyl)-1,3-dithiolane

4-Cyanophenyl-1,5-dithio-β-D-xylopyranoside
See: Thioxyloside
1-(*p*-Cyanophenyl)ethanol
Formed from: *p*-Ethylbenzonitrile
***p*-Cyanophenyl-β-D-glucuronide**
Formed from: *p*-Hydroxybenzonitrile
2-(*p*-Cyanophenyl)-1,3-oxathiolane
→ 2-(*p*-Cyanophenyl)-1,3-oxathiolane sulphoxide
In: Pig F176
Rat F176
2-(*p*-Cyanophenyl)-1,3-oxathiolane sulphone
Probably formed from: 2-(*p*-Cyanophenyl)-1,3-oxathiolane sulphoxide
2-(*p*-Cyanophenyl)-1,3-oxathiolane sulphoxide
Formed from: 2-(*p*-Cyanophenyl)-1,3-oxathiolane
→ 2-(*p*-Cyanophenyl)-1,3-oxathiolane sulphone
Probably in: Pig F176
Rat F176
***o*-Cyanophenyl sulphate**
Formed from: *o*-Hydroxybenzonitrile
***m*-Cyanophenyl sulphate**
Formed from: *m*-Hydroxybenzonitrile
***p*-Cyanophenyl sulphate**
Formed from: *p*-Hydroxybenzonitrile
Cyanophos
See: Cyanox
(5*R*)-3-(2-(3-Cyanopropyl)benzothiazol-6-yl)-5-methoxymethyl-2-oxazolidinone
→ (5*R*)-3-(2-(3-Cyano-1-hydroxypropyl) benzothiazol-6-yl)-5-methoxymethyl-2-oxazolidinone
In: Rat G955
***p*-Cyanosulphinylanisole**
Formed from: *p*-Cyanothioanisole
(+)-N-(1′-(6-Cyano-1,2,3,4-tetrahydro-1-hydroxy-2(*R*)-naphthalenyl)-3,4-dihydro-4(*R*)-hydroxy*spiro*(2*H*-1-benzopyran-2,4′-piperidin)-6-yl) methanesulphonamide
Formed from: (+)-N-(1′-(6-Cyano-1,2,3,4-tetrahydro-2(*R*)-naphthalenyl)-3,4-dihydro-4(*R*)-hydroxy*spiro*(2*H*-1-benzopyran-2,4′-piperidin)-6-yl) methanesulphonamide
(+)-N-(1′-(6-Cyano-1,2,3,4-tetrahydro-4-hydroxy-2(*R*)-naphthalenyl)-3,4-dihydro-4(*R*)-hydroxy*spiro*(2*H*-1-benzopyran-2,4′-piperidin)-6-yl)methanesulphonamide
Formed from: (+)-N-(1′-(6-Cyano-1,2,3,4-tetrahydro-2(*R*)-naphthalenyl)-3,4-dihydro-4(*R*)-hydroxy*spiro*(2*H*-1-benzopyran-2,4′-piperidin)-6-yl)methanesulphonamide
(+)-N-(1′-(6-Cyano-1,2,3,4-tetrahydro-2(*R*)-naphthalenyl)-3,4-dihydro-4(*R*)-hydroxy*spiro*(2*H*-1-benzopyran-2,4′-piperidin)-6-yl)methanesulphonamide
→ 2-(4-Carboxymethyl-4-hydroxypiperidin-1-yl)-6-cyano-1,2,3,4-tetrahydronaphthalene
In: Rat J402
→ 6-Cyano-2-(4-(2-(3-glutathionyl-2-hydroxy-5-methanesulphonamidophenyl)-2-hydroxyethyl)-4-hydroxypiperidin-1-yl)-1,2,3,4-tetrahydronaphthalene
In: Rat J402
→ 6-Cyano-4-(2-(3-glutathionyl-2-hydroxy-5-methanesulphonamidophenyl)-2-hydroxyethyl)-1,2,3,4-tetrahydronaphthalene
In: Rat J402
→ (+)-N-(1′-(6-Cyano-1,2,3,4-tetrahydro-1-hydroxy-2(*R*)-naphthalenyl)-3,4-dihydro-4(*R*)-hydroxy-*spiro*(2*H*-1-benzopyran-2,4′-piperidin-6-yl)methanesulphonamide
In: Rat J402
→ (+)-N-(1′-(6-Cyano-1,2,3,4-tetrahydro-2(*R*)-naphthalenyl)-3,4-dihydro-4(*R*)-hydroxy*spiro*(2*H*-1-benzopyran-2,4′-piperidin-6-yl)methanesulphonamide-*β*-D-glucuronide
In: Dog J402
→ N-(3,4-Dihydro-4(*R*)-hydroxy*spiro*(2*H*-1-benzopyran-2,4′-piperidin)-6-yl) methanesulphonamide
In: Dog J402
Rat J402
(+)-N-(1′-(6-Cyano-1,2,3,4-tetrahydro-2(*R*)-naphthalenyl)-3,4-dihydro-4(*R*)-hydroxy*spiro*(2*H*-1-benzopyran-2,4′-

piperidin-6-yl)methanesulphonamide-β-D-glucuronide
Formed from: (+)-N-(1′-(6-Cyano-1,2,3,4-tetrahydro-2(*R*)-naphthalenyl)-3,4-dihydro-4(*R*)-hydroxy*spiro*(2*H*-1-benzopyran-2,4′-piperidin)-6-yl)methanesulphonamide

***p*-Cyanothioanisole**
→ *p*-Cyanosulphinylanisole
In: Horseradish E403
Mortierella D759

α-Cyano-*o*-tolunitrile
→ *o*-Cyanophenylacetic acid
In: Rhodococcus J825

Cyanox (cyanophos, O-4-cyanophenyl O,O-dimethyl phosphorothioate)
→ Cyanoxon
In: Rat K27
→ Demethylcyanox
In: Rat H461, K27
→ *p*-Hydroxybenzonitrile
In: Rat H461, K27

Cyanoxon (4-cyanophenyl dimethyl phosphate)
Formed from: Cyanox
→ Demethylcyanoxon
Probably in: Rat K27

Cyclandelate
See: 3,3,5-Trimethylcyclohexyl mandelate

Cyclazocine (3-(cyclopropylmethyl)-1,2,3,4,5,6-hexahydro-6,11-dimethyl-2,6-methano-3-benzazocin-8-ol)
→ Cyclazocine-β-D-glucuronide
In: Rat A2134
→ N-Decyclopropylmethycyclazocine
In: Rat C339
See also: A2134
→ Hydroxycyclazocine
In: Rat A2134

Cyclazocine-β-D-glucuronide
Formed from: Cyclazocine

Cyclindole (2,3,4,9-tetrahydro-N,N-dimethyl-1*H*-carbazol-3-amine)
→ Demethylcyclindole
In: Dog A2083
Man A2083
Monkey A2083
Rat A2083
→ 6-Hydroxycyclindole
In: Man A2083
→ 7-Hydroxycyclindole
In: Man A2083
Monkey A2083
Rat A2083

Cyclizine (1-(diphenylmethyl)-4-methoxypiperazine)
→ Cyclizine-N-β-D-glucuronide
In: Man G18

Cyclizine-N-β-D-glucuronide
Formed from: Cyclizine

Cyclo-L-alanyl-L-2-(1,1-dimethylallyl) tryptophanyl
Probably formed from: Cyclo-L-alanyl-L-tryptophanyl

Cyclo-L-alanyl-L-tryptophanyl
→ Cyclo-L-alanyl-L-2-(1,1-dimethylallyl) tryptophanyl
Probably in: Aspergillus A227

Cyclo-(anthraniloyl-phenylalanyl)
Probably formed from: N-Adelylylanthranilic acid
N-Adenylylphenylalanine

Cyclobenzaprine (3-(5*H*-dibenzo[*a*,*d*]cyclohepten-5-ylidene)-N,N-dimethyl-1-propanamide)
Formed from: Cyclobenzaprine-N-oxide
→ Cyclobenzaprine-β-D-glucuronide
In: Dog A2887
Man A3468
→ Cyclobenzaprine-10,11-oxide
In: Dog A2887
Rat A1740, A1797, A2614, A2927
→ Cyclobenzaprine-N-oxide
In: Dog A2887
Rat A1740, A1797, A2614, A2927, A3468
→ Demethylcyclobenzaprine
In: Dog A2887
Man H727
Rat A1740, A1797, A2614, A2927
→ 1,2-Dihydro-1,2-dihydroxycyclobenzaprine
In: Dog A2887
→ 3-Hydroxycyclobenzaprine
In: Dog A2887
Rat A3468
See also: A2927
→ 4-Hydroxycyclobenzaprine

In: Dog A2887
→ 10-Hydroxycyclobenzaprine
Probably in: Rat A2614
Cyclobenzaprine-β-D-glucuronide
Formed from: Cyclobenzaprine
Cyclobenzaprine-10,11-oxide
Formed from: Cyclobenzaprine
→ Cyclobenzaprine-10,11-oxide-N-oxide
Probably in: Rat A2740, A2614, A2927
→ 10,11-Dihydro-10,11-dihydro-N-demethylcyclobenzaprine
Probably in: Dog A2887
Cyclobenzaprine-N-oxide
Formed from: Cyclobenzaprine
→ Cyclobenzaprine
In: Hamster D218
Mouse D218
Pig D218
Rat D218, J876
Cyclobenzaprine-10,11-oxide-N-oxide
Probably formed from: Cyclobenzaprine-10,11-oxide
***l*-N-Cyclobutylmethyl-3,14-dihydroxymorphinan**
See: Butorphanol
Cyclocoumarol (3,4-dihydro-2-methoxy-4-phenyl-2*H*,5*H*-pyrano[3,2-*c*][1]benzopyran-5-one)
→ 4′-Hydroxycyclocoumarol
In: Rat D716
→ 6-Hydroxycyclocoumarol
In: Rat D716
→ 7-Hydroxycyclocoumarol
In: Rat D716
→ 8-Hydroxycyclocoumarol
In: Rat D716
Cyclodopa
See: (2*S*)-2,3-Dihydro-5,6-dihydroxyindole-2-carboxylic acid
1-((1-Cycloheptyl-3-(*p*-dimethylaminophenyl)ureido)methyl)-3-((1-cycloheptyl-3-(*p*-methylaminophenyl)ureido)methyl)benzene
Formed from: 1,3-Bis((1-cycloheptyl-3-(*p*-dimethylaminophenyl)ureido)methyl)benzene
→ 1-((3-(*p*-Aminophenyl)-1-cycloheptyl ureido)methyl)-3-((1-cycloheptyl-3-(*p*-dimethylaminophenyl)ureido)methyl)benzene
In: Dog H130
Rat G581, H130
→ 1,3-Bis((1-cycloheptyl-3-(*p*-methylaminophenyl)ureido)methyl)benzene
In: Dog H130
Rat G581, H130
3,5-Cyclohexadiene-1,2-dione
See: *o*-Quinone
Cyclohexanecarboxylic acid
→ Benzoic acid
In: Guinea pig F750
Mouse F750
Rabbit F750
Rat F750
→ *p*-Hydroxybenzoic acid
In: Trichosporon E502
→ Protocatechuic acid
In: Trichosporon E502
3-Cyclohexanoylmorphine
→ Morphine
In: Man K517
O-Cyclohexanoyl propranolol
→ Propranolol
In: Mouse J222
Cyclohex-1-enecarboxylic acid
Formed from: Benzoic acid
Cyclohex-3-enecarboxylic acid
→ Benzoic acid
In: Alcaligenes C460
2-(Cyclohexylcarbonyl)-1,2,3,6,7,11b-hexahydro-4-pyrazino[2,1-*a*]isoquinolin-4-one
See: Praziquantel
N-Cyclohexyl-4-((1,2-dihydro-2-oxo-6-quinolinyl)oxy)-N-methylbutanamide
See: Cilostamide
N-Cyclohexyl-N′-(4-(β-(5-hydroxymethyl pyrazine-2-carboxamido)ethyl) benzylsulphonyl)urea
Probably formed from: Glipizide
α-Cyclohexylmandelic acid
Formed from: Oxybutynin
→ α-(4-Hydroxycyclohexyl) mandelic acid
Probably in: Rat E294
N-Cyclohexyl-N-methyl-4-(1,2-dihydro-2-oxo-6-quinolyloxy)butyramide
See: Cilostamide

N-Cyclohexyl-N′-(4-(β-(5-methylpyrazine-2-carboxamido)ethyl)benzylsulphonyl)urea
See: Glipizide

4-Cyclohexyl-N-(((4-phenylbutyl)phosphinyl)acetyl)-L-proline
Probably formed from: Fosinopril

1-Cyclohexyl-1-phenyl-3-piperidinopropan-1-ol
See: Benzhexol

4-(Cyclohexyl 1′-piperazinyl-4′-carboxylate)-1-(4′-fluorophenyl)butan-1-one
→ 1-(4′-Fluorophenyl)-4-1′-piperazinyl)butan-1-one
Probably in: Man A2038
Rat A2038
→ 1-(4′-Fluorophenyl)-4-(hydroxycyclohexyl 1′-piperazinyl-4′-carboxylate)butan-1-one
Probably in: Man A2038
Rat A2038

4-(3-Cyclohexylpropionyl)-1-(2-ethoxyphenyl)piperazine
→ 4-(3-Cyclohexylpropionyl)-1-(5-hydroxy-2-ethoxyphenyl)piperazine
In: Rat H430
→ 4-(3-Cyclohexylpropionyl)-1-(2-hydroxyphenyl)piperazine
In: Rat H430
→ 1-(2-Ethoxyphenyl)-4-(3-(hydroxycyclohexyl)propionyl)piperazine
In: Rat H430

4-(3-Cyclohexylpropionyl)-1-(5-hydroxy-2-ethoxyphenyl)piperazine
Formed from: 4-(3-Cyclohexylpropionyl)-1-(2-ethoxyphenyl)piperazine

4-(3-Cyclohexylpropionyl)-1-(2-hydroxyphenyl)piperazine
Formed from: 4-(3-Cyclohexylpropionyl)-1-(2-ethoxyphenyl)piperazine

α-Cyclohexyl-α-phenyl-1-pyrrolidinepropanol
See: Procyclidine

6-(4-(1-Cyclohexyl-1*H*-tetrazol-5-yl)butoxy)-3,4-dihydro-2(1*H*)-quinolinone
See: Cilostazol

Cyclopenine ((2′*R*,3′*S*)-rel-(−)-4-methyl-3′-phenylspiro(3*H*-1,4-benzodiazepine-3,2′-oxirane)-2,5(1*H*,4*H*)-dione)
Formed from: Dehydrocyclopeptine
L-Phenylalanine
→ Cyclopenol
In: Penicillium A1633
See also: A2683

Cyclopenol ((2′*R*,3′*S*)-rel-(−)-3′-(3-hydroxyphenyl)-4-methylspiro(3*H*-1,4-benzodiazepine-3,2′-oxirane)-2,5(1*H*,4*H*)-dione)
Formed from: Cyclopenine

5-(Cyclopentoxycarbonylamino)-1-hydroxymethyl-3-((2-methoxy-4-((2-methylphenyl)sulphonyl)aminocarbonyl)phenylmethyl)indole
Formed from: Zafirlukast
→ N-Demethylzafirlukast
Probably in: Dog J455
Mouse J455
Rat J455

5-(Cyclopentoxycarbonylamino)-3-((4-hydroxy-2-methylphenyl)sulphonyl)aminocarbonyl)-2-methoxyphenylmethyl)-1-methylindole
Formed from: Zafirlukast

3-((Cyclopentylhydroxyphenylacetyl)oxy)-1,1-dimethylpyrrolidinium
See: Glycopyrrolate

4-(3-Cyclopentyloxy-4-methoxyphenyl)-2-pyrrolidinone
See: Rolipram

3-(((4-Cyclopentyl-1-piperazinyl)imino)methyl)rifamycin
See: Ritapentine Cyclopeptine ((3*S*)-3,4-dihydro-4-methyl-3-(phenylmethyl)-1*H*-1,4-benzodiazepine-2,5-dione)
Formed from: Anthranilic acid
L-Phenylalanine
→ 3′-Hydroxucyclopeptine
In: Penicillium A1633

Cyclo-L-prolyl-L-2-(1,1-dimethylallyl)tryptophanyl
Probably formed from: Cyclo-L-prolyl-L-tryptophanyl

Cyclo-L-prolyl-L-tryptophanyl
→ Cyclo-L-prolyl-L-2-(1,1-dimethylallyl) tryptophanyl
Probably in: Aspergillus A227

Cyclopropylanisole
→ 4-Cyclopropylphenol
In: Rat G851
→ 1-(*p*-Methoxyphenyl)cyclopropanol
In: Rat G851

Cyclopropylbenzene
→ Benzyl alcohol
In: Methylococcus F278
→ 3-Cyclopropylcatechol
In: Pseudomonas F283
→ 2-Cyclopropylphenol
In: Rat G851
→ 4-Cyclopropylphenol
In: Methylococcus F278
Rat G851
→ 1-Phenylcyclopropanol
In: Rat G851
→ 3-Phenylprop-2-en-1-ol
In: Methylococcus F278

***p*-Cyclopropylcarbonylphenylacetic acid**
→ *p*-Cyclopropylcarbonylphenylacetyl-β-D-glucuronic acid
In: Monkey A40, A1878
→ *p*-Cyclopropylcarbonylphenylacetylglycine
In: Dog A40, A1878
→ *p*-Cyclopropylcarbonylphenylacetyltaurine
In: Dog A40, A1878
→ *p*-Cyclopropylhydroxymethylphenylacetic acid
In: Rat A40, A1878

***p*-Cyclopropylcarbonylphenylacetyl-β-D-glucuronic acid**
Formed from: *p*-Cyclopropylcarbonylphenylacetic acid

***p*-Cyclopropylcarbonylphenylacetylglycine**
Formed from: *p*-Cyclopropylcarbonylphenylacetic acid

***p*-Cyclopropylcarbonylphenylacetyltaurine**
Formed from: *p*-Cyclopropylcarbonylphenylacetic acid

Cyclopropylcatechol
Formed from: Cyclopropylbenzene

Cyclopropyl-2,3-dihydro-4-phenylpyridinium
Formed from: Cyclopropyl-1,2,3,6-tetrahydro-4-phenylpyridine
→ Cyclopropyl-4-phenylpyridinium
Probably in: Rat J897

(1*S*-(1*R(*R**(*R**))-2*S**,3*R**))-N-Cyclopropyl-2,3-dihydroxypropyl-1-((4-hydroxycyclohexyl)methyl)-α-((2-(((1,1-dimethylethyl) sulphonyl)methyl)-1-oxo-3-phenylpropyl) amino)-1*H*-imidazole-4-propanamide**
Formed from: Remikiren
→ (1*S*-(1*R**(*R**(*R**))-2*S**,3*R**))-N-Cyclopropyl-2,3-dihydroxypropyl-1-((4-oxocyclohexyl)methyl)-α-((2-(((1,1-dimethylethyl)sulphonyl)methyl)-1-oxo-3-phenylpropyl)amino)-1*H*-imidazole-4-propanamide
Probably in: Rat H524

(1*S*-(1*R(*R**(*R**))-2*S**,3*R**))-N-Cyclopropyl-2,3-dihydroxypropyl-1-((4-oxocyclohexyl)methyl)-α-((2-(((1,1-dimethylethyl)sulphonyl)methyl)-1-oxo-3-phenylpropyl)amino)-1*H*-imidazole-4-propanamide**
Formed from: (1*S*-(1*R**(*R**(*R**))-2*S**,3*R**))-N-Cyclopropyl-2,3-dihydroxypropyl-1-((4-hydroxycyclohexyl)methyl)-α-((2-(((1,1-dimethylethyl)sulphonyl)methyl)-1-oxo-3-phenyl propyl)amino)-1*H*-imidazole-4-propanamide

1-Cyclopropyl-6-fluoro-1,4-dihydro-4-oxo-7-(1-piperazinyl)-3-quinolinecarboxylic acid
See: Ciprofloxacin

7-(2-Cyclopropyl-4-(*p*-fluorophenyl)-7,8-dihydro-7,8-dihydroxy-3-quinolinyl)-3,5-dihydroxy-6-heptenoic acid
Formed from: ((3*R*,5*S*,6*E*)-7-(2-Cyclopropyl-4-(*p*-fluorophenyl)-3-quinolinyl)-3,5-dihydroxy-6-heptenoic acid

(3*R*,5*S*,6*E*)-7-(2-Cyclopropyl-4-(*p*-fluorophenyl)-3-quinolinyl)-3,5-dihydroxy-6-heptenoic acid
→ 7-(2-Cyclopropyl-4-(*p*-fluorophenyl)-7,8-dihydro-7,8-dihydroxy-3-quinolinyl)-3,5-dihydroxy-6-heptenoic acid
In: Rabbit K8
→ 7-(2-Cyclopropyl-4-(*p*-fluorophenyl)-3-quinolinyl)-3,5-dihydroxy-6-heptenoic acid lactone

In: Dog K8
Rat K8
→ 7-(2-Cyclopropyl-4-(*p*-fluorophenyl)-3-quinolinyl)-2,4,6-heptatrienoic acid
In: Dog K8
→ 7-(2-Cyclopropyl-4-(*p*-fluorophenyl)-3-quinolinyl)-5-hydroxy-2,6-heptadienoic acid
In: Dog K8
Rabbit K8
Rat K8
→ 7-(2-Cyclopropyl-4-(*p*-fluorophenyl)-3-quinolinyl)-5-hydroxy-6-heptenoic acid
In: Dog K8
→ 7-(2-Cyclopropyl-4-(*p*-fluorophenyl)-3-quinolinyl)-3-hydroxy-5-oxo-6-heptenoic acid
In: Dog K8
Rat K8

7-(2-Cyclopropyl-4-(*p*-fluorophenyl)-3-quinolinyl)-3,5-dihydroxy-6-heptenoic acid lactone
Formed from: ((3*R*,5*S*,6*E*)-7-(2-Cyclopropyl-4-(*p*-fluorophenyl)-3-quinolinyl)-3,5-dihydroxy-6-heptenoic acid

7-(2-Cyclopropyl-4-(*p*-fluorophenyl)-3-quinolinyl)-2,4,6-heptatrienoic acid
Formed from: ((3*R*,5*S*,6*E*)-7-(2-Cyclopropyl-4-(*p*-fluorophenyl)-3-quinolinyl)-3,5-dihydroxy-6-heptenoic acid
→ 7-(2-Cyclopropyl-4-(*p*-fluorophenyl)-3-quinolinyl)-2,4,6-heptatrienoyl-β-D-glucuronic acid
Probably in: Rat K8
→ 7-(2-Cyclopropyl-4-(*p*-fluorophenyl)-3-quinolinyl)-2,4,6-heptatrienoyltaurine
Probably in: Rat K8

7-(2-Cyclopropyl-4-(*p*-fluorophenyl)-3-quinolinyl)-2,4,6-heptatrienoyl-β-D-glucuronic acid
Probably formed from: 7-(2-Cyclopropyl-4-(*p*-fluorophenyl)-3-quinolinyl)-2,4,6-heptatrienoic acid

7-(2-Cyclopropyl-4-(*p*-fluorophenyl)-3-quinolinyl)-2,4,6-heptatrienoyltaurine
Probably formed from: 7-(2-Cyclopropyl-4-(*p*-fluorophenyl)-3-quinolinyl)-2,4,6-heptatrienoic acid

7-(2-Cyclopropyl-4-(*p*-fluorophenyl)-3-quinolinyl)-5-hydroxy-2,6-heptadienoic acid
Formed from: (3*R*,5*S*,6*E*)-7-(2-Cyclopropyl-4-(*p*-fluorophenyl)-3-quinolinyl)-3,5-dihydroxy-6-heptenoic acid

7-(2-Cyclopropyl-4-(*p*-fluorophenyl)-3-quinolinyl)-5-hydroxy-6-heptenoic acid
Formed from: (3*R*,5*S*,6*E*)-7-(2-Cyclopropyl-4-(*p*-fluorophenyl)-3-quinolinyl)-3,5-dihydroxy-6-heptenoic acid

7-(2-Cyclopropyl-4-(*p*-fluorophenyl)-3-quinolinyl)-3-hydroxy-5-oxo-6-heptenoic acid
Formed from: (3*R*,5*S*,6*E*)-7-(2-Cyclopropyl-4-(*p*-fluorophenyl)-3-quinolinyl)-3,5-dihydroxy-6-heptenoic acid

4′-(2-Cyclopropyl-7-hydroxymethyl-3*H*-imidazo[5,4-*b*]pyridin-3-yl)methyl-2-biphenylcarboxylic acid
Formed from: 4′-(2-Cyclopropyl-7-methyl-3*H*-imidazo[5,4-*b*]pyridin-3-yl)methyl-2-biphenylcarboxylic acid

***p*-Cyclopropylhydroxymethylphenylacetic acid**
(α-cyclopropyl-α-hydroxy-*p*-tolylacetic acid)
Formed from: *p*-Cyclopropylcarbonylphenylacetic acid
→ *p*-Cyclopropylhydroxymethylphenylacetylglycine
Probably in: Dog A40
→ *p*-Cyclopropylhydroxymethylphenylacetyltaurine
Probably in: Dog A40

***p*-Cyclopropylhydroxymethylphenylacetylglycine**
Probably formed from: *p*-Cyclopropylhydroxymethylphenylacetic acid

***p*-Cyclopropylhydroxymethylphenylacetyltaurine**
Probably formed from: *p*-Cyclopropylhydroxymethylphenylacetic acid

α-Cyclopropyl-α-hydroxy-*p*-tolylacetic acid
See: *p*-Cyclopropylhydroxymethylphenylacetic acid

2-(((4-Cyclopropylmethoxy-2-pyridinyl)methyl)sulphinyl)-5-fluoro-1*H*-benzimidazole
→ 2-(((4-Cyclopropylmethoxy-2-pyridinyl)methyl)sulphinyl)-5-fluoro-6-hydroxy-1*H*-benzimidazole
In: Man J800

→ 2-(((4-Cyclopropylmethoxy-2-pyridinyl)methyl)sulphonyl)-5-fluoro-1*H*-benzimidazole
In: Man J800
→ 5-Fluoro-2-(((4-pyridon-2-yl)methyl) sulphinyl)-1*H*-benzimidazole
In: Man J800

2-(((4-Cyclopropylmethoxy-2-pyridinyl)methyl) sulphinyl)-5-fluoro-6-hydroxy-1*H*-benzimidazole
Formed from: 2-(((4-Cyclopropylmethoxy-2-pyridinyl)methyl)sulphinyl)-5-fluoro-1*H*-benzimidazole

2-(((4-Cyclopropylmethoxy-2-pyridinyl)methyl) sulphonyl)-5-fluoro-1*H*-benzimidazole
Formed from: 2-(((4-Cyclopropylmethoxy-2-pyridinyl)methyl)sulphinyl)-5-fluoro-1*H*-benzimidazole

(*S*)-(+)-(4-(1-(2-(1-(Cyclopropylmethyl)-3-(3,4-dichlorophenyl)-6-oxo-3-piperidyl)ethyl) azetidin-3-yl)-1-piperazine sulphamide
→ (*S*)-4-(1-(2-(3-(3,4-Dichlorophenyl)-6-oxo-3-piperidyl)ethyl)azetidin-3-yl)-1-piperazine sulphamide
In: Dog K119
Rat K119

N-Cyclopropylmethyl-2,3-dihydroxymorphinan (both (+) and (–) isomers)
Formed from: (+)-N-Cyclopropylmethyl-2-hydroxymorphinan
(–)-N-Cyclopropylmethyl-2-hydroxymorphinan
→ N-Cyclopropylmethyl-3-hydroxy-2-methoxymorphinan
In: Dog A3817

N-Cyclopropylmethyl-2,7β-dihydroxymorphinan
Formed from: (–)-N-Cyclopropylmethyl-2-hydroxymorphinan

N-Cyclopropylmethyl-2,6-dinitro-4-trifluoromethylaniline
Formed from: Profluralin

17-Cyclopropylmethyl-4,5α-epoxy-8β-(2-hydroxyethyl)-3-methoxymorphinan-6-one
Formed from: Codorphone
→ 4,5α-Epoxy-8β-(2-hydroxyethyl)-3-methoxymorphinan-6-one
Probably in: Dog C636
Rat C636

(+)-N-Cyclopropylmethyl-3-hydroxy-2-methoxymorphinan
Formed from: (+)-N-Cyclopropylmethyl-2,3-dihydroxymorphinan
→ 3-Hydroxy-2-methoxymorphinan
In: Dog A3817

(+)-N-Cyclopropylmethyl-2-hydroxymorphinan
→ (+)-N-Cyclopropylmethyl-2,3-dihydroxymorphinan
In: Dog A3817
→ 2-Hydroxymorphinan
In: Dog A3817

(–)-N-Cyclopropylmethyl-2-hydroxymorphinan
→ N-Cyclopropylmethyl-2,7β-dihydroxymorphinan
In: Dog A3817
→ 2-Hydroxymorphinan
In: Dog A3817

4′-(2-Cyclopropyl-7-methyl-3*H*-imidazo[5,4-*b*] pyridin-3-yl)methyl-2-biphenylcarboxylic acid
→ 4′-(2-Cyclopropyl-7-hydroxymethyl-3*H*-imidazo[5,4-*b*]pyridin-3-yl)methyl-2-biphenylcarboxylic acid
In: Cynomolgus monkey H868
→ 4′-(2-Cyclopropyl-7-methyl-3*H*-imidazo[5,4-*b*]pyridin-3-yl)methyl-2-biphenylcarboxylic acid-β-D-glucuronide
In: Rat J167

4′-(2-Cyclopropyl-7-methyl-3*H*-imidazo[5,4-*b*] pyridin-3-yl)methyl-2-biphenylcarboxylic acid-β-D-glucuronide
Formed from: 4′-(2-Cyclopropyl-7-methyl-3*H*-imidazo[5,4-*b*]pyridin-3-yl)methyl-2-biphenylcarboxylic acid

N-(Cyclopropylmethyl)-2,6-dinitro-N-bromo-4-(trifluoromethyl)benzenamine
See: Profluralin

(2*S*,6*R*,11*R*)-3-(Cyclopropylmethyl)-3,4,5,6-tetrahydro-8-hydroxy-6,11-dimethyl-2,6-methano-3-benzazocin-1(2*H*)-one
See: Ketocyclazocine

2-Cyclopropylphenol
Formed from: Cyclopropylbenzene

4-Cyclopropylphenol
Formed from: 4-Cyclopropylanisole
Cyclopropylbenzene
1-Cyclopropyl-4-phenylpyridinium
Probably formed from: 1-Cyclopropyl-2,3-dihydro-4-phenylpyridinium
1-Cyclopropyl-1,2,3,6-tetrahydro-4-phenylpyridine
→ 1-Cyclopropyl-2,3-dihydro-4-phenylpyridinium
In: Rat J897
→ 1,2,3,6-Tetrahydro-4-(*p*-hydroxyphenyl)pyridine
In: Rat J897
→ 1,2,3,6-Tetrahydro-4-phenylpyridine
In: Rat J897
Cyfluthrin (3-(2,2-dichloroethenyl)-2,2-dimethylcyclopropanecarboxylic acid cyano (4-fluoro-3-phenoxyphenyl)methyl ester)
→ 4-Fluoro-3-phenoxymandelonitrile
In: Bacillus G891
See also: H857
L-Cyhalothrin (3-(2-chloro-3,3,3-trifluoro-1-propenyl)-2,2-dimethylcyclopropanecarboxylic acid cyano(3-phenoxyphenyl)methyl ester)
→ *m*-Phenoxymandelonitrile
In: Bacillus G891
***p*-Cymene** (*p*-isopropyltoluene)
→ *p*-Cresol
In: Possum A3934
→ Cumyl alcohol
In: Guinea pig C574
Rabbit C891
Rat C574
See also: A2602, A2692, A2694, A3934
→ *cis*-1,2-Dihydro-1,2-dihydroxy-3-isopropyl-6-methylbenzene
In: Pseudomonas K699
→ *p*-(1-Hydroxy-2-propyl)toluene
In: Guinea pig C574
Rat C574
***cis*-Cypermethrin** (3-(2,2-dichloroethyl)-2,2-dimethylcyclopropanecarboxylic acid cyano (3-phenoxyphenyl)methyl ester)
→ 4′-Hydroxy-*cis*-cypermethrin
In: Rat A3804, C132
Trout E844
→ 5-Hydroxy-*cis*-cypermethrin
In: Rat A3804
→ *tert*-Hydroxy-*cis*-cypermethrin
In: Rat A3804
→ *m*-Phenoxybenzoic acid
In: Mouse A3250, C133
Rat C132
Trout E844
→ *m*-Phenoxymandelonitrile
In: Rat A3804
***trans*-Cypermethrin** (*trans* assumed if not stated; may be a mixture of up to 8 isomers)
→ 4′-Hydroxy-*trans*-cypermethrin
In: Rat A3804, C132
Trout E844
→ 5-Hydroxy-*trans*-cypermethrin
In: Rat A3804
→ *tert*-Hydroxy-*trans*-cypermethrin
In: Rat A3804
→ *m*-Phenoxybenzoic acid
In: Man G377
Mouse A3250, C133, F838
Quail F838
Rat C132, C412
Trout E844
→ *m*-Phenoxybenzyl alcohol
In: Rat C412
→ *m*-Phenoxymandelonitrile
In: Rat A3804
Cyproheptadine (1-methyl-4-(5*H*-dibenzo[*a*,*d*] cycloheptylidene)piperidine)
→ Cyproheptadine-N-β-D-glucuronide
In: Man A3392, G509, J416
Rabbit C30
See also: A1879
→ Demethylcyproheptadine
In: Cunninghamella J90
Dog A1858
Man A3076
Rat A2247, A3076, E885
See also: A1574, A1879
→ 2-Oxocyproheptadine
Probably in: Man A1879
→ Cyproheptadine-10,11-oxide
In: Cunninghamella J90
Dog A1858

Man A3076
Rat A1858, A2247, A3076, E885
→ Cyproheptadine-N-oxide
In: Cat A1858
Cunninghamella J90
Dog A1858
→ *cis*-10,11-Dihydro-10,11-dihydroxycyproheptadine
In: Dog A1858
→ 1-Hydroxycyproheptadine
In: Cunninghamella J90
→ 2-Hydroxycyproheptadine
In: Cunninghamella J90
→ 3-Hydroxycyproheptadine
In: Cunninghamella J90
→ 10-Hydroxycyproheptadine
In: Cunninghamella J90

Cyproheptadine-N-β-D-glucuronide
Probably formed from: Cyproheptadine

Cyproheptadine-10,11-oxide
Formed from: Cyproheptadine
→ *trans*-10,11-Dihydro-10,11-dihydroxycyproheptadine
Probably in: Cat A1858
Dog A1858

Cyproheptadine-N-oxide
Formed from: Cyproheptadine
→ 2-Hydroxycyproheptadine-N-oxide
Probably in: Cunninghamella J90

3-(S-(L-Cysteinyl))-5-(3-(L-alanyl))-*o*-quinone
Probably formed from: 5-(S-(L-Cysteinyl))-3,4-dihydroxy-L-phenylalanine

S-(L-Cysteinyl)carbamazepam
Probably formed from: Carbamazepine

10-(S-(L-Cysteinyl))celiptium
Formed from: Celiptium
→ 10-(N-Acetyl-S-(L-cysteinyl)) celiptium
Probably in: Man D102

3-((4-(L-Cystein-S-yl)-3,4-dihydro-3-hydroxy-5-trifluoromethyl-2-benzothiazolyl)methyl)-3,4-dihydro-4-oxo-1-phthalazineacetic acid
Probably formed from: 3-((4-(L-Glutamyl-L-cystein-S-yl)-3,4-dihydro-3-hydroxy-5-trifluoromethyl-2-benzothiazolyl)methyl)-3,4-dihydro-4-oxo-1-phthalazineacetic acid
→ 3-((4-(N-Acetyl-L-cystein-S-yl)-3,4-dihydro-3-hydroxy-5-trifluoromethyl-2-benzothiazolyl)methyl)-3,4-dihydro-4-oxo-1-phthalazineacetic acid
Probably in: Rat J457

3-((3-(L-Cystein-S-yl)-3,4-dihydro-4-hydroxy-5-trifluoromethyl-2-benzothiazolyl)methyl)-3,4-dihydro-4-oxo-1-phthalazineacetic acid
Probably formed from: 3-((3-(L-Glutamyl-L-cystein-S-yl)-3,4-dihydro-4-hydroxy-5-trifluoromethyl-2-benzothiazolyl)methyl)-3,4-dihydro-4-oxo-1-phthalazineacetic acid
→ 3-((3-(N-Acetyl-L-cystein-S-yl)-3,4-dihydro-4-hydroxy-5-trifluoromethyl-2-benzothiazolyl)methyl)-3,4-dihydro-4-oxo-1-phthalazineacetic acid
Probably in: Rat J457
→ 3-((3-(L-Cystein-S-yl)-4-hydroxy-5-trifluoromethyl-2-benzothiazolyl)methyl)-3,4-dihydro-4-oxo-1-phthalazineacetic acid
Probably in: Rat J457

5-(S-(L-Cysteinyl))-3,4-dihydroxyphenethylamine
→ Phaeomelanin
In: Glycine J252

2-(S-(L-Cysteinyl))-3,4-dihydroxy-L-phenylalanine
Formed from: 3,4-Dihydroxy-L-phenylalanine

5-(S-(L-Cysteinyl))-3,4-dihydroxy-L-phenylalanine
Formed from: 3,4-Dihydroxy-L-phenylalanine
→ 2,5-Bis(S-(L-cysteinyl))-3,4-dihydroxy-L-phenylalanine
Probably in: Horseradish B762
Mushroom A3167
With xanthine oxidase B968
→ 5-(S-(L-Cysteinyl))-3-methoxytyramine
In: Pig B813
→ 5-(S-(L-Cysteinyl))-3-methoxytyrosine
Probably in: Mouse B186
→ 3-(S-(L-Cysteinyl))-5-(3-L-alanyl)-*o*-quinone
Probably in: Man E347
Mushroom E347
→ Phaeomelanin
In: Glycine J252

6-(S-(L-Cysteinyl))-3,4-dihydroxy-L-phenylalanine
Formed from: 3,4-Dihydroxy-L-phenylalanine

3-(S-(L-Cysteinyl))-6-(3-(3,4-dimethoxybenzylamino)-2-hydroxypropoxy)-3,4-dihydro-4-hydroxy-2(1*H*)-quinolinone
Probably formed from: 6-(3,4-Dimethoxybenzylamino)-2-hydroxypropoxy)-3-(S-glutathionyl)-3,4-dihydro-4-hydroxy-2(1*H*)-quinolinone

4-(S-(L-Cysteinyl))-6-(3-(3,4-dimethoxybenzylamino)-2-hydroxypropoxy)-3,4-dihydro-3-hydroxy-2(1*H*)-quinolinone
Formed from: Toborinone

5-(S-(L-Cysteinyl))-6-fluoro-4-hydroxy-3-methoxy-L-phenylalanine
Formed from: 6-Fluoro-4-hydroxy-3-methoxy-L-phenylalanine

13-(S-(L-Cysteinylglycyl))-12,13-dihydro-8,14-dihydroxyefavirenz-8-sulphate
Probably formed from: 13-(S-Glutathionyl)-12,13-dihydro-8,14-dihydroxyefavirenz-8-sulphate

8-(S-(L-Cysteinylglycyl))-8,9-dihydro-9-hydroxyaflatoxin B_1
Probably formed from: 8-(S-Glutathionyl)-8,9-dihydro-9-hydroxyaflatoxin B_1

2-(S-(L-Cysteinylglycyl))-N-isopropyl-N-phenylacetamide
Probably formed from: 2-(S-Glutathionyl)-N-isopropyl-N-phenylacetamide
→ 2-(S-(L-Cysteinyl))-N-isopropyl-N-phenylacetamide
Probably in: Rat F505

2-(S-(L-Cysteinylglycyl))mitoxantrone
Formed from: Mitoxantrone

4-((4-(S-(L-Cysteinylglycyl))-2,3,5,6-tetrafluorophenyl)sulphonamido)-2-fluoro-1-methoxybenzene
Probably formed from: 2-Fluoro-1-methoxy-4-((2,3,5,6-tetrafluoro-4-(S-glutathionyl)phenyl)sulphonamido)benzene
→ 4-((4-(S-(L-Cysteinyl))-2,3,5,6-tetrafluorophenyl)sulphonamido)-2-fluoro-1-methoxybenzene
Probably in: Mouse K62

1-(S-(L-Cysteinyl))-2-hydroxyoestrone
Probably formed from: 1-(S-(Glutathionyl)-2-hydroxyoestrone

4-(S-(L-Cysteinyl))-2-hydroxyoestrone
Probably formed from: 4-(S-(Glutathionyl)-2-hydroxyoestrone

3-((3-(S-(L-Cysteinyl))-4-hydroxy-5-trifluoromethyl-3-benzothiazolyl)methyl)-3,4-dihydro-4-oxo-1-phthalazineacetic acid
Probably formed from: 3-((3-(S-(L-Cysteinyl))-3,4-dihydro-4-hydroxy-5-trifluoromethyl-3-benzothiazolyl)methyl)-3,4-dihydro-4-oxo-1-phthalazineacetic acid

2-(S-(L-Cysteinyl))-N-isopropyl-N-phenylacetamide
Probably formed from: 2-(S-(L-Cysteinylglycyl))-N-isopropyl-N-phenylacetamide
→ 2-(N-Acetyl-S-(L-cysteinyl))-N-isopropyl-N-phenylacetamide
In: Rat C24, G80
See also: A1682, B91, B93, B669, B707, B936, C407, D932, E790, F505
→ 2-(S-(L-Cysteinyl))-N-isopropyl-N-phenylacetamide sulphoxide
Probably in: Chicken D932
Rat B669, B707, B936, F505
→ N-Isopropyl-*o*-acetamidothiophenol
In: Eubacterium D356

2-(S-(L-Cysteinyl))-N-isopropyl-N-phenylacetamide sulphoxide
Probably formed from: 2-(S-(L-Cysteinyl))-N-isopropyl-N-phenylacetamide

(2*R*,4*R*)-3-((S-(S-(L-Cysteinyl))-3-mercaptopropionyl)-2-(*o*-hydroxyphenyl)-4-thiazolidinecarboxylic acid
Probably formed from: (2*R*,4*R*)-3-((S-(S-Glutathionyl))-3-mercaptopropionyl)-2-(*o*-hydroxyphenyl)-4-thiazolidinecarboxylic acid

5-(S-(L-Cysteinyl))-3-methoxytyramine
Probably formed from: 5-(S-(L-Cysteinyl))-3,4-dihydroxy-L-phenylalanine

5-(S-(L-Cysteinyl))-3-methoxytyrosine
Formed from: 5-(S-(L-Cysteinyl))-3,4-dihydroxy-L-phenylalanine

2-(S-(L-Cysteinyl))mitoxantrone
Formed from: Mitoxantrone

2-(S-(L-Cysteinyl))mitoxantrone ethyl ester
Formed from: Mitoxantrone
L-Cysteinyl-β-naphthylamide
→ β-Naphthylamine
In: Bacteria C146
4-((4-(S-(L-Cysteinyl))-2,3,5,6-tetrafluorophenyl)sulphonamido)-2-fluoro-1-methoxybenzene
Probably formed from: 4-((4-(S-(L-Cysteinylglycyl))-2,3,5,6-tetrafluorophenyl)sulphonamido)-2-fluoro-1-methoxybenzene
→ 4-((4-(N-Acetyl-S-(L-cysteinylglycyl))-2,3,5,6-tetrafluorophenyl)sulphonamido)-2-fluoro-1-methoxybenzene
Probably in: Mouse K62
L-Cystinyl-di-β-naphthylamide
→ L-Cystinyl-β-naphthylamide
In: Bothrops G167
→ β-Naphthylamine
In: Bothrops G167
Man A837
L-Cystinyl-β-naphthylamide
Formed from: L-Cystinyl-di-β-naphthylamide
→ β-Naphthylamine
In: Beef A526
Cytemba (*cis*-3-bromo-3-(*p*-methoxybenzoyl)acrylic acid)
→ *cis*-3-Bromo-(*p*-hydroxybenzoyl)acrylic acid
In: Rat A2028
→ 4-Hydroxy-4-(*p*-methoxybenzoyl)butyric acid
In: Rat A2028
→ (*p*-Methoxybenzoyl)propionic acid
In: Rat A2028
Cytenamide (5-carboxamido-5*H*-dibenzo[*a*,*d*] cycloheptene)
→ Cytenamide-10,11-oxide
In: Rat A2758, A2991
Cytenamide-10,11-oxide
Formed from: Cytenamide
→ 10,11-Dihydro-10,11-dihydroxycytenamide
Probably in: A2991
Cytogenin (8-hydroxy-3-(hydroxymethyl)-6-methoxy-1*H*-2-benzopyran-1-one)
Formed from acetate and malonate in Streptovericillium H124

D

2,4-D
See: 2,4-Dichlorophenoxyacetic acid
Daidzein (4′,7-Dihydroxyisoflavone)
Formed from: Daidzin
Formononetin
Liquiritigenin
Puerarin
2,4′,7-Trihydroxyisoflavanone
→ *p*-Coumaric acid
In: Rhizobium H280
→ Coumestrol
In: Medicago A3725
→ Daidzein-β-D-glucoside (position unclear)
In: Rabbit A571
→ Daidzein-4′-β-D-glucuronide
In: Man K18
→ Daidzein-7-β-D-glucuronide
In: Beef E539
Man K18
Rabbit A571
Rat H406
Sheep E539
See also: H408, K531
→ Daidzein-4′-sulphate
In: Hen A1607
Rat H406
See also: H408, K531
→ Daidzin
In: Cicer K738
Pueraria A2373
→ Dihydrodaidzein
In: Microorganism J2
See also: J717
→ 4′,7-Dihydroxyisoflavan
Probably in: Microorganism J717
→ Equol
In: Hen A1607
→ Formononetin
In: Alfalfa H875
→ Isoformononetin
In: Alfalfa H875, G322
Glycyrrhiza K630
→ Phenylacetic acid
In: Rhizobium H280
→ Resorcinol
In: Rhizobium H280
→ 2,4,4′-Trihydroxychalcone
In: Medicago A3725
→ 3′,4′,7-Trihydroxyisoflavan
Probably in: Microorganism J717
→ 4′,7,8-Trihydroxyisoflavone
Probably in: Aspergillus K155
Daidzein-4′,7-disulphate
Formed from: Daidzein-4′-sulphate
Daidzein-β-D-glucoside
Formed from: Daidzein
Daidzein-7-β-D-glucoside
See: Daidzin
Daidzein-4′-β-D-glucuronide
Formed from: Daidzein
Daidzein-7-β-D-glucuronide
Formed from: Daidzein
Daidzein-β-D-glucuronide sulphate
Probably formed from: Daidzein-4′-sulphate
Daidzein-4′-sulphate (4′ assumed if not stated)
Formed from: Daidzein
→ Daidzein-4′,7-disulphate
In: Hen A1607
See also: H408
→ Daidzein-β-D-glucuronide sulphate
Probably in: Rat K531
Daidzin (daidzein-7-β-D-glucoside)
Formed from: Daidzein
→ Daidzein

In: Aspergillus K155
Man J717
Microorganism J525
Rat H406, J717

Dansylamide
See: 5-Dimethylamino-1-naphthalenesulphonamide

Dantrolene (1-(((5-(4-nitrophenyl)-2-furanyl)methylene)amino)-2,4-imidazolidinedione)
→ Aminodantrolene
In: Hamster C538
Mouse C538
Rat C538, G699
→ 5-Hydroxydantrolene
In: Hamster C538
Mouse C538
Rat C538, G699

Daphnetin
See: 7,8-Dihydroxycoumarin

Daphnetin-7-glucoside
See: Daphnin

Daphnin (daphnetin-7-glucoside)
Formed from: 7,8-Dihydroxycoumarin

Dapsone
See: Bis(*p*-aminophenyl)sulphone

Darifenacin ((*S*)-2-(1-(2-(2,3-dihydrobenzofuran-5-yl)ethyl)-3-pyrrolidinyl)-2,2-diphenylacetamide)
→ 5-Carboxymethylbenzofuran
In: Dog H866
Man H866
Mouse H866
Rabbit H866
Rat H866
→ 2-(3-Pyrrolidinyl)-2,2-diphenylacetamide
In: Dog H866
Man H866
Mouse H866
Rabbit H866
Rat H866
→ (*S*)-2-(1-(2-(2,3-Dihydro-3-hydroxybenzofuran-5-yl)ethyl)-3-pyrrolidinyl)-2,2-diphenylacetamide
In: Dog H866
Man H866
Mouse H866
Rabbit H866
Rat H866
→ (*S*)-2-(1-(2-(2,3-Dihydrobenzofuran-5-yl)ethyl)-3-pyrrolidinyl)-2-hydroxyphenyl-2-phenylacetamide
In: Man H866
Mouse H866
Rabbit H866
Rat H866
→ (*S*)-2-(1-(2-(4-Hydroxy-3-(2-hydroxyethyl)phenyl)ethyl)-3-pyrrolidinyl)-2,2-diphenylacetamide
In: Dog H866
Man H866
Mouse H866
Rabbit H866
Rat H866

Datiscetin (2′,3,5,7-tetrahydroxyflavone)
→ Datiscetin-7-β-D-glucoside
In: Parsley K739

Datiscetin-7-β-D-glucoside
Formed from: Datiscetin

Daunomycin
See: Daunorubicin

Daunomycinone
See: Daunorubicin aglycone

Daunomycinone semiquinone radical
Formed from: Daunorubicin aglycone

Daunorubicin (daunomycin, rubidomycin, rubomycin C, 8-acetyl-10-((3-amino-2,3,6-trideoxy-α-L-lyxo-hexopyranosyl)oxy)-7,8,9,10-tetrahydro-6,8,11-trihydroxy-1-methoxy-5,12-naphthacenedione)
Formed from: Rubidazone
→ N-Acetyl-13(*S*)-dihydrodaunorubicin
In Streptomyces D881
→ Adriamycin
In Streptomyces B600
→ Bisanhydro-13-dihydrodaunomycinone
In Streptomyces D881
→ Daunorubicin aglycone
In: Rat A520, D212
→ Daunorubicinol
In: Hamster A1174
Man A37, A138, A406, A764, A1529, B25, B169, B569, C531
Mouse B217, B972, E160

Mucor A2755, A3537
Rabbit A50, A139, A2007, A3268, A3500, A3950, B169, B402, E160
Rat A50, A406, A875, A1327, A2007, A3423
Streptomyces B571, B600, C450
See also: C539, K378
→ Daunorubicin semiquinone radical
In: Beef E116
→ Deoxydaunorubicinone
In: Aeromonas A2670
Cow B370
Rat D212
Vibrio C509
See also: A875
→ Doxorubicin
In Streptomyces C450
See also: C539

Daunorubicin aglycone (daunomycinone, 8-acetyl-10-((3-amino-2,3,6-trideoxy-α-L-lyxo-hexopyranosyl)oxy)-7,8,9,10-tetrahydro-6,8,10,11-tetrahydroxy-1-methoxy-5,12-naphthacenedione)
Formed from: Daunorubicin
Probably formed from: ε-rhodomycinone
→ Daunorubicin
Probably in: Streptomyces B443
→ Daunorubicinol aglycone
Probably in: Rat A520, D212
→ Daunomycinone semiquinone radical
In: Beef E116
→ 13-Dihydrodaunomycinone
In: Streptomyces A2813, B443, B600

Daunorubicinol (daunomycinol, (8*S*,10*S*)-10-((3-amino-2,3,6-trideoxy-α-L-lyxo-hexopyranosyl)oxy)-7,8,9,10-tetrahydro-6,8,11-trihydroxy-8-((1*S*)-1-hydroxyethyl)-1-methoxy-5,12-naphthacenedione)
Formed from: Daunorubicin
Feudomycin A
→ Adriamycin
In Streptomyces B600
→ Daunorubicin
In Streptomyces B600
→ Daunorubicinol aglycone
Probably in: Man A37, A138
→ Deoxydaunorubicinol aglycone
In: Streptomyces A2755, A3537

Daunorubicinol aglycone (7,8,9,10-tetrahydro-6,8,10,11-tetrahydroxy-8-(1-hydroxyethyl)-1-methoxy-5,12-naphthacenedione)
Formed from: Daunorubicin aglycone
→ 4-Demethyldaunorubicinol aglycone
Probably in: Rat D212
→ 7-Deoxydaunorubicinol aglycone
In: Streptomyces B571
See also: A37, A138, A875

Daunorubicinone semiquinone radical
Formed from: Daunorubicin

Dauricine (4-(((1*R*)-1,2,3,4-tetrahydro-6,7-dimethoxy-2-methyl-1-isoquinolinyl)methyl)-2-(4-(((1*R*)-1,2,3,4-tetrahydro-6,7-dimethoxy-2-methyl-1-isoquinolinyl)methyl)phenoxy)phenol)
Formed from: Tyramine

Dazoxiben (4-(2-(1*H*-imidazol-1-yl)ethoxy) benzoic acid)
→ Dazoxiben-β-D-glucuronide
In: Man B843

Dazoxiben-β-D-glucuronide
Formed from: Dazoxiben

DDD
See: 1,1-Dichloro-2,2-bis(*p*-chlorophenyl) ethane

DDE
See: 1,1-Dichloro-2,2-bis(*p*-chlorophenyl) ethylene

DDMI
See: Bisdemethylimipramine

DDMU
See: 1-Chloro-2,2-bis(*p*-chlorophenyl)ethylene

DDT
See: 1,1,1-Trichloro-2,2-bis(*p*-chlorophenyl)ethane

Deacetoxyvindoline ((2β,3β,4β,5α,12*R*,19α)-6,7-didehydro-3-hydroxy-16-methoxy-1-methylaspidospermidine-3-carboxylic acid methyl ester)
→ Deacetylvindoline
In: Catharanthus H5

10-Deacetylbaccatin III
Formed from: Baccatin III
→ Baccatin III
In: Taxus K91

Deacetyl-N,O-bisdemethyldiltiazem
Probably formed from: Deacetyl-N-demethyldiltiazem
Deacetyl-O-demethyldiltiazem
→ Deacetyl-N,O-bisdemethyldiltiazem-β-D-glucuronide
Probably in: Dog F607
Man F607
Rabbit F607
Rat F607
→ Deacetyl-N,O-bisdemethyldiltiazem sulphate
Probably in: Dog F607
Man F607
Rabbit F607
Rat F607

Deacetyl-N,O-bisdemethyldiltiazem-β-D-glucuronide
Probably formed from: Deacetyl-N,O-bisdemethyldiltiazem

Deacetyl-N,O-bisdemethyldiltiazem sulphate
Probably formed from: Deacetyl-N,O-bisdemethyldiltiazem

Deacetyl-N,O-bisnorclentiazem
Probably formed from: Deacetyl-O-norclentiazem

Deacetylclentiazem
Formed from: Clentiazem
→ (+)-(2*S*,3*S*)-8-Chloro-3,4-dihydro-2-(4-hydroxy-3-methoxyphenyl)-4-oxo-1,5-benzothiazepin-5(2*H*)-acetic acid
Probably in: Rat H76
→ Deacetyl-N-norclentiazem
Probably in: Rat H76
→ Deacetyl-O-norclentiazem
Probably in: Rat H76

Deacetylcolchicine
→ Colchicine
Probably in: Colchicum A606

Deacetyl-N-demethyldiltiazem
Probably formed from: Deacetyldiltiazem
N-Demethyldiltiazem
→ Deacetyl-N,O-bisdemethyldiltiazem
Probably in: Dog F607
Man F607
Rabbit F607, G314
Rat F607, G859
→ Deacetyl-N-demethyldiltiazem-β-D-glucuronide
Probably in: Dog F607
Man D126, F607
Rabbit F607
Rat F607

Deacetyl-O-demethyldiltiazem
Probably formed from: Deacetyldiltiazem
O-Demethyldiltiazem
→ Deacetyl-N,O-bisdemethyldiltiazem
Probably in: Man E654
→ Deacetyl-O-demethyldiltiazem-β-D-glucuronide
Probably in: Dog F607
Man D126, F607
Rabbit F607
Rat F607
→ Deacetyl-O-demethyldiltiazem sulphate
Probably in: Dog F607
Man D126, F607
Rabbit F607
Rat F607

Deacetyl-N-demethyldiltiazem-β-D-glucuronide
Probably formed from: Deacetyl-N-demethyldiltiazem

Deacetyl-O-demethyldiltiazem-β-D-glucuronide
Probably formed from: Deacetyl-O-demethyldiltiazem

Deacetyl-O-demethyldiltiazem sulphate
Probably formed from: Deacetyl-O-demethyldiltiazem

N-Deacetyl-16-O-demethyllappaconitine
Probably formed from: N-Deacetyllapppaconitine

Deacetyl-N-demethylthymoxamine
Probably formed from: Deacetylthymoxamine
→ Deacetyl-N-demethylthymoxamine-β-D-glucuronide
Probably in: Dog D788
Man D788, D789
Rat D787, D788
→ Deacetyl-N-demethylthymoxamine sulphate
Probably in: Dog D788
Man D788, D789
Rat D787, D788

Deacetyl-N-demethylthymoxamine-β-D-glucuronide
Probably formed from: Deacetyl-N-demethylthymoxamine

Deacetyl-N-demethylthymoxamine sulphate
Probably formed from: Deacetyl-N-demethylthymoxamine

17-O-Deacetyl-11-O-demethylvindoline
→ 17-O-Deacetylvindoline
In: Catharanthus K729

17-O-Deacetyl-1,2-dihydroraucaffricine
→ 17-O-Deacetyl-1,2-dihydrovomilenine
In: Rauwolfia K939

17-O-Deacetyl-1,2-dihydrovomilenine
Formed from: 17-O-Deacetyl-1,2-dihydroraucaffricine

25-Deacetyl-3′,30-dihydroxy-5′-(4-isobutyl-1-piperazinyl)benzoxazinorifamycin
Probably formed from: 25-Deacetyl-3′-hydroxy-5′-(4-isobutyl-1-piperazinyl)benzoxazinorifamycin

Deacetyldiltiazem
Formed from: Diltiazem
→ Deacetyl-N-demethyldiltiazem
Probably in: Dog F607
Man E654, F607
Rabbit F607
Rat F607, G859
→ Deacetyl-O-demethyldiltiazem
Probably in: Dog F607
Man F607
Rabbit F607, G314
Rat F607, G859
→ Deacetyldiltiazem-N-oxide
Probably in: Dog F607
Man F607
Rabbit F607
Rat F607

Deacetyldiltiazem-N-oxide
Probably formed from: Deacetyldiltiazem

25-Deacetyl-3-formylrifapentine
Probably formed from: 25-Deacetylrifapentine

25-Deacetyl-3′-hydroxy-5′-(4-isobutyl-1-piperazinyl)benzoxazinorifamycin
Formed from: 3′-Hydroxy-5′-(4-isobutyl-1-piperazinyl)benzoxazinorifamycin
→ 25-Deacetyl-3′,30-dihydroxy-5′-(4-isobutyl-1-piperazinyl) benzoxazinorifamycin
Probably in: Dog H527

N-Deacetyl-N-hydroxyketoconazole
Formed from: N-Deacetylketoconazole

25-O-Deacetyl-32-hydroxyrifabutin
Probably formed from: 25-O-Deacetylrifabutin

Deacetylipecoside
Probably formed from: 3,4-Dihydroxyphenethylamine
→ Alangiside
In: Alangium A3154
Cephaelis A3154
→ Deoxytubulosine
In: Alangium C909
→ Ipecoside
In: Alangium A3154
Cephaelis A3154, A3861

Deacetylisoipecoside
→ Cephaeline
In: Alangium A3154
Cephaelis A3154, A3861

N-Deacetylketoconazole
Formed from: Ketoconazole
→ N-Deacetyl-N-hydroxyketoconazole
In: Rat J735, K49

N-Deacetyllappaconitine
Formed from: Lappaconitine
→ N-Deacetyl-16-O-demethyllappaconitine
Probably in: Man F473

Deacetyl-N-methylcolchicine
See: Demecolcine

Deacetyl-N-norclentiazem
Probably formed from: Deacetylclentiazem

Deacetyl-O-norclentiazem
Probably formed from: Deacetylclentiazem
→ Deacetyl-N,O-bisnorclentiazem
Probably in: Rat H76

Deacetylpractolol
Formed from: Practolol

25-O-Deacetylrifabutin
Formed from: Rifabutin
→ 25-O-Deacetyl-32-hydroxyrifabutin
Probably in: Man F453
→ 25-O-Deacetylrifabutin-N-oxide

Probably in: Man F453

25-O-Deacetylrifabutin-N-oxide
Probably formed from: 25-O-Deacetylrifabutin

25-Deacetylrifalazil
Formed from: Rifalazil

25-Deacetylrifampicin
Formed from: Rifampicin

25-O-Deacetylrifamycin LM427
Formed from: Rifamycin LM427

25-Deacetylrifapentine
Formed from: Rifapentine
→ 25-Deacetyl-3-formylrifapentine
In: Cynomolgus monkey J431
Man J430
Mouse J431
Rat J431

Deacetylthymoxamine
Formed from: Thymoxamine
→ Deacetyl-N-demethylthymoxamine
In: Mucor J133
See also: B418, D788, D789, E136, F301
→ Deacetylthymoxamine-β-D-glucuronide
In: Actinomucor J133
Dog D788
Man D788, D789
Mucor J133
Rat B418, D787, D788, J493
→ Deacetylthymoxamine-N-oxide
In: Mortierella J133
→ Deacetylthymoxamine sulphate
In: Cat D788
Dog D788
Man D788, D789, F301
Mortierella J133
Rat B418, D787, D788, J493

Deacetylthymoxamine-β-D-glucuronide
Formed from: Deacetylthymoxamine

Deacetylthymoxamine-N-oxide
Formed from: Deacetylthymoxamine

Deacetylthymoxamine sulphate
Formed from: Deacetylthymoxamine

Deacetylusnic acid
Formed from: Usnic acid

Deacetylvinblastine
Formed from: Vinblastine

4-Deacetylvincristine
Formed from: Vincristine

Deacetylvindoline
Formed from: Deacetoxyvindoline
17-O-Deacetyl-11-O-demethylvindoline
Vindoline
→ Vindoline
In: Catharanthus E580, K729, K931
Plant J701

Deacetylvinorelbine
Formed from: Vinorelbine

Deallylalmitrine
Formed from: Almitrine
→ 1-(4-Amino-6-(2,3-dihydroxypropyl)amino-1,3,5-triazin-2-yl)-4-(bis(*p*-fluorophenyl)methyl)piperazine
Probably in: Man G187
→ Bisdeallylalmitrine
In: Man G187

5-Deazariboflavin (1-deoxy-1-(3,4-dihydro-7,8-dimethyl-2,4-dioxopyrimido[4,5-*b*]quinolin-10(2*H*)-yl)-D-ribitol)
→ Dihydro-5-deazariboflavin
In: Benecka A1270

Debenzylbenidipine
Formed from: Benidipine
→ 4-(*m*-Aminophenyl)-1,4-dihydro-2,6-dimethylpyridine-3,5-dicarboxylic acid 3-(3-(hydroxypiperidinyl)) ester 5-methyl ester
Probably in: Rat H927
→ 1,4-Dihydro-2,6-dimethyl-4-(*m*-nitrophenyl)pyridine-3,5-dicarboxylic acid 3-(3-(hydroxypiperidinyl)) ester 5-methyl ester
Probably in: Rat H927
→ 1,4-Dihydro-2,6-dimethyl-4-(*m*-nitrophenyl)pyridine-3,5-dicarboxylic acid 3-(3-piperidinyl) ester
Probably in: Dog F11
Man J105
Rat F11, H927
→ 2,6-Dimethyl-4-(*m*-nitrophenyl)pyridine-3,5-dicarboxylic acid 3-methyl ester 5-(3-piperidinyl) ester
Probably in: Man J105

Debrisoquine (3,4-dihydro-2(1*H*)-isoquinoline carboximidamide)
Formed from: N-Hydroxydebrisoquine
→ *o*-(2-(4,6-Dimethylpyrimidinylamino) ethyl)benzoic acid
In: Dog A1787
Man A1787
Rat A1787
→ *o*-(2-(4,6-Dimethylpyrimidinylamino) methyl)phenylacetic acid
In: Dog A1787
Man A1787
Rat A1787
→ 1-Hydroxydebrisoquine
In: Man J456
→ 3-Hydroxydebrisoquine
In: Man J456
→ 4-Hydroxydebrisoquine
In: Dog A1787
Man A1653, A1787, A1870, A3411, A3487, B39, B532, B533, B667, C324, C606, D371, D951, E650, J456, K35, K440
Rat A1653, A1787, A1870, D98
→ 5-Hydroxydebrisoquine
In: Man B39, B532, D951, K35
→ 6-Hydroxydebrisoquine
In: Man A1653, A1870, B39, B532, D951, K35
Rat A1653, A1870
→ 7-Hydroxydebrisoquine
In: Man A1870, B39, B532, D951, K35
Rat A1870
→ 8-Hydroxydebrisoquine
In: Man B39, B532, D951, K35
→ N-Hydroxydebrisoquine
In: Rabbit H77
Rat H77

Debromohydroxyzenarestat
Probably formed from: Hydroxyzenarestat

Debutylbucromarone
Formed from: Bucromarone
→ Bisdebutylbucromarone
In: Mouse F715
Rat F715

Decamethrin ((*S*)-α-cyano-3-phenoxybenzyl (1*R*,3*R*)-*cis*-3-(2,2-dibromovinyl)-2,2-dimethylcyclopropanecarboxylate)
→ 2′-Hydroxydecamethrin
In: Rat A3199
→ 4′-Hydroxydecamethrin
In: Rat A3199, A3804
→ 5-Hydroxydecamethrin
In: Rat A3199, A3804
→ *m*-Phenoxymandelonitrile
In: Rat A3199, A3804
See also: A3800

10-Decarbomethoxyaklavinone
Formed from: Aklavinone
→ Daunorubicin
In: Streptomyces B571

Decarboxy-3-hydroxyloflazepate
Probably formed from: 3-Hydroxyloflazepate

Decarboxy-4′-hydroxyloflazepate
Probably formed from: 4′-Hydroxyloflazepate

Decarboxyloflazepate (7-chloro-5-(2-fluorophenyl)-2,3-dihydro-2-oxo-1*H*-1,4-benzodiazepine)
Probably formed from: Loflazepate

Dechlorochloramphenicol
Formed from: Chloramphenicol

Dechlorochloroflavonin
Formed from: L-Phenylalanine
→ Chloroflavonin
In: Aspergillus A154

8-Dechloro-N-demethyl-8-methylthioclozapine
Probably formed from: 8-Dechloro-8-methylthioclozapine

8-Dechloro-8-hydroxyclozapine
Formed from: Clozapine
→ 8-Dechloro-8-hydroxy-N-demethylclozapine
Probably in: Man J153

8-Dechloro-8-hydroxy-N-demethylclozapine
Probably formed from: 8-Dechloro-8-hydroxyclozapine
→ 8-Dechloro-8-hydroxy-N-demethylclozapine-β-D-glucuronide
Probably in: Man J153

8-Dechloro-8-hydroxy-N-demethylclozapine-β-D-glucuronide
Probably formed from: 8-Dechloro-8-hydroxy-N-demethylclozapine

N²-(Dechloromethyl)ifosfamide
Formed from: Ifosfamide

N³-(Dechloromethyl)ifosfamide
Formed from: Ifosfamide

8-Dechloro-8-methylthioclozapine
Formed from: Clozapine
→ 8-Dechloro-N-demethyl-8-methylthioclozapine
Probably in: Man J153
Rat J227

Dechloroticlopidine
Formed from: Ticlopidine

Decyclopropylmethylbuprenorphine
Formed from: Buprenorphine
→ Decyclopropylmethyl-6-O-demethylbuprenorphine
Probably in: Rat D364

N-Decyclopropylmethylcodorphone
Formed from: Codorphone
→ N-Decyclopropylmethyl-demethylcodorphone
In: Dog C636
Man C636
Rat C636

N-Decyclopropylmethylcyclazocine
Formed from: Cyclazocine
→ N-Decyclopropylmethylcyclazocine-β-D-glucuronide
Probably in: Rat A2134
→ N-Decyclopropylmethyl-hydroxycyclazocine
Probably in: Rat A2134

N-Decyclopropylmethylcyclazocine-β-D-glucuronide
Probably formed from: N-Decyclopropyl-methylcyclazocine

Decyclopropylmethyl-6-O-demethylbuprenorphine
Probably formed from: Decyclopropylmethyl-buprenorphine

N-Decyclopropylmethyl-demethylcodorphone
Probably formed from: N-Decyclopropyl-methylcodorphone

Decyclopropylmethyldiprenorphine
Formed from: Diprenorphine

N-Decyclopropylmethyl-hydroxycyclazocine
Probably formed from: N-Decyclopropyl-methylcyclazocine

N-Decyclopropylmethyl-hydroxyvolazocine
Probably formed from: Hydroxyvolazocine

N-Decyclopropylmethylketocyclazocine
Probably formed from: Ketocyclazocine
→ N-Decyclopropylmethylketocyclazocine-β-D-glucuronide
Probably in: Rat A2134

N-Decyclopropylmethylketocyclazocine-β-D-glucuronide
Probably formed from: N-Decyclopropyl-methylketocyclazocine

Decylbenzene
→ Benzoic acid
In: Beauveria D936
Paecilomyces D936
Penicillium D936
Verticillium D936

Decyl salicylate
→ Salicylic acid
In: Guinea pig D144

Deformylformoterol
Formed from: Formoterol

Dehydroabietic acid
→ 2-Hydroxydehydroabietic acid
In: Mortierella E117
→ 7-Hydroxydehydroabietic acid
In: Alcaligenes A932
Pseudomonas A932
→ 15-Hydroxydehydroabietic acid
In: Mortierella E117
→ 16-Hydroxydehydroabietic acid
In: Mortierella E117
Rabbit C275

4,5-Dehydroanhydrolycorine
Formed from: 2-Epilycorine

1,4-Dehydrobenidipine
Formed from: Benidipine

1″,2″-Dehydro-4″,5″-bisnor-Δ⁶-tetrahydrocannabidiol-3″,7-dioic acid
Probably formed from: 4″,5″-Bisnor-Δ⁶-tetrahydrocannabidiol-3″-oic acid

1″,2″-Dehydro-4″,5″-bisnor-Δ^1-tetrahydrocannabinol-3″-oic acid
Probably formed from: 4″,5″-Bisnor-Δ^1-tetrahydrocannabinol-3″-oic acid
4′,5′-Dehydrobrimodinine
Formed from: Brimodinine
Dehydrocanadine
Probably formed from: Tetrahydroberberine
Dehydrochelidonine
Formed from: (+)-Chelidonine
3,4-Dehydrochlorambucil
Formed from: Chlorambucil
→ *p*-(Bis(2-chloroethyl)amino)phenyl)acetic acid
In: Rat A3959
4,5-Dehydrocibenzoline
Formed from: Cibenzoline
3,4-Dehydrocilostazol
Formed from: Cilostazol
4,21-Dehydrocorynantheine
Formed from: Strictosidine
Dehydrocyclopeptine
→ Cyclopenin
In: Penicillium A2734
→ 3′-Hydroxydehydrocyclopeptine
In: Penicillium A1633
6,7-Dehydro-O-demethyldextromethorphan
Probably formed from: O-Demethyldextromethorphan
3-Dehydro-5-deoxy-5-aminoshikimic acid
→ 3-Amino-5-hydroxybenzoic acid
In: Amycolalopsis J643
4,5-Dehydro-desethyl-2,9-dihydro-9-hydroxy-2-oxolisuride
Probably formed from: 4,5-Dehydro-2,9-dihydro-9-hydroxy-2-oxolisuride
Dehydrodicaffeic acid dilactone (3,6-bis(3,4-dihydroxyphenyl)tetrahydro-1*H*,4*H*-furo[3,4-*c*]furan-1,4-dione)
Formed from: Caffeic acid
Dehydrodicatechin A ((2*S*,3*R*,7*R*,14a*S*,14b*R*)-3-(3,4-dihydroxyphenyl)-2,3,8,8a,14a-pentahydro-2,7,10,12,17-pentahydroxy-9*H*-7,14b-methano-1*H*-pyrano[2″,3″:4′,5′]benzofuro[2′,3′:4,5]oxocino[3,2-*b*][1]benzopyran-6(7*H*)-one)
Formed from: (+)-Catechin
Dehydrodicatechin B$_1$
Formed from: (+)-Catechin
Dehydrodicatechin B$_2$
Formed from: (+)-Catechin
Dehydrodicatechin B$_3$
Formed from: (+)-Catechin
Dehydrodicatechin B$_4$
Formed from: (+)-Catechin
Dehydrodiconiferyl alcohol
Formed from: Coniferyl alcohol
→ 5-(2-Carboxyvinyl)-2,3-dihydro-2-(4-hydroxy-3-methoxyphenyl)-3-hydroxymethyl-7-methoxybenzofuran
In: Pseudomonas A3255, F77
→ Coniferyl alcohol
In: Pseudomonas A3255
See also: F77
→ 5-Formylconiferaldehyde
In: Pseudomonas F77
→ Vanillic acid
In: Pseudomonas F77
5,6-Dehydro-7,8-dihydro-dihydroxyyangonin
Probably formed from: 5,6-Dehydro-7,8-dihydro-*p*-hydroxyyangonin
5,6-Dehydrodihydro-*p*-hydroxykawain
Probably formed from: 5,6-Dehydro-*p*-hydroxykawain
4,5-Dehydro-2,9-dihydro-9-hydroxy-2-oxolisuride
Probably formed from: Lisuride
→ 4,5-Dehydro-desethyl-2,9-dihydro-9-hydroxy-2-oxolisuride
Probably in: Rhesus monkey D657
α,β-Dehydro-3,4-dihydroxyphenethylamine
Formed from: N-Acetyldopamine quinone methide
10,11-Dehydro-2,3-dihydroxytrimipramine
Probably formed from: 2,3-Dihydroxytrimipramine
Dehydrodiisoeugenol (4-(2,3-dihydro-7-methoxy-3-methyl-5-(1-propenyl)-2-benzofuranyl)-2-methoxyphenol)
→ 4-Hydroxy-3-methoxybenzaldehyde
In: Microorganism E375
→ Vanillic acid
In: Microorganism E375
See also: G275

4,5-Dehydro-N,N′-dimethylclonidine
Formed from: N,N′-Dimethylclonidine
Dehydrodivanillic acid (6,6′-dihydroxy-5,5′-dimethoxy(1,1′-biphenyl)-3,3′-dicarboxylic acid)
Probably formed from: Dehydrodivanillic semialdehyde
→ 5-Carboxyvanillic acid
Probably in: Microorganism D540
Dehydrodivanillic semialdehyde
Formed from: Dehydrodivanillin
→ Dehydrodivanillic acid
Probably in: Microorganism D540
Dehydrodivanillin (6,6′-dihydroxy-5,5′-dimethoxy(1,1′-biphenyl)-3,3′-dicarboxaldehyde)
→ 5-Carboxyvanillin
Probably in: Microorganism D540
→ Dehydrodivanillic semialdehyde
In: Microorganism D540
Dehydrodivanillyl alcohol (6,6′-dihydroxy-5,5′-dimethoxy(1,1′-biphenyl)-3,3′-dimethanol)
→ Protocatechuic acid
In: Acinetobacter G275
5,6-Dehydrodopazinol
Formed from: Dopazinol
9,11-Dehydro-16-epioestriol
Formed from: 9,11-Dehydrooestrone
→ 9,11-Dehydro-16-oxo-β-oestradiol
Probably in: Streptomyces F810
4,21-Dehydrogeissoschizine
Formed from: Geissoschizine
→ Ajmalicine
In: Catharanthus B371
→ 19-Epiajmalicine
In: Catharanthus B371
→ Tetrahydroalstonine
In: Catharanthus B371
6,7-Dehydrogliclazide
Probably formed from: 6α-Hydroxygliclazide
Dehydrogriseofulvin
Formed from: Griseophenone B
→ Griseofulvin
In: Streptomyces E302, E644
4,5-Dehydrohaloperidol
Formed from: 4,5-Dehydrohaloperidol-N-oxide
4,5-Dehydrohaloperidol-N-oxide
→ 4,5-Dehydrohaloperidol
In: Mouse G569
→ 4-Chlorophenyl-1,2,3,6-tetrahydropyridine
In: Mouse G569
10,11-Dehydro-3-hydroxyamitriptyline
Probably formed from: 3,11-Dihydroxyamitriptyline
3,4-Dehydro-2″-*cis*-hydroxycilostazol
Probably formed from: 2″-*cis*-Hydroxycilostazol
3,4-Dehydro-4″-*trans*-hydroxycilostazol
Probably formed from: 4″-*trans*-Hydroxycilostazol
De(hydrohydroxy)cimetropium
Formed from: Cimetropium
5,6-Dehydro-7,8-dihydro-*p*-hydroxyyangonin
Formed from: 7,8-dihydroyangonin
→ 5,6-Dehydro-7,8-dihydro-dihydroxyyangonin
Probably in: Rat A3485
5,6-Dehydro-*p*-hydroxykawain
Probably formed from: *p*-Hydroxykawain
→ 5,6-Dehydrodihydro-*p*-hydroxykawain
Probably in: Man F862
3,4-Dehydro-6-hydroxymellein
→ 3,4-Dehydro-6-methoxymellein
In: Carrot K730
10,11-Dehydro-2-hydroxynortrimipramine
Probably formed from: 2-Hydroxynortrimipramine
9,11-Dehydro-16α-hydroxyoestrone
Formed from: 9,11-Dehydrooestrone
6,7-Dehydrohyoscyamine
→ Scopolamine
In: Hyoscyamus E350
Dehydroindolidan
Formed from: Indolidan
1,4-Dehydroisradipine
Formed from: Isradipine
5,6-Dehydrokawain
Formed from: Kawain
2,3-Dehydrokievitone (2′-hydroxy-8-prenylgenistein)
→ 2,3-Dehydro-5-O-methylkievitone
In: Lupin E756, F182
→ 2,3-Dihydroxykievitone

In: Aspergillus E634
Botrytis E634
→ Dihydro-2′,4′,5,9-tetrahydroxy-8,8-dimethylfuranoisoflavone
In: Aspergillus E634
Botrytis E634
→ Dihydro-2′,4′,5-trihydroxy-8-(2-hydroxy-2-propyl)pyranoisoflavone
In: Aspergillus E634
Botrytis E634

1,4-Dehydrolacidipine
Probably formed from: Lacidipine
→ 4-(2-(2-Carboxy)ethenyl)phenyl-2,6-dimethyl-3,5-pyridinedicarboxylic acid diethyl ester
Probably in: Dog G371
→ 4-(2-(3-(1,1-Dimethoxy)-3-oxo-1-propenyl)phenyl)-2,6-dimethyl-3,5-pyridinedicarboxylic acid monoethyl ester
Probably in: Dog G371
→ 4-(2-(3-(1,1-Dimethoxy)-3-oxo-1-propenyl)phenyl)-2-hydroxymethyl-6-methyl-3,5-pyridinedicarboxylic acid diethyl ester
Probably in: Dog G371

5,6-Dehydrolevamisole
Formed from: Levamisole

3,4-Dehydro-6-methoxymellein
Formed from: 3,4-Dehydro-6-hydroxymellein

2,3-Dehydro-5-O-methylkievitone
Formed from: 2,3-Dehydrokievitone

1,4-Dehydromodipafant
Formed from: Modipafant
→ 5-Carboxy-4-(*o*-chlorophenyl)-3-ethoxycarbonyl-6-methyl-2-(4-(methylimidazo[4,5,*c*]pyrid-1-yl) phenyl)pyridine
Probably in: Dog H735
Rat H735
→ 4-(*o*-Chlorophenyl)-3-ethoxycarbonyl-5-(N-(3-hydroxy-2-pyridyl)carbamoyl)-6-methyl-2-(4-(methylimidazo[4,5,*c*]pyrid-1-yl)phenyl)pyridine
Probably in: Dog H735
Rat H735

Dehydronilvadipine
Formed from: Nilvadipine
→ 2-Cyano-5-(1-hydroxyisopropoxycarbonyl)-3-methoxycarbonyl-6-methyl-4-(*m*-nitrophenyl)pyridine
Probably in: Dog E818
Rat E818
→ Methyl 5-carboxy-2-cyano-6-methyl-4-(*m*-nitrophenyl)pyridine-3-carboxylate
Probably in: Rat F484

1,4-Dehydronitrendipine
Formed from: Nitrendipine
→ 4-(*m*-Aminophenyl)-2,6-dimethylpyridine-3,5-dicarboxylic acid 3-ethyl ester 3-ethyl ester
Probably in: Dog G398
Mouse G398
Rat G398
→ 2,6-Dimethyl-4-(*m*-nitrophenyl)pyridine-3,5-dicarboxylic acid 3-ethyl ester
Probably in: Dog G398
Mouse G398
Rat G398
→ 2,6-Dimethyl-4-(*m*-nitrophenyl)pyridine-3,5-dicarboxylic acid 3-methyl ester
Probably in: Dog G398
Mouse G398
Rat G398
→ 2-Hydroxymethyl-6-methyl-4-(*m*-nitrophenyl)pyridine-3,5-dicarboxylic acid 3-ethyl ester 5-methyl ester
Probably in: Dog G398
Mouse G398
Rat G398
→ 2-Hydroxymethyl-6-methyl-4-(*m*-nitrophenyl)pyridine-3,5-dicarboxylic acid 5-ethyl ester 3-methyl ester
Probably in: Dog G398
Mouse G398
Rat G398

9,11-Dehydrooestrone
Formed from: Oestrone
→ 9,11-Dehydro-16α-hydroxyoestrone
In: Streptomyces F810
→ 9,11-Dehydro-16-epioestriol
In: Streptomyces F810

9,11-Dehydro-16-oxo-β-oestradiol
Probably formed from: 9,11-Dehydro-16-epioestriol

N-(1,2-Dehydro-5-oxopyrrolidino-2-methyl)-α-hydroxy-α,α-diphenylacetamide
Probably formed from: α-Hydroxy-N-(5-oxopyrrolidino-2-methyl)-α,α-diphenylacetamide

1,4-Dehydropranidipine
Formed from: Pranidipine
→ 2,6-Dimethyl-4-(*m*-nitrophenyl)pyridine-3,5-dicarboxylic acid 3-methyl ester
In: Man J606

1,4-Dehydroproterguride
Formed from: Proterguride

Dehydroreticuline
Formed from: Reticuline

1,2-Dehydroreticulinium
Probably formed from: Reticuline
→ (*R*)-Reticuline
In: Papaver F817, K771

Dehydroscoulerine
Formed from: Reticuline
→ Scoulerine
In: Corydalis G136

4,5-Dehydrosertindole
Formed from: Sertindole

Dehydroshikimic acid
→ Catechol
In: Escherichia G366

Dehydrosinactine
→ Sinactine
In: Corydalis G136

Dehydrosilybin
Probably formed from: Silybin
→ Dehydrosilybin-β-D-glucuronide
Probably in: Rat A2276
→ Dehydrosilybin sulphate
Probably in: Rat A2276

Dehydrosilybin-β-D-glucuronide
Probably formed from: Dehydrosilybin

Dehydrosilybin sulphate
Probably formed from: Dehydrosilybin

4′,5′-Dehydrosirdalud
Formed from: Sirdalud

12-Dehydrotetracycline
Formed from: Anhydrotetracycline
L-Tryptophan

11-N-Dehydrotianeptine
Formed from: Anhydrotetracycline

Dehydroticlopidine
Formed from: Ticlopidine

Dehydrotremetone
Formed from: Tremetone

α,β-Dehydrotryptophan
Formed from: L-Tryptophan

α,β-Dehydrotryptophanamide
Formed from: L-Tryptophanamide

3′,4′-Dehydrovinblastine
→ Catharine
In: Catharanthus B876, E701
→ Leurosine
In: Catharanthus B876, E701

Dehydrowarfarin
Formed from: (*R*)-Warfarin
(*S*)-Warfarin

De(hydroxymethyl)cimetropium
Formed from: Cimetropium

Dehydrozapetine
Formed from: Azapetine
→ Oxazapetine
In: Rabbit A3909
Rat A3909

Deisopropylbidisomide
Formed from: Bidisomide
→ Deisopropylhydroxybidisomide
Probably in: Rat K115

Deisopropyldelaviridine
Formed from: Delaviridine
→ 4-Hydroxydeisopropyldelaviridine
Probably in: Monkey J275

Deisopropyldinorfluvastatin
Probably formed from: Deoxyfluvastatin
→ Deisopropyltetranorfluvastatin
Probably in: Man G585

Deisopropyldisopyramide
Formed from: Disopyramide
→ Bisdeisopropyldisopyramide
Probably in: Man B984

Deisopropylnorfluvastatin
Formed from: Fluvastatin
→ Deisopropyldinorfluvastatin
Probably in: Man G585

Deisopropylhydroxybidisomide
Probably formed from: Deisopropylbidisomide
N-Deisopropylisofenphos (aminoisofenphos)
Formed from: Isofenphos
Deisopropylnilvaldipine
Formed from: Nilvaldipine
→ 5-Carboxy-2-cyano-3-methoxycarbonyl-6-methyl-4-(*m*-nitrophenyl)pyridine
Probably in: Dog E818
Rat E818
Deisopropylrecainam
Formed from: Recainam
Deisopropyltetranorfluvastatin
Probably formed from: Deoxydinorfluvastatin
De(isopropylthiazolylmethyl)ritonavir
Formed from: Ritonavir
Deisopropyltolterone
Formed from: Tolterone
→ Bisdeisopropyltolterone
In: Rat J419
→ Deisopropyltolterone-β-D-glucuronide
Probably in: Dog J419
Mouse J419
→ 3-(2-Hydroxy-5-hydroxymethylphenyl)-N-isopropyl-3-phenylpropylamine
Probably in: Dog J419
Man J375
Mouse J419
Deisopropyltolterone-β-D-glucuronide
Probably formed from: Deisopropyltolterone
Delaviridine (1-(3-((1-methylethyl)amino)-2-pyridinyl)-4-((5-((methylsulphonyl)amino)-1*H*-indol-2-yl)carbonyl)piperazine)
→ N-(β-N-Acetylglucosaminido)delaviridine
In: Monkey J275
Mouse J277
→ 2-Carboxy-5-(S-methylsulphonamido) indole
In: Monkey J275
Mouse J277
→ Deisopropyldelaviridine
Probably in: Dog J424
Man J424
Monkey J275, J424
Mouse J277
Rat J424
→ 4-Hydroxydelaviridine
Probably in: Monkey J275
→ 4′-Hydroxydelaviridine
Probably in: Monkey J275
→ 6′-Hydroxydelaviridine
Probably in: Monkey J275
Mouse J277
Rat J96
→ 1-((5-((Methylsulphonyl)amino)-1*H*-indol-2-yl)carbonyl)piperazine
In: Rat J96
Delphinidin (3,3′,4,5,5′,7-hexahydroxyflavylium)
Formed from: Cyanidin
Dihydromyricetin
→ Delphinidin-3-glucoside
In: Carrot E280
Petunia A3191
Silene B407
See also: A2739
→ Petunidin
Probably in: Petunia A2740
Delphinidin-5-caffeoylglucoside-3-glucoside
Formed from: Delphinidin-3,5-diglucoside
Delphinidin-3-(*p*-coumaroyl)rutinoside
→ Delphinidin-3-(*p*-coumaroyl)rutinoside-5-O-glucoside
In: Petunia D198
Delphinidin-3-(*p*-coumaroyl)rutinoside-5-O-glucoside
Formed from: Delphinidin-3-(*p*-coumaroyl)rutinoside
Delphinidin-3,5-diglucoside
→ Delphinidin-5-caffeoylglucoside-3-glucoside
In: Gentiana K788
Delphinidin-3-glucoside
Formed from: Delphinidin
→ Delphinidin-3-rhamnosylglucoside
In: Silene B409
Delphinidin-3-rhamnoside-5-glucoside
Probably formed from: Petunidin-3-rhamnoside-5-glucoside
→ Delphinidin-3-sambubioside-5-glucoside
Probably in: Pea A194
→ Delphinidin-3-sophoroside-5-glucoside
Probably in: Pea A194

Delphinidin-3-rhamnosylglucoside
Formed from: Delphinidin-3-glucoside
Delphinidin-3-sambubioside-5-glucoside
Probably formed from: Delphinidin-3-rhamnoside-5-glucoside
Delphinidin-3-sophoroside-5-glucoside
Probably formed from: Delphinidin-3-rhamnoside-5-glucoside
Deltamethrin (3-(2,2-dibromoethinyl)-2,2-dimethylcyclopropanecarboxylic acid (*S*)-cyano(3-phenoxyphenyl)methyl ester)
Formed from: Tralomethrin
→ *m*-Phenoxybenzaldehyde
Probably in: Hen F632
→ *m*-Phenoxybenzoic acid
In: Hen E697, H207
Rat C412
→ *m*-Phenoxybenzylalcohol
In: Hen E697
Rat C412
→ *m*-Phenoxymandelonitrile
In: Bacillus G891
Demecolcine (N-deacetyl-N-methylcolchicine)
Probably formed from: O-Methylandrocymbine
→ 2-Demethyldemecolcine
In: Rat E956
→ 3-Demethyldemecolcine
In: Rat E956
3‴-Demethoxycelosianin II
Formed from: Amaranthin
p-Coumaroyl-D-glucose
4-Demethoxydaunorubicin
→ 4-Demethoxydaunorubicinol
In: Mouse E160
Rabbit E160
4-Demethoxydaunorubicinol
Formed from: 4-Demethoxydaunorubicin
4-Demethoxydaunorubicinol aglycone
Probably formed from: Daunorubicinol aglycone
7-Demethoxyencecalin (1-(2,2-dimethyl-2*H*-1-benzopyran-6-yl)ethanone)
→ 6-Acetyl-2-hydroxymethyl-2-methyl-2*H*-benzo[*b*]pyran
In: Melanopus E491
→ 6-(1-Hydroxyethyl)-2,2-dimethyl-2*H*-benzo[*b*]pyran
In: Melanopus E491
5-Demethoxyubiquinone 30
Probably formed from: 2-Hexaprenyl-6-methoxyphenol
→ Ubiquinone 30
Probably in: Saccharomyces A3353
N-Demethylaclacinomycin A
Formed from: Aclacinomycin A
N-Demethylacronine
Formed from: Acronine
O-Demethylacronine
Formed from: Acronine
Demethyladinazolam
Formed from: Adinazolam
Demethylaflatoxin M_1
Probably formed from: Aflatoxin M_1
Demethylalfentanil
Formed from: Alfentanil
Demethylamiflamine
Formed from: Amiflamine
→ Bisdemethylamiflamine
In: Man D287
N^2-Demethylaminopyrine
Formed from: Aminopyrine
Demethylamlodipine
Formed from: Amlodipine
N-Demethylantipyrine
See: 3-Methyl-1-phenyl-3-pyrazolin-5-one
O-Demethylarctigenin
Probably formed from: Arctigenin
→ Arctigenin
In: Rat H48
Demethylastemizole
Formed from: Astemizole
→ *p*-Hydroxyphenylacetic acid
Probably in: Dog E425
Man E425
Rat E425
Demethylazelastine
Formed from: Azelastine
Demethylazonafide
Formed from: Azonafide
→ Bisdemethylazonafide
Probably in: Rat H935

Demethylbambuterol
Formed from: Bambuterol
Demethylbenztropine
Formed from: Benztropine
→ *p*-Hydroxydemethylbenztropine
Probably in: Rat H312
Demethylbenzydamine
Formed from: Benzydamine
→ Bisdemethylbenzydamine
In: Man D678
O-Demethylbosentan
Formed from: Bosentan
Demethylbromazepine
Formed from: Bromazepine
→ Bisdemethylbromazepine
In: Man B556
Demethylbrompheniramine
Formed from: Brompheniramine
9-Demethylbrotizolam
Probably formed from: Brotizolam
Demethylcarbazeran
Formed from: Carbazeran
Demethylcarvedilol
Formed from: Carvedilol
→ Demethylcarvedilol-β-D-glucuronide
In: Dog J443
→ Demethylcarvedilol-O-sulphate
In: Dog J443
Mouse J443
Demethylcarvedilol-β-D-glucuronide
Formed from: Demethylcarvedilol
Demethylcarvedilol-O-sulphate
Formed from: Demethylcarvedilol
Demethylcelosianin II
Formed from: Amaranthin
Caffeoyl-D-glucose
Demethylcentchroman
Formed from: Centchroman
Demethylcerivastatin
Formed from: Cerivastatin
→ (*E*)-(3*R*,5*S*)-7-(4-(*p*-Fluorophenyl)-5-hydroxymethyl-2,6-diisopropylpyrid-3-yl)-2,3,5-trihydroxyhept-6-enoic acid
Probably in: Dog J426
Mouse J426
Rat J426
Demethylcerivastatin lactone
Probably formed from: Cerivastatin lactone
Demethylchlordimeform (N′-(4-chloro-*o*-tolyl)-N-methylformamidine)
Formed from: Chlordimeform
→ 4-Chloro-2-methylformanilide
In: Cabbage looper A1838
Rat A1821
Spider mite A1944
See also: A779, A2047, A2534
→ N′-(4-Chloro-*o*-tolyl)formamidine
In: Cabbage looper A1838
Mouse A2534
Spider mite A1944
See also: A1821
Demethylchlorotrianisene
Formed from: Chlorotrianisene
→ Bisdemethylchlorotrianisene
Probably in: Rabbit A3466
Demethylchlorpheniramine
Formed from: Chlorpheniramine
→ Bisdemethylchlorpheniramine
In: Dog A3931
Man F902
Rabbit E649, E805
Rat F902
→ 3-(*p*-Chlorophenyl)-N-methyl-3-(2-pyridyl)propionamide
Probably in: Rabbit E649
Demethylchlorpromazine
Formed from: Chlorpromazine
N-Hydroxydemethylchlorpromazine
→ Bisdemethylchlorpromazine
In: Guinea pig A901
Man A693
Rabbit A901
Rat A69, J829
See also: A1754, A2881, C619, D119, D860
→ 10-Carboxyethyl-2-chlorophenothiazine
In: Rabbit E822
→ Demethylchlorpromazine sulphoxide
Probably in: Guinea pig A901
Horse A901
Man A693, A901, A1059
Rat F317, G27
Sealion A901

→ 7-Hydroxydemethylchlorpromazine
Probably in: Guinea pig A901
Horse A901
Man A901, A1059, F480
Sealion A901

Demethylchlorpromazine sulphoxide
Probably formed from: Chlorpromazine-N,S-dioxide
Demethylchlorpromazine

Demethylcilostamide
Formed from: Cilostamide
→ 2′-Hydroxydemethylcilostamide
Probably in: Rat D165
→ 4′-Hydroxydemethylcilostamide
Probably in: Rat D165

Demethylcimetropium
Formed from: Cimetropium

Demethylcitalopram
Formed from: Citalopram
→ Bisdemethylcitalopram
In: Man J758, K301

Demethylclobazam
Formed from: Clobazam
→ 3′,4′-Dihydro-3′,4′-dihydroxydemethylclobazam
Probably in: Dog B22
Man B22
Monkey B22
Rat B22

Demethylclozapine
Formed from: Clozapine
→ Demethylclozapine-N-β-D-glucuronide
In: Man H541
→ 7-Hydroxydemethylclozapine
Probably in: Man J153
Rat H702

Demethylclozapine-N-β-D-glucuronide
Formed from: Demethylclozapine

Demethylcodorphone
Formed from: Codorphone

O²-Demethylcolchicine
Formed from: Colchicine
→ O²-Demethylcolchicine-β-D-glucuronide
In: Rat A1540

O³-Demethylcolchicine
Formed from: Colchicine
→ O³-Demethylcolchicine-β-D-glucuronide
In: Rat A1540

O¹⁰-Demethylcolchicine
Formed from: Colchicine

O²-Demethylcolchicine-β-D-glucuronide
Formed from: O²-Demethylcolchicine

O³-Demethylcolchicine-β-D-glucuronide
Formed from: O³-Demethylcolchicine

Demethylcyanox
Formed from: Cyanox
→ Demethylcyanoxon
Probably in: Rat H461

Demethylcyanoxon
Probably formed from: Cyanoxon
Demethylcyanox

Demethylcyclindole
Formed from: Cyclindole
→ 7-Hydroxydemethylcyclindole
Probably in: Dog A2083
Man A2083
Monkey A2083
Rat A2083

Demethylcyclobenzaprine
Formed from: Cyclobenzaprine
→ Demethylcyclobenzaprine-β-D-glucuronide
Probably in: Dog A2887

Demethylcyclobenzaprine-β-D-glucuronide
Probably formed from: Demethylcyclobenzaprine

Demethylcyproheptadine
Formed from: Cyproheptadine
→ 2-Hydroxydemethylcyproheptadine
Probably in: Cunninghamella J90
Man A1879
→ Demethylcyproheptadine-10,11-oxide
Probably in: Cat A1858
Dog A1858
Mouse A1574
Rat A1574, A2247, E885
→ 10-Oxodemethylcyproheptadine
Probably in: Dog A1858

Demethylcyproheptadine-10,11-oxide
Formed from: Demethylcyproheptadine

2-Demethyldemecolcine
Formed from: Demecolcine

3-Demethyldemecolcine
Formed from: Demecolcine

Demethyldeoxyadriamycinol aglycone
Probably formed from: Deoxyadriamycinol aglycone
→ Demethyldeoxyadriamycinol aglycone-β-D-glucuronide
Probably in: Man A1989
→ Demethyldeoxyadriamycinol aglycone sulphate
Probably in: Man A1989

Demethyldeoxyadriamycinol aglycone-β-D-glucuronide
Probably formed from: Demethyldeoxy-adriamycinol aglycone

Demethyldeoxyadriamycinol aglycone sulphate
Probably formed from: Demethyldeoxyadria-mycinol aglycone

Demethyldeoxydaunorubicinol aglycone
Probably formed from: Deoxydaunorubicinol aglycone
→ Demethyldeoxydaunorubicinol aglycone-4-β-D-glucuronide
Probably in: Man A37, A138, C647
→ Demethyldeoxydaunorubicinol aglycone-4-sulphate
Probably in: Man A37, A138

Demethyldeoxydaunorubicinol aglycone-4-β-D-glucuronide
Probably formed from: Demethyldeoxyd-aunorubicinol aglycone

Demethyldeoxydaunorubicinol aglycone-4-sulphate
Probably formed from: Demethyldeoxyd-aunorubicinol aglycone

N-Demethyldeoxynogarol
Formed from: N-Demethylmenogaril

Demethyldeprenyl
Formed from: Deprenyl
→ *l*-Amphetamine
Probably in: Mouse K450

4′-Demethyldeoxypodophyllotoxin
Formed from: Matairesinol
→ Demethylpodophyllotoxin
In: Podophyllum D274
See also: G201
→ α-Peltatin
In: Podophyllum E236
See also: G201

Demethylderamciclane
Formed from: Deramciclane

N-Demethyldextromethorphan (dextrorphan)
Formed from: Dextromethorphan
→ N-Acetyl-N-demethyldextromethorphan
Probably in: Man F956
→ 6-Hydroxy-N-demethyldextromethorphan
Probably in: Man F956

O-Demethyldextromethorphan
Formed from: Dextromethorphan
→ O-Acetyl-O-demethyldextromethorphan
Probably in: Man F956
→ 6,7-Dehydro-O-demethyldextromethorphan
Probably in: Man F956
→ 3-Hydroxymorphinan
Probably in: Man F956, H865
Rat G934

Demethyldihydrocodorphone
Probably formed from: Dihydrocodorphone

N-Demethyldiltiazem
Formed from: Diltiazem
→ N,O-Bisdemethyldiltiazem
Probably in: Man D126
Rat H377
→ Deacetyl-N-demethyldiltiazem
Probably in: Man D126
Rabbit F569

O-Demethyldiltiazem
Formed from: Diltiazem
→ Deacetyl-O-demethyldiltiazem
Probably in: Man D126
Rat H377

Demethyldimethylvinphos
Formed from: Dimethylvinphos

Demethyldimetindene
Formed from: Dimetindene

N-Demethyl-9,9-dioxopromethazine
→ Hydroxy-9,9-dioxo-N-(2-oxopropyl) phenothiazine
In: Rat A2685

Demethyldithiaden
Formed from: Dithiaden

Demethyldithiaden sulphoxide
Probably formed from: Dithiaden sulphoxide

Demethyl-DL-8280
Formed from: DL-8280

Demethyldofetilide
Formed from: Dofetilide

5-O-Demethyldonepezil
Formed from: Donepezil
→ 5,6-Bis-O-demethyldonepezil
Probably in: Rat J789
→ 5-O-Demethyldonepezil-β-D-glucuronide
In: Rat J789

6-O-Demethyldonepezil
Formed from: Donepezil
→ 6-O-Demethyldonepezil-β-D-glucuronide
In: Rat J789

5-O-Demethyldonepezil-β-D-glucuronide
Formed from: 5-O-Demethyldonepezil

6-O-Demethyldonepezil-β-D-glucuronide
Formed from: 6-O-Demethyldonepezil

Demethyldothiepin
See: Northiaden

Demethyldoxepin
Formed from: Doxepin
→ N-Acetyl-N-demethyldoxepin
Probably in: Cunninghamella J782
→ Bisdemethyldoxepin
Probably in: Cunninghamella J782
→ 2-Hydroxydemethyldoxepin
Probably in: Cunninghamella J782
Man F570
→ 3-Hydroxydemethyldoxepin
Probably in: Cunninghamella J782
→ 4-Hydroxydemethyldoxepin
Probably in: Cunninghamella J782
→ 8-Hydroxydemethyldoxepin
Probably in: Cunninghamella J782

N-Demethyldoxycycline
Formed from: Doxycycline

Demethyldoxylamine
Formed from: Doxylamine
→ N-Acetyldemethyldoxylamine
In: Man E785
Rat E11
→ Bisdemethyldoxylamine
In: Man B909, E785
Rat E11, F684
Rhesus monkey E92
Squirrel monkey E11
→ 'Demethyldoxylamine-N-oxide' (identity in doubt)
Probably in: Rat E424
→ *p*-Hydroxydemethyldoxylamine
Probably in: Rat E11, E438
Squirrel monkey E11

'Demethyldoxylamine-N-oxide'
Probably formed from: Demethyldoxylamine

Demethyldroloxifene
Formed from: Droloxifene
→ Demethyldroloxifene-β-D-glucuronide
Probably in: Cynomolgus monkey H924
Mouse H924
→ *trans*-1-(3,4-Dihydroxyphenyl)-1-(*p*-(2-methylaminoethoxy)phenyl)-2-phenylbut-1-ene
Probably in: Cynomolgus monkey H924

Demethyldroloxifene-β-D-glucuronide
Probably formed from: Demethyldroloxifene

Demethylencecalin
Formed from: Encecalin
→ 7-Hydroxy-6-(1-hydroxyethyl)-2,2-dimethyl-2*H*-benzo[*b*]pyran
In: Melanopus E491

2′-Demethylenciprazine
Formed from: Enciprazine

3-Demethylenciprazine
Formed from: Enciprazine

4-Demethylenciprazine
Formed from: Enciprazine

4,5-Demethylenegomisin
Formed from: Gomisin A

3′-Demethyletoposide
Formed from: Etoposide

Demethylfenitrooxon
Probably formed from: Fenitrooxon

Demethylfenitrothion
Formed from: Fenitrothion

Demethylfleroxacin
Formed from: Fleroxacin
→ Demethyl-3-oxofleroxacin
Probably in: Dog G115
Man G115
Rabbit G115

Demethylfluoxetine
Formed from: Fluoxetine

Demethylfluphenazine sulphoxide
Formed from: Fluphenazine
Demethylformoterol
Formed from: Formoterol
→ Demethylformoterol-β-D-glucuronide
Probably in: Man J779
Demethylformoterol-β-D-glucuronide
Probably formed from: Demethylformoterol
N-Demethyl-N-formyldoxepin
Formed from: Doxepin
N-Demethyl-N-formylmesembrinone
Formed from: Mesembrinone
N-Demethyl-N-formylmianserin
Formed from: Mianserin
2-O-Demethylglaucine
Formed from: Glaucine
2-O-Demethylgomisin A
Formed from: Gomisin A
14-O-Demethylgomisin A
Formed from: Gomisin A
Demethylgossic acid (1,3-dihydro-5,6-dihydroxy-7-(1-methylethyl)-1,3-dioxo-4-isobenzofurancarboxylic acid)
Probably formed from: Gossypolonic acid
1-Demethylgranisetron
Formed from: Granisetron
9′-Demethylgranisetron
Formed from: Granisetron
4-Demethylgriseofulvin
Formed from: Griseofulvin
6-Demethylgriseofulvin
Formed from: Griseofulvin
Demethylhomopterocarpin
See: Medicarpin
Demethylhydroxy
See entries under Hydroxydemethyl
Demethylidebenone
Formed from: Idebenone
→ 2-Hydroxy-6-(10-hydroxydecyl)-3-methoxy-5-methylquinol
Probably in: Dog E119
Rat E119
Demethyliloperidone
Formed from: Iloperidone
→ 1-(4-(3-(4-(6-Fluoro-1,2-benzisoazol-3-yl)-1-piperidinyl)propoxy)-3,6-dihydroxyphenyl)ethanone
Probably in: Rat H447
Demethylimipramine (DMI)
Formed from: N-Hydroxydemethyl-imipramine
Imipramine
Imipramine-N-oxide
Lofepramine
→ Bisdemethylimipramine
In: Glycine K359
→ N-(2-Carboxyethyl)iminodibenzyl
In: Rabbit C476
See also: C496
→ Demethylimipramine-N-β-D-glucuronide
Probably in: Dog A979
→ Demethylimipramine-N-sulphate
In: Guinea pig E456
Rabbit J914
Rat E45, E51, H511
→ 2-Hydroxydemethylimipramine
In: Man A1086
See also: A316, A356, A1001, A2029, A3668, A3865, C496, C872, D431, G711, H244, H551
→ 10-Hydroxydemethylimipramine
Probably in: Rabbit C496
→ N-Hydroxydemethylimipramine
In: Guinea pig A888
Rabbit A888, C496
→ Iminodibenzyl
In: Aspergillus A2409
See also: A1001, A2029, A2165, D431
→ Imipramine
In: Man C467, D431
Pig K923
Rabbit E6
Demethylimipramine-N-β-D-glucuronide
Probably formed from: Demethylimipramine
Demethylimipramine-N-sulphate
Formed from: Demethylimipramine
Demethylimipramin-2-yl-β-D-glucuronide
Formed from: 2-Hydroxydemethylimipramine
Demethylindometacinamide-β-D-glucuronide
Formed from: Oxametacin
Demethyliomeglamic acid
Formed from: Iomeglamic acid
Demethyliomeglamic acid methyl ester
→ Probably in: Man A683

Rat A1475

Demethyliomeglamic acid methyl ester
Probably formed from: Demethyliomeglamic acid

Demethyliprindole
Formed from: Iprindole
→ Hydroxydemethyliprindole (alicyclic)
In: Dog A3712
Monkey A3712
Pig A3712
Rat A3712, G23
→ Hydroxydemethyliprindole (aryl)
In: Rat G23

N-Demethylisochlortetracycline
Probably formed from: Isochlortetracycline

3′-Demethylisopapaverine (4-((3-hydroxy-4-methoxyphenyl)methyl)-6,7-dimethoxyisoquinoline)
Formed from: Isopapaverine

4′-Demethylisopapaverine (4-((4-hydroxy-3-methoxyphenyl)methyl)-6,7-dimethoxyisoquinoline)
Formed from: Isopapaverine

6-Demethylisopapaverine (4-((3,4-dimethoxyphenyl)methyl)-6-hydroxy-7-methoxyisoquinoline)
Formed from: Isopapaverine

7-Demethylisopapaverine (4-((3,4-dimethoxyphenyl)methyl)-7-hydroxy-6-methoxyisoquinoline)
Formed from: Isopapaverine

N-Demethylisopropylantipyrine-β-D-glucuronide
Formed from: 4-Isopropyl-3-methyl-1-phenyl-3-pyrazolin-5-one

N-Demethylketamine (2-amino-2-(*o*-chlorophenyl)cyclohexanone)
Formed from: Ketamine
→ 2-Amino-2-(*o*-chlorophenyl)cyclohex-3-en-1-one
Probably in: Man A423
→ 2-Amino-2-(*o*-chlorophenyl)cyclohex-5-en-1-one
Probably in: Dog B565
Man A3297
Monkey A3297

N-Demethyllevomepromazine
Formed from: Levomepromazine
→ 3-Hydroxy-N-demethyllevomepromazine
Probably in: Man H598
→ 7-Hydroxy-N-demethyllevomepromazine
Probably in: Man H598

O-Demethyllevomepromazine
Formed from: Levomepromazine
→ N,O-Bisdemethyllevomepromazine
Probably in: Man H598
→ 3-Hydroxy-O-demethyllevomepromazine
Probably in: Man H598
→ 7-Hydroxy-O-demethyllevomepromazine
Probably in: Man H598

Demethyllevormeloxifene
Formed from: Levormeloxifene
→ C,O-Bisdemethyllevormeloxifene
Probably in: Rat K31
→ Demethyllevormeloxifene-β-D-glucuronide
In: Rat K31

Demethyllevormeloxifene-β-D-glucuronide
Formed from: Demethyllevormeloxifene

5′-Demethyllevosemotiadil
Formed from: Levosemotiadil
→ 4,5′-Bisdemethyllevosemotiadil
Probably in: Rat J130

Demethyllilopristone
Formed from: Demethyllilopristone

Demethyllisuride
Formed from: Lisuride

Demethyllofepramine
Probably formed from: Lofepramine

Demethylloperamide
Formed from: Loperamide
→ Bisdemethylloperamide
In: Rat A3382, B275, C253
→ Hydroxydemethylloperamide
Probably in: Rat B275

Demethyllortalamine
Formed from: Lortalamine

Demethylloxapine
Formed from: Loxapine

Demethylmaprotiline
Formed from: Maprotiline
→ 1′-Hydroxydemethylmaprotiline
Probably in: Mycobacterium D167
→ 2-Hydroxydemethylmaprotiline
Probably in: Mycobacterium D167

→ 3-Hydroxydemethylmaprotiline
Probably in: Mycobacterium D167

***p*-O-Demethylmecinarone**
Formed from: Mecinarone

O-Demethylmeconine
Formed from: Narcotine

Demethylmedicarpin
Formed from: Medicarpin

N-Demethylmenogaril
Formed from: Menogaril
→ N-Demethyldeoxymenogarol
Probably in: Rabbit D897

N-Demethylmenogarol
Formed from: Menogarol

N-Demethylmeptazinol
Formed from: Meptazinol

Demethylmesoridazine
Probably formed from: Mesoridazine

4′-O-Demethylmesembrenol
Formed from: Ferulic acid

4′-O-Demethylmesembrenone
→ Mesembrenol
In: Sceletium A3136

1-Demethylmetergoline
Formed from: Metergoline

Demethylmethindione
Formed from: Methindione

Demethylmethocarbamol
Formed from: Methocarbamol
→ Demethylmethocarbamol-β-D-glucuronide
Probably in: Rat A360

Demethylmethocarbamol-β-D-glucuronide
Formed from: Demethylmethocarbamol

Demethylmianserin
Formed from: Mianserin
→ Demethylmianserin-N-β-D-glucuronide
Probably in: Guinea pig G66
Man G864
→ Demethylmianserin-N-sulphate
Probably in: Guinea pig G66
Man G66, G864
Rat G66, G864
→ 8-Hydroxydemethylmianserin
Probably in: Guinea pig G66
Man G66, G864
Mouse G66
Rat G66, G864, G908, J784

Demethylmianserin-N-β-D-glucuronide
Probably formed from: Demethylmianserin

Demethylmianserin-N-sulphate
Probably formed from: Demethylmianserin

N-Demethylmibefradil
Formed from: Mibefradil
→ 2-(2-(3-(1*H*-Benzimidazol-2-yl)propyl)amino)ethyl-6-fluoro-1,2,3,4-tetrahydro-2-hydroxy-1-isopropylnaphthalene
Probably in: Rat G311
→ Methoxyacetic acid 6-fluoro-2-(2-(3-(4-β-D-glucuronosyloxyl-1*H*-benzimidazol-2-yl)propyl)amino)ethyl-1,2,3,4-tetrahydro-1-isopropyl-2-naphthalenyl ester
Probably in: Rat G311
→ Methoxyacetic acid 6-fluoro-1,2,3,4-tetrahydro-2-(2-(3-(5-hydroxy-1*H*-benzimidazol-2-yl)propyl)amino)ethyl-1-isopropyl-2-naphthalenyl ester
Probably in: Rat G311

O-Demethylmibefradil
Formed from: Mibefradil

Demethylmifepristone
Formed from: Mifepristone
→ Bisdemethylmifepristone
In: Man J752
Monkey J752
Rat J752

Demethylminocycline
Probably formed from: Minocycline

Demethylmirtazepine
Formed from: Mirtazepine

Demethylmitoclomine
Formed from: Mitoclomine

Demethylmociprazine
Formed from: Mociprazine

Demethylnefopam
Formed from: Nefopam

6″-Demethylnefopam
Formed from pentanoic acid in Aspergillus H675

Demethylnortriptyline
Formed from: Nortriptyline
→ 10-Hydroxydemethylnortriptyline
Probably in: Man A1489, A3913, E894
Rat E892

Demethylofloxacin
Formed from: Ofloxacin

4′-Demethylolanzapine
Formed from: Olanzapine

Demethylonapristone
Formed from: Onapristone

Demethylondansetron
Formed from: Ondansetron
→ 7-Hydroxydemethylondansetron
In: Rat H732

Demethyloxaprotiline
Formed from: Oxaprotiline
→ Demethyloxaprotiline-2′-β-D-glucuronide
Probably in: Dog E48
Rat E48
→ 3-Hydroxydemethyloxaprotiline
Probably in: Dog E48
Rat E48

Demethyloxaprotiline-2′-β-D-glucuronide
Probably formed from: Demethyloxaprotiline

Demethyl-3-oxofleroxacin
Probably formed from: Demethylfleroxacin

Demethyloxoiprindole
Probably formed from: Hydroxydemethyliprindole

N-Demethyloxytetracycline
Formed from: Oxytetracycline

3′-Demethylpapaverine
See: Palaudine

4′-Demethylpapaverine (1-((4-hydroxy-3-methoxyphenyl)methyl)-6,7-dimethoxyisoquinoline)
Formed from: Papaverine
→ 4′,6-Bisdemethylpapaverine
In: Rat A305
See also: A309, A3051

6-Demethylpapaverine (1-((3,4-dimethoxyphenyl)methyl)-6-hydroxy-7-methoxyisoquinoline)
Formed from: Papaverine
→ 4′,6-Bisdemethylpapaverine
In: Rat A305
See also: A309

7-Demethylpapaverine (1-((3,4-dimethoxyphenyl) methyl)-7-hydroxy-6 methoxyisoquinoline)
Formed from: Papaverine

Demethylpargyline
Formed from: Pargyline
→ Benzylamine
Probably in: Cunninghamella C68
→ Hydroxydemethylpargyline
Probably in: Rat A3005
→ N-Hydroxy-N-propargylbenzylamine
Probably in: Cunninghamella C68

Demethylperazine
Formed from: Perazine
→ Bisdemethylpromazine
Probably in: Rat A69

Demethylpheniramine
Formed from: Pheniramine

Demethylpicenadol
Formed from: Picenadol

O-Demethylpimobendan
Formed from: Pimobendan
→ O-Demethylpimobendan-N-β-D-glucuronide
Probably in: Man H311
→ O-Demethylpimobendan-O-β-D-glucuronide
In: Man H311

O-Demethylpimobendan-N-β-D-glucuronide
Probably formed from: O-Demethylpimobendan

O-Demethylpimobendan-O-β-D-glucuronide
Formed from: O-Demethylpimobendan

2′-Demethylpiritrexim
Formed from: Piritrexim
→ 2′-Demethylpiritrexim-O-sulphate
In: Dog G44

5′-Demethylpiritrexim
Formed from: Piritrexim
→ 5′-Demethylpiritrexim-β-D-glucuronide
In: Dog G44

5′-Demethylpiritrexim-β-D-glucuronide
Formed from: 5′-Demethylpiritrexim

2′-Demethylpiritrexim-O-sulphate
Formed from: 2′-Demethylpiritrexim

2-Demethylplastoquinone-9
Formed from: Homogentisic caid

4′-Demethylpodophyllotoxin
Formed from: 4′-Demethyldeoxypodophyllotoxin
Ferulic acid
→ Podophyllotoxin

Probably in: Diphylleia G201
Podophyllum G201
Contrast: in Podophyllum D274

Demethylpisatin
See: 3,6a-Dihydroxy-8,9-methylenepterocarpan

6-O-Demethylprazocin
Formed from: Prazocin

7-O-Demethylprazocin
Formed from: Prazocin

Demethylpromazine
Formed from: N-Hydroxynorpromazine
Promazine
→ 10-(2-Carboxyethyl)phenothiazine
In: Rabbit E822
→ Bisdemethylpromazine
Probably in: Rat A69

Demethylpropiverine
Formed from: Propiverine

Demethylpropyphenazone
Formed from: Propyphenazone
→ Demethylpropyphenazone *enol*-β-D-glucuronide
Probably in: Man B10

Demethylpropyphenazone *enol*-β-D-glucuronide
Probably formed from: Demethylpropyphenazone

Demethylprotizinic acid
Formed from: Protizinic acid

O-Demethylpuromycin
N-Acetyl-O-demethylpuromycin
In: Streptomyces D879
→ Puromycin
In: Streptomyces D879

N-Demethylpyrilamine
Formed from: Pyrilamine
→ N,O-Bisdemethylpyrilamine
Probably in: Aspergillus E882
Candida E882
Debaryomyces E882
Mucor E882
Penicillium E882
Saccharomyces E882
Syncephalastrum E882

O-Demethylpyrilamine (*p*-hydroxytripelennamine)
Formed from: Pyrilamine
Tripelennamine
→ N,O-Bisdemethylpyrilamine
Probably in: Rat E907, E930
→ O-Demethylpyrilamine-β-D-glucuronide
In: Man G933
Monkey H100
Mouse H100
Rat E905, H100
→ N-(*p*-Hydroxybenzyl)-N-(2-(5-hydroxypyridyl))-N′,N′-dimethylethylenediamine
Probably in: Rat A379
→ 3-Hydroxy-O-demethylpyrilamine
Probably in: Rat E907, E930
→ N-(3-Hydroxy-4-methoxybenzyl)-N,N′-dimethyl-N-(2-pyridyl)ethylenediamine
Probably in: Man G409
→ N-(4-Hydroxy-3-methoxybenzyl)-N,N′-dimethyl-N-(2-pyridyl)ethylenediamine
Probably in: Man G409

O-Demethylpyrilamine-β-D-glucuronide
Formed from: O-Demethylpyrilamine

6-Demethylquinidine
Formed from: Quinidine

27-Demethylrifabutin
Formed from: Rifabutin

N-Demethylrifampicin
Formed from: Rifampicin

N-Demethylritonavir
Formed from: Ritonavir
→ 2-Hydroxy-N-demethylritonavir
Probably in: Dog J146
Man J146
Rat J146

Demethylrizatriptan
Formed from: Rizatriptan

O-Demethylrolipram
Formed from: Rolipram

Demethylroxifiban
Formed from: Roxifiban
→ N^2-(1-Butyloxycarbonyl)-N^3-(2-(3-(4-formamidinophenyl)-4-hydroxyisoxazolin-5(*R*)-yl)acetyl)-2,3-diaminopropionic acid
Probably in: Man K356
Microorganism K356
Mouse K356

Rat K356
→ N^2-(1-Butyloxycarbonyl)-N^3-(2-(3-(4-formamidinophenyl)-5-hydroxyisoxazolin-5(*R*)-yl)acetyl)-2,3-diaminopropionic acid
Probably in: Man K356
Microorganism K356
Mouse K356
Rat K356
→ N^3-(2-(3-(4-Formamidinophenyl)isoxazolin-5(*R*)-yl)acetyl)-N^2-(3-hydroxy-1-butyloxycarbonyl)-2,3-diaminopropionic acid
Probably in: Man K356
Microorganism K356
Mouse K356
Rat K356

Demethylsalioxon
Probably formed from: Salioxon

Demethylsalithion
Formed from: Salithion

Demethylsemotiadil
Formed from: Semotiadil
→ 4,5′-Bisdemethylsemotiadil
Probably in: Rat H313, J130
→ Demethylsemotiadil-β-D-glucuronide
In: Rat H313

Demethylsemotiadil-β-D-glucuronide
Formed from: Demethylsemotiadil

Demethylsertraline
Formed from: Sertraline

Demethylsterigmatocystin
→ Sterigmatocystin
In: Aspergillus F281, J511

Demethylstobadine
Formed from: Stobadine

Demethylsuberosin (7-hydroxy-6-(3-methyl-2-butenyl)-2*H*-1,4-benzopyran-2-one)
→ Marmesin
In: Ammi E559

Demethylsufentanil
Formed from: Sufentanil

O-Demethylsulphamethomidine
Formed from: Sulphamethomidine

6-Demethylsulphamonomethoxine
Formed from: Sulphamonomethoxine

Demethylsulphoridazine
Formed from: Sulphoridazine
→ 7-Hydroxydemethylthioridazine-2-sulphone
Probably in: Man E342

Demethylsulphoridazine sulphoxide
Probably formed from: Sulphoridazine sulphoxide

Demethylsultopride
Formed from: Sultopride

Demethyltamoxifen
Formed from: Tamoxifen
→ Bisdemethyltamoxifen
Probably in: Man C316, H182
→ 4-Hydroxydemethyltamoxifen
Probably in: Man G844
→ 1-(*p*-Hydroxyphenyl)-1-(*p*-(2-methylaminoethoxy)phenyl)-2-phenyl-1-butene
Probably in: Rat E213

Demethyltamsulosin
Formed from: Tamsulosin
→ Demethyltamsulosin-β-D-glucuronide
In: Dog H528
Rat H528

Demethyltamsulosin-β-D-glucuronide
Formed from: Demethyltamsulosin

6-Demethylterazocin
Formed from: Terazocin

7-Demethylterazocin
Formed from: Terazocin

Demethylterbinafine
Formed from: Terbinafine

Demethylterbinafine dihydrodiol
Probably formed from: Terbinafine dihydrodiol

4′-O-Demethylterritrem B
Formed from: Territrem B

Demethyl tetrachlorovinphos
See: 2-Chloro-1-(2,4,5-trichlorophenyl)vinyl methyl phosphate

N-Demethyltetracycline
Formed from: Tetracycline

O-Demethyltetrandrine
Formed from: Tetrandrine

3-O-Demethylthebaine
See: Oripavine

N-Demethylthioridazine
Formed from: Thioridazine
S-Demethylthioridazine
Formed from: Thioridazine
→ 3-Hydroxy-S-demethylthioridazine
Probably in: Man D366
→ 7-Hydroxy-S-demethylthioridazine
Probably in: Man D366
Demethylthymoxamine
Formed from: Thymoxamine
→ Demethylthymoxamine-β-D-glucuronide
In: Rat J493
→ Demethylthymoxamine-O-sulphate
In: Rat J493
Demethylthymoxamine-β-D-glucuronide
Formed from: Demethylthymoxamine
Demethylthymoxamine-O-sulphate
Formed from: Demethylthymoxamine
Demethyltienocarbine
Formed from: Tienocarbine
3′-Demethyltofizopam
Formed from: Tofizopam
4′-Demethyltofizopam
Formed from: Tofizopam
7-Demethyltofizopam
Formed from: Tofizopam
→ 3′,7-Bisdemethyltofizopam
Probably in: Monkey D548
→ 4′,7-Bisdemethyltofizopam
Probably in: Man D548
Monkey D548
Rat D548
8-Demethyltofizopam
Formed from: Tofizopam
11-Demethyltomaymycin
Formed from: 3,4-Dihydroxy-L-phenylalanine
L-Tryptophan
Demethyltomoxiprole
Formed from: Tomoxiprole
→ Demethyltomoxiprole-β-D-glucuronide
In: Rat D912
Demethyltomoxiprole-β-D-glucuronide
Formed from: Demethyltomoxiprole
Demethyltopotecan
Formed from: Topotecan
→ Bisdemethyltopotecan
In: Rat J704
→ Demethyltopotecan-β-D-glucuronide
Probably in: Man J844
Demethyltopotecan-β-D-glucuronide
Probably formed from: Demethyltopotecan
O-Demethyltrachelogenin
Probably formed from: Trachelogenin
→ Trachelogenin
In: Rat H48
O-Demethyltramadol
Formed from: Tramadol
Demethyltriflubazam
Formed from: Triflubazam
→ 4′-Hydroxydemethyltriflubazam
Probably in: Man A1873
Rat A1864
→ 3′,4′-Dihydro-3′,4′-dihydroxydemethyltriflubazam
Probably in: Dog A1864
Monkey A1864
Rat A1864
Demethyltrifluoperazine
Probably formed from: Fluphenazine
10-(3-(4-Methylpiperazin-1-yl)propyl)trifluoromethylphenothiazine
→ Demethyltrifluoperazine sulphoxide
Probably in: Rat A67
Demethyltrifluoperazine sulphoxide
Formed from: Fluphenazine
Probably formed from: Demethyltrifluoperazine
3′-Demethyltrimethoprim
Formed from: Trimethoprim
→ 3′,5′-Bisdemethyltrimethoprim
Probably in: Beef A2296
→ 3′-Demethyltrimethoprim-O-sulphate
In: Goat A3478
Pig A3458
4′-Demethyltrimethoprim
Formed from: Trimethoprim
3′-Demethyltrimethoprim-O-sulphate
Formed from: 3′-Demethyltrimethoprim
***m*-Demethyltrimetozine**
Formed from: Trimetozine
→ *m*-Demethyltrimetozine-β-D-glucuronide
In: Dog A3302
→ *m*-Demethyltrimetozine-O-sulphate
In: Dog A3302

p-Demethyltrimetozine
Formed from: Trimetozine
→ *p*-Demethyltrimetozine-β-D-glucuronide
In: Dog A3302

m-Demethyltrimetozine-β-D-glucuronide
Formed from: *m*-Demethyltrimetozine

p-Demethyltrimetozine-β-D-glucuronide
Formed from: *p*-Demethyltrimetozine

m-Demethyltrimetozine-O-sulphate
Formed from: *m*-Demethyltrimetozine

Demethyltripelennamine
Formed from: Tripelennamine
→ N-(*p*-Hydroxybenzyl)-N′-methyl-N-(2-pyridyl)ethylenediamine
Probably in: Rat A379

Demethyltropine indole-3-carboxylate
Formed from: Tropine indole-3-carboxylate

5-Demethylubiquinone 45
→ Dihydro-5-demethylubiquinone 45
In: Rat A3084

N-Demethylvenlafaxine
Formed from: Venlafaxine
→ N,N-Bisdemethylvenlafaxine
In: Rat G564
→ N,O-Bisdemethylvenlafaxine
Probably in: Rat G564

O-Demethylvenlafaxine
Formed from: Venlafaxine
→ O-Demethylvenlafaxine-β-D-glucuronide
In: Dog G564
Man G564
Mouse G564
Rat G564
→ O-Demethylvenlafaxine-O-sulphate
In: Man G564

O-Demethylvenlafaxine-β-D-glucuronide
Formed from: O-Demethylvenlafaxine

O-Demethylvenlafaxine-O-sulphate
Formed from: O-Demethylvenlafaxine

Demethylverlukast
Formed from: Verlukast

Demethylzafirlukast
Probably formed from: 5-(Cyclopentoxycarbonylamino)-1-hydroxymethyl-3-((2-methoxy-4-((2-methylphenyl)sulphonyl)aminocarbonyl)phenylmethyl)indole

1-Demethylzolazepam
Formed from: Zolazepam
→ Hydroxy-1-demethylzolazepam
Probably in: Dog A1572

8-Demethylzolazepam
Formed from: Zolazepam

Demethylzolmitriptan
→ (*S*)-4-((3-Formylmethyl)-1*H*-indol-5-yl)methyl)-2-oxazolidine
Probably in: Man J741

Demethylzopiclone
Formed from: Zopiclone

Demethylzotepine
Formed from: Zotepine
→ Demethylzotepine sulphoxide
Probably in: Dog B245
Mouse B245
Rat B245
→ 7-Hydroxydemethylzotepine
Probably in: Dog B245
Mouse B245
Rat B245

Demethylzotepine sulphoxide
Formed from: Demethylzotepine

Demoxepam
See: 7-Chloro-1,3-dihydro-5-phenyl-2*H*-1,4-benzodiazepin-2-one-4-oxide

Denopamine ((−)2-(3,4-dimethoxyphenethyl)amino)-1-(*p*-hydroxyphenyl)ethanol)
→ Denopamine-β-D-glucuronide (both phenolic and side-chain)
In: Dog D839, J869
Rabbit D347
Rat D839
→ Denopamine-O-sulphate
In: Dog D839, J869
→ N-(3,4-Dimethoxyphenethyl) noradrenaline
In: Rat D347
See also: C626, D839, D754, J869
→ N-(3-Hydroxy-4-methoxyphenethyl) octopamine
In: Man C626, D754
Rat D347
→ N-(4-Hydroxy-3-methoxyphenethyl) octopamine
In: Dog D839

Man C626, D754
Rat D347, D839

Denopamine-β-D-glucuronide (both phenolic and side-chain)
Formed from: Denopamine

Denopamine-O-sulphate
Formed from: Denopamine

7-Deoxyadriamycinol aglycone
Formed from: 7,8,9,10-Tetrahydro-6,8,11-trihydroxy-8-hydroxyacetyl-1-methoxy-5,12-naphthacenedione
Probably formed from: Adriamycinol
→ Demethyldeoxyadriamycinol aglycone
Probably in: Man A1989

7-Deoxyadriamycinone
Formed from: 4′-O-Tetrahydropyranyl-adriamycin

7-Deoxyaklavinone
Formed from: Aclacinomycin A
Aklavinone
→ 7-Deoxyaklavinone dimer
Probably in: Rat B154, B229

7-Deoxyaklavinone dimer
Probably formed from: 7-Deoxyaklavinone

7-Deoxyauramycinone
Formed from: Auramycinone

16-Deoxycatharanthine-(15*H*)-15,16-oxide
Probably formed from: Catharanthine

6′-Deoxychalcone
Formed from: *p*-Coumaroyl CoA

7-Deoxydaunorubin aglycone
See: 7-Deoxydaunorubinone

7-Deoxydaunorubinol aglycone (7,8,9,10-tetrahydro-6,8,11-trihydroxy-8-(1-hydroxyethyl)-1-methoxy-5,12-naphthacenedione
Formed from: Daunomycinol
Daunorubinol aglycone
N,N-Dimethyldaunorubin
Probably formed from: 7-Deoxyrubicinone
4′-Epidoxorubicin
→ Demethyldeoxydaunorubinol aglycone
Probably in: Man A37, A138, C647
→ Deoxydaunorubinol aglycone-13-β-D-glucuronide
Probably in: Man A37, A138
→ Deoxydaunorubinol aglycone semiquinone radical
In: Beef E116
→ Deoxydaunorubinol aglycone-4-sulphate
Probably in: Man A138

Deoxydaunorubinol aglycone-13-β-D-glucuronide
Probably formed from: 7-Deoxydaunorubinol aglycone

Deoxydaunorubinol aglycone semiquinone radical
Probably formed from: 7-Deoxydaunorubinol aglycone

Deoxydaunorubinol aglycone-4-sulphate
Probably formed from: 7-Deoxydaunorubinol aglycone

7-Deoxydaunorubinone (7-Deoxydaunorubin aglycone)
Formed from: Daunorubin
N,N-Dibenzyldaunorubin
→ 7-Deoxydaunorubinol aglycone
Probably in: Rat C539, D212
Streptomyces A2757

2-Deoxy-1,5-diethyl-5-(*p*-hydroxyphenyl)barbituric acid
Formed from: 2-Deoxy-N-ethylphenobarbital

2-Deoxy-N,N′-diethylphenobarbital
→ 2-Deoxy-N-ethylphenobarbital
Probably in: Man G74

7-Deoxydihydroadriamycinone
Probably formed from: Dihydro-adriamycinone
13-Dihydro-4′-O-tetrahydropyranyladriamycin

7-Deoxy-13-dihydrodaunomycinone
Formed from: 13-Dihydrodaunomycinone

4-((6-Deoxy-2,4-di-O-methyl-α-L-mannopyranosyl)oxy)-3,4-dihydro-2,5,7-trihydroxy-3,9-dimethoxy-2-methyl-1,6,11(2*H*)-naphthacenetrione
See: Steffimycin B

2-Deoxy-N,N′-dimethylphenobarbital
→ 2-Deoxy-N-methylphenobarbital
Probably in: Man G74

4′-Deoxydoxorubicin
→ 4′-Deoxydoxorubicinol
In: Rat D457

7-Deoxydoxorubicin aglycone
Formed from: Doxorubicin
4′-Deoxydoxorubicinol
Formed from: 4′-Deoxydoxorubicin
7-Deoxydoxorubicinol aglycone
Probably formed from: Doxorubicinol
7-Deoxydoxorubicinone
Probably formed from: Doxorubicin
7-Deoxyepidoxorubicin aglycone
Formed from: Epidoxorubicin
7-Deoxyepidoxorubicinol aglycone
Probably formed from: 4′-Epidoxorubicinol
(3′α,4′α)-4′-Deoxy-3′,4′-epoxyvincaleukablastine
See: Leurosine
2-Deoxy-5-ethyl-5-(*p*-hydroxyphenyl)barbituric acid
Formed from: 2-Deoxyphenobarbital
2-Deoxy-5-ethyl-5-(*p*-hydroxyphenyl)-1-methylbarbituric acid
Formed from: 2-Deoxy-N-methylphenobarbital
2-Deoxy-N-ethylphenobarbital
Probably formed from: 2-Deoxy-N,N′-diethylphenobarbital
→ 2-Deoxyphenobarbital
In: Man G74
→ 2-Deoxy-1,5-diethyl-5-(*p*-hydroxyphenyl)barbituric acid
In: Man G74
→ 2,N-Diethyl-2-phenylmalondiamide
In: Man G74
7-Deoxy-1-hydroxyauramycinone
Formed from: 1-Hydroxyauramycinone
7-Deoxy-11-hydroxyauramycinone
Formed from: 11-Hydroxyauramycinone
3-Deoxy-2-hydroxyoestradiol
Probably formed from: 3-Deoxyoestradiol
→ 2-Hydroxyoestradiol
In: Rat A1915
See also: A1969
3-Deoxy-2-hydroxy-17β-oestradiol-17-sulphate
Formed from: 3-Deoxy-17β-oestradiol-17-sulphate
3-Deoxy-6β-hydroxyoestrone
Formed from: 3-Deoxyoestrone
7-Deoxy-1-hydroxysulfurmycinone
Formed from: 1-Hydroxysulfurmycinone
7-Deoxy-11-hydroxysulfurmycinone
Formed from: 11-Hydroxysulfurmycinone
4′-Deoxy-3′-hydroxyvincristine
Probably formed from: Vincristine
4-((6-Deoxy-2-O-methyl-α-L-mannopyranosyl)oxy)-3,4-dihydro-2,5,7-trihydroxy-3,9-dimethoxy-2-methyl-1,6,11(2*H*)-naphthacenetrione
See: Steffimycin
2-Deoxy-N-methylphenobarbital
Probably formed from: 2-Deoxy-N,N′-dimethylphenobarbital
→ 2-Deoxyphenobarbital
In: Man G74
→ 2-Deoxy-5-ethyl-5-(*p*-hydroxyphenyl)-1-methylbarbituric acid
In: Man G74
→ 2-Ethyl-N-methyl-2-phenylmalondiamide
In: Man G74
7-Deoxynogalamycinone
Formed from: Nogalamycin
7-Deoxynogarol
Formed from: Menogaril
3-Deoxyoestradiol
Probably formed from: 3-Deoxyoestrone
→ 3-Deoxy-2-hydroxyoestradiol
In: Rat A529
3-Deoxy-17β-oestradiol-17-sulphate
→ 3-Deoxy-2-hydroxy-17β-oestradiol-17-sulphate
In: Rat F219
→ 17β-Oestradiol-17-sulphate
In: Rat F219
3-Deoxyoestrone
→ 3-Deoxyoestradiol
In: Rat A529
→ 3-Deoxy-6β-oestrone
In: Rat A529
→ Oestrone
In: Rat A529
2-Deoxyphenobarbital (mysoline, primaclone, primidone)
Formed from: 2-Deoxy-N-ethylphenobarbital
2-Deoxy-N-methylphenobarbital
→ 2-Deoxy-5-ethyl-5-(*p*-hydroxyphenyl) barbituric acid
In: Man C586, G74

Rat C586
→ 2-Ethyl-2-phenylmalondiamide
In: Dog F618
Man G74
Rabbit A2878
See also: A123, A507
→ Phenobarbital
In: Dog F618
Man A507
Mouse A2356, A2357
Rabbit A2878
Rat A123

Deoxypicropodophyllin ((5*R*,5a*S*,8a*R*)-5,8,8a,9-tetrahydro-5-(3,4,5-trimethoxyphenyl)furo[3′,4′:6,7]naphtho[2,3-*d*]-1,3-dioxol-6(5a*H*)-one)
Formed from: Deoxypodophyllotoxin

Deoxypodophyllotoxin
→ Deoxypicropodophyllin
In: Fusarium F156
Penicillium F156
Pseudomonas F156
→ Epipodophyllotoxin
In: Penicillium F156, F157
→ Podophyllotoxin
In: Aspergillus F156
Podophyllum D274
→ β-Peltanin
In: Podophyllum E236

7-Deoxypyrromycinone
Formed from: Marcellomycin

7-Deoxy-ε-rhodomycinone ((1*R*,2*R*)-2-ethyl-2,3,4,6,11-hexahydro-2,4,5,12-tetrahydroxy-6,11-dioxo-1-naphthacenecarboxylic acid methyl ester)
Formed from: Aklanonic acid

10-Deoxysarpagine (normacusine B, sarpagan-17-ol)
Formed from: Strictosidine

Deoxyshikonin
Formed from: L-Phenylalanine
→ Shikonin
Probably in: Lithospermum H556

7-Deoxysteffimycinol
Formed from: Steffimycinol

7-Deoxysteffimycinone
Formed from: Steffimycin
Steffimycinone

7-Deoxysteffimycinone B
Formed from: Steffimycin B

7-Deoxysulfurmycinone
Formed from: Sulfurmycinone

Deoxytubulosine
Formed from: Deacetylipecoside
→ Tubulosine
In: Alangium C909

16-Deoxyvinblastine-(15*H*)-15,16-oxide
Formed from: Vinblastine

4′-Depentyl-Δ^1-tetrahydrocannabinol-4′-acrylic-7-oic acid
Formed from: Δ^1-Tetrahydrocannabinol

Dephosophoferuloyl CoA
→ Feruloyltyramine
In: Nicotiana D558

Deprenyl (selegiline, N-α-dimethyl-N-2-propynylbenzeneethanamide; both isomers)
→ Demethyldeprenyl
In: Man J152, K37, K195, K480
Mouse K450
Rat E781, K581
See also: J241
→ Ephedrine (*l*-isomer assumed)
In: Man J152
→ N-Methylamphetamine
In: Cunninghamella C68
Man A3329, A3961, J152, K37, K195, K480
Mouse K450
Rat D661, E781, K581
See also: J241
→ Pseudoephedrine (*l*-isomer assumed)
In: Man J152

Depropylalpidem
Formed from: Alpidem
→ 6-Chloro-2-(*p*-chlorophenyl)-N-(2-hydroxypropyl)imidazo[1,2-*a*]pyridine-3-acetamide
Probably in: Rat G55

Depropylbesipirdine
Formed from: Besipirdine

Depropyldopazinol
Formed from: Dopazinol
Depropylflavodilol
Formed from: Flavodilol
Depropylpropafenone
Formed from: Propafenone
Depropylropinirole
Formed from: Ropinirole
→ Depropylropinirole carbamyl-β-D-glucuronide
Probably in: Cynomolgus monkey J561
Man J561
Mouse J561
Rat J561
→ 7-Hydroxydepropylropinirole
Probably in: Cynomolgus monkey J561
Man J561
Mouse J561
Depropylropinirole carbamyl-β-D-glucuronide
Probably formed from: Depropylropinirole
Depropyltiropramide
Formed from: Tiropramide
Deramciclane ((1*R*,2*S*,4*R*)-(−)-N,N-dimethyl-((1,7,7-trimethyl-2-phenylbicyclo[2.2.1]hept-2-yl)oxy)ethane)
→ Demethylderamciclane
In: Man K635
Derrone (5-hydroxy-3-(4-hydroxyphenyl)-8,8-dimethyl-4*H*,8*H*-benzo[1,2-*b*:3,4-*b*′]dipyran-4-one)
→ 5-O-Methylderrone
In: Lupin E756, F182
Desethylamiodarone
Formed from: Amiodarone
Desethylamodiaquine
Formed from: Amodiaquine
→ Bisdesethylamodiaquine
Probably in: Rat A278, H159
→ 2-Hydroxydesethylamodiaquine
Probably in: Man D529
Desethylamperozide
Formed from: Amperozide
→ Desethylamperozide-N-oxide
Probably in: Man H745
Desethylamperozide-N-oxide
Probably formed from: Desethylamperozide
Desethylaprindine
Formed from: Aprindine
Desethylbemitradine
Formed from: Bemitradine
→ Desethylbemitradine-β-D-glucuronide
Probably in: Man F314
→ 3′,4′-Dihydro-3′,4′-dihydrodesethylbemitradine
Probably in: Man F314
Desethylbemitradine-β-D-glucuronide
Probably formed from: Desethylbemitradine
Desethylbromerguride
Formed from: Bromerguride
→ Bisdesethylbromerguride
In: Dog F337
Guinea pig F337
Monkey F337
Rat F337
Desethylbromopride
Formed from: Bromopride
Desethylclomiphene
Formed from: Clomiphene
Desethyldexfenfluramine
Formed from: Dexfenfluramine
Desethyldoxapram
Formed from: Doxapram
4′-Desethyldrotaveraldine
Probably formed from: 4′-Desethyldrotaverine
4′-Desethyldrotaverine
Formed from: Drotaverine
→ 4′-Desethyldrotaveraldine
Probably in: Rat B538
6-Desethyldrotaverine
Formed from: Drotaverine
Desethylenclomiphene
Formed from: Enclomiphene
Desethylentacapone
Formed from: Entacapone
→ Bisdesethylentacapone
In: Rat G527
Desethylepicainide
Formed from: Epicainide
→ α-Hydroxy-N-(5-oxopyrrolidino-2-methyl)-α,α-diphenylacetamide
Probably in: Man F323
N-Desethylethacizin
Formed from: Ethacizin

→ N,N-Bisdesethylethacizin
Probably in: Man F452

Desethylfenfluramine
Formed from: Fenfluramine
→ N-Hydroxydesethylfenfluramine
Probably in: Man J870

Desethyllacidipine
Formed from: Lacidipine

Desethyllisuride
Formed from: Lisuride
→ Bisdesethyllisuride
In: Rat B616
→ Desethyl methyllisuride
Probably in: Rat B616

N-Desethylmebeverine
Formed from: Mebeverine

Desethyl methyllisuride
Probably formed from: Desethyllisuride

Desethylmitoclomine
Formed from: Mitoclomine

Desethyl-5-oxozaleplon
Probably formed from: Desethylzaleplon

Desethyloxybutynin
Formed from: Oxybutynin

Desethylpentopril
Formed from: Pentopril

Desethylproterguride
Formed from: Proterguride

Desethylreboxetine
Formed from: Reboxetine

Desethyltamsulosin
Formed from: Tamsulosin
→ Desethyltamsulosin-β-D-glucuronide
In: Dog H528
Man H714
Rat H528
→ Desethyltamsulosin-O-sulphate
In: Dog H528
Man H714
Rat H528

Desethyltamsulosin-β-D-glucuronide
Formed from: Desethyltamsulosin

Desethyltamsulosin-O-sulphate
Formed from: Desethyltamsulosin

Desethylterguride
Formed from: Terguride

Desethylthiapride
Formed from: Thiapride

Desethyltiropramide
Formed from: Tiropramide
→ N-Benzoyl-O-(2-ethylaminoethyl)tyrosyl propylamide
Probably in: Rat F32, F97

Desethylzaleplon
Formed from: Zaleplon
→ Desethyl-5-oxozaleplon
Probably in: Cynomolgus monkey J570

Desoxybenzoin
See: ω-Phenylacetophenone

Desoxymansil (1,2,3,4-tetrahydro-6-methyl-2-(((1-methylethyl)amino)methyl)-7-nitroquinoline)
→ Oxamniquine
In: Mouse A3711
Rabbit A3711
Rat A3711
→ 1,2,3,4-Tetrahydro-2-hydroxymethyl-6-methyl-7-nitroquinoline
Probably in: Mouse A3711
Rabbit A3711
Rat A3711

De-(thiazolylmethylcarbonyl)ritonavir
Formed from: Ritonavir

Detomidine (4-(2,3-dimethylbenzyl)imidazole)
→ 4-(3-Hydroxymethyl-2-methylbenzyl) imidazole
In: Horse F675, G634
Rat E717, E719, G233
→ 4-(3-(S-Glutathionyl)methyl-2-methylbenzyl)imidazole
In: Rat E719
→ N-Methyldetomidine
In: Rat G233

Dexfenfluramine ((+)-fenfluramine, (αS)-N-ethyl-α-methyl-3-(trifluoromethyl) benzeneethanamine)
→ Desethyldexfenfluramine
In: Baboon H459
Cynomolgus monkey H459
Dog H459
Man H459, J870
Mouse H459
Rat H459, J325

Squirrel monkey H459
→ 1-(*m*-Trifluoromethylphenyl)acetone
In: Rat J325

Dexmedetomidine (4-((1*S*)-1-(2,3-dimethylphenyl)ethyl)-1*H*-imidazole)
→ Hydroxydexmedetomidine
In: Rat G233
→ N-Methyldexmedetomidine
In: Rat G233

Dextromethorphan ((+)-3-methoxy-N-methylmorphinan)
→ N-Demethyldextromethorphan
In: Man F956, G925, H262, H864, JH865, J299
Rat G934
→ O-Demethyldextromethorphan
In: Man F956, G925, H864, J299
Rat G934
→ 6-Hydroxydextromethorphan
In: Man F956

Dextrorphan
See: N-Demethyldextromethorphan)

Dezocine ((*R*)-13β-amino-5,6,7,8,9,10,11,12-octahydro-5α-methyl-5,11-methanobenzocyclodecen-3-ol)
→ Dezocine-β-D-glucuronide
In: Monkey B608
Rat B608
→ Dezocine-O-sulphate
In: Monkey B608
Rat B608

Dezocine-β-D-glucuronide
Formed from: Dezocine

Dezocine-O-sulphate
Formed from: Dezocine

DHPG
See: 3,4-Dihydroxyphenylglycol

Dhurrin ((α*S*)-α-(β-D-glucopyranosyloxy)-4-hydroxybenzeneacetonitrile)
Formed from: *p*-Hydroxymandelonitrile
→ *p*-Hydroxymandelonitrile
In: Sorghum F353
→ 6″-O-Malonyldhurrin
In: Petroselinum K725

2,5-Diacetamidoaniline
Formed from: 2,4-Diaminoacetanilide
3,4-Diaminoacetanilide
2-Nitro-*p*-phenylenediamine

2,4-Diacetamidoanisole
Probably formed from: 4-Acetamido-2-aminoanisole
→ 2,4-Diacetamido-5-methoxyphenol
Probably in: Rat A3704, B645
→ 2,4-Diacetamidophenol
Probably in: Rat A3704, B645
→ 5-Glycollamido-2-methoxyacetanilide
Probably in: Rat A3704, B645

3,5-Diacetamidobenzamide
Formed from: 3-Acetamido-5-aminobenzamide

1,3-Diacetamidobenzene
Formed from: *m*-Acetamidoaniline

2,3-Diacetamidobenzoic acid
Probably formed from: 4-Acetamido-anthranilic acid

5,5′-Diacetamido-2,2′-dihydroxybiphenyl
Formed from: *p*-Acetamidophenol
→ 5,5′,5″-Triacetamido-2,2′,2″-trihydroxyterphenyl
With xanthine oxidase F171

5,5′-Diacetamido-3,3′-di-(S-glutathionyl)-2,2′-dihydroxybiphenyl
Formed from: S-(5-Acetamido-2-hydroxyphenyl)glutathione

5,5′-Diacetamido-3-(S-glutathionyl)-2,2′-dihydroxybiphenyl
Formed from: S-(5-Acetamido-2-hydroxyphenyl)glutathione
p-Acetamidophenol

2,4-Diacetamido-5-methoxyphenol
Probably formed from: 2,4-Diacetamidoanisole

2,4-Diacetamido-6-methoxyphenol
Probably formed from: 2,4-Diamino-6-methoxyphenol

2,4-Diacetamidophenol
Probably formed from: 2,4-Diacetamidoanisole
m-Nitroacetanilide

2,4-Diacetamidotoluene
Formed from: 3-Acetamido-4-methylaniline
5-Acetamido-2-methylaniline
→ 3-Acetamido-4-methylaniline
In: Mouse A3949

2,6-Diacetamidotoluene
Formed from: 2-Acetamido-6-aminotoluene

1,8-Diacetoxyanthraquinone
Formed from: 1,8-Diacetoxy-9-anthrone

1,8-Diacetoxy-9-anthrone
Probably formed from: 1,8,9-Triacetoxyanthracene
→ 8-Acetoxy-1-hydroxy-9-anthrone
Probably in: Man B150
→ 1,8-Diacetoxyanthraquinone
Probably in: Man B150

2,3-Diacetoxybenzoic acid
→ Pyrocatechuic acid
In: Acinetobacter A962

2,4-Diacetoxybenzoic acid
→ 2,4-Dihydroxybenzoic acid
In: Acinetobacter A962

2,5-Diacetoxybenzoic acid
→ Gentisic acid
In: Acinetobacter A962

2,6-Diacetoxybenzoic acid
→ 2,6-Dihydroxybenzoic acid
In: Acinetobacter A962

3,4-Diacetoxybenzoic acid
→ Protocatechuic acid
In: Acinetobacter A962

3,5-Diacetoxybenzoic acid
→ 3,5-Dihydroxybenzoic acid
In: Acinetobacter A962

7,8-Diacetoxy-4-methylcoumarin
→ 7,8-Dihydroxy-4-methylcoumarin
In: Rat K400

Diacetyl 2-amino-tetrahydro-6,7-dihydroxy-naphthalene
→ 2-Amino-tetrahydro-6,7-dihydroxy-naphthalene
In: Rat C251

4,4′-Diacetylbenzidine
Formed from: 4,4′-Diacetyl-N-hydroxybenzidine
N-Acetylbenzidine
→ N-Acetylbenzidine
In: Mouse D872
Rat C590
→ 4,4′-Diacetyl-3-hydroxybenzidine
In: Hamster B129
Mouse B129
Rat B129
See also: C590
→ 4,4′-Diacetyl-N-hydroxybenzidine
In: Hamster B129
Mouse B129, D872
Rat B129, D872, J144
See also: C590

Diacetylboldine
→ 9-Acetylboldine
In: Candida D981

N,O-Diacetylcalycotomine
→ N-Acetylcalycotomine
In: Arthrobacter F1

1,3-Diacetylchloramphenicol
Probably formed from: 1-Acetylchloramphenicol

N,N′-Diacetyl-3,3′-dichlorobenzidine
Probably formed from: N-Acetyl-3,3′-dichlorobenzidine

N,N′-Diacetyl-3,3′-dimethoxybenzidine
Probably formed from: N-Acetyl-3,3′-dimethoxybenzidine

Diacetylepinine
→ 3,4-Dihydroxy-N-methylphenethylamine
In: Beef F220

3,8-Diacetylethidium
Probably formed from: 8-Acetylethidium

4,4′-Diacetyl-3-hydroxybenzidine
Formed from: 4,4′-Diacetylbenzidine
4,4′-Diacetyl-3-hydroxybenzidine-β-D-glucuronide
→ N-Acetyl-3-hydroxybenzidine
Probably in: Rat C590

4,4′-Diacetyl-N-hydroxybenzidine
Formed from: N-Acetyl-N-hydroxybenzidine
N-Acetyl-N′-hydroxybenzidine
4,4′-Diacetylbenzidine
4,4′-Diacetyl-N-hydroxybenzidine-β-D-glucuronide
→ 4,4′-Diacetylbenzidine
In: Rat C590
→ 4,4′-Diacetyl-N-hydroxybenzidine-β-D-glucuronide
Probably in: Rat C590

4,4′-Diacetyl-3-hydroxybenzidine-β-D-glucuronide
Probably formed from: 4,4′-Diacetyl-3-hydroxybenzidine

→ 4,4′-Diacetyl-3-hydroxybenzidine
In: Escherichia J766
Man J766

4,4′-Diacetyl-N-hydroxybenzidine-β-D-glucuronide
Probably formed from: 4,4′-Diacetyl-N-hydroxybenzidine
→ 4,4′-Diacetyl-N-hydroxybenzidine
In: Escherichia J766
Man J766

N,N′-Diacetyl-4,4′-(hydroxymethylene)dianiline
Formed from: 4,4′-Methylenedianiline

Diacetylkynurenine
→ L-Kynurenine
In: Aspergillus J489

4,4′-Diacetyl-3-methoxybenzidine
Probably formed from: 4′-N-Acetyl-3-hydroxybenzidine

N,N′-Diacetyl-*p*-phenylenediamine
Formed from: *p*-Acetamidoaniline
Probably formed from: N-Acetyl-*p*-phenylenediamine

6,6′-Diacetylpseudomorphine
Formed from: 6-Acetylmorphine

N^1,N^4-Diacetylsulphanilamide
Formed from: *p*-Acetamidobenzenesulphonamide

N,N′-Diacetyl-*o*-tolidine
Probably formed from: N-Acetyl-*o*-tolidine

Di-(N-acetyl-L-tyrosine)
Formed from: N-Acetyl-L-tyrosine
→ Tri-(N-acetyl-L-tyrosine)
Probably in: Horseradish J327

N,N-Diallylamphetamine
→ N-Allylamphetamine
In: Man K37

N,N′-Diallyl-2,2′-bimorphine
Formed from: Nalorphine

N,N′-Diallyl-7,7′,8,8′-tetrahydro-14,14′-dihydroxy-2,2′-bimorphine
Probably formed from: 6α-Naloxol

2,4-Diaminoacetanilide
→ 2,5-Diacetamidoaniline
In: Rat F828

3,4-Diaminoacetanilide
Probably formed from: 1,2,4-Triaminobenzene
→ 5-Acetamido-2,3-dihydro-2-oxo-1*H*-benzimidazole
Probably in: Rat J586
→ 5-Acetamido-2-methyl-1*H*-benzimidazole
Probably in: Rat J586
→ 2,5-Diacetamidoaniline
In: Rat F828
See also: J586

3,6-Diaminoacridine
See: Proflavine

2,4-Diaminoanisole (1,3-diamino-4-methoxybenzene)
→ 2-Acetamido-4-aminoanisole
In: Rat A3704
→ 4-Acetamido-2-aminoanisole
In: Rat A3740, B645

2,4-Diaminoazobenzene
→ Aniline
In: Pseudomonas E393
→ 1,2,4-Triaminobenzene
In: Pseudomonas E393

4,4′-Diaminoazobenzene
→ 4′-Acetamido-4-aminoazobenzene
In: Pseudomonas E393
→ *p*-Phenylenediamine
In: Pseudomonas E393

3,5-Diaminobenzamide
Formed from: 3-Amino-5-nitrobenzamide
→ 3-Acetamido-5-aminobenzamide
Probably in: Rat B193

1,2-Diaminobenzene
See: *o*-Phenylenediamine

1,3-Diaminobenzene
See: *m*-Phenylenediamine

1,4-Diaminobenzene
See: *p*-Phenylenediamine

2,5-Diaminobenzenesulphonic acid
→ Poly(2,5-diaminobenzenesulphonic acid)
In: Horseradish H946

2,4-Diaminobenzoic acid
Probably formed from: 2-Amino-4-nitrobenzoic acid
2,4-Diaminotoluene
→ 4-Acetamido-2-aminobenzoic acid
Probably in: Rat H515

3,4-Diaminobenzoic acid
→ 3,4-Diaminophenol
In: Agaricus E162

4,4′-Diaminobenzophenone
Probably formed from: 4-Amino-4′-methylaminobenzophenone
→ 4′-Acetamido-4-aminobenzophenone
Probably in: Rat B713

2,4-Diamino-6-benzyloxy-5-nitropyrimidine
Formed from: 2,4-Diamino-6-benzyloxy-5-nitrosopyrimidine

2,4-Diamino-6-benzyloxy-5-nitrosopyrimidine
→ 2,4-Diamino-6-benzyloxy-5-nitropyrimidine
In: Rat J68
→ 2,4,5-Triamino-6-benzyloxypyrimidine
In: Rat J68

2,5-Diamino-N,N′-di-(4-aminophenyl)-2,5-cyclohexadiene-1,4-diimine
Formed from: *p*-Phenylenediamine

4,4′-Diamino-3,3′-dichloro-5-hydroxydiphenylmethane
Formed from: 4,4′-Methylenebis(*o*-chloroaniline)
→ 4,4′-Diamino-3,3′-dichloro-5-hydroxydiphenylmethane-O-sulphate
In: Dog D523, D847

4,4′-Diamino-3,3′-dichloro-5-hydroxydiphenylmethane-O-sulphate
Formed from: 4,4′-Diamino-3,3′-dichloro-5-hydroxydiphenylmethane

2,4-Diamino-6-(2,5-dichloro-3-hydroxyphenyl)-*s*-triazine
Formed from: 2,4-Diamino-6-(2,5-dichlorophenyl)-*s*-triazine

2,4-Diamino-6-(2,5-dichloro-4-hydroxyphenyl)-*s*-triazine
Formed from: 2,4-Diamino-6-(2,5-dichlorophenyl)-*s*-triazine

2,4-Diamino-6-(2,5-dichlorophenyl)-*s*-triazine
→ 2,4-Diamino-6-(2,5-dichloro-3-hydroxyphenyl)-*s*-triazine
In: Dog E159
Guinea pig E159
Monkey E159
Rat E159
→ 2,4-Diamino-6-(2,5-dichloro-4-hydroxyphenyl)-*s*-triazine
In: Dog E159
Guinea pig E159
Monkey E159
Rat E159
→ 2,4-Diamino-6-(2,5-dichlorophenyl)-*s*-triazine-N-oxide
In: Dog E159
Guinea pig E159
Monkey E159
Rat E159
→ 2,4-Diamino-6-(2,5-dichloro-3-sulphinylmethylphenyl)-*s*-triazine
Probably in: Dog E159
Guinea pig E159
Monkey E159
Rat E159
→ 2,4-Diamino-6-(2,5-dichloro-4-sulphinylmethylphenyl)-*s*-triazine
Probably in: Dog E159
Guinea pig E159
Monkey E159
Rat E159

2,4-Diamino-6-(2,5-dichlorophenyl)-*s*-triazine-N-oxide
Formed from: 2,4-Diamino-6-(2,5-dichlorophenyl)-*s*-triazine

2,4-Diamino-6-(2,5-dichloro-3-sulphinylmethylphenyl)-*s*-triazine
Probably formed from: 2,4-Diamino-6-(2,5-dichlorophenyl)-*s*-triazine
→ 2,4-Diamino-6-(2,5-dichloro-3-sulphonylmethylphenyl)-*s*-triazine
Probably in: Dog E159
Guinea pig E159
Monkey E159
Rat E159

2,4-Diamino-6-(2,5-dichloro-4-sulphinylmethylphenyl)-*s*-triazine
Probably formed from: 2,4-Diamino-6-(2,5-dichlorophenyl)-*s*-triazine
→ 2,4-Diamino-6-(2,5-dichloro-4-sulphonylmethylphenyl)-*s*-triazine
Probably in: Dog E159
Guinea pig E159
Monkey E159
Rat E159

2,4-Diamino-6-(2,5-dichloro-3-sulphonylmethylphenyl)-*s*-triazine
Probably formed from: 2,4-Diamino-6-(2,5-dichloro-3-sulphinylmethylphenyl)-*s*-triazine

2,4-Diamino-6-(2,5-dichloro-4-sulphonylmethylphenyl)-*s*-triazine
Probably formed from: 2,4-Diamino-6-(2,5-dichloro-4-sulphinylmethylphenyl)-*s*-triazine

5,7-Diamino-1,2-dihydro-3-(*p*-hydroxyphenyl)-2-methylpyrido[3,4-*b*]pyrazine
Probably formed from: Ethyl 5-amino-1,2-dihydro-3-(*p*-hydroxyphenyl)-2-methylpyrido[3,4-*b*]pyrazine-7-carbamate

2,4-Diamino-6,7-dimethoxyquinazoline
Formed from: Prazocin

2,2′-Diaminodiphenylmethane
Formed from: Dibenzo[*c*,*f*][1,2]diazepine

4,4′-Diaminodiphenylmethane
→ Azodiphenylmethane
In: Rabbit G777
→ 4-Amino-4′-nitrosodiphenylmethane
In: Rabbit G777

3,8-Diamino-5-ethyl-6-phenylphenthridinium
See: Ethidium

1,2-Diamino-4-(*p*-fluorobenzyl)aminobenzene
Probably formed from: Ethyl N-(2-amino-4-(4-fluorobenzylamino)phenyl)carbamate
→ 1-Acetamido-2-amino-4-(*p*-fluorobenzyl)aminobenzene
Probably in: Man J586
Rat J586
→ 1,2,4-Triaminobenzene
Probably in: Rat J586

α,2-Diamino-4-fluoro-γ-oxobenzenebutanoic acid
See: 4-Fluoro-L-kynurenine

α,2-Diamino-5-fluoro-γ-oxobenzenebutanoic acid
See: 5-Fluoro-L-kynurenine

2,4-Diamino-6-((4-hydroxy-3,5-dimethoxyphenyl)amino)methyl-5-methylquinazoline
Formed from: Trimetrexate

2,6-Diamino-5-hydroxy-3-(4-hydroxyphenylazo)pyridine
Probably formed from: 2,6-Diamino-5-hydroxy-3-(phenylazo)pyridine

2,6-Diamino-3-(2-hydroxyphenylazo)pyridine
Formed from: Phenazopyridine

2,6-Diamino-3-(4-hydroxyphenylazo)pyridine
Formed from: Phenazopyridine

2,6-Diamino-5-hydroxy-3-(phenylazo)pyridine
Formed from: Phenazopyridine
→ 2,6-Diamino-5-hydroxy-3-(4-hydroxyphenylazo)pyridine
Probably in: Guinea pig F621
Man F621
Mouse F621
Rat F621, G541

1,3-Diamino-4-methoxybenzene
See: 2,4-Diaminoanisole

2,4-Diamino-3-methylphenol
Formed from: 2,6-Diaminotoluene

2,4-Diamino-5-methylphenol
Formed from: 2,4-Diaminotoluene
→ 2-Acetamido-4-amino-5-methylphenol
Probably in: Guinea pig A1782
Rabbit A1782
Rat A1782

2,4-Diamino-6-methylphenol
Probably formed from: 6-Amino-4-nitro-*o*-cresol
→ 2,4-Diacetamido-6-methylphenol
Probably in: Rat C470

2,6-Diamino-3-methylphenol
Formed from: 2,4-Diaminotoluene
→ 6-Acetamido-2-amino-3-methylphenol
Probably in: Guinea pig A1782
Rabbit A1782
Rat A1782

3,5-Diamino-2-methylphenol
Formed from: 2,4-Diaminotoluene

3,5-Diamino-4-methylphenol
Probably formed from: 2,6-Diaminotoluene

2,4-Diamino-N^{10}-methylpteroic acid
Formed from: Methotrexate

2,4-Diamino-5-methyl-6-quinazolinecarboxylic acid
Formed from: Trimetrexate

5-((4′-((2,6-Diamino-3-methyl-5-sulphophenyl)azo)-3,3′-dimethyl(1,1′-biphenyl)-4-yl)azo)-2-hydroxybenoic acid
See: Direct Orange 6

2,4-Diamino-1-naphthol
Formed from: 2-Amino-1,4-naphthoquinoneimine
2,4-Diamino-6-nitrotoluene
Probably formed from: 4-Amino-2,6-dinitrotoluene
→ 2,4,6-Triaminotoluene
Probably in: Clostridium H810
Desulphovibrio G719
Veillonella A2624
2,6-Diamino-4-nitrotoluene
Formed from: 2-Amino-4,6-dinitrotoluene
3,4-Diamino-5-nitrotoluene
Probably formed from: 2,6-Diamino-*p*-toluidine
2,4-Diaminophenol
Formed from: 2,4-Dinitrophenol
→ Phenol
In: Desulphovibrio G827
3,4-Diaminophenol
Formed from: 3,4-Diaminobenzoic acid
(2,6-Diamino-3-phenylazo)pyridine
See: Phenazopyridine
N-(4-(((2,4-Diamino-6-pteridinyl)methyl)amino) benzoyl)-L-glutamic acid
See: Aminopterin
1,8-Diaminopyrene
Formed from: 1-Amino-8-nitropyrene
4,4′-Diaminostilbene
→ 4-Aminostilben-4′-yl nitroxide radical
In: Rabbit B184
2,4-Diaminotoluene
Formed from: 5-Nitro-*o*-toluidine
Probably formed from: 2-Amino-4-nitrotoluene
4-Amino-2-nitrotoluene
→ 3-Acetamido-4-methylaniline
In: Guinea pig A311
Hamster A311, A461
Mouse A311
Rabbit A311
Rat A311
Contrast (in dog and man) A311
→ 5-Acetamido-2-methylaniline
In: Guinea pig A311
Hamster A311, A461
Mouse A311, B423
Rabbit A311
Rat A311, B423, H3
See also: B885, J196
Contrast (in dog and man) A311
→ 2,4-Diaminobenzoic acid
Probably in: Rat B885
→ 2,4-Diamino-5-methylphenol
In: Guinea pig A1782
Rabbit A1782
Rat A1782
→ 2,6-Diamino-3-methylphenol
In: Guinea pig A1782
Rabbit A1782
Rat A1782
→ 3,5-Diamino-2-methylphenol
In: Guinea pig A1782
Rabbit A1782
Rat A1782
2,5-Diaminotoluene
Formed from: 4-Nitro-*o*-toluidine
2,6-Diaminotoluene
Formed from: 2-Methyl-3-nitroaniline
3-Nitro-*o*-toluidine
→ 2-Acetamido-6-aminotoluene
In: Rat F56
→ 2,4-Diamino-3-methylphenol
In: Rat F56
→ 3,5-Diamino-4-methylphenol
Probably in: Rat F56
3,4-Diaminotoluene
Formed from: 2-Nitro-*p*-toluidine
3-Nitro-*p*-toluidine
2-(4-((4,6-Diamino-1,3,5-triazin-2-yl)amino)phenyl)-1,3,2-dithiarsolane-4,5-dicarboxylic acid
See: Melarsoprol
2,4-Diamino-5-(3,4,5-trimethoxybenzyl) pyrimidine
See: Trimethoprim
Dianisidine
See: 3,3′-Dimethoxybenzidine
Diazepam
See: 7-Chloro-1,3-dihydro-1-methyl-5-phenyl-2*H*-1,4-benzodiazepin-2-one
2,5-Di-(*p*-anisidino)benzoquinone di(*p*-methoxyphenyl)imine
Formed from: *p*-Methoxyaniline

1-(α-Diazoethoxycarbonylmethyl) hydrocotarnine
→ 1,2-Dihydro-8-methoxy-2-methyl-6,7-methylenedioxyisoquinoline
In: Rat A2382

1-(α-Diazo-*p*-methoxy-β-oxophenethyl)hydrocotarnine
→ 1-(α-Diazo-*p*-methoxy-β-oxophenethyl)-4-hydroxyhydrocotarnine
Probably in: Rat A2382

1-(α-Diazo-*p*-methoxy-β-oxophenethyl)-4-hydroxyhydrocotarnine
Probably formed from: 1-(α-Diazo-*p*-methoxy-β-oxophenethyl) hydrocotarnine

Diazoxide (7-chloro-3-methyl-1,2,4-benzothiadiazine-1,1-dioxide)
→ 7-Chloro-3-hydroxymethyl-1,2,4-benzothiadiazine-1,1-dioxide
In: Dog A1232
Man A1232
Monkey A1232

Dibenz[*a*,*j*]acridine
→ Dibenz[*a*,*j*]acridine-5,6-oxide
In: Man H379
→ Dibenz[*a*,*j*]acridine-N-oxide
In: Man H379
→ Dibenz[*a*,*j*]acridin-14-one
In: Man H379
→ 1,2-Dihydro-1,2-dihydroxydibenz[*a*,*j*] acridine
In: Man H379
→ *trans*-3,4-Dihydro-3,4-dihydroxydibenz[*a*,*j*]acridine
In: Man G498, H379
Rat F884
→ *trans*-5,6-Dihydro-5,6-dihydroxydibenz[*a*,*j*]acridine
In: Man H379
Rat F884
→ 2-Hydroxydibenz[*a*,*j*]acridine
In: Mouse G445
Rat G445
→ 3-Hydroxydibenz[*a*,*j*]acridine
In: Man H379
Mouse G445
Rat F884, G445
→ 4-Hydroxydibenz[*a*,*j*]acridine
In: Man H379
Rat F884
→ 5-Hydroxydibenz[*a*,*j*]acridine
In: Mouse G445
Rat G445

Dibenz[*a*,*j*]acridine-5,6-oxide
Formed from: Dibenz[*a*,*j*]acridine

Dibenz[*a*,*j*]acridine-N-oxide
Formed from: Dibenz[*a*,*j*]acridine

Dibenz[*a*,*j*]acridin-14-one
Formed from: Dibenz[*a*,*j*]acridine

Dibenz[*a*,*c*]anthracene
→ 1,2-Dihydro-1,2-dihydroxydibenz[*a*,*c*]anthracene
In: Rat B285
→ 3,4-Dihydro-3,4-dihydroxydibenz[*a*,*c*]anthracene
In: Rat B285
→ 9,10-Dihydro-9,10-dihydroxydibenz[*a*,*c*]anthracene
In: Mouse B376
Rat B285
→ 12,13-Dihydro-12,13-dihydroxydibenz[*a*,*c*]anthracene
In: Rat G442

Dibenzanthracene (dibenz[*a*,*h*]anthracene)
→ Dibenzanthracene-5,6-oxide
In: Rat E556
→ 1,2-Dihydro-1,2-dihydroxydibenzanthracene
In: Man J14
Rat B285, E556
→ 3,4-Dihydro-3,4-dihydroxydibenzanthracene
In: Man J14
Mouse B376
Rat B285, G442
See also: E556
→ 5,6-Dihydro-5,6-dihydroxydibenzanthracene (assumed to be *trans*)
In: Man J14
Rat B285
→ 1-Hydroxydibenzanthracene
In: Rat E556
→ 2-Hydroxydibenzanthracene

In: Rat E556
→ 3-Hydroxydibenzanthracene
In: Rat E556
→ 4-Hydroxydibenzanthracene
In: Rat E556
→ 5-Hydroxydibenzanthracene
In: Rat E556
→ 6-Hydroxydibenzanthracene
In: Rat E556
→ 7-Methyldibenzanthracene
In: Rat E83

Dibenz[*a,j*]anthracene
→ *trans*-3,4-Dihydro-3,4-dihydroxydibenz[*a,j*]anthracene
In: Mouse G611
→ *trans*-5,6-Dihydro-5,6-dihydroxydibenz[*a,j*]anthracene
In: Mouse G611
→ *trans*-3,4,8,9-Tetrahydro-3,4,8,9-tetrahydroxydibenz[*a,j*]anthracene
In: Rat G442

Dibenzanthracene-5,6-oxide
Formed from: Dibenzanthracene
→ *trans*-5,6-Dihydro-5,6-dihydroxydibenzanthracene
In: Man A2486, A3217, B63, B220
Mouse A2486
Rat A2486, A3031, B63, B220

Dibenzanthracen-1-yl-β-D-glucuronide
Formed from: 1-Hydroxydibenzanthracene

Dibenzanthracen-2-yl-β-D-glucuronide
Formed from: 2-Hydroxydibenzanthracene

Dibenzanthracen-4-yl-β-D-glucuronide
Formed from: 4-Hydroxydibenzanthracene

Dibenzanthracen-5-yl-β-D-glucuronide
Formed from: 5-Hydroxydibenzanthracene

Dibenzanthracen-6-yl-β-D-glucuronide
Formed from: 6-Hydroxydibenzanthracene

Dibenzanthracen-7-yl-β-D-glucuronide
Formed from: 7-Hydroxydibenzanthracene

Dibenz[*b,f*]azepine
See: Iminostilbene

5*H*-Dibenz[*b,f*]azepine-5-carboxamide
See: Carbamazepine

4-(3-(5*H*-Dibenz[*b,f*]azepin-5-yl)propyp)-1-piperazineethanol
See: Opipramol

Dibenzepine
See: 10β-Dimethylaminoethyl-10,11-dihydro-5-methyl-11-oxo-5*H*-dibenzo[*b,e*][1,4]diazepine

7*H*-Dibenzo[*c,g*]carbazole
→ 2-Hydroxy-7*H*-dibenzo[*c,g*]carbazole
In: Mouse G445
Rat G445
→ 3-Hydroxy-7*H*-dibenzo[*c,g*]carbazole
In: Mouse G445
Rat G445
→ 5-Hydroxy-7*H*-dibenzo[*c,g*]carbazole
In: Mouse G445
Rat G445

5-(5*H*-Dibenzo[*a,d*]cycloheptan-5-yl)aminopentanoic acid
Formed from: Amineptine
→ 3-(5*H*-Dibenzo[*a,d*]cycloheptan-5-yl)aminopropionic acid
Probably in: Dog G202
Man G202
Rat G202
→ 5-(10-Hydroxy-5*H*-dibenzo[*a,d*]cycloheptan-5-yl)aminopentanoic acid
Probably in: Dog G202
Man G202
Rat G202

3-(5*H*-Dibenzo[*a,d*]cycloheptan-5-yl)aminopropionic acid
Probably formed from: 5-(5*H*-Dibenzo[*a,d*]cycloheptan-5-yl)aminopentanoic acid
→ 3-(10-Hydroxy-5*H*-dibenzo[*a,d*]cycloheptan-5-yl)aminopropionic acid
Probably in: Man G202

5*H*-Dibenzo[*a,d*]cycloheptene
→ 5*H*-Dibenzo[*a,d*]cycloheptene-10,11-oxide
In: Rat A2758, A2851

5*H*-Dibenzo[*a,d*]cycloheptene-10,11-oxide
Formed from: 5*H*-Dibenzo[*a,d*] cycloheptene
→ 10,11-Dihydroxy-5*H*-dibenzo[*a,d*]cycloheptane
In: Rabbit F333

See also: A2851

(5*H*-Dibenzo[*a*,*d*]cyclohepten-5-ylidene)acetic acid

→ (5*H*-Dibenzo[*a*,*d*]cyclohepten-5-ylidene)acetyl-β-D-glucuronic acid
In: Dog E863
Monkey E863
Mouse E863
Rat E863

→ (2-Hydroxy-5*H*-dibenzo[*a*,*d*]cyclohepten-5-ylidene)acetic acid
In: Rat E863

→ (3-Hydroxy-5*H*-dibenzo[*a*,*d*]cyclohepten-5-ylidene)acetic acid
In: Rat E863

(5*H*-Dibenzo[*a*,*d*]cyclohepten-5-ylidene)acetyl-β-D-glucuronic acid

Formed from: (5*H*-Dibenzo[*a*,*d*]cyclohepten-5-ylidene)acetic acid

3-(5*H*-Dibenzo[*a*,*d*]cyclohepten-5-ylidene)-N,N-dimethyl-1-propanamide

See: Cyclobenzaprine

Dibenzo[*c*,*f*][1,2]diazepine

→ 2,2′-Diaminodiphenylmethane
In: Rat B280

→ Dibenzo[*c*,*f*][1,2]diazepine-N-oxide
In: Rat B88, B280

→ 11-Hydroxydibenzo[*c*,*f*][1,2]diazepine
In: Rat B280

See also: B88

Dibenzo[*c*,*f*][1,2]diazepine-N-oxide

Formed from: Dibenzo[*c*,*f*][1,2]diazepine

Dibenzo-*p*-dioxin (dibenzo[1,4]dioxan)

→ *cis*-1,2-Dihydro-1,2-dihydroxydibenzo-*p*-dioxin
In: Beijerinckia B358
Pseudomonas B157

→ 2-Hydroxydibenzo-*p*-dioxin
In: Pseudomonas B157

See also: A3809

→ 2,2′,3-Trihydroxydiphenyl ether
In: Sphingomonas G464

Dibenzo[*a*,*e*]fluoroanthrene

→ *trans*-3,4-Dihydro-3,4-dihydroxydibenzo[*a*,*e*]fluoroanthrene
In: Mouse B453
Rat B453

→ *trans*-12,13-Dihydro-12,13-dihydroxydibenzo[*a*,*e*]fluoroanthrene
In: Mouse B453
Rat B453

→ 3-Hydroxydibenzo[*a*,*e*]fluoroanthrene
In: Mouse B453
Rat B453

→ 4-Hydroxydibenzo[*a*,*e*]fluoroanthrene
In: Mouse B453
Rat B453

→ 7-Hydroxydibenzo[*a*,*e*]fluoroanthrene
In: Mouse B453
Rat B453

Dibenzofuran

→ *cis*-1,2-Dihydro-1,2-dihydroxydibenzofuran
In: Pseudomonas J332

→ *cis*-2,3-Dihydro-2,3-dihydroxydibenzofuran
Probably in: Beijerinckia A3426

→ *trans*-2,3-Dihydro-2,3-dihydroxydibenzofuran
Probably in: Cunninghamella A3426

→ *cis*-3,4-Dihydro-3,4-dihydroxydibenzofuran
In: Pseudomonas J332

→ 1-Hydroxydibenzofuran
In: Pseudomonas F650

→ 2-Hydroxydibenzofuran
In: Cunninghamella A3426
Pseudomonas F650

→ 3-Hydroxydibenzofuran
In: Cunninghamella A3426
Pseudomonas F650

→ 4-Hydroxydibenzofuran
In: Pseudomonas F650

→ Salicylic acid
In: Pseudomonas F650
Staphylococcus G753

→ 2,2′,3-Trihydroxybiphenyl
In: Brevibacterium F460
Pseudomonas H254
Sphingomonas G464, H81

2-Dibenzofuranacetic acid

Formed from: Furobufen

Dibenzo[*b*,*f*]oxepin
→ Dibenzo[*b*,*f*]oxepin-10,11-oxide
In: Rat B865

Dibenzo[*b*,*f*]oxepin-10,11-oxide
Formed from: Dibenzo[*b*,*f*]oxepin
→ 10,11-Dihydro-10,11-dihydroxydibenzo[*b*,*f*]oxepin
Probably in: Rat B865

Dibenzo[*a*,*i*]pyrene
→ 1,2-Dihydro-1,2-dihydroxydibenzo[*a*,*i*]pyrene
Probably in: Mouse C71
→ 3,4-Dihydro-3,4-dihydroxydibenzo[*a*,*i*]pyrene
In: Mouse C71

Dibenzo[*b*,*f*]thiepine
→ Dibenzo[*b*,*f*]thiepine sulphoxide
In: Rat B865

Dibenzo[*b*,*f*]thiepine sulphone
Probably formed from: Dibenzo[*b*,*f*]thiepine sulphoxide

Dibenzo[*b*,*f*]thiepine sulphoxide
Formed from: Dibenzo[*b*,*f*]thiepine
→ Dibenzo[*b*,*f*]thiepine sulphone
Probably in: Rat B865

Dibenzothiophene
→ Benzoic acid
In: Brevibacterium F613
→ Biphenyl
In: Microorganism H692
→ Dibenzothiophene-5-oxide
In: Beijerinckia A3340
Brevibacterium F613
Cunninghamella E46, H60
Mortierella E46
Pseudomonas A1647, F168, H672
Rat D636
Rhizopus E46
Rhodococcus J396
See also: K319, K409
→ *cis*-1,2-Dihydro-1,2-dihydroxydibenzothiophene
In: Beijerinckia A3340
Pseudomonas J332
→ *cis*-3,4-Dihydro-3,4-dihydroxydibenzothiophene
In: Pseudomonas J332
→ 2,2′-Dihydroxybiphenyl
Probably in: Paecilomyces G438
→ Dithiosalicylic acid
In: Microorganism J368
→ 2-Formyl-3-hydroxybenzo[*b*]thiophene
In: Pseudomonas A1647
See also: J303

Dibenzothiophene-5,5-dioxide
Formed from: Dibenzothiophene-5-oxide
→ 2-(*o*-Hydroxyphenyl)benzenesulphinic acid
In: Rhodococcus J396, K315
See also: K319, K409

Dibenzothiophene-5-oxide
Formed from: Dibenzothiophene
→ Dibenzothiophene-5,5-dioxide
In: Brevibacterium F613
Cunninghamella E46, H60
Mortierella E46
Rat D636
Rhizopus E46
Rhodococcus J396
See also: F168, H672, K319, K409

Dibenz[*c*,*e*][1,2]oxathiin
→ Dibenz[*c*,*e*][1,2]oxathiin-6,6-dioxide
Probably in: Rhodococcus K315

Dibenz[*c*,*e*][1,2]oxathiin-6,6-dioxide
Probably formed from: Dibenz[*c*,*e*][1,2]oxathiin
→ 2,2′-Dihydroxybiphenyl
In: Rhodococcus K315

Dibenzo[*b*,*f*]-1,4-oxazepine
→ 2-Amino-2′-hydroxymethyl diphenyl ether
In: Guinea pig C492
Rat C493
→ 10,11-Dihydro-11(10*H*)-oxodibenzo[*b*,*f*]-1,4-oxazepine
In: Guinea pig C492
Monkey C492
Rat C392, C493

Dibenzoyl-2-amino-tetrahydro-6,7-dihydroxynaphthalene
→ 2-Amino-tetrahydro-6,7-dihydroxynaphthalene
In: Rat C251
→ Benzoic acid
In: Rat C251

Dibenzoylbenzene
Formed from: 1,3-Diphenylisobenzofuran
Dibenzoylepinine
→ Benzoic acid
In: Beef F220
→ 3,4-Dihydroxy-N-methylphenethylamine
In: Beef F220
N^2,N^5-Dibenzoyl-L-ornithine
See: Ornithuric acid
Dibenzylamine
Formed from: Tribenzylamine
Probably formed from: N,N-Dibenzylethylamine
N-Methyldibenzylamine
→ Benzaldehyde
Probably in: Rabbit A2331
→ Benzylamine
In: Rabbit A2331
→ N,N-Dibenzylhydroxylamine
In: Rabbit A1731, A2331
Dibenzylaniline
→ Benzaldehyde
Probably in: Rat K296
→ N-Benzylaniline
In: Rat K296
N,N-Dibenzylbenzamide
→ Benzaldehyde
Probably in: Rat K296
→ N-Benzylbenzamide
Probably in: Rat K296
N,N-Dibenzyldaunorubicin
→ N-Benzyldaunorubicin
In: Rat C539, D88
→ N,N-Dibenzyldaunorubicinol
In: Rat C539
→ 7-Deoxydaunorubicinone
In: Rat C539
N,N-Dibenzyldaunorubicinol
Formed from: N,N-Dibenzyldaunorubicin
4,7-Dibenzyl-1-(2,6-dimethylphenoxy)-10-(hexahydro-2,4-dioxopyrimidin-1-yl)-11-methyl-2,9-dioxo-3,8-diazadodecane
Probably formed from: 4,7-Dibenzyl-1-(2,6-dimethylphenoxy)-10-(hexahydro-4-hydroxy-2-oxopyrimidin-1-yl)-11-methyl-2,9-dioxo-3,8-diazadodecane
→ 4-Benzyl-1-(2,6-dimethylphenoxy)-10-(hexahydro-2,4-dioxopyrimidin-1-yl)-7-(*p*-hydroxybenzyl)-11-methyl-2,9-dioxo-3,8-diazadodecane
Probably in: Man J562
→ 7-Benzyl-1-(2,6-dimethylphenoxy)-10-(hexahydro-2,4-dioxopyrimidin-1-yl)-4-(*p*-hydroxybenzyl)-11-methyl-2,9-dioxo-3,8-diazadodecane
Probably in: Man J562
4,7-Dibenzyl-1-(2,6-dimethylphenoxy)-10-(hexahydro-4-hydroxy-2-oxopyrimidin-1-yl)-11-methyl-2,9-dioxo-3,8-diazadodecane
Formed from: 4,7-Dibenzyl-1-(2,6-dimethylphenoxy)-10-(hexahydro-2-oxopyrimidin-1-yl)-11-methyl-2,9-dioxo-3,8-diazadodecane
→ 4-Benzyl-1-(2,6-dimethylphenoxy)-10-(hexahydro-4-hydroxy-2-oxopyrimidin-1-yl)-7-(*p*-hydroxybenzyl)-11-methyl-2,9-dioxo-3,8-diazadodecane
Probably in: Man J562
→ 7-Benzyl-1-(2,6-dimethylphenoxy)-10-(hexahydro-4-hydroxy-2-oxopyrimidin-1-yl)-4-(*p*-hydroxybenzyl)-11-methyl-2,9-dioxo-3,8-diazadodecane
Probably in: Man J562
→ 4,7-Dibenzyl-1-(2,6-dimethylphenoxy)-10-(hexahydro-2,4-dioxopyrimidin-1-yl)-11-methyl-2,9-dioxo-3,8-diazadodecane
Probably in: Man J562
4,7-Dibenzyl-1-(2,6-dimethylphenoxy)-10-(hexahydro-6-hydroxy-2-oxopyrimidin-1-yl)-11-methyl-2,9-dioxo-3,8-diazadodecane
Formed from: 4,7-Dibenzyl-1-(2,6-dimethylphenoxy)-10-(hexahydro-2-oxopyrimidin-1-yl)-11-methyl-2,9-dioxo-3,8-diazadodecane
4,7-Dibenzyl-1-(2,6-dimethylphenoxy)-10-(hexahydro-2-oxopyrimidin-1-yl)-11-methyl-2,9-dioxo-3,8-diazadodecane
→ 4-Benzyl-1-(2,6-dimethylphenoxy)-10-(hexahydro-2-oxopyrimidin-1-yl)-7-(*p*-hydroxybenzyl)-11-methyl-2,9-dioxo-3,8-diazadodecane
Probably in: Man J562

→ 4-Benzyl-1-(2,6-dimethylphenoxy)-10-(hexahydro-2-oxopyrimidin-1-yl)-7-(α-hydroxybenzyl)-11-methyl-2,9-dioxo-3,8-diazadodecane
Probably in: Man J562
→ 7-Benzyl-1-(2,6-dimethylphenoxy)-10-(hexahydro-2-oxopyrimidin-1-yl)-4-(*p*-hydroxybenzyl)-11-methyl-2,9-dioxo-3,8-diazadodecane
Probably in: Man J562
→ 4,7-Dibenzyl-1-(2,6-dimethylphenoxy)-10-(hexahydro-4-hydroxy-2-oxopyrimidin-1-yl)-11-methyl-2,9-dioxo-3,8-diazadodecane
In: Man J562
→ 4,7-Dibenzyl-1-(2,6-dimethylphenoxy)-10-(hexahydro-6-hydroxy-2-oxopyrimidin-1-yl)-11-methyl-2,9-dioxo-3,8-diazadodecane
In: Man J562
→ 4,7-Dibenzyl-10-(hexahydro-2-oxopyrimidin-1-yl)-1-(2-hydroxymethyl-6-methylphenoxy)-11-methyl-2,9-dioxo-3,8-diazadodecane
Probably in: Man J562

Dibenzyl diselenide
Formed from: Benzyl selenocyanate
→ Benzoic acid
In: Rat G282

Dibenzyl disulphide
Formed from: S-Benzyl O-ethyl phenylphosphonothioate
→ Benzyl mercaptan
In: Microorganism G646
→ Toluene
In: Microorganism G646

N,N-Dibenzylethylamine
→ Dibenzylamine
Probably in: Rat K296
→ N,N-Dibenzylethylamine-N-oxide
In: Rat K296

N,N-Dibenzylethylamine-N-oxide
Formed from: N,N-Dibenzylethylamine

4,7-Dibenzyl-10-(hexahydro-2-oxopyrimidin-1-yl)-1-(2-hydroxymethyl-6-methylphenoxy)-11-methyl-2,9-dioxo-3,8-diazadodecane
Probably formed from: 4,7-Dibenzyl-1-(2,6-dimethylphenoxy)-10-(hexahydro-2-oxopyrimidin-1-yl)-11-methyl-2,9-dioxo-3,8-diazadodecane

N,N-Dibenzylhydroxylamine
Formed from: Dibenzylamine

Dibenzyl sulphoxide
→ Benzyl alcohol
In: Rhodococcus H618
→ Toluene
In: Rhodococcus H618

2,4-Dibromoanisole
Formed from: 2,4-Dibromophenol

3,5-Dibromoanthranilic acid
Formed from: Ambroxol
Bromhexine
→ 3,5-Dibromoanthraniloylglycine
In: Dog A763
Mouse A763
Rat A763

3,5-Dibromoanthraniloylglycine
Formed from: 3,5-Dibromoanthranilic acid

3,3′-Dibromoazobenzene
Formed from: 3,3′-Dibromoazoxybenzene

3,3′-Dibromoazoxybenzene
→ 3,3′-Dibromoazobenzene
In: Saccharomyces H894

1,2-Dibromobenzene
→ 2,3-Dibromophenol
In: Rabbit A1825
→ 3,4-Dibromophenol
In: Rabbit A1825

1,3-Dibromobenzene
→ *m*-Bromophenol
Probably in: Rat K263
→ *m*-Bromothiophenol
Probably in: Rat K263
→ 3,5-Dibromophenol
Probably in: Rat K263
→ 3,5-Dibromothiophenol
Probably in: Rat K263

1,4-Dibromobenzene
→ 2,4-Dibromophenol
In: Rabbit A1825
→ 2,5-Dibromophenol
In: Rabbit A1825

4,4′-Dibromobenzilic acid
Formed from: Acarol

4,4′-Dibromobenzoin
Formed from: *p*-Bromobenzaldehyde
4,4′-Dibromobiphenyl
→ 4,4′-Dibromo-3-methoxybiphenyl
In: Rat A3390
6,8-Dibromo-(+)-catechin
→ 6,8-Dibromo-3′-O-methyl-(+)-catechin
In: Pig D112
3,5-Dibromocatechol
Formed from: 2,4-Dibromophenol
4,5-Dibromocatechol
Formed from: 4,5-Dibromo-1,2-methylenedioxybenzene
2,6-Dibromo-4-chloroaniline
Probably formed from: 2-Bromo-4-chloroaniline
3,5-Dibromo-3′-((1,6-dihydro-6-oxopyridazin-3-yl)methyl)thyroacetic acid
Formed from: L-3,5-Dibromo-3′-((1,6-dihydro-6-oxopyridazin-3-yl)methyl)thyronine
3,5-Dibromo-3′-((1,6-dihydro-6-oxopyridazin-3-yl)methyl)thyrolactic acid
Formed from: L-3,5-Dibromo-3′-((1,6-dihydro-6-oxopyridazin-3-yl)methyl) thyronine
L-3,5-Dibromo-3′-((1,6-dihydro-6-oxopyridazin-3-yl)methyl)thyronine
→ 3,5-Dibromo-3′-((1,6-dihydro-6-oxopyridazin-3-yl)methyl)thyroacetic acid
In: Dog F446
Monkey F446
Rat F446
→ 3,5-Dibromo-3′-((1,6-dihydro-6-oxopyridazin-3-yl)methyl)thyrolactic acid
In: Dog F446
Monkey F446
Rat F446
2,3-Dibromo-4,5-dihydroxybenzaldehyde
Probably formed from: 3-Bromo-4-hydroxybenzaldehyde
3,5-Dibromo-α-(((1,1-dimethylethyl)amino)methyl)-4-hydroxylaminobenzenemethanol
Formed from: Bromobuterol
2,2′-Dibromo-4,4′-dimethylhydrazobenzene
Formed from: 2-Bromo-4-methylaniline

3-(2,2-Dibromoethenyl)-2,2-dimethylcyclopropanecarboxylic acid (*S*)-cyano(3-phenoxyphenyl)methyl ester
See: Deltamethrin
N′,N′-Di-(2-bromoethyl)-3-methyl-*p*-phenylenediamine
Formed from: 4-Bis(2-bromoethyl)amino-2-methylazobenzene
2′-Carboxy-4-bis(2-bromoethyl)amino-2-methylazobenzene
N,N-Di-(2-bromoethyl)-*p*-phenylenediamine
Formed from: 4-Bis(2-bromoethyl)aminoazobenzene
2′-Carboxy-4-bis(2-bromoethyl)aminoazobenzene
2,4-Dibromoethynyloestradiol
→ 2,4-Dibromoethynyloestradiol-β-D-glucuronide
In: Rat F476
2,4-Dibromoethynyloestradiol-β-D-glucuronide
Formed from: 2,4-Dibromoethynyloestradiol
3,5-Dibromo-4-hydroxybenzaldehyde
Formed from: 3-Bromo-4-hydroxybenzaldehyde
→ 3,5-Dibromo-4-hydroxybenzoic acid
In: Microorganism E664
→ 3,5-Dibromo-4-hydroxybenzyl alcohol
In: Microorganism E664
3,5-Dibromo-4-hydroxybenzamide
Probably formed from: 3,5-Dibromo-4-hydroxybenzonitrile
→ 3,5-Dibromo-4-hydroxybenzoic acid
Probably in: Microorganism A1251
Wheat A746
3,5-Dibromo-4-hydroxybenzoic acid
Formed from: 3,5-Dibromo-4-hydroxybenzonitrile
3,5-Dibromo-4-hydroxybenzaldehyde
Probably formed from: 3,5-Dibromo-4-hydroxybenzamide
→ 2,6-Dibromophenol
Probably in: Wheat A746
3,5-Dibromo-4-hydroxybenzonitrile (bromoxynil)
Probably formed from: Bromoxynil octanoate
→ 3-Bromo-4,5-dihydroxybenzonitrile
Probably in: Wheat A746
→ 3-Bromo-4-hydroxybenzonitrile

Probably in: Wheat A746
→ 3,5-Dibromo-4-hydroxybenzamide
In: Microorganism A1251
Pseudomonas G609
See also: A746
→ 3,5-Dibromo-4-hydroxybenzoic acid
In: Fusarium A3364
Pseudomonas G609
→ 2,6-Dibromoquinol
In: Flavobacterium G641

3,5-Dibromo-4-hydroxybenzonitrile octanoate ester
See: Bromoxynil octanoate

3,5-Dibromo-4-hydroxybenzyl alcohol
Formed from: 3,5-Dibromo-4-hydroxybenzaldehyde

***cis*-4-(3,5-Dibromo-2-hydroxybenzylamino)cyclohexanol**
Formed from: 4-(3,5-Dibromo-2-hydroxybenzylamino)cyclohexanone

***trans*-4-(3,5-Dibromo-2-hydroxybenzylamino)cyclohexanol**
Formed from: 4-(3,5-Dibromo-2-hydroxybenzylamino)cyclohexanone
→ 4-(3,5-Dibromo-4-hydroxybenzylamino)cyclohexanone
In: Cow E61
Dog E61
Horse E61
Man E61
Rat E61

4-(3,5-Dibromo-2-hydroxybenzylamino) cyclohexanone
Formed from: *trans*-4-(3,5-Dibromo-2-hydroxybenzylamino)cyclohexanol
→ *cis*-4-(3,5-Dibromo-2-hydroxybenzylamino)cyclohexanol
In: Cow E61
Dog E61
Horse E61
Rat E61
→ *trans*-4-(3,5-Dibromo-2-hydroxybenzylamino)cyclohexanol
In: Cow E61
Dog E61
Horse E61
Rat E61

2,4-Dibromo-16α-hydroxyoestrone
Probably formed from: 2,4-Dibromooestrone

3,5-Dibromo-4′-hydroxysalicylanilide
Formed from: 3,5-Dibromosalicylanilide
3,4′,5-Tribromosalicylanilide
→ 3-Bromo-4′5-dihydroxysalicylanilide
Probably in: Rat B708
→ 3,5-Dibromo-4′-hydroxysalicylanilide-β-D-glucuronide
In: Rat B709
→ 3,5-Dibromo-4′-hydroxysalicylanilide sulphate
In: Rat B709

3,4′-Dibromo-5-hydroxysalicylanilide
Formed from: 3,4′,5-Tribromosalicylanilide
→ 3,4′-Dibromo-5-hydroxysalicylanilide-β-D-glucuronide
In: Rat B709
→ 3,4′-Dibromo-5-hydroxysalicylanilide sulphate
In: Rat B709

3,5-Dibromo-4′-hydroxysalicylanilide-β-D-glucuronide
Formed from: 3,5-Dibromo-4′-hydroxysalicylanilide

3,4′-Dibromo-5-hydroxysalicylanilide-β-D-glucuronide
Formed from: 3,4′-Dibromo-5-hydroxysalicylanilide

3,5-Dibromo-4′-hydroxysalicylanilide sulphate
Formed from: 3,5-Dibromo-4′-hydroxysalicylanilide

3,4′-Dibromo-5-hydroxysalicylanilide sulphate
Formed from: 3,4′-Dibromo-5-hydroxysalicylanilide

4,4′-Dibromo-3-methoxybiphenyl
Formed from: 4,4′-Dibromobiphenyl

2-(3,5-Dibromo-2-methoxyphenyl)-1-methylpyrrole
→ 1-Methyl-2-(3,4,5,6-tetrabromo-2-methoxyphenyl)pyrrole
Probably in: Streptomyces F276

6,8-Dibromo-3′-O-methyl-(+)-catechin
Formed from: 6,8-Dibromo-(+)-catechin

4,5-Dibromo-1,2-methylenedioxybenzene
→ 4,5-Dibromocatechol
In: Rat A3947

1,4-Dibromonaphthalene
→ 2,4-Dibromo-α-naphthol
In: Pig A1829
2,4-Dibromo-α-naphthol
Formed from: 1,4-Dibromonaphthalene
2,6-Dibromo-4-nitrophenol
→ 2,6-Dibromo-4-nitroquinol
In: Flavobacterium G641
2,6-Dibromo-4-nitroquinol
Formed from: 2,6-Dibromo-4-nitrophenol
2,4-Dibromo-17β-oestradiol
→ 2,4-Dibromooestriol
In: Rat G579
→ 2,4-Dibromooestrone
In: Rat G579
2,4-Dibromooestriol
Formed from: 2,4-Dibromo-17β-oestradiol
2,4-Dibromooestrone
Formed from: 2,4-Dibromo-17β-oestradiol
→ 2,4-Dibromo-16α-hydroxyoestrone
Probably in: Rat G579
→ 2,4-Dibromooestrone-β-D-glucuronide
Probably in: Rat G579
2,4-Dibromooestrone-β-D-glucuronide
Probably formed from: 2,4-Dibromooestrone
2,3-Dibromophenol
Formed from: 1,2-Dibromobenzene
2,4-Dibromophenol
Formed from: 1,4-Dibromobenzene
Probably formed from: *o*-Bromophenol
p-Bromophenol
→ *o*-Bromophenol
In: Microorganism G331
→ *p*-Bromophenol
In: Microorganism G331
→ 2,4-Dibromoanisole
In: Phanerochete G782
→ 3,5-Dibromocatechol
In: Proteobacteria J466
→ 2,4,6-Tribromophenol
Probably in: Ulva K153
2,5-Dibromophenol
Formed from: 1,4-Dibromobenzene
2,6-Dibromophenol
Probably formed from: *o*-Bromophenol
3,5-Dibromo-4-hydroxybenzoic acid
→ 2,6-Dibromoquinol
In: Flavobacterium G641
→ 2,4,6-Triibromophenol
Probably in: Corallina E597
Ulva K153
3,4-Dibromophenol
Formed from: 1,2-Dibromobenzene
3,5-Dibromophenol
Probably formed from: 1,3-Dibromobenzene
1,2-Dibromo-1-phenylethane
→ S-(2-Hydroxy-1-phenylethyl)glutathione (both *R* and *S*)
In: Rat E870
→ S-(2-Hydroxy-2-phenylethyl)glutathione (both *R* and *S*)
In: Rat E870
N′,N′-Di-(2-bromopropyl)-3-methyl-*p*-phenylenediamine
Formed from: 2′-Carboxy-4-bis(2-bromopropyl)amino-2-methylazobenzene
N,N-Di-(2-bromopropyl)-*p*-phenylenediamine
Formed from: 2-(4′-Bis(2″-bromopropyl)aminophenylazo)benzoic acid
2,6-Dibromoquinol
Formed from: 3,5-Dibromo-4-hydroxybenzonitrile
2,6-Dibromophenol
2,4,6-Tribromophenol
3,5-Dibromosalicylanilide
→ 3,5-Dibromo-4′-hydroxysalicylanilide
In: Rat B708
→ 3-Bromo-5-hydroxysalicylanilide
In: Rat B708
→ 3,5-Dibromosalicylanilide-β-D-glucuronide
In: Rat B708
→ 3,5-Dibromosalicylanilide sulphate
In: Rat B708
4′,5-Dibromosalicylanilide
→ 4′,5-Dibromosalicylanilide-β-D-glucuronide
In: Rat B709
→ 4′,5-Dibromosalicylanilide sulphate
In: Rat B709
3,5-Dibromosalicylanilide-β-D-glucuronide
Formed from: 3,5-Dibromosalicylanilide
4′,5-Dibromosalicylanilide-β-D-glucuronide
Formed from: 4′,5-Dibromosalicylanilide

3,5-Dibromosalicylanilide sulphate
Formed from: 3,5-Dibromosalicylanilide
4′,5-Dibromosalicylanilide sulphate
Formed from: 4′,5-Dibromosalicylanilide
6,8-Dibromo-1,2,3,4-tetrahydro-3-(*trans*-4-hydroxycyclohexyl)quinazoline
Formed from: Ambroxol
3,5-Dibromothiophenol
Probably formed from: 1,3-Dibromobenzene
3,5-Dibromotyramine
Formed from: 3,5-Dibromo-L-tyrosine
3,5-Dibromo-L-tyrosine
→ 3,4′-Bis(3-alanyl)-3′,5,5′-tribromo-4-hydroxydiphenyl ether
In: Horseradish J372
→ 3,5-Dibromotyramine
In: Micrococcus C62
→ Dibromoverongiaquinol
In: Aplysina H696
Dibromoverongiaquinol
Formed from: 3,5-Dibromo-L-tyrosine
Dibucaine (2-butoxy-N-(2-diethylaminoethyl)-4-quinolinecarboxamide)
→ 2-Butoxy-N-(2-ethylamino)ethyl-4-quinolinecarboxamide
In: Man C822
Rabbit C822
Rat C822
→ N-(2-Diethylamino)ethyl-2-(2,3-dihydroxybutoxy)-4-quinolinecarboxamide
In: Man C822
Rabbit C822
Rat C822
→ N-(2-Diethylamino)ethyl-2-(2-hydroxyethoxy)-4-quinolinecarboxamide
In: Man C822
Rabbit C822
Rat C822
3,6-Dibutanoylmorphine (3,6-Dibutyrylmorphine)
→ 6-Butanoylmorphine
In: Man D173
Rat D173, E232
2-(4-(3-(Dibutylamino)propoxy)-3,5-dimethylbenzoyl)-4H-1-benzopyran-4-one
See: Bucromarone
N,N-Di-*sec*-butylaniline
→ N-*sec*-Butylaniline
In: Rabbit A3484
3,3′-Di-*tert*-butyl-2,2′-dihydroxy-5,5′-dimethoxybiphenyl
Formed from: 2-*tert*-Butyl-4-methoxyphenol
2,6-Di-*tert*-butyl-4-hydroperoxy-4-methylcyclohexa-2,5-dienone
Formed from: 2,6-Di-*tert*-butyl-4-methylphenol
→ 2,6-Di-*tert*-butyl-4-hydroxymethylphenol
In: Rat A1602
3,5-Di-*tert*-butyl-4-hydroxybenzaldehyde
Probably formed from: 2,6-Di-*tert*-butyl-4-hydroxymethylphenol
→ 3,5-Di-*tert*-butyl-4-hydroxybenzoic acid
Probably in: Man A2864
Mouse D206, E890
Rat D206, E890
3,5-Di-*tert*-butyl-4-hydroxybenzoic acid
Probably formed from: 3,5-Di-*tert*-butyl-4-hydroxybenzaldehyde
→ 3,5-Di-*tert*-butyl-4-hydroxybenzoic acid-β-D-glucuronide
In: Rat G164
See also: D206
→ 2,6-Di-*tert*-butylphenol
In: Rat G164
3,5-Di-*tert*-butyl-4-hydroxybenzoic acid-β-D-glucuronide
Formed from: 3,5-Di-*tert*-butyl-4-hydroxybenzoic acid
S-(2,6-Di-*tert*-butyl-4-hydroxybenzyl)glutathione
Formed from: 3,5-Di-*tert*-butyl-4-methylphenol
2,6-Di-*tert*-butyl-6-(2-hydroxy-1,1-dimethylethyl)quinol
Probably formed from: 2,6-Di-*tert*-butylquinol
2,6-Di-*tert*-butyl-4-hydroxy-4-methylcyclohexa-2,5-dienone
Formed from: 2,6-Di-*tert*-butyl-4-methylphenol
→ 2,6-Di-*tert*-butyl-4-hydroxymethylphenol
In: Rat A1602
→ 2,6-Di-*tert*-butyl-4-methylphenol
In: Rat A1602

2,6-Di-*tert*-butyl-4-hydroxymethylphenol
Formed from: 2,6-Di-*tert*-butyl-4-methylphenol
2,6-Di-*tert*-butyl-4-hydroperoxy-4-methylcyclohexa-2,5-dienone
2,6-Di-*tert*-butyl-4-hydroxy-4-methylcyclohexa-2,5-dienone
→ 2,6-Di-*tert*-butyl-4-hydroxybenzaldehyde
Probably in: Man A2864
Mouse D206, E890
Rat D206, E890
→ 2,6-Di-*tert*-butyl-4-methylene-2,5-cyclohexadienone
In: Rat C130

3,5-Di-*tert*-butyl-1-hydroxy-4-oxo-1-((thiazolidin-4-on-5-yl)methyl)cyclohexa-2,5-dione
Formed from: Tazofelone

2-(3,5-Di-*tert*-butyl-2-hydroxyphenyl)-2*H*-benzotriazole
Formed from: 2-(3,5-Di-*tert*-butyl-2-hydroxyphenyl)-2*H*-benzotriazole-1-oxide

2-(3,5-Di-tert-butyl-2-hydroxyphenyl)-2*H*-benzotriazole-1-oxide
→ 2-(3,5-Di-*tert*-butyl-2-hydroxyphenyl)-2*H*-benzotriazole
In: Saccharomyces H894

2,6-Di-*tert*-butyl-4-methoxyphenol (topanol 354)
→ 2,6-Di-*tert*-butylquinol
In: Dog A1079
Man A1079
Rat A1079

2,6-Di-*tert*-butyl-4-methylene-2,5-cyclohexadienone
Formed from: 2,6-Di-*tert*-butyl-4-methylphenol

2,6-Di-*tert*-butyl-4-methylphenol (BHT, IONOL, 3,5-di-*tert*-butyl-4-hydroxytoluene)
Formed from: 2-*tert*-Butyl-4-hydroxy-4-methylcyclohexa-2,5-dienone
→ N-Acetyl-S-(3,5-di-*tert*-Butyl-4-hydroxybenzyl)-L-cysteine
Probably in: Man A2864
→ 6-*tert*-Butyl-2-(1-(hydroxymethyl)isopropyl)-4-methylphenol
In: Mouse E890, F738
Rat E890, F738
→ 5-Carboxy-7-(1-carboxy-1-methylethyl)-2,3-dihydro-2-hydroxy-3,3-dimethylbenzofuran
In: Man A2864
→ S-(2,6-Di-*tert*-butyl-4-hydroxybenzyl)glutathione
In: Rat D7
→ 2,6-Di-*tert*-butyl-4-hydroperoxy-4-methylcyclohexa-2,5-dienone
In: Rat A1602, E382
→ 2,6-Di-*tert*-butyl-4-hydroxy-4-methylcyclohexa-2,5-dienone
In: Mouse E890
Rat A1602, E382, E890
→ 2,6-Di-*tert*-butyl-4-hydroxymethylphenol
In: Mouse D206, E890
Rat A1602, D206, E382, E890
See also: A2864
→ 2,6-Di-*tert*-butyl-4-methylene-2,5-cyclohexadienone
In: Mouse F352, F738
Rat B359, C130, F738
→ 2,6-Di-*tert*-butyl-4-methylphenyl-β-D-glucuronide
In: Man J490
→ 2,6-Di-*tert*-butyl-4-((methylthio)methyl)phenol
In: Rat B58
→ 2,6-Di-*tert*-butylquinol
In: Rat B58, D7
See also: E890

2,6-Di-*tert*-butyl-4-methylphenyl-β-D-glucuronide
Formed from: 2,6-Di-*tert*-butyl-4-methylphenyl

2,6-Di-*tert*-butyl-4-((methylthio)methyl)phenol
Formed from: 2,6-Di-*tert*-butyl-4-methylphenyl

2,6-Di-*tert*-butylphenol
Formed from: 2,6-Di-*tert*-butyl-4-hydroxybenzoic acid
→ 2,6-Di-*tert*-butylquinol
In: Rat G164

Di-*n*-butyl phthalate
→ Phthalic acid mono-*n*-butyl ester
In: Baboon A1652
Beef J716
Ferret A1652

Microorganism A2214, K248
Rat A1652, A2376, A3866
See also: C600

2,6-Di-*tert*-butylquinol
Formed from: 2,6-Di-*tert*-butyl-4-methoxyphenol
2,6-Di-*tert*-butylphenol
→ 2,6-Di-*tert*-butylquinol-1-β-D-glucuronide
Probably in: Rat D7, G164
→ 2,6-Di-*tert*-butylquinone
Probably in: Dog E890
Rat B58, G164, E890
→ 2,6-Di-*tert*-butyl-6-(2-hydroxy-1,1-dimethylethyl)quinol
In: Dog A1079
Man A1079
Rat A1079

2,6-Di-*tert*-butylquinol-1-β-D-glucuronide
Probably formed from: 2,6-Di-*tert*-butylquinol

2,6-Di-*tert*-butylquinone
Probably formed from: 2,6-Di-*tert*-butylquinol

3,5-Di-*tert*-butyl-*o*-quinone
Formed from: 3,5-Di-*tert*-butylphenol

3,5-Di-*tert*-butyl-4-hydroxytoluene
See: 2,6-Di-*tert*-butyl-4-methylphenol

Dibutyrylmorphine
See: 3,6-Dibutanoylmorphine

(2*R*,3*R*)-O-Dicaffeoyltartaric acid
See: Cichoric acid

Dicambra
See: 3,6-Dichloro-2-methoxybenzoic acid

3,3′-Dicarboxybiphenyl
Formed from: 4,6-Dimethyldibenzothiophene

3″,7-Dicarboxy-4″,5″-bisnorcannabinol
Probably formed from: 3″-Carboxy-7-hydroxy-4″,5″-bisnorcannabinol

3,3′-Dicarboxy-4-(*o*-chlorophenyl)-2-hydroxymethyl-6-methylpyridine
Probably formed from: 3-Carboxy-4-(*o*-chlorophenyl)-2-hydroxymethyl-5-methoxycarbonyl-6-methylpyridine

3,5-Dicarboxy-2-cyano-6-methyl-4-(3-nitrophenyl)pyridine
Probably formed from: Methyl 5-carboxy-2-cyano-6-methyl-4-(3-nitrophenyl)pyridine-3-carboxylate

2′,3-Dicarboxy-2,6-dichlorodiphenylamine
Probably formed from: 2-Carboxy-2′,6′-dichloro-3′-hydroxymethyldiphenylamine

2,3-Dicarboxy-4-(2,3-dichlorophenyl)-5-ethoxycarbonyl-6-methylpyridine
Probably formed from: 3-Carboxy-4-(2,3-dichlorophenyl)-2-hydroxymethyl-5-ethoxycarbonyl-6-methylpyridine

2,3-Dicarboxy-4-(2,3-dichlorophenyl)-5-methoxycarbonyl-6-methylpyridine
Probably formed from: 3-Carboxy-4-(2,3-dichlorophenyl)-2-hydroxymethyl-5-methoxycarbonyl-6-methylpyridine

5,5′-Dicarboxy-2,2′-dihydroxy-3,3′-dimethoxybiphenyl
Formed from: Vanillic acid
→ 3,3′-Bis(5-carboxy-2-hydroxy-3-methoxyphenyl)-4,4′-dihydroxy-5,5′-dimethoxybiphenyl
In: Coriolus F459
→ 5-Carboxy-3-(5-carboxy-2-hydroxy-3-methoxyphenyl)-*o*-quinone
Probably in: Coriolus F459
→ 5′-Carboxy-5-(5,5′-dicarboxy-2′-hydroxy-3,3′-dimethoxy-2-biphenyloxy)-2,2′-dihydroxy-3,3′-dimethoxybiphenyl
In: Coriolus F459
→ 5-Carboxy-2,2′-dihydroxy-3,3′,5′-trimethoxybiphenyl
In: Coriolus F459
→ 6-Carboxy-3,8-dimethoxybenzofuran-1,4-quinone
In: Coriolus F459
→ 2-(5-Carboxy-2-hydroxy-3-methoxyphenyl)-6-methoxy-*p*-quinone
Probably in: Coriolus F459
→ 5,5′-Dicarboxy-2,2′,3-trihydroxy-3′-methoxybiphenyl
In: Pseudomonas E733
→ 4-Hydroxy-5-methoxyisophthalic acid

In: Pseudomonas E733

3,5-Dicarboxy-2,6-dimethyl-4-(3-nitrophenyl)pyridine
Probably formed from: 2-Nitrooxypropyl 3-nitrooxypropyl 2,6-dimethyl-4-(3-nitrophenyl)pyridine-3,5-dicarboxylate
→ 3,5-Dicarboxy-2-hydroxymethyl-6-methyl-4-(3-nitrophenyl)pyridine
Probably in: Dog G593
Rat G593

3,5-Dicarboxy-2-(4-ethoxy-3-methoxyphenyl)-2,3-dihydro-7-methoxybenzofuran
Formed from: 5-Carboxy-2-(4-ethoxy-3-methoxyphenyl)-3-formyl-2,3-dihydro-7-methoxybenzofuran

2,6-Di-(1-carboxyethyl)naphthalene
Formed from: 2,6-Diisopropylnaphthalene

3,5-Dicarboxy-2-hydroxymethyl-6-methyl-4-(3-nitrophenyl)pyridine
Probably formed from: 3,5-Dicarboxy-2,6-dimethyl-4-(3-nitrophenyl)pyridine

5,5′-Di(carboxymethyl)-2,2′-dihydroxybiphenyl
Formed from: *p*-Hydroxyphenylacetic acid

5,5′-Dicarboxy-2,2′,3-trihydroxy-3′-methoxybiphenyl
Formed from: 5,5′-Dicarboxy-2,2′-dihydroxy-3,3′-dimethoxybiphenyl

Dichlobenil
See: 2,6-Dichlorobenzonitrile

2-(Dichloroacetamido)-1-(*p*-hydroxylaminophenyl)-1,3-propanediol
Formed from: D-*threo*-1-(*p*-Aminophenyl)-2-(dichloroacetamido)-1,3-propanediol
→ 2-(Dichloroacetamido)-1-(*p*-nitrosophenyl)-1,3-propanediol
In: Rat D811

(2-Dichloroacetamido)-3-hydroxy-3-(*p*-nitrophenyl)-1-propanal
Formed from: Chloramphenicol

2-(Dichloroacetamido)-1-(*p*-nitrophenyl)-1,3-propanediol
See: Chloramphenicol

2-(Dichloroacetamido)-1-(*p*-nitrosophenyl)-1,3-propanediol
Formed from: Chloramphenicol
2-(Dichloroacetamido)-1-(*p*-hydroxylaminophenyl)-1,3-propanediol
→ 1-(*p*-Aminophenyl)-2-(dichloroacetamido)-1,3-propanediol
In: Rat D689

2,5-Dichloroacetanilide
→ 2,5-Dichloroaniline
In: Grass A1321

3,4-Dichloroacetanilide
Formed from: 3,4-Dichloroaniline
→ 3,4-Dichloroaniline
In: Dandelion A230

2′,4′-Dichloroacetophenone
Formed from: S-(2,4-Dichlorophenacyl)glutathione
→ 1-(2,4-Dichlorophenyl)ethanol
Probably in: Rat A1752

2′,6′-Dichloroacetophenone imine
→ 2′,6′-Dichloroacetophenone oxime
In: Ferret D955
Guinea pig D955
Hamster D955
Mouse D955
Rabbit C243, D955
Rat D955

2′,6′-Dichloroacetophenone oxime
Formed from: 2′,6-Dichloroacetophenone imine

1-Dichloroacetyl-6-(2-furoyloxy)-1,2,3,4-tetrahydroquinoline
See: Quinfamide

1-Dichloroacetyl-1,2,3,4-tetrahydro-6-hydroxyquinoline
Formed from: Quinfamide

3-(3,3-Dichloroallyl)-2-hydroxy-1,4-naphthoquinone
Formed from: Dichloroallyllawsone
→ *o*-(5,5-Dichloro-2-hydroxy-1-oxo-4-hexen-1-yl)phenylglyoxylic acid
In: Cunninghamella C59
Helicostylum C59
Stemphylium C59
Streptomyces C59

Dichloroallyllawsone
→ 3-(3,3-Dichloroallyl)-2-hydroxy-1,4-naphthoquinone
In: Penicillium C59

2,3-Dichloroaniline
Formed from: 1,2-Dichloro-3-nitrobenzene

2,3-Dichloropropionanilide
→ 2,2′,3,3′-Tetrachloroazobenzene
In: Horseradish A661

2,4-Dichloroaniline
Formed from: *p*-Chloroaniline
1,3-Dichloro-4-nitrobenzene
2,4-Dichloropropionanilide
Diuron
→ 2,2′,4,4′-Tetrachloroazobenzene
In: Geotrichum A661
Horseradish A661

2,5-Dichloroaniline
Formed from: 2,5-Dichloroacetanilide
1,4-Dichloro-2-nitrobenzene
2,5-Dichloropropionanilide
→ 2,2′,5,5′-Tetrachloroazobenzene
In: Horseradish A661

2,6-Dichloroaniline
Formed from: 1,3-Dichloro-2-nitrobenzene

3,4-Dichloroaniline
Formed from: 3′,4′-Dichloroacetanilide
3′,4′-Dichlorobutyranilide
1,2-Dichloro-4-nitrobenzene
3′,4′-Dichloro-2-methacrylanilide
N-(3,4-Dichlorophenyl)-N′-(2,3-dihydrobenzofuran-5-sulphonyl)urea
3′,4′-Dichloropropionanilide
3′,4′-Dichlorovaleranilide
Karsil
3,3′,4,4′-Tetrachloroazobenzene
Probably formed from: Linuron
→ 6-Amino-2,3-dichlorophenol
In: Man C879
Rabbit C879
See also: H535, J156
→ 6-Amino-3,4-dichlorophenol
In: Man C879
Rabbit C879
See also: J156
→ 2-Amino-4,6,7-trichlorophenoxazin-3-one
Probably in: Man C879
→ 1,3-Bis(3,4-dichlorophenyl)triazene
In: Escherichia A3899, C768
→ 3,4-Dichloroacetanilide
In: Chaetomium A3794
→ 3,4-Dichloroaniline-N-β-D-glucoside
In: Carrot H118
Glycine G247, H650
Wheat H650
→ 4,5-Dichlorocatechol
Probably in: Pseudomonas C201
→ 3,4-Dichloro-N-malonylaniline
In: Carrot H118
Glycine G247, H650
→ 1,2-Dichloro-4-nitrobenzene
In: Chaetomium A3794
→ 3,4-Dichloronitrosobenzene
In: Man C879
Pea C797
With chloroperoxidase A3579
→ 3,4-Dichlorophenol
In: Escherichia C768
→ 3,4-Dichlorophenylhydroxylamine
In: Chaetomium A3794
→ N-(3,4-Dichlorophenyl)succinimide
In: Phanerochaete E317
→ 3,3′,4,4′-Tetrachloroazobenzene
In: Arthrobacter A3899
Bacillus A3899
Chaetomium A3794
Corynebacterium A3899
Erwinia A3899
Escherichia A3899, C768
Geotrichum A661
Horseradish A661
→ 3,3′,4,4′-Tetrachloroazoxybenzene
In: Fusarium A487

3,5-Dichloroaniline
Formed from: 1,3-Dichloro-5-nitrobenzene
3,5-Dichloropropionanilide
N-(3,5-Dichlorophenyl)succinimide
→ 3,3′,5,5′-Tetrachloroazobenzene
In: Horseradish A661

3,4-Dichloroaniline-N-β-D-glucoside
Formed from: 3,4-Dichloroaniline
→ 3,4-Dichloroaniline-N-(6-O-malonyl)-β-D-glucoside
Probably in: Daucus H118

3,4-Dichloroaniline-N-(6-O-malonyl)-β-D-glucoside
Probably formed from: 3,4-Dichloroaniline-N-β-D-glucoside

3,5-Dichloroanisic acid (3,5-dichloro-4-methoxybenzoic acid)
Formed from: 3,5-Dichloro-4-methoxybenzyl alcohol
→ 2,6-Dichlorophenol
In: Microorganism J708

3,6-Dichloroanisic acid
Formed from: Disugran

2,4-Dichloroanisole
Formed from: 2,4-Dichlorophenol

3,4-Dichloroanisole
Formed from: 3,4-Dichlorophenol

2,2′-Dichloroazobenzene
Formed from: *o*-Chloroaniline
2,2′-Dichloroazoxybenzene

3,3′-Dichloroazobenzene
Formed from: *m*-Chloroaniline

4,4′-Dichloroazobenzene
Formed from: *p*-Chloroaniline

2,2′-Dichloroazoxybenzene
→ 2,2′-Dichloroazobenzene
In: Saccharomyces H894

2,4-Dichlorobenzaldehyde
Formed from: N-(2,4-Dichlorobenzyl)aniline
→ 2,4-Dichlorobenzyl alcohol
In: Dunaliella K176
Nannochloris K176
Porphyridium K176

2,6-Dichlorobenzaldehyde
Formed from: N-(2,6-Dichlorobenzyl)aniline

3,4-Dichlorobenzaldehyde
→ 3,4-Dichlorobenzoic acid
In: Rat H761
→ 3,4-Dichlorobenzyl alcohol
In: Chlorella K176
Dunaliella K176
Nannochloris K176
Porphyridium K176

3,5-Dichlorobenzaldehyde
→ 3,5-Dichlorobenzoic acid
In: Rat H761

2,6-Dichlorobenzaldoxime
→ 2,6-Dichlorobenzyl-β-D-sulphatoglucoside
In: Brassica F657

2,5-Dichlorobenzamide
→ 2,5-Dichlorobenzoic acid
In: Aspergillus A640

2,6-Dichlorobenzamide
Formed from: 2,6-Dichlorobenzonitrile
→ N-Acetyl-2-chloro-6-(S-(L-cysteinyl))-5-hydroxybenzamide
In: Rat F307

3,5-Dichlorobenzamide
→ 3,5-Dichlorobenzoic acid
In: Aspergillus A640

***o*-Dichlorobenzene**
→ *cis*-3,4-Dichloro-1,2-dihydro-1,2-dihydroxybenzene
In; Pseudomonas E641
→ 1,2-Dichloro-3-(S-glutathionyl)-3,4-dihydro-4-hydroxybenzene
In: Rat H514
→ 1,2-Dichloro-4-(S-glutathionyl)-3,4-dihydro-3-hydroxybenzene
In: Rat H514
→ 3,4-Dichloro-5-(S-glutathionyl)-4,5-dihydro-4-hydroxybenzene
In: Rat H514
→ 2,3-Dichlorophenol
In: Man H653, H814, J337
Pseudomonas C64
Rat G734, H514
→ 2,6-Dichlorophenol
In: Pseudomonas C64
→ 3,4-Dichlorophenol
In: Man H653, H814, J337
Pseudomonas C64
Rat G734, H514

***m*-Dichlorobenzene**
Formed from: Lindane
γ-Pentachlorocyclohex-1-ene
1,2,3-Trichlorobenzene
1,3,5-Trichlorobenzene
→ 3,5-Dichloro-1,2-dihydro-1,2-dihydroxybenzene
In: Alcaigenes E260
→ *trans*-3,5-Dichloro-2-(S-glutathionyl)-1,2-dihydro-1-hydroxybenzene
In: Rat G294
→ *trans*-4,6-Dichloro-2-(S-glutathionyl)-1,2-dihydro-1-hydroxybenzene
In: Rat G294
→ S-(3,5-Dichlorophenyl)cysteinylglycine
In: Rat G294

→ S-(2,4-Dichlorophenyl)glutathione
In: Rat G294
→ 2,4-Dichlorophenyl methyl sulphone
In; Rat E2
→ 3,5-Dichlorophenyl methyl sulphone
In; Rat E2

p-Dichlorobenzene
Formed from: Lindane
1,2,4-Trichlorobenzene
→ 2-(N-Acetyl-S-(L-cysteinyl))-1,4-dichlorobenzene
In: Rat G970
→ 2-(N-Acetyl-S-(L-cysteinyl))-1,4-dichloro-2,3-dihydro-3-hydroxybenzene
In: Rat G970
→ 3,6-Dichloro-1,2-dihydro-1,2-dihydroxybenzene
In; Pseudomonas E731
Xanthobacter H690
→ 2,4-Dichlorophenol
In: Rat G734
→ 2,5-Dichlorophenol
In: Rat A3938, G734, G970, H821
→ 3,6-Dichlorocatechol
In: Alcaligenes E315
Pseudomonas E731
Xanthobacter H690

2,4′-Dichlorobenzhydrol
Probably formed from: 2-(*o*-Chlorophenyl)-2-(*p*-chlorophenyl)acetic acid
2,2,2-Trichloro-1,1-bis(*p*-chlorophenyl)ethanol

4,4′-Dichlorobenzhydrol
See: Bis(*p*-chlorophenyl)carbinol

3,3′-Dichlorobenzidine
Formed from: Direct Red 46
→ N-Acetyl-3,3′-dichlorobenzidine
Probably in: Rat C541
→ 4,4′-Bis(4-amino-3-chlorophenyl)-3,3′-dichloroazobenzene
In: Horseradish F298

4,4′-Dichlorobenzilic acid
Formed from: Chloropropylate

2,3-Dichlorobenzoic acid
Formed from: 2,2′,3,3′-Tetrachlorobiphenyl
o-Chlorobenzoic acid
→ In: Desulfomicrobium J491

2,4-Dichlorobenzoic acid
Formed from: 2,4-Dichlorobiphenyl
2,4,4′-Trichlorobiphenyl
2,2′,4,4′-Tetrachlorobiphenyl
→ *p*-Chlorobenzoic acid
In: Alcaligenes E335
Microorganism K269
→ 4-Chlorocatechol
In: Pseudomonas H212
See also: G434

2,5-Dichlorobenzoic acid
Formed from: Benzoic acid
2,5-Dichlorobenzamide
2,5-Dichlorobiphenyl
2,2′,5-Trichlorobiphenyl
2,3′,5-Trichlorobiphenyl
2,4′,5-Trichlorobiphenyl
→ *o*-Chlorobenzoic acid
In: Desulfomicrobium J491
Desulfomonile G152, G358
→ 4-Chlorocatechol
In: Pseudomonas H357, J173

2,6-Dichlorobenzoic acid
→ *o*-Chlorobenzoic acid
Probably in: Microorganism K269

3,4-Dichlorobenzoic acid
Formed from: 3,4-Dichlorobenzaldehyde
1-(3,4-Dichlorobenzyl)-5-octylbiguanide
3,4-Dichlorobenzyloxyacetic acid
3,4-Dichlorobiphenyl
2′,3,4-Trichlorobiphenyl
3,3′,4,4′-Tetrachlorobiphenyl
→ 4-Carboxy-*o*-quinone
In: Acinetobacter F106

3,5-Dichlorobenzoic acid
Formed from: 3,5-Dichlorobenzaldehyde
3,5-Dichlorobenzamide
3,5-Dichlorobiphenyl
8-Methyl-8-azabicyclo[3.2.1]octan-3-yl 3,5-dichlorobenzoate
→ *m*-Chlorobenzoic acid
In: Desulfomonile G152
→ 3,5-Dichlorocatechol
Probably in: Pseudomonas B11
→ 3,5-Dichlorohippuric acid

In: Dog G386
Monkey G386

4,4′-Dichlorobenzoin
Formed from: *p*-Chlorobenzaldehyde

2,6-Dichlorobenzonitrile (dichlobenil)
Formed from: α-Amino-2,6-dichlorobenzaldoxime
Probably formed from: 2,6-Dichlorobenzamide
→ 2-Chloro-6-(S-(L-cysteinyl))benzonitrile
In: Beef F586
→ 6-Chloro-2,3-dihydroxybenzonitrile
In: Hen F302
See also: F296
→ 6-Chloro-2-(S-glutathionyl)-3-hydroxybenzonitrile
In: Hen F302
See also: F296, F586
→ 2-Chloro-6-sulphydrylbenzonitrile
In: Goat F296
Rat F296
→ 2,6-Dichlorobenzamide
In: Arthrobacter A2367
Rhodococcus H952
→ 2,5-Dichloro-6-hydroxybenzonitrile
Probably in: Goat F296
Rat F296
→ 2,6-Dichloro-3-hydroxybenzonitrile
In: Goat F296
Hen F302
Rat F296
→ 2,6-Dichloro-4-hydroxybenzonitrile
In: Goat F296
Hen F302
Rat F296
→ 2-Sulphinylmethyl-6-thiomethylbenzonitrile
In: Goat F296

4,4′-Dichlorobenzophenone
Formed from: Bis(*p*-chlorophenyl)carbinol
Probably formed from: 3,3-Bis(*p*-chlorophenyl)pyruvic acid
→ *p*-Chlorobenzoic acid
In: Pseudomonas E41
→ Bis(*p*-chlorophenyl)carbinol
In: Trout A3397

2,4-Dichlorobenzoyl CoA
→ *p*-Chlorobenzoyl CoA
In: Corynebacterium J178

2,4-Dichlorobenzyl alcohol
Formed from: 2,4-Dichlorobenzaldehyde

3,4-Dichlorobenzyl alcohol
Formed from: 3,4-Dichlorobenzaldehyde

2,3-Dichlorobenzylamine
→ 2,3-Dichloro-N-methylbenzylamine
In: Rabbit B741

N-(2,4-Dichlorobenzyl)aniline
→ Aniline
In: Hamster G951
→ 2,4-Dichlorobenzaldehyde
In: Hamster G951
→ α-(2,4-Dichlorophenyl)-N-phenylnitrone
In: Hamster G951

N-(2,6-Dichlorobenzyl)aniline
→ Aniline
In: Hamster G951
→ 2,6-Dichlorobenzaldehyde
In: Hamster G951
→ α-(2,6-Dichlorophenyl)-N-phenylnitrone
In: Hamster G951

1-(3,4-Dichlorobenzyl)-5-(6,7-dihydroxyoctyl)biguanide
Probably formed from: 1-(3,4-Dichlorobenzyl)-5-(7-hydroxyoctyl)biguanide
→ 1-(3,4-Dichlorobenzyl)-5-(6-hydroxy-7-oxooctyl)biguanide
Probably in: Dog K61
Rat K61

1-(3,4-Dichlorobenzyl)-5-(7,8-dihydroxyoctyl) biguanide
Probably formed from: 1-(3,4-Dichlorobenzyl)-5-(7-hydroxyoctyl) biguanide

1-(3,4-Dichlorobenzyl)-5-(5-hydroxyoctyl) biguanide
Formed from: 1-(3,4-Dichlorobenzyl)-5-octylbiguanide

1-(3,4-Dichlorobenzyl)-5-(6-hydroxyoctyl) biguanide
Formed from: 1-(3,4-Dichlorobenzyl)-5-octylbiguanide

1-(3,4-Dichlorobenzyl)-5-(7-hydroxyoctyl) biguanide
Formed from: 1-(3,4-Dichlorobenzyl)-5-octylbiguanide
→ 1-(3,4-Dichlorobenzyl)-5-(6,7-dihydroxyoctyl)biguanide
Probably in: Dog K61
Rat K61
→ 1-(3,4-Dichlorobenzyl)-5-(7,8-dihydroxyoctyl)biguanide
Probably in: Dog K61
Rat K61
→ 1-(3,4-Dichlorobenzyl)-5-(7-oxooctyl) biguanide
Probably in: Dog K61
Rat K61

1-(3,4-Dichlorobenzyl)-5-(6-hydroxy-7-oxooctyl)biguanide
Probably formed from: 1-(3,4-Dichlorobenzyl)-5-(6,7-dihydroxyoctyl)biguanide

1-(2,6-Dichlorobenzyl)-4-hydroxyquinolinium
Formed from: 1-(2,6-Dichlorobenzyl)quinolinium

1-(3,4-Dichlorobenzyl)-5-octylbiguanide
→ 5-(4-Carboxybutyl)-1-(3,4-dichlorobenzyl)biguanide
In: Dog J380, K61
Rat K61
→ 1-(5-Carboxypentyl)-5-(3,4-dichlorobenzyl)biguanide
In: Dog J380, K61, K333
Rat K61
→ 3,4-Dichlorobenzoic acid
In: Dog J380, K333
→ 1-(3,4-Dichlorobenzyl)-5-(5-hydroxyoctyl)biguanide
In: Dog K61
Rat K61
→ 1-(3,4-Dichlorobenzyl)-5-(6-hydroxyoctyl)biguanide
In: Dog K61
Rat K61
→ 1-(3,4-Dichlorobenzyl)-5-(7-hydroxyoctyl)biguanide
In: Dog K61, K333
Rat K61

1-(3,4-Dichlorobenzyl)-5-(7-oxooctyl)biguanide
Probably formed from: 1-(3,4-Dichlorobenzyl)-5-(7-hydroxyoctyl)biguanide

3,4-Dichlorobenzyloxyacetic acid
→ 3,4-Dichlorobenzoic acid
In: Rat E866
→ 3,4-Dichlorobenzyloxyacetylglycine
In: Rat E866
→ 3,4-Dichlorobenzyloxyacetyltaurine
In: Rat E866

3,4-Dichlorobenzyloxyacetylglycine
Formed from: 3,4-Dichlorobenzyloxyacetic acid

3,4-Dichlorobenzyloxyacetyltaurine
Formed from: 3,4-Dichlorobenzyloxyacetic acid

1-(2,6-Dichlorobenzylquinolinium
→ 1-(2,6-Dichlorobenzyl)-4-hydroxyquinolinium
In: Rabbit D208

2,6-Dichlorobenzyl-β-D-sulphatoglucoside
Formed from: 2,6-Dichlorobenzaldoxime

2,2′-Dichlorobiphenyl
→ *o*-Chlorobenzoic acid
In: Acinetobacter B125
Alcaligenes A1822, B125
→ 2′-Chloro-2,3-dihydroxybiphenyl
In: Pseudomonas H693
→ 2,2′-Dichloro-3-hydroxybiphenyl
In: Rat C54
→ 2,2′-Dichloro-4-hydroxybiphenyl
In: Rat C54
See also: A1458
→ 2,2′-Dichloro-5-hydroxybiphenyl
In: Rat A2472, C54

2,3-Dichlorobiphenyl
Formed from: 2,2′,3,3′-Tetrachlorobiphenyl
→ 2′,3′-Dichloro-3-hydroxybiphenyl
In: Rat C54
→ 2,3-Dichloro-4-hydroxybiphenyl
In: Rat C54
→ 2′,3′-Dichloro-4-hydroxybiphenyl
In: Rat A1423, A1458, A2470, A2472, C54

2,4-Dichlorobiphenyl
→ 2,4-Dichlorobenzoic acid
In: Acinetobacter B125

Alcaligenes A1822
→ 2′,4′-Dichloro-3-hydroxybiphenyl
In: Rat C54
→ 2′,4′-Dichloro-4-hydroxybiphenyl
In: Rat A2472, C54
See also: A1458
→ 2,4-Dichloro-5-hydroxybiphenyl
Probably in: Rat C54

2,4′-Dichlorobiphenyl
→ *o*-Chlorobenzoic acid
In: Alcaligenes A1822
→ 4′-Chloro-2,3-dihydroxybiphenyl
In: Pseudomonas H693
→ 2,4′-Dichloro-4-hydroxybiphenyl
In: Rat C54

2,5-Dichlorobiphenyl
→ 2,5-Dichlorobenzoic acid
In: Acinetobacter B125
Alcaligenes A1822
→ 2′,5′-Dichloro-3-hydroxybiphenyl
In: Rat C54
→ 2,5-Dichloro-4-hydroxybiphenyl
In: Rat C54
→ 2′,5′-Dichloro-4-hydroxybiphenyl
In: Rat C54

2,6-Dichlorobiphenyl
→ 2′,6′-Dichloro-3-hydroxybiphenyl
In: Rat C54
→ 2′,6′-Dichloro-4-hydroxybiphenyl
In: Rat C54

3,3′-Dichlorobiphenyl
→ *m*-Chlorobenzoic acid
In: Acinetobacter B125
Alcaligenes B125
→ 3,3′-Dichloro-4-hydroxybiphenyl
In: Rat A2472, C54
→ 3,3′-Dichloro-5-hydroxybiphenyl
In: Rat C54

3,4-Dichlorobiphenyl
Formed from: 2′,3,4-Trichlorobiphenyl
→ 3,4-Dichlorobenzoic acid
In: Acinetobacter B125
Alcaligenes A1822, B125
→ 3′,4′-Dichloro-3-hydroxybiphenyl
In: Rat C54
→ 3′,4′-Dichloro-4-hydroxybiphenyl
In: Rat C54
See also: A1458

3,5-Dichlorobiphenyl
→ 3,5-Dichlorobenzoic acid
In: Acinetobacter B125
Alcaligenes B128
→ 3′,5′-Dichloro-3-hydroxybiphenyl
In: Rat C54
→ 3,5-Dichloro-4-hydroxybiphenyl
In: Rat C54
→ 3′,5′-Dichloro-4-hydroxybiphenyl
In: Rat C54

4,4′-Dichlorobiphenyl
→ *p*-Chlorobenzoic acid
In: Achromobacter A706
Acinetobacter B125, F106
Alcaligenes A1822, B125
Bacteria A3413
→ 4′-Chloro-3-hydroxybiphenyl
In: Man D299
→ 4′-Chloro-4-hydroxybiphenyl
In: Frog A2530
Man D299
Rabbit A2566
Rat A2530
See also: A1458
→ 4,4′-Dichloro-2,3-dihydroxybiphenyl
In: Bacteria A3413
→ 3,4′-Dichloro-4-hydroxybiphenyl
In: Man D299
Rabbit A2566
Rat A2530, A2747, C54
→ 4,4′-Dichloro-2-hydroxybiphenyl
Man D299
→ 4,4′-Dichloro-3-hydroxybiphenyl
In: Frog A2530
Goat A1576, A2046
Hen A3893
Man D299
Rabbit A2566
Rat A2168, A2472, A2530, A2747, C54
See also: A701, A3390, G799

1,1-Dichloro-2,2-bis(*p*-chlorophenyl)ethane
(DDD)
Formed from: 1,1,1-Trichloro-2,2-bis(*p*-chlorophenyl)ethane

Probably formed from: 1,1-Dichloro-2,2-bis(*p*-chlorophenyl)ethylene
→ 2,2-Bis(*p*-chlorophenyl)acetic acid
In: Beef A2034
Mouse C439, C675
→ 1-Chloro-2,2-bis(*p*-chlorophenyl)ethylene
In: Mouse C439
Rat A928, J663
Trout A3397
→ 2,2-Dichloro-1,1-bis(*p*-chlorophenyl)ethanol
In: Beef A2055
Mouse C439

2,2-Dichloro-1,1-bis(*p*-chlorophenyl)ethanol
Formed from: 1,1-Dichloro-2,2-bis(*p*-chlorophenyl)ethane
2,2,2-Trichloro-1,1-bis(*p*-chlorophenyl)ethanol

1,1-Dichloro-2,2-bis(*p*-chlorophenyl)ethylene
(DDE)
Formed from: 1,1,1,2-Tetrachloro-2,2-bis(*p*-chlorophenyl)ethane
1,1,1-Trichloro-2,2-bis(*p*-chlorophenyl)ethane
Probably formed from: 1,1,1-Trichloro-2,2-bis(*p*-chlorophenyl)ethanol
→ 1,1-Dichloro-2,2-bis(*p*-chlorophenyl)ethane
Probably in: Phagocata A1395
Contrast in Trout A3397
→ 1,1-Dichloro-2-(3-chloro-4-hydroxyphenyl-2-(*p*-chlorophenyl)ethylene
In: Guillemot A2301
Rat A1935, A2301
Seal A2301
→ 1,1-Dichloro-2-(4-chloro-2-hydroxyphenyl)-2-(*p*-chlorophenyl)ethylene
In: Guillemot A2301
Rat A1935, A2301
Seal A2301
→ 1,1-Dichloro-2-(4-chloro-3-hydroxyphenyl-2-(*p*-chlorophenyl)ethylene
In: Guillemot A2301
Rat A1935, A2301
Seal A2301
→ 1,1-Dichloro-2-(*p*-chlorophenyl)-2-(*p*-hydroxyphenyl)ethylene
In: Rat A1935

1,1-Dichloro-2,2-bis(*p*-hydroxyphenyl)ethane
Probably formed from: 1,1-Dichloro-2-(*p*-hydroxyphenyl)-2-(*p*-methoxyphenyl)ethane
→ 1-Chloro-2,2-bis(*p*-hydroxyphenyl)ethylene
Probably in: Goat C198
→ 1,1-Dichloro-2,2-bis(*p*-hydroxyphenyl)ethane-β-D-glucuronide
Probably in: Goat C198

1,1-Dichloro-2,2-bis(*p*-hydroxyphenyl)ethane-β-D-glucuronide
Probably formed from: 1,1-Dichloro-2,2-bis(*p*-hydroxyphenyl)ethane

1,1-Dichloro-2,2-bis(*p*-hydroxyphenyl)ethylene
Probably formed from: 1,1-Dichloro-2-(*p*-hydroxyphenyl)-2-(*p*-methoxyphenyl)ethylene
→ 1,1-Dichloro-2,2-bis(*p*-hydroxyphenyl)ethylene-β-D-glucuronide
Probably in: Goat C198

1,1-Dichloro-2,2-bis(*p*-hydroxyphenyl)ethylene-β-D-glucuronide
Probably formed from: 1,1-Dichloro-2,2-bis(*p*-hydroxyphenyl)ethylene

1,1-Dichloro-2,2-bis(*p*-methoxyphenyl)ethane
Formed from: 1,1,1-Trichloro-2,2-bis(*p*-methoxyphenyl)ethane
→ 1,1-Dichloro-2-(*p*-hydroxyphenyl)-2-(*p*-methoxyphenyl)ethane
Probably in: Goat C198
Klebsiella C175

1,1-Dichloro-2,2-bis(*p*-methoxyphenyl)ethylene
Formed from: 1,1,1-Trichloro-2,2-bis(*p*-methoxyphenyl)ethylene
→ 1,1-Dichloro-2-(*p*-hydroxyphenyl)-2-(*p*-methoxyphenyl)ethylene
Probably in: Goat C198, D178

2,5-Dichloro-4,6-bis(methylthio)isophthalonitrile
Probably formed from: 2,5,6-Trichloro-4-methylthioisophthalonitrile
→ 2-Chloro-4,5,6-tris(methylthio)-isophthalonitrile
Probably in: Rat J453

1,1-Dichloro-2,2-bis(*p*-nitrophenyl)ethylene
Formed from: 1,1,1-Trichloro-2,2-bis(*p*-nitrophenyl)ethane

3′,4′-Dichlorobutyranilide
→ 3,4-Dichloroaniline

In: Dandelion A230

3,4-Dichlorocatechol

Formed from: 2,3-Dichlorophenol
3,4-Dichlorophenol
Probably formed from: *cis*-3,4-Dichloro-1,2-dihydro-1,2-dihydroxybenzene
3,4,5-Trichlorocatechol
→ 4,5-Dichloro-2-methoxyphenol
Probably in: Phanerochaete J696
→ 2,3-Dichloro-*cis*,*cis*-muconic acid
In: Pseudomonas H506, J702
See also: E641

3,5-Dichlorocatechol

Formed from: 3,5-Dichloro-2-hydroxybenzoic acid
2,4-Dichlorophenol
3,4,5-Trichlorocatechol
Probably formed from: 3,5-Dichlorobenzoic acid
3,5-Dichloro-1,2-dihydro-1,2-dihydroxybenzene
2,4,5-Trichlorophenol
→ 3-Chlorocatechol
In: Bacteria F883
→ 4-Chlorocatechol
Probably in: Pseudomonas B16
→ 2,3-Dichloro-2-hydroxy-*cis*,*cis*-muconic semialdehyde
In: Pseudomonas J515
→ 2,4-Dichloro-*cis*,*cis*-muconic acid
In: Alcaligenes G328
Pseudomonas A3769, H506, J702

3,6-Dichlorocatechol

Formed from: *p*-Dichlorobenzene
3,6-Dichloro-1,2-dihydro-1,2-dihydroxybenzene
2,5-Dichlorophenol
→ 2,5-Dichloro-*cis*,*cis*-muconic acid
In: Alcaligenes E315
Pseudomonas J702
See also: H690

4,5-Dichlorocatechol

Formed from: 4,5-Dichloro-2-methoxyphenol
4,5-Dichloro-1,2-methylenedioxybenzene
3,4-Dichlorophenol
1,4,7,8-Tetrachloro-*p*-dioxin
2,3,7,8-Tetrachloro-*p*-dioxin
Probably formed from: 3,4-Dichloroaniline
→ 4-Chlorocatechol
In: Bacteria F883
→ 4,5-Dichloro-2-hydroxy-*cis*,*cis*-muconic semialdehyde
In: Pseudomonas J515

1,1-Dichloro-2-(2-chloro-3,4-dihydroxyphenyl)-2-(*p*-chlorophenyl)ethane

Probably formed from: 1,1-Dichloro-2-(2-chloro-3-hydroxyphenyl)-2-(*p*-chlorophenyl)ethane
1,1-Dichloro-2-(2-chloro-4-hydroxyphenyl)-2-(*p*-chlorophenyl)ethane
→ 1,1-Dichloro-2-(2-chloro-4-hydroxy-3-methoxyphenyl)-2-(*p*-chlorophenyl)ethane
Probably in: Chicken A2368
Rat A991

1,1-Dichloro-2-(2-chloro-3,4-dihydroxyphenyl)-2-(*p*-chlorophenyl)ethylene

Probably formed from: 1,1-Dichloro-2-(2-chloro-4-hydroxyphenyl)-2-(*p*-chlorophenyl)ethylene
→ 1,1-Dichloro-2-(2-chloro-4-hydroxy-3-methoxyphenyl)-2-(*p*-chlorophenyl)ethylene
Probably in: Chicken A2368

1,1-Dichloro-2-(2-chloro-4-hydroxy-3-methoxyphenyl)-2-(*p*-chlorophenyl)ethane

Probably formed from: 1,1-Dichloro-2-(2-chloro-3,4-dihydroxyphenyl)-2-(*p*-chlorophenyl)ethane
1,1,1-Trichloro-2-(2-chloro-4-hydroxy-3-methoxyphenyl)-2-(*p*-chlorophenyl)ethane

1,1-Dichloro-2-(2-chloro-4-hydroxy-3-methoxyphenyl)-2-(*p*-chlorophenyl)ethylene

Probably formed from: 1,1-Dichloro-2-(2-chloro-3,4-dihydroxyphenyl)-2-(*p*-chlorophenyl)ethylene

1,1-Dichloro-2-(2-chloro-3-hydroxyphenyl)-2-(*p*-chlorophenyl)ethane

Probably formed from: 1,1-Dichloro-2-(*o*-chlorophenyl)-2-(*p*-chlorophenyl)ethane
→ 1,1-Dichloro-2-(2-chloro-3,4-dihydroxyphenyl)-2-(*p*-chlorophenyl)ethane

Probably in: Chicken A2368
Man A1526
Rat A991

1,1-Dichloro-2-(2-chloro-4-hydroxyphenyl)-2-(*p*-chlorophenyl)ethane
Probably formed from: 1,1-Dichloro-2-(*o*-chlorophenyl)-2-(*p*-chlorophenyl)ethane

1,1-Dichloro-2-(2-chloro-3-hydroxyphenyl)-2-(*p*-chlorophenyl)ethylene
Probably formed from: 1,1-Dichloro-2-(*o*-chlorophenyl)-2-(*p*-chlorophenyl)ethylene

1,1-Dichloro-2-(2-chloro-4-hydroxyphenyl)-2-(*p*-chlorophenyl)ethylene
Probably formed from: 1,1-Dichloro-2-(*o*-chlorophenyl)-2-(*p*-chlorophenyl)ethylene
→ 1,1-Dichloro-2-(2-chloro-3,4-dihydroxyphenyl)-2-(*p*-chlorophenyl)ethylene
Probably in: Chicken A2368

1,1-Dichloro-2-(3-chloro-4-hydroxyphenyl)-2-(*p*-chlorophenyl)ethylene
Formed from: 1,1-Dichloro-2,2-bis(*p*-chlorophenyl)ethylene

1,1-Dichloro-2-(4-chloro-2-hydroxyphenyl)-2-(*p*-chlorophenyl)ethylene
Formed from: 1,1-Dichloro-2,2-bis(*p*-chlorophenyl)ethylene

1,1-Dichloro-2-(4-chloro-3-hydroxyphenyl)-2-(*p*-chlorophenyl)ethylene
Formed from: 1,1-Dichloro-2,2-bis(*p*-chlorophenyl)ethylene

1,1-Dichloro-2-(*o*-chlorophenyl)-2-(*p*-chlorophenyl)ethane
Formed from: 1,1,1-Trichloro-2-(*o*-chlorophenyl)-2-(*p*-chlorophenyl)ethane
→ 2-(*o*-Chlorophenyl)-2-(*p*-chlorophenyl)acetic acid
In: Beef A2054
Dog A2419, E876
See also: A1526
→ 1,1-Dichloro-2-(2-chloro-3-hydroxyphenyl)-2-(*p*-chlorophenyl)ethane
In: Man A1526
See also: A991, A2368
→ 1,1-Dichloro-2-(2-chloro-4-hydroxyphenyl)-2-(*p*-chlorophenyl)ethane
Probably in: Chicken A2368

1,1-Dichloro-2-(*o*-chlorophenyl)-2-(*p*-chlorophenyl)ethylene
Formed from: 1,1,1-Trichloro-2-(*o*-chlorophenyl)-2-(*p*-chlorophenyl)ethane
→ 1,1-Dichloro-2-(2-chloro-3-hydroxyphenyl)-2-(*p*-chlorophenyl)ethylene
Probably in: Chicken A2368
→ 1,1-Dichloro-2-(2-chloro-4-hydroxyphenyl)-2-(*p*-chlorophenyl)ethylene
Probably in: Chicken A2368

1,1-Dichloro-2-(*p*-chlorophenyl)-2-(*p*-hydroxyphenyl)ethylene
Formed from: 1,1-Dichloro-2,2-bis(*p*-chlorophenyl)ethylene

2,6-Dichlorocinnamic acid
→ 2,6-Dichlorostyrene
In: Microorganism F390

2-(2,3-Dichloro-4-(3-cyanopropoxy)benzoyl)-5-hydroxythiophene
Formed from: 2-(2,3-Dichloro-4-(3-cyanopropoxy)benzoyl)thiophene

2-(2,3-Dichloro-4-(3-cyanopropoxy)benzoyl)-thiophene
→ 2-(2,3-Dichloro-4-(3-cyanopropoxy)benzoyl)-5-hydroxythiophene
In: Man K165

α,3-Dichloro-4-cyclohexyl)phenylacetic acid
See: Fenclorac

(6,7-Dichloro-2-cyclopentyl-2-methyl-1-oxo-5-indanyloxy)acetic acid
→ (6,7-Dichloro-2-cyclopentyl-2-methyl-1-oxo-5-indanyloxy)acetyl-β-D-glucuronic acid
In: Baboon A2951
Monkey A2951
→ (6,7-Dichloro-2-(1-hydroxycyclopentyl)-2-methyl-1-oxo-5-indanyloxy)acetic acid
In: Baboon A2951
Chimpanzee A2865
Dog A2951
Man A2865
Monkey A2951
Rat A2951
→ (6,7-Dichloro-2-(3-hydroxycyclopentyl)-2-methyl-1-oxo-5-indanyloxy)acetic acid

In: Baboon A2951
Chimpanzee A2865
Man A2865
Monkey A2951
Rat A2951
→ (6,7-Dichloro-2-(4-hydroxycyclopentyl)-2-methyl-1-oxo-5-indanyloxy)acetic acid
In: Baboon A2951
Chimpanzee A2865
Man A2865
Monkey A2951
Rat A2951

(6,7-Dichloro-2-cyclopentyl-2-methyl-1-oxo-5-indanyloxy)acetyl-β-D-glucuronic acid
Formed from: (6,7-Dichloro-2-cyclopentyl-2-methyl-1-oxo-5-indanyloxy)acetic acid

2-(4-(2,2-Dichlorocyclopropyl)phenoxy)-2-methylpropanoic acid (ciprofibrate)
→ 2-(4-(2,2-Dichlorocyclopropyl)phenoxy)-2-methylpropanoyl-β-D-glucuronic acid
In: Man A1868, J416, J783
Monkey H861
Rat A1868

2-(4-(2,2-Dichlorocyclopropyl)phenoxy)-2-methylpropanoyl-β-D-glucuronic acid
Formed from: 2-(4-(2,2-Dichlorocyclopropyl)phenoxy)-2-methylpropanic acid

***trans*-3,5-Dichloro-2-(S-(L-cysteinyl))-1,2-dihydro-1-hydroxybenzene**
Probably formed from: *trans*-3,5-Dichloro-2-(S-glutathionyl)-1,2-dihydro-1-hydroxybenzene
→ *trans*-2-(N-Acetyl-S-(L-cysteinyl))-3,5-dichloro-1,2-dihydro-1-hydroxybenzene
Probably in: Rat G244

***trans*-4,6-Dichloro-2-(S-(L-cysteinyl))-1,2-dihydro-1-hydroxybenzene**
Probably formed from: *trans*-4,6-Dichloro-2-(S-(L-cysteinylglycyl))-1,2-dihydro-1-hydroxybenzene

***trans*-4,6-Dichloro-2-(S-(L-cysteinylglycyl))-1,2-dihydro-1-hydroxybenzene**
Probably formed from: *trans*-4,6-Dichloro-2-(S-glutathionyl)-1,2-dihydro-1-hydroxybenzene
→ *trans*-4,6-Dichloro-2-(S-(L-cysteinyl))-1,2-dihydro-1-hydroxybenzene
Probably in: Rat G294

2′,4′-Dichlorodehydrocyclopeptine
→ 2′,4′-Dichloro-5′-hydroxydehydrocyclopeptine
Probably in: Penicillium A1633

4,4′-Dichlorodiazoaminobenzene
Formed from: *p*-Chloroaniline

2,3-Dichlorodibenzo-*p*-dioxin
→ 2,3-Dichloro-hydroxydibenzo-*p*-dioxin
In: Rat A3809

2,7-Dichlorodibenzo-*p*-dioxin
→ 2,7-Dichloro-hydroxydibenzo-*p*-dioxin
In: Rat A3809

2,3-Dichlorodibenzo-hydroxy-*p*-dioxin
Formed from: 2,3-Dichlorodibenzo-*p*-dioxin

2,7-Dichlorodibenzo-hydroxy-*p*-dioxin
Formed from: 2,7-Dichlorodibenzo-*p*-dioxin

2,4-Dichlorodibenzylamine
Probably formed from: 2,4-Dichlorotribenzylamine

3,5-Dichloro-2,4-difluoroaniline
→ 2-Amino-4,6-dichloro-3,5-difluorophenol
In: Rat J260
→ 4-Amino-2,6-dichloro-3-fluorophenol
In: Rat J260

N-(3,5-Dichloro-2,4-difluorophenyl)urea
→ N-(3,5-Dichloro-2-fluoro-4-hydroxyphenyl)urea
In: Rat J260

***cis*-3,4-Dichloro-1,2-dihydro-1,2-dihydroxybenzene**
Formed from: *o*-Dichlorobenzene
→ 3,4-Dichlorocatechol
Probably in: Pseudomonas E641

3,5-Dichloro-1,2-dihydro-1,2-dihydroxybenzene
Formed from: *m*-Dichlorobenzene
→ 3,5-Dichlorocatechol
Probably in: Alcaligenes E260

3,6-Dichloro-1,2-dihydro-1,2-dihydroxybenzene
Formed from: *p*-Dichlorobenzene
→ 3,6-Dichlorocatechol
In: Xanthobacter J186

5,6-Dichloro-1,2-dihydro-1,2-dihydroxynaphthalene
Formed from: 1,2-Dichloronaphthalene

7,8-Dichloro-3,4-dihydroisoquinoline
Probably formed from: 7,8-Dichloro-1,2,3,4-tetrahydro-2-hydroxyisoquinoline
→ 7,8-Dichloroisoquinoline
Probably in: Rat C649

7,8-Dichloro-3,4-dihydroisoquinoline-N-oxide
Probably formed from: 7,8-Dichloro-1,2,3,4-tetrahydro-2-hydroxyisoquinoline

6,7-Dichloro-2,3-dihydro-5-(N-methylsulphamoyl)benzofuran-2-carboxylic acid
Formed from: 6,7-Dichloro-5-(N,N-dimethylsulphamoyl)-2,3-dihydrobenzofuran-2-carboxylic acid

3,6-Dichloro-2,5-dihydroxybenzoic acid
See: 3,6-Dichlorogentisic acid

2,6-Dichloro-3,4-dihydroxybenzonitrile
Probably formed from: 2,6-Dichloro-4-hydroxybenzonitrile

3,4′-Dichloro-3′,4-dihydroxybiphenyl
Formed from: 4,4′-Dichloro-3-hydroxybiphenyl

4,4′-Dichloro-2,3-dihydroxybiphenyl
Formed from: 4,4′-Dichlorobiphenyl

4,4′-Dichloro-2,5-dihydroxybiphenyl
Formed from: 4,4′-Dichloro-3-hydroxybiphenyl

4,4′-Dichloro-3,3′-dihydroxybiphenyl
Formed from: 4,4′-Dichloro-3-hydroxybiphenyl

(*R*,*R*)-*p*,*p*′-Dichloro-α,β-dihydroxydibenzyl
Formed from: *p*,*p*′-Dichloro-*cis*-stilbene oxide

3,5-Dichloro-4,4′-dihydroxydiphenylamine
Formed from: 2,6-Dichlorophenolindophenol

1,2-Dichloro-4,5-dimethoxybenzene
Formed from: 2,3,7,8-Tetrachlorodibenzo-*p*-dioxin

2,5-Dichloro-1,4-dimethoxybenzene (chloroneb)
Formed from: 2,5-Dichloro-4-methoxyphenol
→ 2,5-Dichloro-4-methoxyphenol
In: Cephalosporium A938
Fusarium A938
Mucor A938
Phaseolus A939
Rhizoctonia A938

2,2′-Dichloro-4,4′-dimethylazobenzene
Formed from: 2-Chloro-4-methylaniline

2,2′-Dichloro-6,6′-dimethylazobenzene
Formed from: 2-Chloro-6-methylaniline

3,3′-Dichloro-2,2′-dimethylazobenzene
Formed from: 3-Chloro-2-methylaniline

3,3′-Dichloro-4,4′-dimethylazobenzene
Formed from: 3-Chloro-4-methylaniline

4,4′-Dichloro-2,2′-dimethylazobenzene
Formed from: 4-Chloro-2-methylaniline

5,5′-Dichloro-2,2′-dimethylazobenzene
Formed from: 5-Chloro-2-methylaniline

3,5-Dichloro-α-(((1,1-dimethylethyl)amino)methyl)-4-hydroxylaminobenzenemethanol
Formed from: Clenbuterol
→ 3,5-Dichloro-α-(((1,1-dimethylethyl)amino)methyl)-4-nitrosobenzenemethanol
Probably in: Beef H930, J591
Rat H855, H930

3,5-Dichloro-α-(((1,1-dimethylethyl)amino)methyl)-4-nitrobenzenemethanol
Probably formed from: 3,5-Dichloro-α-(((1,1-dimethylethyl)amino)methyl)-4-nitrosobenzenemethanol

3,5-Dichloro-α-(((1,1-dimethylethyl)amino)methyl)-4-nitrosobenzenemethanol
Probably formed from: 3,5-Dichloro-α-(((1,1-dimethylethyl)amino)methyl)-4-hydroxylaminobenzenemethanol
→ 3,5-Dichloro-α-(((1,1-dimethylethyl)amino)methyl)-4-nitrobenzenemethanol
Probably in: Beef H930, J591
Rat H855, H930

2,2′-Dichloro-4,4-dimethylhydrazobenzene
Formed from: 2-Chloro-4-methylaniline

6,7-Dichloro-5-(N,N-dimethylsulphamoyl)-2,3-dihydrobenzofuran-2-carboxylic acid
→ 3-(3,4-Dichloro-5-(N,N-dimethylsulphamoyl)-2-hydroxyphenyl)-2-(S-glutathionyl)propionic acid
In: Rat F898, G68
→ 6,7-Dichloro-2,3-dihydro-5-(N-methylsulphamoyl)benzofuran-2-carboxylic acid
In: Rat G81

3-(3,4-Dichloro-5-(N,N-dimethylsulphamoyl)-2-hydroxyphenyl)-2-(S-glutathionyl)propionic acid
Formed from: 6,7-Dichloro-5-(N,N-dimethylsulphamoyl)-2,3-dihydrobenzofuran-2-carboxylic acid

1,1-Dichloro-*cis*-2,3-diphenylcyclopropane (analog 11)
→ 2-Dichloro-3-oxo-1,3-diphenyl-1-propene
In: Rat F573
→ 1,1-Dichloro-*cis*-2-(*p*-hydroxyphenyl)-3-phenylcyclopropane
In: Rat F573

4,4′-Dichlorodiphenyl ether
→ *p*-Chlorophenol
In: Rat A3503
→ 4,4′-Dichloro-2-hydroxydiphenyl ether
In: Rat A3503

4,4′-Dichlorodiphenyl sulphone
→ 4-Chloro-3-hydroxyphenyl 4-chlorophenyl sulphone
In: Rat H707

3-(2,2-Dichloroethenyl)-2,2-dimethylcyclopropanecarboxylic acid cyano(4-fluoro-3-phenoxyphenyl) methyl ester
See: Cyfluthrin

3-(2,2-Dichloroethenyl)-2,2-dimethylcyclopropanecarboxylic acid cyano(3-phenoxyphenyl)methyl ester
See: Cypermethrin

3-(2,2-Dichloroethenyl)-2,2-dimethylcyclopropanecarboxylic acid cyano(3-phenoxy-2-pyridinyl)methyl ester
See: Fenpyrithrin

2,2-Dichloro-N-(2-ethoxyethyl)-N-(*p*-glycylaminophenoxy)benzyl)acetamide
→ N-((*p*-Acetamidophenoxy)benzyl)-2,2-dichloro-N-(2-ethoxyethyl) acetamide
In: Rat C803

Dichlorofluorescein
→ Dichlorofluorescein-β-D-glucuronide
In: Man G524

Dichlorofluorescein-β-D-glucuronide
Formed from: Dichlorofluorescein

N-(3,5-Dichloro-2-fluoro-4-hydroxyphenyl)urea
Formed from: N-(3,5-Dichloro-2,4-difluorophenyl)urea
→ N-(3,5-Dichloro-2-fluoro-4-hydroxyphenyl)urea-β-D-glucuronide
In: Rat J260
→ N-(3,5-Dichloro-2-fluoro-4-hydroxyphenyl)urea sulphate
In: Rat J260

N-(3,5-Dichloro-2-fluoro-4-hydroxyphenyl)urea-β-D-glucuronide
Formed from: N-(3,5-Dichloro-2-fluoro-4-hydroxyphenyl)urea

N-(3,5-Dichloro-2-fluoro-4-hydroxyphenyl) urea sulphate
Formed from: N-(3,5-Dichloro-2-fluoro-4-hydroxyphenyl)urea

3,5-Dichloro-2-fluoro-4-methoxybenzaldehyde
Probably formed from: 3-Chloro-2-fluoro-4-methoxybenzaldehyde

3,5-Dichloro-2-formylmethoxybiphenyl
Formed from: 2-(2-Aminoethoxy)-3,5-dichlorobiphenyl
→ 2-Carboxymethoxy-3,5-dichlorobiphenyl
In: Rat A1890
→ 3,5-Dichloro-2-(2-hydroxyethoxy) biphenyl
In: Rat A1890

3,6-Dichlorogentisic acid (3,6-dichloro-2,5-dihydroxybenzoic acid)
Probably formed from: 2,5-Dichloro-3-hydroxy-6-methoxybenzoic acid

1,2-Dichloro-3-(S-glutathionyl)-3,4-dihydro-4-hydroxybenzene
Formed from: *o*-Dichlorobenzene
→ 3-(N-Acetyl-S-(L-cysteinyl))-1,2-dichloro-3,4-dihydro-4-hydroxybenzene
Probably in: Rat H514

1,2-Dichloro-4-(S-glutathionyl)-3,4-dihydro-3-hydroxybenzene
Formed from: *o*-Dichlorobenzene
→ 4-(N-Acetyl-S-(L-cysteinyl))-1,2-dichloro-3,4-dihydro-3-hydroxybenzene
Probably in: Rat H514

1,2-Dichloro-5-(S-glutathionyl)-4,5-dihydro-4-hydroxybenzene
Formed from: *o*-Dichlorobenzene
→ 5-(N-Acetyl-S-(L-cysteinyl))-1,2-dichloro-4,5-dihydro-4-hydroxybenzene
Probably in: Rat H514

***trans*-3,5-Dichloro-2-(S-glutathionyl)-1,2-dihydro-1-hydroxybenzene**
Formed from: *m*-Dichlorobenzene
→ *trans*-3,5-Dichloro-2-(S-(L-cysteinyl))-1,2-dihydro-1-hydroxybenzene
Probably in: Rat G294

***trans*-4,6-Dichloro-2-(S-glutathionyl)-1,2-dihydro-1-hydroxybenzene**
Formed from: *m*-Dichlorobenzene
→ *trans*-4,6-Dichloro-2-(S-(L-cysteinylglycyl))-1,2-dihydro-1-hydroxybenzene
Probably in: Rat G294

(2,3-Dichloro-4-(2-(S-glutathionyl)-methylenebutyryl)phenoxy)acetic acid
Probably formed from: Ethacrynic acid
→ (4-(2-(N-Acetyl-S-(L-cysteinyl)) methylenebutyryl)-2,3-dichlorophenoxy)acetic acid
Probably in: Rat A136

3,5-Dichloro-2-(S-glutathionyl)quinol
Formed from: Trichloroquinol
→ 2,6-Dichloroquinol
In: Phanerochaete K138

3,5-Dichlorohippuric acid
Formed from: 3,5-Dichlorobenzoic acid

3,5-Dichloro-4-hydroxybenzaldehyde
→ 3,5-Dichloro-4-hydroxybenzoic acid
In: Burkholderia K107
Microorganism E664
→ 3,5-Dichloro-4-hydroxybenzyl alcohol
In: Microorganism E664

3,5-Dichloro-2-hydroxybenzoic acid
→ 3,5-Dichlorocatechol
In: Pseudomonas H506

3,5-Dichloro-4-hydroxybenzoic acid
Formed from: 3,5-Dichloro-4-hydroxybenzaldehyde
→ 3-Chloro-4-hydroxybenzoic acid
In: Desulfitobacterium J232
→ 2,6-Dichlorophenol
Probably in: Pseudomonas G291

3,6-Dichloro-2-hydroxybenzoic acid
Formed from: 3,6-Dichloro-2-methoxybenzoic acid
Methyl 3,6-dichlorosalicylate
→ 3,6-Dichloro-2-hydroxybenzoic acid-β-D-glucuronide
Probably in: Cow B209
→ 2,5-Dichlorophenol
Probably in: Cow B209

3,6-Dichloro-2-hydroxybenzoic acid-β-D-glucuronide
Probably formed from: 3,6-Dichloro-2-hydroxybenzoic acid

2,6-Dichloro-3-hydroxybenzonitrile
Formed from: 2,6-Dichlorobenzonitrile

2,6-Dichloro-4-hydroxybenzonitrile
Formed from: 2,6-Dichlorobenzonitrile
→ 2,6-Dichloro-3,4-dihydroxybenzonitrile
Probably in: Rat F296

2-(2,3-Dichloro-4-hydroxybenzoyl)-5-hydroxythiophene
Formed from: 2-(2,3-Dichloro-4-hydroxybenzoyl)thiophene

2-(2,3-Dichloro-4-hydroxybenzoyl)thiophene
Formed from: 2-(2,3-Dichloro-4-(4-hydroxy-butoxy)benzoylthiophene
→ 2-(2,3-Dichloro-4-hydroxybenzoyl)-5-hydroxythiophene
In: Man K165

3,5-Dichloro-4-hydroxybenzyl alcohol
Formed from: 3,5-Dichloro-4-methoxybenzyl alcohol

2,2′-Dichloro-3-hydroxybiphenyl
Formed from: 2,2′-Dichlorobiphenyl

2,2′-Dichloro-4-hydroxybiphenyl
Formed from: 2,2′-Dichlorobiphenyl

2,2′-Dichloro-5-hydroxybiphenyl
Formed from: 2,2′-Dichlorobiphenyl

2,3-Dichloro-4-hydroxybiphenyl
Formed from: 2,3-Dichlorobiphenyl

2′,3′-Dichloro-3-hydroxybiphenyl
Formed from: 2,3-Dichlorobiphenyl

2′,3′-Dichloro-4-hydroxybiphenyl
Formed from: 2,3-Dichlorobiphenyl

2′,4′-Dichloro-3-hydroxybiphenyl
Formed from: 2,4-Dichlorobiphenyl

2,4′-Dichloro-4-hydroxybiphenyl
Formed from: 2,4′-Dichlorobiphenyl

2′,4′-Dichloro-4-hydroxybiphenyl
Formed from: 2,4-Dichlorobiphenyl

2,4-Dichloro-5-hydroxybiphenyl
Probably formed from: 2,4-Dichlorobiphenyl

2′,5-Dichloro-3-hydroxybiphenyl
Formed from: 2′,5-Dichlorobiphenyl

2,5-Dichloro-4-hydroxybiphenyl
Formed from: 2,5-Dichlorobiphenyl

2′,5′-Dichloro-4-hydroxybiphenyl
Formed from: 2,5-Dichlorobiphenyl

2′,6′-Dichloro-3-hydroxybiphenyl
Formed from: 2,6-Dichlorobiphenyl

2′,6′-Dichloro-4-hydroxybiphenyl
Formed from: 2,6-Dichlorobiphenyl

3,3′-Dichloro-4-hydroxybiphenyl
Formed from: 3,3′-Dichlorobiphenyl

3,3′-Dichloro-5-hydroxybiphenyl
Formed from: 3,3′-Dichlorobiphenyl

3′,4′-Dichloro-3-hydroxybiphenyl
Formed from: 3,4-Dichlorobiphenyl

3,4′-Dichloro-4-hydroxybiphenyl
Formed from: 4,4′-Dichlorobiphenyl

3′,4′-Dichloro-4-hydroxybiphenyl
Formed from: 3,4-Dichlorobiphenyl

3,5-Dichloro-2-hydroxybiphenyl
Formed from: 2-(2-Aminoethoxy)-3,5-dichlorobiphenyl

3′,5′-Dichloro-3-hydroxybiphenyl
Formed from: 3,5-Dichlorobiphenyl

3,5-Dichloro-4-hydroxybiphenyl
Formed from: 3,5-Dichlorobiphenyl

3′,5′-Dichloro-4-hydroxybiphenyl
Formed from: 3,5-Dichlorobiphenyl

4,4′-Dichloro-2-hydroxybiphenyl
Formed from: 4,4′-Dichlorobiphenyl

4,4′-Dichloro-3-hydroxybiphenyl
Formed from: 4,4′-Dichlorobiphenyl
→ 4-Chloro-3,4′-dihydroxybiphenyl
In: Rat A2530
→ 3,4′-Dichloro-3′,4-dihydroxybiphenyl
In: Rat A2530
→ 4,4′-Dichloro-2,5-dihydroxybiphenyl
In: Rat A2530
→ 4,4′-Dichloro-3,3′-dihydroxybiphenyl
In: Rat A2530
See also: D299
→ 4,4′-Dichloro-3-methoxybiphenyl
In: Duckweed A3390

2-(2,3-Dichloro-4-(4-hydroxybutoxy)benzoyl)-5-hydroxythiophene
Formed from: 2-(2,3-Dichloro-4-(4-hydroxybutoxy)benzoyl)thiophene

2-(2,3-Dichloro-4-(4-hydroxybutoxy)benzoyl)thiophene
→ 2-(2,3-Dichloro-4-(4-hydroxybutoxy)benzoyl)-5-hydroxythiophene
In: Man K165
→ 2-(2,3-Dichloro-4-hydroxybenzoyl)thiophene
In: Man K165

(6,7-Dichloro-2-(1-hydroxycyclopentyl)-2-methyl-1-oxo-5-indanyloxy)acetic acid
Formed from: (6,7-Dichloro-2-cyclopentyl-2-methyl-1-oxo-5-indanyloxy)acetic acid

(6,7-Dichloro-2-(3-hydroxycyclopentyl)-2-methyl-1-oxo-5-indanyloxy)acetic acid
Formed from: (6,7-Dichloro-2-cyclopentyl-2-methyl-1-oxo-5-indanyloxy)acetic acid

(6,7-Dichloro-2-(4-hydroxycyclopentyl)-2-methyl-1-oxo-5-indanyloxy)acetic acid
Formed from: (6,7-Dichloro-2-cyclopentyl-2-methyl-1-oxo-5-indanyloxy)acetic acid
→ (6,7-Dichloro-2-methyl-1-oxo-(4-oxocyclopentyl)-5-indanyloxy)acetic acid
Probably in: Baboon A2951
Chimpanzee A2865
Dog A2951
Man A2865
Monkey A2951
Rat A2951

2′,4′-Dichloro-5′-hydroxydehydrocyclopeptine
Formed from: 2′,4′-Dichlorodehydrocyclopeptine

4,4′-Dichloro-2-hydroxydiphenyl ether
Formed from: 4,4′-Dichlorodiphenyl ether

3,5-Dichloro-2-(2-hydroxyethoxy)biphenyl
Formed from: 3,5-Dichloro-2-formylmethoxybiphenyl

2′,5-Dichloro-2-(2-hydroxyethylamino)acetamidobenzophenone
Formed from: Cloxazolam
→ 2-Amino-2′,5-dichlorobenzophenone
Probably in: Mouse A828
Rat A828

2,2-Dichloro-N-((1*R*,2*R*)-2-hydroxy-1-(hydroxymethyl)-2-(4-(methylsulphonyl)phenyl)ethyl)acetamide
See: Thiamphenicol

***o*-((2,6-Dichloro-4-hydroxy-3-hydroxymethylphenyl)amino)benzoic acid**
Probably formed from: *o*-((2,6-Dichloro-3-hydroxymethylphenyl)amino)benzoic acid

Dichloro-7-hydroxymethotrexate
Formed from: Dichloromethotrexate

2,5-Dichloro-3-hydroxy-6-methoxybenzoic acid
Formed from: 3,6-Dichloro-2-methoxybenzoic acid
→ 3,6-Dichlorogentisic acid
In: Bracken A2688

(+)-5,7-Dichloro-4-hydroxy-6′-methoxy-6-methyl-3,4′-dioxospiro(benzofuran-2(3*H*),1′-(2,5)-cyclohexadiene)-2′-carboxylic acid methyl ester
See: Geodin

***o*-((2,6-Dichloro-3-hydroxymethylphenyl)amino)benzoic acid**
Formed from: *o*-((2,6-Dichloro-3-methylphenyl)amino)benzoic acid
→ *o*-((3-Carboxy-2,6-dichlorophenyl)amino)benzoic acid
Probably in: Man E707
→ *o*-((2,6-Dichloro-4-hydroxy-3-hydroxymethylphenyl)amino)benzoic acid
Probably in: Man E707

***o*-((2,6-Dichloro-4-hydroxy-3-methylphenyl)amino)benzoic acid**
Formed from: *o*-((2,6-Dichloro-3-methylphenyl)amino)benzoic acid

(6,7-Dichloro-1α-hydroxy-2-methyl-2-phenyl-5-indanyloxy)acetic acid
Formed from: (6,7-Dichloro-2-methyl-1-oxo-2-phenyl-5-indanyloxy)acetic acid

6,7-Dichloro-5-hydroxy-2-methyl-2-phenyl-5-indanone
Formed from: 6,7-Dichloro-2-methyl-1-oxo-2-phenyl-5-indanyloxy)acetic acid

5,7-Dichloro-8-hydroxy-2-methylquinoline
(sterosan)
→ 5,7-Dichloro-8-hydroxy-2-methylquinoline-β-D-4-O-methylglucoside
In: Sporotrichum A2630

5,7-Dichloro-8-hydroxy-2-methylquinoline-β-D-4-O-methylglucoside
Formed from: 5,7-Dichloro-8-hydroxy-2-methylquinoline

2′,5-Dichloro-2-(3-hydroxymethyl-4*H*-1,2,4-triazol-4-yl)benzophenone
Probably formed from: 2′,5-Dichloro-2-(3-methyl-4*H*-1,2,4-triazol-4-yl)benzophenone

***o*-(5,5-Dichloro-2-hydroxy-1-oxo-4-hexen-1-yl)phenylglyoxylic acid**
Formed from: 3-(3,3-Dichloroallyl)-2-hydroxy-1,4-naphthoquinone

2,3-Dichloro-4-hydroxyphenoxyacetic acid
Formed from: 2,4-Dichlorophenoxyacetic acid
→ 2,3-Dichloro-4-hydroxyphenoxyacetic acid-4-β-D-glucoside
In: Cucumber A3062
See also: A2504

2,4-Dichloro-3-hydroxyphenoxyacetic acid
Probably formed from: 2,4-Dichlorophenoxyacetic acid

2,4-Dichloro-5-hydroxyphenoxyacetic acid
Formed from: 2,4-Dichlorophenoxyacetic acid

2,4-Dichloro-6-hydroxyphenoxyacetic acid
Formed from: 2,4-Dichlorophenoxyacetic acid

2,5-Dichloro-4-hydroxyphenoxyacetic acid
Formed from: 2,4-Dichlorophenoxyacetic acid
→ 2,5-Dichloro-4-hydroxyphenoxyacetic acid-4-β-D-glucoside
In: Cucumber A3062
See also: A2504

2,3-Dichloro-4-hydroxyphenoxyacetic acid-4-β-D-glucoside
Probably formed from: 2,3-Dichloro-4-hydroxyphenoxyacetic acid

2,5-Dichloro-4-hydroxyphenoxyacetic acid-4-β-D-glucoside
Probably formed from: 2,5-Dichloro-4-hydroxyphenoxyacetic acid

N-(2,6-Dichloro-4-hydroxyphenoxyethyl)-N-propylurea
Probably formed from: N-(2,4,6-Trichloro-4-hydroxyphenoxyethyl)-N-propylurea

2-(4-(2′,4′-Dichloro-3′-hydroxyphenoxy)phenoxy) propionic acid
Probably formed from: 2-(4-(2′,4′-Dichlorophenoxy)phenoxy)propionic acid

2-(4-(2′,4′-Dichloro-5′-hydroxyphenoxy)phenoxy) propionic acid
Probably formed from: 2-(4-(2′,4′-Dichlorophenoxy)phenoxy)propionic acid

2-(4-(2′,4′-Dichloro-6′-hydroxyphenoxy)phenoxy) propionic acid
Probably formed from: 2-(4-(2′,4′-Dichlorophenoxy)phenoxy)propionic acid

***o*-((2,6-Dichloro-4-hydroxyphenyl)amino)benzoic acid**
Probably formed from: *o*-((3-Carboxy-2,6-dichloro-4-hydroxyphenyl)amino)benzoic acid

(2-((2,6-Dichloro-4-hydroxyphenyl)amino)phenyl) acetoxyacetic acid
Formed from: Aceclofenac

2,6-Dichloro-N-(*p*-hydroxyphenyl)-*p*-benzoquinoneimine
See: 2,6-Dichlorophenolindophenol

4,6-Dichloro-3-(2-(*p*-hydroxyphenylcarbamoyl) ethenyl)indole-2-carboxylic acid
Formed from: 4,6-Dichloro-3-(2-(phenylcarbamoyl)ethenyl)indole-2-carboxylic acid
→ 4,6-Dichloro-3-(2-(*p*-hydroxyphenylcarbamoyl) ethenyl)indole-2-carboxylic acid-β-D-glucuronide
In: Man K125
→ 4,6-Dichloro-3-(2-(*p*-hydroxyphenylcarbamoyl) ethenyl)indole-2-carboxylic acid-O-sulphate
In: Man K125

4,6-Dichloro-3-(2-(*p*-hydroxyphenylcarbamoyl) ethenyl)indole-2-carboxylic acid-β-D-glucuronide
Formed from: 4,6-Dichloro-3-(2-(*p*-hydroxyphenylcarbamoyl)ethenyl)indole-2-carboxylic acid

4,6-Dichloro-3-(2-(*p*-hydroxyphenylcarbamoyl) ethenyl)indole-2-carboxylic acid-O-sulphate
Formed from: 4,6-Dichloro-3-(2-(*p*-hydroxyphenylcarbamoyl)ethenyl)indole-2-carboxylic acid

N-(2,3-Dichloro-4-phenyl)-1-methylcyclohexanecarboxamide
See: Fenhexamid

1,1-Dichloro-2-(*p*-hydroxyphenyl)-2-(*p*-methoxyphenyl)ethane
Probably formed from: 1,1-Dichloro-2,2-bis(*p*-methoxyphenyl)ethane
→ 1-Chloro-2-(*p*-hydroxyphenyl)-2-(*p*-methoxyphenyl)ethylene
Probably in: Goat C198
→ 1,1-Dichloro-2,2-bis(*p*-hydroxyphenylethane
Probably in: Chicken D178
Goat C198, D178
→ 1,1-Dichloro-2-(*p*-hydroxyphenyl)-2-(*p*-methoxyphenyl)ethane-β-D-glucuronide
Probably in: Chicken D178
Goat D178

1,1-Dichloro-2-(*p*-hydroxyphenyl)-2-(*p*-methoxyphenyl)ethane-β-D-glucuronide
Probably formed from: 1,1-Dichloro-2-(*p*-hydroxyphenyl)-2-(*p*-methoxyphenyl)ethane

1,1-Dichloro-2-(*p*-hydroxyphenyl)-2-(*p*-methoxyphenyl)ethylene
Probably formed from: 1,1-Dichloro-2,2-bis(*p*-methoxyphenyl)ethylene
→ 1,1-Dichloro-2,2-bis(*p*-hydroxyphenylethylene
Probably in: Chicken D178
Goat C198, D178
→ 1,1-Dichloro-2-(*p*-hydroxyphenyl)-2-(*p*-methoxyphenyl)ethylene-β-D-glucuronide
Probably in: Chicken D178

1,1-Dichloro-2-(*p*-hydroxyphenyl)-2-(*p*-methoxyphenyl)ethylene-β-D-glucuronide
Probably formed from: 1,1-Dichloro-2-(*p*-hydroxyphenyl)-2-(*p*-methoxyphenyl)ethylene

(6,7-Dichloro-2-(*p*-hydroxyphenyl)-2-methyl-1-oxo-5-indanyloxy)acetic acid
Formed from: (6,7-Dichloro-2-methyl-1-oxo-2-phenyl-5-indanyloxy)acetic acid
→ (6,7-Dichloro-2-(*p*-methoxyphenyl)-2-methyl-1-oxo-5-indanyloxy)acetic acid
Probably in: Chimpanzee A2122

2,4-Dichloro-5-hydroxyphenyl 4-nitrophenyl ether
Formed from: Nitrofen

1,1-Dichloro-*cis*-2-(*p*-hydroxyphenyl)-3-phenylcyclopropane
Formed from: 1,1-Dichloro-*cis*-2,3-diphenylcyclopropane

N-(3,5-Dichloro-4-hydroxyphenyl)succinamic acid
Probably formed from: N-(3,5-Dichlorophenyl)succinamic acid
N-(3,5-Dichlorophenyl)succinimide

N-(3,4-Dichloro-2-hydroxyphenyl)urea
Probably formed from: N-(3,4-Dichlorophenyl)urea

N-(4,5-Dichloro-2-hydroxyphenyl)urea
Probably formed from: N-(3,4-Dichlorophenyl)urea

5,7-Dichloro-8-hydroxyquinoline
Formed from: 5,7-Dichloroquinolin-8-yl-β-D-glucuronide
Probably formed from: 5,7-Dichloroquinolin-8-yl sulphate
→ 5,7-Dichloroquinolin-8-yl-β-D-glucuronide
In: Rat A3654
→ 5,7-Dichloroquinolin-8-yl sulphate
In: Rat A3654

5,6-Dichloro-4-hydroxy-2-trifluoromethylbenzimidazole
Formed from: 5,6-Dichloro-2-trifluoromethylbenzimidazole

4,4′-Dichloro-2-hydroxy-3′-trifluoromethylcarbanilide
Formed from: 4,4′-Dichloro-3′-trifluoromethylcarbanilide
→ 4,4′-Dichloro-3′-trifluoromethylcarbanilid-2-yl-β-D-glucuronide
In: Rat A2076
→ 4,4′-Dichloro-3′-trifluoromethylcarbanilid-2-yl sulphate
In: Rat A2076

4,4′-Dichloro-3-hydroxy-3′-trifluoromethylcarbanilide
Formed from: 4,4′-Dichloro-3′-trifluoromethylcarbanilide
→ 4,4′-Dichloro-3′-trifluoromethylcarbanilid-3-yl-β-D-glucuronide
In: Rat A2076
→ 4,4′-Dichloro-3′-trifluoromethylcarbanilid-3-yl sulphate
In: Rat A2076

2,6-Dichloroindophenol
See: 2,6-Dichlorophenolindophenol

2,5-Dichloro-4′-isopropylbiphenyl
→ 4-Carboxy-2′,5′-dichlorobiphenyl
In: Cunninghamella A3078
Rat A2531
→ 4-(1-Carboxyethyl)-2′,5′-dichlorobiphenyl
In: Cunninghamella A3078
Frog A3078
Goldfish A3078
Rat A2531

7,8-Dichloroisoquinoline
Probably formed from: 7,8-Dichloro-3,4-dihydroisoquinoline

3,4-Dichloro-N-malonylaniline
Formed from: 3,4-Dichloroaniline

2,4-Dichloromandelic acid
Formed from: Dimethylvinphos

3′,4′-Dichloro-2-methacrylanilide (dicryl)
→ 3,4-Dichloroaniline
In: Bacillus A1006
Dandelion A230

Dichloromethotrexate
→ Dichloro-7-hydroxymethotrexate
In: Rabbit D644

3,6-Dichloro-2-methoxybenzoic acid (dicambra)
→ 3,6-Dichloro-2-hydroxybenzoic acid
In: Beef B209
→ 2,5-Dichloro-3-hydroxy-6-methoxybenzoic acid
In: Bracken A2688

2-(2,3-Dichloro-4-methoxybenzoyl)-5-hydroxythiophene
Formed from: 2-(2,3-Dichloro-4-methoxybenzoyl)thiophene

2-(2,3-Dichloro-4-methoxybenzoyl)thiophene
→ 2-(2,3-Dichloro-4-methoxybenzoyl)-5-hydroxythiophene
In: Man K165
Rat F130

3,5-Dichloro-4-methoxybenzyl alcohol
→ 3,5-Dichloroanisic acid
In: Microorganism J708
→ 3,5-Dichloro-4-hydroxybenzyl alcohol
In: Microorganism J708

4,4′-Dichloro-3-methoxybiphenyl
Probably formed from: 4,4′-Dichloro-3-hydroxybiphenyl

2,5-Dichloro-4-methoxyphenol
Formed from: 2,5-Dichloro-1,4-dimethoxybenzene
→ 2,5-Dichloro-1,4-dimethoxybenzene
In: Mucor A938
Trichoderma A938
→ 2,5-Dichloroquinol
In: Cephalosporium A938
Fusarium A938
Mucor A938
Rhizoctonia A938

2,6-Dichloro-4-methoxyphenol
Formed from: 2,6-Dichloroquinol

4,5-Dichloro-2-methoxyphenol
Probably formed from: 3,4-Dichlorocatechol
→ 4,5-Dichlorocatechol
In: Microorganism H684
→ 4,5-Dichloroveratrole
Probably in: Phanerochaete J696

2,3-Dichloro-4-methoxyphenyl)-2-furylmethanone O-(2-(diethylamino)ethyl) oxime
See: Diclofurime

(6,7-Dichloro-2-(*p*-methoxyphenyl)-2-methyl-1-oxo-5-indanyloxy)acetic acid
Probably formed from: (6,7-Dichloro-2-(*p*-hydroxyphenyl)-2-methyl-1-oxo-5-indanyloxy)acetic acid

2,6-Dichloro-4-methylaniline
→ 4-Amino-2,6-dichlorobenzyl alcohol
In: Rat G580
→ N-(4-Amino-2,6-dichlorobenzyl)-2,6-dichloro-4-methylaniline
In: Rat G580
→ N-(4-Amino-2,6-dichlorobenzylidene)-2,6-dichloro-4-methylaniline
In: Rat G580
→ 2,6-Dichloro-4-methylnitrosobenzene
In: Rat G580
→ 2,2′,6,6′-Tetrachloro-4,4′-dimethylazoxybenzene
In: Rat G580

2,3-Dichloro-N-methylbenzylamine
Formed from: 2,3-Dichlorobenzylamine

(2,3-Dichloro-4-(2-methylenebutyryl)phenoxy) acetic acid
See: Ethacrynic acid

4,5-Dichloro-1,2-methylenedioxybenzene
→ 4,5-Dichlorocatechol
In: Rat A3947

2,6-Dichloro-4-methylnitrosobenzene
Formed from: 2,6-Dichloro-4-methylaniline

(6,7-Dichloro-2-methyl-1-oxo-(4-oxocyclopentyl)-5-indanyloxy)acetic acid
Probably formed from: (6,7-Dichloro-2-(4-hydroxycyclopentyl)-2-methyl-1-oxo-5-indanyloxy)acetic acid

(6,7-Dichloro-2-methyl-1-oxo-2-phenyl-5-indanyloxy)acetic acid
→ 6,7-Dichloro-5-hydroxy-2-methyl-2-phenyl-1-indanone
In: Dog A2124
→ (6,7-Dichloro-1α-hydroxy-2-methyl-2-phenyl-5-indanyloxy)acetic acid
In: Chimpanzee A2122
→ (6,7-Dichloro-2-(*p*-hydroxyphenyl)-2-methyl-1-oxo-5-indanyloxy)acetic acid
In: Chimpanzee A2122
Dog A2124
Man A3254
Monkey A2124
Rat A2124
→ (6,7-Dichloro-2-methyl-1-oxo-2-phenyl-5-indanyloxy)acetyl-β-D-glucuronide
In: Dog A2124
Rat A2124

(6,7-Dichloro-2-methyl-1-oxo-2-phenyl-5-indanyloxy)acetyl-β-D-glucuronide
Formed from: (6,7-Dichloro-2-methyl-1-oxo-2-phenyl-5-indanyloxy)acetic acid

3,4-Dichloro-N-methylphenethylamine
Formed from: 3,4-Dichlorophenylamine

***o*-((2,6-Dichloro-3-methylphenyl)amino)benzoic acid**
→ *o*-((2,6-Dichloro-3-hydroxymethylphenyl) amino)benzoic acid
In: Man E707
→ *o*-((2,6-Dichloro-4-hydroxy-3-methylphenyl)amino)benzoic acid
In: Man E707
→ 2-((2,6-Dichloro-3-methylphenyl)amino)-4-hydroxybenzoic acid
In: Man E707

3,4-Dichloro-N-methylphenylethanolamine
Formed from: 3,4-Dichlorophenylethanolamine

2-((2,6-Dichloro-3-methylphenyl)amino)-4-hydroxybenzoic acid
Formed from: *o*-((2,6-Dichloro-3-methylphenyl)amino)benzoic acid

2′,5-Dichloro-2-(3-methyl-4*H*-1,2,4-triazol-4-yl) benzophenone
Formed from: Triazolam
→ 2′,5-Dichloro-2-(3-hydroxymethyl-4*H*-1,2,4-triazol-4-yl)benzophenone
Probably in: Dog A3702
Rat A3702

3′,4′-Dichloro-2-methylvaleranilide
See: Karsil

1,2-Dichloronaphthalene
→ 5,6-Dichloro-1,2-dihydro-1,2-dihydroxynaphthalene
In: Rat A2540
→ 3,4-Dichloro-α-naphthol
Probably in: Pig A1829

1,4-Dichloronaphthalene
→ 2,4-Dichloro-α-naphthol
In: Pig A1829

2,6-Dichloronaphthalene
→ 2,6-Dichloro-α-naphthol
Probably in: Rat A2540

2,7-Dichloronaphthalene
→ 7-Dichloro-β-naphthol
Probably in: Rat A2540

2,4-Dichloro-α-naphthol
Formed from: 1,4-Dichloronaphthalene

2,6-Dichloro-α-naphthol
Formed from: 2,6-Dichloronaphthalene

3,4-Dichloro-α-naphthol
Probably formed from: 1,2-Dichloronaphthalene

2,6-Dichloro-4-nitroanisole
→ 2,6-Dichloro-4-nitrophenol
In: Guinea pig E221
Mouse E221
Rat A2855, A3495, E221

1,2-Dichloro-3-nitrobenzene
→ 1-Chloro-2-methylthio-3-nitrobenzene
In: Mucor D463
→ 2,3-Dichloroaniline
In: Mucor D463

1,2-Dichloro-4-nitrobenzene
Formed from: 3,4-Dichloroaniline
Probably formed from: 3,4-Dichloronitrosobenzene
→ 1-Chloro-2-methylthio-4-nitrobenzene
In: Mucor B955
→ S-(2-Chloro-4-nitrophenyl)glutathione
In: Rat H516, J583
See also: A1997, A2132, B69, B702, B875, C562, C878, C901, J102
→ 2,3-Dichloroaniline
In: Mucor D463
→ 3,3′,4,4′-Tetrachloroazoxybenzene
In: Escherichia C768

1,3-Dichloro-2-nitrobenzene
→ 2,6-Dichloroaniline
In: Mucor D463

1,3-Dichloro-4-nitrobenzene
→ 1-Chloro-3-methylthio-4-nitrobenzene
In: Aspergillus D463
Mucor D463
→ S-(5-Chloro-2-nitrophenyl) glutathione
In: Mucor F36, F37
→ 2,3-Dichloroaniline
In: Aspergillus D463
Mucor D463

1,3-Dichloro-5-nitrobenzene
→ 3,5-Dichloroaniline
In: Mucor D463

1,4-Dichloro-2-nitrobenzene
→ 2,5-Dichloroaniline
In: Mucor D463

2,6-Dichloro-4-nitrophenol
Formed from: 2,6-Dichloro-4-nitroanisole

2′,5-Dichloro-4′-nitrosalicylanilide
(niclosamide)
→ 4′-Amino-2′,5-dichlorosalicylanilide
In: Rat B295
→ 2′,5-Dichloro-4′-nitrosalicylanilide-β-D-glucuronide
In: Rat B295
Trout A2171

2′,5-Dichloro-4′-nitrosalicylanilide-β-D-glucuronide
Formed from: 2′,5-Dichloro-4′-nitrosalicylanilide

3,4-Dichloronitrosobenzene
Formed from: 3,4-Dichloroaniline
→ 1,2-Dichloro-4-nitrobenzene
Probably in: Man C879

4,6-Dichloro-3-((1*E*)-3-oxo-3-(phenylamino)-1-propenyl)-1*H*-indole-2-carboxylic acid
→ 4,6-Dichloro-3-((1*E*)-3-oxo-3-(phenylamino)-1-propenyl)-1*H*-indole-2-carboxy-β-D-glucuronic acid
In: Man K656

4,6-Dichloro-3-((1*E*)-3-oxo-3-(phenylamino)-1-propenyl)-1*H*-indole-2-carboxy-β-D-glucuronic acid
Formed from: 4,6-Dichloro-3-((1*E*)-3-oxo-3-(phenylamino)-1-propenyl)-1*H*-indole-2-carboxylic acid

Dichlorophen (bis(5-chloro-2-hydroxyphenyl) methane)
→ Dichlorophen-β-D-glucuronide
In: Rat A3750
→ Dichlorophen-O-sulphate
In: Rat A3750

2,4-Dichlorophenacyl chloride
Formed from: Dimethylvinphos
→ S-(2,4-Dichlorophenacyl)glutathione (spontaneous)
In: Dog A1752
Rat A1752, A2023, C184

3,4-Dichlorophenacyl chloride
→ 2-Chloro-1-(3,4-dichlorophenyl)ethanol
In: Geotrichum K454

S-(2,4-Dichlorophenacyl)glutathione
Formed from: 2,4-Dichlorophenacyl chloride
→ 2′,4′-Dichloroacetophenone
In: Rat A1752, A2023, C184

3,4-Dichlorophenethylamine
→ 3,4-Dichloro-N-methylphenethylamine
In: Rabbit B741

Dichlorophen-β-D-glucuronide
Formed from: Dichlorophen

2,3-Dichlorophenol
Formed from: *o*-Dichlorobenzene
2,3,5-Trichlorophenol
2,3,6-Trichlorophenol
→ *o*-Chlorophenol
In: Desulfomonile G471
→ *m*-Chlorophenol
In: Desulfitobacterium J232, K247
See also: H509
→ 3,4-Dichlorocatechol
In: Proteobacteria J466
Rhodococcus F383
See also: G734, H653, J337
→ 2,3-Dichloroquinol
Probably in: Flavobacterium G641

2,4-Dichlorophenol
Formed from: Chlomethoxynil
5-Chloro-(2,4-dichlorophenoxy)phenol
p-Dichlorobenzene
2,4-Dichlorophenoxyacetic acid
2-(2,4-Dichlorophenoxy)propionic acid
3-(2,4-Dichlorophenoxy)propionic acid
Nemacide
Nitrofen
2,2′,4-Trichlorodiphenyl ether
2,3,4-Trichlorophenol
2,4,5-Trichlorophenol
2,4,6-Trichlorophenol
→ *p*-Chlorophenol
In: Desulfitobacterium J232

Microorganism G331, G667
→ Chloroquinol
In: Azotobacter H884
→ 2,4-Dichloroanisole
In: Phanerochaete G782
→ 3,5-Dichlorocatechol
In: Acinetobacter C232
Proteobacteria J466
Pseudomonas G221
See also: G734
→ 2-(2,4-Dichlorophenoxy)-1,4-benzoquinone
In: Rhizoctonia B588
→ 2-(2,4-Dichlorophenoxy)-6-chloro-1,4-benzoquinone
In: Rhizoctonia B588
→ 4-(4-(2,4-Dichlorophenoxy)-2,6-dichlorophenoxy)-2,6-dichlorophenol
In: Rhizoctonia D690
→ 4-(4-(4-(2,4-Dichlorophenoxy)-2,6-dichlorophenoxy)-2,6-dichlorophenoxy)-2,6-dichlorophenol
In: Rhizoctonia D690
→ 4-(4-(4-(4-(2,4-Dichlorophenoxy)-2,6-dichlorophenoxy)-2,6-dichlorophenoxy)-2,6-dichlorophenoxy)-2,6-dichlorophenol
In: Rhizoctonia D690
→ 4-(2,4-Dichlorophenoxy)-2,6-dimethoxyphenol
In: Rhizoctonia D690
→ 5-(2,4-Dichlorophenoxy)-3-methoxy-*o*-quinone
In: Rhizoctonia D690
→ 2,4-Dichlorophenyl-β-D-glucuronide
In: Rat C330, C358
→ 2,4-Dichlorophenyl sulphate
In: Rat C358

2,5-Dichlorophenol
Formed from: *p*-Dichlorobenzene
O-(2,5-Dichlorophenyl) O-hexyl phosphoramidate
2,3,6-Trichlorophenol
Probably formed from: 3,6-Dichloro-2-hydroxybenzoic acid
→ 4-Chlorocatechol
Probably in: Burkholderia J177
→ *o*-Chlorophenol
In: Desulfomonile G471
→ 3,6-Dichlorocatechol
In: Proteobacteria J466
Pseudomonas E731
→ 2,5-Dichlorophenyl-β-D-glucuronide
In: Rat A3938, H821
→ 2,5-Dichlorophenyl sulphate
In: Rat A3938

2,6-Dichlorophenol
Formed from: 3,5-Dichloroanisic acid
o-Dichlorobenzene
Lofexidine
2,3,6-Trichlorophenol
Probably formed from: 3,5-Dichloro-4-hydroxybenzoic acid
→ *o*-Chlorophenol
In: Desulfitobacterium J232, K247
Microorganism G331
→ 4-(2,6-Dichlorophenoxy)-2,6-dimethoxyphenol
In: Rhizoctonia D690
→ 4-(4-(2,4-Dichlorophenoxy)-2,6-dimethoxyphenoxy)-2,6-dimethoxyphenol
In: Rhizoctonia D690
→ 4-(4-(4-(2,4-Dichlorophenoxy)-2,6-dimethoxyphenoxy)-2,6-dimethoxyphenoxy)-2,6-dimethoxyphenol
In: Rhizoctonia D690
→ 4-(4-(4-(4-(2,4-Dichlorophenoxy)-2,6-dimethoxyphenoxy)-2,6-dimethoxyphenoxy)-2,6-dimethoxyphenoxy)-2,6-dimethoxyphenol
In: Rhizoctonia D690
→ 2,6-Dichloroquinol
In: Flavobacterium G641
See also: H884

3,4-Dichlorophenol
Formed from: 3,4-Dichloroaniline
o-Dichlorobenzene
3,3′,4,4′-Tetrachlorodiphenyl ether
2,2′,4-Trichlorodiphenyl ether
2,3,4-Trichlorophenol
2,4,5-Trichlorophenol
→ *p*-Chlorophenol
In: Desulfomonile G471
→ Chloroquinol

In: Azotobacter H884
→ 3,4-Dichloroanisole
In: Phanerochaete J696
→ 3,4-Dichlorocatechol
In: Pseudomonas E731
→ 4,5-Dichlorocatechol
In: Phanerochaete J696
Rhodococcus F383
See also: G734, H653, J337, J466
→ 3,4-Dichlorophenyl sulphate
In: Man E575
Rat E575

3,5-Dichlorophenol
Formed from: 2,3,5-Trichlorophenol
3,4,5-Trichlorophenol
→ *m*-Chlorophenol
In: Desulfomonile G471
Microorganism G331, G667, H509
→ 3,5-Dichlorophenyl sulphate
In: Man E575
Rat E575

Dichlorophenolindophenol (2,6-dichloroindophenol, 2,6-dichloro-N-(*p*-hydroxyphenyl)-*p*-benzoquinoneimine)
→ 3,5-Dichloro-4,4′-dihydroxydiphenylamine
In: Bacillus B725
Beef A1899
Diphyllobothrium A679
Escherichia A1228
Eubacterium B426
Guinea pig K773
Man A3008, B260
Phanerochaete H611
Pseudomonas A2952
Rat A2223
Saccharomyces K229
Thauera K173
→ Dichlorophenolindophenol-β-D-glucuronide
In: Man G524

Dichlorophenolindophenol-β-D-glucuronide
Formed from: Dichlorophenolindophenol

2,4-Dichlorophenoxyacetic acid (2,4-D)
Formed from: 2,4-Dichlorophenoxyacetic acid bisethyleneglycol ester
→ 2-Chloro-4-hydroxyphenoxyacetic acid
In: Wheat A2746
See also: A2504
→ *p*-Chlorophenoxyacetic acid
In: Azotobacter F895
→ 2,3-Dichloro-4-hydroxyphenoxyacetic acid
In: Carrot A2504
Cucumber A3062
Glycine A525, A3200
Jackbean A2504
Nicotiana A2504
Rice A1836
Sunflower A2504
Zea A2504, A3200, A3513
→ 2,4-Dichloro-3-hydroxyphenoxyacetic acid
Probably in: Zea A2504
→ 2,4-Dichloro-5-hydroxyphenoxyacetic acid
In: Glycine A3200
Zea A3200
→ 2,4-Dichloro-6-hydroxyphenoxyacetic acid
In: Wheat A2746
→ 2,5-Dichloro-4-hydroxyphenoxyacetic acid
In: Carrot A2504
Cucumber A3062
Glycine A525, A3200
Jackbean A2504
Nicotiana A2504
Rice A1836
Sunflower A2504
Wheat A2746
Zea A2504, A3200, A3513
See also: G813
→ 2,4-Dichlorophenol
In: Alcaligenes E727, H411, J518
Grass A1268
Legume A1268
Microorganism G173, G183, K584
Sheep A2366
→ 2,4-Dichlorophenoxyacetyl-L-alanine
In: Glycine A3200
Zea A3200
→ 2,4-Dichlorophenoxyacetyl-L-aspartic acid
In: Carrot A2504
Glycine A445, A525, A3200
Jackbean A2504
Nicotiana A2504
Zea A3200

→ 2,4-Dichlorophenoxyacetyl-β-glucoside
In: Carrot A2504
Jackbean A2504
Nicotiana A2504
Sunflower A2504
Zea A2504
See also: G813
→ 2,4-Dichlorophenoxyacetyl-L-glutamic acid
In: Carrot A2504
Glycine A445, A525, A2504, A3200
Jackbean A2504
Nicotiana A2504
Sunflower A2504
→ 2,4-Dichlorophenoxyacetylglycine
In: Rat A1405, E534
→ 2,4-Dichlorophenoxyacetyl-L-leucine
In: Glycine A3200
Zea A3200
→ 2,4-Dichlorophenoxyacetyl-L-phenylalanine
In: Glycine A3200
Zea A3200
→ 2,4-Dichlorophenoxyacetyltaurine
In: Dogfish A1784, A2359, A3935
Flounder A1784
Lobster C38
Rat A1405
→ 2,4-Dichlorophenoxyacetyl-L-tryptophan
In: Glycine A3200
Zea A3200
→ 2,4-Dichlorophenoxyacetyl-L-valine
In: Glycine A3200
Zea A3200
→ Ethyl 2,4-dichlorophenoxyacetate
In: Glycine A3200
Zea A3200
Unchanged in man A298

2,4-Dichlorophenoxyacetic acid bisethyleneglycol ester
→ 2,4-Dichlorophenoxyacetic acid
In: Barley G813

2,4-Dichlorophenoxyacetyl-L-alanine
Formed from: 2,4-Dichlorophenoxyacetic acid

2,4-Dichlorophenoxyacetyl-L-aspartic acid
Formed from: 2,4-Dichlorophenoxyacetic acid

2,4-Dichlorophenoxyacetyl-di-β-glucoside
Probably formed from: 2,4-Dichlorophenoxyacetyl-β-glucoside

2,4-Dichlorophenoxyacetyl-β-glucoside
Formed from: 2,4-Dichlorophenoxyacetic acid
→ 2,4-Dichlorophenoxyacetyl-di-β-glucoside
Probably in: Barley G813

2,4-Dichlorophenoxyacetyl-L-glutamic acid
Formed from: 2,4-Dichlorophenoxyacetic acid

2,4-Dichlorophenoxyacetylglycine
Formed from: 2,4-Dichlorophenoxyacetic acid

2,4-Dichlorophenoxyacetyl-L-leucine
Formed from: 2,4-Dichlorophenoxyacetic acid

2,4-Dichlorophenoxyacetyl-L-phenylalanine
Formed from: 2,4-Dichlorophenoxyacetic acid

2,4-Dichlorophenoxyacetyltaurine
Formed from: 2,4-Dichlorophenoxyacetic acid

2,4-Dichlorophenoxyacetyl-L-tryptophan
Formed from: 2,4-Dichlorophenoxyacetic acid

2,4-Dichlorophenoxyacetyl-L-valine
Formed from: 2,4-Dichlorophenoxyacetic acid

***m*-(3,4-Dichlorophenoxy)benzaldehyde**
→ *m*-(3,4-Dichlorophenoxy)benzoic acid
In: Rat H761

***m*-(3,5-Dichlorophenoxy)benzaldehyde**
→ *m*-(3,5-Dichlorophenoxy)benzoic acid
In: Rat H761

***m*-(3,4-Dichlorophenoxy)benzoic acid**
Formed from: *m*-(3,4-Dichlorophenoxy)benzaldehyde

***m*-(3,5-Dichlorophenoxy)benzoic acid**
Formed from: *m*-(3,5-Dichlorophenoxy)benzaldehyde

2-(2,4-Dichlorophenoxy)-1,4-benzoquinone
Formed from: 2,4-Dichlorophenol

4-(2,4-Dichlorophenoxy)butyric acid
Does not form 2,4-dichlorophenoxyacetic acid in Phytophthora A1828

2-(2,4-Dichlorophenoxy)-6-chloro-1,4-benzoquinone
Formed from: 2,4-Dichlorophenol

4-(2,4-Dichlorophenoxy)-2,6-dimethoxyphenol
Formed from: 2,4-Dichlorophenol
4-Hydroxy-3,5-dimethoxybenzoic acid

4-(2,6-Dichlorophenoxy)-2,6-dimethoxyphenol
Formed from: 2,6-Dichlorophenol

4-Hydroxy-3,5-dimethoxybenzoic acid

4-(4-(2,4-Dichlorophenoxy)-2,6-dimethoxyphenoxy)-2,6-dimethoxyphenol
Formed from: 2,4-Dichlorophenol
4-Hydroxy-3,5-dimethoxybenzoic acid

4-(4-(2,6-Dichlorophenoxy)-2,6-dimethoxyphenoxy)-2,6-dimethoxyphenol
Formed from: 2,6-Dichlorophenol
4-Hydroxy-3,5-dimethoxybenzoic acid

4-(4-(4-(2,4-Dichlorophenoxy)-2,6-dimethoxyphenoxy)-2,6-dimethoxyphenoxy)-2,6-dimethoxyphenol
Formed from: 2,4-Dichlorophenol
4-Hydroxy-3,5-dimethoxybenzoic acid

4-(4-(4-(2,6-Dichlorophenoxy)-2,6-dimethoxyphenoxy)-2,6-dimethoxyphenoxy)-2,6-dimethoxyphenol
Formed from: 2,6-Dichlorophenol
4-Hydroxy-3,5-dimethoxybenzoic acid

4-(4-(4-(4-(2,4-Dichlorophenoxy)-2,6-dimethoxyphenoxy)-2,6-dimethoxyphenoxy)-2,6-dimethoxyphenoxy)-2,6-dimethoxyphenol
Formed from: 2,4-Dichlorophenol
4-Hydroxy-3,5-dimethoxybenzoic acid

4-(4-(4-(4-(2,6-Dichlorophenoxy)-2,6-dimethoxyphenoxy)-2,6-dimethoxyphenoxy)-2,6-dimethoxyphenoxy)-2,6-dimethoxyphenol
Formed from: 2,6-Dichlorophenol
4-Hydroxy-3,5-dimethoxybenzoic acid

2-(1-(2,6-Dichlorophenoxy)ethyl)-4,5-dihydro-1*H*-imidazole
See: Lofexidine

2-(2,4-Dichlorophenoxy)-5-hydroxyphenylacetic acid
Formed from: Fenclofenac

5-(2,4-Dichlorophenoxy)-3-methoxy-*o*-quinone
Formed from: 2,4-Dichlorophenol
4-Hydroxy-3,5-dimethoxybenzoic acid

5-(2,6-Dichlorophenoxy)-3-methoxy-*o*-quinone
Formed from: 2,6-Dichlorophenol
4-Hydroxy-3,5-dimethoxybenzoic acid

2-(4-(2′,4′-Dichlorophenoxy)phenoxy)propionic acid (diclofop)
Formed from: Methyl 2-(4-(2′,4′-Dichlorophenoxy)phenoxy)propionate
→ 2-(4-(2′,4′-Dichloro-3′-hydroxyphenoxy)phenoxy)propionic acid
Probably in: Wheat A1947
→ 2-(4-(2′,4′-Dichloro-5′-hydroxyphenoxy)phenoxy)propionic acid
Probably in: Wheat A1947, A3803
→ 2-(4-(2′,4′-Dichloro-6′-hydroxyphenoxy)phenoxy)propionic acid
Probably in: Wheat A1947

2-(2,4-Dichlorophenoxy)phenylacetic acid
See: Fenclofenac

2-(2,4-Dichlorophenoxy)phenylacetyl-β-D-glucuronic acid
Formed from: Fenclofenac

2-(2,4-Dichlorophenoxy)phenylacetyltaurine
Formed from: Fenclofenac

2-(2,4-Dichlorophenoxy)propionic acid (dichlorprop)
→ 2,4-Dichlorophenol
In: Alcaligenes J911
Flavobacterium F628
Ralstonia J911

3-(2,4-Dichlorophenoxy)propionic acid
→ 2,4-Dichlorophenol
In: Grass A1268
Legume A1268

Dichlorophen-O-sulphate
Formed from: Dichlorophen

2,6-Dichlorophenylacetamide
Formed from: Guanfacine

2,6-Dichlorophenylacetic acid
Formed from: Guanfacine
→ 2,6-Dichlorophenylacetylglycine
Probably in: Rat B889

2,6-Dichlorophenylacetylglycine
Probably formed from: 2,6-Dichlorophenylacetic acid

2,3-Dichloro-L-phenylalanine
→ 2,3-Dichlorophenylpyruvic acid
In: Bean E444

2,4-Dichloro-L-phenylalanine
→ 2,4-Dichlorophenylpyruvic acid

In: Bean E444
Bushbean E640
2,5-Dichloro-L-phenylalanine
→ 2,5-Dichlorophenylpyruvic acid
In: Bean E444
2,6-Dichloro-L-phenylalanine
→ 2,6-Dichlorophenylpyruvic acid
In: Bean E444
Bushbean E640
3,4-Dichloro-L-phenylalanine
→ 3,4-Dichlorophenylpyruvic acid
In: Bean E444
Bushbean E640
3,5-Dichloro-L-phenylalanine
→ 3,5-Dichlorophenylpyruvic acid
In: Bean E444
S-(N-(2,4-Dichlorophenyl)-2-aminoethyl) mercapturic acid
Probably formed from: N-(2-Chloroethyl)-2,4-dinitroaniline
(2-((2,6-Dichlorophenyl)-2-amino)-5-hydroxyphenyl)acetoxyacetic acid
Formed from: Aceclofenac
4,6-Dichloro-3-(2-(phenylcarbamoyl)ethenyl) indole-2-carboxylic acid
→ 4,6-Dichloro-3-(2-(*p*-hydroxyphenylcarbamoyl)ethenyl)indole-2-carboxylic acid
In: Man K125
***o*-((2,6-Dichlorophenyl)amino)phenylacetic acid**
See: Diclofenac
(2-((2,6-Dichlorophenyl)amino)phenyl) acetoxyacetic acid
See: Aceclofenac
***o*-((2,6-Dichlorophenyl)amino) phenylacetyltaurine**
Formed from: Diclofenac
S-(2,4-Dichlorophenyl)-L-cysteine
Probably formed from: S-(2,4-Dichlorophenyl) glutamyl-L-cysteine
→ N-Acetyl-S-(2,4-dichlorophenyl)-L-cysteine
Probably in: Rat G294
S-(3,5-Dichlorophenyl)-L-cysteine
Probably formed from: S-(3,5-Dichlorophenyl)-L-cysteinylglycine
→ N-Acetyl-S-(3,5-dichlorophenyl)-L-cysteine
Probably in: Rat G294
S-(3,5-Dichlorophenyl)-L-cysteinylglycine
Formed from: *m*-Dichlorobenzene
→ S-(3,5-Dichlorophenyl)-L-cysteine
Probably in: Rat G294
O-(2,4-Dichlorophenyl) O,O-diethyl phosphorothioate
See: Nemacide
N-(3,4-Dichlorophenyl)-N′-(2,3-dihydrobenzofuran-5-sulphonyl)urea
→ 3,4-Dichloroaniline
In: Monkey J156
Mouse J156
Rat J156
→ N-(3,4-Dichlorophenyl)-N′-(2,3-dihydro-3-hydroxybenzofuran-5-sulphonyl)urea
In: Monkey J156
Mouse J156
Rat J156
→ N-(3,4-Dichlorophenyl)-N′-((4-hydroxy-3-(2-hydroxyethyl)phenyl)-5-sulphonyl)urea
In: Monkey J156
Mouse J156
Rat J156
4-(2,3-Dichlorophenyl)-1,4-dihydro-2,6-dimethylpyridine-3,5-dicarboxylate 3-ethyl ester 5-methyl ester
See: Felodipine
N-(3,4-Dichlorophenyl)-N′-(2,3-dihydro-3-hydroxybenzofuran-5-sulphonyl)urea
Formed from: N-(3,4-Dichlorophenyl)-N′-(2,3-dihydrobenzofuran-5-sulphonyl)urea
N-(2,6-Dichlorophenyl)-4,5-dihydro-1*H*-imidazol-2-amine
See: Clonidine
N-(2,6-Dichlorophenyl)-4,5-dihydro-N-2-propenyl-1*H*-imidazol-2-amine
See: Alinidine
N-(3,4-Dichlorophenyl)-N-((dimethylamino)-carbonyl)benzamide
See: Phenobenzuron
N-(3,5-Dichlorophenyl)-1,2-dimethylcyclopropane-1,2-dicarboximide
See: Procymidone
N′-(3,4-Dichlorophenyl)-N,N-dimethylurea
See: Diuron

1-(2,4-Dichlorophenyl)ethanol
Formed from: Dimethylvinphos
Probably formed from: 2′,4′-Dichloroacetophenone
→ 1-(2,4-Dichlorophenyl)ethyl-β-D-glucuronide
Probably in: Dog A1752
Rat A1752

3,4-Dichlorophenylethanolamine
→ 3,4-Dichloro-N-methylphenylethanolamine
In: Man A498
Rabbit A498, C190
Rat C190

4-(2,3-Dichlorophenyl)-3-ethoxycarbonyl-2,6-dimethylpyridine
Probably formed from: 3-Carboxy-4-(2,3-dichlorophenyl)-5-ethoxycarbonyl-2,6-dimethylpyridine

3,4-Dichlorophenylethylenediamine
→ N-(3,4-Dichlorophenyl)-N′-methylethylenediamine
In: Man A498
Rabbit A498

1-(2,4-Dichlorophenyl)ethyl-β-D-glucuronide
Formed from: 1-(2,4-Dichlorophenyl)ethanol

2,4-Dichlorophenyl-β-D-glucuronide
Formed from: 2,4-Dichlorophenol

2,5-Dichlorophenyl-β-D-glucuronide
Formed from: 2,5-Dichlorophenol

S-(2,4-Dichlorophenyl)glutamyl-L-cysteine
Probably formed from: S-(2,4-Dichlorophenyl)glutathione
→ S-(2,4-Dichlorophenyl)-L-cysteine
Probably in: Rat G294

S-(2,4-Dichlorophenyl)glutathione
Formed from: *m*-Dichlorobenzene
α-Lindane
β-Lindane
δ-Lindane
→ S-(2,4-Dichlorophenyl)glutamyl-L-cysteine
Probably in: Rat G294

S-(2,5-Dichlorophenyl)glutathione
Formed from: α-Lindane

S-(2,6-Dichlorophenyl)glutathione
Formed from: α-Lindane
β-Lindane
δ-Lindane

S-(3,4-Dichlorophenyl)glutathione
Formed from: α-Lindane
δ-Lindane

2,4-Dichlorophenylglycol
Formed from: Dimethylvinphos
→ 2,4-Dichlorophenylglycol-β-D-glucuronide
Probably in: Dog A1752
Rat A1752

2,4-Dichlorophenylglycol-β-D-glucuronide
Probably formed from: 2,4-Dichlorophenylglycol

O-(2,5-Dichlorophenyl) O-hexyl phosphoramidate
→ 2,5-Dichlorophenol
In: Rabbit J553

N-(2-(2,4-Dichlorophenyl)-2-hydroxyethyl) formamide
Probably formed from: N-(2-(2,4-Dichlorophenyl)-2-hydroxyethyl)urea

1-(2-(2,4-Dichlorophenyl)-2-hydroxyethyl)-4,5-imidazolidinedione
Probably formed from: 1-(2,4-Dichlorophenyl)-2-(1-imidazolyl)ethanol
→ N-(2-(2,4-Dichlorophenyl)-2-hydroxyethyl) oxamic acid
Probably in: Rat B152
→ N-(2-(2,4-Dichlorophenyl)-2-hydroxyethyl) urea
Probably in: Rat B152

N-(2-(3,4-Dichlorophenyl)-2-hydroxyethyl)oxamic acid
Probably formed from: 1-(2-(2,4-Dichlorophenyl)-2-hydroxyethyl)-4,5-imidazolidinedione

N-(3,4-Dichlorophenyl)-N′-((4-hydroxy-3-(2-hydroxyethyl)phenyl)-5-sulphonyl)urea
Formed from: N-(3,4-Dichlorophenyl)-N′-(2,3-dihydrobenzofuran-5-sulphonyl)urea

N-(2-(2,4-Dichlorophenyl)-2-hydroxyethyl)urea
Probably formed from: 1-(2-(2,4-Dichlorophenyl)-2-hydroxyethyl)-4,5-imidazolidinedione
→ N-(2-(2,4-Dichlorophenyl)-2-hydroxyethyl)formamide
Probably in: Rat B152

3,4-Dichlorophenylhydroxylamine
Formed from: 3,4-Dichloroaniline

2,4-Dichlorophenyl 6-hydroxy-3-methoxy-4-nitrophenyl ether
Formed from: Chlomethoxynil
→ 2,4-Dichlorophenyl 6-hydroxy-3-methoxy-4-nitrophenyl ether sulphate
In: Rat C358

2,4-Dichlorophenyl 6-hydroxy-3-methoxy-4-nitrophenyl ether sulphate
Formed from: 2,4-Dichlorophenyl 6-hydroxy-3-methoxy-4-nitrophenyl ether

2,4-Dichlorophenyl 3-hydroxy-4-nitrophenyl ether
Formed from: Chlomethoxynil

N-(3,5-Dichlorophenyl)-2-hydroxysuccinamic acid
Probably formed from: N-(3,5-Dichlorophenyl)-2-hydroxysuccinimide
N-(3,5-Dichlorophenyl)succinamic acid

N-(3,5-Dichlorophenyl)-3-hydroxysuccinamic acid
Probably formed from: N-(3,5-Dichlorophenyl)-2-hydroxysuccinimide
N-(3,5-Dichlorophenyl)succinamic acid

N-(3,5-Dichlorophenyl)-2-hydroxysuccinimide
Formed from: N-(3,5-Dichlorophenyl) succinimide
→ N-(3,5-Dichlorophenyl)-2-hydroxysuccinamic acid
Probably in: Rat H138
→ N-(3,5-Dichlorophenyl)-3-hydroxysuccinamic acid
Probably in: Rat H138

4-(3,4-Dichlorophenyl)-2-hydroxy-1-tetralone
Formed from: Sertraline

1-(2,4-Dichlorophenyl-2-(1-imidazolyl)ethanol
Formed from: Econazole
→ 1-(2-(2,4-Dichlorophenyl)-2-hydroxyethyl)-4,5-imidazolidinedione
In: Rat B152

2-(2,6-Dichlorophenylimino)-1,3-dimethylimidazolidine
See: N,N′-Dimethylclonidine

N-(3,5-Dichlorophenyl)malonamic acid
Formed from: N-(3,5-Dichlorophenyl) succinimide

N-(3,4-Dichlorophenyl)-N′-methylethylenediamine
Probably formed from: 3,4-Dichlorophenylethylenediamine

4-(2,3-Dichlorophenyl)-3-methoxycarbonyl-2,6-dimethylpyridine
Probably formed from: 3-Carboxy-4-(2,3-dichlorophenyl)-3-methoxycarbonyl-2,6-dimethylpyridine

N′-(3,4-Dichlorophenyl)-N-methoxy-N-methylurea
See: Linuron

2,4-Dichlorophenyl 3-methoxy-4-nitrophenyl ether
See: Chlomethoxynil

N′-(3,4-Dichlorophenyl)-N-methoxyurea
Formed from: Linuron
→ N-(3,4-Dichlorophenyl)urea
Probably in: Rabbit G709

2-((2,6-Dichlorophenyl)methylene)-hydrazinecarboximidamide
See: Guanoxabenz

N-(3,4-Dichlorophenyl)-N′-methylethylenediamine
Probably formed from: 3,4-Dichlorophenyl-ethylenediamine

2-(3,4-Dichlorophenyl)-4-methyl-1,2,4-oxadiazolidine-3,5-dione
See: Methazole

2,4-Dichlorophenyl methyl sulphone
Formed from: *m*-Dichlorobenzene

3,5-Dichlorophenyl methyl sulphone
Formed from: *m*-Dichlorobenzene

3-(3,4-Dichlorophenyl)-1-methoxyurea
Formed from: Methazole
→ N-(3,4-Dichlorophenyl)urea
Probably in: Cow A1837
Hen A1837
Rat A1853

2,4-Dichlorophenyl 4-nitrophenyl ether
See: Nitrofen

2,4-Dichloro-6-phenylphenoxyethylamine
See: 2-(2-Aminoethoxy)-3,5-dichlorobiphenyl

(*S*)-4-(1-(2-(3-(3,4-Dichlorophenyl)-6-oxo-3-piperidyl)ethyl)azetidin-3-yl)-1-piperazine sulphamide
Formed from: (*S*)-(+)-4-(1-(2-(1-cyclopropylmethyl)-3-(3,4-dichlorophenyl)-6-oxo-3-piperidyl)ethyl)azetidin-3-yl)-1-piperazine sulphamide

α-(2,4-Dichlorophenyl)-N-phenylnitrone
Formed from: N-(2,4-Dichlorobenzyl)aniline

α-(2,6-Dichlorophenyl)-N-phenylnitrone
Formed from: N-(2,6-Dichlorobenzyl)aniline
2,3-Dichlorophenylpyruvic acid
Formed from: 2,3-Dichloro-L-phenylalanine
2,4-Dichlorophenylpyruvic acid
Formed from: 2,4-Dichloro-L-phenylalanine
2,5-Dichlorophenylpyruvic acid
Formed from: 2,5-Dichloro-L-phenylalanine
2,6-Dichlorophenylpyruvic acid
Formed from: 2,6-Dichloro-L-phenylalanine
3,4-Dichlorophenylpyruvic acid
Formed from: 3,4-Dichloro-L-phenylalanine
3,5-Dichlorophenylpyruvic acid
Formed from: 3,5-Dichloro-L-phenylalanine
N-(3,5-Dichlorophenyl)succinamic acid
Formed from: N-(3,5-Dichlorophenylsuccinimide
→ N-(3,5-Dichloro-4-hydroxyphenyl) succinamic acid
Probably in: Rat H138, J441, J587
→ N-(3,5-Dichlorophenyl)-2-hydroxysuccinamic acid
Probably in: Rat J441, J587
→ N-(3,5-Dichlorophenyl)-3-hydroxysuccinamic acid
Probably in: Rat J441, J587
N-(3,4-Dichlorophenyl)succinimide
Formed from: 3,4-Dichloroaniline
N-(3,5-Dichlorophenyl)succinimide
→ 3,5-Dichloroaniline
In: Rabbit H543
→ N-(3,5-Dichlorophenyl)-2-hydroxysuccinimide
In: Rat H138, H543, J441
→ N-(3,5-Dichlorophenyl)-4-hydroxysuccinamic acid
In: Rat J441, J581
See also: H138
→ N-(3,5-Dichlorophenyl) malonamic acid
In: Rat J441
2,4-Dichlorophenyl sulphate
Formed from: 2,2-Dichlorophenol
2,5-Dichlorophenyl sulphate
Formed from: 2,5-Dichlorophenol
3,4-Dichlorophenyl sulphate
Formed from: 3,4-Dichlorophenol
3,5-Dichlorophenyl sulphate
Formed from: 3,5-Dichlorophenol
6-(2,3-Dichlorophenyl)-1,2,4-triazine-3,5-diamine
See: Lamotrigine
6-(2,5-Dichlorophenyl)-1,2,4-triazine-2,4-diamine
See: Irsogladine
N-(3,4-Dichlorophenyl)urea
Probably formed from: N′-(3,4-Dichlorophenyl)-N-methoxyurea
3-(3,4-Dichlorophenyl)-1-methylurea
→ N-(3,4-Dichloro-2-hydroxyphenyl)urea
Probably in: Cow A1837
Hen A1837
Rat A1853
→ N-(4,5-Dichloro-2-hydroxyphenyl)urea
Probably in: Cow A1837
Hen A1837
Rabbit G709
Rat A1853
2′,3′-Dichloropropionanilide
→ 2,3-Dichloroaniline
In: Dandelion A230
Rice A3773
2′,4′-Dichloropropionanilide
→ 2,4-Dichloroaniline
In: Dandelion A230
Rice A3773
2′,5′-Dichloropropionanilide
→ 2,5-Dichloroaniline
In: Rice A3773
2′,6′-Dichloropropionanilide
Not hydrolysed in dandelion A230
3′,4′-Dichloropropionanilide (propanil)
→ 3,4-Dichloroaniline
In: Aspergillus E322
Bacillus A1006
Dandelion A230
Lactuca A2191
3′,5′-Dichloropropionanilide
→ 3,5-Dichloroaniline
In: Rice A3773
2,3-Dichloroquinol
Formed from: 2,3,4-Trichlorophenol
Trichloroquinol
Probably formed from: 2,3-Dichlorophenol
2,5-Dichloroquinol
Formed from: 2,5-Dichloro-4-methoxyphenol

2,4,5-Trichlorophenol
Trichloroquinol
→ 2-Chloro-5-hydroxyquinol
In: Pseudomonas J177

2,6-Dichloroquinol
Formed from: 4-Amino-2,6-dichlorophenol
3,5-Dichloro-2-(S-glutathionyl)quinol
2,6-Dichlorophenol
2,4,6-Trichlorophenol
3,4,5-Trichlorophenol
Trichloroquinol
→ 2-Chloro-6-(S-glutathionyl)quinol
In: Phanerochaete K138
→ 3,5-Dichloro-4-methoxyphenol
In: Rhodococcus E761

5,7-Dichloroquinolin-8-yl-β-D-glucuronide
Formed from: 5,7-Dichloro-8-hydroxyquinoline
→ 5,7-Dichloro-8-hydroxyquinoline
With β-glucuronidase A878
See also: A3654

5,7-Dichloroquinolin-8-yl sulphate
Formed from: 5,7-Dichloro-8-hydroxyquinoline
→ 5,7-Dichloro-8-hydroxyquinoline
Probably in: Rat A3654

5,6-Dichloro-1-β-D-ribofuranosylbenzimidazole
→ 5,6-Dichloro-1-β-D-ribofuranosylbenzimidazole-5′-phosphate
In: Chironomus B279

5,6-Dichloro-1-β-D-ribofuranosylbenzimidazole-5′-phosphate
Formed from: 5,6-Dichloro-1-β-D-ribofuranosylbenzimidazole

***p,p′*-Dichloro-*cis*-stilbene oxide**
→ (*R,R*)-*p,p′*-Dichloro-α,β-dihydroxydibenzyl
In: Rabbit H568

2,6-Dichlorostyrene
Formed from: 2,6-Dichlorocinnamic acid

8,9-Dichloro-2,3,4,5-tetrahydro-1*H*-2-benzazepine
→ 8,9-Dichloro-2,3,4,5-tetrahydro-N-methyl-1*H*-2-benzazepine
In: Rabbit B741

7,8-Dichloro-1,2,3,4-tetrahydro-1-hydroxyisoquinoline
Formed from: 7,8-Dichloro-1,2,3,4-tetrahydroisoquinoline

7,8-Dichloro-1,2,3,4-tetrahydro-2-hydroxyisoquinoline
Formed from: 7,8-Dichloro-1,2,3,4-tetrahydroisoquinoline
→ 7,8-Dichloro-3,4-dihydroisoquinoline
Probably in: Rat C649
→ 7,8-Dichloro-3,4-dihydroisoquinoline-N-oxide
Probably in: Rat C649

7,8-Dichloro-1,2,3,4-tetrahydro-3-hydroxyisoquinoline
Formed from: 7,8-Dichloro-1,2,3,4-tetrahydroisoquinoline

7,8-Dichloro-1,2,3,4-tetrahydro-4-hydroxyisoquinoline
Formed from: 7,8-Dichloro-1,2,3,4-tetrahydroisoquinoline

7,8-Dichloro-1,2,3,4-tetrahydroisoquinoline
→ 7,8-Dichloro-1,2,3,4-tetrahydro-1-hydroxyisoquinoline
In: Dog B698
Rat B698
→ 7,8-Dichloro-1,2,3,4-tetrahydro-2-hydroxyisoquinoline
In: Rat C649
→ 7,8-Dichloro-1,2,3,4-tetrahydro-3-hydroxyisoquinoline
In: Dog B698
Rat B698
→ 7,8-Dichloro-1,2,3,4-tetrahydro-4-hydroxyisoquinoline
In: Dog B698
Rat B698
→ 7,8-Dichloro-1,2,3,4-tetrahydro-N-methylisoquinoline
In: Dog C657
Rabbit B741, E6
→ 7,8-Dichloro-1,2,3,4-tetrahydro-N-methylisoquinoline-N-oxide
In: Dog B698

8,9-Dichloro-2,3,4,5-tetrahydro-N-methyl-1*H*-2-benzazepine
Formed from: 8,9-Dichloro-2,3,4,5-tetrahydro-1*H*-2-benzazepine

7,8-Dichloro-1,2,3,4-tetrahydro-N-methylisoquinoline
Formed from: 7,8-Dichloro-1,2,3,4-tetrahydroisoquinoline

7,8-Dichloro-1,2,3,4-tetrahydro-N-methylisoquinoline-N-oxide
Formed from: 7,8-Dichloro-1,2,3,4-tetrahydroisoquinoline

2,3-Dichloro-4-(2-thienylcarbonyl) phenoxy)acetic acid
See: Tienilic acid

2,6-Dichlorothiobenzamide (prefix)
→ 2,6-Dichlorobenzonitrile
Probably in: Rat F296

2,6-Dichlorotribenzylamine
→ 2,6-Dichlorodibenzylamine
Probably in: Rat J471

5,6-Dichloro-2-trifluoromethylbenzimidazole
→ 5,6-Dichloro-4-hydroxy-2-trifluoromethylbenzimidazole
In: Rat A2263
→ 5,6-Dichloro-2-trifluoromethylbenzimidazole-β-D-glucuronide
In: Rat A2263

5,6-Dichloro-2-trifluoromethylbenzimidazole-β-D-glucuronide
Formed from: 5,6-Dichloro-2-trifluoromethylbenzimidazole

4,4′-Dichloro-3′-trifluoromethylcarbanilide
→ 4,4′-Dichloro-2-hydroxy-3′-trifluoromethylcarbanilide
In: Rat A2076
→ 4,4′-Dichloro-3-hydroxy-3′-trifluoromethylcarbanilide
In: Rat A2076

4,4′-Dichloro-3′-trifluoromethylcarbanilid-2-yl-β-D-glucuronide
Formed from: 4,4′-Dichloro-2-hydroxy-3′-trifluoromethylcarbanilide

4,4′-Dichloro-3′-trifluoromethylcarbanilid-3-yl-β-D-glucuronide
Formed from: 4,4′-Dichloro-3-hydroxy-3′-trifluoromethylcarbanilide

4,4′-Dichloro-3′-trifluoromethylcarbanilid-2-yl sulphate
Formed from: 4,4′-Dichloro-2-hydroxy-3′-trifluoromethylcarbanilide

4,4′-Dichloro-3′-trifluoromethylcarbanilid-3-yl sulphate
Formed from: 4,4′-Dichloro-3-hydroxy-3′-trifluoromethylcarbanilide

3′,4′-Dichlorovaleranilide
→ 3,4-Dichloroaniline
In: Dandelion A230

4,5-Dichloroveratrole
Probably formed from: 4,5-Dichloro-2-methoxyphenol

Dichlorprop
See: 2-(2,4-dichlorophenoxy)propionic acid

Diclofenac (*o*-((2,6-dichlorophenyl)amino) phenylacetic acid)
Formed from: Aceclofenac
→ *o*-((2,6-Dichlorophenyl)amino) phenylacetyltaurine
In: Baboon A3840
Dog A3840
Man A3840
Rat A3840
→ Diclofenac-β-D-glucuronide
In: Baboon A3840
Dog A3840
Man A3840
Rabbit J107
Rat A3840, J107
→ 3′-(S-Glutathionyl)-4′-hydroxydiclofenac
Man J569
Rat J569
→ 4-(S-Glutathionyl)-5-hydroxydiclofenac
Man J569
Rat J569
→ 6-(S-Glutathionyl)-5-hydroxydiclofenac
Man J569
Rat J569
→ 3′-Hydroxydiclofenac
In: Baboon A3840, A3844
Dog A3840

Man A3840, A3844, F316, H404, K212
Monkey A3844
Rat A3840
→ 4′-Hydroxydiclofenac
In: Baboon A3840, A3844, F320
Dog A3840, A3844
Man A3840, A3844, F316, F320, H404, J47, K22, K212
Monkey A3844, J47
Rat A3840, A3844, J47
See also: F320
→ 5-Hydroxydiclofenac
In: Baboon A3840, A3844
Dog A3840, A3844
Man A3840, A3844, F316, H404, H739, K47, K252
Monkey A3844
Rat A3840, A3844

Diclofenac-β-D-glucuronide
Formed from: Diclofenac

Diclofop
See: 2-(4-(2′,4′-Dichlorophenoxy)phenoxy) propionate

Diclofop-methyl
See: Methyl 2-(4-(2′,4′-dichlorophenoxy) phenoxy)propionate

Diclofurime (2,3-dichloro-4-methoxyphenyl)-2-furylmethanone O-(2-(diethylamino)ethyl) oxime)
→ 3-Carboxy-1-propionyl 1-(2,3-dichloro-4-methoxy)phenyl O-(2-(diethylamino)ethyl) ketoxime
In: Dog D428
Pig D428

Dicofol
See: 2,2,2-Trichloro-1,1-bis(*p*-chlorophenyl) ethanol

Dicoumarol (3,3′-methylenebis(4-hydroxy-2*H*-1-benzopyran-2-one)
→ Dicoumarol-β-D-glucuronide
In: Rat A2257
→ 7-Hydroxydicoumarol
In: Rat A2257, A2668

Dicoumarol-β-D-glucuronide
Formed from: Dicoumarol

1,2-Di-O-(*p*-coumaroyl)glucose
Formed from: *p*-Coumaroyl-β-D-glucose

Di-*o*-cresyl *o*-hydroxymethylphenyl phosphate
Formed from: Tri-*o*-cresyl phosphate
→ *o*-Cresyl di-(*o*-hydroxymethylphenyl) phosphate
In: Cat D375
→ *o*-Cresyl di-(*o*-hydroxymethylphenyl) phosphate
Probably in: Cat D375
→ Salicyl alcohol
Probably in: Cat D375

Di-*o*-cresyl phosphate
Formed from: Tri-*o*-cresyl phosphate
→ *o*-Cresol
In: Cat D375
Rat G129
→ Mono-*o*-cresyl phosphate
In: Cat D375
Rat G129

Dicryl
See: 3,4-Dichloro-2′-methacrylanilide

Dictamnine (4-methylfuro[2,3-*b*]quinoline)
Formed from: 3-Dimethylallyl-2,4-dihydroxyquinoline
Platydesmine
→ Choisyine
In: Choisya A1170, A1484
→ Evoxine
In: Choisya A1170, A1484
→ 8-Methoxydictamnine
In: Ruta A846
→ Skimmianine
In: Choisya A1170, A1484
Skimmia A1170, A1484

Dicumenyl peroxide
Formed from: Cumene hydroperoxide

Dicyclohexyl phthalate
→ Phthalic acid monocyclohexyl ester
In: Baboon A1652
Ferret A1652
Rat A1652

5,8-Dideazaisofolic acid (N-(*p*-(2-amino-4-hydroxy-6-quinazolinyl)amino)methyl)benzoyl)-L-glutamic acid)
→ 5,8-Dideazaisofolylglutamic acid
In: Man F126

5,8-Dideazaisofolylglutamic acid (N-(*p*-(2-amino-4-hydroxy-6-quinazolinyl)amino)methyl) benzoyl)-L-glutamic acid)
Formed from: 5,8-Dideazaisofolic acid
→ 5,8-Dideazaisofolylglutamylglutamic acid
Probably in: Man F126

5,8-Dideazaisofolylglutamylglutamic acid
Probably formed from: 5,8-Dideazaisofolylglutamic acid
→ 5,8-Dideazaisofolylglutamylglutamylglutamic acid
Probably in: Man F126

5,8-Dideazaisofolylglutamylglutamylglutamic acid
Probably formed from: 5,8-Dideazaisofolylglutamylglutamic acid

Di-(9-decenyl) phthalate
→ Phthalic acid mono (9-decenyl) ester
In Rat C869

8,9-Didehydro-6,8-dimethylergoline
See: Agroclavine

(8α)-8,9-Didehydro-6,8-dimethylergolin-8-ol
See: Setoclavine

(5α,6α)-7,8-Didehydro-4,5-epoxymorphinan-3,6-diol
See: Normorphine

Didehydrofelodipine
Formed from: Felodipine
→ 3-Carboxy-4-(2,3-dichlorophenyl)-5-ethoxycarbonyl-2,6-dimethylpyridine
Probably in: Dog E862
Man E862
Rat E846, E862
→ 3-Carboxy-4-(2,3-dichlorophenyl)-5-methoxycarbonyl-2,6-dimethylpyridine
Probably in: Dog E862
Man E862
Rat E846, E862

Didehydroglaucine
Formed from: Glaucine
(6a*R*)-Glaucine

8′,9′-Didehydroisobromocriptine
Probably formed from: 8′α-Hydroxyisobromocriptine

8,9-Didehydro-6-methylergoline-8-carboxylic acid
See: Paspalic acid

N′-((8α)-9,10-Didehydro-6-methylergolin-8-yl)-N,N-diethylurea
See: Lisuride

(8β)-9,10-Didehydro-6-methylergoline-6-methanol
See: Lysergol

16,17-Didehydro-19-methyloxayohimban-1-carboxylic acid methyl ester
See: Ajmalicine

Didehydro-4-methyl-5-phenyl-1,3,4-thiadiazolidine-2-one
Formed from: Didehydro-4-methyl-5-phenyl-1,3,4-thiadiazolidine-2-thione

Didehydro-4-methyl-5-phenyl-1,3,4-thiadiazolidine-2-thione
→ Benzoic acid
In: Rat C438
→ Didehydro-4-methyl-5-phenyl-1,3,4-thiadiazolidine-2-one
In: Rat C438
→ N-Methylbenzamide
In: Rat C438

Didehydro-N-methylcoclaurinium
→ Isotetrandrine
In: Cocculus B673

Didehydronorcoclaurine
→ Norcoclaurine
Probably in: Annona B137

Di-O-demethylnarcotine
Formed from: Narcotine

7,7′-Dideoxy-7,7′-biaklavinone
Formed from: Aclacinomycin A

1,4-Dideoxy-1,4-dihydro-1,4-dioxorifamycin
See: Rifamycin S

7-O-(2,6-Dideoxy-2-fluoro-α-L-talopyranosyl) adriamycinol
Probably formed from: 7-O-(2,6-Dideoxy-2-fluoro-α-L-talopyranosyl) adriamycinone
→ (8*S*,10*S*)-8-(1,2-Dihydroxyethyl)-7,8,9,10-tetrahydro-6,8,10,11-tetrahydroxy-1-methoxy-5,12-naphthacenedione
Probably in: Rat J77

7-O-(2,6-Dideoxy-2-fluoro-α-L-talopyranosyl) adriamycinone
Formed from: (8*S*,10*S*)-8-(3-Aminopropanoyloxyacetyl)-10-((2,6-dideoxy-2-fluoro-α-L-talopyranosyl)oxy)-

7,8,9,10-tetrahydro-6,8,11-trihydroxy-1-methoxy-5,12-naphthacenedione
7-O-(2,6-Dideoxy-2-fluoro-α-L-talopyranosyl)adriamycinone-14-hemipimerate
→ 7-O-(2,6-Dideoxy-2-fluoro-α-L-talopyranosyl)adriamycinol
Probably in: Rat J77
→ (8*S*,10*S*)-7,8,9,10-Tetrahydro-6,8,10,11-tetrahydroxy-8-hydroxyacetyl-1-methoxy-5,12-naphthacenedione
Probably in: Rat J77

7-O-(2,6-Dideoxy-2-fluoro-α-L-talopyranosyl) adriamycinone-14-hemipimerate
→ 7-O-(2,6-Dideoxy-2-fluoro-α-L-talopyranosyl)adriamycinone
In: Man F670
Rat F670

Dienoestrol (3,4-di-(*p*-hydroxyphenyl)hexa-2,4-diene)
Formed from: Stilboestrol
→ Dienoestrol-β-D-glucuronide
In: Man G968

Dienoestrol-β-D-glucuronide
Formed from: Dienoestrol

1,2-Diethenylbenzene
See: 1,2-Divinylbenzene

1,3-Diethenylbenzene
See: 1,3-Divinylbenzene

1,4-Diethenylbenzene
See: 1,4-Divinylbenzene

Diethofenocarb
See: Isopropyl 3,4-diethoxycarbanilate

4,4′-Diethoxyazobenzene
Formed from: *p*-Ethoxyaniline

3,4-Diethoxybenzaldehyde
Formed from: 1-(3,4-Dimethoxyphenyl)-1,3-dihydroxy-2-(*p*-methoxyphenyl)propane
1-(3,4-Dimethoxyphenyl)-1-hydroxy-2-(*p*-methoxyphenyl)ethane

3,4-Diethoxybenzoic acid
→ 3-Ethoxy-4-hydroxybenzoic acid
In: Polyporus A1117

1,6-Diethoxybenzpyrene
→ 6-Ethoxy-1-hydroxybenzpyrene
In: Rat B477

1-(3,4-Diethoxybenzal)-6,7-diethoxy-1,2,3,4-tetrahydroisoquinoline
See: Drotaverine

3-(2′,4′-Diethoxy-5′-hydroxybenzoyl)propionic acid
Formed from: 3-(2′,4′,5′-Triiethoxybenzoyl) propionic acid
→ 3-(2′-Ethoxy-4′,5′-dihydroxybenzoyl) propionic acid
Probably in: Rat A3251

3-(2′,5′-Diethoxy-4′-hydroxybenzoyl)propionic acid
Formed from: 3-(2′,4′,5′-Triiethoxybenzoyl) propionic acid

3-(2′,4′-Diethoxy-5′-(2-hydroxyethoxy)benzoyl) propionic acid
Formed from: 3-(2′,4′,5′-Triethoxybenzoyl) propionic acid

1-(3,4-Diethoxyphenyl)-1,3-dihydroxy-2-(*p*-methoxyphenyl)propane
→ 3,4-Diethoxybenzaldehyde
In: Phanerochaete D181, D467, D655, D758
→ *p*-Methoxyphenylglycol
In: Phanerochaete D181, D467, D655, D758

1-(3,4-Diethoxyphenyl)-1-hydroxy-2-(*p*-methoxyphenyl)ethane
→ Anisyl alcohol
In: Phanerochaete D467
→ 3,4-Diethoxybenzaldehyde
In: Phanerochaete D467

4-((N,N-Diethyl)acetamid-2-oxy)-3-methoxyphenylacetic acid
Formed from: Propanidid

N,N-Diethylalanyl-*o*-toluidide
→ *o*-Toluidine
In: Mouse A830

4-(((Diethylamino)acetyl)amino)-N-(2,6-dimethylphenyl)benzamide
→ 4-Amino-N-(2,6-dimethylphenyl) benzamide
In: Mouse E951
→ 4-(((Ethylamino)acetyl)amino)-N-(2,6-dimethylphenyl) benzamide
In: Mouse E951

4-Diethylaminobutynyl α-cyclohexylmandelate
See: Oxybutynin

4-Diethylamino-1,1-dimethylbut-2-yn-1-yl 2-cyclohexyl-2-hydroxy-2-phenylacetate
→ 4-Diethylamino-1,1-dimethylbut-2-yn-1-yl 2-hydroxy-2-(4-hydroxycyclohexyl)-2-phenylacetate
In: Man J554
Rat J554
→ 4-Ethylamino-1,1-dimethylbut-2-yn-1-yl 2-cyclohexyl-2-hydroxy-2-phenylacetate
In: Man J554
Rat J554

Diethylamino-1,1-dimethylbut-2-yn-1-yl 2-hydroxy-2-(4-hydroxycyclohexyl)-2-phenylacetate
Formed from: 4-Diethylamino-1,1-dimethylbut-2-yn-1-yl 2-cyclohexyl-2-hydroxy-2-phenylacetate

2-Diethylaminoethyl *p*-aminobenzoate (novocaine, procaine)
→ *p*-Aminobenzoic acid
In: Ascaris A324
Baboon A475
Chimpanzee A475
Man A475
Moniezia A324
Monkey A475
Rabbit G517, H563
Rat C677

2-Diethylaminoethyl 4-amino-3-butoxybenzoate
See: Benoxinate

5-((2-(Diethylamino)ethyl)amino)-8-hydroxy-6*H*-imidazo[4,5,1-*de*] acridin-6-one
→ 5-((2-(Diethylamino)ethyl)amino)-8-hydroxy-6*H*-imidazo[4,5,1-*de*] acridin-6-one-β-D-glucuronide
In: Mouse J604

5-((2-(Diethylamino)ethyl)amino)-8-hydroxy-6*H*-imidazo[4,5,1-*de*]acridin-6-one-β-D-glucuronide
Formed from: 5-((2-(Diethylamino)ethyl) amino)-8-hydroxy-6*H*-imidazo[4,5,1-*de*] acridin-6-one

O-(2-Diethylaminoethyl) N-benzoyl tyrosyl dipropylamide
See: Tiropramide

N-(2-Diethylaminoethyl)-2-(2,3-dihydroxybutoxy)-4-quinolinecarboxamide
Formed from: Dibucaine

2-Diethylaminoethyl diphenylvalerate (SKF 525A)
→ 2-Ethylaminoethyl diphenylacetate
In: Pig B817
Rat B817

N-(2-Diethylaminoethyl)-2-*p*-hydroxyphenoxy) acetamide
Formed from: Mefexamide

N-(2-Diethylaminoethyl)-2-(2-hydroxyethoxy)-4-quinolinecarboxamide
Formed from: Dibucaine

2-(Diethylamino)ethyl α-(1-naphthylmethyl)-β-(tetrahydrofuran-2-yl)propionate
→ α-(1-Naphthylmethyl)-β-(tetrahydro-5-oxofuran-2-yl)propionate
In: Rat A2233

N-(2-(Diethylamino)ethyl)-2-methoxy-5-(methylsulphonyl)benzamide
See: Thiapride

N-(2-Diethylaminoethyl)-2-(*p*-methoxyphenoxy) acetamide
See: Mefexamide

2-Diethylaminoethyl β-(1-naphthyl-β′-(2-tetrahydrofuryl)isobutyrate (naftidrofuryl)
→ β-(1-Naphthyl-β′-(2-tetrahydrofuryl) isobutyrate
In: Rat 1260

N-(2-Diethylaminoethyl)-*p*-nitrobenzene
Formed from: *p*-Amino-N-(2-diethylaminoethyl)benzamide

8-(6-Diethylaminohexylamino)-4-hydroxymethyl-6-methoxyquinoline
Formed from: 8-(6-Diethylaminohexylamino)-6-methoxy-4-methylquinoline

8-(6-Diethylaminohexylamino)-6-hydroxy-4-methylquinoline
Formed from: 8-(6-Diethylaminohexylamino)-6-methoxy-4-methylquinoline
→ 8-(6-Diethylaminohexylamino)-4-methylquinolin-6-yl-β-D-glucuronide
In: Rat F477

→ 8-(6-Diethylaminohexylamino)-4-methylquinolin-6-yl sulphate
In: Rat F477
→ 8-(6-Ethylaminohexylamino)-6-hydroxy-4-methylquinoline
Probably in: Rat F477

8-(6-Diethylaminohexylamino)-6-methoxy-4-methylquinoline
→ 8-Amino-6-methoxy-4-methylquinoline
In: Rat F477
→ 8-(5-Carboxypentylamino)-6-methoxy-4-methylquinoline
In: Rat F477
→ 8-(6-Diethylaminohexylamino)-4-hydroxymethyl-6-methoxyquinoline
In: Rat F477
→ 8-(6-Diethylaminohexylamino)-6-hydroxy-4-methylquinoline
In: Rat F477
→ 8-(6-Ethylaminohexylamino)-6-methoxy-4-methylquinoline
In: Rat F477

8-(6-Diethylaminohexylamino)-4-methylquinolin-6-yl-β-D-glucuronide
Formed from: 8-(6-Diethylaminohexylamino)-6-hydroxy-4-methylquinoline

8-(6-Diethylaminohexylamino)-4-methylquinolin-6-yl sulphate
Formed from: 8-(6-Diethylaminohexylamino)-6-hydroxy-4-methylquinoline

3-Diethylamino-*p*-hydroxypropiophenone
Formed from: 3-Diethylamino-O-(*p*-methoxyphenylcarbamoyl)propiophenone oxime

3-Diethylamino-O-(*p*-methoxyphenylcarbamoyl) propiophenone oxime
→ Bis(*p*-methoxyphenyl)urea
In: Rat C203
→ 3-Diethylamino-*p*-hydroxypropiophenone
In: Rat C203

N-(4-(Diethylamino)-1-methylbutyl)-2-(2-(5-nitro-2-furanyl)ethenyl-4-quinolinecarboxamide
See: Quinifuryl

7-Diethylamino-4-methylcoumarin
→ 7-Ethylamino-4-methylcoumarin
Probably in: Man J618

10-(3-(Diethylamino)-1-oxopropyl)-2-(trifluoromethyl)-10*H*-phenothiazine
See: Fluoracizine

2-Diethylamino-1-phenyl-1-propanol (diethylnorephedrine)
Formed from: 2-(Diethylamino) propiophenone
→ 2-Ethylamino-1-phenyl-1-propanol
Probably in: Man A103, A648

10-(3-Diethylaminopropionyl)phenothiazine-2-carbamic acid
Formed from: Ethacizin

2-(Diethylamino)propiophenone (diethylpropion)
→ 2-Diethylamino-1-phenyl-1-propanol
In: Man A103, A648
→ 2-(Ethylamino)propiophenone
In: Man A103, A648

N-(3-Diethylaminopropyl)-2-aminoindane
Formed from: Aprindine

3-Diethylaminopropylaniline
Formed from: Aprindine

2,6-Diethylaniline
Formed from: N-Chloroacetyl-2,6-diethylaniline
2′,6′-Diethyl-2-methylthioacetanilide
→ 4-Amino-3,5-diethylphenol
Probably in: Mouse F528
Rat F528
→ 2,6-Diethylnitrosobenzene
In: Rat D943

N,N-Diethylaniline
→ N,N-Diethylaniline-N-oxide
In: Mouse D446
Pig D446
Rabbit A3484
→ N-Ethylaniline
In: Rabbit A3484

N,N-Diethylaniline-N-oxide
Formed from: N,N-Diethylaniline

N,N-Diethylbenzamide
→ Benzoic acid
In: Rat H475
→ N-Ethylbenzamide
In: Rat G531

1,2-Diethylbenzene
→ 1-(*o*-Ethylphenyl)ethanol

In: Rat J793

1,4-Diethylbenzene
→ 1-(*p*-Ethylphenyl)ethanol
In: Mortierella D969
→ 2-(*p*-Ethylphenyl)ethanol
In: Mortierella D969

3,5-Diethylbenzoquinone-4-imine
Probably formed from: 4-Amino-3,5-diethylphenol

Diethyl 4-((4-bromo-2-cyanophenyl)carbamoyl)benzylphosphonate
→ 2-Amino-5-bromo-3-sulphatoxybenzonitrile
In: Rat H731
→ Diethyl *p*-carboxybenzylphosphonate
In: Rat H731
→ Ethyl 4-((4-bromo-2-cyanophenyl)carbamoyl)benzylphosphonate
In: Rat H731

Diethyl *p*-carboxybenzylphosphonate
Formed from: Diethyl 4-((4-bromo-2-cyanophenyl)carbamoyl)benzylphosphonate

Diethyl 1,4-dihydro-2,6-dimethyl-4-(2-trifluoromethylphenyl)-3,5-pyridinedicarboxylate
→ 1,4-Dihydro-2,6-dimethyl-4-(2-trifluoromethylphenyl)-3,5-pyridinedicarboxylic acid monoethyl ester
In: Dog A715, A840
Monkey A840
Rat A840

(*R*,*R*)-*p*,*p*′-Diethyl-α,β-dihydroxydibenzyl
Formed from: *p*,*p*′-Diethyl-*cis*-stilbene oxide

α,α-Diethyl-3,4-dihydroxyphenylacetamide
Formed from: α,α-Diethyl-3,4-methylenedioxyphenylacetamide
→ α,α-Diethyl-3,4-dihydroxyphenylacetamide-β-D-glucuronide
In: Guinea pig A1014
→ α,α-Diethyl-4-hydroxy-3-methoxyphenylacetamide
Probably in: Guinea pig A1014

α,α-Diethyl-3,4-dihydroxyphenylacetamide-β-D-glucuronide
Formed from: α,α-Diethyl-3,4-dihydroxyphenylacetamide

3,3-Diethyl-N-(1(*R*)-(3,4-dihydroxyphenyl)butyl)-2(*S*)-((4-methyl-1-piperazinyl)carbonyl)phenoxy-4-oxo-1-azetidinecarboxamide
→ 3,3-Diethyl-N-(1(*R*)-(3-hydroxy-4-methoxyphenyl)butyl)-2(*S*)-((4-methyl-1-piperazinyl)carbonyl)phenoxy-4-oxo-1-azetidinecarboxamide
In: Monkey J270
→ 3,3-Diethyl-N-(1(*R*)-(4-hydroxy-3-methoxyphenyl)butyl)-2(*S*)-((4-methyl-1-piperazinyl)carbonyl)phenoxy-4-oxo-1-azetidinecarboxamide
In: Monkey J270

N^3,N^3-Diethyl-2,4-dinitro-6-trifluoromethyl-*m*-phenylenediamine
See: Dinitramine

N,N-Diethyl-N′-(1-ethylergolin-8α-yl)sulphamide
Formed from: N,N-Diethyl-N′-(1-ethyl-6-methylergolin-8α-yl)sulphamide

N,N-Diethyl-N′-(1-ethyl-6-methylergolin-8α-yl)sulphamide
→ N,N-Diethyl-N′-(1-ethylergolin-8α-yl)sulphamide
In: Man G92
→ N,N-Diethyl-N′-(6-methylergolin-8α-yl)sulphamide
In: Man G576
→ N-Ethyl-N′-(1-ethyl-6-methylergolin-8α-yl)sulphamide
In: Man G576

N,N-Diethyl-*m*-formylbenzamide
Formed from: N,N-Diethyl-*m*-hydroxymethylbenzamide
→ N,N-Diethyl-*m*-hydroxymethylbenzamide
In: Rat J572
→ N,N-Diethylisophthamidic acid
In: Rat E858, J572
See also: H530
→ N-Ethyl-*m*-formylbenzamide
In: Rat J572

N,N-Diethylglycine-2′-hydroxymethyl-6′-methylanilide
Formed from: Lignocaine

N,N-Diethylglycine-3′-hydroxy-2′,6-xylidide
Formed from: Lignocaine
N,N-Diethylglycine-4′-hydroxy-2′,6-xylidide
Formed from: Lignocaine
N,N-Diethylglycine-2′,6-xylidide
See: Lignocaine
N,N-Diethylglycyl-*o*-toluidide
→ *o*-Toluidine
In: Mouse A830
Di-(2-ethylhexyl)phthalate
→ Phthalic acid mono (2-ethylhexyl) ester
In: Baboon A1652
Catfish A2239
Earthworm H9
Ferret A1652
Man D360
Minnow A2798
Monkey B579
Rat A1652, A1861, G876
Trout A1990, A2133, A3710
Wheat D259, D886
See also: A39, A842, C887, G91
Di-(2-ethylhexyl)terephthalate
→ Terephthalic acid
In: Rat G876
Di-(5-hexenyl)phthalate
→ Phthalic acid mono 5-hexenyl ester
In: Rat C869
Diethyl 4-hydroxylaminophenyl phosphorothioate
Formed from: Diethyl 4-nitrophenyl phosphorothioate
Diethyl 4-nitrosophenyl phosphorothioate
α,α-Diethyl-4-hydroxy-3-methoxyphenylacetamide
Probably formed from: α,α-Diethyl-3,4-dihydroxyphenylacetamide
→ α,α-Diethyl-4-hydroxy-3-methoxyphenylacetamide-β-D-glucuronide
Probably in: Guinea pig A1014
α,α-Diethyl-4-hydroxy-3-methoxyphenylacetamide-β-D-glucuronide
Probably formed from: α,α-Diethyl-4-hydroxy-3-methoxyphenylacetamide
3,3-Diethyl-N-(1(*R*)-(3-hydroxy-4-methoxyphenyl)butyl)-2(*S*)-((4-methyl-1-piperazinyl)carbonyl)phenoxy-4-oxo-1-azetidinecarboxamide
Probably formed from: 3,3-Diethyl-N-(1(*R*)-(3,4-dihydroxyphenyl)butyl)-2(*S*)-((4-methyl-1-piperazinyl)carbonyl)phenoxy-4-oxo-1-azetidinecarboxamide
3,3-Diethyl-N-(1(*R*)-(4-hydroxy-3-methoxyphenyl)butyl)-2(*S*)-((4-methyl-1-piperazinyl)carbonyl)phenoxy-4-oxo-1-azetidinecarboxamide
Probably formed from: 3,3-Diethyl-N-(1(*R*)-(3,4-dihydroxyphenyl)butyl)-2(*S*)-((4-methyl-1-piperazinyl) carbonyl)phenoxy-4-oxo-1-azetidinecarboxamide
N,N-Diethyl-*m*-hydroxymethylbenzamide
Formed from: N,N-Diethyl-*m*-formylbenzamide
N,N-Diethyl-*m*-toluamide
→ N,N-Diethyl-*m*-formylbenzamide
In: Rat E858, J572
→ N-Ethyl-*m*-hydroxymethylbenzamide
In: Rat J572
6,8-Diethyl-5-hydroxy-4-oxo-4*H*-1-benzopyran-2-carboxylic acid
→ 8-Ethyl-5-hydroxy-6-(1-hydroxyethyl)-4-oxo-4*H*-1-benzopyran-2-carboxylic acid
In: Ferret B973
Hamster B973
Macaque B973
Monkey B973
Mouse B973
Rabbit B973
Rat B973
→ 8-Ethyl-5-hydroxy-6-(2-hydroxyethyl)-4-oxo-4*H*-1-benzopyran-2-carboxylic acid
In: Hamster B973
Monkey B973
Rabbit B973
Rat B973
N,N-Diethylisophthalamidic acid
Formed from: N,N-Diethyl-*m*-formylbenzamide
→ N-Ethylisophthalamidic acid
In: Rat J572
See also: H530

2,6-Diethyl-N-methoxymethylaniline
Formed from: Alachlor
α,α-Diethyl-3,4-methylenedioxyphenylacetamide
→ α,α-Diethyl-3,4-dihydroxyphenylacetamide
In: Guinea pig A1014
N,N-Diethyl-N′-(6-methylergolin-8α-yl) sulphamide
Formed from: N,N-Diethyl-N′-(1-ethyl-6-methylergolin-8α-yl)sulphamide
2′,6′-Diethyl-2-methylthioacetamide
→ 2,6-Diethylaniline
In: Mouse F528
Rat F528
N,N-Diethyl-N′-((8α)-6-methylergolin-8-yl)urea
See: Terguride
Diethyl 4-nitrophenyl phosphate (paraoxon)
Formed from: Diethyl 4-nitrophenyl phosphorothioate
→ Ethyl *p*-nitrophenyl phosphate
In: Man G892
Rat A1831
See also: J198
→ *p*-Nitrophenol
In: Badger K863
Beef K863
Cat K863
Cow A1296
Dog A1296
Goat K863
Guinea pig A1296
Horse A1296
Man A1296, F717, K327
Mouse D804, K863
Pig K863
Rabbit A1296, A1296, J553, K863
Rat A881, A1296, K170, K194, K863
Sheep K863
See also: A798, A2435, E25, G251
Diethyl 4-nitrophenyl phosphorothioate (parathion)
→ 4-Aminophenyl diethyl phosphorothioate
In: Microorganism A2435
Rat A881
Spinach A2050
→ Diethyl 4-hydroxylaminophenyl phosphorothioate
In: Spinach A2050
→ Diethyl 4-nitrophenyl phosphate
In: Microorganism A2435
Mouse A2567, D858
Rabbit A1995
Rat A82, A459, A881, F739
See also: A3100
Contrast (in lobster) A2567
→ Diethyl 4-nitrosophenyl phosphorothioate
In: Spinach A2050
→ *p*-Nitrophenol
In: Carrot H674
Datura H674
Digitalis H674
Flavobacterium A534
Glycine H674
Microorganism A2435
Mouse A2567, D858
Pseudomonas A3100
Rat A881
Spodoptera A1076
Wheat H674
See also: A459
2,6-Diethylnitrosobenzene
Formed from: 2,6-Diethylaniline
Diethyl 4-nitrosophenyl phosphorothioate
Formed from: Diethyl 4-nitrophenyl phosphorothioate
→ 4-Aminophenyl diethyl phosphorothioate
In: Spinach A2050
→ Diethyl 4-hydroxylaminophenyl phosphorothioate
In: Spinach A2050
Diethylnorephedrine
See: 2-Diethylamino-1-phenyl-1-propanol
N,N-Diethylphenylacetamide
→ N-Ethylphenylacetamide
In: Guinea pig F273
Mouse F273
Rat F256, F273
→ Phenylacetic acid
In: Guinea pig F273
Mouse F273
Rat F256, F273

1,5-Diethylphenylbarbituric acid
(1-ethylphenobarbital)
→ 5-Ethyl-5-(4-hydroxy-3-methoxyphenyl) barbituric acid
In: Man E474
→ Phenobarbital
Probably in: Man E474

N,N-Diethyl-1-phenylcyclohexylamine
→ N-Ethyl-1-phenylcyclohexylamine
In: Rat G529
See also: F892

O,O-Diethyl O-(5-phenyl-3-isoxazolyl) phosphate
See: Isoxathion

2,N-Diethyl-2-phenylmalondiamide
Formed from: 2-Deoxy-N-ethylphenobarbital

Diethyl phthalate
→ Phthalic acid monoethyl ester
In: Baboon A1652
Ferret A1652
Micrococcus D686
Rat A1652

Diethylpropion
See: α-Diethylaminopropiophenone

N,N-Diethyl-N′-((8α)-6-propylergolin-8-yl)urea
See: Proterguride

***p,p′*-Diethyl-*cis*-stilbene oxide**
→ (*R,R*)-*p,p′*-Diethyl-α,β-dihydroxydibenzyl
In: Rabbit H568

α,α-Diethylstilbenediol
See: Stilboestrol

Diethylstilboestrol
See: Stilboestrol

N,N-Diethyl-*m*-toluamide
→ N,N-Diethyl-*m*-hydroxymethylbenzamide
In: Rat E858, E934, J572
See also: H530
→ N-Ethyl-*m*-toluamide
In: Rat E858, E934, J572

1,2-Di-O-feruloyl-β-D-glucose
Feruloyl-β-D-glucose

Difloxacin (6-fluoro-1-(4-fluorophenyl)-1,4-dihydro-7-(4-methyl-1-piperazinyl)-4-oxo-3-quinolinecarboxylic acid)
→ 7-(2-Aminoethylamino)-3-carboxy-6-fluoro-N-(*p*-fluorophenyl)-4-quinolone
In: Man E266
→ Difloxacin-β-D-glucuronide
In: Man E266
→ Difloxacin-N-oxide
In: Man E266
→ Nordifloxacin
In: Man E266

Difloxacin-β-D-glucuronide
Formed from: Difloxacin

Difloxacin-N-oxide
Formed from: Difloxacin

Diflubenzuron (N-((*p*-chlorophenyl) amino)carbonyl-2,6-difluorobenzamide)
→ *p*-Chloroaniline
In: Pig C394
Sporoptera E659
→ *p*-Chloroformanilide
In: Spider mite B887
→ N-((4-Chloro-2-hydroxyphenyl)amino) carbonyl-2,6-difluorobenzamide
In: Anthonomus A3432
Beef A2751
Mouse A3386
Rat A3941
Sheep A2751
→ N-((4-Chloro-3-hydroxyphenyl)amino) carbonyl-2,6-difluorobenzamide
In: Beef A2751
Rat A3941
Sheep A2751
→ N-((*p*-Chlorophenyl)amino)carbonyl-2,6-difluoro-3-hydroxybenzamide
In: Anthonomus A3432
Beef A2751
Sheep A2751
See also: A3941
→ *p*-Chlorophenylurea
In: Beef A2751
Fly A2753
Pig C394
Spider mite B887
Sporoptera E659
→ 2,6-Difluorobenzamide
In: Bean B887

Fly A2753
Pig C394
Spider mite B887
→ 2,6-Difluorobenzoic acid
In: Beef A2751
Pig C394
Rat A3941
Sheep A2751
Spider mite B887

Diflunisal (2′,4-difluoro-4-hydroxy-3-biphenylylcarboxylic acid)
→ Diflunisal-β-D-glucuronic acid
In: Man E741, H517, J183
Rat C453, D489, D820, D891, F910, H415, H517, H934
→ Diflunisal-β-D-glucuronide
In: Man E741, H517, J183
Rat C453, D489, D820, F415, F910, J550
See also: H861, J783
→ Diflunisal-O-sulphate
In: Man E741
Rat F415
→ 3-Hydroxydiflunisal
In: Man G28
Rat G28, G30

Diflunisal-β-D-glucuronic acid
Formed from: Diflunisal

Diflunisal-β-D-glucuronide
Formed from: Diflunisal

Diflunisal-O-sulphate
Formed from: Diflunisal
Unchanged in rat F926

ω-Difluoroacetophenone
→ 2,2-Difluoro-1-phenylethanol
In: Candida G826
Rhodotorula G826
Saccharomyces G826
Torulopsis G826
Xanthomonas G826

2,3-Difluoroaniline
→ 4-Amino-2,3-difluorophenol
In: Rat H273

2,5-Difluoroaniline
→ 4-Amino-2,5-difluorophenol
In: Rat H273

2,6-Difluoroaniline
→ 4-Amino-3,5-difluorophenol
In: Rat H273

3,5-Difluoroaniline
→ 4-Amino-2,6-difluorophenol
In: Rat H273

2,2′-Difluoroazobenzene
Formed from: *o*-Fluoroaniline

3,3′-Difluoroazobenzene
Formed from: *m*-Fluoroaniline

4,4′-Difluoroazobenzene
Formed from: *p*-Fluoroaniline

3,4-Difluorobenzaldehyde
→ 3,4-Difluorobenzoic acid
In: Rat H761

2,6-Difluorobenzamide
Formed from: Diflubenzuron
Difluorobenzonitrile
Teflubenzuron
→ 2,6-Difluorobenzoic acid
In: Rat J260
See also: H952

***m*-Difluorobenzene**
→ 2,4-Difluorophenol
In: Rat G953
→ 3-Fluorocatechol
In: Pseudomonas F76

***p*-Difluorobenzene**
→ 3,6-Difluorocatechol
In: Pseudomonas F283

2,4-Difluorobenzoic acid
→ 4-Fluorocatechol
Probably in: Pseudomonas G434

2,6-Difluorobenzoic acid
Formed from: Diflubenzuron
2,6-Difluorobenzamide
Teflubenzuron
→ 2,6-Difluorohippuric acid
In: Beef A2751
Rat J260
Sheep A2751
→ 3-Fluorocatechol
Probably in: Pseudomonas G434

3,4-Difluorobenzoic acid
Formed from: 3,4-Difluorobenzaldehyde
→ *cis*-3,4-Difluoro-1,2-dihydro-1,2-dihydroxybenzoic acid

In: Pseudomonas E543
→ *cis*-3,4-Difluoro-1,6-dihydro-1,6-dihydroxybenzoic acid
In: Pseudomonas E543
→ 4-Fluorocatechol
Probably in: Pseudomonas G434

3,5-Difluorobenzoic acid
→ *cis*-3,5-Difluoro-1,2-dihydro-1,2-dihydroxybenzoic acid
In: Pseudomonas E543

2,2′-Difluorobenzoin
Formed from: *o*-Fluorobenzaldehyde

4,4′-Difluorobenzoin
Formed from: *p*-Fluorobenzaldehyde

2,6-Difluorobenzonitrile
→ 2,6-Difluorobenzamide
In: Rhodococcus H952

4,4′-Difluorobenzophenone
Formed from: Flunarizine
Probably formed from: 1-(Bis(*p*-fluorophenyl)methyl)-4-(2,3,4-trimethoxybenzyl)piperazine

N-(2,6-Difluorobenzoyl)-N′-(2,4-difluoro-3,5-dichlorobenzoyl)urea
See: Teflubenzuron

4,4′-Difluorobiphenyl
→ 4,4′-Difluoro-3-methoxybiphenyl
In: Duckweed A3390

3,4-Difluorocatechol
Formed from: 2,3-Difluorophenol
3,4-Difluorophenol
2,3,6-Trifluorophenol

3,5-Difluorocatechol
Formed from: 2,4-Difluorophenol
3,5-Difluorophenol
2,3,5-Trifluorophenol

3,6-Difluorocatechol
Formed from: *p*-Difluorobenzene
2,5-Difluorophenol
2,3,6-Trifluorophenol

4,5-Difluorocatechol
Formed from: 3,4-Difluorophenol

1-(*trans*-11,11-Difluorodibenzo[*b*,*e*]bicyclo[5.1.0]oct-5-yl)piperazine
Formed from: (2*R*)-1-(4-(*trans*-11,11-Difluorodibenzo[*b*,*e*]bicyclo[5.1.0]oct-5-yl)piperazin-1-yl)-3-(quinolin-5-yl)oxy-2-propanol

(2*R*)-1-(4-(*trans*-11,11-Difluorodibenzo[*b*,*e*]bicyclo[5.1.0]oct-5-yl)piperazin-1-yl)-3-(6-hydroxyquinolin-5-yl)oxy-2-propanol
Formed from: (2*R*)-1-(4-(*trans*-11,11-Difluorodibenzo[*b*,*e*]bicyclo[5.1.0]oct-5-yl)piperazin-1-yl)-3-(quinolin-5-yl)oxy-2-propanol

(2*R*)-1-(4-(*trans*-11,11-Difluorodibenzo[*b*,*e*]bicyclo[5.1.0]oct-5-yl)piperazin-1-yl)-3-(8-hydroxyquinolin-5-yl)oxy-2-propanol
Formed from: (2*R*)-1-(4-(*trans*-11,11-Difluorodibenzo[*b*,*e*]bicyclo[5.1.0]oct-5-yl)piperazin-1-yl)-3-(quinolin-5-yl)oxy-2-propanol

(2*R*)-1-(4-(*trans*-11,11-Difluorodibenzo[*b*,*e*]bicyclo[5.1.0]oct-5-yl)piperazin-1-yl)-3-(quinolin-5-yl)oxy-2-propanol
→ 1-(*trans*-11,11-Difluorodibenzo[*b*,*e*]bicyclo[5.1.0]oct-5-yl)piperazine
In: Man H931
Rat H931
→ (2*R*)-1-(4-(*trans*-11,11-Difluorodibenzo[*b*,*e*]bicyclo[5.1.0]oct-5-yl)piperazin-1-yl)-3-(6-hydroxyquinolin-5-yl)oxy-2-propanol
In: Rat H931
→ (2*R*)-1-(4-(*trans*-11,11-Difluorodibenzo[*b*,*e*]bicyclo[5.1.0]oct-5-yl)piperazin-1-yl)-3-(8-hydroxyquinolin-5-yl)oxy-2-propanol
In: Rat H931
→ (2*R*)-1-(4-(*trans*-11,11-Difluorodibenzo[*b*,*e*]bicyclo[5.1.0]oct-5-yl)piperazin-1-yl)-3-(quinolin-5-yl)oxy-2-propanol-β-D-glucuronide
In: Rat H931
→ (2*R*)-1-(4-(*trans*-11,11-Difluorodibenzo[*b*,*e*]bicyclo[5.1.0]oct-5-yl)piperazin-1-yl)-3-(quinolin-5-yl)oxy-2-propanol-N-oxide
In: Man H931
Rat H931

(2*R*)-1-(4-(*trans*-11,11-Difluorodibenzo[*b*,*e*]bicyclo[5.1.0]oct-5-yl)piperazin-1-yl)-3-(quinolin-5-yl)oxy-2-propanol-β-D-glucuronide
Formed from: (2*R*)-1-(4-(*trans*-11,11-Difluorodibenzo[*b*,*e*]bicyclo[5.1.0]oct-5-

yl)piperazin-1-yl)-3-(quinolin-5-yl)oxy-2-propanol

(2*R*)-1-(4-(*trans*-11,11-Difluorodibenzo[*b,e*]bicyclo[5.1.0]oct-5-yl)piperazin-1-yl)-3-(quinolin-5-yl)oxy-2-propanol-N-oxide
Formed from: (2*R*)-1-(4-(*trans*-11,11-Difluorodibenzo[*b,e*]bicyclo[5.1.0]oct-5-yl)piperazin-1-yl)-3-(quinolin-5-yl)oxy-2-propanol

***cis*-3,4-Difluoro-1,2-dihydro-1,2-dihydroxybenzoic acid**
Formed from: 3,4-Difluorobenzoic acid

***cis*-3,4-Difluoro-1,6-dihydro-1,6-dihydroxybenzoic acid**
Formed from: 3,4-Difluorobenzoic acid

***cis*-3,5-Difluoro-1,2-dihydro-1,2-dihydroxybenzoic acid**
Formed from: 3,5-Difluorobenzoic acid

3,7-Difluoro-1,2-dihydroxy*spiro*(9*H*-fluorene-9,4′-imidazolidine)-2′,5′-dione
Formed from: Imirestat

2,7-Difluoro-4,5-dimethoxy*spiro*(9*H*-fluorene-9,4′-imidazolidine)-2′,5′-dione
→ 2,7-Difluoro-4-hydroxy-5-methoxy*spiro*(9*H*-fluorene-9,4′-imidazolidine)-2′,5′-dione
In: Monkey G513
Rat G513

2,2′-Difluoro-4,4′-dimethylazobenzene
Formed from: 2-Fluoro-4-methylaniline

2,2′-Difluoro-4,4′-dimethylazoxybenzene
Formed from: 2-Fluoro-4-methylaniline

6,8-Difluoro-1-(2-fluoroethyl)-1,4-dihydro-7-(4-methyl-1-piperazinyl)-4-oxo-3-quinolonecarboxylic acid
See: Fleroxacin

2,6-Difluorohippuric acid
Formed from: 2,6-Difluorobenzoic acid

2,5-Difluoro-4-hydroxybenzoic acid
→ 2,5-Difluoroquinol
In: Candida J461

2,6-Difluoro-4-hydroxybenzoic acid
→ 2,6-Difluoroquinol
In: Candida J461

3,5-Difluoro-4-hydroxybenzoic acid
→ 2,6-Difluoroquinol
In: Candida J461

2′,4′-Difluoro-4-hydroxy-3-biphenylylcarboxylic acid
See: Diflunisal

2,7-Difluoro-4-hydroxy-5-methoxy*spiro*(9*H*-fluorene-9,4′-imidazolidine)-2′,5′-dione
Formed from: 2,7-Difluoro-4,5-dimethoxy*spiro*(9*H*-fluorene-9,4′imidazolidine)-2′,5′-dione

2′,4′-Difluoro-γ-hydroxy-α-methyl-1-(1,1′-biphenyl)-4-butanoic acid
Formed from: Flobufen

N-(3,5-Difluoro-4-hydroxyphenyl)succinamic acid
Probably formed from: N-(3,5-Difluorophenyl)succinamic acid

3,7-Difluoro-2-hydroxy*spiro*(9*H*-fluorene-9,4′-imidazolidine)-2′,5′-dione
Formed from: Imirestat

4,4′-Difluoro-3-methoxybiphenyl
Formed from: 4,4′-Difluorobiphenyl

4-(Difluoromethoxy)-α-(1-methylethyl)benzeneacetic acid cyano(3-phenoxyphenyl)methyl ester
See: Flucythrinate

α-Difluoromethyl-3,4-dihydroxy-L-phenylalanine
→ α-Difluoromethyl-4-hydroxy-3-methoxy-L-phenylalanine
In: Pig B813
Rat A3981

α-Difluoromethyl-4-hydroxy-3-methoxy-L-phenylalanine
Formed from: α-Difluoromethyl-3,4-dihydroxy-L-phenylalanine

2,4-Difluoro-17β-oestradiol
→ 4-Fluoro-2-methoxy-17β-oestradiol
In: Hamster D712

2,3-Difluorophenol
→ 3,4-Difluorocatechol
In: Penicillium J843
Trichosporon H420
→ 3-Fluorocatechol
In: Trichosporon H420

2,4-Difluorophenol
Formed from: *m*-Difluorobenzene
→ 2,4-Difluorophenyl-β-D-glucuronide
In: Rat G953, H522
→ 2,4-Difluorophenyl sulphate

In: Rat G953, H522
→ 3,5-Difluorocatechol
In: Penicillium J843
Trichosporon H420
→ 4-Fluorocatechol
In: Trichosporon H420

2,5-Difluorophenol
→ 3,6-Difluorocatechol
In: Penicillium J843
Trichosporon H420
→ 4-Fluorocatechol
In: Trichosporon H420

2,6-Difluorophenol
→ 2,6-Difluorophenyl-β-D-glucuronide
In: Rat H522
→ 2,6-Difluorophenyl sulphate
In: Rat H522
→ 3-Fluorocatechol
In: Trichosporon H420

3,4-Difluorophenol
→ 3,4-Difluorophenyl-β-D-glucuronide
In: Rat H522
→ 3,4-Difluorophenyl sulphate
In: Rat H522
→ 3,4-Difluorocatechol
In: Penicillium J843
Trichosporon H420
→ 4,5-Difluorocatechol
In: Penicillium J843
Trichosporon H420

3,5-Difluorophenol
→ 3,5-Difluorocatechol
In: Trichosporon H420

N-(6-(2,4-Difluorophenoxy)-1-hydroxyindan-5-yl)methanesulphonamide
Formed from: N-(6-(2,4-Difluorophenoxy) indan-5-yl)methanesulphonamide
→ N-(6-(2,4-Difluorophenoxy)-1-oxoindan-5-yl)methanesulphonamide
Probably in: Monkey C485
Rat C485

N-(6-(2,4-Difluorophenoxy)-3-hydroxyindan-5-yl)methanesulphonamide
Probably formed from: N-(6-(2,4-Difluoro)phenoxyindan-5-yl) methanesulphonamide

N-(6-(2,4-Difluorophenoxy)indan-5-yl) methanesulphonamide
→ N-(6-(2,4-Difluorophenoxy)-1-hydroxyindan-5-yl) methanesulphonamide
In: Rat C485
→ N-(6-(2,4-Difluorophenoxy)-3-hydroxyindan-5-yl) methanesulphonamide
In: Rat C485

N-(6-(2,4-Difluorophenoxy)-1-oxoindan-5-yl) methanesulphonamide
Probably formed from: N-(6-(2,4-Difluorophenoxy)-1-hydroxyindan-5-yl)methanesulphonamide

5-(2′,4′-Difluorophenyl)benzoic acid-2-yl-β-D-glucuronide
Formed from: 5-(2′,4′-Difluorophenyl)-2-hydroxybenzoic acid

2,2-Difluoro-1-phenylethanol
Formed from: ω-Difluoroacetophenone

2,4-Difluorophenyl-β-D-glucuronide
Formed from: 2,4-Difluorophenol

2,6-Difluorophenyl-β-D-glucuronide
Formed from: 2,6-Difluorophenol

3,4-Difluorophenyl-β-D-glucuronide
Formed from: 3,4-Difluorophenol

5-(2′,4′-Difluorophenyl)-2-hydroxybenzoic acid
→ 5-(2′,4′-Difluorophenyl)benzoic acid-2-yl-β-D-glucuronide
In: Man A1865
→ 5-(2′,4′-Difluorophenyl)-2-hydroxybenzoyl-β-D-glucuronic acid
In: Dog A1865
Man A1865
Rat A1865

5-(2′,4′-Difluorophenyl)-2-hydroxybenzoyl-β-D-glucuronic acid
Formed from: 5-(2′,4′-Difluorophenyl)-2-hydroxybenzoic acid

N-(3,5-Difluorophenyl)-2-hydroxysuccinamic acid
Probably formed from: N-(3,5-Difluorophenyl)succinamic acid

N-(3,5-Difluorophenyl)-3-hydroxysuccinamic acid
Probably formed from: N-(3,5-Difluorophenyl)succinamic acid

N-(3,5-Difluorophenyl)succinamic acid
Formed from: N-(3,5-Difluorophenyl) succinimide
→ N-(3,5-Difluoro-4-hydroxyphenyl) succinamic acid
Probably in: Rat J587
→ N-(3,5-Difluorophenyl)-2-hydroxysuccinamic acid
Probably in: Rat J587
→ N-(3,5-Difluorophenyl)-3-hydroxysuccinamic acid
Probably in: Rat J587

N-(3,5-Difluorophenyl)succinimide
→ N-(3,5-Difluorophenyl)succinamic acid
In: Rat J587

2,4-Difluorophenyl sulphate
Formed from: 2,4-Difluorophenol

2,6-Difluorophenyl sulphate
Formed from: 2,6-Difluorophenol

3,4-Difluorophenyl sulphate
Formed from: 3,4-Difluorophenol

1-(2,6-Difluorophenyl)-1*H*,3*H*-thiazolo[3,4-*a*] benzimidazole
→ 1-(2,6-Difluorophenyl)-1*H*,3*H*-thiazolo [3,4-*a*]benzimidazole sulphoxide
Probably in: Mouse G554

1-(2,6-Difluorophenyl)-1*H*,3*H*-thiazolo[3,4-*a*] benzimidazole sulphone
Probably formed from: 1-(2,6-Difluorophenyl)-1*H*,3*H*-thiazolo[3,4-*a*] benzimidazole sulphoxide

1-(2,6-Difluorophenyl)-1*H*,3*H*-thiazolo[3,4-*a*] benzimidazole sulphoxide
Probably formed from: 1-(2,6-Difluorophenyl)-1*H*,3*H*-thiazolo[3,4-*a*] benzimidazole
→ 1-(2,6-Difluorophenyl)-1*H*,3*H*-thiazolo [3,4-*a*]benzimidazole sulphone
Probably in: Mouse G554

2,5-Difluoroquinol
Formed from: 2,5-Difluoro-4-hydroxybenzoic acid

2,6-Difluoroquinol
Formed from: 2,6-Difluoro-4-hydroxybenzoic acid
3,5-Difluoro-4-hydroxybenzoic acid

4,4′-Diformamidodiphenyl sulphone
→ 4-Amino-4′-formamidodiphenyl sulphone
In: Dog A521
Man A1722
Monkey A521
→ N-Hydroxy-4,4′-diformamidodiphenyl sulphone
In: Dog A521
Monkey A521

3,5-Diformyl-2-(4-hydroxy-3,5-dimethoxyphenyl)-7-methoxycoumarone
Probably formed from: 5-Formyl-2-(4-hydroxy-3,5-dimethoxyphenyl)-3-hydroxymethyl-7-methoxycoumarone

N′,N$^{\alpha}$-Diformylkynurenine
→ N$^{\alpha}$-Formylkynurenine
Probably in: Neurospora A3374

Diftalone (phthalazino[2,3-*b*]phthalazine-5,12-(7*H*,14*H*)-dione)
→ 7-Hydroxydiftalone
In: Man A113, A2987

***m*-Digallic acid** (digallic acid, 3,4-dihydroxy-5-((3,4,5-trihydroxybenzoyl) oxy)benzoic acid)
Formed from: Tannin

Di-O-galloyl-β-D-glucose
Formed from: Galloyl-β-D-glucose
→ 1,2,6-Tri-O-galloyl-β-D-glucose
In: Rhus K888

Di-(L-glutaminyl-L-glutaminyl-L-methionyl-L-glutaminyl-L-glutaminyl-L-seryl-L-threonyl-L-arginyl-L-prolyl-L-tyrosyl-L-glutaminyl-L-glutaminyl-L-arginyl-L-prolyl-glycyl-L-glutaminyl-L-glutamine)
Formed from: L-Glutaminyl-L-glutaminyl-L-methionyl-L-glutaminyl-L-glutaminyl-L-seryl-L-threonyl-L-arginyl-L-prolyl-L-tyrosyl-L-glutaminyl-L-glutaminyl-L-arginyl-L-prolyl-glycyl-L-glutaminyl-L-glutamine

Di-(5-hexenyl) phthalate
→ Phthalic acid mono 5-hexenyl ester
In: Rat C869

Dihydroacetophenone tricarbonylchromium
Probably formed from: Acetophenone tricarbonylchromium

Dihydroaclacinomycin A (2 isomers)
Formed from: Aclacinomycin A

Dihydroacrivastine
Formed from: Acrivastine
→ 3-(6-(1-(*p*-Carboxyphenyl)-3-(1-pyrrolidinyl)-1-propenyl)-2-pyridinyl) propionic acid
Probably in: Dog G382

Dihydroadriamycinone
Formed from: Adriamycin aglycone
→ 7-Deoxydihydroadriamycinone
Probably in: Streptomyces B571

2,3-Dihydroagroclavine
→ 2,3-Dihydroelymoclavine
In: Claviceps D237

9,10-Dihydroanthracene
→ Anthrone
In: Microorganism D452
Pseudomonas H254
→ *cis*-1,2,9,10-Tetrahydro-1(*R*),2(*S*)-dihydroxyanthracene
In: Pseudomonas J289

Dihydrobenfluron-N-oxide
Formed from: Benfluron

2,3-Dihydrobenz[*d*]isothiazole-1,1-dioxide
→ 2,3-Dihydro-3-hydroxybenz[*d*]isothiazole-1,1-dioxide
Probably in: Rat A2107

2,3-Dihydrobenzisoazol-3-one
Probably formed from: 1,2-Benzisoazole

2-(2,3-Dihydro-1,4-benzodioxin-2-yl)-4,5-dihydro-1*H*-imidazole
See: Idazoxan

(*S*)-2-(1-(2-(2,3-Dihydrobenzofuran-5-yl)ethyl)-3-pyrrolidinyl)-2-hydroxyphenyl-2-phenylacetamide
Formed from: Darifenacin

6,7-Dihydro-5*H*-benzocyclohept-1-ene
→ *cis*-6,7-Dihydro-1(*R*),2(*S*)-dihydroxybenzocycloheptane
In: Pseudomonas J36
→ 6,7-Dihydro-5(*R*)-hydroxy-5*H*-benzocyclohept-1-ene
In: Pseudomonas F629, J36

2-((2,3-Dihydro-1,4-benzodioxin-2-yl)methyl)-1-ethyl-1*H*-imidazole
See: Imiloxan

(*S*)-2-(1-(2-(2,3-Dihydrobenzofuran-5-yl)ethyl)-3-pyrrolidinyl)-2,2-diphenylacetamide
See: Darifenacin

(*S*)-2-(1-(2-(2,3-Dihydrobenzofuran-5-yl)ethyl)-3-pyrrolidinyl)-2-hydroxyphenyl-2-phenylacetamide
Formed from: Darifenacin

1,4-Dihydro-3*H*-2-benzopyran-3-one
See: 3-Isochromanone

Dihydrobenzopyran-4-one
See: 4-Chromanone

1-((3,4-Dihydro-2*H*-1-benzothiopyran-8-yl)oxy)-3-((1,1-dimethylethyl)amino)-2-propanol
See: Tertatolol

2,3-Dihydrobenzoxazol-2-one
Probably formed from: Benzoxazole

7,8-Dihydro-3,4-benzpyrene
→ 7,8-Dihydro-3,4-benzpyrene-9,10-oxide
In: Rat C554
Sheep C554
→ 7,8-Dihydro-7-hydroxy-3,4-benzpyrene
In: Rat D497
→ 7,8-Dihydro-8-hydroxy-3,4-benzpyrene
In: Rat D497
→ *cis*-7,8,9,10-Tetrahydro-9,10-dihydroxy-3,4-benzpyrene
In: Sheep C361
→ *trans*-7,8,9,10-Tetrahydro-9,10-dihydroxy-3,4-benzpyrene
In: Sheep C361

9,10-Dihydro-3,4-benzpyrene
→ 9,10-Dihydro-9-hydroxy-3,4-benzpyrene
In: Rat D497
→ 9,10-Dihydro-10-hydroxy-3,4-benzpyrene
In: Rat D497

7,8-Dihydro-3,4-benzpyrene-9,10-oxide
Formed from: 7,8-Dihydro-3,4-benzpyrene
→ 10-Glutathionyl-7,8,9,10-tetrahydro-9-hydroxy-3,4-benzpyrene
In: Rat D293

9,10-Dihydro-3,4-benzpyrene-7,8-oxide
→ 7-Glutathionyl-7,8,9,10-tetrahydro-8-hydroxy-3,4-benzpyrene
In: Rat D304

7,8-Dihydroberberine
Formed from: Tetrahydroberberine
→ Berberine
Probably in: Coptis D526
Macleaya C696
Contrast (in Berberis) D536
→ Tetrahydroberberine
In: Corydalis G136
Dihydrobupropion
Formed from: Bupropion
Dihydrocaffeic acid (3,4-dihydroxyphenylpropionic acid, hydrocaffeic acid)
Formed from: Caffeic acid
Eriodictyol
Hesperetin
Phloretic acid
Probably formed from: 5-(3,4-Dihydroxyphenyl)pentanoic acid
→ γ-Carboxyethyl-α-hydroxy-*cis*,*cis*-muconic semialdehyde
In: Amanita G119
Arthrobacter A3279
Bacillus A2827
Pseudomonas A1090
→ 4-(2-Carboxyethyl)-*o*-quinone
In: Apple J678
Pear J678
Mushroom K140
See also: E407
→ Dihydroferulic acid
In: Nicotiana A3778, B490
Rat A486
→ Dihydroisoferulic acid
In: Medicago G273
Rat A486
→ 2,3-Dihydro-2,3,4,5-tetrahydroxyphenylpropionic acid
Probably in: Bacterium B871
→ *m*-Hydroxyphenylpropionic acid
In: Rat A557
See also: D894
Dihydrocaffeoyl methylamide
→ Caffeoyl methylamide
In: Sarcophaga F793
→ 3,4-Dihydroxyphenylhydracryloyl methylamide
In: Sarcophaga F793
4,5-Dihydrocanthin-6-one
→ Canthin-6-one
In: Ailanthus E304
Dihydrocapsaicin
→ Dihydro-ω-hydroxycapsaicin
In: Rat H410
→ 4-Hydroxy-3-methoxybenzylamine
In: Rat D574
10,11-Dihydrocarbamazepin-10-yl-β-D-glucuronide
Probably formed from: Carbamazepin-10,11-oxide
1,2-Dihydro-β-carbolin-1-one
Formed from: 1,2,3,4-Tetrahydro-β-carboline
Tetrahydronorharmane
1,2-Dihydrocatechol
See: 1,2-Dihydro-1,2-dihydroxybenzene
Dihydrochelerythrine
→ Chelerythrine
In: Sanguinaria G723
Dihydrochelirubine
Formed from: Dihydro-10-hydroxysanguinarine
→ Chelirubine
Probably in: Eschscholtzia K777
→ Dihydro-12-hydroxychelirubine
In: Thalictrum K776
2,8-Dihydro-4-chloro-6-methyl-9-oxo-9*H*-xanthene-1-carboxylic acid methyl ester
See: 4-Chloropinselin
Dihydrocinnamic acid
See: 3-Phenylpropionic acid
Dihydrocitrinone ((3*R*,4*S*)-3,4-dihydro-6,8-dihydroxy-3,4,5-trimethyl-1-oxo-1*H*-2-benzopyran-7-carboxylic acid)
Formed from: Citrinin
Dihydrocodeine
Formed from: 7,8-Dihydrocodeinone
→ Dihydrocodeine-β-D-glucuronide
In: Dog G935, J629
→ Dihydromorphine
In: Dog G935

→ Dihydronorcodeine
In: Dog G935

Dihydrocodeine-β-D-glucuronide
Formed from: Dihydrocodeine

7,8-Dihydrocodeinone
→ Dihydrocodeine
In: Rat J744
→ 7,8-Dihydroisocodeinone
In: Aureobasidium E666

Dihydrocodorphone
Formed from: Codorphone
→ Demethyldihydrocodorphone
In: Man C636
Rat C636

Dihydroconiferaldehyde
Probably formed from: Coniferaldehyde

Dihydroconiferyl alcohol (3-(4-hydroxy-3-methoxyphenyl)propan-1-ol)
Probably formed from: Coniferyl alcolol

3,4-Dihydrocoumarin
Formed from: Coumarin
Fluorene
→ Melilotic acid
In: Acinetobacter K429
Man K327

Dihydrodaidzein (4′,7-dihydroxyisoflavanone)
Formed from: Daidzein

13-Dihydrodaunomycinone
Formed from: Daunorubicin aglycone
→ 7-Deoxy-13-Dihydrodaunomycinone
In: Streptomyces B443
→ 13-Dihydrodaunomycinone glycoside
In: Streptomyces B443

13-Dihydrodaunomycinone glycoside
Formed from: 13-Dihydrodaunomycinone

Dihydro-5-deazariboflavin
Formed from: 5-Deazariboflavin

Dihydrodemethylsterigmatocystin
→ Dihydrosterigmatocystin
In: Aspergillus F281
See also: J511

Dihydro-5-demethylubiquinone 45
Formed from: 5-Demethylubiquinone 45
→ Dihydroubiquinone 45
In: Rat A3084

Dihydro-11-O-demethylvindoline
Formed from: Dihydrovindoline

Dihydrodeoxymorphine
→ 2,2′-Bi(dihydrodeoxymorphine)
In: Horseradish A1357

***l*-4,5-Dihydrodiazepam**
→ 7-Chloro-1,3-dihydro-1-methyl-5-phenyl-2*H*-1,4-benzodiazepin-2-one
In: Rat A2858, B541
→ 7-Chloro-1,3,4,5-tetrahydro-5-phenyl-2*H*-1,4-benzodiazepin-2-one
In: Rat A2858, B541
→ 1,3,4,5-Tetrahydro-7-hydroxy-5-phenyl-2*H*-1,4-benzodiazepin-2-one
Probably in: Rat A2858

10,11-Dihydro-5*H*-dibenz[*b*,*f*]azepine
See: Iminodibenzyl

3-(10,11-Dihydro-5*H*-dibenzo[*a*,*d*]cyclohepten-5-ylidene)propionic acid
Formed from: Nortriptyline

10,11-Dihydrodibenz[*b*,*f*]-1,4-oxazepine
→ 10,11-Dihydro-11(10*H*)-oxodibenz[*b*,*f*]-1,4-oxazepine
In: Rat C495

7-((10,11-Dihydro-5*H*-dibenzo[*a*,*d*]cyclohepten-5-yl)amino)heptenoic acid
See: Amineptine

5,6-Dihydrodibenzo[*a*,*g*]quinolizinium
See: Protoberberine

1,2-Dihydro-1,2-dihydroxyaceanthrylene
Formed from: Aceanthrylene

(−)-*trans*-1,2-Dihydro-1,2-dihydroxyacenaphthene
→ 1,2-Dihydroxyacenaphthene
In: Rat K817

***cis*-2′,3′-Dihydro-2′,3′-dihydroxyacetophenone**
Formed from: Acetophenone

***cis*-1,2-Dihydro-1,2-dihydroxyacridine**
Formed from: Acridine

***trans*-1,2-Dihydro-1,2-dihydroxyacridine**
Formed from: Acridine

3,4-Dihydro-3,4-dihydroxyacridine
Formed from: Acridine

1,2-Dihydro-1,2-dihydroxyaflatoxin B_1
Formed from: Aflatoxin B_1

8,9-Dihydro-8,9-dihydroxyaflatoxin B_1
Formed from: Aflatoxin B_1-8,9-oxide

→ Aflatoxin B_1-dialcohol
In: Man J662
Rat H14

2′,3′-Dihydro-2′,3′-dihydroxyalphenal
Probably formed from: Alphenal

2′,3′-Dihydro-2′,3′-dihydroxyaminopyrine
Formed from: Aminopyrine
→ 2′,3′-Dihydroxyaminopyrine
Probably in: Phenylobacterium B838

1,2-Dihydro-1,2-dihydroxyamitriptyline
Formed from: Amitriptyline

***cis*-10,11-Dihydro-10,11-dihydroxyamitriptyline**
Formed from: 10-Hydroxyamitriptyline

***trans*-10,11-Dihydro-10,11-dihydroxyamitriptyline**
Formed from: 10-Hydroxyamitriptyline

***cis*-2,3-Dihydro-2,3-dihydroxyanisole**
Formed from: Anisole

***cis*-1,2-Dihydro-1,2-dihydroxyanthracene**
Formed from: Anthracene
→ 1,2-Dihydroxyanthracene
In: Pseudomonas A1298

***trans*-1,2-Dihydro-1,2-dihydroxyanthracene**
Formed from: Anthracene
Anthracene-1,2-oxide

2,3-Dihydro-2,3-dihydroxyantipyrine
Formed from: Antipyrine
→ 2,3-Dihydroxyantipyrine
In: Microorganism A3169

3,4-Dihydro-3,4-dihydroxyantipyrine
Formed from: Antipyrine

***trans*-1,2-Dihydro-1,2-dihydroxybenz[*j*]aceanthrylene**
Formed from: Benz[*j*]aceanthrylene
→ 1,2-Dihydro-1,2,8-trihydroxybenz[*j*]aceanthrylene
In: Man J688
Rat J688
→ 1,2-Dihydro-1,2,10-trihydroxybenz[*j*]aceanthrylene
In: Man J688
Rat J688, J804
→ 1,2,9,10-Tetrahydro-1,2,9,10-tetrahydroxybenz[*j*]aceanthrylene
In: Man J688
Rat J688

1,2-Dihydro-1,2-dihydroxybenz[*l*]aceanthrylene
Formed from: Benz[*l*]aceanthrylene

4,5-Dihydro-4,5-dihydroxybenz[*l*]aceanthrylene
Formed from: Benz[*l*]aceanthrylene

7,8-Dihydro-7,8-dihydroxybenz[*l*]aceanthrylene
Formed from: Benz[*l*]aceanthrylene

9,10-Dihydro-9,10-dihydroxybenz[*j*]aceanthrylene
Formed from: Benz[*j*]aceanthrylene
→ 9,10-Dihydro-9,10-trihydroxybenz[*j*]aceanthrylene-10-sulphate
In: Rat J688

11,12-Dihydro-11,12-dihydroxybenz[*j*]aceanthrylene
Formed from: Benz[*j*]aceanthrylene

9,10-Dihydro-9,10-dihydroxybenz[*j*]aceanthrylene-10-sulphate
Formed from: 9,10-Dihydro-9,10-dihydroxybenz[*j*]aceanthrylene

***cis*-1,2-Dihydro-1,2-dihydroxybenzanthracene**
Formed from: Benzanthracene

***trans*-1,2-Dihydro-1,2-dihydroxybenzanthracene**
Formed from: Benzanthracene
→ 1,2-Dihydroxybenzanthracene
In: Rat G440
→ 1,2,3,4-Tetrahydro-1,2,3,4-tetrahydroxybenzanthracene
In: Man H285

***trans*-3,4-Dihydro-3,4-dihydroxybenzanthracene**
Formed from: Benzanthracene
→ Benzanthracene-3,4-quinone
In: Rat C244, E740
→ 3,4-Dihydroxybenzanthracene
In: Rat F852, G440
→ 1,2,3,4-Tetrahydro-1,2,3,4-tetrahydroxybenzanthracene
In: Rat C244

***trans*-5,6-Dihydro-5,6-dihydroxybenzanthracene**
Formed from: Benzanthracene-5,6-oxide
→ S-(5,6,8,9-Tetrahydro-5,6,9-trihydroxybenzanthr-8-yl)glutathione
Probably in: Rat A306

8,9-Dihydro-8,9-dihydroxybenzanthracene (*cis* and *trans*)
Formed from: Benzanthracene

Benzanthracene-8,9-oxide
→ 8,9-Dihydro-8,9-dihydroxybenzanthracene-10,11-oxide
In: Rat A1422
→ S-(5,6,8,9-Tetrahydro-6,8,9-trihydroxybenzanthr-5-yl)glutathione
Probably in: Rat A306
→ S-(8,9,10,11-Tetrahydro-8,9,10-trihydroxybenzanthr-11-yl)glutathione
Probably in: Rat A306

***trans*-10,11-Dihydro-10,11-dihydroxybenzanthracene**
Formed from: Benzanthracene
Benzanthracene-8,9-oxide
→ 8,9,10,11-Tetrahydro-8,9,10,11-tetrahydroxybenzanthracene
In: Cunninghamella H329
Man H285

***cis*-11,12-Dihydro-11,12-dihydroxybenzanthracene**
Formed from: Benzanthracene

8,9-Dihydro-8,9-dihydroxybenzanthracene-10,11-oxide
Formed from: 8,9-Dihydro-8,9-dihydroxybenzanthracene
→ 8,9,10,11-Tetrahydro-8,9,10,11-tetrahydroxybenzanthracene
Probably in: Rat A1422

***cis*-1,2-Dihydro-1,2-dihydroxybenzene**
Formed from: Benzene
→ Catechol
In: Microoganism J512
Moraxella A638
Mouse F377
Pseudomonas A1298, J338, K763
Xanthobacter J186
See also: B678, G653, H873

***trans*-1,2-Dihydro-1,2-dihydroxybenzene**
(assumed when not stated)
Formed from: Benzene
→ Catechol
In: Beef G750
Guinea pig E67
Hamster F153
Monkey F241, F388, F391
Mouse E712, E821, F377
Pig F389
Rat G440
See also: G655

11,12-Dihydro-11,12-dihydroxybenzo[*g*]chrysene-13,14-oxide
→ 14-(S-Glutathionyl)-11,12,13,14-tetrahydro-11,12,13-trihydroxybenzo[*g*]chrysene
Probably in: Mouse K270

***cis*-6,7-Dihydro-1(*R*),2(*S*)-dihydroxybenzocycloheptane**
Formed from: 6,7-Dihydrobenzocyclohept-1-ene

***cis*-6,7-Dihydro-5,6-dihydroxy-5*H*-benzocyclohept-1-ene**
Probably formed from: 6,7-Dihydro-5-hydroxy-5*H*-benzocyclohept-1-ene

5,6-Dihydro-5,6-dihydroxy-7,8-benzoflavone
Probably formed from: 7,8-Benzoflavone

7,8-Dihydro-7,8-dihydroxy-7,8-benzoflavone
Probably formed from: 7,8-Benzoflavone

1,2-Dihydro-1,2-dihydroxybenzo[*b*]fluoranthene
Formed from: Benzo[*b*]fluoranthene

8,9-Dihydro-8,9-dihydroxybenzo[*k*]fluoranthene
Formed from: Benzo[*k*]fluoranthene

9,10-Dihydro-9,10-dihydroxybenzo[*j*]fluoranthene
Formed from: Benzo[*j*]fluoranthene

10,11-Dihydro-10,11-dihydroxybenzo[*b*]fluoranthene
Formed from: Benzo[*b*]fluoranthene

1,2-Dihydro-1,2-dihydroxybenzoic acid
Formed from: Benzoic acid

***cis*-1,2-Dihydro-1,2-dihydroxybenzo[*b*]naphtho[2,1-*d*]thiophene**
Formed from: Benzo[*b*]naphtho[2,1-*d*]thiophene
→ (+)-(1*R*,2*S*,7*R*,8*S*)-*cis*,*cis*-1,2,7,8-Tetrahydro-1,2,7,8-tetrahydroxybenzo[*b*]naphtho[2,1-*d*]thiophene
In: Sphingomonas K97

***trans*-Dihydro-dihydroxybenzo[*b*]naphtho[1,2-*d*]thiophene**
Formed from: Benzo[*b*]naphtho[1,2-*d*]thiophene

***cis*-2,3-Dihydro-2,3-dihydroxybenzonitrile**
Formed from: Benzonitrile
→ 2,3-Dihydroxybenzonitrile
Probably in: Pseudomonas F283

***trans*-3,4-Dihydro-3,4-dihydroxybenzo[*c*]phenanthrene**
Formed from: Benzo[*c*]phenanthrene
→ *trans*-3,4-Dihydro-3,4-dihydroxybenzo[*c*]phenanthrene-1,2-oxide
In: Mouse E82
Rat E82

5,6-Dihydro-5,6-dihydroxybenzo[*c*]phenanthrene
Formed from: Benzo[*c*]phenanthrene-5,6-oxide

***trans*-3,4-Dihydro-3,4-dihydroxybenzo[*c*]phenanthrene-1,2-oxide**
Formed from: *trans*-3,4-Dihydro-3,4-dihydroxybenzo[*c*]phenanthrene

(+)-(*3S*,*4R*)-*cis*-*3*,4-Dihydro-3,4-dihydroxybenzo[*c*]phenanthridine
Formed from: Benzo[*c*]phenanthridine

(+)-(9*S*,10*R*)-*cis*-10,11-Dihydro-10,11-dihydroxybenzo[*c*]phenanthridine
Formed from: Benzo[*c*]phenanthridine

2,3-Dihydro-2,5-dihydroxybenzopyran-4-one
→ 4,5-Dihydroxy-2*H*-benzopyran-2-one
In: Magnaporthea J522

5,6-Dihydro-5,6-dihydroxybenzo[*f*]quinoline
Formed from: Benzo[*f*]quinoline

5,6-Dihydro-5,6-dihydroxybenzo[*h*]quinoline
Formed from: Benzo[*h*]quinoline

7,8-Dihydro-7,8-dihydroxybenzo[*f*]quinoline
Formed from: Benzo[*f*]quinoline

7,8-Dihydro-7,8-dihydroxybenzo[*h*]quinoline
Formed from: Benzo[*h*]quinoline

9,10-Dihydro-9,10-dihydroxybenzo[*f*]quinoline
Formed from: Benzo[*f*]quinoline

***cis*-4,5-Dihydro-4,5-dihydroxy-3,4-benzpyrene**
Formed from: 3,4-Benzpyrene
→ *cis*-4,5-Dihydro-4,5-dihydroxy-3,4-benzpyrene-β-D-glucuronide
In: Rat A1615, C890

***trans*-4,5-Dihydro-4,5-dihydroxy-3,4-benzpyrene**
Formed from: 3,4-Benzpyrene-4,5-oxide
→ 3,4-Benzpyrene-4,5-quinone
Probably in: Rabbit B286
→ Chrysene
Probably in: Mouse A532
→ *trans*-4,5-Dihydro-4,5-dihydroxy-3,4-benzpyrene-β-D-glucuronide
In: Hamster A2009
Man B387, H92
Rat A1615, A2728, A3902, B33, C890
→ *trans*-4,5-Dihydro-4,5-dihydroxy-3,4-benzpyrene sulphate
In: Rat A2014, A2728
See also: A3732
→ 4,5-Dihydro-4-hydroxy-5-oxo-3,4-benzpyrene
Probably in: Rat B498

4,5-Dihydro-4,5-dihydroxybenz[*e*]pyrene
→ 4,5-Dihydro-4,5-dihydroxybenz[*e*]pyrene-β-D-glucuronide
In: Hamster C340

***cis*-7,8-Dihydro-7,8-dihydroxy-3,4-benzpyrene**
Formed from: 3,4-Benzpyrene

***trans*-7,8-Dihydro-7,8-dihydroxy-3,4-benzpyrene**
Formed from: 3,4-Benzpyrene-7,8-oxide
→ *trans*-7,8-Dihydro-7,8-dihydroxy-3,4-benzpyrene-β-D-glucuronide
In: Hamster A2009, A3625
Man H92
Rat A1615, A3625
See also: K667
→ *trans*-7,8-Dihydro-7,8-dihydroxy-3,4-benzpyrene sulphate
In: Man B387
Rat A2014
See also: A3732
→ 7,8-Dihydro-6,7,8-trihydroxy-3,4-benzpyrene
Probably in: Rat A3093
→ 7,8-Dihydroxy-3,4-benzpyrene
In: Rat E583, G630
→ 7,8,9,10-Tetrahydro-7α,8β-dihydroxy-3,4-benzpyrene-9β,10β-oxide
In: Rat A3525
→ 7,8,9,10-Tetrahydro-7β,8α-dihydroxy-3,4-benzpyrene-9α,10α-oxide
In: Rat A2812
See also: A3093, A3164, A3598, A3605, A3649, B749, E496
→ 7,8,9,10-Tetrahydro-7β,8α-dihydroxy-3,4-benzpyrene-9β,10β-oxide
In: Rat A2812, A3649
See also: A3093, A3164, A3598, A3605, B122, B452, B671, B749, C265, D243, E496, H784

→ 7,8,9,10-Tetrahydro-7,8,9,10-tetrahydroxy-3,4-benzpyrene (stereochemistry unspecified)
In: Hamster A2009
Rat A1661, G101
See also: A1352
→ 7,8,9,10-Tetrahydro-7α,8β,9α,10α-tetrahydroxy-3,4-benzpyrene
In: English sole D305
→ 7,8,9,10-Tetrahydro-7α,8β,9α,10β-tetrahydroxy-3,4-benzpyrene
In: English sole D305
→ 7,8,9,10-Tetrahydro-7α,8β,9β,10α-tetrahydroxy-3,4-benzpyrene
In: English sole D305
→ S-(7,8,9,10-Tetrahydro-7,8,9-trihydroxy-3,4-benzpyren-10-yl)glutathione
Probably in: Rat A1352

***cis*-9,10-Dihydro-9,10-dihydroxy-3,4-benzpyrene**
Formed from: 3,4-Benzpyrene

***trans*-9,10-Dihydro-9,10-dihydroxy-3,4-benzpyrene**
Formed from: 3,4-Benzpyrene-9,10-oxide
→ *trans*-9,10-Dihydro-9,10-dihydroxy-3,4-benzpyrene-β-D-glucuronide
Probably in: Glycine E416
→ *trans*-9,10-Dihydro-9,10-dihydroxy-3,4-benzpyrene-7β,8β-oxide
In: Cunninghamella B570
→ *trans*-9,10-Dihydro-9,10-dihydroxy-3,4-benzpyrene sulphate
In: Man B387
Rat A2014
See also: A3732
→ 9,10-Dihydro-1,9,10-trihydroxy-3,4-benzpyrene
Probably in: Rat A2866
→ 9,10-Dihydroxy-3,4-benzpyrene
In: Hamster A3625
Rat A3625
See also: A2589
→ 7,8,9,10-Tetrahydro-7,8,9,10-tetrahydroxy-3,4-benzpyrene
In: Cunninghamella B570
Hamster A2009
See also: A1352, B387, G147
→ S-(7,8,9,10-Tetrahydro-8,9,10-trihydroxy-3,4-benzpyren-7-yl)glutathione
Probably in: Rat A1352

9,10-Dihydro-9,10-dihydroxybenz[*e*]pyrene
Formed from: Benz[*e*]pyrene
→ 9,10-Dihydro-9,10-dihydroxybenz[*e*]pyrene-11,12-oxide
Probably in: Guinea pig B845
Hamster B845
Man B845
Mouse B845
Rabbit B845
Rat B845
→ 4,5,9,10-Tetrahydro-4,5,9,10-tetrahydroxybenz[*e*]pyrene
Probably in: Guinea pig B845
Hamster B845
Man B845
Mouse B845
Rabbit B845
Rat B845

***cis*-10,11-Dihydro-10,11-dihydroxy-3,4-benzpyrene**
Formed from: 3,4-Benzpyrene

***trans*-11,12-Dihydro-11,12-dihydroxy-3,4-benzpyrene**
Formed from: 3,4-Benzpyrene-11,12-oxide

***cis*-4,5-Dihydro-4,5-dihydroxy-3,4-benzpyrene-β-D-glucuronide**
Formed from: *cis*-4,5-Dihydro-4,5-dihydroxy-3,4-benzpyrene

***trans*-4,5-Dihydro-4,5-dihydroxy-3,4-benzpyrene-β-D-glucuronide**
Formed from: *trans*-4,5-Dihydro-4,5-dihydroxy-3,4-benzpyrene

4,5-Dihydro-4,5-dihydroxybenz[*e*]pyrene-β-D-glucuronide
Formed from: 4,5-Dihydro-4,5-dihydroxybenz[*e*]pyrene

***trans*-7,8-Dihydro-7,8-dihydroxy-3,4-benzpyrene-β-D-glucuronide**
Formed from: *trans*-7,8-Dihydro-7,8-dihydroxy-3,4-benzpyrene

***trans*-9,10-Dihydro-9,10-dihydroxy-3,4-benzpyrene-β-D-glucuronide**
Formed from: *trans*-9,10-Dihydro-9,10-dihydroxy-3,4-benzpyrene

7,8-Dihydro-7β,8α-dihydroxy-3,4-benzpyrene-9α,10α-oxide
→ *trans*-7,8-Dihydro-1,7,8-dihydroxy-3,4-benzpyrene-9,10-oxide
In: Rat F83
See also: E254
→ *trans*-7,8-Dihydro-3,7,8-dihydroxy-3,4-benzpyrene-9,10-oxide
In: Rat F83
See also: E254
→ 7,8,9,10-Tetrahydro-1,7,8,9,10-pentahydroxy-3,4-benzpyrene
Probably in: Rat E254
→ 7,8,9,10-Tetrahydro-3,7,8,9,10-pentahydroxy-3,4-benzpyrene
Probably in: Rat D945, E254

***trans*-9,10-Dihydro-9,10-dihydroxy-3,4-benzpyrene-7β,8β-oxide**
Formed from: *trans*-9,10-Dihydro-9,10-dihydroxy-3,4-benzpyrene

9,10-Dihydro-9,10-dihydroxybenz[*e*]pyrene-11,12-oxide
Probably formed from: 9,10-Dihydro-9,10-dihydroxybenz[*e*]pyrene

***trans*-4,5-Dihydro-4,5-dihydroxy-3,4-benzpyrene sulphate**
Formed from: *trans*-4,5-Dihydro-4,5-dihydroxy-3,4-benzpyrene

***trans*-7,8-Dihydro-7,8-dihydroxy-3,4-benzpyrene sulphate**
Formed from: *trans*-7,8-Dihydro-7,8-dihydroxy-3,4-benzpyrene

9,10-Dihydro-9,10-dihydroxy-3,4-benzpyrene sulphate
Formed from: 9,10-Dihydro-9,10-dihydroxy-3,4-benzpyrene

***cis*-2,3-Dihydro-2,3-dihydroxybenzyl acetate**
Formed from: Benzyl acetate
→ 2,3-Dihydroxybenzyl acetate
Probably in: Pseudomonas J656

***cis*-2,3-Dihydro-2,3-dihydroxybiphenyl**
Formed from: Biphenyl
→ 2,3-Dihydroxybiphenyl
In: Burkholderia J860
Sphingomonas H880
See also: G823

***cis*-3,4-Dihydro-3,4-dihydroxybiphenyl**
Formed from: Biphenyl
→ 3,4-Dihydroxybiphenyl
In: Burkholderia J860
Comamonas K196

3′,4′-Dihydro-3′,4′-dihydroxycannabichromene
Probably formed from: Cannabichromene-3′,4′-oxide

8,9-Dihydro-8,9-dihydroxycannabidiol
Formed from: Cannabidiol
→ 8,9-Dihydro-1″,8,9-trihydroxycannabidiol
Probably in: Rabbit F708
→ 8,9-Dihydro-2″,8,9-trihydroxycannabidiol
Probably in: Rabbit F708
→ 8,9-Dihydro-3″,8,9-trihydroxycannabidiol
Probably in: Rabbit F708
→ 8,9-Dihydro-4″,8,9-trihydroxycannabidiol
Probably in: Rabbit F708
→ 8,9-Dihydro-5″,8,9-trihydroxycannabidiol
Probably in: Rabbit F708
→ 8,9-Dihydro-6,8,9-trihydroxycannabidiol
Probably in: Rabbit F708
→ 8,9-Dihydro-7,8,9-trihydroxycannabidiol
Probably in: Rabbit F708

8,9-Dihydro-8,9-dihydroxycannabidiol-7-oic acid
Probably formed from: Cannabidiol-7-oic acid

2′,3′-Dihydro-2′,3′-dihydroxycannabigerol
Probably formed from: Cannabigerol-2′,3′-oxide

6′,7′-Dihydro-6′,7′-dihydroxycannabigerol
Probably formed from: Cannabigerol-6′,7′-oxide

2,3-Dihydro-2,3-dihydroxycarbamazepine
Formed from: Carbamazepine

10,11-Dihydro-10,11-dihydroxycarbamazepine
(structures unstated)
Formed from: Carbamazepine
→ Dihydrodihydroxycarbamazepine-N-β-D-glucuronide (3 isomers)
Probably in: Man A1967
→ Dihydrodihydroxycarbamazepine-O-β-D-glucuronide (4 isomers)
In: Man A1967

***cis*-10,11-Dihydro-10,11-dihydroxycarbamazepine**
Formed from: Carbamazepine-10,11-oxide

***trans*-10,11-Dihydro-10,11-dihydroxycarbamazepine**
Formed from: Carbamazepine-10,11-oxide

Dihydrodihydroxycarbamazepine-N-β-D-glucuronide
Probably formed from: 10,11-dihydroxyiminodibenzyl-5-carboxamide

Dihydrodihydroxycarbamazepine-O-β-D-glucuronide
Probably formed from: 10,11-dihydroxyiminodibenzyl-5-carboxamide

Dihydro-2′,4-dihydroxychalcone
Probably formed from: Dihydro-2′-hydroxychalcone
→ Dihydro-2′,3,4-trihydroxychalcone
Probably in: Aspergillus F673

Dihydro-2′,5-dihydroxychalcone
Probably formed from: Dihydro-2′-hydroxychalcone

***cis*-2,3-Dihydro-2,3-dihydroxychloridazon**
→ 2,3-Dihydroxychloridazon
In: Bacterium B871

***trans*-1,2-Dihydro-1,2-dihydroxychrysene**
Probably formed from: Chrysene-1,2-oxide
→ 1,2-Dihydro-1,2-dihydroxychrysene-3,4-oxide
In: Rat B772
→ *trans*-1,2-Dihydro-1,2,9-trihydroxychrysene
Probably in: Rat D588

(+)-(3*S*,4*R*)-*cis*-3,4-Dihydro-3,4-dihydroxychrysene
Formed from: Chrysene
→ (+)-(3*S*,4*R*,9*S*,10*R*)-*cis.cis*-3,4,9,10-Tetrahydro-3,4,9,10-tetrahydroxychrysene
In: Sphingomonas K97

***trans*-3,4-Dihydro-3,4-dihydroxychrysene**
Probably formed from: Chrysene-3,4-oxide

***trans*-5,6-Dihydro-5,6-dihydroxychrysene**
Probably formed from: Chrysene
Chrysene-5,6-oxide

1,2-Dihydro-1,2-dihydroxychrysene-3,4-oxide
Formed from: 1,2-Dihydro-1,2-dihydroxychrysene

3′,4′-Dihydro-3′,4′-dihydroxyclemastine
Formed from: Clemastine

3′,4′-Dihydro-3′,4′-dihydroxyclobazam
→ 3′,4′-Dihydro-3′,4′-dihydroxydemethylclobazam
Probably in: Dog B22
Man B22
Monkey B22
Rat B22

***trans*-1,2-Dihydro-1,2-dihydroxycrisnatol**
Formed from: Crisnatol

1,2-Dihydro-1,2-dihydroxycyclobenzaprine
Formed from: Cyclobenzaprine
→ 1,2-Dihydro-1,2-dihydroxycyclobenzaprine-β-D-glucuronide
Probably in: Dog A2887

1,2-Dihydro-1,2-dihydroxycyclobenzaprine-β-D-glucuronide
Probably formed from: 1,2-Dihydro-1,2-dihydroxycyclobenzaprine

***cis*-10,11-Dihydro-10,11-dihydroxycyproheptadine**
Formed from: Cyproheptadine
→ *cis*-10,11-Dihydro-10,11-dihydroxydemethylcyproheptadine
Probably in: Cat A1858

***trans*-10,11-Dihydro-10,11-dihydroxycyproheptadine**
Probably formed from: Cyproheptadine-10,11-oxide

10,11-Dihydro-10,11-dihydroxycytenamide
Probably formed from: Cytenamide-10,11-oxide

3′,4′-Dihydro-3′,4′-dihydroxydemethylclobazam
Probably formed from: Demethylclobazam
3′,4′-Dihydro-3′,4′-dihydroxyclobazam

***cis*-10,11-Dihydro-10,11-dihydroxydemethylcyclobenzaprine**
Probably formed from: *cis*-10,11-Dihydro-10,11-dihydroxycyclobenzaprine

10,11-Dihydro-10,11-dihydroxy-N-demethylcyclobenzaprine (assumed to be *trans*)
Probably formed from: Cyclobenzaprine-10,11-oxide

3′,4′-Dihydro-3′,4′-dihydroxydemethyltriflubazam
Probably formed from: Demethyltriflubazam

3′,4′-Dihydro-3′,4′-dihydroxydesethylbemitradine
Probably formed from: Desethylbemitradine

→ 3′,4′-Dihydro-3′,4′-dihydroxydesethylbemitradine-β-D-glucuronide
Probably in: Man F314

3′,4′-Dihydro-3′,4′-dihydroxydesethylbemitradine-β-D-glucuronide
Probably formed from: 3′,4′-Dihydro-3′,4′-dihydroxydesethylbemitradine

1,2-Dihydro-1,2-dihydroxydibenz[*a*,*j*]acridine
Formed from: Dibenz[*a*,*j*]acridine

***trans*-3,4-Dihydro-3,4-dihydroxydibenz[*a*,*j*]acridine**
Formed from: Dibenz[*a*,*j*]acridine

***trans*-5,6-Dihydro-5,6-dihydroxydibenz[*a*,*j*]acridine**
Formed from: Dibenz[*a*,*j*]acridine

***trans*-3,4-Dihydro-3,4-dihydroxydibenz[*c*,*h*]acridine**
→ 1,2,3,4-Tetrahydro-1,2,3,4-tetrahydroxydibenz[*c*,*h*]acridine
In: Rat K267

1,2-Dihydro-1,2-dihydroxydibenz[*a*,*c*]anthracene
Formed from: Dibenz[*a*,*c*]anthracene

3,4-Dihydro-3,4-dihydroxydibenz[*a*,*c*]anthracene
Formed from: Dibenz[*a*,*c*]anthracene

12,13-Dihydro-12,13-dihydroxydibenz[*a*,*c*]anthracene
Formed from: Dibenz[*a*,*c*]anthracene
→ 1,2,3,4,12,13-Hexahydro-1,2,3,4,12,13-hexahydroxydibenz[*a*,*c*]anthracene
Probably in: Rat G442

1,2-Dihydro-1,2-dihydroxydibenzanthracene
Formed from: Dibenzanthracene

***trans*-3,4-Dihydro-3,4-dihydroxydibenzanthracene**
Formed from: Dibenzanthracene
→ 3,4-Dihydroxydibenzanthracene
In: Rat G440
See also: E556
→ *trans*-1,2,3,4-Tetrahydro-1,2,3,4-tetrahydroxydibenzanthracene
Probably in: Rat E556, G442
→ *trans*-3,4,10,11-Tetrahydro-3,4,10,11-tetrahydroxydibenzanthracene
Probably in: Rat G442
→ *trans*-3,4,12,13-Tetrahydro-3,4,12,13-tetrahydroxydibenzanthracene
Probably in: Rat E556

***trans*-5,6-Dihydro-5,6-dihydroxydibenzanthracene**
Formed from: Dibenzanthracene
Dibenzanthracene-5,6-oxide

***trans*-3,4-Dihydro-3,4-dihydroxydibenz[*a*,*j*]anthracene**
Formed from: Dibenz[*a*,*j*]anthracene
→ *trans*-3,4-Dihydro-3,4-dihydroxydibenz[*a*,*j*]anthracene-β-D-glucuronide
In: Mouse G611

***trans*-5,6-Dihydro-5,6-dihydroxydibenz[*a*,*j*]anthracene**
Formed from: Dibenz[*a*,*j*]anthracene

***trans*-3,4-Dihydro-3,4-dihydroxydibenz[*a*,*j*]anthracene-β-D-glucuronide**
Formed from: *trans*-3,4-Dihydro-3,4-dihydroxydibenz[*a*,*j*]anthracene

***cis*-1,2-Dihydro-1,2-dihydroxydibenzo-*p*-dioxin**
Formed from: Dibenzo-*p*-dioxin
→ 1,2-Dihydroxydibenzo-*p*-dioxin
Probably in: Beijerinckia B358

***trans*-3,4-Dihydro-3,4-dihydroxydibenzo[*a*,*e*]fluoranthrene**
Probably formed from: Dibenzo[*a*,*e*]fluoranthrene

***trans*-12,13-Dihydro-12,13-dihydroxydibenzo[*a*,*e*]fluoranthrene**
Probably formed from: Dibenzo[*a*,*e*]fluoranthrene

***cis*-1,2-Dihydro-1,2-dihydroxydibenzofuran**
Formed from: Dibenzofuran

***cis*-2,3-Dihydro-2,3-dihydroxydibenzofuran**
Formed from: Dibenzofuran

***trans*-2,3-Dihydro-2,3-dihydroxydibenzofuran**
Formed from: Dibenzofuran
→ 2,3-Dihydroxydibenzofuran
In: Cunninghamella A3426

***cis*-3,4-Dihydro-3,4-dihydroxydibenzofuran**
Formed from: Dibenzofuran

10,11-Dihydro-10,11-dihydroxydibenzo[*b*,*f*]oxepin
Probably formed from: Dibenzo[*b*,*f*]oxepin-10,11-oxide

1,2-Dihydro-1,2-dihydroxydibenzo[*a*,*i*]pyrene
Formed from: Dibenzo[*a*,*i*]pyrene

3,4-Dihydro-3,4-dihydroxydibenzo[*a*,*i*]pyrene
Formed from: Dibenzo[*a*,*i*]pyrene

***cis*-1,2-Dihydro-1,2-dihydroxydibenzothiophene**
Formed from: Dibenzothiophene
→ 1,2-Dihydroxydibenzothiophene
Probably in: Beijerinckia A3340

***cis*-3,4-Dihydro-3,4-dihydroxydibenzothiophene**
Formed from: Dibenzothiophene

8,9-Dihydro-8,9-dihydroxy-7,12-di-(hydroxymethyl)benzanthracene
Probably formed from: 7,12-Di-(hydroxymethyl)benzanthracene

2,3-Dihydro-2,3-dihydroxy-2′,3′-dimethoxybiphenyl
Formed from: 2,3-Dimethoxybiphenyl

***cis*-3,4-Dihydro-3,4-dihydroxy-6,7-dimethoxy-2,2-dimethyl-2*H*-benzo[*b*]pyran**
Formed from: Precocene II

***trans*-3,4-Dihydro-3,4-dihydroxy-6,7-dimethoxy-2,2-dimethyl-2*H*-benzo[*b*]pyran**
Formed from: Precocene II

1,2-Dihydro-1,2-dihydroxy-9,10-dimethylanthracene
Formed from: 9,10-Dimethylanthracene

1,2-Dihydro-1,2-dihydroxy-7,12-dimethylbenzanthracene
Formed from: 7,12-Dimethylbenzanthracene

***trans*-3,4-Dihydro-3,4-dihydroxy-7,12-dimethylbenzanthracene**
Formed from: 7,12-Dimethylbenzanthracene
→ 3,4-Dihydro-3,4-dihydroxy-7-hydroxymethyl-12-methylbenzanthracene
Probably in: Cunninghamella C919
→ 3,4-Dihydro-3,4-dihydroxy-12-hydroxymethyl-7-methylbenzanthracene
Probably in: Cunninghamella C919
→ 7,12-Dimethylbenzanthracene-3,4-quinone
In: Rat E740

***trans*-5,6-Dihydro-5,6-dihydroxy-7,12-dimethylbenzanthracene**
Formed from: 7,12-Dimethylbenzanthracene-5,6-oxide

7,12-Dihydro-7,12-dihydroxy-7,12-dimethylbenzanthracene
Formed from: 7,12-Dimethylbenzanthracene-7,12-peroxide

8,9-Dihydro-8,9-dihydroxy-7,12-dimethylbenzanthracene (*cis* and *trans* isomers)
Formed from: 7,12-Dimethylbenzanthracene
→ 8,9-Dihydro-8,9-dihydroxy-7-hydroxymethyl-12-methylbenzanthracene
Probably in: Cunninghamella C919
Rat A2324
→ 8,9-Dihydro-8,9-dihydroxy-12-hydroxymethyl-7-methylbenzanthracene
Probably in: Cunninghamella C919

10,11-Dihydro-10,11-dihydroxy-7,12-dimethylbenzanthracene
Formed from: 7,12-Dimethylbenzanthracene

2,3-Dihydro-3,7-dihydroxy-2,2-dimethylbenzofuran
Probably formed from: 2,3-Dihydro-7-hydroxy-2,2-dimethylbenzofuran
→ 2,3-Dihydro-7-hydroxy-2,2-dimethyl-3-oxobenzofuran
Probably in: Bean D806
Cotton D806
Zea D806

***cis*-3,4-Dihydro-3,4-dihydroxy-6,7-dimethoxy-2,2-dimethyl-2*H*-benzo[*b*]pyran**
Formed from: Precocene II

***trans*-3,4-Dihydro-3,4-dihydroxy-6,7-dimethoxy-2,2-dimethyl-2*H*-benzo[*b*]pyran**
Formed from: Precocene II

3-(2,3-Dihydro-3,5-dihydroxy-2,2-dimethyl-1-benzopyran-6-yl)-5,7-dihydroxy-4*H*-1-benzopyran-4-one
Formed from: Piscerythrone

3-(2,3-Dihydro-3,5-dihydroxy-2,2-dimethyl-1-benzopyran-8-yl)-5,7-dihydroxy-4*H*-1-benzopyran-4-one
Formed from: Piscerythrone

***trans*-3,4-Dihydro-3,4-dihydroxy-2,6-dimethylnaphthalene**
Formed from: 2,6-Dimethylnaphthalene
→ 2,6-Dimethyl-3,4-naphthoquinone
In: Starry flounder B852

9,10-Dihydro-4,5-dihydroxy-9,10-dioxo-2-anthracenecarboxylic acid
See: Rhein

2′,3′-Dihydro-2′,3′-dihydroxyedunol
Formed from: Edunol

α-Dihydro-8′,9′-dihydroxyergocryptine
Probably formed from: α-Dihydro-9′-hydroxyergocryptine

2,3-Dihydro-2,3-dihydroxyestragole
Formed from: Estragole-2,3-oxide

***cis*-2,3-Dihydro-2,3-dihydroxyethylbenzene**
Formed from: Ethylbenzene
→ 3-Ethylcatechol
In: Pseudomonas A914, A1298

2,3-Dihydro-2,3-dihydroxyeugenol
Formed from: Eugenol oxide
→ Vanillactic acid
Probably in: Man F499

***trans*-2,3-Dihydro-2,3-dihydroxyfluoranthene**
Formed from: Fluoranthene
→ Fluoranthene-2,3-quinone
Probably in: Rat E255
→ *trans*-2,3-Dihydro-2,3-dihydroxyfluoranthene-1,10b-oxide
Probably in: Man G672

***trans*-2,3-Dihydro-2,3-dihydroxyfluoranthene-1,10b-oxide**
Probably formed from: *trans*-2,3-Dihydro-2,3-dihydroxyfluoranthene

1,10-Dihydro-1,10-dihydroxyfluoren-9-one
Probably formed from: Fluorenone

***cis*-1,2-Dihydro-1,2-dihydroxy-3-fluorobenzene**
Formed from: Fluorobenzene

6,7-Dihydro-6,7-dihydroxygranisetron
Formed from: Granisetron

3,4-Dihydro-3,4-dihydroxyguanfacine
Formed from: Guanfacine
→ 3,4-Dihydro-3,4-dihydroxyguanfacine-β-D-glucuronide
In: Rat B889

3,4-Dihydro-3,4-dihydroxyguanfacine-β-D-glucuronide
Formed from: 3,4-Dihydro-3,4-dihydroxyguanfacine

3,4-Dihydro-3,4-dihydroxy-7-(α-hydroxyethyl)benz[*a*]anthracene
Probably formed from: 7-(α-Hydroxyethyl)benz[*a*]anthracene

8,9-Dihydro-8,9-dihydroxy-7-(α-hydroxyethyl)benz[*a*]anthracene
Probably formed from: 7-(α-Hydroxyethyl)benz[*a*]anthracene

10,11-Dihydro-10,11-dihydroxy-7-(α-hydroxyethyl)benz[*a*]anthracene
Probably formed from: 7-(α-Hydroxyethyl)benz[*a*]anthracene

3,4-Dihydro-3,4-dihydroxy-8-(2-hydroxy-3-isopropylamino)propoxy-2*H*-1-benzopyran
Probably formed from: 3,4-Dihydro-3-hydroxy-8-(2-hydroxy-3-isopropylamino)propoxy-2*H*-1-benzopyran

3,4-Dihydro-3,5-dihydroxy-8-(2-hydroxy-3-isopropylamino)propoxy-2*H*-1-benzopyran
Probably formed from: 3,4-Dihydro-3-hydroxy-8-(2-hydroxy-3-isopropylamino)propoxy-2*H*-1-benzopyran

3,4-Dihydro-3,4-dihydroxy-9-hydroxymethylanthracene
Formed from: 9-Hydroxymethylanthracene

***trans*-1,2-Dihydro-1,2-dihydroxy-8-hydroxymethylbenzanthracene**
Formed from: 8-Hydroxymethylbenzanthracene

3,4-Dihydro-3,4-dihydroxy-7-hydroxymethylbenzanthracene
Probably formed from: 7-Hydroxymethylbenzanthracene

***trans*-3,4-Dihydro-3,4-dihydroxy-8-hydroxymethylbenzanthracene**
Formed from: 8-Hydroxymethylbenzanthracene

***trans*-5,6-Dihydro-5,6-dihydroxy-8-hydroxymethylbenzanthracene**
Formed from: 8-Hydroxymethylbenzanthracene-5,6-oxide

8,9-Dihydro-8,9-dihydroxy-7-hydroxymethylbenzanthracene
Formed from: 7-Hydroxymethylbenzanthracene

***trans*-8,9-Dihydro-8,9-dihydroxy-8-hydroxymethylbenzanthracene**
Formed from: 8-Hydroxymethylbenzanthracene

9,10-Dihydro-9,10-dihydroxy-7-hydroxymethylbenzanthracene
Formed from: 7-Hydroxymethylbenzanthracene

***trans*-10,11-Dihydro-10,11-dihydroxy-8-hydroxymethylbenzanthracene**
Formed from: 8-Hydroxymethylbenzanthracene

***trans*-7,8-Dihydro-7,8-dihydroxy-3-hydroxymethylcholanthrene**
Formed from: 3-Hydroxymethylcholanthrene

***trans*-9,10-Dihydro-9,10-dihydroxy-3-hydroxymethylcholanthrene**
Formed from: 3-Hydroxymethylcholanthrene

***trans*-11,12-Dihydro-11,12-dihydroxy-3-hydroxymethylcholanthrene**
Formed from: 3-Hydroxymethylcholanthrene

1,2-Dihydro-1,2-dihydroxy-5-hydroxymethylchrysene
Formed from: 5-Hydroxymethylchrysene

7,8-Dihydro-7,8-dihydroxy-5-hydroxymethylchrysene
Formed from: 5-Hydroxymethylchrysene

9,10-Dihydro-9,10-dihydroxy-5-hydroxymethylchrysene
Probably formed from: 5-Hydroxymethylchrysene

3,4-Dihydro-3,4-dihydroxy-7-hydroxymethyl-12-methylbenzanthracene
Formed from: 7-Hydroxymethyl-12-methylbenzanthracene
Probably formed from: 3,4-Dihydro-3,4-dihydroxy-7,12-dimethylbenzanthracene

3,4-Dihydro-3,4-dihydroxy-12-hydroxymethyl-7-methylbenzanthracene
Probably formed from: 3,4-Dihydro-3,4-dihydroxy-7,12-dimethylbenzanthracene

5,6-Dihydro-5,6-dihydroxy-7-hydroxymethyl-12-methylbenzanthracene
Formed from: 7-Hydroxymethyl-12-methylbenzanthracene

5,6-Dihydro-5,6-dihydroxy-12-hydroxymethyl-7-methylbenzanthracene
Probably formed from: 12-Hydroxymethyl-7-methylbenzanthracene

8,9-Dihydro-8,9-dihydroxy-7-hydroxymethyl-12-methylbenzanthracene
Probably formed from: *trans*-8,9-Dihydro-8,9-dihydroxy-7,12-dimethylbenzanthracene

8,9-Dihydro-8,9-dihydroxy-12-hydroxymethyl-7-methylbenzanthracene
Probably formed from: *trans*-8,9-Dihydro-8,9-dihydroxy-7,12-dimethylbenzanthracene

9,10-Dihydro-4,5-dihydroxy-3-(1-hydroxy-3-oxo-1-pentenyl)-9,10-dioxo-2-anthraceneacetic acid
See: Aklanonic acid

6,7-Dihydro-5,7-dihydroxy-2-(*p*-hydroxyphenyl)-3-methoxy-8,8-dimethyl-4*H*,8*H*-[1,2-*b*;5,4-*b′*]benzodipyran-4-one
Formed from: Topazolin

10,11-Dihydro-10,11-dihydroxyiminostilbene
Probably formed from: Iminostilbene-10,11-oxide

***trans*-1,2-Dihydro-1,2-dihydroxyindene[1,2,3-*cd*]pyrene**
Formed from: Indene[1,2,3-*cd*]pyrene
→ *trans*-1,2-Dihydro-1,2,8-trihydroxyindene[1,2,3-*cd*]pyrene
Probably in: Rat D893
→ *trans*-1,2-Dihydro-1,2,9-trihydroxyindene[1,2,3-*cd*]pyrene
Probably in: Rat D893
→ Indene[1,2,3-*cd*]pyrene
Probably in: Rat D893

(2*S*)-2,3-Dihydro-5,6-dihydroxyindole-2-carboxylic acid (cyclodopa)
Formed from: 3,4-Dihydroxy-L-phenylalanine
→ (2*S*)-2,3-Dihydro-5,6-dihydroxyindole-2-carboxylic acid-5-β-D-glucoside
Probably in: Celosia A1325

(2*S*)-2,3-Dihydro-5,6-dihydroxyindole-2-carboxylic acid-5-β-D-glucoside
Probably formed from: (2*S*)-2,3-Dihydro-5,6-dihydroxyindole-2-carboxylic acid
→ Amaranthin
In: Celosia A1325

***cis*-1,2-Dihydro-1,2-dihydroxy-3-iodobenzene**
Formed from: Iodobenzene

2,3-Dihydro-2,3-dihydroxy-4-iodotoluene
→ 3-Iodo-6-methylcatechol
In: Pseudomonas K348
***cis*-1,2-Dihydro-1,2-dihydroxy-3-isopropyl-6-methylbenzene**
Formed from: *p*-Cymene
→ 3-Isopropyl-6-methylcatechol
In: Pseudomonas K699
***cis*-5,6-Dihydro-5,6-dihydroxyisoquinoline**
Formed from: Isoquinoline
***cis*-7,8-Dihydro-7,8-dihydroxyisoquinoline**
Formed from: Isoquinoline
1,2-Dihydro-1,2-dihydroxyisosafrole
Probably formed from: Isosafrole oxide
9,10-Dihydro-9,10-dihydroxykadsurenone
Formed from: Kadsurenone
Dihydro-3′,4′-dihydroxykawain
Probably formed from: 3′,4′-Dihydroxykawain
Dihydro-*p*-hydroxykawain
2″,3″-Dihydro-2″,3″-dihydroxyluteone
Formed from: Luteone
2,3-Dihydro-2,3-dihydroxymaprotiline
Formed from: Maprotiline
→ 3-Hydroxy-2-methoxymaprotiline
Probably in: Man D167
1,2-Dihydro-1,2-dihydroxy-5-methoxybenzoic acid
Probably formed from: *m*-Methoxybenzoic acid
2,3-Dihydro-2,3-dihydroxy-2′-methoxybiphenyl
Formed from: 2-Methoxybiphenyl
3-(2,3-Dihydro-3,8-dihydroxy-7-methoxy-2,2-dimethyl-1-benzopyran-8-yl)-5,7-dihydroxy-4*H*-1-benzopyran-4-one
Formed from: Piscidone
***cis*-2,3-Dihydro-2,3-dihydroxy-1-methoxy-4-methylbenzene**
Formed from: *p*-Methylanisole
→ 3-Methoxy-6-methylcatechol
In: Pseudomonas K699
1,2-Dihydro-1,2-dihydroxy-6-methoxynaphthalene
Formed from: 2-Methoxynaphthalene
***cis*-1,2-Dihydro-1,2-dihydroxy-7-methoxynaphthalene**
Formed from: 2-Methoxynaphthalene
→ 1-Hydroxy-7-methoxynaphthalene
In: Pseudomonas J398
→ 2-Hydroxy-7-methoxynaphthalene
In: Pseudomonas J398
(3*Z*)-2,3-Dihydro-5,7-dihydroxy-3-(4-methoxyphenyl)methylene)-4*H*-1-benzopyran-4-one
See: Eucomin
5,6-Dihydro-5,6-dihydroxy-2-methoxyquinoline
Formed from: 2-Methoxyquinoline
7,8-Dihydro-7,8-dihydroxy-2-methoxyquinoline
Formed from: 2-Methoxyquinoline
1,2-Dihydro-1,2-dihydroxy-9-methylanthracene
Formed from: 9-Methylanthracene
3,4-Dihydro-3,4-dihydroxy-9-methylanthracene
Formed from: 9-Methylanthracene
***trans*-1,2-Dihydro-1,2-dihydroxy-7-methylbenz[*c*]acridine**
Formed from: 7-Methylbenz[*c*]acridine
3,4-Dihydro-3,4-dihydroxy-7-methylbenz[*c*]acridine
Formed from: 7-Methylbenz[*c*]acridine
5,6-Dihydro-5,6-dihydroxy-7-methylbenz[*c*]acridine
Formed from: 7-Methylbenz[*c*]acridine
***trans*-8,9-Dihydro-8,9-dihydroxy-7-methylbenz[*c*]acridine**
Formed from: 7-Methylbenz[*c*]acridine
***trans*-10,11-Dihydro-10,11-dihydroxy-7-methylbenz[*c*]acridine**
Formed from: 7-Methylbenz[*c*]acridine
1,2-Dihydro-1,2-dihydroxy-7-methylbenzanthracene
Formed from: 7-Methylbenzanthracene
3,4-Dihydro-3,4-dihydroxy-7-methylbenzanthracene
Formed from: 7-Methylbenzanthracene
→ 7-Methylbenzanthracene-3,4-quinone
In: Rat E740
3,4-Dihydro-3,4-dihydroxy-11-methylbenz[*a*]anthracene
Formed from: 11-Methylbenz[*a*]anthracene
3,4-Dihydro-3,4-dihydroxy-12-methylbenzanthracene
→ 12-Methylbenzanthracene-3,4-quinone
In: Rat E740
5,6-Dihydro-5,6-dihydroxy-7-methylbenzanthracene
Formed from: 7-Methylbenzanthracene
7-Methylbenzanthracene-5,6-oxide

5,6-Dihydro-5,6-dihydroxy-11-methyl[*a*] benzanthracene
Formed from: 11-Methyl[*a*]benzanthracene

***trans*-5,6-Dihydro-5,6-dihydroxy-12-methylbenzanthracene**
Formed from: 12-Methylbenzanthracene

8,9-Dihydro-8,9-dihydroxy-7-methylbenzanthracene
Formed from: 7-Methylbenzanthracene
→ 8,9,10,11-Tetrahydro-8,9,10,11-tetrahydroxy-7-methylbenzanthracene
Probably in: Cunninghamella C255

8,9-Dihydro-8,9-dihydroxy-8-methylbenzanthracene
Formed from: 8-Methylbenzanthracene

8,9-Dihydro-8,9-dihydroxy-11-methylbenz[*a*] anthracene
Formed from: 11-Methylbenz[*a*]anthracene

10,11-Dihydro-10,11-dihydroxy-7-methylbenzanthracene
Formed from: 7-Methylbenzanthracene

10,11-Dihydro-10,11-dihydroxy-11-methylbenz[*a*]anthracene
Formed from: 11-Methylbenz[*a*]anthracene

1,2-Dihydro-1,2-dihydroxy-4-methylbenzoic acid
Formed from: *p*-Toluic acid

1,2-Dihydro-1,2-dihydroxy-5-methylbenzoic acid
Probably formed from: *m*-Toluic acid

1,2-Dihydro-1,2-dihydroxy-6-methylbenzoic acid
Formed from: *o*-Toluic acid

4,5-Dihydro-4,5-dihydroxy-2-methylbenzotriazole
Formed from: 2-Methylbenzotriazole

4,5-Dihydro-4,5-dihydroxy-7-methyl-3,4-benzpyrene
Formed from: 7-Methyl-3,4-benzpyrene

4,5-Dihydro-4,5-dihydroxy-7-methyl-3,4-benzpyrene
Formed from: 7-Methyl-3,4-benzpyrene

7,8-Dihydro-7,8-dihydroxy-7-methyl-3,4-benzpyrene
Formed from: 7-Methyl-3,4-benzpyrene

9,10-Dihydro-9,10-dihydroxy-7-methyl-3,4-benzpyrene
Formed from: 7-Methyl-3,4-benzpyrene

***trans*-4,5-Dihydro-4,5-dihydroxy-3-methylcholanthrene**
Formed from: 3-Methylcholanthrene

***trans*-9,10-Dihydro-9,10-dihydroxy-3-methylcholanthrene**
Formed from: 3-Methylcholanthrene

***trans*-11,12-Dihydro-11,12-dihydroxy-3-methylcholanthrene**
Formed from: 3-Methylcholanthrene
3-Methylcholanthrene-10,11-oxide

1,2-Dihydro-1,2-dihydroxy-5-methylchrysene
Formed from: 5-Methylchrysene

7,8-Dihydro-7,8-dihydroxy-5-methylchrysene
Formed from: 5-Methylchrysene

9,10-Dihydro-9,10-dihydroxy-5-methylchrysene
Formed from: 5-Methylchrysene

3,4-Dihydro-3,4-dihydroxy-7-methyldibenz[*a*,*j*] anthracene
Formed from: 7-Methyldibenz[*a*,*j*] anthracene

5,6-Dihydro-5,6-dihydroxy-7-methyldibenz[*a*,*j*] anthracene
Formed from: 7-Methyldibenz[*a*,*j*] anthracene

***cis*-1,2-Dihydro-1,2-dihydroxy-3-methylnaphthalene**
Formed from: 2-Methylnaphthalene

***trans*-1,2-Dihydro-1,2-dihydroxy-3-methylnaphthalene**
Formed from: 2-Methylnaphthalene

***trans*-1,2-Dihydro-1,2-dihydroxy-6-methylnaphthalene**
Formed from: 2-Methylnaphthalene

***cis*-1,2-Dihydro-1,2-dihydroxy-7-methylnaphthalene**
Formed from: 2-Methylnaphthalene

***trans*-1,2-Dihydro-1,2-dihydroxy-7-methylnaphthalene**
Formed from: 2-Methylnaphthalene

1,2-Dihydro-1,2-dihydroxy-8-methylnaphthalene
Formed from: 1-Methylnaphthalene

***cis*-2,3-Dihydro-2,3-dihydroxy-α-methylstyrene**
Formed from: α-Methylstyrene

5,6-Dihydro-5,6-dihydroxy-2-(4′-methylsulphonylphenyl)imidazo[1,2-*a*]pyridine
Formed from: Zolimidine

→ 5,6-Dihydro-5,6-dihydroxy-2-(4′-methylsulphonylphenyl)imidazo[1,2-*a*]pyridine-di-β-D-glucuronide
Probably in: Man A1557

7,8-Dihydro-7,8-dihydroxy-2-(4′-methylsulphonylphenyl)imidazo[1,2-*a*]pyridine
Formed from: Zolimidine

5,6-Dihydro-5,6-dihydroxy-2-(4′-methylsulphonylphenyl)imidazo[1,2-*a*]pyridine-di–β-D-glucuronide
Probably formed from: 5,6-Dihydro-5,6-dihydroxy-2-(4′-methylsulphonylphenyl)imidazo[1,2-*a*]pyridine

2,3-Dihydro-3,4-dihydroxy-N(1)-methyltabersonine
Formed from: 2,3-Dihydro-3-hydroxy-N(1)-methyltabersonine

(+)-*cis*-(1*R*,2*S*)-1,2-Dihydro-1,2-dihydroxynaphthalene
Formed from: 1,2-Dihydroxynaphthalene
Naphthalene
→ 1,2-Dihydroxynaphthalene
In: Comamonas K196
Pseudomonas A1298, A1591, J370, J398
Rhodococcus H885
Sphingomonas H880

***trans*-1,2-Dihydro-1,2-dihydroxynaphthalene**
Formed from: Naphthalene-1,2-oxide
→ 1,2-Dihydro-1β,2α-dihydroxynaphthalene-3α,4α-oxide
Probably in: Rat B602
→ 1,2-Dihydro-1-hydroxynaphth-2-yl-β-D-glucuronide
Probably in: Beef F586
→ 1,2-Dihydroxynaphthalene
In: Beef G750
Hamster F153
Monkey F241, F388, F391
Mouse F377
Pig F389
Rat E583
See also: F586
→ Naphthalene-1,2-dione
In: Rat J552

***cis*-1,4-Dihydro-1,4-dihydroxynaphthalene**
Formed from: Naphthalene

3,4-Dihydro-3,4-dihydroxynaphthalene
Formed from: (+)-1,2,3,4-Tetrahydro-*cis*-(1*R*,2*S*)-dihydroxynaphthalene
(−)-1,2,3,4-Tetrahydro-*cis*-(1*R*,2*S*)-dihydroxynaphthalene

1,2-Dihydro-1β,2α-dihydroxynaphthalene-3α,4α-oxide
Probably formed from: *trans*-1,2-Dihydro-1,2-dihydroxynaphthalene
→ 1,2,3,4-Tetrahydro-1β,2α,3α,4β-tetrahydroxynaphthalene
Probably in: Rat B602
→ 1,2,3,4-Tetrahydro-1β,2α,3β,4α-tetrahydroxynaphthalene
Probably in: Rat B602

***cis*-3,4-Dihydro-3,4-dihydroxynaphthalene-2-sulphonic acid**
Formed from: Naphthalene-2-sulphonic acid

5,6-Dihydro-5,6-dihydroxy-α-naphthoflavone
Probably formed from: α-Naphthoflavone-5,6-oxide

7,8-Dihydro-7,8-dihydroxy-α-naphthoflavone
Formed from: α-Naphthoflavone

***cis*-1,2-Dihydro-1,2-dihydroxy-2-naphthoic acid**
Formed from: 2-Naphthoic acid

***cis*-1,2-Dihydro-1,2-dihydroxy-2-naphthylacetic acid**
Formed from: 2-Naphthylacetic acid

S-(1,2-Dihydro-2-hydroxy-1-naphthyl)glutathione
Formed from: Naphthalene
→ N-Acetyl-S-(1,2-dihydro-2-hydroxy-1-naphthyl)-L-cysteine
In: Rat C262

5,6-Dihydro-5,6-dihydroxy-1-naphthyl N-methylcarbamate
Formed from: 1-Naphthyl N-methylcarbamate
→ 5,6-Dihydro-5,6-dihydroxy-1-naphthyl N-methylcarbamate-β-D-glucuronide
Probably in: Rat A602, C482

5,6-Dihydro-5,6-dihydroxy-1-naphthyl N-methylcarbamate-β-D-glucuronide
Probably formed from: 5,6-Dihydro-5,6-dihydroxy-1-naphthyl N-methylcarbamate

N-(3,4-Dihydro-3,4-dihydroxy-1-naphthylmethyl)-N-methyl-3-phenyl-2-propen-1-ylamine
Formed from: Naftifine
→ N-(3,4-Dihydro-3,4-dihydroxy-1-naphthylmethyl)-3-phenyl-2-propen-1-ylamine
Probably in: Man E32

N-(7,8-Dihydro-7,8-dihydroxy-1-naphthylmethyl)-N-methyl-3-phenyl-2-propen-1-ylamine
Formed from: Naftifine
→ N-(7,8-Dihydro-7,8-dihydroxy-1-naphthylmethyl)-3-phenyl-2-propen-1-ylamine
Probably in: Man E32

N-(3,4-Dihydro-3,4-dihydroxy-1-naphthylmethyl)-3-phenyl-2-propen-1-ylamine
Probably formed from: N-(3,4-Dihydro-3,4-dihydroxy-1-naphthylmethyl)-N-methyl-3-phenyl-2-propen-1-ylamine

N-(7,8-Dihydro-7,8-dihydroxy-1-naphthylmethyl)-3-phenyl-2-propen-1-ylamine
Probably formed from: N-(7,8-Dihydro-7,8-dihydroxy-1-naphthylmethyl)-N-methyl-3-phenyl-2-propen-1-ylamine

3,4-Dihydro-3,4-dihydroxy-7-nitrobenzanthracene
Formed from: 7-Nitrobenzanthracene

8,9-Dihydro-8,9-dihydroxy-7-nitrobenzanthracene
Formed from: 7-Nitrobenzanthracene

1,2-Dihydro-1,2-dihydroxy-3-nitrobenzene
Formed from: Nitrobenzene
→ Catechol
In: Pseudomonas H757

***trans*-7,8-Dihydro-7,8-dihydroxy-1-nitro-3,4-benzpyrene**
Formed from: 1-Nitro-3,4-benzpyrene
→ 7,8,9,10-Tetrahydro-7,8,9,10-tetrahydroxy-1-nitro-3,4-benzpyrene
In: Rat E414
See also: C577

7,8-Dihydro-7,8-dihydroxy-6-nitro-3,4-benzpyrene
Formed from: 6-Nitro-3,4-benzpyrene

***trans*-9,10-Dihydro-9,10-dihydroxy-1-nitro-3,4-benzpyrene**
Formed from: 1-Nitro-3,4-benzpyrene
→ 7,8,9,10-Tetrahydro-7,8,9,10-tetrahydroxy-1-nitro-3,4-benzpyrene
In: Rat E415

9,10-Dihydro-9,10-dihydroxy-6-nitro-3,4-benzpyrene
Formed from: 6-Nitro-3,4-benzpyrene

2,3-Dihydro-2,3-dihydroxy-4′-nitrobiphenyl
Formed from: 4-Nitrobiphenyl
→ 2-Hydroxy-6-(*p*-nitrophenyl)-6-oxohexanoic acid
Probably in: Bacterium D911

3,4-Dihydro-3,4-dihydroxy-4′-nitrobiphenyl
Formed from: 4-Nitrobiphenyl

***trans*-1,2-Dihydro-1,2-dihydroxy-6-nitrochrysene**
Formed from: 6-Nitrochrysene

***trans*-9,10-Dihydro-9,10-dihydroxy-6-nitrochrysene**
Formed from: 6-Nitrochrysene

***cis*-1,2-Dihydro-1,2-dihydroxy-7-nitronaphthalene**
→ 1-Hydroxy-7-nitronaphthalene
In: Pseudomonas J398
→ 2-Hydroxy-7-nitronaphthalene
In: Pseudomonas J398

7,8-Dihydro-7,8-dihydroxy-2-nitronaphtho[2,1-*b*]furan
Probably formed from: 7-Methoxy-2-nitronaphtho[2,1-*b*]furan

***cis*-4,5-Dihydro-4,5-dihydroxy-1-nitropyrene**
Formed from: 1-Nitropyrene

***trans*-4,5-Dihydro-4,5-dihydroxy-1-nitropyrene**
Formed from: 1-Nitropyrene
1-Nitropyrene-4,5-oxide
→ 1-Amino-4,5-dihydro-4,5-dihydroxypyrene
In: Rat E689
→ 4,5-Dihydro-4,5-dihydroxy-1-nitropyrene-β-D-glucuronide
In: Rat D571
→ 1-Nitropyrene-4,5-dione
In: Rat E689
→ 4,5,9,10-Tetrahydro-4,5,9,10-tetrahydroxy-1-nitropyrene
In: Rat E689

***trans*-4,5-Dihydro-4,5-dihydroxy-2-nitropyrene**
Formed from: 2-Nitropyrene

***cis*-9,10-Dihydro-9,10-dihydroxy-1-nitropyrene**
Formed from: 1-Nitropyrene

9,10-Dihydro-9,10-dihydroxy-1-nitropyrene (presumably *trans*)
Formed from: 1-Nitropyrene-9,10-oxide

***trans*-9,10-Dihydro-9,10-dihydroxy-4-nitropyrene**
Formed from: 4-Nitropyrene

4,5-Dihydro-4,5-dihydroxy-1-nitropyrene-β-D-glucuronide
Formed from: 4,5-Dihydro-4,5-dihydroxy-1-nitropyrene

1,2-Dihydro-1,2-dihydroxynortriptyline
Probably formed from: Nortriptyline

2′,3′-Dihydro-2′,3′-dihydroxyochratoxin A
Formed from: Ochratoxin A
→ N-(5-Chloro-3,4-dihydro-8-hydroxy-3-methylcoumarin-7-oyl)aspartic acid
Probably in: Phenylbactrium C697

9,10-Dihydro-4,5-dihydroxy-10-oxo-2-anthracenecarboxylic acid
See: Rhein-9-anthrone

4′,5′-Dihydro-4′,aryl-dihydroxy-5-oxoimiloxan
Probably formed from: 4′,5′-Dihydro-4′-hydroxy-5-oxoimiloxan
→ 4′,5′-Dihydro-4′,aryl-dihydroxy-5-oxoimiloxan-β-D-glucuronide
Probably in: Man G67

4′,5′-Dihydro-4′,aryl-dihydroxy-5-oxoimiloxan-β-D-glucuronide
4′,5′-Dihydro-4′,aryl-dihydroxy-5-oxoimiloxan

***cis*-1,2-Dihydro-1,2-dihydroxyphenanthrene**
Formed from: Phenanthrene
→ 1,2-Dihydroxyphenanthrene
In: Pseudomonas A1298

***trans*-1,2-Dihydro-1,2-dihydroxyphenanthrene**
Formed from: Phenanthrene
Phenanthrene-1,2-oxide
→ 1,2-Dihydro-1,2-dihydroxyphenanthrene-3,4-oxide
In: Rat B772
→ 1,2-Dihydroxyphenanthrene
In: Rat G440

(+)-*cis*-3,4-Dihydro-3,4-dihydroxyphenanthrene
Formed from: Phenanthrene

***trans*-3,4-Dihydro-3,4-dihydroxyphenanthrene**
Formed from: Phenanthrene
Phenanthrene-3,4-oxide
→ 3,4-Dihydroxyphenanthrene
In: Rat G440

***cis*-9,10-Dihydro-9,10-dihydroxyphenanthrene**
Formed from: Phenanthrene
→ *cis*-9,10-Dihydro-9,10-dihydroxyphenanthrene-β-D-glucuronide
In: Rat C890

***trans*-9,10-Dihydro-9,10-dihydroxyphenanthrene**
Formed from: Phenanthrene-9,10-oxide
→ *trans*-9,10-Dihydro-9,10-dihydroxyphenanthrene-β-D-glucuronide
In: Rat C890

***cis*-3,4-Dihydro-3,4-dihydroxyphenanthrene-4-carboxylic acid**
Probably formed from: Phenanthrene-4-carboxylic acid

***cis*-9,10-Dihydro-9,10-dihydroxyphenanthrene-β-D-glucuronide**
Formed from: *cis*-9,10-Dihydro-9,10-dihydroxyphenanthrene

***trans*-9,10-Dihydro-9,10-dihydroxyphenanthrene-β-D-glucuronide**
Formed from: *trans*-9,10-Dihydro-9,10-dihydroxyphenanthrene

1,2-Dihydro-1,2-dihydroxyphenanthrene-3,4-oxide
Formed from: *trans*-1,2-Dihydro-1,2-dihydroxyphenanthrene

1,2-Dihydro-1,2-dihydroxyphenanthridine
Formed from: Phenanthridine

9,10-Dihydro-9,10-dihydroxyphenanthridine
Formed from: Phenanthridine

***cis*-2,3-Dihydro-2,3-dihydroxyphenetole**
Formed from: Phenetole

***cis*-2,3-Dihydro-2,3-dihydroxyphenylacetic acid**
Formed from: Phenylacetic acid
Probably formed from: *cis*-2,3-Dihydro-2,3-dihydroxyphenylpyruvic acid
→ 2,3-Dihydroxyphenylacetic acid
In: Bacterium B871, B950

***cis*-2,3-Dihydro-2,3-dihydroxy-L-phenylalanine**
Formed from: L-Phenylalanine
→ 2,3-Dihydroxy-L-phenylalanine
In: Bacterium B950

(2*S*)-2,3-dihydro-5,7-dihydroxy-2-phenyl-4*H*-1-benzopyran-4-one
See: Pinocembrin

***cis*-4-(2,3-Dihydro-2,3-dihydroxyphenyl)butyric acid**
Formed from: 4-Phenylbutyric acid

→ 4-(2,3-Dihydroxyphenyl)butyric acid
In: Bacterium B871, B950

***cis*-2-(2,3-Dihydro-2,3-dihydroxyphenyl) cyclohexanol**
Formed from: 2-Phenylcyclohexanol

***cis*-2-(2,3-Dihydro-2,3-dihydroxyphenyl) cyclohexanone**
Formed from: 2-Phenylcyclohexanone

***cis*-1-(2,3-Dihydro-2,3-dihydroxyphenyl) cyclohexene**
Formed from: 1-Phenylcyclohexene

α-(3,4-Dihydro-3,4-dihydroxyphenyl)-α,N-dimethylsuccinimide
Formed from: Methsuximide

(+)-*cis*-1-(2,3-Dihydro-2,3-dihydroxyphenyl) ethanol
Formed from: 1-Phenylethanol

5-(3,4-Dihydro-3,4-dihydroxyphenyl)-5-ethylbarbituric acid
Formed from: Phenobarbital

2-(3,4-Dihydro-3,4-dihydroxyphenyl)-2-ethylglutarimide
Formed from: Glutethimide

5-(3,4-Dihydro-3,4-dihydroxyphenyl)-3-ethylhydantoin
Probably formed from: 3-Ethyl-5-phenylhydantoin
→ 5-(3,4-Dihydroxyphenyl)-3-ethylhydantoin
Probably in: Man B113

5-(3,4-Dihydro-3,4-dihydroxyphenyl)-5-ethylhydantoin
Probably formed from: 5-Ethyl-5-phenylhydantoin

2-(3,4-Dihydro-3,4-dihydroxyphenyl)-3-hydroxy-N,2-dimethylsuccinimide
Formed from: N-Methyl-α-phenylsuccinimide

2-(3,4-Dihydro-3,4-dihydroxyphenyl)-2-hydroxy-N-methylsuccinimide
Formed from: N-Methyl-α-phenylsuccinimide
→ 2-(3,4-Dihydroxyphenyl)-2-hydroxy-N-methylsuccinimide
Probably in: Man A1709, A2017

***cis*-2,3-Dihydro-2,3-dihydroxyphenyllactic acid**
Formed from: Phenyllactic acid
→ 2,3-Dihydroxyphenyllactic acid
In: Bacterium B871

5-(3,4-Dihydro-3,4-dihydroxyphenyl)-5-phenylhydantoin
Probably formed from: 5,5-Diphenylhydantoin
→ 5-(3,4-Dihydroxyphenyl)-5-phenylhydantoin
In: Mouse D818
See also: A1970, B991, C39, K328

***cis*-2,3-Dihydro-2,3-dihydroxyphenylpropionic acid**
Formed from: 3-Phenylpropionic acid
→ 2,3-Dihydroxyphenylpropionic acid
In: Acinetobacter C817
Bacterium B950

***cis*-2,3-Dihydro-2,3-dihydroxyphenylpyruvic acid**
Formed from: Phenylpyruvic acid
→ *cis*-2,3-Dihydro-2,3-dihydroxyphenylacetic acid
Probably in: Bacterium B950
→ 2,3-Dihydroxyphenylpyruvic acid
Probably in: Bacterium B950

3,4-Dihydro-3,4-dihydroxyphthalic acid
Formed from: Phthalic acid
→ 3,4-Dihydroxyphthalic acid
Probably in: Rhodococcus H789, J211

4,5-Dihydro-4,5-dihydroxyphthalic acid
Formed from: Phthalic acid
Probably formed from: 4-Chlorophthalic acid

***trans*-1,2-Dihydro-1,2-dihydroxypicene**
Formed from: Picene

***trans*-3,4-Dihydro-3,4-dihydroxypicene**
Formed from: Picene

***trans*-5,6-Dihydro-5,6-dihydroxypicene**
Formed from: Picene

Dihydrodihydroxypirprofen
Probably formed from: Pirprofen oxide

4,5-Dihydro-4,5-dihydroxy-2-propylbenzotriazole
Formed from: 2-Propylbenzotriazole

10,11-Dihydro-10,11-dihydroxyprotriptyline
Probably formed from: Protriptyline-10,11-oxide

***cis*-4,5-Dihydro-4,5-dihydroxypyrene**
Formed from: Pyrene
→ Phenanthrene-4,5-dicarboxylic acid
In: Mycobacterium J646

***trans*-4,5-Dihydro-4,5-dihydroxypyrene**
Formed from: Pyrene-4,5-oxide

→ Phenanthrene-4,5-dicarboxylic acid
In: Mycobacterium H779, H844

cis-5,6-Dihydro-5,6-dihydroxyquinazoline
Formed from: Quinazoline

10,11-Dihydro-10,11-dihydroxyquinidine
Formed from: Quinidine

10,11-Dihydro-10,11-dihydroxyquinine
Probably formed from: Quinine-10,11-oxide

cis-5,6-Dihydro-5,6-dihydroxyquinoline
Formed from: Quinoline

trans-5,6-Dihydro-5,6-dihydroxyquinoline
Formed from: Quinoline

cis-5,6-Dihydro-5,6-dihydroxyquinoxaline
Formed from: Quinoxaline

3′,4′-Dihydro-3′,4′-dihydroxyrofecoxib
Formed from: Rofecoxib

1′,2′-Dihydro-1′,2′-dihydroxyrotenone
Formed from: Rotenone

21,22-Dihydro-21α,22α-dihydroxystrychnine
Probably formed from: Strychnine-21,22-oxide

21,22-Dihydro-21α,22β-dihydroxystrychnine
Probably formed from: Strychnine-21,22-oxide

cis-2,3-Dihydro-2,3-dihydroxystyrene
Formed from: Styrene
→ 3-Vinylcatechol
In: Rhodococcus H253
See also: J656

1,2-Dihydro-1,2-dihydroxyterephthalic acid
Formed from: Terephthalic acid
→ Protocatechuic acid
In: Comamonas K762

cis-2,3-Dihydro-2,3-dihydroxythianthrene
Formed from: Thianthrene

cis-2,3-Dihydro-2,3-dihydroxythioanisole
Formed from: Thioanisole

cis-2,3-Dihydro-2,3-dihydroxytoluene
Formed from: Toluene
→ 3-Methylcatechol
In: Bacillus E368
Pseudomonas A1298, J338
See also: G653

1,2-Dihydro-1,2-dihydroxy-3,4,6-trichlorobenzene
Formed from: 1,2,4-Trichlorobenzene
→ 1,2-Dihydro-1,2-dihydroxy-3,4,6-trichlorobenzene-β-D-glucuronide
In: Monkey B988

1,2-Dihydro-1,2-dihydroxy-3,4,6-trichlorobenzene-β-D-glucuronide
Formed from: 1,2-Dihydro-1,2-dihydroxy-3,4,6-trichlorobenzene

3′,4′-Dihydro-3′,4′-dihydroxytriflubazam
Formed from: Triflubazam

cis-1,2-Dihydro-1,2-dihydroxy-3-trifluoromethylbenzene
Formed from: 3-Trifluoromethylbenzene
→ 3-Trifluoromethylcatechol
Probably in: Pseudomonas J656

cis-2,3-Dihydro-2,3-dihydroxy-2-trifluoromethylbenzoic acid
Probably formed from: *o*-Trifluoromethylbenzoic acid

cis-2,3-Dihydro-2,3-dihydroxy-4-trifluoromethylbenzoic acid
Probably formed from: *p*-Trifluoromethylbenzoic acid

1,4-Dihydro-1,4-dihydroxy-6-trifluoromethylquinoxaline-2,3-dione
Probably formed from:1,4-Dihydro-1-hydroxy-6-trifluoromethylquinoxaline-2,3-dione

(3_R_,4_S_)-3,4-Dihydro-6,8-dihydroxy-3,4,5-trimethyl-1-oxo-1_H_-2-benzopyran-7-carboxylic acid
See: Dihydrocitrinone

1,2-Dihydro-1,6-dihydroxy-2,2,4-trimethylquinoline
Probably formed from: 1,2-Dihydro-1-hydroxy-2,2,4-trimethylquinoline

trans-1,2-Dihydro-1,2-dihydroxytriphenylene
Formed from: Triphenylene
Triphenylene-1,2-oxide
→ *trans*-1,2-Dihydro-1,2-dihydroxytriphenylene-3,4-oxide
In: Rat G242

trans-1,2-Dihydro-1,2-dihydroxytriphenylene-3,4-oxide
Formed from: *trans*-1,2-Dihydro-1,2-dihydroxytriphenylene

5″,6″-Dihydro-5″,6″-dihydroxyverlukast
Probably formed from: Verlukast-5″,6″-oxide

2′,3′-Dihydro-2′,3′-dihydroxywighteone
Formed from: Wighteone

→ 2′,3′-Dihydro-3′-hydroxywighteone
Probably in: Aspergillus E8

2,3-Dihydro-2,3-dihydroxy-*p*-xylene
Formed from: *p*-Xylene
→ 2,3-Dihydroxy-*p*-xylene
In: Pseudomonas A1313

4,5-Dihydro-4,5-dihydroxy-*m*-xylene
Probably formed from: *m*-Xylene

3,4-Dihydro-6,7-dimethoxy-2-methylisoquinolinium
Formed from: 1,2,3,4-Tetrahydro-6,7-dimethoxy-N-methylisoquinoline

3,4-Dihydro-6,7-dimethoxy-2-methylisoquinolone
Formed from: 3,4-Dihydro-2′-hydroxymethyl-2-methylpapaverinium
3,4-Dihydro-2-methylpapaverinium
→ 6,7-Dimethoxy-2-methylisoquinolone
Probably in: Rabbit B38

α-(1-((3,4-Dihydro-6,7-dimethoxy-1-methyl-2(1*H*)-isoquinolyl)methyl)-3′-methylbutyl)-3,4-dimethylbenzyl acetate
→ α-(1-((3,4-Dihydro-6,7-dimethoxy-1-methyl-2(1*H*)-isoquinolyl)methyl)-3′-methylbutyl)-3,4-dimethylbenzyl alcohol
In: Monkey A2561
Rat A2561

α-(1-((3,4-Dihydro-6,7-dimethoxy-1-methyl-2(1*H*)-isoquinolyl)methyl)-3′-methylbutyl)-3,4-dimethylbenzyl alcohol
Formed from: α-(1-((3,4-Dihydro-6,7-dimethoxy-1-methyl-2(1*H*)-isoquinolyl) methyl)-3′-methylbutyl)-3,4-dimethylbenzyl acetate

9,10-Dihydro-5,7-dimethoxy-2-phenanthrenol
See: Orchinol

2,3-Dihydro-5,6-dimethoxy-2-((1-(phenylmethyl)-4-piperidinyl)methyl)-1*H*-inden-1-one
See: Donepezil

2,3-Dihydro-2,2-dimethyl-7-benzofuranyl carbamate
Probably formed from: Carbofuran
→ 2,3-Dihydro-3-hydroxy-2,2-dimethyl-7-benzofuranyl carbamate
Probably in: Cotton A3871
Zea A3871

2,3-Dihydro-2,2-dimethyl-7-benzofuranyl N-(di-*n*-butylaminosulphenyl)-N-methylcarbamate
(carbosulfan)
→ Carbofuran
In: Cotton A3871
Housefly C359
Zea A3871
→ Carbosulfan sulphone
In: Rat C359

2,3-Dihydro-2,2-dimethyl-7-benzofuranyl N-hydroxymethylcarbamate
Probably formed from: Carbofuran

2,3-Dihydro-2,2-dimethylbenzofuran-7-yl methylcarbamate
See: Carbofuran

2,3-Dihydro-2,2-dimethyl-7-benzofuranyl N-morpholinosulphenyl-N-methylcarbamate
→ Carbofuran
In: Cotton A3871
Zea A3871

6-((3,4-Dihydro-4,4-dimethyl-2*H*-1-benzothiopyran-6-yl)ethynyl)-3-pyridinecarboxylic acid
See: Tazarotenic acid

6-((3,4-Dihydro-4,4-dimethyl-2*H*-1-benzothiopyran-6-yl)ethynyl)-3-pyridinecarboxylic acid ethyl ester
See: Tazarotene

(3*R*)-3,4-Dihydro-2′,2′-dimethyl-(3,6′-bi-2*H*-1-benzopyran)-5′,7-diol
See: Phaseollinisoflavan

10,11-Dihydro-N,5-dimethyl-5*H*-dibenz[*b*,*f*] azepin-10-amine
See: Metapramine

10,11-Dihydro-N,N-dimethyl-5*H*-dibenz[*b*,*f*] azepine-5-propanamine
See: Imipramine

10,11-Dihydro-N,N-dimethyl-5*H*-dibenzo[*a*,*d*] cycloheptene-$\Delta^{5,\gamma}$-propylamine
See: Amitriptyline

4a,9b-Dihydro-8,9b-dimethyl-3(4*H*)-dibenzofuranone
See: Pummerer's ketone

9,10-Dihydro-2,6-dimethyl-9,10-dioxoellipticinium
Formed from: 9-Hydroxy-2,6-dimethylellipticinium

1,2-Dihydro-1,5-dimethyl-4-(1-methylethyl)-2-phenyl-3*H*-pyrazol-3-one
See: Propyphenazone

1,2-Dihydro-2,6-dimethyl-1-naphthaleneheptanoic acid-δ-lactone
Formed from: Lovastatin

1,4-Dihydro-2,6-dimethyl-4-(*m*-nitrophenyl) pyridine-3-carboxylic acid 3-(3-piperidinyl) ester
Probably formed from: 1,4-Dihydro-2,6-dimethyl-4-(*m*-nitrophenyl)pyridine-3,5-dicarboxylic acid 3-piperidinyl ester

1,4-Dihydro-2,6-dimethyl-4-(*m*-nitrophenyl)pyridine-3,5-dicarboxylic acid ((4*R*) and (4*S*))
Formed from: Barnidipine
Benidipine
→ 2,6-Dimethyl-4-(*m*-nitrophenyl) pyridine-3,5-dicarboxylic acid
In: Dog H518
Rat H518

1,4-Dihydro-2,6-dimethyl-4-(*m*-nitrophenyl) pyridine-3,5-dicarboxylic acid 3-(2-(N-benzyl-N-methylamino))ethyl ester 5-methyl ester
→ 4-(3-Aminophenyl)-1,4-dihydro-2,6-dimethylpyridine-3,5-dicarboxylic acid 3-(2-(N-benzyl-N-methylamino))ethyl ester 5-methyl ester
In: Dog A2096
Rat A2096
→ 1,4-Dihydro-2,6-dimethyl-4-(*m*-nitrophenyl)pyridine-3,5-dicarboxylic acid 3-(2-methylamino)ethyl ester 5-methyl ester
In: Dog A2096
Rat A2096
→ 2,6-Dimethyl-4-(*m*-nitrophenyl)pyridine-3,5-dicarboxylic acid 3-(2-(N-benzyl-N-methylamino))ethyl ester 5-methyl ester
In: Dog A2096
Rat A2096

1,4-Dihydro-2,6-dimethyl-4-(2-nitrophenylpyridine)-3,5-dicarboxylic acid dimethyl ester
See: Nifedipine

1,4-Dihydro-2,6-dimethyl-4-(*m*-nitrophenyl) pyridine-3,5-dicarboxylic acid 3-ethyl ester
Formed from: Nitrendipine
→ 1,4-Dihydro-2-hydroxymethyl-6-methyl-4-(*m*-nitrophenyl)pyridine-3,5-dicarboxylic acid 5-ethyl ester
In: Dog G398
Mouse G398
Rat G398

1,4-Dihydro-2,6-dimethyl-4-(*m*-nitrophenyl) pyridine-3,5-dicarboxylic acid 3-ethyl ester 5-methyl ester
See: Nitrendipine

1,4-Dihydro-2,6-dimethyl-4-(*m*-nitrophenyl) pyridine-3,5-dicarboxylic acid 3-methyl ester 5-(2-methylamino)ethyl ester
→ 4-(3-Aminophenyl)-1,4-dihydro-2,6-dimethylpyridine-3,5-dicarboxylic acid 3-methyl ester 5-(2-methylamino)ethyl ester
In: Dog A2096
Rat A2096

1,4-Dihydro-2,6-dimethyl-4-(*m*-nitrophenyl) pyridine-3,5-dicarboxylic acid 3-(3-(hydroxypiperidinyl)) ester 5-methyl ester
Probably formed from: Debenzylbenidipine
→ 1,4-Dihydro-2,6-dimethyl-4-(*m*-nitrophenyl)pyridine-3,5-dicarboxylic acid 3-(3-(hydroxypiperidinyl)) ester 5-methyl ester-β-D-glucuronide
Probably in: Rat H927

1,4-Dihydro-2,6-dimethyl-4-(*m*-nitrophenyl) pyridine-3,5-dicarboxylic acid 3-(3-(hydroxypiperidinyl)) ester 5-methyl ester-β-D-glucuronide
Probably formed from: 1,4-Dihydro-2,6-dimethyl-4-(*m*-nitrophenyl)pyridine-3,5-dicarboxylic acid 3-(3-(hydroxypiperidinyl)) ester 5-methyl ester

1,4-Dihydro-2,6-dimethyl-4-(3-nitrophenylpyridine)-3,5-dicarboxylic acid 2-methoxyethyl 1-methylethyl ester
See: Nimodipine

1,4-Dihydro-2,6-dimethyl-4-(*m*-nitrophenyl) pyridine-3,5-dicarboxylic acid 3-(2-methylamino)ethyl ester 5-methyl ester
Formed from: 1,4-Dihydro-2,6-dimethyl-4-(*m*-nitrophenyl)pyridine-3,5-dicarboxylic acid 3-(2-(N-benzyl-N-methylamino))ethyl ester 5-methyl ester

1,4-Dihydro-2,6-dimethyl-4-(*m*-nitrophenyl) pyridine-3,5-dicarboxylic acid 3-methyl ester
Formed from: Pranidipine

1,4-Dihydro-2,6-dimethyl-4-(3-nitrophenylpyridine)-3,5-dicarboxylic acid methyl (2*E*)-3-phenyl-2-propenyl ester
See: Pranidipine

1,4-Dihydro-2,6-dimethyl-4-(*m*-nitrophenyl) pyridine-3,5-dicarboxylic acid 3-(3-piperidinyl) ester
Formed from: Debenzylbenidipine
→ 4-(*m*-Aminophenyl)-1,4-dihydro-2,6-dimethylpyridine-3,5-dicarboxylic acid 3-(3-piperidinyl) ester
Probably in: Dog F11
Rat F11
→ 1,4-Dihydro-2,6-dimethyl-4-(*m*-nitrophenyl)pyridine-3-carboxylic acid 3-(3-piperidinyl) ester
Probably in: Dog F11
Rat F11, H927
→ 2,6-Dimethyl-4-(*m*-nitrophenyl)pyridine-3,5-dicarboxylic acid 3-(3-piperidinyl) ester
Probably in: Dog F11
Man J105
Rat F11, H927

((2,3-Dihydro-1,5-dimethyl-3-oxo-2-phenyl-1*H*-pyrazol-4-yl)methylamino)methanesulphonic acid
See: Dipyrone

N-(2,3-Dihydro-1,5-dimethyl-3-oxo-2-phenyl-1*H*-pyrazol-4-yl)-3-pyridinecarboxamide
See: Nifenazone

3-(5,6-Dihydro-6,6-dimethyl-2-oxo-2-pyran-3-yl)-1-(3*H*)-isobenzofuranone
See: Catalpalactone

4,5-Dihydro-1,3-dimethyl-1*H*-pyrazol[3,4-*b*][1,4] benzoxazepine
→ Salicylic acid
In: Dog A876
Rat A876

1,3-Dihydro-3,3-dimethyl-5-(1,4,5,6-tetrahydro-6-oxo-3-pyridazinyl)-2*H*-indol-2-one
See: Indolidan

1,4-Dihydro-2,6-dimethyl-4-(2-trifluoromethyl-phenyl)-3,5-pyridinedicarboxylic acid
Formed from: 1,4-Dihydro-2,6-dimethyl-4-(2-trifluoromethylphenyl)-3,5-pyridinedicarboxylic acid monoethyl ester

1,4-Dihydro-2,6-dimethyl-4-(2-trifluoromethyl-phenyl)-3,5-pyridinedicarboxylic acid monoethyl ester
Formed from: Diethyl 1,4-dihydro-2,6-dimethyl-4-(2-trifluoromethylphenyl)-3,5-pyridinedicarboxylic acid
→ 1,4-Dihydro-2,6-dimethyl-4-(2-trifluoromethylphenyl)-3,5-pyridinedicarboxylic acid
In: Dog A715
→ 1,4-Dihydro-2,6-dimethyl-4-(2-trifluoromethylphenyl)-3,5-pyridinedicarboxylic acid mono-2-hydroxyethyl ester
Probably in: Dog A715
→ 2,6-Dimethyl-4-(2-trifluoromethylphenyl)-3,5-pyridinedicarboxylic acid monoethyl ester
In: Dog A840
Rat A840
Squirrel monkey A840

1,4-Dihydro-2,6-dimethyl-4-(2-trifluoromethylphenyl)-3,5-pyridinedicarboxylic acid mono-2-hydroxyethyl ester
Probably formed from: 1,4-Dihydro-2,6-dimethyl-4-(2-trifluoromethylphenyl)-3,5-pyridinedicarboxylic acid monoethyl ester

3-(2,3-Dihydro-1,1-dioxido-3-oxo-1,2-benzisothiazol-2-yl)-2,6-dioxopiperidine
Probably formed from: 3-(2,3-Dihydro-1,1-dioxido-3-oxo-1,2-benzisothiazol-2-yl)-6-hydroxy-2-oxopiperidine
→ 5-Amido-α-(2,3-dihydro-1,1-dioxido-3-oxo-1,2-benzisothiazol-2-yl)glutaric acid
Probably in: Rat C388
→ 5-Amido-γ-(2,3-dihydro-1,1-dioxido-3-oxo-1,2-benzisothiazol-2-yl)glutaric acid
Probably in: Rat C388
→ 2-Carboxy-N-(2,6-dioxopiperidin-3-yl) benzenesulphonamide
Probably in: Rat C388

→ 3-(2,3-Dihydro-1,1-dioxido-3-oxo-1,2-benzisothiazol-2-yl)-5-hydroxy-2,6-dioxopiperidine
Probably in: Rat C388

γ-(2,3-Dihydro-1,1-dioxido-3-oxo-1,2-benzisothiazol-2-yl)glutaric acid
Probably formed from: 5-Amido-γ-(2,3-dihydro-1,1-dioxido-3-oxo-1,2-benzisothiazol-2-yl)glutaric acid

3-(2,3-Dihydro-1,1-dioxido-3-oxo-1,2-benzisothiazol-2-yl)-5-hydroxy-2,6-dioxopiperidine
Probably formed from: 3-(2,3-Dihydro-1,1-dioxido-3-oxo-1,2-benzisothiazol-2-yl)-2,6-dioxopiperidine

3-(2,3-Dihydro-1,1-dioxido-3-oxo-1,2-benzisothiazol-2-yl)-6-hydroxy-2-oxopiperidine
Formed from: Supidimide
→ 3-(2,3-Dihydro-1,1-dioxido-3-oxo-1,2-benzisothiazol-2-yl)-2,6-dioxopiperidine
Probably in: Rat C388

3-(2,3-Dihydro-1,1-dioxido-3-oxo-1,2-benzisothiazol-2-yl)-2-oxopiperidine
See: Supidimide

4′,5′-Dihydro-4′,5′-dioxoimiloxan
Formed from: Imiloxan

4,5-Dihydro-4,5-dioxo-1*H*-pyrrolo[2,3-*f*]quinoline-2,7,9-tricarboxylic acid
See: Pyrroloquinoline quinone

1,4-Dihydro-2,4-dioxo-3(2*H*)-quinazolineacetic acid
Formed from: Ketanserin

3-Dihydrodolasetron
Formed from: Dolasetron
→ 3-Dihydrodolasetron-N-oxide
In: Man H709
See also: H315
→ ((2α,6α,8α,9α,β)-Octahydro-3-hydroxy-2,6-methano-2*H*-quinolizin-8-yl) 1*H*-5-hydroxyindole-3-carboxylate
In: Man H709
See also: H315
→ ((2α,6α,8α,9α,β)-Octahydro-3-hydroxy-2,6-methano-2*H*-quinolizin-8-yl) 1*H*-6-hydroxyindole-3-carboxylate
In: Man H709
See also: H315
→ ((2α,6α,8α,9α,β)-Octahydro-3-hydroxy-2,6-methano-2*H*-quinolizin-8-yl) 1*H*-7-hydroxyindole-3-carboxylate
In: Man H709

3-Dihydrodolasetron-N-oxide
Formed from: 3-Dihydrodolasetron

2,3-Dihydroelymoclavine
Formed from: 2,3-Dihydroagroclavine

17α-Dihydroequilenin
Probably formed from: Equilenin
→ 17α-Dihydroequilenin-β-D-glucuronide
In: Horse B745
See also: G993
→ 17α-Dihydroequilenin-O-sulphate
In: Horse B745
See also: G993

17β-Dihydroequilenin
Probably formed from: 17β-Dihydroequilin
Equilenin
→ 17β-Dihydroequilenin-β-D-glucuronide
In: Horse B745
→ 17β-Dihydroequilenin-O-sulphate
In: Horse B745

17α-Dihydroequilenin-β-D-glucuronide
Probably formed from: 17α-Dihydroequilenin

17β-Dihydroequilenin-β-D-glucuronide
Probably formed from: 17β-Dihydroequilenin

17α-Dihydroequilenin-O-sulphate
Probably formed from: 17α-Dihydroequilenin

17β-Dihydroequilenin-O-sulphate
Probably formed from: 17β-Dihydroequilenin

17α-Dihydroequilin
Formed from: Equilin
→ 17α-Dihydroequilin-β-D-glucuronide
Probably in: Horse B745
→ 17α-Dihydroequilin-O-sulphate
Probably in: Horse B745

17β-Dihydroequilin
Probably formed from: 17β-Dihydroequilin-O-sulphate
→ 17β-Dihydroequilenin
In: Man H211
→ 17β-Dihydroequilin-β-D-glucuronide
In: Man H211
See also: B745
→ 17β-Dihydroequilin-O-sulphate

Probably in: Horse B745
→ Equilenin
In: Man H211
17α-Dihydroequilin-β-D-glucuronide
Probably formed from: 17α-Dihydroequilin
17β-Dihydroequilin-β-D-glucuronide
Formed from: 17β-Dihydroequilin
17α-Dihydroequilin-O-sulphate
Probably formed from: 17α-Dihydroequilin
17β-Dihydroequilin-O-sulphate
Probably formed from: 17β-Dihydroequilin
→ 17β-Dihydroequilin
Probably in: Man H211
α-Dihydroergocryptine
→ α-Dihydro-9′-hydroxyergocryptine
In: Man K355
Dihydroeriodictyol
See: Luteoforol
Dihydroesculetin quinone
Formed from: 4-(2-Carboxyethyl)-*o*-quinone
Dihydroflavine adenine dinucleotide
Formed from: Flavine adenine dinucleotide
Dihydroferulic acid (3-(4-hydroxy-3-methoxyphenyl)propionic acid)
Formed from: Dihydrocaffeic acid
Eugenol
Ferulic acid
Probably formed from: 3,4-Dimethoxyphenylpropionic acid
4-(4-Hydroxy-3-methoxyphenyl)butan-1,2-diol
→ Dihydroferulic acid-β-D-glucuronide
Probably in: Rat A3492
→ *m*-Methoxyphenylpropionic acid
Probably in: Microorganism D894
Dihydroferulic acid-β-D-glucuronide
Probably formed from: Dihydroferulic acid
Dihydroflavine adenine dinucleotide
Formed from: Flavine adenine dinucleotide
16,21-Dihydroflumioxazin
Formed from: Flumioxazin
→ 16,21-Dihydro-17-hydroxyflumioxazin
Probably in: Rat K152
→ 16,21-Dihydro-18-hydroxyflumioxazin
Probably in: Rat K152
Dihydrofluorescein
See: Fluorescin
1,4-Dihydro-6-fluoro-1-(5-fluoro-2-pyridyl)-7-(4-methyl-1-piperazinyl)-4-oxoquinoline-3-carboxylic acid
→ 1,4-Dihydro-6-fluoro-1-(5-fluoro-2-pyridyl)-7-(4-methyl-1-piperazinyl)-4-oxoquinoline-3-carboxylic acid-β-D-glucuronide
In: Rat J387
1,4-Dihydro-6-fluoro-1-(5-fluoro-2-pyridyl)-7-(4-methyl-1-piperazinyl)-4-oxoquinoline-3-carboxylic acid-β-D-glucuronide
Formed from: 1,4-Dihydro-6-fluoro-1-(5-fluoro-2-pyridyl)-7-(4-methyl-1-piperazinyl)-4-oxoquinoline-3-carboxylic acid
Dihydrofolic acid
Formed from: Dihydropteroic acid
Tetrahydrofolic acid
→ Dihydropteroyldiglutamic acid
In: Pig E277
→ Formyldihydrofolic acid
In: Man K745
→ Formyltetrahydrofolic acid
With 5,10-methylenetetrahydrofolate dehydrogenase B480
→ 5-Methyltetrahydrofolic acid
With dihydrofolate reductase B480
→ Tetrahydrofolic acid
Drosophila K458
Sunflower K461
Dihydrofuranisoflavone
Formed from: Wighteone
Dihydrogallic acid
Formed from: Gallic acid
Dihydrogeodin
→ Geodin
In: Aspergillus E300
Dihydrogossypetin (3,3′,4′,5,7,8-hexahydroxyflavanone)
Formed from: Dihydroquercetin
Dihydrohaloperidol
Formed from: Haloperidol
→ Haloperidol
In: Man F261
Dihydrohomofolic acid
→ Tetrahydrohomofolic acid
In: Mouse A84
Rhesus monkey A1474

3,4-Dihydro-5-hydroxyanthraniloyl CoA
5-Hydroxyanthraniloyl CoA
S-(9,10-Dihydro-9-hydroxy-1-azaphenanthr-10-yl)glutathione
Formed from: 1-Azaphenanthrene-9,10-oxide
S-(9,10-Dihydro-10-hydroxy-1-azaphenanthr-9-yl)glutathione
Formed from: 1-Azaphenanthrene-9,10-oxide
S-(9,10-Dihydro-9-hydroxy-4-azaphenanthr-10-yl)glutathione
Formed from: 4-Azaphenanthrene-9,10-oxide
S-(9,10-Dihydro-10-hydroxy-4-azaphenanthr-10-yl)glutathione
Formed from: 4-Azaphenanthrene-9,10-oxide
S-(5,6-Dihydro-5-hydroxybenzanthr-6-yl) glutathione
Formed from: Benzanthracene-5,6-oxide
S-(5,6-Dihydro-6-hydroxybenzanthr-5-yl) glutathione
Formed from: Benzanthracene-5,6-oxide
1,3-Dihydro-5-hydroxy-2*H*-benzimidazol-2-one
Formed from: Oxatomide
→ 1,3-Dihydro-5-hydroxy-2*H*-benzimidazol-2-one-β-D-glucuronide
Probably in: Dog C870
Man C870
Rat C870
1,3-Dihydro-5-hydroxy-2H-benzimidazol-2-one-β-D-glucuronide
Probably formed from: 1,3-Dihydro-5-hydroxy-2H-benzimidazol-2-one
2,3-Dihydro-3-hydroxybenz[*d*]isothiazole-1,1-dioxide
Probably formed from: 2,3-Dihydrobenz[*d*] isothiazole-1,1-dioxide
→ *o*-Carboxybenzenesulphonamide
Probably in: Rat A2107
→ Saccharin
Probably in: Rat A2107
→ *o*-Sulphamoylbenzyl alcohol
Probably in: Rat A2107
6,7-Dihydro-5-hydroxy-5*H*-benzocyclohept-1-ene
→ *cis*-6,7-Dihydro-5,6-dihydroxy-5*H*-benzocyclohept-1-ene
Probably in: Pseudomonas F629

(*S*)-2-(1-(2-(2,3-Dihydro-3-hydroxybenzofuran-5-yl)ethyl)-3-pyrrolidinyl)-2,2-diphenylacetamide
Formed from: Darifexacin
→ 5-Carboxymethyl-3-hydroxybenzofuran
Probably in: Dog H866
Man H866
Mouse H866
Rabbit H866
Rat H866
(+)-Dihydro-3-hydroxybenzopyran
See: (+)-3-Chromanol
Dihydro-4-hydroxybenzopyran
See: 4-Chromanol
7,8-Dihydro-7-hydroxy-3,4-benzpyrene
Formed from: 7,8-Dihydro-3,4-benzpyrene
7,8-Dihydro-8-hydroxy-3,4-benzpyrene
Formed from: 7,8-Dihydro-3,4-benzpyrene
9,10-Dihydro-9-hydroxy-3,4-benzpyrene
Formed from: 9,10-Dihydro-3,4-benzpyrene
9,10-Dihydro-10-hydroxy-3,4-benzpyrene
Formed from: 9,10-Dihydro-3,4-benzpyrene
S-(4,5-Dihydro-4-hydroxy-3,4-benzpyren-5-yl) glutathione
Formed from: 3,4-benzpyrene-4,5-oxide
S-(4,5-Dihydro-5-hydroxy-3,4-benzpyren-4-yl) glutathione
Formed from: 3,4-benzpyrene-4,5-oxide
S-(7,8-Dihydro-7-hydroxy-3,4-benzpyren-8-yl) glutathione
Formed from: 3,4-benzpyrene-7,8-oxide
5-((3,4-Dihydro-2-(α-hydroxybenzyl)-2*H*-1-benzopyran-6-yl)methyl)thiazolidine-2,4-dione
Formed from: 5-((3,4-Dihydro-2-phenylmethyl-2*H*-1-benzopyran-6-yl)methyl)thiazolidine-2,4-dione
Dihydrohydroxybunolol
Probably formed from: Hydroxybunolol
6′,7′-Dihydro-8′-hydroxycannabigerol
Probably formed from: 8′-Hydroxycannabigerol
Dihydro-ω-hydroxycapsaicin
Formed from: Dihydrocapsaicin
3,4-Dihydro-5-hydroxycarbostyril
Formed from: Carteolol
3,4-Dihydro-6-hydroxycarbostyril
Formed from: Cilostazol

Dihydro-2′-hydroxychalcone
Probably formed from: 2′-Hydroxychalcone
→ Dihydro-2′,4-dihydroxychalcone
Probably in: Streptomyces F673
→ Dihydro-2′,5-dihydroxychalcone
Probably in: Aspergillus F673

Dihydro-10-hydroxychelirubine
→ Dihydro-10-methoxychelirubine
In: Thalictrum K776

Dihydro-12-hydroxychelirubine
Formed from: Dihydrochelirubine
→ Dihydromacarpine
In: Thalictrum K776

Dihydro-6′-hydroxycinchonidine
Probably formed from: Dihydroquinine

3,4-Dihydro-3-hydroxy-6,7-dimethoxy-2,2-dimethyl-2*H*-benzo[*b*]pyran
Formed from: Precocene II

3,4-Dihydro-6-hydroxy-7,8-dimethoxy-2-methylisoquinolinium
See: Tritoqualine

1,4-Dihydro-7-hydroxy-1,6-dimethoxy-4-oxoquinoline-3-carboxylic acid
Formed from: 5,8-Dihydro-5-methoxy-8-oxo-2*H*-1,3-dioxolo[4,5-*g*]quinoline-7-carboxylic acid
→ 1,4-Dihydro-7-hydroxy-1,6-dimethoxy-4-oxoquinoline-3-carboxylic acid-β-D-glucuronide
Probably in: Man A3993

1,4-Dihydro-7-hydroxy-1,6-dimethoxy-4-oxoquinoline-3-carboxylic acid-β-D-glucuronide
Probably formed from: 1,4-Dihydro-7-hydroxy-1,6-dimethoxy-4-oxoquinoline-3-carboxylic acid

3,5-Dihydro-2-(4-hydroxy-3,5-dimethoxyphenyl)-7-methoxy-1-benzopyrylium
See: Hirsutidin

2,3-Dihydro-7-hydroxy-2,2-dimethylbenzofuran
Formed from: Carbofuran
→ 2,3-Dihydro-3,7-dihydroxy-2,2-dimethylbenzofuran
Probably in: Bean D806
Cotton D806
Strawberry A1948
Zea D806

2,3-Dihydro-3-hydroxy-2,2-dimethyl-7-benzofuranyl carbamate
Probably formed from: 2,3-Dihydro-2,2-dimethyl-7-benzofuranyl carbamate

1,4-Dihydro-6-hydroxy-4,4-dimethyl-2*H*-3,1-benzoxazin-2-one
Formed from: Brofoxine

2,3-Dihydro-7-hydroxy-2,2-dimethyl-3-oxobenzofuran
Probably formed from: 2,3-Dihydro-3,7-dihydroxy-2,2-dimethylbenzofuran
3-Oxocarbofuran

1,2-Dihydro-4-hydroxy-1,N-dimethyl-N-phenyl-2-oxoquinoline-3-carboxamide
See; Roquinimex

α-Dihydro-9′-hydroxyergocryptine
Formed from: α-Dihydroergocryptine
→ α-Dihydro-8′,9′-dihydroxyergocryptine
Probably in: Man K355

1,4-Dihydro-3-(2-hydroxyethoxy)carbonyl-5-isopropoxycarbonyl-2,6-dimethyl-4-(*m*-nitrophenyl)pyridine
Formed from: Nimodipine
→ 3-(2-Hydroxyethoxy)carbonyl-5-isopropoxycarbonyl-2,6-dimethyl-4-(*m*-nitrophenyl)pyridine
Probably in: Dog G189
Rat G189

2,3-Dihydro-2-(1-hydroxyethyl)-5-benzofuranylacetic acid
Formed from: 2-Ethyl-2,3-Dihydro-5-benzofuranylacetic acid

2,3-Dihydro-2-(2-hydroxyethyl)-5-benzofuranylacetic acid
Formed from: 2-Ethyl-2,3-Dihydro-5-benzofuranylacetic acid

3,4-Dihydro-2-(1-hydroxyethyl)-4-(3-methylaminopropyl)-3-phenyl-2*H*-1,4-benzothiazine
Formed from: 4-(3-Dimethylaminopropyl)-3,4-dihydro-2-(1-hydroxyethyl)-3-phenyl-2*H*-1,4-benzothiazine
→ 3,4-Dihydro-2-(1-hydroxyethyl)-4-(3-methylaminopropyl)-3-phenyl-2*H*-1,4-benzothiazine sulphoxide

Probably in: Dog A1888
Rat A1769, A1888

3,4-Dihydro-2-(1-hydroxyethyl)-4-(3-methylaminopropyl)-3-phenyl-2*H*-1,4-benzothiazine sulphoxide
Probably formed from: 3,4-Dihydro-2-(1-hydroxyethyl)-4-(3-methylaminopropyl)-3-phenyl-2*H*-1,4-benzothiazine
4-(3-Dimethylaminopropyl)-3,4-dihydro-2-(1-hydroxyethyl)-3-phenyl-2*H*-1,4-benzothiazine sulphoxide

16,21-Dihydro-17-hydroxyflumioxazin
Probably formed from: 16,21-Dihydroflumioxazin

16,21-Dihydro-18-hydroxyflumioxazin
Probably formed from: 16,21-Dihydroflumioxazin

(+)-2,3-Dihydro-7-hydroxy-3-(6-hydroxy-1,3-benzodioxol-5-yl)-4*H*-1-benzopyran-1-one
See: Sophorol

2-((3*S*,4*R*)-3,4-Dihydro-4-hydroxy-3-(*p*-hydroxybenzyl)-2*H*-1-benzopyran-7-yl)-4-trifluoromethylbenzoic acid
Probably formed from: 2-((3*S*,4*R*)-3-Benzyl-3,4-dihydro-4-hydroxy-2*H*-1-benzopyran-7-yl)-4-trifluoromethylbenzoic acid

2-((3*S*,4*R*)-3,4-Dihydro-4-hydroxy-3-(α-hydroxybenzyl)-2*H*-1-benzopyran-7-yl)-4-trifluoromethylbenzoic acid
Probably formed from: 2-((3*S*,4*R*)-3-Benzyl-3,4-dihydro-4-hydroxy-2*H*-1-benzopyran-7-yl)-4-trifluoromethylbenzoic acid

3,4-Dihydro-hydroxy-2-(1-hydroxyethyl)-4-(3-methylaminopropyl)-3-phenyl-2*H*-1,4-benzothiazine
Probably formed from: 4-(3-Dimethylaminopropyl)-3,4-dihydro-hydroxy-2-(1-hydroxyethyl)-3-phenyl-2*H*-1,4-benzothiazine

(3*R*)-3,4-Dihydro-8-hydroxy-3-(3-hydroxy-4-methoxyphenyl)-1*H*-2-benzopyran-1-one
See: Phyllodulcin

5,6-Dihydro-4-hydroxy-6-(4-hydroxy-3-methoxystyryl)-2-pyrone
Probably formed from: Feruloyl CoA

5-((4-((3,4-Dihydro-6-hydroxy-5-hydroxymethyl-2,7,8-trimethyl-2*H*-1-benzopyran-2-yl) methoxy)phenyl)methyl)-2,4-thiazolidinedione
Formed from: Troglitazone

3,4-Dihydro-3-hydroxy-8-(2-hydroxy-3-isopropylamino)propoxy-2*H*-1-benzopyran
Formed from: Nipradilol
→ 3,4-Dihydro-3,4-dihydroxy-8-(2-hydroxy-3-isopropylamino)propoxy-2*H*-1-benzopyran
Probably in: Rat D882
→ 3,4-Dihydro-3,5-dihydroxy-8-(2-hydroxy-3-isopropylamino)propoxy-2*H*-1-benzopyran
Probably in: Rat D882

3,4-Dihydro-8-hydroxy-3-(4-hydroxyphenyl)-1*H*-2-benzopyran-1-one
See: Hydrangenol

1,2-Dihydro-4-hydroxy-N-(*p*-hydroxyphenyl)-1,N-dimethyl-2-oxoquinoline-3-carboxamide
Formed from: Roquinimex

2,3-Dihydro-4-hydroxy-7-(*p*-hydroxyphenyl)-2-(2-hydroxy-2-propyl)-6-methoxy-5*H*-[3,2-*g*][1]furobenzopyran-5-one
Formed from: Topazolin

d-*cis*-2,3-Dihydro-3-hydroxy-2-(*p*-hydroxyphenyl)-5-(2-methylaminoethyl)-1,5-benzothiazepin-4(5*H*)-one
Probably formed from: D-*cis*-2,3-Dihydro-3-hydroxy-2-(*p*-methoxyphenyl)-5-(2-methylaminoethyl)-1,5-benzothiazepin-4(5*H*)-one

3-(2,3-Dihydro-4-hydroxy-2-(2-hydroxy-2-propyl)benzofural-5-yl)-5,7-dihydroxy-4*H*-1-benzopyran-4-one
Formed from: Piscerythrone

3-(2,3-Dihydro-4-hydroxy-2-(2-hydroxy-2-propyl)benzofural-7-yl)-5,7-dihydroxy-4*H*-1-benzopyran-4-one
Formed from: Piscerythrone

5,6-Dihydro-4-hydroxy-6-(4-hydroxystyryl)-2-pyrone
Formed from: *p*-Coumaroyl CoA
Probably formed from: Feruloyl CoA

3,4-Dihydro-8-(2-hydroxy-3-isopropylaminopropoxy)-3-nitroxy-2*H*-1-benzopyran
See: Nipradilol

(4″,5″-Dihydro-5″-(2-hydroxy)isopropylfurano)[2″,3″-*d*]-2′,4,6′,α-tetrahydroxychalcone
Probably formed from: (4″,5″-Dihydro-5″-(2-hydroxy)isopropylfurano)[2″,3″-*h*]-2,4,5-trihydroxyflavanone

(4″,5″-Dihydro-5″-(2-hydroxy)isopropylfurano)[2″,3″-*h*]-3,4,5-trihydroxyflavanone
Probably formed from: 8-(β,γ-Dihydroxyisovaleryl)-3,4′,5,7-tetrahydroxyflavanone
→ (4″,5″-Dihydro-5″-(2-hydroxy)isopropylfurano)[2″,3″-*d*]-2′,4,6″α-tetrahydroxychalcone
Probably in: Aspergillus A3306

9,10-Dihydro-8-hydroxykadsurenone
Formed from: 9,10-Dihydrokadsurenone

9,10-Dihydro-9-hydroxykadsurenone
Formed from: 9,10-Dihydrokadsurenone
→ 9,10-Dihydro-9-hydroxykadsurenone-β-D-glucuronide
In: Monkey E950

9,10-Dihydro-10-hydroxykadsurenone
Formed from: 9,10-Dihydrokadsurenone
→ 9,10-Dihydro-10-hydroxykadsurenone-β-D-glucuronide
In: Monkey E950

9,10-Dihydro-9-hydroxykadsurenone-β-D-glucuronide
Formed from: 9,10-Dihydro-9-hydroxykadsurenone

9,10-Dihydro-10-hydroxykadsurenone-β-D-glucuronide
Formed from: 9,10-Dihydro-10-hydroxykadsurenone

Dihydro-8-hydroxykawain
Probably formed from: Dihydrokawain

Dihydro-*m*-hydroxykawain
Probably formed from: Dihydrokawain

Dihydro-*p*-hydroxykawain
Probably formed from: Dihydrokawain
→ Dihydro-3′,4′-dihydroxykawain
Probably in: Rat A3485
→ 4-Hydroxy-6-(*p*-hydroxyphenyl)hexan-2-one
Probably in: Rat A3485
→ 4-Hydroxy-6-(*p*-hydroxyphenyl)-5-hexen-2-one
Probably in: Rat A3485

2″,3″-Dihydro-3″-hydroxyluteone
Formed from: Luteone

11,12-Dihydro-9-hydroxy-8-methoxybenzo[*a*]-1,3-benzodioxolo[4,5-*g*]quinolizinium
See: Groenlandicine

3,4-Dihydro-6-(4-(4-hydroxy-3-methoxybenzoyl)-1-piperazinyl)-2(1*H*)-quinolone
Formed from: 3,4-Dihydro-6-(4-(3,4-dimethoxybenzoyl)-1-piperazinyl)-2(1*H*)-quinolone
→ 3,4-Dihydro-6-(4-(4-hydroxy-3-methoxybenzoyl)-1-piperazinyl)-2(1*H*)-quinolone-β-D-glucuronide
In: Man F318
Rat F318
→ 3,4-Dihydro-6-(4-(4-hydroxy-3-methoxybenzoyl)-1-piperazinyl)-2(1*H*)-quinolone-O-sulphate
In: Man F318
Rat F318

3,4-Dihydro-6-(4-(4-hydroxy-3-methoxybenzoyl)-1-piperazinyl)-2(1*H*)-quinolone-β-D-glucuronide
Formed from: 3,4-Dihydro-6-(4-(4-hydroxy-3-methoxybenzoyl)-1-piperazinyl)-2(1*H*)-quinolone

3,4-Dihydro-6-(4-(4-hydroxy-3-methoxybenzoyl)-1-piperazinyl)-2(1*H*)-quinolone-O-sulphate
Formed from: 3,4-Dihydro-6-(4-(4-hydroxy-3-methoxybenzoyl)-1-piperazinyl)-2(1*H*)-quinolone

(3*R*)-3,4-Dihydro-8-hydroxy-6-methoxy-3-methyl-1*H*-2-benzopyran-1-one
See: 6-Methoxymellein

3,4-Dihydro-7-hydroxy-6-methoxy-2-methylisoquinolinium
Formed from: N-Methylcoclaurine
Probably formed from: 1,2,3,4-Tetrahydro-7-hydroxy-6-methoxy-2-methylisoquinoline

3,4-Dihydro-3-(4-hydroxy-5-methoxy-2-methylphenyl)-2-methyl-4-oxopyrido[2,3-*d*]pyrimidine
Probably formed from: 3,4-Dihydro-3-(4-hydroxy-2-methylphenyl)-2-methyl-4-oxopyrido[2,3-*d*]pyrimidine

3,4-Dihydro-3-(2-hydroxy-4-methoxyphenyl)-2*H*-1-benzopyran-7-ol
See: Vestitol

d-*cis*-2,3-Dihydro-3-hydroxy-2-(*p*-methoxyphenyl)-5-(2-methylaminoethyl)-1,5-benzothiazepin-4(5*H*)-one
Probably formed from: D-*cis*-5-(2-Dimethylaminoethyl)-2,3-dihydro-3-hydroxy-2-(*p*-methoxyphenyl)-1,5-benzothiazepin-4(5*H*)-one
→ D-*cis*-2,3-Dihydro-3-hydroxy-2-(*p*-hydroxyphenyl)-5-(2-methylaminoethyl)-1,5-benzothiazepin-4(5*H*)-one
Probably in: Rat A391

(3*R*)-3,4-Dihydro-8-hydroxy-3-methyl-1*H*-2-benzopyran-1-one
See: Mellein

(2*R*,4*S*)-3,4-Dihydro-4-hydroxy-3-(3-methyl-2-butenyl-1(2*H*)-naphthalenone
See: Catalponol

10,11-Dihydro-2-hydroxy-5-methyl-5*H*-dibenzo[*a,d*]cyclohepten-5,10-imine
Formed from: 10,11-Dihydro-5-methyl-5*H*-dibenzo[*a,d*]cyclohepten-5,10-imine

10,11-Dihydro-8-hydroxy-5-methyl-5*H*-dibenzo[*a,d*]cyclohepten-5,10-imine
Formed from: 10,11-Dihydro-5-methyl-5*H*-dibenzo[*a,d*]cyclohepten-5,10-imine

10,11-Dihydro-N-hydroxy-5-methyl-5*H*-dibenzo[*a,d*]cyclohepten-5,10-imine
Formed from: 10,11-Dihydro-5-methyl-5*H*-dibenzo[*a,d*]cyclohepten-5,10-imine

2,3-Dihydro-3-hydroxy-3-methyl-9*H*-isoxazolo[3,2-*b*]quinazolin-9-one
Formed from: 2,3-Dihydro-3-methyl-9*H*-isoxazolo[3,2-*b*]quinazolin-9-one

2,3-Dihydro-6-hydroxymethyl-4-(β-methylaminoethyl)-2-phenyl-1,4-benzoxazin-3-one
Probably formed from: 4-(β-Dimethylaminoethyl)-2,3-dihydro-6-hydroxymethyl-2-phenyl-1,4-benzoxazin-3-one

2,3-Dihydro-7-hydroxy-6-methyl-4-(β-methylaminoethyl)-2-phenyl-1,4-benzoxazin-3-one
Probably formed from: 4-(β-Dimethylaminoethyl)-2,3-dihydro-7-hydroxy-6-methyl-2-phenyl-1,4-benzoxazin-3-one

1,4-Dihydro-2-hydroxymethyl-6-methyl-4-(*m*-nitrophenyl)pyridine-3,5-dicarboxylic acid 5-ethyl ester
Probably formed from: 1,4-Dihydro-2,6-dimethyl-4-(*m*-nitrophenyl)pyridine-3,5-dicarboxylic acid 3-ethyl ester

1,4-Dihydro-2-hydroxymethyl-6-methyl-4-(*m*-nitrophenyl)pyridine-3,5-dicarboxylic acid 3-ethyl ester 5-methyl ester
→ 1,4-Dihydro-2-hydroxymethyl-6-methyl-4-(*m*-nitrophenyl)pyridine-3,5-dicarboxylic acid 3-ethyl ester 5-methyl ester-β-D-glucuronide
Probably in: Dog G398
Mouse G398
Rat G398

1,4-Dihydro-2-hydroxymethyl-6-methyl-4-(*m*-nitrophenyl)pyridine-3,5-dicarboxylic acid 3-ethyl ester 5-methyl ester-β-D-glucuronide
Probably formed from: 1,4-Dihydro-2-hydroxymethyl-6-methyl-4-(*m*-nitrophenyl)pyridine-3,5-dicarboxylic acid 3-ethyl ester 5-methyl ester

1,4-Dihydro-2-hydroxymethyl-6-methyl-4-(*m*-nitrophenyl)pyridine-3,5-dicarboxylic acid 5-methyl ester
Probably formed from: 1,4-Dihydro-2,6-dimethyl-4-(*m*-nitrophenyl)pyridine-3,5-dicarboxylic acid 3-methyl ester

3,4-Dihydro-2′-hydroxymethyl-2-methylpapaverinium
→ 3,4-Dihydro-6,7-dimethoxy-2-methylisoquinoline
In: Rabbit B38
→ 4,5-Dimethoxyphthalide
In: Rabbit B38

3,4-Dihydro-2-hydroxymethyl-3-(*o*-methylphenyl)-4-oxopyrido[2,3-*d*]pyrimidine
Formed from: 3,4-Dihydro-2-methyl-3-(*o*-methylphenyl)-4-oxopyrido[2,3-*d*] pyrimidine
→ 2-Carboxy-3,4-dihydro-3-(*o*-methylphenyl)-4-oxopyrido[2,3-*d*] pyrimidine
Probably in: Rat B420

3,7-Dihydro-6-hydroxymethyl-7-methyl-5-phenylpyrrolo[3,4-*e*][1,4]diazepin-2(1*H*)-one
Formed from: Premazepam

1,3-Dihydro-3-(5-hydroxymethyl-3-methyl-1*H*-pyrrol-2-ylmethylene)indol-2-one
Formed from: 3-(3,5-Dimethyl-1*H*-pyrrol-2-ylmethylene)-1,3-dihydroindol-2-one
→ 3-(5-Carboxy-3-methyl-1*H*-pyrrol-2-yl-methylene)-1,3-dihydroindol-2-one
Probably in: Dog K336
Man K336
Monkey K336
Mouse K336
Rat K336
→ 1,3-Dihydro-3-(5-hydroxymethyl-3-methyl-1*H*-pyrrol-2-ylmethylene)indol-2-one-β-D-glucuronide
Probably in: Dog K336
Man K336
Monkey K336
Mouse K336
Rat K336

1,3-Dihydro-3-(5-hydroxymethyl-3-methyl-1*H*-pyrrol-2-ylmethylene)indol-2-one-β-D-glucuronide
Formed from: 1,3-Dihydro-3-(5-hydroxymethyl-3-methyl-1*H*-pyrrol-2-yl-methylene)indol-2-one

S-(1,2-Dihydro-1-hydroxy-1-methylnaphth-2-yl) glutathione
Probably formed from: 1-Methylnaphthalene-1,2-oxide

S-(1,2-Dihydro-2-hydroxy-2-methylnaphth-2-yl) glutathione
Probably formed from: 2-Methylnaphthalene-1,2-oxide

3*H*-1,2-Dihydro-2-(4-hydroxymethylphenylamino)methyl-1-pyrrolizinone
Formed from: 3*H*-1,2-Dihydro-2-(4-methylphenylamino)methyl-1-pyrrolizinone
→ 3*H*-1,2-Dihydro-2-(4-carboxymethylphenylamino)methyl-1-pyrrolizinone
In: Rabbit K555

S-(1,2-Dihydro-2-hydroxy-2-methylphenyl) glutathione
Probably formed from: Toluene-1,2-oxide

3,4-Dihydro-3-(*o*-hydroxymethylphenyl)-2-methyl-4-oxopyrido[2,3-*d*]pyrimidine
Formed from: 3,4-Dihydro-2-methyl-3-(*o*-methylphenyl)-4-oxopyrido[2,3-*d*]pyrimidine
→ 3,4-Dihydro-3-(*o*-hydroxymethylphenyl)-2-methyl-4-oxopyrido[2,3-*d*]pyrimidine-N-oxide
Probably in: Rat B240

3,4-Dihydro-3-(4-hydroxy-2-methylphenyl)-2-methyl-4-oxopyrido[2,3-*d*]pyrimidine
Formed from: 3,4-Dihydro-2-methyl-3-(*o*-methylphenyl)-4-oxopyrido[2,3-*d*]pyrimidine
→ 3,4-Dihydro-3-(4-hydroxy-5-methoxy-2-methylphenyl)-2-methyl-4-oxopyrido[2,3-*d*]pyrimidine
Probably in: Rat B240

3,4-Dihydro-3-(*o*-hydroxymethylphenyl)-2-methyl-4-oxopyrido[2,3-*d*]pyrimidine-N-oxide
Probably formed from: 3,4-Dihydro-3-(*o*-hydroxymethylphenyl)-2-methyl-4-oxopyrido[2,3-*d*]pyrimidine

1,2-Dihydro-4-hydroxy-1-methyl-N-phenyl-2-oxoquinoline-3-carboxamide
Formed from: Roquinimex

1,2-Dihydro-4-hydroxy-N-methyl-N-phenyl-2-oxoquinoline-3-carboxamide
Formed from: Roquinimex

2,3-Dihydro-2-hydroxy-2-methyl-3-phytylnaphthoquinone
Formed from: Vitamin K_1

2,3-Dihydro-3-hydroxy-2-methyl-phytylnaphthoquinone
Formed from: Vitamin K_1

9,10-Dihydro-2-hydroxy-7-methyl-8-propylphenanthrene
→ 9,10-Dihydro-7-methyl-8-propylphenanthr-2-yl sulphate
In: Beef A1211

1,2-Dihydro-1-hydroxy-O-methylsterigmatocystin
→ Aflatoxin G_2
Probably in: Aspergillus E544

2,3-Dihydro-3-hydroxy-N(1)-methyltabersonine
→ 2,3-Dihydro-3,4-dihydroxy-N(1)-methyltabersonine
In: Catharanthus F843

1,2-Dihydro-1-hydroxy-2-methylthionaphthalene
Formed from: N-Acetyl-S-(1,2-dihydro-1-hydroxynaphth-2-yl)-L-cysteine
S-(1,2-Dihydro-1-hydroxynaphth-2-yl)-L-cysteine
→ 1,2,3,4-Tetrahydro-2,4-dihydroxy-1,3-di(methylthio)naphthalene
Probably in: Mouse C634

9,10-Dihydro-9-hydroxy-10-methylthiophenanthrene
Probably formed from: S-(9,10-Dihydro-9-hydroxyphenanthr-10-yl)-L-cysteine
→ 9,10-Dihydro-10-methylthiophenanthr-9-yl-β-D-glucuronide
Probably in: Guinea pig C630
Rat C630

Dihydro-6′-hydroxynabilone
Probably formed from: Dihydronabilone
→ Dihydro-6′-oxonabilone
Probably in: Bacillus A3010

Dihydro-7′-hydroxynabilone
Probably formed from: Dihydronabilone

S-(1,2-Dihydro-1-hydroxynaphth-2-yl)-L-cysteine
→ N-Acetyl-S-(1,2-dihydro-1-hydroxynaphth-2-yl)-L-cysteine
Probably in: Mouse C634
→ 1,2-Dihydro-1-hydroxy-2-methylthionaphthalene
In: Rat D694
See also: C634
→ ((1,2-Dihydro-1-hydroxynaphth-2-yl)thio)acetic acid
Probably in: Mouse C634
→ 3-((1,2-Dihydro-1-hydroxynaphth-2-yl)thio)lactic acid
Probably in: Mouse C634
→ α-Naphthol
In: Rat D694
→ β-Naphthol
In: Rat D694

1,2-Dihydro-1-hydroxynaphth-2-yl-β-D-glucuronide
Probably formed from: *trans*-1,2-Dihydro-1,2-dihydroxynaphthalene

***trans*-S-(1,2-Dihydro-1-hydroxynaphth-2-yl) glutathione**
Formed from: Naphthalene
Naphthalene-1,2-oxide
→ N-Acetyl-S-(1,2-Dihydro-1-hydroxynaphth-2-yl)-L-cysteine
In: Mouse F536

***trans*-S-(1,2-Dihydro-2-hydroxynaphth-1-yl) glutathione**
Probably formed from: Naphthalene
→ N-Acetyl-S-(1,2-Dihydro-2-hydroxynaphth-1-yl)-L-cysteine
In: Mouse F536
→ 1,2,3,4-Tetrahydro-2,3-dihydroxy-1,4-di(methylthio)naphthalene
Probably in: Mouse C634
→ 1,2,3,4-Tetrahydro-1,2,3-trihydroxy-4-methylthionaphthalene
Probably in: Mouse C634

((1,2-Dihydro-1-hydroxynaphth-2-yl)thio)acetic acid
Probably formed from: S-(1,2-Dihydro-1-hydroxynaphth-2-yl)-L-cysteine

3-((1,2-Dihydro-1-hydroxynaphth-2-yl)thio)lactic acid
Probably formed from: S-(1,2-Dihydro-1-hydroxynaphth-2-yl)-L-cysteine

1,3-Dihydro-3-hydroxy-7-nitro-5-phenyl-2*H*-1,4-benzodiazepin-2-one
Formed from: 1,3-Dihydro-7-nitro-5-phenyl-2*H*-1,4-benzodiazepin-2-one

7,8-Dihydro-14-hydroxynormorphine
Probably formed from: 7,8-Dihydro-14-hydroxynormorphinone
6α-Naloxol

7,8-Dihydro-14-hydroxynormorphine
Formed from: α-Naltrexol
Naltrexonium

7,8-Dihydro-14-hydroxynormorphinone
Formed from: Naloxone
Naltrexone
→ 7,8-Dihydro-14-hydroxynormorphine
Probably in: Rat A2000

4,5-Dihydro-4-hydroxy-5-oxo-3,4-benzpyrene
Probably formed from: 4,5-Dihydro-4,5-dihydroxy-3,4-benzpyrene
→ 4,5-Dihydro-4-hydroxy-5-oxo-3,4-benzpyrene-7,8-oxide
In: Rat C69, C225

4,5-Dihydro-4-hydroxy-5-oxo-3,4-benzpyrene-7,8-oxide
Formed from: 4,5-Dihydro-4-hydroxy-5-oxo-3,4-benzpyrene
→ 4,5,7,8-Tetrahydro-4,7,8-trihydroxy-5-oxo-3,4-benzpyrene
In: Rat C69, C225

10,11-Dihydro-4-hydroxy-11(10*H*)-oxodibenz[*b*,*f*]-1,4-oxazepine
Probably formed from: 10,11-Dihydro-11(10*H*)-oxodibenz[*b*,*f*]-1,4-oxazepine
→ 10,11-Dihydro-11(10*H*)-oxodibenz[*b*,*f*]-1,4-oxazepin-4-yl sulphate
In: Rat C495
See also: C492

10,11-Dihydro-7-hydroxy-11(10*H*)-oxodibenz[*b*,*f*]-1,4-oxazepine
Probably formed from: 10,11-Dihydro-11(10*H*)-oxodibenz[*b*,*f*]-1,4-oxazepine
→ 10,11-Dihydro-11(10*H*)-oxodibenz[*b*,*f*]-1,4-oxazepin-7-yl sulphate
In: Rat C495
See also: C492

10,11-Dihydro-9-hydroxy-11(10*H*)-oxodibenz[*b*,*f*]-1,4-oxazepine
Probably formed from: 10,11-Dihydro-11(10*H*)-oxodibenz[*b*,*f*]-1,4-oxazepine
→ 10,11-Dihydro-11(10*H*)-oxodibenz[*b*,*f*]-1,4-oxazepin-9-yl sulphate
In: Rat C495
See also: C492

4′,5′-Dihydro-4′-hydroxy-5-oxoimiloxan
Formed from: Imiloxam
→ 4′,5′-Dihydro-4′,aryl-dihydroxy-5-oxoimiloxan
Probably in: Man G67

2,3-Dihydro-7-hydroxy-2-oxo-1*H*-indole-3-acetic acid
Formed from: 2,3-Dihydro-7-hydroxy-2-oxo-1*H*-indole-3-acetic acid-7-β-D-glucoside
Indole-3-acetic acid
Oxindole-3-acetic acid
→ 2,3-Dihydro-7-hydroxy-2-oxo-1*H*-indole-3-acetic acid-7-β-D-glucoside
Probably in: Zea D543, D619, F947

2,3-Dihydro-7-hydroxy-2-oxo-1*H*-indole-3-acetic acid-7-β-D-glucoside
Probably formed from: 2,3-Dihydro-7-hydroxy-2-oxo-1*H*-indole-3-acetic acid
→ 2,3-Dihydro-7-hydroxy-2-oxo-1*H*-indole-3-acetic acid
With β-glucosidase F947

3,4-Dihydro-5-hydroxy-4-oxo-3-((5-trifluoromethyl-2-benzothiazolyl)methyl)-1-phthalazineacetic acid
Probably formed from: Zopolrestat
→ 3,4-Dihydro-5-hydroxy-4-oxo-3-((5-trifluoromethyl-2-benzothiazolyl)methyl)-1-phthalazineacetic acid-β-D-glucuronide
Probably in: Dog J459
Rat J457

3,4-Dihydro-6-hydroxy-4-oxo-3-((5-trifluoromethyl-2-benzothiazolyl)methyl)-1-phthalazineacetic acid
Probably formed from: Zopolrestat
→ 3,4-Dihydro-6-hydroxy-4-oxo-3-((5-trifluoromethyl-2-benzothiazolyl)methyl)-1-phthalazineacetic acid-β-D-glucuronide
Probably in: Dog J459
Rat J457

3,4-Dihydro-7-hydroxy-4-oxo-3-((5-trifluoromethyl-2-benzothiazolyl)methyl)-1-phthalazineacetic acid
Probably formed from: Zopolrestat
→ 3,4-Dihydro-7-hydroxy-4-oxo-3-((5-trifluoromethyl-2-benzothiazolyl)methyl)-1-phthalazineacetic acid-β-D-glucuronide
Probably in: Dog J459
Rat J457

3,4-Dihydro-5-hydroxy-4-oxo-3-((5-trifluoromethyl-2-benzothiazolyl)methyl)-1-phthalazineacetic acid-β-D-glucuronide
Probably formed from: 3,4-Dihydro-5-hydroxy-4-oxo-3-((5-trifluoromethyl-2-benzothiazolyl)methyl)-1-phthalazineacetic acid

3,4-Dihydro-6-hydroxy-4-oxo-3-((5-trifluoromethyl-2-benzothiazolyl)methyl)-1-phthalazineacetic acid-β-D-glucuronide
Probably formed from: 3,4-Dihydro-6-hydroxy-4-oxo-3-((5-trifluoromethyl-2-benzothiazolyl)methyl)-1-phthalazineacetic acid

3,4-Dihydro-7-hydroxy-4-oxo-3-((5-trifluoromethyl-2-benzothiazolyl)methyl)-1-phthalazineacetic acid-β-D-glucuronide
Probably formed from: 3,4-Dihydro-7-hydroxy-4-oxo-3-((5-trifluoromethyl-2-benzothiazolyl)methyl)-1-phthalazineacetic acid

2,3-Dihydro-8-(2-hydroxy-3-(4-(1-oxo-3-(3,4,5-trimethoxyphenyl)-2-propenyl)-1-piperazinyl)propoxy-1,4-benzodioxin-5-carboxylic acid
Formed from: Isopentyl 2,3-dihydro-8-(2-hydroxy-3-(4-(1-oxo-3-(3,4,5-trimethoxyphenyl)-2-propenyl)-1-piperazinyl)propoxy)-1,4-benzodioxin-5-carboxylate

9,10-Dihydro-9(*S*)-hydroxyphenanthrene
Formed from: 9,10-Dihydrophenanthrene

S-(9,10-Dihydro-9-hydroxyphenanthr-10-yl)-L-cysteine
→ N-Acetyl-S-(9,10-dihydro-9-hydroxyphenanthr-10-yl)-L-cysteine
Probably in: Guinea pig C630
Rat C630
→ 9,10-Dihydro-9-hydroxy-10-methylthiophenanthrene
Probably in: Guinea pig C630
Rat C630
→ S-(9,10-Dihydro-9-hydroxyphenanthr-10-yl)mercaptoacetic acid
Probably in: Guinea pig C630
Rat C630
→ S-(9,10-Dihydro-9-hydroxyphenanthr-10-yl)mercaptolactic acid
Probably in: Guinea pig C630
Rat C630

S-(1,2-Dihydro-2-hydroxyphenanthr-1-yl) glutathione
Probably formed from: Phenanthrene-1,2-oxide

S-(3,4-Dihydro-3-hydroxyphenanthr-4-yl) glutathione
Probably formed from: Phenanthrene-3,4-oxide

S-(9,10-Dihydro-9-hydroxyphenanthr-10-yl) glutathione
Probably formed from: Phenanthrene-9,10-oxide

S-(9,10-Dihydro-9-hydroxyphenanthr-10-yl) mercaptoacetic acid
Probably formed from: S-(9,10-Dihydro-9-hydroxyphenanthr-10-yl)-L-cysteine

S-(9,10-Dihydro-9-hydroxyphenanthr-10-yl) mercaptolactic acid
Probably formed from: S-(9,10-Dihydro-9-hydroxyphenanthr-10-yl)-L-cysteine

1,3-Dihydro-3-hydroxy-5-phenyl-2*H*-1,4-benzodiazepin-2-one
Formed from: 1,3-Dihydro-5-phenyl-2*H*-1,4-benzodiazepin-2-one

(3*S*)-3,4-Dihydro-3-(4-hydroxyphenyl)-2*H*-1-benzopyran-7-ol
See: Equol

(2*R*,3*R*)-3,4-Dihydro-2-(4-hydroxyphenyl)-2*H*-1-benzopyran-3,5,7-triol-7-O-β-D-glucoside
See: Epiafzelechin-7-O-β-D-glucoside

5,6-Dihydro-3-(*p*-hydroxyphenyl)-4,6-dioxo-4*H*-pyran
Formed from: 4′,5,7-trihydroxyisoflavanone
→ 3,4-Dihydroxyphenylacetic acid
In: Fusarium C474

S-(1,2-Dihydro-2-hydroxyphenyl)glutathione
Formed from: Benzene oxide

5-((3,4-Dihydro-2-hydroxy-2-phenylmethyl-2*H*-1-benzopyran-6-yl)methyl)thiazolidine-2,4-dione
Formed from: 5-((3,4-Dihydro-2-phenylmethyl-2*H*-1-benzopyran-6-yl)methyl)thiazolidine-2,4-dione

5-((3,4-Dihydro-2-(*p*-hydroxyphenyl)methyl-2*H*-1-benzopyran-6-yl)methyl)thiazolidine-2,4-dione
Formed from: 5-((3,4-Dihydro-2-phenylmethyl-2*H*-1-benzopyran-6-yl)methyl) thiazolidine-2,4-dione

2,3-Dihydro-2-(*p*-hydroxyphenyl)-6-methyl-4-(β-methylaminoethyl)-1,4-benzoxazin-3-one
Probably formed from: 4-(β-Dimethylaminoethyl)-2,3-dihydro-2-(*p*-hydroxyphenyl)-6-methyl-1,4-benzoxazin-3-one

3*R*-(+)-N-(2,3-Dihydro-5-(*m*-hydroxyphenyl)-1-methyl-2-oxo-1*H*-benzodiazepin-3-yl)-N′-(*m*-methylphenyl)urea
Formed from: 3*R*-(+)-N-(2,3-Dihydro-1-methyl-2-oxo-5-phenyl-1*H*-benzodiazepin-3-yl)-N′-(*m*-methylphenyl)urea

(−)-*cis*-2,3-Dihydro-2-(*p*-hydroxyphenyl)-3-(4-methylpiperazinylmethyl)-1,5-benzothiazepin-4(5*H*)-one
Formed from: (−)-*cis*-2,3-Dihydro-3-(4-methylpiperazinylmethyl)-2-phenyl-1,5-benzothiazepin-4(5*H*)-one

2,3-Dihydro-3-(2-(4-(*p*-hydroxyphenyl) piperazin-1-yl)ethyl)-2-oxooxazolo[4,5-*b*] pyridine
Formed from: 2,3-Dihydro-2-oxo-3-(2-(4-phenylpiperazin-1-yl)ethyl)oxazolo[4,5-*b*] pyridine
→ 2,3-Dihydro-3-(2-(4-(*p*-hydroxyphenyl) piperazin-1-yl)ethyl)-2-oxooxazolo[4,5-*b*] pyridine-β-D-glucuronide
In: Rat H226
→ 2,3-Dihydro-3-(2-(4-(*p*-hydroxyphenyl) piperazin-1-yl)ethyl)-2-oxooxazolo[4,5-*b*] pyridine-O-sulphate
In: Rat H226

2,3-Dihydro-3-(2-(4-(*p*-hydroxyphenyl) piperazin-1-yl)ethyl)-2-oxooxazolo[4,5-*b*] pyridine-β-D-glucuronide
Formed from: 2,3-Dihydro-3-(2-(4-(*p*-hydroxyphenyl)piperazin-1-yl)ethyl)-2-oxooxazolo[4,5-*b*]pyridine

2,3-Dihydro-3-(2-(4-(*p*-hydroxyphenyl) piperazin-1-yl)ethyl)-2-oxooxazolo [4,5-*b*]pyridine-O-sulphate
Formed from: 2,3-Dihydro-3-(2-(4-(*p*-hydroxyphenyl)piperazin-1-yl)ethyl)-2-oxooxazolo[4,5-*b*]pyridine

2,5-Dihydro-2-(*p*-hydroxyphenyl)-3*H*-pyrazolo [4,3-*c*]quinazolin-3-one
Formed from: 2,5-Dihydro-2-phenyl-3*H*-pyrazolo[4,3-*c*]quinazolin-3-one

1,3-Dihydro-5-hydroxy-1-(3-(-1-piperazinyl) propyl)-2*H*-benzimidazol-2-one
Probably formed from: 1,3-Dihydro-1-(3-(1-piperazinyl)propyl)-2*H*-benzimidazol-2-one

1,3-Dihydro-6-hydroxy-1-(3-(-1-piperazinyl) propyl)-2*H*-benzimidazol-2-one
Probably formed from: 1,3-Dihydro-1-(3-(1-piperazinyl)propyl)-2*H*-benzimidazol-2-one

S-(4,5-Dihydro-4-hydroxypyren-5-yl)glutathione
Formed from: Pyrene-4,5-oxide

S-(4,5-Dihydro-5-hydroxypyren-4-yl)glutathione
Formed from: Pyrene-4,5-oxide

Dihydro-3-hydroxyquinine
Formed from: Dihydroquinine

1′,2′-Dihydro-2′-hydroxyrotenone
Formed from: 1′,2′-Dihydrorotenone

Dihydro-10-hydroxysanguinarine
Formed from: Dihydrosanguinarine
→ Dihydrochelirubine
In: Eschscholtzia K777

N-(3,4-Dihydro-4-hydroxy*spiro*(2*H*-1-benzopyr-an-2,4′-piperidin)-6-yl)methanesulphonamide
Formed from: (+)-N-(1′-(6-Cyano-1,2,3,4-tetrahydro-2(*R*)-naphthalenyl)-3,4-dihydro-4(*R*)-hydroxy*spiro*(2*H*-1-benzopyran-2,4′-piperidin-6-yl)methanesulphonamide
Probably formed from: 3,4-Dihydro-4-oxo*spiro*(2*H*-1-benzopyran-2,4′-piperidin-6-yl)methanesulphonamide

16,17-Dihydro-17-hydroxysterigmatocystin
→ 16,17-Dihydro-17-hydroxysterigmatocystin-14,17-hydrate
Probably in: Aspergillus G120

16,17-Dihydro-17-hydroxysterigmatocystin-14,17-hydrate
Probably formed from: 16,17-Dihydro-17-hydroxysterigmatocystin

1,2-Dihydro-2-hydroxy-3-trifluoromethylbenzoic acid
Formed from: *m*-Trifluoromethylbenzoic acid

3,4-Dihydro-3-((4-hydroxy-5-trifluoromethyl-2-benzothiazolyl)methyl)-4-oxo-1-phthalazineacetic acid
Probably formed from: Zopolrestat

1,4-Dihydro-1-hydroxy-6-trifluoromethylquinoxaline-2,3-dione
Formed from: 1,4-Dihydro-6-trifluoromethylquinoxaline-2,3-dione
→ 1,4-Dihydro-1,4-dihydroxy-6-trifluoromethylquinoxaline-2,3-dione
Probably in: Cat A1751

1,4-Dihydro-7-hydroxy-6-trifluoromethylquinoxaline-2,3-dione
Probably formed from: 1,4-Dihydro-6-trifluoromethylquinoxaline-2,3-dione

7,8-Dihydro-7-hydroxy-3,4,12-trimethoxy-10,11-methylenedioxy-5*H*-isoindolo[1,2-*b*][3]benzazepin-5-one
Formed from: 7-Acetoxy-7,8-dihydro-3,4,12-trimethoxy-10,11-methylenedioxy-5*H*-isoindolo[1,2-*b*][3]benzazepin-5-one

1,2-Dihydro-1-hydroxy-2,2,4-trimethylquinoline
Formed from: 1,2-Dihydro-2,2,4-trimethylquinoline
→ 1,2-Dihydro-1,6-dihydroxy-2,2,4-trimethylquinoline
Probably in: Rat E871
→ 1,2-Dihydro-1-hydroxy-2,2,4-trimethylquinoline-β-D-glucuronide
Probably in: Rat E871

1,2-Dihydro-6-hydroxy-2,2,4-trimethylquinoline
Formed from: 1,2-Dihydro-2,2,4-trimethylquinoline
Ethoxyquin
→ 1,2-Dihydro-6-hydroxy-2,2,4-trimethylquinoline-β-D-glucuronide
In: Mouse H712
→ 1,2-Dihydro-6-hydroxy-2,2,4-trimethylquinoline sulphate
In: Mouse H712
Rat H712
See also: E871
→ 2,2,4-Trimethyl-6-quinolone
In: Rat A3846

1,2-Dihydro-1-hydroxy-2,2,4-trimethylquinoline-β-D-glucuronide
Probably formed from: 1,2-Dihydro-1-hydroxy-2,2,4-trimethylquinoline

1,2-Dihydro-6-hydroxy-2,2,4-trimethylquinoline-β-D-glucuronide
Formed from: 1,2-Dihydro-6-hydroxy-2,2,4-trimethylquinoline

1,2-Dihydro-6-hydroxy-2,2,4-trimethylquinoline sulphate
Formed from: 1,2-Dihydro-6-hydroxy-2,2,4-trimethylquinoline

2′,3′-Dihydro-3′-hydroxywighteone
Probably formed from: 2′,3′-Dihydro-2′,3′-dihydroxywighteone

Dihydroiloperidone
Formed from: Iloperidone
→ 1-(4-(3-(4-(6-Fluoro-1,2-benzisoxazol-3-yl)-1-piperidinyl)propoxy)-3-hydroxyphenyl) ethanol
Probably in: Dog H447
Man H447
Rat H447
→ 1-(4-(3-(4-(6-Fluoro-1,2-benzisoxazol-3-yl)-1-piperidinyl)propoxy)-6-hydroxyphenyl) ethanol
Probably in: Dog H447
Man H447
Rat H447

2-(2-(4,5-Dihydro-1*H*-imidazol-2-yl)-1-(3,4-dihydroxyphenyl)ethyl)pyridine
Probably formed from: *p*-Hydroxymidaglizole
→ 2-(2-(4,5-Dihydro-1*H*-imidazol-2-yl)-1-(4-hydroxy-3-methoxyphenyl)ethyl)pyridine
Probably in: Rat F334

2-(2-(4,5-Dihydro-1*H*-imidazol-2-yl)-1-(4-hydroxy-3-methoxyphenyl)ethyl)pyridine
Probably formed from: 2-(2-(4,5-Dihydro-1*H*-imidazol-2-yl)-1-(3,4-dihydroxyphenyl) ethyl)pyridine

2-(2-(4,5-Dihydro-1*H*-imidazol-2-yl)-1-(*p*-hydroxyphenyl)ethyl)pyridine
Probably formed from: *p*-Hydroxymidaglizole

5,6-Dihydro-7-(1*H*-imidazol-1-yl)naphthalene-2-carboxyl-β-D-glucuronic acid
Formed from: 5,6-Dihydro-7-(1*H*-imidazol-1-yl)naphthalene-2-carboxylic acid

5,6-Dihydro-7-(1*H*-imidazol-1-yl)naphthalene-2-carboxylic acid
→ 5,6-Dihydro-7-(1*H*-imidazol-1-yl) naphthalene-2-carboxyl-β-D-glucuronic acid
In: Man G530
Rat G530

2-(2-(4,5-Dihydro-1*H*-imidazol-2-yl)-1-phenylethyl)pyridine
See: Midaglizole

1,2-Dihydro-1-iminoxyl-2-oxonaphthalene radical
Formed from: 1-Nitroso-2-naphthol

1,2-Dihydro-2-iminoxyl-1-oxonaphthalene radical
Formed from: 2-Nitroso-1-naphthol

2,3-Dihydro-1*H*-indole
See: Indoline

7,8-Dihydroisocodeinone
Formed from: 7,8-Dihydrocodeinone

Dihydroisoferulic acid (3-(3-hydroxy-4-methoxyphenyl)propionic acid
Formed from: Dihydrocaffeic acid
Hesperetin

Dihydroisomorphine
Formed from: Hydromorphone

1,4-Dihydro-5-isopropoxycarbonyl-2,6-dimethyl-4-(*m*-nitrophenyl)pyridine-3-carboxylic acid
Formed from: Nimodipine
→ 5-Isopropoxycarbonyl-2,6-dimethyl-4-(*m*-nitrophenyl)pyridine-3-carboxylic acid
Probably in: Dog G189
Monkey G189
Rat G189

2,3-Dihydro-9*H*-isoxazolo[3,2-*b*]quinazolin-9-one
→ 3-(*o*-Carboxyphenylimino)isoxazolidine
In: Rat A3074

9,10-Dihydrokadsurenone (2-(3,4-dimethoxyphenyl)-3,3a,9,10-tetrahydro-3a-methoxy-3-methyl-5-(2-propenyl)-6(2*H*)-benzofuranone)
→ 9,10-Dihydro-8-hydroxykadsurenone
In: Rat E950
→ 9,10-Dihydro-9-hydroxykadsurenone
In: Monkey E950
Rat E950
→ 9,10-Dihydro-10-hydroxykadsurenone
In: Monkey E950
Rat E950

Dihydrokaempferol
See: Aromadendrin

Dihydrokawain ((6*S*)-5,6-dihydro-4-methoxy-6-(2-phenylethyl)-2*H*-pyran-2-one)
→ Benzoic acid
Probably in: Rat A3485
→ Dihydro-*m*-hydroxykawain
Probably in: Rat A3485
→ Dihydro-*p*-hydroxykawain
Probably in: Rat A3485
→ Dihydro-8-hydroxykawain
Probably in: Rat A3485
→ 4-Hydroxy-6-phenylhexan-2-one
In: Rat A3485

Dihydroketotifen
Formed from: Ketotifen
→ Ketotifen
In: Man K341

Dihydrolysergol ((8β)-6-methylergoline-8-methanol)
→ Dihydrolysergol-O-β-D-fructofuranoside
In: Claviceps F512, F516
→ Dihydrolysergol-β-galactoside
In: Aspergillus H11

Dihydrolysergol-O-β-D-fructofuranoside
Formed from: Dihydrolysergol

Dihydrolysergol-β-galactoside
Formed from: Dihydrolysergol

Dihydromaackiain
Formed from: Maackiain

Dihydromacarpine
Formed from: Dihydro-12-hydroxychelirubine
Norcoclaurine
→ Macarpine
In: Eschscholtzia E604
Thalictrum K776

Dihydromagnolol
Formed from: Magnolol
→ 5-Allyl-2,2′-dihydroxy-5′-(1-propen-1(*E*)-yl)biphenyl
Probably in: Rat D549, D625
→ Tetrahydromagnolol
In: Rat D549
See also: D625

Dihydro-10-methoxychelirubine
Formed from: Dihydro-10-hydroxychelirubine

1,2-Dihydro-8-methoxy-2-methyl-6,7-methylenedioxyisoquinoline
Formed from: 1-(α-Diazoethoxycarbonylmethyl)hydrocotarnine
→ 8-Methoxy-2-methyl-6,7-methylenedioxyisoquinoline
Probably in: Rat A2382

4-(2,3-Dihydro-7-methoxy-3-methyl-5-(1-propenyl)-2-benzofuranyl)-2-methoxyphenol
See: Dehydrodiisoeugenol

4,9-Dihydro-7-methoxy-1-methyl-3*H*-pyrido[3,4-*b*]indole
See: Harmaline

5,8-Dihydro-5-methoxy-8-oxo-2*H*-1,3-dioxolo[4,5-*g*]quinoline-7-carboxylic acid
→ 1,4-Dihydro-7-hydroxy-1,6-dimethoxy-4-oxoquinoline-3-carboxylic acid
In: Man A3993
→ 5,8-Dihydro-5-methoxy-8-oxo-2*H*-1,3-dioxolo[4,5-*g*]quinoline-7-carboxyl-β-D-glucuronide
In: Man A3993, A3993a
→ 5,8-Dihydro-8-oxo-2*H*-1,3-dioxolo[4,5-*g*]quinoline-7-carboxylic acid
In: Man A3993, A3993a

5,8-Dihydro-5-methoxy-8-oxo-2*H*-1,3-dioxolo[4,5-*g*]quinoline-7-carboxyl-β-D-glucuronide
Formed from: 5,8-Dihydro-5-methoxy-8-oxo-2*H*-1,3-dioxolo[4,5-*g*]quinoline-7-carboxylic acid

9,10-Dihydro-4-methoxy-2,5-phenanthrenediol
See: Hircinol

4,5-Dihydro-6-(2-(4-methoxyphenyl)-1*H*-benzimidazol-5-yl)-5-methyl-3-(2*H*)-pyridazinone
See: Pimobendan

5,6-Dihydro-3-(*p*-methoxyphenyl)-4,6-dioxo-4*H*-pyran
Formed from: 5,7-Dihydroxy-4′-methoxyisoflavanone
→ 3,4-Dihydroxyphenylacetic acid
In: Fusarium C474
→ *p*-Methoxyphenylacetic acid
Probably in: Fusarium C446

(6*R*)-5,6-Dihydro-4-methoxy-6-((1*E*)-2-phenylethenyl)-2*H*-pyran-2-one
See: Kawain

(6*S*)-5,6-dihydro-4-methoxy-6-(2-phenylethyl)-2*H*-pyran-2-one
See: Dihydrokawain

3,12-Dihydro-6-methoxy-3,3,12-trimethyl-7*H*-pyrano[2,3-*c*]acridan-7-one
See: Acronine

Dihydro-4′-methoxywarfarin
Formed from: 4′-Methoxywarfarin

10,11-Dihydro-10β-methylaminoethyl-11-oxo-5H-dibenzo[*b,e*][1,4]diazepine
Probably formed from: 10,11-Dihydro-5-methyl-10β-methylaminoethyl-11-oxo-5*H*-dibenzo[*b*,*e*][1,4]diazepine

(2*R*)-2,3-Dihydro-2-(3-methyl-2-butenyl)-1,4-naphthalenedione
See: Catalponone

10,11-Dihydro-5-methyl-5*H*-dibenzo[*a*,*d*]cyclohepten-5,10-imine
→ 10,11-Dihydro-2-hydroxy-5-methyl-5*H*-dibenzo[*a*,*d*]cyclohepten-5,10-imine
In: Rat C599
→ 10,11-Dihydro-8-hydroxy-5-methyl-5*H*-dibenzo[*a*,*d*]cyclohepten-5,10-imine
In: Rat C599
→ 10,11-Dihydro-N-hydroxy-5-methyl-5*H*-dibenzo[*a*,*d*]cyclohepten-5,10-imine
In: Rat C599

10,11-Dihydro-N-methyl-5*H*-dibenzo[*a*,*d*]cycloheptene-$\Delta^{5,\gamma}$-propylamine
See: Nortriptyline

1,3-Dihydro-3-methylene-2*H*-indol-2-one
See: 3-Methyleneoxindole

1-((2*R*)-2,3-Dihydro-2-(1-methylethenyl)-5-benzofuranyl)ethanone
See: Tremetone

3,4-Dihydro-1-methylisoquinoline
Formed from: 1,2,3,4-Tetrahydro-1-methylisoquinoline

2,3-Dihydro-3-methyl-9*H*-isoxazolo[3,2-*b*]quinazolin-9-one
→ 2,3-Dihydro-3-hydroxy-3-methyl-9*H*-isoxazolo[3,2-*b*]quinazolin-9-one
In: Dog B215
Rat B215

10,11-Dihydro-5-methyl-10β-methylaminoethyl-11-oxo-5*H*-dibenzo[*b,e*][1,4]diazepine
Formed from: 10β-Dimethylaminoethyl-10,11-dihydro-5-methyl-11-oxo-5*H*-dibenzo[*b,e*][1,4]diazepine
→ 10β-Aminoethyl-10,11-dihydro-5-methyl-11-oxo-5*H*-dibenzo[*b,e*][1,4]diazepine
In: Man A712
→ 10,11-Dihydro-10β-methylaminoethyl-11-oxo-5*H*-dibenzo[*b,e*][1,4]diazepine
Probably in: Man A712

2,3-Dihydro-6-methyl-4-(β-methylaminoethyl)-2-phenyl-1,4-benzoxazin-3-one
Formed from: 4-(β-Dimethylaminoethyl)-2,3-dihydro-6-methyl-2-phenyl-1,4-benzoxazin-3-one

3,4-Dihydro-2-methyl-3-(*o*-methylphenyl)-4-oxopyrido[2,3-*d*]pyrimidine
→ 3,4-Dihydro-2-hydroxymethyl-3-(*o*-methylphenyl)-4-oxopyrido[2,3-*d*] pyrimidine
In: Rat B420
→ 3,4-Dihydro-3-(*o*-hydroxymethylphenyl)-2-methyl-4-oxopyrido[2,3-*d*]pyrimidine
In: Rat B420
→ 3,4-Dihydro-3-(4-hydroxy-2-methylphenyl)-2-methyl-4-oxopyrido [2,3*d*]pyrimidine
In: Rat B420
→ 3,4-Dihydro-3-(*o*-methylphenyl)-4-oxopyrido[2,3*d*]pyrimidine
In: Rat B420

1,3-Dihydro-1-methyl-7-nitro-5-phenyl-2*H*-1,4-benzodiazepin-2-one
See: Nimetazepam

5,6-Dihydro-2-methyl-1,4-oxathiin-3-carboxylic acid
Formed from: 3-Carboxanilido-5,6-dihydro-2-methyl-1,4-oxathiin

3*R*-(+)-N-(2,3-Dihydro-1-methyl-2-oxo-5-phenyl-1*H*-1,4-benzodiazepin-3-yl)-N′-(*m*-methylphenyl)urea
→ 3*R*-(+)-N-(2,3-Dihydro-5-(*m*-hydroxyphenyl)-1-methyl-2-oxo-1*H*-1,4-benzodiazepin-3-yl)-N′-(*m*-methylphenyl)urea
In: Dog G85
→ 3*R*-(+)-N-(2,3-Dihydro-1-methyl-2-oxo-5-phenyl-1*H*-1,4-benzodiazepin-3-yl)-N′-(4-hydroxy-3-methylphenyl)urea
In: Dog G85

3*R*-(+)-N-(2,3-Dihydro-1-methyl-2-oxo-5-phenyl-1*H*-1,4-benzodiazepin-3-yl)-N′-(4-hydroxy-3-methylphenyl)urea
Formed from: 3*R*-(+)-N-(2,3-Dihydro-1-methyl-2-oxo-5-phenyl-1*H*-1,4-benzodiazepin-3-yl)-N′-(*m*-methylphenyl)urea

3,4-Dihydro-2-methylpapaverinium
→ 3,4-Dihydro-6,7-dimethoxy-2-methylisoquinolone
In: Rabbit B38
→ Veratraldehyde
In: Rabbit B38

3*H*-1,2-Dihydro-2-(4-methylphenylamino)methyl-1-pyrrolizinone
→ 3*H*-1,2-Dihydro-2-(4-hydroxymethylphenylamino)methyl-1-pyrrolizinone
In: Rabbit K555

(3*S*)-3,4-Dihydro-4-methyl-3-(phenylmethyl)-1*H*-1,4-benzodiazepine-2,5-dione
See: Cyclopeptine

3,4-Dihydro-3-(*o*-methylphenyl)-4-oxopyrido [2,3-*d*]pyrimidine
Formed from: 3,4-Dihydro-2-methyl-3-(*o*-methylphenyl)-4-oxopyrido[2,3-*d*]pyrimidine

2,3-Dihydro-1-methyl-4-phenylpyridinium
Formed from: 1,2,3,6-Tetrahydro-1-methyl-4-phenylpyridine
→ 5,6-Dihydro-1-methyl-4-phenyl-2(1*H*)-pyridone
In: Mouse E684
→ 1-Methyl-4-phenylpyridinium
In: Mouse E684

1,3-Dihydro-4-methyl-5-(4-(methylthio)benzoyl)-2*H*-imidazol-2-one
See: Fenoximone

5,6-Dihydro-1-methyl-4-phenyl-2(1*H*)-pyridone
Formed from: 2,3-Dihydro-1-methyl-4-phenylpyridinium
1,2,3,6-Tetrahydro-1-methyl-4-phenylpyridine
→ 1-Methyl-4-phenyl-2(1*H*)-pyridone
In: Mouse E684

See also: E711, K32

3,7-Dihydro-6-methyl-5-phenylpyrrolo[3,4-*e*][1,4]diazepin-2(1*H*)-one
Formed from: Premazepam

(−)-*cis*-2,3-Dihydro-3-(4-methylpiperazinylmethyl)-2-phenyl-1,5-benzothiazepin-4(5*H*)-one
→ (−)-*cis*-2,3-Dihydro-2-(*p*-hydroxyphenyl)-3-(4-methylpiperazinylmethyl)-1,5-benzothiazepin-4(5*H*)-one
In: Rabbit G270
Rat G270
→ (−)-*cis*-2,3-Dihydro-3-(4-methylpiperazinylmethyl)-2-phenyl-1,5-benzothiazepin-4(5*H*)-one sulphoxide
In: Rabbit G270
Rat G270
→ (−)-*cis*-2,3-Dihydro-2-phenyl-3-piperazinylmethyl-1,5-benzothiazepin-4(5*H*)-one
In: Rabbit G270
Rat G270, G271

(−)-*cis*-2,3-Dihydro-3-(4-methylpiperazinylmethyl)-2-phenyl-1,5-benzothiazepin-4(5*H*)-one sulphoxide
Formed from: (−)-*cis*-2,3-Dihydro-3-(4-methylpiperazinylmethyl)-2-phenyl-1,5-benzothiazepin-4(5*H*)-one

10,11-Dihydro-5-(1-methyl-4-piperidinylidene)-5*H*-dibenzo[*a*,*d*]cycloheptene-3-carboxylic acid
→ 10,11-Dihydro-5-(4-piperidinylidene)-5*H*-dibenzo[*a*,*d*]cycloheptene-3-carboxylic acid
In: Dog C17
Rat C17

2-(*p*-((2,3-Dihydro-2-methylpropyl)amino)phenyl)propionic acid
Formed from: 2-(*p*-((2-Methyl-2-propenyl)amino)phenyl)propionic acid

9,10-Dihydro-7-methyl-8-propylphenanthr-2-yl sulphate
Formed from: 9,10-Dihydro-2-hydroxy-7-methyl-8-propylphenanthrene

4,9-Dihydro-1-methyl-3*H*-pyrido[3,4-*b*]indole
See: Harmalan

4,9-Dihydro-1-methyl-3*H*-pyrido[3,4-*b*]indol-7-ol
See: Harmalol

3,4-Dihydro-6-methyl-2-quinolone
Probably formed from: 2-Hydroxy-6-methylquinoline

3,4-Dihydro-8-methyl-2-quinolone
Probably formed from: 2-Hydroxy-8-methylquinoline

1,2-Dihydro-1-methylraucaffricine
→ 1,2-Dihydro-1-methylvomilenine
In: Rauwolfia K939

Dihydro-O-methylsterigmatocystin
Probably formed from: Dihydrosterigmatocystin
→ Aflatoxin B_2
In: Aspergillus E767, F398
See also: F281
→ Aflatoxin G_2
In: Aspergillus E767, F398, K237
See also: F281

Dihydromorphinone
See: Hydromorphone

9,10-Dihydro-10-methylthiophenthr-9-yl-β-D-glucuronide
Probably formed from: 9,10-Dihydro-9-hydroxy-10-methylthiophenanthrene

1,2-Dihydro-1-methylvomilenine
Formed from: 1,2-Dihydro-1-methylraucaffricine

Dihydro-S-methylziprasidone
Formed from: Ziprasidone
→ Dihydro-S-methylziprasidone sulphoxide
Probably in: Man J279
Rat J281

Dihydro-S-methylziprasidone sulphoxide
Probably formed from: Dihydro-S-methylziprasidone

Dihydromorphine
Formed from: Dihydrocodeine
Hydromorphone
Morphine
→ Dihydromorphine-3-β-D-glucuronide
In: Rat A2350
See also: A823, H861, H933
→ Dihydronormorphine
In: Rat A2350

Dihydromorphine-β-D-glucuronide
Probably formed from: Dihydromorphine

S-(7,8-Dihydromorphinon-8-yl)glutathione
Probably formed from: Morphinone

Dihydromyricetin (3,3′,4′,5,5′,7-hexahydroxyflavanone)
Formed from: 3′,4′,5,5′,7-Pentahydroxyflavanone
→ Delphinidin
In: Petunia A2740
→ (+)-Gallocatechin
In: Ginkgo D617
→ Leucodelphinidin
In: Fir D617
Ginkgo D617
Matthiola D782
Petunia E557

Dihydronabilone
Formed from: Nabilone
→ Dihydro-6′-hydroxynabilone
In: Bacillus A3010
→ Dihydro-7′-hydroxynabilone
In: Nocardia B37

Dihydronafimidone
Formed from: Nafimidone
→ Dihydronafimidone-N-β-D-glucuronide
Probably in: Baboon F483
Cynomolgus monkey F483
Man F483
→ Dihydronafimidone-O-β-D-glucuronide
Probably in: Baboon F483
Cynomolgus monkey F483
Dog F483
Man F483
Rat E908
→ 1-(2-Hydroxy-2-(β-naphthyl)ethyl) hydantoin
Probably in: Baboon F483
Cynomolgus monkey F483
Dog F483
Man F483

Dihydronafimidone-N-β-D-glucuronide
Probably formed from: Dihydronafimidone

Dihydronafimidone-O-β-D-glucuronide
Formed from: Dihydronafimidone

1,2-Dihydronaphthalene
→ (+)-1,2-Dihydro-*cis*-(1*R*,2*S*)-dihydroxynaphthalene
In: Pseudomonas H678
→ (*1R*)-1,2-Dihydro-α-naphthol
In: Pseudomonas G108, H678
→ (*1R*)-1,2-Dihydro-β-naphthol
In: Sphingomonas H880
→ Naphthalene
In: Sphingomonas H880
→ (+)-1,2,3,4-Tetrahydro-*cis*-(1*S*,2*R*)-dihydroxynaphthalene (stereochemistry uncertain)
In: Pseudomonas H678
See also: K167
→ (−)-1,2,3,4-Tetrahydro-*cis*-(1*R*,2*S*)-dihydroxynaphthalene
In: Pseudomonas H678
Sphingomonas H880
→ Tetralone
In: Sphingomonas H880

1,4-Dihydronaphthalene
→ 1,2-Dihydro-β-naphthol
In: Pseudomonas G108
→ 1,4-Dihydro-α-naphthol
In: Pseudomonas G108

1,2-Dihydronaphthalene-3,4-oxide
→ 1-(S-Glutathionyl)-1,2,3,4-tetrahydro-2-hydroxynaphthalene
In: Rat D293

(*1R*)-1,2-Dihydro-α-naphthol
Formed from: 1,2-Dihydronaphthalene

(*1R*)-1,2-Dihydro-α-naphthol
Formed from: 1,2-Dihydronaphthalene
1,4-Dihydronaphthalene

1,4-Dihydro-α-naphthol
Formed from: 1,4-Dihydronaphthalene

7,8-Dihydro-2-naphthol
Probably formed from: 5,6,7,8-Tetrahydro-2-naphthol

Dihydronaringenin
See: Apiforol

1,3-Dihydro-7-nitro-5-phenyl-2*H*-1,4-benzodiazepin-2-one (mogadon, nitrazepam)
Formed from: Nimetazepam
→ 7-Amino-1,3-dihydro-5-phenyl-2*H*-1,4-benzodiazepin-2-one
In: Clostridium J388
Dog A1763
Man B104

Mouse G903
Rat A297, A352, D113, G903
See also: A2466, A3876
→ 2-Amino-3-hydroxy-5-nitrobenzophenone
In: Rabbit A2647
→ 2-Amino-5-nitrobenzophenone
Species unclear A3843
→ 2′-Benzoyl-2-hydroxy-4′-nitroacetanilide
In: Rabbit A3876
→ 1,3-Dihydro-3-hydroxy-7-nitro-5-phenyl-2*H*-1,4-benzodiazepin-2-one
In: Rat A352
See also: A1763
→ 2-(2-Hydroxyacetamido)-5-nitrobenzophenone
Probably in: Rabbit A2647

Dihydro-4′-nitrowarfarin
Formed from: 4′-Nitrowarfarin

Dihydronorcodeine
Formed from: Dihydrocodeine

Dihydronormorphine
Formed from: Dihydromorphine
Normorphine

2,3-Dihydro-2-oxo-1*H*-benzimidazole-1-propionic acid
Formed from: Oxatomide

2,3-Dihydro-2-oxobrimodinine
Formed from: Brimodinine

3,4-Dihydro-3-oxobrimodinine
Formed from: Brimodinine

10,11-Dihydro-11(10*H*)-oxodibenz[*b*,*f*]-1,4-oxazepine
Formed from: Dibenz[*b*,*f*]-1,4-oxazepine
10,11-Dihydrodibenz[*b*,*f*]-1,4-oxazepine
Probably formed from: 2-Amino-2′-carboxydiphenyl ether
→ 10,11-Dihydro-4-hydroxy-11(10*H*)-oxodibenz[*b*,*f*]-1,4-oxazepine
In: Rat A3596, C495
See also: C492
→ 10,11-Dihydro-7-hydroxy-11(10*H*)-oxodibenz[*b*,*f*]-1,4-oxazepine
In: Rat A3596, C495
See also: C492
→ 10,11-Dihydro-9-hydroxy-11(10*H*)-oxodibenz[*b*,*f*]-1,4-oxazepine
In: Rat C495
See also: C492

6,11-Dihydro-1-oxodibenz[*b*,*e*]oxepin-3-acetic acid
See: Oxepinac

10,11-Dihydro-11(10*H*)-oxodibenz[*b*,*f*]-1,4-oxazepin-4-yl sulphate
Formed from: 10,11-Dihydro-4-hydroxy-11(10*H*)-oxodibenz[*b*,*f*]-1,4-oxazepine

10,11-Dihydro-11(10*H*)-oxodibenz[*b*,*f*]-1,4-oxazepin-7-yl sulphate
Formed from: 10,11-Dihydro-7-hydroxy-11(10*H*)-oxodibenz[*b*,*f*]-1,4-oxazepine

10,11-Dihydro-11(10*H*)-oxodibenz[*b*,*f*]-1,4-oxazepin-9-yl sulphate
Formed from: 10,11-Dihydro-9-hydroxy-11(10*H*)-oxodibenz[*b*,*f*]-1,4-oxazepine

6,11-Dihydro-11-oxodibenz[*b*,*e*]oxepin-2-acetyl-β-D-glucuronic acid
Formed from: Isoxepac

6,11-Dihydro-11-oxodibenz[*b*,*e*]oxepin-2-acetylglycine
Formed from: Isoxepac

6,11-Dihydro-11-oxodibenz[*b*,*e*]oxepin-2-acetyltaurine
Formed from: Isoxepac

6,11-Dihydro-11-oxodibenz[*b*,*e*]oxepin-3-acetyltaurine
Formed from: Isoxepac

5,8-Dihydro-8-oxo-2*H*-1,3-dioxolo[4,5-*g*]quinoline-7-carboxylic acid
→ 5,8-Dihydro-8-oxo-2*H*-1,3-dioxolo[4,5-*g*]quinoline-7-carboxyl-β-D-glucuronic acid
Probably in: Man A3993, A3993a

5,8-Dihydro-8-oxo-2*H*-1,3-dioxolo[4,5-*g*]quinoline-7-carboxyl-β-D-glucuronic acid
Probably formed from: 5,8-Dihydro-8-oxo-2*H*-1,3-dioxolo[4,5-*g*]quinoline-7-carboxylic acid

3-(1,3-Dihydro-3-oxo-3*H*-indol-2-ylidene)-1,3-dihydro-2*H*-indol-2-one
See: Indirubin

4-(1,3-Dihydro-1-oxo-2*H*-isoindol-2-yl)-α-methylbenzeneacetic acid
See: (*R*)-Indoprofen

2,3-Dihydro-2-oxolysergic diethylamide
Formed from: Lysergic diethylamide

Dihydro-6′-oxonabilone
Probably formed from: Dihydro-6′-hydroxynabilone

Dihydro-7′-oxonabilone
Probably formed from: Dihydro-7′-hydroxynabilone
→ 3-(6-Carboxy-1,1-dimethylhexyl)-6,6α,7,8,10,10α-hexahydro-1,9-dihydroxy-6,6-dimethyl-9*H*-dibenzo[*b*,*d*]pyran
Probably in: Nocardia B37

2,5-Dihydro-5-oxo-4-phenyl-2-furanacetic acid
See: 4-Phenylmuconolactone

5-((3,4-Dihydro-4-oxo-2-phenylmethyl-2*H*-1-benzopyran-6-yl)methyl)thiazolidine-2,4-dione
Formed from: 5-((3,4-Dihydro-2-phenylmethyl-2*H*-1-benzopyran-6-yl)methyl)thiazolidine-2,4-dione

2,3-Dihydro-2-oxo-3-(2-(4-phenylpiperazin-1-yl)ethyl)oxazolo[4,5-*b*]pyridine
→ 2,3-Dihydro-3-(2-(4-(*p*-hydroxyphenyl)piperazin-1-yl)ethyl)-2-oxooxazolo[4,5-*b*]pyridine
In: Rat H226
→ 1-Phenylpiperazine
In: Rat H226

2,3-Dihydro-5-(2-oxopropyl)-1,4-benzoxazepine
Probably formed from: 4-(2-Isopropylaminoethoxy)-3-oxobutyrophenone

4-(1,2-Dihydro-2-oxo-6-quinolyloxy)-N-methylbutyramide
Formed from: Cilostamide

N-**(1′-(3,4-Dihydro-4-oxo*spiro*((2*H*)-1-benzopyran-2,4′-piperidin)-6-yl)methanesulphonamide**
Formed from: N-(1′-(2-(5-Benzofurazanyl)ethyl)-3,4-dihydro-4-oxo*spiro*(2*H*-1-benzopyran-2,4′-piperidin)-6-yl)methanesulphonamide
→ N-(3,4-Dihydro-4-hydroxy*spiro*(2*H*-1-benzopyran-2,4′-piperidin)-6-yl)methanesulphonamide
Probably in: Rat G577

3,4-Dihydro-4-oxo-3-((5-(trifluoromethyl)-2-benzothiazolyl)methyl)-1-phthalazineacetic acid
See: Zopolrestat

3,4-Dihydropapaveroline (1-((3,4-dihydroxyphenyl)methyl)-3,4-dihydro-6,7-isoquinolinediol)
Formed from: Tetrahydropapaveroline
→ Tetrahydropapaveroline
Probably in: Litsea A3222

9,10-Dihydrophenanthrene
→ 9,10-Dihydro-9(*S*)-hydroxyphenanthrene
In: Pseudomonas J289
→ *cis*-3,4,9,10-Tetrahydro-3(*S*),4(*R*)-dihydroxyphenanthrene
In: Pseudomonas J289

3,4-Dihydrophenanthrene-1,2-oxide
→ 1-(S-Glutathionyl)-1,2,3,4-tetrahydro-2-hydroxyphenanthrene
In: Rat D293

5-(4,5-Dihydro-2-phenylbenz[*e*]indol-3-yl)salicylic acid
See: Fendosal

1,3-Dihydro-5-phenyl-2*H*-1,4-benzodiazepin-2-one
→ 1,3-Dihydro-3-hydroxy-5-phenyl-2*H*-1,4-benzodiazepin-2-one
In: Mouse A3269
Rat A3269

2,3-Dihydro-2-phenyl-4*H*-1-benzopyran-4-one
See: Flavanone

2,3-Dihydro-3-phenyl-4*H*-1-benzopyran-4-one
See: Isoflavanone

5-((3,4-Dihydro-2-phenylmethyl-2*H*-1-benzopyran-6-yl)methyl)-5-hydroxythiazolidine-2,4-dione
Probably formed from: 5-((3,4-Dihydro-2-phenylmethyl-2*H*-1-benzopyran-6-yl)methyl)thiazolidine-2,4-dione

5-((3,4-Dihydro-2-phenylmethyl-2*H*-1-benzopyran-6-yl)methyl)thiazolidine-2,4-dione
→ 5-((3,4-Dihydro-2-(α-hydroxybenzyl)-2*H*-1-benzopyran-6-yl)methyl)thiazolidine-2,4-dione
In: Rat G6
→ 5-((3,4-Dihydro-2-hydroxy-2-phenylmethyl-2*H*-1-benzopyran-6-yl)methyl)thiazolidine-2,4-dione
In: Rat G6
→ 5-((3,4-Dihydro-2-(*p*-hydroxyphenyl)methyl-2*H*-1-benzopyran-6-yl)methyl)thiazolidine-2,4-dione

In: Rat G6
→ 5-((3,4-Dihydro-4-oxo-2-phenylmethyl-2*H*-1-benzopyran-6-yl)methyl)thiazolidine-2,4-dione
In: Rat G6
→ 5-((3,4-Dihydro-2-phenylmethyl-2*H*-1-benzopyran-6-yl)methyl)-5-hydroxythiazolidine-2,4-dione
Probably in: Rat G6

4,5-Dihydro-4-(((4-(phenylmethyl)-1-piperazinyl)acetyl)-7-(3-trifluoromethylphenyl))pyrazolo[1,5-*a*]pyrimidine-3-carbonitrile
→ 4,5-Dihydro-4-(((piperazin-1-yl)acetyl)-7-(3-trifluoromethylphenyl))pyrazolo[1,5-*a*]pyrimidine-3-carbonitrile
In: Rat G700

(−)-*cis*-2,3-Dihydro-2-phenyl-3-piperazinylmethyl-1,5-benzothiazepin-4(5*H*)-one
Formed from: (−)-*cis*-2,3-Dihydro-3-(4-methylpiperazinylmethyl)-2-phenyl-1,5-benzothiazepin-4(5*H*)-one

2,5-Dihydro-2-phenyl-3*H*-pyrazolo[4,3-*c*]quinazolin-3-one
→ 2,5-Dihydro-2-(*p*-hydroxyphenyl)-3*H*-pyrazolo[4,3-*c*]quinazolin-3-one
In Man E129
Rat E129

13,14-Dihydro-17-phenyl-18,19,20-trisnorprostaglandin-$F_{2\alpha}$
Formed from: Latanoprost

Dihydrophloroglucinol
Formed from: Phloroglucinol
→ Phloroglucinol
In: Coprococcus C99
Eubacterium F75

Dihydropinosylvin
See: 3,5-Dihydroxybibenzyl

Dihydropipamperone
Formed from: Pipamperone

4,5-Dihydro-4-(((piperazin-1-yl)acetyl)-7-(3-trifluoromethylphenyl))pyrazolo[1,5-*a*]pyrimidine-3-carbonitrile
Formed from: 4,5-Dihydro-4-(((4-(phenylmethyl)-1-piperazinyl)acetyl)-7-(3-trifluoromethylphenyl))pyrazolo[1,5-*a*]pyrimidine-3-carbonitrile

N-((3,4-Dihydro-7-(piperazin-1-yl)-1(1*H*)-isoquinolinon-2-yl)acetyl)-3(*S*)-(ethynyl)-β-alanine
Formed from: N-((3,4-Dihydro-7-(piperazin-1-yl)-1(1*H*)-isoquinolinon-2-yl)acetyl)-3(*S*)-(ethynyl)-β-alanine benzyl ester
N-((3,4-Dihydro-7-(piperazin-1-yl)-1(1*H*)-isoquinolinon-2-yl)acetyl)-3(*S*)-(ethynyl)-β-alanine ethyl ester
N-((3,4-Dihydro-7-(piperazin-1-yl)-1(1*H*)-isoquinolinon-2-yl)acetyl)-3(*S*)-(ethynyl)-β-alanine 1-(ethyl)propyl ester
N-((3,4-Dihydro-7-(piperazin-1-yl)-1(1*H*)-isoquinolinon-2-yl)acetyl)-3(*S*)-(ethynyl)-β-alanine isopropyl ester
N-((3,4-Dihydro-7-(piperazin-1-yl)-1(1*H*)-isoquinolinon-2-yl)acetyl)-3(*S*)-(ethynyl)-β-alanine methyl ester
N-((3,4-Dihydro-7-(piperazin-1-yl)-1(1*H*)-isoquinolinon-2-yl)acetyl)-3(*S*)-(ethynyl)-β-alanine phenethyl ester

N-((3,4-Dihydro-7-(piperazin-1-yl)-1(1*H*)-isoquinolinon-2-yl)acetyl)-3(*S*)-ethynyl)-β-alanine benzyl ester
→ Benzyl alcohol
In: Dog J274
Man J274
Rhesus monkey J274
→ N-((3,4-Dihydro-7-(piperazin-1-yl)-1(1*H*)-isoquinolinon-2-yl)acetyl)-3(*S*)-(ethynyl)-β-alanine
In: Dog J274
Man J274
Rat J418
Rhesus monkey J274

N-((3,4-Dihydro-7-(piperazin-1-yl)-1(1*H*)-isoquinolinon-2-yl)acetyl)-3(*S*)-ethynyl)-β-alanine ethyl ester
→ N-((3,4-Dihydro-7-(piperazin-1-yl)-1(1*H*)-isoquinolinon-2-yl)acetyl)-3(*S*)-(ethynyl)-β-alanine
In: Dog J274
Man J274
Rhesus monkey J274

N-((3,4-Dihydro-7-(piperazin-1-yl)-1(1*H*)-isoquinolinon-2-yl)acetyl)-3(*S*)-(ethynyl)-β-alanine 1-(ethyl)propyl ester
→ N-((3,4-Dihydro-7-(piperazin-1-yl)-1(1*H*)-isoquinolinon-2-yl)acetyl)-3(*S*)-(ethynyl)-β-alanine
In: Dog J274
Man J274
Rhesus monkey J274

N-((3,4-Dihydro-7-(piperazin-1-yl)-1(1*H*)-isoquinolinon-2-yl)acetyl)-3(*S*)-(ethynyl)-β-alanine isopropyl ester
→ N-((3,4-Dihydro-7-(piperazin-1-yl)-1(1*H*)-isoquinolinon-2-yl)acetyl)-3(*S*)-(ethynyl)-β-alanine
In: Dog J274
Man J274
Rhesus monkey J274

N-((3,4-Dihydro-7-(piperazin-1-yl)-1(1*H*)-isoquinolinon-2-yl)acetyl)-3(*S*)-(ethynyl)-β-alanine methyl ester
→ N-((3,4-Dihydro-7-(piperazin-1-yl)-1(1*H*)-isoquinolinon-2-yl)acetyl)-3(*S*)-(ethynyl)-β-alanine
In: Dog J274
Man J274
Rhesus monkey J274

N-((3,4-Dihydro-7-(piperazin-1-yl)-1(1*H*)-isoquinolinon-2-yl)acetyl)-3(*S*)-(ethynyl)-β-alanine phenethyl ester
→ N-((3,4-Dihydro-7-(piperazin-1-yl)-1(1*H*)-isoquinolinon-2-yl)acetyl)-3(*S*)-(ethynyl)-β-alanine
In: Dog J274
Man J274
Rhesus monkey J274
→ 1-Phenylethanol
In: Dog J274
Man J274
Rhesus monkey J274

1,3-Dihydro-1-(3-(1-piperazinyl)propyl)-2*H*-benzimidazol-2-one
Formed from: Oxatomide
→ 1,3-Dihydro-5-hydroxy-1-(3-(1-piperazinyl) propyl)-2*H*-benzimidazol-2-one
Probably in: Dog C870
Man C870
Rat C870
→ 1,3-Dihydro-6-hydroxy-1-(3-(1-piperazinyl) propyl)-2*H*-benzimidazol-2-one
Probably in: Dog C870
Man C870
Rat C870

10,11-Dihydro-5-(4-piperidinylidene)-5*H*-dibenzo[*a*,*d*]cyclopentene-3-carboxylic acid
Formed from: 10,11-Dihydro-5-(1-methyl-4-piperidinylidene)-5*H*-dibenzo[*a*,*d*] cyclopentene-3-carboxylic acid

2,3-Dihydro-1-(4-piperidinyl)-2-thioxo-1*H*-benzimidazole
Formed from: Timiperone

Dihydropratensein (3′,5,7-trihydroxy-4′-methoxyisoflavanone)
Formed from: Pratensein

6,7-Dihydro-6-(2-propenyl)-5*H*-dibenz[*c*,*e*]azepine
See: Azapetine

6,7-Dihydro-6-(2-propenyl)-5*H*-dibenz[*c*,*e*]azepin-5-one
See: Oxapetine

2,3-Dihydro-5-propyl-1,4-benzoxazepine
Probably formed from: *o*-(2-Isopropylaminoethoxy)butyrophenone

Dihydropteroic acid
→ Dihydrofolic acid
In: Escherichia A1362
Pea A1363, A2240
Pig K455
Serratia A2836

Dihydropteroyldiglutamic acid
Formed from: Dihydrofolic acid
→ Dihydropteroyltriglutamic acid
In: Pig K455
See also: E277

Dihydropteroylheptaglutamic acid
Formed from: Tetrahydropteroylhepta-glutamic acid

Dihydropteroylhexaglutamic acid
Formed from: Dihydropteroylpentaglutamic acid

Dihydropteroylpentaglutamic acid
Formed from: Tetrahydropteroylpenta-glutamic acid
→ Dihydropteroylhexaglutamic acid
In: Pig K455

Dihydropteroyltetraglutamic acid
Probably formed from: Dihydropteroyltriglutamic acid

Dihydropteroyltriglutamic acid
Formed from: Dihydropteroyldiglutamic acid
Tetrahydropteroyltriglutamic acid
→ Dihydropteroyltetraglutamic acid
Probably in: Pig E277

Dihydropyranisoflavone
Formed from: Wighteone

4′-O-Dihydropyranyldaunomycin
Formed from: 4′-O-Tetrahydropyranyldaunomycin

Dihydroquercetin (2*R*,3*R*-3,3′,4′,5,7-pentahydroxyflavanone, taxifolin)
Formed from: Aromadendrin
Eriodictyol
Probably formed from: Dihydroquercetin-7-glucoside
4′-O-Methyldihydroquercetin
→ Cyanidin
In: Haplopappus A200
Petunia A2740
See also: A3088, B404
→ Dihydrogossypetin
In: Pseudomonas K917
→ 2,3-*trans*-3,4-*cis*-3,3′,4,4′,5,7-Hexahydroxyflavan
In: Cryptomeria E758
Fir D455, D617, D702, D863, D866
Ginkgo D617
Matthiola D782
Petunia E557
Zea E545
See also: G368
→ Quercetin
In: Dianthus G761
Parsley B957
Petunia D950
→ Silybin
In: Horseradish A3221
Silybum A3221

Dihydroquercetin-7-glucoside
→ Dihydroquercetin
Probably in: Petunia A2740

Dihydroquinine
Formed from: Quinine
→ Dihydro-6′-hydroxycinchonidine
Probably in: Man B399
→ Dihydro-3-hydroxyquinine
Probably in: Man B399

3,4-Dihydro-2(1*H*)-quinoxalinone
Probably formed from: 2-Hydroxyquinoxaline

1,2-Dihydrorraucaffricine
→ 1,2-Dihydrovomilenine
In: Rauwolfia K939

Dihydroriboflavin
Formed from: Riboflavin

Dihydroriboflavin-5′-phosphate
Formed from: Riboflavin-5′-phosphate

Dihydrorobinetin (3,3′,4′,5′,7-pentahydroxyflavanone)
→ Robinetinidin
In: Petunia A2740

***cis*-3,4-Dihydrorofecoxib**
Formed from: Rofecoxib

***trans*-3,4-Dihydrorofecoxib**
Formed from: Rofecoxib

1′,2′-Dihydrorotenone
→ 1′,2′-Dihydro-2′-hydroxyrotenone
In: Cunninghamella C416

Dihydroruine (dihydro-7-methoxy-1-methyl-9*H*-pyrido[3,4-*b*]indol-8-yl-β-D-glucoside)
Formed from: Harmaline

Dihydrosafrole (1,2-methylenedioxy-4-propylbenzene)
→ 4-*n*-Propylcatechol
In: Rat C550

Dihydrosanguinarine
Formed from: Chelidonine
Sanguinarine
→ Dihydro-10-hydroxysanguinarine
In: Eschscholtzia K777
See also: F699
→ Sanguinarine
In: Papaver G615
Sanguinaria G615, G723

Dihydrosinapic acid (3-(4-hydroxy-3,5-dimethoxyphenyl)propionic acid)
Probably formed from: 3,4,5-Trimethoxyphenylpropionic acid
→ 3,4-Dihydroxy-5-methoxyphenylpropionic acid
Probably in: Microorganism A261, A744

→ 3,5-Dimethoxyphenylpropionic acid
Probably in: Microorganism D894

1,3-Dihydrosirdalud-2,2-dioxide
Formed from: Sirdalud

Dihydrosterigmatocystin
Formed from: Dihydrodemethylsterigmatocystin
→ Aflatoxin B_2
In: Aspergillus F398
→ Aflatoxin G_2
In: Aspergillus F398
→ Dihydro-O-methylsterigmatocystin
In: Aspergillus F281, F398
See also: E767

13-Dihydro-4′-O-tetrahydropyranyladriamycin
Formed from: 4′-O-Tetrahydropyranyladriamycin
→ 7-Deoxy-13-dihydroadriamycinone
Probably in: Mouse D785

8,9-Dihydro-4″,6,8,9-tetrahydroxycannabidiol
Probably formed from: 4″,6β-Dihydroxycannabidiol

8,9-Dihydro-4″,7,8,9-tetrahydroxycannabidiol
Probably formed from: 8,9-Dihydro-7,8,9-trihydroxycannabidiol
4″,7-Dihydroxycannabidiol

8,9-Dihydro-5″,6,8,9-tetrahydroxycannabidiol
Probably formed from: 5″,6β-Dihydroxycannabidiol

8,9-Dihydro-5″,7,8,9-tetrahydroxycannabidiol
Probably formed from: 8,9-Dihydro-7,8,9-trihydroxycannabidiol
5″,7-Dihydroxycannabidiol

8,9-Dihydro-6,7,8,9-tetrahydroxycannabidiol
Probably formed from: 8,9-Dihydro-7,8,9-trihydroxycannabidiol
6,7-Dihydroxycannabidiol

Dihydro-2′,3,4,6′-tetrahydroxychalcone
Formed from: Dihydro-2′,3,6′-trihydroxychalcone
Dihydro-2′,4,6′-trihydroxychalcone
→ Dihydro-2′,4,6′-trihydroxy-3-methoxychalcone
Probably in: Rat D650

3,4-Dihydro-2,4,5,7-tetrahydroxy-3,9-dimethoxy-2-methyl-1,6,11(2*H*)-naphthacenetrione
See: Steffimycin

Dihydro-2′,4′,5,9-tetrahydroxy-8,8-dimethylfuranoisoflavone
Formed from: 2,3-Dehydrokievitone

2,3-Dihydro-2,3,4,5-tetrahydroxyphenylpropionic acid
Probably formed from: Dihydrocinnamic acid
→ 2,3,4,5-Tetrahydroxyphenylpropionic acid
In: Bacterium B871

5,6-Dihydro-2,3,9,10-tetramethoxydibenzo[*a*,*g*]quinolizinium
See: Palmatine

4-(4-(2,3-Dihydro-2-thioxo-1*H*-benzimidazol-1-yl)-1-piperidinyl)-1-(4-fluorophenyl)-1-butanol
Formed from: Timiperone

Dihydrotienilic acid
Formed from: Tienilic acid
→ O-Methyldihydrotienilic acid
In: Dog A3554
Pig A3554
Rat A3554

13-Dihydro-N-trifluoroacetyldoxorubicin
Probably formed from: N-Trifluoroacetyldoxorubicin

1,4-Dihydro-6-trifluoromethylquinoxaline-2,3-dione
→ 1,4-Dihydro-1-hydroxy-6-trifluoromethylquinoxaline-2,3-dione
In: Cat A1751
Rat A1751
→ 1,4-Dihydro-7-hydroxy-6-trifluoromethylquinoxaline-2,3-dione
In: Cat A1751
Rat A1751

1,2-Dihydro-1,2,8-trihydroxybenz[*j*]aceanthrylene
Probably formed from: *trans*-1,2-Dihydro-1,2-dihydroxybenz[*j*]aceanthrylene
→ 1,2-Dihydro-1,2,8-trihydroxybenz[*j*]aceanthrylene-β-D-glucuronide
Probably in: Rat J804

1,2-Dihydro-1,2,10-trihydroxybenz[*j*]aceanthrylene
Probably formed from: *trans*-1,2-Dihydro-1,2-dihydroxybenz[*j*]aceanthrylene
→ 1,2-Dihydro-1,2,10-trihydroxybenz[*j*]aceanthrylene-10-β-D-glucuronide
Probably in: Man J688
Rat J804

→ 1,2-Dihydro-1,2,10-trihydroxybenz[*j*]aceanthrylene-10-sulphate
Probably in: Man J688
Rat J804

1,2-Dihydro-1,2,8-trihydroxybenz[*j*]aceanthrylene-β-D-glucuronide
Probably formed from: 1,2-Dihydro-1,2,8-trihydroxybenz[*j*]aceanthrylene

1,2-Dihydro-1,2,10-trihydroxybenz[*j*] aceanthrylene-10-β-D-glucuronide
Probably formed from: 1,2-Dihydro-1,2,10-trihydroxybenz[*j*]aceanthrylene

1,2-Dihydro-1,2,10-trihydroxybenz[*j*] aceanthrylene-10-sulphate
Probably formed from: 1,2-Dihydro-1,2,10-trihydroxybenz[*j*]aceanthrylene

1,2-Dihydro-3,4,6-trihydroxybenzene
Formed from: Hydroxyquinol

2,3-Dihydro-2,3,4-trihydroxybenzoic acid
Formed from: *p*-Hydroxybenzoic acid

2,3-Dihydro-2,3,5-trihydroxybenzoic acid
Probably formed from: *m*-Hydroxybenzoic acid

2,3-Dihydro-2,3,6-trihydroxybenzoic acid
Probably formed from: Salicylic acid

***trans*-7,8-Dihydro-3,7,8-trihydroxy-3,4-benzpyrene**
Probably formed from: 3-Hydroxy-3,4-benzpyrene

7,8-Dihydro-6,7,8-trihydroxy-3,4-benzpyrene
Probably formed from: *trans*-7,8-Dihydro-7,8-dihydroxy-3,4-benzpyrene

9,10-Dihydro-1,9,10-trihydroxy-3,4-benzpyrene
Probably formed from: 9,10-Dihydro-9,10-dihydroxy-3,4-benzpyrene

***trans*-7,8-Dihydro-1,7,8-trihydroxy-3,4-benzpyrene-9,10-oxide**
Formed from: *trans*-7,8-Dihydro-7,8-dihydroxy-3,4-benzpyrene-9,10-oxide

***trans*-7,8-Dihydro-3,7,8-trihydroxy-3,4-benzpyrene-9,10-oxide**
Formed from: *trans*-7,8-Dihydro-7β,8α-dihydroxy-3,4-benzpyrene-9α,10α-oxide

8,9-Dihydro-1″,8,9-trihydroxycannabidiol
Probably formed from: 8,9-Dihydro-8,9-dihydroxycannabidiol

8,9-Dihydro-2″,8,9-trihydroxycannabidiol
Probably formed from: 8,9-Dihydro-8,9-dihydroxycannabidiol

8,9-Dihydro-3″,8,9-trihydroxycannabidiol
Probably formed from: 8,9-Dihydro-8,9-dihydroxycannabidiol

8,9-Dihydro-4″,8,9-trihydroxycannabidiol
Probably formed from: 8,9-Dihydro-8,9-dihydroxycannabidiol

8,9-Dihydro-5″,8,9-trihydroxycannabidiol
Probably formed from: 8,9-Dihydro-8,9-dihydroxycannabidiol

8,9-Dihydro-6,8,9-trihydroxycannabidiol
Probably formed from: 8,9-Dihydro-8,9-dihydroxycannabidiol

8,9-Dihydro-7,8,9-trihydroxycannabidiol
Probably formed from: 8,9-Dihydro-8,9-dihydroxycannabidiol
7-Hydroxycannabidiol
→ 8,9-Dihydro-4″,7,8,9-tetrahydroxycannabidiol
Probably in: Dog G113
→ 8,9-Dihydro-5″,7,8,9-tetrahydroxycannabidiol
Probably in: Dog G113
→ 8,9-Dihydro-6,7,8,9-tetrahydroxycannabidiol
Probably in: Dog G113

Dihydro-2′,3,4-trihydroxychalcone
Probably formed from: Dihydro-2′,4-dihydroxychalcone

Dihydro-2′,3,6′-trihydroxychalcone
→ Dihydro-2′,3,4,6′-tetrahydroxychalcone
In: Rat D650

Dihydro-2′,4,4′-trihydroxychalcone
→ Phloretic acid
In: Rat D650
→ Resorcinol
In: Rat D650

Dihydro-2′,4,6′-trihydroxychalcone
→ Dihydro-2′,3,4,6′-tetrahydroxychalcone
In: Rat D650
→ Phloretic acid
In: Rat D650
→ Resorcinol
In: Rat D650

1,2-Dihydro-1,2,3-trihydroxychrysene
Probably formed from: 3-Hydroxychrysene

***trans*-1,2-Dihydro-1,2,9-trihydroxychrysene**
Formed from: 3-Hydroxychrysene
Probably formed from: 1,2-Dihydro-1,2-dihydroxychrysene
→ 1,2,3,4-Tetrahydro-1,2,3,4,9-pentahydroxychrysene
Probably in: Rat F652

10,11-Dihydro-6′,10,11-trihydroxycinchonidine
Probably formed from: 6′-Hydroxycinchonidine

3′,4′-Dihydro-3′,4′,9-trihydroxydemethylclobazam
Probably formed from: 3′,4′-Dihydro-3′,4′-dihydroxydemethylclobazam

9,10-Dihydro-4,5-trihydroxy-9,10-dioxo-2-anthracenecarboxylic acid
See: Emodic acid

Dihydro-2′,4′,5-trihydroxy-8-(2-hydroxy-2-propyl)pyranoisoflavone
Formed from: 2,3-Dehydrokievitone

***trans*-1,2-Dihydro-1,2,8-trihydroxyindene [1,2,3-*cd*]pyrene**
Probably formed from: *trans*-1,2-Dihydro-1,2-dihydroxyindene[1,2,3-*cd*]pyrene

***trans*-1,2-Dihydro-1,2,9-trihydroxyindene [1,2,3-*cd*]pyrene**
Probably formed from: *trans*-1,2-Dihydro-1,2-dihydroxyindene[1,2,3-*cd*]pyrene

Dihydro-2′,4,6′-trihydroxy-3-methoxychalcone
Probably formed from: Dihydro-2′,3,4,6′-tetrahydroxychalcone

***trans*-9,10-Dihydro-1,9,10-trihydroxy-3-methylcholanthrene**
Formed from: 1-Hydroxy-3-methylcholanthrene

9,10-Dihydro-3,6,8-trihydroxy-1-methyl-9,10-dioxo-2-anthracenecarboxylic acid methyl ester
See: Laccaic acid methyl ester

5,6-Dihydro-2,5,6-trihydroxy-3-methylquinoline
Formed from: 2-Hydroxy-3-methylquinoline

3,4-Dihydro-2-(3,4,5-trihydroxyphenyl)-2*H*-1-benzopyran-3,4,5,7-tetrol
See: Leucodelphinidin

10,11-Dihydro-2′,10,11-trihydroxyquinine
Probably formed from: 2′-Hydroxyquinine

5,6-Dihydro-2,5,6-trihydroxyquinoline
Formed from: 2-Hydroxyquinoline

1,2-Dihydro-2,2,4-trimethylquinoline
→ 1,2-Dihydro-1-hydroxy-2,2,4-trimethylquinoline
In: Rat E871
→ 1,2-Dihydro-6-hydroxy-2,2,4-trimethylquinoline
In: Rat E871

Dihydroubiquinone 45
Formed from: Dihydro-5-demethylubiquinone 45
→ Ubiquinone 45
In: Rat A3084

Dihydrovindoline
Formed from: Vindoline
→ Dihydro-11-O-demethylvindoline
In: Streptomyces C349

Dihydrovindoline ether
Formed from: Vindoline

1,2-Dihydrovomilenine
Formed from: 1,2-Raucaffricine

***R*,*R*-Dihydrowarfarin**
Formed from: *R*-Warfarin

***R*,*S*-Dihydrowarfarin**
Formed from: *R*-Warfarin

***S*,*R*-Dihydrowarfarin**
Formed from: *S*-Warfarin

***S*,*S*-Dihydrowarfarin**
Formed from: *S*-Warfarin

Dihydrowarfarin (stereochemistry unclear)
Formed from: Warfarin

Dihydroxanthommatin
Formed from: Xanthommatin

1,2-Dihydroxyacenaphthene
Formed from: 1,2-Dihydro-1,2-dihydroxyacenaphthene

2′,3′-Dihydroxyacetanilide
Formed from: *m*-Acetamidophenol

2′,5′-Dihydroxyacetanilide
Formed from: *o*-Acetamidophenol
m-Acetamidophenol
2-Acetamido-*p*-quinone
Probably formed from: Aminoquinol
→ 2-Acetamido-*p*-quinone
In: Mouse D814
See also: F31, F190

3′,4′-Dihydroxyacetanilide
Formed from: *m*-Acetamidophenol
→ 4-Acetamido-*o*-quinone
Probably in: Mouse F31
→ 3′-Hydroxy-4′-methoxyacetanilide
With catechol O-methyltransferase A597
→ 4′-Hydroxy-3′-methoxyacetanilide
With catechol O-methyltransferase A597

2′,3′-Dihydroxyacetophenone
Probably formed from: 8-Hydroxychromone

2′,4′-Dihydroxyacetophenone (resacetophenone)
Formed from: 2′-Hydroxy-4′-methoxyacetophenone
→ 2′,4′-Dihydroxyacetophenone-β-D-glucuronide
Probably in: Guinea pig B506
Man B920
Mouse B506
Rabbit B506
Rat B506
→ 2′,4′-Dihydroxyacetophenone-O-sulphate
Probably in: Guinea pig B506
Mouse B506
Rabbit B506
Rat B506, K209
→ 2′,4′,5′-Trihydroxyacetophenone
Probably in: Rat K209

3′,4′-Dihydroxyacetophenone
Probably formed from: 4′-Hydroxy-3′-methoxyacetophenone
→ 3′-Hydroxy-4′-methoxyacetophenone
In: Rat A486, A2817
With catechol O-methyltransferase A597
See also: E849
→ 4′-Hydroxy-3′-methoxyacetophenone
In: Rat A486, A2817
With catechol O-methyltransferase A597

***o*,ω-Dihydroxyacetophenone** (*o*-hydroxybenzoylmethanol)
Formed from: Flavone
→ *o*-Hydroxyphenylglycol
Probably in: Absidia F504

***p*,ω-Dihydroxyacetophenone** (*p*-hydroxybenzoylmethanol)
Probably formed from: *p*-Hydroxyacetophenone
→ *p*-Hydroxybenzoic acid
In: Alcaligenes D761, E173
See also: A813

2′,4′-Dihydroxyacetophenone-β-D-glucuronide
Probably formed from: 2′,4′-Dihydroxyacetophenone

2′,4′-Dihydroxyacetophenone-O-sulphate
Probably formed from: 2′,4′-Dihydroxyacetophenone

9,11-Dihydroxyacronine
Probably formed from: 9-Hydroxyacronine

9,11-Dihydroxyacronycine
Probably formed from: 9-Hydroxyacronycine

2′,3′-Dihydroxyaminopyrine
Probably formed from: 2′,3′-Dihydro-2′,3′-dihydroxyaminopyrine
→ 6-(1,2-Dihydro-1,5-dimethyl-3*H*-pyrazol-3-on-2-yl)-*cis,cis*-muconic semialdehyde
In: Phenylobacterium B838

2,3-Dihydroxyamitriptyline
Probably formed from: 2-Hydroxy-amitriptyline
→ 2-Hydroxy-3-methoxyamitriptyline
Probably in: Rat E892, E931

2,10-Dihydroxyamitriptyline
Probably formed from: 2-Hydroxy-amitriptyline
→ 2,10,11-Triydroxyamitriptyline
Probably in: Rat E892

2,11-Dihydroxyamitriptyline
Probably formed from: 2-Hydroxy-amitriptyline
→ 2-Hydroxy-11-oxoamitriptyline
Probably in: Rat E892

3,11-Dihydroxyamitriptyline
Probably formed from: 3-Hydroxy-amitriptyline
→ 10,11-Dehydro-3-hydroxyamitriptyline
Probably in: Rat E892

***cis*-10,11-Dihydroxyamitriptyline**
Probably formed from: 10-Hydroxyamitriptyline
→ *cis*-10,11-Dihydroxyamitriptyline-β-D-glucuronide
Probably in: Rat E931

***trans*-10,11-Dihydroxyamitriptyline**
Probably formed from: 10-Hydroxyamitriptyline

→ *trans*-10,11-Dihydroxyamitriptyline-β-D-glucuronide
In: Rat E931

***cis*-10,11-Dihydroxyamitriptyline-β-D-glucuronide**
Probably formed from: *cis*-10,11-Dihydroxyamitriptyline

***trans*-10,11-Dihydroxyamitriptyline-β-D-glucuronide**
Probably formed from: *trans*-10,11-Dihydroxyamitriptyline

***trans*-10,11-Dihydroxyamitriptyline-N-β-D-glucuronide**
Formed from: 10-Hydroxyamitriptyline-N-β-D-glucuronide
→ 2,11,12-Trihydroxyamitriptyline
Probably in: Rat E892

3,4-Dihydroxyamphetamine (3,4-dihydroxy-α-methylphenethylamine)
Formed from: 3,4-Dihydroxy-α-methyl-L-phenylalanine
L-α-Hydrazino-3,4-dihydroxy-α-methylphenylpropionic acid
p-Hydroxyamphetamine
3,4-Methylenedioxyamphetamine
Probably formed from: N-Ethyl-3,4-dihydroxyamphetamine
m-Hydroxyamphetamine
p-Hydroxyamphetamine
4-Hydroxy-3-methoxyamphetamine
N-Methyl-3,4-methylenedioxyamphetamine
→ 4-(2-Aminopropyl)-*o*-quinone
Probably in: Man A63
→ 3,4-Dihydroxyamphetamine-β-D-glucuronide
Probably in: Man G944
→ 3,4-Dihydroxyamphetamine-O-sulphate
In: Man C345, D35
See also: D656, G944
→ 3,4-Dihydroxyphenylacetone
Probably in: Dog A3465, A3697
→ 3-Hydroxy-4-methoxyamphetamine
In: Rat A486
See also: B403
→ 4-Hydroxy-3-methoxyamphetamine
In: Dog A2077
Monkey A2077
Rat A486
See also: A2964, A3465, B403, B632, H736, J544
→ α-Methylnoradrenaline
Probably in: Rat C758, D424, D766

3,4-Dihydroxyamphetamine-β-D-glucuronide
Probably formed from: 3,4-Dihydroxyamphetamine

3,4-Dihydroxyamphetamine-O-sulphate
Formed from: 3,4-Dihydroxyamphetamine

2,4-Dihydroxyaniline
Formed from: Acid Orange 6

1,2-Dihydroxyanthracene
Formed from: *cis*-1,2-Dihydro-1,2-dihydroxyanthracene

2,3-Dihydroxyanthracene
Formed from: Anthracene

1,8-Dihydroxy-9,10-anthracenedione diacetate
See: Chrysazin diacetate

3,4-Dihydroxyanthranilic acid (2-amino-3,4-dihydroxybenzoic acid)
Formed from: 3-Hydroxyanthranilic acid

3,6-Dihydroxyanthranilic acid (2-amino-3,6-dihydroxybenzoic acid)
Formed from: 3-Hydroxyanthranilic acid

1,2-Dihydroxyanthraquinone
See: Alizarin

1,3-Dihydroxyanthraquinone
→ 1,3-Dihydroxyanthraquinone-β-D-glucoside
In: Streptomyces B462

1,4-Dihydroxyanthraquinone (Pigment Violet 12, quinizarin)
→ 4-β-D-Glucopyranosyloxy)-1-hydroxyanthraquinone
In: Cinchona K889
Streptomyces A1497
→ Phthalic acid
In: Coriolus J631

1,5-Dihydroxyanthraquinone
See: Anthrarufin

1,8-Dihydroxyanthraquinone (chrysazin)
Probably formed from: 1,8-Dihydroxy-9-anthrone
→ 1,8-Dihydroxyanthraquinone-β-D-glucuronide
In: Rat A2828

→ 8-β-D-Glucopyranosyloxy-1-hydroxyanthraquinone
In: Streptomyces A1497

2,6-Dihydroxyanthraquinone
See: Anthraflavin

1,3-Dihydroxyanthraquinone-β-D-glucoside
Formed from: 1,3-Dihydroxyanthraquinone

1,8-Dihydroxyanthraquinone-β-D-glucuronide
Formed from: 1,8-Dihydroxyanthraquinone

1,8-Dihydroxy-9-anthrone
Probably formed from: 8-Acetoxy-1-hydroxy-9-anthrone
→ 1,8-Dihydroxyanthraquinone
Probably in: Man B150
→ 1,1′,8,8′-Tetrahydroxybisanthrone
Probably in: Man B150

2,3-Dihydroxyantipyrine
Probably formed from: 2,3-Dihydro-2,3-dihydroxyantipyrine

4,4′-Dihydroxyantipyrine
Probably formed from: 4-Hydroxyantipyrine
→ 4,4′-Dihydroxyantipyrine-β-D-glucuronide
In: Rat B305
See also: D632
→ 4,4′-Dihydroxyantipyrine sulphate
In: Rat B305
See also: C415, C418, J769

4,4′-Dihydroxyantipyrine-β-D-glucuronide
Formed from: 4,4′-Dihydroxyantipyrine

4,4′-Dihydroxyantipyrine sulphate
Formed from: 4,4′-Dihydroxyantipyrine

Dihydroxyapovincamic acid
Probably formed from: Apovincamic acid

2,3-Dihydroxybenzaldehyde
Formed from: 2,3-Dimethoxybenzaldehyde
Probably formed from: 2,3-Dihydroxybenzylamine
2,3-Dihydroxyphenylpyruvic acid
→ Pyrocatechuic acid
Probably in: Rat G865

2,4-Dihydroxybenzaldehyde
→ 2,4,5-Trihydroxybenzaldehyde
In: Pseudomonas A2391

2,5-Dihydroxybenzaldehyde
See: Gentisaldehyde

3,4-Dihydroxybenzaldehyde
See: Protocatechualdehyde

3,5-Dihydroxybenzaldehyde
Formed from: Astringin
Isorhapontin
Piceid

2,5-Dihydroxybenzamide
See: Gentisamide

2,6-Dihydroxybenzamide
See: 2,6-Dihydroxybenzoic

2,4′-Dihydroxybenzanilide
Formed from: Salicylanilide

2,5-Dihydroxybenzanilide
Formed from: Salicylanilide

3′,4′-Dihydroxybenzanilide
Formed from: Salicylanilide
→ 3′-Hydroxy-4′-methoxybenzanilide
In: Rat A486
→ 4′-Hydroxy-3′-methoxybenzanilide
In: Rat A486

1,2-Dihydroxybenzanthracene
Formed from: *trans*-1,2-Dihydro-1,2-dihydroxybenzanthracene

3,4-Dihydroxybenzanthracene
Formed from: 3,4-Dihydro-3,4-dihydroxybenzanthracene

1′,6-Dihydroxybenzbromarone
Probably formed from: 1′-Hydroxybenzbromarone

1,2-Dihydroxybenzene
See: Catechol

1,3-Dihydroxybenzene
See: Resorcinol

1,4-Dihydroxybenzene
See: Quinol

2,3-Dihydroxybenzenesulphonic acid (3-sulphocatechol)
Formed from: Orthanilic acid
Probably formed from: *m*-Aminobenzenesulphonic acid
Benzenesulphonic acid
o-Nitrobenzenesulphonic acid

3,4-Dihydroxybenzenesulphonic acid (catechol-4-sulphonic acid, 4-sulphocatechol)
Formed from: Sulphanilic acid
Probably formed from: *p*-Chlorobenzenesulphonic acid
p-Hydroxybenzenesulphonic acid
5-Sulphosalicylic acid

→ 3-Sulphomuconic acid
In: Alcaligenes K775
Hydrogenophaga G785, J218

2,3-Dihydroxybenzoic acid
See: Pyrocatechuic acid

2,4-Dihydroxybenzoic acid (β-resorcylic acid)
Formed from: 2,4-Diacetoxybenzoic acid
7-Hydroxynaphthalene-1-sulphonic acid
Medicarpin
Morin
Salicylic acid
→ 4-Carboxy-3-hydroxyphenyl-D-glucoside
In: Cucumis A611
→ 2,4-Dihydroxybenzoyladenylic acid
In: Escherichia K830
→ Hydroxyquinol
In: Candida J461
Pseudomonas H18
Sporotrichum B73, C57
Trichosporon D232
→ Resorcinol
In: Microorganism F844
→ 2,3,4-Trihydroxybenzoic acid
In: Pseudomonas K109
→ 2,4,5-Trihydroxybenzoic acid
In: Pseudomonas K109
See also: A2554

2,5-Dihydroxybenzoic acid
See: Gentisic acid

2,6-Dihydroxybenzoic acid (γ-resorcylic acid)
Formed from: 2,6-Diacetoxybenzoic acid
5-Hydroxynaphthalene-1-sulphonic acid
→ 2,6-Dihydroxybenzamide
In: Streptomyces E279
→ Pyrogallol
In: Trichosporon D232
→ Resorcinol
In: Microorganism F844

3,4-Dihydroxybenzoic acid
See: Protocatechuic acid

3,5-Dihydroxybenzoic acid (α-resorcylic acid)
Formed from: 3,5-Diacetoxybenzoic acid
→ 3-Carboxy-5-hydroxyphenyl sulphate
In: Eubacterium H439
Haemophilus H439
Klebsiella H439
→ 3,5-Dihydroxybenzoyl-D-glucose
In: Bacillus H234
Cucumis A611
→ Gallic acid
In: Aspergillus A1130
Pseudomonas A327, A1229
→ 3-Hydroxy-5-methoxybenzoic acid
In: Phanerochaete J339
→ 2,3,5-Trihydroxybenzoic acid
In: Pseudomonas A326

2,3-Dihydroxybenzonitrile
Formed from: *m*-Fluorobenzonitrile
Probably formed from: *cis*-2,3-Dihydro-2,3-dihydroxybenzonitrile

3,4-Dihydroxybenzonitrile
Formed from: 3,4-Methylenedioxybenzonitrile
p-Hydroxybenzonitrile
→ 3-Hydroxy-4-methoxybenzonitrile
In: Rat A486
With catechol O-methyltransferase A597
→ 4-Hydroxy-3-methoxybenzonitrile
In: Rat A486
With catechol O-methyltransferase A597

2,3′-Dihydroxybenzophenone
Formed from: *o*-Hydroxybenzophenone

2,4-Dihydroxybenzophenone
Formed from: 2-Hydroxy-4-methoxybenzophenone
→ 2,3′,4-Trihydroxybenzophenone
In: Hypericum J509

3,4-Dihydroxybenzophenone
→ 3-Hydroxy-4-methoxybenzophenone
In: Rat A486
With catechol O-methyltransferase A597
→ 4-Hydroxy-3-methoxybenzophenone
In: Rat A486
With catechol O-methyltransferase A597

3,4′-Dihydroxybenzophenone
Formed from: *p*-Hydroxybenzophenone

2,3-Dihydroxybenzo[*b*]thiophene
Probably formed from: 2-Formyl-3-hydroxybenzo[*b*]thiophene
→ Benzo[*b*]thiophene-2,3-dione
Probably in: Microorganism J303

4,5-Dihydroxybenzopyran-2-one
Formed from: 2,3-Dihydro-2,5-dihydroxybenzopyran-4-one

2,4-Dihydroxy-1,4-benzoxazin-3-one
Formed from: 2,4-Dihydroxy-1,4-benzoxazin-3-one-β-D-glucoside
2-Hydroxy-1,4-benzoxazin-3-one

2,4-Dihydroxy-1,4-benzoxazin-3-one-β-D-glucoside
→ 2,4-Dihydroxy-1,4-benzoxazin-3-one
In: Zea K149

2,3-Dihydroxybenzoyladenylic acid
Formed from: Pyrocatechuic acid
→ N-(2,3-Dihydroxybenzoyl)-1,7-diaminoheptane
In: Vibrio K666
→ N-(2,3-Dihydroxybenzoyl)hexylamine
In: Vibrio K666
→ N-(2,3-Dihydroxybenzoyl)norspermidine
In: Vibrio K666
→ N-(2,3-Dihydroxybenzoyl)octylamine
In: Vibrio K666

2,4-Dihydroxybenzoyladenylic acid
Formed from: 2,4-Dihydroxybenzoic acid

2,5-Dihydroxybenzoyladenylic acid
Formed from: Gentisic acid

2,5-Dihydroxybenzoyl CoA
Formed from: *m*-Hydroxybenzoyl CoA

N-(2,3-Dihydroxybenzoyl)-1,7-diaminoheptane
Formed from: 2,3-Dihydroxybenzoyladenylic acid

3,4-Dihydroxybenzoyl-D-glucose
Formed from: Protocatechuic acid

3,5-Dihydroxybenzoyl-D-glucose
Formed from: 3,5-Dihydroxybenzoic acid

N-(2,3-Dihydroxybenzoyl)hexylamine
Formed from: 2,3-Dihydroxybenzoyladenylic acid

N-(2,3-Dihydroxybenzoyl)norspermidine
Formed from: 2,3-Dihydroxybenzoyladenylic acid

N-(2,3-Dihydroxybenzoyl)octylamine
Formed from: 2,3-Dihydroxybenzoyladenylic acid

1,6-Dihydroxy-3,4-benzpyrene
Probably formed from: 3,4-Benzpyrene-1,6-quinone
→ 1,6-Dihydroxy-3,4-benzpyrene-di-β-D-glucuronide
Probably in: Rat G601
→ 1,6-Dihydroxy-3,4-benzpyrene-O-sulphate
Probably in: Rat A3762

3,5-Dihydroxy-3,4-benzpyrene
Probably formed from: 3-Hydroxy-3,4-benzpyrene

3,6-Dihydroxy-3,4-benzpyrene
Formed from: 3,4-Benzpyrene-3,6-quinone
Probably formed from: 3-Hydroxy-3,4-benzpyrene
→ 3,4-Benzpyrene-3,6-quinone
In: Rat A3637
See also: C540
→ 3,6-Dihydroxy-3,4-benzpyrene-3-β-D-glucuronide
In: Rat D713, G715
→ 3,6-Dihydroxy-3,4-benzpyrene-6-β-D-glucuronide
In: Rat D707, D713, G715, H575
See also: B674, B691
→ 1,6-Dihydroxy-3,4-benzpyrene-O-sulphate
Probably in: Rat A3762

3,9-Dihydroxy-3,4-benzpyrene
Probably formed from: 3-Hydroxy-3,4-benzpyrene
9-Hydroxy-3,4-benzpyrene

6,12-Dihydroxy-3,4-benzpyrene
Probably formed from: 3,4-Benzpyrene-6,12-quinone
→ 6,12-Dihydroxy-3,4-benzpyrene-di-β-D-glucuronide
Probably in: Rat G601

7,8-Dihydroxy-3,4-benzpyrene
Formed from: *trans*-7,8-Dihydro-7,8-dihydroxy-3,4-benzpyrene

9,10-Dihydroxy-3,4-benzpyrene
Formed from: *trans*-9,10-Dihydro-9,10-dihydroxy-3,4-benzpyrene

1,6-Dihydroxy-3,4-benzpyrene-di-β-D-glucuronide
Probably formed from: 1,6-Dihydroxy-3,4-benzpyrene

3,6-Dihydroxy-3,4-benzpyrene-di-β-D-glucuronide
Formed from: 3,6-Dihydroxy-3,4-benzpyrene-3-β-D-glucuronide
3,6-Dihydroxy-3,4-benzpyrene-6-β-D-glucuronide

6,12-Dihydroxy-3,4-benzpyrene-di-β-D-glucuronide
Probably formed from: 6,12-Dihydroxy-3,4-benzpyrene

3,6-Dihydroxy-3,4-benzpyrene-3-β-D-glucuronide
Formed from: 3,6-Dihydroxy-3,4-benzpyrene
→ 3,6-Dihydroxy-3,4-benzpyrene-di-β-D-glucuronide
In: Rat D713
See also: G601, G715

3,6-Dihydroxy-3,4-benzpyrene-6-β-D-glucuronide
Formed from: 3,6-Dihydroxy-3,4-benzpyrene
→ 3,6-Dihydroxy-3,4-benzpyrene-di-β-D-glucuronide
In: Rat D707, D713
See also: H575

1,6-Dihydroxy-3,4-benzpyrene-O-sulphate
Probably formed from: 1,6-Dihydroxy-3,4-benzpyrene

3,6-Dihydroxy-3,4-benzpyrene-O-sulphate
Probably formed from: 3,6-Dihydroxy-3,4-benzpyrene

2,3-Dihydroxybenzyl acetate
Probably formed from: *cis*-2,3-Dihydro-2,3-dihydroxybenzyl acetate

2,3-Dihydroxybenzyl alcohol
Formed from: *m*-Fluorobenzyl alcohol

2,5-Dihydroxybenzyl alcohol
See: Gentisyl alcohol

3,4-Dihydroxybenzyl alcohol
Formed from: *m*-Hydroxybenzyl alcohol
p-Hydroxybenzyl alcohol
4-Methylcatechol
Protocatechualdehyde
Probably formed from: Vanillyl alcohol
→ 4-Hydroxymethyl-*o*-quinone
In: Mushroom H907
→ Isovanillyl alcohol
With catechol O-methyltransferase A597
→ Protocatechualdehyde
Probably in: Sarcophaga C692
→ Vanillyl alcohol
With catechol O-methyltransferase A597

2,3-Dihydroxybenzylamine
→ 2,3-Dihydroxybenzaldehyde
Probably in: Rat G865

3,4-Dihydroxybenzylamine
→ 4-Aminomethyl-*o*-quinone
Probably in: Man A63
→ 3,4-Dihydroxybenzylamine-O-sulphate
In: Man D583
Rat E962
→ 3-Hydroxy-4-methoxybenzylamine
In: Rat A486
→ 4-Hydroxy-3-methoxybenzylamine
In: Rat A486
→ Protocatechualdehyde
In: Penicillium J16
See also: Rat G865

3,4-Dihydroxybenzylamine-O-sulphate
Formed from: 3,4-Dihydroxybenzylamine

3,4-Dihydroxybenzyl cyanide
→ 4-Cyanomethyl-*o*-quinone
In: Mushroom K834

5-(3,4-Dihydroxybenzyl)hydantoin
→ 3,4-Dihydroxy-D-phenylalanine
In: Pseudomonas E450
→ 3,4-Dihydroxy-L-phenylalanine
In: Flavobacterium F940

N-(3,4-Dihydroxybenzyl)-*p*-hydoxyphenylpropylamine
See: Norbelladine

5-(3,4-Dihydroxybenzyl)-2-((5-methyl-4-imidazolylmethylthio)ethylamino)-4-pyrimidone
Formed from: Oxmetidine
→ 5-(4-Hydroxy-3-methoxybenzyl)-2-(5-methyl-4-imidazolylmethylthio) ethylamino-4-pyrimidone
Probably in: Dog C875
→ 5-(3-Hydroxy-4-sulphatoxybenzyl)-2-(5-methyl-4-imidazolylmethylthio) ethylamino-4-pyrimidone
Probably in: Dog C875
→ 5-(4-Hydroxy-3-sulphatoxybenzyl)-2-(5-methyl-4-imidazolylmethylthio) ethylamino-4-pyrimidone
Probably in: Dog C875

1-(3,4-Dihydroxybenzyl)-4-(2-pyrimidinyl) piperazine
Formed from: Piribedil

1-(3,4-Dihydroxybenzyl)-1,2,3,4-tetrahydro-6-hydroxy-7-methoxyisoquinoline
Formed from: Tetrahydropapaveroline

→ 1-(3,4-Dihydroxybenzyl)-1,2,3,4-tetrahydro-6-hydroxy-7-methoxy-2-methylisoquinoline
In: Berberis K780

1-(3,4-Dihydroxybenzyl)-1,2,3,4-tetrahydro-7-hydroxy-6-methoxyisoquinoline
Formed from: Coclaurine
Tetrahydropapaveroline
→ 1-(3,4-Dihydroxybenzyl)-1,2,3,4-tetrahydro-7-hydroxy-6-methoxy-2-methylisoquinoline
In: Berberis K780
→ Nororientaline
In: Argemone D76

1-(3,4-Dihydroxybenzyl)-1,2,3,4-tetrahydro-6-hydroxy-7-methoxy-2-methylisoquinoline
Formed from: 1-(3,4-Dihydroxybenzyl)-1,2,3,4-tetrahydro-6-hydroxy-7-methoxyisoquinoline

1-(3,4-Dihydroxybenzyl)-1,2,3,4-tetrahydro-7-hydroxy-6-methoxy-2-methylisoquinoline
Formed from: N-Methylcoclaurine
→ (*S*)-Reticuline
In: Berberis K933
Eschscholtzia K933
Peumus K933

2,3-Dihydroxybibenzyl
Probably formed from: Bibenzyl
→ 2,2′3,3′-Tetrahydroxybibenzyl
Probably in: Pseudomonas J10

3,4-Dihydroxybibenzyl
Probably formed from: 3,4-Dihydroxy-*trans*-stilbene

3,4′-Dihydroxybibenzyl (lunularin)
Formed from: Lunularic acid

3,5-Dihydroxybibenzyl (dihydropinosylvin)
Formed from: Phenylpropionyl CoA

4,4′-Dihydroxybibenzyl
Probably formed from: 4,4′-Dihydroxy-*trans*-stilbene

Dihydroxybidisomide
Probably formed from: Hydroxybidisomide

2,2′-Dihydroxy-1,1′-binaphthyl
Formed from: β-Naphthol

2,2′-Dihydroxy-1,1′-binaphthyl-3,3′-dicarboxylic acid dimethyl ether
Probably formed from: 3-Hydroxy-2-naphthoate

2,2′-Dihydroxybiphenyl
Formed from: Dibenz[*c*,*e*][1,2]oxathiin-6,6-dioxide
Phenol
Probably formed from: Dibenzothiophene
2-Hydroxybiphenyl
→ 2-Hydroxybiphenyl-2′-yl-β-D-glucuronide
Probably in: Rainbow trout J582
→ 2,2′,3-Trihydroxybiphenyl
In: Pseudomonas E958, E960, H22, J392

2,3-Dihydroxybiphenyl
Formed from: 2-Hydroxybiphenyl
3-Hydroxybiphenyl
cis-2,3-Dihydro-2,3-dihydroxybiphenyl
→ 2-Hydroxy-6-oxo-6-phenylhexa-2,4-dienoic acid
In: Microorganism H816
Pseudomonas A1599, E329, G771
Rhodococcus J221
Sphingomonas J654
See also: B546, E958, E960, F161, G478, G823, H272, J653
→ 2-Phenyl-*cis*-*cis*-muconic acid
Probably in: Pseudomonas G478

2,4-Dihydroxybiphenyl
Formed from: 4-Hydroxybiphenyl

2,4′-Dihydroxybiphenyl
Formed from: 2-Hydroxybiphenyl
Probably formed from: 4-Hydroxybiphenyl
→ 2,4′-Dihydroxybiphenyl-O-sulphate
Probably in: Man J404
Rat J404

2,5-Dihydroxybiphenyl
Formed from: 2-Hydroxybiphenyl
3-Hydroxybiphenyl
→ Biphenyl-2,5-quinone
With P450 F850
See also: K682
→ 2,3,5-Trihydroxybiphenyl
In: Pseudomonas E958, J392

3,4-Dihydroxybiphenyl
Formed from: *cis*-3,4-Dihydro-3,4-dihydroxybiphenyl

3-Hydroxybiphenyl
4-Hydroxybiphenyl
4-Hydroxy-3-nitrobiphenyl
→ 3-Hydroxy-4-methoxybiphenyl
Probably in: Guinea pig A2911
Rabbit A613, A2911
→ 4-Hydroxy-3-methoxybiphenyl
Probably in: Candida B775
Guinea pig A2911
Mouse B615
Rabbit A613, A2911
Rat A3808, B615
→ 4-Phenylmuconolactone
Probably in: Debaryomyces J824
→ 3,4,4′-Trihydroxybiphenyl
Probably in: Guinea pig A2911
Mouse B615
Pig A2650
Rabbit A2911
Rat A2660, A2724, A3808, B615
Trichosporon K682

3,4′-Dihydroxybiphenyl
Probably formed from: 4-Hydroxybiphenyl
→ 2,3,4′-Trihydroxybiphenyl
Probably in: Pseudomonas F107

4,4′-Dihydroxybiphenyl
Formed from: 4-Hydroxybiphenyl
Phenol
→ 4,4′-Dihydroxy-3-methylthiobiphenyl
Probably in: Rat A3808
→ *p*-Diphenoquinone
In: Horseradish C468, D433
→ 4-Hydroxybiphenyl-4′-yl-β-D-glucuronide
In: Rat B962, G968
See also: A2678, J582
→ 4-Hydroxybiphenyl-4′-yl sulphate
Probably in: Aspergillus C860

2,4′-Dihydroxybiphenyl-O-sulphate
Probably formed from: 2,4′-Dihydroxybiphenyl

3-(3′,4′-Dihydroxy-4-biphenylyl)-3-hydroxybutyronitrile
Formed from: 3-(4-Biphenylyl)butyronitrile

6,6′-Dihydroxy-4,4′-biquinoline
Probably formed from: 6-Hydroxyquinoline

1,3-Dihydroxy-1,2-bis(4-hydroxy-3,5-dimethoxyphenyl)propane
→ 4-Hydroxy-3,5-dimethoxybenzaldehyde
In: Coriolus E547
→ Syringylglycerol
In: Coriolus E547

1,4-Dihydroxy-5,8-bis((2-((2-hydroxyethyl)amino)ethyl)amino)-9,10-anthracenedione
See: Mitoxantrone

2,2′-Dihydroxy-4,5;4′,5′-bis(methylenedioxy)biphenyl
Formed from: 3,4-Methylenedioxyphenol
→ 4,5;4′,5′-Bis(methylenedioxy)biphenyl-2,2′-quinone
In: Horseradish B948

2,2′-Dihydroxy-5,5′-bis(1-propen-(1*E*)-yl)biphenyl
See: Magnolol

8′,9′-Dihydroxybromocryptine
Probably formed from: 8′α-Hydroxybromocryptine

1′,6-Dihydroxybufurolol
Probably formed from: 1′-Hydroxybufurolol
→ 6-Hydroxy-1′-oxobufurolol
Probably in: Dog A2806
Man A2806
Rat A2806

5,6-Dihydroxybufurolol
Probably formed from: 6-Hydroxybufurolol
→ 6-Hydroxy-5-oxobufurolol
Probably in: Dog A2806
Man A2806
Rat A2806

ω,ω-Dihydroxybupropion
Probably formed from: ω-Hydroxybupropion

6,7-Dihydroxybutamoxane
Formed from: 6-Hydroxybutamoxane
7-Hydroxybutamoxane

3-((2,3-Dihydroxy-*n*-butyl)amino)-4-phenoxy-5-sulphamoylbenzoic acid
Probably formed from: 3-((3-Hydroxy-*n*-butyl)amino)-4-phenoxy-5-sulphamoylbenzoic acid

3′,4′-Dihydroxybutyrophenone
→ 3′-Hydroxy-4′-methoxybutyrophenone
In: Rat A486
With catechol O-methyltransferase A597

→ 4′-Hydroxy-3′-methoxybutyrophenone
In: Rat A486
With catechol O-methyltransferase A597

1″,5′-Dihydroxycannabichromene
Probably formed from: 5′-Hydroxycannabichromene

1″,6′-Dihydroxycannabichromene
Probably formed from: 6′-Hydroxycannabichromene

3″,5′-Dihydroxycannabichromene
Probably formed from: 5′-Hydroxycannabichromene

3″,6′-Dihydroxycannabichromene
Probably formed from: 6′-Hydroxycannabichromene

4″,5′-Dihydroxycannabichromene
Probably formed from: 5′-Hydroxycannabichromene

4″,6′-Dihydroxycannabichromene
Probably formed from: 6′-Hydroxycannabichromene

5′,5″-Dihydroxycannabichromene
Probably formed from: 5′-Hydroxycannabichromene

5″,6′-Dihydroxycannabichromene
Probably formed from: 6′-Hydroxycannabichromene

1″,6β-Dihydroxycannabidiol
Probably formed from: 6β-Hydroxycannabidiol

1″,7-Dihydroxycannabidiol
Probably formed from: 7-Hydroxycannabidiol

2″,6β-Dihydroxycannabidiol
Probably formed from: 6β-Hydroxycannabidiol

2″,7-Dihydroxycannabidiol
Probably formed from: 7-Hydroxycannabidiol

3″,6β-Dihydroxycannabidiol
Probably formed from: 6β-Hydroxycannabidiol
→ 3″-Hydroxy-6-oxocannabidiol
Probably in: Rat A2777

3″,7-Dihydroxycannabidiol
Probably formed from: 7-Hydroxycannabidiol
→ 3″,6,7-Trihydroxycannabidiol
Probably in: Rabbit F708

4″,6β-Dihydroxycannabidiol
Probably formed from: 6β-Hydroxycannabidiol
→ 8,9-Dihydro-4″,6,8,9-tetrahydroxycannabidiol
Probably in: Dog F557, G113
→ 4″-Hydroxy-6-oxocannabidiol
Probably in: Rat A2777

4″,7-Dihydroxycannabidiol
Probably formed from: 7-Hydroxycannabidiol
→ 8,9-Dihydro-4″,7,8,9-tetrahydroxycannabidiol
Probably in: Dog F557
→ 4″,6,7-Trihydroxycannabidiol
Probably in: Rabbit F708

5″,6β-Dihydroxycannabidiol
Probably formed from: 6β-Hydroxycannabidiol
→ 8,9-Dihydro-5″,6,8,9-tetrahydroxycannabidiol
Probably in: Dog F557, G113
→ 6β-Hydroxycannabidiol-5″-oic acid
Probably in: Dog F557, G113
Man G113
→ 5″-Hydroxy-6-oxocannabidiol
Probably in: Rat F557

5″,7-Dihydroxycannabidiol
Probably formed from: 7-Hydroxycannabidiol
→ 8,9-Dihydro-5″,7,8,9-tetrahydroxycannabidiol
Probably in: Dog F557
→ 7-Hydroxycannabidiol-5″-oic acid
Probably in: Dog F557, G113
Man G113
Rat G608
→ 5″,6,7-Trihydroxycannabidiol
Probably in: Rabbit F708

6β,7-Dihydroxycannabidiol (β assumed unless otherwise stated)
Probably formed from: 7-Hydroxycannabidiol
→ 8,9-Dihydro-5″,7,8,9-tetrahydroxyxycannabidiol
Probably in: Dog F557
→ 2″,6,7-Trihydroxycannabidiol
Probably in: Dog F557, G113
→ 4″,6,7-Trihydroxycannabidiol
Probably in: Dog F557, G113

Rat G113
→ 5″,6,7-Trihydroxycannabidiol
Probably in: Dog F557, G113
Rat G113

6,7-Dihydroxycannabidiol-5″-oic acid
Probably formed from: 7-Hydroxycannabidiol-5″-oic acid
5″,6,7-Trihydroxycannabidiol
→ 4″,5″-Bisnor-6,7-dihydroxycannabidiol-3″-oic acid
Probably in: Dog F557

1″,7-Dihydroxycannabinol
Probably formed from: 7-Hydroxycannabinol
→ 1″-Hydroxycannabinol-7-oic acid
Probably in: Mouse A3405
Rabbit A440

2″,7-Dihydroxycannabinol
Probably formed from: 7-Hydroxycannabinol
→ 2″-Hydroxycannabinol-7-oic acid
Probably in: Mouse A3405
Rabbit A440

3″,7-Dihydroxycannabinol
Probably formed from: 7-Hydroxycannabinol
→ 3″-Hydroxycannabinol-7-oic acid
Probably in: Mouse A3405

4″,7-Dihydroxycannabinol
Probably formed from: 7-Hydroxycannabinol
→ 4″-Hydroxycannabinol-7-oic acid
Probably in: Mouse A3405

2,3-Dihydroxycarbamazepine
Probably formed from: 2-Hydroxycarbamazepine
→ 2-Hydroxy-3-methoxycarbamazepine
Probably in: Man A1967, A3453, J101

10,11-Dihydroxycarbamazepine
Formed from: Oxcarbamazepine

2,2″-Dihydroxycarpipramine
Probably formed from: 2-Hydroxycarpipramine

2,10-Dihydroxycarpipramine
Probably formed from: 2-Hydroxycarpipramine

2″,10-Dihydroxycarpipramine
Probably formed from: 2″-Hydroxycarpipramine

2′,4-Dihydroxychalcone
Probably formed from: 4′-Hydroxyflavone

2′,4′-Dihydroxychalcone
Probably formed from: 2′-Hydroxychalcone

2,3-Dihydroxychloridazon
Formed from: *cis*-2,3-Dihydro-2,3-dihydroxychloridazon
→ 4-Amino-5-chloro-1-(5-hydroxy-*cis,cis*-mucon-1-amido)-6(1*H*)-oxopyridazine
In: Phenylobacterium C332

3,7-Dihydroxychlorpromazine
Probably formed from: 7-Hydroxychlorpromazine
→ 3,7-Dimethoxychlorpromazine
Probably in: Man A693

7,8-Dihydroxychlorpromazine
Formed from: 7-Hydroxychlorpromazine
→ Chlorpromazine-7,8-quinone
In: Mushroom A1290
→ 7,8-Dihydroxychlorpromazine-β-D-glucuronide
In: Guinea pig A2092
Rat A2092
→ 7-Hydroxy-8-methoxychlorpromazine
Probably in: Man A693
→ 8-Hydroxy-7-methoxychlorpromazine
Probably in: Man A693

7,8-Dihydroxychlorpromazine-β-D-glucuronide
Formed from: 7,8-Dihydroxychlorpromazine

5,7-Dihydroxychromone-7-O-glucoside
Formed from: 3′,4′,5,7-Tetrahydroxyflavanone-7-O-glucoside

3,6-Dihydroxychrysene
→ 3,6-Dihydroxychrysene-3-β-D-glucuronide
In: Man H575
Rat G715
→ 3,6-Dihydroxychrysene-6-β-D-glucuronide
In: Rat G715

3,6-Dihydroxychrysene-di-β-D-glucuronide
Probably formed from: 3,6-Dihydroxychrysene-6-β-D-glucuronide

3,6-Dihydroxychrysene-3-β-D-glucuronide
Formed from: 3,6-Dihydroxychrysene

3,6-Dihydroxychrysene-6-β-D-glucuronide
Formed from: 3,6-Dihydroxychrysene
→ 3,6-Dihydroxychrysene-di-β-D-glucuronide
Probably in: Rat G715

Dihydroxycimetropium
Probably formed from: *p*-Hydroxycimetropium
→ Trihydroxycimetropium
Probably in: Guinea pig G234
Hamster G234
Mouse G234
Rat G234

2′,6′-Dihydroxycinchonidine
Probably formed from: 2′-Hydroxyquinine

3,6′-Dihydroxycinchonidine
Probably formed from: 6′-Hydroxycinchonidine

α,6-Dihydroxycinmethylin
Probably formed from: α-Hydroxycinmethylin
→ 2-((2-Carboxyphenyl)methoxy)-6-hydroxy-1-methyl-4-(1-methylethyl)-7-oxabicyclo[2.2.1]heptane
Probably in: Goat F98

α,8-Dihydroxycinmethylin
Probably formed from: α-Hydroxycinmethylin
→ 2-((2-Carboxyphenyl)methoxy)-4-(1-hydroxy-1-methylethyl)-1-methyl-7-oxabicyclo[2.2.1]heptane
Probably in: Goat F98
→ α,6,8-Trihydroxycinmethylin
Probably in: Goat F98

α,9-Dihydroxycinmethylin
Probably formed from: α-Hydroxycinmethylin
→ 2-((2-Carboxyphenyl)methoxy)-4-(1-(hydroxymethyl)ethyl)-1-methyl-7-oxabicyclo[2.2.1]heptane
Probably in: Goat F98
→ α,6,9-Trihydroxycinmethylin
Probably in: Goat F98

3,4-Dihydroxycinnamaldehyde
See: Caffealdehyde

2,3-Dihydroxycinnamic acid
Formed from: *o*-Coumaric acid

2,4-Dihydroxycinnamic acid (umbellic acid)
→ 2,4-Dihydroxystyrene
In: Klebsiella J853

3,4-Dihydroxycinnamic acid
See: Caffeic acid

3,5-Dihydroxycinnamic acid
Formed from: 3-Hydroxy-5-methoxycinnamic acid

α-(3,5-Dihydroxycinnamido)isobutyric acid
Formed from: N-(3,3-Dimethylpropynyl)-3,4,5-trimethoxycinnamide

4,6-Dihydroxyclausenamide
Probably formed from: 6-Hydroxyclausenamide

3′,4′-Dihydroxyclobazam
Probably formed from: *p*-Hydroxyclobazam
→ 3′,4′-Dihydroxyclobazam-β-D-glucuronide
Probably in: Rat G572
→ 3′,4′-Dihydroxyclobazam-O-sulphate
Probably in: Rat G572
→ 3′-Hydroxy-4′-methoxyclobazam
Probably in: Rat G572
→ 4′-Hydroxy-3′-methoxyclobazam
Probably in: Rat G572

3′,4′-Dihydroxyclobazam-β-D-glucuronide
Probably formed from: 3′,4′-Dihydroxyclobazam

3′,4′-Dihydroxyclobazam-O-sulphate
Probably formed from: 3′,4′-Dihydroxyclobazam

3,7-Dihydroxycoumarin
Formed from: Umbelliferone
Probably formed from: 7-Ethoxy-3-hydroxycoumarin

4,5-Dihydroxycoumarin
Formed from: 2,5-Dihydroxychroman-4-one

5,7-Dihydroxycoumarin
Formed from: Naringenin

6,7-Dihydroxycoumarin (esculetin)
Formed from: *cis*-Caffeic acid
7-Hydroxycoumarin-6-yl-β-D-glucoside
Umbelliferone
Probably formed from: Scopoletin
→ 6-Hydroxycoumarin-7-yl-β-D-glucoside
In: Nicotiana B442, B819, K556
See also: E681, G897
→ 7-Hydroxycoumarin-6-yl-β-D-glucuronide
In: Gardenia D87
Lithospermum D87
Nicotiana D87
See also: G968, J416
→ 7-Hydroxycoumarin-7-yl sulphate
In: Eubacterium H354
→ Isoscopoletin
In: Nicotiana A3437, A3757, B490

Streptomyces C774
→ Scopoletin
In: Apple D258
Nicotiana A3437, A3757, B490
Streptomyces C774
See also: A3778, K652

7,8-Dihydroxycoumarin (daphnetin)
Formed from: Umbelliferone
→ Daphnin
In: Nicotiana B442, B819
See also: E681
→ 8-Hydroxy-7-methoxycoumarin
Probably in: Nicotiana A3757

6,7-Dihydroxycoumarin-di-β-D-glucoside
Formed from: 7-Hydroxycoumarin-6-yl-β-D-glucoside

2,3-Dihydroxycumic acid
Formed from: Cumic acid
→ 3-Carboxy-2-hydroxy-6-isopropyl-*cis,cis*-muconic semialdehyde
In: Bacillus G766
See also: A2694, G768

1,2-Dihydroxycyclohexa-3,5-diene
See: 1,2-Dihydrocatechol

1-(*cis*-3,*cis*-4-Dihydroxycyclohexyl)-1-phenyl-3-(1-pyrrolidinyl)-1-propanol
Probably formed from: 1-(*cis*-4-Hydroxycyclohexyl)-1-phenyl-3-(1-pyrrolidinyl)-1-propanol

1-(*cis*-3,*trans*-4-Dihydroxycyclohexyl)-1-phenyl-3-(1-pyrrolidinyl)-1-propanol
Probably formed from: 1-(*trans*-4-Hydroxycyclohexyl)-1-phenyl-3-(1-pyrrolidinyl)-1-propanol

1-(*trans*-3,*trans*-4-Dihydroxycyclohexyl)-1-phenyl-3-(1-pyrrolidinyl)-1-propanol
Probably formed from: 1-(*trans*-4-Hydroxycyclohexyl)-1-phenyl-3-(1-pyrrolidinyl)-1-propanol

4-(Dihydroxycyclohexyl 1′-piperazinyl-4′-carboxylate)-1-(4′-fluorophenyl)butan-1-ol
Probably formed from: 4-(Dihydroxycyclohexyl 1′-piperazinyl-4′-carboxylate)-1-(4′-fluorophenyl)butan-1-one

4-(Dihydroxycyclohexyl 1′-piperazinyl-4′-carboxylate)-1-(4′-fluorophenyl)butan-1-one
Probably formed from: 1-(4′-Fluorophenyl)-4-(hydroxycyclohexyl 1′-piperazinyl-4′-carboxylate)butan-1-one
→ 4-(Dihydroxycyclohexyl 1′-piperazinyl-4′-carboxylate)-1-(4′-fluorophenyl)butan-1-ol
Probably in: Man A2038
Rat A2038

2-(*p*-((2,3-Dihydroxycyclopentyl)methyl)phenyl)propionic acid
Formed from: 2-((*p*-(3-Hydroxy-2-oxocyclopentyl)methyl)phenyl)propionic acid

2-(*p*-((2,4-Dihydroxycyclopentyl)methyl)phenyl) propionic acid
Formed from: 2-(*p*-((2-Hydroxycyclopentyl) methyl)phenyl)propionic acid

3′,10-Dihydroxycyclopeptine
Formed from: 10-Hydroxycyclopeptine

4,6-Dihydroxydebrisoquine
Formed from: 4-Hydroxydebrisoquine

10,11-Dihydroxydemethylnortriptyline
Probably formed from: 10-Hydroxydemethylnortriptyline

7,8-Dihydroxy-2-demethylschizandrin
Formed from: 7,8-Dihydroxyschizandrin

7,8-Dihydroxy-3-demethylschizandrin
Formed from: 7,8-Dihydroxyschizandrin

3′,4′-Dihydroxydemethyltriflubazam
Probably formed from: 4′-Hydroxydemethyltriflubazam
→ 4′-Hydroxy-3′-methoxydemethyltriflubazam
Probably in: Man A1873
Monkey A1864

3,4-Dihydroxydibenzanthracene
Formed from: *trans*-3,4-Dihydro-3,4-dihydroxydibenzanthracene

10,11-Dihydroxy-5*H*-dibenzo[*a*,*d*]cycloheptane
Probably formed from: 5*H*-Dibenzo[*a*,*d*] cycloheptane-10,11-oxide

1,2-Dihydroxydibenzo-*p*-dioxin
Probably formed from: *cis*-1,2-Dihydro-1,2-dihydroxydibenzo-*p*-dioxin

1,2-Dihydroxydibenzofuran
→ 2-Hydroxy-4-(3-oxo-3*H*-benzofuran-2-yliden)but-2-enoic acid

In: Ralstonia K614

2,3-Dihydroxydibenzofuran
Formed from: *trans*-2,3-Dihydro-2,3-dihydroxydibenzofuran

3,8-Dihydroxydibenzo[*b*,*d*]pyran-6-one
Formed from: Ellagic acid

1,2-Dihydroxydibenzothiophene
Probably formed from: *cis*-1,2-Dihydro-1,2-dihydroxydibenzothiophene
→ *trans*-4-(2-(3-hydroxy)thianaphthenyl)-2-oxo-3-butenoic acid
In: Beijerinckia A3340
See also: A811

4′,5-Dihydroxydiclofenac
Formed from: 5-Hydroxydiclofenac
Probably formed from: 4′-Hydroxydiclofenac

N,5-Dihydroxydiclofenac
Formed from: 5-Hydroxydiclofenac

7,14-Dihydroxydiftalone
Probably formed from: 7-Hydroxydiftalone

3,6-Dihydroxy-2,4-dimethoxybenzaldehyde
Formed from: 4-Hydroxy-3,5-dimethoxybenzaldehyde

1,3-Dihydroxy-5,6-dimethoxybiphenyl
Probably formed from: 1-(2,6-Dihydroxy-3,4-dimethoxyphenyl)-3-(*p*-hydroxyphenyl) propan-1-one

4,4′-Dihydroxy-3,3′-dimethoxybiphenyl
Formed from: Guaiacol
→ 3,3′-Dimethoxy-4,4′-biphenylquinone
With lactoperoxidase J352

6,6′-Dihydroxy-5,5′-dimethoxy(1,1′-biphenyl)-3,3′-dicarboxaldehyde
See: Dehydrodivanillin

6,6′-Dihydroxy-5,5′-dimethoxy(1,1′-biphenyl)-3,3′-dicarboxylic acid
See: Dehydrodivanillic acid

6,6′-Dihydroxy-5,5′-dimethoxy(1,1′-biphenyl)-3,3′-dimethanol
See: Dehydrodivanillyl alcohol

2′,4′-Dihydroxy-4,6′-dimethoxychalcone
→ Eucomin
In: Eucomis A228

2,2′-Dihydroxy-3,3′-dimethoxy-5,5′-di-(2-propenyl)biphenyl
Formed from: Eugenol

1,4-Dihydroxy-2,3-dimethoxynaphthalene
Formed from: 2,3-Dimethoxy-1,4-naphthoquinone

3′,5-Dihydroxy-4′,7-dimethoxyflavone
Probably formed from: Diosmetin

4′,5-Dihydroxy-3′,7-dimethoxyflavone
Probably formed from: Chrysoeriol

5,7-Dihydroxy-4′,8-dimethoxyflavone
Probably formed from: 5,7,8-Trihydroxy-4′-methoxyflavone

5,8-Dihydroxy-4′,7-dimethoxyflavone
Probably formed from: 5,7,8-Trihydroxy-4′-methoxyflavone

1-(2,6-Dihydroxy-3,4-dimethoxyphenyl)-3-(*p*-hydroxyphenyl)propan-1-one
Probably formed from: Abrusin
→ 1,3-Dihydroxy-5,6-dimethoxybenzene
Probably in: Microorganism K574
→ Phloretic acid
Probably in: Microorganism K574

2,3-Dihydroxy-3,4-dimethoxyprotoberberine
→ Jatrorrhizine
In: Agemone D77

3,6-Dihydroxy-2,4-dimethylbenzaldehyde
Formed from: 4-Hydroxy-3,5-dimethylbenzaldehyde

4,6-Dihydroxy-2,3-dimethylbenzaldehyde
→ 4,6-Dihydroxy-2,3-dimethylbenzyl alcohol
In: Penicillium B62

2,5-Dihydroxy-3,4-dimethylbenzoic acid
→ 6-Carboxy-4-hydroxy-2,3-dimethyl-*cis,cis*-muconic semialdehyde
In: Pseudomonas F649

2,5-Dihydroxy-3,4-dimethylbenzoic acid
(5-methylorsellinic acid)
→ Mycophenolic acid
In: Penicillium A814

4,6-Dihydroxy-2,3-dimethylbenzyl alcohol
Formed from: 4,6-Dihydroxy-2,3-dimethylbenzaldehyde
→ 4,5,6-Trimethylresorcinol
In: Penicillium B62

2,2′-Dihydroxy-5,5′-dimethylbiphenyl
Formed from: *p*-Cresol

6,7-Dihydroxy-1″-1″-dimethylcannabidiol-5″-oic acid
Probably formed from: 7-Hydroxy-1″-1″-dimethylcannabidiol-5″-oic acid

4,5-Dihydroxy-N,N′-dimethylclonidine
Probably formed from: 4-Hydroxy-N,N′-dimethylclonidine

(*R,R*)-α,β-Dihydroxy-*p,p′*-dimethyldibenzyl
Formed from: *p,p′*-Dimethyl-*cis*-stilbene oxide

3,5-Dihydroxy-4,4-dimethyl-1-(3,4-methylenedioxyphenyl)-1-pentene
Formed from: Stiripentol

1,4-Dihydroxy-2,3-dimethylnaphthalene
Formed from: 2,3-Dimethyl-1,4-naphthoquinone

3,4-Dihydroxy-N,N-dimethylphenethylamine
Formed from: 3,4-Dihydroxy-N-methylphenethylamine
Hordenine
→ 3-Hydroxy-4-methoxy-N,N-dimethylphenethylamine
In: Rat A486
→ 4-Hydroxy-3-methoxy-N,N-dimethylphenethylamine
In: Rat A486
→ 4-(2-Dimethylaminoethyl)-*o*-quinone
Probably in: Man A63

4,4′-Dihydroxydiphenoxybiphenyl
Formed from: 2-Hydroxyphenyl phenyl ether

2,4′-Dihydroxydiphenylamine
Probably formed from: 4-Hydroxydiphenylamine

Dihydroxydiprenylpulvinone
Formed from: Dihydroxypulvinone

(*R,R*)-α,β-Dihydroxy-*p,p′*-dipropyldibenzyl
Formed from: *p,p′*-Dipropyl-*cis*-stilbene oxide

3,4-Dihydroxy-N,N-dipropylphenethylamine
→ 4-(2-((N,N-Dipropyl)amino)ethyl)-*o*-quinone
In: Agaricus J460

7,14-Dihydroxyefavirenz
Probably formed from: 7-Hydroxyefavirenz
→ 7,14-Dihydroxyefavirenz-7-β-D-glucuronide
Probably in: Cynomolgus monkey J775
Guinea pig J775
Man J775
→ 7,14-Dihydroxyefavirenz-7-sulphate
Probably in: Guinea pig J775

8,14-Dihydroxyefavirenz
Probably formed from: 8-Hydroxyefavirenz
→ 8,14-Dihydroxyefavirenz-8-β-D-glucuronide
Probably in: Guinea pig J775
Rat J775
→ 8,14-Dihydroxyefavirenz-8-sulphate
Probably in: Cynomolgus monkey J775
Guinea pig J775
Hamster J775
Rat J765, J775

7,14-Dihydroxyefavirenz-7-β-D-glucuronide
Probably formed from: 7,14-Dihydroxyefavirenz

8,14-Dihydroxyefavirenz-8-β-D-glucuronide
Probably formed from: 8,14-Dihydroxyefavirenz

8,14-Dihydroxyefavirenz-14-β-D-glucuronide-8-sulphate
Probably formed from: 8,14-Dihydroxyefavirenz-8-sulphate

7,14-Dihydroxyefavirenz-7-sulphate
Probably formed from: 7,14-Dihydroxyefavirenz

8,14-Dihydroxyefavirenz-8-sulphate
Probably formed from: 8,14-Dihydroxyefavirenz
→ 8,14-Dihydroxyefavirenz-14-β-D-glucuronide-8-sulphate
Probably in: Guinea pig J775
Rat J765, J775
→ 13-(S-Glutathionyl)-12,13-dihydro-8,14-dihydroxyefavirenz-8-sulphate
Probably in: Rat J765

8,9-Dihydroxyergotamine
Probably formed from: 9-Hydroxyergotamine

4-(1,2-Dihydroxyethyl)benzoic acid
Probably formed from: *p*-Ethenylphenylglycol

4-(1,2-Dihydroxyethyl)biphenyl
Formed from: *p*-Phenylstyrene

2-(1,2-Dihydroxyethyl)-3-(*p*-hydroxybenzoyl)benzofuran
Probably formed from: 3-(*p*-Hydroxybenzoyl)-2-(1-hydroxyethyl)benzofuran

1-(1,2-Dihydroxyethyl)phenoxathiin-10,10-dioxide
Probably formed from: 1-(1-Hydroxyethyl) phenoxathiin-10,10-dioxide

3-(4-(1,2-Dihydroxyethyl)phenoxy)-1-isopropylaminopropan-2-ol
Probably formed from: 3-(4-(1-Hydroxy-2-methoxyethyl)phenoxy)-1-isopropylaminopropan-2-ol
→ 3-(4-(1-Carboxy-1-hydroxymethyl) phenoxy)-1-isopropylaminopropan-2-ol
Probably in: Rat A1750

3,4-Dihydroxy-N,N-dimethylphenylethanolamine
See: N-Methyladrenaline

4-(1,2-Dihydroxyethyl)-*o*-quinone
Probably formed from: 3,4-Dihydroxyphenylglycol
→ ω,3′,4′-Trihydroxyacetophenone
In: Sarcophaga F792, F793, G184

4′,7-Di-O-(β-hydroxyethyl)rutin
Probably formed from: 7-O-(β-Hydroxyethyl)rutin
→ 4′,7-Di-O-(β-hydroxyethyl)rutin aglycone
In: Microorganism A283
Rat A271
→ 4′,7-Di-O-(β-hydroxyethyl)rutin-β-D-glucuronide
In: Rat A271
→ 3′,4′,7-Tri-O-(β-hydroxyethyl)rutin
Probably in: Rat A271

4′,7-Di-O-(β-hydroxyethyl)rutin aglycone
Formed from: 4′,7-Di-O-(β-hydroxyethyl) rutin

4′,7-Di-O-(β-hydroxyethyl)rutin-β-D-glucuronide
Formed from: 4′,7-Di-O-(β-hydroxyethyl) rutin

(8*S*,10*S*)-8-(1,2-Dihydroxyethyl)-7,8,9,10-tetrahydro-6,8,10,11-tetrahydroxy-1-methoxy-5,12-naphthacenedione
Probably formed from: 7-O-(2,6-Dideoxy-2-fluoro-α–L-talopyranosyl)adriamycinol

8-(1,2-Dihydroxyethyl)-7,8,9,10-tetrahydro-6,8,11-trihydroxy-1-methoxy-5,12-naphthacenedione
(adriamycinol aglycone)
Formed from: Adriamycin
Probably formed from: Adriamycinol

6-(1,2-Dihydroxyethyl)-3-(1*H*-tetrazol-5-yl) chromone
Probably formed from:6-(1-Hydroxyethyl)-3-(1*H*-tetrazol-5-yl)chromone

α-(1,2-Dihydroxyethyl)-3-trifluoromethylbenzhydrol
Formed from: α-Ethyl-3-trifluoromethylbenzhydrol

4α,4′-Dihydroxyflavan
Probably formed from: 4-Hydroxyflavone

3,4′-Dihydroxyflavan-4α-ol
Probably formed from: *trans*-3-Hydroxyflavan-4α-ol

3,4′-Dihydroxyflavan-4β-ol
Probably formed from: *trans*-3-Hydroxyflavan-4β-ol

3,6-Dihydroxyflavan-4α-ol
Probably formed from: *trans*-3-Hydroxyflavan-4α-ol

3,6-Dihydroxyflavan-4β-ol
Probably formed from: *trans*-3-Hydroxyflavan-4β-ol

3′,4′-Dihydroxyflavanone
Probably formed from: 4′-Hydroxyflavanone

4′,7-Dihydroxyflavanone
See: Liquiritigenin

3′,4′-Dihydroxyflavone
Formed from: 4′-Hydroxyflavone

4′,5-Dihydroxyflavone
Formed from: 5-Hydroxyflavone
→ 3′,4′,5-Trihydroxyflavone
Probably in: Aspergillus G102

4′,6-Dihydroxyflavone
Formed from: 6-Hydroxyflavone

4′,7-Dihydroxyflavone
Formed from: 7-Hydroxyflavone
→ 4′,7-Dihydroxyflavone-7-β-D-glucoside
In: Parsley K739
→ *p*-Hydroxybenzoic acid
In: Rhizobium H280
→ Resorcinol
In: Rhizobium H280

5,6-Dihydroxyflavone
→ 5-Hydroxy-6-methoxyflavone
In: Calamondin orange A3987
→ 6-Hydroxy-5-methoxyflavone
In: Calamondin orange A3987

5,7-Dihydroxyflavone
→ 5,7-Dihydroxyflavone-7-β-D-glucoside
In: Parsley K739

7,8-Dihydroxyflavone
Formed from: 7-Hydroxyflavone
→ 7-Hydroxy-8-methoxyflavone
In: Calamondin orange A3987
→ 8-Hydroxy-7-methoxyflavone
In: Calamondin orange A3987

4′,7-Dihydroxyflavone-7-β-D-glucoside
Formed from: 4′,7-Dihydroxyflavone

5,7-Dihydroxyflavone-7-β-D-glucoside
Formed from: 5,7-Dihydroxyflavone

4′,5-Dihydroxyflufenamic acid
Probably formed from: 4′-Hydroxyflufenamic acid

1,1a-Dihydroxy-1*H*-fluoren-9-one
Formed from: Fluorene
→ 1-Hydroxy-1′-oxalylbiphenyl
Probably in: Pseudomonas H254

3′,4′-Dihydroxyflurbiprofen
Probably formed from: 4′-Hydroxyflurbiprofen
→ 3′-Hydroxy-4′-methoxyflurbiprofen
Probably in: Mouse A3393
Pig C224
Rat A3393

2,2′-Dihydroxy-3,3′,4,4′,6,6′-hexamethylbiphenyl
Probably formed from: 2,3,5-Trimethylbiphenyl

2,5-Dihydroxyhippuric acid
Formed from: Gentisic acid
o-Hydroxyhippuric acid
→ Gentisamide
In: Man K660

3,4-Dihydroxyhippuric acid
Formed from: Protocatechuic acid

4′,6-Dihydroxy-2-(α-hydroxybenzyl)coumaranone
Formed from: 2′,4,4′-Trihydroxychalcone

3,14-Dihydroxy-N-(3-hydroxycyclobutylmethyl) morphinan
Formed from: Butorphanol

7,10-Dihydroxy-2-(2-((2-hydroxyethyl)amino)ethyl)-5-((2-(methylamino)ethyl)amino)anthra[1,9-*c*,*d*]pyrazol-6(2*H*)-one
→ 7,10-Dihydroxy-2-(2-((2-hydroxyethyl) amino)ethyl)-5-((2-(methylamino)ethyl) amino)anthra[1,9-*c*,*d*]pyrazol-6(2*H*)-one-β-D-glucuronide
In: Rat H139
→ 8-(S-Glutathionyl)-7,10-dihydroxy-2-(2-((2-hydroxyethyl)amino)ethyl)-5-((2-(methylamino)ethyl)amino)anthra [1,9-*c*,*d*]pyrazol-6(2*H*)-one
In: Rat H139
→ 9-(S-Glutathionyl)-7,10-dihydroxy-2-(2-((2-hydroxyethyl)amino)ethyl)-5-((2-(methylamino)ethyl)amino)anthra [1,9-*c*,*d*]pyrazol-6(2*H*)-one
In: Rat H139

7,10-Dihydroxy-2-(2-((2-hydroxyethyl)amino) ethyl)-5-((2-(methylamino)ethyl)amino) anthra[1,9-*c*,*d*]pyrazol-6(2*H*)-one-β-D-glucuronide
Formed from: 7,10-Dihydroxy-2-(2-((2-hydroxyethyl)amino)ethyl)-5-((2-(methylamino)ethyl)amino)anthra[1,9-*c*,*d*] pyrazol-6(2*H*)-one

1,4-Dihydroxy-6-(2-hydroxyethyl)-2-hydroxymethyl-2,5,7-trimethylindane
Formed from: Illudin S

3,5-Dihydroxy-4-(6-hydroxy-4-isopropenyl-1-methylcyclohexen-3-yl)benzoic acid
Probably formed from: 3,5-Dihydroxy-4-(4-isopropenyl-1-methylcyclohexen-3-yl) benzoic acid

3,5-Dihydroxy-4-(6-hydroxy-4-isopropenyl-1-methylcyclohexen-3-yl)phenylbutyric acid
Probably formed from: 3,5-Dihydroxy-4-(4-isopropenyl-1-methylcyclohexen-3-yl) phenylbutyric acid

3,5-Dihydroxy-4-(6-hydroxy-4-isopropenyl-1-methylcyclohexen-3-yl)phenylpropionic acid
Probably formed from: 3,5-Dihydroxy-4-(4-isopropenyl-1-methylcyclohexen-3-yl) phenylpropionic acid

2,6-Dihydroxy-8-hydroxylamino-4-hydroxymethyl-5-(3-trifluoromethylphenoxy) quinoline
Probably formed from: 8-Amino-2,6-dimethoxy-4-methyl-5-(3-trifluoromethylphenoxy)quinoline

5,6-Dihydroxy-8-hydroxylaminoquinoline
Probably formed from: 8-Amino-6-hydroxyquinoline

3,5-Dihydroxy-1-(3-hydroxy-4-methoxyphenyl)-4,4-dimethyl-1-pentene
Probably formed from: 3-Hydroxy-1-(3-hydroxy-4-methoxyphenyl)-4,4-dimethyl-1-pentene
→ 4-Carboxy-3-hydroxy-1-(3-hydroxy-4-methoxyphenyl)-4,4-dimethyl-1-butene
Probably in: Rat F574

3,5-Dihydroxy-1-(4-hydroxy-3-methoxyphenyl)-4,4-dimethyl-1-pentene
Probably formed from: 3-Hydroxy-1-(4-hydroxy-3-methoxyphenyl)-4,4-dimethyl-1-pentene

1,3-Dihydroxy-5-((1*E*)-2-(3-hydroxy-4-methoxyphenyl)ethenyl)benzene
See: Rhapontigenin

1,8-Dihydroxy-3-hydroxymethylanthraquinone
See: Aloe-emodin

1,8-Dihydroxy-3-hydroxymethylanthrone
Formed from: Barbaloin

3,5-Dihydroxy-4-(1-hydroxymethyl-4-isopropenylcyclohexen-3-yl)benzoic acid
Probably formed from: 3,5-Dihydroxy-4-(4-isopropenyl-1-methylcyclohexen-3-yl) benzoic acid

3,5-Dihydroxy-4-(1-hydroxymethyl-4-isopropenylcyclohexen-3-yl)phenylbutyric acid
Probably formed from: 3,5-Dihydroxy-4-(4-isopropenyl-1-methylcyclohexen-3-yl) phenylbutyric acid

3,5-Dihydroxy-4-(1-hydroxymethyl-4-isopropenylcyclohexen-3-yl)phenylpropionic acid
Probably formed from: 3,5-Dihydroxy-4-(4-isopropenyl-1-methylcyclohexen-3-yl) phenylpropionic acid

5,6-Dihydroxy-8-(3-hydroxy-5-methyl-2-oxopyrrolidin-1-yl)quinoline
Probably formed from: 8-((4-Amino-1-methyl-3-oxobutyl)amino)-5-hydroxy-6-methoxyquinoline

5,8-Dihydroxy-2-((1*R*)-1-hydroxy-4-methyl-3-pentenyl)-1,4-naphthalenedione
See: Shikonin

2,4-Dihydroxy-6-(10-hydroxy-6-oxo-*trans*-1-undecenyl)benzoic acid γ-lactone
See: Zearalenone

6,7-Dihydroxy-1-(*p*-hydroxyphenoxy)-3,7-dimethyl-2-octene
Formed from: *trans*-6,7-Epoxy-1-(4′-ethylphenoxy)-3,7-dimethyl-2-octene
→ 6,7-Dihydroxy-1-(*p*-hydroxyphenoxy)-3,7-dimethyl-2-octene-β-D-glucuronide
Probably in: Rat A813

6,7-Dihydroxy-1-(*p*-hydroxyphenoxy)-3,7-dimethyl-2-octene-β-D-glucuronide
Probably formed from: 6,7-Dihydroxy-1-(*p*-hydroxyphenoxy)-3,7-dimethyl-2-octene

2,6-Dihydroxy-1-(*p*-hydroxyphenylazo) naphthalene
Formed from: 2-Hydroxy-1-(*p*-hydroxyphenylazo)naphthalene
1-(4-Hydroxyphenylazo)naphthalene

***meso*-3,4-Di-(*p*-hydroxyphenyl)hexane**
See: Hexoestrol

5,7-Dihydroxy-3-(4-hydroxyphenyl)-6-(3-methyl-2-butenyl)-4-*H*-1-benzopyran-4-one
See: Wighteone

3,5-Dihydroxy-N-(2-(3-(*p*-hydroxyphenyl)propyl) phenylethanolamine
See: Fenoterol

2,2-Dihydroxy-3-(*p*-hydroxyphenyl)-1-(2,4,6-trihydroxyphenyl)-1,3-propanedione
Formed from: Kaempferol

3,4-Dihydroxy-N-(3-hydroxypropyl)amphetamine
Probably formed from: *p*-Hydroxy-N-(3-hydroxypropyl)amphetamine
→ 4-Hydroxy-N-(3-hydroxypropyl)-3-methoxyamphetamine
Probably in: Man J544

2,3-Dihydroxyiminodibenzyl
Probably formed from: 2-Hydroxyiminodibenzyl

***cis*-1,2-Dihydroxyindane**
Formed from: *cis*-2-Amino-1-hydroxyindane
trans-2-Amino-1-hydroxyindane
1-Hydroxyindane
Indan-2-one
Indene
→ 2-Hydroxyindan-1-one
Probably in: Pseudomonas K207, K348
Does not form *trans*-1,2-dihydroxyindane in Rhodococcus H885

***trans*-1,2-Dihydroxyindane**
Formed from: *cis*-2-Amino-1-hydroxyindane
trans-2-Amino-1-hydroxyindane
1-Hydroxyindane
Indan-2-one
Indene
Indene-1,2-oxide

1,5-Dihydroxyindane
Probably formed from: 5-Hydroxyindane
→ 5-Hydroxyindan-1-one
Probably in: Penicillium J836

N-(2,3-Dihydroxy-1-indanyl)-5(*S*)-(1,1-dimethylethoxycarbonylamino)-4(*S*)-hydroxy-2(*R*)-(4-(2-(4-morpholinyl)ethoxy)phenyl) methyl-6-phenylhexamide
Formed from: 5(*S*)-(1,1-Dimethylethoxycarbonylamino)-4(*S*)-hydroxy-N-(2(*R*)-hydroxy-1(*S*)-indanyl)-2(*R*)-(4-(2-(4-morpholinyl)ethoxy)phenyl)methyl-6-phenylhexamide

N-(2,4-Dihydroxy-1-indanyl)-5(*S*)-(1,1-dimethylethoxycarbonylamino)-4(*S*)-hydroxy-2(*R*)-(4-(2-(4-morpholinyl)ethoxy)phenyl) methyl-6-phenylhexamide
Formed from: 5(*S*)-(1,1-Dimethylethoxycarbonylamino)-4(*S*)-hydroxy-N-(2(*R*)-hydroxy-1(*S*)-indanyl)-2(*R*)-(4-(2-(4-morpholinyl)ethoxy)phenyl)methyl-6-phenylhexamide

N-(2(*R*),3-Dihydroxy-1(*S*)-indanyl)-5-((2(*S*)-(1,1-dimethylethylaminocarbonyl)piperazin-1-yl)-4(*S*)-hydroxy-2(*R*)-phenylmethyl) pentanamide
Probably formed from: 5-((2(*S*)-(1,1-Dimethylethylaminocarbonyl)piperazin-1-yl)-4(*S*)-hydroxy-N-(2(*R*)-hydroxy-1(*S*)-indanyl)-2(*R*)-phenylmethyl)pentanamide

N-(2(*R*),3-Dihydroxy-1(*S*)-indanyl)-5-((2(*S*)-(1,1-dimethylethylaminocarbonyl)-4-((pyridine-N-oxide-3-yl)methyl)piperazin-1-yl)-4(*S*)-hydroxy-2(*R*)-phenylmethyl) pentanamide
Probably formed from: N-(2(*R*),3-Dihydroxy-1(*S*)-indanyl)-5-((2(*S*)-(1,1-dimethylethyl-aminocarbonyl)-4-((pyridin-3-yl)methyl)piperazin-1-yl)-4(*S*)-hydroxy-2(*R*)-phenyl-methyl)pentanamide

N-(2(*R*),3-Dihydroxy-1(*S*)-indanyl)-5-((2(*S*)-(1,1-dimethylethylaminocarbonyl)-4-((pyridin-3-yl)methyl)piperazin-1-yl)-4(*S*)-hydroxy-2(*R*)-phenylmethyl)pentanamide
Formed from: 5-((2(*S*)-(1,1-Dimethylethyl aminocarbonyl)-4-((pyridin-3-yl)methyl) piperazin-1-yl)-4(*S*)-hydroxy-N-(2(*R*)-hydroxy-1(*S*)-indanyl-2(*R*)-phenylmethyl) pentanamide
→ N-(2(*R*),3-Dihydroxy-1(*S*)-indanyl)-5-((2(*S*)-(1,1-dimethylethylaminocarbonyl)-4-((pyridine-N-oxide-3-yl)methyl) piperazin-1-yl)-4(*S*)-hydroxy-2(*R*)-phenylmethyl)pentanamide
Probably in: Man H151

2,3-Dihydroxyindole
→ N-Formylanthranilic acid
In: Microorganism K821

5,6-Dihydroxyindole
Formed from: 4-(3-L-alanyl)-*o*-quinone
D-Dopachrome
Probably formed from: 5,6-Dihydroxyindole-3-acetic acid

5,6-Dihydroxyindole-3-acetic acid
Probably formed from: 5-Hydroxyindole-3-acetic acid
→ 5,6-Dihydroxyindole
Probably in: Man G762
→ 8′,9′-Dihydroxyisobromocryptine
Probably formed from: 8′α-Hydroxyisobromocryptine
→ 4′,7-Dihydroxyisoflavan

5,6-Dihydroxyindole-2-carboxylic acid
Formed from: L-Dopachrome
→ Indole-5,6-quinone-2-carboxylic acid
Probably in: Mouse H235

8′,9′-Dihydroxyisobromocryptine
Probably formed from: 8′α-Hydroxyisobromocryptine

3′,30-Dihydroxy-5′-(4-isobutyl-1-piperazinyl) benzoxazinorifamycin
Formed from: 3′-Hydroxy-5′-(4-isobutyl-1-piperazinyl)benzoxazinorifamycin

4′,7-Dihydroxyisoflavan
Probably formed from: Daidzein

4′,7-Dihydroxyisoflavanone
See: Dihydrodaidzein

3′,4′-Dihydroxyisoflavone
Probably formed from: 4′-Hydroxyisoflavone
Isoflavone

4′,7-Dihydroxyisoflavone
See: Daidzein

4,5-Dihydroxyisophthalic acid
Formed from: 5-Hydroxyisophthalic acid
4-Hydroxy-5-methoxyisophthalic acid

3,5-Dihydroxy-4-(4-isopropenyl-1-methylcyclohexen-3-yl)phenylpropionic acid
Probably formed from: 3″-Carboxy-4″,5″-bisnorcannabinol
→ 3,5-Dihydroxy-4-(6-hydroxy-4-isopropenyl-1-methylcyclohexen-3-yl)phenylpropionic acid
Probably in: Man F503
→ 3,5-Dihydroxy-4-(1-hydroxymethyl-4-isopropenyl-cyclohexen-3-yl) phenylpropionic acid
Probably in: Man F503
→ 3,5-Dihydroxy-4-(4-isopropenyl-1-methylcyclohexen-3-yl) phenylacetic acid
Probably in: Man F503

3,5-Dihydroxy-4-(4-isopropenyl-1-methylcyclohexen-3-yl)benzoic acid
Probably formed from: 3,5-Dihydroxy-4-(4-isopropenyl-1-methylcyclohexen-3-yl) phenylacetic acid
→ 3,5-Dihydroxy-4-(6-hydroxy-4-isopropenyl-1-methylcyclohexen-3-yl)benzoic acid
Probably in: Man F503
→ 3,5-Dihydroxy-4-(1-hydroxymethyl-4-isopropenylcyclohexen-3-yl)benzoic acid
Probably in: Man F503

3,5-Dihydroxy-4-(4-isopropenyl-1-methylcyclohexen-3-yl)phenylacetic acid
Probably formed from: 3,5-Dihydroxy-4-(4-isopropenyl-1-methylcyclohexen-3-yl) phenylpropionic acid
→ 3,5-Dihydroxy-4-(4-isopropenyl-1-methylcyclohexen-3-yl)benzoic acid
Probably in: Man F503

3,5-Dihydroxy-4-(4-isopropenyl-1-methylcyclohexen-3-yl)phenylbutyric acid
Formed from: Cannabidiol
→ 3,5-Dihydroxy-4-(6-hydroxy-4-isopropenyl-1-methylcyclohexen-3-yl) phenylbutyric acid
Probably in: Man F503
→ 3,5-Dihydroxy-4-(1-hydroxymethyl-4-isopropenylcyclohexen-3-yl)phenylbutyric acid
Probably in: Man F503
→ 3,5-Dihydroxy-4-(4-isopropenyl-1-methylcyclohexen-3-yl)phenypropionic acid
Probably in: Man F503

2,3-Dihydroxy-4-isopropylbenzoic acid
→ 2-Carboxy-5-isopropyl-*cis,cis*-muconic acid
In: Tecoma A2241

2,5-Dihydroxy-3-isopropylbenzoic acid
→ 6-Carboxy-4-hydroxy-2-isopropyl-*cis,cis*-muconic semialdehyde
In: Pseudomonas F649

3,5-Dihydroxy-N-isopropylphenylethanolamine
(metaproterenol, orciprenaline)
→ 3-Hydroxy-N-isopropylphenylethanolamine-5-yl-β-D-glucuronide
In: Rat D344
→ 3-Hydroxy-N-isopropylphenylethanolamine-5-yl-sulphate
In: Man C583

1,2-Dihydroxy-1-(4-isopropylsulphonylphenyl) propane
Probably formed from: 2-Amino-1-(4-isopropylsulphonylphenyl)propanol

3,4-Dihydroxy-2(1*H*)-isoquinoline carboxamidine
See: Debrisoquine

8-(β,γ-Dihydroxyisovaleryl)-3,4′,5,7-tetrahydroxyflavanone
Probably formed from: Neophellamuretin
→ (4″,5″-Dihydro-5″-(2-hydroxy)isopropylfurano)[2″,3″-*h*]-3,4′,5-trihydroxyflavanone
Probably in: Aspergillus A3306

3′,4′-Dihydroxykawain
Formed from: Methystictin
Probably formed from: *p*-Hydroxykawain
→ Dihydro-3′,4′-dihydroxykawain
Probably in: Rat A3485

4,9-Dihydroxy-α-lapachone
Formed from: 2-Carboxy-4-hydroxy-α-tetralone

6,6a-Dihydroxymaackiain
Probably formed from: 6a-Hydroxymaackiain

3,4-Dihydroxymandelaldehyde
Formed from: D-Adrenaline
3,4-Dihydroxyphenylglycol
Isoprenaline
D-Noradrenaline

2,5-Dihydroxymandelic acid
Probably formed from: Homogentisic acid
→ 2,5-Dihydroxyphenylglyoxylic acid
In: Aspergillus A1204

3,4-Dihydroxymandelic acid (DOMA)
Formed from: 3,4-Dihydroxyphenylacetic acid
2-(3,4-Dihydroxyphenyl)-2-oxoethylamine
4-Hydroxy-3-methoxymandelic acid
→ δ-(Carboxy-α-hydroxymethyl)-α-hydroxy-*cis,cis*-muconic semialdehyde
In: Arthrobacter A3279
Bacillus A2827
Pseudomonas A1090
→ 3,4-Dihydroxymandelic acid-O-sulphate
In: Beef F196
→ 3,4-Dihydroxyphenylglyoxylic acid
In: Pseudomonas G794
→ 3-Hydroxy-4-methoxymandelic acid
In: Rat A486, A1072
→ 4-Hydroxy-3-methoxymandelic acid
In: Rat A486, A1072
→ Protocatechualdehyde
In: Mushroom D865, E963

3,4-Dihydroxymandelic acid-O-sulphate
Formed from: 3,4-Dihydroxymandelic acid

3,4-Dihydroxymandelonitrile
Probably formed from: *p*-Hydroxymandelonitrile
→ 3,4-Dihydroxymandelonitrile-2-glucoside
In: Triglochin D58
→ Triglochinin
In: Thalictrum B492
Triglochin B492

3,4-Dihydroxymandelonitrile-2-glucoside
Formed from: 3,4-Dihydroxymandelonitrile

3,4-Dihydroxymefenorex
Probably formed from: *p*-Hydroxymefenorex
→ 4-Hydroxy-3-methoxymefenorex
Probably in: Man J544

3′,4′-Dihydroxymeperidine
Formed from: *p*-Hydroxymeperidine
Meperidine
→ 3′,4′-Dihydroxynormeperidine
Probably in: Cat D546
Dog D546
Guinea pig D546
Rabbit D546
Rat D546
→ 4′-Hydroxy-3′-methoxymeperidine
Probably in: Cat D546
Dog D546
Guinea pig D546
Rabbit D546
Rat D546

2′,3′-Dihydroxy-4′-methoxyacetophenone
Probably formed from: 2′-Hydroxy-4′-methoxyacetophenone

2′,3′-Dihydroxy-6′-methoxyacetophenone
Probably formed from: 2′-Hydroxy-6′-methoxyacetophenone

2′,4′-Dihydroxy-3′-methoxyacetophenone
Formed from: 4′-Hydroxy-3′-methoxyacetophenone

2′,4′-Dihydroxy-5′-methoxyacetophenone
Formed from: 4′-Hydroxy-3′-methoxyacetophenone

2′,5′-Dihydroxy-4′-methoxyacetophenone
Formed from: 2′-Hydroxy-4′-methoxyacetophenone
→ 2′,5′-Dihydroxy-4′-methoxyacetophenone-β-D-glucuronide
Probably in: Guinea pig B506
Man B920
Mouse B506
Rabbit B506
Rat B506
→ 2′,5′-Dihydroxy-4′-methoxyacetophenone-O-sulphate
Probably in: Guinea pig B506
Mouse B506
Rabbit B506
Rat B506

2′,6′-Dihydroxy-4′-methoxyacetophenone
Probably formed from: 2′-Hydroxy-4′-methoxyacetophenone

2′,ω-Dihydroxy-4′-methoxyacetophenone
Formed from: 2′-Hydroxy-4′-methoxyacetophenone

3′,4′-Dihydroxy-5′-methoxyacetophenone
Formed from: 4′-Hydroxy-3′-methoxyacetophenone
→ 4′-Hydroxy-3′,5′-dimethoxyacetophenone
Probably in: Rat E849

3′,6′-Dihydroxy-2′-methoxyacetophenone
Formed from: 2′-Hydroxy-6′-methoxyacetophenone

4′,ω-Dihydroxy-3′-methoxyacetophenone
Formed from: 4′-Hydroxy-3′-methoxyacetophenone
4-Hydroxy-3-methoxyphenylglycol

2′,5′-Dihydroxy-4′-methoxyacetophenone-β-D-glucuronide
Probably formed from: 2′,5′-Dihydroxy-4′-methoxyacetophenone

2′,5′-Dihydroxy-4′-methoxyacetophenone-O-sulphate
Probably formed from: 2′,5′-Dihydroxy-4′-methoxyacetophenone

1,3-Dihydroxy-2-methoxybenzene
Probably formed from: Pyrogallol
→ 1,3-Dihydroxy-2-methoxybenzene-β-D-glucuronide
Probably in: Rat K575

1,4-Dihydroxy-2-methoxybenzene
See: Methoxyquinol

1,3-Dihydroxy-2-methoxybenzene-β-D-glucuronide
Probably formed from: 1,3-Dihydroxy-2-methoxybenzene

2,3-Dihydroxy-4-methoxybenzoic acid
→ 3-Carboxy-2-hydroxy-*cis,cis*-muconic semialdehyde
In: Pseudomonas A2694

2,5-Dihydroxy-3-methoxybenzoic acid
→ 6-Carboxy-4-hydroxy-2-methoxy-*cis,cis*-muconic semialdehyde
In: Pseudomonas F649

2,5-Dihydroxy-4-methoxybenzoic acid
→ 6-Carboxy-4-hydroxy-3-methoxy-*cis,cis*-muconic semialdehyde
In: Pseudomonas F649

3,4-Dihydroxy-5-methoxybenzoic acid
Formed from: Gallic acid
4-Hydroxy-3,5-dimethoxybenzoic acid
3-Hydroxy-5-methoxybenzoic acid
4-Hydroxy-5-methoxyisophthalic acid
Probably formed from: Vanillic acid
→ 4-Carboxy-2-hydroxymuconic acid 6-methyl ester
In: Pseudomonas A2399
→ Gallic acid
In: Clostridium G367
Pseudomonas D16
→ 3-Hydroxy-4,5-dimethoxybenzoic acid
Probably in: Mouse A647
→ 4-Hydroxy-3,5-dimethoxybenzoic acid
Probably in: Phanerochaete J339

3,5-Dihydroxy-4-methoxybenzoic acid
Formed from: Bergenin
Gallic acid
→ 3,5-Dihydroxy-4-methoxybenzoic acid sulphate
Probably in: Rat K575
→ 3,5-Dihydroxy-4-methoxybenzoyl-D-glucose
In: Cucumis A611
→ Gallic acid
In: Erwinia A466
→ 1,3,5-Trihydroxy-2-methoxybenzene
In: Sporotrichum C57

3,6-Dihydroxy-2-methoxybenzoic acid
See: Leiocarpic acid

3,5-Dihydroxy-4-methoxybenzoic acid sulphate
Probably formed from: 3,5-Dihydroxy-4-methoxybenzoic acid

2,2′-Dihydroxy-4-methoxybenzophenone
Formed from: 2-Hydroxy-4-methoxybenzophenone
→ 2,2′,4-Trihydroxybenzophenone
Probably in: Rat G692

7,8-Dihydroxy-6-methoxy-2*H*-benzopyran-2-one
See: Fraxetin

2,4-Dihydroxy-7-methoxy-1,4-benzoxazin-3-one
Formed from: 2,4-Dihydroxy-7-methoxy-1,4-benzoxazin-3-one-β-D-glucoside
→ 2,4-Dihydroxy-7-methoxy-1,4-benzoxazin-3-one-β-D-glucoside

In: Zea K892

2,4-Dihydroxy-7-methoxy-1,4-benzoxazin-3-one-β-D-glucoside
Formed from: 2,4-Dihydroxy-7-methoxy-1,4-benzoxazin-3-one
→ 2,4-Dihydroxy-7-methoxy-1,4-benzoxazin-3-one
In: Secale K512
Zea K149

3,5-Dihydroxy-4-methoxybenzoyl-D-glucose
Formed from: 3,5-Dihydroxy-4-methoxybenzoic acid

1-(3,5-Dihydroxy-4-methoxybenzoyl)-1,2,3,4-tetrahydro-6,7-dimethoxy-2-methylisoquinoline
Formed from: Laudanosine

1-(2,4-Dihydroxy-3-methoxybenzyl)hydrazine
Probably formed from: 1-(2,3,4-Trihydroxybenzyl)hydrazine

N-(4,5-Dihydroxy-3-methoxybenzyl)-(*E*)-6-nonenamide
Formed from: Capsaicin

2-(2,4-Dihydroxy-3-methoxybenzyl)-1-serylhydrazine
Probably formed from: Benserazide

3,4′-Dihydroxy-4-methoxybiphenyl
Probably formed from: 3,4,4′-Trihydroxybiphenyl

4,4′-Dihydroxy-3-methoxybiphenyl
Probably formed from: 3,4,4′-Trihydroxybiphenyl

1′,6-Dihydroxy-5-methoxybufuralol
Probably formed from: 6-Hydroxy-5-methoxybufuralol
→ 6-Hydroxy-5-methoxy-1′-oxobufuralol
Probably in: Dog A2806
Man A2806
Rat A2806

2′,4-Dihydroxy-3-methoxychalcone
Formed from: 2′,3,4′-Trihydroxychalcone

4,4′-Dihydroxy-2′-methoxychalcone
Formed from: 2′,4,4′-Trihydroxychalcone
Probably formed from: 4-Hydroxy-2′,4′-dimethoxychalcone

3,4-Dihydroxy-5-methoxycinnamaldehyde
Formed from: Coniferaldehyde
5-Methoxycaffeoyl CoA
→ Sinapaldehyde
In: Escherichia K217

4,6-Dihydroxy-7-methoxy-5-coumaranoyl-β-acrylic acid
Formed from: 8-Methoxypsoralen

3,4-Dihydroxy-7-methoxy-2,2-dimethylchromane
Formed from: Precocene I

4′,5-Dihydroxy-7-methoxyflavanone
See: Sakuranetin

4′,7-Dihydroxy-5-methoxyflavanone
→ 3′,4′,7-Trihydroxy-5-methoxyflavanone
In: Horseradish A3777

5,7-Dihydroxy-4′-methoxyflavanone
See: Isosakuranetin

4′,7-Dihydroxy-5-methoxyflavanone-7-neohesperidoside
Probably formed from: 2′,4,4′-Trihydroxy-6′-methoxychalcone-4′-neohesperidoside

3,5-Dihydroxy-5-methoxyflavone
Formed from: Galangin

4′,5-Dihydroxy-7-methoxyflavone
See: Genkwanin

5,7-Dihydroxy-3-methoxyflavone
Formed from: Galangin
→ 5-Hydroxy-3,7-dimethoxyflavone
Probably in: Citrus A3987

5,7-Dihydroxy-4′-methoxyflavone
See: Acacetin

5,7-Dihydroxy-8-methoxyflavone
See: Wogonin

2′,7-Dihydroxy-4′-methoxyisoflavanone
Formed from: Formononetin
→ Medicarpin
In: Medicago A3723
Trifolium A1929

5,7-Dihydroxy-4′-methoxyisoflavanone
Formed from: Biochanin A
2′,7-Dihydroxy-4′-methoxyisoflavone
Probably formed from: Naringenin
→ 5,6-Dihydro-3-(*p*-methoxyphenyl)-4,6-dioxo-4*H*-pyran
In: Fusarium C474
See also: C446

2′,7-Dihydroxy-4′-methoxyisoflavone
Formed from: Formononetin
→ 2′,7-Dihydroxy-4′-methoxyisoflavanone
In: Glycine K702
→ Medicarpin

In: Medicago A3723
Trifolium A1929

3′,7-Dihydroxy-4′-methoxyisoflavone
See: Calycosin

4′,5-Dihydroxy-7-methoxyisoflavone
See: Prunetin

4′,6-Dihydroxy-7-methoxyisoflavone
Formed from: 4′,6,7-Trihydroxyisoflavone

4′,7-Dihydroxy-5-methoxyisoflavone
Formed from: Genistein

4′,7-Dihydroxy-6-methoxyisoflavone (glycitein)
Formed from: 4′,6,7-Trihydroxyisoflavone
→ 7-Hydroxy-4′,6-dimethoxyisoflavone
Probably in: Citrus A3987

5,7-Dihydroxy-4′-methoxyisoflavone
See: Biochanin A

6,7-Dihydroxy-4′-methoxyisoflavone (texasin)
Formed from: 4′,6,7-Trihydroxyisoflavone
→ 6-Hydroxy-4′,7-dimethoxyisoflavone
Probably in: Citrus A3987
→ 7-Hydroxy-4′,6-dimethoxyisoflavone
Probably in: Citrus A3987
→ 4′,6,7-Trihydroxyisoflavone
In: Eubacterium K627

1,6-Dihydroxy-8-methoxy-3-methyl-9,10-anthracenedione
See: Questin

1,7-Dihydroxy-3-methoxy-6-methyl-9,10-anthracenedione
See: Macrosporin

1,8-Dihydroxy-3-methoxy-6-methyl-9,10-anthracenedione
See: Physcion

3,4′-Dihydroxy-5-methoxy-6-methyldibenzo-α-pyrone
Formed from: Alternariol

5,6-Dihydroxy-8-methoxy-2-methyl-4*H*-naphtho[2,3-*b*]pyran-4-one
See: Rubrofusarin

3,4-Dihydroxy-5-methoxy-N-methylphenethylamine
→ 3-Hydroxy-4,5-dimethoxy-N-methylphenethylamine
In: Lophophora A3600
→ 4-Hydroxy-3,5-dimethoxy-N-methylphenethylamine
In: Lophophora A3600

3,5-Dihydroxy-4-methoxy-N-methylphenethylamine
Formed from: 3,4,5-Trihydroxy-N-methylphenethylamine

16β,17β-Dihydroxy-3-methoxy-1,3,5(10),9(11)-oestratetraene
Formed from: 3-Methoxy-17-oxo-1,3,5(10),9(11)-oestratetraene
→ 17β-Hydroxy-3-methoxy-16-oxo-1,3,5(10),9(11)-oestratetraene
Probably in: Streptomyces F810

7,α-Dihydroxy-8-methoxy-2-oxo-2*H*-1-benzopyran-6-acetic acid
Probably formed from: 6-Carboxymethyl-7-hydroxy-8-methoxycoumarin

2,4-Dihydroxy-5-methoxyphenethylamine
Formed from: 2,4,5-Trihydroxyphenethylamine

4-(3,4-Dihydroxy-5-methoxyphenoxy)-3,5-dimethoxybenzoic acid
Probably formed from: 4-Hydroxy-3,5-dimethoxybenzoic acid

2,5-Dihydroxy-4-methoxyphenylacetic acid
Probably formed from: 4-Hydroxy-3-methoxyphenylacetic acid

3,4-Dihydroxy-5-methoxyphenylacetic acid
Probably formed from: 3,4,5-Trimethoxyphenylacetic acid
→ γ-Carboxy-α-hydroxy-*cis*,*cis*-muconic acid-ε-methyl ester
Probably in: Arthrobacter B931

1,2-Dihydroxy-1-(*p*-methoxyphenyl)propane
Probably formed from: Anethole oxide

3,4-Dihydroxy-5-methoxyphenylpropionic acid
Formed from: Dihydrosinapic acid
3,4,5-Trihydroxyphenylpropionic acid
Probably formed from: 3-(3-Hydroxy-4,5-dimethoxyphenyl)propionic acid
→ 3-Hydroxy-4,5-dimethoxyphenylpropionic acid
Probably in: Rat A745
→ 3-Hydroxy-5-methoxyphenylpropionic acid
In: Microorganism A744
→ 3,4,5-Trihydroxyphenylpropionic acid
In: Microorganism A744

3,5-Dihydroxy-4-methoxyphenylpropionic acid
Formed from: 3,4,5-Trihydroxyphenylpropionic acid
→ 3,4,5-Trihydroxyphenylpropionic acid
In: Microorganism A744

3,4′-Dihydroxy-5-methoxystilbene
Probably formed from: Resveratol

2-(3,4-Dihydroxy-5-methoxyphenyl)-3,5,7-trihydroxy-1-benzopyrilium
See: Petunidin

2,5-Dihydroxy-4-methylamphetamine
Probably formed from: 5-Hydroxy-2-methoxy-4-methylamphetamine

3,4-Dihydroxy-N-methylamphetamine
Formed from: N-Methyl-3,4-methylenedioxyamphetamine
→ 3-Hydroxy-4-methoxy-N-methylamphetamine
Probably in: Man F187
Rat E988, G145
→ 4-Hydroxy-3-methoxy-N-methylamphetamine
Probably in: Man F187
Rat E988, G145
→ 4-(2-Methylaminopropyl)-*o*-quinone
Probably in: Rat F805
→ 2,4,5-Trihydroxy-N-methylamphetamine
Probably in: Rat G354

3,6-Dihydroxy-N-methylamphetamine
→ 2-(2-Aminopropyl)-4-methyl-*p*-quinone
Probably in: Man A63

9,10-Di-(hydroxymethyl)anthracene
Probably formed from: 9-Hydroxymethyl-10-methylanthracene

3,8-Dihydroxy-1-methyl-9,10-anthracenedione
See: Aloesaponarin 1

1,8-Dihydroxy-3-methyl-9(10*H*)-anthracenone
See: Chrysophanolanthrone

1,8-Dihydroxy-3-methylanthraquinone
See: Chrysophanol

7,12-Di-(hydroxymethyl)benzanthracene
Formed from: 7-Hydroxymethyl-12-methylbenzanthracene
→ 8,9-Dihydro-8,9-dihydroxy-7,12-di-(hydroxymethyl)benzanthracene
Probably in: Rat A2324
→ 7,12-Di-(hydroxymethyl)benzanthracene-7-O-sulphate
In: Rat D769
See also: J683

7,12-Di-(hydroxymethyl)benzanthracene-7-O-sulphate
Formed from: 7,12-Di-(hydroxymethyl)benzanthracene

2,3-Dihydroxy-4-methylbenzoic acid
→ 3-Carboxy-2-hydroxy-6-methyl-*cis,cis*-muconic semialdehyde
In: Pseudomonas A2694
See also: G766
→ α-Carboxy-δ-methyl-*cis,cis*-muconic acid
In: Tecoma A2241

2,4-Dihydroxy-6-methylbenzoic acid
See: Orsellinic acid

2,5-Dihydroxy-3-methylbenzoic acid
Formed from: 3-Hydroxy-5-methylbenzoic acid
→ 6-Carboxy-4-hydroxy-2-methyl-*cis,cis*-muconic semialdehyde
In: Pseudomonas F649
Rhodococcus H441

2,5-Dihydroxy-4-methylbenzoic acid
Formed from: 3-Hydroxy-5-methylbenzoic acid
→ 6-Carboxy-4-hydroxy-3-methyl-*cis,cis*-muconic semialdehyde
In: Pseudomonas F649
Rhodococcus H441

2,5-Dihydroxy-6-methylbenzoic acid
→ 6-Carboxy-4-hydroxy-5-methyl-*cis,cis*-muconic semialdehyde
In: Rhodococcus H441

3,4-Dihydroxy-5-methylbenzoic acid
→ 2,3,5-Trihydroxytoluene
In: Trichosporon B34

6-(2,3-Dihydroxy-3-methylbutyl)-4′,5,7-trihydroxy-3-methoxyflavone
Formed from: Topazolin
→ 6-(3-Hydroxy-3-methylbutyl)-4′,5,7-trihydroxy-3-methoxyflavone
Probably in: Aspergillus G264
Botrytis G264

3,4-Dihydroxy-N-(N-(*m*-(N-methylcarboxamido)phenyl)-β-alanyl)phenethylamine
Probably formed from: 4-Hydroxy-3-methoxy-N-(N-(*m*-(N-methylcarboxamido)phenyl)-β-alanyl)phenethylamine

***cis*-1,2-Dihydroxy-3-methylcholanthrene**
Formed from: 3-Methylcholanthrene

***trans*-1,2-Dihydroxy-3-methylcholanthrene**
Formed from: 3-Methylcholanthrene

***cis*-9,10-Dihydroxy-3-methylcholanthrene**
Formed from: 3-Methylcholanthrene

6,7-Dihydroxy-4-methylcoumarin
Formed from: 4-Methylumbelliferone
→ 6,7-Dihydroxy-4-methylcoumarin-β-D-glucuronide
In: Rat A773
→ 6-Hydroxy-7-methoxy-4-methylcoumarin
In: Rat A486, A773
→ 6-Methoxy-4-methylumbelliferone
In: Rat A486, A773

7,8-Dihydroxy-4-methylcoumarin
Formed from: 7,8-Diacetoxy-4-methylcoumarin
→ 8-Hydroxy-7-methoxy-4-methylcoumarin
In: Rat A486
→ 8-Methoxy-4-methylumbelliferone
In: Rat A486

6,7-Dihydroxy-4-methylcoumarin-β-D-glucuronide
Formed from: 6,7-Dihydroxy-4-methylcoumarin

2′,7-Dihydroxy-4′,5′-methylenedioxyisoflavone
→ Maackiain
In: Trifolium A3474
→ Sophorol
In: Pisum F860

3,6α-Dihydroxy-8,9-methylenedioxypterocarpan
(demethylpisatin, 6a-hydroxyinermin)
Formed from: Pisatin
→ 3-Hydroxyinerminisoflavan
In: Fusarium B382

2-(1,2-Dihydroxy-1-methylethyl)-6-(1-hydroxy-1-methylethyl)naphthalene
Probably formed from: 2,6-Di-(1-hydroxy-1-methylethyl)naphthalene
→ 2-(1-Carboxy-1-hydroxyethyl)-6-(1-hydroxy-1-methylethyl)naphthalene
Probably in: Rat C11

2,6-Di-(1-hydroxy-1-methylethyl)naphthalene
Formed from: 2,6-Diisopropylnaphthalene
→ 2-(1,2-Dihydroxy-1-methylethyl)-6-(1-hydroxy-1-methylethyl)naphthalene
Probably in: Rat C11

5,7-Dihydroxy-4′-methoxyflavanone
See: Isosakuranetin

7,9-Dihydroxy-α-methylfluorene-2-acetic acid
Probably formed from: 7-Hydroxy-α-methylfluorene-2-acetic acid

5,6-Dihydroxy-3-methylindole-2-carboxylic acid
Probably formed from: L-Methyldopachrome

2′,3′-Dihydroxy-7-O-methylkievitone
Probably formed from: 7-O-Methylkievitone-2′,3′-epoxide

2,6-Dihydroxy-4-methyl-8-(5-methyl-2,3-dioxopyrrolidin-5-yl)amino-5-(3-trifluoromethylphenoxy)quinoline
Probably formed from: 8-(4-Amino-3-hydroxy-1-methylbutyl)amino-2,6-dihydroxy-4-methyl-5-(3-trifluoromethylphenoxy)quinoline

2,4-Dihydroxy-3-((1*R*,6*R*)-3-methyl-6-(1-methylethenyl)-2-cyclohexen-1-yl)-6-pentylbenzoic acid
See: Cannabidiolic acid

1,4-Dihydroxy-2-methylnaphthalene
Formed from: 2-Methyl-1,4-naphthoquinone
→ 1,4-Dihydroxy-2-methylnaphthalene-β-D-glucuronide
In: Rat A662
→ 2-Methyl-1,4-naphthoquinone
In: Arum E87

1,4-Dihydroxy-2-methylnaphthalene-β-D-glucuronide
Formed from: 1,4-Dihydroxy-2-methylnaphthalene

2,4-Dihydroxy-3-O-methyloestrone
Formed from: 2,4-Dihydroxyoestrone

2,16α-Dihydroxy-3-O-methyloestrone
Probably formed from: 2,16α-Dihydroxyoestrone

5,6-Dihydroxy-8-(5-methyl-2-oxopyrrolidin-1-yl)quinoline-N-oxide
Probably formed from: 8-((4-Amino-1-methyl-3-oxobutyl)amino)-5-hydroxy-6-methoxyquinoline

3,4-Dihydroxy-N-methylphenethylamine (epinine)
Formed from: Diacetylepinine
Dibenzoylepinine
3,4-Dihydroxyphenethylamine
Dipivaloylepinine
Epinine-4-O-phosphate
Probably formed from: Epinine monoisobutyrate
→ 3,4-Dihydroxy-N,N-dimethylphenethylamine
In: Rat A80, C657
→ 3,4-Dihydroxyphenylacetaldehyde
Probably in: Beef F220
Man D970, H34
Rat D990, G865
See also: E940
→ Epinine-4-β-D-glucuronide
Probably in: Rat D990, E940
→ Epinine-3-O-sulphate
Probably in: Dog E940
Rat D990, E940
→ 3-Hydroxy-4-methoxy-N-methylphenethylamine
In: Rat A486
→ 4-Hydroxy-3-methoxy-N-methylphenethylamine
In: Lophophora A3600
Rat A486, G865
See also: D990, E940
→ 4-(2-Methylaminoethyl)-*o*-quinone
Probably in: Beef A624
→ 1,2,3,4-Tetrahydro-6,7-dihydroxy-2-methylisoquinoline
In: Pig A121

3,4-Dihydroxy-α-methylphenethylamine
See: 3,4-Dihydroxyamphetamine

2,5-Dihydroxy-3-methylphenylacetic acid
Formed from: 4-Hydroxy-2-methylphenylacetic acid

2,4-Dihydroxy-6-methyl-L-phenylalanine (orcylalanine)
Formed from: Orsellinic acid
→ 2,4-Dihydroxy-6-methylphenylpyruvic acid
With L-aminoacid oxidase A641

3,4-Dihydroxy-α-methyl-D-phenylalanine
Does not yield 3,4-dihydroxyamphetamine in rat A2012

3,4-Dihydroxy-α-methyl-L-phenylalanine (α-methyldopa)
Formed from: 3,4-Dihydroxy-α-methyl-L-phenylalanine pivaloyloxyethyl ester
3,4-Dihydroxy-α-methyl-L-phenylalanine succinimidoethyl ester
→ 3,4-Dihydroxyamphetamine
In: Man B325, C835, D656, G944
Rat A2012, C758, D766
→ 3,4-Dihydroxyphenylacetone
In: Pig B780
With dopa decarboxylase J594
→ 3,4-Dihydroxy-α-methyl-L-phenylalanine-β-D-glucuronide
In: Man G944
→ 4-Hydroxy-3-methoxy-α-methyl-L-phenylalanine
In: Man C835, D35, D656, G944
Pig B813
Rat A2667
→ 4-Hydroxy-α-methyl-3-sulphatoxy-L-phenylalanine
In: Man B325
See also: C345, C835, D35, F196, G944
→ 4-(3-(L-(α-Methyl)alanyl))-*o*-quinone
In: Apple J678
Pear J678
See also: A2556, D984

3,4-Dihydroxy-N-methylphenylalanine
→ 3,4-Dihydroxy-L-phenylalanine
Probably in: Rat A2961
→ 4-(N-Methyl-3-(L-alanyl))-*o*-quinone
In: Mushroom B897
Does not yield 3,4-Dihydroxy-N-methylphenethylamine in rat A2961

3,4-Dihydroxy-α-methyl-L-phenylalanine-β-D-glucuronide
Formed from: 3,4-Dihydroxy-α-methyl-L-phenylalanine

3,4-Dihydroxy-α-methyl-L-phenylalanine pivaloyloxyethyl ester
→ 3,4-Dihydroxy-α-methyl-L-phenylalanine
In: Man A3467, C835
Rat A3467

3,4-Dihydroxy-α-methyl-L-phenylalanine succinimidoethyl ester
→ 3,4-Dihydroxy-α-methyl-L-phenylalanine

In: Man A3467
Rat A3467

6-(2,4-Dihydroxy-6-methylphenyl)-4-methoxy-2-pyrone
Formed from: Aloenin

3,4-Dihydroxy-α-methylphenylpropionic acid
Formed from: L-α-Hydrazino-3,4-dihydroxy-α-methylphenylpropionic acid
→ 3-Hydroxy-α-methylphenylpropionic acid
Probably in: Man A1502
Monkey A1502
Rat A1502
→ 4-Hydroxy-3-methoxy-α-methylphenylpropionic acid
Probably in: Dog A1436
Man A1436
Monkey A1436

2,4-Dihydroxy-6-methylphenylpyruvic acid
Formed from: 2,4-Dihydroxy-6-methyl-L-phenylalanine

3,4-Dihydroxy-N-methyl-L-*threo*-phenylserine
→ 4-Hydroxy-3-methoxy-N-methyl-L-*threo*-phenylserine
In: Mouse E415
Rat E415

1,4-Dihydroxy-2-methyl-3-phytylnaphthalene
(dihydrovitamin K_1)
Probably formed from: Vitamin K_1
→ 1,4-Dihydroxy-2-methyl-3-phytylnaphthalene epoxide
In: Rat B338

1,4-Dihydroxy-2-methyl-3-phytylnaphthalene epoxide
Formed from: 1,4-Dihydroxy-2-methyl-3-phytylnaphthalene

3,4-Dihydroxy-N-(2-(2-(1-methyl-2-piperidinyl)ethyl)phenyl)benzamide
Probably formed from: 4-Hydroxy-N-(2-(2-(1-methyl-2-piperidinyl)ethyl)phenyl)benzamide
→ 3-Hydroxy-4-methoxy-N-(2-(2-(1-methyl-2-piperidinyl)ethyl)phenyl)benzamide
Probably in: Man F532
→ 4-Hydroxy-3-methoxy-N-(2-(2-(1-methyl-2-piperidinyl)ethyl)phenyl)benzamide
Probably in: Man F532

3,4-Dihydroxy-α-methyl-N-propylphenethylamine
→ 4-Hydroxy-3-methoxy-α-methyl-N-propylphenethylamine
In: Rat C299

3′,4′-Dihydroxy-2-methylselenobenzanilide
Probably formed from: 4′-Hydroxy-2-methylselenobenzanilide
→ 4′-Hydroxy-3′-methoxy-2-methylselenobenzanilide
Probably in: Man F235
Pig F235
Rat F235

4,4′-Dihydroxy-α-methylstilbene
Formed from: 1,2-Bis(*p*-hydroxyphenyl)-2-propanol

4,4′-Dihydroxy-3-methylthiobiphenyl
Probably formed from: 4,4′-Dihydroxybiphenyl

3,14-Dihydroxymorphinan
Formed from: Butorphanol

1,2-Dihydroxynaphthalene
Formed from: *cis*-1,2-Dihydro-1,2-dihydroxynaphthalene
trans-1,2-Dihydro-1,2-dihydroxynaphthalene
α-Naphthol
β-Naphthol
1,2-Naphthoquinone
→ *o*-Carboxy-*cis*-cinnamic acid
In: Pseudomonas G478
See also: K421
→ *cis*-*o*-Hydroxybenzylidenepyruvic acid
In: Pseudomonas B527
→ 1,2-Naphthoquinone
Probably in: Claviceps A3430
Cunninghamella A3218, A3430
Mucor A3430
Mushroom D418
Neurospora A3430
Psilocybe A3430
Syncephalastrum A3430
→ 1,2,4-Trihydroxynaphthalene
Probably in: Rat B602

1,3-Dihydroxynaphthalene
→ 1,3-Dihydroxynaphthalene-β-D-glucuronide
In: Rat B962

→ 3-Hydroxy-1-tetralone
In: Verticillium K593

1,4-Dihydroxynaphthalene
Formed from: 1-Hydroxy-4-methoxynaphthalene
1,4-Naphthoquinone
Probably formed from: α-Naphthol
→ 1,4-Naphthoquinone
Probably in: Arthrobacter K540
Claviceps A3430
Cunninghamella A3218
Mouse G117
Mushroom D418
Neurospora A3430
→ 4-Hydroxy-1-naphthalenyl-β-D-glucuronide
In: Juglans A3470

1,5-Dihydroxynaphthalene
Formed from: α-Naphthol

1,7-Dihydroxynaphthalene
Formed from: α-Naphthol
Probably formed from: β-Naphthol

2,3-Dihydroxynaphthalene
Formed from: 2,3-Methylenedioxynaphthalene
Naphthalene

2,6-Dihydroxynaphthalene
Formed from: 2-Hydroxy-6-methoxynaphthalene
β-Naphthol

2,7-Dihydroxynaphthalene
Formed from: 2-Hydroxy-7-methoxynaphthalene
β-Naphthol

1,3-Dihydroxynaphthalene-β-D-glucuronide
Formed from: 1,3-Dihydroxynaphthalene

4,8-Dihydroxy-1-naphthalenyl-β-D-glucuronide
Probably formed from: 1,4,5-Trihydroxynaphthalene

1,4-Dihydroxy-2-naphthoic acid
Formed from: 4-*o*-Carboxyphenylsuccinyl CoA
o-Succinylbenzoic acid
o-Succinylbenzoyl CoA
→ Juglone
In: Juglans A1639
→ Lawsone
In: Impatiens A1639
→ Menaquinone 3
In: Escherichia A2774

1,5-Dihydroxy-2-naphthoic acid
Probably formed from: 1-Hydroxy-2-naphthoic acid

7,8-Dihydroxy-1-naphthoic acid
Formed from: 1-Naphthoic acid
→ 2-Hydroxyisophthaldehydic acid
Probably in: Pseudomonas H415

2,5-Dihydroxy-1,4-naphthoquinone
Formed from: 5-Hydroxyscytalone
→ 5-Hydroxy-1,4-naphthoquinone
In: Wangiella D760
→ 3,4,8-Trihydroxy-1-tetralone
In: Wangiella D760

3,5-Dihydroxy-1,4-naphthoquinone
Formed from: 5-Hydroxy-1,4-naphthoquinone
→ 2,4,8-Trihydroxy-1-tetralone
In: Wangiella D760

5,8-Dihydroxy-1,4-naphthoquinone
(naphthazarin)
→ 1,4,5,8-Tetrahydroxynaphthalene
In: Beef G401
→ 2,5,8-Trihydroxy-1,4-naphthoquinone
In: Pseudomonas F46, K897

1-(3,4-Dihydroxy-1-naphthoxy)-2,3-propanediol
Probably formed from: 3,4-Dihydroxypropanolol
→ 1-(3-Hydroxy-4-methoxy-1-naphthoxy)-2,3-propanediol
Probably in: Rat E916
→ 1-(4-Hydroxy-3-methoxy-1-naphthoxy)-2,3-propanediol
Probably in: Rat E916

1,2-Dihydroxy-1-(2-naphthyl)ethane
Formed from: 2-Isopropylamino-1-(2-naphthyl)ethanol

3,4-Dihydroxy-5-nitrobenzaldehyde
Formed from: 3-(3,4-Dihydroxy-5-nitrobenzylidene)-2,4-pentanedione
→ 3,4-Dihydroxy-5-nitrobenzaldehyde-β-D-glucuronide
Probably in: Man F721
→ 3,4-Dihydroxy-5-nitrobenzoic acid
Probably in: Dog G560
Rat G560

3,4-Dihydroxy-5-nitrobenzaldehyde-β-D-glucuronide
Probably formed from: 3,4-Dihydroxy-5-nitrobenzaldehyde

3,4-Dihydroxy-5-nitrobenzoic acid
Probably formed from: 3,4-Dihydroxy-5-nitrobenzaldehyde

1,9-Dihydroxy-6-nitro-3,4-benzpyrene
Formed from: 1-Hydroxy-6-nitro-3,4-benzpyrene

3,9-Dihydroxy-6-nitro-3,4-benzpyrene
Formed from: 3-Hydroxy-6-nitro-3,4-benzpyrene

3-(3,4-Dihydroxy-5-nitrobenzyl)-2-hydroxypentan-4-one
Probably formed from: 3-(3,4-Dihydroxy-5-nitrobenzylidene)-2-hydroxypentan-4-one
3-(3,4-Dihydroxy-5-nitrobenzyl)pentan-2,4-dione
→ 3-(3,4-Dihydroxy-5-nitrobenzyl)-2-hydroxypentan-4-one-β-D-glucuronide
Probably in: Man F721
→ 3-(3,4-Dihydroxy-5-nitrobenzyl)pentan-2,4-diol
Probably in: Dog G560
Man F721
Rat G560

3-(3,4-Dihydroxy-5-nitrobenzyl)-2-hydroxypentan-4-one-β-D-glucuronide
Probably formed from: 3-(3,4-Dihydroxy-5-nitrobenzyl)-2-hydroxypentan-4-one

3-(3,4-Dihydroxy-5-nitrobenzylidene)-2-hydroxypentan-4-one
Formed from: 3-(3,4-Dihydroxy-5-nitrobenzylidene)-2,4-pentanedione
→ 3-(3,4-Dihydroxy-5-nitrobenzyl)-2-hydroxypentan-4-one
Probably in: Man F271
→ 3-(3,4-Dihydroxy-5-nitrobenzylidene)-2-hydroxypentan-4-one-β-D-glucuronide
Probably in: Man F271

3-(3,4-Dihydroxy-5-nitrobenzylidene)-2-hydroxypentan-4-one-β-D-glucuronide
Probably formed from: 3-(3,4-Dihydroxy-5-nitrobenzylidene)-2-hydroxypentan-4-one

3-(3,4-Dihydroxy-5-nitrobenzylidene)-2,4-pentanedione (nitecapone)
→ 3-Acetyl-4-(3,4-dihydroxy-5-nitrophenyl)butan-2-one-3-sulphonic acid
Probably in: Dog G560
Rat G560
→ 3,4-Dihydroxy-5-nitrobenzaldehyde
In: Dog G560
Man F271
Rat G560
→ 3-(3,4-Dihydroxy-5-nitrobenzylidene)-2-hydroxypentan-4-one
In: Dog G560
Man F271
Rat G560
→ 3-(3,4-Dihydroxy-5-nitrobenzylidene)-2,4-pentanedione-β-D-glucuronide
In: Man F271
→ 3-(3,4-Dihydroxy-5-nitrobenzyl)-2,4-pentanedione
In: Dog G560
Man F271
Rat G560
→ 4-(3,4-Dihydroxy-5-nitrophenyl)but-3-en-2-one
In: Dog G560
Rat G560
→ 3-(3,4-Dihydroxy-5-nitro-α-sulphonylbenzyl)-2,4-pentanedione
Probably in: Dog G560
Rat G560
→ 3-(4-Hydroxy-3-methoxy-5-nitrobenzylidene)-2,4-pentanedione
In: Dog G560
Rat G560

3-(3,4-Dihydroxy-5-nitrobenzylidene)-2,4-pentanedione-β-D-glucuronide
Probably formed from: 3-(3,4-Dihydroxy-5-nitrobenzylidene)-2,4-pentanedione

3-(3,4-Dihydroxy-5-nitrobenzyl)pentane-2,4-diol
Probably formed from: 3-(3,4-Dihydroxy-5-nitrobenzyl)-2-hydroxypentan-4-one
→ 3-(3,4-Dihydroxy-5-nitrobenzyl)pentan-2,4-diol-β-D-glucuronide
Probably in: Man F721

3-(3,4-Dihydroxy-5-nitrobenzyl)pentane-2,4-diol-β-D-glucuronide
Probably formed from: 3-(3,4-Dihydroxy-5-nitrobenzyl)pentan-2,4-diol

3-(3,4-Dihydroxy-5-nitrobenzyl)-2,4-pentanedione
Probably in: 3-(3,4-Dihydroxy-5-nitrobenzylidene)-2,4-pentanedione
→ 3-(3,4-Dihydroxy-5-nitrobenzyl)-2-hydroxypentan-4-one
Probably in: Dog G560
Rat G560
→ 3-(3,4-Dihydroxy-5-nitrobenzyl)-2,4-pentanedione-β-D-glucuronide
Probably in: Man F721

3-(3,4-Dihydroxy-5-nitrobenzyl)-2,4-pentanedione-β-D-glucuronide
Probably formed from: 3-(3,4-Dihydroxy-5-nitrobenzyl)-2,4-pentanedione

1,3-Dihydroxy-1-(*p*-nitrophenyl)acetone
Formed from: *p*-Nitrophenylserinol

4-(3,4-Dihydroxy-5-nitrophenyl)but-3-en-2-one
Formed from: 3-(3,4-Dihydroxy-5-nitrobenzylidene)-2,4-pentanedione

3-(3,4-Dihydroxy-5-nitro-α-sulphonylbenzyl)-2,4-pentanedione
Formed from: 3-(3,4-Dihydroxy-5-nitrobenzylidene)-2,4-pentanedione

3′,4′-Dihydroxynomifensine
Probably formed from: 4′-Hydroxynomifensine
→ 3′-Hydroxy-4′-methoxynomifensine
Probably in: Man A3348, B373, E37
Monkey A3348
→ 4′-Hydroxy-3′-methoxynomifensine
Probably in: Man A3348, B373, E37
Monkey A3348

3,4-Dihydroxynorephedrine
See: α-Methylnoradrenaline

3′,4′-Dihydroxynormeperidine
Probably formed from: 3′,4′-Dihydroxymeperidine

2,11-Dihydroxynortriptyline
Probably formed from: 2-Hydroxynortriptyline
→ 2-Hydroxy-11-oxonortriptyline
Probably in: Rat E892

***cis*-10,11-Dihydroxynortriptyline**
Formed from: 10-Hydroxynortriptyline
Nortriptyline
→ *cis*-10,11-Dihydroxynortriptyline-β-D-glucuronide
In: Rat E931

***trans*-10,11-Dihydroxynortriptyline**
Formed from: 10-Hydroxynortriptyline
Nortriptyline
→ *trans*-10,11-Dihydroxynortriptyline-β-D-glucuronide
In: Rat E931

***cis*-10,11-Dihydroxynortriptyline-β-D-glucuronide**
Formed from: *cis*-10,11-Dihydroxynortriptyline

***trans*-10,11-Dihydroxynortriptyline-β-D-glucuronide**
Formed from: *trans*-10,11-Dihydroxynortriptyline

3,16α-Dihydroxyoestra-1,3,5(10)-triene
→ 16α-Hydroxyoestra-1,3,5(10)-trien-3-yl sulphate
In: Beef A1211

3,17β-Dihydroxyoestra-1,3,5(10),9-tetraen-12-one
→ 17β-Hydroxyoestra-1,3,5(10),9-tetraen-12-on-3-yl sulphate
In: Beef A1211

2,4-Dihydroxyoestrone
→ 2,4-Dihydroxy-3-O-methyloestrone
In: Rat A2439
→ 2-Hydroxy-4-methoxyoestrone
In: Rat A2439
→ 4-Hydroxy-2-methoxyoestrone
In: Rat A2439

2,16α-Dihydroxyoestrone
Probably formed from: 16α-Hydroxyoestrone
→ 2,16α-Dihydroxy-3-O-methyloestrone
Probably in: Rat A950, A1280
→ 16α-Hydroxy-2-methoxyoestrone
Probably in: Rat A950, A1280, A1487

3,6-Dihydroxyoxaprotiline
Probably formed from: 3-Hydroxyoxaprotiline

11,12-Dihydroxy-7-oxodehydroabietic acid
Probably formed from: 7-Oxodehydroabietic acid

3,5-Dihydroxy-4-(6-oxo-*p*-mentha-1,8(9)-dien-3-yl)phenylacetic acid
Probably formed from: 4-(3′,5′-Dihydroxy-4′-(*p*-mentha-1,8(9)-dien-3-yl)phenyl)butyric acid

3-(3′,5′-Dihydroxy-4′-(6-oxo-*p*-mentha-1,8(9)-dien-3-yl)phenyl)propionic acid
Probably formed from: 4-(3′,5′-Dihydroxy-4′-(*p*-mentha-1,8(9)-dien-3-yl)phenyl)propionic acid

2,4-Dihydroxy-6-oxo-6-phenylhexa-2,4-dienoic acid
→ Benzoic acid
In: Burkholderia K522

5,5′-Dihydroxy-2′,3,4′,6,7-pentamethoxyflavone
→ 5,5′-Dihydroxy-2′,3,4′,6,7-pentamethoxyflavone-β-D-glucuronide
In: Chrysosplenium C755, G262

5,5′-Dihydroxy-2′,3,4′,6,7-pentamethoxyflavone-β-D-glucuronide
Formed from: 5,5′-Dihydroxy-2′,3,4′,6,7-pentamethoxyflavone

N,N′-Dihydroxypentamidine
Formed from: N′-Hydroxypentamidine

5-(4-(3,4-Dihydroxypentyloxy)-3-ethoxyphenyl)-2,4-thiazolidinedione
Probably formed from: 5-(3-Ethoxy-4-(4-hydroxypentyloxy)phenyl)-2,4-thiazolidinedione
→ 5-(4-(2-Carboxyethoxy)-3-ethoxyphenyl)-2,4-thiazolidinedione
Probably in: Dog D250
Rat D250

4′,*t*-Dihydroxy-*cis*-permethrin
Probably formed from: 4′-Hydroxy-*cis*-permethrin

4′,*t*-Dihydroxy-*trans*-permethrin
Probably formed from: 4′-Hydroxy-*trans*-permethrin

6a,7-Dihydroxyphaseollin
Formed from: Phaseollin

1,2-Dihydroxyphenanthrene
Formed from: *cis*-1,2-Dihydro-1,2-dihydroxyphenanthrene
trans-1,2-Dihydro-1,2-dihydroxyphenanthrene

3,4-Dihydroxyphenanthrene
Formed from: *trans*-3,4-Dihydro-3,4-dihydroxyphenanthrene

9,10-Dihydroxyphenanthrene
Formed from: Phenanthrene-9,10-quinone

5,6-Dihydroxy-1,10-phenanthroline
Formed from: 1,10-Phenanthroline-5,6-dione

1,8-Dihydroxyphenazine
→ 1,8-Dihydroxyphenazine-10-oxide
In: Pseudomonas A3284

1,6-Dihydroxyphenazine-6,10-dioxide (iodinin)
Formed from: 6-Hydroxyphenazine-1-carboxylic acid
Shikimic acid

1,8-Dihydroxyphenazine-10-oxide
Formed from: 1,8-Dihydroxyphenazine

2,3-Dihydroxyphenethyl acetate
Formed from: Phenethyl acetate

2,3-Dihydroxyphenethylamine
→ 2-Hydroxy-3-methoxyphenethylamine
In: Rat A486

2,4-Dihydroxyphenethylamine
Formed from: 2,4-Dihydroxy-L-phenylalanine

2,5-Dihydroxyphenethylamine
Probably formed from: 2,5-Dihydroxy-L-phenylalanine

3,4-Dihydroxyphenethylamine (dopamine)
Formed from: N-Acetyl-3,4-dihydroxyphenethylamine
N-(L-Alanyl)-3,4-dihydroxyphenethylamine
Amyl L-γ-glutamyl dopamide
Benzyl L-γ-glutamyl dopamide
n-Butyl L-γ-glutamyl dopamide
3,4-Dihydroxy-L-phenylalanine
Docarbamine
L-γ-Glutamyl-3,4-dihydroxyphenethylamine
Heptyl L-γ-glutamyl dopamide
Hexyl L-γ-glutamyl dopamide
3-Hydroxyphenethylamin-4-yl sulphate
4-Hydroxyphenethylamin-3-yl sulphate
Isobutyl L-γ-glutamyl dopamide
Lauryl L-γ-glutamyl dopamide
3-Methoxytyramine
3,4-Methylenedioxyphenethylamine
D-Noradrenaline
Octyl L-γ-glutamyl dopamide

L-Phenylalanyl-3,4-dihydroxyphenethylamine
L-Phenylalanyl-L-phenylalanyl-3,4-dihydroxyphenethylamine
L-Phenylalanyl-L-prolyl-3,4-dihydroxyphenethylamine
L-Prolyl-3,4-dihydroxyphenethylamine
L-Prolyl-L-phenylalanyl-3,4-dihydroxyphenethylamine
L-Prolyl-L-prolyl-3,4-dihydroxyphenethylamine
n-Propyl-L-γ-glutamyl dopamide
m-Tyramine
Tyramine

Probably formed from: N-(N-Acetyl-L-methionyl)-3,4-dihydroxyphenethylamine

→ N-Acetyl-3,4-dihydroxyphenethylamine
In: Aedes F193
Cockroach H835
Drosophila A643, H758
Periplaneta F514, G103
Rhizoglyphus F669
Schistocerca A2804

See also: B765, F791

→ γ-(2-Aminoethyl)-α-hydroxy-*cis*,*cis*-muconic semialdehyde
In: Amanita G119

→ 4-(2-Aminoethyl)-*o*-quinone
In: Apple J678
Banana K628
Mushroom H907
Pear J678

See also: A63, A218, A624, A677, A2360, A2556, A3114, A3975, B897, D530, E407, F944, G620

→ *p*-Coumaroyl-3,4-dihydroxyphenethylamine
In: Nicotiana F72

→ (−)-Cryptostyline I
In: Cryptostylis A1515

→ Deacetylipecoside
Probably in: Alangium C909

→ 3,4-Dihydroxy-N-methylphenethylamine
In: Dog C657
Rat A80, E867, F397, F623

→ 3,4-Dihydroxyphenethylamine-O-β-D-glucuronide
In: Dog A1887
Monkey A1887
Mouse A1887
Rabbit A1887
Rat A105, A1887, C185, K595

See also: A1026, A1275, A2856

→ 3,4-Dihydroxy-N-methylphenethylamine
In: Dog C657
Rat A80, E867, F397, F623

→ 3,4-Dihydroxyphenylacetaldehyde
In: Alcaligenes H367, H682
Amblyomma J4
Arthrobacter F626, H912
Bacteria B225
Beef A1285, A3885
Dog A806
Guinea pig A3958
Helix A657
Hymenolepis A3815
Man A2010, A2967, F754
Micrococcus K420
Pig A72
Planorbis A2055
Pseudomonas C517
Rat A1285, A1309, A1324, A1455, A2003, A2641, E100, E339
Rhiziglyphus F669
Tetrahymena A3131

See also: A321, A371, A451, A652, A658, A801, A850, A1026, A1716, A1887, A1909, A2138, A2568, A2584, A2655, A2771, A2961, A3079, A3574, A3594, A3603, A3722, A3862, A3966, A3969, B892, B977, C785, F380, G902, G926, H432

→ Feruloyl-3,4-dihydroxyphenethylamine
In: Nicotiana D558, F72

→ 3-Hydroxy-4-methoxyphenethylamine
In: Rat A486, A2902, D297

→ 3-Hydroxyphenethylamin-4-yl sulphate
In: Eubacterium E432, G602
Man E186

See also: A153, A378, A1716

→ 4-Hydroxyphenethylamin-3-yl sulphate

In: Dog A806
Eubacterium G602
Man E186
See also: A378, A1343, A1716, A1887, A3533, A3814, B834, B941, C185, C322, C441, C528, D233, D583, D586, D607, D736, D741, D748, D977, D979, E194, E195, E198, E211, E695, E962, F116, F205, F464, F770, H261, H276, H628, H888, J292, J535, J644, K344, K588
Contrast in dog: E804
→ 3-Methoxytyramine
In: Cat A1567
Catfish F282
Dog A849, A1567
Man A1567, A2411, A3831, A3862, B118, D55
Pig C346
Planorbis A2055
Rabbit B892
Rat A486, A1257, A1337, A1455, A2961, A3529, A3862, D297, D704, E100
Streptomyces K652
See also: A652, A1716, A2902, A3966, A3969, G926, J363
→ D-Noradrenaline
In: Beef K156
Mouse A2558
Planorbis A2055
Rat E100
See also: A321, A898, A1092, A1155, A1257, A2572, A3966, B858
→ Norcoclaurine
Probably in: Annona B137
→ Norsalsolinol
In: Pig A2346
→ Salsolinol
In: Rat A2641
See also: A795, A1820
→ Tetrahydropapaveroline
In: Berberis B919, K932
Corydalis B919
Dicentra B919
Eschscholzia B919, C859
Fumaria B919
Glaucium B919
Meconopsis B919
Papaver B919
Ranunculus B919
Rat A2641
Thalictrum B919
See also: A795, A1189
→ Tetrahydropapaveroline-1-carboxylic acid
In: Papaver A1513
See also: A3222
→ *m*-Tyramine
In: Rat A2961, C456, D6
See also: B162, B907
→ Tyramine
In: Rat A1257, A1337, C465, D6
See also: B162

3,4-Dihydroxyphenethylamine-O-β-D-glucuronide
Formed from: 3,4-Dihydroxyphenethylamine

3,4-Dihydroxyphenethyl trimethyl ammonium
→ 4-(2-Trimethylammonioethyl)-*o*-quinone
Probably in: Man A63

2,5-Dihydroxyphenoxyacetic acid
Probably formed from: *p*-Hydroxyphenoxyacetic acid

(+)-(*R*)-2-(2-((N-(2-(3,4-Dihydroxyphenoxy)ethyl)-N-methylamino)propoxy)phenyl)-4-methyl-2*H*-1,4-benzothiazin-3-(4*H*)-one
Formed from: Semotiadil
→ (+)-(*R*)-2-(2-((N-(2-(4-Hydroxy-3-methoxyphenoxy)ethyl)-N-methylamino)propoxy)phenyl)-4-methyl-2*H*-1,4-benzothiazin-3-(4*H*)-one
Probably in: Rat H313, J130

(*S*)-2-(2-((N-(2-(3,4-Dihydroxyphenoxy)ethyl)-N-methylamino)propoxy)phenyl)-4-methyl-2*H*-1,4-benzothiazin-3-(4*H*)-one
Formed from: Levosemotiadil
→ (*S*)-2-(2-((N-(2-(4-Hydroxy-3-methoxyphenoxy)ethyl)-N-methylamino)propoxy)phenyl)-4-methyl-2*H*-1,4-benzothiazin-3-(4*H*)-one
Probably in: Rat J130

3,4-Dihydroxyphenylacetaldehyde
Formed from: 3,4-Dihydroxy-N-methylphenethylamine
3,4-Dihydroxyphenethylamine
3,4-Dihydroxy-L-phenylalanine
2-(3,4-Dihydroxyphenyl)ethanol
→ 3,4-Dihydroxyphenylacetic acid

In: Horse A239
Man E206
Pig A1287
Planorbis A2055
Rat A1455, A2003, A2641, A2961, B659, E257
See also: A2700, A2771, A3722, G865
→ 2-(3,4-Dihydroxyphenyl)ethanol
In: Man E206
Monkey A1908
Pig A1287
Rat A2641, A2961, B659
See also: A2700, A2771, A3722
→ Tetrahydropapaveroline
In: Berberis B919, K932
Corydalis B919
Dicentra B919
Eschscholtzia B919, C859
Fumaria B919
Glaucium B919
Meconopsis B919
Papaver B919
Ranunculus B919
Thalictrum B919

3,4-Dihydroxyphenylacetamide
→ 3-Hydroxy-4-methoxyphenylacetamide
In: Rat A486
With catechol O-methyltransferase A597
→ 4-Hydroxy-3-methoxyphenylacetamide
In: Rat A486
With catechol O-methyltransferase A597

2,3-Dihydroxyphenylacetic acid
Formed from: *cis*-2,3-Dihydro-2,3-dihydroxyphenylacetic acid
→ Catechol
In: Bacteria J469

2,4-Dihydroxyphenylacetic acid
Formed from: *p*-Hydroxyphenylacetic acid

2,6-Dihydroxyphenylacetic acid
Formed from: *o*-Hydroxyphenylacetic acid
Phenylacetic acid

3,4-Dihydroxyphenylacetic acid (dopac)
Formed from: *p*-Chlorophenylacetic acid
5,6-Dihydro-3-(*p*-hydroxyphenyl)-4,6-dioxo-4*H*-pyran
5,6-Dihydro-3-(*p*-methoxyphenyl)-4,6-dioxo-4*H*-pyran
3,4-Dihydroxyphenylacetaldehyde
3,4-Dihydroxyphenylpyruvic acid
3-Hydroxy-4-methoxyphenylacetic acid
4-Hydroxy-3-methoxyphenylacetic acid
m-Hydroxyphenylacetic acid
p-Hydroxyphenylacetic acid
Phenylacetic acid
Quercetin
Rutin
Probably formed from: 4-(3,4-Dihydroxyphenyl)butyric acid
m-Hydroxyphenylacetic acid
→ γ-Carboxymethyl-α-hydroxy-*cis*,*cis*-muconic semialdehyde
In: Bacillus A1464
See also: J653
→ δ-Carboxymethyl-α-hydroxy-*cis*,*cis*-muconic semialdehyde
In: Acinetobacter A1449
Arthrobacter A3279, G664
Bacillus A2827
Klebsiella H232
Pseudomonas A1090
See also: A2651, A3208, H387
→ 4-Carboxymethyl-2-hydroxyphenyl-β-D-glucuronide
Probably in: Rat A105, K595
→ 4-Carboxymethyl-2-hydroxyphenyl sulphate
Probably in: Man D613
Rat A153, A1343,
→ 3-Carboxymethyl-*cis*,*cis*-muconic acid
In: Hydrogenophaga J218
→ 4-Carboxymethyl-*o*-quinone
In: Apple J678
Mushroom H907
Pear J678
See also: E407, F944, G620
→ 3,4-Dihydroxymandelic acid
In: Sarcophaga C692
→ 2-(3,4-Dihydroxyphenyl)ethanol
In: Rat H798
→ 2,3,6,7,9,10-Hexahydro-2,3,7,8-tetraoxoanthracene

In: Klebsiella H748
→ 3-Hydroxy-4-methoxyphenylacetic acid
In: Rat A486, A1608, A2961
See also: A801
→ 4-Hydroxy-3-methoxyphenylacetic acid
In: Nicotiana A3778
Pig A3614, B813, C346
Rat A486, A1608, A2620, A2817, A2961
See also: A371, A801, A1455, A2568, C211, D422
→ *m*-Hydroxyphenylacetic acid
In: Rat A2961
See also: B162, B918
→ *p*-Hydroxyphenylacetic acid
Probably in: Rat B162
→ 2,4,5-Trihydroxyphenylacetic acid
In: Flavobacterium E679, F780, G404, K779
→ 3,4,5-Trihydroxyphenylacetic acid
Probably in: Escherichia G817
See also: A489, A1480

3,5-Dihydroxyphenylacetic acid
Probably formed from: 6-(Carboxypentyl)-2,4-dihydroxybenzoic acid

3,4-Dihydroxyphenylacetone
Formed from: 4-Hydroxy-3-methoxyphenylacetone
3,4-Dihydroxy-α-methyl-L-phenylalanine
L-α-Hydrazino-3,4-dihydroxy-α-methylphenylpropionic acid
Probably formed from: 3,4-Methylenedioxy-phenylacetone
→ 4-Hydroxy-3-methoxyphenylacetone
Probably in: Rat E988

2,3-Dihydroxy-L-phenylalanine
Probably formed from: *cis*-2,3-Dihydro-2,3-dihydroxy-L-phenylalanine
L-*o*-Tyrosine
L-Tyrosine
→ Pyrocatechuic acid
In: Aspergillus A3377

2,4-Dihydroxy-L-phenylalanine
Formed from: Resorcinol
L-Tyrosine
→ 2,4-Dihydroxyphenethylamine
In: Micrococcus C62
Mouse E409
→ 2,4,5-Trihydroxy-L-phenylalanine
Probably in: Aeromonas D677

2,5-Dihydroxy-L-phenylalanine
→ 2,5-Dihydroxyphenethylamine
Probably in: Aspergillus A3377

3,4-Dihydroxy-D-phenylalanine
Formed from: 5-(3,4-Dihydroxybenzyl) hydrantoin
D-Tyrosine
→ 4-(3-(D-Alanyl))-*o*-quinone
Probably in: Avocado A3975
Probably with caeruloplasmin A677
→ 3,4-Dihydroxyphenylpyruvic acid
In: Rat A1393
See also: A2961
→ 3-Hydroxy-4-methoxy-D-phenylalanine
In: Rat A486
→ 3-Methoxy-D-tyrosine
In: Pig B813
Rat A486, G635

3,4-Dihydroxy-L-phenylalanine (L-dopa)
Formed from: Catechol
5-(3,4-Dihydroxybenzyl)hydrantoin
3,4-Dihydroxyphenylpyruvic acid
γ-Glutamyl-3,4-dihydroxy-L-phenylalanine
L-*m*-Tyrosine
L-Tyrosine
Probably formed from: 3,4-Dihydroxy-L-phenylalanine ethyl ester
3-Hydroxy-4-methoxy-L-phenylalanine
→ γ-(3-Alanyl)-α-hydroxy-*cis*,*cis*-muconic semialdehyde
In: Amanita G119, J872
Stizolobium K706, K708
→ δ-(3-Alanyl)-α-hydroxy-*cis*,*cis*-muconic semialdehyde
In: Pseudomonas A1090
Stizolobium K706, K708
See also: D451, J642
→ 4-(3-(L-Alanyl))-*o*-quinone
In: Amasya H567
Apple J678
Mushroom H907
Pear J678
Potato A2997

See also: A7, A63, A218, A594, A677, A702, A1208, A1636, A1930, A2360, A2556, A3114, A3135, A3975, B897, E407, G449, G620, J313

→ Amaranthin
In: Amaranthus A235

→ Anthramycin
In: Streptomyces A1677

→ Berberine
In: Coptis G749

→ Betacyanin
In: Amaranthus D4

→ Betalamic acid
In: Amanita J872
Portulaca A184

→ Betanidin
In: Portulaca J872

→ Caffeic acid
Probably in: Rat A1035

→ Catechol
In: Erwinia A708

→ Corydaline
In: Corydalis A3684

→ 2-(S-(L-Cysteinyl))-3,4-dihydroxy-L-phenylalanine
In: Horseradish B762
Mushroom A3167, C442

→ 5-(S-(L-Cysteinyl))-3,4-dihydroxy-L-phenylalanine
In: Horseradish B762
Mushroom A3167, C442
With xanthine oxidase B968

→ 6-(S-(L-Cysteinyl))-3,4-dihydroxy-L-phenylalanine
In: Mushroom A3167

→ 11-Demethyltomaymycin
In: Streptomyces A2154

→ (*2S*)-2,3-Dihydro-5,6-dihydroxyindole-2-carboxylic acid
In: Portulaca J872

→ 3,4-Dihydroxyphenethylamine
In: Barley A1210
Dog G902, H592
Drosophila A643
Helix A1386
Locust A2804, B907
Man A371, A451, A801, A898, A1107, A1505, D793, F783, G902
Micrococcus G918
Microorganism E762
Mouse A2558
Mycobacterium F189
Opossum J463
Papaver H557
Periplaneta E448
Pig A1906
Pigtail monkey 1716
Planorbis A2055
Pseudomonas F191
Rat A652, A669, A801, A1026, A1092, A1155, A1288, A1393, A1505, A1781, A2138, A2378, A2570, A2584, A2654, A2655, A2856, A2961, A2981, A3966, A3969, B344, D256, E735, F197, G192, G635, G902
Sheep A2568
Streptococcus A537, F197

See also: A1035, A1900, A2572, A2700, A2771, B162

→ 3,4-Dihydroxyphenylacetaldehyde
In: Pig A2425

→ 3,4-Dihydroxy-L-phenylalanine-β-D-glucuronide
In: Rat A105, K595

→ 3,4-Dihydroxyphenylpyruvic acid
In: Hansenula D978
Man A801
Rat A801, A2268, D196

See also: A1609, A2961

→ γ-Glutamyl-3,4-dihydroxy-L-phenylalanine
In: Sheep A2981

→ 3-Hydroxy-4-methoxy-L-phenylalanine
In: Rat 486, B513

→ 3-Hydroxy-L-phenylalanin-4-yl sulphate
In: Eubacterium E432

See also: G602
Contrast (in man) J837

→ 4-Hydroxy-L-phenylalanin-3-yl sulphate
In: Man J837

See also: E804, F196, H888

→ 3-Methoxy-L-tyrosine
In: Man C883

Pig B813, C346
Planorbis A2055
Rat A321, A486, A1035, A1288, A3325, A3966, A3969, B513, F748, G635
See also: A3239
→ Ochotensimine
In: Corydalis A3684
→ Protocatechualdehyde
In: Aspergillus A596
→ Protopine
In: Corydalis A3684
→ Rosmarinic acid
Probably in: Coleus A3313
→ Stizolobic acid
In: Mucuna A1640
→ Tetrahydropapaveroline-1-carboxylic acid
In: Papaver A1513
→ 2,4,5-Trihydroxy-L-phenylalanine
In: Sea anemone D245
See also: A1364
→ 3,4,5-Trihydroxy-L-phenylalanine
In: Mushroom C442

3,4-Dihydroxy-L-phenylalanine ethyl ester
→ 3,4-Dihydroxy-L-phenylalanine
Probably in: Cat A1567
Dog A1567
Man A1567

3,4-Dihydroxy-L-phenylalanine-β-D-glucuronide
Formed from: 3,4-Dihydroxy-L-phenylalanine

3,4-Dihydroxy-L-phenylalanine methyl ester
Formed from: L-Tyrosine methyl ester
→ 4-(3-L-Alanyl)-*o*-quinone methyl ester
In: Apple J678
Mushroom H907
Pear J678

1-(3,4-Dihydroxyphenylazo)-2-hydroxynaphthalene
Formed from: 2-Hydroxy-1-(*m*-hydroxyphenylazo)naphthalene
2-Hydroxy-1-(*p*-hydroxyphenylazo)naphthalene
Probably formed from: 1-Phenylazo-2-naphthol

2,6-Dihydroxy-1-phenylazonaphthalene
Formed from: 1-Phenylazo-2-naphthol

N-(3,4-Dihydroxyphenyl)benzenesulphonamide
→ N-(3-Hydroxy-4-methoxyphenyl)benzenesulphonamide
In: Rat A486
→ N-(4-Hydroxy-3-methoxyphenyl)benzenesulphonamide
In: Rat A486

(*R*,*R*)-*threo*-1,2-Dihydroxy-1-phenylbutane
Formed from: β-Ethylstyrene oxide

4-(3,4-Dihydroxyphenyl)butan-2-ol
Probably formed from: 4-(3,4-Dihydroxyphenyl)butan-2-one

4-(3,4-Dihydroxyphenyl)butan-2-one
Formed from: 4-(*p*-Hydroxyphenyl)butan-2-one
Zingerone
→ 4-(3,4-Dihydroxyphenyl)butan-2-ol
Probably in: Rat A1786

***p*,γ-Dihydroxyphenylbutazone**
Probably formed from: γ-Hydroxyphenylbutazone
Oxyphenylbutazone
→ *p*,γ-Dihydroxyphenylbutazone-β-D-glucuronide
Probably in: Rat A296

***p*,γ-Dihydroxyphenylbutazone-β-D-glucuronide**
Probably formed from: *p*,γ-Dihydroxyphenylbutazone

4-(3,4-Dihydroxyphenyl)butylamine
→ 4-(4-Aminobutyl)-*o*-quinone
Probably in: Man A63
→ 4-(3,4-Dihydroxyphenyl)butyric acid
In: Rat G865

4-(2,3-Dihydroxyphenyl)butyric acid
Formed from: *cis*-4-(2,3-Dihydro-2,3-dihydroxyphenyl)butyric acid

4-(3,4-Dihydroxyphenyl)butyric acid
Formed from: 4-(3,4-Dihydroxyphenyl)butylamine
6-(3,4-Dihydroxyphenyl)hexylamine
Probably formed from: 6-(3,4-Dihydroxyphenyl)hexanoic acid
→ 3,4-Dihydroxyphenylacetic acid
Probably in: Rat G865

1,2-Dihydroxy-1-phenylcyclohexane
Formed from: 1-Phenyl-1-cyclohexene

3,4-Dihydroxy-1-phenyl-1-cyclohexene
Probably formed from: 3-Hydroxy-1-phenyl-1-cyclohexene

3,6-Dihydroxy-1-phenyl-1-cyclohexene
Probably formed from: 3-Hydroxy-1-phenyl-1-cyclohexene
→ 6-Hydroxy-3-oxo-1-phenyl-1-cyclohexene
Probably in: Rat C829

S-(2,5-Dihydroxyphenyl)-L-cysteine
Formed from: *p*-Acetamidophenol
N-Acetyl-S-(2,5-dihydroxyphenyl)-L-cysteine
Probably formed from: Quinol-2-S-glutathione
→ N-Acetyl-S-(2,5-Dihydroxyphenyl)-L-cysteine
In: Mouse J321
Rat J321
See also: F54, F600

1-(3,4-Dihydroxyphenyl)-3,5-dihydroxy-4,4-dimethyl-1-pentene
Probably formed from: 1-(3,4-Dihydroxyphenyl)-3-hydroxy-4,4-dimethyl-1-pentene
→ 1-(3,4-Dihydroxyphenyl)-5-hydroxy-4,4-dimethyl-3-oxo-1-pentene
Probably in: Rat E852, F574

1-(3,4-Dihydroxyphenyl)-4,4-dimethyl-4-oxo-1-pentene
Probably formed from: 1-(3,4-Dihydroxyphenyl)-3-hydroxy-4,4-dimethyl-1-pentene

1-(2,3-Dihydroxyphenyl)ethanol
Formed from: 1-Phenylethanol

1-(3,4-Dihydroxyphenyl)ethanol
Probably formed from: 1-(4-Hydroxy-3-methoxyphenyl)ethanol

2-(3,4-Dihydroxyphenyl)ethanol
Formed from: 3,4-Dihydroxyphenylacetaldehyde
3,4-Dihydroxyphenylacetic acid
2-(3-Hydroxy-4-methoxyphenyl)ethanol
p-Tyrosol
→ 4-(2-Hydroxyethyl)-*o*-quinone
In: Mushroom H907

3,4-Dihydroxyphenylethanolamine
See: Noradrenaline

3-((1*E*)-2-(3,4-Dihydroxyphenyl)ethenyl)-5-hydroxyphenyl-β-D-glucopyranoside
See: Astringin

5-(3,4-Dihydroxyphenyl)-5-ethylbarbituric acid
Probably formed from: 5-(*p*-hydroxyphenyl) barbituric acid
→ 5-Ethyl-5-(4-Hydroxy-3-methoxyphenyl) barbituric acid
Probably in: Man B642, E474

1-(2-(3,4-Dihydroxyphenyl)ethyl)-3-diphenylmethoxypiperidine
Formed from: Zamifenacin
→ 1-(2-(3,4-Dihydroxyphenyl)ethyl)-3-diphenylmethoxypiperidine-β-D-glucuronide
In: Man H547
→ 1-(2-(3-Hydroxy-4-methoxyphenyl)ethyl)-3-diphenylmethoxypiperidine
Probably in: Man H547
→ 1-(2-(4-Hydroxy-3-methoxyphenyl)ethyl)-3-diphenylmethoxypiperidine
Probably in: Man H547

1-(2-(3,4-Dihydroxyphenyl)ethyl)-3-diphenylmethoxypiperidine-β-D-glucuronide
Formed from: 1-(2-(3,4-Dihydroxyphenyl) ethyl)-3-diphenylmethoxypiperidine

2-(3,4-Dihydroxyphenyl)-2-ethylglutarimide
Probably formed from: 2-Ethyl-2-(*p*-hydroxyphenyl)glutarimide
→ 2-Ethyl-2-(4-hydroxy-3-methoxyphenyl)glutarimide
Probably in: Man A2200, A3438, A3991

5-(3,4-Dihydroxyphenyl)-3-ethylhydantoin
Probably formed from: 5-(3,4-Dihydro-3,4-dihydroxyphenyl)-3-ethylhydantoin
→ 3-Ethyl-5-(4-hydroxy-3-methoxyphenyl) hydantoin
Probably in: Man B113

5-(3,4-Dihydroxyphenyl)-3-(2-ethylphenyl)-1*H*-1,2,4-triazole
Probably formed from: 3-(2-Ethylphenyl)-5-(3-hydroxyphenyl)-1*H*-1,2,4-triazole
→ 3-(2-Ethylphenyl)-5-(4-hydroxy-3-methoxyphenyl)-1*H*-1,2,4-triazole
Probably in: Rat C871

3,5-Dihydroxyphenyl-α-D-glucoside
Probably formed from: Phloroglucinol

3,5-Dihydroxyphenyl-β-D-glucuronide
Probably formed from: Phloroglucinol

D-3,4-Dihydroxyphenylglycine
Formed from: D-*p*-Hydroxyphenylglycine

L-3,4-Dihydroxyphenylglycine (L-assumed unless otherwise stated)
→ Protocatechualdehyde
In: Mushroom J176
→ 2-(*o*-Quinon-4-yl)glycine
In: Mushroom J176

3,4-Dihydroxyphenylglycol (DHPG)
Formed from: 3,4-Dihydroxyphenylethanol
4-(2-Hydroxyethyl)-*o*-quinone
4-Hydroxy-3-methoxyphenylglycol
→ 4-(1,2-Dihydroxyethyl)-*o*-quinone
Probably in: Drosophila G620
→ 3,4-Dihydroxymandelaldehyde
In: Man D411, F87
→ 3,4-Dihydroxyphenylglycol-β-D-glucuronide
In: Rat K595
→ 3,4-Dihydroxyphenylglycol sulphate
In: Beef F196
Rat A153, A1240, A1343
→ ω,3′,4′-Trihydroxyacetophenone
In: Sarcophaga F793

3,4-Dihydroxyphenylglycol-β-D-glucuronide
Formed from: 3,4-Dihydroxyphenylglycol

3,4-Dihydroxyphenylglycol sulphate
Formed from: 3,4-Dihydroxyphenylglycol

2,5-Dihydroxyphenylglyoxylic acid (2,5-dihydroxybenzoylformic acid)
Formed from: 2,5-Dihydroxymandelic acid

3,4-Dihydroxyphenylglyoxylic acid
Formed from: 3,4-Dihydroxymandelic acid
→ Protocatechualdehyde
Probably in: Horse A254

7-(3,4-Dihydroxyphenyl)heptanoic acid
Formed from: 7-(3,4-Dihydroxyphenyl)heptylamine
→ 5-(3,4-Dihydroxyphenyl)pentanoic acid
Probably in: Rat G865

7-(3,4-Dihydroxyphenyl)heptylamine
→ 7-(3,4-Dihydroxyphenyl)heptanoic acid
In: Rat G865

3,4-Di-(*p*-hydroxyphenyl)hexa-2,4-diene
See: Dienoestrol

(*R*,*R*)-*threo*-1,2-Dihydroxy-1-phenylhexane
Formed from: β-*n*-Butylstyrene oxide

6-(3,4-Dihydroxyphenyl)hexanoic acid
Formed from: 8-(3,4-Dihydroxyphenyl)octylamine
→ 4-(3,4-Dihydroxyphenyl)butyric acid
Probably in: Rat G865

6-(3,4-Dihydroxyphenyl)hexylamine
→ 4-(3,4-Dihydroxyphenyl)butyric acid
In: Rat G865

3,4-Dihydroxyphenylhydracryloyl methylamide
Formed from: Dihydrocaffeoyl methylamide

1-(3,4-Dihydroxyphenyl)-5-hydroxy-4,4-dimethyl-3-oxo-1-pentene
Probably formed from: 1-(3,4-Dihydroxyphenyl)-3,5-dihydroxy-4,4-dimethyl-1-pentene

1-(3,4-Dihydroxyphenyl)-3-hydroxy-4,4-dimethyl-1-pentene
Formed from: Stiripentol
→ 1-(3,4-Dihydroxyphenyl)-3,5-dihydroxy-4,4-dimethyl-1-pentene
Probably in: Man E852
Rat F574
→ 1-(3,4-Dihydroxyphenyl)-4,4-dimethyl-3-oxo-1-pentene
Probably in: Rat F574
→ 3-Hydroxy-1-(3-hydroxy-4-methoxyphenyl)-4,4-dimethyl-1-pentene
Probably in: Man D264, E852
Rat F574, G831
→ 3-Hydroxy-1-(4-hydroxy-3-methoxyphenyl)-4,4-dimethyl-1-pentene
Probably in: Man D264, E852
Rat F574, G831

(*R*)-4-(3-(2-(3,4-Dihydroxyphenyl)-2-hydroxyethyl)amino-3-methylbutyl)benzamide
Probably formed from: (*R*)-4-(3-(2-Hydroxy-2-(*m*-hydroxyphenyl)ethyl)amino-3-methylbutyl)benzamide
→ (*R*)-4-(3-(2-Hydroxy-2-(4-hydroxy-3-methoxyphenyl)ethyl)amino-3-methylbutyl)benzamide
Probably in: Mouse F407
Rat F407

1-(2,4-Dihydroxyphenyl)-3-(4-hydroxy-2-methoxyphenyl)-1,3-propanedione
Formed from: 1,3-Bis(2,4-dihydroxyphenyl)-1,3-propanedione

3-(3,4-Dihydroxyphenyl)-3-(2-hydroxy-5-methylphenyl)propylamine
Probably formed from: 3-(2-Hydroxy-5-methylphenyl)-3-(*p*-hydroxyphenyl) propylamine

2-(3,4-Dihydroxyphenyl)-2-hydroxy-N-methylsuccinimide
Probably formed from: 2-(3,4-Dihydro-3,4-dihydroxyphenyl)-2-hydroxy-N-methylsuccinimide

6-(2,3-Dihydroxyphenyl)-2-hydroxy-6-oxomuconic acid
Formed from: 2,2′,3,3′-Tetrahydroxybiphenyl

5-(3,4-Dihydroxyphenyl)-5-(*p*-hydroxyphenyl) hydantoin
Probably formed from: 5,5-Bis(*p*-hydroxyphenyl)hydantoin
→ 5-(3,4-Dihydroxyphenyl)-5-(*p*-hydroxyphenyl)hydantoin-β-D-glucuronide
Probably in: Man A2110
Rat A2110
→ 5-(4-Hydroxy-3-methoxyphenyl)-5-(*p*-hydroxyphenyl)hydantoin
Probably in: Man A2110
Rat A2110

5-(3,4-Dihydroxyphenyl)-5-(*p*-hydroxyphenyl) hydantoin-β-D-glucuronide
Probably formed from: 5-(3,4-Dihydroxyphenyl)-5-(*p*-hydroxyphenyl) hydantoin

3-(3,4-Dihydroxyphenyl)-3-(*p*-hydroxyphenyl)-1(3*H*)-isobenzofuran
Formed from: Phenolphthalein

1-(2,4-Dihydroxyphenyl)-3-(*p*-hydroxyphenyl)-1,3-propanedione
See: Licodione

1-(3,4-Dihydroxyphenyl)-2-iminoethanol
Formed from: D-Noradrenaline

1-(3,4-Dihydroxyphenyl)-2-isopropylaminobutan-1-ol
See: Isoetharine

2,3-Dihydroxyphenyllactic acid
Formed from: *cis*-2,3-Dihydro-2,3-dihydroxyphenyllactic acid

3,4-Dihydroxyphenyllactic acid
Formed from: 3,4-Dihydroxyphenylpyruvic acid
p-Hydroxyphenyllactic acid
Rosmarinic acid
Vanillactic acid
Probably formed from: 3,4-Methylenedioxyphenyllactic acid
→ *p*-Coumaroyl 3,4-dihydroxyphenyllactic acid ester
In: Coleus G420, J349
→ 3,4-Dihydroxyphenylpyruvic acid
Probably in: Rat A2961
→ Rosmarinic acid
In: Coleus E660, G257, G420, H17
→ Vanillactic acid
Probably in: Rat A2961

***trans*-1-(3,4-Dihydroxyphenyl)-1-(*p*-(2-methylaminoethoxy)phenyl)-2-phenylbut-1-ene**
Probably formed from: Demethyldroloxifene
→ *trans*-1-(*p*-(2-Methylaminoethoxy)phenyl)-1-(3-hydroxy-4-methoxyphenyl)-2-phenylbut-1-ene
Probably in: Mouse H924
→ *trans*-1-(*p*-(2-Methylaminoethoxy)phenyl)-1-(4-hydroxy-3-methoxyphenyl)-2-phenylbut-1-ene
Probably in: Mouse H924

1-((3,4-Dihydroxyphenyl)methyl)-3,4-dihydro-6,7-isoquinolinediol
See: 3,4-Dihydropapaveroline

1-(3,4-Dihydroxyphenyl)-2-methyliminoethanol
Formed from: D-Adrenaline

3-(3,4-Dihydroxyphenyl)-2-methyllactic acid
Formed from: L-α-Hydrazino-3,4-dihydroxy-α-methylphenylpropionic acid
→ 3-(4-Hydroxy-3-methoxyphenyl)-2-methyllactic acid
Probably in: Dog A1502

3-(3,4-Dihydroxyphenyl)-5-methyl-3-phenyl-1-pyrroline
Probably formed from: 3-(*p*-Hydroxyphenyl)-5-methyl-3-phenyl-1-pyrroline

→ 3-(3,4-Dihydroxyphenyl)-5-methyl-3-phenyl-1-pyrroline-β-D-glucuronide
Probably in: Rat A2779
→ 3-(4-Hydroxy-3-methoxyphenyl)-5-methyl-3-phenyl-1-pyrroline
Probably in: Rat A2779

3-(3,4-Dihydroxyphenyl)-5-methyl-3-phenyl-1-pyrroline-β-D-glucuronide
Probably formed from: 3-(3,4-Dihydroxyphenyl)-5-methyl-3-phenyl-1-pyrroline

1-(4-(3,4-Dihydroxyphenyl)-1-methyl-4-piperidyl)-1-propanone
Formed from: Ketobemidone
→ 1-(4-(4-Hydroxy-3-methoxyphenyl)-1-methyl-4-piperidyl)-1-propanone
In: Man B643

(1*S*)-1-((3,4-Dihydroxyphenyl)methyl)-1,2,3,4-tetrahydro-6,7-dimethoxyisoquinoline
See: Norlaudanosine

1-((3,4-Dihydroxyphenyl)methyl)-1,2,3,4-tetrahydro-7-hydroxy-6-methoxyisoquinoline
See: 6-O-Methyllaudanosoline

1-((3,4-Dihydroxyphenyl)methyl)-1,2,3,4-tetrahydro-6,7-isoquinolinediol
See: Laudanosoline

3-(3,4-Dihydroxyphenyl)-N-(1-naphthylmethyl)-2-propen-1-ylamine
Probably formed from: *p*-Hydroxynornaftifine

(*R*,*R*)-*threo*-1,2-Dihydroxy-1-phenyloctane
Formed from: β-*n*-Hexylstyrene oxide

8-(3,4-Dihydroxyphenyl)octylamine
→ 6-(3,4-Dihydroxyphenyl)hexanoic acid
In: Rat G865

2-(3,4-Dihydroxyphenyl)-2-oxoethylamine
→ 3,4-Dihydroxymandelic acid
In: Rat J287

(α*R*)-α-(((2*E*)-3-(3,4-Dihydroxyphenyl)-1-oxo-2-propenyl)oxy)-3,4-dihydroxybenzenepropanoic acid
See: Rosmarinic acid

(*R*,*R*)-*threo*-1,2-Dihydroxy-1-phenylpentane
Formed from: β-*n*-Propylstyrene oxide

5-(3,4-Dihydroxyphenyl)pentanoic acid
Formed from: 5-(3,4-Dihydroxyphenyl) pentylamine
Probably formed from: 7-(3,4-Dihydroxyphenyl)heptanoic acid
→ Dihydrocaffeic acid
Probably in: Rat G865

5-(3,4-Dihydroxyphenyl)pentylamine
→ 4-(5-Aminopentyl)-*o*-quinone
Probably in: Man A63
→ 5-(3,4-Dihydroxyphenyl)pentanoic acid
Probably in: Rat G865

N-(1-(3,4-Dihydroxyphenyl)pent-2-yl)pyrrolidine
Probably formed from: *p*-Hydroxyprolintane

7-(3,4-Dihydroxyphenyl)-1-phenylhepta-1,3-dien-5-one
→ Angiorufine
In: Angiozanthos H157

5-(3,4-Dihydroxyphenyl)-5-phenylhydantoin
Formed from: 5-(3,4-Dihydro-3,4-dihydroxyphenyl)-5-phenylhydantoin
5-(*p*-Hydroxyphenyl)-5-phenylhydantoin
→ 5-(3,4-Dihydroxyphenyl)-5-phenylhydantoin-β-D-glucuronide
Probably in: Rat A1018
→ 5-(4-Hydroxy-3-methoxyphenyl)-5-phenylhydantoin
In: Rat A567
See also: A1018, A1970, A3305, C39

5-(3,4-Dihydroxyphenyl)-5-phenylhydantoin-β-D-glucuronide
Probably formed from: 5-(3,4-Dihydroxyphenyl)-5-phenylhydantoin

4-((3,4-Dihydroxyphenyl)phenyl)methyl-1-((octylimino)methyl)piperidine
Formed from: *p*-Hydroxyfenoctimine
→ 4-((4-Hydroxy-3-methoxyphenyl)phenyl) methyl-1-((octylimino)methyl)piperidine
Probably in: Dog H62

3-(3,4-Dihydroxyphenyl)propanol
Formed from: 3-(*p*-Hydroxyphenyl)propanol
→ 4-(3-Hydroxypropyl)-*o*-quinone
In: Mushroom H907

2,3-Dihydroxyphenylpropionic acid
Formed from: *cis*-2,3-Dihydro-2,3-dihydroxyphenylpropionic acid
Melilotic acid
→ *cis*,*cis*-(*S*)-Carboxyocta-2,4-dien-1,8-dioic acid
In: Bacteria B950

→ 2-Hydroxy-6-oxo-*cis*,*cis*-nona-2,4-dien-1,9-dioic acid
In: Achromobacter K819
Acinetobacter C817
Escherichia H383, H606

2-(2,5-Dihydroxyphenyl)propionic acid
Formed from: 2-(*p*-Hydroxyphenyl)propionic acid

3-(2,5-Dihydroxyphenyl)propionic acid
Formed from: Phloretic acid

3-(3,4-Dihydroxyphenyl)propionic acid
See: Dihydrocaffeic acid

3-(3,5-Dihydroxyphenyl)propionic acid
Formed from: 3-Hydroxy-4,5-dimethoxyphenylpropionic acid
3-Hydroxy-5-methoxyphenylpropionic acid
3-(3,4,5-Trihydroxyphenyl)propionic acid

β,γ-Dihydroxy-γ-phenylpropyl carbamate
Probably formed from: γ-Hydroxy-γ-phenylpropyl carbamate

γ,*p*-Dihydroxy-γ-phenylpropyl carbamate
Probably formed from: γ-Hydroxy-γ-phenylpropyl carbamate

3,4-Dihydroxyphenyl-N-*n*-propylpiperidine
→ 4-Hydroxy-3-methoxyphenyl-N-*n*-propylpiperidine
Probably in: Beef D422

3-(3,5-Dihydroxyphenyl)propanal
Formed from: Olivetol
→ 3-(3,5-Dihydroxyphenyl)propionic acid
Probably in: Syncephalastrum D450

2,3-Dihydroxyphenylpropionic acid
Formed from: Melilotic acid
Quinoline

3-(3,5-Dihydroxyphenyl)propionic acid
Probably formed from: 3-(3,5-Dihydroxphenyl)propanal

3-(3,4-Dihydroxyphenyl)propylamine
→ 4-(3-Aminopropyl)-*o*-quinone
Probably in: Man A63

β,γ-Dihydroxy-γ-phenylpropyl carbamate
Probably formed from: γ-Hydroxy-γ-phenylpropyl carbamate

γ,*p*-Dihydroxy-γ-phenylpropyl carbamate
Probably formed from: γ-Hydroxy-γ-phenylpropyl carbamate

3-(3,4-Dihydroxyphenyl)propyl trimethyl ammonium
Formed from: 3-(*p*-Hydroxyphenyl)propyl trimethyl ammonium
→ Trimethyl 3-(*o*-quinon-4-yl)propyl ammonium
In: Mushroom H907

2,3-Dihydroxyphenylpyruvic acid
Probably formed from: *cis*-2,3-Dihydro-2,3-dihydroxyphenylpyruvic acid
→ 2,3-Dihydroxybenzaldehyde
Probably in: Bacteria B950

2,5-Dihydroxyphenylpyruvic acid
Probably formed from: *p*-Hydroxyphenylpyruvic acid

3,4-Dihydroxyphenylpyruvic acid
Formed from: 3,4-Dihydroxy-D-phenylalanine
3,4-Dihydroxy-L-phenylalanine
p-Hydroxyphenylpyruvic acid
Probably formed from: 3,4-Dihydroxyphenyllactic acid
→ δ-(2-Carboxy-2-oxoethyl)-α-hydroxy-*cis*,*cis*-muconic semialdehyde
In: Pseudomonas A1090
→ 3,4-Dihydroxy-L-phenylalanine
In: Alcaligenes A956
Enterobacter A1616
Rat A1393, A2138, A2139, A2983
See also: A2961
→ 3,4-Dihydroxyphenylacetic acid
In: Rat C504
→ 3,4-Dihydroxyphenyllactic acid
In: Coleus E660, G289
Rat A2961, A2983
→ Tetrahydropapaveroline-1-carboxylic acid
Probably in: Litsea A3222
→ 2,4,5-Trihydroxyphenylacetic acid
In: Chicken A1099

***erythro*-3,4-Dihydroxy-D-phenylserine**
→ L-Noradrenaline
In: Rat A125

***erythro*-3,4-Dihydroxy-L-phenylserine**
→ *erythro*-3-Hydroxy-4-methoxy-L-phenylserine
In: Rat A486
→ *erythro*-4-Hydroxy-3-methoxy-L-phenylserine

In: Rat A486, H231
→ L-Noradrenaline
In: Mouse A581
Rat A125, A581, H231
See also: A3177, F951
→ L-Phenylserine-3,4-quinone
See also: Mushroom B897
→ Protocatechualdehyde
In: Man E371
Rat H231

threo-3,4-Dihydroxy-L-phenylserine
→ *threo*-3-Hydroxy-4-methoxy-L-phenylserine
In: Rat A486
→ *threo*-4-Hydroxy-3-methoxy-L-phenylserine
In: Mouse F946
Rat A486, H231
→ D-Noradrenaline
In: Man C566, D99, D793
Pig A2593
Rat A2621, A3882, A3888, C566, G810, H231
See also: A3177, E348
→ Protocatechualdehyde
In: Man E371
Rat H231

2,3-Dihydroxyphenyl sulphate
Formed from: Pyrogallol

2,6-Dihydroxyphenyl sulphate
Formed from: Pyrogallol

3,5-Dihydroxyphenyl sulphate
Formed from: Phloroglucinol

1-(3,4-Dihydroxyphenyl)-1-(m-trifluoromethylphenyl)-1-propanol
Probably formed from: 1-(*p*-Hydroxyphenyl)-1-(*m*-trifluoromethylphenyl)-1-propanol

(3,4-Dihydroxyphenyl)(2,4,6-trihydroxyphenyl) methanone
See: Maclurin

1-(3,4-Dihydroxyphenyl)-3-(2,4,6-trihydroxyphenyl)propan-2-ol
Formed from: (+)-Catechin
→ 1-(*m*-Hydroxyphenyl)-3-(2,4,6-trihydroxyphenyl)propan-2-ol
In: Microorganism C880

d-δ-(3,4-Dihydroxyphenyl)-γ-valerolactone
→ *d*-δ-(4-Hydroxy-3-methoxyphenyl)-γ-valerolactone
Probably in: Rat C781
→ *d*-δ-(*m*-Hydroxyphenyl)-γ-valerolactone
Probably in: Rat C781

3,4-Dihydroxyphthalic acid
Probably formed from: 3,4-Dihydro-3,4-dihydroxyphthalic acid
→ Protocatechuic acid
In: Micrococcus K753

4,5-Dihydroxyphthalic acid
Formed from: Phthalic acid
→ Protocatechuic acid
In: Bacillus D31
Pseudomonas A3096, A3332, D538
See also: A2653, A3528, B2

2,3-Dihydroxypromazine
Formed from: Chlorpromazine
3-Hydroxychlorpromazine
→ 2,3-Dihydroxypromazine-β-D-glucuronide
In: Rat A2092

2,3-Dihydroxypromazine-β-D-glucuronide
Formed from: 2,3-Dihydroxypromazine

3′,4′-Dihydroxypropafenone
Probably formed from: 4′-Hydroxypropafenone
→ 4′-Hydroxy-3′-methoxypropafenone
Probably in: Dog E18

4,5-Dihydroxypropafenone
Probably formed from: 5-Hydroxypropafenone
→ 5-Hydroxy-4-methoxypropafenone
Probably in: Dog E18
Rat J538

2,2′-Dihydroxy-5-(1-propen-1(E)-yl)-5′-propylbiphenyl
Probably formed from: Tetrahydromagnolol
→ Isomagnolol
Probably in: Rat D549, D625

2,m-Dihydroxypropiophenone
Formed from: *m*-Hydroxybenzaldehyde

2,p-Dihydroxypropiophenone
Formed from: *p*-Hydroxybenzaldehyde

3′,4′-Dihydroxypropiophenone
Formed from: *p*-Hydroxypropiophenone
→ 3′-Hydroxy-4′-methoxypropiophenone

In: Rat A486
With catechol O-methyltransferase A597
→ 4′-Hydroxy-3′-methoxypropiophenone
In: Rat A486
With catechol O-methyltransferase A597

3,*p*-Dihydroxypropiophenone
→ 3-β-D-Glucosidoxy-*p*-hydroxypropiophenone
In: Sulfolobus H573

4-(2′(2″,3″-Dihydroxypropoxycarbonyl)phenylamino)-7-chloroquinoline
See: Glifanan

7-(1′,2′-Dihydroxy-3′-propoxy)coumarin
Formed from: 7-Glycidoxycoumarin

2-(*p*-(2,3-Dihydroxypropoxy)phenyl)-2-(*p*-hydroxyphenyl)propane
Formed from: 2,2-Bis(*p*-(2,3-epoxypropoxy)phenyl)propane
→ 2-(*p*-(2-Carboxy-2-hydroxyethoxy)phenyl)-2-(*p*-(2,3-dihydroxypropoxy)phenyl)propane
Probably in: Mouse B704

***o*-(2,3-Dihydroxypropoxy)-ω-phenylpropiophenone**
Probably formed from: 2-Hydroxy-3-propionaldoxy)-ω-phenylpropiophenone

2-(2,3-Dihydroxypropoxy)-1-(2-propen-1-oxy)benzene
Formed from: Oxprenolol

6-(2,3-Dihydroxypropoxy)-2(1*H*)-quinolinone
Formed from: Toborinone

2,3-Dihydroxypropranolol
Formed from: 2-Hydroxypropranolol
3-Hydroxypropranolol

3,4-Dihydroxypropranolol
Formed from: 3-Hydroxypropranolol
Probably formed from: 4-Hydroxypropranolol
→ 1-((3,4-Dihydroxy)-1-naphthoxy)-2,3-propanediol
Probably in: Man E916
→ 3-Hydroxy-4-methoxypropranolol
Probably in: Rat E916
→ 4-Hydroxy-3-methoxypropranolol
Probably in: Man E916
Rat E916

3,7-Dihydroxypropranolol
Formed from: 3-Hydroxypropranolol
7-Hydroxypropranolol

4,6-Dihydroxypropranolol
Formed from: 6-Hydroxypropranolol

4,8-Dihydroxypropranolol
Formed from: 8-Hydroxypropranolol
Probably formed from: 4-Hydroxypropranolol

5,6-Dihydroxypropranolol
Formed from: 5-Hydroxypropranolol
6-Hydroxypropranolol

7,8-Dihydroxypropranolol
Formed from: 7-Hydroxypropranolol
8-Hydroxypropranolol

4-(1,2-Dihydroxy-2-propyl)antipyrine
Formed from: 4-Isopropylantipyrine

2,3-Dihydroxypropyl 3-carboxy-2,6-dimethyl-4-(3-nitrophenyl)pyridine-5-carboxylate
Probably formed from: 2-Hydroxy-3-nitrooxypropyl 3-carboxy-2,6-dimethyl-4-(3-nitrophenyl)pyridine-5-carboxylate

4-(2,3-Dihydroxypropyl)catechol
Probably formed from: Allylcatechol-2,3-oxide
Safrole glycol

10-(2,3-Dihydroxypropyl)-9,9-dioxophenothiazine
Formed from: Prothanon

1-(2,3-Dihydroxypropyl)-5-ethyl-5-phenyl-(1*H*,3*H*,5*H*)-pyrimidine-2,4,6-trione
Probably formed from: 3-(5-Ethyl-5-phenyl-(1*H*,3*H*,5*H*)-pyrimidine-2,4,6-trion-1-yl)-2-carbamoylpropanol

N-(2,3-Dihydroxypropyl)-2-(*o*-hydroxyphenoxy)ethylamine
Formed from: Carvedilol
N-(2,3-Dihydroxypropyl)-2-(*o*-methoxyphenoxy)ethylamine

N-(2,3-Dihydroxypropyl)-2-(*o*-methoxyphenoxy)ethylamine
Formed from: Carvedilol
→ N-(2,3-Dihydroxypropyl)-2-(*o*-hydroxyphenoxy)ethylamine
Probably in: Dog J443

4-(2,3-Dihydroxypropyl)-1,2-methylenedioxybenzene
See: Safrole glycol

p-(2,3-Dihydroxypropyl)phenol
Probably formed from: 4-Allylphenol-2,3-oxide

5,6-Dihydroxy-2-(N-propyl-N-(2-thienyl)ethylamino)tetralin
Formed from: 5-Hydroxy-2-(N-propyl-N-(2-thienyl)ethylamino)tetralin
→ 5-Hydroxy-2-(N-propyl-N-(2-thienyl)ethylamino)tetralin-6-yl-β-D-glucuronide
In: Rat F888
→ 6-Hydroxy-2-(N-propyl-N-(2-thienyl)ethylamino)tetralin-5-yl-β-D-glucuronide
Probably in: Rat F888

p-(1,2-Dihydroxy-2-propyl)toluene
Probably formed from: *p*-(1-Hydroxy-2-propyl)toluene

2,3-Dihydroxypropyl N-(8-trifluoromethyl-4-quinolinyl)anthranilate
See: Floctafenin

5,8-Dihydroxypsoralen
Probably formed from: 5-Hydroxy-8-methoxypsoralen
→ 5,8-Dioxopsoralen
Probably in: Rat B355, D966
→ 8-Methoxypsoralen
In: Ruta D28

3,9-Dihydroxypterocarpan
→ 3,6a,9-Trihydroxypterocarpan
In: Glycine F242

Dihydroxypulvinone
→ Dihydroxydiprenylpulvinone
In: Aspergillus A2311

1,6-Dihydroxypyrene
Probably formed from: 1-Hydroxypyrene
→ 1,6-Dihydroxypyrene-β-D-glucuronide
Probably in: Cunninghamella E181
→ 6-Hydroxypyren-1-yl sulphate
Probably in: Crinipellis H879
Marasmiellus H879
Marasmius H879
→ Pyrene-1,6-quinone
Probably in: Aspergillus H388
Crinipellis H879
Cunninghamella E181
Marasmiellus H879
Marasmius H879
Pseudomonas J530

1,8-Dihydroxypyrene
Probably formed from: 1-Hydroxypyrene
→ 1,8-Dihydroxypyrene-β-D-glucuronide
Probably in: Cunninghamella E181
→ 1-Hydroxypyren-8-yl sulphate
Probably in: Aspergillus H388
Marasmiellus H879
Marasmius H879
→ Pyrene-1,8-quinone
Probably in: Aspergillus H388
Crinipellis H879
Cunninghamella E181
Marasmiellus H879
Marasmius H879
Pseudomonas J530

1,6-Dihydroxypyrene-β-D-glucuronide
Probably formed from: 1,6-Dihydroxypyrene

1,8-Dihydroxypyrene-β-D-glucuronide
Probably formed from: 1,8-Dihydroxypyrene

5,6-Dihydroxy-N-(4-pyridinyl)-1-indolamine
Probably formed from: 5-Hydroxy-N-(4-pyridinyl)-1-indolamine

N′-(2,6-Dihydroxy-4-pyrimidinyl)sulphanilamide
Probably formed from: N′-(2-Hydroxy-6-methoxy-4-pyrimidinyl)sulphanilamide

1-(2,x-Dihydroxypyrrolidine)-N-(1-methyl-5-oxo-2-pyrrolidinylidene)-N′-phenylcarboximidamide
Probably formed from: 1-(2-Hydroxypyrrolidine)-N-(1-methyl-2-pyrrolidinylidene)-N′-phenylcarboximidamide

N-(p-((2,4-Dihydroxyquinazolin-6-yl)aminomethyl)benzoyl)diglutamylglutamic acid
Probably formed from: N-(*p*-((2,4-Dihydroxyquinazolin-6-yl)aminomethyl)benzoyl)glutamylglutamic acid

N-(p-((2,4-Dihydroxyquinazolin-6-yl)aminomethyl)benzoyl)glutamic acid
→ N-(*p*-((2,4-Dihydroxyquinazolin-6-yl)aminomethyl)benzoyl)glutamylglutamic acid
In: Pig F760

N-(p-((2,4-Dihydroxyquinazolin-6-yl)aminomethyl)benzoyl)glutamylglutamic acid
Formed from: N-(*p*-((2,4-Dihydroxyquinazolin-6-yl)aminomethyl)benzoyl)glutamic acid

→ N-(*p*-((2,4-Dihydroxyquinazolin-6-yl) aminomethyl)benzoyl)diglutamylglutamic acid
Probably in: Pig F760

N-(*p*-((2,4-Dihydroxyquinazolin-6-yl) formamidomethyl)benzoyl)diglutamylglutamic acid
Probably formed from: N-(*p*-((2,4-Dihydroxyquinazolin-6-yl)formamidomethyl) benzoyl)glutamylglutamic acid

N-(*p*-((2,4-Dihydroxyquinazolin-6-yl) formamidomethyl)benzoyl)glutamic acid
→ N-(*p*-((2,4-Dihydroxyquinazolin-6-yl) formamidomethyl)benzoyl) glutamylglutamic acid
In: Pig F760

N-(*p*-((2,4-Dihydroxyquinazolin-6-yl) formamidomethyl)benzoyl)glutamylglutamic acid
Formed from: N-(*p*-((2,4-Dihydroxyquinazolin-6-yl)formamidomethyl) benzoyl)glutamic acid
→ N-(*p*-((2,4-Dihydroxyquinazolin-6-yl) formamidomethyl)benzoyl) diglutamylglutamic acid
Probably in: Pig F760

N-(*p*-((2,4-Dihydroxyquinazolin-6-yl)methylamino)benzoyl) diglutamylglutamic acid
Probably formed from: N-(*p*-((2,4-Dihydroxyquinazolin-6-yl)methylamino) benzoyl)glutamylglutamic acid

N-(*p*-((2,4-Dihydroxyquinazolin-6-yl) methylamino)benzoyl)glutamic acid
→ N-(*p*-((2,4-Dihydroxyquinazolin-6-yl)methylamino)benzoyl) glutamylglutamic acid
In: Pig F760

N-(*p*-((2,4-Dihydroxyquinazolin-6-yl) methylamino)benzoyl)glutamylglutamic acid
Formed from: N-(*p*-((2,4-Dihydroxyquinazolin-6-yl)methylamino) benzoyl)glutamic acid
→ N-(*p*-((2,4-Dihydroxyquinazolin-6-yl)methylamino)benzoyl) diglutamylglutamic acid
Probably in: Pig F760

N-(*p*-(((2,4-Dihydroxyquinazolin-6-yl)-N-methylamino)methyl)benzoyl) diglutamylglutamic acid
Probably formed from: N-(*p*-(((2,4-Dihydroxyquinazolin-6-yl)-N-methylamino)methyl)benzoyl) glutamylglutamic acid

N-(*p*-(((2,4-Dihydroxyquinazolin-6-yl)-N-methylamino)methyl)benzoyl)glutamic acid
→ N-(*p*-(((2,4-Dihydroxyquinazolin-6-yl)-N-methylamino)methyl)benzoyl) glutamylglutamic acid
In: Pig F760

N-(*p*-(((2,4-Dihydroxyquinazolin-6-yl)-N-methylamino)methyl)benzoyl)glutamylglutamic acid
Formed from: N-(*p*-(((2,4-Dihydroxyquinazolin-6-yl)-N-methylamino)methyl)benzoyl)glutamic acid
→ N-(*p*-(((2,4-Dihydroxyquinazolin-6-yl)-N-methylamino)methyl)benzoyl) diglutamylglutamic acid
Probably in: Pig F760

N-(*p*-((2,4-Dihydroxyquinazolin-6-yl) methylformamido)benzoyl)diglutamylglutamic acid
Probably formed from: N-(*p*-((2,4-Dihydroxyquinazolin-6-yl)methylformamido) benzoyl)glutamylglutamic acid

N-(*p*-((2,4-Dihydroxyquinazolin-6-yl) methylformamido)benzoyl)glutamic acid
→ N-(*p*-((2,4-Dihydroxyquinazolin-6-yl)methylformamido)benzoyl) glutamylglutamic acid
In: Pig F760

N-(*p*-((2,4-Dihydroxyquinazolin-6-yl) methylformamido)benzoyl)glutamylglutamic acid
Formed from: N-(*p*-((2,4-Dihydroxyquinazolin-6-yl) methylformamido)benzoyl) glutamic acid
→ N-(*p*-((2,4-Dihydroxyquinazolin-6-yl)methylformamido)benzoyl) diglutamylglutamic acid
Probably in: Pig F760

N-(*p*-(((2,4-Dihydroxyquinazolin-6-yl)methyl)-N-methylamino)benzoyl) diglutamylglutamic acid
Probably formed from: N-(*p*-(((2,4-Dihydroxyquinazolin-6-yl)methyl)-N-methylamino)benzoyl)glutamylglutamic acid

N-(*p*-(((2,4-Dihydroxyquinazolin-6-yl)-N-methyl)methylamino)benzoyl)glutamic acid
→ N-(*p*-(((2,4-Dihydroxyquinazolin-6-yl)methyl)-N-methylamino)benzoyl) glutamylglutamic acid
In: Pig F760

N-(*p*-(((2,4-Dihydroxyquinazolin-6-yl)methyl)-N-methylamino)benzoyl) glutamylglutamic acid
Formed from: N-(*p*-(((2,4-Dihydroxyquinazolin-6-yl)methyl)-N-methylamino)benzoyl)glutamic acid
→ N-(*p*-(((2,4-Dihydroxyquinazolin-6-yl)methyl)-N-methylamino)benzoyl) diglutamylglutamic acid
Probably in: Pig F760

2′,3-Dihydroxyquinoline
Probably formed from: 2′-Hydroxyquinoline

2,4-Dihydroxyquinoline (4-hydroxy-2-quinolone)
→ 3-Dimethylallyl-2,4-dihydroxyquinoline
Probably in: Skimmia A1674

2,5-Dihydroxyquinoline
Formed from: 5-Hydroxyquinoline

2,6-Dihydroxyquinoline
Formed from: Cilostamide
2-Hydroxyquinoline
6-Hydroxyquinoline
Probably formed from: 6-Amino-2-quinolone
→ 2,6-Dihydroxyquinoline-β-D-glucuronide
Probably in: Man F318
→ 2,6-Dihydroxyquinoline-O-sulphate
Probably in: Man F318

2,7-Dihydroxyquinoline
Formed from: 7-Hydroxyquinoline

2,8-Dihydroxyquinoline (8-hydroxy-2-quinolone)
Formed from: 2-Hydroxyquinoline
8-Hydroxyquinoline
→ 8-Hydroxycoumarin
Probably in: Alcaligenes J32

4,6-Dihydroxyquinoline
Probably formed from: 6-Hydroxykynurenic acid

4,8-Dihydroxyquinoline-2-carboxylic acid
See: Xanthurenic acid

2,6-Dihydroxyquinoline-β-D-glucuronide
Probably formed from: 2,6-Dihydroxyquinoline

2,6-Dihydroxyquinoline-O-sulphate
Probably formed from: 2,6-Dihydroxyquinoline

2,5-Dihydroxyquinone
→ 1,2,4,5-Tetrahydroxybenzene
In: Sporotrichum B301

2,6-Dihydroxyquinone
Formed from: Hydroxyquinol

2,3-Dihydroxyquinoxaline
Probably formed from: 2-Hydroxyquinoxaline

7,8-Dihydroxyschizandrin
Formed from: Schizandrin
→ 7,8-Dihydroxy-2-demethylschizandrin
Probably in: Rat H87
→ 7,8-Dihydroxy-3-demethylschizandrin
Probably in: Rat H87

3′,4′-Dihydroxyscopolamine
Probably formed from: *p*-Hydroxyscopolamine

6,7-Dihydroxyshikonin
Formed from: 7-Hydroxyshikonin

3,4-Dihydroxy-*trans*-stilbene
Probably formed from: 4-Hydroxy-*trans*-stilbene
→ 3,4-Dihydroxybibenzyl
Probably in: Rabbit A1991
Rat A1991
→ 3-Hydroxy-4-methoxy-*trans*-stilbene
In: Rat A486
See also: A1991
→ 4-Hydroxy-3-methoxy-*trans*-stilbene
In: Rat A486
→ 3,4,4′-Trihydroxy-*trans*-stilbene
Probably in: Rabbit A1991
Rat A1991

3,5-Dihydroxystilbene
See: Pinosylvine

4,4′-Dihydroxy-*trans*-stilbene
Probably formed from: 4-Hydroxy-*trans*-stilbene
→ 4,4′-Dihydroxybibenzyl
Probably in: Rabbit A1991
Rat A1991
→ 3,4,4′-Trihydroxy-*trans*-stilbene
Probably in: Rat A1421

2,4-Dihydroxystyrene
Formed from: 2,4-Dihydroxycinnamic acid

3,4-Dihydroxystyrene
Formed from: Caffeic acid

1,2-Dihydroxysulofenur
Probably formed from: 1-Hydroxysulofenur

1,3-Dihydroxysulofenur
Probably formed from: 1-Hydroxysulofenur

3,4-Dihydroxytamoxifen
Formed from: *trans*-1-(*p*-Dimethylaminoethoxyphenyl)-1-(*p*-hydroxyphenyl)-2-phenylbut-1-ene
Droloxifene
→ 3,4-Dihydroxytamoxifen-β-D-glucuronide
Probably in: Dog A253
Mouse H924
Rat A253, H924
→ *trans*-1-(*p*-(2-Dimethylaminoethoxy) phenyl)-1-(4-hydroxy-3-methoxyphenyl)-2-phenylbut-1-ene
In: Mouse H924
Rat H924
See also: A253
→ 3-Hydroxy-4-methoxytamoxifen
In: Mouse H924
Rat H924

3,5-Dihydroxytamoxifen
Probably formed from: Droloxifene

3,4-Dihydroxytamoxifen-di-β-D-glucuronide
Probably formed from: 3,4-Dihydroxytamoxifen-β-D-glucuronide

3,4-Dihydroxytamoxifen-β-D-glucuronide
Formed from: 3,4-Dihydroxytamoxifen
→ 3,4-Dihydroxytamoxifen-di-β-D-glucuronide
Probably in: Mouse H924

6,7-Dihydroxytargetrin
Probably formed from: 6-Hydroxytargetrin
→ 7-Hydroxy-6-oxotargetrin
Probably in: Man J290
→ 4-(1-(5,6,7,8-Tetrahydro-6,7-dihydroxy-3,5,5,8,8-pentamethyl-2-naphthalenyl)-1-ethenyl) benzoyltaurine
Probably in: Man J290

(6α,3′-*p*-Dihydroxy)taxol
Formed from: (6α-Hydroxy)taxol

3′,4′-Dihydroxytenidap
Probably formed from: 3′-Hydroxytenidap
→ 3′-Hydroxy-4′-methoxytenidap
Probably in: Rat J92

3,8-Dihydroxy-1-tetralone
See: Vermelone

4,8-Dihydroxy-1-tetralone
Formed from: 5-Hydroxynaphthoquinone
Probably formed from: 2,4,8-Trihydroxy-1-tetralone

2′,5-Dihydroxy-3,4′,5′,7-tetramethoxyflavone
→ 2′,5-Dihydroxy-3,4′,5′,7-tetramethoxyflavone-2′-glucoside
In: Chrysosplenium C755, G262

3′,4′-Dihydroxy-5,6,7,8-tetramethoxyflavone
Probably formed from: 4′-Hydroxy-5,6,7,8-tetramethoxyflavone

4′,5-Dihydroxy-3,3′,6,7-tetramethoxyflavone
Formed from: 4′,5,6-Trihydroxy-3,3′,7-trimethoxyflavone

5,5′-Dihydroxy-3,4′,6,7-tetramethoxyflavone
→ 5,5′-Dihydroxy-3,4′,6,7-tetramethoxyflavone-5′-glucoside
In: Chrysosplenium C755

2′,5-Dihydroxy-3,4′,5′,7-tetramethoxyflavone-2′-glucoside
Formed from: 2′,5-Dihydroxy-3,4′,5′,7-tetramethoxyflavone

5,5′-Dihydroxy-3,4′,6,7-tetramethoxyflavone-5′-glucoside
Formed from: 5,5′-Dihydroxy-3,4′,6,7-tetramethoxyflavone

2′,5-Dihydroxy-3,4′,6,7′-tetramethoxyflavone-5′-O-glucoside
→ 5-Hydroxy-2′,3,4′,6,7′-pentamethoxyflavone-5′-O-glucoside
In: Chrysosplenium E351

2,4-Dihydroxytoluene
See: 4-Methylresorcinol

2,5-Dihydroxytoluene
See: Methylquinol
2,6-Dihydroxytoluene
See: 2-Methylresorcinol
3,5-Dihydroxytoluene
See: Orcinol
3′,4′-Dihydroxy-2-trifluoromethylbenzanilide
Probably formed from: 3′-Hydroxy-2-trifluoromethylbenzanilide
→ 4′-Hydroxy-3′-methoxy-2-trifluoromethylbenzanilide
Probably in: Cucumber D191
Oryza D191
Rat D190
3,4-Dihydroxy-5-((3,4,5-trihydroxybenzoyl)oxy) benzoic acid
See: *m*-Digallic acid
3,7-Dihydroxy-2-(3,4,5-trihydroxyphenyl)-1-benzopyrilium
See: Robinetinidin
3,5-Dihydroxy-3′,4′,7-trimethoxyflavone
Formed from: 3,5,7-Trihydroxy-3′,4′-dimethoxyflavone
3′,5-Dihydroxy-3,4′,7-trimethoxyflavone
Formed from: 3′,4′,5-Trihydroxy-3,7-dimethoxyflavone
→ 3′,5,6-Trihydroxy-3,4′,7-trimethoxyflavone
In: Chrysosplenium K658
4′,5-Dihydroxy-3,3′,7-trimethoxyflavone
Probably formed from: 3′,4′,5-Trihydroxy-3,7-dimethoxyflavone
4′,6-Dihydroxy-5,7,8-trimethoxyflavone
Probably formed from: 4′-Hydroxy-5,6,7,8-tetramethoxyflavone
→ 4′,6,7-Trihydroxy-5,8-dimethoxyflavone
Probably in: Rat K582
4′,7-Dihydroxy-5,6,8-trimethoxyflavone
Probably formed from: 4′-Hydroxy-5,6,7,8-tetramethoxyflavone
5,6-Dihydroxy-4′,7,8-trimethoxyflavone
Probably formed from: 6-Hydroxy-4′,5,7,8-tetramethoxyflavone
5,7-Dihydroxy-3,3′,4′-trimethoxyflavone
Formed from: 3,5,7-Trihydroxy-3′,4′-dimethoxyflavone
5,5′-Dihydroxy-3,4′,7-trimethoxyflavone-2′-O-glucoside
→ 5-Hydroxy-3,4′,5′,7-tetramethoxyflavone-2′-O-glucoside
In: Chrysosplenium E351
2,3-Dihydroxy-9,10,11-trimethoxytetrahydroprotoberberine
Formed from: Trimetoquinol
5-((4-(2,5-Dihydroxy-3,4,6-trimethylphenyl)-2-hydroxy-2-methylbutoxy)benzyl)-2,4-thiazolidinedione
→ 5-((4-(2,5-Dihydroxy-3,4,6-trimethylphenyl)-2-hydroxy-2-methylbutoxy)benzyl)-2,4-thiazolidinedione-β-D-glucuronide
In: Cynomolgus monkey K601
5-((4-(2,5-Dihydroxy-3,4,6-trimethylphenyl)-2-hydroxy-2-methylbutoxy)benzyl)-2,4-thiazolidinedione-β-D-glucuronide
Formed from: 5-((4-(2,5-Dihydroxy-3,4,6-trimethylphenyl)-2-hydroxy-2-methylbutoxy)benzyl)-2,4-thiazolidinedione
2,3-Dihydroxytrimipramine
Probably formed from: 2-Hydroxy-trimipramine
→ 10,11-Dehydro-2,3-dihydroxytrimipramine
Probably in: Rat G269
→ 2-Hydroxy-3-methoxytrimipramine
Probably in: Rat G269
2,10-Dihydroxytrimipramine
Probably formed from: 10-Hydroxytrimipramine
3,4-Dihydroxy-5-triprenylbenzoic acid
→ 4-Hydroxy-3-methoxy-5-triprenylbenzoic acid
In: Saccharomyces J30
β,5-Dihydroxytryptamine
Formed from: 5-Hydroxytryptamine
5,6-Dihydroxytryptamine
→ N-Acetyl-5,6-dihydroxytryptamine
In: Quail H685
→ 5-Hydroxy-6-methoxytryptamine
Probably in: Rat A93
β,5-Dihydroxy-L-tryptophan
Probably formed from: L-Tryptophan
3,4-Dihydroxytulobuterol
Probably formed from: 4-Hydroxytulobuterol

→ 4-Hydroxy-3-methoxytulobuterol
In: Dog C43
Rabbit C43
Rat C43

4,5-Dihydroxytulobuterol
Probably formed from: 4-Hydroxytulobuterol
→ 4-Hydroxy-5-methoxytulobuterol
In: Dog C43
Rat C43

Dihydroxyvolazocine
Probably formed from: Hydroxyvolazocine
→ Hydroxymethoxyvolazocine
Probably in: Rat A2134

3′,4′-Dihydroxywarfarin
Probably formed from: 4′-Hydroxywarfarin
→ 3′,4′-Dihydroxywarfarin-3′-(4-methoxy) glucoside
Probably in: Beauvaria J82

3′,4′-Dihydroxywarfarin-3′-(4-methoxy)glucoside
Probably formed from: 3′,4′-Dihydroxywarfarin

2,3-Dihydroxy-*p*-xylene
Formed from: 2,3-Dihydro-2,3-dihydroxy-*p*-xylene

6′,8′-Dihydroxyzearalenone
Probably formed from: 8′-Hydroxyzearalenone

7,8-Dihydroyangonin
→ 5,6-Dehydro-7,8-dihydro-*p*-hydroxyyangonin
In: Rat A3485

3,3′-Diindolylmethane
Formed from: Indole-3-methanol
→ 3-Formylindole
Probably in: Zea A727

2,2′-Diiodoazobenzene
Formed from: *o*-Iodoaniline

3,3′-Diiodoazobenzene
Formed from: 3,3′-Diiodoazoxybenzene
m-Iodoaniline

4,4′-Diiodoazobenzene
Formed from: *p*-Iodoaniline

3,3′-Diiodoazoxybenzene
→ 3,3′-Diiodoazobenzene
In: Saccharomyces H894

3,5-Diiodo-4-hydroxybenzonitrile
→ 2,6-Diiodoquinol
In: Flavobacterium G641

2,6-Diiodo-4-nitrophenol (disophenol)
→ 4-Amino-2,6-diiodophenol
In: Ascaris A3694
Moniezia A3694
Mouse A3694
Sheep A3694

3,5-Diiodophenyllactic acid
Formed from: 3,5-Diiodophenylpyruvic acid

3,5-Diiodophenylpyruvic acid
→ 3,5-Diiodophenyllactic acid
In: Dog A2917

2,6-Diiodoquinol
Formed from: 3,5-Diiodo-4-hydroxybenzonitrile
2,4,6-Triiodophenol

3,3′-Diiodothyroacetic acid
Formed from: 3,3′-Diiodo-L-thyronine
3,3′,5-Triiodothyroacetic acid
→ 3-Iodothyroacetic acid
In: Rat F218
→ 3′-Iodothyroacetic acid
In: Rat F218

3,3′-Diiodothyroacetic acid-4′-sulphate
Formed from: 3,3′,5-Triiodothyroacetic acid-4′-sulphate

3,3′-Diiodo-L-thyronine
Formed from: 3,3′-Diiodo-L-thyronine-O-sulphate
L-Thyroxine
3,3′,5-Triiodo-L-thyronine
3,3′,5′-Triiodo-L-thyronine
→ 3,3′-Diiodothyroacetic acid
Probably in: Rat F807
→ 3,3′-Diiodo-L-thyronine-β-D-glucuronide
In: Rat B771
→ 3,3′-Diiodo-L-thyronine-O-sulphate
In: Man K90, K620
Monkey B776
Rat B770, B771, C769, D531, E363, J495, K649
→ 3-Iodo-L-thyronine
In: Guinea pig D910
Rat C724, G337
→ 3′-Iodo-L-thyronine
Probably in: Rat C109

3,5-Diiodo-L-thyronine
Formed from: 3,3′,5-Triiodo-L-thyronine
→ 3,5-Diiodothyropyruvic acid
In: Rabbit K713
→ 3-Iodo-L-thyronine
In: Rat C108
See also: A3889
→ 3,3′,5-Triiodo-L-thyronine
In: Rat C163

3′,5′-Diiodo-L-thyronine
→ 3′-Iodo-L-thyronine
In: Rat B833, B868

3,3′-Diiodo-L-thyronine-β-D-glucuronide
Formed from: 3,3′-Diiodo-L-thyronine

3,3′-Diiodo-L-thyronine sulphamate
Formed from: 3,3′,5-Triiodo-L-thyronine sulphamate
3,3′,5′-Triiodo-L-thyronine sulphamate
→ 3-Iodo-L-thyronine sulphamate
In: Rat G337

3,3′-Diiodo-L-thyronine-O-sulphate
Formed from: 3,3′-Diiodo-L-thyronine
3,3′,5-Triiodo-L-thyronine-O-sulphate
3,3′,5′-Triiodo-L-thyronine-O-sulphate
→ 3,3′-Diiodo-L-thyronine
In: Rat C769
→ 3-Iodo-L-thyronine-O-sulphate
In: Dog H882
Man H882
Rat C724, G337

3,5-Diiodothyropyruvic acid
Formed from: 3,5-Diiodo-L-thyronine

3,5-Diiodo-L-tyrosine
Formed from: 3-Iodo-L-tyrosine
L-Thyroxine

1-(3,5-Diisobutyroxyphenyl)-2-(*tert*-butylamino) ethanol
See: Ibuterol

Diisobutyryl 2-amino-tetrahydro-6,7-dihydroxynaphthalene
→ 2-Amino-tetrahydro-6,7-dihydroxynaphthalene
In: Rat C251

Diisobutyryl apomorphine
→ Apomorphine
In: Rat A2013

***o*-(2-Diisopropylaminoethoxy)acetophenone**
Probably formed from: *o*-(2-Diisopropylaminoethoxy)-3-hydroxybutyrophenone
→ *o*-(2-Isopropylaminoethoxy)acetophenone
Probably in: Rat A3324

***o*-(2-Diisopropylaminoethoxy)butyrophenone**
See: Ketocaine

2′-(2-Diisopropylaminoethoxy)-3,5′-dihydroxybutyrophenone
Probably formed from: *o*-(2-Diisopropylaminoethoxy)-3-hydroxybutyrophenone
→ 2′-(2-Diisopropylaminoethoxy)-3,5′-dihydroxybutyrophenone-β-D-glucuronide
Probably in: Rat A3324

2′-(2-Diisopropylaminoethoxy)-3,5′-dihydroxybutyrophenone-β-D-glucuronide
Probably formed from: 2′-(2-Diisopropylaminoethoxy)-3,5′-dihydroxybutyrophenone

***o*-(2-Diisopropylaminoethoxy)-3-hydroxybutyrophenone**
Formed from: Ketocaine
→ 2′-(2-Diisopropylaminoethoxy)-3,5′-dihydroxybutyrophenone
Probably in: Rat A3324
→ *o*-(2-Diisopropylaminoethoxy) acetophenone
Probably in: Rat A3324

***o*-(2-Diisopropylaminoethoxy)phenylacetic acid**
Formed from: Ketocaine

1-(*o*-(2-Diisopropylaminoethoxy)phenyl)-1-butanol
→ *o*-(1-Hydroxybutyl)phenol
In: Rat A3075
→ 1-(*o*-(2-Isopropylaminoethoxy)phenyl)-1-butanol
In: Rat A3075

N,N-Diisopropylaniline
→ N-Isopropylaniline
In: Rabbit A3484

N,N-Diisopropyl-3-(2-hydroxy-5-methylphenyl)phenylpropanamine
See: Tolterodine

2,6-Diisopropylnaphthalene
→ 2,6-Di-(1-Carboxyethyl)naphthalene

In: Rat C11
→ 2,6-Di-(1-Hydroxy-1-methyl)naphthalene
In: Rat C11

2,6-Diisopropylphenol (propofol)
→ 2,6-Diisopropylphenyl-β-D-glucuronide
In: Dog G16
Man F293, G61, G524, G975, J183, J617, J783
Rabbit G16
→ 2,6-Diisopropylphenyl sulphate
In: Rabbit G16
→ 2,6-Diisopropylquinol
In: Dog G16. J773, K116
Man F293, G975, H956
Rabbit G16
→ 2-(2-Hydroxy-1-methylethyl)-6-isopropylphenol
In: Dog G16

2,6-Diisopropylphenyl-β-D-glucuronide
Formed from: 2,6-Diisopropylphenol
→ 2-(1-Hydroxy-2-propyl)-6-isopropylphenyl-β-D-glucuronide
Probably in: Dog G496
Rat G496
Does not yield 2,6-diisopropylphenol in rat or dog G496

2,6-Diisopropylphenyl sulphate
Formed from: 2,6-Diisopropylphenol

2,6-Diisopropylquinol
Formed from: 2,6-Diisopropylphenol
→ 2,6-Diisopropylquinol-1-β-D-glucuronide
Probably formed from: Dog G16
Man F293, G975
Rabbit G16
→ 2,6-Diisopropylquinol-4-β-D-glucuronide
Probably formed from: Dog G16
Man F293, G975
Rabbit G16
Rat G16
→ 2,6-Diisopropylquinol-4-sulphate
Probably formed from: Dog G16
Man F293, G975
Rabbit G16
Rat G16
→ 2-(2-Hydroxy-1-methylethyl)-6-isopropylquinol
In: Dog G16

2,6-Diisopropylquinol-1-β-D-glucuronide
Probably formed from: 2,6-Diisopropylquinol

2,6-Diisopropylquinol-4-β-D-glucuronide
Probably formed from: 2,6-Diisopropylquinol

2,6-Diisopropylquinol-4-sulphate
Probably formed from: 2,6-Diisopropylquinol

Diisovaleryl adrenaline
Probably formed from: Diisovaleryl adrenalone

Diisovaleryl adrenalone
→ Diisovaleryl adrenaline
In: Rabbit D268

Dilazep (1,4-bis(3-(3,4,5-trimethoxybenzoyl) propyl)piperazine)
→ 1-(3-Hydroxypropyl)-,4-(3-(3,4,5-trimethoxybenzoyl)propyl)piperazine
In: Rat A646
→ 3,4,5-Trimethoxybenzoic acid
In: Rat A646

Dilevalol (2-hydroxy-5-((1*R*)-1-hydroxy-2-(((1*R*)-1-methyl-3-phenylpropyl) amino)ethyl)benzamide)
→ 7-Carboxamido-5,6-dihydroxy-1-(4-phenylbut-2-yl)indole
In: Man H55
Monkey H55
→ Dilevalol-1′-β-D-glucuronide
In: Dog H55
Man H55
Monkey H55
Rat H55
→ Dilevalol-2-β-D-glucuronide
In: Rat H55
→ 3-Hydroxydilevalol
In: Dog H55
Man H55
Monkey H55
Rat H55
→ 4″-Hydroxydilevalol
In: Rat H55

Dilevalol-1′-β-D-glucuronide
Formed from: Dilevalol

Dilevalol-2-β-D-glucuronide
Formed from: Dilevalol

Diltiazem (*cis*-3-acetyloxy-5-(2-dimethylaminoethyl)-2,3-dihydro-2-(4-

methoxyphenyl)-1,5-benzothiazepin-4(5*H*)-one)
→ 3-Acetoxy-N-formylmethyl-2,3-dihydro-2-(*p*-methoxyphenyl)-1,5-benzothiazepin-4(5*H*)-one
In: Rat F835
→ Deacetyldiltiazem
In: Dog F607
Man D126, E654, F607
Rabbit F569, F607, G314, J18
Rat F607, G849, H377, J807
→ N-Demethyldiltiazem
In: Dog F607
Glycine K359
Man D126, E654, F607
Rabbit F569, F607, G314, J18
Rat F607, G859, H377, J807
→ O-Demethyldiltiazem
In: Man D126, E654
Rat H377

Dimetacrine (N,N,9,9-tetramethyl-10(9*H*)-acridinepropanamine)
→ Dimetacrine-N-oxide
In: Rat A3819
→ 2-Hydroxydimetacrine
In: Rat A3819
→ Nordimetacrine
In: Rat A3819

Dimetacrine-N-oxide
Formed from: Dimetacrine

Dimethophrine
See: 5-Methoxymetanephrine

2′,4′-Dimethoxyacetophenone
Formed from: 2′,4,4′-Trimethoxychalcone

2′,6′-Dimethoxyacetophenone
→ 3′-Hydroxy-2′,6′-dimethoxyacetophenone
In: Rat A357
→ 2′-Hydroxy-6′-methoxyacetophenone
In: Rat A357

3′,4′-Dimethoxyacetophenone
Formed from: 4′-Hydroxy-3′-methoxyacetophenone
Probably formed from: 3,4-Dimethoxyphenylhydracrylic acid
→ 4′-Hydroxy-3′-methoxyacetophenone
Probably in: Rat A1765

3,4-Dimethoxyallylbenzene (eugenol methyl ether)
→ 1-(3,4-Dihydroxyphenyl)propane
In: Rat A1765
→ 3,4-Dimethoxyallylbenzene-2,3-oxide
In: Rat A1765
→ 3,4-Dimethoxyphenylacetaldehyde
In: Rat A1765
→ 3-(3,4-Dimethoxyphenyl)propan-1-ol
In: Rat A1765
→ 1-(3,4-Dimethoxyphenyl)-2-propen-1-ol
In: Rat A1765
→ Eugenol
In: Rat A1765
See also: B528
→ 2-Hydroxy-4,5-dimethoxyallylbenzene
In: Rat A1765
→ 3-Hydroxy-4-methoxyallylbenzene
In: Rat A1765
→ Veratraldehyde
In: Rat A1765

3,4-Dimethoxyallylbenzene-2,3-oxide
Formed from: 3,4-Dimethoxyallylbenzene
→ 3-(3,4-Dimethoxyphenyl)propane-1,2-diol
Probably in: Rat B258

3,4-Dimethoxyamphetamine
Formed from: *p*-Chloroamphetamine
→ 4-Hydroxy-3-methoxyamphetamine
In: Dog A3697
Monkey A3697

3,4-Dimethoxyaniline
→ 4-Amino-2-methoxyphenol
Probably in: Sporomusa J199

10,11-Dimethoxyaporphine
→ Apocodeine
In: Cunninghamella A2769
Streptomyces A2769
→ Isoapocodeine
In: Cunninghamella A2769
Microsporum A2769
Mucor A2769
Penicillium A2769
Streptomyces A2769

4,4′-Dimethoxyazobenzene
Formed from: 4,4′-Dimethoxyazoxybenzene

4,4′-Dimethoxyazoxybenzene
→ 4,4′-Dimethoxyazobenzene

In: Saccharomyces H894
2,3-Dimethoxybenzaldehyde
Formed from: 2,3-Dimethoxybenzyl alcohol
Probably formed from: 2,3-Dimethoxybenzylamine
→ 2,3-Dimethoxybenzoic acid
In: Acetobacterium F655
See also: G865
2,4-Dimethoxybenzaldehyde
Formed from: 2,4-Dimethoxybenzyl alcohol
2,5-Dimethoxybenzaldehyde
Probably formed from: 2,5-Dimethoxybenzyl alcohol
2,5-Dimethoxybenzylamine
→ 2,5-Dimethoxybenzoic acid
Probably in: Rat G865
→ 2,5-Dimethoxybenzyl alcohol
Probably in: Nocardia F266
3,4-Dimethoxybenzaldehyde
See: Veratraldehyde
3,5-Dimethoxybenzaldehyde
Formed from: 3,5-Dimethoxybenzyl alcohol
Probably formed from: 3.5-Dimethoxybenzylamine
→ 3,5-Dimethoxybenzoic acid
In: Phanerochaete F866
See also: G865
1,2-Dimethoxybenzene
See: Veratrole
1,3-Dimethoxybenzene (*m*-methoxyanisole)
→ Resorcinol
In: Clostridium G367
Rat J201
Sporomusa G931
1,4-Dimethoxybenzene
→ *p*-Methoxyphenol
In: Rat J201
Sporomusa J199
→ Quinol
In: Clostridium G367
3,3′-Dimethoxybenzidine (dianisidine)
Formed from: Direct Black 114
Direct Blue 1
Direct Blue 8
Direct Blue 10
Direct Blue 15
Direct Violet 32
→ N-Acetyl-3,3′-dimethoxybenzidine
In: Rat C623
See also: C541
→ 3-Hydroxy-3′-methoxybenzidine
In: Rat C623
2,3-Dimethoxybenzoic acid
Probably formed from: 2,3-Dimethoxybenzaldehyde
→ Pyrocatechuic acid
In: Acetobacterium F655
2,5-Dimethoxybenzoic acid
Probably formed from: 2,5-Dimethoxybenzaldehyde
Gentisic acid
→ 2,5-Dimethoxybenzaldehyde
Probably in: Nocardia F266
3,4-Dimethoxybenzoic acid
See: Veratric acid
3,5-Dimethoxybenzoic acid
Formed from: 3,5-Dimethoxybenzaldehyde
3,5-Dimethoxycinnamic acid
3-Hydroxy-5-methoxybenzoic acid
Probably formed from: Gentisic acid
→ 3-Hydroxy-5-methoxybenzoic acid
In: Pseudomonas D16
3,4-Dimethoxybenzoic acid 4-(ethyl(2-(4-methoxyphenyl)-1-methylethyl)amino)butyl ester
See: Mebeverine
3,3′-Dimethoxybenzoin
Formed from: *m*-Methoxybenzaldehyde
4,4′-Dimethoxybenzoin
See: Anisoin
6-(4-(3,4-Dimethoxybenzoyl)-2,3-dihydroxy-1-piperazinyl)-3,4-dihydro-2(1*H*)-quinolone
Formed from: 6-(4-(3,4-Dimethoxybenzoyl)-1-piperazinyl)-3,4-dihydro-2(1*H*)-quinolone
→ N-(3,4-Dimethoxybenzoyl)-N′-(2-quinolon-6-yl)ethylenediamine
Probably in: Dog F318
Man F318
Monkey F318
Mouse F318
Rat F318
6-(4-(3,4-Dimethoxybenzoyl)-1-piperazinyl)-3,4-dihydro-2(1*H*)-quinolone
→ 6-Amino-2-quinolone

In: Monkey F318
Mouse F318
Rat F318
→ 6-(4-(3,4-Dimethoxybenzoyl)-2,3-dihydroxy-1-piperazinyl)-3,4-dihydro-2(1*H*)-quinolone
In: Dog F318
Man F318
Monkey F318
Mouse F318
Rat F318
→ 6-(4-(4-Hydroxy-3-methoxybenzoyl)-1-piperazinyl)-3,4-dihydro-2(1*H*)-quinolone
In: Dog F318
Man F318
Monkey F318
Mouse F318
Rat F318
→ 6-(1-Piperazinyl)-2-quinolone
In: Dog F318
Man F318
Monkey F318
Mouse F318
Rat F318

N-(3,4-Dimethoxybenzoyl)-N′-(2-quinolon-6-yl)ethylenediamine
Probably formed from: 6-(4-(3,4-Dimethoxybenzoyl)-2,3-dihydroxy-1-piperazinyl)-3,4-dihydro-2(1*H*)-quinolone

1,6-Dimethoxybenzpyrene
→ 1-Hydroxy-6-methoxybenzpyrene
In: Rat B477

3,6-Dimethoxybenzpyrene
→ 3-Hydroxy-6-methoxybenzpyrene
In: Rat B477

6,12-Dimethoxybenzpyrene
→ 3-Hydroxy-6,12-dimethoxybenzpyrene
In: Rat B477

2,3-Dimethoxybenzyl alcohol
→ 2,3-Dimethoxybenzaldehyde
In: Pleurotus F867, G668
→ 2,3-Dimethoxybenzyl-β-D-glucoside
In: Verbesina K98

2,4-Dimethoxybenzyl alcohol
→ 2,4-Dimethoxybenzaldehyde
In: Pleurotus F867, G668
→ 2,4-Dimethoxybenzyl-β-D-glucoside
In: Verbesina K98

2,5-Dimethoxybenzyl alcohol
Probably formed from: 2,5-Dimethoxybenzaldehyde
→ 2,5-Dimethoxybenzyl-β-D-glucoside
In: Verbesina K98

3,4-Dimethoxybenzyl alcohol
See: Veratryl alcohol

3,5-Dimethoxybenzyl alcohol
Formed from: 3,5-Dimethoxybenzaldehyde
→ 3,5-Dimethoxybenzaldehyde
In: Arion G804
Pleurotus F867

2,3-Dimethoxybenzylamine
→ 2,3-Dimethoxybenzaldehyde
Probably in: Rat G865

2,5-Dimethoxybenzylamine
→ 2,5-Dimethoxybenzaldehyde
Probably in: Rat G865

3,4-Dimethoxybenzylamine
See: Veratrylamine

3,5-Dimethoxybenzylamine
→ 3,5-Dimethoxybenzaldehyde
Probably in: Rat G865

6-(3-(3,4-Dimethoxybenzylamino)-2-hydroxypropoxy)-8-hydroxy-2(1*H*)-quinolinone
Formed from: Toborinone
→ 6-(3-(3,4-Dimethoxybenzylamino)-2-hydroxypropoxy)-8-hydroxy-2(1*H*)-quinolinone-β-D-glucuronide
Probably in: Dog J154
Rat J154

6-(3-(3,4-Dimethoxybenzylamino)-2-hydroxypropoxy)-8-hydroxy-2(1*H*)-quinolinone-β-D-glucuronide
Probably formed from: 6-(3-(3,4-Dimethoxybenzylamino)-2-hydroxypropoxy)-8-hydroxy-2(1*H*)-quinolinone

6-(3-(3,4-Dimethoxybenzylamino)-2-hydroxypropoxy)-3-(S-glutathionyl)-3,4-dihydro-4-hydroxy-2(1*H*)-quinolinone
Formed from: Toborinone
→ 3-(S-(L-Cysteinyl))-6-(3-(3,4-dimethoxybenzylamino)-2-hydroxypropoxy)-4-hydroxy-2(1*H*)-quinolinone
Probably in: Dog J154

Rat J154

6-(3-(3,4-Dimethoxybenzylamino)-2-hydroxypropoxy)-2(1*H*)-quinolinone
See: Toborinone

4-(3,4-Dimethoxybenzyl)-6,7-dimethoxyisocoumarin
Formed from: Papaverine

4-(3,4-Dimethoxybenzyl)-6,7-dimethoxyisoquinoline
See: Isopapaverine

2,3-Dimethoxybenzyl-β-D-glucoside
Formed from: 2,3-Dimethoxybenzyl alcohol

2,4-Dimethoxybenzyl-β-D-glucoside
Formed from: 2,4-Dimethoxybenzyl alcohol

2,5-Dimethoxybenzyl-β-D-glucoside
Formed from: 2,5-Dimethoxybenzyl alcohol

3,4-Dimethoxybenzyl-β-D-glucoside
Formed from: Veratryl alcohol

5-(3,4-Dimethoxybenzyl)hydantoin (both stereoisomers)
Formed from: N-Carbamyl-3,4-dimethoxy-D-phenylalanine
N-Carbamyl-3,4-dimethoxy-L-phenylalanine
→ N-Carbamyl-3,4-dimethoxy-D-phenylalanine
In: Flavobacterium F938
→ N-Carbamyl-3,4-dimethoxy-L-phenylalanine
In: Flavobacterium F938
→ 3,4-Dimethoxy-D-phenylalanine
In: Pseudomonas E450
→ 3,4-Dimethoxy-L-phenylalanine
In: Flavobacterium F940

3-(3,4-Dimethoxybenzyl)-2-hydroxy-5,6-dimethoxy-1-indenone
Formed from: Papaverine

(*E*)-3-(2,4-Dimethoxybenzylidene)-3,4,5,6-tetrahydro-2,3′-bipyridine
→ (*E*)-3,4,5,6-Tetrahydro-3-(2-hydroxy-4-methoxybenzylidene)-2,3′-bipyridine
In: Rat J728
→ (*E*)-3,4,5,6-Tetrahydro-3-(4-hydroxy-2-methoxybenzylidene)-2,3′-bipyridine
In: Rat J728

2,3-Dimethoxybiphenyl
→ 2,3-Dihydro-2,3-dihydroxy-2′,3′-dimethoxybiphenyl
In: Escherichia H951

6,6′-((3,3′-Dimethoxy(1,1′-biphenyl)-4,4′-diyl) bis(azo)bis(4-amino-5-hydroxy-1,3-naphthalenedisulphonic acid)
See: Direct Blue 1

3,3′-((3,3′-Dimethoxy(1,1′-biphenyl)-4,4′-diyl) bis(azo)bis(4,5-dihydroxy-2,7-naphthalenedisulphonic acid)
See: Direct Blue 10

3,3′-((3,3′-Dimethoxy(1,1′-biphenyl)-4,4′-diyl) bis(azo)bis(4-hydroxy-1-naphthalenesulphonic acid)
See: Direct Blue 8

3,3′-Dimethoxy-4,4′-biphenylquinone
Probably formed from: 4,4′-Dihydroxy-3,3′-dimethoxybiphenyl

3,5-Dimethoxycarbonyl-2,6-dimethyl-4-(*o*-difluoromethoxyphenyl)-1,4-dihydropyridine
See: Ryosidine

3,5-Di-(methoxycarbonyl)-2,6-dimethyl-4-(2-nitrophenyl)pyridine
Formed from: Nifedipine
→ 3-Carboxy-5-methoxycarbonyl-2,6-dimethyl-4-(2-nitrophenyl)pyridine
Probably in: Man F919

4,5-Dimethoxycatechol
Formed from: 4,5-Dimethoxy-*o*-quinone

3,7-Dimethoxychlorpromazine
Probably formed from: 3,7-Dihydroxychlorpromazine

7,8-Dimethoxychlorpromazine
Probably formed from: 7-Hydroxy-8-methoxychlorpromazine

3,4-Dimethoxycinnamaldehyde
Formed from: 3,4-Dimethoxycinnamic acid
3,4-Dimethoxycinnamyl alcohol
→ 3,4-Dimethoxycinnamyl alcohol
In: Coriolus D335
Glycine A2325
See also: A2744

2,4-Dimethoxycinnamic acid
→ 2,4-Dimethoxycinnamoyl-β-D-glucose

In: Verbesina K98

2,5-Dimethoxycinnamic acid
→ 2,5-Dimethoxycinnamoyl-β-D-glucose
In: Verbesina K98

2,6-Dimethoxycinnamic acid
→ 2,6-Dimethoxycinnamoyl-β-D-glucose
In: Verbesina K98

3,4-Dimethoxycinnamic acid
Probably formed from: 3-(3,4-Dimethoxyphenyl)-2-propen-1-ol
Ferulic acid
→ *trans*-ω-Bromo-3,4-dimethoxystyrene
In: Glycine K439
→ 3,4-Dimethoxycinnamaldehyde
In: Neurospora A750
See also: A2744, B758
→ 3,4-Dimethoxycinnamoyl CoA
In: Glycine A2148
→ 3,4-Dimethoxycinnamoyl-β-D-glucose
In: Verbesina K98
→ 3,4-Dimethoxycinnamoylglycine
In: Rat A1765
→ 3,4-Dimethoxycinnamyl alcohol
In: Coriolus G821
→ 3,4-Dimethoxyphenylhydracrylic acid
In: Rat A1765
See also: A2744, B758
→ 3-(3,4-Dimethoxyphenyl)propionic acid
In: Phanerochaete B758
Rat A1765
→ 3,4-Dimethoxystyrene
In: Hansenula F390
Saccharomyces A2157
→ Ferulic acid
In: Rat A1765
See also: A261, K896
→ Veratric acid
In: Vanilla F824

3,5-Dimethoxycinnamic acid
→ 3,5-Dimethoxybenzoic acid
In: Aspergillus G821
→ 3,5-Dimethoxycinnamoyl-β-D-glucose
In: Raphanus B440
Verbesina K98
→ 3-(3,5-Dimethoxyphenyl)propionic acid
In: Microorganism A744
→ 3-Hydroxy-5-methoxycinnamic acid
In: Microorganism A261

3,4-Dimethoxycinnamoyl CoA
Formed from: 3,4-Dimethoxycinnamic acid

2,4-Dimethoxycinnamoyl-β-D-glucose
Formed from: 2,4-Dimethoxycinnamic acid

2,5-Dimethoxycinnamoyl-β-D-glucose
Formed from: 2,5-Dimethoxycinnamic acid

2,6-Dimethoxycinnamoyl-β-D-glucose
Formed from: 2,6-Dimethoxycinnamic acid

3,4-Dimethoxycinnamoyl-β-D-glucose
Formed from: 3,4-Dimethoxycinnamic acid

3,5-Dimethoxycinnamoyl-β-D-glucose
Formed from: 3,5-Dimethoxycinnamic acid

3,4-Dimethoxycinnamoylglycine
Formed from: 3,4-Dimethoxycinnamic acid

3,4-Dimethoxycinnamyl alcohol
Formed from: 3,4-Dimethoxycinnamaldehyde
3,4-Dimethoxycinnamic acid
→ 3,4-Dimethoxycinnamaldehyde
In: Glycine A2325
→ 3-(3,4-Dimethoxyphenyl)propan-1-ol
In: Coriolus D335
See also: B758, G821
→ Veratrylglycerol
In: Coriolus D335

5,7-Dimethoxycoumarin
Formed from: 4′-Hydroxy-5,7-dimethoxyflavanone

6,7-Dimethoxycoumarin (scoparone)
→ Isoscopoletin
In: Hamster F744
Monkey F744
Mouse A3623
Rat D895, F52, F744, G566
→ Scopoletin
In: Hamster F744
Monkey F744
Mouse A3623
Rat D895, F52, F744, G566

4-(2,6-Dimethoxy-4-(2,6-dimethoxy-4-(pentachlorophenoxy)phenoxy)phenoxy)-2,6-dimethoxyphenol
Formed from: 4-Hydroxy-3,5-dimethoxybenzoic acid
Pentachlorophenol

4-(2,6-Dimethoxy-4-(2,6-dimethoxy-4-(2,6-dimethoxy-4-(pentachlorophenoxy)phenoxy) phenoxy)phenoxy)-2,6-dimethoxyphenol
Formed from: 4-Hydroxy-3,5-dimethoxybenzoic acid
Pentachlorophenol

6,7-Dimethoxy-2,2-dimethyl-2*H*-benzo[*b*]pyran
See: Precocene II

3,4-Dimethoxy-N,N-dimethylphenethylamine
Formed from: 3,4-Dimethoxy-N-methylphenethylamine

5,6-Dimethoxyflavone
Probably formed from: 5-Hydroxy-6-methoxyflavone

7,8-Dimethoxyflavone
Probably formed from: 7-Hydroxy-8-methoxyflavone

3,5-Dimethoxy-4-hydroxybenzoic acid
→ 2,6-Dimethoxyphenol
In: Microorganism G632

3,5-Dimethoxy-4-hydroxycinnamaldehyde
See: Sinapic acid

3,5-Dimethoxy-4-hydroxycinnamic acid
See: Sinapaldehyde

3,5-Dimethoxy-4-hydroxycinnamyl alcohol
See: Sinapyl alcohol

(6,7-Dimethoxy-1-isoquinolinyl)(3,4-dimethoxyphenyl)methanone
See: Papaveraldine

2,5-Dimethoxymandelic acid
Formed from: 2-(2,5-Dimethoxyphenyl)-N,N-dimethyl-2-oxoethylamine

3,4-Dimethoxymandelic acid
Formed from: 2-(3,4-Dimethoxyphenyl)-N,N-dimethyl-2-oxoethylamine
3,4-Dimethoxyphenylethanolamine
3,4-Dimethoxyphenylglyoxal

3,4-Dimethoxymandelonitrile
Formed from: Veratraldehyde
→ O-Acetyl-3,4-dimethoxymandelonitrile
In: Pseudomonas H303

2,5-Dimethoxy-4-methylamphetamine (2-amino-1-(2,5-dimethoxy-4-methylphenyl)propane)
→ 2-Amino-1-(4-hydroxymethyl-2,5-dimethoxyphenyl)propane
Probably in: Rabbit A1401
→ 1-(2,5-Dimethoxy-4-methylphenyl)-2-hydroxylaminopropane
In: Rabbit A2251
→ 2,5-Dimethoxy-4-methylpropan-2-one
In: Guinea pig A3228
Rabbit A3228
→ 4-Hydroxymethyl-2,5-dimethoxyamphetamine
In: Guinea pig A3228
Rabbit A3228
→ 2-Hydroxy-5-methoxy-4-methylamphetamine
In: Rabbit A2427, A2622, A2829
→ 5-Hydroxy-2-methoxy-4-methylamphetamine
In: Rabbit A2427, A2622, A2829

2,5-Dimethoxy-4-methylbenzoic acid
Probably formed from: 2,5-Dimethoxy-4-methylpropan-2-one

6,8-Dimethoxy-7-((3-methyl-2-butenyl)oxy)-2*H*-1-benzopyran-2-one
See: Puberulin

1-(2-(5,6-Dimethoxy-2-methyl-3-indolyl)ethyl)-4-phenylpiperazine (oxypertine)
→ 1-Phenylpiperazine
In: Rat D308

6,7-Dimethoxy-2-methylisoquinoline
Formed from: 2-Methylpapaverinium

6,7-Dimethoxy-2-methylisoquinolone
Probably formed from: 3,4-Dihydro-6,7-dimethoxy-2-methylisoquinolone

7,8-Dimethoxy-4-methyl-3-(γ-morpholino-β-(3′,4′,5′-trimethoxybenzoyl)propyl)coumarin
→ 3-(β-Hydroxy-γ-morpholinopropyl)-7,8-dimethoxy-4-methylcoumarin
In: Rabbit A275

3,4-Dimethoxy-N-methylphenethylamine
Formed from: 3,4-Dimethoxyphenethylamine
5-(N-(3,4-Dimethoxyphenethyl) methylamino)-2-(3,4-dimethoxyphenyl)-2-isopropylvaleronitrile
3-Hydroxy-4-methoxy-N-methylphenethylamine
4-Hydroxy-3-methoxy-N-methylphenethylamine

N-Methyl-*p*-tyramine
Normetanephrine
→ 3,4-Dimethoxy-N,N-dimethylphenethylamine
In: Rat A80
→ 3,4-Dimethoxyphenylacetaldehyde
Probably in: Rat G865

N,N′-Dimethoxymethylphenobarbital
→ N-Methoxymethylphenobarbital
In: Rat A805, A1233

3,4-Dimethoxy-N-methylphenylethanolamine
(normacromerine)
Formed from: N-Methyl-*p*-tyramine
D-Normetanephrine

1-(2,5-Dimethoxy-4-methylphenyl)-2-hydroxylaminopropane
Formed from: 2,5-Dimethoxy-4-methylamphetamine

4,5-Dimethoxy-2-methylphenyl methyl sulphoxide
See: Tolmesoxide

2,5-Dimethoxy-4-methylphenylpropan-2-ol
Probably formed from: 2,5-Dimethoxy-4-methylphenylpropan-2-one

2,5-Dimethoxy-4-methylphenylpropan-2-one
Formed from: 2,5-Dimethoxy-4-methylamphetamine
→ 2,5-Dimethoxy-4-methylbenzoic acid
Probably in: Guinea pig A3228
Rabbit A3228
→ 2,5-Dimethoxy-4-methylphenylpropan-2-ol
Probably in: Guinea pig A3228
Rabbit A3228

5,6-Dimethoxy-3-methyl-2-prenylquinol
Formed from: Ubiquinone-5

2,3-Dimethoxy-5-methyl-*p*-quinone
(ubiquinone 0)
→ Ubiquinone 30
In: Man A774
→ Ubiquinone 45
In: Man A774
→ Ubiquinone 50
In: Man A774

2,5-Dimethoxy-4-methylsulphinylamphetamine
Formed from: 2,5-Dimethoxy-4-methylthioamphetamine

2,5-Dimethoxy-4-methylthioamphetamine
→ 2,5-Dimethoxy-4-methylsulphinylamphetamine
In: Cunninghamella G516

2,3-Dimethoxy-1,4-naphthoquinone
→ 1,4-Dihydroxy-2,3-dimethoxynaphthalene
In: Guinea pig H790
Rat H790

4,5-Dimethoxy-2-nitrobenzaldehyde
Formed from: Veratryl alcohol

1,2-Dimethoxy-4-nitrobenzene
See: 4-Nitroveratrole

3,4-Dimethoxy-1-nitrobenzene
See: 4-Nitroveratrole

2,4-Dimethoxyoestrone
Formed from: 4-Hydroxy-2-methoxyoestrone
2-Hydroxy-4-methoxyoestrone

3,4-Dimethoxy-N-oxalylphenethylamine
Formed from: Ecabapide
→ 4-Hydroxy-3-methoxy-N-oxalylphenethylamine
Probably in: Rat H737

2,6-Dimethoxy-4-(pentachlorophenoxy)phenol
Formed from: 4-Hydroxy-3,5-dimethoxybenzoic acid
Pentachlorophenol

4-(2,6-Dimethoxy-4-(pentachlorophenoxy)phenoxy)-2,6-dimethoxyphenol
Formed from: 4-Hydroxy-3,5-dimethoxybenzoic acid
Pentachlorophenol

2,3-Dimethoxyphenethylamine
→ 2,3-Dimethoxyphenylacetaldehyde
Probably in: Rat G865

2,4-Dimethoxyphenethylamine
→ 2,4-Dimethoxyphenylacetaldehyde
Probably in: Rat G865

2,5-Dimethoxyphenethylamine
→ 2,5-Dimethoxyphenylacetaldehyde
Probably in: Rat G865

3,4-Dimethoxyphenethylamine
(homoveratrylamine)
Formed from: 5-((3,4-Dimethoxyphenethyl)amino)-2-isopropyl-2-(3,4,5-trimethoxyphenyl)valeronitrile
Probably formed from: Norgallopamil

→ (−)-Cryptostyline I
In: Cryptostylis A1515
→ 3,4-Dimethoxy-N-methylphenethylamine
In: Phalaris A3891
→ 3,4-Dimethoxyphenylacetaldehyde
Probably in: Rat G865

3,5-Dimethoxyphenethylamine
→ 3,5-Dimethoxyphenylacetaldehyde
Probably in: Rat G865

5-(N-(3,4-Dimethoxyphenethyl)amino)-2-(3,4-dimethoxyphenyl)-2-isopropylvaleronitrile
Formed from: 5-(N-(3,4-Dimethoxyphenethyl)methylamino)-2-(3,4-dimethoxyphenyl)-2-isopropylvaleronitrile
Probably formed from: 5-(N-(3,4-Dimethoxyphenethyl)-N-formylamino)-2-(3,4-dimethoxyphenyl)-2-isopropylvaleronitrile
→ 5-(N-(3,4-Dimethoxyphenethyl)amino)-2-(3,4-dimethoxyphenyl)-2-isopropylvaleronitrile-β-D-glucuronide
Probably in: Rat F474

5-(N-(3,4-Dimethoxyphenethyl)amino)-2-(3,4-dimethoxyphenyl)-2-isopropylvaleronitrile-β-D-glucuronide
Probably formed from: 5-(N-(3,4-Dimethoxyphenethyl)amino)-2-(3,4-dimethoxyphenyl)-2-isopropylvaleronitrile

5-((3,4-Dimethoxyphenethyl)amino)-2-(3-hydroxy-4,5-dimethoxyphenyl)-2-isopropylvaleronitrile
Formed from: 5-((3,4-Dimethoxyphenethyl)methylamino)-2-(3-hydroxy-4,5-dimethoxyphenyl)-2-isopropylvaleronitrile

(−)-2-(3,4-Dimethoxyphenethyl)amino)-1-(*p*-hydroxyphenyl)ethanol
See: Denopamine

5-((3,4-Dimethoxyphenethyl)amino)-2-isopropyl-2-(3,4,5-trimethoxyphenyl) valeronitrile
→ 3,4-Dimethoxyphenethylamine
In: Man F524
Rat F524
→ 2-Isopropyl-2-(3,4,5-trimethoxyphenyl) valeronitrile
In: Man F524
Rat F524

5-(N-(3,4-Dimethoxyphenethyl)-N-formylamino)-2-(3,4-dimethoxyphenyl)-2-isopropylvaleronitrile
Formed from: 5-(N-(3,4-Dimethoxyphenethyl)methylamino)-2-(3,4-dimethoxyphenyl)-2-isopropylvaleronitrile
→ 5-(N-(3,4-Dimethoxyphenethyl)amino)-2-(3,4-dimethoxyphenyl)-2-isopropylvaleronitrile
Probably in: Man A3674
→ 5-(N-(3,4-Dimethoxyphenethyl)-N-formylamino)-2-(4-hydroxy-3-methoxyphenyl)-2-isopropylvaleronitrile
Probably in: Man A3674

5-(N-(3,4-Dimethoxyphenethyl)-N-formylamino)-2-(4-hydroxy-3-methoxyphenyl)-2-isopropylvaleronitrile
Probably formed from: 5-(N-(3,4-Dimethoxyphenethyl)-N-formylamino)-2-(3,4-dimethoxyphenyl)-2-isopropylvaleronitrile

N-(3,4-Dimethoxyphenethyl)-3-hydroxy-4-methoxyphenylethanolamine
Probably formed from: N-(3,4-Dimethoxyphenethyl)noradrenaline
→ N-(3,4-Dimethoxyphenethyl)noradrenaline
Probably in: Rat D347
→ N-(3-Hydroxy-4-methoxyphenethyl)-3-hydroxy-4-methoxyphenylethanolamine
Probably in: Man D754
→ N-(4-Hydroxy-3-methoxyphenethyl)-3-hydroxy-4-methoxyphenylethanolamine
Probably in: Man D754

5-(N-(3,4-Dimethoxyphenethyl)methylamino)-2-(3,4-dimethoxyphenyl)-2-isopropylvaleronitrile
(verapamil)
→ 4-Cyano-4-(3,4-dimethoxyphenyl)-N,5-dimethylhexylamine
In: Man E943
Rat E943
→ 3,4-Dimethoxy-N-methylphenethylamine
In: Man E943
Rat E943

→ 5-(N-(3,4-Dimethoxyphenethyl)amino)-2-(3,4-dimethoxyphenyl)-2-isopropylvaleronitrile
In: Man E943, G977, K54
Rat E943, F474
→ 5-(N-(3,4-Dimethoxyphenethyl)-N-formylamino)-2-(3,4-dimethoxyphenyl)-2-isopropylvaleronitrile
In: Man A3674
Rat E943
→ 5-(N-(3,4-Dimethoxyphenethyl) methylamino)-2-(3,4-dimethoxyphenyl)-2-isopropylvaleronitrile-N-oxide
In: Rat G392
Pig G392
→ 5-(N-(3,4-Dimethoxyphenethyl) methylamino)-2-(3-hydroxy-4-methoxyphenyl)-2-isopropylvaleronitrile
In: Man E912, E943
Rat E912, E943
→ 5-(N-(3,4-Dimethoxyphenethyl) methylamino)-2-(4-hydroxy-3-methoxyphenyl)-2-isopropylvaleronitrile
In: Man A3674, E912, E943, K54
Rat E912, E943
→ 2-(3,4-Dimethoxyphenyl)-5-(N-(4-hydroxy-3-methoxyphenethyl) methylamino)-2-isopropylvaleronitrile
In: Man E912, K54
Rat E912
→ 2-(3,4-Dimethoxyphenyl)-2-isopropy-5-methylaminolvaleronitrile
In: Man A3674, G977, K54

5-(N-(3,4-Dimethoxyphenethyl)methylamino)-2-(3,4-dimethoxyphenyl)-2-isopropylvaleronitrile-N-oxide
Formed from: 5-(N-(3,4-Dimethoxyphenethyl)methylamino)-2-(3,4-dimethoxyphenyl)-2-isopropylvaleronitrile

5-((3,4-Dimethoxyphenethyl)methylamino)-2-(3-hydroxy-4,5-dimethoxyphenyl)-2-isopropylvaleronitrile
Formed from: Gallopamil
→ 5-((3,4-Dimethoxyphenethyl)amino)-2-(3-hydroxy-4,5-dimethoxyphenyl)-2-isopropylvaleronitrile
In: Man F524
Rat F524

5-((3,4-Dimethoxyphenethyl)methylamino)-2-(4-hydroxy-3,5-dimethoxyphenyl)-2-isopropylvaleronitrile
Formed from: Gallopamil

5-(N-(3,4-Dimethoxyphenethyl)methylamino)-2-(3-hydroxy-4-methoxyphenyl)-2-isopropylvaleronitrile
Formed from: 5-(N-(3,4-Dimethoxyphenethyl)methylamino)-2-(3,4-dimethoxyphenyl)-2-isopropylvaleronitrile

5-(N-(3,4-Dimethoxyphenethyl)methylamino)-2-(4-hydroxy-3-methoxyphenyl)-2-isopropylvaleronitrile
Formed from: 5-(N-(3,4-Dimethoxyphenethyl)methylamino)-2-(3,4-dimethoxyphenyl)-2-isopropylvaleronitrile

N-(3,4-Dimethoxyphenethyl)noradrenaline
Formed from: Denopamine
Probably formed from: N-(3,4-Dimethoxyphenethyl)-3-hydroxy-4-methoxyphenethylamine
→ N-(3,4-Dimethoxyphenethyl)-3-hydroxy-4-methoxyphenylethanolamine
Probably in: Man C626, D754
Rat D347
→ N-(3,4-Dimethoxyphenethyl) normetanephrine
Probably in: Dog D839, J869
Man C626, D754
Rabbit D347
Rat D347, D839

N-(3,4-Dimethoxyphenethyl)normetanephrine
Probably formed from: N-(3,4-Dimethoxyphenethyl)noradrenaline
→ N-(3,4-Dimethoxyphenethyl) normetanephrine-β-D-glucuronide
In: Rabbit D347
See also: J869
→ N-(3-Hydroxy-4-methoxyphenethyl) normetanephrine
Probably in: Man D754
→ N-(4-Hydroxy-3-methoxyphenethyl) normetanephrine
Probably in: Dog D839
Man C626, D754
Rat D839

N-(3,4-Dimethoxyphenethyl)normetanephrine-β-D-glucuronide
Formed from: N-(3,4-Dimethoxyphenethyl) normetanephrine

2,3-Dimethoxyphenol
Formed from: *o*-Xylene
→ 3-Methoxycatechol
In: Clostridium G367

2,4-Dimethoxyphenol
Probably formed from: 4-Methoxycatechol
m-Xylene

2,5-Dimethoxyphenol
Formed from: *p*-Xylene
Probably formed from: 4-Methoxycatechol

2,6-Dimethoxyphenol
Formed from: 3,5-Dimethoxy-4-hydroxybenzoic acid
→ 2,6-Dimethoxyphenyl-β-D-glucuronide
In: Rat A293
→ 2,6-Dimethoxyphenyl sulphate
In: Rat A293
→ 2,6-Dimethoxyquinol
In: Cat A2686
See also: H612
→ 3-Methoxycatechol
In: Clostridium G367
→ 3,3′,5,5′-Tetramethoxydiphenoquinone
In: Rhizoctonia A3836

3,4-Dimethoxyphenol
Formed from: 3,4-Dimethoxyphenyl-β-D-glucuronide
Veratric acid
o-Xylene
→ Methoxyquinol
In: Sporomusa J199

2,3-Dimethoxyphenylacetaldehyde
Probably formed from: 2,3-Dimethoxyphenethylamine
→ 2,3-Dimethoxyphenylacetic acid
Probably in: Rat G865

2,4-Dimethoxyphenylacetaldehyde
Probably formed from: 2,3-Dimethoxyphenethylamine
→ 2,4-Dimethoxyphenylacetic acid
Probably in: Rat G865

2,5-Dimethoxyphenylacetaldehyde
Probably formed from: 2,3-Dimethoxyphenethylamine
→ 2,5-Dimethoxyphenylacetic acid
Probably in: Rat G865

3,4-Dimethoxyphenylacetaldehyde
Formed from: 3,4-Dimethoxyallylbenzene
3,4-Dimethoxy-N-methylphenethylamine
3,4-Dimethoxypropenylbenzene
Probably formed from: 3,4-Dimethoxyphenethylamine
Gallopamil
→ 3,4-Dimethoxyphenylacetic acid
Probably in: Man F527
Rat A1765, G865
→ 3,4-Dimethoxyphenylethanol
Probably in: Man F527

3,5-Dimethoxyphenylacetaldehyde
Probably formed from: 2,3-Dimethoxyphenethylamine
→ 3,5-Dimethoxyphenylacetic acid
Probably in: Rat G865

2,3-Dimethoxyphenylacetic acid
Probably formed from: 2,3-Dimethoxyphenylacetaldehyde

2,4-Dimethoxyphenylacetic acid
Probably formed from: 2,4-Dimethoxyphenylacetaldehyde

2,5-Dimethoxyphenylacetic acid
Probably formed from: 2,5-Dimethoxyphenylacetaldehyde

3,4-Dimethoxyphenylacetic acid
Formed from: 4-(3,4-Dimethoxyphenyl) butylamine
N-(2-(3,4-Dimethoxyphenyl)ethyl)-5-methoxy-N,α-dimethyl-2-(phenylethynyl) benzenepropanamine
Probably formed from: 3,4-Dimethoxyphenylacetaldehyde
→ Veratraldehyde
In: Bjerkandera F610
Phanerochaete F610

3,5-Dimethoxyphenylacetic acid
Probably formed from: 3,5-Dimethoxyphenylacetaldehyde
3,5-Dimethoxyphenylpropionic acid

3,4-Dimethoxy-D-phenylalanine
Formed from: 5-(3,4-Dimethoxybenzyl) hydantoin

3,4-Dimethoxy-L-phenylalanine
Formed from: N-Carbamyl-3,4-dimethoxy-L-phenylalanine
5-(3,4-Dimethoxybenzyl)hydantoin

4-(3,4-Dimethoxyphenyl)butylamine
→ 3,4-Dimethoxyphenylacetic acid
In: Rat G865

4-(3,4-Dimethoxyphenyl)butyric acid
Formed from: 6-(3,4-Dimethoxyphenyl) hexylamine

(3*R*)-3-(2,4-Dimethoxyphenyl)-3,4-dihydro-2*H*-1-benzopyran-7-ol
See: Sativan

2-(3,4-Dimethoxyphenyl)-2,3-dihydro-3,5-di-(hydroxymethyl)-7-methoxybenzofuran
Formed from: 2-(3,4-Dimethoxyphenyl)-5-formyl-2,3-dihydro-3-hydroxymethyl-7-methoxybenzofuran
→ 2-(3,4-Dimethoxyphenyl)-5-formyl-2,3-dihydro-3-hydroxymethyl-7-methoxybenzofuran
In: Phanerochaete B796

***threo*-2-(3,4-Dimethoxyphenyl)-2,3-dihydro-3-hydroxymethyl-7-methoxy-5-(1,2,3-trihydroxypropyl)benzofuran**
Formed from: Methyl dehydrodiconiferyl alcohol
→ 2-(3,4-Dimethoxyphenyl)-5-formyl-2,3-dihydro-3-hydroxymethyl-7-methoxybenzofuran
In: Phanerochaete B796

(2*S*,3*R*,3a*S*)-2-(3,4-Dimethoxyphenyl)-3,3a-dihydro-3a-methoxy-3-methyl-5-(2-propenyl)-6(2*H*)-benzofuranone
See: Kadsurenone

α-((3,4-Dimethoxyphenyl)-6,7-dimethoxy-1-isoquinolinemethanol
See: Papaverinol

2-(2,5-Dimethoxyphenyl)-N,N-dimethyl-2-oxoethylamine
→ 2,5-Dimethoxymandelic acid
In: Rat J287

2-(3,4-Dimethoxyphenyl)-N,N-dimethyl-2-oxoethylamine
→ 3,4-Dimethoxymandelic acid
In: Rat J287

3,4-Dimethoxyphenylethanol
Probably formed from: 3,4-Dimethoxyphenylacetaldehyde

3,4-Dimethoxyphenylethanolamine
→ 3,4-Dimethoxymandelic acid
In: Rat J287

***p*-(3-(N-(1-(3,4-Dimethoxyphenyl)ethyl)amino)-2-hydroxypropoxy)phenylpropionic acid**
Formed from: Methyl *p*-(3-(N-(1-(3,4-dimethoxyphenyl)ethyl)amino)-2-hydroxypropoxy)phenylpropionate

3,4-Dimethoxy-6-(2-phenylethyl)-1,2-benzenediol
See: Batatasin I

3-(2-(3,4-Dimethoxyphenyl)ethylcarbamoylmethyl)amino-N-methylbenzamide
See: Ecabapide

N-(2-(3,4-Dimethoxyphenyl)ethyl)-4,5-dihydroxy-N,α-dimethyl-2-(phenylethynyl)benzenepropanamine
Probably formed from: N-(2-(3,4-Dimethoxyphenyl)ethyl)-4-hydroxy-5-methoxy-N,α-dimethyl-2-(phenylethynyl)benzenepropanamine

1-(3,4-Dimethoxyphenyl)-5-ethyl-7,8-dimethoxy-4-methyl-5*H*-2,3-benzodiazepine
See: Tofizopam

N-(2-(3,4-Dimethoxyphenyl)ethyl)-5-hydroxy-N,α-dimethyl-2-(phenylethynyl)benzenepropanamine
Probably formed from: N-(2-(3,4-Dimethoxyphenyl)ethyl)-5-methoxy-N,α-dimethyl-2-(phenylethynyl)benzenepropanamine
→ N-(2-(3,4-Dimethoxyphenyl)ethyl)-5-hydroxy-α-methyl-2-(phenylethynyl)benzenepropanamine
Probably in: Dog H937
Rat J444

N-(2-(3,4-Dimethoxyphenyl)ethyl)-4-hydroxy-5-methoxy-N,α-dimethyl-2-(phenylethynyl) benzenepropanamine
Formed from: N-(2-(3,4-Dimethoxyphenyl) ethyl)-5-methoxy-N,α-dimethyl-2-(phenylethynyl)benzenepropanamine
→ N-(2-(3,4-Dimethoxyphenyl) ethyl)-4,5-dihydroxy-N,α-dimethyl-2-(phenylethynyl)benzenepropanamine
Probably in: Rat J444
→ N-(2-(3,4-Dimethoxyphenyl)ethyl)-4-hydroxy-5-methoxy-α-methyl-2-(phenylethynyl)benzenepropanamine
Probably in: Dog H937
Rat J444

N-(2-(3,4-Dimethoxyphenyl)ethyl)-4-hydroxy-5-methoxy-α-methyl-2-(phenylethynyl) benzenepropanamine
Probably formed from: N-(2-(3,4-Dimethoxyphenyl)ethyl)-4-hydroxy-5-methoxy-N,α-dimethyl-2-(phenylethynyl) benzenepropanamine

N-(2-(3,4-Dimethoxyphenyl)ethyl)-5-hydroxy-α-methyl-2-(phenylethynyl)benzenepropanamine
Probably formed from: N-(2-(3,4-Dimethoxyphenyl)ethyl)-5-hydroxy-N,α-dimethyl-2-(phenylethynyl) benzenepropanamine

N-(2-(3,4-Dimethoxyphenyl)ethyl)-2-((*p*-hydroxyphenyl)ethynyl)-5-methoxy-N,α-dimethylbenzenepropanamine
Formed from: N-(2-(3,4-Dimethoxyphenyl) ethyl)-5-methoxy-N,α-dimethyl-2-(phenylethynyl)benzenepropanamine
→ N-(2-(3,4-Dimethoxyphenyl)ethyl)-2-((*p*-hydroxyphenyl)ethynyl)-5-methoxy-α-methylbenzenepropanamine
Probably in: Dog H937

N-(2-(3,4-Dimethoxyphenyl)ethyl)-2-((*p*-hydroxyphenyl)ethynyl)-5-methoxy-α-methylbenzenepropanamine
Probably formed from: N-(2-(3,4-Dimethoxyphenyl)ethyl)-2-((*p*-hydroxyphenyl)ethynyl)-5-methoxy-N,α-dimethylbenzenepropanamine

N-(2-(3,4-Dimethoxyphenyl)ethyl)-5-methoxy-N,α-dimethyl-2-(phenylethynyl) benzenepropanamine
→ 3,4-Dimethoxyphenylacetic acid
In: Dog H937
Rat J444
→ N-(2-(3,4-Dimethoxyphenyl)ethyl)-5-hydroxy-N,α-dimethyl-2-(phenylethynyl) benzenepropanamine
In: Dog H937
Rat J444
→ N-(2-(3,4-Dimethoxyphenyl)ethyl)-4-hydroxy-5-methoxy-N,α-dimethyl-2-(phenylethynyl)benzenepropanamine
In: Dog H937
Rat J444
→ N-(2-(3,4-Dimethoxyphenyl)ethyl)-2-((*p*-hydroxyphenyl)ethynyl)-5-methoxy-N,α-dimethylbenzenepropanamine
In: Dog H937
Rat J444
→ N-(2-(3,4-Dimethoxyphenyl)ethyl)-5-methoxy-α-methyl-2-(phenylethynyl) benzenepropanamine
In: Dog H937
Rat J444
→ N-(2-(4-Hydroxy-3-methoxyphenyl)ethyl)-5-methoxy-N,α-dimethyl-2-(phenylethynyl) benzenepropanamine
In: Dog H937
Rat J444
→ 5-Methoxy-N,α-dimethyl-2-(phenylethynyl) benzenepropanamine
In: Dog H937
→ 4-(5-Methoxy-2-(2-phenylethynyl)phenyl) butan-2-one
In: Rat J444

N-(2-(3,4-Dimethoxyphenyl)ethyl)-5-methoxy-α-methyl-2-(phenylethynyl)benzenepropanamine
Formed from: N-(2-(3,4-Dimethoxyphenyl) ethyl)-5-methoxy-N,α-dimethyl-2-(phenylethynyl)benzenepropanamine

2-(3,4-Dimethoxyphenyl)-5-formyl-2,3-dihydro-3-hydroxymethyl-7-methoxybenzofuran
Formed from: 2-(3,4-Dimethoxyphenyl)-2,3-dihydro-3,5-di-(hydroxymethyl)-7-methoxybenzofuran

threo-2-(3,4-Dimethoxyphenyl)-2,3-dihydro-3-hydroxymethyl-7-methoxy-5-(1,2,3-trihydroxypropyl)benzofuran
→ 5-Carboxy-2-(3,4-dimethoxyphenyl)-2,3-dihydro-3-hydroxymethyl-7-methoxybenzofuran
In: Phanerochaete B796
→ 2-(3,4-Dimethoxyphenyl)-2,3-dihydro-3,5-di-(hydroxymethyl)-7-methoxybenzofuran
In: Phanerochaete B796

2,6-Dimethoxyphenyl-β-D-glucuronide
Formed from: 2,6-Dimethoxyphenol

3,4-Dimethoxyphenyl-β-D-glucuronide
→ 2,6-Dimethoxyphenol
In: Beef A1029
Escherichia A1029

3,4-Dimethoxyphenylglycol
Formed from: 1,2-Bis(3,4-dimethoxyphenyl)-1,3-propanediol
→ ω-Hydroxy-3′,4′-dimethoxyacetophenone
In: Phanerochaete C401
→ Veratraldehyde
In: Phanerochaete C401

3,4-Dimethoxyphenylglyoxal
→ 3,4-Dimethoxyphenylmandelic acid
In: Rat J287

7-(3,4-Dimethoxyphenyl)heptanoic acid
Formed from: 7-(3,4-Dimethoxyphenyl)heptylamine
→ 3-(3,4-Dimethoxyphenyl)propionic acid
Probably in: Rat G865

7-(3,4-Dimethoxyphenyl)heptylamine
→ 7-(3,4-Dimethoxyphenyl)heptanoic acid
In: Rat G865

6-(3,4-Dimethoxyphenyl)hexylamine
→ 4-(3,4-Dimethoxyphenyl)butyric acid
In: Rat G865

3,4-Dimethoxyphenylhydracrylic acid
Formed from: 3,4-Dimethoxycinnamic acid
→ 3′,4′-Dimethoxyacetophenone
Probably in: Rat A1765
→ Veratric acid
Probably in: Rat A1765

2-(3,4-Dimethoxyphenyl)-5-(N-(4-hydroxy-3-methoxyphenethyl)methylamino)-2-isopropylvaleronitrile
Formed from: 5-(N-(3,4-Dimethoxyphenethyl)methylamino)-2-(3,4-dimethoxyphenyl)-2-isopropylvaleronitrile

3-(3,4-Dimethoxyphenyl)-3-hydroxy-2-phenoxypropionic acid
Formed from: Veratrylglycerol β-phenyl ether

2-(3,4-Dimethoxyphenyl)-2-isopropyl-5-methylaminovaleronitrile
Formed from: 5-N-(3,4-Dimethoxyphenethyl)methylamino)-2-(3,4-dimethoxyphenyl)-2-isopropylvaleronitrile
→ 5-Amino-2-(3,4-dimethoxyphenyl)-2-isopropylvaleronitrile
Probably in: Man K54
→ 2-(3-Hydroxy-4-methoxyphenyl)-2-isopropyl-5-methylaminovaleronitrile
Probably in: Man K54

3,4-Dimethoxyphenyllactic acid
Probably formed from: 3-(3,4-Dimethoxyphenyl)propane-1,2-diol
→ 3,4-Dimethoxyphenylpyruvic acid
Probably in: Rat A1765

(3*S*,4*S*)-4-((3,4-Dimethoxyphenyl)methyl)dihydro-3-hydroxy-3-((4-hydroxy-3-methoxyphenyl)methyl)-2(3*H*)-furanone
See: Trachelogenin

(3*S*,4*S*)-4-((3,4-Dimethoxyphenyl)methyl)-3-((4-(β-D-glucopyranosyloxy) 3-Methoxyphenyl)methyl)dihydro-3-hydroxy-2(3*H*)-furanone
See: Tracheloside

1-((3,4-Dimethoxyphenyl)methyl)-6-methoxy-7-isoquinolinol
See: 7-Demethylpapaverine

1-((3,4-Dimethoxyphenyl)methyl)-7-methoxy-6-isoquinolinol
See: 6-Demethylpapaverine

4-((3,4-Dimethoxyphenyl)methyl)-6-methoxy-7-isoquinolinol
See: 7-Demethylisopapaverine

4-((3,4-Dimethoxyphenyl)methyl)-7-methoxy-6-isoquinolinol
See: 6-Demethylisopapaverine

1-((3,4-Dimethoxyphenyl)methyl)-1,2,3,4-tetrahydro-6,7-dimethoxyisoquinoline
See: Tetrahydropapaverine

(1*S*)-1-((3,4-Dimethoxyphenyl)methyl)-1,2,3,4-tetrahydro-6,7-dimethoxy-2-methylisoquinoline
See: Laudanosine

(1*S*)-1-((3,4-Dimethoxyphenyl)methyl)-1,2,3,4-tetrahydro-6-methoxy-2-methyl-7-isoquinolinol
See: Codamine

1-((3,4-Dimethoxyphenyl)methyl)-1,2,3,4-tetrahydro-7-methoxy-2-methyl-6-isoquinolinol
See: Pseudolaudanine

1-(3,4-Dimethoxyphenyl)-1-oxopropane-1,2-diol
Formed from: Veratrylglycerol
→ Veratraldehyde
In: Coriolus D335
→ Veratrylglycerol
In: Coriolus D335

1-(3,4-Dimethoxyphenyl)propane-1,2-diol
Formed from: 3,4-Dimethoxypropenylbenzene

1-(3,4-Dimethoxyphenyl)propane-1,3-diol
Formed from: 3-(3,4-Dimethoxyphenyl) propan-1-ol

3-(3,4-Dimethoxyphenyl)propane-1,2-diol
Probably formed from: 3,4-Dimethoxyallylbenzene-2,3-oxide
→ 3,4-Dimethoxyphenyllactic acid
Probably in: Rat A1765

3-(3,4-Dimethoxyphenyl)propan-1-ol
Formed from: 3,4-Dimethoxyallylbenzene
3,4-Dimethoxycinnamyl alcohol
Probably formed from: 3-(3,4-Dimethoxyphenyl)-2-propen-1-ol
→ 3-(3,4-Dimethoxyphenyl)propane-1,3-diol
In: Coriolus D335

1-(3,4-Dimethoxyphenyl)-2-propen-1-ol
Formed from: 3,4-Dimethoxyallylbenzene
→ 3-(3,4-Dimethoxyphenyl)-2-propen-1-ol
Probably in: Rat A1765
→ 1-(3,4-Dimethoxyphenyl)-2-propen-1-one
Probably in: Rat A1765

3-(3,4-Dimethoxyphenyl)-2-propen-1-ol
Formed from: 3,4-Dimethoxypropenylbenzene
Probably formed from: 1-(3,4-Dimethoxyphenyl)-2-propen-1-ol
→ 3,4-Dimethoxycinnamic acid
Probably in: Rat A1765
→ 1-(3,4-Dimethoxyphenyl)propan-1-ol
Probably in: Rat A1765
→ 3-(4-Hydroxy-3-methoxyphenyl)-2-propen-1-ol
Probably in: Rat A1765

1-(3,4-Dimethoxyphenyl)-2-propen-1-one
Probably formed from: 1-(3,4-Dimethoxyphenyl)-2-propen-1-ol

3,4-Dimethoxyphenylpropionic acid
Formed from: 3,4-Dimethoxycinnamic acid
Probably formed from: 7-(3,4-Dimethoxyphenyl)heptanoic acid
→ Dihydroferulic acid
Probably in: Rat A1765
→ Veratric acid
Probably in: Rat G865

3,5-Dimethoxyphenylpropionic acid
Formed from: 3,5-Dimethoxycinnamic acid
3-(3-Hydroxy-4,5-dimethoxyphenyl) propionic acid
Probably formed from: Dihydrosinapic acid
→ 3,5-Dimethoxyphenylacetic acid
Probably in: Microorganism D894
→ 3-Hydroxy-5-methoxyphenylpropionic acid
Probably in: Microorganism A261, A744

3,4-Dimethoxyphenylpyruvic acid
Probably formed from: 3,4-Dimethoxyphenyllactic acid

2,6-Dimethoxyphenyl sulphate
Formed from: 2,6-Dimethoxyphenol

4-((1*S*,3a*R*,4*R*,6a*R*)-4-(3,4-dimethoxyphenyl)tetrahydro-1*H*,3*H*-furo[3,4-*c*]furan-1-yl)-2-methoxyphenol
See: Phillygenin

4,5-Dimethoxyphthalide
Formed from: 3,4-Dihydro-2′-hydroxymethyl-2-methylpapaverinium

6,7-Dimethoxyphthalide
See: Meconine

5,6-Dimethoxy-2-((piperidin-4-yl)methyl)-α-hydrindone
Formed from: Donepezil
→ 3-Hydroxy-5,6-dimethoxy-2-((piperidin-4-yl)methyl)-α-hydrindone
Probably in: Rat J789

3,4-Dimethoxypropenylbenzene
→ 3,4-Dimethoxyphenylacetaldehyde
In: Rat A1765
→ 1-(3,4-Dimethoxyphenyl)propane-1,2-diol
In: Rat A1765
→ 3-(3,4-Dimethoxyphenyl)-2-propen-1-ol
In: Rat A1765
→ 4-Hydroxy-3-methoxyphenylacetone
In: Rat A1765
→ 3-Hydroxy-4-methoxypropenylbenzene
In: Rat A1765
→ Isoeugenol
In: Rat A1765

2,5-Dimethoxy-4-propoxyamphetamine
→ N-Acetyl-2,5-dimethoxy-4-propoxyamphetamine
In: Cunninghamella G516
→ 4-Hydroxy-2,5-dimethoxyamphetamine
In: Cunninghamella G516
→ 4-(2-Hydroxypropoxy)-2,5-dimethoxyamphetamine
Probably in: Cunninghamella G516
→ 4-(3-Hydroxypropoxy)-2,5-dimethoxyamphetamine
Probably in: Cunninghamella G516

10,11-Dimethoxy-N-propylnorapomorphine
→ 10-Hydroxy-11-methoxy-N-propylnorapomorphine
In: Cunninghamella E282

N′-(2,6-Dimethoxy-4-pyrimidinyl) sulphanilamide
See: Sulphadimethoxine

N-(4,7-Dimethoxy-6-(2-(1-(pyrrolidinyl)ethoxy)-5-benzofuranyl)-N′-methylurea
See: Carocainide

2,6-Dimethoxyquinol
Formed from: 2,6-Dimethoxyphenol
3,5-Dimethoxyquinone
4-Hydroxy-3,5-dimethoxybenzoic acid
Probably formed from: 4-Hydroxy-3,5-dimethoxybenzaldehyde
→ 3,5-Dimethoxyquinone
Probably in: Coriolus E572
Pleurotus H612
Rhodotorula E262

3,5-Dimethoxyquinone (2,6-dimethoxy-*p*-quinone)
Formed from: Syringylglycerol β-vanillin ether
Probably formed from: 2,6-Dimethoxyquinol
→ 2,6-Dimethoxyquinol
In: Phanerochaete H611
→ 1,2,3,5-Tetrahydroxybenzene
In: Sporotrichum B301

4,5-Dimethoxy-*o*-quinone
→ 4,5-Dimethoxycatechol
In: Sporotrichum B301

2,6-Dimethoxy-*p*-quinone
See: 3,5-Dimethoxyquinone

3,4-Dimethoxystyrene
Formed from: 3,4-Dimethoxycinnamic acid

5,6-Dimethoxy-4-sulphanilamidopyrimidine
See: Sulphadoxine

2,6-Dimethoxy-4-(2,3,5,6-tetrachlorophenoxy) phenol
Formed from: 4-Hydroxy-3,5-dimethoxybenzoic acid
2,3,5,6-Tetrachlorophenol

4-(2,6-Dimethoxy-4-(2,3,5,6-tetrachlorophenoxy)phenoxy)-2,6-dimethoxyphenol
Formed from: 4-Hydroxy-3,5-dimethoxybenzoic acid
2,3,5,6-Tetrachlorophenol

4-(4-(2,6-Dimethoxy-4-(2,3,5,6-tetrachlorophenoxy)phenoxy)-2,6-dimethoxyphenoxy)-2,6-dimethoxyphenol
Formed from: 4-Hydroxy-3,5-dimethoxybenzoic acid
2,3,5,6-Tetrachlorophenol

4-(4-(4-(2,6-Dimethoxy-4-(2,3,5,6-tetrachlorophenoxy)phenoxy)-2,6-dimethoxyphenoxy)-2,6-dimethoxyphenoxy)-2,6-dimethoxyphenol
Formed from: 4-Hydroxy-3,5-dimethoxybenzoic acid
2,3,5,6-Tetrachlorophenol

2,5-Dimethoxythioanisole
Formed from: Bromobenzene
o-Bromophenol
m-Bromophenol
p-Bromophenol

Bromoquinol

3,4-Dimethoxytoluene
→ 2-Methoxy-4-methylphenol
Probably in: Sporomusa J199

2,6-Dimethoxy-4-(2,3,6-trichlorophenoxy)phenol
Formed from: 4-Hydroxy-3,5-dimethoxybenzoic acid
2,3,6-Trichlorophenol

2,6-Dimethoxy-4-(2,4,5-trichlorophenoxy)phenol
Formed from: 4-Hydroxy-3,5-dimethoxybenzoic acid
2,4,5-Trichlorophenol

4-(2,6-Dimethoxy-4-(2,3,6-trichlorophenoxy)phenoxy)-2,6-dimethoxyphenol
Formed from: 4-Hydroxy-3,5-dimethoxybenzoic acid
2,3,6-Trichlorophenol

4-(2,6-Dimethoxy-4-(2,4,5-trichlorophenoxy)phenoxy)-2,6-dimethoxyphenol
Formed from: 4-Hydroxy-3,5-dimethoxybenzoic acid
2,4,5-Trichlorophenol

4-(4-(2,6-Dimethoxy-4-(2,3,6-trichlorophenoxy)phenoxy)-2,6-dimethoxyphenoxy)-2,6-dimethoxyphenol
Formed from: 4-Hydroxy-3,5-dimethoxybenzoic acid
2,3,6-Trichlorophenol

4-(4-(2,6-Dimethoxy-4-(2,4,5-trichlorophenoxy)phenoxy)-2,6-dimethoxyphenoxy)-2,6-dimethoxyphenol
Formed from: 4-Hydroxy-3,5-dimethoxybenzoic acid
2,4,5-Trichlorophenol

4-(4-(4-(2,6-Dimethoxy-4-(2,3,6-trichlorophenoxy)phenoxy)-2,6-dimethoxyphenoxy)-2,6-dimethoxyphenoxy)-2,6-dimethoxyphenol
Formed from: 4-Hydroxy-3,5-dimethoxybenzoic acid
2,3,6-Trichlorophenol

4-(4-(4-(2,6-Dimethoxy-4-(2,4,5-trichlorophenoxy)phenoxy)-2,6-dimethoxyphenoxy)-2,6-dimethoxyphenoxy)-2,6-dimethoxyphenol
Formed from: 4-Hydroxy-3,5-dimethoxybenzoic acid
2,4,5-Trichlorophenol

1,2-Dimethoxy-4-trifluoromethylbenzene
See: 4-Trifluoromethylveratrole

6,7-Dimethoxy-1-veratrylisoquinoline
See: Papaverine

2,3-Dimethylacetanilide
Formed from: 2,3-Dimethylaniline

2,4-Dimethylacetanilide
Formed from: 2,4-Dimethylaniline

2,5-Dimethylacetanilide
Formed from: 2,5-Dimethylaniline

3,4-Dimethylacetanilide
Formed from: 3,4-Dimethylaniline

3,5-Dimethylacetanilide
Formed from: 3,5-Dimethylaniline

3,5-Dimethylacetanilide-4-phenoxyl radical
Formed from: 4-Hydroxy-3,5-dimethylacetanilide

2,6-Dimethylacetophenone imine
→ 2,6-Dimethylacetophenone oxime
In: Ferret D955
Guinea pig D955
Hamster D955
Mouse D955
Rabbit C243, D955
Rat D955

2,6-Dimethylacetophenone oxime
Formed from: 2,6-Dimethylacetophenone imine

9,9-Dimethylacridane-10-carboxylic acid S-(dimethylamino)thiolethyl ester
→ 9,9-Dimethylacridane-10-carboxylic acid S-(dimethylamino)thiolethyl ester-N-oxide
In: Dog A3003
→ 9,9-Dimethylacridane-10-carboxylic acid S-(methylamino)thiolethyl ester
In: Dog A3003

9,9-Dimethylacridane-10-carboxylic acid S-(dimethylamino)thiolethyl ester-N-oxide
Formed from: 9,9-Dimethylacridane-10-carboxylic acid S-(dimethylamino)thiolethyl ester

9,9-Dimethylacridane-10-carboxylic acid S-(methylamino)thiolethyl ester
Formed from: 9,9-Dimethylacridane-10-carboxylic acid S-(dimethylamino)thiolethyl ester

O-β,β-Dimethylacrylylshikonin
Probably formed from: Shikonin

N,N-Dimethyladriamycin
→ N,N-Dimethyladriamycinol
Probably in: Mouse B109
Rat B109

N,N-Dimethyladriamycinol
Probably formed from: N,N-Dimethyladriamycin

3-Dimethylallyl-2,4-dihydroxyquinoline
Probably formed from: 2,4-Dihydroxyquinoline
→ Dictamnine
In: Skimmia A1674
→ Platydesminium
In: Skimmia A1674

4-Dimethylallyl-N-methyl-L-tryptophan
Probably formed from: 4-Dimethylallyl-L-tryptophan
→ Agroclavine
In: Claviceps A2412
→ Elymoclavine
In: Claviceps B250
See also: B415

3′-(3,3-Dimethylallyl)-2′,4′,5,7-tetrahydroxy-5′-methoxyisoflavone
See: Piscerythrone

6′-(3,3-Dimethylallyl)-4′,5,5′,7-tetrahydroxy-2′-methoxyisoflavone
See: Piscidone

4-Dimethylallyl-L-tryptophan
Probably formed from: L-Tryptophan
→ Chanoclavine I
Probably in: Claviceps A752
→ (−)-Chanoclavine II
Probably in: Claviceps A752
→ Clavicipitic acid
In: Claviceps A1731
See also: B329
→ 4-Dimethylallyl-N-methyl-L-tryptophan
Probably in: Claviceps A2412, B415
→ 4-(4-Hydroxy-3-methyl-Δ^2-butenyl)-L-tryptophan
In: Claviceps A3139, A3599

4-Dimethylaminoantipyrine
See: Aminopyrine

4-Dimethylaminoazobenzene (butter yellow)
Formed from: 4-Dimethylaminoazobenzene-N-oxide
→ Aniline
In: Ascaris A304, A324
Boophilus A1024
Microorganism D257
Moniezia A324, A1785
Rabbit C884
Rat C615, D257, G133, G380, G511
Spider A1024
→ 4′-Dimethylamino-2-hydroxyazobenzene
In: Rat A1981
→ 4′-Dimethylamino-4-hydroxyazobenzene
In: Rat A1979, A1981, A3898, B111, B985, C615, D68, D849, D851, E240
→ N,N-Dimethyl-*p*-phenylenediamine
In: Ascaris A304, A324
Microorganism D257
Moniezia A324, A1785
Rabbit C884
Rat D257, D851, G133, G380, G511
→ N-(Glutathion-S-methylene)-4-aminoazobenzene
In: Rat C142
See also: J409
→ N-(Glutathion-S-methylene)-4-methylaminoazobenzene
In: Rat D68
→ 4-Methylaminoazobenzene
Boophilus A1024
Rat A1975, A1979, A1981, A3898, B111, B985, D849, D851
Spider A1024

4-Dimethylaminoazobenzene-4′-arsonic acid
→ *p*-Arsanilic acid
In: Rat G133, G380, G511
→ N,N-Dimethyl-*p*-phenylenediamine
In: Rat G133, G380, G511

4-Dimethylaminoazobenzene-N-oxide
→ 4-Dimethylaminoazobenzene

In: Rat E187

4′-Dimethylaminoazobenzene-4-sulphonic acid
See: Methyl orange

4′-Dimethylaminoazobenzen-4-yl-β-D-glucuronide
Formed from: 4′-Dimethylamino-4-hydroxyazobenzene

4′-Dimethylaminoazobenzen-4-yl sulphate
Formed from: 4′-Dimethylamino-4-hydroxyazobenzene

***p*-Dimethylaminobenzaldehyde**
→ *p*-Dimethylaminobenzoic acid
In: Man J24

***p*-Dimethylaminobenzoic acid**
Formed from: *p*-Dimethylaminobenzaldehyde
→ *p*-Methylaminobenzoic acid
Probably in: Pseudomonas A903
Rat A114

***p*-Dimethylaminobenzonitrile**
→ 4-Dimethylamino-3-hydroxybenzonitrile
In: Rat D160
p-Methylaminobenzonitrile
→ In: Rabbit F839
Rat D160

***p*-Dimethylaminobenzonitril-3-yl sulphate**
Probably formed from: 4-Dimethylamino-3-hydroxybenzonitrile

4-Dimethylaminobiphenyl
→ N-Methylbiphenyl-4-yl nitroxide radical
Probably in: Rabbit B184

***p*-Dimethylaminocinnamaldehyde**
→ *p*-Dimethylaminocinnamic acid
In: Man J24

***p*-Dimethylaminocinnamic acid**
Formed from: *p*-Dimethylaminocinnamaldehyde

4-Dimethylamino-2,α-dimethylphenethylamine
See: Amiflamine

4-Dimethylamino-3,5-dimethylphenol
Formed from: Mexacarbate

4-Dimethylamino-3,5-dimethylphenyl N-hydroxymethylcarbamate
Formed from: 4-Dimethylamino-3,5-dimethylphenyl methylcarbamate

4-Dimethylamino-3,5-dimethylphenyl methylcarbamate (zectran)
→ 4-Dimethylamino-3,5-dimethylphenyl N-hydroxymethylcarbamate
In: Rat A1545
→ 3,5-Dimethyl-4-methylaminophenyl methylcarbamate
In: Rat A1545
→ 3,5-Dimethyl-4-methylformamidophenyl methylcarbamate
In: Rat A1545

3-Dimethylamino-1,1-diphenylbutane (recipavrin)
→ 1,1-Diphenylbutane-3-one
In: Rat F582
→ 1,1-Diphenylbutane-3-one oxime
In: Rat F582
→ 3-Methylamino-1,1-diphenylbutane
In: Rat F582

***d*-3-Dimethylamino-1,2-diphenyl-2-butyl propionate** (*d*-propoxyphene)
→ *d*-3-Methylamino-1,2-diphenyl-2-butyl propionate
In: Rat A2421

***l*-3-Dimethylamino-1,2-diphenyl-2-butyl propionate** (*l*-propoxyphene)
Formed from: *l*-3-Dimethylamino-1,2-diphenyl-2-butyl propionate-N-oxide

***l*-3-Dimethylamino-1,2-diphenyl-2-butyl propionate-N-oxide**
→ *l*-3-Dimethylamino-1,2-diphenyl-2-butyl propionate
In: Dog A2033
Rat A2033

2-Dimethylamino-4,4-diphenyl-5-heptanol (*d* and *l*-isomers; methadol)
l-α-Acetyl-2-dimethylamino-4,4-diphenyl-5-heptanol
2-Dimethylamino-4,4-diphenyl-5-heptanone
→ 2-Dimethylamino-4,4-diphenyl-5-heptanone
Probably in: Rat A2900
→ 2-Methylamino-4,4-diphenyl-5-heptanol
In: Man A1123
Rat A1123
See also: A1000, A2088, A2384, A2979

2-Dimethylamino-4,4-diphenyl-5-heptanone (methadone)
Probably formed from: 2-Dimethylamino-4,4-diphenyl-5-heptanol
→ 2-Dimethylamino-4,4-diphenyl-5-heptanol

In: Man A1123
Rabbit A3500
Rat A1123, A2384
See also: A133
→ 2-Dimethylamino-4,4-diphenyl-5-heptanone-N-oxide
In: Man A409
Rhesus monkey A3299
See also: A133
Contrast (man) A1000
→ 4-Dimethylamino-2,2-diphenylvaleric acid
In: Man 1000
→ 2-Dimethylamino-4-(*p*-hydroxyphenyl)-4-phenyl-5-heptanone
In: Rat E864
→ 2-(4′,4′-Diphenylheptan-5′-on-2′-yl)oxaziridine
In: Rat D352
→ 2-Ethylidene-1,5-dimethyl-3,3-diphenyl-1-pyrrolidine
In: Dog D899
Man A760, A1000, A1175, A2803
Mouse A1693
Rat A81, A1912, A1973
See also: A133
→ 3-(*p*-Hydroxyphenyl)-5-methyl-3-phenyl-1-pyrroline
In: Rat A1912
See also: A2779
→ 2-Methylamino-4,4-diphenyl-5-heptanone
In: Macaque A2117
Man A2125, A2803
Rat A1983
→ Normethadone
In: Man H859

2-Dimethylamino-4,4-diphenyl-5-heptanone-N-oxide
Probably formed from: 2-Dimethylamino-4,4-diphenyl-5-heptanone

4-Dimethylamino-2,2-diphenylvaleric acid
Formed from: 2-Dimethylamino-4,4-diphenyl-5-heptanone
→ 1,5-Dimethyl-3,3-diphenyl-1-pyrrolidone
Probably in: Man A1000

5-(α-(2-Dimethylaminoethoxy)benzyl)-1*H*-pyrazole
Probably formed from: Cizolirtine

1-(6-(2-(Dimethylamino)ethoxy)-4,7-dimethoxy-5-benzofuranyl)-3-(4-methoxyphenyl)-2-propen-1-one
See: Mecinarone

5-(2-(Dimethylamino)ethoxy)-7-hydroxy-7*H*-benzo[*c*]fluorene
Formed from: Benfluron

4-(2-Dimethylaminoethoxy)-5-isopropyl-2-methylphenyl acetate (ester)
See: Thymoxamine

***trans*-1-(*p*-β-2-Dimethylaminoethoxyphenyl)-1,2-diphenylbut-1-ene**
See: Tamoxifen

1-(*p*-(2-Dimethylaminoethoxy)phenyl)-1,2-diphenyl-1-buten-3-ol-N-oxide
Formed from: Tamoxifen
Tamoxifen-N-oxide

1-(*p*-β-Dimethylaminoethoxyphenyl)-1-hydroxy-1,2-diphenylbutane
Formed from: Tamoxifen

***trans*-1-(*p*-Dimethylaminoethoxyphenyl)-3-hydroxy-1-(*p*-hydroxyphenyl)-2-phenylbut-1-ene**
Probably formed from: *trans*-1-(*p*-Dimethylaminoethoxyphenyl)-1-(*p*-hydroxyphenyl)-2-phenylbut-1-ene
→ *trans*-1-(*p*-Dimethylaminoethoxyphenyl)-3-hydroxy-1-(*p*-hydroxyphenyl)-2-phenylbut-1-ene-β-D-glucuronide
Probably in: Rat A253

***trans*-1-(*p*-Dimethylaminoethoxyphenyl)-3-hydroxy-1-(*p*-hydroxyphenyl)-2-phenylbut-1-ene-β-D-glucuronide**
Probably formed from: *trans*-1-(*p*-Dimethylaminoethoxyphenyl)-3-hydroxy-1-(*p*-hydroxyphenyl)-2-phenylbut-1-ene

***trans*-1-(*p*-β-Dimethylaminoethoxyphenyl)-3-hydroxy-1-(*p*-iodophenyl)-2-phenylbut-1-ene**
Formed from: *trans*-1-(*p*-β-Dimethylaminoethoxyphenyl)-1-(*p*-iodophenyl)-2-phenylbut-1-ene

***trans*-1-(*p*-(2-Dimethylaminoethoxy)phenyl)-1-(3-hydroxy-4-methoxyphenyl)-2-(hydroxyphenyl)but-1-ene**
Probably formed from: 3-Hydroxy-4-methoxytamoxifen
→ *trans*-1-(*p*-(2-Dimethylaminoethoxy)phenyl)-1-(3-hydroxy-4-methoxyphenyl)-2-(hydroxyphenyl)but-1-ene-β-D-glucuronide
Probably in: Cynomolgus monkey H924
Mouse H924
Rat H924

***trans*-1-(*p*-(2-Dimethylaminoethoxy)phenyl)-1-(4-hydroxy-3-methoxyphenyl)-2-(hydroxyphenyl)but-1-ene**
Probably formed from: *trans*-1-(*p*-(2-Dimethylaminoethoxy)phenyl)-1-(4-hydroxy-3-methoxyphenyl)-2-phenylbut-1-ene
→ *trans*-1-(*p*-(2-Dimethylaminoethoxy)phenyl)-1-(4-hydroxy-3-methoxyphenyl)-2-(hydroxyphenyl)but-1-ene-β-D-glucuronide
Probably in: Cynomolgus monkey H924
Mouse H924
Rat H924

***trans*-1-(*p*-(2-Dimethylaminoethoxy)phenyl)-1-(3-hydroxy-4-methoxyphenyl)-2-(hydroxyphenyl)but-1-ene-β-D-glucuronide**
Probably formed from: *trans*-1-(*p*-(2-Dimethylaminoethoxy)phenyl)-1-(3-hydroxy-4-methoxyphenyl)-2-(hydroxyphenyl)but-1-ene

***trans*-1-(*p*-(2-Dimethylaminoethoxy)phenyl)-1-(4-hydroxy-3-methoxyphenyl)-2-(hydroxyphenyl)but-1-ene-β-D-glucuronide**
Probably formed from: *trans*-1-(*p*-(2-Dimethylaminoethoxy)phenyl)-1-(4-hydroxy-3-methoxyphenyl)-2-(hydroxyphenyl)but-1-ene

***trans*-1-(*p*-(2-Dimethylaminoethoxy)phenyl)-1-(4-hydroxy-3-methoxyphenyl)-2-phenylbut-1-ene**
Formed from: 3,4-Dihydroxytamoxifen
→ *trans*-1-(*p*-(2-Dimethylaminoethoxy)phenyl)-1-(4-hydroxy-3-methoxyphenyl)-2-(hydroxyphenyl)but-1-ene
Probably in: Mouse H924
Rat H924
→ *trans*-1-(*p*-(2-Dimethylaminoethoxy)phenyl)-1-(4-hydroxy-3-methoxyphenyl)-2-phenylbut-1-ene-β-D-glucuronide
Probably in: Cynomolgus monkey H924
Mouse H924
Rat H924
→ *trans*-1-(*p*-(2-Dimethylaminoethoxy)phenyl)-1-(4-hydroxy-3-methoxyphenyl)-2-phenylbut-1-ene sulphate
Probably in: Cynomolgus monkey H924

***trans*-1-(*p*-(2-Dimethylaminoethoxy)phenyl)-1-(4-hydroxy-3-methoxyphenyl)-2-phenylbut-1-ene-β-D-glucuronide**
Probably formed from: *trans*-1-(*p*-(2-Dimethylaminoethoxy)phenyl)-1-(4-hydroxy-3-methoxyphenyl)-2-phenylbut-1-ene

***trans*-1-(*p*-(2-Dimethylaminoethoxy)phenyl)-1-(4-hydroxy-3-methoxyphenyl)-2-phenylbut-1-ene sulphate**
Probably formed from: *trans*-1-(*p*-(2-Dimethylaminoethoxy)phenyl)-1-(4-hydroxy-3-methoxyphenyl)-2-phenylbut-1-ene

***trans*-1-(*p*-(2-Dimethylaminoethoxy)phenyl)-1-(*m*-hydroxyphenyl)-2-(hydroxyphenyl)but-1-ene**
Formed from: Droloxifene

***trans*-1-(*p*-β-Dimethylaminoethoxyphenyl)-2-(*p*-hydroxyphenyl)-1-(*p*-iodophenyl)but-1-ene**
Formed from: *trans*-1-(*p*-β-Dimethylaminoethoxyphenyl)-1-(*p*-iodophenyl)-2-phenylbut-1-ene

***cis*-1-(*p*-Dimethylaminoethoxyphenyl)-1-(*p*-hydroxyphenyl)-2-phenylbut-1-ene**
Formed from: *trans*-1-(*p*-Dimethylaminoethoxyphenyl)-1-(*p*-hydroxyphenyl)-2-phenylbut-1-ene
→ *trans*-1-(*p*-Dimethylaminoethoxyphenyl)-1-(*p*-hydroxyphenyl)-2-phenylbut-1-ene
In: Man H360

***trans*-1-(*p*-Dimethylaminoethoxyphenyl)-1-(*p*-hydroxyphenyl)-2-phenylbut-1-ene** (4-hydroxytamoxifen)
Formed from: *cis*-1-(*p*-Dimethylaminoethoxyphenyl)-1-(*p*-hydroxyphenyl)-2-phenylbut-1-ene
Tamoxifen
→ 3,4-Dihydroxytamoxifen
In: Man J589

Rat J589
→ *trans*-1-(*p*-Dimethylaminoethoxyphenyl)-3-hydroxy-1-(*p*-hydroxyphenyl)-2-phenylbut-1-ene
Probably in: Dog A253
Man A252
Rhesus monkey A253
→ *cis*-1-(*p*-Dimethylaminoethoxyphenyl)-1-(*p*-hydroxyphenyl)-2-phenylbut-1-ene
In: Man H360
→ *trans*-1-(*p*-Dimethylaminoethoxyphenyl)-1-(*p*-hydroxyphenyl)-2-phenylbut-1-ene-β-D-glucuronide
In: Man G844
See also: A253, F131, F772

***trans*-1-(*p*-Dimethylaminoethoxyphenyl)-1-(*p*-hydroxyphenyl)-2-phenylbut-1-ene-β-D-glucuronide**
Formed from: *trans*-1-(*p*-Dimethylaminoethoxyphenyl)-1-(*p*-hydroxyphenyl)-2-phenylbut-1-ene

1-(*p*-(2-Dimethylaminoethoxy)phenyl)-1-(*p*-hydroxyphenyl)-2-phenylbut-1-ene-N-oxide
Probably formed from: Tamoxifen-N-oxide

***trans*-1-(*p*-(2-Dimethylaminoethoxy)phenyl)-1-(*m*-hydroxyphenyl)-2-phenylbut-1-en-2-one**
Formed from: Droloxifene
→ *trans*-1-(*p*-(2-Dimethylaminoethoxy)phenyl)-1-(*m*-hydroxyphenyl)-2-phenylbut-1-en-2-one-β-D-glucuronide
Probably in: Mouse H924

***trans*-1-(*p*-(2-Dimethylaminoethoxy)phenyl)-1-(*m*-hydroxyphenyl)-2-phenylbut-1-en-2-one-β-D-glucuronide**
Probably formed from: *trans*-1-(*p*-(2-Dimethylaminoethoxy)phenyl)-1-(*m*-hydroxyphenyl)-2-phenylbut-1-en-2-one

***trans*-1-(*p*-β-Dimethylaminoethoxyphenyl)-1-(*p*-iodophenyl)-2-phenylbut-1-ene** (4-iodotamoxifen)
→ *trans*-1-(*p*-β-Dimethylaminoethoxyphenyl)-3-hydroxy-1-(*p*-iodophenyl)-2-phenylbut-1-ene
In: Rat F750
→ *trans*-1-(*p*-β-Dimethylaminoethoxyphenyl)-2-(*p*-hydroxyphenyl)-1-(*p*-iodophenyl)but-1-ene
In: Rat F750
→ *trans*-1-(*p*-β-Dimethylaminoethoxyphenyl)-1-(*p*-iodophenyl)-2-phenylbut-1-ene-N-oxide
In: Rat F750
→ *trans*-1-(*p*-Iodophenyl)-1-(*p*-β-methylaminoethoxyphenyl)-2-phenylbut-1-ene
In: Rat F750

***trans*-1-(*p*-β-Dimethylaminoethoxyphenyl)-1-(*p*-iodophenyl)-2-phenylbut-1-ene-N-oxide**
Formed from: *trans*-1-(*p*-β-Dimethylaminoethoxyphenyl)-1-(*p*-iodophenyl)-2-phenylbut-1-ene

N-((4-(2-(Dimethylamino)ethoxy)phenyl)methyl)-3,4-dimethoxybenzamide
See: Itopride

N-(2-(Dimethylamino)ethyl)acridine-4-carboxamide
→ N-(2-(Dimethylamino)ethyl)acridine-4-carboxamide-N-oxide
In: Man J323, K142
Mouse G583
Rat G565, J323
→ N-(2-(Dimethylamino)ethyl)-9(10*H*)-acridone-4-carboxamide
In: Man K142
Mouse G583
Rat G565
→ N-(2-(Methylamino)ethyl)acridine-4-carboxamide
In: Man K142
Mouse G583
Rat G565

N-(2-(Dimethylamino)ethyl)acridine-4-carboxamide-N-oxide
Formed from: N-(2-(Dimethylamino)ethyl)acridine-4-carboxamide

N-(2-(Dimethylamino)ethyl)-9(10*H*)-acridone-4-carboxamide
Formed from: N-(2-(Dimethylamino)ethyl)acridine-4-carboxamide
→ N-(2-(Dimethylamino)ethyl)-9(10*H*)-acridone-4-carboxamide-N-oxide
Probably in: Man K142
Rat G565

→ N-(2-(Dimethylamino)ethyl)-7-hydroxy-9(10*H*)-acridone-4-carboxamide
Probably in: Mouse G583
→ N-(2-(Methylamino)ethyl)-9(10*H*)-acridone-4-carboxamide
Probably in: Man K142
Mouse G583

N-(2-(Dimethylamino)ethyl)-9(10*H*)-acridone-4-carboxamide-N-oxide
Probably formed from: N-(2-(Dimethylamino)ethyl)-9(10*H*)-acridone-4-carboxamide

1-(2-Dimethylaminoethyl)amino-4-(2-(dimethylamino-N-oxide)ethyl)amino-5,8-dihydroxyanthraquinone
Formed from: 1,4-Bis(2-(dimethylamino-N-oxide)ethyl)amino-5,8-dihydroxyanthraquinone
→ 1,4-Bis(2-dimethylaminoethyl)amino-5,8-dihydroxyanthraquinone
In: Rat J753

3-(2-Dimethylaminoethyl)benzo[*b*]thiophene
→ Benzo[*b*]thiophen-3-ylacetic acid
In: Rat A1293
→ 3-(2-Dimethylaminoethyl)-6-hydroxybenzo[*b*]thiophene
In: Rat A1293
→ 3-(2-Methylaminoethyl)benzo[*b*]thiophene
In: Rat A1293

***d-cis*-5-(2-Dimethylaminoethyl)-2,3-dihydro-3-hydroxy-2-(*p*-hydroxyphenyl)-1,5-benzothiazepin-4(5*H*)-one**
Probably formed from: *d-cis*-5-(2-Dimethylaminoethyl)-2,3-dihydro-3-hydroxy-2-(*p*-methoxyphenyl)-1,5-benzothiazepin-4(5*H*)-one

***d-cis*-5-(2-Dimethylaminoethyl)-2,3-dihydro-3-hydroxy-2-(*p*-methoxyphenyl)-1,5-benzothiazepin-4(5*H*)-one**
Formed from: *d-cis*-3-Acetoxy-5-(2-dimethylaminoethyl)-2,3-dihydro-2-(*p*-methoxyphenyl)-1,5-benzothiazepin-4(5*H*)-one
→ *d-cis*-5-(2-Dimethylaminoethyl)-2,3-dihydro-3-hydroxy-2-(*p*-hydroxyphenyl)-1,5-benzothiazepin-4(5*H*)-one
Probably in: Rat A391
→ *d-cis*-2,3-Dihydro-3-hydroxy-2-(*p*-methoxyphenyl)-5-(2-methylaminoethyl)-1,5-benzothiazepin-4(5*H*)-one
Probably in: Rat A391

4-(β-Dimethylaminoethyl)-2,3-dihydro-6-hydroxymethyl-2-phenyl-1,4-benzoxazin-3-one
Formed from: 4-(β-Dimethylaminoethyl)-2,3-dihydro-6-methyl-2-phenyl-1,4-benzoxazin-3-one
→ 2,3-Dihydro-6-hydroxymethyl-4-(β-methylaminoethyl)-2-phenyl-1,4-benzoxazin-3-one
Probably in: Rat A3358

4-(β-Dimethylaminoethyl)-2,3-dihydro-7-hydroxy-6-methyl-2-phenyl-1,4-benzoxazin-3-one
Formed from: 4-(β-Dimethylaminoethyl)-2,3-dihydro-6-methyl-2-phenyl-1,4-benzoxazin-3-one
→ 2,3-Dihydro-7-hydroxy-6-methyl-4-(β-methylaminoethyl)-2-phenyl-1,4-benzoxazin-3-one
Probably in: Rat A3358

4-(β-Dimethylaminoethyl)-2,3-dihydro-2-(*p*-hydroxyphenyl)-6-methyl-1,4-benzoxazin-3-one
Formed from: 4-(β-Dimethylaminoethyl)-2,3-dihydro-6-methyl-2-phenyl-1,4-benzoxazin-3-one
→ 2,3-Dihydro-2-(*p*-hydroxyphenyl)-6-methyl-4-(β-methylaminoethyl)-1,4-benzoxazin-3-one
Probably in: Rat A3358

10β-Dimethylaminoethyl-10,11-dihydro-5-methyl-11-oxo-5*H*-dibenzo[*b*,*e*][1,4]diazepine
(dibenzepine)
→ 10,11-dihydro-5-methyl-10β-methylaminoethyl-11-oxo-5*H*-dibenzo[*b*,*e*][1,4]diazepine
In: Man A712
→ 10β-Dimethylaminoethyl-10,11-dihydro-11-oxo-5*H*-dibenzo[*b*,*e*][1,4]diazepine
In: Man A712

4-(β-Dimethylaminoethyl)-2,3-dihydro-6-methyl-2-phenyl-1,4-benzoxazin-3-one
→ 2,3-Dihydro-6-methyl-4-(β-methylaminoethyl)-2-phenyl-1,4-benzoxazin-3-one

In: Rat A3358
→ 4-(β-Dimethylaminoethyl)-2,3-dihydro-6-hydroxymethyl-2-phenyl-1,4-benzoxazin-3-one
In: Rat A3358
→ 4-(β-Dimethylaminoethyl)-2,3-dihydro-7-hydroxy-6-methyl-2-phenyl-1,4-benzoxazin-3-one
In: Rat A3358
→ 4-(β-Dimethylaminoethyl)-2,3-dihydro-2-(*p*-hydroxyphenyl)-6-methyl-1,4-benzoxazin-3-one
In: Rat A3358
→ 4-(β-Dimethylaminoethyl)-2,3-dihydro-6-methyl-2-phenyl-1,4-benzoxazin-3-one-N-oxide
In: Rat A3358

4-(β-Dimethylaminoethyl)-2,3-dihydro-6-methyl-2-phenyl-1,4-benzoxazin-3-one-N-oxide
Formed from: 4-(β-Dimethylaminoethyl)-2,3-dihydro-6-methyl-2-phenyl-1,4-benzoxazin-3-one

10β-Dimethylaminoethyl-10,11-dihydro-11-oxo-5*H*-dibenzo[*b*,*e*][1,4]diazepine
Formed from: 10β-Dimethylaminoethyl-10,11-dihydro-5-methyl-11-oxo-5*H*-dibenzo[*b*,*e*][1,4]diazepine

N-(2-(Dimethylamino)ethyl)-7-hydroxy-9(10*H*)-acridine-4-carboxamide
Probably formed from: N-(2-(Dimethylamino)ethyl)-9(10*H*)-acridine-4-carboxamide

3-(2-Dimethylaminoethyl)-6-hydroxybenzo[*b*]thiepine
Formed from: 3-(2-Dimethylaminoethyl)benzo[*b*]thiepine

N-(2-Dimethylaminoethyl)-*p*-hydroxybenzylamine
Formed from: Pyrilamine

(1*S*)-4-((3-(2-Dimethylaminoethyl)-1*H*-indol-5-yl)methyl)-2-oxazolidine
See: Zolmitriptan

3-(2-(Dimethylamino)ethyl)-N-methyl-1*H*-indole-5-methanesulphonamide
See: Sumatriptan

4-(2-Dimethylaminoethyl)-*o*-quinone
Probably formed from: 3,4-Dihydroxy-N,N-dimethylphenethylamine

4-Dimethylamino-3′-fluoroazobenzene
→ 4′-Dimethylamino-3-fluoro-4-hydroxyazobenzene
In: Rat A776
→ 3′-Fluoro-4-methylaminoazobenzene
Probably in: Rat A776

4-Dimethylamino-4′-fluoroazobenzene
→ 4′-Dimethylamino-4-hydroxyazobenzene
In: Rat A1979
→ 4′-Fluoro-4-methylaminoazobenzene
In: Rat A1979

4′-Dimethylamino-3-fluoro-4-hydroxyazobenzene
Formed from: 4-Dimethylamino-3′-fluoroazobenzene
→ 3-Fluoro-4-hydroxy-4′-methylaminoazobenzene
Probably in: Rat A776

4′-Dimethylamino-3-formylazobenzene
Probably formed from: 4′-Dimethylamino-3-hydroxymethylazobenzene
→ 3-Carboxy-4′-dimethylaminoazobenzene
Probably in: Rat G9

6-Dimethylamino-4-formylcarbostyril
→ 4-Carboxy-6-dimethylaminocarbostyril
In: Man J24

6-Dimethylamino-4-formylcoumarin
→ 4-Carboxy-6-dimethylaminocoumarin
In: Man J24

4′-Dimethylamino-2-hydroxyazobenzene
Formed from: 4-Dimethylaminoazobenzene

4′-Dimethylamino-4-hydroxyazobenzene
Formed from: 4′-Bromo-4-dimethylaminoazobenzene
4′-Chloro-4-dimethylaminoazobenzene
4-Dimethylaminoazobenzene
4-Dimethylamino-4′-fluoroazobenzene
4-Dimethylamino-4′-iodoazobenzene
4-Dimethylamino-4-methoxyazobenzene
→ *p*-Aminophenol
In: Rat G133, G511
→ 4′-Dimethylaminoazobenzen-4-yl-β-D-glucuronide
In: Rat E240
→ 4′-Dimethylaminoazobenzen-4-yl sulphate
Probably in: Rat D68
→ N,N-Dimethyl-*p*-phenylenediamine

In: Rat G133, G511
→ 4-Hydroxy-4′-methylaminoazobenzene
In: Rat B985
See also: A1979, A1981, D68

4-Dimethylamino-3-hydroxybenzonitrile
Formed from: *p*-Dimethylaminobenzonitrile
→ 4-Dimethylamino-3-hydroxybenzonitril-3-yl sulphate
Probably in: Rat D160

(−)-*cis*-2-(α-Dimethylamino-*m*-hydroxybenzyl) cyclohexanol
See: Ciramadol

4′-Dimethylamino-4-hydroxydiphenylamine
Formed from: 2-Acetamido-N,N-dimethylindoaniline

4′-Dimethylamino-4-hydroxy-3-methoxyazobenzene
Formed from: 4-Dimethylamino-3′-methoxyazobenzene

4-Dimethylamino-3′-hydroxymethylazobenzene
Formed from: 4-Dimethylamino-3′-methylazobenzene
→ 4′-Dimethylamino-3-formylazobenzene
Probably in: Rat A3783, A3835, G9
→ 3′-Hydroxymethyl-4-methylaminoazobenzene
Probably in: Rat E872

4′-Dimethylamino-4-hydroxy-3-methylazobenzene
→ 4-Amino-*o*-cresol
Probably in: Rat A3835
→ 4-Hydroxy-3-methyl-4′-methylaminoazobenzene
Probably in: Rat A3835, B9

4-Dimethylamino-3-hydroxymethyl-2-methyl-1-phenyl-3-pyrazolin-5-one
Formed from: Aminopyrine

S,S,S-(2-Dimethylamino-5-hydroxy-1,3,4-phenylene)tris-L-cysteine
Probably formed from: S,S,S-(2-Dimethylamino-5-hydroxy-1,3,4-phenylene)tris-glutathione

S,S-(2-Dimethylamino-5-hydroxy-1,3,4-phenylene)bis-glutathione
Formed from: 4-Dimethylaminophenol
→ S,S,S-(2-Dimethylamino-5-hydroxy-1,3,4-phenylene)tris-glutathione
Probably in: Dog A2970
Man D642
Rat B182

S,S,S-(2-Dimethylamino-5-hydroxy-1,3,4-phenylene)tris-glutathione
Formed from: 4-Dimethylaminophenol
→ S,S,S-(2-Dimethylamino-5-hydroxy-1,3,4-phenylene)tris-L-cysteine
Probably in: Rat B182

2-Dimethylamino-4-(*p*-hydroxyphenyl)-4-phenyl-5-heptanone
Formed from: 2-Dimethylamino-4,4-diphenyl-5-heptanone

4-Dimethylamino-3′-iodoazobenzene
→ 3′-Iodo-4-methylaminoazobenzene
In: Rat A776

4-Dimethylamino-4′-iodoazobenzene
→ 4′-Dimethylamino-4-hydroxyazobenzene
In: Rat A1797
→ 4′-Iodo-4-methylaminoazobenzene
In: Rat A1797

4-Dimethylamino-3′-methoxyazobenzene
→ 4′-Dimethylamino-4-hydroxy-3-methoxyazobenzene
In: Rat A776
→ 3′-Methoxy-4-methylaminoazobenzene
In: Rat A776

4-Dimethylamino-4′-methoxyazobenzene
→ 4′-Dimethylamino-4-hydroxyazobenzene
In: Rat A1797
→ 4′-Methoxy-4-methylaminoazobenzene
In: Rat A1797

1-(*p*-Dimethylaminomethoxyphenyl)-1-(*p*-hydroxyphenyl)-2-phenylbut-1-ene
Formed from: Tamoxifen
→ 3,4-Dihydroxytamoxifen
In: Man J589
Rat J589

1-(2-Dimethylamino-1-(*p*-methoxyphenyl) ethyl)cyclohexane-1,4-diol
Formed from: Venlafaxine

1-(2-Dimethylamino-1-(4-methoxyphenyl)ethyl) cyclohexanol
See: Venlafaxine

4-Dimethylamino-4′-(N-methylacetamido) azobenzene
→ 4′-Acetamido-4-dimethylaminoazobenzene
In: Rat A2833
→ 4′-(N-Methylacetamido)-4-methylaminoazobenzene
In: Rat A2833

4′-Dimethylamino-4-methylaminobenzophenone
Formed from: Michler's ketone
→ 4-Amino-4′-dimethylaminobenzophenone
Probably in: Rat B713

4-Dimethylamino-2-methylazobenzene
→ Aniline
In: Rat G133, G511
→ N,N,2-Trimethyl-*p*-phenylenediamine
In: Rat G133, G511

4-Dimethylamino-3′-methylazobenzene
→ 4-Amino-*o*-cresol
Probably in: Rat A1756
→ 4′-Dimethylamino-4-hydroxy-3-methylazobenzene
In: Rat A1114, A1748, A1756, A3835, B9, G9
→ 4-Dimethylamino-3′-hydroxymethylazobenzene
In: Rat A3835, E872, G9
See also: A3783
→ N,N-Dimethyl-*p*-phenylenediamine
In: Microorganism D257
Rat G133, G511, D257
→ 3′-Methyl-4-methylaminoazobenzene
In: Rat A3835, E872, G9
See also: A673, A3783
→ *m*-Toluidine
In: Microorganism D257
Rat A673, A3783, G133, G511, D257

4-Dimethylamino-4′-methylazobenzene
→ 4′-Methyl-4-methylaminoazobenzene
In: Rat A1979

***p*-Dimethylaminomethylbenzaldehyde**
Probably formed from: *p*-Dimethylaminomethylbenzylamine

***p*-Dimethylaminomethylbenzylamine**
→ *p*-Dimethylaminomethylbenzaldehyde
Probably in: Man A1686
Pig A3693

3-Dimethylaminomethyl-1,2-diphenyl-2-butyl propionate
→ 3-Dimethylaminomethyl-1-(hydroxyphenyl)-2-phenyl-2-butyl propionate
In: Rat A1683
→ 3-Methylaminomethyl-1,2-diphenyl-2-butyl propionate
In: Dog A1683
Man A1168, A1683
Rat H803

α-Dimethylaminomethyl-2-(3-ethyl-5-methyl-4-isoxazolyl)-6-hydroxy-1*H*-indole-3-methanol
Formed from: α-Dimethylaminomethyl-2-(3-ethyl-5-methyl-4-isoxazolyl)-1*H*-indole-3-methanol

α-Dimethylaminomethyl-2-(3-ethyl-5-methyl-4-isoxazolyl)-1*H*-indole-3-methanol
→ α-Dimethylaminomethyl-2-(3-ethyl-5-methyl-4-isoxazolyl)-6-hydroxy-1*H*-indole-3-methanol
In: Man E815
→ α-Dimethylaminomethyl-2-(3-ethyl-5-methyl-4-isoxazolyl)-1*H*-indole-3-methyl-β-D-glucuronide
In: Man E815

α-Dimethylaminomethyl-2-(3-ethyl-5-methyl-4-isoxazolyl)-1*H*-indole-3-methyl-β-D-glucuronide
Formed from: α-Dimethylaminomethyl-2-(3-ethyl-5-methyl-4-isoxazolyl)-1*H*-indole-3-methanol

α-(2-Dimethylamino-1-methylethyl)-α-phenylbenzeneethanol (propoxyphene)
→ α-(1-Methyl-2-methylaminoethyl)-α-phenylbenzeneethanol
In: Man A1168

4-Dimethylaminomethyl-7-hydroxycoumarin
Formed from: 4-Dimethylaminomethyl-7-methoxycoumarin

3-Dimethylaminomethyl-1-hydroxyphenyl-2-phenyl-2-butyl propionate
Formed from: 3-Dimethylaminomethyl-1,2-diphenyl-2-butyl propionate

3-Dimethylaminomethylindole
See: Gramine

4-Dimethylaminomethyl-7-methoxycoumarin
→ 4-Dimethylaminomethyl-7-hydroxycoumarin
In: Man K339

4-Dimethylamino-3-methyl-1-phenyl-3-pyrazolin-5-one
Formed from: Aminopyrine

6-Dimethylaminomethyl-β-naphthaldehyde
→ 6-Dimethylaminomethyl-β-naphthoic acid
In: Man J24

6-Dimethylaminomethyl-β-naphthoic acid
Formed from: 6-Dimethylaminomethyl-β-naphthaldehyde

5-Dimethylamino-1-naphthalenesulphonamide (dansylamide)
→ 5-Methylamino-1-naphthalenesulphonamide
In: Rat A1365

4-Dimethylamino-3′-nitroazobenzene
→ 4-Methylamino-3′-nitroazobenzene
In: Rat A776

5-(5-(Dimethyamino)penta-1-en-3-yn-1-yl)-5*H*-dibenzo[*a*,*d*]cycloheptene
See: Intriptyline

10-(N,N-Dimethylaminopentyl)-2-trifluoromethylphenothiazine
→ 10-(N,N-Dimethylaminopentyl)-2-trifluoromethylphenothiazine-N-oxide
In: Man G579
→ 10-(N,N-Dimethylaminopentyl)-2-trifluoromethylphenothiazine sulphoxide
In: Man G579

10-(N,N-Dimethylaminopentyl)-2-trifluoromethylphenothiazine-N-oxide
Formed from: 10-(N,N-Dimethylaminopentyl)-2-trifluoromethylphenothiazine

10-(N,N-Dimethylaminopentyl)-2-trifluoromethylphenothiazine sulphoxide
Formed from: 10-(N,N-Dimethylaminopentyl)-2-trifluoromethylphenothiazine

***o*-Dimethylaminophenol**
Formed from: N,N-Dimethylaniline

***m*-Dimethylaminophenol**
Formed from: *m*-Hydroxyphenyl trimethylammonium

***p*-Dimethylaminophenol**
Formed from: N,N-Dimethylaniline
→ 2,3-Bis(CoA-S-yl)-4-dimethylamino-5-(S-glutathionyl)phenol
Probably in: Man F138
→ 2-(CoA-S-yl)-4-dimethylamino-3,5-bis(S-glutathionyl)phenol
Probably in: Man F138
→ S,S-(2-Dimethylamino-5-hydroxy-1,3-phenylene)bis-glutathione
In: Man D642
See also: A2970, B182
→ *p*-Dimethylaminophenyl-β-D-glucuronide
In: Dog A2970
Rat A2971, B182, B183
→ *p*-Dimethylaminophenyl sulphate
In: Dog A2970
Rat A2971, B182, B183
→ 4-Dimethylamino-2,3,5-tris(S-glutathionyl)phenol
In: Man F138
→ *p*-Methylaminophenol
In: Rabbit C89
→ 2,3,5-Tris(CoA-S-yl)-4-dimethylaminophenol
Probably in: Man D642

***p*-Dimethylaminophenyl acetate**
→ *p*-Methylaminophenyl acetate
In: Rat D578

6-(*p*-Dimethylaminophenyl)azobenzothiazole
→ 6-Aminobenzothiazole
In: Microorganism D257
Rat D257
→ N,N-Dimethyl-*p*-phenylenediamine
In: Microorganism D257
Rat D257

5-(*p*-Dimethylaminophenyl)azoindazole
→ 6-Aminoindazole
In: Microorganism D257
Rat D257
→ N,N-Dimethyl-*p*-phenylenediamine

In: Microorganism D257
Rat D257

2-Dimethylamino-2-phenylbutan-1-ol
Formed from: Trimebutine
→ 2-Methylamino-2-phenylbutan-1-ol
Probably in: Dog F28
Rat F28

((1*R*,2*S*)-rel-2-(Dimethylamino)-1-phenyl-3-cyclohexane-1-carboxylic acid ethyl ester)
See: Tilidine

***p*-Dimethylaminophenyl-β-D-glucuronide**
Formed from: *p*-Dimethylaminophenol

(11β,13α,17α)-11-(4-(Dimethylamino)phenyl)-17-hydroxy-17-(3-hydroxypropyl)oestra-4,9-dien-3-one
See: Onapristone

11-β-(*p*-Dimethylaminophenyl)-17β-hydroxy-17α-(1-(3-hydroxy)propynyl) oestra-4,9-dien-3-one
Formed from: Mifepristone
→ 17β-Hydroxy-17α-(1-(3-hydroxy) propynyl)-11-β-(*p*-methylaminophenyl) oestra-4,9-dien-3-one
Probably in: Man J752
Monkey J752
Rat J752

2-(Dimethylamino)-1-phenyl-1-propanone
See: Dimethylpropion

***p*-Dimethylaminophenyl sulphate**
Formed from: *p*-Dimethylaminophenol

2-(Dimethylamino)propiophenone
→ N-Methylephedrine
In: Man A504

3-(3-Dimethylaminopropoxy)-1-(*p*-hydroxybenzyl)-1*H*-indazole
Formed from: Benzydamine

3-(3-Dimethylaminopropoxy)-1*H*-indazole
Formed from: Benzydamine

9-((3-Dimethylaminopropyl)amino)-1-nitroacridine
See: Nitracrine

9-((3-Dimethylaminopropyl)amino)-1-(S-glutathionyl)acridine-N(10)-oxide
Formed from: Nitracrine
Nitracrine-N(10)-oxide

4-(3-Dimethylaminopropyl)-3,4-dihydro-2-(1-hydroxyethyl)-3-phenyl-2*H*-1,4-benzothiazine
→ 3,4-Dihydro-2-(1-hydroxyethyl)-4-(3-methylaminopropyl)-3-phenyl-2*H*-1,4-benzothiazine
In: Dog A1888
Monkey A1888
Rat A1769, A1888
→ 4-(3-Dimethylaminopropyl)-3,4-dihydro-2-(1-hydroxyethyl)-3-phenyl-2*H*-1,4-benzothiazine-N-oxide
In: Dog A1888
Monkey A1888
→ 4-(3-Dimethylaminopropyl)-3,4-dihydro-2-(1-hydroxyethyl)-3-phenyl-2*H*-1,4-benzothiazine sulphoxide
In: Dog A1888
Monkey A1888
Rat A1769
→ 4-(3-Dimethylaminopropyl)-3,4-dihydro-hydroxy-2-(1-hydroxyethyl)-3-phenyl-2*H*-1,4-benzothiazine
In: Dog A1888
Monkey A1888
Rat A1769

4-(3-Dimethylaminopropyl)-3,4-dihydro-2-(1-hydroxyethyl)-3-phenyl-2*H*-1,4-benzothiazine-N-oxide
Formed from: 4-(3-Dimethylaminopropyl)-3,4-dihydro-2-(1-hydroxyethyl)-3-phenyl-2*H*-1,4-benzothiazine

4-(3-Dimethylaminopropyl)-3,4-dihydro-2-(1-hydroxyethyl)-3-phenyl-2*H*-1,4-benzothiazine sulphoxide
Formed from: 4-(3-Dimethylaminopropyl)-3,4-dihydro-2-(1-hydroxyethyl)-3-phenyl-2*H*-1,4-benzothiazine
→ 3,4-Dihydro-2-(1-hydroxyethyl)-4-(3-methylaminopropyl)-3-phenyl-2*H*-1,4-benzothiazine sulphoxide
Probably in: Rat A1769

4-(3-Dimethylaminopropyl)-3,4-dihydro-hydroxy-2-(1-hydroxyethyl)-3-phenyl-2*H*-1,4-benzothiazine
Formed from: 4-(3-Dimethylaminopropyl)-3,4-dihydro-2-(1-hydroxyethyl)-3-phenyl-2*H*-1,4-benzothiazine

→ 3,4-Dihydro-hydroxy-2-(1-hydroxyethyl)-4-(3-methylaminopropyl)-3-phenyl-2*H*-1,4-benzothiazine
Probably in: Rat A1769

1-((3-Dimethylamino)propyl)-1-(ergolin-8β-yl) carbonyl-3-ethylurea
Formed from: Cabergoline

4-(3-Dimethylaminopropylidene)-4,9-dihydrothieno[2,3-*c*][2]benzothiepin
See: Dithiadene

10-(3-Dimethylaminopropyl)phenothiazine
See: Promazine

1-(*o*-((3-Dimethylamino)propyl)sulphinyl)phenyl-3-methylurea
Formed from: 1-(*o*-((3-Dimethylamino) propyl)thio)phenyl-3-methylurea

1-(*o*-((3-Dimethylamino)propyl)sulphinyl)phenyl-3-methylurea-N-oxide
Probably formed from: 1-(*o*-((3-Dimethylamino)propyl)thio)phenyl-3-methylurea-N-oxide

1-(*o*-((3-Dimethylamino)propyl)thio)phenyl-3-methylurea
→ 1-(*o*-((3-Dimethylamino)propyl)sulphinyl) phenyl-3-methylurea
In: Dog A436
→ 1-(*o*-((3-Dimethylamino)propyl)thio) phenyl-3-methylurea-N-oxide
In: Dog A436
→ 3-Methyl-1-(*o*-((3-methylamino)propyl) thio)phenylurea
In: Dog A436

1-(*o*-((3-Dimethylamino)propyl)thio)phenyl-3-methylurea-N-oxide
Formed from: 1-(*o*-((3-Dimethylamino) propyl)thio)phenyl-3-methylurea
→ 1-(*o*-((3-Dimethylamino)propyl) sulphinyl)phenyl-3-methylurea-N-oxide
Probably in: Dog A436

10-(3-Dimethylaminopropyl)-2-trifluoromethylphenothiazine
(triflupromazine)
→ 10-(3-Methylaminopropyl)-2-trifluoromethylphenothiazine
Probably in: Rat A69
→ Triflupromazine-N-β-D-glucuronide
In: Man H541

***cis*-Dimethylaminostilbene**
→ *threo*-1-(4-Acetamido)phenyl-2-phenyl-1,2-ethanediol
In: Rat A2026

***trans*-Dimethylaminostilbene**
→ *threo*-1-(4-Acetamido)phenyl-2-phenyl-1,2-ethanediol
In: Rat A2026

4-Dimethylamino-*m*-tolyl N-methylcarbamate
See: Aminocarb

4-Dimethylamino-2,3,5-tris(S-glutathionyl)phenol
Formed from: *p*-Dimethylaminophenol

N,N-Dimethylamphetamine
N-Methylamphetamine
→ N,N-Dimethylamphetamine-N-oxide
In: Man E783
Rat E783
→ *p*-Hydroxy-N,N-dimethylamphetamine
In: Man E783
Rat E783
→ N-Methylamphetamine
In: Man E783
Rat A300, E783

4-Dimethylamino-3,5-xylyl N-methylcarbamate
See: Mexacarbate

N,N-Dimethylamphetamine-N-oxide
Formed from: N,N-Dimethylamphetamine

N,N-Dimethylaniline
Formed from: N,N-Dimethylaniline-N-oxide
→ N,N-Dimethylaniline-N-oxide
In: Man A1143
Mouse C877
Pig C877, D976, K160
Rabbit A1144, A3161, A3484, C89, C126, C196, C877, G928
Rat A1091, A1746, C126, C196, C751, C877, F336
Turbot H155
→ *o*-Dimethylaminophenol
In: Guinea pig C89
Rabbit C89
→ *p*-Dimethylaminophenol
In: Guinea pig C89
Rabbit C89
Rat A1091
See also: B185
→ N-Methylaniline

In: Beef C814
Dog A1250
Glycine K419
Horseradish C814
Man K419
Rabbit A3484, C89, D40, E828, F839
Rat C751, C669, C814, D578, F336
Whale C814

2,3-Dimethylaniline
→ 2,3-Dimethylacetanilide
In: Rabbit E399

2,4-Dimethylaniline
Probably formed from: Amitraz
→ 2-Amino-3,5-dimethylphenol
In: Dog F240
→ 4-Amino-3-methylbenzoic acid
In: Dog F240
→ 2,4-Dimethylacetanilide
In: Rabbit E399

2,5-Dimethylaniline
→ 2,5-Dimethylacetanilide
In: Rabbit E399

2,6-Dimethylaniline
Formed from: N-Ethylglycine-2,6-xylidide
N-Isopropylglycyl-2,6-dimethylanilide
Lignocaine
N-Methylglycyl-2,6-dimethylanilide
N-(*n*-Propyl)glycyl-2,6-dimethylanilide
Probably formed from: Bupivacaine
Etidocaine
→ 4-Amino-3,5-dimethylphenol
Probably in: Man A1744, J67
→ 2-Amino-3-methylbenzoic acid
Probably in: Rabbit E53

3,4-Dimethylaniline (4-amino-*o*-xylene)
→ 3,4-Dimethylacetanilide
In: Rabbit E399
→ N-Hydroxy-3,4-dimethylaniline
In: Rat C192

3,5-Dimethylaniline
→ 3,5-Dimethylacetanilide
In: Rabbit E399

N,N-Dimethylaniline-N-oxide
Formed from: N,N-Dimethylaniline
→ N,N-Dimethylaniline
In: Rat A1913
With cytochrome c B185
→ N-Methylaniline
In: Rabbit D40
Rat A1847
With cytochrome c B185
See also: A1091, E828

2,4-Dimethylaniline-5-sulphonic acid
Formed from: 1-Hydroxy-2-(5-sulpho-2,4-xylylazo)naphthalene-4-sulphonic acid

N,N-Dimethyl-*p*-anisidine
→ N-Methyl-*p*-anisidine
In: Rat D578
→ *p*-Quinone
In: Beef F500

2,3-Dimethylanisole
Formed from: 2,3-Dimethylphenol

2,4-Dimethylanisole
Formed from: 2,4-Dimethylphenol

2,5-Dimethylanisole
Formed from: 2,5-Dimethylphenol

2,6-Dimethylanisole
Formed from: 2,6-Dimethylphenol

3,4-Dimethylanisole
Formed from: 3,4-Dimethylphenol

3,5-Dimethylanisole
Formed from: 3,5-Dimethylphenol

9,10-Dimethylanthracene
Probably formed from: 9-Methylanthracene
→ 1,2-Dihydro-1,2-dihydroxy-9,10-dimethylanthracene
In: Salmonella D805
→ 9-Hydroxymethyl-10-methylanthracene
In: Horseradish H796
Salmonella D805
With cytochrome P450 H796
See also: E612

N,N-Dimethylanthranilic acid
→ Pyrocatechuic acid
In: Trichosporon E270

4,4′-Dimethylazobenzene
Formed from: 4,4′-Dimethylazoxybenzene

4,4′-Dimethylazoxybenzene
→ 4,4′-Dimethylazobenzene
In: Saccharomyces H896

3,4-Dimethylbenzaldehyde
Probably formed from: 3,4-Dimethylbenzyl alcohol

→ 3,4-Dimethylbenzoic acid
Probably in: Pseudomonas D236

N,N-Dimethylbenzamide
→ N-Hydroxymethyl-N-methylbenzamide
In: Mouse C319
Rat G683

3,5-Dimethylbenzamide
Formed from: 3,5-Dimethyl-N-2-picolylbenzamide
3,5-Dimethyl-N-4-picolylbenzamide

7,12-Dimethylbenzanthracene (7,12-dimethylbenz[*a*]anthracene)
Formed from: 7,12-Dimethylbenzanthracene-5,6-oxide
7-Methylbenzanthracene
12-Methylbenzanthracene
→ 1,2-Dihydro-1,2-dihydroxy-7,12-dimethylbenzanthracene
In: Man C889
Mouse C889
Rat C889
→ *trans*-3,4-Dihydro-3,4-dihydroxy-7,12-dimethylbenzanthracene
In: Cunninghamella C919
Man G340
Mouse B376
Rainbow trout G889
Rat B784, G340
Syncephalastrum E514
→ *trans*-5,6-Dihydro-5,6-dihydroxy-7,12-dimethylbenzanthracene
In: Rat F374
→ 8,9-Dihydro-8,9-dihydroxy-7,12-dimethylbenzanthracene (assumed to be *trans*)
In: Cunninghamella C919
Hamster A267, A2040, A3745
Man D339, G340
Mouse B376
Rabbit A2040
Rat A2040, B784, D53, G340
Syncephalastrum E514
See also: A2324
→ 10,11-Dihydro-10,11-dihydroxy-7,12-dimethylbenzanthracene
Mouse B376
Syncephalastrum E514
→ 7,12-Dimethylbenzanthracene-5,6-oxide
In: Rat E364
See also: A267, A1034, A3930, B376, B784, D442, D483, G340, E514
→ 7,12-Dimethylbenzanthracene-5,6-oxide
In: Rat E364
See also: A267, A1530, A1034, A3930, B376, B784, D442, D483, E514, G340
→ 7,12-Dimethylbenzanthracene-7,12-peroxide
In: Rat A54
See also: B784
→ 2-Hydroxy-7,12-dimethylbenzanthracene
In: Rat B60, B784
→ 3-Hydroxy-7,12-dimethylbenzanthracene
In: Hamster A267, A3745
Rat B784
Trout G889
See also: A2324
→ 4-Hydroxy-7,12-dimethylbenzanthracene
In: Hamster A267, A3745
Rat A3930, B784
See also: A2324
→ 5-Hydroxy-7,12-dimethylbenzanthracene
Probably in: Rat A1530, B880
→ 8-Hydroxy-7,12-dimethylbenzanthracene
Probably in: Rat A1530
→ 9-Hydroxy-7,12-dimethylbenzanthracene
Probably in: Rat A3930
→ 7-Hydroxymethyl-12-methylbenzanthracene
In: Hamster A267, A2040, A3745
Man G340, J381
Mouse A2087
Penicillium B779
Pseudomonas B779
Rabbit A2040
Rat A339, A1180, A2040, A3930, B784, B913, D53, D442, F136, F742, G491
Syncephalastrum E514
See also: A2324, A2863, B60, D339
→ 12-Hydroxymethyl-7-methylbenzanthracene
In: Hamster A267, A2040, A3745
Man G340, J381
Mouse A2087

Penicillium B779
Pseudomonas B779
Rabbit A2040
Rat A339, A1180, A2040, B784, B913, F136
See also: A2324, A2863, B60
Contrast (in rat) A3930
→ 7-Methylbenzanthracene
In: Rat G491

7,12-Dimethylbenzanthracene-5,6-oxide
→ *trans*-5,6-Dihydro-5,6-dihydroxy-7,12-dimethylbenzanthracene
In: Rat F374
→ S-(5,6-Dihydro-6-hydroxy-7,12-dimethylbenzanthr-5-yl)glutathione
Probably in: Rat A1034
Sheep A2284
→ 7,12-Dimethylbenzanthracene
In: Rat A350, F374

7,12-Dimethylbenzanthracene-7,12-peroxide
Formed from: 7,12-Dimethylbenzanthracene
→ 7,12-Dihydro-7,12-dihydroxy-7,12-dimethylbenzanthracene
In: Rat A54
See also: B784

7,12-Dimethylbenzanthracene-3,4-quinone
Formed from: 3,4-Dihydro-3,4-dihydroxy-7,12-dimethylbenzanthracene

1,2-Dimethylbenzene
See: *o*-Xylene

1,3-Dimethylbenzene
See: *m*-Xylene

1,4-Dimethylbenzene
See: *p*-Xylene

3,3′-Dimethylbenzidine
See: *o*-Tolidine

5,6-Dimethylbenzimidazole
Formed from:10-Formylmethylisoalloxazine
Riboflavine-5′-phosphate
Formed from erythrose in Eubacterium G461

2,2-Dimethylbenzo-1,3-dioxol-4-ol
Formed from: Bendiocarb

2,2-Dimethylbenzo-1,3-dioxol-4-ol N-hydroxymethylcarbamate
Formed from: Bendiocarb

2,2-Dimethylbenzo-1,3-dioxol-4-ol methylcarbamate
See: Bendiocarb

2,3-Dimethylbenzoic acid
Probably formed from: 1,2,3-Trimethylbenzene
→ 3,4-Dimethylcatechol
In: Pseudomonas G448

2,4-Dimethylbenzoic acid
Probably formed from: 2,4-Dimethylbenzyl alcohol
→ 2,4-Dimethylhippuric acid
In: Rat A1561
See also: F328

2,5-Dimethylbenzoic acid
Probably formed from: 2,5-Dimethylbenzyl alcohol
→ 2,5-Dimethylhippuric acid
In: Rat A1561
See also: F328

3,4-Dimethylbenzoic acid
Probably formed from: 3,4-Dimethylbenzaldehyde
→ 3,4-Dimethylhippuric acid
In: Rat A1561
See also: F328
→ 3,4-Dimethylphenol
Probably in: Pseudomonas D236

3,5-Dimethylbenzoic acid
Probably formed from: Mesitylene
→ 3,5-Dimethylhippuric acid
In: Rabbit F375
Rat A1561
→ 5-Methylisophthalic acid
In: Rabbit F375

4,4′-Dimethylbenzoin
Formed from: Tolualdehyde

1-(2,2-Dimethyl-2*H*-1-benzopyran-6-yl)ethanone
See: 7-Demethoxyencecalin

N,N-Dimethyl-*p*-benzoquinonediiminium cation
Formed from: N,N-Dimethyl-4-nitrosoaniline

2,4-Dimethylbenzyl alcohol
Formed from: 1,2,4-Trimethylbenzene
→ 2,4-Dimethylbenzoic acid
Probably in: Rat F328

2,5-Dimethylbenzyl alcohol
Formed from: 1,2,4-Trimethylbenzene
→ 2,5-Dimethylbenzoic acid
Probably in: Rat F328

3,4-Dimethylbenzyl alcohol
Formed from: 1,2,4-Trimethylbenzene
→ 3,4-Dimethylbenzaldehyde
Probably in: Pseudomonas D236

α,α-Dimethylbenzyl hydroperoxide
→ 2-Phenyl-2-propanol
In: Rat E382

4-(2,3-Dimethylbenzyl)imidazole
See: Detomidine

3,3′-((3,3′-Dimethyl(1,1′-biphenyl)-4,4′-diyl)bis (azo))bis(5-amino-4-hydroxy-2,7-naphthalenedisulphonic acid)
See: Direct Blue 14

6,6′-((3,3′-Dimethyl(1,1′-biphenyl)-4,4′-diyl)bis (azo))bis(4-amino-5-hydroxy-1,3-naphthalenedisulphonic acid)
See: Direct Blue 53

3,3′-(3,3′-Dimethyl(1,1′-biphenyl)-4,4′-diyl)bis (azo))bis(4-amino-1-naphthalenesulphonic acid)
See: Direct Red 2

3,3′-((3,3′-Dimethyl(1,1′-biphenyl)-4,4′-diyl)bis (azo))bis(4,5-dihydroxy-2,7-naphthalenedisulphonic acid)
See: Direct Blue 25

(+)-3,3′-Dimethyl-5,5′-bis(1-methylethyl)-(2,2′-binaphthalene)-1,1′,6,6′,7,7′-hexol
See: Apogossypol

1-(2-(1,3-Dimethyl-2-butenylidene)hydrazino) phthalazine
See: Budralazine

Dimethylcarbamic acid 5-(2-(1,1-dimethylethyl) amino)-1-hydroxyethyl)-1,3-phenylene ester
See: Bambuterol

N,N-Dimethylcarbamoylmethyl 4-(4-guanidinobenzoyloxy)phenylacetate
See: Camostat

1,3-Dimethylcarbazole
Formed from: N-Acetyl-1,3-dimethylcarbazole

3,3′-Di-O-methylcatechin
See: *cis*-4′5,7-Trihydroxy-3,3′-dimethoxyflavan

3,4-Dimethylcatechol
Formed from: 2,3-Dimethylbenzoic acid
2,3-Dimethylphenol
→ α-Hydroxy-δ,ε-dimethylmuconic semialdehyde
In: Pseudomonas B849

3,6-Dimethylcatechol
Formed from: 2,5-Dimethylphenol

4,5-Dimethylcatechol
Formed from: 3,4-Dimethylbenzoic acid
Probably formed from: 3,4-Dimethylphenol

5,6-Dimethylchrysene
→ 5-Methylchrysene
In: Rat G118
→ 6-Methylchrysene
In: Rat G118

N,N′-Dimethylclonidine (2-(2,6-dichlorophenylimino)-1,3-dimethylimidazolidine)
→ 4,5-Dehydro-N,N′-dimethylclonidine
In: Rat B214
→ *p*-Hydroxy-N,N′-dimethylclonidine
In: Rat B214
→ 4-Hydroxy-N,N′-dimethylclonidine
In: Rat B214

N,N-Dimethyldaunorubicin
→ Deoxydaunorubicinol aglycone
In: Mouse B109
Rabbit B109
→ N-Methyldaunorubicinol
In: Mouse B109
Rabbit B109

7,14-Dimethyldibenzanthracene
Probably formed from: 7-Methyl-dibenzanthracene
→ 7-Hydroxymethyl-14-methyldibenzanthracene
Probably in: Rat E83

4,6-Dimethyldibenzothiophene
→ 3,3′-Dicarboxybiphenyl
In: Microorganism J368
→ 2-Formyl-3-hydroxy-7-methylbenzothiophene
Probably in: Sphingomonas K304
→ 7-Methylbenzothiophene-2,3-dione
Probably in: Sphingomonas K304

Dimethyl 4,4′-dimethoxy-5,6,5′,6′-dimethylenedioxybiphenyl-2,2′-dicarboxylate
→ Dimethyl 4-hydroxy-4′-methoxy-5, 6,5′,6′-dimethylenedioxybiphenyl-2,2′-dicarboxylate
In: Rat D189

2,2-Dimethyl-2-(3,5-dimethyl-2,4,6-trinitrophenyl)ethanol
Formed from: Musk xylene

4-(5,5-Dimethyl-1,3,2-dioxaphosphorinan-2-yl)-*m*-fluoro-4-(5-hydroxymethyl-5-methyl-1,3,2-dioxaphosphorinan-2-yl)butyrophenone
Formed from: 4,4-Bis(5,5-dimethyl-1,3,2-dioxaphosphorinan-2-yl)-*m*-fluorobutyrophenone

N,N-Dimethyldiphenylacetamide (dimid, diphenamid)
→ N-Hydroxymethyl-N-methyl-2,2-diphenylacetamide
In: Glycine A3772
Tomato A1113
→ N-Methyldiphenylacetamide
In: Glycine A3772
Plant A1205

1,5-Dimethyl-3,3-diphenyl-2-ethylidenepyrrolidone
Formed from: L-α-Acetyl-2-dimethylamino-4,4-diphenyl-5-heptanol

N,N-Dimethyl-2-diphenylmethoxyethylamine
See: Diphenhydramine

1,5-Dimethyl-3,3-diphenyl-2-pyrrolidone
Probably formed from: 4-Dimethylamino-2,2-diphenylvaleric acid

(((8β)-1,6-Dimethylergolin-8-yl)methyl)carbamic acid phenylmethyl ester
See: Metergoline

5(*S*)-(1,1-Dimethylethoxycarbonylamino)-4(*S*)-hydroxy-N-(2(*R*)-hydroxy-1(*S*)-indanyl)-2(*R*)-(4-(2-(4-morpholinyl)ethoxy)phenyl)methyl-6-phenylhexamide
→ 2(*R*)-((4-Carboxymethoxy)phenyl)methyl-5(*S*)-(1,1-dimethylethoxycarbonylamino)-4(*S*)-hydroxy-N-(2(*R*)-hydroxy-1(*S*)-indanyl)hexamide
In: Rat H145
→ N-(2,3-Dihydroxy-1-indanyl)-5(*S*)-(1,1-dimethylethoxycarbonylamino)-4(*S*)-hydroxy-2(*R*)-(4-(2-(4-morpholinyl)ethoxy)phenyl)methyl-6-phenylhexamide
In: Rat H145
→ 5(*S*)-(1,1-Dimethylethoxycarbonylamino)-4(*S*)-hydroxy-N-(2(*R*)-hydroxy-1(*S*)-indanyl)-2(*R*)-(4-(2-(4-morpholinyl)ethoxy)phenyl)methyl-6-phenylhexamide-β-D-glucuronide
In: Rat H145
→ 5(*S*)-(1,1-Dimethylethoxycarbonylamino)-4(*S*)-hydroxy-N-(2(*R*)-hydroxy-1(*S*)-indanyl)-2(*R*)-(4-(2-(3-oxo-4-morpholinyl)ethoxy)phenyl)methyl-6-phenylhexamide
In: Rat H145

5(*S*)-(1,1-Dimethylethoxycarbonylamino)-4(*S*)-hydroxy-N-(2(*R*)-hydroxy-1(*S*)-indanyl)-2(*R*)-(4-(2-(4-morpholinyl)ethoxy)phenyl)methyl-6-phenylhexamide-β-D-glucuronide
Formed from: 5(*S*)-(1,1-Dimethylethoxycarbonylamino)-4(*S*)-hydroxy-N-(2(*R*)-hydroxy-1(*S*)-indanyl)-2(*R*)-(4-(2-(4-morpholinyl)ethoxy)phenyl)methyl-6-phenylhexamide

5(*S*)-(1,1-Dimethylethoxycarbonylamino)-4(*S*)-hydroxy-N-(2(*R*)-hydroxy-1(*S*)-indanyl)-2(*R*)-(4-(2-(3-oxo-4-morpholinyl)ethoxy)phenyl)methyl-6-phenylhexamide
Formed from: 5(*S*)-(1,1-Dimethylethoxycarbonylamino)-4(*S*)-hydroxy-N-(2(*R*)-hydroxy-1(*S*)-indanyl)-2(*R*)-(4-(2-(4-morpholinyl)ethoxy)phenyl)methyl-6-phenylhexamide

4-(2-(3-(1,1-Dimethylethoxy)-3-oxo-1-propenyl)phenyl)-2,6-dimethyl-3,5-pyridinedicarboxylic acid monoethyl ester
Probably formed from: 1,4-Dehydrolacidipine

4-(2-(3-(1,1-Dimethylethoxy)-3-oxo-1-propenyl)phenyl)-2-hydroxymethyl-6-methyl-3,5-pyridinedicarboxylic acid diethyl ester
Probably formed from: 1,4-Dehydrolacidipine

4-(2-(3-(1,1-Dimethylethoxy)-3-oxo-1-propenyl)phenyl)-1,4-dihydro-2,6-dimethyl-3,5-pyridinedicarboxylic acid diethyl ester
See: Lacidipine

N′-(((1,1-Dimethylethyl)amino)carbonyl)octahydrohydroxy-2(1*H*)isoquinolinyl-2-

hydroxy-1-(phenylmethyl)propyl)-2-((2-quinolinylcarbonyl)aminobutandiamide
Formed from: Saquinavir

5-((2(*S*)-(1,1-Dimethylethyl)aminocarbonyl)piperazin-1-yl)-4-(*S*)-hydroxy-N-(2(*R*)-hydroxy-1(*S*)-indanyl)-2(*R*)-(phenylmethyl)pentanamide
Formed from: 5-((2(*S*)-(1,1-Dimethylethyl)aminocarbonyl)-4-((pyridin-3-yl)methyl)piperazin-1-yl)-4-(*S*)-hydroxy-N-(2(*R*)-hydroxy-1(*S*)-indanyl)-2(*R*)-(phenylmethyl)pentanamide
→ N-(2(*R*),3-Dihydroxy-1(*S*)-indanyl)-5-((2(*S*)-(1,1-dimethylethyl)aminocarbonyl)piperazin-1-yl)-4-(*S*)-hydroxy-2(*R*)-(phenylmethyl)pentanamide
Probably in: Man H151

5-((2(*S*)-(1,1-Dimethylethyl)aminocarbonyl)-4-((pyridine-N-β-D-glucuronide-3-yl)methyl)piperazin-1-yl)-4-(*S*)-hydroxy-N-(2(*R*)-hydroxy-1(*S*)-indanyl)-2(*R*)-(phenylmethyl)pentanamide
Formed from: 5-((2(*S*)-(1,1-Dimethylethyl)aminocarbonyl)-4-((pyrid-3-yl)methyl)piperazin-1-yl)-4(*S*)-hydroxy-N-(2(*R*)-hydroxy-1(*S*)-indanyl)-2(*R*)-(phenylmethyl)pentanamide

5-((2(*S*)-(1,1-Dimethylethyl)aminocarbonyl)-4-((pyridine-N-oxide-3-yl)methyl)piperazin-1-yl)-4(*S*)-hydroxy-N-(2(*R*)-hydroxy-1(*S*)-indanyl)-2(*R*)-(phenylmethyl)pentanamide
Formed from: 5-((2(*S*)-(1,1-Dimethylethyl)aminocarbonyl)-4-((pyrid-3-yl)methyl)piperazin-1-yl)-4(*S*)-hydroxy-N-(2(*R*)-hydroxy-1(*S*)-indanyl)-2(*R*)-(phenylmethyl)pentanamide

5-((2(*S*)-(1,1-Dimethylethyl)aminocarbonyl)-4-((pyrid-3-yl)methyl)piperazin-1-yl)-4-(*S*)-hydroxy-N-(2(*R*)-hydroxy-1(*S*)-indanyl)-2(*R*)-((*p*-hydroxyphenylmethyl)pentanamide
Formed from: 5-((2(*S*)-(1,1-Dimethylethyl)aminocarbonyl)-4-((pyrid-3-yl)methyl)piperazin-1-yl)-4(*S*)-hydroxy-N-(2(*R*)-hydroxy-1(*S*)-indanyl)-2(*R*)-(phenylmethyl)pentanamide

5-((2(*S*)-(1,1-Dimethylethyl)aminocarbonyl)-4-((pyrid-3-yl)methyl)piperazin-1-yl)-4-(*S*)-hydroxy-N-(2(*R*)-hydroxy-1(*S*)-indanyl)-2(*R*)-(phenylmethyl)pentanamide
→ N-(2(*R*),3-Dihydroxy-1(*S*)-indanyl)-5-((2(*S*)-(1,1-dimethylethyl)aminocarbonyl)-4-((pyridin-3-yl)methyl)piperazin-1-yl)-4(*S*)-hydroxy-2(*R*)-(phenylmethyl)pentanamide
In: Man H151
→ 5-((2(*S*)-(1,1-Dimethylethyl)aminocarbonyl)piperazin-1-yl)-4-(*S*)-hydroxy-N-(2(*R*)-hydroxy-1(*S*)-indanyl)-2(*R*)-(phenylmethyl)pentanamide
In: Man H151
→ 5-((2(*S*)-(1,1-Dimethylethyl)aminocarbonyl)-4-((pyridine-N-β-D-glucuronide-3-yl)methyl)piperazin-1-yl)-4-(*S*)-hydroxy-N-(2(*R*)-hydroxy-1(*S*)-indanyl)-2(*R*)-(phenylmethyl)pentanamide
In: Man H151
→ 5-((2(*S*)-(1,1-Dimethylethyl)aminocarbonyl)-4-((pyridine-N-oxide-3-yl)methyl)piperazin-1-yl)-4-(*S*)-hydroxy-N-(2(*R*)-hydroxy-1(*S*)-indanyl)-2(*R*)-(phenylmethyl)pentanamide
In: Man H151
→ 5-((2(*S*)-(1,1-Dimethylethyl)aminocarbonyl)-4-((pyridin-3-yl)methyl)piperazin-1-yl)-4-(*S*)-hydroxy-N-(2(*R*)-hydroxy-1(*S*)-indanyl)-2(*R*)-((*p*-hydroxyphenylmethyl)pentanamide
In: Man H151

2-(2-((1,1-Dimethylethyl)amino)-1-hydroxyethyl)-7-ethylbenzofuran
See: Bifuralol

2-(3-((1,1-Dimethylethyl)amino)-2-hydroxypropoxy)benzonitrile
See: Bunitrolol

5-(3-((1,1-Dimethylethyl)amino)-2-hydroxypropoxy)-1,2,3,4-tetrahydro-2,3-naphthalenediol
See: *trans*-Nadolol

1-(1,1-Dimethylethyl)-4,4-diphenylpiperidine
See: Budipine

4-(4-(4-(1,1-Dimethylethyl)phenyl)-2-hydroxybutoxy)benzoic acid
See: Lifibrol

α-(4-(1,1-Dimethylethyl)phenyl)-2-(hydroxydiphenylmethyl)-1-piperidinebutanol
See: Terfenadine

2,5-Dimethylfuran-3-carboxanilide
→ Aniline
In: Nocardia A2663
See also: A3263

N-((2*E*)-6,6-Dimethyl-2-hepten-4-ynyl)-N-methyl-1-naphthalenemethanamine
See: Terbinafine

(−)-*cis*-3-(4-(1,1-Dimethylheptyl)-2-hydroxyphenyl)-*trans*-4-(3-hydroxypropyl) cyclohexanol
→ (−)-*cis*-3-(2-Hydroxy-4-(2-hydroxy-1,1-dimethylheptyl)phenyl)-*trans*-4-(3-hydroxypropyl)cyclohexanol
In: Mouse F605
→ (−)-*cis*-3-(2-Hydroxy-4-(3-hydroxy-1,1-dimethylheptyl)phenyl)-*trans*-4-(3-hydroxypropyl)cyclohexanol
In: Mouse F605
→ (−)-*cis*-3-(2-Hydroxy-4-(4-hydroxy-1,1-dimethylheptyl)phenyl)-*trans*-4-(3-hydroxypropyl)cyclohexanol
In: Mouse F605
→ (−)-*cis*-3-(2-Hydroxy-4-(5-hydroxy-1,1-dimethylheptyl)phenyl)-*trans*-4-(3-hydroxypropyl)cyclohexanol
In: Mouse F605
→ (−)-*cis*-3-(2-Hydroxy-4-(6-hydroxy-1,1-dimethylheptyl)phenyl)-*trans*-4-(3-hydroxypropyl)cyclohexanol
In: Mouse F605

(−)-*cis*-2,8-Dimethyl-2,3,4,4a,5,9b-hexahydro-1*H*-pyrido[4,3-*b*]indole
See: Stobadine

2,4-Dimethylhippuric acid
Formed from: 2,4-Dimethylbenzoic acid

2,5-Dimethylhippuric acid
Formed from: 2,5-Dimethylbenzoic acid

3,4-Dimethylhippuric acid
Formed from: 3,4-Dimethylbenzoic acid

3,5-Dimethylhippuric acid
Formed from: 3,5-Dimethylbenzoic acid

Dimethyl 4-hydroxy-4′-methoxy-5, 6,5′,6′-dimethylenedioxybiphenyl-2,2′-dicarboxylate
Formed from: Dimethyl 4,4′-dimethoxy-5, 6,5′,6′-dimethylenedioxybiphenyl-2,2′-dicarboxylate
→ Dimethyl 4-hydroxy-4′-methoxy-5, 6,5′,6′-dimethylenedioxybiphenyl-2,2′-dicarboxylate-β-D-glucuronide
In: Rat D189

Dimethyl 4-hydroxy-4′-methoxy-5, 6,5′,6′-dimethylenedioxybiphenyl-2,2′-dicarboxylate-β-D-glucuronide
Formed from: Dimethyl 4-hydroxy-4′-methoxy-5, 6,5′,6′-dimethylenedioxybiphenyl-2,2′-dicarboxylate

N,N-Dimethylindoaniline
→ N-(*p*-Hydroxyphenyl)-N′,N′-dimethyl-*p*-phenylenediamine
In: Rat E214

2,3-Dimethylindole
→ 2-Methylindole-3-methanol
In: Pseudomonas A3734

N,N-Dimethylindoliumolate
Formed from: Hordenine

N′-(3,4-Dimethyl-5-isoxazolyl)sulphanilamide
See: Sulphisoxazole

3,5-Dimethyl-4-methylaminophenyl methylcarbamate
Formed from: 4-Dimethylamino-3,5-dimethylphenyl methylcarbamate
→ 4-Amino-3,5-dimethylphenyl methylcarbamate
Probably in: Rat A1545

N,N-Dimethyl-4,4′-methylenedianiline
Formed from: 4,4′-Methylenebis((N,N-dimethyl)aniline)

N,N′-Dimethyl-4,4′-methylenedianiline
Formed from: 4,4′-Methylenebis((N,N-dimethyl)aniline)
→ N,N′-Diacetyl-4,4′-(hydroxymethylene) dianiline
Probably in: Rat C362
→ N-Methyl-4,4′-methylenedianiline
Probably in: Rat C362, C364

3,4-((Dimethyl)methylenedioxy)benzaldehyde
→ 3,4-((Dimethyl)methylenedioxy) mandelonitrile (both isomers)
In: Almond J194, J302
Apple J302
Sorghum J194

3,4-((Dimethyl)methylenedioxy)mandelonitrile
Formed from: 3,4-((Dimethyl) methylenedioxy)benzaldehyde

N,N-Dimethyl-2-(3,4-methylenedioxyphenyl)-2-oxoethylamine
→ 3,4-Methylenedioxymandelic acid
In: Rat J287

3,5-Dimethyl-4-methylformamidophenyl methylcarbamate
Formed from: 4-Dimethylamino-3,5-dimethylphenyl methylcarbamate

2,6-Dimethyl-N-(5-methyl-3-isoxazolyl) benzamide
→ 2-Hydroxymethyl-6-methyl-N-(5-methyl-3-isoxazolyl)benzamide
Probably in: Rat J76

O,O-Dimethyl O-(3-methyl-4-methylsulphinylphenyl) phosphate
Probably formed from: O,O-Dimethyl O-(3-methyl-4-methylsulphinylphenyl) thiophosphate
→ O,O-Dimethyl O-(3-methyl-4-methylsulphonylphenyl) phosphate
Probably in: Beef A3802

O,O-Dimethyl O-(3-methyl-4-methylsulphinylphenyl) thiophosphate
Formed from: Fenthion
→ O,O-Dimethyl O-(3-methyl-4-methylsulphinylphenyl) phosphate
Probably in: Beef A3802
→ O,O-Dimethyl O-(3-methyl-4-methylsulphonylphenyl) thiophosphate
Probably in: Beef A3802

O,O-Dimethyl O-(3-methyl-4-methylsulphonylphenyl) phosphate
Probably formed from: O,O-Dimethyl O-(3-methyl-4-methylsulphinylphenyl) phosphate

O,O-Dimethyl O-(3-methyl-4-methylsulphonylphenyl) thiophosphate
Probably formed from: O,O-Dimethyl O-(3-methyl-4-methylsulphinylphenyl) thiophosphate

O,O-Dimethyl O-(3-methyl-4-methylthiophenyl) phosphate
Formed from: Fenthion

O,O-Dimethyl O-(3-methyl-4-methylthiophenyl) thiophosphate
See: Fenthion

Dimethyl 3-methyl-4-nitrophenyl phosphate
See: Fenitrooxon

O,O-Dimethyl 3-methyl-4-nitrophenyl phosphorothioate
See: Fenitrothion

O,S-Dimethyl O-(3-methyl-4-nitrophenyl) phosphorothioate
→ 3-Methyl-4-nitrophenol
In: Rat A2609

3,3-Dimethyl-N-(1-methyl-1-phenylethyl) butyramide
Formed from: Bromobutide
→ N-(2-Hydroxy-1-methyl-1-phenylethyl)-3,3-dimethylbutyramide
In: Mouse D417
See also: D193
→ 4-Hydroxy-3,3-dimethyl-N-(1-methyl-1-phenylethyl)butyramide
In: Mouse D417
See also: D193, D215
→ N-(1-(*p*-Hydroxyphenyl)-1-methylethyl)-3,3-dimethylbutyramide
In: Mouse D417
See also: D193, D215

2,3-Dimethyl-4-(3′-methyl-2′-phenylmorpholinomethyl)-1-phenylpyrazolone
See: Tarugan

8-((3,3′-Dimethyl-4′-((4-(((4-methylphenyl) sulphonyl)oxy)phenyl)azo)(1,1′-biphenyl)-4-yl)azo)-7-hydroxy-1,3-naphthalenedisulphonic acid
See: Acid Red 114

3,3-Dimethyl-1-(*p*-methylphenyl)triazene
→ 3-Methyl-1-(*p*-methylphenyl)triazene
In: Rat A112

N,N-Dimethyl-10-(3-(4-methyl-1-piperazinyl) propyl)-10*H*-phenothiazine-2-sulphonamide
See: Thioproperazine

N,N-Dimethyl-5-(N-methylsulphonamidomethyl) tryptamine
→ 5-(N-Methylsulphonamidomethyl)indole-3-acetic acid
In: Dog G691
Man G691
Rabbit G691
Rat G691

3,5-Dimethyl-4-methylsulphinylphenyl methylcarbamate
Formed from: 3,5-Dimethyl-4-methylthiophenyl methylcarbamate

3,5-Dimethyl-4-methylthiophenyl N-hydroxymethylcarbamate
Formed from: Methiocarb

3,5-Dimethyl-4-methylthiophenyl methylcarbamate
See: Methiocarb

***d*-N,3-Dimethylmorphinan** (dinemorphan)
→ *d*-3-Methylmorphinan
In: Mouse D61

N-(1,1-Dimethyl-2-(((morpholino)carbonyl) amino)ethyl)-2-hydroxy-3-(*p*-hydroxybenzoyloxy)propylamine
→ *p*-Hydroxybenzoic acid
In: Rabbit H563

1,5-Dimethylnaphthalene
→ 5-Methyl-1-naphthoic acid
In: Microorganism A1116

1,8-Dimethylnaphthalene
→ 1-Hydroxymethyl-8-methylnaphthalene
In: Pseudomonas A1482

2,6-Dimethylnaphthalene
→ *trans*-3,4-Dihydro-3,4-dihydroxy-2,6-dimethylnaphthalene
In: Starry flounder B852
→ 3-Hydroxy-2,6-dimethylnaphthalene
In: Starry flounder B852
→ 2-Hydroxymethyl-6-methylnaphthalene
In: Starry flounder B852

2,6-Dimethyl-3,4-naphthoquinone
Probably formed from: *trans*-3,4-Dihydro-3,4-dihydroxy-2,6-dimethylnaphthalene

3,5-Dimethyl-4-nitroacetanilide
Formed from: 3,5-Dimethyl-4-nitroaniline

N,N-Dimethyl-*m*-nitroaniline
→ N-Methyl-*m*-nitroaniline
In: Glycine K419
Man K419

3,5-Dimethyl-4-nitroaniline
→ 3,5-Dimethyl-4-nitroacetanilide
In: Arthrobacter E276
Bacillus E276
Cordyceps E276

N,N-Dimethyl-*p*-nitrobenzamide
→ N-Methyl-*p*-nitrobenzamide
In: Rat G683

Dimethyl *p*-nitrophenyl phosphate
Formed from: Dimethyl *p*-nitrophenyl phosphorothioate

Dimethyl *p*-nitrophenyl phosphorothioate (methylparathion)
→ Dimethyl *p*-nitrophenyl phosphate
In: Rat F739, K599
→ Methyl *p*-nitrophenyl phosphorothioate
In: Crayfish E816
Man E64, E225
Prawn E816
→ *p*-Nitrophenol
In: Crayfish E816
Prawn E816
Pseudomonas H386
Rat K599
→ S-(*p*-Nitrophenyl)glutathione
In: Man E225
Rat A3509

O,O-Dimethyl O-(4-nitrophenyl) phosphorothioate
See: Methylparathion

2,6-Dimethyl-4-(*m*-nitrophenyl)pyridine-3,5-dicarboxylic acid
Probably formed from: (4S)-1,4-Dihydro-2,6-dimethyl-4-(*m*-nitrophenyl)pyridine-3,5-dicarboxylic acid
2,6-Dimethyl-4-(*m*-nitrophenyl)pyridine-3,5-dicarboxylic acid 3-methyl ester
5-Isopropoxycarbonyl-2,6-dimethyl-4-(*m*-nitrophenyl)pyridine-3-carboxylic acid
Probably formed from: 2,6-Dimethyl-4-(*m*-nitrophenyl)pyridine-3,5-dicarboxylic acid 3-(2-(N-benzyl-N-methylamino))ethyl ester

→ 2-Hydroxymethyl-6-methyl-4-(*m*-nitrophenyl)pyridine-3,5-dicarboxylic acid
Probably in: Dog G398
Mouse G398
Rat G398

2,6-Dimethyl-4-(3-nitrophenyl)pyridine-3,5-dicarboxylic acid 3-(2-(N-benzyl-N-methylamino))ethyl ester
Probably formed from: 2,6-Dimethyl-4-(3-nitrophenyl)pyridine-3,5-dicarboxylic acid 3-(2-(N-benzyl-N-methylamino))ethyl ester 5-methyl ester
→ 2,6-Dimethyl-4-(3-nitrophenyl)pyridine-3,5-dicarboxylic acid
Probably in: Dog A2096
Rat A2096

2,6-Dimethyl-4-(3-nitrophenyl)pyridine-3,5-dicarboxylic acid 3-(2-(N-benzyl-N-methylamino))ethyl ester 5-methyl ester
Formed from: 1,4-Dihydro-2,6-dimethyl-4-(3-nitrophenyl)pyridine-3,5-dicarboxylic acid 3-(2-(N-benzyl-N-methylamino)) ethyl ester 5-methyl ester
→ 2,6-Dimethyl-4-(3-nitrophenyl)pyridine-3,5-dicarboxylic acid 3-(2-(N-benzyl-N-methylamino))ethyl ester
Probably in: Dog A2096
Rat A2096
→ 2,6-Dimethyl-4-(3-nitrophenyl)pyridine-3,5-dicarboxylic acid 3-carboxymethyl ester 5-methyl ester
Probably in: Dog A2096
Rat A2096

2,6-Dimethyl-4-(3-nitrophenyl)pyridine-3,5-dicarboxylic acid 3-carboxymethyl ester 5-methyl ester
Probably formed from: 2,6-Dimethyl-4-(3-nitrophenyl)pyridine-3,5-dicarboxylic acid 3-(2-(N-benzyl-N-methylamino))ethyl ester 5-methyl ester

2,6-Dimethyl-4-(*o*-nitrophenyl)pyridine-3,5-dicarboxylic acid dimethyl ester
→ 2,6-Dimethyl-4-(*o*-nitrophenyl)pyridine-3,5-dicarboxylic acid methyl ester
In: Rat F540

2,6-Dimethyl-4-(*m*-nitrophenyl)pyridine-3,5-dicarboxylic acid 3-ethyl ester
Probably formed from: 1,4-Dehydronitrendipine

2,6-Dimethyl-4-(*o*-nitrophenyl)pyridine-3,5-dicarboxylic acid 3-methyl ester
Formed from: 2,6-Dimethyl-4-(*o*-nitrophenyl)pyridine-3,5-dicarboxylic acid dimethyl ester

2,6-Dimethyl-4-(*m*-nitrophenyl)pyridine-3,5-dicarboxylic acid 3-methyl ester
Formed from: Nicardipine
Probably formed from: 1,4-Dehydronitrendipine
1,4-Dehydropranidipine
1,4-Dihydro-2,6-dimethyl-4-(*m*-nitrophenyl)pyridine-3,5-dicarboxylic acid 3-methyl ester
→ 2,6-Dimethyl-4-(*m*-nitrophenyl)pyridine-3,5-dicarboxylic acid
In: Dog G398, H518
Mouse G398
Rat G398, H518
→ 2,6-Dimethyl-4-(*m*-nitrophenyl)pyridine-3,5-dicarboxylic acid 3-methyl ester β-D-glucuronide
In: Dog G398
Mouse G398
Rat G398
→ 6-Hydroxymethyl-2-methyl-4-(*m*-nitrophenyl)pyridine-3,5-dicarboxylic acid 3-methyl ester
Probably in: Rat F11, H927
→ Methyl 2,6-dimethyl-4-(*m*-nitrophenyl) pyridine-3-carboxylate
Probably in: Man D889
→ Methyl 3-hydroxymethyl-2,6-dimethyl-4-(*m*-nitrophenyl)pyridine-5-carboxylate
Probably in: Man D889

2,6-Dimethyl-4-(*m*-nitrophenyl)pyridine-3,5-dicarboxylic acid 3-methyl ester β-D-glucuronide
Probably formed from: 2,6-Dimethyl-4-(*m*-nitrophenyl)pyridine-3,5-dicarboxylic acid 3-methyl ester

2,6-Dimethyl-4-(*m*-nitrophenyl)pyridine-3,5-dicarboxylic acid 3-methyl ester 5-(3-piperidinyl) ester
Probably formed from: Debenzylbenidipine
→ 2-Hydroxymethyl-6-methyl-4-(*m*-nitrophenyl)pyridine-3,5-dicarboxylic acid 3-methyl ester 5-(3-piperidinyl) ester
Probably in: Man J105

2,6-Dimethyl-4-(*m*-nitrophenyl)pyridine-3,5-dicarboxylic acid 3-(3-piperidinyl) ester
Probably formed from: 1,4-Dihydro-2,6-dimethyl-4-(*m*-nitrophenyl)pyridine-3,5-dicarboxylic acid 3-(3-piperidinyl) ester
→ 2-Hydroxymethyl-6-methyl-4-(*m*-nitrophenyl)pyridine-3,5-dicarboxylic acid 3-(3-piperidinyl) ester
Probably in: Rat H927
→ 2-Hydroxymethyl-6-methyl-4-(*m*-nitrophenyl)pyridine-3,5-dicarboxylic acid 5-(3-piperidinyl) ester
Probably in: Dog F11
Rat F11, H927

N,N-Dimethyl-4-nitrosoaniline
→ N,N-Dimethyl-*p*-benzoquinonediiminium cation
In: Saccharomyces K229

α,α-Dimethyl-α-nitro-β-phenethylamine
Formed from: N-Hydroxyphentermine

α,α-Dimethyl-α-nitroso-β-phenylethane
See: Nitrosophentermine

2-((2*E*)-3,7-Dimethyl-2,6-octadienyl)-1,4-benzenediol
See: Geranylhydroquinone

2-((2*E*)-3,7-Dimethyl-2,6-octadienyl)-2,4-dihydroxy-6-pentylbenzoic acid
See: Cannabigerolic acid

2-((2*E*)-3,7-Dimethyl-2,6-octadienyl)-5-pentyl-1,3-benzenediol
See: Cannabigerol

O,O-Dimethyl S-((4-oxo-1,2,3-benzotriazin-3(4*H*)-yl)methyl) phosphorodithioate
See: Azinphos-methyl

O,O-Dimethyl S-((4-oxo-1,2,3-benzotriazin-3(4*H*)-yl)methyl) phosphorothioate
See: Gutoxon

N,N′-Dimethyl-4-oxoclonidine
Probably formed from: 4-Hydroxy-N,N′-dimethylclonidine

5-(5,5-Dimethyl-2-oxo-1,3,2-dioxaphosphorinan-2-yl)-1,4-dihydro-2,6-dimethyl-4-(*m*-nitrophenyl) pyridine-3-carboxylic acid
Formed from: Efonidipine

3-((2*E*,4*E*)-4,6-Dimethyl-1-oxo-2,4-octadienyl)-1,4-dihydroxy-5-(4-hydroxyphenyl)-2(1*H*)-pyridinone
See: Tenellin

4-(4,4-Dimethyl-3-oxopentyl)-3,5-dioxo-1,2-diphenylpyrazolidine
See: Trimetazon

4-(4,4-Dimethyl-3-oxopentyl)-3,5-dioxo-1-(*p*-hydroxyphenyl)-2-phenylpyrazolidine
Formed from: Trimetazon

4-(4,4-Dimethyl-3-oxopentyl)-4-hydroxy-3,5-dioxo-1,2-diphenylpyrazolidine
Formed from: Trimetazon

N,N-Dimethyl-2-oxo-2-phenylethylamine
→ Mandelic acid
In: Rat J287

5-(2,5-Dimethylphenoxy)-2,2-dimethylpentanoic acid
See: Gemfibrozil

N-(2,6-Dimethylphenyl)-2-(2-oxo-1-pyrrolidinyl) acetamide
See: Nefiracetam

11,11-Dimethyl-12-oxo-12-((2,4,6-trimethoxyphenyl)amino)dodecanoic acid
Formed from: 2,2-Dimethyl-N-((2,4,6-trimethoxyphenyl) dodecanamide

6,6-Dimethyl-7-oxo-7-((2,4,6-trimethoxyphenyl) amino)heptanoic acid
Formed from: 2,2-Dimethyl-N-((2,4,6-trimethoxyphenyl) dodecanamide

5,5-Dimethyl-6-oxo-6-((2,4,6-trimethoxyphenyl) amino)hexanoic acid
Formed from: 2,2-Dimethyl-N-((2,4,6-trimethoxyphenyl) dodecanamide

N,N-Dimethyl-2-oxo-2-(2,3,4-trimethoxyphenyl) ethylamine
→ 2,3,4-Trimethoxymandelic acid
In: Rat J287

N,N-Dimethyl-2-oxo-2-(3,4,5-trimethoxyphenyl) ethylamine
→ 3,4,5-Trimethoxymandelic acid
In: Rat J287

Dimethyl phenazine-1,6-dicarboxylate
→ Phenazine-1,6-dicarboxylic acid
Probably in: Pseudomonas A2805

N,N-Dimethylphenethylamine
Formed from: N-Methylphenethylamine
→ N-Methylphenethylamine
In: Rat A300

α,α-Dimethylphenethylamine
See: Phentermine

2,3-Dimethylphenol
Formed from: 1-*tert*-Butylamino-3-(2,3-dimethylphenoxy)-2-propanol
2,3-Dimethylbenzoic acid
o-Xylene
→ 2,3-Dimethylanisole
In: Guinea pig A115
→ 3,4-Dimethylcatechol
Probably in: Pseudomonas E463

2,4-Dimethylphenol
Formed from: *m*-Xylene
→ 2,4-Dimethylanisole
In: Guinea pig A115
→ 2,4-Dimethylphenyl-β-D-glucuronide
In: Rat C166
→ 4-Hydroxy-3-methylbenzyl alcohol
Probably in: Microorganism F89
Pseudomonas D696, F381

2,5-Dimethylphenol
Formed from: *p*-Xylene
→ 2,5-Dimethylanisole
In: Guinea pig A115
→ 3,6-Dimethylcatechol
Probably in: Pseudomonas A2391

2,6-Dimethylphenol
Formed from: 2-Amino-1-(2,6-dimethylphenoxy)propane
m-Xylene
→ 2,6-Dimethylanisole
In: Guinea pig A115
→ 2,6-Dimethylphenyl sulphate
In: Mouse H959

3,4-Dimethylphenol
Probably formed from: 3,4-Dimethylbenzoic acid
o-Xylene
→ 3,4-Dimethylanisole
In: Guinea pig A115
→ 4,5-Dimethylcatechol
Probably in: Botrytis A1930
→ 4-Hydroxy-2-methylbenzyl alcohol
Probably in: Microorganism F89

3,5-Dimethylphenol
Probably formed from: 3,5-Dimethylphenyl N-methylcarbamate
→ 3,5-Dimethylanisole
In: Guinea pig A115
→ 3,5-Dimethylphenyl sulphate
In: Man E575
Rat E575
→ 3-Hydroxy-5-methylbenzyl alcohol
Probably in: Microorganism A3870

2,3-Dimethylphenoxyacetic acid
Probably formed from: 2,3-Dimethylphenoxypropane-2,3-diol

(α*S*)-N-((1*S*,3*S*,4*S*)-4-(((2,6-Dimethylphenoxy) acetyl)amino)-3-hydroxy-5-phenyl-1-(phenylmethyl)pentyl)tetrahydro-α-isopropyl-2,4-dioxo-1(2*H*)-pyrimidineacetamide
Probably formed from: (αS)-N-((1S,3S,4S)-4-(((2,6-Dimethylphenoxy)acetyl)amino)-3-hydroxy-5-phenyl-1-(phenylmethyl)pentyl) tetrahydro-4-hydroxy-α-isopropyl-2-oxo-1(2*H*)-pyrimidineacetamide

(αS)-N-((1S,3S,4S)-4-(((2,6-Dimethylphenoxy) acetyl)amino)-3-hydroxy-5-phenyl-1-(phenylmethyl)pentyl)tetrahydro-4-hydroxy-α-isopropyl-2-oxo-1(2*H*)-pyrimidineacetamide
Formed from: (αS)-N-((1S,3S,4S)-4-(((2,6-Dimethylphenoxy)acetyl)amino)-3-hydroxy-5-phenyl-1-(phenylmethyl)pentyl)tetrahydro-α-isopropyl-2-oxo-1(2*H*)-pyrimidineacetamide
→ (αS)-N-((1S,3S,4S)-4-(((2,6-Dimethylphenoxy)acetyl)amino)-3-hydroxy-5-phenyl-1-(phenylmethyl) pentyl)tetrahydro-α-isopropyl-2,4-dioxo-1(2*H*)-pyrimidineacetamide
Probably in: Man J734

(αS)-N-((1S,3S,4S)-4-(((2,6-Dimethylphenoxy)acetyl)amino)-3-hydroxy-5-phenyl-1-(phenylmethyl)pentyl)tetrahydro-α-isopropyl-2-oxo-1(2H)-pyrimidineacetamide
→ (αS)-N-((1S,3S,4S)-4-(((2,6-Dimethylphenoxy)acetyl)amino)-3-hydroxy-5-phenyl-1-(phenylmethyl)pentyl)tetrahydro-4-hydroxy-α-isopropyl-2-oxo-1(2H)-pyrimidineacetamide
In: Man J734

N,N-Dimethylphenoxyethylamine
→ N-Methylphenoxyethylamine
In: Rat A2421

1-(2,6-Dimethylphenoxy)-2-oximinopropane
Formed from: Mexiletine
→ 1-(2,6-Dimethylphenoxy)propan-2-one
In: Cunninghamella J155

N,N-Dimethyl-N′-(p-phenoxyphenyl)sulphamide
→ N′-(p-(o-Hydroxyphenoxy)phenyl)-N,N-dimethylsulphamide
In: Dog A1263
Rat A1263
→ N′-(p-(p-Hydroxyphenoxy)phenyl)-N,N-dimethylsulphamide
In: Dog A1263
Man A1263
Rat A1263
→ N-Methyl-N′-(p-phenoxyphenyl)sulphamide
In: Dog A1263
Man A1263
Rat A1263

1-(2,6-Dimethylphenoxy)-2-propanamine
See: Mexiletine

2,3-Dimethylphenoxypropan-2,3-diol
Formed from: 1-*tert*-Butylamino-3-(2,3-dimethylphenoxy)-2-propanol
→ 2,3-Dimethylphenoxyacetic acid
Probably in: Monkey A2369
Rabbit A2369
Rat A2369
→ 4-Hydroxy-2,3-dimethylphenoxypropane-2,3-diol
Probably in: Monkey A2369
Rabbit A2369
Rat A2369

1-(2,6-Dimethylphenoxy)propan-2-ol
Formed from: Mexiletine
Probably formed from: 1-(2,6-Dimethylphenoxy)propan-2-one
→ 1-(2,6-Dimethylphenoxy)propan-2-ol-β-D-glucuronide
Probably in: Man G385
Rat G385
→ 1-(4-Hydroxy-2,6-dimethylphenoxy)propan-2-ol
Probably in: Cunninghamella J155
→ 1-(2-Hydroxymethyl-6-methylphenoxy)propan-2-ol
Probably in: Cunninghamella J155

1-(2,6-Dimethylphenoxy)propan-2-ol-β-D-glucuronide
Probably formed from: 1-(2,6-Dimethylphenoxy)propan-2-ol

1-(2,6-Dimethylphenoxy)propan-2-one
Probably formed from: 1-(2,6-Dimethylphenoxy)-2-oximinopropane
→ 1-(2,6-Dimethylphenoxy)propan-2-ol
Probably in: Cunninghamella J155

Dimethylphentermine
→ Methylphentermine
In: Rabbit F839
Rat A300

2-((2,3-Dimethylphenyl)amino)benzoic acid
See: Mefenamic acid

2-(2,6-Dimethylphenylamino)-5,6-dihydro-4H-1,3-thiazine
See: Xylazine

N-((2,6-Dimethylphenylcarbamoyl)methyl)succinamic acid
Formed from: Nefiracetam

1-(2,6-Dimethylphenyl)-2,4-diethyl-2-imidazolin-5-one
Formed from: Etidocaine

N′-(2,4-Dimethylphenyl)-N-(((2,4-dimethylphenyl)imino)methyl)-N-methylmethanimidamide
See: Amitraz

N,N-Dimethyl-p-phenylenediamine
Formed from: 2-Amino-4′-dimethylaminoazobenzene
3-Amino-4′-dimethylaminoazobenzene
4-Amino-4′-dimethylaminoazobenzene

4-Dimethylaminoazobenzene
4-Dimethylaminoazobenzene-4′-arsonic acid
4′-Dimethylamino-4-hydroxyazobenzene
4-Dimethylamino-3-methylazobenzene
4-Dimethylamino-3′-methylazobenzene
6-(*p*-Dimethylaminophenyl) azobenzothiazole
5-(*p*-Dimethylaminophenyl)azoindazole
N,N-Dimethyl-4-(2-pyridylazo)aniline
Methyl orange
Methyl red
p-Methyl red
Methyl red methyl ester
p-Methyl red methyl ester
→ N-Methyl-*p*-phenylenediamine
In: Glycine K419
Man K419
See also: D851
→ Free radical
In: Botrytis A1930

N,N-Dimethylphenylethanolamine
→ Mandelaldehyde
Probably in: Rat J287

N-(2,6-Dimethylphenyl)-2-ethylaminobutanamide
Formed from: Etidocaine
→ 2-Amino-N-(2,6-dimethylphenyl)-butanamide
Probably in: Man A1744

N,N-Dimethyl-N′-phenylethylenediamine
Formed from: Methaphenilene

4-(1-(2,3-Dimethylphenyl)ethyl)-1*H*-imidazole
See: Medetomidine

4-((1*S*)-1-(2,3-Dimethylphenyl)ethyl)-1*H*-imidazole
See: Dexmedetomidine

3-(2,6-Dimethylphenyl)-5-ethyl-2,4-imidazolinedione
Formed from: Etidocaine

1-(2,6-Dimethylphenyl)-4-ethyl-2-methyl-2-imidazolin-5-one
Formed from: Etidocaine

N-(2,6-Dimethylphenyl)-2-(ethylpropylamino) butanamide, 2-(N-ethyl-N-propylamino)-2′,6′-butyroxylidide
See: Etidocaine

2,4-Dimethylphenyl-β-D-glucuronide
Formed from: 2,4-Dimethylphenol

N-(2,6-Dimethylphenyl)-2-hydroxyacetamide
Formed from: Lidoflazine

2,6-Dimethylphenylhydroxylamine
Formed from: Lignocaine

N-(2,6-Dimethylphenyl)-2-(3-hydroxy-2-oxo-1-pyrrolidinyl)acetamide
Formed from: N-(2,6-Dimethylphenyl)-2-(2-oxo-1-pyrrolidinyl)acetamide

N-(2,6-Dimethylphenyl)-2-(4-hydroxy-2-oxo-1-pyrrolidinyl)acetamide
Formed from: N-(2,6-Dimethylphenyl)-2-(2-oxo-1-pyrrolidinyl)acetamide

N-(2,6-Dimethylphenyl)-2-(5-hydroxy-2-oxo-1-pyrrolidinyl)acetamide
Formed from: N-(2,6-Dimethylphenyl)-2-(2-oxo-1-pyrrolidinyl)acetamide
→ N-(3-Hydroxy-2,6-dimethylphenyl)-2-(5-hydroxy-2-oxo-1-pyrrolidinyl)acetamide
Probably in: Man F587

1-(2,6-Dimethylphenyl)-3-methylamidinourea
See: Lidamidine

3,5-Dimethylphenyl N-methylcarbamate
→ 3-Carboxy-5-methylphenyl N-methylcarbamate
In: Rabbit A1477
→ 3,5-Dimethylphenol
In: Rabbit A1477

N-(2,6-Dimethylphenyl)-N′-(3-((1-methylethyl) amino)phenyl)urea
See: Recainam

4-(2,6-Dimethylphenyl)-1-methylpyridinium
Probably formed from: 4-(2,6-Dimethylphenyl)-1,2,3,6-tetrahydro-1-methylpyridine

1-(2,6-Dimethylphenyl)-3-(1-methyl-2-pyrrolidinylidene)urea
See: Xilopam

N-(2,6-Dimethylphenyl)-N′-(N-methylsuccinamido)urea
Probably formed from: 5-Oxoxilobam
→ N-(2-Carboxyethylcarbonyl)-N′-(2,6-dimethylphenyl)urea
Probably in: Dog G938
Mouse G938

(2*S*,3*S*)-3,4-Dimethyl-2-phenylmorpholine
See: Phendimetrazine

N-(2,6-Dimethylphenyl)-2-(2-oxo-1-pyrrolidinyl) acetamide
→ N-(2,6-Dimethylphenyl)-2-(3-hydroxy-2-oxo-1-pyrrolidinyl)acetamide
In: Man F587
→ N-(2,6-Dimethylphenyl)-2-(4-hydroxy-2-oxo-1-pyrrolidinyl)acetamide
In: Man F587
→ N-(2,6-Dimethylphenyl)-2-(5-hydroxy-2-oxo-1-pyrrolidinyl)acetamide
In: Man F587
→ N-(3-Hydroxy-2,6-dimethylphenyl)-2-(2-oxo-1-pyrrolidinyl)acetamide
In: Man F587
→ N-(4-Hydroxy-2,6-dimethylphenyl)-2-(2-oxo-1-pyrrolidinyl)acetamide
In: Man F587
→ N-(2-Hydroxymethyl-6-methylphenyl)-2-(2-oxo-1-pyrrolidinyl)acetamide
In: Man F587
→ N-(4-Hydroxybutyrylglycyl)-2,6-dimethylaniline
In: Man F587

2,2-Dimethyl-1-phenylpropanol
Formed from: *tert*-Butyl phenyl ketone

N-(2,6-Dimethylphenyl)-2-propylaminobutanamide
Formed from: Etidocaine

2,3-Dimethyl-1-phenyl-3-pyrazolin-5-one
See: Antipyrine

N,N-Dimethyl-2-(1-phenyl-1-(2-pyridinyl) ethoxy)ethanamine
See: Doxylamine

2,6-Dimethylphenyl sulphate
Formed from: 2,6-Dimethylphenol

3,5-Dimethylphenyl sulphate
Formed from: 3,5-Dimethylphenol

4-(2,6-Dimethylphenyl)-1,2,3,6-tetrahydro-1-methylpyridine
→ 4-(2,6-Dimethylphenyl)-1-methylpyridinium
Probably in: Beef F395
Man F395

α,N-Dimethyl-α-phenylsuccinimide
See: Methsuximide

N,N-Dimethyl-N′-phenyl-N′-(2-thienylmethyl)-1,2-ethanediamine
See: Methaphenilene

N-(2,6-Dimethylphenyl)thiourea
Probably formed from: 4-Oxoxylazine

3,3-Dimethyl-1-phenyltriazene
→ Aniline
In: Rat A2019, A2648, A2714
→ 3-Hydroxymethyl-3-methyl-1-phenyltriazene
Probably in: Rat A2019
→ 1-(*o*-Hydroxyphenyl)-3,3-dimethyltriazene
Probably in: Rat A2648, A2704
→ 1-(*p*-Hydroxyphenyl)-3,3-dimethyltriazene
Probably in: Rat A2019, A2648, A2704
→ 3-Methyl-1-phenyltriazene
In: Rat A112

N-(2,6-Dimethylphenyl)urea
Formed from: Recainam
Xilobam
→ N-(4-Hydroxy-2,6-dimethylphenyl)urea
Probably in: Man G938
Mouse G938
→ N-(2-Hydroxymethyl-6-methylphenyl)urea
Probably in: Dog G938
Man G938
Mouse G938

Dimethyl phthalate
→ Phthalic acid monomethyl ester
In: Aspergillus K622
Baboon A1662
Ferret A1662
Microorganism K282
Rat A1662

2,3-Dimethyl-5-phytylquinol (phytylplastoquinol)
Formed from: Phytyltoluquinol
→ 2,3-Dimethyl-5-phytylquinone
Probably in: Spinach A3972
→ γ-Tocopherol
In: Anabaena G905
Spinach A3972

2,3-Dimethyl-5-phytylquinone
Formed from: 2-Methyl-6-phytylquinone
Probably formed from: 2,3-Dimethyl-5-phytylquinol

3,5-Dimethyl-N-2-picolylbenzamide
→ 3,5-Dimethylbenzamide

In: Rat E231
→ 3,5-Dimethyl-N-2-picolylbenzamide-N-oxide
In: Rat E231
→ 4-Hydroxy-3,5-dimethyl-N-2-picolylbenzamide
In: Rat E231
→ 3-Hydroxymethyl-5-methyl-N-2-picolylbenzamide
In: Rat E231

3,5-Dimethyl-N-3-picolylbenzamide
→ 3,5-Dimethyl-N-3-picolylbenzamide-N-oxide
In: Rat E231
→ 3-Hydroxymethyl-5-methyl-N-3-picolylbenzamide
In: Rat E231

3,5-Dimethyl-N-4-picolylbenzamide
→ 3,5-Dimethylbenzamide
In: Rat E231
→ 3,5-Dimethyl-N-4-picolylbenzamide-N-oxide
In: Rat E231
→ 3-Hydroxymethyl-5-methyl-N-4-picolylbenzamide
In: Rat E231

3,5-Dimethyl-N-2-picolylbenzamide-N-oxide
Formed from: 3,5-Dimethyl-N-2-picolylbenzamide

3,5-Dimethyl-N-3-picolylbenzamide-N-oxide
Formed from: 3,5-Dimethyl-N-3-picolylbenzamide

3,5-Dimethyl-N-4-picolylbenzamide-N-oxide
Formed from: 3,5-Dimethyl-N-4-picolylbenzamide

2,4-Dimethyl-3-pipecolamidophenol
→ 2,4-Dimethyl-3-pipecolamidophenyl sulphate
In: Man J764

2,4-Dimethyl-3-pipecolamidophenyl sulphate
Formed from: 2,4-Dimethyl-3-pipecolamidophenol

(2*E*)-3-(5-(1,1-dimethyl-2-propenyl)-4-hydroxy-2-methoxyphenyl)-1-(4-hydroxyphenyl-2-propen-1-one
See: Licochalcone A

Dimethylpropion (2-(dimethylamino)-1-phenyl-1-propanone)
→ Methylpropion
In: Guinea pig F249
Mouse F249
Rabbit F249
→ N-Methylpseudoephedrine
In: Guinea pig F249
Mouse F249
Rabbit F249

(3*S*,6*S*)-3-((2-(1,1-Dimethylpropyl)-5,7-bis(3-methylbutyl)-1*H*-indol-3-yl)methyl)-6-methyl-2,5-piperazinedione
See: Echinuline

2,4-Dimethyl-3-(*n*-propyl)pipecolamidophenol
→ 2,4-Dimethyl-3-(*n*-propyl) pipecolamidophenyl sulphate
In: Man J764

2,4-Dimethyl-4-(*n*-propyl)pipecolamidophenol
→ 2,4-Dimethyl-4-(*n*-propyl) pipecolamidophenyl sulphate
In: Man J764

2,4-Dimethyl-3-(*n*-propyl)pipecolamidophenyl sulphate
Formed from: 2,4-Dimethyl-3-(*n*-propyl) pipecolamidophenol

3,5-Dimethyl-4-(*n*-propyl)pipecolamidophenyl sulphate
Formed from: 3,5-Dimethyl-4-(*n*-propyl) pipecolamidophenol

rel-3-((3*R*,4*S*)-1,3-Dimethyl-4-propyl-4-piperidinyl)phenol
See: Picenadol

N-(3,3-Dimethylpropynyl)-3,4,5-trimethoxycinnamide
→ α-(3,5-Dihydroxycinnamido) isobutyric acid
In: Microorganism A568

β,β-Dimethyl-α-pyridinyl-3-pyridineethanol
See: Metyrapone

5,11-Dimethyl-6*H*-pyrido[4,3-*b*]carbazole
See: Ellipticine

N,N-Dimethyl-4-(2-pyridylazo)aniline
→ N,N-Dimethyl-*p*-phenylenediamine
In: Rat G133, G380, G511
3,5-Dimethyl-N-((4-pyridyl)methyl)benzamide
See: Picobenzide
***o*-(2-(4,6-Dimethylpyrimidinylamino)ethyl)benzoic acid**
Formed from: Debrisoquine
***o*-(2-(4,6-Dimethylpyrimidinylamino)methyl) benzoic acid**
Formed from: Debrisoquine
N-(4,6-Dimethyl-2-pyrimidinyl) benzenesulphonamide
Formed from: Sulphamethazine
→ N-(4,6-Dimethyl-2-pyrimidinyl)-*p*-hydroxybenzenesulphonamide
In: Rat E902
→ N-(5-Hydroxy-4,6-dimethyl-2-pyrimidinyl)benzenesulphonamide
In: Rat E902
→ N-(4-Hydroxymethyl-6-methyl-2-pyrimidinyl)benzenesulphonamide
In: Rat E902
N-(4,6-Dimethyl-2-pyrimidinyl)-*p*-hydroxybenzenesulphonamide
Formed from: N-(4,6-Dimethyl-2-pyrimidinyl)benzenesulphonamide
3-(3,5-Dimethyl-1*H*-pyrrol-2-ylmethylene)-1,3-dihydro-5-hydroxyindol-2-one
Formed from: 3-(3,5-Dimethyl-1*H*-pyrrol-2-ylmethylene)-1,3-dihydroindol-2-one
3-(3,5-Dimethyl-1*H*-pyrrol-2-ylmethylene)-1,3-dihydroindol-2-one
→ 1,3-Dihydro-3-(5-hydroxymethyl-3-methyl-1*H*-pyrrol-2-ylmethylene)-indol-2-one
In: Dog K336
Man K336
Monkey K336
Mouse K336
Rat K336
→ 3-(3,5-Dimethyl-1*H*-pyrrol-2-ylmethylene)-1,3-dihydro-5-hydroxyindol-2-one
In: Dog K336
Man K336
Monkey K336
Mouse K336
Rat K336

2,3-Dimethylquinol
Probably formed from: 1-*tert*-Butylamino-3-(4-hydroxy-2,3-dimethylphenoxy)-2-propanol
2,5-Dimethylquinol
Formed from: 2,5-Dimethyl-*p*-quinone
6,7-Dimethylquinoline
→ 6-Hydroxymethyl-7-methylquinoline
In: Rainbow trout F451
→ 7-Hydroxymethyl-6-methylquinoline
In: Rainbow trout F451
6,8-Dimethylquinoline
→ 3-Hydroxy-6,8-dimethylquinoline
In: Rainbow trout F451
→ 5-Hydroxy-6,8-dimethylquinoline
In: Rainbow trout F451
→ 7-Hydroxy-6,8-dimethylquinoline
In: Rainbow trout F451
→ 6-Hydroxymethyl-8-methylquinoline
In: Rainbow trout F451
1,2-Dimethylquinolinium
→ 4-Hydroxy-1,2-dimethylquinolinium
In: Rabbit D208
1,4-Dimethylquinolinium
→ 2-Hydroxy-1,4-dimethylquinolinium
In: Rabbit D208
2,5-Dimethyl-*p*-quinone
→ 2,5-Dimethylquinol
In: Beef G401
Guinea pig H790
Rat H790
6,7-Dimethyl-8-ribityllumazine
Probably formed from guanine in Aerobacter A1032
Ribitol in Eremothecium A740
→ Riboflavin
In: Eremothecium A3341
With riboflavin synthase E447
See also: A1032
***p*-Dimethylsulphoniumbenzaldehyde**
Formed from: *p*-Dimethylsulphoniumbenzylamine
***p,p'*-Dimethyl-*cis*-stilbene oxide**
→ (*R,R*)-α,β-Dihydroxy-*p,p'*-dimethyldibenzyl
In: Rabbit H568
Dimethylstilboestrol
→ Dimethylstilboestrol quinone

In: Horseradish D892

Dimethylstilboestrol quinone
Formed from: Dimethylstilboestrol

2,6-Dimethyl-N-succinylglycylaniline
Probably formed from: N-(4-Hydroxybutyrylglycyl)-2,6-dimethylaniline

***p*-Dimethylsulphoniumbenzaldehyde**
Formed from: *p*-Dimethylsulphoniumbenzylamine

***p*-Dimethylsulphoniumbenzylamine**
→ *p*-Dimethylsulphoniumbenzaldehyde
In: Pig A3693

Dimethyl terephthalate
→ Monomethyl terephthalate
In: Aspergillus H424
Bacillus F403

2,2-Dimethyl-3-(1,2,2,2-tetrabromoethyl)cyclopropanecarboxylic acid cyano(3-phenoxyphenyl) methyl ester
See: Tralomethrin

α,α-Dimethyl-4-(α,α,β,β-tetrafluorophenethyl) benzylamine
→ α,α-Dimethyl-4-(α,α,β,β-tetrafluorophenethyl)benzylamine-β-D-glucuronide
In: Baboon A2116, A3021
Dog A2116, A3021
Man A2116, A3021
Monkey A2116, A3021
Rat A2116
→ 2-Hydroxylamino-2-(4-(α,α,β,β-tetrafluorophenethyl)phenyl)propane
In: Baboon A2116, A3021
Dog A2116, A3021
Man A2116, A3021
Monkey A2116, A3021
Rat A2116
→ 2-Nitro-2-(4-(α,α,β,β-tetrafluorophenethyl)phenyl)propane
In: Baboon A2116, A3021
Dog A2116, A3021
Man A2116, A3021
Monkey A2116, A3021
Rat A2116
→ 4′-(α,α,β,β-Tetrafluorophenethyl) acetophenone
In: Baboon A2116, A3021
Dog A2116, A3021
Man A2116, A3021
Monkey A2116, A3021
Rat A2116
→ 2-(4-(α,α,β,β-Tetrafluorophenethyl) phenyl)-2-propanol
In: Baboon A2116, A3021
Dog A2116, A3021
Man A2116, A3021
Monkey A2116, A3021
Rat A2116

α,α-Dimethyl-4-(α,α,β,β-tetrafluorophenethyl) benzylamine-β-D-glucuronide
Formed from: α,α-Dimethyl-4-(α,α,β,β-tetrafluorophenethyl)benzylamine

α,α-Dimethyl-4-(α,α,β,β-tetrafluorophenethyl) benzyl methyl ether
2-(4-(α,α,β,β-Tetrafluorophenethyl)phenyl)-2-propanol

4′,7-Di-O-methyltetrahydropapaveroline
See: Norprotosinomenine

N,N-Dimethyl-*p*-toluidine
→ N-Methyl-*p*-toluidine
In: Glycine K419
Man K419
Rabbit F839
Rat D578

***p*-(3,3-Dimethyl-1-triazeno)benzoic acid**
Formed from: 1-(*p*-Carboxamidophenyl)-3,3-dimethyltriazene
→ *p*-(3,3-Dimethyl-1-triazeno)benzoyl-β-D-glucuronic acid
In: Mouse F385
→ *p*-(3,3-Dimethyl-1-triazeno)benzoylglycine
In: Mouse F385

***p*-(3,3-Dimethyl-1-triazeno)benzoyl-β-D-glucuronic acid**
Formed from: *p*-(3,3-Dimethyl-1-triazeno)benzoic acid

***p*-(3,3-Dimethyl-1-triazeno)benzoylglycine**
Formed from: *p*-(3,3-Dimethyl-1-triazeno)benzoic acid

3,3-Dimethyl-1-(2,4,6-trichlorophenyl)triazene
→ 3-Hydroxymethyl-3-methyl-1-(2,4,6-trichlorophenyl)triazene-β-D-glucuronide
In: Rat B241

N,N-Dimethyl-*p*-trifluoroacetamidoaniline
→ N-Methyl-*p*-trifluoroacetamidoaniline
In: Rat D578

2,6-Dimethyl-4-(2-trifluoromethylphenyl)-3,5-pyridinedicarboxylic acid
Probably formed from: 2,6-Dimethyl-4-(2-trifluoromethylphenyl)-3,5-pyridinedicarboxylic acid monoethyl ester
→ 2-Hydroxymethyl-6-methyl-4-(2-trifluoromethylphenyl)-3,5-pyridinedicarboxylic acid
Probably in: Dog A840
Rat A840
Squirrel monkey A840

2,6-Dimethyl-4-(2-trifluoromethylphenyl)-3,5-pyridinedicarboxylic acid monoethyl ester
Probably formed from: 1,4-Dihydro-2,6-dimethyl-4-(2-trifluoromethylphenyl)-3,5-pyridinedicarboxylic acid monoethyl ester
→ 2,6-Dimethyl-4-(2-trifluoromethylphenyl)-3,5-pyridinedicarboxylic acid
Probably in: Dog A840
Rat A840
Squirrel monkey A840

2,2-Dimethyl-N-(2,4,6-trimethoxyphenyl) dodecanamide
→ 11,11-Dimethyl-12-oxo-12-((2,4,6-trimethoxyphenyl)amino)dodecanoic acid
In: Rat J83
→ 6,6-Dimethyl-7-oxo-7-((2,4,6-trimethoxyphenyl)amino)heptanoic acid
In: Rat J83
→ 5,5-Dimethyl-6-oxo-6-((2,4,6-trimethoxyphenyl)amino)hexanoic acid
In: Monkey F766
Rat F766, J83

α,N-Dimethyltryptamine
Formed from: α-Methyltryptamine

N,N-Dimethyltryptamine
Formed from: N,N-Dimethyltryptamine-N-oxide
N-Methyltryptamine
→ N,N-Dimethyltryptamine-N-oxide
In: Rat A3946, E210, E212, E506
→ Indole-3-acetaldehyde
In: Rat A3946, B653, E210
→ N-Methyltryptamine
In: Rat A3946, C679, E210, E212, E506
→ 1,N,N-Trimethyltryptamine
In: Rabbit C333

1,N-Dimethyltryptamine
Formed from: 1-Methyltryptamine
N-Methyltryptamine
→ 1,N,N-Trimethyltryptamine
In: Rabbit A106

5,N-Dimethyltryptamine
Formed from: 5-Methyltryptamine

N,N-Dimethyltryptamine-N-oxide
→ N,N-Dimethyltryptamine
In: Rat A3946
Formed from: 5-Methyltryptamine

N,N-Dimethyl-*m*-tyramine
→ *m*-Hydroxyphenylacetaldehyde
In: Rat A2961

N,N-Dimethyltyramine
See: Hordenine

Di-(α-methyltyrosine)
Formed from: α-Methyltyrosine

N-(1,1-Dimethyl-2-ureidoethyl)-2-hydroxy-3-(*o*-fluorobenzoyloxy)propylamine
See: Flestolol

1,3-Di-O-methyl-ε-viniferin
Formed from: *trans*-Pterostilbene

Dimethylvinphos
→ Demethyldimethylvinphos
In: Dog A1752
Rat A1752
→ 2,4-Dichloromandelic acid
In: Dog A1752
Rat A1752
→ 2,4-Dichlorophenacyl chloride
In: Dog A1752
Rat A1752
→ 1-(2,4-Dichlorophenyl)ethanol
In: Dog A1752
Rat A1752
→ 2,4-Dichlorophenylglycol
In: Dog A1752
Rat A1752

5,6-Dimethylxanthenone-4-acetic acid
→ 5,6-Dimethylxanthenone-4-acetyl-β-D-glucuronic acid
In: Rat H183

→ 6-Hydroxymethyl-5-methylxanthenone-4-acetic acid
In: Man K335
See also: H183

5,6-Dimethylxanthenone-4-acetyl-β-D-glucuronic acid
Formed from: 5,6-Dimethylxanthenone-4-acetic acid

Dimetindene
→ Demethyldimetindene
In: Rat H782
→ 6-Hydroxydimetindene
In: Rat H782

Dimid
See: N,N-Dimethyldiphenylacetamide

Dimorphine
Formed from: Morphine

N,N′-Di-(α-(1-naphthyl)propionyloxy-2-ethyl) piperazine
See: Nafiverine

Dinemorphan
See: *d*-N,3-Dimethylmorphinan

3,6-Dinicotinoylmorphine
See: Nicomorphine

Dinitramine (N^3,N^3-diethyl-2,4-dinitro-6-trifluoromethyl-*m*-phenylenediamine)
→ 6-Amino-1-ethyl-2-methyl-7-nitro-5-trifluoromethylbenzimidazole
In: Aspergillus A1002
→ N^3-Ethyl-2,4-dinitro-6-trifluoromethyl-*m*-phenylenediamine
In: Aspergillus A1002
Carp A1951

2,4-Dinitroaniline
Formed from: 1-(1-Aziridinyl)-2,4-dinitrobenzene
→ 2,4-Dinitrophenylhydroxylamine
In: Rat D649
→ 2-Nitro-*p*-phenylenediamine
Probably in: Veillonella A2624
→ 1,2,4-Triaminobenzene
In: Haloanaerobium G369
Sporohalobacter G369
See also: G291

3,5-Dinitroaniline
Probably formed from: 1,3,5-Trinitrobenzene
→ *m*-Dinitrobenzene
Probably in: Pseudomonas H288

2,2′-Dinitro-4,4′-azoxytoluene
Probably formed from: 4-Amino-2-nitrotoluene

4,4′-Dinitro-2,2′-azoxytoluene
Probably formed from: 2-Amino-4-nitrotoluene

2,4-Dinitrobenzaldehyde
Formed from: 2,4-Dinitrobenzyl alcohol
→ 2,4-Dinitrobenzyl alcohol
In: Rat F252, F324
→ 2,4-Dinitrobenzoic acid
In: Man J24
Rat F252
See also: C903, F324, H515, H854

2,6-Dinitrobenzaldehyde
Formed from: 2,6-Dinitrobenzyl alcohol
→ 2,6-Dinitrobenzyl alcohol
In: Rat F252, F450
→ 2,4-Dinitrobenzoic acid
In: Rat C33, F252, F450
See also: H515

3,5-Dinitrobenzamide (nitromide)
→ 3-Amino-5-nitrobenzamide
In: Rat B193

***o*-Dinitrobenzene**
→ *o*-Nitroaniline
In: Rat D826, E888
Saccharomyces F201
See also: A2624
→ S-(*o*-Nitrophenyl)glutathione
In: Japanese monkey E429, F139
Man E429
Pig C901
Rat D826, E429
Rhesus monkey E429
See also E888
→ *o*-Nitrosonitrobenzene
In: Rat E888

***m*-Dinitrobenzene**
Probably formed from: 3,5-Dinitroaniline
→ *m*-Nitroaniline
In: Beef H449
Hamster F764
Methanococcus H328
Mycobacterium J916

Rat D826, E440, E888, F764, H449, H815
Saccharomyces F201
See also: A2624, H288

p-Dinitrobenzene
→ *p*-Nitroaniline
In: Rat D826
Saccharomyces F201
See also: A2624, E888
→ S-(*p*-Nitrophenyl)glutathione
In: Japanese monkey E429, F139
Man E429
Pig C901
Rat B429
Rhesus monkey F429
See also: E888

2,4-Dinitrobenzoic acid
Probably formed from: 2,4-Dinitrobenzaldehyde
→ 2-Amino-4-nitrobenzoic acid
Probably in: Rat B578, H515
→ 4-Amino-2-nitrobenzoic acid
Probably in: Rat C903, H515
Veillonella A2624
→ S-(4-Carboxy-3-nitrophenyl)glutathione
Probably in: Japanese monkey F139
Pig C901
Rhesus monkey F139

2,5-Dinitrobenzoic acid
→ S-(3-Carboxy-4-nitrophenyl)glutathione
Probably in: Pig C901

2,6-Dinitrobenzoic acid
Probably formed from: 2,6-Dinitrobenzaldehyde
→ 2-Amino-6-nitrobenzoic acid
Probably in: Rat C33

3,4-Dinitrobenzoic acid
→ S-(4-Carboxy-2-nitrophenyl)glutathione
Probably in: Japanese monkey F139
Pig C901
Rhesus monkey F139

3,5-Dinitrobenzoic acid
→ 3-Amino-5-nitrobenzoic acid
Probably in: Veillonella A2624

2,4-Dinitrobenzyl alcohol
Formed from: 2,4-Dinitrobenzaldehyde
2,4-Dinitrobenzyl-β-D-glucuronide
2,4-Dinitrotoluene
→ 2-Amino-4-nitrobenzyl alcohol
Probably in: Mouse D319
Rat B885
→ 4-Amino-2-nitrobenzyl alcohol
Probably in: Rat B885
→ 2,4-Dinitrobenzaldehyde
In: Rat F252, F324
See also: B578, C903, H515, H854
→ 2,4-Dinitrobenzyl-β-D-glucuronide
In: Rat B578, C903, F324, H515, H854

2,6-Dinitrobenzyl alcohol
Formed from: 2,6-Dinitrobenzaldehyde
2,6-Dinitrobenzyl-β-D-glucuronide
2,6-Dinitrotoluene
→ 2-Amino-6-nitrobenzyl alcohol
Probably in: Rat F450
→ 2,6-Dinitrobenzaldehyde
In: Rat C33, F252, F450
See also: H854
→ 2,6-Dinitrobenzyl-β-D-glucuronide
In: Rat B578, C33, H515, H854

2,4-Dinitrobenzyl-β-D-glucuronide
Formed from: 2,4-Dinitrobenzyl alcohol
→ 2,4-Dinitrobenzyl alcohol
In: Rat H854

2,6-Dinitrobenzyl-β-D-glucuronide
Formed from: 2,6-Dinitrobenzyl alcohol
→ 2,6-Dinitrobenzyl alcohol
In: Rat H854

4,4′-Dinitrobiphenyl
→ 4-Amino-3-chloro-4′-nitrobiphenyl
In: Rat A721
→ Benzidine
In: Guinea pig A721

3,5-Dinitrocatechol
→ 3,5-Dinitrocatechol-β-D-glucuronide
In: Rat J546

3,5-Dinitrocatechol-β-D-glucuronide
Formed from: 3,5-Dinitrocatechol

4,6-Dinitro-o-cresol
→ 4-Amino-6-nitro-*o*-cresol
Probably in: Rat C470
Veillonella A2624
→ 6-Amino-4-nitro-*o*-cresol
In: Rat C470

→ 2-Hydroxy-3,5-dinitrobenzyl alcohol
In: Rat C470

2,6-Dinitro-N,N-dipropyl-4-trifluoromethylaniline
(trifluralin, α,α,α-trifluoro-2,6-dinitro-N,N-di-*n*-propyl-*p*-toluidine)
→ 2,6-Dinitro-N-propyl-4-trifluoromethylaniline
In: Rat A2912
→ 2-Ethyl-7-nitro-1-propyl-5-trifluoromethylbenzimidazole
In: Rat A2912
→ N-(2-Hydroxypropyl)-2,6-dinitro-N-propyl-4-trifluoromethylaniline
In: Rat A2912
→ N-(3-Hydroxypropyl)-2,6-dinitro-N-propyl-4-trifluoromethylaniline
In: Rat A2912

3,9-Dinitrofluoranthene
→ 3-Amino-9-nitrofluoranthene
In: Rat G887

2,4-Dinitrophenol
Formed from: Formed from: N-Acetyl-2,4-dinitrophenyl-β-D-glucosaminide
2,4-Dinitrophenyl-β-D-xyloside
→ 2-Amino-4-nitrophenol
In: Rat A1065
→ 4-Amino-2-nitrophenol
In: Rat A1065
See also: A2624, G921, H328
→ 2,4-Diaminophenol
In: Haloanaerobium G369
Sporohalobacter G369
See also: G827
→ 2-(S-Glutathionyl)-4-nitrophenol
Probably in: Japanese monkey F139

2,5-Dinitrophenol
→ 3-(S-Glutathionyl)-2-nitrophenol
Probably in: Pig C901

3,4-Dinitrophenol
Formed from: 3,4-Dinitrophenyl tetra-N-acetylchitotetraoside

S-(2,4-Dinitrophenyl)-L-cysteine
Probably formed from: S-(2,4-Dinitrophenyl)cysteinylglycine
→ N-Acetyl-S-(2,4-dinitrophenyl)-L-cysteine
Probably in: Guinea pig G352
Rat G352

S-(2,4-Dinitrophenyl)cysteinylglycine
Probably formed from: S-(2,4-Dinitrophenyl)glutathione
→ S-(2,4-Dinitrophenyl)-L-cysteine
Probably in: Guinea pig G352
Rat G352

S-(2,4-Dinitrophenyl)glutathione
Formed from: 1-Chloro-2,4-dinitrobenzene
→ S-(2,4-Dinitrophenyl)cysteinylglycine
Probably in: Guinea pig G352
Rat G352

2,4-Dinitrophenylhydroxylamine
Formed from: 2,4-Dinitroaniline
→ 2,4-Dinitrophenylhydroxylamine-β-D-glucuronide
In: Rat D649
→ 2,4-Dinitrophenylhydroxylamine sulphate
In: Rat D649

2,4-Dinitrophenylhydroxylamine-β-D-glucuronide
Formed from: 2,4-Dinitrophenylhydroxylamine

2,4-Dinitrophenylhydroxylamine sulphate
Formed from: 2,4-Dinitrophenylhydroxylamine

3,4-Dinitrophenyl tetra-N-acetylchitotetraoside
→ 3,4-Dinitrophenol
In: Vibrio A2722

2,4-Dinitrophenyl-β-D-xyloside
→ 2,4-Dinitrophenol
In: Penicillium A3240

2,6-Dinitro-N-propyl-4-trifluoromethylaniline
Formed from: 2,6-Dinitro-N,N-dipropyl-4-trifluoromethylaniline
Fluchloralin
Profluralin
→ 2,6-Dinitro-4-trifluoromethylaniline
In: Rat A2912
→ N-(2-Hydroxypropyl)-2,6-dinitro-4-trifluoromethylaniline
Probably in: Rat A2912
→ N-(3-Hydroxypropyl)-2,6-dinitro-4-trifluoromethylaniline
Probably in: Rat A2912

1,8-Dinitropyrene
→ 1-Amino-8-nitropyrene
In: Rabbit E953
Salmonella F439

2,4-Dinitrotoluene
Formed from: 2,4,6-Trinitrotoluene
→ 2-Amino-4-nitrotoluene
In: Dog A1703
Microorganism C435, D151, D319, G542
Microsporium A2716
Monkey A1703
Mouse D319
Phanerochaete G436
Pseudomonas J196
Rabbit A1703
Rat A1703 B885, C131, F324, H515, H854
See also: B297, G827
→ 4-Amino-2-nitrotoluene
In: Dog A1703
Microorganism C435, D151, D319, G542
Microsporium A2716
Monkey A1703
Mouse D319
Pseudomonas J196
Rabbit A1703
Rat A1703 B885, C131, H854
Veillonella A2624
See also: B297, C903
→ 2,4-Dinitrobenzyl alcohol
In: Dog A1703, B297
Microorganism D319
Monkey A1703, B297
Mouse B297, D319
Rabbit A1703, B297
Rat A1703, B297, B578, B607, B885, C131, C903, D718, F324, H515, H854
→ 4-Methyl-5-nitrocatechol
In: Pseudomonas G407, J524
→ 2-Nitro-4-nitrosotoluene
In: Microorganism C263, D151, G542
→ 4-Nitro-2-nitrosotoluene
In: Microorganism C263, D151

2,6-Dinitrotoluene
→ 2-Amino-6-nitrotoluene
In: Man G378
Methanococcus H328
Rat F450, G378, H515, H854
See also: G827
→ 2,6-Dinitrobenzyl alcohol
In: Man G378
Rat C33, F450, G378, F252, H515, H894
→ 4-Nitro-6-nitrosotoluene
In: Microorganism G542
Salmonella G235

2,6-Dinitro-*p*-toluidine
→ 3,4-Diamino-5-nitrotoluene
Probably in: Veillonella A2624

2,6-Dinitro-4-trifluoromethylaniline
Formed from: 2,6-Dinitro-N-propyl-4-trifluoromethylaniline
→ 3-Nitro-5-trifluoromethy-*o*-phenylenediamine
In: Streptomyces B13
See also: A2912

2,6-Dinitro-6-trifluoromethyl-*m*-phenylenediamine
Formed from: N^3-Ethyl-2,6-dinitro-6-trifluoromethyl-*m*-phenylenediamine

Dinorcloprostenol
Formed from: Cloprostenol

Dinoseb
See: 2-*sec*-Butyl-4,6-dinitrophenol

Di-*n*-octyl phthalate
→ Phthalic acid mono-*n*-octyl ester
In: Baboon A1652
Ferret A1652
Rat A1652

Diosmetin (3′,5,7-trihydroxy-4′-methoxyflavone)
→ Diosmetin-3′-β-D-glucuronide
In: Rat G848
→ Diosmetin-O-sulphate
In: Rat G848

Diosmetin-3′,7-di-β-D-glucuronide
Probably formed from: Diosmetin-3′-β-D-glucuronide

Diosmetin-7-glucoside
→ Diosmetin-7-malonylglucoside
In: Parsley D684

Diosmetin-3′-β-D-glucuronide
Formed from: Diosmetin
→ Diosmetin-3′,7-di-β-D-glucuronide
Probably in: Rat G848

Diosmetin-7-malonylglucoside
Formed from: Diosmetin-7-glucoside

Diosmetin-7-rutoside
See: Diosmin

Diosmetin-O-sulphate
Formed from: Diosmetin

Diosmin (diosmetin-7-rutoside)
→ Diosmin-β-D-glucuronide
In: Rat G830
→ Diosmin-O-sulphate
In: Rat G830

Diosmin-β-D-glucuronide
Formed from: Diosmin

Diosmin-O-sulphate
Formed from: Diosmin

Dioxidine (2,6-quinoxalinedimethanol-1,4-dioxide)
Formed from: Quinoxidine

4,4′-(1,1′-Dioxido-3*H*-2,1-benzoxathiol-3-ylidene)bisphenol
See: Phenol red

Dioxindole (3-hydroxyoxindole)
Formed from: Isatin
Probably formed from: Dioxindole-3-acetic acid
Oxindole

Dioxindole-3-acetic acid
Formed from: Indole-3-acetic acid
→ Dioxindole
Probably in: Bradyrhizobium H679

Dioxindole-3-acetylaspartic acid
Formed from: Indole-3-acetylaspartic acid
→ Dioxindole-3-acetylaspartic acid-3-O-β-D-glucoside
In: Vicia E122, E437

Dioxindole-3-acetylaspartic acid-3-O-β-D-glucoside
Formed from: Dioxindole-3-acetylaspartic acid

3,5-Dioxo-1,2-diphenyl-4-(2′-phenylsulphonyl-ethyl)pyrazolidine
Formed from: Sulphinpyrazone

3,5-Dioxo-1,2-diphenyl-4-(2′-phenylthioethyl)pyrazolidine
Formed from: Sulphinpyrazole

2,3-Dioxoindole
See: Isatin

2-(1,3-Dioxolan-2-yl)-2-formyl-6-nitro-4-(2-oxopyrrolidin-1-yl)-2*H*-1-benzopyran
Probably formed from: 2-(1,3-Dioxolan-2-yl)-2-hydroxymethyl-6-nitro-4-(2-oxopyrrolidin-1-yl)-2*H*-1-benzopyran
→ 2-Carboxy-2-(1,3-dioxolan-2-yl)-6-nitro-4-(2-oxopyrrolidin-1-yl)-2*H*-1-benzopyran
Probably in: Rat J578

2-(1,3-Dioxolan-2-yl)-2-hydroxymethyl-6-nitro-4-(2-oxopyrrolidin-1-yl)-2*H*-1-benzopyran
Formed from: 2-(1,3-Dioxolan-2-yl)-2-methyl-6-nitro-4-(2-oxopyrrolidin-1-yl)-2*H*-1-benzopyran
→ 2-(1,3-Dioxolan-2-yl)-2-formyl-6-nitro-4-(2-oxopyrrolidin-1-yl)-2*H*-1-benzopyran
Probably in: Rat J578
→ 2-(1,3-Dioxolan-2-yl)-2-hydroxymethyl-6-nitro-4-(2-oxopyrrolidin-1-yl)-2*H*-1-benzopyran-β-D-glucuronide
In: Rat J578

2-(1,3-Dioxolan-2-yl)-2-hydroxymethyl-6-nitro-4-(2-oxopyrrolidin-1-yl)-2*H*-1-benzopyran-β-D-glucuronide
Formed from: 2-(1,3-Dioxolan-2-yl)-2-hydroxymethyl-6-nitro-4-(2-oxopyrrolidin-1-yl)-2*H*-1-benzopyran

2-(1,3-Dioxolan-2-yl)-2-methyl-6-nitro-4-(2-oxopyrrolidin-1-yl)-2*H*-1-benzopyran
→ 2-(1,3-Dioxolan-2-yl)-2-hydroxymethyl-6-nitro-4-(2-oxopyrrolidin-1-yl)-2*H*-1-benzopyran
In: Rat J578

8-(2,3-Dioxo-5-methylpyrrolidin-5-yl)amino-2,6-dihydroxy-4-methyl-5-(3-trifluoromethylphenoxy)quinoline
Probably formed from: 8-(4-Amino-3-hydroxy-1-methylbutyl)amino-2,6-dihydroxy-4-methyl-5-(3-trifluoromethylphenoxy) quinoline

3,5-Dioxo-4-(3-oxobutyl)-1,2-diphenylpyrazolidine
See: Kebuzon

9,9-Dioxophenothiazine
Formed from: Prothanon
→ Hydroxy-9,9-dioxophenothiazine
Probably in: Rat A3301

2,6-Dioxo-6-phenylhexanoic acid
Formed from: 2-Hydroxy-6-oxo-6-phenyl-hexa-2,4-dienoic acid
2-Methoxy-6-oxo-6-phenylhexa-2,4-dienoic acid
→ 2-Hydroxy-6-oxo-6-phenylhexanoic acid
Probably in: Pseudomonas E616

N-(2,6-Dioxo-3-piperidyl)phthalimide
See: Phthalidomide
9,9-Dioxopromethazine
See: Prothanon
5,8-Dioxopsoralen
Probably formed from: 5,8-Dihydroxypsoralen
1,2-Dipalmitoyl-3-(*m*-phenoxybenzoyl)glycerol
Formed from: *m*-Phenoxybenzoyl CoA
→ *m*-Phenoxybenzoic acid
In: Pig J875
See also: J877
1,3-Dipalmitoyl-2-(*m*-phenoxybenzoyl)glycerol
→ *m*-Phenoxybenzoic acid
Probably in: Rat J877
Diphacinone (2-diphenylacetyl-1,3-indandione)
→ Hydroxydiphacinone
In: Mouse C637
Rat C637
Diphenamid
See: N,N-Dimethyldiphenylacetamide
Diphenhydramine (benadryl, N,N-dimethyl-2-diphenylmethoxyethylamine)
→ Benzhydryl methylaminoethyl ether
In: Man A1554
Rat C669
→ Diphenhydramine-β-D-glucuronide
In: Man G18, H541, J137
→ Diphenylmethoxyacetic acid
In: Man A1554
Sheep J574, J607
Diphenhydramine-β-D-glucuronide
Formed from: Diphenhydramine
Diphenic acid ((1,1′-biphenyl)-2,2′-dicarboxylic acid)
Formed from: Phenanthrene
Phenanthrene-9,10-quinone
Diphenidol (α,α-diphenyl-1-piperidinebutanol)
→ Diphenidol-β-D-glucuronide
In: Dog A619
Man A619
→ N-(4-Hydroxy-4,4-diphenylbutyl)-δ-aminovaleric acid
In: Dog A619
Man A619
Diphenidol-β-D-glucuronide
Formed from: Diphenidol
***p*-Diphenoquinone**
Formed from: 4,4′-Dihydroxybiphenyl
Diphenoxylate
→ Diphenoxylic acid
In: Man A452
Diphenoxylic acid
Formed from: Diphenoxylate
→ Hydroxydiphenoxylic acid
Probably in: Man A452
2,2-Diphenylacetamide
Formed from: N-Methyldiphenylacetamide
→ Diphenylacetic acid
In: Plant A1205
N,N′-Diphenyl-2-acetamidoxyacetamide
→ Glycollanilide
In: Rat C45
Diphenylacetic acid
Formed from: Adiphenine
Piperidolate
Probably formed from: 2,2-Diphenylacetamide
→ Diphenylacetyl-β-D-glucuronic acid
In: Bushbaby A2527
Capuchin A2527
Cat A2527
Ferret A394, A3046
Fruit bat A2527
Monkey A2527
Mouse B401
Pigeon A395
Rabbit A2527
Rat A384, A2527, B401, B526, C313
→ Diphenylacetylglutamine
In: Rat C313
→ Diphenylacetylglycine
In: Bushbaby A2527
Cat A2527
Ferret A3046
Marmoset A2527
Pigeon A395
Rat A2527, C313
→ Diphenylacetyltaurine
In: Cat A2527
Ferret A3046
Pigeon A395
1,3-Diphenylacetone
Formed from: 1,3-Diphenylpropan-2-ol

Diphenylacetyl-β-D-glucuronic acid
Formed from: Diphenylacetic acid
Diphenylacetylglutamine
Formed from: Diphenylacetic acid
Diphenylacetylglycine
Formed from: Diphenylacetic acid
2-Diphenylacetyl-1,3-indandione
See: Diphacinone
Diphenylacetyltaurine
Formed from: Diphenylacetic acid
Diphenylamine
Formed from: Diphenylnitrosamine
→ Aniline
In: Microorganism C124, J474
→ Diphenylamine-N-β-D-glucuronide
In: Man H541, J416
→ 2-Hydroxydiphenylamine
In: Apple J494
→ 3-Hydroxydiphenylamine
In: Apple J494
→ 4-Hydroxydiphenylamine
In: Apple J494
Microorganism C124
See also: E383
→ Indole
In: Microorganism C124
Diphenylamine-N-β-D-glucuronide
Formed from: Diphenylamine
1,1-Diphenylbutan-3-one
Formed from: 3-Dimethylamino-1,1-diphenylbutane
→ 1-(*p*-Hydroxyphenyl)-1-phenylbutan-3-one
Probably in: Rat F582
1,1-Diphenylbutan-3-one oxime
Formed from: 3-Dimethylamino-1,1-diphenylbutane
→ 1-(*p*-Hydroxyphenyl)-1-phenylbutan-3-one oxime
Probably in: Rat F582
(*Z*)-2-(4-(1,2-Diphenyl-1-butenyl)phenoxy)acetaldehyde
Formed from: (*Z*)-2-(4-(1,2-Diphenyl-1-butenyl)phenoxy)ethanol
4-(1,2-Diphenyl-1-butenyl)phenoxyacetic acid
Formed from: Tamoxifen
→ 4-(1-(*p*-Hydroxyphenyl)-2-phenyl-1-butenyl)phenoxyacetic acid
Probably in: Rat F560, H450
(*Z*)-2-(4-(1,2-Diphenyl-1-butenyl)phenoxy)ethanol
→ (*Z*)-2-(4-(1,2-Diphenyl-1-butenyl)phenoxy)acetaldehyde
In: Rat H450
Diphenylcarbinol
See: Benzhydrol
4,4-Diphenylcyclohexanol
Formed from: Pramiverine
4,4-Diphenylcyclohexylamine
Formed from: Pramiverine
2-(2,2-Diphenylcyclopropyl)-4,5-dihydro-1*H*-imidazole
See: Cibenzoline
Diphenyldiazene-1-oxide
See: Fenazox
1,2-Diphenyl-3,5-dioxo-4-prenylpyrazolidine
See: Feprazone
Diphenyl disulphide
See: Thianthrene
Diphenylenedioxide-2,3-quinone
Formed from: Catechol
Diphenylethanedione
See: Benzil
1,2-Diphenylethanol
Formed from: Bibenzyl
ω-Phenylacetophenone
2,2-Diphenylethanol
Formed from: 2,2-Diphenylethylene
Diphenyl ether
→ 2-Hydroxyphenyl phenyl ether
In: Guinea pig E10
Rat C673
Trout E10
→ 3-Hydroxyphenyl phenyl ether
Probably in: Trout E10
→ 4-Hydroxyphenyl phenyl ether
In: Guinea pig E10
Rat C673
Trout B756, E10
→ Phenol
In: Pseudomonas E41
→ 2-Phenoxymuconic acid
In: Pseudomonas E41
2,2-Diphenylethylene
→ Atropic acid
In: Microorganism A1193

→ 2,2-Diphenylethanol
In: Microorganism A1193

N,N′-Diphenylguanidine
→ N-(*p*-Hydroxyphenyl)-N′-phenylguanidine
In: Rabbit G547
Rat G547
→ N″-Hydroxy-N,N′-diphenylguanidine
In: Rabbit G547
Rat G547

2-(4′,4′-Diphenylheptan-5′-on-2′-yl)oxaziridine
Formed from: 2-Dimethylamino-4,4-diphenyl-5-heptanone

2,2-Diphenylhydantoic acid
Formed from: 5,5-Diphenylhydantin

5,5-Diphenylhydantoin (dilantin, epanutin, phenytoin, 5,5-diphenyl-2,4-imidazolidinedione)
→ N,N-Dichloro-5,5-diphenylhydantoin
With myeloperoxidase E682
→ 5-(3,4-Dihydro-3,4-dihydroxyphenyl)-5-phenylhydantoin
In: Cat D459
Dog B57
Frog C39
Man A387, B57, D757, K328
Monkey A947
Mouse B991, C597, D818
Rat A626, A1970, B57, C286
→ 2,2-Diphenylhydantoic acid
In: Dog A2760
Mouse B991
→ 5,5-Diphenylhydantoin-N^3-β-D-glucuronide
In: Cat D459
Man A3402
Rat A3402
→ 5,5-Diphenylhydantoin-3′,4′-oxide
Probably in: Rabbit B471
Rat B471
→ 5-(*o*-Hydroxyphenyl)-5-phenylhydantoin
In: Man H635
→ 5-(*m*-Hydroxyphenyl)-5-phenylhydantoin
In: Dog A61, A1719, A2760, A3349, C582
Frog C39
Man A1719, C288, H635, K328
Rat C286, C288
→ 5-(*p*-Hydroxyphenyl)-5-phenylhydantoin
In: Cat D459, K201
Chicken A3431
Dog A61, A1719, A2760, A3349, C582
Frog C39
Man A629, A922, A1377, A1555, A1719, A2760, C288, C638, H635, H846, J59, K63, K201, K328
Mouse B991, C547, D818
Rat A626, A1970, A2160, A2182, A2854, A2879, A2976, C286, C288, C638, F830, K201
See also: A494, A506, A567, A947, A1018, A1971, A2110, A3305

5,5-Diphenylhydantoin-N^3-β-D-glucuronide
Probably formed from: 5,5-Diphenylhydantoin

5,5-Diphenylhydantoin-3′,4′-oxide
Formed from: 5,5-Diphenylhydantoin

1,2-Diphenylhydrazine
→ Aniline
In: Microorganism A3616

1,3-Diphenylisobenzofuran
→ *o*-Dibenzoylbenzene
In: Rabbit A3806

Diphenylmethane
→ Benzhydrol
In: Pseudomonas A2831
Rat B40, B706
→ *p*-Hydroxyphenyl phenylmethane
In: Pseudomonas H156
Rat B706
→ Phenylacetic acid
In: Pseudomonas A2831
→ 1,1,1′,1′-Tetraphenyldimethyl ether
In: Pseudomonas A2831

Diphenylmethanone
See: Benzophenone

Diphenylmethan-4-yl-β-D-glucuronide
Formed from: 4-Hydroxydiphenylmethane

Diphenylmethoxyacetic acid
Formed from: Diphenhydramine

4-(Diphenylmethoxy)piperidine
Formed from: Ebastine

4-Diphenylmethyl-1-formylpiperidine
Formed from: Fenoctimine
→ 1-Formyl-4-((*p*-hydroxyphenyl)phenyl) methylpiperidine

Probably in: Mouse H62

4-Diphenylmethyl-1-(((6-hydroxy)hexylimino) methyl)piperidine
Probably formed from: 4-Diphenylmethyl-1-(((8-hydroxy)octylimino)methyl)piperidine

4-Diphenylmethyl-1-(((7-hydroxy)octylimino) methyl)piperidine
→ 4-Diphenylmethyl-1-(((7-oxo)octylimino) methyl)piperidine
Probably in: Rat H228

4-Diphenylmethyl-1-(((8-hydroxy)octylimino) methyl)piperidine
Formed from: Fenoctimine
→ 4-Diphenylmethyl-1-(((6-hydroxy) hexylimino)methyl)piperidine
In: Mouse H62

1-(Diphenylmethyl)-4-(3-(*p*-hydroxyphenyl)-2-propenyl)piperazine
Formed from: Cinnarizine

(2*S*,3*S*,4*S*)-2-Diphenylmethyl-3-(5-(2-hydroxy-2-propyl)-2-methoxybenzylamino)-1-azabicyclo [2.2.2]octane
Formed from: Ezlopitant
→ (2*S*,3*S*,4*S*)-2-Diphenylmethyl-3-(2-methoxy-5-(2-propenyl)benzylamino)-1-azabicyclo[2.2.2]octane
In: Cynomolgus monkey K48
Dog K48
Guinea pig K48
Man K48
Rat K48

(3*R*)-(+)-Diphenylmethoxy-1-(3,4-methylenedioxyphenethyl)piperidine
See: Zamifenacin

(2*S*,3*S*,4*S*)-2-Diphenylmethyl-3-(2-methoxy-5-(2-propenyl)benzylamino)-1-azabicyclo[2.2.2] octane
Probably formed from: (2*S*,3*S*,4*S*)-2-Diphenylmethyl-3-(5-(2-hydroxy-2-propyl)-2-methoxybenzylamino)-1-azabicyclo[2.2.2] octane

1-(Diphenylmethyl)-4-methylpiperazine
See: Cyclizine

1-(3-(4-(Diphenylmethyl)-1-piperazinyl)propyl)-1,3-dihydro-2*H*-benzimidazol-2-one
See: Oxatomide

4-Diphenylmethyl-1-methylpiperidine
Formed from: Fenoctimine

4-(Diphenylmethyl)-1-((octylimino)methyl) piperidine
See: Fenoctimine

4-Diphenylmethyl-1-(((7-oxo)octylimino) methyl)piperidine
Probably formed from: 4-Diphenylmethyl-1-(((7-hydroxy)octylimino)methyl)piperidine

((*E*)-1-diphenylmethyl)-4-(3-phenyl-2-propyl) piperazine
See: Cinnarizine

1-(Diphenylmethyl)piperazine
Formed from: Cinnarizine

1-(3-(4-(Diphenylmethyl)-1-piperazinyl)propyl)-1,3-dihydro-5-hydroxy-2*H*-benzimidazol-2-one
Formed from: Oxatomide

1-(3-(4-(Diphenylmethyl)-1-piperazinyl)propyl)-1,3-dihydro-6-hydroxy-2*H*-benzimidazol-2-one
Formed from: Oxatomide

4-Diphenylmethylpiperidine
Formed from: Fenoctimine
→ 4-((*p*-Hydroxyphenyl)phenyl) methylpiperidine
Probably in: Dog H62

α,N-Diphenylnitrone
Formed from: N-Benzylaniline

Diphenylnitrosamine (N-nitrosodiphenylamine)
→ Diphenylamine
In: Mouse E383, F942
Rat D537
→ 4-Hydroxydiphenylamine
In: Mouse F942
See also: D537

2,5-Diphenyloxazole
→ 2-(*o*-Hydroxyphenyl)-5-phenyloxazole
In: Man E520
Rat E520
→ 2-(*p*-Hydroxyphenyl)-5-phenyloxazole
In: Man E520
Rat E520
→ 5-(*p*-Hydroxyphenyl)-2-phenyloxazole
In: Man E520
Rat E520

4,5-Diphenyl-4-oxazolin-2-one
Formed from: Ditazole

3,3-Diphenyl-N-(1-phenylethyl)propylamine
→ Benzoic acid
In: Rat A1068
→ Mandelic acid
In: Rat A1068
→ 1-Phenylethylamine
In: Rat A1068

1,2-Diphenyl-4-(2-(phenylsulphinyl)ethyl)-3,5-pyrazolidinedione
See: Sulphinpyrazole

1,2-Diphenyl-4-(2-(phenylsulphonyl)ethyl)-3,5-pyrazolidinedione
Formed from: Sulphinpyrazole
→ 1-(*p*-Hydroxyphenyl-2-phenyl-4-(2-(phenylsulphonyl)ethyl)-3,5-pyrazolidinedione
Probably in: Man B510

1,2-Diphenyl-4-(2-(phenylthio)ethyl)-3,5-pyrazolidinedione
Formed from: Sulphinpyrazole
→ 1-(*p*-Hydroxyphenyl-2-phenyl-4-(2-(phenylthio)ethyl)-3,5-pyrazolidinedione
In: Rabbit D114
See also: B510, B856, E152
→ Sulphinpyrazole
In: Rabbit D114

Diphenyl phosphate
Formed from: *p*-*tert*-butylphenyl diphenyl phosphate
Triphenyl phosphate
Probably formed from: Bisphenyl 9-((2-phosphonomethoxy)ethyl)adenine

S,S-Diphenyl phosphorodithioate
Formed from: O-Ethyl S,S-diphenyl phosphorodithioate
→ S-Phenyl phosphorothioate
Probably in: Goat A2322

α,α-Diphenyl-1-piperidinebutanol
See: Diphenidol

Diphenyl-4-piperidinylmethanol (azacyclonol)
Formed from: 4-(4-(Hydroxy(diphenyl)methyl)piperidin-1-yl)-1-(4-(1-hydroxymethyl)-1-(methylethyl)phenyl)-1-butanol
Terfenadine

1,3-Diphenylpropane
→ 1,3-Diphenylpropan-1-ol
In: Rat A3018
→ 1,3-Diphenylpropan-2-ol
In: Rat A3018

1,3-Diphenylpropan-1-ol
Formed from: 1,3-Diphenylpropane
→ 1,3-Diphenylpropan-1-one
In: Rat A3018

1,3-Diphenylpropan-2-ol
Formed from: 1,3-Diphenylpropane
→ 1,3-Diphenylacetone
In: Rat A3018

3,3-Diphenylpropan-1-ol
Probably formed from: 3,3-Diphenylpropylamine

1,3-Diphenylpropan-1-one
Formed from: 1,3-Diphenylpropan-1-ol

2,2-Diphenylpropionic acid
Formed from: Aprophen

2,2-Diphenylpropionic acid 2-(N-ethyl-N-hydroxyethyl))aminoethyl ester
Formed from: Aprophen

3,3-Diphenylpropylamine
Formed from: Fendiline
→ N-Acetyl-3,3-diphenylpropylamine
In: Man C778
→ 3,3-Diphenylpropan-1-ol
Probably in: Man C778
→ 3-(*p*-Hydroxyphenyl)-3-phenylpropylamine
Probably in: Man C778

N-(3,3-Diphenylpropyl)-1-(4-hydroxy-3-methoxyphenyl)ethylamine
Probably formed from: N-(3,3-Diphenylpropyl)-1-(*p*-hydroxyphenyl) ethylamine

N-(3,3-Diphenylpropyl)-1-(*p*-hydroxyphenyl) ethylamine
Probably formed from: Fendiline
→ N-(3,3-Diphenylpropyl)-1-(4-hydroxy-3-methoxyphenyl)ethylamine
Probably in: Man C778
→ *p*-Hydroxyacetophenone
Probably in: Man C778

N-(3,3-Diphenylpropyl)-α-methylphenethylamine
See: Prenylamine

N-(3,3-Diphenylpropyl)-1-phenylethylamine
See: Fendiline

Diphenyl sulphide (1,1′-thiobisbenzene)
Formed from: Diphenyl sulphoxide
→ Diphenyl sulphoxide
In: Guinea pig F592
Pig D446
Pseudomonas H165
See also: G438

Diphenyl sulphone
Formed from: Diphenyl sulphoxide
→ Hydroxydiphenyl sulphone
Probably in: Rat J413

Diphenyl sulphoxide
Formed from: Diphenyl sulphide
→ Diphenyl sulphide
In: Escherichia H560
Rabbit H561
Rat H561
→ Diphenyl sulphone
In: Guinea pig F592
Rat J413
See also: G438

Diphenyltin
→ Phenyltin
In: Rat A1932

1,3-Diphenyl-1-triazene
→ Aniline
In: Mouse J795
Rat J795
→ Benzene
In: Mouse J795
Rat J795

Dipivaloylapomorphine
→ Apomorphine
In: Rat A2013

Dipivaloyladrenaline
→ Adrenaline
In: Man A3562

Dipivaloyl 2-amino-tetrahydro-6,7-dihydroxynaphthalene
→ 2-Amino-tetrahydro-6,7-dihydroxynaphthalene
In: Rat C251

Dipivaloylepinine
→ 3,4-Dihydroxy-N-methylphenethylamine
In: Beef F220

Diprenorphine ((5α,7α)-17-(cyclopropylmethyl)-4,5-epoxy-18,19-dihydro-3-hyroxy-6-methoxy-α,α-dimethyl-6,14-ethenomorphinan-7-methanol
→ Decyclopropylmethyldiprenorphine
In: Rat A2350
→ Diprenorphine-β-D-glucuronide
In: Man J416
Monkey H861
Rat A2350, J97a

Diprenorphine-β-D-glucuronide
Formed from: Diprenorphine

Di-(L-prolyl-L-glutaminyl-L-glutaminyl-L-prolyl-L-tyrosine)
Formed from: L-Prolyl-L-glutaminyl-L-glutaminyl-L-prolyl-L-tyrosine
→ Tri-(L-prolyl-L-glutaminyl-L-glutaminyl-L-prolyl-L-tyrosine)
In: Horseradish J327

Di-(L-prolyl-L-glutaminyl-L-glutaminyl-L-prolyl-L-tyrosyl-L-prolyl-L-glutaminyl-L-glutaminyl-L-prolyl-L-alanine)
Formed from: L-Prolyl-L-glutaminyl-L-glutaminyl-L-prolyl-L-tyrosyl-L-prolyl-L-glutaminyl-L-glutaminyl-L-prolyl-L-alanine

3,5′-Di(1*E*)-1-propenyl(1,1′-biphenyl)-2,2′-diol
See: Isomagnolol

Dipropionyl apomorphine
→ Apomorphine
In: Rat A2013

Dipropionyl morphine
→ 6-Propionyl morphine
In: Man D173
Rat D173

1,6-Dipropoxybenzpyrene
→ 1-Hydroxy-6-propoxybenzpyrene
In: Rat B477

4-(2-(Dipropylamino)ethyl)-1,3-dihydro-2*H*-indol-2-one
See: Ropinirole

4-(2-Di-*n*-propylaminoethyl)-7-hydroxy-2(3*H*)-indolone
Formed from: 4-(2-Di-*n*-propylaminoethyl)-2(3*H*)-indolone
→ 4-(2-Di-*n*-propylaminoethyl)-7-hydroxy-2(3*H*)-indolone-β-D-glucuronide
In: Dog E295
Rat E295

4-(2-Di-*n*-propylaminoethyl)-7-hydroxy-2(3*H*)-indolone-β-D-glucuronide
Formed from: 4-(2-Di-*n*-propylaminoethyl)-7-hydroxy-2(3*H*)-indolone

N-(5-(2-Di-*n*-propylaminoethyl)-2-hydroxyphenyl) methanesulphonamide
→ N-(5-(2-Di-*n*-propylaminoethyl)-2-hydroxyphenyl)methanesulphonamide-β-D-glucuronide
In: Dog E295
Rat E295

N-(5-(2-Di-*n*-propylaminoethyl)-2-hydroxyphenyl)methanesulphonamide-β-D-glucuronide
Formed from: N-(5-(2-Di-*n*-propylaminoethyl)-2-hydroxyphenyl)methanesulphonamide

4-(2-Di-*n*-propylaminoethyl)-2(3*H*)-indolone
→ 4-(2-Di-*n*-propylaminoethyl)-7-hydroxy-2(3*H*)-indolone
In: Dog E295
Rat E295
→ 4-(2-*n*-Propylaminoethyl)-2(3*H*)-indolone
In: Dog E295
Rat E295

4-(2-((N,N-Dipropyl)amino)ethyl)-*o*-quinone
Formed from: 3,4-Dihydroxy-N,N-dipropylphenethylamine
→ N,N-Dipropylindoliumolate
In: Agaricus J460

2-((Di-*n*-propyl)amino)-5-hydroxy-6-methyltetralin
→ 6-Carboxy-2-((di-*n*-propyl)amino)-5-hydroxytetralin
In: Rat D819
→ 5-Hydroxy-6-methyl-2-((*n*-propyl) amino)tetralin
In: Rat D819

2-((Di-*n*-propyl)amino)-8-hydroxytetralin
→ 2-((Di-*n*-propyl)amino)-8-hydroxytetralin-β-D-glucuronide
In: Rat H132, H472
→ 2-(*n*-Propylamino)-8-hydroxytetralin
In: Rat H132, H472

2-((Di-*n*-propyl)amino)-8-hydroxytetralin-β-D-glucuronide
Formed from: 2-((Di-*n*-propyl)amino)-8-hydroxytetralin

(−)-4-Dipropylamino-1,3,4,5-tetrahydrobenz [*c*,*d*]indole-6-carboxamide
→ (−)-1,3,4,5-Tetrahydro-4-propylaminobenz[*c*,*d*]indole-6-carboxamide
In: Monkey G96
Rat G96

N,N-Dipropylaniline
→ N-Propylaniline
In: Rabbit A3486

N,N-Dipropylindoliumolate
Formed from: 4-(2-((N,N-Dipropyl)amino) ethyl)-*o*-quinone

N,N-Dipropylphenethylamine
→ N-Propylphenethylamine
Probably in: Cunninghamella F345

***p*,*p*′-Dipropyl-*cis*-stilbene oxide**
→ (*R*,*R*)-α,β-Dihydroxy-*p*,*p*′-dipropyldibenzyl
In: Rabbit H568

***p*-Dipropylsulphamoylbenzoic acid**
(probenecid)
→ *p*-Dipropylsulphamoylbenzoyl-β-D-glucuronic acid
In: Dog A1886
Man H66
Rat A1441
→ *p*-(N-(2-Hydroxypropyl)-N-propylsulphamoyl)benzoic acid
In: Monkey A1886
Rat A1441, A2081
→ *p*-(N-(3-Hydroxypropyl)-N-propylsulphamoyl)benzoic acid
In: Monkey A1886
Rat A1441, A2081
→ *p*-Propylsulphamoylbenzoic acid
In: Dog A1886
Monkey A1886
Rat A2081

***p*-Dipropylsulphamoylbenzoyl-β-D-glucuronic acid**
Formed from: *p*-Dipropylsulphamoylbenzoic acid

Dipyrone (metamizol, sulpyrine, ((2,3-dihydro-1,5-dimethyl-3-oxo-2-phenyl-1*H*-pyrazol-4-yl)methylamino)methanesulphonic acid)
→ 4-Formamidoantipyrine
In: Man B10
→ 4-Methylaminoantipyrine
In: Man A1161, B10, F467
Rat A1124, A1140
See also: A1141

Direct Black 4 (4-amino-3-((4′-((2,4-diamino-5-methylphenyl)azo)(1,1′-biphenyl)-4-yl)azo)-5-hydroxy-6-(phenylazo)-2,7-naphthalenedisulphonic acid)
→ Benzidine
In: Dog B739
Rat B739

Direct Black 38 (Direct Deep Black EX, 4-amino-3-((4′-((2,4-diaminophenyl)azo)(1,1′-biphenyl)-4-yl)azo)-5-hydroxy-6-(phenylazo)-2,7-naphthalenedisulphonic acid)
→ Benzidine
In: Microorganism C154, C254
Mouse A1182
Rat A1182

Direct Black 114 (4-hydroxy-3-((2-hydroxy-4-nitrophenyl)azo)-7-((8-hydroxy-5-quinolinyl)azo)-2-naphthalenesulphonic acid)
→ 3,3′-Dimethylbenzidine
In: Rat C541

Direct Blue 1 (6,6′-((3,3′-dimethoxy(1,1′-biphenyl)-4,4′-diyl)bis(azo))bis(4-amino-5-hydroxy-1,3-naphthalenedisulphonic acid))
→ 3,3′-Dimethylbenzidine
In: Dog B739
Rat B739

Direct Blue 2 (5-amino-3-((4′-((7-amino-1-hydroxy-3-sulpho-2-naphthalenyl)azo)(1,1′-biphenyl)-4-yl)azo)-4-hydroxy-2,7-naphthalenedisulphonic acid)
→ Benzidine
In: Dog B739
Rat B739

Direct Blue 8 (3,3′-((3,3′-dimethoxy(1,1′-biphenyl)-4,4′-diyl)bis(azo))bis(4-hydroxy-1-naphthalenesulphonic acid))
→ 3,3′-Dimethylbenzidine
In: Rat C541

Direct Blue 10 (3,3′-((3,3′-dimethoxy(1,1′-biphenyl)-4,4′-diyl)bis(azo))bis(4,5-dihydroxy-2,7-naphthalenedisulphonic acid))
→ 3,3′-Dimethylbenzidine
In: Rat C541

Direct Blue 14 (3,3′-((3,3′-dimethyl-(1,1′-biphenyl)-4,4-diyl)bis(azo))bis(5-amino-4-hydroxy-2,7-naphthalenedisulphonic acid))
→ 4-Amino-3,3′-dimethylbiphenyl
In: Staphylococcus K288
→ *o*-Tolidine
In: Rat C541
Staphylococcus K288

Direct Blue 15 (3,3-dimethyl-4,4-bis(azo(5-amino-4-hydroxynaphth-2-yl-2,7-disulphonic acid)biphenyl)
→ 3,3′-Dimethylbenzidine
In: Dog B739
Microorganism C154, C254
Rat B739

Direct Blue 25 (3,3′-((3,3′-dimethyl(1,1′-biphenyl)-4,4′-diyl)bis(azo))bis(4,5-dihydroxy-2,7-naphthalenedisulphonic acid)
→ *o*-Tolidine
In: Dog B739
Rat B739

Direct Blue 53 (Evans Blue, 6,6′-((3,3′-dimethyl(1,1′-biphenyl)-4,4′-diyl)bis(azo)) bis(4-amino-5-hydroxy-1,3-naphthalenedisulphonic acid))
→ *o*-Tolidine
In: Rat C541
See also: C471

Direct Brown 2 (5-((4′-((7-amino-1-hydroxy-3-sulpho-2-naphthalenyl)azo)(1,1′-biphenyl)-4-yl)azo)-2-hydroxybenzoic acid)
→ Benzidine
In: Dog B739
Rat B739

Direct Deep Black EX
See: Direct Black 38

Direct Green 1 (4-amino-5-hydroxy-3-((4′-((4-hydroxyphenyl)azo)(1,1′-biphenyl)-4-yl)azo)-6-(phenylazo)-2,7-naphthalenedisulphonic acid)
→ Benzidine
In: Dog B739

Rat B739

Direct Orange 1 (a mixture of three components)
→ Benzidine
In: Dog B739
Rat B739

Direct Orange 6 (5-((4′-((2,6-diamino-3-methyl-5-sulphophenyl)azo)-3,3′-dimethyl(1,1′-biphenyl)-4-yl)azo)-2-hydroxybenzoic acid)
→ *o*-Tolidine
In: Rat C541

Direct Orange 8 (5-((4′-((1-amino-4-sulpho-2-naphthalenyl)azo)(1,1′-biphenyl)-4-yl)azo)-2-hydroxybenzoic acid)
→ Benzidine
In: Dog B739
Rat B739

Direct Red 2 (3,3′-((3,3′-dimethyl(1,1′-biphenyl)-4,4′-diyl)bis(azo))bis(4-amino-1-naphthalenesulphonic acid))
→ *o*-Tolidine
In: Microorganism C154, C254

Direct Red 28 (3,3′-((1,1′-biphenyl)-4,4′-diylbis(azo))bis(4-amino-1-naphthalenesulphonic acid))
→ Benzidine
In: Dog B739
Rat B739

Direct Red 39 (8-((4′-((4-ethoxyphenyl)azo)-3,3′-dimethyl((1,1′-biphenyl)-4-yl)azo)-7-hydroxy-1,3-naphthalenedisulphonic acid))
→ *o*-Tolidine
In: Rat C541

Direct Red 46 (2-(4-((1-hydroxy-4-sulpho-2-naphalenyl)azo)phenyl)-6-methyl-7-benzothiazolesulphonic acid)
→ 3,3′-Dichlorobenzidine
In: Rat C541

Direct Violet 32 (4-amino-3-((4′-((1-hydroxy-4-sulpho-2-naphthalenyl)azo)-3,3′-dimethoxy(1,1′-biphenyl)-4-yl)azo-1-naphthalenesulphonic acid)
→ 3,3′-Dimethoxybenzidine
In: Rat C541

2,2′-Diselenylbisbenzanilide
Probably formed from: 2-Selenylbisbenzanilide

1,2-Di-O-sinapoyl-β-D-glucose
Formed from: Sinapoyl-β-D-glucose

Disophenol
See: 2,6-Diiodo-4-nitrophenol

Disopyramide (α-(2-(bis(1-methylethyl)amino)ethyl)-α-phenyl-2-pyridineacetamide)
→ Deisopropyldisopyramide
In: Dog B984
Man B984
Rat B984
→ *p*-Hydroxydisopyramide
In: Rat B984

Disugran (methyl 3,6-dichloro-*o*-anisate)
→ 3,6-Dichloro-*o*-anisic acid
In: Microorganism A1317
→ Methyl 3,6-dichlorosalicylate
In: Microorganism A1317

1,6-Disulphatoxypyrene
Probably formed from: 6-Hydroxypyren-1-yl sulphate

Ditazole (2-bis(2-hydroxyethyl)amino-4,5-diphenyloxazole)
→ Benzil
In: Man F30
→ 4,5-Diphenyl-4-oxazolin-2-one
In: Man A1056
Rabbit A1056
Rat A1056
→ Hydroxyditazole
In: Man F30
→ 2-(2-Hydroxyethyl)amino-4,5-diphenyloxazole
In: Man A1056, F30
Rabbit A1056
Rat A1056

Dithiaden (4-(3-dimethylaminopropylidene)-4,9-dihydrothieno[2,3-*c*][2]benzothiepin)
→ Demethyldithiaden
In: Rat A397
→ Dithiaden sulphoxide
In: Rat A397

Dithiaden sulphoxide
Formed from: Dithiaden
→ Demethyldithiaden sulphoxide
Probably in: Rat A397

5,5′-Dithiobis(2-nitrobenzoic acid)
→ 5-Mercapto-2-nitrobenzoic acid

In: Rat D611

3,3′-(Dithiobis(1-oxo-3,1-propanediyl))bis(2-(*o*-hydroxyphenyl))-(2*R*-(2α,3(2′*R,4′*R**),4α))-4-thiazolidinecarboxylic acid**
Formed from: (2*R*,4*R*)-(*o*-Hydroxyphenyl)-3-mercaptopropionyl-4-thiazolidinecarboxylic acid

Dithiosalicylic acid (2-hydroxybenzene-carbodithioic acid)
Formed from: Dibenzothiophene

Ditropan
See: Oxybutynin

L-Dityrosine (5,5′-bis(3-L-alanyl)-2,2′-dihydroxybiphenyl)
Formed from: L-Tyrosine

Diuron (N′-(3,4-dichlorophenyl)-N,N-dimethylurea)
Formed from: Phenobenzuron
→ 2,4-Dichloroaniline (a misprint?)
In: Arthrobacter A634
Pseudomonas A634

1,2-Divinylbenzene (1,2-diethenylbenzene)
→ N-Acetyl-S-(2-(2-ethenylphenyl)-2-hydroxyethyl)-L-cysteine
In: Rat J60
→ *o*-Ethenylphenylacetic acid
In: Rat J60
→ *o*-Ethenylphenylglycol
In: Rat J60

1,3-Divinylbenzene (1,3-diethenylbenzene)
→ N-Acetyl-S-(1-(3-ethenylphenyl)-2-hydroxyethyl)-L-cysteine
In: Rat G70
→ N-Acetyl-S-(2-(3-ethenylphenyl)-2-hydroxyethyl)-L-cysteine
In: Rat G70
→ *m*-Ethenylmandelic acid
In: Rat G70

1,4-Divinylbenzene (1,4-diethenylbenzene)
→ N-Acetyl-S-(1-(4-ethenylphenyl)-2-hydroxyethyl)-L-cysteine
In: Rat F349
→ N-Acetyl-S-(2-(4-ethenylphenyl)-2-hydroxyethyl)-L-cysteine
In: Rat F349
→ N-Acetyl-S-(1-(4-formylphenyl)-2-hydroxyethyl)-L-cysteine
In: Rat F349
→ 1,4-Bis(carboxymethyl)benzene
In: Rat F881
→ 1-Ethenyl-4-(1-hydroxyethyl)benzene
In: Rat F881
→ *p*-Ethenylphenylacetic acid
In: Rat F349
→ *p*-Ethenylphenylglycol
In: Rat F349, F881

DMI
See: Demethylimipramine

Dobutamine (4-(2-((3-(4-hydroxyphenyl)-1-methylpropyl)amino)ethyl)-1,2-benzenediol)
→ Dobutamine-β-D-glucuronide
In: Dog A1850
→ 3-O-Methyldobutamine
In: Dog A1850

Dobutamine-β-D-glucuronide
Formed from: Dobutamine

Docarpamine (4-(2-(((2*S*)-2-(acetylamino)-4-(methylthio)-1-oxobutyl)amino)ethyl)-1,2-phenylene bis(ethylcarbonate) ester)
→ 3,4-Dihydroxyphenethylamine
In: Dog H636
Rat H636

Docetaxel (taxotere)
→ (α-Hydroxy-N-((1-(hydroxymethyl)-1-methylethoxy)carbonyl-β-phenyl-β-alanyl)-10-deacetylbaccatin III
In: Man H256, H721

***n*-Dodecylbenzene**
→ Benzoic acid
In: Beauveria D936
Paecilomyces D936
Penicillium D936
Verticillium D936
→ 4-(*o*-Hydroxyphenyl)butyric acid
In: Candida A1206
→ Phenylacetic acid
In: Candida A1206

5-(2-Dodecylphenyl)-4,6-dithianonanedioic acid
→ 5-(2-Dodecylphenyl)-4,6-dithianonanedioic acid-β-D-glucuronide
In: Guinea pig E873
→ 5-(2-(7-Hydroxydodecylphenyl)-4,6-dithianonanedioic acid
In: Guinea pig E873

→ 5-(2-(8-Hydroxydodecylphenyl)-4,6-dithianonanedioic acid
In: Guinea pig E873
→ 5-(2-(9-Hydroxydodecylphenyl)-4,6-dithianonanedioic acid
In: Guinea pig E873
→ 5-(2-(10-Hydroxydodecylphenyl)-4,6-dithianonanedioic acid
In: Guinea pig E873
In: Rat E871b
→ 5-(2-(11-Hydroxydodecylphenyl)-4,6-dithianonanedioic acid
In: Guinea pig E873
Rat E871b
→ 5-(2-(12-Hydroxydodecylphenyl)-4,6-dithianonanedioic acid
In: Rat E871b

5-(2-Dodecylphenyl)-4,6-dithianonanedioic acid-di-β-D-glucuronide
Formed from: 5-(2-Dodecylphenyl)-4,6-dithianonanedioic acid-β-D-glucuronide

5-(2-Dodecylphenyl)-4,6-dithianonanedioic acid-β-D-glucuronide
Formed from: 5-(2-Dodecylphenyl)-4,6-dithianonanedioic acid
→ 5-(2-Dodecylphenyl)-4,6-dithianonanedioic acid-di-β-D-glucuronide
In: Guinea pig E873

Dofetilide (1-(4-methanesulphonamidophenoxy)-2-(N-(4-methanesulphonamidophenethyl)-N-methylamino)ethane)
→ Demethyldofetilide
In: Dog G240, H699, H742
Man G240, H699
Mouse G240
Rat G240, H699
→ Dofetilide-N-oxide
In: Man G240
→ *p*-Methanesulphonamidophenoxyacetic acid
In: Dog G240, H699
Man G240, H699
Mouse G240
Rat G240, H699
→ *p*-Methanesulphonamidophenoxy-N-methylethylamine
In: Dog G240, H699
Man G240, H699
Mouse G240
Rat G240, H699
→ *p*-Methanesulphonamidophenylacetic acid
In: Dog G240, H699
Man G240, H699
Mouse G240
Rat G240, H699
→ *p*-Methanesulphonamidophenyl-N-methylethylamine
In: Dog G240, H699
Man G240, H699
Mouse G240
Rat G240, H699

Dofetilide-N-oxide
Formed from: Dofetilide

Dolasetron ((2α,6α,9aβ)-octahydro-3-oxo-2,6-methano-2*H*-quinolizin-8-yl 1*H*-indol-3-carboxylate)
→ 3-Hydrodolasetron
In: Man H315

Dolichantoside ((2*S*,3*R*,4*S*)-3-Ethenyl-2-(β-D-glucopyranosyloxy)-3,4-dihydro-4-(((1*S*)-2,3,4,9-tetrahydro-2-methyl-1*H*-pyrido[3,4-*b*]indol-1-yl)methyl-2*H*-pyran-5-carboxylic acid methyl ester, palicoside methyl ester)
→ Dolichantoside aglycone
In: Catharanthus K428
Strychnos K428

Dolichantoside aglycone
Formed from: Dolichantoside

DOMA
See: 3,4-Dihydroxymandelic acid

Donepezil (aricept, 2,3-dihydro-5,6-dimethoxy-2-((1-(phenylmethyl)-4-piperidinyl)methyl)-1*H*-inden-1-one)
→ Benzaldehyde
Probably in: Dog J751
Man J751
Rat J751, J789
→ 5-O-Demethyldonepezil
In: Dog J751
Man J751, J873
Rat J751, J789
→ 6-O-Demethyldonepezil
In: Dog J751
Man J751, J873

Rat J751, J789
→ 5,6-Dimethyl-2-((piperidin-4-yl)methyl)-α-hydrindone
In: Dog J751
Man J751, J873
Rat J751, J789
→ Donepezil-N-oxide (2 stereoisomers)
In: Man J873
Rat J751, J789
→ *p*-Hydroxydonepezil
In: Rat J751, J789

Donepezil-N-oxide
Formed from: Donepezil

L-Dopa
See: 3,4-Dihydroxy-L-phenylalanine

Dopac
See: 3,4-Dihydroxyphenylacetic acid

D-Dopachrome (D-2,3,5,6-tetrahydro-5,6-dioxo-1*H*-indole-2-carboxylic acid)
→ 5,6-Dihydroxyindole
In: Man H872
Rat J514

L-Dopachrome (L-2,3,5,6-tetrahydro-5,6-dioxo-1*H*-indole-2-carboxylic acid)
→ 5,6-Dihydroxyindole-2-carboxylic acid
In: Mouse H338
See also: J842

L-Dopachrome methyl ester
→ Methyl 5,6-dihydroxyindole-2-carboxylate
Probably in: Locusta J842

Dopamine
See: 3,4-Dihydroxyphenethylamine

Dopazinol ((+)-9-hydroxy-4-propylnaphthoxazine)
→ 5,6-Dehydrodopazinol
In: Rat F602
→ Depropyldopazinol
In: Rat F602
→ 5-Hydroxydopazinol
Probably in: Rat F602
→ 8-Hydroxydopazinol
In: Rat F602

Dosulepin (prothiaden, 3-dibenzo[*b*,*e*]thiepin-11(6*H*)-ylidene-N,N-dimethyl-1,1-propanamine)
→ Dosulepin-N-oxide
In: Man A1500, A2797
→ Dosulepin sulphoxide
In: Man A1500, A2797
→ Northiaden
In: Man A2797

Dosulepin-N-oxide
Formed from: Dosulepin

Dosulepin sulphoxide
Formed from: Dosulepin

Dothiepin (3-dibenzo[*b*,*e*]thiepin-11(6*H*)-ylidene-N,N-dimethyl-1-propanamine)
→ Demethyldothiepin
In: Man E141
→ Dothiepin-β-D-glucuronide
In: Man E141
→ Dothiepin sulphoxide
In: Man E141

Dothiepin-β-D-glucuronide
Formed from: Dothiepin

Dothiepin sulphoxide
Formed from: Dothiepin

Dothistromin (2,3,3a,12a-tetrahydro-2,3a,4,6,9-pentahydroxyanthra[2,3-*b*]furo[3,2-*d*]furan-5,10-dione)
Formed from acetate in Dothistroma A3473

Doxaminol (1-((2-(6,11-dihydrodibenz[*b*,*e*]oxepin-11-yl)ethyl)methylamino)-3-phenoxy-2-propanol)
→ Doxaminol-16-β-D-glucuronide
In: Dog F447, F550
→ 4′-Hydroxydoxaminol
In: Dog F447, F550
→ 3-Phenoxylactic acid
In: Dog F447, F550
→ 1-O-Phenylglycerol
In: Dog F447, F550

Doxaminol-16-β-D-glucuronide
Formed from: Doxaminol

Doxapram (1-ethyl-4-(2-morpholinoethyl)-3,3-diphenyl-2-pyrrolidinone)
→ Desethyldoxapram
In: Man G25
→ 4-Ethenyl-1-ethyl-3,3-diphenyl-2-pyrrolidinone
In: Man G25
→ 1-Ethyl-4-formylmethyl-3,3-diphenyl-2-pyrrolidinone

In: Man G25
→ 1-Ethyl-4-(N,N-di-(2-hydroxyethyl)amino-ethyl)-3,3-diphenylpyrrolidin-2-one
In: Rat A1389
→ 2-Oxodoxapram
In: Man G25
→ 3-Oxodoxapram
In: Man G25

Doxepin (*cis* and *trans,* 3-dibenzo[*b*,*e*]oxepin-11(6*H*)-ylidene-N,N-dimethyl-1,1-propanamine)
→ Demethyldoxepin
In: Cunninghamella J782
Dog A785, H856
Guinea pig H856
Man A785, F570, F755, H856
Rabbit H856
Rat A785, H856
→ N-Demethyl-N-formyldoxepin
In: Cunninghamella J782
Doxepin-β-D-glucuronide
In: Man F769, G509
→ Doxepin-N-oxide
In: Cunninghamella J782
Dog A785
Man A785, F570
Rat A785
→ 2-Hydroxydoxepin
In: Cunninghamella J782
Dog A785
Man A785, F570, F609
Rat A785, F609
→ 3-Hydroxydoxepin
In: Cunninghamella J782
Man F609
Rat F609

Doxepin-β-D-glucuronide
Formed from: Doxepin

Doxepin-N-oxide
Formed from: Doxepin

Doxorubicin ((8*S*,10*S*)-10-((3-amino-2,3,6-trideoxy-α-L-lyxo-hexopyranosyl)oxy)-7,8,9,10-tetrahydro-6,8,11-trihydroxy-8-(hydroxyacetyl)-1-methoxy-5,12-naphthacenedione)
Formed from: Daunorubicin
Doxorubicin-β-D-glucuronide
Probably formed from: N-Trifluoroacetyl-doxorubicin
→ 7-Deoxydoxorubicin aglycone
In: Mouse F634
→ 7-Deoxydoxorubicinone
Probably in: Rat C539
→ Doxorubicinol
In: Man G144
Mouse F634
Rat D457
→ 13-Hydroxydoxorubicin
In: Mouse K287

Doxorubicin-β-D-glucuronide
→ Doxorubicin
In: Man J333

Doxorubicinol ((8*S*,10*S*)-10-((3-amino-2,3,6-trideoxy-α-L-lyxo-hexopyranosyl)oxy)-7,8,9,10-tetrahydro-6,8,11-trihydroxy-8-((1*S*)-1,2-dihydroxyethyl)-1-methoxy-5,12-naphthacenedione)
Formed from: Doxorubicin
→ 7-Deoxydoxorubicinol aglycone
Probably in: Mouse F634

Doxycycline (4*S*,4a*R*,5*S*,5a*R*,6*R*,12a*S*)-4-(dimethylamino)-1,4,4a,5,5a,6,11,12a-octahydro-3,5,10,12,12a-pentahydroxy-6-methyl-1,11-dioxo-2-naphthacenecarboxamide)
→ N-Demethyldoxycycline
In: Mouse D47

Doxylamine (N,N-dimethyl-2-(1-phenyl-1-(2-pyridinyl)ethoxy)ethanamine)
→ Demethyldoxylamine
In: Man B909, E785
Monkey E11, E92
Rat E11, E424, F684
→ Doxylamine-β-D-glucuronide
In: Man G18, H541
Rat F684
→ Doxylamine-N-oxide (ring)
Probably in: Rat E11
→ Doxylamine-N-oxide (side chain)
In: Rat E11
See also: E92, E424
→ *p*-Hydroxydoxylamine
In: Rat E438

See also: E11
→ 1-Phenyl-1-(2-pyridyl)ethanol
In: Man B909
See also: E11
→ 2-(1-Phenyl-1-pyridylethoxy)ethanol
In: Monkey E11
Rat E11

Doxylamine-β-D-glucuronide
Formed from: Doxylamine

Doxylamine-N-oxide (ring)
Formed from: Doxylamine

Doxylamine-N-oxide (side chain)
Formed from: Doxylamine

Drazoxolon
See: 4-(2-Chlorophenylhydrazono)-3-methyloxazol-5-one

Drepamon (S-benzyl N,N-di-*sec*-butylthiolcarbamate)
→ Benzyl N,N-di-*sec*-butylsulphinylcarbamate
In: Grass A1827
Oryza A1827

Drobuline (1-isopropylamino-4,4-diphenyl-2-butanol)
→ 1-Amino-4,4-diphenyl-2-butanol
In: Dog A3726
→ Drobuline-β-D-glucuronide
In: Dog A3726
→ Hydroxydrobuline
In: Dog A3726

Drobuline-β-D-glucuronide
Formed from: Drobuline

Droloxifene (3-hydroxytamoxifen)
Probably formed from: Tamoxifen
→ Demethyldroloxifene
In: Cynomolgus monkey H924
Mouse H924
Rat H924
→ 3,4-Dihydroxytamoxifen
In: Cynomolgus monkey H924
Mouse H924
Rat H924
→ 3,5-Dihydroxytamoxifen
In: Cynomolgus monkey H924
Rat H924
→ *trans*-1-(*p*-(2-Dimethylaminoethoxy)phenyl)-1-(*m*-hydroxyphenyl)-2-(hydroxyphenyl)but-1-ene
In: Mouse H924
→ *trans*-1-(*p*-(2-Dimethylaminoethoxy)phenyl)-1-(*m*-hydroxyphenyl)-2-phenylbut-1-en-3-one
In: Mouse H924
→ Droloxifene-β-D-glucuronide
In: Mouse H924
→ *trans*-1-(*p*-(2-Hydroxyethoxy)phenyl)-1-(*m*-hydroxyphenyl)-2-phenylbut-1-ene
In: Rat H924

Droloxifene-β-D-glucuronide
Formed from: Droloxifene

Droperidol (1-(1-(4-(4-fluorophenyl)-4-oxobutyl)-1,2,3,6-hexahydro-4-pyridinyl)-1,3-dihydro-2*H*-benzimidazol-2-one)
→ 2(3*H*)-Benzimidazolinone
In: Man A3593

Drotaverine (1-(3,4-diethoxybenzal)-6,7-diethoxy-1,2,3,4-tetrahydroisoquinoline)
→ 4′-Desethyldrotaverine
In: Rat B538
→ 6-Desethyldrotaverine
In: Rat B538
→ Drotaverine-β-D-glucuronide
In: Mouse A9816

Drotaverine-β-D-glucuronide
Formed from: Drotaverine

Duroquinol
See: Tetramethylquinol

Duroquinone
See: Tetramethylquinone

Duspatal
See: Mebeverine

E

Ebastine (4′-*tert*-butyl-4-(4-(diphenylmethoxy)piperidino)butyrophenone)
→ 4-(Diphenylmethoxy)piperidine
In: Man J422
→ 4′-(1-Hydroxymethyl-1-methylethyl)-4-(diphenylmethoxy)piperidino) butyrophenone
In: Man J422

Ebrotidine ((N(*E*)N-(((2-(((2-((aminoimino-methyl)amino)-4-thiazolyl)methyl)thio)ethyl) amino)methylene)-4-bromobenzenesulphonamide)
→ *p*-Bromobenzenesulphonamide
In: Man H74
→ 4-Bromo-N-(2-methylsulphinylethylamino-methylene)benzenesulphonamide
In: Man H448
→ Ebrotidine sulphoxide
In: Man H74, H448

Ebrotidine sulphone
Formed from: Ebrotidine sulphoxide

Ebrotidine sulphoxide
Formed from: Ebrotidine
→ Ebrotidine sulphone
In: Man H74, H448

Ebselen (2-phenyl-1,2-benzoisoselenazol-3(2*H*)-one)
Formed from: 2-(Methylseleno)benzanilide
→ N-(*p*-Hydroxyphenyl)-2-(methylseleno)benzanilide
In: Rat F311
→ 2-(Methylseleninyl)benzanilide
In: Guinea pig F311
Pig F311
Rat F311
→ 2-(Methylseleno)benzanilide
In: Guinea pig H376
Pig F311
Rat F311
→ 2-Selenobenzanilide
In: Guinea pig H376
Man F311
Pig F311, H376
Rat E808, F127, F311, H376
See also: F235

Ecabapide (3-(2-(3,4-dimethoxyphenyl)ethylcarbamoylmethyl)amino-N-methylbenzamide)
→ 3-Amino-N-methylbenzamide
In: Man J806
Rat H170
→ 3,4-Dimethoxy-N-oxalylphenethylamine
In: Rat H170, H737
→ N-Glycollyl-3,4-dimethoxyphenethylamine
In: Rat H737
→ N-Glyoxylyl-3,4-dimethoxyphenethylamine
In: Rat H737
→ 3-Hydroxy-4-methoxy-N-(N-(*m*-(N-methylcarboxamido)phenyl)-β-alanyl)phenethylamine
In: Man J806
See also: H170
→ 4-Hydroxy-3-methoxy-N-(N-(*m*-(N-methylcarboxamido)phenyl)-β-alanyl)phenethylamine
In: Man J806
→ N-(N-(2-Hydroxy-5-(N-methylcarboxamido)phenyl)-β-alanyl)-3,4-dimethoxyphenethylamine
In: Man J806

Echinuline (3*S*,6*S*)-3-((2-(1,1-dimethylpropyl)-5,7-bis(3-methylbutyl)-1*H*-indol-3-yl)methyl)-6-methyl-2,5-piperazinedione)
Formed from: L-Tryptophan

Econazole (1-(2-((4-chlorophenyl)methoxy)-2-(2,4-dichlorophenyl)ethyl)-1*H*-imidazole)
→ 1-(2-(4-Chlorobenzyloxy)-2-(2,4-dichlorophenyl)ethyl)-tetrahydro-2,4-dioxoimidazole
In: Monkey Man B715
C780
Rat C780
→ 1-(2-(4-Chlorobenzyloxy)-2-(2,4-dichlorophenyl)ethyl)-tetrahydro-2-hydroxy-5-oxoimidazole
In: Man B715
Monkey C780
Rat C780
→ 1-(2,4-Dichlorophenyl)-2-(1-imidazolyl)ethanol
In: Rat B153
→ Econazole-N-oxide
In: Man B715
Monkey C780
Rat C780
Econazole-N-oxide
Formed from: Econazole
Ecstasy
See: N-Methyl-3,4-methylenedioxyamphetamine
Edifenphos
See: O-Ethyl S,S-diphenyl phosphorodithioate
Edrophonium
See: Ethyl *m*-hydroxyphenyl dimethylammonium
Edunol (6a*S*,12a*S*)-6a,12a-dihydro-2-(3-methyl-2-butenyl)-6*H*-[1,3]dioxolo[5,6]benzofuro[3,2-*c*][1]benzopyran-3-ol)
→ 2′,3-Dihydro-2′,3-dihydroxyedunol
In: Aspergillus E526
→ Neopranol
In: Aspergillus E526
→ Neorautanol
In: Aspergillus E526
Efavirenz (sustiva, (*S*)-6-chloro-4-(cyclopropylethynyl)-1,4-dihydro-4-trifluoromethyl-2*H*-3,1-benzoxazin-2-one)
→ (*S*)-6-Chloro-4-(2-(S-glutathionyl)-4-hydroxy-3-oxo-1-buten-1-yl)-1,4-dihydro-8-hydroxy-4-trifluoromethyl-2*H*-3,1-benzoxazin-2-one-8-sulphate
In: Rat J765
→ Efavirenz-N-β-D-glucuronide
In: Cynomolgus monkey J775
Guinea pig J775
Hamster J775
Man J775
Rat J775
→ 7-Hydroxyefavirenz
In: Cynomolgus monkey J775
Guinea pig J775
Hamster J775
Man J775
→ 8-Hydroxyefavirenz
In: Cynomolgus monkey J775
Guinea pig J775
Hamster J775
Man J775
Rat J775
Efavirenz-β-D-glucuronide
Formed from: Efavirenz
Efonidipine (2-(Benzyl(phenyl)amino)ethyl-1,4-dihydro-2,6-dimethyl-5-(5,5-dimethyl-2-oxo-1,3,2-dioxaphosphorinan-2-yl)-4-(*m*-nitrophenyl)-3-pyridinecarboxylate)
→ 2-(Benzylamino)ethyl 5-(5,5-dimethyl-2-oxo-1,3,2-dioxaphosphorinan-2-yl)-1,4-dihydro-2,6-dimethyl-4-(*m*-nitrophenyl)-3-pyridinecarboxylate
In: Man H523
Rat H625
→ Carboxymethyl 5-(5,5-dimethyl-2-oxo-1,3,2-dioxaphosphorinan-2-yl)-1,4-dihydro-2,6-dimethyl-4-(*m*-nitrophenyl)-3-pyridinecarboxylate
In: Rat H625
→ 5-(5,5-Dimethyl-2-oxo-1,3,2-dioxaphosphorinan-2-yl)-1,4-dihydro-2,6-dimethyl-4-(*m*-nitrophenyl)pyridine-3-carboxylic acid
In: Man H523
Rat H625
→ 2-Hydroxyethyl 5-(5,5-dimethyl-2-oxo-1,3,2-dioxaphosphorinan-2-yl)-1,4-dihydro-2,6-dimethyl-4-(*m*-nitrophenyl)-3-pyridinecarboxylate
In: Rat H625

→ 2-(Phenylamino)ethyl 5-(5,5-dimethyl-2-oxo-1,3,2-dioxaphosphorinan-2-yl)-1,4-dihydro-2,6-dimethyl-4-(*m*-nitrophenyl)-3-pyridinecarboxylate
In: Man H523
Rat H625

Eleagnine (2,3,4,9-tetrahydro-1-methyl-1*H*-pyrido[3,4-*b*]indole)
Formed from: 1,2,3,4-Tetrahydro-1-methyl-β-carboline-1-carboxylic acid

Elemicin (3,4,5-trimethoxyallylbenzene)
→ Elemicin-2,3-oxide
In: Rat B198
→ 3-Hydroxy-4,5-dimethoxyallylbenzene
In: Rat B198
→ 1-(3,4,5-Trimethoxyphenyl)-2-propen-1-ol
In: Rat B198
→ 3-(3,4,5-Trimethoxyphenyl)-2-propen-1-ol
In: Rat B198

Elemicin-2,3-oxide
Formed from: Elemicin
→ 3-(3,4,5-Trimethoxyphenyl)-2-propan-1,2-diol
Probably in: Rat B198

Ellagic acid (2,3,7,8-tetrahydroxy[1]benzopyrano[5,4,3-*cde*][1]benzopyran-5,10-dione)
→ 3,8-Dihydroxy-6*H*-dibenzo[*b*,*d*]pyran-6-one
In: Microorganism B189
→ Ellagic acid-β-D-glucuronide
In: Mouse F290
→ Ellagic acid-O-sulphate
In: Eubacterium G475
Mouse F290

Ellagic acid-β-D-glucuronide
Formed from: Ellagic acid

Ellagic acid-O-sulphate
Formed from: Ellagic acid

Ellipticine (4,11-dimethyl-6*H*-pyrido[4,3-*b*]carbazole)
→ 7-Hydroxyellipticine
In: Rat A3558
→ 8-Hydroxyellipticine
In: Aspergillus A3908
→ 9-Hydroxyellipticine
In: Aspergillus A3908
Rat A2176, A2988, A3458, A3558

Ellipticin-9-yl-β-D-glucuronide
Formed from: 9-Hydroxyellipticine

Ellipticin-9-yl sulphate
Formed from: 9-Hydroxyellipticine

Elliptinium
See: 9-Hydroxy-2-methylellipticinium

Elymoclavine (8,9-didehydro-6-methylergoline-8-methanol)
Formed from: Agroclavine
Chanoclavine I aldehyde
4-Dimethylallyl-N-methyl-L-tryptophan
4-(3-Hydroxy-3-methylbut-1-enyl)-N-methyl-L-tryptophan
N-Methyl-4-(3-methylbut-1,3-dienyl)-L-tryptophan
→ Elymoclavine-O-β-D-fructofuranoside
In: Claviceps F512, F516
→ Elymoclavine-β-D-galactoside
In: Aspergillus H11
→ Ergocornine
In: Claviceps B572
→ Ergocryptine
In: Claviceps B572
→ Ergometrine
In: Claviceps B572
→ Ergotoxine (a mixture)
In: Claviceps A887
→ 10-Hydroxyelymoclavine
In: Plant H750
→ Paspalic acid
In: Claviceps C796

Elymoclavine-O-acetate
→ Ergotoxine (a mixture)
In: Claviceps A887

Elymoclavine-O-β-D-fructofuranoside
Formed from: Elymoclavine
→ Elymoclavine-O-β-D-fructofuranosyl(2→1)-O-β-D-fructofuranoside
In: Claviceps F512, F516

Elymoclavine-O-β-D-fructofuranosyl(2→1)-O-β-D-fructofuranoside
Probably formed from: Elymoclavine-O-β-D-fructofuranoside

Elymoclavine-β-D-galactoside
Formed from: Elymoclavine

Emetine (2*S*,3*R*,11b*S*)-3-ethyl-1,3,4,6,7,11b-hexahydro-9,10-dimethoxy-2-(((1*R*)-1,2,3,4-

tetrahydro-6,7-dimethoxy-1-isoquinolinyl)methyl-2*H*-benzo[*a*]quinolizine)
Probably formed from: Cephaeline
→ O-Methylpsychotrine
In: Cunninghamella H903

Emodic acid (9,10-dihydro-4,5,7-trihydroxy-9,10-dioxo-2-anthracenecarboxylic acid)
Formed from: Emodin

Emodin (1,3,8-trihydroxy-6-methylanthraquinone)
Formed from: Emodin anthrone
Physcion
→ Chrysophanol
In: Mouse E151
Pyrenochaeta D948
→ Emodic acid
In: Rat B587
→ Emodin-β-D-glucoside
In: Cinchona K889
→ Emodin-β-D-glucuronide
In: Man G542, G968, J416, J617, J783
→ Physcion
In: Aspergillus E75
→ Questin
In: Aspergillus C334, G633
→ Skyrin
In: Penicillium A1046
→ 1,2,3,8-Tetrahydroxy-6-methylanthraquinone
In: Rat J421
→ 1,6,8-Trihydroxy-3-hydroxymethylanthraquinone
In: Mouse E151
Rat J521
See also: E273

Emodin anthrone (1,3,8-trihydroxy-6-methyl-9(10*H*)-anthracenone)
→ Chrysophanol
In: Pyrenochaeta D948
→ Chrysophanolanthrone
In: Pyrenochaeta D948
→ Emodin
In: Aspergillus G479
Claviceps G479
Myrothecium G479
Penicillium G479
Pyrenochaeta G479
→ Iridoskyrin
In: Penicillium A1046
→ Islandicin
In: Penicillium A1046
→ Rubroskyrin
In: Penicillium A1046
→ Skyrin
In: Penicillium A1046

Emodin-β-D-glucoside
Formed from: Emodin

Emodin-β-D-glucuronide
Formed from: Emodin

Enalapril (N-((1*S*)-1-(ethoxycarbonyl)-3-phenylpropyl)-L-alanyl-L-proline)
→ Enalaprilat
In: Rat H489

Enalaprilat (N-((1*S*)-1-carboxy-3-phenylpropyl)-L-alanyl-L-proline)
Formed from: Enalapril

Encainide (4-methoxy-N-(2-(2-(1-methyl-2-piperidinyl)ethyl)phenyl)benzamide)
→ 4-Hydroxy-N-(2-(2-(1-methyl-2-piperidinyl)ethyl)phenyl)benzamide
In: Man D89, F532
Rat D98
→ 4-Methoxy-N-(2-(2-(2-piperidinyl)ethyl)phenyl)benzamide
In: Man D89
Rat D98

Encecalin (1-(7-methoxy-2,2-dimethyl-2*H*-1-benzopyran-6-yl)ethanone)
→ 6-Acetyl-2-hydroxymethyl-7-methoxy-2-methyl-2*H*-benzo[*b*]pyran
In: Melanopus E491
→ Demethylencecalin
In: Melanopus E491
→ 6-(1-Hydroxyethyl)-7-methoxy-2,2-dimethyl-2*H*-benzo[*b*]pyran
In: Melanopus E491

Enciprazine (4-(2-methoxyphenyl)-α-((3,4,5-trimethoxyphenoxy)methyl)-1-piperazineethanol)
→ 2′-Demethylenciprazine
In: Man G31
→ 3-Demethylenciprazine
In: Dog G31

Man G31
Rat G31
→ 4-Demethylenciprazine
In: Dog G31
Man G31
Rat G31
→ 3-(1-(*o*-Methoxyphenyl)piperazin-4-yl)propan-1,2-diol
In: Dog G31
Man G31
Rat G31
→ 3,4,5-Trimethoxyphenol
Probably in: Dog G31
Man G31
Rat G31

Enclomiphene (2-(4-((1*E*)-2-chloro-1,2-diphenylethenyl)phenoxy)-N,N-diethylethanamine)
→ Desethylenclomiphene
In: Rabbit C18
→ 4-Hydroxyenclomiphene
In: Rabbit C18

Enpiprazole (1-(2-chlorophenyl)-4-(2-(1-methyl-1*H*-pyrazol-4-yl)ethyl)piperazine)
→ 1-(*o*-Chlorophenyl)piperazine
In: Rat D308

Entacapone ((*E*)-2-cyano-N,N-diethyl-3-(3,4-dihydroxy-5-nitrophenyl)propenamide)
→ 3-(5-Acetamido-3,4-dihydroxyphenyl)-2-cyano-N,N-diethylpropenamide
In: Rat G527
→ (*Z*)-Entacapone
In: Dog H283
Man G527
Rat G527
→ (*E*)-2-Cyano-N,N-diethyl-3-(3-hydroxy-4-methoxy-5-nitrophenyl)propenamide
In: Rat G527
Contrast (in man) G527
→ 2-Cyano-N-ethyl-N-(2-hydroxyethyl)-3-(3,4-dihydroxy-5-nitrophenyl)propenamide
In: Man G527
→ Desethylentacapone
In: Rat G527
→ Entacapone-3-β-D-glucuronide
In: Man G527, K331
Rat G527, J546, J886

(*Z*)-Entacapone
Formed from: Entacapone
→ (*Z*)-Entacapone-3-β-D-glucuronide
In: Man G527
Rat J546, J886

Entacapone-3-β-D-glucuronide
Formed from: Entacapone

(*Z*)-Entacapone-3-β-D-glucuronide
Formed from: (*Z*)-Entacapone

Enterovioform
See: 5-Chloro-8-hydroxy-7-iodoquinoline

Epanutin
See: 5,5-Diphenylhydrantoin

D-Ephedrine (2-methylamino-1-phenyl-1-propanol)
→ Benzoic acid
In: Guinea pig A1153
Rabbit A996, A2135
→ D-Ephedrine-β-D-glucuronide
In: Rat A697
→ *p*-Hydroxy-D-ephedrine
In: Rat A697
→ 2-Methylimino-1-phenylpropan-1-ol
In: Arthrobacter B360
→ D-Norephedrine
In: Guinea pig A1153
Rabbit A996, A2135
Rat A697
→ 2-Oxo-1-phenyl-1-propanol
In: Rabbit A1552
See also: A2135

L-Ephedrine ((1*R*,2*S*)-2-methylamino-1-phenyl-1-propanol; this isomer is assumed unless otherwise stated)
Formed from: O-Acetylephedrine
Benzaldehyde
Benzoic acid
Cinnamic acid
Deprenyl
N-Methylephedrine
→ Benzoic acid
In: Guinea pig A717
Man A687, A1814, A2431
Rabbit A717, A996, A2135
→ Catechol

In: Pseudomonas B360
→ L-Ephedrine-β-D-glucuronide
In: Rat A697
→ *p*-Hydroxy-L-ephedrine
In: Man A2071
Mouse A717
Rat A697, A717
See also: F481
→ 2-Methylimino-1-phenylpropan-1-ol
In: Arthrobacter B360
→ L-Norephedrine
In: Guinea pig A717
Man A687, A1814, A2431
Mouse A717
Rabbit A996, A2135
Rat A697
See also: F481
→ 1-Oxo-1-phenyl-2-propanol
In: Pseudomonas B360
→ 2-Oxo-1-phenyl-1-propanol
Probably in: Man A2431
Mouse A717
Rabbit A1552, A2135

D-Ephedrine-β-D-glucuronide
Formed from: D-Ephedrine

L-Ephedrine-β-D-glucuronide
Formed from: L-Ephedrine

(−)-Epiafzelechin-7-O-β-D-glucoside ((2*R*,3*R*)-3,4-dihydro-2-(4-hydroxyphenyl)-2*H*-1-benzopyran-3,5,7-trione-7-O-β-D-glucoside)
Formed from: (−)-Epicatechin

19-Epiajmalicine ((19β)-16,17-didehydro-19-methyloxayohimban-16-carboxylic acid methyl ester)
Formed from: Cathenamine
4,21-Dehydrogeissoschizine
Geissoschizine
Tryptamine

Epicainide (N-((1-ethyl-2-pyrrolidinyl)methyl)-α-hydroxy-α-phenylbenzeneacetamide)
→ Desethylepicainide
In: Man F323
→ *o*-Hydroxyepicainide
Probably in: Man F323
→ *m*-Hydroxyepicainide
Probably in: Man F323
→ *p*-Hydroxyepicainide
Probably in: Man F323
→ α-Hydroxydiphenylacetamide
In: Man F323
→ α-Hydroxydiphenylacetic acid
In: Man F323

(+)-Epicatechin (*trans*-3,3′,4′,5,7-pentahydroxyflavan)
→ 3,4′,5,7-Tetrahydroxy-3′-methoxyflavan
Probably in: Man K45

(−)-Epicatechin (*trans*-3,3′,4′,5,7-pentahydroxyflavan)
Formed from: (+)-Catechin
Cinnamic acid
→ (−)-Epiafzelechin-7-O-β-D-glucoside
In: Lobelia H686
→ (−)-Epicatechin-7-O-β-D-glucoside
In: Lobelia H686
→ (−)-Epicatechin-5-β-D-glucuronide
In: Rat J608
See also: K645
→ (−)-Epicatechin quinone
Probably in: Acacia J187
Tea A972
→ (−)-Epicatechin-O-sulphate
In: Eubacterium G475
→ 3,3′,5,7-Tetrahydroxy-4′-methoxyflavan
In: Man K45
Rat J608
→ 3,4′,5,7-Tetrahydroxy-3′-methoxyflavan
In: Man K45
Rat J608, K645
See also: J655

(−)-Epicatechin gallate
Probably formed from: Gallic acid
→ (−)-Epicatechin gallate-O-sulphate
In: Eubacterium G475
→ Gallic acid
In: Tea A972
→ O-Methylepicatechin gallate
In: Man K45

(−)-Epicatechin gallate-O-sulphate
Formed from: (−)-Epicatechin gallate

(−)-Epicatechin-7-O-β-D-glucoside
Formed from: (−)-Epicatechin

(–)-Epicatechin-5-O-β-D-glucuronide
Formed from: (–)-Epicatechin

(–)-Epicatechin-O-sulphate
Formed from: (–)-Epicatechin

(–)-Epicatechin quinone
Probably formed from: (–)-Epicatechin

4-Epichlortetracycline ((4*R*,4a*S*,5a*S*,6*S*,12a*S*)-7-chloro-4-(dimethylamino-1,4,4a,5,5a,6,11,12a,-octahydro-3,6,10,12,12a-pentahydroxy-6-methyl-1,11-dioxo-2-naphthacenecarboxamide)
Formed from: Chlortetracycline

4′-Epidoxorubicin ((8*S*,10*S*)-10-((3-amino-2,3,6-trideoxy-α-L-arabino-hexopyranosyl)oxy)-7,8,9,10-tetrahydro-6,8,11-trihydroxy-8-(hydroxyacetyl)-1-methoxy-5,12-naphthacenedione)
→ 7-Deoxydaunorubicinol aglycone
Probably in: Animal E548
Man C647
→ 7-Deoxyepidoxorubicin aglycone
In: Mouse F634
→ 4′-Epidoxorubicinol
In: Man C647
Mouse F634
Rat D457
See also: E548
→ 4′-Epidoxorubicin-β-D-glucuronide
In: Man C647
Rat D457

4′-Epidoxorubicin-β-D-glucuronide
Formed from: 4′-Epidoxorubicin

4′-Epidoxorubicinol
Formed from: 4′-Epidoxorubicin
→ 7-Deoxyepidoxorubicin aglycone
Probably in: Mouse F634
→ 4′-Epidoxorubicinol-β-D-glucuronide
Probably in: Rat D457

4′-Epidoxorubicinol-β-D-glucuronide
Probably formed from: 4′-Epidoxorubicinol

Epigalanthamine ((4a*S*,6*S*,8a*S*)-4a,5,9,10,11,12-hexahydro-3-methoxy-11-methyl-6*H*-benzofuro[3a,3,2-*ef*]benzazepin-6-ol)
Formed from: Galanthaminone
→ Galanthaminone
In: Rabbit E427

(–)-Epigallocatechin (3,3′,4′,5,5′,7-hexahydroxyflavan)
Probably formed from: (–)-Epigallocatechin gallate
→ (–)-Epigallocatechin quinone
Probably in: Tea A972
→ (–)-Epigallocatechin-O-sulphate
In: Eubacterium G475
→ (–)-Gallocatechin
In: Musa K423
→ 2-Hydroxy-3-(2,4,6-trihydroxyphenyl)-1-(3,4,5-trihydroxyphenyl)-1-propanone
In: Musa K423
→ 3,3′,5,5′,7-Pentahydroxy-4′-methoxyflavan
In: Rat K10
See also: K45

(–)-Epigallocatechin gallate
Formed from: Tannin
Probably formed from: Gallic acid
→ (–)-Epigallocatechin
Probably in: Mouse A647
→ Gallic acid
In: Tea A972
→ (–)-Epigallocatechin gallate-4′-α-D-glucoside
In: Leuconostoc J5
→ (–)-Epigallocatechin gallate-O-sulphate
In: Eubacterium G475
→ 4″-O-Methylepigallocatechin gallate
In: Rat K10
See also: K45

(–)-Epigallocatechin gallate-4′,4″-di-α-D-glucoside
Probably formed from: (–)-Epigallocatechin gallate-4′-α-D-glucoside

(–)-Epigallocatechin gallate-4′-α-D-glucoside
Formed from: (–)-Epigallocatechin gallate

(–)-Epigallocatechin gallate-4′,4″-di-α-D-glucoside
Probably in: Leuconostoc J5

(–)-Epigallocatechin gallate-O-sulphate
Formed from: (–)-Epigallocatechin gallate

(–)-Epigallocatechin quinone
Formed from: (–)-Epigallocatechin

(–)-Epigallocatechin-O-sulphate
Formed from: (–)-Epigallocatechin

4-Epiisochlortetracycline
Probably formed from: Isochlortetracycline
2-Epilycorine (1*S*,2*R*,12b*S*,12c*S*)-2,4,5,7,12b, 12c-hexahydro-1*H*-[1,3]dioxolo[4,5-*j*] pyrrolo[3,2,1-*de*]phenanthridine-1,2-diol)
→ 4,5-Dehydroanhydrolycorine
In: Crinum F630
→ 2-Epipancrassidine
In: Crinum F630
→ Hippadine
In: Crinum F630
Epinine
See: 3,4-Dihydroxy-N-methylphenethylamine
Epinine diisobutyryl ester
See: Ibopamine
Epinine-4-β-D-glucuronide
Probably formed from: 3,4-Dihydroxy-N-methylphenethylamine
Epinine monoisobutyrate
Formed from: Ibopamine
→ 3,4-Dihydroxy-N-methylphenethylamine
Probably in: Beef F220
Dog E940
Man D970
Rat D990, E940
Epinine-4-phosphate
→ 3,4-Dihydroxy-N-methylphenethylamine
In: Man H34
Epinine-3-O-sulphate
Probably formed from: 3,4-Dihydroxy-N-methylphenethylamine
16-Epioestriol (16β-hydroxy-17β-oestradiol)
Formed from: 17β-oestradiol
1,3,5(10),16-oestratetraen-3-ol-16α,17α-oxide
1,3,5(10),16-oestratetraen-3-ol-16β,17β-oxide
16-Oxo-17β-oestradiol
→ 16-Epioestriol-3-β-D-glucuronide
Probably in: Rat A3102
→ 16-Epioestriol-16-β-D-glucuronide
In: Man A1188
See also: A951, A3652
→ 16-Epioestriol-3-sulphate
In: Beef A1211
See also: A3652
→ 2-Hydroxy-16-epioestriol
Probably in: Rat A3102
→ 16-Oxo-17β-oestradiol
Probably in: Hen A3652
Man A1188
16-Epioestriol-3,16-di-β-D-glucuronide
Formed from: 16-Epioestriol-16-β-D-glucuronide
16-Epioestriol-3-β-D-glucuronide
Formed from: 16-Epioestriol
16-Epioestriol-16-β-D-glucuronide
Formed from: 16-Epioestriol
→ 16-Epioestriol-3,16-di-β-D-glucuronide
Probably in: Man A1188
→ 16-Epioestriol-16-β-D-glucuronide-3-sulphate
Probably in: Man A1188
16-Epioestriol-16-β-D-glucuronide-3-sulphate
Probably formed from: 16-Epioestriol-16-β-D-glucuronide
16-Epioestriol-3-sulphate
Formed from: 16-Epioestriol
4-Epioxytetracycline (4*S*,4a*R*,5*S*,,5a*R*,6*S*,12a*S*)-4-(dimethylamino)-1,4,4a,5,5a,6,11,12a-octahydro-3,5,6,10,12,12a-hexahydroxy-6-methyl-1,11-dioxo-2-naphthacenecarboxamide)
Formed from: Oxytetracycline
2-Epipancrassidine
Formed from: 2-Epilycorine
Epipinoresinol (4-((1*S*,3a*R*,4*R*,6a*R*)-tetrahydro-1*H*,3*H*-furo[3,4-*c*]furan-1,4-diyl)bis(2-methoxyphenol))
Formed from: Coniferyl alcohol
Epipodophyllotoxin (5*R*,5a*R*,8a*R*,9*S*)-5,8,8a,9-tetrahydro-9-hydroxy-5-(3,4,5-trimethoxyphenyl)furo[3′,4′:6,7]naphtho[2,3-*d*]-1,3-dioxol-6(5a*H*)-one)
Formed from: Deoxypodophyllotoxin
7-Epitaxol
Formed from: Taxol
Epoxyaurapten (7-(((2*E*)-5-((2*R*)-3,3-dimethyloxiranyl)-3-methyl-2-pentenyl)oxy)-2*H*-1-benzopyran-2-one)
→ 7-(7-Hydroxy-3,7-dimethyl-6-oxo-2-octenyl)coumarin
In: Pseudomonas G467

(5α,6α)-4,5-Epoxy-17-(cyclopropylmethyl)morphinan-3,14-diol-6-one
See: Naltrexone
(5α)-4,5-Epoxy-3,14-dihydroxy-17-methylmorphinan-6-one
See: Oxymorphone
***trans*-6,7-Epoxy-1-(4′-ethylphenoxy)-3,7-dimethyl-2-octene**
→ 1-(*p*-Acetylphenoxy)-2,6,7-trihydroxy-3,7-dimethyl-2-octane
Probably in: Rat A813
→ 1-(*p*-Acetylphenoxy)-3,6,7-trihydroxy-3,7-dimethyl-2-octane
Probably in: Rat A813
→ 6,7-Dihydroxy-1-(*p*-hydroxyphenoxy)-3,7-dimethyl-2-octene
In: Rat A813
→ 1-(*p*-Ethylphenoxy)-6,7-dihydroxy-3,7-dimethyl-*trans*-2-octene
In: Mouse A3955
→ *p*-Hydroxyacetophenone
In: Rat A813
→ *p*-Hydroxyphenylacetic acid
Probably in: Rat A813
1α,2α-Epoxyhexahydrocannabinol
See: Δ^1-Tetrahydrocannabinol-1α,2α-oxide
4,5α-Epoxy-8β-(2-hydroxyethyl)-3-methoxymorphinan-6-one
Probably formed from: 17-Cyclopropylmethyl-4,5α-epoxy-8β-(2-hydroxyethyl)-3-methoxymorphinan-6-one
Epoxyhydroxysafrole
Formed from: Safrole
15,16-Epoxy-10′-hydroxyvinorelbine
Probably formed from: 15,16-Epoxyvinorelbine
(5α)-4,5-Epoxy-3-methoxy-17-methylmorphinan-6-one
See: Hydrocodone
1,2-Epoxy-3-(*p*-nitrophenyl)propane
See: 3-(*p*-Nitrophenoxy)-1-propene oxide
1,2-Epoxy-3-phenoxypropane
→ N-Acetyl-S-(2-hydroxy-3-phenoxypropyl)-L-cysteine
In: Rabbit A3042
Rat A3042
→ 3-Phenoxylactic acid
In: Rabbit A3042
Rat A3042
Epoxypseudoisoeugenol 2-methylbutyrate
Formed from: *p*-(1-Propenyl)phenol
3′,4′-Epoxyvincristine-N-oxide
Probably formed from: Vincristine
15,16-Epoxyvinorelbine
Formed from: Vinorelbine
→ 15,16-Epoxy-10′-hydroxyvinorelbine
Probably in: Rat H941
Eprazinone (4-(2-benzoylpropyl)-1-(2-ethoxy-2-phenylethyl)piperazine)
→ 1-(2-Hydroxy-2-phenylethyl)piperazine
In: Man A2332
Rabbit A2332
Rat A2332
Equilenin (3-hydroxyoestra-1,3,5,7,9-pentaen-17-one)
Formed from: 17β-Dihydroequilin
Equilin
3-Hydroxy-3,5,7-androstatrien-17-one
→ 17α-Dihydroequilenin
Probably in: Man G217, G993
→ 17β-Dihydroequilenin
Probably in: Horse B745
Man C385
→ Equilenin-β-D-glucuronide
Probably in: Man H211
→ Equilenin-O-sulphate
Probably in: Horse B745
Man G993, H211
→ 2-Hydroxyequilenin
In: Hamster D514
Rat D514
→ 4-Hydroxyequilenin
In: Hamster D514, J310
Rat D514
Equilenin-β-D-glucuronide
Probably formed from: Equilenin
Equilenin-O-sulphate
Probably formed from: Equilenin
Equilin (3-hydroxyoestra-1,3,5(10),7-tetraen-17-one)
Formed from: 3-Hydroxy-3,5,7-androstatrien-17-one
19-Hydroxyandrosta-4,17-diene-3,17-dione
→ Equilenin

In: Horse B745
Man G217, G993
→ Equilin-β-D-glucuronide
In: Horse B745
→ Equilin sulphate
In: Beef A1211
Horse B745
→ 17α-Dihydroxyequilin
In: Horse B745
→ 17β-Dihydroxyequilin
In: Horse B745
Man C385, G217, G993

Equilin-β-D-glucuronide
Formed from: Equilin

Equilin sulphate
Formed from: Equilin

Equol (4′,7-dihydroxyisoflavanone)
Formed from: Daidzein
→ Equol-β-D-glucoside
In: Rabbit A571
→ Equol-β-D-glucuronide
In: Rabbit A571
→ Equol sulphate
Probably in: Hen A1607

Equol disulphate
Probably formed from: Equol sulphate

Equol-β-D-glucoside
Formed from: Equol

Equol-β-D-glucuronide
Formed from: Equol

Equol sulphate
Probably formed from: Equol
→ Equol disulphate
Probably in: Hen A1607

Eraldin
See: Practolol

Ergocornine ((5′α)-12′-hydroxy-2′,5′-bis(1-methylethyl)ergotaman-3′,6′,18-trione)
Formed from: Elymoclavine
D-Lysergic acid

Ergocryptine ((5′α)-12′-hydroxy-2′-(1-methylethyl)-5′-(2-methylpropyl)ergotaman-3′,6′,18-trione)
Formed from: Elymoclavine
D-Lysergic acid

Ergoline-8β-carboxylic acid
Probably formed from: 6-Allylergoline-8β-carboxylic acid

Ergometrine ((8β(*S*))-9,10-didehydro-N-(2-hydroxy-1-methylethyl)-6-methylergoline-8-carboxamide)
Formed from: Elymoclavine
D-Lysergic acid
→ Ergometrine-β-D-galactoside
In: Aspergillus H11

Ergometrine-β-D-galactoside
Formed from: Ergometrine

Ergosine ((5′α)-12′-hydroxy-2′-methyl-5′-(2-methylpropyl)ergotaman-3′,6′,18-trione)
Formed from: Agroclavine

Ergotamine ((5′α)-12′-hydroxy-2′-methyl-5′-(phenylmethyl)ergotaman-3′,6′,18-trione)
Formed from: L-Phenylalanine
→ 9-Hydroxyergotamine
In: Beef K587

Ergotoxine (a mixture)
Formed from: 8-Acetoxymethylene-6-methyl-9-ergoline
Elymoclavine
Elymoclavine-O-acetate

Eriocitrin (eriodictyol-7-rutinoside)
→ Eriodictyol
In: Bacteroides J362
Rat K533

Eriodictyol (3′,4′,5,7-tetrahydroxyflavanone)
Formed from: Caffeoyl CoA
Eriocitrin
Hesperetin
Naringenin
Naringin
Probably formed from: 2′,3,4,4′,6′-Pentahydroxychalcone
→ Columnidin
In: Columnea E702
→ Dihydrocaffeic acid
In: Clostridium J362
→ Dihydroquercetin
In: Matthiola B679
Parsley B957
Petunia D772, E72
See also: B404, K533
→ Eriodictyol-7-O-glucoside

In: Chrysanthemum J320
→ Hesperetin
Probably in: Rat K533
→ Homoeriodictyol
In: Zea F173, G223
See also: K533
→ Luteoforol
In: Sinningia E703
→ Luteolin
In: Antirrhinum C65
Parsley B957
Sinningia E570
Verbena D471
→ 3′,4′,5,5′,7-Pentahydroxyflavanone
In: Zea D947
→ Phloroglucinol
In: Clostridium J362

Eriodictyol-7-O-glucoside
Formed from: Eriodictyol

Eriodictyol-7-rutinoside
See: Eriocitrin

Esculetin
See: 6,7-Dihydroxycoumarin

Esculin
See: 7-Hydroxycoumarin-6-yl-β-D-glucoside

Esmolol (methyl 3-(4-(2-hydroxy-3-isopropylaminopropoxy)phenyl)propionate)
→ 3-(4-(2-Hydroxy-3-isopropylamino-propoxy)phenyl)propionic acid
In: Dog E929
Guinea pig E929
Man C911, D956, D961, E929
Monkey E929
Rabbit H563
Rat E929

Estazolam (8-chloro-6-phenyl-4*H*-[1,2,4]triazolo[4,3-*a*][1,4]benzodiazepine)
→ 8-Chloro-2-(4*H*-1,2,4-*s*-triazol-4-yl)benzophenone
In: Man D651
Dog D651
→ *keto*-1-Hydroxyestazolam
In: Mouse D362
→ 4-Hydroxyestazolam
In: Man D651
Dog D651
→ *p*-Hydroxyestazolam
In: Man D651
Dog D651

Estragole (*p*-methoxyallylbenzene)
Formed from: L-Phenylalanine
→ 4-Allylphenol
Probably in: Rat B528
→ Estragole-2,3-oxide
In: Rat B528
→ 1-Hydroxy-1-(*p*-methoxyphenyl)prop-2-ene
In: Mouse B961
Rat B961
→ *p*-Methoxycinnamyl alcohol
Probably in: Man E812
→ *p*-Methoxyphenylacetaldehyde
Probably in: Man E812

Estragole-2,3-oxide
Formed from: Estragole
→ 2,3-Dihydro-2,3-dihydroxyestragole
In: Guinea pig G86, G963
Mouse G86, G963
Rat G963
Rabbit G963
See also: B528
→ 2-(S-Glutathionyl)-2,3-dihydro-3-hydroxyestragole
Probably in: Guinea pig G963
Mouse G963
Rat G963
Rabbit G963

Estramustine ((17β)-oestra-1,3,5(10)-triene-3,17-diol 3-(bis(2-chloroethyl)carbamate ester)ester)
Formed from: Estramustine phosphate
→ Estromustine
In: Rat F81
→ 17β-Oestradiol
In: Rat F81
See also: D83

Estramustine phosphate
→ Estramustine
In: Dog D83
Man D83
Mouse D83

Estromustine (3-(((bis(2-chloroethyl)amino)carbonyl)oxy)oestra-1,3,5(10)-trien-17-one)
Formed from: Estramustine

Eterylate (2-(4-Acetamidophenoxy)ethyl 2-acetoxybenzoate)
→ *p*-Acetamidophenoxyacetic acid
In: Dog C662
Man C662
Rat C662
→ *p*-Acetamidophenol
In: Dog C662
Man C662, C783
Rat C662
→ Salicylic acid
In: Dog C662
Man C662, C783
Rat C662

Ethacizin (ethyl 10-(3-diethylaminopropionyl)phenothiazine-2-carbamate)
→ N-Desethylethacizin
In: Man F452
→ 10-(3-Diethylaminopropionyl) phenothiazine-2-carbamic acid
In: Man F452
→ Ethacizin sulphoxide
In: Man F452

Ethacizin sulphoxide
Formed from: Ethacizin
→ Ethyl phenothiazine-2-carbamate sulphoxide
Probably in: Man F452

Ethacrynic acid ((2,3-dichloro-4-(2-methylenebutyryl)phenoxy)acetic acid)
→ (2,3-Dichloro-4-(2-(S-glutathionyl)methylenebutyryl)phenoxy)acetic acid
Probably in: Antechinus J102
Cat C562
Dog C562
Guinea pig C562
Mouse C562
Quail C562
Rabbit C562
Rat A136, C562
Trout C562

1,1′-(1,2-Ethanediyl)bisbenzene
See: Bibenzyl

9,10-Ethanoanthracene-9(10*H*)-2′-hydroxypropionaldehyde
Probably formed from: Oxaprotiline
→ 9,10-Ethanoanthracene-9(10*H*)-lactic acid
Probably in: Dog E48
Rat E48
→ 9,10-Ethanoanthracene-9(10*H*)-2′,3′-propanediol
Probably in: Dog E48
Rat E48

9,10-Ethanoanthracene-9(10*H*)-lactic acid
Probably formed from: 9,10-Ethanoanthracene-9(10*H*)-2′-hydroxypropionaldehyde

9,10-Ethanoanthracene-9(10*H*)-2′,3′-propanediol
Probably formed from: 9,10-Ethanoanthracene-9(10*H*)-2′-hydroxypropionaldehyde
→ 9,10-Ethanoanthracene-9(10*H*)-2′,3′-propanediol-β-D-glucuronide
Probably in: Dog E48

9,10-Ethanoanthracene-9(10*H*)-2′,3′-propanediol-β-D-glucuronide
Formed from: 9,10-Ethanoanthracene-9(10*H*)-2′,3′-propanediol

5-(Ethanolamino)-2-hydroxy-3-(10′-(*Z*)-pentadecenyl)-1,4-benzoquinone
Formed from: Maesanin

Ethenylbenzene
See: Styrene

***o*-Ethenylbenzoic acid**
Probably formed from: *o*-Ethenylphenylglyoxylic acid

***p*-Ethenylbenzoic acid**
Probably formed from: *p*-Ethenylphenylglyoxylic acid
→ *p*-Ethenylhippuric acid
In: Rat F349

4-((1*E*,3*S*)-3-Ethenyl-3,7-dimethyl-1,6-octadienyl)phenol
See: Bakuchiol

4-Ethenyl-1-ethyl-3,3-diphenyl-2-pyrrolidinone
Formed from: Doxapram

***p*-Ethenylhippuric acid**
Formed from: *p*-Ethenylbenzoic acid

1-Ethenyl-4-(1-hydroxyethyl)benzene
Formed from: 1,4-Divinylbenzene

2′-Ethenyl-4-hydroxylaminobiphenyl
Formed from: N-Hydroxy-4-acetamido-2′-ethenylbiphenyl

***o*-Ethenylmandelic acid**
Probably formed from: *o*-Ethenylphenylglycol
→ *o*-Ethenylphenylglyoxylic acid

Probably in: Rat J60

m-Ethenylmandelic acid
Probably formed from: 1,3-Divinylbenzene
→ *m*-Ethenylphenylglyoxylic acid
Probably in: Rat G70

p-Ethenylmandelic acid
Probably formed from: *p*-Ethenylphenylglycol
→ *p*-Ethenylphenylglyoxylic acid
Probably in: Rat F349

Ethenyloxybenzene
Formed from: Phenetole

p-Ethenylphenol
See: *p*-Hydroxystyrene

o-Ethenylphenylacetic acid
Formed from: 1,2-Divinylbenzene
→ *o*-Ethenylphenylacetylglycine
In: Rat J60

p-Ethenylphenylacetic acid
Formed from: 1,4-Divinylbenzene

o-Ethenylphenylacetylglycine
Formed from: *o*-Ethenylphenylacetic acid

o-Ethenylphenylglycol
Formed from: 1,2-Divinylbenzene
→ *o*-Ethenylmandelic acid
Probably in: Rat J60

p-Ethenylphenylglycol
Formed from: 1,4-Divinylbenzene
→ 4-(1,2-Dihydroxyethyl)benzoic acid
Probably in: Rat F881
→ *p*-Ethenylmandelic acid
Probably in: Rat F349
→ 1-(*p*-Ethenylphenyl)-2-hydroxyethyl-β-D-glucuronide
Probably in: Rat F881
→ 2-(*p*-Ethenylphenyl)-2-hydroxyethyl-β-D-glucuronide
Probably in: Rat F881

o-Ethenylphenylglyoxylic acid
Probably formed from: *o*-Ethenylmandelic acid
→ *o*-Ethenylbenzoic acid
Probably in: Rat J60

m-Ethenylphenylglyoxylic acid
Probably formed from: *m*-Ethenylmandelic acid

p-Ethenylphenylglyoxylic acid
Probably formed from: *p*-Ethenylmandelic acid
→ *p*-Ethenylbenzoic acid
Probably in: Rat F349

1-(p-Ethenylphenyl)-2-hydroxyethyl-β-D-glucuronide
Probably formed from: *p*-Ethenylphenylglycol

2-(p-Ethenylphenyl)-2-hydroxyethyl-β-D-glucuronide
Probably formed from: *p*-Ethenylphenylglycol

Ethenylsulphinylbenzene
Formed from: Ethenylthiobenzene

Ethenylthiobenzene
→ Ethenylsulphinylbenzene
In: Pseudomonas H165

Ethidium (3,8-diamino-5-ethyl-6-phenylphenanthridinium)
→ 8-Acetylethidium
In: Rat B625
See also: J573
→ 3-Amino-5-ethyl-8-β-D-glucuronosyl-amino-6-phenylphenanthridinium
In: Rat J573
→ 8-Amino-5-ethyl-3-β-D-glucuronosyl-amino-6-phenylphenanthridinium
In: Rat J573

17α-Ethinyl-2-hydroxyoestradiol
Formed from: 17α-Ethinyloestradiol

17α-Ethinyl-6α-hydroxyoestradiol
Formed from: 17α-Ethinyloestradiol

17α-Ethinyl-6β-hydroxyoestradiol
Formed from: 17α-Ethinyloestradiol

17α-Ethinyl-16-hydroxyoestradiol
Formed from: 17α-Ethinyloestradiol

17α-Ethinyl-6α-hydroxyoestradiol-3-methyl ether
Formed from: 17α-Ethinyloestradiol-3-methyl ether

17α-Ethinyl-6β-hydroxyoestradiol-3-methyl ether
Formed from: 17α-Ethinyloestradiol-3-methyl ether

17α-Ethinyloestradiol
→ 17α-Ethinyl-2-hydroxyoestradiol
In: Man E412, F627
Rat F627
→ 17α-Ethinyl-6α-hydroxyoestradiol
In: Penicillium A2225

→ 17α-Ethinyl-6β-hydroxyoestradiol
In: Penicillium A2225
→ 17α-Ethinyl-16-hydroxyoestradiol
In: Man E412
→ 17α-Ethinyloestradiol-β-D-glucuronide
In: Man E412
→ 17α-Ethinyloestradiol-O-sulphate
In: Man E412

17α-Ethinyloestradiol-β-D-glucuronide
Formed from: 17α-Ethinyloestradiol

17α-Ethinyloestradiol-3-methyl ether
→ 17α-Ethinyl-6α-hydroxyoestradiol-3-methyl ether
In: Penicillium A2225
→ 17α-Ethinyl-6β-hydroxyoestradiol-3-methyl ether
In: Penicillium A2225

17α-Ethinyloestradiol-O-sulphate
Formed from: 17α-Ethinyloestradiol

Ethotoin
See: 3-Ethyl-5-phenylhydantoin

***o*-Ethoxyacetanilide**
→ *o*-Ethoxyaniline
In: Rat D491

***m*-Ethoxyacetanilide**
→ *m*-Ethoxyaniline
In: Rat D491

***p*-Ethoxyacetanilide** (phenacetin)
Formed from: *p*-Ethoxyaniline
p-Ethoxy-N-hydroxyacetanilide
→ 2-Acetamido-5-ethoxyphenol
In: Ferret A263
Guinea pig A263
Man A263
Rabbit A263
Rat A262, A263
See also: D631
→ 5-Acetamido-2-ethoxyphenol
Probably in: Rat D631
→ *p*-Acetamidophenol
In: Cunninghamella F590
Dog A1223, A1559
Ferret A263
Guinea pig A263
Hamster A3446
Man A263, B646, H465, J269, J733, J742
Microorganism A261
Mouse A3502
Rabbit A263, D355, E208
Rat A19, A122, A262, A263, A1780, A2106, A3287, B646, D98, E112, H801
See also: A3739, D630, D631
→ N-Acetyl-*p*-ethoxyphenyl nitroxide radical
Probably in: Rabbit B184
→ *p*-Ethoxyaniline
In: Aspergillus E112
Dog A1223
Man H465
Mouse C609
Penicillium A2341
Rat C677, D491
→ *p*-Ethoxyglycollanilide
In: Dog A1559
Rabbit D355
→ *p*-Ethoxy-N-hydroxyacetanilide
In: Hamster A2119
Man D630
Mouse A3502
Rabbit D355
Rat B633, D630
→ N-(4-(2-Hydroxyethoxy)phenyl)acetamide
In: Rabbit D355
See also: A3024, D631
→ Phenacetin glutathione (structure uncertain)
In: Hamster B20

***p*-Ethoxyamphetamine**
→ N-Acetyl-*p*-ethoxyamphetamine
In: Candida F893
→ *p*-Hydroxyamphetamine
In: Candida F893
Cunninghamella F893

***o*-Ethoxyaniline**
Formed from: *o*-Ethoxyacetanilide

***m*-Ethoxyaniline**
Formed from: *m*-Ethoxyacetanilide

***p*-Ethoxyaniline** (*p*-phenetidine)
Formed from: *p*-Ethoxyacetanilide
p-Ethoxyglycollanilide
→ *p*-Aminophenol
In: Microorganism A261
See also: A2811

→ 4,4′-Diethoxyazobenzene
In: Horseradish D307
With prostaglandin synthase D307
→ *p*-Ethoxyacetanilide
In: Hen F694
Rat D282, F874
Sheep D282
→ *p*-Ethoxyaniline-N-β-D-glucuronide
In: Man H541
→ *p*-Ethoxyaniline radical
In: Beef E203
Horseradish E203
→ N-(*p*-Ethoxyphenyl)-*p*-benzoquinoneimine
Probably in: Horseradish D573

p-Ethoxyaniline-N-β-D-glucuronide
Formed from: *p*-Ethoxyaniline

p-Ethoxybenzaldehyde
→ *p*-Ethoxybenzoic acid
In: Beef C508

o-Ethoxybenzamide
→ Salicylamide
In: Rat D528

Ethoxybenzene
See: Phenetole

p-Ethoxybenzoic acid
Formed from: *p*-Ethoxybenzaldehyde
→ *p*-Ethoxybenzoyl CoA
In: Beef H740

p-Ethoxybenzoyl CoA
Formed from: *p*-Ethoxybenzoic acid

N-(4-(Ethoxycarbonyl)butyl)-2-hydroxy-3-(o-tolyloxy)propylamine
→ N-(4-Carboxybutyl)-2-hydroxy-3-(*o*-tolyloxy)propylamine
In: Rabbit H563

N-(2-(Ethoxycarbonyl)ethyl)-2-hydroxy-3-(α-naphthyloxy)propylamine
→ N-(2-Carboxyethyl)-2-hydroxy-3-(α-naphthyloxy)propylamine
In: Rabbit H563

***trans*-2-(Ethoxycarbonylethyl)-3-(p-hydroxyphenyl)-N-phenyllactam**
→ *trans*-2-(2-Carboxyethyl)-3-(*p*-hydroxyphenyl)-N-phenyllactam
In: Microorganism K110

N-(2-(Ethoxycarbonyl)ethyl)-2-hydroxy-3-(o-tolyloxy)propylamine
→ N-(2-Carboxyethyl)-2-hydroxy-3-(*o*-tolyloxy)propylamine
In: Rabbit H563

3-(4-(1-Ethoxycarbonyl-1-methylethoxy)phenyl)-5-(3-pyridyl)-1,2,4-oxadiazole
→ 3-(4-(1-Carboxy-1-methylethoxy)phenyl)-5-(3-pyridyl)-1,2,4-oxadiazole
In: Rat A3267
→ 3-(*p*-Hydroxyphenyl)-5-(3-pyridyl)-1,2,4-oxadiazole
In: Rat A3267

1-(4-Ethoxycarbonyl-2-methyl-2(*R*),4(*R*)-pentanoyl)-2,3-dihydro-2(*S*)-indole-2-carboxylic acid
See: Pentopril

N-Ethoxycarbonylphenethylamine
→ Phenethylamine
In: Rat D436

N-((1*S*)-1-(Ethoxycarbonyl)-3-phenylpropyl)-L-alanyl-L-proline
See: Enalapril

(4*S*-(4α,7α,(*R),12bβ))-7-(*S*-(1-Ethoxycarbonyl-3-phenylpropyl)amino)-1,2,3,4,6,7,8,12b-octahydro)-6-oxopyrido[2,1-*a*][2]benzazepine-4-carboxylic acid**
→ (4*S*-(4α,7α,(*R**),12bβ))-7-(*S*-(1-Carboxy-3-phenylpropyl)amino)-1,2,3,4,6,7,8,12b-octahydro)-6-oxopyrido[2,1-*a*][2]benzazepine-4-carboxylic acid
In: Dog F722
Monkey F722

4-Ethoxycarbonyl-tetrahydro-N-hydroxy-4-phenylpyridine
Probably formed from: 4-Ethoxycarbonyl-tetrahydro-4-phenylpyridine

4-Ethoxycarbonyl-tetrahydro-4-phenylpyridine
Formed from: Meperidine
→ 4-Ethoxycarbonyl-tetrahydro-N-hydroxy-4-phenylpyridine
Probably in: Cat D546
Dog D546
Guinea pig D546
Rabbit D546
Rat D546

2-(3-Ethoxycarbonyl-2-thioureido)aniline
Formed from: 1,2-Bis(3-ethoxycarbonyl-2-thioureido)benzene

2-(3-Ethoxycarbonyl-2-thioureido)-1-(3-ethoxycarbonyl-2-ureido)benzene
→ 2-(3-Ethoxycarbonyl-2-ureido)aniline
In: Mouse A260

2-(3-Ethoxycarbonyl-2-thioureido)-1-(3-ethoxycarbonyl-2-ureido)benzene
Formed from: 1,2-Bis(3-ethoxycarbonyl-2-thioureido)benzene
→ 1,2-Bis(3-ethoxycarbonyl-2-ureido)benzene
In: Mouse A260
→ 2-(3-Ethoxycarbonyl-2-thioureido)-1-(2-ureido)benzene
In: Mouse A260
→ 2-(3-Ethoxycarbonyl-2-thioureido)aniline
In: Mouse A260
→ 2-(3-Ethoxycarbonyl-2-ureido)aniline
In: Mouse A260

2-(3-Ethoxycarbonyl-2-thioureido)-1-(2-thioureido)benzene
→ 2-(3-Ethoxycarbonyl-2-thioureido)-1-(2-ureido)benzene
In: Mouse A260
→ 2-(3-Ethoxycarbonyl-2-ureido)-1-(2-thioureido)benzene
In: Mouse A260

2-(3-Ethoxycarbonyl-2-thioureido)-1-(2-ureido)benzene
Formed from: 1,2-Bis(3-ethoxycarbonyl-2-ureido)benzene
2-(3-Ethoxycarbonyl-2-thioureido)-1-(3-ethoxycarbonyl-2-ureido)benzene
2-(3-Ethoxycarbonyl-2-thioureido)-1-(2-thioureido)benzene
→ 2-(3-Ethoxycarbonyl-2-ureido)-1-(2-ureido)benzene
In: Mouse A260

2-(3-Ethoxycarbonyl-2-ureido)aniline
Formed from: 2-(3-Ethoxycarbonyl-2-thioureido)aniline
2-(3-Ethoxycarbonyl-2-thioureido)-1-(3-ethoxycarbonyl-2-ureido)benzene
→ 2-Aminobenzimidazole
In: Mouse A260
Sheep A260
→ Ethyl benzimidazol-2-ylcarbamate
In: Mouse A260
Sheep A260

2-(3-Ethoxycarbonyl-2-ureido)-1-(2-thioureido)benzene
Formed from: 2-(3-Ethoxycarbonyl-2-thioureido)-1-(2-thioureido)benzene
→ 2-(3-Ethoxycarbonyl-2-ureido)-1-(2-ureido)benzene
In: Mouse A260

2-(3-Ethoxycarbonyl-2-ureido)-1-(2-ureido)benzene
Formed from: 1,2-Bis(3-ethoxycarbonyl-2-ureido)benzene
2-(3-Ethoxycarbonyl-2-thioureido)-1-(2-ureido)benzene
2-(3-Ethoxycarbonyl-2-ureido)-1-(2-thioureido)benzene
→ 2-Aminobenzimidazole
In: Mouse A260
Sheep A260
→ Ethyl benzimidazol-2-ylcarbamate
In: Mouse A260
Sheep A260

4-Ethoxycatechol
Formed from: *p*-Ethoxyphenol
→ 4-Ethoxy-*o*-quinone
In: Mushroom H907

7-Ethoxycoumarin
→ 7-Ethoxy-3-hydroxycoumarin
In: Mouse D560
Pig D560
Rabbit D560
Rat D560
→ 7-Ethoxy-6-hydroxycoumarin
In: Mouse D560
Pig D560
Rat D560
→ Umbelliferone
In: Dog C160
Guinea pig A3971, B303, C160, G956
Man B646, F300, G956
Monkey C160, G956
Mouse A770, A1965, A2120, A2180, A2982, C160, D560, G956
Rabbit C160, D560

Rat A114, A2127, A3932, B646, C160, D98, D155, D560, G956, H697, J610, K75
Streptomyces C774
Trout C610
See also: B655

1-(4-Ethoxy-3-(6,7-dihydro-3-(hydroxypropyl)-1-methyl-7-oxo-1*H*-pyrazolo[4,3-*d*]pyrimidin-5-yl)phenylsulphonyl)-4-methylpiperazine
Formed from: Sildenafil

6-Ethoxy-1,2-dihydro-8-hydroxy-2,2,4-trimethylquinoline
Formed from: Ethoxyquin
→ 6-Ethoxy-2,2,4-trimethyl-8-quinolone
Probably in: Rat A3848

N-(4-Ethoxy-3-(6,7-dihydro-1-methyl-7-oxo-3-propyl-1*H*-pyrazolo[4,3-*d*]pyrimidin-5-yl)phenylsulphonyl)ethylenediamine
Probably formed from: N^1-(4-Ethoxy-3-(6,7-dihydro-1-methyl-7-oxo-3-propyl-1*H*-pyrazolo[4,3-*d*]pyrimidin-5-yl)phenylsulphonyl)-N^2-methylethylenediamine

1-(4-Ethoxy-3-(6,7-dihydro-1-methyl-7-oxo-3-propyl-1*H*-pyrazolo[4,3-*d*]pyrimidin-5-yl)phenylsulphonyl)hydroxy-4-methylpiperazine
Formed from: Sildenafil
→ N^1-(4-Ethoxy-3-(6,7-dihydro-1-methyl-7-oxo-3-propyl-1*H*-pyrazolo[4,3-*d*]pyrimidin-5-yl)phenylsulphonyl)-N^2-methylethylenediamine
Probably in: Dog J557
Man J557
Mouse J557
Rabbit J557
Rat J557

N^1-(4-Ethoxy-3-(6,7-dihydro-1-methyl-7-oxo-3-propyl-1*H*-pyrazolo[4,3-*d*]pyrimidin-5-yl)phenylsulphonyl)-N^2-methylethylenediamine
Probably formed from: 1-(4-Ethoxy-3-(6,7-dihydro-1-methyl-7-oxo-3-propyl-1*H*-pyrazolo[4,3-*d*]pyrimidin-5-yl)phenylsulphonyl)-hydroxy-4-methylpiperazine
→ N-(4-Ethoxy-3-(6,7-dihydro-1-methyl-7-oxo-3-propyl-1*H*-pyrazolo[4,3-*d*]pyrimidin-5-yl)phenylsulphonyl)ethylenediamine
Probably in: Dog J557
Man J557
Mouse J557
Rabbit J557
Rat J557

1-(4-Ethoxy-3-(6,7-dihydro-1-methyl-7-oxo-3-propyl-1*H*-pyrazolo[4,3-*d*]pyrimidin-5-yl)phenylsulphonyl)piperazine
Formed from: Sildenafil

1-(4-Ethoxy-3-(6,7-dihydro-7-oxo-3-propyl-1*H*-pyrazolo[4,3-*d*]pyrimidin-5-yl)phenylsulphonyl)-4-methylpiperazine
Formed from: Sildenafil
→ 1-(4-Ethoxy-3-(6,7-dihydro-7-oxo-3-propyl-1*H*-pyrazolo[4,3-*d*]pyrimidin-5-yl)phenylsulphonyl)piperazine
Probably in: Dog J557
Rat J557

1-(4-Ethoxy-3-(6,7-dihydro-7-oxo-3-propyl-1*H*-pyrazolo[4,3-*d*]pyrimidin-5-yl)phenylsulphonyl)piperazine
Probably formed from: 1-(4-Ethoxy-3-(6,7-dihydro-7-oxo-3-propyl-1*H*-pyrazolo[4,3-*d*]pyrimidin-5-yl)phenylsulphonyl)-4-methylpiperazine

6-Ethoxy-1,2-dihydro-2,2,4-trimethylquinoline
See: Ethoxyquin

3-(2′-Ethoxy-4′,5′-dihydroxybenzoyl)propionic acid
Probably formed from: 3-(2′,4′-Diethoxy-5′-hydroxybenzoyl)propionic acid
→ 3-(2′-Ethoxy-4′-hydroxy-5′-methoxybenzoyl)propionic acid
Probably in: Rat A3251

4-Ethoxy-3,5-dimethoxybenzaldehyde
Formed from: 1-(4-Ethoxy-3,5-dimethoxyphenyl)-2-(4-ethoxy-3-methoxyphenyl)propane-1,3-diol

4-Ethoxy-3,5-dimethoxybenzyl alcohol
→ Probably in: Coriolus D124

4-Ethoxy-3,5-dimethoxybenzyl alcohol
Probably formed from: 4-Ethoxy-3,5-dimethoxybenzaldehyde

1-(4-Ethoxy-3,5-dimethoxyphenyl)-2-(4-ethoxy-3-methoxyphenyl)propane-1,3-diol
→ 4-Ethoxy-3,5-dimethoxybenzaldehyde
In: Coriolus D124
→ 4-Ethoxy-3-methoxyphenylglycol
In: Coriolus D124

7-Ethoxy-3,4-dimethylcoumarin
→ 7-Hydroxy-3,4-dimethylcoumarin
In: Rat J229

6-Ethoxy-2,4-dimethoxyquinoline
Formed from: Ethoxyquin

17-O-(2-Ethoxyethyl)-7-ethylcamptothecin-21-(2-dimethylamino)ethylamide
→ 7-Ethylcamptothecin-21-(2-dimethylamino)ethylamide
In: Rat H632

***p*-Ethoxyglycollanilide**
Formed from: *p*-Ethoxyacetanilide
→ *p*-Ethoxyaniline
In: Microorganism A261

***p*-Ethoxy-N-hydroxyacetanilide** (N-hydroxyphenacetin)
Formed from: *p*-Ethoxyacetanilide
→ *p*-Ethoxyacetanilide
In: Rabbit B178, K111
Rat B178, D631
→ *p*-Ethoxy-N-hydroxyacetanilide-β-D-glucuronide
In: Rabbit C588
Rat A1840
→ *p*-Ethoxy-N-hydroxyacetanilide-O-sulphate
In: Rat A1840
→ *p*-Ethoxyphenylhydroxylamine
In: Mouse C609
→ *p*-Nitrosophenetole
In: Rabbit B178
Rat B178, D631

3-Ethoxy-4-hydroxyacetanilide
Probably formed from: 3-Ethoxy-4-hydroxyaniline
→ 3-Ethoxy-4-hydroxyacetanilide-β-D-glucuronide
Probably in: Rat F922
→ 3-Ethoxy-4-hydroxyacetanilide-O-sulphate
Probably in: Rat F922

***p*-Ethoxy-N-hydroxyacetanilide-β-D-glucuronide**
Probably formed from: *p*-Ethoxy-N-hydroxyacetanilide

3-Ethoxy-4-hydroxyacetanilide-β-D-glucuronide
Probably formed from: 3-Ethoxy-4-hydroxyacetanilide

***p*-Ethoxy-N-hydroxyacetanilide-O-sulphate**
Probably formed from: *p*-Ethoxy-N-hydroxyacetanilide

3-Ethoxy-4-hydroxyacetanilide-O-sulphate
Probably formed from: 3-Ethoxy-4-hydroxyacetanilide

3-Ethoxy-4-hydroxyaniline
Probably formed from: Isopropyl 3-ethoxy-4-hydroxycarbanilate
→ 3-Ethoxy-4-hydroxyacetanilide
Probably in: F922

3-Ethoxy-4-hydroxybenzaldehyde (bourbonal, ethylvanillin)
Probably formed from: 3-Ethoxy-4-hydroxybenzylamine
→ 3-Ethoxy-4-hydroxybenzoic acid
In: Rhodopseudomonas G246
See also: G865
→ 3-Ethoxy-4-hydroxybenzyl alcohol
In: Dunaliella G459
Euglena G422

4-Ethoxy-3-hydroxybenzaldehyde
Probably formed from: 4-Ethoxy-3-hydroxybenzylamine
→ 4-Ethoxy-3-hydroxybenzoic acid
Probably in: Rat G865

3-Ethoxy-4-hydroxybenzoic acid
Formed from: 3,4-Diethoxybenzoic acid
3-Ethoxy-4-hydroxybenzaldehyde
3-Ethoxy-4-methoxybenzoic acid
3-Ethoxy-4-*n*-propoxybenzoic acid

4-Ethoxy-3-hydroxybenzoic acid
Probably formed from: 4-Ethoxy-3-hydroxybenzaldehyde

6-Ethoxy-1-hydroxybenzpyrene
Formed from: 1,6-Diethoxybenzpyrene

3-Ethoxy-4-hydroxybenzyl alcohol
Formed from: 3-Ethoxy-4-hydroxybenzaldehyde

3-Ethoxy-4-hydroxybenzylamine
→ 3-Ethoxy-4-hydroxybenzaldehyde
Probably in: Rat G865

4-Ethoxy-3-hydroxybenzylamine
→ 4-Ethoxy-3-hydroxybenzaldehyde
Probably in: Rat G865

7-Ethoxy-3-hydroxycoumarin
Formed from: 7-Ethoxycoumarin

→ 3,7-Dihydroxycoumarin
Probably in: Mouse D560

7-Ethoxy-6-hydroxycoumarin
Formed from: 7-Ethoxycoumarin

3-(2′-Ethoxy-4′-hydroxy-5′-methoxybenzoyl) propionic acid
Probably formed from: 3-(2′-Ethoxy-4′,5′-dihydroxybenzoyl)propionic acid

5-(3-Ethoxy-4-(4-hydroxypentyloxy)phenyl)-2,4-thiazolidinedione
Formed from: 5-(3-Ethoxy-4-pentyloxyphenyl)-2,4-thiazolidinedione
→ 5-(4-(3,4-Dihydroxypentyloxy)-3-ethoxyphenyl)-2,4-thiazolidinedione
Probably in: Dog D250
Rat D250
→ 5-(3-Ethoxy-4-(4-oxopentyloxy)phenyl)-2,4-thiazolidinedione
Probably in: Dog D250
Rat D250

3-Ethoxy-4-hydroxyphenethylamine
→ 3-Ethoxy-4-hydroxyphenylacetaldehyde
Probably in: Rat G865

2-(α-(2-Ethoxy-4-hydroxyphenoxy)benzyl) morpholine
Formed from: Reboxetine

2-(α-(2-Ethoxy-5-hydroxyphenoxy)benzyl) morpholine
Formed from: Reboxetine

2-(α-(2-Ethoxy-4-hydroxyphenoxy)benzyl)-5-oxomorpholine
Probably formed from: 2-(α-(2-Ethoxyphenoxy)benzyl)-5-oxomorpholine

2-(α-(2-Ethoxy-5-hydroxyphenoxy)benzyl)-5-oxomorpholine
Probably formed from: 2-(α-(2-Ethoxyphenoxy)benzyl)-5-oxomorpholine

(−)-(*R*)-5-(2-((2-(2-Ethoxy-4-hydroxyphenoxy)ethyl)amino)propyl)-2-methoxybenzenesulphonamide
Formed from: Tamsulan
→ (−)-(*R*)-5-(2-((2-(2-Ethoxy-4-hydroxyphenoxy)ethyl)amino)propyl)-2-methoxybenzenesulphonamide-β-D-glucuronide
In: Dog H528
Man H714
Rat H528

(−)-(*R*)-5-(2-((2-(2-Ethoxy-5-hydroxyphenoxy)ethyl)amino)propyl)-2-methoxybenzenesulphonamide
Formed from: Tamsulan
→ (−)-(*R*)-5-(2-((2-(2-Ethoxy-5-hydroxyphenoxy)ethyl)amino)propyl)-2-methoxybenzenesulphonamide-β-D-glucuronide
In: Dog H528
Man H714
Rat H528
→ (−)-(*R*)-5-(2-((2-(2-Ethoxy-5-hydroxyphenoxy)ethyl)amino)propyl)-2-methoxybenzenesulphonamide-O-sulphate
In: Dog H528
Man H714
Rat H528

(−)-(*R*)-5-(2-((2-(2-Ethoxy-4-hydroxyphenoxy)ethyl)amino)propyl)-2-methoxybenzenesulphonamide-β-D-glucuronide
Formed from: (−)-(*R*)-5-(2-((2-(2-Ethoxy-4-hydroxyphenoxy)ethyl)amino)propyl)-2-methoxybenzenesulphonamide

(−)-(*R*)-5-(2-((2-(2-Ethoxy-5-hydroxyphenoxy)ethyl)amino)propyl)-2-methoxybenzenesulphonamide-β-D-glucuronide
Formed from: (−)-(*R*)-5-(2-((2-(2-Ethoxy-5-hydroxyphenoxy)ethyl)amino)propyl)-2-methoxybenzenesulphonamide

(−)-(*R*)-5-(2-((2-(2-Ethoxy-5-hydroxyphenoxy)ethyl)amino)propyl)-2-methoxybenzenesulphonamide-O-sulphate
Formed from: (−)-(*R*)-5-(2-((2-(2-Ethoxy-5-hydroxyphenoxy)ethyl)amino)propyl)-2-methoxybenzenesulphonamide

3-Ethoxy-4-hydroxyphenylacetaldehyde
Probably formed from: 3-Ethoxy-4-hydroxyphenethylamine
→ 3-Ethoxy-4-hydroxyphenylacetic acid
Probably in: Rat G865

3-Ethoxy-4-hydroxyphenylacetic acid
Probably formed from: 3-Ethoxy-4-hydroxyphenylacetaldehyde

***m*-Ethoxy-2-hydroxypropiophenone**
Formed from: *m*-Ethoxybenzaldehyde

3-Ethoxy-4-(1-hydroxy-2-propoxy)benzoic acid
Formed from: 3-Ethoxy-4-isopropoxybenzoic acid

6-Ethoxy-8-hydroxy-2,2,4-trimethylquinoline
Probably formed from: 6-Ethoxy-1,2-dihydro-8-hydroxy-2,2,4-trimethylquinoline

3-Ethoxy-4-isopropoxybenzaldehyde
Formed from: 3-Ethoxy-4-isopropoxybenzoic acid
→ 3-Ethoxy-4-isopropoxybenzyl alcohol
In: Polyporus A1117

3-Ethoxy-4-isopropoxybenzoic acid
→ 3-Ethoxy-4-(1-hydroxy-2-propoxy)benzoic acid
In: Polyporus A1117
→ 3-Ethoxy-4-isopropoxybenzaldehyde
In: Polyporus A1117

3-Ethoxy-4-isopropoxybenzyl alcohol
Formed from: 3-Ethoxy-4-isopropoxybenzaldehyde

3-Ethoxy-4-methoxybenzaldehyde
Formed from: 3-Ethoxy-4-methoxybenzyl alcohol

4-Ethoxy-3-methoxybenzaldehyde
Formed from: 4-Ethoxy-3-methoxybenzyl alcohol
1-(4-Ethoxy-3-methoxyphenyl)-1,2-dihydroxypropane
1-(4-Ethoxy-3-methoxyphenyl)propene
1-(4-Ethoxy-3-methoxyphenyl)-2-(*o*-methoxyphenoxy)propane-1,3-diol
Probably formed from: 1-(4-Ethoxy-3-methoxyphenyl)glycerol
2-(4-Ethoxy-3-methoxyphenyl)-2-oxoethanol
→ 4-Ethoxy-3-methoxybenzoic acid
Probably in: Coriolus D124
→ 4-Ethoxy-3-methoxybenzyl alcohol
Probably in: Coriolus D124

3-Ethoxy-4-methoxybenzoic acid
→ 3-Ethoxy-4-hydroxybenzoic acid
In: Polyporus A1117

4-Ethoxy-3-methoxybenzoic acid
Formed from: 5-Carboxy-2-(4-ethoxy-3-methoxyphenyl)-3-formyl-2,3-dihydro-7-methoxybenzofuran
Probably formed from: 4-Ethoxy-3-methoxybenzaldehyde
→ Vanillic acid
In: Polyporus A1117

3-Ethoxy-4-methoxybenzyl alcohol
→ 3-Ethoxy-4-methoxybenzaldehyde
In: Arion G804

4-Ethoxy-3-methoxybenzyl alcohol
Formed from: 3-O-Benzyl-1-(4-ethoxy-3-methoxyphenyl)glycerol β-(4-formyl-2-methoxyphenoxy) ether
Probably formed from: 4-Ethoxy-3-methoxybenzaldehyde
→ 4-Ethoxy-3-methoxybenzaldehyde
In: Arion G804
→ Vanillyl alcohol
Probably in: Phanerochaete D19

1-(4-Ethoxy-3-methoxyphenyl)-1,3-dihydroxy-2-(*o*-methoxyphenoxy)propane
→ 1-(4-Ethoxy-3-methoxyphenyl)glycerol
In: Phanerochaete D467, D655, D758

1-(4-Ethoxy-3-methoxyphenyl)-1,2-dihydroxypropane
Formed from: 1-(4-Ethoxy-3-methoxyphenyl)-1-propene
→ 4-Ethoxy-3-methoxybenzaldehyde
In: Phanerochaete D467, D655

1-(4-Ethoxy-3-methoxyphenyl)-1,3-dihydroxypropane
Formed from: 1-(4-Ethoxy-3-methoxyphenyl)-2-(4-hydroxymethyl-2-methoxyphenoxy) propane-1,3-diol
→ 3-(4-Ethoxy-3-methoxyphenyl)-3-oxopropan-1-ol
In: Phanerochaete F257

1-(4-Ethoxy-3-methoxyphenyl)glycerol
Formed from: 1-(4-Ethoxy-3-methoxyphenyl)-1,3-dihydroxy-2-(*o*-methoxyphenoxy)propane
→ 1-(4-Ethoxy-3-methoxyphenyl)benzaldehyde
In: Phanerochaete D467, D655, D758

4-Ethoxy-3-methoxyphenylglycol
Formed from: 1-(4-Ethoxy-3,5-dimethoxyphenyl)-2-(4-ethoxy-3-methoxyphenyl)propane-1,3-diol
4-Ethoxy-3-methoxyphenylglycol β-guaiacyl ether
4-Ethoxy-3-methoxyphenylglycol β-vanillic acid ether
→ 2-(4-Ethoxy-3-methoxyphenyl)-2-oxoethanol

Probably in: Coriolus D124
Phanerochaete B872

4-Ethoxy-3-methoxyphenylglycol β-guaiacyl ether
→ 4-Ethoxy-3-methoxyphenylglycol
In: Phanerochaete B872
→ Guaiacol
In: Phanerochaete B872

4-Ethoxy-3-methoxyphenylglycol β-vanillic acid ether
→ 4-Ethoxy-3-methoxyphenylglycol
In: Phanerochaete B872
→ 4-Ethoxy-3-methoxyphenylglycol β-vanillyl alcohol ether
In: Phanerochaete B872
→ Vanillic acid
In: Phanerochaete B872

4-Ethoxy-3-methoxyphenylglycol β-vanillyl alcohol ether
Formed from: 4-Ethoxy-3-methoxyphenylglycol β-vanillic acid ether

1-(4-Ethoxy-3-methoxyphenyl)-2-(4-hydroxymethyl-2-methoxyphenoxy)propane-1,3-diol
→ 1-(4-Ethoxy-3-methoxyphenyl)-1,3-dihydroxypropane
In: Phanerochaete F257
→ Vanillin
In: Phanerochaete F257

1-(4-Ethoxy-3-methoxyphenyl)-1-hydroxypropane
Formed from: 4-Ethoxy-3-methoxyphenylpropane

1-(4-Ethoxy-3-methoxyphenyl)-2-(*o*-methoxyphenoxy)propane-1,3-diol
→ 4-Ethoxy-3-methoxybenzaldehyde
In: Phanerochaete C401

2-(4-Ethoxy-3-methoxyphenyl)-2-oxoethanol
Probably formed from: 4-Ethoxy-3-methoxyphenylglycol
→ 4-Ethoxy-3-methoxybenzaldehyde
Probably in: Coriolus D124

3-(4-Ethoxy-3-methoxyphenyl)-3-oxopropan-1-ol
Probably formed from: 1-(4-Ethoxy-3-methoxyphenyl)-1,3-dihydroxypropane

4-Ethoxy-3-methoxyphenylpropane
→ 1-(4-Ethoxy-3-methoxyphenyl)-1-hydroxypropane
In: Phanerochaete D655, D758

1-(4-Ethoxy-3-methoxyphenyl)-1-propene
→ 1-(4-Ethoxy-3-methoxyphenyl)-1,2-dihydroxypropane
In: Phanerochaete D467, D655, D758
→ 4-Ethoxy-3-methoxybenzaldehyde
In: Phanerochaete D758

2-((Ethoxy((1-methylethyl)amino)phosphinothioyl)oxy)benzoic acid 1-methylethyl ester
See: Isofenphos

1-Ethoxynaphthalene
→ α-Naphthol
In: Rat B582

***p*-Ethoxynitrobenzene**
→ *p*-Nitrophenol
In: Rabbit A2441

***p*-Ethoxynitrosobenzene**
→ N-(*p*-Ethoxyphenyl)isobutyrohydroxamic acid
In: Rat J429

5-(3-Ethoxy-4-(4-oxopentyloxy)phenyl-2,4-thiazolidinedione
Probably formed from: 5-(3-Ethoxy-4-(4-hydroxypentyloxy)phenyl-2,4-thiazolidinedione

5-(3-Ethoxy-4-pentyloxyphenyl)-2,4-thiazolidinedione
→ 5-(3-Ethoxy-4-(4-hydroxypentyloxy)phenyl)-2,4-thiazolidinedione
In: Dog D250
Rat D250

***o*-Ethoxyphenol**
Formed from: Phenetole

***m*-Ethoxyphenol**
Formed from: Phenetole

***p*-Ethoxyphenol**
Formed from: Phenetole
→ 4-Ethoxycatechol
In: Apple J678
Mushroom H907
Pear J678
See also: B897

7-Ethoxyphenoxazone
See: 7-Ethoxyresorufin

***o*-Ethoxyphenoxyacetic acid**
Formed from: Tamsulosin

2-(α-(2-Ethoxyphenoxy)benzyl)morpholine
See: Reboxetine

2-(α-(2-Ethoxyphenoxy)benzyl)-3-oxomorpholine
Formed from: Reboxetine

2-(α-(2-Ethoxyphenoxy)benzyl)-5-oxomorpholine
Formed from: Reboxetine
→ 2-(α-(2-Ethoxy-4-hydroxyphenoxy)benzyl)-5-oxomorpholine
Probably in: Dog G462
Man G462
Monkey G462
Rat G462
→ 2-(α-(2-Ethoxy-5-hydroxyphenoxy)benzyl)-5-oxomorpholine
Probably in: Dog G462
Man G462
Monkey G462
Rat G462

8-((4′-((4-Ethoxyphenyl)azo)-3,3′-dimethyl(1,1′-biphenyl)-4-yl)azo)-7-hydroxy-1,3-naphthalenedisulphonic acid
See: Direct Red 39

N-(*p*-Ethoxyphenyl)-*p*-benzoquinoneimine
Probably formed from: *p*-Ethoxyaniline

(−)-(*R*)-5-(2-((2-(*o*-Ethoxyphenoxy)ethyl)amino)propyl)-2-methoxybenzenesulphonamide
See: Tamsulosin

N-(*p*-Ethoxyphenyl)-2-hydroxybenzamide
See: Etofesalamide

1-(2-Ethoxyphenyl)-4-(3-(hydroxycyclohexyl)propionyl)piperazine
Formed from: 4-(3-(Cyclohexyl)propionyl)-1-(2-ethoxyphenyl)piperazine

***p*-Ethoxyphenylhydroxylamine**
Formed from: *p*-Ethoxy-N-hydroxyacetanilide

N-(*p*-Ethoxyphenyl)isobutyrohydroxamic acid
Formed from: *p*-Ethoxynitrosobenzene

3-Ethoxy-4-propoxybenzoic acid
→ 1-(4-Carboxy-2-ethoxyphenoxy)propan-2-ol
In: Polyporus A1117
→ 1-(4-Carboxy-2-ethoxyphenoxy)propan-3-ol
In: Polyporus A1117
→ 3-Ethoxy-4-hydroxybenzoic acid
In: Polyporus A1117

Ethoxyquin (6-ethoxy-1,2-dihydro-2,2,4-trimethylquinoline)
→ 1,2-Dihydro-6-hydroxy-2,2,4-trimethylquinoline
In: Mouse H712
Rat A3846, H712
See also: A3848
→ 6-Ethoxy-1,2-dihydro-8-hydroxy-2,2,4-trimethylquinoline
In: Rat A3846
→ 6-Ethoxy-2,4-dimethylquinoline
In: Rat A3846

6-Ethoxyquinoline-4-carboxylic acid hydrazide
Formed from: 5-(6-Ethoxyquinolin-4-yl)-1,3,4-oxadiazol-2(3*H*)-one

5-(6-Ethoxyquinolin-4-yl)-1,3,4-oxadiazol-2(3*H*)-one
→ 6-Ethoxyquinoline-4-carboxylic acid hydrazide
In: Rat A505

4-Ethoxy-*o*-quinone
Formed from: 4-Ethoxycatechol

7-Ethoxyresorufin (7-ethoxyphenoxazone, resorufin ethyl ether)
→ Resorufin
In: Alligator J855
Cat C562
Channel catfish E209
Dog C562
Guinea pig C562
Hamster A2129
Man J245, J299
Mouse C562
Quail C562
Rabbit C562
Rat A2127, A2129, A2371, C562, C748, D421, J855, J904
Trout C562, K436
See also: B655

7-Ethoxy-4-trifluoromethylcoumarin
→ 7-Hydroxy-4-trifluoromethylcoumarin
In: Rat E825

6-Ethoxy-2,2,4-trimethyl-8-quinolone
Probably formed from: 6-Ethoxy-1,2-dihydro-8-hydroxy-2,2,4-trimethylquinoline

Ethyl *m*-acetamidobenzoate
Formed from: Ethyl *m*-aminobenzoate

Ethyl *p*-acetamidobenzoate
Formed from: Ethyl *p*-aminobenzoate

***m*-Ethylacetanilide**
Formed from: *m*-Ethylaniline

***p*-Ethylacetanilide**
Formed from: *p*-Ethylaniline

Ethyl N-alanyl-*p*-aminobenzoate
→ Ethyl *p*-aminobenzoate
In: Rat C305

N-Ethylalanyl-*o*-toluidide
→ *o*-Toluidine
In: Mouse A830

4-(((Ethylamino)acetyl)amino)-N-(2,6-dimethylphenyl)benzamide
Formed from: 4-(((Diethylamino)acetyl)amino)-N-(2,6-dimethylphenyl)benzamide
→ 4-((Aminoacetyl)amino)-N-(2,6-dimethylphenyl)benzamide
In: Mouse E951

Ethyl *m*-aminobenzoate
Formed from: Ethyl *m*-nitrobenzoate
→ *m*-Aminobenzoic acid
In: Squalus A1388
→ Ethyl *m*-acetamidobenzoate
In: Squalus A1388

Ethyl *p*-aminobenzoate (benzocaine)
Formed from: Ethyl N-alanyl-*p*-aminobenzoate
Ethyl N-dimethylalanyl-*p*-aminobenzoate
Ethyl N-dimethylglycyl-*p*-aminobenzoate
Ethyl N-glycyl-*p*-aminobenzoate
Ethyl N-leucyl-*p*-aminobenzoate
Ethyl N-γ-methylglutamyl-*p*-aminobenzoate
Ethyl *p*-nitrobenzoate
Ethyl N-valyl-*p*-aminobenzoate
→ *p*-Aminobenzoic acid
In: Rabbit G517, H246
Rat A3992
→ Ethyl *p*-acetamidobenzoate
In: Guinea pig J203
Pig F916
→ Ethyl *p*-sulphamidobenzoate
In: Eubacterium H439
Haemophilus H439
Klebsiella H439

Ethyl 5-amino-1,2-dihydro-3-(*p*-hydroxyphenyl)-2-methylpyrido[3,4-*b*]pyrazin-7-carbamate
Probably formed from: Ethyl 5-amino-1,2-dihydro-2-methyl-3-phenylpyrido[3,4-*b*]pyrazin-7-carbamate
→ 5,7-Diamino-1,2-dihydro-3-(*p*-hydroxyphenyl)-2-methylpyrido[3,4-*b*]pyrazine
Probably in: Mouse D828

Ethyl 5-amino-1,2-dihydro-2-methyl-3-phenylpyrido[3,4-*b*]pyrazin-7-carbamate
→ Ethyl 5-amino-1,2-dihydro-3-(*p*-hydroxyphenyl)-2-methylpyrido[3,4-*b*]pyrazin-7-carbamate
Probably in: Mouse D828

4-Ethylamino-1,1-dimethylbut-2-yn-1-yl 2-cyclohexyl-2-hydroxy-2-phenylacetate
Formed from: 4-Diethylamino-1,1-dimethylbut-2-yn-1-yl 2-cyclohexyl-2-hydroxy-2-phenylacetate

4-Ethylaminoethoxyhexoestrol
Formed from: 4,4′-Bis(diethylaminoethoxy)hexoestrol

2-Ethylaminoethyl 2,2-diphenylvalerate
Formed from: 2-Diethylaminoethyl 2,2-diphenylvalerate

Ethyl N-(2-amino-4-(*p*-fluorobenzylamino)phenyl)carbamate (retigabine)
→ 1,2-Diamino-4-(4-fluorobenzylamino)benzene
In: Man J586
Rat J586
→ Ethyl N-(2-amino-4-(N-(*p*-fluorobenzyl)-N-(β-D-glucuronyl)amino)phenyl)carbamate
In: Dog J586
Man J108, J586
Rat J586
→ Ethyl N-(4-(*p*-fluorobenzylamino)-2-β-D-glucuronamido)phenyl)carbamate
In: Dog J586
Man J108, J586
Rat J586
→ Retigabine-β-D-glucoside
In: Dog J586

Ethyl N-(2-amino-4-(N-(*p*-fluorobenzyl)-N-(β-D-glucuronyl)amino)phenyl)carbamate
Formed from: Ethyl N-(2-amino-4-(*p*-fluorobenzylamino)phenyl)carbamate

8-(6-Ethylaminohexylamino)-6-hydroxy-4-methylquinoline
Probably formed from: 8-(6-Diethylaminohexylamino)-6-hydroxy-4-methylquinoline

8-(6-Ethylaminohexylamino)-6-hydroxy-4-methylquinoline
Probably formed from: 8-(6-Ethylaminohexylamino)-6-methoxy-4-methylquinoline
→ 8-(6-Ethylaminohexylamino)-4-methylquinolin-6-yl sulphate
Probably in: Rat F477

8-(6-Ethylaminohexylamino)-6-methoxy-4-methylquinoline
Formed from: 8-(6-Diethylaminohexylamino)-6-methoxy-4-methylquinoline
→ 8-(6-Aminohexylamino)-6-methoxy-4-methylquinoline
Probably in: Rat F477
→ 8-(6-Diethylaminohexylamino)-4-hydroxymethyl-6-methoxyquinoline
Probably in: Rat F477
→ 8-(6-Ethylaminohexylamino)-6-hydroxy-4-methylquinoline
Probably in: Rat F477

8-(6-Ethylaminohexylamino)-4-methylquinolin-6-yl sulphate
Probably formed from: 8-(6-Ethylaminohexylamino)-6-hydroxy-4-methylquinoline

2-Ethylamino-N-(3-hydroxy-2,6-dimethylphenyl)butanamide
Probably formed from: 2-(Ethylpropylamino)-N-(3-hydroxy-2,6-dimethylphenyl)butanamide
→ 2-Amino-N-(3-hydroxy-2,6-dimethylphenyl)butanamide
Probably in: Man A3247

2-Ethylamino-N-(4-hydroxy-2,6-dimethylphenyl)butanamide
Probably formed from: 2-(Ethylpropylamino)-N-(4-hydroxy-2,6-dimethylphenyl)butanamide
→ 2-Amino-N-(4-hydroxy-2,6-dimethylphenyl)butanamide
Probably in: Man A3247

1-(3-(Ethylamino)-4-hydroxypyridinyl)-4-((5-methoxy-1*H*-indol-2-yl)carbonyl)piperazine
Probably formed from: Ateviridine

7-Ethylamino-4-methylcoumarin
Probably formed from: 7-Diethylamino-4-methylcoumarin
→ 7-Amino-4-methylcoumarin
In: Mouse J618

2-Ethylamino-2-methyl-1-phenylethanol
Formed from: 2-Ethylaminopropiophenone

***p*-Ethylaminophenol**
Formed from: N-Ethylaniline

2-Ethylamino-1-phenyl-1-propanol
Probably formed from: 2-Diethylamino-1-propanol
→ 2-Amino-1-phenyl-1-propanol
Probably in: Man A103

Ethyl (10-(3-aminopropionyl)phenothiazin-2-yl) carbamate
Probably formed from: Ethyl (10-(N-(2-hydroxyethyl)-3-aminopropionyl)phenothiazin-2-yl)carbamate

2-Ethylaminopropiophenone
Formed from: α-Diethylaminopropiophenone
→ Cathinone
In: Man A103
→ 2-Ethylamino-2-methyl-1-phenylethanol
In: Man A1167

1-(3-(Ethylamino)-2-pyridinyl)-4-((5-hydroxy-1*H*-indol-2-yl)carbonyl)piperazine
Formed from: Ateviridine
→ 1-(3-(Ethylamino)-2-pyridinyl)-4-((5-hydroxy-1*H*-indol-2-yl)carbonyl)piperazine-β-D-glucuronide
In: Rat J449
→ 1-(3-(Ethylamino)-2-pyridinyl)-4-((5-hydroxy-1*H*-indol-2-yl)carbonyl)piperazine-O-sulphate
In: Rat J449

1-(3-(Ethylamino)-2-pyridinyl)-4-((6-hydroxy-5-methoxy-1*H*-indol-2-yl)carbonyl)piperazine
Probably formed from: Ateviridine

1-(3-(Ethylamino)-2-pyridinyl)-4-((7-hydroxy-5-methoxy-1*H*-indol-2-yl)carbonyl)piperazine
Probably formed from: Ateviridine

1-(3-(Ethylamino)-2-pyridinyl)-4-((5-hydroxy-1*H*-indol-2-yl)carbonyl)piperazine-β-D-glucuronide
Formed from: (3-(Ethylamino)-2-pyridinyl)-4-((5-hydroxy-1*H*-indol-2-yl)carbonyl)piperazine

1-(3-(Ethylamino)-2-pyridinyl)-4-((5-hydroxy-1*H*-indol-2-yl)carbonyl)piperazine-O-sulphate
Formed from: (3-(Ethylamino)-2-pyridinyl)-4-((5-hydroxy-1*H*-indol-2-yl)carbonyl) piperazine

1-(3-(Ethylamino)-2-pyridinyl)-4-((5-methoxy-1*H*-indol-2-yl)carbonyl)piperazine
See: Ateviridine

2-Ethylamino-1-(*m*-trifluoromethylthiophenyl) propane
See: Tiflorex

N-Ethylamphetamine
Formed from: N-Ethyl-N-methylamphetamine
→ Amphetamine
In: Man A408
Mycobacterium B380
Rabbit A2719, A3013
→ N-Ethyl-N-hydroxyamphetamine
In: Rabbit A2719, A3013
→ N-Ethyl-*p*-hydroxyamphetamine
In: Rat B7
→ α-Methyl-N-(1′-phenylprop-2′-yl)nitrone
In: Rabbit A2719, A3013

N-Ethylaniline
Formed from: N,N-Diethylaniline
N-Ethyl-N-methylaniline
Probably formed from: N-Ethyl-N-methylaniline-N-oxide
→ Aniline
Probably in: Rabbit A307
→ *p*-Ethylaminophenol
Probably in: Rabbit B181
→ N-Ethylphenylhydroxylamine
In: Pig K160

***m*-Ethylaniline**
→ *m*-Ethylacetanilide
In: Rabbit E399

***p*-Ethylaniline**
Probably formed from: N-Ethyl-N-methylaniline-N-oxide
→ *p*-Ethylacetanilide
In: Rabbit E399
→ *p*-Ethylnitrosobenzene
With chloroperoxidase A3579

***p*-Ethylanisole**
→ 1-(*p*-Methoxyphenyl)ethanol
In: Mortierella D969

Ethyl anthranilate
Formed from: Ethyl *o*-nitrobenzoate

Ethyl apovincaminate (vinpocetine)
→ Apovincaminic acid
In: Dog D252
Rat A2810, B81, D252
→ Ethyl hydroxyapovincaminate
In: Rat A2810
→ Ethyl vincaminate
In: Rat A2810

N-Ethylbenzamide
Formed from: N,N-Diethylbenzamide
→ Benzoic acid
In: Rat H475

7-Ethylbenz[*a*]anthracene
→ 7-Ethyl-1,2-dihydro-1,2-dihydroxybenz[*a*]anthracene
In: Rat E523, E903
→ 7-Ethyl-3,4-dihydro-3,4-dihydroxybenz[*a*]anthracene
In: Rat E523, E903
→ 7-Ethyl-5,6-dihydro-5,6-dihydroxybenz[*a*]anthracene
In: Rat E523, E903
→ 7-Ethyl-8,9-dihydro-8,9-dihydroxybenz[*a*]anthracene
In: Rat E523, E903
→ 7-Ethyl-10,11-dihydro-10,11-dihydroxybenz[*a*]anthracene
In: Rat E523, E903
→ 7-(α-Hydroxyethyl)benz[*a*]anthracene
In: Rat E903

Ethylbenzene
Formed from: Ferulic acid
→ (+)-*cis*-2,3-Dihydro-2,3-dihydroxyethylbenzene
In: Pseudomonas A914, J656
→ *p*-Ethylphenol
In: Man D482
Nitrosomonas H219
→ Mandelic acid
In: Man F330
Pseudomonas F870
→ 1-Phenylethanol (both isomers)
In: Azoarcus K241
Chinook salmon E710

Desulfobacula J847
Escherichia G740
Man D482
Mortierella D969
Nitrosomonas H219, H631
Nocardia B144
Pseudomonas F870, J219
Rabbit A287
Rat C670, D223, D365
See also: A914, A1743
→ 2-Phenylethanol
In: Nitrosomonas H631
Pseudomonas F870, H334
See also: D223
→ Salicylic acid
In: Pseudomonas F870
→ Styrene
In: Pseudomonas J219

Ethylbenzene hydroperoxide
→ 1-Phenylethanol
Probably in: Rat C670

***p*-Ethylbenzenesulphonyl-4-hydroxy-2-methylpiperidine**
Formed from: *p*-Ethylbenzenesulphonyl-2-methylpiperidine

***p*-Ethylbenzenesulphonyl-2-methylpiperidine**
→ *p*-Ethylbenzenesulphonyl-4-hydroxy-2-methylpiperidine
In: Rat C23
→ *p*-(1-Hydroxyethyl)benzenesulphonyl-2-methylpiperidine
In: Rat C23

Ethyl benzimidazol-2-ylcarbamate
Formed from: 2-(3-Ethoxycarbonyl-2-ureido)aniline
2-(3-Ethoxycarbonyl-2-ureido)-1-(2-ureido)benzene
Methyl 1-(butylcarbamoyl)benzimidazol-2-ylcarbamate
→ Ethyl 5-hydroxybenzimidazol-2-ylcarbamate
Probably in: Mouse A260

2-Ethyl-5-benzofuranylacetic acid
Formed from: 2-Ethyl-2,3-dihydro-5-benzofuranylacetic acid

***o*-Ethylbenzoic acid**
→ (*S*)-3-methylphthalide
In: Aspergillus J486
Pseudomonas J486

***p*-Ethylbenzoic acid**
→ *p*-Ethylbenzoyl-β-D-glucuronic acid
In: Rat J549
→ *p*-Ethylhippuric acid
In: Rat J549

***p*-Ethylbenzonitrile**
→ 1-(*p*-Cyanophenyl)ethanol
In: Mortierella D969

2-(5-Ethyl-3-benzothienyl)ethylamine
→ N-Acetyl-2-(5-ethyl-3-benzothienyl)ethylamine
In: Man K490

1-Ethyl-1*H*-benzotriazole
→ 1*H*-Benzotriazole
In: Rat C240

2-Ethylbenzotriazole
→ 2-Ethyl-4,5-dihydro-4,5-dihydroxybenzotriazole
In: Rat C419
→ 2-Ethyl-4-hydroxybenzotriazole
In: Rat C419
→ 2-Ethyl-5-hydroxybenzotriazole
In: Rat C419

Ethylbenzoylglycyl-L-lysine
→ Hippuric acid
With carboxypeptidase B A243

***p*-Ethylbenzoyl-β-D-glucuronic acid**
Formed from: *p*-Ethylbenzoic acid

***p*-Ethylbenzoyl-3-hydroxy-N-methylbutylamine**
Formed from: *p*-Ethylbenzoyl-N-methylbutylamine

***p*-Ethylbenzoyl-N-methylbutylamine**
→ *p*-Ethylbenzoyl-3-hydroxy-N-methylbutylamine
In: Rat C23
→ *p*-(1-Hydroxyethyl)benzoyl-N-methylbutylamine
In: Rat C23

***p*-Ethylbenzoyl-1,2,3,4-tetrahydroisoquinoline**
→ 1,2,3,4-Tetrahydro-N-(*p*-(1-hydroxyethyl)benzoyl)isoquinoline
In: Cunninghamella D180

3-(α-Ethylbenzyl)-4-hydroxycoumarin
See: Phenprocoumon

Ethyl 4-((4-bromo-2-cyanophenyl)carbamoyl) benzylphosphonate
Formed from: Diethyl 4-((4-bromo-2-cyanophenyl)carbamoyl)benzylphosphonate

N-(2-Ethylbut-1-en-1-yl)-N′-(1-(3,4-methylenedioxyphenyl)but-1-yl)urea
Formed from: N-(1(*R*)-(1,3-Benzodioxol-5-yl)butyl)-3,3-diethyl-2(*S*)-((4-methyl-1-piperazinyl)carbonyl)phenoxy-4-oxo-1-azetidinecarboxamide

7-Ethylcamptothecin-21-(2-dimethylamino) ethylamide
Formed from: 17-O-Butyl-7-ethylcamptothecin-21-(2-dimethylamino)ethylamide
17-O-*tert*-Butyl-7-ethylcamptothecin-21-(2-dimethylamino)ethylamide
17-O-(2-Ethoxyethyl)-7-ethylcamptothecin-21-(2-dimethylamino)ethylamide
7-Ethyl-17-O-ethylcamptothecin-21-(2-dimethylamino)ethylamide
7-Ethyl-17-O-(*p*-fluorophenyl)camptothecin-21-(2-dimethylamino)ethylamide
7-Ethyl-17-O-isopropylcamptothecin-21-(2-dimethylamino)ethylamide
7-Ethyl-17-O-(*p*-methoxyphenyl)camptothecin-21-(2-dimethylamino)ethylamide
7-Ethyl-17-O-methylcamptothecin-21-(2-dimethylamino)ethylamide
7-Ethyl-17-O-(methylthio)ethylcamptothecin-21-(2-dimethylamino)ethylamide
7-Ethyl-17-O-phenylcamptothecin-21-(2-dimethylamino)ethylamide
7-Ethyl-17-O-propylcamptothecin-21-(2-dimethylamino)ethylamide

Ethyl 2-carbamoyloxybenzoate
→ Benzoxazine-2,4(3*H*)-dione
In: Rat F759
→ Salicylamide
In: Rat F759
→ Salicylic acid
In: Rat F759

1-(4-(Ethylcarbamoyloxy)hydroxypiperidino)-6,7-dimethoxyphthalazine
Formed from: Carbazeran

1-(4-Ethylcarbamoyloxy)piperidino)-6,7-dimethoxy)phthalazine
See: Carbazeran

3-Ethylcatechol
Formed from: (+)-*cis*-2,3-dihydro-2,3-dihydroxyethylbenzene
o-Ethylphenol
→ ε-Ethyl-α-hydroxymuconic semialdehyde
In: Pseudomonas B849

4-Ethylcatechol
Formed from: Caffeic acid
Phenylethane-*p*-sulphonic acid
Probably formed from: *m*-Ethylphenol
→ 5-Ethyl-2-hydroxy-*cis*,*cis*-muconic semialdehyde
Probably in: Alcaligenes A1323
→ 4-Ethyl-2-methoxyphenol
In: Rat A486
With catechol O-methyltransferase A597
→ 5-Ethyl-2-methoxyphenol
In: Rat A486
With catechol O-methyltransferase A597

Ethyl 2-(*p*-chlorophenyl)-5-(3-formyl-1-oxoprop-2-en-1-yl)oxazole-4-acetate
→ Ethyl 2-(*p*-chlorophenyl)-5-(4-hydroxy-1-oxobutyl)oxazole-4-acetate
In: Rat E893

Ethyl 2-(*p*-chlorophenyl)-5-(2-furyl)oxazole-4-acetate
→ 2-(*p*-Chlorophenyl)-5-(2-furyl)oxazole-4-acetic acid
In: Dog E877
Rabbit E877
Rat E877
→ Ethyl 2-(*p*-chlorophenyl)-5-(4-hydroxy-1-oxobutyl)oxazole-4-acetate
In: Rat E893

Ethyl 2-(*p*-chlorophenyl)-5-(4-hydroxy-1-oxobutyl)oxazole-4-acetate
Formed from: Ethyl 2-(*p*-chlorophenyl-5-(2-furyl)oxazole-4-acetate
Ethyl 2-(*p*-chlorophenyl)-5-(3-formyl-1-oxoprop-2-en-1-yl)oxazole-4-acetate

Ethyl cinnamate
→ Cinnamic acid
In: Rat K911

Ethylcocaine
Formed from: Cocaine

Ethyl 2,4-dichlorophenoxyacetate
Formed from: 2,4-Dichlorophenoxyacetic acid

Ethyl 10-(diethylaminopropionyl)phenothiazine-2-carbamate
See: Ethacizin

***erythro*-α′-Ethyl-3,3′-difluoro-4,4′-dihydroxy-α-methylbibenzyl**
See: Bifluranol

2-Ethyl-2,3-dihydro-5-benzofuranylacetic acid
→ 2,3-Dihydro-2-(1-hydroxyethyl)-5-benzofuranylacetic acid
In: Rat B276
→ 2,3-Dihydro-2-(2-hydroxyethyl)-5-benzofuranylacetic acid
In: Rat B276
→ 2-Ethyl-5-benzofuranylacetic acid
In: Rat B276
→ 2-Ethyl-2,3-dihydro-3-hydroxy-5-benzofuranylacetic acid
In: Rat B276

7-Ethyl-1,2-dihydro-1,2-dihydroxybenz[*a*]anthracene
Formed from: 7-Ethylbenz[*a*]anthracene

7-Ethyl-3,4-dihydro-3,4-dihydroxybenz[*a*]anthracene
Formed from: 7-Ethylbenz[*a*]anthracene

7-Ethyl-5,6-dihydro-5,6-dihydroxybenz[*a*]anthracene
Formed from: 7-Ethylbenz[*a*]anthracene

7-Ethyl-8,9-dihydro-8,9-dihydroxybenz[*a*]anthracene
Formed from: 7-Ethylbenz[*a*]anthracene

7-Ethyl-10,11-dihydro-10,11-dihydroxybenz[*a*]anthracene
Formed from: 7-Ethylbenz[*a*]anthracene

2-Ethyl-4,5-dihydro-4,5-dihydroxybenzotriazole
Formed from: 2-Ethylbenzotriazole

***cis*-7-Ethyl-1,2-dihydro-1,2-dihydroxynaphthalene**
Formed from: 2-Ethylnaphthalene
→ 7-Ethyl-1,2-dihydroxynaphthalene
In: Pseudomonas J370

***cis*-8-Ethyl-1,2-dihydro-1,2-dihydroxynaphthalene**
Formed from: 1-Ethylnaphthalene
→ 8-Ethyl-1,2-dihydroxynaphthalene
In: Pseudomonas J370

2-Ethyl-2,3-dihydro-3-hydroxy-5-benzofuranylacetic acid
Formed from: 2-Ethyl-2,3-dihydro-5-benzofuranylacetic acid

5-Ethyl-2,4-dihydro-2-(3-hydroxypropyl)-4-(2-phenoxyethyl)-1,2,4-triazol-3-one
Formed from: Nefazodone

N-(1-(2-(4-Ethyl-4,5-dihydro-5-oxo-1*H*-tetrazol-1-yl)ethyl)-4-(methoxymethyl)-4-piperidinyl)-N-phenylpropanamide
See: Alfentanil

N-Ethyl-3,4-dihydroxyamphetamine
Formed from: N-Ethyl-3,4-methylenedioxyamphetamine
→ 3,4-Dihydroxyamphetamine
Probably in: Man H736
→ N-Ethyl-4-hydroxy-3-methoxyamphetamine
Probably in: Man H736

3-Ethyl-2,5-dihydroxybenzoic acid
→ 6-Carboxy-2-ethyl-4-hydroxy-*cis*,*cis*-muconic semialdehyde
In: Pseudomonas F649

β-Ethyl-α,β-dihydroxydibenzyl
Formed from: β-Ethylstilbene oxide

1-Ethyl-4-(N,N-di-(2-hydroxyethyl)aminoethyl)-3,3-diphenylpyrrolidin-2-one
Formed from: Doxapram
→ 1-Ethyl-4-(2-(2-hydroxyethylamino)ethyl)-3,3-diphenylpyrrolidin-2-one
Probably in: Rat A1389

7-Ethyl-1,2-dihydroxynaphthalene
Formed from: *cis*-7-Ethyl-1,2-dihydro-1,2-dihydroxynaphthalene

8-Ethyl-1,2-dihydroxynaphthalene
Formed from: *cis*-8-Ethyl-1,2-dihydro-1,2-dihydroxynaphthalene

4-(N-Ethyl-N-(2-(3,4-dihydroxyphenyl)-1-methylethyl)amino)butanol
Probably formed from: 4-(N-Ethyl-N-(2-(*p*-hydroxyphenyl)-1-methylethyl)amino)butanol
→ 4-(N-Ethyl-N-(2-(4-hydroxy-3-methoxyphenyl)-1-methylethyl)amino)butanol
Probably in: Man K21
Rat K21

4-Ethyl-2,6-dimethoxyphenol (4-ethylsyringol)
Probably formed from: 4-Hydroxy-3,5-dimethoxstyrene

Ethyl 4-(3,4-dimethoxyphenyl)-6,7-dimethoxy-2-(1,2,4-triazol-1-ylmethyl)quinoline-3-carboxylate
→ Ethyl 4-(4-hydroxy-3-methoxyphenyl)-6,7-dimethoxy-2-(1,2,4-triazol-1-ylmethyl) quinoline-3-carboxylate
In: Rat J579

4-Ethyl-6,7-dimethoxyquinazoline
See: Quazodine

Ethyl N-dimethylalanyl-*p*-aminobenzoate
→ Ethyl *p*-aminobenzoate
In: Rat C305

***p*-Ethyl-N,N-dimethylaniline**
→ *p*-Ethyl-N-methylaniline
In: Rat D578

5″-Ethyl-1″,1″-dimethylcannabidiol
→ 5″-Ethyl-6-hydroxy-1″,1″-dimethylcannabidiol
In: Guinea pig F853
Mouse F853
Rabbit F853
Rat F853, F885
→ 5″-Ethyl-7-hydroxy-1″,1″-dimethylcannabidiol
In: Guinea pig F853
Mouse F853
Rabbit F853
Rat F853, F885
→ 6-Hydroxy-1″,1″-dimethylcannabidiol-5″-oic acid
In: Dog F885
→ 7-Hydroxy-1″,1″-dimethylcannabidiol-5″-oic acid
In: Dog F885

Ethyl N-dimethylglycyl-*p*-aminobenzoate
→ Ethyl *p*-aminobenzoate
In: Rat C305

4-((2-Ethyl-5,7-dimethyl-3*H*-imidazol[4,5-*b*]pyridin-3-yl)methyl)-2′-(1*H*-tetrazol-5-yl)biphenyl
→ 4-((2-Ethyl-5,7-dimethyl-3*H*-imidazol[4,5-*b*]pyridin-3-yl)methyl)-2′-(2-β-D-glucuronido-1*H*-tetrazol-5-yl)biphenyl
In: Monkey G866

4-((2-Ethyl-5,7-dimethyl-3*H*-imidazol[4,5-*b*] pyridin-3-yl)methyl)-2′-(2-β-D-glucuronido-1*H*-tetrazol-5-yl)biphenyl
Formed from: 4-((2-Ethyl-5,7-dimethyl-3*H*-imidazol[4,5-*b*]pyridin-3-yl)methyl)-2′-(1*H*-tetrazol-5-yl)biphenyl

5-Ethyl-1,3-dimethyl-5-phenylbarbituric acid
→ 5-Ethyl-5-(4-hydroxy-3-methoxyphenyl)-1,3-dimethylbarbituric acid
In: Man E474

N[3]-Ethyl-2,4-dinitro-6-trifluoromethyl-*m*-phenylenediamine
Formed from: Dinitramine
→ 2,4-Dinitro-6-trifluoromethyl-*m*-phenylenediamine
In: Aspergillus A1002
Carp A1951

O-Ethyl S,S-diphenyl phosphorodithioate (edifenphos)
→ S,S-Diphenyl phosphorodithioate
In: Goat A2322
→ O-Ethyl S-phenyl phosphorodithioate
In: Goat A2322
→ Methyl phenyl sulphone
In: Goat A2322

1-Ethyl-6-fluoro-1,4-dihydro-4-oxo-7-(4-methyl-1-piperazinyl)quinoline-3-carboxylic acid
See: Pefloxacin

1-Ethyl-6-fluoro-1,4-dihydro-4-oxo-7-(1-piperazinyl)quinoline-3-carboxylic acid
See: Norfloxacin

Ethylene glycol monovanillic acid ether
Formed from: Glycerol-2-vanillic acid ether

7-Ethyl-17-O-ethylcamptothecin-21-(2-dimethylamino)ethylamide
→ 7-Ethylcamptothecin-21-(2-dimethylamino)ethylamide
In: Rat H632

N-Ethyl-N′-(1-ethyl-6-methylergolin-8-yl) sulphamide
Formed from: N,N-Diethyl-N′-(1-ethyl-6-methylergolin-8-yl)sulphamide
→ N-Ethyl-N′-(6-methylergolin-8-yl) sulphamide
Probably in: Man G576

1-Ethyl-4-fluorobenzene
→ 1-(*p*-Fluorophenyl)ethanol

In: Mortierella D969
→ 2-(*p*-Fluorophenyl)ethanol
In: Mortierella D969

Ethyl N-(4-(*p*-fluorobenzylamino)-2-β-D-glucuronamido)phenyl)carbamate
Formed from: Ethyl N-(2-amino-4-(*p*-fluorobenzylamino)phenyl)carbamate

Ethyl 8-fluoro-5,6-dihydro-5-methyl-6-oxo-4*H*-imidazo[1,5-*a*][1,4]benzodiazepine-3-carboxylate
→ 8-Fluoro-5,6-dihydro-5-methyl-6-oxo-4*H*-imidazo[1,5-*a*][1,4]benzodiazepine-3-carboxylic acid
In: Man F406

1-Ethyl-6-fluoro-1,4-dihydro-4-oxo-7-(3-oxo-1-piperazinyl)-3-quinolinecarboxylic acid
Formed from: 1-Ethyl-6-fluoro-1,4-dihydro-4-oxo-7-(1-piperazinyl)-3-quinolinecarboxylic acid
→ 7-(2-Aminoethylamino)-1-ethyl-6-fluoro-1,4-dihydro-4-oxo-3-quinolinecarboxylic acid
Probably in: Dog C104
Man C106
Monkey C104
Rat C104

1-Ethyl-6-fluoro-1,4-dihydro-4-oxo-7-(1-piperazinyl)-3-quinolinecarboxylic acid
→ 7-(4-Acetyl-1-piperazinyl)-1-ethyl-6-fluoro-1,4-dihydro-4-oxo-3-quinolinecarboxylic acid
In: Dog C104
Monkey C104
Rat C104
→ 1-Ethyl-6-fluoro-1,4-dihydro-4-oxo-7-(3-oxo-1-piperazinyl)-3-quinolinecarboxylic acid
In: Dog C104
Man C106
Monkey C104
Rat C104
→ 1-Ethyl-6-fluoro-7-(4-formyl-1-piperazinyl)-1,4-dihydro-4-oxo-3-quinolinecarboxylic acid
In: Dog C104
Monkey C104
Rat C104

2-Ethyl-7-fluoro-3,4-dihydro-3-oxo-2*H*-quinoxaline
Formed from: Isopropyl (*S*)-2-ethyl-7-fluoro-3,4-dihydro-3-oxo-2*H*-quinoxaline-1-carboxylate

(*S*)-2-Ethyl-7-fluoro-3,4-dihydro-3-oxo-2*H*-quinoxalinecarboxylic acid 1-hydroxyisopropyl ester
Formed from: Isopropyl (*S*)-2-ethyl-7-fluoro-3,4-dihydro-3-oxo-2*H*-quinoxaline-1-carboxylate
→ (*S*)-2-Ethyl-7-fluoro-3,4-dihydro-3-oxo-2*H*-quinoxalinecarboxylic acid 1-hydroxyisopropyl ester-β-D-glucuronide
Probably in: Pig J749
→ (*S*)-2-Ethyl-7-fluoro-3,4-dihydro-3-oxo-2*H*-quinoxalinecarboxylic acid lactic acid ester
Probably in: Pig J749

(*S*)-2-Ethyl-7-fluoro-3,4-dihydro-3-oxo-2*H*-quinoxalinecarboxylic acid 1-hydroxyisopropyl ester-β-D-glucuronide
Probably formed from: (*S*)-2-Ethyl-7-fluoro-3,4-dihydro-3-oxo-2*H*-quinoxalinecarboxylic acid 1-hydroxyisopropyl ester

(*S*)-2-Ethyl-7-fluoro-3,4-dihydro-3-oxo-2*H*-quinoxalinecarboxylic acid lactic acid ester
Probably formed from: (*S*)-2-Ethyl-7-fluoro-3,4-dihydro-3-oxo-2*H*-quinoxalinecarboxylic acid 1-hydroxyisopropyl ester

1-Ethyl-6-fluoro-7-(4-formyl-1-piperazinyl)-1,4-dihydro-4-oxo-3-quinolinecarboxylic acid
Formed from: 1-Ethyl-6-fluoro-1,4-dihydro-4-oxo-7-(1-piperazinyl)-3-quinolinecarboxylic acid

7-Ethyl-17-O-(*p*-fluorophenyl)camptothecin-21-(2-dimethylamino)ethylamide
→ 7-Ethylcamptothecin-21-(2-dimethylamino)ethylamide
In: Rat H632

4-Ethyl-5-(*p*-fluorophenyl)-2,4-dihydro-3*H*-1,2,4-triazole-3-thione
→ 5-(*p*-Fluorophenyl)-2,4-dihydro-3*H*-1,2,4-triazole-3-thione
In: Rat K297

N-Ethyl-*m*-formylbenzamide
Formed from: N,N-Diethyl-*m*-formylbenzamide

N-Ethyl-*m*-hydroxymethylbenzamide
→ N-Ethyl-*m*-hydroxymethylbenzamide
In: Rat J572
→ N-Ethylisophthalamidic acid
In: Rat J572
→ *m*-Formylbenzamide
In: Rat J572

1-Ethyl-4-(N-formyl-N-(2-hydroxyethyl) aminoethyl)-3,3-diphenyl-2-pyrrolidinone
Formed from: 2-Oxodoxapram

1-Ethyl-4-(2-(2-formylmethoxyethylamino)ethyl)-3,3-diphenyl-2-pyrrolidinone
Formed from: 3-Oxodoxapram

1-Ethyl-4-formylmethyl-3,3-diphenyl-2-pyrrolidinone
Formed from: Doxapram

Ethyl gallate
→ Ethyl gallate-O-sulphate
In: Eubacterium G485

Ethyl gallate-O-sulphate
Formed from: Ethyl gallate

N-Ethylglycine-2,6-xylidide
Formed from: Lignocaine
→ 2,6-dimethylaniline
In: Man J67
Mouse A830
→ Glycine-2,6-xylidide
Probably in: Rabbit E53, H723
Rat E546, F135

Ethyl N-glycyl-*p*-aminobenzoate
→ Ethyl *p*-aminobenzoate
In: Rat C305

Ethyl *p*-(6-guanidinohexanoyloxy)benzoate
→ Ethyl *p*-hydroxybenzoate
In: Rat A2297

3-(3-Ethylhexahydro-1-methyl-1*H*-azepin-3-yl)phenol
See: Meptazinol

6-Ethylhexahydro-6-(*m*-hydroxyphenyl)-1-methylazepin-2(2*H*)-one
Formed from: Meptazinol

(1*R*,2*R*,4*S*)-2-Ethyl-1,2,3,4,6,11-hexahydro-2,4,5,7,12-pentahydroxy-6,11-dioxo-1-naphthacenecarboxylic acid methyl ester
See: ε-Rhodomycinone

2-Ethyl-1,2,3,4,6,11-hexahydro-2,4,5,7-tetrahydroxy-6,11-dioxo-1-naphthacenecarboxylic acid methyl ester
See: Aklavinone

(1*R*,2*R*)-2-Ethyl-1,2,3,4,6,11-hexahydro-2,5,7,10-tetrahydroxy-6,11-dioxo-1-naphthacenecarboxylic acid methyl ester
See: ζ-Pyrromycinone

(1*R*,2*R*)-2-Ethyl-1,2,3,4,6,11-hexahydro-2,5,7,12-tetrahydroxy-6,11-dioxo-1-naphthacenecarboxylic acid methyl ester
See: ζ-Rhodomycinone

(1*R*,2*R*)-2-Ethyl-1,2,3,4,6,11-hexahydro-2,4,5,12-tetrahydroxy-6,11-dioxo-1-naphthacenecarboxylic acid methyl ester
See: 7-Deoxy-ε-rhodomycinone

(1*R*,2*R*)-2-Ethyl-1,2,3,4,6,11-hexahydro-2,5,7,12-tetrahydroxy-4,6,11-trioxo-1-naphthacenecarboxylic acid methyl ester
See: Maggiemycin

(1*R*,2*R*)-2-Ethyl-1,2,3,4,6,11-hexahydro-2,5,7-trihydroxy-4,6,11-trioxo-1-naphthacenecarboxylic acid methyl ester
See: Aklaviketone

N-Ethyl-1-hexyl-N-methyl-4-phenyl-4-piperidinecarboxamide
See: Sameridine

***p*-Ethylhippuric acid**
Formed from: *p*-Ethylbenzoic acid

α-Ethyl-α-(*p*-hydroxylaminophenyl)glutarimide
Formed from: Aminoglutethimide

N-Ethyl-*p*-hydroxyamphetamine
Formed from: N-Ethylamphetamine
N-Ethyl-*p*-methoxyamphetamine
→ N-Ethyl-4-hydroxy-3-methoxyamphetamine
Probably in: Rat B7

N-Ethyl-N-hydroxyamphetamine
Formed from: N-Ethylamphetamine

Ethyl hydroxyapovincaminate
Formed from: Ethyl apovincaminate

Ethyl 5-hydroxybenzimidazol-2-ylcarbamate
Probably formed from: Ethyl benzimidazol-2-ylcarbamate

Ethyl *p*-hydroxybenzoate (ethylparaben)
Formed from: Ethyl *p*-(6-guanidinohexanoyloxy)benzoate
→ Ethyl *p*-hydroxybenzoate-β-D-glucuronide
In: Man G524
Rat A2297
→ Ethyl *p*-hydroxybenzoate-O-sulphate
In: Eubacterium E432
→ *p*-Hydroxybenzoate
In: Cat A3387
See also: A2229

Ethyl *p*-hydroxybenzoate-β-D-glucuronide
Formed from: Ethyl *p*-hydroxybenzoate

Ethyl *p*-hydroxybenzoate-O-sulphate
Formed from: Ethyl *p*-hydroxybenzoate

3-Ethyl-2-hydroxybenzoic acid
See: 3-Ethylsalicylic acid

2-Ethyl-4-hydroxybenzotriazole
Formed from: 2-Ethylbenzotriazole

2-Ethyl-5-hydroxybenzotriazole
Formed from: 2-Ethylbenzotriazole

5-Ethyl-3-(2-(6-hydroxybenzoxazol-2-yl)ethyl)-6-methylpyridin-2(1*H*)-one
Formed from: 3-(2-(Benzoxazol-2-yl)ethyl)-5-ethyl-6-methylpyridin-2(1*H*)-one
→ 5-Ethyl-3-(2-(6-hydroxybenzoxazol-2-yl)ethyl)-6-methylpyridin-2(1*H*)-one-O-sulphate
In: Rat G505

5-Ethyl-3-(2-(7-hydroxybenzoxazol-2-yl)ethyl)-6-methylpyridin-2(1*H*)-one
Formed from: 3-(2-(Benzoxazol-2-yl)ethyl)-5-ethyl-6-methylpyridin-2(1*H*)-one

5-Ethyl-3-(2-(6-hydroxybenzoxazol-2-yl)ethyl)-6-methylpyridin-2(1*H*)-one-O-sulphate
Formed from: 5-Ethyl-3-(2-(6-hydroxybenzoxazol-2-yl)ethyl)-6-methylpyridin-2(1*H*)-one

2-Ethyl-3-(*p*-hydroxybenzoyl)benzofuran
See: Benzarone

7-Ethyl-10-hydroxycamptothecin (11-ethyl-9-hydroxycamptothecin)
Formed from: 7-Ethyl-10-hydroxycamptothecin-β-D-glucuronide
Irinotecan
→ 7-Ethyl-10-hydroxycamptothecin-β-D-glucuronide
In: Man K180, K532

7-Ethyl-10-hydroxycamptothecin-β-D-glucuronide
Formed from: 7-Ethyl-10-hydroxycamptothecin
→ 7-Ethyl-10-hydroxycamptothecin
In: Escherichia G534

Ethyl *m*-hydroxycarbanilate carbanilate
→ Ethyl N-(*m*-hydroxyphenyl)carbamate
In: Sugar beet A633

5″-Ethyl-6-hydroxy-1″,1″-dimethylcannabidiol
Formed from: 5″-Ethyl-1″,1″-dimethylcannabidiol

5″-Ethyl-7-hydroxy-1″,1″-dimethylcannabidiol
Formed from: 5″-Ethyl-1″,1″-dimethylcannabidiol

1-Ethyl-4-(2-(2-hydroxyethylamino)ethyl)-3,3-diphenylpyrrolidin-2-one
Probably formed from: 1-Ethyl-4-(N,N-di-(2-hydroxyethyl)aminoethyl)-3,3-diphenylpyrrolidin-2-one
3-Oxodoxapram

Ethyl (10-(N-(2-hydroxyethyl)-3-aminopropionyl)phenothiazin-2-yl)carbamate
Formed from: Moricizine
→ Ethyl (10-(3-aminopropionyl)phenothiazin-2-yl)carbamate
Probably in: Man J747

2-Ethyl-5-hydroxy-3-(*p*-hydroxybenzoyl) benzofuran
Probably formed from: Benzarone

8-Ethyl-5-hydroxy-6-(1-hydroxyethyl)-4-oxo-4*H*-1-benzopyran-2-carboxylic acid
Formed from: 6,8-Diethyl-5-hydroxy-4-oxo-4*H*-1-benzopyran-2-carboxylic acid

8-Ethyl-5-hydroxy-6-(2-hydroxyethyl)-4-oxo-4*H*-1-benzopyran-2-carboxylic acid
Formed from: 6,8-Diethyl-5-hydroxy-4-oxo-4*H*-1-benzopyran-2-carboxylic acid

2-Ethyl-4-hydroxy-2-(*p*-hydroxyphenyl) glutarimide
Probably formed from: 4-Hydroxyglutethimide

3-Ethyl-5-hydroxy-5-(*p*-hydroxyphenyl)hydantoin
Probably formed from: 3-Ethyl-5-hydroxy-5-phenylhydantoin

2-Ethyl-4-hydroxy-2-(*p*-hydroxyphenyl) glutarimide
Probably formed from: 4-Hydroxy-glutethimide

α-Ethyl-3-hydroxy-N-isopropyl-4-methoxyphenylethanolamine
Formed from: α-Ethyl-N-isopropylnoradrenaline

Ethyl *p*-hydroxymandelate
→ *p*-Hydroxymandelic acid
In: Pig K911

N-Ethyl-4-hydroxy-3-methoxyamphetamine
Probably formed from: N-Ethyl-3,4-dihydroxyamphetamine
N-Ethyl-*p*-hydroxyamphetamine

5-Ethyl-5-(4-hydroxy-3-methoxyphenyl)barbituric acid
Probably formed from: 1,5-Diethyl-5-phenylbarbituric acid
5-(3,4-Dihydroxyphenyl)-5-ethylbarbituric acid
5-Ethyl-5-(4-hydroxy-3-methoxyphenyl)-1-methylbarbituric acid

Ethyl 4-(4-hydroxy-3-methoxyphenyl)-6,7-dimethoxy-2-(1,2,4-triazol-1-ylmethyl)quinoline-3-carboxylate
Formed from: Ethyl 4-(3,4-dimethoxyphenyl)-6,7-dimethoxy-2-(1,2,4-triazol-1-ylmethyl) quinoline-3-carboxylate

5-Ethyl-5-(4-hydroxy-3-methoxyphenyl)-1,3-dimethylbarbituric acid
Formed from: 5-Ethyl-1,3-dimethyl-5-phenyl-barbituric acid
→ 5-Ethyl-5-(4-hydroxy-3-methoxyphenyl)-1-methylbarbituric acid
Probably in: Man E474

α-Ethyl-3-hydroxy-4-methoxyphenylethanolamine
Formed from: α-Ethylnoradrenaline

N-Ethyl-3-hydroxy-4-methoxyphenylethanolamine
Formed from: N-Ethylnoradrenaline

2-Ethyl-2-(4-hydroxy-3-methoxyphenyl) glutarimide
Probably formed from: 2-(3,4-Dihydroxyphenyl)-2-ethylglutarimide

3-Ethyl-5-(4-hydroxy-3-methoxyphenyl)hydantoin
Probably formed from: 5-(3,4-Dihydroxyphenyl)-3-ethylhydantoin

5-Ethyl-5-(4-hydroxy-3-methoxyphenyl)-1-methylbarbituric acid
Probably formed from: 5-Ethyl-5-(4-hydroxy-3-methoxyphenyl)-1,3-dimethylbarbituric acid
5-Ethyl-5-(*p*-hydroxyphenyl)-1-methylbarbituric acid
→ 5-Ethyl-5-(4-hydroxy-3-methoxyphenyl)barbituric acid
Probably in: Man E474

4-(N-Ethyl-N-(2-(4-hydroxy-3-methoxyphenyl)-1-methylethyl)amino)butanol
Probably formed from: 4-(N-Ethyl-N-(2-(3,4-dihydroxyphenyl)-1-methylethyl)amino) butanol

5-Ethyl-5-(4-hydroxy-3-methoxyphenyl)-3-methylhydantoin
Probably formed from: 5-Ethyl-5-(*p*-hydroxyphenyl)-3-methylhydantoin

4-Ethyl-7-hydroxy-6-methoxyquinazoline
Formed from: Quazodine
→ 4-Ethyl-6-methoxyquinazolin-7-yl-β-D-glucuronide
In: Dog A3212
Man A3212
Rat A3212
→ 4-Ethyl-6-methoxyquinazolin-7-yl sulphate
In: Dog A3212
Man A3212
Rat A3212

N-Ethyl-*m*-hydroxymethylbenzamide
Formed from: N,N-Diethyl-*m*-hydroxymethylbenzamide
N-Ethyl-*m*-formylmethylbenzamide
N-Ethyl-*m*-toluamide
→ N-Ethyl-*m*-formylbenzamide
In: Rat J572
→ *m*-Hydroxymethylbenzamide
In: Rat J572

2-Ethyl-N-(*p*-hydroxymethylbenzoyl)anilide
Formed from: 2-Ethyl-N-(*p*-toluoyl)anilide

5-Ethyl-5-(*p*-hydroxyphenyl)barbituric acid
Formed from: Phenobarbital
→ 5-(3,4-Dihydroxyphenyl)-5-ethylbarbituric acid
Probably in: Guinea pig A470
Man A470, B642, E474
Rat A470

→ Phenobarbital-4′-yl-β-D-glucuronide
In: Man A1706
Rat A1959
See also: A2184
→ Phenobarbital-4′-yl sulphate
In: Rat A1959

Ethyl N-(*m*-hydroxyphenylcarbamate)
Formed from: Ethyl *m*-hydroxycarbanilate carbanilate
→ *m*-Aminophenol
In: Sugar beet A633

α-Ethyl-1-hydroxy-4-phenylcyclohexaneacetic acid
See: Fencibuterol

Ethyl *m*-hydroxyphenyl dimethylammonium (edrophonium)
→ Ethyl *m*-hydroxyphenyl dimethylammonium-3-β-D-glucuronide
In: Rat A426

Ethyl *m*-hydroxyphenyl dimethylammonium-3-β-D-glucuronide
Formed from: Ethyl *m*-hydroxyphenyl dimethylammonium

2-Ethyl-2-(*p*-hydroxyphenyl)glutarimide
Formed from: Glutethimide
→ 2-(3,4-Dihydroxyphenyl)-2-ethylglutarimide
Probably in: Guinea pig A1021
Man A2200, A2313, A3991
Rat A1021

3-Ethyl-5-(*o*-hydroxyphenyl)hydantoin
Formed from: 3-Ethyl-5-phenylhydantoin

3-Ethyl-5-(*m*-hydroxyphenyl)hydantoin
Formed from: 3-Ethyl-5-phenylhydantoin

3-Ethyl-5-(*p*-hydroxyphenyl)hydantoin
Formed from: 3-Ethyl-5-phenylhydantoin

3-Ethyl-5-hydroxy-5-phenylhydantoin
Formed from: 3-Ethyl-5-phenylhydantoin
→ 3-Ethyl-5-hydroxy-5-(*p*-hydroxyphenyl)hydantoin
Probably in: Man B113

3-(2-Ethyl-3-hydroxyphenyl)-5-(4-hydroxy-3-methylphenyl)-1*H*-1,2,4-triazole
Probably formed from: 3-(2-Ethylphenyl)-5-(4-hydroxy-3-methylphenyl)-1*H*-1,2,4-triazole
→ 3-(2-Ethyl-3-hydroxyphenyl)-5-(4-hydroxy-3-methylphenyl)-1*H*-1,2,4-triazole-β-D-glucuronide
Probably in: Rat C871

3-(2-Ethyl-4-hydroxyphenyl)-5-(4-hydroxy-3-methylphenyl)-1*H*-1,2,4-triazole
Probably formed from: 3-(2-Ethylphenyl)-5-(4-hydroxy-3-methylphenyl)-1*H*-1,2,4-triazole

3-(2-Ethyl-3-hydroxyphenyl)-5-(4-hydroxy-3-methylphenyl)-1*H*-1,2,4-triazole-β-D-glucuronide
Probably formed from: 3-(2-Ethyl-3-hydroxyphenyl)-5-(4-hydroxy-3-methylphenyl)-1*H*-1,2,4-triazole

5-Ethyl-5-(*p*-hydroxyphenyl)-1-methylbarbituric acid
Formed from: Mephobarbital
→ 5-Ethyl-5-(4-hydroxy-3-methoxyphenyl)-1-methylbarbituric acid
Probably in: Man B642, E474

4-(N-Ethyl-N-(2-(*p*-hydroxyphenyl)-1-methylethyl)amino)butanol
Probably formed from: 4-(N-Ethyl-N-(2-(*p*-methoxyphenyl)-1-methylethyl)amino)butanol
→ 4-(N-Ethyl-N-(2-(3,4-dihydroxyphenyl)-1-methylethyl)amino)butanol
Probably in: Man K21
Rat K21

5-Ethyl-5-(*m*-hydroxyphenyl)-3-methylhydantoin
Probably formed from: 5-Ethyl-3-methylhydantoin

5-Ethyl-5-(*p*-hydroxyphenyl)-3-methylhydantoin
Probably formed from: 5-Ethyl-3-methyl-5-phenylhydantoin
→ 5-Ethyl-5-(4-hydroxy-3-methoxyphenyl)-3-methylhydantoin
Probably in: Man A3675
→ 5-Ethyl-5-(*p*-hydroxyphenyl)-3-methylhydantoin-β-D-glucuronide
Probably in: Man A3675, A3929

5-Ethyl-5-(*p*-hydroxyphenyl)-3-methylhydantoin-β-D-glucuronide
Probably formed from: 5-Ethyl-5-(*p*-hydroxyphenyl)-3-methylhydantoin

3-Ethyl-2-(4-hydroxyphenyl)-1-methyl-1*H*-inden-6-ol
See: Indenoestrol A

2-Ethyl-3-(*p*-hydroxyphenyl)-5-methyl-3-phenylpyrroline
Formed from: 2-Ethyl-5-methyl-3,3-diphenylpyrroline
→ 2-Ethyl-3-(*p*-hydroxyphenyl)-5-methyl-3-phenylpyrroline-β-D-glucuronide
In: Rat A1718

2-Ethyl-3-(*p*-hydroxyphenyl)-5-methyl-3-phenylpyrroline-β-D-glucuronide
Formed from: 2-Ethyl-3-(*p*-hydroxyphenyl)-5-methyl-3-phenylpyrroline

Ethyl *m*-hydroxyphenylphosphonic acid
Probably formed from: Ethyl phenylphosphonic acid

Ethyl *p*-hydroxyphenylphosphonic acid
Probably formed from: Ethyl phenylphosphonic acid

Ethyl 2-(*p*-hydroxyphenyl)propionate
→ Ethyl 2-(*p*-hydroxyphenyl)propionate-β-D-glucuronide
In: Panax F255

Ethyl 2-(*p*-hydroxyphenyl)propionate-β-D-glucuronide
Formed from: Ethyl 2-(*p*-hydroxyphenyl) propionate

3-(5-Ethyl-5-(*p*-hydroxyphenyl)-(1*H*,3*H*,5*H*)-pyrimidine-2,4,6-trion-1-yl)-2-carbamoyloxypropanol
Probably formed from: 3-(5-Ethyl-5-phenyl-(1*H*,3*H*,5*H*)-pyrimidine-2,4,6-trion-1-yl)-2-carbamoyloxypropanol
→ 1-(2-Carbamoyloxy-3-hydroxypropyl)-5-ethyl-5-(4-hydroxy-3-methoxyphenyl)-(1*H*,3*H*,5*H*)-pyrimidine-2,4,6-trione
Probably in: Man D26

3-(5-Ethyl-5-(*p*-hydroxyphenyl)-(1*H*,3*H*,5*H*)-pyrimidine-2,4,6-trion-1-yl)-2-carbamoyloxypropoxybutyric acid
Probably formed from: 3-(5-Ethyl-5-phenyl-(1*H*,3*H*,5*H*)-pyrimidine-2,4,6-trion-1-yl)-2-carbamoyloxypropoxybutyric acid

5-Ethyl-2-hydroxyquinol
Formed from: 4-Ethylresorcinol

α-Ethyl-4-hydroxy-3′-trifluoromethylbenzhydrol
Formed from: α-Ethyl-3-trifluoromethylbenzhydrol

1,1′-Ethylidenebis(tryptophan)
→ Tetrahydro-1-methyl-β-carboline-3-carboxylic acid
In: Rat H26

2-Ethylidene-dihydroxy-1,5-dimethyl-3,3-diphenyl-1-pyrrolidine
Probably formed from: 2-Ethylidene-1,5-dimethyl-3,3-diphenyl-1-pyrrolidine

2-Ethylidene-hydroxy-1,5-dimethyl-3,3-diphenyl-1-pyrrolidine
Probably formed from: 2-Ethylidene-1,5-dimethyl-3,3-diphenyl-1-pyrrolidine

2-Ethylidene-1,5-dimethyl-3,3-diphenyl-1-pyrrolidine
Formed from: 2-Dimethylamino-4,4-diphenyl-5-heptanone
2-Ethyl-5-methyl-3,3-diphenyl-1-pyrroline
→ 2-Ethylidenedihydroxy-1,5-dimethyl-3,3-diphenyl-1-pyrrolidine
Probably in: Rat A1912
→ 2-Ethylidenehydroxy-1,5-dimethyl-3,3-diphenyl-1-pyrrolidine
Probably in: Rat A1912
→ 2-Ethylidene-3-(hydroxyphenyl)-1,5-dimethyl-3-phenyl-1-pyrrolidine
Probably in: Man A1000
→ 2-Ethyl-5-methyl-3,3-diphenyl-1-pyrroline
In: Rat A1718
See also: A760, A1000, H859

2-Ethylidene-3-(hydroxyphenyl)-1,5-dimethyl-3-phenyl-1-pyrrolidine
Probably formed from: 2-Ethylidene-1,5-dimethyl-3,3-diphenyl-1-pyrrolidine

Ethyl indole-3-propionylglycinate
→ 3-(*o*-aminobenzoyl)propionic acid
In: Wheat K823

N-Ethylisophthalamidic acid
Formed from: N,N-Diethylisophthalamidic acid
N-Ethyl-*m*-formylbenzamide
→ Isophthalamidic acid
In: Rat J572

O-Ethyl O-(2-isopropoxycarbonylphenyl) N-isopropylphosphoramidate (isofenphos oxon)
Formed from: Isofenphos

O-Ethyl O-(2-isopropoxycarbonylphenyl) phosphate
Formed from: Isofenphos

O-Ethyl O-(2-isopropoxycarbonylphenyl) phosphorothioate
Formed from: Isofenphos

7-Ethyl-17-O-isopropylcamptothecin-21-(2-dimethylamino)ethylamide
→ 7-Ethylcamptothecin-21-(2-dimethylamino)ethylamide
In: Rat H632

α-Ethyl-N-isopropylnoradrenaline
→ α-Ethyl-3-hydroxy-N-isopropyl-4-methoxyphenylethanolamine
In: Rat A486
→ α-Ethyl-N-isopropylnormetanephrine
In: Rat A486

α-Ethyl-N-isopropylnormetanephrine
Formed from: α-Ethyl-N-isopropylnoradrenaline

Ethyl isovanillate
Formed from: Ethyl protocatechuate

Ethyl N-leucyl-*p*-aminobenzoate
→ Ethyl *p*-aminobenzoate
In: Rat C305

Ethyl loflazepate
→ 2-Amino-5-chloro-2′-fluorobenzophenone
In: Baboon D752
Dog D752
Man D752
→ 6-Chloro-(*o*-fluorophenyl)-2(1*H*)-quinazolinone
In: Dog D752
Man D752
→ Loflazepate
In: Baboon D752
Dog D752
Man D752
Rat D752

Ethyl mandelate
Formed from: Phenothioate

N-Ethyl-*p*-methoxyamphetamine
Formed from: Mebeverine
→ N-Ethyl-*p*-hydroxyamphetamine
In: Man J727
See also: K21
→ *p*-Methoxyamphetamine
Probably in: Man K21
Rat K21

4-Ethyl-2-methoxyphenol
Formed from: 4-Ethylcatechol

5-Ethyl-2-methoxyphenol
Formed from: 4-Ethylcatechol
Ferulic acid

7-Ethyl-17-O-(*p*-methoxyphenyl)camptothecin-21-(2-dimethylamino)ethylamide
→ 7-Ethylcamptothecin-21-(2-dimethylamino)ethylamide
In: Rat H632

4-(N-Ethyl-N-(2-(*p*-methoxyphenyl)-1-methylethyl)amino)butanol
Formed from: Mebeverine
→ 4-(N-Ethyl-N-(2-(*p*-hydroxyphenyl)-1-methylethyl)amino)butanol
Probably in: Man K21
Rat K21

4-(N-Ethyl-N-(2-(*p*-methoxyphenyl)-1-methylethyl)amino)butyl protocatechuate
Probably formed from: 4-(N-Ethyl-N-(2-(*p*-methoxyphenyl)-1-methylethyl)amino)butyl vanillate

4-(N-Ethyl-N-(2-(*p*-methoxyphenyl)-1-methylethyl)amino)butyl vanillate
Probably formed from: Mebeverine
→ 4-(N-Ethyl-N-(2-(*p*-methoxyphenyl)-1-methylethyl)amino)butyl protocatechuate
Probably in: Man K21
Rat K21

2-Ethyl-N-(1-methoxy-2-propyl)-6-methylaniline
Formed from: Metolachlor

4-Ethyl-6-methoxyquinazolin-7-yl-β-D-glucuronide
Formed from: 4-Ethyl-7-hydroxy-6-methoxyquinazoline

4-Ethyl-6-methoxyquinazolin-7-yl sulphate
Formed from: 4-Ethyl-7-hydroxy-6-methoxyquinazoline

N-Ethyl-β-methoxy-β-(*m*-trifluoromethylphenyl) ethylamine
→ N-Ethyl-β-methoxy-β-(*m*-trifluoromethylphenyl)ethylhydroxylamine
In: Rabbit A2352
→ 2-Methoxy-2-(*m*-trifluoromethylphenyl) ethanol
In: Rabbit A2352

→ β-Methoxy-β-(*m*-trifluoromethylphenyl) ethylamine
In: Rabbit A2352

N-Ethyl-β-methoxy-β-(*m*-trifluoromethylphenyl) ethylhydroxylamine
Formed from: N-Ethyl-β-methoxy-β-(*m*-trifluoromethylphenyl)ethylamine

2-Ethyl-2-(methylamino)-1*H*-indene-1,3(2*H*)-dione
See: Methindione

N-Ethyl-N-methylamphetamine
→ N-Ethylamphetamine
In: Rat A300
→ N-Methylamphetamine
In: Rat A300

2-Ethyl-6-methylaniline
→ 2-Ethyl-6-methylnitrosobenzene
In: Rat D943

***p*-Ethyl-N-methylaniline**
Formed from: *p*-Ethyl-N,N-dimethylaniline

N-Ethyl-N-methylaniline
Formed from: N-Ethyl-N-methylaniline-N-oxide
→ N-Ethylaniline
In: Dog H848
Guinea pig C569, H848
Hamster C569
Mouse C569, H848
Rabbit A310, A3484, C569, H848
Rat C569, H848
See also: B181
→ *p*-(N-Ethyl-N-methyl)aminophenol
In: Rabbit B181
→ N-Ethyl-N-methylaniline-N-oxide
In: Dog H848
Guinea pig C569, H848
Hamster C569
Mouse C569, H848
Pig H317
Rabbit A307, A310, A3484, C569, H848
Rat C569, H848
→ N-Methylaniline
In: Dog H848
Guinea pig C569, H848
Hamster C569
Mouse C569, H848
Rabbit A310, A3484, C569, H848
Rat C569, H848
See also: B181

N-Ethyl-N-methylaniline-N-oxide
Formed from: N-Ethyl-N-methylaniline
→ N-Ethylaniline
Probably in: Rabbit A307, A310
→ N-Ethyl-N-methylaniline
In: Rabbit A310
→ N-Methylaniline
Probably in: Rabbit A307, A310

7-Ethyl-17-O-methylcamptothecin-21-(2-dimethylamino)ethylamide
→ 7-Ethylcamptothecin-21-(2-dimethylamino)ethylamide
In: Rat H632

2-Ethyl-5-methyl-3,3-diphenylpyrroline
Formed from: 2-Ethylidene-1,5-dimethyl-3,3-diphenyl-1-pyrrolidine
Probably formed from: 2-Methylamino-4,4-diphenyl-5-heptanol
→ 2-Ethylidene-1,5-dimethyl-3,3-diphenylpyrrolidine
In: Rat A1718
→ 2-Ethyl-5-methyl-3-(*p*-hydroxyphenyl)-3-phenylpyrroline
In: Rat A1718
See also: A1000, A2979

N-Ethyl-3,4-methylenedioxyamphetamine (eve)
→ N-Ethyl-3,4-dihydroxyamphetamine
In: Man H736, K475
→ 3,4-Methylenedioxyamphetamine
In: Man K475
Rat H671
→ 3,4-Methylenedioxyphenylacetone
In: Rat H671

N-Ethyl-N′-(6-methylergolin-8-yl)sulphamide
Probably formed from: N-Ethyl-N′-(1-ethyl-6-methylergolin-8-yl)sulphamide

Ethyl N-γ-methylglutamyl-*p*-aminobenzoate
→ Ethyl *p*-aminobenzoate
In: Rat C305

4′-Ethyl-6-methyllaudanosoline
→ Morphine
In: Papaver C443

6-Ethyl-4′-methyllaudanosoline
→ Morphine

In: Papaver C443

Ethyl *trans*-β-methyl-*p*-(1-methyl-2-propynyloxy)cinnamate
→ *trans*-β-Methyl-*p*-(1-methyl-2-propynyloxy)cinnamate
In: Baboon C409
Dog C409
Man C409
Marmoset C409
Mouse C409
Rabbit C409
Rat C409

2-Ethyl-6-methylnitrosobenzene
Formed from: 2-Ethyl-6-methylaniline

α-Ethyl-N-methylphenethylamine
→ α-Ethylphenethylamine
In: Rat C595

5-Ethyl-1-methyl-5-phenylbarbituric acid
See: Mephobarbital

5-Ethyl-3-methyl-5-phenylhydantoin
(mephenytoin, 3-methylnirvanol)
→ 5-Ethyl-5-(*m*-hydroxyphenyl)-3-methylhydantoin
In: Dog B51
→ 5-Ethyl-5-(*p*-hydroxyphenyl)-3-methylhydantoin
In: Dog B51
Man A3929, B822, D342, G381, H244, H577, H596, K22
See also: A3675
→ 5-Ethyl-5-phenylhydantoin
In: Dog A3577, B51
Man B822, D342, J42, J436
Mouse A1541
See also: A3675
→ 5-Hydroxyethyl-3-methyl-5-phenylhydantoin
In: Man A3675

N-(2-Ethyl-6-methylphenyl)-N-(2-hydroxy-1-methylethyl)glycollamide
Probably formed from: N-(2-Ethyl-6-methylphenyl)-N-(2-methoxy-1-methylethyl)glycollamide

2-Ethyl-N-methyl-2-phenylmalondiamide
Formed from: 2-Deoxy-N-methylphenobarbital

N-(2-Ethyl-6-methylphenyl)-N-(2-methoxy-1-methylethyl)glycollamide
Formed from: Metolachlor
→ N-(2-N-(2-Ethyl-6-methylphenyl)-N-(2-hydroxy-1-methylethyl)glycollamide
Probably in: Phanerochaete F871
Rhizoctonia F871
Streptomyces F871
Syncephalastrum F871

O-Ethyl O-(4-(methylsulphinyl)phenyl) S-propylphosphorodithioate
Formed from: O-Ethyl O-(4-(methylthio) phenyl) S-propylphosphorodithioate
→ O-Ethyl O-(4-(methylsulphonyl)phenyl) S-propylphosphorodithioate
Probably in: Cow A1818
Rat A1819

O-Ethyl O-(4-(methylsulphinyl)phenyl) S-propylphosphorothioate
Probably formed from: O-Ethyl O-(4-(methylthio)phenyl) S-propylphosphorothioate
→ O-Ethyl O-(4-(methylsulphonyl)phenyl) S-propylphosphorothioate
Probably in: Rat A1819

O-Ethyl O-(4-(methylsulphonyl)phenyl) S-propylphosphorodithioate
Probably formed from: O-Ethyl O-(4-(methyl-sulphinyl)phenyl) S-propylphosphorodithioate
→ O-Ethyl O-(4-(methylsulphonyl)phenyl) S-propylphosphorothioate
Probably in: Cotton A1832

O-Ethyl O-(4-(methylsulphonyl)phenyl) S-propylphosphorothioate
Probably formed from: O-Ethyl O-(4-(methyl-sulphinyl)phenyl) S-propylphosphorothioate
O-Ethyl O-(4-(methylsulphonyl)phenyl) S-propylphosphorodithioate

7-Ethyl-17-O-(methylthio)ethylcamptothecin-21-(2-dimethylamino)ethylamide
→ 7-Ethylcamptothecin-21-(2-dimethylamino)ethylamide
In: Rat H632

O-Ethyl O-(4-(methylthio)phenyl) S-propylphosphorodithioate
→ O-Ethyl O-(4-(methylsulphinyl)phenyl) S-propylphosphorodithioate

In: Cotton A1819
See also: A1818
→ O-Ethyl O-(4-(methylthio)phenyl) S-propylphosphorothioate
In: Rat A1819
→ *p*-(Methylthio)phenol
In: Beef A1818
Rat A1819
See also: A1832

O-Ethyl O-(4-(methylthio)phenyl) S-propylphosphorothioate
Formed from: O-Ethyl O-(4-(methylthio) phenyl) S-propylphosphorodithioate
→ O-Ethyl O-(4-(methylsulphinyl)phenyl) S-propylphosphorothioate
Probably in: Rat A1819

N-Ethyl-N-methyl-*p*-toluidine
→ N-Ethyl-N-methyl-*p*-toluidine-N-oxide
In: Guinea pig C569
Hamster C569
Mouse C569
Rabbit C569
Rat C569
→ N-Ethyl-*p*-toluidine
In: Guinea pig C569
Hamster C569
Mouse C569
Rabbit C569
Rat C569
→ N-Methyl-*p*-toluidine
In: Guinea pig C569
Hamster C569
Mouse C569
Rabbit C569
Rat C569

N-Ethyl-N-methyl-*p*-toluidine-N-oxide
Formed from: N-Ethyl-N-methyl-*p*-toluidine

N-Ethyl-α-methyl-3-trifluoromethylphenethylamine (fenfluramine)
→ α-Methyl-3-trifluoromethylphenethylamine
In: Dog G494
Guinea pig E709
Man F347, G126, G494
Mouse G494
Rat F287, G494
See also: A762, A3123, A3252
→ *m*-Trifluoromethylbenzoic acid
In: Dog G494
Man G126
In: Rat A3123, A3730
→ 1-(*m*-Trifluoromethylphenyl)acetone
Probably in: Dog G494
Man C410, G126
Rat G494

3-Ethylmorphine
Formed from: Oripavine 3-ethyl ether
→ 3-Ethylmorphine-6-β-D-glucuronide
In: Man H595, H652
→ 3-Ethylmorphine-N-oxide
In: Mouse D446
Pig D446
→ Ethylmorphinone
In: Rabbit H920
→ 3-Ethylnormorphine
In: Cat C562
Dog C562
Guinea pig C562
Hamster A1658
Macaque A2117
Man A572, H595
Monkey A1997
Mouse A2022, A3246, C562
Pig A1997
Quail C562
Rabbit A1996, C562
Rat A572, A1859, A1983, A1997, A2131 A2882, B388, C562, C669, D98, E538, G909
Tree shrew A1997
Trout C562
→ Morphine
In: Man H595
Rat A2131, A2882, B388, G909

3-Ethylmorphine-6-β-D-glucuronide
Formed from: 3-Ethylmorphine

3-Ethylmorphine-N-oxide
Formed from: 3-Ethylmorphine

Ethylmorphinone
Formed from: 3-Ethylmorphine

1-Ethyl-4-(2-morpholinoethyl)-3,3-diphenylpyrrolidin-2-one
See: Doxapram

1-Ethylnaphthalene
→ *cis*-8-Ethyl-1,2-dihydro-1,2-dihydroxynaphthalene
In: Pseudomonas J370

2-Ethylnaphthalene
→ *cis*-7-Ethyl-1,2-dihydro-1,2-dihydroxynaphthalene
In: Pseudomonas J370

1-Ethyl-4-nitrobenzene
→ 1-Nitrophenyl)ethanol
In: Mortierella D969

Ethyl *o*-nitrobenzoate
→ Ethyl anthranilate
In: Saccharomyces F201

Ethyl *m*-nitrobenzoate
→ Ethyl *m*-aminobenzoate
In: Saccharomyces F201

Ethyl *p*-nitrobenzoate
→ Ethyl *p*-aminobenzoate
In: Saccharomyces F201

α-Ethyl-α-(*p*-nitrophenyl)glutarimide
Formed from: Aminoglutethimide

Ethyl *p*-nitrophenyl phosphate
Formed from: Diethyl *p*-nitrophenyl phosphate

Ethyl *p*-nitrophenyl phosphorothioate
Formed from: Ethyl *p*-nitrophenyl phenylphosphorothioate
→ *p*-Nitrophenol
In: Housefly B441
Mouse A3996
→ S-(*p*-Nitrophenyl)glutathione
In: Housefly B441
→ *p*-Nitrophenyl phenylphosphoric acid
In: Housefly B441

Ethyl *p*-nitrophenyl phenylphosphorothioate
→ Ethyl *p*-nitrophenyl phenylphosphate
In: Hen B873
Housefly B441
Rat B873
→ O-Ethyl phenylphosphorothioic acid
In: Cat C596
Hen B590
Housefly B441
Mouse A3996
Rat B590
See also: C200
→ S-(Ethyl phenylphosphorothionyl) glutathione
In: Housefly B441
→ *p*-Nitrophenol
In: Hen B590, B873
Housefly B441
Mouse A3996
Rat B590, B873
→ S-(*p*-Nitrophenyl)glutathione
In: Housefly B441

2-Ethyl-7-nitro-1-propyl-5-trifluoromethylbenzimidazole
Formed from: 2,6-Dinitro-N,N-dipropyl-4-trifluoromethylaniline

***p*-Ethylnitrosobenzene**
Formed from: *p*-Ethylaniline

α-Ethylnoradrenaline
→ α-Ethyl-3-hydroxy-4-methoxyphenylethanolamine
In: Rat A486
→ α-Ethylnormetanephrine
In: Rat A486

N-Ethylnoradrenaline
→ N-Ethyl-3-hydroxy-4-methoxyphenylethanolamine
In: Rat A486
→ N-Ethylnormetanephrine
In: Rat A486

6-Ethyl-6-noragroclavine
→ 6-Ethyl-6-norelymoclavine
In: Claviceps D237

6-Ethyl-6-norelymoclavine
Formed from: 6-Ethyl-6-noragroclavine

α-Ethylnormetanephrine
Formed from: α-Ethylnoradrenaline

N-Ethylnormetanephrine
Formed from: N-Ethylnoradrenaline

N-Ethylnormorphine
Formed from: N-Ethylnorreticuline

3-Ethylnormorphine
Formed from: 3-Ethylmorphine
→ 3-Ethylnormorphine-6-β-D-glucuronide
Probably in: Man H595

3-Ethylnormorphine-6-β-D-glucuronide
Probably formed from: 3-Ethylnormorphine

N-Ethylnorreticuline
→ N-Ethylnormorphine

In: Papaver C443

N-((1-Ethyl-5-oxo-2-pyrrolidinyl)methyl)-2-hydroxy-5-sulphamoylbenzamide
Probably formed from: N-((1-Ethyl-5-oxo-2-pyrrolidinyl)methyl)-5-sulphamoyl-2-anisamide

N-((1-Ethyl-5-oxo-2-pyrrolidinyl)methyl)-5-sulphamoyl-2-anisamide
Formed from: Sulpiride
→ N-((1-Ethyl-5-oxo-2-pyrrolidinyl)methyl)-2-hydroxy-5-sulphamoylbenzamide
Probably in: Rat A3564

N-(1-Ethyl-5-oxo-2-pyrrolidylmethyl)-2-hydroxy-5-sulphonylethylbenzamide
Formed from: N-(1-Ethyl-5-oxo-2-pyrrolidylmethyl)-2-methoxy-5-sulphonylethylbenzamide

N-(1-Ethyl-5-oxo-2-pyrrolidylmethyl)-2-methoxy-5-sulphonylethylbenzamide
Formed from: Sultopride
→ N-(1-Ethyl-5-oxo-2-pyrrolidylmethyl)-2-hydroxy-5-sulphonylethylbenzamide
Probably in: Rat D174

Ethylparaben
See: Ethyl *p*-hydroxybenzoate

α-Ethylphenethylamine
Formed from: α-Ethyl-N-methylphenethylamine

Ethylphenidate (α-phenyl-2-piperidineacetic acid ethyl ester)
Probably formed from: α-Phenyl-2-piperidineacetic acid

1-Ethylphenobarbital
See: 1,5-Diethyl-5-phenylbarbituric acid

***o*-Ethylphenol**
Formed from: Ferulic acid
→ 3-Ethylcatechol
In: Pseudomonas E958, J392
→ *o*-Ethylphenyl-β-D-glucuronide
In: Man F903
→ *o*-Ethylphenyl sulphate
In: Man F903, K405

***m*-Ethylphenol**
Formed from: Ferulic acid
→ 4-Ethylcatechol
Probably in: Pseudomonas A2391
→ *m*-Ethylphenyl-β-D-glucuronide
In: Man F903, J490
→ *m*-Ethylphenyl sulphate
In: Man F903, K405

***p*-Ethylphenol**
Formed from: *p*-Coumaric acid
Probably formed from: *p*-Hydroxystyrene
→ *p*-Ethylphenyl-β-D-glucuronide
In: Man F903, G524, J490
→ *p*-Ethylphenyl sulphate
In: Man E575, F903, K405
Rat E575
→ 1-(*p*-Hydroxyphenyl)ethanol (both stereoisomers)
In: Aspergillus H236
Byssochlamys K164
Penicillium J664, J680, J836, K596
Pseudomonas D469, E361, E396, F88, F381, F695, K177
→ *p*-Hydroxystyrene
In: Penicillium J664

Ethyl phenothiazine-2-carbamate sulphoxide
Probably formed from: Ethacizin sulphoxide
Moricizine sulphoxide

1-Ethylphenoxathiin-10,10-dioxide
→ 1-(1-Hydroxyethyl)phenoxathiin-10,10-dioxide
In: Man H513
→ 1-(2-Hydroxyethyl)phenoxathiin-10,10-dioxide
In: Man H513

1-(*p*-Ethylphenoxy)-6,7-dihydroxy-3,7-dimethyl-*trans*-2-octene
Formed from: 6,7-Epoxy-1-(*p*-ethylphenoxy)-3,7-dimethyl-*trans*-2-octene

N-Ethylphenylacetamide
Formed from: N,N-Diethylphenylacetamide
→ Phenylacetamide
In: Guinea pig F273
Mouse F273
Rat F256, F273

β-Ethyl-L-phenylalanine
See: L-2-Amino-3-phenylvaleric acid

5-Ethyl-5-phenylbarbituric acid
See: Phenobarbital

7-Ethyl-17-O-phenylcamptothecin-21-(2-dimethylamino)ethylamide
→ 7-Ethylcamptothecin-21-(2-

dimethylamino)ethylamide
In: Rat H632

N-Ethyl-1-phenylcyclohexylamine
Formed from: N,N-Diethyl-1-phenylcyclohexylamine
→ 1-Phenylcyclohexylamine
In: Rat G529
See also: F892

O-(*o*-Ethylphenyl) O,O-diphenyl phosphate
→ *o*-(α-Hydroxyethyl)phenol
In: Mouse D244
Rat D244

1-(*o*-Ethylphenyl)ethanol
Formed from: 1,2-Diethylbenzene
→ 1-(*o*-Ethylphenyl)ethyl-β-D-glucuronide
In: Rat J793

1-(*p*-Ethylphenyl)ethanol
Formed from: 1,4-Diethylbenzene

2-(*p*-Ethylphenyl)ethanol
Formed from: 1,4-Diethylbenzene

1-(*o*-Ethylphenyl)ethyl-β-D-glucuronide
Formed from: 1-(*o*-Ethylphenyl)ethanol

Ethyl 1-(1-phenylethyl)-1*H*-imidazole-5-carboxylate
See: (+)- and (−)-Etomidate

O-Ethyl S-phenyl ethylphosphonodithioate (dyfonate, fonofos)
→ O-Ethyl S-phenyl ethylphosphonothioate
In: Housefly A3613
Microorganism A1462
Mouse A1672, A3613
Rhizopus A2539
→ 3-Hydroxyphenyl methyl sulphone
In: Housefly A3613
Mouse A3613
Rat A481
→ 4-Hydroxyphenyl methyl sulphone
In: Housefly A3613
Mouse A3613
Rat A481
→ Thioanisole sulphoxide
In: Housefly A3613
Microorganism A1462
Mouse A3613
See also: A2539
→ Thiophenol
In: Rhizopus A2539
See also: A3613

O-Ethyl S-phenyl ethylphosphonothioate (oxon)
Formed from: O-Ethyl S-phenyl ethylphosphonodithioate

***o*-Ethylphenyl-β-D-glucuronide**
Formed from: *o*-Ethylphenol

***m*-Ethylphenyl-β-D-glucuronide**
Formed from: *m*-Ethylphenol

***p*-Ethylphenyl-β-D-glucuronide**
Formed from: *p*-Ethylphenol

2-Ethyl-2-phenylglutaconimide
Formed from: Glutethimide

2-Ethyl-2-phenylglutarimide
See: Glutethimide

3-Ethyl-5-phenylhydantoin (ethotoin)
→ 5-(3,4-Dihydro-3,4-dihydroxyphenyl)-3-ethylhydantoin
Probably in: Man B113
→ 3-Ethyl-5-hydroxy-5-phenylhydantoin
In: Man B113
→ 3-Ethyl-5-(*o*-hydroxyphenyl)hydantoin
In: Man B113
→ 3-Ethyl-5-(*m*-hydroxyphenyl)hydantoin
In: Man B113
→ 3-Ethyl-5-(*p*-hydroxyphenyl)hydantoin
In: Man B113
→ 5-Phenylhydantoin
In: Man B113

5-Ethyl-5-phenylhydantoin (nirvanol; *R* and *S* enantiomers)
Formed from: 5-Ethyl-3-methyl-5-phenylhydantoin
→ 5-(3,4-Dihydro-3,4-dihydroxyphenyl)-5-ethylhydantoin
In: Dog C13
→ 5-Ethyl-5-(*m*-hydroxyphenyl)hydantoin
In: Dog C13
→ 5-Ethyl-5-(*p*-hydroxyphenyl)hydantoin
In: Dog A2474, C13
→ 5-Ethyl-5-phenylhydantoin-β-D-glucuronide
In: Dog C13, C628

5-Ethyl-5-phenylhydantoin-β-D-glucuronide
Probably formed from: 5-Ethyl-5-phenylhydantoin

N-Ethylphenylhydroxylamine
Formed from: N-Ethylaniline
3-(2-Ethylphenyl)-5-(4-hydroxy-3-methoxyphenyl)-1*H*-1,2,4-triazole
Probably formed from: 5-(3,4-Dihydroxyphenyl)-3-(2-ethylphenyl)-1*H*-1,2,4-triazole
→ 3-(2-Ethylphenyl)-5-(4-hydroxy-3-methoxyphenyl)-1*H*-1,2,4-triazole-β-D-glucuronide
Probably in: Rat C871
→ 3-(2-Ethyl-3-hydroxyphenyl)-5-(4-hydroxy-3-methoxyphenyl)-1*H*-1,2,4-triazole
Probably in: Rat C871
→ 3-(2-Ethyl-4-hydroxyphenyl)-5-(4-hydroxy-3-methoxyphenyl)-1*H*-1,2,4-triazole
Probably in: Rat C871
→ 3-(2-(1-Hydroxyethyl)phenyl)-5-(4-hydroxy-3-methoxyphenyl)-1*H*-1,2,4-triazole
Probably in: Rat C871
3-(2-Ethylphenyl)-5-(4-hydroxy-3-methoxyphenyl)-1*H*-1,2,4-triazole-β-D-glucuronide
Probably formed from: 3-(2-Ethylphenyl)-5-(4-hydroxy-3-methoxyphenyl)-1*H*-1,2,4-triazole
3-(2-Ethylphenyl)-5-(*m*-hydroxyphenyl)-1*H*-1,2,4-triazole
Formed from: 3-(2-Ethylphenyl)-5-(*m*-methoxyphenyl)-1*H*-1,2,4-triazole
→ 5-(3,4-Dihydroxyphenyl)-3-(2-ethylphenyl)-1*H*-1,2,4-triazole
Probably in: Rat C871
→ 3-(2-(1-Hydroxyethyl)phenyl)-5-(*m*-hydroxyphenyl)-1*H*-1,2,4-triazole
Probably in: Rat C871
2-Ethyl-2-phenylmalonamide
Probably formed from: 2-Ethyl-2-phenylmalondiamide
2-Ethyl-2-phenylmalondiamide
Formed from: 2-Deoxyphenobarbital
→ 2-Ethyl-2-phenylmalonamide
Probably in: Man A507
3-(2-Ethylphenyl)-5-(*m*-methoxyphenyl)-1*H*-1,2,4-triazole
→ 3-(2-Ethylphenyl)-5-(*m*-hydroxyphenyl)-1*H*-1,2,4-triazole
In: Rat C871
4-(*o*-Ethylphenyl)-1-methylpyridinium
Probably formed from: 4-(*o*-Ethylphenyl)-1,2,3,6-tetrahydro-1-methylpyridine
1-Ethyl-4-phenylpyridinium
Probably formed from: 1-Ethyl-1,2,3,6-tetrahydro-4-phenylpyridine
Ethyl phenylphosphonic acid
Probably formed from: O-Ethyl phenylphosphorothioic acid
→ Ethyl *m*-hydroxyphenylphosphonic acid
Probably in: Hen B590
Rat B590
→ Ethyl *p*-hydroxyphenylphosphonic acid
Probably in: Hen B590
Rat B590
→ Phenylphosphonic acid
Probably in: Hen B590
Rat B590
O-Ethyl S-phenyl phosphorodithioate
Formed from: O-Ethyl S,S-diphenyl phosphorodithioate
O-Ethyl phenylphosphorothioic acid
Formed from: S-Benzyl O-ethyl phenylphosphonothioate
Ethyl *p*-nitrophenyl phenylphosphorothioate
→ Ethyl phenylphosphonic acid
Probably in: Cat C596
Hen B590
Piricularia A431
Rat B590
→ Phenylphosphorothioic acid
Probably in: Hen B590
Rat B590
S-(Ethyl phenylphosphorothionyl)glutathione
Formed from: Ethyl *p*-nitrophenyl phenylphosphorothioate
2-(*p*-Ethylphenyl)propane-1,2-diol
Formed from: 2-(*p*-Ethylphenyl)propylene oxide
2-(*p*-Ethylphenyl)propylene oxide
→ 2-(*p*-Ethylphenyl)propan-1,2-diol
In: Pig H753
3-(5-Ethyl-5-phenyl-(1*H*,3*H*,5*H*)-pyrimidine-2,4,6-trion-yl)-2-carbamoyloxypropanol
Formed from: Febarbamate

→ 1-(2,3-Dihydroxypropyl)-5-ethyl-5-phenyl-(1*H*,3*H*,5*H*)-pyrimidine-2,4,6-trione
Probably in: Man D26
→ 3-(5-Ethyl-5-(*p*-hydroxyphenyl)-(1*H*,3*H*,5*H*)-pyrimidine-2,4,6-trion-yl)-2-carbamoyloxypropanol
Probably in: Man D26
→ 3-(5-Ethyl-5-phenyl-(1*H*,3*H*,5*H*)-pyrimidine-2,4,6-trion-yl)-2-carbamoyloxypropionic acid
Probably in: Man D26

3-(5-Ethyl-5-phenyl-(1*H*,3*H*,5*H*)-pyrimidine-2,4,6-trion-yl)-2-carbamoyloxypropionic acid
Probably formed from: 3-(5-Ethyl-5-phenyl-(1*H*,3*H*,5*H*)-pyrimidine-2,4,6-trion-yl)-2-carbamoyloxypropanol
3-(5-Ethyl-5-phenyl-(1*H*,3*H*,5*H*)-pyrimidine-2,4,6-trion-yl)-2-carbamoyloxypropoxybutyric acid

3-(5-Ethyl-5-phenyl-(1*H*,3*H*,5*H*)-pyrimidine-2,4,6-trion-yl)-2-carbamoyloxypropoxybutyric acid
Probably formed from: 1-(2-Carbamoyloxy-3-(4-hydroxybutoxy)propyl)-5-ethyl-5-phenyl-(1*H*,3*H*,5*H*)-pyrimidine-2,4,6-trione
→ 3-(5-Ethyl-5-(*p*-hydroxyphenyl)-(1*H*,3*H*,5*H*)-pyrimidine-2,4,6-trion-yl)-2-carbamoyloxypropoxybutyric acid
Probably in: Man D26
→ 3-(5-Ethyl-5-phenyl-(1*H*,3*H*,5*H*)-pyrimidine-2,4,6-trion-yl)-2-carbamoyloxypropionic acid
Probably in: Man D26

***o*-Ethylphenyl sulphate**
Formed from: *o*-Ethylphenol

***m*-Ethylphenyl sulphate**
Formed from: *m*-Ethylphenol

***p*-Ethylphenyl sulphate**
Formed from: *p*-Ethylphenol

4-(*o*-Ethylphenyl)-1,2,3,6-tetrahydro-1-methylpyridine
→ 4-(*o*-Ethylphenyl)-1-methylpyridinium
Probably in: Beef F395
Man F395

α-Ethyl-α-phenyl-3-(trifluoromethyl)benzene-methanol
See: Flumecinol

7-Ethyl-17-O-propylcamptothecin-21-(2-dimethylamino)ethylamide
→ 7-Ethylcamptothecin-21-(2-dimethylamino)ethylamide
In: Rat H632

2-(Ethylpropylamino)-N-(3-hydroxy-2,6-dimethylphenyl)butanamide
Formed from: Etidocaine
→ 2-Ethylamino-N-(3-hydroxy-2,6-dimethylphenyl)butanamide
Probably in: Man A3247
→ N-(3-Hydroxy-2,6-dimethylphenyl)-2-propylaminobutanamide
Probably in: Man A3247

2-(Ethylpropylamino)-N-(4-hydroxy-2,6-dimethylphenyl)butanamide
Formed from: Etidocaine
→ 2-Ethylamino-N-(4-hydroxy-2,6-dimethylphenyl)butanamide
Probably in: Man A3247
→ N-(4-Hydroxy-2,6-dimethylphenyl)-2-propylaminobutanamide
Probably in: Man A3247

N-(1-Ethylpropyl)-3,4-dimethyl-2,6-dinitro-N-nitrosoaniline
→ 6-Amino-N-(1-ethylpropyl)-3,4-dimethyl-2-nitro-N-nitrosoaniline
In: Streptomyces B589
→ N-(1-Ethylpropyl)-3-hydroxymethyl-4-methyl-2,6-dinitro-N-nitrosoaniline
Probably in: Streptomyces B589
→ N-(1-Ethylpropyl)-4-hydroxymethyl-3-methyl-2,6-dinitro-N-nitrosoaniline
In: Streptomyces B589

N-(1-Ethylpropyl)-3,4-dimethyl-N,2,6-trinitroaniline
→ N-(1-Ethylpropyl)-3-hydroxymethyl-4-methyl-N,2,6-trinitroaniline
In: Streptomyces B589
→ N-(1-Ethylpropyl)-4-hydroxymethyl-3-methyl-N,2,6-trinitroaniline
In: Streptomyces B589

N-(1-Ethylpropyl)-3-hydroxymethyl-4-methyl-2,6-dinitro-N-nitrosoaniline
Probably formed from: N-(1-Ethylpropyl)-3,4-dimethyl-2,6-dinitro-N-nitrosoaniline

N-(1-Ethylpropyl)-4-hydroxymethyl-3-methyl-2,6-dinitro-N-nitrosoaniline
Formed from: N-(1-Ethylpropyl)-3,4-dimethyl-2,6-dinitro-N-nitrosoaniline
N-(1-Ethylpropyl)-3-hydroxymethyl-4-methyl-N,2,6-trinitroaniline
Formed from: N-(1-Ethylpropyl)-3,4-dimethyl-N,2,6-trinitroaniline
N-(1-Ethylpropyl)-4-hydroxymethyl-3-methyl-N,2,6-trinitroaniline
Formed from: N-(1-Ethylpropyl)-3,4-dimethyl-N,2,6-trinitroaniline
Ethyl protocatechuate
→ Ethyl isovanillate
In: Rat A486
With catechol O-methyltransferase A597
→ Ethyl vanillate
In: Rat A486
With catechol O-methyltransferase A597
→ Protocatechuic acid
In: Bacillus A1525
2-(5-Ethylpyridin-2-yl)benzimidazole
→ 2-(5-(1-Hydroxyethyl)pyridin-2-yl) benzimidazole
In: Rat E716, F620
5-((4-(2-(5-Ethyl-2-pyridinyl)ethoxy) phenyl)methyl)-2,4-thiazolidinedione
See: Pioglitazone
N-((1-Ethyl-2-pyrrolidinyl)methyl)-α-hydroxy-α-phenylbenzeneacetamide
See: Epicainide
N-((1-Ethyl-2-pyrrolidinyl)methyl)-2-hydroxy-5-sulphamoylbenzamide
Formed from: Sulpiride
N-((1-Ethyl-2-pyrrolidinyl)methyl)-2-methoxy-5-sulphobenzamide
Formed from: Sultopride
N-(1-Ethyl-2-pyrrolidinylmethyl)-2-methoxy-5-sulphonylethylbenzamide
See: Sultopride
N-((1-Ethyl-2-pyrrolidinyl)methyl)-5-sulphamoylanisamide
See: Sulpiride
4-Ethylresorcinol
Formed from: Ferulic acid
→ 2-(2-(N-Acetamido)ethyl)-3′-ethyl-4,4′,5,6′-tetrahydroxybiphenyl
In: Locust G346
→ 5-Ethyl-2-hydroxyquinol
In: Pseudomonas A2391
Ethyl salicylate
→ Salicylic acid
In: Guinea pig D144
3-Ethylsalicylic acid (3-ethyl-2-hydroxybenzoic acid)
Formed from: 3-Ethyltoluene
O-Ethylsterigmatocystin
→ Aflatoxin B_1
In: Aspergillus K249
β-Ethylstilbene oxide
→ β-Ethyl-α,β-dihydroxydibenzyl
In: Mouse B648
β-Ethylstyrene oxide
→ (*R,R*)-*threo*-1,2-Dihydroxy-1-phenylbutane
In: Mouse C861
Rabbit J17
Rat E759
Ethyl *p*-sulphamidobenzoate
Formed from: Ethyl *p*-aminobenzoate
Ethylsulphinylbenzene
Formed from: Ethylthiobenzene
→ Ethylsulphonylbenzene
Probably in: Paecilomyces G438
2-Ethylsulphinylmethylphenol
Probably formed from: 2-Ethylsulphinylmethylphenyl N-methylcarbamate
→ 2-Ethylsulphonylmethylphenol
Probably in: Rat A1830
2-Ethylsulphinylmethylphenyl N-methylcarbamate
Formed from: 2-Ethylthiomethylphenyl N-methylcarbamate
→ 2-Ethylsulphinylmethylphenol
Probably in: Rat A1830
→ 2-Ethylsulphonylmethylphenyl N-methylcarbamate
Probably in: Rat A1830
2-Ethylsulphinylnaphthalene
Formed from: 2-Ethylthionaphthalene
***p*-Ethylsulphinyltoluene**
Formed from: *p*-Ethylthiotoluene
3-(3-Ethyl-6-sulphobenzothiazol-2-ylidenehydrazono)-3-oxo-1,4-cyclohexadiene-1-carboxylic acid
Formed from: 2,2′-Azinodi-(3-ethylbenzothiazoline-6-sulphonic acid)

m-Hydroxybenzoic acid

3-(3-Ethyl-6-sulphobenzothiazol-2-ylidenehydrazono)-4-oxo-1,5-cyclohexadiene-1-carboxylic acid
Formed from: 2,2′-Azinodi-(3-ethylbenzothiazoline-6-sulphonic acid)
p-Hydroxybenzoic acid

Ethylsulphonylbenzene
Probably formed from: Ethylsulphinylbenzene

5-Ethylsulphonyl-2-methoxybenzoic acid
Formed from: Sultopride

2-Ethylsulphonylmethylphenol
Probably formed from: 2-Ethylsulphinylmethylphenol

2-Ethylsulphonylmethylphenyl N-methylcarbamate
Probably formed from: 2-Ethylsulphinylmethylphenyl N-methylcarbamate

4-Ethylsyringol
See: 4-Ethyl-2,6-dimethoxyphenol

Ethyl-Δ^1-tetrahydrocannabinol
→ Ethyl-Δ^1-tetrahydro-7-hydroxycannabinol
In: Mouse G210

Ethyl-Δ^6-tetrahydrocannabinol
→ Ethyl-Δ^6-tetrahydrocannabinol-7-oic acid
In: Mouse G210
→ Ethyl-Δ^6-tetrahydro-5α,7-dihydroxycannabinol
In: Mouse G210

Ethyl-Δ^1-tetrahydrocannabinol-7-oic acid
Formed from: Ethyl-Δ^1-tetrahydro-7-hydroxycannabinol

Ethyl-Δ^6-tetrahydrocannabinol-7-oic acid
Formed from: Ethyl-Δ^6-tetrahydrocannabinol

Ethyl-Δ^6-tetrahydro-5α,7-dihydroxycannabinol
Formed from: Ethyl-Δ^6-tetrahydrocannabinol
→ Ethyl-Δ^6-tetrahydro-5α-hydroxycannabinol-7-oic acid
Probably in: Mouse G210

Ethyl-Δ^1-tetrahydro-6α,7-dihydroxycannabinol
Formed from: Ethyl-Δ^1-tetrahydro-7-hydroxycannabinol

2-Ethyl-1,2,3,4-tetrahydro-4,6-dihydroxyisoquinoline
Probably formed from: Etilefrine
→ 2-Ethyl-1,2,3,4-tetrahydro-4,6-dihydroxyisoquinoline sulphate
Probably in: Man B446

2-Ethyl-1,2,3,4-tetrahydro-4,8-dihydroxyisoquinoline
Probably formed from: Etilefrine
→ 2-Ethyl-1,2,3,4-tetrahydro-4,8-dihydroxyisoquinoline sulphate
Probably in: Man B446

2-Ethyl-1,2,3,4-tetrahydro-4,6-dihydroxyisoquinoline sulphate
Probably formed from: 2-Ethyl-1,2,3,4-tetrahydro-4,6-dihydroxyisoquinoline

2-Ethyl-1,2,3,4-tetrahydro-4,8-dihydroxyisoquinoline sulphate
Probably formed from: 2-Ethyl-1,2,3,4-tetrahydro-4,8-dihydroxyisoquinoline

Ethyl-Δ^1-tetrahydro-7-hydroxycannabinol
Formed from: Ethyl-Δ^1-tetrahydrocannabinol
→ Ethyl-Δ^1-tetrahydrocannabinol-7-oic acid
Probably in: Mouse G210
→ Ethyl-Δ^1-tetrahydro-6α,7-dihydroxycannabinol
Probably in: Mouse G210

Ethyl-Δ^6-tetrahydro-5α-hydroxycannabinol-7-oic acid
Probably formed from: Ethyl-Δ^6-tetrahydro-5α,7-dihydroxycannabinol

1-Ethyl-1,3,4,9-tetrahydro-8-(1-hydroxyethyl)pyrano[3,4-*b*]indole-1-acetic acid
Formed from: Etodolac

1-Ethyl-1,2,3,6-tetrahydro-4-phenylpyridine
→ 1-Ethyl-4-phenylpyridinium
Probably in: Beef F395
Man F395

6-Ethyl-3-(1*H*-tetrazol-5-yl)chromone
→ 6-Carboxymethyl-3-(1*H*-tetrazol-5-yl)chromone
In: Rabbit A3486
→ 6-Ethyl-3-(1*H*-tetrazol-5-yl)chromone-β-D-glucuronide
In: Dog A3486
Rabbit A3486
→ 6-(1-Hydroxyethyl)-3-(1*H*-tetrazol-5-yl)chromone
In: Dog A3486
Monkey A3486
Rabbit A3486
Rat A3486

6-Ethyl-3-(1*H*-tetrazol-5-yl)chromone-β-D-glucuronide
Formed from: 6-Ethyl-3-(1*H*-tetrazol-5-yl)chromone

Ethylthiobenzene
→ Ethylsulphinylbenzene
In: Mortierella D759
Pseudomonas H165
See also: G438

2-Ethylthiomethylphenyl N-methylcarbamate
(croneton)
→ 2-Ethylsulphinylmethylphenyl N-methylcarbamate
In: Rat A1830

2-Ethylthionaphthalene
→ 2-Ethylsulphinylnaphthalene
In: Rabbit H484

***p*-Ethylthiotoluene**
→ 4-Ethylsulphinyltoluene
In: Acinetobacter C179
Pig A179
Rabbit H484

N-Ethyl-*m*-toluamide
Formed from: N,N-Diethyl-*m*-toluamide
→ N-Ethyl-*m*-hydroxymethylbenzamide
In: Rat J572
→ *m*-Toluamide
In: Rat J572

3-Ethyltoluene
→ 3-Ethylsalicylic acid
In: Pseudomonas A1190

4-Ethyltoluene
→ 1-(*p*-Methylphenyl)ethanol
In: Mortierella D969
→ 2-(*p*-Methylphenyl)ethanol
In: Mortierella D969

N-Ethyl-*p*-toluidine
Formed from: N-Ethyl-N-methyl-*p*-toluidine

2-Ethyl-N-(*p*-toluoyl)anilide
→ 2-Ethyl-N-(*p*-(hydroxymethyl)benzoyl) anilide
In: Cunninghamella D810

5-Ethyl-5-(*p*-tolyl)barbituric acid
→ 5-(*p*-Carboxyphenyl)-5-ethylbarbituric acid
In: Rat A2402

Ethyl tolyl sulphide
See: *p*-Ethylthiotoluene

α-Ethyl-3-trifluoromethylbenzhydrol
→ α-(1,2-Dihydroxyethyl)-3-trifluoromethylbenzhydrol
In: Rat A3556
→ α-Ethyl-4-hydroxy-3′-trifluoromethylbenzhydrol
In: Rat A3556

N-Ethyl-3-trifluoromethylphenethylamine
See: Norfenfluramine

Ethyl 4-(3,4,5-trimethoxycinnamoyl)-1-piperazinylacetate
→ 4-(3,4,5-Trimethoxycinnamoyl)-1-piperazinylacetic acid
In: Man A664

Ethyl 3,5,6-tris-O-(phenylmethyl)-D-glucofuranoside
See: Tribenoside

Ethyl N-valyl-*p*-aminobenzoate
→ Ethyl *p*-aminobenzoate
In: Rat C305

Ethyl vanillate
Formed from: Ethyl protocatechuate

Ethylvanillin
See: 3-Ethoxy-4-hydroxybenzaldehyde

Ethyl vincaminate
Formed from: Ethyl apovincaminate

4-Ethynylbiphenyl
→ 4-Biphenylylacetic acid
In: Rabbit A3631
Rat A3631, B786

α-(((1-Ethynylcyclohexyl)oxy)methyl)-4-(5-hydroxy-2-methoxyphenyl)-1-piperazineethanol
Formed from: Mociprazine
→ α-(((1-Ethynyl-hydroxycyclohexyl) oxy)methyl)-4-(5-hydroxy-2-methoxyphenyl)-1-piperazineethanol
Probably in: Dog G433

α-(((1-Ethynylcyclohexyl)oxy)methyl)-4-(2-methoxyphenyl)-1-piperazineethanol
See: Mociprazine

α-(((1-Ethynylcyclohexyl)oxy)methyl)-1-piperazineethanol
Formed from: Mociprazine

4′-Ethynyl-2-fluorobiphenyl
→ *p*-(*o*-fluorophenyl)phenylacetic acid
In: Rat A3670, B786

α-(((1-Ethynyl-hydroxycyclohexyl)oxy)methyl)-4-(2-hydroxyphenyl)-1-piperazineethanol
Probably formed from: α-(((1-Ethynyl-hydroxycyclohexyl)oxy)methyl)-4-(2-methoxyphenyl)-1-piperazineethanol

α-(((1-Ethynyl-hydroxycyclohexyl)oxy)methyl)-4-(5-hydroxy-2-methoxyphenyl)-1-piperazineethanol
Probably formed from: α-(((1-Ethynylcyclohexyl)oxy)methyl)-4-(5-hydroxy-2-methoxyphenyl)-1-piperazineethanol

α-(((1-Ethynyl-hydroxycyclohexyl)oxy)methyl)-4-(2-methoxyphenyl)-1-piperazineethanol
→ α-(((1-Ethynyl-hydroxycyclohexyl)oxy)methyl)-4-(2-hydroxyphenyl)-1-piperazineethanol
Probably in: Dog G433

17α-Ethynyl-2-hydroxy-3-O-methyl-17β-oestradiol
Formed from: 17α-Ethynyl-2-hydroxy-17β-oestradiol

17α-Ethynyl-2-hydroxy-17β-oestradiol
Formed from: 17α-Ethynyl-17β-oestradiol
→ 17α-Ethynyl-2-hydroxy-3-O-methyl-17β-oestradiol
In: Man A1575
See also: B566
→ 17α-Ethynyl-2-hydroxy-17β-oestradiol-β-D-glucuronide
Probably in: Rat C378, C671
→ 17α-Ethynyl-2-hydroxy-17β-oestradiol sulphate
Probably in: Rat C378
→ 17α-Ethynyl-2-methoxy-17β-oestradiol
In: Man A1575, C498
See also: A3295, B566, C378, C671

17α-Ethynyl-4-hydroxy-17β-oestradiol
Formed from: 17α-Ethynyl-17β-oestradiol

17α-Ethynyl-6α-hydroxy-17β-oestradiol
Formed from: 17α-Ethynyl-17β-oestradiol
→ 17α-Ethynyl-6-oxo-17β-oestradiol
Probably in: Man A1575

17α-Ethynyl-16α-hydroxy-17β-oestradiol
Formed from: 17α-Ethynyl-17β-oestradiol
→ 17α-Ethynyl-16α-hydroxy-17β-oestradiol-β-D-glucuronide
Probably in: Rat C378
→ 17α-Ethynyl-16α-hydroxy-17β-oestradiol sulphate
Probably in: Rat C378

17α-Ethynyl-16β-hydroxy-17β-oestradiol
Formed from: 17α-Ethynyl-17β-oestradiol

17α-Ethynyl-2-hydroxy-17β-oestradiol-β-D-glucuronide
Probably formed from: 17α-Ethynyl-2-hydroxy-17β-oestradiol

17α-Ethynyl-16α-hydroxy-17β-oestradiol-β-D-glucuronide
Probably formed from: 17α-Ethynyl-16α-hydroxy-17β-oestradiol

17α-Ethynyl-2-hydroxy-17β-oestradiol sulphate
Probably formed from: 17α-Ethynyl-2-hydroxy-17β-oestradiol

17α-Ethynyl-16α-hydroxy-17β-oestradiol sulphate
Probably formed from: 17α-Ethynyl-16α-hydroxy-17β-oestradiol

17α-Ethynyl-2-methoxy-17β-oestradiol
Formed from: 17α-Ethynyl-2-hydroxy-17β-oestradiol
→ 17α-Ethynyl-2-methoxy-17β-oestradiol-β-D-glucuronide
Probably in: Rat C378, C671
→ 17α-Ethynyl-2-methoxy-17β-oestradiol sulphate
Probably in: Rat C378

17α-Ethynyl-2-methoxy-17β-oestradiol-β-D-glucuronide
Probably formed from: 17α-Ethynyl-2-methoxy-17β-oestradiol

17α-Ethynyl-2-methoxy-17β-oestradiol sulphate
Probably formed from: 17α-Ethynyl-2-methoxy-17β-oestradiol

17α-Ethynyl-17β-oestradiol (17α-ethynyl-1,3,5(10)-oestratrien-3,17β-diol)
Formed from: Mestranol
→ 17α-Ethynyl-2-hydroxy-17β-oestradiol
In: Hamster D514
Man A1575, C498
Rat A257, A3295, C378, C671, D514
See also: A1577, B566
→ 17α-Ethynyl-4-hydroxy-17β-oestradiol
In: Hamster D514
Rat D514
→ 17α-Ethynyl-16α-hydroxy-17β-oestradiol

In: Man A1575
→ 17α-Ethynyl-16β-hydroxy-17β-oestradiol
In: Man A1575, B566
See also: C378
→ 17α-Ethynyl-17β-oestradiol-β-D-glucuronide
In: Man C498, J783
Rat C378
→ 17α-Ethynyl-17β-oestradiol-3-sulphate
In: Beef A1211
Rat C378
See also: C498, K344
→ D-Homooestradiol-17aβ
In: Man A1575
→ D-Homooestrone
In: Monkey C578
→ 2-Hydroxymestranol
In: Rat C378
→ 17β-Oestradiol
In: Baboon A3290
Man A1575
Monkey B621

17α-Ethynyl-17β-oestradiol-β-D-glucuronide
Formed from: 17α-Ethynyl-17β-oestradiol

17α-Ethynyl-17β-oestradiol-3-sulphate
Formed from: 17α-Ethynyl-17β-oestradiol

17α-Ethynyl-6-oxo-17β-oestradiol
Probably formed from: 17α-Ethynyl-6α-hydroxy-17β-oestradiol

Etidocaine (N-(2,6-dimethylphenyl)-2-(ethylpropylamino)butanamide, 2-(N-ethyl-N-propylamino)-2′,6′-butyroxylidide)
→ 2,6-Dimethylaniline
In: Man A1744
→ 1-(2,6-Dimethylphenyl)-2,4-diethyl-2-imidazolin-5-one
In: Man A2031
→ N-(2,6-Dimethylphenyl)-2-ethylamino butanamide
In: Man A1744
→ 3-(2,6-Dimethylphenyl)-5-ethyl-2,4-imidazolinedione
In: Man A2031
See also: A1744
→ 1-(2,6-Dimethylphenyl)-4-ethyl-2-methyl-2-imidazolin-5-one
In: Man A2031
→ N-(2,6-Dimethylphenyl)-2-propylaminobutanamide
In: Man A1744
→ 2-(Ethylpropylamino)-N-(3-hydroxy-2,6-dimethylphenyl)butanamide
In: Man A3247
→ 2-(Ethylpropylamino)-N-(4-hydroxy-2,6-dimethylphenyl)butanamide
In: Man A3247

Etilefrine (α-(*m*-hydroxyphenyl-β-ethylaminoethanol)
Formed from: Etilefrine-3-stearate
→ Etilefrine-3-sulphate
In: Man A2429, B446
→ 2-Ethyl-1,2,3,4-tetrahydro-4,6-dihydroxyisoquinoline
Probably in: Man B446
→ 2-Ethyl-1,2,3,4-tetrahydro-4,8-dihydroxyisoquinoline
Probably in: Man B446
→ *m*-Hydroxymandelic acid
In: Man A2429

Etilefrine-3-stearate
→ Etilefrine
In: Man B446

Etilefrine-3-sulphate
Formed from: Etilefrine

Etodilic acid
See: Etodolac

Etodolac (etodilic acid)
→ 1-Ethyl-1,3,4,9-tetrahydro-8-(1-hydroxyethyl)pyrano[3,4-*b*]indole-1-acetic acid
In: Man D662, K637
→ Etodolac-β-D-glucuronic acid
In: Dog C51
Man D662
→ 3-Hydroxyetodolac
Probably in: Dog C51
→ 4-Hydroxyetodolac
Probably in: Dog C51
→ 5-Hydroxyetodolac
In: Man K637
→ 6-Hydroxyetodolac
In: Man D662, K637
→ 7-Hydroxyetodolac
In: Man D662, K637

→ 4-Ureidoetodolac
In: Dog E896
Man E896, K637
Mouse E896
Rat E896

Etodolac-β-D-glucuronic acid
Formed from: Etodolac

Etofenamate (2-((3-(trifluoromethyl)phenyl) amino)benzoic acid 2-(2-hydroxyethoxy)ethyl ester)
→ Flufenamic acid
In: Rat C445
→ 4′-Hydroxyflufenamic acid
Probably in: Rat C445
→ 5-Hydroxyflufenamic acid
Probably in: Rat C445
→ β-Hydroxyethylflufenamate
Probably in: Dog C238

Etofesalamide (N-(*p*-ethoxyphenyl)-2-hydroxybenzamide)
→ Etofesalamide-β-D-glucuronide
In: Rabbit K478
→ Oxaphenamide
In: Rabbit K478

Etofesalamide-β-D-glucuronide
Formed from: Etofesalamide

(+)-Etomidate (ethyl 1-(1-phenylethyl)-1*H*-imidazole-5-carboxylate)
→ Benzaldehyde
Probably in: Rat A2633
→ (+)-1-(1-Phenylethyl)-1*H*-imidazole-5-carboxylic acid
In: Rat A2633

(−)-Etomidate
→ Acetophenone
In: Rat A2633
→ Benzaldehyde
Probably in: Rat A2633
→ (−)-1-(1-Phenylethyl)-1*H*-imidazole-5-carboxylic acid
In: Rat A2633

Etoperidone (2-(3-(4-(3-chlorophenyl)-1-piperazinyl)propyl)-4,5-diethyl-2,4-dihydro-3*H*-1,2,4-triazol-3-one)
→ 1-(*m*-Chlorophenyl)piperazine
In: Rat D308

Etoposide ((5*R*,5a*R*,8a*R*,9*S*)-9-((4,6-O-((1*R*)-ethylidene-β-D-glucopyranosyl)oxy)-5,8,8a,9-tetrahydro-5-(4-hydroxy-3,5-dimethoxyphenyl)furo[3′,4′:6,7]naphtho[2,3-*d*]-1,3-dioxol-6(5a*H*)-one)
Formed from: Etoposide-4′-sulphate
→ 3′-Demethyletoposide
In: Man J719
Rat E479
→ Etoposide-β-D-glucuronide
In: Rabbit E622
Rat E622, E652
→ Free radical
In: Horseradish E74
Sheep E74

Etoposide-β-D-glucuronide
Formed from: Etoposide

Etoposide-4′-sulphate
→ Etoposide
In: Streptomyces H588, H603, H627

Etorphine ((α*R*,5α,7α)-4,5-epoxy-3-hydroxy-6-methoxy-α,17-dimethyl-α-propyl-6,14-ethenomorphinan-7-methanol)
→ Etorphine-β-D-glucuronide
In: Rat A2350

Etorphine-β-D-glucuronide
Formed from: Etorphine

Etretinate (9-(4-methoxy-2,3,6-trimethylphenyl)-3,7-dimethyl-2,4,6,8-nonatetraenoic acid)
Formed from: Acitretin
Atretinoyl CoA
→ Acitretin
In: Dog F58
Man F118
Rat G858

Eucomin ((3*Z*)-2,3-dihydro-5,7-dihydroxy-3-((4-methoxyphenyl)methylene)-4*H*-1-benzopyran-4-one)
Formed from: 2′,4′-Dihydroxy-4,6′-dimethoxychalcone
L-Phenylalanine
2′,4,4′-Trihydroxy-6′-methoxychalcone

Eugenol (4-allyl-2-methoxyphenol)
Formed from: Cinnamic acid
L-Phenylalanine
Probably formed from: 4-Allylcatechol
→ 4-Allylcatechol

Probably in: Rat B528, D312
→ 4-Allylveratrole
In: Pimpinella J188
→ Coniferaldehyde
In: Pseudomonas D3
→ Coniferyl alcohol
In: Byssochlamys K164
Penicillium H389, K761
Pseudomonas J890
→ Dihydroferulic acid
In: Man F499
→ 2,2′-Dihydroxy-3,3′-dimethoxy-5,5′-di-(2-propenyl)biphenyl
In: Horseradish F213
→ Eugenol-β-D-glucuronide
In: Man G524, G968, J416, J783
Mouse C80
Rat B962, C172, D312
→ Eugenol oxide
In: Man F499
Pseudomonas D3
Rat B528
→ Eugenol-O-sulphate
In: Mouse H959
Rat D312
→ Ferulic acid
In: Microorganism A3068
→ S-Glutathionyleugenol (3 isomers)
In: Rat F774
→ 1-(4-Hydroxy-3-methoxyphenyl)-2-propen-1-ol
In: Man F499
Pseudomonas F381
→ 4-Hydroxy-3-methoxystyrene
In: Fusarium E115
→ Isoeugenol
In: Man F499
→ 2-Methoxy-4-propylphenol
In: Man F499
Rat D312

Eugenol-β-D-glucuronide
Formed from: Eugenol

Eugenol methyl ester
See: 3,4-Dimethoxyallylbenzene

Eugenol oxide
Formed from: Eugenol
→ 2,3-Dihydro-2,3-dihydroxyeugenol
In: Guinea pig G86
Rat G86
See also: B528, D3, F499

Eugenol-O-sulphate
Formed from: Eugenol

Eupalitin-3,4′-disulphate (3,5-dihydroxy-2-(4-hydroxyphenyl)-6,7-dimethoxy-4*H*-1-benzopyran-4-one-3,4′-disulphate)
→ Eupalitin-3-sulphate
In: Helix F102

Eupalitin-3-sulphate
Formed from: Eupalitin-3,4′-disulphate

Eupatolitin-3,4′-disulphate (2-(3,4-dihydroxyphenyl-3,5-dihydroxy-6,7-dimethoxy-4*H*-1-benzopyran-4-one-3,4′-disulphate)
→ Eupatolitin-3-sulphate
In: Helix F102

Eupatolitin-3-sulphate
Formed from: Eupatolitin-3,4′-disulphate

Eutypine (4-hydroxy-3-(3-methyl-3-buten-1-ynyl)benzaldehyde)
→ Eutypinol
In: Mung bean J910

Eutypinol
Formed from: Eutypine

Eve
See: N-Ethyl-3,4-methylenedioxyamphetamine

Evoxine ((+)-1-((4,8-dimethoxyfuro[2,3-*b*]quinolin-7-yl)oxy)-3-methyl-2,3-butanediol)
Formed from: Dictamnine
4-Methoxy-3-prenylquinol-2-one

Ezlopitant ((2*S*,3*S*,4*S*)-2-diphenylmethyl-3-(5-isopropyl-2-methoxybenzylamino)-1-azabicyclo[2.2.1]octane)
→ (2*S*,3*S*,4*S*)-2-Diphenylmethyl-3-(5-(2-hydroxy-2-propyl)-2-methoxybenzylamino)-1-azabicyclo[2.2.1]octane
In: Cynomolgus monkey K48
Dog K48
Guinea pig K48
Man K48
Rat K48

F

FAD
See: Flavine adenine dinucleotide
β-Fagarine
See: Skimmianine
Famprofazone (1,2-dihydro-1-methyl-4-(1-methylethyl)-5-((methyl(1-methyl-2-phenylethyl)amino)methyl)-2-phenyl-3*H*-pyrazol-3-one)
→ N-Methylamphetamine
In: Man G75
Fastac (1*S*,3*S*)-rel-3-(2,2-dichloroethenyl)-2,2-dimethylcyclopropanecarboxylic acid (*R*)-cyano(3-(phenoxyphenyl)methyl ester)
→ *m*-Phenoxybenzonitrile
In: Bacillus G891
Febantel (((2-((methoxyacetyl)amino)-4-(phenylthio)phenyl)carbonimidoyl)bis (carbamic acid) dimethyl ester)
→ Febantel sulphoxide
In: Beef E647, F325
Chicken E647, F325
Horse E647, F325
Pig E647, F325
Rat E647, F325
Sheep E647, F325
Trout E647, F325
→ Febendzole
In: Microorganism E647
Febantel sulphoxide
Formed from: Febantel
Febarbamate (1-(3-butoxy-2-(carbamoyloxy)propyl)-5-ethyl-5-phenyl)-(1*H*,3*H*,5*H*)-pyrimidine-2,4,6-trione)
→ 1-(2-Carbamoyloxy-3-(2-hydroxybutoxy)propyl)-5-ethyl-5-phenyl-(1*H*,3*H*,5*H*)-pyrimidine-2,4,6-trione
In: Man D26
Rat C368
→ 1-(2-Carbamoyloxy-3-(3-hydroxybutoxy)propyl)-5-ethyl-5-phenyl-(1*H*,3*H*,5*H*)-pyrimidine-2,4,6-trione
In: Man D26
Rat C368
→ 1-(2-Carbamoyloxy-3-(4-hydroxybutoxy)propyl)-5-ethyl-5-phenyl-(1*H*,3*H*,5*H*)-pyrimidine-2,4,6-trione
In: Man D26
→ 1-(2-Carbamoyloxy-3-(2-hydroxyethoxy)propyl)-5-ethyl-5-phenyl-(1*H*,3*H*,5*H*)-pyrimidine-2,4,6-trione
In: Rat C368
→ 1-(2-Carbamoyloxy-3-hydroxypropyl)-5-ethyl-5-phenyl-(1*H*,3*H*,5*H*)-pyrimidine-2,4,6-trione
In: Rat C368
→ 3-(5-Ethyl-5-phenyl-(1*H*,3*H*,5*H*)-pyrimidine-2,4,6-trion-4-yl)-2-carbamoyloxypropanol
In: Man D26
→ *p*-Hydroxyfebarbamate
In: Man D26
Rat C368
Febuprol (3-butoxy-1-phenoxy-2-propanol)
→ *p*-Hydroxyfebuprol
In: Mouse A2656
Rat A2656
Felbamate (2-phenyl-1,3-propanediol dicarbamate)
→ 3-Carbamoyl-2-phenylpropionaldehyde
In: Man K357
→ α-Hydroxyfelbamate
In: Dog G43, G94
Man J454
Rabbit G43

Rat G42, G43
→ *p*-Hydroxyfelbamate
In: Dog G43, G94
Man J454
Rabbit G43
Rat G42, G43
→ 2-Phenyl-1,3-propanediol carbamate
In: Dog G43, G94
Man J454
Rabbit G43
Rat G42, G43
See also: G591

Felodipine (4-(2,3-dichlorophenyl)-1,4-dihydro-2,6-dimethylpyridine-3,5-dicarboxylic acid 3-ethyl ester 5-methyl ester)
→ 3-Carboxy-4-(2,3-dichlorophenyl)-1,4-dihydro-2-hydroxymethyl-5-methoxycarbonyl-6-methylpyridine
In: Dog E862
Man E862
Rat E862
→ 3-Carboxy-4-(2,3-dichlorophenyl)-5-ethoxycarbonyl-1,4-dihydro-2-hydroxymethyl-6-methylpyridine
In: Dog E862
Man E862
Rat E862
→ Didehydrofelodipine
In: Dog D5, F901, G49
Man D5, G49
Mouse D5
Rat C882, D5, E846, G49
See also: E862

Femoxetine (3-((*p*-methoxyphenoxy)methyl)-1-methyl-4-phenylpiperidine)
→ Femoxetine-N-oxide
In: Dog B866
Man B866
Monkey B866
Rat B866
→ 3-((*p*-Hydroxyphenoxy)methyl)-1-methyl-4-phenylpiperidine
In: Dog B866
Man B866
Monkey B866
Rat B866
→ 4-(*p*-Hydroxyphenyl)-3-((*p*-methoxyphenoxy)methyl)-1-methylpiperidine
In: Dog B866
Man B866
Monkey B866
Rat B866
→ Norfemoxetine
In: Dog B866
Man B866
Monkey B866
Rat B866

Femoxetine-N-oxide
Formed from: Femoxetine

Fenazox (diphenyldiazene-1-oxide)
→ 2-Hydroxyazobenzene
In: Tomato D729

Fenbendazole (5-(phenylthio)-1*H*-benzimidazol-2-ylcarbamic acid methyl ester)
Formed from: Febantel
Fenbendazole sulphoxide
→ *p*-Hydroxyfenbendazole
In: Beef E637
Catfish E637
Chicken E637
Duck E637
Goat E637
Rabbit E637
Turkey E637
→ Oxfendazole
In: Ascaris A3849
Beef E637, F325
Catfish E637
Chicken E637, F325
Duck E637
Goat E637
Horse F325
Moniezia A3849
Mouse A3849
Pig F325, K506
Rabbit E637, E704
Rat F325
Sheep E637, F325, G163
Trout F325
Turkey E637
See also: E647

Fenbendazole sulphone
Probably formed from: Oxfendazole
Fenbendazole sulphoxide
See: Oxfendazole
Fenbufen (4-(biphenyl-4-yl)-4-oxobutyric acid)
→ 4-Biphenylylacetic acid
In: Man C253
→ 4-(Biphenyl-4-yl)-4-hydroxybutyric acid
In: Man C253
→ Fenbufenoylglycerol
In: Adipocyte H795
→ Fenbufenoyl phosphatidylcholine
In: Adipocyte H795
Fenbufenoylglycerol
Formed from: Fenbufen
Fenbufenoyl phosphatidylcholine
Formed from: Fenbufen
Fencibuterol (α-ethyl-1-hydroxy-4-phenylcyclohexaneacetic acid)
→ *p*-Hydroxyfencibuterol
In: Aspergillus C273
Fenclofenac (2-(2,4-dichlorophenoxy)phenylacetic acid)
→ 2,4-Dichlorophenoxy)-5-hydroxyphenylacetic acid
In: Baboon B529
→ 2-(2,4-Dichlorophenoxy)phenylacetyl-β-D-glucuronic acid
In: Baboon B529
Guinea pig B529
Horse C432
Man B529
Rabbit B529
Rat B529
→ 2-(2,4-Dichlorophenoxy)phenyl-acetyltaurine
In: Dog A1358
Fenclorac (α,3-dichloro-4-cyclohexyl) phenylacetic acid)
→ 3-Chloro-4-cyclohexylphenylglyoxylic acid
In: Dog A2074
Man A2074
Monkey A2074
Rat A2074
Fenclozic acid (2-(*p*-chlorophenyl)thiazol-4-ylacetic acid)
→ 2-(*p*-Chlorophenyl)thiazol-4-ylacetyl-β-D-glucuronic acid
In: Rat B803
→ 2-(*p*-Chlorophenyl)thiazol-4-ylacetyltaurine
In: Rat B803
Fendiline (N-(3,3-diphenylpropyl)-1-phenylethylamine)
→ 3,3-Diphenylpropylamine
In: Man C778
→ N-(3,3-Diphenylpropyl)-1-(*p*-hydroxyphenyl)ethylamine
Probably in: Man C778
→ N-(3-(*p*-Hydroxyphenyl)-3-phenylpropyl)-1-phenylethylamine
Probably in: Man C778
Fendosal (5-(4,5-dihydro-2-phenylbenz[*e*]indol-3-yl)salicylic acid)
→ 7-Hydroxyfendosal
In: Dog B624
Man B624
Mouse B624
Rabbit B624
Rat B624
Rhesus monkey B624
Fenetylline (7-(2-(α-methylphenylethylamino)ethyl)theophylline)
→ Amphetamine
In: Man F303
Rat F303
Fenfluramine
See: N-Ethyl-α-methyl-3-trifluoromethylphenethylamine
(+)-Fenfluramine
See: Dexfenfluramine
Fenhexamid (N-(2,3-dichloro-4-hydroxyphenyl)-1-methylcyclohexanecarboxamide)
→ Fenhexamid glucoside
In: Apple K256
Grape K256
Lettuce K256
Pea K256
Tomato K256
Fenhexamid glucoside
Formed from: Fenhexamid
Fenitrooxon (dimethyl 3-methyl-4-nitrophenyl phosphate)
Formed from: Fenitrothion
→ Demethylfenitrooxon

Probably in: Cipangopaludina E639
Daphnia E340
Oryzias E638
Palaemon E340
Physa E639

Fenitrothion (sumithion, O,O-dimethyl O-(3-methyl-4-nitrophenyl) phosphorothioate)
→ O-(4-Amino-3-methylphenyl) O,O-dimethylphosphorothioate
In: Cipangopaludina E639
Physa E639
→ Demethylfenitrothion
In: Cipangopaludina E639
Daphnia E340
Oryzias E638
Palaemon E340
Physa E639
Trout A2830
→ Fenitrooxon
In: Apple A2170
Cipangopaludina E639
Daphnia E340
Oryzias E638
Palaemon E340
Physa E639
Trout A2830
→ 3-Methyl-4-nitrophenol
In: Apple A2170
Cipangopaludina E639
Oryzias E638
Physa E639
Rat A2609
Trout A2830

Fenobam (N-(3-chlorophenyl)-N′-(4,5-dihydro-1-methyl-4-oxo-1*H*-imidazol-2-yl)urea)
→ N-(3-Chloro-4-hydroxyphenyl)-N′-(4,5-dihydro-1-methyl-4-oxo-1*H*-imidazol-2-yl)urea
In: Rat H350
→ N-(*m*-Chlorophenyl)-N′-(4,5-dihydro-5-hydroxy-1-methyl-4-oxo-1*H*-imidazol-2-yl)urea
In: Rat H350
→ N-(*m*-Chlorophenyl)-N′-(imino(methylamino)methyl)urea
In: Rat H350

Fenoctimine (4-(diphenylmethyl)-1-((octylimino)methyl)piperidine)
→ 4-Diphenylmethyl-1-formylpiperidine
In: Man H62
Rat H228
→ 4-Diphenylmethyl-1-(((7-hydroxy)octylimino)methyl)piperidine
In: Rat H228
→ 4-Diphenylmethyl-1-(((8-hydroxy)octylimino)methyl)piperidine
In: Dog H62
→ 4-Diphenylmethyl-1-methylpiperidine
In: Dog H62
→ 4-Diphenylmethylpiperidine
In: Man H62
Rat H228
→ *p*-Hydroxyfenoctimine
In: Dog H62

Fenofibrate (procetofene, 2-(4-chlorobenzoyl)phenoxy)-2-methylpropanoic acid 1-methylethyl ester)
→ 2-(4-(*p*-Chlorobenzoyl)phenoxy)-2-methylpropionic acid
In: Dog A2612, E920
Guinea pig E920
Man A2612, A2636, F535
Rat A2612, E920
See also: F262
→ 2-(4-(*p*-Chlorophenyl) hydroxymethylphenoxy)-2-methylpropionic acid
In: Dog F262
Guinea pig F262
Man F262
Rat F262
→ Fenofibrate-β-D-glucuronide
In: Man G349

Fenofibrate-β-D-glucuronide
Formed from: Fenofibrate

Fenofibric acid
See: 2-(4′-(*p*-Chlorobenzoyl)phenoxy)-2-methylpropionic acid

Fenoldopam (6-chloro-2,3,4,5-tetrahydro-7,8-dihydroxy-1-(*p*-hydroxyphenyl)-1*H*-3-benzazepine)
→ Fenoldopam-8-O-sulphate

In: Man J383
→ 7-O-Methylfenoldopam
In: Man J383
→ 8-O-Methylfenoldopam
In: Man J383

Fenoldopam-8-O-sulphate
Formed from: Fenoldopam

(*R*)-Fenoprofen (2-(3-phenoxyphenyl)propionic acid; *R*-isomer assumed unless otherwise stated)
→ (*S*)-Fenoprofen
In: Rat H597
Sheep H597
→ (*R*)-Fenoprofen-CoA
In: Rat J756, K15
→ (*R*)-Fenoprofen-β-D-glucuronide
In: Man A726, G348, J416
Monkey H861
Rabbit G348
Rat G348
Sheep G348
→ 4′-Hydroxy-(*R*)-fenoprofen
In: Man A726

(*S*)-Fenoprofen
Formed from: (*R*)-Fenoprofen
(*S*)-Fenoprofen-CoA
→ (*S*)-Fenoprofen-CoA
In: Rat K15

(*R*)-Fenoprofen-CoA
Formed from: (*R*)-Fenoprofen
(*S*)-Fenoprofen-CoA
→ (*R*)-Fenoprofen
In: Rat J756
→ (*S*)-Fenoprofen-CoA
In: Rat J756

(*S*)-Fenoprofen-CoA
Formed from: (*S*)-Fenoprofen
(*R*)-Fenoprofen-CoA
→ (*S*)-Fenoprofen
In: Rat J756
→ (*R*)-Fenoprofen-CoA
In: Rat J756

(*R*)-Fenoprofen-β-D-glucuronide
Formed from: (*R*)-Fenoprofen

Fenoterol (1-(3,5-dihydroxyphenyl)-2-(4-hydroxy-α-methylphenethylamio)ethanol)
→ Fenoterol-*m*-β-D-glucuronide
In: Man D871
Rat G855
→ Fenoterol-*p*′-β-D-glucuronide
In: Man D871
Rat G855
See also: D344
→ Fenoterol-O-sulphate
In: Man G855

Fenoterol-*m*-β-D-glucuronide
Formed from: Fenoterol

Fenoterol-*p*′-β-D-glucuronide
Formed from: Fenoterol

Fenoterol-O-sulphate
Formed from: Fenoterol

Fenoximone (1,3-dihydro-4-methyl-5-(4-(methylthio)benzoyl)-2-*H*-imidazol-2-one)
→ Fenoximone sulphoxide
In: Man D253

Fenoximone sulphoxide
Formed from: Fenoximone

Fenproporex
See: N-(2-Cyanoethyl)amphetamine

Fenpyrithrin (3-(2,2-dichloroethenyl)-2,2-dimethylcyclopropanecarboxylic acid cyano(6-phenoxy-2-pyridinyl)methyl ester)
→ *m*-Phenoxymandelonitrile
In: Bacillus G891

Fentanyl (N-(1-(2-phenethyl-4-piperidinyl)) propionanilide)
→ N-(1-(2-(*p*-Hydroxyphenyl)ethyl)-4-piperidinyl)propionanilide
In: Guinea pig D906
Man C81
Rat C79, C81, C137, D906
→ Hydroxy-1-(2-phenethyl)-4-(N-propionylanilino)piperidine
In: Rat C42
→ N-(1-((2-(2-Hydroxy-2-pheny)lethyl)-4-piperidinyl))propionanilide
In: Guinea pig D906
Rat D906
→ 4-(N-(N-(3-Hydroxypropionyl)anilino))-1-(2-phenethyl)piperidine
In: Rat C42

Man J266
→ Hydroxy-4-(N-propionylanilino)piperidine
In: Rat C42
→ N-(1-(2-Phenethyl-4-piperidinyl)) aniline
In: Man J266
→ N-(1-(2-Phenethyl-4-piperidinyl)) lactanilide
In: Guinea pig D906
Rat D906
→ N-(1-(2-Phenethyl-4-piperidinyl)) malonanilic acid
In: Horse B605, B884
Man B884
→ Phenylacetic acid
In: Man C81
Mouse C53
Rat C79, C81, C137
→ N-(1-(2-Piperidinyl))propionanilide
In: Rat C137
→ 4-(N-Propionylanilino)piperidine
In: Guinea pig D906
Man C81, J41, J266
Mouse C53
Rat C42, C79, C81, D906

Fenthion (O,O-dimethyl O-(3-methyl-4-methylthiophenyl) thiophosphate)
→ O,O-Dimethyl O-(3-methyl-4-methylsulphinylphenyl) thiophosphate
In: Beef A3802
Rhizopus A3571
→ O,O-Dimethyl O-(3-methyl-4-methylthiophenyl) phosphate
In: Beef A3802
Rhizopus A3571

Fentiazac (4-(*p*-chlorophenyl)-2-phenylthiazol-5-ylacetic acid)
→ 4-(*p*-Chlorophenyl)-2-(*p*-hydroxyphenyl)thiazol-5-ylacetic acid
In: Man D162
Rat B521

Fenvalerate (sumicidin, α-cyano-3-phenoxybenzyl 2-(*p*-chlorophenyl)isovalerate)
→ α-Amido-3-phenoxybenzyl 2-(*p*-chlorophenyl)isovalerate
In: Bean B722
→ (α-Carbamoyl-3-phenoxybenzyl 2-(*p*-chlorophenyl)-3-methylbutyrate
In: Cabbage D565
Japanese quail E286
Rat E286
→ 2-(*p*-Chlorophenyl)isovaleric acid
In: Bean B722
Cabbage D565
Chicken C766
Dog D443
Japanese quail E286
Mouse C127
Rat B233, C127, E286, H828
See also: E823, G891
→ 2′-Hydroxyfenvalerate
In: Mouse C127
Rat B233, C127
→ 4′-Hydroxyfenvalerate
In: Dog D443
Japanese quail E286
Mouse C127
Rat B233, C127, E286
→ *m*-Phenoxybenzaldehyde
In: Bacillus G891
Chicken C766
Mouse C127
Rat C127
See also: D443, D565, F632, H828
→ *m*-Phenoxybenzoic acid
In: Rat B233
→ *m*-Phenoxymandelonitrile
In: Bacillus G891

Feprazone (prenazone, 3,5-dioxo-1,2-diphenyl-4-prenylpyrazolidine)
→ Feprazone-β-D-glucuronide
In: Man A3916, B103, B256, F306
→ Feprazone-2′,3′-oxide
Probably in: Man B256
→ *cis*-4-(3-Hydroxymethylbut-2-enyl)-3,5-dioxo-1,2-diphenylpyrazolidine
In: Dog B256
Rat B256
→ *trans*-4-(3-Hydroxymethylbut-2-enyl)-3,5-dioxo-1,2-diphenylpyrazolidine

In: Dog B256
Man A3916, B103, B256
Rat A1739, B256
→ 1-(*p*-Hydroxyphenyl)-3,5-dioxo-2-phenyl-4-prenylpyrazolidine
In: Man F306
Rat B256

Feprazone-β-D-glucuronide
Formed from: Feprazone

Feprazone-2′,3′-oxide
Formed from: Feprazone

Ferulaldehyde
See: Coniferaldehyde

Ferulic acid (4-hydroxy-3-methoxycinnamic acid)
Formed from: O-(α-L-Arabinofuranosyl)(1→3)-O-(2-O-feruloyl-α-L-arabinofuranosyl)(1→5)-L-arabinofuranose
Caffeic acid
Curcumin
3,4-Dimethoxycinnamic acid
Eugenol
Ferulic acid-4-glucoside
O-(5-O-Feruloyl)-L-arabinofuranose
O-(2-O-Feruloyl-α-L-arabinofuranosyl)(1→5-L-arabinofuranose
O-(5-O-Feruloyl-α-L-arabinofuranosyl)(1→3)-O-β-D-xylopyranosyl(1→4)-D-xylopyranose
Feruloylgalactarate
O-(6-O-Feruloyl-β-D-galactopyranosyl)(1→4)-β-D-galactopyranose
Feruloylglucarate
Feruloylglucarolactone
Feruloyl-β-D-glucose
Feruloylshikimate
Isoeugenol
γ-Oryzanol
O-β-D-Xylopyranosyl(1→4)-O-(5-O-feruloyl-α-L-arabinofuranosyl)(1→3)-O-β-D-xylopyranosyl(1→4)-O-D-xylopyranose
Probably formed from: 3-(4-Hydroxy-3-methoxyphenyl)-2-propen-1-ol
→ Benzene
In: Enterobacter E325
→ 2,3-Bis(α,4-Dihydroxy-3-methoxybenzyl)succinic dilactone
In: Sorghum A1641
→ Caffeic acid
In: Acetobacterium F655
Microorganism D737, G432
Pseudomonas D16
Sporomusa G931
See also: D710
→ Catechol
Microorganism E133, G432
→ Cinnamic acid
In: Bacterium B363
→ Coniferaldehyde
In: Mycobacterium G821
See also: H309
→ *o*-Cresol
In: Enterobacter E325
→ *p*-Cresol
In: Microorganism E133
→ Curcumin
In: Curcuma A1147
→ Demethylmesembrenol
In: Sceletium A3136
→ 4′-Demethylpodophyllotoxin
In: Podophyllum D275
→ Dihydroferulic acid
In: Microorganism D894, H838
Mycobacterium G821
Pseudomonas D503, G895, H426
Saccharomyces G895, G911
See also: A1765, A3492, C849
→ 3,4-Dimethoxycinnamic acid
Probably in: Populus A3477
→ Ethylbenzene
In: Enterobacter E325
Microorganism E133
→ 5-Ethyl-2-methoxyphenol
In: Brettanomyces E263, H356
→ *o*-Ethylphenol
In: Enterobacter E325
→ *m*-Ethylphenol
In: Microorganism E133
→ 4-Ethylresorcinol
Microorganism E133
→ Ferulic acid-4-glucoside
In: Forsythia A3365
Nicotiana B442, K556
Rose K809

Tomato B846
→ Ferulic acid-O-sulphate
Probably in: Rat J686
→ Feruloyl CoA
In: Aubergine A2514
Cucumber A2514
Forsythia A848
Glycine A2148, A2514
Leek A2514
Marrow A2514
Pea A2514
Pepper A2514
Petunia A2845
Phaseolus A2514
Polyporus A226
Potato A204
Pseudomonas B298, J650
Tomato A3149
Willow D466
→ Feruloyl-β-D-glucose
In: Chenopodium G255
Quercus K737
Raphanus B440
Sweet potato D70
Tomato B846
See also: E28, G897
→ Free radical
In: Coprinus J346
Coriolus K677
→ 4′-Hydroxy-3′-methoxyacetophenone
Probably in: Pseudomonas G910
→ 4-Hydroxy-3-methoxyphenylacetic acid
In: Microorganism D894
→ 4-Hydroxy-3-methoxyphenylhydracrylic acid
Probably in: Aspergillus K72
Corynebacterium A3068
Escherichia D929
Microorganism H838
Nocardia B445
Pseudomonas G910, H426
Pycnoporus H309
Rat C849
Streptopmyces C818
Trametes A2744
→ 4-Hydroxy-3-methoxystyrene
In: Aspergillus F390
Bacillus F390, G895, H346, J623, J697
Brettanomyces E263, H356
Candida K70, F390
Catharanthus G895, K244
Cladosporium A2675
Corynespora F390
Curvularia F390, G895
Daucus K244
Fusarium E115, F3908
Hansenula F390, G895
Klebsiella J853
Lactobacillus K567
Mycobacterium G895
Nicotiana K244
Nocardia G895
Paecilomyces F199
Penicillium G895
Pestalotia F199
Pseudomonas D503, G895, G911, H364
Rhizopus G895
Rhodotorula G895, K70
Saccharomyces F390, G895, G911, G924
Streptpmyces G895
→ 5-Methoxycaffeic acid
In: Poplar D179
See also: A1930
→ Methoxyquinol
In: Trametes A2744
→ Methyl ferulate
In: Sceletium A3136
→ Phenol
In: Enterobacter E325
Microorganism E133
→ Phenylacetic acid
In: Enterobacter E325
→ Propylbenzene
In: Enterobacter E325
→ Salicyl alcohol
In: Enterobacter E325
→ Toluene
In: Enterobacter E325
Microorganism E133

Ferulic acid-4-glucoside
Formed from: Ferulic acid
Ferulic acid-O-sulphate

→ Ferulic acid
In: Cicer K832
O-(5-O-Feruloyl-L-arabinofuranose
→ Ferulic acid
In: Aspergillus H381
See also: J638
O-(2-O-Feruloyl-α-L-arabinofuranosyl)(1→5)-L-arabinofuranose
→ Ferulic acid
In: Aspergillus H381
O-(5-O-Feruloyl-α-L-arabinofuranosyl)(1→3)-O-β-D-xylopyranosyl(1→4)-D-xylopyranose
→ Ferulic acid
In: Aspergillus H381, H751
Neocallimastix G759
Penicillium J499
Streptomyces G999
Feruloylcholine
Formed from: Feruloyl-β-D-glucose
Feruloyl CoA
Formed from: Coniferaldehyde
Caffeoyl CoA
Ferulic acid
→ Coniferaldehyde
In: Eucalyptus H330
Forsythia A848
Glycine A2576, B922
Populus D54
Spruce B922
See also: A216, A2744, E558
→ 5,6-Dihydro-4-hydroxy-6-(4-hydroxy-3-methoxystyryl)-2-pyrone
Probably in: Haplopappus A3362
→ Feruloyl-3,4-dihydroxyphenethylamine
In: Nicotiana D558, F72
→ Feruloylepilupinine
In: Lupinus H43
→ Feruloylgalactarate
In: Secale K722
→ Feruloylglucarate
In: Secale K722
→ Feruloylglucarolactone
In: Secale K722
→ Feruloyl-4-hydroxy-3-methoxyphenethylamine
In: Nicotiana F72
→ Feruloylnoradrenaline
In: Nicotiana D558, F72
→ Feruloyloctopamine
In: Nicotiana D558, F72
→ Feruloylphenethylamine
In: Nicotiana F72
→ Feruloylquinic acid
In: Potato A3755, K722
→ Feruloylshikimate
In: Secale K722
→ Feruloyltyramine
In: Nicotiana D558, F72
→ Homoeriodictyol
In: Haplopappus A3362
Spinach D596, D867
Tulip A3792
→ 4-Hydroxy-3-methoxybenzalacetone
In: Haplopappus A3362
→ 2′,4,4′,6′-Tetrahydroxy-3-methoxychalcone
In: Buckwheat E180
Tulip B377
→ Vanillin
In: Pseudomonas K92
See also: J650
Feruloyl-3′-dephospho CoA
Formed from: Caffeoyl-3′-dephospho CoA
Feruloyl-3,4-dihydroxyphenethylamine
Formed from: Feruloyl CoA
3,4-Dihydroxyphenethylamine
Feruloylepilupinine
Formed from: Feruloyl CoA
O-(6-O-Feruloyl-β-D-galactopyranosyl)(1→4)-β-D-galactopyranose
→ Ferulic acid
In: Aspergillus H381
Feruloyl-β-D-glucose
Formed from: Ferulic acid
→ Celosianin II
In: Chenopodium G255
→ 1,2-Di-O-feruloyl-β-D-glucose
In: Raphanus K724
→ Ferulic acid
In: Raphanus K724
→ Feruloylcholine
In: Raphanus K724
→ Feruloylglucuronosylglucose

In: Chenopodium G255
Feruloylglucuronosylglucose
Formed from: Feruloyl-β-D-glucose
Feruloyl-4-hydroxy-3-methoxyphenethylamine
Formed from: Feruloyl CoA
3-Methoxytyramine
Feruloylnoradrenaline
Formed from: Feruloyl CoA
D-Noradrenaline
Feruloyloctopamine
Formed from: Feruloyl CoA
Octopamine
Feruloylphenethylamine
Formed from: Feruloyl CoA
Phenethylamine
Feruloyl-4′-phosphopantetheine
→ Feruloyltyramine
In: Nicotiana D558
Feruloylquinic acid
Formed from: Chlorogenic acid
Feruloyl CoA
Feruloyltyramine
Formed from: Dephosphoferuloyl CoA
Feruloyl CoA
Feruloyl-4′-phosphopantetheine
Tyramine
Feudomycin A ((8*S*,10*S*)-10-((3-amino-2,3,6-trideoxy-α-L-lyxo-hexopyranosyl)oxy)-7,8,9,10-tetrahydro-6,8,11-trihydroxy-1-methoxy-8-(2-oxopropyl)-5,12-naphthacenedione)
→ Daunorubicinol
In: Streptomyces B600
Fisetin (3,3′,4′,7-tetrahydroxyflavone)
→ Fisetin-3-glucoside
In: Cabbage A1733
→ Fisetin-7-β-D-glucoside
In: Parsley K739
→ Fisetin-β-D-glucuronide
In: Man G524, J416, J783
→ Fisetin quinone
In: Vicia J880
→ Fisetin-3-D-xyloside
In: Euonymus G444
→ 3,4′,7-Trihydroxy-3′-methoxyflavone
In: Streptomyces K652
Fisetin-3-glucoside
Formed from: Fisetin
Fisetin-7-β-D-glucoside
Formed from: Fisetin
Fisetin-β-D-glucuronide
Formed from: Fisetin
Fisetin quinone
Formed from: Fisetin
Fisetin-3-D-xylidide
Formed from: Fisetin
Flavan-4α-ol (2-phenyl-4*H*-1-benzopyran-4α-ol)
Formed from: Flavanone
→ *trans*-3-Hydroxyflavan-4α-ol
Probably in: Rat B222
→ 4′-Hydroxyflavan-4α-ol
Probably in: Rat B222
→ 6-Hydroxyflavan-4α-ol
Probably in: Rat B222
Flavan-4β-ol (2-phenyl-4*H*-1-benzopyran-4β-ol)
Formed from: Flavanone
→ *trans*-3-Hydroxyflavan-4β-ol
Probably in: Rat B222
→ 4′-Hydroxyflavan-4β-ol
Probably in: Rat B222
→ 6-Hydroxyflavan-4β-ol
Probably in: Rat B222
→ 4′-Hydroxy-3′-methoxyflavan-4β-ol
Probably in: Rat B222
Flavanone (2,3-dihydro-2-phenyl-4*H*-1-benzopyran-4-one)
→ Flavan-4α-ol
In: Gibberella A1512
Rat B222
→ Flavan-4β-ol
In: Gibberella A1512
Rat B222
→ Flavone
In: Gibberella A1512
See also: F673
→ Flavonol
In: Gibberella A1512
Rat B222
→ 2′-Hydroxychalcone
In: Gibberella A1512
See also: F673
→ *o*-Hydroxydibenzoylmethane

In: Aspergillus F673
→ *cis*-3-Hydroxyflavanone
In: Rat B222
→ *trans*-3-Hydroxyflavanone
In: Rat B222
→ 4′-Hydroxyflavanone
In: Absidia G14
Gibberella A1512
Rat B222
Streptomyces F673
→ 6-Hydroxyflavanone
In: Rat B222

Flavine adenine dinucleotide (FAD)
Formed from: Riboflavine-5′-phosphate
→ Dihydroflavine adenine dinucleotide
In: Beneckea A3209
→ Riboflavine
In: Mouse A3780
→ Riboflavine-5′-phosphate
In: Mouse A3780
See also: B428

Flavine mononucleotide
See: Riboflavine-5′-phosphate

Flaviolin (2,5,7-trihydroxy-1,4-naphthalenedione)
→ 5-Hydroxyscytalone
In: Wangiella D760

Flavodilol (7-(2-hydroxy-3-propylaminopropoxy)-2-phenyl-4*H*-1-benzopyran-4-one)
→ Depropylflavodilol
In: Rat F26
→ 3-(4-Oxo-2-phenyl-4*H*-benzopyran-7-oxy)lactic acid
In: Rat F26

Flavone (2-phenyl-4*H*-1-benzopyran-4-one)
Formed from: Flavanone
→ *o*,ω-Dihydroxyacetophenone
In: Rhizopus F504
→ *o*-Hydroxydibenzoylmethane
Probably in: Gibberella A1512
→ 3-Hydroxyflavone
Probably in: Aspergillus F673
Gibberella A1512
→ 4′-Hydroxyflavone
In: Absidia F504
Aspergillus F504
Cunninghamella F504
Curvularia F504
Gliocladium F504
Gongronella F504
Gymnascella F504
Helicostylum F504
Linderina F504
Manascus F504
Penicillium F504
Sepedonium F504
Streptomyces F504
See also: A1512
→ 4′-Hydroxyisoflavone
In: Gongronella F504

Flavoneacetic acid (4-oxo-2-phenyl-4*H*-1-benzopyran-8-acetic acid)
→ Flavoneacetic acid-β-D-glucuronide
In: Man F215
→ Flavoneacetyl-β-D-glucuronic acid
In: Man F215

Flavoneacetic acid-β-D-glucuronide
Formed from: Flavoneacetic acid

Flavoneacetyl-β-D-glucuronic acid
Formed from: Flavoneacetic acid

Flavonol
See: 3-Hydroxyflavone

Flavopyridol (*cis*-(−)-2-(2-chlorophenyl)-5,7-dihydroxy-8-(3-hydroxy-1-methyl-4-piperidinyl)-4*H*-1-benzopyran-1-one)
→ Flavopyridol-5-β-D-glucuronide
In: Rat J597
→ Flavopyridol-7-β-D-glucuronide
In: Rat J597

Flavopyridol-5-β-D-glucuronide
Formed from: Flavopyridol

Flavopyridol-7-β-D-glucuronide
Formed from: Flavopyridol

Flavoxate (2-piperidinoethyl 3-methylflavone-8-carboxylate)
→ 3-Methylflavone-8-carboxylic acid
In: Man A2199, A2632
Rat A3395, G110

Flecainide (N-(2-piperidinylmethyl)-2,5-bis(2,2,2-trifluoroethoxy)benzamide)
→ 2-Hydroxy-N-(2-piperidinylmethyl)-5-(2,2,2-trifluoroethoxy)benzamide
In: Man D388

Fleroxacin (6,8-difluoro-1-(2-fluoroethyl)-1,4-dihydro-7-(4-methyl-1-piperazinyl)-4-oxo-3-quinolonecarboxylic acid)
→ Demethylfleroxacin
In: Dog G115
Man E985, G115
Rabbit G115
→ Fleroxacin-N-oxide
In: Dog G115
Man E985, G115
Rabbit G115
→ 4-Formyldemethylfleroxacin
In: Man G115
Rabbit G115

Fleroxacin-N-oxide
Formed from: Fleroxacin

Flestolol (N-(1,1-dimethyl-2-ureidoethyl)-2-hydroxy-3-(2-fluorobenzoyloxy)propylamine)
→ *o*-Fluorobenzoic acid
In: Rabbit H563

Flexirubin (7-(4-hydroxy-3-methylphenyl)-2,4,6,8,10,12,14,16-heptadecaoctaenoic acid 2-dodecyl-3-hydroxy-5-methylphenyl ester)
Formed from: L-Tyrosine

Flobufen (4-(2′,4′-difluoro-α-methyl-γ-oxo-(1,1′-biphenyl-4-yl))butyric acid)
→ 3-Carboxymethyl-2′,4′-difluorobiphenyl
In: Rat F606
→ 2′,4′-Difluoro-γ-hydroxy-α-methyl-(1,1′-biphenyl)-4-butanoic acid
In: Mouse K211
Rat K211

Flocoumafen (4-hydroxy-3-(1,2,3,4-tetrahydro-3-(4-((4-(trifluoromethyl)phenyl)methoxy)phenyl)-1-naphthalenyl)-2*H*-1-benzopyran-2-one)
→ Flocoumafen-β-D-glucuronide
In: Quail F322

Flocoumafen-β-D-glucuronide
Formed from: Flocoumafen

Floctafenin (2,3-dihydroxypropyl N-(8-trifluoromethyl-4-quinolinyl)anthranilate)
→ Hydroxyfloctafenin
In: Rat A3798
→ N-(8-Trifluoromethyl-4-quinolinyl) anthranilate
In: Man A1707, A3798
Mouse A3798

Flosequinan (7-fluoro-1-methyl-3-methylsulphinyl-4-quinolone; *R* and *S* isomers)
Formed from: 7-Fluoro-1-methyl-3-methylthio-4-quinolone
→ 7-Fluoro-1-methyl-3-methylsulphonyl-4-quinolone
In: Rat G877, J738
See also: J159
→ 7-Fluoro-1-methyl-3-methylthio-4-quinolone
In: Rat G877, J738

Fluchloralin (N-(2-chloroethyl)-2,6-dinitro-N-propyl-4-(trifluoromethyl)benzenamine)
→ 2-Amino-N-(2-chloroethyl)-6-nitro-N-propyl-4-trifluoromethylaniline
In: Glycine A3895, A3985
→ 1-(2-Chloroethyl)-2-ethyl-7-nitro-5-trifluoromethylbenzimidazole
In: Glycine A3895, A3985
→ 2,6-Dinitro-N-propyl-4-trifluoromethylaniline
In: Rat A2912
→ 5-Nitro-7-trifluoromethylquinoxaline
In: Rat A2912

Fluconazole (α-(2,4-difluorophenyl)-α-(1*H*-1,2,4-triazol-1-ylmethyl)-1*H*-1,2,4-triazole-1-ethanol)
→ Fluconazole-β-D-glucuronide
In: Man G35
→ Fluconazole-N-oxide
In: Man G35

Fluconazole-β-D-glucuronide
Formed from: Fluconazole

Fluconazole-N-oxide
Formed from: Fluconazole

Flucythrinate (4-(difluoromethoxy)-α-(1-methylethyl)benzeneacetic acid cyano(3-phenoxyphenyl)methyl ester)
→ *m*-Phenoxymandelonitrile
In: Bacillus G891

Flufenamic acid (2-((3-(trifluoromethyl)phenyl)amino)benzoic acid)
Formed from: Butyl 2-((*m*-trifluoromethyl-phenyl)amino)benzoate

Flumazenil (8-fluoro-5,6-dihydro-5-methyl-6-oxo-4*H*-imidazo[1,5-*a*][1,4]benzodiazepine-3-carboxylic acid ethyl ester)
→ Flumazenil acid
In: Man K294

Flumazenil acid (8-fluoro-5,6-dihydro-5-methyl-6-oxo-4*H*-imidazo[1,5-*a*][1,4]benzodiazepine-3-carboxylic acid)
Formed from: Flumazenil
→ Flumazenil acid methyl ester
In: Man K294

Flumazenil acid methyl ester
Formed from: Flumazenil acid

Flumecinol
See: α-Ethyl-α-phenyl-3-(trifluoromethyl)benzenemethanol

Flumequine (9-fluoro-6,7-dihydro-5-methyl-1-oxobenzo[*ij*]quinolizine-2-carboxylic acid)
→ 7-Hydroxyflumequine
In: Dog E860
Rat E860

Flumethrin (3-(2-chloro-2-(4-chlorophenyl) ethenyl)-2,2-dimethylcyclopropanecarboxylic acid cyano(4-fluoro-3-phenoxyphenyl)methyl ester)
→ 1-Carboxy-2-(2-chloro-2-phenylethenyl)-3,3-dimethylcyclopropane
In: Bacillus G891
→ 4-Fluoro-3-phenoxymandelonitrile
In: Bacillus G891

Flumioxazin (7-fluoro-6-(3,4,5,6-tetrahydrophthalimido)-4-(2-propynyl)-2*H*-1,4-benzoxazin-3(4*H*)-one)
→ 16,21-Dihydroflumioxazin
In: Rat K152
→ 17-Hydroxyflumioxazin
In: Rat K152
→ 18-Hydroxyflumioxazin
In: Rat K152

Flunarizine ((*E*-)-1-(bis(4-fluorophenyl)methyl)-4-(3-phenyl-2-propenyl)piperazine)
→ Bis(*p*-fluorophenyl)methanol
In: Dog C773
Man G310
Rat C773, G310
See also: H852
→ 1-(Bis(*p*-fluorophenyl)methyl)-4-((2,3-epoxy-3-phenyl)propyl)piperazine
Probably in: Man G310
Rat G310
→ 1-(Bis(*p*-fluorophenyl)methyl)piperazine
In: Dog C773, G310
Man H852
Rat C773, G310, H35
→ Cinnamaldehyde
Probably in: Man H852
→ 1-Cinnamylpiperazine
Probably in: Man H852
→ 4,4′-Difluorobenzophenone
In: Rat H35
→ Flunarizine-N-oxide
In: Dog C773
→ 4-Hydroxyflunarizine
In: Dog C773, G310
Man G310, H852
Rat C773, G310, H35

Flunarizine-N-oxide
Formed from: Flunarizine

Flunitrazepam (5-(*o*-fluorophenyl)-1,3-dihydro-1-methyl-7-nitro-2*H*-1,4-benzodiazepin-2-one)
→ 7-Amino-5-(*o*-fluorophenyl)-1,3-dihydro-1-methyl-2*H*-1,4-benzodiazepin-2-one
In: Man A3604, A3841
→ 5-(2-Fluoro-4-hydroxyphenyl)-1,3-dihydro-1-methyl-7-nitro-2*H*-1,4-benzodiazepin-2-one
In: Rat B496
→ 5-(*o*-Fluorophenyl)-1,3-dihydro-7-nitro-2*H*-1,4-benzodiazepin-2-one
In: Man A3604, J757
→ 3-Hydroxyflunitrazepam
In: Man A3604, J757
Rat B496

Flunixin (2-((2-methyl-3-(trifluoromethyl)phenyl)amino)-3-pyridinecarboxylic acid)
→ Flunixin-β-D-glucuronic acid
In: Dog J374

Flunixin-β-D-glucuronic acid
Formed from: Flunixin

Flunoxaprofen ((α*S*)-2-(4-fluorophenyl)-α-methyl-5-benzoxazoleacetic acid)
→ Flunoxaprofen-β-D-glucuronide
In: Rat G46

(*R*)-Flunoxaprofen → (*S*)-Flunoxaprofen
In: Rat G46
(*S*)-Flunoxaprofen → (*R*)-Flunoxaprofen
In: Rat G46

Flunoxaprofen-β-D-glucuronide
Formed from: Flunoxaprofen

Fluoracizine (10-(3-(diethylamino)-1-oxopropyl)-2-(trifluoromethyl)-10*H*-phenothiazine)
→ 10-(3-Aminopropyl)-2-trifluoromethylphenothiazine
In: Rat A829
→ 2-Trifluoromethylphenothiazine
In: Rat A829

Fluoranthene
→ Acenaphthenone
In: Alcaligenes G182
Mycobacterium G818
→ *trans*-2,3-Dihydro-2,3-dihydroxyfluoranthene
In: Man G672
Rat E255, G672
→ 9-Fluorenone-1-carboxylic acid
In: Mycobacterium G122, G818, J483
Pasteurella J856, K312
→ *o*-Formylbenzoic acid
In: Mycobacterium G818
Pasteurella J856, K312
→ 1-Hydroxyfluoranthene
In: Rat E255
→ 3-Hydroxyfluoranthene
In: Rat E255
→ 8-Hydroxyfluoranthene
In: Rat E255
→ 8-Hydroxy-7-methoxyfluoranthene
In: Mycobacterium G818
→ 3-Hydroxymethyl-4,5-benzocoumarin
In: Alcaligenes G182
→ Phenylacetic acid
In: Pasteurella J856, K312

Fluoranthene-2,3-quinone
Probably formed from: *trans*-2,3-Dihydro-2,3-dihydroxyfluoranthene

1-Fluorenamine
See: 1-Aminofluoene

2-Fluorenamine
See: 2-Aminofluoene

Fluorene
→ Dihydrocoumarin
In: Arthrobacter G669
→ 1,1a-Dihydroxy-1*H*-fluoren-9-one
In: Pseudomonas H37, H254
→ Fluorene-9-hydroperoxide
In: Rat A1955
→ Fluorenone
In: Arthrobacter G669, H901
Brevibacterium F460
Coreolopsis K589
Cunninghamella H47
Pseudomonas H254
→ Indan-1-one
In: Arthrobacter H901
→ Indan-2-one
In: Arthrobacter H901
→ Phthalic acid
In: Brevibacterium F460
Pseudomonas H254

Fluorene-1-carboxylic acid
Formed from: 1-Methylfluorene
→ 9-Hydroxyfluorene-1-carboxylic acid
In: Pseudomonas H672

Fluorene-2-carboxylic acid
Formed from: 2-Formylfluorene
→ 1,2,4-Tricarboxybenzene
In: Pseudomonas H672

Fluorene-9-hydroperoxide
Formed from: Fluorene
→ 9-Hydroxyfluorene
Probably in: Rat A1955

Fluorenone (9-oxofluorene)
Formed from: Fluorene
9-Hydroxyfluorene
→ 1,10-Dihydro-1,10-dihydroxyfluoren-9-one
Probably in: Brevibacterium F460
Staphylococcus G753
→ 9-Hydroxyfluorene
In: Rat D390
See also: G818
→ 1-Hydroxyfluorenone
Probably in: Brevibacterium F460
Staphylococcus G753
→ 2-Hydroxyfluorenone
In: Cunninghamella H47

→ 4-Hydroxyfluorenone
Probably in: Arthrobacter H901
Staphylococcus G753

9-Fluorenone-1-carboxylic acid
Formed from: Fluoranthene
→ Fluorenone
Probably in: Pasteurella J856, K312
→ 9-Hydroxyfluorene-1-carboxylic acid
Probably in: Mycobacterium G818
Pasteurella J856, K312

9-Fluorenone oxime
→ 9-Fluorenone oxime-O-sulphate
In: Rat E261
→ Phenanthridinone
In: Rat A3230

9-Fluorenone oxime-O-sulphate
Formed from: 9-Fluorenone oxime

9-Fluorenon-2-yl sulphate
Formed from: 2-Hydroxyfluorenone

1-(N-(2′-Fluorenyl)acetamido)-2-acetamidofluorene
Formed from: N-Hydroxy-2-acetamidofluorene

2-Fluorenyl nitroxide radical
Formed from: 2-Aminofluorene
Probably formed from: 2-Hydroxylaminofluorene

Fluoren-1-yl sulphate
Formed from: 1-Hydroxyfluorene

Fluorescein (3′,6′-dihydroxyspiro(isobenzofuran-1(3*H*),9′-(9*H*)-xanthen)-3-one)
Formed from: Fluorescein mono-β-D-galactoside
→ Fluorescein-β-D-glucuronide
In: Man B512, G524, G968
Rabbit B512
→ Fluorescin
In: Man B512
Rabbit B512

Fluorescein di-β-D-galactoside
→ Fluorescein mono-β-D-galactoside
In: Escherichia H237

Fluorescein mono-β-D-galactoside
Formed from: Fluorescein di-β-D-galactoside
→ Fluorescein
In: Escherichia H237

Fluorescein-β-D-glucuronide
Formed from: Fluorescein

Fluorescin (dihydrofluorescein)
Formed from: Fluorescein
→ Fluorescin-β-D-glucuronide
Probably in: Man B512
Rabbit B512

Fluorescin-β-D-glucuronide
Probably formed from: Fluorescin

***o*-Fluoroacetanilide**
Formed from: *o*-Fluoroaniline
→ *o*-Fluoroaniline
In: Nocardia G343

***m*-Fluoroacetanilide**
→ *m*-Fluoroaniline
In: Nocardia G343

***p*-Fluoroacetanilide**
Formed from: *p*-Fluoroaniline
→ 2-Acetamido-5-fluorophenol
In: Rat J603
→ *p*-Acetamidophenol
In: Rat J558
→ *p*-Fluoroaniline
Probably in: Rat J603

***o*-Fluoroacetophenone**
→ 1-(*o*-Fluorophenyl)ethanol
In: Rabbit C25

***p*-Fluoroacetophenone**
→ 1-(*p*-Fluorophenyl)ethanol
In: Man G89

ω-Fluoroacetophenone
→ 2-Fluoro-1-phenylethanol
In: Candida G826
Rhodotorula G826
Saccharomyces G826
Torulopsis G826
Xanthomonas G826

10-Fluoroajmalicine
Probably formed from: 5-Fluorotryptamine

11-Fluoroajmalicine
Probably formed from: 6-Fluorotryptamine

***o*-Fluoroaniline**
Formed from: *o*-Fluoroacetanilide
→ 2-Amino-3-fluorophenol
In: Rat G790
→ 4-Amino-3-fluorophenol
In: Man D952

Marmoset D952
Rabbit D952
Rat D952, F489, F518, G790, H273
→ 2,2′-Difluoroazobenzene
In: Horseradish A661
→ *o*-Fluoroacetanilide
In: Rat F518, G790

***m*-Fluoroaniline**
Formed from: *m*-Fluoroacetanilide
→ 2-Amino-4-fluorophenol
In: Rat G790
→ 4-Amino-2-fluorophenol
In: Rat F489, G790, H273
→ 3,3′-Difluoroazobenzene
In: Horseradish A661

***p*-Fluoroaniline**
Formed from: *p*-Fluoro-N-methylaniline
Probably formed from: *p*-Fluoroacetanilide
→ 2-Amino-5-fluorophenol
In: Rat F489, G219, G790, J558, J603
→ *p*-Aminophenol
In: Rat J603
→ 4,4′-Difluoroazobenzene
In: Geotrichum A661
Horseradish A661
→ *p*-Fluoroacetanilide
In: Rabbit E399
Rat J558, J603
→ *p*-Fluoronitrosobenzene
With chloroperoxidase A3579
→ *p*-Fluorophenylhydroxylamine
In: Rabbit F839
Rat G219

***o*-Fluoroanisole**
→ *o*-Fluorophenol
In: Sporomusa J199

***m*-Fluoroanisole**
→ 3-Methoxycatechol
In: Pseudomonas F76

***p*-Fluoroanisole**
→ *p*-Fluorophenol
In: Methylosinus D513
Sporomusa J199

4-Fluoroanthranilic acid (2-amino-4-fluorobenzoic acid)
Formed from: 4-Fluoro-L-kynurenine

5-Fluoroanthranilic acid (2-amino-5-fluorobenzoic acid)
Formed from: 5-Fluoro-L-kynurenine

***o*-Fluorobenzaldehyde**
Formed from: *o*-Fluorobenzoic acid
o-Fluoro-L-phenylalanine
→ N-Acetyl-S-(*o*-fluorobenzyl)-L-cysteine
In: Rat C553
→ 2,2′-Difluorobenzoin
In: Pseudomonas K424
→ *o*-Fluorobenzoic acid
In: Acinetobacter F366
Yeast C508
→ *o*-Fluorobenzyl alcohol
In: Horse C508
Mung bean J910
Yeast C508, H425
See also: E743
→ *o*-Fluoro-2-hydroxypropiophenone
In: Pseudomonas K524
→ 2-Fluoro-4-methoxybenzaldehyde
Probably in: Bjerkandera H914
→ (1*R*,2*S*)-1-(*o*-Fluorophenyl)propane-1,2-diol
In: Saccharomyces E487

***m*-Fluorobenzaldehyde**
Formed from: *m*-Fluorobenzoic acid
m-Fluorobenzyl alcohol
m-Fluorobenzylamine
m-Fluoro-L-phenylalanine
→ *m*-Fluorobenzoic acid
In: Acinetobacter E663, F366
Rat H761
→ *m*-Fluorobenzyl alcohol
In: Mung bean J910
Saccharomyces H425
See also: E743
→ *m*-Fluoro-2-hydroxypropiophenone
In: Pseudomonas K524
→ 3-Fluoro-4-methoxybenzaldehyde
Probably in: Bjerkandera H914
→ (1*R*,2*S*)-1-(*m*-Fluorophenyl)propane-1,2-diol
In: Saccharomyces E487

***p*-Fluorobenzaldehyde**
Formed from: *p*-Fluorobenzoic acid

p-Fluorobenzylamine
→ 4,4′-Difluorobenzoin
In: Pseudomonas K424
→ *p*-Fluorobenzoic acid
In: Acinetobacter E663, F366
Cow C508
Rat H761
Yeast C508
→ *p*-Fluorobenzyl alcohol
In: Horse C508
Mung bean J910
Yeast C508, H425
→ *p*-Fluoro-2-hydroxypropiophenone
In: Pseudomonas K524
→ (1*R*,2*S*)-1-(*p*-Fluorophenyl)propane-1,2-diol
In: Saccharomyces E487

***o*-Fluorobenzaldoxime**
→ *o*-Fluorobenzyl-β-D-sulphatoglucoside
In: Brassica F657

***m*-Fluorobenzaldoxime**
→ *m*-Fluorobenzyl-β-D-sulphatoglucoside
In: Brassica F657

7-Fluorobenzanthracene
→ 7-Fluoro-3,4-dihydro-3,4-dihydroxybenzanthracene
In: Rat C331
→ 7-Fluoro-5,6-dihydro-5,6-dihydroxybenzanthracene
In: Rat C331
→ 7-Fluoro-8,9-dihydro-8,9-dihydroxybenzanthracene
In: Rat C331
→ 7-Fluoro-10,11-dihydro-10,11-dihydroxybenzanthracene
In: Rat C331

Fluorobenzene
→ *cis*-1,2-Dihydro-1,2-dihydroxy-3-fluorobenzene
In: Pseudomonas J656
→ *o*-Fluorophenol
In: Rat A2160, C899
See also: F278
→ *m*-Fluorophenol
In: Nitrosomonas H631
Rat A2160
→ *p*-Fluorophenol
In: Nitrosomonas H631
Rat A2160, C899

4-(6-Fluoro-1,2-benzisoxazol-3-yl)-1-(3-(4-hydroxy-2-methoxyphenoxy)propyl)piperidine
Formed from: Iloperidone

1-(4-(3-(4-(6-Fluoro-1,2-benzisoxazol-3-yl)-1-piperidinyl)propoxy)-3,6-dihydroxyphenyl)ethanone
Probably formed from: Demethyliloperidone

1-(4-(3-(4-(6-Fluoro-1,2-benzisoxazol-3-yl)-1-piperidinyl)propoxy)-6-hydroxy-3-methoxyphenyl)ethanol
Probably formed from: Dihydroiloperidone

1-(4-(3-(4-(6-Fluoro-1,2-benzisoxazol-3-yl)-1-piperidinyl)propoxy)-6-hydroxy-3-methoxyphenyl)ethanone
Probably formed from: Iloperidone

1-(4-(3-(4-(6-Fluoro-1,2-benzisoxazol-3-yl)-1-piperidinyl)propoxy)-3-hydroxyphenyl)ethanol
Probably formed from: Dihydroiloperidone

1-(4-(3-(4-(6-Fluoro-1,2-benzisoxazol-3-yl)-1-piperidinyl)propoxy)-3-methoxyphenyl)-2-hydroxyethanone
Probably formed from: Iloperidone

***o*-Fluorobenzoic acid**
Formed from: N-(*tert*-Butyl)-3-(*o*-fluorobenzoyloxy)-2-hydroxypropylamine
Flestolol
o-Fluorobenzaldehyde
o-Fluorobenzonitrile
3-Fluorophthalic acid
o-Fluorotoluene
→ Catechol
In: Acinetobacter A1361
Alcaligenes B339
Pseudomonas B339, F247, G434
→ *o*-Fluorobenzaldehyde
In: Nocardia G207
See also: E743
→ *o*-Fluorobenzoyl CoA
In: Microorganism H500
Pseudomonas F65
Rhodopseudomonas E605
→ *o*-Fluorobenzoyl-β-D-glucuronic acid
In: Rabbit H519
→ 3-Fluorocatechol
In: Acinetobacter A1361

See also: B336
→ 6-Fluoro-1,2-dihydro-1,2-dihydroxybenzoic acid
In: Alcaligenes B339
See also: B465
→ *o*-Fluorohippuric acid
In: Rabbit H519
Rat H519
→ Fluoroquinol
In: Candida J461

m-Fluorobenzoic acid
Formed from: *m*-Fluorobenzaldehyde
m-Fluorobenzonitrile
o-Fluorophenol
3-Fluorophthalic acid
m-Fluorotoluene
→ *m*-Fluorobenzaldehyde
In: Clostridium G416
Nocardia G207
See also: E743
→ *m*-Fluorobenzoyl CoA
In: Microorganism H500
Pseudomonas F65
→ *m*-Fluorobenzoyl-β-D-glucuronic acid
In: Rabbit H519
→ *m*-Fluorobenzyl alcohol
In: Desulfovibrio K615
→ 3-Fluorocatechol
In: Acinetobacter A1361
Pseudomonas B336
→ 4-Fluorocatechol
In: Acinetobacter A1361
Pseudomonas B336
→ 5-Fluoro-1,2-dihydro-1,2-dihydroxybenzoic acid
Probably in: Pseudomonas B465
→ *m*-Fluorohippuric acid
In: Rabbit H519
Rat J549
→ 2-Fluoro-*cis*,*cis*-muconic acid
In: Acinetobacter D516
Pseudomonas D516
→ Fluoroquinol
In: Candida J461

p-Fluorobenzoic acid
Formed from: Astemizole
2-Chloro-4-fluorobenzoic acid
p-Fluorobenzaldehyde
p-Fluorobenzonitrile
p-Fluorotoluene
Mosapride
→ *p*-Fluorobenzaldehyde
In: Clostridium G416
Nocardia G207
See also: K615
→ *p*-Fluorobenzoyl CoA
In: Microorganism H500
Pseudomonas F65, G729
Rhodopseudomonas E605
→ *o*-Fluorobenzoyl-β-D-glucuronic acid
In: Rabbit H519
→ 4-Fluorocatechol
In: Acinetobacter A1361
Alcaligenes F200
See also: B336
→ 4-Fluoro-1,2-dihydro-1,2-dihydroxybenzoic acid
In: Pseudomonas B465
→ *p*-Fluorohippuric acid
In: Rabbit H519
Rat J549
→ *p*-Hydroxybenzoic acid
In: Arthrobacter D544
Aureobacterium F200

o-Fluorobenzonitrile
→ *o*-Fluorobenzoic acid
In: Fusarium A3364
Nocardia A3420, D725

m-Fluorobenzonitrile
→ 2,3-Dihydroxybenzonitrile
In: Pseudomonas F76
→ *m*-Fluorobenzoic acid
In: Fusarium A3364
Nocardia A3420, D725

p-Fluorobenzonitrile
→ 4-Fluoro-2,3-dihydroxybenzonitrile
In: Pseudomonas F283
→ *p*-Fluorobenzoic acid
In: Fusarium A3364
Nocardia A3420, D725
Rhodococcus H952

1-(6-Fluorobenzo[*b*]thien-3-yl)piperazine
Formed from: 3-(4-(4-(6-Fluorobenzo[*b*]thien-3-yl)piperazinyl)butyl)-2,5,5-trimethyl-4-thiazolidinone

3-(4-(4-(6-Fluorobenzo[*b*]thien-3-yl) piperazinyl)butyl)2,5,5-trimethyl-4-thiazolidinone
→ 1-(4-((N-Acetyl-N-butyryl)amino)butyl)-4-(6-fluorobenzo[*b*]thien-3-yl)piperazine
In: Rat J53
→ 3-(4-(4-(6-Fluorobenzo[*b*]thien-3-yl) piperazinyl)butyl)-2,5,5-trimethyl-4-thiazolidinone-1-oxide
In: Rat J53
→ 3-(4-(4-(6-Fluoro-1-oxobenzo[*b*]thien-3-yl)piperazinyl)butyl)-2,5,5-trimethyl-4-thiazolidinone
In: Rat J53
→ 1-(6-Fluorobenzo[*b*]thien-3-yl)piperazine
In: Rat J53

3-(4-(4-(6-Fluorobenzo[*b*]thien-3-yl) piperazinyl)butyl)-2,5,5-trimethyl-4-thiazolidinone-1-oxide
Formed from: 3-(4-(4-(6-Fluorobenzo[*b*] thien-3-yl)piperazinyl)butyl)-2,5,5-trimethyl-4-thiazolidinone

***o*-Fluorobenzoyl CoA**
Formed from: *o*-Fluorobenzoic acid
→ 2-Fluorocyclohexa-1,5-diene-1-carboxyl CoA
Probably in: Thauera J175

***m*-Fluorobenzoyl CoA**
Formed from: *m*-Fluorobenzoic acid
→ 3-Fluorocyclohexa-1,5-diene-1-carboxyl CoA
Probably in: Thauera J175

***p*-Fluorobenzoyl CoA**
Formed from: *p*-Fluorobenzoic acid
→ 4-Fluorocyclohexa-1,5-diene-1-carboxyl CoA
Probably in: Thauera J175
→ *p*-Hydroxybenzoyl CoA
In: Arthrobacter H279

***o*-Fluorobenzoyl-β-D-glucuronic acid**
Formed from: *o*-Fluorobenzoic acid

***m*-Fluorobenzoyl-β-D-glucuronic acid**
Formed from: *m*-Fluorobenzoic acid

***p*-Fluorobenzoyl-β-D-glucuronic acid**
Formed from: *p*-Fluorobenzoic acid

***p*-Fluorobenzoylpropionic acid**
Formed from: Bromperidol
Haloperidol
→ 4-(*p*-Fluorophenyl)-4-hydroxybutyric acid
Probably in: Rat E16

3-(2-(4-(*p*-Fluorobenzoyl)-2-oxo-1-piperidinyl)ethyl)-hydroxy-2,4-quinazolinedione
Formed from: Ketanserin

3-(2-(4-(*p*-Fluorobenzoyl)-1-piperidinyl)ethyl)-6-hydroxy-2,4(1*H*,3*H*)-quinazolinedione
Formed from: Ketanserin

3-(2-(4-(*p*-Fluorobenzoyl)-1-piperidinyl)ethyl)-2,4(1*H*,3*H*)-quinazolinedione
See: Ketanserin

1-(3-(*p*-Fluorobenzoyl)propyl)-1,4′-bipiperidine-4-carboxamide
See: Pipamperone

1-(3-(*p*-Fluorobenzoyl)propyl)-hydroxy-1,4′-bipiperidine-4′-carboxamide
Formed from: Pipamperone

8-Fluoro-3,4-benzpyrene
→ 8-Fluoro-4,5-dihydro-4,5-dihydroxy-3,4-benzpyrene
In: Rat D511
→ 8-Fluoro-9,10-dihydro-9,10-dihydroxy-3,4-benzpyrene
In: Rat D511
→ 8-Fluoro-3-hydroxy-3,4-benzpyrene
In: Rat D511

9-Fluoro-3,4-benzpyrene
→ 9-Fluoro-4,5-dihydro-4,5-dihydroxy-3,4-benzpyrene
In: Rat D511
→ 9-Fluoro-7,8-dihydro-7,8-dihydroxy-3,4-benzpyrene
In: Rat D511
→ 9-Fluoro-3-hydroxy-3,4-benzpyrene
In: Rat D511

8-Fluoro-3,4-benzpyrene-3,6-quinone
Probably formed from: 8-Fluoro-3-hydroxy-3,4-benzpyrene

9-Fluoro-3,4-benzpyrene-3,6-quinone
Probably formed from: 9-Fluoro-3-hydroxy-3,4-benzpyrene

***o*-Fluorobenzyl alcohol**
Formed from: *o*-Fluorobenzaldehyde
***m*-Fluorobenzyl alcohol**
Formed from: *m*-Fluorobenzaldehyde
m-Fluorobenzoic acid
→ 2,3-Dihydroxybenzyl alcohol
In: Pseudomonas F76
→ *m*-Fluorobenzaldehyde
In: Arion G804
***p*-Fluorobenzyl alcohol**
Formed from: *p*-Fluorobenzaldehyde
p-Fluorotoluene
***m*-Fluorobenzylamine**
→ *m*-Fluorobenzaldehyde
In: Pig A72
***p*-Fluorobenzylamine**
→ *p*-Fluorobenzaldehyde
In: Pig A72
(2-Fluorobenzylidene)malononitrile
→ (2-Fluorobenzyl)malononitrile
In: Man B260
(2-Fluorobenzyl)malononitrile
Formed from: (2-Fluorobenzylidene) malononitrile
***o*-Fluorobenzyl-β-D-sulphatoglucoside**
Formed from: *o*-Fluorobenzaldehyde
***m*-Fluorobenzyl-β-D-sulphatoglucoside**
Formed from: *m*-Fluorobenzaldoxime
(*R*)-(+)-N-(3-(5-(*p*-Fluorobenzyl)thiophen-2-yl)-1-methyl-2-propynyl)-N-hydroxyurea
→ (*R*)-(+)-N-(3-(5-(*p*-Fluorobenzyl) thiophen-2-yl)-1-methyl-2-propynyl)urea
In: Man J446
→ (*R*)-(+)-N-(3-(5-(*p*-Fluoro-α-hydroxybenzyl)thiophen-2-yl)-1-methyl-2-propynyl)-N-hydroxyurea
In: Man J446
(*R*)-(+)-N-(3-(5-(*p*-Fluorobenzyl)thiophen-2-yl)-1-methyl-2-propynyl)urea
Formed from: (*R*)-(+)-N-(3-(5-(*p*-Fluorobenzyl)thiophen-2-yl)-1-methyl-2-propynyl)-N-hydroxyurea
→ (*R*)-(+)-N-(3-(5-(*p*-Fluoro-α-hydroxybenzyl)thiophen-2-yl)-1-methyl-2-propynyl)urea
In: Man J446

4-Fluorobiphenyl
→ 4′-Fluoro-4-hydroxybiphenyl
In: Rat B835
(*S*)-(2-Fluorobiphenyl-4-yl)propanol
Formed from: (*S*)-(2-Fluorobiphenyl-4-yl)propanol acetyl ester
(*S*)-(2-Fluorobiphenyl-4-yl)propanol acetyl ester
→ (*S*)-(2-Fluorobiphenyl-4-yl)propanol
In: Pig K560
4-(1-(2-Fluorobiphenyl-4-yl)ethyl)-2-methylaminothiazole
→ 2-Amino-4-(1-(2-fluorobiphenyl-4-yl)ethyl)thiazole
In: Rat J343
→ 4-(1-(2-Fluorobiphenyl-4-yl)ethyl)-2-methylaminothiazole-β-D-glucuronide
In: Rat J343
→ 4-(1-(2-Fluoro-4′-hydroxybiphenyl-4-yl)ethyl)-2-methylaminothiazole
In: Rat J343
4-(1-(2-Fluorobiphenyl-4-yl)ethyl)-2-methylaminothiazole-β-D-glucuronide
Formed from: 4-(1-(2-Fluorobiphenyl-4-yl)ethyl)-2-methylaminothiazole
3-Fluorocatechol
Formed from: *m*-Difluorobenzene
2,3-Difluorophenol
2,6-Difluorophenol
o-Fluorobenzoic acid
m-Fluorobenzoic acid
o-Fluorophenol
m-Fluorophenol
Probably formed from: 2,6-Difluorobenzoic acid
→ 3-Fluoro-2-hydroxy-*cis*,*cis*-muconic semialdehyde
Probably in: Pseudomonas G355
→ 2-Fluoromuconic acid
In: Pseudomonas A3769, B754
See also: B336
→ 4-Fluoropyrogallol
In: Trichosporon A902
4-Fluorocatechol
Formed from: 2,4-Difluorophenol
2,5-Difluorophenol
m-Fluorobenzoic acid
p-Fluorobenzoic acid

5-Fluoro-2-nitrophenol
m-Fluorophenol
p-Fluorophenol
5-Fluorosalicylic acid
Probably formed from: 2-Chloro-4-fluorobenzoic acid
3,4-Difluorobenzoic acid
→ 5-Fluoro-2-hydroxy-*cis*,*cis*-mucomic semialdehyde
In: Pseudomonas J515
→ 4-Fluoro-2-methoxyphenol
In: Rat A486
With catechol O-methyltransferase A597
→ 5-Fluoro-2-methoxyphenol
In: Rat A486
With catechol O-methyltransferase A597
→ 3-Fluoromuconic acid
In: Alcaligenes A3769
Pseudomonas A3769, B754
See also: B336
→ 6-Fluoro-3,4-dihydroxyphenylalanine
With β-tyrosinase K99

***p*-Fluorodemethyldeprenyl**
Formed from: *p*-Fluorodeprenyl

***p*-Fluorodeprenyl**
→ *p*-Fluorodemethyldeprenyl
Species unclear J241
→ *p*-Fluoro-N-methylamphetamine
Species unclear J241

Fluorodifen
See: *p*-Nitrophenyl 2-nitro-4-trifluoromethylphenyl ether

7-Fluoro-3,4-dihydro-3,4-dihydroxybenzanthracene
Formed from: 7-Fluorobenzanthracene

7-Fluoro-5,6-dihydro-5,6-dihydroxybenzanthracene
Formed from: 7-Fluorobenzanthracene

7-Fluoro-8,9-dihydro-8,9-dihydroxybenzanthracene
Formed from: 7-Fluorobenzanthracene

7-Fluoro-10,11-dihydro-10,11-dihydroxybenzanthracene
Formed from: 7-Fluorobenzanthracene

4-Fluoro-1,2-dihydro-1,2-dihydroxybenzoic acid
Formed from: *p*-Fluorobenzoic acid

5-Fluoro-1,2-dihydro-1,2-dihydroxybenzoic acid
Probably formed from: *m*-Fluorobenzoic acid

6-Fluoro-1,2-dihydro-1,2-dihydroxybenzoic acid
Probably formed from: *o*-Fluorobenzoic acid

***trans*-6-Fluoro-7,8-dihydro-7,8-dihydroxy-3,4-benzpyrene**
→ *trans*-6-Fluoro-7,8-dihydro-7,8-dihydroxy-3,4-benzpyrene-4,5-oxide
In: Rat D331
→ *trans*-6-Fluoro-7,8-dihydro-7,8-dihydroxy-3,4-benzpyrene-9,10-oxide
In: Rat D331

***trans*-6-Fluoro-7,8-dihydro-7,8-dihydroxy-3,4-benzpyrene-4,5-oxide**
Probably formed from: *trans*-6-Fluoro-7,8-dihydro-7,8-dihydroxy-3,4-benzpyrene

***trans*-6-Fluoro-7,8-dihydro-7,8-dihydroxy-3,4-benzpyrene-9,10-oxide**
Probably formed from: *trans*-6-Fluoro-7,8-dihydro-7,8-dihydroxy-3,4-benzpyrene

8-Fluoro-4,5-dihydro-4,5-dihydroxy-3,4-benzpyrene
Formed from: 8-Fluoro-3,4-benzpyrene

8-Fluoro-9,10-dihydro-9,10-dihydroxy-3,4-benzpyrene
Probably formed from: 8-Fluoro-3,4-benzpyrene

9-Fluoro-4,5-dihydro-4,5-dihydroxy-3,4-benzpyrene
Formed from: 9-Fluoro-3,4-benzpyrene

9-Fluoro-7,8-dihydro-7,8-dihydroxy-3,4-benzpyrene
Formed from: 9-Fluoro-3,4-benzpyrene

3-Fluoro-1,2-dihydro-1,2-dihydroxy-5-hydroxymethylchrysene
Probably formed from: 3-Fluoro-5-hydroxymethylchrysene

3-Fluoro-7,8-dihydro-7,8-dihydroxy-5-hydroxymethylchrysene
Probably formed from: 3-Fluoro-5-hydroxymethylchrysene
3-Fluoro-5-methylchrysene

3-Fluoro-9,10-dihydro-9,10-dihydroxy-5-hydroxymethylchrysene
Probably formed from: 3-Fluoro-5-hydroxymethylchrysene

7-Fluoro-1,2-dihydro-1,2-dihydroxy-5-hydroxymethylchrysene
Probably formed from: 7-Fluoro-5-hydroxymethylchrysene

7-Fluoro-9,10-dihydro-9,10-dihydroxy-5-hydroxymethylchrysene
Probably formed from: 7-Fluoro-5-hydroxymethylchrysene

1-Fluoro-1,2-dihydro-1,2-dihydroxy-5-methylchrysene
Probably formed from: 1-Fluoro-5-methylchrysene

1-Fluoro-7,8-dihydro-7,8-dihydroxy-5-methylchrysene
Probably formed from: 1-Fluoro-5-methylchrysene

3-Fluoro-1,2-dihydro-1,2-dihydroxy-5-methylchrysene
Probably formed from: 3-Fluoro-5-methylchrysene

3-Fluoro-7,8-dihydro-7,8-dihydroxy-5-methylchrysene
Probably formed from: 3-Fluoro-5-methylchrysene

6-Fluoro-1,2-dihydro-1,2-dihydroxy-5-methylchrysene
Probably formed from: 6-Fluoro-5-methylchrysene

6-Fluoro-7,8-dihydro-7,8-dihydroxy-5-methylchrysene
Probably formed from: 6-Fluoro-5-methylchrysene

7-Fluoro-1,2-dihydro-1,2-dihydroxy-5-methylchrysene
Probably formed from: 7-Fluoro-5-methylchrysene

7-Fluoro-7,8-dihydro-7,8-dihydroxy-5-methylchrysene
Probably formed from: 7-Fluoro-5-methylchrysene

12-Fluoro-1,2-dihydro-1,2-dihydroxy-5-methylchrysene
Probably formed from: 12-Fluoro-5-methylchrysene

12-Fluoro-7,8-dihydro-7,8-dihydroxy-5-methylchrysene
Probably formed from: 12-Fluoro-5-methylchrysene

***trans*-1-Fluoro-3,4-dihydro-3,4-dihydroxynaphthalene**
Formed from: 1-Fluoronaphthalene

***trans*-1-Fluoro-5,6-dihydro-5,6-dihydroxynaphthalene**
Formed from: 1-Fluoronaphthalene

4-Fluoro-2,3-dihydro-2,3-dihydroxytoluene
→ 3-Fluoro-6-methylcatechol
In: Pseudomonas K348

6-Fluoro-1,4-dihydro-1-methylamino-7-(4-methyl-1-piperazinyl)-4-oxo-3-quinolinecarboxylic acid
See: Amifloxacin

6-Fluoro-1,4-dihydro-1-methylamino-4-oxo-7-(1-piperazinyl)-3-quinolinecarboxylic acid
Formed from: Amifloxacin

9-Fluoro-2,3-dihydro-3-methyl-1-(4-methyl-1-piperazinyl)-7-oxo-7*H*-pyrido[1,2,3-*de*]-1,4-benzoxazine-3-carboxylic acid (DL-8280)
→ 9-Fluoro-2,3-dihydro-3-methyl-1-(1-piperazinyl)-7-oxo-7*H*-pyrido[1,2,3-*de*]-1,4-benzoxazine-3-carboxylic acid
In: Dog D119
Monkey D119
Rat D119
→ 9-Fluoro-2,3-dihydro-3-methyl-1-(4-methyl-1-piperazinyl)-7-oxo-7*H*-pyrido[1,2,3-*de*]-1,4-benzoxazine-3-carboxylic acid-N-oxide
In: Dog D119
Monkey D119
Rat D119

9-Fluoro-2,3-dihydro-3-methyl-1-(4-methyl-1-piperazinyl)-7-oxo-7*H*-pyrido[1,2,3-*de*]-1,4-benzoxazine-3-carboxylic acid-N-oxide
Formed from: 9-Fluoro-2,3-dihydro-3-methyl-1-(4-methyl-1-piperazinyl)-7-oxo-7*H*-pyrido[1,2,3-*de*]-1,4-benzoxazine-3-carboxylic acid

8-Fluoro-5,6-dihydro-5-methyl-6-oxo-4*H*-imidazo[1,5-*a*][1,4]benzodiazepine-3-carboxylic acid
Formed from: Ethyl 8-fluoro-5,6-dihydro-5-methyl-6-oxo-4*H*-imidazo[1,5-*a*][1,4]benzodiazepine-3-carboxylate

9-Fluoro-2,3-dihydro-3-methyl-1-(1-piperazinyl)-7-oxo-7*H*-pyrido[1,2,3-*de*]-1,4-benzoxazine-3-carboxylic acid
Formed from: 9-Fluoro-2,3-dihydro-3-methyl-1-(4-methyl-1-piperazinyl)-7-oxo-7*H*-pyrido[1,2,3-*de*]-1,4-benzoxazine-3-carboxylic acid

2-Fluoro-4,5-dihydroxybenzoic acid
Formed from: 2-Fluoro-5-hydroxybenzoic acid

3-Fluoro-2,5-dihydroxybenzoic acid
See: 3-Fluorogentisic acid

4-Fluoro-2,5-dihydroxybenzoic acid
Formed from: 4-Fluoro-3-hydroxybenzoic acid
→ 6-Carboxy-3-fluoro-4-hydroxy-*cis*,*cis*-muconic semialdehyde
In: Pseudomonas F649

4-Fluoro-2,3-dihydroxybenzonitrile
Formed from: *p*-Fluorobenzonitrile

2-Fluoro-3,4-dihydroxyphenethylamine
Formed from: 2-Fluoro-3,4-dihydroxy-L-phenylalanine

6-Fluoro-3,4-dihydroxyphenethylamine
Formed from: 6-Fluoro-3,4-dihydroxy-D-phenylalanine
6-Fluoro-3,4-dihydroxy-L-phenylalanine
→ 6-Fluoro-3,4-dihydroxyphenylacetic acid
Probably in: Man E216
Rat E216, G457, G635
→ 6-Fluoronoradrenaline
With dopamine β-hydroxylase J660

3-Fluoro-2,5-dihydroxyphenylacetic acid
Formed from: 2-Fluoro-4-hydroxyphenylpyruvic acid

6-Fluoro-3,4-dihydroxyphenylacetic acid
Probably formed from: 6-Fluoro-3,4-dihydroxyphenethylamine

2-Fluoro-3,4-dihydroxy-L-phenylalanine
→ 2-Fluoro-3,4-dihydroxyphenethylamine
In: Rat E810
→ 2-Fluoro-4-hydroxy-3-methoxy-L-phenylalanine
Probably in: Rat E810

5-Fluoro-3,4-dihydroxy-D-phenylalanine
→ 5-Fluoro-4-hydroxy-3-methoxy-D-phenylalanine
In: Rat G635
Does not form 6-fluoro-3,4-dihydroxyphenethylamine
In: Rat G635

5-Fluoro-3,4-dihydroxy-L-phenylalanine
→ 5-Fluoro-4-hydroxy-3-methoxy-L-phenylalanine
In: Rat G457

6-Fluoro-3,4-dihydroxy-D-phenylalanine
→ 6-Fluoro-3,4-dihydroxyphenethylamine
In: Rat G635
→ 6-Fluoro-4-hydroxy-3-methoxy-D-phenylalanine
In: Rat G635

6-Fluoro-3,4-dihydroxy-L-phenylalanine
Formed from: 4-Fluorocatechol
→ 6-Fluoro-3,4-dihydroxyphenethylamine
In: Man E216
Rat E216, E810, G457, G635
→ 6-Fluoro-4-hydroxy-3-methoxy-L-phenylalanine
In: Man E216
Rat E216, E810, G457, G635

***p*-Fluoro-N,N-dimethylaniline**
→ *p*-Fluoro-N-methylaniline
In: Rabbit F839
Rat D578

3-(4-(4-(6-Fluoro-1,1-dioxobenzo[*b*]thien-3-yl)-1-piperazinyl)butyl)-2,5,5-trimethyl-4-thiazolidinone
Formed from: 3-(4-(4-(6-Fluoro-1-oxobenzo[*b*]thien-3-yl)-1-piperazinyl)butyl)-2,5,5-trimethyl-4-thiazolidinone

2-Fluorofluorene
→ 4-Fluorophthalic acid
In: Pseudomonas H672

(β*E*)-4-Fluoro-β-(fluoromethylene)-benzenebutanamide
See: Mofegiline

6-Fluoro-1-(4-fluorophenyl)-1,4-dihydro-7-(4-methyl-1-piperazinyl)-4-oxo-3-quinolinecarboxylic acid
See: Difloxacin

4-Fluoro-N-formyl-L-kynurenine
Formed from: 6-Fluoro-L-tryptophan
→ 4-Fluoro-L-kynurenine

In: Rat A886

5-Fluoro-N-formyl-L-kynurenine
Formed from: 5-Fluoro-L-tryptophan

3-Fluorogentisic acid (3-fluoro-2,5-dihydroxybenzoic acid)
→ 6-Fluoro-4-hydroxy-2-oxohepta-3,5-dienoic acid
In: Rhodococcus H441

4-Fluorogentisic acid
See: 4-Fluoro-2,5-dihydroxybenzoic acid

6-Fluoro-2-(2-(3-(4-(β-glucuronosyloxy)-1*H* benzimidazol-2-yl)propyl)amino)ethyl-1,2,3,4-tetrahydro-2-hydroxy-1-isopropylnaphthalene
Probably formed from: Methoxyacetic acid 6-fluoro-2-(2-(3-(4-β-glucuronosyloxy-1*H*-benzimidazol-2-yl)propyl)amino)ethyl-1,2,3,4-tetrahydro-1-isopropyl-2-naphthalenyl ester

6-Fluoro-2-(2-(3-(5-β-glucuronosyloxy-1*H*-benzimidazol-2-yl)propyl)amino)ethyl-1,2,3,4-tetrahydro-2-hydroxy-1-isopropylnaphthalene
Probably formed from: 2-(2-(3-(1*H*-Benzimidazol-2-yl)propyl)amino)ethyl-6-fluoro-1,2,3,4-tetrahydro-2-hydroxy-1-isopropylnaphthalene

1-(3-Fluoro-5-(S-glutathionyl)-4,5-dihydro-4-hydroxyphenyl)-4,4-bis(5,5-dimethyl-1,3,2-dioxaphosophorinan-2-yl)-1-butanone
Probably formed from: *m*-Fluoro-4,4-bis(5,5-dimethyl-1,3,2-dioxaphosophorinan-2-yl) butyrophenone

4-Fluoro-4-glycollamidostilbene
Formed from: 4-Acetamido-4′-fluorostilbene

***o*-Fluorohippuric acid**
Formed from: *o*-Fluorobenzoic acid

***p*-Fluorohippuric acid**
Formed from: *p*-Fluorobenzoic acid
Timiperone

2-Fluoro-5-hydroxybenzoic acid
→ 2-Fluoro-4,5-dihydroxybenzoic acid
In: Pseudomonas A327

4-Fluoro-3-hydroxybenzoic acid
→ 4-Fluoro-2,5-dihydroxybenzoic acid
Probably in: Pseudomonas A326
→ Protocatechuic acid
Probably in: Pseudomonas A327

5-Fluoro-2-hydroxybenzoic acid
See: 5-Fluorosalicylic acid

4-(4-Fluoro-2-hydroxybenzoyl)-1-(3-(4-(1-hydroxyethyl)-2-methoxyphenoxy)propyl) piperidine
Probably formed from: 1-(3-(4-Acetyl-2-methoxyphenoxy)propyl)-4-(4-fluoro-2-hydroxybenzoyl)piperidine

8-Fluoro-3-hydroxy-3,4-benzpyrene
Formed from: 8-Fluoro-3,4-benzpyrene
→ 8-Fluoro-3,4-benzpyrene-3,6-quinone
Probably in: Rat D511

9-Fluoro-3-hydroxy-3,4-benzpyrene
Formed from: 9-Fluoro-3,4-benzpyrene
→ 9-Fluoro-3,4-benzpyrene-3,6-quinone
Probably in: Rat D511

3-(2-(4-(*p*-Fluoro-α-hydroxybenzyl)-1-piperidinyl)ethyl)-6-hydroxy-2,4(1*H*,3*H*)-quinazolinedione
Probably formed from: 3-(2-(4-(*p*-Fluoro-α-hydroxybenzyl)-1-piperidinyl)ethyl)-2,4(1*H*,3*H*)-quinazolinedione

3-(2-(4-(*p*-Fluoro-α-hydroxybenzyl)-1-piperidinyl)ethyl)-2,4(1*H*,3*H*)-quinazolinedione
Formed from: Ketanserin
→ 3-(2-(4-(*p*-Fluoro-α-hydroxybenzyl)-1-piperidinyl)ethyl)-6-hydroxy-2,4(1*H*,3*H*)-quinazolinedione
Probably in: Dog D379
Rat D379

(*R*)-(+)-N-(3-(5-(*p*-Fluoro-α-hydroxybenzyl) thiophen-2-yl)-1-methyl-2-propynyl)-N-hydroxyurea
Formed from: (*R*)-(+)-N-(3-(5-(*p*-Fluorobenzyl)thiophen-2-yl)-1-methyl-2-propynyl)-N-hydroxyurea

(*R*)-(+)-N-(3-(5-(*p*-Fluoro-α-hydroxybenzyl) thiophen-2-yl)-1-methyl-2-propynyl)urea
Formed from: (*R*)-(+)-N-(3-(5-(*p*-Fluorobenzyl)thiophen-2-yl)-1-methyl-2-propynyl)urea

4′-Fluoro-4-hydroxybiphenyl
Formed from: 4′-Fluorobiphenyl

4-(1-(2-Fluoro-4′-hydroxybiphenyl-4-yl)ethyl)-2-methylaminothiazole
Formed from: 4-(1-(2-Fluorobiphenyl-4-yl)ethyl)-2-methylaminothiazole

→ 2-Amino-4-(1-(2-fluoro-4′-hydroxybiphenyl-4-yl)ethyl)thiazole
Probably in: Rat J343
→ 4-(1-(2-Fluoro-4′-hydroxy-3′-methoxybiphenyl-4-yl)ethyl)-2-methylaminothiazole
Probably in: Rat J343
→ 4-(1-(2-Fluoro-4′-hydroxybiphenyl-4-yl)ethyl)-2-methylaminothiazole-N-β-D-glucuronide
Probably in: Rat J343
→ 4-(1-(2-Fluoro-4′-hydroxybiphenyl-4-yl)ethyl)-2-methylaminothiazole-O-β-D-glucuronide
Probably in: Rat J343

4-(1-(2-Fluoro-4′-hydroxybiphenyl-4-yl)ethyl)-2-methylaminothiazole-N-β-D-glucuronide
Probably formed from: 4-(1-(2-Fluoro-4′-hydroxybiphenyl-4-yl)ethyl)-2-methylaminothiazole

4-(1-(2-Fluoro-4′-hydroxybiphenyl-4-yl)ethyl)-2-methylaminothiazole-O-β-D-glucuronide
Probably formed from: 4-(1-(2-Fluoro-4′-hydroxybiphenyl-4-yl)ethyl)-2-methylaminothiazole

3-Fluoro-7-hydroxy-5-hydroxymethylchrysene
Probably formed from: 3-Fluoro-5-hydroxymethylchrysene

7-Fluoro-1-hydroxy-5-hydroxymethylchrysene
Probably formed from: 7-Fluoro-5-hydroxymethylchrysene

2-(((7-Fluoro-1-hydroxyindan-4-yl)oxy)methyl)morpholine
Formed from: (*S*)-2-(((7-Fluoroindan-4-yl)oxy)methyl)morpholine
→ 2-(((7-Fluoro-1-oxoindan-4-yl)oxy)methyl)morpholine
Probably in: Man K114

2-(((7-Fluoro-2-hydroxyindan-4-yl)oxy)methyl)morpholine
Formed from: (*S*)-2-(((7-Fluoroindan-4-yl)oxy)methyl)morpholine

4-(1-(2-Fluoro-4′-hydroxy-3′-methoxybiphenyl-4-yl)ethyl)-2-methylaminothiazole
Probably formed from: 4-(1-(2-Fluoro-4′-hydroxybiphenyl-4-yl)ethyl)-2-methylaminothiazole
→ 4-(1-(2-Fluoro-4′-hydroxy-3′-methoxybiphenyl-4-yl)ethyl)-2-methylaminothiazole-β-D-glucuronide
Probably in: Rat J343

4-(1-(2-Fluoro-4′-hydroxy-3′-methoxybiphenyl-4-yl)ethyl)-2-methylaminothiazole-β-D-glucuronide
Probably formed from: 4-(1-(2-Fluoro-4′-hydroxy-3′-methoxybiphenyl-4-yl)ethyl)-2-methylaminothiazole

16β-Fluoro-15α-hydroxy-2-methoxy-17β-oestradiol
Probably formed from: 16β-Fluoro-2-methoxy-17β-oestradiol
→ 16β-Fluoro-15α-hydroxy-2-methoxy-17β-oestradiol-β-D-glucuronide
Probably in: Rat J498

16β-Fluoro-15α-hydroxy-2-methoxy-17β-oestradiol-β-D-glucuronide
Probably formed from: 16β-Fluoro-15α-hydroxy-2-methoxy-17β-oestradiol

2-Fluoro-4-hydroxy-3-methoxy-L-phenylalanine
Probably formed from: 2-Fluoro-3,4-dihydroxy-L-phenylalanine

5-Fluoro-4-hydroxy-3-methoxy-D-phenylalanine
Probably formed from: 5-Fluoro-3,4-dihydroxy-D-phenylalanine

5-Fluoro-4-hydroxy-3-methoxy-L-phenylalanine
Probably formed from: 5-Fluoro-3,4-dihydroxy-L-phenylalanine

6-Fluoro-4-hydroxy-3-methoxy-D-phenylalanine
Probably formed from: 6-Fluoro-3,4-dihydroxy-D-phenylalanine

6-Fluoro-4-hydroxy-3-methoxy-L-phenylalanine
Probably formed from: 6-Fluoro-3,4-dihydroxy-L-phenylalanine
→ 5-(S-(L-Cysteinyl))-6-fluoro-4-hydroxy-3-methoxy-L-phenylalanine
In: Rat K427
→ 6-Fluoro-4-hydroxy-3-methoxy-L-phenylalanine-β-D-glucuronide
In: Rat K427

6-Fluoro-4-hydroxy-3-methoxy-L-phenylalanine-β-D-glucuronide
Formed from: 6-Fluoro-4-hydroxy-3-methoxy-L-phenylalanine

3-Fluoro-4-hydroxy-4′-methylaminoazobenzene
Probably formed from: 4′-Dimethylamino-3-fluoro-4-hydroxyazobenzene
→ 4′-Amino-3-fluoro-4-hydroxyazobenzene
Probably in: Rat A776

1-Fluoro-5-hydroxymethylchrysene
Probably formed from: 1-Fluoro-5-methylchrysene

3-Fluoro-5-hydroxymethylchrysene
Probably formed from: 3-Fluoro-5-methylchrysene
→ 3-Fluoro-1,2-dihydro-1,2-dihydroxy-5-hydroxymethylchrysene
Probably in: Rat C83
→ 3-Fluoro-7,8-dihydro-7,8-dihydroxy-5-hydroxymethylchrysene
Probably in: Rat C83
→ 3-Fluoro-9,10-dihydro-9,10-dihydroxy-5-hydroxymethylchrysene
Probably in: Rat C83
→ 3-Fluoro-7-hydroxy-5-hydroxymethylchrysene
Probably in: Rat C83

6-Fluoro-5-hydroxymethylchrysene
Probably formed from: 6-Fluoro-5-methylchrysene

7-Fluoro-5-hydroxymethylchrysene
Probably formed from: 7-Fluoro-5-methylchrysene
→ 7-Fluoro-1,2-dihydro-1,2-dihydroxy-5-hydroxymethylchrysene
Probably in: Rat C83
→ 7-Fluoro-9,10-dihydro-9,10-dihydroxy-5-hydroxymethylchrysene
Probably in: Rat C83
→ 7-Fluoro-1-hydroxy-5-hydroxymethylchrysene
Probably in: Rat C83

12-Fluoro-5-hydroxymethylchrysene
Probably formed from: 12-Fluoro-5-methylchrysene

***m*-Fluoro-4-(5-hydroxymethyl-5-methyl-1,3,2-dioxaphosphorinan-2-yl)-4-(5,5-dimethyl-1,3,2-dioxaphosphorinan-2-yl)butyrophenone**
Formed from: 4,4-Bis(5,5-dimethyl-1,3,2-dioxaphosphorinan-2-yl)-*m*-fluorobutyrophenone
→ 4-(5-Carboxy-5-methyl-1,3,2-dioxaphosphorinan-2-yl)-4-(5,5-dimethyl-1,3,2-dioxaphosphorinan-2-yl)-*m*-fluorobutyrophenone
Probably in: Rat J285

***cis*-5-Fluoro-2-hydroxymethyl-1-((*p*-methylsulphinyl)-α-hydroxybenzyl)indene-3-acetic acid**
Probably formed from: *cis*-5-Fluoro-2-methyl-1-((*p*-methylsulphinyl)benzylidenyl)indene-3-acetic acid

16α-Fluoro-2-hydroxy-3-O-methyl-17α-oestradiol
Probably formed from: 16α-Fluoro-2-hydroxy-17α-oestradiol
→ 16α-Fluoro-2-hydroxy-3-O-methyl-17α-oestradiol-2-β-D-glucuronide
Probably in: Rat J498

16β-Fluoro-2-hydroxy-3-O-methyl-17β-oestradiol
Probably formed from: 16β-Fluoro-2-hydroxy-17β-oestradiol
→ 16β-Fluoro-2-hydroxy-3-O-methyl-17β-oestradiol-2-β-D-glucuronide
Probably in: Rat J498
→ 16β-Fluoro-2-hydroxy-3-O-methyl-17β-oestradiol-17-β-D-glucuronide
Probably in: Rat J498

16α-Fluoro-2-hydroxy-3-O-methyl-17α-oestradiol-2-β-D-glucuronide
Probably formed from: 16α-Fluoro-2-hydroxy-3-O-methyl-17α-oestradiol

16β-Fluoro-2-hydroxy-3-O-methyl-17β-oestradiol-2-β-D-glucuronide
Probably formed from: 16β-Fluoro-2-hydroxy-3-O-methyl-17β-oestradiol

16β-Fluoro-2-hydroxy-3-O-methyl-17β-oestradiol-17-β-D-glucuronide
Probably formed from: 16β-Fluoro-2-hydroxy-3-O-methyl-17β-oestradiol

***p*-Fluoro-4-(2-hydroxy-4-methylpiperidin-1-yl)butyrophenone**
Formed from: Melperone

***p*-Fluoro-4-(3-hydroxy-4-methylpiperidin-1-yl)butyrophenone**
Formed from: Melperone

1-Fluoro-4-hydroxynaphthalene
Formed from: 1-Fluoronaphthalene

1-Fluoro-5-hydroxynaphthalene
Formed from: 1-Fluoronaphthalene
→ 1-Fluoro-8-hydroxy-5-tetralone
Probably in: Cunninghamella D109

2-Fluoro-4-hydroxy-17β-oestradiol
Formed from: 2-Fluoro-17β-oestradiol
→ 2-Fluoro-4-methoxy-17β-oestradiol
Probably in: Hamster H1

4-Fluoro-2-hydroxy-17β-oestradiol
Formed from: 4-Fluoro-17β-oestradiol
→ 4-Fluoro-2-methoxy-17β-oestradiol
Probably in: Hamster H1

16α-Fluoro-2-hydroxy-17α-oestradiol
Formed from: 16α-Fluoro-17α-oestradiol
→ 16α-Fluoro-2-hydroxy-17α-oestradiol-2-β-D-glucuronide
Probably in: Rat J498
→ 16α-Fluoro-2-methoxy-17α-oestradiol
Probably in: Rat J498
→ 16α-Fluoro-2-hydroxy-3-O-methyl-17α-oestradiol
Probably in: Rat J498

16α-Fluoro-2-hydroxy-17β-oestradiol
Formed from: 16α-Fluoro-17β-oestradiol
→ 16α-Fluoro-2-hydroxy-17β-oestradiol-2-β-D-glucuronide
Probably in: Rat J498
→ 16α-Fluoro-2-methoxy-17β-oestradiol
Probably in: Rat J498

16β-Fluoro-2-hydroxy-17β-oestradiol
Formed from: 16β-Fluoro-17β-oestradiol
→ 16β-Fluoro-2-hydroxy-17β-oestradiol-2-β-D-glucuronide
Probably in: Rat J498
→ 16β-Fluoro-2-methoxy-17β-oestradiol
Probably in: Rat J498
→ 16β-Fluoro-2-hydroxy-3-O-methyl-17β-oestradiol
Probably in: Rat J498

16α-Fluoro-2-hydroxy-17α-oestradiol-2-β-D-glucuronide
Probably formed from: 16α-Fluoro-2-hydroxy-17α-oestradiol

16α-Fluoro-2-hydroxy-17β-oestradiol-2-β-D-glucuronide
Probably formed from: 16α-Fluoro-2-hydroxy-17β-oestradiol

16β-Fluoro-2-hydroxy-17β-oestradiol-2-β-D-glucuronide
Probably formed from: 16β-Fluoro-2-hydroxy-17β-oestradiol

4-Fluoro-2-hydroxyoestrone
Formed from: 4-Fluoro-17β-oestradiol

4-(4-Fluoro-3-hydroxyphenoxy)benzaldehyde semicarbazone
Probably formed from: 4-(4-Fluorophenoxy)benzaldehyde semicarbazone

5-(2-Fluoro-4-hydroxyphenyl)-1,3-dihydro-3-hydroxy-1-methyl-7-nitro-2*H*-1,4-benzodiazepin-2-one
Probably formed from: 3-Hydroxyflunitrazepam

5-(2-Fluoro-4-hydroxyphenyl)-1,3-dihydro-1-methyl-7-nitro-2*H*-1,4-benzodiazepin-2-one
Probably formed from: Flunitrazepam

***p*-(2-Fluoro-4-hydroxyphenyl)phenylacetic acid**
Probably formed from: *p*-(*o*-Fluorophenyl)phenylacetic acid

2-Fluoro-4-hydroxyphenylpyruvic acid
→ 3-Fluoro-2,5-dihydroxyphenylacetic acid
In: Pseudomonas D688

***o*-Fluoro-2-hydroxypropiophenone**
Formed from: *o*-Fluorobenzaldehyde

***m*-Fluoro-2-hydroxypropiophenone**
Formed from: *m*-Fluorobenzaldehyde

***p*-Fluoro-2-hydroxypropiophenone**
Formed from: *p*-Fluorobenzaldehyde

6-Fluoro-3-(1-(3-hydroxypropyl)-4-piperidinyl) benzisoxazole
Formed from: Iloperidone

7-Fluoro-2-hydroxy*spiro*(9*H*-fluorene-9,4′-imidazolidine)-2′,5′-dione
Formed from: Imirestat

1-Fluoro-8-hydroxy-5-tetralone
Probably formed from: 1-Fluoro-5-hydroxynaphthalene

6-Fluoro-5-hydroxytryptamine
Formed from: 6-Fluorotryptamine

5-Fluoro-β-hydroxy-L-tryptophan
Probably formed from: 5-Fluoro-L-tryptophan

(*S*)-2-(((7-Fluoroindan-4-yl)oxy)methyl)-N-hydroxymorpholine
Formed from: (*S*)-2-(((7-Fluoroindan-4-yl) oxy)methyl)morpholine

(*S*)-2-(((7-Fluoroindan-4-yl)oxy)methyl) morpholine
→ (*S*)-2-(((7-Fluoro-1-hydroxyindan-4-yl) oxy)methyl)morpholine
In: Man K114
→ (*S*)-2-(((7-Fluoro-2-hydroxyindan-4-yl) oxy)methyl)morpholine
In: Man K114
→ (*S*)-2-(((7-Fluoroindan-4-yl)oxy)methyl)-N-hydroxymorpholine
In: Man K114

4-Fluoroindole
→ 4-Fluoro-L-tryptophan
In: Carrot B874
Nicotiana B874

5-Fluoroindole
Probably formed from: 5-Fluoro-L-tryptophan
→ 5-Fluoro-L-tryptophan
In: Carrot B874
Nicotiana B874

6-Fluoroindole
→ 6-Fluoro-L-tryptophan
In: Carrot B874
Nicotiana B874

5-Fluoroindole-3-pyruvic acid
Formed from: 5-Fluoro-L-tryptophan

***m*-Fluoroiodobenzene**
→ 3-Iodocatechol
In: Pseudomonas F76

4-Fluoro-L-kynurenine (α,2-diamino-4-fluoro-γ-oxobenzenebutanoic acid)
Formed from: 4-Fluoro-N-formyl-L-kynurenine
→ 4-Fluoroanthranilic acid
In: Pseudomonas B248

5-Fluoro-L-kynurenine (α,2-diamino-5-fluoro-γ-oxobenzenebutanoic acid)
→ 5-Fluoroanthranilic acid
In: Pseudomonas B248

***p*-Fluoromandelic acid**
→ *p*-Fluorophenylglyoxylic acid
In: Rhodotorula H90

2-Fluoro-4-methoxybenzaldehyde
Probably formed from: *o*-Fluorobenzaldehyde
→ 3-Chloro-2-fluoro-4-methoxybenzaldehyde
Probably in: Bjerkandera H914

3-Fluoro-4-methoxybenzaldehyde
Probably formed from: *m*-Fluorobenzaldehyde

3-Fluoro-L-tyrosine
→ 3-Chloro-5-fluoro-4-methoxybenzaldehyde
Probably in: Bjerkandera H914

2-Fluoro-4-methoxy-17β-oestradiol
Probably formed from: 2-Fluoro-4-hydroxy-17β-oestradiol

4-Fluoro-2-methoxy-17β-oestradiol
Formed from: 2,4-Difluoro-17β-oestradiol
4-Fluoro-17β-oestradiol
Probably formed from: 4-Fluoro-2-hydroxy-17β-oestradiol

16α-Fluoro-2-methoxy-17α-oestradiol
Probably formed from: 16α-Fluoro-2-hydroxy-17α-oestradiol
→ 16α-Fluoro-2-methoxy-17α-oestradiol-3-β-D-glucuronide
Probably in: Rat J498

16α-Fluoro-2-methoxy-17β-oestradiol
Probably formed from: 16α-Fluoro-2-hydroxy-17β-oestradiol
→ 16α-Fluoro-2-methoxy-17β-oestradiol-3-β-D-glucuronide
Probably in: Rat J498

16β-Fluoro-2-methoxy-17β-oestradiol
Probably formed from: 16β-Fluoro-2-hydroxy-17β-oestradiol
→ 16β-Fluoro-15α-hydroxy-2-methoxy-17β-oestradiol
Probably in: Rat J498
→ 16β-Fluoro-2-methoxy-17β-oestradiol-3-β-D-glucuronide
Probably in: Rat J498
→ 16β-Fluoro-2-methoxy-17β-oestradiol-17-β-D-glucuronide
Probably in: Rat J498

16α-Fluoro-2-methoxy-17α-oestradiol-3-β-D-glucuronide
Probably formed from: 16α-Fluoro-2-methoxy-17α-oestradiol

16α-Fluoro-2-methoxy-17β-oestradiol-3-β-D-glucuronide
Probably formed from: 16α-Fluoro-2-methoxy-17β-oestradiol

16β-Fluoro-2-methoxy-17β-oestradiol-3-β-D-glucuronide
Probably formed from: 16β-Fluoro-2-methoxy-17β-oestradiol

16β-Fluoro-2-methoxy-17β-oestradiol-17-β-D-glucuronide
Probably formed from: 16β-Fluoro-2-methoxy-17β-oestradiol

2-Fluoro-1-methoxy-4-((pentafluorophenyl)sulphonamido)benzene
→ 2-Fluoro-1-methoxy-4-((2,3,5,6-tetrafluoro-4-(S-glutathionyl)phenyl)sulphonamido)benzene
In: Mouse K62

4-Fluoro-2-methoxyphenol
Formed from: 4-Fluorocatechol

5-Fluoro-2-methoxyphenol
Formed from: 4-Fluorocatechol

2-Fluoro-1-methoxy-4-((2,3,5,6-tetrafluoro-4-(S-glutathionyl)phenyl)sulphonamido)benzene
Formed from: 2-Fluoro-1-methoxy-4-((pentafluorophenyl)sulphonamido)benzene
→ 4-((4-(S-(L-Cysteinylglycyl))-2,3,5,6-tetrafluorophenyl)sulphonamido)-2-fluoro-1-methoxybenzene
Probably in: Mouse K62
→ 2-Fluoro-1-methoxy-4-((2,3,5,6-tetrafluoro-4-mercaptophenyl)sulphonamido)benzene
Probably in: Mouse K62

2-Fluoro-1-methoxy-4-((2,3,5,6-tetrafluoro-4-mercaptophenyl)sulphonamido)benzene
Probably formed from: 2-Fluoro-1-methoxy-4-((2,3,5,6-tetrafluoro-4-(S-glutathionyl)phenyl)sulphonamido)benzene

3′-Fluoro-4-methylaminoazobenzene
Probably formed from: 4-Dimethylamino-3′-fluoroazobenzene

4′-Fluoro-4-methylaminoazobenzene
Probably formed from: 4-Dimethylamino-4′-fluoroazobenzene
→ 4-Amino-4′-fluoroazobenzene
In: Rat A1979

***p*-Fluoro-N-methylamphetamine**
Formed from: *p*-Fluorodeprenyl

2-Fluoro-4-methylaniline
→ N-(4-Amino-3-fluorobenzyl)-2-fluoro-4-methylaniline
In: Rat G518
→ 2,2′-Difluoro-4,4′-dimethylazobenzene
In: Rat G518
→ 2,2′-Difluoro-4,4′-dimethylazoxybenzene
In: Rat G518

***p*-Fluoro-N-methylaniline**
Formed from: *p*-Fluoro-N,N-dimethylaniline
→ *p*-Fluoroaniline
In: Rat G555
→ *p*-Fluoro-N-methylphenylhydroxylamine
In: Rat G555
→ *p*-Methylaminophenol
In: Rat G555

***p*-Fluoromethylbenzaldehyde**
Formed from: *p*-Fluoromethylphenylglyoxylic acid

3-Fluoro-6-methylcatechol
Formed from: 4-Fluoro-2,3-dihydro-2,3-dihydroxytoluene
Probably formed from: 2-Fluoro-5-methylphenol

5-Fluoro-4-methylcatechol
Probably formed from: 4-Fluoro-3-methylphenol

1-Fluoro-5-methylchrysene
→ 1-Fluoro-1,2-dihydro-1,2-dihydroxy-5-methylchrysene
Probably in: Rat B234
→ 1-Fluoro-7,8-dihydro-7,8-dihydroxy-5-methylchrysene
Probably in: Rat B234
→ 1-Fluoro-5-hydroxymethylchrysene
Probably in: Rat B234

3-Fluoro-5-methylchrysene
→ 3-Fluoro-1,2-dihydro-1,2-dihydroxy-5-methylchrysene
Probably in: Rat B234
→ 3-Fluoro-7,8-dihydro-7,8-dihydroxy-5-methylchrysene
Probably in: Rat B234
→ 3-Fluoro-5-hydroxymethylchrysene
Probably in: Rat B234

6-Fluoro-5-methylchrysene
→ 6-Fluoro-1,2-dihydro-1,2-dihydroxy-5-methylchrysene
Probably in: Rat B234
→ 6-Fluoro-7,8-dihydro-7,8-dihydroxy-5-methylchrysene
Probably in: Rat B234
→ 6-Fluoro-5-hydroxymethylchrysene
Probably in: Rat B234

7-Fluoro-5-methylchrysene
→ 7-Fluoro-1,2-dihydro-1,2-dihydroxy-5-methylchrysene
Probably in: Rat B234
→ 7-Fluoro-7,8-dihydro-7,8-dihydroxy-5-methylchrysene
Probably in: Rat B234
→ 7-Fluoro-5-hydroxymethylchrysene
Probably in: Rat B234

12-Fluoro-5-methylchrysene
→ 12-Fluoro-1,2-dihydro-1,2-dihydroxy-5-methylchrysene
Probably in: Rat B234
→ 12-Fluoro-7,8-dihydro-7,8-dihydroxy-5-methylchrysene
Probably in: Rat B234
→ 12-Fluoro-5-hydroxymethylchrysene
Probably in: Rat B234

α-Fluoromethyl-3,4-dihydroxy-L-phenylalanine
→ α-Fluoromethyl-4-hydroxy-3-methoxy-L-phenylalanine
In: Pig B813

2-Fluoromethyl-6-glycollamido-3-(*o*-hydroxymethylphenyl)-4(3*H*)-quinazolinone
Probably formed from: 6-Acetamido-2-fluoromethyl-3-(*o*-hydroxymethylphenyl)-4(3*H*)-quinazolinone

α-Fluoromethyl-4-hydroxy-3-methoxy-L-phenylalanine
Formed from: α-Fluoromethyl-3,4-dihydroxy-L-phenylalanine

6-Fluoro-1-methyl-7-(4-((5-methyl-2-oxo-1,3-dioxol-4-yl)methyl)-1-piperazinyl)-4-oxo-1*H*,4*H*-[1,3]thiazeto[3,2-*a*]quinoline-3-carboxylic acid (prulifloxacin)
→ 6-Fluoro-1-methyl-7-(4-((5-methyl-2-oxo-1,3-dioxol-4-yl)methyl)-1-piperazinyl)-4-oxo-1*H*,4*H*-[1,3]thiazeto[3,2-*a*]quinoline-3-carboxy-β-D-glucuronic acid
In: Man H764
→ 6-Fluoro-1-methyl-4-oxo-7-(1-piperazinyl)-4*H*-[1,3]thiazeto[3,2-*a*]quinoline-3-carboxylic acid
In: Rat J378

6-Fluoro-1-methyl-7-(4-((5-methyl-2-oxo-1,3-dioxol-4-yl)methyl)-1-piperazinyl)-4-oxo-1*H*,4*H*-[1,3]thiazeto[3,2-*a*]quinoline-3-carboxyl-β-D-glucuronic acid
Formed from: 6-Fluoro-1-methyl-7-(4-((5-methyl-2-oxo-1,3-dioxol-4-yl)methyl)-1-piperazinyl)-4-oxo-1*H*,4*H*-[1,3]thiazeto[3,2-*a*]quinoline-3-carboxylic acid

***cis*-5-Fluoro-2-methyl-1-((*p*-methylsulphinyl)benzylidenyl)indene-3-acetic acid** (sulindac)
Formed from: Sulindac sulphide
Probably formed from: *cis*-5-Fluoro-2-methyl-1-((*p*-methylthio)benzylidenyl)indene-3-acetic acid
→ *cis*-5-Fluoro-2-hydroxymethyl-1-((*p*-methylsulphinyl)-α-hydroxybenzyl)indene-3-acetic acid
Probably in: Man A1885
→ *cis*-5-Fluoro-2-methyl-1-((*p*-methylsulphinyl)benzylidenyl)indene-3-acetyl-β-D-glucuronic acid
In: Man A1885
→ *cis*-5-Fluoro-2-methyl-1-((*p*-methylsulphonyl)benzylidenyl)indene-3-acetic acid
In: Aspergillus D432
Dog A1885, C449
Guinea pig B114, C449
Man A1885, C518
Monkey A1885
Nocardia D432
Rabbit C449, D115
Rat A1885, B114, F836
→ *cis*-5-Fluoro-2-methyl-1-((*p*-methylthio)benzylidenyl)indene-3-acetic acid
In: Dog A1885
Escherichia H560
Guinea pig B114
Man C518

Microorganism E241
Monkey A1885
Nocardia D432
Rabbit H561
Rat A1885, B114, B612, B637, D836, H561
Sporobolomyces D432

***cis*-5-Fluoro-2-methyl-1-((*p*-methylsulphinyl)benzylidenyl)indene-3-acetyl-β-D-glucuronic acid**
Formed from: *cis*-5-Fluoro-2-methyl-1-((*p*-methylsulphinyl)benzylidenyl)indene-3-acetic acid

7-Fluoro-1-methyl-3-methylsulphinyl-4-quinolone
See: Flosequinan

***cis*-5-Fluoro-2-methyl-1-((*p*-methylsulphonyl)benzylidenyl)indene-3-acetic acid**
Formed from: *cis*-5-Fluoro-2-methyl-1-((*p*-methylsulphinyl)benzylidenyl)indene-3-acetic acid
→ *cis*-5-Fluoro-2-methyl-1-((*p*-methylsulphonyl)benzylidenyl)indene-3-acety-β-D-glucuronic acid
Probably in: Man A1885

***cis*-5-Fluoro-2-methyl-1-((*p*-methylsulphonyl)benzylidenyl)indene-3-acetyl-β-D-glucuronic acid**
Probably formed from: *cis*-5-Fluoro-2-methyl-1-((*p*-methylsulphonyl)benzylidenyl)indene-3-acetic acid

7-Fluoro-1-methyl-3-methylsulphonyl-4-quinolone
Formed from: Flosequinan

***cis*-5-Fluoro-2-methyl-1-((*p*-methylthio)benzylidenyl)indene-3-acetic acid**
Formed from: *cis*-5-Fluoro-2-methyl-1-((*p*-methylsulphinyl)benzylidenyl)indene-3-acetic acid
→ *cis*-5-Fluoro-2-methyl-1-((*p*-methylsulphinyl)benzylidenyl)indene-3-acetic acid
Probably in: Dog C449
Guinea pig C449
Rabbit C449, H561
Rat H561

7-Fluoro-1-methyl-3-methylthio-4-quinolone
Formed from: Flosequinan
→ Flosequinan
In: Man J159
Rat J159, J738

6-Fluoro-1-methyl-4-oxo-7-(1-piperazinyl)-4*H*-[1,3]thiazeto[3,2-*a*]quinoline-3-carboxylic acid
Formed from: 6-Fluoro-1-methyl-7-(4-((5-methyl-2-oxo-1,3-dioxol-4-yl)methyl)-1-piperazinyl)-4-oxo-1*H*,4*H*-[1,3]thiazeto[3,2-*a*]quinoline-3-carboxylic acid

2-Fluoro-5-methylphenol
→ 3-Fluoro-6-methylcatechol
Probably in: Pseudomonas A2391

4-Fluoro-3-methylphenol
→ 4-Fluoro-3-methylphenyl-β-D-glucuronide
In: Rat H468, H522
→ 4-Fluoro-3-methylphenyl sulphate
In: Rat H468, H522
→ 5-Fluoro-4-methylcatechol
Probably in: Pseudomonas A2391

4-Fluoro-3-methylphenyl-β-D-glucuronide
Formed from: 4-Fluoro-3-methylphenol

***p*-Fluoromethylphenylglyoxylic acid**
→ *p*-Fluoromethylbenzaldehyde
In: Pseudomonas E714

***p*-Fluoro-N-methylphenylhydroxylamine**
Formed from: *p*-Fluoro-N-methylaniline

4-Fluoro-3-methylphenyl sulphate
Formed from: 4-Fluoro-3-methylphenol

1-Fluoronaphthalene
→ *trans*-1-Fluoro-3,4-dihydro-3,4-dihydroxynaphthalene
In: Cunninghamella D109
→ *trans*-1-Fluoro-5,6-dihydro-5,6-dihydroxynaphthalene
In: Cunninghamella D109
→ 1-Fluoro-4-hydroxynaphthalene
In: Cunninghamella D109
→ 1-Fluoro-5-hydroxynaphthalene
In: Cunninghamella D109

3-Fluoro-6-(4-methyl-1-piperazinyl)-11*H*-dibenzo[*b*,*e*]azepine
See: Fluperlapine

***p*-Fluoro-4-(4-methylpiperidin-1-yl)butyrophenone**
See: Melperone

(*S*)-N-((3-(3-Fluoro-4-(4-morpholinyl)phenyl)-2-oxo-5-oxazolidinyl)methyl)acetamide
See: Linezolid

3-Fluoro-4-nitrocatechol
Formed from: 3-Fluoro-4-nitrophenol
3-Fluoro-5-nitrocatechol
Formed from: 2-Fluoro-4-nitrophenol
4-Fluoro-5-nitrocatechol
Formed from: 3-Fluoro-4-nitrophenol
2-Fluoro-4-nitrophenol
→ 3-Fluoro-5-nitrocatechol
In: Trichosporon H420
3-Fluoro-4-nitrophenol
→ 3-Fluoro-4-nitrocatechol
In: Trichosporon H420
→ 4-Fluoro-5-nitrocatechol
In: Trichosporon H420
5-Fluoro-2-nitrophenol
→ 4-Fluorocatechol
In: Pseudomonas E607, H65
***p*-Fluoronitrosobenzene**
Formed from: *p*-Fluoroaniline
6-Fluoronoradrenaline
Formed from: 6-Fluoro-3,4-dihydroxyphenethylamine
6′-Fluoronorsolorinic acid (2-(6-fluoro-1-oxohexyl)-1,3,6,8-tetrahydroxy-9,10-anthracenedione)
Formed from 6-fluorohexanoic acid in Aspergillus H675
2-Fluoro-17β-oestradiol
→ 2-Fluorooestrone
In: Rat G488
→ 2-Fluoro-4-hydroxy-17β-oestradiol
In: Hamster H1
→ 2-Hydroxy-17β-oestradiol
In: Hamster H1
See also: D712
4-Fluoro-17β-oestradiol
→ 4-Fluoro-2-hydroxyoestrone
In: Hamster H1
Rat G488
→ 4-Fluoro-2-methoxy-17β-oestradiol
In: Hamster D712
→ 4-Fluorooestrone
In: Rat G488
16α-Fluoro-17α-oestradiol
→ 16α-Fluoro-2-hydroxy-17α-oestradiol
In: Rat J498
→ 16α-Fluoro-17α-oestradiol-3-β-D-glucuronide
In: Rat J498
16α-Fluoro-17β-oestradiol
→ 16α-Fluoro-2-hydroxy-17β-oestradiol
In: Rat J498
→ 16α-Fluoro-17β-oestradiol-3-β-D-glucuronide
In: Rat J498
16β-Fluoro-17β-oestradiol
→ 16β-Fluoro-2-hydroxy-17β-oestradiol
In: Rat J498
→ 16β-Fluoro-17β-oestradiol-3-β-D-glucuronide
In: Rat J498
16α-Fluoro-17α-oestradiol-3-β-D-glucuronide
Formed from: 16α-Fluoro-17α-oestradiol
16α-Fluoro-17β-oestradiol-3-β-D-glucuronide
Formed from: 16α-Fluoro-17β-oestradiol
16β-Fluoro-17β-oestradiol-3-β-D-glucuronide
Formed from: 16β-Fluoro-17β-oestradiol
2-Fluorooestrone
Formed from: 2-Fluoro-17β-oestradiol
→ 2-Fluorooestrone-β-D-glucuronide
Probably in: Rat G488
4-Fluorooestrone
Formed from: 4-Fluoro-17β-oestradiol
2-Fluorooestrone-β-D-glucuronide
Formed from: 2-Fluorooestrone
3-(4-(4-(6-Fluoro-1-oxobenzo[*b*]thien-3-yl)piperazinyl)butyl)-2,5,5-trimethyl-4-thiazolidinone
Formed from: 3-(4-(4-(6-Fluorobenzo[*b*]thien-3-yl)piperazinyl)butyl)-2,5,5-trimethyl-4-thiazolidinone
→ 3-(4-(4-(6-Fluoro-1,1-dioxobenzo[*b*]thien-3-yl)piperazinyl)butyl)-2,5,5-trimethyl-4-thiazolidinone
In: Rat J53
2-(6-Fluoro-1-oxohexyl)-1,3,6,8-tetrahydroxy-9,10-anthracenedione
See: 6′-Fluoronorsolorinic acid
2-(((7-Fluoro-1-oxoindan-4-yl)oxy)methyl)morpholine
Probably formed from: 2-(((7-Fluoro-2-hydroxyindan-4-yl)oxy)methyl)morpholine

***o*-Fluorophenethylamine**
→ *o*-Fluorophenylacetaldehyde
In: Pig A72

***m*-Fluorophenethylamine**
→ *m*-Fluorophenylacetaldehyde
In: Pig A72

***p*-Fluorophenethylamine**
→ *p*-Fluorophenylacetaldehyde
In: Alcaligenes H367
Pig A72

***o*-Fluorophenol**
Formed from: *o*-Fluoroanisole
Fluorobenzene
3-Fluoro-L-tyrosine
→ *m*-Fluorobenzoic acid
In: Microorganism G971
→ 3-Fluorocatechol
In: Bacillus J340
Candida K653
Trichosporon A902, H420
→ *o*-Fluorophenyl-β-D-glucuronide
In: Rat H468, H522
→ *o*-Fluorophenyl sulphate
In: Man E575
Rat E575, H468, H522
→ 3-Fluoro-L-tyrosine
In: Erwinia F4

***m*-Fluorophenol**
Formed from: Fluorobenzene
→ 3-Fluorocatechol
In: Trichosporon H420
→ 4-Fluorocatechol
In: Candida K653
Trichosporon A902, H420
→ *m*-Fluorophenyl-β-D-glucuronide
In: Rat H468, H522
→ *m*-Fluorophenyl sulphate
In: Rat H468, H522
→ 2-Fluoro-L-tyrosine
In: Erwinia F4

***p*-Fluorophenol**
Formed from: Cisapride
p-Fluoroanisole
Fluorobenzene
p-Fluorophenyl nicotinate
p-Fluorophenyl-β-D-xyloside
→ 4-Fluorocatechol
In: Candida K653
Trichosporon A902, H420
→ *p*-Fluorophenyl-β-D-glucuronide
In: Rat H468, H522
→ *p*-Fluorophenyl sulphate
In: Rat H468, H522
See also: E926
→ Quinol
In: Rat J81

4-(*p*-Fluorophenoxy)benzaldehyde semicarbazone
→ 4-(4-Fluoro-3-hydroxyphenoxy) benzaldehyde semicarbazide
Probably in: Rat K131
→ 4-(*p*-Fluorophenoxy)benzoic acid
In: Rat K131
→ 4-(*p*-Fluorophenoxy)benzoylsemicarbazide
In: Rat K131

4-Fluoro-3-phenoxybenzoic acid
Probably formed from: 4-Fluoro-3-phenoxymandelonitrile

4-(*p*-Fluorophenoxy)benzoic acid
Formed from: 4-(*p*-Fluorophenoxy) benzaldehyde semicarbazone
→ 4-(*p*-Fluorophenoxy)benzoyl-β-D-glucuronic acid
In: Rat K131
→ 4-(*p*-Fluorophenoxy)benzoylglycine
In: Rat K131

4-(*p*-Fluorophenoxy)benzoyl-β-D-glucuronic acid
Formed from: 4-(*p*-Fluorophenoxy)benzoic acid

4-(*p*-Fluorophenoxy)benzoylglycine
Formed from: 4-(*p*-Fluorophenoxy)benzoic acid

4-(*p*-Fluorophenoxy)benzoylsemicarbazide
Formed from: 4-(*p*-Fluorophenoxy) benzaldehyde semicarbazone

2-(*p*-Fluorophenoxy)-3-(4-hydroxy-3-methoxyphenyl)-3-oxopropanol
→ Guaiacylglycerol β-(*p*-fluorophenyl) ether
With guaiacylglycerol β-guaiacyl ether dehydrogenase G357

2-(*p*-Fluorophenoxy)-1-(*o*-hydroxyphenyl)ethane
→ 2-(*p*-Fluorophenoxy)-1-(*o*-hydroxyphenyl)ethane-β-D-glucuronide

In: Pig F849
→ 2-(*p*-Fluorophenoxy)-1-(*o*-hydroxyphenyl)ethane-O-sulphate
In: Dog F849
Rat F849

2-(*p*-Fluorophenoxy)-1-(*o*-hydroxyphenyl)ethane-β-D-glucuronide
Formed from: 2-(*p*-Fluorophenoxy)-1-(*o*-hydroxyphenyl)ethane

2-(*p*-Fluorophenoxy)-1-(*o*-hydroxyphenyl)ethane-O-sulphate
Formed from: 2-(*p*-Fluorophenoxy)-1-(*o*-hydroxyphenyl)ethane

4-Fluoro-3-phenoxymandelonitrile
Formed from: Cyfluthrin
Flumethrin
→ 4-Fluoro-3-phenoxybenzoic acid
Probably in: Man H857

3-(*p*-Fluorophenoxy)propionaldehyde
Probably formed from: Cisapride

***o*-Fluorophenylacetaldehyde**
Formed from: *o*-Fluorophenethylamine

***m*-Fluorophenylacetaldehyde**
Formed from: *m*-Fluorophenethylamine

***p*-Fluorophenylacetaldehyde**
Formed from: *p*-Fluorophenethylamine

***p*-Fluorophenylacetamide**
Probably formed from: *p*-Fluorophenylalanine

(*R*)-α-Fluorophenylacetic acid
Formed from: α-Fluorophenylmalonic acid

***p*-Fluorophenylacetic acid**
Formed from: Bromperidol
Haloperidol
→ *p*-Fluorophenylacetylglycine
In: Dog C343
Man A3090, C343
Rat C343
See also: E18

***p*-Fluorophenylacetylglycine**
Formed from: *p*-Fluorophenylacetic acid
Timiperone

***o*-Fluoro-L-phenylalanine**
→ *o*-Fluorobenzaldehyde
In: Bjerkandera H914
→ *o*-Fluorophenylpyruvic acid
In: Bushbean E500

***m*-Fluoro-L-phenylalanine**
→ *m*-Fluorobenzaldehyde
In: Bjerkandera H914
→ *m*-Fluorophenylpyruvic acid
In: Bushbean E500

***p*-Fluoro-L-phenylalanine**
→ *p*-Fluorophenylacetamide
Probably in: Pseudomonas D246
→ *p*-Fluorophenylpyruvic acid
In: Bushbean E500
Pseudomonas D246

3-(*m*-Fluorophenyl)-2,5-dihydro-5,5-dimethyl-4-((4-methylsulphonyl)phenyl)-2-furanol
→ 3-(*m*-Fluorophenyl)-2,5-dihydro-5,5-dimethyl-4-((4-methylsulphonyl)phenyl)-2-furanol-β-D-glucuronide
In: Rat J568
→ 3-(*m*-Fluorophenyl)-2,5-dihydro-5,5-dimethyl-4-((4-methylsulphonyl)phenyl)-2-furanone
In: Man J568
Rat J568

3-(*m*-Fluorophenyl)-2,5-dihydro-5,5-dimethyl-4-((4-methylsulphonyl)phenyl)-2-furanol-β-D-glucuronide
Formed from: 3-(*m*-Fluorophenyl)-2,5-dihydro-5,5-dimethyl-4-((4-methylsulphonyl)phenyl)-2-furanol

3-(*m*-Fluorophenyl)-2,5-dihydro-5,5-dimethyl-4-((4-methylsulphonyl)phenyl)-2-furanone
Formed from: 3-(*m*-Fluorophenyl)-2,5-dihydro-5,5-dimethyl-4-((4-methylsulphonyl)phenyl)-2-furanol

5-(*o*-Fluorophenyl)-1,3-dihydro-1-methyl-7-nitro-2*H*-1,4-benzodiazepin-2-one
See: Flunitrazepam

5-(*o*-Fluorophenyl)-1,3-dihydro-7-nitro-2*H*-1,4-benzodiazepin-2-one
Formed from: Flunitrazepam

2-(*p*-Fluorophenyl)-4,5-dihydro-1*H*-pyrrole-3-acetic acid
Formed from: Ketanserin

6-(*p*-Fluorophenyl)-2,3-dihydro-5-(4-pyridyl)imidazo[2,1-*b*]thiazole
→ 6-(*p*-Fluorophenyl)-2,3-dihydro-5-(4-pyridyl)imidazo[2,1-*b*]thiazole-1-oxide

In: Rat F20, F559

6-(*p*-Fluorophenyl)-2,3-dihydro-5-(4-pyridyl) imidazo[2,1-*b*]thiazole-1,1-dioxide
Formed from: 6-(*p*-Fluorophenyl)-2,3-dihydro-5-(4-pyridyl)imidazo[2,1-*b*]thiazole-1-oxide
→ 6-(*p*-Fluorophenyl)-2,3-dihydro-5-(4-pyridyl-N-oxide)imidazo[2,1-*b*]thiazole-1,1-dioxide
In: Rat F559

6-(*p*-Fluorophenyl)-2,3-dihydro-5-(4-pyridyl) imidazo[2,1-*b*]thiazole-1-oxide
Formed from: 6-(*p*-Fluorophenyl)-2,3-dihydro-5-(4-pyridyl)imidazo[2,1-*b*]thiazole
→ 6-(*p*-Fluorophenyl)-2,3-dihydro-5-(4-pyridyl)imidazo[2,1-*b*]thiazole-1,1-dioxide
In: Rat F20, F559

6-(*p*-Fluorophenyl)-2,3-dihydro-5-(4-pyridyl-N-oxide)imidazo[2,1-*b*]thiazole-1,1-dioxide
Formed from: 6-(*p*-Fluorophenyl)-2,3-dihydro-5-(4-pyridyl)imidazo[2,1-*b*]thiazole-1,1-dioxide

2-(*p*-Fluorophenyl)-4,5-dihydro-1*H*-pyrrole-3-acetic acid
Formed from: Ketanserin

5-(*p*-Fluorophenyl)-2,4-dihydro-3*H*-1,2,4-triazolo-3-thione
Formed from: 4-Ethyl-5-(*p*-fluorophenyl)-2,4-dihydro-3*H*-1,2,4-triazolo-3-thione

2-(*p*-Fluorophenyl)-β,δ-dihydroxy-4-(((*o*-hydroxyphenyl)amino)carbonyl)-5-(1-methylethyl)-3-phenyl-1*H*-pyrrole-1-heptanoic acid
Formed from: Atorvastatin
→ 2-(*p*-Fluorophenyl)-β,δ-dihydroxy-4-(((*o*-hydroxyphenyl)amino)carbonyl)-5-(1-methylethyl)-3-phenyl-1*H*-pyrrole-1-heptanoic acid-β-D-glucuronide
Probably in: Dog J736
Rat J736

2-(*p*-Fluorophenyl)-β,δ-dihydroxy-4-(((*p*-hydroxyphenyl)amino)carbonyl)-5-(1-methylethyl)-3-phenyl-1*H*-pyrrole-1-heptanoic acid
Formed from: Atorvastatin

2-(*p*-Fluorophenyl)-β,δ-dihydroxy-4-(((*o*-hydroxyphenyl)amino)carbonyl)-5-(1-methylethyl)-3-phenyl-1*H*-pyrrole-1-heptanoic acid-β-D-glucuronide
Probably formed from: 2-(*p*-Fluorophenyl)-β,δ-dihydroxy-4-(((*o*-hydroxyphenyl)amino)carbonyl)-5-(1-methylethyl)-3-phenyl-1*H*-pyrrole-1-heptanoic acid

2-(*p*-Fluorophenyl)-β,δ-dihydroxy-5-(1-methylethenyl)-3-phenyl-4-((phenyl)amino)carbonyl)-1*H*-pyrrole-1-heptanoic acid
Formed from: Atorvastatin

(β*R*,δ*R*))-2-(4-Fluorophenyl)-β,δ-dihydroxy-5-(1-methylethyl)-3-phenyl-4-((phenylamino)carbonyl)-1*H*-pyrrole-1-heptanoic acid
See: Atorvasatin

(*E*)-(5*S*)-7-(4-(*p*-Fluorophenyl)-2,6-diisopropyl-5-(methoxymethyl)pyrid-3-yl)-5-hydroxyhept-6-enoic acid
Formed from: Cerivastatin
→ (*E*)-(5*S*)-7-(4-(*p*-Fluorophenyl)-2,6-diisopropyl-5-(methoxymethyl)pyrid-3-yl)-5-hydroxyhept-6-enoyltaurine
Probably in: Mouse J426
Rat J426
→ (*E*)-(5*S*)-7-(4-(*p*-Fluorophenyl)-5-hydroxymethyl-2,6-diisopropylpyrid-3-yl)-5-hydroxyhept-6-enoic acid
Probably in: Mouse J426
Rat J426

(*E*)-(5*S*)-7-(4-(*p*-Fluorophenyl)-2,6-diisopropyl-5-(methoxymethyl)pyrid-3-yl)-5-hydroxyhept-6-enoyltaurine
Probably formed from: (*E*)-(5*S*)-7-(4-(*p*-Fluorophenyl)-2,6-diisopropyl-5-(methoxymethyl)pyrid-3-yl)-5-hydroxyhept-6-enoic acid

5-(4-(*p*-Fluorophenyl)-2,6-diisopropyl-5-(methoxymethyl)pyrid-3-yl)-3-hydroxypent-4-enoic acid
Formed from: Cerivastatin
→ 5-(4-(*p*-Fluorophenyl)-2,6-diisopropyl-5-(methoxymethyl)pyrid-3-yl)penta-2,4-dienoic acid
Probably in: Mouse J426
Rat J426

5-(4-(*p*-Fluorophenyl)-2,6-diisopropyl-5-(methoxymethyl)pyrid-3-yl)penta-2,4-dienoic acid
Probably formed from: 5-(4-(*p*-Fluorophenyl)-2,6-diisopropyl-5-(methoxymethyl)pyrid-3-yl)-3-hydroxypent-4-enoic acid
→ 5-(4-(*p*-Fluorophenyl)-2,6-diisopropyl-5-(methoxymethyl)pyrid-3-yl)pent-4-enoic acid
Probably in: Mouse J426
Rat J426
→ 5-(4-(*p*-Fluorophenyl)-2,6-diisopropyl-5-(methoxymethyl)pyrid-3-yl)penta-2,4-dienoyltaurine
Probably in: Mouse J426
Rat J426

5-(4-(*p*-Fluorophenyl)-2,6-diisopropyl-5-(methoxymethyl)pyrid-3-yl)penta-2,4-dienoyltaurine
Probably formed from: 5-(4-(*p*-Fluorophenyl)-2,6-diisopropyl-5-(methoxymethyl)pyrid-3-yl)penta-2,4-dienoic acid

5-(4-(*p*-Fluorophenyl)-2,6-diisopropyl-5-(methoxymethyl)pyrid-3-ylpent-4-enoic acid
Probably formed from: 5-(4-(*p*-Fluorophenyl)-2,6-diisopropyl-5-(methoxymethyl)pyrid-3-yl)penta-2,4-dienoic acid
→ 5-(4-(*p*-Fluorophenyl)-2,6-diisopropyl-5-(methoxymethyl)pyrid-3-yl)pent-4-enoyltaurine
Probably in: Mouse J426
Rat J426

5-(4-(*p*-Fluorophenyl)-2,6-diisopropyl-5-(methoxymethyl)pyrid-3-yl)pent-4-enoyltaurine
Probably formed from: 5-(4-(*p*-Fluorophenyl)-2,6-diisopropyl-5-(methoxymethyl)pyrid-3-yl)pent-4-enoic acid

1-(*o*-Fluorophenyl)ethanol
Formed from: *o*-Fluoroacetophenone

1-(*p*-Fluorophenyl)ethanol
Formed from: 1-Ethyl-4-fluorobenzene
p-Fluoroacetophenone

2-(*p*-Fluorophenyl)ethanol
Formed from: 1-Ethyl-4-fluorobenzene

2-Fluoro-1-phenylethanol
Formed from: ω-Fluoroacetophenone

1-Fluoro-2-phenylethan-2-one
See: ω-Fluoroacetophenone

1-(*o*-Fluorophenyl)ethylamine
Formed from: Octyl 1-(*o*-Fluorophenyl) ethylamine oxalamate

3-(3-(4-(2-(*m*-Fluorophenyl)ethyl)piperazine-1-oxide-1-yl)propyl)-5-(1,2,4-triazol-4-yl)-1*H*-indole
Formed from: 3-(3-(4-(2-(*m*-Fluorophenyl)ethyl)piperazin-1-yl)propyl)-5-(1,2,4-triazol-4-yl)-1*H*-indole

3-(3-(4-(2-(*m*-Fluorophenyl)ethyl)piperazine-4-oxide-1-yl)propyl)-5-(1,2,4-triazol-4-yl)-1*H*-indole
Formed from: 3-(3-(4-(2-(*m*-Fluorophenyl)ethyl)piperazin-1-yl)propyl)-5-(1,2,4-triazol-4-yl)-1*H*-indole

3-(3-(4-(2-(*m*-Fluorophenyl)ethyl)piperazin-1-yl)-2-hydroxypropyl)-5-(1,2,4-triazol-4-yl)-1*H*-indole
Formed from: 3-(3-(4-(2-(*m*-Fluorophenyl)ethyl)piperazin-1-yl)propyl)-5-(1,2,4-triazol-4-yl)-1*H*-indole

3-(3-(4-(2-(*m*-Fluorophenyl)ethyl)piperazin -1-yl)propyl)-5-(1,2,4-triazol-4-yl)-1*H*-indole
→ 3-(3-(4-(2-(*m*-Fluorophenyl)ethyl) piperazine-1-oxide-1-yl)propyl)-5-(1,2,4-triazol-4-yl)-1*H*-indole
In: Dog J792
Man J792
Rat J792
Rhesus monkey J792
→ 3-(3-(4-(2-(*m*-Fluorophenyl)ethyl) piperazine-4-oxide-1-yl)propyl)-5-(1,2,4-triazol-4-yl)-1*H*-indole
In: Dog J792
Man J792
Rat J792
Rhesus monkey J792
→ 3-(3-(4-(2-(*m*-Fluorophenyl)ethyl) piperazin-1-yl)-2-hydroxypropyl)-5-(1,2,4-triazol-4-yl)-1*H*-indole
In: Dog J792
Man J792
Rat J792
Rhesus monkey J792
→ 3-(*m*-Fluorophenyl)propionic acid

Probably in: Dog J792
Man J792
Rat J792
Rhesus monkey J792
→ 3-(3-(Piperazin-1-yl)propyl)-5-(1,2,4-triazol-4-yl)-1*H*-indole
In: Dog J792
Man J792
Rhesus monkey J792

(3*R*,4*S*)-1-(*p*-Fluorophenyl)-3-((3*S*)-3-(*p*-fluorophenyl)-3-hydroxypropyl)-4-(*p*-hydroxyphenyl)-2-azetidinone
→ (3*R*,4*S*)-1-(*p*-Fluorophenyl)-3-((3*S*)-3-(*p*-fluorophenyl)-3-hydroxypropyl)-4-(*p*-hydroxyphenyl)-2-azetidinone-β-D-glucuronide
In: Beef J814

(3*R*,4*S*)-1-(*p*-Fluorophenyl)-3-((3*S*)-3-(*p*-fluorophenyl)-3-hydroxypropyl)-4-(*p*-hydroxyphenyl)-2-azetidinone-β-D-glucuronide
Formed from: (3*R*,4*S*)-1-(*p*-Fluorophenyl)-3-((3*S*)-3-(*p*-fluorophenyl)-3-hydroxypropyl)-4-(*p*-hydroxyphenyl)-2-azetidinone

N-((4-(Fluorophenyl)-2-formyl)but-1-enyl)urea
Formed from: Mofegiline

(*E*)-(3*R*,5*S*)-7-(4-(*p*-Fluorophenyl)-5-(β-D-glucosidomethyl)-2,6-diisopropylpyrid-3-yl)-2,3,5-trihydroxyhept-6-enoic acid
Probably formed from: (*E*)-(3*R*,5*S*)-7-(4-(*p*-Fluorophenyl)-5-hydroxymethyl-2,6-diisopropylpyrid-3-yl)-2,3,5-trihydroxyhept-6-enoic acid

(*E*)-(3*R*,5*S*)-7-(4-(*p*-Fluorophenyl)-5-(β-D-glucuronidomethyl)-2,6-diisopropylpyrid-3-yl)-2,3,5-trihydroxyhept-6-enoic acid
Probably formed from: (*E*)-(3*R*,5*S*)-7-(4-(*p*-Fluorophenyl)-5-hydroxymethyl-2,6-diisopropylpyrid-3-yl)-2,3,5-trihydroxyhept-6-enoic acid

***o*-Fluorophenyl-β-D-glucuronide**
Formed from: *o*-Fluorophenol

***m*-Fluorophenyl-β-D-glucuronide**
Formed from: *m*-Fluorophenol

***p*-Fluorophenyl-β-D-glucuronide**
Formed from: *p*-Fluorophenol

***p*-Fluorophenylglyoxylic acid**
Formed from: *p*-Fluoromandelic acid

4-(*p*-Fluorophenyl)-4-hydroxybutyric acid
Probably formed from: *p*-Fluorobenzoylpropionic acid
→ 4-(*p*-Fluorophenyl)-4-hydroxybutyric acid-β-D-glucuronide
Probably in: Rat E16

4-(*p*-Fluorophenyl)-4-hydroxybutyric acid-β-D-glucuronide
Probably formed from: 4-(*p*-Fluorophenyl)-4-hydroxybutyric acid

1-(4′-Fluorophenyl)-4-(hydroxycyclohexyl 1′-piperazinyl-4′-carboxylate)butan-1-ol
Probably formed from: 1-(4′-Fluorophenyl)-4-(hydroxycyclohexyl 1′-piperazinyl-4′-carboxylate)butan-1-one

1-(4′-Fluorophenyl)-4-(hydroxycyclohexyl 1′-piperazinyl-4′-carboxylate)butan-1-one
Probably formed from: 4-(Cyclohexyl 1′-piperazinyl-4′-carboxylate)-1-(4′-fluorophenyl)butan-1-one
→ 4-(Dihydroxycyclohexyl 1′-piperazinyl-4′-carboxylate)-1-(4′-fluorophenyl)butan-1-one
Probably in: Man A2038
Rat A2038
→ 1-(4′-Fluorophenyl)-4-(hydroxycyclohexyl 1′-piperazinyl-4′-carboxylate)butan-1-ol
Probably in: Man A2038
Rat A2038

***p*-Fluorophenylhydroxylamine**
Formed from: *p*-Fluoroaniline

(*E*)-(5*S*)-7-(4-(*p*-Fluorophenyl)-5-hydroxymethyl-2,6-diisopropylpyrid-3-yl)-5-hydroxyhept-6-enoic acid
Probably formed from: (*E*)-(5*S*)-7-(4-(*p*-Fluorophenyl)-2,6-diisopropyl-5-methoxymethyl)pyrid-3-yl)-5-hydroxyhept-6-enoic acid
→ (*E*)-(5*S*)-7-(4-(*p*-Fluorophenyl)-5-hydroxymethyl-2,6-diisopropylpyrid-3-yl)-5-hydroxyhept-6-enoyltaurine
Probably in: Mouse J426
Rat J426

(*E*)-(5*S*)-7-(4-(*p*-Fluorophenyl)-5-hydroxymethyl-2,6-diisopropylpyrid-3-yl)-5-hydroxyhept-6-enoyltaurine
Probably formed from: (*E*)-(5*S*)-7-(4-(*p*-Fluorophenyl)-5-hydroxymethyl-2,6-diisopropylpyrid-3-yl)-5-hydroxyhept-6-enoic acid

(*E*)-(3*R*,5*S*)-7-(4-(*p*-Fluorophenyl)-5-hydroxymethyl-2,6-diisopropylpyrid-3-yl)-2,3,5-trihydroxyhept-6-enoic acid
Probably formed from: Demethylcerivastatin
→ (*E*)-(3*R*,5*S*)-7-(4-(*p*-Fluorophenyl)-5-(β-D-glucosidomethyl)-2,6-diisopropylpyrid-3-yl)-2,3,5-trihydroxyhept-6-enoic acid
Probably in: Mouse J426
Rat J426
→ (*E*)-(3*R*,5*S*)-7-(4-(*p*-Fluorophenyl)-5-(β-D-glucuronidomethyl)-2,6-diisopropylpyrid-3-yl)-2,3,5-trihydroxyhept-6-enoic acid
Probably in: Mouse J426
Rat J426

4-(*p*-Fluorophenyl)-7-hydroxymethyl-1-isopropyl-2(1*H*)-quinazolinone
Formed from: Fluproquazone
→ 7-Carboxy-4-(*p*-fluorophenyl)-1-isopropyl-2(1*H*)-quinazolinone
In: Dog B891
Man B891
Mouse B891
Rabbit B891
Rat B891

2-(*p*-Fluorophenyl)-4-(((*p*-hydroxyphenyl)amino)carbonyl)-5-(1-methylethyl)-3-phenyl-1*H*-pyrrole-1-pentanoic acid
Probably formed from: 2-(*p*-Fluorophenyl)-5-(1-methylethyl)-3-phenyl-4-((phenylamino)carbonyl)-1*H*-pyrrole-1-pentanoic acid
→ 2-(*p*-Fluorophenyl)-4-(((*p*-hydroxyphenyl)amino)carbonyl)-5-(1-methylethyl)-3-phenyl-1*H*-pyrrole-1-pentanoic acid
Probably in: Mouse J433

2-(*p*-Fluorophenyl)-4-(((*p*-hydroxyphenyl)amino)carbonyl)-5-(1-methylethenyl)-3-phenyl-1*H*-pyrrole-1-pentanoic acid
Probably formed from: 2-(*p*-Fluorophenyl)-4-(((*p*-hydroxyphenyl)amino)carbonyl)-5-(1-methylethyl)-3-phenyl-1*H*-pyrrole-1-pentanoic acid

5-(*p*-Fluorophenyl)-5-(*p*-hydroxyphenyl)hydantoin
Formed from: 5-(*p*-Fluorophenyl)-5-phenylhydantoin

(*E*)-(+)-(3*R*,5*S*)-7-(4-(*p*-Fluorophenyl)-6-(1-hydroxy-2-propyl)-2-isopropyl-5-(methoxymethyl)pyrid-3-yl)-3,5-dihydroxyhept-6-enoic acid
Probably formed from: Cerivastatin

4-(*p*-Fluorophenyl)-1-isopropyl-7-methyl-2(1*H*)-quinazolinone
See: Fluproquazone

α-Fluorophenylmalonic acid
→ (*R*)-α-Fluorophenylacetic acid
In: Escherichia K232

(α*S*)-2-(4-Fluorophenyl)-α-methyl-5-benzoxazoleacetic acid
See: Flunoxaprofen

2-(*p*-Fluorophenyl)-5-(1-methylethenyl)-3-phenyl-4-((phenylamino)carbonyl)-1*H*-pyrrole-1-pentanoic acid
Probably formed from: 2-(*p*-Fluorophenyl)-5-(1-methylethyl)-3-phenyl-4-((phenylamino)carbonyl)-1*H*-pyrrole-1-pentanoic acid

2-(*p*-Fluorophenyl)-5-(1-methylethy)-3-phenyl-4-((phenylamino)carbonyl)-1*H*-pyrrole-1-pentanoic acid
Formed from: Atorvastatin
→ 2-(*p*-Fluorophenyl)-4-(((*p*-hydroxyphenyl)amino)carbonyl)-5-(1-methylethyl)-3-phenyl-1*H*-pyrrole-1-pentanoic acid
Probably in: Mouse J433
→ 2-(*p*-Fluorophenyl)-5-(1-methylethenyl)-3-phenyl-4-((phenylamino)carbonyl)-1*H*-pyrrole-1-pentanoic acid
Probably in: Mouse J433

α-(*o*-Fluorophenyl)-α-methylmalonic acid
→ 2-(*o*-Fluorophenyl)propionic acid
In: Alcaligenes G716

α-(*m*-Fluorophenyl)-α-methylmalonic acid
→ 2-(*m*-Fluorophenyl)propionic acid
In: Alcaligenes G716

α-(*p*-Fluorophenyl)-α-methylmalonic acid
→ 2-(*p*-Fluorophenyl)propionic acid
In: Alcaligenes G716

1-(*p*-Fluorophenyl)-4-(4-methylpiperidin-1-yl)-1-butanol
Formed from: Melperone

4-(*m*-Fluorophenyl)-1-methylpyridinium
Probably formed from: 4-(*m*-Fluorophenyl)-1,2,3,6-tetrahydro-1-methylpyridine

4-(*p*-Fluorophenyl)-1-methylpyridinium
Probably formed from: 4-(*p*-Fluorophenyl)-1,2,3,6-tetrahydro-1-methylpyridine

***p*-Fluorophenyl nicotinate**
→ *p*-Fluorophenol
In: Man J141

1-(4′-Fluorophenyl)-4-(oxopiperazin-1′-yl)butan-1-ol
Probably formed from: 1-(4′-Fluorophenyl)-4-(1′-piperazinyl)butan-1-ol

3-(*p*-Fluorophenyl)-3-oxovaleric acid
→ L-2-Amino-3-(*p*-fluorophenyl)valeric acid
In: Bacillus G168

***p*-(*o*-Fluorophenyl)phenylacetic acid**
Formed from: 4′-Ethynyl-2-fluorobiphenyl
→ *p*-(2-Fluoro-4-phenyl)phenylacetic acid
Probably in: Rat A3670

5-(*p*-Fluorophenyl)-5-phenylhydantoin
→ 5-(*p*-Fluorophenyl)-5-(*p*-hydroxyphenyl)hydantoin
In: Rat B80

1-(*p*-Fluorophenyl)piperazine
Formed from: Niaprazine

1-(4′-Fluorophenyl)-4-(1′-piperazinyl)butan-1-ol
Probably formed from: 1-(4′-Fluorophenyl)-4-(1′-piperazinyl)butan-1-one
→ 1-(4′-Fluorophenyl)-4-(oxopiperazin-1′-yl)butan-1-ol
Probably in: Man A2038
Rat A2038

1-(4′-Fluorophenyl)-4-(1′-piperazinyl)butan-1-one
Probably formed from: 4-(Cyclohexyl 1′-piperazinyl-4′-carboxylate)-1-(4′-fluorophenyl)butan-1-one
→ 1-(4′-Fluorophenyl)-4-(1′-piperazinyl)butan-1-ol
Probably in: Man A2038
Rat A2038

N-(3-(4-(*p*-Fluorophenyl)-1-piperazinyl)-1-methylpropyl)-3-pyridinecarboxamide
See: Niaprazine

(1*R*,2*S*)-1-(*o*-Fluorophenyl)propane-1,2-diol
Formed from: *o*-Fluorobenzaldehyde

(1*R*,2*S*)-1-(*m*-Fluorophenyl)propane-1,2-diol
Formed from: *m*-Fluorobenzaldehyde

(1*R*,2*S*)-1-(*p*-Fluorophenyl)propane-1,2-diol
Formed from: *p*-Fluorobenzaldehyde

1-(*p*-Fluorophenyl)-1-propanol
Formed from: *p*-Fluoropropiophenone

2-(*o*-Fluorophenyl)propionic acid
Formed from: α-(*o*-Fluorophenyl)-α-methylmalonic acid

2-(*m*-Fluorophenyl)propionic acid
Formed from: α-(*m*-Fluorophenyl)-α-methylmalonic acid

2-(*p*-Fluorophenyl)propionic acid
Formed from: α-(*p*-Fluorophenyl)-α-methylmalonic acid

3-(*m*-Fluorophenyl)propionic acid
Probably formed from: 3-(3-(4-(2-(*m*-Fluorophenyl)ethyl)piperazin-1-yl)propyl)-5-(1,2,4-triazol-4-yl)-1*H*-indole

α-(4-Fluorophenyl)-4-(2-pyridinyl)-1-piperazinebutanol
See: Azaperol

1-(4-Fluorophenyl)-4-(4-(2-pyridinyl)-1-piperazinyl)-1-butanone
See: Azaperone

***o*-Fluorophenylpyruvic acid**
Formed from: *o*-Fluoro-L-phenylalanine

***m*-Fluorophenylpyruvic acid**
Formed from: *m*-Fluoro-L-phenylalanine

***p*-Fluorophenylpyruvic acid**
Formed from: *p*-Fluoro-L-phenylalanine
→ *p*-Fluoro-L-phenylalanine
In: Bacillus G168

4-(*p*-Fluorophenyl)succindialdehyde
Formed from: Melperone

***o*-Fluorophenyl sulphate**
Formed from: *o*-Fluorophenol

***m*-Fluorophenyl sulphate**
Formed from: *m*-Fluorophenol

***p*-Fluorophenyl sulphate**
Formed from: *p*-Fluorophenol

3-(*p*-Fluorophenylsulphonyl)-2-methyllactic acid
Formed from: Casodex

4-(*m*-Fluorophenyl)-1,2,3,6-tetrahydro-1-methylpyridine
→ 4-(*m*-Fluorophenyl)-1-methylpyridinium
Probably in: Beef F395
Man F395

4-(*p*-Fluorophenyl)-1,2,3,6-tetrahydro-1-methylpyridine
→ 4-(*p*-Fluorophenyl)-1-methylpyridinium
Probably in: Beef F395
Man F395

***p*-Fluorophenyl-β-D-xylidide**
→ *p*-Fluorophenol
In: Bacillus A2061

3-Fluorophthalic acid
Formed from: N-(1(1*H*)-(1-Carboxy-1-(*n*-heptyl))imidazol-4-yl)-6-fluorophthalamidic acid
→ *o*-Fluorobenzoic acid
In: Bacillus D31
→ *m*-Fluorobenzoic acid
In: Bacillus D31

4-Fluorophthalic acid
Formed from: 2-Fluorofluorene

5-Fluoro-1-phthalidyluracil
→ *o*-Formylbenzoic acid
In: Rat E244

6-Fluoro-3-(4-piperidinyl)benzisoxazole
Formed from: Iloperidone

***p*-Fluoropropiophenone**
→ 1-(*p*-Fluorophenyl)-1-propanol
In: Man G89

5-Fluoro-2-(((4-pyridon-2-yl)methyl)sulphinyl)-1*H*-benzimidazole
Formed from: 2-(((4-Cyclopropylmethoxy-2-pyridin-2-yl)methyl)sulphinyl)-5-fluoro-1*H*-benzimidazole

4-Fluoropyrogallol
Probably formed from: 3-Fluorocatechol

Fluoroquinol
Formed from: *o*-Fluorobenzoic acid
m-Fluorobenzoic acid
2-Fluoro-4-hydroxybenzoic acid
3-Fluoro-4-hydroxybenzoic acid

5-Fluorosalicylic acid (5-fluoro-2-hydroxybenzoic acid)
→ 4-Fluorocatechol
In: Trichosporon D232
See also: E146

***p*-Fluorosulphinylanisole**
Formed from: *p*-Fluorothioanisole

6-Fluoro-1,2,3,4-tetrahydro-2-hydroxy-2-(2-((3-(5-hydroxy-1*H*-benzimidazol-2-yl)propyl)amino)ethyl)-1-isopropylnaphthalene
Probably formed from: 6-Fluoro-1,2,3,4-tetrahydro-2-hydroxy-2-(2-((3-(5-hydroxy-1*H*-benzimidazol-2-yl)propyl)methylamino)ethyl)-1-isopropylnaphthalene

6-Fluoro-1,2,3,4-tetrahydro-2-hydroxy-2-(2-((3-(4-hydroxy-1*H*-benzimidazol-2-yl)propyl)methylamino)ethyl)-1-isopropylnaphthalene
Formed from: 2-(2-((1*H*-Benzimidazol-2-yl)propyl)methylamino)ethyl)-6-fluoro-1,2,3,4-tetrahydro-2-hydroxy-1-isopropylnaphthalene
6-Fluoro-1,2,3,4-tetrahydro-2-hydroxy-2-(2-((3-(4-hydroxy-1*H*-benzimidazol-2-yl)propyl)methylamino)ethyl)-1-isopropylnaphthalene
→ 6-Fluoro-1,2,3,4-tetrahydro-2-hydroxy-2-(2-((3-(4-hydroxy-1*H*-benzimidazol-2-yl)propyl)methylamino)ethyl)-1-isopropylnaphthalene-β-D-glucuronide
In: Rat G311

6-Fluoro-1,2,3,4-tetrahydro-2-hydroxy-2-(2-((3-(5-hydroxy-1*H*-benzimidazol-2-yl)propyl)methylamino)ethyl)-1-isopropylnaphthalene
Formed from: 2-(2-((1*H*-Benzimidazol-2-yl)propyl)methylamino)ethyl)-6-fluoro-1,2,3,4-tetrahydro-2-hydroxy-1-isopropylnaphthalene
6-Fluoro-1,2,3,4-tetrahydro-2-hydroxy-2-(2-((3-(5-hydroxy-1*H*-benzimidazol-2-

yl)propyl)methylamino)ethyl)-1-isopropylnaphthalene
→ 6-Fluoro-1,2,3,4-tetrahydro-2-hydroxy-2-(2-((3-(5-hydroxy-1*H*-benzimidazol-2-yl)propyl)amino)ethyl)-1-isopropylnaphthalene
Probably in: Cynomolgus monkey J103
Marmoset J103
Rabbit J103
Rat J103
→ 6-Fluoro-1,2,3,4-tetrahydro-2-hydroxy-2-(2-((3-(5-hydroxy-1*H*-benzimidazol-2-yl)propyl)methylamino)ethyl)-1-isopropylnaphthalene-β-D-glucuronide
Probably in: Rat J103

6-Fluoro-1,2,3,4-tetrahydro-2-hydroxy-2-(2-((3-(4-hydroxy-1*H*-benzimidazol-2-yl)propyl) methylamino)ethyl)-1-isopropylnaphthalene-β-D-glucuronide
Formed from:6-Fluoro-1,2,3,4-tetrahydro-2-hydroxy-2-(2-((3-(4-hydroxy-1*H*-benzimidazol-2-yl)propyl)methylamino)ethyl)-1-isopropylnaphthalene

6-Fluoro-1,2,3,4-tetrahydro-2-hydroxy-2-(2-((3-(5-hydroxy-1*H*-benzimidazol-2-yl)propyl) methylamino)ethyl)-1-isopropylnaphthalene-β-D-glucuronide
Probably formed from: 6-Fluoro-1,2,3,4-tetrahydro-2-hydroxy-2-(2-((3-(5-hydroxy-1*H*-benzimidazol-2-yl)propyl)methylamino)ethyl)-1-isopropylnaphthalene

6-Fluoro-1,2,3,4-tetrahydro-2-hydroxy-1-isopropyl-2-((N-methylamino)ethyl)naphthalene
Formed from: 2-(2-((3-(1*H*-Benzimidazol-2-yl)propyl)methylamino)ethyl)-6-fluoro-1,2,3,4-tetrahydro-2-hydroxy-1-isopropylnaphthalene

***p*-Fluorothioanisole**
Formed from: *p*-Fluorothiophenol
→ *p*-Fluorosulphinylanisole
In: Mortierella D759

***p*-Fluorothiophenol**
→ *p*-Fluorothioanisole
In: Man D926

4′-Fluoro-4-(4-(2-thioxo-1-benzimidazolinyl) piperidino)butyrophenone
See: Timiperone

***o*-Fluorotoluene**
→ *o*-Fluorobenzoic acid
In: Thauera H616

***m*-Fluorotoluene**
→ *m*-Fluorobenzoic acid
In: Thauera H616
→ 3-Methylcatechol
In: Pseudomonas F76

***p*-Fluorotoluene**
→ *p*-Fluorobenzoic acid
In: Rabbit C891
Thauera H616
→ *p*-Fluorobenzyl alcohol
In: Mortierella F823

1-Fluoro-3-(*p*-tolylsulphenyl)propan-2-ol
Formed from: 1-Fluoro-3-(*p*-tolylsulphenyl)propan-2-one

1-Fluoro-3-(*p*-tolylsulphenyl)propan-2-one
→ 1-Fluoro-3-(*p*-tolylsulphenyl)propan-2-ol
In: Aspergillus E667
Aureobasidium E667
Fusarium E667
Geotrichum E667
Kloeckera E667
Rhizoctonia E667
Streptomyces E667
Yeast E667

1-Fluoro-3-(*p*-tolylsulphonyl)propan-2-ol
Formed from: 1-Fluoro-3-(*p*-tolylsulphonyl)propan-2-one

1-Fluoro-3-(*p*-tolylsulphonyl)propan-2-one
→ 1-Fluoro-3-(*p*-tolylsulphonyl)propan-2-ol
In: Aspergillus E667
Aureobasidium E667
Fusarium E667
Geotrichum E667
Kloeckera E667
Rhizoctonia E667
Streptomyces E667
Yeast E667

***m*-Fluorotrifluoromethylbenzene**
→ 3-Trifluoromethylcatechol
In: Pseudomonas F76

2-Fluoro-4-trifluoromethylbenzoic acid
→ 2-Fluoro-4-trifluoromethylbenzoyl-β-D-glucuronic acid
In: Rat J549

4-Fluoro-2-trifluoromethylbenzoic acid
→ 4-Fluoro-2-trifluoromethylbenzoyl-β-D-glucuronic acid
In: Rat J549

4-Fluoro-3-trifluoromethylbenzoic acid
→ 4-Fluoro-3-trifluoromethylbenzoyl-β-D-glucuronic acid
In: Rat J549
→ 4-Fluoro-3-trifluoromethylhippuric acid
In: Rat J549

2-Fluoro-4-trifluoromethylbenzoyl-β-D-glucuronic acid
Formed from: 2-Fluoro-4-trifluoromethylbenzoic acid

4-Fluoro-2-trifluoromethylbenzoyl-β-D-glucuronic acid
Formed from: 4-Fluoro-2-trifluoromethylbenzoic acid

4-Fluoro-3-trifluoromethylbenzoyl-β-D-glucuronic acid
Formed from: 4-Fluoro-3-trifluoromethylbenzoic acid

4-Fluoro-3-trifluoromethylhippuric acid
Formed from: 4-Fluoro-3-trifluoromethylbenzoic acid

2-Fluoro-5-trifluoromethylphenol
→ 2-Fluoro-5-trifluoromethylphenyl-β-D-glucuronide
In: Rat H468, H522
→ 2-Fluoro-5-trifluoromethylphenyl sulphate
In: Rat H468, H522

3-Fluoro-5-trifluoromethylphenol
→ 3-Fluoro-5-trifluoromethylphenyl-β-D-glucuronide
In: Rat H468, H522
→ 3-Fluoro-5-trifluoromethylphenyl sulphate
In: Rat H468, H522

4-Fluoro-2-trifluoromethylphenol
→ 4-Fluoro-2-trifluoromethylphenyl-β-D-glucuronide
In: Rat H468, H522
→ 4-Fluoro-2-trifluoromethylphenyl sulphate
In: Rat H468, H522

2-Fluoro-5-trifluoromethylphenyl-β-D-glucuronide
Formed from: 2-Fluoro-5-trifluoromethylphenol

3-Fluoro-5-trifluoromethylphenyl-β-D-glucuronide
Formed from: 3-Fluoro-5-trifluoromethylphenol

4-Fluoro-2-trifluoromethylphenyl-β-D-glucuronide
Formed from: 2-Fluoro-2-trifluoromethylphenol

2-Fluoro-5-trifluoromethylphenyl sulphate
Formed from: 2-Fluoro-5-trifluoromethylphenol

3-Fluoro-5-trifluoromethylphenyl sulphate
Formed from: 3-Fluoro-5-trifluoromethylphenol

4-Fluoro-2-trifluoromethylphenyl sulphate
Formed from: 4-Fluoro-2-trifluoromethylphenol

5-Fluorotryptamine
Formed from: 5-Fluoro-L-tryptophan
→ 10-Fluoroajmalicine
Probably in: Catharanthus A2741

6-Fluorotryptamine
→ N-Acetyl-6-fluorotryptamine
In: Rat D282
Sheep D282
→ 11-Fluoroajmalicine
Probably in: Catharanthus A2741
→ 6-Fluoro-5-hydroxytryptamine
In: Peganum F9

4-Fluoro-L-tryptophan
Formed from: 4-Fluoroindole

5-Fluoro-L-tryptophan
Formed from: 5-Fluoroindole
→ 5-Fluoro-N-formyl-L-kynurenine
In: Streptomyces E589
Wheat E24
→ 5-Fluoro-β-hydroxy-L-tryptophan
Probably in: Pseudomonas A3007
→ 5-Fluoroindole
Probably in: Enterobacter A781
Erwinia A781
Escherichia A781
Kluyvera A781
Proteus A781
→ 5-Fluoroindole-3-pyruvic acid
In: Azospirillum J22
→ 5-Fluorotryptamine
In: Micrococcus C62

6-Fluoro-L-tryptophan
Formed from: 6-Fluoroindole
→ 4-Fluoro-N-formyl-L-kynurenine
In: Rat A886
Wheat E24

2-Fluoro-L-tyrosine
Formed from: *m*-Fluorophenol

3-Fluoro-L-tyrosine
Formed from: *o*-Fluorophenol
→ *o*-Fluorophenol
In: Citrobacter F186
→ 3-Fluoro-4-methoxybenzaldehyde
In: Bjerkandera H914

Fluoxetine (N-methyl-γ-(4-(trifluoromethyl) phenoxy)benzenepropanamide, both isomers)
→ Demethylfluoxetine
In: Man K133, K163
Mouse H928

Flupentixol (4-(3-(2-(trifluoromethyl)-9*H*-thioxanthen-9-ylidene)propyl)-1-piperazineethanol)
→ Flupentixol-β-D-glucuronide
In: Rat F719
→ Flupentixol sulphoxide
In: Rat F719
→ 1′-(5-Glutathionyl)-1′,10-dihydroflupentixol
In: Rat F719
→ 7-Hydroxyflupentixol
Probably in: Rat F719

Flupentixol-β-D-glucuronide
Formed from: Flupentixol

Flupentixol sulphoxide
Formed from: Flupentixol
→ Flupentixol sulphoxide sulphate
Probably in: Rat F719

Flupentixol sulphoxide sulphate
Probably formed from: Flupentixol sulphoxide

Fluperlapine (3-fluoro-6-(4-methyl-1-piperazinyl)-11*H*-dibenzo[*b*,*e*]azepime)
→ Fluperlapine-N-β-D-glucuronide
In: Man E835
→ Fluperlapine-N-oxide
In: Man E835
Rat E918, F771
→ 2-Hydroxyfluperlapine
In: Rat E918
→ Norfluperlapine
In: Man E835
Rat E918

Fluperlapine-N-β-D-glucuronide
Formed from: Fluperlapine

Fluperlapine-N-oxide
Formed from: Fluperlapine
→ 2-Hydroxyfluperlapine-N-oxide
Probably in: Man E835
Rat E918

Fluphenazine (4-(3-(2-(trifluoromethyl)-10*H*-phenothiazin-10-yl)propyl)-1-piperazineethanol)
→ Demethylfluphenazine sulphoxide
In: Rat A67
→ Demethyltrifluoperazine
In: Rat A1865
→ Fluphenazine-β-D-glucuronide
In: Rabbit F909
Rat F720
→ Fluphenazine-N-oxide
In: Pig A1865
→ Fluphenazine sulphoxide
In: Rat F720
→ 7-Hydroxyfluphenazine
In: Rabbit F909
Rat F720
→ N-Hydroxymethyl-N′-(γ-(-(2-(trifluoromethylphenothiazin-10-yl)propyl)ethylenediamine
In: Rat A1865
→ N-(γ-(2-(Trifluoromethylphenothiazin-10-yl)propyl)ethylenediamine
In: Rat A576, A1865

Fluphenazine-β-D-glucuronide
Formed from: Fluphenazine

Fluphenazine-N-oxide
Formed from: Fluphenazine

Fluphenazine sulphoxide
Formed from: Fluphenazine

Fluproquazone (4-(*p*-fluorophenyl)-1-isopropyl-7-methyl-2(1*H*)-quinazolinone)
→ 4-(*p*-Fluorophenyl)-7-hydroxymethyl-1-isopropyl-2(1*H*)-quinazolinone
In: Dog B891
Man B891

Mouse B891
Rabbit B891
Rat B891

Flupyrazofos (O,O-diethyl O-(1-phenyl-3-(trifluoromethyl)-1*H*-pyrazol-5-yl) phosphorothioic ester)
→ Flupyrazofos oxon
In: Rat J106
→ 5-Hydroxy-1-phenyl-3-trifluoromethylpyrazole
In: Rat J106

Flupyrazofos oxon (O,O-diethyl O-(1-phenyl-3-(trifluoromethyl)-1*H*-pyrazol-5-yl) phosphate ester)
Formed from: Flupyrazofos
→ 5-Hydroxy-1-phenyl-3-trifluoromethylpyrazole
In: Rat J106

Flurazepam
See: 7-Chloro-1-(2-diethylaminoethyl)-5-(2-fluorophenyl)-1,3-dihydro-2*H*-1,4-benzodiazepin-2-one

(*R*)-Flurbiprofen (4-(1-carboxyethyl)-2-fluorobiphenyl; *R* assumed unless otherwise stated)
Formed from: (*R*)-Flurbiprofen-β-D-glucuronide
→ (*S*)-Flurbiprofen
In: Rat J307
→ (*R*)-Flurbiprofen-β-D-glucuronide
In: Guinea pig H180
Man F727, H180, J617
Monkey H180
Rat F564, H180
→ 3′-Hydroxyflurbiprofen
In: Baboon A3482
Dog A3482
Man A3482
Mouse A3482
Rat A3482
See also: A3393
→ 4′-Hydroxyflurbiprofen
In: Baboon A3482
Dog A3482
Man A3482
Mouse A3482
Rat A3482

(*S*)-Flurbiprofen
Formed from: (*R*)-Flurbiprofen
(*S*)-Flurbiprofen-β-D-glucuronide
→ (*S*)-Flurbiprofen-β-D-glucuronide
In: Guinea pig H180
Man F727, H180
Monkey H180
Rat H180

(*R*)-Flurbiprofen-β-D-glucuronide
Formed from: (*R*)-Flurbiprofen
→ (*R*)-Flurbiprofen
In: Helix F727

(*S*)-Flurbiprofen-β-D-glucuronide
Formed from: (*S*)-Flurbiprofen
→ (*S*)-Flurbiprofen
In: Helix F727

Flusilazole ((bis(*p*-fluorophenyl)methyl-(1*H*)-1,2,4-triazol-1-ylmethyl)silane)
→ Bis(*p*-fluorophenyl)(hydroxymethyl)(methyl)silane
In: Goat K144
→ Bis(*p*-fluorophenyl)(methyl)silanol
In: Goat K144
→ Oxy(bis(bis(*p*-fluorophenyl)(methyl)(silane))
In: Goat K144

Flutamide (2-methyl-N-(4-nitro-3-(trifluoromethyl)phenyl)propanamide)
→ 2-Hydroxyflutamide
In: Man H805

Flutolanil (3′-isopropoxy-2-trifluoromethylbenzanilide)
→ 4′-Hydroxyflutolanil
In: Rat D190
→ 3′-(1-Hydroxy-2-propoxy)-2-trifluoromethylbenzanilide
In: Cucumber D191
Rice D191
→ 3′-Hydroxy-2-trifluoromethylbenzanilide
In: Cucumber D191
Rat D190
Rice D191

Flutoprazepam (7-chloro-1-(cyclopropylmethyl)-5-(2-fluorophenyl)-1,3-dihydro-2*H*-1,4-benzodiazepin-2-one)
→ 7-Chloro-5-(*o*-fluorophenyl)-1,3-dihydro-2*H*-1,4-benzodiazepin-2-one

In: Dog C545
Man F637
→ 3-Hydroxyflutoprazepam
In: Man F637

Fluvalinate (N-(2-chloro-4-(trifluoromethyl) valine cyano(3-phenoxyphenyl)methyl ester)
→ *m*-Phenoxybenzoic acid
In: Hen E941
Rat D224
→ *m*-Phenoxybenzyl alcohol
In: Rat D224
→ *m*-Phenoxymandelonitrile
In: Bacillus G891

Fluvastatin ((3*R*,5*S*,6*E*)-rel-7-(3-(4-fluorophenyl)-1-(-1-methylethyl)-1*H*-indol-2-yl)-3,5-dihydroxy-6-heptenoic acid)
→ Deisopropylfluvastatin
In: Man G585
→ 5-Hydroxyfluvastatin
In: Man G585
→ 6-Hydroxyfluvastatin
In: Man G585

Fluvoxamine (5-methoxy-4′-(trifluoromethyl) valerophenone (*E*)-O-(2-aminoethyl) oxime)
→ 5-Hydroxy-*p*-trifluoromethyl valerophenone (*E*)-O-(2-aminoethyl) oxime
In: Dog C652
Hamster C652
Mouse C652
Rat C652
→ 5-Methoxy-*p*-trifluoromethyl valerophenone oxime
In: Dog C652
Hamster C652
Man D24
Mouse C652
Rat C652
→ 5-Methoxy-*p*-trifluoromethyl valerophenone O-(carboxymethyl) oxime
In: Man D24
→ 5-Methoxy-*p*-trifluoromethyl valerophenone O-(2-hydroxyethyl) oxime
In: Man D24
→ Methyl 5-(2-aminoethyl)oximino)-5-(*p*-trifluoromethylphenyl)valerate
In: Dog C652
Hamster C652
Mouse C652
Rat C652
→ *p*-Trifluoromethylbenzoic acid
In: Dog C652
Hamster C652
Mouse C652
Rat C652

Folic acid (pteroylglutamic acid)
Formed from: Pteroyldiglutamic acid
→ *p*-Aminobenzoyl-L-glutamic acid
In: Pseudomonas A971, A1061
Rat A2103, B74
→ 10-Formylpteroylpentaglutamic acid
In: Rat A711
→ 10-Formyltetrahydrofolic acid
In: Rat A2103
→ 5-Formyltetrahydropteroylpentaglutamic acid
In: Rat A711
→ 10-Formyltetrahydropteroylpentaglutamic acid
In: Rat A711
→ 5,10-Methylenetetrahydrofolic acid
In: Pediococcus A867
→ 5-Methyltetrahydrofolic acid
In: Rat A2103
→ Pteroyldiglutamic acid
In: Pig E277, K455
See also: A1098
→ Pteroylheptaglutamic acid
Probably in: Monkey A1098
→ Pteroylhexaglutamic acid
In: Pig E277
See also: A1098, A1217
→ Pteroylpentaglutamic acid
In: Pig E277
See also: A1098, A1217
→ Pteroylpentaglutamic acid
In: Pig E277
See also: A1098, A1217
→ Pteroyltetraglutamic acid
In: Pig E277
See also: A1098
→ Pteroyltriglutamic acid
In: Pig E277

See also: A1098
→ Tetrahydro-5-methylpteroylpentaglutamic acid
In: Rat A711
→ Tetrahydropteroylpentaglutamic acid
In: Rat A711

Folinic acid
See: 5-Formyltetrahydrofolic acid

Fominoben (3′-chloro-2′-(N-methyl-N-(morpholinocarbonylmethyl)aminomethyl) benzanilide)
→ N-(2-Amino-6-chlorobenzyl)-N-methylglycinemorpholide
In: Man A1109

Fomocaine (1-morpholino-3-((*p*-phenoxymethyl)phenyl)propane)
Formed from: Fomocaine-N-oxide
→ Fomocaine-N-oxide
In: Dog A348
Rat A1558, A3303
→ 3-(*p*-((*p*-Hydroxyphenoxy)methyl)phenyl)-1-morpholinopropane
In: Dog A348
Rabbit A3303
Rat A2687
→ *p*-(γ-Morpholinopropyl)benzoic acid
In: Dog A348
Rat A2687
→ Phenol
In: Rat A2687

***o*-Fomocaine** (4-(3-(2-(phenoxymethyl) phenyl)propyl)morpholine)
→ *o*-Fomocaine-N-oxide
In: Rat A155, B21
→ *p*-Hydroxy-*o*-fomocaine
In: Rat B21
→ Phenol
In: Rat B21

Fomocaine-N-oxide
Formed from: Fomocaine
→ Fomocaine
In: Rat A3303

***o*-Fomocaine-N-oxide**
Formed from: *o*-Fomocaine

Fonofos
See: O-Ethyl S-phenyl ethylphosphonodithioate

N^3^-(2-(3-(4-Formamidinophenyl)isoxazolin-5(*R*)-yl)acetyl)-N^2^-(3-hydroxy-1-butyloxycarbonyl)-2,3-diaminopropionic acid
Probably formed from: Demethylroxifiban

***o*-Formamidoacetophenone**
Formed from: Skatole
→ *o*-Aminoacetophenone
Probably in: Pig J780

2-Formamidoanthraquinone
Formed from: 2-Aminoanthraquinone

4-Formamidoantipyrine
Formed from: Aminopyrine
Dipyrone
4-Methylaminoantipyrine
→ 4-Aminoantipyrine
Probably in: Man F467

***o*-Formamidobenzaldehyde**
Formed from: Indole
→ Anthranilaldehyde
Probably in: Tecoma G659

4-Formamidobiphenyl
→ 4-Aminobiphenyl
In: Rat F109

6-Formamido-11,13-bis(β-D-glucopyranosyl)-12,13-dihydro-1-hydroxy-5*H*-indolo[2,3-*a*]pyrrolo[3,4-*c*]carbazole-5,7(6*H*)-dione
Formed from: 6-Formamido-13-(β-D-glucopyranosyl)-12,13-dihydro-1,11-dihydroxy-5*H*-indolo[2,3-*a*]pyrrolo[3,4-*c*]carbazole-5,7(6*H*)-dione

2-Formamido-4,6-dinitrotoluene
Formed from: 2-Amino-4,6-dinitrotoluene

4-Formamido-2,6-dinitrotoluene
Formed from: 4-Amino-2,6-dinitrotoluene

2-Formamidofluorene
Formed from: 2-Aminofluorene
→ 2-Aminofluorene
In: Rat F109

6-Formamido-13-(β-D-glucopyranosyl)-12,13-dihydro-1,11-dihydroxy-5*H*-indolo[2,3-*a*]pyrrolo[3,4-*c*]carbazole-5,7(6*H*)-dione
→ 6-Amino-13-(β-D-glucopyranosyl)-12,13-dihydro-1,11-dihydroxy-5*H*-indolo[2,3-*a*]pyrrolo[3,4-*c*]carbazole-5,7(6*H*)-dione

In: Mouse J602
Rat J600, J602
→ 6-Formamido-11,13-bis(β-D-glucopyranosyl)-12,13-dihydro-1-hydroxy-5*H*-indolo[2,3-*a*]pyrrolo[3,4-*c*]carbazole-5,7(6*H*)-dione
In: Man J602
Mouse J602
Rat J600, J602
→ 13-(β-D-Glucopyranosyl)-12,13-dihydro-1,11-dihydroxy-5*H*-indolo[2,3-*a*]-furano[3,4-*c*]carbazole-5,7(6*H*)-dione
In: Dog J602
Man J602

α-(N-Formamido)-N-glyoxalylbenzamide
Formed from: 2-Acetyl-tetrahydro-3-phenyl-1,2,4-oxadiazin-5-one
→ N-Glyoxalylbenzamide
Probably in: Rat D762

5-Formamido-2-hydroxybenzoic acid
Formed from: 5-Amino-2-hydroxybenzoic acid

1-Formamido-2-(3-methoxycarbonyl-2-thioureido)benzene
→ 1-Formamido-2-(3-methoxycarbonyl-2-ureido)benzene
In: Mouse A260
→ 2-(3-Methoxycarbonyl-2-thioureido)aniline
In: Mouse A260

1-Formamido-2-(3-methoxycarbonyl-2-ureido)benzene
Formed from: 1-Formamido-2-(3-methoxycarbonyl-2-thioureido)benzene
2-(3-Methoxycarbonyl-2-thioureido)-1-thioformamidobenzene
→ 2-(3-Methoxycarbonyl-2-ureido)aniline
In: Mouse A260

1-Formamidonaphthalene
Formed from: α-Naphthylamine
→ α-Naphthylamine
In: Rat F109

2-Formamidonaphthalene
Formed from: β-Naphthylamine
→ β-Naphthylamine
In: Rat F109

2-Formamido-5-nitrobenzophenone
Probably formed from: 2-Amino-5-nitrobenzophenone

***p*-Formamidophenol**
Formed from: *p*-Aminophenol

1-(*p*-Formamidophenyl)-3,3-dimethyltriazene
→ 1-(*p*-Formamidophenyl)-3-methyltriazene
In: Rat A112

1-(*p*-Formamidophenyl)-3-methyltriazene
Formed from: 1-(*p*-Formamidophenyl)-3,3-dimethyltriazene

***p*-Formamidophenyl 2,4,6-trichlorophenyl ether**
Formed from: *p*-Aminophenyl 2,4,6-trichlorophenyl ether

1-Formamidopyrene
Formed from: 1-Aminopyrene
→ 1-Aminopyrene
In: Rat F109

4-Formamido-3,5-xylyl methylcarbamate
Probably formed from: 4-Methylamino-3,5-xylyl methylcarbamate

Formanilide
→ Aniline
In: Microorganism A261

Formononetin (biochanin B, 7-hydroxy-4′-methoxyisoflavone)
Formed from: Daidzein
Formononetin-7-β-D-glucoside
Liquiritigenin
2,4′,7-trihydroxyflavanone
→ Afrormosin
In: Onobrychis A2742
→ Calycosin
In: Chickpea E433, H4 K751, K799
→ Daidzein
In: Eubacterium K627
→ 2′,7-Dihydroxy-4′-methoxyisoflavone
In: Chickpea E433, H4, K751, K799
→ Formononetin glucoside (structure uncertain)
In: Rabbit A571
→ Formononetin-7-β-D-glucoside
In: Cicer K738
Nicotiana K550
→ Formononetin-β-D-glucuronide
In: Beef E539
Rabbit A571

Sheep E539
→ Maackiain
In: Trifolium A229, A1929, A3474
→ Medicarpin
In: Medicago A3725
Trifolium A229
→ 9-O-Methylcoumestrol
In: Medicago A3725
→ Pisatin
In: Pea C235

Formononetin glucoside (structure unclear)
Formed from: Formononetin

Formononetin-7-β-D-glucoside
Formed from: Formononetin
→ 7-O-(β-D-Apiofuranosyl(1→2)-β-D-glucosyl)formononetin
In: Parsley K827
→ Formononetin
In: Secale K512
→ Formononetin-7-β-D-glucoside-6″-malonate
In: Cicer D445

Formononetin-7-β-D-glucoside-6″-malonate
Formed from: Formononetin-7-O-β-D-glucoside

Formononetin-β-D-glucuronide
Formed from: Formononetin

Formoterol ((*R*,*R* and *S*,*S*)-rel-N-(2-hydroxy-5-((1*R*)-1-hydroxy-2-(((1*R*)-2-(4-methoxyphenyl)-1-methylethyl)amino)ethyl)phenylformamide)
→ Deformylformoterol
In: Man J779
→ Demethylformoterol
In: Man J779
→ Formoterol-2-β-D-glucuronide
In: Dog C380
Guinea pig K365
Man J779, K468
Rat C380

Formoterol-2-β-D-glucuronide
Formed from: Formoterol

N-Formoxy-2-aminofluorene
Formed from: N-Hydroxy-2-formamidofluorene

N-Formylacetylkynurenine
Formed from: N-Acetyl-L-tryptophan

9-Formylanthracene (9-anthraldehyde)
Probably formed from: 9-Hydroxymethylanthracene
→ 9-Carboxyanthracene
In: Mouse F432, H67

N-Formylanthranilic acid
Formed from: 2,3-Dihydroxyindole
N-Formyl-L-kynurenine
1*H*-3-Hydroxy-4-oxoquinoline
4-Hydroxyquinoline
Indole
→ Anthranilic acid
In: Hansenula A3776
See also: D158, F417

***o*-Formylbenzaldehyde**
See: *o*-Phthalaldehyde

***m*-Formylbenzaldehyde**
See: Isophthalaldehyde

***p*-Formylbenzaldehyde**
See: Terephthalaldehyde

***m*-Formylbenzamide**
Formed from: N-Ethyl-*m*-formylbenzamide
m-Hydroxymethylbenzamide
→ *m*-Hydroxymethylbenzamide
In: Rat J572
→ Isophthalamidic acid
In: Rat J572

N-Formylbenzamide
Probably formed from: N-Hydroxymethylbenzamide

7-Formylbenzanthracene
Probably formed from: 7-Hydroxymethylbenzanthracene

8-Formylbenzanthracene
Probably formed from: 8-Hydroxymethylbenzanthracene
→ 8-Carboxybenzanthracene
Probably in: Rat D389

***p*-Formylbenzenesulphonamide**
Formed from: *p*-Hydroxymethylbenzenesulphonamide
→ *p*-Carboxybenzenesulphonamide
Probably in: Guinea pig A559
Rat A559, A2021

***p*-Formylbenzenesulphonic acid**
Formed from: *p*-Hydroxymethylbenzene-sulphonic acid

→ *p*-Carboxybenzenesulphonic acid
In: Pseudomonas F188
→ *p*-Hydroxymethylbenzenesulphonic acid
In: Pseudomonas F188

***o*-Formylbenzoic acid** (phthaldehydic acid)
Formed from: *o*-Carboxybenzalpyruvic acid
5-Fluoro-1-phthalidyluracil
Phthalidyl theophylline
Talampicillin
→ 2-Formyl-3,4-dihydroxybenzoic acid
In: Micrococcus C247
→ *o*-Hydroxymethylbenzoic acid
In: Dog A2094, A3716
Hamster F153
Man A2094, J662
Monkey A3716
Mouse A2094, A3716
Rabbit A2094, A3716
Rat A2094, A3716, G239, K546
See also: A3058, E244
→ Phthalic acid
Probably in: Burkholderia H813
Dog A3058
Man A3058
Mycobacterium G818
Rat A3058
→ Phthalide
In: Euglena G422

***m*-Formylbenzoic acid** (isophthalaldehydic acid)
Formed from: Isophthalaldehyde
→ *m*-Hydroxymethylbenzoic acid
In: Rabbit A3716
→ Isophthalic acid
In: Dog A2094, A3716
Monkey A3716
Mouse A3716
Rabbit A3716
Rat A2094, A3716

***p*-Formylbenzoic acid** (terephthalaldehydic acid)
Formed from: *p*-Hydroxymethylbenzoic acid
Terephthalaldehyde
→ *p*-Hydroxymethylbenzoic acid
In: Dog A2094, A3716
Monkey A3716
Mouse A3716
Pig A1679
Rabbit A3716
Rat A2007, A2094, A3716
→ Terephthalic acid
In: Bacteria K74
Dog A3716
Monkey A3716
Mouse A3716
Rabbit A3716
Rat A3716, H759
See also: G405

***o*-Formylbenzonitrile** (*o*-cyanobenzaldehyde)
→ *o*-Hydroxymethylbenzonitrile
In: Dunaliella G459
Euglena G422
→ Phthalide
In: Dunaliella G459
Euglena G422

***m*-Formylbenzonitrile** (*m*-cyanobenzaldehyde)
→ *m*-Cyanobenzoic acid
In: Rat H761
→ *m*-Cyano-2-hydroxypropiophenone
In: Pseudomonas K524
→ *m*-Hydroxymethylbenzonitrile
In: Dunaliella G459
Euglena G422

***p*-Formylbenzonitrile** (*p*-cyanobenzaldehyde)
→ *p*-Cyanobenzoic acid
In: Rat H761
→ *p*-Cyano-2-hydroxypropiophenone
In: Pseudomonas K524

4-Formylbiphenyl
Formed from: 4-Hydroxymethylbiphenyl
→ 4-Carboxybiphenyl
In: Rat H761

2-Formyl-2,3-bis(hydroxymethyl)-3-methylbutyl benzoate
Probably formed from: 2-Formyl-2-hydroxymethyl-3,3-dimethylbutyl benzoate

N-(4-Formylbutyl)-3-((1-piperazinyl)methyl) benzamide
Probably formed from: 1-(3-((1-Piperazinyl)methyl)benzoyl)piperidine
→ N-(5-Hydroxypentyl)-3-((1-piperazinyl)methyl)benzamide
Probably in: Dog J581

N-Formylcarbamoylaspartame
→ L-Phenylalanine

Probably in: Man H223
Pig H223

1-Formylcarbazole
Formed from: 1-Methylcarbazole
→ 1-Carboxycarbazole
In: Pseudomonas J180

3-Formylcarbazole
Formed from: 3-Methylcarbazole
→ 3-Carboxycarbazole
In: Pseudomonas J180

4-Formylcarbostyril
→ 4-Carboxycarbostyril
In: Man J24

N-Formylcatharinine
Formed from: Vincristine

N-Formyl-5-chloroanthranilic acid
Probably formed from: 4-Chloro-2-methylformanilide
→ 5-Chloroanthranilic acid
Probably in: Mouse A2534
Rat A1821

N-Formylciprofloxacin
Formed from: Ciprofloxacin

5-Formylconiferaldehyde
Formed from: Dehydrodiconiferyl alcohol
→ 5-Formylferulic acid
Probably in: Pseudomonas F77

5-Formyldemethylfleroxacin
Formed from: Fleroxacin

10-Formyldihydrofolic acid
Formed from: Dihydrofolic acid

2-Formyl-3,4-dihydroxybenzoic acid
Formed from: 2-Formylbenzoic acid

N-Formyl-3,4-dihydroxy-L-phenylalanine
Formed from: N-Formyl-L-tyrosine
→ δ-(3-(N-Formyl)alanyl)-α-hydroxy-*cis*,*cis*-muconic semialdehyde
In: Pseudomonas A1090

4-Formyl-6,7-dimethoxycoumarin
→ 4-Carboxy-6,7-dimethoxycoumarin
In: Man J24

5-Formylferulic acid
Probably formed from: 5-Formylconiferaldehyde
→ 5-Hydroxymethylferulic acid
Probably in: Pseudomonas F77

2-Formylfluorene
→ Fluorene-2-carboxylic acid
In: Man J24

10-Formylfolic acid
→ *p*-Aminobenzoic acid
In: Rat B74

10-Formylfolic tetraglutamate
→ *p*-Aminobenzoic acid
In: Rat B74

4-Formyl-2-hydroxybenzoic acid
Formed from: 2-Methylnaphthalene

2-Formyl-3-hydroxybenzo[*b*]thiophene
Formed from: Dibenzothiophene
→ 2-Carboxy-3-hydroxybenzo[*b*]thiophene
Probably in: Microorganism J303
→ 2,3-Dihydroxybenzo[*b*]thiophene
Probably in: Microorganism J303
→ 2-(3′-Hydroxythionaphthenyl[2]methylene)-3-oxodihydrothionaphthene
Probably in: Pseudomonas A1647

5-Formyl-8-hydroxycarbostyril
Formed from: Procaterol
→ 8-Hydroxycarbostyril
Probably in: Dog A3483
Man A3483
Rat A3483

4-Formyl-6-hydroxycoumarin
→ 4-Carboxy-6-hydroxycoumarin
In: Man J24

5-Formyl-2-(4-hydroxy-3,5-dimethoxyphenyl)-3-hydroxymethyl-7-methoxycoumarin
→ 5-Carboxy-2-(4-hydroxy-3,5-dimethoxyphenyl)-7-methoxycoumarin
In: Fusarium E174
→ 5-Formyl-2-(4-hydroxy-3,5-dimethoxyphenyl)-3-hydroxymethyl-7-methoxycoumarone
In: Fusarium E174
→ 2-(5-Formyl-2-hydroxy-3-methoxyphenyl)-3-hydroxy-1-(4-hydroxy-3,5-dimethoxyphenyl)-1-propanone
In: Fusarium E174
→ 2-(4-Hydroxy-3,5-dimethoxyphenyl)-3,5-bis(hydroxymethyl)-7-methoxycoumarin
In: Fusarium E174

5-Formyl-2-(4-hydroxy-3,5-dimethoxyphenyl)-3-hydroxymethyl-7-methoxycoumarone
Formed from: 5-Formyl-2-(4-hydroxy-3,5-dimethoxyphenyl)-3-hydroxymethyl-7-methoxycoumarin
→ 3,5-Diformyl-2-(4-hydroxy-3,5-dimethoxyphenyl)-7-methoxycoumarone
Probably in: Fusarium E174

Formyl-N-(1*R*,2*R*)-2-hydroxy-1-(hydroxymethyl)-2-(4-(methylsulphonyl)phenyl)ethyl)formamide
See: Thiamphenicol aldehyde

N-Formyl-5-hydroxy-L-kynurenine
Formed from: 5-Hydroxy-L-tryptophan
→ 5-Hydroxy-L-kynurenine
In: Rat A1284

5-Formyl-2-hydroxy-3-methoxybenzoic acid
See: 5-Carboxyvanillin

5-Formyl-2-hydroxy-3-methoxyphenylacetic acid
See: 5-Carboxymethylvanillin

2-(5-Formyl-2-hydroxy-3-methoxyphenyl)-3-hydroxy-1-(4-hydroxy-3,5-dimethoxyphenyl)-1-propanone
Probably formed from: 5-Formyl-2-(4-hydroxy-3,5-dimethoxyphenyl)-3-hydroxymethyl-7-methoxycoumarin
→ 2-(3-Carboxy-6-hydroxy-5-methoxyphenyl)ethanol
Probably in: Fusarium E174
→ 4-Hydroxy-3,5-dimethoxybenzoic acid
Probably in: Fusarium E174
→ 4-Hydroxy-5-methoxyisophthalic acid
Probably in: Fusarium E174

2-Formyl-3-hydroxy-7-methylbenzothiophene
Probably formed from: 4,6-Dimethyldibenzothiophene

2-Formyl-2-hydroxymethyl-3,3-dimethylbutyl benzoate
Formed from: 2,2-Bis(hydroxymethyl)-3,3-dimethylbutyl benzoate
tert-Butylbicycloorthobenzoate
→ 2-Formyl-2,3-bis(hydroxymethyl)-3-methylbutyl benzoate
Probably in: Mouse E803

1-Formyl-8β-hydroxymethyl-10α-methoxy-6-methylergoline
Probably formed from: 1,8β-Bis(hydroxymethyl)-10α-methoxy-6-methylergoline
→ 8β-Hydroxymethyl-10α-methoxy-6-methylergoline
Probably in: Dog E167
Rat E167

1-Formyl-4-((*p*-hydroxyphenyl)phenyl) methylpiperidine
Probably formed from: 4-Diphenylmethyl-1-formylpiperidine

2-Formyl-3-hydroxythianaphthene
Formed from: *trans*-4-(2-(3-hydroxy)thianaphthenyl)-2-oxo-3-butanoic acid

N-Formyl-4-hydroxy-2-((thiazol-2-yl)carbonyl)aniline
Formed from: Camalexine

N-Formylindecainide
Formed from: Indecainide

3-Formylindole
Formed from: Gramine
Indole-3-acetaldehyde
Indole-3-acetic acid
Indole-3-lactic acid
Indole-3-methanol
D-Tryptophan
Probably formed from: 3,3′-Diindolylmethane
→ Indole-3-carboxylic acid
In: Citrus H80
Orobanche C181
Pea G446, G520
Wheat D150
See also: A993, A2555, E306, G139, G702
→ Indole-3-methanol
In: Orobanche C181
See also: C503, D234, E306

3-Formylisochroman
→ 2-Hydroxy-2-(isochroman-3-yl)acetonitrile
In: Almond K320

p-Formyl-N-isopropylbenzamide
Formed from: Azo-procarbazine
→ *p*-Hydroxymethyl-N-isopropylbenzamide
Probably in: Rat C35

N-Formyl-D-kynurenine (D-α-amino-2-(formylamino)-γ-oxobenzenebutanoic acid)
Formed from: D-Tryptophan
→ D-Kynurenine
In: Rabbit A845
Rat A845

N-Formyl-L-kynurenine (L-α-amino-2-(formylamino)-γ-oxobenzenebutanoic acid)
Formed from: L-Tryptophan
→ L-Kynurenine
In: Hansenula A3776
Rabbit A845
Rat A845, A1284, A2893
See also: A1195
→ N-Formylanthranilic acid
In: Neurospora A2766
Pseudomonas A2766, B248

N^{α}-Formylkynurenine
Probably formed from: N',N^{α}-Diformylkynurenine

N-Formyl-L-kynurenine methyl ester
Formed from: L-Tryptophan methyl ester

N-Formyl-L-kynureninyl-L-alanine
Formed from: L-Tryptophanyl-L-alanine

N-Formyl-L-kynureninylglycine
Formed from: L-Tryptophanylglycine

N-Formyl-L-kynureninyl-L-leucne
Formed from: L-Tryptophanyl-L-leucine

N-Formyl-L-kynureninyl-L-phenylalanine
Formed from: L-Tryptophanyl-L-phenylalanine

N-Formyl-L-kynureninyl-L-tyrosine
Formed from: L-Tryptophanyl-L-tyrosine

4-Formyl-5-methoxycarbostyril
→ 4-Carboxy-5-methoxycarbostyril
In: Man J24

4-Formyl-6-methoxycarbostyril
→ 4-Carboxy-6-methoxycarbostyril
In: Man J24

4-Formyl-6-methoxycoumarin
→ 4-Carboxy-6-methoxycoumarin
In: Man J24

1-(Formylmethoxy)naphthalene
→ 1-Hydroxy-3-(1-naphthyloxy)propionitrile
In: Almond K6

4-Formyl-2-methoxyphenoxyacetic acid
→ Vanillin
In: Acinetobacter G637
See also: G275

2-(4-Formyl-2-methoxyphenoxy)-1,3-dihydroxy-1-vanillylpropane
→ 2-(4-Formyl-2-methoxyphenoxy)-3-hydroxypropiovanillone
Probably in: Pseudomonas E832

2-(4-Formyl-2-methoxyphenoxy)-3-hydroxypropiovanillone
Probably formed from: 2-(4-Formyl-2-methoxyphenoxy)-1,3-dihydroxy-1-vanillylpropane
→ β-Hydroxypropiovanillone
In: Pseudomonas E832

2-Formyl-6-methoxyphenyl-β-D-glucuronide
Formed from: 2-Hydroxy-3-methoxybenzaldehyde

4-Formyl-6-methoxy-*o*-quinone
Probably formed from: 4-Hydroxy-3,5-dimethoxybenzaldehyde

3-Formyl-4′-methylaminoazobenzene
Probably formed from: 3′-Hydroxymethyl-4-methylaminoazobenzene
→ 3-Carboxy-4′-methylaminoazobenzene
Probably in: Rat A3835, B27, G9

7-Formyl-12-methylbenzanthracene
Probably formed from: 7-Hydroxymethyl-12-methylbenzanthracene

N-Formylmethyl-2,3-dihydro-3-hydroxy-2-(*p*-hydroxyphenyl)-1,5-benzothiazepin-4(5*H*)-one
Probably formed from: N-Formylmethyl-2,3-dihydro-3-hydroxy-2-(*p*-methoxyphenyl)-1,5-benzothiazepin-4(5*H*)-one
→ N-Carboxymethyl-2,3-dihydro-3-hydroxy-2-(*p*-hydroxyphenyl)-1,5-benzothiazepin-4(5*H*)-one
Probably in: Rat F835

N-Formylmethyl-2,3-dihydro-3-hydroxy-2-(*p*-methoxyphenyl)-1,5-benzothiazepin-4(5*H*)-one
Probably formed from: 3-Acetoxy-N-formylmethyl-2,3-dihydro-2-(*p*-

methoxyphenyl)-1,5-benzothiazepin-4(5*H*)-one
→ N-Carboxymethyl-2,3-dihydro-3-hydroxy-2-(*p*-methoxyphenyl)-1,5-benzothiazepin-4(5*H*)-one
Probably in: Rat F835
→ N-Formylmethyl-2,3-dihydro-3-hydroxy-2-(*p*-hydroxyphenyl)-1,5-benzothiazepin-4(5*H*)-one
Probably in: Rat F835

N-((1-Formyl-1-methylethoxy)carbonyl-α-hydroxy-β-phenyl-β-alanyl)-10-deacetylbaccatin III
Probably formed from: (α-Hydroxy-N-((1-(hydroxymethyl)-1-methylethoxy)carbonyl-β-phenyl-β-alanyl)-10-deacetylbaccatin III
→ (N-((1-Carboxy-1-methylethoxy)carbonyl-α-hydroxy-β-phenyl-β-alanyl)-10-deacetylbaccatin III
Probably in: Man H256

(*S*)-4-((3-Formylmethyl-1*H*-indol-5-yl)methyl)-2-oxazolidine
Probably formed from: Demethylzolmitriptan
Zolmitriptan
→ (*S*)-4-((3-Carboxymethyl-1*H*-indol-5-yl)methyl)-2-oxazolidine
Probably in: Man J741
→ (*S*)-4-((3-(2-Hydroxyethyl)-1*H*-indol-5-yl)methyl)-2-oxazolidine
Probably in: Man J741

10-Formylmethylisoalloxazine
→ 5,6-Dimethylbenzimidazole
In: Propiobacterium A3415

N-Formyl-3-methyl-L-kynurenine
Formed from: 7-Methyl-L-tryptophan

N-Formyl-4-methyl-L-kynurenine
Formed from: 6-Methyl-L-tryptophan

N-Formyl-5-methyl-L-kynurenine
Formed from: 5-Methyl-L-tryptophan

N-Formyl-α-methyl-L-kynurenine
Formed from: α-Methyl-L-tryptophan

8-(3-Formyl-1-methylpropylamino)-6-methoxyquinoline
Formed from: Primaquine
→ 8-(3-Carboxy-1-methylpropylamino)-6-methoxyquinoline
In: Yeast G230
See also: C279, D117, D131, E150, E705, F321, F437
→ 8-(3-Hydroxymethyl-1-methylpropylamino)-6-methoxyquinoline
In: Horse G230

9-Formylphenanthrene
→ Phenanthrene-9-carboxylic acid
In: Man J24

***o*-Formylphenoxyacetic acid**
→ *o*-Carboxyphenoxyacetic acid
In: Mouse A2094
→ *o*-Hydroxymethylphenoxyacetic acid
In: Dog A2094, A3716
Monkey A3716
Mouse A2094, A3716
Rabbit A2094, A3716
Rat A2094, A3716

***m*-Formylphenoxyacetic acid**
→ *m*-Carboxyphenoxyacetic acid
In: Rat A2094
→ *m*-Hydroxymethylphenoxyacetic acid
In: Rat A2094

***p*-Formylphenoxyacetic acid**
→ *p*-Carboxyphenoxyacetic acid
In: Rat A2094
→ *p*-Hydroxymethylphenoxyacetic acid
In: Rat A2094

***o*-Formylphenoxypropionic acid**
→ *o*-Carboxyphenoxyacetic acid
In: Mouse A2094
Rabbit A2094
Rat A2094
→ *o*-Hydroxymethylphenoxyacetic acid
In: Dog A2094
Mouse A2094
Rabbit A2094
Rat A2094

10-Formylpteroylpentaglutamic acid
Formed from: Folic acid

3-Formylquinoline
→ Quinoline-3-carboxylic acid
In: Man J24

4-Formylquinoline
→ 4-Hydroxymethylquinoline
Probably in: Cinchona E357
→ Quinoline-4-carboxylic acid

In: Man J24

3-Formylrifampicin
Formed from: Rifampicin

3-Formylrifapentine
Formed from: Rifapentine

5-Formyltetrahydrofolic acid (folinic acid)
→ 5-Formyltetrahydropteroyldiglutamic acid
In: Pig K445
→ 5,10-Methenyltetrahydrofolic acid
In: Rabbit H657
→ 5-Methyltetrahydrofolic acid
In: Man A1681

10-Formyltetrahydrofolic acid
Formed from: Dihydrofolic acid
Folic acid
5,10-Methenyltetrahydrofolic acid
5,10-Methylenetetrahydrofolic acid
Tetrahydrofolic acid
→ 10-Formyltetrahydropteroyldiglutamic acid
In: Pig K445
→ Tetrahydrofolic acid
In: Clostridium E360
Escherichia H380
With 10-formyltetrahydrofolate dehydrogenase K316

5-Formyl-2,3,4,5-tetrahydro-1-(phenylcyclohexyl)pyridinium
Formed from: 2,3,4,5-Tetrahydro-1-(phenylcyclohexyl)pyridinium

5-Formylpteroyldiglutamic acid
Formed from: 5-Formyltetrahydrofolic acid

10-Formylpteroyldiglutamic acid
Formed from: 10-Formyltetrahydrofolic acid

5-Formyltetrahydropteroylpentaglutamic acid
Formed from: Folic acid

10-Formyltetrahydropteroylpentaglutamic acid
Formed from: Folic acid

3-Formyl-1,6,8-trihydroxyanthraquinone
Probably formed from: 1,6,8-trihydroxy-3-hydroxymethylanthraquinone
→ 3-Carboxy-1,6,8-trihydroxyanthraquinone
Probably in: Mouse E151

10-(4-Formyl-2,3,4-trihydroxybutyl)-7,8-dimethylisoalloxazine
See: Vitamin B_2 aldehyde

N-Formyltrimetazidine
Formed from: Trimetazidine

N-Formyl-L-tyrosine
→ N-Formyl-3,4-dihydroxy-L-phenylalanine
In: Aspergillus A2771
Mucuna E769
Streptomyces A1433

3-Formylureido-4-hydroxy-2-methyl-2*H*-1,2-benzothiazine-1,1-dioxide
Probably formed from: 3-(N′-Carboxymethyl)formylureido-4-hydroxy-2-methyl-2*H*-1,2-benzothiazine-1,1-dioxide

N-Formylviloxazine
Formed from: N-Methylviloxazine

Fosinopril ((4*S*)-4-cyclohexyl-1-(((*R*)-((1*S*)-2-methyl-1-(1-oxopropoxy)propoxy)(4-phenylbutyl)phosphinyl)acetyl)-L-proline)
→ 4-Cyclohexyl-N-(((4-phenylbutyl)phosphinyl)acetyl)-L-proline
Probably in: Dog F538

Fraxetin (8-(β-D-glucopyranosyloxy)-7-hydroxy-6-methoxy-2*H*-1-benzopyran-2-one)
→ Fraxin
Probably in: Bapleura E681
Datura E681
Gardenia E681
Mallotus E681
Manduca G897
Perilla E681

Fraxin (7,8-dihydroxy-6-methoxy-2*H*-1-benzopyran-2-one)
Formed from: Fraxetin

Fructose-phenylalanine
→ Phenylacetic acid
In: Rat C125

Furfenorex (furfurylmethylamphetamine)
→ *p*-Hydroxyfurfenorex
In: Rat D658
→ N-Methylamphetamine
In: Man A589, E52
Rat D658
→ N-Methyl-N-(γ-valerolactonyl)amphetamine
In: Man E52
Rat D658

N-Furfuryl-*p*-hydroxyamphetamine
Probably formed from: *p*-Hydroxyfurfenorex

Furfurylmethylamphetamine
See: Furfenorex

Furoben (α-oxo-2-dibenzofuranbutyric acid)
→ 2-Dibenzofuranacetic acid
In: Dog B619
Man A1978, B619
Mouse B619
Rat B619

***7H*-Furo[3,2-*g*][1]benzopyran-7-one**
See: Psoralen

Furosemide (5-(aminosulphonyl)-4-chloro-2-((2-furanylmethyl)amino)benzoic acid)
→ Furosemide-β-D-glucoside
In: Cunninghamella G557
→ Furosemide-β-D-glucuronide
In: Man J783
Monkey H861
→ Saluamine
In: Cunninghamella G506
Dog A2761
Man A3920
Monkey A2761

Furosemide-β-D-glucoside
Formed from: Furosemide

Furosemide-β-D-glucuronide
Formed from: Furosemide

G

Galactosyl(β1 → 3)-6-O-acetylgalactosyl (β1 → 4)xylyl(β)-4-methylumbelliferone
Formed from: 6-O-Acetylgalactosyl (β1 → 4)xylyl(β)-4-methylumbelliferone

Galactosyl(β1 → 3)xylyl(β)-4-methylumbelliferone
Formed from: 4-Methylumbelliferyl-β-D-xyloside

Galactosyl(β1 → 4)xylyl(β)-4-methylumbelliferone
Formed from: 4-Methylumbelliferyl-β-D-xyloside

Galangin (3,5,7-trihydroxyflavone)
→ 3,5-Dihydroxy-7-methoxyflavone
In: Calamondin orange A3987
→ 5,7-Dihydroxy-3-methoxyflavone
In: Calamondin orange A3987, B830
Nicotiana B830
→ Galangin-β-D-glucoside
In: Parsley K739
→ Galangin-β-D-glucuronide
In: Man G524, G968, J416
→ Kaempferol
In: Rat J300

Galangin-β-D-glucoside
Formed from: Galangin

Galangin-β-D-glucuronide
Formed from: Galangin

Galanthamine ((4a*S*,6*R*,8a*S*)-4a,5,9,10,11,12-hexahydro-3-methoxy-11-methyl-6*H*-benzofuro[3a,3,2-*ef*][2]benzazepin-6-ol)
Formed from: Galanthaminone
→ Galanthaminone
In: Rabbit E427

Galanthaminone ((4a*S*,8a*S*)-4a,5,9,10,11,12-hexahydro-3-methoxy-11-methyl-6*H*-benzofuro[3a,3,2-*ef*][2]benzazepin-6-one)
Formed from: Epigalanthamine
Galanthamine
→ Epigalanthamine
In: Rabbit E427
→ Galanthamine
In: Rabbit E427

Gallic acid (3,4,5-trihydroxybenzoic acid)
Formed from: Cinnamic acid
3,5-Dihydroxybenzoic acid
3,4-Dihydroxy-5-methoxybenzoic acid
3,5-Dihydroxy-4-methoxybenzoic acid
(−)-Epicatechin gallate
(−)-Epigallocatechin gallate
4-Hydroxy-3,5-dimethoxybenzoic acid
Methyl gallate
Propyl gallate
Quinic acid
Tannin
3,4,5-Trihydroxybenzaldehyde
Formed from glucose in Escherichia K591
Probably formed from: Galloyl-β-D-glucose
→ γ-Carboxy-α-hydroxy-*cis*,*cis*-muconic acid
In: Aspergillus E343
Pseudomonas A2399
See also: A698, C709
→ 5-Carboxy-3-hydroxy-*o*-quinone
Probably in: Botrytis A1930
Coriolus E369, F274
Mushroom B897
→ Dihydrogallic acid
In: Penicillium G784
→ 3,4-Dihydroxy-5-methoxybenzoic acid
Probably in: Mouse A647
Nicotiana A3757
Phanerochaete J339
→ 3,5-Dihydroxy-4-methoxybenzoic acid
In: Rat K71, K575, K697
→ Ellagic acid
In: Turnip A1926

→ (−)-Epicatechin-3-gallate
Probably in: Tea C548
→ (−)-Epigallocatechin-3-gallate
Probably in: Tea C548
→ Gallic acid-O-sulphate
In: Eubacterium G475
→ Galloyl-β-D-glucose
In: Quercus K736, K737
→ Protocatechuic acid
In: Penicillium G784
See also: A647
→ Pyrogallol
In: Bacterium C396
Enterobacter E113, J859
Escherichia K591
Eubacterium E388, G772
Microorganism G632
Pelobacter E189, G795
Rat K575, K697
Streptomyces E113
→ 1,2,3,5-Tetrahydroxybenzene
In: Sporotrichum B73, C57

Gallic acid-β-D-glucoside
→ Gallic acid-β-D-glucoside-6′-sulphate
In: Mimosa J473

Gallic acid-β-D-glucoside-6′-sulphate
Formed from: Gallic acid-β-D-glucoside

Gallic acid-O-sulphate
Formed from: Gallic acid

(+)-Gallocatechin ((+)-3,3′,4′,5,5′,7-hexahydroxyflavan)
Formed from: Dihydromyricetin

(−)-Gallocatechin ((−)-3,3′,4′,5.5′,7-hexahydroxyflavan)
Formed from: (−)-Epigallocatechin

(−)-Gallocatechin gallate
Formed from: Tannin

Gallopamil (α-(3-((2-(3,4-dimethoxyphenyl)ethyl)methylamino)propyl)-3,4,5-trimethoxy-α-(1-methylethyl)benzeneacetonitrile)
→ N-(3-Cyano-2-methyl-3-(3,4,5-trimethoxyphenyl)-6-hexyl)methylamine
In: Man F527
Rat F527
→ 5-((3,4-Dimethoxyphenethyl)methylamino)-2-(3-hydroxy-4,5-dimethoxyphenyl)-2-isopropylvaleronitrile
In: Man F527
Rat F527
→ 5-((3,4-Dimethoxyphenethyl)methylamino)-2-(4-hydroxy-3,5-dimethoxyphenyl)-2-isopropylvaleronitrile
In: Man F527
Rat F527
→ 3,4-Dimethoxyphenylacetaldehyde
Probably in: Man F527
Rat F527
→ 5-((3-Hydroxy-4-methoxyphenethyl)methylamino)-2-isopropyl-2-(3,4,5-trimethoxyphenyl)valeronitrile
In: Man F527
Rat F527
→ 5-((4-Hydroxy-3-methoxyphenethyl)methylamino)-2-isopropyl-2-(3,4,5-trimethoxyphenyl)valeronitrile
In: Man F527
Rat F527
→ Norgallopamil
In: Man F527
Rat F474, F527

Gallotannin
See: Tannin

Galloyl-β-D-glucoside (β-glucogallin)
Formed from: Gallic acid
→ 1,6-Di-O-galloyl-β-D-glucoside
In: Quercus K894
→ Gallic acid
Probably in: Quercus K894
→ 1,2,6-Tri-O-galloyl-β-D-glucoside
In: Rhus K888

Garbanzol (3,4′,7-trihydroxyflavanone)
Formed from: 2′4,4′-Trihydroxychalcone

Gardona (tetrachlorvinphos, (*Z*)-2-chloro-1-(2,4,5-trichlorophenyl)vinyl dimethyl phosphate)
→ 2-Chloro-1-(2,4,5-trichlorophenyl)ethanol
In: Chicken A2532
Goose B208
Moth A777
Turkey B208

→ (*E*)-2-Chloro-1-(2,4,5-trichlorophenyl)vinyl dimethyl phosphate
In: Goose B208
Turkey B208
→ 2-Chloro-1-(2,4,5-trichlorophenyl)vinyl methyl phosphate
In: Beef A3807
Chicken A2532
Pig A3807
Sheep A3807
→ 2′,4′,5′-Trichloroacetophenone
In: Chicken A2532
Goose B208
Moth A777
Turkey B208
→ 2,4,5-Trichlorophenacyl chloride
In: Goose B208
Moth A777
Turkey B208
→ 1-(2,4,5-Trichlorophenyl)ethanol
In: Chicken A2532
Goose B208
Moth A777
Turkey B208
→ 2,4,5-Trichlorophenylglycol
In: Goose B208

Geissoschizine ((αZ,2S,3E,12bS)-3-ethylidene-1,2,3,4,6,7,12,12b-octahydro-α-(hydroxymethylene)indolo[2,3-*a*]quinolizine-2-acetic acid methyl ester)
Formed from: Tryptamine
→ Ajmalicine
In: Catharanthus A3808, B46
Corynathe A3256
→ 4,21-Dehydrogeissoschizine
In: Cathanthus K700
→ Epiajmalicine
In: Corynathe A3256

Gemfibrozil (5-(2,5-dimethylphenoxy)-2,2-dimethylpentanoic acid)
Formed from: Gemfibrozil-β-D-glucuronide
→ Gemfibrozil-β-D-glucuronide
In: Hamster J565
Rat J46, J565
→ 4′-Hydroxygemfibrozil
In: Hamster J565
Rat J564, J565
See also: A3033
→ 5-(2-Hydroxymethyl-5-methylphenoxy)-2,2-dimethylpentanoic acid
In: Hamster J565
Rat J565
→ 5-(5-Hydroxymethyl-2-methylphenoxy)-2,2-dimethylpentanoic acid
In: Hamster J565
Man A3033
Rat J564, J565

Gemfibrozil-β-D-glucuronide
Formed from: Gemfibrozil
→ Gemfibrozil
In: Rat J906

Genistein (4′,5′,7-trihydroxyisoflavone)
Formed from: Biochanin A
Genistein glucoside
Genistin
Naringenin
→ *p*-Coumaric acid
In: Rhizobium H280
See also: K363
→ 4′,7-Dihydroxy-5-methoxyisoflavone
In: Lupin E756, F182
→ Genistein-4′-β-D-glucuronide
In: Man K18
→ Genistein-7-β-D-glucuronide
In: Man J416, J783, K18
Rabbit A571
Rat J169
See also K363, K453, K531
→ Genistein-4′-sulphate
In: Man J169
See also: K531
→ Genistein-7-sulphate
In: Man J865
See also: K18, K363
→ Genistin
In: Cicer K738
Rabbit A571
→ Homogentisic acid
In: Armillaria H831
→ 6′-Hydroxy-O-demethylangolensin
In: Rat K363
→ Phloroglucinol

In: Armillaria H831
→ Prunetin
In: Alfalfa G322, H875
Prunus G982
→ 4′,5,6,7-Tetrahydroxyisoflavone
In: Microorganism J627
→ 4′,5,7,8-Tetrahydroxyisoflavone
In: Microorganism J627
See also: K155
→ 2,4,6-Trihydroxybenzoic acid
In: Rhizobium H280
→ 4′,5′,7-Trihydroxyisoflavanone
In: Fusarium C474, E724
Rat K363

Genistein glucoside (structure uncertain)
→ Genistein
In: Secale K512

Genistein-7-glucoside
See: Genistin

Genistein-7-glucoside-6″-malonate
Formed from: Genistin

Genistein-4′-β-D-glucuronide
Formed from: Genistein

Genistein-7-β-D-glucuronide
Formed from: Genistein

Genistein-7-β-D-glucuronide sulphate
Probably in: Rat K531

Genistein-7-β-D-glucuronide sulphate
Probably formed from: Genistein-7-β-D-glucuronide
Genistein-4′-sulphate

Genistein-4′-sulphate
Formed from: Genistein
→ Genistein-7-β-D-glucuronide sulphate
Probably in: Rat K531

Genistein-7-sulphate
Formed from: Genistein

Genistin (genistein-7-glucoside)
Formed from: Genistein
→ Genistein
In: Ascochyta D744
Aspergillus K155
Man J821
Microorganism J394
Pig K309
See also: J169
→ Genistein-7-glucoside-6″-malonate
In: Cicer D445

Genkwanin (4′,5-dihydroxy-7-methoxyflavone)
Formed from: Apigenin

Gentian violet (N-(4-bis(4-dimethylamino) phenyl)methylene)-2,5-cyclohexadien-1-ylidene-N-methanaminium)
→ Free radical
In: Rat F960
→ Leucogentian violet
In: Bacteroides C838
Bifidobacterium C838
Citrobacter C838
Clostridium C838
Escherichia C838
Lactobacillus C838
Peptococcus C838
Peptostreptococcus C838
Salmonella C838
→ Pentamethylpararosaniline
In: Microorganism C838

Gentisaldehyde (2,5-dihydroxybenzaldehyde)
→ Gentisic acid
Probably in: Aspergillus C60

Gentisamide (2,5-dihydroxybenzamide)
Formed from: 2,5-Dihydroxyhippuric acid
Salicylamide
→ Gentisamide-5-β-D-glucuronide
In: Rat E706, F556, F558, H738
See also: A1199, E683, F671
→ Gentisamide-2-O-sulphate
In: Rat E706, F556, F558, H738
→ Gentisamide-5-O-sulphate
In: Rat E706, F556, F558, H738
See also: E683
Contrast in rat F671

Gentisamide-5-β-D-glucuronide
Formed from: Gentisamide

Gentisamide-2-O-sulphate
Formed from: Gentisamide

Gentisamide-5-O-sulphate
Formed from: Gentisamide

Gentisic acid (2,5-dihydroxybenzoic acid)
Formed from: 5-Amino-2-hydroxybenzoic acid
2,5-Diacetoxybenzoic acid

m-Hydroxybenzoic acid
p-Hydroxybenzoic acid
1-Naphthylsulphonic acid
Quinol
Salicylic acid
5-Sulphosalicylic acid
Probably formed from: Gentisaldehyde
2-Hydroxy-5-methoxybenzoic acid
→ 3-Carboxy-4-hydroxyphenyl-D-glucoside
In: Cucumis A611
Fagopyrum H31
→ Carboxy-*p*-quinone
Probably in: Sarcophaga E407
Sheep blowfly F944
→ 2,5-Dihydroxybenzoyladenylic acid
In: Escherichia K830
→ 2,5-Dihydroxyhippuric acid
In: Beef A3602
Rat A3602
See also: F588, G203
→ 2,5-Dimethoxybenzoic acid
Probably in: Nocardia F266
→ Gentisoyl CoA
In: Bacterium H110
Desulfococcus H293
→ Hydroxyquinol
In: Sporotrichum B73
Trichosporon B368, D232
→ Maleylpyruvic acid
In: Bacillus A2316
Klebsiella F822
Nocardia B2
Pseudomonas D771, K4
Rhodococcus H441
Streptomyces B877
See also: B161, J27
→ 2,4,5-Trihydroxybenzoic acid
In: Comamonas J519
Pseudomonas A327, A1229

Gentisoyl CoA
Formed from: Gentisic acid
→ Benzoyl CoA
In: Bacterium H110

Gentisyl alcohol (2,5-dihydroxybenzyl alcohol)
Formed from: *m*-Hydroxybenzyl alcohol
→ Patulin
In: Penicillium E161

Geodin ((+)-5,7-dichloro-4-hydroxy-6′-methoxy-6-methyl-3,4′-dioxospiro (benzofuran-2(3*H*),1′-(2,5)-cyclohexadiene)-2′-carboxylic acid methyl ester)
Formed from: Dihydrogeodin

Geraniin
→ Geraniin-O-sulphate
In: Eubacterium G475

Geraniin-O-sulphate
Formed from: Geraniin

Geranylhydroquinone (2-((2*E*)-3,7-dimethyl-2,6-octadienyl)-1,4-benzenediol)
Probably formed from: 3-Geranyl-4-hydroxybenzoic acid
→ Shikonin
In: Lithospermum A3763

3-Geranyl-4-hydroxybenzoic acid (4-carboxy-2-diprenylphenol)
Formed from: *p*-Hydroxybenzoic acid
→ Geranylhydroquinone
Probably in: Lithospermum A3763

Glafenic acid (2-((7-chloro-4-quinolinyl) amino)benzoic acid)
Formed from: Glafenine

Glafenine (2-((7-chloro-4-quinolinyl)amino) benzoic acid 2,3-dihydroxypropyl ester)
→ Glafenic acid
In: Man B216
→ Glafenine-N-oxide
In: Man B216
Rat B216
→ 5-Hydroxyglafenine
In: Man B216

Glafenine-N-oxide
Formed from: Glafenine
→ Glafenic acid
In: Man B216

(6a*R*)-Glaucine
→ Didehydroglaucine
In: Aspergillus B831

(*S*)-Glaucine ((6a*S*)-5,6,6a,7-tetrahydro-1,2,9,10-tetramethoxy-6-methyl-4*H*-dibenzo[*de*,*g*] quinoline)
→ 2-O-Demethylglaucine
In: Streptomyces A2919
→ Didehydroglaucine

In: Fusarium A2919, B789
→ Norglaucine
In: Streptomyces A2919

Glaziovine ((8′a*S*)-2′,3′,8′,8′a-tetrahydro-6′-hydroxy-5′-methoxy-1′-methylspiro (2,5 cyclohexadiene-1,7′(1′*H*)-cyclopent [*i*,*j*]isoquinolin)-4-one)
Formed from: N-Methylcoclaurine

Glibornuride (N-((((1*S*,2*S*,3*R*,4*R*)-3-hydroxy-4,7,7-trimethylbicyclo[2.2.1] hept-2-yl)amino)carbonyl)-4-methylbenzenesulphonamide)
→ 1-((1*R*)-2-*endo*)-hydroxy-3-*endo*-bornyl)-3-(*p*-(α-hydroxy)toluenesulphonyl)urea
In: Man A759
Rat A874
→ (1*S*)-(*endo*,*endo*)-N-(((3-Hydroxy-4-hydroxymethyl-7,7-dimethyl bicyclo[2.2.1]hept-2-yl)amino)carbonyl)-4-methylbenzenesulphonamide
In: Man A759
→ (1*S*)-(*endo*,*endo*)-N-(((3-Hydroxy-7-hydroxymethyl-4,7-dimethyl bicyclo[2.2.1]hept-2-yl)amino)carbonyl)-4-methylbenzenesulphonamide
In: Man A759

Gliclazide (N′-(3-azabicyclo[3.3.0]oct-3-yl)-N-(4-methylbenzenesulphonyl)urea)
→ 1-(3-Azabicyclo[3.3.0]oct-3-yl)-3-(4-hydroxymethylbenzenesulphonyl)urea
In: Man D169, H488
Rat C782
→ 4β-Hydroxygliclazide
In: Man H488
→ 6α-Hydroxygliclazide
In: Man D169, H488
Rat H473
→ 6β-Hydroxygliclazide
In: Man D169, H488
See also: C782
→ 7α-Hydroxygliclazide
In: Man D169, H488
→ 7β-Hydroxygliclazide
In: Man H488
See also: C782

Glifanan (4-(2′-(2″,3″-dihydroxypropoxy carbonyl)phenylamino)-7-chloroquinoline)
→ 4-(2′-(Carboxymethyl)phenylamino)-7-chloroquinoline
In: Rat A1498
→ 4-(2′-Carboxyphenylamino)-7-chloroquinoline
In: Rat A1498

Glipizide (N-cyclohexyl-N′-(4-(β-(5-methyl pyrazine-2-carboxamido)ethyl)benzyl sulphonyl)urea)
→ N-Cyclohexyl-N′-(4-(β-(5-hydroxy methylpyrazine-2-carboxamido)ethyl) benzylsulphonyl)urea
Probably in: Cat A2234
Dog A2234
Guinea pig A2234
Monkey A2234
Mouse A2234
Rabbit A2234
Rat A2234
→ N-(3-*cis*-Hydroxycyclohexyl)-N′-(4-(β-(5-methylpyrazine-2-carboxamido)ethyl) benzylsulphonyl)urea
In: Cat A2234
Dog A2234
Guinea pig A2234
Monkey A2234
Mouse A2234
Rabbit A2234
Rat A2234
→ N-(4-*trans*-hydroxycyclohexyl)-N′-(4-(β-(5-methylpyrazine-2-carboxamido)ethyl) benzylsulphonyl)urea
In: Cat A2234
Dog A2234
Guinea pig A2234
Monkey A2234
Mouse A2234
Rabbit A2234
Rat A2234
→ 4-(β-(5-Methylpyrazine-2-carboxamido)ethyl)benzenesulphonyl urea
In: Cat A2234
Dog A2234
Guinea pig A2234
Monkey A2234

Mouse A2234
Rabbit A2234
Rat A2234

Glucobrassicin (1-(N-(sulphooxy)-1*H*-indole-3-ethanimidate) 1-thio-β-glucopyranose)
Formed from: L-Tryptophan

β-Glucogallin
See: Galloyl-β-D-glucoside

Gluconasturtiin (1-(N-(sulphooxy)-1*H*-indole-3-propanimidate) 1-thio-β-glucopyranose)
Probably formed from: 3-Phenylpropionaldoxime

13-(β-D-Glucopyranosyl)-12,13-dihydro-1,11-dihydroxy-5*H*-indolo[2,3-*a*]furano[3,4-*c*]carbazole-5,7(6*H*)-dione
Formed from: 6-Formamido-13-(β-D-glucopyranosyl)-12,13-dihydro-1,11-dihydroxy-5*H*- indolo[2,3-*a*]furano[3,4-*c*]carbazole-5,7(6*H*)-dione

(10*S*)-10-β-D-Glucopyranosyl-1,8-dihydroxy-3-(hydroxymethyl)-9(10*H*)-anthracenone
See: Barbaloin

8-β-D-Glucopyranosyl-5,7-dihydroxy-2-(4-hydroxyphenyl)-4*H*-1-benzopyran-4-one
See: Vitexin

8-β-D-Glucopyranosyl-7-hydroxy-3-(4-hydroxyphenyl)-4*H*-1-benzopyran-4-one
See: Puerarin

8-β-D-Glucopyranosyl-5-hydroxy-2-(4-hydroxyphenyl)-6,7-dimethoxy-4*H*-1-benzopyran-4-one
See: Abrusin

8-β-D-Glucopyranosyl-7-hydroxy-5-methyl-2-(2-oxopropyl)-4*H*-1-benzopyran-4-one
See: Aloeresin B

7-(β-D-Glucopyranosyloxy)-2*H*-1-benzopyran-2-one
See: Skimmin

7-(β-D-Glucopyranosyloxy)-3,5-dihydroxy-2-(4-methoxyphenyl)-8-(3-methyl-2-butenyl)-4*H*-1-benzopyran-4-one
See: Icariside I

2-(β-D-Glucopyranosyloxy)-1-hydroxyanthraquinone (alizarin-4-β-D-glucoside)
Formed from: Alizarin

4-(β-D-Glucopyranosyloxy)-1-hydroxyanthraquinone (quinizarin-2-β-D-glucoside)
Formed from: 1,4-Dihydroxyanthraquinone

6-(β-D-Glucopyranosyloxy)-1-hydroxyanthraquinone (anthraflavin-6-β-D-glucoside)
Formed from: Anthraflavin

8-(β-D-Glucopyranosyloxy)-1-hydroxyanthraquinone (chrysazin-8-β-D-glucoside)
Formed from: 1,8-Dihydroxyanthraquinone

(α*R*)-α-(β-D-Glucopyranosyloxy)-4-hydroxybenzeneacetonitrile
See: Taxiphyllin

(α*S*)-α-(β-D-Glucopyranosyloxy)-4-hydroxybenzeneacetonitrile
See: Dhurrin

8-(β-D-Glucopyranosyloxy)-7-hydroxy-6-methoxy-2*H*-1-benzopyran-2-one
See: Fraxin

(6*S*)-6-(4-(β-D-Glucopyranosyloxy)phenyl)-5,6-dihydro-2*H*-pyran-2-one
See: Psilotin

3-(4-(β-D-Glucopyranosyloxy)phenyl)-5,7-dihydroxy-4*H*-1-benzopyran-4-one
See: Sophoricoside

8-β-D-Glucopyranosyl-3′,4′,5,7-tetrahydroxyflavone
See: Orientin

2-β-D-Glucopyranosyl-1,3,6,7-tetrahydroxy-9*H*-xanthen-9-one
See: Mangiferin

4-β-D-Glucopyranosyl-1,3,6,7-tetrahydroxy-9*H*-xanthen-9-one
See: Isomangiferin

6-β-D-Glucopyranosyl-4′,5,7-trihydroxy-3′-methoxyflavone
See: Isoscoparine

3-(β-D-Glucosidoxy)-*p*-hydroxypropiophenone
Formed from: 3,*p*-Dihydroxypropiophenone

***p*-(D-Glucosylamino)benzoic acid**
Formed from: *p*-Aminobenzoic acid

4-O-Glucosylconiferyl alcohol
See: Coniferin

7-Glucosyloxy-8-methoxycoumarin (hydrangin)
Formed from: 7-Hydroxy-8-methoxycoumarin

8-Glucosylrhein
Formed from: Sennoside A
→ Rhein
Probably in: Bifidobacterium B696
Clostridium B696
Eubacterium B696
Peptostreptococcus B696

8-Glucosylrheinanthrone
Formed from: Sennoside A

5′-(α-Glucosyl)riboflavine
Formed from: Riboflavine

β-Glucosylriboflavine
Formed from: Riboflavine

***o*-(D-Glucuronamino)benzoic acid**
Probably formed from: Anthranilic acid

***p*-(D-Glucuronamino)benzoic acid**
Probably formed from: *p*-Aminobenzoic acid

***o*-(D-Glucuronamino)phenol**
Probably formed from: *o*-Aminophenol

5-O-β-D-Glucuronyl-2-phenylbenzimidazole
Formed from: Phenzidole

L-Glutaminyl-L-glutaminyl-L-methionyl-L-glutaminyl-L-glutaminyl-Lseryl-L-threonyl-L-arginyl-L-proly-L-tyrosyl-L-glutaminyl-L-glutaminyl-L-arginyl-L-prolyl-glycyl-L-glutaminyl-L-glutamine
→ Di-(L-Glutaminyl-L-glutaminyl-L-methionyl-L-glutaminyl-L-glutaminyl-Lseryl-L-threonyl-L-arginyl-L-proly-L-tyrosyl-L-glutaminyl-L-glutaminyl-L-arginyl-L-prolyl-glycyl-L-glutaminyl-L-glutamine)
In: Horseradish J327

γ-Glutaminyl-4-hydroxybenzene
Formed from: Shikimic acid

L-Glutaminyl-*p*-nitroaniline
→ *p*-Nitroaniline
In: Man A1219

γ-Glutamylaniline
Formed from: Aniline

3-((3-(L-Glutamyl-L-cystein-S-yl)-3,4-dihydro-4-hydroxy-5-trifluoromethyl-2-benzothiazolyl)methyl)-3,4-dihydro-4-oxo-1-phthalazineacetic acid
Probably formed from: 3-((3-(S-Glutathionyl)-3,4-dihydro-4-hydroxy-5-trifluoromethyl-2-benzothiazolyl)methyl)-3,4-dihydro-4-oxo-1-phthalazineacetic acid
→ 3-((3-(L-Cystein-S-yl)-3,4-dihydro-4-hydroxy-5-trifluoromethyl-2-benzothiazolyl)methyl)-3,4-dihydro-4-oxo-1-phthalazineacetic acid
Probably in: Rat J457

3-((4-(L-Glutamyl-L-cystein-S-yl)-3,4-dihydro-3-hydroxy-5-trifluoromethyl-2-benzothiazolyl)methyl)-3,4-dihydro-4-oxo-1-phthalazineacetic acid
Probably formed from: 3-((4-(S-Glutathionyl)-3,4-dihydro-3-hydroxy-5-trifluoromethyl-2-benzothiazolyl)methyl)-3,4-dihydro-4-oxo-1-phthalazineacetic acid
→ 3-((4-(L-Cystein-S-yl)-3,4-dihydro-3-hydroxy-5-trifluoromethyl-2-benzothiazolyl)methyl)-3,4-dihydro-4-oxo-1-phthalazineacetic acid
Probably in: Rat J457

γ-Glutamyl-3,4-dihydroxyphenethylamine
Formed from: γ-Glutamyl-3,4-dihydroxy-L-phenylalanine
→ 3,4-Dihydroxyphenethylamine
In: Dog A3945

γ-Glutamyl-3,4-dihydroxy-L-phenylalanine
Formed from: 3,4-Dihydroxy-L-phenylalanine
→ 3,4-Dihydroxy-L-phenylalanine
In: Man F663
Rat F653, F663
See also: A2981, F361, G192
→ γ-Glutamyl-3,4-dihydroxyphenethylamine
In: Rat F361, F663

γ-Glutamyl-γ-glutamyltetrahydrofolic acid
Probably formed from: γ-Glutamyltetrahydrofolic acid

α-L-Glutamyl-β-naphthylamide
→ β-Naphthylamine
In: Beef A526
Crithidia A1102
Rabbit A1589, A2347, B334
Rat A523

γ-L-Glutamyl-β-naphthylamide
→ β-Naphthylamine
In: Rat A523

γ-L-Glutamyl-*p*-nitroanilide
→ *p*-Nitroaniline
In: Rat A3043
See also: C146

γ-Glutamyltetrahydrofolic acid
Formed from: Tetrahydrofolic acid
→ γ-Glutamyl-γ-glutamyltetrahydrofolic acid
In: Escherichia A1362
See also: E277

L-Glutamyl-S-(trichlorophenyl)-L-cysteine
Probably formed from: S-(Trichlorophenyl)glutathione

L-Glutamyl-L-tyrosine
→ L-Tyrosine
In: Aspergillus A3658

Glutathione sulphinanilide
→ Aniline
In: Man B177

N-(Glutathion-S-methylene)-4-aminoazobenzene
Formed from: 4-Dimethylaminoazobenzene
→ 4-(N-(Glutathion-S-methylene)amino-4′-sulphonyloxyazobenzene
Probably in: Rat J409

4-(N-(Glutathion-S-methylene)amino-4′-sulphonyloxyazobenzene
Probably formed from: N-(Glutathion-S-methylene)-4-aminoazobenzene

N-(Glutathion-S-methylene)-4-methylamino-azobenzene
Formed from: 4-Dimethylaminoazobenzene

S-Glutathionyl-*m*-acetamidophenol
Probably formed from: *m*-Acetamidophenol

5′-(S-Glutathionyl)amodiaquine
Formed from: Amodiaquine

5′-(S-Glutathionyl)amsacrine
Formed from: Amsacrine

6′-(S-Glutathionyl)amsacrine
Formed from: Amsacrine

3-(S-Glutathionyl)benzidine
Formed from: Benzidine

***p*-(S-Glutathionyl)benzyl alcohol**
Formed from: *p*-Nitrobenzyl alcohol

S-Glutathionylcaffeoylquinone tartrate
Probably formed from: S-Glutathionylcaffeoyl tartrate

S-Glutathionylcaffeoyl tartrate
→ S-Glutathionylcaffeoylquinone tartrate
Probably in: Grape F701

4-Glutathionylcambendazole
Formed from: Cambendazole

13-(S-Glutathionyl)-12,13-dihydro-8,14-dihydroxyefavirenz-8-sulphate
Probably formed from: 8,14-Dihydroxyefavirenz-8-sulphate
→ 13-(S-(L-Cysteinyl))-12,13-dihydro-8,14-dihydroxyefavirenz-8-sulphate
Probably in: Rat J765

1-(S-Glutathionyl)-1′,10-dihydroflupentixol
Formed from: Flupentixol

2-(S-Glutathionyl)-2,3-dihydro-3-hydroxy-aflatoxin B_1
Formed from: Aflatoxin B_1

8-(S-Glutathionyl)-8,9-dihydro-9-hydroxyaflatoxin B_1
Formed from: Aflatoxin B_1
Aflatoxin B_1-8,9-oxide
→ 8-(S-Cysteinylglycylyl)-8,9-dihydro-9-hydroxyaflatoxin B_1
Probably in: Rat C852

3-(S-Glutathionyl)-2,3-dihydro-2-hydroxy-carbamazepine
Formed from: Carbamazepine

2-(S-Glutathionyl)-2,3-dihydro-3-hydroxyestragole
Probably formed from: Estragole-2,3-oxide

2-(S-Glutathionyl)-1,2-dihydro-1-hydroxy-5-nitronaphthalene
Formed from: 1-Nitronaphthalene

2-(S-Glutathionyl)-1,2-dihydro-1-hydroxy-8-nitronaphthalene
Formed from: 1-Nitronaphthalene

4-(S-Glutathionyl)-4,5-dihydro-5-hydroxy-1-nitropyrene
Formed from: 1-Nitropyrene
1-Nitropyrene-4,5-oxide

5-(S-Glutathionyl)-4,5-dihydro-4-hydroxy-1-nitropyrene
Formed from: 1-Nitropyrene
1-Nitropyrene-4,5-oxide

9-(S-Glutathionyl)-9,10-dihydro-10-hydroxy-1-nitropyrene
Formed from: 1-Nitropyrene
1-Nitropyrene-9,10-oxide

10-(S-Glutathionyl)-9,10-dihydro-9-hydroxy-1-nitropyrene
Formed from: 1-Nitropyrene
1-Nitropyrene-9,10-oxide

3-((3-(S-Glutathionyl)-3,4-dihydro-3-hydroxy-5-trifluoromethyl-2-benzothiazolyl)methyl)-3,4-dihydro-4-oxo-1-phthalazineacetic acid
Probably formed from: Zopolrestat
→ 3-((3-(L-Glutamyl-L-cystein-S-yl)-3,4-dihydro-3-hydroxy-5-trifluoromethyl-2-benzothiazolyl)methyl)-3,4-dihydro-4-oxo-1-phthalazineacetic acid
Probably in: Rat J457

3-((4-(S-Glutathionyl)-3,4-dihydro-3-hydroxy-5-trifluoromethyl-2-benzothiazolyl)methyl)-3,4-dihydro-4-oxo-1-phthalazineacetic acid
Probably formed from: Zopolrestat
→ 3-((4-(L-Glutamyl-L-cystein-S-yl)-3,4-dihydro-3-hydroxy-5-trifluoromethyl-2-benzothiazolyl)methyl)-3,4-dihydro-4-oxo-1-phthalazineacetic acid
Probably in: Rat J457

(8*S*)-(S-Glutathionyl)dihydromorphinone
Formed from: Morphine

3′-(S-Glutathionyl)-2′,3′-dihydro-1′-oxosafrole
Formed from: 1′-Oxosafrole
→ 3′-(N-Acetyl-S-(L-cysteinyl))-2′,3′-dihydro-1′-oxosafrole
Probably in: Mouse D242
Rat D242

12-(S-Glutathionyl-11,12-dihydroverlukast
Formed from: Verlukast

3-(S-Glutathionyl)-4,4′-dihydroxybiphenyl
Formed from: Phenol

8-(S-Glutathionyl)-7,10-dihydroxy-2-(2-((2-hydroxyethyl)amino)ethyl)-5-(2-(methylamino)ethyl)amino)anthra[1,9-*c*,*d*]pyrazol-6(2*H*)-one
Formed from: 7,10-Dihydroxy-2-(2-((2-hydroxyethyl)amino)ethyl)-5-(2-(methylamino)ethyl)amino)anthra[1,9-*c*,*d*]pyrazol-6(2*H*)-one

9-(S-Glutathionyl)-7,10-dihydroxy-2-(2-((2-hydroxyethyl)amino)ethyl)-5-(2-(methylamino)ethyl)amino)anthra[1,9-*c*,*d*]pyrazol-6(2*H*)-one
Formed from: 7,10-Dihydroxy-2-(2-((2-hydroxyethyl)amino)ethyl)-5-(2-(methylamino)ethyl)amino)anthra[1,9-*c*,*d*]pyrazol-6(2*H*)-one
→ 8,9-Bis(S-glutathionyl)-7,10-dihydroxy-2-(2-((2-hydroxyethyl)amino)ethyl)-5-(2-(methylamino)ethyl)amino)anthra[1,9-*c*,*d*]pyrazol-6(2*H*)-one
Probably in: Rat H139

5-(S-Glutathionyl)-3,4-dihydroxy-L-phenylalanine
→ 5-(3-(L-Alanyl))-5-(S-glutathionyl)-*o*-quinone
Probably in: Man E347
Mushroom E347

2-(S-Glutathionyl)-3,5-dimethylquinol
Formed from: 4-Acetamido-2,6-dimethylphenol

S-Glutathionyleugenol
Formed from: Eugenol

2-(S-Glutathionyl)-1-hydroxydibenzyl
Probably formed from: *cis*-stilbene oxide

3′-(S-Glutathionyl)-4′-hydroxydiclofenac
Formed from: Diclofenac

4-(S-Glutathionyl)-5-hydroxydiclofenac
Formed from: Diclofenac

6-(S-Glutathionyl)-5-hydroxydiclofenac
Formed from: Diclofenac

1-(S-Glutathionyl)-2-hydroxy-3-(*p*-nitrophenoxy)propane
Formed from: 3-(*p*-Nitrophenoxy)-1-propane oxide
→ S-(2-Hydroxy-3-(*p*-nitrophenoxy)propyl)-L-cysteine
Probably in: Rabbit A3042
Rat A3042, A3043

1-(S-Glutathionyl)-2-hydroxyoestrone
Formed from: Oestrone
→ 1-(S-(L-Cysteinyl))-2-hydroxyoestrone
Probably in: Rat A1396

4-(S-Glutathionyl)-2-hydroxyoestrone
Formed from: Oestrone
→ 4-(S-(L-Cysteinyl))-2-hydroxyoestrone
Probably in: Rat A1396

1-(S-Glutathionyl)-1-(*p*-hydroxyphenyl)acetone
Formed from: *p*-Hydroxyphenylacetone

2-(S-Glutathionyl)-N-isopropyl-N-phenylacetamide
Formed from: Propachlor
→ 2-(S-(L-Cysteinylglycyl))-N-isopropyl-N-phenylacetamide
Probably in: Chicken D932
Microorganism C407
Rat B669, B707, F505

(2*R*,4*R*)-3-((S-(S-Glutathionyl))-3-mercaptopropionyl)-2-(*o*-hydroxyphenyl)-4-thiazolidinecarboxylic acid
Formed from: (2*R*,4*R*)- 2-(*o*-Hydroxyphenyl)-3-(3-mercaptopropionyl)-4-thiazolidinecarboxylic acid
→ (2*R*,4*R*)-3-((S-(S-(L-Cysteinyl))-3-mercaptopropionyl)-2-(*o*-hydroxyphenyl)-4-thiazolidinecarboxylic acid
Probably in: Dog E137
Rat E137

3-(S-Glutathionyl)-methylindole
Formed from: Skatole

4-(3-(S-Glutathionyl)methyl-2-methylbenzyl) imidazole
Formed from: Detomidine

6-(S-Glutathionyl)mitoxantrone
Formed from: Mitoxantrone
→ 6,7-Bis(S-glutathionyl)mitoxantrone
In: Horseradish G48

2-(S-Glutathionyl)-5-nitrobenzaldehyde
Formed from: 2-Chloro-5-nitrobenzaldehyde

5′-(S-Glutathionyl)-4-methyl-5-N-methylcarboxamido-9-anilinoacridine
Probably formed from: 4-Methyl-5-N-methylcarboxamido-9-anilinoacridine

6′-(S-Glutathionyl)-4-methyl-5-N-methylcarboxamido-9-anilinoacridine
Probably formed from: 4-Methyl-5-N-methylcarboxamido-9-anilinoacridine

9-(S-Glutathionyl)-4-methyl-5-N-methylcarboxamido-9-anilinoacridine
Probably formed from: 4-Methyl-5-N-methylcarboxamido-9-anilinoacridine

1-(S-Glutathionyl)-3-(*o*-methylphenoxy) propan-2-ol
Formed from: *o*-Cresol glycidyl ether

1-(S-Glutathionyl)-2-methyl-1-phenyl-2-propanol
Probably formed from: 2-Methyl-1-phenyl-1-propene-1,2-oxide

2-(S-Glutathionyl)-5-nitrobenzonitrile
Formed from: 2-Chloro-5-nitrobenzonitrile

2-(S-Glutathionyl)-5-nitrobenzophenone
Formed from: 2-Chloro-5-nitrobenzophenone

2-(S-Glutathionyl)-4-nitrophenol
Formed from: 2,4-Dinitrophenol

1-(S-Glutathionyl)-2-(*m*-nitrophenyl)ethanol
Probably formed from: *m*-Nitrostyrene oxide

1-(S-Glutathionyl)-2-(*p*-nitrophenyl)ethanol
Probably formed from: *p*-Nitrostyrene oxide

2-(S-Glutathionyl)-5-nitro-1-trifluoromethylbenzene
Formed from: 2-Chloro-5-nitro-1-trifluoromethylbenzene

6-(S-Glutathionyl)-4-oxo-10-propyl-4*H*-pyrano[3,2-*g*]quinoline-2,8-dicarboxylic acid
Formed from: 6-Chloro-4-oxo-10-propyl-4*H*-pyrano[3,2-*g*]quinoline-2,8-dicarboxylic acid

4-(S-Glutathionyl)phenyl acetate
Formed from: *p*-Nitrophenyl acetate

2-(S-Glutathionyl)-2-phenyl-3-butanol
Probably formed from: *cis*-2-Phenyl-2-butene-2,3-oxide
trans-2-Phenyl-2-butene-2,3-oxide

3-(S-Glutathionyl)-4-phenylbutan-2-one
Probably formed from: *trans*-4-Phenyl-3-buten-2-one

4-(S-Glutathionyl)-4-phenylbutan-2-one
Probably formed from: *trans*-4-Phenyl-3-buten-2-one
→ 4-(N-Acetyl-S-(L-cysteinyl))-4-phenylbutan-2-one
Probably in: Mouse J293

2-(S-Glutathionyl)-2-phenylethanethiol
Probably formed from: Styrene episulphide

2-(S-Glutathionyl)-1-phenylethanol
Formed from: Styrene oxide
→ S-(2-Hydroxy-2-phenylethyl)-L-cysteine
In: Winter flounder D387
See also: B714, C605

2-(S-Glutathionyl)-2-phenylethanol
Formed from: Styrene oxide

→ S-(2-Hydroxy-1-phenylethyl)cysteinylglycine
In: Rabbit A1713
See also: C605, D387

1-(S-Glutathionyl)-1-phenyl-2-propanol
Formed from: *trans*-1-phenyl-2-propene oxide
Probably formed from: *cis*-1-phenyl-2-propene oxide

2-(S-Glutathionyl)-3-phenyl-1-propanol
Formed from: Allylbenzene oxide

2-(S-Glutathionyl)-3-phenylpropionaldehyde
Probably formed from: Cinnamaldehyde

S-Glutathionyl-2-phenylquinol
Formed from: 2-Hydroxybiphenyl

4-(S-Glutathionyl)quinoline-N-oxide
Formed from: 4-Nitroquinoline-N-oxide

4-(S-Glutathionyl)-7-sulphobenzofuran
Probably formed from: 4-Chloro-7-sulphobenzofuran

7-(S-Glutathionyl)-7,8,9,10-tetrahydro-8-hydroxy-3,4-benzpyrene
Formed from: 9,10-Dihydro-3,4-benzpyrene-7,8-oxide

10-(S-Glutathionyl)-7,8,9,10-tetrahydro-9-hydroxy-3,4-benzpyrene
Formed from: 7,8-Dihydro-3,4-benzpyrene-9,10-oxide

1-(S-Glutathionyl)-1,2,3,4-tetrahydro-2-hydroxy-naphthalene
Formed from: 1,2-Dihydronaphthalene-3,4-oxide

1-(S-Glutathionyl)-1,2,3,4-tetrahydro-2-hydroxy-phenanthrene
Formed from: 3,4-Dihydrophenanthrene-1,2-oxide

14-(S-Glutathionyl)-11,12,13,14-tetrahydro-11,12,13-trihydroxybenzo[*g*]chrysene
Probably formed from: 11,12-Dihydro-11,12-dihydroxybenzo[*g*]chrysene-13,14-oxide

1α-(S-Glutathionyl)-1,2,3,4-tetrahydro-2β,3β,4α-trihydroxybenzo[*c*]phenanthrene
Formed from: 1,2,3,4-Tetrahydro-3β,4α-trihydroxybenzo[*c*]phenanthrene-1β,2β-oxide

10β-(S-Glutathionyl)-7,8,9,10-tetrahydro-7β,8α,9α-trihydroxy-3,4-benzpyrene
Formed from: 7,8,9,10-Tetrahydro-7β,8α-diihydroxy-3,4-benzpyrene-9α,10α-oxide

4-(S-Glutathionyl)-1,2,3,4-tetrahydro-1,2,3-2,3-trihydroxychrysene chrysene
Formed from: 1,2,3,4-Tetrahydro-r-1,t-2,3-trihydroxychrysene-3,4-oxide

1-(S-Glutathionyl)-1,2,3,4-tetrahydro-2,c-3,t-4-trihydroxydibenz[*a*,*h*]anthracene
Formed from: 1,2,3,4-Tetrahydro-3,4-dihydroxydibenz[*a*,*h*]anthracene-1,2-oxide

Glutethimide (2-ethyl-2-phenylglutarimide)
→ 2-(3,4-Dihydro-3,4-dihydroxyphenyl)-2-ethylglutarimide
In: Monkey A1021
Rat A1021
→ 2-Ethyl-2-(4-hydroxy-3-methoxy-phenyl)glutarimide
Probably in: Man A2392
→ 2-Ethyl-2-(*p*-hydroxyphenyl)glutarimide
In: Guinea pig A1021
Man A1710, A2200, A2313, A2392
Monkey A1710
Rat A1021
→ 2-Ethyl-2-phenylglutaconimide
In: Guinea pig A1021
Man A1710
Monkey A1710
Rat A1021
→ 2-(1-Hydroxyethyl)-2-phenylglutarimide
In: Man A2313, A3883
Monkey A1710
→ 2-(2-Hydroxyethyl)-2-phenylglutarimide
In: Man A2313
→ 3-Hydroxyglutethimide
In: Monkey A1021
Rat A1021
→ 4-Hydroxyglutethimide
In: Dog A1435
Guinea pig A1021
Man A1710, A2313, A3883, A3991
Rat A1021
→ 2-Phenylglutarimide
In: Guinea pig A1021
Man A1710
Monkey A1710
Rat A1021

Glyceric acid 2-vanillic acid ether
Formed from: Glycerol 2-vanillic acid ether

Glycerol 2-vanillic acid ether
Formed from: Guaiacylglycerol β-vanillic acid ether
Probably formed from: Syringylglycerol β-vanillic acid ether
→ Ethylene glycol monovanillic acid ether
In: Fusarium C117
→ Glyceric acid 2-vanillic acid ether
In: Fusarium C117
→ Glycerol 2-vanillic acid ether monoacetate
In: Fusarium C117

Glycerol 2-vanillic acid ether monoacetate
Formed from: Glycerol 2-vanillic acid ether

Glyceryl guaiacolate
→ β-(2-Methoxyphenoxy)lactic acid
In: Man A668

7-Glycidoxycoumarin
→ S-(3-(7′-Coumaroxy)-2-hydroxy-1-propyl)glutathione
In: Rat C684
See also: F142
→ 7-(1′,2′-Dihydroxy-3′-propoxy)coumarin
In: Rat C684

Glycine-2,6-xylidide
Probably formed from: N-Ethylglycine-2,6-xylidide

Glycitein
See: 4′,7-Dihydroxy-6-methoxyisoflavone

2-Glycollamidofluorene
Formed from: 2-Acetamidofluorene
→ N-Hydroxy-2-glycollamidofluorene
In: Rat F638

6-Glycollamido-2-hydroxymethyl-3-(*o*-tolyl)-4(3*H*)-quinazolinone
Probably formed from: 6-Acetamido-2-hydroxymethyl-3-(*o*-tolyl)-4(3*H*)-quinazolinone

5-Glycollamido-2-methoxyacetanilide
Probably formed from: 2,4-Diacetamidoanisole

6-Glycollamido-2-methylsulphonylmethyl-3-(*o*-tolyl)-4(3*H*)-quinazolinone
Probably formed from: N-Acetyl-2-methylsulphonylafloqualone

2-Glycollamido-1-(*p*-nitrophenyl)-1,3-propanediol
Formed from: Chloramphenicol
Probably formed from: Chloramphenicol-2-oxamaldehyde

2-Glycollamido-9-oxofluorene
Formed from: 2-Acetamido-9-oxofluorene

***p*-Glycollamidophenol**
See: *p*-Hydroxyglycollanilide

Glycollanilide (2-hydroxyacetanilide)
Formed from: 1,2′-Bis(N-phenylacetamidoxy)ethoxyethane
N,N′-Diphenyl-2-acetamidoxyacetamide
→ Aniline
In: Microorganism A261

N-Glycollylafloqualone
Formed from: Afloqualone
Probably formed from: N-Acetylafloqualone

N-Glycollyl-3,4-dimethoxyphenethylamine
Formed from: Ecabapide
→ N-Glycollyl-4-hydroxy-3-methoxyphenethylamine
Probably in: Rat H737

N-Glycollyl-4-hydroxy-3-methoxyphenethylamine
Probably formed from: N-Glycollyl-3,4-dimethoxyphenethylamine

Glycopyrrolate (3-((cyclopentylhydroxyphenylacetyl)oxy)-1,1-dimethylpyrrolidinium)
→ 3-Hydroxy-1,1-dimethylpyrrolidinium benzoylformate
In: Mouse A1074, A1075
Rat A1075
→ 3-Hydroxy-1,1-dimethylpyrrolidinium α-(2-hydroxycyclopentyl)mandelate
In: Mouse A1074, A1075
Rat A1075
→ 3-Hydroxy-1,1-dimethylpyrrolidinium α-(3-hydroxycyclopentyl)mandelate
In: Mouse A1074

Glycyl-L-arginyl-β-naphthylamide
→ β-Naphthylamine
In: Man A2149

Glycyl-N-formyl-L-kynurenine
Formed from: Glycyl-L-tryptophan

Glycylglycyl-β-naphthylamide
→ β-Naphthylamine
In: Pseudomonas G643

Glycyl-L-leucyl-β-naphthylamide
→ β-Naphthylamine

In: Rabbit A1589

Glycyl-β-naphthylamide
→ β-Naphthylamine
In: Agave A2745
Aspergillus A3607
Beef A526
Glycine A1596
Man A1219, A1397
Rabbit A1589

Glycyl-*p*-nitroanilide
→ *p*-Nitroaniline
In: Agave A2745
Glycine A1596
Nocardia G343
Streptococcus C277

Glycyl-L-phenylalaninamide
→ Glycyl-L-phenylalanine
In: Aspergillus A3607

Glycyl-L-phenylalanine
Formed from: Glycyl-L-phenylalaninamide
Glycyl-L-phenylalanyl-β-naphthylamide
→ L-Phenylalanine
In: Brevibacterium E587

Glycyl-L-phenylalanyl-β-naphthylamide
→ Glycyl-L-phenylalanine
In: Beef A526
Man A2149
→ β-Naphthylamine
In: Beef A526
Man A2149
Pseudomonas G643

Glycyl-L-prolyl-β-naphthylamide
→ β-Naphthylamine
In: Aspergillus A3607
Beef A526
Man A1308, A2226

Glycyl-L-tryptophan
→ Glycyl-N-formyl-L-kynurenine
In: Wheat C559, E29

Glycyl-L-tyrosine
→ L-Tyrosine
In: Aspergillus A3658

Glyoxalylbenzamide
Probably formed from: α-(N-Formamido)-N-glyoxalylbenzamide

N-Glyoxylyl-3,4-dimethoxyphenethylamine
Formed from: Ecabapide

Gomisin A (6*S*,7*S*,13a*R*)-5,6,7,8-tetrahydro-1,2,3,13-tetramethoxy-6,7-dimethylbenzo[3,4]cycloocta[1,2-*f*][1,3]benzodioxol-6-ol)
→ 4,5-Demethylenegomisin A
In: Rat E750
→ 2-O-Demethylgomisin A
In: Rat E750
→ 14-O-Demethylgomisin A
In: Rat E750
→ 8-Hydroxygomisin A
In: Rat E750
→ 9-Hydroxygomisin A
In: Rat E750

Gomphrenin I ((2*S*)-4-(2-((2*S*)-2-carboxy-6-(β-D-glucopyranosyloxy)-2,3-dihydro-5-hydroxy-1*H*-indol-1-yl)ethenyl)-2,3-dihydro-2,6-pyridinedicarboxylic acid)
Formed from: Betanidin

Gossypetin (3,3′,4′,5,7,8-hexahydroxyflavone)
→ Gossypetin-7-glucoside
In: Chrysanthemum J320
→ 8-Hydroxy-3-O-methylquercetin
In: Lotus C905, D904
→ 8-Methoxyquercetin
In: Lotus C905

Gossypetin-7-glucoside
Formed from: Gossypetin

Gossypol (1,1′,6,6′,7,7′-hexahydroxy-3,3′-dimethyl-5,5′-bis(1-methylethyl)(2,2′-binaphthalene)-8,8′-dicarboxaldehyde)
Formed from acetate in cotton A1394, E29,
mevalonate in cotton D971
→ Apogossypol
Probably in: Pig A2146
→ Gossypol hexaglucoside
In: Heliothis G825
→ Gossypolone
In: Pig A2146

Gossypol hexaglucoside
Formed from: Gossypol

Gossypolone (1,1′,4,4′-tetrahydro-6,6′,7,7′-tetrahydroxy-3,3′-dimethyl-5,5′-bis(1-methylethyl)-1,1′-,4,4′-tetraoxo-(2,2′-binaphthalene)-8,8′-dicarboxaldehyde)
Formed from: Gossypol

→ Gossypolonic acid
Probably in: Pig A2146

Gossypolonic acid (1,1′,4,4′-tetrahydro-6,6′,7,7′-tetrahydroxy-3,3′-dimethyl-5,5′-bis(1-methylethyl)-1,1′-,4,4′-tetraoxo-(2,2′-binaphthalene)-8,8′-dicarboxylic acid)
Probably formed from: Gossypolone
→ Demethylgossic acid
Probably in: Pig A2146

Gramine (3-dimethylaminomethylindole)
Formed from: 3-Methylaminomethylindole
N-Methyl-L-tryptophan
→ 3-Formylindole
In: Hordeum G139
Lupinus A233

Granisetron (1-methyl-N-((3-*endo*)-9-methyl-9-azabicyclo[3.3.1]non-3-yl)-1*H*-indazole-3-carboxamide)
→ 1-Demethylgranisetron
In: Dog H63
Man H63
Rat H63
→ 9′-Demethylgranisetron
In: Dog H63
Man H63
Rat H63
→ 6,7-Dihydro-6,7-dihydroxygranisetron
In: Dog H63
Man H63
Rat H63
→ 5-Hydroxygranisetron
In: Dog H63
Man H63
Rat H63
→ 7-Hydroxygranisetron
In: Dog H63
Man H63
Rat H63

Griseofulvic acid
Formed from: Griseofulvin

Griseofulvin (1′*S*,6′*R*)-7-chloro-2′,4,6-trimethoxy-6′-methylspiro(benzofuran-2(3*H*),1′-[2]cyclohexene)-3,4′-dione)
Formed from: Dehydrogriseofulvin
→ 4-Demethylgriseofulvin
In: Man A1094
Mouse A618, A920, A1889
Rat A58, A920, A1867, A1889
→ 6-Demethylgriseofulvin
In: Man A1094
Mouse A618, A920, A1889
Rat A58, A920, A1867, A1889
→ Griseofulvic acid
In: Man B219
→ *ax*-5′-Hydroxygriseofulvin
In: Cephalosporium A839
→ 6′-Hydroxygriseofulvin
In: Cephalosporium A839

Griseophenone B (5-chloro-2,4′,6-trihydroxy-2′,4-dimethoxy-6′-methylbenzophenone)
→ Dehydrogriseofulvin
In: Aspergillus C556

Griseophenone C (2,4′,6-trihydroxy-2′,4-dimethoxy-6′-methylbenzophenone)
Probably formed from acetate in Penicillium A1931, A2448

Groenlandicine (11,12-dihydro-9-hydroxy-8-methoxybenzo[*a*]-1,3-benzodioxolo[4,5-*g*] quinolizinium)
→ N-Methylgroenlandicine
In: Eschscholtzia K782

Guaiacol (*o*-methoxyphenol)
Formed from: Anisole
Catechol
4-Ethoxy-3-methoxyphenylglycol β-guaiacyl ether
Guaiacyl glycerol ether
o-Methoxyphenoxyacetic acid
3-(*o*-Methoxyphenoxy)-1,2-propanediol
o-Methoxyphenyl nicotinate
Syringylglycerol β-guaiacyl ether
Vanillic acid
Vanillyl alcohol
Veratrole
Veratrylglycerol β-guaiacyl ether
Probably formed from: β-Hydroxy-propiovanillone
Isovanillic acid
Isovanillyl alcohol
Mociprazine
→ Catechol
In: Acetobacterium J500, J667
Clostridium G367

Microorganism G917
Paecilomyces F199
Pseudomonas D16
Rat J201
Sporomusa J199
Streptomyces B877
See also: F101, F859
→ 4,4′-Dihydroxy-3,3′-dimethoxybiphenyl
With lactoperoxidase J352
→ Free radical
In: Coriolus K677
Horseradish A248
See also: A1926, A3764, J318
→ *o*-Methoxyphenyl-β-D-glucoside
In: Agrotis G816
Manduca G816, G897
Melanopus G816
Periplaneta G816
Tenebrio G816
Trilobium G816
→ *o*-Methoxyphenyl sulphate
In: Man D979, E575
Mouse H959
Rat E575
→ Polyguaiacol
In: Bjerkandera K565
→ 2,3,4′-Trihydroxy-3′-methoxybiphenyl
With lactoperoxidase J352
→ Veratrole
In: Phanerochaete G782

Guaiacylglycerol β-coniferaldehyde ether
Formed from: Guaiacylglycerol β-coniferyl ether
→ Guaiacylglycerol β-ferulic acid ether
Probably in: Fusarium B482

Guaiacylglycerol β-coniferyl ether
→ Coniferyl alcohol
In: Pseudomonas A3030
→ Guaiacylglycerol β-coniferaldehyde ether
In: Fusarium B482
→ β-Hydroxypropiovanillone
In: Pseudomonas A3030

Guaiacyl glycerol ether
→ Guaiacol
In: Bacillus E404

Guaiacylglycerol β-ferulic acid ether
Probably formed from: Guaiacylglycerol β-coniferaldehyde ether
→ Guaiacylglycerol β-vanillin ether
Probably in: Fusarium B482

Guaiacylglycerol β-(*p*-fluorophenyl) ether
Formed from: 2-(*p*-Fluorophenoxy)-3-(4-hydroxy-3-methoxyphenyl)-3-oxopropanol

Guaiacylglycerol β-guaiacyl ether
→ 3-Hydroxy-1-(4-hydroxy-3-methoxyphenyl)-2-(2-methoxyphenoxy)-1-propanone
In: Pseudomonas D967
→ β-Hydroxypropiovanillone
In: Pseudomonas D967

Guaiacylglycerol β-vanillic acid ether
Probably formed from: Guaiacylglycerol β-vanillin ether
→ Glycerol 2-vanillic acid ether
In: Fusarium C117

Guaiacylglycerol β-vanillin ether
Probably formed from: Guaiacylglycerol β-ferulic acid ether
→ Guaiacylglycerol β-vanillic acid ether
Probably in: Fusarium B482

Guaiphenesin
See: 3-(2-Methoxyphenoxy)-1,2-propanediol

Guaiphenesin-α-D-glucoside
Formed from: 3-(2-Methoxyphenoxy)-1,2-propanediol

Guanabenz (2-((2,6-dichlorophenyl)-methylene)hydrazinecarboximidamide)
Formed from: Guanoxabenz
→ Guanoxabenz
In: Man J170, H802
Pig J170
Rabbit J170

Guanfacine (N-amidino-2-(2,6-dichlorophenyl)-acetamide)
Formed from: Guanfacine-N-oxide
→ 2,6-Dichlorophenylacetamide
In: Rat B889
→ 2,6-Dichlorophenylacetic acid
In: Rat B889
→ 3,4-Dihydro-3,4-dihydroxyguanfacine
In: Rat B889
→ Guanfacine-N-oxide

In: Rat B889
→ 3-Hydroxyguanfacine
In: Rat B889
→ 4-Hydroxyguanfacine
In: Rat B889

Guanfacine-N-oxide
Formed from: Guanfacine
→ Guanfacine
In: Man J548
Pig J548
Rabbit J548

4-Guanidinobenzoic acid
Formed from: Camostat

4-(4-Guanidinobenzoyloxy)phenylacetic acid
Formed from: Camostat

***o*-((2-Guanidino)ethyl)benzoic acid**
Probably formed from: 1-Hydroxy-debrisoquine

***o*-(Guanidinomethyl)phenylacetic acid**
Probably formed from: 3-Hydroxy-debrisoquine

Guanoxabenz (2-((2,6-dichlorophenyl)-methylene)-N-hydroxyhydrazine-carboximidamide)
Formed from: Guanabenz
→ Guanabenz
In: Man J170
Pig J170
Rabbit J170

2-(N^7-Guanyl)-2,3-dihydro-3-hydroxyaflatoxin B_1
Formed from: Aflatoxin B_1

Guattegaumerine
Formed from: N-Methylcoclaurine

Guthion
See: Azinophos-methyl

Gutoxon (O,O-dimethyl S-(4-oxo-1,2,3-benzotriazin-3(4*H*)-yl)methyl) phosphorothioate)
→ Benzazimide
In: Rat B283

H

Haemanthamine ((3*S*,4a*S*,5*S*,11b*S*,12*R*)-2,4a-dihydro-3-methoxy-3*H*,6*H*-5,11b-ethano[1,3]dioxolo[4,5-*j*] phenanthridin-12-ol)
Formed from: O-Methylnorbelladine

Halazepam (7-chloro-1,3-dihydro-5-phenyl-1-(2,2,2-trifluoroethyl)-2*H*-1,4-benzodiazepin-2-one)
→ 7-Chloro-1,3-dihydro-5-phenyl-2H-1,4-benzodiazepin-2-one
In: Man F765
Rat F643
→ 3-Hydroxyhalazepam
In: Man F765
Rat F643

Haloperidol (4-(4-(*p*-chlorophenyl)-4-hydroxypiperidino)-4′-fluorobutyrophenone)
Formed from: Dihydrohaloperidol
Haloperidol decanoate
→ 4-(*p*-Chlorophenyl)-1-(4-(*p*-fluorophenyl)-4-oxobutyl)pyridine
In: Mouse G569
→ 4-(*p*-Chlorophenyl)-4-hydroxypiperidine
In: Mouse G569
→ 4-(4-(*p*-Chlorophenyl)-1,2,3,6-tetrahydropyridin-1-yl)-1-(*p*-fluorophenyl)-1-butanone
In: Mouse G451
See also: K545
→ Dihydrohaloperidol
In: Guinea pig F122
Man A3013, F122, F261, G89, J846
Mouse G569
→ *p*-Fluorobenzoylpropionic acid
In: Man A3090
Mouse G569
Rat E17
See also: E16
→ *p*-Fluorophenylacetic acid
In: Man A3090
Rat E17
See also: E16
→ Haloperidol-β-D-glucuronide
In: Rat E17
See also: E16
→ Haloperidol-O-sulphate
In: Rat E17
See also: E16

Haloperidol decanoate
→ Haloperidol
In: Rat E16

Haloperidol-β-D-glucuronide
Formed from: Haloperidol

Haloperidol-O-sulphate
Formed from: Haloperidol

Haloprogin (2,4,5-trichlorophenyl γ-iodopropargyl ether)
→ 2,4,5-Trichlorophenol
In: Dog A477
Mouse A477
Rabbit A477
Rat A477

(*R*)-Haloxyfop (2-(4-((3-chloro-5-(trifluoromethyl)-2-pyridinyl)oxy) phenoxy)propanoic acid)
Formed from: (*S*)-Haloxyfop

(*S*)-Haloxyfop
→ (*R*)-Haloxyfop
In: Rat F120

Harmalan (4,9-dihydro-1-methyl-3*H*-pyrido[3,4-*b*]indole)
Formed from: 1,2,3,4-Tetrahydro-1-methyl-β-carboline-1-carboxylic acid

Harmaline (4,9-dihydro-7-methoxy-1-methyl-3*H*-pyrido[3,4-*b*]indole)
→ Dihydroruine
In: Peganum A1330
→ Harmine
In: Peganum A1330

Harmalol (4,9-dihydro-1-methyl-3*H*-pyrido[3,4-*b*]indol-7-ol)
→ Harmalol-β-D-glucuronide
In: Rat A90
→ Harmalol sulphate
In: Rat A90

Harmalol-β-D-glucuronide
Formed from: Harmalol

Harmalol sulphate
Formed from: Harmalol

Harman (1-methyl-3*H*-pyrido[3,4-*b*]indole)
Formed from: 1,2,3,4-Tetrahydro-1-methyl-β-carboline-1-carboxylic acid
L-Tryptophan

Harmine (7-methoxy-1-methyl-9*H*-pyrido[3,4-*b*]indole)
Formed from: Harmaline
→ Harmol
In: Cat A1777
Cow A1777
Guinea pig A1777
Mouse A1777, E880
Rabbit A1777
Rat A1777, B287
→ 3-Hydroxyharmine
Probably in: Mouse E880
→ 6-Hydroxyharmine
In: Mouse E880

Harmol (1-methyl-9*H*-pyrido[3,4-*b*]indol-7-ol)
Formed from: Harmine
→ Harmol-β-D-glucuronide
In: Cat A107
Guinea pig A107
Man G516
Mouse A107
Rabbit A107
Rat A90, A107, A1610, A3464, A3714, B287, B644, B824, C180, C602, F728
→ Harmol sulphate
In: Cat A107
Dog C187
Guinea pig A107
Man G516, D736
Mouse A107, C187
Monkey C187
Rabbit C187
Rat A90, A107, A1610, A3464, A3714, B644, B824, C180, C187, C320, C602, F728

Harmol-β-D-glucuronide
Formed from: Harmol

Harmol sulphate
Formed from: Harmol

Hawkinsin (6-((2-amino-2-carboxyethyl)thio)-1,4-dihydroxy-2-cyclohexene-1-acetic acid)
Probably formed from: *p*-Hydroxyphenylpyruvic acid

(*S*,*R*)-Hayatidine (13a*S*,25a*R*)-2,3,13a,14,15,16,25,25a-octahydro-9,18,29-trimethoxy-1,14-dimethyl-13*H*-4,6:21,24-dietheno-8,12-metheno-1*H*-pyrido[3′,2′:14,15][1,11]dioxacycloeicosino[2,3,4-*ij*]isoquinolin-19-ol)
Formed from: (*R*)-N-Methylcoclaurine
(*S*)-N-Methylcoclaurine

Helianthin
See: Methyl orange

Helicin
See: Salicylaldehyde-β-D-glucoside

Hemimellitene
See: 1,2,3-Trimethylbenzene

1,2,3,4,6,7,8-Heptachlorodibenzo-*p*-dioxin
Formed from: Octachlorodibenzo-*p*-dioxin
→ 1,2,3,4,6,7-Hexachlorodibenzo-*p*-dioxin
Probably in: Microorganism H874
→ 1,2,3,4,6,8-Hexachlorodibenzo-*p*-dioxin
Probably in: Microorganism H874
→ 1,2,3,4,7,8-Hexachlorodibenzo-*p*-dioxin
Probably in: Microorganism H874
→ 1,2,3,6,7,8-Hexachlorodibenzo-*p*-dioxin
Probably in: Microorganism H874

1,2,3,4,6,7,9-Heptachlorodibenzo-*p*-dioxin
Formed from: Octachlorodibenzo-*p*-dioxin
→ 1,2,3,4,6,7-Hexachlorodibenzo-*p*-dioxin
Probably in: Microorganism H874
→ 1,2,3,6,7,9-Hexachlorodibenzo-*p*-dioxin
Probably in: Microorganism H874

Heptachlorophenylacetic acid
Formed from: Octachlorostyrene

Heptachlorostyrene
Formed from: Octachlorostyrene
2,3,3,3′,4′,5,7-Heptahydroxyflavan-4-one (quercetin dihydrodiol)
Formed from: Quercetin
***trans*-2-(Heptoxycarbonylethyl)-3-(*p*-hydroxyphenyl)-N-phenyllactam**
→ *trans*-2-(2-Carboxyethyl)-3-(*p*-hydroxyphenyl)-N-phenyllactam
In: Microorganism K110
***p*-*n*-Heptylbenzoic acid**
→ *p*-*n*-Heptylbenzoyl CoA
In: Beef H740
***p*-*n*-Heptylbenzoyl CoA**
Formed from: *p*-*n*-Heptylbenzoic acid
Heptyl L-γ-glutamyl dopamide
→ 3,4-Dihydroxyphenethylamine
In: Dog A3945
2-Heptyloxyaniline
Formed from: Heptacaine
***p*-*n*-Heptylphenol**
→ 1-(*p*-Hydroxyphenyl)-1-heptanol
In: Pseudomonas F381
Hernandalinol (4,5-dimethoxy-2-(((6a*S*)-5,6,6a,7-tetrahydro-1,2,10-trimethoxy-6-methyl-4*H*-dibenzo[*de*,*g*]quinolin-9-yl)oxy)benzenemethanol)
Formed from: Thalicarpine
Herniarin
See: 7-Methoxycoumarin
Heroin (3,6-diacetylmorphine)
→ 6-Acetylmorphine
In: Man A798, A1347, A2907, B213, D173, H890
Rat D173
Hesperetin (3′,5,7-trihydroxy-4′-methoxyflavanone)
Formed from: Hesperidin
Rutin
Probably formed from: Eriodictyol
→ Dihydrocaffeic acid
In: Rat A1835
→ Dihydroisoferulic acid
In: Rat A1835
→ Eriodictyol
In: Rat G912
→ Hesperetin-3′,7-disulphate
In: Rat G912
→ Hesperetin-7-β-D-glucoside
In: Calamondin F419
Grapefruit K810, K891
Petunia K303
→ Hesperetin-3′-β-D-glucuronide
In: Rat G912
See also: A3696, B244
→ Hesperetin-7-β-D-glucuronide
In: Rat G912
→ *m*-Hydroxyphenylpropionic acid
In: Rat A1835
Hesperetin-3′,7-disulphate
Formed from: Hesperetin
Hesperetin-7-β-D-glucoside
Formed from: Hesperetin
Hesperetin-3′-β-D-glucuronide
Formed from: Hesperetin
→ Hesperetin-3′-β-D-glucuronide-7-sulphate
Probably in: Rat G912
Hesperetin-7-β-D-glucuronide
Formed from: Hesperetin
Hesperetin-3′-β-D-glucuronide-7-sulphate
Probably formed from: Hesperetin-3′-β-D-glucuronide
Hesperidin (hesperetin-7-(6-O-rhamnosyl)-β-D-glucoside)
→ Hesperetin
In: Microorganism J525, K643
→ Hesperidin-β-D-glucuronide
In: Rat A3696, B244
Hesperidin-β-D-glucuronide
Formed from: Hesperidin
Hesperetin-7-(2-O-rhamnosyl)-β-D-glucoside
See: Neohesperidin
Hesperetin-7-(6-O-rhamnosyl)-β-D-glucoside
See: Hesperidin
Heteroauxin
See: Indole-3-acetic acid
2,2′,4,4′,5,5′-Hexabromobiphenyl
→ Hexabromohydroxybiphenyl
In: Rat H264
Hexabromohydroxybiphenyl
Formed from: 2,2′,4,4′,5,5′-Hexabromobiphenyl
2,2′,3,3′,4,4′-Hexachloroazobenzene
Formed from: 2,3,4-Trichloroaniline

Hexachlorobenzene
→ N-Acetyl-S-(pentachlorophenyl)-L-cysteine
In: Rat A3742, H205, H769
→ Pentachlorobenzene
In: Monkey A2377, A2993, A3201
Rat A1563, A2377, A2789, B841, B923
See also: A1950
→ Pentachlorophenol
In: Monkey A2377, A2993, A3201
Rat A2167, A2789, A2872, A3822, B841, B923, C44, D552, D916, E54, H205, H769
→ Pentachlorothiophenol
In: Man G663
Rat A2784, A2872, B42, B923, C44, D916
→ 1,2,3,4-Tetrachlorobenzene
In: Rat A3822
See also: A2377, A2789
→ 1,2,4,5-Tetrachlorobenzene
In: Rat A3822
→ Tetrachloromercaptothioanisole
In: Rat H205
→ 1,3,5-Trichlorobenzene
In: Rat A3822
→ 2,3,4-Trichlorophenol
In: Rat A2789
→ 2,4,5-Trichlorophenol
In: Rat A3822

2,2′,3,3′,4,4′-Hexachlorobiphenyl
→ 2,2′,3,3′,4,4′-Hexachloro-5-hydroxybiphenyl
In: Rat K200
→ 2,2′,3,3′,4,4′-Hexachloro-5-methylthiobiphenyl
In: Rat K200
→ 2,2′,3,3′,4,4′-Hexachloro-6-methylthiobiphenyl
In: Rat K200

2,2′,3,3′,5,5′-Hexachlorobiphenyl
→ 2,2′,3,3′,5,5′-Hexachloro-4-hydroxybiphenyl
In: Rat B450
→ 2,2′,3,3′,5,5′-Hexachloro-6-hydroxybiphenyl
In: Rat B450
→ 2,2′,3′,4,5,5′-Hexachloro-3-hydroxybiphenyl
In: Rat B450
→ 2,2′,3,5,5′-Pentachloro-4-hydroxybiphenyl
In: Rat B450

2,2′,3,3′,6,6′-Hexachlorobiphenyl
→ 2,2′,3,3′,6,6′-Hexachloro-4-hydroxybiphenyl
In: Rat B450
→ 2,2′,3,3′,6,6′-Hexachloro-5-hydroxybiphenyl
In: Rat C660

2,2′,3,4,5,5′-Hexachlorobiphenyl
→ 2,2′,3′,4′,5,5′-Hexachloro-3-hydroxybiphenyl
In: Rat J278
→ 2,2′,3′,4′,5,5′-Hexachloro-4-hydroxybiphenyl
In: Rat J278
→ 2,2′,3,4,5,5′-Hexachloro-3′-methylthiobiphenyl
In: Rat J278
→ 2,2′,3,4,5,5′-Hexachloro-4′-methylthiobiphenyl
In: Rat J278

2,2′,4,4′,5,5′-Hexachlorobiphenyl
→ 2,2′,3,4′,5,5′-Hexachloro-4-hydroxybiphenyl
In: Rat B450
→ 2,2′,4,4′,5,5′-Hexachloro-3-hydroxybiphenyl
In: Dog G947, H601
Guinea pig H601, J258
Man H601
Monkey B954
Mouse A2533
Rabbit A2533
Rat A2533, B450
→ 2,2′,4,4′,5,5′-Hexachloro-6-hydroxybiphenyl
In: Rat B450
→ 2′,3,4,4′,5,5′-Hexachloro-2-hydroxybiphenyl
In: Dog G497, H601
Guinea pig H601, J258
→ 2,2′,4′,5,5′-Pentachloro-3-hydroxybiphenyl
In: Guinea pig J258
Rabbit A2533
Rat B450

→ 2,2′,4′,5,5′-Pentachloro-4-hydroxybiphenyl
In: Guinea pig J258
Rabbit A2533
→ 2′,4,4′,5,5′-Pentachloro-2-hydroxybiphenyl
In: Dog H601
Guinea pig H601, J258

2,2′,4,4′,6,6′-Hexachlorobiphenyl
→ 2,2′,4,4′,6,6′-Hexachloro-3-hydroxybiphenyl
In: Rat B450

α-1,2,3,4,5,6-Hexachlorocyclohexane
See: α-Lindane

β-1,2,3,4,5,6-Hexachlorocyclohexane
See: β-Lindane

γ-1,2,3,4,5,6-Hexachlorocyclohexane
See: Lindane

δ-1,2,3,4,5,6-Hexachlorocyclohexane
See: δ-Lindane

ε-1,2,3,4,5,6-Hexachlorocyclohexane
See: ε-Lindane

1,2,3,4,6,7-Hexachlorodibenzo-*p*-dioxin
Formed from: 2,4,5-Trichlorophenol
Probably formed from: 1,2,3,4,6,7,8-Heptachlorodibenzo-*p*-dioxin
1,2,3,4,6,7,9-Heptachlorodibenzo-*p*-dioxin

1,2,3,4,6,8-Hexachlorodibenzo-*p*-dioxin
Probably formed from: 1,2,3,4,6,7,8-Heptachlorodibenzo-*p*-dioxin

1,2,3,4,7,8-Hexachlorodibenzo-*p*-dioxin
Probably formed from: 1,2,3,4,6,7,8-Heptachlorodibenzo-*p*-dioxin
→ 2,3,7,8-Tetrachlorodibenzo-*p*-dioxin
Probably in: Microorganism H874

1,2,3,6,7,8-Hexachlorodibenzo-*p*-dioxin
Probably formed from: 1,2,3,4,6,7,8-Heptachlorodibenzo-*p*-dioxin
→ 2,3,7,8-Tetrachlorodibenzo-*p*-dioxin
Probably in: Microorganism H874

1,2,3,6,7,9-Hexachlorodibenzo-*p*-dioxin
Formed from: 2,4,5-Trichlorophenol

1,2,3,6,8,9-Hexachlorodibenzo-*p*-dioxin
Formed from: 2,4,5-Trichlorophenol

1,2,4,6,7,9-Hexachlorodibenzo-*p*-dioxin
Formed from: 2,4,5-Trichlorophenol
Probably formed from: 1,2,3,4,6,7,9-Heptachlorodibenzo-*p*-dioxin

1,2,4,6,8,9-Hexachlorodibenzo-*p*-dioxin
Formed from: 2,4,5-Trichlorophenol

2,2′,3,3′,4,4′-Hexachloro-5-hydroxybiphenyl
Formed from: 2,2′,3,3′,4,4′-Hexachlorobiphenyl

2,2′,3,3′,5,5′-Hexachloro-4-hydroxybiphenyl
Formed from: 2,2′,3,3′,5,5′-Hexachlorobiphenyl

2,2′,3,3′,5,5′-Hexachloro-6-hydroxybiphenyl
Formed from: 2,2′,3,3′,5,5′-Hexachlorobiphenyl

2,2′,3,3′,6,6′-Hexachloro-4-hydroxybiphenyl
Formed from: 2,2′,3,3′,6,6′-Hexachlorobiphenyl

2,2′,3,3′,6,6′-Hexachloro-5-hydroxybiphenyl
Formed from: 2,2′,3,3′,6,6′-Hexachlorobiphenyl

2,2′,3′,4′,5,5′-Hexachloro-3-hydroxybiphenyl
Formed from: 2,2′,3,4,5,5′-Hexachlorobiphenyl

2,2′,3,4′,5,5′-Hexachloro-4-hydroxybiphenyl
Formed from: 2,2′,4,4′,5,5′-Hexachlorobiphenyl

2,2′,3′,4,5,5′-Hexachloro-3-hydroxybiphenyl
Formed from: 2,2′,3,3′,5,5′-Hexachlorobiphenyl

2,2′,3′,4′,5,5′-Hexachloro-4-hydroxybiphenyl
Formed from: 2,2′,3,4′,5,5′-Hexachlorobiphenyl

2,2′,3′,4,5,5′-Hexachloro-6-hydroxybiphenyl
Formed from: 2,2′,3,3′,5,5′-Hexachlorobiphenyl

2,2′,4,4′,5,5′-Hexachloro-3-hydroxybiphenyl
Formed from: 2,2′,4,4′,5,5′-Hexachlorobiphenyl

2,2′,4,4′,5,5′-Hexachloro-6-hydroxybiphenyl
Formed from: 2,2′,4,4′,5,5′-Hexachlorobiphenyl

2,2′,4,4′,6,6′-Hexachloro-3-hydroxybiphenyl
Formed from: 2,2′,4,4′,6,6′-Hexachlorobiphenyl

2′,3,4,4′,5,5′-Hexachloro-2-hydroxybiphenyl
Formed from: 2,2′,4,4′,5,5′-Hexachlorobiphenyl
→ 2′,4,4′,5,5′-Pentachloro-2-hydroxybiphenyl
In: Dog G497

2,2′,3,4,5,5′-Hexachloro-3′-methylsulphonylbiphenyl
Probably formed from: 2,2′,3,4,5,5′-Hexachloro-3′-methylthiobiphenyl

2,2′,3,4,5,5′-Hexachloro-4′-methylsulphonylbiphenyl
Probably formed from: 2,2′,3,4,5,5′-Hexachloro-4′-methylthiobiphenyl

2,2′,3,3′,4,4′-Hexachloro-5-methylthiobiphenyl
Formed from: 2,2′,3,3′,4,4′-Hexachlorobiphenyl

2,2′,3,3′,4,4′-Hexachloro-6-methylthiobiphenyl
Formed from: 2,2′,3,3′,4,4′-Hexachlorobiphenyl

2,2′,3,4,5,5′-Hexachloro-3′-methylthiobiphenyl
Formed from: 2,2′,3,4,5,5′-Hexachlorobiphenyl
→ 2,2′,3,4,5,5′-Hexachloro-3′-methylsulphonylbiphenyl
Probably in: Rat J278

2,2′,3,4,5,5′-Hexachloro-4′-methylthiobiphenyl
Formed from: 2,2′,3,4,5,5′-Hexachlorobiphenyl
→ 2,2′,3,4,5,5′-Hexachloro-4′-methylsulphonylbiphenyl
Probably in: Rat J278

Hexachlorophene (2,2′-methylenebis(3,4,6-trichlorophenol))
→ Hexachlorophene-β-D-glucuronide
In: Rabbit A1937
Rat A281

Hexachlorophene-β-D-glucuronide
Formed from: Hexachlorophene

Hexadecyl-1-naphthylacetylglycerol
Formed from: 1-Naphthylacetyl CoA

2-Hexadecyl-3-palmitoyl-1-(*m*-phenoxybenzoyl)glycerol
Probably formed from: 2-Hexadecyl-1-(*m*-phenoxybenzoyl)glycerol

2-Hexadecyl-1-(*m*-phenoxybenzoyl)glycerol
Formed from: *m*-Phenoxybenzoyl CoA
→ 2-Hexadecyl-3-palmitoyl-1-(*m*-phenoxybenzoyl)glycerol
Probably in: Rat D416

Hexafluorobenzene
→ Pentafluorophenol
In: Rat H769

1-(Hexahydroazepin-1-yl)-3-(*p*-hydroxymethylphenyl)sulphonylurea
Formed from: Tolazamide
→ 3-(*p*-Carboxyphenyl)sulphonyl-1-(hexahydroazepin-1-yl)urea
Probably in: Man A1416, A3563

1-(Hexahydroazepin-1-yl)-1-phenylcyclohexane
→ 1-Phenylcyclohexylamine
In: Rat F892

1-(Hexahydroazepin-1-yl)-3-(*p*-tolyl)sulphonylurea
See: Tolazamide

***cis*-Hexahydrocannabinol** (*cis* assumed unless otherwise stated)
Formed from: $\Delta^{1(7)}$-Tetrahydrocannabinol
→ Hexahydro-1″-hydroxycannabinol
In: Guinea pig F767
Hamster F767
Rabbit F767
Rat F767
→ Hexahydro-3″-hydroxycannabinol
In: Guinea pig F767
Hamster F767
Rabbit F767
→ Hexahydro-4″-hydroxycannabinol
In: Guinea pig F767
Hamster F767
Rabbit F767
Rat F767
→ Hexahydro-5″-hydroxycannabinol
In: Guinea pig F767
Hamster F767
Rabbit F767
Rat F767
→ Hexahydro-6α-hydroxycannabinol
In: Guinea pig F767
Hamster F767
Mouse F767
Rabbit F767
Rat F767
→ Hexahydro-6β-hydroxycannabinol
In: Guinea pig F767
Hamster F767
Mouse F767
Rabbit F767
Rat F767

→ Hexahydro-7-hydroxycannabinol
In: Guinea pig F767
Hamster F767
Mouse F767
Rabbit F767
Rat F767
See also: B604

***trans*-Hexahydrocannabinol**
Formed from: $\Delta^{1(7)}$-Tetrahydrocannabinol
→ *trans*-Hexahydro-7-hydroxycannabinol
Probably in: Mouse B604

***cis*-Hexahydrocannabinol-7-oic acid**
Formed from: *cis*-Hexahydro-7-hydroxycannabinol
Δ^{1}-Tetrahydrocannabinol
Δ^{6}-Tetrahydrocannabinol
→ *cis*-Hexahydro-1α-hydroxycannabinol-7-oic acid
Probably in: Mouse B604
→ *cis*-Hexahydro-2″-hydroxycannabinol-7-oic acid
Probably in: Mouse B604
→ *cis*-Hexahydro-3″-hydroxycannabinol-7-oic acid
Probably in: Mouse B604
→ *cis*-Hexahydro-4″-hydroxycannabinol-7-oic acid
Probably in: Mouse B604

***trans*-Hexahydrocannabinol-7-oic acid**
Formed from: *trans*-Hexahydro-7-hydroxycannabinol
Δ^{1}-Tetrahydrocannabinol
Δ^{6}-Tetrahydrocannabinol
→ *trans*-Hexahydro-2″-hydroxycannabinol-7-oic acid
Probably in: Mouse B604
→ *trans*-Hexahydro-3″-hydroxycannabinol-7-oic acid
Probably in: Mouse B604
→ *trans*-Hexahydro-4″-hydroxycannabinol-7-oic acid
Probably in: Mouse B604

Hexahydrocurcumin
Probably formed from: Tetrahydrocurcumin
→ Hexahydrocurcumin-β-D-glucuronide
Probably in: Rat A3492

Hexahydrocurcumin-β-D-glucuronide
Probably formed from: Hexahydrocurcumin

Hexahydro-1α,2β-dihydroxycannabinol
Formed from: Δ^{1}-Tetrahydrocannabinol-1α,2α-oxide

Hexahydro-1β,2α-dihydroxycannabinol
Formed from: Δ^{1}-Tetrahydrocannabinol-1α,2α-oxide
Δ^{1}-Tetrahydrocannabinol-1β,2β-oxide

Hexahydro-1,6-dihydroxycannabinol
(stereochemistry unclear)
Formed from: Δ^{6}-Tetrahydrocannabinol

Hexahydro-1α,6β-dihydroxycannabinol
Formed from: Δ^{6}-Tetrahydrocannabinol-1,6-oxide
Probably formed from: Δ^{6}-Tetrahydrocannabinol
→ Hexahydro-1α,3″,6β-trihydroxycannabinol
Probably in: Mouse A3425
→ Hexahydro-1α,4″,6β-trihydroxycannabinol
Probably in: Mouse A3425

Hexahydro-1β,6α-dihydroxycannabinol
Formed from: Δ^{6}-Tetrahydrocannabinol-1,6-oxide

Hexahydro-1α,7-dihydroxycannabinol
Probably formed from: *cis*-Hexahydro-7-hydroxycannabinol
$\Delta^{1(7)}$-Tetrahydrocannabinol-1β,7-oxide
→ Hexahydro-1α,2″,7-trihydroxycannabinol
Probably in: Mouse B604
Rat B367
→ Hexahydro-1α,3″,7-trihydroxycannabinol
Probably in: Mouse B604
→ Hexahydro-1α,4″,7-trihydroxycannabinol
Probably in: Mouse B604
Rat B367
→ Hexahydro-1α,6α,7-trihydroxycannabinol
Probably in: B367

Hexahydro-1β,7-dihydroxycannabinol
Probably formed from: *cis*-Hexahydro-7-hydroxycannabinol
$\Delta^{1(7)}$-Tetrahydrocannabinol-1β,7-oxide
→ Hexahydro-1β,2″,7-trihydroxycannabinol
Probably in: Mouse B604
→ Hexahydro-1β,3″,7-trihydroxycannabinol
Probably in: Mouse B604

→ Hexahydro-1β,4″,7-trihydroxycannabinol
Probably in: Mouse B604

Hexahydro-1α,2″-dihydroxycannabinol-7-oic acid
Probably formed from: *cis*-Hexahydro-1α-hydroxycannabinol-7-oic acid

Hexahydro-1α,3″-dihydroxycannabinol-7-oic acid
Probably formed from: *cis*-Hexahydro-1α-hydroxycannabinol-7-oic acid

Hexahydro-1α,4″-dihydroxycannabinol-7-oic acid
Probably formed from: *cis*-Hexahydro-1α-hydroxycannabinol-7-oic acid

6,7,7a,8,9,11a-Hexahydro-1,7-dihydroxy-7,10-dimethyldibenzo[*b*,*d*]oxepin
Formed from: Cannabidiol

1,2,3,4,7,12-Hexahydro-8,11-dihydroxy-4-(2-hydroxyethyl)-6-(2-(2-hydroxyethyl)aminoethyl)amino-7,12-dioxonaphtho[2,3-*f*]quinoxaline
Formed from: Mitoxantrone

1,2,3,4,8,9-Hexahydro-1,2,3,4,8,9-hexahydroxydibenzanthracene
Probably formed from: *trans*-1,2,3,4-Tetrahydro-1,2,3,4-tetrahydroxydibenzanthracene

1,2,3,4,10,11-Hexahydro-1,2,3,4,10,11-hexahydroxydibenzanthracene
Probably formed from: 3,4,10,11-Tetrahydro-3,4,10,11-tetrahydroxydibenzanthracene

1,2,3,4,12,13-Hexahydro-1,2,3,4,12,13-hexahydroxydibenz[*a*,*c*]anthracene
Probably formed from: 12,13-Dihydroxydibenz[*a*,*c*]anthracene

1-(Hexahydro-4-hydroxyazepin-1-yl)-3-(*p*-tolyl)sulphonylurea
Formed from: Tolazamide

Hexahydro-1″-hydroxycannabinol
Formed from: *cis*-Hexahydrocannabinol

Hexahydro-3″-hydroxycannabinol
Formed from: *cis*-Hexahydrocannabinol

Hexahydro-4″-hydroxycannabinol
Formed from: *cis*-Hexahydrocannabinol

Hexahydro-5″-hydroxycannabinol
Formed from: *cis*-Hexahydrocannabinol

Hexahydro-6α-hydroxycannabinol
Formed from: *cis*-Hexahydrocannabinol

Hexahydro-6β-hydroxycannabinol
Formed from: *cis*-Hexahydrocannabinol

***cis*-Hexahydro-7-hydroxycannabinol**
Formed from: *cis*-Hexahydrocannabinol
→ *cis*-Hexahydrocannabinol-7-oic acid
Probably in: Mouse B604
→ Hexahydro-1α,7-dihydroxycannabinol
Probably in: Mouse B604
→ Hexahydro-1β,7-dihydroxycannabinol
Probably in: Mouse B604

***trans*-Hexahydro-7-hydroxycannabinol**
Probably formed from: *trans*-Hexahydrocannabinol
→ *trans*-Hexahydrocannabinol-7-oic acid
Probably in: Mouse B604

***cis*-Hexahydro-1α-hydroxycannabinol-7-oic acid**
Probably formed from: *cis*-Hexahydrocannabinol-7-oic acid
→ Hexahydro-1α,2″-dihydroxycannabinol-7-oic acid
Probably in: Mouse B604
→ Hexahydro-1α,3″-dihydroxycannabinol-7-oic acid
Probably in: Mouse B604
→ Hexahydro-1α,4″-dihydroxycannabinol-7-oic acid
Probably in: Mouse B604

***cis*-Hexahydro-2″-hydroxycannabinol-7-oic acid**
Probably formed from: *cis*-Hexahydrocannabinol-7-oic acid

***trans*-Hexahydro-2″-hydroxycannabinol-7-oic acid**
Probably formed from: *trans*-Hexahydrocannabinol-7-oic acid

***cis*-Hexahydro-3″-hydroxycannabinol-7-oic acid**
Probably formed from: *cis*-Hexahydrocannabinol-7-oic acid

***trans*-Hexahydro-3″-hydroxycannabinol-7-oic acid**
Probably formed from: *trans*-Hexahydrocannabinol-7-oic acid

***cis*-Hexahydro-4″-hydroxycannabinol-7-oic acid**
Probably formed from: *cis*-Hexahydrocannabinol-7-oic acid

***trans*-Hexahydro-4″-hydroxycannabinol-7-oic acid**
Probably formed from: *trans*-Hexahydrocannabinol-7-oic acid

1,2,3,4,5,6-Hexahydro-8-hydroxy-6-(*eq*),11(*ax*)-dimethyl-2,6-methano-3-benzazocine-3-(*cis*-3-carboxy-2-butene)
Formed from: 1,2,3,4,5,6-Hexahydro-8-hydroxy-α,6-(*eq*),11(*ax*)-trimethyl-2,6-methano-3-benzazocine-3-(*trans*-3-carboxy-2-butene)

3-(Hexahydro-3-(2-hydroxyethyl)-1-methyl-1*H*-azepin-3-yl)phenol
Probably formed from: Meptazinol

1,2,3,4,10,14b-Hexahydro-8-hydroxy-2-methylpyrazino[2,1-*a*]pyrido[2,3-*c*][2]benzazepine
Formed from: 1,2,3,4,10,14b-Hexahydro-2-methylpyrazino[2,1-*a*]pyrido[2,3-*c*][2]benzazepine
→ 1,2,3,4,10,14b-Hexahydro-8-hydroxy-2-methylpyrazino[2,1-*a*]pyrido[2,3-*c*][2]benzazepine-β-D-glucuronide
In: Man G864
Rat G864
→ 1,2,3,4,10,14b-Hexahydro-8-hydroxy-2-methylpyrazino[2,1-*a*]pyrido[2,3-*c*][2]benzazepine sulphate
In: Man G864
Rat G864

1,2,3,4,10,14b-Hexahydro-8-hydroxy-2-methylpyrazino[2,1-*a*]pyrido[2,3-*c*][2]benzazepine-β-D-glucuronide
Formed from: 1,2,3,4,10,14b-Hexahydro-8-hydroxy-2-methylpyrazino[2,1-*a*]pyrido[2,3-*c*][2]benzazepine

1,2,3,4,10,14b-Hexahydro-8-hydroxy-2-methylpyrazino[2,1-*a*]pyrido[2,3-*c*][2]benzazepine sulphate
Formed from: 1,2,3,4,10,14b-Hexahydro-8-hydroxy-2-methylpyrazino[2,1-*a*]pyrido[2,3-*c*][2]benzazepine

(3a*R*,7a*S*)-1,2,3,3a,7,7a-Hexahydro-3a-(4-hydroxyphenyl)-1-methyl-6*H*-indol-6-one
See: Sceletenone

1,2,3,4,5,6-Hexahydro-8-hydroxy-α,6-(*eq*),11(*ax*)-trimethyl-2,6-methano-3-benzazocine-3-(*trans*-3-carboxy-2-butene)
Formed from: Pentazocine
→ 1,2,3,4,5,6-Hexahydro-8-hydroxy-6-(*eq*),11(*ax*)-dimethyl-2,6-methano-3-benzazocine-3-(*cis*-3-carboxy-2-butene)
In: Mouse A1603, B30
See also: A2134

1,2,3,4,10,14b-Hexahydro-2-methylpyrazino[2,1-*a*]pyrido[2,3-*c*][2]benzazepine
→ 1,2,3,4,10,14b-Hexahydro-8-hydroxy-2-methylpyrazino[2,1-*a*]pyrido[2,3-*c*][2]benzazepine
In: Man G864
Rat G864
→ 1,2,3,4,10,14b-Hexahydro-2-methylpyrazino[2,1-*a*]pyrido[2,3-*c*][2]benzazepine-N-β-D-glucuronide
In: Man G864
→ 1,2,3,4,10,14b-Hexahydro-2-methylpyrazino[2,1-*a*]pyrido[2,3-*c*][2]benzazepine-N-oxide
In: Man G864
Rat G864
→ 1,2,3,4,10,14b-Hexahydropyrazino[2,1-*a*]pyrido[2,3-*c*][2]benzazepine
In: Man G864
Rat G864

1,2,3,4,10,14b-Hexahydro-2-methylpyrazino[2,1-*a*]pyrido[2,3-*c*][2]benzazepine-N-β-D-glucuronide
Formed from: 1,2,3,4,10,14b-Hexahydro-2-methylpyrazino[2,1-*a*]pyrido[2,3-*c*][2]benzazepine

1,2,3,4,10,14b-Hexahydro-2-methylpyrazino[2,1-*a*]pyrido[2,3-*c*][2]benzazepine-N-oxide
Formed from: 1,2,3,4,10,14b-Hexahydro-2-methylpyrazino[2,1-*a*]pyrido[2,3-*c*][2]benzazepine

1,2,3,4,10,14b-Hexahydropyrazino[2,1-*a*]pyrido[2,3-*c*][2]benzazepine
Formed from: 1,2,3,4,10,14b-Hexahydro-2-methylpyrazino[2,1-*a*]pyrido[2,3-*c*][2]benzazepine
→ 1,2,3,4,10,14b-Hexahydropyrazino[2,1-*a*]pyrido[2,3-*c*][2]benzazepine-N-sulphate
Probably in: Man G864
Rat G864

1,2,3,4,10,14b-Hexahydropyrazino[2,1-*a*]pyrido[2,3-*c*][2]benzazepine-N-sulphate
Probably formed from: 1,2,3,4,10,14b-Hexahydropyrazino[2,1-*a*]pyrido[2,3-*c*][2]benzazepine

(1*R*,2*R*,4*S*)-1,2,3,4,6,11-Hexahydro-2,4,5,7-tetrahydroxy-6,11-dioxo-2-(2-oxopropyl)-1-naphthacenecarboxylic acid methyl ester
See: Sulfurmycinone

(1*R*,2*R*,4*S*)-1,2,3,4,6,11-Hexahydro-2,4,5,7-tetrahydroxy-2-methyl-6,11-dioxo-1-naphthacenecarboxylic acid methyl ester
See: Auramycinone

2,3,6,7,9,10-Hexahydro-2,3,7,8-tetraoxoanthracene
Formed from: 3,4-Dihydroxyphenylacetic acid
Protocatechuic acid

Hexahydro-1α,2″,7-trihydroxycannabinol
Probably formed from: Hexahydro-1α,7-dihydroxycannabinol

Hexahydro-1β,2″,7-trihydroxycannabinol
Probably formed from: Hexahydro-1β,7-dihydroxycannabinol

Hexahydro-1α,3″,6β-trihydroxycannabinol
Probably formed from: Hexahydro-1α,6β-dihydroxycannabinol
Δ^6-Tetrahydro-3″-hydroxycannabinol

Hexahydro-1α,3″,7-trihydroxycannabinol
Probably formed from: Hexahydro-1α,7-dihydroxycannabinol

Hexahydro-1β,3″,7-trihydroxycannabinol
Probably formed from: Hexahydro-1β,7-dihydroxycannabinol

Hexahydro-1α,4″,6β-trihydroxycannabinol
Probably formed from: Hexahydro-1α,6β-dihydroxycannabinol
Δ^6-Tetrahydro-4″-hydroxycannabinol

Hexahydro-1α,4″,7-trihydroxycannabinol
Probably formed from: Hexahydro-1α,7-dihydroxycannabinol

Hexahydro-1β,4″,7-trihydroxycannabinol
Probably formed from: Hexahydro-1β,7-dihydroxycannabinol

Hexahydro-1α,6α,7-trihydroxycannabinol
Probably formed from: Hexahydro-1α,7-dihydroxycannabinol

2,2′,4,4′,5,5′-Hexahydroxy-7,7′-dimethyl-1,1′-bianthraquinone
See: Skyrin

2,3-*trans*-3,4-*cis*-3,3′,4,4′,5,7-Hexahydroxyflavan (leucocyanidin)
Formed from: Dihydroquercetin
→ (+)-Catechin
In: Douglas fir D455, D702, D863
→ Cyanidin
Probably in: Matthiola D575

3,3′,4′,5,5′,7-Hexahydroxyflavan
See: (−)-Epigallocatechin

3,3′,4′,5,5′,7-Hexahydroxyflavanone
See: Dihydromyricetin

3,3′,4′,5,7,8-Hexahydroxyflavanone (dihydrogossypetin)
Formed from: Dihydroquercetin

2′,3,4′,5,5′,7-Hexahydroxyflavone
Probably formed from: Morin

3,3′,4′,5,5′,7-Hexahydroxyflavone
See: Myricetin

3,3′,4′,5,6,7-Hexahydroxyflavone
See: Quercetagetin

3,3′,4′,5,7,8-Hexahydroxyflavone
See: Gossypetin

3,3′,4,5,5′,7-Hexahydroxyflavylium
See: Delphinidin

2,2′,4,4′,5,5′-Hexamethylazobenzene
Formed from: 2,4,5-Trimethylaniline

2,2′,4,4′,6,6′-Hexamethylazobenzene
Formed from: 2,4,6-Trimethylaniline

***o*-Hexanoylmercaptobenzoic acid**
→ *o*-Mercaptobenzoic acid
In: Guinea pig H214

5-Hexaprenyl-3,4-dihydroxybenzoic acid
Formed from: *p*-Hydroxybenzoic acid
→ 5-Hexaprenyl-4-hydroxy-3-methoxybenzoic acid
Probably in: Saccharomyces A3261

5-Hexaprenyl-4-hydroxy-3-methoxybenzoic acid
Formed from: 5-Hexaprenyl-3,4-dihydroxybenzoic acid

2-Hexaprenyl-6-methoxyphenol
Formed from: 4-Carboxy-2-hexaprenylphenol
→ 5-Demethylubiquinone 30
Probably in: Saccharomyces A3353

Hexoestrol (*meso*-3,4-di-(*p*-hydroxyphenyl) hexane)
→ 2-Hydroxyhexoestrol
Probably in: Hamster D514
Rat D514

***p*-Hexoxybenzaldehyde**
→ *p*-Hexoxybenzoic acid
In: Beef C508

***p*-Hexoxybenzoic acid**
Formed from: *p*-Hexoxybenzaldehyde

***trans*-2-(Hexoxycarbonylethyl)-3-(*p*-hydroxyphenyl)-N-phenyllactam**
→ *trans*-2-(2-Carboxyethyl)-3-(*p*-hydroxyphenyl)-N-phenyllactam
In: Microorganism K110

Hexyl acitretinate
Formed from: Acitretinoyl CoA

Hexylbenzene
→ Benzoic acid
In: Paecilomyces D936

Hexyl L-γ-glutamyl dopamide
→ 3,4-Dihydroxyphenethylamine
In: Dog A3945

Hexyl salicylate
→ Salicylic acid
In: Guinea pig D144

β-*n*-Hexylstyrene oxide
→ (*R*,*R*)-*threo*-1,2-Dihydroxy-1-phenyloctane
In: Rabbit J17

Higenamine (1,2,3,4-tetrahydro-1-((4-hydroxyphenyl)methyl)-6,7-isoquinolinediol)
→ Coclaurine
In: Rat H97
→ Isococlaurine
In: Rat H97

Hippadine (7*H*-[1,3]dioxolo[4,5-*j*]pyrrolo[3,2,1-*de*]phenanthridin-7-one)
Formed from: 2-Epilycorine

Hippeastrine ((5*S*,5a*S*,12b*S*,12c*S*)-2,3,5,5a,12b,12c-hexahydro-5-hydroxy-1-methyl[1,3]dioxolo[6,7][2]benzopyrano[3,4-*g*]indol-7(1*H*)-one)
Formed from: Lycorine

Hippuric acid (benzoylglycine)
Formed from: Benzoic acid
Benzoyl CoA
Benzoylglycyl-L-lysine
Ethyl benzoylglycyl-L-lysine
Methyl benzoylglycyl-L-lysine
→ Benzoic acid
In: Corynebacterium E145
Streptococcus A824
Streptomyces A435
Contrast (in Streptomyces) A435

Hircinol (9,10-dihydro-4-methoxy-2,5-phenanthrenediol)
Formed from: Cinnamic acid
m-Hydroxyphenylpropionic acid

Hirsutidin (3,5-dihydro-2-(4-hydroxy-3,5-dimethoxyphenyl)-7-methoxy-1-benzopyrilium)
Formed from: Cinnamic acid

Hispidin (6-(3,4-dihydroxystyryl)-4-hydroxypyrone)
Formed from: Bisnoryangonin

Hispidol-6-β-D-glucoside ((2*Z*)-6-hydroxy-2-((4-hydroxyphenyl)methylene)-3-((2*H*)-benzofuranone-6-β-D-glucoside)
→ *p*-Hydroxybenzoic acid
In: Cicer A188
Glycine A188
Petroselinum A188
Phaseolus A188

L-Histidyl-β-naphthylamide
→ β-Naphthylamine
In: Bacteria C146
Beef A526
Glycine A1596
Rabbit B334
Rat A523

L-Histidyl-L-phenylalanine
→ L-Phenylalanine
In: Brevibacterium E587

HMPG
See: 4-Hydroxy-3-methoxyphenylglycol

Homatropine (α-hydroxybenzeneacetic acid (3-*endo*)-8-methyl-8-azabicyclo[3.2.1]oct-3-yl ester)
→ 6β-Hydroxyhomatropine
In: Hyoscyamus E350

Homocatechol
See: 4-Methylcatechol

2-(S-(L-(Homocysteinyl)))-N-isopropyl-N-phenylacetamide
→ 2-(S-(L-(N-Acetylhomocysteinyl)))-N-isopropyl-N-phenylacetamide
In: Rat G80

Homoeriodictyol (3′-methoxynaringenin, 4′,5,7-trihydroxy-3′-methoxyflavanone)
Formed from: Eriodictyol
Feruloyl CoA
→ Homoeriodictyol-D-glucoside
In: Rose K809

Homoeriodictyol-D-glucoside
Formed from: Homoeriodictyol

Homogentisic acid (2,5-dihydroxyphenylacetic acid)
Formed from: Biochanin A
Genistein
Homogentisic acid lactone
o-Hydroxyphenylacetic acid
m-Hydroxyphenylacetic acid
p-Hydroxyphenylacetic acid
p-Hydroxyphenylpyruvic acid
Phenylacetic acid
→ δ-Carboxymethyl-α-hydroxy-*cis*,*cis*-muconic semialdehyde
In: Pseudomonas A2399
→ 2-Demethylplastoquinone-9
In: Lettuce D902
Spinach D902
→ 2,5-Dihydroxymandelic acid
Probably in: Aspergillus C60, D711
→ Maleylacetoacetic acid
In: Bacillus A2463
Flavobacterium E679, F780
Moraxella A2463
Xanthobacter E303
See also: B368
→ 2-Methyl-6-phytylquinone
In: Lettuce D902
Spinach D902
→ 3-Nonaprenyltoluquinol
In: Euglena A201, A1584
Sugar beet A1584
→ 3-Octaprenyltoluquinol
In: Euglena A201, A1584
Sugar beet A1584
→ Phytyltoluquinone
In: Euglena A1584
Sugar beet A1584
→ Plastoquinone
In: Lettuce D902
Spinach A2522, D902
See also: A1724
→ α-Tocopherol
In: Lettuce D902
Spinach A2522, D902
→ γ-Tocopherol
In: Euglena G754
→ 2,4,5-Trihydroxyphenylacetic acid
Probably in: Escherichia G817

Homogentisic acid lactone
→ Homogentisic acid
In: Man K327

D-Homooestradiol-17aβ
Formed from: 17α-Ethynyl-17β-oestradiol

18-Homooestradiol
→ 18-Homooestradiol-β-D-glucuronide
In: Dog A1892
→ 18-Homooestradiol sulphate
In: Rat A1892
→ 2-Hydroxy-18-homooestradiol
In: Rat A1892
→ 16α-Hydroxy-18-homooestradiol
In: Rat A1892

18-Homooestradiol-β-D-glucuronide
Formed from: 18-Homooestradiol

18-Homooestradiol sulphate
Formed from: 18-Homooestradiol

D-Homooestrone ((4a*S*,4b*R*,10b*S*,12a*S*)-3,4,4a,4b,5,6,10b,11,12,12a-decahydro-8-hydroxy-12a-methyl-1(2*H*)-chrysenone)
Formed from: 17α-Ethynyloestradiol

Homophenylalanine
See: L-2-Amino-4-phenylbutyric acid

Homophenylalanine aldoxime
Formed from: L-2-Amino-4-phenylbutyric acid

Homophthalic acid (2-carboxyphenylacetic acid)
→ *o*-Hydroxyphenylacetic acid
In: Aspergillus H108
→ Phenylacetic acid
In: Pseudomonas H49

Homotryptophan (α-amino-1*H*-indole-3-butanoic acid)
→ β-Hydroxyhomotryptophan
In: Pseudomonas B84
Homovanillic acid
See: 4-Hydroxy-3-methoxyphenylacetic acid
Homoveratrylamine
See: 3,4-Dimethoxyphenethylamine
Homozimeldine (*E* and *Z* isomers)
→ Homozimeldine-N-oxide
In: Pig E936, G392
Rat E936, G392
Homozimeldine-N-oxide
Formed from: Homozimeldine
Hordenine (N,N-dimethyltyramine)
Probably formed from: N-Methyltyramine
→ 3,4-Dihydroxy-N,N-dimethylphenethylamine
In: Agaricus J460
→ 3,4-Dimethylindoliumolate
In: Mushroom K374
→ *p*-Hydroxyphenylacetaldehyde
In: Pig A72
Rat F217
Willia A1157
See also: G865
→ N-Methyl-*p*-tyramine
In: Barley A3594
Parsley A3594
Wheat A3594
5HT
See: 5-Hydroxytryptamine
5HTP
See: 5-Hydroxytryptophan
HVA
See: Homovanillic acid
Hydnocarpine (rel-2-((2*R*,3*R*)-2,3-dihydro-2-(4-hydroxy-3-methoxyphenyl)-3-(hydroxymethyl)-1,4-benzodioxin-6-yl)-5,7-4*H*-1-benzopyran-4-one)
Formed from: Coniferyl alcohol
Luteolin
Hydralazine (1-hydrazinophthalazine)
Formed from: Apiracohl
Budralazine
3-Methyl-*s*-triazolo[3,4-*a*]phthalazine
→ 4-(2-Acetylhydrazino)phthalazinone
In: Man A3371, B908
See also: C613
→ Bis(phthalazin-1-yl)amine
In: Rat E218
See also: E861
→ 1-Hydrazinophthalazin-4-one
In: Rat C613
→ Hydroxyhydralazine
Probably in: Rat A1138
→ 3-Methyl-*s*-triazolo[3,4-*s*]phthalazine
In: Man A1016, A1058, A2299, A3371, A3592, B908
Microorganism H172
Rat C613, E861
→ Phthalazine
In: Man A1150, A3592
Rat D349, E218, E861
→ 1(2*H*)-Phthalazinone
In: Man A1187, A2299, A3371, A3592, B908
Rat C613, D349, E218, E861
→ *s*-Triazolo[3,4-*s*]phthalazine
In: Man A1187, A2299, A3371, B908
Rat C613, E861
Hydralazinepyruvic acid hydrazone
→ 3-Methyl-*s*-triazolo[3,4-*s*]phthalazine
In: Rabbit A3903
Hydrangea glucoside A
Probably formed from: *p*-Coumaric acid
→ Hydrangenol
In: Hydrangea A1927
Hydrangenol (3,4-dihydro-8-hydroxy-3-(4-hydroxyphenyl)-1*H*-2-benzopyran-1-one)
Formed from: L-Phenylalanine
L-Tyrosine
Probably formed from: Hydrangea glucoside A
→ Lunularic acid
Probably in: Conocephalum A1924
→ Phyllodulcin
Probably in: Hydrangea A1927
Hydrangetin
See: 7-Hydroxy-8-methoxycoumarin
Hydrangin
See: 7-Glucosyloxy-8-methoxycoumarin

Hydrastine ((3*S*)-6,7-dimethoxy-3-((5*R*)-5,6,7,8-tetrahydro-6-methyl-1,3-dioxolo[4,5-*g*]isoquinolin-5-yl)-1(3*H*)-isobenzofuranone)
Formed from: Tetrahydropapaveroline

Hydratropic acid
See: 2-Phenylpropionic acid

L-α-Hydrazino-3,4-dihydroxy-α-methylphenylpropionic acid (carbidopa)
→ 3,4-Dihydroxyamphetamine
In: Rat D424
→ 3,4-Dihydroxy-α-methylphenylpropionic acid
In: Dog A1436, A1502
Man A1436, A1502
Monkey A1436, A1502
→ 3,4-Dihydroxyphenylacetone
In: Dog A1436, A1502
Man A1436, A1502
→ 3-(3,4-Dihydroxyphenyl)-2-methyllactic acid
In: Dog A1502
→ L-α-Hydrazino-4-hydroxy-3-methoxy-α-methylphenylpropionic acid
In: Dog A1502
Man A1502
Monkey A1502
Pig B117

L-α-Hydrazino-4-hydroxy-3-methoxy-α-methylphenylpropionic acid
Formed from: L-α-Hydrazino-3,4-dihydroxy-α-methylphenylpropionic acid

1-Hydrazinophthalazine
See: Hydralazine

1-Hydrazinophthalazin-4-one
Probably formed from: Hydralazine

Hydrazobenzene (1,2-diphenylhydrazine)
Formed from: Azobenzene
→ Aniline
In: Rat A1129
→ Benzidine
In: Rat A1129

***meso*-Hydrobenzoin** ((1*R*,2*S*)-rel-1,2-diphenyl-1,2-ethanediol)
Formed from: Benzoin
cis-Stilbene
trans-Stilbene oxide

***threo*-Hydrobenzoin**
Formed from: Benzoin
cis-Stilbene oxide

Hydrocaffeic acid
See: Dihydrocaffeic acid

Hydrochlorothiazide (6-chloro-3,4-dihydro-7-sulphamoyl-2*H*-1,2,4-benzothiadiazine-1,1-dioxide)
→ 2-Amino-4-chloro-1,5-benzenedisulphonamide
In: Man E486

6α-Hydrocodol ((5α,6α)-4,5-epoxy-3-methoxy-17-methylmorphinan-6-ol)
Formed from: Hydrocodone
→ 6α-Hydromorphol
Probably in: Guinea pig A3434
Man A3434

6β-Hydrocodol ((5α,6β)-4,5-epoxy-3-methoxy-17-methylmorphinan-6-ol)
Formed from: Hydrocodone
→ 6β-Hydromorphol
Probably in: Guinea pig A3434
Man A3434

Hydrocodone ((5α)-4,5-epoxy-3-methoxy-17-methylmorphinan-6-one)
Formed from: Codeinone
→ 6α-Hydrocodol
In: Guinea pig A3434
Man A3434
See also K756, K757
→ 6β-Hydrocodol
In: Guinea pig A3434
Man A3434
Rat A3434
→ Norhydrocodone
In: Dog A3434
Man A3434

Hydrocotarnine (5,6,7,8-tetrahydro-4-methoxy-6-methyl-1,3-dioxolo[4,5-*g*]isoquinoline)
Formed from: Narcotine
Noscapine

3-Hydrodolasetron
Formed from: Dolasetron

Hydrojuglone-β-D-glucuronide
→ 5-Hydroxynaphthoquinone
In: Juglans J695

6α-Hydromorphol (4,5-epoxy-3-hydroxy-17-methylmorphinan-6α-ol)
Probably formed from: 6α-Hydrocodol
6β-Hydromorphol (4,5-epoxy-3-hydroxy-17-methylmorphinan-6β-ol)
Probably formed from: 6β-Hydrocodol
Hydromorphone (dihydromorphinone)
Formed from: Morphine
→ Dihydroisomorphine
In: Dog A3359
Guinea pig A3359
Man A3359
Rabbit A3359
Rat A3359
→ Dihydromorphine
In: Dog A3359
Guinea pig A3359
Man A3359
Papaver K756, K757
Rabbit A3359
Rat A3359, J744
See also: A823
→ Hydromorphone-3-β-D-glucuronide
In: Rabbit A3359
Rat J675
See also: H861, H933, J416, J783
Hydromorphone-3-β-D-glucuronide
Formed from: Hydromorphone
Hydroquinone
See: Quinol
Hydroxocobalamin
Formed from: Vitamin B_{12}
1-Hydroxyacenaphthene
Formed from: Acenaphthene
→ Acenaphthene-1,2-diol
Probably in: Cunninghamella G725
Microorganism J596
→ Acenaphthene-1,5-diol
Probably in: Cunninghamella G725
→ Acenaphthenone
In: Beijerinckia D248
Guinea pig E67
Hamster F153, K443
Microorganism J596
Monkey F241, F388
Mouse F377
Rabbit A2032
See also: G725, H28
6-Hydroxyacenaphthene
Formed from: Acenaphthene
6-Hydroxyacenocoumarol
Formed from: Acenocoumarol
7-Hydroxyacenocoumarol
Formed from: Acenocoumarol
8-Hydroxyacenocoumarol
Formed from: Acenocoumarol
N-Hydroxy-2-acetamido-7-acetylfluorene
→ N-Acetoxy-7-acetamido-2-aminofluorene
In: Hamster H106
N-Hydroxy-4-acetamidobenzophenone
→ 4-Hydroxylaminobenzophenone
In: Hamster F522
N-Hydroxy-4-acetamidobiphenyl
Formed from: N-Hydroxy-4-acetamido-biphenyl-O-β-D-glucuronide
Probably formed from: 4-Hydroxylamino-biphenyl
→ 4-Acetamidobiphenyl
In: Rabbit B178
Rat B178
→ 4-Acetamidobiphenyl nitroxide radical
Probably in: Horseradish A517
→ 4-Amino-3-methylmercaptobiphenyl
In: Rat A729
→ 4,4′-Bisazoxybiphenyl
In: Rabbit B178
Rat B178
→ 4-Hydroxylaminobiphenyl
In: Animal C812
Hamster F522
Rat K747
→ 4-Nitrosobiphenyl
In: Rabbit B178
Rat B178
→ N-Sulphatoxy-4-acetamidobiphenyl
In: Man H304
N-Hydroxy-4-acetamidobiphenyl-O-β-D-glucuronide
→ N-Hydroxy-4-acetamidobiphenyl
In: Escherichia J766
Man J766

N-Hydroxy-2-acetamido-7-bromofluorene
→ N-Acetoxy-2-amino-7-bromofluorene
In: Hamster H106
N-Hydroxy-4-acetamido-4′-bromostilbene
Formed from: 4-Acetamido-4′-bromostilbene
N-Hydroxy-4-acetamido-4′-butoxystilbene
Formed from: 4-Acetamido-4′-butoxystilbene
N-Hydroxy-4-acetamido-2′-ethenylbiphenyl
→ 2′-Ethenyl-4-hydroxylaminobiphenyl
In: Hamster F522
N-Hydroxy-2-acetamido-7-ethoxyfluorene
→ N-Acetoxy-2-amino-7-ethoxyfluorene
In: Hamster H106
N-Hydroxy-4-acetamido-4′-ethoxystilbene
Formed from: 4-Acetamido-4′-ethoxystilbene
N-Hydroxy-2-acetamido-7-ethylfluorene
→ N-Acetoxy-2-amino-7-ethylfluorene
In: Hamster H106
N-Hydroxy-2-acetamidofluorene
Formed from: 2-Acetamidofluorene
Probably formed from: 2-Hydroxylamino-fluorene
→ 2-Acetamidofluorene
In: Guinea pig A3965
Hamster A3965
Horse D32
Horseradish D32
Mouse A703, A3965
Rabbit A3965, B178, K111
Rat A3965, B178
See also: H16
→ 2-Acetamidofluorene nitroxide radical
Probably in: Horseradish A517
→ 2-Acetamido-7-hydroxyfluorene
In: Guinea pig C440
See also: B179
→ N-Acetoxy-2-acetamidofluorene
In: Hamster E110, G757
Rat C70, G811
See also: D467
→ N-Acetoxy-2-aminofluorene
In: Hamster G744, H106
Rat G735
→ 4-Amino-3-methylmercaptofluorene
In: Rat A729
→ 2,2′-Bisazoxyfluorene
In: Rabbit B178
Rat B178
→ 1-(N-2′-Fluorenylacetamido)-2-acetamidofluorene
In: Rat A3912
→ N-Hydroxy-2-acetamidofluorene-β-D-glucuronide
In: Man H92
Rabbit C588
Rat A1840, D723, E219
→ N-Hydroxy-2-acetamidofluorene-O-phosphate
In: Man A2086
Monkey A2086
Rat A2086
→ N-Hydroxy-2-acetamidofluorene-O-sulphate
In: Man A2086, K34
Monkey A2086
Mouse D756
Rat A1840, A2086, A2206, C320, G317, G675
→ 2-Hydroxylaminofluorene
In: Animal C812
Guinea pig D721
Rat G735, H96, K747
See also: G125
→ 2-Nitrosofluorene
In: Horse D32
Horseradish D32
Rat C230, G811
N-Hydroxy-3-acetamidofluorene
→ 3-Acetamidofluorene
In: Horse D32
Horseradish D32
→ 3-Acetamidofluorene nitroxide radical
Probably in: Horseradish A517
→ 3-Nitrosofluorene
In: Horse D32
Horseradish D32
N-Hydroxy-2-acetamidofluorene-O-β-D-glucuronide
→ 2-Hydroxyaminofluorene-O-β-D-glucuronide
In: Guinea pig A1354

N-Hydroxy-2-acetamidofluorene-O-phosphate
Formed from: N-Hydroxy-2-acetamidofluorene

N-Hydroxy-2-acetamidofluorene-O-sulphate
Formed from: N-Hydroxy-2-acetamidofluorene

N-Hydroxy-4-acetamido-4′-fluorostilbene
Formed from: 4-Acetamido-4′-fluorostilbene

N-Hydroxy-4-acetamido-4′-methoxystilbene
Formed from: 4-Acetamido-4′-methoxystilbene

N-Hydroxy-1-acetamidonaphthalene
→ 1-Acetamidonaphthalene nitroxide radical
In: Horseradish A517

N-Hydroxy-2-acetamidonaphthalene
Formed from: 2-Acetamidonaphthalene
→ 2-Acetamidonaphthalene nitroxide radical
In: Horseradish A517
→ N-Hydroxy-2-acetamidonaphthalene-β-D-glucuronide
In: Rabbit C588
Rat A1840
→ N-Hydroxy-2-acetamidonaphthalene-O-sulphate
In: Rat A1840

N-Hydroxy-2-acetamidonaphthalene-β-D-glucuronide
Formed from: N-Hydroxy-2-acetamidonaphthalene

N-Hydroxy-2-acetamidonaphthalene-O-sulphate
Formed from: N-Hydroxy-2-acetamidonaphthalene

2-(2-Hydroxyacetamido)-5-nitrobenzophenone
Formed from: 1,3-Dihydro-7-nitro-5-phenyl-2*H*-1,4-benzodiazepin-2-one

N-Hydroxy-2-acetamido-9-oxofluorene
Formed from: 2-Acetamido-9-oxofluorene

N-Hydroxy-2-acetamidophenanthrene
→ 2-Acetamidophenanthrene nitroxide radical
In: Horseradish A517

N-Hydroxy-*m*-acetamidophenol
Formed from: *m*-Acetamidophenol

N-Hydroxy-4-acetamido-4′-propoxystilbene
Formed from: 4-Acetamido-4′-propoxystilbene

N-Hydroxy-4-acetamidostilbene
Formed from: 4-Acetamidostilbene
→ 2-Acetamidostilbene nitroxide radical
In: Horseradish A517
→ 4-Hydroxylaminostilbene
In: Animal C812

2-Hydroxyacetanilide
See: Glycollanilide

N-Hydroxyacetanilide
→ N-Hydroxyacetanilide-β-D-glucuronide
In: Rabbit C588
Rat A1840
→ N-Hydroxyacetanilide-O-sulphate
In: Rat A1840

N-Hydroxyacetanilide-β-D-glucuronide
Formed from: N-Hydroxyacetanilide

N-Hydroxyacetanilide-O-sulphate
Formed from: N-Hydroxyacetanilide

4-(Hydroxyacetate)rifamycin
See: Rifamycin

Hydroxyacetic acid 2-(2-((3-(1*H*-benzimidazol-2-yl)propyl)amino)ethyl)-6-fluoro-1,2,3,4-tetrahydro-4-hydroxy-1-isopropyl-2-naphthalenyl ester
Probably formed from: Methoxyacetic acid 2-(2-((3-(1*H*-benzimidazol-2-yl)propyl)amino)ethyl)-6-fluoro-1,2,3,4-tetrahydro-4-hydroxy-1-isopropyl-2-naphthalenyl ester

Hydroxyacetic acid 2-(2-((3-(1*H*-benzimidazol-2-yl)propyl)methylamino)ethyl)-6-fluoro-1,2,3,4-tetrahydro-4-hydroxy-1-isopropyl-2-naphthalenyl ester
Probably formed from: Methoxyacetic acid 2-(2-((3-(1*H*-benzimidazol-2-yl)propyl)methylamino)ethyl)-6-fluoro-1,2,3,4-tetrahydro-4-hydroxy-1-isopropyl-2-naphthalenyl ester

***o*-Hydroxyacetophenone**
Probably formed from: *o*-(*m*-Chlorobenzyloxy)acetophenone
→ Acetophenon-2′-yl sulphate
Probably in: Rat F507

***m*-Hydroxyacetophenone**
Formed from: Acetophenone

p-Hydroxyacetophenone
Formed from: Acetophenone
Acetophenon-4′-yl sulphate
2,2-Bis(p-hydroxyphenyl)propane
1,2-Bis(p-hydroxyphenyl)-2-propanol
1-(p-Hydroxyphenyl)ethanol
p-Methoxyacetophenone
Picein
Probably formed from: N-(3,3-Diphenylpropyl)-1-(p-hydroxyphenyl)ethylamine
p-Hydroxy-β-oxophenylpropionaldehyde
→ Acetophenon-4′-yl-β-D-glucuronide
In: Man G524
See also: A813
→ Acetophenon-4′-yl sulphate
Probably in: Rat A813
→ p,ω-Dihydroxyacetophenone
Probably in: Alcaligenes D761
→ p-Hydroxyphenyl acetate
In: Aspergillus H236
Pseudomonas K698
→ Quinol
In: Pseudomonas E484
Rat J81

p-Hydroxyacetophenone-β-D-glucoside
See: Picein

ω-Hydroxyacetosyringone
Probably formed from: Syringylglycerol

2-Hydroxyacridine
Formed from: Acridine

9-Hydroxyacronine
Formed from: Acronine
→ 9,11-Dihydroxyacronine
Probably in: Man A1646
Mouse A1646
Rat A1646

11-Hydroxyacronine
Formed from: Acronine
→ 11-Hydroxy-O-demethylacronine
Probably in: Pig A1646
→ 11-Hydroxy-3-hydroxymethyl-3-demethylacronine
Probably in: Dog A1646
Man A1646
Mouse A1646
Rat A1646

2-Hydroxyaflatoxin B_1
Formed from: Aflatoxin B_1

3-Hydroxyaflatoxin B_1
Formed from: Aflatoxin B_1

4-Hydroxyaflatoxin B_1
Formed from: Aflatoxin B_1

2-Hydroxyaflatoxin B_2
Formed from: Aflatoxin B_2

4-Hydroxyaflatoxin B_2
Formed from: Aflatoxin B_2

2-Hydroxyaflatoxin G_1
Formed from: Aflatoxin G_1

4-Hydroxyaflatoxin G_1
Formed from: Aflatoxin G_1

3′-Hydroxyafloqualone
Formed from: Afloqualone

4′-Hydroxyafloqualone
Formed from: Afloqualone

10-Hydroxyajmalicine
Probably formed from: 5-Hydroxytryptamine

(1-Hydroxyallyl)benzene
Formed from: Allylbenzene
→ Phenyl vinyl ketone
In: Rat A2998

5-(1-Hydroxyallyl)-1-methoxy-2,3-methylenedioxybenzene
Formed from: Myristicin

4-(1-Hydroxyallyl)-1,2-methylenedioxybenzene
Formed from: Safrole

4-(3-Hydroxyallyl)-1,2-methylenedioxybenzene
Formed from: Safrole

p-Hydroxyalphenal
Formed from: Alphenal
→ 5-Allyl-5-(3,4-dihydroxyphenyl)barbituric acid
Probably in: Rat A2423

α-Hydroxyalprazolam
Formed from: Alprazolam

4-Hydroxyalprazolam
Formed from: Alprazolam

p-Hydroxyalprenolol
Formed from: Alprenolol
→ p-Hydroxyalprenolol-p-β-D-glucuronide
Probably in: Dog D376
→ p-Hydroxyalprenolol-α-β-D-glucuronide

Probably in: Dog D376

p-Hydroxyalprenolol-p-β-D-glucuronide
Probably formed from: *p*-Hydroxyalprenolol

p-Hydroxyalprenolol-α-β-D-glucuronide
Probably formed from: *p*-Hydroxyalprenolol

3-Hydroxy-5-(2-aminoethyl)-o-quinone
Probably formed from: 3,4,5-Trihydroxy-phenethylamine

4-Hydroxy-5-(2-aminoethyl)-o-quinone
Probably formed from: 2,4,5-Trihydroxy-phenethylamine

7-Hydroxyaminopterin
Formed from: Aminopterin

2-Hydroxyamitriptyline
Formed from: Amitriptyline
→ 2,3-Dihydroxyamitriptyline
Probably in: Rat E892, E931
→ 2,10-Dihydroxyamitriptyline
Probably in: Rat E892
→ 2,11-Dihydroxyamitriptyline
Probably in: Rat E892
→ 2-Hydroxyamitriptyline-β-D-glucuronide
In: Rat E931
→ 2-Hydroxynortriptyline
Probably in: Rat E931

3-Hydroxyamitriptyline
Formed from: Amitriptyline
→ 3,11-Dihydroxyamitriptyline
Probably in: Rat E892

10-Hydroxyamitriptyline (*E* and *Z* isomers)
Formed from: Amitriptyline
Amitriptyline-N-oxide
→ *cis*-10,11-Dihydro-10,11-dihydroxyamitriptyline
In: Rat E931
See also: E892, E894
→ *trans*-10,11-Dihydro-10,11-dihydroxyamitriptyline
Probably in: Man E894, F523
Rat E892
→ 10-Hydroxyamitriptyline-N-β-D-glucuronide
In: Man F523
→ 10-Hydroxyamitriptyline-β-D-glucuronide (presumably O-)
In: Man F523, G472
Rat E931
→ 10-Hydroxyamitriptyline-N-oxide
Probably in: Rat E892
→ 10-Hydroxynortriptyline
Probably in: Man C26, J70
Rat E931
See also: A3768
→ 10-Oxoamitriptyline
Probably in: Man E894

11-Hydroxyamitriptyline
Formed from: Amitriptyline
→ 11-Hydroxyamitriptyline-N-oxide
Probably in: Cunninghamella H485
→ 11-Hydroxynortriptyline
Probably in: Cunninghamella H485

2-Hydroxyamitriptyline-β-D-glucuronide
Formed from: 2-Hydroxyamitriptyline

10-Hydroxyamitriptyline-N-β-D-glucuronide
Formed from: Amitriptyline-N-β-D-glucuronide
10-Hydroxyamitriptyline
→ *trans*-10,11-Dihydroxyamitriptyline-N-β-D-glucuronide
Probably in: Man F523

10-Hydroxyamitriptyline-β-D-glucuronide (presumably O-)
Formed from: 10-Hydroxyamitriptyline

10-Hydroxyamitriptyline-N-oxide
Formed from: Amitriptyline-N-oxide
Probably formed from: 10-Hydroxy-amitriptyline

11-Hydroxyamitriptyline-N-oxide
Probably formed from: 11-Hydroxy-amitriptyline

7-Hydroxyamoxapine
Formed from: Amoxapine
→ 7-Hydroxyamoxapine-β-D-glucuronide
In: Rat B953

7-Hydroxyamoxapine-β-D-glucuronide
Formed from: 7-Hydroxyamoxapine

o-Hydroxyamphetamine
Formed from: *o*-Hydroxy-N-methyl-amphetamine
Probably formed from: *o*-Methoxy-amphetamine

m-Hydroxyamphetamine
Formed from: Amphetamine
m-Methoxyamphetamine

→ 3,4-Dihydroxyamphetamine
Probably in: Dog B403
Man B403
Monkey B403
→ *m*-Hydroxyphenylacetone
Probably in: Dog B403
Man B403
Monkey B403

***p*-Hydroxyamphetamine** (D and L isomers; α-methyltyramine)
Formed from: D-Amphetamine
L-Amphetamine
p-Benzyloxyamphetamine
p-Ethoxyamphetamine
p-Hydroxy-D-amphetamine
p-Hydroxy-L-amphetamine
p-Hydroxy-N-methylamphetamine
p-Methoxyamphetamine
p-Propoxyamphetamine
→ Amphetamin-4-yl-β-D-glucuronide
In: Guinea pig A1771
Rat A4, A1771
See also: E783
→ Amphetamin-4-yl sulphate
In: Man A1771, C523
→ 3,4-Dihydroxyamphetamine
Probably in: Dog B632
Man J544
Monkey B632
Mushroom B897
→ *p*-Hydroxybenzoic acid
In: Guinea pig A1771
Rat A1771
See also: A2090
→ *p*-Hydroxynorephedrine
In: Guinea pig A1771
Man A1771
Rat A1492, A1771, E76
See also: A3289
→ *p*-Hydroxyphenylacetone
Probably in: Dog B632
Monkey B632

N-Hydroxyamphetamine
See: α-Methyl-β-phenethylhydroxylamine

α-Hydroxy-D-amphetamine
See: D-Norephedrine

α-Hydroxy-L-amphetamine
See: L-Norephedrine

2-Hydroxyamylthiobenzthiazole
Formed from: Butylcaptax

19-Hydroxyandrosta-4,7-diene-3,17-dione
→ Equilin
In: Fusarium E617

3-Hydroxy-3,5,7-androstatrien-17-one
→ Equilenin
In: Horse E337
→ Equilin
In: Horse E337
Man E337

19-Hydroxyandrostene-3,17-dione
→ Oestrone
Probably in: Chimpanzee A784
Escherichia A625
Macaque A645
Man A645, A3011, B289
Monkey A897
Rabbit A645
Rat A645
Sheep A593

Hydroxyangiorufone (2-hydroxy-9-(4-hydroxyphenyl)-1*H*-phenalen-1-one)
Formed from: 3-Phenylpropionic acid

***p*-Hydroxyaniline mustard** (N,N-bis(2-chloroethyl)-*p*-hydroxybenzenamine)
Formed from: *p*-Hydroxyaniline mustard-β-D-glucuronide
p-Hydroxyaniline mustard-O-phosphate

***p*-Hydroxyaniline mustard-β-D-glucuronide**
→ *p*-Hydroxyaniline mustard
In: Beef A99

***p*-Hydroxyaniline mustard-O-phosphate**
→ *p*-Hydroxyaniline mustard
In: Beef A99
Mouse A2492
Wheat A99
HeLa cells A1612

5-(*p*-Hydroxyanilino)-1,2,3,4-thiatriazole
→ *p*-Aminophenol
In: Dog A42
→ 5-(*p*-Hydroxyanilino)-1,2,3,4-thiatriazole-β-D-glucuronide
In: Rat A994

→ 5-(*p*-Hydroxyanilino)-1,2,3,4-thiatriazole sulphate
In: Rat A994
→ *p*-Hydroxyphenylthiourea
In: Dog A42
→ *p*-Hydroxyphenylurea
In: Dog A42

5-(*p*-Hydroxyanilino)-1,2,3,4-thiatriazole-β-D-glucuronide
Formed from: 5-(*p*-Hydroxyanilino)-1,2,3,4-thiatriazole

5-(*p*-Hydroxyanilino)-1,2,3,4-thiatriazole sulphate
Formed from: 5-(*p*-Hydroxyanilino)-1,2,3,4-thiatriazole

3-Hydroxyanthranilamide
Formed from: Anthranilamide
→ Anthranilamid-3-yl sulphate
In: Rat A1411, A2891

5-Hydroxyanthranilamide
Formed from: Anthranilamide
→ Anthranilamid-5-yl sulphate
In: Rat A1411, A2891

3-Hydroxyanthranilic acid (2-amino-3-hydroxybenzoic acid)
Formed from: Anthranilic acid
3-Hydroxy-L-kynurenine
→ 2-Amino-3-carboxyphenyl-β-D-glucuronide
Probably in: Guinea pig A674
Man A674
Rat A674
→ 2-Amino-3-carboxymuconic semialdehyde
In: Rat E472
→ 2-Amino-3-carboxyphenyl sulphate
Probably in: Guinea pig A674
Man A674
Rat A674
→ *o*-Aminophenol
In: Streptomyces F190
→ Cinnabarinic acid
In: Drosophila A2574
Man D276
Streptomyces B924
→ 3,4-Dihydroxyanthranilic acid
In: Aspergillus A1130
Pseudomonas A327, A1229
→ 3-Hydroxy-4-methylanthranilic acid
In: Streptomyces E542, E676, E725, F934
→ 3-Methoxyanthranilic acid
In: Rat A1427, A1506

5-Hydroxyanthranilic acid (2-amino-5-hydroxybenzoic acid)
Formed from: Anthranilic acid
5-Hydroxyindole

5-Hydroxyanthraniloyl CoA
Formed from: Anthraniloyl CoA
→ 3,4-Dihydro-5-hydroxyanthraniloyl CoA
In: Pseudomonas H605

2-Hydroxyanthraquinone
→ 2-Hydroxyanthraquinone-β-D-glucoside
In: Streptomyces B462

2-Hydroxyanthraquinone-β-D-glucoside
Formed from: 2-Hydroxyanthraquinone

2-Hydroxyanthraquinon-1-yl sulphate
Probably formed from: Alizarin

4-Hydroxyantipyrine
Formed from: 4-Aminoantipyrine
Antipyrine
→ 4,4′-Dihydroxyantipyrine
Probably in: Guinea pig B305
Man B305, C415, C418, D632
Rabbit B305, C418
Rat B305, B395, C415, C418, J769
→ 4-Hydroxyantipyrine-β-D-glucuronide
In: Man C415
Rat C415
See also: B394, D632, J769
→ 4-Hydroxyantipyrine sulphate
In: Man C415
Rat C415
See also: B394, C418, G546, J769

4-Hydroxyantipyrine-β-D-glucuronide
Formed from: 4-Hydroxyantipyrine

4-Hydroxyantipyrine sulphate
Formed from: 4-Hydroxyantipyrine

30-Hydroxyalphelandrine
Formed from: Alphelandrine
→ Alphelandrine-30,31-quinone
In: Alphelandra H351

6β-Hydroxyapoatropine
Formed from: Apoatropine

10-Hydroxyapomorphine
Formed from: 10-Methoxyapomorphine

Hydroxyaprindine
Formed from: Aprindine
8-Hydroxyarteflene
Formed from: Arteflene
→ 8-Hydroxyarteflene-β-D-glucuronide
In: Rat K76
8-Hydroxyarteflene-β-D-glucuronide
Formed from: 8-Hydroxyarteflene
5-Hydroxyastemizole
Formed from: Astemizole
→ 5-Hydroxydemethylastemizole
Probably in: Dog E425
Rat E425
6-Hydroxyastemizole
Formed from: Astemizole
→ 6-Hydroxydemethylastemizole
Probably in: Dog E425
Rat E425
Hydroxyatenolol
Formed from: Atenolol
***o*-Hydroxyatorvastatin lactone**
Formed from: Atorvastatin lactone
***p*-Hydroxyatorvastatin lactone**
Formed from: Atorvastatin lactone
1-Hydroxyauramycin A
Formed from: 1-Hydroxyauramycin B
1-Hydroxyauramycinone
1-Hydroxyauramycin B
Formed from: 1-Hydroxyauramycinone
→ 1-Hydroxyauramycin A
In: Streptomyces D17
11-Hydroxyauramycin A
Formed from: 11-Hydroxyauramycinone
11-Hydroxyauramycin B
Formed from: 11-Hydroxyauramycinone
1-Hydroxyauramycinone
→ 1-Hydroxyauramycin A
In: Streptomyces D170
→ 1-Hydroxyauramycin B
In: Streptomyces D170
11-Hydroxyauramycinone
Formed from: Auramycinone
→ 7-Deoxy-11-hydroxyauramycinone
In: Streptomyces D170
→ 11-Hydroxyauramycin A
In: Streptomyces D170
→ 11-Hydroxyauramycin B
In: Streptomyces D170
5′-Hydroxyazaperol
Probably formed from: 5′-Hydroxyazaperone
5′-Hydroxyazaperone
Formed from: Azaperone
→ 5′-Hydroxyazaperol
Probably in: Horse H754
6-Hydroxyazapropazone
Formed from: Azapropazone
8-Hydroxyazapropazone
Formed from: Azapropazone
2-Hydroxyazobenzene
Formed from: Fenazox
→ *o*-Aminophenol
Probably in: Tomato D729
→ Aniline
Probably in: Tomato D729
4-Hydroxyazobenzene
→ *p*-Aminophenol
In: Rat G380
→ Aniline
In: Rat G380
4-Hydroxybefunolol
Formed from: Befunolol
***p*-Hydroxybemitradine**
Formed from: Bemitradine
→ *p*-Hydroxybemitradine-β-D-glucuronide
Probably in: Man F314
***p*-Hydroxybemitradine-β-D-glucuronide**
Probably formed from: *p*-Hydroxybemitradine
9-Hydroxybenfluron
Formed from: Benfluron
→ 9-Hydroxybenfluron-β-D-glucuronide
Probably in: Rat K387
9-Hydroxybenfluron-β-D-glucuronide
Probably formed from: 9-Hydroxybenfluron
3-Hydroxybenfuracarb
Formed from: Benfuracarb
→ 3-Oxobenfuracarb
Probably in: Bean D806
Cotton D806
Zea D806
10-Hydroxybenz[*j*]aceanthrylene
Formed from: Benz[*j*]aceanthrylene
→ 10-Hydroxybenz[*j*]aceanthrylene-β-D-glucuronide
In: Rat J688

→ 10-Hydroxybenz[*j*]aceanthrylene-O-sulphate
In: Rat J688

10-Hydroxybenz[*j*]aceanthrylene-β-D-glucuronide
Formed from: 10-Hydroxybenz[*j*]aceanthrylene

10-Hydroxybenz[*j*]aceanthrylene-O-sulphate
Formed from: 10-Hydroxybenz[*j*]aceanthrylene

8-Hydroxybenz[*j*]aceanthrylen-1-one
→ 8-Hydroxybenz[*j*]aceanthrylen-1-one-O-sulphate
Probably in: Rat J688

8-Hydroxybenz[*j*]aceanthrylen-1-one-O-sulphate
Probably formed from: 8-Hydroxybenz[*j*]aceanthrylen-1-one

***p*-Hydroxybenzalacetone**
Formed from: *p*-Coumaroyl CoA

***o*-Hydroxybenzaldehyde**
See: Salicylaldehyde

***m*-Hydroxybenzaldehyde**
Formed from: *m*-Hydroxybenzoic acid
m-Hydroxybenzyl alcohol
m-Hydroxybenzylidenepyruvic acid
m-Hydroxymandelonitrile
m-Hydroxyphenylglyoxylic acid
Probably formed from: *m*-Hydroxybenzylamine
m-Phenoxybenzaldehyde
→ 2,*m*-Dihydroxypropiophenone
In: Pseudomonas K524
→ *m*-Hydroxybenzoic acid
In: Acinetobacter E663, F366
Arion G804
Helix G804
Limax G804
Lithospermum G229
Man J24
Pseudomonas A30, E771
Rhodopseudomonas G246
Streptomyces C284
See also: A707, A2155, G865
→ *m*-Hydroxybenzyl alcohol
In: Dunaliella G459
Euglena G422
→ *m*-Hydroxymandelonitrile
With (*S*)-oxynitrilase F680
→ Protocatechualdehyde
In: Trichosporon A902

***p*-Hydroxybenzaldehyde**
Formed from: *p*-Benzyloxyphenol
1,2-Bis(*p*-hydroxyphenyl)-2-propanol
p-Coumaric acid
p-Coumaroyl CoA
p-Hydroxybenzoic acid
p-Hydroxybenzyl alcohol
p-Hydroxymandelic acid
p-Hydroxymandelonitrile
p-Hydroxyphenylacetaldoxime
p-Hydroxyphenylacetonitrile
p-Hydroxy-L-phenylglycine
p-Hydroxyphenylglyoxylic acid
enol-*p*-Hydroxyphenylpyruvic acid
p-Methoxymethylphenol
Piceid
L-Tyrosine
Probably formed from: *p*-Hydroxybenzylamine
4-Hydroxy-β-oxophenylpropionic acid
p-Hydroxyphenylpyruvic acid
→ 3-Bromo-4-hydroxybenzaldehyde
In: Odonthalia A3989
See also: B300
→ *p*-Bromophenol
In: Ulva K153
→ 2,*p*-Dihydroxypropiophenone
In: Pseudomonas K524
→ *p*-Hydroxybenzoic acid
In: Acinetobacter E663, F366
Aspergillus A329, A2691
Burkholderia K655
Clostridium H102, J864
Desulphovibrio F666
Glucuronobacter G161
Hyphomicrobium C123
Lithospermum G229
Microorganism E176
Pseudomonas A30, E669, E771, K203
Rhodopseudomonas G246
Serratia G197
Streptomyces C284
See also: A126, A2336, A2573, A2744, A2889, F51, F89, G865, H616, K210

→ *p*-Hydroxybenzyl alcohol
In: Dunaliella G459
Euglena G422
Glucuronobacter G161
Phanerochaete F866
Rat A2308
See also: A2366, A2573, A2744, D569
→ *p*-Hydroxymandelonitrile (*R* and *S* isomers)
In: Almond J194, J302
Apple J302
Cherry J302
Sorghum J194
With (*S*)-oxynitrilase F680
→ Quinol
Probably in: Naja J684

p-Hydroxybenzaldehyde-4-sulphate
→ *p*-Hydroxybenzyl alcohol-4-sulphate
In: Rat A2308

o-Hydroxybenzamide
See: Salicylamide

p-Hydroxybenzamide
→ *p*-Carboxamidophenyl sulphate
In: Man E575
Rat E575

N-Hydroxybenzamidine
See: Benzamidoxime

N-Hydroxy-2-benzamidofluorene
→ Benzoic acid
In: Rat H96
→ 2-Hydroxylaminofluorene
In: Rat H96

4-Hydroxybenzanilide
Formed from: Benzanilide

1-Hydroxybenzanthracene
Formed from: Benzanthracene

3-Hydroxybenzanthracene
→ Benzanthr-3-yl-β-D-glucuronide
In: Rat E471

9-Hydroxybenzanthracene
Formed from: Benzanthracene

11-Hydroxybenzanthracene
Formed from: Benzanthracene

1′-Hydroxybenzbromarone
Formed from: Benzbromarone
→ 1′,6-Dihydroxybenzbromarone
Probably in: Man G835
→ 1′-Oxobenzbromarone
Probably in: Man F619, G835

2′-Hydroxybenzbromarone
Formed from: Benzbromarone

6-Hydroxybenzbromarone
Formed from: Benzbromarone

α-Hydroxybenzeneacetic acid
See: (*R*)-Mandelic acid

2-Hydroxybenzenecarbodithioic acid
See: Dithiosalicylic acid

p-Hydroxybenzenesulphonic acid
→ 3,4-Dihydroxybenzenesulphonic acid
Probably in: Alcaligenes K293

3-Hydroxybenzidine
Formed from: 3-Hydroxybenzidine-O-β-D-glucuronide

3-Hydroxybenzidine-O-β-D-glucuronide
→ 3-Hydroxybenzidine
In: Escherichia J766
Man J766

Hydroxybenzil
Probably formed from: Benzil

5-Hydroxybenzimidazole
Formed from: Benzimidazole
→ 5-Hydroxybenzimidazolylcobamide
In: Clostridium A2272
Eubacterium G919
→ 5-Methoxybenzimidazole
In: Clostridium A2272
→ 5-Methoxy-6-methylbenzimidazolylcobamide
In: Eubacterium G919
→ Vitamin B_{12}
In: Eubacterium G919

5-Hydroxybenzimidazolylcobamide
Formed from: 5-Hydroxybenzimidazole

1-(5-Hydroxy-2-benzimidazolyl)-2-(5-methoxy-2-benzimidazolyl)-1,2-ethanediol
Formed from: 1,2-Bis(5-methoxy-2-benzimidazolyl)-1,2-ethanediol

6-Hydroxy-7,8-benzoflavone
Formed from: 7,8-Benzoflavone

7-Hydroxy-7,8-benzoflavone
Formed from: 7,8-Benzoflavone

4-Hydroxybenzo[*b*]fluoranthene
Probably formed from: Benzo[*b*]fluoranthene

5-Hydroxybenzo[*b*]fluoranthene
Probably formed from: Benzo[*b*]fluoranthene
6-Hydroxybenzo[*b*]fluoranthene
Probably formed from: Benzo[*b*]fluoranthene
***o*-Hydroxybenzoic acid**
See: Salicylic acid
***m*-Hydroxybenzoic acid**
Formed from: *m*-Acetoxybenzoic acid
Benzoic acid
Catechol
m-Chlorobenzoic acid
2-Chloro-5-hydroxybenzoic acid
m-Hydroxybenzaldehyde
m-Hydroxybenzonitrile
m-(*p*-Hydroxyphenoxy)benzoic acid
4-Hydroxyphthalic acid
m-Methoxybenzoic acid
m-Phenoxybenzoic acid
Protocatechuic acid
Probably formed from: *m*-Carboxyphenyl N-methylcarbamate
m-(*p*-Chlorophenoxy)benzoic acid
m-(*p*-Methylphenoxy)benzoic acid
→ Benzoic acid
In: Sporotomaculum J670
See also: B321, H665
→ *m*-Carboxyphenyl sulphate
Probably in: Chicken F632
→ 2,3-Dihydro-2,3,5-trihydroxybenzoic acid
Probably in: Pseudomonas B465
→ 6-(3-Ethyl-6-sulphobenzthiazol-2-ylidenehydrazono)-3-oxo-1,4-cyclohexadiene-1-carboxylic acid
In: Polyporus E134
→ Gentisic acid
In: Arthrobacter A672
Arxula G773
Bacillus A2316
Candida G773
Flavobacterium E355
Klebsiella F822, G199, H399
Microorganism C269, J669
Nocardia F266
Pseudomonas A326, D771, F649, G434
Rhodococcus H341
Salmonella G718
Streptomyces B763, B877, F615
See also: A488, A2155
→ *m*-Hydroxybenzaldehyde
In: Clostridium G416
Neurospora A750
Nocardia G207
See also: E743
→ *m*-Hydroxybenzoyl CoA
In: Bacterium G431
Centaurium J306
Desulfitomaculum K210
Hypericum J509
Pseudomonas F65, G983
Sporomaculum K493
→ *m*-Hydroxybenzoyl-β-D-glucose
In: Cucumis A611
→ *m*-Hydroxybenzoyl-β-D-glucuronic acid
In: Rat J549
See also: F632
→ *m*-Hydroxyhippuric acid
In: Rabbit H519
Rat J549
→ *m*-Methoxybenzoic acid
In: Phanerochaete J339
→ Phenol
In: Bacterium C396
→ Protocatechuic acid
In: Amycolatopsis F615
Aspergillus A1130
Bacillus J37
Comamonas J519
Nocardia F266
Pseudomonas A327, A903, A1229, D771, F649, K716
See also: A707
→ Pyrocatechuic acid
In: Pseudomonas B361, B593, D771, K748
***p*-Hydroxybenzoic acid**
Formed from: *p*-Acetoxybenzoic acid
Anisic acid
Apigenin
Benzoic acid
2,2-Bis(*p*-hydroxyphenyl)propane
p-Bromobenzoic acid
p-Carboxyphenyl-β-D-glucoside
p-Chlorobenzoic acid

p-Chlorobenzoyl CoA
3-Chloro-4-hydroxybenzoic acid
Cyclohexanecarboxylic acid
p,ω-Dihydroxyacetophenone
4′,7-Dihydroxyflavone
N-(1,1-Dimethyl-2-(((morpholino) carbonyl)amino)ethyl)-2-hydroxy-3-(p-hydroxybenzoyloxy)propylamine
Ethyl p-hydroxybenzoate
p-Fluorobenzoic acid
Hispidol-6-β-D-glucoside
p-Hydroxyamphetamine
p-Hydroxybenzaldehyde
p-Hydroxybenzonitrile
p-Hydroxybenzoyl CoA
p-Hydroxybenzoyl-L-alanine
p-Hydroxyhippuric acid
2-(p-Hydroxyphenyl)propionic acid
4-Hydroxyphthalic acid
p-Iodobenzoic acid
Kaempferol
Phenol
Phenylacetic acid
n-Propyl p-hydroxybenzoate
Shikimic acid
Tyramine
L-Tyrosine

Probably formed from: p-Aminobenzoic acid
3-Chloro-N-(p-hydroxybenzyl) propionamide
Ethyl p-hydroxybenzoate
p-Hydroxyacetophenone
p-Hydroxybenzoylphloroglucinolcarboxylic acid
4-Hydroxy-*trans*-stilbene
p-Phenoxybenzoic acid

→ Anisic acid
In: Nocardia F266

→ Benzoic acid
In: Desulfitomaculum K210
Microorganism G167, G981

See also: D654, F266

→ p-Bromophenol
In: Ulva K153

→ 4-Carboxy-2-decaprenylphenol
In: Rat A147
Yeast A192

→ 4-Carboxy-2-hexaprenyl-6-methoxyphenol
In: Saccharomyces B870

→ 4-Carboxy-2-hexaprenylphenol
In: Saccharomyces A3353

→ 4-Carboxy-2-nonaprenylphenol
In: Escherichia A1524
Rat A147
Yeast A192

→ 4-Carboxy-2-octaprenylphenol
In: Escherichia A1524
Rat A147
Yeast A192

→ 4-Carboxy-2-pentaprenylphenol
In: Yeast A192

→ p-Carboxyphenyl-β-D-glucoside
In: Lithospermum F369, G341, J15, J365
Pinus F108

See also: A1725

→ p-Carboxyphenyl-β-D-glucuronide
Probably in: Rat A813

→ p-Carboxyphenyl sulphate
In: Eubacterium H439
Haemophilus H439
Klebsiella H439
Rat B304

See also: A813

→ 4-Carboxy-2-tetaprenylphenol
In: Yeast A192

→ 4-Carboxy-2-triprenylphenol
In: Yeast A192

→ 2,3-Dihydro-2,3,4-trihydroxybenzoic acid
In: Pseudomonas B465

→ 3-(3-Ethyl-6-sulphobenzthiazol-2-ylidenehydrazono)-4-oxo-1,5-cyclohexadiene-1-carboxylic acid
In: Polyporus E134

→ Gentisic acid
In: Bacillus B161
Bacterium A489

→ 3-Geranyl-4-hydroxybenzoic acid
In: Lithospermum A3763, F369, J15
Yeast A192

→ 5-Hexaprenyl-3,4-dihydroxybenzoic acid
In: Saccharomyces A3097

See also: A3261

→ *p*-Hydroxybenzaldehyde
In: Clostridium G416
Neurospora A750
Nocardia G207
Pycnoporus D569
See also: E743
→ *p*-Hydroxybenzoyl CoA
In: Desulfitomaculum K210
Hypericum J509
Man K220
Pseudomonas F65, G225
Quercus K737
Rhodopseudomonas F48
See also: H827
→ *p*-Hydroxybenzoyl-D-glucose
In: Cucumis A611
Raphanus B490
See also: A1725, G897
→ *p*-Hydroxybenzoyl-β-D-glucuronic acid
In: Rabbit H519
Rat B304, J549
→ *p*-Hydroxyhippuric acid
In: Cat A3165
Rabbit H519
Rat F693, J549
See also: A1991, A3833
→ Lignin
In: Poplar A213
→ Phenol
In: Clostridium F402, H576
Klebsiella A2229
Microorganism A2767, F177, F515, G167, G632, H842, K573
Pseudomonas G225
See also: A653, B321, G291
→ Protocatechuic acid
In: Acinetobacter F106, K318
Alcaligenes D761
Amycolatopsis F615
Arxula G773
Aspergillus A329, A640, A2691
Bacillus A1464
Candida G773
Corynebacterium D554, J178
Klebsiella H662
Micrococcus D56
Moraxella G793
Mucuna E769
Nocardia B2, B226, F266
Pseudomonas A563, A903, A2554, E669, E733, G156, K109, K391, K410, K716
Rhizobium C500
Streptomyces B763, B877, F615
Trichosporon B34, B368, E288
See also: A1218, A1466, A1991, A2790, A3219, B695, B927, C76, C817, D542, D544, E335, F51, J482
→ Quinol
In: Candida H277, J461
Pseudomonas G434
Pycnoporus D569
Sporotrichum B73, C57

2-Hydroxybenzoic acid 4-(acetylamino)phenyl ester
See: Phenetsal

2-Hydroxybenzoic acid 2-carboxyphenyl ester
See: Salsalate

4-Hydroxybenzoin
→ Benzoic acid
In: Pseudomonas E737

Hydroxybenzo[*b*]naphtho[1,2-*d*]thiophene sulphone
Probably formed from: Benzo[*b*]naphtho[1,2-*d*]thiophene sulphone

Hydroxybenzo[*b*]naphtho[2,1-*d*]thiophene sulphone
Probably formed from: Benzo[*b*]naphtho[2,1-*d*]thiophene sulphone

***o*-Hydroxybenzonitrile** (*o*-cyanophenol)
Formed from: Benzonitrile
Bunitrolol
→ *o*-Cyanophenyl sulphate
In: Man E575
Rat E575
→ Salicylamide
In: Rhodococcus H952
→ Salicylic acid
In: Fusarium A3364
Nocardia D725
Rhodococcus G172, H952

***m*-Hydroxybenzonitrile**
Formed from: Benzonitrile
→ *m*-Cyanophenyl sulphate
In: Man E575
Rat E575
→ *m*-Hydroxybenzoic acid
In: Fusarium A3364
Nocardia A3420, D725
Rhodococcus G172, H952

***p*-Hydroxybenzonitrile**
Formed from: Benzonitrile
Bis(*p*-cyanophenyl) phosphate
Cyanox
p-Methoxybenzonitrile
Probably formed from: 3-Bromo-4-hydroxybenzonitrile
→ *p*-Cyanophenyl-β-D-glucuronide
In: Rat K27
→ *p*-Cyanophenyl sulphate
In: Man E575
Rat E575, H461, K27
→ 3,4-Dihydroxybenzonitrile
In: Rat H305
→ *p*-Hydroxybenzoic acid
In: Fusarium A3364
Nocardia A3420, D725
Rhodococcus H952
→ Quinol
In: Rat J81

***o*-Hydroxybenzophenone**
→ 2,3′-Dihydroxybenzophenone
In: Hypericum J509

***p*-Hydroxybenzophenone**
Formed from: Benzophenone
Probably formed from: *p*-Bromobenzophenone
p-Hydroxycibenzoline
→ 3,4′-Dihydroxybenzophenone
In: Hypericum J509
→ Quinol
In: Rat J81

7-Hydroxybenzo[*f*]quinoline
Formed from: Benzo[*f*]quinoline

9-Hydroxybenzo[*f*]quinoline
Formed from: Benzo[*f*]quinoline

7-Hydroxybenzo[*h*]quinoline
Formed from: Benzo[*h*]quinoline

2-Hydroxybenzothiazole
Formed from: Benzothiazole
→ 2-Methylmercaptoaniline
In: Guinea pig J9

4-Hydroxy-1*H*-benzotriazole
Formed from: 1*H*-Benzotriazole

5-Hydroxy-1*H*-benzotriazole
Formed from: 1*H*-Benzotriazole

D-2-(6′-Hydroxy-2′-benzothiazolyl)-Δ^2-thiazoline-4-carboxylic acid
See: Luciferin (firefly)

1-(6-Hydroxy-2-benzothiazolyl)-1,3-dimethylurea
Formed from: 1-(2-Benzothiazolyl)-1,3-dimethylurea

2-Hydroxy-1,4-benzoxazin-3-one
→ 2,4-Dihydroxy-1,4-benzoxazin-3-one
In: Wheat K251
Zea G245

6-Hydroxybenzoxazolin-2-one
Formed from: Chlorzoxazone

2-Hydroxy-1,4-benzoxazin-3-one
Formed from: 2-Hydroxy-1,4-benzoxazin-3-one-β-D-glucoside
→ 2,4-Dihydroxy-1,4-benzoxazin-3-one
In: Wheat K251
Zea G245

2-Hydroxy-1,4-benzoxazin-3-one-β-D-glucoside
→ 2-Hydroxy-1,4-benzoxazin-3-one
In: Secale K512

***p*-Hydroxybenzoyl-L-alanine**
→ *p*-Hydroxybenzoic acid
In: Corynebacterium E145

***m*-Hydroxybenzoyl CoA**
Formed from: Benzoyl CoA
m-Hydroxybenzoic acid
m-Hydroxybenzylsuccinyl CoA
→ 2,5-Dihydroxybenzoyl CoA
In: Pseudomonas H397
→ 3-Hydroxycyclohexa-1,5-diene-1-carboxyl CoA
Probably in: Thauera J175
→ 2,3′,4,6-Tetrahydroxybenzophenone
In: Centaurium J35
Hypericum J509

p-Hydroxybenzoyl CoA
Formed from: p-Bromobenzoyl CoA
p-Chlorobenzoyl CoA
p-Fluorobenzoyl CoA
p-Hydroxybenzoic acid
p-Iodobenzoyl CoA
→ Benzoyl CoA
Probably in: Rhodopseudomonas H827
→ p-Hydroxybenzoic acid
In: Corynebacterium J178
Pseudomonas H645, K191
o-Hydroxybenzoylformic acid
See: o-Hydroxyphenylglyoxylic acid
m-Hydroxybenzoyl-D-glucose
Formed from: m-Hydroxybenzoic acid
p-Hydroxybenzoyl-D-glucose
Formed from: p-Hydroxybenzoic acid
m-Hydroxybenzoyl-β-D-glucuronic acid
Formed from: m-Hydroxybenzoic acid
p-Hydroxybenzoyl-β-D-glucuronic acid
Formed from: p-Hydroxybenzoic acid
3-(p-Hydroxybenzoyl)-2-(1-hydroxyethyl)benzoylfuran
Formed from: Benzarone
→ 2-(1,2-Dihydroxyethyl)-3-(p-hydroxybenzoyl)benzoylfuran
Probably in: Dog E789
Man E789
Rat E789
o-Hydroxybenzoylmethanol
See: o,ω-Dihydroxyacetophenone
p-Hydroxybenzoylmethanol
See: p,ω-Dihydroxyacetophenone
N-Hydroxybenzoylnorecgonine
Formed from: N-Hydroxynorcocaine
N-(p-Hydroxybenzoyl)-N′-methylpiperazine
Formed from: N-(1(R)-(1,3-Benzodioxol-5-yl)butyl)-3,3-diethyl-2(S)-((4-methyl-1-piperazinyl)carbonyl)phenoxy-4-oxo-1-azetidinecarboxamide
o-Hydroxybenzoyl-L-phenylalanine
Formed from: o-Hydroxybenzoyl-L-phenylalanine ethyl ester
o-Hydroxybenzoyl-L-phenylalanine ethyl ester
→ o-Hydroxybenzoyl-L-phenylalanine
In: Man D775
With carboxypeptidase D775
With α-chymotrypsin D775
p-Hydroxybenzoylphloroglucinolcarboxylic acid
Probably formed from: Kaempferol
→ p-Hydroxybenzoic acid
Probably in: Horseradish A231
1-Hydroxy-3,4-benzpyrene
Formed from: 3,4-Benzpyrene
Probably formed from: 1-Acetoxyy-3,4-benzpyrene
→ 3,4-Benzpyrene-1,6-quinone
In: Rat B929, C540
→ 3,4-Benzpyren-1-yl-β-D-glucuronide
In: Man H92
Mouse C459
Rat A1615, B746, H290
→ 3,4-Benzpyren-1-yl sulphate
In: Rat A2014, A3732, H290
→ Free radical
In: Rat B476
2-Hydroxy-3,4-benzpyrene
→ 3,4-Benzpyren-2-yl-β-D-glucuronide
In: Man H92
Mouse C459
Rat A1615, B746
→ 3,4-Benzpyren-2-yl sulphate
In: Rat A2014, A3732
→ Free radical
In: Rat B476
3-Hydroxy-3,4-benzpyrene
Formed from: 3,4-Benzpyrene
3,4-Benzpyren-3-yl-β-D-glucuronide
3,4-Benzpyren-3-yl sulphate
→ 3,4-Benzpyren-3-yl-β-D-glucuronide
In: Bluegill E480
Guinea pig B261
Hamster A2009, A3262, E480
Hen D840
Man E480
Mouse C155, C459, D315, E480
Rabbit B269, B286
Rat A1615, B261, B746, C155, C672, D444, D840, E471, E480, G715
→ 3,4-Benzpyren-3-yl sulphate
In: Beef D841
Channel catfish K497
Hamster A1644, A2009, A3625

Hen D841
Lobster H45
Rabbit B286
Rat A1644, A1897, A2014, A2589, A3625, A3732, D841
→ *trans*-7,8-Dihydro-3,7,8-trihydroxy-3,4-benzpyrene
In: Rat D877, E70, E367
→ 3,5-Dihydroxy-3,4-benzpyrene
Probably in: Rat D877, E70
→ 3,6-Dihydroxy-3,4-benzpyrene
Probably in: Rat B944, C540, C668
→ 3,9-Dihydroxy-3,4-benzpyrene
Probably in: Rat B944

4-Hydroxy-3,4-benzpyrene
Formed from: 3,4-Benzpyrene
→ 3,4-Benzpyren-4-yl-β-D-glucuronide
In: Man H92
Mouse C459
Rat A1615, B746
→ 3,4-Benzpyren-4-yl sulphate
In: Rat A2014, A3732
→ Free radical
In: Rat B476

5-Hydroxy-3,4-benzpyrene
→ 3,4-Benzpyren-5-yl-β-D-glucuronide
In: Man H92
Mouse C459
Rat B746
→ Free radical
In: Rat B476

6-Hydroxy-3,4-benzpyrene
Formed from: 6-Nitro-3,4-benzpyrene
→ 6-Acetoxy-3,4-benzpyrene
In: Rat E252
→ 3,4-Benzpyren-6-yl-β-D-glucuronide
In: Man H92
Mouse C459
Rat A1615, B746
→ 3,4-Benzpyren-6-yl sulphate
In: Rat A2014, A3732
→ 6-Oxo-3,4-benzpyrene radical
In: Rat A2298
See also: B476

7-Hydroxy-3,4-benzpyrene
Formed from: 3,4-Benzpyrene
3,4-Benzpyren-7-yl sulphate
→ 3,4-Benzpyren-7-yl-β-D-glucuronide
In: Mouse C459
Rat A1615, B746, E471
→ 3,4-Benzpyren-7-yl sulphate
In: Channel catfish K497
Rat A1848, A2014, A3732
→ Free radical
In: Rat B476

8-Hydroxy-3,4-benzpyrene
Formed from: 3,4-Benzpyrene
→ 3,4-Benzpyren-8-yl-β-D-glucuronide
In: Man H92
Mouse C459
Rat A1615, B746, E471
→ 3,4-Benzpyren-8-yl sulphate
In: Rat A2014, A3732
→ Free radical
In: Rat B476

9-Hydroxy-3,4-benzpyrene
Formed from: 3,4-Benzpyrene
3,4-Benzpyren-9-yl-β-D-glucuronide
3,4-Benzpyren-9-yl sulphate
→ 3,4-Benzpyren-9-yl-β-D-glucuronide
In: Hamster A3262
Man H92
Mouse C155, C459, D315
Rabbit B286
Rat A1615, B746, C155, E471
→ 3,4-Benzpyren-9-yl sulphate
In: Channel catfish K497
Glycine E416
Man B387
Rabbit B286
Rat A1848, A2014, A3732
→ 3,9-Dihydroxy-3,4-benzpyrene
Probably in: Rat B944
→ Free radical
In: Rat B476

10-Hydroxy-3,4-benzpyrene
→ 3,4-Benzpyren-10-yl-β-D-glucuronide
In: Man H92
Rabbit C459
Rat A1615, B746
→ 3,4-Benzpyren-10-yl sulphate
In: Rat A2014, A3732

→ Free radical
In: Rat B476

11-Hydroxy-3,4-benzpyrene
→ 3,4-Benzpyren-11-yl-β-D-glucuronide
In: Mouse C459
→ Free radical
In: Rat B476

12-Hydroxy-3,4-benzpyrene
→ 3,4-Benzpyren-12-yl-β-D-glucuronide
In: Man H92
Mouse C459
Rat A1615, B746
→ 3,4-Benzpyren-12-yl sulphate
In: Rat A2014, A3732

1-Hydroxybenz[*e*]pyrene
Formed from: Benz[*e*]pyrene

3-Hydroxybenz[*e*]pyrene
Formed from: Benz[*e*]pyrene
→ 3-Hydroxybenz[*e*]pyrene-β-D-glucuronide
In: Hamster C340

3-Hydroxybenz[*e*]pyrene-β-D-glucuronide
Formed from: 3-Hydroxybenz[*e*]pyrene

***p*-Hydroxybenztropine**
Formed from: Benztropine
→ *p*-Hydroxybenztropine-β-D-glucuronide
In: Rat H312
→ 4′-Hydroxy-3′-methoxybenztropine
Probably in: H312

***p*-Hydroxybenztropine-β-D-glucuronide**
Formed from: *p*-Hydroxybenztropine

***p*-Hydroxybenztropine-N-oxide**
Formed from: Benztropine-N-oxide
→ *p*-Hydroxybenztropine-N-oxide-β-D-glucuronide
In: Rat H312

***p*-Hydroxybenztropine-N-oxide-β-D-glucuronide**
Formed from: *p*-Hydroxybenztropine-N-oxide

N-Hydroxy-*p*-benzylacetanilide
→ *p*-Benzylphenylhydroxylamine
In: Hamster F522

***o*-Hydroxybenzyl alcohol**
See: Salicyl alcohol

***m*-Hydroxybenzyl alcohol**
Formed from: *m*-Cresol
m-Hydroxybenzaldehyde
→ 3,4-Dihydroxybenzyl alcohol
In: Trichosporon A902
See also: A2391
→ Gentisyl alcohol
In: Penicillium A2314
→ *m*-Hydroxybenzaldehyde
In: Penicillium K946
Pseudomonas A1366, A3837
Rhodopseudomonas D197
See also: A707, A2155
→ *m*-Hydroxymethylphenyl-α-D-glucoside
Probably in: Bacillus H234
→ *m*-Hydroxymethylphenyl sulphate
Probably in: Man E575
Rat E575

***p*-Hydroxybenzyl alcohol**
Formed from: Armepavine
p-Cresol
p-Hydroxybenzaldehyde
N-Methylcoclaurine
→ Anisyl alcohol
In: Phanerochaete G782
→ *p*-Bromophenol
In: Ulva K153
→ 3,4-Dihydroxybenzyl alcohol
In: Mucuna E769
→ *p*-Hydroxybenzaldehyde
In: Acinetobacter E663
Desulfitomaculum K210
Microorganism F89
Penicillium G662
Pseudomonas A1366, A3829, A3837, D696, F381, K203, K289, K410, K841
Rhodopseudomonas D197
Sheep A57
See also: E176, H20, H616
→ *p*-Hydroxybenzyl alcohol 4-sulphate
In: Rat A2308
→ *p*-Hydroxymethylphenyl-α-D-glucoside
Probably in: Bacillus H234
→ *p*-Hydroxymethylphenyl-α-D-glucoside
Probably in: Triglochin D58

***p*-Hydroxybenzyl alcohol 4-sulphate**
Formed from: *p*-Hydroxybenzaldehye 4-sulphate
p-Hydroxybenzyl alcohol

***o*-Hydroxybenzylamine**
Formed from: *o*-Methoxybenzylamine

→ Salicylaldehyde
Probably in: Rat G865

_m_-Hydroxybenzylamine
Formed from: *m*-Methoxybenzylamine
→ *m*-Hydroxybenzaldehye
Probably in: Rat G865

_p_-Hydroxybenzylamine
Formed from: *p*-Methoxybenzylamine
→ *p*-Hydroxybenzaldehye
Probably in: Rat G865

2-(α-Hydroxybenzyl)aminopyridine
Probably formed from: α-Hydroxytripelennamine

N-(Hydroxybenzyl)-N′,N″-dimethylguanidine
Formed from: Bethanidine
→ N-(Hydroxybenzyl)-N′-methylguanidine
Probably in: Dog A2001
Rat A2001

5-(α-Hydroxybenzyl)-3-furoic acid
Probably formed from: 5-Benzyl-3-furoic acid

5-(β-Hydroxybenzyl)-3-furoic acid
Probably formed from: 5-Benzyl-3-furoic acid

5-(β-Hydroxybenzyl)-3-furylcarboxylic acid
Probably formed from: 5-Benzyl-3-furylcarboxylic acid
4′-Hydroxykadethrin

_o_-Hydroxybenzyl-β-D-glucoside
Formed from: Salicyl alcohol

_p_-Hydroxybenzyl-β-D-glucoside
Formed from: *p*-Hydroxybenzyl alcohol

_p_-Hydroxybenzylglucosinolate
Formed from: L-Tyrosine

5-(_m_-Hydroxybenzyl)hydantoin
→ D-*m*-Tyrosine
In: Pseudomonas E450

5-(_p_-Hydroxybenzyl)hydantoin
(D and L isomers)
Formed from: N-Carbamyl-D-tyrosine
N-Carbamyl-L-tyrosine
→ N-Carbamyl-D-tyrosine
In: Flavobacterium F938
→ N-Carbamyl-L-tyrosine
In: Flavobacterium F938
→ D-Tyrosine
In: Pseudomonas E450
→ L-Tyrosine
In: Flavobacterium F940

5-(α-Hydroxybenzyl)-3-hydroxymethylfuran
Probably formed from: 5-Benzyl-3-furylmethanol
→ 3-Hydroxymethyl-5-(α-oxobenzyl)furan
Probably in: Rat B14

N-(_p_-Hydroxybenzyl)-N-(2-(5-hydroxypyridyl))-N′,N′-dimethylethylenediamine
Probably formed from: O-Demethylpyrilamine

cis-_o_-Hydroxybenzylidenepyruvic acid (4-(*o*-hydroxyphenyl)-2-oxobut-3-enoic acid)
Formed from: 1,2-Dihydroxynaphthalene
2-Hydroxychromene-2-carboxylic acid
→ Salicylaldehyde
In: Escherichia K537

_m_-Hydroxybenzylidenepyruvic acid
→ *m*-Hydroxybenzaldehyde
In: Escherichia K537

N-(Hydroxybenzyl)-N′-methylguanidine
Probably formed from: N-(Hydroxybenzyl)-N′,N″-dimethylguanidine

α-Hydroxybenzyl methyl ketoxime
Formed from: Benzyl methyl ketoxime

3-(α-Hydroxybenzyl)-1-methylpyridinium
Probably formed from: 3-Benzoyl-1-methylpyridinium

4-(_p_-Hydroxybenzyl)-1-methylpyridinium
Probably formed from: Tetrahydro-4-(*p*-hydroxybenzyl)-1-methylpyridine

N-(_p_-Hydroxybenzyl)-N′-methyl-N-(2-pyridyl)ethylenediamine
Probably formed from: Demethyltripelennamine

4-(_o_-(_p_-Hydroxybenzyl)phenoxy)butyric acid
Formed from: 4-(*o*-(*p*-Hydroxybenzyl) phenoxy)-N-methylbutylamine

4-(_o_-(_p_-Hydroxybenzyl)phenoxy)-N-methylbutylamine
Probably formed from: 4-(*o*-Benzylphenoxy)-N-methylbutylamine
→ 4-(*o*-(*p*-Hydroxybenzyl)phenoxy) butyric acid
Probably in: Dog D989
Mouse D989
Rabbit D989
Rat D989

O-(_o_-Hydroxy)benzyl phosphorothioate
Formed from: Salithion

N-(*m*-Hydroxybenzyl)piperidine
Formed from: 2-(2-Hydroxyethyl-1-thio)-N-(3-(3-(piperidinomethyl)phenoxy)propyl) acetamide

***erythro*-2-(4-(*p*-Hydroxybenzyl)piperidino)-1-(4-hydroxy-3-methoxyphenyl)-1-propanol**
Probably formed from: *erythro*-2-(4-Benzylpiperidino)-1-(4-hydroxy-3-methoxyphenyl)-1-propanol

***erythro*-2-(4-(*p*-Hydroxybenzyl)piperidino)-1-(*p*-hydroxyphenyl)-1-propanol**
Formed from: *erythro*-2-(4-Benzylpiperidino)-1-(*p*-hydroxyphenyl)-1-propanol
→ *erythro*-2-(4-(*p*-Hydroxybenzyl) piperidino)-1-(*p*-hydroxyphenyl)-1-propanol-β-D-glucuronide
Probably in: Rat A3408

***erythro*-2-(4-(*p*-Hydroxybenzyl)piperidino)-1-(*p*-hydroxyphenyl)-1-propanol-β-D-glucuronide**
Probably formed from: *erythro*-2-(4-(*p*-Hydroxybenzyl)piperidino)-1-(*p*-hydroxyphenyl)-1-propanol

2-(α-Hydroxybenzyl)pyridine
Formed from: 2-Benzoylpyridine

3-(α-Hydroxybenzyl)pyridine
Formed from: 3-Benzoylpyridine
→ 3-(α-Hydroxybenzyl)pyridine-β-D-glucuronide
Probably in: Rat C664
→ 3-(α-Hydroxybenzyl)pyridine-N-oxide
In: Rat C664

4-(α-Hydroxybenzyl)pyridine
Formed from: 4-Benzoylpyridine

3-(α-Hydroxybenzyl)pyridine-β-D-glucuronide
Probably formed from: 3-(α-Hydroxybenzyl) pyridine

3-(α-Hydroxybenzyl)pyridine-N-oxide
Formed from: 3-Benzoylpyridine-N-oxide
3-(α-Hydroxybenzyl)pyridine

***m*-Hydroxybenzylsuccinic acid**
Formed from: *m*-Cresol
→ *m*-Hydroxybenzylsuccinyl CoA
In: Desulfobacterium K292

***m*-Hydroxybenzylsuccinyl CoA**
Formed from: *m*-Hydroxybenzylsuccinic acid
→ *m*-Hydroxybenzoyl CoA
In: Desulfobacterium K292

α-Hydroxybenzyl trimethyl germane
Formed from: Benzoyl trimethyl germane

α-Hydroxybenzyl trimethyl silane
Formed from: Benzoyl trimethyl silane

8-Hydroxybergapten
→ Isopimpinellin
In: Ruta A2582, A3572

5-Hydroxybesipirdine
Formed from: Besipirdine
→ 5-Hydroxy-N-(4-pyridinyl)-1-indolamine
Probably in: Cunninghamella J158

Hydroxybezafibrate
Formed from: Bezafibrate

Hydroxybidisomide
Formed from: Bidisomide
→ Dihydroxybidisomide
Probably in: Man H321

1′-Hydroxybifuralol
Formed from: Bifuralol

2-Hydroxybiphenyl
Formed from: Biphenyl
2-(*o*-Hydroxyphenyl)benzene sulphinate
→ Biphenyl-2-yl-β-D-glucuronide
In: Cat B785
Dog B785
Guinea pig A3971, B303
Man J416, J783
Mouse D315
Rabbit J107
Rainbow trout J582
→ Biphenyl-2-yl sulphate
In: Beef A1211
Cat B785
Dog B785
Guinea pig A3971, B303
Man G10, J404
Mouse J404
Rat J404
→ 2,2′-Dihydroxybiphenyl
In: Rainbow trout J582
See also: A323
→ 2,3-Dihydroxybiphenyl
In: Mouse B61
Pseudomonas E958, E960, F107, J392
Rat B617
See also: B546, B615, F161, K682

→ 2,4′-Dihydroxybiphenyl
In: Man J404
Rat J404
See also: A3427, B41, B615
→ 2,5-Dihydroxybiphenyl
In: Cunninghamella A3427, B41
Debaryomyces J824
Man K353
Mouse B617
Mucor A832, D602
Rat B617, G595, K353
See also: A2660, A3808, B549, B615, K682
→ S-Glutathionyl-2-phenylquinol
In: Rat F343
See also: J582

3-Hydroxybiphenyl
Formed from: Biphenyl
→ Biphenyl-3-yl-β-D-glucuronide
In: Man J416
→ Biphenyl-3-yl sulphate
In: Man G10
→ 2,3-Dihydroxybiphenyl
In: Pseudomonas F107
→ 2,5-Dihydroxybiphenyl
In: Mouse B617
Rat B617
→ 3,4-Dihydroxybiphenyl
In: Mouse B617
Rat B617
See also: A603

4-Hydroxybiphenyl
Formed from: Biphenyl
4-Biphenylyl N-methylcarbamate
4-Methoxybiphenyl
→ Biphenyl-4-yl-β-D-glucuronide
In: Dog B786
Guinea pig A3971, B303
Man C654, G524, G968, J416, J783
Mouse A2678, C266, C80, D315
Rabbit J107
Rainbow trout J582
Rat A3231, B962, C672, D413, E242, H539, J107
→ Biphenyl-4-yl sulphate
In: Aspergillus C860
Beef A1211
Guinea pig A3971, B303
Man E575, G10
Rainbow trout J582
Rat A3231, D413, E242, E575
→ 2,4-Dihydroxybiphenyl
In: Mouse B617
Rat B617
→ 2,4′-Dihydroxybiphenyl
Probably in: Cunninghamella A3427, B41
Trichosporon K682
→ 3,4-Dihydroxybiphenyl
In: Debaryomyces J824
Mouse B617
Rat B617
See also: A2650, A2660, A2724, A2911, A3808, B615, B775, G885
→ 3,4′-Dihydroxybiphenyl
Probably in: Mouse B615
Rabbit A2911
Rat A2724, A2660, A2808, B615
Pig A2650
Pseudomonas F107
Trichosporon K682
→ 4,4′-Dihydroxybiphenyl
In: Absidia B434
Buccinium A2913
Candida B775
Cirolana A2913
Cunninghamella A1218, A3427, B41, B434, B549
Guinea pig A2911
Mouse A2678, B615, B617
Ophiocomina A2913
Pig A2650
Rabbit A2911
Rainbow trout J582
Rat A1753, A2160, A2660, A3808, B615, B617
Trichosporon K682
See also: A1323, A2724, C860

4-Hydroxybiphenyl-4-ylacetic acid
Probably formed from: 4-Biphenylylacetic acid
→ 4-Hydroxybiphenyl-4-ylacetic acid-β-D-glucuronide
Probably in: Rabbit A3631
Rat A3631

4′-Hydroxybiphenyl-4-ylacetic acid-β-D-glucuronide
Probably formed from: 4′-Hydroxybiphenyl-4-ylacetic acid

3-(4′-Hydroxy-4-biphenylyl)butyric acid
Formed from: 3-(4-Biphenyl)butyronitrile

2-Hydroxybiphenyl-2′-yl-β-D-glucuronide
Probably formed from: 2,2′-Dihydroxy-biphenyl

4-Hydroxybiphenyl-4′-yl-β-D-glucuronide
Probably formed from: 4,4′-Dihydroxy-biphenyl

2-(4′-Hydroxybiphenyl-4-yl)-2-propanol
Probably formed from: 2-(Biphenyl-4-yl)-2-propanol

2-(4′-Hydroxybiphenyl-4-yl)-2-propionic acid
Probably formed from: 2-(Biphenyl-4-yl)-2-propionic acid

4-Hydroxybiphenyl-4-yl sulphate
Probably formed from: 4,4′-Dihydroxy-biphenyl

2′-Hydroxybiphenyl-2-sulphinic acid
See: 2-(*o*-Hydroxyphenyl)benzenesulphinic acid

7-Hydroxybisdemethylchlorpromazine
Probably formed from: Bisdemethylchlorpromazine
→ 7-Hydroxybisdemethylchlorpromazine sulphoxide
Probably in: Man A693

7-Hydroxybisdemethylchlorpromazine sulphoxide
Probably formed from: 7-Hydroxybisdemethylchlorpromazine

8-Hydroxybisdemethylclomipramine
Formed from: 8-Hydroxydemethylclomipramine

***p*-Hydroxy-5,6-bis-O-demethyldonezepil**
Probably formed from: *p*-Hydroxy-5-O-demethyldonezepil

Hydroxybisdemethyliprindole
Probably formed from: Demethyl-hydroxyiprindole

7-Hydroxy-N,O-bisdemethyllevomepromazine
Probably formed from: 7-Hydroxy-N-demethyllevomepromazine

Hydroxybisdemethylloperamide
Probably formed from: Bisdemethylloperamide

7-Hydroxybisdemethyl-8-methoxychlorpromazine
Probably formed from: 7-Hydroxydemethyl-8-methoxychlorpromazine

10-Hydroxybisdemethyltrimipramine
Probably formed from: trimipramine

β-Hydroxybishomotryptophan
Formed from: Bishomotryptophan

3-Hydroxy-1,2-bis(4-hydroxy-3-methoxyphenyl)-1-propanone
Formed from: *erythro*-1,2-bis(4-hydroxy-3-methoxyphenyl)-1,3-propanediol

1-Hydroxy-2,2-bis(4-hydroxy-3-methylphenyl) propane
Formed from: 2,2-Bis(4-hydroxy-3-methylphenyl)propane

5-Hydroxy-3,4-bis(4-hydroxyphenyl)-7-phenyl-2,6-benzofurandione
See: Peniophorin

1-Hydroxy-2,2-bis(*p*-hydroxyphenyl)propane
Formed from: 2,2-Bis(*p*-hydroxyphenyl) propane
→ 2,2-Bis(*p*-hydroxyphenyl)-1,2-propanediol
In: Bacillus G623
Bacterium H119
→ 2,2-Bis(*p*-hydroxyphenyl)propionic acid
In: Bacterium H119

2-Hydroxy-1,2-bis(methoxyphenyl)ethanone
See: Anisoin

N-Hydroxybisnorpromazine
→ Bisdemethylpromazine
In: Rabbit E822

N-Hydroxybisnorpromethazine
Formed from: N-Hydroxynorpromethazine

2-Hydroxybisnortrimipramine
Probably formed from: 2-Hydroxynortrimipramine
→ 2,10-Dihydroxybisnortrimipramine
Probably in: Rat G269

10-Hydroxybisnortriptyline
Probably formed from: 10-Hydroxynortriptyline

1-((1*R*)-2-*endo*-Hydroxy-3-*endo*-bornyl)-3-(*p*-(α-hydroxy)toluenesulphonyl)urea
Formed from: Glibornuride

→ 3-(*p*-Carboxybenzenesulphonyl)-1-(2-*endo*-hydroxy-3-*endo*-bornyl)urea
In: Man A759

8-Hydroxybrofoxine
Formed from: Brofoxine

5′-Hydroxybromazepam
Formed from: Bromazepam

8-Hydroxybromazepam
Formed from: Bromazepam

9-Hydroxybromazepam
Formed from: Bromazepam

8′α-Hydroxybromocriptine
Formed from: Bromocriptine
→ 8′,9′-Dihydroxybromocriptine
Probably in: Rat C607

8′β-Hydroxybromocriptine
Formed from: Bromocriptine

***m*-Hydroxybropirimine**
Formed from: Bropirimine

***p*-Hydroxybropirimine**
Formed from: Bropirimine

6-Hydroxybrotizolam
Formed from: Brotizolam
→ 2-Bromo-4-(*o*-chlorophenyl)-6-hydroxy-9-hydroxymethyl-6*H*-thieno[3,2-*f*][1,2,4]triazolo[4,3-*a*][1,4]diazepine
Probably in: Mouse G186
→ 6-Oxobrotizolam
Probably in: Dog E91

***p*-Hydroxybrotizolam**
Formed from: Brotizolam
→ 4′-Hydroxy-5′-methoxybrotizolam
Probably in: Dog E91
Man E91
Rat E91

α-Hydroxybrotizolam
See: 2-Bromo-4-(*o*-chlorophenyl)-9-hydroxymethyl-6*H*-thieno[3,2-*f*][1,2,4]triazolo[4,3-*a*][1,4]diazepine

16-Hydroxybrucine
Formed from: Brucine

5-Hydroxybucindolol
Formed from: Bucindolol

Hydroxybupidipine
Formed from: Bupidipine
→ Hydroxybupidipine-O-sulphate
In: Rat G415

Hydroxybupidipine-N-oxide
Probably formed from: Bupidipine-N-oxide

Hydroxybupidipine-O-sulphate
Formed from: Hydroxybupidipine

1′-Hydroxybufuralol
Formed from: Bufuralol
→ 1′,6-Dihydroxybufuralol
Probably in: Dog A2806
Man A2806
Rat A2806
→ 1′-Hydroxybufuralol-β-D-glucuronide
Probably in: Man E283
→ 1′-Oxobufuralol
In: Dog A2806
Man A2806
Rat A2806

2′-Hydroxybufuralol
Formed from: Bufuralol

4-Hydroxybufuralol
Formed from: Bufuralol

6-Hydroxybufuralol
Formed from: Bufuralol
→ 5,6-Dihydroxybufuralol
Probably in: Dog A2806
Man A2806
Rat A2806

1′-Hydroxybufuralol-β-D-glucuronide
Probably formed from: 1′-Hydroxybufuralol

4-Hydroxybunitrolol
Formed from: Bunitrolol

Hydroxybunolol
Formed from: 5-(3-*tert*-Butylamino-2-hydroxypropoxy)-1,2,3,4-tetrahydro-1-oxonaphthalene
→ Dihydrohydroxybunolol
Probably in: Dog A3309
Man A3369
Rat A3404

Hydroxybupranolol
Formed from: Bupranolol

3-Hydroxybupropion
Formed from: Bupropion

***p*-Hydroxybupropion**
Probably formed from: Bupropion

ω-Hydroxybupropion
Formed from: Bupropion
→ ω,ω-Dihydroxybupropion

Probably in: Man H640

4-Hydroxybutachlor
Formed from: Butachlor

6-Hydroxybutamoxane
Formed from: Butamoxane
→ 6,7-Dihydroxybutamoxane
In: Rat A1390

7-Hydroxybutamoxane
Formed from: Butamoxane
→ 6,7-Dihydroxybutamoxane
In: Rat A1390

8-Hydroxybutamoxane
Formed from: 8-Methoxybutamoxane

Hydroxybutaperazine
Formed from: Butaperazine
→ Hydroxydemethylbutaperazine
Probably in: Dog A295
Rat A295

***p*-(2-Hydroxybutoxy)acetanilide**
Formed from: *p*-Butoxyacetanilide
→ *p*-(Carboxymethoxy)acetanilide
Probably in: Cunninghamella F590
→ *p*-(2-Hydroxyethoxy)acetanilide
Probably in: Cunninghamella F590

1-(2-Hydroxybutoxy)naphthalene
Formed from: 1-Butoxynaphthalene

1-(3-Hydroxybutoxy)naphthalene
Formed from: 1-Butoxynaphthalene

3-((3-Hydroxy-*n*-butyl)amino)-4-phenoxy-5-sulphamoylbenzoic acid
Formed from: Bumetanide
→ 3-((2,3-Dihydroxy-*n*-butyl)amino)-4-phenoxy-5-sulphamoylbenzoic acid
Probably in: Rat A1992

3-((4-Hydroxy-*n*-butyl)amino)-4-phenoxy-5-sulphamoylbenzoic acid
Formed from: Bumetanide
→ 3-((3-Carboxypropyl)amino)-4-phenoxy-5-sulphamoylbenzoic acid
Probably in: Man A2335
Rat A1992, C611

1-(3-Hydroxybut-1-yl)-1*H*-benzotriazole
Formed from: 1-Butyl-1*H*-benzotriazole

4-(3-Hydroxybutyl)-3,5-dioxo-1,2-diphenylpyrazolidine
Formed from: Kebuzon

3-Hydroxybutyl 9-hydroxyfluorenyl-9-carboxylate
Formed from:
Butyl 9-hydroxyfluorenyl-9-carboxylate

4-(3-Hydroxybutyl)-1-(*p*-hydroxyphenyl)-3,5-dioxo-2-phenylpyrazolidine
Probably formed from: 1-(*p*-Hydroxyphenyl)-3,5-dioxo-4-(3-oxobutyl)-2-phenylpyrazolidine

2-(1-Hydroxybutyl)-10-(3-(4-methylpiperazin-1-yl)propyl)phenothiazine
Formed from: Butaperazine

3-Hydroxybutyl *p*-nitrophenyl ether
Formed from: Butyl *p*-nitrophenyl ether
→ *p*-Nitrophenyl 3-oxobutyl ether
In: Rabbit A3047

4-Hydroxybutyl *p*-nitrophenyl ether
Formed from: Butyl *p*-nitrophenyl ether
→ *p*-Nitrophenoxybutyric acid
In: Rabbit A3047

N-(β-Hydroxy-*tert*-butyl)norchlorcyclizine
Formed from: N-*tert*-butylnorchlorcyclizine

***o*-(1-Hydroxybutyl)phenol**
Formed from: 1-(*o*-(2-Diisopropylamino)ethoxy)phenyl-1-butanol

2-(3-Hydroxybutyl)-3-((2-(1*H*-tetrazol-5-yl)(1,1′-biphenyl)-4-yl)methyl)-1,3-diaza*spiro*[4,4]non-1-en-4-one
Formed from: Irbesartan
→ 2-(3-Oxobutyl)-3-((2-(1*H*-tetrazol-5-yl)(1,1′-biphenyl)-4-yl)methyl)-1,3-diaza*spiro*[4,4]non-1-en-4-one
Probably in: Man J403

N-(4-Hydroxybutyrylglycyl)-2,6-dimethylaniline
Formed from: N-(2,6-Dimethylphenyl)-2-(2-oxo-1-pyrrolidinyl)acetamide
→ 2,6-Dimethyl-N-succinylglycylaniline
Probably in: Man F587

1-(1-Hydroxybutyryl)-4-(*p*-hydroxycinnamyl)piperazine
Probably formed from: 4-Cinnamyl-1-(1-hydroxybutyryl)piperazine

1-(2-Hydroxybutyryl)-4-(*p*-hydroxycinnamyl)piperazine
Probably formed from: 4-Cinnamyl-1-(2-hydroxybutyryl)piperazine

5-Hydroxycamalexine
Formed from: Camalexine
→ 3-Cyano-5-hydroxyindole

Probably in: Rhizoctonia K392
1′-Hydroxycannabichromene
Formed from: Cannabichromene
1″-Hydroxycannabichromene
Formed from: Cannabichromene
2′-Hydroxycannabichromene
Formed from: Cannabichromene
2″-Hydroxycannabichromene
Formed from: Cannabichromene
3″-Hydroxycannabichromene
Formed from: Cannabichromene
4″-Hydroxycannabichromene
Formed from: Cannabichromene
5′-Hydroxycannabichromene
Formed from: Cannabichromene
→ 1″,5′-Dihydroxycannabichromene
Probably in: Rabbit G335
→ 3″,5′-Dihydroxycannabichromene
Probably in: Rabbit G335
→ 4″,5′-Dihydroxycannabichromene
Probably in: Rabbit G335
→ 5′,5″-Dihydroxycannabichromene
Probably in: Rabbit G335
5″-Hydroxycannabichromene
Formed from: Cannabichromene
6′-Hydroxycannabichromene
Formed from: Cannabichromene
→ 1″,6′-Dihydroxycannabichromene
Probably in: Rabbit G335
→ 3″,6′-Dihydroxycannabichromene
Probably in: Rabbit G335
→ 4″,6′-Dihydroxycannabichromene
Probably in: Rabbit G335
→ 5″,6′-Dihydroxycannabichromene
Probably in: Rabbit G335
1″-Hydroxycannabidiol
Formed from: Cannabidiol
2″-Hydroxycannabidiol
Formed from: Cannabidiol
3″-Hydroxycannabidiol
Formed from: Cannabidiol
4″-Hydroxycannabidiol
Formed from: Cannabidiol
→ 4″-Hydroxycannabidiol-β-D-glucoside
Probably in: Dog F497, G460
5″-Hydroxycannabidiol
Formed from: Cannabidiol
→ 5″-Hydroxycannabidiol-β-D-glucoside
Probably in: Dog F497, G460
→ 5″-Norcannabidiol-4″-oic acid
In: Dog G113
Man G113
Rat G113
See also: A2085, B458, F503
6α-Hydroxycannabidiol
Formed from: Cannabidiol
→ 6α-Hydroxycannabidiol-β-D-glucuronide
Probably in: Mouse A2085
6β-Hyoxycannabidiol (this isomer assumed unless otherwise stated)
Formed from: Cannabidiol
→ 4″,5″-Bisnor-6-hydroxycannabidiol-3″-oic acid
Probably in: Rat G608
→ 1″,6β-Dihydroxycannabidiol
Probably in: Rabbit F708
→ 2″,6β-Dihydroxycannabidiol
Probably in: Rat A2777
→ 3″,6β-Dihydroxycannabidiol
Probably in: Rabbit F708
Rat A2777
→ 4″,6β-Dihydroxycannabidiol
Probably in: Dog F557, G113
Guinea pig B458
Rabbit F708
Rat A2777
→ 5″,6β-Dihydroxycannabidiol
Probably in: Dog F497, G113
Rabbit F708
→ 6-Hydroxycannabidiol-5″-oic acid
Probably in: Dog G113
→ 6-Oxocannabidiol
Probably in: Dog F497, G113
7-Hydroxycannabidiol
Formed from: Cannabidiol
→ 4″,5″-Bisnor-7-hydroxycannabidiol-3″-oic acid
Probably in: Rat G608
→ Cannabidiol-7-oic acid
In: Mouse A2085
See also: B458, F503, F557, G113
→ 8,9-Dihydro-7,8,9-trihydroxycannabidiol
Probably in: Dog G113

→ 1″,7-Dihydroxycannabidiol
Probably in: Guinea pig B458
Rabbit F708
Rat A2777
→ 2″,7-Dihydroxycannabidiol
Probably in: Rabbit F708
→ 3″,7-Dihydroxycannabidiol
Probably in: Guinea pig B458
Rabbit F708
Rat A2777
→ 4″,7-Dihydroxycannabidiol
Probably in: Dog F557, G113
Guinea pig B458
Rabbit F708
Rat A2777
→ 5″,7-Dihydroxycannabidiol
Probably in: Dog F557, G113
Rabbit F708
Rat A2777, G608
→ 6β,7-Dihydroxycannabidiol
Probably in: Dog F557, G113
Guinea pig B458
Man F503
Rat A2777
→ 7-Hydroxycannabidiol-β-D-glucuronide
In: Rabbit B666
See also: A1660, A2085, A2989
→ 2-(7-Hydroxy-*p*-mentha-1,8(9)-dien-3-yl)-5-(1-hydroxypent-1-yl)resorcinol
Probably in: Mouse A2085
→ 2-(7-Hydroxy-*p*-mentha-1,8(9)-dien-3-yl)-5-(4-hydroxypent-1-yl)resorcinol
Probably in: Mouse A2085

10-Hydroxycannabidiol
Formed from: Cannabidiol

4″-Hydroxycannabidiol-β-D-glucuronide
Probably formed from: 4″-Hydroxycannabidiol

5″-Hydroxycannabidiol-β-D-glucuronide
Probably formed from: 5″-Hydroxycannabidiol

6α-Hydroxycannabidiol-β-D-glucuronide
Probably formed from: 6α-Hydroxycannabidiol

7-Hydroxycannabidiol-β-D-glucuronide
Probably formed from: 7-Hydroxycannabidiol

2″-Hydroxycannabidiol-7-oic acid
Formed from: Cannabidiol-7-oic acid

3″-Hydroxycannabidiol-7-oic acid
Formed from: Cannabidiol-7-oic acid

4″-Hydroxycannabidiol-7-oic acid
Formed from: Cannabidiol-7-oic acid

5″-Hydroxycannabidiol-7-oic acid
Probably formed from: Cannabidiol-7-oic acid
→ Cannabidiol-5″,7-dioic acid
Probably in: Dog G113

6-Hydroxycannabidiol-5″-oic acid
Probably formed from: 5″,6-Dihydroxycannabidiol
6-Hydroxycannabidiol
→ 4″,5″-Bisnor-6-hydroxycannabidiol-3″-oic acid
Probably in: Dog F557
→ 6-Oxocannabidiol-5″-oic acid
Probably in: Dog F557

6β-Hydroxycannabidiol-7-oic acid
Probably formed from: 6β,7-Dihydroxycannabidiol

7-Hydroxycannabidiol-5″-oic acid
Probably formed from: 5″,7-Dihydroxycannabidiol
→ 4″,5″-Bisnor-7-hydroxycannabidiol-3″-oic acid
Probably in: Dog F557
→ Cannabidiol-5″,7-dioic acid
Probably in: Rat G608
→ 6,7-Dihydroxycannabidiol-5″-oic acid
Probably in: Dog G113

10-Hydroxycannabidiol-7-oic acid
Formed from: Cannabidiol-7-oic acid

1″-Hydroxycannabigerol
Formed from: Cannabigerol

2″-Hydroxycannabigerol
Formed from: Cannabigerol

3″-Hydroxycannabigerol
Formed from: Cannabigerol

4′-Hydroxycannabigerol
Formed from: Cannabigerol

4″-Hydroxycannabigerol
Formed from: Cannabigerol

5″-Hydroxycannabigerol
Formed from: Cannabigerol

8′-Hydroxycannabigerol
Formed from: Cannabigerol
→ 6′,7′-Dihydro-8′-hydroxycannabigerol
Probably in: Cat F704
Gerbil F704
Guinea pig F704
Hamster F704
Mouse F704
Rabbit F704
Rat F704

9′-Hydroxycannabigerol
Formed from: Cannabigerol

1″-Hydroxycannabinol
Formed from: Cannabinol

2″-Hydroxycannabinol
Formed from: Cannabinol

3″-Hydroxycannabinol
Formed from: Cannabinol

4″-Hydroxycannabinol
Formed from: Cannabinol
→ 4″-Hydroxycannabinol oleate
Probably in: Rat A3767
→ 4″-Hydroxycannabinol palmitate
Probably in: Rat A3767

5″-Hydroxycannabinol
Formed from: Cannabinol
→ 5″-Hydroxycannabinol oleate
Probably in: Rat A3767
→ 5″-Hydroxycannabinol palmitate
Probably in: Rat A3767

7-Hydroxycannabinol
Formed from: Cannabinol
→ 1″,7-Dihydroxycannabinol
Probably in: Guinea pig B458
Mouse A3405
Rat A2994, A3436
→ 2″,7-Dihydroxycannabinol
Probably in: Mouse A3405
Rat A2994, A3436
→ 3″,7-Dihydroxycannabinol
Probably in: Mouse A3405
Rat A2994
→ 4″,7-Dihydroxycannabinol
Probably in: Mouse A3405
Rat A2994, A3436
→ 7-Hydroxycannabinol-β-D-glucuronide
Probably in: Mouse A2989, A3405
→ 7-Hydroxycannabinol oleate
Probably in: Rat A3767
→ 7-Hydroxycannabinol palmitate
Probably in: Rat A3767
→ 7-Oxocannabinol
In: Mouse H67
See also: A2256, A2989, A3405, A3436, B458

9′-Hydroxycannabinol
Formed from: Cannabinol
→ 9′-Hydroxycannabinol-β-D-glucuronide
Probably in: Pinellia H72

7-Hydroxycannabinol-β-D-glucuronide
Probably formed from: 7-Hydroxycannabinol

9′-Hydroxycannabinol-β-D-glucuronide
Probably formed from: 9′-Hydroxycannabinol

1″-Hydroxycannabinol-7-oic acid
Probably formed from: 1″,7-Dihydroxycannabinol

2″-Hydroxycannabinol-7-oic acid
Probably formed from: 2″,7-Dihydroxycannabinol

3″-Hydroxycannabinol-7-oic acid
Probably formed from: 3″,7-Dihydroxycannabinol

4″-Hydroxycannabinol-7-oic acid
Probably formed from: 4″,7-Dihydroxycannabinol

4″-Hydroxycannabinol oleate
Probably formed from: 4″-Hydroxycannabinol

5″-Hydroxycannabinol oleate
Probably formed from: 5″-Hydroxycannabinol

7-Hydroxycannabinol oleate
Probably formed from: 7-Hydroxycannabinol

4″-Hydroxycannabinol palmitate
Probably formed from: 4″-Hydroxycannabinol

5″-Hydroxycannabinol palmitate
Probably formed from: 5″-Hydroxycannabinol

7-Hydroxycannabinol palmitate
Probably formed from: 7-Hydroxycannabinol

1-Hydroxycanthin-6-one
Formed from: Canthin-6-one
→ 1-Methoxycanthin-6-one
Probably in: Ailanthus E304

ω-Hydroxycapsaicin
Formed from: Capsaicin

2-Hydroxycarbamazepine
Formed from: Carbamazepine
→ 2,3-Dihydroxycarbamazepine
Probably in: Man A3453, C3, J101
→ 2-Hydroxycarbamazepine-β-D-glucuronide
In: Man J101

3-Hydroxycarbamazepine
Formed from: Carbamazepine
→ 3-Hydroxycarbamazepine-N-β-D-glucuronide
Probably in: Man A1967
→ 3-Hydroxycarbamazepine-O-β-D-glucuronide
In: Man A1967, J101
→ Hydroxycarbamazepine-10,11-oxide
Probably in: Man A1967

10-Hydroxycarbamazepine (both isomers)
Formed from: Carbamazepine
Oxcarbazepine

3-Hydroxycarbamazepine-N-β-D-glucuronide
Probably formed from: 3-Hydroxycarbamazepine

2-Hydroxycarbamazepine-O-β-D-glucuronide
Probably formed from: 2-Hydroxycarbamazepine

3-Hydroxycarbamazepine-O-β-D-glucuronide
Probably formed from: 3-Hydroxycarbamazepine

Hydroxycarbamazepine-10,11-oxide
Probably formed from: 3-Hydroxycarbamazepine

4-Hydroxycarbazeran
Formed from: Carbazeran
→ 1-(4-Carbamoyloxypiperidino)-4-hydroxy-6,7-dimethoxyphthalazine
Probably in: Man D161
→ 4-Hydroxy-1-(4-hydroxypiperidino)-6,7-dimethoxyphthalazine
Probably in: Man D161

3-Hydroxycarbazole
Formed from: Carbazole

3-Hydroxycarbofuran
Formed from: Carbofuran
3-Hydroxycarbofuran glucoside
→ 3-Hydroxycarbofuran-β-D-glucuronide
In: Rat D176
→ 3-Oxocarbofuran
Probably in: Aspergillus J641
Cotton A3871
Fusarium J641
Nicotiana A402
Pine A1315
Strawberry A1948
Trichogaster B731
Zea A952, A3871

3-Hydroxycarbofuran glucoside
→ 3-Hydroxycarbofuran
Probably in: Rat D176

3-Hydroxycarbofuran-β-D-glucuronide
Formed from: 3-Hydroxycarbofuran

8-Hydroxycarbostyril
Probably formed from: 5-Formyl-8-hydroxycarbostyril

2-Hydroxycarpipramine
Formed from: Carpipramine
→ 2,2″-Dihydroxycarpipramine
Probably in: Man D368
→ 2,10-Dihydroxycarpipramine
Probably in: Man D368

2″-Hydroxycarpipramine
Formed from: Carpipramine
→ 2″,10-Dihydroxycarpipramine
Probably in: Man D368

10-Hydroxycarpipramine
Formed from: Carpipramine

α-Hydroxycarprofen
Formed from: Carprofen

7-Hydroxycarprofen
Formed from: Carprofen

8-Hydroxycarprofen
Formed from: Carprofen

8-Hydroxycarteolol
Formed from: Carteolol

1-Hydroxycarvedilol
Formed from: Carvedilol
→ 1-Hydroxycarvedilol-β-D-glucuronide
Probably in: Dog J443
Mouse J443
Rat F584, G33

3-Hydroxycarvedilol
Formed from: Carvedilol

4′-Hydroxycarvedilol
Formed from: Carvedilol
→ 4′-Hydroxycarvedilol-β-D-glucuronide

In: Mouse J443

5′-Hydroxycarvedilol
Formed from: Carvedilol
→ 5′-Hydroxycarvedilol-β-D-glucuronide
In: Mouse J443

8-Hydroxycarvedilol
Formed from: Carvedilol
→ 8-Hydroxycarvedilol-β-D-glucuronide
Probably in: Dog J443
Mouse J443
Rat F584, G33

1-Hydroxycarvedilol-β-D-glucuronide
Probably formed from: 1-Hydroxycarvedilol

4′-Hydroxycarvedilol-β-D-glucuronide
Formed from: 4′-Hydroxycarvedilol

5′-Hydroxycarvedilol-β-D-glucuronide
Formed from: 5′-Hydroxycarvedilol

8-Hydroxycarvedilol-β-D-glucuronide
Probably formed from: 8-Hydroxycarvedilol

8-Hydroxycatechin
Formed from: Catechin

ω-Hydroxycatenarin
Formed from: ω-Hydroxyemodin

15′-Hydroxycatharinine
Formed from: Leurosine

Hydroxycelecoxib
Formed from: Celecoxib

6-Hydroxycentpiperalone
Formed from: Centpiperalone

2′-Hydroxychalcone
Formed from: Flavanone
→ Dihydro-2′-hydroxychalcone
Probably in: Gibberella A1512
Streptomyces F673
→ 2′,4′-Dihydroxychalcone
Probably in: Gibberella A1512

***m*-Hydroxychloridazon**
Formed from: Chloridazon

***p*-Hydroxychloridazon**
Formed from: Chloridazon

N-Hydroxychlorphentermine
Formed from: Chlorphentermine
Nitrochlorphentermine
→ N-Hydroxychlorphentermine-β-D-glucuronide
In: Rat A3951

N-Hydroxychlorphentermine-β-D-glucuronide
Formed from: N-Hydroxychlorphentermine

Hydroxy-3-(3-chlorophenoxy)pyrrolidine-1-carboxamide
Probably formed from: 3-(3-Chlorophenoxy) pyrrolidine-1-carboxamide

3-Hydroxychlorpromazine
→ 2,3-Dihydroxypromazine
In: Man A693
Probably in: Mushroom A1290

6-Hydroxychlorpromazine
→ Chlorpromazin-6-yl-β-D-glucuronide
In: Guinea pig A2092
Rat A2092

7-Hydroxychlorpromazine
Formed from: Chlorpromazine
→ Chlorpromazin-7-yl-β-D-glucuronide
In: Guinea pig A2092
Rat A2092
See also: G27
→ 3,7-Dihydroxychlorpromazine
Probably in: Man A693
→ 7,8-Dihydroxychlorpromazine
Probably in: Man A693
→ 7-Hydroxychlorpromazine sulphoxide
Probably in: Man A693
→ 7-Methoxychlorpromazine
In: Man A693

8-Hydroxychlorpromazine
Formed from: Chlorpromazine
→ Chlorpromazin-8-yl-β-D-glucuronide
In: Guinea pig A2092
Rat A2092

7-Hydroxychlorpromazine sulphoxide
Formed from: 7-Hydroxychlorpromazine

4-Hydroxychlorprophan
Formed from: Chlorprophan

6-Hydroxychlorzoxazone
Formed from: Chlorzoxazone

2-Hydroxychromene-2-carboxylic acid
→ *o*-Hydroxybenzylidenepyruvic acid
In: Pseudomonas J711

8-Hydroxychromone
Probably formed from: 3-Carboxy-8-hydroxy-chromone
→ 2′,3′-Dihydroxyacetophenone

Probably in: Dog A1761

1-Hydroxychrysene
Formed from: Chrysene

2-Hydroxychrysene
Formed from: Chrysene

3-Hydroxychrysene
Formed from: Chrysene
→ 1,2-Dihydro-1,2,3-trihydroxychrysene
Probably in: Hamster F449
Man F449
→ *trans*-1,2-Dihydro-1,2,9-trihydroxychrysene
In: Rat F652
→ 3-Hydroxychrysene-β-D-glucuronide
In: Rat G715

4-Hydroxychrysene
Formed from: Chrysene

6-Hydroxychrysene
Formed from: Chrysene
→ 6-Hydroxychrysene-β-D-glucuronide
In: Rat G715

3-Hydroxychrysene-β-D-glucuronide
Formed from: 3-Hydroxychrysene

6-Hydroxychrysene-β-D-glucuronide
Formed from: 6-Hydroxychrysene

***p*-Hydroxycibenzoline**
Formed from: Cibenzoline
→ *p*-Hydroxybenzophenone
Probably in: Dog D845
Man D653
→ 4-Hydroxy-3-methoxycibenzoline
Probably in: Dog D845

6-Hydroxyciclobendazole
Formed from: Ciclobendazole

2′-Hydroxyciglitazone
Formed from: Ciglitazone
→ 2′-Oxociglitazone
Probably in: Rat C753

***cis*-3′-Hydroxyciglitazone**
Formed from: Ciglitazone
→ 3′-Oxociglitazone
Probably in: Rat C753

***trans*-3′-Hydroxyciglitazone**
Formed from: Ciglitazone

***cis*-4′-Hydroxyciglatizone**
Formed from: Ciglatizone
→ 4′-Oxociglitazone
Probably in: Rat C753

***trans*-4′-Hydroxyciglatizone**
Formed from: Ciglatizone

2′-Hydroxycilostamide
Formed from: Cilostamide

3′-Hydroxycilostamide
Formed from: Cilostamide

4′-Hydroxycilostamide
Formed from: Cilostamide

2″-*cis*-Hydroxycilostazol
Formed from: Cilostazol
→ 3,4-Dehydro-2″-*cis*-hydroxycilostazol
Probably in: Man D755
→ 2″-*cis*-Hydroxycilostazol-β-D-glucuronide
In: Dog D755
Rat D755

2″-*trans*-Hydroxycilostazol
Formed from: Cilostazol

3″-*cis*-Hydroxycilostazol
Formed from: Cilostazol

3″-*trans*-Hydroxycilostazol
Formed from: Cilostazol

4-Hydroxycilostazol
Formed from: Cilostazol

4″-*cis*-Hydroxycilostazol
Formed from: Cilostazol

4″-*trans*-Hydroxycilostazol
Formed from: Cilostazol
→ 3,4-Dehydro-4″-*trans*-hydroxycilostazol
Probably in: Man D755

2″-*cis*-Hydroxycilostazol-β-D-glucuronide
Formed from: 2″-*cis*-Hydroxycilostazol

***p*-Hydroxycimetropium**
Formed from: Cimetropium
→ Dihydroxycimetropium
Probably in: Guinea pig G234
Hamster G234
Mouse G234
Rat G234
→ *p*-Hydroxy-de(hydroxymethyl)cimetropium
Probably in: Guinea pig G234
Hamster G234
Mouse G234
Rat G234

2′-Hydroxycinchonidine
Formed from: Cinchonidine

6′-Hydroxycinchonidine
Formed from: Quinine

→ 10,11-Dihydro-6′,10,11-trihydroxycinchonidine
Probably in: Man J826
→ 3,6′-Dihydroxycinchonidine
Probably in: Man J826

6′-Hydroxycinchonidinone
Formed from: Cupreine

2′-Hydroxycinchonine
Formed from: Cinchonine

α-Hydroxycinmethylin
Formed from: Cinmethylin
→ α,6-Dihydroxycinmethylin
Probably in: Goat F98
→ α,8-Dihydroxycinmethylin
Probably in: Goat F98
→ α,9-Dihydroxycinmethylin
Probably in: Goat F98

6-Hydroxycinmethylin
Formed from: Cinmethylin

8-Hydroxycinmethylin
Formed from: Cinmethylin
→ 4-(1-Hydroxy-1-methylethyl)-2-((3-hydroxy-2-methylphenyl)methoxy)-1-methyl-7-oxabicyclo[2.2.1]heptane
Probably in: Goat F98
→ 4-(1-Hydroxy-1-methylethyl)-2-((4-hydroxy-2-methylphenyl)methoxy)-1-methyl-7-oxabicyclo[2.2.1]heptane
Probably in: Goat F98

9-Hydroxycinmethylin
Formed from: Cinmethylin

***o*-Hydroxycinnamaldehyde**
See: *o*-Coumaraldehyde

***m*-Hydroxycinnamaldehyde**
See: *m*-Coumaraldehyde

***p*-Hydroxycinnamaldehyde**
See: *p*-Coumaraldehyde

***o*-Hydroxycinnamic acid**
See: *o*-Coumaric acid

***m*-Hydroxycinnamic acid**
See: *m*-Coumaric acid

***p*-Hydroxycinnamic acid**
See: *p*-Coumaric acid

(−)-(*trans*-*p*-Hydroxycinnamoyl)lupinine
Formed from: *p*-Coumaric acid

6β-Hydroxycinnamoyltropine
Formed from: Cinnamoyltropine

1-(*p*-Hydroxycinnamoyl)piperazine
Probably formed from: 1-*n*-Butyryl-4-(*p*-hydroxycinnamoyl)piperazine

***p*-Hydroxycinnamyl alcihol**
See: *p*-Coumaryl alcohol

4-Hydroxycinnoline
Formed from: Cinnoline

Hydroxyciramadol
Formed from: Ciramadol

2-Hydroxycisapride
Formed from: Cisapride

3-Hydroxycisapride
Formed from: Cisapride

3′-Hydroxycisapride
Formed from: Cisapride
→ 3′-Hydroxycisapride-O-sulphate
In: Rat E620

3′-Hydroxycisapride-O-sulphate
Formed from: 3′-Hydroxycisapride

4-Hydroxyclausenamide
Formed from: (+)-Clausenamide

6-Hydroxyclausenamide
Formed from: (+)-Clausenamide
→ 4,6-Dihydroxyclausenamide
Probably in: Rat K198

3′-Hydroxyclemastine
Formed from: Clemastine

Hydroxyclenbuterol
Probably formed from: Clenbuterol

***p*-Hydroxyclobazam**
Formed from: Clobazam
→ 3′,4′-Dihydroxyclobazam
Probably in: Rat G572
→ *p*-Hydroxyclobazam-β-D-glucuronide
Probably in: Rat G572
→ *p*-Hydroxyclobazam-O-sulphate
Probably in: Rat G572
→ 4′-Hydroxy-3′-methoxyclobazam
Probably in: Man B22
Monkey B22
Rat B22
→ *p*-Hydroxynorclobazam
Probably in: Rat G572

***p*-Hydroxyclobazam-β-D-glucuronide**
Probably formed from: *p*-Hydroxyclobazam

***p*-Hydroxyclobazam-O-sulphate**
Probably formed from: *p*-Hydroxyclobazam

4-Hydroxyclomiphene
Formed from: Clomiphene
4′-Hydroxyclomiphene
Formed from: Clomiphene
→ 4′-Hydroxy-3′-methoxyclomiphene
Probably in: Rat E476
3-Hydroxyclonazepam
Formed from: Clonazepam
***p*-Hydroxyclonidine**
Formed from: Clonidine
4′-Hydroxyclonixin
Formed from: Clonixin
5-Hydroxyclonixin
Formed from: Clonixin
7-Hydroxyclozapine
Formed from: Clozapine
→ 7-Hydroxyclozapine-β-D-glucuronide
In: Man J153
Rat H702
→ 7-Hydroxyclozapine sulphate
In: Rat H702
See also: J153
7-Hydroxyclozapine-β-D-glucuronide
Formed from: 7-Hydroxyclozapine
7-Hydroxyclozapine sulphate
Formed from: 7-Hydroxyclozapine
Hydroxycobalamin
See: Vitamin B_{12}
Hydroxycocaine
Formed from: Cocaine
14β-Hydroxycodeine
Formed from: Codeine
14β-Hydroxycodeinone
Formed from: Codeinone
→ Oxycodone
In: Pseudomonas K311
5-Hydroxyconiferaldehyde
See: 3,4-Dihydroxy-5-methoxycinnamaldehyde
3-Hydroxycoumarin
Formed from: Coumarin
→ *o*-Hydroxyphenyllactic acid
Probably in: Beef H154
Rat G605
Does not yield *o*-hydroxyphenylacetaldehyde in hamster (J809) or mouse (H807)
4-Hydroxycoumarin
Formed from: Coumarin
→ Coumarin-4-yl-β-D-glucoside
In: Manduca G897
Melanopus G816
→ Coumarin-4-yl-β-D-glucuronide
In: Chicken B468
5-Hydroxycoumarin
Formed from: Coumarin
6-Hydroxycoumarin
Formed from: Coumarin
7-Hydroxycoumarin
See: Umbelliferone
8-Hydroxycoumarin
Formed from: Coumarin
Probably formed from: 2,8-Dihydroxyquinoline
7-Hydroxycoumarin-6-yl-β-D-glucoside (esculin)
→ 6,7-Hydroxycoumarin
In: Guinea pig A1517
Rat A1517
Secale K512
→ 6,7-Hydroxycoumarin diglucoside
In: Mucor A493
6-Hydroxycoumarin-7-yl-β-D-glucuronide
Formed from: 6,7-Hydroxycoumarin
7-Hydroxycoumarin-6-yl-β-D-glucuronide
Formed from: 6,7-Hydroxycoumarin
N-((7-Hydroxycoumarin-4-yl)methyl)pyridinium
Formed from: N-((7-Methoxycoumarin-4-yl)methyl)pyridinium
1-(4-Hydroxy-3-coumarinyl)-1-phenyl-2-butanone
See: Warfarin
7-Hydroxycoumarin-6-yl sulphate
Formed from: 6,7-Hydroxycoumarin
1-Hydroxycrisnatol
Formed from: Crisnatol
→ 1-Hydroxycrisnatol-β-D-glucuronide
Probably in: Rat F741
→ 1-Hydroxy-6-hydroxymethylchrysene
Probably in: Rat F741
1-Hydroxycrisnatol-β-D-glucuronide
Probably formed from: 1-Hydroxycrisnatol
3-Hydroxycromakalin
Formed from: Cromakalin
4-Hydroxycromakalin
Formed from: Cromakalin

Hydroxycyclazocine
Formed from: Cyclazocine
→ Methoxycyclazocine
In: Rat A2134
6-Hydroxycyclindole
Formed from: Cyclindole
7-Hydroxycyclindole
Formed from: Cyclindole
3-Hydroxycyclobenzaprine
Formed from: Cyclobenzaprine
→ 3-Hydroxydemethylcyclobenzaprine
Probably in: Dog A2887
4-Hydroxycyclobenzaprine
Formed from: Cyclobenzaprine
10-Hydroxycyclobenzaprine
Probably formed from: Cyclobenzaprine
→ 10-Hydroxycyclobenzaprine-N-oxide
Probably in: Rat A2614
→ 10-Hydroxydemethylcyclobenzaprine
Probably in: Rat A2614
10-Hydroxycyclobenzaprine-N-oxide
Probably formed from: 10-Hydroxycyclobenzaprine
4′-Hydroxycyclocoumarol
Formed from: Cyclocoumarol
6-Hydroxycyclocoumarol
Formed from: Cyclocoumarol
7-Hydroxycyclocoumarol
Formed from: Cyclocoumarol
8-Hydroxycyclocoumarol
Formed from: Cyclocoumarol
2-(4-(2-Hydroxycyclohexylidenemethyl)phenyl)propionyltaurine
Formed from: 2-(4-(2-Oxocyclohexylidenemethyl)phenyl)propionic acid
α-(4-Hydroxycyclohexyl)mandelic acid
Probably formed from: α-Cyclohexylmandelic acid
2-(4-(2-Hydroxycyclohexylmethyl)phenyl)propionic acid
Formed from: 2-(4-(2-Oxocyclohexylmethyl)phenyl)propionic acid
N-(3-*cis*-Hydroxycyclohexyl)-N′-(4-(β-(5-methylpyrazine-2-carboxamido)ethyl)benzylsulphonyl)urea
Formed from: Glipizide
N-(3-*trans*-Hydroxycyclohexyl)-N′-(4-(β-(5-methylpyrazine-2-carboxamido)ethyl)benzylsulphonyl)urea
Formed from: Glipizide
1-(2-Hydroxy)cyclohexyl-1-phenyl-3-piperidinopropan-1-ol
Formed from: Benzhexol
1-(3-Hydroxy)cyclohexyl-1-phenyl-3-piperidinopropan-1-ol
Formed from: Benzhexol
1-(4-Hydroxy)cyclohexyl-1-phenyl-3-piperidinopropan-1-ol
Formed from: Benzhexol
2-(4-(3-Hydroxycyclohexyl)phenyl)propionic acid
Probably formed from: 2-(4-(3-Oxocyclohexyl)phenyl)propionic acid
1-(3-Hydroxycyclohexyl)-1-phenyl-3-(1-pyrrolidinyl)-1-propanol
Formed from: Procyclidine
1-(*cis*-4-Hydroxycyclohexyl)-1-phenyl-3-(1-pyrrolidinyl)-1-propanol
Formed from: Procyclidine
→ 1-(*cis*-3,*cis*-4-Dihydroxycyclohexyl)-1-phenyl-3-(1-pyrrolidinyl)-1-propanol
Probably in: Rat D240, E788
→ 1-(4-Oxocyclohexyl)-1-phenyl-3-(1-pyrrolidinyl)-1-propanol
Probably in: Rat D240
1-(*trans*-4-Hydroxycyclohexyl)-1-phenyl-3-(1-pyrrolidinyl)-1-propanol
Formed from: Procyclidine
→ 1-(*cis*-3,*trans*-4-Dihydroxycyclohexyl)-1-phenyl-3-(1-pyrrolidinyl)-1-propanol
Probably in: Rat D240
→ 1-(*trans*-3,*trans*-4-Dihydroxycyclohexyl)-1-phenyl-3-(1-pyrrolidinyl)-1-propanol
Probably in: Rat D240, E788
2-(4-(2-Hydroxycyclopent-3-en-1-ylmethyl)phenyl)propionic acid
Probably formed from: 2-(4-(2-Oxocyclopent-3-en-1-ylmethyl)phenyl)propionic acid
5-((2-Hydroxycyclopentoxy)carbonylamino)-3-((2-methoxy-4-((2-methylphenyl)sulphonyl)aminocarbonyl)phenylmethyl)-1-methylindole
Formed from: Zafirlukast

5-((3-Hydroxycyclopentoxy)carbonylamino)-3-((2-methoxy-4-((2-methylphenyl)sulphonyl)aminocarbonyl)phenylmethyl)-1-methylindole
Formed from: Zafirlukast

2-(*p*-((2-Hydroxycyclopentyl)methyl)phenyl)propionic acid
Formed from: Loxoprofen
→ 2-(*p*-((2,4-Dihydroxycyclopentyl)methyl)phenyl)propionic acid
Probably in: Rat C914
→ 2-(*p*-((2-Hydroxycyclopentyl)methyl)phenyl)propionic acid-β-D-glucuronide
Probably in: Dog C914
Monkey C914
Mouse C914
→ 2-(*p*-((2-Hydroxycyclopentyl)methyl)phenyl)propionyltaurine
Probably in: Dog C914

2-(*p*-((2-Hydroxycyclopentyl)methyl)phenyl)propionic acid-β-D-glucuronide
Probably formed from: 2-(*p*-((2-Hydroxycyclopentyl)methyl)phenyl)propionic acid

2-(*p*-((2-Hydroxycyclopentyl)methyl)phenyl)propionyltaurine
Probably formed from: 2-(*p*-((2-Hydroxycyclopentyl)methyl)phenyl)propionic acid

3′-Hydroxycyclopeptine
Formed from: Cyclopeptine

10-Hydroxycyclopeptine
→ 3′,10-Dihydroxycyclopeptine
In: Penicillium A1633

4′-Hydroxy-*cis*-cypermethrin
Formed from: *cis*-Cypermethrin

4′-Hydroxy-*trans*-cypermethrin
Formed from: *trans*-Cypermethrin

5-Hydroxy-*cis*-cypermethrin
Formed from: *cis*-Cypermethrin
→ 3-Hydroxy-5-phenoxybenzoic acid
Probably in: Rat A3804

5-Hydroxy-*trans*-cypermethrin
Formed from: *trans*-Cypermethrin

***t*-Hydroxy-*cis*-cypermethrin**
Formed from: *cis*-Cypermethrin

***t*-Hydroxy-*trans*-cypermethrin**
Formed from: *trans*-Cypermethrin

1-Hydroxycyproheptadine
Formed from: Cyproheptadine

2-Hydroxycyproheptadine
Formed from: Cyproheptadine

3-Hydroxycyproheptadine
Formed from: Cyproheptadine

10-Hydroxycyproheptadine
Formed from: Cyproheptadine

2-Hydroxycyproheptadine-N-oxide
Formed from: Cyproheptadine-N-oxide

5-Hydroxydantrolene
Formed from: Dantrolene

1-Hydroxydebrisoquine
Formed from: Debrisoquine
→ *o*-((2-Guanidino)ethyl)benzoic acid
Probably in: Man J456

3-Hydroxydebrisoquine
Formed from: Debrisoquine
→ *o*-((2-Guanidino)methyl)phenylacetic acid
Probably in: Man J456

4-Hydroxydebrisoquine
Formed from: Debrisoquine
→ 4,6-Dihydroxydebrisoquine
Probably in: Man A1653
Rat A1653

5-Hydroxydebrisoquine
Formed from: Debrisoquine

6-Hydroxydebrisoquine
Formed from: Debrisoquine

7-Hydroxydebrisoquine
Formed from: Debrisoquine

8-Hydroxydebrisoquine
Formed from: Debrisoquine

N-Hydroxydebrisoquine
Formed from: Debrisoquine
→ 2-Carboxamido-1,2,3,4-tetrahydroisoquinoline
In: Rabbit H77
→ Debrisoquine
In: Rabbit H77
Rat H77

2′-Hydroxydecamethrin
Formed from: Decamethrin

4′-Hydroxydecamethrin
Formed from: Decamethrin

5-Hydroxydecamethrin
Formed from: Decamethrin

→ 3-Hydroxy-5-phenoxybenzoic acid
Probably in: Rat A3804

6-(10-Hydroxydecyl)-2,3-dimethoxy-5-methylquinol
Probably formed from: Idebenone
→ 6-(10-Hydroxydecyl)-2,3-dimethoxy-5-methylquinol-β-D-glucuronide
Probably in: Dog D764
Rat E119
→ 6-(10-Hydroxydecyl)-2,3-dimethoxy-5-methylquinol sulphate
Probably in: Dog D764
Rat E119

6-(10-Hydroxydecyl)-2,3-dimethoxy-5-methylquinol-β-D-glucuronide
Probably formed from: 6-(10-Hydroxydecyl)-2,3-dimethoxy-5-methylquinol

6-(10-Hydroxydecyl)-2,3-dimethoxy-5-methylquinol sulphate
Probably formed from: 6-(10-Hydroxydecyl)-2,3-dimethoxy-5-methylquinol

6-(10-Hydroxydecyl)-2,3-dimethoxy-5-methylquinone
See: Idebenone

2-Hydroxydehydroabietic acid
Formed from: Dehydroabietic acid

7-Hydroxydehydroabietic acid
Formed from: Dehydroabietic acid
→ 7-Oxodehydroabietic acid
Probably in: Alcaligenes A932
Pseudomonas A932

15-Hydroxydehydroabietic acid
Formed from: Dehydroabietic acid

16-Hydroxydehydroabietic acid
Formed from: Dehydroabietic acid

3′-Hydroxydehydrocyclopeptine
Formed from: Dehydrocyclopeptine

3-Hydroxydehydro-*iso*-α-lapachone
Formed from: *o*-Succinylbenzoic acid

***p*-Hydroxy-de(hydroxymethyl)cimetropium**
Probably formed from: *p*-Hydroxycimetropium

4-Hydroxydeisopropyldelaviridine
Probably formed from: Deisopropyldelaviridine
→ 4-Hydroxydeisopropyldelaviridine-O-β-D-glucuronide
Probably in: Monkey J275
→ 4-Hydroxydeisopropyldelaviridine-N-sulphate
Probably in: Monkey J275
Mouse J277
→ 4-Hydroxydeisopropyldelaviridine-O-sulphate
Probably in: Monkey J275

4-Hydroxydeisopropyldelaviridine-O-β-D-glucuronide
Probably formed from: 4-Hydroxydeisopropyldelaviridine

4-Hydroxydeisopropyldelaviridine-N-sulphate
Probably formed from: 4-Hydroxydeisopropyldelaviridine

4-Hydroxydeisopropyldelaviridine-O-sulphate
Probably formed from: 4-Hydroxydeisopropyldelaviridine

4-Hydroxydelaviridine
Probably formed from: Delaviridine
→ 4-Hydroxydelaviridine-β-D-glucuronide
Probably in: Monkey J275

4′-Hydroxydelaviridine
Probably formed from: Delaviridine

6′-Hydroxydelaviridine
Probably formed from: Delaviridine
→ 6′-Hydroxydelaviridine-β-D-glucuronide
In: Rat J96
→ 6′-Hydroxydelaviridine-O-sulphate
In: Rat J96

4-Hydroxydelaviridine-β-D-glucuronide
Probably formed from: 4-Hydroxydelaviridine

6′-Hydroxydelaviridine-β-D-glucuronide
Probably formed from: 6′-Hydroxydelaviridine

6′-Hydroxydelaviridine-O-sulphate
Probably formed from: 6′-Hydroxydelaviridine

11-Hydroxy-O-demethylacronine
Probably formed from: 11-Hydroxyacronine

6-Hydroxy-O-demethylangolensin
Formed from: Genistein

5-Hydroxydemethylastemizole
Probably formed from: 5-Hydroxyastemizole

6-Hydroxydemethylastemizole
Probably formed from: 6-Hydroxyastemizole

***p*-Hydroxydemethylbenztropine**
Probably formed from: Demethylbenztropine

→ *p*-Hydroxydemethylbenztropine-β-D-glucuronide
Probably in: Rat H312

p-Hydroxydemethylbenztropine-β-D-glucuronide
Probably formed from: *p*-Hydroxydemethylbenztropine

Hydroxydemethylbutaperazine
Probably formed from: Hydroxybutaperazine

7-Hydroxydemethylchlorpromazine
Probably formed from: Demethylchlorpromazine
→ N-Acetyl-7-hydroxydemethylchlorpromazine
Probably in: Man A693
→ 7-Hydroxydemethylchlorpromazine sulphoxide
Probably in: Guinea pig A901
Man A901

N-Hydroxydemethylchlorpromazine
→ Demethylchlorpromazine
In: Rabbit E822

7-Hydroxydemethylchlorpromazine sulphoxide
Formed from: 7-Hydroxydemethylchlorpromazine

2′-Hydroxydemethylcilostamide
Probably formed from: Demethylcilostamide

4′-Hydroxydemethylcilostamide
Probably formed from: Demethylcilostamide

7-Hydroxydemethylcyclindole
Probably formed from: Demethylcyclindole

3-Hydroxydemethylcyclobenzaprine
Probably formed from: 3-Hydroxycyclobenzaprine

10-Hydroxydemethylcyclobenzaprine
Probably formed from: 10-Hydroxycyclobenzaprine

2-Hydroxydemethylcyproheptadine
Probably formed from: Cyproheptadine

6-Hydroxy-N-demethyldextromethorphan
Probably formed from: N-Demethyldextromethorphan
→ 6-Oxo-N-demethyldextromethorphan
Probably in: Man F956

p-Hydroxy-5-O-demethyldonezepil
Probably formed from: *p*-Hydroxydonezepil
→ *p*-Hydroxy-5,6-bis-O-demethyldonezepil
Probably in: Rat J789
→ *p*-Hydroxy-5-O-demethyldonezepil sulphate
Probably in: Rat J789

p-Hydroxy-5-O-demethyldonezepil sulphate
Probably formed from: *p*-Hydroxy-5-O-demethyldonezepil

2-Hydroxydemethyldoxepin
Probably formed from: Demethyldoxepin

3-Hydroxydemethyldoxepin
Probably formed from: Demethyldoxepin

4-Hydroxydemethyldoxepin
Probably formed from: Demethyldoxepin

8-Hydroxydemethyldoxepin
Probably formed from: Demethyldoxepin

p-Hydroxydemethyldoxylamine
Probably formed from: Demethyldoxylamine

5-Hydroxy-1-demethylgranisetron
Probably formed from: 5-Hydroxygranisetron

2-Hydroxydemethylimipramine
Formed from: Demethylimipramine
→ Demethylimipramin-2-yl-β-D-glucuronide
In: Guinea pig A2092
Rat A2092
See also: C872

10-Hydroxydemethylimipramine
Formed from: Demethylimipramine

N-Hydroxydemethylimipramine
Formed from: Demethylimipramine
→ Demethylimipramine
In: Rabbit C496

Hydroxydemethyliprindole (alicyclic)
Probably formed from: Demethyliprindole
→ Demethyloxoiprindole
Probably in: Dog A3712
Rat A3712, G23
→ Hydroxybisdemethyliprindole
Probably in: Dog A3712
Monkey A3712
Pig A3712
Rat A3712

Hydroxydemethyliprindole (aryl)
Probably formed from: Demethyliprindole

6-Hydroxy-N-demethylketamine
Formed from: 6-Hydroxyketamine

3-Hydroxy-N-demethyllevomepromazine
Probably formed from: N-Demethyllevomepromazine

3-Hydroxy-O-demethyllevomepromazine
Probably formed from: O-Demethyllevomepromazine

7-Hydroxy-N-demethyllevomepromazine
Probably formed from: N-Demethyllevomepromazine
→ 7-Hydroxy-N,O-bisdemethyllevomepromazine
Probably in: Man H598

Hydroxydemethylloperamide
Probably formed from: Demethylloperamide

1′-Hydroxydemethylmaprotiline
Probably formed from: Demethylmaprotiline

2-Hydroxydemethylmaprotiline
Probably formed from: Demethylmaprotiline

3-Hydroxydemethylmaprotiline
Probably formed from: Demethylmaprotiline

8-Hydroxydemethylmianserin
Probably formed from: Demethylmianserin
→ 8-Hydroxydemethylmianserin-O-β-D-glucuronide
In: Guinea pig G66
Man G66, G864
Mouse G66
Rabbit G66
Rat G66, G864, G908
→ 8-Hydroxydemethylmianserin-N-sulphate
In: Mouse G66
Rat G864
→ 8-Hydroxydemethylmianserin-O-sulphate
In: Mouse G66
Rat G66, G864, G908
→ 8-Hydroxydemethyl-4-oxomianserin-O-β-D-glucuronide
In: Guinea pig G66
Man G66, G864
Mouse G66
Rabbit G66
Rat G66, G864

8-Hydroxydemethylmianserin-O-β-D-glucuronide
Probably formed from: 8-Hydroxydemethylmianserin

8-Hydroxydemethylmianserin-N-sulphate
Probably formed from: 8-Hydroxydemethylmianserin

8-Hydroxydemethylmianserin-O-sulphate
Probably formed from: 8-Hydroxydemethylmianserin

10-Hydroxydemethylnortriptyline
Probably formed from: Demethylnortriptyline
→ 10,11-Dihydroxydemethylnortriptyline
Probably in: Man E894

7-Hydroxydemethylondansetron
Formed from: Demethylondansetron

3-Hydroxydemethyloxaprotiline
Probably formed from: Demethyloxaprotiline
→ 3-Hydroxydemethyloxaprotiline-β-D-glucuronide
Probably in: Dog E48
Rat E48

3-Hydroxydemethyloxaprotiline-β-D-glucuronide
Probably formed from: 3-Hydroxydemethyloxaprotiline

8-Hydroxydemethyl-4-oxomianserin-O-β-D-glucuronide
Probably formed from: 8-Hydroxydemethylmianserin

Hydroxydemethylpargyline
Probably formed from: Demethylpargyline

3-Hydroxy-O-demethylpyrilamine
Probably formed from: O-Demethylpyrilamine
→ 3-Hydroxypyrilamine
Probably in: Rat E907, E930
→ 3-Methoxy-O-demethylpyrilamine
Probably in: Rat E907, E930

2-Hydroxy-N-demethylritonavir
Probably formed from: N-Demethylritonavir

4-Hydroxydemethyltamoxifen
Probably formed from: Demethyltamoxifen
→ 4-Hydroxydemethyltamoxifen-β-D-glucuronide
Probably in: Man G844

4-Hydroxydemethyltamoxifen-β-D-glucuronide
Probably formed from: 4-Hydroxydemethyltamoxifen

3-Hydroxy-S-demethylthioridazine
Probably formed from: S-Demethylthioridazine

7-Hydroxy-S-demethylthioridazine
Probably formed from: S-Demethylthioridazine

7-Hydroxydemethylthioridazine-2-sulphone
Probably formed from: Demethylsulphoridazine

4′-Hydroxydemethyltriflubazam
Probably formed from: Demethyltriflubazam
→ 3′,4′-Dihydroxydemethyltriflubazam
Probably in: Man A1873

Hydroxy-1-demethylzolazepam
Probably formed from: 1-Demethylzolazepam

6-Hydroxy-8-demethylzolazepam
Probably formed from: 6-Hydroxyzolazepam

7-Hydroxydemethylzotepine
Probably formed from: Demethylzotepine
→ 7-Hydroxydemethylzotepine sulphate
Probably in: Mammal B245

7-Hydroxydemethylzotepine sulphate
Probably formed from: 7-Hydroxydemethylzotepine

7-Hydroxydepropylropinirole
→ 7-Hydroxydepropylropinirole-β-D-glucuronide
In: Cynomolgus monkey J561
Man J561
Mouse J561

7-Hydroxydepropylropinirole-β-D-glucuronide
Formed from: 7-Hydroxydepropylropinirole

2-Hydroxydesethylamodiaquine
Probably formed from: Desethylamodiaquine

N-Hydroxydesethylfenfluramine
Probably formed from: Desethylfenfluramine

3-Hydroxydesethyl-2-oxolisuride
Probably formed from: 3-Hydroxy-2-oxolisuride

Hydroxydexmedetomidine
Formed from: Dexmedetomidine

6-Hydroxydextromethorphan
Formed from: Dextromethorphan
→ 6-Oxodextromethorphan
Probably in: Man F956

N-Hydroxy-4,4′-diaminodiphenyl sulphone
Probably formed from: N-Hydroxy-4,4′-diformamidodiphenyl sulphone

3-Hydroxydiazepam
See: 7-Chloro-1,3-dihydro-3-hydroxy-1-methyl-5-phenyl-2*H*-1,4-benzodiazepin-2-one

2-Hydroxydibenz[*a,j*]acridine
Formed from: Dibenz[*a,j*]acridine

3-Hydroxydibenz[*a,j*]acridine
Formed from: Dibenz[*a,j*]acridine

4-Hydroxydibenz[*a,j*]acridine
Formed from: Dibenz[*a,j*]acridine

5-Hydroxydibenz[*a,j*]acridine
Formed from: Dibenz[*a,j*]acridine

1-Hydroxydibenzanthracene
Formed from: Dibenzanthracene
→ Dibenzanthracen-1-yl-β-D-glucuronide
In: Rat E471

2-Hydroxydibenzanthracene
Formed from: Dibenzanthracene
→ Dibenzanthracen-2-yl-β-D-glucuronide
In: Rat E471

3-Hydroxydibenzanthracene
Formed from: Dibenzanthracene

4-Hydroxydibenzanthracene
Formed from: Dibenzanthracene
→ Dibenzanthracen-4-yl-β-D-glucuronide
In: Rat E471

5-Hydroxydibenzanthracene
Formed from: Dibenzanthracene
→ Dibenzanthracen-5-yl-β-D-glucuronide
In: Rat E471

6-Hydroxydibenzanthracene
Formed from: Dibenzanthracene
→ Dibenzanthracen-6-yl-β-D-glucuronide
In: Rat E471

7-Hydroxydibenzanthracene
Formed from: Dibenzanthracene
→ Dibenzanthracen-7-yl-β-D-glucuronide
In: Rat E471

2-Hydroxy-7*H*-dibenzo[*c,g*]carbazole
Formed from: 7*H*-Dibenzo[*c,g*]carbazole

3-Hydroxy-7*H*-dibenzo[*c,g*]carbazole
Formed from: 7*H*-Dibenzo[*c,g*]carbazole

5-Hydroxy-7*H*-dibenzo[*c,g*]carbazole
Formed from: 7*H*-Dibenzo[*c,g*]carbazole

3-(10-Hydroxy-5*H*-dibenzo[*a,d*]cycloheptan-5-yl)aminopropionic acid
Probably formed from: 3-(5*H*-Dibenzo[*a,d*]cycloheptan-5-yl)aminopropionic acid

5-(10-Hydroxy-5*H*-dibenzo[*a*,*d*]cycloheptan-5-yl)aminopropionic acid
Probably formed from: 3-(5*H*-Dibenzo[*a*,*d*]cycloheptan-5-yl)aminopropionic acid

(2-Hydroxy-5*H*-dibenzo[*a*,*d*]cyclohepten-5-ylidene)acetic acid
Formed from: (5*H*-Dibenzo[*a*,*d*]cyclohepten-5-ylidene)acetic acid

(3-Hydroxy-5*H*-dibenzo[*a*,*d*]cyclohepten-5-ylidene)acetic acid
Formed from: (5*H*-Dibenzo[*a*,*d*]cyclohepten-5-ylidene)acetic acid

11-Hydroxydibenzo[*c*,*f*][1,2]diazepine
Formed from: Dibenzo[*c*,*f*][1,2]diazepine
→ 11-Oxodibenzo[*c*,*f*][1,2]diazepine
Probably in: Rat B280

2-Hydroxydibenzo-*p*-dioxin
Formed from: Dibenzo-*p*-dioxin

3-Hydroxydibenzo[*a*,*e*]fluoranthene
Formed from: Dibenzo[*a*,*e*]fluoranthene

4-Hydroxydibenzo[*a*,*e*]fluoranthene
Formed from: Dibenzo[*a*,*e*]fluoranthene

7-Hydroxydibenzo[*a*,*e*]fluoranthene
Formed from: Dibenzo[*a*,*e*]fluoranthene

1-Hydroxydibenzofuran
Formed from: Dibenzofuran

2-Hydroxydibenzofuran
Formed from: Dibenzofuran

3-Hydroxydibenzofuran
Formed from: Dibenzofuran

4-Hydroxydibenzofuran
Formed from: Dibenzofuran

***o*-Hydroxydibenzoylmethane**
Formed from: Flavanone
Probably formed from: Flavone

3′-Hydroxydiclofenac
Formed from: Diclofenac

4′-Hydroxydiclofenac
Formed from: Diclofenac
→ 4′,5-Dihydroxydiclofenac
Probably in: Baboon A3840, A3844
Dog A3840
Man A3840, A3844, F316
Rat A3840
Rhesus monkey A3844
→ 4′-Hydroxydiclofenac sulphate
Probably in: Baboon A3840
Dog A3840
Man A3840
Rat A3840
→ 3′-Hydroxy-4′-methoxydiclofenac
Probably in: Baboon F320
Man F316, F320, H414

5-Hydroxydiclofenac
Formed from: Diclofenac
→ 4′,5-Dihydroxydiclofenac
In: Man K212
→ N,5-Dihydroxydiclofenac
In: Man K212

4′-Hydroxydiclofenac sulphate
Formed from: 4′-Hydroxydiclofenac

7-Hydroxydicoumarol
Formed from: Dicoumarol
→ 7-Hydroxydicoumarol-β-D-glucuronide
Probably in: Rat A2257

7-Hydroxydicoumarol-β-D-glucuronide
Probably formed from: 7-Hydroxydicoumarol

ω-Hydroxydienoestrol
Formed from: Stilboestrol

3-Hydroxydiflunisal
Formed from: Diflunisal

N-Hydroxy-4,4′-diformamidodiphenyl sulphone
Formed from: 4,4′-Diformamidodiphenyl sulphone
→ N-Hydroxy-4,4′-diaminodiphenyl sulphone
Probably in: Dog 6521
Monkey 6521

7-Hydroxydiftalone
Formed from: Diftalone
→ 7,14-Dihydroxydiftalone
Probably in: Man A113

3-Hydroxydiflunisal
Formed from: Diflunisal

14-Hydroxydihydrocodeine
Formed from: Oxycodone
→ 14-Hydroxydihydrocodeine-N-oxide
Probably in: Rabbit A3677

14-Hydroxydihydrocodeine-N-oxide
Probably formed from: 14-Hydroxydihydrocodeine

14-Hydroxydihydroisocodeine
Formed from: Oxycodone

14-Hydroxydihydromorphine
Formed from: Oxycodone

2-Hydroxy-3,5-diiodobenzoic acid
Formed from: 2-Acetoxy-3,5-diiodobenzoic acid
4-Hydroxy-3,5-diiodobenzoic acid
Formed from: 4-Hydroxy-3,5-diiodobenzonitrile
4-Hydroxy-3,5-diiodobenzonitrile
→ 4-Hydroxy-3,5-diiodobenzoic acid
In: Fusarium A3364
***p*-(4-Hydroxy-3,5-diiodophenoxy)-3,5-diiodo-D-phenylalanine**
See: D-Thyroxine
***p*-(4-Hydroxy-3,5-diiodophenoxy)-3,5-diiodo-L-phenylalanine**
See: L-Thyroxine
3-Hydroxydilevalol
Formed from: Dilevalol
→ 3-Hydroxydilevalol-1′-β-D-glucuronide
Probably in: Cynomolgus monkey H55
Dog H55
Man H55
Rat H55
→ 3-Hydroxydilevalol-3-β-D-glucuronide
Probably in: Cynomolgus monkey H55
Dog H55
Man H55
Rat H55
4″-Hydroxydilevalol
Formed from: Dilevalol
→ 4″-Hydroxydilevalol-β-D-glucuronide
Probably in: Rat H55
3-Hydroxydilevalol-1′-β-D-glucuronide
Probably formed from: 3-Hydroxydilevalol
3-Hydroxydilevalol-3-β-D-glucuronide
Probably formed from: 3-Hydroxydilevalol
4″-Hydroxydilevalol-β-D-glucuronide
Probably formed from: 4″-Hydroxydilevalol
2-Hydroxydimetacrine
Formed from: Dimetacrine
3′-Hydroxy-2′,6′-dimethoxyacetophenone
Formed from: 2′,6′-Dimethoxyacetophenone
4′-Hydroxy-3′,5′-dimethoxyacetophenone
(acetosyringone)
Probably formed from: 3′,4′-Dimethoxy-5′-methoxyacetophenone
→ Free radical
In: Coriolus K677
Pleurotus H378
→ 3′,4′,5′-Trimethoxyacetophenone
In: Phanerochaete G782
ω-Hydroxy-3′,4′-dimethoxyacetophenone
Formed from: 3,4-Dimethoxyphenylglycol
2-Hydroxy-4,5-dimethoxyallylbenzene
Formed from: 3,4-Dimethoxyallylbenzene
3-Hydroxy-4,5-dimethoxyallylbenzene
Formed from: Elemicin
4-Hydroxy-2,5-dimethoxyamphetamine
Formed from: 2,5-Dimethoxy-4-propoxyamphetamine
4-Hydroxy-3,5-dimethoxybenzaldehyde
(syringaldehyde)
Formed from: 1,3-Dihydroxy-1,2-bis(4-hydroxy-3,5-dimethoxyphenyl)propane
Syringylglycerol β-guaiacyl ether
Probably formed from: 4-Hydroxy-3,5-dimethoxy-β-oxophenylpropionic acid
→ 3,6-Dihydroxy-2,4-dimethoxybenzaldehyde
In: Burkholderia K107
→ 2,6-Dimethoxyquinol
Probably in: Coriolus E547, E572
→ 4-Formyl-6-methoxy-*o*-quinone
Probably in: Coriolus E572
→ 4-Hydroxy-3,5-dimethoxybenzoic acid
In: Burkholderia K107, K655
Desulphovibrio F666
Serratia G197
Streptomyces C284
→ 3,4,5-Trimethoxybenzaldehyde
In: Phanerochaete G782
4-Hydroxy-3,5-dimethoxybenzaldehyde ((4-hydroxy-3,5-dimethoxyphenyl)methylene)hydrazone
See: Syringaldazine
3-(3-Hydroxy-4,5-dimethoxybenzamido)piperidine
Formed from: 3-(3,4,5-Trimethoxybenzamido) piperidine
3-(4-Hydroxy-3,5-dimethoxybenzamido)piperidine
Formed from: 3-(3,4,5-Trimethoxybenzamido) piperidine
→ 4-Hydroxy-3,5-dimethoxybenzoic acid
Probably in: Dog B357
Guinea pig B357
Mouse B357
Rat B357

3-Hydroxy-4,5-dimethoxybenzoic acid
Probably formed from: 3,4-Dihydroxy-5-methoxybenzoic acid

4-Hydroxy-3,5-dimethoxybenzoic acid
(syringic acid)
Formed from: 4-Hydroxy-3,5-dimethoxybenzaldehyde
3,4,5-Trimethoxybenzoic acid
Probably formed from: 3,4-Dihydroxy-5-methoxybenzoic acid
2-(5-Formyl-2-hydroxy-3-methoxyphenyl)-3-hydroxy-1-(4-hydroxy-3,5-dimethoxyphenyl)-1-propanone
3-(4-Hydroxy-3,5-dimethoxybenzamido) piperidine
→ 4-(4-Bromo-2-chlorophenoxy)-2,6-dimethoxyphenol
In: Rhizoctonia D690
→ 4-(4-(4-Bromo-2-chlorophenoxy)-2,6-dimethoxyphenoxy)-2,6-dimethoxyphenol
In: Rhizoctonia D690
→ 5-(4-Bromo-2-chlorophenoxy)-3-methoxy-*o*-quinone
In: Rhizoctonia D690
→ 4-(*p*-Chlorophenoxy)-2,6-dimethoxyphenol
In: Rhizoctonia D690
→ 4-(4-(*p*-Chlorophenoxy)-2,6-dimethoxyphenoxy)-2,6-dimethoxyphenol
In: Rhizoctonia D690
→ 4-(4-(4-(*p*-Chlorophenoxy)-2,6-dimethoxyphenoxy)-2,6-dimethoxyphenoxy)-2,6-dimethoxyphenol
In: Rhizoctonia D690
→ 4-(4-(4-(4-(*p*-Chlorophenoxy)-2,6-dimethoxyphenoxy)-2,6-dimethoxyphenoxy)-2,6-dimethoxyphenoxy)-2,6-dimethoxyphenol
In: Rhizoctonia D690
→ 5-(*p*-Chlorophenoxy)-3-methoxy-*o*-quinone
In: Rhizoctonia D690
→ 4-(2,4-Dichlorophenoxy)-2,6-dimethoxyphenol
In: Rhizoctonia D690
→ 4-(2,6-Dichlorophenoxy)-2,6-dimethoxyphenol
In: Rhizoctonia D690
→ 4-(4-(2,4-Dichlorophenoxy)-2,6-dimethoxyphenoxy)-2,6-dimethoxyphenol
In: Rhizoctonia D690
→ 4-(4-(2,6-Dichlorophenoxy)-2,6-dimethoxyphenoxy)-2,6-dimethoxyphenol
In: Rhizoctonia D690
→ 4-(4-(4-(2,4-Dichlorophenoxy)-2,6-dimethoxyphenoxy)-2,6-dimethoxyphenoxy)-2,6-dimethoxyphenol
In: Rhizoctonia D690
→ 4-(4-(4-(2,6-Dichlorophenoxy)-2,6-dimethoxyphenoxy)-2,6-dimethoxyphenoxy)-2,6-dimethoxyphenol
In: Rhizoctonia D690
→ 4-(4-(4-(4-(2,4-Dichlorophenoxy)-2,6-dimethoxyphenoxy)-2,6-dimethoxyphenoxy)-2,6-dimethoxyphenoxy)-2,6-dimethoxyphenol
In: Rhizoctonia D690
→ 4-(4-(4-(4-(2,6-Dichlorophenoxy)-2,6-dimethoxyphenoxy)-2,6-dimethoxyphenoxy)-2,6-dimethoxyphenoxy)-2,6-dimethoxyphenol
In: Rhizoctonia D690
→ 5-(2,4-Dichlorophenoxy)-3-methoxy-*o*-quinone
In: Rhizoctonia D690
→ 5-(2,6-Dichlorophenoxy)-3-methoxy-*o*-quinone
In: Rhizoctonia D690
→ 3,4-Dihydroxy-5-methoxybenzoic acid
In: Acetobacterium J667
Azotobacter A3308
Fomes D453
Microorganism H838
Sporotrichum D453
Pseudomonas A2399, D16, E733
See also: B501, E262
→ 4-(3,4-Dihydroxy-5-methoxyphenoxy)-3,5-dimethoxybenzoic acid
Probably in: Rhizoctonia C143
→ 2,6-Dimethoxy-4-(pentachlorophenoxy) phenol
In: Rhizoctonia D690

→ 4-(2,6-Dimethoxy-4-(pentachlorophenoxy) phenoxy)-2,6-dimethoxyphenol
In: Rhizoctonia D690
→ 4-(2,6-Dimethoxy-4-(2,6-dimethoxy-4-(pentachlorophenoxy)phenoxy)phenoxy)-2,6-dimethoxyphenol
In: Rhizoctonia D690
→ 4-(2,6-Dimethoxy-4-(2,6-dimethoxy-4-(2,6-dimethoxy-4-(pentachlorophenoxy) phenoxy)phenoxy)phenoxy)-2,6-dimethoxyphenol
In: Rhizoctonia D690
→ 2,6-Dimethoxyphenol
In: Pleurotus H612
→ 2,6-Dimethoxyquinol
In: Sporotrichum B73
See also: C143, E174, E262
→ 2,6-Dimethoxy-4-(2,3,5,6-tetrachlorophenoxy)phenol
In: Rhizoctonia D690
→ 4-(2,6-Dimethoxy-4-(2,3,5,6-tetrachlorophenoxy)phenoxy)-2,6-dimethoxyphenol
In: Rhizoctonia D690
→ 4-(4-(2,6-Dimethoxy-4-(2,3,5,6-tetrachlorophenoxy)phenoxy)-2,6-dimethoxyphenoxy)-2,6-dimethoxyphenol
In: Rhizoctonia D690
→ 4-(4-(4-(2,6-Dimethoxy-4-(2,3,5,6-tetrachlorophenoxy)phenoxy)-2,6-dimethoxyphenoxy)- 2,6-dimethoxy phenoxy)-2,6-dimethoxyphenol
In: Rhizoctonia D690
→ 2,6-Dimethoxy-4-(2,3,6-trichlorophenoxy) phenol
In: Rhizoctonia D690
→ 2,6-Dimethoxy-4-(2,4,5-trichlorophenoxy) phenol
In: Rhizoctonia D690
→ 4-(2,6-Dimethoxy-4-(2,3,6-trichlorophenoxy)phenoxy)-2,6-dimethoxyphenol
In: Rhizoctonia D690
→ 4-(2,6-Dimethoxy-4-(2,4,5-trichlorophenoxy)phenoxy)-2,6-dimethoxyphenol
In: Rhizoctonia D690
→ 4-(4-(2,6-Dimethoxy-4-(2,3,6-trichlorophenoxy)phenoxy)-2,6-dimethoxyphenoxy)-2,6-dimethoxyphenol
In: Rhizoctonia D690
→ 4-(4-(2,6-Dimethoxy-4-(2,4,5-trichlorophenoxy)phenoxy)-2,6-dimethoxyphenoxy)-2,6-dimethoxyphenol
In: Rhizoctonia D690
→ 4-(4-(4-(2,6-Dimethoxy-4-(2,3,6-trichlorophenoxy)phenoxy)-2,6-dimethoxyphenoxy)-2,6-dimethoxyphenoxy)-2,6-dimethoxyphenol
In: Rhizoctonia D690
→ 4-(4-(4-(2,6-Dimethoxy-4-(2,4,5-trichlorophenoxy)phenoxy)-2,6-dimethoxyphenoxy)- 2,6-dimethoxyphenoxy)-2,6-dimethoxyphenol
In: Rhizoctonia D690
→ 4-(4-(4-(4-(2,6-Dimethoxy-4-(2,4,5-trichlorophenoxy)phenoxy)-2,6-dimethoxyphenoxy)- 2,6-dimethoxyphenoxy)-2,6-dimethoxyphenoxy)-2,6-dimethoxyphenol
In: Rhizoctonia D690
→ Gallic acid
In: Clostridium G367
Microorganism H838
Sporomusa G931
→ 4-Hydroxy-3,5-dimethoxybenzoyl-β-D-glucose
In: Raphanus B440
→ 4-(4-Hydroxy-3,5-dimethoxyphenoxy)-3,5-dimethoxybenzoic acid
In: Rhizoctonia C143
→ 3-Methoxy-5-(2,3,6-trichlorophenoxy)-*o*-quinone
In: Rhizoctonia D690
→ 3-Methoxy-5-(2,4,5-trichlorophenoxy)-*o*-quinone
In: Rhizoctonia D690
→ Poly-4-hydroxy-3,5-dimethoxybenzoic acid
In:Bjerkandera K565
→ Syringic acid-4-β-D-glucoside
In: Nicotiana B442
→ 3,4,5-Trimethoxybenzoic acid
In: Acetobacterium F655
Petriellidium D453

Phanerochaete G782
Phialophora D453
Sporotrichum D453

4-Hydroxy-3,5-dimethoxybenzoyl-β-D-glucose
Formed from: 4-Hydroxy-3,5-dimethoxybenzoic acid

3-Hydroxy-6,12-dimethoxybenzpyrene
Formed from: 6,12-Dimethoxybenzpyrene

4-Hydroxy-3,5-dimethoxybenzyl alcohol
Formed from: Syringin

4-Hydroxy-3,5-dimethoxybenzyl alcohol-4-β-D-glucoside
See: Syringin

3-Hydroxy-2,4-dimethoxybenzylpiperazine
Formed from: Trimetazidine

4-Hydroxy-2′,4′-dimethoxychalcone
Formed from: 2′,4,4′-Trimethoxychalcone
→ 4,4′-Dihydroxy-2′-methoxychalcone
Probably in: Rat A1171

3-Hydroxy-4,5-dimethoxycinnamic acid
Formed from: 3,4,5-Trimethoxycinnamic acid
→ 3-Hydroxy-4,5-dimethoxyphenylpropionic acid
Probably in: Rat A745, B198

2-(4-(4-Hydroxy-3,5-dimethoxycinnamoyl)-1-piperazinyl)-N-isopropyl)acetamide
Formed from: Cinpropazide
→ 2-(4-(4-Hydroxy-3,5-dimethoxyphenylpropionyl)-1-piperazinyl)-N-isopropyl) acetamide
Probably in: Dog A1400

4-(3-Hydroxy-4,5-dimethoxycinnamoyl)-1-(N-pyrrolidinocarbonylmethyl)piperazine
Formed from: Cinepazide
→ 4-(3-Hydroxy-4,5-dimethoxycinnamoyl)-1-(N-pyrrolidinocarbonylmethyl) piperazine-β-D-glucuronide
Probably in: Dog A1789
Man A1789
Rat A1789

4-(4-Hydroxy-3,5-dimethoxycinnamoyl)-1-(N-pyrrolidinocarbonylmethyl)piperazine
Formed from: Cinepazide
→ 4-(4-Hydroxy-3,5-dimethoxycinnamoyl)-1-(N-pyrrolidinocarbonylmethyl) piperazine-β-D-glucuronide
Probably in: Dog A1789
Man A1789
Rat A1789
→ 1-(N-Pyrrolidinocarbonylmethyl)-4-(3,4,5-trihydroxycinnamoyl)piperazine
Probably in: Dog A1789
Man A1789
Rat A1789

4-(3-Hydroxy-4,5-dimethoxycinnamoyl)-1-(N-pyrrolidinocarbonylmethyl)piperazine-β-D-glucuronide
Probably formed from: 4-(3-Hydroxy-4,5-dimethoxycinnamoyl)-1-(N-pyrrolidinocarbonylmethyl)piperazine

4-(4-Hydroxy-3,5-dimethoxycinnamoyl)-1-(N-pyrrolidinocarbonylmethyl)piperazine-β-D-glucuronide
Probably formed from: 4-(4-Hydroxy-3,5-dimethoxycinnamoyl)-1-(N-pyrrolidinocarbonylmethyl)piperazine

4′-Hydroxy-5,7-dimethoxyflavanone
→ 5,7-Dimethoxycoumarin
In: Horseradish A3777

5-Hydroxy-3,7-dimethoxyflavone
Probably formed from: 5,7-Dihydroxy-3-methoxyflavone

5-Hydroxy-4′,7-dimethoxyflavone
Probably formed from: Acacetin

5-Hydroxy-7,8-dimethoxyflavone
Probably formed from: Wogonin

5-Hydroxy-4′,7-dimethoxyisoflavone
Formed from: Biochanin A

6-Hydroxy-4′,7-dimethoxyisoflavone
Formed from: 6,7-Dihydroxy-4′-methoxyisoflavanone

7-Hydroxy-4′,6-dimethoxyisoflavone
(afrormosin)
Formed from: 6,7-Dihydroxy-4′-methoxyisoflavanone
Formononetin
Probably formed from: 4′,7-Dihydroxy-6-methoxyisoflavanone

3-Hydroxy-4,5-dimethoxy-N-methylphenethylamine
Formed from: 3,4-Dihydroxy-5-methoxy-N-methylphenethylamine

4-Hydroxy-3,5-dimethoxy-N-methylphenethylamine
Formed from: 3,4-Dihydroxy-5-methoxy-N-methylphenethylamine
→ 3,4,5-Trimethoxy-N-methylphenethylamine
In: Lophophora A3600

4-Hydroxy-3,5-dimethoxy-N-methylphenylethanolamine
See: 5-Methoxymetanephrine

4-Hydroxy-3,5-dimethoxy-β-oxophenylpropionic acid
Probably formed from: 4-Hydroxy-3,5-dimethoxyphenylhydracrylic acid
→ 4-Hydroxy-3,5-dimethoxybenzaldehyde
Probably in: Trametes A2744

4-Hydroxy-3,5-dimethoxyphenethylamine
→ Mescaline
In: Lophophora A195

4-(4-Hydroxy-3,5-dimethoxyphenoxy)-3,5-dimethoxybenzoic acid
Formed from: 4-Hydroxy-3,5-dimethoxybenzoic acid

2-Hydroxy-4,5-dimethoxyphenylacetaldehyde
Probably formed from: Papaverine
→ 2-Hydroxy-4,5-dimethoxyphenylacetic acid
Probably in: Nocardia C268
→ 2-Hydroxy-4,5-dimethoxyphenylethanol
Probably in: Nocardia C268

2-Hydroxy-4,5-dimethoxyphenylacetic acid
Probably formed from: 2-Hydroxy-4,5-dimethoxyphenylacetaldehyde

2-(4-Hydroxy-3,5-dimethoxyphenyl)-3,5-bis(hydroxymethyl)-7-methoxycoumarin
Formed from: 5-Formyl-2-(4-hydroxy-3,5-dimethoxyphenyl)-3-hydroxymethyl-7-methoxycoumarin

2-Hydroxy-4,5-dimethoxyphenylethanol
Probably formed from: 2-Hydroxy-4,5-dimethoxyphenylacetaldehyde

4-Hydroxy-3,5-dimethoxyphenylhydracrylic acid
Probably formed from: Sinapic acid
→ 4-Hydroxy-3,5-dimethoxy-β-oxophenylpropionic acid
Probably in: Schizophyllum C447

3-(3-Hydroxy-4,5-dimethoxyphenyl)propan-1,2-diol
Probably formed from: 3-(3,4,5-Trimethoxyphenyl)propane-1,2-diol

3-(3-Hydroxy-4,5-dimethoxyphenyl)propionic acid
Formed from: 3,4,5-Trimethoxyphenylpropionic acid
Probably formed from: 3-(3,4-Dihydroxy-5-methoxyphenyl)propionic acid
3-Hydroxy-4,5-dimethoxycinnamic acid
→ 3-(3,4-Dihydroxy-5-methoxyphenyl) propionic acid
Probably in: Microorganism A744
→ 3,5-Dimethoxyphenylpropionic acid
In: Rat A745

3-(4-Hydroxy-3,5-dimethoxyphenyl)propionic acid
See: Dihydrosinapic acid

2-(4-(4-Hydroxy-3,5-dimethoxyphenylpropionyl)-1-piperazinyl)-N-isopropylacetamide
Probably formed from: 2-(4-(4-Hydroxy-3,5-dimethoxycinnamoyl)-1-piperazinyl)-N-isopropylacetamide

3-Hydroxy-5,6-dimethoxy-2-((piperidin-4-yl)methyl)-α-hydrindone
Probably formed from: 5,6-Dimethoxy-2-((piperidin-4-yl)methyl)-α-hydrindone

3-Hydroxy-4,5-dimethoxypropenylbenzene
Formed from: Isoelemicin

6a-Hydroxy-3,9-dimethoxypterocarpan
See: Variabilin

4-Hydroxy-3′,5′-dimethoxystilbene
See: Pterostilbene

4-Hydroxy-3,5-dimethoxystyrene (4-vinylsyringol)
Formed from: Sinapic acid
→ 4-Ethyl-2,6-dimethoxyphenol
Probably in: Brettanomyces E263

4′-Hydroxy-3′,5′-dimethylacetanilide
→ N-Acetyl-3,5-dimethyl-*p*-benzoquinone imine
In: Horseradish D267
→ 3′,5′-Dimethylacetanilide 4′-phenoxyl radical
In: Horseradish D267

***p*-Hydroxy-N,N-dimethylamphetamine**
Formed from: N,N-Dimethylamphetamine
→ *p*-Hydroxy-N,N-dimethylamphetamine-β-D-glucuronide

In: Man E783
Rat E783

_p_-Hydroxy-N,N-dimethylamphetamine-β-D-glucuronide
Formed from: *p*-Hydroxy-N,N-dimethylamphetamine

N-Hydroxy-3,4-dimethylaniline
Formed from: 3,4-Dimethylaniline

4-Hydroxy-3,5-dimethylbenzaldehyde
→ 3,6-Dihydroxy-2,4-dimethylbenzaldehyde
In: Burkholderia K107
→ 4-Hydroxy-3,5-dimethylbenzoic acid
In: Burkholderia K107

2-Hydroxy-7,12-dimethylbenzanthracene
Formed from: 7,12-Dimethylbenzanthracene

3-Hydroxy-7,12-dimethylbenzanthracene
Formed from: 7,12-Dimethylbenzanthracene

4-Hydroxy-7,12-dimethylbenzanthracene
Formed from: 7,12-Dimethylbenzanthracene

5-Hydroxy-7,12-dimethylbenzanthracene
Formed from: 7,12-Dimethylbenzanthracene

8-Hydroxy-7,12-dimethylbenzanthracene
Formed from: 7,12-Dimethylbenzanthracene

9-Hydroxy-7,12-dimethylbenzanthracene
Formed from: 7,12-Dimethylbenzanthracene

4-Hydroxy-3,5-dimethylbenzoic acid
Formed from: 4-Hydroxy-3,5-dimethylbenzaldehyde

2-(4-Hydroxy-3,5-dimethylbenzoyl)chromone
Formed from: Bucromarone

4-Hydroxy-N,N-dimethylbutyramide 2-(_p_-chlorophenoxy)isobutyrate
See: Clofibride

6-Hydroxy-1″,1″-dimethylcannabidiol-5″-oic acid
Formed from: 5″-Ethyl-1″,1″-dimethylcannabidiol
→ 6-Hydroxy-1″,1″-dimethylnorcannabidiol-4″-oic acid
Probably in: Dog F885
Rat F885

7-Hydroxy-1″,1″-dimethylcannabidiol-5″-oic acid
Formed from: 5″-Ethyl-1″,1″-dimethylcannabidiol
→ 6,7-Dihydroxy-1″,1″-dimethylcannabidiol-5″-oic acid
Probably in: Dog F885
Rat F885
→ 7-Hydroxy-1″,1″-dimethylnorcannabidiol-4″-oic acid
Probably in: Dog F885
Rat F885

7-Hydroxy-2,5-dimethylchromone
Formed from: Aloenin

4-Hydroxy-N,N′-dimethylclonidine
Formed from: N,N′-Dimethylclonidine
→ 4,5-Dihydroxy-N,N′-dimethylclonidine
Probably in: Rat B214
→ N,N′-Dimethyl-4-oxoclonidine
Probably in: Rat B214

_p_-Hydroxy-N,N′-dimethylclonidine
Formed from: N,N′-Dimethylclonidine

7-Hydroxy-3,4-dimethylcoumarin
Formed from: 7-Ethoxy-3,4-dimethylcoumarin

9-Hydroxy-2,6-dimethylellipticinium
→ 9,10-Dihydro-2,6-dimethyl-9,10-dioxoellipticinium
In: Horseradish E217

N-(_p_-(2-Hydroxy-1,1-dimethylethyl)phenyl)urea
Formed from: 4-*tert*-Butyl-(3-(2-chloroethyl)ureido)benzene
→ N-(*p*-(2-Hydroxy-1,1-dimethylethyl)phenyl)urea-β-D-glucuronide
Probably in: Mouse H936

N-(_p_-(2-Hydroxy-1,1-dimethylethyl)phenyl)urea-β-D-glucuronide
Probably formed from: N-(*p*-(2-Hydroxy-1,1-dimethylethyl)phenyl)urea

3-Hydroxy-4,4-dimethyl-1-(3,4-methylenedioxyphenyl)-1-pentene
See: Stiripentol

4-Hydroxy-3,3-dimethyl-N-(1-methyl-1-phenylethyl)butyramide
Formed from: Debromobutide
→ 3-Carboxy-3-methyl-N-(1-methyl-1-phenylethyl)butyramide
Probably in: Rat D215
→ 4-Hydroxy-3,3-dimethyl-N-(1-methyl-1-phenylethyl)butyramide-β-D-glucuronide
Probably in: Mouse D193
Rat D193

4-Hydroxy-3,3-dimethyl-N-(1-methyl-1-phenylethyl)butyramide-β-D-glucuronide
Probably formed from: 4-Hydroxy-3,3-dimethyl-N-(1-methyl-1-phenylethyl) butyramide

3-Hydroxy-2,6-dimethylnaphthalene
Formed from: 2,6-Dimethylnaphthalene

6-Hydroxy-1″,1″-dimethylnorcannabidiol-4″-oic acid
Probably formed from: 6-Hydroxy-1″,1″-dimethylcannabidiol-5″-oic acid

7-Hydroxy-1″,1″-dimethylnorcannabidiol-4″-oic acid
Probably formed from: 7-Hydroxy-1″,1″-dimethylcannabidiol-5″-oic acid

7-(7-Hydroxy-3,7-dimethyl-6-oxo-2-octenyl) coumarin
Formed from: Epoxyaurapten

4-(3-Hydroxy-4,4-dimethylpentyl)-3,5-dioxo-1,2-diphenylpyrazolidine
Formed from: Trimetazon
→ 4-(3-Hydroxy-4,4-dimethylpentyl)-1-(*p*-hydroxyphenyl)-3,5-dioxo-2-phenylpyrazolidine
Probably in: Rat A2284

4-(3-Hydroxy-4,4-dimethylpentyl)-1-(*p*-hydroxyphenyl)-3,5-dioxo-2-phenylpyrazolidine
Probably formed from: 4-(3-Hydroxy-4,4-dimethylpentyl)-3,5-dioxo-1,2-diphenyl-pyrazolidine

3-(4-Hydroxy-2,3-dimethylphenoxy)-1-(2-hydroxymethyl-2-propylamino)-2-propanol
Probably formed from: 1-*tert*-Butylamino-3-(4-hydroxy-2,3-dimethylphenoxy)-2-propanol

2-(4-Hydroxy-2,6-dimethylphenoxy)-1-methylethylamine
Formed from: 1-Methyl-2-(2,6-xylyloxy)ethylamine

4-Hydroxy-2,3-dimethylphenoxypropan-2,3-diol
Probably formed from: 2,3-Dimethyl-phenoxypropane-2,3-diol

1-(4-Hydroxy-2,3-dimethylphenoxy)propan-2-ol
Probably formed from: 1-(2,3-Dimethyl-phenoxy)propan-2-ol

N-(3-Hydroxy-2,6-dimethylphenyl)-2-(5-hydroxy-2-oxo-1-pyrrolidinyl)acetamide
Probably formed from: N-(2,6-Dimethyl-phenyl)-2-(5-hydroxy-2-oxo-1-pyrrolidinyl) acetamide
N-(2,6-Dimethylphenyl)-2-(2-oxo-1-pyrrolidinyl)acetamide

N-(3-Hydroxy-2,6-dimethylphenyl)-2-(2-oxo-1-pyrrolidinyl)acetamide
Probably formed from: N-(2,6-Dimethyl-phenyl)-2-(2-oxo-1-pyrrolidinyl)acetamide

N-(4-Hydroxy-2,6-dimethylphenyl)-2-(2-oxo-1-pyrrolidinyl)acetamide
Probably formed from: N-(2,6-Dimethyl-phenyl)-2-(2-oxo-1-pyrrolidinyl)acetamide

N-(3-Hydroxy-2,6-dimethylphenyl)-2-propylaminobutanamide
Probably formed from: 2-(Ethylpropylamino)-N-(3-hydroxy-2,6-dimethylphenyl)butanamide
→ 2-Amino-N-(3-hydroxy-2,6-dimethylphe-nyl)butanamide
Probably in: Man A3247

N-(4-Hydroxy-2,6-dimethylphenyl)-2-propylaminobutanamide
Probably formed from: 2-(Ethylpropylamino)-N-(4-hydroxy-2,6-dimethylphenyl)butanamide
→ 2-Amino-N-(4-hydroxy-2,6-dimethylphenyl)butanamide
Probably in: Man A3247

β-Hydroxy-α,N-dimethyl-α-phenylsuccinimide
Formed from: Methsuximide

N-(4-Hydroxy-2,6-dimethylphenyl)urea
Probably formed from: N-(2,6-Dimethyl-phenyl)urea

4-Hydroxy-3,5-dimethyl-N-2-picolylbenzamide
Formed from: 3,5-Dimethyl-N-2-picolylbenzamide

2-(((4-Hydroxy-3,5-dimethyl-2-pyridinyl) methyl)sulphinyl)-5-methoxy-1*H*-benzimidazole
Formed from: Omeprazole

N-(5-Hydroxy-4,6-dimethyl-2-pyrimidinyl) benzenesulphonamide
Formed from: N-(4,6-Dimethyl-2-pyrimidinyl)benzenesulphonamide

3-Hydroxy-1,1-dimethylpyrrolidinium benzoylformate
Formed from: Glycopyrrolate

3-Hydroxy-1,1-dimethylpyrrolidinium α-(2-hydroxycyclopentyl)mandelate
Formed from: Glycopyrrolate

3-Hydroxy-1,1-dimethylpyrrolidinium α-(3-hydroxycyclopentyl)mandelate
Formed from: Glycopyrrolate

3-Hydroxy-6,8-dimethylquinoline
Formed from: 6,8-Dimethylquinoline

5-Hydroxy-6,8-dimethylquinoline
Formed from: 6,8-Dimethylquinoline

7-Hydroxy-6,8-dimethylquinoline
Formed from: 6,8-Dimethylquinoline

2-Hydroxy-1,4-dimethylquinolinium
Formed from: 1,4-Dimethylquinolinium

4-Hydroxy-1,2-dimethylquinolinium
Formed from: 1,2-Dimethylquinolinium

3′-(((Hydroxydimethylsilyl)methylamino)methyl)phenol
Formed from: 4-((3-Methoxyphenyl)methyl)-2,2,6,6-tetramethyl-1-oxa-4-aza-2,6-disilacyclohexane

5-Hydroxy-N,N-dimethyltryptamine (bufotenine)
Formed from: 5-Hydroxy-N-methyltryptamine
5-Methoxy-N,N-dimethyltryptamine
→ 5-Hydroxyindole-3-acetaldehyde
In: Man A60, A2786
Rat B211, B653
→ 5-Methoxy-N,N-dimethyltryptamine
In: Beef C522
Chicken C522

5-Hydroxy-1,N-dimethyltryptamine
Formed from: 5-Hydroxy-N-methyltryptamine

Hydroxydimetindene
Formed from: Dimetindene

6-Hydroxy-2,4-dinitrobenzaldehyde
Probably formed from: 6-Hydroxy-2,4-dinitrotoluene

2-Hydroxy-3,5-dinitrobenzoic acid
→ 5-Amino-2-hydroxy-3-nitrobenzoic acid
Probably in: Veillonella A2624

2-Hydroxy-3,5-dinitrobenzyl alcohol
Formed from: 4,6-Dinitro-*o*-cresol

6-Hydroxy-2,4-dinitrotoluene
Formed from: 2,4,6-Trinitrotoluene
→ 6-Hydroxy-2,4-dinitrobenzaldehyde
Probably in: Myriophyllum K416

Hydroxy-9,9-dioxo-N-(2-oxopropyl)phenothiazine
Formed from: N-Demethyl-9,9-dioxopromethazine
Probably formed from: Hydroxypromethanon
→ Hydroxy-N-(2-hydroxypropyl)-9,9-dioxophenothiazine
Probably in: Rat A2685

Hydroxy-9,9-dioxophenothiazine
Probably formed from: 9,9-Dioxophenothiazine

Hydroxydiphacinone
Formed from: Diphacinone

Hydroxydiphenoxylic acid
Formed from: Diphenoxylic acid

α-Hydroxydiphenylacetamide
Formed from: Epicainide

α-Hydroxydiphenylacetic acid
Formed from: Epicainide

2-Hydroxydiphenylamine
Formed from: Diphenylamine

3-Hydroxydiphenylamine
Formed from: Diphenylamine

4-Hydroxydiphenylamine
Formed from: Diphenylamine
Diphenylnitrosamine
→ 2,4′-Dihydroxydiphenylamine
Probably in: Apple J494
→ 4-Hydroxydiphenylamine glucoside
In: Apple J494
→ N-Phenylquinoneimine
Probably in: Mouse F942

4-Hydroxydiphenylamine glucoside
Formed from: 4-Hydroxydiphenylamine

N-(4-Hydroxy-4,4-diphenylbutyl)-δ-aminovaleric acid
Formed from: Diphenidol

2-Hydroxy-1,2-diphenylethanone
See: Benzoin

4-Hydroxydiphenyl ether
See: 4-Hydroxyphenyl phenyl ether

2-Hydroxy-2,2-diphenylethylmethylnitrosamine
→ Benzophenone
Probably in: Rat C739
N″-Hydroxy-N,N′-diphenylguanidine
Formed from: N,N′-Diphenylguanidine
4-Hydroxydiphenylmethane
→ 4-Hydroxydiphenylmethane-β-D-glucuronide
In: Man G968
4-(4-(Hydroxy(diphenyl)methyl)piperidin-1-yl)-1-(4-(1-(hydroxymethyl)-1-methylethyl)phenyl)-1-butanol
Formed from: Terfenadine
→ 1-(4-(1-Carboxy-1-methylethyl) phenyl)-4-(4-(hydroxy(diphenyl) methyl)piperidin-1-yl)-1-butanol
In: Man H196
See also: H53, H342, J616
→ Diphenyl 4-piperidinylmethanol
In: Man H196
Hydroxydiphenylnitrosamine
Formed from: Diphenylnitrosamine
Hydroxydiphenyl sulphone
Formed from: Diphenyl sulphone
***p*-Hydroxydisopyramide**
Formed from: Disopyramide
→ 4-Hydroxy-3-methoxydisopyramide
Probably in: Rat B984
Hydroxyditazole
Formed from: Ditazole
5-(2-(7-Hydroxydodecyl)phenyl)-4,6-dithianonanedioic acid
Formed from: 5-(2-Dodecylphenyl)-4,6-dithianonanedioic acid
5-(2-(8-Hydroxydodecyl)phenyl)-4,6-dithianonanedioic acid
Formed from: 5-(2-Dodecylphenyl)-4,6-dithianonanedioic acid
5-(2-(9-Hydroxydodecyl)phenyl)-4,6-dithianonanedioic acid
Formed from: 5-(2-Dodecylphenyl)-4,6-dithianonanedioic acid
5-(2-(10-Hydroxydodecyl)phenyl)-4,6-dithianonanedioic acid
Formed from: 5-(2-Dodecylphenyl)-4,6-dithianonanedioic acid
5-(2-(11-Hydroxydodecyl)phenyl)-4,6-dithianonanedioic acid
Formed from: 5-(2-Dodecylphenyl)-4,6-dithianonanedioic acid
→ 5-(2-(11-Oxododecyl)phenyl)-4,6-dithianonanedioic acid
Probably in: Guinea pig E873
5-(2-(12-Hydroxydodecyl)phenyl)-4,6-dithianonanedioic acid
Formed from: 5-(2-Dodecylphenyl)-4,6-dithianonanedioic acid
→ 5-(2-(9-Carboxynonyl)phenyl)-4,6-dithianonanedioic acid
Probably in: Guinea pig E873
***p*-Hydroxydonepezil**
Formed from: Donepezil
→ *p*-Hydroxy-5-O-demethyldonepezil
Probably in: Rat J789
6-Hydroxydopa
See: 2,4,5-Trihydroxy-L-phenylalanine
5-Hydroxydopazinol
Probably formed from: Dopazinol
8-Hydroxydopazinol
Probably formed from: Dopazinol
4′-Hydroxydoxaminol
Formed from: Doxaminol
→ 4′-Hydroxydoxaminol-β-D-glucuronide
Probably in: Dog F551
4′-Hydroxydoxaminol-β-D-glucuronide
Probably formed from: 4′-Hydroxydoxaminol
2-Hydroxydoxepin
Formed from: Doxepin
→ 2-Hydroxydoxepin-β-D-glucuronide
In: Man F609
Rat F609
3-Hydroxydoxepin
Formed from: Doxepin
→ 3-Hydroxydoxepin-β-D-glucuronide
In: Man F609
Rat F609
2-Hydroxydoxepin-β-D-glucuronide
Formed from: 2-Hydroxydoxepin
3-Hydroxydoxepin-β-D-glucuronide
Formed from: 3-Hydroxydoxepin
13-Hydroxydoxorubicin
Formed from: Doxorubicin

***p*-Hydroxydoxylamine**
Formed from: Doxylamine
→ *p*-Hydroxydoxylamine-β-D-glucuronide
Probably in: Rat E11
→ 1-(Hydroxyphenyl)-1-(2-pyridyl)ethanol
Probably in: Rat E11

***p*-Hydroxydoxylamine-β-D-glucuronide**
Probably formed from: *p*-Hydroxydoxylamine

Hydroxydrobuline
Formed from: Drobuline
→ Hydroxymethoxydrobuline
Probably in: Dog A3726

7-Hydroxyefavirenz
Formed from: Efavirenz
→ 7,14-Dihydroxyefavirenz
Probably in: Cynomolgus monkey J775
Guinea pig J775
Man J775
→ 7-Hydroxyefavirenz-7-β-D-glucuronide
In: Cynomolgus monkey J775
Guinea pig J775
Hamster J775
Man J775
→ 7-Hydroxyefavirenz-7-sulphate
In: Cynomolgus monkey J775
Guinea pig J775
Hamster J775
Man J775

8-Hydroxyefavirenz
Formed from: Efavirenz
→ 8,14-Dihydroxyefavirenz
Probably in: Cynomolgus monkey J775
Hamster J775
Guinea pig J775
Man J775
Rat J765, J775
→ 8-Hydroxyefavirenz-8-β-D-glucuronide
In: Cynomolgus monkey J775
Guinea pig J775
Hamster J775
Man J775
Rat J775
→ 8-Hydroxyefavirenz-8-sulphate
In: Cynomolgus monkey J775
Guinea pig J775
Hamster J775
Man J775
Rat J775

7-Hydroxyefavirenz-7-β-D-glucuronide
Formed from: 7-Hydroxyefavirenz

8-Hydroxyefavirenz-8-β-D-glucuronide
Formed from: 8-Hydroxyefavirenz

7-Hydroxyefavirenz-7-sulphate
Formed from: 7-Hydroxyefavirenz

8-Hydroxyefavirenz-8-sulphate
Formed from: 8-Hydroxyefavirenz

7-Hydroxyellipticine
Formed from: Ellipticine

8-Hydroxyellipticine
Formed from: Ellipticine

9-Hydroxyellipticine
Formed from: Ellipticine
9-Methoxyellipticine
→ Ellipticin-9-yl-β-D-glucuronide
In: Mouse B688
Rat A3458, E3459
→ Ellipticin-9-yl sulphate
In: Rat A3458, E3459

10-Hydroxyelymoclavine
Formed from: Elymoclavine

ω-Hydroxyemodin
→ ω-Hydroxycatenarin
In: Aspergillus C334

4-Hydroxyenclomiphene
Formed from: Enclomiphene

***p*-Hydroxy-D-ephedrine**
→ *p*-Hydroxy-D-ephedrine-β-D-glucuronide
Probably in: Rat A697

***p*-Hydroxy-L-ephedrine** (4-hydroxy-α-((1*R*)-1-(methylamino)ethyl)benzenemethanol)
Formed from: L-Ephedrine
→ *p*-Hydroxy-L-ephedrine-β-D-glucuronide
Probably in: Rat A697, A2071

***p*-Hydroxy-D-ephedrine-β-D-glucuronide**
Probably formed from: *p*-Hydroxy-D-ephedrine

***p*-Hydroxy-L-ephedrine-β-D-glucuronide**
Probably formed from: *p*-Hydroxy-L-ephedrine

17-Hydroxy-19-*epi*-ajmalicine
Formed from: 17-Hydroxy-19-*epi*-cathenamine

***o*-Hydroxyepicainide**
Probably formed from: Epicainide

***m*-Hydroxyepicainide**
Probably formed from: Epicainide

***p*-Hydroxyepicainide**
Probably formed from: Epicainide

17-Hydroxy-19-*epi*-cathenamine
→ 17-Hydroxy-19-*epi*-ajmalicine
In: Catharanthus B485

2-Hydroxy-16-epioestriol
Probably formed from: Epioestriol
→ 2-Hydroxy-16-epioestriol-2-sulphate
Probably in: Rat A3102
→ 2-Methoxy-16-epioestriol
Probably in: Rat A3102

2-Hydroxy-16-epioestriol-2-sulphate
Probably formed from: 2-Hydroxy-16-epioestriol

2-Hydroxyequilenin
Formed from: Equilenin

4-Hydroxyequilenin
Formed from: Equilenin

9-Hydroxyergotamine
Formed from: Ergotamine
→ 8,9-Dihydroxyergotamine
Probably in: Beef K587

(keto)-1-Hydroxyestalozam
Formed from: Estalozam

4-Hydroxyestalozam
Formed from: Estalozam
→ 4-Oxoestalozam
Probably in: Dog D651
Man D661

***p*-Hydroxyestalozam**
Formed from: Estalozam

1′-Hydroxyestragole
See: 1-Hydroxy-1-(*p*-methoxyphenyl)prop-2-ene

***p*-(2-Hydroxyethoxy)acetanilide**
Probably formed from: *p*-(2-Hydroxybutoxy)acetanilide
→ *p*-(Carboxymethoxy)acetanilide
In: Cunninghamella F590

2-(((10-(2-Hydroxyethoxy)-9-anthracenyl)methyl)amino)-2-hydroxymethylacetic acid
Formed from: 2-(((10-(2-Hydroxyethoxy)-9-anthracenyl)methyl)amino)-1,3-propanediol
→ 2-(((10-(2-Hydroxyethoxy)-9-anthracenyl)methyl)amino)malonic acid
Probably in: Rat G522

2-(((10-(2-Hydroxyethoxy)-9-anthracenyl)methyl)amino)malonic acid
Probably formed from: 2-(((10-(2-Hydroxyethoxy)-9-anthracenyl)methyl)amino)-2-hydroxymethylacetic acid

2-(((10-(2-Hydroxyethoxy)-9-anthracenyl)methyl)amino)-1,3-propanediol
→ 9-Carboxy-10-(2-hydroxyethoxy)anthracene
In: Rat G522
→ 2-(((10-Carboxymethoxy)-9-anthracenyl)methyl)amino)-1,3-propanediol
In: Rat G522
→ 2-(((10-(2-Hydroxyethoxy)-9-anthracenyl)methyl)amino)-2-hydroxymethylacetic acid
In: Rat G522
→ 2-(((10-(2-Hydroxyethoxy)-9-anthracenyl)methyl)amino)-1,3-propanediol-β-D-glucuronide
In: Rat G522

2-(((10-(2-Hydroxyethoxy)-9-anthracenyl)methyl)amino)-1,3-propanediol-β-D-glucuronide
Formed from: 2-(((10-(2-Hydroxyethoxy)-9-anthracenyl)methyl)amino)-1,3-propanediol

3-(2-Hydroxyethoxy)carbonyl-6-hydroxymethyl-5-isopropoxycarbonyl-2-methyl-4-(*m*-nitrophenyl)pyridine
Probably formed from: 3-(2-Hydroxyethoxy)carbonyl-5-isopropoxycarbonyl-2,6-dimethyl-4-(*m*-nitrophenyl)pyridine
→ 4-(*m*-Aminophenyl)-3-(2-hydroxyethoxy)carbonyl-6-hydroxymethyl-5-isopropoxycarbonyl-2-methylpyridine
Probably in: Dog G189
Monkey G189

3-(2-Hydroxyethoxy)carbonyl-5-isopropoxycarbonyl-2,6-dimethyl-4-(*m*-nitrophenyl)pyridine
Probably formed from: 1,4-Dihydro-3-(2-hydroxyethoxy)carbonyl-5-isopropoxycarbonyl-2,6-dimethyl-4-(*m*-nitrophenyl)pyridine

→ 4-(*m*-Aminophenyl)-3-(2-hydroxyethoxy) carbonyl-5-isopropoxycarbonyl-2,6-dimethylpyridine
Probably in: Dog G189
Monkey G189
→ 3-Carboxy-5-(2-hydroxyethoxy)carbonyl-2,6-dimethyl-4-(*m*-nitrophenyl)pyridine
Probably in: Dog G189
Monkey G189
→ 3-(2-Hydroxyethoxy)carbonyl-6-hydroxymethyl-5-isopropoxycarbonyl-2-methyl-4-(*m*-nitrophenyl)pyridine
Probably in: Dog G189
Monkey G189

4-(2-Hydroxyethoxy)catechol
Formed from: *p*-(2-Hydroxyethoxy)phenol
→ 4-(2-Hydroxyethoxy)-*o*-quinone
In: Mushroom H907

N-(6-(2-Hydroxyethoxy)-5-(2-hydroxyphenoxy)-(2,2′-bipyrimidin)-4-yl)-4-(2-hydroxymethyl-2-propyl)benzenesulphonamide
Probably formed from: N-(6-(2-Hydroxyethoxy)-5-(2-methoxyphenoxy)-(2,2′-bipyrimidin)-4-yl)-4-(2-hydroxymethyl-2-propyl)benzenesulphonamide

N-(6-(2-Hydroxyethoxy)-5-(2-methoxyphenoxy)-(2,2′-bipyrimidin)-4-yl)-4-(2-hydroxymethyl-2-propyl)benzenesulphonamide
Formed from: Bosentan
→ N-(6-(2-Hydroxyethoxy)-5-(2-hydroxyphenoxy)-(2,2′-bipyrimidin)-4-yl)-4-(2-hydroxymethyl-2-propyl)benzenesulphonamide
Probably in: Man J732

***p*-(Hydroxyethoxy)phenol**
→ 4-(2-Hydroxyethoxy)catechol
In: Mushroom H907

N-(4-(2-Hydroxyethoxy)phenyl)acetamide
Formed from: *p*-Ethoxyacetanilide
→ *p*-Acetamidophenoxyacetic acid
In: Rabbit D355
See also: A3024

***trans*-1-(*p*-(*β*-Hydroxyethoxy)phenyl)-1,2-diphenylbut-1-ene**
Formed from: Tamoxifen

***trans*-1-(*p*-(2-Hydroxyethoxy)phenyl)-1-(*m*-hydroxyphenyl)-2-phenylbut-1-ene**
Formed from: Droloxifene
→ *trans*-1-(*p*-(2-Hydroxyethoxy)phenyl)-1-(*m*-hydroxyphenyl)-2-phenylbut-1-ene-β-D-glucuronide
Probably in: Rat H924

***trans*-1-(*p*-(β-2-Hydroxyethoxy)phenyl)-1-(*m*-hydroxyphenyl)-2-phenylbut-1-ene-β-D-glucuronide**
Probably formed from: *trans*-1-(*p*-(2-Hydroxyethoxy)phenyl)-1-(*m*-hydroxyphenyl)-2-phenylbut-1-ene

3-(3-(β-Hydroxyethoxy)propionamido)-2,4,6-triiodobenzoic acid
Formed from: Iodoxamic acid

4-(2-Hydroxyethoxy)-*o*-quinone
Formed from: 4-(2-Hydroxyethoxy)catechol

5-(2-Hydroxyethyl)amino-2,4-dinitrobenzamide
Formed from: 5-(1-Aziridinyl)-2,4-dinitrobenzamide

2-(2-Hydroxyethyl)amino-4,5-diphenyloxazole
Formed from: Ditazole
→ 2-Amino-4,5-diphenyloxazole
In: Man F30
→ Hydroxy-2-(2-hydroxyethyl)amino-4,5-diphenyloxazole
Probably in: Man F30

3-(2-(2-Hydroxyethyl)amino)ethylamino-6-(*p*-hydroxyphenyl)-4-methylpyridazine
Probably formed from: 3-(2-(2-Hydroxyethyl)amino)ethylamino-4-methyl-6-phenylpyridazine

3-(2-(2-Hydroxyethyl)amino)ethylamino)-4-methyl-6-phenylpyridazine
Formed from: Minaprine
→ 3-(2-(2-Hydroxyethyl)amino)ethylamino-6-(*p*-hydroxyphenyl)-4-methylpyridazine
Probably in: Baboon B839
Dog B839
Macaca B839
Man B839
Marmoset B839
Rat B839

***o*-(2-β-Hydroxyethylamino)phenyl *p*-hydroxyphenyl ether**
Probably formed from: *o*-(2-(N-2,3-Dioxomorpholino)ethoxy)phenyl phenyl ether

7-(α-Hydroxyethyl)benz[*a*]anthracene
Formed from: 7-Ethylbenz[*a*]anthracene
→ 3,4-Dihydro-3,4-dihydroxy-7-(α-hydroxyethyl)benz[*a*]anthracene
Probably in: Rat E903
→ 8,9-Dihydro-8,9-dihydroxy-7-(α-hydroxyethyl)benz[*a*]anthracene
Probably in: Rat E903
→ 10,11-Dihydro-10,11-dihydroxy-7-(α-hydroxyethyl)benz[*a*]anthracene
Probably in: Rat E903

***p*-(1-Hydroxyethyl)benzenesulphonyl-2-methylpiperidine**
Formed from: *p*-Ethylbenzenesulphonyl-2-methylpiperidine

***p*-(1-Hydroxyethyl)benzoyl-N-methylbutylamine**
Formed from: *p*-Ethylbenzoyl-N-methylbutylamine

N-(2-Hydroxyethyl)cinnamide
→ Cinnamic acid
In: Man A1539
Rat A2198
→ Cinnamide
In: Man A1539
Rat A2198

2-(1-Hydroxyethyl)chromone
Formed from: 2-Acetylchromone

3-(1-Hydroxyethyl)coumarin
Formed from: 3-Acetylcoumarin

6-(1-Hydroxyethyl)-2,2-dimethyl-2*H*-benzo[*b*]pyran
Formed from: 7-Demethoxyencecalin
→ 6-(1-Hydroxyethyl)-2-hydroxymethyl-2-methyl-2*H*-benzo[*b*]pyran
Probably in: Melanopus E491

2-Hydroxyethyl 5-(5,5-dimethyl-2-oxo-1,3,2-dioxaphorinan-2-yl)-1,4-dihydro-2,6-dimethyl-4-(*m*-nitrophenyl)-3-pyridinecarboxylate
Formed from: Efonidipine
Probably formed from: 2-Aminoethyl 5-(5,5-dimethyl-2-oxo-1,3,2-dioxaphorinan-2-yl)-1,4-dihydro-2,6-dimethyl-4-(*m*-nitrophenyl)-3-pyridinecarboxylate
→ 2-Hydroxyethyl 5-(5,5-dimethyl-2-oxo-1,3,2-dioxaphorinan-2-yl)-2,6-dimethyl-4-(*m*-nitrophenyl)-3-pyridinecarboxylate
Probably in: Man H523
Rat H625

2-Hydroxyethyl 5-(5,5-dimethyl-2-oxo-1,3,2-dioxaphorinan-2-yl)-2,6-dimethyl-4-(*m*-nitrophenyl)-3-pyridinecarboxylate
Probably formed from: 2-Hydroxyethyl 5-(5,5-dimethyl-2-oxo-1,3,2-dioxaphorinan-2-yl)-1,4-dihydro-2,6-dimethyl-4-(*m*-nitrophenyl)-3-pyridinecarboxylate

N-(2-Hydroxyethyl)-2,4-dinitroaniline
Formed from: N-(1-Aziridinyl)-2,4-dinitrobenzene

β-Hydroxyethylflufenamate
Probably formed from: Etofenamate
→ 5-Hydroxy-β-hydroxyethylflufenamate
Probably in: Dog C238

2-(1-Hydroxyethyl)-7-(2-hydroxy-3-isopropylaminopropoxy)benzofuran
Formed from: Befunolol

6-(1-Hydroxyethyl)-2-hydroxymethyl-7-methoxy-2-methyl-2*H*-benzo[*b*]pyran
Probably formed from: Encecalin
6-(1-Hydroxyethyl)-7-methoxy-2,2-dimethyl-2*H*-benzo[*b*]pyran

6-(1-Hydroxyethyl)-2-hydroxymethyl-2-methyl-2*H*-benzo[*b*]pyran
Probably formed from: 6-(1-Hydroxyethyl)-2,2-dimethyl-2*H*-benzo[*b*]pyran

***p*-((2-Hydroxyethylidene)amino)phenylacetic acid**
Probably formed from: *p*-(N,N-Bis(2-hydroxyethyl)amino)phenylacetic acid

4-(*p*-((2-Hydroxyethylidene)amino)phenyl)butyric acid
Probably formed from: 4-(*p*-(N,N-Bis(2-hydroxyethyl)amino)phenyl)butyric acid

N-(2-Hydroxyethyl)indole-3-carboxamide
Formed from: Camalexine

(*S*)-4-((3-(2-Hydroxyethyl)-1*H*-indol-5-yl)methyl)-2-oxazolidine
Probably formed from: (*S*)-4-((3-Formylmethyl-1*H*-indol-5-yl)methyl)-2-oxazolidine

10-(2-Hydroxyethyl)isoalloxazine
Formed from: Riboflavine

6-(1-Hydroxyethyl)-7-methoxy-2,2-dimethyl-2*H*-benzo[*b*]pyran
Probably formed from: Encecalin

→ 6-(1-Hydroxyethyl)-2-hydroxymethyl-7-methoxy-2-methyl-2*H*-benzo[*b*]pyran
Probably in: Melanopus E491

2-Hydroxyethyl methyl 1,4-dihydro-2,6-dimethyl-4-(*m*-nitrophenyl)pyridine-3,5-dicarboxylate
Probably formed from: Methyl 2-oxoethyl 1,4-dihydro-2,6-dimethyl-4-(*m*-nitrophenyl) pyridine-3,5-dicarboxylate
→ 2-Hydroxyethyl methyl 2,6-dimethyl-4-(*m*-nitrophenyl)pyridine-3,5-dicarboxylate
Probably in: Man D889

2-Hydroxyethyl methyl 2,6-dimethyl-4-(*m*-nitrophenyl)pyridine-3,5-dicarboxylate
Probably formed from: 2-Hydroxyethyl methyl 1,4-dihydro-2,6-dimethyl-4-(*m*-nitrophenyl)pyridine-3,5-dicarboxylate

5-Hydroxyethyl-3-methyl-5-phenylhydantoin
Formed from: 5-Ethyl-3-methyl-5-phenylhydantoin

4-(1-Hydroxyethyl)-3-methyl-1-phenyl-3-pyrazolin-5-one
Probably formed from: 4-Isopropyl-3-methyl-1-phenyl-3-pyrazolin-5-one
2-(3-Methyl-1-phenyl-3-pyrazolin-5-on-4-yl)propionic acid

3-O-(2-Hydroxyethyl)morphine
Formed from: Pholcodine
→ 3-O-(2-Hydroxyethyl)normorphine
Probably in: Man F682

N-(2-Hydroxyethyl)-N′-(α-(1-naphthyl)propionyloxy-2-ethyl)piperazine
Formed from: Nafiverine

N-(2-Hydroxyethyl)norfenfluramine
Formed from: N-(2-Benzoyloxyethyl)norfenfluramine
→ N-(2-Hydroxyethyl)norfenfluramine-β-D-glucuronide
In: Man A427
→ N-(2-Hydroxyethyl)norfenfluramine-O-sulphate
In: Man A427

N-(2-Hydroxyethyl)norfenfluramine-β-D-glucuronide
Formed from: N-(2-Hydroxyethyl)norfenfluramine

N-(2-Hydroxyethyl)norfenfluramine-O-sulphate
Formed from: N-(2-Hydroxyethyl)norfenfluramine

3-O-(2-Hydroxyethyl)normorphine
Probably formed from: 3-O-(2-Hydroxyethyl)morphine

5-(2-Hydroxyethyl)oximino-5-(*p*-trifluoromethylphenyl)valeric acid
Probably formed from: 5-(2-Aminoethyl)oximino-5-(*p*-trifluoromethylphenyl)valeric acid

***o*-(α-Hydroxyethyl)phenol** (1-(*o*-hydroxyphenyl)ethanol)
Formed from: O,O-Bis(*o*-ethylphenyl) O-phenyl phosphate
O-(*o*-Ethylphenyl) O,O-diphenyl phosphate

1-(1-Hydroxyethyl)phenoxathiin-10,10-dioxide
Formed from: 1-Ethylphenoxathiin-10,10-dioxide
→ 1-Carboxymethylphenoxathiin-10,10-dioxide
Probably in: Man H513
→ 1-(1,2-Dihydroxyethyl)phenoxathiin-10,10-dioxide
Probably in: Man H513

1-(2-Hydroxyethyl)phenoxathiin-10,10-dioxide
Formed from: 1-Ethylphenoxathiin-10,10-dioxide

3-(4-(2-Hydroxyethyl)phenoxy)-1-isopropylaminopropan-2-ol
Formed from: Metoprolol
→ 3-(4-Carboxymethylphenoxy)-1-isopropylaminopropan-2-ol
Probably in: Dog A2255, G41
Rat A1750, A2255
Man A2255, F794

1-(*p*-(1-Hydroxyethyl)phenyl)-3,3-dimethyltriazene
Formed from: 1-(*p*-Acetylphenyl)-3,3-dimethyltriazene

2-(1-Hydroxyethyl)-2-phenylglutarimide
Formed from: Glutethimide
→ 4-Hydroxy-2-(1-hydroxyethyl)-2-phenylglutarimide
Probably in: Man A2313

2-(2-Hydroxyethyl)-2-phenylglutarimide
Formed from: Glutethimide

3-(2-(1-Hydroxyethyl)phenyl)-5-(4-hydroxy-3-methoxyphenyl)-1*H*-1,2,4-triazole
Probably formed from: 3-(2-Ethylphenyl)-5-(4-hydroxy-3-methoxyphenyl)-1*H*-1,2,4-triazole
→ 3-(2-(1-Hydroxyethyl)phenyl)-5-(4-hydroxy-3-methoxyphenyl)-1*H*-1,2,4-triazole-β-D-glucuronide
Probably in: Rat C871

3-(2-(1-Hydroxyethyl)phenyl)-5-(4-hydroxy-3-methoxyphenyl)-1*H*-1,2,4-triazole-β-D-glucuronide
Probably formed from: 3-(2-(1-Hydroxyethyl)phenyl)-5-(4-hydroxy-3-methoxyphenyl)-1*H*-1,2,4-triazole

3-(2-(1-Hydroxyethyl)phenyl)-5-(*m*-hydroxyphenyl)-1*H*-1,2,4-triazole
Probably formed from: 3-(2-Ethylphenyl)-5-(*m*-hydroxyphenyl)-1*H*-1,2,4-triazole

O-(β-Hydroxyethyl)phloroglucinol
Probably formed from: 7-O-(β-Hydroxyethyl)rutin aglycone

2-(1-Hydroxyethyl)promazine
Formed from: 2-Acetylpromazine
→ 2-(1-Hydroxyethyl)promazine sulphoxide
Probably in: Horse B609

2-(1-Hydroxyethyl)promazine sulphoxide
Probably formed from: 2-(1-Hydroxyethyl)promazine

2-(5-(1-Hydroxyethyl)pyridin-2-yl)benzimidazole
Formed from: 2-(5-Ethylpyridin-2-yl)benzimidazole
→ 2-(5-Acetylpyridin-2-yl)benzimidazole
Probably in: Rat E716, F620

5-((4-(2-(5-(2-Hydroxyethyl)-2-pyridinyl)ethoxy)phenyl)methyl)-2,4-thiazolidinedione
Formed from: Pioglitazone
→ 5-((4-(2-(5-Carboxymethyl-2-pyridinyl)ethoxy)phenyl)methyl)-2,4-thiazolidinedione
Probably in: Rat G946
→ 5-((4-(2-(5-(2-Hydroxyethyl)-2-pyridinyl)ethoxy)phenyl)methyl)-2,4-thiazolidinedione-β-D-glucuronide
In: Rat G946

5-((4-(2-Hydroxy-2-(5-ethyl-2-pyridinyl)ethoxy)phenyl)methyl)-2,4-thiazolidinedione
Formed from: Pioglitazone
→ 5-((4-(2-Hydroxy-2-(5-ethyl-2-pyridinyl)ethoxy)phenyl)methyl)-2,4-thiazolidinedione-β-D-glucuronide
In: Rat G946

5-((4-(2-(5-(2-Hydroxyethyl)-2-pyridinyl)ethoxy)phenyl)methyl)-2,4-thiazolidinedione-β-D-glucuronide
Formed from: 5-((4-(2-(5-(2-Hydroxyethyl)-2-pyridinyl)ethoxy)phenyl)methyl)-2,4-thiazolidinedione

5-((4-(2-Hydroxy-2-(5-ethyl-2-pyridinyl)ethoxy)phenyl)methyl)-2,4-thiazolidinedione-β-D-glucuronide
Formed from: 5-((4-(2-Hydroxy-2-(5-ethyl-2-pyridinyl)ethoxy)phenyl)methyl)-2,4-thiazolidinedione

7-O-(β-Hydroxyethyl)quercetin
See: 7-O-(β-Hydroxyethyl)rutin aglycone

3-(2-Hydroxyethyl)quinazoline-2,4-dione
→ 6-Hydroxy-3-(2-hydroxyethyl)quinazoline-2,4-dione
Species unclear E202

3-(2-Hydroxyethyl)quinazolin-4-one
Formed from: 3-(2-(2-(2-Benzimidazolyl)benzyloxy)ethyl)quinazolin-4-one

4-(2-Hydroxyethyl)-*o*-quinone
Formed from: 2-(3,4-Dihydroxyphenyl)ethanol
→ 3,4-Dihydroxyphenylglycol
In: Sarcophaga F792, F793, G184
→ 4-(2-Hydroxyethyl)-*o*-quinone methide
In: Sarcophaga F793

4-(2-Hydroxyethyl)-*o*-quinone methide
Formed from: 4-(2-Hydroxyethyl)-*o*-quinone

7-O-(β-Hydroxyethyl)rutin
Formed from: Rutin
→ 4′,7-Di-O-(β-hydroxyethyl)rutin
Probably in: Rat A271
→ 7-O-(β-Hydroxyethyl)rutin aglycone
In: Microorganism A283
Mouse A2358
Rat A271
→ 7-O-(β-Hydroxyethyl)rutin-β-D-glucuronide

In: Rat A271

7-O-(β-Hydroxyethyl)rutin aglycone
Formed from: 7-O-(β-Hydroxyethyl)rutin
→ O-(β-Hydroxyethyl)phloroglucinol
Probably in: Microorganism A283
→ *m*-Hydroxyphenylacetic acid
In: Microorganism A283

7-O-(β-Hydroxyethyl)rutin-β-D-glucuronide
Formed from: 7-O-(β-Hydroxyethyl)rutin

2-(2-Hydroxyethyl-1-sulphinyl)-N-(3-(3-(piperidinomethyl)phenoxy)propyl)acetamide
Formed from: 2-(2-Hydroxyethyl-1-thio)-N-(3-(3-(piperidinomethyl)phenoxy)propyl)acetamide
→ 2-(2-Hydroxyethyl-1-sulphonyl)-N-(3-(3-(piperidinomethyl)phenoxy)propyl)acetamide
In: Rat J739

2-(2-Hydroxyethyl-1-sulphonyl)-N-(3-(3-(piperidinomethyl)phenoxy)propyl)acetamide
Formed from: 2-(2-Hydroxyethyl-1-sulphinyl)-N-(3-(3-(piperidinomethyl)phenoxy)propyl)acetamide

6-(1-Hydroxyethyl)-3-(1*H*-tetrazol-5-yl)chromone
Formed from: 6-Ethyl-3-(1*H*-tetrazol-5-yl)chromone
→ 6-Acetyl-3-(1*H*-tetrazol-5-yl)chromone
Probably in: Rat A3486
→ 6-(1,2-Dihydroxyethyl)-3-(1*H*-tetrazol-5-yl)chromone
Probably in: Rat A3486

4-(2-Hydroxyethylthio)catechol
Formed from: *p*-(2-Hydroxyethylthio)phenol
→ 4-(2-Hydroxyethylthio)-*o*-quinone
In: Mushroom H907

4-(*p*-(N-(2-Hydroxyethyl)-N-(2-thiocyano)ethyl)amino)phenyl)butyric acid
Formed from: Chlorambucil

***p*-(2-Hydroxyethylthio)phenol**
→ 4-(2-Hydroxyethylthio)catechol
In: Mushroom H907

2-(2-Hydroxyethyl-1-thio)-N-(3-(3-(piperidinomethyl)phenoxy)propyl)acetamide
→ 2-((Carboxymethyl)thio)-N-(3-(3-(piperidinomethyl)phenoxy)propyl)acetamide
In: Rat J739
→ N-(*m*-Hydroxybenzyl)piperidine
In: Rat J739
→ 2-(2-Hydroxyethyl-1-sulphinyl)-N-(3-(3-(piperidinomethyl)phenoxy)propyl)acetamide
In: Rat J739
→ 3-(*m*-((1-Piperidinyl)methyl)phenoxy)propionic acid
In: Rat J739

4-(2-Hydroxyethylthio)-*o*-quinone
Formed from: 4-(2-Hydroxyethylthio)catechol

3-(2-Hydroxyethyl)-1-(3-trifluoromethylphenyl)quinazoline-2,4(1*H*,3*H*)-dione
→ 3-Carboxymethyl-1-(3-trifluoromethylphenyl)quinazoline-2,4(1*H*,3*H*)-dione
In: Guinea pig A359
Hamster A359
Man A359
Monkey A359
Mouse A359
Rabbit A359
Rat A319, A359
→ 3-(2-Hydroxyethyl)-1-(3-trifluoromethylphenyl)quinazoline-2,4(1*H*,3*H*)-dione-β-D-glucuronide
In: Guinea pig A359
Hamster A359
Man A359
Monkey A359
Mouse A359
Rabbit A359
Rat A319, A359
→ 6-Hydroxy-3-(2-hydroxyethyl)-1-(3-trifluoromethylphenyl)quinazoline-2,4(1*H*,3*H*)-dione
In: Guinea pig A359
Hamster A359
Man A359
Monkey A359
Mouse A359
Rabbit A359
Rat A359

3-(2-Hydroxyethyl)-1-(3-trifluoromethylphenyl) quinazoline-2,4(1*H*,3*H*)-dione-β-D-glucuronide
Formed from: 3-(2-Hydroxyethyl)-1-(3-trifluoromethylphenyl)quinazoline-2,4(1*H*,3*H*)-dione

N-(2-Hydroxyethyl)-N-(3,4,5-trimethoxybenzoyl)glycine
Probably formed from: N-(3,4,5-Trimethoxybenzoyl)diethanolamine

2-Hydroxyethynyloestradiol
Formed from: Ethynyloestradiol
→ 2-Methoxyethynyloestradiol
Probably in: Pig C346

3-Hydroxyetodolac
Formed from: Etodolac

4-Hydroxyetodolac
Formed from: Etodolac

5-Hydroxyetodolac
Formed from: Etodolac

6-Hydroxyetodolac
Formed from: Etodolac

7-Hydroxyetodolac
Formed from: Etodolac
→ 7-Hydroxyetodolac-β-D-glucuronic acid
Probably in: Man D662

7-Hydroxyetodolac-β-D-glucuronic acid
Probably formed from: 7-Hydroxyetodolac

***p*-Hydroxyfebarbamate**
Formed from: Febarbamate
→ 1-(3-Butoxy-2-carbamoyloxypropyl)-5-ethyl-5-(4-hydroxy-3-methoxyphenyl)-(1*H*,3*H*,5*H*)-pyrimidine-2,4,6-trione
Probably in: Man D26
→ 1-(2-Carbamoyloxy-3-(3-hydroxybutoxy) propyl)-5-ethyl-5-(*p*-hydroxyphenyl)-(1*H*,3*H*,5*H*)-pyrimidine-2,4,6-trione
Probably in: Rat C368
→ 1-(2-Carbamoyloxy-3-hydroxypropyl)-5-ethyl-5-(*p*-hydroxyphenyl)-(1*H*,3*H*,5*H*)-pyrimidine-2,4,6-trione
Probably in: Rat C368

***p*-Hydroxyfenbendazole**
Formed from: Fenbendazole

***p*-Hydroxyfebuprol**
Formed from: Febuprol
→ *p*-Hydroxyfebuprol-O-sulphate
Probably in: Mouse A2656
Rat A2656

***p*-Hydroxyfebuprol-O-sulphate**
Probably formed from: *p*-Hydroxyfebuprol

α-Hydroxyfelbamate
Formed from: Felbamate

***p*-Hydroxyfelbamate**
Formed from: Felbamate

***p*-Hydroxyfencibuterol**
Formed from: Fencibuterol

7-Hydroxyfendosal
Formed from: Fendosal
→ 7-Hydroxyfendosal-β-D-glucuronide
In: Dog B624
Man B624
Monkey B624
Mouse B624
Rabbit B624
Rat B624

7-Hydroxyfendosal-β-D-glucuronide
Formed from: 7-Hydroxyfendosal

***p*-Hydroxyfenoctimine**
Formed from: Fenoctimine
→ 4-((3,4-Dihydroxyphenyl) phenyl)methyl-1-((octylimino) methyl)piperidine
Probably in: Dog H62

4′-Hydroxy-(*R*)-fenoprofen
Formed from: (*R*)-Fenoprofen
→ 4′-Hydroxy-(*R*)-fenoprofen-β-D-glucuronide
Probably in: Man A726

4′-Hydroxy-(*R*)-fenoprofen-β-D-glucuronide
Probably formed from: 4′-Hydroxy-(*R*)-fenoprofen

2′-Hydroxyfenvalerate
Formed from: Fenvalerate

4′-Hydroxyfenvalerate
Formed from: Fenvalerate

5-Hydroxyferulic acid
See: 5-Methoxycaffeic acid

***trans*-3-Hydroxyflavan-4α-ol**
Probably formed from: Flavan-4α-ol
→ 3,4′-Dihydroxyflavan-4α-ol
Probably in: Rat B222
→ 3,6-Dihydroxyflavan-4α-ol
Probably in: Rat B222

***trans*-3-Hydroxyflavan-4β-ol**
Probably formed from: Flavan-4β-ol
→ 3,4′-Dihydroxyflavan-4β-ol
Probably in: Rat B222
→ 3,6-Dihydroxyflavan-4β-ol
Probably in: Rat B222

4′-Hydroxyflavan-4α-ol
Probably formed from: Flavan-4α-ol
4′-Hydroxyflavanone
→ 4′-Hydroxy-3′-methoxyflavan-4α-ol
Probably in: Rat B222

4′-Hydroxyflavan-4β-ol
Probably formed from: Flavan-4β-ol

6-Hydroxyflavan-4α-ol
Probably formed from: Flavan-4α-ol

6-Hydroxyflavan-4β-ol
Probably formed from: Flavan-4β-ol

***cis*-3-Hydroxyflavanone**
Probably formed from: Flavanone

***trans*-3-Hydroxyflavanone**
Probably formed from: Flavanone

4′-Hydroxyflavanone
Probably formed from: Flavanone
→ 3′,4′-Dihydroxyflavanone
Probably in: Gibberella A1512
Rat B222
→ 4′-Hydroxyflavan-4α-ol
Probably in: Gibberella A1512

6-Hydroxyflavanone
Probably formed from: Flavanone

3-Hydroxyflavone (flavonol)
Formed from: Flavanone
Probably formed from: Flavone
→ 3-Hydroxyflavone-β-D-glucoside
In: Manduca G897
→ Phenylglyoxylic acid
In: Horseradish A343
→ Salicylic acid
In: Horseradish A343

4′-Hydroxyflavone
Formed from: Flavone
→ 2′,4-Dihydroxychalcone
Probably in: Streptomyces F673
→ 4α,4′-Dihydroxyflavan
Probably in: Absidia F673
→ 3′,4′-Dihydroxyflavone
Probably in: Absidia F504
Aspergillus F504
Cunninghamella F504
Curvularia F504
Gliocladium F504
Helicostylum F504
Linderina F504
Penicillium F504
Rhizopus F504
Sepedonium F504
Streptomyces F504

5-Hydroxyflavone
→ 4′,5-Dihydroxyflavone
Probably in: Aspergillus G102

6-Hydroxyflavone
→ 4′,6-Dihydroxyflavone
Probably in: Aspergillus G102

7-Hydroxyflavone
→ 4′,7-Dihydroxyflavone
Probably in: Aspergillus G102
→ 7,8-Dihydroxyflavone
In: Pseudomonas A1328
→ 7-Hydroxyflavone-β-D-glucoside
In: Manduca G897
→ 7-Hydroxyflavone-β-D-glucuronide
In: Man G524, G968, J416, J783

3-Hydroxyflavone-β-D-glucoside
Formed from: 3-Hydroxyflavone

7-Hydroxyflavone-β-D-glucoside
Formed from: 7-Hydroxyflavone

7-Hydroxyflavone-β-D-glucuronide
Formed from: 7-Hydroxyflavone

Hydroxyfloctafenin
Formed from: Foctafenin

4′-Hydroxyflufenamic acid
Probably formed from: Flufenamic acid
→ 4′,5-Dihydroxyflufenamic acid
Probably in: Rat C445

5-Hydroxyflufenamic acid
Probably formed from: Flufenamic acid

7-Hydroxyflumequine
Formed from: Flumequine

17-Hydroxyflumioxazin
Formed from: Flumioxazin

18-Hydroxyflumioxazin
Formed from: Flumioxazin

4-Hydroxyflunarizine
Formed from: Flunarizine
→ 4-Hydroxyflunarizine-β-D-glucuronide
In: Rat G310
→ 4-Hydroxyflunarizine-N-oxide
Probably in: Dog C773

4-Hydroxyflunarizine-β-D-glucuronide
Formed from: 4-Hydroxyflunarizine

4-Hydroxyflunarizine-N-oxide
Formed from: 4-Hydroxyflunarizine

3-Hydroxyflunitrazepam
Formed from: Flunitrazepam
→ 5-(2-Fluoro-4-hydroxyphenyl)-1,3-dihydro-3-hydroxy-1-methyl-7-nitro-2*H*-1,4-benzodiazepin-2-one
Probably in: Rat B496

1-Hydroxyfluoranthene
Formed from: Fluoranthene

3-Hydroxyfluoranthene
Formed from: Fluoranthene

8-Hydroxyfluoranthene
Formed from: Fluoranthene

Hydroxyfluoranthenic acid
Formed from: Benzo[*k*]fluoranthene

1-Hydroxyfluorene
→ Fluoren-1-yl sulphate
In: Beef A1211

9-Hydroxyfluorene
Formed from: Fluorene
Fluorene-9-hydroperoxide
Fluorenone
→ Fluorenone
In: Beijerinckia D248
Brevibacterium H112
Rabbit A2032
Rat D390
See also: G753

9-Hydroxyfluorene-1-carboxylic acid
Formed from: Fluorene-1-carboxylic acid
9-Fluorenone-1-carboxylic acid

1-Hydroxyfluorenone
Probably formed from: Fluorenone

2-Hydroxyfluorenone
Probably formed from: Fluorenone
→ 9-Fluorenon-2-yl sulphate
In: Beef A1211

4-Hydroxyfluorenone
Probably formed from: Fluorenone

9-Hydroxyfluorenyl-9-carboxylic acid
Formed from: Butyl 9-hydroxyfluorenyl-9-carboxylic acid

7-Hydroxyflupentixol
Probably formed from: Flupentixol

2-Hydroxyfluperlapine
Formed from: Fluperlapine
→ 2-Hydroxyfluperlapine-O-sulphate
Probably in: Rat E918

2-Hydroxyfluperlapine-N-oxide
Formed from: Fluperlapine-N-oxide
→ 2-Hydroxyfluperlapine-N-oxide-2-β-D-glucuronide
Probably in: Rat E918

2-Hydroxyfluperlapine-N-oxide-2-β-D-glucuronide
Probably formed from: 2-Hydroxy-fluperlapine-N-oxide

2-Hydroxyfluperlapine-O-sulphate
Probably formed from: 2-Hydroxyfluperlapine

7-Hydroxyfluphenazine
Formed from: Fluphenazine
→ 7-Hydroxyfluphenazine-β-D-glucuronide
Probably in: Rabbit F909
Rat F720

7-Hydroxyfluphenazine-β-D-glucuronide
Formed from: 7-Hydroxyfluphenazine

3′-Hydroxyflurbiprofen
Formed from: Flurbiprofen

4′-Hydroxyflurbiprofen
Formed from: Flurbiprofen
→ 3′,4′-Dihydroxyflurbiprofen
Probably in: Baboon A3482
Dog A3482
Man A3482
Mouse A3482, A3482
Pig C224
Rat A3393, A3482

2-Hydroxyflutamide
Formed from: Flutamide

4′-Hydroxyflutolanil
Formed from: Flutolanil

3-Hydroxyflutoprazepam
Formed from: Flutoprazepam

5-Hydroxyfluvastatin
Formed from: Fluvastatin
6-Hydroxyfluvastatin
Formed from: Fluvastatin
***p*-Hydroxy-*o*-fomocaine**
Formed from: *o*-Fomocaine
→ *p*-Hydroxy-*o*-fomocaine-N-oxide
Probably in: Rat B21
***p*-Hydroxy-*o*-fomocaine-N-oxide**
Probably formed from: *p*-Hydroxy-*o*-fomocaine
N-Hydroxy-2-formamidofluorene
→ N-Formoxy-2-aminofluorene
In: Dog H247
Rat G735
***p*-Hydroxyfurfenorex**
Formed from: Furfenorex
→ N-Furfuryl-*p*-hydroxyamphetamine
Probably in: Rat D658
4-Hydroxy-7*H*-furo[3,2-*g*][1]benzopyran-7-one
See: Bergaptol
4′-Hydroxygemfibrozil
Formed from: Gemfibrozil
→ 4′-Hydroxygemfibrozil-β-D-glucuronide
In: Hamster J565
Rat J564, J565
→ 4′-Hydroxygemfibrozil-O-sulphate
In: Hamster J565
Rat J564, J565
4′-Hydroxygemfibrozil-β-D-glucuronide
Formed from: 4′-Hydroxygemfibrozil
4′-Hydroxygemfibrozil-O-sulphate
Formed from: 4′-Hydroxygemfibrozil
5-Hydroxyglafenic acid
Probably formed from: 5-Hydroxyglafenine
5-Hydroxyglafenine
Formed from: Glafenine
→ 5-Hydroxyglafenic acid
Probably in: Man B216
4β-Hydroxygliclazide
Formed from: Gliclazide
6α-Hydroxygliclazide
Formed from: Gliclazide
→ 6,7-Dehydrogliclazide
Probably in: Man H488
6β-Hydroxygliclazide
Formed from: Gliclazide
→ 6β-Hydroxygliclazide-β-D-glucuronide
In: Man D169
7α-Hydroxygliclazide
Formed from: Gliclazide
7β-Hydroxygliclazide
Formed from: Gliclazide
→ 7β-Hydroxygliclazide-β-D-glucuronide
In: Man D169
6β-Hydroxygliclazide-β-D-glucuronide
Formed from: 6β-Hydroxygliclazide
7β-Hydroxyglilcazide-β-D-glucuronide
Formed from: 7β-Hydroxygliclazide
6-(4-Hydroxy-2-β-D-glucoyloxy-6-methylphenyl)-4-methoxy-2-pyrone
See: Aloenin
3-Hydroxyglutethimide
Formed from: Glutethimide
4-Hydroxyglutethimide
Formed from: Glutethimide
→ 2-Ethyl-4-hydroxy-2-(*p*-hydroxyphenyl)glutarimide
Probably in: Man A2313
N-Hydroxy-2-glycollamidifluorene
Formed from: 2-Glycollamidifluorene
***p*-Hydroxyglycollanilide** (*p*-glycollamidophenol)
→ *p*-Aminophenol
In: Microorganism A261
8-Hydroxygomisin A
Formed from: Gomisin A
9-Hydroxygomisin A
Formed from: Gomisin A
5-Hydroxygranisetron
Formed from: Granisetron
→ 5-Hydroxy-1-demethylgranisetron
Probably in: Rat H63
7-Hydroxygranisetron
Formed from: Granisetron
***ax*-5′-Hydroxygriseofulvin**
Formed from: Griseofulvin
6′-Hydroxygriseofulvin
Formed from: Griseofulvin
3-Hydroxyguanfacine
Formed from: Guanfacine
→ 3-Hydroxyguanfacine-β-D-glucuronide
In: Rat B889
→ 3-Hydroxyguanfacine-O-sulphate
In: Rat B889

4-Hydroxyguanfacine
Formed from: Guanfacine

3-Hydroxyguanfacine-β-D-glucuronide
Formed from: 3-Hydroxyguanfacine

3-Hydroxyguanfacine-O-sulphate
Formed from: 3-Hydroxyguanfacine

3-Hydroxyhalazepam
Formed from: Halazepam
→ 7-Chloro-1,3-dihydro-3-hydroxy-5-phenyl-2*H*-1,4-benzodiazepin-2-one
In: Rat F643

3-Hydroxyharmine
Probably formed from: Harmine

6-Hydroxyharmine
Probably formed from: Harmine

2-Hydroxyhexoestrol
Probably formed from: Hexoestrol

***o*-Hydroxyhippuric acid** (salicyluric acid)
Formed from: Salicylic acid
→ 2,5-Dihydroxyhippuric acid
In: Rat A3602
→ *o*-Hydroxyhippuric acid-β-D-glucuronide
Probably in: Man C374, D668
Rat F588
→ *o*-Hydroxyhippuric acid-O-sulphate
Probably in: Man A1590
→ Salicylic acid
In: Microorganism D616, E987, F543, F544, F546
Rabbit E200
Rat B56, E987, F545

***m*-Hydroxyhippuric acid**
Formed from: *m*-Hydroxybenzoic acid

***p*-Hydroxyhippuric acid**
Formed from: *p*-Hydroxybenzoic acid
→ *p*-Hydroxybenzoic acid
In: Corynebacterium E145

***o*-Hydroxyhippuric acid-β-D-glucuronide**
Probably formed from: *o*-Hydroxyhippuric acid

***o*-Hydroxyhippuric acid-O-sulphate**
Probably formed from: *o*-Hydroxyhippuric acid

6β-Hydroxyhomatropine
Formed from: Homatropine

2-Hydroxy-18-homooestriol
Formed from: 18-Homooestriol

16α-Hydroxy-18-homooestrone
Formed from: 18-Homooestriol

β-Hydroxyhomotryptophan
Formed from: Homotryptophan

Hydroxyhydralazine
Formed from: Hydralazine

***p*-Hydroxyhydratropic acid**
See: 2-(*p*-Hydroxyphenyl)propionic acid

2-Hydroxy-2-(4′-hydroxy-4-biphenylyl)propionic acid
Formed from: 3-(4-Biphenylyl)butyronitrile

1-Hydroxy-1-(4′-hydroxy-3′-coumarinyl)-1-phenyl-3-butanone
Formed from: *R*-Warfarin
S-Warfarin

2-Hydroxy-6-(10-hydroxydecyl)-3-methoxy-5-methylquinol
Formed from: Demethylidebenone
→ 2-Hydroxy-6-(10-hydroxydecyl)-3-methoxy-5-methylquinol-β-D-glucuronide
Probably in: Dog E119
Rat E119

2-Hydroxy-6-(10-hydroxydecyl)-3-methoxy-5-methylquinol-β-D-glucuronide
Probably formed from: 2-Hydroxy-6-(10-hydroxydecyl)-3-methoxy-5-methylquinol

7-Hydroxy-6-(2-hydroxyethyl)-5-methoxycoumarin
Formed from: 5-Methoxypsoralen

(—)-*cis*-3-(2-Hydroxy-4-(2-hydroxy-1,1-dimethylheptyl)phenyl)-*trans*-4-(3-hydroxypropyl)cyclohexanol
Formed from: (—)-*cis*-3-(4-(1,1-Dimethylheptyl)-2-hydroxyphenyl)-*trans*-4-(3-hydroxypropyl)cyclohexanol

(—)-*cis*-3-(2-Hydroxy-4-(3-hydroxy-1,1-dimethylheptyl)phenyl)-*trans*-4-(3-hydroxypropyl)cyclohexanol
Formed from: (—)-*cis*-3-(4-(1,1-Dimethylheptyl)-2-hydroxyphenyl)-*trans*-4-(3-hydroxypropyl)cyclohexanol

(−)-*cis*-3-(2-Hydroxy-4-(4-hydroxy-1,1-dimethylheptyl)phenyl)-*trans*-4-(3-hydroxypropyl)cyclohexanol
Formed from: (−)-*cis*-3-(4-(1,1-Dimethylheptyl)-2-hydroxyphenyl)-*trans*-4-(3-hydroxypropyl)cyclohexanol

(−)-*cis*-3-(2-Hydroxy-4-(5-hydroxy-1,1-dimethylheptyl)phenyl)-*trans*-4-(3-hydroxypropyl)cyclohexanol
Formed from: (−)-*cis*-3-(4-(1,1-Dimethylheptyl)-2-hydroxyphenyl)-*trans*-4-(3-hydroxypropyl)cyclohexanol

(−)-*cis*-3-(2-Hydroxy-4-(6-hydroxy-1,1-dimethylheptyl)phenyl)-*trans*-4-(3-hydroxypropyl)cyclohexanol
Formed from: (−)-*cis*-3-(4-(1,1-Dimethylheptyl)-2-hydroxyphenyl)-*trans*-4-(3-hydroxypropyl)cyclohexanol

Hydroxy-2-(2-hydroxyethyl)amino-4,5-diphenyloxazole
Probably formed from: 2-(2-Hydroxyethyl)amino-4,5-diphenyloxazole

7-Hydroxy-6-(1-hydroxyethyl)-2,2-dimethyl-2*H*-benzo[*b*]pyran
Formed from: Demethylencecalin

5-Hydroxy-β-hydroxyethylflufenamate
Probably formed from: β-Hydroxyethylflufenamate

5-Hydroxy-3-(2-hydroxyethyl)indole
See: 5-Hydroxytryptophol

(*S*)-2-(1-(2-(4-Hydroxy-3-(2-hydroxyethyl)phenyl)ethyl)-3-pyrrolidinyl)-2,2-diphenylacetamide
Formed from: Darifenacin
→ (*S*)-2-(1-(2-(3-Carboxymethyl-4-hydroxyphenyl)ethyl)-3-pyrrolidinyl)-2,2-diphenylacetamide
Probably in: Dog H866
Man H866
Mouse H866
Rabbit H866
Rat H866

4-Hydroxy-2-(1-hydroxyethyl)-2-phenylglutarimide
Probably formed from: 2-(1-Hydroxyethyl)-2-phenylglutarimide

6-Hydroxy-3-(2-hydroxyethyl)quinazolin-2,4-dione
Probably formed from: 3-(2-Hydroxyethyl)quinazoline-2,4-dione

6-Hydroxy-3-(2-hydroxyethyl)-1-(3-trifluoromethylphenyl)quinazolin-2,4(1*H*,3*H*)-dione
Formed from: 3-(2-Hydroxyethyl)-1-(3-trifluoromethylphenyl)quinazolin-2,4(1*H*,3*H*)-dione

4-Hydroxy-α-(((6-hydroxyhexyl)amino)methyl)-1,3-benzenedimethanol
Formed from: Salmeterol
→ α-(((5-Carboxypentyl)amino)methyl)-4-hydroxy-1,3-benzenedimethanol
Probably in: Dog G838
Man G838
Rat G838

3-Hydroxy-5-(hydroxy-(*m*-hydroxyphenyl)methyl)-1-methyl-4-phenyl-2-pyrrolidinone
Probably formed from: (+)-Clausenamide

8-Hydroxy-5-(1-hydroxy-2-isopropylaminobutyl)carbostyril
See: Procaterol

3-Hydroxy-2-(4-(1-hydroxyisopropyl)cyclohexyl)-1,4-naphthoquinone
Formed from: 2-(4-*tert*-Butylcyclohexyl)-3-hydroxy-1,4-naphthoquinone

2-Hydroxy-2-(7-hydroxy-8-methoxycoumarin-6-yl)acetic acid
Probably formed from: 2-(7-Hydroxy-8-methoxycoumarin-6-yl)acetic acid

(+)-(*R*)-2-(5-Hydroxy-2-((N-(2-((4-hydroxy-3-methoxy)phenoxy)ethyl)-N-methylamino)propoxy)phenyl)-4-methyl-2*H*-1,4-benzothiazin-3-(4*H*)-one
Probably formed from: (+)-(*R*)-2-(2-((N-(2-((4-Hydroxy-3-methoxy)phenoxy)ethyl)-N-methylamino)propoxy)phenyl)-4-methyl-2*H*-1,4-benzothiazin-3(4*H*)-one

5-(1-Hydroxy-2-((2-(4-hydroxy-2-methoxyphenoxy)ethyl)amino)ethyl)-2-methylbenzenesulphonamide
Formed from: Amosulalol

5-(1-Hydroxy-2-((2-(5-hydroxy-2-methoxyphenoxy)ethyl)amino)ethyl)-2-methylbenzenesulphonamide
Formed from: Amosulalol
→ 5-(1-Hydroxy-2-((2-(5-hydroxy-2-methoxyphenoxy)ethyl)amino)ethyl)-2-methylbenzenesulphonamide-O-sulphate
Probably in: Man D369
Monkey D369

5-(1-Hydroxy-2-((2-(5-hydroxy-2-methoxyphenoxy)ethyl)amino)ethyl)-2-methylbenzenesulphonamide-O-sulphate
Probably formed from: 5-(1-Hydroxy-2-((2-(5-hydroxy-2-methoxyphenoxy)ethyl)amino) ethyl)-2-methylbenzenesulphonamide

1-Hydroxy-4-(4-hydroxy-3-methoxyphenyl) butan-2-one
Formed from: Zingerone

4-Hydroxy-4-(4-hydroxy-3-methoxyphenyl) butan-2-one
Formed from: Zingerone

3-Hydroxy-1-(3-hydroxy-4-methoxyphenyl)-4,4-dimethyl-1-pentene
Probably formed from: 1-(3,4-Dihydroxyphenyl)-3-hydroxy-4,4-dimethyl-1-pentene
→ 3,5-Dihydroxy-1-(3-hydroxy-4-methoxyphenyl)-4,4-dimethyl-1-pentene
Probably in: Rat F574

3-Hydroxy-1-(4-hydroxy-3-methoxyphenyl)-4,4-dimethyl-1-pentene
Probably formed from: 1-(3,4-Dihydroxyphenyl)-3-hydroxy-4,4-dimethyl-1-pentene
→ 3,5-Dihydroxy-1-(4-hydroxy-3-methoxyphenyl)-4,4-dimethyl-1-pentene
Probably in: Rat F574

3-Hydroxy-5-((1*E*)-2-(3-hydroxy-4-methoxyphenyl)ethenyl)phenyl-β-D-glucuronide
See: Rhaponticin

(*R*)-4-(3-(2-Hydroxy-2-(4-hydroxy-3-methoxyphenyl)ethyl)amino-3-methylbutyl)benzamide
Probably formed from: (*R*)-4-(3-(2-(3,4-Dihydroxyphenyl)-2-hydroxyethyl)amino-3-methylbutyl)benzamide

4-Hydroxy-1-(hydroxymethoxyphenyl)-4-isopropyl-2,3-dimethyl-3-pyrazolin-5-one
Probably formed from: 4-Hydroxy-1-(hydroxyphenyl)-4-isopropyl-2,3-dimethyl-3-pyrazolin-5-one

3-Hydroxy-1-(4-hydroxy-3-methoxyphenyl)-2-(2-methoxyphenoxy)-1-propanone
Formed from: Guaiacylglycerol β-guaiacyl ether

2-Hydroxy-5-(1-hydroxy-2-((2-(4-methoxyphenyl)-1-methylethyl)amino)ethyl)formanilide
See: Formoterol

3-Hydroxy-1-(4-hydroxy-3-methoxyphenyl)-2-(2-methoxyphenoxy)-1-propanone
Formed from: Guaiacylglycerol β-guaiacyl ether

4-Hydroxy-N-(1-(4-hydroxy-3-methoxyphenyl)-1-methylethyl)-3,3-dimethylbutyramide
Probably formed from: 2-Bromo-4-hydroxy-N-(1-(4-hydroxy-3-methoxyphenyl)-1-methylethyl)-3,3-dimethylbutyramide

3-Hydroxy-1-(4-hydroxy-3-methoxyphenyl)-1-propanone
See: β-Hydroxypropiovanillone

4-Hydroxy-4′-(N-hydroxy-N-methyl) aminoazobenzene
Probably formed from: 4-Hydroxy-4′-methylaminoazobenzene

2-Hydroxy-4-hydroxymethylbenzaldehyde
(4-hydroxymethylsalicylaldehyde)
Formed from: 2-Methylnaphthalene

1-Hydroxy-8-hydroxymethylbenzanthracene
Formed from: 8-Hydroxymethylbenzanthracene

2-Hydroxy-8-hydroxymethylbenzanthracene
Formed from: 8-Hydroxymethylbenzanthracene

3-Hydroxy-8-hydroxymethylbenzanthracene
Formed from: 8-Hydroxymethylbenzanthracene

4-Hydroxy-8-hydroxymethylbenzanthracene
Formed from: 8-Hydroxymethylbenzanthracene

2-Hydroxy-5-(*p*-hydroxymethylbenzenesulphonyloxy)benzenesulphonic acid
Formed from: 2-Hydroxy-5-(*p*-toluenesulphonyloxy)benzenesulphonic acid

→ 5-(*p*-Carboxybenzenesulphonyloxy)-2-hydroxybenzenesulphonic acid
In: Man B975
Rat B975

4-Hydroxy-3-hydroxymethylbenzoic acid
Formed from: 4-Hydroxy-3-methylbenzoic acid

3-Hydroxy-6-hydroxymethyl-3,4-benzpyrene
Formed from: 6-Hydroxymethyl-3,4-benzpyrene

3-Hydroxy-7-hydroxymethyl-3,4-benzpyrene
Formed from: 7-Hydroxymethyl-3,4-benzpyrene

3-Hydroxy-N-hydroxymethylcarbazole
Probably formed from: N-Hydroxymethylcarbazole

1-Hydroxy-3-hydroxymethylcholanthrene
Formed from: 3-Hydroxymethylcholanthrene
→ 3-Hydroxymethyl-1-oxocholanthrene
Probably in: Rat H708

2-Hydroxy-3-hydroxymethylcholanthrene
Formed from: 3-Hydroxymethylcholanthrene
→ 3-Hydroxymethyl-2-oxocholanthrene
Probably in: Rat H708

8-Hydroxy-3-hydroxymethylcholanthrene
Formed from: 3-Hydroxymethylcholanthrene

2-Hydroxy-7-hydroxymethylchromene-2-carboxylic acid
Formed from: 2-Methylnaphthalene

8-Hydroxy-3-hydroxymethylchromone
Formed from: 3-Hydroxymethyl-8-methoxychromone
→ 3-Carboxy-8-methoxychromone
Probably in: Rat A1761

1-Hydroxy-5-hydroxymethylchrysene
Formed from: 5-Hydroxymethylchrysene

1-Hydroxy-6-hydroxymethylchrysene
Formed from: 1-Hydroxycrisnatol

7-Hydroxy-5-hydroxymethylchrysene
Formed from: 5-Hydroxymethylchrysene

11-Hydroxy-3-hydroxymethyl-3-demethylacronine
Probably formed from: 11-Hydroxyacronine

1*S*-(*endo*,*endo*)-N-(((3-Hydroxy-4-hydroxymethyl-7,7-dimethylbicyclo[2.2.1]hept-2-yl)amino)carbonyl)-4-methylbenzenesulphonamide
Formed from: Glibornuride

1*S*-(*endo*,*endo*)-N-(((3-Hydroxy-7-hydroxymethyl-4,7-dimethylbicyclo[2.2.1]hept-2-yl)amino)carbonyl)-4-methylbenzenesulphonamide
Formed from: Glibornuride

Hydroxy-1-(3-((4-(4-hydroxy-2-(1-methylethoxy)phenyl)-1-piperazinyl)methyl)benzoyl)piperidine
Probably formed from: 1-(3-((4-(4-Hydroxy-2-(1-methylethoxy)phenyl)-1-piperazinyl)methyl)benzoyl)piperidine
→ 1-(3-((4-(4-Hydroxy-2-(1-methylethoxy)phenyl)-1-piperazinyl)methyl)benzoyl)oxopiperidine
Probably in: Dog J581

6-Hydroxy-4-(1-hydroxy-1-methylethyl)-2-((4-hydroxy-2-methylphenyl)methoxy)-1-methyl-7-oxabicyclo[2.2.1]heptane
Probably formed from: 4-(1-Hydroxy-1-methylethyl)-2-((4-hydroxy-2-methylphenyl)methoxy)-1-methyl-7-oxabicyclo[2.2.1]heptane

[5*S*-(5*R,8*R**,10*R**,11*R**)]-10-Hydroxy-1-(2-(1-hydroxy-1-methylethyl)-4-thiazolyl)-2-methyl-5-(1-methylethyl)3,6-dioxo-8,11-bis(phenylmethyl)-2,4,7,12-tetraazatridecan-13-oic acid 5-thiazolylmethyl ester**
Formed from: Ritonavir
→ 11-Amino-10-hydroxy-1-(2-(1-hydroxy-1-methylethyl)-4-thiazolyl)-2-methyl-3,6-dioxo-11-phenyl-8-phenylmethyl-2,4,7-triazadodecan
Probably in: Dog J146
Rat J146

3-Hydroxy-2-(4-(1-hydroxymethylisopropyl)cyclohexyl)-1,4-naphthoquinone
Formed from: 2-(4-*tert*-Butylcyclohexyl)-3-hydroxy-1,4-naphthoquinone

4-Hydroxy-N-(5-hydroxymethyl-3-isoxazolyl)-2*H*-1,2-benzothiazine-3-carboxamide-1,1-dioxide
Formed from: Isoxicam

8-Hydroxy-3-(hydroxymethyl)-6-methoxy-1*H*-2-benzopyran-1-one
See: Cytogenin

4-Hydroxy-5-hydroxymethyl-3-methoxycinnamic acid
See: 5-Hydroxymethylferulic acid

6-Hydroxy-4-hydroxymethyl-2-methoxy-8-(5-methyl-2,3-dioxopyrrolidin-5-yl)amino-5-(3-trifluoromethylphenoxy)quinoline
Probably formed from: 8-(4-Amino-1-methyl-3-oxobutyl)amino-6-hydroxy-4-hydroxymethyl-2-methoxy-5-(3-trifluoromethylphenoxy)quinoline

6-Hydroxy-2-hydroxymethyl-7-methoxy-2-methyl-2*H*-benzo[*b*]pyran
Probably formed from: 2-Hydroxymethyl-6,7-dimethoxy-2-methyl-2*H*-benzo[*b*]pyran

2-Hydroxy-7-hydroxymethyl-12-methylbenzanthracene
Formed from: 7-Hydroxymethyl-12-methylbenzanthracene

2-Hydroxy-12-hydroxymethyl-7-methylbenzanthracene
Formed from: 12-Hydroxymethyl-7-methylbenzanthracene

3-Hydroxy-7-hydroxymethyl-12-methylbenzanthracene
Probably formed from: 7-Hydroxymethyl-12-methylbenzanthracene

4-Hydroxy-7-hydroxymethyl-12-methylbenzanthracene
Formed from: 7-Hydroxymethyl-12-methylbenzanthracene

4-Hydroxy-12-hydroxymethyl-7-methylbenzanthracene
Probably formed from: 12-Hydroxymethyl-7-methylbenzanthracene

9-Hydroxy-7-hydroxymethyl-12-methylbenzanthracene
Probably formed from: 7-Hydroxymethyl-12-methylbenzanthracene

(α-Hydroxy-N-((1-(hydroxymethyl)-1-methylethoxy)carbonyl-β-phenyl-β-alanyl)-10-deacetylbaccatin III
Formed from: Docetaxel
→ (N-((1-Formyl-1-methylethoxy)carbonyl-α-hydroxy-β-phenyl-β-alanyl)-10-deacetylbaccatin III
Probably in: Man H256

(1*S*, 2*R*-*S*-(3*S*,4α*S*,8α*S*)-N′-(2-Hydroxy-3-(3-(((1-hydroxymethyl-1-methylethyl)amino)carbonyl)octahydro-2(1*H*)-isoquinolinyl)-1-(phenylmethyl)propyl)-2-(2-quinolinylcarbonyl)aminobutandiamide
Formed from: Saquinavir

***p*-(2-Hydroxy-4-(*p*-((1-hydroxymethyl-1-methyl)ethyl)phenyl)butoxy)benzoic acid**
Formed from: Lifibrol
→ *p*-(4-(*p*-((1-Carboxy-1-methyl)ethyl)phenyl)-2-hydroxy)butoxy)benzoic acid
Probably in: Man H533

[5*S*-(5*R,8*R**,10*R**,11*R**)]-10-hydroxy-2-hydroxymethyl-1-(2-(1-methylethyl)-4-thiazolyl)-5-(1-methylethyl)3,6-dioxo-8,11-bis(phenylmethyl)-2,4,7,12-tetraazatridecan-13-oic acid 5-thiazolylmethyl ester**
Formed from: Ritonavir

4-Hydroxy-N-(5-hydroxymethyl-3-oxazolyl)-2-methyl-2*H*-1,2-benzothiazine-3-carboxamide-1,1-dioxide
Formed from: Isoxicam

2-Hydroxy-3-(4-hydroxy-3-methylphenoxy)-N-isopropylpropylamine
Formed from: 2-Hydroxy-N-isopropyl-3-(*m*-methylphenoxy)propylamine

3-(2-Hydroxy-5-hydroxymethylphenyl)-N,N-diisopropyl-3-phenylpropylamine
Formed from: Tolterone
→ 3-(5-Carboxy-2-hydroxyphenyl)-N,N-diisopropyl-3-phenylpropylamine
In: Dog J419
Man J419
Mouse J419
Rat J419
→ 3-(2-Hydroxy-5-hydroxymethylphenyl)-N,N-diisopropyl-3-phenylpropylamine-β-D-glucuronide
Probably in: Dog J419
Mouse J419
→ 3-(2-Hydroxy-5-hydroxymethylphenyl)-N-isopropyl-3-phenylpropylamine
Probably in: Dog J419
Mouse J419

3-(2-Hydroxy-5-hydroxymethylphenyl)-N,N-diisopropyl-3-phenylpropylamine-β-D-glucuronide
Probably formed from: 3-(2-Hydroxy-5-hydroxymethylphenyl)-N,N-diisopropyl-3-phenylpropylamine

3-(2-Hydroxy-5-hydroxymethylphenyl)-3-(*p*-hydroxyphenyl)-N,N-diisopropyl-propylamine
Probably formed from: 3-(2-Hydroxy-5-methylphenyl)-3-(*p*-hydroxyphenyl)-N,N-diisopropylpropylamine
→ 3-(5-Carboxy-2-hydroxyphenyl)-3-(*p*-hydroxyphenyl)-N,N-diisopropylpropylamine
Probably in: Rat J419

3-(2-Hydroxy-5-hydroxymethylphenyl)-N-isopropyl-3-phenylpropylamine
Probably formed from: Deisopropyltolterodine

3-(2-Hydroxy-5-hydroxymethylphenyl)-N,N-diisopropyl-3-phenylpropylamine
→ 3-(2-Hydroxy-5-hydroxymethylphenyl)-N-isopropyl-3-phenylpropylamine-β-D-glucuronide
Probably in: Dog J419
Mouse J419

3-(2-Hydroxy-5-hydroxymethylphenyl)-N-isopropyl-3-phenylpropylamine-β-D-glucuronide
Probably formed from: 3-(2-Hydroxy-5-hydroxymethylphenyl)-N-isopropyl-3-phenylpropylamine

2-Hydroxy-5-(1-hydroxy-2-((1-methyl-3-phenylpropyl)amino)ethyl)benzamide
See: Labetalol

2-Hydroxy-5-((1*R*)-1-hydroxy-2-(((1*R*)-1-methyl-3-phenylpropyl)amino)ethyl) benzamide
See: Dilevalol

4-Hydroxy-N-(5-hydroxymethyl-2-thiazolyl)-2-methyl-2*H*-1,2-benzothiazine-3-carboxamide-1,1-dioxide
→ N-(5-Carboxy-2-thiazolyl)-4-hydroxy-2-methyl-2*H*-1,2-benzothiazine-3-carboxamide-1,1-dioxide
In: Man H477, H867, J425
Pig J425
Rat J425

7-Hydroxy-3-(β-hydroxy-γ-morpholinopropyl)-8-methoxy-4-methylcoumarin
Probably formed from: 3-(β-Hydroxy-γ-morpholinopropyl)-7,8-dimethoxy-4-methylcoumarin

2-Hydroxy-3-(4-hydroxy-1-naphthyloxy) propylamine
Probably formed from: 4-Hydroxypropranolol
→ 2-Hydroxy-3-(4-hydroxy-1-naphthyloxy) propylamine-β-D-glucuronide
Probably in: Man D343
→ 2-Hydroxy-3-(4-hydroxy-1-naphthyloxy)propylamine sulphate
Probably in: Man D343

2-Hydroxy-3-(4-hydroxy-1-naphthyloxy) propylamine-β-D-glucuronide
Probably formed from: 2-Hydroxy-3-(4-hydroxy-1-naphthyloxy)propylamine

2-Hydroxy-3-(4-hydroxy-1-naphthyloxy) propylamine sulphate
Probably formed from: 2-Hydroxy-3-(4-hydroxy-1-naphthyloxy)propylamine

4-Hydroxy-3-((2-hydroxy-4-nitrophenyl)azo)-7-((8-hydroxy-5-quinolinyl)azo)-2-naphthalenesulphonic acid
See: Direct Black 114

4-Hydroxy-3-(γ-hydroxy-(*p*-nitrophenyl)butyl) coumarin
Formed from: Acenocoumarol

7-(3-Hydroxy-2-(3-hydroxy-4-phenoxy-1-butenyl)-5-oxocyclopentyl)-5-heptenamide
Formed from: Sulprostone
→ 7-(2-(3-Hydroxy-4-phenoxy-1-butenyl)-5-oxocyclopentyl)-5-heptenamide
Probably in: Guinea pig G50
Man G50

7-(3-Hydroxy-2-(3-hydroxy-4-phenoxy-1-butenyl)-5-oxocyclopentyl)-N-(methylsulphonyl)-5-heptenamide
See: Sulprostone

5-(1-Hydroxy-2-((2-(*o*-hydroxyphenoxy) ethyl)amino)ethyl)-2-hydroxymethylbenzene-sulphonamide
Probably formed from: 5-(1-Hydroxy-2-((2-(*o*-methoxyphenoxy)ethyl)amino)ethyl)-2-hydroxymethylbenzenesulphonamide
→ 5-(1-Hydroxy-2-((2-(*o*-hydroxyphenoxy) ethyl)amino)ethyl)-2-hydroxymethylbenzenesulphonamide sulphate
Probably in: Dog D1
Rat D1

5-(1-Hydroxy-2-((2-(*o*-hydroxyphenoxy) ethyl)amino)ethyl)-2-hydroxymethylbenzenesulphonamide sulphate
Probably formed from: 5-(1-Hydroxy-2-((2-(*o*-hydroxyphenoxy)ethyl)amino)ethyl)-2-hydroxymethylbenzenesulphonamide

5-(1-Hydroxy-2-((2-(*o*-hydroxyphenoxy) ethyl)amino)ethyl)-2-methylbenzenesulphonamide
Formed from: Amosulalol

2-Hydroxy-3-(4-hydroxyphenoxy)-N-(isopropyl)propylamine
See: Prenalterol

N-(2-((2-Hydroxy-3-(4-hydroxyphenoxy)propyl) amino)ethyl)-4-morpholine carboxamide
See: Xamoterol

1-Hydroxy-1-(*m*-hydroxyphenyl)acetone
Probably formed from: *m*-Hydroxyphenylpropiophenone

2-Hydroxy-1-(*m*-hydroxyphenylazo)naphthalene
→ (3,4-Dihydroxyphenylazo)-2-hydroxynaphthalene
In: Rat H362

2-Hydroxy-1-(*p*-hydroxyphenylazo)naphthalene
Formed from: 1-Phenylazo-2-naphthol
→ 2,6-Dihydroxy-1-(*p*-hydroxyphenylazo)-2-hydroxynaphthalene
In: Rat H362
See also: K539
→ (3,4-Dihydroxyphenylazo)-2-hydroxynaphthalene
In: Rat H362

2-Hydroxy-1-(*p*-hydroxyphenylazo)naphthalene-6-sulphonic acid
Formed from: Acid Orange 12

2-Hydroxy-N-(4-hydroxyphenyl)benzamide
See: Oxaphenamide

4-Hydroxy-α-(((6-(4-hydroxy-4-phenylbutoxy)hexyl)amino)methyl)-1,3-benzenedimethanol
Formed from: Salmeterol

4-Hydroxy-1-(1-(*p*-hydroxyphenyl)cyclohexyl) piperidine
Probably formed from: 4-Hydroxy-1-(1-phenylcyclohexyl)piperidine

β-Hydroxy-α-(*p*-hydroxyphenyl)-α,N-dimethylsuccinimide
Probably formed from: α-(*p*-Hydroxyphenyl)-α,N-dimethylsuccinimide

5-Hydroxy-3-(4-hydroxyphenyl)-4,7-diphenyl-2,6-benzofurandione
See: Xylerythrin

3-Hydroxy-4-methoxyphenylethanolamine
See: Norparanephrine

4-Hydroxy-3-methoxyphenylethanolamine
See: Normetanephrine

3-Hydroxy-5-((1*E*)-2-(4-hydroxyphenyl)ethenyl) phenyl-β-D-glucopyranoside
See: Piceid

(*R*)-4-(3-(2-Hydroxy-2-(*m*-hydroxyphenyl)ethyl) amino-3-methylbutyl)benzamide
Probably formed from: (*R*)-4-(3-(2-(2-hydroxy-2-phenyl)ethyl)amino-3-methylbutyl) benzamide
→ (*R*)-4-(3-(2-(3,4-Dihydroxyphenyl)-2-hydroxy-ethyl)amino-3-methylbutyl) benzamide
Probably in: Mouse F407
Rat F407

(*R*)-4-(3-(2-Hydroxy-2-(*p*-hydroxyphenyl)ethyl) amino-3-methylbutyl)benzamide
Probably formed from: (*R*)-4-(3-(2-hydroxy-2-phenylethyl)amino-3-methylbutyl)benzamide

2-Hydroxy-6-(2-(4-hydroxyphenyl)ethyl) benzoic acid
See: Lunularic acid

p-Hydroxy-N-(2-(p-hydroxyphenyl)ethyl) norephedrine (ritodrine)
→ *p*-Hydroxy-N-(2-(*p*-hydroxyphenyl-β-D-glucuronide)ethyl)norephedrine
In: Rat F578
→ *p*-Hydroxy-N-(2-(*p*-hydroxyphenyl-O-sulphate)ethyl)norephedrine
In: Rat F578
→ N-(2-(*p*-Hydroxyphenyl)ethyl)norephedrin-4-yl-β-D-glucuronide
In: Rat F578
See also: D344, D562
→ N-(2-(*p*-Hydroxyphenyl)ethyl)norephedrin-4-yl sulphate
In: Rat F578

p-Hydroxy-N-(2-(p-hydroxyphenyl-β-D-glucuronide)ethyl)norephedrine
Formed from: *p*-Hydroxy-N-(2-(*p*-hydroxyphenyl)ethyl)norephedrine

4-Hydroxy-6-(p-hydroxyphenyl)hexan-2-one
Probably formed from: Dihydro-*p*-hydroxy-kawain

4-Hydroxy-6-(p-hydroxyphenyl)-5-hexen-2-one
Probably formed from: Dihydro-*p*-hydroxy-kawain

3-Hydroxy-4-(o-hydroxyphenyl)-5-(hydroxy-phenylmethyl)-1-methyl-2-pyrrolidinone
Probably formed from: (+)-Clausenamide

4-Hydroxy-1-(hydroxyphenyl)-4-isopropyl-2,3-dimethyl-3-pyrazolin-5-one
Probably formed from: 1-(Hydroxyphenyl)-4-isopropyl-2,3-dimethyl-3-pyrazolin-5-one
→ 4-Hydroxy-1-(hydroxymethoxyphenyl)-4-isopropyl-2,3-dimethyl-3-pyrazolin-5-one
Probably in: Rat D285

(2Z)-6-Hydroxy-2-((4-hydroxyphenyl) methylene)-3-((2H)-benzofuranone-6-β-D-glucoside
See: Hispidol 6-β-D-glucoside

4-Hydroxy-N-(1-(p-hydroxyphenyl)-1-methylethyl)-3,3-dimethylbutyramide
Probably formed from: 2-Bromo-4-hydroxy-N-(1-(*p*-hydroxyphenyl)-1-methylethyl)-3,3-dimethylbutyramide
N-(1-(*p*-Hydroxyphenyl)-1-methylethyl)-3,3-dimethylbutyramide
→ 4-Hydroxy-N-(1-(*p*-hydroxyphenyl)-1-methylethyl)-3,3-dimethylbutyramide-β-D-glucuronide
Probably in: Mouse D193
Rat D193

4-Hydroxy-N-(1-(p-hydroxyphenyl)-1-methylethyl)-3,3-dimethylbutyramide-β-D-glucuronide
Probably formed from: 4-Hydroxy-N-(1-(*p*-hydroxyphenyl)-1-methylethyl)-3,3-dimethylbutyramide

5-(1-Hydroxy-2-((3-(p-hydroxyphenyl)-1-methyl-propyl)amino)ethyl)salicylamide
Formed from: Labetalol
→ 5-(1-Hydroxy-2-((3-(*p*-hydroxyphenyl)-1-methylpropyl)amino)ethyl)salicylamide-β-D-glucuronide
In: Rabbit A1655
See also: A2831

5-(1-Hydroxy-2-((3-(p-hydroxyphenyl)-1-methylpropyl)amino)ethyl)salicylamide-β-D-glucuronide
Formed from: 5-(1-Hydroxy-2-((3-(*p*-hydroxyphenyl)-1-methylpropyl)amino) ethyl)salicylamide

4-Hydroxy-α-(((3-(4-hydroxyphenyl)-1-methylpropyl)amino)methyl)benzenemethanol
See: Ractopamine

2-Hydroxy-6-(o-hydroxyphenyl)-6-oxohexa-2,4-dienoic acid
→ Salicylic acid
In: Sphingomonas H217

2-Hydroxy-6-(o-hydroxyphenyl)-6-oxomuconic acid
Formed from: 2,2′,3-Trihydroxybiphenyl

Hydroxy-1-(3-((4-(o-hydroxyphenyl)-1-piperazinyl)methyl)benzoyl)piperidine
Probably formed from: 1-(3-((4-(*o*-Hydroxy-phenyl)-1-piperazinyl) methyl)benzoyl)piperidine

2-Hydroxy-1-(p-hydroxyphenyl)propan-1-one
Probably formed from: *p*-Hydroxy-propiophenone
→ 1-(*p*-Hydroxyphenyl)propane-1,2-dione
Probably in: Horseradish D166
Rat D166

6β-Hydroxy-(3-(*o*-hydroxyphenyl)propionyl) tropine
Formed from: 3-(*o*-Hydroxyphenyl) propionyltropine

6β-Hydroxy-(3-(*m*-hydroxyphenyl)propionyl) tropine
Formed from: 3-(*m*-Hydroxyphenyl) propionyltropine

***p*-Hydroxy-N-(2-(*p*-hydroxyphenyl-O-sulphate) ethyl)norephedrine**
Formed from: *p*-Hydroxy-N-(2-(*p*-hydroxyphenyl)ethyl)norephedrine

4(*S*)-Hydroxy-4-(*p*-hydroxyphenyl)-L-threonine
Probably formed from: N-Acetyl-4(*S*)-hydroxy-4-(*p*-hydroxyphenyl)-L-threonine

4-Hydroxy-1-(4-hydroxypiperidino)-6,7-dimethoxyphthalazine
Probably formed from: 4-Hydroxycarbazeran

3-(1-Hydroxy-2-(3-hydroxypiperidino)ethyl)-5-phenylisoxazole
Formed from: 3-(1-Hydroxy-2-piperidinoethyl)-5-phenylisoxazole

3-(1-Hydroxy-2-(4-hydroxypiperidino)ethyl)-5-phenylisoxazole
Formed from: 3-(1-Hydroxy-2-piperidinoethyl)-5-phenylisoxazole

(*S*,*S*)-2-Hydroxy-*m*-(2-hydroxypropionyl)propiophenone
Probably formed from: (*S*)-3-(2-Hydroxypropionyl)benzaldehyde

(*S*,*S*)-2-Hydroxy-*p*-(2-hydroxypropionyl) propiophenone
Probably formed from: (*S*)-4-(2-Hydroxypropionyl)benzaldehyde

4-Hydroxy-N-((2-hydroxypropyl) aminothioxomethyl)-2-methyl-2*H*-1,2-benzothiazine-3-carboxamide-1,1-dioxide
Formed from: Meloxicam
→ N-(Aminothioxomethyl)-4-hydroxy-2-methyl-2*H*-1,2-benzothiazine-3-carboxamide-1,1-dioxide
Probably in: Rat H462

***p*-Hydroxy-N-(3-hydroxypropyl)amphetamine**
Probably formed from: N-(3-Hydroxypropyl) amphetamine
→ 3,4-Dihydroxy-N-(3-hydroxypropyl) amphetamine
Probably in: Man J544

Hydroxy-N-(2-hydroxypropyl)-9,9-dioxophenothiazine
Probably formed from: Hydroxy-9,9-dioxo-N-(2-oxopropyl)phenothiazine

4-Hydroxy-N-(3-hydroxypropyl)-3-methoxyamphetamine
Probably formed from: 3,4-Dihydroxy-N-(3-hydroxypropyl)amphetamine

7-Hydroxy-2-(2-hydroxypropyl)-5-methylchromone
Formed from: Aloeresin B

17β-Hydroxy-17α-(1-(3-hydroxy)propynyl)-11β-(*p*-methylaminophenyl)oestra-4,9-dien-3-one
Probably formed from: 11β-(*p*-Dimethylaminophenyl)-17β-hydroxy-17α-(1-(3-hydroxy)propynyl)oestra-4,9-dien-3-one
→ 11β-(*p*-Aminophenyl)-17β-hydroxy-17α-(1-(3-hydroxy)propynyl)oestra-4,9-dien-3-one
Probably in: Man J752
Monkey J752
Rat J752

6β-Hydroxyhyoscyamine
Formed from: Hyoscyamine
→ Scopolamine
In: Hyoscyamus K887

12-Hydroxyibogaine
Formed from: Ibogaine

5-Hydroxyidazoxan
Formed from: Idazoxan
→ 5-Hydroxyidazoxan-β-D-glucuronide
In: Rat F284
→ 5-Hydroxyidazoxan-O-sulphate
In: Rat F284

6-Hydroxyidazoxan
Formed from: Idazoxan
→ 6-Hydroxyidazoxan-β-D-glucuronide
In: Rat F284
→ 6-Hydroxyidazoxan-O-sulphate
In: Rat F284

7-Hydroxyidazoxan
Formed from: Idazoxan

5-Hydroxyidazoxan-β-D-glucuronide
Formed from: 5-Hydroxyidazoxan

6-Hydroxyidazoxan-β-D-glucuronide
Formed from: 6-Hydroxyidazoxan
5-Hydroxyidazoxan-O-sulphate
Formed from: 5-Hydroxyidazoxan
6-Hydroxyidazoxan-O-sulphate
Formed from: 6-Hydroxyidazoxan
2-Hydroxy-4-((1*H*)-imidazol-4-yl)aniline
Probably formed from: 4-((1*H*)-Imidazol-4-yl)aniline
4′-Hydroxyimiloxan
Probably formed from: Imiloxan
Hydroxyimiloxan (aryl)
Probably formed from: Imiloxan
→ Hydroxyimiloxan-β-D-glucuronide (aryl)
In: Man G67
Hydroxyimiloxan-β-D-glucuronide (aryl)
Formed from: Hydroxyimiloxan
2-Hydroxyiminodibenzyl
Probably formed from: Iminodibenzyl
→ 2,3-Dihydroxyiminodibenzyl
Probably in: Rat G269
10-Hydroxyiminodibenzyl
Probably formed from: Iminodibenzyl
2-Hydroxyimipramine
Formed from: Imipramine
→ Imipramin-2-yl-β-D-glucuronide
In: Guinea pig A2092
Rat A2092, C872
See also: D431
10-Hydroxyimipramine
Formed from: Imipramine
→ Imipramin-10-yl-β-D-glucuronide
In: Rat C872
3-Hydroxyimirestat
Formed from: Imirestat
4-Hydroxyimirestat
Formed from: Imirestat
3-Hydroxyinabenfide
Probably formed from: Inabenfide
4-Hydroxyinabenfide
Probably formed from: Inabenfide
1-Hydroxyindane (1-indanol)
Formed from: Indane
Indan-1-one
→ 1,2-Dihydroxyindane (*cis* and *trans*)
In: Pseudomonas K207
See also: E533, F617
→ Indan-1-one
In: Guinea pig E67
Hamster F153, K443
Man D415
Monkey F241, F388
Mouse E821, F377
Pseudomonas K207, K348
Rabbit A2032, H920
See also: E533, F617
2-Hydroxyindane (2-indanol)
Formed from: 2-Aminoindane
Indane
Indan-2-one
→ Indan-2-one
In: Monkey F388
Rabbit A449
Rat A449
See also: E533, F617
5-Hydroxyindane
Formed from: Indane
Probably formed from: Candroxatril
→ 1,5-Dihydroxyindane
Probably in: Penicillium J836
Pseudomonas F381
→ 5-Hydroxyindene
Probably in: Penicillium J836
2-(2-Hydroxy-3-((2-(1*H*-indol-3-yl)-1,1-dimethylethyl)amino)propoxy)benzonitrile
See: Bucindolol
2-Hydroxyindan-1-one
Formed from: Indan-1-one
Indan-2-one
Probably formed from: *cis*-1,2-Dihydroxyindane
3-Hydroxyindan-1-one
Probably formed from: Indan-1-one
5-Hydroxyindan-1-one
Probably formed from: 1,5-Dihydroxyindane
S-(1-Hydroxyindan-2-yl)glutathione
Probably formed from: Indene-1,2-oxide
→ N-Acetyl-S-(1-hydroxyindan-2-yl)-L-cysteine
Probably in: Guinea pig D855
Rat A2886, D855
(1*S*)-Hydroxyindene
Formed from: Indene

5-Hydroxyindene
Probably formed from: 5-Hydroxyindane
2-Hydroxyindenoestrol A
Formed from: Indenoestrol A
8-Hydroxyindene[1,2,3-*cd*]pyrene
Formed from: Indene[1,2,3-*cd*]pyrene
9-Hydroxyindene[1,2,3-*cd*]pyrene
Formed from: Indene[1,2,3-*cd*]pyrene
10-Hydroxyindene[1,2,3-*cd*]pyrene
Formed from: Indene[1,2,3-*cd*]pyrene
3-Hydroxyindinavir
Formed from: Indinavir
→ 2(*R*)-Benzyl-5-((2(*S*)-*tert*-butylaminocarbonyl)piperazino)-N-(2,3-dihydroxy-1(*S*)-indanyl)-4(*S*)-hydroxypentanamide
Probably in: Dog J50
Man J58
Monkey J50
Rat J50
***p*-Hydroxyindinavir**
Formed from: Indinavir
3-Hydroxyindinavir-N-oxide
Formed from: Indinavir-N-oxide
3-Hydroxyindole
See: Indoxyl
4-Hydroxyindole
Formed from: Indole
→ 4-Hydroxyindole-β-D-glucoside
In: Baphicacanthus K347
5-Hydroxyindole
Formed from: Indole
→ 5-Hydroxyanthranilic acid
In: Jasminum B70
Tecoma C170
→ 5-Hydroxyindole-β-D-glucoside
In: Baphicacanthus K347
→ 5-Hydroxyindole sulphate
In: Rat B61
→ 5-Hydroxy-L-tryptophan
In: Carrot B874
Enterobacter A781
Erwinia A781
Escherichia A781
Kluyvera A781
Nicotiana B874
Proteus A781
Contrast: In Acetobacter A781
Aeromonas A781
Agrobacterium A781
Alcaligenes A781
Azobacter A781
Clostridium A781
Enterobacter A781
Erwinia A781
Escherichia A781
Flavobabterium A781
Kluyvera A781
Mycoplana A781
Proteus A781
Pseudomonas A781
Rhizobium A781
Salmonella A781
Serratia A781
Vibrio A781
Xanthomonas A781
6-Hydroxyindole
Formed from: Indole
→ 6-Hydroxyindole-β-D-glucoside
In: Baphicacanthus K347
7-Hydroxyindole
→ 7-Hydroxyindole-β-D-glucoside
In: Baphicacanthus K347
α-Hydroxyindole-3-acetaldehyde
Probably formed from: Indole-3-acetaldehyde
5-Hydroxyindole-3-acetaldehyde
Formed from: 5-Hydroxy-N,N-dimethyltryptamine
5-Hydroxy-N-methyltryptamine
5-Hydroxytryptamine
5-Hydroxy-L-tryptophan
Probably formed from: 5-Hydroxyindole-3-pyruvic acid
→ 5-Hydroxyindole-3-acetic acid
In: Horse A239
Man A60, E206
Pig A1287
See also: A923, A2894, A3875
→ 5-Hydroxytryptophol
In: Man A60, E206
Monkey A1908
Pig A1287
See also: A923, A3875

α-Hydroxyindole-3-acetic acid
Probably formed from: Indole-3-acetic acid
4-Hydroxyindole-3-acetic acid
Probably formed from: Indole-3-acetic acid
5-Hydroxyindole-3-acetic acid
Formed from: 5-Hydroxyindole-3-acetaldehyde
(4*R*)-2-(3-(5-Hydroxyindolyl)methyl)-1,3-thiazolidine-4-carboxylic acid
Indole-3-acetic acid
→ 5,6-Dihydroxyindole-3-acetic acid
Probably in: Man G762
→ 5-Hydroxyindole-3-acetic acid-O-sulphate
In: Eubacterium G602
6-Hydroxyindole-3-acetic acid
Probably formed from: Indole-3-acetic acid
5-Hydroxyindole-3-acetic acid-O-sulphate
Formed from: 5-Hydroxyindole-3-acetic acid
α-Hydroxyindole-3-acetonitrile
Probably formed from: Indole-3-acetonitrile
Hydroxyindole-3-acetylaspartic acid
Formed from: Indole-3-acetylaspartic acid
4-Hydroxyindole-3-butyric acid
Formed from: Indole-3-butyric acid
γ-Hydroxyindole-3-butyric acid
Formed from: Indole-3-butyric acid
1-Hydroxyindole-3-carboxylic acid
Probably formed from: 1-Methoxyindole-3-carboxylic acid
→ Indole-3-carboxylic acid
In: Rat D158
4-Hydroxyindole-β-D-glucoside
Formed from: 4-Hydroxyindole
5-Hydroxyindole-β-D-glucoside
Formed from: 5-Hydroxyindole
6-Hydroxyindole-β-D-glucoside
Formed from: 6-Hydroxyindole
7-Hydroxyindole-β-D-glucoside
Formed from: 7-Hydroxyindole
5-Hydroxyindole-3-glyoxal
Formed from: 5-Hydroxy-L-tryptophan
Probably formed from: 5-Hydroxytryptamine
1-Hydroxyindole-3-glyoxylic acid
→ Indole-3-glyoxylic acid
In: Rat D158
β-Hydroxyindole-3-propionic acid
Formed from: Indole-3-propionic acid
5-Hydroxyindole-3-pyruvic acid
Formed from: 5-Hydroxy-L-tryptophan
→ 5-Hydroxyindole-3-acetaldehyde
Probably in: Rat A2894
5-Hydroxyindole sulphate
Formed from: 5-Hydroxyindole
5′-Hydroxyindolidan
Formed from: Indolidan
7-Hydroxy-2-indolinone-3-acetic acid
→ 7-Hydroxy-2-indolinone-3-acetic acid-7-glucoside
In: Zea E421
7-Hydroxy-2-indolinone-3-acetic acid-7-glucoside
Formed from: 7-Hydroxy-2-indolinone-3-acetic acid
4-Hydroxyindolo[2,3-*a*]quinolizidine
Formed from: Indolo[2,3-*a*]quinolizidine
6-Hydroxyindolo[2,3-*a*]quinolizidine
Formed from: Indolo[2,3-*a*]quinolizidine
7-Hydroxyindolo[2,3-*a*]quinolizidine
Formed from: Indolo[2,3-*a*]quinolizidine
5-(5-Hydroxyindolylmethyl)hydantoin
→ 5-Hydroxy-L-tryptophan
In: Flavobacterium F940
(4*R*)-2-(3-(5-Hydroxyindolyl)methyl)-1,3-thiazolidine-4-carboxylic acid
→ 5-Hydroxyindole-3-acetic acid
In: Rat F158
***p*-Hydroxyindoramin**
Formed from: Indoramin
5-Hydroxyindoramin
Formed from: Indoramin
6-Hydroxyindoramin
Formed from: Indoramin
6a-Hydroxyinermin
See: 3,6a-Dihydroxy-8,9-methylenedioxypterocarpan
3-Hydroxyinerminisoflavin
Formed from: 3,6a-Dihydroxy-8,9-methylenedioxypterocarpan
4′-Hydroxy-2-iodobenzanilide
Formed from: 2-Iodobenzanilide
→ 2-Iodobenzanilid-4′-yl-β-D-glucuronide
In: Rat B878
→ 2-Iodobenzanilid-4′-yl sulphate
In: Rat B878

2-Hydroxy-5-iodobenzoic acid
Formed from: 2-Acetoxy-5-iodobenzoic acid
4-Hydroxy-3-iodobenzylguanidine
Formed from: *m*-Iodobenzylguanidine
4-Hydroxy-4′-iodobiphenyl
Formed from: 4-Iodobiphenyl
4-Hydroxy-3-iodo-5-nitrobenzamide
Formed from: 4-Hydroxy-3-iodo-5-nitrobenzonitrile
→ 4-Hydroxy-3-iodo-5-nitrobenzoic acid
Probably in: Mouse A3694
Rat H305
Sheep A3694
4-Hydroxy-3-iodo-5-nitrobenzoic acid
Probably formed from: 4-Hydroxy-3-iodo-5-nitrobenzamide
4-Hydroxy-3-iodo-5-nitrobenzonitrile (nitroxynil)
→ 5-Amino-4-hydroxy-3-iodobenzonitrile
In: Ascaris A3694
Moniezia A3694
Mouse A3694
Rat C297
Sheep A3694
→ 4-Hydroxy-3-iodo-5-nitrobenzamide
In: Mouse A3694
Rat H305
Sheep A3694
4-Hydroxy-3-iodophenylpyruvic acid
Formed from: 3-Iodo-L-tyrosine
Hydroxyiprindole (alicyclic)
Formed from: Iprindole
→ Oxoiprindole
Probably in: Rat G23
Hydroxyiprindole (aryl)
Formed from: Iprindole
2-Hydroxyirbesartan
Formed from: Irbesartan
3-Hydroxyirbesartan
Formed from: Irbesartan
Hydroxyirinotecan
Formed from: Irinotecan
Hydroxyirsogladine
Formed from: Irsogladine
8′α-Hydroxyisobromocryptine
Formed from: Bromocryptine
→ 8′,9′-Didehydroisobromocryptine
Probably in: Rat C607
→ 8′,9-Dihydroxyisobromocryptine
Probably in: Rat C607
8′β-Hydroxyisobromocryptine
Formed from: Bromocryptine
5-(2-Hydroxyisobutyl)-1-naphthylacetic acid
Formed from: 5-Isobutyl-1-naphthylacetic acid
3′-Hydroxy-5′-(4-isobutyl-1-piperazinyl) benzoxazinorifamycin
→ 25-Deacetyl-3′-hydroxy-5′-(4-isobutyl-1-piperazinyl) benzoxazinorifamycin
In: Dog H729
Man H729
→ 3′,30-Dihydroxy-5′-(4-isobutyl-1-piperazinyl)benzoxazinorifamycin
In: Dog H527, H729
Man H729
Mouse H729
5-Hydroxyisocarbostyril
Formed from: 5-Hydroxyisoquinoline
2-Hydroxy-2-(isochroman-3-yl)acetonitrile
Formed from: 3-Formylisochroman
9-Hydroxyisoergotamine
Formed from: Isoergotamine
2-Hydroxyisoflavanone
Formed from: Isoflavanone
4′-Hydroxyisoflavanone
Formed from: Isoflavanone
4′-Hydroxyisoflavone
Formed from: Flavone
→ 3′,4′-Dihydroxyisoflavone
Probably in: Absidia F504
Aspergillus F504
7-Hydroxyisoflavone
Formed from: 7-Methoxyisoflavone
13-Hydroxyisolysergic diethylamide
Formed from: Isolysergic diethylamide
14-Hydroxyisolysergic diethylamide
Formed from: Isolysergic diethylamide
2-Hydroxyisophthaldehydic acid
Formed from: 7,8-Dihydroxy-1-naphthoic acid
→ 2-Hydroxyisophthalic acid
Probably in: Pseudomonas H415

2-Hydroxyisophthalic acid
Probably formed from: 2-Hydroxyisophthaldehydic acid
→ Salicylic acid
Probably in: Pseudomonas H415

4-Hydroxyisophthalic acid
→ Protocatechuic acid
In: Pseudomonas A3066, B709

5-Hydroxyisophthalic acid
→ 4,5-Dihydroxyisophthalic acid
In: Bacterium A2985

5-Hydroxy-2-isopropoxyphenyl N-methylcarbamate
Probably formed from: *o*-Isopropoxyphenyl N-methylcarbamate

2-Hydroxy-N-isopropylacetanilide
See: N-Isopropylglycollanilide

***p*-Hydroxy-N-isopropylacetanilide**
Formed from: N-Isopropylacetanilide

3-Hydroxy-*o*-(2-isopropylaminoethoxy) butyrophenone
Probably formed from: *o*-(2-Isopropylaminoethoxy)butyrophenone
→ *o*-(2-Isopropylaminoethoxy) acetophenone
Probably in: Rat A3324
→ *o*-(2-Isopropylaminoethoxy)-3-oxobutyrophenone
Probably in: Rat A3324

5-Hydroxy-*o*-(2-isopropylaminoethoxy) butyrophenone
Probably formed from: *o*-(2-Isopropylaminoethoxy)butyrophenone
→ 2′-(2-Isopropylaminoethoxy) butyrophenon-5′-yl-β-D-glucuronide
Probably in: Rat A3324

6-(2-Hydroxy-3-isopropylaminopropoxy) anthranilic acid
Formed from: Pindolol

2-(2-Hydroxy-3-isopropylamino-1-propoxy)-1-(2-propen-1-oxy)benzene
See: Oxprenolol

***o*-(2-Hydroxy-3-((N-isopropyl)amino)propoxy) phenylpropionic acid**
Formed from: Methyl *o*-(2-hydroxy-3-((N-isopropyl)amino)propoxy) phenylpropionate

***p*-(2-Hydroxy-3-isopropylaminopropoxy)benzoic acid**
Formed from: Methyl *p*-(2-hydroxy-3-((N-isopropyl)amino)propoxy)benzoate
Pamatolol

***o*-(2-Hydroxy-3-((N-isopropyl)amino)propoxy) benzoic acid**
Formed from: Methyl *o*-(2-hydroxy-3-((N-isopropyl)amino)propoxy)benzoate

4-(2-Hydroxy-3-isopropylaminopropoxy)indole
See: Pindolol

4-(2-Hydroxy-3-isopropylaminopropoxy) indolin-2-one
Formed from: Pindolol

***p*-(2-Hydroxy-3-isopropylaminopropoxy) phenylacetic acid**
Formed from: Pamatolol

3-(4-(2-Hydroxy-3-isopropylaminopropoxy) phenyl)propionic acid
Formed from: Esmolol

4-(2-Hydroxy-3-isopropylamino-1-propoxy)-3-(2-propen-1-oxy)phenol
Formed from: Oxprenolol

***p*-Hydroxy-N-isopropylamphetamine**
Formed from: N-Isopropylamphetamine

N-Hydroxy-N-isopropylamphetamine
Formed from: N-Isopropylamphetamine

N-(1-Hydroxyisopropyl)aniline
Probably formed from: N-(1-Hydroxyisoprop-1-yl)-2-methylsulphonyl-N-phenylacetamide
→ N-(1-Hydroxyisopropyl)aniline-β-D-glucuronide
Probably in: Rat B93

N-(1-Hydroxyisopropyl)aniline-β-D-glucuronide
Probably formed from: N-(1-Hydroxyisopropyl)aniline

1-Hydroxyisopropyl N-benzoyl-N-(3-chloro-4-fluorophenyl)-2-aminopropionate
Formed from: Barnon
→ 1-Carboxy-1-ethyl N-benzoyl-N-(3-chloro-4-fluorophenyl)-2-aminopropionate
Probably in: Dog A2035, A3057
Rat A2035, A3057

4-Hydroxy-4′-isopropylbiphenyl
Probably formed from: 4-Chloro-4′-isopropylbiphenyl

3-Hydroxy-N-isopropyl-4-methoxyphenylethanolamine
Formed from: Isoprenaline

L-3-Hydroxy-N-isopropyl-4-methoxyphenylethanolamine
Formed from: L-Isoprenaline

3-(2-Hydroxyisopropyl) 5-methyl 2-carbamoyloxymethyl-4-(2,3-dichlorophenyl-1,4-dihydro-6-methylpyridine-3,5-dicarboxylate
Formed from: 3-Isopropyl 5-methyl 2-carbamoyloxymethyl-4-(2,3-dichlorophenyl-1,4-dihydro-6-methylpyridine-3,5-dicarboxylate

3-(N-(4-Hydroxy-2-isopropyl-5-methylphenoxy)ethyl)-N-methylaminomethyl)-4-methoxy-2,5,6-trimethylphenol
→ 3-(N-(4-Hydroxy-2-isopropyl-5-methylphenoxy)ethyl)-N-methylaminomethyl)-4-methoxy-2,5,6-trimethylphenol-β-D-glucuronide
In: Rat K334
→ 3-(N-(4-Hydroxy-2-isopropyl-5-methylphenoxy)ethyl)-N-methylaminomethyl)-4-methoxy-2,5,6-trimethylphenol sulphate
In: Rat K334

3-(N-(4-Hydroxy-2-isopropyl-5-methylphenoxy)ethyl)-N-methylaminomethyl)-4-methoxy-2,5,6-trimethylphenol-β-D-glucuronide
Formed from: 3-(N-(4-Hydroxy-2-isopropyl-5-methylphenoxy)ethyl)-N-methylamino-methyl)-4-methoxy-2,5,6-trimethylphenol

3-(N-(4-Hydroxy-2-isopropyl-5-methylphenoxy)ethyl)-N-methylaminomethyl)-4-methoxy-2,5,6-trimethylphenol sulphate
Formed from: 3-(N-(4-Hydroxy-2-isopropyl-5-methylphenoxy)ethyl)-N-methylaminomethyl)-4-methoxy-2,5,6-trimethylphenol

2-Hydroxy-N-isopropyl-3-(*m*-methyphenoxy)propylamine
→ 2-Hydroxy-3-(4-hydroxy-3-methyphenoxy)-N-isopropylpropylamine
In: Mammal A853
→ 2-Hydroxy-N-isopropyl-3-(*m*-methyphenoxy)propylamine-β-D-glucuronide
In: Man A853
→ 2-Hydroxy-N-isopropyl-3-(*m*-methyphenoxy)propylamine-O-sulphate
In: Dog A853
→ 3-(*m*-Methyphenoxy)lactic acid
In: Mammal A853

2-Hydroxy-N-isopropyl-3-(*m*-methyphenoxy)propylamine-β-D-glucuronide
Formed from: 2-Hydroxy-N-isopropyl-3-(*m*-methyphenoxy)propylamine

2-Hydroxy-N-isopropyl-3-(*m*-methyphenoxy)propylamine-O-sulphate
Formed from: 2-Hydroxy-N-isopropyl-3-(*m*-methyphenoxy)propylamine

N-(1-Hydroxyisopropyl)-2-methylsulphonylacetamide-β-D-glucuronide
Probably formed from: N-Isopropyl-2-mercaptoacetanilide

***p*-Hydroxy-N-isopropyl-2-methylsulphonylacetanilide**
Probably formed from: N-Isopropyl-2-methylsulphonylacetanilide
→ *p*-Hydroxy-N-isopropyl-2-methylsulphonylacetanilide-β-D-glucuronide
Probably in: Rat C24
→ *p*-Hydroxy-2-methylsulphonylacetanilide
Probably in: Rat C24

***p*-Hydroxy-N-isopropyl-2-methylsulphonylacetanilide-β-D-glucuronide**
Probably formed from: *p*-Hydroxy-N-isopropyl-2-methylsulphonylacetanilide N-Isopropyl-2-mercaptoacetanilide

***p*-Hydroxy-N-isopropylphentermine**
Formed from: N-Isopropylphentermine

3-Hydroxy-N-isopropylphenylethanolamin-5-yl-β-D-glucuronide
Formed from: 3,5-Dihydroxy-N-isopropylphenylethanolamine

3-Hydroxy-N-isopropylphenylethanolamin-5-yl sulphate
Formed from: 3,5-Dihydroxy-N-isopropylphenylethanolamine

2-Hydroxy-*m*-isopropylpropiophenone
Formed from: *m*-Isopropylbenzaldehyde

1-Hydroxyisopropyl 2-(4-thiazolyl)-5-benzimidazolecarbamate
Formed from: Cambendazole

1(2*H*)-Hydroxyisoquinoline
See: Isocarbostyril

5-Hydroxyisoquinoline
Formed from: Isoquinoline
→ 5-Hydroxyisocarbostyril
In: Pseudomonas K765

3′-Hydroxyisosafrole
→ 3′-Hydroxyisosafrole-O-sulphate
In: Mouse E235
→ 3,4-Methylenedioxycinnamic acid
In: Mouse E235

3′-Hydroxyisosafrole-O-sulphate
Formed from: 3′-Hydroxyisosafrole

O-β-Hydroxyisovalerylshikonin
Probably formed from: Shikonin

Hydroxyitraconazole
Formed from: Itraconazole

4′-Hydroxykadethrin
Formed from: Kadethrin
→ 5-(*p*-Hydroxybenzyl)-3-furylcarboxylic acid
Probably in: Rat B14

α-Hydroxykadethrin
Formed from: Kadethrin

8-Hydroxykaempferol
(3,4′,5,7,8-pentahydroxyflavone)
→ 8-Hydroxy-3-O-methylkaempferol
In: Lotus C905
→ 8-Methoxykaempferol
In: Lotus C905, D904

***p*-Hydroxykawain**
Probably formed from: Kawain
Yangonin
→ 5,6-Dehydro-*p*-hydroxykawain
Probably in: Man F862
Rat A3485
→ 3′,4′-Dihydroxykawain
Probably in: Rat A3485
→ 4-Hydroxy-6-(*p*-hydroxyphenyl)-5-hexen-2-one
Probably in: Rat A3485

5-Hydroxykawain
Probably formed from: Kawain

4-Hydroxyketamine
Formed from: Ketamine

5-Hydroxyketamine
Formed from: Ketamine

6-Hydroxyketamine
Formed from: Ketamine
→ 6-Hydroxy-N-demethylketamine
In: Man E786
Rabbit E786
Rat E786

Hydroxyketanserin
Formed from: Ketanserin
→ Hydroxyketanserinol
Probably in: Man E728

Hydroxyketanserinol
Probably formed from: Hydroxyketanserin

Hydroxyketocyclazocine
Probably formed from: Ketocyclazocine
→ Hydroxyketocyclazocine-β-D-glucuronide
Probably in: Rat A2134

Hydroxyketocyclazocine-β-D-glucuronide
Probably formed from: Hydroxyketocyclazocine

'Hydroxyketoprofen'
Formed from: Ketoprofen

***p*-Hydroxyketorolac**
Formed from: Ketorolac

3-Hydroxykynuramine
Formed from: 3-Hydroxy-L-kynurenine

6-Hydroxykynurenic acid
Probably formed from: 5-Hydroxy-L-kynurenine
→ 4,6-Dihydroxyquinoline
Probably in: Rat A1469

3-Hydroxy-L-kynurenine
Formed from: L-Kynurenine
→ 3-Hydroxyanthranilic acid
In: Aspergillus A2150, A3659
Man D272
Neurospora A2766
Penicillium A2150, A3659
Pig A3872
Pseudomonas A2150, A2766, A3659, B248
Rat A2644, F676
Rhizopus A2150, A3659
Streptomyces B335
Suncus F175

Xanthomonas A3615
See also: A780, A854, H762
→ 3-Hydroxykynuramine
In: Micrococcus C62
→ 3-Hydroxy-L-kynurenine-3-β-D-glucoside
In: Man G787, H271
→ Ommochrome (probably a mixture)
In: Cerura A1430
→ Xanthommatin
In: Drosophila A2574
→ Xanthurenic acid
In: Aedes J510, J668
Drosophila F636
Hansenula D978
Rat A2342, C557
See also: A1195

5-Hydroxy-L-kynurenine
Formed from: N-Formyl-5-hydroxy-L-kynurenine
→ 6-Hydroxykynurenic acid
Probably in: Rat A1469

3-Hydroxy-L-kynurenine-3-β-D-glucoside
Probably formed from: 3-Hydroxy-L-kynurenine

***o*-Hydroxylaminobenzoic acid**
Formed from: *o*-Nitrobenzoic acid

***p*-Hydroxylaminobenzoic acid**
Formed from: *p*-Nitrobenzoic acid
→ Protocatechuic acid
In: Pseudomonas G980

***p*-Hydroxylaminobenzonitrile**
Formed from: *p*-Aminobenzonitrile

4-Hydroxylaminobenzophenone
Formed from: N-Hydroxy-4-acetamidobenzophenone

4-Hydroxylaminobiphenyl
Formed from: 4-Aminobiphenyl
N-Hydroxy-4-acetamidobiphenyl
4-Hydroxylaminobiphenyl-N-β-D-glucuronide
4-Nitrobiphenyl
Probably formed from: 4-Nitrosobiphenyl
→ 4-(N-Acetoxylamino)biphenyl
In: Rat K747
→ N-Hydroxy-4-acetamidobiphenyl
Probably in: Monkey A880
→ 4-Hydroxylaminobiphenyl-N-β-D-glucuronide
In: Rat D314, D600
See also: A1972
→ 4-Nitrosobiphenyl
In: Rat G5
→ 4-(N-Sulphatoxyamino)biphenyl
In: Man H304, H434

4-Hydroxylaminobiphenyl-N-β-D-glucuronide
Formed from: 4-Hydroxylaminobiphenyl
→ 4-Hydroxylaminobiphenyl
In: Escherichia J766
Man J766

2-Hydroxylamino-α-carboline
Probably formed from: 2-Amino-α-carboline

4-Hydroxylamino-2′,3-dimethylbiphenyl
→ N-Acetoxy-4-amino-2′,3-dimethylbiphenyl
In: Man F172

2-Hydroxylamino-3,8-dimethylimidazo[4,5-*f*]quinoxaline
Formed from: 2-Amino-3,8-dimethylimidazo[4,5-*f*]quinoxaline
→ 2-Hydroxylamino-3,8-dimethylimidazo[4,5-*f*]quinoxaline-β-D-glucuronide
Probably in: Rat G414

2-Hydroxylamino-3,8-dimethylimidazo[4,5-*f*]quinoxaline-β-D-glucuronide
Probably formed from: 2-Hydroxylamino-3,8-dimethylimidazo[4,5-*f*]quinoxaline

2-Hydroxylamino-4,6-dinitrotoluene
Formed from: 2,4,6-Trinitrotoluene

4-Hydroxylamino-2,6-dinitrotoluene
Formed from: 2,4,6-Trinitrotoluene
→ 2,4-Bis(hydroxylamino)-6-nitrotoluene
In: Pseudomonas H957

3-Hydroxylaminofluoranthene
Probably formed from: 3-Nitrosofluoranthene
→ 3-Aminofluoranthene
Probably in: Rabbit G361

2-Hydroxylaminofluorene
Formed from: 2-Aminofluorene
N-Hydroxy-2-acetamidofluorene
N-Hydroxy-2-benzamidofluorene
2-Nitrofluorene
2-Nitrosofluorene
→ N-Acetoxylaminofluorene
In: Hamster G757

Rat K747
→ 2-Aminofluorene
Probably in: Horse H291
→ 2-Amino-1-hydroxyfluorene
Probably in: Horse H291
→ 2-Amino-3-hydroxyfluorene
Probably in: Horse H291
→ 2-Fluorenyl nitroxide radical
Probably in: Rabbit B184
→ N-Hydroxy-2-acetamidofluorene
Probably in: Rat F953
→ 2-Hydroxylaminofluorene-β-D-glucuronide
In: Rat D314, D600
→ 2-Hydroxylaminofluorene-O-sulphate
In: Mouse D756

2-Hydroxylaminofluorene-β-D-glucuronide
Formed from: N-Hydroxy-2-acetamidofluorene-β-D-glucuronide
2-Hydroxylaminofluorene

2-Hydroxylaminofluorene-O-sulphate
Formed from: 2-Hydroxylaminofluorene

2-Hydroxylamino-3-methyimidazo[4,5-*f*]quinoline
Formed from: 2-Amino-3-methyimidazo[4,5-*f*]quinoline
→ 2-Hydroxylamino-3-methyimidazo[4,5-*f*]quinoline-N^2-β-D-glucuronide
Probably in: Cynomolgus monkey G673

2-Hydroxylamino-3-methyimidazo[4,5-*f*]quinoline-N^2-β-D-glucuronide
Probably formed from: 2-Hydroxylamino-3-methyimidazo[4,5-*f*]quinoline

5-Hydroxylamino-2-methyl-7-methylsulphonylquinoxaline
Probably formed from: 2-Methyl-7-methylsulphonyl-5-nitrosoquinoxaline
→ 5-Acetoxylamino-2-methyl-7-methylsulphonylquinoxaline
Probably in: Rat C750

2-Hydroxylamino-1-methyl-6-phenylimidazo[4,5-*b*]pyridine
Formed from: 2-Amino-1-methyl-6-phenylimidazo[4,5-*b*]pyridine
→ 2-Hydroxylamino-1-methyl-6-phenylimidazo[4,5-*b*]pyridine-β-D-glucuronide
In: Guinea pig H345
Monkey H88
Rat G443, H345, H417

2-Hydroxylamino-1-methyl-6-phenylimidazo[4,5-*b*]pyridine-β-D-glucuronide
Formed from: 2-Hydroxylamino-1-methyl-6-phenylimidazo[4,5-*b*]pyridine

3-Hydroxylamino-1-methyl-5*H*-pyrido[4,3-*b*]indole
Formed from: 3-Amino-1-methyl-5*H*-pyrido[4,3-*b*]indole

1-Hydroxylaminonaphthalene
See: 1-Naphthylhydroxylamine

2-Hydroxylaminonaphthalene
See: 2-Naphthylhydroxylamine

2-Hydroxylamino-6-nitrotoluene
Probably formed from: 2-Nitro-6-nitrosotoluene
→ 2-Methyl-3-nitroaniline
Probably in: Microorganism G542
Salmonella G235

4-Hydroxylaminoquinoline-N-oxide
Formed from: 4-Nitroquinoline-N-oxide
→ 4-Aminoquinoline-N-oxide
In: Chick D67
Mouse A3227
See also: A75

4-Hydroxylaminostilbene
Formed from: N-Hydroxy-4-acetamidostilbene

2-Hydroxylamino-2-(4-(α,α,β,β-tetrafluorophenethyl)phenyl)propane
Probably formed from: α,α-Dimethyl-4-(α,α,β,β-tetrafluorophenethyl)benzylamine

5-Hydroxylansoprazone
Formed from: Lansoprazone

13-Hydroxylergotrile
Formed from: Lergotrile

10′-Hydroxyleurosine
Formed from: Leurosine

21′-Hydroxyleurosine
Formed from: Leurosine

4-Hydroxylevamisole
Formed from: Levamisole

3-Hydroxylevomepromazine
Formed from: Levomepromazine

7-Hydroxylevomepromazine
Formed from: Levomepromazine

Hydroxylevormeloxifene
Formed from: Levormeloxifene
β-Hydroxylicochalcone A
Formed from: Licochalcone A
→ β-Hydroxylicochalcone A-β-D-glucuronide
Probably in: Pig J166
Rabbit J166
β-Hydroxylicochalcone A-β-D-glucuronide
Formed from: β-Hydroxylicochalcone A
***p*-Hydroxylidamidine**
Formed from: Lidamidine
***p*-Hydroxylidoflazine**
Formed from: Lidoflazine
5-Hydroxylinogliride
Formed from: Linogliride
2-Hydroxylisuride
Formed from: Lisuride
12-Hydroxylisuride
Probably formed from: Lisuride
13-Hydroxylisuride
Probably formed from: Lisuride
2-Hydroxylofepramine
Probably formed from: Lofepramine
3-Hydroxyloflazepate
Probably formed from: Loflazepate
→ Decarboxy-3-hydroxyloflazepate
Probably in: Baboon D752
Dog D752
Man D752
4′-Hydroxyloflazepate
Probably formed from: Loflazepate
→ Decarboxy-4′-hydroxyflolazepate
Probably in: Rat D752
Hydroxylonazolac
Formed from: Lonazolac
4-Hydroxyloprazolam
Formed from: Loprazolam
→ 4-Hydroxyloprazolam-β-D-glucuronide
Probably in: Dog C568, C570
Man C570
Rat C568, C570
4-Hydroxyloprazolam-β-D-glucuronide
Probably formed from: 4-Hydroxyloprazolam
Hydroxylorazepam
Formed from: 7-Chloro-5-(*o*-chlorophenyl)-1,3-dihydro-3-hydroxy-2*H*-1,4-benzodiazepin-2-one
5′-Hydroxylosigamone
Formed from: Losigamone
Hydroxyloxapine
Formed from: Loxapine
2′-Hydroxylupalbigenin
→ Lupinisoflavone F
In: Aspergillus E95
4″-Hydroxyluteone
Formed from: Luteone
→ 6-(*cis*-3-Carboxy-2-buten-1-yl)-2′,4′,5,7-tetrahydroxyisoflavone
Probably in: Rat G339
5″-Hydroxyluteone
Formed from: Luteone
13-Hydroxylysergic diethylamide
Formed from: Lysergic diethylamide
→ 13-Hydroxylysergic diethylamide-β-D-glucuronide
Probably in: Rat A3720
14-Hydroxylysergic diethylamide
Formed from: Lysergic diethylamide
→ 14-Hydroxylysergic diethylamide-β-D-glucuronide
Probably in: Rat A3720
13-Hydroxylysergic diethylamide-β-D-glucuronide
Probably formed from: 13-Hydroxylysergic diethylamide
14-Hydroxylysergic diethylamide-β-D-glucuronide
Probably formed from: 14-Hydroxylysergic diethylamide
6a-Hydroxymaackiain
Formed from: Maackiain
→ 6,6a-Dihydroxymaackiain
Probably in: Colletorichum J345
→ Pisatin
In: Pea F396, F860, G345
***p*-Hydroxymalonanilic acid**
Probably formed from: Malonanilic acid
→ *p*-Hydroxymalonanilic acid-β-D-glucoside
Probably in: Arachis D52
***p*-Hydroxymalonanilic acid-β-D-glucoside**
Probably formed from: *p*-Hydroxymalonanilic acid

o-Hydroxymandelaldehyde
Probably formed from: *o*-Octopamine
→ *o*-Hydroxymandelic acid
Probably in: Rat C823
→ *o*-Hydroxyphenylglycol
Probably in: Rat C823

m-Hydroxymandelaldehyde
Probably formed from: *m*-Octopamine

p-Hydroxymandelaldehyde
Formed from: Octopamine
Probably formed from: Naringenin
p-Sympatol
→ *p*-Hydroxymandelic acid
In: Man A1372, A3791
Periplaneta A2336
Pig A1287
See also: A3777
→ *p*-Hydroxyphenylglycol
In: Man B569
Monkey A1908
Pig A1287, A1679
Sheep A57

o-Hydroxymandelic acid
Formed from: 2-(*o*-Hydroxyphenyl)-N,N-dimethyl-2-oxoethylamine
2-(*o*-Hydroxyphenyl)-N-methyl-2-oxoethylamine
2-(*o*-Hydroxyphenyl)-2-oxoethylamine
Probably formed from: *o*-Hydroxymandelaldehyde

m-Hydroxymandelic acid
Formed from: Etilefrine
2-(*m*-Hydroxyphenyl)-N,N-dimethyl-2-oxoethylamine
2-(*m*-Hydroxyphenyl)-N-methyl-2-oxoethylamine
2-(*m*-Hydroxyphenyl)-2-oxoethylamine
m-Sympatol
L-*m*-Tyrosine
→ *m*-Hydroxyphenylglyoxylic acid
In: Aspergillus A1204
Pseudomonas G794
Rhodotorula H90

p-Hydroxymandelic acid
Formed from: *p*-Hydroxymandelaldehyde
Ethyl *p*-hydroxymandelate
p-Hydroxyphenylacetic acid
2-(*p*-Hydroxyphenyl)-N,N-dimethyl-2-oxoethylamine
p-Hydroxyphenylglyoxylic acid
2-(*p*-Hydroxyphenyl)-N-methyl-2-oxoethylamine
2-(*p*-Hydroxyphenyl)-2-oxoethylamine
p-Hydroxyphenylpyruvic acid
p-Hydroxystyrene
L-Mandelic acid
Phellamurin
→ *p*-Hydroxyphenylglyoxylic acid
In: Maycoletopsis K688
Aspergillus A1204
Pseudomonas G794
Rhodotorula H90
See also: A329, A2336, A3306, A3594, G159
→ *p*-Hydroxybenzaldehyde
In: Aspergillus A2691
Coriolus G656
Pseudomonas A2681
See also: A2674, A3833

m-Hydroxymandelonitrile
Formed from: *m*-Hydroxybenzaldehyde
→ *m*-Hydroxybenzaldehyde
With hydroxynitrile lyase H944

p-Hydroxymandelonitrile
Formed from: Dhurrin
p-Hydroxyphenylacetaldoxime
p-Hydroxyphenylacetonitrile
N-Hydroxy-L-tyrosine
Taxiphyllin
→ Dhurrin
In: Sorghum F353, G228, K313
Triglochin D58
→ 3,4-Dihydroxymandelonitrile
Probably in: Thalictrum B492
Triglochin B492
→ *p*-Hydroxybenzaldehyde
In: Sorghum F353

1′-Hydroxymaprotiline
Formed from: Maprotiline

2-Hydroxymaprotiline
Formed from: Maprotiline

3-Hydroxymaprotiline
Formed from: Maprotiline

5-Hydroxymarmesin
Probably formed from: Marmesin
→ 5-Methoxypsoralen
In: Ficus B757
Ruta B757
2-Hydroxymecloqualone
Formed from: Mecloqualone
6-Hydroxymecloqualone
Formed from: Mecloqualone
7-Hydroxymecloqualone
Formed from: Mecloqualone
8-Hydroxymecloqualone
Formed from: Mecloqualone
Hydroxymedicarpin
Formed from: Medicarpin
***p*-Hydroxymefenorex**
Probably formed from: Mefenorex
→ 3,4-Dihydroxymefenorex
Probably in: Man J544
Hydroxymefloquine
Formed from: Mefloquine
6-Hydroxymelatonin
Formed from: Melatonin
→ Melatonin-6-yl-β-D-glucuronide
In: Man E441
→ Melatonin-6-yl sulphate
In: Man E194, E441, J31
Rat H413
6-Hydroxymellein
→ 6-Methoxymellein
In: Carrot J13, K730
2-(7-Hydroxy-*p*-mentha-1,8(9)-dien-3-yl)-5-(1-hydroxypent-1-yl)resorcinol
Probably formed from: 7-Hydroxycannabidiol
2-(7-Hydroxy-*p*-mentha-1,8(9)-dien-3-yl)-5-(4-hydroxypent-1-yl)resorcinol
Probably formed from: 7-Hydroxycannabidiol
***p*-Hydroxymeperidine**
→ 3′,4′-Dihydroxymeperidine
In: Cat D546
Dog D546
Guinea pig D546
Rabbit D546
Rat D546
→ *p*-Hydroxynormeperidine
In: Cat D546
Dog D546
Guinea pig D546
Rabbit D546
Rat D546
N-Hydroxymephentermine
Formed from: Mephentermine
→ N-Hydroxyphentermine
In: Rat G79, G536
→ Mephentermine
In: Microorganism G238
***p*-Hydroxymephentermine**
Formed from: Mephentermine
→ *p*-Hydroxyphentermine
In: Rat G79, G238, G536
5-Hydroxy-3-mercapto-4-methoxybenzoic acid
→ 5-Hydroxy-4-methoxy-3-methylthiobenzoic acid
In: Man C917
7-Hydroxymesoridazine
Formed from: Mesoridazine
Probably formed from: 7-Hydroxythioridazine
2-Hydroxymestranol
Formed from: 17α-Ethynyl-17β-oestradiol
→ 2-Hydroxymestranol-β-D-glucuronide
Probably in: Rat C378
→ 2-Hydroxymestranol sulphate
Probably in: Rat C378
2-Hydroxymestranol-β-D-glucuronide
Probably formed from: 2-Hydroxymestranol
2-Hydroxymestranol sulphate
Probably formed from: 2-Hydroxymestranol
***p*-Hydroxymethaphenilene**
Formed from: Methaphenilene
→ *p*-Aminophenol
Probably in: Rat F297
12-Hydroxymetergoline
Formed from: Metergoline
Hydroxymethocarbamol
Formed from: Methocarbamol
7-Hydroxymethotrexate
Formed from: Methotrexate
3′-Hydroxy-4′-methoxyacetanilide
Formed from: 3′,4′-Dihydroxyacetanilide
4′-Hydroxy-3′-methoxyacetanilide
Formed from: 3′,4′-Dihydroxyacetanilide
N-Hydroxy-*p*-methoxyacetanilide
Formed from: *p*-Methoxyacetanilide

→ *p*-Methoxyphenylhydroxylamine
In: Mouse C609

2-Hydroxy-*p*-methoxyacetophenone
Probably formed from: *p*-Methoxyacetophenone

2′-Hydroxy-3′-methoxyacetophenone
Probably formed from: 8-Methoxychromone

2′-Hydroxy-4′-methoxyacetophenone (paeonol)
→ 2′,4′- Dihydroxyacetophenone
In: Guinea pig B506
Man B920, C915
Mouse B506
Rabbit B506
Rat A3336, B506, E849, K209
→ 2′,3′- Dihydroxy-4′-methoxyacetophenone
Probably in: Rat E849
→ 2′,5′- Dihydroxy-4′-methoxyacetophenone
In: Guinea pig B506
Man B920, C915
Mouse B506
Rabbit B506
Rat A3336, B506, E849, K209
→ 2′,6′- Dihydroxy-4′-methoxyacetophenone
Probably in: Rat E849
→ 2′,ω- Dihydroxy-4′-methoxyacetophenone
In: Rat E849
→ 2′-Hydroxy-4′-methoxyacetophenone-β-D-glucuronide
In: Man B920
→ 2′-Hydroxy-4′-methoxyacetophenone sulphate
In: Rat K209

2′-Hydroxy-6′-methoxyacetophenone
Formed from: 2′,6′- Dihydroxyacetophenone
→ 2′,3′- Dihydroxy-6′-methoxyacetophenone
In: Rat A357
→ 3′,6′- Dihydroxy-2′-methoxyacetophenone
In: Rat A357

3′-Hydroxy-4′-methoxyacetophenone
(acetoisovanillone)
Formed from: 3,′4′- Dihydroxyacetophenone

4′-Hydroxy-3′-methoxyacetophenone
(acetovanillone)
Formed from: ω-(4-Carboxy-2-methoxyphenoxy)acetovanillone
3,′4′- Dihydroxyacetophenone
Ferulic acid
γ-Oryzanol
→ 3′,4′- Dihydroxyacetophenone
Probably in: Rat E849
→ 2′,4′- Dihydroxy-3′-methoxyacetophenone
Probably in: Rat E849
→ 2′,4′- Dihydroxy-5′-methoxyacetophenone
In: Rat E849
→ 3′,4′- Dihydroxy-5′-methoxyacetophenone
In: Rat E849
→ 4′,ω- Dihydroxy-3′-methoxyacetophenone
In: Solanum H906
→ 3′,4′-Dimethoxyacetophenone
In: Phanerochaete G782
→ 1-(4-Hydroxy-3-methoxyphenyl)ethanol
In: Rat E849
→ Poly-4′-hydroxy-3′-methoxyacetophenone
In: Bjerkandera K565

2′-Hydroxy-4′-methoxyacetophenone-β-D-glucuronide
Formed from: 2′-Hydroxy-4′-methoxyacetophenone

2′-Hydroxy-4′-methoxyacetophenone sulphate
Formed from: 2′-Hydroxy-4′-methoxyacetophenone

3-Hydroxy-4-methoxyallylbenzene
Formed from: 3,4-Dimethoxyallylbenzene

2-Hydroxy-3-methoxyamitriptyline
Probably formed from: 2,3-Dihydroxyamitriptyline

N-Hydroxy-*p*-methoxyamphetamine
Formed from: *p*-Methoxyamphetamine

3-Hydroxy-4-methoxyamphetamine
Probably formed from:
3,4-Dihydroxyamphetamine
3-Hydroxy-4-methoxy-N-methylamphetamine

4-Hydroxy-3-methoxyamphetamine
(3-methoxy-α-methyltyramine)
Formed from: 3,4-Dihydroxyamphetamine
3,4-Dimethoxyamphetamine
→ 3,4-Dihydroxyamphetamine
Probably in: Dog A3697
Monkey A3697
→ 4-Hydroxy-3-methoxyphenylacetone
Probably in: Dog A3697, B403, B362
Man A3697, B403
Monkey A3697, B403, B362

5-Hydroxy-2-methoxyamphetamine
Probably formed from: 1-(5-Hydroxy-2-methoxyphenyl)2-methylaminopropane
→ 5-Hydroxy-2-methoxyphenylacetone
Probably in: Man C434

4-Hydroxy-3-methoxybenzalacetone
Formed from: Feruloyl CoA

2-Hydroxy-3-methoxybenzaldehyde (*o*-vanillin)
Probably formed from: 2-Hydroxy-3-methoxybenzylamine
→ 2-Formyl-6-methoxyphenyl-β-D-glucoside
In: Manduca G897
→ 2-Hydroxy-3-methoxybenzyl alcohol
In: Dunaliella G459
→ *o*-Vanillic acid
In: Acetobacterium F655
Desulphovibrio F666
Rhodopseudomonas G246
See also: G865

2-Hydroxy-5-methoxybenzaldehyde
Probably formed from: 2-Hydroxy-5-methoxybenzylamine
→ 2-Hydroxy-5-methoxybenzoic acid
Probably in: Rat G865

3-Hydroxy-4-methoxybenzaldehyde
See: Isovanillin

4-Hydroxy-3-methoxybenzaldehyde
See: Vanillin

3′-Hydroxy-4′-methoxybenzanilide
Formed from: 3′,4′-Dihydroxybenzanilide

4′-Hydroxy-3′-methoxybenzanilide
Formed from: 3′,4′-Dihydroxybenzanilide

3-Hydroxy-3′-methoxybenzidine
Formed from: 3,3′-Dimethoxybenzidine
→ N-Acetyl-3-hydroxy-3′-methoxybenzidine
Probably in: Rat C623

(*Z*)-3-(6-Hydroxy-7-methoxybenzofuran-5-yl)-2-propenoic acid
Formed from: 8-Methoxpsoralen

3-(5-(6-Hydroxy-7-methoxybenzofuryl)) propionic acid
Formed from: 8-Methoxypsoralen

2-Hydroxy-3-methoxybenzoic acid
See: *o*-Vanillic acid

2-Hydroxy-4-methoxybenzoic acid
Formed from: 2-Methoxynaphthalene

2-Hydroxy-5-methoxybenzoic acid
(5-methoxysalicylic acid)
Probably formed from: 2-Hydroxy-5-methoxybenzaldehyde
→ Gentisic acid
Probably in: Rat G865
→ 4-Methoxycatechol
In: Trichosporon D232

3-Hydroxy-4-methoxybenzoic acid
See: Isovanillic acid

3-Hydroxy-5-methoxybenzoic acid
Formed from: 3,5-Dihydroxybenzoic acid
→ 3,4-Dihydroxy-5-methoxybenzoic acid
In: Aspergillus A1130
→ 3,5-Dimethoxybenzoic acid
In: Phanerochaete J339

4-Hydroxy-3-methoxybenzoic acid
See: Vanillic acid

4-Hydroxy-4′-methoxybenzoin
→ Anisaldehyde
In: Pseudomonas E737

3-Hydroxy-4-methoxybenzonitrile
Formed from: 3,4-Dihydroxybenzonitrile

4-Hydroxy-3-methoxybenzonitrile
Formed from: 3,4-Dihydroxybenzonitrile

2-Hydroxy-4-methoxybenzophenone
→ 2,4-Dihydroxybenzophenone
In: Rat G692
→ 2,2′-Dihydroxy-4-methoxybenzophenone
In: Rat G692

3-Hydroxy-4-methoxybenzophenone
Formed from: 3,4-Dihydroxybenzophenone

4-Hydroxy-3-methoxybenzophenone
Formed from: 3,4-Dihydroxybenzophenone

4-Hydroxy-4′-methoxybenzophenone
Probably formed from: 1,1,1-Trichloro-2-(*p*-hydroxyphenyl)-2-(*p*-methoxyphenyl)ethane

6-Hydroxy-7-methoxy-2*H*-1-benzopyran-2-one
See: Isoscopoletin

7-Hydroxy-8-methoxy-2*H*-1-benzopyran-2-one
See: Hydrangetin

2-Hydroxy-7-methoxy-1,4-benzoxazin-3-one
Formed from: 2-Hydroxy-7-methoxy-1,4-benzoxazin-3-one-β-D-glucuronide

2-Hydroxy-7-methoxy-1,4-benzoxazin-3-one-β-D-glucuronide
→ 2-Hydroxy-7-methoxy-1,4-benzoxazin-3-one
In: Secale K512

4-Hydroxy-4-(*p*-methoxybenzoyl)butyric acid
Formed from: Cytemba

4-Hydroxy-3-methoxybenzoylformic acid
See: 4-Hydroxy-3-methoxyphenylglyoxylic acid

N-(4-Hydroxy-3-methoxybenzoyl)-*o*-(2-(piperidin-2-yl)ethyl)aniline
Probably formed from: O-Demethylencainide

1-Hydroxy-6-methoxybenzpyrene
Formed from: 1,6-Dimethoxybenzpyrene

3-Hydroxy-6-methoxybenzpyrene
Formed from: 3,6-Dimethoxybenzpyrene

4′-Hydroxy-3′-methoxybenztropine
Probably formed from: *p*-Hydroxybenztropine

2-Hydroxy-3-methoxybenzyl alcohol
Formed from: 2-Hydroxy-3-methoxybenzaldehyde

3-Hydroxy-4-methoxybenzyl alcohol
See: Isovanillyl alcohol

4-Hydroxy-3-methoxybenzyl alcohol
See: Vanillyl alcohol

2-Hydroxy-3-methoxybenzylamine
→ 2-Hydroxy-3-methoxybenzaldehyde
Probably in: Rat G865

2-Hydroxy-5-methoxybenzylamine
→ 2-Hydroxy-5-methoxybenzaldehyde
Probably in: Rat G865

3-Hydroxy-4-methoxybenzylamine
Formed from: 3,4-Dihydroxybenzylamine
→ Isovanillin
Probably in: Rat G865

4-Hydroxy-3-methoxybenzylamine
Formed from: Capsaicin
Dihydrocapsaicin
3,4-Dihydroxybenzylamine
3,4-Dimethoxybenzylamine
Veratrylamine
→ Vanillin
In: Pseudomonas F3, J16
See also: D574, G865, K761

6-(3-(3-Hydroxy-4-methoxybenzylamino)-2-hydroxypropoxy)-2(1*H*)-quinolinone
Formed from: Toborinone
→ 6-(3-(3-Hydroxy-4-methoxybenzylamino)-2-hydroxypropoxy)-2(1*H*)-quinolinone-β-D-glucuronide
Probably in: Dog J154
Rat J154

6-(3-(4-Hydroxy-3-methoxybenzylamino)-2-hydroxypropoxy)-2(1*H*)-quinolinone
Formed from: Toborinone
→ 6-(3-(4-Hydroxy-3-methoxybenzylamino)-2-hydroxypropoxy)-2(1*H*)-quinolinone-β-D-glucuronide
Probably in: Dog J154
Rat J154

6-(3-(3-Hydroxy-4-methoxybenzylamino)-2-hydroxypropoxy)-2(1*H*)-quinolinone-β-D-glucuronide
Probably formed from: 6-(3-(3-Hydroxy-4-methoxybenzylamino)-2-hydroxypropoxy)-2(1*H*)-quinolinone

6-(3-(4-Hydroxy-3-methoxybenzylamino)-2-hydroxypropoxy)-2(1*H*)-quinolinone-β-D-glucuronide
Probably formed from: 6-(3-(4-Hydroxy-3-methoxybenzylamino)-2-hydroxypropoxy)-2(1*H*)-quinolinone

N-(3-Hydroxy-4-methoxybenzyl)-N′,N′-dimethyl-N-(2-pyridyl)ethylenediamine
Probably formed from: O-Demethylpyrilamine

N-(4-Hydroxy-3-methoxybenzyl)-N′,N′-dimethyl-N-(2-pyridyl)ethylenediamine
Probably formed from: O-Demethylpyrilamine

5-(4-Hydroxy-3-methoxybenzyl)-2-(5-methyl-4-imidazolylmethylthio)ethylamino-4-pyrimidone
Formed from: 5-(3,4-Dihydroxybenzyl)-2-(5-methyl-4-imidazolylmethylthio) ethylamino-4-pyrimidone
→ 5-(3-Methoxy-4-sulphatoxybenzyl)-2-(5-methyl-4-imidazolylmethylthio) ethylamino-4-pyrimidone
Probably in: Dog C875

3-Hydroxy-4-methoxybiphenyl
Probably formed from: 3,4-Dihydroxybiphenyl

4-Hydroxy-3-methoxybiphenyl
Probably formed from: 3,4-Dihydroxybiphenyl

7-Hydroxy-8-methoxybisdemethylchlorpromazine
Probably formed from: 7-Hydroxy-8-methoxydemethylchlorpromazine

4′-Hydroxy-5′-methoxybrotizolam
Probably formed from: *p*-Hydroxybrotizolam
→ 2-Bromo-4-(2-chloro-4-hydroxy-5-methoxyphenyl)-9-hydroxymethyl-6*H*-thieno[3,2-*f*]-1,2,4-triazolo[4,3-*a*]-1,4-diazepine
Probably in: Dog E91
Man E91
Rat E91

6-Hydroxy-5-methoxybufuralol
Probably formed from: 5,6-Dihydroxy-bufuralol
→ 1′,6-Dihydroxy-5-methoxybufuralol
Probably in: Dog A2806
Man A2806
Rat A2806

5-Hydroxy-8-methoxybutamoxane
Probably formed from: 8-Methoxy-butamoxane

3′-Hydroxy-4′-methoxybutyrophenone
Formed from: 3′,4′-Dihydroxybutyrophenone

4′-Hydroxy-3′-methoxybutyrophenone
Formed from: 3′,4′-Dihydroxybutyrophenone

2-Hydroxy-3-methoxycarbamazepine
Probably formed from: 2,3-Dihydroxy-carbamazepine
→ 2-Hydroxy-3-methoxycarbamazepine-N-β-D-glucuronide
Probably in: Man A1967
→ 2-Hydroxy-3-methoxycarbamazepine-O-β-D-glucuronide
Probably in: Man A1967, J101

2-Hydroxy-3-methoxycarbamazepine-N-β-D-glucuronide
Probably formed from: 2-Hydroxy-3-methoxy-carbamazepine

2-Hydroxy-3-methoxycarbamazepine-O-β-D-glucuronide
Probably formed from: 2-Hydroxy-3-methoxy-carbamazepine

***o*-Hydroxy-N′-methoxycarbonylphenylhydrazone**
Probably formed from: *o*-Methoxy-N′-methoxycarbonylphenylhydrazone
→ N′-Carboxy-*o*-hydroxyphenylhydrazone
Probably in: Rat E599
→ *o*-Hydroxy-N′-methoxycarbonylphenylhydrazone-O-sulphate
Probably in: Rat E599

***o*-Hydroxy-N′-methoxycarbonylphenylhydrazone-O-sulphate**
Probably formed from: *o*-Hydroxy-N′-methoxycarbonylphenylhydrazone

7-Hydroxy-8-methoxychlorpromazine
Probably formed from: 7,8-Dihydroxy-chlorpromazine
→ 2-Chloro-7-hydroxy-8-methoxyphenothiazine
Probably in: Man A693
→ 7,8-Dimethoxychlorpromazine
Probably in: Man A693
→ 7-Hydroxy-8-methoxydemethylchlorpromazine
Probably in: Man A693

8-Hydroxy-7-methoxychlorpromazine
Probably formed from: 7,8-Dihydroxy-chlorpromazine

4-Hydroxy-3-methoxycibenzoline
Probably formed from: *p*-Hydroxycibenzoline

4-Hydroxy-3-methoxycinnamaldehyde
See: Coniferaldehyde

3-Hydroxy-4-methoxycinnamic acid
See: Isoferulic acid

3-Hydroxy-5-methoxycinnamic acid
Formed from: 3,5-Dimethoxycinnamic acid
→ 3,5-Dihydroxycinnamic acid
Probably in: Microorganism A261

4-Hydroxy-3-methoxycinnamic acid
See: Ferulic acid

4-Hydroxy-3-methoxycinnamyl alcohol
See: Coniferyl alcohol

Hydroxymethoxyciramadol
Formed from: Ciramadol

3′-Hydroxy-4′-methoxyclobazam
Probably formed from: 3′,4′-Dihydroxy-clobazam
→ 3′-Hydroxy-4′-methoxyclobazam-β-D-glucuronide
Probably in: Rat G572
→ 3′-Hydroxy-4′-methoxyclobazam-O-sulphate

Probably in: Rat G572

4′-Hydroxy-3′-methoxyclobazam
Probably formed from: 3′,4′-Dihydroxy-clobazam
p-Hydroxyclobazam
→ 4′-Hydroxy-3′-methoxyclobazam-β-D-glucuronide
Probably in: Rat G572
→ 4′-Hydroxy-3′-methoxyclobazam-O-sulphate
Probably in: Rat G572
→ 4′-Hydroxy-3′-methoxynorclobazam
Probably in: Man B22
Monkey B22
Rat B22, G572

3′-Hydroxy-4′-methoxyclobazam-β-D-glucuronide
Probably formed from: 3′-Hydroxy-4′-methoxyclobazam

4′-Hydroxy-3′-methoxyclobazam-β-D-glucuronide
Probably formed from: 4′-Hydroxy-3′-methoxyclobazam

3′-Hydroxy-4′-methoxyclobazam-O-sulphate
Probably formed from: 3′-Hydroxy-4′-methoxyclobazam

4′-Hydroxy-3′-methoxyclobazam-O-sulphate
Probably formed from: 4′-Hydroxy-3′-methoxyclobazam

4′-Hydroxy-3′-methoxyclomipramine
Probably formed from: 4′-Hydroxy-clomipramine

7-Hydroxy-6-methoxycoumarin
See: Scopoletin

7-Hydroxy-8-methoxycoumarin (hydrangetin)
→ 7-Glucosyloxy-8-methoxycoumarin
In: Forsythia A3365
Nicotiana B442, B819

8-Hydroxy-7-methoxycoumarin
Probably formed from: 7,8-Dihydroxy-coumarin

2-(7-Hydroxy-8-methoxycoumarin-6-yl)acetic acid
Formed from: 8-Methoxypsoralen
→ 2-Hydroxy-2-(7-hydroxy-8-methoxycoumarin-6-yl)acetic acid
Probably in: Goat G456
Hen G456

2-(7-Hydroxy-8-methoxycoumarin-6-yl)ethanol
Formed from: 8-Methoxypsoralen

(7-Hydroxy-8-methoxycoumarin-6-yl)glycol
Formed from: 8-Methoxypsoralen

3′-Hydroxy-5′-methoxydehydrocyclopeptine
Probably formed from: 3′-Methoxydehydro-cyclopeptine

7-Hydroxy-8-methoxydemethylchlorpromazine
Probably formed from: 7-Hydroxy-8-methoxy-chlorpromazine
→ 7-Hydroxy-8-methoxybisdemethylchlor-promazine
Probably in: Man A693

4′-Hydroxy-3′-methoxydemethyltriflubazam
Probably formed from:
3′,4′-Dihydroxydemethyltriflubazam

3′-Hydroxy-4′-methoxydiclofenac
Probably formed from: 4′-Hydroxydiclofenac

6-Hydroxy-4-methoxy-2,3-dimethylbenzaldehyde
→ 5-Methoxy-2,3,4-trimethylphenol
In: Penicillium B62

6-Hydroxy-7-methoxy-2,2-dimethyl-2*H*-benzo[*b*]pyran
Formed from: Precocene II
→ 6-Hydroxy-7-methoxy-2,2-dimethyl-2*H*-benzo[*b*]pyran-β-D-glucoside
In: Grasshopper B439

7-Hydroxy-6-methoxy-2,2-dimethyl-2*H*-benzo[*b*]pyran
Formed from: Precocene II
→ 7-Hydroxy-6-methoxy-2,2-dimethyl-2*H*-benzo[*b*]pyran-β-D-glucoside
In: Grasshopper B439

6-Hydroxy-7-methoxy-2,2-dimethyl-2*H*-benzo[*b*]pyran-β-D-glucoside
Formed from: 6-Hydroxy-7-methoxy-2,2-dimethyl-2*H*-benzo[*b*]pyran

7-Hydroxy-6-methoxy-2,2-dimethyl-2*H*-benzo[*b*]pyran-β-D-glucoside
Formed from: 7-Hydroxy-6-methoxy-2,2-dimethyl-2*H*-benzo[*b*]pyran

3-Hydroxy-4-methoxy-N,N-dimethylphenethylamine
Formed from: 3,4-Dihydroxy-N,N-dimethylphenethylamine

4-Hydroxy-3-methoxy-N,N-dimethylphenethylamine
Formed from: 3,4-Dihydroxy-N,N-dimethylphenethylamine

3-Hydroxy-4-methoxy-N,N-dimethylphenylethanolamine
Formed from: N-Methyladrenaline

5-Hydroxy-2-(((4-methoxy-3,5-dimethyl-2-pyridinyl)methyl)sulphinyl)-1*H*-benzimidazole
Formed from: Omeprazole

4-Hydroxy-3-methoxydisopyramide
Probably formed from: *p*-Hydroxy-disopyramide

Hydroxymethoxydrobuline
Probably formed from: Hydroxydrobuline

3-(4-(1-Hydroxy-2-methoxyethyl)phenoxy)-1-isopropylaminopropan-2-ol
Formed from: Metoprolol
→ 3-(4-(1,2-Dihydroxyethyl)phenoxy)-1-isopropylaminopropan-2-ol
Probably in: Rat A1750

4′-Hydroxy-3′-methoxyflavan-4α-ol
Probably formed from: 4′-Hydroxyflavan-4α-ol

4′-Hydroxy-3′-methoxyflavan-4β-ol
Probably formed from: 4′-Hydroxyflavan-4β-ol

5-Hydroxy-6-methoxyflavone
Formed from: 5,6-Dimethoxyflavone
→ 5,6-Dimethoxyflavone
Probably in: Citrus A3987

6-Hydroxy-5-methoxyflavone
Formed from: 5,6-Dihydroxyflavone

7-Hydroxy-8-methoxyflavone
Formed from: 7,8-Dihydroxyflavone
→ 7,8-Dimethoxyflavone
Probably in: Citrus A3987

8-Hydroxy-7-methoxyflavone
Formed from: 7,8-Dimethoxyflavone

3′-Hydroxy-4′methoxyflurbiprofen
Probably formed from: 3′,4′-Dihydroxy-flurbiprofen

8-Hydroxy-7-methoxyfluoranthene
Formed from: Fluoranthene

1-Hydroxy-2-methoxyindane (*cis* and *trans* isomers)
Formed from: 2-Methoxyindane

4′-Hydroxy-7-methoxyisoflavone
See: Isoformononetin

7-Hydroxy-4′-methoxyisoflavone
See: Formononetin

4-Hydroxy-5-methoxyisophthalic acid
Probably formed from: 5,5′-Dicarboxy-2,2′-dihydroxy-3,3′-dimethoxybiphenyl
2-(5-Formyl-2-hydroxy-3-methoxyphenyl)-3-hydroxy-1-(4-hydroxy-3,5-dimethoxyphenyl)-1-propanone
→ 4,5-Dihydroxyisophthalic acid
In: Pseudomonas E733
→ 3,4-Dihydroxy-5-methoxybenzoic acid
In: Pseudomonas E733

Hydroxymethoxylorazepam
Formed from: 7-Chloro-5-(*o*-chlorophenyl)-1,3-dihydro-3-hydroxy-2*H*-1,4-benzodiazepin-2-one

4-Hydroxy-3-methoxymandelaldehyde
Formed from: 4-Hydroxy-3-methoxyphenyl-glycol
D-Normetanephrine
Probably formed from: D-Metanephrine

3-Hydroxy-4-methoxymandelic acid
Formed from: 3,4-Dihydroxymandelic acid
2-(3-Hydroxy-4-methoxyphenyl)-N,N-dimethyl-2-oxoethylamine
2-(3-Hydroxy-4-methoxyphenyl)-2-oxoethylamine

4-Hydroxy-3-methoxymandelic acid
(vanilmandelic acid, VMA)
Formed from: 3,4-Dihydroxymandelic acid
2-(4-Hydroxy-3-methoxyphenyl)-N,N-dimethyl-2-oxoethylamine
2-(4-Hydroxy-3-methoxyphenyl)-2-oxoethylamine
→ 3,4-Dihydroxymandelic acid
In: Rat A3063
→ 4-Hydroxy-3-methoxymandelic acid-4-sulphate
In Beef F196
Man D613
Rat A153, A1343, C441
→ 4-Hydroxy-3-methoxyphenylglyoxylic acid
In: Aspergillus A1204
Microorganism J256
Pseudomonas G794

4-Hydroxy-3-methoxymandelic acid-4-sulphate
Probably formed from: 4-Hydroxy-3-methoxy-mandelic acid

3-Hydroxy-4-methoxymaprotiline
Probably formed from: 2,3-Dihydro-2,3-dihydroxymaprotiline

4-Hydroxy-3-methoxymefenorex
Probably formed from: 3,4-Dihydroxy-mefenorex

4′-Hydroxy-3′-methoxymeperidine
Probably formed from: 3′,4′-Dihydroxy-meperidine

6-Hydroxy-5-methoxy-2-(((4-methoxy-3,5-dimethyl-2-pyridinyl)methyl)sulphinyl)-1*H*-benzimidazole
Formed from: Omeprazole

4-Hydroxy-3-methoxy-4′-methylamino-azobenzene
Probably formed from: 3′-Methoxy-4-methyl-aminoazobenzene

2-Hydroxy-5-methoxy-4-methylamphetamine
Formed from: 2,5-Dimethoxy-4-methyl-amphetamine

3-Hydroxy-4-methoxy-N-methylamphetamine
Formed from: 3,4-Dimethoxy-N-methyl-amphetamine
→ 3-Hydroxy-4-methoxyamphetamine
Probably in: Man F187

4-Hydroxy-3-methoxy-N-methylamphetamine
Formed from: 3,4-Dimethoxy-N-methyl-amphetamine
→ 4-Hydroxy-3-methoxyphenylacetone
Probably in: Man F187
Rat E988

5-Hydroxy-2-methoxy-4-methylamphetamine
Formed from: 2,5-Dimethoxy-4-methyl-amphetamine
→ 2,5-Dimethoxy-4-methylamphetamine
Probably in: Rabbit A2427, A2889

2-Hydroxy-4-methoxy-6-methylbenzaldehyde
→ 2-Hydroxy-4-methoxy-6-methylbenzyl alcohol
In: Penicillium B62

2-Hydroxy-4-methoxy-6-methylbenzyl alcohol
Formed from: 2-Hydroxy-4-methoxy-6-methylbenzaldehyde
→ 5-Methoxy-2,3-dimethylphenol
In: Penicillium B62

3-Hydroxy-4-methoxy-N-(N-(*m*-(N-methylcarboxamido)phenyl)-β-alanyl)phenethylamine
Formed from: Ecabapide

4-Hydroxy-3-methoxy-N-(N-(*m*-(N-methylcarboxamido)phenyl)-β-alanyl)phenethylamine
Formed from: Ecabapide
→ 3,4-Dihydroxy-N-(N-(*m*-(N-methylcarboxamido)phenyl)-β-alanyl)phenethylamine
Probably in: Man J806

6-Hydroxy-7-methoxy-4-methylcoumarin
Formed from: 6,7-Dihydroxy-4-methyl-coumarin
→ 7-Methoxy-4-methylcoumarin-6-yl-β-D-glucuronide
Probably in: Rat A773

8-Hydroxy-7-methoxy-4-methylcoumarin
Formed from: 7,8-Dihydroxy-4-methyl-coumarin

3-Hydroxy-4-methoxy-N-(1-(2,3-methylenedioxyphenyl)isopropyl)phenylethanolamine
Formed from: N-(1-(2,3-Methylenedioxy-phenyl)isopropyl)noradrenaline

9-Hydroxy-9-methoxy-α-methylfluorene-2-acetic acid
Formed from: d-α-Methylfluorene-2-acetic acid
l-α-Methylfluorene-2-acetic acid

6-Hydroxy-2-methoxy-4-methyl-8-nitroso-5-(3-trifluoromethylphenoxy)quinoline
Probably formed from: 8-Amino-2,6-dimethoxy-4-methyl-5-(3-trifluoromethylphenoxy)quinoline

2-Hydroxy-4-methoxy-3-O-methyloestrone
Formed from: 2-Hydroxy-4-methoxyloestrone

4-Hydroxy-2-methoxy-3-O-methyloestrone
Formed from: 4-Hydroxy-2-methoxyloestrone

4-Hydroxy-3-methoxy-N-methyl-α-oxophenethylamine
See: Metanephrone

6-(4-Hydroxy-6-methoxy-7-methyl-4-oxo-5-phthalanyl)-4-methyl-4-hexanoic acid
See: Mycophenolic acid

D-3-Hydroxy-4-methoxy-N-methylphenethylamine
See: D-Paranephrine

4-Hydroxy-3-methoxy-N-methylphenethylamine
Formed from: 3,4-Dihydroxy-N-methylphenethylamine
3-Methoxytyramine
→ 3,4-Dimethoxy-N-methylphenethylamine
In: Lophophora A3600
→ 4-Hydroxy-3-methoxy-N-methylphenethylamine-β-D-glucuronide
Probably in: Rat D990, E940
→ 4-Hydroxy-3-methoxy-N-methylphenylacetaldehyde
Probably in: Rat G865

3-Hydroxy-4-methoxy-α-methylphenethylamine
Formed from: 3,4-Dihydroxy-α-methylphenethylamine

4-Hydroxy-3-methoxy-α-methylphenethylamine
Formed from: 3,4-Dihydroxy-α-methylphenethylamine
→ 4-Hydroxy-3-methoxy-α-methylphenethylamine-β-D-glucuronide
Probably in: Rat A2964

3-Hydroxy-4-methoxy-N-methylphenethylamine
Formed from: 3-Hydroxy-4-methoxyphenethylamine
→ 3,4-Dimethoxy-N-methylphenethylamine
In: Lophophora A3600

4-Hydroxy-3-methoxy-N-methylphenethylamine-β-D-glucuronide
Probably formed from: 4-Hydroxy-3-methoxy-N-methylphenethylamine

4-Hydroxy-3-methoxy-α-methylphenethylamine-β-D-glucuronide
Probably formed from: 4-Hydroxy-3-methoxy-α-methylphenethylamine

4-Hydroxy-3-methoxy-α-methyl-L-phenylalanine
Formed from: 3,4-Dihydroxy-α-methyl-L-phenylalanine
→ 4-Hydroxy-3-methoxy-α-methyl-L-phenylalanine-O-sulphate
Probably in: Man G944

4-Hydroxy-3-methoxy-α-methyl-L-phenylalanine-O-sulphate
Formed from: 4-Hydroxy-3-methoxy-α-methyl-L-phenylalanine

3-Hydroxy-4-methoxy-6-methylphenylethanolamine
Formed from: 6-Methylnoradrenaline

D-*erythro*-3-Hydroxy-4-methoxy-α-methylphenylethanolamine
Formed from: D-*erythro*-3-Hydroxy-4-methoxy-α-methylnoradrenaline

L-*erythro*-3-Hydroxy-4-methoxy-α-methylphenylethanolamine
Formed from: L-*erythro*-3-Hydroxy-4-methoxy-α-methylnoradrenaline

L-*threo*-3-Hydroxy-4-methoxy-α-methylphenylethanolamine
Formed from: L-*threo*-3-Hydroxy-4-methoxy-α-methylnoradrenaline

4-Hydroxy-3-methoxy-N-methylphenylethanolamine
See: Metanephrine

3-(4-Hydroxy-5-methoxy-2-methylphenyl)-2-methyl-4(3*H*)-quinazolinone
Probably formed from: 3-(4-Hydroxy-2-methylphenyl)-2-methyl-4(3*H*)-quinazolinone

4-Hydroxy-5-methoxy-2-methylphenyl methyl sulphone
Probably formed from: 4-Hydroxy-5-methoxy-2-methylphenyl methyl sulphoxide

5-Hydroxy-4-methoxy-2-methylphenyl methyl sulphone
Formed from: Tolmesoxide

4-Hydroxy-5-methoxy-2-methylphenyl methyl sulphoxide
Formed from: Tolmesoxide
→ 4-Hydroxy-5-methoxy-2-methylphenyl methyl sulphone
Probably in: Dog B400
Man B400
Rat B400

4-Hydroxy-3-methoxy-α-methylphenylpropionic acid
Probably formed from: 3,4-Dihydroxy-α-methylphenylpropionic acid

4-Hydroxy-3-methoxy-N-methyl-L-*threo*-phenylserine
Formed from: 3,4-Dihydroxy-N-methyl-L-*threo*-phenylserine

3-Hydroxy-4-methoxy-N-(2-(2-(1-methyl-2-piperidinyl)ethyl)phenyl)benzamide
Probably formed from: 3,4-Dihydroxy-N-(2-(2-(1-methyl-2-piperidinyl)ethyl)phenyl) benzamide

4-Hydroxy-3-methoxy-N-(2-(2-(1-methyl-2-piperidinyl)ethyl)phenyl)benzamide
Probably formed from: 3,4-Dihydroxy-N-(2-(2-(1-methyl-2-piperidinyl)ethyl) phenyl)benzamide

4-Hydroxy-3-methoxy-α-methyl-N-propylphenethylamine
Formed from: 3,4-Hydroxy-α-methyl-N-propylphenethylamine

4′-Hydroxy-3′-methoxy-2-methylselenobenzanilide
Probably formed from: 3′,4′-Dihydroxy-2-methylselenobenzanilide

***trans*-2-(4-Hydroxy-3-methoxy-5-methylsulphonylphenyl)-5-(3,4,5-trimethoxyphenyl)tetrahydrofuran**
Formed from: *trans*-2-(3-Methoxy-5-methylsulphonyl-4-propoxyphenyl)-5-(3,4,5-trimethoxyphenyl)tetrahydrofuran
→ *trans*-2-(4-Hydroxy-3-methoxy-5-methylsulphonylphenyl)-5-(3,4,5-trimethoxyphenyl)tetrahydrofuran-β-D-glucuronide
In: Monkey F925

***trans*-2-(4-Hydroxy-3-methoxy-5-methylsulphonylphenyl)-5-(3,4,5-trimethoxyphenyl)tetrahydrofuran-β-D-glucuronide**
Formed from: *trans*-2-(4-Hydroxy-3-methoxy-5-methylsulphonylphenyl)-5-(3,4,5-trimethoxyphenyl)tetrahydrofuran

5-Hydroxy-4-methoxy-3-methylthiobenzoic acid
Formed from: 5-Hydroxy-3-mercapto-4-methoxybenzoic acid

3-Hydroxy-2-methoxymorphinan
Probably formed from: (+)-N-Cyclopropylmethyl-3-hydroxy-2-methoxymorphinan

2-Hydroxy-3-methoxy-6β-naltrexol
Probably formed from: 6β-Naltrexol

2-Hydroxy-3-methoxynaltrexone
Probably formed from: Naltrexone

1-Hydroxy-4-methoxynaphthalene
→ 1,4-Dihydroxynaphthalene
In: Sporomusa J199

1-Hydroxy-7-methoxynaphthalene
Formed from: *cis*-1,2-Dihydro-1,2-dihydroxy-7-methoxynaphthalene

2-Hydroxy-6-methoxynaphthalene
→ 2,6-Dihydroxynaphthalene
In: Sporomusa J199

2-Hydroxy-7-methoxynaphthalene
→ 2,7-Dihydroxynaphthalene
In: Sporomusa J199

1-(3-Hydroxy-4-methoxy-1-naphthoxy)-2,3-propanediol
Probably formed from: 1-(3,4-Dihydroxy-1-naphthoxy)-2,3-propanediol

1-(4-Hydroxy-3-methoxy-1-naphthoxy)-2,3-propanediol
Probably formed from: 1-(3,4-Dihydroxy-1-naphthoxy)-2,3-propanediol

2-Hydroxy-2-(6-methoxy-2-naphthyl)propionitrile
Formed from: N-Acetyl-6-methoxy-naphthalene

3-(4-Hydroxy-3-methoxy-5-nitrobenzylidene)-2,4-pentanedione
Formed from: 3-(3,4-Dihydroxy-5-nitrobenzylidene)-2,4-pentanedione

3′-Hydroxy-4′-methoxynomifensine
Probably formed from: 3′,4′-Dihydroxy-nomifensine

4′-Hydroxy-3′-methoxynomifensine
Probably formed from: 3′,4′-Dihydroxy-nomifensine

4′-Hydroxy-3′-methoxynorclobazam
Probably formed from: 4′-Hydroxy-3′-methoxyclobazam
→ 4′-Hydroxy-3′-methoxynorclobazam-β-D-glucuronide
Probably in: Rat G572
→ 4′-Hydroxy-3′-methoxynorclobazam-O-sulphate
Probably in: Rat G572

4′-Hydroxy-3′-methoxynorclobazam-β-D-glucuronide
Probably formed from: 4′-Hydroxy-3′-methoxynorclobazam

4′-Hydroxy-3′-methoxynorclobazam-O-sulphate
Probably formed from: 4′-Hydroxy-3′-methoxynorclobazam

2-Hydroxy-4-methoxy-17β-oestradiol
Probably formed from: 4-Methoxy-17β--oestradiol

3-Hydroxy-17-methoxy-1,3,5(10)-oestradiene
Formed from: 4-Androstene-3,17-dione

2-Hydroxy-4-methoxyoestrone
Formed from: 2,4-Dihydroxyoestrone
→ 2,4-Dimethoxyoestrone
In: Rat A2439
→ 2-Hydroxy-4-methoxy-3-O-methyloestrone
In: Rat A2439

4-Hydroxy-2-methoxyoestrone
Formed from: 2,4-Dihydroxyoestrone
2-Methoxyoestrone
→ 2,4-Dimethoxyoestrone
In: Rat A2439
→ 4-Hydroxy-2-methoxy-3-O-methyloestrone
In: Rat A2439

16α-Hydroxy-2-methoxyoestrone
Formed from: 2,16α-Dihydroxyoestrone
→ 16α-Hydroxy-2-methoxyoestrone-3-β-D-glucuronide
Probably in: Rat A1280, A1487
→ 16α-Hydroxy-2-methoxyoestrone sulphate
Probably in: Rat A1487

16α-Hydroxy-2-methoxyoestrone-3-β-D-glucuronide
Probably formed from: 16α-Hydroxy-2-methoxyoestrone

16α-Hydroxy-2-methoxyoestrone sulphate
Probably formed from: 16α-Hydroxy-2-methoxyoestrone

4-Hydroxy-3-methoxy-N-oxalylphenethylamine
Probably formed from: 3,4-Dihydroxy-N-oxalylphenethylamine

3-Hydroxy-2-methoxyoxaprotiline
Probably formed from: 2-Hydroxyoxaprotiline
→ 3-Hydroxy-2-methoxyoxaprotiline-β-D-glucuronide
Probably in: Dog E48
Rat E48

3-Hydroxy-2-methoxyoxaprotiline-β-D-glucuronide
Probably formed from: 3-Hydroxy-2-methoxyoxaprotiline

7-Hydroxy-8-methoxy-2-oxo-2*H*-1-benzopyran-6-acetic acid
Formed from: 8-Methoxypsoralen
→ 7-Hydroxy-8-methoxy-2-oxo-2*H*-1-benzopyran-6-(α-hydroxy)acetic acid
Probably in: Dog A3911

7-Hydroxy-8-methoxy-2-oxo-2*H*-1-benzopyran-6-(α-hydroxy)acetic acid
Probably formed from: 7-Hydroxy-8-methoxy-2-oxo-2*H*-1-benzopyran-6-acetic acid

6-Hydroxy-5-methoxy-1′-oxobufuralol
Probably formed from: 1′,6-Dihydroxy-5-methoxybufuralol

16α-Hydroxy-3-methoxy-17-oxo-1,3,5(10),9(11)-oestratetraene
Formed from: 3-Methoxy-17-oxo-1,3,5(10),9(11)-oestratetraene

16β-Hydroxy-3-methoxy-16-oxo-1,3,5(10),9(11)-oestratetraene
Probably formed from: 16β,17β-Dihydroxy-3-methoxy-1,3,5(10),9(11)-oestratetraene

4-Hydroxy-3-methoxy-α-oxophenethylamine
See: Normetanephrone

4-Hydroxy-3-methoxy-β-oxophenylpropionic acid
Probably formed from: 4-Hydroxy-3-methoxyphenylhydracrylic acid
→ Vanillic acid
Probably in: Pseudomonas G920

2-Hydroxy-3-methoxyphenethylamine
Formed from: 2,3-Dihydroxyphenethylamine

3-Hydroxy-4-methoxyphenethylamine
Formed from: 3,4-Dihydroxyphenethylamine
3-Hydroxy-4-methoxy-L-phenylalanine
→ 3-Hydroxy-4-methoxy-N-methylphenethylamine
In: Phalaris A3891
→ 3-Hydroxy-4-methoxyphenylacetaldehyde
In: Rat A1608, A2961
See also: A801
→ 4-Methoxyphenethylamin-3-yl sulphate
In: Man D583
→ *m*-Tyramine
In: Rat A2961

4-Hydroxy-3-methoxyphenethylamine
See: 3-Methoxytyramine

N-(3-Hydroxy-4-methoxyphenethyl)-3-hydroxy-4-methoxyphenylethanolamine
Probably formed from: N-(3,4-Dihydroxyphenethyl)-3-hydroxy-4-methoxyphenylethanolamine

N-(4-Hydroxy-3-methoxyphenethyl)-3-hydroxy-4-methoxyphenylethanolamine
Probably formed from: N-(3,4-Dihydroxyphenethyl)-3-hydroxy-4-methoxyphenylethanolamine

5-((3-Hydroxy-4-methoxyphenethyl)methylamino)-2-isopropyl-2-(3,4,5-trimethoxyphenyl)valeronitrile
Formed from: Gallopamil

5-((4-Hydroxy-3-methoxyphenethyl)methylamino)-2-isopropyl-2-(3,4,5-trimethoxyphenyl)valeronitrile
Formed from: Gallopamil

N-(3-Hydroxy-4-methoxyphenethyl)normetanephrine
Probably formed from: N-(3,4-Dihydroxyphenethyl)normetanephrine

N-(4-Hydroxy-3-methoxyphenethyl)normetanephrine
Probably formed from: N-(3,4-Dimethoxyphenethyl)normetanephrine
→ N-(4-Hydroxy-3-methoxyphenethyl)normetanephrine-β-D-glucuronide
In: Rabbit D347

N-(4-Hydroxy-3-methoxyphenethyl)normetanephrine-β-D-glucuronide
Probably formed from: N-(4-Hydroxy-3-methoxyphenethyl)normetanephrine

N-(3-Hydroxy-4-methoxyphenethyl)octopamine
Formed from: Denopamine

N-(4-Hydroxy-3-methoxyphenethyl)octopamine
Formed from: Denopamine
→ N-(4-Hydroxy-3-methoxyphenethyl)octopamine-β-D-glucuronide
In: Rabbit D347

N-(4-Hydroxy-3-methoxyphenethyl)octopamine-β-D-glucuronide
Formed from: N-(4-Hydroxy-3-methoxyphenethyl)octopamine

5-(1-Hydroxy-2-((2-(*o*-methoxyphenoxy)ethyl)amino)ethyl)-2-methylbenzenesulphonamide
See: Amosulalol

(+)-(*R*)-2-(2-((N-(2-((4-Hydroxy-3-methoxy)phenoxy)ethyl)-N-methylamino)propoxy)-5-methoxyphenyl)-4-methyl-2*H*-1,4-benzothiazin-3(4*H*)-one
Formed from: Semotiadil
→ (+)-(*R*)-2-(5-Hydroxy-2-((N-(2-((4-hydroxy-3-methoxy)phenoxy)ethyl)-N-methylamino)propoxy)phenyl)-4-methyl-2*H*-1,4-benzothiazin-3(4*H*)-one
Probably in: Rat H313
→ (+)-(*R*)-2-(2-((N-(2-((4-Hydroxy-3-methoxy)phenoxy)ethyl)-N-methylamino)propoxy)-5-methoxyphenyl)-4-methyl-2*H*-1,4-benzothiazin-3(4*H*)-one-β-D-glucuronide
Probably in: Rat H313

(+)-(*R*)-2-(2-((N-(2-((4-Hydroxy-3-methoxy)phenoxy)ethyl)-N-methylamino)propoxy)-5-methoxyphenyl)-4-methyl-2*H*-1,4-benzothiazin-3(4*H*)-one-β-D-glucuronide
Probably formed from: (+)-(R)-2-(2-((N-(2-((4-Hydroxy-3-methoxy)phenoxy)ethyl)-N-methylamino)propoxy)-5-methoxyphenyl)-4-methyl-2*H*-1,4-benzothiazin-3(4*H*)-one

(+)-(*R*)-2-(2-((N-(2-((4-Hydroxy-3-methoxy)phenoxy)ethyl)-N-methylamino)propoxy)phenyl)-4-methyl-2*H*-1,4-benzothiazin-3(4*H*)-one
Probably formed from: (+)-(*R*)-(2-((N-(2-(3,4-Dihydroxyphenoxy)ethyl)-N-methylamino)propoxy)phenyl)-4-methyl-2*H*-1,4-benzothiazin-3(4*H*)-one

(*S*)-2-(2-((N-(2-((4-Hydroxy-3-methoxy)phenoxy)ethyl)-N-methylamino)propoxy)phenyl)-4-methyl-2*H*-1,4-benzothiazin-3(4*H*)-one
Probably formed from: (*S*)-2-(2-((N-(2-(3,4-Dihydroxyphenoxy)ethyl)-N-methylamino)propoxy)phenyl)-4-methyl-2*H*-1,4-benzothiazin-3(4*H*)-one

5-(1-Hydroxy-2-((2-(*o*-methoxyphenoxy) ethyl)amino)ethyl)-2-hydroxymethylbenzenesulphonamide
Formed from: Amosulalol
→ 5-(1-Hydroxy-2-((2-(*o*-hydroxyphenoxy) ethyl)amino)ethyl)-2-hydroxymethylbenzenesulphonamide
Probably in: Dog D1
Rat D1

3-(4-Hydroxy-3-methoxyphenoxy)-N-methylmorphinan
Probably formed from: 3-(*p*-Hydroxyphenoxy)-N-methylmorphinan

N′-(*p*-(4-Hydroxy-3-methoxyphenoxy)phenyl)-N,N-dimethylsulphamide
Probably formed from: N′-(*p*-(*p*-Hydroxyphenoxy)phenyl)-N,N-dimethylsulphamide
→ N′-(*p*-(4-Hydroxy-3-methoxyphenoxy) phenyl)-N-methylsulphamide
Probably in: Dog A1263
Man A1263
Rat A1263

N′-(*p*-(4-Hydroxy-3-methoxyphenoxy)phenyl)-N-methylsulphamide
Probably formed from: N′-(*p*-(4-Hydroxy-3-methoxyphenoxy)phenyl)-N,N-dimethylsulphamide

3-Hydroxy-2-(*o*-methoxyphenoxy)propionic acid
Formed from: Syringylglyceryl β-guaiacyl ether

3-Hydroxy-4-methoxyphenylacetaldehyde
Probably formed from: 3-Hydroxy-4-methoxyphenethylamine
→ 3-Hydroxy-4-methoxyphenylacetic acid
In: Rat A2961
See also: A801
→ 2-(3-Hydroxy-4-methoxyphenyl)ethanol
In: Rat A2961

4-Hydroxy-3-methoxyphenylacetaldehyde
Formed from: 3-Methoxytyramine
Probably formed from: 4-Hydroxy-3-methoxy-N-methylphenethylamine
4-(4-Hydroxy-3-methoxyphenyl)butan-2,3-diol
→ 4-Hydroxy-3-methoxyphenylacetic acid
In: Horse A239
Pig A1287
Rat A41, A2961
See also: G865
→ 2-(4-Hydroxy-3-methoxyphenyl)ethanol
In: Pig A1287
Rat A2961

3-Hydroxy-4-methoxyphenylacetamide
Formed from: 3,4-Dihydroxyphenylacetamide

4-Hydroxy-3-methoxyphenylacetamide
Formed from: 3,4-Dihydroxyphenylacetamide

3-Hydroxy-4-methoxyphenylacetic acid
Formed from: 3,4-Dihydroxyphenylacetic acid
3-Hydroxy-4-methoxyphenylacetaldehyde
3-Hydroxy-4-methoxyphenylacetylglycine
→ 3,4-Dihydroxyphenylacetic acid
Probably in: Rat A2961
→ 3-Hydroxy-4-methoxyphenylacetylglycine
In: Rat A2961

4-Hydroxy-3-methoxyphenylacetic acid
(homovanillic acid)
Formed from: (+)-Catechin
3,4-Dihydroxyphenylacetic acid
Ferulic acid
4-Hydroxy-3-methoxyphenylacetaldehyde
Rutin
Vanilpyruvic acid
→ 3,4-Dihydroxyphenylacetic acid
In: Rat B162
Pseudomonas H387
→ 2,5-Dihydroxy-4-methoxyphenylacetic acid
In: Pseudomonas A1480
Variovax H760
→ 4-Hydroxy-3-methoxyphenylacetic acid-4-sulphate
In: Beef F196
Man D613
Rat A153, A1343, C441
→ 4-Hydroxy-3-methoxyphenylacetylglycine
In: Rat A2961
→ *m*-Methoxyphenylacetic acid
In: Rat B162

4-Hydroxy-3-methoxyphenylacetic acid-4-sulphate
Formed from: 4-Hydroxy-3-methoxyphenylacetic acid

4-Hydroxy-3-methoxyphenylacetone
Formed from: 3,4-Dimethoxypropenylbenzene

4-Hydroxy-3-methoxy-N-methylamphetamine
Probably formed from:
3,4-Dihydroxyphenylacetone
4-Hydroxy-3-methoxyamphetamine
→ 3,4-Dihydroxyphenylacetone
In: Rat G571
→ 1-(4-Hydroxy-3-methoxyphenyl)propan-2-ol
In: Rat G571
See also: B403

5-Hydroxy-2-methoxyphenylacetone
Probably formed from: 5-Hydroxy-2-methoxyamphetamine

3-Hydroxy-4-methoxyphenylacetylglycine
Formed from: 3-Hydroxy-4-methoxyphenylacetic acid
→ 3-Hydroxy-4-methoxyphenylacetic acid
In: Rat A2961

4-Hydroxy-3-methoxyphenylacetylglycine
Formed from: 4-Hydroxy-3-methoxyphenylacetic acid

3-Hydroxy-4-methoxy-D-phenylalanine
Formed from: 3,4-Dihydroxy-D-phenylalanine

3-Hydroxy-4-methoxy-L-phenylalanine
Formed from: 3,4-Dihydroxy-L-phenylalanine
Probably formed from:
N-Acetyl-3-hydroxy-4-methoxy-L-phenylalanine
3-Hydroxy-4-methoxyphenylpyruvic acid
→ N-Acetyl-3-hydroxy-4-methoxy-L-phenylalanine
In: Man A801
Rat A801
→ 3,4-Dihydroxy-L-phenylalanine
Probably in: Rat A2961
→ 3-Hydroxy-4-methoxyphenethylamine
In: Man A801
Rat A801, A2961
See also: A1608
→ 3-Hydroxy-4-methoxyphenylpyruvic acid
In: Man A801
Rat A801, A2961

4-Hydroxy-3-methoxy-L-phenylalanine
See: 3-Methoxy-L-tyrosine

N-(3-Hydroxy-4-methoxyphenyl) benzenesulphonamide
Formed from: N-(3,4-Dihydroxyphenyl) benzenesulphonamide

N-(4-Hydroxy-3-methoxyphenyl) benzenesulphonamide
Formed from: N-(3,4-Dihydroxyphenyl) benzenesulphonamide

4-(4-Hydroxy-3-methoxyphenyl)butane-1,2-diol
Formed from: Zingerone
→ Dihydroferulic acid
Probably in: Rat A1786

4-(4-Hydroxy-3-methoxyphenyl)butane-2,3-diol
Formed from: Zingerone
→ 4-Hydroxy-3-methoxyphenylacetaldehyde
Probably in: Rat A1786

4-(4-Hydroxy-3-methoxyphenyl)butan-2-ol
See: Zingerol

4-(4-Hydroxy-3-methoxyphenyl)butan-2-one
See: Zingerone

2-(4-Hydroxy-3-methoxyphenyl)butylurea
Probably formed from: 2-(*p*-Hydroxyphenyl) butylurea

2-Hydroxy-3-methoxy-4-phenylbutyronitrile
Formed from: 2-Methoxy-3-phenylpropionaldehyde

2-Hydroxy-4-methoxy-4-phenylbutyronitrile
Formed from: 3-Methoxy-3-phenylpropionaldehyde

2-(4-Hydroxy-3-methoxyphenyl)butyrylurea
Probably formed from: 2-(*p*-Hydroxyphenyl) butyrylurea

2-(3-Hydroxy-4-methoxyphenyl)-N,N-dimethyl-2-oxoethylamine
→ 3-Hydroxy-4-methoxymandelic acid
In: Rat J287

2-(4-Hydroxy-3-methoxyphenyl)-N,N-dimethyl-2-oxoethylamine
→ 4-Hydroxy-3-methoxymandelic acid
In: Rat J287

1-(4-Hydroxy-3-methoxyphenyl)ethanol
Formed from: 4′-Hydroxy-3′-methoxyacetophenone
→ 1-(3,4-Dihydroxyphenyl)ethanol
Probably in: Rat E849

2-(3-Hydroxy-4-methoxyphenyl)ethanol
Formed from: 3-Hydroxy-4-methoxyphenylacetaldehyde
→ 2-(3,4-Dihydroxyphenyl)ethanol
In: Rat A2961
→ 3-Hydroxy-4-methoxyphenylacetaldehyde
In: Rat A2961

2-(4-Hydroxy-3-methoxyphenyl)ethanol (HMPE)
Formed from: 4-Hydroxy-3-methoxyphenylacetaldehyde
→ 3-Methoxyphenylethanol-4-yl sulphate
In: Rat D807

3-Hydroxy-4-methoxyphenylethanolamine
See: L-Norparanephrine

4-Hydroxy-3-methoxyphenylethanolamine
See: Normetanephrine

1-(2-(3-Hydroxy-4-methoxyphenyl)ethyl)-3-diphenylmethoxypiperidine
Probably formed from: 1-(2-(3,4-Dihydroxyphenyl)ethyl)-3-diphenylmethoxypiperidine

1-(2-(4-Hydroxy-3-methoxyphenyl)ethyl)-3-diphenylmethoxypiperidine
Probably formed from: 1-(2-(3,4-Dihydroxyphenyl)ethyl)-3-diphenylmethoxypiperidine

N-(2-(4-Hydroxy-3-methoxyphenyl)ethyl)-5-methoxy-N,α-dimethyl-2-(phenylethynyl) benzenepropanamine
Formed from: N-(2-(3,4-Dihydroxyphenyl) ethyl)-5-methoxy-N,α-dimethyl-2-(phenylethynyl)benzenepropanamine
→ N-(2-(4-Hydroxy-3-methoxyphenyl)ethyl)-5-methoxy-α-methyl-2-(phenylethynyl) benzenepropanamine
Probably in: Dog H937
Rat J444

N-(2-(4-Hydroxy-3-methoxyphenyl)ethyl)-5-methoxy-α-methyl-2-(phenylethynyl) benzenepropanamine
Probably formed from: N-(2-(4-Hydroxy-3-methoxyphenyl)ethyl)-5-methoxy-N,α-dimethyl-2-(phenylethynyl) benzenepropanamine

4-Hydroxy-3-methoxyphenylglycol
→ 4′,ω-Dihydroxy-3′-methylacetophenone
In: Penicillium J16
→ 4-Hydroxy-3-methoxymandelaldehyde
In: Man F87
See also: K761
→ 3-Methoxyphenylglycol-4-yl sulphate
In: Beef F196
Rat D807

4-Hydroxy-3-methoxyphenylglyoxylic acid
(4-hydroxy-3-methoxybenzoylformic acid)
Formed from: 4-Hydroxy-3-methoxymandelic acid
→ Vanillin
Probably in: Horse A254

4-Hydroxy-3-methoxyphenylhydracrylic acid
Probably formed from: Ferulic acid
→ 4-Hydroxy-3-methoxy-β-oxophenylpropionic acid
Probably in: Pseudomonas G910

4-Hydroxy-3-methoxyphenyl *p*-hydroxyphenyl ether
Probably formed from: Bis(4-hydroxyphenyl) ether

5-(4-Hydroxy-3-methoxyphenyl)-5-(*p*-hydroxyphenyl)hydantoin
Probably formed from: 5-(3,4-Dihydroxyphenyl)-5-(*p*-hydroxyphenyl)hydantoin

1-(Hydroxymethoxyphenyl)-4-isopropyl-2,3-dimethyl-3-pyrazolin-5-one
Probably formed from: 1-(Hydroxyphenyl)-4-isopropyl-2,3-dimethyl-3-pyrazolin-5-one

2-(3-Hydroxy-4-methoxyphenyl)-2-isopropyl-5-methylaminovaleronitrile
Probably formed from: 2-(3,4-Dimethoxyphenyl)-2-isopropyl-5-methylaminovaleronitrile

1-(Hydroxymethoxyphenyl)-4-isopropyl-3-methyl-3-pyrazolin-5-one
Probably formed from: 1-(Hydroxyphenyl)-4-isopropyl-3-methyl-3-pyrazolin-5-one

3-Hydroxy-4-methoxyphenyllactic acid
Formed from: 3-Hydroxy-4-methoxyphenylpyruvic acid
→ 3-Hydroxy-4-methoxyphenylpyruvic acid
In: Rat A2961

4-Hydroxy-3-methoxyphenyllactic acid
See: Vanillactic acid

2-(4-Hydroxy-3-methoxyphenyl)-7-methoxy-3-methyl-5-(1-propenyl)benzofuran
Formed from: Isoeugenol

***trans*-1-(3-Hydroxy-4-methoxyphenyl)-1-(*p*-(2-methylaminoethoxy)phenyl)-2-phenylbut-1-ene**
Probably formed from: *trans*-1-(3,4-Dihydroxyphenyl)-1-(*p*-(2-methylamino-ethoxy)phenyl)-2-phenylbut-1-ene

***trans*-1-(4-Hydroxy-3-methoxyphenyl)-1-(*p*-(2-methylaminoethoxy)phenyl)-2-phenylbut-1-ene**
Probably formed from: *trans*-1-(3,4-Dihydroxyphenyl)-1-(*p*-(2-methylamino-ethoxy)phenyl)-2-phenylbut-1-ene

1-(5-Hydroxy-2-methoxyphenyl)-2-methylaminopropane
Formed from: Methoxyphenamine
→ 5-Hydroxy-2-methoxyamphetamine
Probably in Cunninghamella F340

1-((4-Hydroxy-3-methoxyphenyl)methyl)-6,7-dimethoxyisoquinoline
See: 4′-Demethylisopapaverine

4-((3-Hydroxy-4-methoxyphenyl)methyl)-6,7-dimethoxyisoquinoline
See: 3′-Demethylisopapaverine

4-((4-Hydroxy-3-methoxyphenyl)methyl)-6,7-dimethoxyisoquinoline
See: 4′-Demethylisopapaverine

N-(2-((4-Hydroxy-3-methoxyphenyl)methylene)cyclohexyl)cyclopropylamine
Probably formed from: *p*-Hydroxytazadolene
→ N-(2-((4-Hydroxy-3-methoxyphenyl)methylene)cyclohexyl)cyclopropylamine-β-D-glucuronide
Probably in: Dog G375

N-(2-((4-Hydroxy-3-methoxyphenyl)methylene)cyclohexyl)cyclopropylamine-β-D-glucuronide
Probably formed from: N-(2-((4-Hydroxy-3-methoxyphenyl)methylene)cyclohexyl)cyclopropylamine

4-Hydroxy-α′-(((2-(4-methoxyphenyl)-1-methylethyl)amino)methyl)-1,3-benzenedimethanol
See: Salmeterol

N-(1-(4-Hydroxy-3-methoxyphenyl)-1-methylethyl)-3,3-dimethylbutyramide
Probably formed from: 2-Bromo-N-(1-(4-Hydroxy-3-methoxyphenyl)-1-methylethyl)-3,3-dimethylbutyramide

3-(4-Hydroxy-3-methoxyphenyl)-2-methyllactic acid
Probably formed from: 3-(3,4-Dihydroxyphenyl)-2-methyllactic acid

(6*E*)-N-((4-Hydroxy-3-methoxyphenyl)methyl)-8-methyl-6-nonenamide
See: Capsaicin

3-(4-Hydroxy-3-methoxyphenyl)-N-methyl-N-(1-naphthylmethyl)-2-propen-1-ylamine
Probably formed from: *p*-Hydroxynaftifine

N-((4-Hydroxy-3-methoxyphenyl)methyl)nonanamide
See: Nonivamide

3-(4-Hydroxy-3-methoxyphenyl)-5-methyl-3-phenyl-1-pyrroline
Probably formed from: 3-(3,4-Dihydroxyphenyl)-5-methyl-3-phenyl-1-pyrroline

1-(4-(4-Hydroxy-3-methoxyphenyl)-1-methyl-4-piperidyl)-1-propanone
Probably formed from: 1-(4-(3,4-Dihydroxyphenyl)-1-methyl-4-piperidyl)-1-propanone

1-(1-Hydroxy-1-(*m*-methoxyphenyl)-2-methylprop-2-yl)imidazole
Formed from: 1-(*m*-Methoxyphenyl)-2-methyl-1-oxoprop-2-yl)imidazole

4-((4-Hydroxy-3-methoxyphenyl)methyl)-2,2,6,6-tetramethyl-1-oxa-4-aza-2,6-disilacyclohexane
Formed from: 4-((3-Methoxyphenyl)methyl)-2,2,6,6-tetramethyl-1-oxa-4-aza-2,6-disilacyclohexane

2-(3-Hydroxy-4-methoxyphenyl)-2-oxoethylamine
→ 3-Hydroxy-4-methoxymandelic acid
In: Rat J287

2-(4-Hydroxy-3-methoxyphenyl)-2-oxoethylamine
→ 4-Hydroxy-3-methoxymandelic acid
In: Rat J287

1-(4-Hydroxy-3-methoxyphenyl)-1-phenylbutan-3-one
Probably formed from: 1-(*p*-Hydroxyphenyl)-1-phenylbutan-3-one

1-(4-Hydroxy-3-methoxyphenyl)-1-phenylbutan-3-one oxime
Probably formed from: 1-(*p*-Hydroxyphenyl)-1-phenylbutan-3-one oxime

3-Hydroxy-4-methoxyphenyl phenyl ether
Probably formed from: 4-Hydroxyphenyl phenyl ether

5-(4-Hydroxy-3-methoxyphenyl)-5-phenylhydantoin
Formed from: 5-(3,4-Dihydroxyphenyl)-5-phenylhydantoin
→ 5-(4-Hydroxy-3-methoxyphenyl)-5-phenylhydantoin-β-D-glucuronide
Probably in: Rat A1018

5-(4-Hydroxy-3-methoxyphenyl)-5-phenylhydantoin-β-D-glucuronide
Probably formed from: 5-(4-Hydroxy-3-methoxyphenyl)-5-phenylhydantoin

4-((4-Hydroxy-3-methoxyphenyl)phenyl)methyl-1-((octylimino)methyl)piperidine
Probably formed from: 4-((3,4-Dihydroxyphenyl)phenyl)methyl-1-((octylimino) methyl)piperidine

3-(4-Hydroxy-3-methoxyphenyl)-3-phenylpropylamine
Probably formed from: 3-(*p*-Hydroxyphenyl)-3-phenylpropylamine

N-(3-(4-Hydroxy-3-methoxyphenyl)-3-phenylpropyl)-1-phenylethylamine
Probably formed from: N-(3-(*p*-Hydroxyphenyl)-3-phenylpropyl)-1-phenylethylamine

3-(1-(4-Hydroxy-2-methoxyphenyl)piperazin-4-yl)-1-naphthyloxypropan-2-ol
Formed from: Naftopidil

6-(3-(4-(4-Hydroxy-2-methoxyphenyl)piperazinyl) propylamino)-1,3-dimethyluracil
Formed from: Urapidil

1-(4-Hydroxy-3-methoxyphenyl)propan-2-ol
Formed from: 4-Hydroxy-3-methoxyphenylacetone

3-(4-Hydroxy-3-methoxyphenyl)propan-1-ol
See: Dihydroconiferyl alcohol

1-Hydroxy-1-(*p*-methoxyphenyl)prop-2-ene
(1′-hydroxyestragole)
Formed from: Estragole

1-(4-Hydroxy-3-methoxyphenyl)-2-propen-1-ol
Formed from: Eugenol

3-(4-Hydroxy-3-methoxyphenyl)-2-propen-1-ol
Probably formed from: 3-(3,4-Dihydroxyphenyl)-2-propen-1-ol
→ Ferulic acid
Probably in: Rat A1765

3-(4-Hydroxy-3-methoxyphenyl)-2-propenyl-β-D-glucopyranoside
See: Isoconiferin

3-(4-Hydroxy-3-methoxyphenyl)propionaldehyde
See: Dihydroconiferaldehyde

3-(3-Hydroxy-4-methoxyphenyl)propionic acid
See: Dihydroisoferulic acid

3-(4-Hydroxy-3-methoxyphenyl)propionic acid
See: Dihydroferulic acid

3-Hydroxy-5-methoxyphenylpropionic acid
Formed from: 3,4-Dihydroxy-5-methoxyphenylpropionic acid
3,5-Dimethoxyphenylpropionic acid
→ 3-(3,5-Dihydroxyphenyl)propionic acid
Probably in: Microorganism A261, A744

1-(2-Hydroxy-3-methoxy-3-phenylpropyl)-4-(2-methoxy-2-phenylethyl)piperazine
See: Zipeprol

1-(2-Hydroxy-3-methoxy-3-phenylpropyl) piperazine
Formed from: Zipeprol
→ 1-(2-Hydroxy-3-methoxy-3-phenylpropyl)-4-methylpiperazine
Probably in: Man B766

1-(2-Hydroxy-3-methoxy-3-phenylpropyl)-4-methylpiperazine
Probably formed from: 1-(2-Hydroxy-3-methoxy-3-phenylpropyl)piperazine

4-Hydroxy-3-methoxyphenyl-N-*n*-propylpiperidine
Probably formed from: 3,4-Dihydroxyphenyl-N-*n*-propylpiperidine

4-(3-Hydroxy-4-methoxyphenyl)-2-pyrrolidone
Formed from: Rolipram
→ 4-(3-Hydroxy-4-methoxyphenyl)-2-pyrrolidone-O-sulphate
In: Cynomolgus monkey G710

Man G590, G710
Rabbit G710
Rat G590, G710
Rhesus monkey G710

4-(3-Hydroxy-4-methoxyphenyl)-2-pyrrolidone-O-sulphate
Formed from: 4-(3-Hydroxy-4-methoxyphenyl)-2-pyrrolidone

3-Hydroxy-4-methoxyphenylpyruvic acid
Formed from: 3-Hydroxy-4-methoxy-L-phenylalanine
3-Hydroxy-4-methoxyphenyllactic acid
→ 3-Hydroxy-4-methoxy-L-phenylalanine
Probably in: Rat A801
→ 3-Hydroxy-4-methoxyphenyllactic acid
In: Man A801
Rat A801, A2961

4-Hydroxy-3-methoxyphenylpyruvic acid
See: Vanilpyruvic acid

***erythro*-3-Hydroxy-4-methoxyphenylserine**
Formed from: *erythro*-3,4-Dihydroxyphenylserine

***threo*-3-Hydroxy-4-methoxyphenylserine**
Formed from: *threo*-3,4-Dihydroxyphenylserine

***erythro*-4-Hydroxy-3-methoxyphenylserine**
Formed from: *erythro*-3,4-Dihydroxyphenylserine

***threo*-4-Hydroxy-3-methoxyphenylserine**
Formed from: *threo*-3,4-Dihydroxyphenylserine

***d*-δ-(4-Hydroxy-4-methoxyphenyl)-γ-valerolactone**
Probably formed from: *d*-δ-(3,4-Dihydroxyphenyl)-γ-valerolactone

5-(2-(4-Hydroxy-3-methoxyphenyl)vinyl)ferulic acid
Probably formed from: 5-(2-Carboxyvinyl)-2,3-dihydro-2-(4-hydroxy-3-methoxyphenyl)-3-hydroxymethyl-7-methoxybenzofuran

6-Hydroxy-7-methoxyphthalide
Formed from: Meconine

7-Hydroxy-6-methoxyphthalide
Formed from: Meconine

4′-Hydroxy-3′-methoxypropafenone
Probably formed from: 3′,4′-Dihydroxypropafenone

5-Hydroxy-4-methoxypropafenone
Probably formed from: 4,5-Dihydroxypropafenone

3-Hydroxy-4-methoxypropenylbenzene
Formed from: 3,4-Dihydroxypropenylbenzene

4-Hydroxy-3-methoxypropenylbenzene
See: Isoeugenol

2-Hydroxy-*m*-methoxypropiophenone
Formed from: *m*-Methoxybenzaldehyde

2-Hydroxy-*p*-methoxypropiophenone
Formed from: Anisaldehyde

3′-Hydroxy-4′-methoxypropiophenone
Formed from: 3′,4′-Dihydroxypropiophenone

4′-Hydroxy-3′-methoxypropiophenone
Formed from: 3′,4′-Dihydroxypropiophenone

3-Hydroxy-4-methoxypropranolol
Probably formed from: 3,4-Dihydroxypropranolol

4-Hydroxy-3-methoxypropranolol
Probably formed from: 3,4-Dihydroxypropranolol

4-Hydroxy-3-methoxy-N-propylamphetamine
Formed from: *p*-Hydroxy-N-propylamphetamine

10-Hydroxy-11-methoxy-N-propylnorapomorphine
Formed from: 10,11-Dihydroxy-N-propylnorapomorphine

3-Hydroxy-4-methoxy-N-propylphenylethanolamine
Formed from: N-Propylnoradrenaline

5-Hydroxy-8-methoxypsoralen
Formed from: 8-Methoxypsoralen
→ 5,8-Dihydroxypsoralen
Probably in: Rat B355, D966
→ 5-Hydroxy-8-methoxypsoralen-O-sulphate
In: Rat E868

5-Hydroxy-8-methoxypsoralen-O-sulphate
Formed from: 5-Hydroxy-8-methoxypsoralen

N[1]-(2-Hydroxy-6-methoxy-4-pyrimidinyl)sulphanilamide
Formed from: Sulphadimethoxine
Sulphamonomethoxine
→ N[4]-Acetyl-N[1]-(2-hydroxy-6-methoxy-4-pyrimidinyl)sulphanilamide
Probably in: Pseudemys F541

→ N^1-(2,6-Dihydroxy-4-pyrimidinyl) sulphanilamide
Probably in: Chicken H352
Pseudemys F551

N^1-(6-Hydroxy-2-methoxy-4-pyrimidinyl) sulphanilamide
Formed from: Sulphadimethoxine
→ N^4-Acetyl-N^1-(6-hydroxy-2-methoxy-4-pyrimidinyl)sulphanilamide
Probably in: Pseudemys F541

4-Hydroxy-8-methoxyquinoline-2-carboxylic acid
See: 8-Methoxykynurenic acid

3-Hydroxy-4-methoxy-*trans*-stilbene
Formed from: 3,4-Dihydroxy-*trans*-stilbene

4-Hydroxy-3-methoxy-*trans*-stilbene
Formed from: 3,4-Dihydroxy-*trans*-stilbene

3-Hydroxy-4-methoxystyrene
Formed from: Isoferulic acid

4-Hydroxy-3-methoxystyrene (4-vinylguaiacol)
Formed from: Eugenol
Ferulic acid
→ Vanillin
In: Fusarium E115
Paecilomyces F199

3-Hydroxy-4-methoxytamoxifen
Probably formed from: 3,4-Dihydroxy-tamoxifen
→ *trans*-1-(*p*-(2-Dimethylaminoethoxy) phenyl)-1-(3-hydroxy-4-methoxyphenyl)-2-(hydroxyphenyl)but-1-ene
Probably in: Mouse H924
Rat H924
→ 3-Hydroxy-4-methoxytamoxifen-β-D-glucuronide
Probably in: Cynomolgus monkey H924
Mouse H924
Rat H924
→ 3-Hydroxy-4-methoxytamoxifen-O-sulphate
Probably in: Cynomolgus monkey H924

4-Hydroxy-3-methoxytamoxifen
Probably formed from: 3,4-Dihydroxy-tamoxifen
→ *trans*-1-(*p*-(2-Dimethylaminoethoxy) phenyl)-1-(4-hydroxy-3-methoxyphenyl)-2-(hydroxyphenyl)but-1-ene
Probably in: Mouse H924
Rat H924
→ 4-Hydroxy-3-methoxytamoxifen-β-D-glucuronide
Probably in: Cynomolgus monkey H924
Mouse H924
Rat H924
→ 4-Hydroxy-3-methoxytamoxifen-O-sulphate
Probably in: Cynomolgus monkey H924

3-Hydroxy-4-methoxytamoxifen-β-D-glucuronide
Probably formed from: 3-Hydroxy-4-methoxytamoxifen

4-Hydroxy-3-methoxytamoxifen-β-D-glucuronide
Probably formed from: 4-Hydroxy-3-methoxytamoxifen

3-Hydroxy-4-methoxytamoxifen-O-sulphate
Probably formed from: 3-Hydroxy-4-methoxytamoxifen

4-Hydroxy-3-methoxytamoxifen-O-sulphate
Probably formed from: 4-Hydroxy-3-methoxytamoxifen

3′-Hydroxy-4′-methoxytenidap
Probably formed from: 3′,4′-Dihydroxy-tenidap

4′-Hydroxy-3′-methoxy-2-trifluoromethylbenzanilide
Probably formed from: 3′,4′-Dihydroxy-2-trifluoromethylbenzanilide

2-Hydroxy-3-methoxytrimipramine
Formed from: 2,3-Dihydroxytrimipramine

4-Hydroxy-3-methoxy-5-triprenylbenzoic acid
Formed from: 3,4-Dihydroxy-5-triprenylbenzoic acid

5-Hydroxy-6-methoxytryptamine
Formed from: N-Acetyl-5-hydroxy-6-methoxytryptamine
Probably formed from: 5,6-Dihydroxy-tryptamine

4-Hydroxy-3-methoxytulobuterol
Formed from: 3,4-Dihydroxytulobuterol

4-Hydroxy-5-methoxytulobuterol
Formed from: 4,5-Dihydroxytulobuterol

Hydroxymethoxyvolazocine
Probably formed from: Volazocine

4′-Hydroxy-11-methoxywarfarin
Probably formed from: 4′-Hydroxywarfarin

4′-Hydroxy-3′-methylacetophenone
Formed from: 2,2-Bis(4-hydroxy-3-methylphenyl)propane

1-Hydroxymethylagroclavine
→ 1-Hydroxymethylelymoclavine
In: Claviceps D237

N-Hydroxy-4-methylaminoazobenzene
Formed from: 4-Methylaminoazobenzene

2-Hydroxy-4′-methylaminoazobenzene
Formed from: 4-Methylaminoazobenzene

4-Hydroxy-4′-methylaminoazobenzene
Formed from: 4′-Dimethylamino-4-hydroxyazobenzene
4-Methylaminoazobenzene
→ 4′-(N-Acetyl-N-methyl)amino-4-hydroxyazobenzene
Probably in: Rat B111
→ *p*-Aminophenol
In: Rat G511
→ 4′-Amino-4-hydroxyazobenzene
Probably in: Rat A1979
→ 4-Hydroxy-4′-(N-hydroxy-N-methyl)aminoazobenzene
Probably in: Rat B111
→ 4′-Methylaminoazobenzen-4-yl-β-D-glucuronide
Probably in: Rat A1975
→ 4′-Methylaminoazobenzen-4-yl sulphate
Probably in: Rat A1975, D68
→ N-Methyl-*p*-phenylenediamine
In: Rat G511

3-Hydroxy-4-methylaminobenzonitrile
Probably formed from: *p*-Methylaminobenzonitrile
→ 4-Methylaminobenzonitril-3-yl sulphate
Probably in: Rat F305

9-Hydroxy-5-(2-(methylamino)ethoxy)-7*H*-benzo[*c*]fluoren-7-one
Probably formed from: 5-(2-(Methylamino)ethoxy)-7*H*-benzo[*c*]fluoren-7-one

7-Hydroxy-N-(2-(methylamino)ethyl)-9(10*H*)-acridone-4-carboxamide
Probably formed from: N-(2-(Methylamino)ethyl)-9(10*H*)-acridone-4-carboxamide

4-((1*R*)-1-Hydroxy-2-(methylamino)ethyl)-1,2-benzenediol
See: D-Adrenaline

4-Hydroxy-α-(1-(methylamino)ethyl) benzenemethanol
See: *p*-Hydroxyephedrine

7-Hydroxy-4-methylaminomethylcoumarin
Formed from: 7-Methoxy-4-methylaminomethylcoumarin

2-(2-Hydroxy-3-methylamino-1-propoxy)-1-(2-propen-1-oxy)benzene
Probably formed from: 2-(3-Amino-2-hydroxy-1-propoxy)-1-(2-propen-1-oxy) benzene

N-Hydroxy-N-methylamphetamine
Formed from: N-Methylamphetamine
→ Amphetamine
Probably in: Guinea pig E806
→ N-Methylamphetamine
In: Man K608

***o*-Hydroxy-N-methylamphetamine**
Formed from: Methoxyphenamine
→ *o*-Hydroxyamphetamine
Probably in: Cunninghamella F340

***p*-Hydroxy-N-methylamphetamine**
Formed from: N-Methylamphetamine
→ *p*-Hydroxyamphetamine
In: Rat D156
→ N-Methylamphetamin-4-yl-β-D-glucuronide
Probably in: Rat E783

N-Hydroxy-N-methylaniline
See: N-Methyl-N-phenylhydroxylamine

9-Hydroxymethylanthracene
Formed from: Anthracene
9-Methylanthracene
9-(Trimethylsilyl)methylanthracene
→ Anthraquinone
With cytochrome P450 H796
→ 3,4-Dihydro-3,4-dihydroxy-9-hydroxymethylanthracene
In: Rat D884
→ 9-Formylanthracene
Probably in: Rat E612
→ 9-Hydroxymethylanthracene-O-sulphate
Probably in: Rat H339

9-Hydroxymethylanthracene-O-sulphate
Probably formed from: 9-Hydroxymethylanthracene

3-Hydroxy-4-methylanthranilic acid
Formed from: 3-Hydroxyanthranilic acid
→ Actinocin
In: Streptomyces E676
→ 3-Hydroxy-4-methylanthranilic acid-3-adenylate
In: Streptomyces G791

3-Hydroxy-4-methylanthranilic acid-3-adenylate
Formed from: 3-Hydroxy-4-methylanthranilic acid

7-Hydroxymethylbenz[*c*]acridine
Formed from: 7-Methylbenz[*c*]acridine
→ Benz[*c*]acridine-7-carboxylic acid
Probably in: Rat D163, D430
→ 7-Hydroxymethylbenz[*c*]acridine-5,6-oxide
Probably in: Rat D34, D163

5-Hydroxy-7-methylbenz[*c*]acridine
Formed from: 7-Methylbenz[*c*]acridine

9-Hydroxy-7-methylbenz[*c*]acridine
Formed from: 7-Methylbenz[*c*]acridine

11-Hydroxy-7-methylbenz[*c*]acridine
Formed from: 7-Methylbenz[*c*]acridine

7-Hydroxymethylbenz[*c*]acridine-5,6-oxide
Probably formed from: 7-Hydroxymethylbenz[*c*]acridine

2-Hydroxy-3-methylbenzaldehyde
Formed from: 1-Methylnaphthalene

4-Hydroxy-3-methylbenzaldehyde
Formed from: Dehydrodiisoeugenol

***m*-Hydroxymethylbenzaldehyde**
Formed from: Isophthalaldehyde
→ 1,3-Bis(hydroxymethyl)benzene
In: Dunaliella G459

***p*-Hydroxymethylbenzaldehyde**
Formed from: *p*-Hydroxymethylbenzylamine
Terephthalaldehyde
→ 1,4-Bis(hydroxymethyl)benzene
In: Dunaliella G459

***m*-Hydroxymethylbenzamide**
Formed from: N-Ethyl-*m*-hydroxymethylbenzamide
m-Formylbenzamide
m-Toluamide
→ *m*-Formylbenzamide
In: Rat J572

2-Hydroxy-3-methylbenzamide
Formed from: 2-Hydroxy-3-methylbenzoic acid

2-Hydroxy-4-methylbenzamide
Formed from: 2-Hydroxy-4-methylbenzoic acid

N-Hydroxymethylbenzamide
Formed from: N-Methylbenzamide
→ N-Formylbenzamide
Probably in: Mouse C319

N-Hydroxy-*m*-methylbenzamidine
See: *m*-Methylbenzamidoxime

N-Hydroxy-*p*-methylbenzamidine
See: *p*-Methylbenzamidoxime

4′-Hydroxy-2-methylbenzanilide
Formed from: 2-Methylbenzanilide

7-Hydroxymethylbenzanthracene
Formed from: 7-Hydroxymethylbenzanthracene-5,6-oxide
→ 3,4-Dihydro-3,4-dihydroxy-7-hydroxymethylbenzanthracene
Probably in: Cunninghamella C255
→ 8,9-Dihydro-8,9-dihydroxy-7-hydroxymethylbenzanthracene
Probably in: Cunninghamella C255
→ 7-Formylbenzanthracene
Probably in: Rat F534, G491
→ 7-Hydroxymethyl-1,2-methylbenzanthracene
Probably in: Rat F534

8-Hydroxymethylbenzanthracene
→ *trans*-1,2-Dihydro-1,2-dihydroxy-7-hydroxymethylbenzanthracene
In: Rat D389
→ *trans*-3,4-Dihydro-3,4-dihydroxy-7-hydroxymethylbenzanthracene
In: Rat D389
→ *trans*-8,9-Dihydro-8,9-dihydroxy-7-hydroxymethylbenzanthracene
In: Rat D389
→ *trans*-10,11-Dihydro-10,11-dihydroxy-7-hydroxymethylbenzanthracene
In: Rat D389
→ 8-Formylbenzanthracene
In: Rat D389

→ 1-Hydroxy-8-hydroxymethylbenzanthracene
In: Rat D389
→ 2-Hydroxy-8-hydroxymethylbenzanthracene
In: Rat D389
→ 3-Hydroxy-8-hydroxymethylbenzanthracene
In: Rat D389
→ 4-Hydroxy-8-hydroxymethylbenzanthracene
In: Rat D389
→ 8-Hydroxymethylbenzanthracene-5,6-oxide
In: Rat D389

7-Hydroxymethylbenzanthracene-5,6-oxide
→ 7-Hydroxymethylbenzanthracene
In: Rat A350

8-Hydroxymethylbenzanthracene-5,6-oxide
Formed from:
8-Hydroxymethylbenzanthracene
→ *trans*-5,6-Dihydro-5,6-dihydroxy-8-hydroxymethylbenzanthracene
Probably in: Rat D389

(α*S*)-α-(Hydroxymethyl)benzeneacetic acid (3-*endo*)-8-methyl-8-azabicyclo[3.2.1]oct-3-yl ester
See: Hyoscyamine

***p*-Hydroxymethylbenzenesulphonamide**
Formed from: *p*-Toluenesulphonamide
→ *p*-Formylbenzenesulphonamide
In: Rat A3041
See also: A559, A2021

***p*-Hydroxymethylbenzenesulphonic acid**
Formed from:
p-Formylbenzenesulphonic acid
p-Toluenesulphonic acid
→ *p*-Formylbenzenesulphonic acid
In: Pseudomonas F188

***p*-Hydroxymethylbenzenesulphonylpiperidine**
Formed from:
p-Toluenesulphonylpiperidine

5-Hydroxy-6-methylbenzimidazole
→ 5-Hydroxy-6-methylbenzimidazolylcobamide
In: Eubacterium G919

3-Hydroxymethyl-4,5-benzocoumarin
Formed from: Fluoranthene

2-Hydroxy-3-methylbenzoic acid
(3-methylsalicylic acid)
Formed from: 1-Methylnaphthalene
Probably formed from: *m*-Toluic acid
→ 2-Hydroxy-3-methylbenzamide
In: Streptomyces E279
→ 3-Methylcatechol
Probably in: Pseudomonas A1301

2-Hydroxy-4-methylbenzoic acid
Formed from: 2-Methylnaphthalene
→ 2-Hydroxy-4-methylbenzamide
In: Streptomyces E279
→ 4-Methylcatechol
In: Trichosporon D232

2-Hydroxy-6-methylbenzoic acid
(6-methylsalicylic acid)
Formed from acetic acid in Penicillium C448
acetyl CoA in Penicillium G751
malonyl CoA in Penicillium F470
→ *m*-Cresol
In: Valsa A1049

3-Hydroxy-4-methylbenzoic acid
→ 3-Methoxy-4-methylbenzoic acid
In: Phanerochaete J339

3-Hydroxy-5-methylbenzoic acid
Probably formed from:
3-Carboxy-5-methylphenyl N-methylcarbamate
→ 2,5-Hydroxy-3-methylbenzoic acid
In: Pseudomonas A326

4-Hydroxy-2-methylbenzoic acid
Formed from: *m*-Cresol

4-Hydroxy-3-methylbenzoic acid
Formed from: 2,2-Bis(4-hydroxy-3-methylphenyl)propane
o-Cresol
→ *o*-Cresol
Probably in: Pseudomonas G291
→ 4-Hydroxy-3-hydroxymethylbenzoic acid
In: Pseudomonas F223
See also: F624

***o*-Hydroxymethylbenzoic acid**
Formed from: Cinmethylin
o-Formylbenzoic acid
o-Toluic acid
→ Phthalic acid
Probably in: Rat D942

_m_-Hydroxymethylbenzoic acid
Formed from: *m*-Formylbenzoic acid

_p_-Hydroxymethylbenzoic acid
Formed from: *p*-Formylbenzoic acid
p-Toluic acid
Probably formed from: β-Artelinic acid
→ *p*-Formylbenzoic acid
Probably in: Comamonas G405

_o_-Hydroxymethylbenzonitrile
Formed from: *o*-Formylbenzonitrile
o-Tolunitrile

_m_-Hydroxymethylbenzonitrile
Formed from: *m*-Formylbenzonitrile

_p_-Hydroxymethylbenzonitrile
Formed from: *p*-Tolunitrile

4-Hydroxy-2-methylbenzotriazole
Formed from: 2-Methylbenzotriazole

5-Hydroxy-2-methylbenzotriazole
Formed from: 2-Methylbenzotriazole

N-(_p_-Hydroxymethylbenzoyl)-N-methylanilide
Formed from: N-Methyl-*p*-toluanilide

_p_-Hydroxymethylbenzoylpiperidine
Formed from: *p*-Toluoylpiperidine

N-(4-Hydroxy-3-methylbenzoyl)-_o_-(2-(piperidin-2-yl)ethyl)aniline
Formed from: 4-Hydroxy-N-(2-(2-(1-methyl-2-piperidinyl)ethyl)phenyl)benzamide

3-Hydroxy-7-methyl-3,4-benzpyrene
Formed from: 7-Methyl-3,4-benzpyrene

6-Hydroxymethyl-3,4-benzpyrene
Formed from: 3,4-Benzpyrene
6-Methyl-3,4-benzpyrene
→ 3-Hydroxy-6-hydroxymethyl-3,4-benzpyrene
In: Rat B477

7-Hydroxymethyl-3,4-benzpyrene
Formed from: 7-Methyl-3,4-benzpyrene
→ 9,10-Dihydro-9,10-dihydroxy-7-hydroxymethyl-3,4-benzpyrene
Probably in: Rat C427, C521
→ 3-Hydroxy-7-hydroxymethyl-3,4-benzpyrene
Probably in: Rat C521

3-Hydroxy-2-methylbenzyl alcohol
Probably formed from: 1-*tert*-Butylamino-3-(3-hydroxymethyl-2-methylphenoxy)-2-propanol

3-Hydroxy-5-methylbenzyl alcohol
Formed from: 3,5-Dimethylphenol

4-Hydroxy-2-methylbenzyl alcohol
Formed from: 3,4-Dimethylphenol

4-Hydroxy-3-methylbenzyl alcohol
Formed from: 2,4-Dimethylphenol

_p_-Hydroxymethylbenzbenzylamine
→ *p*-Hydroxymethylbenzbenzaldehyde
In: Pig A3693

4-Hydroxymethylbiphenyl
→ 4-Formylmethylbiphenyl
In: Arion G804

3-Hydroxymethyl-5,6-bis(_p_-hydroxyphenyl)-1,2,4-triazine
Probably formed from: 5,6-Bis(*p*-hydroxyphenyl)-3-methyl-1,2,4-triazine

4-Hydroxymethyl-2,8-bis(trifluoromethyl)quinoline
Formed from: Mefloquine

7-Hydroxy-6-(3-methyl-2-butenyl)-2_H_-1,4-benzopyran-2-one
See: Demethylsuberosin

cis-4-(3-Hydroxymethylbut-2-enyl)-3,5-dioxo-1,2-diphenylpyrazolidine
Formed from: Feprazone

trans-4-(3-Hydroxymethylbut-2-enyl)-3,5-dioxo-1,2-diphenylpyrazolidine
Formed from: Feprazone
→ *trans*-4-(3-Hydroxymethylbut-2-enyl)-1-(*p*-hydroxyphenyl)-3,5-dioxo-2-phenylpyrazolidine
Probably in: Rat A1739

4-(3-Hydroxy-3-methylbut-1-enyl)-N-methyl-L-tryptophan
→ Elymoclavine
In: Claviceps H7

3-Hydroxy-3-methylbutenylpaspalinine
Formed from: L-Tryptophan

9-Hydroxy-4-(3-methyl-2-butenyl-7_H_-furo[3,2-_g_][1]benzopyran-7-one
See: Alloimperatorin

cis-4-(3-Hydroxymethylbut-2-enyl)-1-(_p_-hydroxyphenyl)-3,5-dioxo-2-phenylpyrazolidine
Formed from: 1-(*p*-Hydroxyphenyl)-3,5-dioxo-2-phenyl-4-prenylpyrazolidine

***trans*-4-(3-Hydroxymethylbut-2-enyl)-(*p*-hydroxyphenyl)-3,5-dioxo-2-phenylpyrazolidine**
Probably formed from: *trans*-4-(3-Hydroxymethylbut-2-enyl)-3,5-dioxo-1,2-diphenylpyrazolidine
1-(*p*-Hydroxyphenyl)-3,5-dioxo-2-phenyl-4-prenylpyrazolidine

2-Hydroxy-3-(3-methyl-2-butenyl)-1,4-naphthoquinone
See: Lapachol

4-(4-Hydroxy-3-methyl-Δ^2-butenyl)-L-tryptophan
Formed from: 4-Dimethylallyl-L-tryptophan

4-Hydroxy-3-(3-methyl-3-buten-1-ynyl) benzaldehyde
See: Eutypine

***m*-(1-Hydroxy-1-methylbutyl)phenyl N-methylcarbamate**
Formed from: *m*-(1-Methylbutyl)phenyl N-methylcarbamate

6-(3-Hydroxy-3-methylbutyl)-4′,5,7-trihydroxy-3-methoxyflavone
Probably formed from: 6-(2,3-Dihydroxy-3-methylbutyl)-4′,5,7-trihydroxy-3-methoxyflavone

1-Hydroxy-N-methylcarbazole
Formed from: N-Methylcarbazole

2-Hydroxy-N-methylcarbazole
Formed from: N-Methylcarbazole

3-Hydroxy-N-methylcarbazole
Formed from: N-Methylcarbazole

N-Hydroxy-N-methylcarbazole
Formed from: N-Methylcarbazole
→ 3-Hydroxy-N-hydroxymethylcarbazole
Probably in: Cunninghamella G90, G701

N-(N-(2-Hydroxy-5-(N-methylcarboxamido) phenyl)-β-alanyl)-3,4-dimethoxyphenethylamine
Formed from: Ecabapide

1-Hydroxy-3-methylcholanthrene
Formed from: 3-Methylcholanthrene
→ *trans*-9,10-Dihydro-1,9,10-trihydroxy-3-methylcholanthrene
In: Cunninghamella B837
→ 3-Methylcholanthren-1-one
Probably in: Cunninghamella B837
Man F562
Rat B391, F382

2-Hydroxy-3-methylcholanthrene
Formed from: 3-Methylcholanthrene
→ 3-Methylcholanthren-2-one
Probably in: Cunninghamella B837
Mouse A444, A632
Rat B391

3-Hydroxymethylcholanthrene
Formed from: 3-Methylcholanthrene
→ *trans*-7,8-Dihydro-7,8-dihydroxy-3-hydroxymethylcholanthrene
In: Rat H708
→ *trans*-9,10-Dihydro-9,10-dihydroxy-3-hydroxymethylcholanthrene
In: Rat H708
→ *trans*-11,12-Dihydro-11,12-dihydroxy-3-hydroxymethylcholanthrene
In: Rat H708
→ 1-Hydroxy-3-hydroxymethylcholanthrene
In: Rat H708
→ 2-Hydroxy-3-hydroxymethylcholanthrene
In: Rat H708
→ 8-Hydroxy-3-hydroxymethylcholanthrene
In: Rat H708

2-Hydroxy-3-methylchromene-2-carboxylic acid
Formed from: 1-Methylnaphthalene

1-Hydroxy-5-methylchrysene
Formed from: 5-Methylchrysene

1-Hydroxy-6-methylchrysene
Formed from: 6-Methylchrysene

5-Hydroxymethylchrysene
Formed from: 5-Methylchrysene
→ 1,2-Dihydro-1,2-dihydroxy-5-hydroxymethylchrysene
In: Rat C83
→ 7,8-Dihydro-7,8-dihydroxy-5-hydroxymethylchrysene
In: Rat C83
→ 9,10-Dihydro-9,10-dihydroxy-5-hydroxymethylchrysene
Probably in: Rat C83
→ 1-Hydroxy-5-hydroxymethylchrysene
In: Rat C83
→ 7-Hydroxy-5-hydroxymethylchrysene
In: Rat C83
→ 5-Hydroxymethylchrysene sulphate

In: Guinea pig F141
Hamster F141
Mouse E811, F141
Rat D610, E68, F141

6-Hydroxymethylchrysene
Formed from: 6-Methylchrysene

7-Hydroxy-5-methylchrysene
Formed from: 5-Methylchrysene

7-Hydroxy-6-methylchrysene
Formed from: 6-Methylchrysene

9-Hydroxy-5-methylchrysene
Formed from: 5-Methylchrysene

5-Hydroxymethylchrysene sulphate
Formed from: 5-Hydroxymethylchrysene
→ S-(Chrysen-5-yl)methyl-L-cysteine
In: Rat E68

3′-Hydroxy-N-methyl-(*S*)-coclaurine
→ Reticuline
In: Berberis F854
Coptis K572

7-Hydroxy-4-methylcoumarin
See: 4-Methylumbelliferone

1-(2-Hydroxymethylcyclohexyl)-3-phenylurea
Formed from: 1-(2-Methylcyclohexyl)-3-phenylurea

1-(4-Hydroxy-2-methylcyclohexyl)-3-phenylurea
Formed from: 1-(2-Methylcyclohexyl)-3-phenylurea

10-Hydroxy-9-methyldaunomycin
Formed from: Auramycinone

3-Hydroxymethyl-3-demethylacronine
Formed from: Acronine

8-Hydroxymethyl-8-demethyltaxol
Formed from: Taxol

7-Hydroxymethyldibenzanthracene
Probably formed from: 7-Methyldibenzanthracene

4-Hydroxymethyl-2,5-dimethoxyamphetamine
Formed from: 2,5-Dimethoxy-4-methylamphetamine
→ 4-Carboxy-2,5-dimethoxyamphetamine
In: Guinea pig A3228
Rabbit A3228

2-Hydroxymethyl-6,7-dimethoxy-2-methyl-2*H*-benzo[*b*]pyran
Formed from: Precocene II
→ 6-Hydroxy-2-hydroxymethyl-7-methoxy-2-methyl-2*H*-benzo[*b*]pyran
Probably in: Melanopus E491

2-Hydroxymethyl-4,5-dimethoxyphenyl methyl sulphone
Formed from: Tolmesoxide

N-Hydroxymethyl-N,6-dimethyl-2-(4-methylphenyl)imidazo[1,2-*a*]pyridine-3-acetamide
Formed from: Zolpiden

6-Hydroxymethyl-N,N-dimethyl-2-(4-methylphenyl)imidazo[1,2-*a*]pyridine-3-acetamide
Formed from: Zolpiden

5′-Hydroxymethyl-4,8-dimethylpsoralen
Probably formed from: 4,5′,8-Trimethylpsoralen

7-(5-Hydroxymethyl-3,6-dimethyl-*p*-quinon-2-yl)-7-phenylheptanoic acid
Formed from: Seratrodast

9-Hydroxy-2-methylellipticinium (ellipticinium)
Probably formed from: 2-Methylellipticinium
→ S-(9-Hydroxy-2-methylellipticinium-10-yl)glutathione
In: Rat C655, D309, E530, E590
→ 2-Methylellipticinium-9-yl-β-D-glucuronide
In: Rat C491, C655, D309

S-(9-Hydroxy-2-methylellipticinium-10-yl)-L-cysteine
Probably formed from: S-(9-Hydroxy-2-methylellipticinium-10-yl)glutathione
→ N-Acetyl-S-(9-Hydroxy-2-methylellipticinium-10-yl)-L-cysteine
Probably in: Rat D309

S-(9-Hydroxy-2-methylellipticinium-10-yl) glutathione
Formed from: 9-Hydroxy-2-methylellipticinium
→ S-(9-Hydroxy-2-methylellipticinium-10-yl)-L-cysteine
Probably in: Rat D309, E590

1-Hydroxymethylelymoclavine
Formed from: 1-Hydroxymethylagroclavine

N-Hydroxy-4,4′-methylenebis(*o*-chloroaniline)
Formed from: 4,4′-Methylenebis(*o*-chloroaniline)
→ 4-Amino-3,3′-dichloro-4′-nitrosodiphenylmethane
Probably in: Dog F25
Guinea pig F25
Rat F25

N-Hydroxy-3,4-methylenedioxyamphetamine
→ 3,4-Methylenedioxyamphetamine
In: Rat H224

7-Hydroxy-3′,4′-methylenedioxyisoflavone
See: Pseudobaptigenin

8-Hydroxy-3,4-methylenedioxy-1-phenanthrenecarboxylic acid
Probably formed from: Aristolic acid I

***p*-Hydroxy-N-methylephedrine**
Formed from: N-Methylephedrine
→ N-Methylephedrin-4-yl-β-D-glucuronide
In: Rat F481

***p*-Hydroxy-N-methylephedrine-N-oxide**
Probably formed from: N-Methylephedrine-N-oxide

1-(3-((4-(4-Hydroxy-2-(1-methylethoxy)phenyl)-1-piperazinyl)methyl)benzoyl)oxopiperidine
Probably formed from: Hydroxy-1-(3-((4-(4-hydroxy-2-(1-methylethoxy)phenyl)-1-piperazinyl)methyl)benzoyl)piperidine

4-Hydroxy-1-(3-((4-(2-(1-methylethoxy)phenyl)-1-piperazinyl)methyl)benzoyl)piperidine
Formed from: Mazapertine

1-(3-((4-(4-Hydroxy-2-(1-methylethoxy)phenyl)-1-piperazinyl)methyl)benzoyl)piperidine
Formed from: Mazapertine
→ Hydroxy-1-(3-((4-(4-hydroxy-2-(1-methylethoxy)phenyl)-1-piperazinyl)methyl)benzoyl)piperidine
Probably in: Dog J581

N-(2-(4-(2-Hydroxy-3-((1-methylethyl)amino)propoxy)phenyl)ethyl)-N′-(1-methylethyl)urea
See: Pafenolol

4-(1-Hydroxy-1-methylethyl)-2-((3-hydroxy-2-methylphenyl)methoxy)-1-methyl-7-oxabicyclo[2.2.1]heptane
Probably formed from: 8-Hydroxycinmethylin

4-(1-Hydroxy-1-methylethyl)-2-((4-hydroxy-2-methylphenyl)methoxy)-1-methyl-7-oxabicyclo[2.2.1]heptane
Probably formed from: 8-Hydroxycinmethylin
→ 6-Hydroxy-4-(1-hydroxy-1-methylethyl)-2-((4-hydroxy-2-methylphenyl)methoxy)-1-methyl-7-oxabicyclo[2.2.1]heptane
Probably in: Goat F98

2-(2-Hydroxy-1-methylethyl)-6-isopropylphenol
Formed from: 2,6-Diisopropylphenol
→ 2-(2-Hydroxy-1-methylethyl)-6-isopropylphenyl-β-D-glucuronide
Probably in: Rabbit G16

2-(2-Hydroxy-1-methylethyl)-6-isopropylphenyl-β-D-glucuronide
Probably formed from: 2-(2-Hydroxy-1-methylethyl)-6-isopropylphenyl

2-(2-Hydroxy-1-methylethyl)-6-isopropylquinol
Probably formed from: 2,6-Diisopropylquinol
→ 2-(2-Hydroxy-1-methylethyl)-6-isopropylquinol-4-sulphate
Probably in: Dog G16
Rabbit G16
Rat G16

2-(2-Hydroxy-1-methylethyl)-6-isopropylquinol-4-sulphate
Probably formed from: 2-(2-Hydroxy-1-methylethyl)-6-isopropylquinol

N-(2-Hydroxy-1-methylethyl)-2-methylsulphinyl-N-phenylacetamide
Probably formed from: N-(2-Methoxy-1-methylethyl)-2-methylsulphinyl-N-phenylacetamide

5-Hydroxymethylferulic acid (4-hydroxy-5-hydroxymethyl-3-methoxycinnamic acid)
Probably formed from: 5-Formylferulic acid

9-Hydroxy-9-methylfluorene
Probably formed from: 9-Methylfluorene-9-hydroperoxide

7-Hydroxy-α-methylfluorene-2-acetic acid
Formed from: *d*-α-Methylfluorene-2-acetic acid
l-α-Methylfluorene-2-acetic acid
→ 7,9-Dihydroxy-α-methylfluorene-2-acetic acid
Probably in: Rat A2353, A2718

9-Hydroxy-α-methylfluorene-2-acetic acid
Formed from: *d*-α-Methylfluorene-2-acetic acid
l-α-Methylfluorene-2-acetic acid

3-Hydroxymethyl-1-(hydroxymethoxyphenyl)-4-isopropyl-2-methyl-3-pyrazolin-5-one
Probably formed from: 3-Hydroxymethyl-1-(hydroxyphenyl)-4-isopropyl-2-methyl-3-pyrazolin-5-one

3-Hydroxymethyl-1-(hydroxyphenyl)-4-isopropyl-2-methyl-3-pyrazolin-5-one
Probably formed from: 3-Hydroxymethyl-4-isopropyl-2-methyl-1-phenyl-3-pyrazolin-5-one
→ 3-Hydroxymethyl-1-(hydroxymethoxyphenyl)-4-isopropyl-2-methyl-3-pyrazolin-5-one
Probably in: Rat D285

2-Hydroxymethyl-5-(1-hydroxy-2-propoxy) carbonyl-3-(2-methyoxyethoxy)carbonyl-6-methyl-4-(*m*-nitrophenyl)pyridine
Probably formed from: 2-Hydroxymethyl-5-isopropoxycarbonyl-3-(2-methyoxyethoxy) carbonyl-6-methyl-4-(*m*-nitrophenyl)pyridine

1-Hydroxy-2-methylindane
Formed from: 2-Methylindane
→ 2-Methylindan-1-one
Probably in: Pseudomonas K207

1-Hydroxy-2-methylindan-3-one
Formed from: 2-Methylindan-1-one

3-Hydroxy-3-methylindolenine
Formed from: Skatole

1-(3-Hydroxymethyl)indolyl-3-indolylmethanol
Formed from: Indole-3-methanol

6-Hydroxymethyl-5-isopropoxycarbonyl-2-methyl-4-(*m*-nitrophenyl)pyridine-3-carboxylic acid
Probably formed from: 5-Isopropoxycarbonyl-2,6-dimethyl-4-(*m*-nitrophenyl)pyridine-3-carboxylic acid

2-Hydroxymethyl-5-isopropoxycarbonyl-3-(2-methyoxyethoxy)carbonyl-6-methyl-4-(*m*-nitrophenyl)pyridine
Probably formed from: 5-Isopropoxycarbonyl-3-(2-methyoxyethoxy)carbonyl-2,6-dimethyl-4-(*m*-nitrophenyl)pyridine
→ 4-(*m*-Aminophenyl)-2-hydroxymethyl-5-isopropoxycarbonyl-3-(2-methyoxy-ethoxy)carbonyl-6-methylpyridine
Probably in: Dog G189
Monkey G189
→ 2-Hydroxymethyl-5-(1-hydroxy-2-propoxy)carbonyl-3-(2-methyoxyethoxy) carbonyl-6-methyl-4-(*m*-nitrophenyl) pyridine
Probably in: Monkey G189

***p*-Hydroxymethyl-N-isopropylbenzamide**
Formed from: Azoprocarbazine
Probably formed from: *p*-Formyl-N-isopropylbenzamide

3-Hydroxymethyl-4-isopropyl-2-methyl-1-phenyl-3-pyrazolin-5-one
Formed from: 4-Isopropylantipyrine
→ 3-Hydroxymethyl-1-(hydroxyphenyl)-4-isopropyl-2-methyl-3-pyrazolin-5-one
Probably in: Rat D285

3-Hydroxymethyl-4-isopropyl-1-phenyl-3-pyrazolin-5-one
Formed from: 4-Isopropyl-3-methyl-1-phenyl-3-pyrazolin-5-one

N^1-(5-Hydroxymethyl-3-isoxazolyl)sulphanilamide
Formed from: Sulphamethoxazole
→ N^4-Acetyl-N^1-(5-hydroxymethyl-3-isoxazolyl)sulphanilamide
Probably in: Man H502
Pseudemys E553

8-Hydroxy-3-O-methylkaempferol (4′,5,7,8-tetrahydroxy-3-methoxyflavone)
Formed from: 8-Hydroxykaempferol

N^1-Hydroxymethylmethergine
Formed from: Methysergide

3-Hydroxymethyl-8-methoxychromone
→ 3-Carboxy-8-methoxychromone
In: Rat A2113
See also: A1761
→ 8-Hydroxy-3-hydroxymethylchromone
Probably in: Dog A1761
Rat A1761

8β-Hydroxymethyl-10α-methoxy-1,6-dimethylergoline
Formed from: Methergoline
Nicergoline

→ 8β-Bis(hydroxymethyl)-10α-methoxy-6-methylergoline
Probably in: Dog E167
Rat E167
→ 8β-Hydroxymethyl-10α-methoxy-6-methylergoline
Probably in: Mammal A425

9-(3-Hydroxymethyl-4-methoxy-2,6-dimethylphenyl)-3,7-dimethylnona-2,4,6,8-tetraenoic acid
Probably formed from: Acitretin

8β-Hydroxymethyl-10α-methoxy-6-methylergoline
Formed from: Nicergoline
Probably formed from: 1-Formyl-8β-hydroxymethyl-10α-methoxy-6-methylergoline
8β-Hydroxymethyl-10α-methoxy-1,6-dimethylergoline

8β-Hydroxymethyl-10α-methoxy-6-methylergoline 5-bromo-3-pyridinecarboxylate
Formed from: Nicergoline

2-(2-(3-Hydroxymethyl-4-methoxy-5-methyl)pyridinyl)methylsulphinyl-5-methoxybenzimidazole
Formed from: Omeprazole
→ 2-(2-(3-Carboxyl-4-methoxy-5-methyl) pyridinyl)methylsulphinyl-5-methoxybenzimidazole
Probably in: Man F19

2-(2-(5-Hydroxymethyl-4-methoxy-3-methyl)pyridinyl)methylsulphinyl-5-methoxybenzimidazole
Formed from: Omeprazole
→ 2-(2-(5-Carboxyl-4-methoxy-3-methyl) pyridinyl)methylsulphinyl-5-methoxybenzimidazole
In: Dog D844
Rat D844

2-Hydroxymethyl-3-(2-methoxy-4-nitrophenyl)-4(3*H*)-quinazolinone
Formed from: Nitromethaqualone
→ 3-(4-Amino-2-methoxyphenyl)-2-hydroxymethyl-4(3*H*)-quinazolinone
Probably in: Rat C437
→ 2-Hydroxymethyl-3-(2-methoxy-4-nitrophenyl)-4(3*H*)-quinazolinone-β-D-glucuronide
In: Rat C437

2-Hydroxymethyl-3-(2-methoxy-4-nitrophenyl)-4(3*H*)-quinazolinone-β-D-glucuronide
Formed from: 2-Hydroxymethyl-3-(2-methoxy-4-nitrophenyl)-4(3*H*)-quinazolinone

3′-Hydroxymethyl-4-methylaminoazobenzene
Formed from: 3′-Methyl-4-methylaminoazobenzene
Probably formed from: 4-Dimethylamino-3′-hydroxymethylazobenzene
→ 3-Formyl-4′-methylaminoazobenzene
Probably in: Rat A3835, B27, G9

4-Hydroxy-3-methyl-4′-methylaminoazobenzene
Formed from: 3′-Methyl-4-methylaminoazobenzene
Probably formed from: 4′-Dimethylamino-4-hydroxy-3-methylazobenzene
→ 4′-Amino-4-hydroxy-3-methylazobenzene
Probably in: Rat A3835, B9, B27, G9

9-Hydroxymethyl-10-methylanthracene
Formed from: 9,10-Dimethylanthracene
→ 9,10-Di(hydroxymethyl)anthracene
Probably in: Rat E612

N-Hydroxymethyl-N-methylbenzamide
Formed from: N,N-Dimethylbenzamide
→ N-Methylbenzamide
In: Rat G683

7-Hydroxymethyl-12-methylbenzanthracene
Formed from: 7,12-Dimethylbenzanthracene
Probably formed from: 7-Hydroxymethylbenzanthracene
→ 7,12-Di-(hydroxymethyl)benzanthracene
In: Rat B60
See also: A267, A1180, A2040, A2087, A2324, A2863, F136, F534, F742, G491, J381
→ 3,4-Dihydro-3,4-dihydroxy-7-hydroxymethyl-12-methylbenzanthracene
In: Mouse B128
Rat B128, B784, D53
→ 5,6-Dihydro-5,6-dihydroxy-7-hydroxymethyl-12-methylbenzanthracene
In: Rat B128, B784

→ 8,9-Dihydro-8,9-dihydroxy-7-hydroxymethyl-12-methylbenzanthracene
In: Mouse B128
Rat B128, B784
See also: A267
→ 10,11-Dihydro-10,11-dihydroxy-7-hydroxymethyl-12-methylbenzanthracene
In: Mouse B128
Rat B128, B784, D53
See also: A267
→ 7-Formyl-12-methylbenzanthracene
Probably in: Rat F534
→ 2-Hydroxy-7-hydroxymethyl-12-methylbenzanthracene
In: Rat B60
→ 3-Hydroxy-7-hydroxymethyl-12-methylbenzanthracene
Probably in: Rat A3324
→ 4-Hydroxy-7-hydroxymethyl-12-methylbenzanthracene
Probably in: Rat A2324, A3930
→ 9-Hydroxy-7-hydroxymethyl-12-methylbenzanthracene
Probably in: Rat A3930
→ 7-Hydroxymethyl-12-methylbenzanthracene sulphate
In: Channel catfish K497
Man J683
→ 7-Hydroxymethyl-12-methylbenzanthracene-5,6-oxide
Probably in: Rat B913
→ 12-Methylbenzanthracene-7-carboxylic acid
Probably in: Hamster A267

12-Hydroxymethyl-7-methylbenzanthracene
Formed from: 7,12-Dimethylbenzanthracene
→ 7,12-Di(hydroxymethyl)benzanthracene
In: Rat B60
See also: A267, A1180, A2040, A2087, A2324, A2863, D339, F136, J381
→ 5,6-Dihydro-5,6-dihydroxy-12-hydroxymethyl-7-methylbenzanthracene
Probably in: Rat A2324
→ 2-Hydroxy-12-hydroxymethyl-7-methylbenzanthracene
In: Rat B60
→ 4-Hydroxy-12-hydroxymethyl-7-methylbenzanthracene
Probably in: Rat A2324
→ 12-Hydroxymethyl-7-methylbenzanthracene sulphate
In: Man J683

7-Hydroxymethyl-12-methylbenzanthracene-5,6-oxide
Probably formed from: 7-Hydroxymethyl-12-methylbenzanthracene

7-Hydroxymethyl-12-methylbenzanthracene sulphate
Formed from: 7-Hydroxymethyl-12-methylbenzanthracene
→ S-((12-Methylbenzanthracenyl)-7-methyl)glutathione
In: Rat B965

12-Hydroxymethyl-7-methylbenzanthracene sulphate
Formed from: 12-Hydroxymethyl-7-methylbenzanthracene

4-(3-Hydroxymethyl-2-methylbenzyl)imidazole
Formed from: Detomidine
→ 4-(3-Carboxy-2-methylbenzyl)imidazole
In: Rat E719
See also: F675, G634
→ 4-(3-Hydroxymethyl-2-methylbenzyl) imidazole-β-D-glucuronide
Probably in: Horse G634
Rat E717, E719

4-(3-Hydroxymethyl-2-methylbenzyl)imidazole-β-D-glucuronide
Probably formed from: 4-(3-Hydroxymethyl-2-methylbenzyl)imidazole

4-Hydroxymethyl-5-N-methylcarboxamido-9-anilinoacridine
Probably formed from: 4-Methyl-5-(N-methyl)carboxamido-9-anilinoacridine

7-Hydroxymethyl-14-methyldibenzanthracene
Probably formed from: 7,14-Dimethyldibenzanthracene

N-Hydroxymethyl-N-methyl-2,2-diphenylacetamide
Formed from: N,N-Dimethyldiphenylacetamide
→ N-Hydroxymethyl-N-methyl-2,2-diphenylacetamide-β-D-glucoside

In: Tomato A1113

N-Hydroxymethyl-N-methyl-2,2-diphenylacetamide-β-D-gentiobioside
Probably formed from: N-Hydroxymethyl-N-methyl-2,2-diphenylacetamide-β-D-glucoside

N-Hydroxymethyl-N-methyl-2,2-diphenylacetamide-β-D-glucoside
Formed from: N-Hydroxymethyl-N-methyl-2,2-diphenylacetamide
→ N-Hydroxymethyl-N-methyl-2,2-diphenylacetamide-β-D-gentiobioside
Probably in: Tomato A1113

N-Hydroxymethyl-N-(1-methyl-3,3-diphenylpropyl)formamide
Formed from: N-Methyl-N-(1-methyl-3,3-diphenylpropyl)formamide
→ N-Hydroxymethyl-N-(1-methyl-3,3-diphenylpropyl)formamide-β-D-glucuronide
In: Rat G88

N-Hydroxymethyl-N-(1-methyl-3,3-diphenylpropyl)formamide-β-D-glucuronide
Formed from: N-Hydroxymethyl-N-(1-methyl-3,3-diphenylpropyl)formamide

Hydroxy-N-methyl-3,4-methylenedioxyamphetamine
Formed from: N-Methyl-3,4-methylenedioxyamphetamine

4′-(1-Hydroxymethyl-1-methylethyl)-4-(4-(diphenylmethoxy)piperidino)butyrophenone
Formed from: Ebastine
→ Carebastine
In: Man J422

1-(N-((6-Hydroxymethyl)-6-methylhept-2-en-4-yn-1-yl)-N-((methylamino)methyl))naphthalene
Formed from: Terbinafine

2-Hydroxymethyl-6-methyl-N-(5-methyl-3-isoxazolyl)benzamide
Probably formed from: 2,6-Dimethyl-N-(5-methyl-3-isoxazolyl)benzamide

1-Hydroxymethyl-8-methylnaphthalene
Formed from: 1,8-Dimethylnaphthalene
→ 1,8-Bis(hydroxymethyl)naphthalene
In: Pseudomonas A1482
→ 8-Methyl-1-naphthoic acid
In: Pseudomonas A1482

2-Hydroxymethyl-6-methylnaphthalene
Formed from: 2,6-Dimethylnaphthalene

2-Hydroxymethyl-6-methyl-4-(*m*-nitrophenyl)pyridine-3,5-dicarboxylic acid
Probably formed from: 2,6-Dimethyl-4-(*m*-nitrophenyl)pyridine-3,5-dicarboxylic acid
2,6-Dimethyl-4-(*m*-nitrophenyl)pyridine-3,5-dicarboxylic acid 3-methyl ester 5-(3-piperidinyl) ester
6-Hydroxymethyl-2-methyl-4-(*m*-nitrophenyl)pyridine-3,5-dicarboxylic acid 3-methyl ester
2-Hydroxymethyl-6-methyl-4-(*m*-nitrophenyl)pyridine-3,5-dicarboxylic acid 5-(3-piperidinyl) ester

2-Hydroxymethyl-6-methyl-4-(*m*-nitrophenyl)pyridine-3,5-dicarboxylic acid 3-ethyl ester 5-methyl ester
Probably formed from: 1,4-Dehydro-nitrendipine
→ 2-Hydroxymethyl-6-methyl-4-(*m*-nitrophenyl)pyridine-3,5-dicarboxylic acid 3-ethyl ester 5-(3-piperidinyl) ester-β-D-glucuronide
Probably in: Dog G398
Mouse G398
Rat G398

2-Hydroxymethyl-6-methyl-4-(*m*-nitrophenyl)pyridine-3,5-dicarboxylic acid 5-ethyl ester 3-methyl ester
Probably formed from: 1,4-Dehydro-nitrendipine

2-Hydroxymethyl-6-methyl-4-(*m*-nitrophenyl)pyridine-3,5-dicarboxylic acid 3-ethyl ester 5-methyl ester-β-D-glucuronide
Probably formed from: 2-Hydroxymethyl-6-methyl-4-(*m*-nitrophenyl)pyridine-3,5-dicarboxylic acid 3-ethyl ester 5-methyl ester

6-Hydroxymethyl-2-methyl-4-(*m*-nitrophenyl)pyridine-3,5-dicarboxylic acid 3-methyl ester
Probably formed from: 2,6-Dimethyl-4-(*m*-nitrophenyl)pyridine-3,5-dicarboxylic acid 3-methyl ester
→ 2-Hydroxymethyl-6-methyl-4-(*m*-nitrophenyl)pyridine-3,5-dicarboxylic acid
Probably in: Rat H927

2-Hydroxymethyl-6-methyl-4-(*m*-nitrophenyl) pyridine-3,5-dicarboxylic acid 3-methyl ester 5-(3-piperidinyl) ester
Probably formed from: 2,6-Dimethyl-4-(*m*-nitrophenyl)pyridine-3,5-dicarboxylic acid 3-methyl ester 5-(3-piperidinyl) ester

2-Hydroxymethyl-6-methyl-4-(*m*-nitrophenyl) pyridine-3,5-dicarboxylic acid 3-(3-piperidinyl) ester
Probably formed from: 2,6-Dimethyl-4-(*m*-nitrophenyl)pyridine-3,5-dicarboxylic acid 3-(3-piperidinyl) ester

2-Hydroxymethyl-6-methyl-4-(*m*-nitrophenyl) pyridine-3,5-dicarboxylic acid 5-(3-piperidinyl) ester
Probably formed from: 2,6-Dimethyl-4-(*m*-nitrophenyl)pyridine-3,5-dicarboxylic acid 3-(3-piperidinyl) ester
→ 2-Hydroxymethyl-6-methyl-4-(*m*-nitrophenyl)pyridine-3,5-dicarboxylic acid
Probably in: Rat F11

(+)-5-Hydroxy-2-methyl-2-(4-methyl-3-pentenyl)-7-pentyl-2*H*-1-benzopyran-6-carboxylic acid
See: Cannabichromenic acid

5-(2-Hydroxymethyl-5-methylphenoxy)-2,2-dimethylpentanoic acid
Formed from: Gemfibrozil
→ 5-(2,5-Bis(hydroxymethyl)phenoxy)-2,2-dimethylpentanoic acid
Probably in: Hamster J565
Rat J564, J565
→ 5-(2-Hydroxymethyl-5-methylphenoxy)-2,2-dimethylpentanoic acid-β-D-glucuronide
Probably in: Hamster J565
Rat J565

5-(5-Hydroxymethyl-2-methylphenoxy)-2,2-dimethylpentanoic acid
Formed from: Gemfibrozil
→ 5-(2,5-Bis(hydroxymethyl)phenoxy)-2,2-dimethylpentanoic acid
Probably in: Rat J564
→ 5-(5-Carboxy-2-methylphenoxy)-2,2-dimethylpentanoic acid
In: Man A3033
Rat J565
→ 5-(5-Hydroxymethyl-2-methylphenoxy)-2,2-dimethylpentanoic acid-β-D-glucuronide
Probably in: Hamster J565
Rat J564, J565

5-(2-Hydroxymethyl-5-methylphenoxy)-2,2-dimethylpentanoic acid-β-D-glucuronide
Probably formed from: 5-(2-Hydroxymethyl-5-methylphenoxy)-2,2-dimethylpentanoic acid

5-(5-Hydroxymethyl-2-methylphenoxy)-2,2-dimethylpentanoic acid-β-D-glucuronide
Probably formed from: 5-(5-Hydroxymethyl-2-methylphenoxy)-2,2-dimethylpentanoic acid

2-(2-Hydroxymethyl-6-methylphenoxy)-1-methylethylamine
Formed from: 1-Methyl-2-(2,6-xylyloxy) ethylamine

1-(2-Hydroxymethyl-6-methylphenoxy)propan-2-ol
Probably formed from: 1-(2,6-Dimethylphenoxy)propan-2-ol

N-((2-Hydroxymethyl-6-methylphenylcarbamoyl) methyl)pyrrolidone
Formed from: Nefiracetam

4-(1-(3-Hydroxymethyl-2-methylphenyl)ethyl)-1*H*-imidazole
Formed from: Medetomidine
→ 4-(1-(3-Carboxy-2-methylphenyl)ethyl)-1*H*-imidazole
Probably in: Rat F886
→ 4-(1-(3-Hydroxymethyl-2-methylphenyl) ethyl)-1*H*-imidazole-β-D-glucuronide
Probably in: Rat F886

4-(1-(3-Hydroxymethyl-2-methylphenyl)ethyl)-1*H*-imidazole-β-D-glucuronide
Probably formed from: 4-(1-(3-Hydroxymethyl-2-methylphenyl)ethyl)-1*H*-imidazole

4-Hydroxymethyl-3-methylphenyl N-methylcarbamate
Formed from: 3,4-Xylyl N-methylcarbamate

1-(2-Hydroxymethyl-6-methylphenyl)-3-(1-methyl-5-oxopyrrolidinylidene)urea
Probably formed from: 5′-Oxoxilobam

5-Hydroxymethyl-3-(3-methylphenyl)-2-oxazolidinone
See: Toloxatone

N-(2-Hydroxymethyl-6-methylphenyl)-2-(2-oxo-1-pyrrolidinyl)acetamide
Formed from: N-(2,6-Dimethylphenyl)-2-(2-oxo-1-pyrrolidinyl)acetamide

α-Hydroxymethyl-N-methyl-α-phenylsuccinimide
Formed from: Methsuximide

3-Hydroxymethyl-2-methyl-1-phenyl-3-pyrazolin-5-one
Formed from: Antipyrine
→ 3-Carboxy-2-methyl-1-phenyl-3-pyrazolin-5-one-β-D-glucuronide
In: Man B394
Rat A3645, A3850
→ 3-Hydroxymethyl-2-methyl-1-phenyl-3-pyrazolin-5-one-β-D-glucuronide
In: Man B394, C415
Rat C415
See also: D322, J769

3-Hydroxymethyl-2-methyl-1-phenyl-3-pyrazolin-5-one-β-D-glucuronide
Formed from: 3-Hydroxymethyl-2-methyl-1-phenyl-3-pyrazolin-5-one

N-((2-Hydroxymethyl-6-methylphenylcarbamoyl)methyl)pyrrolidone
Formed from: Nefiracetam

4-(1-(3-Hydroxymethyl-2-methylphenyl)ethyl)-1*H*-imidazole
Formed from: Medetomidine
→ 4-(1-(3-Carboxy-2-methylphenyl)ethyl)-1*H*-imidazole
Probably in: Rat F886
→ 4-(1-(3-Hydroxymethyl-2-methylphenyl) ethyl)-1*H*-imidazole-β-D-glucuronide
Probably in: Rat F886

4-(1-(3-Hydroxymethyl-2-methylphenyl)ethyl)-1*H*-imidazole-β-D-glucuronide
Probably formed from: 4-(1-(3-Hydroxymethyl-2-methylphenyl)ethyl)-1*H*-imidazole

4-Hydroxymethyl-3-methylphenyl N-methylcarbamate
Formed from: 3,4-Xylyl N-methylcarbamate

1-(2-Hydroxymethyl-6-methylphenyl)-3-(1-methyl-5-oxopyrrolidinylidene)urea
Probably formed from: 5′-Oxoxilobam

N-(2-Hydroxymethyl-6-methylphenyl)-2-(2-oxo-1-pyrrolidinyl)acetamide
Formed from: N-(2,6-Dimethylphenyl)-2-(2-oxo-1-pyrrolidinyl)acetamide

3-Hydroxymethyl-2-methyl-1-phenyl-3-pyrazolin-5-one
Formed from: Antipyrine
→ 3-Carboxy-2-methyl-1-phenyl-3-pyrazolin-5-one
In: Man B394
Rat A3645, A3850
→ 3-Hydroxymethyl-2-methyl-1-phenyl-3-pyrazolin-5-one-β-D-glucuronide
In: Man B394, C415, D352
Rat C415, J769
See also: D322

3-Hydroxymethyl-2-methyl-1-phenyl-3-pyrazolin-5-one-β-D-glucuronide
Formed from: 3-Hydroxymethyl-2-methyl-1-phenyl-3-pyrazolin-5-one

α-Hydroxymethyl-N-methyl-α-phenylsuccinimide
Formed from: Methsuximide

3-Hydroxymethyl-3-methyl-1-phenyltriazene
Probably formed from: 3,3-Dimethyl-1-phenyltriazene

N-(2-Hydroxymethyl-6-methylphenyl)urea
Probably formed from: N-(2,6-Dimethylphenyl)urea

3-Hydroxymethyl-5-methyl-N-2-picolylbenzamide
Formed from: 3,5-Dimethyl-N-2-picolylbenzamide

3-Hydroxymethyl-5-methyl-N-3-picolylbenzamide
Formed from: 3,5-Dimethyl-N-3-picolylbenzamide

3-Hydroxymethyl-5-methyl-N-4-picolylbenzamide
Formed from: 3,5-Dimethyl-N-4-picolylbenzamide

2-Hydroxymethyl-4-(4-methyl-1-piperazinyl)-10*H*-thieno[2,3-*b*][1,5]benzodiazepine
Formed from: Olanzapine
→ 2-Carboxy-4-(4-methyl-1-piperazinyl)-10*H*-thieno[2,3-*b*][1,5]benzodiazepine
Probably in: Dog J150
Man J150
Rat J96
Rhesus monkey J150

→ 2-Hydroxymethyl-4-(4-methyl-1-piperazinyl)-10*H*-thieno[2,3-*b*][1,5]benzodiazepine-β-D-glucuronide
Probably in: Rhesus monkey J150
→ 2-Hydroxymethyl-4-(4-methyl-1-piperazinyl)-10*H*-thieno[2,3-*b*][1,5]benzodiazepine-N-oxide
Probably in: Rhesus monkey J150

2-Hydroxymethyl-4-(4-methyl-1-piperazinyl)-10*H*-thieno[2,3-*b*][1,5]benzodiazepine-β-D-glucuronide
Probably formed from: 2-Hydroxymethyl-4-(4-methyl-1-piperazinyl)-10*H*-thieno[2,3-*b*][1,5]benzodiazepine

2-Hydroxymethyl-4-(4-methyl-1-piperazinyl)-10*H*-thieno[2,3-*b*][1,5]benzodiazepine-N-oxide
Probably formed from: 2-Hydroxymethyl-4-(4-methyl-1-piperazinyl)-10*H*-thieno[2,3-*b*][1,5]benzodiazepine

8-(3-Hydroxymethyl-1-methylpropylamino)-6-methoxyquinoline
Formed from: 8-(3-Formyl-1-methylpropylamino)-6-methoxyquinoline

2′-Hydroxymethyl-6′-methyl-N-propylpipecoloxylidide
Formed from: Ropivacaine

N-(4-Hydroxymethyl-6-methyl-2-pyrimidinyl)benzenesulphonamide
Formed from: N-(4,6-Dimethyl-2-pyrimidinyl)benzenesulphonamide

N^1-(4-Hydroxymethyl-6-methyl-2-pyrimidinyl)sulphanilamide
Formed from: Sulphamethazine

6-Hydroxymethyl-7-methylquinoline
Formed from: 6,7-Dimethylquinoline

6-Hydroxymethyl-8-methylquinoline
Formed from: 6,8-Dimethylquinoline

7-Hydroxymethyl-6-methylquinoline
Formed from: 6,7-Dimethylquinoline

4-Hydroxy-2-methyl-N-(5-methyl-2-thiazolyl)-2*H*-1,2-benzothiazine-3-carboxamide-1,1-dioxide
See: Meloxicam

3-Hydroxymethyl-3-methyl-1-(2,4,6-trichlorophenyl)triazene-β-D-glucuronide
Formed from: 3,3-Dimethyl-1-(2,4,6-trichlorophenyl)triazene

2-Hydroxymethyl-6-methyl-4-(2-trifluoromethylphenyl)-3,5-pyridinedicarboxylic acid
Probably formed from: 2,6-Dimethyl-4-(2-trifluoromethylphenyl)-3,5-pyridinedicarboxylic acid

6-Hydroxymethyl-5-methylxanthenone-4-acetic acid
Formed from: 5,6-Dimethylxanthenone-4-acetic acid

Hydroxymethylmexiletine
Formed from: Mexiletine

3-Hydroxy-N-methylmorphinan
Formed from: N-Methyl-3-phenoxymorphinan
→ 3-Hydroxymorphinan
Probably in: Dog B634

1-Hydroxymethylnaphthalene
(1-naphthylmethanol)
Formed from: 1-Methylnaphthalene
Probably formed from: 1-Naphthaldehyde
→ 1-Naphthaldehyde
Probably in: Cunninghamella D33
→ 1-Naphthylmethyl sulphate
In: Rat E330, E630

2-Hydroxymethylnaphthalene
(2-naphthylmethanol)
Formed from: 2-Methylnaphthalene
→ 2-Naphthaldehyde
In: Arion G804
See also: D33
→ 2-Naphthylmethyl sulphate
In: Rat E630

5-Hydroxy-2-methyl-1,4-naphthalenedione
See: Plumbagin

5-Hydroxymethyl-1-naphthoic acid
Probably formed from: 5-Methyl-1-naphthoic acid

7-Hydroxymethyl-1,2-naphthoquinone
Formed from: 2-Methylnaphthalene

2-Hydroxy-3-methyl-3-(2-naphthyl)propionitrile
Formed from: 2-(2-Naphthyl)propanal

2-Hydroxymethyl-3-(2-nitrooxypropoxycarbonyl)-5-(*m*-nitrooxypropoxycarbonyl)-6-methyl-4-(*m*-nitrophenyl)pyridine
Probably formed from: 2-Nitrooxypropyl 3-nitrooxypropyl 2,6-dimethyl-4-(*m*-nitrophenyl)pyridine-3,5-dicarboxylate
→ 3-Nitrooxypropyl 3-carboxy-2-hydroxymethyl-6-methyl-4-(*m*-nitrophenyl)pyridine-5-carboxylate
Probably in: Dog G593
Rat G593

2-Hydroxymethyl-5-(2-nitrooxypropoxycarbonyl)-3-(*m*-nitrooxypropoxycarbonyl)-6-methyl-4-(*m*-nitrophenyl)pyridine
Probably formed from: 2-Nitrooxypropyl 3-nitrooxypropyl 2,6-dimethyl-4-(*m*-nitrophenyl)pyridine-3,5-dicarboxylate
→ 2-Nitrooxypropyl 3-carboxy-2-hydroxymethyl-6-methyl-4-(*m*-nitrophenyl)pyridine-5-carboxylate
Probably in: Dog G593
Rat G593

17α-Hydroxymethyloestradiol
Formed from: 17α-Azidomethyloestradiol

2-Hydroxy-3-O-methyl-17β-oestradiol
Formed from: 2-Hydroxy-17β-oestradiol
→ 2-Hydroxy-3-O-methyl-17β-oestradiol-2-β-D-glucuronide
Probably in: Rat J498

16β-Hydroxy-3-O-methyl-17β-oestradiol
Formed from: 3-O-Methyloestrone
→ 3-O-Methyl-16-oxo-17β-oestradiol
Probably in: Streptomyces F810

2-Hydroxy-3-O-methyl-17β-oestradiol-2-β-D-glucuronide
Probably formed from: 2-Hydroxy-3-O-methyl-17β-oestradiol

S-(2-Hydroxy-3-O-methyl-17β-oestradiol-1-yl)-L-cysteine
Formed from: S-(2-Hydroxy-17β-oestradiol-1-yl)-L-cysteine

2-Hydroxy-3-O-methyloestriol
Formed from: 2-Hydroxyoestriol
→ 2-Hydroxy-3-O-methyloestriol-β-D-glucuronide
Probably in: Rat A1487
→ 2-Hydroxy-3-O-methyloestriol sulphate
Probably in: Rat A1487

2-Hydroxy-3-O-methyloestriol-β-D-glucuronide
Probably formed from: 2-Hydroxy-3-O-methyloestriol

2-Hydroxy-3-O-methyloestriol sulphate
Probably formed from: 2-Hydroxy-3-O-methyloestriol

2-Hydroxy-3-O-methyloestrone
Formed from: 2-Hydroxyoestrone
→ 2-Hydroxy-3-O-methyloestrone-2-β-D-glucuronide
In: Rat A1280

4-Hydroxy-3-O-methyloestrone
Formed from: 4-Hydroxyoestrone

16α-Hydroxy-3-O-methyloestrone
Formed from: 3-O-Methyloestrone

2-Hydroxy-3-O-methyloestrone-2-β-D-glucuronide
Formed from: 2-Hydroxy-3-O-methyloestrone

9-Hydroxy-2-methylolivacinium
(1,2,5-trimethyl-6*H*-pyrido[4,3-*b*]carbazole)
→ S-(9-Hydroxy-2-methylolivacinium-10-yl)glutathione
In: Rat C655
→ 2-Methylolivacinium-9-yl-β-D-glucuronide
In: Rat C491, C655

S-(9-Hydroxy-2-methylolivacinium-10-yl) glutathione
Formed from: 9-Hydroxy-2-methylolivacinium

2-(3-(5-Hydroxymethyl-1,3,4-oxadiazol-2-yl)-3,3-diphenylpropyl)-2-azabicyclo[2.2.1]octane
Formed from: Nufenoxole

5-Hydroxy-2-(3-(5-methyl-1,3,4-oxadiazol-2-yl)-3,3-diphenylpropyl)-2-azabicyclo[2.2.1]octane
Formed from: Nufenoxole

3-Hydroxy-3-methyloxindole
Probably formed from: 3-Methyloxindole

4-Hydroxy-3-methyloxindole
Probably formed from: 3-Methyloxindole

3-Hydroxymethyloxindole
Formed from: Indole-3-acetic acid
Indole-3-methanol
Probably formed from: 3-Methyleneindolenine
→ 3-Methyleneoxindole
In: Wheat D150
See also: A727

3-Hydroxymethyl-5-(α-oxobenzyl)furan
Probably formed from: 5-(α-Hydroxybenzyl)-3-hydroxymethylfuran

3-Hydroxymethyl-1-oxocholanthrene
Probably formed from: 1-Hydroxy-3-hydroxymethylcholanthrene

3-Hydroxymethyl-2-oxocholanthrene
Probably formed from: 2-Hydroxy-3-hydroxymethylcholanthrene

***o*-(2-Hydroxy-5-methyl-1-oxo-4-hexen-1-yl) phenylglyoxylic acid**
Formed from: Lapachol

2-Hydroxy-3-O-methyl-6-oxo-17β-oestradiol
Formed from: 2-Hydroxy-6-oxo-17β-oestradiol

2-Hydroxy-3-O-methyl-16-oxo-17β-oestradiol
Formed from: 2-Hydroxy-16-oxo-17β-oestradiol

4-(4-Hydroxy-4-methyl-3-oxopentyl)-3,5-dioxo-1,2-diphenylpyrazolidine
Formed from: Trimetazon

7-Hydroxy-5-methyl-2-(2-oxopropyl)-4*H*-1-benzopyran-4-one
See: Aloesone

4-Hydroxymethyloxybutazone-β-D-glucuronide
Formed from: Suxibuzone

4-(3-Hydroxy-4-methylpentyl)-3,5-dioxo-1,2-diphenylpyrazolidine
Formed from: Trimetazon

***o*-Hydroxymethylphenoxyacetic acid**
Formed from: *o*-Formylphenoxyacetic acid
o-Formylphenoxypropionic acid

***m*-Hydroxymethylphenoxyacetic acid**
Formed from: *m*-Formylphenoxyacetic acid

***p*-Hydroxymethylphenoxyacetic acid**
Formed from: *p*-Formylphenoxyacetic acid

3-(4-Hydroxymethylphenoxy)-1-isopropylaminopropan-2-ol
Probably formed from: 3-(4-(1-Carboxy-1-hydroxymethyl)phenoxy)-1-isopropylaminopropan-2-ol

(*E*)-2-(2-((*o*-Hydroxymethylphenoxy)methyl) phenyl)-2-methoxyiminoacetic acid
Probably formed from: (*E*)-2-Methoxyimino-2-((*o*-tolyloxymethyl)phenyl)acetic acid

(*E*)-2-(2-((4-Hydroxy-2-methylphenoxy) methyl)phenyl)-2-methoxyiminoacetic acid
Probably formed from: (*E*)-2-Methoxyimino-2-((*o*-tolyloxymethyl)phenyl)acetic acid

3-(4-Hydroxy-2-methylphenoxy)-1,2-propanediol
Formed from: Mephenesin

***o*-Hydroxymethylphenoxypropionic acid**
Formed from: *o*-Formylphenoxypropionic acid

3-(*o*-Hydroxymethylphenoxy)pyridazine
Formed from: Credazine

3-(4-Hydroxy-2-methylphenoxy)pyridazine
Formed from: Credazine
→ 3-(4-Hydroxy-2-methylphenoxy)pyridazine sulphate
Probably in: Rat A3372

3-(4-Hydroxy-2-methylphenoxy)pyridazine sulphate
Probably formed from: 3-(4-Hydroxy-2-methylphenoxy)pyridazine

4-Hydroxy-2-methylphenylacetic acid
→ 2,5-Dihydroxy-3-methylphenylacetic acid
In: Pseudomonas A1480

***m*-Hydroxymethyl-L-phenylalanine**
Formed from: 3-Carboxy-L-phenylalanine

***p*-Hydroxymethyl-L-phenylalanine**
Formed from: *p*-Methyl-L-phenylalanine

4-Hydroxymethylphenylbutazone-β-D-glucuronide
Formed from: Suxibuzone

***m*-Hydroxy-N-methylphenylethanolamine**
See: *m*-Sympatol

***p*-Hydroxy-N-methylphenylethanolamine**
See: *p*-Sympatol

N-(2-Hydroxy-1-methyl-1-phenylethyl)-3,3-dimethylbutyramide
Formed from: 3,3-Dimethyl-N-(1-methyl-1-phenylethyl)butyramide

3-(2-Hydroxy-1-methyl-2-phenylethyl)-N-(*p*-hydroxyphenylaminocarbonyl) sydnone imine
Probably formed from: N-(*p*-Hydroxyphenylaminocarbonyl)-3-(1-methyl-2-phenylethyl) sydnone imine

***o*-Hydroxymethylphenyl-α-D-glucoside**
Probably formed from: Salicyl alcohol

***o*-Hydroxymethylphenyl-β-D-glucoside**
See: Salicin

***m*-Hydroxymethylphenyl-α-D-glucoside**
Probably formed from: *m*-Hydroxybenzyl alcohol

***p*-Hydroxymethylphenyl-α-D-glucoside**
Probably formed from: *p*-Hydroxybenzyl alcohol

***m*-Hydroxymethylphenylglycol**
Probably formed from: *m*-Methylphenylglycol
→ *m*-Carboxyphenylglycol
Probably in: Rat C153
→ *m*-Hydroxymethylphenylglyoxylic acid
Probably in: Rat C153

***m*-Hydroxymethylphenylglyoxylic acid**
Probably formed from: *m*-Hydroxymethylphenylglycol

3-(2-Hydroxy-5-methylphenyl)-3-(*p*-hydroxyphenyl)-N,N-diisopropylpropylamine
Probably formed from: Tolterodine
→ 3-(2-Hydroxy-5-hydroxymethylphenyl)-3-(*p*-hydroxyphenyl)-N,N-diisopropylpropylamine
Probably in: Rat J419

3-(2-Hydroxy-5-methylphenyl)-3-(*p*-hydroxyphenyl)propylamine
Probably formed from: Bisdeisopropyltolterodine
→ 3-(3,4-Dihydroxyphenyl)-3-(2-hydroxy-5-hydroxymethylphenyl)propylamine
Probably in: Rat J419

(5-Hydroxymethyl-3-phenyl-4-isoxazolyl) penicillin
Formed from: Oxacillin

2-((*o*-Hydroxymethyl)phenyl)-2-methoxyiminoacetic acid
Probably formed from: (*E*)-2-Methoxyimino-2-((*o*-tolyloxymethyl)phenyl)acetic acid

2-((3-Hydroxy-2-methylphenyl)methoxy)-1-methyl-4-(1-methylethyl)-7-oxabicyclo [2.2.1]heptane
Formed from: Cinmethylin

2-((4-Hydroxy-2-methylphenyl)methoxy)-1-methyl-4-(1-methylethyl)-7-oxabicyclo [2.2.1]heptane
Formed from: Cinmethylin

2-((5-Hydroxy-2-methylphenyl)methoxy)-1-methyl-4-(1-methylethyl)-7-oxabicyclo [2.2.1]heptane
Formed from: Cinmethylin

***m*-Hydroxymethylphenyl N-methylcarbamate**
Formed from: *m*-Cresyl N-methylcarbamate
→ *m*-Carboxyphenyl N-methylcarbamate
In: Housefly A1209
Rat A1209
See also: A2666

4-Hydroxy-3-methylphenyl N-methylcarbamate
Formed from: *m*-Cresyl N-methylcarbamate

5-(*o*-Hydroxymethylphenyl)-1-methyl-3-(3-pyridyl-N-oxide)-1*H*-1,2,4-triazole
Formed from: 5-(*o*-Hydroxymethylphenyl)-1-methyl-3-(3-pyridyl)-1*H*-1,2,4-triazole
→ 5-(*o*-Hydroxymethylphenyl)-1-methyl-3-(3-pyridyl-N-oxide)-1*H*-1,2,4-triazole-β-D-glucuronide
Probably in: Cat B900
Dog B900
Rat B900
→ 5-(*o*-Hydroxymethylphenyl)-1-methyl-3-(3-pyridyl-N-oxide)-1*H*-1,2,4-triazole sulphate
Probably in: Cat B900
Dog B900
Rat B900

5-(*o*-Hydroxymethylphenyl)-1-methyl-3-(3-pyridyl-N-oxide)-1*H*-1,2,4-triazole-β-D-glucuronide
Probably formed from: 5-(*o*-Hydroxymethylphenyl)-1-methyl-3-(3-pyridyl-N-oxide)-1*H*-1,2,4-triazole

5-(*o*-Hydroxymethylphenyl)-1-methyl-3-(3-pyridyl-N-oxide)-1*H*-1,2,4-triazole sulphate
Probably formed from: 5-(*o*-Hydroxymethylphenyl)-1-methyl-3-(3-pyridyl-N-oxide)-1*H*-1,2,4-triazole

5-(*o*-Hydroxymethylphenyl)-1-methyl-3-(3-pyridyl)-1*H*-1,2,4-triazole
→ 5-(*o*-Carboxyphenyl)-1-methyl-3-(3-pyridyl)-1*H*-1,2,4-triazole
In: Cat B900
→ 5-(*o*-Hydroxymethylphenyl)-1-methyl-3-(3-pyridyl-N-oxide)-1*H*-1,2,4-triazole

In: Man B900
Rat B900
→ 5-(*o*-Hydroxymethylphenyl)-1-methyl-3-(3-pyridyl)-1*H*-1,2,4-triazole-β-D-glucuronide
Probably in: Cat B900
Rat B900
→ 5-(*o*-Hydroxymethylphenyl)-1-methyl-3-(3-pyridyl)-1*H*-1,2,4-triazole sulphate
Probably in: Cat B900
Rat B900

5-(*o*-Hydroxymethylphenyl)-1-methyl-3-(3-pyridyl)-1*H*-1,2,4-triazole-β-D-glucuronide
Probably formed from: 5-(*o*-Hydroxymethylphenyl)-1-methyl-3-(3-pyridyl)-1*H*-1,2,4-triazole

5-(*o*-Hydroxymethylphenyl)-1-methyl-3-(3-pyridyl)-1*H*-1,2,4-triazole sulphate
Probably formed from: 5-(*o*-Hydroxymethylphenyl)-1-methyl-3-(3-pyridyl)-1*H*-1,2,4-triazole

5-(*p*-Hydroxymethylphenyl)-1-methylpyrrole-2-acetic acid
Formed from: Tolmetin
→ 5-(*p*-Carboxyphenyl)-1-methylpyrrole-2-acetic acid
In: Rat A3666
See also: A32, A1875, A2121, A2986, B105

3-(*o*-Hydroxymethylphenyl)-2-methyl-4(3*H*)-quinazolinone
Formed from: 2-Methyl-3-(*o*-tolyl)-4(3*H*)-quinazolinone

3-(3-Hydroxy-2-methylphenyl)-2-methyl-4(3*H*)-quinazolinone
Formed from: 2-Methyl-3-(*o*-tolyl)-4(3*H*)-quinazolinone

3-(4-Hydroxy-2-methylphenyl)-2-methyl-4(3*H*)-quinazolinone
Formed from: 2-Methyl-3-(*o*-tolyl)-4(3*H*)-quinazolinone
→ 3-(4-Hydroxy-5-methoxy-2-methylphenyl)-2-methyl-4(3*H*)-quinazolinone
Probably in: Man A2418

5-Hydroxymethyl-3-phenyl-1,2,4-oxadiazole
Formed from: 5-Methyl-3-phenyl-1,2,4-oxadiazole

5-(*p*-(Hydroxymethyl)phenyl)-5-phenylhydantoin
Formed from: 5-Phenyl-5-(*p*-tolyl)hydantoin
→ 5-(*p*-Carboxyphenyl)-5-phenylhydantoin
In: Rat A2402
→ 5-(*p*-(Hydroxymethyl)phenyl)-5-phenylhydantoin-β-D-glucuronide
Probably in: Rat A2402

5-(*p*-(Hydroxymethyl)phenyl)-5-phenylhydantoin-β-D-glucuronide
Probably formed from: 5-(*p*-(Hydroxymethyl)phenyl)-5-phenylhydantoin

4-(4-(4-Hydroxymethylphenyl)phenylmethoxy-1-piperidinyl)butyric acid
Formed from: 4-(4-(4-Methylphenyl)phenylmethoxy-1-piperidinyl)butyric acid

3-Hydroxy-α-methylphenylpropionic acid
Probably formed from: 3,4-Dihydroxy-α-methylphenylpropionic acid

2-(*p*-Hydroxymethylphenyl)propionic acid
Probably formed from: 2-(*p*-Methylphenyl)propionic acid

2-Hydroxy-3-methyl-3-phenylpropionitrile
Formed from: 2-Phenylpropionaldehyde

1-(1-Hydroxy-2-methyl-1-phenylprop-2-yl)imidazole
Formed from: 1-(2-Methyl-1-oxo-1-phenylprop-2-yl)imidazole

***o*-Hydroxymethylphenyl sulphate**
Probably formed from: Salicyl alcohol

***m*-Hydroxymethylphenyl sulphate**
Probably formed from: *m*-Hydroxybenzyl alcohol

N-(*p*-Hydroxymethylphenylsulphonyl)-5-methyl-2-pyrazoline-1-carboxamide
Formed from: N-(*p*-toluenesulphonyl)-2-pyrazoline-1-carboxamide
→ N-(*p*-Carboxyphenylsulphonyl)-5-methyl-2-pyrazoline-1-carboxamide
In: Man B684
Rat B684

N-(*p*-Hydroxymethylphenylsulphonyl)-N′-2-(1-phenylpropyl)urea
Formed from: Tosifen
N-(*p*-Carboxyphenylsulphonyl)-N′-2-(1-phenylpropyl)urea
In: Dog A2877
Rat A2877

4-(5-((*p*-Hydroxymethyl)phenyl)-3-trifluoromethyl-1*H*-pyrazol-1-yl) benzenesulphonamide
Formed from: Celecoxib
→ 4-(5-(*p*-Carboxyphenyl)-3-trifluoromethyl-1*H*-pyrazol-1-yl)benzenesulphonamide
In: Cynomolgus monkey K123
Dog J781
Man K19
Mouse K123
Rabbit K123
Rat K30
Rhesus monkey K123
→ 4-(5-((*p*-Hydroxymethyl)phenyl)-3-trifluoromethyl-1*H*-pyrazol-1-yl) benzenesulphonamide-β-D-glucuronide
In: Rat K19

4-(5-((*p*-Hydroxymethyl)phenyl)-3-trifluoromethyl-1*H*-pyrazol-1-yl) benzenesulphonamide-β-D-glucuronide
Formed from: 4-(5-((*p*-Hydroxymethyl) phenyl)-3-trifluoromethyl-1*H*-pyrazol-1-yl) benzenesulphonamide

2-((*p*-Hydroxymethyl)phenyl)-N,N,6-trimethylimidazo[1,2-*a*]pyridine-3-acetamide
Formed from: Zolpidem

4-Hydroxy-N-(2-(2-(1-methyl-2-piperidinyl) ethyl)phenyl)benzamide
Formed from: Encainide
→ 3,4-Dihydroxy-N-(2-(2-(1-methyl-2-piperidinyl)ethyl)phenyl)benzamide
Probably in: Man F532
→ N-(4-Hydroxy-3-methylbenzoyl)-*o*-(2-(piperidin-2-yl)ethyl)aniline
Probably in: Man D89
→ 4-Hydroxy-N-(2-(2-(2-piperidinyl) ethyl)phenyl)benzamide
Probably in: Man D89, F532
Rat D98

N-(4-Hydroxymethyl-4-piperidinyl)-N-phenylpropanamide
Probably formed from: N-(4-Methoxymethyl-4-piperidinyl)-N-phenylpropanamide

3-(2-Hydroxymethyl-1-propenyl)-2,2-dimethylcyclopropanecarboxylic acid (3-phenoxyphenyl)methyl ester
Formed from: *cis*-Phenothrin
→ 3-(2-Carboxy-1-propenyl)-2,2-dimethylcyclopropanecarboxylic acid (3-phenoxyphenyl)methyl ester
In: Rat C63

N-(3-(2-Hydroxy-2-methyl-1-propoxy)-2-(1-pyrrolidino)propyl)aniline
Probably formed from: N-Benzyl-N-(3-(2-hydroxy-2-methyl-1-propoxy)-2-(1-pyrrolidino)propyl)aniline

5-Hydroxy-6-methyl-2-((*n*-propyl)amino)tetralin
Formed from: 2-((Di-*n*-propyl)amino)-5-hydroxy-6-methyltetralin
→ 6-Methyl-2-((*n*-propyl)amino)tetralin-5-yl sulphate
Probably in: Rat D819

N-(2-Hydroxymethyl-2-propyl)-4,4-diphenyl-2-butylamine
Formed from: Terodiline
→ N-(2-Hydroxymethyl-2-propyl)-4-(*p*-hydroxyphenyl)-4-phenyl-2-butylamine
Probably in: Dog D784

N-(2-Hydroxymethyl-2-propyl)-4-(*p*-hydroxyphenyl)-4-phenyl-2-butylamine
Probably formed from: N-(2-Hydroxy-methyl-2-propyl)-4,4-diphenyl-2-butylamine

***o*-(1-Hydroxy-1-methylpropyl)phenol**
Probably formed from: *o*-(1-Hydroxy-1-methylpropyl)phenyl N-methyl-carbamate

***erythro-o*-(2-Hydroxy-1-methylpropyl)phenol**
Probably formed from: *erythro-o*-(2-Hydroxy-1-methylpropyl)phenyl N-methylcarbamate

***threo-o*-(2-Hydroxy-1-methylpropyl)phenol**
Probably formed from: *threo-o*-(2-Hydroxy-1-methylpropyl)phenyl N-methylcarbamate

***o*-(3-Hydroxy-1-methylpropyl)phenol**
Probably formed from: *o*-(3-Hydroxy-1-methylpropyl)phenyl N-methylcarbamate

***p*-(1-Hydroxy-2-methylpropyl)phenylacetic acid**
Formed from: *p*-Isobutylphenylacetic acid

***p*-(2-Hydroxy-2-methylpropyl)phenylacetic acid**
Formed from: *p*-Isobutylphenylacetic acid

***o*-(1-Hydroxymethylpropyl)phenyl N-methylcarbamate**
Formed from: *o*-*sec*-Butylphenyl N-methylcarbamate

***o*-(1-Hydroxy-1-methylpropyl)phenyl N-methylcarbamate**
Formed from: *o*-*sec*-Butylphenyl N-methylcarbamate
→ *o*-(1-Hydroxy-1-methylpropyl)phenol
Probably in: Cladosporium A1815

***erythro-o*-(2-Hydroxy-1-methylpropyl)phenyl N-methylcarbamate**
Formed from: *o*-*sec*-Butylphenyl N-methylcarbamate
→ *erythro-o*-(2-Hydroxy-1-methylpropyl) phenol
Probably in: Cladosporium A1815
→ *o*-(1-Methylacetonyl)phenyl N-methylcarbamate
Probably in: Aspergillus A1811, A1813
Cladosporium A1813
Coniothyrium A1813
Fusarium A1813
Penicillium A1813

***threo-o*-(1-Hydroxy-1-methylpropyl)phenyl N-methylcarbamate**
Formed from: *o*-*sec*-Butylphenyl N-methylcarbamate
→ *threo-o*-(2-Hydroxy-1-methylpropyl)phenol
Probably in: Cladosporium A1815
→ *o*-(1-Methylacetonyl)phenyl)phenyl N-methylcarbamate
Probably in: Aspergillus A1811, A1813
Cladosporium A1813
Coniothyrium A1813
Fusarium A1813
Penicillium A1813

***o*-(3-Hydroxy-1-methylpropyl)phenyl N-methylcarbamate**
Formed from: *o*-*sec*-Butylphenyl N-methylcarbamate
→ *o*-(2-Carboxy-1-methylethyl)phenyl N-methylcarbamate
Probably in: Aspergillus A1813, A1815
Cladosporium A1813
Coniothyrium A1813
Penicillium A1813
→ *o*-(3-Hydroxy-1-methylpropyl)phenol
Probably in: Cladosporium A1815

(*R*)-2-(4-(2-Hydroxy-2-methylpropyl)phenyl) propionic acid
Formed from: (*R*)-2-(4-Isobutylphenyl) propionic acid
→ 2-(4-(2-Carboxypropyl)phenyl)propionic acid
Probably in: Man A2428, A3560
Rabbit B470
→ (*S*)-2-(4-(2-Hydroxy-2-methylpropyl) phenyl)propionic acid
In: Man A2142

(*S*)-2-(4-(2-Hydroxy-2-methylpropyl)phenyl) propionic acid
Formed from: (*R*)-2-(4-(2-Hydroxy-2-methylpropyl)phenyl)propionic acid
(*S*)-2-(4-Isobutylphenyl)propionic acid

6-Hydroxy-7-methyl-2-propyl-3-((2′-(1*H*-tetrazol-5-yl)(1,1′-biphenyl)-4-yl)methyl)-3*H*-imidazo[4,5-*b*]pyridine
Formed from: 7-Methyl-2-propyl-3-((2′-(1*H*-tetrazol-5-yl)(1,1′-biphenyl)-4-yl) methyl)-3*H*-imidazo[4,5-*b*]pyridine

7-Hydroxymethyl-2-propyl-3-((2′-(1*H*-tetrazol-5-yl)(1,1′-biphenyl)-4-yl)methyl)-3*H*-imidazo [4,5-*b*]pyridine
Formed from: 7-Methyl-2-propyl-3-((2′-(1*H*-tetrazol-5-yl)(1,1′-biphenyl)-4-yl)methyl)-3*H*-imidazo[4,5-*b*]pyridine

1-Hydroxymethylpyrene
→ 1-Hydroxymethylpyrene-O-sulphate
In: Rat G194

1-Hydroxymethylpyrene-O-sulphate
Formed from: 1-Hydroxymethylpyrene

4-Hydroxy-2-methyl-N-(2-pyridyl)-2*H*-1,2-benzothiazine-3-carboxamide-1,1-dioxide
See: Piroxicam

***o*-Hydroxy-2-methyl-2-(3-pyridyl) propiophenone**
Formed from: *o*-Methoxy-2-methyl-2-(3-pyridyl)propiophenone
→ *o*-Hydroxy-2-methyl-2-(3-pyridyl) propiophenone-N-oxide
Probably in: Rat J86

o-Hydroxy-2-methyl-2-(3-pyridyl)propiophenone-N-oxide
Probably formed from: *o*-Hydroxy-2-methyl-2-(3-pyridyl)propiophenone

N-(5-Hydroxy-1-methyl-2-pyrrolidinylidene)-1-(2-hydroxypyrrolidine)-N′-phenyl-carboximidamide
Probably formed from: N-(5-Hydroxy-1-methyl-2-pyrrolidinylidene)-N′-phenyl-1-pyrrolidinecarboximidamide

N-(5-Hydroxy-1-methyl-2-pyrrolidinylidene)-N′-phenyl-1-pyrrolidinecarboximidamide
Formed from: Piroglide
→ N-(5-Hydroxy-1-methyl-2-pyrrolidinylidene)-1-(2-hydroxypyrrolidine)-N′-phenyl-carboximidamide
Probably in: Dog J693
Man J693
Monkey J693
Rat J693
→ N-(1-Methyl-5-oxo-2-pyrrolidinylidene)-N′-phenyl-1-pyrrolidinecarboximidamide
Probably in: Dog J693
Man J693
Monkey J693
Rat J693

8-Hydroxy-3-O-methylquercetin
(3′,4′,5,7,8-pentahydoxy-3-methoxyflavone)
Formed from: Gossypetin

2-Hydroxy-3-methylquinoline
Formed from: 3-Methylquinoline
→ 5,6-Dihydro-2,5,6-trihydroxy-3-methylquinoline
In: Comamonas K769

2-Hydroxy-4-methylquinoline
Formed from: 4-Methylquinoline

2-Hydroxy-6-methylquinoline
Formed from: 6-Methylquinoline
→ 3,4-Dihydro-6-methyl-2-quinolone
Probably in: Microorganism J367

2-Hydroxy-8-methylquinoline
Formed from: 8-Methylquinoline
→ 3,4-Dihydro-8-methyl-2-quinolone
Probably in: Microorganism J367

4-Hydroxymethylquinoline
Probably formed from: 4-Formylquinoline

2-Hydroxy-5-methylquinone
Formed from: 4-Methyl-5-nitrocatechol
→ 2,4,5-Trihydroxytoluene
In: Pseudomonas H238

4-Hydroxymethyl-*o*-quinone
Formed from: 3,4-Dihydroxybenzyl alcohol

4′-Hydroxy-2-(methylseleno)benzanilide
Formed from: 2-(Methylseleno)benzanilide
→ 3′,4′-Dihydroxy-2-(methylseleno) benzanilide
Probably in: Man F235
Pig F235
Rat F235
→ 4′-Hydroxy-2-(methylseleno)-3′-methylthiobenzanilide
Probably in: Man F235
Pig F235
Rat F235
→ 2-(Methylseleno)benzanilid-4′-yl-β-D-glucuronide
Probably in: Man F235
Pig F235
Rat E808, F127, F235

4′-Hydroxy-2-(methylseleno)-3′methylthiobenzanilide
Probably formed from: 4′-Hydroxy-2-methylselenobenzanilide

4-Hydroxy-α-methyl-3-sulphatoxyphenethylamine
Probably formed from: 3,4-Dihydroxy-α-methylphenethylamine

4-Hydroxy-α-methyl-3-sulphatoxy-L-phenylalanine
Formed from: 3,4-Dihydroxy-α-methyl-L-phenylalanine

Hydroxy-methylsulphinylcarbamazepine
Formed from: Carbamazepine

Hydroxy-3-(2-methylsulphinylethyl)quinazoline-2,4(1*H*,3*H*)-dione
Probably formed from: Hydroxy-3-(2-methylthioethyl)quinazoline-2,4(1*H*,3*H*)-dione
→ Hydroxy-3-(2-methylsulphonylethyl) quinazoline-2,4(1*H*,3*H*)-dione
Probably in: Rat H122

2-Hydroxy-3-(4-methylsulphinyl-α-naphthyloxy)propylamine
Probably formed from: 2-Hydroxy-3-(4-methylthio-α-naphthyloxy)propylamine
→ 2-Hydroxy-3-(4-methylsulphonyl-α-naphthyloxy)propylamine
Probably in: Dog F60

***p*-Hydroxy-2-methylsulphonylacetanilide**
Probably formed from: *p*-Hydroxy-N-isopropyl-2-methylsulphonylacetanilide
→ *p*-Hydroxy-2-methylsulphonylacetanilide-β-D-glucuronide
Probably in: Rat C24

***p*-Hydroxy-2-methylsulphonylacetanilide-β-D-glucuronide**
Probably formed from: *p*-Hydroxy-2-methylsulphonylacetanilide

Hydroxy-3-(2-methylsulphonylethyl)quinazoline-2,4(1*H*,3*H*)-dione
Probably formed from: Hydroxy-3-(2-methylsulphinylethyl)quinazoline-2,4(1*H*,3*H*)-dione

2-Hydroxy-3-(4-methylsulphonyl-α-naphthyloxy) propylamine
Probably formed from: 2-Hydroxy-3-(4-methylsulphinyl-α-naphthyloxy)propylamine

4-Hydroxy-2-methyl-N-(2-thiazolyl)-2*H*-1,2-benzothiazine-3-carboxamide-1,1-dioxide
See: Sudoxicam

(*S*)-2-(4-(5-Hydroxy-3-methyl-2-thienyl) phenyl)propionic acid
Formed from: (*S*)-2-(4-(3-Methyl-2-thienyl)phenyl)propionic acid

Hydroxy-3-(2-methylthioethyl)quinazoline-2,4(1*H*,3*H*)-dione
Probably formed from: 3-(2-Methylthioethyl)quinazoline-2,4(1*H*,3*H*)-dione
→ Hydroxy-3-(2-methylsulphinylethyl) quinazoline-2,4(1*H*,3*H*)-dione
Probably in: Rat H122

2-Hydroxy-1-methylthioindane
Formed from: Indene

2-Hydroxy-3-(4-methylthio-α-naphthyloxy) propylamine
Formed from: 4′-Methylthiopropranolol
→ 2-Hydroxy-3-(4-methylsulphinyl-α-naphthyloxy)propylamine
Probably in: Dog F60

1-Hydroxy-2-methylthio-1-phenylethane
Probably formed from: S-(2-Hydroxy-2-phenylethyl)-L-cysteine

1-Hydroxymethylthioxanthone
Formed from: 1-Methylthioxanthone

3-Hydroxymethylthioxanthone
Formed from: 3-Methylthioxanthone

2-Hydroxymethyl-3-(*o*-tolyl)-4(3*H*)-quinazolinone
Formed from: 2-Methyl-3-(*o*-tolyl)-4(3*H*)-quinazolinone

6-Hydroxy-2-methyl-3-(*o*-tolyl)-4(3*H*)-quinazolinone
Formed from: 2-Methyl-3-(*o*-tolyl)-4(3*H*)-quinazolinone

8-Hydroxy-2-methyl-3-(*o*-tolyl)-4(3*H*)-quinazolinone
Formed from: 2-Methyl-3-(*o*-tolyl)-4(3*H*)-quinazolinone

3-Hydroxymethyl-*s*-triazolo[3,4-*a*]phthalazine
Formed from: 3-Methyl-*s*-triazolo[3,4-*a*] phthalazine
→ 3-Carboxy-*s*-triazolo[3,4-*a*]phthalazine
In: Rat A2544
See also: A1503, A1816, A3592

4-Hydroxy-3-methyl-*s*-triazolo[3,4-*a*]phthalazine
Formed from: 3-Methyl-*s*-triazolo[3,4-*a*] phthalazine

8-Hydroxy-3-methyl-*s*-triazolo[3,4-*a*]phthalazine
Formed from: 3-Methyl-*s*-triazolo[3,4-*a*] phthalazine

9-Hydroxy-3-methyl-*s*-triazolo[3,4-*a*]phthalazine
Formed from: 3-Methyl-*s*-triazolo[3,4-*a*] phthalazine

N-Hydroxymethyl-N′-(γ-(2-trifluoromethyl-10-phenothiazinyl)propyl)ethylenediamine
Formed from: Fluphenazine

5-(4-(2-Hydroxy-2-methyl-4-(3,5,6-trimethyl-1,4-benzoquinon-2-yl)butoxy)benzyl)-2,4-thiazolidinedione
Formed from: Troglitazone

4-(2-Hydroxymethyl-2,4,4-trimethylbut-4-yl) phenol
Formed from: 4-*tert*-Octylphenol

→ 4-(2-Hydroxymethyl-2,4,4-trimethylbut-4-yl)phenyl-β-D-glucuronide
Probably in: Rainbow trout K126

4-(2-Hydroxymethyl-2,4,4-trimethylbut-4-yl) phenyl-β-D-glucuronide
Probably formed from: 4-(2-Hydroxymethyl-2,4,4-trimethylbut-4-yl)phenol

5-Hydroxy-1-methyltryptamine
Formed from: 5-Hydroxytryptamine

5-Hydroxy-N-methyltryptamine
Formed from: 5-Hydroxytryptamine
N-Methyltryptamine
→ 5-Hydroxy-1,N-dimethyltryptamine
In: Rabbit C333
→ 5-Hydroxy-N,N-dimethyltryptamine
In: Rabbit A106, A111
Rat A111
→ 5-Hydroxyindole-3-acetaldehyde
In: Rat B653
→ 1,2,3,4-Tetrahydro-6-hydroxy-2-methyl-β-carboline
In: Rabbit A111
Rat A111

5-Hydroxy-α-methyltryptamine
Formed from: α-Methyltryptamine

6-Hydroxy-N-methyltryptamine
Formed from: 6-Hydroxytryptamine

β-Hydroxy-5-methyl-L-tryptophan
Probably formed from: 5-Methyl-L-tryptophan

***m*-Hydroxymexiletine**
Formed from: Mexiletine

***p*-Hydroxymexiletine**
Formed from: Mexiletine

N-Hydroxymexiletine
Formed from: Mexiletine
→ N-Hydroxymexiletine-N-β-D-glucuronide
In: Man G385, G976
Rabbit G385

N-Hydroxymexiletine-N-β-D-glucuronide
Formed from: N-Hydroxymexiletine

8-Hydroxymianserin
Formed from: Mianserin
→ 8-Hydroxymianserin-O-β-D-glucuronide
In: Man G66
Mouse G66
Rat G66, G864, G908
→ 8-Hydroxymianserin-N-oxide
Probably in: Rat G66, G864
→ 8-Hydroxymianserin-O-sulphate
In: Mouse G66
Rat G66, G864, G908

8-Hydroxymianserin-O-β-D-glucuronide
Formed from: 8-Hydroxymianserin

8-Hydroxymianserin-N-oxide
Probably formed from: 8-Hydroxymianserin

8-Hydroxymianserin-O-sulphate
Formed from: 8-Hydroxymianserin

***p*-Hydroxymidaglizole**
Formed from: Midaglizole
→ 2-(2-(4,5-Dihydro-1*H*-imidazol-2-yl)-1-(3,4-dihydroxyphenyl)ethyl)pyridine
Probably in: Rat F334
→ 2-(2-(4,5-Dihydro-1*H*-imidazol-2-yl)-1-(*p*-hydroxyphenyl)ethyl)pyridine
Probably in: Rat F334
→ *p*-Hydroxymidaglizole-β-D-glucuronide
Probably in: Rat F334
→ 3-(*p*-Hydroxyphenyl)-3-(2-pyridyl) propioimidamide
Probably in: Rat F334

***p*-Hydroxymidaglizole-β-D-glucuronide**
Probably formed from: *p*-Hydroxymidaglizole

1′-Hydroxymidazolam
See: 8-Chloro-6-(*o*-fluorophenyl)-1-hydroxymethyl-4*H*-imidazo[1,5-*a*][1,4] benzodiazepine

4-Hydroxymidazolam
Formed from: Midazolam
→ 8-Chloro-6-(*o*-fluorophenyl)-4-hydroxy-4*H*-1-imidazo[1,5-*a*][1,4]benzodiazepine
In: Man J43
Rat J43
See also: B705

1′-Hydroxymidazolam-β-D-glucuronide
Probably formed from: 8-Chloro-6-(*o*-fluorophenyl)-1-hydroxymethyl-4*H*-1-imidazo[1,5-*a*][1,4]benzodiazepine

***p*-Hydroxyminaprine**
Formed from: Minaprine
→ 6-(*p*-Hydroxyphenyl)-4-methyl-3-(3-oxo-2-morpholinoethyamino)pyridazine
Probably in: Dog G65

9-Hydroxyminocycline
Probably formed from: Minocycline
8-Hydroxymirtazapine
Formed from: Mirtazapine
21-Hydroxymontelukast
Formed from: Montelukast
25-Hydroxymontelukast
Formed from: Montelukast
33-Hydroxymontelukast
Formed from: Montelukast
2-Hydroxymorphinan
Formed from: (+)-N-Cyclopropylmethyl-2-hydroxymorphinan
(−)-N-Cyclopropylmethyl-2-hydroxymorphinan
3-Hydroxymorphinan (N,O-bisdemethyldextromethorphan)
Probably formed from: O-Demethyldextromethorphan
3-Hydroxy-N-methylmorphinan
3-O-(2-(3-Hydroxymorpholino)ethyl)morphine
Probably formed from: Pholcodine
→ 3-O-(2-(3-Oxomorpholino)ethyl)morphine
Probably in: Man F682
3-(β-Hydroxy-γ-morpholinopropyl)-7,8-dimethoxy-4-methylcoumarin
Formed from: 7,8-Dimethoxy-4-methyl-3-(γ-morpholino-β-3′,4′,5′-trimethoxybenzoxypropyl)coumarin
→ 7-Hydroxy-3-(β-hydroxy-γ-morpholinopropyl)-8-methoxy-4-methylcoumarin
Probably in: Rabbit A275
HMPE
See: 2-(4-Hydroxy-3-methoxyphenyl)ethanol
6′-Hydroxynabilone
Formed from: Nabilone
→ 6′-Oxonabilone
In: Bacillus A3010
7′-Hydroxynabilone
Formed from: Nabilone
→ 7′-Oxonabilone
Probably in: Nocardia B37
***m*-Hydroxynaftifine**
Formed from: Naftifine
***p*-Hydroxynaftifine**
Formed from: Naftifine
→ 3-(4-Hydroxy-3-methoxyphenyl)-N-methyl-N-(1-naphthylmethyl)-2-propen-1-ylamine
Probably in: Man E32
1-Hydroxy-2-naphthaldehyde
→ 1-Hydroxy-2-naphthoic acid
Probably in: Pseudomonas E776
1-Hydroxynaphthalene
See: α-Naphthol
2-Hydroxynaphthalene
See: β-Naphthol
2-Hydroxynaphthalene-1-sulphonic acid
→ Salicylic acid
In: Pseudomonas F185
5-Hydroxynaphthalene-1-sulphonic acid
→ 2,6-Dihydroxybenzoic acid
In: Pseudomonas F185
7-Hydroxynaphthalene-1-sulphonic acid
→ 2,4-Dihydroxybenzoic acid
In: Pseudomonas F185
4-((2-Hydroxy-1-naphthalenyl)azo)benzenesulphonic acid
See: Acid Orange 7
4-Hydroxy-1-naphthalenyl-β-D-glucuronide
Formed from: 1,4-Dihydroxynaphthalene
4-((2-Hydroxy-1-naphthalenyl)azo)-1-naphthalenesulphonic acid
See: Acid Red 88
4-((4-(2-Hydroxy-1-naphthalenyl)azo)phenyl)azobenzenesulphonic acid
See: Acid Red 151
6-Hydroxy-α-naphthoflavone
Formed from: α-Naphthoflavone
9-Hydroxy-α-naphthoflavone
Formed from: α-Naphthoflavone
1-Hydroxy-2-naphthoic acid
Probably formed from: 1-Hydroxy-2-naphthaldehyde
2-Naphthoic acid
Phenanthrene
→ *o*-Carboxybenzalpyruvic acid
In: Coccus C730
Nocardioides J598
See also: H813, J674
→ 1,5-Dihydroxy-2-naphthoic acid
Probably in: Sphingomonas K610
→ Phthalic acid

In: Bacillus K528
2-Hydroxy-1-naphthoic acid
Formed from: Phenanthrene
3-Hydroxy-2-naphthoic acid
Formed from: Anthracene
5-Hydroxy-1-naphthoic acid
Probably formed from: 1-Naphthoic acid
5-Hydroxy-2-naphthoic acid
Probably formed from: 2-Naphthoic acid
6-Hydroxy-2-naphthoic acid
Probably formed from: 2-Naphthoic acid
2-Hydroxy-1,4-naphthquinone (lawsone)
Formed from: 1,4-Dihydroxy-2-naphthoic acid
1,4-Naphthquinone
o-Succinylbenzoic acid
→ Salicylic acid
In: Pseudomonas D918
5-Hydroxy-1,4-naphthquinone (juglone)
Formed from: 1,4-Dihydroxy-2-naphthoic acid
2,5-Dihydroxy-1,4-naphthquinone
Hydrojuglone-β-D-glucuronide
1,4-Naphthquinone
o-Succinylbenzoic acid
→ 3,5-Dihydroxy-1,4-naphthquinone
In: Pseudomonas D918, K897
→ 4,8-Dihydroxy-1-tetralone
In: Wangiella D760
→ 1,4,5-Trihydroxynaphthalene
Probably in: Beef G401
Guinea pig K773
Juglans A3470
6-Hydroxy-2-naphthylacetic acid
Formed from: 6-Methoxy-2-naphthylacetic acid
→ 6-Hydroxy-2-naphthylacetic acid sulphate
Probably in: Rat C867
6-Hydroxy-2-naphthylacetic acid sulphate
Probably formed from: 6-Hydroxy-2-naphthylacetic acid
2-Hydroxy-2-(2-naphthyl)acetonitrile
Formed from: 2-Naphthaldehyde
***p*-(2-Hydroxy-1-naphthylazo) benzenesulphonic acid**
See: Acid Orange 7

4-(6-Hydroxy-2-naphthyl)butan-2-one
Formed from: Nabumetone
1-(2-Hydroxy-2-(β-naphthyl)ethyl)hydantoin
Probably formed from: Dihydronafimidone
→ 2,3,4,5-Tetrahydro-4-hydroxy-1-(2-hydroxy-2-(β-naphthyl)ethyl)-2-oxoimidazole
Probably in: Baboon F483
Cynomolgus monkey F483
Dog F483
Man F483
1-(6-Hydroxy-2-naphthyl)-2-isopropylaminoethanol
Formed from: 2-Isopropylamino-1-(2-naphthyl)ethanol
1-(7-Hydroxy-2-naphthyl)-2-isopropylaminoethanol
Formed from: 2-Isopropylamino-1-(2-naphthyl)ethanol
2-Hydroxy-1-naphthyl methylcarbamate
Formed from: 1-Naphthyl methylcarbamate
4-Hydroxy-1-naphthyl methylcarbamate
Formed from: 1-Naphthyl methylcarbamate
(1-Hydroxy-2-naphthyloxy)acetic acid
Probably formed from: (2-Naphthyloxy)acetic acid
3-(4-Hydroxy-1-naphthyloxy)lactic acid
Probably formed from: 4-Hydroxypropranolol
3-(α-Naphthyloxy)lactic acid
1-(4-Hydroxy-1-naphthyloxy)-3-(1-(*o*-methoxyphenyl)piperazin-4-yl)propan-2-ol
Formed from: Naftopidil
2-Hydroxy-3-(α-naphthyloxy)propionaldehyde
Probably formed from: 2-Hydroxy-3-(α-naphthyloxy)propylamine
Propranolol
→ 1-(α-Naphthyloxy)-2,3-propylene glycol
Probably in: Dog B55, C581, C620, C625
Man D343
Rat A620, B55, C840
→ 3-(α-Naphthyloxy)lactic acid
Probably in: Dog C581, C620
Man D343
Rat A620, C840
1-Hydroxy-3-(1-naphthyloxy)propionitrile
Formed from: 1-(Formylmethoxy)naphthalene

2-Hydroxy-3-(α-naphthyloxy)propylamine
Formed from: Propranolol
→ 1-Acetamido-3-(α-naphthyloxy)-2-propanol
In: Rat H409
→ 2-Hydroxy-3-(α-naphthyloxy) propionaldehyde
Probably in: Dog B55
Man D343
Rat B55
→ 2-Hydroxy-3-(α-naphthyloxy) propylamine-β-D-glucuronide
Probably in: Dog C581
Man D343
→ α-Naphthol
In: Dog B55

2-Hydroxy-3-(α-naphthyloxy)propylamine-β-D-glucuronide
Probably formed from: 2-Hydroxy-3-(α-naphthyloxy)propylamine

1-(4-Hydroxy-α-naphthyloxy)-2,3-propylene glycol
Probably formed from: 1-(α-Naphthyloxy)-2,3-propylene glycol
→ 1-(4-Hydroxy-α-naphthyloxy)-2,3-propylene glycol-β-D-glucuronide
Probably in: Dog C581
Man D343
→ 1-(4-Hydroxy-α-naphthyloxy)-2,3-propylene glycol-O-sulphate
Probably in: Dog C581
Man D343

1-(4-Hydroxy-α-naphthyloxy)-2,3-propylene glycol-β-D-glucuronide
Probably formed from: 1-(4-Hydroxy-α-naphthyloxy)-2,3-propylene glycol

1-(4-Hydroxy-α-naphthyloxy)-2,3-propylene glycol-O-sulphate
Probably formed from: 1-(4-Hydroxy-α-naphthyloxy)-2,3-propylene glycol

2-(6-Hydroxynaphth-2-yl)propionic acid
Formed from: Naproxen
→ 2-(6-Hydroxynaphth-2-yl)propionic acid-β-D-glucuronide
In: Dog A3960
Guinea pig A3960
Mouse A3960
Pig A3960
Rabbit A3960
Rat A3960
→ 2-(6-Hydroxynaphth-2-yl)propionic acid-O-sulphate
In: Dog A3960
Guinea pig A3960
Man F14
Mouse A3960
Pig A3960
Rabbit A3960
Rat A3960

2-(6-Hydroxynaphth-2-yl)propionic acid-β-D-glucuronide
Probably formed from: 2-(6-Hydroxynaphth-2-yl)propionic acid

2-(6-Hydroxynaphth-2-yl)propionic acid-O-sulphate
Formed from: 2-(6-Hydroxynaphth-2-yl) propionic acid

2-Hydroxy-3-(2-naphthyl)propionitrile
Formed from: 2-Naphthylacetaldehyde

(1-Hydroxy-2-naphthyl)semicarbazide
Formed from: Naftazone
→ (1-Hydroxy-2-naphthyl)semicarbazide-β-D-glucuronide
In: Man H480
Rat H480

(1-Hydroxy-2-naphthyl)semicarbazide-β-D-glucuronide
Formed from: (1-Hydroxy-2-naphthyl) semicarbazide

3-Hydroxynaringenin
See: Aromadendrin

3-Hydroxynefiracetam
Formed from: Nefiracetam

3′-Hydroxynefiracetam
Formed from: Nefiracetam

4-Hydroxynefiracetam
Formed from: Nefiracetam

4′-Hydroxynefiracetam
Formed from: Nefiracetam

5-Hydroxynefiracetam
Formed from: Nefiracetam

4′-Hydroxynifenazone
Formed from: Nifenazone

4′-Hydroxyniflumic acid
Formed from: Niflumic acid

→ 4′-Hydroxyniflumic acid-β-D-glucuronide
In: Dog A973
Man A973
→ 4′-Hydroxyniflumic acid-4′-sulphate
In: Dog A973
Man A973

5-Hydroxyniflumic acid
Formed from: Niflumic acid
→ 5-Hydroxyniflumic acid-β-D-glucuronide
In: Dog A973
Man A973
→ 5-Hydroxyniflumic acid-4′-sulphate
In: Dog A973
Man A973

4′-Hydroxyniflumic acid-β-D-glucuronide
Formed from: 4′-Hydroxyniflumic acid

5-Hydroxyniflumic acid-β-D-glucuronide
Formed from: 5-Hydroxyniflumic acid

4′-Hydroxyniflumic acid-4′-sulphate
Formed from: 4′-Hydroxyniflumic acid

5-Hydroxyniflumic acid-4′-sulphate
Formed from: 5-Hydroxyniflumic acid

***p*-Hydroxynimesulide**
Formed from: Nimesulide

4-Hydroxynipradolol
Probably formed from: Nipradolol

2-Hydroxynitracrine
Formed from: Nitracrine
→ 2-Hydroxynitracrine-β-D-glucuronide
In: Rat H704

2-Hydroxynitracrine-β-D-glucuronide
Formed from: 2-Hydroxynitracrine

2-Hydroxy-3-nitrobenzaldehyde
→ 2-Hydroxy-3-nitrobenzyl alcohol
In: Dunaliella G459

4-Hydroxy-3-nitrobenzaldehyde
→ Protocatechualdehyde
In: Pseudomonas E607

5-Hydroxy-2-nitrobenzaldehyde
Formed from: 3-Methyl-4-nitrophenol

N-Hydroxy-*m*-nitrobenzamidine
Formed from: *m*-Nitrobenzamidine

4-Hydroxy-3-nitrobenzoic acid
Formed from: 4-Methoxy-3-nitrobenzoic acid
→ Nitroquinol
In: Candida J461

1-Hydroxy-6-nitro-3,4-benzpyrene
Formed from: 6-Nitro-3,4-benzpyrene
→ 1,9-Dihydroxy-6-nitro-3,4-benzpyrene
In: Rat C540
See also: C231
→ 6-Nitro-3,4-benzpyren-1-yl-β-D-glucuronide
In: Cunninghamella E298

3-Hydroxy-6-nitro-3,4-benzpyrene
Formed from: 6-Nitro-3,4-benzpyrene
→ 3,4-Benzpyrene-3,6-quinone
Probably in: Rat C231, C540
→ 3,9-Dihydroxy-6-nitro-3,4-benzpyrene
In: Rat C540
See also: C231, C383, C788
→ 6-Nitrobenzpyren-1-yl-β-D-glucuronide
In: Cunninghamella E298

2-Hydroxy-3-nitrobenzyl alcohol
Formed from: 2-Hydroxy-3-nitrobenzaldehyde

2-Hydroxy-4′-nitrobiphenyl
Formed from: 4′-Nitrobiphenyl

4-Hydroxy-3-nitrobiphenyl
→ 3,4-Dihydroxybiphenyl
In: Pseudomonas E607, H65

4-Hydroxy-4′-nitrobiphenyl
Formed from: 4-Nitrobiphenyl

8-Hydroxy-3-nitrofluoranthene
Formed from: 3-Nitrofluoranthene

9-Hydroxy-2-nitrofluorene
Formed from: 2-Nitrofluorene

1-Hydroxy-7-nitronaphthalene
Formed from: *cis*-1,2-Dihydro-1,2-dihydroxy-7-nitronaphthalene

2-Hydroxy-7-nitronaphthalene
Formed from: *cis*-1,2-Dihydro-1,2-dihydroxy-7-nitronaphthalene

7-Hydroxy-2-nitronaphtho[2,1-*b*]furan
Formed from: 7-Methoxy-2-nitronaphtho[2,1-*b*]furan
→ 7-Hydroxy-2-nitronaphtho[2,1-*b*]furan-β-D-glucuronide
In: Rat D354

7-Hydroxy-2-nitronaphtho[2,1-*b*]furan-β-D-glucuronide
Formed from: 7-Hydroxy-2-nitronaphtho[2,1-*b*]furan

3-Hydroxy-2-nitrooestrone
Formed from: 3-Acetyl-2-nitrooestrone

2-Hydroxy-3-nitrooxypropyl 3-carboxy-2,6-dimethyl-4-(*m*-nitrophenyl)pyridine-5-carboxylate
Probably formed from: 2-Nitrooxypropyl-3-nitrooxypropyl 2,6-dimethyl-4-(*m*-nitrophenyl)pyridine-3,5-dicarboxylate
→ 2,3-Dihydroxypropyl 3-carboxy-2,6-dimethyl-4-(*m*-nitrophenyl)pyridine-5-carboxylate
Probably in: Dog G593
Rat G593

S-(2-Hydroxy-3-(*p*-nitrophenoxy)propyl)-L-cysteine
Probably formed from: 1-(S-Glutathionyl)-2-hydroxy-3-(*p*-nitrophenoxy)propane
→ N-Acetyl-S-(2-hydroxy-3-(*p*-nitrophenoxy) propyl)-L-cysteine
Probably in: Rat A3043

2-Hydroxy-5-((4-nitrophenyl)azo)benzoic acid
See: Mordant Orange 1

2-Hydroxy-4-nitrophenyl-β-D-glucuronide
Formed from: 4-Nitrocatechol

2-Hydroxy-5-nitrophenyl-β-D-glucuronide
Formed from: 4-Nitrocatechol

S-(3-Hydroxy-4-nitrophenyl)glutathione
Formed from: 2,5-Dinitrophenol

2-Hydroxy-6-(*p*-nitrophenyl)-6-oxohexanoic acid
Probably formed from: 2,3-Dihydro-2,3-dihydroxy-4′-nitrobiphenyl

β-Hydroxy-*p*-nitrophenylpyruvic acid
Formed from: β-(*p*-Nitrophenyl)serine

4-Hydroxy-3-nitrophenylpyruvic acid
Formed from: 3-Nitro-L-tyrosine

2-Hydroxy-5-nitrophenyl sulphate
(nitrocatechol sulphate)
Probably formed from: 4-Nitrocatechol
→ 4-Nitrocatechol
In: Ascaris A324
Cyanobacterium E179
Man A595, A2457, A3839, E177
Moniezia A324
Peptococcus E449
Rabbit A1145, A2945, B481
See also: A653

3-Hydroxy-1-nitropyrene
Formed from: 1-Nitropyrene
→ 1-Nitropyren-3-yl-β-D-glucuronide
In: Rat D571

6-Hydroxy-1-nitropyrene
Formed from: 1-Nitropyrene
→ 1-Nitropyren-6-yl-β-D-glucoside
In: Cunninghamella D624
→ 1-Nitropyren-6-yl-β-D-glucuronide
In: Rat D571

6-Hydroxy-2-nitropyrene
Formed from: 2-Nitropyrene

6-Hydroxy-4-nitropyrene
Probably formed from: 4-Nitropyrene

8-Hydroxy-1-nitropyrene
Formed from: 1-Nitropyrene
→ 1-Nitropyren-8-yl-β-D-glucoside
In: Cunninghamella D624
→ 1-Nitropyren-8-yl-β-D-glucuronide
In: Rat D571

9-Hydroxy-4-nitropyrene
Formed from: 4-Nitropyrene

10-Hydroxy-1-nitropyrene
Formed from: 1-Nitropyrene

2-Hydroxy-5-nitroquinone
Formed from: 4-Methyl-5-nitrocatechol

4′-Hydroxynomifensine
Formed from: Nomifensine
→ 3′,4′-Dihydroxynomifensine
Probably in: Man A3348, B373, E37
Monkey A3348
→ 4′-Hydroxynomifensine-β-D-glucuronide
In: Man B373

4′-Hydroxynomifensine-β-D-glucuronide
Formed from: 4′-Hydroxynomifensine

ω-Hydroxynonivamide
Formed from: Nonivamide

6-Hydroxy-5″-norcannabidiol-4″-oic acid
Probably formed from: 5″-Norcannabidiol-4″-oic acid

7-Hydroxy-5″-norcannabidiol-4″-oic acid
Probably formed from: 5″-Norcannabidiol-4″-oic acid

***p*-Hydroxynorclobazam**
Probably formed from: *p*-Hydroxyclobazam
→ *p*-Hydroxynorclobazam-β-D-glucuronide
Probably in: Rat G572

→ *p*-Hydroxynorclobazam-O-sulphate
Probably in: Rat G572

p-Hydroxynorclobazam-β-D-glucuronide
Probably formed from: *p*-Hydroxynorclobazam

p-Hydroxynorclobazam-O-sulphate
Probably formed from: *p*-Hydroxynorclobazam

p-Hydroxynorephedrine (both isomers)
Formed from: *p*-Hydroxyamphetamine
D-Norephedrine
Probably formed from: L-Norephedrine
→ Norephedrin-4-yl-β-D-glucuronide
Probably in: Rat B553

N-Hydroxynorcocaine
Formed from: Norcocaine
→ N-Hydroxybenzoylnorecgonine
In: Rat A3721

N-Hydroxynorephedrine
Formed from: Norephedrine

6β-Hydroxynorhyoscyamine
Formed from: Norhyoscyamine

p-Hydroxynormeperidine
Formed from: *p*-Hydroxymeperidine

p-Hydroxynornaftifine
Probably formed from: Nornaftifine
→ 3-(3,4-Dihydroxyphenyl)-N-(1-naphthylmethyl)-2-propen-1-ylamine
Probably in: Man E32

N-Hydroxynorpromazine
→ Demethylpromazine
In: Rabbit E822

Hydroxynorpromethazine (aryl hydroxyl?)
Formed from: Norpromethazine

N-Hydroxynorpromethazine
→ N-Hydroxybisnorpromethazine
In: Pig G525
→ 10-(2′-Propanoneoxime)phenothiazine
In: Pig G525

Hydroxynorprothanon
Probably formed from: Norprothanon

4-Hydroxynorsalsolinol (1,2,3,4-tetrahydro-4,6,7-trihydroxyisoquinoline)
→ 1,2,3,4-Tetrahydro-4,6-dihydroxy-7-methoxyisoquinoline
In: Rat D27
→ 1,2,3,4-Tetrahydro-4,7-dihydroxy-6-methoxyisoquinoline
In: Rat D27

N-Hydroxynorscopolamine
Formed from: Norscopolamine

2-Hydroxynortrimipramine
Probably formed from: 2-Hydroxytrimipramine
→ 10,11-Dehydro-2-hydroxynortrimipramine
Probably in: Rat G269
→ 2-Hydroxybisnortrimipramine
Probably in: Rat G269

10-Hydroxynortrimipramine
Probably formed from: Nortrimipramine

2-Hydroxynortriptyline
Formed from: Nortriptyline
→ 2,11-Dihydroxynortriptyline
Probably in: Man E894
Rat E892
→ 2-Hydroxynortriptyline-β-D-glucuronide
In: Rat E931

10-Hydroxynortriptyline (both *E* and *Z* isomers)
Formed from: 10-Hydroxyamitriptyline
Nortriptyline
10-Oxonortriptyline
→ *cis*-10,11-Dihydroxynortriptyline
In: Rat E931
→ *trans*-10,11-Dihydroxynortriptyline
In: Rat E931
→ 10-Hydroxybisnortriptyline
Probably in: Man G472
→ 10-Hydroxynortriptyline-β-D-glucuronide
In: Man E564, F27, F207
Rat E564, E931
→ 10-Oxonortriptyline
In: Man H470, K341

11-Hydroxynortriptyline
Probably formed from: 11-Hydroxyamitriptyline

N-Hydroxynortriptyline
Formed from: Nortriptyline

2-Hydroxynortriptyline-β-D-glucuronide
Formed from: 2-Hydroxynortriptyline

10-Hydroxynortriptyline-β-D-glucuronide
Formed from: 10-Hydroxynortriptyline

***m*-Hydroxyochratoxin A**
Formed from: Ochratoxin A

(4*R*)-Hydroxyochratoxin A
Formed from: Ochratoxin A
(4*S*)-Hydroxyochratoxin A
Formed from: Ochratoxin A
10-Hydroxyochratoxin A
Formed from: Ochratoxin A
4-Hydroxy-3-octaprenylbenzoic acid
See: 4-Carboxy-2-octaprenylphenol
2-Hydroxy-17α-oestradiol
Formed from: 17α-Oestradiol
2-Hydroxy-17β-oestradiol
Formed from: 3-Deoxy-2-hydroxy-17β-oestradiol
2-Fluoro-17β-oestradiol
2-Hydroxyoestrone
2-Methoxy-17β-oestradiol
17β-Oestradiol
Probably formed from: 2-Bromo-17β-oestradiol
→ 2-Hydroxy-3-O-methyl-17β-oestradiol
In: Man A447, A2465, E519
Rat A2175, B319
See also: A486, D712, J498
→ 2-Hydroxy-17β-oestradiol-2-β-D-glucuronide
Probably in: Monkey A951
Rat A1487, J498
→ 2-Hydroxy-17β-oestradiol-3-β-D-glucuronide-2-sulphate
Probably in: Rat A1280
→ S-(2-Hydroxy-17β-oestradiol-1-yl) glutathione
In: Rat A2969
See also: A1915, J171
→ S-(2-Hydroxy-17β-oestradiol-4-yl) glutathione
In: Rat A2969
See also: A1915, J171
→ 2-Hydroxy-17β-oestradiol-3-sulphate
In: Beef 1211
See also: A605, A951, A1487
→ 2-Hydroxyoestriol
In: Rat A2467
→ 2-Hydroxyoestrone
In: Man A2465
Microorganism F469
See also: A3344
→ 2-Methoxy-17β-oestradiol
In: Man A447, A2465, E519
Pig E346
Rat A1487, A2175, B319
See also: A486, A1575, A2408, A3295, A3344, C346, D712, E267, H1, H36, H706, J171, J498
4-Hydroxy-17α-oestradiol
Formed from: 17α-Oestradiol
4-Hydroxy-17β-oestradiol
Formed from: 4-Methoxy-17β-oestradiol
17β-Oestradiol
→ 4-Hydroxyoestradiol-β-D-glucuronide
In: Rat K595
→ S-(4-Hydroxy-17β-oestradiol-1-yl) glutathione
Probably in: Hamster J171
→ 4-Hydroxyoestrone
In: Man E519
See also: A3344
→ 4-Methoxy-17β-oestradiol
Probably in: Hamster D712, H706, J171
Pig H36
6α-Hydroxy-17β-oestradiol
Formed from: 17β-Oestradiol
→ 6α-Hydroxyoestriol
Probably in: Rat H342
→ 6-Oxo-17β-oestradiol
Probably in: Rat E525
6β-Hydroxy-17β-oestradiol
Formed from: 17β-Oestradiol
7α-Hydroxy-17β-oestradiol
Formed from: 17β-Oestradiol
7β-Hydroxy-17β-oestradiol
Formed from: 17β-Oestradiol
11β-Hydroxy-17β-oestradiol
Formed from: 17β-Oestradiol
→ 11β-Hydroxy-17β-oestradiol-3-sulphate
In: Beef A1211
14α-Hydroxy-17β-oestradiol
Formed from: 17β-Oestradiol
→ 14α-Hydroxyoestrone
Probably in: Rat H342
15α-Hydroxy-17α-oestradiol
Probably formed from: 15α-Oestradiol
→ 15α-Hydroxy-17α-oestradiol-β-D-glucuronide

Probably in: Man A670

15α-Hydroxy-17β-oestradiol

Formed from: 17β-Oestradiol
15α-Hydroxyoestrone
→ 15α-Hydroxy-17β-oestradiol-15-N-acetylglucosaminide
In: Man A670, A1151
→ 15α-Hydroxy-17β-oestradiol-3-β-D-glucuronide
Probably in: Guinea pig A467
Man A670
See also: A545, J498
→ 15α-Hydroxy-17β-oestradiol-15-β-D-glucuronide
Probably in: Guinea pig A467
→ 15α-Hydroxyoestrone
Probably in: Guinea pig A467
Man A3241
See also: A670
→ Oestetrol
In: Man A670, A1151, A3241

16α-Hydroxy-17β-oestradiol

See: Oestriol

16β-Hydroxy-17β-oestradiol

See: 16-Epioestriol

18-Hydroxy-17β-oestradiol

Formed from: 18-Hydroxyoestrone

15α-Hydroxy-17α-oestradiol-15-β-D-N-acetylglucosaminide

Probably formed from: 15α-Hydroxyoestrone
→ 15α-Hydroxy-17α-oestradiol-15-β-N-acetylglucosaminide-3-sulphate
Probably in: Man A670

15α-Hydroxy-17β-oestradiol-15-N-acetylglucosaminide

Formed from: 15α-Hydroxy-17β-oestradiol
→ 15α-Hydroxy-17β-oestradiol-15-N-acetylglucosaminide-3-sulphate
Probably in: Man A670

15α-Hydroxy-17α-oestradiol-15-β-N-acetylglucosaminide-3-sulphate

Probably formed from: 15α-Hydroxy-17α-oestradiol-15-β-N-acetylglucosaminide

15α-Hydroxy-17β-oestradiol-15-N-acetylglucosaminide-3-sulphate

Probably formed from: 15α-Hydroxy-17β-oestradiol-15-N-acetylglucosaminide

2-Hydroxy-17β-oestradiol-2-β-D-glucuronide

Probably formed from: 2-Hydroxy-17β-oestradiol

2-Hydroxy-17β-oestradiol-3-β-D-glucuronide

→ 2-Hydroxy-17β-oestradiol
Probably in: Rat A2175

2-Hydroxy-17β-oestradiol-17-β-D-glucuronide

Formed from: 17β-Oestradiol-17-β-D-glucuronide

4-Hydroxyoestradiol-β-D-glucuronide

Formed from: 4-Hydroxyoestradiol

15α-Hydroxy-17α-oestradiol-β-D-glucuronide

Formed from: 15α-Hydroxy-17α-oestradiol

15α-Hydroxy-17β-oestradiol-3-β-D-glucuronide

Formed from: 15α-Hydroxy-17β-oestradiol

15α-Hydroxy-17β-oestradiol-15-β-D-glucuronide

Formed from: 15α-Hydroxy-17β-oestradiol

2-Hydroxy-17β-oestradiol-3-β-D-glucuronide-2-sulphate

Probably formed from: 2-Hydroxy-17β-oestradiol

2-Hydroxy-17β-oestradiol-17-β-D-glucuronide-2-sulphate

Probably formed from: 2-Hydroxy-17β-oestradiol

16α-Hydroxyoestradiol methyl ether

Formed from: Oestradiol methyl ether

2-Hydroxy-17β-oestradiol-3-sulphate

Formed from: 2-Hydroxy-17β-oestradiol
Probably formed from: 17β-Oestradiol-3-sulphate
→ 2-Methoxy-17β-oestradiol-3-sulphate
Probably in: Rat A1844, A2175

2-Hydroxy-17β-oestradiol-17-sulphate

Formed from: 17β-Oestradiol-17-sulphate

4-Hydroxy-17β-oestradiol-17-sulphate

Formed from: 17β-Oestradiol-17-sulphate

6β-Hydroxy-17β-oestradiol-17-sulphate

Formed from: 17β-Oestradiol-17-sulphate

7α-Hydroxy-17β-oestradiol-17-sulphate

Formed from: 17β-Oestradiol-17-sulphate

7β-Hydroxy-17β-oestradiol-17-sulphate

Formed from: 17β-Oestradiol-17-sulphate

11β-Hydroxy-17β-oestradiol-3-sulphate

Formed from: 11β-Hydroxy-17β-oestradiol

15α-Hydroxy-17β-oestradiol-3-sulphate

Formed from: 17β-Oestradiol-3-sulphate

S-(2-Hydroxy-17β-oestradiol-1-yl)-L-cysteine
→ S-(2-Hydroxy-3-O-methyl-17β-oestradiol-1-yl)-L-cysteine
In: Man A420
Rat A420

S-(2-Hydroxy-17β-oestradiol-4-yl)-L-cysteine
→ S-(2-Methoxy-17β-oestradiol-4-yl)-L-cysteine
In: Man A420
Rat A420

S-(2-Hydroxy-17β-oestradiol-1-yl)glutathione
Formed from: 2-Hydroxy-17β-oestradiol

S-(2-Hydroxy-17β-oestradiol-4-yl)glutathione
Formed from: 2-Hydroxy-17β-oestradiol
→ 1,4-Bis(S-glutathionyl)-2-hydroxy-17β-oestradiol
Probably in: Hamster J171

S-(4-Hydroxy-17β-oestradiol-1-yl)glutathione
Probably formed from: 4-Hydroxy-17β-oestradiol

3-Hydroxyoestra-1,3,5,7,9-pentaene-17-one
See: Equilenin

3-Hydroxyoestra-1,3,5(10),9-tetraene-12,17-dione
→ Oestra-1,3,5(10),9-tetraene-12,17-dion-3-yl sulphate
In: Beef A1211

3-Hydroxyoestra-1,3,5(10),6-tetraene-17-one
→ Oestra-1,3,5(10),6-tetraen-17-on-3-yl sulphate
In: Beef A1211

3-Hydroxyoestra-1,3,5(10),7-tetraene-17-one
See: Equilin

3-Hydroxyoestra-1,3,5(10),9-tetraene-17-one
→ Oestra-1,3,5(10),9-tetraen-17-on-3-yl sulphate
In: Beef A1211

17β-Hydroxyoestra-1,3,5(10),9-tetraene-12-on-3-yl sulphate
Formed from: 3,17β-Dihydroxyoestra-1,3,5(10),9-tetraen-12-one

3-Hydroxyoestra-1,3,5(10)-triene
→ Oestra-1,3,5(10)-trien-3-yl sulphate
In: Beef A1211

3-Hydroxyoestra-1,3,5(10)-trien-16-one
→ Oestra-1,3,5(10)-trien-16-on-3-yl sulphate
In: Beef A1211

3-Hydroxyoestra-1,3,5(10)-trien-17-one
See: Oestrone

16α-Hydroxyoestra-1,3,5(10)-trien-3-yl sulphate
Formed from: 3,16α-Dihydroxyoestra-1,3,5(10)-triene

2-Hydroxyoestriol
Formed from: 2-Hydroxy-17β-oestriol
Oestriol
→ 2-Hydroxy-3-O-methyloestriol
In: Rat A1255
See also: A1487, A3102
→ 2-Hydroxyoestriol-2-β-D-glucuronide
In: Dog F806
Guinea pig F806
Rat F806
→ 2-Hydroxyoestriol-3-β-D-glucuronide
In: Guinea pig F806
Rat F806
→ 2-Hydroxyoestriol-16-β-D-glucuronide
Probably in: Monkey A951
Rat A605, A1487, A3102
→ 2-Hydroxyoestriol-3-sulphate
In: Rat C261
See also: A951, A1487, A3102
→ 2-Methoxyoestriol
In: Rat A1255, C242
See also: A1280, A1487, A2408, A3102, E267

6α-Hydroxyoestriol
Probably formed from: 6α-Hydroxy-17β-oestradiol

6ξ-Hydroxyoestriol
Formed from: Oestriol

15α-Hydroxyoestriol
See: Oestetrol

2-Hydroxyoestriol-2-β-D-glucuronide
Formed from: 2-Hydroxyoestriol

2-Hydroxyoestriol-3-β-D-glucuronide
Formed from: 2-Hydroxyoestriol

2-Hydroxyoestriol-16-β-D-glucuronide
Probably formed from: 2-Hydroxyoestriol

2-Hydroxyoestriol-16-β-D-glucuronide-3-sulphate
Formed from: 2-Hydroxyoestriol-3-sulphate

2-Hydroxyoestriol-3-sulphate
Formed from: 2-Hydroxyoestriol
→ 2-Hydroxyoestriol-16-β-D-glucuronide-3-sulphate
In: Rat C261

See also: A605

2-Hydroxyoestrone

Formed from: 2-Hydroxy-17β-oestradiol
2-Methoxyoestrone
Oestrone
→ 2-Hydroxy-3-O-methyloestrone
In: Dog G988
Guinea pig G988
Man A3225, G988
Monkey G988
Mouse G988
Rabbit G988
Rat A1280, A1281, D177, G988, K603
See also: A1457, A2175, A2408, A2465
→ 2-Hydroxy-17β-oestradiol
In: Microorganism F469
See also: A951, A1280, A3225
→ S-(2-Hydroxyoestron-1-yl)glutathione
Probably in: Rat A1281
→ 2-Hydroxyoestrone-2-β-D-glucuronide
In: Guinea pig D177
Rat A1280, D177
See also: A951, A1487, G968, J783
→ 2-Hydroxyoestrone-3-β-D-glucuronide
In: Guinea pig D177
→ 2-Hydroxyoestrone-2-sulphate
In: Man D177
Rat D177
See also: A605, A1487
→ 2-Methoxyoestrone
In: Dog G988
Guinea pig G988
Man A3292, G988
Monkey G988
Mouse G988
Rabbit G988
Rat A1280, A1281, D177, G988, K603
See also: A490, A545, A1457, A1487, A2175, A2408, A2465, A3295, A3344, C346, F282

4-Hydroxyoestrone

Formed from: 4-Hydroxy-17β-oestradiol
4-Methoxyoestrone
Oestrone
Probably formed from: 4-Hydroxyoestrone-3-sulphate
→ 4-Hydroxy-3-O-methyloestrone
In: Dog G988
Guinea pig G988
Man G988
Monkey G988
Mouse G988
Rabbit G988
Rat C329, D177, G988, K603
→ 4-Hydroxyoestrone-3-β-D-glucuronide
In: Guinea pig D177
See also: G968, J783
→ 4-Hydroxyoestrone-4-β-D-glucuronide
In: Guinea pig D177
Rat D177
See also: G524
→ 4-Hydroxyoestrone-4-sulphate
In: Man D177
Rat D177
→ 4-Methoxyoestrone
In: Dog G988
Guinea pig G988
Man G988
Monkey G988
Mouse G988
Rabbit G988
Rat C329, D177, G988, K603
See also: C346

6α-Hydroxyoestrone

Formed from: Oestrone

6β-Hydroxyoestrone

Formed from: Oestrone
→ 6-Oxooestrone
Probably in: Rat H342

7α-Hydroxyoestrone

Formed from: Oestrone

7β-Hydroxyoestrone

Formed from: Oestrone

14α-Hydroxyoestrone

Probably formed from: 14α-Hydroxy-17β-oestradiol

15α-Hydroxyoestrone

Formed from: 15α-Hydroxy-17β-oestradiol
Oestrone
→ 15α-Hydroxy-17α-oestradiol
Probably in: Man A670
→ 15α-Hydroxy-17β-oestradiol
In: Guinea pig A467

Microorganism B454
→ 15α-Hydroxy-17α-oestradiol-15-β-D-N-acetylglucosaminide
Probably in: Man A670
→ 15α-Hydroxyoestrone-3-β-D-glucuronide
In: Guinea pig A467
→ 15α-Hydroxyoestrone-15-β-D-glucuronide
In: Guinea pig A467

15β-Hydroxyoestrone
Formed from: Oestrone

16α-Hydroxyoestrone
Formed from: Oestrone
Probably formed from: Oestriol
→ 2,16α-Dihydroxyoestrone
Probably in: Rat A1280, A1487
→ 16α-Hydroxyoestrone-β-D-glucuronide
In: Man G968
See also: A545, A1487
→ 16α-Hydroxyoestrone-3-sulphate
In: Beef A1211
See also: A1487
→ Oestriol
In: MicroorganismB454

16β-Hydroxyoestrone
Formed from: Oestrone
→ 16β-Hydroxyoestrone-3-sulphate
Probably in: Guinea pig A1936

16β-Hydroxyoestrone
Formed from: Oestrone
→ 16β-Hydroxyoestrone-3-sulphate
Probably in: Guinea pig A1936

18-Hydroxyoestrone
→ 18-Hydroxy-17β-oestradiol
In: Rabbit A1215
Rat A1215

16α-Hydroxyoestrone disulphate
Probably formed from: 16α-Hydroxyoestrone-3-sulphate

16β-Hydroxyoestrone disulphate
Probably formed from: 16β-Hydroxyoestrone-3-sulphate

2-Hydroxyoestrone-2-β-D-glucuronide
Formed from: 2-Hydroxyoestrone
→ 2-Hydroxyoestrone-2-β-D-glucuronide-3-methyl ether
In: Rat A2175, A2364

2-Hydroxyoestrone-3-β-D-glucuronide
Formed from: 2-Hydroxyoestrone
→ 2-Methoxyoestrone-3-β-D-glucuronide
In: Rat A2364

4-Hydroxyoestrone-3-β-D-glucuronide
Formed from: 4-Hydroxyoestrone

4-Hydroxyoestrone-4-β-D-glucuronide
Formed from: 4-Hydroxyoestrone

15α-Hydroxyoestrone-3-β-D-glucuronide
Formed from: 15α-Hydroxyoestrone

15α-Hydroxyoestrone-15-β-D-glucuronide
Formed from: 15α-Hydroxyoestrone

16α-Hydroxyoestrone-β-D-glucuronide
Probably formed from: 16α-Hydroxyoestrone

2-Hydroxyoestrone-2-β-D-glucuronide-3-methyl ether
Formed from: 2-Hydroxyoestrone-2-β-D-glucuronide

2-Hydroxyoestrone-2-sulphate
Formed from: 2-Hydroxyoestrone
Probably formed from: 2-Hydroxyoestrone-3-sulphate

2-Hydroxyoestrone-3-sulphate
Formed from: Oestrone-3-sulphate
→ 2-Hydroxyoestrone-2-sulphate
Probably in: Man A1022
→ 2-Methoxyoestrone-3-sulphate
In: Man A1022
See also: A1844

4-Hydroxyoestrone-3-sulphate
→ 4-Hydroxyoestrone
Probably in: Rat C329

4-Hydroxyoestrone-4-sulphate
Formed from: 4-Hydroxyoestrone

15α-Hydroxyoestrone-3-sulphate
Formed from: Oestrone-3-sulphate

16α-Hydroxyoestrone-3-sulphate
Formed from: 16α-Hydroxyoestrone
Oestrone-3-sulphate
→ 16α-Hydroxyoestrone disulphate
Probably in: Guinea pig A3754, B683

16β-Hydroxyoestrone-3-sulphate
Formed from: Oestrone-3-sulphate
Probably formed from: 16β-Hydroxyoestrone
→ 16β-Hydroxyoestrone-3,16-disulphate
Probably in: Guinea pig A1936, B683

18-Hydroxyoestrone-3-sulphate
Formed from: Oestrone-3-sulphate
S-(2-Hydroxyoestron-1-yl)glutathione
Formed from: 2-Hydroxyoestrone
7-Hydroxyolanzapine
Formed from: Olanzapine
→ 7-Hydroxyolanzapine-β-D-glucuronide
In: Dog J150
Mouse J150
Rhesus monkey J150
→ 7-Hydroxyolanzapine-N-oxide
Probably in: Dog J150
7-Hydroxyolanzapine-β-D-glucuronide
Formed from: 7-Hydroxyolanzapine
7-Hydroxyolanzapine-4′-N-oxide
Probably formed from: 7-Hydroxyolanzapine
Olanzapine-4′-N-oxide
6-Hydroxyondansetron
Formed from: Ondansetron
7-Hydroxyondansetron
Formed from: Ondansetron
8-Hydroxyondansetron
Formed from: Ondansetron
N-Hydroxyorbifloxacin
Formed from: Orbifloxacin
1-Hydroxy-1′-oxalylbiphenyl
Probably formed from: 1,1a-Dihydroxy-1*H*-fluoren-9-one
***p*-Hydroxyoxanilic acid**
→ *p*-Aminophenol
In: Microorganism A261
7-Hydroxyoxapadol
Probably formed from: Oxapadol
8-Hydroxyoxapadol
Probably formed from: Oxapadol
9-Hydroxyoxapadol
Probably formed from: Oxapadol
2-Hydroxyoxaprotiline
Formed from: Oxaprotiline
→ 3-Hydroxy-2-methoxyoxaprotiline
Probably in: Dog E48
Rat E48
→ 2-Hydroxyoxaprotiline-β-D-glucuronide
Probably in: Dog E48
Rat E48
3-Hydroxyoxaprotiline
Formed from: Oxaprotiline
→ 3,6-Dihydroxyoxaprotiline
Probably in: Dog E48
→ 3-Hydroxyoxaprotiline-β-D-glucuronide
Probably in: Dog E48
Man C865
Rat E48
2-Hydroxyoxaprotiline-β-D-glucuronide
Probably formed from: 2-Hydroxyoxaprotiline
3-Hydroxyoxaprotiline-β-D-glucuronide
Probably formed from: 3-Hydroxyoxaprotiline
Hydroxyoxepinac
Formed from: Oxepinac
3-Hydroxyoxindole
See: Dioxindole
5-Hydroxyoxindole
Formed from: Oxindole
→ Oxindol-5-yl sulphate
Probably in: Rat D158
2-Hydroxy-11-oxoamitriptyline
Probably formed from: 2, 11-Dihydroxy-amitriptyline
2-Hydroxy-4-(3-oxo-3*H*-benzofuran-2-yliden) but-2-enoic acid
Formed from: 1,2-Dihydroxybenzofuran
→ Salicylic acid
In: Ralstonia K614
6-Hydroxy-1′-oxobufuralol
Probably formed from: 1′,6-Dihydroxy-bufuralol
4-(3-Hydroxy-1-oxobutyl)-3-methoxy-5-methylbenzoic acid
Formed from: 5-(3-Hydroxy-1-oxobutyl)-4-methoxy-6-methyl-2-pyrone
5-(3-Hydroxy-1-oxobutyl)-4-methoxy-6-methyl-2-pyrone
→ 4-(3-Hydroxy-1-oxobutyl)-3-methoxy-5-methylbenzoic acid
In: Macrophora E766
4-(2-Hydroxy-3-oxobutylthio)benzenesulphonic acid
Probably formed from: 4-((5-Methyl-2-oxo-1,3-dioxol-4-yl)methylthio) benzenesulphonic acid
3″-Hydroxy-6-oxocannabidiol
Probably formed from: 3″,6β-Dihydroxy-cannabidiol

4″-Hydroxy-6-oxocannabidiol
Probably formed from: 4″,6β-Dihydroxy-cannabidiol

5″-Hydroxy-6-oxocannabidiol
Probably formed from: 5″,6β-Dihydroxy-cannabidiol

2-(*p*-((3-Hydroxy-2-oxocyclopentyl)methyl) phenyl)propionic acid
Formed from: Loxoprofen
→ 2-(*p*-((2,3-Dihydroxycyclopentyl)methyl)phenyl)propionic acid
In: Rabbit D125

Hydroxy-2-oxocyproheptadine
Probably formed from: 2-Oxocyproheptadine

α-(4-(5-Hydroxy-1-oxo-2-isoindolinyl)phenyl) propionic acid
Formed from: α-(4-(1-Oxo-2-isoindolinyl) phenyl)propionic acid

α-(4-(6-Hydroxy-1-oxo-2-isoindolinyl)phenyl) propionic acid
Formed from: α-(4-(1-Oxo-2-isoindolinyl) phenyl)propionic acid

Hydroxy-11-oxo-11*H*-isoindolo[2,1-*a*] benzimidazole
Probably formed from: 11-Oxo-11*H*-isoindolo[2,1-*a*]benzimidazole

3-Hydroxy-2-oxolisuride
Probably formed from: Lisuride
→ 3-Hydroxy-desethyl-2-oxolisuride
Probably in: Rhesus monkey D657

***o*-(2-N-Hydroxyoxomorpholinoethoxy)phenyl phenyl ether**
Formed from: *o*-(2-N-Morpholinoethoxy) phenyl phenyl ether
→ *o*-(2-(N-2,3-Dioxomorpholino)ethoxy) phenyl phenyl ether
Probably in: Rabbit A325
Rat A325

2-Hydroxy-10-oxonortrimipramine
Probably formed from: 2-Hydroxy-10-oxotrimipramine

2-Hydroxy-11-oxonortriptyline
Probably formed from: 2,11-Dihydroxy-nortriptyline

2-Hydroxy-6-oxo-17β-oestradiol
→ 2-Hydroxy-3-O-methyl-6-oxo-17β-oestradiol
In: Rat B319
→ 2-Methoxy-6-oxo-17β-oestradiol
In: Rat B319

2-Hydroxy-16-oxo-17β-oestradiol
Probably formed from: 16-Oxo-17β-oestradiol
→ 2-Hydroxy-3-O-methyl-16-oxo-17β-oestradiol
Probably in: Rat A1255
→ 2-Methoxy-16-oxo-17β-oestradiol
Probably in: Rat A1255

9-Hydroxy-10-oxophenanthrene
Formed from: Phenanthrene-9,10-quinone

4-Hydroxy-2-oxo-4-phenylbutane
Probably formed from: 2,4-Dioxo-4-phenylbutane

6-Hydroxy-3-oxo-1-phenyl-1-cyclohexene
Probably formed from: 3,6-Dihydroxy-1-phenyl-1-cyclohexene

(α*E*)-α-(3-Hydroxy-5-oxo-4-phenyl-2(5*H*)-furanylidene)benzeneacetic acid
See: Pulvinic acid

2-Hydroxy-6-oxo-6-phenylhexa-2,4-dienoic acid
Formed from: 2,3-Dihydroxybiphenyl
→ Benzoic acid
In: Burkholderia J710, K522
Pseudomonas E85
Sphingomonas H217
See also: F161
→ 2,6-Dioxo-6-phenylhexanoic acid
In: Pseudomonas E156

2-Hydroxy-6-oxo-6-phenylhexanoic acid
Probably formed from: 2,6-Dioxo-6-phenylhexanoic acid

5-Hydroxy-6-oxo-α-phenyl-2-piperidineacetic acid
Probably formed from: 6-Oxo-α-phenyl-2-piperidineacetic acid

1-Hydroxy-2-oxo-1-phenylpropane oxime
Formed from: Norephedrine

***p*-Hydroxy-β-oxophenylpropionaldehyde**
Formed from: Naringenin
→ *p*-Hydroxyacetophenone
Probably in: Horseradish A3777

***p*-Hydroxy-β-oxophenylpropionic acid**
Probably formed from: *p*-Hydroxy-phenylhydracrylic acid
→ *p*-Hydroxybenzaldehyde
Probably in: Man A126

Trichosporon B34

3-(1-Hydroxy-2-(2-oxopiperidino)ethyl)-5-phenylisoxazole
Formed from: 3-(1-Hydroxy-2-piperidinoethyl)-5-phenylisoxazole

2-Hydroxy-N-(6-oxo-2-piperidinymethyl)-5-(2,2,2-trifluoroethoxy)benzamide
Probably formed from: 2-Hydroxy-N-(2-piperidinymethyl)-5-(2,2,2-trifluoroethoxy)benzamide

3-Hydroxy-2-oxoproterguride
Probably formed from: 2-Oxoproterguride

α-Hydroxy-N-(5-oxopyrrolidino-2-methyl)-α,α-diphenylacetamide
Probably formed from: Desethylepicainide
→ N-(1,2-Dehydro-5-oxopyrrolidino-2-methyl)-α-hydroxy-α,α-diphenylacetamide
Probably in: Man F323
Rat F323

4-Hydroxy-2-(2-oxopyrrolidino)-1-phenylpentane
Probably formed from: *p*-Hydroxyprolintane

N-(3-Hydroxy-2-(2-oxo-1-pyrrolidino)propyl) aniline
Probably formed from: N-Benzyl-N-(3-(2-methyl-1-propoxy)-2-(2-oxo-1-pyrrolidino)propyl)aniline

1*H*-3-Hydroxy-4-oxoquinaldine
→ N-Acetylanthranilic acid
In: Arthrobacter K774
Pseudomonas H368, K774

1*H*-3-Hydroxy-4-oxoquinoline
Formed from: 4-Hydroxyquinoline
→ N-Formylanthranilic acid
In: Arthrobacter K774
Pseudomonas H368, K774

7-Hydroxy-6-oxotargetrin
Probably formed from: 6,7-Dihydroxytargetrin
→ 7-Hydroxy-6-oxotargetrin-O-sulphate
Probably in: Man J290

7-Hydroxy-6-oxotargetrin-O-sulphate
Probably formed from: 7-Hydroxy-6-oxotargetrin

16α-Hydroxy-19-oxotestosterone
→ Oestriol
In: Man E665

2-Hydroxy-10-oxotrimipramine
Probably formed from: 2-Hydroxy-trimipramine
→ 2-Hydroxy-10-oxonortrimipramine
Probably in: Rat G269

5′-Hydroxy-5-oxoviloxazine
Probably formed from: 5-Oxoviloxazine

4-Hydroxy-5′-oxoxilobam
Probably formed from: 5′-Oxoxilobam

7β-Hydroxy-6β-oxycodol
Probably formed from: 6β-Oxycodol

4-Hydroxyoxyphenbutazone
Formed from: Oxyphenbutazone

α-Hydroxypafenolol
Formed from: Pafenolol

4-Hydroxypanomifene
Formed from: Panomifene

Hydroxypenfluridol
Formed from: Penfluridol

5-Hydroxy-2′,3,4′,6,7′-pentamethoxyflavone-5′-O-glucoside
Formed from: 2′,5-Dihydroxy-,3,4′,6,7′-tetramethoxyflavone-5′-O-glucoside

N-Hydroxypentamidine
Formed from: Pentamidine
→ N,N′-Dihydroxypentamidine
In: Rat F706, J410

N-(4-Hydroxypentanoyl)-N-(2′-(1*H*-tetrazol-5-yl)(1,1′-biphenyl-4-yl)methyl)valine
Formed from: Valsartan

Hydroxypentazocine
Probably formed from: Pentazocine
→ Hydroxypentazocine-β-D-glucuronide
Probably in: Rat A2134

Hydroxypentazocine-β-D-glucuronide
Probably formed from: Hydroxypentazocine

5-Hydroxypentyl N-(2-carboxyethyl) laudanosinium
Formed from: Atracurium

2-Hydroxypent-1-yl *p*-nitrophenyl ether
Probably formed from: *p*-Nitrophenyl pent-1-yl ether

4-Hydroxypent-1-yl *p*-nitrophenyl ether
Probably formed from: *p*-Nitrophenyl pent-1-yl ether

N-(5-Hydroxypentyl)-3-((1-piperazinyl)methyl)benzamide
Probably formed from: N-(4-Formylbutyl)-3-((1-piperazinyl)methyl)benzamide
5-(4-Hydroxypentyl)resorcinol
Formed from: Olivetol
3-Hydroxyperazine
Formed from: Perazine
→ Perazin-3-yl-β-D-glucuronide
In: Rat A2183
***m*-Hydroxyperisoxal**
Formed from: Perisoxal
→ *m*-Hydroxyperisoxal-β-D-glucuronide
Probably in: Man B554
Rabbit Man B554
Rat Man B554
***p*-Hydroxyperisoxal**
Formed from: Perisoxal
→ *p*-Hydroxyperisoxal-β-D-glucuronide
Probably in: Man B554
Rabbit Man B554
Rat Man B554
***m*-Hydroxyperisoxal-β-D-glucuronide**
Probably formed from: *m*-Hydroxyperisoxal
***p*-Hydroxyperisoxal-β-D-glucuronide**
Probably formed from: *p*-Hydroxyperisoxal
2′-Hydroxy-*cis*-permethrin
Formed from: *cis*-Permethrin
2′-Hydroxy-*trans*-permethrin
Formed from: *trans*-Permethrin
2′-Hydroxy-[1*RS,cis*]-permethrin
Formed from: [1*RS,cis*]-Permethrin
4′-Hydroxy-*cis*-permethrin
Formed from: *cis*-Permethrin
→ 4′,*t*-Dihydroxy-*cis*-permethrin
Probably in: Carp A3853
Cotton A3195
Cow A3194
Hen A3511
Trout A3853
→ *m*-(*p*-Hydroxyphenoxy)benzyl alcohol
Probably in: Carp A3853
Trout A3853
4′-Hydroxy-*trans*-permethrin
Formed from: *trans*-Permethrin
→ 4′,*t*-Dihydroxy-*trans*-permethrin
Probably in: Carp A3853
Cotton A3195
Cow A3194
Hen A3511
Trout A3853
→ *m*-(*p*-Hydroxyphenoxy)benzyl alcohol
Probably in: Carp A3853
Trout A3853
4′-Hydroxy-[1*RS,cis*]-permethrin
Formed from: [1*RS,cis*]-Permethrin
6-Hydroxy-*cis*-permethrin
Formed from: *cis*-Permethrin
6-Hydroxy-*trans*-permethrin
Formed from: *trans*-Permethrin
***c*-Hydroxy-*cis*-permethrin**
Formed from: *cis*-Permethrin
***c*-Hydroxy-*trans*-permethrin**
Formed from: *trans*-Permethrin
***t*-Hydroxy-*cis*-permethrin**
Formed from: *cis*-Permethrin
***t*-Hydroxy-*trans*-permethrin**
Formed from: *trans*-Permethrin
6α-Hydroxyphaseollin
Formed from: Phaseollin
1α-Hydroxyphaseollone
Formed from: Phaseollin
N-Hydroxyphenacetin
See: *p*-Ethyl-N-hydroxyacetanilide
1-Hydroxyphenanthrene
Formed from: Phenanthrene
→ 1-Hydroxyphenanthrene-β-D-glucoside
In: Cunninghamella F198
→ 1-Methoxyphenanthrene
Probably in: Agmenellum G474
2-Hydroxyphenanthrene
Formed from: Phenanthrene
3-Hydroxyphenanthrene
Formed from: Phenanthrene
4-Hydroxyphenanthrene
Formed from: Phenanthrene
9-Hydroxyphenanthrene
Formed from: Phenanthrene
→ 9-Hydroxyphenanthrene-β-D-glucoside
In: Phanerochaete G406, H495
1-Hydroxyphenanthrene-β-D-glucoside
Formed from: 1-Hydroxyphenanthrene
9-Hydroxyphenanthrene-β-D-glucoside
Formed from: 9-Hydroxyphenanthrene

2-Hydroxyphenanthridine
Formed from: Phenanthridine
6-Hydroxyphenazine-1-carboxylic acid
→ 1,6-Dihydroxyphenazine-5,10-dioxide
In: Brevibacterium A419
Pseudomonas A3284
→ Phenazine-1-carboxylic acid
Pseudomonas A3284
***o*-Hydroxyphenethylamine**
See: *o*-Tyramine
***m*-Hydroxyphenethylamine**
See: *m*-Tyramine
***p*-Hydroxyphenethylamine**
See: Tyramine
N-Hydroxyphenethylamine
(phenethylhydroxylamine)
Formed from: Phenethylamine
→ Phenylacetaldoxime
Probably in: Rabbit A1166
3-Hydroxyphenethylamin-4-yl sulphate
Formed from: 3,4-Dihydroxyphenethylamine
→ 3,4-Dihydroxyphenethylamine
In: Dog E829
4-Hydroxyphenethylamin-3-yl sulphate
Probably formed from: 3,4-Dihydroxyphenethylamine
4-Hydroxy-L-phenylalanin-3-yl sulphate
→ N-Acetyl-4-hydroxyphenethylamin-3-yl sulphate
In: Periplaneta A1239
→ 3,4-Dihydroxyphenethylamine
In: Dog A806, G732
Guinea pig A806
Man E177
Rat A806
Contrast (dog) E829
→ Noradrenaline
In: Beef A3334
Hydroxy-1-(2-phenethyl)-4-N-(N-propionylanilino)piperidine
Formed from: Fentanyl
N^3-(α-Hydroxy-β-phenethyl)uridine
Formed from: N^3-Phenacyluridine
4-Hydroxyphenformin
Formed from: Phenformin
→ 4-Hydroxyphenformin-β-D-glucuronide
In: Dog A3854
Guinea pig A3851, A3854
Rat A3851, A3854
4-Hydroxyphenformin-β-D-glucuronide
Formed from: 4-Hydroxyphenformin
Hydroxypheniramidol
Formed from: Pheniramidol
4′-Hydroxyphenmetrazine
Formed from: Phenmetrazine
4-Hydroxyphenobarbital
→ 4-Hydroxyphenobarbital-β-D-glucuronide
In: Mouse C80
Rat B962
N-Hydroxyphenobarbital
Formed from: Phenobarbital
4-Hydroxyphenobarbital-β-D-glucuronide
Formed from: 4-Hydroxyphenobarbital
3-Hydroxyphenothiazine (leucophenothiazone)
→ 3-Hydroxyphenothiazin-10-yl-L-alanyl-L-prolyl-L-prolyl-L-alanyl-L-glutamic acid
Probably in: Beef D410
→ 3-Hydroxyphenothiazin-10-ylglycyl-L-prolylglycine
Probably in: Beef D410
→ 3-Hydroxyphenothiazin-10-yl-L-threonyl-L-lysylglycyl-L-seryl-L-serylglycine
Probably in: Beef D410
→ Phenothiazin-3-yl sulphate
In: Gerbil B735
Guinea pig A3923
Hamster B735
Mouse B735
Rat B735
→ Phenothiazone
Probably in: Guinea pig A3923
7-Hydroxyphenothiazine-2-sulphonic acid
Probably formed from: 7-Hydroxysulphoridazine
3-Hydroxyphenothiazine
Formed from: Phenothiazine
7-Hydroxy-3*H*-phenothiazine
See: Thionol
N-Hydroxyphenothiazine (identity challenged)
Formed from: Phenothiazine
3-Hydroxyphenothiazin-10-yl-L-alanyl-L-prolyl-L-prolyl-L-alanyl-L-glutamic acid
Probably formed from: 3-Hydroxyphenothiazine

3-Hydroxyphenothiazin-10-ylglycyl-L-prolylglycine
Probably formed from: 3-Hydroxyphenothiazine

3-Hydroxyphenothiazin-10-yl-L-threonyl-L-lysylglycyl-L-seryl-L-serylglycine
Probably formed from: 3-Hydroxyphenothiazine

N-Hydroxy-*p*-phenoxyacetanilide
→ *p*-Phenoxyphenylhydroxylamine
In: Hamster F522

***o*-Hydroxyphenoxyacetic acid**
Formed from: Phenoxyacetic acid
Probably formed from: 2-(Carboxymethoxy)-1-(2-propen-1-oxy)benzene

***o*-Hydroxyphenoxyacetic acid**
Formed from: Phenoxyacetic acid
→ 2,5-Dihydroxyphenoxyacetic acid
Probably in: Pseudomonas A1480
→ Quinol
In: Bacillus A3639

7-Hydroxyphenoxazone
See: Resorufin

2-Hydroxy-5-phenoxybenzoic acid
Probably formed from: *m*-Phenoxybenzoic acid

3-Hydroxy-5-phenoxybenzoic acid
Probably formed from: 5-Hydroxy-*cis*-cypermethrin
5-Hydroxydecamethrin
m-Phenoxybenzoic acid
→ 3-Hydroxy-5-phenoxybenzoyltaurine
Probably in: Mouse A3800

4-Hydroxy-3-phenoxybenzoic acid
Probably formed from: *m*-Phenoxybenzoic acid

***m*-(*o*-Hydroxyphenoxy)benzoic acid**
Probably formed from: *m*-Phenoxybenzoic acid
→ *m*-(*o*-Sulphatoxyphenoxy)benzoic acid
Probably in: Rat A1934, A3199

***m*-(*p*-Hydroxyphenoxy)benzoic acid**
Formed from: *m*-Phenoxybenzoic acid
Probably formed from: *m*-(*p*-Hydroxyphenoxy)benzyl alcohol
→ *m*-Hydroxybenzoic acid
In: Hen G83
See also: G206
→ *m*-(*p*-Hydroxyphenoxy)benzoic acid-4′-β-D-glucuronide
Probably in: Rainbow trout E582
Rat B638, B640, D671, E582, E844
→ *m*-(*p*-Hydroxyphenoxy)benzoyl-β-D-glucuronic acid
In: Rat C167
See also: B638, B640, D671, E582, E844
→ *m*-(*p*-Hydroxyphenoxy)hippuric acid
Probably in: Dog D443
Rat A3199, D224
→ Quinol
In: Pseudomonas G206
See also: G979
→ *m*-(*p*-Sulphatoxyphenoxy)benzoic acid
In: Duck B638
Ferret B638
Gerbil B638
Guinea pig B638
Hamster B638
Marmoset B638
Mouse B638
Rabbit B638
Rat B199, B638, B640, C167, C836
See also: A1823, A1934, A3194, C133, D443, D671, E844, J877

***m*-(*p*-Hydroxyphenoxy)benzoic acid-4′-β-D-glucuronide**
Probably formed from: *m*-(*p*-Hydroxyphenoxy)benzoic acid

***m*-(*p*-Hydroxyphenoxy)benzoyl-β-D-glucuronic acid**
Formed from: *m*-(*p*-Hydroxyphenoxy) benzoic acid

3-Hydroxy-5-phenoxybenzoyltaurine
Probably formed from: 3-Hydroxy-5-phenoxybenzoic acid

***m*-(*p*-Hydroxyphenoxy)benzyl alcohol**
Formed from: *m*-Phenoxybenzyl alcohol
Probably formed from: 4′-Hydroxy-*cis*-permethrin
4′-Hydroxy-*trans*-permethrin
→ *m*-(*p*-Hydroxyphenoxy)benzoic acid
Probably in: Bollworm A3515
Hen A3511

→ *m*-(*p*-Hydroxyphenoxy)benzyl alcohol-β-D-glucuronide
Probably in: Mouse A3800
Rat D671
→ *m*-(*p*-Hydroxyphenoxy)benzyl alcohol-O-sulphate
Probably in: Bollworm A3515
Dog D443
Rat D671

m-(p-Hydroxyphenoxy)benzyl alcohol-β-D-glucuronide
Probably formed from: *m*-(*p*-Hydroxyphenoxy)benzyl alcohol

m-(p-Hydroxyphenoxy)benzyl alcohol-O-sulphate
Probably formed from: *m*-(*p*-Hydroxyphenoxy)benzyl alcohol

7-(2-(3-Hydroxy-4-phenoxy-1-butenyl)-5-oxocyclopentyl)-5-heptenamide
Probably formed from: 7-(3-Hydroxy-2-(3-hydroxy-4-phenoxy-1-butenyl)-5-oxocyclopentyl)-5-heptenamide

7-(2-(3-Hydroxy-4-phenoxy-1-butenyl)-5-oxocyclopentyl)-N-(methylsulphonyl)-5-heptenamide
Formed from: Sulprostone

m-(p-Hydroxyphenoxy)hippuric acid
Probably formed from: *m*-(*p*-Hydroxyphenoxy)benzoic acid

N-(1-Hydroxy-6-phenoxyindan-5-yl) methanesulphonamide
Formed from: N-(6-Phenoxyindan-5-yl) methanesulphonamide
→ N-(1-Oxo-6-phenoxyindan-5-yl) methanesulphonamide
Probably in: Rat C485

N-(3-Hydroxy-6-phenoxyindan-5-yl) methanesulphonamide
Formed from: N-(6-Phenoxyindan-5-yl) methanesulphonamide
→ N-(3-Oxo-6-phenoxyindan-5-yl) methanesulphonamide
Probably in: Rat C485

β-(p-Hydroxyphenoxy)lactic acid
Formed from: Prenalterol

3-((p-Hydroxyphenoxy)methyl)-1-methyl-4-phenylpiperidine
Formed from: Femoxetine

3-(p-Hydroxyphenoxy)-N-methylmorphinan
Formed from: N-Methyl-3-phenoxymorphinan
→ 3-(4-Hydroxy-3-methoxyphenoxy)-N-methylmorphinan
Probably in: Dog B634

3-(p-((p-Hydroxyphenoxy)methyl)phenyl)-1-morpholinopropane
Formed from: Fomocaine
→ 3-(*p*-((*p*-Hydroxyphenoxy)methyl)phenyl)-1-morpholinopropane-β-D-glucuronide
In: Rat A2687
→ 3-(*p*-((*p*-Hydroxyphenoxy)methyl)phenyl)-1-morpholinopropane-N-oxide
Probably in: Dog A348
Rat A2687

3-(p-((p-Hydroxyphenoxy)methyl)phenyl)-1-morpholinopropane-β-D-glucuronide
Formed from: 3-(*p*-((*p*-Hydroxyphenoxy) methyl)phenyl)-1-morpholinopropane

3-(p-((p-Hydroxyphenoxy)methyl)phenyl)-1-morpholinopropane-N-oxide
Probably formed from: 3-(*p*-((*p*-Hydroxyphenoxy)methyl)phenyl)-1-morpholinopropane

p-(4-Hydroxyphenoxy)phenol methanesulphonate
Formed from: *p*-Phenoxyphenol methanesulphonate

4-(p-Hydroxyphenoxy)phenylacetic acid
See: Thyroacetic acid

4-(4-Hydroxyphenoxy)-D-phenylalanine
See: D-Thyronine

4-(4-Hydroxyphenoxy)-L-phenylalanine
See: L-Thyronine

N′-(p-(o-Hydroxyphenoxy)phenyl)-N,N-dimethylsulphamide
Formed from: N,N-Dimethyl-N′-(*p*-phenoxyphenyl)sulphamide
→ N′-(*p*-(*o*-Hydroxyphenoxy)phenyl)-N-methylsulphamide
Probably in: Dog A1263
Rat A1263

N′-(p-(p-Hydroxyphenoxy)phenyl)-N,N-dimethylsulphamide
Formed from: N,N-dimethyl-N′-(*p*-phenoxyphenyl)sulphamide

→ N′-(*p*-(4-Hydroxy-3-methoxyphenoxy) phenyl)-N,N-dimethylsulphamide
Probably in: Dog A1263
Man A1263
Rat A1263
→ N′-(*p*-(*p*-Hydroxyphenoxy)phenyl)-N-methylsulphamide
Probably in: Dog A1263
Man A1263
Rat A1263

N′-(*p*-(*o*-Hydroxyphenoxy)phenyl)-N-methylsulphamide
Probably formed from: N′-(*p*-(*o*-Hydroxyphenoxy)phenyl)-N,N-dimethylsulphamide

N′-(*p*-(*p*-Hydroxyphenoxy)phenyl)-N-methylsulphamide
Probably formed from: N′-(*p*-(*p*-Hydroxyphenoxy)phenyl)-N,N-dimethylsulphamide

2-Hydroxy-3-phenoxypropionitrile
Formed from: 2-Phenoxyacetaldehyde

2-Hydroxy-*m*-phenoxypropiophenone
Formed from: *m*-Phenoxypropiophenone

1-Hydroxy-3-phenoxypropyl sulphate
Probably formed from: 1-O-Phenylglycerol

4′-Hydroxyphenprocoumon
Formed from: Phenprocoumon

6-Hydroxyphenprocoumon
Formed from: Phenprocoumon

7-Hydroxyphenprocoumon
Formed from: Phenprocoumon

8-Hydroxyphenprocoumon
Formed from: Phenprocoumon

N-Hydroxyphentermine
Formed from: N-Hydroxymephentermine
Phentermine
→ α,α-Dimethyl-α-nitroso-β-phenylethane
In: Rat A3671
See also: A264
→ 2-Methyl-2-nitro-1-phenylpropane
In: Rabbit B654
Rat A3671, B654
→ Phentermine
In: Guinea pig A2118
Rabbit A2118, A2721
Rat A2118, G79, G570

***p*-Hydroxyphentermine**
Formed from: *p*-Hydroxymephentermine
Phentermine

***o*-Hydroxyphenylacetaldehyde**
Formed from: Coumarin
o-Tyramine
Probably formed from: N-Methyl-*o*-tyramine
→ *o*-Hydroxyphenylacetic acid
Probably in: Man G954
Rat G605, G865
→ *o*-Tyrosol
Probably in: Rat G338
See also: G605, H154

***m*-Hydroxyphenylacetaldehyde**
Formed from: N,N-Dimethyl-*m*-tyramine
N-Methyl-*m*-tyramine
L-*m*-Tyrosine
m-Tyramine
m-Tyrosol
→ *m*-Hydroxyphenylacetic acid
In: Rat A2961
→ *m*-Tyrosol
In: Rat A2961

***p*-Hydroxyphenylacetaldehyde**
Formed from: Hordenine
N-Methyltyramine
Octopamine
p-Sympatol
Tyramine
Probably formed from: *p*-Hydroxyphenylpyruvic acid
→ *p*-Hydroxyphenylacetic acid
In: Aspergillus A2888
Cat C135
Enterobacteria A2889
Epidermophyton A2888
Fungus A2889
Hormodendrum A2888
Klebsiella A2888
Man A410, A561
Microsporum A2888
Mucor A2888
Mycobacteria A2888, A2889
Penicillium A2888
Proteus A2888

Rat A411, A1331
Streptococcus A2888
Trichophyton A2888
See also:A2209, A2651, D75, E97, G865
→ *p*-Tyrosol
In: Aspergillus A2888
Cat C135
Enterobacteria A2889
Fungus A2889
Hormodendrum A2888
Man B569
Monkey A1908
Mycobacteria A2889
Rat A411, A1331
Willia A1157
See also: A2209, D75, E97

α-Hydroxyphenylacetaldehyde
See: Mandelaldehyde

***p*-Hydroxyphenylacetaldoxime**
Formed from: 2-(*p*-Hydroxyphenyl)-1-nitroethane
Tyramine
L-Tyrosine
→ *p*-Hydroxyphenylacetonitrile
In: Sorghum A2914, J504
See also: A3331, G228, K216
E-Isomer → *Z*-Isomer
In: Sorghum F547
Z-Isomer → *E*-Isomer
In: Sorghum F547

2-Hydroxy-2-phenylacetaldoxime
Probably formed from: Phenylacetaldoxime
→ Mandelonitrile
Probably in: Millipede A2262

***o*-Hydroxyphenylacetamide**
Probably formed from: *o*-Tyrosine

***m*-Hydroxyphenylacetamide**
Probably formed from: L-*m*-Tyrosine

***p*-Hydroxyphenylacetamide**
Probably formed from: L-Tyrosine

***p*-Hydroxyphenyl acetate**
Formed from: 3-Chloro-4-hydroxyphenyl acetate
p-Hydroxyacetophenone
→ Quinol
In: Aspergillus H236
Pseudomonas E484

***o*-Hydroxyphenylacetic acid**
Formed from: 2-Coumaranone
Coumarin
Homophthalic acid
4-(*o*-Hydroxyphenyl)butyric acid
Phenylacetic acid
o-Tyrosol
Probably formed from: *o*-Hydroxyphenylacetaldehyde
o-Methoxyphenylacetic acid
Phenylpyruvic acid
→ 2,6-Dihydroxyphenylacetic acid
In: Aspergillus A3210, B727, C484
Trichosporon E332
See also: E746
→ Homogentisic acid
In: Aspergillus C484
Rhodococcus J23
Trichosporon E332
See also: A819, C532, E746, H108

***m*-Hydroxyphenylacetic acid**
Formed from 3,4-Dihydroxyphenylacetic acid
7-O-(β-Hydroxyethyl)rutin aglycone
m-Hydroxyphenylacetaldehyde
m-Methoxyphenylacetic acid
Phenylacetic acid
Quercetin
→ *m*-Cresol
In: Aspergillus A707
→ 3,4-Dihydroxyphenylacetic acid
In: Escherichia B517
Klebsiella G214
Trichosporon E332
See also: A2961, A3208, E746, G817
→ Homogentisic acid
In: Aspergillus C484
Bacteria B950
Flavobacterium E679, F780, G404, K779
Rhodococcus J23
Trichosporon E332
See also: B368
→ *m*-Hydroxyphenylacetylglycine
In: Rat A2961

***p*-Hydroxyphenylacetic acid**
Formed from: Astemizole

Biochanin A
Caffeic acid
3-Chloro-4-hydroxyphenylacetic acid
trans-6,7-Epoxy-1-(4′-ethylphenoxy)-3,7-dimethyl-2-octene
p-Hydroxyphenylacetaldehyde
p-Hydroxyphenylacetonitrile
4-(*p*-Hydroxyphenyl)butyric acid
p-Hydroxyphenylpyruvic acid
Kaempferol
p-Methoxyphenylacetic acid
Phenylacetic acid
Proanthocyanidin B-3
4-(2,4,6-Trihydroxyphenyl)flavan-3-ol
Probably formed from: Demethylastimazole
3,4-Dihydroxyphenylacetic acid
→ *p*-Bromophenol
In: Ulva K153
→ *p*-Carboxymethylphenyl sulphate
Probably in: Rat A411
→ *p*-Cresol
In: Clostridium D917, E528
Lactobacillus B750
Microorganism A2767, E291
→ 5,5′-Di(carboxymethyl)-2,2′-dihydroxybiphenyl
With peroxidase E44
→ 2,4-Dihydroxyphenylacetic acid
In: Flavobacterium G404, K779
Pseudomonas J467
→ 3,4-Dihydroxyphenylacetic acid
In: Alnus A3343
Arthrobacter A3208
Aspergillus C454
Bacillus A1464
Bacteria B950
Barley A3594
Escherichia B517, G817
Glycine A3594
Klebsiella G214
Mucuna E769, F957
Mushroom H907
Parsley A3594
Potato D964
Trichosporon E332
Wheat A3594
See also: A1011, A1449, A2209, A2651
→ Homogentisic acid
In: Bacterium A489
Flavobacterium E679
Pseudomonas A52, A1480, B798
Xanthobacter E303
See also: J23
→ *p*-Hydroxymandelic acid
In: Aspergillus A329, A2691
Barley A3594
Glycine A3594
Parsley A3594
Wheat A3594
See also: B300, G159, G929
→ *p*-Hydroxyphenylacetic acid-4-β-D-glucoside
In: Panax F255
→ *p*-Hydroxyphenylacetyl CoA
In: Pseudomonas G225, G929
→ *p*-Hydroxyphenylacetylglycine
In: Rat A411
→ Phenylacetic acid
Probably in: Microorganism A1164
→ L-Tyrosine
In: Microorganism A1164

α-Hydroxyphenylacetic acid
See: Mandelic acid

***p*-Hydroxyphenylacetic acid-4-β-D-glucoside**
Formed from: *p*-Hydroxyphenylacetic acid

***m*-Hydroxyphenylacetone**
Formed from: α-Methyl-*m*-tyrosine
Probably formed from: *m*-Hydroxyamphetamine

***p*-Hydroxyphenylacetone**
Probably formed from: *p*-Hydroxyamphetamine
→ 1-(S-Glutathionyl)-1-(*p*-hydroxyphenyl) acetone
In: Rat H741

***p*-Hydroxyphenylacetonitrile**
Formed from: *p*-Hydroxyphenylacetaldehyde
→ *p*-Hydroxybenzaldehyde
In: Triglochin D58
→ *p*-Hydroxyphenylacetic acid
In: Rhodococcus G172
→ *p*-Hydroxymandelonitrile

In: Sorghum K711
See also: A2914, A3471
→ Triglochinin
In: Thalictrum B492
Triglochin B492
See also: F164, G228, K216

α-Hydroxyphenylacetonitrile
See: Mandelonitrile

***p*-Hydroxyphenylacetoxime**
Formed from: N-Hydroxytyramine
Tyramine

***p*-Hydroxyphenylacetyl CoA**
Formed from: *p*-Hydroxyphenylacetic acid

***m*-Hydroxyphenylacetylglycine**
Formed from: *m*-Hydroxyphenylacetic acid

***p*-Hydroxyphenylacetylglycine**
→ *p*-Hydroxyphenylacetylglycine-O-sulphate
Probably in: Rat A411

***p*-Hydroxyphenylacetylglycine-O-sulphate**
Probably formed from: *p*-Hydroxyphenylacetylglycine

6β-Hydroxyphenylacetyltropine
Formed from: Phenylacetyltropine

N-Hydroxy-L-phenylalanine
Probably formed from: L-Phenylalanine
→ Phenylacetaldoxime
Probably in: Millipede A2262, A3389

3-Hydroxy-L-phenylalanin-4-yl sulphate
Formed from: 3,4-Dihydroxy-L-phenylalanine

4-Hydroxy-L-phenylalanin-3-yl sulphate
Formed from: 3,4-Dihydroxy-L-phenylalanine
→ 4-Hydroxyphenethylamin-3-yl sulphate
Probably in: Dog E804

N-(*p*-Hydroxyphenylaminocarbonyl)-3-(1-methyl-2-phenylethyl)sydnone imine
Formed from: Mesocarb
→ 3-(2-Hydroxy-1-methyl-2-phenylethyl)-N-(*p*-hydroxyphenylaminocarbonyl) sydnone imine
Probably in: Rat A3699

***p*-Hydroxyphenyl azide**
Formed from: *p*-Methoxyphenyl azide
Phenyl azide

2-Hydroxy-1-phenylazonaphthalene-6-sulphonic acid
See: Acid Orange 12

1-(4-Hydroxyphenylazo)-2-naphthol
Formed from: 1-Phenylazo-2-naphthol
→ 2,6-Dihydroxy-1-(*p*-hydroxyphenylazo) naphthalene
Probably in: Tulipa K439

2-Hydroxy-N-phenylbenzamide
See: Salicylanilide

α-Hydroxy-α-phenylbenzeneacetic acid
See: Benzilic acid

2-(*o*-Hydroxyphenyl)benzenesulphinic acid
(2′-hydroxybiphenyl-2-sulphinic acid)
Formed from: Dibenzothiophene-5,5-dioxide
→ 2-Hydroxybiphenyl
In: Rhodococcus J396
See also: K319, K409

N-(*p*-Hydroxyphenyl)benzenesulphonamide
Formed from: Benzenesulphonanilide

2-Hydroxy-3-phenylbenzoic acid
Formed from: 2-Methoxy-3-phenylbenzoic acid
Methyl 2-hydroxy-3-phenylbenzoate

2-Hydroxy-5-phenylbenzoyl-β-D-glucuronic acid
Formed from: 3-Carboxy-5-hydroxybiphenyl

N-(*p*-Hydroxyphenyl)benzylamine
Formed from: Bepridil
→ *p*-Aminophenol
Probably in: Man G62
Mouse G62
Rabbit G62
Rat G62
Rhesus monkey G62

***exo*-5-Hydroxy-*endo*-2-phenylbicyclo[2.2.1] heptane-2-carboxylic acid 3-aminopropyl ester**
Probably formed from: *exo*-5-Hydroxy-*endo*-2-phenylbicyclo[2.2.1]heptane-2-carboxylic acid 3-ethylaminopropyl ester
→ *exo*-5-Hydroxy-*endo*-2-phenylbicyclo [2.2.1]heptane-2-carboxylic acid 3-aminopropyl ester O-sulphate
Probably in: Rat F5

***exo*-5-Hydroxy-*endo*-2-phenylbicyclo[2.2.1] heptane-2-carboxylic acid 3-aminopropyl ester O-sulphate**
Probably formed from: *exo*-5-Hydroxy-*endo*-2-phenylbicyclo[2.2.1]heptane-2-carboxylic acid 3-aminopropyl ester

***endo*-5-Hydroxy-*endo*-2-phenylbicyclo[2.2.1] heptane-2-carboxylic acid 3-diethylaminopropyl ester**
Formed from: *endo*-2-Phenylbicyclo [2.2.1]heptane-2-carboxylic acid 3-diethylaminopropyl ester
→ *endo*-5-Hydroxy-*endo*-2-phenylbicyclo [2.2.1]heptane-2-carboxylic acid 3-diethylaminopropyl ester O-sulphate
Probably in: Rat F5

***exo*-5-Hydroxy-*endo*-2-phenylbicyclo[2.2.1] heptane-2-carboxylic acid 3-diethylaminopropyl ester**
Formed from: *endo*-2-Phenylbicyclo [2.2.1]heptane-2-carboxylic acid 3-diethylaminopropyl ester
→ *exo*-5-Hydroxy-*endo*-2-phenylbicyclo [2.2.1]heptane-2-carboxylic acid 3-diethylaminopropyl ester O-sulphate
Probably in: Rat F5
→ *exo*-5-Hydroxy-*endo*-2-phenylbicyclo[2.2.1] heptane-2-carboxylic acid 3-Ethylamino-propyl ester
Probably in: Rat F5

***exo*-6-Hydroxy-*endo*-2-phenylbicyclo[2.2.1] heptane-2-carboxylic acid 3-diethylaminopropyl ester**
Formed from: *endo*-2-Phenylbicyclo [2.2.1]heptane-2-carboxylic acid 3-diethylaminopropyl ester
→ *exo*-6-Hydroxy-*endo*-2-phenylbicyclo [2.2.1]heptane-2-carboxylic acid 3-diethylaminopropyl ester O-sulphate
Probably in: Rat F5

***endo*-5-Hydroxy-*endo*-2-phenylbicyclo[2.2.1] heptane-2-carboxylic acid 3-diethylaminopropyl ester O-sulphate**
Formed from: *endo*-5-Hydroxy-*endo*-2-phenylbicyclo[2.2.1]heptane-2-carboxylic acid 3-diethylaminopropyl ester

***exo*-5-Hydroxy-*endo*-2-phenylbicyclo[2.2.1] heptane-2-carboxylic acid 3-diethylaminopropyl ester O-sulphate**
Formed from: *exo*-5-Hydroxy-*endo*-2-phenylbicyclo[2.2.1]heptane-2-carboxylic acid 3-diethylaminopropyl ester

***exo*-6-Hydroxy-*endo*-2-phenylbicyclo[2.2.1] heptane-2-carboxylic acid 3-diethylaminopropyl ester O-sulphate**
Formed from: *exo*-6-Hydroxy-*endo*-2-phenylbicyclo[2.2.1]heptane-2-carboxylic acid 3-diethylaminopropyl ester

***exo*-5-Hydroxy-*endo*-2-phenylbicyclo[2.2.1] heptane-2-carboxylic acid 3-ethylaminopropyl ester**
Probably formed from: *exo*-5-Hydroxy-*endo*-2-phenylbicyclo[2.2.1]heptane-2-carboxylic acid 3-diethylaminopropyl ester
→ *exo*-5-Hydroxy-*endo*-2-phenylbicyclo [2.2.1]heptane-2-carboxylic acid 3-aminopropyl ester
Probably in: Rat F5
→ *exo*-5-Hydroxy-*endo*-2-phenylbicy-clo[2.2.1]heptane-2-carboxylic acid 3-ethylaminopropyl ester O-sulphate
Probably in: Rat F5

***exo*-5-Hydroxy-*endo*-2-phenylbicyclo[2.2.1] heptane-2-carboxylic acid 3-ethylaminopropyl ester O-sulphate**
Probably formed from: *exo*-5-Hydroxy-*endo*-2-phenylbicyclo[2.2.1]heptane-2-carboxylic acid 3-ethylaminopropyl ester

4-(*p*-Hydroxyphenyl)butane-2,3-diol
Formed from: 4-(*p*-Hydroxyphenyl)butan-2-one

4-(*o*-Hydroxyphenyl)-2-butanol
Formed from: 3-Acetylcoumarin

1-(*p*-Hydroxyphenyl)-*n*-butanol
Formed from: *p*-*n*-Butylphenol

4-(*p*-Hydroxyphenyl)butan-2-ol (rhododendrol)
Formed from: 4-(*p*-Hydroxyphenyl)butan-2-one
→ Rhododendrin
In: Acer H651
→ 4-(*p*-Hydroxyphenyl)but-2-yl-β-D-xylopyranosyl(1→6)-β-D-glucopyranoside
In: Acer H651

3-Hydroxy-4-phenylbutan-2-one
Formed from: Phenylacetaldehyde

4-Hydroxy-4-phenylbutan-2-one
Formed from: *trans*-4-Phenyl-3-buten-2-one

4-(*p*-Hydroxyphenyl)butan-2-one (raspberry ketone)
Formed from: Benzalacetone
→ 4-(3,4-Dihydroxyphenyl)butan-2-one
In: Guinea pig C303
Rabbit C303
Rat C303
See also: H907
→ 4-(*p*-Hydroxyphenyl)butane-2,3-diol
In: Rabbit C303
Rat C303
→ 4-(*p*-Hydroxyphenyl)butan-2-ol
In: Guinea pig C303
Rabbit C303
Rat C303

4-Hydroxyphenylbutazone
Formed from: Phenylbutazone

γ-Hydroxyphenylbutazone
Formed from: Phenylbutazone
→ *p*,γ-Dihydroxyphenylbutazone
Probably in: Horse D310
Man A3158, B419
→ γ-Hydroxyphenylbutazone-β-D-glucuronide
Probably in: Rat A296
→ γ-Oxophenylbutazone
In: Horse D310

γ-Hydroxyphenylbutazone-β-D-glucuronide
Probably formed from: γ-Hydroxyphenylbutazone

2-(*p*-Hydroxyphenyl)but-1-ene
Formed from: *p-sec*-Butylphenol

2-(*p*-Hydroxyphenyl)but-2-ene
Formed from: *p-sec*-Butylphenol

4-(*o*-Hydroxyphenyl)but-3-enoic acid
Probably formed from: 4-Phenylbut-3-enoic acid

2-Hydroxy-4-phenylbut-3-enonitrile
Formed from: Cinnamaldehyde
Cinnamonitrile

4-(*o*-Hydroxyphenyl)butylamine
Formed from: 4-(*o*-Methoxyphenyl)butylamine
→ 4-(*o*-Hydroxyphenyl)butyric acid
In: Rat G865

4-(*m*-Hydroxyphenyl)butylamine
Formed from: 4-(*m*-Methoxyphenyl)butylamine
→ 4-(*m*-Hydroxyphenyl)butyric acid
In: Rat G865

4-(*p*-Hydroxyphenyl)butylamine
Formed from: 4-(*p*-Methoxyphenyl)butylamine
→ 4-(*p*-Hydroxyphenyl)butyric acid
In: Rat G865

4-(*p*-Hydroxyphenyl)but-2-yl-β-D-xylopyranosyl(1→6)-β-D-glucopyranoside
Formed from: 4-(*p*-Hydroxyphenyl)butan-2-ol

3-Hydroxy-2-phenylbutyric acid
Probably formed from: 3-Hydroxy-2-phenylbutyrylurea

4-(*o*-Hydroxyphenyl)butyric acid
Formed from: *n*-Dodecylamine
4-(*o*-Hydroxyphenyl)butylamine
4-(*o*-Methoxyphenyl)butyric acid
Probably formed from: 4-Phenylbutyric acid
→ *o*-Hydroxyphenylacetic acid
In: Rat G865
See also: A1206

4-(*m*-Hydroxyphenyl)butyric acid
Formed from: 4-(*m*-Hydroxyphenyl)butylamine
4-(*m*-Methoxyphenyl)butyric acid

4-(*p*-Hydroxyphenyl)butyric acid
Formed from: 4-(*p*-Hydroxyphenyl)butylamine
6-(*p*-Hydroxyphenyl)hexylamine
4-(*p*-Methoxyphenyl)butyric acid
→ *p*-Hydroxyphenylacetic acid
In: Rat G865

(*S*)-α-Hydroxy-4-phenylbutyronitrile
Formed from: 3-Phenylpropionaldehyde

3-Hydroxy-2-phenylbutyrylurea
Formed from: 2-Phenylbutyrylurea
→ 3-Hydroxy-2-phenylbutyric acid
Probably in: Man E320
→ 2-Phenyl-2-butenoylurea
Probably in: Man E320

2-(*p*-Hydroxyphenyl)butyrylurea
Formed from: 2-Phenylbutyrylurea

→ 2-(4-Hydroxy-3-methoxyphenyl) butyrylurea
Probably in: Man E320

5-(N-(*p*-Hydroxyphenyl)carboxamido)barbituric acid
Probably formed from: 5-(N-Phenyl-carboxamido)barbituric acid

5-(N-(*p*-Hydroxyphenyl)carboxamido)-2-thiobarbituric acid
Formed from: Merbarone

2-(*p*-Hydroxyphenyl)chroman
Formed from: 2-Phenylchroman

3-(*p*-Hydroxyphenyl)chroman
Formed from: 2-Phenylchroman

3-Hydroxy-1-phenyl-1-cyclohexene
Formed from: 1-Phenyl-1-cyclohexene
→ 3,4-Dihydroxy-1-phenyl-1-cyclohexene
Probably in: Rat C829
→ 3,6-Dihydroxy-1-phenyl-1-cyclohexene
Probably in: Rat C829
→ 3-Oxo-1-phenyl-1-cyclohexene
Probably in: Mouse C641
Rat C829

6-Hydroxy-1-phenyl-1-cyclohexene
Formed from: 1-Phenyl-1-cyclohexene
→ 6-Oxo-1-phenyl-1-cyclohexene
Probably in: Mouse C641

3-Hydroxy-1-(1-phenylcyclohexyl)piperidine
Formed from: Phencyclidine

4-Hydroxy-1-(1-phenylcyclohexyl)piperidine
Formed from: Phencyclidine
→ 4-Hydroxy-1-(1-(*p*-hydroxyphenylcyclohexyl)piperidine
Probably in: Rabbit G29
→ 4-Oxo-1-(1-phenylcyclohexyl)piperidine
Probably in: Rabbit G29

1-(4-Hydroxy-1-phenylcyclohexyl)piperidine-N-oxide
Formed from: Phencyclidine-N-oxide

1-(4-Hydroxyphenyl)-2,3-diisocyano-4-(4-methoxyphenyl)-1,3-butadiene
See: Xanthocillin

2-(*o*-Hydroxyphenyl)-N,N-dimethyl-2-oxoethylamine
→ *o*-Hydroxymandelic acid
In: Rat J287

2-(*m*-Hydroxyphenyl)-N,N-dimethyl-2-oxoethylamine
→ *m*-Hydroxymandelic acid
In: Rat J287

2-(*p*-Hydroxyphenyl)-N,N-dimethyl-2-oxoethylamine
→ *p*-Hydroxymandelic acid
In: Rat J287

N-(*p*-Hydroxyphenyl)-N′,N′-dimethyl-*p*-phenylenediamine
Formed from: N,N-Dimethylindoaniline

1-(*o*-Hydroxyphenyl)-2,3-dimethyl-3-pyrazolin-5-one
Formed from: Antipyrine

1-(*p*-Hydroxyphenyl)-2,3-dimethyl-3-pyrazolin-5-one
Formed from: Antipyrine
→ 1-(*p*-Hydroxyphenyl)-3-methyl-3-pyrazolin-5-one
Probably in: Rat B395

α-(*m*-Hydroxyphenyl)-α,N-dimethylsuccinimide
Formed from: Methsuximide

α-(*p*-Hydroxyphenyl)-α,N-dimethylsuccinimide
Formed from: Methsuximide
→ β-Hydroxy-α-(*p*-hydroxyphenyl)-α, N-dimethylsuccinimide
Probably in: Guinea pig A1020
Man A1020
Rat A1020
→ α-(*p*-Hydroxyphenyl)-α,N-dimethylsuccinimide-β-D-glucuronide
In: Dog A1434

α-(*p*-Hydroxyphenyl)-α,N-dimethylsuccinimide-β-D-glucuronide
Formed from: α-(*p*-Hydroxyphenyl)-α,N-dimethylsuccinimide

1-(*o*-Hydroxyphenyl)-3,3-dimethyltriazene
Probably formed from: 3,3-Dimethyl-1-phenyltriazene

1-(*p*-Hydroxyphenyl)-3,3-dimethyltriazene
Probably formed from: 3,3-Dimethyl-1-phenyltriazene

1-(*p*-Hydroxyphenyl)-3,5-dioxo-4-(3-oxobutyl)-2-phenylpyrazolidine
Formed from: Kebuzon
→ 4-(3-Hydroxybutyl)-1-(*p*-hydroxyphenyl)-3,5-dioxo-2-phenylpyrazolidine

Probably in: Man A710
Rat A710

1-(*p*-Hydroxyphenyl)-3,5-dioxo-2-phenyl-4-(2′-phenylsulphonylethyl)pyrazolidine
Probably formed from: *p*-Hydroxysulphinpyrazole

1-(*p*-Hydroxyphenyl)-3,5-dioxo-2-phenyl-4-(2′-phenylthioethyl)pyrazolidine
Probably formed from: *p*-Hydroxysulphinpyrazole

1-(*p*-Hydroxyphenyl)-3,5-dioxo-2-phenyl-4-prenylpyrazolidine
Formed from: Feprazone
→ *cis*-4-(3-Hydroxymethylbut-2-enyl)-1-(*p*-hydroxyphenyl)-3,5-dioxo-2-phenyl-4-prenylpyrazolidine
Probably in: Dog B256
Rat B256
→ *trans*-4-(3-Hydroxymethylbut-2-enyl)-1-(*p*-hydroxyphenyl)-3,5-dioxo-2-phenyl-4-prenylpyrazolidine
Probably in: Dog B256
Man B256
Rat B256
→ 1-(*p*-Hydroxyphenyl)-3,5-dioxo-2-phenyl-4-prenylpyrazolidine-β-D-glucuronide
Probably in: Man B256

1-(*p*-Hydroxyphenyl)-3,5-dioxo-2-phenyl-4-prenylpyrazolidine-β-D-glucuronide
Probably formed from: 1-(*p*-Hydroxyphenyl)-3,5-dioxo-2-phenyl-4-prenylpyrazolidine

1-(*p*-Hydroxyphenyl)-1,2-diphenylbut-1-ene
Formed from: Tamoxifen
→ 1-(*p*-Hydroxyphenyl)-1,2-diphenylbut-1-ene-β-D-glucuronide
In: Dog A252

1-(*p*-Hydroxyphenyl)-1,2-diphenylbut-1-ene-β-D-glucuronide
Formed from: 1-(*p*-Hydroxyphenyl)-1,2-diphenylbut-1-ene

1-(*p*-Hydroxyphenyl)ethanol
Formed from: *p*-Ethylphenol
→ *p*-Hydroxyacetophenone
In: Pseudomonas F381, F695
See also: E396, E484, H236, J664, J680

2-(*o*-Hydroxyphenyl)ethanol
See: *o*-Tyrosol

2-(*m*-Hydroxyphenyl)ethanol
See: *m*-Tyrosol

2-(*p*-Hydroxyphenyl)ethanol
See: Tyrosol

***o*-Hydroxyphenylethanolamine**
See: *o*-Octopamine

***m*-Hydroxyphenylethanolamine**
See: *m*-Octopamine

***p*-Hydroxyphenylethanolamine**
See: Octopamine

***l*-N-Hydroxy-1-phenylethylamine**
Formed from: Acetophenone oxime

α-(*m*-Hydroxyphenyl-β-ethylaminoethanol
See: Etilefrine

4-(((2-(4-Hydroxyphenyl)ethyl)amino)methyl)-1,2-benzenediol
See: Norbelladine

(*R*)-4-(3-(2-Hydroxy-2-phenylethyl)amino-3-methylbutyl)benzamide
→ (*R*)-4-(3-(2-Hydroxy-2-(*m*-hydroxyphenyl)ethyl)amino-3-methylbutyl)benzamide
Probably in: Mouse F407
Rat F407
→ (*R*)-4-(3-(2-Hydroxy-2-(*p*-hydroxyphenyl)ethyl)amino-3-methylbutyl)benzamide
Probably in: Mouse F407
Rat F407

S-(2-Hydroxy-1-phenylethyl)-L-cysteine
→ N-Acetyl-S-(2-hydroxy-1-phenylethyl)-L-cysteine
Probably in: Guinea pig C605
Rat B714, C605
Winter flounder D387
→ 1-Hydroxy-2-methylthio-2-phenylethane
Probably in: Guinea pig C605
Rat C605
→ S-(2-Hydroxy-1-phenylethyl)mercaptopyruvic acid
Probably in: Guinea pig C605
Rat C605

S-(2-Hydroxy-2-phenylethyl)-L-cysteine
Formed from: 2-(S-Glutathionyl)-1-phenylethanol
→ N-Acetyl-S-(2-hydroxy-2-phenylethyl)-L-cysteine
Probably in: Guinea pig C605
Rat C605

→ 1-Hydroxy-2-methylthio-2-phenylethane
Probably in: Guinea pig C605
Rat C605
→ S-(2-Hydroxy-2-phenylethyl) mercaptopyruvic acid
Probably in: Guinea pig C605
Rat C605

S-(2-Hydroxy-1-phenylethyl)cysteinylglycine
Formed from: 2-(S-Glutathionyl)-2-phenylethanol
→ N-Acetyl-S-(2-hydroxy-1-phenylethyl)-L-cysteine
Probably in: Guinea pig C605
Rat B714, C605
Winter flounder D387

2-(*p*-Hydroxyphenyl)ethyl-β-D-glucoside
See: Salidroside

(*R*)-S-(2-Hydroxy-1-phenylethyl)glutathione
Formed from: 1,2-Dibromo-1-phenylethane
Styrene

(*S*)-S-(2-Hydroxy-1-phenylethyl)glutathione
Formed from: 1,2-Dibromo-1-phenylethane
Styrene
→ N-Acetyl-S-(2-hydroxy-1-phenylethyl)-L-cysteine
Probably in: Rat E870

(*R*)-S-(2-Hydroxy-2-phenylethyl)glutathione
Formed from: 1,2-Dibromo-1-phenylethane
Styrene

(*S*)-S-(2-Hydroxy-2-phenylethyl)glutathione
Formed from: 1,2-Dibromo-1-phenylethane
Styrene

N-(1-(2-Hydroxy-2-phenylethyl)-4-(hydroxypiperidinyl))propionanilide
Probably formed from: N-(1-(2-Hydroxy-2-phenylethyl)-4-piperidinyl)propionanilide

1-((2-Hydroxy-2-phenyl)ethyl)imidazole
Formed from: 1-((2-Oxo-2-phenyl)ethyl) imidazole

S-(2-Hydroxy-1-phenylethyl)mercaptoacetic acid
Probably formed from: S-(2-Hydroxy-1-phenylethyl)mercaptopyruvic acid

S-(2-Hydroxy-2-phenylethyl)mercaptoacetic acid
Probably formed from: S-(2-Hydroxy-2-phenylethyl)mercaptopyruvic acid

S-(2-Hydroxy-1-phenylethyl)mercaptolactic acid
Probably formed from: S-(2-Hydroxy-1-phenylethyl)mercaptopyruvic acid

S-(2-Hydroxy-2-phenylethyl)mercaptolactic acid
Probably formed from: S-(2-Hydroxy-2-phenylethyl)mercaptopyruvic acid

S-(2-Hydroxy-1-phenylethyl)mercaptopyruvic acid
Probably formed from: S-(2-Hydroxy-1-phenylethyl)-L-cysteine
→ S-(2-Hydroxy-1-phenylethyl) mercaptoacetic acid
Probably in: Guinea pig C605
Rat C605
→ S-(2-Hydroxy-1-phenylethyl)mercaptolactic acid
Probably in: Guinea pig C605

S-(2-Hydroxy-2-phenylethyl)mercaptopyruvic acid
Probably formed from: S-(2-Hydroxy-2-phenylethyl)-L-cysteine
→ S-(2-Hydroxy-2-phenylethyl) mercaptoacetic acid
Probably in: Guinea pig C605
Rat C605
→ S-(2-Hydroxy-2-phenylethyl)mercaptolactic acid
Probably in: Mouse C605

3-(2-(3-Hydroxyphenyl)ethyl)-5-methoxyphenol
See: Batatasin III

N-(2-Hydroxy-2-phenylethyl)-N-methylnitrosamine
→ Benzaldehyde
In: Rat D731
See also: C739
→ N-(2-Oxo-2-phenylethyl)-N-methylnitrosamine
In: Rat D731

N-(2-(*p*-Hydroxyphenyl)ethyl)norephedrin-4-yl-β-D-glucuronide
Formed from: *p*-Hydroxy-N-(2-(*p*-hydroxyphenyl)ethyl)norephedrine

N-(2-(*p*-Hydroxyphenyl)ethyl)norephedrin-4-yl sulphate
Formed from: *p*-Hydroxy-N-(2-(*p*-hydroxyphenyl)ethyl)norephedrine

1-(2-Hydroxy-2-phenylethyl)piperazine
Formed from: Eprazinone

Probably formed from: 1-(2-Methoxy-2-phenylethyl)piperazine

N-(1-(2-Hydroxy-2-phenylethyl)-4-piperidinyl) lactanilide
Probably formed from: N-(1-(2-Hydroxy-2-phenylethyl)-4-piperidinyl)propionanilide

N-(1-(2-Hydroxy-2-phenylethyl)-4-piperidinyl) propionanilide
Formed from: Fentanyl
→ N-(1-(2-Hydroxy-2-phenylethyl)-4-piperidinyl)lactanilide
Probably in: Guinea pig D906
Rat D906
→ N-(1-(2-Hydroxy-2-phenylethyl)-4-(hydroxypiperidinyl))propionanilide
Probably in: Guinea pig D906
Rat D906

N-(1-(2-(*p*-Hydroxyphenyl)ethyl)-4-piperidinyl) propionanilide
Formed from: Fentanyl
→ 4-N-(N-Propionylanilino)piperidine
In: Guinea pig D906
Rat D906

***o*-Hydroxyphenyl-α-D-glucoside**
Probably formed from: Catechol

***o*-Hydroxyphenyl-β-D-glucoside**
Formed from: Catechol

***m*-Hydroxyphenyl-β-D-glucoside**
Formed from: Resorcinol

***p*-Hydroxyphenyl-α-D-glucoside**
Formed from: Quinol

***p*-Hydroxyphenyl-β-D-glucoside**
See: Arbutin

***o*-Hydroxyphenyl-β-D-glucuronide**
Formed from: Catechol

***m*-Hydroxyphenyl-β-D-glucuronide**
Formed from: Resorcinol
→ *m*-Sulphatoxyphenyl-β-D-glucuronide
Probably in: Rat E532

***p*-Hydroxyphenyl-β-D-glucuronide**
Formed from: Quinol
→ Quinol
In: Escherichia J364

1-O-(*p*-Hydroxyphenyl)glycerol
Probably formed from: 1-O-Phenylglycerol

***o*-Hydroxy-D-phenylglycine**
Formed from: 5-(*o*-Hydroxyphenyl)hydantoin

***m*-Hydroxy-D-phenylglycine**
Formed from: 5-(*m*-Hydroxyphenyl) hydantoin
→ *m*-Hydroxyphenylglyoxylic acid
In: Pseudomonas E771, K716

***p*-Hydroxy-D-phenylglycine**
Formed from: *p*-Hydroxyphenylglyoxylic acid
5-(*o*-Hydroxyphenyl)hydantoin
D-Phenylglycine
→ Amoxicillin
In: Pseudomonas C920
→ 3,4-Dihydroxy-D-phenylglycine
In: Horseradish B925
→ *p*-Hydroxyphenylglyoxylic acid
In: Pseudomonas E669, E771, J353, K716

***p*-Hydroxy-L-phenylglycine**
Formed from: *p*-Hydroxyphenylglyoxylic acid
→ *p*-Hydroxybenzaldehyde
In: Coriolus G656
→ *p*-Hydroxyphenylglyoxylic acid
In: Pseudomonas E669

***o*-Hydroxyphenylglycol**
Probably formed from: *o*,ω-Dihydroxyacetophenone
o-Hydroxymandelaldehyde

***m*-Hydroxyphenylglycol**
Formed from: *m*-Sympatol
→ *m*-Hydroxyphenylglycol-O-sulphate
Probably in: Man C894
Rat C894

***p*-Hydroxyphenylglycol**
Formed from: *p*-Hydroxymandelaldehyde
Probably formed from: *p*-Hydroxystyrene oxide

***m*-Hydroxyphenylglycol-O-sulphate**
Probably formed from: *m*-Hydroxyphenylglycol

***o*-Hydroxyphenylglyoxylic acid**
(*o*-hydroxybenzoylformic acid)
→ Salicylaldehyde
In: Moraxella G356

***m*-Hydroxyphenylglyoxylic acid**
Formed from: *m*-Hydroxymandelic acid
m-Hydroxy-D-phenylglycine
→ *m*-Hydroxybenzaldehyde
In: Pseudomonas E771
See also: K716

p-Hydroxyphenylglyoxylic acid
Formed from: *p*-Hydroxymandelic acid
p-Hydroxy-D-phenylglycine
p-Hydroxy-L-phenylglycine
→ *p*-Hydroxybenzaldehyde
In: Pseudomonas E669, E771
See also: A2336, K716
→ *p*-Hydroxymandelic acid
In: Streptococcus E271
→ *p*-Hydroxy-D-phenylglycine
In: Pseudomonas K716
→ *p*-Hydroxy-L-phenylglycine
In: Amycoletopsis H688

7-(p-Hydroxyphenyl)heptanoic acid
Probably formed from: 9-(*p*-Hydroxyphenyl)nonanoic acid
7-(*p*-Methoxyphenyl)heptanoic acid
→ 5-(*p*-Hydroxyphenyl)pentanoic acid
Probably in: Rainbow trout J412
Rat G865

1-(p-Hydroxyphenyl)-1-heptanol
Formed from: *p*-*n*-Heptylphenol

7-(p-Hydroxyphenyl)heptylamine
→ 5-(*p*-Hydroxyphenyl)pentanoic acid
In: Rat G865

4-Hydroxy-6-phenylhexan-2-one
Formed from: Dihydrokawain

4-Hydroxy-6-phenyl-5-hexen-2-one
Probably formed from: Kawain
→ 6-Phenyl-5-hexene-2,4-dione
Probably in: Man F862
→ 6-Phenyl-5-hexen-2-one
Probably in: Man F862

3-(p-Hydroxyphenyl)hex-2-en-4-one
Formed from: Stilboestrol

6-(p-Hydroxyphenyl)hexylamine
→ 4-(*p*-Hydroxyphenyl)butyric acid
In: Rat G865

5-(o-Hydroxyphenyl)hydantoin
→ *o*-Hydroxy-D-phenylglycine
In: Pseudomonas E450

5-(m-Hydroxyphenyl)hydantoin
→ *m*-Hydroxy-D-phenylglycine
In: Pseudomonas E450

5-(p-Hydroxyphenyl)hydantoin
→ *p*-Hydroxy-D-phenylglycine
In: Pseudomonas E391, E450, F943

p-Hydroxyphenylhydracrylaldehyde
Formed from: Naringenin
→ *p*-Hydroxyphenylhydracrylic acid
Probably in: Horseradish A3777

m-Hydroxyphenylhydracrylic acid
Formed from: (+)-Catechin
Rutin
Probably formed from: *m*-Coumaric acid

p-Hydroxyphenylhydracrylic acid
Probably formed from: *p*-Hydroxyphenylhydracrylaldehyde
→ 4-Hydroxy-β-oxophenylpropionic acid
Probably in: Man A126
Polyporus A2573
Trichosporon B34

5-(m-Hydroxyphenyl)-3-(1-hydroxy-2-piperidinoethyl)isoxazole
Formed from: 3-(1-Hydroxy-2-piperidinoethyl)-5-phenylisoxazole

5-(p-Hydroxyphenyl)-3-(1-hydroxy-2-piperidinoethyl)isoxazole
Formed from: 3-(1-Hydroxy-2-piperidinoethyl)-5-phenylisoxazole

1-(o-Hydroxyphenyl)-1-(1H-imidazol-1-yl)ethylene
Formed from: Croconazole

2-(p-Hydroxyphenyl)indole
Formed from: 2-Phenylindole

5-Hydroxy-2-phenylindole
Formed from: 2-Phenylindole

6-Hydroxy-2-phenylindole
Formed from: 2-Phenylindole

(−)-ω-(1-Hydroxy-1-phenylisopropylamino)-3′-methoxypropiophenone
See: Oxyfedrine

1-(Hydroxyphenyl)-4-isopropyl-2,3-dimethyl-3-pyrazolin-5-one
Formed from: 4-Isopropylantipyrine
→ 4-Hydroxy-1-(hydroxyphenyl)-4-isopropyl-2,3-dimethyl-3-pyrazolin-5-one
Probably in: Rat D285

1-(Hydroxyphenyl)-4-isopropyl-3-methyl-3-pyrazolin-5-one
Probably formed from: 4-Isopropyl-3-methyl-1-phenyl-3-pyrazolin-5-one
→ 1-(Hydroxymethoxyphenyl)-4-isopropyl-3-methyl-3-pyrazolin-5-one

Probably in: Rat D285

1-(Hydroxyphenyl)-4-isopropyl-3-methyl-3-pyrazolin-5-one-N-oxide
Probably formed from: 4-Isopropyl-3-methyl-1-phenyl-3-pyrazolin-5-one-N-oxide

N-(*p*-Hydroxyphenyl)-N-isopropyl-2-methylsulphonylacetamide
Formed from: Propachlor
→ N-(*p*-Hydroxyphenyl)-N-isopropyl-2-methylsulphonylacetamide-β-D-glucuronide
In: Rat B93
Sheep B91

N-(*p*-Hydroxyphenyl)-N-isopropyl-2-methylsulphonylacetamide-β-D-glucuronide
Formed from: N-(*p*-Hydroxyphenyl)-N-isopropyl-2-methylsulphonylacetamide

3-Hydroxy-5-phenylisoxazole
Formed from: Isoxathion
→ 5-Phenylisoxazol-3-yl-β-D-glucuronide
In: Rat A2287
→ 5-Phenylisoxazol-3-yl sulphate
In: Rat A2287

***o*-Hydroxyphenyllactic acid**
Formed from: 3-Hydroxycoumarin

***m*-Hydroxyphenyllactic acid**
Formed from: *m*-Hydroxyphenylpyruvic acid
→ *m*-Hydroxyphenylpyruvic acid
In: Rat A2961

***p*-Hydroxyphenyllactic acid**
Formed from: *p*-hydroxyphenylpyruvic acid
Methyl *p*-hydroxyphenyllactate
Proanthocyanidin B 3
4-(2,4,6-Trihydroxyphenyl)flavan-3-ol
→ Caffeoyl *p*-hydroxyphenyllactic acid ester
In: Coleus G420, J349
→ *p*-Coumaric acid
Probably in: Microorganism A2767, A3990
→ *p*-Coumaroyl *p*-hydroxyphenyllactic acid ester
In: Coleus G420, H17, J349
→ 3,4-Dihydroxyphenyllactic acid
In: Mucuna E769
Potato D964
→ *p*-Hydroxyphenylpyruvic acid
In: Candida D975
Neisseria F92
See also: A2767

9-Hydroxy-N-phenyllinoleamide
Formed from: N-Phenyllinoleamide

13-Hydroxy-N-phenyllinoleamide
Formed from: N-Phenyllinoleamide

(2*R*,4*R*)-2-(*o*-Hydroxyphenyl)-3-(3-mercaptopropionyl)-4-thiazolidinecarboxylic acid
→ 3,3′-(Dithiobis(1-oxo-3,1-propanediyl))bis(2-(*o*-hydroxyphenyl))-(2*R*-(2α,3(2′*R**,4′*R**),4α))-4-thiazolidinecarboxylic acid
In: Rat E137
→ (2*R*,4*R*)- 3-(S-(S-Glutathionyl)-3-mercaptopropionyl)-2-(*o*-hydroxyphenyl)-4-thiazolidinecarboxylic acid
In: Dog E137
→ (2*R*,4*R*)-2-(*o*-Hydroxyphenyl)-3-(3-methylmercaptopropionyl)-4-thiazolidinecarboxylic acid
In: Dog E137
Rat E137

***trans*-3-(*p*-Hydroxyphenyl)-2-(methoxycarbonylethyl)-N-phenyllactam**
→ *trans*-2-(2-Carboxyethyl)-3-(*p*-hydroxyphenyl)-N-phenyllactam
In: Microorganism K110

***trans*-3-(*p*-Hydroxyphenyl)-2-(methoxyethoxyethoxycarbonylethyl)-N-phenyllactam**
→ *trans*-2-(2-Carboxyethyl)-3-(*p*-hydroxyphenyl)-N-phenyllactam
In: Microorganism K110

5-(*p*-Hydroxyphenyl)-6-(*p*-methoxyphenyl)-3-methyl-1,2,4-triazine
Formed from: 5,6-Bis(*p*-methoxyphenyl)-3-methyl-1,2,4-triazine
→ 5,6-Bis(*p*-hydroxyphenyl)-3-methyl-1,2,4-triazine
Probably in: Rat B396

4-(*p*-Hydroxyphenyl)-3-((*p*-methoxyphenoxy)methyl)-1-methylpiperidine
Formed from: Femoxetine

6-(*p*-Hydroxyphenyl)-5-(*p*-methoxyphenyl)-3-methyl-1,2,4-triazine
Formed from: 5,6-Bis(*p*-methoxyphenyl)-3-methyl-1,2,4-triazine

1-(*p*-Hydroxyphenyl)-1-(*p*-(2-methylaminoethoxy) phenyl)-2-phenyl-1-butene
Probably formed from: Demethyltamoxifen

1-Hydroxyphenyl-3-methylaminomethyl-2-phenyl-2-butanol
Probably formed from: 1-Hydroxyphenyl-3-methylaminomethyl-2-phenyl-2-butyl propionate

1-Hydroxyphenyl-3-methylaminomethyl-2-phenyl-2-butyl propionate
Probably formed from: 3-Methylaminomethyl-1,2-diphenyl-2-butyl propionate
→ N-Acetyl-1-hydroxyphenyl-3-methylaminomethyl-2-phenyl-2-butyl propionate
Probably in: Rat A1683
→ 3-Aminomethyl-1-hydroxyphenyl-2-phenyl-2-butyl propionate
Probably in: Rat A1683
→ 1-Hydroxyphenyl-3-methylaminomethyl-2-phenyl-2-butanol
Probably in: Rabbit A1683
Rat A1683

4-(*p*-Hydroxyphenyl)-3-methyl-3-butenoic diethylamide
Probably formed from: 3-Methyl-4-phenyl-3-butenoic diethylamide
→ 4-(*p*-Hydroxyphenyl)-3-methyl-3-butenoic ethylamide
Probably in: Rabbit B435

4-(*p*-Hydroxyphenyl)-3-methyl-3-butenoic ethylamide
Probably formed from: 4-(*p*-Hydroxyphenyl)-3-methyl-3-butenoic diethylamide

***m*-Hydroxyphenyl N-methylcarbamate**
Formed from: *m*-Cresyl N-methylcarbamate

1-(*p*-Hydroxyphenyl)-3-(2-methylcyclohexyl)urea
Formed from: 1-(2-methylcyclohexyl)-3-phenylurea

N-(1-(*p*-Hydroxyphenyl)-1-methylethyl)-3,3-dimethylbutyramide
Formed from: 3,3-Dimethyl-N-(1-methyl-1-phenylethyl)butyramide
→ 4-Hydroxy-N-(1-(*p*-hydroxyphenyl)-1-methylethyl)-3,3-dimethylbutyramide
Probably in: Mouse D417
Rat D215
→ N-(1-(*p*-Hydroxyphenyl)-1-methylethyl)-3,3-dimethylbutyramide-β-D-glucuronide
Probably in: Mouse D193
Rat D193

N-(1-(*p*-Hydroxyphenyl)-1-methylethyl)-3,3-dimethylbutyramide-β-D-glucuronide
Probably formed from: N-(1-(*p*-Hydroxyphenyl)-1-methylethyl)-3,3-dimethylbutyramide

1-(*p*-Hydroxyphenyl)-2-methyliminoethanol
Formed from: *p*-Sympatol

1-(*p*-Hydroxyphenyl)-2-methyliminopropan-1-ol
Formed from: *p*-Hydroxyephedrine

(2*R*,4*R*)-2-(*o*-Hydroxyphenyl)-3-(3-methylmercaptopropionyl)-4-thiazolidinecarboxylic acid
Formed from: (2*R*,4*R*)-2-(*o*-Hydroxyphenyl)-3-(3-mercaptopropionyl)-4-thiazolidinecarboxylic acid
→ (2*R*,4*R*)-2-(*o*-Hydroxyphenyl)-3-(3-methylsulphinylpropionyl)-4-thiazolidinecarboxylic acid
Probably in: Rat E137

2-(*o*-Hydroxyphenyl)-N-methyl-2-oxoethylamine
→ *o*-Hydroxymandelic acid
In: Rat J287

2-(*m*-Hydroxyphenyl)-N-methyl-2-oxoethylamine
→ *m*-Hydroxymandelic acid
In: Rat J287

2-(*p*-Hydroxyphenyl)-N-methyl-2-oxoethylamine
→ *p*-Hydroxymandelic acid
In: Rat J287

6-(*p*-Hydroxyphenyl)-4-methyl-3-(3-oxo-2-morpholinoethylamino)pyridazine
Probably formed from: *p*-Hydroxyminaprine

N-(3-(4-(N-(*p*-Hydroxyphenyl)methyl)phenoxy) propyl)-N-methyl-β-alanine
Formed from: N-Methyl-N-(3-(4-(N-phenylmethyl)phenoxy)propyl)-β-alanine

4-(4-(4-Hydroxyphenyl)(4-methylphenyl)methoxy-1-piperidinyl)butyric acid
Formed from: 4-(4-(4-methylphenyl) phenylmethoxy-1-piperidinyl)butyric acid

N-(3-(*m*-Hydroxyphenyl)-1-methyl-3-phenylpropyl)-N-methylformamide
Formed from: N-Methyl-N-(1-methyl-3,3-diphenylpropyl)formamide

→ N-(3-(*m*-Hydroxyphenyl)-1-methyl-3-phenylpropyl)-N-methylformamide-β-D-glucuronide
In: Rat G88

N-(3-(*p*-Hydroxyphenyl)-1-methyl-3-phenylpropyl)-N-methylformamide
Formed from: N-Methyl-N-(1-methyl-3,3-diphenylpropyl)formamide
→ N-(3-(*p*-Hydroxyphenyl)-1-methyl-3-phenylpropyl)-N-methylformamide-β-D-glucuronide
In: Rat G88

N-(3-(*m*-Hydroxyphenyl)-1-methyl-3-phenylpropyl)-N-methylformamide-β-D-glucuronide
Formed from: N-(3-(*m*-Hydroxyphenyl)-1-methyl-3-phenylpropyl)-N-methylformamide

N-(3-(*p*-Hydroxyphenyl)-1-methyl-3-phenylpropyl)-N-methylformamide-β-D-glucuronide
Formed from: N-(3-(*p*-Hydroxyphenyl)-1-methyl-3-phenylpropyl)-N-methylformamide

3-(*p*-Hydroxyphenyl)-5-methyl-3-phenyl-1-pyrroline
Formed from: 2-Dimethylamino-4,4-diphenyl-5-heptanone
→ 3-(3,4-Dihydroxyphenyl)-5-methyl-3-phenyl-1-pyrroline
Probably in: Rat A2779
→ 3-(*p*-Hydroxyphenyl)-5-methyl-3-phenyl-1-pyrroline-β-D-glucuronide
Probably in: Rat A1912, A2779

3-(*p*-Hydroxyphenyl)-5-methyl-3-phenyl-1-pyrroline-β-D-glucuronide
Probably formed from: 3-(*p*-Hydroxyphenyl)-5-methyl-3-phenyl-1-pyrroline

1-(4-(3-Hydroxyphenyl)-1-methyl-4-piperidinyl)-1-propanone
See: Ketobemidone

2-(*p*-Hydroxyphenyl)-2-methylpropionic acid
Probably formed from: *p*-(2-Carboxy-2-propyl)phenyl diphenyl phosphate

4-(2-((3-(4-Hydroxyphenyl)-1-methylpropyl)amino)ethyl)-1,2-benzenediol
See: Dobutamine

(1*R*)-3-(4-Hydroxyphenyl)-1-methylpropyl-β-D-glucopyranoside
See: Rhododendrin

1-(*p*-Hydroxyphenyl)-3-methyl-3-pyrazolin-5-one
Probably formed from: 1-(*p*-Hydroxyphenyl)-2,3-dimethyl-3-pyrazolin-5-one

1-(2-Hydroxyphenyl)-2-methyl-2-(3-pyridyl)-1-propanone
Formed from: 1-(2-Methoxyphenyl)-2-methyl-2-(3-pyridyl)-1-propanone

N′-(*p*-Hydroxyphenyl)-N-(1-methyl-2-pyrrolidinyidene)-1-pyrrolidinecarboxamidamide
Formed from: Pirogliride
→ N′-(*p*-Hydroxyphenyl)-N-(1-methyl-2-pyrrolidinyidene)-1-pyrrolidinecarboxamidamide-β-D-glucuronide
Probably in: Dog J693
Man J693
Monkey J693
Rat J693

N′-(*p*-Hydroxyphenyl)-N-(1-methyl-2-pyrrolidinyidene)-1-pyrrolidinecarboxamidamide-β-D-glucuronide
Probably formed from: N′-(*p*-Hydroxyphenyl)-N-(1-methyl-2-pyrrolidinyidene)-1-pyrrolidinecarboxamidamide

N-(*p*-Hydroxyphenyl)-2-(methylseleno)benzanilide
Formed from: Ebselen
→ N-(*p*-Hydroxyphenyl)-2-(methylseleno)benzanilide-β-D-glucuronide
In: Rat F311

N-(*p*-Hydroxyphenyl)-2-(methylseleno)benzanilide-β-D-glucuronide
Formed from: N-(*p*-Hydroxyphenyl)-2-(methylseleno)benzanilide

α-(*p*-Hydroxyphenyl)-α-methylsuccinimide
Probably formed from: α-Methyl-α-phenylsuccinimide
→ α-(*p*-Hydroxyphenyl)-α-methylsuccinimide-β-D-glucuronide
Probably in: Dog A1434

α-(*p*-Hydroxyphenyl)-α-methylsuccinimide-β-D-glucuronide
Probably formed from: α-(*p*-Hydroxyphenyl)-α-methylsuccinimide

(2*R*,4*R*)-2-(*o*-Hydroxyphenyl)-3-(3-methylsulphinylpropionyl)-4-thiazolidinecarboxylic acid
Probably formed from: (2*R*,4*R*)-2-(*o*-Hydroxyphenyl)-3-(3-methylmercapto-propionyl)-4-thiazolidinecarboxylic acid

3-Hydroxyphenyl methyl sulphone
Formed from: O-Ethyl S-phenyl ethylphosphonodithioate
Methyl phenyl sulphone

4-Hydroxyphenyl methyl sulphone
Formed from: O-Ethyl S-phenyl ethylphosphonodithioate
Methyl phenyl sulphone

N-(*p*-Hydroxyphenyl)-2-methylsulphonylacetamide
Formed from: Propachlor
Probably formed from: 2-Methylsulphonyl-N-phenylacetamide
→ N-(*p*-Hydroxyphenyl)-2-methylsulphonylacetamide-β-D-glucuronide
Probably in: Rat B93, C421

N-(*p*-Hydroxyphenyl)-2-methylsulphonylacetamide-β-D-glucuronide
Probably formed from: N-(*p*-Hydroxyphenyl)-2-methylsulphonylacetamide

4-((3-Hydroxyphenyl)methyl)-2,2,6,6-tetramethyl-1-oxa-4-aza-2,6-disilacyclohexane
Formed from: 4-((3-Methoxyphenyl) methyl)-2,2,6,6-tetramethyl-1-oxa-4-aza-2,6-disilacyclohexane

***p*-Hydroxyphenyl *o*-(2-N-morpholinoethoxy) phenyl ether**
Formed from: *o*-(2-N-Morpholinoethoxy) phenyl phenyl ether

3-(*p*-Hydroxyphenyl)muconolactone
Formed from: Biphenyl

6-Hydroxy-N-phenyl-β-naphthylamine
Formed from: N-Phenyl-β-naphthylamine

N-(*p*-Hydroxyphenyl)-β-naphthylamine
Formed from: N-Phenyl-β-naphthylamine

2-(*p*-Hydroxyphenyl)-1-nitroethane
→ (*E*)-*p*-Hydroxyphenylacetaldoxime
In: Sorghum K711

2-(*p*-Hydroxyphenyl)nonane
→ 2-(*p*-Hydroxyphenyl)nonan-8-ol
In: Rainbow trout J57

3-(*p*-Hydroxyphenyl)nonane
→ 7-(*p*-Hydroxyphenyl)nonan-2-ol
In: Rainbow trout J57

4-(*p*-Hydroxyphenyl)nonane
→ 6-(*p*-Hydroxyphenyl)nonan-2-ol
In: Rainbow trout J57

9-(*p*-Hydroxyphenyl)nonanoic acid
Probably formed from: 9-(*p*-Hydroxy-phenyl)nonan-1-ol
→ 7-(*p*-Hydroxyphenyl)heptanoic acid
Probably in: Rainbow trout J412

1-(*p*-Hydroxyphenyl)nonan-1-ol
Formed from: *p*-*n*-Nonylphenol

2-(*p*-Hydroxyphenyl)nonan-8-ol
Formed from: 2-(*p*-Hydroxyphenyl)nonane
→ 2-(*p*-Hydroxyphenyl)nonan-8-ol-β-D-glucuronide
In: Rainbow trout J57

6-(*p*-Hydroxyphenyl)nonan-2-ol
Formed from: 4-(*p*-Hydroxyphenyl)nonane
→ 6-(*p*-Hydroxyphenyl)nonan-2-ol-β-D-glucuronide
In: Rainbow trout J57

7-(*p*-Hydroxyphenyl)nonan-2-ol
Formed from: 3-(*p*-Hydroxyphenyl)nonane
→ 7-(*p*-Hydroxyphenyl)nonan-2-ol-β-D-glucuronide
In: Rainbow trout J57

9-(*p*-Hydroxyphenyl)nonan-1-ol
Formed from: *p*-*n*-Nonylphenol
→ 9-(*p*-Hydroxyphenyl)nonanoic acid
In: Rainbow trout J412

9-(*p*-Hydroxyphenyl)nonan-2-ol
Formed from: *p*-*n*-Nonylphenol

2-(*p*-Hydroxyphenyl)nonan-8-ol-β-D-glucuronide
Formed from: 2-(*p*-Hydroxyphenyl) nonan-8-ol

6-(*p*-Hydroxyphenyl)nonan-2-ol-β-D-glucuronide
Formed from: 6-(*p*-Hydroxyphenyl) nonan-2-ol

7-(*p*-Hydroxyphenyl)nonan-2-ol-β-D-glucuronide
Formed from: 7-(*p*-Hydroxyphenyl)nonan-2-ol

9-(*p*-Hydroxyphenyl)nonylamine
→ Phloretic acid
In: Rat G865

4-(*o*-Hydroxyphenyl)-2-oxobut-3-enoic acid
See: *o*-Hydroxybenzylidenepyruvic acid

2-(*o*-Hydroxyphenyl)-2-oxoethylamine
→ *o*-Hydroxymandelic acid
In: Rat J287

2-(*m*-Hydroxyphenyl)-2-oxoethylamine
→ *m*-Hydroxymandelic acid
In: Rat J287

2-(*p*-Hydroxyphenyl)-2-oxoethylamine
→ *p*-Hydroxymandelic acid
In: Rat J287

1-(*p*-Hydroxyphenyl)-2-(2-oxopyrrolidino) pentane
Probably formed from: 1-(*p*-Hydroxyphenyl)-2-pyrrolidinopentane
→ 4-Hydroxy-2-(2-oxopyrrolidino)-1-phenylpentane
Probably in: Rabbit A1274

1-Hydroxy-1-phenyl-*n*-pentane
See: 1-Phenylpentan-1-ol

5-(*p*-Hydroxyphenyl)pentanoic acid
Formed from: 5-(*p*-Hydroxyphenyl) pentylamine
7-(*p*-Methoxyphenyl)heptylamine
1-Phenylnonane
Probably formed from: 7-(*p*-Hydroxyphenyl)heptanoic acid
7-(*p*-Methoxyphenyl)heptanoic acid
→ Phloretic acid
Probably in: Rainbow trout J412

1-(*p*-Hydroxyphenyl)-1-pentanol
Formed from: *p*-*n*-Pentylphenol

(2*E*)-3-(4-Hydroxyphenyl)-2-pentenedioic acid
See: *trans*-Sphagnum acid

***trans*-3-(*p*-Hydroxyphenyl)-2-(pentoxycarbonylethyl)-N-phenyllactam**
→ *trans*-2-(2-Carboxyethyl)-3-(*p*-hydroxyphenyl)-N-phenyllactam
In: Microorganism K110

5-(*p*-Hydroxyphenyl)pentylamine
→ *p*-Hydroxyphenylpentanoic acid
In: Rat G865

N-(4-Hydroxy-1-phenylpent-2-yl)pyrrolidine
Formed from: Prolintane

N-(5-Hydroxy-1-phenylpent-2-yl)pyrrolidine
Formed from: Prolintane

2-Hydroxy-9-phenyl-1*H*-phenalen-1-one
See: Angiorufone

1-(*p*-Hydroxyphenyl)-1-phenyl-3-butanol
Probably formed from: 1-(*p*-Hydroxyphenyl)-1-phenyl-3-butanone

1-(*p*-Hydroxyphenyl)-1-phenylbutan-3-one
Probably formed from: N-(*tert*-Butyl)-4-(*p*-hydroxyphenyl)-4-phenyl-2-butylamine
1,1-Diphenyl-3-butanone
→ 1-(4-Hydroxy-3-methoxyphenyl)-1-phenylbutan-3-one
Probably in: Dog D784
Rat F582
→ 1-(*p*-Hydroxyphenyl)-1-phenyl-3-butanol
Probably in: Dog D784

1-(*p*-Hydroxyphenyl)-1-phenylbutan-3-one oxime
Probably formed from: 1,1-Diphenyl-3-butanone oxime
→ 1-(4-Hydroxy-3-methoxyphenyl)-1-phenylbutan-3-one oxime
Probably in: Rat F582

1-(4-(*p*-Hydroxyphenyl)-2-phenyl-1-butenyl) phenoxyacetic acid
Probably formed from: 4-(1, 2-Diphenyl-1-butenyl)phenoxyacetic acid

2-Hydroxyphenyl phenyl ether
Formed from: Diphenyl ether
o-(2-N-Morpholinoethoxy)phenyl phenyl ether

3-Hydroxyphenyl phenyl ether
Probably formed from: Diphenyl ether

4-Hydroxyphenyl phenyl ether
(4-hydroxydiphenyl ether)
Formed from: Diphenyl ether
→ Bis(4-hydroxyphenyl) ether
In: Guinea pig E10
Rat C673
Sphingomonas J882
Trout E10
→ 4,4′-Dihydroxy-3,3′-diphenoxybiphenyl
In: Pycnoporus K689
→ 3-Hydroxy-4-methoxyphenyl phenyl ether

In: Guinea pig E10
Trout E10
→ Phenol
In: Sphingomonas J882

N-(*p*-Hydroxyphenyl)-N′-phenylguanidine
Formed from: N,N′-Diphenylguanidine

1-(*o*-Hydroxyphenyl)-3-phenylhexane-1,2,5-trione
Formed from: Warfarin

5-(*o*-Hydroxyphenyl)-5-phenylhydantoin
Formed from: 5,5-Diphenylhydantoin

5-(*m*-Hydroxyphenyl)-5-phenylhydantoin
Formed from: 5,5-Diphenylhydantoin

5-(*p*-Hydroxyphenyl)-5-phenylhydantoin
Formed from: 5,5-Diphenylhydantoin
→ 5,5-Bis(*p*-hydroxyphenyl)hydantoin
Probably in Man A2110
Rat A2110
→ 5-(3,4-Dihydroxyphenyl)-5-phenylhydantoin
In: Mouse D818
See also: A494, A506, A567, A1018, A3305
→ 5-(*p*-Hydroxyphenyl)-5-phenylhydantoin-4′-β-D-glucuronide
In: Frog C39
Rat A2184, A2879, A2976, B962, C286

5-(*p*-Hydroxyphenyl)-5-phenylhydantoin-4′-β-D-glucuronide
Formed from: 5-(*p*-Hydroxyphenyl)-5-phenylhydantoin

***p*-Hydroxyphenyl phenylmethane**
Formed from: Diphenylmethane

1-((*p*-Hydroxyphenyl)phenylmethyl)-4-(3-phenyl-2-propenyl)piperazine
Formed from: Cinnarizine

4-((*p*-Hydroxyphenyl)phenyl)methylpiperidine
Probably formed from: 4-Diphenylmethyl-piperidine

***m*-(*p*-Hydroxyphenyl)phenylmuconolactone**
Formed from: *m*-Terphenyl

2-(*o*-Hydroxyphenyl)-5-phenyloxazole
Formed from: 2,5-Diphenyloxazole

2-(*o*-Hydroxyphenyl)-5-phenyloxazole
Formed from: 2,5-Diphenyloxazole

2-(*p*-Hydroxyphenyl)-5-phenyloxazole
Formed from: 2,5-Diphenyloxazole

5-(*p*-Hydroxyphenyl)-2-phenyloxazole
Formed from: 2,5-Diphenyloxazole

4-(*p*-Hydroxyphenyl)-5-phenyl-2-oxazolepropionic acid
Formed from: Oxaprozin
→ 4-(*p*-Hydroxyphenyl)-5-phenyl-2-oxazolepropionic acid-β-D-glucuronic acid
Probably in: Man B375
→ 4-(*p*-Hydroxyphenyl)-5-phenyl-2-oxazolepropionyl-β-D-glucuronide
In: Man B375

4-(*p*-Hydroxyphenyl)-5-phenyl-2-oxazolepropionic acid-β-D-glucuronic acid
Formed from: 4-(*p*-Hydroxyphenyl)-5-phenyl-2-oxazolepropionic acid

4-(*p*-Hydroxyphenyl)-5-phenyl-2-oxazolepropionyl-β-D-glucuronide
Formed from: 4-(*p*-Hydroxyphenyl)-5-phenyl-2-oxazolepropionic acid

1-(*o*-Hydroxyphenyl)-3-phenylpentan-1,2-dione
Formed from: Phenprocoumon

1-(*p*-Hydroxyphenyl)-2-phenyl-4-(2-phenylsulphonyl)ethyl)-3,5-pyrazolidinedione
Probably formed from: 1,2-Diphenyl-4-(2-phenylsulphonyl)ethyl)-3,5-pyrazolidinedione

1-(*p*-Hydroxyphenyl)-2-phenyl-4-((2-phenylthio)ethyl)-3,5-pyrazolidinedione
Formed from: 1,2-Diphenyl-4-((2-phenylthio)ethyl)-3,5-pyrazolidinedione
→ 1-(*p*-Hydroxyphenyl)-2-phenyl-4-(2-phenylthio)ethyl)-3,5-pyrazolidinedione-β-D-glucuronide
Probably in: Man B510

1-(*p*-Hydroxyphenyl)-2-phenyl-4-((2-phenylthio)ethyl)-3,5-pyrazolidinedione-β-D-glucuronide
Formed from: 1-(*p*-Hydroxyphenyl)-2-phenyl-4-((2-phenylthio)ethyl)-3,5-pyrazolidinedione

***trans*-3-(*p*-Hydroxyphenyl)-N-phenyl-2-(propoxycarbonylethyl)lactam**
→ *trans*-2-(2-Carboxyethyl)-3-(*p*-hydroxyphenyl)-N-phenyllactam
In: Microorganism K110

(*R*)-2-Hydroxy-2-phenylpropionitrile
→ Phenylacetone
In: Almond K617

3-(*p*-Hydroxyphenyl)-3-phenylpropylamine
Probably formed from: 3,3-Diphenylpropylamine

→ 3-(4-Hydroxy-3-methoxyphenyl)-3-phenylpropylamine
Probably in: Man C778

N-(3-(*p*-Hydroxyphenyl)-3-phenylpropyl)-1-phenylethylamine
Probably formed from: Fendiline
→ N-(3-(4-Hydroxy-3-methoxyphenyl)-3-phenylpropyl)-1-phenylethylamine
Probably in: Man C778

***trans*-3-(*p*-Hydroxyphenyl)-N-phenyl-2-(2,2,2-trifluoroethoxycarbonylethyl) lactam**
→ *trans*-2-(2-Carboxyethyl)-3-(*p*-hydroxyphenyl)-N-phenyllactam
In: Microorganism K110

***p*-Hydroxyphenyl phosphate**
Probably formed from: Quinol

1-(3-((4-(*o*-Hydroxyphenyl)-1-piperazinyl) methyl)benzoyl)piperidine
Formed from: Mazapertine
→ Hydroxy-1-(3-((4-(*o*-hydroxyphenyl)-1-piperazinyl)methyl)benzoyl) piperidine
Probably in: Dog J581

3-(1-(*o*-Hydroxyphenyl)piperazin-4-yl)-1-naphthyloxypropan-2-ol
Formed from: Naftopidil

6-(3-(4-(*o*-Hydroxyphenyl)piperazinyl) propylamino)-1,3-dimethyluracil
Formed from: Urapidil

α-(*p*-Hydroxyphenyl)-2-piperidineacetic acid
Probably formed from: α-Phenyl-2-piperidineacetic acid
→ α-(*p*-Hydroxyphenyl)-2-piperidineacetic acid-β-D-glucuronide
Probably in: Dog C14

α-(*p*-Hydroxyphenyl)-2-piperidineacetic acid-β-D-glucuronide
Probably formed from: α-(*p*-Hydroxyphenyl)-2-piperidineacetic acid

4-(*p*-Hydroxyphenyl)-4-piperidinocyclohexanol
Probably formed from: 4-Phenyl-4-piperidinocyclohexanol

***p*-Hydroxyphenyl 2-(3′-piperidinopropyl)benzyl ether**
Probably formed from: Phenyl 2-(3′-piperidinopropyl)benzyl ether

1-(*m*-Hydroxyphenyl)-1-(1-piperidinyl) cyclohexane
Formed from: Phencyclidine

α-(*p*-Hydroxyphenyl)-α-(2-piperidyl)acetic acid
Probably formed from: α-Phenyl-α-(2-piperidyl)acetic acid
→ α-(*p*-Hydroxyphenyl)-α-(2-piperidyl)acetic acid-β-D-glucuronide
Probably in: Rat A137

α-(*p*-Hydroxyphenyl)-α-(2-piperidyl)acetic acid-β-D-glucuronide
Probably formed from: α-(*p*-Hydroxyphenyl)-α-(2-piperidyl)acetic acid

1-(4-(*m*-Hydroxyphenyl)-4-piperidyl)-1-propanone
Formed from: Ketobemidone

1-(*p*-Hydroxyphenyl)propan-1,2-dione
Probably formed from: 2-Hydroxy-1-(*p*-hydroxyphenyl)propan-1,2-dione

1-(*m*-Hydroxyphenyl)propan-2-ol
Probably formed from: *m*-Hydroxyphenylpropiophenone

2-(*p*-Hydroxyphenyl)-2-propanol
Probably formed from: 2-Phenyl-2-propanol

3-(*p*-Hydroxyphenyl)propanol
Probably formed from: Phloretic acid
p-Propylphenol
→ *p*-Coumaraldehyde
Probably in: Penicillium J836
→ 3-(3,4-Dihydroxyphenyl)propanol
In: Mucuna E769
Mushroom H907
→ Rhododendrin
In: Alnus A3144
Betula A3144

2-Hydroxy-1-phenyl-1-propanone
See: 2-Oxo-1-phenyl-1-propanol

***p*-Hydroxyphenylpropan-2-one**
See: *p*-Hydroxyphenylacetone

***p*-Hydroxyphenyl propionate**
Formed from: *p*-Hydroxypropiophenone

3-(*o*-Hydroxyphenyl)propionic acid
See: Melilotic acid

***m*-Hydroxyphenylpropionic acid**
Formed from: (+)-Catechin
m-Coumaric acid
Dihydrocaffeic acid
Hesperetin

Phloretic acid
Proanthocyanidin B_3
Rutin
4-(2,4,6-Trihydroxyphenyl)flavan-3-ol
→ Batatasin III
In: Dioscorea D136
→ Hircinol
In: Dioscorea D136

2-(*p*-Hydroxyphenyl)propionic acid
(*p*-hydroxyhydratropic acid)
→ 2-(2,5-Dihydroxyphenyl)propionic acid
In: Pseudomonas A52, A1480
→ *p*-Hydroxybenzoic acid
In: Pseudomonas H426
→ 2-(*p*-Hydroxyphenyl)propionic acid-*p*-*β*-D-glucoside
In: Panax F255
→ 2-(*p*-Hydroxyphenyl)propionyl-β-D-glucoside
In: Panax F255

3-(*p*-Hydroxyphenyl)propionic acid
See: Phloretic acid

3-Hydroxy-2-phenylpropionic acid
See: Tropic acid

3-Hydroxy-3-phenylpropionic acid
See: β-Phenylhydracrylic acid

2-(*p*-Hydroxyphenyl)propionic acid-*p*-β-D-glucoside
Formed from: 2-(*p*-Hydroxyphenyl) propionic acid

2-Hydroxy-3-phenylpropionitrile
See: Phenyllactonitrile

***m*-Hydroxyphenylpropionyl CoA**
→ 3,3′,5′-Trihydroxybibenzyl
In: Barlia C734
Bletilla H101
Dioscorea D136
Epipactis G138
Orchid H582

***p*-Hydroxyphenylpropionyl CoA**
→ 3,4′,5′-Trihydroxybibenzyl
In: Barlia C734

2-(*p*-Hydroxyphenyl)propionyl-β-D-glucoside
Formed from: 2-(*p*-Hydroxyphenyl)propionic acid

3-(*o*-Hydroxyphenyl)propionyltropine
→ 6β-Hydroxy-(3-(*o*-hydroxyphenyl) propionyl)tropine
In: Hyoscyamus E350

3-(*m*-Hydroxyphenyl)propionyltropine
→ 6β-Hydroxy-(3-(*m*-hydroxyphenyl) propionyl)tropine
In: Hyoscyamus E350

***o*-Hydroxy-ω-phenylpropiophenone**
Formed from: Propafenone

***m*-Hydroxyphenylpropiophenone**
Probably formed from: *m*-Hydroxy-amphetamine
→ 1-(*m*-Hydroxyphenyl)propan-2-ol
Probably in: Dog B403
Man B403
Monkey B403
→ 1-Hydroxy-1-(*m*-hydroxyphenyl) acetone
Probably in: Dog B403
Man B403
Monkey B403

***o*-Hydroxyphenylpropylamine**
→ Melilotic acid
In: Rat G865

***p*-Hydroxyphenylpropylamine**
→ Phloretic acid
In: Rat G865

γ-Hydroxy-γ-phenylpropyl carbamate
Formed from: γ-Phenylpropyl carbamate
→ Benzoic acid
Probably in: Man A3226
→ β,γ-Dihydroxy-γ-phenylpropyl carbamate
Probably in: Man A3226
→ γ,β-Dihydroxy-γ-phenylpropyl carbamate
Probably in: Man A3226

3-(*p*-Hydroxyphenyl)propyl trimethyl ammonium
→ 3-(3,4-Dihydroxyphenyl)propyl trimethyl ammonium
In: Mushroom H907

1-(Hydroxyphenyl)-1-(2-pyridyl)ethane
Probably formed from: 1-(Hydroxyphenyl)-1-(2-pyridyl)ethanol

1-(Hydroxyphenyl)-1-(2-pyridyl)ethanol
Probably formed from: Hydroxydoxylamine
→ 1-(Hydroxyphenyl)-1-(2-pyridyl)ethane
Probably in: Rat E11

4-Hydroxy-2-(2-phenyl-2-(2-pyridyl)ethyl) imidazole
Probably formed from: 2-(2-Phenyl-2-(2-pyridyl)ethyl)imidazole
→ 3-Phenyl-3-(2-pyridyl) propioimidazolylaminoacetic acid
Probably in: Dog E842
→ 3-Phenyl-3-(2-pyridyl) propionylaminoacetamide
Probably in: Dog E842

3-(*p*-Hydroxyphenyl)-5-(3-pyridyl)-1,2,4-oxadiazole
Formed from: 3-(4-(1-Ethoxycarbony-1-methylethoxy)phenyl)-5-(3-pyridyl)-1,2,4-oxadiazole

3-(*p*-Hydroxyphenyl)-3-(2-pyridyl) propioimidamide
Probably formed from: *p*-Hydroxy-midaglizole

***m*-Hydroxyphenylpyruvic acid**
Formed from: *m*-Hydroxyphenyllactic acid
D-*m*-Tyrosine
L-*m*-Tyrosine
→ *m*-Hydroxyphenyllactic acid
In: Coleus G289
Rat A2961
→ L-*m*-Tyrosine
Probably in: Rat A2961, A3493

***p*-Hydroxyphenylpyruvic acid**
Formed from: *p*-Hydroxyphenyllactic acid
enol-p-Hydroxyphenylpyruvic acid
D-Tyrosine
L-Tyrosine
Probably formed from: Prephenic acid
→ 2,5-Dihydroxyphenylpyruvic acid
Probably in: Bacterium A489
→ 3,4-Dihydroxyphenylpyruvic acid
In: Mucuna F112
Potato D964
→ Hawkinsin
Probably in: Man A3187, A3259
→ Homogentisic acid
In: Aspergillus C484
Drosophyllum A181
See also: B455
→ *p*-Hydroxybenzaldehyde
Probably in: Flavobacterium B455
→ *cis*-4-Hydroxycyclohexylacetic acid
Probably in: Man A3259
→ *trans*-4-Hydroxycyclohexylacetic acid
Probably in: Man A3259
→ *p*-Hydroxymandelic acid
In: Amycolatopsis K499, K688
Rat E334
See also: A3368
→ *p*-Hydroxyphenylacetaldehyde
Probably in: Bacterium A2651
Candida E97
Cat A1197
Clostridium A2571
Microorganism A2767
Nocardia E606
→ *p*-Hydroxyphenylacetic acid
In: Arthrobacter A3208
Naja J684
Rat E485
See also: B455, E97
→ *p*-Hydroxyphenylacetyl CoA
In: Pyrococcus H306
→ *p*-Hydroxyphenyllactic acid
In: Candida D975
Cat A1197
Coleus E660, G289
Man A3368
Rat A3026
Streptococcus E271
Trypanosoma A1619
With malate dehydrogenase E561
See also: E97
→ *enol-p*-Hydroxyphenylpyruvic acid
In: Man J475
Naja J684
Pig G916
→ L-Tyrosine
In: Bacillus E445, G168
Brevibacterium B266, E642
Corynebacterium B266
Escherichia A3548
Hansenula D978
Methanophilus H114
Microtetraspora F222
Mung bean B290, H826
Neisseria F92

Nocardia F394
Rat K718
Rhodococcus E451
Sporosarcina E423
See also: E485

***enol-p*-Hydroxyphenylpyruvic acid**
Formed from: *p*-Hydroxyphenylpyruvic acid
→ *p*-Hydroxybenzaldehyde
In: Naja J684
→ *p*-Hydroxyphenylpyruvic acid
In: Pig G916

***p*-Hydroxyphenylpyruvic acid oxime**
Formed from: L-Tyrosine

N-(*p*-Hydroxyphenyl)-*all-trans*-retinamide
Formed from: N-(*p*-Methoxyphenyl)-*all-trans*-retinamide
→ N-(*p*-Hydroxyphenyl)-*cis*-retinamide
In: Mouse E857
Rat E857
→ N-(*p*-Methoxyphenyl)-*all-trans*-retinamide
In: Mouse E857
Rat E857

N-(*p*-Hydroxyphenyl)-*cis*-retinamide
Formed from: N-(*p*-Hydroxyphenyl)-*all-trans*-retinamide

***threo*-*p*-Hydroxyphenylserine**
→ Novobiocin
In: Streptomyces A945

***o*-Hydroxyphenyl sulphate**
Formed from: Catechol

***m*-Hydroxyphenyl sulphate**
Formed from: Resorcinol

***p*-Hydroxyphenyl sulphate**
Formed from: Quinol
→ Quinol
In: Escherichia J364

2-(*p*-Hydroxyphenyl)-2-(*p*-sulphatoxyphenyl) propane
Formed from: 2,2-Bis(*p*-hydroxyphenyl) propane

3-((*p*-Hydroxyphenyl)sulphinyl)acetic acid
Formed from: 3-((*p*-Hydroxyphenyl) thio)pyruvic acid

8-(4-((*p*-Hydroxyphenyl)sulphinyl)-3-methoxyphenyl)pyrazolo)[1,5-*a*]-1,3,5-triazine-2,4(1*H*)-dione
Formed from: 8-(4-((*p*-Hydroxyphenyl) sulphinyl)-3-methoxyphenyl)pyrazolo[1,5-*a*]-1,3,5-triazine-2,4(1*H*)-dione

8-(4-((*p*-Hydroxyphenyl)sulphinyl)-3-methoxyphenyl)pyrazolo)[1,5-*a*]-1,3,5-triazin-4(1*H*)-one
Formed from: 8-(4-((*p*-Hydroxyphenyl) sulphinyl)-3-methoxyphenyl)pyrazolo[1,5-*a*]-1,3,5-triazin-4(1*H*)-one

7-(5-Hydroxy-3-phenylsulphonylaminobicyclo [2.2.1]hept-2-yl)heptenoic acid
Formed from: 5-(+)-(*Z*)-7-((1*R*,2*S*,3*S*,4*S*)-3-Phenylsulphonylaminobicyclo[2.2.1]hept-2-yl)heptenoic acid

7-(6-Hydroxy-3-phenylsulphonylaminobicyclo [2.2.1]hept-2-yl)heptenoic acid
Formed from: 5-(+)-(*Z*)-7-((1*R*,2*S*,3*S*,4*S*)-3-Phenylsulphonylaminobicyclo[2.2.1]hept-2-yl)heptenoic acid

5-(5-Hydroxy-3-phenylsulphonylaminobicyclo [2.2.1]hept-2-yl)pentanoic acid
Formed from: 5-((1*R*,2*S*,3*S*,4*S*)-3-Phenylsulphonylaminobicyclo[2.2.1]hept-2-yl)pentanoic acid

5-(6-Hydroxy-3-phenylsulphonylaminobicyclo [2.2.1]hept-2-yl)pentanoic acid
Formed from: 5-((1*R*,2*S*,3*S*,4*S*)-3-Phenylsulphonylaminobicyclo[2.2.1]hept-2-yl)pentanoic acid

5-(5-Hydroxy-3-phenylsulphonylaminobicyclo[2.2.1]hept-2-yl)pentenoic acid
Formed from: 5-((1*R*,2*S*,3*S*,4*S*)-3-Phenylsulphonylaminobicyclo[2.2.1]hept-2-yl)pentenoic acid

5-(6-Hydroxy-3-phenylsulphonylaminobicyclo [2.2.1]hept-2-yl)pentenoic acid
Formed from: 5-((1*R*,2*S*,3*S*,4*S*)-3-Phenylsulphonylaminobicyclo[2.2.1]hept-2-yl)pentenoic acid

3-(5-Hydroxy-3-phenylsulphonylaminobicyclo [2.2.1]hept-2-yl)propionic acid
Formed from: 3-((1*R*,2*S*,3*S*,4*S*)-3-Phenylsulphonylaminobicyclo[2.2.1]hept-2-yl)propionic acid

3-(6-Hydroxy-3-phenylsulphonylaminobicyclo [2.2.1]hept-2-yl)propionic acid
Formed from: 3-((1*R*,2*S*,3*S*,4*S*)-3-Phenylsulphonylaminobicyclo[2.2.1]hept-2-yl)propionic acid

2-(*p*-Hydroxyphenyl)-1-tetralone
Formed from: 2-Phenyl-1-tetralone

3-(*p*-Hydroxyphenyl)-1-tetralone
Formed from: 3-Phenyl-1-tetralone

8-(4-(*p*-Hydroxyphenyl)thio-3-methoxyphenyl) pyrazolo[1,5-*a*]-1,3,5-triazin-4(1*H*)-one
Formed from: 8-(3-Methoxy-4-phenylthiophenyl)pyrazolo[1,5-*a*]-1,3,5-triazin-4(1*H*)-one
→ 8-(4-(*p*-Hydroxyphenyl)thio-3-methoxyphenyl)pyrazolo[1,5-*a*]-1,3,5-triazin-4(1*H*)-one sulphate
In: Rat J797

8-(4-(*p*-Hydroxyphenyl)thio-3-methoxyphenyl) pyrazolo[1,5-*a*]-1,3,5-triazin-4(1*H*)-one sulphate
Formed from: 8-(4-(*p*-Hydroxyphenyl)thio-3-methoxyphenyl)pyrazolo[1,5-*a*]-1,3,5-triazin-4(1*H*)-one

8-(3-Hydroxy-4-phenylthiophenyl) pyrazolo[1,5-*a*]-1,3,5-triazin-4(1*H*)-one
Formed from: 8-(3-Methoxy-4-phenylthiophenyl)pyrazolo[1,5-*a*]-1,3,5-triazin-4(1*H*)-one
→ 8-(3-Hydroxy-4-phenylthiophenyl) pyrazolo[1,5-*a*]-1,3,5-triazin-4(1*H*)-one-β-D-glucuronide
In: Rat J797

8-(3-Hydroxy-4-phenylthiophenyl)pyrazolo [1,5-*a*]-1,3,5-triazin-4(1*H*)-one-β-D-glucuronide
Formed from: 8-(3-Hydroxy-4-phenylthiophenyl)pyrazolo[1,5-*a*]-1,3,5-triazin-4(1*H*)-one

3-((*p*-Hydroxyphenyl)thio)pyruvic acid
→ 3-((*p*-Hydroxyphenyl)sulphinyl) acetic acid
In: Pseudomonas D688

***p*-Hydroxyphenylthiourea**
Formed from: 5-(*p*-Hydroxyanilino)-1,2,3,4-thiatriazole

1-(*p*-Hydroxyphenyl)-1-(*m*-trifluoromethylphenyl) propane-1,2-diol
Probably formed from: 1-(*p*-Hydroxyphenyl)-1-(*m*-trifluoromethylphenyl)propanol

1-(*p*-Hydroxyphenyl)-1-(*m*-trifluoromethylphenyl) propane-1,3-diol
Probably formed from: 1-Phenyl-1-(*m*-trifluoromethylphenyl)propane-1,3-diol-1-(*p*-Hydroxyphenyl)-1-(*m*-trifluoromethylphenyl)propanol
Formed from: 1-Phenyl-1-(*m*-trifluoromethylphenyl)propanol
→ 1-(3,4-Dihydroxyphenyl)-1-(*m*-trifluoromethylphenyl)propanol
Probably in: Man E814
→ 1-(*p*-Hydroxyphenyl)-1-(*m*-trifluoromethylphenyl)propane-1,2-diol
Probably in: Man E814

5-Hydroxy-1-phenyl-3-trifluoromethylpyrazole
Formed from: Flupyrazofos
Flupyrazofos oxon

(4-Hydroxyphenyl)(2,4,6-trihydroxyphenyl) methanone
See: Iriflophenone

1-(*m*-Hydroxyphenyl)-3-(2,4,6-trihydroxyphenyl) propan-2-ol
Formed from: 4-(2,4,6-trihydroxyphenyl) flavan-3-ol
Proanthocyanidin B_3
Probably formed from: 1-(3,4-Dihydroxyphenyl)-3-(2,4,6-trihydroxyphenyl)propan-2-ol

***m*-Hydroxyphenyl trimethylammonium**
(trimophonium)
Formed from: Neostigmine
→ *m*-Dimethylaminophenol
Probably in: Rat A616
→ *m*-Hydroxyphenyl trimethylammonium-β-D-glucuronide
In: Rat A2130, A2393

***m*-Hydroxyphenyl trimethylammonium-β-D-glucuronide**
Formed from: *m*-Hydroxyphenyl trimethylammonium

***p*-Hydroxyphenylurea**
Formed from: 5-(*p*-Hydroxyanilino)-1,2,3,4-thiatriazole

δ-(*m*-Hydroxyphenyl)-γ-valerolactone
Formed from: (+)-Catechin
d-δ-(3,4-Dihydroxyphenyl)-γ-valerolactone
Hydroxyphloroglucinol
(1,2,4,5-tetrahydroxybenzene)
Probably formed from: Phloroglucinol-carboxylic acid
6-Hydroxy-1(2*H*)-phthalazinone
Probably formed from: 1(2*H*)-Phthalazinone
3-Hydroxyphthalic acid
Formed from: Phthalic acid
→ Protocatechuic acid
Probably in: Corynebacterium A2999
3-Hydroxyphthalic acid
→ *m*-Hydroxybenzoic acid
In: Bacillus D31
Pseudomonas A3332, D31
→ *p*-Hydroxybenzoic acid
In: Bacillus D31
2-Hydroxypicene
Formed from: Picene
4-Hydroxypicene
Formed from: Picene
Hydroxypinaverium
Formed from: Pinaverium
3-Hydroxypindolol
Formed from: Pindolol
→ 3-Hydroxypindolol sulphate
In: Monkey B398
5-Hydroxypindolol
Formed from: Pindolol
→ 5-Hydroxypindolol-β-D-glucuronide
In: Man E835
→ 5-Hydroxypindolol sulphate
In: Man E835
Monkey B398
6-Hydroxypindolol
Formed from: Pindolol
→ 6-Hydroxypindolol sulphate
In: Man E835
5-Hydroxypindolol-β-D-glucuronide
Formed from: 5-Hydroxypindolol
3-Hydroxypindolol sulphate
Formed from: 3-Hydroxypindolol
5-Hydroxypindolol sulphate
Formed from: 5-Hydroxypindolol
6-Hydroxypindolol sulphate
Formed from: 6-Hydroxypindolol
3′-Hydroxypipecolo-2′,6′-xylidide
Probably formed from: Pipecolo-2′,6′-xylidide
4′-Hydroxypipecolo-2′,6′-xylidide
Probably formed from: Pipecolo-2′,6′-xylidide
1-(4-Hydroxypiperidino)-6,7-dimethoxyphthalazine
Formed from: Carbazeran
3-(1-Hydroxy-2-piperidinoethyl)-5-phenylisoxazole (perisoxal)
→ 3-(1-Hydroxy-2-(3-hydroxypiperidino) ethyl)-5-phenylisoxazole
In: Rabbit A1809
→ 3-(1-Hydroxy-2-(4-hydroxypiperidino) ethyl)-5-phenylisoxazole
In: Rabbit A1809
→ 3-(1-Hydroxy-2-(2-oxopiperidino)ethyl)-5-phenylisoxazole
In: Rabbit A1809
→ 5-(*m*-Hydroxyphenyl)-3-(1-hydroxy-2-piperidinoethyl)isoxazole
In: Man B554
Rabbit A1809, B554
Rat B554
→ 5-(*p*-Hydroxyphenyl-3-(1-hydroxy-2-piperidinoethyl)isoxazole
In: Man B554
Rabbit A1809, B554
Rat B554
→ 3-(1-Hydroxy-2-piperidinoethyl)-5-phenylisoxazole-N-oxide
In: Rabbit A1809
→ 5-Phenyl-3-isoxazolylglycollic acid
In: Rabbit A1809
→ Perisoxal-β-D-glucuronide
In: Man B554
Rabbit B554
Rat B554
3-(1-Hydroxy-2-piperidinoethyl)-5-phenylisoxazole-N-oxide
Formed from: 3-(1-Hydroxy-2-piperidino) ethyl)-5-phenylisoxazole
4-(4-Hydroxypiperidino)-4-phenylcyclohexanol
Probably formed from: 4-Phenyl-4-piperidino-cyclohexanol

4-Hydroxy-N-(2-(2-(2-piperidinyl)ethyl)phenyl) benzamide
Probably formed from: 4-Hydroxy-N-(2-(2-(1-methyl-2-piperidinyl)ethyl)phenyl) benzamide

2-Hydroxy-N-(3-(*m*-(1-piperidinylmethyl) phenoxy)propyl)acetamide
See: Roxatidine
Formed from: 2-Acetoxy-N-(3-(*m*-(1-piperidinylmethyl)phenoxy) propyl)acetamide

2-Hydroxy-N-(2-piperidinylmethyl)-5-(2,2,2-trifluoroethoxy)benzamide
Formed from: Flecainide
→ 2-Hydroxy-N-(6-oxo-2-piperidinylmethyl)-5-(2,2,2-trifluoroethoxy)benzamide
Probably in: Man D388

3′-Hydroxypiritrexim
Formed from: Piritrexim
→ 3′-Hydroxypiritrexim-O-sulphate
In: Rat F757

3′-Hydroxypiritrexim-O-sulphate
Formed from: 3′-Hydroxypiritrexim

2-Hydroxypiroheptine
Formed from: Piroheptine

5′-Hydroxypiroxicam
Formed from: Piroxicam

3-Hydroxyplumbagin
Formed from: Plumbagin

3-Hydroxypractolol
See: 1-(4-Acetamido-3-hydroxyphenoxy)-3-isopropylaminopropan-2-ol

4-*cis*-Hydroxypraziquantel
Formed from: Praziquantel
→ 4-*cis*-Hydroxypraziquantel-β-D-glucuronide
In: Man K638

4-*trans*-Hydroxypraziquantel
Formed from: Praziquantel
→ 4-*trans*-Hydroxypraziquantel-β-D-glucuronide
In: Man K638

6-Hydroxypraziquantel
Formed from: Praziquantel
→ 6-Hydroxypraziquantel-β-D-glucuronide
In: Man K638

4-*cis*-Hydroxypraziquantel-β-D-glucuronide
Formed from: 4-*cis*-Hydroxypraziquantel

4-*trans*-Hydroxypraziquantel-β-D-glucuronide
Formed from: 4-*trans*-Hydroxypraziquantel

6-Hydroxypraziquantel-β-D-glucuronide
Formed from: 6-Hydroxypraziquantel

Hydroxyoxprenolol
Formed from: Oxprenolol

5-Hydroxyprimaquine
Formed from: Primaquine
→ 8-Amino-5-hydroxy-6-methoxyquinoline
Probably in: Dog C66
→ 8-((4-Amino-3-hydroxy-1-methylbutyl) amino)-5-hydroxy-6-methoxyquinoline
Probably in: Rat H136
→ 8-((4-Amino-1-methylbutyl)amino)-5,6-dihydroxyquinoline
Probably in: Rat H136

2-Hydroxyprimidone
→ Phenobarbital
In: Dog F618

3-Hydroxyprocymidone
Formed from: Procymidone
→ 3-Hydroxyprocymidone-β-D-glucuronide
In: Rabbit J488

3-Hydroxyprocymidone-β-D-glucuronide
Formed from: 3-Hydroxyprocymidone

Hydroxyprodolic acid
Formed from: Prodolic acid

***p*-Hydroxyprolintane**
Formed from: Prolintane
→ N-(1-(3,4-Dihydroxyphenyl)pent-2-yl) pyrrolidine
Probably in: Man G63
→ 1-(*p*-Hydroxyphenyl)-2-(2-oxopyrrolidino) pentane
Probably in: Rabbit A1274

L-Hydroxyprolyl-β-naphthylamide
→ β-Naphthylamine
Beef A526

2-Hydroxypromazine
Probably formed from: Promazine
→ Promazin-2-yl-β-D-glucuronide
In: Guinea pig A2092
Rat A2092

3-Hydroxypromazine
Probably formed from: Promazine

→ Promazin-3-yl-β-D-glucuronide
In: Guinea pig A2092
Rat A2092

Hydroxypromethazine
Formed from: Promethazine

4′-Hydroxypropafenone
Formed from: Propafenone
→ 3′,4′-Dihydroxypropafenone
Probably in: Dog D439, E18
→ 4′-Hydroxypropafenone-4′-yl-β-D-glucuronide
Probably in: Dog E18

5-Hydroxypropafenone
Formed from: Propafenone
→ 4,5-Dihydroxypropafenone
Probably in: Dog E18
Rat J538
→ 5-Hydroxypropafenone-5-yl-β-D-glucuronide
Probably in: Dog E18
Man D200

4′-Hydroxypropafenone-4′-yl-β-D-glucuronide
Probably formed from: 4′-Hydroxypropafenone

5-Hydroxypropafenone-5-yl-β-D-glucuronide
Probably formed from: 5-Hydroxypropafenone

Hydroxypropantheline
Probably formed from: Propantheline

N-Hydroxy-N-propargylbenzylamine
Probably formed from: Demethylpargyline

***o*-(2-Hydroxy-3-propionaldoxy)-ω-phenylpropiophenone**
Probably formed from: Propafenone
→ *o*-(2-Carboxy-2-hydroxyethoxy)-ω-phenylpropiophenone
Probably in: Dog D439
Man D200
→ *o*-(2,3-Dihydroxypropoxy)-ω-phenylpropiophenone
Probably in: Dog D439
Man D200

(3-Hydroxypropion)anilide
Formed from: 1,2′-Bis(N-phenyl-3-propionamidoxy)ethoxyethane

2-Hydroxypropiono-2′,6′-xylidide
Formed from: 2-Oxopropiono-2′,6′-xylidide

4-(N-(N-(3-Hydroxypropionylanilino))-1-(2-phenethyl)piperidine
Formed from: Fentanyl

Hydroxy-4-N-(N-propionylanilino)piperidine
Formed from: Fentanyl

(*S*)-3-(2-Hydroxypropionyl)benzaldehyde
Formed from: Isophthaldehyde
→ (*S*,*S*)-2-Hydroxy-*m*-(2-hydroxypropionyl)propiophenone
Probably with benzoylformate decarboxylase K657

(*S*)-4-(2-Hydroxypropionyl)benzaldehyde
Formed from: Isophthaldehyde
→ (*S*,*S*)-2-Hydroxy-*p*-(2-hydroxypropionyl)propiophenone
Probably with benzoylformate decarboxylase K657

2-Hydroxypropiophenone
Formed from: Mandelic acid
Phenylglyoxylic acid

***p*-Hydroxypropiophenone**
Formed from: Stilboestrol
Stilboestrol epoxide
Probably formed from: *p*-(1-Hydroxy-*n*-propyl)phenol
→ 3,4-Dihydroxypropiophenone
In: Rat A1154
→ 2-Hydroxy-1-(*p*-hydroxyphenyl)propan-1-one
Probably in: Horseradish D166
Mouse D166
→ *p*-Hydroxyphenyl propionate
In: Pseudomonas K698

β-Hydroxypropiovanillone (3-hydroxy-1-(4-hydroxy-3-methoxyphenyl)-1-propanone)
Formed from: Guaiacylglycerol β-coniferyl ether
2-(4-Formyl-2-methoxyphenoxy)-3-hydroxypropiovanillone
Formed from: Guaiacylglycerol β-guaiacyl ether
→ Guaiacol
Probably in: Pseudomonas D967

***p*-(2-Hydroxypropoxy)acetanilide**
Formed from: *p*-Propoxyacetanilide

1-Hydroxy-6-propoxybenzpyrene
Formed from: 1,6-Dipropoxybenzpyrene

4-(2-Hydroxypropoxy)-2,5-dimethoxyamphetamine
Probably formed from: 2,5-Dimethoxy-4-propoxyamphetamine

4-(3-Hydroxypropoxy)-2,5-dimethoxyamphetamine
Probably formed from: 2,5-Dimethoxy-4-propoxyamphetamine

4-(1-Hydroxy-2-propoxy)-3,5-dimethoxybenzoic acid
Formed from: 4-Isopropoxy-3,5-dimethoxybenzoic acid

4-(1-Hydroxy-2-propoxy)-3-methoxybenzoic acid
Formed from: 4-Isopropoxy-3-methoxybenzoic acid

***trans*-2-(4-(2-Hydroxypropoxy)-3-methoxy-5-methylsulphonylphenyl)-5-(3,4,5-trimethoxyphenyl)tetrahydrofuran**
Formed from: *trans*-2-(3-Methoxy-5-methylsulphonyl-4-propoxyphenyl)-5-(3,4,5-trimethoxyphenyl)tetrahydrofuran

1-(1-Hydroxy-2-propoxy)naphthalene
Formed from: 1-(2-Propoxy)naphthalene

1-(2-Hydroxypropoxy)naphthalene
Formed from: 1-Propoxynaphthalene

3′-(1-Hydroxy-2-propoxy)-2-trifluoromethylbenzanilide
Formed from: Flutolanil
→ 3′-(1-Carboxy-1-hydroxyethoxy)-2-trifluoromethylbenzanilide
Probably in: Rat D190

7-(2-Hydroxy-3-propylaminopropoxy)-2-phenyl-4*H*-1-benzopyran-4-one
See: Flavodilol

2-Hydroxypropranolol
Formed from: Propranolol
→ 2,3-Dihydroxypropranolol
In: Rat E915

3-Hydroxypropranolol
Formed from: Propranolol
→ 2,3-Dihydroxypropranolol
In: Rat E915
→ 3,4-Dihydroxypropranolol
In: Rat E915
→ 3,7-Dihydroxypropranolol
In: Rat E915

4-Hydroxypropranolol
Formed from: 4-Hydroxypropranolol-O-sulphate
Propranolol
→ 3,4-Dihydroxypropranolol
Probably in: Man E917
Rat E916, E917
→ 4,8-Dihydroxypropranolol
Probably in: Man E917
Rat E917
→ 2-Hydroxy-3-(4-hydroxy-1-naphthyloxy)propylamine
Probably in: Man D343
→ 3-(4-Hydroxy-1-naphthyloxy)lactic acid
Probably in: Man D859
Rat D859
→ 4-Hydroxypropranolol-β-D-glucuronide (side chain)
Probably in: Dog C581
Man D343, D527
Rat C19
→ 4-Hydroxypropranolol-4-β-D-glucuronide
In: Rat C19
See also: D343, D376, F531
→ 4-Hydroxypropranolol-4-sulphate
In: Cat D824
Dog D824, F408, F531
Hamster D824
Man F735, G36, G307
Rat D824
See also: C581, C625, D343, D414, D527

5-Hydroxypropranolol
Formed from: Propranolol
→ 5,6-Dihydroxypropranolol
In: Rat E915

6-Hydroxypropranolol
Formed from: Propranolol
→ 4,6-Dihydroxypropranolol
In: Rat E915
→ 5,6-Dihydroxypropranolol
In: Rat E915

7-Hydroxypropranolol
Formed from: Propranolol
→ 3,7-Dihydroxypropranolol
In: Rat E915
→ 7,8-Dihydroxypropranolol

In: Rat E915

8-Hydroxypropranolol
Formed from: Propranolol
→ 4,8-Dihydroxypropranolol
In: Rat E915
→ 7,8-Dihydroxypropranolol
In: Rat E915

4-Hydroxypropranolol-β-D-glucuronide (side chain)
Probably formed from: 4-Hydroxypropranolol

4-Hydroxypropranolol-4-β-D-glucuronide
Formed from: 4-Hydroxypropranolol

4-Hydroxypropranolol-4-sulphate
Formed from: 4-Hydroxypropranolol
→ 4-Hydroxypropranolol
By arylsulphatase D585

N-(1-Hydroxy-2-propyl)acetanilide
Formed from: N-Isopropylacetanilide

7-Hydroxy-4-propylaminomethylcoumarin
Formed from: 7-Methoxy-4-propylaminomethylcoumarin

1-(1-Hydroxy-2-propylamino)-3-(1-naphthyloxy)-2-propanol
Formed from: Propranolol

N-(2-Hydroxypropyl)amphetamine
Formed from: N-Propylamphetamine

N-(3-Hydroxypropyl)amphetamine
Formed from: Mefenorex
→ *p*-Hydroxy-N-(3-hydroxypropyl) amphetamine
Probably in: Man J544

N-Hydroxy-N-propylamphetamine
Formed from: N-Propylamphetamine
→ Amphetamine
In: Mycobacterium B380

***p*-Hydroxy-N-propylamphetamine**
Formed from: N-Propylamphetamine
→ 4-Hydroxy-3-methoxy-N-propylamphetamine
In: Rat A1998

***p*-(1-Hydroxy-2-propyl)benzoic acid**
Probably formed from: Cumic acid

***p*-(2-Hydroxy-2-propyl)benzoic acid**
Probably formed from: Cumic acid
→ *p*-(Propen-2-yl)benzoic acid
Probably in: Rat C574

1-(2-Hydroxypropyl)-1*H*-benzotriazole
Formed from: 1-Propyl-1*H*-benzotriazole

4-Hydroxy-2-propylbenzotriazole
Formed from: 2-Propylbenzotriazole

5-Hydroxy-2-propylbenzotriazole
Formed from: 2-Propylbenzotriazole

2-Hydroxypropyl 3-carboxy-2-hydroxymethyl-6-methyl-4-(*m*-nitrophenyl)pyridine-5-carboxylate
Probably formed from: 2-Nitrooxypropyl 3-carboxy-2-hydroxymethyl-6-methyl-4-(*m*-nitrophenyl)pyridine-5-carboxylate

3-Hydroxypropyl 3-carboxy-2-hydroxymethyl-6-methyl-4-(*m*-nitrophenyl)pyridine-5-carboxylate
Probably formed from: 2-Nitrooxypropyl 3-carboxy-2-hydroxymethyl-6-methyl-4-(*m*-nitrophenyl)pyridine-5-carboxylate

1-Hydroxy-2-propyl 3-chlorocarbanilate
Formed from: Isopropyl 3-chlorocarbanilate

6-(1-Hydroxypropyl)chromone-2-carboxylic acid
Formed from: 6-*n*-Propylchromone-2-carboxylic acid

6-(2-Hydroxypropyl)chromone-2-carboxylic acid
Formed from: 6-*n*-Propylchromone-2-carboxylic acid

N-(2-Hydroxypropyl)-2,6-dinitro-N-propyl-4-trifluoromethylaniline
Formed from: 2,6-Dinitro-N,N-dipropyl-4-trifluoromethylaniline

N-(3-Hydroxypropyl)-2,6-dinitro-N-propyl-4-trifluoromethylaniline
Formed from: 2,6-Dinitro-N,N-dipropyl-4-trifluoromethylaniline

N-(2-Hydroxypropyl)-2,6-dinitro-4-trifluoromethylaniline
Probably formed from: 2,6-Dinitro-N-propyl-4-trifluoromethylaniline

N-(3-Hydroxypropyl)-2,6-dinitro-4-trifluoromethylaniline
Probably formed from: 2,6-Dinitro-N-propyl-4-trifluoromethylaniline

1-Hydroxy-2-propyl (*S*)-2-ethyl-7-fluoro-3,4-dihydro-5-hydroxy-3-oxo-2*H*-quinoxalinecarboxylate
Probably formed from: Isopropyl (*S*)-2-ethyl-7-fluoro-3,4-dihydro-5-hydroxy-3-oxo-2*H*-quinoxalinecarboxylate

1-Hydroxy-2-propyl (*S*)-2-ethyl-7-fluoro-3,4-dihydro-3-oxo-2*H*-quinoxalinecarboxylate
Probably formed from: Isopropyl (*S*)-2-ethyl-7-fluoro-3,4-dihydro-3-oxo-2*H*-quinoxaline-1-carboxylate
→ 1-Carboxyethyl (*S*)-2-ethyl-7-fluoro-3,4-dihydro-3-oxo-2*H*-quinoxaline-1-carboxylate
Probably in: Cynomolgus monkey J749, K29
Dog J749, K29
Man K29
Mouse K29
Pig J749, K29
Rabbit K29
→ 1-Hydroxy-2-propyl (*S*)-2-ethyl-7-fluoro-3,4-dihydro-3-oxo-2*H*-quinoxaline-1-carboxylate-β-D-glucuronide
Probably in: Cynomolgus monkey J749, K29
Dog J749, K29
Man K29
Pig J749, K29
Rabbit K29

1-Hydroxy-2-propyl (*S*)-2-ethyl-7-fluoro-3,4-dihydro-3-oxo-2*H*-quinoxaline-1-carboxylate-β-D-glucuronide
Probably formed from: 1-Hydroxy-2-propyl (*S*)-2-ethyl-7-fluoro-3,4-dihydro-3-oxo-2*H*-quinoxaline-1-carboxylate

2-Hydroxypropyl 3-hydroxypropyl 1,4-dihydro-2,6-dimethyl-4-(*m*-nitrophenyl)pyridine-3,5-dicarboxylate
Probably formed from: 3-Hydroxypropyl 2-nitrooxypropyl 1,4-dihydro-2,6-dimethyl-4-(*m*-nitrophenyl)pyridine-3,5-dicarboxylate

2-Hydroxypropyl 3-hydroxypropyl 2,6-dimethyl-4-(*m*-nitrophenyl)pyridine-3,5-dicarboxylate
Probably formed from: 3-Hydroxypropyl 2-nitrooxypropyl 2,6-dimethyl-4-(*m*-nitrophenyl)pyridine-3,5-dicarboxylate
→ 2-Carboxyethyl 2-hydroxypropyl 2,6-dimethyl-4-(*m*-nitrophenyl)pyridine-3,5-dicarboxylate
Probably in: Dog G593
Rat G593

2-(1-Hydroxy-2-propyl)-6-isopropylphenyl-β-D-glucuronide
Probably formed from: 2,6-Diisopropylphenyl-β-D-glucuronide

6-(2-Hydroxypropyl)-2-methoxybenzonitrile
Formed from: 2-Methoxy-6-(2-oxopropyl) benzonitrile

4-(1-Hydroxypropyl)-2-methoxyphenol
Formed from: 2-Methoxy-4-propylphenol

3-(1-Hydroxy-2-propyl) 5-methyl 2-carbamoyloxymethyl-4-(2,3-dichlorophenyl)-1,4-dihydro-6-methyl-3,5-pyridinedicarboxylate
Formed from: 3-Isopropyl 5-methyl 2-carbamoyloxymethyl-4-(2,3-dichlorophenyl)-1,4-dihydro-6-methyl-3,5-pyridinedicarboxylate

4-(1-Hydroxy-2-propyl)-3-methyl-1-phenyl-3-pyrazolin-5-one
Probably formed from: 4-Isopropyl-3-methyl-1-phenyl-3-pyrazolin-5-one

N-(1-Hydroxyprop-2-yl)-2-methylsulphonyl-N-phenylacetamide
Formed from: N-Isopropyl-2-methylsulphonylacetanilide
Propachlor
→ N-(1-Hydroxyisopropyl)aniline
Probably in: Rat B93
→ N-(1-Hydroxyprop-2-yl)-2-methylsulphonyl-N-phenylacetamide-β-D-glucuronide
In: Rat B93
Sheep B91
See also: C24, C421

N-(1-Hydroxyprop-2-yl)-2-methylsulphonyl-N-phenylacetamide-β-D-glucuronide
Formed from: N-(1-Hydroxyprop-2-yl)-2-methylsulphonyl-N-phenylacetamide

(+)-9-Hydroxy-4-propylnaphthoxazine
See: Dopazinol

3-Hydroxypropyl 2-nitrooxypropyl 1,4-dihydro-2,6-dimethyl-4-(*m*-nitrophenyl)pyridine-3,5-dicarboxylate
Formed from: 2-Nitrooxypropyl 3-nitrooxypropyl 1,4-dihydro-2,6-dimethyl-4-(*m*-nitrophenyl)pyridine-3,5-dicarboxylate
→ 2-Hydroxypropyl 3-hydroxypropyl 1,4-dihydro-2,6-dimethyl-4-(*m*-nitrophenyl) pyridine-3,5-dicarboxylate
Probably in: Dog G593
Rabbit H163
Rat G593

2-Hydroxypropyl 3-nitrooxypropyl 2,6-dimethyl-4-(*m*-nitrophenyl)pyridine-3,5-dicarboxylate
Formed from: 2-Nitrooxypropyl 3-nitrooxypropyl 2,6-dimethyl-4-(*m*-nitrophenyl)pyridine-3,5-dicarboxylate
→ 2-Hydroxypropyl 3-hydroxypropyl 2,6-dimethyl-4-(*m*-nitrophenyl)pyridine-3,5-dicarboxylate
Probably in: Rabbit H163

3-Hydroxypropyl 2-nitrooxypropyl 2,6-dimethyl-4-(*m*-nitrophenyl)pyridine-3,5-dicarboxylate
Formed from: 2-Nitrooxypropyl 3-nitrooxypropyl 2,6-dimethyl-4-(*m*-nitrophenyl)pyridine-3,5-dicarboxylate
→ 2-Hydroxypropyl 3-hydroxypropyl 2,6-dimethyl-4-(*m*-nitrophenyl)pyridine-3,5-dicarboxylate
Probably in: Dog G593
Rabbit H163
Rat G593

***p*-(1-Hydroxy-*n*-propyl)phenol**
Formed from: *p*-(*n*-Propyl)phenol
→ *p*-Hydroxypropiophenone
Probably in: Penicillium J664, J680, J836

***p*-(2-Hydroxy-2-propyl)phenol**
Formed from: *p*-Isopropylphenol

***p*-Hydroxy-N-propylphentermine**
Formed from: N-Propylphentermine

2-(4-(3-Hydroxypropyl)piperazin-1-yl)ethyl (1-(*p*-chlorobenzoyl)-5-methoxy-2-methylindol-3-yl)acetate
Formed from: Proglumetacin

***p*-(N-(2-Hydroxypropyl)-N-propylsulphamoyl) benzoic acid**
Formed from: *p*-Dipropylsulphamoylbenzoic acid
→ *p*-(N-(2-Hydroxypropyl)-N-propylsulphamoyl)benzoic acid-β-D-glucuronide
Probably in: Rat A1441

***p*-(N-(3-Hydroxypropyl)-N-propylsulphamoyl) benzoic acid**
Formed from: *p*-Dipropylsulphamoylbenzoic acid
→ *p*-(N-(2-Carboxyethyl)-N-propylsulphamoyl)benzoic acid
In: Monkey A1886
Rat A1441, A2081

***p*-(N-(2-Hydroxypropyl)-N-propylsulphamoyl) benzoic acid-β-D-glucuronide**
Probably formed from: *p*-(N-(2-Hydroxypropyl)-N-propylsulphamoyl)benzoic acid

4-(3-Hydroxypropyl)-*o*-quinone
Formed from: 3-(3,4-Dihydroxyphenyl) propanol

5-Hydroxy-2-(N-propyl-N-(2-thienyl)ethylamino) tetralin
→ 2-(N-Propyl-N-(2-thienyl)ethylamino) tetralin-5-yl-β-D-glucuronide
In: Rat F888, F889
→ 5,6-Dihydroxy-2-(N-propyl-N-(2-thienyl) ethylamino)tetralin
In: Rat F888

5-Hydroxy-2-(N-propyl-N-(2-thienyl)ethylamino) tetralin-6-yl-β-D-glucuronide
Formed from: 5,6-Dihydroxy-2-(N-propyl-N-(2-thienyl)ethylamino)tetralin

6-Hydroxy-2-(N-propyl-N-(2-thienyl)ethylamino) tetralin-5-yl-β-D-glucuronide
Formed from: 5,6-Dihydroxy-2-(N-propyl-N-(2-thienyl)ethylamino)tetralin

***p*-(1-Hydroxy-2-propyl)toluene**
Probably formed from: *p*-Cymene
→ *p*-(1,2-Dihydroxy-2-propyl)toluene
Probably in: Guinea pig C574
Rat C574
→ 2-(*p*-Methylphenyl)propionic acid
Probably in: Guinea pig C574
Rat C574

1-(3-Hydroxypropyl)-4-(3-(3,4,5-trimethoxybenzoyl)propyl)piperazine
Formed from: Dilazap
→ 3,4,5-Trimethoxybenzoic acid
In: Rat A646

Hydroxyprothanon
Formed from: Prothanon
→ Hydroxy-9,9-dioxo-10-(2-oxopropyl) phenothiazine
Probably in: Rat A3301

2-Hydroxyprotriptyline
Formed from: Protriptyline
→ N-Acetoxy-2-hydroxyprotriptyline
Probably in: Cunninghamella J726

10-Hydroxyprotriptyline
Formed from: Protriptyline

***p*-Hydroxypseudoephedrine**
Formed from: Pseudoephedrine

***cis*-ω-Hydroxypseudostilboestrol**
Formed from: Stilboestrol

***trans*-ω-Hydroxypseudostilboestrol**
Formed from: Stilboestrol

8-Hydroxypsoralen (xanthotoxol)
Formed from: 8-Methoxypsoralen
→ 8-Methoxypsoralen
In: Parsley D983
Ruta A3572
→ 4,6,7-Trihydroxy-5-coumaranoyl-β-acrylic acid
Probably in: Rat B355

1α-Hydroxypterocarp-1,4-dien-3-one
Formed from: Medicarpin

4-Hydroxypulvinic acid
Probably formed from: Pulvinic acid
→ 5-O-Methylxylerythrin
In: Peniophora A2738
→ Peniophorin
In: Peniophora A2738
→ Peniophoronin
In: Peniophora A2738
→ Xylerythrin
In: Peniophora A2738

1-Hydroxypyrene
Formed from: Pyrene
→ 1,6-Dihydroxypyrene
Probably in: Aspergillus H388
Crinipellis H879
Cunninghamella E181
Marasmius H879
Marasmiellus H879
Pseudomonas J530
Rainbow trout G874
Rat B903
→ 1,8-Dihydroxypyrene
Probably in: Aspergillus H388
Crinipellis H879
Cunninghamella E181
Marasmius H879
Marasmiellus H879
Pseudomonas J530
Rainbow trout G874
→ 1-Methoxypyrene
Probably in: Datura H876
Digitalis H876
Glycine H876
Wheat H876
→ Pyren-1-yl glucoside
Probably in: Porcellio K260
→ Pyren-1-yl-β-D-glucuronide
Probably in: Cunninghamella E181
→ Pyren-1-yl sulphate
In: Aspergillus H388
Crinipellis H879
Marasmius H879
Marasmiellus H879
Porcellio K260
→ 1,4,5-Trihydroxypyrene
Probably in: Rat B903

6-Hydroxypyrene
Formed from: Pyrene
→ Pyren-6-yl sulphate
In: Crinipellis H879
Marasmius H879
Marasmiellus H879

1-Hydroxypyren-8-yl sulphate
Probably formed from: 1,8-Dihydroxypyrene

6-Hydroxypyren-1-yl sulphate
Probably formed from: 1,6-Dihydroxypyrene
→ 1,6-Disulphatoxypyrene
Probably in: Crinipellis H879
Marasmius H879
Marasmiellus H879

5-((4-(2-(3-Hydroxy-2-pyridinylamino)ethoxy)phenyl)methyl)-2,4-thiazolidinedione
Probably formed from: 5-((4-(((3-Hydroxy-2-pyridinyl)methylamino)ethoxy)phenyl)methyl)-2,4-thiazolidinedione
5-((4-(2-(2-Pyridinylamino)ethoxy)phenyl)methyl)-2,4-thiazolidinedione
→ 5-((4-(((3-Hydroxy-2-pyridinyl)amino)ethoxy)phenyl)methyl)-2,4-thiazolidinedione-O-sulphate
Probably in: Man K55

5-((4-(2-(5-Hydroxy-2-pyridinylamino)ethoxy)phenyl)methyl)-2,4-thiazolidinedione
Probably formed from: 5-((4-(2-(5-Carboxy-2-pyridinyl)ethoxy)phenyl)methyl)-2,4-thiazolidinedione
5-((4-(((5-Hydroxy-2-pyridinyl)methylamino)ethoxy)phenyl)methyl)-2,4-thiazolidinedione
5-((4-(2-(2-Pyridinylamino)ethoxy)phenyl)methyl)-2,4-thiazolidinedione
→ 5-((4-(2-(5-Hydroxy-2-pyridinylamino)ethoxy)phenyl)methyl)-2,4-thiazolidinedione-O-sulphate
In: Rat G946
See also: H715

5-((4-(((3-Hydroxy-2-pyridinyl)amino)ethoxy)phenyl)methyl)-2,4-thiazolidinedione-O-sulphate
Probably formed from: 5-((4-(((3-Hydroxy-2-pyridinyl)amino)ethoxy)phenyl)methyl)-2,4-thiazolidinedione

5-((4-(2-(5-Hydroxy-2-pyridinylamino)ethoxy)phenyl)methyl)-2,4-thiazolidinedione-O-sulphate
Probably formed from: 5-((4-(2-(5-Hydroxy-2-pyridinylamino)ethoxy)phenyl)methyl)-2,4-thiazolidinedione

5-((4-(2-(5-Hydroxy-2-pyridinyl)ethoxy)phenyl)methyl)-2,4-thiazolidinedione
Probably formed from: 5-((4-(2-(5-Carboxy-2-pyridinyl)ethoxy)phenyl)methyl)-2,4-thiazolidinedione
→ 5-((4-(2-(5-Hydroxy-2-pyridinyl)ethoxy)phenyl)methyl)-2,4-thiazolidinedione-O-sulphate
In: Rat G946

5-((4-(2-(5-Hydroxy-2-pyridinyl)ethoxy)phenyl)methyl)-2,4-thiazolidinedione-O-sulphate
Formed from: 5-((4-(2-(5-Hydroxy-2-pyridinyl)ethoxy)phenyl)methyl)-2,4-thiazolidinedione

5-Hydroxy-N-(4-pyridinyl)-1-indolamine
Probably formed from: 5-Hydroxybesipirdine
→ 5,6-Dihydroxy-N-(4-pyridinyl)-1-indolamine
Probably in: Cunninghamella J158

5-((4-(((3-Hydroxy-2-pyridinyl)methylamino)ethoxy)phenyl)methyl)-2,4-thiazolidinedione
Formed from: 5-((4-(2-(Methyl-2-pyridinylamino)ethoxy)phenyl)methyl)-2,4-thiazolidinedione
→ 5-((4-(((3-Hydroxy-2-pyridinyl)amino)ethoxy)phenyl)methyl)-2,4-thiazolidinedione
Probably in: Man K55
→ 5-((4-(((3-Hydroxy-2-pyridinyl)methylamino)ethoxy)phenyl)methyl)-2,4-thiazolidinedione-β-D-glucuronide
Probably in: Man K55
Rat H715
→ 5-((4-(((3-Hydroxy-2-pyridinyl)methylamino)ethoxy)phenyl)methyl)-2,4-thiazolidinedione-O-sulphate
Probably in: Dog H715
Man K55
Rat H715

5-((4-(((5-Hydroxy-2-pyridinyl)methylamino)ethoxy)phenyl)methyl)-2,4-thiazolidinedione
Probably formed from: 5-((4-(2-(Methyl-2-pyridinylamino)ethoxy)phenyl)methyl)-2,4-thiazolidinedione
→ 5-((4-(((5-Hydroxy-2-pyridinyl)amino)ethoxy)phenyl)methyl)-2,4-thiazolidinedione
Probably in: Man K55
→ 5-((4-(((5-Hydroxy-2-pyridinyl)methylamino)ethoxy)phenyl)methyl)-2,4-thiazolidinedione-β-D-glucuronide
Probably in: Man K55
Rat H715
→ 5-((4-(((5-Hydroxy-2-pyridinyl)methylamino)ethoxy)phenyl)methyl)-2,4-thiazolidinedione-O-sulphate

Probably in: Dog H715
Man K55

5-((4-(((3-Hydroxy-2-pyridinyl)methylamino)ethoxy)phenyl)methyl)-2,4-thiazolidinedione-β-D-glucuronide
Formed from: 5-((4-(((5-Hydroxy-2-pyridinyl)methylamino)ethoxy)phenyl)methyl)-2,4-thiazolidinedione

5-((4-(((5-Hydroxy-2-pyridinyl)methylamino)ethoxy)phenyl)methyl)-2,4-thiazolidinedione-β-D-glucuronide
Formed from: 5-((4-(((5-Hydroxy-2-pyridinyl)methylamino)ethoxy)phenyl)methyl)-2,4-thiazolidinedione

5-((4-(((3-Hydroxy-2-pyridinyl)methylamino)ethoxy)phenyl)methyl)-2,4-thiazolidinedione-O-sulphate
Formed from: 5-((4-(((3-Hydroxy-2-pyridinyl)methylamino)ethoxy)phenyl)methyl)-2,4-thiazolidinedione

5-((4-(((5-Hydroxy-2-pyridinyl)methylamino)ethoxy)phenyl)methyl)-2,4-thiazolidinedione-O-sulphate
Formed from: 5-((4-(((5-Hydroxy-2-pyridinyl)methylamino)ethoxy)phenyl)methyl)-2,4-thiazolidinedione

Hydroxypyrilamine (position uncertain)
Formed from: Pyrilamine

3-Hydroxypyrilamine
Probably formed from: 3-Hydroxy-O-demethylpyrilamine

4-(5-Hydroxy-2-pyrimidinyl)-1-(3,4-methylenedioxybenzyl)piperazine
Formed from: Piribedil

1-(2-Hydroxypyrrolidine)-N-(1-methyl-5-oxo-2-pyrrolidinylidene)-N′-phenylcarboximidamide
Probably formed from: N-(1-Methyl-5-oxo-2-pyrrolidinylidene)-N′-phenyl-1-pyrrolidine carboximidamide
→ 1-(2,x-Dihydroxypyrrolidine)-N-(1-methyl-5-oxo-2-pyrrolidinylidene)-N′-phenylcarboximidamide
Probably in: Dog J693
Man J693
Monkey J693
Rat J693

N-(2-(2-Hydroxy-1-pyrrolidino)-3-(2-methyl-1-propoxy)propyl)aniline
Probably formed from: N-Benzyl-N-(2-(2-hydroxy-1-pyrrolidino)-3-(2-methyl-1-propoxy)propyl)aniline

(*E*)-4-Hydroxy-2-3-((1-pyrrolidinyl)-1-(*p*-tolyl)propenyl)pyridine
Formed from: Triprolidine

5-Hydroxyquestin
Formed from: Catenarin

4-Hydroxyquinaldic acid
See: Kynurenic acid

4-Hydroxyquinazoline (4(3*H*)-quinazolinone)
Formed from: Quinazoline
→ 2,4(1*H*,3*H*)-Quinazolinedione
Probably in: Streptomyces J648

2′-Hydroxyquinidine
Formed from: Quinidine

3-Hydroxyquinidine
Formed from: Quinidine

5′-Hydroxyquinidine
Formed from: Quinidine

2′-Hydroxyquinine (2′-quinone)
Formed from: Quinine
→ 2′,6′-Dihydroxycinchonidine
Probably in: Rabbit A543
→ 2′,3-Dihydroxyquinine
Probably in: Rabbit A543
→ 10,11-Dihydro-2′,10,11-trihydroxyquinine
Probably in: Man J826

3-Hydroxyquinine
Formed from: Quinine

Hydroxyquinol (1,2,4-trihydroxybenzene)
Formed from: 2,4-Dihydroxybenzoic acid
Gentisic acid
4-Hydroxy-*o*-quinone
4-Hydroxy-*p*-quinone
Methoxyquinol
3,4-Methylenedioxyphenol
m-Nitrophenol
4-Nitroquinol
4-Nitroresorcinol
Phloroglucinol
Protocatechuic acid
Resorcinol
Tetrabromoquinol
Tetrachloroquinol

Tetrafluoroquinol
1,2,3,5-Tetrahydroxybenzene
1,2,4,5-Tetrahydroxybenzene
3′,4′,5,7-Tetrahydroxyflavanone-7-O-glucoside
2,3,5-Trihydroxybenzoic acid
2,3,6-Trihydroxybenzoic acid
→ 1,2-Dihydro-3,4,6-trihydroxybenzene
In: Desulfovibrio K483
→ 2,6-Dihydroxyquinone
In: Eubacterium E507
→ 4-Hydroxy-*cis*,*cis*-muconic semialdehyde
Probably in: Burkholderia J903
→ 4-Hydroxy-*o*-quinone
Probably in: Thauera K446
→ Maleylacetic acid
In: Arthrobacter H270
Aspergillus H236, H570
Azotobacter H602
Bradyrhizobium J484
Phanerochaete H284
Pseudomonas A2697, H386, J705
Sprorotrichum B124
Streptomyces H347
Trichosporon A3826, B34, B386, C84
See also: C475, H270, K445
→ Maleylpyruvic acid
In: Pseudomonas H18
→ Quinol
In: Burkholderia K553
→ Resorcinol
In: Eubacterium E507
Pelobacter F434, G198, G621
→ 1,2,4,5-Tetrahydroxybenzene
In: Pelobacter G621
See also: F434

2-Hydroxyquinoline (carbostyril, 2(1*H*)-quinolone)
Formed from: Quinoline
→ 2,6-Dihydroxyquinoline
Probably in: Nocardia E694
Pseudomonas E694
→ 5,6-Dihydro-2,5,6-trihydroxyquinoline
In: Comamonas K769
→ 2,8-Dihydroxyquinoline
In: Alcaligenes J32
Pseudomonas K778
See also: E694, F454

3-Hydroxyquinoline
Formed from: Quinoline

4-Hydroxyquinoline ((1*H*)-quinolone, γ-quinolone)
Formed from: Carbazole
Indole-3-acetaldehyde
Kynuramine
Quinoline
→ N-Formylanthranilic acid
In: Pseudomonas F880
→ 1*H*-3-Hydroxy-4-oxoquinoline
In: Pseudomonas F880, G812

5-Hydroxyquinoline
→ 2,5-Dihydroxyquinoline
In: Pseudomonas F856

6-Hydroxyquinoline
Formed from: Quinoline
→ 6,6′-Dihydroxy-4,4′-biquinoline
Probably in: Horseradish J592
→ 2,6-Dihydroxyquinoline
In: Pseudomonas F856
Rhodococcus K767
→ Quinolin-6-yl-β-D-glucuronide
In: Man G968

7-Hydroxyquinoline
→ 2,7-Dihydroxyquinoline
In: Pseudomonas F856
Rhodococcus K767

8-Hydroxyquinoline
Formed from: Quinoline
→ 2,8-Dihydroxyquinoline
In: Pseudomonas F856
Rhodococcus K767
→ Quinolin-6-yl glucoside-6-phosphate
Probably in: Mealworm C494
→ Quinolin-8-yl-β-D-glucuronide
In: Man A3206, A3654, G968
→ Quinolin-8-yl sulphate
In: Rat A3206, A3654

4-Hydroxyquinoline-2-carboxylic acid
Formed from: Quinaldic acid

4-Hydroxy-2-quinolone
See: 2,4-Dihydroxyquinoline

8-Hydroxy-2-quinolone
See: 2,8-Dihydroxyquinoline
2-Hydroxy-*p*-quinone
Formed from: 2-Chloro-5-hydroxyquinol
4-Nitrocatechol
→ Hydroxyquinol
In: Phanerochaete G436
4-Hydroxy-*o*-quinone
Formed from: Methoxyquinol
Probably formed from: Hydroxyquinol
→ Hydroxyquinol
In: Burkholderia J705
Phanerochaete G436
N-Hydroxy-3-(*o*-quinon-4-yl) acryloylsuccinamide
Formed from: Caffeoyl-N-hydroxysuccinamide
2-Hydroxyquinoxaline (2(1*H*)-quinoxalinone)
Formed from: Quinoxaline
→ 3,4-Dihydro-2(1*H*)-quinoxalinone
Probably in: Streptomyces J648
→ 2,3-Dihydroxyquinoxaline
Probably in: Rabbit A3695
Rat A3695
→ 1-Methyl-2(1*H*)-quinoxalinone
Probably in: Streptomyces J648
5-Hydroxyquinoxaline
Formed from: Quinoxaline
***m*-Hydroxyrecainam**
Formed from: Recainam
***p*-Hydroxyrecainam**
Formed from: Recainam
4-Hydroxyreticuline
Formed from: Reticuline
8-Hydroxyrhein
Probably formed from: Rhein
→ 2-Carboxy-4-hydroxy-5,8-dioxoanthrequinone
Probably in: Rabbit J143
7α-Hydroxyriboflavine
Formed from: Riboflavine
8α-Hydroxyriboflavine
Formed from: Riboflavine
20-Hydroxyrifabutin
Formed from: Rifabutin
31-Hydroxyrifabutin
Formed from: Rifabutin
32-Hydroxyrifabutin
Formed from: Rifabutin
32-Hydroxyrifalazil
Formed from: Rifalazil
30-Hydroxyrifamycin LM 427
Probably formed from: Rifamycin LM 427
31-Hydroxyrifamycin LM 427
Probably formed from: Rifamycin LM 427
4-Hydroxyriluzole
Formed from: Riluzole
5-Hydroxyriluzole
Formed from: Riluzole
7-Hydroxyriluzole
Formed from: Riluzole
N-Hydroxyriluzole
Formed from: Riluzole
7-Hydroxyrisperidone
Formed from: Risperidone
9-Hydroxyrisperidone
Formed from: Risperidone
6-Hydroxyrizatriptan
Formed from: Rizatriptan
→ 6-Hydroxyrizatriptan sulphate
In: Man J802
6-Hydroxyrizatriptan sulphate
Formed from: 6-Hydroxyrizatriptan
4-Hydroxyrofecoxib
Formed from: Rofecoxib
→ 4-Hydroxyrofecoxib-O-sulphate
In: Dog K139
Rat K139
5-Hydroxyrofecoxib
Formed from: Rofecoxib
→ 5-Hydroxyrofecoxib-β-D-glucuronide
In: Dog K139
Rat K139
5-Hydroxyrofecoxib-β-D-glucuronide
Formed from: 5-Hydroxyrofecoxib
4-Hydroxyrofecoxib-O-sulphate
Formed from: 4-Hydroxyrofecoxib
2′-Hydroxyrolipram
Formed from: Rolipram
3′-Hydroxyrolipram
Formed from: Rolipram
→ 3′-Hydroxyrolipram-O-sulphate
In: Cynomolgus monkey G710
Man G590, G710

Rabbit G710
Rat G590, G710
Rhesus monkey G710

5″-Hydroxyrolipram
Formed from: Rolipram

3′-Hydroxyrolipram-O-sulphate
Formed from: 3′-Hydroxyrolipram

7-Hydroxyropinirole
Formed from: Ropinrole
→ 7-Hydroxyropinirole-β-D-glucuronide
In: Cynomolgus monkey J561
Man J561
Mouse J561
Rat J561

7-Hydroxyropinirole-β-D-glucuronide
Formed from: 7-Hydroxyropinirole

3-Hydroxyropivacaine
Formed from: Ropivacaine

4-Hydroxyropivacaine
Formed from: Ropivacaine

6-Hydroxyroquinimex
Formed from: Roquinimex

7-Hydroxyroquinimex
Formed from: Roquinimex

8-Hydroxyroquinimex
Formed from: Roquinimex

2′-Hydroxyrotenone
Formed from: Rotenone

3-Hydroxyroxatidine acetate
Formed from: Roxatidine acetate

1′-Hydroxysafrole
→ 1′-Hydroxysafrole-β-D-glucuronide
In: Mouse D242
Rat D242
→ 1′-Hydroxysafrole-O-sulphate
In: Mouse E235

1′-Hydroxysafrole-β-D-glucuronide
Formed from: 1′-Hydroxysafrole

1′-Hydroxysafrole-O-sulphate
Formed from: 1′-Hydroxysafrole

5′-Hydroxysameridine
Formed from: Sameridine
→ 1-(4-Carboxybutyl)-N-ethyl-N-methyl-4-phenyl-4-piperidinecarboxamide
Probably in: Rat K43
→ 5′-Oxosameridine
Probably in: Rat K43

6′-Hydroxysameridine
Formed from: Sameridine
→ 1-(5-Carboxypentyl)-N-ethyl-N-methyl-4-phenyl-4-piperidinecarboxamide
Probably in: Rat K43

***m*-Hydroxyscopolamine**
Formed from: Scopolamine

***p*-Hydroxyscopolamine**
Formed from: Scopolamine
→ 3′,4′-Dihydroxyscopolamine
Probably in: Rat G19

4′-Hydroxyscopolammonium
→ 4′-Methoxyscopolammonium
Probably in: Rat A1201

5-Hydroxyscytalone
Formed from: Flaviolin
→ 2,5-Dihydroxy-1,4-naphthoquinone
In: Wangiella D760

Hydroxysemotiadil
Formed from: Semotiadil

4′-Hydroxyseratrodast
Formed from: Seratrodast

5-Hydroxysertindole
Formed from: Sertindole
→ 5-Oxosertindole
Probably in: Dog H471
Rat H471

N-Hydroxysertraline
Formed from: Sertraline
→ N-Hydroxysertraline-β-D-glucuronide
Probably in: Dog F82
Rat F82

N-Hydroxysertraline-β-D-glucuronide
Probably formed from: N-Hydroxysertraline

1′-Hydroxyshikonin
Formed from: Shikonin

6-Hydroxyshikonin
Formed from: Shikonin

7-Hydroxyshikonin
Formed from: Shikonin
→ 6,7-Dihydroxyshikonin
Probably in: Rat K178

4′-Hydroxysirdalud
Formed from: Sirdalud
→ 4′-Oxosirdalud
Probably in: Dog F487
Man F487

Mouse F487
Rabbit F487
Rat F487

7-Hydroxysirdalud
Formed from: Sirdalud
→ 7-Hydroxysirdalud-β-D-glucuronide
In: Dog F487
Man F487
Mouse F487
Rabbit F487
Rat F487
→ 7-Hydroxysirdalud-O-sulphate
In: Dog F487
Man F487
Mouse F487
Rabbit F487
Rat F487

7-Hydroxysirdalud-β-D-glucuronide
Formed from: 7-Hydroxysirdalud

7-Hydroxysirdalud-O-sulphate
Formed from: 7-Hydroxysirdalud

4-Hydroxyskatole
Probably formed from: Skatole

5-Hydroxyskatole
Formed from: Skatole

6-Hydroxyskatole
Formed from: Skatole

6-Hydroxy*spiro*(9*H*-fluorene-3′,9-piperidine)-2′-one
Probably formed from: 9-(3-Aminopropyl)-9*H*-fluorene-9-carboxamide
→ *spiro*(9*H*-Fluorene-3′,9-piperidine)-2′,6′-dione
Probably in: Dog D372
Rat D372

4-Hydroxy-*trans*-stilbene
Formed from: *trans*-Stilbene
→ Benzoic acid
Probably in: Rabbit A1991
Rat A1991
→ 3,4-Dihydroxy-*trans*-stilbene
Probably in: Rabbit A1991
Rat A1991
→ 4,4′-Dihydroxy-*trans*-stilbene
In: Cunninghamella A1218
Rabbit A1991
Rat A1421, A1991
Streptomyces A1218
→ *p*-Hydroxybenzoic acid
Probably in: Rabbit A1991
Rat A1991

1-Hydroxystilboestriol
Formed from: Stilboestriol

3-Hydroxystilboestriol
Formed from: Stilboestriol
→ 3-Methoxystilboestriol
Probably in: Rat A3295, B324

2-Hydroxystrychnine
Formed from: Strychnine

16-Hydroxystrychnine
Formed from: Strychnine

22-Hydroxystrychnine
Formed from: Strychnine

***o*-Hydroxystyrene** (2-vinylphenol)
Formed from: *o*-Coumaric acid

***m*-Hydroxystyrene**
Formed from: *m*-Coumaric acid

***p*-Hydroxystyrene** (*p*-ethenylphenol, 4-vinylphenol)
Formed from: *cis*-*p*-Coumaric acid
p-Coumaric acid
p-Ethylphenol
→ *p*-Ethylphenol
Probably in: Brettanomyces E263
→ *p*-Hydroxymandelic acid
Probably in: Rat A3833
→ *p*-Hydroxystyrene oxide
Probably in: Rat D295

***p*-Hydroxystyrene oxide**
Probably formed from: *p*-Hydroxystyrene
→ *p*-Hydroxyphenylglycol
Probably in: Rat D295

2-Hydroxy-6-succinyl-2,4-cyclohexadiene-1-carboxylic acid
→ *o*-Succinylbenzoic acid
Probably in: Escherichia D780

1-Hydroxysulfurmycin A
Formed from: 1-Hydroxysulfurmycin B
1-Hydroxysulfurmycinone

1-Hydroxysulfurmycin B
Formed from: 1-Hydroxysulfurmycinone
→ 1-Hydroxysulfurmycin A
In: Streptomyces D17

11-Hydroxysulfurmycin A
Formed from: 11-Hydroxysulfurmycinone

11-Hydroxysulfurmycin B
Formed from: 11-Hydroxysulfurmycinone

1-Hydroxysulfurmycinone
→ 7-Deoxy-1-hydroxysulfurmycinone
In: Streptomyces D170
→ 1-Hydroxysulfurmycin A
In: Streptomyces D170
→ 1-Hydroxysulfurmycin B
In: Streptomyces D170

11-Hydroxysulfurmycinone
→ 7-Deoxy-11-hydroxysulfurmycinone
In: Streptomyces D170
→ 11-Hydroxysulfurmycin A
In: Streptomyces D170
→ 11-Hydroxysulfurmycin B
In: Streptomyces D170

1-Hydroxysulofenur
Formed from: Sulofenur
→ 1,2-Dihydroxysulofenur
Probably in: Man F731
Monkey F731
Mouse F731
Rat F731
→ 1,3-Dihydroxysulofenur
Probably in: Man F731
Monkey F731
Mouse F731
Rat F731
→ 1-Oxosulofenur
Probably in: Man F731
Monkey F731
Mouse F731
Rat F731, G898

3-Hydroxysulofenur
Formed from: Sulofenur
→ 3-Oxosulofenur
Probably in: Mouse F731
Rat F731

3-Hydroxysulphadimethoxine
Formed from: Sulphadimethoxine
→ 3-Hydroxysulphadimethoxine sulphate
In: Pig H734

3-Hydroxysulphadimethoxine sulphate
Formed from: 3-Hydroxysulphadimethoxine

5-Hydroxysulphamerazine
Formed from: Sulphadimidine
→ 5-Hydroxysulphamerazine-β-D-glucuronide
Probably in: Horse F928

5-Hydroxysulphamerazine-β-D-glucuronide
Probably formed from: 5-Hydroxysulphamerazine

4-Hydroxysulphamethazine
Formed from: Sulphamethazine
→ 4-Hydroxysulphamethazine-β-D-glucuronide
Probably in: Rhesus monkey H637
→ 4-Hydroxysulphamethiazine sulphate
Probably in: Rhesus monkey H637

5-Hydroxysulphamethazine
Formed from: Sulphamethazine
→ 5-Hydroxysulphamethazine-β-D-glucuronide
Probably in: Beef D869
Horse F928
Rhesus monkey H637
→ 5-Hydroxysulphamethazine sulphate
Probably in: Rhesus monkey H637

4-Hydroxysulphamethazine-β-D-glucuronide
Probably formed from: 4-Hydroxysulphamethazine

5-Hydroxysulphamethazine-β-D-glucuronide
Probably formed from: 5-Hydroxysulphamethazine

4-Hydroxysulphamethazine sulphate
Probably formed from: 4-Hydroxysulphamethazine

5-Hydroxysulphamethazine sulphate
Probably formed from: 5-Hydroxysulphamethazine

N-Hydroxysulphamethoxazole
Formed from: Sulphamethoxazole

6-Hydroxy-2-(4-sulphamoylbenzylamino)-4,5,7-trimethylaminobenzothiazole
→ 6-Hydroxy-2-(4-sulphamoylbenzylamino)-4,5,7-trimethylaminobenzothiazole-β-D-glucuronide
In: Dog F927
Guinea pig F927
Monkey F927
Rat F927

→ 6-Hydroxy-2-(4-sulphamoylbenzylamino)-4,5,7-trimethylaminobenzothiazole sulphate
In: Dog F927
Rat F927

6-Hydroxy-2-(4-sulphamoylbenzylamino)-4,5,7-trimethylaminobenzothiazole-β-D-glucuronide
Formed from: 6-Hydroxy-2-(4-sulphamoylbenzylamino)-4,5,7-trimethylaminobenzothiazole

6-Hydroxy-2-(4-sulphamoylbenzylamino)-4,5,7-trimethylaminobenzothiazole sulphate
Formed from: 6-Hydroxy-2-(4-sulphamoylbenzylamino)-4,5,7-trimethylaminobenzothiazole

4-Hydroxy-2-sulphanilamidopyrimidine
Formed from: Sulphadiazine
→ N^4-Acetyl-4-hydroxy-2-sulphanilamidopyrimidine
Probably in: Rat A3672

5-Hydroxy-2-sulphanilamidopyrimidine
Formed from: Sulphadiazine

4′-Hydroxysulphapyridine
Formed from: Sulphapyridine

5′-Hydroxysulphapyridine
Formed from: Sulphapyridine
→ 5′-Hydroxysulphapyridine-β-D-glucuronide
Probably in: Man A1023

5′-Hydroxysulphapyridine-β-D-glucuronide
Probably formed from: 5′-Hydroxysulphapyridine

5-(3-Hydroxy-4-sulphatoxybenzyl)-2-(5-methyl-4-imidazolylmethylthio)ethylamino-4-pyrimidone
Probably formed from: 5-(3,4-Dihydroxybenzyl)-2-(5-methyl-4-imidazolylmethylthio)ethylamino-4-pyrimidone

5-(4-Hydroxy-3-sulphatoxybenzyl)-2-(5-methyl-4-imidazolylmethylthio)ethylamino-4-pyrimidone
Probably formed from: 5-(3,4-Dihydroxybenzyl)-2-(5-methyl-4-imidazolylmethylthio)ethylamino-4-pyrimidone

4-Hydroxysulphinpyrazole
Formed from: Sulphinpyrazole

***p*-Hydroxysulphinpyrazole**
Formed from: Sulphinpyrazone
→ 1-(*p*-Hydroxyphenyl)-3,5-dioxo-2-phenyl-4-(2′-phenylsulphonylethyl)pyrazolidine
Probably in: Pig B711
Rat B711
Rhesus monkey B711
→ 1-(*p*-Hydroxyphenyl)-3,5-dioxo-2-phenyl-4-(2′-phenylthioethyl)pyrazolidine
Probably in: Guinea pig B711
Pig B711
Rabbit B711
Rat B711
Rhesus monkey B711

2-Hydroxy-5-sulphobenzoic acid
See: 5-Sulphosalicylic acid

2-(4-((1-Hydroxy-4-sulpho-2-naphthalenyl)azo)phenyl)-6-methyl-7-benzothiazolesulphonic acid
See: Direct Red 46

2-Hydroxy-1-(4-sulpho-1-naphthylazo)naphthalene-3,6-disulphonic acid (amaranth)
→ 1-Amino-2-hydroxynaphthalene-3,6-disulphonic acid
In: Proteus A317
Rabbit C884
Rat A1701, D601, G133
Shigella F152, G767
→ 1-Naphthylamine-4-sulphonic acid
In: Proteus A317
Rabbit C884
Rat A1701, D601, G133
Shigella F152, G767

2-Hydroxy-5-((4-sulphophenyl)azo)benzoic acid
See: Mordant Yellow 10

2-Hydroxy-1-(4-sulphophenylazo)naphthalene-6-sulphonic acid (sunset yellow)
→ 1-Amino-2-hydroxynaphthalene-6-sulphonic acid
In: Proteus A317
Rat A1701
→ Sunset Yellow-O-sulphate
In: Rat C526
→ Sulphanilic acid
In: Proteus A317
Rat A1701

7-Hydroxysulphoridazine
Formed from: Sulphoridazine

→ 7-Hydroxyphenothiazine-2-sulphonic acid
Probably in: Man D366

1-Hydroxy-2-(5-sulpho-2,4-xylylazo) naphthalene-4-sulphonic acid (ponceau SX)
→ 2-Amino-1-hydroxynaphthalene-4-sulphonic acid
In: Proteus A317
Shigella F152, G767
→ 2,4-Dimethylaniline-5-sulphonic acid
In: Proteus A317
Shigella F152, G767

1-Hydroxytacrine
Formed from: Tacrine

2-Hydroxytacrine
Formed from: Tacrine

3-Hydroxytacrine
Formed from: Tacrine

4-Hydroxytacrine
Formed from: Tacrine

7-Hydroxytacrine
Formed from: Tacrine

3-Hydroxytamoxifen
See: Droloxifene

4-Hydroxytamoxifen
See: *trans*-1-(*p*-Dimethylaminoethoxyphenyl)-1-(*p*-hydroxyphenyl)-2-phenylbut-1-ene

α-Hydroxytamoxifen
Formed from: Tamoxifen
→ α-Hydroxytamoxifen-β-D-glucuronide
In: Man J900
Mouse J900
Rat J900
→ α-Hydroxytamoxifen-N-oxide
Probably in: Rat H572

α-Hydroxytamoxifen-β-D-glucuronide
Formed from: α-Hydroxytamoxifen

α-Hydroxytamoxifen-N-oxide
Formed from: α-Hydroxytamoxifen

6-Hydroxytargetrin
Formed from: Targetrin
→ 6,7-Dihydroxytargetrin
Probably in: Man J290
→ 6-Oxotargetrin
Probably in: Man J290

7-Hydroxytargetrin
Formed from: Targetrin
→ 7-Oxotargetrin
Probably in: Man J290
→ 4-(1-(5,6,7,8-Tetrahydro-7-hydroxy-3,5,5,8,8-pentamethyl-2-naphthalenyl)-1-ethenyl)benzoyltaurine
Probably in: Man J290

2-(*m*-Hydroxy)taxol
Formed from: Taxol

3′-(*p*-Hydroxy)taxol
Formed from: Taxol

6α-Hydroxytaxol
Formed from: Taxol
→ (6α,3′-*p*-Dihydroxy)taxol
Probably in: Man H504, H654

***m*-Hydroxytazadolene**
Formed from: Tazadolene
→ *m*-Hydroxytazadolene-β-D-glucuronide
In: Dog G375

***p*-Hydroxytazadolene**
Formed from: Tazadolene
→ N-(2-((4-Hydroxy-3-methoxyphenyl) methylene)cyclohexyl)cyclopropylamine
Probably in: Dog G375
→ *p*-Hydroxytazadolene-β-D-glucuronide
Probably in: Dog G375

***m*-Hydroxytazadolene-β-D-glucuronide**
Formed from: *m*-Hydroxytazadolene

***p*-Hydroxytazadolene-β-D-glucuronide**
Formed from: *p*-Hydroxytazadolene

***p*-Hydroxytemazepam-β-D-glucuronide**
Probably formed from: 7-Chloro-1,3-dihydro-3-hydroxy-5-(*p*-hydroxyphenyl)-1-methyl-2*H*-1,4-benzodiazepin-2-one

***p*-Hydroxytemazepam**
See: 7-Chloro-1,3-dihydro-3-hydroxy-5-(*p*-hydroxyphenyl)-1-methyl-2*H*-1,4-benzodiazepin-2-one

3′-Hydroxytenidap
Formed from: Tenidap
→ 3′,4′-Dihydroxytenidap
Probably in: Rat J92

4-Hydroxytenidap
Formed from: Tenidap

5′-Hydroxytenidap
Formed from: Tenidap
→ 5′-Hydroxytenidap-β-D-glucuronide
Probably in: Rat J92

6-Hydroxytenidap
Formed from: Tenidap
5′-Hydroxytenidap-β-D-glucuronide
Probably formed from: 5′-Hydroxytenidap
4(methyl)-Hydroxyterritrem A
Formed from: Territrem A
4(methyl)-Hydroxyterritrem B
Formed from: Territrem B
4-Hydroxytertatolol
Formed from: Tertatolol
16-Hydroxytestosterone
→ Oestriol
In: Horse G395
3-Hydroxy-1-tetralone
Formed from: 1,3-Dihydroxynaphthalene
4-Hydroxy-1-tetralone
Formed from: α-Naphthol
1,4-Naphthoquinone
6-Hydroxy-1-tetralone
→ 6-Hydroxy-1-tetralone-β-D-4-O-methylglucuronide
In: Sporotrichum A2630
6-Hydroxy-1-tetralone-β-D-4-O-methylglucuronide
Formed from: 6-Hydroxy-1-tetralone
4′-Hydroxy-5,6,7,8-tetramethoxyflavone
Formed from: Tangeretin
→ 3′,4′-Dihydroxy-5,6,7,8-tetramethoxyflavone
Probably in: Rat K582
→ 4′,6-Dihydroxy-5,7,8-trimethoxyflavone
Probably in: Rat K582
→ 4′,7-Dihydroxy-5,6,8-trimethoxyflavone
Probably in: Rat K582
5-Hydroxy-3,3′,4′,7-tetramethoxyflavone
Formed from: 3′,4′,5,7-Tetrahydroxy-3-methoxyflavone
3,3′,5-Trihydroxy-4′,7-dimethoxyflavone
6-Hydroxy-4′,5,7,8-tetramethoxyflavone
Formed from: Tangeretin
→ 5,6-Dihydroxy-4′,7,8-trimethoxyflavone
Probably in: Rat K582
5-Hydroxy-3,4′,5′,7-tetramethoxyflavone-2′-O-glucoside
Formed from: 5,5′-Dihydroxy-3,4′,7-trimethoxyflavone-2′-O-glucoside
4-(3-Hydroxy-2,2,4,4-tetramethylbut-4-yl)phenol
Formed from: 4-*tert*-Octylphenol
→ 4-(3-Hydroxy-2,2,4,4-tetramethylbut-4-yl)phenol-β-D-glucuronide
Probably in: Rainbow trout K126
4-(3-Hydroxy-2,2,4,4-tetramethylbut-4-yl)phenol-β-D-glucuronide
Formed from: 4-(3-Hydroxy-2,2,4,4-tetramethylbut-4-yl)phenol
5-Hydroxy-(*S*)-thalidomide
Formed from: (*S*)-Thalidomide
5′-Hydroxy-(*R*)-thalidomide
Formed from: (*R*)-Thalidomide
4-Hydroxythiabendazole
Formed from: Thiabendazole
5-Hydroxythiabendazole
Formed from: Thiabendazole
***trans*-4-(2-(3-Hydroxy)thianaphthenyl)-2-oxo-3-butenoic acid**
Formed from: 1,2-Dihydroxy-dibenzothiophene
→ 2-Formyl-3-hydroxythianaphthene
In: Beijerinckia A3340
1-Hydroxythianthrene
Probably formed from: Thianthrene
2-Hydroxythianthrene
Probably formed from: Thianthrene
***p*-Hydroxythidiazuron**
Formed from: Thidiazuron
3-Hydroxy-2-(4-(2-thienylcarbonyl)phenyl)propionic acid
Formed from: Suprofen
→ 2-(4-(2-Thienylcarbonyl)phenyl)malonic acid
Probably in: Man H229
2-(4-(5-Hydroxy-2-thienylcarbonyl)phenyl)propionic acid
Formed from: Suprofen
1-Hydroxy-2-thioethyl-2-phenylethane
Probably formed from: S-(2-Hydroxy-1-phenylethyl)-L-cysteine

2-(3′-Hydroxythionaphthenyl[2]methylene)-3-oxodihydrothionaphthene
Probably formed from: 2-Formyl-3-hydroxy-benzo[*b*]thiophene

N-Hydroxy-*p*-(S-thiophenyl)acetanilide
→ *p*-(S-Thiophenyl)phenylhydroxylamine
In: Hamster F522

3-Hydroxythioridazine
Formed from: Thioridazine

7-Hydroxythioridazine
Formed from: Thioridazine
→ 7-Hydroxymesoridazine
Probably in: Man D366
→ 7-Hydroxythioridazine-5-sulphoxide
Probably in: Man D366

7-Hydroxythioridazine-5-sulphoxide
Probably formed from: 7-Hydroxythioridazine

9-Hydroxythioxanthene
Formed from: Thioxanthene
→ Thioxanthene
Probably in: Pseudomonas H672

9-Hydroxythioxanthene sulphone
Formed from: Thioxanthene sulphone
→ Thioxanthene sulphone
Probably in: Pseudomonas H672

Hydroxytiazuril
Formed from: Tiazuril

Hydroxytiazuril sulphone
Probably formed from: Tiazuril sulphone

Hydroxytiazuril sulphoxide
Probably formed from: Tiazuril sulphoxide

5-Hydroxytienilic acid
Formed from: Tienilic acid

3′-Hydroxytiospirone
Formed from: Tiospirone
→ 3′-Hydroxytiospirone sulphone
Probably in: Man F732

6′-Hydroxytiospirone
Formed from: Tiospirone
→ 6′-Hydroxytiospirone sulphone
Probably in: Man F732

3′-Hydroxytiospirone sulphone
Probably formed from: 3′-Hydroxytiospirone

6′-Hydroxytiospirone sulphone
Probably formed from: 6′-Hydroxytiospirone

Hydroxytoloxatone
Formed from: Toloxatone
→ Hydroxytoloxatone-β-D-glucuronide
Probably in: Man A3709

Hydroxytoloxatone-β-D-glucuronide
Probably formed from: Hydroxytoloxatone

2-Hydroxy-5-(*p*-toluenesulphonyloxy) benzenesulphonic acid (sultosilic acid)
→ 2-Hydroxy-5-(*p*-hydroxymethylbenzenesulphonyloxy) benzenesulphonic acid
In: Man B975
Ray B975

N-Hydroxy-*o*-toluidine
Formed from: *o*-Nitrosotoluene
o-Toluidine

***p*-Hydroxytranylcypromine**
Formed from: Tranylcypromine

***p*-Hydroxytrazodone**
Formed from: Trazodone
→ *p*-Hydroxytrazodone-β-D-glucuronide
In: Rabbit A1438
Rat A285

***p*-Hydroxytrazodone-β-D-glucuronide**
Formed from: *p*-Hydroxytrazodone

***p*-Hydroxytriamterene**
Formed from: *p*-Hydroxytriamterene sulphate
Triamterene
→ *p*-Hydroxytriamterene sulphate
In: Man C528
See also: A1871

***p*-Hydroxytriamterene sulphate**
Formed from: *p*-Hydroxytriamterene
→ *p*-Hydroxytriamterene
In: Rat F351

4-Hydroxytriazolam
Formed from: Triazolam

α-Hydroxytriazolam
Formed from: Triazolam

8-Hydroxy-*s*-triazolo[3,4-*a*]phthalazine
Probably formed from: *s*-Triazolo[3,4-*a*] phthalazine

4′-Hydroxytriflubazam
Formed from: Triflubazam

7-Hydroxytrifluoperazine
Formed from: 10-(3-(4-Methylpiperazin-1-yl) propyl)trifluoromethylphenothiazine
→ 7-Hydroxytrifluoperazine-β-D-glucuronide
In: Rat A2720

7-Hydroxytrifluoperazine-β-D-glucuronide
Formed from: 7-Hydroxytrifluoperazine

3′-Hydroxy-2-trifluoromethylbenzanilide
→ 3′,4′-Dihydroxy-2-trifluoromethylbenzanilide
Probably in: Cucumber D191
Rat D190
Rice D191
→ 3′-Methoxy-2-trifluoromethylbenzanilide
Probably in: Cucumber D191
Rice D191

7-Hydroxy-4-trifluoromethylcoumarin
Formed from: 7-Benzyloxy-4-trifluoromethylcoumarin
7-Ethoxy-4-trifluoromethylcoumarin

5-Hydroxy-*p*-trifluoromethylvalerophenone (*E*)-O-2-(aminoethyl)oxime
Formed from: Fluvoxamine
→ 5-(2-Aminoethyl)oximino-5-(*p*-trifluoromethylphenyl)valeric acid
Probably in: Man D24
→ 5-Hydroxy-*p*-trifluoromethylvalerophenone (*E*)-O-2-(aminoethyl)oxime-β-D-glucuronide
In: Dog C652
Hamster C652
Mouse C652
Rat C652
→ 5-Hydroxy-*p*-trifluoromethylvalerophenone (*E*)-O-2-(carboxymethyl)oxime
Probably in: Dog C652
Hamster C652
Mouse C652
Rat C652

5-Hydroxy-*p*-trifluoromethylvalerophenone (*E*)-O-2-(aminoethyl)oxime-β-D-glucuronide
Formed from: 5-Hydroxy-*p*-trifluoromethylvalerophenone (*E*)-O-2-(aminoethyl)oxime

5-Hydroxy-*p*-trifluoromethylvalerophenone (*E*)-O-2-(carboxymethyl)oxime
Probably formed from: 5-Hydroxy-*p*-trifluoromethylvalerophenone (*E*)-O-2-(aminoethyl) oxime
→ 5-(Carboxymethyl)oximino-5-(*p*-trifluoromethylphenyl)valeric acid
Probably in: Dog C652
Hamster C652
Mouse C652
Rat C652

2-Hydroxy-3-(2,4,6-trihydroxyphenyl)-1-(3,4,5-trihydroxyphenyl)-1-propanone
Formed from: (−)-Epigallocatechin

α-Hydroxytrimethoprim
Formed from: Trimethoprim

2-Hydroxy-N-(3,4,5-trimethoxybenzoyl) morpholine
Formed from: Trimetozine
→ 2-Hydroxy-N-(3,4,5-trimethoxybenzoyl) morpholine-β-D-glucuronide
In: Dog A3302

2-Hydroxy-N-(3,4,5-trimethoxybenzoyl) morpholine-β-D-glucuronide
Formed from: 2-Hydroxy-N-(3,4,5-trimethoxybenzoyl)morpholine

7-Hydroxy-2,4,6-trimethoxyphenanthrene (isobatatasin I)
Probably formed from: Resveratrol

2-Hydroxy-1-(2,4,5-trimethylphenylazo) naphthalene-3,6-disulphonic acid (ponceau 3R)
→ 1-Amino-2-hydroxynaphthalene-3,6-disulphonic acid
In: Proteus A317
→ 2,4,5-Trimethylaniline
In: Proteus A317

8-(4-Hydroxy-2,3,6-trimethylphenyl)-2,6-dimethyl-*trans*,*trans*-octa-5,7-dienoic acid
Probably formed from: 8-(4-Methoxy-2,3,6-trimethylphenyl)-2,6-dimethyl-*trans*,*trans*-octa-5,7-dienoic acid

2-Hydroxytrimipramine
Formed from: Trimipramine
→ 2,3-Dihydroxytrimipramine
Probably in: Rat G269
→ 2-Hydroxynortrimipramine
Probably in: Rat G269
→ 2-Hydroxy-10-oxotrimipramine
Probably in: Rat G269

10-Hydroxytrimipramine
Formed from: Trimipramine
→ 2,10-Dihydroxytrimipramine
Probably in: Man G991
Rat G269

→ 10-Hydroxybisdemethyltrimipramine
Probably in: Rat G269
→ 10-Oxotrimipramine
In: Rat G269

3-Hydroxytripamide
Formed from: Tripamide

8-Hydroxytripamide
Formed from: Tripamide

9-Hydroxytripamide
Formed from: Tripamide

***p*-Hydroxytripelennamine**
See: O-Demethylpyrilamine

α-Hydroxytripelennamine
Formed from: Tripelennamine
→ 2-(α-Hydroxybenzyl)aminopyridine
Probably in: Rat E907

1-Hydroxytriphenylene
Formed from: Triphenylene

2-Hydroxytriphenylene
Formed from: Triphenylene

5-Hydroxytropisetron
Formed from: Tropisetron
→ 5-Hydroxytropisetron-β-D-glucuronide
In: Man G388
→ 5-Hydroxytropisetron-O-sulphate
In: Man G388

6-Hydroxytropisetron
→ 6-Hydroxytropisetron-O-sulphate
In: Man G388

7-Hydroxytropisetron
→ 7-Hydroxytropisetron-O-sulphate
In: Man G388

5-Hydroxytropisetron-β-D-glucuronide
Formed from: 5-Hydroxytropisetron

5-Hydroxytropisetron-O-sulphate
Formed from: 5-Hydroxytropisetron

6-Hydroxytropisetron-O-sulphate
Formed from: 6-Hydroxytropisetron

7-Hydroxytropisetron-O-sulphate
Formed from: 7-Hydroxytropisetron

5-Hydroxytryptamine (serotonin)
Formed from: N-Acetyl-5-hydroxytryptamine
5-Hydroxy-L-tryptophan
Tryptamine
→ N-Acetyl-5-hydroxytryptamine
In: Ascaridia G997
Cockroach H835
Drosophila A643, E569, H758
Hen A1626
Man K490
Nematode H40
Periplaneta F514, G103, J632, J633
Quail H685
Rabbit G324
Rat A1350, D282
Rhizoglyphus F669
Sheep D282, J850
See also: G393, K550
→ β,5-Dihydroxytryptamine
In: Pseudomonas A3007
→ 10-Hydroxyajmalicine
Probably in: Catharanthus A2741
→ 5-Hydroxyindole-3-acetaldehyde
In: Amblyomma J4
Beef A1285
Guinea pig A3958
Hen A1626
Man A2967, A3957, E55
Mouse A946, A1895
Octopus J88
Pig A2587, B206
Pseudomonas C517
Rabbit A94
Rat A1053, A1227, A1285, A1324, A1348, A1355, A1613, A2575, A3863, A3948, B653, E339
Rhizoglyphus F669
Sheep A535
Tetrahymena A3131
See also: A658, A1533, A1909, A2390, A3079, A3875, D105, F158
→ 5-Hydroxyindole-3-glyoxal
Probably in: Pseudomonas A3009
→ 5-Hydroxy-1-methyltryptamine
In: Rabbit C333
→ 5-Hydroxy-N-methyltryptamine
In: Animal A1310
Rabbit A111, A1311, A1320, B741, C156, J884
Rat A80, A111, A1311
→ 10-Methoxyharmalan
In: Rat A908
→ 5-Methoxytryptamine

In: Beef C522
Chicken C522
→ Tetrahydro-6-hydroxy-β-carboline
In: Rabbit A111
Rat A111
→ 1,2,3,4-Tetrahydro-6-hydroxy-1-methyl-β-carboline
In: Rat A908
→ Tryptamin-5-yl-β-D-glucuronide
In: Man K108
→ Tryptamin-5-yl sulphate
In: Dog J292
Eubacterium G602
Guinea pig E695
Man A3814, C322, C523, D583, D586
Rat B61, E695
See also: A604

6-Hydroxytryptamine
→ 6-Hydroxy-N-methyltryptamine
In: Phalaris A3891

β-Hydroxytryptamine
Formed from: Tryptamine
Probably formed from: β-Hydroxy-L-tryptophan

5-Hydroxy-L-tryptophan (5HTP)
Formed from: 5-Hydroxyindole
5-(5-Hydroxyindolylmethyl)hydantoin
L-Tryptophan
→ β,5-Dihydroxy-L-tryptophan
Probably in: Pseudomonas A3007
→ N-Formyl-5-hydroxy-L-kynurenine
In: Wheat E24
See also: A3875
→ 5-Hydroxyindole-3-glyoxal
In: Pseudomonas A3009
→ 5-Hydroxyindole-3-acetaldehyde
In: Pig B780
→ 5-Hydroxyindole-3-pyruvic acid
In: Hansenula D978
Monkey A3570, K717
Rat A2339, A2342, A2847, B379, K717
See also: A923, A2894
→ 5-Hydroxytryptamine
In: Drosophila E569
Helix A1051, A1386
Hymenolepis C524
Man C566, D793, G393, H666, K501
Micrococcus C62
Mouse A946
Nematode A2572
Periplaneta E448
Pig A1906
Rabbit A3875
Rat A604, A1533, A2378, A2390, C566, E735
Tomato A536
Tuna H683
See also: D105, E269, E967

β-Hydroxy-D-tryptophan
Probably formed from: D-Tryptophan

β-Hydroxy-L-tryptophan
Probably formed from: L-Tryptophan
→ β-Hydroxytryptamine
Probably in: Pseudomonas A3007

β-Hydroxy-L-tryptophanamide
Probably formed from: L-Tryptophanamide

β-Hydroxy-L-tryptophan methyl ester
Probably formed from: L-Tryptophan methyl ester

***erythro*-β-Hydroxy-L-tryptophanyl-L-leucine**
Formed from: L-Tryptophanyl-L-leucine

***threo*-β-Hydroxy-L-tryptophanyl-L-leucine**
Formed from: L-Tryptophanyl-L-leucine
→ β-Oxo-L-tryptophanyl-L-leucine
Probably in: Pseudomonas B917

β-Hydroxy-L-tryptophanyl-L-phenylalanine
Probably formed from: L-Tryptophanyl-L-phenylalanine

β-Hydroxy-L-tryptophanyl-L-tryptophan
Probably formed from: L-Tryptophanyl-L-tryptophan

5-Hydroxytryptophol (5-hydroxy-3-(2-hydroxyethyl)indole)
Formed from: 5-Hydroxyindole-3-acetaldehyde
5-Hydroxyindoleacetic acid
→ 5-Hydroxytryptophol-O-sulphate
In: Rat B941
→ 5-Methoxytryptophol
In: Rabbit H84

β-Hydroxytryptophol
Formed from: Indole-3-glycolaldehyde
Tryptophol

5-Hydroxytryptophol-O-sulphate
Formed from: 5-Hydroxytryptophol
3-Hydroxytulobuterol
Formed from: Tulobuterol
4-Hydroxytulobuterol
Formed from: Tulobuterol
→ 3,4-Dihydroxytulobuterol
Probably in: Dog C43
Guinea pig C43
Rabbit C43
Rat C43
→ 4,5-Dihydroxytulobuterol
Probably in: Dog C43
Rabbit C43
Rat C43
5-Hydroxytulobuterol
Formed from: Tulobuterol
N-Hydroxytyramine
→ *p*-Hydroxyphenylacetoxime
In: Man J336
Pig J336
→ Tyramine
In: Man J336
Pig J336
N-Hydroxy-L-tyrosine
Formed from: L-Tyrosine
→ *p*-Hydroxymandelonitrile
In: Sorghum A3335
See also: G228, J504
***cis*-3-Hydroxyvelnacrine**
Formed from: Velnacrine
***cis*-4-Hydroxyvelnacrine**
Formed from: Velnacrine
***trans*-4-Hydroxyvelnacrine**
Formed from: Velnacrine
7-Hydroxyvelnacrine
Formed from: Velnacrine
6″-Hydroxyverlukast
Formed from: Verlukast
4′-Hydroxyviloxazine
Formed from: Viloxazine
→ 4′-Hydroxyviloxazine-β-D-glucuronide
In: Man A354
5′-Hydroxyviloxazine
Formed from: Viloxazine
→ 5′-Hydroxyviloxazine-β-D-glucuronide
In: Man A354
4′-Hydroxyviloxazine-β-D-glucuronide
Formed from: 4′-Hydroxyviloxazine
5′-Hydroxyviloxazine-β-D-glucuronide
Formed from: 5′-Hydroxyviloxazine
Hydroxyvinamidine
Formed from: 3′,4′-Anhydrovinblastine
10′-Hydroxyvinblastine
Formed from: Vinblastine
6α-Hydroxyvincamine
Formed from: Vincamine
→ 6-Oxovincamine
Probably in: Rat B4, B773
6β-Hydroxyvincamine
Formed from: Vincamine
10′-Hydroxyvinorelbine
Formed from: Vinorelbine
11′-Hydroxyvinorelbine
Formed from: Vinorelbine
19′-Hydroxyvinorelbine
Formed from: Vinorelbine
11-Hydroxyvittatine
Formed from: O-Methylnorbelladine
→ Narciclasine
In: Pancratium A1322
Hydroxyvolazocine
Formed from: Volazocine
→ N-Decyclopropylmethyl-hydroxyvolazocine
Probably in: Rat A2134
→ Dihydroxyvolazocine
Probably in: Rat A2134
→ Hydroxyvolazocine-β-D-glucuronide
In: Rat A2134
Hydroxyvolazocine-β-D-glucuronide
Formed from: Hydroxyvolazocine
3′-Hydroxywarfarin
Formed from: Warfarin
4′-Hydroxywarfarin
Formed from: Warfarin
→ 3′,4′-Dihydroxywarfarin
Probably in: Beauveria J82
Streptomyces J82
→ 4′-Hydroxy-11-methoxywarfarin
Probably in: Streptomyces J82
6-Hydroxywarfarin
Formed from: Warfarin

7-Hydroxywarfarin
Formed from: Warfarin
8-Hydroxywarfarin
Formed from: Warfarin
10-Hydroxywarfarin (α-hydroxywarfarin)
Formed from: Warfarin
12-Hydroxywarfarin
Formed from: Warfarin
4′-Hydroxy-*R*-warfarin
Formed from: *R*-Warfarin
6-Hydroxy-*R*-warfarin
Formed from: *R*-Warfarin
7-Hydroxy-*R*-warfarin
Formed from: *R*-Warfarin
8-Hydroxy-*R*-warfarin
Formed from: *R*-Warfarin
10-Hydroxy-*R*-warfarin
Formed from: *R*-Warfarin
4′-Hydroxy-*S*-warfarin
Formed from: *S*-Warfarin
6-Hydroxy-*S*-warfarin
Formed from: *S*-Warfarin
7-Hydroxy-*S*-warfarin
Formed from: *S*-Warfarin
8-Hydroxy-*S*-warfarin
Formed from: *S*-Warfarin
10-Hydroxy-*S*-warfarin
Formed from: *S*-Warfarin
11-Hydroxy-*S*-warfarin
Formed from: *S*-Warfarin
α-Hydroxywarfarin
See: 10-Hydroxywarfarin
Hydroxyxanthanoic acid
Probably formed from: Xanthanoic acid
2-Hydroxyxanthone
Formed from: Xanthone
3-Hydroxyxanthone
Formed from: Xanthone
4-Hydroxyxanthone
Formed from: Xanthone
3-Hydroxyxanthone-1-acetic acid
Formed from: Xanthone-1-acetic acid
5-Hydroxyxanthotoxin
→ Isopimpinellin
In: Parsley D983
Ruta A2582, A3572
5′-Hydroxyxilobam
Formed from: Xilobam
→ 5′-Oxoxilobam
In: Man G938
3′-Hydroxyxylazine
Formed from: Xylazine
4′-Hydroxyxylazine
Formed from: Xylazine
10-Hydroxyohimbine
Formed from: Yohimbine
11-Hydroxyohimbine
Formed from: Yohimbine
Hydroxyzamifenacin
Formed from: Zamifenacin
3-Hydroxyzatosetron
Formed from: Zatosetron
3′-Hydroxyzearalenone
Formed from: Zearalenone
8′-Hydroxyzearalenone
Formed from: Zearalenone
→ 6′,8′-Dihydroxyzearalenone
Probably in: Fusarium A1817
Hydroxyzenarestat
Formed from: Zenarestat
→ Debromohydroxyzenarestat
Probably in: Rat G58
Hydroxyzileuton
Formed from: Zileuton
Hydroxyzileuton
Formed from: Zileuton
Hydroxyzine (2-(2-(4-((4-chlorophenyl) phenylmethyl)-1-piperazinyl)ethoxy) ethanol)
→ *p*-Chlorobenzophenone
In: Rat A2197
6-Hydroxyzolazepam
Formed from: Zolazepam
→ 6-Hydroxy-8-demethylzolazepam
Probably in: Rat A1572
2-Hydroxyzotepine
Formed from: Zotepine
3-Hydroxyzotepine
Formed from: Zotepine
Hyoscine
See: Scopolamine

Hyoscyamine (atropine, (a*S*)-α-(hydroxymethyl) benzeneacetic acid (3-*endo*)-8-methyl-8-azabicyclo[3.2.1]oct-3-yl ester)
Formed from: Phenyllactic acid
Tropic acid
→ Apoatropine
In: Guinea pig G957
→ 6β-Hydroxyhyoscyamine
In: Atropa K885
Datura K885
Duboisia K885
Hyoscyamus E350, G135, K885
→ Hyoscyamine-N-oxide
In: Atropa A1727
Guinea pig A2816
→ Norhyoscyamine
In: Guinea pig A2816
→ Scopolamine
Probably in: Datura A1638
→ (*S*)-Tropic acid
In: Pseudomonas G752
Rabbit H563

Hyoscyamine-N-oxide
Formed from: Hyoscyamine

Hypaphorine (lenticin, N,N,N-trimethyl-L-tryptophan)
Formed from: L-Tryptophan
→ Indole-3-acrylic acid
In: Lens A232
See also: A234

Hypoxoside ((1*E*)-1-penten-4-yne-1,5-diylbis(2-hydroxy-4,1-phenylene)bis-β-D-glucuronide)
Formed from: Caffeic acid

I

IAA
See: Indole-3-acetic acid
Ibogaine (12-methoxyibogamine)
→ 12-Hydroxyibogaine
In: Man H669, J435
Ibopamine (epinine diisobutyryl ester)
→ Epinine monoisobutyrate
In: Rat D990
See also: D970, E940, F220
Ibuprofen
See: 2-(4-Isobutylphenyl)propionic acid
Ibuprofenac
See: *p*-Isobutylphenylacetic acid
(*R*)-Ibuprofen adenylate
→ (*S*)-2-(4-Isobutylphenyl)propionic acid
In: Rat H289
Ibuprofen amide (α-methyl-4-(2-methylpropyl) benzeneacetamide)
→ (*S*)-2-(4-Isobutylphenyl)propionic acid
In: Rhodococcus J885
Ibuterol (1-(3,5-diisobutyroxyphenyl)-2-(*tert*-butylamino)ethanol)
→ Terbutaline
In: Guinea pig A2494
Icariin (3-(((6-deoxy-α-L-mannopyranosyl)oxy)-7-(β-D-glucopyranosyloxy)-5-hydroxy-2-(4-methoxyphenyl)-8-(3-methyl-2-butenyl)-4*H*-1-benzopyran-4-one)
→ Icariside I
In: Rat K226
→ Icariside II
In: Rat K150, K192, K213, K226
→ Icaritin
In: Rat K150, K213, K226
Icariside I (7-(β-D-glucopyranosyloxy)-3,5-dihydroxy-2-(4-methoxyphenyl)-8-(3-methyl-2-butenyl)-4*H*-1-benzopyran-4-one)
Formed from: Icariin
Icariside II (3-((6-deoxy-α-L-mannopyranosyl)oxy)-5,7-dihydroxy-2-(4-methoxyphenyl)-8-(3-methyl-2-butenyl)-4*H*-1-benzopyran-4-one)
Formed from: Icariin
Icaritin (3,5,7-trihydroxy-2-(4-methoxyphenyl)-8-(3-methyl-2-butenyl)-4*H*-1-benzopyran-4-one)
Formed from: Icariin
Idarubin ((7*S*,9*S*)-9-acetyl-7-((3-amino-2,3,6-trideoxy-α-L-lyxo-hexopyranosyl)oxy)-7,8,9,10-tetrahydro-6,9,11-trihydroxy-5,12-naphthacenedione)
→ Idarubicinol
In: Man F915
Idarubicinol ((7*S*,9*S*)-7-((3-amino-2,3,6-trideoxy-α-L-lyxo-hexopyranosyl)oxy)-7,8,9,10-tetrahydro-6,9,11-trihydroxy-9-((1*S*)-1-hydroxyethyl)-5,12-naphthacenedione)
Formed from: Idarubicin
Idazoxan (2-(2,3-dihydro-1,4-benzodioxin-2-yl)-4,5-dihydro-1*H*-imidazole)
→ N-(2-Aminoethyl)-2-carboxamido-1,4-benzodioxan
In: Rat F284
→ 5-Hydroxyidazoxan
In: Rat F284
→ 6-Hydroxyidazoxan
In: Rat F284
→ 7-Hydroxyidazoxan
In: Rat F284
Idebenone (6-(10-hydroxydecyl)-2,3-dimethoxy-5-methylquinoline)
→ 6-(9-Carboxynonyl)-2,3-dimethoxy-5-methylquinol
In: Dog D764, E119
Rat D764, E119
→ Demethylidebenone

In: Dog E119
Rat E119
→ 6-(10-Hydroxydecyl)-2,3-dimethoxy-5-methylquinol
In: Dog D764, E119
Rat D764, E119

Idein
See: Cyanidin-3-galactoside

Ifenprodil
See: *erythro*-2-(4-Benzylpiperidino)-1-(*p*-hydroxyphenyl)-1-propanol

Ifetroban (2-(((1*S*,2*R*,3*S*,4*R*)-3-(4-((pentylamino)carbonyl)-2-oxazolyl)-7-oxabicyclo[2.2.1]hept-2-yl)methyl) benzenepropanoic acid)
→ Ifetroban-β-D-glucuronide
In: Dog K67
Man K67
Monkey K67
Rat K67

Ifetroban-β-D-glucuronide
Formed from: Ifetroban

Ifosfamide (N,3-bis(2-chloroethyl)tetrahydro-2*H*-1,3,2-oxaphosphorin-2-amine-2-oxide)
→ N^2-(Dechloromethyl)ifosfamide
In: Man J438
Rat J267
→ N^3-(Dechloromethyl)ifosfamide
In: Man J438
Rat J267

Illudin S ((2′*S*,3′*R*,6′*R*)-2′,3′-dihydro-3′,6′-dihydroxy-2′-(hydroxymethyl)-2′,4′,6′-trimethylspiro(cyclopropane-1,5′-(5*H*) inden)-7′(6′*H*)-one)
→ 1-(N-Acetyl-S-(L-cysteinyl))-1,7-dihydroxy-2-hydroxymethyl-5-(3-hydroxypropyl)-2,4,6-trimethylindane
In: Rat H526
→ 6-(2-Chloroethyl)-1,4-dihydroxy-2-hydroxymethyl-2,5,7-trimethylindane
In: Rat F520
→ 1,4-Dihydroxy-6-(2-hydroxyethyl)-2-hydroxymethyl-2,5,7-trimethylindane
In: Rat F520

Iloperidone (1-(4-(3-(4-(6-fluoro-1,2-benzisoxazol-3-yl)-1-piperidinyl)propoxy)-3-methoxyphenyl)ethanone)
→ 1-(3-(4-Acetyl-2-methoxyphenoxy)propyl)-4-(4-fluoro-2-hydroxybenzoyl)piperidine
In: Man H447
Rat H447
→ 4′-((2-Carboxy)ethoxy)-2′-methoxyacetophenone
In: Rat H447
→ Demethyliloperidone
In: Man H447, J748
Rat H447
→ Dihydroiloperidone
In: Dog H447
Man H447, J748
Rat H447
→ 4-(6-Fluoro-1,2-benzisoxazol-3-yl)-1-(3-(4-hydroxy-2-methoxyphenoxy)propyl) piperidine
In: Man J748
→ 1-(4-(3-(4-(6-Fluoro-1,2-benzisoxazol-3-yl)-1-piperidinyl)propoxy)-6-hydroxy-3-methoxyphenyl)ethanone
In: Dog H447
Man H447
Rat H447
→ 1-(4-(3-(4-(6-Fluoro-1,2-benzisoxazol-3-yl)-1-piperidinyl)propoxy)-3-methoxyphenyl)-2-hydroxyethanone
In: Dog H447
Man J748
Rat H447
→ 6-Fluoro-3-(1-(3-hydroxypropyl)-4-piperidinyl)benzisoxazole
In: Rat H447
→ 6-Fluoro-3-(4-piperidinyl)benzisoxazole
In: Rat H447

6*H*-Imidazo[4,5-*e*]benzothiazole
See: Thiazolo[4,5-*g*]benzimidazole

***p*-(1*H*-Imidazol-4-yl)aniline**
Formed from: *p*-(1*H*-Imidazol-4-yl) formanilide
Mifentidine
→ 2-Hydroxy-4-(1*H*-imidazol-4-yl)aniline
Probably in: Rat G411

4-(2-(1*H*-Imidazol-1-yl)ethoxy)benzoic acid
See: Dazoxiben
***p*-(1*H*-Imidazol-4-yl)formanilide**
Formed from: Mifentidine
→ *p*-(1*H*-Imidazol-4-yl)aniline
In: Guinea pig G411
Hamster G411
Mouse G411
Rat G411
***p*-(1-Imidazolylmethyl)benzoic acid**
Formed from: (*E*)-3-(4-(1-Imidazolylmethyl) phenyl)-2-propenoyl CoA
(*E*)-3-(4-(1-Imidazolylmethyl)phenyl)-2-propenoic acid
→ (*E*)-3-(4-(1-Imidazolylmethyl)phenyl)-2-propenoyl CoA
In: Rat D506
(*E*)-3-(4-(1-Imidazolylmethyl)phenyl)-2-propenoyl CoA
Formed from: (*E*)-3-(4-(1-Imidazolylmethyl) phenyl)-2-propenoic acid
→ *p*-(1-Imidazolylmethyl)benzoic acid
In: Rat D506
2-(1*H*-Imidazol-1-yl)-1-(2-naphthalenyl) ethanone
See: Nafimidone
1-(4-(1*H*)-Imidazol-4-yl)phenyl)-3-isopropylurea
Formed from: Mifentidine
Imiloxan (2-((2,3-dihydro-1,4-benzodioxin-2-yl) methyl)-1-ethyl-1*H*-imidazole)
→ 4′,5′-Dihydro-4′,5′-dihydroxyimiloxan
In: Man G67
→ 4′,5′-Dihydro-4′-hydroxy-5-oxoimiloxan
In: Man G67
→ Hydroxyimiloxan (aryl)
In: Man G67
→ 4′-Hydroxyimiloxan
Probably in: Man G67
→ Imiloxan-3′-β-D-glucuronide
In: Man G67
Imiloxan-3′-β-D-glucuronide
Formed from: Imiloxan
5-Iminodaunorubicin
→ Free radical
In: Horseradish E831

Iminodibenzyl (10,11-dihydro-5*H*-dibenz[*b*,*f*] azepine)
Formed from: Carpipramine
Imipramine
Trimipramine
→ 2-Hydroxyiminodibenzyl
Probably in: Man A1001
Rat G269
→ 10-Hydroxyiminodibenzyl
Probably in: Man D368
Rat G269
→ Iminostilbene
Probably in: Rat G269
2-Imino-1-phenylpropan-1-ol
Formed from: D-Norpseudoephedrine
Iminostilbene (dibenz[*b*,*f*]azepine)
Formed from: Carbamazepine
Probably formed from: Iminodibenzyl
→ Iminostilbene-10,11-oxide
In: Man A2990
Rat A2758
Iminostilbene-10,11-oxide
Formed from: Carbamazepine-10,11-oxide
Iminostilbene
→ 10,11-Dihydro-10,11-dihydroxyiminostilbene
Probably in: Man A2990
Imipramine (10,11-dihydro-N,N-dimethyl-5*H*-dibenz[*b*,*f*]azepine-5-propanamide)
Formed from: Demethylimipramine
Imipramine-N-oxide
→ Demethylimipramine
In: Arthrobacter C904
Bacillus C904
Beauveria B736
Chaetomium B736
Cunninghamella B736
Enterobacter C904
Escherichia C904
Glycine K359
Guinea pig C735
Klebsiella C904
Man A1001, A3270, A3274, A3280, A3337, F410, G711, H33, H244, H551, J774
Monkey C315

Nocardia C904
Pseudomonas C904
Rat A2165, A3178, A3668, C315, C669, D431, F410, H932, K168
Rhizopus B736
Staphylococcus C904
Streptococcus C904
See also: A888, C872
→ 2-Hydroxyimipramine
In: Guinea pig C735
Man A1001, C315, F410, G711, G948, H551
Monkey C315
Rat A316, A3668, C872, D431, F410, G932, G948
→ 10-Hydroxyimipramine
In: Chaetomium B736
Cunninghamella B736
Man A1001, F410, H33
Mucor B736
Penicillium B736
Rat A3668, C872, F410, H932
Rhizopus B736
→ Imipramine-β-D-glucuronide
In: Man C614, G492, H162, H541, J416
Rabbit G492
Rat C614
→ Imipramine-N-oxide
In: Aspergillus B736
Beauveria B736
Chaetomium B736
Cunninghamella B736
Fusarium B736
Guinea pig C735
Man A1001, C315, F410
Monkey C315
Penicillium B736
Pig D446
Rabbit A3161, B720
Rat A2165, B720, B829, C126, D431, D446, F410, H932, J174, K168
Rhizopus B736

Imipramine-β-D-glucuronide
Formed from: Imipramine

Imipramine-N-oxide
Formed from: Imipramine
→ Demethylimipramine
In: Rat A3656
See also: C496
→ Imipramine
In: Hamster D218
Mouse D218
Pig D218
Rat A1847, A1913, A3656, D218, J563, J876

Imipramin-2-yl-β-D-glucuronide
Formed from: 2-Hydroxyimipramine

Imipramin-10-yl-β-D-glucuronide
Formed from: 10-Hydroxyimipramine

Imirestat (2,7-difluoro*spiro*(9*H*-fluorene-9,4′-imidazolidine)-2′,5′-dione)
→ 3,7-Difluoro-1,2-dihydroxy*spiro*(9*H*-fluorene-9,4′-imidazolidine)-2′,5′-dione
Probably in: Dog G309
→ 3,7-Difluoro-2-hydroxy*spiro*(9*H*-fluorene-9,4′-imidazolidine)-2′,5′-dione
In: Dog G309
→ 7-Fluoro-2-hydroxy*spiro*(9*H*-fluorene-9,4′-imidazolidine)-2′,5′-dione
In: Dog G309
→ 3-Hydroxyimirestat
In: Dog G309
→ 4-Hydroxyimirestat
In: Dog G309

Inabenfide (4′-chloro-2′-(α-hydroxybenzyl) isonicotinanilide)
→ 2-Amino-5-chlorobenzhydrol
In: Rat E780
→ 2′-Benzoyl-4′-chloroisonicotinanilide
In: Rat E780
→ 3-Hydroxyinabenfide
Probably in: Rat E780
→ 4-Hydroxyinabenfide
Probably in: Rat E780
→ Inabenfide-N-oxide
In: Rat E780

Inabenfide-N-oxide
Formed from: Inabenfide

Indane
→ 1-Hydroxyindane

In: Pseudomonas E567, H580, K207
Rat E533, F617
→ 2-Hydroxyindane
In: Rat E533, F617
→ 5-Hydroxyindane
In: Rat E533, F617
→ Indene
In: Pseudomonas H580

1-Indanol
See: 1-Hydroxyindane

2-Indanol
See: 2-Hydroxyindane

Indan-1-one
Formed from: 1-Aminoindane
Fluorene
1-Hydroxyindane
Indene
→ 3-Hydroxyindan-1-one
Probably in: Pseudomonas K207

Indan-2-one
Formed from: 2-Aminoindane
Fluorene
2-Hydroxyindane
→ *cis*-1,2-Dihydroxyindane
In: Rabbit A449
Rat A449
→ *trans*-1,2-Dihydroxyindane
In: Rabbit A449
Rat A449
→ 2-Hydroxyindane
In: Rabbit F295
Rat F295
→ 2-Hydroxyindan-1-one
In: Pseudomonas H375
→ 3-Isochromanone
In: Arthrobacter H901

Indecainide (9-(3-isopropylamino)-9*H*-fluorene-9-carboxamide)
→ 9-(3-Aminopropyl)-9*H*-fluorene-9-carboxamide
In: Dog D352
Rat D372
→ N-Formylindecainide
In: Dog D352
Rat D372

Indeloxazine (2-((inden-7-yloxy)methyl) morpholine)
→ Inden-7-ol
In: Rat E239
→ 6-((Inden-7-yloxy)methyl)-3-morpholinone
In: Rat E239
→ *trans*-4-(2-Morpholinylmethoxy)-1,2-indandiol
In: Rat E239, E837
→ 7-(2-Morpholinylmethoxy)inden-4-ol
In: Rat E239

Indene
Formed from: Indane
→ *trans*-(2*S*, 1*S*)-Bromoindanol
In: Curvularia J851
→ *cis*-1,2-Dihydroxyindan (1*S*,2*R* and 1*R*,2*S*)
In: Pseudomonas E567, H580, K197
Rhodococcus J896, K305
→ *trans*-(1*R*,2*R*)-1,2-Dihydroxyindane
In: Rhodococcus H885
→ (1*S*)-Hydroxyindene
In: Pseudomonas H580
→ 2-Hydroxy-1-methylthioindane
In: Guinea pig D855
Rat D855
→ Indan-1-one
In: Pseudomonas E567
See also: H580
→ Indene-1,2-oxide
In: Curvularia J851
Rat A380
See also: A2886, D855

Indene-1,2-oxide
Formed from: Indene
Probably formed from: 2-Bromoindene
1-Chloroindane
→ *trans*-1,2-Dihydroxyindane
In: Guinea pig D855
Rat D855
See also: A380
→ S-(1-Hydroxyindan-2-yl)glutathione
Probably in: Guinea pig D855
Rat A380, A2886, D855

Indene[1,2,3-*cd*]pyrene
→ *trans*-1,2-Dihydro-1,2-dihydroxyindene [1,2,3-*cd*]pyrene
In: Rat D893
→ 8-Hydroxyindene[1,2,3-*cd*]pyrene
In: Rat D893
→ 9-Hydroxyindene[1,2,3-*cd*]pyrene
In: Rat D893
→ 10-Hydroxyindene[1,2,3-*cd*]pyrene
In: Rat D893

Indene[1,2,3-*cd*]pyrene-1,2-quinone
Probably formed from: *trans*-1,2-Dihydro-1,2-dihydroxyindene[1,2,3-*cd*]pyrene

Indenoestrol A (3-ethyl-2-(4-hydroxyphenyl)-1-methyl-1*H*-inden-6-ol)
Formed from: Stilboestrol
→ 2-Hydroxyindenoestrol A
In: Hamster D514
Rat D514

Inden-7-ol
Formed from: Indeloxazine

3-(Inden-7-yloxy)lactic acid
Probably formed from: 6-((Inden-7-yloxy)methyl)-3-morpholinone

2-((Inden-7-yloxy)methyl)morpholine
See: Indeloxazine

6-((Inden-7-yloxy)methyl)-3-morpholinone
Formed from: Indeloxazine
→ 3-(Inden-7-yloxy)lactic acid
Probably in: Rat E239
→ *trans*-4-(5-Oxo-2-morpholinylmethoxy)-1,2-indandiol
In: Rat E837
See also: E239

Inderal
See: Propranolol

Indican
See: Indoxyl-β-D-glucoside

Indigo (2-(1,3-dihydro-3-oxo-2*H*-indol-2-ylidene)-1,2-dihydro-3*H*-indol-3-one)
Formed from: Indole
Probably formed from: Indoxyl

Indinavir (crixivan, N-(2(*R*)-hydroxy-1(*S*)-indanyl)-5-(2(*S*)-*tert*-butylaminocarbonyl)-4-(3-pyridylmethyl)piperazino)-4(*S*)-hydroxy-2(*R*)-phenylmethylpentanamide)
→ 2(*R*)-Benzyl-5((2(*S*)-*tert*-butylaminocarbonyl)piperazino)-4(*S*)-hydroxy-N-(2(*R*)-hydroxy-1(*S*)-indanyl) pentanamide
In: Chimpanzee J867
Dog J50
Man J58, J297, J867
Monkey J50, J867
Rat J50
→ 3-Hydroxyindinavir
In: Chimpanzee J867
Dog J50
Man J58, J297, J867
Monkey J50, J867
Rat J50
→ *p*-Hydroxyindinavir
In: Dog J50
Man J58
Monkey J50
Rat J50
→ Indinavir-N-β-D-glucuronide
In: Dog J50
Man J58
Monkey J50
Rat J50
→ Indinavir-N-oxide
In: Dog J50
Man J58, J297
Monkey J50
Rat J50

Indinavir-N-β-D-glucuronide
Formed from: Indinavir

Indinavir-N-oxide
Formed from: Indinavir
→ 3-Hydroxyindinavir-N-oxide
Probably in: Dog J50
Man J58
Monkey J50
Rat J50

Indirubin (3-(1,3-dihydro-3-oxo-3*H*-indol-2-ylidene)-1,3-dihydro-2*H*-indol-2-one)
Formed from: Indole
Probably formed from: Indoxyl

Indole
Formed from: Diphenylamine
Indole-3-acetic acid

D-Tryptophan
L-Tryptophan
Probably formed from: 1-Methoxyindole
→ Anthranil
In: Tecoma G659
Zea A3747
→ 2,2-Bis(3-indolyl)indoxyl
In: Horse chestnut A3125
Oak A3125
Rhus A3125
→ 5-Bromoindole
In: Glycine K439
→ *o*-Formamidobenzaldehyde
In: Tecoma G659
→ N-Formylanthranilic acid
In: Aspergillus F417
Rat D158
See also: A3747, B70, C170, C506, D334, G639
→ 4-Hydroxyindole
In: Pumpkin A3530
→ 5-Hydroxyindole
In: Glycine K439
Pumpkin A3530
Rat D158
Seterasia A3538
Tradescantia A3538
Zebrina A3538
→ 6-Hydroxyindole
In: Man K623
Rat D158
Seterasia A3538
Tradescantia A3538
Zebrina A3538
→ Indigo
In: Rhodococcus H885
Pseudomonas J479, K607
→ Indole-3-acetic acid
In: Crown gall A1527
Heibeloma F371
Zea H126
→ Indirubin
In: Pseudomonas K607
→ Indole-3-β-D-glucuronide
Probably in: Man G762
→ Indoxyl
In: Aspergillus F417
Man K623
Rat D158
See also: C472, G762
→ 6*H*-Oxazolo[2,3-*a*:4,5-*b*′]diindole
In: Man K623
→ Oxindole
In: Microorganism E293, F15, F781, G330
Rat D158, J628
Wheat K251
See also: J314
→ L-Tryptophan
In: Carrot B874
Enterobacter A781
Erwinia A781
Escherichia A781, F642
Kluyvera A781
Nicotiana B874
Proteus A781, A2890
Salmonella K275
Vibrio B85
Zea H126
Contrast (in many genuses of microorganism) A781

Indole-3-acetaldehyde
Formed from: N,N-Dimethyltryptamine
Indole-3-acetaldoxime
Indole-3-acetic acid
Indole-3-pyruvic acid
N-Methyltryptamine
Tryptamine
L-Tryptophan
Tryptophol
→ 3-Formylindole
In: Orobanche C181
Pseudomonas C706
See also: A556, A993, B863
→ α-Hydroxyindole-3-acetaldehyde
Probably in: Pseudomonas A3007
→ 4-Hydroxyquinoline
In: Horseradish A1036
→ Indole-3-acetic acid
In: Achromobacter K869
Azospirillum F49
Bradyrhizobium J189
Crown gall A1527

Cucumber A3378
Man J24
Orobanche C181
Pea G446, G520
Rhizobium E405
Ustilago H902
Zygosaccharomyces A923
See also: A446, A3551, B98, E97, E210, E352, E578, G320
→ Tryptophol
In: Cabbage F931
Cucumber F462
Man B569
Mung bean J207
Orobanche C181
Pea A755
Phycomyces F846
Pig A1679
Rat A2007
Zygosaccharomyces A923
See also: A446, A993, A1922, A3551, B98, B863, C533, D221, E97, G320

Indole-3-acetaldoxime (*E* and *Z* isomers)
Formed from: L-Tryptophan
→ Indole-3-acetaldehyde
In: Cabbage E352
Zea D896
See also: K882
→ Indole-3-acetonitrile
In: Aspergillus K882
Bacillus K411
Banana K882
Cabbage D896
Helianthus D896
Isatis D896
Zea D896
→ Indole-3-desulphomethylglucosinolate
In: Cabbage D896
→ Indole-3-methylglucosinolate
In: Cabbage D896

Indole-3-acetamide
Formed from: Indole-3-acetonitrile
L-Tryptophan
Probably formed from: 7-Chloroindole-3-acetamide
→ Indole-3-acetic acid
In: Agrobacterium H366
Azospirillum G776
Bradyrhizobium E754, F788
Poncirus G760
See also: G169

Indole-3-acetic acid (heteroauxin, IAA)
Formed from: Carbazole
Indole
Indole-3-acetaldehyde
Indole-3-acetamide
Indole-3-acetonitrile
Indole-3-acetylaspartic acid
Indole-3-acetyl-β-D-glucuronic acid
Indole-3-glycerol phosphate
Indole-3-pyruvic acid
D-Tryptophan
L-Tryptophan
→ *o*-Aminophenylglyoxylic acid
In: Bradyrhizobium H679
→ 3,3′-Bisindolylmethaneaspartic acid
In: Lupinus F847
→ 2,3-Dihydro-7-hydroxy-2-oxindole-3-acetic acid
In: Zea F947
→ Dioxindole-3-acetic acid
In: Bradyrhizobium H679
→ 3-Formylindole
In: Bean A635
Chromobacterium A2555
Horseradish C503, D234, F961
Lupinus F847
Orobanche C181
Pea A635
See also: E306, D747, K541
→ 4-Hydroxyindole-3-acetic acid
In: Aspergillus A2224, A2932
→ 5-Hydroxyindole-3-acetic acid
In: Aspergillus A2932
Man G762
Sedum D41
See also: H349
→ 6-Hydroxyindole-3-acetic acid
In: Aspergillus A2932, A3070
→ α-Hydroxyindole-3-acetic acid
Probably in: Pseudomonas A3007
→ 3-Hydroxymethyloxindole

In: Horseradish C525, D234, F961
Lupinus F847
→ Indole
In: Man G762
See also: H274
→ Indole-3-acetaldehyde
In: Nectria A556
→ Indole-3-acetyl-L-alanine
In: Crown gall A2843
→ Indole-3-acetylaspartic acid
In: Arabidopsis J713
Bean A635, E437
Crown gall A2843
Dalbergia E729
Pea A635
Populus H349
Sycamore A3220
Teucrium E307
See also: E88
→ Indole-3-acetyl-β-D-glucose
In: Lupinus F847
Orobanche C181
Zea B433, G487, K746
→ Indole-3-acetyl-β-D-glucuronic acid
In: Man A274
→ Indole-3-acetyl-L-glutamic acid
In: Arabidopsis J713
Crown gall A2843
Teucrium E307
→ Indole-3-acetylglutamine
In: Aotus A274
Cebus A274
Cercopithecus A274
Macaca A274
Man A274
Papio A274
Saimiri A274
→ Indole-3-acetylglycine
In: Aotus A274
Canis A274
Cebus A274
Crown gall A2843
Felis A274
Ferret A394, A3046
Galago A274
Mesocricetus A274
Mus A274
Mustela A274
Nycticebus A274
Oryctolagus A274
Rattus A274
Saimiri A274
→ Indole-3-acetyltaurine
In: Aotus A274
Cebus A274
Cercopithecus A274
Ferret A394, A3046
Macaca A274
Mustela A274
Papio A274
Pigeon A274, A395
Saimiri A274
→ Indole-3-acetyl-L-valine
In: Crown gall A2843
→ Indole-3-butyric acid
In: Zea G758, H694
→ Indole-3-methanol
In: Horseradish C525, F961
Lupinus F847
Pinus D597
Rhizobium E405
Wheat D150
Zea F798
→ Isatin
In: Bradyrhizobium H679
→ 3-Methyleneindolenine
In: Zea A727
See also: A635
→ 3-Methyleneoxindole
In: Horseradish D234
Lupinus F847
→ Oxindole-3-acetic acid
In: Arabidopsis J713
Populus H349
Zea D545, D618
→ Skatole
In: Lactobacillus B750
→ Skatole free radical
In: Horseradish E118
→ L-Tryptophan
In: Microorganism A1164
Ruminococcus A1070

Indole-3-acetonitrile
Formed from: Indole-3-acetaldoxime
→ α-Hydroxyindole-3-acetonitrile
Probably in: Pseudomonas A3007
→ Indole-3-acetamide
In: Agrobacterium H366
→ Indole-3-acetic acid
In: Brassica A3986
Fusarium A3364
Rhodococcus G649

Indole-3-acetyl-L-alanine
Formed from: Indole-3-acetic acid

Indole-3-acetylaspartic acid
Formed from: Indole-3-acetic acid
→ Dioxindole-3-acetylaspartic acid
In: Phaseolus E122, E437
See also: E88
→ Hydroxyindole-3-acetylaspartic acid
In: Arabidopsis J713
→ Indole-3-acetic acid
In: Dalbergia E729
Enterobacter J254
→ Oxindole-3-acetylaspartic acid
In: Arabidopsis J713

Indole-3-acetyl CoA
Formed from: Indole-3-pyruvic acid

Indole-3-acetyl-β-D-glucose
Formed from: Indole-3-acetic acid
→ Indole-3-acetyl-*myo*-inositol
In: Zea B433, K746

Indole-3-acetyl-β-D-glucuronic acid
Formed from: Indole-3-acetic acid
→ Indole-3-acetic acid
In: Beef A1029
Escherichia A1029

Indole-3-acetyl-L-glutamic acid
Formed from: Indole-3-acetic acid

Indole-3-acetylglutamine
Formed from: Indole-3-acetic acid

Indole-3-acetylglycine
Formed from: Indole-3-acetic acid

Indole-3-acetyl-*myo*-inositol
Formed from: Indole-3-acetyl-β-D-glucose
→ Indole-3-acetyl-*myo*-inositol arabinoside
In: Zea K734
→ Indole-3-acetyl-*myo*-inositol galactoside
In: Zea K734

Indole-3-acetyl-*myo*-inositol arabinoside
Formed from: Indole-3-acetyl-*myo*-inositol

Indole-3-acetyl-*myo*-inositol galactoside
Formed from: Indole-3-acetyl-*myo*-inositol

Indole-3-acetyltaurine
Formed from: Indole-3-acetic acid

Indole-3-acetyl-L-valine
Formed from: Indole-3-acetic acid

Indole-3-acrylic acid
Formed from: Hypaphorine
Indole-3-propionic acid
Probably formed from: L-Tryptophan
→ Indole-3-propionic acid
Probably in: Clostridium A2571

Indole-3-butyric acid
Formed from: Indole-3-acetic acid
→ 4-Hydroxyindole-3-butyric acid
In: Bapleurum J838
Phytolacca J838
→ γ-Hydroxyindole-3-butyric acid
In: Pseudomonas A3007
→ Indole-3-butyrylglucoside
Probably in: Petunia H46

Indole-3-butyrylaspartic acid
Probably formed from: Indole-3-butyrylglucoside

Indole-3-butyrylglucoside
Probably formed from: Indole-3-butyric acid
→ Indole-3-butyrylaspartic acid
Probably in: Petunia H46

Indole-2-carboxylic acid
Probably formed from: Indole-2,3-dicarboxylic acid

Indole-3-carboxylic acid
Formed from: 3-Formylindole
1-Hydroxyindole-3-carboxylic acid
Indole-3-nitrile
Probably formed from: 1-Methoxyindole-3-carboxylic acid
Indole-2,3-dicarboxylic acid
→ Anthranilic acid
Probably in: Hordeum G139
→ Indole-3-carboxyl-β-D-glucuronic acid
In: Rat D158, E306, G702

Indole-3-desulphomethylglucosinolate
Formed from: Indole-3-acetaldoxime

Indole-2,3-dicarboxylic acid
Formed from: Carbazole
→ Indole-2-carboxylic acid
Probably in Ralstonia K442
→ Indole-3-carboxylic acid
Probably in Ralstonia K442
Indole-β-D-glucuronide
Probably formed from: Indole
Indole-3-glycolaldehyde
Formed from: L-Tryptophan
→ β-Hydroxytryptophol
Probably in: Pseudomonas B98
→ Indole-3-glycollic acid
Probably in: Pseudomonas B98
Indole-3-glycollic acid
Probably formed from: Indole-3-glycolaldehyde
→ Indole-3-glyoxylic acid
In: Pseudomonas K95
Indole-3-glyoxal
Formed from: N-Acetyl-L-tryptophan
N-Acetyl-L-tryptophanamide
Indole-3-lactic acid
Indole-3-propionic acid
D-Tryptophan
L-Tryptophan
Probably formed from: Tryptamine
Tryptophol
→ Indole-3-glyoxylic acid
Probably in: Pseudomonas B98
→ β-Oxotryptophol
Probably in: Pseudomonas B98
Indole-3-glyoxylic acid
Formed from: 1-Hydroxyindole-3-glyoxylic acid
Indole-3-glycollic acid
Probably formed from: Indole-3-glyoxal
Indole-3-glycerol phosphate
→ Indole-3-acetic acid
In: Arabidopsis K684
Indole-3-lactic acid
Formed from: Indole-3-pyruvic acid
→ 3-Formylindole
In: Orobanche C181
See also: E435
→ Indole-3-glyoxal
In: Pseudomonas A3009
→ Indole-3-pyruvic acid
In: Candida D985
Pseudomonas K95
Indole-3-methanol (indole-3-carbinol)
Formed from: 3-Formylindole
Indole-3-acetic acid
Skatole
→ 3,3′-Diindolylmethane
In: Rat H446
See also: A727
→ 3-Formylindole
In: Horseradish C503
Orobanche C181
Pseudomonas A3734
See also: D150, E306
→ (1-(3-Hydroxymethyl)indolyl-3-indolylmethane
In: Rat H446
→ 3-Hydroxymethyloxindole
In: Zea A727
See also: C503, C525
→ Indole-3-methanol-β-D-glucoside
In: Wheat D150, E306
→ (2-(Indol-3-ylmethyl)indol-3-yl)indol-3-ylmethane
In: Rat H446
Indole-3-methanol-β-D-glucoside
Formed from: Indole-3-methanol
Indole-3-methylglucosinolate
Formed from: Indole-3-acetaldoxime
Indole-3-nitrile
→ Indole-3-carboxylic acid
In: Rhodococcus G174, G649
Indole-3-propionic acid
Probably formed from: Indole-3-acrylic acid
→ Indole-3-acrylic acid
In: Chromobacterium H248, H664
→ Indole-3-glyoxal
In: Pseudomonas A3009
→ β-Hydroxyindole-3-propionic acid
In: Pseudomonas A3007
Indole-3-pyruvic acid
Formed from: Indole-3-lactic acid
D-Tryptophan
L-Tryptophan

→ Indole-3-acetaldehyde
In: Azospirillum F49
Bradyrhizobium J189
Enterobacter H348
See also: A923, A2571, A3551, B483, B863, B855, E97, E435, G320
→ Indole-3-acetic acid
In: Pyrococcus J312
→ Indole-3-acetyl CoA
In: Pyrococcus H306
→ Indole-3-lactic acid
In: Candida D975
Taphrina A405
See also: A446, B483, B863, E97, E435
→ 3-(3-Indolyl)-3-methylpyruvic acid
In: Streptomyces A2281, A2307, D148
→ L-Tryptophan
In: Bevibacterium E642
Enterobacter H348
Escherichia E278
Hansenula D978
Microorganism K382
Mung bean H826
Nocardia F394
Pseudomonas B672
Rhodococcus E451
Sporosarcina E423
See also: A446, K259

Indole-5,6-quinone-2-carboxylic acid
Probably formed from: 5,6-Dihydroxyindole-2-carboxylic acid

Indolidan (1,3-dihydro-3,3-dimethyl-5-(1,4,5,6-tetrahydro-6-oxo-3-pyridazinyl)-2*H*-indol-2-one)
→ Dehydroindolidan
In: Dog F310
Man E627
Monkey F310
Mouse F310
Rat F310
→ 5′-Hydroxyindolidan
In: Dog F310
Monkey F310
Mouse F310
Rat F310

Indoline (2,3-dihydroindole)
Formed from: N-Benzylindoline
N-(*p*-Toluoyl)indoline

Indolin-2-one
See: Oxindole

Indolo[2,3-*a*]quinolizidine
→ 4-Hydroxyindolo[2,3-*a*]quinolizidine
In: Rat J359
→ 6-Hydroxyindolo[2,3-*a*]quinolizidine
In: Rat J359
→ 7-Hydroxyindolo[2,3-*a*]quinolizidine
In: Rat J359

3-Indolyl acetate
→ Indoxyl
In: Microorganism G330

2-(3-Indolyl)ethanol
See: Tryptophol

(1*H*-Indol-3-ylmethyl)carbamodithioic acid methyl ester
See: Brassinine

5-Indolylmethylhydantoin (both D and L)
Formed from: N-Carbamyl-L-tryptophan
→ D-Tryptophan
In: Pseudomonas E391, E450
→ L-Tryptophan
In: Flavobacterium A3000, F940, F941

(2-(Indol-3-ylmethyl)indol-3-yl)indol-3-ylmethane
Formed from: Indole-3-methanol

5-(3-Indolyl)-2-methyl-2-oxazoline-4-carboxamide
Formed from: N-Acetyl-L-tryptophanamide

3-(3-Indolyl)-3-methylpyruvic acid
Formed from: Indole-3-pyruvic acid
β-Methyl-L-tryptophan
→ β-Methyl-L-tryptophan
In: Streptomyces D148

1-(3-Indolyl)-1-oxoethanol
See: β-Oxotryptophol

3-(4-Indolyloxy)lactic acid
Formed from: Pindolol

Indomethacin
See: 1-(*p*-Chlorobenzoyl)-5-methoxy-2-methyl-1*H*-indole-3-acetic acid

Indomethacin farnesil (1-(4-chlorobenzoyl)-5-methoxy-2-methyl-1*H*-indole-3-acetic acid 3,7,11-trimethyl-2,6,10-dodecatrienyl ester)
→ 1-(*p*-Chlorobenzoyl)-5-methoxy-2-methyl-indole-2-acetic acid

In: Rat F494
Indoprofen (stereochemistry unspecified)
See: α-(4-(1-Oxo-2-isoindolinyl)phenyl) propionic acid
(*R*)-Indoprofen (4-(1,3-dihydro-1-oxo-2*H*-isoindol-2-yl)-α-methylbenzeneacetic acid)
→ (*S*)-Indoprofen
Probably in: Rat G763
(*S*)-Indoprofen
Probably formed from: (*R*)-Indoprofen
Indoramin (3-(2-(4-benzamidopiperidino) ethyl)indole)
→ 4-Benzamidopiperazine
In: Patas monkey B840
→ 5-Hydroxyindoramin
In: Patas monkey B840
→ 6-Hydroxyindoramin
In: Man C723, E609
Patas monkey B840
→ *p*-Hydroxyindoramin
In: Patas monkey B840
Indoramin-6-yl sulphate
Formed from: 6-Hydroxyindoramin
Indoxyl (3-hydroxyindole)
Formed from: Indole
3-Indolyl acetate
Indoxyl phosphate
Indoxyl-3-sulphate
→ Indigo
Probably in: Man K623
→ Indirubin
Probably in: Man K623
→ Indoxyl-β-D-glucoside
In: Baphicacanthus K347
Polygonum K375
→ Indoxyl-3-β-D-glucuronide
Probably in: Rat D158
→ Indoxyl-3-sulphate
In: Rat B61
See also: D158
→ Isatin
Probably in: Alcaligenes C742
Man K623
Indoxyl-β-D-glucoside (indican)
Formed from: Indoxyl
→ Indoxyl-β-D-glucoside-6′-malonate
In: Cicer D445
Indoxyl-β-D-glucoside-6′-malonate
Formed from: Indoxyl-β-D-glucoside
Indoxyl-3-β-D-glucuronide
Formed from: Indoxyl
Indoxyl phosphate
→ Indoxyl
In: Polygonum K375
Indoxyl-3-sulphate
Formed from: Indoxyl
→ Indoxyl
In: Klebsiella E668
Providencia E668
Inezin
See: S-Benzyl O-ethyl phenylphosphonothioate
Intriptyline (5-(5-(dimethylamino)penta-1-en-3-yn-1-yl)-5*H*-dibenzo[*a*,*d*]cycloheptene)
→ Intriptyline-10,11-oxide
In: Rat A2758
Intriptyline-10,11-oxide
Formed from: Intriptyline
Iodinin
See: 1,6-Dihydroxyphenazine-6,10-dioxide
***endo*-Iodixanol** (5,5′-((2-hydroxy-1,3-propanediyl)bis(acetylamino))bis(N,N′-bis(2,3-dihydroxypropyl)-2,4,6-triiodo-1,3-benzenedicarboxamide)
→ *endo*-Iodixanol-α-D-glucoside
In: Cynomolgus monkey J786
Dog J786
Rat J786
***exo*-Iodixanol**
→ *exo*-Iodixanol-α-D-glucoside
In: Cynomolgus monkey J786
Dog J786
Rat J786
***endo*-Iodixanol-di-α-D-glucoside**
Probably formed from: *endo*-Iodixanol-α-D-glucoside
***exo*-Iodixanol-di-α-D-glucoside**
Probably formed from: *exo*-Iodixanol-α-D-glucoside
***endo*-Iodixanol-α-D-glucoside**
Formed from: *endo*-Iodixanol
→ *endo*-Iodixanol-di-α-D-glucoside
Probably in: Cynomolgus monkey J786
Dog J786

Rat J786

***exo*-Iodixanol-α-D-glucoside**
Formed from: *exo*-Iodixanol
→ *exo*-Iodixanol-di-α-D-glucoside
Probably in: Cynomolgus monkey J786
Dog J786
Rat J786

***p*-Iodoacetanilide**
Formed from: *p*-Iodoaniline

***o*-Iodoaniline**
→ 4-Amino-3-iodophenol
In: Rat H273
→ 2,2′-Diiodoazobenzene
In: Horseradish A661

***m*-Iodoaniline**
→ 4-Amino-2-iodophenol
In: Rat H273
→ 3,3′-Diiodoazobenzene
In: Horseradish A661

***p*-Iodoaniline**
→ 4,4′-Diiodoazobenzene
In: Geotrichum A661
Horseradish A661
→ *p*-Iodoacetanilide
In: Rabbit E399
→ *p*-Iodonitrosobenzene
With chloroperoxidase A3579
→ *p*-Iodophenylhydroxylamine
In: Rabbit F839

***p*-Iodoanisole**
→ 2-Iodo-5-methoxyphenol
In: Rat H158
→ 5-Iodo-2-methoxyphenol
In: Rat H158
→ *p*-Iodophenol
In: Methylosinus D513
Rat H158
→ *m*-Methoxyphenol
In: Rat H158
→ *p*-Methoxyphenol
In: Rat H158

***m*-Iodobenzaldehyde**
Formed from: *m*-Iodobenzoic acid
→ *m*-Iodobenzyl alcohol
Probably in: Nocardia E743

***p*-Iodobenzaldehyde**
Probably formed from: *p*-Iodobenzoic acid
→ *p*-Iodobenzoic acid
In: Beef C508
Yeast C508
→ *p*-Iodobenzyl alcohol
In: Horse C508
Yeast C508
See also: E743

2-Iodobenzanilide (benodanil)
→ 4′-Hydroxy-2-iodobenzanilide
In: Rat B878

2-Iodobenzanilid-4′-yl-β-D-glucuronide
Formed from: 4′-Hydroxy-2-iodobenzanilide

2-Iodobenzanilid-4′-yl sulphate
Formed from: 4′-Hydroxy-2-iodobenzanilide

Iodobenzene
→ *cis*-1,2-Dihydro-1,2-dihydroxy-3-iodobenzene
In: Pseudomonas J656, K699
→ *p*-Iodophenol
In: Rat C899

***o*-Iodobenzoic acid**
→ Catechol
In: Pseudomonas G434

***m*-Iodobenzoic acid**
Formed from: *m*-Iodobenzylguanidine
→ *m*-Iodobenzaldehyde
In: Nocardia G207
See also: E743, K615

***p*-Iodobenzoic acid**
Formed from: *p*-Iodobenzaldehyde
→ *p*-Hydroxybenzoic acid
In: Alcaligenes E335
Arthrobacter E752
Pseudomonas E359
→ *p*-Iodobenzaldehyde
Probably in: Nocardia E743
→ *p*-Iodobenzoyl CoA
In: Pseudomonas G729, K191
→ *p*-Iodobenzoyl-β-D-glucuronic acid
In: Rabbit H519

***p*-Iodobenzoyl CoA**
Formed from: *p*-Iodobenzoic acid
→ *p*-Hydroxybenzoyl CoA
In: Arthrobacter H279

Pseudomonas G920, H645

p-Iodobenzoyl-β-D-glucuronic acid
Formed from: *p*-Iodobenzoic acid

***m*-Iodobenzyl alcohol**
Probably formed from: *m*-Iodobenzaldehyde

***p*-Iodobenzyl alcohol**
Formed from: *p*-Iodobenzaldehyde

***m*-Iodobenzylguanidine**
→ 4-Hydroxy-3-iodobenzylguanidine
In: Man E508
→ *m*-Iodobenzoic acid
In: Man E508

4-(*p*-Iodobenzyl)-1-methylpyridinium
Probably formed from: Tetrahydro-4-(*p*-iodobenzyl)-1-methylpyridine

4-Iodobiphenyl
→ 4-Hydroxy-4′-iodobiphenyl
In: Rat C297

3-Iodocatechol
Formed from: *m*-Fluoroiodobenzene

Iodochlorhydroxyquin
See: 5-Chloro-8-hydroxy-7-iodoquinoline

Iododoxorubicin ((8*S*,10*S*)-10-(((2*R*,4*S*,5*S*,6*S*)-4-amino-tetrahydro-5-iodo-6-methyl-2*H*-pyran-2-yl)oxy)-7,8,9,10-tetrahydro-6,8,11-trihydroxy-8-(hydroxyacetyl)-1-methoxy-5,12-naphthacenedione)
→ Iododoxorubicinol
In: Dog G53
Man G144, G543
Monkey G53, G543
Rat G53

Iododoxorubicinol
Formed from: Iododoxorubicin
→ Iododoxorubicinol aglycone
Probably in: Man G144

Iododoxorubicinol aglycone
Probably formed from: Iododoxorubicinol

3-Iodo-4-(3-iodophenoxy)phenylacetic acid
See: 3,3′-Diiodothyroacetic acid

2-Iodo-5-methoxyphenol
Formed from: *p*-Iodoanisole

5-Iodo-2-methoxyphenol
Formed from: *p*-Iodoanisole

3′-Iodo-4-methylaminoazobenzene
Formed from: 4-Dimethylamino-3′-iodoazobenzene
→ 4-Amino-3′-iodoazobenzene
In: Rat A776

4′-Iodo-4-methylaminoazobenzene
Formed from: 4-Dimethylamino-4′-iodoazobenzene
→ 4-Amino-4′-iodoazobenzene
In: Rat A1979

3-Iodo-6-methylcatechol
Formed from: 2,3-Dihydro-2,3-dihydroxy-4-iodotoluene

***p*-Iodonitrosobenzene**
Formed from: *p*-Iodoaniline

***o*-Iodophenol**
→ *o*-Iodophenyl sulphate
In: Man E575
Rat E575

***m*-Iodophenol**
→ *m*-Iodophenyl-β-D-glucuronide
In: Man J490
→ *m*-Iodophenyl sulphate
In: Man E575
Rat E575

***p*-Iodophenol**
Formed from: *p*-Iodobenzene

***p*-Iodophenylacetic acid**
Formed from: 15-(*p*-Iodophenyl)-3-methylpentadecanoic acid

***p*-Iodo-L-phenylalanine**
→ *p*-Iodophenylpyruvic acid
In: Bushbean E500

***m*-Iodophenyl-β-D-glucuronide**
Formed from: *m*-Iodophenol

***p*-Iodophenylhydroxylamine**
Formed from: *p*-Iodoaniline

***trans*-1-(*p*-Iodophenyl)-1-(*p*-β-methylaminoethoxyphenyl)-2-phenylbut-1-ene**
Formed from: *trans*-1-(*p*-β-Dimethylaminoethoxyphenyl)-1-(*p*-iodophenyl)-2-phenylbut-1-ene

15-(*p*-Iodophenyl)-3-methylpentadecanoic acid
→ 14-(*p*-Iodophenyl)-2-methyltetradecanoic acid
In: Dog H811
→ *p*-Iodophenylacetic acid
In: Dog H811

14-(*p*-Iodophenyl)-2-methyltetradecanoic acid
Formed from: 15-(*p*-Iodophenyl)-3-methylpentadecanoic acid

***p*-Iodophenylpyruvic acid**
Formed from: *p*-Iodo-L-phenylalanine

***o*-Iodophenyl sulphate**
Formed from: *o*-Iodophenol

***m*-Iodophenyl sulphate**
Formed from: *m*-Iodophenol

4-Iodotamoxifen
See: *trans*-1-(*p*-β-Dimethylaminoethoxyphenyl)-1-(*p*-iodophenyl)-2-phenylbut-1-ene

4-Iodothymol
→ 4-Iodothymol-β-D-glucuronide
In: Man A3633
→ Thymoquinone
In: Man A3633

4-Iodothymol-β-D-glucuronide
Formed from: 4-Iodothymol

3-Iodothyroacetic acid
Formed from: 3,3′-Diiodothyroacetic acid

3′-Iodothyroacetic acid
Formed from: 3,3′-Diiodothyroacetic acid

3-Iodo-L-thyronine (T_1)
Formed from: 3,3′-Diiodo-L-thyronine
3,5-Diiodo-L-thyronine
→ 3-Iodo-L-thyronine-O-sulphate
In: Monkey B770
Rat B770
→ 3-Iodothyropyruvic acid
In: Rabbit K713

3′-Iodo-L-thyronine (rT_1)
Formed from: 3,3′-Diiodo-L-thyronine
3′,5′-Diiodo-L-thyronine
→ 3′-Iodo-L-thyronine-O-β-D-glucuronide
In: Rat B770
→ 3′-Iodo-L-thyronine-O-sulphate
In: Monkey B770
Rat B770, B771

3′-Iodo-L-thyronine-O-β-D-glucuronide
Formed from: 3′-Iodo-L-thyronine

3-Iodo-L-thyronine sulphamate
Formed from: 3,3′-Diiodo-L-thyronine sulphamate

3-Iodo-L-thyronine-O-sulphate
Formed from: 3,3′-Diiodo-L-thyronine-O-sulphate
3-Iodo-L-thyronine

3′-Iodo-L-thyronine-O-sulphate
Formed from: 3′-Iodo-L-thyronine

3-Iodothyropyruvic acid
Formed from: 3-Iodo-L-thyronine

3-Iodo-L-tyrosine
Formed from: L-Tyrosine
→ 3,5-Diiodo-L-tyrosine
In: Goat A997
Pigeon A432
→ 4-Hydroxy-3-iodophenylpyruvic acid
In: Rat A1383
→ L-Tyrosine
In: Bungarus C698
Ctenopharyngodon C698
Elaphe C698
Gallus C698
Gekko C698
Geoclemys C698
Hydropis C698
Naja C698
Ptyas C698
Scarops C698
Trionyx C698
Triturus C698
Varantus C698
Xenopus C698

Iodoxamic acid ((3,3′-((1,16-dioxo-4,7,10,13-tetraoxahexadecane-1,16-diyl)diimino) bis(2,4,6-triiodobenzoic acid))
→ 3-(3-(β-Hydroxyethoxy)propionamido)-2,4,6-triiodobenzoic acid
In: Man A1456

Iomeglamic acid (N-(3-amino-2,4,6-triiodophenyl)-N-methylglutaramidic acid)
→ N-Acetyliomeglamic acid
In: Man A683
→ N-(3-Amino-2,6-diiodophenyl)-N-methylglutaramidic acid
In: Man A683
→ N-(3-Amino-4,5-diiodophenyl)-N-methylglutaramidic acid
In: Rat A1475

→ N-(3-Amino-4,6-diiodophenyl)-N-methylglutaramidic acid
In: Man A683
→ Demethyliomeglamic acid
In: Man A683
Rat A1475
→ Iomeglamic acid methyl ester
In: Man A683
Rat A1475

Iomeglamic acid methyl ester
Formed from: Iomeglamic acid

Iopanoic acid (3-amino-α-ethyl-2,4,6-triiodobenzenepropanoic acid)
→ Iopanoic acid-β-D-glucuronide
In: Rat C687

Iopanoic acid-β-D-glucuronide
Formed from: Iopanoic acid

Iopronic acid (2-(2-(3-acetamido-2,4,6-triiodophenoxy)ethoxy)methyl)butyric acid)
→ 2-(3-Acetamido-2,4,6-triiodophenoxy) ethanol
In: Rat A2780, A3381
See also: A2778, A2781, A2783
→ Iopronic acid-β-D-glucuronide
In: Dog A2778
Man A2781
Rat A2780

Iopronic acid-β-D-glucuronide
Formed from: Iopronic acid

Ipecoside
Formed from: Deacetylipecoside

Iprindole (6,7,8,9,10,11-hexahydro-N,N-dimethyl-5*H*-cyclooct[*b*]indole-5-propanamine)
→ Demethyliprindole
In: Dog A3712, G23
Rat A3712, G23
→ Hydroxyiprindole (alicyclic)
In: Rat G23
→ Hydroxyiprindole (aryl)
In: Rat G23

Iproclozide (1-(*p*-chlorophenoxyacetyl)-2-isopropylhydrazine)
→ *p*-Chlorophenoxyacetamide
In: Man A1863

Irbesartan (2-butyl-3-((2′-(1*H*-tetrazol-5-yl)(1′,1′-biphenyl-4-yl)methyl)-1,3-diazaspiro[4,4] non-1-en-4-one)
→ 3-(((2-Butyramido)cyclopentyl) formamidomethyl)-2′-(1*H*-tetrazol-5-yl)biphenyl
In: Macaque J803
Mouse J803
Rabbit J803
Rat J803
→ 2-(3-Carboxyethyl)-3-((2-(1*H*-tetrazol-5-yl)(1,1′-biphenyl)-4-yl)methyl)-1,3-diaza*spiro*[4,4]non-1-en-4-one
In: Macaque J803
Man J403
Mouse J803
Rabbit J803
Rat J803
→ 2-(3-Hydroxybutyl)-3-((2-(1*H*-tetrazol-5-yl)(1,1′-biphenyl)-4-yl)methyl)-1,3-diaza*spiro*[4,4]non-1-en-4-one
In: Man J403
See also: J605
→ 2-Hydroxyirbesartan
In: Man J403
See also: J605
→ 3-Hydroxyirbesartan
In: Man J403
See also: J605
→ Irbesartan-N-β-D-glucuronide
In: Macaque J803
Man J403
Mouse J803
Rabbit J803
Rat J803

Irbesartan-N-β-D-glucuronide
Formed from: Irbesartan

Iridoskyrin ((+)-4,4′,5,5′,8,8′-hexahydro-7,7′-dimethyl(1,1′-bianthracene)-9,9′10,10′-tetrone)
Formed from: Emodinanthrone
→ Skyrin
In: Penicillium A1046

Iriflophenone ((4-hydroxyphenyl)(2,4,6-trihydroxyphenyl)methanone)
→ Maclurin
Probably in: Anemarrhena B806

Irinotecan (camptosar, 1,4′-bipiperidine-1′-carboxylic acid (4*S*)-4,11-diethyl-3,4,12,14-tetrahydro-4*H*-3,14-dioxo-1*H*-pyrano [3′,4′:6,7]indolizino[1,2-*b*]quinolin-9-yl ester)
→ 10-(4-N-(5-Aminopentanoic acid)-1-piperidino)carbonyloxycamptothecin
In: Man J223, K17
→ 10-((4-Amino-1-piperidino)carbonyloxy)-7-ethylcamptothecin
In: Man J434, K17
→ 7-Ethyl-10-hydroxycamptothecin
In: Man J240, J291
Mouse K83
Rat G12, H641, K17
→ Hydroxyirinotecan
In: Man K17

Irsogladine (6-(2,5-dichlorophenyl)-1,3,5-triazine-2,4-diamine)
→ Hydroxyirsogladine
In: Dog J280
Man J280
Rat J280
Rhesus monkey J280
→ Irsogladine-N-oxide
In: Dog J280
Man J280
Rat J280
Rhesus monkey J280

Irsogladine-N-oxide
Formed from: Irsogladine

Isacitretin
Formed from: Acitretin

Isatin (2,3-dioxoindole)
Formed from: 3-Acetonyl-3-hydroxyoxindole
Indole-3-acetic acid
Isatin hydrazone
Isatin oxime
Isatin phenylhydrazone
Isatin semicarbazone
Isatin thiocarbazone
Oxindole
Probably formed from: Indoxyl
→ Anthranilic acid
In: Microorganism F15
Rabbit F803
→ Dioxindole
In: Bradyrhizibium H820
Microorganism F15
Pea H820
Rhizobium H820
→ Isatinic acid
In: Bradyrhizibium H820
Rhizobium H820
→ L-Tryptophan
In: Rabbit F803
See also: C742, F829

Isatin hydrazone
→ Isatin
In: Rabbit F829

Isatinic acid (2-amino-α-oxobenzeneacetic acid)
Formed from: Isatin

Isatin oxime
→ Isatin
In: Rabbit F829

Isatin phenylhydrazone
→ Isatin
In: Rabbit F829

Isatin semicarbazone
→ Isatin
In: Rabbit F829

Isatin thiocarbazone
→ Isatin
In: Rabbit F829

Islandicin (1,4,5-trihydroxy-2-methyl-9,10-anthracenedione)
Formed from: 2-Acetyl-1,8-dihydroxy-3-(2-oxopropyl)naphthalene
Catenarin
Chrysophanol
Emodinanthrone

Isoacitretin ((2*Z*,4*E*,6*E*,8*E*)-9-(4-methoxy-2,3,6-trimethoxyphenyl)-3,7-dimethyl-2,4,6,8-nonatetraenoic acid)
Formed from: Acitretin
→ Acitretin
In: Dog F58
→ Isoacitretin-β-D-glucuronide
In: Rat G495
→ 8-(4-Methoxy-2,3,6-trimethylphenyl)-2,6-dimethyl-*trans*,*trans*-octa-5,7-dienoic acid
In: Rat G495

Isoacitretin-β-D-glucuronide
Formed from: Isoacitretin

Isoamyl gallate
→ Isoamyl gallate-O-sulphate
In: Eubacterium G475
Isoamyl gallate-O-sulphate
Formed from: Isoamyl gallate
Isoandrocymbine ((7*S*,11a*R*)-6,7-dihydro-3-hydroxy-1,2,10-trimethoxy-14-methyl-7,11a-(iminoethano)-11a*H*-dibenzo[*a*,*c*] cyclohepten-9(5*H*)-one)
→ Colchicine
Probably in: Colchicum J458
Isoapocodeine ((6a*R*)-5,6,6a,7-tetrahydro-11-methoxy-6-methyl-4*H*-dibenzo[*de*,*g*]quinolin-10-ol)
Formed from: 10,11-Dimethoxyaporphine
→ Isoapomorphine
In: Rat A1673
Isoapomorphine ((6a*R*)-5,6,6a,7-tetrahydro-6-methyl-4*H*-dibenzo[*de*,*g*]quinoline-9,10-diol)
Formed from: Isoapocodeine
Isobatatasin I
See: 7-Hydroxy-2,4,6-trimethoxyphenanthrene
(+)-Isoboldine (6a*S*)-5,6,6a,7-tetrahydro-2,10-dimethoxy-6-methyl-4*H*-dibenzo[*de*,*g*] quinoline-1,9-diol)
Formed from: Reticuline
→ Boldine
In: Litsea A1547, A3089
→ Thalicarpine
In: Thalictrum F229
4-Isobutoxycatechol
Formed from: *p*-Isobutoxyphenol
→ 4-Isobutoxy-*o*-quinone
In: Mushroom H907
***p*-Isobutoxyphenol**
→ 4-Isobutoxycatechol
In: Mushroom H907
4-Isobutoxy-*o*-quinone
Formed from: 4-Isobutoxycatechol
N-Isobutylalanyl-*o*-toluidide
→ *o*-Toluidine
In: Mouse A830
N-Isobutylaniline
Formed from: N-Isobutyl-N-methylaniline
4-Isobutylcatechol
Formed from: *p*-Isobutylphenol
→ 4-Isobutyl-*o*-quinone
Probably in: Mushroom B897
Isobutyl L-γ-glutamyl dopamide
→ 3,4-Dihydroxyphenethylamine
In: Dog A3945
N-Isobutyl-N-methylaniline
→ N-Isobutylaniline
In: Rabbit A3484
→ N-Isobutyl-N-methylaniline-N-oxide
In: Rabbit A3484
→ N-Methylaniline
In: Rabbit A3484
N-Isobutyl-N-methylaniline-N-oxide
Formed from: N-Isobutyl-N-methylaniline
α-Isobutyl-N-methylphenethylamine
→ α-Isobutylphenethylamine
In: Rat C595
5-Isobutyl-1-naphthylacetic acid
→ 5-(2-Carboxypropyl)-1-naphthylacetic acid
In: Rat F214, F635
→ 5-(2-Hydroxyisobutyl)-1-naphthylacetic acid
In: Rat F214, F635
α-Isobutylphenethylamine
Formed from: α-Isobutyl-N-methylphenethylamine
***p*-Isobutylphenylacetic acid** (ibufenac)
→ *p*-(2-Carboxypropyl)phenylacetic acid
In: Man A3812
→ *p*-(1-Hydroxy-2-methylpropyl)phenylacetic acid
In: Man A3812
→ *p*-(2-Hydroxy-2-methylpropyl)phenylacetic acid
In: Man A3812
2-(*p*-Isobutylphenyl)propan-1,2-diol
Formed from: 2-(*p*-Isobutylphenyl)propylene oxide
2-(*p*-Isobutylphenyl)propionamide
Formed from: 2-(*p*-Isobutylphenyl) propionitrile
→ 2-(4-Isobutylphenyl)propionic acid
Probably in: Rhodococcus G172
(*R*)-2-(4-Isobutylphenyl)propionic acid ((*R*)-ibuprofen; *R* assumed where not stated)
Formed from: (*R*)-2-(4-Isobutylphenyl) propionyl CoA

Probably formed from: 2-(*p*-Isobutylphenyl)propionamide
→ 2-(4-(2-Carboxypropyl)phenyl)propionic acid
In: Man G722
→ (*R*)-2-(4-(2-Hydroxy-2-methylpropyl) phenyl)propionic acid
In: Dog B470
Man A2142, A2428
Rabbit B470
Rat F581
See also: A3560, G722
→ (*S*)-2-(4-Isobutylphenyl)propionic acid
In: Man A2142, A2428
Rat F471, F581, F733, H289, H710
→ (*R*)-2-(4-Isobutylphenyl)propionyl CoA
In: Man G528
Rat G528, J234, K15
→ (*R*)-2-(4-Isobutylphenyl)propionyl-β-D-glucuronic acid
In: Man E886, J416, J617
Monkey E886, H861
Rabbit E886
Rat F564, H539
→ (*R*)-2-(4-Isobutylphenyl)propionyltaurine
In: Man H258
Rat H258
→ 2-(4-Isobutylphenyl)propionyltriglyceride
In: Rat F745

(*S*)-2-(4-Isobutylphenyl)propionic acid
Formed from: (*R*)-Ibuprofen adenylate
Ibuprofen amide
(*R*)-2-(4-Isobutylphenyl)propionic acid
(*S*)-2-(4-Isobutylphenyl)propionyl CoA
→ (*S*)-2-(4-(2-Hydroxy-2-methylpropyl) phenyl)propionic acid
In: Man A2428
Rat F581
→ (*S*)-2-(4-Isobutylphenyl)propionyl CoA
In: Rat K15
→ (*S*)-2-(4-Isobutylphenyl)propionyl-β-D-glucuronic acid
In: Man E886, J416
Monkey E886, H861
Rabbit E886

2-(*p*-Isobutylphenyl)propionitrile
→ 2-(*p*-Isobutylphenyl)propionamide
In: Rhodococcus G172

(*R*)-2-(4-Isobutylphenyl)propionyl CoA
Formed from: (*R*)-2-(4-Isobutylphenyl) propionic acid
(*S*)-2-(4-Isobutylphenyl)propionyl CoA
→ (*R*)-2-(4-Isobutylphenyl)propionic acid
In: Man G304
Rat G304
→ (*S*)-2-(4-Isobutylphenyl)propionyl CoA
In: Man G304
Rat G304

(*S*)-2-(4-Isobutylphenyl)propionyl CoA
Formed from: (*S*)-2-(4-Isobutylphenyl) propionic acid
(*R*)-2-(4-Isobutylphenyl)propionyl CoA
→ (*S*)-2-(4-Isobutylphenyl)propionic acid
In: Man G304
Rat G304
→ (*R*)-2-(4-Isobutylphenyl)propionyl CoA
In: Man G304
Rat G304

(*R*)-2-(4-Isobutylphenyl)propionyl-β-D-glucuronic acid
Formed from: (*R*)-2-(4-Isobutylphenyl) propionic acid

(*S*)-2-(4-Isobutylphenyl)propionyl-β-D-glucuronic acid
Formed from: (*S*)-2-(4-Isobutylphenyl) propionic acid

(*R*)-2-(4-Isobutylphenyl)propionyltaurine
Formed from: (*R*)-2-(4-Isobutylphenyl) propionic acid

2-(4-Isobutylphenyl)propionyltriglyceride
Formed from: 2-(4-Isobutylphenyl)propionic acid

2-(*p*-Isobutylphenyl)propylene oxide
→ 2-(*p*-Isobutylphenyl)propan-1,2-diol
In: Pig H753

4-Isobutyl-*o*-quinone
Probably formed from: 4-Isobutylcatechol

3-Isobutyrylchloramphenicol
Formed from: Chloramphenicol

Isobutyrylsalicylic acid
→ Salicylic acid

In: Acinetobacter A962

O-Isobutylshikonin
Probably formed from: Shikonin

Isocarbostyril (1-(2*H*)-hydroxyisoquinoline)
Formed from: Isoquinoline

Isocarboxazide (5-methyl-3-isoxazolecarboxylic acid 2-(phenylmethyl)hydrazide)
→ Benzylhydrazine
In: Man H465

Isochlorogenic acid
Formed from: Chlorogenic acid

Isochlortetracycline ((4*S*,4a*S*,6*S*,8a*S*)-6-((1*S*)-7-chloro-1,3-dihydro-4-hydroxy-1-methyl-3-oxo-1-isobenzofuranyl)-4-(dimethylamino)-1,4,4a,5,6,7,8,8a-octahydro-3,8a-dihydroxy-1,8-dioxo-2-naphthalenecarboxamide)
Formed from: Isochlortetracycline
→ N-Demethylisochlortetracycline
Probably in: Hen K669
→ 4-Epiisochlortetracycline
Probably in: Hen K669

(*R*,*R*)-Isochondrodendrine ((12a*R*,24a*R*)-2,3,12a,13,14,15,24,24a-octahydro-5,17-dimethoxy-1,13-dimethyl-8,11:20,23-dietheno-1*H*,12*H*-[1,10]dioxacyclooctadecino[2,3,4-*ij*:11,12,13-*i′j′*]diisoquinoline-6,18-dione)
Formed from: (*R*)-N-Methylcoclaurine

Isochorismic acid ((5*S*,6*S*)-5-((1-carboxyethenyl)oxy)-6-hydroxy-1,3-cyclohexadiene-1-carboxylic acid)
→ *p*-Aminobenzoic acid
In: Enterobacter F111
Streptomyces F111
→ *o*-Succinylbenzoic acid
In: Aerobacter E430
Escherichia D682, E430, E457

3-Isochromanone (1,4-dihydro-3*H*-2-benzopyran-3-one)
Formed from: Indan-2-one

Isococculidine ((3β)-1,2-didehydro-3-methoxyerythrinan-15-ol)
Formed from: Cocculidine
→ Cocculidine
In: Erythrina A3588

Isococlaurine ((1*R*)-1,2,3,4-tetrahydro-1-((4-hydroxyphenyl)methyl)-7-methoxy-6-isoquinolinol)
Formed from: Higenamine

Isoconiferin (3-(4-hydroxy-3-methoxyphenyl)-2-propenyl-β-D-glucopyranoside)
Formed from: Coniferyl alcohol

Isoelemicin (3,4,5-trimethoxypropenylbenzene)
→ 3-Hydroxy-4,5-dimethoxypropenylbenzene
In: Rat B198
→ 1-(3,4,5-Trimethoxyphenyl)propan-1,2-diol
In: Rat B198
→ 1-(3,4,5-Trimethoxyphenyl)-2-propen-1-ol
In: Rat B198

'Isoergotamine'
→ 9-Hydroxyisoergotamine
In: Beef K587

Isoetharine (1-(3,4-dihydroxyphenyl)-2-isopropylaminobutan-1-ol)
→ Isoetharine-β-D-glucuronide
In: Dog A286
Man A286
→ Isoetharine-O-sulphate
In: Dog A286
Man A286
→ 3-O-Methylisoetharine
In: Dog A1222
See also: A286, D652

Isoetharine-β-D-glucuronide
Formed from: Isoetharine

Isoetharine-O-sulphate
Formed from: Isoetharine

Isoeugenol (2-methoxy-4-(1-propenyl)phenol)
Formed from: 3,4-Dimethoxypropenylbenzene
Eugenol
2-Methoxy-4-propylphenol
→ Capsaicin
In: Capsicum K295
→ Ferulic acid
In: Capsicum K295
→ 2-(4-Hydroxy-3-methoxyphenyl)-7-methoxy-3-methyl-5-(1-propenyl)benzofuran
In: Horseradish F213

Isofenphos (2-((ethoxy((1-methylethyl)amino) phosphinothioyl)oxy)benzoic acid 1-methylethyl ester)
→ Aminoisofenphos
In: Rat E5, E466
→ N-Deisopropylisofenphos
In: Dog H39
Guinea pig H39
Man H39
Monkey H39
Rat H39
→ O-Ethyl O-(2-isopropoxycarbonylphenyl) N-isopropylphosphoramidate
In: Dog H39
Guinea pig H39
Man H39
Monkey H39
Rat E5, E454, E466, H39
→ O-Ethyl O-(2-isopropoxycarbonylphenyl) phosphate
In: Rat E5, E454, E466
→ O-Ethyl O-(2-isopropoxycarbonylphenyl) phosphorothioate
In: Rat E454
→ O-(2-Isopropoxycarbonylphenyl) phosphorothioate
In: Rat E5

Isofenphos oxon
See: O-Ethyl O-(2-isopropoxycarbonylphenyl) N-isopropylphosphoramidate

Isoferulic acid (3-hydroxy-4-methoxycinnamic acid)
Formed from: Caffeic acid
→ 3-Hydroxy-4-methoxystyrene
In: Candida F390
Fusarium F390
→ Isoferuloyl CoA
In: Aubergine A2514
Cucumber A2514
Glycine A2148, A2514
Leek A2514
Marrow A2514
Pea A2514
Pepper A2514
Phaseolus A2514
Tomato A3149

Isoferuloyl CoA
Formed from: Isoferulic acid

Isoflavanone (2,3-dihydro-3-phenyl-4*H*-1-benzopyran-4-one)
→ 2-Hydroxyisoflavanone
In: Aspergillus F673
→ 4′-Hydroxyisoflavanone
In: Absidia F673, G14
→ Isoflavone
In: Aspergillus F673

Isoflav-3-ene (3-phenyl-2*H*-1-benzopyran)
→ 9-O-Methylcoumestrol
In: Medicago A3640

Isoflavone (3-phenyl-4*H*-1-benzopyran-4-one)
Formed from: Isoflavanone
→ 3′,4′-Dihydroxyisoflavone
In: Aspergillus F673

Isoformononetin (4′-hydroxy-7-methoxyisoflavone)
Formed from: Daidzein

Isoimperatorin (4-((3-methyl-2-butenyl)oxy)-7*H*-furo[3,2-*g*][1]benzopyran-7-one)
Formed from: L-Phenylalanine
Formed from mevalonic acid in Thamnosma A1013

L-Isoleucyl-β-naphthylamide
→ β-Naphthylamine
In: Agave A2745
Glycine A1596
Rat A523

L-Isoleucyl-*p*-nitroanilide
Formed from: Succinyl-L-alanyl-L-alanyl-L-isoleucyl-*p*-nitroanilide

Isoliquiritigenin
See: 2′,4,4′-Trihydroxychalcone

D-Isolysergamide ((8α)-9,10-didehydro-6-methylergoline-8-carboxamide)
→ D-Norisolysergamide
In: Rat A98

Isolysergic diethylamide
→ 13-Hydroxyisolysergic diethylamide
In: Rat B390
→ 14-Hydroxyisolysergic diethylamide
In: Rat B390
→ Norisolysergic diethylamide
In: Rat A98
→ 2-Oxoisolysergic diethylamide

In: Rat B390
D-Isolysergic dimethylamide
→ D-Norisolysergic dimethylamide
In: Rat A98
Isomagnolol (5,5′-di(1*E*)-1-propenyl(1,1′-biphenyl)-2,2′-diol)
Probably formed from: 2,2′-Dihydroxy-5-(1-propen-1(*E*)-yl)-5′-propylbiphenyl
Isomangiferin (4-β-D-glucopyranosyl-1,3,6,7-tetrahydroxy-9*H*-xanthen-9-one)
Formed from: *p*-Coumaric acid
Maclurin
Isomethergine ((8α)-9,10-didehydro-N-(1-(hydroxymethyl)propyl)-6-methylergoline-8-carboxamide)
→ Norisomethergine
In: Rat A98
Isoorientin (2-(3,4-dihydroxyphenyl)-6-β-D-glucopyranosyl-5,7-dihydroxy-4*H*-1-benzopyran-4-one)
Formed from: Cinnamic acid
→ Isoorientin-2″-D-glucoside
In: Silene B385
→ Isoscoparin
In: Silene C351
Isoorientin-2″-D-glucoside
Formed from: Isoorientin
Isopapaverine (4-(3,4-dimethoxybenzyl)-6,7-dimethoxyisoquinoline)
→ 3′-Demethylisopapaverine
In: Silene E560
→ 4′-Demethylisopapaverine
In: Silene E560
→ 6-Demethylisopapaverine
In: Silene E560
→ 7-Demethylisopapaverine
In: Silene E560
Isopentyl 2,3-dihydro-8-(2-hydroxy-3-(4-(1-oxo-3-(3,4,5-trimethoxyphenyl)-2-propenyl)-1-piperazinyl)propoxy)-1,4-benzodioxin-5-carboxylate
→ 2,3-Dihydro-8-(2-hydroxy-3-(4-(1-oxo-3-(3,4,5-trimethoxyphenyl)-2-propenyl)-1-piperazinyl)propoxy)-1,4-benzodioxin-5-carboxylic acid
In: Dog C377
Man C377
Rat C377
Isophthalaldehyde (3-formylbenzaldehyde)
→ *m*-Formylbenzoic acid
In: Euglena G422
See also: A2094, G197
→ *m*-Hydroxymethylbenzaldehyde
In: Dunaliella G459
Euglena G422
→ Isophthalic acid
In: Streptomyces C284
Isophthalaldehydic acid
See: *m*-Formylbenzoic acid
Isophthalamidic acid (isophthalic acid monoamide)
Formed from: N-Ethylisophthalamidic acid
m-Formylbenzamide
Isophthalic acid (1,3-benzenedicarboxylic acid)
Formed from: *m*-Cyanobenzoic acid
m-Formylbenzoic acid
Isophthalaldehyde
Salicylic acid
→ Protocatechuic acid
In: Microorganism C269
Pseudomonas C91
Isophthalic acid monoamide
See: Isophthalamidic acid
Isophthalonitrile (1,3-benzodinitrile)
→ *m*-Cyanobenzoic acid
In: Rhododcoous J825
See also: A3364, A3420, D725
Isopimpinellin (4,9-dimethoxy-7*H*-furo[3,2-*c*][1]benzopyran-7-one)
Formed from: Bergapten
8-Hydroxybergapten
5-Hydroxyxanthotoxin
8-Methoxypsoralen
L-Phenylalanine
Formed from acetate and glycine in Thamnosma A1012
Probably formed from: Marmesin
Isoprenaline (N-isopropylnoradrenaline, isoproterenol)
→ 3,4-Dihydroxymandelaldehyde
In: Rat A724
Contrast (in man) A724
→ 3-Hydroxy-N-isopropyl-4-methoxyphenylethanolamine

In: Man A724
Rat A486, A724
→ Isoprenaline quinone
In: Apple J678
See also: A63, A624, B897
→ Isoprenaline sulphate
In: Dog A588, A1404, A1839, B665, C187
Guinea pig B665, C187
Man A588, A1404, A1839, C523, D736
Monkey B665, C187
Mouse B665, C187
Rabbit B665, C187
Rat B665, C187
→ N-Isopropylnormetanephrine
In: Cat A3452
Dog A588, A684, A1222, A1404, A1839
Guinea pig A115
Man A38, A588, A684, A1404, A1839, A3831
Rabbit A115, D728
Rat A115, A314, A486, A1471, B168, D704

L-Isoprenaline
→ L-3-Hydroxy-N-isopropyl-4-methoxyphenylethanolamine
In: Rat A486
→ L-N-Isopropylnormetanephrine
In: Pig D652
Rat A486

Isoprenaline quinone
Formed from: Isoprenaline

Isoprenaline sulphate
Formed from: Isoprenaline

Isopropenylbenzene
See: α-Methylstyrene

***p*-Isopropoxyacetanilide**
→ *p*-Acetamidophenol
In: Cunninghamella F590

Isopropoxybenzene
→ *o*-Isopropoxyphenol
In: Rat G500
→ *p*-Isopropoxyphenol
In: Rat G500
→ Phenol
In: Rat G500

5-Isopropoxycarbonyl-2,6-dimethyl-4-(*m*-nitrophenyl)pyridine-3-carboxylic acid
Probably formed from: 1,4-Dihydro-5-isopropoxycarbonyl-2,6-dimethyl-4-(*m*-nitrophenyl)pyridine-3-carboxylic acid
→ 2,6-Dimethyl-4-(*m*-nitrophenyl)pyridine-3,5-dicarboxylic acid
Probably in: Rat G189
→ 6-Hydroxymethyl-5-isopropoxycarbonyl-2-methyl-4-(*m*-nitrophenyl)pyridine-3-carboxylic acid
Probably in: Rat G189

5-Isopropoxycarbonyl-3-(2-methoxyethoxy)carbonyl-2,6-dimethyl-4-(*m*-nitrophenyl)pyridine
Formed from: Nimodipine
→ 2-Hydroxymethyl-5-isopropoxycarbonyl-3-(2-methoxyethoxy)carbonyl-6-methyl-4-(*m*-nitrophenyl)pyridine
Probably in: Dog G189
Monkey G189
Rat G189

O-(2-Isopropoxycarbonylphenyl) phosphorothioate
Formed from: Isofenphos

4-Isopropoxy-3,5-dimethoxybenzoic acid
→ 4-(1-Hydroxy-2-propoxy)-3,5-dimethoxybenzoic acid
In: Polyporus A1117

4-Isopropoxy-3-methoxybenzaldehyde
Formed from: 4-Isopropoxy-3-methoxybenzoic acid
→ 4-Isopropoxy-3-methoxybenzyl alcohol
In: Polyporus A1117

4-Isopropoxy-3-methoxybenzoic acid
→ 4-(1-Hydroxy-2-propoxy)-3-methoxybenzoic acid
In: Polyporus A1117
→ 4-Isopropoxy-3-methoxybenzaldehyde
In: Polyporus A1117

4-Isopropoxy-3-methoxybenzyl alcohol
Formed from: 4-Isopropoxy-3-methoxybenzaldehyde

***o*-Isopropoxyphenol**
Formed from: Isopropoxybenzene
Propoxur

***p*-Isopropoxyphenol**
Formed from: Isopropoxybenzene

o-Isopropoxyphenyl N-methylcarbamate
(baygon)
→ 5-Hydroxy-2-isopropoxyphenyl N-methylcarbamate
Probably in: Anopheles D195

1-(o-Isopropoxyphenyl)piperazine
Formed from: Mazapertine

3′-Isopropoxy-2-trifluoromethylbenzanilide
See: Flutolanil

N-Isopropyl-o-acetamidothiophenol
Formed from: 2-(S-(L-Cysteinyl))-N-isopropyl-N-phenylacetamide

N-Isopropylacetanilide
Formed from: Propachlor
→ Acetanilide
In: Rat K340
→ p-Hydroxy-N-isopropylacetanilide
In: Rat K340
→ N-(1-Hydroxy-2-propyl)acetanilide
In: Rat K340
→ N-Isopropylglycollanilide
In: Rat K340

Isopropyl 2-acetyl-5-benzimidazolecarbamate
Probably formed from: Isopropyl 2-(1-hydroxyethyl)-5-benzimidazolecarbamate

1-Isopropylamino-4,4-diphenyl-2-butanol
See: Drobuline

o-(2-Isopropylaminoethoxy)acetophenone
Probably formed from: o-(2-Diisopropylaminoethoxy)acetophenone
3-Hydroxy-o-(2-isopropylaminoethoxy) butyrophenone

o-(2-Isopropylaminoethoxy)butyrophenone
Formed from: Ketocaine
→ 1-(o-(2-Aminoethoxy)phenyl)butanol
Probably in: Rat A3324
→ 2,3-Dihydro-5-propyl-1,4-benzoxazepine
Probably in: Rat A3324
→ 3-Hydroxy-o-(2-isopropylaminoethoxy) butyrophenone
Probably in: Rat A3324
→ 5′-Hydroxy-2′-(2-isopropylaminoethoxy) butyrophenone
Probably in: Rat A3324

2′-(2-Isopropylaminoethoxy)butyrophenon-5′-yl-β-D-glucuronide
Probably formed from: 5′-Hydroxy-2′-(2-isopropylaminoethoxy)butyrophenone

o-(2-Isopropylaminoethoxy)-3-oxobutyrophenone
Probably formed from: 3-Hydroxy-o-(2-isopropylaminoethoxy)butyrophenone
→ 2,3-Dihydro-5-(2-oxopropyl)-1,4-benzoxazepine
Probably in: Rat A3324

1-(o-(2-Isopropylamino)ethoxy)phenyl-1-butanol
Formed from: 1-(o-(2-Diisopropylamino) ethoxy)phenyl-1-butanol

9-(3-Isopropylamino)-9H-fluorene-9-carboxamide
See: Indecainide

1-Isopropylamino-3-(4-(2-methoxyethyl)phenoxy) propan-2-ol
See: Metoprolol

2-Isopropylamino-1-(2-naphthyl)ethanol
(pronethalol)
→ 1,2-Dihydroxy-1-(2-naphthyl)ethane
In: Mouse A1536
Rat A1536
→ 1-(6-Hydroxy-2-naphthyl)-2-isopropylaminoethanol
In: Mouse A1536
→ 1-(7-Hydroxy-2-naphthyl)-2-isopropylaminoethanol
In: Mouse A1536

N-Isopropylamphetamine
Formed from: N-Isopropy-N-methylamphetamine
→ Amphetamine
In: Cunninghamella F345
Rat A300
→ N-Hydroxy-N-isopropylamphetamine
In: Cunninghamella F345
→ p-Hydroxy-N-isopropylamphetamine
In: Cunninghamella F345
→ Phenylacetone
In: Rat A300

N-Isopropylaniline
Formed from: N,N-Diisopropylaniline
N-Isopropyl-N-methylaniline

p-Isopropylaniline
→ p-Isopropylnitrosobenzene

With chloroperoxidase A3579
4-Isopropylantipyrine
→ 4-(1,2-Dihydroxy-2-propyl)antipyrine
Probably in: Rat D285
→ 3-Hydroxymethyl-4-isopropyl-2-methyl-1-phenyl-3-pyrazolin-5-one
In: Rat D285
→ 1-(Hydroxypropyl)-4-isopropyl-2,3-dimethyl-3-pyrazolin-5-one
In: Rat D285
→ 4-Isopropyl-3-methyl-1-phenyl-3-pyrazolin-5-one
In: Man A1778
Rat A1778, D285
→ 2-(3-Methyl-1-phenylpyrazolin-5-on-4-yl) propionic acid
In: Rat B818
***o*-Isopropylbenzaldehyde**
→ *o*-Isopropylbenzyl alcohol
In: Rabbit F457
***p*-Isopropylbenzaldehyde**
See: Cumaldehyde
N-Isopropylbenzamide radical
Formed from: Procarbazine
Isopropylbenzene (cumene)
→ 1-Phenyl-1-propanol
In: Rabbit C891
→ 2-Phenyl-1-propanol
In: Rabbit G499
Rhodococcus H213
Isopropyl 5-benzimidazolecarbamate
Probably formed from: Isopropyl 2-carboxamido-5-benzimidazolecarbamate
***p*-Isopropylbenzoic acid**
See: Cumic acid
3-Isopropyl-1*H*-2,1,3-benzothiadiazin-4(3*H*)-one-2,2-dioxide
See: Bentazon
Isopropyl N-benzoyl-N-(3-chloro-4-fluorophenyl)-2-aminopropionate
See: Barnon
***o*-Isopropylbenzyl alcohol**
Formed from: *o*-Isopropylbenzaldehyde
***p*-Isopropylbenzyl alcohol**
See: Cumyl alcohol
Isopropyl 6-benzyloxy-4-methoxynethyl-β-carboline-3-carboxylate
See: Abecarnil
4-Isopropylbiphenyl
→ 2-(Biphenyl-4-yl)-1-propanol
In: Rat A3055
→ 2-(Biphenyl-4-yl)-2-propanol
In: Rat A3055
1-(N-Isopropylcarbamoylmethyl)-4-(3,4,5-trimethoxycinnamoyl)piperazine
→ 1-Carbamoylmethyl-4-(3,4,5-trimethoxycinnamoyl)piperazine
In: Dog A259
Man A259
Rat A259
→ N-(3,4,5-Trimethoxycinnamoyl)piperazine
In: Dog A259
Man A259
Rat A259
Isopropyl carbanilate (propham, isopropyl N-phenylcarbamate)
→ Aniline
In: Bacillus A1006
→ Isopropyl *m*-hydroxycarbanilate
In: Chick A508
→ Isopropyl *p*-hydroxycarbanilate
In: Chick A508, A1691
Goat A1690
Isopropyl 2-carboxamido-5-benzimidazolecarbamate
Formed from: Cambendazole
→ Isopropyl 5-benzimidazolecarbamate
Probably in: Pig A3510
Sheep A3510
3-Isopropylcatechol
Formed from: *o*-Isopropylphenol
→ 2-Hydroxy-6-isopropyl-*cis*,*cis*-muconic semialdehyde
In: Pseudomonas H337
See also: A1323, A2694, G766
4-Isopropylcatechol
Formed from: 2-Phenylpropanesulphonic acid
→ 4-Isopropyl-2-methoxyphenol
In: Rat A486
→ 5-Isopropyl-2-methoxyphenol
In: Rat A486

Isopropyl (4′-(*p*-chlorobenzoyl)-2-methyl-2-phenoxy)propionate
See: Fenofibrate

Isopropyl 3-chlorocarbanilate
→ 1-Hydroxy-2-propyl 3-chlorocarbanilate
In: Glycine A1833

Isopropyl N-(3-chloro-4-fluorophenyl)-N-(*m*-hydroxybenzoyl)-2-aminopropionate
Formed from: Barnon
→ N-(3-Chloro-4-fluorophenyl)-N-(*m*-hydroxybenzoyl)-2-aminopropionic acid
Probably in: Dog A2035
Rat A2035

Isopropyl N-(3-chloro-4-fluorophenyl)-N-(*p*-hydroxybenzoyl)-2-aminopropionate
Formed from: Barnon
→ N-(3-Chloro-4-fluorophenyl)-N-(*p*-hydroxybenzoyl)-2-aminopropionic acid
Probably in: Dog A2035
Rat A2035

Isopropyl N-(3-chloro-4-hydroxyphenyl)carbamate
→ Isopropyl N-(3-chloro-4-hydroxyphenyl) carbamate-β-D-glucuronide
In: Rat G748

Isopropyl N-(3-chloro-4-hydroxyphenyl) carbamate-β-D-glucuronide
Formed from: Isopropyl N-(3-chloro-4-hydroxyphenyl)carbamate

Isopropyl 6-(S-(L-cysteinyl))-2-ethyl-3,4-dihydro-7-hydroxy-3-oxo-2*H*-quinoxaline-1-carboxylate
Formed from: Isopropyl (*S*)-2-ethyl-7-fluoro-3,4-dihydro-3-oxo-2*H*-quinoxaline-1-carboxylate

Isopropyl 3,4-diethoxycarbanilate (diethofenocarb)
→ Isopropyl 3-ethoxy-4-hydroxycarbanilate
In: Rat F922

Isopropyl 3,4-dihydroxycarbanilate
Formed from: Isopropyl *p*-hydroxycarbanilate
→ Isopropyl 3,4-dihydroxycarbanilate-3-sulphate
Probably in: Chick A508
→ Isopropyl 4-hydroxy-3-methoxycarbanilate
In: Chick A508

Isopropyl 3,4-dihydroxycarbanilate-3-sulphate
Probably formed from: Isopropyl 3,4-dihydroxycarbanilate

Isopropyl 2-(1,2-dihydroxyethyl)-5-benzimidazolecarbamate
Probably formed from: Isopropyl 2-(1-formamido-2-hydroxyethyl)-5-benzimidazolecarbamate

***p*-Isopropyl-N,N-dimethylaniline**
→ *p*-Isopropyl-N-methylaniline
In: Rat D578

N-Isopropyl-4,4-diphenylcyclohexylamine
See: Pramiverine

Isopropyl 3-ethoxy-4-hydroxycarbanilate
Formed from: Isopropyl 3,4-diethoxycarbanilate
→ 3-Ethoxy-4-hydroxycarbanilate
Probably in: Rat F922
→ Isopropyl 3-ethoxy-4-hydroxycarbanilate-β-D-glucuronide
Probably in: Rat F922
→ Isopropyl 3-ethoxy-4-hydroxycarbanilate sulphate
Probably in: Rat F922

Isopropyl 3-ethoxy-4-hydroxycarbanilate-β-D-glucuronide
Probably formed from: Isopropyl 3-ethoxy-4-hydroxycarbanilate

Isopropyl 3-ethoxy-4-hydroxycarbanilate sulphate
Probably formed from: Isopropyl 3-ethoxy-4-hydroxycarbanilate

Isopropyl 2-ethyl-3,4-dihydro-7-hydroxy-3-oxo-2*H*-quinoxaline-1-carboxylate
Formed from: Isopropyl 2-ethyl-7-fluoro-3,4-dihydro-3-oxo-2*H*-quinoxaline-1-carboxylate
→ Isopropyl 2-ethyl-3,4-dihydro-7-hydroxy-3-oxo-2*H*-quinoxaline-1-carboxylate-β-D-glucuronide
Probably in: Pig J749

Isopropyl 2-ethyl-3,4-dihydro-7-hydroxy-3-oxo-2*H*-quinoxaline-1-carboxylate-β-D-glucuronide
Probably formed from: Isopropyl 2-ethyl-3,4-dihydro-7-hydroxy-3-oxo-2*H*-quinoxaline-1-carboxylate

Isopropyl (*S*)-2-ethyl-6-fluoro-3,4-dihydro-7-hydroxy-3-oxo-2*H*-quinoxaline-1-carboxylate
Formed from: Isopropyl (*S*)-2-ethyl-7-fluoro-3,4-dihydro-3-oxo-2*H*-quinoxaline-1-carboxylate
→ Isopropyl (*S*)-2-ethyl-6-fluoro-3,4-dihydro-7-hydroxy-3-oxo-2*H*-quinoxaline-1-carboxylate-β-D-glucuronide
Probably in: Cynomolgus monkey J749, K29
Dog J749, K29
Man K29
Mouse K29
Pig J749, K29
Rabbit K29

Isopropyl (*S*)-2-ethyl-7-fluoro-3,4-dihydro-5-hydroxy-3-oxo-2*H*-quinoxaline-1-carboxylate
Formed from: Isopropyl (*S*)-2-ethyl-7-fluoro-3,4-dihydro-3-oxo-2*H*-quinoxaline-1-carboxylate
→ 1-Hydroxy-2-propyl (*S*)-2-ethyl-7-fluoro-3,4-dihydro-5-hydroxy-3-oxo-2*H*-quinoxaline-1-carboxylate
Probably in: Cynomolgus monkey K29
Dog K29
Man K29
Mouse K29
Pig K29
Rabbit K29
→ Isopropyl (*S*)-2-ethyl-7-fluoro-3,4-dihydro-5-hydroxy-3-oxo-2*H*-quinoxaline-1-carboxylate-5-β-D-glucuronide
In: Cynomolgus monkey J749, K29
Dog J749, K29
Man K29
Mouse K29
Pig J749, K29
Rabbit K29

Isopropyl (*S*)-2-ethyl-7-fluoro-3,4-dihydro-6-hydroxy-3-oxo-2*H*-quinoxaline-1-carboxylate
Formed from: Isopropyl (*S*)-2-ethyl-7-fluoro-3,4-dihydro-3-oxo-2*H*-quinoxaline-1-carboxylate
→ Isopropyl (*S*)-2-ethyl-7-fluoro-3,4-dihydro-6-hydroxy-3-oxo-2*H*-quinoxaline-1-carboxylate-6-β-D-glucuronide
In: Cynomolgus monkey J749, K29
Dog J749, K29
Man K29
Mouse K29
Pig J749, K29
Rabbit K29

Isopropyl (*S*)-2-ethyl-8-fluoro-3,4-dihydro-7-hydroxy-3-oxo-2*H*-quinoxaline-1-carboxylate
Formed from: Isopropyl (*S*)-2-ethyl-7-fluoro-3,4-dihydro-3-oxo-2*H*-quinoxaline-1-carboxylate

Isopropyl (*S*)-2-ethyl-6-fluoro-3,4-dihydro-7-hydroxy-3-oxo-2*H*-quinoxaline-1-carboxylate-5-β-D-glucuronide
Probably formed from: Isopropyl (*S*)-2-ethyl-6-fluoro-3,4-dihydro-7-hydroxy-3-oxo-2*H*-quinoxaline-1-carboxylate

Isopropyl (*S*)-2-ethyl-7-fluoro-3,4-dihydro-5-hydroxy-3-oxo-2*H*-quinoxaline-1-carboxylate-5-β-D-glucuronide
Probably formed from: Isopropyl (*S*)-2-ethyl-7-fluoro-3,4-dihydro-5-hydroxy-3-oxo-2*H*-quinoxaline-1-carboxylate

Isopropyl (*S*)-2-ethyl-7-fluoro-3,4-dihydro-6-hydroxy-3-oxo-2*H*-quinoxaline-1-carboxylate-6-β-D-glucuronide
Probably formed from: Isopropyl (*S*)-2-ethyl-7-fluoro-3,4-dihydro-6-hydroxy-3-oxo-2*H*-quinoxaline-1-carboxylate

Isopropyl (*S*)-2-ethyl-7-fluoro-3,4-dihydro-3-oxo-2*H*-quinoxaline-1-carboxylate
→ 2-Ethyl-7-fluoro-3,4-dihydro-3-oxo-2*H*-quinoxaline
In: Rabbit K29
→ 1-Hydroxy-2-propyl (*S*)-2-ethyl-7-fluoro-3,4-dihydro-3-oxo-2*H*-quinoxaline-1-carboxylate
In: Cynomolgus monkey J749, K29
Dog J749, K29
Man K29
Pig J749, K29
Rabbit K29
→ Isopropyl 6-(S-(L-cysteinyl))-2-ethyl-3,4-dihydro-7-hydroxy-3-oxo-2*H*-quinoxaline-1-carboxylate
In: Rabbit K29

→ Isopropyl (*S*)-2-ethyl-6-fluoro-3,4-dihydro-7-hydroxy-3-oxo-2*H*-quinoxaline-1-carboxylate
In: Man K29
Rabbit K29
→ Isopropyl (*S*)-2-ethyl-7-fluoro-3,4-dihydro-5-hydroxy-3-oxo-2*H*-quinoxaline-1-carboxylate
In: Cynomolgus monkey J749, K29
Dog J749, K29
Man K29
Mouse K29
Pig J749, K29
Rabbit K29
→ Isopropyl (*S*)-2-ethyl-7-fluoro-3,4-dihydro-6-hydroxy-3-oxo-2*H*-quinoxaline-1-carboxylate
In: Cynomolgus monkey J749, K29
Dog J749, K29
Man K29
Mouse K29
Pig J749, K29
Rabbit K29
→ Isopropyl (*S*)-2-ethyl-8-fluoro-3,4-dihydro-7-hydroxy-3-oxo-2*H*-quinoxaline-1-carboxylate
In: Man K29
Mouse K29
Rabbit K29
→ Isopropyl (*S*)-7-fluoro-3,4-dihydro-2-(2-hydroxyethyl)-3-oxo-2*H*-quinoxaline-1-carboxylate
In: Cynomolgus monkey J749, K29
Dog J749, K29
Man K29
Mouse K29
Pig J749, K29
Rabbit K29
→ Isopropyl 2-ethyl-3,4-dihydro-7-hydroxy-3-oxo-2*H*-quinoxaline-1-carboxylate
In: Cynomolgus monkey J749, K29
Dog J749, K29
Man K29
Mouse K29
Pig J749, K29
Rabbit K29

Isopropyl (*S*)-7-fluoro-3,4-dihydro-2-(2-hydroxyethyl)-3-oxo-2*H*-quinoxaline-1-carboxylate
Formed from: Isopropyl (*S*)-2-ethyl-7-fluoro-3,4-dihydro-3-oxo-2*H*-quinoxaline-1-carboxylate
→ Isopropyl (*S*)-7-fluoro-3,4-dihydro-2-(2-hydroxyethyl)-3-oxo-2*H*-quinoxaline-1-carboxylate-β-D-glucuronide
In: Cynomolgus monkey J749, K29
Dog J749, K29
Man K29
Mouse K29
Pig J749, K29
Rabbit K29

Isopropyl (*S*)-7-fluoro-3,4-dihydro-2-(2-hydroxyethyl)-3-oxo-2*H*-quinoxaline-1-carboxylate-β-D-glucuronide
Probably formed from: Isopropyl (*S*)-7-fluoro-3,4-dihydro-2-(2-hydroxyethyl)-3-oxo-2*H*-quinoxaline-1-carboxylate

Isopropyl 2-(1-formamido-2-hydroxyethyl)-5-benzimidazolecarbamate
Formed from: Cambendazole
→ Isopropyl 2-(1,2-dihydroxyethyl)-5-benzimidazolecarbamate
Probably in: Pig A3510
Sheep A3510

Isopropyl 2-formamidomethyl-5-benzimidazolecarbamate
Formed from: Cambendazole
→ Isopropyl 2-hydroxymethyl-5-benzimidazolecarbamate
Probably in: Pig A3510
Sheep A3510

N-Isopropylglycollanilide (2-hydroxy-N-isopropylacetanilide)
Formed from: N-Isopropylacetanilide
→ N-Isopropyloxanilic acid
Probably in: Rat K340
→ N-Isopropylglycollanilide-β-D-glucuronide
In: Rat K340

N-Isopropylglycollanilide-β-D-glucuronide
Formed from: N-Isopropylglycollanilide

N-Isopropylglycyl-2,6-dimethylanilide
→ 2,6-Dimethylaniline
In: Mouse A830

Isopropyl *m*-hydroxycarbanilate
Formed from: Isopropyl carbanilate
→ Isopropyl *m*-hydroxycarbanilate sulphate
In: Chick A508

Isopropyl *p*-hydroxycarbanilate
Formed from: Isopropyl carbanilate
→ *p*-Aminophenol
Probably in: Chick A508
→ Isopropyl 3,4-dihydroxycarbanilate
In: Chick A508
→ Isopropyl *p*-hydroxycarbanilate-β-D-glucuronide
In: Chicken A1691
See also: A508
→ Isopropyl *p*-hydroxycarbanilate sulphate
In: Chicken A1691
See also: A508

Isopropyl *p*-hydroxycarbanilate-β-D-glucuronide
Formed from: Isopropyl *p*-hydroxycarbanilate

Isopropyl *m*-hydroxycarbanilate sulphate
Formed from: Isopropyl *m*-hydroxycarbanilate

Isopropyl *p*-hydroxycarbanilate sulphate
Formed from: Isopropyl *p*-hydroxycarbanilate

Isopropyl 2-(1-hydroxyethyl)-5-benzimidazolecarbamate
Formed from: Cambendazole
→ Isopropyl 2-acetyl-5-benzimidazolecarbamate
Probably in: Pig A3510
Sheep A3510

Isopropyl 4-hydroxy-3-methoxycarbanilate
Formed from: Isopropyl 3,4-dihydroxycarbanilate
→ Isopropyl 4-hydroxy-3-methoxycarbanilate sulphate
Probably in: Chick A508

Isopropyl 4-hydroxy-3-methoxycarbanilate sulphate
Probably formed from: Isopropyl 4-hydroxy-3-methoxycarbanilate

Isopropyl 6-hydroxy-4-methoxymethyl-β-carboline-3-carboxylate
Formed from: Abecarnil
→ Isopropyl 6-hydroxy-4-methoxymethyl-β-carboline-3-carboxylate-β-D-glucuronide
In: Man G2
→ Isopropyl 6-hydroxy-4-methoxymethyl-β-carboline-3-carboxylate-O-sulphate
In: Man G2

Isopropyl 6-hydroxy-4-methoxymethyl-β-carboline-3-carboxylate-β-D-glucuronide
Formed from: Isopropyl 6-hydroxy-4-methoxymethyl-β-carboline-3-carboxylate

Isopropyl 6-hydroxy-4-methoxymethyl-β-carboline-3-carboxylate-O-sulphate
Formed from: Isopropyl 6-hydroxy-4-methoxymethyl-β-carboline-3-carboxylate

Isopropyl 2-hydroxymethyl-5-benzimidazolecarbamate
Probably formed from: Isopropyl 2-formamidomethyl-5-benzimidazolecarbamate

Isopropyl 2-(2-hydroxy-2-methylsulphonylethyl)-5-benzimidazolecarbamate
Formed from: Cambendazole

4,4′-Isopropylidenebis(2,6-dibromophenol)
See: Tetrabromobisphenol A

***p*-Isopropylmandelonitrile**
Formed from: Cumaldehyde

N-Isopropyl-2-mercaptoacetanilide
Formed from: Propachlor
→ N-(1-Hydroxyisopropyl)-2-methylsulphonylacetanilide-β-D-glucuronide
Probably in: Rat B707
→ *p*-Hydroxy-N-isopropyl-2-methylsulphonylacetanilide-β-D-glucuronide
Probably in: Rat B707

4-Isopropyl-2-methoxyphenol
Formed from: 4-Isopropylcatechol

5-Isopropyl-2-methoxyphenol
Formed from: 4-Isopropylcatechol

N-Isopropyl-N-methylamphetamine
→ N-Isopropylamphetamine
In: Rat A300
→ N-Methylamphetamine
In: Rat A300

***p*-Isopropyl-N-methylaniline**
Formed from: *p*-Isopropyl-N,N-dimethylaniline
→ N-Isopropylaniline
In: Rabbit A3484
→ N-Isopropyl-N-methylaniline-N-oxide

In: Rabbit A3484
→ N-Methylaniline
In: Rabbit A3484

N-Isopropyl-N-methylaniline-N-oxide
Formed from: N-Isopropyl-N-methylaniline

N-Isopropyl-α-(2-methylazo)-*p*-toluamide
See: Azo-procarbazine

N-Isopropyl-α-(2-methyl-ONN-azoxy)-*p*-toluamide
Formed from: Azo-procarbazine

3-Isopropyl 5-methyl 2-carbamoyloxymethyl-6-carboxy-4-(2,3-dichlorophenyl)-3,5-pyridinedicarboxylate
Probably formed from: 3-Isopropyl 5-methyl 2-carbamoyloxymethyl-4-(2,3-dichlorophenyl)-6-methyl-3,5-pyridinedicarboxylate

3-Isopropyl 5-methyl 2-carbamoyloxymethyl-4-(2,3-dichlorophenyl)-1,4-dihydro-6-methyl-3,5-pyridinedicarboxylate
→ 2-Carbamoyloxymethyl-5-carboxy-4-(2,3-dichlorophenyl)-1,4-dihydro-3-isopropoxycarbonyl-6-methylpyridine
In: Dog F493
Rat F493, F921
→ 2-Carbamoyloxymethyl-3-carboxy-4-(2,3-dichlorophenyl)-1,4-dihydro-5-methoxycarbonyl-6-methylpyridine
In: Dog F493
Rat F493, F921
→ 3-(2-Hydroxyisopropyl) 5-methyl 2-carbamoyloxymethyl-4-(2,3-dichlorophenyl)-1,4-dihydro-6-methyl-3,5-pyridinedicarboxylate
In: Rat F921
→ 3-(1-Hydroxy-2-propyl) 5-methyl 2-carbamoyloxymethyl-4-(2,3-dichlorophenyl)-1,4-dihydro-6-methyl-3,5-pyridinedicarboxylate
In: Dog F493
Rat F493
→ 3-Isopropyl 5-methyl 2-carbamoyloxymethyl-4-(2,3-dichlorophenyl)-6-methyl-3,5-pyridinedicarboxylate
In: Dog F493
Rat F493, F921

3-Isopropyl 5-methyl 2-carbamoyloxymethyl-4-(2,3-dichlorophenyl)-6-methyl-3,5-pyridinedicarboxylate
Formed from: 3-Isopropyl 5-methyl 2-carbamoyloxymethyl-4-(2,3-dichlorophenyl)-1,4-dihydro-6-methyl-3,5-pyridinedicarboxylate
→ 2-Carbamoyloxymethyl-5-carboxy-4-(2,3-dichlorophenyl)-3-isopropoxycarbonyl-6-methylpyridine
Probably in: Dog F493
Rat F493
→ 2-Carbamoyloxymethyl-3-carboxy-4-(2,3-dichlorophenyl)-5-methoxycarbonyl-6-methylpyridine
Probably in: Rat F493
→ 3-Isopropyl 5-methyl 2-carbamoyloxymethyl-6-carboxy-4-(2,3-dichlorophenyl)-3,5-pyridinedicarboxylate
Probably in: Rat F493

3-Isopropyl-6-methylcatechol
Formed from: 1,2-Dihydro-1,2-dihydroxy-3-isopropyl-6-methylbenzene

5-Isopropyl 3-methyl 2-cyano-1,4-dihydro-6-methyl-4-(3-nitrophenyl)-3,5-pyridinedicarboxylate
See: Nilvadipine

N-Isopropyl-α-(2-methylhydrazino)-*p*-toluamide
See: Procarbazine

α-Isopropyl-N-methylphenethylamine
→ α-Isopropylphenethylamine
In: Rat C595

2-Isopropyl-5-methylphenol
See: Thymol

5-Isopropyl-2-methylphenol
See: Carvacrol

4-Isopropyl-3-methyl-1-phenyl-3-pyrazolin-5-one
Formed from: 4-Isopropylantipyrine
→ 4-(1-Carboxyethyl)-3-methyl-1-phenyl-3-pyrazolin-5-one
Probably in: Rat A3979
→ N-Demethylisopropylantipyrine-β-D-glucuronide
In: Man A1778
Rat A1778
→ 4-(1-Hydroxyethyl)-3-methyl-1-phenyl-3-pyrazolin-5-one

Probably in: Rat A3979
→ 3-Hydroxymethyl-4-isopropyl-1-phenyl-3-pyrazolin-5-one
Probably in: Rat D285
→ 4-(1-Hydroxy-2-propyl)-3-methyl-1-phenyl-3-pyrazolin-5-one
Probably in: Rat D285
→ 1-(Hydroxyphenyl)-4-isopropyl-3-methyl-3-pyrazolin-5-one
Probably in: Rat D285
→ 4-Isopropyl-3-methyl-1-phenyl-3-pyrazolin-5-one-N-oxide
Probably in: Rat D285
→ 3-Methyl-1-phenyl-3-pyrazolin-5-one
Probably in: Rat A3979, B818

4-Isopropyl-3-methyl-1-phenyl-3-pyrazolin-5-one-N-oxide
Formed from: 4-Isopropyl-3-methyl-1-phenyl-3-pyrazolin-5-one
→ 1-(Hydroxyphenyl)-4-isopropyl-3-methyl-3-pyrazolin-5-one-N-oxide
Probably in: Rat D285

5-Isopropyl-2-methyl-4-(piperazin-1-yl) ethoxyphenol
Formed from: 3-(Piperazin-1-yl)ethoxy-*p*-cymene

2-Isopropyl-5-methylquinol (thymohydroquinone)
Formed from: Carvacrol
Thymol
→ Thymohydroquinone glucoside
In: Mucor A832

N-Isopropyl-2-methylsulphonylacetanilide
Formed from: N-Isopropyl-2-methylthioacetanilide
Propachlor
→ *p*-Hydroxy-N-isopropyl-2-methylsulphonylacetanilide
Probably in: Rat C24
→ N-(1-Hydroxyprop-2-yl)-2-methylsulphonyl-N-phenylacetamide
Probably in: Rat C24

N-Isopropyl-2-methylsulphonyl-N-(*p*-hydroxyphenyl)acetamide
Formed from: Propachlor

N-Isopropyl-2-methylthioacetanilide
→ N-Isopropyl-2-methylsulphonylacetanilide
In: Rat C24

2-Isopropylnaphthalene
→ 2-(2-Naphthyl)-2-propanol
In: Rat B606, D214
→ 2-(2-Naphthyl)propionic acid
In: Rat B606, D214

Isopropyl *p*-nitrophenyl ether
→ *p*-Nitrophenol
In: Rat A119

***p*-Isopropylnitrosobenzene**
Formed from: *p*-Isopropylaniline

N-Isopropylnoradrenaline
See: Isoprenaline

D-Isopropylnorephedrine
→ Benzoic acid
In: Rabbit A996

D-N-Isopropylnormetanephrine
Formed from: Isoprenaline
→ N-Isopropylnormetanephrine-4-sulphate
Probably in: Dog A1404
Man A1404

L-N-Isopropylnormetanephrine
Formed from: L-Isoprenaline

N-Isopropylnormetanephrine-4-sulphate
Probably formed from: N-Isopropylnormetanephrine

N-Isopropyloxanilic acid
Formed from: Propachlor
Probably formed from: N-Isopropylglycollanilide

α-Isopropylphenethylamine
Formed from: α-Isopropyl-N-methylphenethylamine

***o*-Isopropylphenol**
→ 3-Isopropylcatechol
In: Pseudomonas H337

***p*-Isopropylphenol**
→ *p*-(2-Hydroxy-2-propyl)phenol
In: Penicillium J836
See also: D696
→ *p*-(1-Propen-2-yl)phenol
In: Penicillium J836

N-Isopropylphentermine
→ *p*-Hydroxy-N-isopropylphentermine
In: Cunninghamella F345
→ Phentermine
In: Cunninghamella F345
Rat A300

Isopropyl N-phenylcarbamate
See: Isopropyl carbanilate

4-(*o*-Isopropylphenyl)-1-methylpyridinium
Probably formed from: 1,2,3,6-Tetrahydro-4-(*o*-isopropylphenyl)-1-methylpyridine

N-Isopropylphthalamidic acid
→ Phthalic acid
In: Xanthobacter J253

4-Isopropyl-3-(2-(piperazin-1-yl)ethoxy)benzyl alcohol
Formed from: 3-(Piperazin-1-yl)ethoxy-*p*-cymene

***p*-Isopropylsulphinylanisole**
Formed from: *p*-Isopropylthioanisole

***p*-Isopropylsulphinylbenzene**
Formed from: *p*-Isopropylthiobenzene

1-(4-Isopropylsulphonylphenyl)-2-aminopropanol
Formed from: Suloctidil (artifact)

Isopropyl 2-(4-thiazolyl)-5-benzimidazolecarbamate
See: Cambendazole

***p*-Isopropylthioanisole**
→ *p*-Isopropylsulphinylanisole
In: Mortierella D759

***p*-Isopropylthiobenzene**
→ *p*-Isopropylsulphinylbenzene
In: Mortierella D759
Pseudomonas H165

1-(4-Isopropylthiophenyl)-2-N-octylaminopropanol
See: Suloctidil

***p*-Isopropyltoluene**
See: *p*-Cymene

N-Isopropyl-2-(4-(3,4,5-trimethoxycinnamoyl)-1-piperazinyl)acetamide
See: Cinpropazide

2-Isopropyl-2-(3,4,5-trimethoxyphenyl) valeronitrile
Formed from: 5-((3,4-Dimethoxyphenyl) amino)-2-isopropyl-2-(3,4,5-trimethoxyphenyl)valeronitrile

Isoproterenol
See: Isoprenaline

Isoquercitrin
See: Quercetin-3-β-D-glucoside

Isoquinoline
Formed from: Isoquinoline-N-oxide
1,2,3,4-Tetrahydroisoquinoline
→ *cis*-5,6-Dihydro-5,6-dihydroxyisoquinoline
In: Pseudomonas E541
→ *cis*-7,8-Dihydro-7,8-dihydroxyisoquinoline
In: Pseudomonas H50
→ 5-Hydroxyisoquinoline
In: Pseudomonas E541, H50
→ Isocarbostyril
In: Acinetobacter F399
Arthrobacter J33
Microorganism G277
Pseudomonas K765
Rabbit A3695
Rat A3695
Streptomyces J648
→ Isoquinoline-N-oxide
In: Cunninghamella H332
→ N-Methylisoquinolinium
In: Rabbit E824

Isoquinoline-N-oxide
Formed from: Isoquinoline
→ Isoquinoline
In: Saccharomyces H894

Isoreserpiline ((19a,20a)-16,17-didehydro-10,11-dimethoxy-19-methyloxayohimban-16-carboxylic acid methyl ester)
Formed from: Strictosidine

Isoreserpinine ((3β,19α,20α)-16,17-didehydro-11-methoxy-19-methyloxayohimban-16-carboxylic acid methyl ester)
Formed from: Strictosidine

Isorhamnetin ((3,4′,5,7-tetrahydroxy-3′-methoxyflavone)
Formed from: Isorhamnetin-3-malonylglucoside
Quercetin
→ Isorhamnetin-3-β-D-glucoside
In: Norway spruce G215
Red cabbage A1733
→ Isorhamnetin-4′-glucoside
Probably in: Forsythia A3365
Rose K809
→ Isorhamnetin-β-D-glucuronide(s)
Probably in: Rat J690
→ Isorhamnetin-3-sulphate
In: Flaveria F210, F243, G769
→ Rhamnazin

In: Citrus A3435

Isorhamnetin-3,4′-di-β-D-glucoside
Probably formed from: Isorhamnetin-3-β-D-glucoside

Isorhamnetin-3,7-di-β-D-glucoside
Probably formed from: Isorhamnetin-3-β-D-glucoside

Isorhamnetin-3,4′-disulphate
Formed from: Isorhamnetin-3-sulphate

Isorhamnetin-3,7-disulphate
Formed from: Isorhamnetin-3-sulphate

Isorhamnetin-3-β-D-glucoside
Formed from: Isorhamnetin
→ Isorhamnetin-3,4′-di-β-D-glucoside
Probably in: Vitis G137
→ Isorhamnetin-3,7-di-β-D-glucoside
Probably in: Vitis G137
→ Isorhamnetin-3-malonylglucoside
In: Parsley D684

Isorhamnetin-4′-glucoside
Probably formed from: Isorhamnetin

Isorhamnetin-β-D-glucuronide(s)
Probably formed from: Isorhamnetin
→ Isorhamnetin-β-D-glucuronide sulphate
Probably in: Rat J690

Isorhamnetin-β-D-glucuronide sulphate
Probably formed from: Isorhamnetin-β-D-glucuronide

Isorhamnetin-3-malonylglucoside
Formed from: Isorhamnetin-3-β-D-glucoside
→ Isorhamnetin
In: Parsley D684

Isorhamnetin-3-sulphate
Formed from: Isorhamnetin
→ Isorhamnetin-3,4′-disulphate
In: Flaveria F210, F243
→ Isorhamnetin-3,7-disulphate
In: Flaveria F210

Isorhapontin (3-hydroxy-5-((1*E*)-2-(4-hydroxy-3-methoxyphenyl)ethenyl)phenyl-β-D-glucopyranoside)
→ 3,5-Dihydroxybenzaldehyde
In: Pseudomonas F370
→ Vanillin
In: Pseudomonas F370

ε-Isorhodomycinone ((1*R*,2*R*)-2-ethyl-1,2,3,4,6,11-hexahydro-2,5,7,10,12-pentahydroxy-6,11-dioxo-1-naphthacenecarboxylic acid methyl ester)
→ ε-Isorhodomycinone glycoside
In: Streptomyces B443

ε-Isorhodomycinone glycoside
Formed from: ε-Isorhodomycinone

Isosafrole (1,2-methylenedioxy-4-propenylbenzene)
Probably formed from: Safrole
→ Isosafrole oxide
Probably in: Rat B528
→ 4-Propenylcatechol
In: Rat C550

Isosafrole oxide
Probably formed from: Isosafrole
→ 1,2-Dihydro-1,2-dihydroxyisosafrole
Probably in: Rat B528

Isosakuranetin (4′,5,7-trihydroxyflavanone)
Formed from: Poncirin

Isosalsoline
See: (1*S*)-1,2,3,4-Tetrahydro-6-methoxy-1-methyl-7-isoquinolinol

Isoscoparine (6-β-glucopyranosyl-4′,5,7-trihydroxy-3′-methoxyflavone)
Formed from: Isoorientin

Isoscopoletin (6-hydroxy-7-methoxy-2*H*-benzopyran-2-one)
Formed from: 6,7-Dihydroxycoumarin
6,7-Dimethoxycoumarin

Isosetoclavine (8β)-9.10-didehydro-6,8-dimethylergolin-8-ol)
Formed from: Agroclavine

Isotetrandrine
Formed from: Didehydro-N-methylcoclaurinium
N-Methylcoclaurine

***p*-(S-Isothiuronium)benzaldehyde**
Formed from: *p*-(S-Isothiuronium) benzylamine

***p*-(S-Isothiuronium)benzylamine**
→ *p*-(S-Isothiuronium)benzaldehyde
In: Pig A3693

Iso-TR-2
Formed from: Verruculogen

3-Isovalerylchloramphenicol
Formed from: Chloramphenicol

O-Isovalerylshikonin
Probably formed from: Shikonin
Isovanillamide
Formed from: Isovanillic acid
Isovanillic acid (3-hydroxy-4-methoxybenzoic acid)
Formed from: Anisic acid
Isovanillin
Protocatechuic acid
Veratric acid
→ Anisic acid
In: Nocardia E374
→ Guaiacol
Probably in: Clostridium G367
Rat A361
→ Isovanillamide
In: Streptomyces E279
→ Isovanillic acid-3-sulphate
In: Rat B854
→ Isovanillin
In: Nocardia F266
See also: J237
→ Isovanilloylglycine
In: Rat A361
→ Protocatechuic acid
In: Acetobacterium J667
Nocardia A960, E374
Sporomusa J199
See also: A361, A1241
→ Veratric acid
In: Nocardia F266, E374
Phanerochaete J339
Isovanillic acid-3-sulphate
Formed from: Isovanillic acid
Isovanillin (3-hydroxy-4-methoxybenzaldehyde)
Formed from: Isovanillic acid
Isovanillyl alcohol
Protocatechualdehyde
Probably formed from: 3-Hydroxy-4-methoxybenzylamine
→ Isovanillic acid
In: Acinetobacter F366
Rat A361
Rhodopseudomonas G246
Serratia G197
See also: G865
→ Isovanillyl alcohol
In: Dunaliella G459
Euglena G422
Rat A361
See also: F266
→ Protocatechualdehyde
Probably in: Rat A361
Isovanilloylglycine
Formed from: Isovanillic acid
Isovanillyl alcohol (3-hydroxy-4-methoxybenzyl alcohol)
Formed from: 3,4-Dihydroxybenzyl alcohol
Isovanillin
→ Guaiacol
Probably in: Rat A361
→ Isovanillin
In: Pleurotus F867, G668
Rat A361
→ 2-Methoxy-5-methylphenol
Probably in: Rat A361
Isovincoside
See: Strictosidine
Isovincoside lactam
See: Strictosamide
Isovitexin (6-β-D-glucopyranosyl-5,7-dihydroxy-2-(4-hydroxyphenyl)-4*H*-1-benzopyran-4-one)
→ Isovitexin-2″-O-arabinoside
In: Silene A3759
→ Isovitexin-2″-D-glucoside
In: Silene B203, B385
→ Isovitexin-7-O-xyloside
In: Melandrium A183
Isovitexin-2″-O-arabinoside
Formed from: Isovitexin
Isovitexin-2″-O-arabinoside-7-O-xyloside
Formed from: Isovitexin-7-O-xyloside
Isovitexin-2″-D-glucoside
Formed from: Isovitexin
Isovitexin-7-O-xyloside
Formed from: Isovitexin
→ Isovitexin-2″-O-arabinoside-7-O-xyloside
In: Silene A3759
Isoxathion (diethyl 5-phenyl-3-isoxazolyl phosphate)
→ Benzoic acid

In: Rat A2287
→ 3-Hydroxy-5-phenylisoxazole
In: Rat A2287

Isoxepac (6,7-dihydro-11-oxodibenz[*b*,*e*]oxepin-2-acetic acid)
→ 6,11-Dihydro-11-oxodibenz[*b*,*e*]oxepin-2-acetyl-β-D-glucuronide
In: Man A3456
Monkey A3456
Rat A3456
→ 6,11-Dihydro-11-oxodibenz[*b*,*e*]oxepin-2-acetylglycine
In: Rabbit A3456
Rat A3456
→ 6,11-Dihydro-11-oxodibenz[*b*,*e*]oxepin-2-acetyltaurine
In: Dog A3456
Rabbit A3456

Isoxicam (4-hydroxy-2-methyl-N-(5-methyl-3-oxazolyl)-2*H*-1,2-benzothiazine-3-carboxamide-1,1-dioxide)
→ 4-Hydroxy-N-(5-hydroxymethyl-3-oxazolyl)-2-methyl-2*H*-1,2-benzothiazine-3-carboxamide-1,1-dioxide
In: Dog D393, F59
Man D393, G651
Monkey D393, F59
Rat D393, F59
→ N-Methylsaccharin
In: Dog F59
Horseradish F491
Man G651
Monkey F59
Rat F59

Isoyohimbine
See: α-Yohimbine

Isradipine (isopropyl 4-(2,1,3-benzoxadiazol-4-yl)-1,4-dihydro-5-methoxycarbonyl-2,6-dimethyl-3-pyridinecarboxylate)
→ 1,4-Dehydroisradipine
In: Man F50

Itanoxone (2′-chloro-α-methylene-γ-oxo-(1,1′-biphenyl)-4-butanoic acid)
→ 2′-Chlorobiphenyl-4-acetic acid
In: Crab D491
Man D491
Rat D491
→ 2′-Chlorobiphenyl-4-carboxylic acid
In: Crab D491
Man D491
Rat D491

Itopride (N-((4-(2-(dimethylamino)ethoxy) phenyl)methyl)-3,4-dimethoxybenzamide)
→ Itopride-N-oxide
In: Man K137

Itopride-N-oxide
Formed from: Itopride

Itraconazole
→ Hydroxyitraconazole
In: Rat K513

J

Jatrorrhizine (5,6-dihydro-3-hydroxy-2,9,10-trimethoxy-6*H*-dibenzo[]quinolizin-3-ol)
Formed from: 2,3-dihydroxy-1,10-dimethoxy-protoberberine
Norcoclaurine
Reticuline
Probably formed from: Tetrahydro-jatrorrhizine
→ Tetrahydrojatrorrhizine
In: Corydalis G136

Juglone
See: 5-Hydroxy-1,4-naphthoquinone

K

Kadethrin (5-benzyl-3-furylmethyl (1*R*,*cis*(*E*))-2,2-dimethyl-3-(2′-oxo-3′-thiocyclopentylidenemethyl)-1-cyclopropanecarboxylate)
→ 5-Benzyl-3-furylmethyl (1*R*,*cis*(*E*))-3-(2′-carboxy-4-sulphydrylbut-1-en-1-yl)-2,2-dimethyl-1-cyclopropanecarboxylate
In: Rat B14
→ 5-Benzyl-3-hydroxymethylfuran
In: Rat B14
→ 4′-Hydroxykadethrin
In: Rat B14
→ α-Hydroxykadethrin
In: Rat B14

Kadsurenone ((2*S*,3*R*,3a*S*)-2-(3,4-dimethoxyphenyl)-3,3a-dihydro-3a-methoxy-3-methyl-5-(2-propenyl)-6(2*H*)-benzofuranone)
→ 9,10-Dihydro-9,10-dihydroxykadsurenone
In: Monkey E950
Rat E950

Kaempferide (3,5,7-trihydroxy-4′-methoxyflavone)
Formed from: Kaempferol
→ Kaempferide-7-β-D-glucoside
In: Parsley K739
→ Kaempferide-3-O-glucosylgalactoside
In: Petunia H658

Kaempferide-7-β-D-glucoside
Formed from: Kaempferide

Kaempferide-3-O-glucosylgalactoside
Formed from: Kaempferide

Kaempferol (3,4′,5,7-tetrahydroxyflavone)
Formed from: Aromadendrin
Cinnamic acid
Galangin
Kaempferol-3-β-D-glucoside
Quercetin
Robinin
→ Acacetin
With peroxidase D751
→ 2,2-Dihydroxy-3-(*p*-hydroxyphenyl)-1-(2,4,6-trihydroxyphenyl)-1,3-propanedione
With peroxidase D751
→ *p*-Hydroxybenzoic acid
With peroxidase D751
→ *p*-Hydroxybenzoylphloroglucinol-carboxylic acid
Probably in: Horseradish A231
→ *p*-Hydroxyphenylacetic acid
In: Clostridium F203
→ Kaempferide
In: Calamondin orange A3987
→ Kaempferol-3-β-D-glucoside
In: Cabbage A1733
Carrot E280
Hippeastrum E529, E636
Narcissus B680
Pisum C90
Quercus B680
Secale B680
Spruce G215
Tulipa B680
See also: K889
→ Kaempferol-4′-β-D-glucoside
In: Onion G248
→ Kaempferol-7-O-glucoside
In: Chrysanthemum J320
Petunia K303
See also: K556
→ Kaempferol-3-O-(2″-O-β-D-glucopyranosyl)-β-D-galactopyranoside
In: Petunia H658
→ Kaempferol-3-β-D-glucuronide

In: Man K467
→ Kaempferol-4′-β-D-glucuronide
In: Man K467
→ Kaempferol-5-β-D-glucuronide
In: Man K467
→ Kaempferol-7-β-D-glucuronide
In: Man K467
→ Kaempferol-3-sulphate
In: Flaveria F210, F243, G769
→ Kaempferol-3-D-xyloside
In: Euonymus G444
→ 2,3,4′,5,7-Pentahydroxyflavanone
In: Horseradish A231
→ Quercetin
In: Parsley C772
Rat J300
Zea D947, E125
→ Rhamnocitrin
In: Oryza J190
→ 2,4,6-Trihydroxybenzoic acid
With peroxidase D751
→ 4′,5,7-Trihydroxy-3-methoxyflavone
In: Calamondin orange A3987

Kaempferol-3-diglucoside
Formed from: Kaempferol-3-β-D-glucoside
→ Kaempferol-3-triglucoside
In: Pisum C90

Kaempferol-3,4′-disulphate
Formed from: Kaempferol-3-sulphate

Kaempferol-3-β-D-galactopyranoside
→ Kaempferol-3-O-(2″-O-β-D-glucopyranosyl)-β-D-galactopyranoside
In: Petunia H658

Kaempferol-3-O-(2″-O-β-D-glucopyranosyl)-β-D-galactopyranoside
Formed from: Kaempferol
Kaempferol-3-β-D-galactopyranoside

Kaempferol-3-β-D-glucoside
Formed from: Kaempferol
→ Kaempferol
In: Spruce G215
→ Kaempferol-3-diglucoside
In: Petunia H658
Pisum C90
→ Kaempferol-3-malonylglucoside
In: Parsley D684

Kaempferol-4′-β-D-glucoside
Formed from: Kaempferol

Kaempferol-7-O-glucoside
Formed from: Kaempferol
→ Kaempferol-7-O-glucoside-6″-malonate
In: Cicer D445

Kaempferol-7-O-glucoside-6″-malonate
Formed from: Kaempferol-7-O-glucoside

Kaempferol-3-glucosylgalactoside-7-rhamnoglucoside
Formed from: Kaempferol-7-rhamnoglucoside

Kaempferol-3-β-D-glucuronide
Formed from: Kaempferol

Kaempferol-4′-β-D-glucuronide
Formed from: Kaempferol

Kaempferol-5-β-D-glucuronide
Formed from: Kaempferol

Kaempferol-7-β-D-glucuronide
Formed from: Kaempferol

Kaempferol-3-malonylglucoside
Formed from: Kaempferol-3-glucoside

Kaempferol-7-rhamnoglucoside
→ Kaempferol-3-glucosylgalactoside-7-rhamnoglucoside
In: Petunia H658

Kaempferol-7-rhamnoside-3-robinoside
See: Robinin

Kaempferol-3-sulphate
Formed from: Kaempferol

Kaempferol-3,4′,7-trisulphate
→ Kaempferol-3,4′-disulphate
In: Flaveria F243

Kaempferol-3-triglucoside
Formed from: Kaempferol-3-diglucoside

Kaempferol-3,4′,7-trisulphate
→ Kaempferol-3-sulphate
In: Helix F102

Kaempferol-3-D-xyloside
Formed from: Kaempferol

Karsil (3′,4′-dichloro-2-methylvaleranilide)
→ 3,4-Dichloroaniline
In: Bacillus A1006

Katurenin
See: Aromadendrin

Kawain ((6*R*)-5,6-dihydro-4-methoxy-6-((1*E*)-2-phenylethenyl)-2*H*-pyran-2-one)
→ Benzoic acid

Probably in: Rat A3485
→ 5,6-Dehydrokawain
In: Man F862
→ 5-Hydroxykawain
In: Man F862
→ *p*-Hydroxykawain
Probably in: Man F862
Rat A3485
→ 4-Hydroxy-6-phenyl-5-hexen-2-one
Probably in: Man F862
Rat A3485
→ 3-Methoxy-7-phenylhepta-2,4,6-trienoic acid
In: Man F862

Kebuzon (3,5-dioxo-4-(3-oxobutyl)-1,2-diphenylpyrazolidine)
→ 1-(*p*-Hydroxyphenyl)-3,5-dioxo-4-(3-oxobutyl)-2-phenylpyrazolidine
In: Man A710
Rat A710
→ 4-(3-Hydroxybutyl)-3,5-dioxo-1,2-diphenylpyrazolidine
In: Man A710
Rat A710

Kelthane
See: 2,2,2-Trichloro-1,1-bis (*p*-chlorophenyl)ethanol

Ketamine (2-(*o*-chlorophenyl)-2-methylaminocyclohexanone; both stereoisomers)
→ N-Demethylketamine
In: Dog A3297, B565
Horse G285
Man A423, A3297, E786
Monkey A3297
Rabbit E786
Rat A135, A1052, A1345, C148, C669, E786, F159
→ 4′-Hydroxyketamine
In: Man E786
Rabbit E786
Rat E786
See also: C148
→ 5-Hydroxyketamine
In: Man E786
Rabbit E786
Rat E786
→ 6-Hydroxyketamine
In: Man E786
Rabbit E786
Rat E786

Ketanserin (3-(2-(4-(*p*-fluorobenzoyl)-1-piperidinyl)ethyl)-2,4(1*H*,3*H*)-quinazolinedione)
→ 1,4-Dihydro-2,4-dioxo-3(2*H*)-quinazoline acetic acid
In: Dog D379
Man E728
Rat D379
→ 3-(2-(4-(*p*-Fluorobenzoyl)-1-piperidinyl)ethyl)-hydroxy-2,4(1*H*,3*H*)-quinazolinedione
In: Dog D379
Rat D379
→ 3-(2-(4-(*p*-Fluorobenzoyl)-1-piperidinyl)ethyl)-6-hydroxy-2,4(1*H*,3*H*)-quinazolinedione
Probably in: Dog D379
Rat D379
→ 3-(2-(4-(*p*-Fluoro-α-hydroxybenzyl)-1-piperidinyl)ethyl)-2,4(1*H*,3*H*)-quinazolinedione
In: Dog D379
Man E728
Rat D379
→ 2-(*p*-Fluorophenyl)-4,5-dihydro-1*H*-pyrrole-3-acetic acid
In: Dog D379
Rat D379
→ (aryl)-Hydroxyketanserin
In: Man E728

Ketobemidone (1-(4-(3-hydroxyphenyl)-1-methyl-4-piperidinyl)-1-propanone)
→ 1-(4-(3,4-Dihydroxyphenyl)-1-methyl-4-piperidyl)-1-propanone
In: Man B643
→ 1-(4-(*m*-Hydroxyphenyl)-4-piperidyl)-1-propanone
In: Man B643

Ketocaine (*o*-(2-diisopropylaminoethoxy) butyrophenone)
→ *o*-(2-Diisopropylaminoethoxy)-3-hydroxy-butyrophenone
In: Rat A3324

→ *o*-(2-Diisopropylaminoethoxy)phenylacetic acid
In: Rat A3324
→ *o*-(2-Isopropylaminoethoxy) butyrophenone
In: Rat A3324

Ketoconazole (rel-1-acetyl-4-(4-(((2*R*,4*S*)-2-(2,4-dichlorophenyl)-2-(1*H*-imidazol-1-ylmethyl)-1,3-dioxolan-4-yl)methoxy)phenyl)piperazine)
→ N-Deacetylketoconazole
In: Rat J163

Ketocyclazocine ((2*S*,6*R*,11*R*)-3-(cyclopropyl-methyl)-3,4,5,6-tetrahydro-8-hydroxy-6,11-di-methyl-2,6-methano-3-benzazocin-1(2*H*)-one)
→ N-Decyclopropylmethylketocyclazocine
Probably in: Rat A2134
→ Hydroxyketocyclazocine
Probably in: Rat A2134
→ Ketocyclazocine-β-D-glucuronide
In: Rat A2134

Ketocyclazocine-β-D-glucuronide
Formed from: Ketocyclazocine

(*R*)-Ketoprofen (*R* assumed when not stated; 3-benzoyl-α-methylbenzeneacetic acid)
→ 'Hydroxyketoprofen'
In: Camel K181
→ (*S*)-Ketoprofen
In: Dog H670
Gerbil H670
Man J202
Mouse H670
Rat H670
See also: E36
→ (*R*)-Ketoprofen-β-D-glucuronide
In: Man J202, J617
Rat J550
See also: H861

(*S*)-Ketoprofen
Formed from: (*R*)-Ketoprofen
→ (*S*)-Ketoprofen-β-D-glucuronide
In: Rat J550
See also: J202, J416

(*R*)-Ketoprofen-β-D-glucuronide
Formed from: (*R*)-Ketoprofen

(*S*)-Ketoprofen-β-D-glucuronide
Formed from: (*S*)-Ketoprofen

Ketorolac (5-benzyl-2,3-dihydro-1*H*-pyrrolizine-1-carboxylic acid)
→ *p*-Hydroxyketorolac
In: Man E901
Mouse E901
Monkey E901
Rabbit E901
Rat E901

Ketotifen (4,9-dihydro-4-(1-methyl-4-piperidinylidene)-10*H*-benzo[4,5]cyclohepta[1,2-*b*]thiophen-10-one)
Formed from: Dihydroketotifen
Ketotifen-N-β-D-glucuronide
→ Dihydroketotifen
In: Man C585, K341, K358
→ Ketotifen-N-β-D-glucuronide
In: Man C585, J771, K58
→ Ketotifen-N-oxide
In: Man C585
→ Norketotifen
In: Man C585

Ketotifen-N-β-D-glucuronide
Formed from: Ketotifen
→ Ketotifen
In: Escherichia J771

Ketotifen-N-oxide
Formed from: Ketotifen

Kievitone (3-(2,4-dihydroxyphenyl)-2,3-dihydro-5,7-dihydroxy-8-(3-methyl-2-butenyl)-4*H*-1-benzopyran-4-one)
Formed from: L-Phenylalanine
→ Kievitone hydrate
In: Fusarium A1933, C58, C74, F687

Kievitone hydrate
Formed from: Kievitone

Kotanin (4,4′,7,7′-tetramethoxy-5,5′-dimethyl-(8,8′-bi-2*H*-1-benzopyran)-2,2′-dione)
Formed from acetate in Aspergillus E774, F785

Kynuramine (2-(3-aminopropionyl)aniline)
Formed from: L-Kynurenine
→ 4-Hydroxyquinoline
In: Beef A1285
Rat A1285
See also: F395

Kynurenic acid (4-hydroxyquinaldic acid)
Formed from: L-Kynurenine

Quinaldic acid

D-Kynurenine

N-Formyl-D-kynurenine

L-Kynurenine (L-3-anthraniloylalanine)

Formed from: Diacetylkynurenine

N-Formyl-L-kynurenine

→ 4-(*o*-Aminophenyl)-2,4-dioxobutyric acid

In: Monkey K717

Rat K717

→ Anthranilic acid

In: Aspergillus A2150, A3659

Man D272

Mouse A854

Neurospora A2766

Penicillium A2150, A3659

Pig A3872

Pseudomonas A2150, A2766, A3659, B248

Rat A2664

Rhizopus A2150, A3659

Suncus F175

Trichosporon B790

Viper A1044

Xanthomonas A3615

See also: A780, A1195, B677

→ 3-Hydroxy-L-kynurenine

In: Drosophila A1194

Rat H762

See also: A1195, G787

→ Kynuramine

In: Micrococcus C62, G918

→ Kynurenic acid

In: Hansenula D978

Rat A2339, A2340, F800, H608, H762, G423, J470

See also: A3570, H792

Kyotorphin (L-tyrosyl-L-arginine)

Formed from: L-Tyrosine

L

Labetalol (2-hydroxy-5-(1-hydroxy-2-((1-methyl-3-phenylpropyl)amino)ethyl)benzamide)
→ 3-Amino-1-phenylbutane
In: Man E965, H433
Sheep G559
→ 7-Carboxamido-2,3-dihydro-5,6-dihydroxy-1-(4-phenylbut-2-yl)indole
In: Monkey G533
→ 5-(1-Hydroxy-2-((3-(*p*-hydroxyphenyl)-1-methylpropyl)amino)ethyl)salicylamide
In: Dog A2818
Man E965
Rabbit A2818
Rat A2818
→ Labetalol-β-D-glucuronide (both isomers)
In: Beef F713
Dog A2818
Man A2818
Mouse A2818
Rabbit A2818
Rat A2818
Sheep G559
→ Labetalol-O-sulphate
In: Sheep G559

Labetalol-β-D-glucuronide (both α- and *p*-isomers)
Formed from: Labetalol

Labetalol-O-sulphate
Formed from: Labetalol

Laccaic acid D methyl ester (9,10-dihydro-3,6,8-trihydroxy-1-methyl-9,10-dioxo-2-anthracenecarboxylic acid methyl ester)
Probably formed from: Aloesaponol 1

Lacidipine (4-(2-(3-(1,1-dimethylethoxy)-3-oxo-1-propenyl)phenyl)-1,4-dihydro-2,6-dimethyl-3,5-pyridinedicarboxylic acid diethyl ester)
→ 1,4-Dehydrolacidipine
Probably in: Dog G371
→ Desethyllacidipine
In: Rat G371

4-N(N-Lactanilino)piperidine
Formed from: N-(1-(2-Phenethyl4-piperidinyl)lactanilide

Lamotrigine (6-(2,3-dichlorophenyl)-1,2,4-triazine-3,5-diamine)
→ Lamotrigine-2-β-D-glucuronide
In: Guinea pig F763
Man F718

Lamotrigine-2-β-D-glucuronide
Formed from: Lamotrigine

Lansoprazole (2-(((3-methyl-4-(2,2,2-trifluoroethoxy)-2-pyridinyl)methyl)sulphinyl)-1*H*-benzimidazole)
→ 5-Hydroxylansoprazole
In: Man H431, J182
→ Lansoprazole sulphone
In: Man H431, J182, J308

Lansoprazole sulphone
Formed from: Lansoprazole

Lapachol (2-hydroxy-3-(3-methylbutenyl)-1,4-naphthalenedione)
→ *o*-(2-Hydroxy-5-methyl-1-oxo-4-hexen-1-yl)phenylglyoxylic acid
In: Penicillium A2772, C59

α-Lapachone (3,4-dihydro-2,2-dimethyl-2*H*-naphtho[2,3-*b*]pyran-5,10-dione)
Formed from: *o*-Succinylbenzoic acid

Lappaconitine ((1α,14α,16β)-20-ethyl-1,14,16-trimethoxyaconitane-4,8,9-triol 4-(2-(acetylamino)benzoate))
→ N-Deacetyllappaconitine
In: Man F473

(+)-Lariciresinol ((2*S*,3*R*,4*R*)-tetrahydro-2-(4-hydroxy-3-methoxyphenyl)-4-((4-hydroxy-3-methoxyphenyl)methyl)-3-furanmethanol)
Formed from: Pinoresinol
→ (−)-Secoisolariciresinol
In: Forsythia H883

Latanoprost (13,14-dihydro-17-phenyl-18,19,20-trisnorprostaglandin $F_{2\alpha}$ 1-isopropyl ester)
→ 13,14-Dihydro-17-phenyl-18,19,20-trisnorprostaglandin $F_{2\alpha}$
In: Rabbit J432

Laudanine (2-methoxy-5-((1,2,3,4-tetrahydro-6,7-dimethoxy-2-methyl-1-isoquinolinyl)methyl)phenol)
Formed from: Laudanosine

Laudanosine ((1*S*)-1-((3,4-dimethoxyphenyl)methyl)-1,2,3,4-tetrahydro-6,7-dimethoxy-2-methylisoquinoline)
Formed from: Tetrahydropapaverine
→ Codamine
In: Dog E855
Man E855
Rabbit E855
→ 1-(3,5-Dihydroxy-4-methylbenzoyl)-1,2,3,4-tetrahydro-6,7-dimethoxy-2-methylisoquinoline
In: Pseudomonas C902
→ Laudanine
In: Dog E855
Man E855
Rabbit E855
→ O-Methylcoripalline
In: Pseudomonas C902
→ Norlaudanosine
In: Dog E855
Man E855
Rabbit E855
→ Pseudocodamine
In: Cunninghamella A2608
Dog E855
Man E855
Rabbit E855
→ Pseudolaudanine
In: Dog E855
Man E855
Rabbit E855

Laudanosoline (N-methyltetrahydropapaveroline, 1-((3,4-dihydroxyphenyl)methyl)-1,2,3,4-tetrahydro-2-methyl-6,7-isoquinolinediol)
Probably formed from: Tetrahydropapaveroline
→ 1,2,3,4-Tetrahydro-6,7-dihydroxy-1-(3-hydroxy-4-methoxybenzyl)-2-methylisoquinoline
In: Coptis K572
→ Tetrahydro-2,3,9,10-tetrahydroxyprotoberberine
In: Berberis D615, D878
Eschscholtzia H680

Lauryl gallate
→ Lauryl gallate-O-sulphate
In: Eubacterium G475

Lauryl gallate-O-sulphate
Formed from: Lauryl gallate

Lauryl L-γ-glutamyldopamine
→ 3,4-Dihydroxyphenethylamine
In: Dog A3945

1-Lauryl-1-phenyl-2-(*p*-toluoyl)hydrazine
Probably formed from: Toluic acid phenylhydrazide

Lawsone
See: 2-Hydroxy-1,4-naphthoquinone

Leicocarpic acid (3,6-dihydroxy-2-methoxybenzoic acid)
Formed from: Leicocarposide

Leicocarposide (2-(((3-(β-D-glucopyranosyloxy)-6-hydroxy-2-methoxybenzoyl)oxy)methyl)-phenyl-β-D-glucopyranoside)
→ Leicocarpic acid
In: Rat F263
→ Salicin
Probably in: Rat F263

Lenticin
See: Hypaphorine

Leptophos (O-(4-bromo-2,5-dichlorophenyl) O-methyl phenylphosphonothioate)
→ 4-Bromo-2,5-dichlorophenol
In: Mouse A1063
Rat A2903
→ O-(4-Bromo-2,5-dichlorophenyl) O-methyl phenylphosphonate
In: Mouse A1063
→ O-Methyl phenylphosphonothioic acid

In: Mouse A1063
Rat A2903
Lergotrile (2-chloro-6-methylergoline-8β-acetonitrile)
→ 8β-Acetamido-2-chloro-6-methylergoline
In: Guinea pig A62
→ 2-Chloro-6-methylergoline-8β-acetic acid
In: Man A1708
→ 13-Hydroxylergotrile
In: Guinea pig A2857
See also: A62
→ Norlergotrile
In: Cunninghamella B268
Guinea pig A62
Man A1708
Streptomyces B268
Leucoapigeninidin
See: Apiforol
Leucocyanidin
See: 2,3-*trans*-3,4-*cis*-3,3′,4,4′,5,7-Hexahydroxyflavan
Leucodelphinidin (3,4-dihydro-2-(3,4,5-trihydroxyphenyl)-2*H*-1-benzopyran-3,4,5,7-tetrol)
Formed from: Dihydromyricetin
Leucogentian violet (4,4′,4″-methylidynetris(N,N-dimethylbenzenamine))
Formed from: Gentian violet
Leucoluteolinidin
See: Luteoforol
Leucomethylene blue (N,N,N′,N′-tetramethyl-10*H*-phenothiazine-3,7-diamine)
Formed from: Methylene blue
3,4-*cis*-Leucopelargonidin (*cis*-3,4,4′,5,7-pentahydroxyflavan)
Formed from: Aromadendrin
→ Pelargonidin
Probably in: Matthiola D575
Leucophenothiazone
See: 3-Hydroxyphenothiazine
L-Leucyl-L-alanyl-β-naphthylamide
→ β-Naphthylamine
In: Rabbit A1589
L-Leucyl-4-aminoantipyrine
→ 4-Aminoantipyrine
In: Man B658
Rat B240
L-Leucyl-α,β-dehydrotryptophan
Formed from: L-Leucyl-L-tryptophan
L-Leucyl-N-formyl-L-kynurenine
Formed from: L-Leucyl-L-tryptophan
L-Leucylglycyl-β-naphthylamide
→ β-Naphthylamine
In: Rabbit A1589
L-Leucyl-β-naphthylamide
→ β-Naphthylamine
In: Agave A2745
Aspergillus A3607
Beef A526
Crithidia A1102
Escherichia E586
Euonymus D216
Flavobacterium C73
Glycine A1596
Hen G654
Man A837, A1219, A1397, A3067
Mytilus A1514
Nemertean worm A1622
Pea A3213
Pig A3067
Pseudomonas G643
Rabbit A1589, A2347, B334, B524, B787
Rat A523
Streptococcus C277
L-Leucyl-*p*-nitroanilide
Formed from: Succinyl-L-alanyl-L-alanyl-L-leucyl-*p*-nitroanilide
→ *p*-Nitroaniline
In: Agave A2745
Aspergillus A3607
Bacteria C146
Euonymus C907, D216
Glycine A1596
Man A1219, A2450
Nocardia G343
Rat A3126, C173
Streptococcus C277
L-Leucyl-L-phenylalanine
Formed from: L-Leucyl-L-phenylalanyl-β-naphthylamide
→ L-Phenylalanine
In: Brevibacterium E587

L-Leucyl-L-phenylalanyl-β-naphthylamide
→ L-Leucyl-L-phenylalanine
In: Glycine A1596
→ β-Naphthylamine
In: Glycine A1596

L-Leucyl-L-tryptophan
→ L-Leucyl-α,β-dehydrotryptophan
In: Pseudomonas B917
→ L-Leucyl-N-formyl-L-kynurenine
In: Wheat E24
→ L-Tryptophan
In: Brevibacterium E587

L-Leucyl-L-tyrosine
Formed from: L-Leucyl-L-tyrosyl-β-naphthylamide
→ L-Tyrosine
In: Aspergillus A3658
Brevibacterium E587

L-Leucyl-L-tyrosyl-β-naphthylamide
→ L-Leucyl-L-tyrosine
In: Glycine A1596
→ β-Naphthylamine
In: Glycine A1596

Leurosine ((3′α,4′α)-3′,4′-epoxyvincaleukoblastine)
Formed from: 3′,4′-Dehydrovinblastine
→ 15′-Hydroxycatharinine
In: Horseradish E723
→ 10′-Hydroxyleurosine
In: Streptomyces B932
→ 21′-Hydroxyleurosine
In: Aspergillus C471
Polyporus C417

Levallorphan ((−)-N-allyl-3-hydroxymorphinan)
→ (−)-N-Allyl-3,6β-dihydroxymorphinan
In: Rat A855

Levametabol I
Formed from: Levamisole

Levametabol II
Formed from: Levamisole

Levametabol III
Formed from: Levamisole

Levamisole ((6*S*)-2,3,5,6-tetrahydro-6-phenylimidazo[2,1-*b*]thiazole)
→ 1-Carboxymethyl-4-phenyl-2-thioimidazilidine
In: Beef H733
→ 5,6-Dehydrolevamisole
In: Beef H733
→ 4-Hydroxylevamisole
In: Rat C497
→ Levametabol I
In: Microorganism G3
→ Levametabol II
In: Bacteroides G3
Clostridium G3
Peptostreptococcus G3
→ Levametabol III
In: Microorganism G3

Levofloxacin ((*S*)-9-fluoro-2,3-dihydro-3-methyl-10-(4-methyl-1-piperazinyl)-7-oxo-7*H*-pyrido[1,2,3-*de*]-1,4-benzoxazine-6-carboxylic acid)
→ Levofloxacin-β-D-glucuronide
In: Man H764

Levofloxacin-β-D-glucuronide
Formed from: Levofloxacin

Levomepromazine ((*βR*)-2-methoxy-N,N,β-trimethyl-10*H*-phenothiazine-10-propanamine)
→ N-Demethyllevomepromazine
In: Man H598
→ O-Demethyllevomepromazine
In: Man H598
→ 3-Hydroxylevomepromazine
In: Man H598
→ 7-Hydroxylevomepromazine
In: Man H598
→ Levomepromazine sulphoxide
In: Man H598

Levomepromazine sulphoxide
Formed from: Levomepromazine

Levorin
Formed from: *p*-Aminoacetophenone

Levormeloxifene ((−)-3,4-*trans*-7-methoxy-2,2-dimethyl-3-phenyl-4-(4-((2-pyrrolidin-1-yl)ethoxy)phenyl)chromanone)
→ Demethyllevormeloxifene
In: Cynomolgus monkey J866
Man J866
Rat J866, K31
→ Hydroxylevormeloxifene (2 compounds)
In: Cynomolgus monkey J866
Man J866
Rat J866, K31

→ Levormeloxifene-β-D-glucuronide
In: Rat K31
→ 3,4-*trans*-7-Methoxy-2,2-dimethyl-4-(4-((2-oxopyrrolidin-1-yl)ethoxy)phenyl)-3-phenylchromane
In: Cynomolgus monkey J866
Man J866
Rat J866
See also: K31

Levormeloxifene-β-D-glucuronide
Formed from: Levormeloxifene

Levorphanol (17-methylmorphinan-3-ol)
→ Levorphanol-β-D-glucuronide
In: Man H933
Rat J97a

Levorphanol-β-D-glucuronide
Formed from: Levorphanol

Levosemotiadil ((2*S*)-2-(-2-(3-((2-(1,3-benzodioxol-5-yloxy)ethyl)methylamino)-propoxy)-5-methoxyphenyl)-4-methyl-2*H*-1,4-benzothiazin-3-(4*H*)-one)
→ 5′-Demethyllevosemotiadil
In: Rat J130
→ (*S*)-2-(2-((N-(2-((3,4-Dihydroxy)phenoxy)ethyl)-N-methylamino)propoxy)phenyl)-4-methyl-2*H*-1,4-benzothiazin-3-(4*H*)-one
In: Rat J130

Librium
See: Chlordiazepoxide

Licochalcone A ((2*E*)-3-(5-(1,1-dimethyl-2-propenyl)-4-hydroxy-2-methoxyphenyl)-1-(4-hydroxyphenyl)-2-propen-1-one)
→ β-Hydroxylicochalcone A
In: Pig J166
Rabbit J166
→ Licochalcone A-4-β-D-glucuronide
In: Pig J166
Rabbit J166
→ Licochalcone A-4′-β-D-glucuronide
In: Pig J166
Rabbit J166

Licochalcone A-4-β-D-glucuronide
Formed from: Licochalcone A

Licochalcone A-4′-β-D-glucuronide
Formed from: Licochalcone A

Licodione (1-(2,4-dihydroxyphenyl)-3-(4-hydroxyphenyl)-1,3-propanedione)
→ 2′-O-Methyllicodione
In: Glycyrrhiza H908, K886

Lidamidine (1-(2,6-dimethylphenyl)-3-methylamidinourea)
→ 1-Amidino-3-(2,6-dimethylphenyl)urea
In: Rat B626
→ p-Hydroxylidamidine
In: Rat B626

Lidocaine
See: Lignocaine

Lidoflazine (4-(4,4-bis(4-fluorophenyl)butyl)-N-(2,6-dimethylphenyl)-1-piperazineacetamide)
→ 4,4-Bis(*p*-fluorophenyl)butanoic acid
In: Dog A2699
Rat A2699
→ 1-(4,4-Bis(*p*-fluorophenyl)butyl)-1-(2-oxopiperazine)aceto-2′6′-xylidide
In: Dog A2699
Rat A2699
→ 1-(4,4-Bis(*p*-fluorophenyl)butyl)piperazine
In: Dog A2699
Rat A2699
→ N-(2,6-Dimethylphenyl)-2-hydroxyacetamide
In: Dog A2699
Rat A2699
→ *p*-Hydroxylidoflazine
In: Dog A2699
Rat A2699

Lifibrol (4-(4-(4-(1,1-dimethylethylphenyl)-2-hydroxybutoxy)benzoic acid)
→ *p*-(2-Hydroxy-4-(*p*-((1-hydroxymethyl-1-methyl)ethyl)phenyl)butoxy)benzoic acid
In: Man H533
→ Lifibrol acyl glucuronide
In: Man H533
→ Lifibrol ether glucuronide
In: Man H533
→ Lifibrol triglycerides
In: Man H533

Lifibrol acyl glucuronide
Formed from: Lifibrol

Lifibrol ether glucuronide
Formed from: Lifibrol

Lifibrol triglycerides
Formed from: Lifibrol
Lignin
Formed from: *p*-Hydroxybenzoic acid
Lignocaine (lidocaine, N,N-diethylglycine-2′,6′-xylidide)
→ N,N-Diethylglycine-2′-hydroxymethyl-6′-methylanilide
In: Rat D861, F135
→ N,N-Diethylglycine-3′-hydroxy-2′,6′-xylidide
In: Man A610, B472, F848, K64
Rabbit E53, H720
Rat A610, D861, E546, F135, F848, H123, J131
→ N,N-Diethylglycine-4′-hydroxy-2′,6′-xylidide
In: Man A610
Rabbit E53
Rat A610
→ 2,6-Dimethylaniline
In: Man J67
Rabbit E53
Rat E546
→ 2,6-Dimethylphenylhydroxylamine
In: Man A3618
→ N-Ethylglycine-2′,6′-xylidide
In: Dog A415
Man A2794, B472, F848, J67, K64
Rabbit E53, H720, H723
Rat D861, E546, F135, H123, J131
→ Lignocaine-N-oxide
In: Rat E79
See also: A617
Lignocaine-N-oxide
Formed from: Lignocaine
Lilopristone ((11β,17β)-11-(dimethylamino)phenyl)-17-hydroxy-17-((1*Z*)-3-hydroxy-1-propenyl)oestra-4,9-dien-3-one)
→ Demethyllilopristone
In: Man J288
α-Lindane (α-1,2,3,4,5,6-hexachlorocyclohexane)
→ S-(2,4-Dichlorophenyl)glutathione
In: Rat A3707
→ S-(2,5-Dichlorophenyl)glutathione
In: Rat A3707
→ S-(2,6-Dichlorophenyl)glutathione
In: Rat A3707
→ S-(3,4-Dichlorophenyl)glutathione
In: Rat A3707
→ 2,4,6-Trichlorophenol
In: Rat A3101
β-Lindane
→ Chlorobenzene
In: Microorganism J700
→ 2,3,4,5-Tetrachlorophenol
In: Rat A3101
→ 2,3,4,6-Tetrachlorophenol
In: Rat A3101
See also: A3101
Lindane (assumed to be γ-)
→ Chlorobenzene
In: Desulfococcus K219
Desulfovibrio K219
→ *o*-Chlorophenol
In: Arthrobacter K598
→ *m*-Dichlorobenzene
In: Wheat A1476
→ *p*-Dichlorobenzene
In: Wheat A1476
→ S-(2,4-Dichlorophenyl)glutathione
In: Rat A3707
→ S-(2,6-Dichlorophenyl)glutathione
In: Rat A3707
→ Pentachlorobenzene
In: Lettuce A2444
Wheat A1476
See also: A3648, B974
→ 1,2,3,4-Tetrachlorobenzene
In: Wheat A1476
→ 1,2,3,5-Tetrachlorobenzene
In: Wheat A1476
See also: A2444, A3101, A3648, B974
→ 1,2,4,5-Tetrachlorobenzene
In: Wheat A1476
See also: A3648
→ 1,2,3-Trichlorobenzene
In: Lettuce A2444
Wheat A1476
Zea A766
→ 1,2,4-Trichlorobenzene
In: Bean A766

Lettuce A2444
Wheat A1476
Zea A766
→ 1,3,5-Trichlorobenzene
In: Wheat A1476
See also: A3101, A3648, B974

δ-Lindane
→ S-(2,4-Dichlorophenyl)glutathione
In: Rat A3707
→ S-(2,6-Dichlorophenyl)glutathione
In: Rat A3707
→ S-(3,4-Dichlorophenyl)glutathione
In: Rat A3707

ε-Lindane
→ 2,4,6-Trichlorophenol
In: Rat A3101

Linezolid (zyvox, (*S*)-N-((3-(3-fluoro-4-(4-morpholinyl)phenyl)-2-oxo-5-oxazolidinyl)methyl)acetamide)
→ (*S*)-4-(5-Acetamidomethyl-2-oxooxazolidin-3-yl)-N-carboxymethyl-2-fluoro-N-(2-hydroxyethyl)aniline
In: Man K44

Linogliride (N-(1-methyl-2-pyrrolidinylidene)-N′-phenyl-4-morpholinecarboximidamide)
→ 5-Hydroxylinogliride
In: Dog H655
Man H655
Rat H655

Linomide
See: Roquinimex

Linuron (N′-(3,4-dichlorophenyl)-N-methoxy-N-methylurea)
→ 3,4-Dichloroaniline
Probably in: Bacillus A1006
→ N′-(3,4-Dichlorophenyl)-N-methoxyurea
In: Rabbit G709

Liquiritigenin (4′,7-dihydroxyflavanone)
Formed from: *p*-Coumaroyl CoA
Liquiritin
→ Daidzein
In: Glycine D468
→ Formononetin
In: Glycyrrhiza K630
→ Liquiritigenin-4′,7-disulphate
Probably in: Rat G211
→ Liquiritigenin-β-D-glucoside
In: Bupleurum E681
Catharanthus E681
Datura E681
Gardenia E681
Lithospermum E681
Perilla E681
→ Liquiritigenin-4′-β-D-glucuronide
Probably in: Rat G211
→ Liquiritigenin-7-β-D-glucuronide
Probably in: Rat G211
→ 2,4′,7-Trihydroxyisoflavanone
In: Glycyrrhiza K308
Pueraria F845

Liquiritigenin-4′,7-disulphate
Formed from: Liquiritigenin

Liquiritigenin-β-D-glucoside (structure unspecified)
Probably formed from: Liquiritigenin

Liquiritigenin-4′-β-D-glucoside
See: Liquiritin

Liquiritigenin-4′-β-D-glucuronide
Probably formed from: Liquiritigenin
→ Liquiritigenin-4′-β-D-glucuronide-7-sulphate
Probably in: Rat G211

Liquiritigenin-7-β-D-glucuronide
Formed from: Liquiritigenin
→ Liquiritigenin-7-β-D-glucuronide-4′-sulphate
Probably in: Rat G211

Liquiritigenin-4′-β-D-glucuronide-7-sulphate
Probably formed from: Liquiritigenin-4′-β-D-glucuronide

Liquiritigenin-7-β-D-glucuronide-4-sulphate
Probably formed from: Liquiritigenin-7-β-D-glucuronide

Liquiritin (liquiritigenin-β-D-glucoside)
→ Liquiritigenin
In: Rat G211

Lisuride (N′-((8α)-9,10-didehydro-6-methylergolin-8-yl)-N,N-diethylurea)
→ 4,5-Dehydro-2,9-dihydro-9-hydroxy-2-oxolisuride
Probably in: Rhesus monkey D657
→ Demethyllisuride
In: Rat B616
→ Desethyllisuride

In: Dog F337
Guinea pig F337
Monkey F337
Rat B616, F337
→ 2-Hydroxylisuride
In: Rat B616
→ 12-Hydroxylisuride
Probably in: Rat B616
→ 13-Hydroxylisuride
Probably in: Rat B616
→ 3-Hydroxy-2-oxolisuride
Probably in: Rhesus monkey D657

Littorine ((αR)-α-hydroxybenzenepropanoic acid (3-*endo*)-8-methyl-8-azabicyclo[3.2.1]oct-3-yl ester)
Formed from: L-Phenylalanine

Lofepraprine (*p*-chlorobenzoylmethyl demethylimipramine)
→ *p*-Chlorobenzoic acid
In: Dog A2029
Man A2029
Rat A356, A2029, B971
See also: G948
→ Demethylimipramine
In: Dog A2029
Man A2029, A3865, G948
Rat A356, A2029, B971, G948
→ Demethyllofepraprine
Probably in: Man G948
Rat G948
→ 2-Hydroxylofepraprine
Probably in: Man G948
Rat G948

Lofexidine (2-(1-(2,6-dichlorophenoxy)ethyl)-4,5-dihydro-1*H*-imidazole)
→ 2,4-Dichlorophenol
In: Man C405

Loflazepate (7-chloro-5-(2-fluorophenyl)-2,3-dihydro-2-oxo-1*H*-1,4-benzodiazepine-3-carboxylic acid)
Formed from: Ethylloflazepate
→ Decarboxyloflazepate
Probably in: Baboon D752
Dog D752
Man D752
Rat D752
→ 3-Hydroxyloflazepate
Probably in: Baboon D752
Dog D752
Man D752
→ 4′-Hydroxyloflazepate
Probably in: Rat D752

Lonapalene (6-chloro-2,3-dimethoxy-1,4-naphthalenediol diacetate)
→ 6-Chloro-2,3-dimethoxy-1,4-naphthoquinone
In: Rat G175

Lonazolac (3-(4-chlorophenyl)-1-phenyl-1*H*-pyrazole-4-acetic acid)
→ Hydroxylonazolac
In: Dog F337
Guinea pig F337
Monkey F337
Rat F337

Loperamide (4-(4-chlorophenyl)-4-hydroxy-N,N-dimethyl-α,α-diphenyl-1-piperidinebutanamide)
Formed from: Loperamide-N-oxide
→ 4-(*p*-Chlorophenyl)-4-hydroxypiperidine
In: Rat A3382
→ Demethylloperamide
In: Rat A3382, B275, C253

Loperamide-N-oxide
→ Loperamide
In: Dog H147
Microorganism H181

Loprazolam ((2*Z*)-6-(2-chlorophenyl)-2,4-dihydro-2-((4-methyl-1-piperazinyl)-methylene)-8-nitro-1*H*-imidazo[1,2-*a*][1,4]benzodiazepin-1-one)
→ 8-Amino-6-(*o*-chlorophenyl)-1,2-dihydro-2-(N-methylpiperazinylmethyl)-1*H*,4*H*-imidazo[1,2-*a*][1,4]benzodiazepin-1-one
In: Dog C568
Rat C568, C570
→ 4-Hydroxyloprazolam
In: Rat C568, C570
→ Loprazolam-N-oxide
In: Dog C568, C570
Man C570
Rat C568
→ Norloprazolam
In: Rat C568

Loprazolam-N-oxide
Formed from: Loprazolam
Lorazepam
See: 7-Chloro-5-(*o*-chlorophenyl)-1,3-dihydro-3-hydroxy-2*H*-1,4-benzodiazepin-2-one
Lorazepam-3-acetate
→ 7-Chloro-5-(*o*-chlorophenyl)-1,3-dihydro-3-hydroxy-2*H*-1,4-benzodiazepin-2-one
In: Man F756
Rat F756
Lorazepam dihydrodiol
Formed from: 7-Chloro-5-(*o*-chlorophenyl)-1,3-dihydro-3-hydroxy-2*H*-1,4-benzodiazepin-2-one
Lorcainide (N-(*p*-chlorophenyl)-N-(1-(1-methylethyl)-4-piperidinyl)benzeneacetamide)
→ N-(*p*-Chlorophenyl)-*p*-hydroxy-N-(1-(1-methylethyl)-4-piperidinyl)benzene-acetamide
In: Dog D94
Man D94
Rat D94
→ N-(*p*-Chlorophenyl)-1-(1-methylethyl)-4-piperidinamine
In: Dog D94
Man D94
Rat D94
→ N-(*p*-Chlorophenyl)-N-(4-piperidinyl)benzeneacetamide
In: Dog D94
Man D94
Rat D94
Lormetazepam (7-chloro-5-(2-chlorophenyl)-1,3-dihydro-3-hydroxy-1-methyl-2*H*-1,4-benzodiazepin-2-one)
→ 7-Chloro-5-(*o*-chlorophenyl)-1,3-dihydro-3-hydroxy-2*H*-1,4-benzodiazepin-2-one
In: Dog B200
Man B437
Monkey B200
→ Lormetazepam-β-D-glucuronide
In: Man H225
Lormetazepam-β-D-glucuronide
Formed from: Lormetazepam
Lortalamine ((4a*R*,10*R*,10a*S*)-rel-8-chloro-1,2,3,4,10,10a-hexahydro-2-methyl-4a,10-(iminoethano)-4a*H*-[1]benzopyrano[3,2-*c*]pyridin-12-one)
→ Demethyllortalamine
In: Man D958
→ Oxolortalamine
In: Man D958
Losartan
See: 2-*n*-Butyl-4-chloro-5-hydroxymethyl-1-((2′-(1*H*-tetrazol-5-yl)-1,1′-biphenyl-4-yl)methyl)-1*H*-imidazole
Losigamone ((5*S*)-rel-5-((*R*)-(2-chlorophenyl)hydroxymethyl)-4-methoxy-2(5*H*)-furanone)
→ 5′-Hydroxylosigamone
In: Man J49
Losoxantrone
See: Biantrazole
Lovastatin ((2*S*)-2-methylbutanoic acid (1*S*,3*R*,7*S*,8*S*,8a*R*)-1,2,3,7,8,8a-hexahydro-3,7-dimethyl-8-(2-((2*R*,4*R*)-tetrahydro-4-hydroxy-6-oxo-2*H*-pyran-2-yl)ethyl)-1-naphthalenyl ester)
→ 1,2-Dihydro-2,6-dimethyl-1-naphthalene-heptanoic acid-δ-lactone
In: Man E946
Loxapine (2-chloro-11-(4-methyl-1-piperaziny)dibenz[*b*,*f*][1,4]oxazepine)
→ Demethylloxapine
In: Man A3700
→ Hydroxyloxapine
In: Man A3700
→ Loxapine-N-β-D-glucuronide
In: Man H162, H541
→ Loxapine-N-oxide
In: Man A3700
Loxapine-N-β-D-glucuronide
Formed from: Loxapine
Loxapine-N-oxide
Formed from: Loxapine
Loxoprofen (2-(*p*-((2-oxocyclopentyl)methyl)-phenyl)propionic acid; (2*R*) assumed)
→ 2-(*p*-((2-Hydroxycyclopentyl)methyl)-phenyl)propionic acid
In: Dog C914
Monkey C914
Mouse C914
Rabbit D125

Rat C914
→ 2-(*p*-((3-Hydroxy-2-oxocyclopentyl)methyl)phenyl)propionic acid
In: Rat C914
→ (2*S*)-Loxoprofen
In: Rat D80

(2*S*)-Loxoprofen
Formed from: (2*R*)-Loxoprofen

Loxtidine (1-methyl-5-(3-(3-((1-piperidinyl)methyl)phenoxy)propyl)amino-1*H*-1,2,4-triazole-3-methanol)
→ N-(3-Carboxy-1-methyl-1,2,4-triazol-5-yl)-3-(*m*-(piperidin-1-ylmethyl)phenoxy)-propylamine
In: Dog D115, J556
Man J556
Rat D115, J556
→ Loxtidine-O-β-D-glucuronide
In: Dog D115, J556
Man J556
Rat D115, J556
→ *m*-(Piperidin-1-ylmethyl)phenol
In: Dog D115, J556
Man J556
Rat D115, J556
→ 3-(*m*-(1-Piperidin-1-ylmethyl)phenoxy)-propionic acid
In: Dog D115
Rat D115

Loxtidine-O-β-D-glucuronide
Formed from: Loxtidine

LSD
See: Lysergic diethylamide

Lucenin
Formed from: Cinnamic acid

Luciferin (Cypridina) (3-(3,7-dihydro-6-(1*H*-indol-3-yl))-2-((*S*)-1-methyl-6-propyl-3-(3-oxoimidazo[1,2-*a*]pyrazin-8-yl)propyl)guanidine)
→ Luciferin (Cypridina) oxidized
In: Cypridina K815

Luciferin (firefly) ((4*S*)-4,5-dihydro-2-(6-hydroxy-2-benzothiazolyl)-4-thiazolecarboxylic acid)
Formed from: *p*-Quinone

L-Luciferin (firefly)
L-Luciferyl adenylate
→ In: Firefly J204

Luciferin (Renilla) (6-(4-hydroxyphenyl)-2,8-bis(phenylmethyl)imidazo[1,2-*a*]pyrazin-3(7*H*)-one)
→ Luciferin (Renilla) sulphate
In: Renilla K831
→ Oxyluciferin (Renilla)
In: Renilla K813, K926

Luciferin (Watasenia) ((phenylmethyl)-6-(4-(sulphooxy)phenyl)-2-((4-sulphooxy)phenyl)-methyl)imidazo[1,2-*a*]pyrazin-3(7*H*)-one)
→ Oxyluciferin (Watasenia)
In: Watasenia K899

Luciferin (Renilla) sulphate
Formed from: Luciferin (Renilla)

L-Luciferyl adenylate
Formed from: L-Luciferin (firefly)

Lumazine (2,4(1*H*,3*H*)-pteridinedione)
→ Riboflavine
In: Eremothecium A1546

Lumiflavin (7,8,10-trimethylbenzo[*g*]pteridine-2,4(3*H*,10*H*)-dione)
Formed from: Riboflavine

Luminol (3-aminophthalhydrazide)
→ Luminol-8-β-D-glucuronide
In: Rat K36
→ Luminol-8-sulphate
In: Rat K36

Luminol-8-β-D-glucuronide
Formed from: Luminol

Luminol-8-sulphate
Formed from: Luminol

Lunarine ((1*E*,15*E*,20a*R*,24a*S*)-4,5,6,7,8,9,10,11,12,13,20a,21,23,24-tetradecahydro-17,19-etheno-22*H*-benzofuro[3a,3-*n*][1,5,10]triazacycloeicosine-3,14,22-trione)
Formed from: N^1,N^{10}-Bis(*p*-coumaroyl)-spermidine

Lunularic acid (2-hydroxy-6-(2-(4-hydroxyphenyl)ethyl)benzoic acid)
Probably formed from: Hydrangenol
→ 3,4′-Dihydroxybibenzyl
In: Conocephalum A1924

Lunularin
See: 3,4′-Dihydroxybibenzyl

Lupinisoflavone F (6-(2,3-dihydro-4-hydroxy-2-(1-hydroxy-1-methylethyl)-5-benzofuranyl)-2,3-dihydro-4-hydroxy-2-(1-hydroxy-1-methylethyl)-5*H*-furo[3,2-*g*][1]benzopyran-5-one)
Formed from: 2′-Hydroxylupalbigenin

Lupiwighteone (5,7-dihydroxy-3-(4-hydroxyphenyl)-8-(3-methyl-2-butenyl)-4*H*-benzopyran-4-one)
→ 5-O-Methyllupiwighteone
In: Lupinus E756, F182

Luteoforol (dihydroeriodictyol, leucoluteolinidin, 3′,4,4′,5,7-pentahydroxyflavan)
Formed from: Eriodictyol

Luteolin (3′,4′,5,7-tetrahydroxyflavone)
Formed from: Eriodictyol
Probably formed from: Apigenin
Luteolin-7-O-glucoside
Luteolin-4′-sulphate
Naringenin
→ Caffeic acid
In: Rhizobium H280
→ Chrysoeriol
In: Calamondin orange A3987
Parsley K805
Zea F173, G223
See also: A2434, A2629, A3757, C905
→ Diosmetin
In: Calamondin orange A3987
→ Hydnocarpin
In: Silybum A3221
→ Luteolin-4′-D-glucoside
In: Onion G248
→ Luteolin-7-O-glucoside
In: Chrysanthemum J320
Parsley K739
Petunia K303
Spirodela A1635
See also: A1030
→ Luteolin-7-β-D-glucuronide
In: Rye K890
→ Orientin
Probably in: Spirodela A1030
→ Phenylacetic acid
In: Rhizobium H280
→ 2,4,6-Trihydroxybenzoic acid
In: Rhizobium H280
→ 3′,4′,5-Trihydroxy-7-methoxyflavone
In: Rice J190

Luteolin-4′,7-disulphate
→ Luteolin-4′-sulphate
In: Helix F102
→ Luteolin-7-sulphate
In: Helix F102

Luteolin-4′-D-glucoside
Formed from: Luteolin

Luteolin-7-O-glucoside
Formed from: Luteolin
→ 7-O-(β-D-apiofuranosyl(1→2)-β-D-glucosyl)luteolin
In: Parsley K827
→ Luteolin
Probably in: Secale K512
→ Luteolin-7-O-glucoside-6″-malonate
In: Cicer D445
Parsley D684
→ 4′,5,7-Trihydroxy-3′-methoxyflavone-7-O-glucoside
In: Parsley K805

Luteolin-7-O-glucoside-6″-malonate
Formed from: Luteolin-7-O-glucoside

Luteolin-7-β-D-glucuronide
Formed from: Luteolin

Luteolin-7-O-(β-D-glucuronosyl(1→2)-β-D-glucuronide
In: Rye K890

Luteolin-7-O-(β-D-glucuronosyl(1→2)-β-D-glucuronide
Formed from: Luteolin-7-β-D-glucuronide
→ Luteolin-7-O-(β-D-glucuronosyl(1→2)-β-D-glucuronide)-4′-O-β-D-glucuronide
In: Rye G600. K890

Luteolin-7-O-(β-D-glucuronosyl(1→2)-β-D-glucuronide)-4′-O-β-D-glucuronide
Formed from: Luteolin-7-O-(β-D-glucuronosyl(1→2)-β-D-glucuronide
→ Luteolin-7-O-(β-D-glucuronosyl(1→2)-β-D-glucuronide)-4′-O-β-D-glucuronide)-β-D-glucuronide
Probably in: Rye G600

Luteolin-7-O-(β-D-glucuronosyl(1 → 2)-β-D-glucuronide)-4′-O-β-D-glucuronide)-β-D-glucuronide
Probably formed from: Luteolin-7-O-(β-D-glucuronosyl(1 → 2)-β-D-glucuronide)-4′-O-β-D-glucuronide
Luteolin-4′-sulphate
Formed from: Luteolin-4′,7-disulphate
→ Luteolin
In: Helix F102
Luteolin-7-sulphate
Formed from: Luteolin-4′,7-disulphate
Luteone (2′,4′,5,7-tetrahydroxy-6-(3,3-dimethylallyl)isoflavone)
→ 4″-Hydroxyluteone
In: Rat G339
→ 5″-Hydroxyluteone
In: Rat G339
→ 2″,3″-Dihydro-2″,3″-dihydroxyluteone
In: Aspergillus D219
Botrytis D219
→ 2″,3″-Dihydro-3″-hydroxyluteone
In: Aspergillus D219
Lycorine ((1*S*,2*S*,12b*S*,12c*S*)-2,4,5,7,12b,12c-hexahydro-1*H*-[1,3]dioxolo[4,5-*j*]pyrrolo[3,2,1-*de*]phenanthridine-1,2-diol)
→ Hippeastrine
In: Crinum F630
→ Lycorin-2-one
In: Crinum F630
→ Ungeremine
In: Crinum F630
Lycorin-2-one
Formed from: Lycorine
D-Lysergamide ((8β)-9,10-didehydro-6-methylergoline-8-carboxamide)
→ D-Norlysergamide
In: Rat A98
D-Lysergic acid (9,10-didehydro-6-methylergoline-8-carboxylic acid)
Formed from: D-Lysergyl-L-alanine methyl ester
→ Ergocornine
In: Claviceps B572
→ Ergocryptine
In: Claviceps B572
→ Ergometrine
In: Claviceps B572
Lysergic allylamide
Formed from: Lysergic diallylamide
Lysergic butanolamide
Formed from: Methergine
Lysergic butyl 3-hydroxybutylamide
Formed from: Lysergic dibutylamide
→ Lysergic butyl 3-oxoxybutylamide
Probably in: Streptomyces B595
***epi*-Lysergic butyl 3-hydroxybutylamide**
Formed from: Lysergic dibutylamide
Lysergic butyl 3-oxoxybutylamide
Probably formed from: Lysergic butyl 3-hydroxybutylamide
Lysergic diallylamide
→ Lysergic allylamide
In: Streptomyces B595
Lysergic dibutylamide
→ Lysergic butyl 3-hydroxybutylamide
In: Streptomyces B595
→ *epi*-Lysergic butyl 3-hydroxybutylamide
In: Streptomyces B595
Lysergic diethylamide (LSD)
→ 2,3-Dihydro-2-oxolysergic diethylamide
In: Rat A3720
See also: A10
→ 3-Hydroxy-2-oxolysergic diethylamide
In: Man K619
→ 13-Hydroxylysergic diethylamide
In: Guinea pig B390
Rat A3720, B197, B390, B536
Rhesus monkey B390
See also: A10
→ 14-Hydroxylysergic diethylamide
In: Guinea pig B390
Rat A3720, B390
Rhesus monkey B390
See also: A10
→ Lysergic ethyl vinylamide
In: Gliocladium B596
Rat B197, B536
Streptomyces B516
→ Lysergic ethylamide
In: Aspergillus B596
Cunninghamella B596
Gliocladium B596

Guinea pig B390
Mycobacterium B596
Penicillium B596
Rat A76, A3720, B197, B390, B536
Rhesus monkey B390
Rhizopus B596
Streptomyces B516, B596
Trametes B596
See also: A10
→ Lysergic ethyl 2-hydroxyethylamide
In: Gliocladium B596
Streptomyces B516
→ Norlysergic diethylamide
In: Cunninghamella B596
Mycobacterium B596
Penicillium B596
Rat A76, A3720, B197, B536
Rhizopus B596
Streptomyces B596
→ 2-Oxolysergic diethylamide
In: Guinea pig B390
Rat B390

D-Lysergic dimethylamide
→ D-Lysergic methylamide
In: Rat A98
Streptomyces B595
→ D-Norlysergic dimethylamide
In: Rat A98

Lysergic dipropylamide
→ Lysergic 2-hydroxypropyl propylamide
In: Streptomyces B595
→ *epi*-Lysergic 2-hydroxypropyl propylamide
In: Streptomyces B595

Lysergic ethylamide
Formed from: Lysergic diethylamide

Lysergic ethyl 2-hydroxyethylamide
Formed from: Lysergic diethylamide

Lysergic ethyl vinylamide
Formed from: Lysergic diethylamide

Lysergic 2-hydroxypropyl propylamide
Formed from: Lysergic dipropylamide
→ Lysergic 2-oxopropyl propylamide
Probably in: Streptomyces B595

***epi*-Lysergic 2-hydroxypropyl propylamide**
Formed from: Lysergic dipropylamide

D-Lysergic methylamide
Formed from: D-Lysergic dimethylamide

Lysergic 2-oxopropyl propylamide
Probably formed from: Lysergic 2-hydroxypropyl propylamide

Lysergol ((8β)-9,10-didehydro-6-methylergoline-6-methanol)
→ Lysergol-O-β-D-fructofuranoside
In: Claviceps F512, F516
→ Lysergol-β-galactoside
In: Aspergillus H11

Lysergol-O-β-D-fructofuranoside
Formed from: Lysergol

Lysergol-β-galactoside
Formed from: Lysergol

D-Lysergyl-L-alanine
Formed from: D-Lysergyl-L-alanine methyl ester

D-Lysergyl-L-alanine methyl ester
→ D-Lysergic acid
In: Claviceps A1254
→ D-Lysergyl-L-alanine
In: Claviceps A1254

L-Lysyl-L-alanyl-β-naphthylamide
→ β-Naphthylamine
In: Man A2149

L-Lysyl-β-naphthylamide
→ β-Naphthylamine
In: Agave A2745
Beef A526
Bacteria C146
Crithidia A1102
Glycine A1596
Man A1219, A1397
Pseudomonas G643
Rabbit A1589, B334
Rat A523

L-Lysyl-*p*-nitroanilide
→ *p*-Nitroaniline
In: Agave A2745
Glycine A1596
Man A1219
Nocardia G343
Rat C173
Streptococcus C277

M

Maackiain ((6aR,12aR)-6a,12a-dihydro-6H-[1,3]dioxolo[5,6]benzofuro[3,2-c][1]benzopyran-3-ol)
Formed from: Calycosin
2′,7-Dihydroxy-4′,5′-methylenedioxyisoflavone
Formononetin
Pseudobaptigenin
Probably formed from: Sophorol
→ Dihydromaackiain
In: Stemphylium A2283
→ 6a-Hydroxymaackiain
In: Nectria E319
See also: J345
→ Maackiain-3-O-glucoside-6″-O-malonate
In: Chickpea H71
→ 1α,3,6α,11α-Tetrahydro-1α-hydroxy-9-methoxy-3-oxo-6H-[1,3]dioxolo[5,6]benzofuro[3,2-c][1]benzopyran
In: Ascochyta G213

Maackiain-3-O-glucoside
→ Maackiain-3-O-glucoside-6″-O-malonate
In: Cicer D445

Maackiain-3-O-glucoside-6″-O-malonate
Formed from: Maackiain
Maackiain-3-O-glucoside

Macarpine (5,7-dimethoxy-13-methyl(1,3-benzodioxolo[5,6-c]-1,3-dioxolo[4,5-i]phenanthridinium)
Formed from: Chelirubine
Dihydromacarpine

Maclurin ((3,4-dihydroxyphenyl)(2,4,6-trihydroxyphenyl)methanone)
Probably formed from: Caffeic acid
Iriflophenone
L-Phenylalanine
→ Isomangiferin
In: Anemarrhena B563
See also: B806
→ Mangiferin
In: Anemarrhena B563
See also: B806

Macrosporin (1,7-dihydroxy-3-methoxy-6-methyl-9,10-anthracenedione)
→ Alterporriol A
In: Alternaria G466

Maesanin (5-carboxy-2-hydroxy-3-(10′-(Z)-pentadecenyl)-1,4-benzoquinone)
→ 5-(Ethanolamino)-2-hydroxy-3-(10′-(Z)-pentadecenyl)-1,4-benzoquinone
In: Debaryomyces K388

Maggiemycin ((1R,2R)-2-ethyl-1,2,3,4,6,11-hexahydro-2,5,7,12-tetrahydroxy-4,6,11-trioxo-1-naphthacenecarboxylic acid methyl ester)
Formed from: Aklaviketone

Magnolol (2,2′-dihydroxy-5,5′-bis(1-propen-1(E)-yl)biphenyl)
→ Dihydromagnolol
In: Rat D549, D625
→ Magnolol-β-D-glucuronide
In: Rat D625

Magnolol-β-D-glucuronide
Formed from: Magnolol

Malonanilic acid (3-oxo-3-(phenylamino)propanoic acid)
Formed from: 3-Carboxanilido-5,6-dihydro-2-methyl-1,4-oxathiin
→ p-Hydroxymalonanilic acid
Probably in: Arachis D52

N-Malonylanthranilic acid
Formed from: Anthranilic acid

Malonylapiin
Formed from: Apiin

6″-O-Malonyldhurrin
Formed from: Dhurrin

6′-O-Malonyl-*p*-nitrophenyl-β-D-glucoside
Formed from: 6′-O-Malonyl-*p*-nitrophenyl-β-D-glucoside methyl ester
Probably formed from: *p*-Nitrophenyl-β-D-glucoside

6′-O-Malonyl-*p*-nitrophenyl-β-D-glucoside methyl ester
→ 6′-O-Malonyl-*p*-nitrophenyl-β-D-glucoside
In: Pig H660

N-Malonyl-S-(pentachlorophenyl)-L-cysteine
Probably formed from: S-(Pentachlorophenyl)-L-cysteine

Malonylprunin
Formed from: Prunin

N-Malonyl-S-(tetrachloronitrophenyl)-L-cysteine
Probably formed from: S-(Tetrachloronitrophenyl)-L-cysteine

N-Malonyl-D-tryptophan
Formed from: D-Tryptophan

N-Malonyl-L-tryptophan
→ L-Tryptophan
Probably in: Chinese cabbage F372

Maltol acetylsalicylate
See: Aspalatone

Maltol salicylate
Formed from: Aspalatone
→ Salicylic acid
Probably in: Rat H777

Malvidin (3,5,7-trihydroxy-4-(4-hydroxy-3,5-dimethoxyphenyl)-1-benzopyrilium)
Formed from: Cinnamic acid
Probably formed from: Petunidin
→ Malvidin-3-glucoside
In: Red cabbage A1733

Malvidin-3-glucoside
Formed from: Malvidin

Malvidin-3-rhamnoside
Probably formed from: Malvidin-3-rhamnoside-5-glucoside

Malvidin-3-rhamnoside-5-glucoside
→ Malvidin-3-rhamnoside
Probably in: Pea A194
→ Petunidin-3-rhamnoside-5-glucoside
Probably in: Pea A194

Mandelaldehyde (α-hydroxyphenylacetaldehyde)
Formed from: N-Methylphenylethanolamine
Phenylethanolamine
Probably formed from: N,N-Dimethylphenylethanolamine
→ Mandelic acid
In: Rat A2950
See also: A1713, A1745, A2093, G665
→ Phenylglycol
In: Rat A2950, A3258

(*R*)-Mandelic acid (α-hydroxybenzeneacetic acid)
Formed from: Mandelonitrile
Styrene
→ D-O-Acetylmandelic acid
With lipase K310
→ (*S*)-Mandelic acid
In: Pseudomonas K310
→ Phenylglycol
In: Nocardia F651
→ Phenylglyoxylic acid
In: Acinetobacter C801, D487
Aspergillus A1204
Micrococcus H668
Pseudomonas G794, H668

(*S*)-Mandelic acid (assumed unless otherwise stated)
Formed from: Benzhydrol
Benzyl mandelate
β-Bromostyrene
N,N-Dimethyl-2-oxo-2-phenylethylamine
3,3-Diphenyl-N-(1-phenylethyl)propylamine
Ethylbenzene
Mandelaldehyde
(*R*)-Mandelic acid
Mandelonitrile
N-Methyl-2-oxo-2-phenylethylamine
2-Oxo-2-phenylethylamine
Phenthoate
Phenylacetic acid
Phenylglyoxal
Phenylglyoxylic acid
Phenylpyruvic acid
Phenyramidol
Styrene
3,3,5-Trimethylcyclohexyl mandelate

Zipeprol
→ *p*-Hydroxymandelic acid
In: Pseudomonas A2674
→ 2-Hydroxypropiophenone
In: Pseudomonas G640
→ Phenylglycol
In: Nocardia F651
→ Phenylglyoxylic acid
In: Acinetobacter D487
Neurospora A3219
Pseudomonas K95
Rat F496
Rhodotorula G774, G904, H90
With mandelate dehydrogenase H648
See also: A569, A688, A3172, A3394, B282, B927, C76, D482, G333, G640, G929, H825

Mandelonitrile (α-hydroxyphenylacetonitrile)
Formed from: Amygdalin
Benzaldehyde
Phenylacetonitrile
Prunasin
Vicianin
Probably formed from:
O-Acetylmandelonitrile
Benzyl cyanide
2-Hydroxy-2-phenylacetaldoxime
L-Phenylalanine
→ Benzaldehyde
In: Cherry H695
Prunus H343
With hydroxynitrile lyase H944, K795
See also: A2262, D323, D815
→ Mandelic acid
In: Rat D323, D815
→ Prunasin
In: Cherry C756
Sorghum K313

Mandelonitrile-β-gentiobioside
See: Amygdalin

Mandelonitrile-β-D-glucoside
See: Prunasin

N-Mandelyl-2-aminoethanol
Formed from: Cefcanel daloxate

Mandelylglycine
Formed from: Cefcanel daloxate

Mangiferin (2-β-D-glucopyranosyl-1,3,6,7-tetrahydroxy-9*H*-xanthen-9-one)
Formed from: *p*-Coumaric acid
Maclurin

Mangostin (1,3,6-trihydroxy-7-methoxy-2,8-bis(3-methyl-2-butenyl)-9*H*-xanthen-9-one)
Formed from: 2,3′,4,6-Tetrahydroxybenzophenone

Maprotiline (N-methyl-9,10-ethanoanthracene-9(1*H*)-propanamine)
→ Demethylmaprotiline
In: Dog D167
Man D167
→ 2,3-Dihydro-2,3-dihydroxymaprotiline
In: Dog D167
Man D167
→ 1′-Hydroxymaprotiline
In: Man D167
→ 2-Hydroxymaprotiline
In: Dog D167
Man D167
→ 3-Hydroxymaprotiline
In: Dog D167
Man D167
Rabbit D167
Rat D167

Marcellomycin
→ 7-Deoxypyrromycinone
In: Mouse C832

Marchantin A (7,8,19,20-tetrahydro-15,18-etheno-2,6:9,13-dimetheno-1,14-benzodioxacyclodocosin-11,12,24-triol)
Formed from: Phloretic acid

Marmesin ((2*S*)-2,3-dihydro-2-(1-hydroxy-1-methylethyl)-7*H*-furo[3,2-*g*][1]benzopyran-7-one)
Formed from: Demethylsuberosin
Rutaretin
Umbelliferone
→ 5-Hydroxymarmesin
Probably in: Ficus B757
Ruta B757
→ Isopimpinellin
Probably in: Thamnosma A1013
→ Psoralen
In: Ammi E559
Ficus A548

→ Rutaretin
In: Ruta A547

Matairesinol ((3*R*,4*R*)-dihydro-3,4-bis((4-hydroxy-3-methoxyphenyl)methyl)-2(3*H*)-furanone)
→ Arctigenin
In: Forsythia F661
→ 4′-Demethyl-desoxypodophyllotoxin
In: Diphylleira G201

Mazapertine (1-(3-((4-(2-(1-methylethoxy)phenyl)-1-piperazinyl)methyl)benzoyl)piperidine)
→ 1-(3-((4-(4-Hydroxy-2-(1-methylethoxy)phenyl)-1-piperazinyl)methyl)benzoyl)piperidine
In: Dog J581
→ 4-Hydroxy-1-(3-((4-(2-(1-methylethoxy)phenyl)-1-piperazinyl)methyl)benzoyl)piperidine
In: Dog J581
→ 1-(3-((4-(*o*-Hydroxyphenyl)-1-piperazinyl)methyl)benzoyl)piperidine
In: Dog J581
→ 1-(*o*-Isopropoxyphenyl)piperazine
In: Dog J581
→ 3-((4-(2-(1-Methylethoxy)phenyl)-1-piperazinyl)methyl)benzamide
In: Dog J581
→ 3-((4-(2-(1-Methylethoxy)phenyl)-1-piperazinyl)methyl)benzoic acid
In: Dog J581
→ 1-(3-((1-Piperazinyl)methyl)benzoyl)piperidine
In: Dog J581

Mazindol (5-(4-chlrophenyl)-2,5-dihydro-3*H*-imidazo[2,1-*a*]isoindol-5-ol)
→ 2-(2-Aminoethyl)-3-(*p*-chlorophenyl)-3-hydroxyphthalimidine
In: Dog A3673
Man A3673
→ 5-(*p*-Chlorophenyl)-2,3-dihydro-2,5-dihydroxy-5*H*-imidazo[2,1-*a*]isoindole
In: Dog A3673
Man A3673
Rat A3673
→ 5-(*p*-Chlorophenyl)-2,5-dihydro-5-hydroxy-3*H*-imidazo[2,1-*a*]isoindol-3-one
In: Dog A3673
Man A3673
Rat A1999, A3673
→ 5-(*p*-Chlorophenyl)-5-hydroxy-5*H*-imidazo[2,1-*a*]isoindole
In: Rat A3673

Mebanazine
See: α-Methylbenzylhydrazine

Mebendazole (methyl (5-benzoyl-1*H*-benzimidazol-2-yl) carbamate)
→ (2-Amino-1*H*-benzimidazol-5-yl)phenylmethanone
In: Pig A2634
Rat G24
→ 2-Amino-5(6)-(α-hydroxybenzyl)imidazole
In: Rat C369
See also: J134
→ Methyl (5-(α-hydroxybenzyl)-1*H*-benzimidazol-2-yl)carbamate
In: Dog A2634
Eel J134
Pig A2634
Rat A2634. D97, G24

Mebeverine (duspatal, 3,4-dimethoxybenzoic acid 4-(ethyl(2-(4-methoxyphenyl)-1-methylethyl)amino)butyl ester)
→ N-Desethylmebeverine
In: Man K21
Rat K21
→ N-Ethyl-*p*-methoxyamphetamine
In: Man K21
Rat K21
→ 4-(N-Ethyl-N-(2-(*p*-methoxyphenyl)-1-methylethyl)amino)butanol
In: Man K21
Rat K21
→ 4-(N-Ethyl-N-(2-(*p*-methoxyphenyl)-1-methylethyl)amino)butyl vanillate
Probably in: Man K21
Rat K21

Mecinarone (1-(6-(2-(dimethylamino)ethoxy)-4,7-dimethoxy-5-benzofuranyl)-3-(4-methoxyphenyl)-2-propen-1-one)
→ *p*-O-Demethylmecinarone
In: Man B767

Meclofenamic acid (2-carboxy-2′,6′-dichloro-3′-methyldiphenylamine, meclomen)
→ 2-Carboxy-2′,6′-dichloro-3′-hydroxymethyldiphenylamine
In: Man A3561
→ 2-Carboxy-2′,6′-dichloro-4-hydroxy-3′-methyldiphenylamine
In: Man A3561
→ 2-Carboxy-2′,6′-dichloro-4′-hydroxy-3′-methyldiphenylamine
In: Man A3561
→ Meclofenamic acid-β-D-glucuronide
In: Monkey H861

Meclofenamic acid-β-D-glucuronide
Formed from: Meclofenamic acid

Meclomen
See: Meclofenamic acid

Mecloqualone (3-(*o*-chlorophenyl)-2-methyl-4(3*H*)-quinazolinone)
→ 2-Amino-N-(*o*-chlorophenyl)benzamide
In: Man A1247
→ 3-(2-Chloro-3-hydroxyphenyl)-2-methyl-4(3*H*)-quinazolinone
In: Man A1247
→ 3-(2-Chloro-4-hydroxyphenyl)-2-methyl-4(3*H*)-quinazolinone
In: Dog A428
Man A428, A1247, A1447
Mouse A428
Rat A428
→ 3-(*o*-Chlorophenyl)quinazolinone
In: Man A1247
→ 2-Hydroxymecloqualone
In: Dog A428
Man A428
Mouse A428
Rat A428
→ 6-Hydroxymecloqualone
In: Dog A428
Man A428, A1247
Mouse A428
Rat A428
→ 7-Hydroxymecloqualone
In: Man A1247, A1447
→ 8-Hydroxymecloqualone
In: Man A1247, A1447

Meconine (6,7-dimethoxyphthalide)
Formed from: Noscapine
→ 6-Hydroxy-7-methoxyphthalide
Probably in: Man B397
Rabbit B397
Rat B397
→ 7-Hydroxy-6-methoxyphthalide
Probably in: Man B397
Rabbit B397
Rat B397

Mecoprop
See: 2-(4-Chloro-2-methylphenoxy)propionic acid

Medazepam
See: 7-Chloro-2,3-dihydro-1-methyl-5-phenyl-1*H*-1,4-benzodiazepine

Medetomidine (4-(1-(2,3-dimethylphenyl)ethyl)-1*H*-imidazole)
→ 4-(1-(3-Hydroxymethyl-2-methylphenyl) ethyl)-1*H*-imidazole
In: Rat F886, G233
→ N-Methylmedetomidine
In: Rat G233

Medicarpin ((6a*R*,11a*R*)-6a,11a-dihydro-9-methoxy-6*H*-benzofuro[3,2-*c*][1] benzopyran-3-ol)
Formed from: 2′,7-Dihydroxy-4′-methoxyisoflavanone
2′,7-Dihydroxy-4′-methoxyisoflavone
Formonometin
Vestitol
→ Demethylmedicarpin
In: Ascochyta E370
Colletotrichum B958
→ 2,4-Dihydroxybenzoic acid
In: Ascochyta E370
→ Hydroxymedicarpin
In: Ascochyta E370
→ 1α-Hydroxypterocarp-1,4-dien-3-one
In: Ascochyta E370, G213
→ Medicarpin-3-O-glucoside-6′-O-malonate
In: Alfalfa H830
Chickpea H71
→ Sativan
In: Alfalfa H830
→ Variabilin
In: Alfalfa H830

→ Vestitol
In: Medicago A3723
Stemphylium A2283

Medicarpin-3-O-glucoside-6′-O-malonate
Formed from: Medicarpin

Mefenamic acid (2-((2,3-dimethylphenyl) amino)benzoic acid)
→ Mefenamic acid-β-D-glucuronide
In: Man J617

Mefenamic acid-β-D-glucuronide
Formed from: Mefenamic acid

Mefexamide (N-(2-diethylaminoethyl)-2-(*p*-methoxyphenoxy)acetamide)
→ N-(2-Diethylaminoethyl)-2-(*p*-hydroxyphenoxy)acetamide
In: Man B952
→ *p*-Methoxyphenol
In: Man B952
→ *p*-Methoxyphenoxyacetic acid
In: Man B952

Mefenorex (N-(3-chloropropyl)amphetamine, N-(3-chloropropyl)-α-methylphenethylamine)
→ Amphetamine
In: Man J544
→ *p*-Hydroxymefenorex
Probably in: Man J544
→ N-(3-Hydroxypropyl)amphetamine
In: Man J544

Mefloquine ((α*S*)-rel-α-(2*R*)-2-piperidinyl-2,8-bis(trifluoromethyl)-4-quinolinemethanol)
→ 2,8-Bis(trifluoromethyl)quinoline-4-carboxylic acid
In: Man B343, E413
→ Hydroxymefloquine
In: Man B343
→ 4-Hydroxymethyl-2,8-bis(trifluoromethyl) quinoline
In: Rat B343

Mefruside (baycaron, 2-(N-(4-chloro-3-sulphamoylbenzenesulphonyl)-N-methylaminomethyl)-2-methyltetrahydrofuran)
→ 5-Oxomefruside
In: Man A3073
Rat A1645

Melarsen oxide (N-(4-arsenosophenyl)-1,3,5-triazine-2,4,6-triamine)
Formed from: Melarsoprol

Melarsoprol (2-(4-((4,6-diamino-1,3,5-triazin-2-yl)amino)phenyl)-1,3,2-dithiarsolane-4,5-dicarboxylic acid)
→ Melarsen oxide
In: Man K568

Melatonin (N-acetyl-5-methoxytryptamine)
Formed from: N-Acetyl-5-hydroxytryptamine
5-Methoxytryptamine
→ N^{γ}-Acetyl-N^2-formyl-5-methoxykynurenamine
In: Rabbit A1212
→ 1-Acetyl-1,2,3,3a,8,8a,hexahydro-8a-hydroxy-5-methoxypyrrolo[2,3-*b*]indole
In: Man E380
→ N-Acetyl-5-hydroxytryptamine
In: Man D508, E441, G393
→ 6-Hydroxymelatonin
In: Man D508, E441
Rat B100
→ 5-Methoxyindole-3-glyoxal
Probably in: Pseudomonas A3009
→ 5-Methoxytryptamine
In: Rat B100, B864
Xenopus F39
→ 1-Methylmelatonin
In: Rabbit C333

Melatonin-6-yl-β-D-glucuronide
Formed from: 6-Hydroxymelatonin

Melatonin-6-yl sulphate
Formed from: 6-Hydroxymelatonin

Melilotic acid (3-(*o*-hydroxyphenyl)propionic acid)
Formed from: 3,4-Dihydrocoumarin
o-Hydroxyphenylpropylamine
Proanthocyanidin B_3
4-(2,4,6-Trihydroxyphenyl)flavan-3-ol
Probably formed from: Indan-1-one
→ 2,3-Dihydroxyphenylpropionic acid
In: Pseudomonas A934

Mellein ((3*R*)-3,4-dihydro-8-hydroxy-3-methyl-1*H*-2-benzopyran-1-one)
Formed from acetic acid in Aspergillus C448

Meloxicam (4-hydroxy-2-methyl-N-(5-methyl-2-thiazolyl)-2*H*-1,2-benzothiazine-3-carboxamide-1,1-dioxide)
→ 4-Hydroxy-N-(5-hydroxymethyl-2-thiazolyl)-2-methyl-2*H*-1,2-benzothiazine-3-carboxamide-1,1-dioxide
In: Man H477, H867, J425
Pig J425
Rat H462, J425
→ 4-Hydroxy-N-(2-hydroxypropyl) aminothioxomethyl)-2-methyl-2*H*-1,2-benzothiazine-3-carboxamide-1,1-dioxide
In: Rat H462

Melperone (*p*-fluoro-4-(4-methylpiperidin-1-yl) butyrophenone)
→ *p*-Fluoro-4-(2-hydroxy-4-methylpiperidin-1-yl)butyrophenone
In: Man E697
→ *p*-Fluoro-4-(3-hydroxy-4-methylpiperidin-1-yl)butyrophenone
In: Man E697
→ 1-(*p*-Fluorophenyl)-4-(4-methylpiperidin-1-yl)-1-butanol
In: Man E697
→ 4-(*p*-Fluorophenyl)succinaldehyde
In: Man E697

Menadiol diphosphate (1,4-dihydroxy-2-methylnaphthalene-1,4-diphosphate, menadione diphosphate)
→ Vitamin $K_{2(20)}$
In: Man A774
→ Vitamin $K_{2(45)}$
In: Man A774
→ Vitamin $K_{2(50)}$
In: Man A774

Menadione
See: 2-Methyl-1,4-naphthoquinone

Menaquinone-2 (vitamin $K_{2(10)}$)
→ Menaquinone-2-oxide
In: Rat A3441

Menaquinone-3 (vitamin $K_{2(15)}$)
Probably formed from: 1,4-Dihydroxy-2-naphthoic acid
→ Menaquinone-3-oxide
In: Rat A3441

Menaquinone-4
See: Vitamin $K_{2(20)}$

Menaquinone MK-9(II-H_2)
Formed from: Shikimic acid

Menaquinone-2-oxide
Formed from: Menaquinone-2

Menaquinone-3-oxide
Formed from: Menaquinone-3

Mendon (7-chloro-1,3-dihydro-2-oxo-5-phenyl-1,4-benzodiazepine-3-carboxylic acid)
→ 7-Chloro-1,3-dihydro-3-hydroxy-5-phenyl-2*H*-1,4-benzodiazepin-2-one
In: Rabbit A1538

Menogaril ((2*R*,3*S*,4*R*,5*R*,6*R*,11*R*,13*R*)-4-(dimethylamino)-3,4,5,6,11,12,13,14-octahydro-3,5,8,10,13-pentahydroxy-11-methoxy-6,13-dimethyl-2,6-epoxy-2*H*-naphthaceno[1,2-*b*]oxocin-9,16-dione)
→ N-Demethylmenogaril
In: Rabbit D897
→ 7-Deoxynogarol
In: Rabbit D897

Menogarol ((13*S*)-menogaril)
→ N-Demethylmenogarol
In: Mouse C841, D397

Meobal
See: 3,4-Xylyl N-methylcarbamate

Meperidine (4-carbethoxy-N-methyl-4-phenylpiperidine, pethidine)
→ 4-Carbethoxy-4-(*p*-hydroxyphenyl)-N-methylpiperidine
In: Cat B890
Dog B890
Guinea pig B890
Rabbit B890
Rat A2659, B890
→ 3′,4′-Dihydroxymeperidine
In: Cat D546
Dog D546
Guinea pig D546
Rabbit D546
Rat D546
→ 4-Ethoxycarbonyl-tetrahydro-4-phenylpyridine
In: Cat D546
Dog D546
Guinea pig D546
Rabbit D546
Rat D546

→ Meperidine-N-β-D-glucuronide
In: Man H541
→ Meperidine-N-oxide
In: Guinea pig A2659
Man A713, A2231, A2659
Rat A2659, B613
→ N-Methyl-4-phenylpiperidine-4-carboxylic acid
In: Cat B890
Dog B890
Guinea pig B890
Man K179
Mouse B890
Rabbit B890
Rat B890, C8
→ Normeperidine
In: Cat B890
Dog B890
Guinea pig A2659, B890
Macaque A2117
Man A2231, A2659
Mouse B890
Rabbit B890
Rat A1983, A2513, A2659, B890, D374

Meperidine-N-β-D-glucuronide
Formed from: Meperidine

Meperidine-N-oxide
Formed from: Meperidine

Mephenesin (3-(*o*-tolyloxy)-1,2-propanediol)
→ N-Acetyl-β-(*o*-tolyloxy)alanine
In: Rat A2924
→ 3-(4-Hydroxy-2-methylphenoxy)-1,2-propanediol
In: Rat A1437
→ 2-(*o*-Tolyloxy)acetic acid
In: Rat A1437
→ 3-(*o*-Tolyloxy)lactic acid
In: Rat A1437

Mephentermine (N,α,α-trimethylbenzeneethanamide)
Formed from: N-Hydroxymephentermine
→ N-Hydroxymephentermine
In: Cat A2008
Chick A2008
Guinea pig A2008, G21
Hamster A2008
Mouse A2008, G21
Rabbit A2008, A2438
Rat A2008, F335, G79
→ *p*-Hydroxymephentermine
In: Guinea pig G21
Mouse G21
Rat F335, F519, G79
→ Phentermine
In: Cat A2008
Chick A2008
Guinea pig A2008, G21
Hamster A2008
Mouse G21
Rabbit A2438
Rat A2008, C669, F335, G79, G536
See also: F519

Mephenytoin
See: 5-Ethyl-3-methyl-5-phenylhydantoin

Mephobarbital (5-ethyl-1-methyl-5-phenylbarbituric acid, prominal)
→ 5-Ethyl-5-(*p*-hydroxyphenyl)-1-methylbarbituric acid
In: Man B642, F22, J790
→ Phenobarbital
In: Man B642, J790
See also: A470, E474

Mepivacaine
See: 1-Methylpipecolo-2′,6′-xylidide

Meptazinol (3-(3-ethylhexahydro-1-methyl-1*H*-azepin-3-yl)phenol)
→ Demethylmeptazinol
In: Rat A1790
See also: F350
→ 6-Ethylhexahydro-6-(*m*-hydroxyphenyl)-1-methylazepin-2(2*H*)-one
In: Man A2548
See also: A2698, F350
→ 3-(Hexahydro-3-(2-hydroxyethyl)-1-methyl-1*H*-azepin-3-yl)phenol
Probably in: Man F350
→ Meptazinol-β-D-glucuronide
In: Man F350
Monkey A1790
Rat A1790
See also: A2698

Meptazinol-β-D-glucuronide
Formed from: Meptazinol

Merbarone (5-(N-phenylcarboxamido)-2-thiobarbituric acid)
→ 5-(N-(*p*-Hydroxyphenyl)carboxamido)-2-thiobarbituric acid
In: Man F724
→ 5-(N-Phenylcarboxamido)barbituric acid
In: Man F724

***o*-Mercaptoacetanilide**
→ *o*-Methylthioacetanilide
In: Man C917
Rat C121

***p*-Mercaptoacetanilide**
→ *p*-Methylthioacetanilide
In: Man D926

2-Mercaptobenzimidazole
→ 2-Sulphinylbenzimidazole
In: Mouse D446
Pig D446

2-Mercaptobenzaldehyde
→ (*R*)-2,2′-Bismercaptobenzoin
In: Pseudomonas K424

***o*-Mercaptobenzoic acid** (thiosalicylic acid)
Formed from: *o*-Acetylmercaptobenzoic acid
o-Hexanoylmercaptobenzoic acid
o-Propionylmercaptobenzoic acid
→ 2-Methylthiosalicylic acid
In: Brassica K492, K594
Man C917, D926
See also: E96

2-Mercaptobenzothiazole
Formed from: S-Aminocyclohexyl-2-benzothiazolyl sulphenamide
S-(2-Benzothiazolyl)-L-cysteine
S-Morpholino-2-benzothiazolyl sulphenamide
2-(Thiocyanomethylthio)benzothiazole
→ 1,2-Bis(2-benzothiazolyl)disulphide
Probably in: Rat H767
→ 2-Mercaptobenzothiazole-β-D-glucuronide
In: Carp A3566
Rat F579, H767
→ 2-Mercaptobenzothiazole sulphate
In: Rat H767

2-Mercaptobenzothiazole-β-D-glucuronide
Formed from: 2-Mercaptobenzothiazole

2-Mercaptobenzothiazole sulphate
Formed from: 2-Mercaptobenzothiazole

3-(2-Mercaptoethyl)quinazoline-2,4(1*H*,3*H*)-dione
→ 3-(2-Methylthioethyl)quinazoline-2,4-(1*H*,3*H*)-dione
In: Rat H122

1-((*S*)-3-Mercapto-2-methylpropanoyl)-L-prolyl-L-phenylalanine
Formed from: 1-((*S*)-3-Acetylthio-2-methylpropanoyl)-L-prolyl-L-phenylalanine

2-(Mercaptomethylthio)benzothiazole
Formed from: 2-(Thiocyanomethylthio) benzothiazole

5-Mercapto-2-nitrobenzoic acid
Formed from: 5,5′-Dithiobis(2-nitrobenzoic acid)

Mescaline (3,4,5-trimethoxyphenethylamine)
Probably formed from: 4-Hydroxy-3,5-dimethoxyphenethylamine
→ N-Acetylmescaline
In: Monkey A877
Mouse A66
Rat A925
→ 3,4,5-Trimethoxybenzoic acid
In: Man A3454
Mouse A66, A117
→ 3,4,5-Trimethoxy-N-methylphenethylamine
In: Rat A80
→ 3,4,5-Trimethoxyphenylacetaldehyde
In: Mouse A117
Rat A925

Meseclazone (7-chloro-3,3a-dihydro-2-methyl-2*H*,9*H*-isoazolo[3,2-*b*][1,3]benzoxazin-9-one)
→ 5-Chlorosalicylic acid
In: Man A2880

Mesembrenol (3a-(3,4-dimethoxyphenyl)-2,3,3a,6,7,7a-hexahydro-1-methyl-1*H*-indol-6-ol)
Formed from: 4′-O-Demethylmesembrenone
Mesembrenone
3′-O-Methylnorbelladine
4′-O-Methylnorbelladine
Norbelladine
Sceletone

Mesembrenone (3a-(3,4-dimethoxyphenyl)-2,3,3a,6,7,7a-hexahydro-1-methyl-1*H*-indol-6-one)
Formed from: Norbelladine
→ N-Demethyl-N-formylmesembrenone
In: Sceletium A3136
→ Mesembrine
In: Sceletium A3136
→ Mesembrenol
In: Sceletium A3136

Mesembrine ((3a*S*,7a*S*)-3a-(3,4-dimethoxyphenyl)octahydro-1-methyl-1*H*-indol-6-one)
Formed from: Mesembrenone
Norbelladine

Mesidine
See: 2,4,6-Trimethylaniline

Mesitylene (1,3,5-trimethylbenzene)
→ 3,5-Dimethylbenzoic acid
In: Rabbit F375
Rat A1561, A2160
→ 2,4,6-Trimethylphenol
In: Rat A2160

Mesocarb (3-(1-methyl-2-phenylethyl)-N-(phenylaminocarbonyl)sydnone imine)
→ Amphetamine
In: Rat A3699
→ N-(*p*-Hydroxyphenylaminocarbonyl)-3-(1-methyl-2-phenylethyl)sydnone imine
In: Rat A3699

Mesocorydaline ((13*R*,13a*R*)-5,8,13,13a-tetrahydro-2,3,9,10-tetramethoxy-13-methyl-6*H*-dibenzo[*a*,*g*]quinolizine)
Formed from: Corydaline
→ Corydaline
In: Corydalis H585
→ 13-Methylpalmatine
In: Corydalis H585

Mesoridazine (thioridazine-2-sulphoxide, 10-(2-(1-methyl-2-piperidinyl)ethyl)-2-(methylsulphinyl)-10*H*-phenothiazine)
Formed from: Thioridazine
→ Demethylmesoridazine
Probably in: Man A3186, D366
→ 7-Hydroxymesoridazine (position assumed)
In: Dog G538
Man G538
→ Mesoridazine-N-oxide
In: Dog G538
Man G538
Rat G538
→ Mesoridazine sulphoxide
In: Beef A3445
Dog G538, G705, J208
Man G538, G705
Rat G538, G705, J208
See also: A3186
→ 6′-Oxomesoridazine
In: Dog G538, J208
Man G538
Rat G538, J208
→ Sulphoridazine
In: Beef A3445
Dog G538, J208
Man A1573, A3121, A3186, A3188, F436, G538
Rat G538, J208
See also: D366, H166
→ Thioridazine
Probably in: Man A3188

Mesoridazine-N-oxide
Formed from: Mesoridazine

Mesoridazine sulphoxide
Formed from: Mesoridazine
Thioridazine-5-sulphoxide
→ 6′-Oxomesoridazine sulphoxide
Probably in: Dog G538
Man G538
Rat G538, J208

Mesothalictricavine ((13*R*,13a*R*)-5,8,13,13a-tetrahydro-9,10-dimethoxy-13-methyl-6*H*-benzo[*g*]-1,3-benzodioxolo[5,6-*a*]quinolizine)
Formed from: 13-Methylberberine
→ 13-Methylallocryptopine
In: Dicentra J496
→ 13-Methylberberine
In: Corydalis H585
Dicentra J496
→ N-Methylmesothalictricavine
In: Dicentra J496
→ Thalictricavine
In: Corydalis H585

Mestranol ((17α)-3-methoxy-19-norpregna-1,3,5(10)-trien-20-yn-17-ol)
→ 17α-Ethynyl-17β-oestradiol
In: Man A1577, B566

Mesurol
See: Methiocarb

Metacetamol
See: *m*-Acetamidophenol

Metaclazepam
See: 7-Bromo-5-(*o*-chlorophenyl)-2,3-dihydro-2-methoxymethyl-1-methyl-1*H*-benzodiazepine

Metadrenaline
See: D-Metanephrine

Metamizol
See: Dipyrone

D-Metanephrine (4-hydroxy-3-methoxy-N-methylphenylethanolamine, metadrenaline)
Formed from: D-Adrenaline
D-Normetanephrine
→ D-Adrenaline
In: Rat A3063
→ 4-Hydroxy-3-methoxymandelaldehyde
Probably in: Horse A254
Rat A1054
→ Metanephrine-O-β-D-glucuronide
In: Rabbit A2497
→ Metanephrine-O-sulphate
In: Beef F196
→ Metanephrone
In: Penicillium J16, K761
→ N-Methylmetanephrine
In: Beef A1463

L-Metanephrine
Formed from: L-Adrenaline

Metanephrine-O-β-D-glucuronide
Formed from: Metanephrine

Metanephrine-O-sulphate
Formed from: Metanephrine

Metanephrone (4-hydroxy-3-methoxy-N-methyl-β-oxophenylethylamine)
Formed from: Metanephrine

Metanil Yellow
See: Acid Yellow 36

Metanilic acid
See: *m*-Aminobenzenesulphonic acid

Metapramine (10,11-dihydro-N,5-dimethyl-5*H*-dibenz[*b*,*f*]azepin-10-amine)
→ 5-Normetapramine
In: Dog C356
Man C356, E251
Rat C356
→ 10-Normetapramine
In: Man E250

Metergoline (methergoline, (((8β)-1,6-dimethylergolin-8-yl)methyl)carbamic acid phenylmethyl ester)
→ 8β-Aminomethyl-1,6-dimethylergoline
In: Rat C436
→ 1-Demethylmetergoline
In: Mammalia A425
Rat C436
→ 12-Hydroxymetergoline
In: Mammalia A425
→ 8β-Hydroxymethyl-10α-methoxy-1,6-dimethylergoline
In: Mammalia A425

N-((5-Methanesulphonamido)benzofuran-2-yl) methyl)-N-(2-(4-methanesulphonamidophenyl) ethylamine
Formed from: N-(((5-Methanesulphonamido) benzofuran-2-yl)methyl)-N-(2-(4-methanesulphonamidophenyl)ethyl) methylamine

N-((5-Methanesulphonamido)benzofuran-2-yl) methyl)-N-(2-(4-methanesulphonamidophenyl) ethylmethylamine
→ N-((5-Methanesulphonamido) benzofuran-2-yl)methyl)-N-(2-(4-methanesulphonamidophenyl)ethylamine
In: Dog H742

5-(Methanesulphonamido)indole-3-acetic acid
Formed from: Sumitriptan

***p*-Methanesulphonamidophenoxyacetic acid**
Formed from: Dofetilide

***p*-Methanesulphonamidophenoxy-N-methylethylamine**
Formed from: Dofetilide

***p*-Methanesulphonamidophenylacetic acid**
Formed from: Dofetilide

***p*-Methanesulphonamidophenyl-N-methylethylamine**
Formed from: Dofetilide

Methaphenilene (N,N-dimethyl-N′-phenyl-N′-(2-thienylmethyl)-1,2-ethanediamine)
→ Aniline
In: Rat F297
→ N,N-Dimethyl-N′-phenylethylenediamine
In: Rat F297
→ *p*-Hydroxymethapheniline
In: Rat F297
→ N-Methyl-N′-phenyl-N′-((2-thiophenyl) methyl)ethylenediamine
In: Rat F297

Methaqualone
See: 2-Methyl-3-(*o*-tolyl)-4(3*H*)-quinazolinone

Methazole (2-(3,4-dichlorophenyl)-4-methyl-1,2,3-oxadiazolidine-3,5-dione)
→ 3-(3,4-Dichlorophenyl)-1-methylurea
In: Cow A1837
Hen A1837
See also: A1853

5,10-Methenyltetrahydrofolic acid
Formed from: 5-Formyltetrahydrofolic acid
5,10-Methylenetetrahydrofolic acid
→ 10-Formyltetrahydrofolic acid
In: Man H646

Methergine ((8β)-9,10-didehydro-N-((1*S*)-1-hydroxymethyl)propyl)-6-methylergoline-8-carboxamide)
Formed from: Methysergide
→ Lysergic butanolmide
In: Rat A98
→ Normethergine
In: Rat A98

Methergoline
See: Metergoline

Methindione (2-ethyl-2-(methylamino)-1*H*-indene-1,3(2*H*)-dione)
→ Demethylmethindione
In: Rat A3077

Methiocarb (3,5-dimethyl-4-(methylthio)phenol methylcarbamate)
→ 3,5-Dimethyl-4-methylthiophenyl N-hydroxymethylcarbamate
In: Rat A1545
→ Methiocarb sulphoxide
In: Glycine D306
Rat A1545

Methiocarb sulphoxide
Formed from: Methiocarb

L-Methionyl-β-naphthylamide
→ β-Naphthylamine
In: Beef A526
Glycine A1596
Man A1397
Rabbit B334

L-Methionyl-*p*-nitroanilide
→ *p*-Nitroaniline
In: Nocardia G343

Methocarbamol (robaxin, 3-(2-methoxyphenoxy)-1,2-propanediol-1-carbamate)
→ Demethylmethocarbamol
In: Rat A360
→ Hydroxymethocarbamol
In: Rat A360
→ Methocarbamol-β-D-glucuronide
In: Rat A360

Methocarbamol-β-D-glucuronide
Formed from: Methocarbamol

Methotrexate (amethopterin, N-(4-(((2,4-diamino-6-pteridinyl)methyl)methylamino) benzoyl)-L-glutamic acid)
→ *p*-Aminobenzoyl-L-glutamic acid
Probably in: Rat A667
→ 4-Amino-10-methylpteroic acid
In: Lactobacillus A1594
Microorganism A396
→ 2,4-Diamino-N^{10}-methylpteroic acid
In: Man B87
→ 7-Hydroxymethotrexate
In: Hamster K381
Man A2426, A3099, B507, K381
Monkey A2426, A3099, K381
Rabbit C193, D644, K381
Rat F104, K381
Contrast (in dog) K381
→ Methotrexate-γ-glutamic acid
In: Man C680
Rat B141
→ Methotrexate-γ-glutamyl-γ-glutamic acid
In: Man C680
Rat B141

→ Methotrexate-γ-glutamyl-γ-glutamyl-γ-glutamic acid
In: Man C680
→ Methotrexate-γ-glutamyl-γ-glutamyl-γ-glutamyl-γ-glutamic acid
In: Man C680
→ Methotrexate polyglutamic acid(s)
In: Rat A1594

Methotrexate-γ-glutamic acid
Formed from: Methotrexate

Methotrexate-γ-glutamyl-γ-glutamic acid
Formed from: Methotrexate

Methotrexate-γ-glutamyl-γ-glutamyl-γ-glutamic acid
Formed from: Methotrexate

Methotrexate-γ-glutamyl-γ-glutamyl-γ-glutamyl-γ-glutamic acid
Formed from: Methotrexate

Methotrexate polyglutamic acid(s)
Formed from: Methotrexate

Methotrimeprazine ((βR)-2-methoxy-N,N,β-trimethyl-10*H*-phenothiazine-10-propanamide)
→ Methotrimeprazine sulphoxide
In: Man A2550, A2904, C292
→ Normethotrimeprazine
In: Man A2904, C292

Methotrimeprazine sulphoxide
Formed from: Methotrimeprazine
→ Normethotrimeprazine sulphoxide
Probably in: Man A2904

***o*-Methoxyacetanilide**
→ *o*-Methoxyaniline
In: Ascaris A2036
Moniezia A1755

***m*-Methoxyacetanilide**
→ *m*-Methoxyaniline
In: Ascaris A2036
Moniezia A1755

***p*-Methoxyacetanilide** (methacetin)
Formed from: *p*-Acetamidophenol
p-Methoxyaniline
→ *p*-Acetamidophenol
In: Mouse A3502
Rat A1780
→ N-Hydroxy-*p*-methoxyacetanilide
In: Mouse A3502
→ *p*-Methoxyaniline
In: Ascaris A2036
Microorganism A261
Moniezia A1755
Mouse C609

Methoxyacetic acid 2-(2-((3-(1*H*-benzimidazol-2-yl)propyl)amino)ethyl)-6-fluoro-1,2,3,4-tetrahydro-4-hydroxy-1-isopropyl-2-naphthalenyl ester
Probably formed from: Methoxyacetic acid 2-(2-((3-(1*H*-benzimidazol-2-yl)propyl)methylamino)ethyl)-6-fluoro-1,2,3,4-tetrahydro-4-hydroxy-1-isopropyl-2-naphthalenyl ester
→ Hydroxyacetic acid 2-(2-((3-(1*H*-benzimidazol-2-yl)propyl)amino)ethyl)-6-fluoro-1,2,3,4-tetrahydro-4-hydroxy-1-isopropyl-2-naphthalenyl ester
Probably in: Cynomolgus monkey J103
Man J103
Marmoset J103
Rabbit J103
Rat J103

Methoxyacetic acid 2-(2-((3-(1*H*-benzimidazol-2-yl)propyl)-N-(carboxyglucuronyl)amino)ethyl)-6-fluoro-1,2,3,4-tetrahydro-1-isopropyl-2-naphthalenyl ester
Formed from: Mibefradil
Probably formed from: Methoxyacetic acid 2-(2-((3-(1*H*-benzimidazol-2-yl)propyl)methylamino)ethyl)-6-fluoro-1,2,3,4-tetrahydro-1-isopropyl-2-naphthalenyl ester

Methoxyacetic acid 2-(2-((3-(1*H*-benzimidazol-2-yl)propyl)methylamino)ethyl)-6-fluoro-4-(β-D-glucuronyloxy)-1,2,3,4-tetrahydro-1-isopropyl-2-naphthalenyl ester
Formed from: Methoxyacetic acid 2-(2-((3-(1*H*-benzimidazol-2-yl)propyl)methylamino)ethyl)-6-fluoro-4-(β-D-glucuronyloxy)-1,2,3,4-tetrahydro-1-isopropyl-2-naphthalenyl ester

Methoxyacetic acid 2-(2-((3-(1*H*-benzimidazol-2-yl)propyl)methylamino)ethyl)-6-fluoro-1,2,3,4-tetrahydro-4-hydroxy-1-isopropyl-2-naphthalenyl ester
Formed from: Mibefradil
→ Hydroxyacetic acid 2-(2-((3-(1*H*-benzimidazol-2-yl)propyl)methylamino)

ethyl)-6-fluoro-1,2,3,4-tetrahydro-4-hydroxy-1-isopropyl-2-naphthalenyl ester
Probably in: Cynomolgus monkey J103
Man J103
Marmoset J103
Rabbit J103
Rat J103
→ Methoxyacetic acid 2-(2-((3-(1*H*-benzimidazol-2-yl)propyl)amino)ethyl)-6-fluoro-1,2,3,4-tetrahydro-4-hydroxy-1-isopropyl-2-naphthalenyl ester
Probably in: Cynomolgus monkey J103
Man J103
Marmoset J103
Rabbit J103
Rat J103
→ Methoxyacetic acid 2-(2-((3-(1*H*-benzimidazol-2-yl)propyl)methylamino)ethyl)-6-fluoro-4-(β-D-glucuronyloxy)-1,2,3,4-tetrahydro-1-isopropyl-2-naphthalenyl ester
In: Rat J103

Methoxyacetic acid 6-fluoro-2-(2-(3-(4-(β-D-glucuronosyloxy)-1*H*-benzimidazol-2-yl)propyl)amino)ethyl)-1,2,3,4-tetrahydro-1-isopropyl-2-naphthalenyl ester
Probably formed from: N-Demethylmibefradil
→ 6-Fluoro-2-(2-(3-(4-(β-D-glucuronosyloxy)-1*H*-benzimidazol-2-yl)propyl)amino)ethyl-1,2,3,4-tetrahydro-2-hydroxy-1-isopropylnaphthalene
Probably in: Rat G311

Methoxyacetic acid 6-fluoro-1,2,3,4-tetrahydro-2-(2-((3-(4-hydroxy-1*H*-benzimidazol-2-yl)propyl)amino)ethyl)-1-isopropyl-2-naphthalenyl ester
Probably formed from: Methoxyacetic acid 6-fluoro-1,2,3,4-tetrahydro-2-(2-((3-(4-hydroxy-1*H*-benzimidazol-2-yl)propyl)methylamino)ethyl)-1-isopropyl-2-naphthalenyl ester

Methoxyacetic acid 6-fluoro-1,2,3,4-tetrahydro-2-(2-((3-(5-hydroxy-1*H*-benzimidazol-2-yl)propyl)amino)ethyl)-1-isopropyl-2-naphthalenyl ester
Probably formed from: N-Demethylmibefradil

Methoxyacetic acid 6-fluoro-1,2,3,4-tetrahydro-2-(2-((3-(4-hydroxy-1*H*-benzimidazol-2-yl)propyl)methylamino)ethyl)-1-isopropyl-2-naphthalenyl ester
Formed from: Mibefradil
→ Methoxyacetic acid 6-fluoro-1,2,3,4-tetrahydro-2-(2-((3-(4-hydroxy-1*H*-benzimidazol-2-yl)propyl)amino)ethyl)-1-isopropyl-2-naphthalenyl ester
Probably in: Rat G311
→ Methoxyacetic acid 6-fluoro-1,2,3,4-tetrahydro-2-(2-((3-(4-hydroxy-1*H*-benzimidazol-2-yl)propyl)methylamino)ethyl)-1-isopropyl-2-naphthalenyl ester-β-D-glucuronide
Probably in: Rat G311

Methoxyacetic acid 6-fluoro-1,2,3,4-tetrahydro-2-(2-((3-(5-hydroxy-1*H*-benzimidazol-2-yl)propyl)methylamino)ethyl)-1-isopropyl-2-naphthalenyl ester
Formed from: Mibefradil
→ Methoxyacetic acid 6-fluoro-1,2,3,4-tetrahydro-2-(2-((3-(5-hydroxy-1*H*-benzimidazol-2-yl)propyl)methylamino)ethyl)-1-isopropyl-2-naphthalenyl ester-β-D-glucuronide
In: Rat G311

Methoxyacetic acid 6-fluoro-1,2,3,4-tetrahydro-2-(2-((3-(4-hydroxy-1*H*-benzimidazol-2-yl)propyl)methylamino)ethyl)-1-isopropyl-2-naphthalenyl ester-β-D-glucuronide
Probably formed from: Methoxyacetic acid 6-fluoro-1,2,3,4-tetrahydro-2-(2-((3-(4-hydroxy-1*H*-benzimidazol-2-yl)propyl)methylamino)ethyl)-1-isopropyl-2-naphthalenyl ester

Methoxyacetic acid 6-fluoro-1,2,3,4-tetrahydro-2-(2-((3-(5-hydroxy-1*H*-benzimidazol-2-yl)propyl)methylamino)ethyl)-1-isopropyl-2-naphthalenyl ester-β-D-glucuronide
Probably formed from: Methoxyacetic acid 6-fluoro-1,2,3,4-tetrahydro-2-(2-((3-(5-hydroxy-1*H*-benzimidazol-2-yl)propyl)methylamino)ethyl)-1-isopropyl-2-naphthalenyl ester

Methoxyacetic acid 6-fluoro-1,2,3,4-tetrahydro-1-isopropyl-2-(2-(N-methylamino)ethyl)-2-naphthalenyl ester
Formed from: Mibefradil

***o*-Methoxyacetophenone**
→ 1-(*o*-Methoxyphenyl)ethanol

In: Rabbit C25

***p*-Methoxyacetophenone**
Formed from: 1-(*p*-Methoxyphenyl)ethanol
Probably formed from:
p-Methoxycinnamic acid
trans-p-Methoxypropenylbenzene
→ *p*-Hydroxyacetophenone
In: Sporomusa J199
→ 2-Hydroxy-*p*-methoxyacetophenone
Probably in: Man E812
→ (*R*)-1-(*p*-Methoxyphenyl)ethanol
In: Geotrichum H749

10-Methoxyajmalicine
Probably formed from: 5-Methoxytryptamine

4-Methoxy-1-allylbenzene
See: Estragole

***o*-Methoxyamphetamine**
Formed from: Methoxyphenamine
→ N-Acetyl-*o*-methoxyamphetamine
Probably in: Cunninghamella F340
→ *o*-Hydroxyamphetamine
Probably in: Rat A3781
→ *o*-Methoxyphenylacetone
Probably in: Man C434

***m*-Methoxyamphetamine**
Formed from: *m*-Methoxyphenylacetone
→ *m*-Hydroxyamphetamine
In: Cunninghamella G20
Dog B403
Man B403
Monkey B403
→ 1-(*m*-Methoxyphenyl)propan-2-ol
In: Dog B403
Man B403
Monkey B403

***p*-Methoxyamphetamine**
Formed from: *p*-Methoxyphenylacetone
Probably formed from: N-Ethyl-*p*-methoxyamphetamine
→ N-Acetyl-*p*-methoxyamphetamine
In: Cunninghamella F893
→ *p*-Hydroxyamphetamine
In: Cunninghamella F893, G20
Dog A2090, B632
Guinea pig A3701
Man A3701, J727
Monkey A2090, B632
Rat A3701
→ N-Hydroxy-*p*-methoxyamphetamine
In: Guinea pig A292
Rabbit A292
Rat A292
→ *p*-Methoxyphenylacetone oxime
In: Man A3701
Rat A3701

***o*-Methoxyaniline**
Formed from: *o*-Methoxyacetanilide
o-Nitroanisole
→ Poly-*o*-methoxyaniline
In: Bjerkandera K565

***m*-Methoxyaniline**
Formed from: *m*-Methoxyacetanilide
m-Nitroanisole

***p*-Methoxyaniline** (*p*-anisidine)
Formed from: *p*-Methoxyacetanilide
p-Methoxyglycollanilide
p-Methoxyoxanilic acid
p-Nitroanisole
→ 2-Amino-5-(*p*-anisidino)benzoquinone di(*p*-methoxyphenyl)imine
In: Horseradish D675
Rhizoctonia D675
Trametes D675
→ *p*-Aminophenol
In: Man D135
Sporomusa J199
See also: A2811
→ 2,5-Di-(*p*-anisidino)benzoquinone di(*p*-methoxyphenyl)imine
In: Trametes D675
→ *p*-Methoxyacetanilide
In: Guinea pig F109
Rabbit E399, F109
Rat F109
Sea urchin B703, B895
Sheep A2788
→ *p*-Methoxyformanilide
In: Guinea pig F109
Rabbit F109
Rat F109
→ *p*-Methoxy-N-methylaniline
In: Rabbit C333

3-Methoxyanthranilic acid
Formed from: 3-Hydroxyanthranilic acid

10-Methoxyapomorphine
→ 10-Hydroxyapomorphine
In: Rat A1673

***o*-Methoxybenzaldehyde**
Formed from: *o*-Methoxybenzyl alcohol
Probably formed from: *o*-Methoxybenzyl-amine
→ *o*-Methoxybenzoic acid
In: Acinetobacter F366
Desulphovibrio F666
Man J24
Streptomyces E653
See also: G865
→ *o*-Methoxybenzyl alcohol
In: Dunaliella G459
Euglena G422
Mung bean J912

***m*-Methoxybenzaldehyde**
Formed from: *m*-Methoxybenzyl alcohol
m-Methoxybenzoic acid
Probably formed from: *m*-Methoxybenzylamine
→ 3,3′-Dimethoxybenzoin
In: Pseudomonas K424
→ 2-Hydroxy-*m*-methoxypropiophenone
In: Pseudomonas K524
→ *m*-Methoxybenzoic acid
In: Acinetobacter E663, F366
Arion G804
Helix G804
Limax G804
Man J24
Mouse G804
Rat H761
Rhodopseudomonas G246
Streptomyces E653
See also: G865
→ *m*-Methoxybenzyl alcohol
In: Dunaliella G459
Euglena G422
Mung bean J912
→ *m*-Methoxymandelonitrile
With (*S*)-oxynitrilase F680

***o*-Methoxybenzamide**
Formed from: *o*-Methoxybenzonitrile
→ *o*-Methoxybenzoic acid
Probably in: Rhodococcus H952

***p*-Methoxybenzamidine**
→ *p*-Methoxybenzamidoxime
In: Rabbit C573, E830, F768

***p*-Methoxybenzamidoxime**
Formed from: *p*-Methoxybenzamidine

3-Methoxybenzidine
→ 4′-Acetyl-3-methoxybenzidine
In: Rat A1579

5-Methoxybenzimidazole
Formed from: 5-Hydroxybenzimidazole
→ 5-Methoxybenzimidazolylcobamide
Probably in: Clostridium A1272

5-Methoxybenzimidazolylcobamide
Probably formed from: 5-Methoxy-benzimidazole

2-(5-Methoxybenzofuryl)ethylamine
→ N-Acetyl-2-(5-methoxybenzofuryl) ethylamine
In: Man K490

***o*-Methoxybenzoic acid**
Formed from: *o*-Methoxybenzaldehyde
Probably formed from: *o*-Methoxy-benzamide
→ Catechol
In: Pseudomonas G434
→ Salicylic acid
In: Acetobacterium F655
Arthrobacter + Pantoea K696
Microorganism B321
Sporomusa G931

***m*-Methoxybenzoic acid**
Formed from: *m*-Hydroxybenzoic acid
m-Methoxybenzaldehyde
m-Methoxybenzonitrile
4-((3-Methoxyphenyl)methyl)-2,2,6,6-tetramethyl-1-oxa-4-aza-2,6-disilacyclohexane
m-Methoxyphenylpropionic acid
m-Methoxyphenylpropylamine
→ 1,2-Dihydro-1,2-dihydroxybenzoic acid
Probably in: Pseudomonas B465
→ *m*-Hydroxybenzoic acid
In: Acetobacterium F655

Arthrobacter A672
Chaetomium K896
Microorganism B321, E305, E700
Nocardia F266
Pseudomonas A903, D16, E957
Xerocomus K896
See also: G865
→ *m*-Methoxybenzaldehyde
In: Clostridium G416
→ *m*-Methoxybenzyl alcohol
In: Aspergillus G821
Corynespora G821

o-Methoxybenzonitrile
→ *o*-Methoxybenzamide
In: Rhodococcus H952

m-Methoxybenzonitrile
→ *m*-Methoxybenzoic acid
In: Rhodococcus H952

p-Methoxybenzonitrile
→ Anisic acid
In: Rhodococcus H952
→ *p*-Hydroxybenzonitrile
In: Methylosinus D513

8-Methoxy-4H-1-benzopyran-4-one
See: 8-Methoxychromone

p-Methoxybenzoyl-L-alanine
→ Anisic acid
In: Corynebacterium E145

3-(p-Methoxybenzoyl)propionic acid
Formed from: Cytemba

o-Methoxybenzyl alcohol
Formed from: *o*-Methoxybenzaldehyde
→ *o*-Methoxybenzaldehyde
In: Acinetobacter E663
Helix G804
Limax G804
→ Salicyl alcohol
In: Clostridium G367

m-Methoxybenzyl alcohol
Formed from: *m*-Methoxybenzaldehyde
m-Methoxybenzoic acid
→ *m*-Methoxybenzaldehyde
In: Acinetobacter E663
Arion G804
Botrytis G613
Helix G804
Limax G804
Phanerochaete H613
Pleurotus F867, G668, K373
Pseudomonas A3837
Rhodopseudomonas D197

o-Methoxybenzylamine
→ *o*-Hydroxybenzylamine
In: Rat G865
→ *o*-Methoxybenzaldehyde
Probably in: Rat G865

m-Methoxybenzylamine
→ *m*-Hydroxybenzylamine
In: Rat G865
→ *m*-Methoxybenzaldehyde
Probably in: Rat G865

p-Methoxybenzylamine
→ *p*-Hydroxybenzylamine
In: Rat G865
→ *p*-Methoxybenzaldehyde
Probably in: Rat G865

2-(p-Methoxybenzylamino)pyridine
Formed from: Pyrilamine

(4-Methoxybenzylidene)malononitrile
→ (4-Methoxybenzyl)malononitrile
In: Man B260

(4-Methoxybenzyl)malononitrile
Formed from: (4-Methoxybenzylidene) malononitrile

4-(o-Methoxybenzyl)-1-methylpyridinium
Probably formed from: Tetrahydro-4-(*o*-methoxybenzyl)-1-methylpyridine

4-(p-Methoxybenzyl)-1-methylpyridinium
Probably formed from: Tetrahydro-4-(*p*-methoxybenzyl)-1-methylpyridine

p-Methoxybenzylpiperazine
Formed from: *erythro*-2-(*o*-Chlorophenyl)-2-(4-(*p*-methoxybenzyl)-1-piperazinyl)-1-phenylethanol
→ *p*-Methoxybenzylpiperazine-β-D-glucuronide
In: Rat A3927

p-Methoxybenzylpiperazine-β-D-glucuronide
Formed from: *p*-Methoxybenzylpiperazine

2-Methoxybiphenyl
→ 2,3-Dihydro-2,3-dihydroxy-2′-methoxybiphenyl
In: Escherichia H951

4-Methoxybiphenyl
→ 4-Hydroxybiphenyl
In: Mucor D602
Rat B651, D413, E242

8-Methoxybutamoxane (2-butylaminomethyl-8-methoxy-1,4-benzodioxane)
→ 8-Hydroxybutamoxane
In: Rat A2421
→ 5-Hydroxy-8-methoxybutamoxane
Probably in: Rat A2421

5-Methoxycaffeic acid (5-hydroxyferulic acid)
Formed from: Ferulic acid
Sinapic acid
3,4,5-Trihydroxycinnamic acid
→ 5-Methoxycaffeic acid glucoside
In: Nicotiana B442
→ 5-Methoxycaffeoyl CoA
In: Glycine A2148
→ Sinapic acid
In: Alfalfa B487
Apple D258
Aspen G624
Bamboo A169
Brassica F367
Glycine A2434
Japanese black pine A1634
Loblolly pine J304
Nicotiana A3757, A3778, B490
Wheat J39
See also: C100

5-Methoxycaffeic acid glucoside
Formed from: 5-Methoxycaffeic acid

5-Methoxycaffeoyl CoA
Formed from: 5-Methoxycaffeic acid
→ 3,4-Dihydroxy-5-methoxycinnamaldehyde
In: Glycine A2576
→ Sinapoyl CoA
Loblolly pine J304

1-Methoxycanthin-6-one
Probably formed from: 1-Hydroxycanthin-6-one

5-Methoxycarbonyl-1,4-dihydro-2,6-dimethyl-4-(2-nitrophenyl)pyridine-3-carboxylic acid
Formed from: Nifedipine
→ 5-Methoxycarbonyl-2,6-dimethyl-4-(2-nitrophenyl)pyridine-3-carboxylic acid
Probably in: Man C744

5-Methoxycarbonyl-1,4-dihydro-2-hydroxymethyl-6-methyl-4-(2-nitrophenyl)pyridine-3-carboxylic acid
Formed from: Nifedipine
→ 5-Methoxycarbonyl-2-hydroxymethyl-6-methyl-4-(2-nitrophenyl)pyridine-3-carboxylic acid
Probably in: Man C744

5-Methoxycarbonyl-2,6-dimethyl-4-(2-nitrophenyl)pyridine-3-carboxylic acid
Probably formed from: 5-Methoxycarbonyl-1,4-dihydro-2,6-dimethyl-4-(2-nitrophenyl) pyridine-3-carboxylic acid

5-Methoxycarbonyl-2-hydroxymethyl-6-methyl-4-(2-nitrophenyl)pyridine-3-carboxylic acid
Probably formed from: 5-Methoxycarbonyl-1,4-dihydro-2-hydroxymethyl-6-methyl-4-(2-nitrophenyl)pyridine-3-carboxylic acid

2-(3-Methoxycarbonyl-2-thioureido)aniline
Formed from: 1,2-Bis(3-methoxycarbonyl-2-thioureido)benzene
1-Formamido-2-(3-methoxycarbonyl-2-thioureido)benzene

2-(3-Methoxycarbonyl-2-thioureido)-1-thioformamidobenzene
→ 2-(3-Methoxycarbonyl-2-ureido)aniline
In: Mouse A260

2-(3-Methoxycarbonyl-2-thioureido)-1-(3-methoxycarbonyl-2-ureido)benzene
Formed from: 1,2-Bis(3-methoxycarbonyl-2-thioureido)benzene
→ 1,2-Bis(3-methoxycarbonyl-2-ureido) benzene
In: Mouse A260
→ 2-(3-Methoxycarbonyl-2-thioureido)-1-(2-thioureido)benzene
In: Mouse A260

2-(3-Methoxycarbonyl-2-thioureido)-1-thioformamidobenzene
→ 1-Formamido-2-(3-methoxycarbonyl-2-ureido)benzene
In: Mouse A260
→ 2-(3-Methoxycarbonyl-2-thioureido)aniline
In: Mouse A260

2-(3-Methoxycarbonyl-2-thioureido)-1-(2-thioureido)benzene
Formed from: 1,2-Bis(3-methoxycarbonyl-2-thioureido)benzene
2-(3-Methoxycarbonyl-2-thioureido)-1-(3-methoxycarbonyl-2-ureido)benzene
→ 2-(3-Methoxycarbonyl-2-thioureido)-1-(2-ureido)benzene
In: Mouse A260

2-(3-Methoxycarbonyl-2-thioureido)-1-(2-ureido)benzene
Formed from: 2-(3-Methoxycarbonyl-2-thioureido)-1-(2-thioureido)benzene
→ 2-(3-Methoxycarbonyl-2-ureido)-1-(2-ureido)benzene
In: Mouse A260

2-(3-Methoxycarbonyl-2-ureido)aniline
Formed from: 1,2-Bis(3-methoxycarbonyl-2-ureido)benzene
1-Formamido-2-(3-methoxycarbonyl-2-ureido)benzene
2-(3-Methoxycarbonyl-2-thioureido)aniline
2-(3-Methoxycarbonyl-2-thioureido)-1-thioformamidobenzene
→ 2-Aminobenzimidazole
In: Mouse A260
Sheep A260
→ 2-Methylbenzimidazol-2-ylcarbamate
In: Mouse A260
Sheep A260

2-(3-Methoxycarbonyl-2-ureido)-1-thioformamidobenzene
→ 2-(3-Methoxycarbonyl-2-ureido)aniline
In: Mouse A260

2-(3-Methoxycarbonyl-2-ureido)-1-(2-thioureido)benzene
→ 2-(3-Methoxycarbonyl-2-ureido)-1-(2-ureido)aniline
In: Mouse A260

2-(3-Methoxycarbonyl-2-ureido)-1-(2-ureido) aniline
Formed from: 1,2-Bis(3-methoxycarbonyl-2-ureido)benzene
2-(3-Methoxycarbonyl-2-thioureido)-1-(2-ureido)benzene
2-(3-Methoxycarbonyl-2-ureido)-1-(2-thioureido)benzene
→ 2-Aminobenzimidazole
In: Mouse A260
→ Methyl benzimidazol-2-ylcarbamate
Probably in: Mouse A260

3-Methoxycatechol
Formed from: Anisole
2,3-Dimethoxyphenol
2,6-Dimethoxyphenol
m-Fluoroanisole
Probably formed from: Pyrogallol
→ 2-Methyl-*cis,cis*-muconic acid
In: Pseudomonas A3769
→ Pyrogallol
In: Clostridium G367

4-Methoxycatechol
Formed from: 2-Hydroxy-5-methoxybenzoic acid
4-Methoxy-2-nitrophenol
p-Methoxyphenol
3,4-Methylenedioxyanisole
→ 2,4-Dimethoxyphenol
Probably in: Man G359, H501
→ 2,5-Dimethoxyphenol
Probably in: Man G359, H501

5‴-Methoxycelosianin II
Formed from: Amaranthin
Sinapoyl-β-D-glucose

7-Methoxychlorpromazine
Formed from: 7-Hydroxychlorpromazine

8-Methoxychromone (8-methoxy-4*H*-1-benzopyran-4-one)
Probably formed from: 3-Carboxy-8-methoxychromone
→ 2′-Hydroxy-3′-methoxyacetophenone
Probably in: Dog A1761
Rat A2113

(8α,9*R*)-6′-Methoxycinchonan-9-ol
See: Quinine

***o*-Methoxycinnamaldehyde**
→ *o*-Methoxycinnamic acid
In: Rat E13

***p*-Methoxycinnamaldehyde**
Probably formed from: *p*-Methoxycinnamic acid
p-Methoxycinnamyl alcohol
→ *p*-Methoxycinnamic acid
Probably in: Man E812

→ *p*-Methoxycinnamyl alcohol
Probably in: Trametes A2744

o-Methoxycinnamic acid
Formed from: *o*-Methoxycinnamaldehyde
→ *o*-Coumaric acid
Probably in: Rat E13
→ *o*-Methoxycinnamoyl-β-D-glucose
In: Verbesina K98
→ *o*-Methoxycinnamoylglycine
In: Rat E13
→ *o*-Methoxyphenylhydracrylic acid
Probably in: Rat E13
→ *o*-Methoxyphenylpropionic acid
Probably in: Rat E13
→ *o*-Methoxystyrene
In: Microorganism F390

m-Methoxycinnamic acid
→ *m*-Methoxycinnamoyl CoA
In: Potato A204
Tomato A3149
→ *m*-Methoxystyrene
In: Microorganism F390

p-Methoxycinnamic acid
Probably formed from: *p*-Methoxycinnamaldehyde
trans-*p*-Methoxypropenylbenzene
→ Anisaldehyde
Probably in: Trametes A2744
→ Anisic acid
Probably in: Mouse H725
Rat H725
Vanilla F824
→ *p*-Methoxyacetophenone
Probably in: Mouse H725
Rat H725
→ *p*-Methoxycinnamaldehyde
Probably in: Trametes A2744
→ *p*-Methoxycinnamoyl CoA
In: Cactus H633
Glycine A2148
Leek A2514
Phaseolus A2514
Potato A204
Pseudomonas B298
Tomato A3149
→ *p*-Methoxycinnamoyl-β-D-glucose
In: Vanilla F824
Verbesina K98
→ *p*-Methoxycinnamoylglycine
In: Man E812
See also: H725
→ *p*-Methoxyphenyllactic acid
Probably in: Man E812
→ *p*-Methoxystyrene
In: Candida F390
Catharanthus K244

m-Methoxycinnamoyl CoA
Formed from: *m*-Methoxycinnamic acid

p-Methoxycinnamoyl CoA
Formed from: *p*-Methoxycinnamic acid

o-Methoxycinnamoyl-β-D-glucose
Formed from: *o*-Methoxycinnamic acid

p-Methoxycinnamoyl-β-D-glucose
Formed from: *p*-Methoxycinnamic acid

o-Methoxycinnamoylglycine
Formed from: *o*-Methoxycinnamic acid

p-Methoxycinnamoylglycine
Formed from: *p*-Methoxycinnamic acid

p-Methoxycinnamyl alcohol
Formed from: *trans-p*-Methoxypropenylbenzene
Probably formed from: Estragole
p-Methoxycinnamaldehyde
→ *p*-Methoxycinnamaldehyde
Probably in: Man E812

5-Methoxyconiferaldehyde
Formed from: Sinapoyl CoA

7-Methoxycoumarin
→ Umbelliferone
In: Dog C160
Guinea pig C160
Mouse C160
Monkey C160
Rabbit C160
Rat C160, D155

N-((7-Methoxycoumarin-4-yl)methyl)pyridinium
→ N-((7-Hydroxycoumarin-4-yl)methyl) pyridinium
In: Man K339

Methoxycyclazocine
Formed from: Hydroxycyclazocine

3′-Methoxydehydrocyclopeptine
→ 3′-Hydroxy-5′-methoxydehydrocyclopeptine
Probably in: Penicillium A1633

3-Methoxy-O-demethylpyrilamine
Probably formed from: 3-Hydroxy-O-demethylpyrilamine

8-Methoxydictamnine
Formed from: Dictamnine

***p*-Methoxy-N,N-dimethylbenzamide**
→ *p*-Methoxy-N-methylbenzamide
In: Rat G683

7-Methoxy-2,2-dimethyl-2*H*-benzo[*b*]pyran
See: Precocene I

3,4-*trans*-7-Methoxy-2,2-dimethyl-4-(4-((2-oxopyrrolidin-1-yl)ethoxy)phenyl-3-phenylchromane
Formed from: Levormeloxifene
→ 3,4-*trans*-4-(4-(((3-Carboxypropyl)amino)ethoxy)phenyl)-7-methoxy-2,2-dimethyl-3-phenylchromane
Probably in: Cynomolgus monkey J866
Man J866
Rat J866, K31

5-Methoxy-2,3-dimethylphenol
Formed from: 2-Hydroxy-4-methoxy-6-methylbenzyl alcohol

***o*-Methoxy-N,N-dimethylphenylethanolamine**
→ *o*-Methoxymandelic acid
In: Rat J287

***m*-Methoxy-N,N-dimethylphenylethanolamine**
→ *m*-Methoxymandelic acid
In: Rat J287

***p*-Methoxy-N,N-dimethylphenylethanolamine**
→ *p*-Methoxymandelic acid
In: Rat J287

5-Methoxy-N,α-dimethyl-2-(phenylethynyl)benzenepropanamine
Formed from: N-(2-(3,4-Dimethoxyphenyl)ethyl)-5-methoxy-N,α-dimethyl-2-(phenylethynyl)benzenepropanamine

5-Methoxy-N,N-dimethyltryptamine
Formed from: 5-Hydroxy-N,N-dimethyltryptamine
5-Methoxy-N-methyltryptamine
→ 5-Hydroxy-N,N-dimethyltryptamine
In: Rat E210, E212, E506
→ 5-Methoxy-N,N-dimethyltryptamine-N-oxide
In: Rat E210, E212, E506
→ 5-Methoxy-N-methyltryptamine
In: Rat E210, E212, E506
→ 5-Methoxyindole-3-acetaldehyde
In: Rat B653, E210
See also: A2278

5-Methoxy-N,N-dimethyltryptamine-N-oxide
Formed from: 5-Methoxy-N,N-dimethyltryptamine

9-Methoxyellipticine
→ 9-Hydroxyellipticine
In: Botrytis B306
Cunninghamella B306
Penicillium B306
Rat C737

2-Methoxy-16-epioestriol
Probably formed from: 2-Hydroxy-16-epioestriol
→ 2-Methoxy-16-epioestriol-3-β-D-glucuronide
Probably in: Rat A3102

2-Methoxy-16-epioestriol-3-β-D-glucuronide
Probably formed from: 2-Methoxy-16-epioestriol

3-(4-(2-Methoxyethyl)phenoxy)lactic acid
Probably formed from: 3-(4-(2-Methoxyethyl)phenoxy)propan-1,2-diol

3-(4-(2-Methoxyethyl)phenoxy)propan-1,2-diol
Formed from: Metoprolol
→ 3-(4-(2-Methoxyethyl)phenoxy)lactic acid
Probably in: Dog A2255, G41
Man A2255
Rat A1750, A2255

2-Methoxyethynyloestradiol
Probably formed from: 2-Hydroxyethynyloestradiol

***p*-Methoxyformanilide**
Formed from: *p*-Methoxyaniline

4-Methoxygentisic acid
Formed from: 2,4,5-Trimethoxybenzoic acid
→ 6-Carboxy-4-hydroxy-3-methoxy-*cis,cis*-muconic semialdehyde
In: Arthrobacter A2773

***p*-Methoxyglycollanilide**
→ *p*-Methoxyaniline

In: Microorganism A261
10-Methoxyharmalan
Formed from: 5-Hydroxytryptamine
***p*-Methoxyhippuric acid**
Formed from: Anisic acid
→ Anisic acid
In: Corynebacterium E145
(*E*)-2-Methoxyimino-2-(2-(*o*-tolyloxymethyl) phenyl)acetic acid
Formed from: Methyl (*E*)-2-methoxyimino-2-(2-(*o*-tolyloxymethyl)phenyl)acetate
→ (*E*)-2-(2-((*o*-Hydroxymethylphenoxy) methyl)phenyl)-2-methoxyiminoacetic acid
Probably in: Goat H728
Hen H728
Pig H728
Rat H728
→ (*E*)-2-(2-((4-Hydroxy-2-methylphenoxy) methyl)phenyl)-2-methoxyiminoacetic acid
Probably in: Goat H728
Hen H728
Pig H728
Rat H728
→ 2-((*o*-Hydroxymethyl)phenyl)-2-methoxyiminoacetic acid
Probably in: Goat H728
Hen H728
Pig H728
Rat H728
2-Methoxyindane
→ 1-Hydroxy-2-methoxyindane (*cis* and *trans* isomers)
In: Pseudomonas K207
1-Methoxyindole
→ Indole
Probably in: Rat D158
5-Methoxyindole
→ 5-Methoxy-L-tryptophan
In: Carrot B874
Nicotiana B874
6-Methoxyindole
→ 6-Methoxy-L-tryptophan
In: Carrot B874
Nicotiana B874
7-Methoxyindole
→ 7-Methoxy-L-tryptophan
In: Carrot B874
Nicotiana B874
5-Methoxyindole-3-acetaldehyde
Formed from: 5-Methoxy-N,N-dimethyltryptamine
5-Methoxytryptamine
Probably formed from: 5-Methoxy-tryptophol
→ 5-Methoxyindole-3-acetic acid
In: Dog A1184
Monkey A1184
Mouse A1184
See also: A2278, C659, E210
5-Methoxyindole-3-acetic acid
Formed from: 5-Methoxyindole-3-acetaldehyde
1-Methoxyindole-3-carboxylic acid
→ 1-Hydroxyindole-3-carboxylic acid
Probably in: Rat F920
→ Indole-3-carboxylic acid
Probably in: Rat D158
→ 1-Methoxyindole-3-carboxyl-β-D-glucuronic acid
In: Rat D158
1-Methoxyindole-3-carboxyl-β-D-glucuronic acid
Formed from: 1-Methoxyindole-3-carboxylic acid
5-Methoxyindole-3-glyoxal
Probably formed from: Melatonin
7-Methoxyisoflavone
→ 7-Hydroxyisoflavone
In: Penicillium A817
8-Methoxykaempferol (3,4′,5,7-tetrahydroxy-8-methoxyflavone)
Formed from: 8-Hydroxykaempferol
8-Methoxykynurenic acid (4-hydroxy-8-methoxyquinoline-2-carboxylic acid)
Formed from: Xanthurenic acid
→ Xanthurenic acid
In: Rat A88, A1427
***o*-Methoxymandelic acid**
Formed from: 2-(*o*-Methoxyphenyl)-N,N-dimethyl-2-oxoethylamine
o-Methoxy-N,N-dimethylphenyl-ethanolamine
o-Methoxyphenylethanolamine
o-Methoxy-N-methylphenylethanolamine

o-Methoxyphenylglyoxylic acid
2-(*o*-Methoxyphenyl)-N-methyl-2-oxoethylamine
2-(*o*-Methoxyphenyl)-2-oxoethylamine

***m*-Methoxymandelic acid**
Formed from: 2-(*m*-Methoxyphenyl)-N,N-dimethyl-2-oxoethylamine
m-Methoxy-N,N-dimethylphenylethanolamine
m-Methoxy-N-methylphenylethanolamine
m-Methoxyphenylethanolamine
m-Methoxyphenylglyoxal
m-Methoxyphenylglyoxylic acid
2-(*m*-Methoxyphenyl)-N-methyl-2-oxoethylamine
2-(*m*-Methoxyphenyl)-2-oxoethylamine
→ *m*-Methoxyphenylglyoxylic acid
In: Rhodotorula H90

***p*-Methoxymandelic acid**
Formed from: 2-(*p*-Methoxyphenyl)-N,N-dimethyl-2-oxoethylamine
p-Methoxy-N,N-dimethylphenylethanolamine
p-Methoxyphenylethanolamine
p-Methoxyphenylglyoxal
p-Methoxyphenylglyoxylic acid
2-(*p*-Methoxyphenyl-N-methyl-2-oxoethylamine
2-(*p*-Methoxyphenyl-2-oxoethylamine
→ *p*-Methoxyphenylglyoxylic acid
In: Rhodotorula H90

***m*-Methoxymandelonitrile**
Formed from: *m*-Methoxybenzaldehyde
→ O-Acetyl-*m*-methoxymandelonitrile
In: Pseudomonas H303

***p*-Methoxymandelonitrile**
Formed from: Anisaldehyde

6-Methoxymellein (3,4-dihydro-8-hydroxy-6-methoxy-3-methyl-1*H*-2-benzopyran-1-one)
Formed from: 6-Hydroxymellein

5-Methoxymetanephrine (dimethophrine)
Unchanged in man A2842

***o*-Methoxy-N′-methoxycarbonylphenylhydrazone**
Probably formed from: Metoxadiazone
→ *o*-Hydroxy-N′-methoxycarbonylphenylhydrazone
Probably in: Rat E599

3′-Methoxy-4-methylaminoazobenzene
Formed from: 4-Dimethylamino-3′-methoxyazobenzene
→ 4-Amino-3′-methoxyazobenzene
In: Rat A776
→ 4-Hydroxy-3-methoxy-4′-methylaminoazobenzene
Probably in: Rat A776

4′-Methoxy-4-methylaminoazobenzene
Formed from: 4-Dimethylamino-4′-methoxyazobenzene
→ 4-Amino-4′-methoxyazobenzene
In: Rat A1979

7-Methoxy-4-methylaminomethylcoumarin
→ 7-Hydroxy-4-methylaminomethylcoumarin
In: Man K339

***p*-Methoxy-N-methylaniline**
Formed from: *p*-Methoxyaniline

***p*-Methoxy-N-methylbenzamide**
Formed from: *p*-Methoxy-N,N-dimethylbenzamide

5-Methoxy-6-methylbenzimidazolylcobamide
Formed from: 5-Hydroxybenzimidazole
5-Hydroxy-6-methylbenzimidazole

3-Methoxy-4-methylbenzoic acid
Formed from: 3-Hydroxy-4-methylbenzoic acid

4-Methoxy-3-methylbenzoic acid
Formed from: 5-Methoxy-6-methyl-2-pyrone

9-Methoxy-4-(3-methyl-2-butenyl-7*H*-furo[3,2-*g*][1]benzopyran-7-one
See: Alloimperatorin methyl ether

3-Methoxy-6-methylcatechol
Formed from: *cis*-2,3-dihydro-2,3-dihydroxy-1-methoxy-4-methylbenzene

7-Methoxy-4-methylcoumarin-6-yl-β-D-glucuronide
Probably formed from: 6-Hydroxy-7-methoxy-4-methylcoumarin

3-Methoxy-4,5-methylenedioxyamphetamine
Formed from: Myristicin

N-(2-Methoxy-1-methylethyl)-2-methylsulphinyl-N-phenylacetamide
→ N-(2-Hydroxy-1-methylethyl)-2-methylsulphinyl-N-phenylacetamide
Probably in: Rat G973

5-Methoxy-2-methylindole-3-acetic acid
Formed from: 1-(*p*-Chlorobenzoyl)-5-methoxy-2-methylindole-3-acetic acid

8-Methoxy-2-methyl-6,7-methylenedioxyisoquinoline
Probably formed from: 1,2-Dihydro-8-methoxy-2-methyl-6,7-methylenedioxyisoquinoline

***o*-Methoxy-N-methylphenethylamine**
→ *o*-Methoxyphenylacetaldehyde
Probably in: Rat G865

***m*-Methoxy-N-methylphenethylamine**
→ *m*-Methoxyphenylacetaldehyde
Probably in: Rat G865
→ N-Methyl-*m*-tyramine
In: Rat G865

***p*-Methoxy-N-methylphenethylamine**
→ *p*-Methoxyphenylacetaldehyde
Probably in: Rat G865
→ N-Methyltyramine
In: Rat G865

N-Methoxymethylphenobarbital
Formed from: N,N′-Dimethoxymethyl-phenobarbital
→ Phenobarbital
In: Rat A1233
See also: A805

2-Methoxy-3-methylphenol
Formed from: 3-Methylcatechol

2-Methoxy-4-methylphenol
Formed from: 4-Methylcatechol
Probably formed from: 3,4-Dimethoxytoluene
Vanillyl alcohol

2-Methoxy-5-methylphenol
Formed from: *p*-Methylanisole
4-Methylcatechol
Probably formed from: Isovanillyl alcohol

2-Methoxy-6-methylphenol
Formed from: 3-Methylcatechol

4-Methoxy-2-methylphenol
Formed from: *p*-Methylanisole

5-Methoxy-2-methylphenol
Formed from: *p*-Methylanisole

***p*-Methoxymethylphenol**
→ *p*-Hydroxybenzaldehyde
In: Penicillium J354
See also: K761

***o*-Methoxy-N-methylphenylethanolamine**
→ *o*-Methoxymandelic acid
In: Rat J287

***m*-Methoxy-N-methylphenylethanolamine**
→ *m*-Methoxymandelic acid
In: Rat J287

N-(4-(Methoxymethyl)-4-piperidinyl)-N-phenylpropanamide
Formed from: Alfentanil
Sufentanil
→ N-(4-(Hydroxymethyl)-4-piperidinyl)-N-phenylpropanamide
Probably in: Dog E838, F567, E895
Man F567
Rat E838, F567, E895

***o*-Methoxy-2-methyl-2-(3-pyridyl)propiophenone**
→ *o*-Hydroxy-2-methyl-2-(3-pyridyl) propiophenone
In: Rat J86

5-Methoxy-6-methyl-2-pyrone
→ 4-Methoxy-3-methylbenzoic acid
In: Macrophora B766

4-Methoxy-6-methyl-2-pyrone-5-carboxylic acid ethyl ester
→ 3-Methoxy-5-methylterephthalic acid 4-ethyl ester
In: Macrophora B766

4-Methoxy-6-methyl-2-pyrone-5-carboxylic acid methyl ester
→ 3-Methoxy-5-methylterephthalic acid 4-methyl ester
In: Macrophora B766

***trans*-2-(3-Methoxy-5-methylsulphonyl-4-propoxyphenyl)-5-(3,4,5-trimethoxyphenyl)tetrahydrofuran**
→ *trans*-2-(4-Hydroxy-3-methoxy-5-methylsulphonylphenyl)-5-(3,4,5-trimethoxyphenyl)tetrahydrofuran
In: Rhesus monkey F925
→ *trans*-2-(4-(2-Hydroxypropoxy)-3-methoxy-5-methylsulphonylphenyl)-5-(3,4,5-trimethoxyphenyl)tetrahydrofuran
In: Rhesus monkey F925

3-Methoxy-5-methylterephthalic acid 4-ethyl ester
Formed from: 4-Methoxy-6-methyl-2-pyrone-5-carboxylic acid ethyl ester

3-Methoxy-5-methylterephthalic acid 4-methyl ester
Formed from: 4-Methoxy-6-methyl-2-pyrone-5-carboxylic acid methyl ester

5-Methoxy-N-methyltryptamine
Formed from: 5-Methoxy-N,N-dimethyltryptamine
5-Methoxytryptamine
→ 5-Methoxy-N,N-dimethyltryptamine
In: Man A1483
Rabbit A1320, A1483

6-Methoxy-N-methyltryptamine
Formed from: 6-Methoxytryptamine

6-Methoxy-4-methylumbelliferone
Formed from: 6,7-Dihydroxy-4-methylcoumarin
→ 6-Methoxy-4-methylumbelliferyl-β-D-glucuronide
Probably in: Rat A773

8-Methoxy-4-methylumbelliferone
Formed from: 7,8-Dihydroxy-4-methylcoumarin

6-Methoxy-4-methylumbelliferyl-β-D-glucuronide
Probably formed from: 6-Methoxy-4-methylumbelliferone

1-Methoxynaphthalene
→ 4-Chloro-1-methoxynaphthalene
In: Corallina E59
→ α-Naphthol
In: Rat B582

2-Methoxynaphthalene
→ 1,2-Dihydro-1,2-dihydroxy-6-methoxynaphthalene
In: Beijerinckia H358
Escherichia H358
Pseudomonas H358
→ 1,2-Dihydro-1,2-dihydroxy-7-methoxynaphthalene
In: Beijerinckia H358
Pseudomonas H358
→ 2-Hydroxy-4-methoxybenzoic acid
In: Pseudomonas B899, B905
→ β-Naphthol
In: Sporomusa J199

2-(6-Methoxynaphth-2-yl)propionylglycine
Formed from: Naproxen

6-Methoxy-2-naphthylacetic acid
Formed from: 4-(6-Methoxy-2-naphthyl)butan-2-ol
Nabumetone
→ 6-Hydroxy-2-naphthylacetic acid
In: Rat C867

4-(6-Methoxy-2-naphthyl)butan-2-ol
Formed from: Nabumetone
→ 4-(6-Hydroxy-2-naphthyl)butan-2-ol
Probably in: Rat C867
→ 6-Methoxy-2-naphthylacetic acid
In: Rat C867

(*S*)-2-(6-Methoxynaphth-2-yl)propanol
Formed from: (*S*)-2-(6-Methoxynaphth-2-yl)propanol acetyl ester

(*S*)-2-(6-Methoxynaphth-2-yl)propanol acetyl ester
→ (*S*)-2-(6-Methoxynaphth-2-yl)propanol
In: Pig K560

2-(6-Methoxynaphth-2-yl)propionylglycine
Formed from: Naproxen

2-(6-Methoxynaphth-2-yl)propionyl-β-D-glucoside
Formed from: Naproxen

2-(6-Methoxynaphth-2-yl)propionylsophoroside
Formed from: Naproxen

2-Methoxy-4-nitroaniline
Formed from: Nitromethaqualone

4-Methoxy-3-nitrobenzoic acid
→ 4-Hydroxy-3-nitrobenzoic acid
In: Pseudomonas A2953, K801

4-Methoxy-5-nitrocatechol
Formed from: 4,5-Methylenedioxy-2-nitroanisole

7-Methoxy-2-nitronaphtho[2,1-*b*]furan
→ 7,8-Dihydro-7,8-dihydroxy-2-nitronaphtho[2,1-*b*]furan
Probably in: Rat E63
→ 7-Hydroxy-2-nitronaphtho[2,1-*b*]furan
In: Rat D354, E63

2-Methoxy-4-nitrophenol
Formed from: 4-Nitrocatechol

2-Methoxy-5-nitrophenol
Formed from: 4-Nitrocatechol

4-Methoxy-2-nitrophenol
→ 4-Methoxycatechol
In: Pseudomonas E607, H65

2-Methoxy-4-(2′-nitrovinyl)phenol
Formed from: N-Acetyl-2-methoxy-4-(2′-nitrovinyl)phenyl-β-D-galactosaminide
2-Methoxy-4-(2′-nitrovinyl)phenyl-β-D-galactoside

2-Methoxy-4-(2′-nitrovinyl)phenyl-β-D-galactoside
→ 2-Methoxy-4-(2′-nitrovinyl)phenol
In: Man C7

2-Methoxy-17β-oestradiol
Formed from: 2-Hydroxy-17β-oestradiol
2-Methoxyoestrone
→ 2-Hydroxy-17β-oestradiol
In: Man K621
Rat B668
See also: A2410
→ 2-Methoxy-17β-oestradiol-3-β-D-glucuronide
Probably in: Rat A1487, J498
→ 2-Methoxy-17β-oestradiol-3-sulphate
In: Beef A1211
See also: A1487
→ 2-Methoxyoestriol
Probably in: Man A3225
Rat A2410

4-Methoxy-17β-oestradiol
Formed from: 4-Hydroxy-17β-oestradiol
4-Methoxyoestrone
→ 2-Hydroxy-4-methoxy-17β-oestradiol
Probably in: Rat A2410
→ 4-Hydroxy-17β-oestradiol
In: Man K621
See also: A2410, A3225

2-Methoxy-17β-oestradiol-3-β-D-glucuronide
Probably formed from: 2-Methoxy-17β-oestradiol

2-Methoxy-17β-oestradiol-3-sulphate
Formed from: 2-Hydroxy-17β-oestradiol-3-sulphate
2-Methoxy-17β-oestradiol

S-(2-Methoxy-17β-oestradiol-4-yl)-L-cysteine
Formed from: S-(2-Hydroxy-17β-oestradiol-4-yl)-L-cysteine

2-Methoxyoestriol
Formed from: 2-Hydroxyoestriol
Probably formed from: 2-Methoxy-17β-oestradiol
→ 2-Methoxyoestriol-16α-β-D-glucuronide
In: Rat C242
See also: A1487
→ 2-Methoxyoestriol sulphate
In: Rat C242
See also: A1487

2-Methoxyoestriol-16α-β-D-glucuronide
Formed from: 2-Methoxyoestriol

2-Methoxyoestriol sulphate
Formed from: 2-Methoxyoestriol

2-Methoxyoestrone
Formed from: 2-Hydroxyoestrone
→ 4-Hydroxy-2-methoxyoestrone
In: Rat A2410
→ 2-Hydroxyoestrone
In: Man A3225
Rat A2410, B668
→ 2-Methoxy-17β-oestradiol
In: Man A3225
Rat A2410
→ 2-Methoxyoestrone-3-β-D-glucuronide
In: Rat A1280
See also: A490, A1487
→ 2-Methoxyoestrone-3-sulphate
In: Beef A1211
See also: A1487

4-Methoxyoestrone
Formed from: 4-Hydroxyoestrone
→ 4-Hydroxyoestrone
In: Man A3225
Rat A2410
→ 4-Methoxy-17β-oestradiol
In: Man A3225
Rat A2410

2-Methoxyoestrone-3-β-D-glucuronide
Formed from: 2-Hydroxyoestrone-3-β-D-glucuronide
2-Methoxyoestrone

2-Methoxyoestrone-3-sulphate
Formed from: 2-Hydroxyoestrone-3-sulphate
2-Methoxyoestrone

***p*-Methoxyoxanilic acid**
→ *p*-Methoxyaniline
In: Microorganism A261

(*Z*)-3-((5-Methoxy-3-oxo-2(3*H*)-benzofuranylidene)methyl)benzoic acid
→ 3-((5-Methoxy-3-oxo-2(3*H*)-

benzofuranyl)methyl)benzoic acid
In: Man C874
Rat C651, C874

3-((5-Methoxy-3-oxo-2(3*H*)-benzofuranyl)methyl)benzoic acid
Formed from: (*Z*)-3-((5-Methoxy-3-oxo-2(3*H*)-benzofuranylidene)methyl) benzoic acid

3-Methoxy-4-(1-oxo-2-butenyl)-5-methylbenzoic acid
Formed from: 4-Methoxy-5-(1-oxo-2-butenyl)-6-methyl-2-pyrone

4-Methoxy-5-(1-oxo-2-butenyl)-6-methyl-2-pyrone
→ 3-Methoxy-4-(1-oxo-2-butenyl)-5-methylbenzoic acid
In: Macrophora E766

2-Methoxy-6-oxo-17β-oestradiol
Formed from: 2-Hydroxy-6-oxo-17β-oestradiol

2-Methoxy-16-oxo-17β-oestradiol
Probably formed from: 2-Hydroxy-16-oxo-17β-oestradiol

3-Methoxy-17-oxo-1,3,5(10),9(11)-oestratetraene
→ 16α,17β-Dihydroxy-3-methoxy-1,3,5(10),9(11)-oestratetraene
In: Streptomyces F810
→ 16α-Hydroxy-3-methoxy-17-oxo-1,3,5(10),9(11)-oestratetraene
In: Streptomyces F810

2-Methoxy-6-oxo-6-phenylhexa-2,4-dienoic acid
→ γ-Benzoylbutyraldehyde
In: Pseudomonas E616
→ 2-,6-Dioxo-6-phenylhexanoic acid
In: Pseudomonas E616

2-Methoxy-6-oxo-6-phenylhexa-2,4-dienoic acid methyl ester
→ 2-Methoxy-6-oxo-6-phenylhexa-2-enoic acid methyl ester
In: Pseudomonas E616

2-Methoxy-6-oxo-6-phenylhexa-2-enoic acid methyl ester
Formed from: 2-Methoxy-6-oxo-6-phenyl-hexa-2,4-dienoic acid methyl ester

2-Methoxy-6-(2-oxopropyl)benzonitrile
→ 6-(2-Hydroxypropyl)-2-methoxybenzonitrile
In: Candida K366

1-Methoxy-4-((pentafluorophenyl)sulphonamido) benzene
→ 1-Methoxy-4-((2,3,5,6-tetrafluoro-4-(S-glutathionyl)phenyl)sulphonamido) benzene
In: Mouse K62

Methoxyphenamine (1-(*o*-methoxyphenyl)-2-methylaminopropane)
→ 1-(5-Hydroxy-2-methoxyphenyl)-2-methylaminopropane
In: Cunninghamella F340
Man A2925, C434, D327, D907, G965
Monkey A2925
Rat D767, F331, F478
See also: A2128
→ *o*-Hydroxy-N-methylamphetamine
In: Cunninghamella F340
Man A2128, A2567, A2925, D327, G965
Monkey A2128, A2925
Rat A3781, D767, F331, F478
→ *o*-Methoxyamphetamine
In: Cunninghamella F340
Man A2128, A2551, A2925, C434, D327, D907
Monkey A2128, A2925
Rat A3781, D767, F331, F478

1-Methoxyphenanthrene
Probably formed from: 1-Hydroxy-phenanthrene

***o*-Methoxyphenethylamine**
→ *o*-Methoxyphenylacetaldehyde
Probably in: Rat G865

***m*-Methoxyphenethylamine**
→ *m*-Methoxyphenylacetaldehyde
Probably in: Rat G865

***p*-Methoxyphenethylamine**
p-Methoxyphenylacetaldehyde
→ In: Pig A72
See also: G865

4-Methoxyphenethylamin-3-yl sulphate
Formed from: 3-Hydroxy-4-methoxyphenethylamine

***m*-Methoxyphenol**
Formed from: Anisole
p-Iodophenol
→ Resorcinol
In: Acetobacterium F655

Clostridium G367
Rat J201
Sporomusa J199

***p*-Methoxyphenol**
Formed from: N-Acetyl-*p*-methoxyphenyl-β-D-glucosaminide
Anisole
1,4-Dimethoxybenzene
p-Iodoanisole
Mefexamide
→ 4-Methoxycatechol
In: Man G359, H501
Mushroom H907
Rainbow trout J901
See also: B897
→ *p*-Methoxyphenyl-α-D-glucoside
Probably in: Bacillus H234
→ *p*-Methoxyphenyl-β-D-glucoside
In: Manduca G816
Trilobium G816
→ *p*-Methoxyphenyl sulphate
In: Man E575
Rat B941, E575, K735
→ Quinol
In: Man D135, G359, H501
Rainbow trout J901
Rat J201
Sporomusa J199

Methoxyphenoxazone (7-methoxyresorufin, resorufin methyl ether)
→ Resorufin
In: Alligator J855
Channel catfish E209
Man J63
Rat J63, J855, J904

***o*-Methoxyphenoxyacetic acid**
Formed from: Amosulalol
Probably formed from: 2-(*o*-Methoxyphenoxy)ethanol
→ Guaiacol
In: Microorganism D920
→ Phenoxyacetic acid
In: Microorganism D920

***p*-Methoxyphenoxyacetic acid**
Formed from: Astemizole
Mefexamide

2-(*o*-Methoxyphenoxy)ethanol
Formed from: Syringylglycerol β-guaiacyl ether
Veratrylglycerol β-guaiacyl ether
→ Guaiacol
In: Acinetobacter F859
→ *o*-Methoxyphenoxyacetic acid
Probably in: Microorganism D920

β-(2-Methoxyphenoxy)lactic acid
Formed from: Glyceryl guaiacolate

3-(*o*-Methoxyphenoxy)-1,2-propanediol
→ Guaiacol
In: Acinetobacter F101

3-(2-Methoxyphenoxy)-1,2-propanediol-1-carbamate
See: Methocarbamol

***o*-Methoxyphenylacetaldehyde**
Probably formed from: *o*-Methoxy-N-methylphenethylamine
o-Methoxyphenethylamine
→ *o*-Methoxyphenylacetic acid
Probably in: Rat G865

***m*-Methoxyphenylacetaldehyde**
Probably formed from: *m*-Methoxy-N-methylphenethylamine
m-Methoxyphenethylamine
→ *m*-Methoxyphenylacetic acid
Probably in: Rat G865

***p*-Methoxyphenylacetaldehyde**
Probably formed from: Estragole
p-Methoxy-N-methylphenethylamine
p-Methoxyphenethylamine
→ *p*-Methoxyphenylacetic acid
Probably in: Rat G865

***p*-Methoxyphenylacetaldoxime**
→ *p*-Methoxyphenylacetonitrile
In: Bacillus K411

***o*-Methoxyphenylacetic acid**
Formed from: 4-(*o*-Methoxyphenyl)butyric acid
Probably formed from: *o*-Methoxyphenylacetaldehyde
→ *o*-Hydroxyphenylacetic acid
Probably in: Rat G865

***m*-Methoxyphenylacetic acid**
Formed from: 4-Hydroxy-3-methoxyphenylacetic acid

m-Methoxyphenylpyruvic acid
Probably formed from: *m*-Methoxyphenyl-acetaldehyde
→ *m*-Hydroxyphenylacetic acid
In: Acetobacterium F655
See also: G865

p-Methoxyphenylacetic acid
Formed from: Astemizole
p-Methoxyphenylacetonitrile
4-(*p*-Methoxyphenyl)butylamine
4-(*p*-Methoxyphenyl)butyric acid
p-Methoxyphenylpyruvic acid
Probably formed from: 5,6-Dihydro-3-(*p*-methoxyphenyl)-4,6-dioxo-4*H*-pyran
p-Methoxyphenylacetaldehyde
→ *p*-Hydroxyphenylacetic acid
Probably in: Rat G865

o-Methoxyphenylacetone
Probably formed from: *o*-Methoxy-amphetamine

m-Methoxyphenylacetone
→ (*S*)-(+)-*m*-Methoxyamphetamine
In: Brevibacterium F813

p-Methoxyphenylacetone
Probably formed from: Anethole oxide
→ (*S*)-(+)-*p*-Methoxyamphetamine
In: Brevibacterium F813
Chromobacterium F813
Flavobacterium F813
Mycobacterium F813
Pseudomonas F813
Sarcina F813
→ 1-(*p*-Methoxyphenyl)propan-2-ol
In: Guinea pig A292
Rabbit A292
Rat A292

p-Methoxyphenylacetone oxime
Formed from: *p*-Methoxyamphetamine

p-Methoxyphenylacetonitrile
Formed from: *p*-Methoxyphenylacetaldoxime
→ *p*-Methoxyphenylacetic acid
In: Rhodococcus G172

p-Methoxyphenyl azide
→ *p*-Hydroxyphenyl azide
In: Mouse G8

2-Methoxy-3-phenylbenzoic acid
→ 2-Hydroxy-3-phenylbenzoic acid
In: Phanerochaete D768

4-(p-Methoxyphenyl)-2-butanone
→ (*S*)-(+)-2-Amino-4-(*p*-methoxyphenyl)-2-butanone
In: Brevibacterium F813

4-(o-Methoxyphenyl)butylamine
→ 4-(*o*-Hydroxyphenyl)butylamine
In: Rat G865
→ 4-(*o*-Methoxyphenyl)butyric acid
In: Rat G865

4-(m-Methoxyphenyl)butylamine
→ 4-(*m*-Hydroxyphenyl)butylamine
In: Rat G865

4-(p-Methoxyphenyl)butylamine
→ 4-(*p*-Hydroxyphenyl)butylamine
In: Rat G865
→ 4-(*p*-Methoxyphenyl)butyric acid
In: Rat G865

4-(o-Methoxyphenyl)butyric acid
Formed from: 4-(*o*-Methoxyphenyl) butylamine
→ 4-(*o*-Hydroxyphenyl)butyric acid
In: Rat G865
→ *o*-Methoxyphenylacetic acid
In: Rat G865

4-(m-Methoxyphenyl)butyric acid
→ 4-(*m*-Hydroxyphenyl)butyric acid
In: Rat G865

4-(p-Methoxyphenyl)butyric acid
→ 4-(*p*-Hydroxyphenyl)butyric acid
In: Rat G865
→ *p*-Methoxyphenylacetic acid
In: Rat G865

1-(p-Methoxyphenyl)cyclopropanol
Formed from: 4-Cyclopropylanisole

2-(o-Methoxyphenyl)-N,N-dimethyl-2-oxoethylamine
→ *o*-Methoxymandelic acid
In: Rat J287

2-(m-Methoxyphenyl)-N,N-dimethyl-2-oxoethylamine
→ *m*-Methoxymandelic acid
In: Rat J287

2-(*p*-Methoxyphenyl)-N,N-dimethyl-2-oxoethylamine
→ *p*-Methoxymandelic acid
In: Rat J287

1-(*p*-Methoxyphenyl-2,2-dimethyl-1-propanol
→ 1-(*p*-Methoxyphenyl-2,2-dimethyl-1-propanone
In: Rat J529

1-(*p*-Methoxyphenyl-2,2-dimethyl-1-propanone
Formed from: 1-(*p*-Methoxyphenyl-2,2-dimethyl-1-propanol

1-(*p*-Methoxyphenyl)-3,3-dimethyltriazene
→ 1-(*p*-Methoxyphenyl)-3-methyltriazene
In: Rat A112

2-(*p*-Methoxyphenyl)-1,3-dithiolane
→ 2-(*p*-Methoxyphenyl)-1,3-dithiolane-1-oxide
In: Rabbit F355

2-(*p*-Methoxyphenyl)-1,3-dithiolane-1,3-dioxide
Formed from: 2-(*p*-Methoxyphenyl)-1,3-dithiolane-1-oxide

2-(*p*-Methoxyphenyl)-1,3-dithiolane-1-oxide
Formed from: 2-(*p*-Methoxyphenyl)-1,3-dithiolane
→ 2-(*p*-Methoxyphenyl)-1,3-dithiolane-1,3-dioxide
In: Rabbit F355

1-(*o*-Methoxyphenyl)ethanol
Formed from: *o*-Methoxyacetophenone

1-*p*-Methoxyphenyl)ethanol
Formed from: *p*-Ethylanisole
p-Methoxyacetophenone
→ *p*-Methoxyacetophenone
In: Geotrichum H749

***o*-Methoxyphenylethanolamine**
→ *o*-Methoxymandelic acid
In: Rat J287

***m*-Methoxyphenylethanolamine**
→ *m*-Methoxymandelic acid
In: Rat J287

***p*-Methoxyphenylethanolamine**
→ *p*-Methoxymandelic acid
In: Rat J287

3-Methoxyphenylethanol-4-yl sulphate
Formed from: 2-(4-Hydroxy-3-methoxyphenyl)ethanol

1-(*o*-Methoxyphenyl)ethylamine
Formed from: Octyl 1-(*o*-methoxyphenyl)ethylamine oxalamate

1-(*m*-Methoxyphenyl)ethylamine
Formed from: Octyl 1-(*m*-methoxyphenyl)ethylamine oxalamate

1-(*p*-Methoxyphenyl)ethylamine
Formed from: Octyl 1-(*p*-methoxyphenyl)ethylamine oxalamate

1-(2-Methoxy-2-phenylethyl)piperazine
Formed from: Zipeprol
→ 1-(2-Hydroxy-2-phenylethyl)piperazine
Probably in: Man A3300

2-(1-(2-(*p*-Methoxyphenyl)ethyl)piperidin-4-yl)aminobenzimidazole
Formed from: Astemizole

4-(5-Methoxy-2-(2-phenylethynyl)phenyl)butan-2-one
Formed from: N-(2-(3,4-Dimethoxyphenyl)ethyl)-5-methoxy-N,α-dimethyl-2-phenylethynyl)benzenepropanamine

***o*-Methoxyphenyl-β-D-glucoside**
Formed from: Guaiacol

***p*-Methoxyphenyl-α-D-glucoside**
Formed from: *p*-Methoxyphenol

***p*-Methoxyphenyl-β-D-glucoside**
Formed from: *p*-Methoxyphenol

***p*-Methoxyphenylglycol**
Formed from: 1-(3,4-Diethoxyphenyl)-1,3-dihydroxy-2-(*p*-methoxyphenyl)propane
p-Methoxystyrene
→ Anisaldehyde
In: Phanerochaete D467, D655, D758
→ 2-(*p*-Methoxyphenyl)-2-oxoethanol
In: Phanerochaete D467
See also: D758

3-Methoxyphenylglycol-4-yl sulphate
Formed from: 4-Hydroxy-3-methoxyphenylglycol

***m*-Methoxyphenylglyoxal** (1-(3-methoxyphenyl)-1,2-dioxoethane)
→ *m*-Methoxymandelic acid
In: Rat J287

***p*-Methoxyphenylglyoxal** (1-(4-methoxyphenyl)-1,2-dioxoethane)
→ *p*-Methoxymandelic acid
In: Rat J287

_o_-Methoxyphenylglyoxylic acid
→ *o*-Methoxymandelic acid
In: Rat J287
_m_-Methoxyphenylglyoxylic acid
Formed from: *m*-Methoxymandelic acid
→ *m*-Methoxymandelic acid
In: Rat J287
_p_-Methoxyphenylglyoxylic acid
Formed from: *p*-Methoxymandelic acid
→ *p*-Methoxymandelic acid
In: Rat J287
3-Methoxy-7-phenylhepta-2,4,6-trienoic acid
Formed from: Kawain
7-(_p_-Methoxyphenyl)heptanoic acid
Formed from: 9-(*p*-Methoxyphenyl) nonylamine
→ 7-(*p*-Hydroxyphenyl)heptanoic acid
Probably in: Rat G865
→ 5-(*p*-Methoxyphenyl)pentanoic acid
Probably in: Rat G865
7-(_p_-Methoxyphenyl)heptylamine
→ 5-(*p*-Hydroxyphenyl)pentanoic acid
In: Rat G865
→ 3-(*p*-Methoxyphenyl)propionic acid
In: Rat G865
_o_-Methoxyphenylhydracrylic acid
Probably formed from: *o*-Methoxycinnamic acid
_p_-Methoxyphenylhydroxylamine
Formed from: N-Hydroxy-*p*-methoxyacetanilide
_m_-Methoxyphenyllactic acid
Formed from: *m*-Methoxyphenylpyruvic acid
_p_-Methoxyphenyllactic acid
Probably formed from: *p*-Methoxycinnamic acid
(2_Z_)-N-((4-Methoxyphenyl)methyl)-N′,N′-dimethyl-N-2-pyridinyl-1,2-ethanediamine
See: Pyrilamine
N-((4-Methoxyphenyl)methyl)-N′,N′-dimethyl-N-2-pyrimidinyl-1,2-ethanediamine
See: Thonzylamine
α-(_p_-Methoxyphenyl)-α-methylmalonic acid
→ 2-(*p*-Methoxyphenyl)propionic acid
In: Alcaligenes G716
5-(_p_-Methoxyphenyl)-3-methyl-1-oxa-3-aza_spiro_[5,5]undecan-9-ol
Formed from: Venlafaxine
2-(_o_-Methoxyphenyl)-N-methyl-2-oxoethylamine
→ *o*-Methoxymandelic acid
In: Rat J287
2-(_m_-Methoxyphenyl)-N-methyl-2-oxoethylamine
→ *m*-Methoxymandelic acid
In: Rat J287
2-(_p_-Methoxyphenyl)-N-methyl-2-oxoethylamine
→ *p*-Methoxymandelic acid
In: Rat J287
1-(1-(_m_-Methoxyphenyl)-2-methyl-1-oxoprop-2-yl)imidazole
→ 1-(1-Hydroxy-1-(*m*-methoxyphenyl)-2-methylprop-2-yl)imidazole
In: Man J61
1-(_p_-Methoxyphenyl)-2-methyl-1-propanol
→ 1-(*p*-Methoxyphenyl)-2-methyl-1-propanone
In: Rat J529
1-(_p_-Methoxyphenyl)-2-methyl-1-propanone
Formed from: 1-(*p*-Methoxyphenyl)-2-methyl-1-propanol
4-(_o_-Methoxyphenyl)-1-methylpyridinium
Probably formed from: 1,2,3,6-Tetrahydro-4-(*o*-methoxyphenyl)-1-methylpyridine
4-(_m_-Methoxyphenyl)-1-methylpyridinium
Probably formed from: 1,2,3,6-Tetrahydro-4-(*m*-methoxyphenyl)-1-methylpyridine
1-(2-Methoxyphenyl)-2-methyl-2-(3-pyridyl)-1-propanone
→ 1-(2-Hydroxyphenyl)-2-methyl-2-(3-pyridyl)-1-propanone
In: Rat H520
→ 1-(2-Methoxyphenyl)-2-methyl-2-(3-pyridyl)-1-propanone-N-oxide
In: Rat H520
1-(2-Methoxyphenyl)-2-methyl-2-(3-pyridyl)-1-propanone-N-oxide
Formed from: 1-(2-Methoxyphenyl)-2-methyl-2-(3-pyridyl)-1-propanone
4-((3-Methoxyphenyl)methyl)-2,2,6,6-tetramethyl-1-oxa-4-aza-2,6-disilacyclohexane
→ 3′-(((Hydroxydimethylsilyl) methylamino)methyl)phenol
In: Man H316

Rat H316
→ 4-((4-Hydroxy-3-methoxyphenyl)methyl)-2,2,6,6-tetramethyl-1-oxa-4-aza-2,6-disilacyclohexane
In: Man H316
Rat H316
→ 4-((3-Hydroxyphenyl)methyl)-2,2,6,6-tetramethyl-1-oxa-4-aza-2,6-disilacyclohexane
In: Man H316
Rat H316
→ *m*-Methoxybenzoic acid
In: Dog H316

2-(*p*-Methoxyphenyl)-4-methylthiazole
→ *p*-Methoxythiobenzamide
In: Mouse G964

1-(*p*-Methoxyphenyl)-1-methylthiopropan-2-ol
Formed from: *trans-p*-Methoxypropenylbenzene

1-(*p*-Methoxyphenyl)-3-methyltriazene
Formed from: 1-(*p*-Methoxyphenyl)-3,3-dimethyltriazene

4-(2-Methoxyphenyl)-α-((1-naphthalenyloxy) methyl)-1-piperazineethanol
See: Naftopidil

***o*-Methoxyphenyl nicotinate**
→ Guaiacol
In: Man J141

9-(*p*-Methoxyphenyl)nonylamine
→ 7-(*p*-Methoxyphenyl)heptanoic acid
In: Rat G865

2-(*p*-Methoxyphenyl)-1,3-oxathiolane
→ 2-(*p*-Methoxyphenyl)-1,3-oxathiolane sulphoxide
In: Pig F176
Rat F176

2-(*p*-Methoxyphenyl)-1,3-oxathiolane sulphone
Probably formed from: 2-(*p*-Methoxyphenyl)-1,3-oxathiolane sulphoxide

2-(*p*-Methoxyphenyl)-1,3-oxathiolane sulphoxide
Probably formed from: 2-(*p*-Methoxyphenyl)-1,3-oxathiolane
→ 2-(*p*-Methoxyphenyl)-1,3-oxathiolane sulphone
Probably in: Pig F176
Rat F176

2-(*p*-Methoxyphenyl)-2-oxoethanol
Formed from: *p*-Methoxyphenylglycol

2-(*o*-Methoxyphenyl)-2-oxoethylamine
→ *o*-Methoxymandelic acid
In: Rat J287

2-(*m*-Methoxyphenyl)-2-oxoethylamine
→ *m*-Methoxymandelic acid
In: Rat J287

2-(*p*-Methoxyphenyl)-2-oxoethylamine
→ *p*-Methoxymandelic acid
In: Rat J287

5-(*p*-Methoxyphenyl)pentanoic acid
Probably formed from: 7-(*p*-Methoxyphenyl)heptanoic acid
→ 3-(*p*-Methoxyphenyl)propionic acid
Probably in: Rat G865

1-(*o*-Methoxyphenyl)piperazine
Formed from: Millipertine

3-(1-(*o*-Methoxyphenyl)piperazin-4-yl) propane-1,2-diol
Formed from: Enciprazine
Naftopidil

6-(3-(4-(*o*-Methoxyphenyl)piperazinyl) propylamino)-1,3-dimethyluracil
See: Urapidil

6-(3-(4-(*o*-Methoxyphenyl)piperazinyl) propylamino)-1-methyluracil
Formed from: Urapidil

1-(*m*-Methoxyphenyl)propane-1,2-diol
Probably formed from: 1-(*m*-Methoxyphenyl)propan-2-ol

(1*R*,2*S*)-1-(*p*-Methoxyphenyl)propane-1,2-diol
Formed from: Anisaldehyde
trans-p-Methoxypropenylbenzene
Probably formed from: 1-(*p*-Methoxyphenyl)propan-1-ol

1-(*m*-Methoxyphenyl)propan-2-ol
Formed from: *m*-Methoxyamphetamine
→ 1-(*m*-Methoxyphenyl)propane-1,2-diol
Probably in: Dog B403
Man B403
Monkey B403

1-(*p*-Methoxyphenyl)propan-1-ol
Formed from: *p*-Methoxypropylbenzene
→ 1-(*p*-Methoxyphenyl)propane-1,2-diol
Probably in: Man E812

1-(*p*-Methoxyphenyl)propan-2-ol
Formed from: *p*-Methoxyphenylacetone
p-Methoxypropylbenzene

2-Methoxy-3-phenylpropionaldehyde
→ 2-Hydroxy-3-methoxy-4-phenylbutyronitrile
In: Almond K320

3-Methoxy-3-phenylpropionaldehyde
→ 2-Hydroxy-4-methoxy-4-phenylbutyronitrile
In: Almond K320

2-(*p*-Methoxyphenyl)propionamide
Formed from: 2-(*p*-Methoxyphenyl)propionitrile
→ 2-(*p*-Methoxyphenyl)propionic acid
Probably in: Rhodococcus G172

***o*-Methoxyphenylpropionic acid**
Formed from: *o*-Methoxyphenylpropylamine
Probably formed from: *o*-Methoxycinnamic acid
→ *o*-Methoxyphenylpropionylglycine
Probably in: Rat E13

***m*-Methoxyphenylpropionic acid**
Probably formed from: Dihydroferulic acid
→ *m*-Methoxybenzoic acid
In: Rat G865

2-(*p*-Methoxyphenyl)propionic acid
Formed from: α-(*p*-Methoxyphenyl)-α-methylmalonic acid
Probably formed from: 2-(*p*-Methoxyphenyl)propionamide

3-(*p*-Methoxyphenyl)propionic acid
Formed from: 7-(*p*-Methoxyphenyl)heptylamine
Probably formed from: 5-(*p*-Methoxyphenyl)pentanoic acid
→ Anisic acid
Probably in: Rat G865

2-(*p*-Methoxyphenyl)propionitrile
→ 2-(*p*-Methoxyphenyl)propionamide
In: Rhodococcus G172

***o*-Methoxyphenylpropionylglycine**
Probably formed from: *o*-Methoxyphenylpropionic acid

***o*-Methoxyphenylpropylamine**
→ *o*-Methoxyphenylpropionic acid
In: Rat G865

***m*-Methoxyphenylpropylamine**
→ *m*-Methoxybenzoic acid
In: Rat G865

***p*-Methoxyphenylpropylamine**
→ Anisic acid
In: Rat G865

***m*-Methoxyphenylpyruvic acid**
→ *m*-Methoxyphenylacetic acid
In: Rat J411
→ *m*-Methoxyphenyllactic acid
In: Rat J411

***p*-Methoxyphenylpyruvic acid**
→ *p*-Methoxyphenylacetic acid
In: Rat J411

N-(*p*-Methoxyphenyl)-*all-trans*-retinamide
Formed from: N-(*p*-Hydroxyphenyl)-*all-trans*-retinamide
→ N-(*p*-Hydroxyphenyl)-*all-trans*-retinamide
In: Mouse F537

***o*-Methoxyphenyl sulphate**
Formed from: Guaiacol

***p*-Methoxyphenyl sulphate**
Formed from: *p*-Methoxyphenol

8-(3-Methoxy-4-phenylsulphinylphenyl)pyrazolo[1,5-*a*]-1,3,5-triazine-2,4(1*H*)-dione
Formed from: 8-(3-Methoxy-4-phenylthiophenyl)pyrazolo[1,5-*a*]-1,3,5-triazin-4(1*H*)-one
→ 8-(4-((*p*-Hydroxyphenyl)sulphinyl)-3-methoxyphenyl)pyrazolo[1,5-*a*]-1,3,5-triazine-2,4(1*H*)-dione
Probably in: Cynomolgus monkey K426
→ 8-(3-Methoxy-4-phenylthiophenyl)pyrazolo[1,5-*a*]-1,3,5-triazine-2,4(1*H*)-dione
In: Microorganism J797

8-(3-Methoxy-4-phenylsulphinylphenyl)pyrazolo[1,5-*a*]-1,3,5-triazin-4(1*H*)-one
Formed from: 8-(3-Methoxy-4-phenylthiophenyl)pyrazolo[1,5-*a*]-1,3,5-triazin-2(1*H*)-one
→ 8-(4-((*p*-Hydroxyphenyl)sulphinyl-3-methoxyphenyl)pyrazolo[1,5-*a*]-1,3,5-triazin-4(1*H*)-one
Probably in: Cynomolgus monkey K426
Rat K477
→ 8-(3-Hydroxy-4-phenylsulphinylphenyl)pyrazolo[1,5-*a*]-1,3,5-triazin-4(1*H*)-one
In: Cynomolgus monkey K426
Rat K477

→ 8-(3-Methoxy-4-phenylsulphinylphenyl) pyrazolo[1,5-*a*]-1,3,5-triazine-2,4(1*H*)-dione
In: Rat J797
→ 8-(3-Methoxy-4-phenylsulphonylphenyl) pyrazolo[1,5-*a*]-1,3,5-triazin-4(1*H*)-one
In: Cynomolgus monkey K426
Rat J797
→ 8-(3-Methoxy-4-phenylthiophenyl) pyrazolo[1,5-*a*]-1,3,5-triazin-4(1*H*)-one
In: Rat J797, K477

8-(3-Methoxy-4-phenylsulphonylphenyl) pyrazolo[1,5-*a*]-1,3,5-triazine-4(1*H*)-one
Formed from: 8-(3-Methoxy-4-phenylsulphinylphenyl)pyrazolo[1,5-*a*]-1,3,5-triazin-4(1*H*)-one

8-(3-Methoxy-4-phenylthiophenyl)pyrazolo[1,5-*a*]-1,3,5-triazine-2,4(1*H*)-dione
Formed from: 8-(3-Methoxy-4-phenylsulphinylphenyl)pyrazolo[1,5-*a*]-1,3,5-triazine-2,4(1*H*)-dione
8-(3-Methoxy-4-phenylthiophenyl) pyrazolo[1,5-*a*]-1,3,5-triazin-4(1*H*)-one

8-(3-Methoxy-4-phenylthiophenyl)pyrazolo[1,5-*a*]-1,3,5-triazine-4(1*H*)-one
Formed from: 8-(3-Methoxy-4-phenylsulphinylphenyl)pyrazolo[1,5-*a*]-1,3,5-triazine-2,4(1*H*)-dione
→ 8-(4-(*p*-Hydroxyphenyl)thio-3-methoxyphenyl)pyrazolo[1,5-*a*]-1,3,5-triazin-4(1*H*)-one
In: Rat J797
→ 8-(3-Methoxy-4-phenylsulphinylphenyl) pyrazolo[1,5-*a*]-1,3,5-triazin-4(1*H*)-one
→ 8-(3-Methoxy-4-phenylsulphinylphenyl) pyrazolo[1,5-*a*]-1,3,5-triazin-4(1*H*)-one
In: Rat J797
→ 8-(3-Methoxy-4-phenylthiophenyl) pyrazolo[1,5-*a*]-1,3,5-triazin-4(1*H*)-one
In: Microorganism J797
→ 8-(3-Methoxy-4-phenylthiophenyl) pyrazolo[1,5-*a*]-1,3,5-triazine-2,4(1*H*)-dione
In: Rat J797

4-(*o*-Methoxyphenyl)-α-((3,4,5-trimethoxyphenoxy)methyl)-1-piperazineethanol
See: Enciprazine

4-Methoxy-N-(2-(2-(2-piperidinyl)ethyl)phenyl) benzamide
Formed from: Encainide

5-Methoxypodophyllotoxin
Formed from: L-Phenylalanine

4-Methoxy-3-prenylquinol-2-one
→ Choisyine
In: Choisya A1170
→ Evoxine
In: Choisya A1170
→ Skimmianine
In: Choisya A1170

Methoxypropachlor (2-chloro-N-((2-methoxy-1-methyl)ethyl)acetanilide)
→ N-(2-Methoxy-1-methylethyl)-2-methysulphinyl-N-phenylacetamide
In: Rat G973

***trans*-*p*-Methoxypropenylbenzene** (anethole)
Formed from: L-Phenylalanine
→ Anethole oxide
Probably in: Mouse H725
Rat H725
→ Anisic acid
In: Man E812
Rat A989
→ *p*-Methoxyacetophenone
In: Man E812
→ *p*-Methoxycinnamic acid
In: Mouse H725
Rat H725
→ *p*-Methoxycinnamyl alcohol
In: Man E812
→ 1-(*p*-Methoxyphenyl)-1-methylthiopropan-1,2-diol
In: Man E812
→ 1-(*p*-Methoxyphenyl)propane-1,2-diol
In: Man E812
→ *p*-(1-Propenyl)phenol
In: Mouse H725
Rat H725

2-Methoxy-4-(1-propenyl)phenol
See: Isoeugenol

7-Methoxy-4-propylaminomethylcoumarin
→ 7-Hydroxy-4-propylaminomethylcoumarin
In: Man K339

***p*-Methoxypropylbenzene** (*p*-propylanisole)
Formed from: *p*-(Propyl)phenol

→ Anisic acid
In: Man E812
→ 1-(*p*-Methoxyphenyl)propan-1-ol
In: Man E812
→ 1-(*p*-Methoxyphenyl)propan-2-ol
In: Man E812

2-Methoxy-4-propylphenol
Formed from: Eugenol
→ 4-(1-Hydroxypropyl)-2-methoxyphenol (both isomers)
In: Penicillium J664, J836
→ Isoeugenol
In: Penicillium J836

5-Methoxypsoralen
Formed from: 5-Hydroxymarmesin
→ 5-(2-Carboxyethyl)-6-hydroxy-4-methoxybenzofuran
In: Dog G501
Man G501
Rat G501
→ 6-(α-Carboxy-α-hydroxymethyl)-7-hydroxy-5-methoxycoumarin
In: Dog G501
Man G501
Rat G501
→ 6-Carboxymethyl-7-hydroxy-5-methoxycoumarin
In: Dog G501
Man G501
Rat G501
→ 7-Hydroxy-6-(2-hydroxyethyl)-5-methoxycoumarin
In: Dog G501
Man G501
Rat G501

8-Methoxypsoralen (methoxsalen, xanthotoxin, 9-methoxy-7*H*-furo[3,2-*g*][1]benzopyran-7-one)
Formed from: 5,8-Dihydroxypsoralen
8-Hydroxypsoralen
→ 6-Carboxymethyl-7,8-dimethoxycoumarin methyl ester
In: Papilio D220
Spodoptera D220
→ 6-Carboxymethyl-7-hydroxy-8-methoxycoumarin
In: Dog A3911
Papilio D220
Spodoptera D220
→ 4,6-Dihydroxy-7-methoxy-5-coumaranoyl-β-acrylic acid
In: Rat B355
→ (*Z*)-3-(6-Hydroxy-7-methoxybenzofuran-5-yl)-2-propenoic acid
In: Dog A3911
→ 3-(5-(6-Hydroxy-7-methoxybenzofuryl))propionic acid
In: Goat G456
→ 2-(7-Hydroxy-8-methoxycoumarin-6-yl) acetic acid
In: Goat G456
Hen G456
Man B699
Rat D966, E868
→ 2-(7-Hydroxy-8-methoxycoumarin-6-yl)ethanol
In: Man B699
→ 2-(7-Hydroxy-8-methoxycoumarin-6-yl) glycol
In: Man B699
→ 7-Hydroxy-8-methoxy-2-oxo-2*H*-1-benzopyran-6-acetic acid
In: Dog A3911
→ 5-Hydroxy-8-methoxypsoralen
In: Goat G456
Rat B355, D966, E868
→ 8-Hydroxypsoralen
In: Goat G456
Hen G456
Man A3642
Rat B355, D966, E868
→ Isopimpinellin
In: Heracleum A2582
Ruta D28
→ Psoralenquinone
In: Goat G456

1-Methoxypyrene
Probably formed from: 1-Hydroxypyrene

2-Methoxy-N-(2-pyrrolidylmethyl)-5-sulphonylethylbenzamide
Formed from: Sultopride

8-Methoxyquercetin (3,3′,4′,5,7-pentahydroxy-8-methoxyflavone)
Formed from: Gossypetin

Methoxyquinol (1,4-dihydroxy-2-methoxybenzene)
Formed from: Ferulic acid
Methoxy-*p*-quinone
Vanillic acid
Probably formed from: 3,4-Dimethoxyphenol
→ Hydroxyquinol
In: Paecilomyces F199
→ 4-Hydroxy-*o*-quinone
In: Phanerochaete G436
→ Methoxy-*p*-quinone
Probably in: Trametes A2744

2-Methoxyquinoline
→ 5,6-Dihydro-5,6-dihydroxy-2-methoxyquinoline
In: Pseudomonas J537
→ 7,8-Dihydro-7,8-dihydroxy-2-methoxyquinoline
In: Pseudomonas J537

6-Methoxyquinoline
Formed from: 6-Methoxyquinoline-N-oxide

6-Methoxyquinoline-N-oxide
→ 6-Methoxyquinoline
In: Saccharomyces H894

4-((6-Methoxyquinolin-8-yl)amino)-*n*-pentanol
Formed from: Primaquine

Methoxy-*p*-quinone
Formed from: 4-Nitroveratrole
Probably formed from: Methoxyquinol
→ Methoxyquinol
In: Phanerochaete G436, H611
Sporotrichum B301

7-Methoxyresorufin
See: 7-Methoxyphenoxazone

5-Methoxysalicylic acid
See: 2-Hydroxy-5-methoxybenzoic acid

4′-Methoxyscopolammonium
Probably formed from: 4′-Hydroxyscopolammonium

3-Methoxystilboestrol
Probably formed from: 3-Hydroxystilboestrol

***o*-Methoxystyrene**
Formed from: *o*-Methoxycinnamic acid

***m*-Methoxystyrene**
Formed from: *m*-Methoxycinnamic acid

***p*-Methoxystyrene**
Formed from: *p*-Methoxycinnamic acid
→ *p*-Methoxyphenylglycol
In: Rat A2859

4-Methoxy-6-sulphanilamidopyrimidine
See: Sulphamonomethoxine

5-(3-Methoxy-4-sulphatoxybenzyl)-2-(5-methyl-4-imidazolylmethylthio)ethylamino-4-pyrimidone
Probably formed from: 5-(4-Hydroxy-3-methoxybenzyl)-2-(5-methyl-4-imidazolyl-methylthio)ethylamino-4-pyrimidone

***p*-Methoxysulphinylanisole**
Formed from: *p*-Methoxythioanisole

***p*-Methoxysulphinylbenzamide**
Formed from: *p*-Methoxythiobenzamide

1-Methoxy-4-((2,3,5,6-tetrafluoro-4-(S-glutathionyl)phenyl)sulphonamido)benzene
Formed from: 1-Methoxy-4-((pentafluorophenyl)sulphonamido)benzene

2-Methoxy-4-((1,2,3,4-tetrahydro-6,7-dimethoxy-2-methyl-1-isoquinolinyl)methyl)phenol
See: Pseudocodamine

2-Methoxy-5-(((1*R*)-1,2,3,4-tetrahydro-6,7-dimethoxy-1-isoquinolinyl)methyl)phenol
See: Norlaudanidine

2-Methoxy-5-((1,2,3,4-tetrahydro-6,7-dimethoxy-2-methyl-1-isoquinolinyl)methyl)phenol
See: Laudanosine

***o*-Methoxythioanisole**
Formed from: *o*-Methoxythiophenol

***m*-Methoxythioanisole**
Formed from: *m*-Methoxythiophenol

***p*-Methoxythioanisole**
Formed from: *p*-Methoxythiophenol
→ *p*-Methoxysulphinylanisole
In: Mortierella D759

***p*-Methoxythiobenzamide**
Formed from: 2-(*p*-Methoxyphenyl)-4-methylthiazole
→ *p*-Methoxysulphinylbenzamide
In: Mouse G964

***o*-Methoxythiophenol**
→ *o*-Methoxythioanisole
In: Man D926

***m*-Methoxythiophenol**
→ *m*-Methoxythioanisole
In: Man D926
***p*-Methoxythiophenol**
→ *p*-Methoxythioanisole
In: Man D926
***p*-Methoxythiotoluene**
See: *p*-Methylthioanisole
***o*-Methoxytoluene**
See: *o*-Methylanisole
***m*-Methoxytoluene**
See: *m*-Methylanisole
***p*-Methoxytoluene**
See: *p*-Methylanisole
3-Methoxy-5-(2,3,6-trichlorophenoxy)-*o*-quinone
Formed from: 4-Hydroxy-3,5-dimethoxybenzoic acid
2,3,6-Trichlorophenol
3-Methoxy-5-(2,4,5-trichlorophenoxy)-*o*-quinone
Formed from: 4-Hydroxy-3,5-dimethoxybenzoic acid
2,4,5-Trichlorophenol
β-Methoxy-3-trifluoromethylphenethylamine
Formed from: *l*-N-Benzyl-β-methoxy-3-trifluoromethylphenethylamine
2-Methoxy-4-trifluoromethylphenol
Formed from: 4-Trifluoromethycatechol
2-Methoxy-5-trifluoromethylphenol
Formed from: 4-Trifluoromethylcatechol
Probably formed from: 4-Trifluoromethylveratrole
β-Methoxy-3-trifluoromethylphenylacetic acid
Formed from: *l*-N-Benzyl-β-methoxy-3-trifluoromethylphenethylamine
→ *m*-Trifluoromethylmandelic acid
Probably in: Monkey A631
***anti*-2-Methoxy-2-(*m*-trifluoromethylphenyl) acetaldoxime**
Probably formed from: β-Methoxy-β-(*m*-trifluoromethylphenyl)ethylhydroxylamine
***syn*-2-Methoxy-2-(*m*-trifluoromethylphenyl) acetaldoxime**
Probably formed from: β-Methoxy-β-(*m*-trifluoromethylphenyl)ethylhydroxylamine
2-Methoxy-2-(*m*-trifluoromethylphenyl)ethanol
Probably formed from: N-Ethyl-β-methoxy-β-(*m*-trifluoromethylphenyl)ethylamine
β-Methoxy-β-(*m*-trifluoromethylphenyl) ethylamine
Formed from: N-Ethyl-β-methoxy-β-(*m*-trifluoromethylphenyl)ethylamine
→ β-Methoxy-β-(*m*-trifluoromethylphenyl) ethylhydroxylamine
Probably in: Rabbit A2352
β-Methoxy-β-(*m*-trifluoromethylphenyl) ethylhydroxylamine
Probably formed from: β-Methoxy-β-(*m*-trifluoromethylphenyl)ethylamine
→ *anti*-2-Methoxy-2-(*m*-trifluoromethylphenyl)acetaldoxime
Probably in: Rabbit A2352
→ *syn*-2-Methoxy-2-(*m*-trifluoromethylphenyl)acetaldoxime
Probably in: Rabbit A2352
5-Methoxy-*p*-trifluoromethylvalerophenone (*E*)-O-(2-aminoethyl) oxime
See: Fluvoxamine
5-Methoxy-*p*-trifluoromethylvalerophenone O-(carboxymethyl) oxime
Formed from: Fluvoxamine
5-Methoxy-*p*-trifluoromethylvalerophenone O-(2-hydroxyethyl) oxime
Formed from: Fluvoxamine
5-Methoxy-*p*-trifluoromethylvalerophenone oxime
Formed from: Fluvoxamine
→ 5-Methoxy-*p*-trifluoromethylvalerophenone oxime-β-D-glucuronide
In: Dog C652
Hamster C652
Mouse C652
Rat C652
5-Methoxy-*p*-trifluoromethylvalerophenone oxime-β-D-glucuronide
Formed from: 5-Methoxy-*p*-trifluoromethylvalerophenone oxime
5-Methoxy-2,3,4-trimethylphenol
Formed from: 6-Hydroxy-4-methoxy-2,3-dimethylbenzamide
(β*R*)-2-Methoxy-N,N,β-trimethyl-10*H*-phenothiazine-10-propanamide
See: Methotrimeprazine
(2*E*,4*E*,6*E*,8*E*)-9-(4-Methoxy-2,3,6-trimethylphenyl)-3,7-dimethyl-2,4,6,8-nonatetraenoic acid
See: Acitretin

(2*E*,4*E*,6*E*,8*E*)-9-(4-Methoxy-2,3,6-trimethylphenyl)-3,7-dimethyl-2,4,6,8-nonatetraenoic acid ethyl ester
See: Etretinate

8-(4-Methoxy-2,3,6-trimethylphenyl)-2,6-dimethyl-*trans*,*trans*-octa-5,7-dienoic acid
Formed from: Isoacitretin
Probably formed from: Acitretin
→ 8-(4-Hydroxy-2,3,6-trimethylphenyl)-2,6-dimethyl-*trans*,*trans*-octa-5,7-dienoic acid
Probably in: Man F118, G869
Rat G495

5-Methoxytryptamine
Formed from: 5-Hydroxytryptamine
Melatonin
→ Melatonin
In: Hamster G287
Hen F694
Man K490
Prawn G824
Rat D282
Sheep D282
See also: H835
→ 10-Methoxyajmalicine
Probably in: Catharanthus A2741
→ 5-Methoxyindole-3-acetaldehyde
In: Dog A1184
Monkey A1184
Mouse A1184
Rat B653
See also: B100, F39, G287
→ 5-Methoxy-N-methyltryptamine
In: Man A1483
Phalaris A3891, E974
Rabbit A1320, A1483
Rat A1311, D282
Sheep D282
See also: A1310
→ 1,2,3,4-Tetrahydro-6-methoxy-β-carboline
In: Rat A2499

6-Methoxytryptamine
→ 6-Methoxy-N-methyltryptamine
In: Phalaris A3891

5-Methoxy-L-tryptophan
Formed from: 5-Methoxyindole

6-Methoxy-L-tryptophan
Formed from: 6-Methoxyindole

7-Methoxy-L-tryptophan
Formed from: 7-Methoxyindole

5-Methoxytryptophol
Formed from: 5-Hydroxytryptophol
Probably formed from: O-Acetyl-5-methoxytryptophol
→ 5-Methoxyindole-3-acetaldehyde
Probably in: Rat C659

3-Methoxytyramine (4-hydroxy-3-methoxyphenethylamine)
Formed from: 3,4-Dihydroxyphenethylamine
→ N-Acetyl-3-methoxytyramine
Probably in: Rat A652
→ 5,5′-Bis(2-aminoethyl-2,2′-dihydroxy-3,3′-dimethoxybiphenyl
In: Horseradish F213
→ 3,4-Dihydroxyphenethylamine
In: Rat A2961, A3063
→ Feruloyl-4-hydroxy-3-methoxyphenethylamine
In: Nictiana F72
→ 4-Hydroxy-3-methoxy-N-methylphenethylamine
In: Rat A80
See also: A1310
→ 4-Hydroxy-3-methoxyphenylacetaldehyde
In: Man A2967
Rat A1608, A2961, E339
See also: A801, A1716, A2055, A2411, A3862
→ 3-Methoxytyramine-4-β-D-glucuronide
Probably in: Rat A1275
→ 3-Methoxytyramine-4-sulphate
In: Man D583
Rat E962
See also: A1716, A3862
→ *m*-Tyramine
In: Rat A2961

3-Methoxytyramine-4-β-D-glucuronide
Formed from: 3-Methoxytyramine

3-Methoxytyramine-4-sulphate
Formed from: 3-Methoxytyramine

3-Methoxy-D-tyrosine
Formed from: 3,4-Dihydroxy-D-phenylalanine

3-Methoxy-L-tyrosine (4-hydroxy-3-methoxy-L-phenylalanine)
Formed from: 3,4-Dihydroxy-L-phenylalanine
→ 3-Methoxy-L-tyrosine-β-D-glucuronide
In: Man J490
→ Vanilpyruvic acid
In: Rat A3229

3-Methoxy-L-tyrosine-β-D-glucuronide
Formed from: 3-Methoxy-L-tyrosine

6-Methoxy-α-(5-vinyl-2-quinuclidinyl)-4-quinolinemethanol
See: Quinidine

4′-Methoxywarfarin
→ Dihydro-4′-methoxywarfarin
In: Rabbit G276

Methsuximide (α,N-dimethyl-α-phenylsuccinimide)
→ α-(3,4-Dihydro-3,4-dihydroxyphenyl)-α,N-dimethylsuccinimide
In: Guinea pig A1020
Man A1020
Rat A1020
See also; A1709
→ β-Hydroxy-α,N-dimethyl-α-phenylsuccinimide
In: Guinea pig A1020
Man A1020
Rat A1020
→ α-Hydroxymethyl-N-methyl-α-phenylsuccinimide
In: Guinea pig A1020
Man A1020
Rat A1020
→ α-(*m*-Hydroxyphenyl)-α,N-dimethylsuccinimide
In: Guinea pig A1020
Man A1020
Rat A1020
→ α-(*p*-Hydroxyphenyl)-α,N-dimethylsuccinimide
In: Dog A1434
Guinea pig A1020
Man A1020
Rat A1020
→ α-Methyl-α-phenylsuccinimide
In: Dog A1434
Man A474
Rat A474

Methtryptoline
See: 1,2,3,4-Tetrahydro-1-methyl-β-carboline

Methyl *p*-acetamidobenzenesulphonylcarbamate
Formed from: Asulam

Methyl α-acetamido-*p*-coumarate
→ Novobiocin
In: Streptomyces A945

Methyl 5-(2-acetamidoethyl)oximino-5-(*p*-trifluoromethylphenyl)valerate
Probably formed from: Methyl 5-(2-aminoethyl)oximino-5-(*p*-trifluoromethylphenyl)valerate
→ 5-(2-Acetamidoethyl)oximino-5-(*p*-trifluoromethylphenyl)valeric acid
Probably in: Dog C652
Hamster C652
Mouse C652
Rat C652

4′-(N-Methylacetamido)-4-methylaminoazobenzene
Formed from: 4-Dimethylamino-4′-(N-methylacetamido)azobenzene
→ 4-Amino-4′-(N-methylacetamido)azobenzene
In: Rat A2833

***o*-Methylacetanilide** (*o*-acetotoluidide)
Formed from: *o*-Toluidine
→ *o*-Toluidine
In: Nocardia G343

***m*-Methylacetanilide**
Formed from: *m*-Toluidine
→ 4-Acetamido-2-methylphenol
In: Rat A1748, A1756
→ *m*-Toluidine
In: Nocardia G343

***p*-Methylacetanilide**
Formed from: *p*-Toluidine
→ *p*-Toluidine
In: Nocardia G343

***p*-Methylacetanilide**
Formed from: *p*-Toluidine
→ *p*-Toluidine
In: Nocardia G343

***o*-(1-Methylacetonyl)phenyl N-methylcarbamate**
Probably formed from: *erythro*-*o*-(2-Hydroxy-1-methylpropyl)phenyl N-methylcarbamate

threo-*o*-(2-Hydroxy-1-methylpropyl)phenyl N-methylcarbamate

m-Methylacetophenone
Formed from: (*S*)-1-(*m*-Methylphenyl) ethanol
→ (*R*)-1-(*m*-Methylphenyl)ethanol
In: Geotrichum H749

p-Methylacetophenone
Formed from: 1-(*p*-Methylphenyl)ethanol
→ 1-(*p*-Methylphenyl)-2-propanol
In: Geotrichum H749
Rabbit C25

Methyl o-acetylbenzoate
→ (*S*)-Methylphthalide
In: Endomyces J486
Geotrichum J486
Mucor J486
Saccharomyces J486

Methyl acetylsalicylate
→ Methyl salicylate
In: Acinetobacter A962

Methyl acitretinate
Formed from: Acitretinoyl CoA

N-Methyladrenaline
Formed from: D-Adrenaline
→ 3-Hydroxy-4-methoxy-N,N-dimethylphenylethanolamine
In: Rat A486
→ N-Methylmetanephrine
In: Rat A486

α-Methyladrenaline
Formed from: α-Methylnoradrenaline

12-Methylajmalicine
Probably formed from: 7-Methyltryptamine

4-(3-(L-(α-Methyl)alanyl))-o-quinone
Formed from: 3,4-Dihydroxy-α-methylphenylalanine

4-(N-Methyl-3-L-alanyl))-o-quinone
Formed from: 3,4-Dihydroxy-N-methylphenylalanine

13-Methylallocryptopine
Formed from: Mesothalictricavine

p-Methylaminoacetophenone
Formed from: *p*-Aminoacetophenone

1-Methylaminoanthraquinone
→ 1-Aminoanthraquinone
In: Sheep C808
→ 1-Methylaminoanthraquinone-β-D-glucuronide
In: Sheep C808

1-Methylaminoanthraquinone-β-D-glucuronide
Formed from: 1-Methylaminoanthraquinone

4-Methylaminoantipyrine
Formed from: Aminopyrine
Dipyrone
→ 4-Aminoantipyrine
Probably in: Goat A1110
Man A910, A911, A1161, A1265, B10, F467, H167
Rat A1124, A1140, A1141, A3163, E488
→ 4-Formamidoantipyrine
Probably in: Man F467

4-Methylaminoazobenzene
Formed from: 4-Dimethylaminoazobenzene
→ 4-Aminoazobenzene
In: Boophilus A1024
Ram C360
Rat A1975, A1979, A1981, A3898, B111, B985, D849, D851
Spider A1024
See also: C615
→ Aniline
In: Rat G133, G380, G511
→ 2-Hydroxy-4′-methylaminoazobenzene
Probably in: Rat A1981
→ 4-Hydroxy-4′-methylaminoazobenzene
In: Rat B985
See also: A1975, A1979, A1981, A3898, B111
→ N-Hydroxy-4-methylaminoazobenzene
In: Rat A2481
→ N-Methyl-*p*-phenylenediamine
In: Rat G133, G380, G511

4-Methylaminoazobenzene-N-β-D-glucuronide-4-yl sulphate
Probably formed from: 4-Methylaminoazobenzene-4-yl sulphate

Methyl p-aminobenzenesulphonylcarbamate
See: Asulam

4′-Methylaminoazobenzen-4-yl-β-D-glucuronide
Formed from: 4-Hydroxy-4′-methylaminoazobenzene

4-Methylaminoazobenzen-4-yl sulphate
Formed from: 4-Hydroxy-4′-methylaminoazobenzene

→ 4-Methylaminoazobenzene-N-β-D-glucuronid-4-yl sulphate
Probably in: Rat D68

Methyl *p*-aminobenzoate (methylparaben)
Formed from: *p*-Methyl red methyl ester
→ Methyl *p*-sulphamidobenzoate
In: Eubacterium H439
Haemophilus H439
Klebsiella H439

***p*-Methylaminobenzoic acid**
Probably formed from: *p*-Dimethylaminobenzoic acid
→ *p*-Aminobenzoic acid
In: Man A114
Pseudomonas A903
Rat A114

***p*-Methylaminobenzonitrile**
Formed from: *p*-Dimethylaminobenzonitrile
Probably formed from: *p*-Aminobenzonitrile
→ *p*-Aminobenzonitrile
Probably in: Rat D160
→ 3-Hydroxy-4-methylaminobenzonitrile
Probably in: Rat D160, F305

4-Methylaminobenzonitril-3-yl sulphate
Probably formed from: 3-Hydroxy-4-methylaminobenzonitrile

2-Methylamino-2-benzothiazole
Formed from: 1-(2-Benzothiazolyl)-1,3-dimethylurea

4-Methylaminobiphenyl
Formed from: 4-Aminobiphenyl

4-Methylaminocatechol
Formed from: *p*-Methylaminophenol

Methyl N-(3-amino-2,6-diiodophenyl)-N-methylglutaramidate
Probably formed from: N-(3-Amino-2,6-diiodophenyl)-N-methylglutaramidic acid

Methyl N-(3-amino-4,6-diiodophenyl)-N-methylglutaramidate
Probably formed from: N-(3-Amino-4,6-diiodophenyl)-N-methylglutaramidic acid

3-Methylamino-1,1-diphenylbutane
Formed from: 3-Dimethylamino-1,1-diphenylbutane

d-3-Methylamino-1,2-diphenyl-2-butyl propionate
Formed from: *d*-3-Dimethylamino-1,2-diphenyl-2-butyl propionate

2-Methylamino-4,4-diphenyl-5-heptanol
Formed from: 2-Dimethylamino-4,4-diphenyl-5-heptanol
→ 2-Amino-4,4-diphenyl-5-heptanol
In: Man A1123
Rat A1123
See also: A1000, A2384, A3180
→ 2-Ethyl-5-methy-3,3-diphenyl-1-pyrroline
Probably in: Monkey A2979

2-Methylamino-4,4-diphenyl-5-heptanone
Formed from: 2-Dimethylamino-4,4-diphenyl-5-heptanone

5-(2-(Methylamino)ethoxy)-7*H*-benzo[*c*]fluoren-7-one
Formed from: Benfluron
→ 9-Hydroxy-5-(2-(methylamino)ethoxy)-7*H*-benzo[*c*]fluoren-7-one
Probably in: Rat K387

5-(α-(2-Methylaminoethoxy)benzyl)-1-methyl-1*H*-pyrazole
Formed from: Cizolirtine

5-((4-((Methylamino)ethoxy)phenyl)methyl)-2,4-thiazolidinedione
Formed from: 5-((4-(2-(Methyl-2-pyridinyl-amino)
ethoxy)phenyl)methyl)-2,4-thiazolidinedione

N-(2-(Methylamino)ethyl)acridine-4-carboxamide
Formed from: N-(2-(Dimethylamino)ethyl) acridine-4-carboxamide
→ N-(2-(Amino)ethyl)acridine-4-carboxamide
In: Rat G565

N-(2-(Methylamino)ethyl)-9(10*H*)-acridone-4-carboxamide
Probably formed from: N-(2-(Dimethylamino) ethyl)-9(10*H*)-acridone-4-carboxamide
→ 7-Hydroxy-N-(2-(methylamino)ethyl)-9(10*H*)-acridone-4-carboxamide
Probably in: Mouse G583

(αS)-α-((1*S*)-1-(methylamino)ethyl) benzenemethanol
See: D-Pseudoephedrine

3-(2-Methylaminoethyl)benzo[*b*]thiophene
Formed from: 3-(2-Dimethylaminoethyl) benzo[*b*]thiophene

2-(Methylamino)ethyl methyl 4-(*m*-aminophenyl)-1,4-dihydro-2,6-dimethylpyridine-3,5-dicarboxylate
Probably formed from: 2-(Methylamino)ethyl methyl 1,4-dihydro-2,6-dimethyl-4-(*m*-nitrophenyl)pyridine-3,5-dicarboxylate

2-(Methylamino)ethyl methyl 1,4-dihydro-2,6-dimethyl-4-(*m*-nitrophenyl)pyridine-3,5-dicarboxylate
→ 2-(Methylamino)ethyl methyl 4-(*m*-aminophenyl)-1,4-dihydro-2,6-dimethylpyridine-3,5-dicarboxylate
Probably in: Dog B535

Methyl 5-(2-aminoethyl)oximino-5-(*p*-trifluoromethylphenyl)valerate
Formed from: Fluvoxamine
→ 5-(2-Aminoethyl)oximino-5-(*p*-trifluoromethylphenyl)valeric acid
Probably in: Dog C652
Hamster C652
Mouse C652
Rat C652
→ Methyl 5-(2-acetamidoethyl)oximino-5-(*p*-trifluoromethylphenyl)valerate
Probably in: Dog C652
Hamster C652
Mouse C652
Rat C652

4-(2-Methylaminoethyl)-*o*-quinone
Probably formed from: 3,4-Dihydroxy-N-methylphenethylamine

***p*-Methylaminomethylbenzaldehyde**
Probably formed from: *p*-Methylaminomethylbenzylamine

***p*-Methylaminomethylbenzylamine**
→ *p*-Methylaminomethylbenzaldehyde
Probably in: Pig A3693

3-Methylaminomethyl-1,2-diphenyl-2-butyl propionate
Formed from: 3-Dimethylaminomethyl-1,2-diphenyl-2-butyl propionate
→ N-Acetyl-3-methylaminomethyl-1,2-diphenyl-2-butyl propionate
In: Man A1683
Rat A1683
→ 3-Aminomethyl-1,2-diphenyl-2-butyl propionate
In: Man A1683
→ 1-Hydroxyphenyl-3-methylaminomethyl-2-phenyl-2-butyl propionate
In: Rabbit A1683

α-((Methylamino)methyl)-9,10-ethanoanthracene-9(10*H*)-ethanol
See: Oxaprotiline

3-Methylaminomethylindole
→ Gramine
Probably in: Lupinus A233
Phalaris A1928

1-((Methylamino)methyl)naphthalene
Formed from: Terbinafine

5-Methylamino-1-naphthalenesulphonamide
Formed from: 5-Dimethylamino-1-naphthalenesulphonamide

4-Methylamino-3′-nitroazobenzene
Formed from: 4-Dimethylamino-3′-nitroazobenzene
→ 4-Amino-3′-nitroazobenzene
In: Rat A776

***o*-Methylaminophenol**
Probably formed from: N-Methylaniline

***p*-Methylaminophenol**
Formed from: *p*-Dimethylaminophenol
p-Fluoro-N-methylaniline

***p*-Methylaminophenol** (metol)
Formed from: *p*-Fluoro-N-methylaniline
N-Methylaniline
Probably formed from: *p*-Dimethylaminophenol
→ *p*-Aminophenol
Probably in: Guinea pig C89
Rabbit C89
Rat G555
→ 4-Methylaminocatechol
In: Candida E204

***p*-Methylaminophenyl acetate**
Formed from: *p*-Dimethylaminophenyl acetate

2-Methylamino-2-phenylbutan-1-ol
Probably formed from: 2-Dimethylamino-2-phenylbutan-1-ol

→ 2-Amino-2-phenylbutan-1-ol
Probably in: Dog F28
Rat F28

Methyl 4-(3-aminophenyl)-5-carboxy-2-cyano-6-methylpyridine-3-carboxylate
Probably formed from: Methyl 5-carboxy-2-cyano-6-methyl-4-(3-nitrophenyl)pyridine-3-carboxylate

2-Methylamino-1-phenyl-1-propanol
See: Ephedrine

2-(Methylamino)-1-phenyl-1-propanone
See: Methylpropion

2-(Methylamino)propiophenone
Formed from: 2-(Dimethylamino) propiophenone
→ 2-Aminopropiophenone
In: Man A504
→ Ephedrine
Probably in: Man A504

4-(2-Methylaminopropyl)-*o*-quinone
Probably formed from: 3,4-Dihydroxy-N-methylamphetamine

10-(3-Methylaminopropyl)-2-trifluoromethyl-10*H*-phenothiazine
Probably formed from: 10-(3-Dimethylamino-propyl)-2-trifluoromethyl-10*H*-phenothiazine
→ 10-(3-Aminopropyl)-2-trifluoromethyl-10*H*-phenothiazine
Probably in: Rat A69

4-Methylamino-*m*-tolyl N-methylcarbamate
Formed from: Aminocarb
→ 4-Amino-*m*-tolyl N-methylcarbamate
In: Crayfish B232

4-Methylamino-3,5-xylyl methylcarbamate
Formed from: Mexacarbate
→ 4-Amino-3,5-xylyl methylcarbamate
Probably in: Bacterium A1207
Trichoderma A1207
→ 4-Formamido-3,5-xylyl methylcarbamate
Probably in: Bacterium A1207
Trichoderma A1207

N-Methylamphetamine (methamphetamine)
Formed from: N-Allyl-N-methylamphetamine
Amphetamine
N-Benzyl-N,α-dimethylphenethylamine
Deprenyl
N,N-Dimethylamphetamine
N-Ethyl-N-methylamphetamine
Famprofazone
Furfenorex
N-Hydroxy-N-methylamphetamine
N-Isopropyl-N-methylamphetamine
N-Methyl-N-propylamphetamine
→ Amphetamine
In: Bacteroides A907
Clostridium A907
Enterobacterium A907
Enterococcus A907
Guinea pig D157
Lactobacillus A907
Man J264
Mycobacterium B380
Rat A300, A1917, B7, C595, C663, C669, D156, D157, D660, D661, D800
See also: A3961, C68, C300, C376, D658, E52, E781, E783, J152, J692, K195, K581
→ N,N-Dimethylamphetamine
In: Rat A80
→ N-Hydroxy-N-methylamphetamine
In: Guinea pig E806
Rat A1917
→ *p*-Hydroxy-N-methylamphetamine
In: Man J264
Rat B7, C663, D156, D164, D660, D800, H197
See also: C376, D658, D661, E52, E783
→ N-((1-Methyl-2-phenyl)ethyl)ethanamine-N-oxide
In: Guinea pig E791
Rat E791
→ Phenylacetone
In: Rat A300

N-Methylamphetamin-4-yl-β-D-glucuronide
Formed from: *p*-Hydroxy-N-methylamphetamine

O-Methylandrocymbine ((7*S*)-6,7-dihydro-2-methoxy-1,3,10-trimethoxy-14-methyl-7,11a-(iminoethano)-11a*H*-dibenzo[*a*,*c*]cyclohepten-9(5*H*)-one)
→ Demecolcine
Probably in: Colchicum A606

***o*-Methylaniline**
See: *o*-Toluidine

m-Methylaniline
See: *m*-Toluidine
***p*-Methylaniline**
See: *p*-Toluidine
N-Methylaniline
Formed from: Aniline
N-*n*-Butyl-N-methylaniline
N,N-Dimethylaniline
N,N-Dimethylaniline-N-oxide
N-Ethyl-N-methylaniline
N-Isobutyl-N-methylaniline
N-Isopropyl-N-methylaniline
N-Methyl-N-propylaniline
N-Methyl-N-nitrosoaniline
Probably formed from: N-Ethyl-N-methylaniline-N-oxide
→ Aniline
In: Glycine K419
Man D135, K419
Mouse G116
Rat C669
Tulipa K539
See also: A1250, C89, F336
→ *o*-Methylaminophenol
Probably in: Dog A1250
→ *p*-Methylaminophenol
In: Man D135
Mouse G116
Tulipa K539
See also: A1250, B181
→ N-Methylaniline-β-D-glucuronide
In: Rat F336
→ N-Methyl-N-phenylhydroxylamine
In: Pig K160
→ N-Methylphenyl nitroxide radical
Probably in: Rabbit B184
N-Methylaniline-β-D-glucuronide
Formed from: N-Methylaniline
Methyl *p*-anisate
→ Methyl *p*-hydroxybenzoate
In: Sporomusa J199
N-Methylanisidine
Formed from: N,N-Dimethylanisidine
***o*-Methylanisole** (*o*-methoxytoluene)
Formed from: *o*-Cresol
→ *o*-Cresol
In: Sporomusa J199
***m*-Methylanisole** (*m*-methoxytoluene)
Formed from: *m*-Cresol
→ *m*-Cresol
In: Sporomusa J199
***p*-Methylanisole** (*p*-methoxytoluene)
Formed from: *p*-Cresol
→ Anisyl alcohol
In: Mortierella F823
Rabbit A931
→ *p*-Cresol
In: Methylosinus D513
Rabbit A931
Sporomusa J199
→ *cis*-2,3-Dihydro-2,3-dihydroxy-1-methoxy-4-methylbenzene
In: Pseudomonas K699
→ 2-Methoxy-5-methylphenol
In: Rabbit A931
→ 4-Methoxy-2-methylphenol
In: Rabbit A931
→ 5-Methoxy-2-methylphenol
In: Rabbit A931
9-Methylanthracene
Formed from: Anthracene
→ 1,2-Dihydro-1,2-dihydroxy-9-methylanthracene
In: Rat D884
Salmonella D805
→ 3,4-Dihydro-3,4-dihydroxy-9-methylanthracene
In: Rat D884
Salmonella D805
→ 9,10-Dimethylanthracene
Probably in: Rat E612
→ 9-Hydroxymethylanthracene
In: Horseradish H796
Rat D884
With cytochrome P450 H796
See also: E612
→ 10-Methyl-9-anthrone
In: Horseradish H796
With cytochrome P450 H796
Methyl anthranilate
Formed from: Methyl red methyl ester
p-Methyl red methyl ester

5-Methylanthranilic acid
Formed from: 5-Methylindole
N-Methylanthranilic acid
Formed from: Anthranilic acid
→ Pyrocatechuic acid
In: Trichosporon C393, E270
10-Methyl-9-anthrone
Formed from: 10-Methylanthracene
4′-Methylapigenin
See: Acacetin
O-Methylapomorphine
Formed from: Apomorphine
Methylaposcopinium
Formed from: Methylscopolammonium
Methylaposcopolammonium
Formed from: Methylscopolammonium
8-Methyl-8-azabicyclo[3.2.1]octan-N-oxide-3-yl 3,5-dichlorobenzoate
Formed from: 8-Methyl-8-azabicyclo[3.2.1] octan-3-yl 3,5-dichlorobenzoate
8-Methyl-8-azabicyclo[3.2.1]octan-3-yl 3,5-dichlorobenzoate
→ 8-Azabicyclo[3.2.1]octan-3-yl 3,5-dichlorobenzoate
In: Dog G386
→ 3,5-Dichlorobenzoic acid
In: Dog G386
Monkey G386
→ 8-Methyl-8-azabicyclo[3.2.1]octan-N-oxide-3-yl 3,5-dichlorobenzoate
In: Dog G386
7-Methylbenz[*c*]acridine
→ *trans*-1,2-Dihydro-1,2-dihydroxy-7-methylbenz[*c*]acridine
In: Rat D34, D430
→ *trans*-3,4-Dihydro-3,4-dihydroxy-7-methylbenz[*c*]acridine
In: Man G498, H379
Rat D430, D962
→ *trans*-8,9-Dihydro-8,9-dihydroxy-7-methylbenz[*c*]acridine
In: Man H379
Rat D34, D163, D430
→ *trans*-10,11-Dihydro-10,11-dihydroxy-7-methylbenz[*c*]acridine
In: Rat D34, D430
→ 5-Hydroxy-7-methylbenz[*c*]acridine
In: Man H379
→ 7-Hydroxymethylbenz[*c*]acridine
In: Man H379
Rat D34, D163, D430
→ 9-Hydroxy-7-methylbenz[*c*]acridine
In: Man H379
→ 11-Hydroxy-7-methylbenz[*c*]acridine
In: Rat D430
→ 7-Methylbenz[*c*]acridine-5,6-oxide
In: Man H379
Rat D34, D430
See also: D163
7-Methylbenz[*c*]acridine-5,6-oxide
Formed from: 7-Methylbenz[*c*]acridine
→ *trans*-5,6-Dihydro-5,6-dihydroxy-7-methylbenz[*c*]acridine
Probably in: Man H379
Rat D430
***o*-Methylbenzaldehyde**
See: *o*-Tolualdehyde
***m*-Methylbenzaldehyde**
See: *m*-Tolualdehyde
***p*-Methylbenzaldehyde**
See: *p*-Tolualdehyde
N-Methylbenzamide
Formed from: Didehydro-4-methyl-5-phenyl-2,3,4-thiadiazolidine-2-thione
N-Hydroxymethyl-N-methylbenzamide
N-Methylthiobenzamide
→ N-Hydroxymethylbenzamide
In: Mouse C319
***o*-Methylbenzamide**
Formed from: *o*-Tolunitrile
→ *o*-Toluic acid
Probably in: Rhodococcus H952
***m*-Methylbenzamide**
See: *m*-Toluamide
***m*-Methylbenzamidine**
→ *m*-Methylbenzamidoxime
In: Rabbit C573
***p*-Methylbenzamidine**
→ *p*-Methylbenzamidoxime
In: Rabbit C573, E830, F768
N-Methylbenzamidine
→ Benzamidine

In: Rabbit H168

***m*-Methylbenzamidoxime** (N-hydroxy-*m*-methylbenzamidine)
Formed from: *m*-Methylbenzamidine

***p*-Methylbenzamidoxime** (N-hydroxy-*p*-methylbenzamidine)
Formed from: *p*-Methylbenzamidine

2-Methylbenzanilide
→ 4′-Hydroxy-2-methylbenzanilide
In: Rhizopus A485
→ *o*-Toluic acid
In: Nocardia A2623

7-Methylbenzanthracene (7-methylbenz[*a*] anthracene)
Formed from: Benzanthracene
7,12-Dimethylbenzanthracene
7-Methylbenzanthracene-5,6-oxide
→ Benzanthracene
In: Rat G491
→ 1,2-Dihydro-1,2-dihydroxy-7-methylbenzanthracene
In: Mouse B376
Rat D209
→ 3,4-Dihydro-3,4-dihydroxy-7-methylbenzanthracene
In: Cunninghamella C255
Mouse B376
Rat A68, D209
→ *trans*-5,6-Dihydro-5,6-dihydroxy-7-methylbenzanthracene
In: Hamster A13
Man A13, E523
Mouse B376
Rat A13, A68, D209
→ 8,9-Dihydro-8,9-dihydroxy-7-methylbenzanthracene
In: Cunninghamella C255
Hamster A13
Man A13, E523
Mouse B376
Rat A13, A68, D209
→ 10,11-Dihydro-10,11-dihydroxy-7-methylbenzanthracene
In: Rat D209
→ 7,12-Dimethylbenzanthracene
In: Rat F742, G491
→ 7-Hydroxymethylbenzanthracene
In: Cunninghamella C255
Hamster A13
Man A13
Rat A13, A68, A339, F742, G491
See also: F534
→ 7-Methylbenzanthracene-5,6-oxide
In: Rat A70

8-Methylbenzanthracene
→ 8,9-Dihydro-8,9-dihydroxy-8-methylbenzanthracene
In: Rat B135

11-Methylbenz[*a*]anthracene
→ 3,4-Dihydro-3,4-dihydroxy-11-methylbenz[*a*]anthracene
In: Rat B979
→ 5,6-Dihydro-5,6-dihydroxy-11-methylbenz[*a*]anthracene
In: Rat B979
→ 8,9-Dihydro-8,9-dihydroxy-11-methylbenz[*a*]anthracene
In: Rat B979
→ 11,12-Dihydro-10,11-dihydroxy-11-methylbenz[*a*]anthracene
In: Rat B979

12-Methylbenzanthracene (12-methylbenz[*a*] anthracene)
→ *trans*-5,6-Dihydro-5,6-dihydroxy-12-methylbenz[*a*]anthracene
In: Rat C267, E749, F935
→ 7,12-Dimethylbenzanthracene
In: Rat F742

12-Methylbenzanthracene-7-carboxylic acid
Probably formed from: 7-Hydroxymethyl-12-methylbenzanthracene

7-Methylbenzanthracene-5,6-oxide
Formed from: 7-Methylbenzanthracene
→ *trans*-5,6-Dihydro-5,6-dihydroxy-7-methylbenz[*a*]anthracene
In Man A2486
Mouse A2486
Rat A2486, E394
→ 7-Methylbenzanthracene
In: Rat A350

7-Methylbenzanthracene-3,4-quinone
Formed from: 3,4-Dihydro-3,4-dihydroxy-7-methylbenzanthracene

12-Methylbenzanthracene-3,4-quinone
Formed from: 3,4-Dihydro-3,4-dihydroxy-12-methylbenzanthracene

S-((12-Methylbenzanthracenyl)-7-methyl) glutathione
Formed from: 7-Hydroxymethyl-12-methylbenzanthracene sulphate

N-(4-Methylbenzenesulphonyl)-N′-(3-azabicyclo[3.3.0]oct-3-yl)urea
See: Gliclazide

N-Methylbenzidine
Formed from: Benzidine

1-Methylbenzimidazole
Formed from: Benzimidazole

Methyl benzimidazol-2-ylcarbamate
(carbendazim)
Formed from: 2-(3-Methoxycarbonyl-2-ureido)aniline
Methyl 1-(butylcarbamoyl)benzimidazol-2-ylcarbamate
Probably formed from: 2-(3-Methoxy-carbonyl-2-ureido)-1-(2-ureido)aniline
→ 2-Aminobenzimidazole
In: Rat D927, E12
Strawberry A974
See also: A3064, A3065, A3821
→ Methyl 5-hydroxybenzimidazol-2-ylcarbamate
In: Rat D927, E12
See also: A260
→ *o*-Phenylenediamine
Probably in: Apple A3064
Carrot A3064
Strawberry A3064

Methyl benzoate
Formed from: Benzoic acid
→ Benzoic acid
In: Nocardia G343

***o*-Methylbenzoic acid**
See: *o*-Toluic acid

***m*-Methylbenzoic acid**
See: *m*-Toluic acid

***p*-Methylbenzoic acid**
See: *p*-Toluic acid

***o*-Methylbenzonitrile**
See: *o*-Tolunitrile

***m*-Methylbenzonitrile**
See: *m*-Tolunitrile

***p*-Methylbenzonitrile**
See: *p*-Tolunitrile

Methylbenzoquinone
See: Toluquinone

7-Methylbenzothiophene-2,3-dione
Probably formed from: 4,6-Dimethyldibenzothiophene

1-Methyl-1*H*-benzotriazole
→ 1*H*-Benzotriazole
In: Rat C240

2-Methylbenzotriazole
→ 4,5-Dihydro-4,5-dihydroxy-2-methylbenzotriazole
In: Rat C419
→ 4-Hydroxy-2-methylbenzotriazole
In: Rat C419
→ 5-Hydroxy-2-methylbenzotriazole
In: Rat C419

Methyl (5-benzoyl-*H*-benzimidazol-2-yl) carbamate
See: Mebendazole

***o*-Methylbenzoyl CoA**
Formed from: *o*-Toluic acid

***m*-Methylbenzoyl CoA**
→ 3-Methylcyclohexa-1,5-diene-1-carboxyl CoA
Probably in: Thauera J175

***p*-Methylbenzoyl CoA**
Formed from: *p*-Toluic acid

***o*-Methylbenzoyl-β-D-glucuronic acid**
Formed from: *o*-Toluic acid

***m*-Methylbenzoyl-β-D-glucuronic acid**
Formed from: *m*-Toluic acid

***p*-Methylbenzoyl-β-D-glucuronic acid**
Formed from: *p*-Toluic acid

Methyl benzoylglycyl-L-lysine
→ Hippuric acid
With carboxypeptidase B A243

***p*-Methylbenzoylthiamine pyrophosphoric acid**
Formed from: *p*-Bromomethylphenylglyoxylic acid
→ *p*-Toluic acid
In: Pseudomonas E714

6-Methyl-3,4-benzpyrene
→ 6-Hydroxymethyl-3,4-benzpyrene

In: Rat A815

7-Methyl-3,4-benzpyrene
→ 4,5-Dihydro-4,5-dihydroxy-7-methyl-3,4-benzpyrene
In: Rat C390, C427, C521
→ 7,8-Dihydro-7,8-dihydroxy-7-methyl-3,4-benzpyrene
In: Rat C390, C427, C521
→ 9,10-Dihydro-9,10-dihydroxy-7-methyl-3,4-benzpyrene
In: Rat C390, C427, C521
→ 3-Hydroxymethyl-3,4-benzpyrene
In: Rat C521
→ 7-Hydroxymethyl-3,4-benzpyrene
In: Rat C390, C427, C521
→ 7-Methyl-3,4-benzpyrene-1,6-quinone
In: Rat C521
→ 7-Methyl-3,4-benzpyrene-3,6-quinone
In: Rat C521

7-Methyl-3,4-benzpyrene-1,6-quinone
Formed from: 7-Methyl-3,4-benzpyrene

7-Methyl-3,4-benzpyrene-3,6-quinone
Formed from: 7-Methyl-3,4-benzpyrene

N-(α-Methylbenzyl)adipic acid monoamide
Formed from: N-(α-Methylbenzyl)linoleamide

***o*-Methylbenzyl alcohol**
Formed from: *o*-Tolualdehyde
o-Toluic acid
o-Xylene
→ N-Acetyl-S-(*o*-methylbenzyl)-L-cysteine
In: Rat C98
→ *o*-Tolualdehyde
In: Acinetobacter J621
See also: J219

***m*-Methylbenzyl alcohol**
Formed from: *m*-Tolualdehyde
m-Xylene
→ *m*-Tolualdehyde
In: Acinetobacter J621
Botrytis G613
Rhodopseudomonas D197
See also: J219

***p*-Methylbenzyl alcohol**
Formed from: *p*-Methylbenzyl benzoate
p-Tolualdehyde
p-Xylene
→ 1,4-Benzenedimethanol
In: Nitrosomonas H219
→ *p*-Methylbenzyl sulphate
In: Rat E630
→ *p*-Tolualdehyde
In: Acinetobacter J621
Botrytis G613
Nitrosomonas H219
Rhodopseudomonas D197
See also: A59, A1301, A1996, A3448, H616, J219

N-Methylbenzylamine
Formed from: Benzylamine
N-Benzyl-1,N-dimethyl-2-propynylamine
Pargyline
→ Benzaldehyde
In: Beef A574
Pig A72
→ Benzylamine
Probably in: Rat A3005, E643
→ N-Methylbenzylamine-N-sulphate
In: Rat E51

***p*-Methylbenzylamine**
→ *p*-Tolualdehyde
In: Pig A3693
See also: G865

N-Methylbenzylamine-N-sulphate
Formed from: N-Methylbenzylamine

***p*-Methylbenzyl benzoate**
→ Benzoic acid
In: Pseudomonas K206
→ *p*-Methylbenzyl alcohol
In: Pseudomonas K206

N-(α-Methylbenzyl)glutaric acid monoamide
Formed from: N-(α-Methylbenzyl)linoleamide

α-Methylbenzylhydrazine (mebenazine)
→ α-Methylbenzylhydrazine-β-D-glucuronide
Probably in: Rat A2104

α-Methylbenzylhydrazine-β-D-glucuronide
Probably formed from: α-Methylbenzylhydrazine

N-(α-Methylbenzyl)linoleamide
→ N-(α-Methylbenzyl)adipic acid monoamide
In: Man A1768
Rat A1768

→ N-(α-Methylbenzyl)glutaric acid monoamide
In: Man A1768
Rat A1768
→ N-(α-Methylbenzyl)succinic acid monoamide
In: Man A1768
Rat A1768

Methyl 3-(3-((N-benzyl-N-methyl)aminomethyl)-1,2,4-oxadiazol-5-yl)-1,4-dihydro-2,6-dimethyl-4-(*m*-nitrophenyl)pyridine-5-carboxylate
→ Methyl 3-(N-cyanocarboxamido)-2,6-dimethyl-4-(*m*-nitrophenyl)pyridine-5-carboxylate
In: Rat G849
→ Methyl 1,4-dihydro-2,6-dimethyl-4-(*m*-nitrophenyl)-3-(1,2,4-oxadiazol-5-yl) pyridine-5-carboxylate
In: Rat G849

4-(*o*-Methylbenzyl)-1-methylpyridinium
Probably formed from: Tetrahydro-4-(*o*-methylbenzyl)-1-methylpyridine

4-(*m*-Methylbenzyl)-1-methylpyridinium
Probably formed from: Tetrahydro-4-(*m*-methylbenzyl)-1-methylpyridine

4-(*p*-Methylbenzyl)-1-methylpyridinium
Probably formed from: Tetrahydro-4-(*p*-methylbenzyl)-1-methylpyridine

4-(α-Methylbenzyl)-1-methylpyridinium
Probably formed from: Tetrahydro-4-(α-methylbenzyl)-1-methylpyridine

(3-Methylbenzyl)succinic acid
Formed from: *m*-Xylene
→ (*E*)-(3-Methylphenyl)itaconic acid
Probably in: Azoarcus K271
→ *m*-Toluic acid
In: Azoarcus K271

N-(α-Methylbenzyl)succinic acid monoamide
Formed from: N-(α-Methylbenzyl)linoleamide

***p*-Methylbenzyl sulphate**
Formed from: *p*-Methylbenzyl alcohol

13-Methylberberine
Formed from: Mesothalictricavine
Thalictricavine
→ Mesothalictricavine
In: Corydalis H585
Dicentra J496
→ Thalictricavine
In: Corydalis H585
Dicentra J496

4-(4′-Methylbiphenyl-2-yl)-1,3-imidazole
→ 4-(4′-Methylbiphenyl-2-yl)-1,3-imidazole-1-β-D-glucuronide
In: Monkey G949

5-(4′-Methylbiphenyl-2-yl)-1,3-imidazole
→ 5-(4′-Methylbiphenyl-2-yl)-1,3-imidazole-1-β-D-glucuronide
In: Monkey G949

4-(4′-Methylbiphenyl-2-yl)-1,3-imidazole-1-β-D-glucuronide
Formed from: 4-(4′-Methylbiphenyl-2-yl)-1,3-imidazole

5-(4′-Methylbiphenyl-2-yl)-1,3-imidazole-1-β-D-glucuronide
Formed from: 5-(4′-Methylbiphenyl-2-yl)-1,3-imidazole

N-Methylbiphenyl-4-yl nitroxide radical
Probably formed from: 4-Dimethylaminobiphenyl

5-(4′-Methylbiphenyl-2-yl)tetrazole
→ 5-(4′-Methylbiphenyl-2-yl)tetrazole-β-D-glucuronide
In: Man G693, G949
Monkey G949
Rat G949

5-(4′-Methylbiphenyl-2-yl)tetrazole-β-D-glucuronide
Formed from: 5-(4′-Methylbiphenyl-2-yl) tetrazole

5-(3′-Methylbiphenyl-2-yl)-1,2,4-triazole
Formed from: 5-(3′-Methylbiphenyl-2-yl)-1,2,4-triazole-1-β-D-glucuronide
→ 5-(3′-Methylbiphenyl-2-yl)-1,2,4-triazole-1-β-D-glucuronide
In: Man G949
Monkey G949
→ 5-(3′-Methylbiphenyl-2-yl)-1,2,4-triazole-2-β-D-glucuronide
In: Man G949

5-(4′-Methylbiphenyl-2-yl)-1,2,3-triazole
Formed from: 5-(4′-Methylbiphenyl-2-yl)-1,2,3-triazole-1-β-D-glucuronide
→ 5-(4′-Methylbiphenyl-2-yl)-1,2,3-triazole-1-β-D-glucuronide

In: Man G949
Monkey G949
→ 5-(4′-Methylbiphenyl-2-yl)-1,2,3-triazole-2-β-D-glucuronide
In: Man G949

5-(3′-Methylbiphenyl-2-yl)-1,2,4-triazole-1-β-D-glucuronide
Formed from: 5-(3′-Methylbiphenyl-2-yl)-1,2,4-triazole
→ 5-(3′-Methylbiphenyl-2-yl)-1,2,4-triazole
In: Beef G949
Escherichia G949
Limpet G949

5-(3′-Methylbiphenyl-2-yl)-1,2,4-triazole-2-β-D-glucuronide
Formed from: 5-(3′-Methylbiphenyl-2-yl)-1,2,4-triazole

5-(4′-Methylbiphenyl-2-yl)-1,2,3-triazole-1-β-D-glucuronide
Formed from: 5-(4′-Methylbiphenyl-2-yl)-1,2,3-triazole
→ 5-(4′-Methylbiphenyl-2-yl)-1,2,3-triazole
In: Beef G949
Escherichia G949
Helix G949
Limpet G949

5-(4′-Methylbiphenyl-2-yl)-1,2,3-triazole-2-β-D-glucuronide
Formed from: 5-(4′-Methylbiphenyl-2-yl)-1,2,3-triazole

Methyl 5(6)-butyl-2-benzimidazolecarbamate (parbendazole)
→ Methyl 5(6)-(1-hydroxybutyl)-2-benzimidazolecarbamate
In: Beef A1066
Cunninghamella A1066
Sheep A1066
→ Methyl 5(6)-(1-hydroxybutyl)-2-benzimidazolecarbamate
In: Beef A1066
Cunninghamella A1066
Sheep A1066
→ Methyl 5(6)-(2-hydroxybutyl)-2-benzimidazolecarbamate
In: Beef A1066
Sheep A1066
→ Methyl 5(6)-(4-hydroxybutyl)-2-benzimidazolecarbamate
In: Beef A1066
Cunninghamella A1066
Sheep A1066
→ Methyl 5(6)-(3-oxobutyl)-2-benzimidazolecarbamate
In: Paecilomyces A1066

Methyl 1-(butylcarbamoyl)benzimidazol-2-yl carbamate (benomyl)
→ N-(2-Acetamidophenyl)-N′-ethoxycarbonyl-S-methylisothiourea
In: Tomato A2479
→ N-(2-Acetamidophenyl)-N′-methoxycarbonyl-S-methylisothiourea
In: Tomato A2479
→ N-(2-Benzamidophenyl)-N′-ethoxycarbonyl-S-methylisothiourea
In: Tomato A2479
→ N-(2-Benzamidophenyl)-N′-methoxycarbonyl-S-methylisothiourea
In: Tomato A2479
→ 1,2-Bis(3-ethoxycarbonyl)-2-thioureidobenzene
In: Tomato A2479
→ 1,2-Bis(3-methoxycarbonyl)-2-thioureidobenzene
In: Tomato A2479
→ Ethyl benzimidazol-2-yl carbamate
In: Tomato A2479
→ Methyl benzimidazol-2-yl carbamate
In: Apple A3064, B592
Carrot A3064
Melon A3065
Pea A502
Pseudomonas A3821
Strawberry A3064
Tomato A2479

Methyl 5(6)-butyl-6-hydroxy-2-benzimidazolecarbamate
Formed from: Methyl 5(6)-butyl-2-benzimidazolecarbamate

Methyl N[2]-(1-butyloxycarbonyl)-N[3]-(2-(3-formamidinophenyl)isoxazolin-5(*S*)-yl)acetyl)-2,3-diaminopropionate
Formed from: Roxifibam

***m*-(1-Methylbutyl)phenyl N-methylcarbamate**
→ *m*-(1-Hydroxy-1-methylbutyl)phenyl N-methylcarbamate
In: Microorganism A403

O-α-Methyl-*n*-butyrylshikonin
Probably formed from: Shikonin

Methyl caffeate
→ Caffeic acid
In: Aspergillus G915
→ Methyl ferulate
In: Parsley F160
→ Methyl isoferulate
In: Lentinus A214

N-Methylcalycotomine
Formed from: O-Acetyl-N-methylcalycotomine

(*S*)-*cis*-N-Methylcanadine (tetrahydroberbinium)
Formed from: Canadine
→ Allocryptopine
In: Corydalis E577

Methyl 1-carbamoyl-α-phenyl-2-piperidineacetate
Formed from: Methylphenidate
→ 1-Carbamoyl-α-phenyl-2-piperidineacetic acid
Probably in: Dog C14

1-Methylcarbazole
→ 1-Formylcarbazole
In: Aspergillus J180
Pseudomonas J180

3-Methylcarbazole
Formed from: N-Acetyl-3-methylcarbazole
→ 3-Formylcarbazole
In: Aspergillus J180
Pseudomonas J180

N-Methylcarbazole
→ Carbazole
In: Animal A2725
Cunninghamella G90
See also: A1773
→ 1-Hydroxy-N-methylcarbazole
In: Rabbit B425
→ 2-Hydroxy-N-methylcarbazole
In: Rabbit B425
→ 3-Hydroxy-N-methylcarbazole
In: Animal A2725
Rabbit B425
→ N-Hydroxymethylcarbazole
In: Animal A2725
Cunninghamella G90, G701
Rabbit A1773, B425
Rat H266

2-Methyl-β-carbolinium
Formed from: Norharman

9-Methyl-β-carbolinium
Formed from: Norharman

Methyl 5-carboxy-2-cyano-6-hydroxymethyl-4-(3-nitrophenyl)pyridine-3-carboxylate
Probably formed from: Methyl 5-carboxy-2-cyano-6-methyl-4-(3-nitrophenyl)pyridine-3-carboxylate

Methyl 5-carboxy-2-cyano-6-methyl-4-(3-nitrophenyl)pyridine-3-carboxylate
Probably formed from: Dehydronilvadipine
→ 3,5-Dicarboxy-2-cyano-6-methyl-4-(3-nitrophenyl)pyridine
Probably in: Rat F484
→ Methyl 4-(3-aminophenyl)-5-carboxy-2-cyano-6-methylpyridine-3-carboxylate
Probably in: Rat F484
→ Methyl 5-carboxy-2-cyano-6-hydroxymethyl-4-(3-nitrophenyl)pyridine-3-carboxylate
Probably in: Rat F484

Methyl 3-carboxy-1,4-dihydro-2,6-dimethyl-4-(*m*-nitrophenyl)pyridine-5-carboxylate
Formed from: Barnidipine
→ Methyl 3-carboxy-2,6-dimethyl-4-(*m*-nitrophenyl)pyridine-5-carboxylate
Probably in: Man J84, J257

Methyl 3-carboxy-2,6-dimethyl-4-(*m*-nitrophenyl)pyridine-5-carboxylate
Probably formed from: Methyl 3-carboxy-1,4-dihydro-2,6-dimethyl-4-(*m*-nitrophenyl)pyridine-5-carboxylate

Methyl 2-(4-(2-carboxy-2-hydroxyethoxy)phenyl) ethylcarbamate
Formed from: Pamatolol

Methyl 5(6)-(3-carboxypropyl)-2-benzimidazolecarbamate
Formed from: Methyl 5(6)-(4-hydroxybutyl)-2-benzimidazolecarbamate

3-O-Methylcatechin
See: *cis*-3′,4′-5,7-Tetrahydroxy-3-methoxyflavan

3-Methylcatechol
Formed from: *o*-Cresol
cis-2,3-Dihydro-2,3-dihydroxytoluene
m-Fluorotoluene
3-Methyl-2-nitrophenol
6-Methyl-2-nitrophenol
o-Nitrotoluene
m-Nitrotoluene
o-Toluidine
Probably formed from: 2-Hydroxy-3-methylbenzoic acid
→ 2-Hydroxy-6-methyl-*cis*,*cis*-muconic semialdehyde
In: Bacillus A2397
Microorganism E31
Sphingomonas J654, J221
Pseudomonas J515, K5
See also: A546, A768, A2694, A2695, D484, G355, G647, G766, H272, J221, J649, J653, J903, K204, K445
→ 2-Methoxy-3-methylphenol
In: Rat A486
→ 2-Methoxy-6-methylphenol
In: Rat A486
→ 3-Methyl-*o*-quinone
Probably in: Botrytis A1930
→ 2-Methyl-*cis*,*cis*-muconic acid
In: Acinetobacter J390
Alcaligenes A3769, K775
Hydrogenophaga J218
Microorganism E31
Pseudomonas A3769, B754
Rhodococcus D247, J703
Trichosporon C84
See also: A2765, G647

4-Methylcatechol (homocatechol)
Formed from: *m*-Cresol
p-Cresol
2-Hydroxy-4-methylbenzoic acid
3,4-Methylenedioxytoluene
4-Methyl-2-nitrophenol
5-Methyl-2-nitrophenol
m-Nitrotoluene
p-Nitrotoluene
Rutin
p-Toluenesulphonic acid
m-Toluidine
p-Toluidine
Probably formed from: *m*-Toluic acid
→ 3,4-Dihydroxybenzyl alcohol
In: Sarcophaga C692
See also: E175, F89, H20
→ 2-Hydroxy-4-methyl-*cis*,*cis*-muconic semialdehyde
In: Candida A943
→ 2-Hydroxy-5-methyl-*cis*,*cis*-muconic semialdehyde
In: Bacillus A2397
See also: A546, A768, A1323, D484, G355, G647, H272, J221, J515, J649, J653, J903, K5, K204, K445
→ 2-Methoxy-4-methylphenol
In: Nicotiana A3778, B490
Rat A486
With catechol O-methyltransferase A597
→ 2-Methoxy-5-methylphenol
In: Rat A486
With catechol O-methyltransferase A597
→ 3-Methyl-*cis*,*cis*-muconic acid
In: Acinetobacter J390
Alcaligenes A3769, K775
Candida E204
Hydrogenophaga J218
Pseudomonas A3769, B754, G647
Rhizobium D672
Rhodococcus D247, J703
Trichosporon C84
See also: A2765
→ 4-Methyl-*o*-quinone
In: Amasya H567
Mushroom H907
See also: A218, A1636, A1930, A2360, A3190, A3138, A3974, A3975, D726, E369, E407, E458, F944, G620, J187, K96

Methyl 2-(*o*-chlorophenyl)-2-(*p*-chlorophenyl) acetate
Probably formed from: 2-(*o*-Chlorophenyl)-2-(*p*-chlorophenyl)acetic acid

Methyl 2-chloro-5-nitrobenzoate
→ Methyl 2-(S-glutathionyl)-5-nitrobenzoate
In: Rat H516

3-Methylcholanthrene
Formed from: 3-Methylcholanthrene-11,12-oxide
→ *trans*-4,5-Dihydro-4,5-dihydroxy-3-methylcholanthrene
In: Rat B391
→ *trans*-9,10-Dihydro-9,10-dihydroxy-3-methylcholanthrene
In: Rat B284, B391
→ *trans*-11,12-Dihydro-11,12-dihydroxy-3-methylcholanthrene
In: Mouse A632
Rat B284, B391
→ *cis*-1,2-Dihydroxy-3-methylcholanthrene
In: Mouse A444
Rat B391
→ *trans*-1,2-Dihydroxy-3-methylcholanthrene
In: Mouse A444
→ *cis*-9,10-Dihydroxy-3-methylcholanthrene
In: Rat B391
→ 1-Hydroxy-3-methylcholanthrene
In: Cunninghamella B837
Man F562
Mouse A632
Rat B391, B847, F382, G243
→ 2-Hydroxy-3-methylcholanthrene
In: Cunninghamella B837
Mouse A444, A632
Rat B284, B391, B847, G243
→ 3-Hydroxymethylcholanthrene
In: Rat H708

3-Methylcholanthren-1-one
Formed from: 1-Hydroxy-3-methylcholanthrene

3-Methylcholanthren-2-one
Formed from: 2-Hydroxy-3-methylcholanthrene

3-Methylcholanthrene-11,12-oxide
→ *trans*-11,12-Dihydro-11,12-dihydroxy-3-methylcholanthrene
In: Man A2486, A3217, B63, B220
Mouse A2486
Rat A2486, A3031, B63, B142, B220
→ 3-Methylcholanthrene
In: Rat A350

5-Methylchrysene
Formed from: 5,6-Dimethylchrysene
→ Chrysene
In: Rat G118
→ 1,2-Dihydro-1,2-dihydroxy-5-methylchrysene
In: Man H792
Mouse C916
Rat B234
→ 7,8-Dihydro-7,8-dihydroxy-5-methylchrysene
In: Man H792
Mouse C916
Rat B234
→ 9,10-Dihydro-9,10-dihydroxy-5-methylchrysene
In: Rat B234
→ 1-Hydroxy-5-methylchrysene
In: Man H792
Rat B234
→ 7-Hydroxy-5-methylchrysene
In: Rat B234
→ 9-Hydroxy-5-methylchrysene
In: Man H792
Rat B234
→ 5-Hydroxymethylchrysene
In: Rat B234, G118
→ 4,5-Methylenechrysene
In: Rat G118

6-Methylchrysene
Formed from: Chrysene
5,6-Dimethylchrysene
→ Chrysene
In: Rat G118
→ 1-Hydroxy-6-methylchrysene
In: Man H792
→ 7-Hydroxy-6-methylchrysene
In: Man H792
→ 6-Hydroxymethylchrysene
In: Man H792
Rat G118

***p*-Methylcinnamaldehyde**
Formed from: 2-Oxo-4-(*p*-tolyl)-3-butenoic acid

α-Methylcinnamaldehyde
→ α-Methylcinnamyl alcohol
In: Dunaliella G459
Euglena G422

Methyl cinnamate
Formed from: Cinnamic acid
→ Methyl *p*-coumarate
In: Lentinus G477

***p*-Methylcinnamic acid**
Formed from: (*E*)-4-(*p*-Bromomethylphenyl)-2-oxo-3-butenoic acid
(*E*)-4-(*p*-Chloromethylphenyl)-2-oxo-3-butenoic acid

α-Methylcinnamyl alcohol
Formed from: α-Methylcinnamaldehyde
→ 2-Methyl-3-phenylpropan-1-ol
Probably in: Dunaliella G459
Euglena G422

N-Methylcoclaurine (1,2,3,4-tetrahydro-1-((4-hydroxyphenyl)methyl)-6-methoxy-2-methyl-7-isoquinolinol; both (*R*) and (*S*) isomers)
Formed from: Coclaurine
→ (*R*,*R*)-Bebeerine
In: Cissampelos E581
→ Berbamunine
In: Berberis G736
→ Coclaurine
In: Eschscholtzia H680
→ 3,4-Dihydro-7-hydroxy-6-methoxy-2-methylisoquinolinium
In: Nelumbo D566
With ascorbate oxidase C335
→ 1-(3,4-Dihydroxybenzyl)-1,2,3,4-tetrahydro-7-hydroxy-6-methoxy-2-methylisoquinoline
In: Berberis F855, K933
Eschscholtzia K933
Peumus K933
→ Glaziovine
In: Nelumbo D566
With ascorbate oxidase C335
→ Guattegaumerine
In: Berberis G736
→ (*S*,*R*)-Hayatidine
In: Cissampelos E517, E581
→ *p*-Hydroxybenzyl alcohol
In: Nelumbo D566
With ascorbate oxidase C335
→ (*R*,*R*)-Isochondrodendrine
In: Cissampelos E581
→ Isotetrandrine
In: Cocculus B673
→ 2′-Norberbamunine
In: Berberis G736
→ 1,2,3,4-Tetrahydro-7-hydroxy-6-methoxy-2-methylisoquinoline
In: Nelumbo D566
With ascorbate oxidase C335
Peroxidase C335
→ 5,6,6a,7-Tetrahydro-1,10,11-trihydroxy-2-methoxy-6-methyl-1*H*-dibenzo[*de*,*g*]quinoline
With ascorbate oxidase C335
Peroxidase C335

O-Methylcoripalline
Formed from: Laudanosine

Methyl *p*-coumarate
Formed from: Methyl cinnamate
→ *p*-Coumaric acid
In: Aspergillus G915
Penicillium G805, G650, J499
→ Methyl *p*-methoxycinnamate
In: Lentinus A214

9-O-Methylcoumestrol
Formed from: Formononetin
Isoflav-3-ene

N-Methylcrotsparine
Probably formed from: Crotsparine
→ Nornuciferine
In: Croton B432

N-Methylcrotsparinol
→ Nornuciferine
In: Croton B432

Methyl *p*-cyanobenzoate
→ Monomethyl terephthalate
In: Rhodococcus H952

Methyl 3-(N-cyanocarboxamido)-2,6-dimethyl-4-(*m*-nitrophenyl)pyridine-5-carboxylate
Formed from: Methyl 3-(3-((N-benzyl-N-methyl)aminomethyl)-1,2,4-oxadiazol-5-yl)-1,4-dihydro-2,6-dimethyl-4-(*m*-nitrophenyl)pyridine-5-carboxylate

3-Methylcyclohexanecarboxylic acid
→ *m*-Toluic acid
In: Guinea pig D750
Mouse D750
Rabbit D750

Rat D750

4-Methylcyclohexanecarboxylic acid
→ *p*-Toluic acid
In: Guinea pig D750
Mouse D750
Rabbit D750
Rat D750

5-(4-(1-Methylcyclohexylmethoxy)benzyl)-2,4-thiazolidinedione
See: Ciglitazone

1-(2-Methylcyclohexyl)-3-phenylurea (siduron)
→ 1-(2-Hydroxymethylcyclohexyl)-3-phenylurea
In: Kentucky bluegrass A2051
→ 1-(4-Hydroxy-2-methylcyclohexyl)-3-phenylurea
In: Kentucky bluegrass A2051
→ 1-(*p*-Hydroxyphenyl)-3-(2-methylcyclohexyl)urea
In: Kentucky bluegrass A2051

1-Methylcyclopenine
→ 1-Methylcyclopenol
In: Penicillium A1633

1-Methylcyclopenol
Formed from: 1-Methylcyclopenine

Methyl 5-cyclopropylcarbonyl-2-benzimidazolecarbamate
See: Ciclobendazole

N-Methyldaunorubicinol
Formed from: N,N-Dimethyldaunorubicin

Methyl dehydrodiconiferyl alcohol (2,3-dihydro-2-(4-hydroxy-3-methoxyphenyl)-3-hydroxymethyl-5-(3-hydroxy-1-propenyl)-7-methoxybenzofuran)
→ *threo*-2-(3,4-Dimethoxyphenyl)-2,3-dihydro-3-hydroxymethyl-7-methoxy-5-(1,2,3-trihydroxypropyl)benzofuran
In: Phanerochaete B796

7-O-Methyl-2′,3′-dehydrokievitone
→ 7-O-Methylkievitone-2′,3′-epoxide
In: Botrytis F246

5-Methylderrone
Formed from: Derrone

N-Methyldetomidine
Formed from: Detomidine

N-Methyldexmedetomidine
Formed from: Dexmedetomidine

3-Methyldiazepam
See: 7-Chloro-1,3-dihydro-1,3-dimethyl-5-phenyl-2*H*-1,4-benzodiazepin-2-one

7-Methyldibenzanthracene
Formed from: Dibenzanthracene
→ 7,14-Dimethyldibenzanthracene
Probably in: Rat E83
→ 7-Hydroxymethyldibenzanthracene
Probably in: Rat E83

7-Methyldibenz[*a*,*j*]anthracene
→ 3,4-Dihydro-3,4-dihydroxy-7-methyldibenz[*a*,*j*]anthracene
In: Mouse G611
→ 5,6-Dihydro-5,6-dihydroxy-7-methyldibenz[*a*,*j*]anthracene
In: Mouse G611

N-Methyl-5*H*-dibenzo[*a*,*d*]cycloheptene-5-propylamine
See: Protriptyline

1-Methyl-4-(5*H*-dibenzo[*a*,*d*]cycloheptylidene)piperidine
See: Cyproheptadine

N-Methyldibenzylamine
→ Dibenzylamine
Probably in: Rat J471
→ N-Methyldibenzylamine-N-oxide
In: Rat J471

N-Methyldibenzylamine-N-oxide
Formed from: N-Methyldibenzylamine

Methyl 2-(4-(2′,4′-dichlorophenoxy)phenoxy)propionate (diclofop-methyl)
→ 2-(4-(2′,4′-Dichlorophenoxy)phenoxy)propionic acid
In: Wheat A1947, A3803

Methyl 3,6-dichloro-*o*-anisate
See: Disugran

(1*S*,4*S*)-N-Methyl-4-(3,4-dichlorophenyl)-1,2,3,4-tetrahydro-α-naphthylamine
See: Sertraline

Methyl 3,6-dichlorosalicylate
Formed from: Disugran
→ 3,6-Dichloro-2-hydroxybenzoic acid
In: Microorganism A1317

Methyl *cis*-7,8-dihydro-7,8-dihydroxy-1-naphthoate
Formed from: Methyl 1-naphthoate
→ Methyl 7,8-dihydroxy-1-naphthoate

In: Pseudomonas J370

Methyl *cis*-7,8-dihydro-7,8-dihydroxy-2-naphthoate
Formed from: Methyl 2-naphthoate
→ Methyl 7,8-dihydroxy-2-naphthoate
In: Pseudomonas J370

Methyl 1,4-dihydro-2,6-dimethyl-4-(*m*-nitrophenyl)-3-(1,2,4-oxadiazol-5-yl)pyridine-5-carboxylate
Formed from: Methyl 3-(3-((N-benzyl-N-methyl)aminomethyl)-1,2,4-oxadiazol-5-yl)-1,4-dihydro-2,6-dimethyl-4-(*m*-nitrophenyl) pyridine-5-carboxylate

4′-O-Methyldihydroquercetin
→ Dihydroquercetin
Probably in: Petunia A2740

O-Methyldihydrotienilic acid
Formed from: Dihydrotienilic acid

Methyl 5(6)-(1,2-dihydroxybutyl)-2-benzimidazolecarbamate
Probably formed from: Methyl 5(6)-(*threo*-1-hydroxybutyl)-2-benzimidazolecarbamate

Methyl 5,6-dihydroxyindole-2-carboxylate
Probably formed from: L-Dopachrome methyl ester

Methyl 7,8-dihydroxy-1-naphthoate
Formed from: Methyl *cis*-7,8-dihydro-7,8-dihydroxy-1-naphthoate

Methyl 7,8-dihydroxy-2-naphthoate
Formed from: Methyl *cis*-7,8-dihydro-7,8-dihydroxy-2-naphthoate

Methyl 3,4-dihydroxyphenylacetate
→ Methyl 3-hydroxy-4-methoxyphenylacetate
In: Rat A486
With catechol O-methyltransferase A597
→ Methyl 4-hydroxy-3-methoxyphenylacetate
In: Rat A486
With catechol O-methyltransferase A597

N-(*p*-((5-Methyl-2,4-dihydroxyquinazolin-6-yl) methylamino)benzoyl)diglutamylglutamic acid
Probably formed from: N-(*p*-((5-Methyl-2,4-dihydroxyquinazolin-6-yl)methylamino) benzoyl)glutamylglutamic acid
→ N-(*p*-((5-Methyl-2,4-dihydroxyquinazolin-6-yl)methylamino)benzoyl) triglutamylglutamic acid
Probably in: Pig F760

N-(*p*-((5-Methyl-2,4-dihydroxyquinazolin-6-yl)methylamino)benzoyl)glutamic acid
→ N-(*p*-((5-Methyl-2,4-dihydroxyquinazolin-6-yl)methylamino)benzoyl) glutamylglutamic acid
Probably in: Pig F760

N-(*p*-((5-Methyl-2,4-dihydroxyquinazolin-6-yl)methylamino)benzoyl)glutamylglutamic acid
Probably formed from: N-(*p*-((5-Methyl-2,4-dihydroxyquinazolin-6-yl)methylamino) benzoyl)glutamic acid
→ N-(*p*-((5-Methyl-2,4-dihydroxyquinazolin-6-yl)methylamino)benzoyl) diiglutamylglutamic acid
Probably in: Pig F760

N-(*p*-((5-Methyl-2,4-dihydroxyquinazolin-6-yl)methylamino)benzoyl)triglutamylglutamic acid
Probably formed from: N-(*p*-((5-Methyl-2,4-dihydroxyquinazolin-6-yl)methylamino) benzoyl)diglutamylglutamic acid

Methyl 3,4-dimethoxycinnamate
Formed from: Methyl ferulate

Methyl 1-(3,4-dimethoxyphenyl)-3-(3-ethylvaleryl)-4,7-dihydroxy-6,8-dimethoxy-2-naphthoate
Formed from: Methyl 1-(3,4-dimethoxyphenyl)-3-(3-ethylvaleryl)-4-hydroxy-6,7,8-trimethoxy-2-naphthoate
→ Methyl 1-(3,4-dimethoxyphenyl)-3-(3-ethylvaleryl)-4,7-dihydroxy-6,8-dimethoxy-2-naphthoate-β-D-glucuronide
Probably in: Dog J839
Rat J839

Methyl 1-(3,4-dimethoxyphenyl)-3-(3-ethylvaleryl)-4,7-dihydroxy-6,8-dimethoxy-2-naphthoate-β-D-glucuronide
Probably formed from: Methyl 1-(3,4-dimethoxyphenyl)-3-(3-ethylvaleryl)-4,7-dihydroxy-6,8-dimethoxy-2-naphthoate

Methyl 1-(3,4-dimethoxyphenyl)-3-(3-ethylvaleryl)-4-hydroxy-6,7,8-trimethoxy-2-naphthoate
→ Methyl 1-(3,4-dimethoxyphenyl)-3-(3-ethylvaleryl)-4,7-dihydroxy-6,8-dimethoxy-2-naphthoate
In: Dog J839

Rat J839
→ Methyl 1-(3,4-dimethoxyphenyl)-3-(3-ethylvaleryl)-4-hydroxy-6,7,8-trimethoxy-2-naphthoate-β-D-glucuronide
In: Dog J839
Rat J839
→ Methyl 3-(3-ethylvaleryl)-4-hydroxy-1-(4-hydroxy-3-methoxyphenyl)-6,7,8-trimethoxy-2-naphthoate
In: Dog J839
Rat J839

Methyl 1-(3,4-dimethoxyphenyl)-3-(3-ethylvaleryl)-4-hydroxy-6,7,8-trimethoxy-2-naphthoate-β-D-glucuronide
Formed from: Methyl 1-(3,4-dimethoxyphenyl)-3-(3-ethylvaleryl)-4-hydroxy-6,7,8-trimethoxy-2-naphthoate

Methyl *p*-(3-(N-(1-(3,4-dimethoxyphenyl)ethyl)amino)-2-hydroxypropoxy)phenylpropionate
→ *p*-(3-(N-(1-(3,4-Dimethoxyphenyl)ethyl)amino)-2-hydroxypropoxy)phenylpropionic acid
In: Rabbit H563

Methyl *p*-dimethylaminobenzoate
→ Methyl *p*-methylaminobenzoate
In: Rat D578

Methyl 2,6-dimethyl-4-(*m*-nitrophenyl)pyridine-3-carboxylate
Probably formed from: 2,6-Dimethyl-4-(*m*-nitrophenyl)pyridine-3,5-dicarboxylic acid 3-methyl ester

2-Methyl-4,6-dinitrophenol
→ 6-Amino-2-methyl-4-nitrophenol
In: Azotobacter A3824

Methyl 2,6-dioxo-6-phenylhexanoate
Formed from: Methyl 2-hydroxy-6-oxo-6-phenylhexa-2,4-dienoate

8-(5-Methyl-2,3-dioxopyrrolidin-5-yl)amino-6-hydroxy-4-hydroxymethyl-2-methoxy-5-(3-trifluoromethylphenoxy)quinoline
Probably formed from: 8-(4-Amino-1-methyl-3-oxobutyl)amino-6-hydroxy-4-hydroxymethyl-2-methoxy-5-(3-trifluoromethylphenoxy)quinoline

N-Methyldiphenylacetamide
Formed from: N,N-Dimethyldiphenylacetamide
→ 2,2-Diphenylacetamide
In: Glycine A3772
Plant A1205

5-Methyl-2,2-diphenyl-1,4-dioxan-3-one
O-(*n*-Propyl)benzilic acid

S-(N-(1-Methyl-3,3-diphenylpropylcarbamoyl)glutathione
Formed from: N-(1-Methyl-3,3-diphenylpropyl)formamide
→ N-Acetyl-S-(N-(1-methyl-3,3-diphenylpropylcarbamoyl)-L-cysteine
Probably in: Rat F604

N-(1-Methyl-3,3-diphenylpropyl)formamide
→ S-(N-(1-Methyl-3,3-diphenylpropylcarbamoyl)glutathione
In: Rat F604

2-Methyl-1,2-di-3-pyridinyl-1-propanone
See: Metyrapone

3-O-Methyldobutamine
Formed from: Dobutamine
→ 3-O-Methyldobutamine-β-D-glucuronide
Probably in: Dog A1850

3-O-Methyldobutamine-β-D-glucuronide
Probably formed from: 3-O-Methyldobutamine

α-Methyldopa
See: 3,4-Dihydroxy-α-methyl-L-phenylalanine

L-Methyldopachrome
→ 5,6-Dihydroxy-3-methylindole-2-carboxylic acid
Probably in: Locusta J842

2-Methylellipticinium
→ 9-Hydroxy-2-methylellipticinium
Probably in: Rat B557

2-Methylellipticinium-9-yl-β-D-glucuronide
Formed from: 9-Hydroxy-2-methylellipticinium

10-Methylene-9-anthrone
→ Anthraquinone
With cytochrome P450 H796

α-Methylenebenzeneacetic acid
See: Atropic acid

4,4′-Methylenebis(*o*-chloroaniline)
→ N-Acetyl-4,4′-methylenebis(*o*-chloroaniline)
In: Man A3253

→ 4,4′-Diamino-3,3′-dichloro-5-hydroxydiphenylmethane
In: Dog D523, D847, F25
Guinea pig F25
Rat F25
→ N-Hydroxy-4,4′-methylenebis(*o*-chloroaniline)
In: Dog F25
Guinea pig F25
Rat F25
→ 4,4′-Methylenebis(*o*-chloroaniline) glucoside
In: Dog D847

4,4′-Methylenebis(*o*-chloroaniline) glucoside
Formed from: 4,4′-Methylenebis(*o*-chloroaniline)

4,4′-Methylenebis((N,N-dimethyl)aniline) (reduced Michler's ketone)
→ N,N-Dimethyl-4,4′ methylenedianiline
In: Rat C362, C364
→ N,N′-Dimethyl-4,4′ methylenedianiline
In: Rat C362, C364

3,3′-Methylenebis(4-hydroxy-2*H*-1-benzopyran-2-one
See: Dicoumarol

2,2′-Methylenebis(3,4,6-trichlorophenol)
See: Hexachlorophene

Methylene blue (3,7-bis(dimethylamino) phenothiazin-5-ium)
→ Leucomethylene blue
In: Clostridium D260
Eubacterium B426
Man A421, A530, A3008, B260

4,5-Methylenechrysene
Formed from: 5-Methylchrysene

4,4′-Methylenedianiline
Probably formed from: M-Methyl-4,4′-methylenedianiline
→ N,N′-Diacetyl-4,4′-(hydroxymethylene) dianiline
Probably in: Rat C632, C634

3,4-Methylenedioxyamphetamine
Formed from: N-Ethyl-3,4-methylenedioxyamphetamine
N-Hydroxy-3,4-methylenedioxyamphetamine
N-Methyl-3,4-methylenedioxyamphetamine
→ 3,4-Dihydroxyamphetamine
In: Dog A2077, A3465
Man K475
Monkey A2077, A3465
Rabbit F761, G268
Rat A1570, A2964
→ 3,4-Methylenedioxyphenylacetone
In: Monkey A3465
→ 3,4-Methylenedioxybenzyl methyl ketoxime
In: Guinea pig A1490
Rabbit A1490

3,4-Methylenedioxyanisole
→ 4-Methoxycatechol
In: Rat A3947

3,4-Methylenedioxybenzaldehyde
See: Piperonal

1,2-Methylenedioxybenzene
→ Catechol
In: Rabbit F761, G268
Rat A3947
→ 3,4-Methylenedioxyphenol
In: Rabbit G76

3,4-Methylenedioxybenzoic acid
See: Piperonylic acid

3,4-Methylenedioxybenzonitrile
→ 3,4-Dihydroxybenzonitrile
In: Rat A3947
→ Piperonylic acid
In: Rhodococcus H952

3,4-Methylenedioxybenzyl alcohol
See: Piperonyl alcohol

3,4-Methylenedioxybenzylamine
→ Piperonal
Probably in: Rat G865

5-(3,4-Methylenedioxybenzyl)hydantoin (D and L isomers)
Formed from: N-Carbamoyl-3,4-methylenedioxy-D-phenylalanine
N-Carbamoyl-3,4-methylenedioxy-L-phenylalanine
→ N-Carbamoyl-3,4-methylenedioxy-D-phenylalanine
In: Flavobacterium F938
→ N-Carbamoyl-3,4-methylenedioxy-L-phenylalanine

In: Flavobacterium F938
→ 3,4-Methylenedioxy-D-phenylalanine
In: Pseudomonas E391, E450
→ 3,4-Methylenedioxy-L-phenylalanine
In: Pseudomonas F940

3,4-Methylenedioxybenzyl methyl ketoxime
Formed from: 3,4-Methylenedioxyamphetamine
→ 3,4-Methylenedioxyphenylacetone
Probably in: Guinea pig A1490
Rabbit A1490

1-(3,4-Methylenedioxybenzyl)-4-(2-pyrimidinyl) piperazine
See: Piribedil

3,4-Methylenedioxycinnamaldehyde
Formed from: 3,4-Methylenedioxycinnamic acid

3,4-Methylenedioxycinnamic acid
Formed from: 3′-Hydroxyisosafrole
3,4-Methylenedioxyphenylpropionic acid
→ 3,4-Methylenedioxycinnamaldehyde
In: Neurospora A750
→ 3,4-Methylenedioxycinnamoyl CoA
In: Pseudomonas B298
→ 3,4-Methylenedioxyphenylhydracrylic acid
In: Rat C164
→ 3,4-Methylenedioxyphenylpropionic acid
In: Rat C164
→ Piperonylic acid
Probably in: Mouse E235

3,4-Methylenedioxycinnamoyl CoA
Formed from: 3,4-Methylenedioxycinnamic acid

3,4-Methylenedioxymandelic acid
Formed from: N,N-Dimethyl-2-(3,4-methylenedioxyphenyl)-2-oxoethylamine
2-(3,4-Methylenedioxyphenyl)-2-oxoethylamine
→ 3,4-Methylenedioxyphenylglyoxylic acid
In: Rat C164

3,4-Methylenedioxymandelonitrile
Formed from: Piperonal

N-Methyl-3,4-methylenedioxyamphetamine
→ 3,4-Dihydroxyamphetamine
In: Man J264
Rat F565
→ 3,4-Dihydroxy-N-methylamphetamine
In: Man F187, J264
Rat F565
→ 3,4-Methylenedioxyphenylacetone
In: Man F187

2,3-Methylenedioxynaphthalene
→ 2,3-Dihydroxynaphthalene
In: Rat A3947

4,5-Methylenedioxy-2-nitroanisole
→ 4-Methoxy-5-nitrocatechol
In: Rat A3947

3,4-Methylenedioxynitrobenzene
→ 4-Nitrocatechol
In: Rat A3947, C674

3,4-Methylenedioxy-1-phenanthrenecarboxylic acid
Formed from: Aristolochic acid

3,4-Methylenedioxyphenethylamine
→ 3,4-Dihydroxyphenethylamine
In: Rat A3603
→ 3,4-Methylenedioxyphenylacetic acid
In: Rat A3603, G865

3,4-Methylenedioxyphenol (sesamol)
Formed from: 1,2-Methylenedioxybenzene
→ 2,2′-Dihydroxy-4,5;4′,5′-bis (methylenedioxy)biphenyl
In: Horseradish B948
→ Hydroxyquinol
In: Rat A3947

3,4-Methylenedioxyphenylacetic acid
Formed from: 3,4-Methylenedioxyphenethylamine
→ 3,4-Methylenedioxyphenylacetylglycine
In: Rat C164

3,4-Methylenedioxyphenylacetone
Formed from: N-Ethyl-3,4-methylenedioxyamphetamine
3,4-Methylenedioxybenzyl methyl ketone
N-Methyl-3,4-methylenedioxyamphetamine
→ 3,4-Dihydroxyphenylacetone
Probably in: Rat E988
→ 4-(3,4-Methylenedioxyphenyl)butan-2-ol
In: Rabbit B188
→ Piperonylic acid
Probably in: Man H736

3,4-Methylenedioxyphenylacetylglycine
Formed from: 3,4-Methylenedioxyphenylacetic acid
3,4-Methylenedioxy-D-phenylalanine
Formed from: 5-(3,4-Methylenedioxybenzyl)hydantoin
3,4-Methylenedioxy-L-phenylalanine
Formed from: N-Carbamoyl-3,4-methylenedioxy-L-phenylalanine
5-(3,4-Methylenedioxybenzyl)hydantoin
4-(3,4-Methylenedioxyphenyl)butan-2-ol
Formed from: 3,4-Methylenedioxyphenylacetone
3,4-Methylenedioxyphenylglyoxylic acid
Formed from: 3,4-Methylenedioxymandelic acid
→ Piperonylic acid
Probably in: Rat C164
N-(1-(2,3-Methylenedioxyphenyl)isopropyl) noradrenaline
→ 3-Hydroxy-4-methoxy-N-(1-(2,3-methylenedioxyphenyl)isopropyl) phenylethanolamine
In: Rat A486
→ N-(1-(2,3-Methylenedioxyphenyl) isopropyl)normetanephrine
In: Rat A486
1-(3,4-Methylenedioxyphenyl)glycerol
Probably formed from: Safrole glycol
3,4-Methylenedioxyphenylhydracrylic acid
Formed from: 3,4-Methylenedioxycinnamic acid
→ Piperonylic acid
Probably in: Rat C164
N-(1-(2,3-Methylenedioxyphenyl)isopropyl) normetanephrine
Formed from: N-(1-(2,3-Methylenedioxyphenyl)isopropyl)noradrenaline
3,4-Methylenedioxyphenyllactic acid
Probably formed from: Safrole glycol
→ 3,4-Dihydroxyphenyllactic acid
Probably in: Guinea pig A1544
Rat A1544
3,4-Methylenedioxyphenylpropionic acid
Formed from: 3,4-Methylenedioxycinnamic acid
→ 3,4-Methylenedioxycinnamic acid
In: Rat C164
2-(3,4-Methylenedioxyphenyl)-2-oxoethylamine
→ 3,4-Methylenedioxymandelic acid
In: Rat J287
1,2-Methylenedioxy-4-propenylbenzene
See: Isosafrole
1,2-Methylenedioxy-4-propylbenzene
See: Dihydrosafrole
3,4-Methylenedioxytoluene
→ 4-Methylcatechol
In: Rat A3947
3-Methylene-3*H*-indole
See: 3-Methyleneindolenine
3-Methyleneindole-2-carboxylic acid
Formed from: Carbazole
3-Methyleneindolenine (3-methylene-3*H*-indole)
Formed from: 3-Hydroxymethylindole
Indole-3-acetic acid
Skatole
→ 3-Hydroxymethylindole
In: Zea A727
→ 3-Hydroxymethyloxindole
Probably in: Bean A635
Pea A635
Zea A727
3-Methyleneoxindole (1,3-dihydro-3-methylene-2*H*-indol-2-one)
Probably formed from:
3-Hydroxymethyloxindole
Indole-3-acetic acid
5,10-Methylenetetrahydrofolic acid
Formed from: Folic acid
5-Methyltetrahydrofolic acid
→ 10-Formyltetrahydrofolic acid
In: Saccharomyces H794
→ 5,10-Methenyltetrahydrofolic acid
In: Clostridium K928
Man H646
Pea A1429
Peptostreptococcus F869
→ 5,10-Methylenetetrahydropteroyldiglutamic acid
In: Pig K455
→ 5-Methyltetrahydrofolic acid
In: Clostridium D260
→ Tetrahydrofolic acid
In: Streptococcus K837

5,10-Methylenetetrahydropteroyldiglutamic acid
Formed from: 5,10-Methylenetetrahydrofolic acid

5,10-Methylenetetrahydropteroylheptaglutamic acid
Formed from: 5-Methyltetrahydropteroyl-heptaglutamic acid

5,10-Methylenetetrahydropteroylpentaglutamic acid
Formed from: 5-Methyltetrahydropteroyl-pentaglutamic acid

5,10-Methylenetetrahydropteroyltriglutamic acid
Formed from: 5-Methyltetrahydropteroyl-triglutamic acid

N-Methylephedrine
Formed from: 2-(Dimethylamino) propiophenone
→ Ephedrine
In: Man A504
Rat F481, F501
→ *p*-Hydroxy-N-methylephedrine
In: Rat F481
→ N-Methylephedrine-N-oxide
In: Man F481
Rat F481, F501

N-Methylephedrine-N-oxide
Formed from: N-Methylephedrine
→ *p*-Hydroxy-N-methylephedrine-N-oxide
Probably in: Rat F481

N-Methylephedrin-4-yl-β-D-glucuronide
Formed from: *p*-Hydroxy-N-methylephedrine

4″-O-Methylepigallocatechin gallate
Probably formed from: (−)-Epigallocatechin gallate

1-(((8β)-6-Methylergolin-8-yl)methyl)-2,4-imidazolidinedione
Formed from: 1-(((8β)-6-Methylergolin-8-yl)methyl)-2,4-imidazolidinedione-N-oxide

1-(((8β)-6-Methylergolin-8-yl)methyl)-2,4-imidazolidinedione-N-oxide
→ 1-(((8β)-6-Methylergolin-8-yl)methyl)-2,4-imidazolidinedione
In: Rat G807

N-Methyl-9,10-ethanoanthracene-9(10*H*)-propanamine
See: Maprotiline

2-(1-Methylethoxy)phenol methylcarbamate
See: Propoxur

3-((4-(2-(1-Methylethoxy)phenyl)-1-piperazinyl)methyl)benzamide
Formed from: Mazapertine

3-((4-(2-(1-Methylethoxy)phenyl)-1-piperazinyl)methyl)benzoic acid
Formed from: Mazapertine

1-((1-Methylethyl)amino)-3-(1-naphthalenyloxy)-2-propanol
See: Propranolol

Methyl 3-(3-ethylvaleryl)-4-hydroxy-1-(4-hydroxy-3-methoxyphenyl)-6,7,8-trimethoxy-2-naphthoate
Formed from: Methyl 1-(3,4-dimethoxyphenyl)-3-(3-ethylvaleryl)-4-hydroxy-6,7,8-trimethoxy-2-naphthoate
→ Methyl 3-(3-ethylvaleryl)-4-hydroxy-1-(4-hydroxy-3-methoxyphenyl)-6,7,8-trimethoxy-2-naphthoate-β-D-glucuronide
Probably in: Dog J839
Rat J839

Methyl 3-(3-ethylvaleryl)-4-hydroxy-1-(4-hydroxy-3-methoxyphenyl)-6,7,8-trimethoxy-2-naphthoate-β-D-glucuronide
Probably formed from: Methyl 3-(3-ethylvaleryl)-4-hydroxy-1-(4-hydroxy-3-methoxyphenyl)-6,7,8-trimethoxy-2-naphthoate

7-O-Methylfenoldopam
Formed from: Fenoldopam

8-O-Methylfenoldopam
Formed from: Fenoldopam

Methyl ferulate
Formed from: Ferulic acid
Methyl caffeate
→ 3-Carboxymethyl-5-(2-carboxymethylethenyl)-2-(4-hydroxy-3-methoxyphenyl)-7-methoxybenzofuran
In: Horseradish F123
→ Ferulic acid
In: Aspergillus G915, H381, K158
Microorganism H589
Penicillium G650, G805, J499
Streptomyces G363
→ Methyl 3,4-dihydroxycinnamate
In: Lentinus A214

3-Methylflavone-8-carboxylic acid
Formed from: Flavoxate
1-Methylfluorene
→ Fluorene-1-carboxylic acid
In: Pseudomonas H672
9-Methylfluorene
→ 9-Methylfluorene-9-hydroperoxide
In: Rat A1955
***d*-α-Methylfluorene-2-acetic acid**
Formed from: *l*-α-Methylfluorene-2-acetic acid
→ 9-Hydroxy-9-methoxy-α-methylfluorene-2-acetic acid
In: Rat A2353
→ 7-Hydroxy-α-methylfluorene-2-acetic acid
In: Rat A2353
→ 9-Hydroxy-α-methylfluorene-2-acetic acid
In: Rat A2353
***l*-α-Methylfluorene-2-acetic acid** (cicloprofen; this isomer is assumed unless otherwise stated)
→ 9-Hydroxy-9-methoxy-α-methylfluorene-2-acetic acid
In: Rat A2353, A2718
→ 7-Hydroxy-α-methylfluorene-2-acetic acid
In: Rat A2353, A2718
→ 9-Hydroxy-α-methylfluorene-2-acetic acid
In: Rat A2353, A2718
→ *d*-α-Methylfluorene-2-acetic acid
In: Dog A1696, A1767
Man A1767
Monkey A1696, A1767, A2111
Rat A1696, A1767, A2111
9-Methylfluorene-9-hydroperoxide
Formed from: 9-Methylfluorene
→ 9-Hydroxy-9-methylfluorene
Probably in: Rat A1955
4-Methylformamido-3,5-xylyl methylcarbamate
Formed from: Mexacarbate
***p*-Methylformanilide**
Formed from: *p*-Toluidine
Methyl gallate
→ Gallic acid
With tannase K911
→ Methyl gallate O-sulphate
In: Eubacterium G475
Methyl gallate O-sulphate
Formed from: Methyl gallate
Methyl 2-(S-glutathionyl)-5-nitrobenzoate
Formed from: Methyl 2-chloro-5-nitrobenzoate
2-Methylglycyl-2,6-dimethylanilide
→ 2,6-Dimethylaniline
In: Mouse A830
N-Methylglycyl-*o*-toluidide
→ *o*-Toluidide
In: Mouse A830
N-Methylgroenlandicine
Formed from: Groenlandicine
9-Methylharman
Formed from: 9-Methylnorharman
***o*-Methylhippuric acid**
Formed from: *o*-Toluic acid
***m*-Methylhippuric acid**
Formed from: *m*-Toluic acid
***p*-Methylhippuric acid**
Formed from: *p*-Toluic acid
Methyl 5-hydroxybenzimidazol-2-ylcarbamate
Probably formed from: Methyl benzimidazol-2-ylcarbamate
Methyl *p*-hydroxybenzoate
Formed from: Methyl *p*-anisate
→ Methyl *p*-hydroxybenzoate-β-D-glucuronide
In: Man G524
Methyl *p*-hydroxybenzoate-β-D-glucuronide
Formed from: Methyl *p*-hydroxybenzoate
Methyl (5-(α-hydroxybenzyl)-1*H*-benzimidazol-2-yl)carbamate
Formed from: Mebendazole
Methyl 5(6)-(*threo*-1-hydroxybutyl)-2-benzimidazolecarbamate
Formed from: Methyl 5(6)-butyl-2-benzimidazolecarbamate
→ Methyl 5(6)-(1,2-dihydroxybutyl)-2-benzimidazolecarbamate
Probably in: Beef A1066
Sheep A1066
Methyl 5(6)-(2-hydroxybutyl)-2-benzimidazolecarbamate
Formed from: Methyl 5(6)-butyl-2-benzimidazolecarbamate
Methyl 5(6)-(4-hydroxybutyl)-2-benzimidazolecarbamate
→ Formed from: Methyl 5(6)-butyl-2-

benzimidazolecarbamate
In: Beef A1066
Cunninghamella A1066
Sheep A1066

Methyl 3-hydroxy-4,5-dimethoxycinnamate
Probably formed from: Methyl 5-methoxycaffeate

Methyl 2-hydroxy-2-(4-(2-hydroxy-3-isopropylamonopropoxy)phenyl)ethylcarbamate
Formed from: Pamatolol

Methyl 3-(4-(2-hydroxy-3-isopropylaminopropoxy) phenyl)propionate
See: Esmolol

Methyl *o*-(2-hydroxy-3-((N-isopropyl)amino) propoxy)benzoate
→ *o*-(2-Hydroxy-3-((N-isopropyl)amino) propoxy)benzoic acid
In: Rabbit H563

Methyl *p*-(2-hydroxy-3-((N-isopropyl)amino) propoxy)benzoate
→ *p*-(2-Hydroxy-3-((N-isopropyl)amino) propoxy)benzoic acid
In: Rabbit H563

Methyl *o*-(2-hydroxy-3-((N-isopropyl)amino) propoxy)phenylpropionate
→ *o*-(2-Hydroxy-3-((N-isopropyl)amino) propoxy)phenylpropionic acid
In: Rabbit H563

Methyl 3-hydroxy-4-methoxyphenylacetate
Formed from: Methyl 3,4-dihydroxyphenylacetate

Methyl 4-hydroxy-3-methoxyphenylacetate
Formed from: Methyl 3,4-dihydroxyphenylacetate

Methyl 3-hydroxymethyl-2,6-dimethyl-4-(*m*-nitrophenyl)pyridine-5-carboxylate
Probably formed from: 2,6-Dimethyl-4-(*m*-nitrophenyl)pyridine-3,5-dicarboxylic acid 3-methyl ester
Methyl 3-hydroxy-2-naphthoate
→ 2,2′-Dihydroxy-1,1′-binaphthyl-3,3′-dicarboxylic acid dimethyl ester
Probably in: Horseradish J592

Methyl 2-hydroxy-6-oxo-6-phenylhexa-2,4-dienoate
→ Methyl 2,6-dioxo-6-phenylhexanoate
In: Pseudomonas E156

Methyl 5-hydroxy-6-oxo-α-phenyl-2-piperidineacetate
Probably formed from: Methyl 6-oxo-α-phenyl-2-piperidineacetate
→ Methyl 5-hydroxy-6-oxo-α-phenyl-2-piperidineacetate-β-D-glucuronide
In: Dog C14

Methyl 5-hydroxy-6-oxo-α-phenyl-2-piperidineacetate-β-D-glucuronide
Formed from: Methyl 5-hydroxy-6-oxo-α-phenyl-2-piperidineacetate

Methyl 2-hydroxy-3-phenylbenzoate
→ 2-Hydroxy-3-phenylbenzoic acid
In: Phanerochaete D768

Methyl 2-hydroxy-3-phenylbutyrate
Formed from: 2-Oxo-3-phenylbutyrate

Methyl *p*-hydroxyphenyllactate
→ *p*-Hydroxyphenyllactic acid
In: Rat H904

Methyl α-(*p*-hydroxyphenyl)-6-oxo-2-piperidineacetate
Probably formed from: Methyl 6-oxo-α-phenyl-2-piperidineacetate
→ Methyl α-(*p*-hydroxyphenyl)-6-oxo-2-piperidineacetate-β-D-glucuronide
Probably in: Dog C14

Methyl α-(*p*-hydroxyphenyl)-6-oxo-2-piperidineacetate-β-D-glucuronide
Probably formed from: Methyl α-(*p*-hydroxyphenyl)-6-oxo-2-piperidineacetate
Formed from: Oxfendazole

Methyl α-(*p*-hydroxyphenyl)-2-piperidineacetate
Formed from: Methylphenidate

Methyl 5(6)-(4′-hydroxyphenylthio) benzimidazole-2-carbamate
Formed from: Oxfendazole
→ Methyl 5(6)-(4′-hydroxyphenylthio) benzimidazole-2-carbamate-β-D-glucuronide
In: Beef A1702
→ Methyl 5(6)-(4′-hydroxyphenylthio) benzimidazole-2-carbamate-O-sulphate
In: Beef A1702

Methyl 5(6)-(4′-hydroxyphenylthio) benzimidazole-2-carbamate-β-D-glucuronide
Formed from: Methyl 5(6)-(4′-hydroxyphenylthio)benzimidazole-2-carbamate

Methyl 5(6)-(4′-hydroxyphenylthio) benzimidazole-2-carbamate-O-sulphate
Formed from: Methyl 5(6)-(4′-hydroxyphenylthio)benzimidazole-2-carbamate

Methyl 5-((2-hydroxypropyl)sulphonyl)-1*H*-benzimidazol-2-yl)carbamate
Probably formed from: Albendazole sulphone

Methyl 5-((3-hydroxypropyl)sulphonyl)-1*H*-benzimidazol-2-yl)carbamate
Probably formed from: Albendazole sulphone

Methyl (6-hydroxy-5-(propylsulphonyl)-1*H*-benzimidazol-2-yl)carbamate
Probably formed from: Albendazole sulphone

4,4′,4″-Methylidynetris(N,N-dimethylbenzenamine)
See: Leucogentian violet

1,2-(2-(5-Methyl-4-imidazolylmethylthio) ethylamino)-5-(3,4-methylenedioxybenzyl)-4-pyrimidone
See: Oxmetidine

2-Methylimino-1-phenylpropan-1-ol
Formed from: D-Ephedrine
L-Ephedrine
D-Pseudoephedrine
L-Pseudoephedrine

3-Methylimino-5-phenyl-1,2,4-thiazole
Formed from: 1-Methyl-3-thiobenzoylthiourea

2′-Methylimipramine
See: Trimipramine

2-Methylindane
→ 1-Hydroxy-2-methylindane
In: Pseudomonas K207

2-Methylindan-1-one
Probably formed from: 1-Hydroxy-2-methylindane
→ 1-Hydroxy-2-methylindan-3-one
Probably in: Pseudomonas K207

2-Methylindole
Probably formed from: 2-Methyl-L-tryptophan
→ 2-Methyl-L-tryptophan
In: Escherichia H617

3-Methylindole
See: Skatole

5-Methylindole
→ 5-Methylanthranilic acid
In: Tecoma C170
→ 5-Methyl-L-tryptophan
In: Carrot B874
Enterobacter A781
Erwinia A781
Escherichia A781
Kluyvera A781
Nicotiana B874
Proteus A781
See also: A781

7-Methylindole
→ 7-Methyl-L-tryptophan
Probably in: Carrot B874
Nicotiana B874

5-Methylindole-3-acetaldehyde
Formed from: 5-Methyltryptamine

7-Methylindole-3-acetaldehyde
Formed from: 7-Methyltryptamine

7-Methylindole-3-acetic acid
Formed from: 7-Methyltryptamine

2-Methylindole-3-methanol
Formed from: 2,3-Dimethylindole

2-Methylindole-3-pyruvic acid
Formed from: Indole-3-pyruvic acid

5-Methylindole-3-pyruvic acid
Formed from: 5-Methyl-L-tryptophan

Methyl (3-indolylmethyl)dithiocarbamate sulphoxide
Formed from: Brassinine

5-Methyl-3-isoazolecarboxylic acid 2-(phenylmethyl)hydrazide
See: Isocarboxazide

N-Methylisococlaurine
Formed from: Isococlaurine

3-O-Methylisoetharine
Formed from: Isoetharine
→ 3-O-Methylisoetharine-β-D-glucuronide
Probably in: Dog A286
Man A286
→ 3-O-Methylisoetharine-O-sulphate
Probably in: Dog A286
Man A286

3-O-Methylisoetharine-β-D-glucuronide
Probably formed from: 3-O-Methylisoetharine

3-O-Methylisoetharine-O-sulphate
Probably formed from: 3-O-Methylisoetharine

Methyl isoferulate
Formed from: Methyl caffeate
Methyl isophthalic acid
Formed from: 3,5-Dimethylbenzoic acid
N-Methylisoquinolinium
Formed from: Isoquinoline
N-Methylisosalsoline (1,2,3,4-tetrahydro-7-hydroxy-6-methoxy-1,2-dimethylisoquinoline)
Formed from: 1,2,3,4-Tetrahydro-7-hydroxy-6-methoxy-1-methylisoquinoline
N[1]-(5-Methyl-3-isoxazolyl)sulphanilamide
See: Sulphamethoxazole
7-O-Methylkievitone-2′,3′-epoxide
Formed from: 7-O-Methyl-2′,3′-dehydrokievitone
→ 2′,3′-Dihydroxy-7-O-methylkievitone
Probably in: Botrytis F246
β-Methyl-L-kynurenine
→ Anthranilic acid
In: Pseudomonas K233
6-O-Methyllaudanosoline (1-((3,4-dihydroxyphenyl)methyl)-1,2,3,4-tetrahydro-7-hydroxy-6-methoxyisoquinoline)
→ Tetrahydro-2,9,10-trihydroxy-3-methoxy-protoberberine
In: Eschscholtzia H680
2′-O-Methyllicodione
Formed from: Licodione
5-O-Methyllupiwighteone
Formed from: Lupiwighteone
7-O-Methylluteone (3-(2,4-dihydroxyphenyl)-5-hydroxy-7-methoxy-6-(3-methyl-2-butenyl)-4*H*-1-benzopyran-4-one)
→ 7-O-Methylluteone oxide
In: Botrytis K456
7-O-Methylluteone oxide
Formed from: 7-O-Methylluteone
***o*-Methylmandelic acid**
Formed from: *o*-Methylphenylglycol
→ *o*-Methylphenylglyoxylic acid
Probably in: Rat C153
***m*-Methylmandelic acid**
Formed from: *m*-Methylphenylglycol
→ *m*-Methylphenylglyoxylic acid
Probably in: Rat C153
***p*-Methylmandelic acid**
Formed from: *p*-Methylphenylglycol
→ *p*-Methylphenylglyoxylic acid
In: Rhodotorula H90
See also: C153, C681
***m*-Methylmandelonitrile**
→ *m*-Tolualdehyde
With hydroxynitrile lyase H944
***p*-Methylmandelonitrile**
Formed from: *p*-Tolualdehyde
→ *p*-Tolualdehyde
With hydroxynitrile lyase H944
α-Methylmandelonitrile
Formed from: Acetophenone
N-Methylmedetomidine
Formed from: Medetomidine
1-Methylmelatonin
Formed from: Melatonin
2-Methylmercaptoaniline
Formed from: Benzothiazole
2-Hydroxybenzothiazole
→ 2-Methylsulphinylaniline
Probably in: Guinea pig G15, J9
2-Methylmercaptobenzimidazole
→ 2-Methylsulphinylbenzimidazole
In: Mouse D446
Pig D446
α-Methylmercaptophenylacetic acid
Formed from: Phenthoate
N-Methylmesothalictricavine
Formed from: Mesothalictricavine
N-Methylmetanephrine
Formed from: D-Metanephrine
N-Methyladrenaline
Methyl 5-methoxycaffeate
→ Methyl 3-hydroxy-4,5-dimethoxycinnamate
Probably in: Lentinus A214
Methyl *p*-methoxycinnamate
Formed from: Methyl *p*-coumarate
Methyl (*E*)-2-methoxyimino-2-(2-(*o*-tolyloxymethyl)phenyl)acetate
→ (*E*)-2-Methoxyimino-2-(2-(*o*-tolyloxymethyl)phenyl)acetic acid
In: Goat H728
Hen H728
Pig H728
Rat H728
3′-Methyl-4-methylaminoazobenzene
Formed from: 4-Dimethylamino-3′-

methylazobenzene
→ N-Acetyl-3′-methyl-4-methylaminoazobenzene
Probably in: Rat E872, G9
→ 4-Amino-3′-methylazobenzene
In: Rat A3835, B27, E872
See also: A673
→ 3′-Hydroxymethyl-4-methylaminoazobenzene
In: Rat B9, B27
See also: A3835, G9
→ 4-Hydroxy-3-methyl-4′-methylaminoazobenzene
In: Rat B9, B27
See also: G9
→ *m*-Toluidine
In: Rat B27
See also: A3963

4′-Methyl-4-methylaminoazobenzene
Formed from: 4′-Dimethyl-4-methylaminoazobenzene
→ 4-Amino-4′-methylazobenzene
In: Rat A1979

Methyl *p*-methylaminobenzoate
Formed from: Methyl *p*-dimethylaminobenzoate

α-(1-Methyl-2-methylaminoethyl)-α-phenylbenzeneethanol
Formed from: α-(2-Dimethylamino-1-methylethyl)-α-phenylbenzeneethanol

3-Methyl-1-(*o*-((3-methylamino)propyl)thio) phenylurea
Formed from: 1-(*o*-((3-Dimethylamino) propyl)thio)phenyl-3-methylurea
→ 3-Methyl-1-(*o*-((3-methylhydroxylamino) propyl)thio)phenylurea
Probably in: Dog A436

N-Methyl-4-(3-methylbuta-1,3-dienyl)-L-tryptophan
→ Elymoclavine
In: Claviceps H7

4-Methyl-5-(N-methyl)carboxamido-9-anilinoacridine
Formed from: Amsacrine
→ 5′-(S-Glutathionyl)-4-methyl-5-(N-methyl)carboxamido-9-anilinoacridine
Probably in: Mouse G236
Rat G236
→ 6′-(S-Glutathionyl)-4-methyl-5-(N-methyl)carboxamido-9-anilinoacridine
Probably in: Mouse G236
Rat G236
→ 9-(S-Glutathionyl)-4-methyl-5-(N-methyl)carboxamido-9-anilinoacridine
Probably in: Mouse G236
Rat G236
→ 4-Hydroxymethyl-5-(N-methyl) carboxamido-9-anilinoacridine
In: Mouse G236
Rat G236

N-Methyl-N-(1-methyl-3,3-diphenylpropyl) formamide
→ N-Hydroxymethyl-N-(1-methyl-3,3-diphenylpropyl)formamide
In: Rat G88
→ N-(3-(*m*-Hydroxyphenyl)-1-methyl-3-phenylpropyl)-N-methylformamide
In: Rat G88
→ N-(3-(*p*-Hydroxyphenyl)-1-methyl-3-phenylpropyl)-N-methylformamide
In: Rat G88

N-Methyl-4,4′-methylenedianiline
Probably formed from: N,N′-Dimethyl-4,4′-methylenedianiline
→ 4,4′-Methylenedianiline
Probably in: Rat C362, C364

N-Methyl-3,4-methylenedioxyamphetamine (ecstasy)
→ 3,4-Dihydroxy-N-methylamphetamine
In: Man F187, K475
Rabbit F761, G268
Rat E988, F565, F805
See also: G145, G354
→ Hydroxy-N-methyl-3,4-methylenedioxyamphetamine
In: Rabbit G76
→ 3,4-Methylenedioxyamphetamine
In: Man K475
Rat E988, F565
→ 3,4-Methylenedioxyphenylacetone
In: Man F187
Rat E988

2-((1*R*,6*R*)-3-Methyl-6-(1-methylethenyl)-2-cyclohexen-1-yl)-5-pentyl-1,3-benzenediol
See: Cannabidiol

2-Methyl-5-(1-methylethyl)-2,5-cyclohexadiene-1,4-dione
See: Thymoquinone

3-Methyl-6-(1-methylethylidene)-2-cyclohexen-1-one
See: Piperitenone

3-Methyl-1-(*o*-((3-methylhydroxylamino)propyl)thio)phenylurea
Probably formed from: 3-Methyl-1-(*o*-((3-methylamino)propyl)thio)phenylurea

Methyl 3-methyl-4-nitrophenyl phosphorothioate
Formed from: Fenitrothion

2-Methyl-2-(4′-methyl-3′-pentenyl)-7-pentyl-2*H*-1-benzopyran-5-ol
See: Cannabichromene

α-Methyl-α-(*p*-methylphenyl)malonic acid
→ 2-(*p*-Methylphenyl)propionic acid
In: Alcaligenes G716

1-Methyl-4-(*o*-methylphenyl)pyridinium
Probably formed from: 1,2,3,4-Tetrahydro-1-methyl-4-(*o*-methylphenyl)pyridine

1-Methyl-4-(*m*-methylphenyl)pyridinium
Probably formed from: 1,2,3,4-Tetrahydro-1-methyl-4-(*m*-methylphenyl)pyridine

1-Methyl-4-(*p*-methylphenyl)pyridinium
Probably formed from: 1,2,3,4-Tetrahydro-1-methyl-4-(*p*-methylphenyl)pyridine

α-Methyl-N-(2-methylphenyl)-1-pyrrolidineacetamide
See: Aptocaine

3-Methyl-1-(*p*-methylphenyl)triazene
Formed from: 3,3-Dimethyl-1-(*p*-methylphenyl)triazene

α-Methyl-4-(2-methylpropyl)benzeneacetamide
See: Ibuprofen amide

***trans*-β-Methyl-*p*-(1-methyl-2-propynyloxy)cinnamic acid**
Formed from: Ethyl *trans*-β-methyl-*p*-(1-methyl-2-propynyloxy)cinnamate
→ β-Methyl-*p*-(1-methyl-2-propynyloxy)propionic acid
Probably in: Baboon C409
Dog C409
Man C409
Marmoset C409
Mouse C409
Rabbit C409
Rat C409
→ 2-(*p*-(1-Methyl-2-propynyloxy)propane
Probably in: Baboon C409
Dog C409
Man C409
Marmoset C409
Mouse C409
Rabbit C409
Rat C409

β-Methyl-*p*-(1-methyl-2-propynyloxy)propionic acid
Probably formed from: *trans*-β-Methyl-*p*-(1-methyl-2-propynyloxy)cinnamic acid

Methyl (5-(methylsulphinyl)-1*H*-benzimidazol-2-yl)carbamate
Formed from: Albendazole

2-Methyl-5-methylsulphonyl-7-nitro-1-propylbenzimidazole
Probably formed from: Nitralin

2-Methyl-7-methylsulphonyl-5-nitroquinoxaline
Probably formed from: Nitralin
→ 2-Methyl-7-methylsulphonyl-5-nitrosoquinoxaline
Probably in: Rat C750

2-Methyl-7-methylsulphonyl-5-nitrosoquinoxaline
Probably formed from: 2-Methyl-7-methylsulphonyl-5-nitroquinoxaline
→ 5-Hydroxylamino-2-methyl-7-methylsulphonylquinoxaline
Probably in: Rat C750

3-Methyl-4-(methylthio)catechol
Formed from: 3-Methyl-4-(methylthio)phenol

4-Methyl-5-(methylthio)catechol
Formed from: 3-Methyl-4-(methylthio)phenol

3-Methyl-4-(methylthio)phenol
→ 3-Methyl-4-(methylthio)catechol
In: Nocardia A3799
→ 4-Methyl-5-(methylthio)catechol
In: Nocardia A3799

N-Methylmexiletine
Formed from: Mexiletine

N-Methylmogadon
See: Nimetazepam

d-3-Methylmorphinan
Formed from: d-3,N-Dimethylmorphinan
17-Methylmorphinan-3-ol
See: Levorphanol
3-O-Methylmorphine
→ 3-O-Methylmorphine-6-β-D-glucuronide
In: Man K517
3-O-Methylmorphine-6-β-D-glucuronide
Formed from: 3-O-Methylmorphine
N-Methylnaltrexone
See: Naltrexonium
N-Methylnandinine
Formed from: Nandinine
1-Methylnaphthalene
→ 1,2-Dihydro-1,2-dihydroxy-8-methylnaphthalene
In: Pseudomonas H252
→ 2-Hydroxy-3-methylbenzaldehyde
In: Pseudomonas H252
→ 2-Hydroxy-3-methylbenzoic acid
In: Pseudomonas K496
→ 2-Hydroxy-3-methylchromene-2-carboxylic acid
In: Pseudomonas H252
→ 1-Hydroxymethylnaphthalene
In: Agmenellum D29, D297
Anabaena D29
Cunninghamella D33
Oscillatoria D29
2-Methylnaphthalene
→ *cis*-1,2-Dihydro-1,2-dihydroxy-3-methylnaphthalene
In: Pseudomonas F641
→ *trans*-1,2-Dihydro-1,2-dihydroxy-3-methylnaphthalene
In: Rat C616, D833
Trout D833
→ *trans*-1,2-Dihydro-1,2-dihydroxy-6-methylnaphthalene
In: Rat C616, D833
Trout D833
→ *cis*-1,2-Dihydro-1,2-dihydroxy-7-methylnaphthalene
In: Pseudomonas F641
→ *trans*-1,2-Dihydro-1,2-dihydroxy-7-methylnaphthalene
In: Rat C616, D833
Trout D833
See also: B988, B905
→ 4-Formyl-2-hydroxybenzoic acid
In: Pseudomonas H252
→ 2-Hydroxy-4-hydroxymethylbenzaldehyde
In: Pseudomonas H252
→ 2-Hydroxy-7-hydroxymethylchromene-2-carboxylic acid
In: Pseudomonas H252
→ 2-Hydroxy-4-methylbenzoic acid
In: Pseudomonas K496
→ 2-Hydroxymethylnaphthalene
In: Agmenellum D29, D297
Anabaena D29
Cunninghamella D33
Oscillatoria D29
Rat C616, D833
Trout D833
→ 7-Hydroxymethyl-1,2-naphthoquinone
In: Pseudomonas H252
→ 7-Methyl-α-naphthol
In: Guinea pig C592
→ S-(7-Methyl-1-naphthyl)glutathione
Probably in: Guinea pig C592
→ 2-Naphthoic acid
In: Guinea pig C592
Rat B986
Trout D534
1-Methylnaphthalene-1,2-oxide
→ S-(1,2-Dihydro-1-hydroxy-1-methylnaphth-2-yl)glutathione
Probably in: Sheep A2282
2-Methylnaphthalene-1,2-oxide
→ S-(1,2-Dihydro-2-hydroxy-2-methylnaphth-2-yl)glutathione
Probably in: Sheep A2282
Methyl 1-naphthoate
→ Methyl *cis*-7,8-dihydro-7,8-dihydroxy-1-naphthoate
In: Pseudomonas J370
Methyl 2-naphthoate
→ Methyl *cis*-7,8-dihydro-7,8-dihydroxy-2-naphthoate
In: Pseudomonas J370
5-Methyl-1-naphthoic acid
Formed from: 1,5-Dimethylnaphthalene

→ 5-Hydroxymethyl-1-naphthoic acid
Probably in: Microorganism A1116

8-Methyl-1-naphthoic acid
Formed from: 1-Hydroxymethyl-8-methylnaphthalene

7-Methyl-α-naphthol
Formed from: 2-Methylnaphthalene
→ 7-Methyl-α-naphthyl-β-D-glucuronide
In: Guinea pig C592
→ 7-Methyl-α-naphthyl sulphate
In: Guinea pig C592

2-Methyl-1,4-naphthoquinone (menadione)
Formed from: 1,4-Dihydroxy-2-methylnaphthalene
→ 1,4-Dihydroxy-2-methylnaphthalene
In: Beef G401
Clostridium D260
Cochliobolus K378
Escherichia A1228
Man B569
Rat A662, J744
Sporotrichum B301
Thauera K173
See also: E821, K569

N-Methyl-β-naphthylamine
Formed from: β-Naphthylamine

S-(7-Methyl-1-naphthyl)-L-cysteine
Probably formed from: S-(7-Methyl-1-naphthyl)glutathione

S-(7-Methyl-1-naphthyl)glutathione
Probably formed from: 2-Methylnaphthalene
→ S-(7-Methyl-1-naphthyl)-L-cysteine
Probably in: Guinea pig C592

7-Methyl-α-naphthyl-β-D-glucuronide
Formed from: 7-Methyl-α-naphthol

Methyl β-naphthyl ketone
See: 2-Acetylnaphthalene

α-Methyl-α-(1-naphthyl)malonic acid
→ 2-(1-Naphthyl)propionic acid
In: Alcaligenes G716

α-Methyl-α-(2-naphthyl)malonic acid
→ 2-(2-Naphthyl)propionic acid
In: Alcaligenes G716

(1*E*)-N-Methyl-N-(1-naphthylmethyl)-3-phenyl-2-propen-1-amine
See: Naftifine

1-Methyl-4-(1-naphthylmethyl)pyridinium
Probably formed from: Tetrahydro-1-methyl-4-(1-naphthylmethyl)pyridine

7-Methyl-α-naphthyl sulphate
Formed from: 7-Methyl-α-naphthol

3-Methylnirvanol
See: 5-Ethyl-3-methyl-5-phenylhydantoin

4-Methyl-3-nitroacetanilide
Formed from: 4-Amino-2-nitrotoluene

2-Methyl-3-nitroaniline
Probably formed from: 2-Hydroxylamino-3-nitrotoluene
→ 2,6-Diaminotoluene
In: Microorganism G542

N-Methyl-*m*-nitroaniline
Formed from: N,N-Dimethyl-*m*-nitroaniline

N-Methyl-*p*-nitroaniline
Formed from: *p*-Nitroaniline

N-Methyl-*p*-nitrobenzamide
Formed from: N,N-Dimethyl-*p*-nitrobenzamide

3-Methyl-4-nitrocatechol
Formed from: 2,6-Dinitrotoluene

3-Methyl-6-nitrocatechol
Probably formed from: 2-Methyl-5-nitrophenol

4-Methyl-5-nitrocatechol
Formed from: 2,4-Dinitrotoluene
→ 2-Hydroxy-5-nitroquinone
In: Pseudomonas H238, J249

2-Methyl-5-nitrophenol
Formed from: *p*-Nitrotoluene
→ 5-Amino-2-methylphenol
In: Veillonella A2624
→ 3-Methyl-6-nitrocatechol
Probably in: Pseudomonas G661

3-Methyl-2-nitrophenol
→ 3-Methylcatechol
In: Pseudomonas E607, H65

3-Methyl-4-nitrophenol
Formed from: N-Acetyl-3-methyl-4-nitrophenyl-β-D-glucosaminide
O,S-Dimethyl-O-(3-methyl-4-nitrophenyl) phosphorothioate
Fenitrothion
→ 5-Hydroxy-2-nitrobenzaldehyde
In: Crayfish E816

→ 3-Methyl-4-nitrophenyl-β-D-glucoside
In: Cipangopaludina E639
Crayfish E816
Physa E639
→ 3-Methyl-4-nitrophenyl sulphate
In: Cipangopaludina E639
Physa E639
Prawn E816

4-Methyl-2-nitrophenol
→ 4-Methylcatechol
In: Pseudomonas E607, H65

4-Methyl-3-nitrophenol
→ 3-Amino-*p*-cresol
In: Veillonella A2624

5-Methyl-2-nitrophenol
→ 4-Methylcatechol
In: Pseudomonas E607, H65

6-Methyl-2-nitrophenol
→ 3-Methylcatechol
In: Pseudomonas H65

3-Methyl-5-(*p*-nitrophenylazo)rhodanine
(nitrodan)
→ 5-(*p*-Aminophenylazo)-3-methylrhodanine
In: Ascaris A3694
Moniezia A3694
Mouse A3694
Sheep A3694

3-Methyl-4-nitrophenyl-β-D-glucoside
Formed from: 3-Methyl-4-nitrophenol

3-Methyl-4-nitrophenyl sulphate
Formed from: 3-Methyl-4-nitrophenol

Methyl *p*-nitrophenyl phosphorothioate
Formed from: Dimethyl *p*-nitrophenyl phosphorothioate

N-Methyl-N-nitrosoaniline
→ Aniline
In: Mouse G116
Rat E366, H819
→ N-Methylaniline
In: Mouse G116
Rat E366
→ Phenol
In: Rat E366

***p*-Methylnitrosobenzene**
Formed from: *p*-Toluidine
→ N-(*p*-Methylphenyl)isobutyrohydroxamic acid
In: Rat J429

N-Methylnitrosobenzylamine
→ Benzaldehyde
In: Mouse B131

6-Methylnoradrenaline
→ 3-Hydroxy-4-methoxy-6-methylphenylethanolamine
In: Rat A486
→ 6-Methylnormetanephrine
In: Rat A486

D-*erythro*-α-Methylnoradrenaline
→ D-*erythro*-3-Hydroxy-4-methoxy-α-methylphenylethanolamine
In: Rat A486
→ D-*erythro*-α-Methylnormetanephrine
In: Rat A486

L-*erythro*-α-Methylnoradrenaline
→ L-*erythro*-3-Hydroxy-4-methoxy-α-methylphenylethanolamine
In: Rat A486
→ L-*erythro*-α-Methylnormetanephrine
In: Rat A486

***threo*-α-Methylnoradrenaline**
→ *threo*-3-Hydroxy-4-methoxy-N-methylphenylethanolamine
In: Rat A486
→ *threo*-α-Methylnormetanephrine
In: Rat A486

α-Methylnoradrenaline
(3,4-dihydroxynorephedrine; stereochemistry unclear)
Probably formed from: 3,4-Dihydroxyamphetamine
→ 4-(2-Amino-1-hydroxyprop-1-yl)-*o*-quinone
Probably in: Beef A624
Mushroom D796
3,4-Dihydroxyephedrine
Probably in: Rat D766
→ α-Methyladrenaline
Probably in: Rabbit C182
Rat C182, D766
→ α-Methylnoradrenaline-O-sulphate
In: Beef F196
Man C345

α-Methylnoradrenaline-O-sulphate
Formed from: α-Methylnoradrenaline

3′-O-Methylnorbelladine
→ Mesembrenol
In: Sceletium A210
4′-O-Methylnorbelladine
→ Haemanthamine
In: Narcissus A992
→ 11-Hydroxyvittatine
In: Pancratium A1322
→ Mesembrenol
In: Sceletium A210
→ Narcissidine
In: Narcissus A1252
→ Oduline
In: Narcissus A992
N-Methylnorcoclaurine
Formed from: Norcoclaurine
9-Methylnorharman
→ 9-Methylharman
In: Beef H949
D-*erythro*-α-Methylnormetanephrine
Formed from: D-*erythro*-α-Methylnoradrenaline
L-*erythro*-α-Methylnormetanephrine
Formed from: L-*erythro*-α-Methylnoradrenaline
***threo*-α-Methylnormetanephrine**
Formed from: *threo*-α-Methylnoradrenaline
6-Methylnormetanephrine
Formed from: 6-Methylnoradrenaline
N-Methylnororientaline
Formed from: Nororientaline
N-Methylnorsalsolinol
See: 1,2,3,4-Tetrahydro-6,7-dihydroxy-2-methylisoquinoline
3-O-Methyl-17β-oestradiol
Formed from: 3-O-Methyloestrone
17β-Oestradiol
→ 3-O-Methyloestrone
In: Man A3086
7α-Methyl-17β-oestradiol
→ 7α-Methyl-17β-oestradiol-3-sulphate
In: Beef A1211
7α-Methyl-17β-oestradiol-3-sulphate
Formed from: 7α-Methyl-17β-oestradiol
1-Methyloestrone
→ 1-Methyloestrone-3-sulphate
In: Beef A1211
3-O-Methyloestrone
Formed from: 3-O-Methyl-17β-oestradiol
→ 16α-Hydroxy-3-O-methyloestrone
In: Streptomyces A2225, F810
→ 16β-Hydroxy-3-O-methyloestrone
In: Streptomyces F810
→ 3-O-Methyl-17β-oestradiol
In: Man A3086
→ Oestrone
In: Microorganism F469
7α-Methyloestrone
→ 7α-Methyloestrone-3-sulphate
In: Beef A1211
1-Methyloestrone-3-sulphate
Formed from: 1-Methyloestrone
7α-Methyloestrone-3-sulphate
Formed from: 7α-Methyloestrone
2-Methylolivacinium-9-yl-β-D-glucuronide
Formed from: 9-Hydroxy-2-methylolivacinium
Methyl orange
(helianthin, 4′-dimethylaminoazobenzene-4-sulphonic acid)
→ N,N-Dimethyl-*p*-phenylenediamine
In: Rat G133, G380, G511
→ Sulphanilic acid
In: Rat G133, G380, G511
5-Methylorsellinic acid
See: 4,6-Dihydroxy-2,3-dimethylbenzoic acid
N-Methyloxazepam
See: 7-Chloro-1,3-dihydro-3-hydroxy-1-methyl-5-phenyl-2*H*-1,4-benzodiazepin-2-one
3-Methyloxindole
Formed from: Skatole
→ 3-Hydroxy-3-methyloxindole
Probably in: Goat G702
→ 4-Hydroxy-3-methyloxindole
Probably in: Goat G702
O-Methyl S-(4-oxo-1,2,3-benzotriazin-3(4*H*)-ylethyl phosphorodithioate
Formed from: Azinphos-methyl
Methyl 5(6)-(3-oxobutyl)-2-benzimidazolecarbamate
Formed from: Methyl 5(6)-butyl-2-benzimidazolecarbamate

3-Methyl-1-oxocholanthrene
See: 3-Methylcholanthren-1-one
3-Methyl-2-oxocholanthrene
See: 3-Methylcholanthren-2-one
4-((5-Methyl-2-oxo-1,3-dioxol-4-yl)methylthio) benzenesulphonic acid
→ 4-(2-Hydroxy-3-oxobutylthio) benzenesulphonic acid
Probably in: Man G892
Methyl 2-oxoethyl 1,4-dihydro-2,6-dimethyl-4-(*m*-nitrophenyl)pyridine-3,5-dicarboxylate
Formed from: Nicardipine
→ Carboxymethyl methyl 1,4-dihydro-2,6-dimethyl-4-(*m*-nitrophenyl)pyridine-3,5-dicarboxylate
Probably in: Man D889
→ 2-Hydroxyethyl methyl 1,4-dihydro-2,6-dimethyl-4-(*m*-nitrophenyl)pyridine-3,5-dicarboxylate
Probably in: Man D889
4-Methyl-3-(2-(3-oxo-2-morpholino)ethylamino)-6-phenylpyridazine
Formed from: Minaprine
3-O-Methyl-16-oxo-17β-oestradiol
Probably formed from: 16β-Hydroxy-3-O-methyl-17β-oestradiol
Methyl 2-oxo-3-phenylbutyrate
→ Methyl 2-hydroxy-3-phenylbutyrate
In: Candida K883
Endomycopsis K883
Hansenula K883
Kloeckera K883
Lipomyces K883
Pichia K883
Rhodotorula K883
Saccharomyces K883
N-Methyl-2-oxo-2-phenylethylamine
→ Mandelic acid
In: Rat J287
Methyl 6-oxo-α-phenyl-2-piperidineacetate
Formed from: Methylphenidate
→ Methyl 5-hydroxy-6-oxo-α-phenyl-2-piperidineacetate
Probably in: Dog C14
Rat C14
→ Methyl α-(*p*-hydroxyphenyl)-6-oxo-2-piperidineacetate
Probably in: Rat C14
→ 6-Oxo-α-phenyl-2-piperidineacetic acid
Probably in: Dog C14
Rat C14
1-(2-Methyl-1-oxo-1-phenylprop-2-yl)Imidazole
→ 1-(1-Hydroxy-2-methyl-1-phenylprop-2-yl)Imidazole
In: Man J61
6-Methyl-6*H*-7-oxopyrido[1,2-*a*]pyrimidino[5,4-*c*] benzothiazine-5,5-oxide
Formed from: Piroxicam
N-(1-Methyl-5-oxo-2-pyrrolidinylidene)-N′-phenyl-1-pyrrolidinecarboximidamide
Probably formed from: N-(5-Hydroxy-1-methyl-2-pyrrolidinylidene)-N′-phenyl-1-pyrrolidinecarboximidamide
→ 1-(2-Hydroxypyrrolidine)-N-(1-methyl-5-oxo-2-pyrrolidinylidene)-N′-phenyl-carboximidamide
Probably in: Dog J693
Man J693
Monkey J693
Rat J693
13-Methylpalmatine
Formed from: Corydaline
Mesocorydaline
Palmatine
→ Corydaline
In: Corydalis H585
2-Methylpapaverinium
→ 6,7-Dimethoxy-2-methylisoquinoline
In: Rabbit B38
→ Papaverine
In: Rabbit B38
→ Veratraldehyde
In: Rabbit B38
Methyl paraben
See: Methyl *p*-aminobenzoate
Methylparaoxon
See: O,O-Dimethyl O-(4-nitrophenyl) phosphate
Methylparathion
See: O,O-Dimethyl O-(4-nitrophenyl) phosphorothioate
Methyl pentachlorophenyl sulphide
See: Pentachlorothioanisole

N-Methylphenethylamine
Formed from: N,N-Dimethylphenethylamine
Phenethylamine
→ N,N-Dimethylphenethylamine
In: Phalaris A3891
→ Phenylacetaldehyde
In: Rat A3978
***p*-Methylphenethylamine**
→ *p*-Methylphenylacetaldehyde
In: Alcaligenes H367
α-Methylphenethylamine
See: Amphetamine
N′-Methyl-N-(β-phenethyl)hydrazine
Formed from: β-Phenethylhydrazine
N-(α-Methylphenethyl)-2-phenylglycinonitrile
See: Aponeuron
α-Methyl-β-phenethylhydroxylamine
(N-hydroxyamphetamine)
Formed from: D-Amphetamine
L-Amphetamine
→ Amphetamine
In: Man K608
Mouse A3160
Mycobacterium B380
Rat A3160
→ Benzyl methyl ketoxime
In: Mycobacterium B380
Rat B664
→ 2-Nitro-1-phenylpropane
In: Rabbit A3621
→ Phenylacetone
In: Rabbit A3621
Methylphenidate (methyl α-phenyl-2-piperidineacetate)
→ Methyl 1-carbamoyl-α-phenyl-2-piperidineacetate
In: Dog C14
Rat C14
→ Methyl α-(*p*-hydroxyphenyl)-2-piperidineacetate
In: Rat C14
→ Methyl 6-oxo-α-phenyl-2-piperidineacetate
In: Dog C14
Rat C14
→ α-Phenyl-2-piperidineacetic acid
In: Dog C14
Man A137, C643, K38
Rat A137, C14
2-Methylphenol
See: *o*-Cresol
3-Methylphenol
See: *m*-Cresol
4-Methylphenol
See: *p*-Cresol
10-Methyl-10*H*-phenothiazine-2-acetic acid
See: Metiazinic acid
***o*-Methylphenoxyacetic acid**
→ *o*-Cresol
In: Alcaligenes E727
***p*-Methylphenoxyacetyl CoA**
→ *p*-Methylphenoxymethylpenicillin
In: Penicillium D12
***m*-(*p*-Methylphenoxy)benzoic acid**
→ *p*-Cresol
In: Pseudomonas F696, G206
→ *m*-Hydroxybenzoic acid
Probably in: Pseudomonas G206
N-Methylphenoxyethylamine
Formed from: N,N-Dimethylphenoxy-ethylamine
3-(*m*-Methylphenoxy)lactic acid
Formed from: 2-Hydroxy-N-isopropyl-3-(*m*-methylphenoxy)propylamine
***p*-Methylphenoxymethylpenicillin**
Formed from: *p*-Methylphenoxyacetyl CoA
N-Methyl-3-phenoxymorphinan
→ 3-(*p*-Hydroxyphenoxy)-N-methylmorphinan
In: Dog B634
→ 3-Hydroxy-N-methylmorphinan
In: Dog B634
N-Methyl-N′-(*p*-phenoxyphenyl)sulphamide
Formed from: N,N-Dimethyl-N′-(*p*-phenoxyphenyl)sulphamide
3-(*o*-Methylphenoxy)pyridazine
See: Credazine
N-Methylphentermine
Formed from: N,N-Dimethylphentermine
***p*-Methylphenylacetaldehyde**
Formed from: *p*-Methylphenethylamine
***o*-Methylphenylacetic acid**
Formed from: *o*-Methylstyrene
o-Methylphenylacetylene
→ *o*-Methylphenylacetyl CoA

In: Penicillium C483
→ *o*-Methylphenylacetylglycine
In: Rat C153

m-Methylphenylacetic acid
Formed from: *m*-Methylstyrene
→ *m*-Methylphenylacetylglycine
In: Rat C153

p-Methylphenylacetic acid
Formed from: *p*-Methylphenylacetylene
p-Methylstyrene
→ *p*-Methylphenylacetyl CoA
In: Penicillium C483
→ *p*-Methylphenylacetylglycine
In: Rat C153

1-(p-Methylphenyl)acetone
Formed from: Anethole oxide
→ *p*-Methylphenylacetone cyanhydrin
In: Almond J334
Apple J334

p-Methylphenylacetone cyanhydrin
Formed from: 1-(*p*-Methylphenyl)acetone

o-Methylphenylacetyl CoA
Formed from: *o*-Methylphenylacetic acid

p-Methylphenylacetyl CoA
Formed from: *p*-Methylphenylacetic acid

o-Methylphenylacetylene
→ *o*-Methylphenylacetic acid
In: Rat E461

p-Methylphenylacetylene
→ *p*-Methylphenylacetic acid
In: Rat E461

o-Methylphenylacetylglycine
Formed from: *o*-Methylphenylacetic acid

m-Methylphenylacetylglycine
Formed from: *m*-Methylphenylacetic acid

p-Methylphenylacetylglycine
Formed from: *p*-Methylphenylacetic acid

N-Methyl-p-phenylenediamine
Formed from: N,N-Dimethyl-*p*-phenylenediamine
4-Hydroxy-4′-methylaminoazobenzene
4-Methylaminoazobenzene

p-Methyl-L-phenylalanine
→ *p*-Hydroxymethyl-L-phenylalanine
In: Rat G150
→ 3-Methyl-L-tyrosine
In: Rat G150

α-Methyl-L-phenylalanine
→ Amphetamine
In: Micrococcus D539

N-Methyl-4-(phenylazo)aniline
→ N-Methyl-4-(phenylazo)benzene nitroxide radical
Probably in: Rabbit B184

N-Methyl-4-(phenylazo)benzene nitroxide radical
Probably formed from: N-Methyl-4-(phenylazo)aniline

α-Methyl-α-phenylbenzeneacetic acid 2-(diethylamino)ethyl ester
See: Aprophen

4-Methyl-2-(4-phenylbenzyl)-2-oxazoline-4-methanol
→ N-(1,1-Bis(hydroxymethyl)ethyl)-4-biphenylylacetamide
In: Dog A1108

3-Methyl-1-phenyl-2-butanol
Formed from: 3-Methyl-1-phenyl-2-butanone

3-Methyl-1-phenyl-2-butanone
→ 3-Methyl-1-phenyl-2-butanol
In: Rabbit A3451

3-Methyl-4-phenyl-3-butenoic diethylamide
→ 3-Benzyl-4-hydroxybutanoic acid
In: Rabbit B435
→ 4-(*p*-Hydroxyphenyl)-3-methyl-3-butenoic diethylamide
Probably in: Rabbit B435
→ 3-Methyl-4-phenyl-3-butenoic ethylamide
In: Rabbit B435

3-Methyl-4-phenyl-3-butenoic ethylamide
Formed from: 3-Methyl-4-phenyl-3-butenoic diethylamide

3-Methyl-1-phenyl-2-butylamine
→ 3-Methyl-1-phenyl-2-butylhydroxylamine
In: Rabbit D91

3-Methyl-1-phenyl-2-butylhydroxylamine
Formed from: 3-Methyl-1-phenyl-2-butylamine

N-Methyl-p-phenylenediamine
Formed from: N,N-Dimethyl-*p*-phenylenediamine
4-Hydroxy-4′-methylaminoazobenzene
4-Methylaminoazobenzene

(R)-1-(m-Methylphenyl)ethanol
Formed from: *m*-Methylacetophenone

(*S*)-1-(*m*-Methylphenyl)ethanol
→ *m*-Methylacetophenone
In: Geotrichum H749

1-(*p*-Methylphenyl)ethanol
Formed from: 4-Ethyltoluene
→ *p*-Methylacetophenone
In: Geotrichum H749

2-(*p*-Methylphenyl)ethanol
Formed from: 4-Ethyltoluene

N-Methylphenylethanolamine
Formed from: Phenylethanolamine
→ Mandelaldehyde
In: Rat A3978
See also: J287

N-Methyl-1-phenylethylamine
Formed from: N-Methyl-N-(1-phenylethyl)-2-propynylamine

N-((1-Methyl-2-phenyl)ethyl)ethanimine-N-oxide
Formed from: N-Methylamphetamine

Methyl phenyl ether
See: Anisole

3-(1-Methyl-2-phenylethyl)-5-(((phenylamino) carbonyl)amino)-1,2,3-oxazolium inner salt
See: Sydnocarb

α-((1-Methyl-2-phenylethyl)amino) benzeneacetonitrile
See: Amphetaminil

7-(2-(α-Methylphenylethylamino)ethyl) theophylline
See: Fenetylline

1-Methyl-1-phenylethyl hydroperoxide
See: Cumene hydroperoxide

N-((1-Methyl-2-phenyl)ethyl)propanimine-N-oxide
Formed from: N-Propylamphetamine

N-Methyl-N-(1-phenylethyl)-2-propynylamine
→ N-Methyl-1-phenylethylamine
In: Rat E645
→ N-(1-Phenylethyl)-2-propynylamine
In: Rat E645

1-Methyl-4-(2-phenylethyl)pyridinium
Probably formed from: 1,2,3,4-Tetrahydro-1-methyl-4-(2-phenylethyl)pyridine

***o*-Methylphenylglycol**
Formed from: *o*-Methylstyrene
→ *o*-Methylmandelic acid
Probably in: Rat C153
→ *o*-Methylphenylglycol-β-D-glucuronide
In: Rat C153

***m*-Methylphenylglycol**
Formed from: *m*-Methylstyrene
→ *m*-Hydroxymethylphenylglycol
Probably in: Rat C153
→ *m*-Methylmandelic acid
Probably in: Rat C153

***p*-Methylphenylglycol**
Formed from: *p*-Methylstyrene
→ *p*-Methylphenylglycol-β-D-glucuronide
In: Rat C153
→ *p*-Methylmandelic acid
Probably in: Rat C153, C681

***o*-Methylphenylglycol-β-D-glucuronide**
Formed from: *o*-Methylphenylglycol

***p*-Methylphenylglycol-β-D-glucuronide**
Formed from: *p*-Methylphenylglycol

***o*-Methylphenylglyoxylic acid**
Probably formed from: *o*-Methylmandelic acid

***m*-Methylphenylglyoxylic acid**
Probably formed from: *m*-Methylmandelic acid

***p*-Methylphenylglyoxylic acid**
Formed from: *p*-Methylmandelic acid
→ *p*-Tolualdehyde
In: Pseudomonas E714
See also: C153, C681
→ *p*-Toluic acid
Probably in: Rat C153, C681

N-Methyl-N-phenylhydroxylamine (N-hydroxy-N-methylaniline)
Formed from: N-Methylaniline

N-(*p*-Methylphenyl)isobutyrohydroxamic acid
Formed from: *p*-Methylnitrosobenzene

5-Methyl-3-phenyl-4-isoxazolylpenicillin
See: Oxacillin

(*E*)-(3-Methylphenyl)itaconic acid
Probably formed from: (3-Methylbenzyl) succinic acid

α-Methyl-α-phenylmalonic acid
→ 2-Phenylpropionic acid
In: Alcaligenes G716

N-Methyl-N-(3-(4-(N-phenylmethyl)phenoxy) propyl)-β-alanine
→ N-(3-(4-(N-(*p*-Hydroxyphenyl)methyl) phenoxy)propyl)-N-methyl-β-alanine
In: Rat J52

→ N-Methyl-N-(3-(4-(N-phenylmethyl)phenoxy)propyl)-β-alanine-N-oxide
In: Rat J52

N-Methyl-N-(3-(4-(N-phenylmethyl)phenoxy)propyl)-β-alanine-N-oxide
Formed from: N-Methyl-N-(3-(4-(N-phenylmethyl)phenoxy)propyl)-β-alanine

***threo*-3-Methyl-2-phenylmorpholine**
See: Phenmetrazine

N-Methylphenyl nitroxide radical
Probably formed from: N-Methylaniline

1-Methyl-2-phenylmethyl)-4(1*H*)-quinazolinone
See: Arborine

N-Methylphenyl nitroxide radical
Probably formed from: N-Methylaniline

5-Methyl-3-phenyl-1,2,4-oxadiazole
→ 5-Hydroxymethyl-3-phenyl-1,2,4-oxadiazole
In: Rat A1834
→ S-((3-Phenyl-1,2,4-oxadiazol-5-yl)methyl)glutathione
In: Rat A1834

2-(*p*-Methylphenyl)-1,3-oxathiolane
→ 2-(*p*-Methylphenyl)-1,3-oxathiolane sulphoxide
In: Pig F176
Rat F176

2-(*p*-Methylphenyl)-1,3-oxathiolane sulphone
Probably formed from: 2-(*p*-Methylphenyl)-1,3-oxathiolane sulphoxide

2-(*p*-Methylphenyl)-1,3-oxathiolane sulphoxide
Formed from: 2-(*p*-Methylphenyl)-1,3-oxathiolane
→ 2-(*p*-Methylphenyl)-1,3-oxathiolane sulphone
Probably in: Pig F176
Rat F176

4-(4-Methylphenyl)-4-oxo-2-thiomethylbutanoic acid
Formed from: 2-Acetylthiomethyl-4-(4-methylphenyl)-4-oxobutanoic acid
→ 4-(4-Methylphenyl)-2-methylthiomethyl-4-oxobutanoic acid
Probably in: Rat J168

4-(4-(4-Methylphenyl)phenylmethoxy-1-piperidinyl)butyric acid
→ 4-(4-(4-Hydroxymethylphenyl)phenylmethoxy-1-piperidinyl)butyric acid
In: Rat G967
→ 4-(4-(4-Hydroxyphenyl)(4-methylphenyl)methoxy-1-piperidinyl)butyric acid
In: Rat G967

1-Methyl-N-phenyl-N-(phenylmethyl)-4-piperidinamine
See: Bamipine

N-(*p*-Methylphenyl-α-phenylnitrone
Formed from: N-Benzyl-*p*-toluidine

O-Methyl phenylphosphonic acid
Probably formed from: O-Methyl phenylphosphonothioic acid
→ Benzene
In: Klebsiella B337
→ Phenylphosphonic acid
In: Mouse A1063
Rat A2903

O-Methyl phenylphosphonothioic acid
Formed from: Leptophos
→ O-Methyl phenylphosphonic acid
Probably in: Mouse A1063
Rat A2903

N-Methyl-4-phenylpiperidine-4-carboxylic acid
Formed from: Meperidine
→ 4-Phenylpiperidine-4-carboxylic acid
Probably in: Cat B890
Dog B890
Guinea pig B890
Mouse B890
Rabbit B890
Rat B890

Methyl α-phenyl-2-piperidineacetate
See: Methylphenidate

1-Methyl-4-phenylpiperidinol propionate
Formed from: 1-Methyl-4-phenylpiperidinol propionate N-oxide

1-Methyl-4-phenylpiperidinol propionate N-oxide
→ 1-Methyl-4-phenylpiperidinol propionate
In: Mouse A1118
Rabbit A1118
Rat A1118

1-Methyl-4-phenyl-2-piperidinone
Formed from: 1,2,3,6-Tetrahydro-1-methyl-4-phenylpiperidine

Methyl α-phenyl-α-(2-piperidyl)acetate
See: Methylphenidate

(1*R*,2*S*)-1-(*o*-Methylphenyl)propane-1,2-diol
Formed from: *o*-Tolualdehyde

(1*R*,2*S*)-1-(*p*-Methylphenyl)propane-1,2-diol
Formed from: *p*-Tolualdehyde

2-(*p*-Methylphenyl)propane-1,2-diol
Formed from: 2-(*p*-Methylphenyl)propylene oxide

1-(*p*-Methylphenyl)-2-propanol
Formed from: *p*-Methylacetophenone

2-Methyl-3-phenylpropan-1-ol
Probably formed from: α-Methylcinnamyl alcohol

2-Methyl-1-phenyl-1-propene-1,2-oxide
→ 1-(S-Glutathionyl)-2-methyl-1-phenyl-2-propanol
Probably in: Sheep A2282

N-Methyl-3-phenyl-2-propen-1-ylamine
Formed from: Naftifine

α-Methyl-β-phenylpropionic acid
Formed from: Oxazepam α-methyl-β-phenylpropionate

2-(*p*-Methylphenyl)propionic acid
Formed from: α-Methyl-α-(*p*-methylphenyl)malonic acid
Probably formed from: *p*-(1-Hydroxy-2-propyl)toluene
→ 2-(*p*-Hydroxymethylphenyl)propionic acid
Probably in: Rat C574

2-(*p*-Methylphenyl)propylene oxide
→ 2-(*p*-Methylphenyl)propane-1,2-diol
In: Pig H753

α-Methyl-N-(1′-phenylprop-2′-yl)nitrone
Formed from: Ethylamphetamine

3-Methyl-1-phenyl-3-pyrazolin-5-one
Formed from: Antipyrine
4-Isopropyl-3-methyl-1-phenyl-3-pyrazolin-5-one
→ 3-Methyl-1-phenyl-3-pyrazolin-5-one-β-D-glucuronide
In: Man A2862, B394, C415, C481
Rat C415
See also: D632, J769
→ 3-Methyl-1-phenyl-3-pyrazolin-5-one sulphate
In: Man B394
Rat C415, C481
See also: B818, C418, J769

3-Methyl-1-phenyl-3-pyrazolin-5-one-β-D-glucuronide
Formed from: 3-Methyl-1-phenyl-3-pyrazolin-5-one

3-Methyl-1-phenyl-3-pyrazolin-5-one sulphate
Formed from: 3-Methyl-1-phenyl-3-pyrazolin-5-one

2-(3-Methyl-1-phenyl-3-pyrazolin-5-on-4-yl) propionic acid
Formed from: 4-Isopropylantipyrine
→ 4-(1-Hydroxyethyl)-3-methyl-1-phenyl-3-pyrazolin-5-one
Probably in: Rat B818

N-(4-Methyl-6-phenyl-3-pyridazinyl)-4-morpholineethanamine
See: Minaprine

N-Methyl-2-phenylpyridinium
Formed from: 2-Phenylpyridine

N-Methyl-3-phenylpyridinium
Formed from: 3-Phenylpyridine

N-Methyl-4-phenylpyridinium
Formed from: 2,3-Dihydro-1-methyl-4-phenylpyridinium
4-Phenylpyridine
1,2,3,6-Tetrahydro-1-methyl-4-phenylpyridine

1-Methyl-4-phenyl-2(1*H*)-pyridone
Formed from: 2,3-Dihydro-1-methyl-4-phenylpyridinium
5,6-Dihydro-1-methyl-4-phenyl-2(1*H*)-pyridone

Methyl phenylpyruvate
Formed from: L-Phenylalanine methyl ester

N-Methyl-α-phenylsuccinimide (phensuximide)
→ 2-(3,4-Dihydro-3,4-dihydroxyphenyl)-3-hydroxy-N,2-dimethylsuccinimide
In: Man A1709, A2017
→ 2-(3,4-Dihydro-3,4-dihydroxyphenyl)-2-hydroxy-N-methylsuccinimide
In: Man A1709, A2017

α-Methyl-α-phenylsuccinimide
Formed from: Methsuximide
→ α-(*p*-Hydroxyphenyl)-α-methylsuccinimide
Probably in: Dog A1434

Methyl 5(6)-phenylsulphinyl-2-benzimidazolecarbamate
See: Oxfendazole

Methyl phenyl sulphone
Formed from: O-Ethyl S,S-diphenyl phosphorodithioate
Probably formed from: Thioanisole sulphoxide
→ 3-Hydroxyphenyl methyl sulphone
In: Rat A2160
See also: A2322
→ 4-Hydroxyphenyl methyl sulphone
In: Rat A2160
See also: A2322
Methyl phenyl sulphoxide
See: Thioanisole sulphoxide
2-(*p*-Methylphenyl)thioethan-1-ol
Formed from: 3-(2-(*p*-Methylphenyl)thioethyl)sydnone
3-(2-(*p*-Methylphenyl)thioethyl)sydnone
→ 2-(*p*-Methylphenyl)thioethan-1-ol
In: Rat D676
→ *p*-Methylthiophenol
In: Rat D676
4-(4-Methylphenyl)-2-thiomethyl-4-oxobutanoic acid
Probably formed from: 2-Acetylthiomethyl-4-(4-methylphenyl)-4-oxobutanoic acid
N-Methyl-N′-phenyl-N′-((2-thiophenyl) methyl)ethylenediamine
Formed from: Methaphenilene
3-Methyl-1-phenyltriazene
Formed from: 3,3-Dimethyl-1-phenyltriazene
N-Methyl-3-phenyl-3-(*p*-trifluoromethylphenoxy) propylamine
→ 3-Phenyl-3-(*p*-trifluoromethylphenoxy) propylamine
In: Dog A47
Rat A47
4-(5-(*p*-Methylphenyl)-3-trifluoromethyl-1*H*-pyrazol-1-yl)benzenesulphonamide
See: Celecoxib
Methyl phosphoramidate isopropyl salicylate ester
Formed from: O-Methyl phosphoramidothioate isopropyl salicylate-O-ester
O-Methyl phosphoramidothioate isopropyl salicylate-O-ester
→ Methyl phosphoramidate isopropyl salicylate ester
In: Cotton A560
(*S*)-Methylphthalide
Formed from: *o*-Ethylbenzoic acid
Methyl *o*-acetylbenzoate
2-Methyl-3-phytyl-1,4-naphthoquinone
See: Vitamin K_1
2-Methyl-6-phytylquinol
See: Phytyltoluquinol
2-Methyl-6-phytylquinone
Formed from: Homogentisic acid
→ 2,3-Dimethyl-5-phytylquinone
In: Spinach A3972, B575
See also: A902
1-Methylpipecolo-3′-hydroxy-2′,6′-xylidide
Formed from: 1-Methylpipecolo-2′,6′-xylidide
1-Methylpipecolo-4′-hydroxy-2′,6′-xylidide
Formed from: 1-Methylpipecolo-2′,6′-xylidide
1-Methylpipecolo-2′,6′-xylidide (mepivacaine)
→ 1-Methylpipecolo-3′-hydroxy-2′,6′-xylidide
In: Man A610
Rat A610
→ 1-Methylpipecolo-4′-hydroxy-2′,6′-xylidide
In: Man A610
Rat A610
N-Methylpiperazine-L-phenylalanine-homophenylalanine-vinylsulphone-phenyl
→ N-Methylpiperazine-L-phenylalanine-homophenylalanine-vinylsulphone-phenyl-N-oxide
In: Man K326
N-Methylpiperazine-L-phenylalanine-homophenylalanine-vinylsulphone-phenyl-N-oxide
Formed from: N-Methylpiperazine-L-phenylalanine-homophenylalanine-vinylsulphone-phenyl
3-(((4-Methyl-1-piperazinyl)imino)methyl) rifamycin
See: Rifampicin
10-(3-(4-Methylpiperazinyl)propyl)phenothiazine
See: Perazine
10-(3-(4-Methylpiperazin-1-yl)propyl)-2-trifluoromethyl-10*H*-phenothiazine (trifluoperazine)
→ 10-(3-Aminopropyl)-2-trifluoromethyl-10*H*-phenothiazine
Probably in: Rat A69

→ Demethyltrifluoperazine
In: Rat A1865, A2072, A2073, A2720
See also: A67
→ 7-Hydroxytrifluoperazine
In: Rat A2072, A2073, A2702
→ 10-(3-(4-Methylpiperazin-1-yl)propyl)-2-trifluoromethyl-10*H*-phenothiazine sulphoxide
In: Rat A67, A2073
→ Trifluoperazine-N-β-D-glucuronide
In: Man H541
→ Trifluoperazine-1″-oxide
In: Rabbit A1202
→ Trifluoperazine-4″-oxide
In: Pig A2861
Rabbit A1202
→ N-(γ-(2-Trifluoromethyl-10-phenothiazinyl)propyl)ethylenediamine
In: Rat A576
→ N′-(γ-(2-Trifluoromethyl-10-phenothiazinyl)propyl)ethylenediamine
In: Rat A576

10-(3-(4-Methylpiperazin-1-yl)propyl)-2-trifluoromethyl-10*H*-phenothiazine sulphoxide
Formed from: 10-(3-(4-Methylpiperazin-1-yl)propyl)-2-trifluoromethyl-10*H*-phenothiazine

N-Methylpiperidinyl-4-benzilate
Formed from: Propiverine
→ N-Methylpiperidinyl-4-benzilate-N-oxide
Probably in: Rat E628, H552

N-Methylpiperidinyl-4-benzilate-N-oxide
Probably formed from: N-Methylpiperidinyl-4-benzilate

N-Methylpiperidinyl α-(2-carboxyethoxy)-α,α-diphenylacetate
Formed from: Propiverine

10-(2-(1-Methyl-2-piperidinyl)ethyl)-2-(methylsulphinyl)-10*H*-phenothiazine
See: Mesoridazine

10-(2-(1-Methyl-2-piperidinyl)ethyl)-2-(methylthio)-10*H*-phenothiazine
See: Thioridazine

N-Methylplatydesmine
See: Platydesminium

2-(*p*-((2-Methyl-2-propenyl)amino)phenyl) propionic acid
→ 2-(*p*-((2,3-Dihydro-2-methyl-2-propenyl) amino)phenyl)propionic acid
In: Animal E135

Methylpropion (2-(methylamino)-1-phenyl-1-propanone)
Formed from: Dimethylpropion

β-((2-Methylpropoxy)methyl)-N-phenyl-N-(phenylmethyl)-1-pyrrolidineethanamine
See: Bepridil

N-(3-(2-Methyl-1-propoxy)-2-(2-oxo-1-pyrrolidino)propyl)aniline
Probably formed from: N-Benzyl-N-(3-(2-methyl-1-propoxy)-2-(2-oxo-1-pyrrolidino)propyl)aniline

N-(3-(2-Methyl-1-propoxy)-2-(1-pyrrolidino) propyl)aniline
Formed from: Bepridil

6-Methyl-2-((*n*-propyl)amino)tetralin-5-yl sulphate
Probably formed from: 5-Hydroxy-6-methyl-2-((*n*-propyl)amino)tetralin

N-Methyl-N-propylamphetamine
→ N-Methylamphetamine
In: Man K37
→ N-Propylamphetamine
In: Man K37

N-Methyl-N-propylaniline
→ N-Methylaniline
In: Rabbit A3484
→ N-Methyl-N-propylaniline-N-oxide
In: Rabbit A3484
→ N-Propylaniline
In: Rabbit A3484

N-Methyl-N-propylaniline-N-oxide
Formed from: N-Methyl-N-propylaniline

α-Methyl-N-(*n*-propylidene)benzenemethanamine-N-oxide
→ Amphetamine
In: Mycobacterium B380

***p*-(2-Methylpropyl)phenol**
Formed from: *p*-(2-Methylpropyl)phenyl acetate

***p*-(2-Methylpropyl)phenyl acetate**
→ *p*-(2-Methylpropyl)phenol
In: Pig K560

1-Methyl-4-(*o*-propylphenyl)pyridinium
Probably formed from: 1,2,3,6-Tetrahydro-1-methyl-4-(*o*-propylphenyl)pyridine

7-Methyl-2-propyl-3-((2′-(1*H*-tetrazol-5-yl)(1,1′-biphenyl)-4-yl)methyl)-3*H*-imidazo[4,5-*b*] pyridine
→ 6-Hydroxy-7-methyl-2-propyl-3-((2′-(1*H*-tetrazol-5-yl)(1,1′-biphenyl)-4-yl)methyl)-3*H*-imidazo[4,5-*b*]pyridine
In: Rat G588
→ 7-Hydroxymethyl-2-propyl-3-((2′-(1*H*-tetrazol-5-yl)(1,1′-biphenyl)-4-yl)methyl)-3*H*-imidazo[4,5-*b*]pyridine
In: Rat G588

2-(*p*-(1-Methyl-2-propynyloxy)propane
Probably formed from: *trans*-β-Methyl-*p*-(1-methyl-2-propynyloxy)cinnamic acid

Methyl protocatechuate
→ Protocatechuic acid
In: Bacillus A1525

N-Methylpseudoephedrine
Formed from: Dimethylpropion

O-Methylpsychotrine
Formed from: Emetine

Methyl purple I
Formed from: Methyl purple I-β-D-galactoside

Methyl purple II
Formed from: Methyl purple II-β-D-galactoside

Methyl purple I-β-D-galactoside
→ Methyl purple I
With β-galactosidase H30

Methyl purple II-β-D-galactoside
→ Methyl purple II
With β-galactosidase H30

4-(β-(5-Methylpyrazine-2-carboxamido)ethyl) benzenesulphonylurea
Formed from: Glipizide

5-((4-(2-(Methyl-2-pyridinylamino)ethoxy) phenyl)methyl)-2,4-thiazolidinedione
(rosiglitazone)
→ 5-((4-(2-((3-Hydroxy-2-pyridinyl) methylamino)ethoxy)phenyl)methyl)-2,4-thiazolidinedione
In: Dog H715
Man K55
Rat H715
→ 5-((4-(2-((5-Hydroxy-2-pyridinyl) methylamino)ethoxy)phenyl)methyl)-2,4-thiazolidinedione
In: Dog H715
Man K55
→ 5-((4-((Methylamino)ethoxy) phenyl)methyl)-2,4-thiazolidinedione
In: Man K55
→ 5-((4-(2-(2-Pyridinylamino)ethoxy) phenyl)methyl)-2,4-thiazolidinedione
In: Man K55
Rat H715

1-Methyl-9*H*-pyrido[3,4-*b*]indole
See: Harman

1-Methyl-9*H*-pyrido[3,4-*b*]indol-7-ol
See: Harmol

N[1]-(4-Methyl-2-pyrimidinyl)sulphanilamide
See: Sulphamerazine

(+)-(3′*S*,4*S*)-Methyl 3′-pyrrolidinyl 4-(*m*-aminophenyl)-1,4-dihydro-2,6-dimethylpyridine-3,5-dicarboxylate
Probably formed from: (+)-(3′*S*,4*S*)-Methyl 3′-pyrrolidinyl 1,4-dihydro-2,6-dimethyl-4-(*m*-nitrophenyl)pyridine-3,5-dicarboxylate

(+)-(3′*S*,4*S*)-Methyl 3′-pyrrolidinyl 1,4-dihydro-2,6-dimethyl-4-(*m*-nitrophenyl)pyridine-3,5-dicarboxylate
Formed from: Barnidipine
→ (+)-(3′*S*,4*S*)-Methyl 3′-pyrrolidinyl 4-(*m*-aminophenyl)-1,4-dihydro-2,6-dimethylpyridine-3,5-dicarboxylate
Probably in: Dog H518
Man J84
Rat H518
→ (+)-(3′*S*,4*S*)-Methyl 3′-pyrrolidinyl 1,4-dihydro-2,6-dimethyl-4-(*m*-nitrophenyl) pyridine-3,5-dicarboxylate-β-D-glucuronide
Probably in: Dog H518
Rat H518
→ Methyl 3′-pyrrolidinyl 2,6-dimethyl-4-(*m*-nitrophenyl)pyridine-3,5-dicarboxylate
Probably in: Dog H518
Man J84, J257
Rat H518

(+)-(3′S,4S)-Methyl 3′-pyrrolidinyl 1,4-dihydro-2,6-dimethyl-4-(m-nitrophenyl)pyridine-3,5-dicarboxylate-β-D-glucuronide
Probably formed from: (+)-(3′S,4S)-Methyl 3′-pyrrolidinyl 1,4-dihydro-2,6-dimethyl-4-(m-nitrophenyl)pyridine-3,5-dicarboxylate

Methyl 3′-pyrrolidinyl 2,6-dimethyl-4-(m-nitrophenyl)pyridine-3,5-dicarboxylate
Probably formed from: (+)-(3′S,4S)-Methyl 3′-pyrrolidinyl 1,4-dihydro-2,6-dimethyl-4-(m-nitrophenyl)pyridine-3,5-dicarboxylate
→ 3′-Pyrrolidinyl 3-carboxy-2,6-dimethyl-4-(m-nitrophenyl)pyridine-5-carboxylate
Probably in: Man J84

N-(1-Methyl-2-pyrrolidinylidene)-N′-phenyl-4-morpholinecarboximidamide
See: Linogliride

N-(1-Methyl-2-pyrrolidinylidene)-N′-phenyl-1-pyrrolidinecarboximidamide
See: Pirogliride

O-Methylquercetin
Formed from: Quercetin

Methylquinol (2,5-dihydroxytoluene)
Formed from: o-Cresol
m-Cresol
Toluquinone
→ Methylquinol-β-D-glucoside
In: Manduca G897
→ Toluquinone
Probably in: Sarcophaga E407
Sheep blowfly F944

Methylquinol-β-D-glucoside
Formed from: Methylquinol

2-Methylquinoline
See: Quinaldine

3-Methylquinoline
→ 2-Hydroxy-3-methylquinoline
In: Comamonas K769
Desulfobacterium H917
Microorganism J367
Rhodococcus K767

4-Methylquinoline
Formed from: 4-Methylquinoline-N-oxide
→ 2-Hydroxy-4-methylquinoline
In: Agrobacterium G778
Desulfobacterium H917
Microorganism J367
Rhodococcus K767

6-Methylquinoline
→ 2-Hydroxy-6-methylquinoline
In: Desulfobacterium H917
Microorganism J367

8-Methylquinoline
→ 2-Hydroxy-8-methylquinoline
In: Desulfobacterium H917
Microorganism J367
Rhodococcus K767

4-Methylquinoline-N-oxide
→ 4-Methylquinoline
In: Saccharomyces H894

N-Methylquinolinium
Formed from: Quinoline
→ 1-Methyl-2-quinolone
In: Guinea pig D147
Rabbit D147, D208
→ 1-Methyl-4-quinolone
In: Guinea pig D147
Rabbit D147

1-Methyl-2-quinolone
Formed from: N-Methylquinolinium

1-Methyl-4-quinolone
Formed from: N-Methylquinolinium

3-Methyl-o-quinone
Formed from: 3-Methylcatechol

4-Methyl-o-quinone
Formed from: 4-Methylcatechol

N-Methylquinoxalinium
Formed from: Quinoxaline

1-Methyl-2(1H)-quinoxalinone
Probably formed from: 2-Hydroxyquinoxaline

Methyl red (2-carboxy-4′-dimethylaminoazobenzene)
→ Anthranilic acid
In: Acetobacter F707
Aerobacter A653
Bacillus A653
Bacteroides A653
Cautobacter K314
Clostridium A653
Escherichia A653
Lactobacillus A653
Proteus A653
Pseudomonas A653
Rabbit C884

Rat A2730, G380, G511
Streptococcus A653
→ N,N-Dimethyl-*p*-phenylenediamine
In: Acetobacter F707
Aerobacter A653
Bacillus A653
Bacteroides A653
Cautobacter K314
Clostridium A653
Escherichia A653
Lactobacillus A653
Proteus A653
Pseudomonas A653
Rabbit C884
Rat A2730, G380, G511
Streptococcus A653

***p*-Methyl red** (4-carboxy-4′-dimethylaminoazobenzene)
→ *p*-Aminobenzoic acid
In: Rat G380, G511
→ 4-Carboxy-4′-methylaminoazobenzene
In: Rat A1979
→ N,N-Dimethyl-*p*-phenylenediamine
In: Rat G380, G511

Methyl red methyl ester
→ Methyl anthranilate
In: Rat G133, G380, G511
→ N,N-Dimethyl-*p*-phenylenediamine
In: Rat G133, G380, G511

***p*-Methyl red methyl ester**
Methyl *p*-aminobenzoate
In: Rat G133, G380, G511
→ N,N-Dimethyl-*p*-phenylenediamine
In: Rat G133, G380, G511

Methyl reserpate
Probably formed from: Syringomethyl reserpate

2-Methylresorcinol (2,6-dihydroxytoluene)
Formed from: *o*-Cresol
→ 2,3,6-Trihydroxytoluene
In: Pseudomonas A2391

4-Methylresorcinol (2,4-dihydroxytoluene)
Formed from: *o*-Cresol
→ 2-Hydroxy-5-methylquinol
In: Pseudomonas A2172, A2391

3-O-Methylrimiterol
Formed from: Rimiterol
→ 3-O-Methylrimiterol-β-D-glucuronide
Probably in: Rat A284
→ 3-O-Methylrimiterol-O-sulphate
Probably in: Man A279

3-O-Methylrimiterol-β-D-glucuronide
Probably formed from: 3-O-Methylrimiterol

3-O-Methylrimiterol-O-sulphate
Probably formed from: 3-O-Methylrimiterol

O-Methylrosmarinic acid
Formed from: Rosmarinic acid

4-Methylrubazoic acid
Formed from: Aminopyrine
→ Rubazoic acid
Probably in: Man A910, A911

O-Methylrutin
Formed from: Rutin

N-Methylsaccharin
Formed from: Isoxicam
Piroxicam
→ Saccharin
Probably in: Dog F59
Monkey F59
Rat F59

Methyl salicylate
Formed from: Methyl acetylsalicylate
Salicylic acid
→ Methyl salicylate-β-D-glucuronide
In: Man J490
→ Salicylic acid
In: Rabbit F327, G517

Methyl salicylate-β-D-glucuronide
Formed from: Methyl salicylate

3-Methylsalicylic acid
See: 2-Hydroxy-3-methylbenzoic acid

6-Methylsalicylic acid
See: 2-Hydroxy-6-methylbenzoic acid

N-Methylsalsolidine
Formed from: Salsolidine

N-Methyl-(*R*)-salsolinol
Formed from: Salsolinol

Methylscopolammonium
→ 4′-Hydroxymethylscopolammonium
In: Rat A1201

→ Methylaposcopinium
In: Rat A1201
→ Methylaposcopolammonium
In: Rat A1201
→ Methylscopolammonium-β-D-glucuronide
In: Rat A1201
→ (*S*)-Tropic acid
In: Rabbit A2330

Methylscopolammonium-β-D-glucuronide
Formed from: Methylscopolammonium

2-(Methylseleninyl)benzanilide
Probably formed from: Ebselen

2-(Methylseleno)benzanilide
Probably formed from: 2-Selenobenzanilide
→ 4′-Hydroxy-2-(methylseleno)benzanilide
In: Rat F616
See also: E808, F127, F235
→ 2-(Methylseleno)benzoic acid
Probably in: Man F235
Pig F235
Rat F235

2-(Methylseleno)benzanilid-4′-yl-β-D-glucuronide
Probably formed from: 4′-Hydroxy-2-(methylseleno)benzanilide

2-(Methylseleno)benzoic acid
Probably formed from: 2-(Methylseleno) benzanilide

4-(Methylselenyl)catechol
Formed from: *p*-(Methylselenyl)phenol
→ 4-(Methylselenyl)-*o*-quinone
In: Mushroom H907

***p*-(Methylselenyl)phenol**
→ 4-(Methylselenyl)catechol
In: Mushroom H907

4-(Methylselenyl)-*o*-quinone
Formed from: 4-(Methylselenyl)catechol

Methyl sinapate
→ Methyl 3,4,5-trimethoxycinnamate
In: Lentinus A214
→ Sinapic acid
In: Aspergillus G915

7-O-Methylsophoricoside
Formed from: Sophoricoside

O-Methylsterigmatocystin
Formed from: Sterigmatocystin
→ Aflatoxin B_1
In: Aspergillus E373, E419, E512, E544, E763, E767, F79, F398, F782, G154, K249, K732
See also: D745, F281
→ Aflatoxin G_1
In: Aspergillus E373, E544, E767, F398, K237, K732
See also: F281

4′-O-Methylstilboestrol
Formed from: Stilboestrol

N-Methylstylopine
Formed from: Stylopine
→ Protopine
In: Corydalis E577

***o*-Methylstyrene**
→ N-Acetyl-S-(2-hydroxy-1-(*o*-methylphenyl) ethyl)-L-cysteine
In: Rat C868
→ N-Acetyl-S-(2-hydroxy-2-(*o*-methylphenyl) ethyl)-L-cysteine
In: Rat C153, C868
→ *o*-Methylphenylacetic acid
In: Rat C153
→ *o*-Methylphenylglycol
Probably in: Rat C153

***m*-Methylstyrene**
→ N-Acetyl-S-(2-hydroxy-1-(*m*-methylphenyl) ethyl)-L-cysteine
In: Rat C868
→ N-Acetyl-S-(2-hydroxy-2-(*m*-methylphenyl) ethyl)-L-cysteine
In: Rat C153, C868
→ *m*-Methylphenylacetic acid
In: Rat C153
→ *m*-Methylphenylglycol
Probably in: Rat C153

***p*-Methylstyrene**
→ N-Acetyl-S-(2-hydroxy-1-(*p*-methylphenyl) ethyl)-L-cysteine
In: Rat C868
→ N-Acetyl-S-(2-hydroxy-2-(*p*-methylphenyl) ethyl)-L-cysteine
In: Rat C153, C868
→ *p*-Methylphenylacetic acid
In: Rat C153
→ *p*-Methylphenylglycol
Probably in: Rat A2859, C153

→ *p*-Vinylbenzoic acid
In: Rat C153, C681

α-Methylstyrene (isopropenylbenzene)
→ *cis*-2,3-Dihydro-2,3-dihydroxy-α-methylstyrene
In: Bacteria A1648

β-Methylstyrene
→ Benzoic acid
In: Bacteria A1648
→ 3-Phenylpropionaldehyde
In: Bacteria A1648

***o*-Methylstyrene oxide**
→ N-Acetyl-S-(2-hydroxy-1-(*o*-methylphenyl) ethyl)-L-cysteine
In: Rat C868
→ N-Acetyl-S-(2-hydroxy-2-(*o*-methylphenyl) ethyl)-L-cysteine
In: Rat C868

***m*-Methylstyrene oxide**
→ N-Acetyl-S-(2-hydroxy-1-(*m*-methylphenyl) ethyl)-L-cysteine
In: Rat C868
→ N-Acetyl-S-(2-hydroxy-2-(*m*-methylphenyl) ethyl)-L-cysteine
In: Rat C868

***p*-Methylstyrene oxide**
→ N-Acetyl-S-(2-hydroxy-1-(*p*-methylphenyl) ethyl)-L-cysteine
In: Rat C868
→ N-Acetyl-S-(2-hydroxy-2-(*p*-methylphenyl) ethyl)-L-cysteine
In: Rat C868

Methyl styryl ketone
See: *trans*-4-Phenyl-3-buten-2-one

Methyl *p*-sulphamidobenzoate
Formed from: Methyl *p*-aminobenzoate

***p*-N-Methylsulphinylacetamidobenzonitrile**
Formed from: *p*-Acetamidobenzonitrile
→ *p*-N-Methylsulphonylacetamidobenzonitrile
Probably in: Rat D160, F305

2-Methylsulphinylafloqualone
Probably formed from: 2-Methylthioafloqualone
→ 2-Methylsulphonylafloqualone
Probably in: Rat C804

2-Methylsulphinylaniline
Probably formed from: 2-Methylmercaptoaniline
→ 2-Methylsulphinylphenylhydroxylamine
Probably in: Guinea pig G15
→ 2-Methylsulphonylaniline
Probably in: Guinea pig G15, J9

***p*-Methylsulphinylanisole**
Formed from: *p*-Methylthioanisole

2-Methylsulphinylbenzanilide
Formed from: 2-Methylthiobenzanilide

N-Methylsulphinylbenzamide
Formed from: N-Methylthiobenzamide
→ N-Methylthiobenzamide
In: Rat D705

2-Methylsulphinylbenzimidazole
Formed from: 2-Methylmercaptobenzimidazole

Methylsulphinylbromazepam
Probably formed from: Methylthiobromazepam
→ Methylsulphonylbromazepam
Probably in: Rat A1788

4-(Methylsulphinyl)catechol
Formed from: 4-(Methylsulphinyl)phenol

3-(2-Methylsulphinylethyl)quinazoline-2,4(1*H*,3*H*)-dione
Probably formed from: 3-(2-Methylthioethyl) quinazoline-2,4(1*H*,3*H*)-dione
→ 3-(2-Methylsulphonylethyl)quinazoline-2,4(1*H*,3*H*)-dione
Probably in: Rat H122

2-Methylsulphinylnaphthalene
Formed from: 2-Methylthionaphthalene

4-(Methylsulphinyl)phenol
Formed from: 4-(Methylthio)phenol
→ 2-Hydroxy-5-methylsulphinylmuconic semialdehyde
Probably in: Nocardia A2701
→ 4-(Methylsulphinyl)catechol
In: Nocardia A2701
→ 4-(Methylsulphonyl)phenol
Probably in: Cotton A1832
Cow A1818
Rat A1819

2-Methylsulphinylphenylhydroxylamine
Probably formed from: 2-Methylsulphinyl-aniline

4′-Methylsulphinylpropranolol
Formed from: 4′-Methylthiopropranolol
→ 4′-Methylsulphonylpropranolol-β-D-glucuronide
Probably in: Dog F60
→ 4′-Methylsulphinylpropranolol
Probably in: Dog F60
Man F502

4′-Methylsulphinylpropranolol-β-D-glucuronide
Probably formed from: 4′-Methylsulphinyl-propranolol

5-(N-Methylsulphonamidomethyl)indole-3-acetic acid
Formed from: N,N-Dimethyl-5-(N-methylsulphonamidomethyl)tryptamine

***p*-N-Methylsulphonylacetamidobenzonitrile**
Probably formed from: *p*-N-Methylsulphinyl-acetamidobenzonitrile

2-Methylsulphonylafloqualone
Probably formed from: 2-Methylsulphinyl-afloqualone

1-((5((Methylsulphonyl)amino)-1*H*-indol-2-yl)carbonyl)piperazine
Formed from: Delaviridine

2-Methylsulphonylaniline
Probably formed from: 2-Methylsulphinyl-aniline
→ 2-Methylsulphonylphenylhydroxylamine
Probably in: Guinea pig G15

Methylsulphonylbromazepam
Probably formed from: Methylsulphinyl-bromazepam

Methylsulphonylcarbamazepine
Probably formed from: Methylsulphonyl-carbamazepine

4-Methylsulphonyl-2,6-dinitro-N,N-dipropylaniline
See: Nitralin

3-(2-Methylsulphonylethyl)quinazoline-2,4(1*H*,3*H*)-dione
Probably formed from: 3-(2-Methylsulphinyl-ethyl)quinazoline-2,4(1*H*,3*H*)-dione

2-Methylsulphonylphenylhydroxylamine
Probably formed from: 2-Methylsulphonyl-aniline

4-(Methylsulphonyl)phenol
Probably formed from: 4-(Methylsulphinyl)-phenol

2-Methylsulphonyl-N-phenylacetamide
Formed from: Propachlor
→ N-(*p*-Hydroxyphenyl)-2-methylsulphonylacetamide
Probably in: Rat A1682, B93, C421
Sheep A1682

4-(4-(Methylsulphonyl)phenyl)-3-phenyl-2(5*H*)-furanone
See: Rofecoxib

4′-Methylsulphonylpropranolol
Probably formed from: 4′-Methylsulphinyl-propranolol
→ 4′-Methylsulphonylpropranolol-β-D-glucuronide
Probably in: Dog F60

4′-Methylsulphonylpropranolol-β-D-glucuronide
Probably formed from: 4′-Methylsulphonyl-propranolol

4-Methylsulphonyltoluene
Formed from: 4-Methylthiotoluene

N-Methyl-*p*-sympatol (*p*-hydroxy-N,N-dimethyl-phenylethanolamine)
Formed from: *p*-Sympatol

1-Methyl-2-(3,4,5,6-tetrabromo-2-methoxyphenyl)pyrrole
Probably formed from: 2-(3,5,-Dibromo-2-methoxyphenyl)-1-methylpyrrole

Methyl 2,3,5,6-tetrachlorophenyl disulphide
Probably formed from: 2,3,5,6-Tetrachloro-thiophenol

5-Methyltetrahydrofolic acid
Formed from: Dihydrofolic acid
Folic acid
5-Formyltetrahydrofolic acid
5,10-Methylenetetrahydrofolic acid
Tetrahydrofolic acid
→ 5,10-Methylenetetrahydrofolic acid
In: Lactobacillus A3376
See also: A2502
→ 5-Methyltetrahydropteroyldiglutamic acid
In: Pig K455

→ Tetrahydrofolic acid
In: Escherichia K807
Rat A2502
See also: A2446, A2946

Methyl 3,3a,8,8a-tetrahydro-3a-hydroxypyrrolo [2,3-*b*]indol-1(2*H*)-yl carbodithioate
Formed from: Methyl tryptamine dithiocarbamate

4′-O-Methyltetrahydropapaveroline
See: 1,2,3,4-Tetrahydro-6,7-dihydroxy-1-(3-hydroxy-4-methoxybenzyl)isoquinoline

6-O-Methyltetrahydropapaveroline
See: 1,2,3,4-Tetrahydro-1-(3,4-dihydroxybenzyl)-7-hydroxy-6-methoxyisoquinoline

N-Methyltetrahydropapaveroline
See: Laudanosoline

O-Methyltetrahydropapaveroline-1-carboxylic acid
Formed from: Tetrahydropapaveroline-1-carboxylic acid

5-Methyltetrahydropteroyldiglutamic acid
Formed from: 5-Methyltetrahydrofolic acid

5-Methyltetrahydropteroylheptaglutamic acid
→ 5,10-Methylenetetrahydropteroylheptaglutamic acid
In: Lactobacillus A3376
→ Tetrahydropteroylheptaglutamic acid
Probably in: Man A2946

5-Methyltetrahydropteroylpentaglutamic acid
→ 5,10-Methylenetetrahydropteroylpentaglutamic acid
In: Lactobacillus A3376
See also: A2502
→ Tetrahydropteroylpentaglutamic acid
Probably in: Rat A2502
See also: A2946

5-Methyltetrahydropteroyltriglutamic acid
→ 5,10-Methylenetetrahydropteroyltriglutamic acid
In: Lactobacillus A3376
→ Tetrahydropteroyltriglutamic acid
Probably in: Beef A2946

α-N-Methylthalictricavine
Formed from: Thalictricavine

(*R*)-2-(4-(3-Methyl-2-thienyl)phenyl)propionic acid
→ (*S*)-2-(4-(3-Methyl-2-thienyl)phenyl) propionic acid
In: Dog J566
Rat J566
→ (*R*)-2-(4-(3-Methyl-2-thienyl)phenyl) propionyltaurine
In: Dog J566

(*S*)-2-(4-(3-Methyl-2-thienyl)phenyl)propionic acid
Formed from: (*R*)-2-(4-(3-Methyl-2-thienyl) phenyl)propionic acid
→ (*S*)-2-(4-(5-Methyl-3-thienyl)phenyl) propionic acid
In: Man J745
→ (*S*)-2-(4-(3-Methyl-2-thienyl)phenyl) propionyltaurine
In: Dog J566

(*R*)-2-(4-(3-Methyl-2-thienyl)phenyl) propionyltaurine
Formed from: (*R*)-2-(4-(3-Methyl-2-thienyl) phenyl)propionic acid

(*S*)-2-(4-(3-Methyl-2-thienyl)phenyl) propionyltaurine
Formed from: (*S*)-2-(4-(3-Methyl-2-thienyl) phenyl)propionic acid

***o*-Methylthioacetanilide**
Formed from: *o*-Mercaptoacetanilide

***p*-Methylthioacetanilide**
Formed from: *p*-Mercaptoacetanilide

2-Methylthioafloqualone
Probably formed from: S-(2-Afloqualonyl)-L-cysteine
→ N-Acetyl-2-methylthioafloqualone
Probably in: Monkey C912
Rat C793a
→ 2-Methylsulphinylafloqualone
Probably in: Monkey C912

***p*-(Methylthio)anisole** (*p*-methoxythiotoluene)
Formed from: *p*-(Methylthio)phenol
→ *p*-(Methylsulphinyl)anisole
In: Horseradish E403
Man K50
Mortierella D759
Rabbitt H484

N-Methylthiobenzamide
Formed from: N-Methylsulphinylbenzamide
→ N-Methylbenzamide
In: Rat D705
→ N-Methylsulphinylbenzamide
In: Rat D705

2-Methylthiobenzanilide
→ 2-Methylsulphinylbenzanilide
In: Rat F616

2-Methylthiobenzoic acid
Formed from: *o*-Mercaptobenzoic acid

2-Methylthiobenzothiazole
→ S-(2-Benzothiazolyl)glutathione
In: Rat E853

3-Methyl-4,4′-thiobis(benzenethiol)
Formed from: 4,4′-Thiobis(benzenethiol)

Methylthiobromazepam
Formed from: Bromazepam
→ Methylsulphinylbromazepam
Probably in: Rat A1788

4-(Methylthio)catechol
Formed from: 4-(Methylthio)phenol
→ 4-(Methylthio)-*o*-quinone
In: Mushroom H907

7-Methylthioclozapine
Formed from: Clozapine

3-(2-Methylthioethyl)quinazoline-2,4(1*H*,3*H*)-dione
Formed from: 3-(2-Mercaptoethyl)quinazoline-2,4(1*H*,3*H*)-dione
→ Hydroxy-3-(2-methylthioethyl)quinazoline-2,4(1*H*,3*H*)-dione
Probably in: Rat H122
→ 3-(2-Methylsulphinylethyl)quinazoline-2,4(1*H*,3*H*)-dione
Probably in: Rat H122

Methylthiomethylsulphinylbenzene
Formed from: Methylthiomethylthiobenzene

Methylthiomethylthiobenzene
→ Methylthiomethylsulphinylbenzene
In: Pseudomonas H165

1-Methylthionaphthalene
Formed from: Naphthalene

2-Methylthionaphthalene
→ 1-Methylsulphinylnaphthalene
In: Rabbit H484

2-(7-Methylthionaphth-1-yl)ethylamine
→ N-Acetyl-2-(7-methylthionaphth-1-yl)ethylamine
In: Man K490

3-(4-Methylthio-α-naphthyloxy)lactic acid
Formed from: 4′-Methylthiopropranolol

4-(Methylthio)phenol
Formed from: O-Ethyl O-(4-(methylthio)phenyl) S-propylphosphorodithioate
3-(2-(*p*-Methylphenyl)thioethyl)sydnone
→ 2-Hydroxy-5-methylmercaptomuconic semialdehyde
In: Nocardia A2701
→ 4-(Methylsulphinyl)phenol
In: Nocardia A2701
See also: A1818, A1819, A1832
→ 4-(Methylthio)anisole
In: Man D926
→ 4-(Methylthio)catechol
In: Mushroom H907

4′-Methylthiopropranolol
→ 2-Hydroxy-3-(4-methylthio-α-naphthyloxy)propylamine
In: Dog F60
→ 4′-Methylsulphinylpropranolol
Probably in: Dog F60
Man F502
→ 3-(4-Methylthio-α-naphthyloxy)lactic acid
In: Dog F60
→ 4′-Methylthiopropranolol-β-D-glucuronide
Probably in: Dog F60
Man F502

4′-Methylthiopropranolol-β-D-glucuronide
Formed from: 4′-Methylthiopropranolol

4-(Methylthio)-*o*-quinone
Formed from: 4-(Methylthio)catechol

1-Methylthio-3-thiobenzoylthiourea
→ Benzoic acid
In: Pythium F154
→ 3-Benzoyl-1-methylthiourea
In: Pythium F154
→ 3-Methylimino-5-phenyl-1,2,4-thiazole
In: Pythium F154

4-Methylthiotoluene
→ 4-Methylsulphonyltoluene
In: Man G202

1-Methylthioxanthone
→ 1-Hydroxymethylthioxanthone
In: Cunninghamella E46
→ 1-Methylthioxanthone sulphone
In: Cunninghamella E46

2-Methylthioxanthone
→ 2-Methylthioxanthone sulphone
In: Cunninghamella E46

3-Methylthioxanthone
→ 3-Hydroxymethylthioxanthone
In: Cunninghamella E46
→ 3-Methylthioxanthone sulphone
In: Cunninghamella E46

4-Methylthioxanthone
→ 4-Methylthioxanthone sulphone
In: Cunninghamella E46

1-Methylthioxanthone sulphone
Formed from: 1-Methylthioxanthone

2-Methylthioxanthone sulphone
Formed from: 2-Methylthioxanthone

3-Methylthioxanthone sulphone
Formed from: 3-Methylthioxanthone

4-Methylthioxanthone sulphone
Formed from: 4-Methylthioxanthone

3-O-Methyltolcapone
Formed from: Tolcapone

N-Methyl-*p*-toluanilide
→ N-*p*-Hydroxymethylbenzoyl)-N-methylanilide
In: Cunninghamella D810

5-Methyl-N-(*p*-toluenesulphonyl)-2-pyrazoline-1-carboxamide
→ N-(*p*-Hydroxymethylphenylsulphonyl)-5-methyl-2-pyrazoline-1-carboxamide
In: Man B684
Rat B684

N-Methyl-*p*-toluidine
Formed from: N,N-Dimethyl-*p*-toluidine
N-Ethyl-N-methyl-*p*-toluidine
p-Toluidine

2-Methyl-2-(*p*-tolyl)propan-1-ol
Formed from: *p*-*tert*-Butyltoluene
→ 2-Methyl-2-(*p*-tolyl)propionic acid
Probably in: Rat C764

2-Methyl-2-(*p*-tolyl)propionic acid
Probably formed from: 2-Methyl-2-(*p*-tolyl)propan-1-ol

2-Methyl-3-(*o*-tolyl)-4(3*H*)-quinazolinone (methaqualone)
→ 3-(*o*-Hydroxymethylphenyl)-2-methyl-4(3*H*)-quinazolinone
In: Man A1764, A2658, A3311, A3498, A3632, A3634, C215
Rat A1877, A3360
→ 3-(3-Hydroxy-2-methylphenyl)-2-methyl-4(3*H*)-quinazolinone
In: Man A1764, A3311, A3498, A3632, A3634, C215
Rat A1877
→ 3-(4-Hydroxy-2-methylphenyl)-2-methyl-4(3*H*)-quinazolinone
In: Man A1764, A2188, A2418, A3311, A3498, A3632, A3634, C215
Rat A1877
→ 2-Hydroxymethyl-3-(*o*-tolyl)-4(3*H*)-quinazolinone
In: Man A1764, A3311, A3498, A3632, A3634, C215
Rat A1877
→ 6-Hydroxy-2-methyl-3-(*o*-tolyl)-4(3*H*)-quinazolinone
In: Man A1764, A3311, A3498, A3632, A3634, C215
→ 8-Hydroxy-2-methyl-3-(*o*-tolyl)-4(3*H*)-quinazolinone
In: Man A1764
→ 2-Methyl-3-(*o*-tolyl)-4(3*H*)-quinazolinone-N-oxide
In: Man A1762, A2418, A3632, A3634, C177, C215

2-Methyl-3-(*o*-tolyl)-4(3*H*)-quinazolinone-N-oxide
Formed from: 2-Methyl-3-(*o*-tolyl)-4(3*H*)-quinazolinone

3-Methyl-*s*-triazolo[3,4-*a*]phthalazine
Formed from: Budralazine
Hydralazine
Hydralazine pyruvic acid hydrazone
→ Hydralazine
In: Rabbit A3903
→ 3-Hydroxymethyl-*s*-triazolo[3,4-*a*]phthalazine
In: Rat A2544

See also: A1503, A1816, A2299, A3371, A3592, B908
→ 4-Hydroxy-3-methyl-*s*-triazolo[3,4-*a*] phthalazine
Probably in: Man A3592
→ 8-Hydroxy-3-methyl-*s*-triazolo[3,4-*a*] phthalazine
Probably in: Rat A1503, A1816, A2544
→ 9-Hydroxy-3-methyl-*s*-triazolo[3,4-*a*] phthalazine
Probably in: Man A3371

N-Methyl-*p*-trifluoroacetamidoaniline
Formed from: N,N-Dimethyl-*p*-trifluoroacetamidoaniline

2-(((3-Methyl-4-(2,2,2-trifluoroethoxy)-2-pyridinyl)methyl)sulphinyl)-1*H*-benzimidazole
See: Lansoprazole

3-Methyl-4-trifluoromethylacetanilide
Formed from: 3-Methyl-4-trifluoromethylaniline
→ 5-Acetamido-2-trifluoromethylbenzoic acid
Probably in: Rat J785
→ 3-Methyl-4-trifluoromethylglycollanilide
Probably in: Rat J785

3-Methyl-4-trifluoromethylaniline
→ 2-Amino-4-methyl-5-trifluoromethylphenol
In: Rat J785
→ 6-Amino-2-methyl-3-trifluoromethylphenol
In: Rat J785
→ 3-Methyl-4-trifluoromethylacetanilide
In: Rat J785
→ 3-Methyl-4-trifluoromethylaniline-β-D-glucuronide
In: Rat J785

3-Methyl-4-trifluoromethylaniline-β-D-glucuronide
Formed from: 3-Methyl-4-trifluoromethylaniline

3-Methyl-4-trifluoromethylglycollanilide
Probably formed from: 3-Methyl-4-trifluoromethylacetanilide

α-Methyl-3-trifluoromethylphenethylamine
Formed from: N-Ethyl-α-methyl-3-trifluoromethylphenethylamine
→ N-Acetyl-α-methyl-3-trifluoromethylphenethylamine
Probably in: Rat A3123

N-Methyl-N′-(γ-(2-trifluoromethyl-10-phenothiazinyl)propyl) ethylenediamine
Formed from: 10-(3-(4-Methylpiperazin-1-yl)propyl)trifluoromethylphenothiazine

N-Methyl-γ-(4-(trifluoromethyl)phenoxy) benzenepropanamide
See: Fluoxetine

2-((2-Methyl-3-(trifluoromethyl)phenyl)amino)-3-pyridinecarboxylic acid
See: Flunixin

N-Methyltrimetazidine
Formed from Trimetazidine

Methyl 3,4,5-trimethoxycinnamate
Formed from: Methyl sinapate

5-Methyl-6-(((3,4,5-trimethoxyphenyl)amino) methyl)-2,4-quinazolinediamine
See: Trimetrexate

N-Methyltryptamine
Formed from: N,N-Dimethyltryptamine
Tryptamine
→ N-Acetyl-N-methyltryptamine
In: Sheep J850
→ 1,N-Dimethyltryptamine
In: Rabbit C333
→ N,N-Dimethyltryptamine
In: Cat A1567
Dog A1567
Man A1567, A1662
Rabbit A106, A111, A1320, A1662, B741, D646, J884
Rat A80, A111, A968, A1311, A1667, E6
→ 5-Hydroxy-N-methyltryptamine
In: Peganum F9
→ Indole-3-acetaldehyde
In: Rat B653
→ 1,2,3,4-Tetrahydro-2-methyl-β-carboline
In: Chick A1428
Rabbit A111
Rat A104, A111, A1428
See also: A3946, C679
→ Tryptamine
Probably in: Rat A3946, C679

1-Methyltryptamine
Formed from: Tryptamine
→ 1,N-Dimethyltryptamine
In: Rabbit A106, E6

See also: A1310
5-Methyltryptamine
Formed from: 5-Methyl-L-tryptophan
→ 5,N-Dimethyltryptamine
In: Phalaris A3891
Rabbit E6
→ 5-Methylindole-3-acetaldehyde
In: Guinea pig A3958
7-Methyltryptamine
→ 12-Methylajmalicine
Probably in: Catharanthus A2741
→ 7-Methylindole-3-acetic acid
In: Guinea pig A3958
α-Methyltryptamine
→ N-Acetyl-α-methyltryptamine
In: Sheep J850
→ α,N-Dimethyltryptamine
In: Rabbit E6, J884
→ 5-Hydroxy-α-methyltryptamine
In: Peganum F9
Methyl tryptamine dithiocarbamate
→ N-Acetyltryptamine
In: Alternaria K541
Phoma K541
→ Methyl 3,3a,8,8a-tetrahydro-3a-hydroxypyrrolo[2,3-*b*]indol-1(2*H*)-yl carbodithioate
In: Phoma K541
2-Methyl-D-tryptophan
Formed from: D-Tryptophan
2-Methyl-L-tryptophan
Formed from: 2-Methylindole
L-Tryptophan
→ 2-Methylindole
Probably in: Escherichia H617
5-Methyl-L-tryptophan
Formed from: 5-Methylindole
→ N-Formyl-5-methyl-L-kynurenine
In: Streptomyces E589
Wheat E24
→ β-Hydroxy-5-methyl-L-tryptophan
Probably in: Pseudomonas A3007
→ 5-Methylindole-3-pyruvic acid
In: Azospirillum J23
→ 5-Methyltryptamine
In: Micrococcus C62
6-Methyl-L-tryptophan
→ N-Formyl-4-methyl-L-kynurenine
In: Wheat E24
7-Methyl-L-tryptophan
Formed from: 7-Methylindole
→ N-Formyl-3-methyl-L-kynurenine
In: Wheat E24
N-Methyl-L-tryptophan (abrine)
→ Gramine
In: Hordeum A1385
α-Methyl-L-tryptophan
→ N-Formyl-α-methyl-L-kynurenine
In: Streptomyces E589
β-Methyl-L-tryptophan
Formed from: 3-(3-Indolyl)-3-methylpyruvic acid
L-Tryptophan
→ 3-(3-Indolyl)-3-methylpyruvic acid
In: Streptomyces A2307
1-Methyl-L-tryptophan methyl ester
Formed from: L-Tryptophan methyl ester
N-Methyl-L-tryptophan methyl ester
Formed from: L-Tryptophan methyl ester
N-Methyl-*o*-tyramine
→ *o*-Hydroxyphenylacetaldehyde
Probably in: Rat G865
2-Methyltyramine
Formed from: 2-Methyl-L-tyrosine
3-Methyltyramine
Formed from: 3-Methyl-L-tyrosine
α-Methyltyramine
See: *p*-Hydroxyamphetamine
N-Methyl-*m*-tyramine
Formed from: *m*-Methoxy-N-methylphenethylamine
→ *m*-Hydroxyphenylacetaldehyde
In: Rat A2961
N-Methyl-*p*-tyramine
Formed from: Hordenine
Tyramine
→ 3,4-Dimethoxy-N-methylphenethylamine
In: Coryphantha B577
→ 3,4-Dimethoxy-N-methylphenylethanolamine
In: Coryphantha A3977
→ Hordenine
Probably in: Barley A3594

Chickpea A3594
Glycine A3594
Mung bean A3594
Parsley A3594
Wheat A3594
→ *p*-Hydroxyphenylacetaldehyde
Probably in: Rat G865
→ Tyramine
Probably in: Barley A3594
Parsley A3594
Wheat A3594

2-Methyl-L-tyrosine
Formed from: *m*-Cresol
→ 2-Methyltyramine
In: Micrococcus C62

3-Methyl-L-tyrosine
Formed from: *o*-Cresol
p-Methyl-L-phenylalanine
→ 3-Methyltyramine
In: Micrococcus C62

α-Methyl-*m*-tyrosine
→ *m*-Hydroxyphenylacetone
In: Pig A2425

α-Methyltyrosine
→ Di-(α-methyltyrosine)
In: Rat G616

4-Methylumbelliferone (7-hydroxy-4-methylcoumarin)
Formed from: 4-Methylumbelliferyl acetate
4-Methylumbelliferyl-N-acetyl-β-D-galactosaminide
4-Methylumbelliferyl-N-acetyl-β-D-glucosaminide
4-Methylumbelliferyl-α-L-fucoside
4-Methylumbelliferyl-α-D-galactoside
4-Methylumbelliferyl-β-D-galactoside
4-Methylumbelliferyl-β-D-glucosaminide
4-Methylumbelliferyl-α-D-glucoside
4-Methylumbelliferyl-β-D-glucoside
4-Methylumbelliferyl-β-D-glucuronide
4-Methylumbelliferyl-α-D-mannoside
4-Methylumbelliferyl palmitate
4-Methylumbelliferyl sulphate
Probably formed from:
4-Methylumbelliferyl-β-D-xyloside
→ 6,7-Dihydroxy-4-methylcoumarin
In: Rat A773
→ 4-Methylumbelliferyl-β-D-glucoside
In: Nicotiana K556
→ 4-Methylumbelliferyl glucoside-6-phosphate
Probably in: Mealworm C494
→ 4-Methylumbelliferyl-β-D-glucuronide
In: Man A91, B800, F309, G524, G968, H575, J416, J490, J617
Mouse D315
Rabbit H720, J107
Rat A1531, A1668, A3464, B740, C672, E106, F69, J107
See also: D816
→ 4-Methylumbelliferyl sulphate
In: Eubacterium E432
Rabbit H720
Rat A3464, F69

4-Methylumbelliferyl acetate
→ 4-Methylumbelliferone
In: Ascaris A324, A3040
Moniezia A324, A3040
Rat F753

4-Methylumbelliferyl-N-acetyl-β-D-galactosaminide
→ 4-Methylumbelliferone
In: Man A551, A642
Physarium A23
Rabbit A1221
Rat A512

4-Methylumbelliferyl-N-acetyl-β-D-glucosaminide
→ 4-Methylumbelliferone
In: Man A551
Physarium A23
Rabbit A1221
Rat A512

4-Methylumbelliferyl-α-L-fucoside
→ 4-Methylumbelliferone
In: Man A1845

4-Methylumbelliferyl-α-D-galactoside
→ 4-Methylumbelliferone
In: Man A551, A642
Poterioochromonas B486
Rabbit A1221
Rat A512

4-Methylumbelliferyl-β-D-galactoside
→ 4-Methylumbelliferone
In: Man A551, A642
Rabbit A1221
Rat A512
4-Methylumbelliferyl-β-D-glucosaminide
→ 4-Methylumbelliferone
In: Cicer K832
4-Methylumbelliferyl-α-D-glucoside
→ 4-Methylumbelliferone
In: Man A551, A642
4-Methylumbelliferyl-β-D-glucoside
Formed from: 4-Methylumbelliferone
→ 4-Methylumbelliferone
In: Man A642
Picea K833
Rat A512
Triglochin A3756
→ 4-Methylumbelliferyl-6″-O-malonyl-β-D-glucoside
In: Petroselinum K725
4-Methylumbelliferyl glucoside-6-phosphate
Probably formed from: 4-Methylumbelliferone
4-Methylumbelliferyl-β-D-glucuronide
Formed from: 4-Methylumbelliferone
→ 4-Methylumbelliferone
In: Man A642, A682
Microorganism G567
Rat A512
With β-glucuronidase J585
4-Methylumbelliferyl-6″-O-malonyl-β-D-glucoside
Formed from: 4-Methylumbelliferyl-β-D-glucoside
4-Methylumbelliferyl-α-D-mannoside
→ 4-Methylumbelliferone
In: Man A551, A642, A1842
Rat A512, A1842
4-Methylumbelliferyl palmitate
→ 4-Methylumbelliferone
In: Man J233
4-Methylumbelliferyl sulphate
Formed from: 4-Methylumbelliferone
→ 4-Methylumbelliferone
In: Ascaris A324
Eubacterium E432
Haemophilus H756
Man A942
Moniezia A324
Rabbit B481
Rat A512, A1050, E106, E426
4-Methylumbelliferyl-β-D-xyloside
→ Galactosyl(β1→3)xylyl(β)-4-methylumbelliferone
In: Escherichia J399
→ Galactosyl(β1→4)xylyl(β)-4-methylumbelliferone
In: Escherichia J399
→ 4-Methylumbelliferone
In: Mouse C77
N-Methyl-N-(γ-valerolactonyl)amphetamine
Formed from: Furfenorex
α-Methylvanilalanine
See: 4-Hydroxy-3-methoxy-α-methyl-L-phenylalanine
N-Methylviloxazine
Formed from: Viloxazine
→ N-Formylviloxazine
Probably in: Dog A355
→ N-Methylviloxazine-N-oxide
In: Dog A355
N-Methylviloxazine-N-oxide
Formed from: N-Methylviloxazine
5-O-Methylxylerythrin
Formed from: 4-Hydroxypulvinic acid
L-Phenylalanine
D-Tyrosine
1-Methyl-2-(2,6-xylyloxy)ethylamine
→ 2-(4-Hydroxy-2,6-dimethylphenoxy)-1-methylethylamine
In: Man A1893
→ 2-(2-Hydroxymethyl-6-methylphenoxy)-1-methylethylamine
In: Man A1893
Methysergide ((8β)-9,10-didehydro-N-((1*S*)-1-(hydroxymethyl)propyl)-1,6-dimethylergoline-8-carboxamide)
→ N^1-Hydroxymethylmethergine
In: Rat A98
→ Methergine
In: Rat A98

Methystictin ((6*R*)-6-((1*E*)-2-(1,3-benzodioxol-5-yl)ethenyl)-5,6-dihydro-4-methoxy-2*H*-pyran-2-one)
→ 3′,4′-Dihydroxykawain
In: Rat A3485
Metiazinic acid (10-methyl-10*H*-phenothiazine-2-acetic acid)
→ Metiazinic acid sulphoxide
In: Rabbit A2216
Rat A2216
Metiazinic acid sulphoxide
Formed from: Metiazinic acid
Metoclopramide (4-amino-5-chloro-N-diethylaminoethyl-2-methoxybenzamide)
→ 4-Amino-5-chloro-N-diethylaminoethyl-2-hydroxybenzamide
In: Rabbit A1799
Rat A3346
→ 4-Amino-5-chloro-N-ethylaminoethyl-2-methoxybenzamide
In: Dog A3346
Rabbit A1799
Rat A3346
See also: A2400
→ 4-Amino-5-chloro-N-(2′-hydroxyethyl)-2-methoxybenzamide
In: Dog A3346
Rabbit A1799
Rat A3346
→ 4-Amino-5-chloro-2-methoxybenzamide
In: Dog A3346
Rat A3346
→ 2-((4-Amino-5-chloro-2-methoxybenzoyl)amino)acetic acid
In: Dog A3346
Man A3346
Rat A3346
Metol
See: *p*-Methylaminophenol
Metolachlor (2-chloro-N-(2-ethyl-6-methylphenyl)-N-(2-methoxy-1-methylethyl)acetamide)
→ 2-(N-Acetyl-S-(L-cysteinyl))-N-(2-ethyl-6-methylphenyl)-N-(2-methoxy-1-methylethyl)acetamide
In: Man H839
Sunfish G779
→ 2-Chloro-N-(2-ethyl-6-hydroxymethylphenyl)-N-(2-methoxy-1-methylethyl)acetamide
In: Cunninghamella H847
Phanerochaete F871
Rhizoctonia F871
Streptomyces F871
Sunfish G779
Syncephalastrum F871
→ 2-Chloro-N-(2-ethyl-6-methylphenyl)acetanilide
In: Phanerochaete F871
Rhizoctonia F871
Streptomyces F871
Syncephalastrum F871
→ 2-Chloro-N-(2-ethyl-6-methylphenyl)-N-(2-hydroxy-1-methylethyl)acetamide
In: Cunninghamella H847
Phanerochaete F871
Rhizoctonia F871
Streptomyces F871
Sunfish G779
Syncephalastrum F871
→ 2-Chloro-N-(2-(1-hydroxyethyl)-6-methylphenyl)-N-(2-methoxy-1-methylethyl)acetamide
See also: Cunninghamella H847
→ 2-Ethyl-N-(1-methoxy-2-propyl)-6-methylaniline
In: Rat D943
→ N-(2-Ethyl-6-methylphenyl)-N-(2-methoxy-1-methylethyl)glycollamide
In: Phanerochaete F871
Rhizoctonia F871
Streptomyces F871
Syncephalastrum F871
Metoprolol (1-isopropylamino-3-(4-(2-methoxyethyl)phenoxy)propan-2-ol)
→ 3-(4-(2-Hydroxyethyl)phenoxy)-1-isopropylaminopropan-2-ol
In: Dog G41
Man F794, G558, H788, J673
Rat A1750, F794, J590
See also: A2255
→ 3-(4-(1-Hydroxy-2-methoxyethyl)phenoxy)-1-isopropylaminopropan-2-ol
In: Dog A2255, G41

Man A2255, F794, H788
Rat A1750, A2255, F794
→ 3-(4-(2-Methoxyethyl)phenoxy)propan-2-ol
In: Man G558, J673
Rat A1750, J590

Metoxadiazone (5-methoxy-3-(2-methoxyphenyl)-1,3,4-oxadiazol-2(3*H*)-one)
→ *o*-Methoxy-N′-methoxycarbonylphenylhydrazone
Probably in: Rat E599
→ Resorcinol
In: Rat E599

Metyrapol (β,β-dimethyl-α-3-pyridinyl-3-pyridineethanol)
Formed from: Metyrapone

Metyrapone (2-methyl-1,2-di-3-pyridinyl-1-propanone)
→ Metyrapol
In: Mouse H230
Rabbit K569
Rat G390
→ Metyrapone-N-oxide
In: Rat G390

Metyrapone-N-oxide
Formed from: Metyrapone

Mexacarbate (4-dimethylamino-3,5-xylyl methylcarbamate)
→ 4-Dimethylamino-3,5-dimethylphenol
In: Bacterium A1207
→ 4-Methylamino-3,5-xylyl methylcarbamate
In: Rat B8
Trichoderma A1207
→ 4-Methylformamido-3,5-xylyl methylcarbamate
In: Bacterium A1207

Mexiletine (1-(2,6-dimethylphenoxy)-2-propanamine)
→ 1-(2,6-Dimethylphenoxy)-2-oximinopropane
In: Cunninghamella J155
→ 1-(2,6-Dimethylphenoxy)propan-2-ol
In: Man G385
Rabbit G385
→ Hydroxymethylmexiletine
In: Cunninghamella J155
Man F529
Rabbit F734
→ *m*-Hydroxymexiletine
In: Cunninghamella J155
Guinea pig F734
Hamster F734
Man F734
Mouse F734
Rabbit F734
Rat F734
→ *p*-Hydroxymexiletine
In: Cunninghamella J155
Man F529
Rabbit F734
→ N-Hydroxymexiletine
In: Cunninghamella J155
Man G385, G976
Rabbit G385
→ N-Methylmexiletine
In: Cunninghamella J155

Mianserin (1,2,3,4,10,14b-hexahydro-2-methyldibenzo[*c*,*f*]pyrazino[1,2-*a*]azepine)
Probably formed from: Mianserin-N-oxide
→ 2-Demethyl-2-formylmianserin
In: Man B610
Rabbit B610
Rat B610
→ Demethylmianserin
In: Man A2920, B610, F140, G66, G864
Mouse G66
Rabbit B610
Rat B610, G864, G908, J784
See also: A2018
→ 8-Hydroxymianserin
In: Guinea pig G66
Man B610, F140, G864
Mouse G66
Rabbit B610, G66
Rat B610, G66, G864, G908, J784
→ Mianserin-N-β-D-glucuronide
In: Man G66, G864
→ Mianserin-N-oxide
In: Guinea pig G66
Man A2920, B610, F140, G66, G864
Mouse G66
Rabbit B610
Rat B610, G864, J784

Mianserin-N-β-D-glucuronide
Formed from: Mianserin
Mianserin-N-oxide
Formed from: Mianserin
→ 8-Hydroxymianserin-N-oxide
Probably in: Man G864
Mouse G66
→ Mianserin
Probably in: Rat A2018
Mibefradil (methoxyacetic acid (1*S*,2*S*)-2-(2-((3-(1*H*-benzimidazol-2-yl)propyl)methylamino)ethyl)-6-fluoro-1,2,3,4-tetrahydro-1-isopropyl-2-naphthyl ester)
→ 2-(2-((3-(1*H*-Benzimidazol-2-yl)propyl)methylamino)ethyl)-6-fluoro-1,2,3,4-tetrahydro-2-hydroxy-1-isopropylnaphthalene
In: Cynomolgus monkey J103
Man J103
Marmoset J103
Rabbit J103
Rat G311, J103
→ N-Demethylmibefradil
In: Cynomolgus monkey J103
Man J103
Marmoset J103
Rabbit J103
Rat G311, J103
→ O-Demethylmibefradil
In: Cynomolgus monkey J103
Man J103
Marmoset J103
Rabbit J103
Rat J103
→ Methoxyacetic acid 2-(2-((3-(1*H*-benzimidazol-2-yl)propyl)-N-(carboxyglucuronyl)amino)ethyl)-6-fluoro-1,2,3,4-tetrahydro-1-isopropyl-2-naphthalenyl ester
In: Rat G311
→ Methoxyacetic acid 2-(2-((3-(1*H*-benzimidazol-2-yl)propyl)methylamino)ethyl)-6-fluoro-1,2,3,4-tetrahydro-4-hydroxy-1-isopropyl-2-naphthalenyl ester
In: Cynomolgus monkey J103
Man J103
Marmoset J103
Rabbit J103
Rat G311, J103
→ Methoxyacetic acid 6-fluoro-1,2,3,4-tetrahydro-2-(2-((3-(4-hydroxy-1*H*-benzimidazol-2-yl)propyl)methylamino)ethyl)-1-isopropyl-2-naphthalenyl ester
In: Cynomolgus monkey J103
Man J103
Marmoset J103
Rabbit J103
Rat G311, J103
→ Methoxyacetic acid 6-fluoro-1,2,3,4-tetrahydro-2-(2-((3-(5-hydroxy-1*H*-benzimidazol-2-yl)propyl)methylamino)ethyl)-1-isopropyl-2-naphthalenyl ester
In: Cynomolgus monkey J103
Man J103
Marmoset J103
Rabbit J103
Rat G311, J103
→ Methoxyacetic acid 6-fluoro-1,2,3,4-tetrahydro-1-isopropyl-2-(2-(N-methylamino)ethyl)-2-naphthalenyl ester
In: Cynomolgus monkey J103
Man J103
Marmoset J103
Rabbit J103
Rat J103
Michler's ketone (4,4′-bis(dimethylamino)benzophenone)
→ 4′-Dimethylamino-4-methylaminobenzophenone
Probably in: Rat B713
Midaglizole (2-(2-(4,5-dihydro-1*H*-imidazol-2-yl)-1-phenylethyl)pyridine)
→ *p*-Hydroxymidaglizole
In: Rat F334
→ 2-(2-Phenyl-2-(pyridyl)ethyl)imidazole
In: Dog E842
Midazolam (8-chloro-6-(2-fluorophenyl)-1-methyl-4*H*-imidazo[1,5*a*][1,4]benzodiazepine)
→ 8-Chloro-6-(*o*-fluorophenyl)-1-hydroxymethyl-4*H*-imidazo[1,5-*a*][1,4]benzodiazepine
In: Man C129, J43, J225, J672, K254
Mouse K441, K161
Rat B705, J43

See also: K84
→ 4-Hydroxymidazolam
In: Man J43, J672
Mouse K161, K441
Rat B705, J43

Mifentidine (N-(4-(1*H*-imidazol-4-yl)phenyl)-N′-(1-methylethyl)methamidate)
→ *p*-(1*H*-Imidazol-4-yl)aniline
In: Guinea pig G411
Hamster G411
Mouse G411
Rat G411
→ *p*-(1*H*-Imidazol-4-yl)formanilide
In: Guinea pig G411
Hamster G411
Mouse G411
Rat G411
→ 1-(4-(1*H*-Imidazol-4-yl)phenyl)-3-isopropylurea
In: Guinea pig G411
Hamster G411
Mouse G411
Rat G411

Mifepristone ((11β,17β)-11-(4-(dimethylamino)phenyl)-17-hydroxy-17-(1-propynyl)oestra-4,9-dien-3-one)
→ Demethylmifepristone
In: Man J752
Monkey J752
Rat J752
→ 11β-(*p*-Dimethylaminophenyl)-17β-hydroxy-17α-(1-(3-hydroxy)propynyl)oestra-4,9-dien-3-one
In: Man J752
Monkey J752
Rat J752

Millipertine
→ 1-(*o*-Methoxyphenyl)piperazine
In: Rat D308

Minaprine (N-(4-methyl-6-phenyl-3-pyridazinyl)-4-morpholineethanamine)
→ 3-Amino-4-methyl-6-phenylpyridazine
In: Baboon B839, D348
Dog B839
Macaca B839, G65
Marmoset B839
Rat B839, F906, G65
→ 3-(2-(2-Hydroxyethylamino)ethylamino)-4-methyl-6-phenylpyridazine
In: Baboon B839, D348
Dog B839, G65
Macaca B839, G65
Man B839, G65
Marmoset B839
Rat B839, F906, G65
→ *p*-Hydroxyminaprine
In: Baboon B839, D348, F906
Dog B839, F906, G65
Macaca B839, G65
Man B839, G65, G302
Marmoset B839
Rabbit F906
Rat B839, D640, F906, G65
→ 4-Methyl-3-(2-(3-oxo-2-morpholino)ethylamino)-6-phenylpyridazine
In: Baboon D348
Dog G65
Rat G65
→ Minaprine-N-oxide
In: Baboon B839, D348
Dog B839, G65
Macaca B839, G65
Man B839, G65
Marmoset B839
Rat B839, D640, F906, G65

Minaprine-N-oxide
Formed from: Minaprine

Minocycline ((4*S*,4a*S*,5a*R*,12a*S*)-4,7-bis(dimethylamino)-1,4,4a,5,5a,6,11,12a-octahydro-3,10,12,12a-tetrahydroxy-1,11-dioxo-2-naphthacenecarboxamide)
→ Demethylminocycline
Probably in: Man B989
→ 9-Hydroxyminocycline
Probably in: Man B989

Mirtazapine (1,2,3,4,10,14b-hexahydro-2-methylpyrazino[2,1-*a*]pyrido[2,3-*c*][2]benzazepine)
→ 8-Hydroxymirtazapine
In: Man K130
→ Mirtazapine-N-oxide
In: Man K130

Mirtazapine-N-oxide
Formed from: Mirtazapine
Mitoclomine (4-amino-N,N-bis(2-chloroethyl)-1-methoxy-2-methylnaphthalene)
→ Demethylmitoclomine
Probably in: Rat A3248
→ Desethylmitoclomine
Probably in: Rat A3248
Mitoxantrone (1,4-dihydroxy-5,8-bis((2-((2-hydroxyethyl)amino)ethyl)amino)-9,10-anthracenedione)
→ 1-(2-(Carboxymethylamino)ethylamino)-5,8-dihydroxy-4-(2-(2-hydroxyethylamino) ethylamino)anthraquinone
In: Man E153, G48
Pig G48, G231
Rat G48
→ 1,2,3,4,7,12-Hexahydro-8,11-dihydroxy-4-(2-hydroxyethyl)-6-(2-(2-hydroxyethyl) aminoethyl)amino-7,12-dioxonaphtho [2,3-*f*]quinoxaline
In: Man H353
In: Pig G231
Rat D647
→ 2-(S-(L-Cysteinylglycyl))mitoxantrone
In: Horseradish G875
→ 2-(S-(L-Cysteinyl))mitoxantrone
In: Horseradish G875
→ 2-(S-(L-Cysteinyl))mitoxantrone ethyl ester
In: Horseradish G875
→ 6-(S-Glutathionyl)mitoxantrone
In: Horseradish G48
→ Mitoxantrone-β-D-glucuronide
In: Pig G231
Rat D647
Mitoxantrone-β-D-glucuronide
Formed from: Mitoxantrone
Mitragynine ((α*E*,2*S*,3*S*,12b*S*)-3-ethyl-1,2,3,4,6,7,12,12b-octahydro-8-methoxy-α-(methoxymethylene)indolo[2,3-*a*]quinolizine-2-acetic acid methyl ester)
Formed from: Strictosidine
Mociprazine (α-(((1-ethynylcyclohexyl)oxy) methyl)-4-(2-methoxyphenyl)-1-piperazineethanol)
→ Demethylmociprazine
In: Dog G433
→ α-(((1-Ethynylcyclohexyl)oxy)methyl)-4-(5-hydroxy-2-methoxyphenyl)-1-piperazineethanol
In: Dog G433
→ α-(((1-Ethynylcyclohexyl)oxy)methyl)-1-piperazineethanol
In: Dog G433
→ α-(((1-Ethynyl-hydroxycyclohexyl)oxy) methyl)-4-(2-methoxyphenyl)-1-piperazineethanol
In: Dog G433
→ Guaiacol
Probably in: Dog D708, G433
Moclobemide (4-chloro-N-(2-(4-morpholinyl) ethyl)benzamide)
→ N-(*p*-Chlorobenzoyl)-N′-(2-hydroxyethyl) ethylenediamine
In: Rat F831
→ *p*-Chloro-N-(2-(N-(2-hydroxyethyl)amino) ethyl)benzamide
In: Man G707
Rat G707
Modipafant ((+)-4-(*o*-chlorophenyl)-3-ethoxycarbonyl-1,4-dihydro-6-methyl-2-(4-(methylimidazo[4,5-*c*]pyrid-1-yl)phenyl)-5-(N-(2-pyridyl)carbamoyl)pyridine)
→ 4-(*o*-Chlorophenyl)-3-ethoxycarbonyl-1,4-dihydro-5-(N-(3-hydroxy-2-pyridyl) carbamoyl)-6-methyl-2-(4-(methylimidazo [4,5-*c*]pyrid-1-yl)phenyl)pyridine
In: Dog H735
Rat H735
→ 1,4-Dehydromodipafant
In: Dog H735
Rat H735
→ Modipafant-N-oxide
In: Dog H735
Rat H735
Modipafant-N-oxide
Formed from: Modipafant
→ 4-(*o*-Chlorophenyl)-3-ethoxycarbonyl-1,4-dihydro-5-(N-(3-hydroxy-2-pyridyl) carbamoyl)-6-methyl-2-(4-(methylimidazo [4,5-*c*]pyrid-1-yl)phenyl)pyridine-N-oxide
Probably in: Dog H735
Rat H735

Mofegiline ((βE)-4-fluoro-β-(fluoromethylene) benzenebutanamide)
→ N-((4-(*p*-Fluorophenyl)-2-formyl)but-1-enyl)urea
In: Dog G962
Man G962
→ Mofegiline carbamate
In: Dog G962
Man G962
→ N-Succinylmofegiline
In: Dog G962
Man G962

Mofegiline carbamate
Formed from: Mofegiline

Mogadon
See: 1,3-Dihydro-7-nitro-5-phenyl-2*H*-1,4-benzodiazepin-2-one

Moniliphenone (2′,6,6′-trihydroxy-2-methoxycarbonyl-4′-methylbenzophenone)
→ Chloromonilicin
In: Monilinia E190
→ 4-Chloropinselin
In: Monilinia E190

Mono-*o*-cresyl phosphate
Formed from: Di-*o*-cresyl phosphate
→ Cyclic-*o*-tolyl phosphate
Probably in: Rat G129

Monolinuron (3-(*p*-chlorophenyl)-1-methoxy-1-methylurea)
→ *p*-Chloroaniline
In: Fusarium B237
→ 3-(*p*-Chlorophenyl)-1-methylurea
In: Fusarium B237

Monomethyl terephthalate
Formed from: Dimethyl terephthalate
Methyl *p*-cyanobenzoate
→ Terephthalic acid
In: Aspergillus H424
Bacillus F403

Montelukast ((*R*)-1-((1-(3-(2-(7-chloro-2-quinolinyl)ethenyl)phenyl)-3-(2-(2-hydroxy-2-propyl)phenyl)propyl)thiomethyl) cyclopropaneacetic acid)
→ (*R*)-1-((1-(3-(2-(7-chloro-2-quinolinyl) ethenyl)phenyl)-3-(2-(1,2-dihydroxy-2-propyl)phenyl)propyl)thiomethyl) cyclopropaneacetic acid
In: Man J265
See also: H804
→ 21-Hydroxymontelukast
In: Man H804, J265
→ 25-Hydroxymontelukast
In: Man H804
→ 33-Hydroxymontelukast
In: Man H804
→ Montelukast-β-D-glucuronide
In: Man J265, H804
→ Montelukast sulphoxide
In: Man J265, H804

Montelukast-β-D-glucuronide
Formed from: Montelukast

Montelukast sulphoxide
Formed from: Montelukast

Monuron
See: 3-(*p*-Chlorophenyl)-1,1-dimethylurea

Mordant Orange 1 (2-hydroxy-5-((4-nitrophenyl) azo)benzoic acid)
→ 5-Amino-2-hydroxybenzoic acid
In: Microorganism J476
→ *p*-Phenylenediamine
In: Microorganism J476

Morforex (N-morpholinoethylamphetamine)
→ Amphetamine
In: Man A2453

Moricizine (10-(3-(4-morpholinyl)-1-oxopropyl)-10*H*-phenothiazin-2-yl)carbamic acid ethyl ester)
→ 2-Amino-10-(3-morpholinopropionyl) phenothiazine
In: Man J85, J747
→ Ethyl (10-(N-(2-hydroxyethyl)-3-aminopropionyl)phenothiazin-2-yl) carbamate
In: Man J85, J747
→ Moricizine sulphoxide
In: Man J85, J747

Moricizine sulphoxide
Formed from: Moricizine
→ Ethyl phenothiazine-2-carbamate sulphoxide
Probably in: Man J85, J747

Morin (2′,3,4′,5,7-pentahydroxyflavone)
→ 2,4-Dihydroxybenzoic acid
In: Horseradish A1183

→ 2′,3,4′,5,5′,7-Hexahydroxyflavone
Probably in: Horseradish A1183
→ 2,4,4′-Trihydroxychalcone
In: Horseradish A1183

Morphine ((−) isomer assumed unless otherwise stated)
Formed from: 6-Acetylmorphine
3-Benzoylmorphine
6-Butanoylmorphine
Codeine
3-Cyclohexanoylmorphine
4′-Ethyl-6-methyllaudanosoline
6-Ethyl-4′-methyllaudanosoline
3-Ethylmorphine
Morphine-3-β-D-glucuronide
Morphine-6-β-D-glucuronide
Morphinone
Pholcodine
3-Pivaloylmorphine
Probably formed from: 6-Nicotinoylmorphine
Normorphine
→ Codeine
In: Man A1453
Mouse G567
Rabbit G567
Rat G567
Contrast (in man) A1160
→ Dihydromorphine
In: Cat A2091
Guinea pig A2091
Mouse A2091
Rabbit A2091
Rat A1857, A2091
→ Dimorphine
In: Rat G616
→ (8*S*)-(S-Glutathionyl)dihydromorphinone
In: Guinea pig F18
Rat J165
→ Hydromorphone
In: Pseudomonas G907, H892
→ Morphine-3-β-D-glucuronide
In: Dog H786, K41
Guinea pig G521
Man A2370, A3296, G789, H292, H909, H933, J74, J183, K33
Monkey D130, H861
Mouse C80
Ochotona F438
Rabbit A1961, A2084
Rat A858, A1374, A1862, A1919, B526, B962, C221, D344, D423, D799, F364, J400
Sheep G846, H178
See also: A134, A804, A2312, A2907, A3107, A3827, C28, C562, C654, C672, D315, D495, D659, E62, F521, G909, H537, H539, H595, H639, H766, J97a, J416, J783
→ Morphine-6-β-D-glucuronide
In: Dog K41
Guinea pig G521
Man A3296, G789, H292, H766, H909, H933, J74, K33, K517
Monkey D130, H861
Sheep G846, H178
See also: F521, H595, H639
Contrast (in rat) D799
→ Morphine-N-oxide
In: Guinea pig A3729
Horseradish C414, D104
Papaver A2743, D104
→ Morphine-3-sulphate
In: Cat A858
Man A2086, A3296
Monkey A2086
Rat A2086
→ Morphinone
In: Guinea pig C357, F876
Hamster K443
Papaver K756
Pseudomonas F815, G181
Rabbit H920
Rat E62, J165
See also: G51
→ S-(10α-Morphinyl)glutathione
Probably in: Rat D266, E127
→ Normorphine
In: Cat A2091
Man A1453, A2091, A2370, A3296
Mouse A2091
Rabbit A2091
Rat A1857, A1983, A2091, C669, D495, E62

See also: A1347, A2236, A2907, D419, G909, H595
→ Pseudomorphine
In: Horseradish A1357, C414, D104
Papaver D104

(+)-Morphine
→ (+)-Morphine-3-β-D-glucuronide
In: Rat D799, F364
→ (+)-Morphine-6-β-D-glucuronide
In: Rat D799, F364

Morphine-3,6-di-β-D-glucuronide
Probably formed from: Morphine-3-β-D-glucuronide
→ Morphine-6-β-D-glucuronide
In: Abalone H428
Helix H428
Limpet H428

Morphine-3-β-D-glucuronide
Formed from: Morphine
→ Morphine
In: Abalone H419
Beef H419
Helix H419
Microorganism G567
Patella G879
Rat A2236, G861
See also: C102
→ Morphine-3,6-di-β-D-glucuronide
Probably in: Man A3296

(+)-Morphine-3-β-D-glucuronide
Formed from: (+)-Morphine

Morphine-6-β-D-glucuronide
Formed from: Morphine
Morphine-3,6-di-β-D-glucuronide
→ Morphine
In: Abalone H419
Beef H419
Helix H419
Microorganism J400
Rat A2236

(+)-Morphine-6-β-D-glucuronide
Formed from: (+)-Morphine

Morphine-N-oxide
Formed from: Morphine

Morphine-3-sulphate
Formed from: Morphine

Morphinone ((5α)-7,8-didehydro-4,5-epoxy-3-hydroxy-17-methylmorphinan-6-one)
Formed from: Morphine
Oripavine
→ S-(7,8-Dihydromorphinon-8-yl)glutathione
Probably in: Guinea pig F876
Rat E62
→ Morphine
In: Papaver K756, K757
Pseudomonas H785
Rat J744
See also D572

S-(10α-Morphinyl)glutathione
Probably formed from: Morphine
→ N-Acetyl-S-(10α-morphinyl)-L-cysteine
Probably in: Man E127
Rat E127

S-Morpholino-2-benzothiazolyl sulphenamide
→ N-Acetyl-S-(2-benzothiazolyl)-L-cysteine
In: Rat H767
→ 2-Mercaptobenzothiazole
In: Rat H767

***o*-(2-N-Morpholinoethoxy)phenyl phenyl ether**
→ *o*-(2-N-Hydroxyoxomorpholinoethoxy) phenyl phenyl ether
In: Rabbit A325
Rat A325
→ *p*-Hydroxyphenyl *o*-(2-N-morpholinoethoxy)phenyl ether
In: Rabbit A325
Rat A325
→ 2-Hydroxyphenyl phenyl ether
In: Rabbit A325
Rat A325

N-Morpholinoethylamphetamine
See: Morforex

3-O-Morpholinoethylmorphine
See: Pholcodine

3-O-(2-Morpholinoethyl)normorphine
Formed from: Pholcodine

1-Morpholino-3-((*p*-phenoxymethyl)phenyl) propane
See: Fomocaine

***p*-Morpholinophenylacetic acid**
Probably formed from: Chlorambucil
p-(γ-Morpholinopropyl)benzoic acid
Formed from: Fomocaine

cis-4-(2-Morpholinylmethoxy)-1,2-indandiol
→ *cis*-4-(2-Morpholinylmethoxy)-1,2-indandiol-α-D-glucoside
In: Rat G303

trans-4-(2-Morpholinylmethoxy)-1,2-indandiol
Formed from: Indeloxazine
→ *trans*-4-(2-Morpholinylmethoxy)-1,2-indandiol-1-α-D-glucoside
In: Rat E837
→ *trans*-4-(5-Oxo-2-morpholinylmethoxy)-1,2-indandiol
In: Rat E837

cis-4-(2-Morpholinylmethoxy)-1,2-indandiol-α-D-glucoside
Formed from: *cis*-4-(2-Morpholinylmethoxy)-1,2-indandiol

trans-4-(2-Morpholinylmethoxy)-1,2-indandiol-1-α-D-glucoside
Formed from: *trans*-4-(2-Morpholinylmethoxy)-1,2-indandiol

10-(3-(4-Morpholinyl)-1-oxopropyl)-10*H*-phenothiazin-2-yl)carbamic acid ethyl ester
See: Moricizine

7-(2-Morpholinylmethoxy)inden-4-ol
Formed from: Indeloxazine

Mosapride (4-amino-5-chloro-2-ethoxy-N-((4-((4-fluorophenyl)methyl)-2-morpholinyl)methyl)benzamide)
→ 4-Amino-5-chloro-2-ethoxy-N-((1,4-oxazin-2-yl)methyl)benzamide
In: Man K137
→ *p*-Fluorobenzoic acid
Probably in: Man K137

Moxifloxacin (1-cyclopropyl-6-fluoro-1,4-dihydro-8-methoxy-7-(4a*S*,7a*S*)-octahydro-6*H*-pyrrolo[3,4-*b*]pyridin-6-yl)-4-oxo-3-quinolinecarboxylic acid)
→ Moxifloxacin-β-D-glucuronide
In: Man K530
→ Moxifloxacin-N-sulphate
In: Man K530

Moxifloxacin-β-D-glucuronide
Formed from: Moxifloxacin

Moxifloxacin-N-sulphate
Formed from: Moxifloxacin

MPP$^+$
See: 1-Methyl-4-phenylpyridinium

MPTP
See: 1,2,3,6-Tetrahydro-1-methyl-4-phenylpyridine

Muramine (5,7,8,14-tetrahydro-3,4,10,11-tetramethoxy-6-methyldibenz[*c*,*g*]azecin-13(6*H*)-one)
Formed from: Tetrahydropalmatine
→ Alpinigenine
In: Papaver A1923

Musk xylene (5-*tert*-butyl-2,4,6-trinitro-*m*-xylene)
→ 4-*tert*-Butyl-2,6-dimethyl-3,5-dinitroaniline
In: Rat K13
→ 3-*tert*-Butyl-5-methyl-2,4,6-trinitrobenzyl alcohol
In: Rat K13
→ 2,2-Dimethyl-2-(3,5-dimethyl-2,4,6-trinitrophenyl)ethanol
In: Rat K13

Mycophenolic acid (6-(4-hydroxy-6-methoxy-7-methyl-4-oxo-5-phthalanyl)-4-methyl-4-hexanoic acid)
Formed from: 4,6-Dihydroxy-2,3-dimethylbenzoic acid
O-(N-(*p*-Carboxyphenyl)carbamoyl)ethyl mycophenolate
→ Mycophenolic acid-β-D-glucuronide
In: Man J138, J783
See also: D518

Mycophenolic acid-β-D-glucuronide
Formed from: Mycophenolic acid

Myricetin (3,3′,4′,5,5′,7-hexahydroxyflavone)
→ Myricetin-3-β-D-glucoside
In: Pisum C90
Red cabbage A1733
Spruce G215
See also: H658
→ Myricetin-4′-O-glucoside
In: Onion G248
→ Myricetin-7-O-glucoside
In: Chrysanthemum J320
→ 3,4′,5,5′,7-Pentahydroxy-3′-methoxyflavone
Probably in: Lotus C905
→ 3′,4′,5,5′,7-Pentahydroxy-3-methoxyflavone
In: Chrysosplenium C792

Myricetin-3-diglucoside
Formed from: Myricetin-3-β-D-glucoside
→ Myricetin-3-triglucoside
In: Pisum C90

Myricetin-3-β-D-glucoside
Formed from: Myricetin
→ Myricetin-3-diglucoside
In: Pisum C90
See also: H658

Myricetin-4′-O-glucoside
Formed from: Myricetin

Myricetin-7-O-glucoside
Formed from: Myricetin

Myricetin-3-triglucoside
Formed from: Myricetin-3-diglucoside

Myristicin (1-allyl-3-methoxy-4,5-methylenedioxybenzene)
→ 5-Allyl-3-methoxycatechol
In: Rat J543
→ 5-(1-Hydroxyallyl)-1-methoxy-2,3-methylenedioxybenzene
In: Rat J543
→ 3-Methoxy-4,5-methylenedioxyamphetamine
In: Rat A699, A975

1-Myristyl-1-phenyl-2-(*p*-toluoyl)hydrazine
Probably formed from: Toluic acid phenylhydrazide

Mysoline
See: 2-Deoxyphenobarbital

N

Nabilone ((6a*R*,10a*R*)-rel-3-(1,1-dimethyl)heptyl)-6,6a,7,8,10,10a-hexahydro-1-hydroxy-6,6-dimethyl-9*H*-dibenzo[*b*,*d*]pyran-9-one)
→ Dihydronabilone (*R*,*R* and *S*,*S* isomers)
In: Bacillus A3010, B37
Bacterium A3010, B37
Dog E247
Monkey E247
Rat A3936
→ 6′-Hydroxynabilone
In: Bacillus A3010
→ 7′-Hydroxynabilone
In: Nocardia B37

Nabumetone (4-(6-methoxy-2-naphthyl)butan-2-one)
→ 4-(6-Hydroxy-2-naphthyl)butan-2-one
In: Dog C867
Man C867
Monkey C867
Mouse C867
Rabbit C867
Rat C867
→ 6-Methoxy-2-naphthylacetic acid
In: Man C867
Monkey C867
Mouse C867
Rabbit C867
Rat C867
→ 4-(6-Methoxy-2-naphthyl)butan-2-ol
In: Monkey C867
Mouse C867
Rabbit C867

***trans*-Nadolol** (5-(3-((1,1-dimethylethyl)amino)-2-hydroxypropoxy)-1,2,3,4-tetrahydro-2,3-naphthalenediol)
→ *trans*-Nadolol-α-D-glucoside
In: Rat G303

***trans*-Nadolol-α-D-glucoside**
Formed from: *trans*-Nadolol

Nafarelin (L-glutamyl-L-histidyl-L-tryptophanyl-L-seryl-L-tyrosyl-(3-(2-naphthyl-D-alanyl)-L-leucyl-L-arginyl-L-prolylglycinamide)
→ L-Tyrosyl-(3-(2-naphthyl-D-alanyl)-L-leucyl-L-arginyl-L-prolylglycinamide
In: Monkey D835
→ 2-Naphthyl-D-alanine
In: Mammal E657

Nafimidone (2-(1*H*-imidazol-1-yl)-1-(2-naphthalenyl)ethanone)
→ Dihydronafimidone
In: Baboon F483
Cynomolgus monkey F483
Man F483
Rat E908
→ Nafimidone-5,6-dihydrodiol
In: Baboon F483
Cynomolgus monkey F483
Man F483
Rat E908

Nafimidone-5,6-dihydrodiol
Formed from: Nafimidone

Nafiverine (N,N′-di-(α-(1-naphthyl)propionyloxy-2-ethyl)piperazine)
→ N-(2-Hydroxyethyl)-N′-(α-(1-naphthyl)propionyloxy-2-ethyl)piperazine
In: Rabbit A1556
→ α-(1-Naphthyl)propionic acid
In: Rabbit A1556

Nafronyl (tetrahydro-α-(1-naphthalenylmethyl)-2-furanpropionic acid 2-(diethylamino) ethyl ester)
→ Tetrahydro-β-(1-naphthylmethyl)-2-furanpropionic acid

In: Man F384
→ Tetrahydro-3-(1-naphthylmethyl)furan-5-ylpropionic acid
In: Man F384

Naftazone ((1-oxo-2-naphthyl)semicarbazone)
→ (1-Hydroxy-2-naphthyl)semicarbazide
In: Man H480
Rat H480

Naftifine ((*E*)-N-methyl-N-(1-naphthylmethyl)-3-phenyl-2-propen-1-amine)
→ Cinnamic acid
In: Man E32
→ N-(3,4-Dihydro-3,4-dihydroxy-1-naphthylmethyl)-N-methyl-3-phenyl-2-propen-1-ylamine
In: Man E32
→ N-(7,8-Dihydro-7,8-dihydroxy-1-naphthylmethyl)-N-methyl-3-phenyl-2-propen-1-ylamine
In: Man E32
→ *m*-Hydroxynaftifine
In: Man E32
→ *p*-Hydroxynaftifine
In: Man E32
→ N-Methyl-3-phenyl-2-propen-1-ylamine
In: Man E32
→ 1-Naphthaldehyde
Probably in: Man E32
→ Nornaftifine
In: Man E32

Naftopidil (4-(2-methoxyphenyl)-α-((1-naphthalenyloxy)methyl)-1-piperazineethanol)
→ 3-(1-(4-Hydroxy-2-methoxyphenyl)piperazin-4-yl)-1-naphthyloxypropan-2-ol
In: Dog G400
Man G400
Mouse G400
Rat G400
→ 1-(4-Hydroxy-1-naphthyloxy)-3-(1-(*o*-methoxyphenyl)piperazin-4-yl)propan-2-ol
In: Dog G400
Man G400
Mouse G400
Rat G400
→ 3-(1-(*o*-Hydroxyphenyl)piperazin-4-yl)-1-naphthyloxypropan-2-ol
In: Dog G400
Man G400
Mouse G400
Rat G400
→ 3-(1-(*o*-Methoxyphenyl)piperazin-4-yl)propane-1,2-diol
In: Dog G400
Man G400
Mouse G400
Rat G400

Nalbuphine ((−)-17-(cyclobutylmethyl)-4,5α-epoxymorphinan-3,6α,14-triol)
Formed from: Nalbuphine acetylsalicylate
→ Nalbuphine-β-D-glucuronide
In: Man H933

Nalbuphine acetylsalicylate
→ Acetylsalicylic acid
In: Dog F308
→ Nalbuphine
In: Dog F308

Nalbuphine-β-D-glucuronide
Formed from: Nalbuphine

Nalmefene ((5α)-17-(cyclopropylmethyl)-4,5-epoxy-6-methylenemorphinan-3,14-diol)
→ Nalmefene-β-D-glucuronide
In: Dog H719
Man H933
Rat H719
→ Nornalmefene
In: Dog H719
Rat H719

Nalmefene-di-β-D-glucuronide
Probably formed from: Nalmefene-β-D-glucuronide

Nalmefene-β-D-glucuronide
Formed from: Nalmefene
→ Nalmefene-di-β-D-glucuronide
Probably in: Rat H719

Nalorphine (N-allylnormorphine)
→ N,N′-Diallyl-2,2′-binormorphine
In: Horseradish A1357
→ Nalorphine-β-D-glucuronide
In: Man H933, J416, J783
Monkey H861
Rat J74, J97a
→ Normorphine

Probably in: Rat A2475

Nalorphine-β-D-glucuronide
Formed from: Nalorphine

6α-Naloxol ((5α,6α)-4,5-epoxy-17-(2-propenyl) morphinan-3,6,14-triol)
Formed from: Naloxone
→ N,N′-Diallyl-7,7′,8,8′-tetrahydro-14,14′-dihydroxy-2,2′-bimorphine
Probably in: Horseradish A1357
→ 7,8-Dihydro-14-hydroxynormorphine
Probably in: Rat A1351

6β-Naloxol ((5α,6β)-4,5-epoxy-17-(2-propenyl)morphinan-3,6,14-triol)
Formed from: Naloxone

Naloxone (N-allyl-7,8-dihydro-14-hydroxynormorphinone)
→ 7,8-Dihydro-14-hydroxynormorphinone
In: Man A1351
Rat A1351, A2350
→ 6α-Naloxol
In: Guinea pig A1704, A2416, F579
Man A3500
Rabbit A3500
See also: A1351, A2290, A3950, K756, K757
→ 6β-Naloxol
In: Chicken A1985
Mammalia A1704
Man A3500
Rabbit A1985, A3500
Rat J744
→ Naloxone-3-β-D-glucuronide
In: Rat A1351, A2350
See also: H933, J74, J97a, J416, J783
→ Naloxone-3-sulphate
In: Eubacterium G602
Rat A1351
→ Noroxymorphone
In: Man A3162

Naloxone-3-β-D-glucuronide
Formed from: Naloxone

Naloxone-3-sulphate
Formed from: Naloxone

6α-Naltrexol ((5α,6α)-4,5-epoxy-17-(cyclopropylmethyl)morphinan-3,6,14-triol)
Formed from: Naltrexone

6β-Naltrexol ((5α,6β)-4,5-epoxy-17-(cyclopropylmethyl)morphinan-3,6,14-triol)
Formed from: Naltrexone
→ 2-Hydroxy-3-methoxy-6β-naltrexol
Probably in: Man A2319, A3657, A3830
Rat A3657

Naltrexone ((5α,6β)-4,5-epoxy-17-(cyclopropylmethyl)morphinan-3,14-diol-6-one)
Formed from: Naltrexonium
→ 7,8-Dihydro-14-hydroxynormorphinone
In: Rabbit A3962
Rat A2000
→ 2-Hydroxy-3-methoxynaltrexone
Probably in: Man A3657, B641
Rat A3657
→ 6α-Naltrexol
In: Animal A1704
Guinea pig A2310, A2625, A2867, B52
Man B641
Monkey A2310
Rabbit A3500
→ 6β-Naltrexol
In: Animal A1704
Chicken A1985
Guinea pig A2164, A2310, A2625, A2867, B52
Man A1488, A1519, A2123, A2310, A2677, D373
Monkey A2123, A2310
Rabbit A1985, A2123, A3500, A2985
Rat A2000, A2123, J744
See also: A2290, A2319, A3657, A3830, A3950, K756, K757
→ Naltrexone-β-D-glucuronide
In: Man H933, J74, J783
Monkey H861
Rabbit A3962
Rat B603, J97a, J416
→ Noroxymorphone
In: Man A3162
Monkey A3162

Naltrexone-β-D-glucuronide
Formed from: Naltrexone

Naltrexonium (N-methylnaltrexone)
→ 7,8-Dihydro-14-hydroxynormorphine

In: Rat E328
→ Naltrexone
In: Dog F486
Man F486
Mouse F486
Rat E328, F486

Naltriben ((4b*S*,8*R*,8a*S*,14b*R*)-7-(cyclopropylmethyl)-5,6,7,8,9,14b-hexahydro-4,8-methano-8a*H*-bisbenzofuro[3,2-*e*:2′,3′-*g*] isoquinoline-1,8-diol)
→ Naltriben-β-D-glucuronide
In: Man H933

Naltriben-β-D-glucuronide
Formed from: Naltriben

Naltrindole ((4b*S*,8*R*,8a*S*,14b*R*)-7-(cyclopropylmethyl)-5,6,7,8,9,14b-hexahydro-4,8-methanobenzofuro[2,3-*a*]pyrido[4,3-*b*] carbazole-1,8a(9*H*)-diol)
→ Naltrindole-β-D-glucuronide
In: Man H933
Rat J97a

Naltrindole-β-D-glucuronide
Formed from: Naltrindole

Nandinine ((13a*S*)-5,8,13,13a-tetrahydro-10-methoxy-6*H*-benzo[*g*]-1,3-benzodioxolo [5,6-*a*]quinolizin-9-ol)
→ N-Methylnandinine
In: Eschscholtzia K782

1-Naphthaldehyde
Formed from: Naftifine
Terbinafine
Probably formed from: 1-Hydroxymethylnaphthalene
→ 1-Hydroxymethylnaphthalene
Probably in: Man J763
→ 1-Naphthoic acid
Probably in: Cunninghamella D33
Man J763

2-Naphthaldehyde
Formed from: 2-Hydroxymethylnaphthalene
→ 2-Hydroxy-2-(2-naphthyl)acetonitrile
In: Almond K320
→ 2-Naphthoic acid
Probably in: Cunninghamella D33

1,8-Naphthaldehydic acid
→ Naphthalene-1,8-dicarboxylic acid
In: Rat A2094

Naphthalene
Formed from: 1,2-Dihydronaphthalene
Probably formed from: Naphthalene-1-sulphonic acid
Naphthalene-2-sulphonic acid
→ *cis*-(1*R*,2*S*)-1,2-Dihydro-1,2-dihydroxynaphthalene
In: Corynebacterium B949, C714
Mycobacterium G267
Oscillatoria B429
Pseudomonas A1591, B365, C144, J216, J332, J370, J524, K246, K376, K613
Rat B602
Rhodococcus H885
→ *cis*-1,4-Dihydro-1,4-dihydroxynaphthalene
In: Rat B602
→ *trans*-S-(1,2-Dihydro-1-hydroxynaphth-2-yl)glutathione
In: Mouse E906, G117
Rat E906
See also: F294, H140
→ *trans*-S-(1,2-Dihydro-2-hydroxynaphth-1-yl)glutathione
In: Mouse E906, G117
Rat C262, E906
→ 2,3-Dihydroxynaphthalene
In: Bacillus K421
→ 1-Methylthionaphthalene
In: Mouse C634
Rat A2957
→ Naphthalene-1,2-oxide
In: Mouse E478
Pigeon B236
Rat D261, D792
See also: A859, A1543, A1663, A2160, A2731, A2957, A3174, A3218, A3352, A3430, A3823, A3879, B294, B414, B602, B775, C256, C634, C716, F294, F555, F586, G117, G267, H129, H140, H199
→ 2-Naphthoic acid
In: Microorganism K253, K564
→ α-Naphthol
In: Absidia A3430
Agmenellum B3
Aspergillus A1218, A3430

Bacillus D434, H240, K421
Basidiobolus A3430
Candida B775, E27
Choanephora A3430
Circinella A3430
Claviceps A3430
Cokermyces A3430
Conidiobolus A3430
Cunninghamella A1218, A3218, A3352, A3430, C256, C716
Emericellopsis A3430
Epicoccum A3430
Gilbertella A3430
Gliocladium A3430
Helicostylum A1218
Linderina A3430Maja A1543
Microorganism G680
Mucor A3430
Neurospora A3430
Nitrosomonas H631
Oscillatoria B429, G476
Panacolus A3430
Penicillium A3430
Phlyctochytrium A3430
Phytophthora A3430
Pigeon B236
Pseudomonas J530
Psilocybe A3430
Rat A1663, A2160, A2957, B602, D603, F299, H140
Rhizopus A1218, A3430
Saccharomyces A3430,E27
Salmon A2731, A3174, A3879
Saprolegnia A3430
Sordaria A3430
Streptomyces A1218
Syncephalastrum A3430
Thamnidium A3430
Thraustochytrium A3430
Trout A3823
Zygorhynchus A3430

→ β-Naphthol
In: Aspergillus A1218, A3430
Bacillus H240, K421
Candida B775
Claviceps A3430
Cunninghamella A1218, A3218, A3352, C256, C716
Epicoccum A3430
Gliocladium A3430
Helicostylum A1218
Mucor A3430
Neurospora A3430
Nitrosomonas H631
Penicillium A3430
Pigeon B236
Pseudomonas J530
Rat A2957, B602

→ S-(1-Naphthyl)glutathione
Probably in: Maja A1543
Mouse C634

Naphthalene-1-carboxylic acid
See: 1-Naphthoic acid

Naphthalene-2-carboxylic acid
See: 2-Naphthoic acid

Naphthalene-1,8-dicarboxylic acid
(naphthalic acid)
Formed from: Acenaphthylene
1,8-Naphthaldehydic acid
Probably formed from: *trans*-Acenaphthene-1,2-diol
1,8-Bis(hydroxymethyl)naphthalene

Naphthalene-1,2-dione
Formed from: *trans*-1,2-Dihydro-1,2-dihydroxynaphthalene

Naphthalene-1,6-disulphonic acid
→ 5-Sulphosalicylic acid
In: Pseudomonas E980

Naphthalene-2,6-disulphonic acid
→ 5-Sulphosalicylic acid
In: Pseudomonas E980

Naphthalene-1,2-oxide
Formed from: Naphthalene
→ *trans*-1,2-Dihydro-1,2-dihydroxynaphthalene
In: Man A1460, A3217, C386
Monkey A1460
Rat A3031, C386, D261
See also: A1543
→ S-(1,2-Dihydro-1-hydroxynaphth-2-yl) glutathione
In: Rat J583

See also: A143, A1460, A2282, C366
→ α-Naphthol
In: Mouse G117
Naphthalene-1-sulphonic acid
→ Naphthalene
Probably in: Rhodococcus H618
→ α-Naphthol
In: Pseudomonas E408
→ Salicylic acid
In: Pseudomonas F185
Naphthalene-2-sulphonic acid
→ *cis*-3,4-Dihydro-3,4-dihydroxynaphthalene-2-sulphonic acid
In: Pseudomonas F185
→ Naphthalene
Probably in: Rhodococcus H618
→ β-Naphthol
In: Pseudomonas E408
→ Salicylic acid
In: Burkholderia K369
Naphthalic acid
See: Naphthalene-1,8-dicarboxylic acid
Naphthazarin
See: 5,8-Dihydroxy-1,4-naphthoquinone
α-Naphthoflavone (2-phenyl-4*H*-naphtho[1,2-*b*]pyran-4-one)
→ 7,8-Dihydro-7,8-dihydroxy-α-naphthoflavone
In: Rat F670
→ 6-Hydroxy-α-naphthoflavone
In: Rat B986, F679
→ 9-Hydroxy-α-naphthoflavone
In Rat B896
→ α-Naphthoflavone-5,6-oxide
In: Rat B896, F679
α-Naphthoflavone-5,6-oxide
Formed from: α-Naphthoflavone
→ 5,6-Dihydro-5,6-dihydroxy-α-naphthoflavone
Probably in: Rat B896, F679
1-Naphthoic acid (naphthalene-1-carboxylic acid)
Formed from: β-(1-Naphthyl)alanine
1-Naphthonitrile
Probably formed from: 1-Naphthaldehyde
→ 7,8-Dihydroxy-1-naphthoic acid
In: Pseudomonas H415
→ 5-Hydroxy-1-naphthoic acid
Probably in: Cunninghamella D33
→ 1-Naphthoylglucose
In: Glycine A3679
2-Naphthoic acid (naphthalene-2-carboxylic acid)
Formed from: 2-Methylnaphthalene
Naphthalene
2-Naphthonitrile
Probably formed from: 2-Naphthaldehyde
→ *cis*-1,2-Dihydro-1,2-dihydroxy-2-naphthoic acid
In: Pseudomonas B899, B905
→ 1-Hydroxy-2-naphthoic acid
In: Burkholderia H813
→ 5-Hydroxy-2-naphthoic acid
Probably in: Cunninghamella D33
→ 6-Hydroxy-2-naphthoic acid
Probably in: Cunninghamella D33
→ 2-Naphthoyl-β-D-glucuronic acid
In: Guinea pig C592
→ 2-Naphthoylglycine
In: Guinea pig C592
Rat B986
→ 1,2,3,4-Tetrahydro-2-naphthoic acid
Probably in: Microorganism K564
→ 5,6,7,8-Tetrahydro-2-naphthoic acid
Probably in: Microorganism K564
α-Naphthol (1-hydroxynaphthalene)
Formed from: N-Acetyl-S-(1,2-dihydro-1-hydroxynaphth-2-yl)-L-cysteine
1-Butoxynaphthalene
S-(1,2-Dihydro-1-hydroxynaphth-2-yl)-L-cysteine
1-Ethoxynaphthalene
2-Hydroxy-3-(α-naphthyloxy)propylamine
1-Methoxynaphthalene
Naphthalene
Naphthalene-1,2-oxide
Naphthalene-1-sulphonic acid
1-Naphthyl acetate
1-Naphthyl butyrate
1-Naphthyl decanoate
1-Naphthyl-β-D-glucoside
1-Naphthyl hexanoate
1-Naphthyl methylcarbamate
1-Naphthyl octanoate

3-(α-Naphthyloxy)lactic acid
1-Naphthyl palmitate
1-Naphthyl pentanoate
1-Naphthyl propionate
1-Naphthyl sulphate
Pranolium
1-(1-Propoxy)naphthalene
1-(2-Propoxy)naphthalene
Propranolol
→ 1,2-Dihydroxynaphthalene
In: Pseudomonas E958
See also: A586, A2957, A3218, B602, D418, E138
→ 1,4-Dihydroxynaphthalene
Probably in: Arthrobacter K540
Coriolus E369
Cunninghamella A3218
Man A2045
Mouse G117
Mushroom D418
Rat A3801, D298, D480, D493
→ 1,5-Dihydroxynaphthalene
In: Pseudomonas J398
→ 1,7-Dihydroxynaphthalene
In: Pseudomonas J398
→ 4-Hydroxy-1-tetralone
In: Microorganism A1542
See also: A3218, A3430, B429, B775, C256, C716
→ 1-Naphthyl galacturonide
In: Man H575
→ 1-Naphthyl-β-D-glucoside
In: Man H575
Salmon A2731, A3879
See also: A1604
→ 1-Naphthyl-D-glucoside-6-phosphate
In: Housefly A1604
See also: C494
→ 1-Naphthyl phosphate
In: Housefly A1604
→ 1-Naphthyl-β-D-glucuronide
In: Bitterling C406
Bluegill A3896
Bream C406
Bushbaby A3061
Capuchin A3061
Carp C406
Cat A273, C562
Catfish A3896, G93
Dog A3897, C562
Goldfish A3896, C406
Gudgeon C406
Guinea pig C562, D960
Guppy C406
Kissing gourami A3896
Maja A1543
Man A1304, B800, C654, E109, F309, F648, G524, G968, H575, J416, J490, J783
Minnow C406
Monkey A3061
Mouse C317, C562, C567
Perch A3896, C406
Pig A273
Quail C562
Rabbit C562, F309, J107
Rat A118, A273, A372, A479, A804, A1314, A1663, A1919, A3827, B165, B287, B392, B740, B962, C195, C562, C672, D103, D288, D344, D423, D807, E365, F86, F309, F688, F728, H140, J99, J107, K379, K595
Roach C406
Rudd C406
Salmon A2731, A3174, A3879
Stickleback C406
Tamarin A3061
Tench C406
Tree shrew A3061
Trout A1874, A3823, C562, H487
See also: A3801
→ 1-Naphthyl sulphate
In: Bitterling C406
Bluegill A3896
Bream C406
Bushbaby A3061
Carp C406
Cat A2, A273
Catfish A3896
Dog A3897, J292
Eubacterium E432
Goldfish A3896, C406
Gudgeon C406

Guinea pig D960
Guppy C406
Haemophilus H756
Kissing gourami A3896
Maja A1543
Man A1304, B392, F648
Minnow C406
Monkey A3061
Mouse C317, C567
Mytilus A3417
Perch A3896, C406
Pig A273
Rat A273, A1314, C441, D103, D807, F728, H140, J99, K379
Roach C406
Rudd C406
Stickleback C406
Salmon A2731, A3879
Tamarin A3061
Tench C406
Tree shrew A3061
Trout H487
See also: A1604, A3801
→ 1,2,3,4-Tetrahydro-3,4-dihydroxy-1-oxo-naphthalene
In: Pseudomonas A2460

β-Naphthol (2-hydroxynaphthalene)
Formed from: N-Acetyl-S-(1,2-dihydro-1-hydroxynaphth-2-yl)-L-cysteine
S-(1,2-Dihydro-1-hydroxynaphth-2-yl)-L-cysteine
2-Hydroxy-3-(α-naphthyloxy)propylamine
2-Methoxynaphthalene
Naphthalene
Naphthalene-2-sulphonic acid
2-Naphthyl acetate
2-Naphthyl butyrate
2-Naphthyl decanoate
2-Naphthyl hexanoate
2-Naphthyl octanoate
2-Naphthyl pentanoate
2-Naphthyl propionate
2-Naphthyl sulphate
→ 2,2′-Dihydroxy-1,1′-binaphthyl
In: Horseradish J572
→ 1,2-Dihydroxynaphthalene
In: Hamster J310
→ 1,7-Dihydroxynaphthalene
Probably in: Rat A2957, B602
→ 2,6-Dihydroxynaphthalene
In: Pseudomonas J398
See also: A2957, B602
→ 2,7-Dihydroxynaphthalene
In: Pseudomonas J398
See also: B602
→ 2-Naphthyl-β-D-glucuronide
In: Cat A273
Guinea pig H41
Lobster J104
Man G524, G968
Pig A273
Rabbit J107
Rat A273, A3464, B287, D288, J107
→ 2-Naphthyl sulphate
In: Beef D607, E800
Cat A2, A273
Channel catfish K497
Dog H455
Eubacterium E432
Guinea pig E456, E695, H41, H455
Lobster H45, J104
Man D607, D977, D979, E575, F464, F770, J571, K46
Mouse H455, H959
Monkey B770, H455
Pig A273
Rabbit H455
Rat A273, A3464, B61, B770, B941, C441, D807, E45, E261, E575, E695, H455

1-Naphthonitrile (1-cyanonaphthalene)
→ 1-Naphthoic acid
In: Rhodococcus H952

2-Naphthonitrile (2-cyanonaphthalene)
→ 2-Naphthoic acid
In: Rhodococcus H952

1,2-Naphthoquinone
Probably formed from: 1,2-Dihydroxynaphthalene
→ 1,2-Dihydroxynaphthalene
In: Cochliobolus K378
Guinea pig K773
Rabbit H920

1,4-Naphthoquinone
Formed from: *o*-Succinylbenzoic acid
Probably formed from: 1,4-Dihydroxynaphthalene
→ 1,4-Dihydroxynaphthalene
In: Beef G401
Guinea pig K773
Juglans A3470
Sporotrichum B301
→ 2-Hydroxy-1,4-naphthoquinone
In: Pseudomonas D918, F46, K897
→ 5-Hydroxy-1,4-naphthoquinone
In: Juglans A3470
→ 4-Hydroxy-1-tetralone
In: Cunninghamella A3218

1-Naphthoylglucose
Formed from: 1-Naphthoic acid

2-Naphthoyl-β-D-glucuronic acid
Formed from: 2-Naphthoic acid

2-Naphthoylglycine
Formed from: 2-Naphthoic acid

2-Naphthylacetaldehyde
2-Hydroxy-3-(2-naphthyl)propionitrile
In: Almond K320

1-Naphthylacetamide
→ 1-Naphthylacetic acid
In: Bradyrhizobium E754, F788

1-Naphthyl acetate
→ α-Naphthol
In: Acinetobacter A962
Aspergillus J831
Cat A1598
Cave roach A2485
Guinea pig A1186
Mouse D582
Parsley C721
Pig K911
Rat F753
Yeast E258

2-Naphthyl acetate
→ β-Naphthol
In: Acinetobacter A962
Ascaris A324, A3040
Moniezia A324, A3040
Rat F753
Yeast E258

1-Naphthylacetic acid
Formed from: 1-Naphthylacetamide
1-Naphthylacetonitrile
β-(1-Naphthyl)alanine
→ 1-Naphthylacetyl-L-aspartic acid
In: Actinidia K281
→ 1-Naphthylacetyl-β-D-glucuronic acid
In: Bat A17, A2524
Bushbaby A17, A2524
Capuchin A2524
Civet A1312, A2387
Dog A2387
Ferret A394, A2056, A3046, A3249
Genet A1312, A2387
Gerbil A3501, E787
Guinea pig A3501, E787
Hamster A3501, E787
Hyaena A2387
Lion A1312
Man A17, A2524
Marmoset A17, A2524
Monkey A17, A2524
Mouse A3501, E787
Pigeon A395
Rabbit A17, A2056, A2524, A3249
Rat A17, A2056, A2524, A3249, E937
→ 1-Naphthylacetylglucose
In: Glycine A3679, K746
→ 1-Naphthylacetylglutamine
In: Capuchin A2524
Marmoset A2524
Monkey A17, A2524
→ 1-Naphthylacetylglycine
In: Bushbaby A17, A2524
Capuchin A2524
Cat A17, A2387, A2524
Civet A1312, A2387
Dog A2387
Ferret A394, A2056, A2387, A3046
Genet A1312, A2387
Gerbil A3501, E787
Guinea pig A3501, E787
Hamster A3501, E787
Hyaena A2387
Lion A1312
Marmoset A17, A2524

Monkey A2524
Mouse A3501, E787
Pigeon A395
Rabbit A2056, A2524
Rat A17, A2056, A2524
→ 1-Naphthylacetylornithine
In: Hen A395
→ 1-Naphthylacetyltaurine
In: Bushbaby A17, A2524
Capuchin A2524
Cat A17, A2387, A2524
Civet A1312, A2387
Dog A2387
Ferret A394, A2387, A3046
Genet A1312, A2387
Gerbil E787
Hamster E787
Hen A395
Hyaena A2387
Lion A1312
Man A17, A2524
Marmoset A17, A2524
Monkey A17, A2524
Mouse E787
Pigeon A395
Rat A394
→ 1-Naphthylacetylglucose
In: Glycine A3679

2-Naphthylacetic acid
Formed from: 2-Naphthylacetonitrile
Probably formed from: 3-(2-Naphthyl)-D-alanyl-L-leucine
→ *cis*-1,2-Dihydro-1,2-dihydroxy-2-naphthylacetic acid
In: Pseudomonas B899, B905
→ 2-Naphthylacetyl-β-D-glucuronic acid
In: Ferret A394, A2056, A3046, A3249
Gerbil A3501, E787
Guinea pig A3501, E787
Hamster A3501, E787
Horse B801
Mouse A3501, E787
Rabbit A2056, A3249
Rat A2056, A3249, E937
→ 2-Naphthylacetylglutamine
In: Ferret A2056, A3249
Hamster A3501, E787
Rabbit A2056, A3249
Rat A2056, A3249
→ 2-Naphthylacetylglycine
In: Ferret A394, A2056, A3046, A3249
Gerbil A3501, E787
Guinea pig A3501, E787
Hamster A3501, E787
Horse B801
Mouse A3501, E787
Rabbit A2056, A3249
Rat A2056, A3249
→ 2-Naphthylacetyltaurine
In: Ferret A394, A2056, A3046, A3249, C420
Gerbil A3501, E787
Hamster A3501, E787
Horse B801
Mouse A3501, E787
Rabbit A2056, A3249
Rat A2056, A3249

1-Naphthylacetonitrile
→ 1-Naphthylacetic acid
In: Rhodococcus G172

2-Naphthylacetonitrile
→ 2-Naphthylacetic acid
In: Rhodococcus G172

1-Naphthylacetylaspartic acid
Formed from: 1-Naphthylacetic acid
Probably formed from: 1-Naphthylacetylglucose

1-Naphthylacetyl CoA
→ 2-Hexadecyl-1-naphthylacetylglycerol
In: Rat F749
→ 1-Naphthylacetyl-2,3-dipalmitoylglycerol
In: Rat F749

1-Naphthylacetyl-2,3-dipalmitoylglycerol
Formed from: 1-Naphthylacetyl CoA

1-Naphthylacetylglucose
Formed from: 1-Naphthylacetic acid
→ 1-Naphthylacetyl-L-aspartic acid
Probably in: Glycine A3679

1-Naphthylacetyl-β-D-glucuronic acid
Formed from: 1-Naphthylacetic acid

2-Naphthylacetyl-β-D-glucuronic acid
Formed from: 2-Naphthylacetic acid

1-Naphthylacetylglutamine
Formed from: 1-Naphthylacetic acid
2-Naphthylacetylglutamine
Formed from: 2-Naphthylacetic acid
1-Naphthylacetylglycine
Formed from: 1-Naphthylacetic acid
2-Naphthylacetylglycine
Formed from: 2-Naphthylacetic acid
1-Naphthylacetylornithine
Formed from: 1-Naphthylacetic acid
1-Naphthylacetyltaurine
Formed from: 1-Naphthylacetic acid
2-Naphthylacetyltaurine
Formed from: 2-Naphthylacetic acid
β-(1-Naphthyl)alanine
→ 1-Naphthoic acid
In: Rat A2020
→ 1-Naphthylacetic acid
In: Rat A2020
→ 1-Naphthyllactic acid
In: Rat A2020
2-Naphthyl-D-alanine
Formed from: Nafarelin
3-(2-Naphthyl)-D-alanyl-L-leucine
Probably formed from: L-Tyrosyl-3-(2-naphthyl)-D-alanyl-L-leucine
→ 2-Naphthylacetic acid
Probably in: Rhesus monkey D835
3-(2-Naphthyl)-D-alanyl-L-leucyl-L-arginyl-L-prolylglycinamide
Probably formed from: L-Tyrosyl-3-(2-naphthyl)-D-alanyl-L-leucyl-L-arginyl-L-prolylglycinamide
α-Naphthylamine (1-aminonaphthalene)
Formed from: 1-Formamidonaphthalene
1-Naphthylhydroxylamine
1-Nitronaphthalene
→ 1-Acetamidonaphthalene
In: Guinea pig F109
Man A3253
Rabbit A3253, F109
Rat F109, G743
→ 1-Amino-2-naphthol
In: Dog D738
Man C798, D738
Rat D738
→ 1-Formamidonaphthalene
In: Guinea pig F109
Rabbit F109
Rat F109
→ α-Naphthylamine-N-β-D-glucuronide
In: Man F309, J416
Rat D288, F309, G743
→ 1-Naphthylhydroxylamine
In: Beef C727
Dog C727
See also: G743
β-Naphthylamine (2-aminonaphthalene)
Formed from: 2-Acetamidonaphthalene
L-Alanyl-L-alanyl-β-naphthylamide
L-Alanyl-L-leucyl-β-naphthylamide
L-Alanyl-β-naphthylamide
L-Arginyl-L-arginyl-β-naphthylamide
L-Arginyl-β-naphthylamide
L-Asparaginyl-β-naphthylamide
L-Aspartyl-β-naphthylamide
Benzoyl-L-arginyl-β-naphthylamide
L-Cysteinyl-β-naphthylamide
L-Cystinyl-di-β-naphthylamide
L-Cystinyl-β-naphthylamide
2-Formamidonaphthalene
α-L-Glutamyl-β-naphthylamide
γ-L-Glutamyl-β-naphthylamide
Glycyl-L-arginyl-β-naphthylamide
Glycylglycyl-β-naphthylamide
Glycyl-L-leucyl-β-naphthylamide
Glycyl-β-naphthylamide
Glycyl-L-phenylalanyl-β-naphthylamide
Glycyl-L-prolyl-β-naphthylamide
L-Histidyl-β-naphthylamide
L-Hydroxyprolyl-β-naphthylamide
L-Isoleucyl-β-naphthylamide
L-Leucyl-L-alanyl-β-naphthylamide
L-Leucyl-glycyl-β-naphthylamide
L-Leucyl-β-naphthylamide
L-Leucyl-L-phenylalanyl-β-naphthylamide
L-Leucyl-L-tyrosyl-β-naphthylamide
L-Lysyl-L-alanyl-β-naphthylamide
L-Lysyl-β-naphthylamide
L-Methionyl-β-naphthylamide
2-Naphthylhydroxylamine
N-(1-Naphthyl)phthalamic acid
2-Nitronaphthalene

L-Norleucyl-β-naphthylamide
L-Norvalyl-β-naphthylamide
L-Ornithyl-β-naphthylamide
L-Phenylalanyl-β-naphthylamide
L-Prolyl-β-naphthylamide
L-Pyrrolidinyl-β-naphthylamide
L-Seryl-β-naphthylamide
L-Seryl-L-tyrosyl-β-naphthylamide
L-Threonyl-β-naphthylamide
L-Tryptophanyl-β-naphthylamide
L-Tyrosyl-β-naphthylamide
L-Valyl-β-naphthylamide
→ 2-Acetamidonaphthalene
In: Guinea pig F109
Hamster F115
Man A3253, D229, F172
Rabbit A3253, F109
Rat D229, F109, G743
→ 2-Amino-1-naphthol
In: Rat B773
→ 5-Amino-2-naphthol
In: Dog D738
Man D738
Rat D738
→ 7-Amino-2-naphthol
In: Dog D738
Man D738
Rat D738
→ 2-Formamidonaphthalene
In: Guinea pig F109
Rabbit F109
Rat F109
→ N-Methyl-β-naphthylamine
In: Pig K923
→ β-Naphthylamine-N-β-D-glucuronide
In: Man J416
Rat D288, F309, G743
→ 2-Naphthylhydroxylamine
In: Dog D738
Man D738
Monkey A880
Pig A1791
Rat B773, D738
→ β-Naphthyl nitroxide radical
In: Rabbit B184
→ 2-Naphthyl sulphamate
In: Guinea pig B967, E456
Man G40
Rat B967
→ 2-Nitrosonaphthalene
In: Ram C360

α-Naphthylamine-N-β-D-glucuronide
Formed from: α-Naphthylamine

β-Naphthylamine-N-β-D-glucuronide
Formed from: β-Naphthylamine

1-Naphthylamine-4-sulphonic acid
Formed from: 2-Hydroxy-1-(4-sulpho-1-naphthylazo)naphthalene-3,6-disulphonic acid

1-Naphthyl butyrate
→ α-Naphthol
In: Aspergillus J831
Yeast E258

2-Naphthyl butyrate
→ β-Naphthol
In: Yeast E258

S-(1-Naphthyl)-L-cysteine
Probably formed from: S-(1-Naphthyl) glutathione
→ N-Acetyl-S-(1-naphthyl)-L-cysteine
Probably in: Mouse C634
→ ((1-Naphthyl)thio)acetic acid
Probably in: Mouse C634
→ ((1-Naphthyl)thio)lactic acid
Probably in: Mouse C634

1-Naphthyl decanoate
→ α-Naphthol
In: Yeast E258

2-Naphthyl decanoate
→ β-Naphthol
In: Yeast E258

1-(1-Naphthyl)ethanol
Formed from: 1-(1-Naphthyl)ethyl acetate

2-(1-Naphthyl)ethanol
Formed from: 2-Acetylnaphthalene
2-(1-Naphthyl)ethyl acetate

1-(1-Naphthyl)ethyl acetate
→ 1-(1-Naphthyl)ethanol
In: Helianthus J661
Solanum J661

2-(1-Naphthyl)ethyl acetate
→ 2-(1-Naphthyl)ethanol
In: Helianthus J661
Solanum J661

1-Naphthyl galacturonide
Formed from: α-Naphthol
1-Naphthyl-β-D-glucoside
Formed from: α-Naphthol
→ α-Naphthol
In: Rat A1314
→ 1-Naphthyl-β-D-glucuronide
In: Mouse C567
See also: A1314
1-Naphthyl-D-glucoside-6-phosphate
Formed from: α-Naphthol
1-Naphthyl-β-D-glucoside
1-Naphthyl-β-D-glucuronide
Formed from: α-Naphthol
α-Naphthyl-β-D-glucoside
2-Naphthyl-β-D-glucuronide
Formed from: β-Naphthol
S-(1-Naphthyl)glutathione
Formed from: Naphthalene
→ S-(1-Naphthyl)-L-cysteine
Probably in: Mouse C634
1-Naphthyl hexanoate
→ α-Naphthol
In: Yeast E258
2-Naphthyl hexanoate
→ β-Naphthol
In: Yeast E258
1-Naphthylhydroxylamine (1-hydroxylaminonaphthalene)
Formed from: α-Naphthylamine
1-Nitronaphthalene
1-Nitrosonaphthalene
→ α-Naphthylamine
In: Rabbit A2173
→ 1-Naphthylhydroxylamine-β-D-glucuronide
In: Rat G743
2-Naphthylhydroxylamine (2-hydroxylaminonaphthalene)
Formed from: β-Naphthylamine
Probably formed from: 2-Nitrosonaphthalene
→ 2-Amino-1-naphthol
Probably in: Rat B773
→ N-Hydroxy-2-acetamidonaphthalene
Probably in: Dog A880
→ β-Naphthylamine
In: Rat A75
→ 2-Naphthylhydroxylamine-β-D-glucuronide
In: Rat B773, D314, D600
1-Naphthylhydroxylamine-β-D-glucuronide
Formed from: 1-Naphthylhydroxylamine
2-Naphthylhydroxylamine-β-D-glucuronide
Formed from: 2-Naphthylhydroxylamine
1-Naphthyl hydroxymethylcarbamate
Formed from: 1-Naphthyl methylcarbamate
1-Naphthyl N-hydroxy-N-methylcarbamate
Formed from: 1-Naphthyl methylcarbamate
1-Naphthyllactic acid
Formed from: β-(1-Naphthyl)alanine
2-(2-Naphthyl)lactic acid
Probably formed from: 2-(2-Naphthyl)-1,2-propanediol
1-Naphthylmethanol
See: 1-Hydroxymethylnaphthalene
2-Naphthylmethanol
See: 2-Hydroxymethylnaphthalene
1-Naphthyl methylcarbamate (carbaryl, sevin)
→ 5,6-Dihydro-5,6-dihydroxy-1-naphthyl methylcarbamate
In: Bee A2048
Man A2045
Rat A602
See also: A1874, A3801, A3896, A3897, C482
→ 2-Hydroxy-1-naphthyl methylcarbamate
In: Silkworm A549, A583
→ 4-Hydroxy-1-naphthyl methylcarbamate
In: Bee A2048
Man A2045
Rabbit A2672
Rat A1545
See also: A3601, A3896, A3897
→ 5-Hydroxy-1-naphthyl methylcarbamate
In: Bee A2048
Man A2045
→ α-Naphthol
In: Arthrobacter K540
Bluegill A3896
Catfish A3896
Dog A3897
Goldfish A3896
Kissing gourami A3896
Man A1304, A2045

Perch A3896
Pseudomonas E138
Rabbit A2672
Rainbow trout A1874
Rat A580
Sporoptera A1076
See also: A443, A2048, A3801
→ 1-Naphthyl hydroxymethylcarbamate
In: Bee A2048
Ostrinia A2288
Rat A1545
→ 1-Naphthyl N-hydroxy-N-methylcarbamate
In: Bee A2048

1-Naphthylmethyl sulphate
Formed from: 1-Hydroxymethylnaphthalene

2-Naphthylmethyl sulphate
Formed from: 2-Hydroxymethylnaphthalene

α-(1-Naphthylmethyl)-β-(tetrahydro-5-oxofuran) propionic acid
Formed from: 2-(Diethylamino)ethyl α-(1-naphthylmethyl)-β-(tetrahydrofuran-2-yl) propionate

β-Naphthyl nitroxide radical
Formed from: β-Naphthylamine

1-Naphthyl octanoate
→ α-Naphthol
In: Yeast E258

2-Naphthyl octanoate
→ β-Naphthol
In: Yeast E258

(1-Naphthyloxy)acetic acid
Formed from: 3-(α-Naphthyloxy)lactic acid

(2-Naphthyloxy)acetic acid
Formed from: Propranolol
→ (1-Hydroxy-2-naphthyloxy)acetic acid
Probably in: Aspergillus A2849

3-(1-Naphthyloxy)lactic acid
Probably formed from: 2-Hydroxy-3-(α-naphthyloxy)propionaldehyde
→ 3-(4-Hydroxy-1-naphthyloxy)lactic acid
Probably in: Man D343
→ α-Naphthol
In: Dog B55
Rat B55, C840
→ (1-Naphthyloxy)acetic acid
In: Dog C840
Rat C840
See also: A620, D343

(*R*)-2-(1-Naphthyloxy)propanol
Formed from: (*R*)-2-(1-Naphthyloxy) propionic acid

(*S*)-2-(1-Naphthyloxy)propanol
Formed from: (*S*)-2-(1-Naphthyloxy) propionic acid

(*R*)-2-(1-Naphthyloxy)propionic acid
→ (*R*)-2-(1-Naphthyloxy)propanol
In: Gloesporium F948
Glomerella F948

(*S*)-2-(1-Naphthyloxy)propionic acid
→ (*S*)-2-(1-Naphthyloxy)propanol
In: Gloesporium F948
Glomerella F948

1-(α-Naphthyloxy)-2,3-propylene glycol
Probably formed from: 2-Hydroxy-3-(α-naphthyloxy)propionaldehyde
→ 1-(4-Hydroxy-α-naphthyloxy)-2,3-propylene glycol
Probably in: Dog B55, C581, C620, D9
Man D9
Rat B55, D9
→ 1-(α-Naphthyloxy)-2,3-propylene glycol-β-D-glucuronide
Probably in: Dog C581
Man D343
→ 1-(α-Naphthyloxy)-2,3-propylene glycol sulphate
Probably in: Dog C625

1-(α-Naphthyloxy)-2,3-propylene glycol-β-D-glucuronide
Probably formed from: 1-(α-Naphthyloxy)-2,3-propylene glycol

1-(α-Naphthyloxy)-2,3-propylene glycol sulphate
Probably formed from: 1-(α-Naphthyloxy)-2,3-propylene glycol

1-Naphthyl palmitate
→ α-Naphthol
In: Cave roach A2485

1-Naphthyl pentanoate
→ α-Naphthol
In: Saccharomyces E258

2-Naphthyl pentanoate
→ β-Naphthol
In: Saccharomyces E258

1-Naphthyl phosphate
Formed from: α-Naphthol
2-(2-Naphthyl)-1,2-propanediol
Probably formed from: 2-(2-Naphthyl)-2-propanol
→ 2-(2-Naphthyl)lactic acid
Probably in: Rat B606, D214
2-(2-Naphthyl)propanal
→ 2-Hydroxy-3-methyl-3-(2-naphthyl) propionitrile
In: Almond K320
2-(2-Naphthyl)-2-propanol
Formed from: 2-Isopropylnaphthalene
→ 2-(2-Naphthyl)-1,2-propanediol
Probably in: Rat B606, D214
1-Naphthyl propionate
→ α-Naphthol
In: Aspergillus J831
Dog A868
Parsley C721
Saccharomyces E258
2-Naphthyl propionate
→ β-Naphthol
In: Saccharomyces E258
2-(1-Naphthyl)propionic acid
Formed from: α-Methyl-α-(1-naphthyl) malonic acid
Nafiverine
2-(2-Naphthyl)propionic acid
Formed from: α-Methyl-α-(2-naphthyl) malonic acid
2-Isopropylnaphthalene
2-Naphthyl sulphamate
Formed from: β-Naphthylamine
1-Naphthyl sulphate
Formed from: α-Naphthol
→ α-Naphthol
In: Haemophilus H756
Rabbit B481
See also: C567
2-Naphthyl sulphate
Formed from: β-Naphthol
→ β-Naphthol
In: Haemophilus H756
1-Naphthylsulphonic acid
→ Gentisic acid
In: Pseudomonas G280

N-α-(2-Naphthylsulphonylglycyl)-4-oxamidinophenylalanine piperidide
→ 3-Amidino-N-α-(2-naphthylsulphonyl-glycyl)phenylalanine piperidide
In: Rat E986
β-(1-Naphthyl)-β′-(2-tetrahydrofuryl)isobutyric acid
Formed from: 2-Diethylaminoethyl β-(1-naphthyl)-β′-(2-tetrahydrofuryl)isobutyrate
((1-Naphthyl)thio)acetic acid
Probably formed from: S-(1-Naphthyl)-L-cysteine
((1-Naphthyl)thio)lactic acid
Probably formed from: S-(1-Naphthyl)-L-cysteine
Naproxen (2-(6-Methoxynaphth-2-yl)propionic acid)
→ 2-(6-Hydroxynaphth-2-yl)propionic acid
In: Dog A3960
Guinea pig A3960
Man A3960, F14, G974, H493
Mouse A1262, A3960
Pig A3960
Rabbit A3960
Rat A1262, A3960
→ 2-(6-Methoxynaphth-2-yl)propionyl-β-D-glucoside
In: Panax F255
→ 2-(6-Methoxynaphth-2-yl)propionylglycine
In: Dog A3960
Guinea pig A3960
Mouse A3960
Pig A3960
Rabbit A3960
Rat A3960
→ 2-(6-Methoxynaphth-2-yl) propionylsophoroside
In: Panax F255
→ Naproxen-β-D-glucuronide
In: Dog A3960
Guinea pig A3960, G37
Man E886, G974, J617
Monkey E886, H861
Mouse A3960
Panax F255
Pig A3960
Rabbit A3960, E886

Rat A3960, F726, G46, H466

Naproxen-β-D-glucuronide
Formed from: Naproxen

Narciclasine ((2*S*,3*R*,4*S*,4a*R*)-3,4,4a,5-tetrahydro-2,3,4,7-tetrahydroxy-[1,3]dioxolo[4,5-*j*]phenanthridin-6(2*H*)-one)
Formed from: 11-Hydroxyvittatine

Narcissidine ((1*S*,2*R*,3*R*,11b*S*,11c*S*)-2,3,5,7,11b,11c-hexahydro-2,9,10-trimethoxy-1*H*-pyrrolo[3,2,1-*de*]phenanthridine-1,3-diol)
Formed from: O-Methylnorbelladine

Narcotine ((3*S*)-6,7-dimethoxy-3-((5*R*)-5,6,7,8-tetrahydro-4-methoxy-6-methyl-1,3-dioxolo [4,5-*g*]isoquinolin-5-yl-1(3*H*)-isobenzofuranone)
→ Cotarnine
In: Rat A3316
→ O-Demethylmeconine
In: Rat A3316
→ Di-O-demethylnarcotine
In: Rat A3316
→ Hydroxycotarnine
In: Rat A3316
→ Oxocotarnine
In: Rat A3316

Nardil
See: β-Phenethylhydrazine

Naringenin (4′,5,7-trihydroxyflavanone)
Formed from: *p*-Coumaroyl CoA
Naringenin-β-D-glucuronide
Naringin
Rutin
2′,4,4′,6′-Tetrahydroxychalcone
→ Apiforol
In: Sinningia E703
→ Apigenin
In: Antirrrinum C65, K290
Parsley B957
Sinningia E570
Torenia K290
Verbena D471
See also: A1030, A1635
→ Aromadendrin
In: Haplopappus A200
Matthiola A3890, B679
Parsley B957
Petunia D772, E77
Tulipa F162
See also: B404
→ *p*-Coumaric acid
In: Horseradish A3777
→ Cyanidin
In: Haplopappus A200
→ Eriodictyol
In: Dianthus C392
Haplopappus A200
Horseradish A3777
Matthiola B679
Parsley C772
Petunia H310
Rat J300
Sinningia E570
Verbena D471
Zea D947
→ 5,7-Dihydroxycoumarin
In: Horseradish A3777
→ 5,7-Dihydroxy-4′-methoxyflavanone
Probably in: Robinia C100
→ Genistein
In: Glycine D468, E23
→ *p*-Hydroxymandelaldehyde
In: Horseradish A3777
Rhizobium H280
→ *p*-Hydroxyphenylhydracrylaldehyde
In: Horseradish A3777
→ *p*-Hydroxy-β-oxophenylpropionaldehyde
In: Horseradish A3777
→ Luteolin
In: Verbena D471
See also: A1030, A1635
→ Naringenin-β-D-glucuronide
In: Man G968, J416, J783
Rat A3696, B244, J714, J849
→ Orientin
In: Spirodela A1635
→ Phenylacetic acid
In: Clostridium F203
→ Phloroglucinol
In: Rhizobium H280
→ Prunin
In: Calamondin F419
Chrysanthemum J320
Citrus K810, K891

Matthiola A3890
Parsley K739
Petunia K303
See also: E681
→ Sakuranetin
In: Rice J190, H895
→ 2,4′,5,7-Tetrahydroxyisoflavanone
In: Glycyrrhiza K209, K308
→ 2,4,6-Trihydroxybenzoylformic acid
In: Horseradish A3777
→ Vitexin
In: Spirodela A1635

Naringenin-chalcone
See: 2′,4,4′,6′-Tetrahydroxychalcone

Naringenin-7-glucorhamnoside
See: Naringin

Naringenin-7-β-D-glucoside
See: Prunin

Naringenin-β-D-glucuronide
Formed from: Naringenin
→ Naringenin
In: Helix J714

Naringin (naringenin-7-glucorhamnoside)
Formed from: L-Phenylalanine
→ Eriodictyol
In: Columnea E702
→ Naringenin
In: Microorganism J525, K643
→ Naringin-β-D-glucuronide
In: Rat A3696, B244
→ Prunin
With naringinase D454
See also: F227

Naringin-β-D-glucuronide
Formed from: Naringin

Nefazodone (2-(3-(4-(3-chlorophenyl)piperazin-1-yl)propyl)-5-ethyl-2,4-dihydro-4-(2-phenoxyethyl)-3*H*-1,2,4-triazol-3-one)
→ 2-(3-(4-(3-chloro-4-hydroxyphenyl) piperazin-1-yl)propyl)-5-ethyl-2,4-dihydro-4-(2-phenoxyethyl)-3*H*-1,2,4-triazol-3-one
In: Dog G578
Man K473
See also: H490
→ 1-(*m*-Chlorophenyl)piperazine
In: Dog G578
Man H490, K473
→ 2-(3-(4-(*m*-Chlorophenyl)piperazin-1-yl)propyl)-5-(1-hydroxyethyl)-2,4-dihydro-4-(2-phenoxyethyl)-3*H*-1,2,4-triazol-3-one
In: Dog G578
Man G872, K473
→ 5-Ethyl-2,4-dihydro-2-(3-hydroxypropyl)-4-(2-phenoxyethyl)-1,2,4-triazol-3-one
In: Man G872

Nefiracetam (N-(2,6-Dimethylphenyl)-2-(2-oxo-1-pyrrolidinyl)acetamide)
→ N-((2,6-Dimethylphenylcarbamoyl)methyl)succinamic acid
In: Man G540
→ N-((2-Hydroxymethyl-6-methylphenylcarbamoyl)methyl)pyrrolidine
In: Rat H730
→ 3-Hydroxynefiracetam
In: Dog H638
Monkey H638
Rat H638, H730
→ 3′-Hydroxynefiracetam
In: Dog H638
Monkey H638
Rat H638, H730
→ 4-Hydroxynefiracetam
In: Dog H638
Man G540
Monkey H638
Rat H638, H730
→ 4′-Hydroxynefiracetam
In: Dog H638
Monkey H638
Rat H638, H730
→ 5-Hydroxynefiracetam
In: Dog H638
Man G540
Monkey H638
Rat H638, H730

Nefopam (3,4,5,6-tetrahydro-5-methyl-1-phenyl-1*H*-2,5-benzoxazocine)
→ Demethylnefopam
In: Man A289
Rabbit A289
Rat A289

Nemacide (O-(2,4-dichlorophenyl) O,O-diethyl phosphorothioate)
→ 2,4-Dichlorophenol
In: Hen A462

Neohesperidin (hesperetin-7-(2-O-rhamnosyl)-β-D-glucoside)
Formed from: Prunin

Neophellamuretin (3,4′,5,7-tetrahydroxy-8-isoprenylflavanone)
Formed from: Phellamurin
→ 8-(β,γ-Dihydroxyisovaleryl)-3,4′,5,7-tetrahydroxyflavanone
Probably in: Aspergillus A3306

Neopranol (1,2,6a,12a-tetrahydro-α,α-dimethyl-6*H*-[1,3]dioxolo[5,6]benzofuro[3,2-*c*]furo [3,2-*g*][1]benzopyran-2-methanol)
Formed from: Edunol

Neoprontosil
See: 7-Acetamido-1-hydroxy-2-(*p*-sulphamoylphenylazo)naphthalene-3,6-disulphonic acid

Neorautanol (2,3,7a,13a-tetrahydro-3,3-dimethyl-1*H*,7*H*-[1,3]dioxolo[5,6]benzofuro[3,2-*c*] pyrano[3,2-*g*][1]benzopyran-2-ol)
Formed from: Edunol

Neostigmine ((*m*-hydroxyphenyl) trimethyl ammonium dimethylcarbamate)
→ *m*-Hydroxyphenyl trimethyl ammonium
In: Rat A616, A909

Netobimin (2-((((methoxycarbonyl)amino) ((2-nitro-5-(propythio)phenyl)amino) methylene)amino)ethanesulphonic acid)
Probably formed from: Netobimin sulphoxide
→ Albendazole
In: Microorganism K86
→ Netobimin sulphoxide
In: Microorganism K86

Netobimin sulphoxide
Formed from: Netobimin
→ Netobimin
Probably in: Microorganism K86

Niaprazine (N-(3-(4-(*p*-fluorophenyl)-1-piperazinyl)-1-methylpropyl)-3-pyridinecarboxamide)
Formed from: Niaprazine-N-oxide
→ 1-(*p*-Fluorophenyl)piperazine
In: Rat D308
→ Niaprazine-N-oxide
In: Rat D429
→ 6-Oxoniaprazine
In: Rat D429

Niaprazine-N-oxide
Formed from: Niaprazine
→ Niaprazine
In: Rat D429

Nicardipine (2-(N-benzyl-N-methylamino)ethyl methyl 1,4-dihydro-2,6-dimethyl-4-(*m*-nitrophenyl)pyridine-3,5-dicarboxylate)
→ 2-(N-Benzyl-N-methylamino)ethyl methyl 4-(*m*-aminophenyl)-1,4-dihydro-2,6-dimethylpyridine-3,5-dicarboxylate
In: Man B535
→ 2-(N-Benzyl-N-methylamino)ethyl methyl 2,6-dimethyl-4-(*m*-nitroohenyl) pyridine-3,5-dicarboxylate
In: Man D889
→ 2,6-Dimethyl-4-(*m*-nitroohenyl) pyridine-3,5-dicarboxylate 3-methyl ester
In: Dog B535
Man B535, D889
Monkey B535
Rat B535
→ 2-(Methylamino)ethyl methyl 1,4-dihydro-2,6-dimethyl-4-(*m*-nitrophenyl) pyridine-3,5-dicarboxylate
In: Dog B535
Man B535, D889
Monkey B535
Rat B535
→ Methyl 2-oxoethyl 1,4-dihydro-2,6-dimethyl-4-(*m*-nitrophenyl)pyridine-3,5-dicarboxylate
In: Man D889

Nicergoline (8β-(5-bromonicotinoylhydroxy-methyl)-10α-methoxy-1,6-dimethylergoline)
→ 8β-Hydroxymethyl-10α-methoxy-1,6-dimethylergoline
Probably in: Dog E167
Rat E167
→ 8β-Hydroxymethyl-10α-methoxy-6-methyl-ergoline
In: Dog E167
Rat E167

→ 8β-Hydroxymethyl-10α-methoxy-6-methyl-ergoline 5-bromo-3-pyridinecarboxylate
In: Dog E167
Rat E167

Niclosamide
See: 2′,5-Dichloro-4′-nitrosalicylanilide

Nicomorphine (3,6-dinicotinoylmorphine)
→ 6-Nicotinoylmorphine
In: Man G350, H494, H639, H766

6-Nicotinoylmorphine
Formed from: Nicomorphine
→ Morphine
Probably in: Man G350, H494, H639, H766

Nifedipine (1,4-dihydro-2,6-dimethyl-4-(2-nitrophenyl)-3,5-pyridinedicarboxylic acid dimethyl ester)
→ 3,5-Di-(methoxycarbonyl)-2,6-dimethyl-4-(2-nitrophenyl)pyridine
In: Man F919
→ 5-Methoxycarbonyl-1,4-dihydro-2,6-dimethyl-4-(2-nitrophenyl)pyridine-3-carboxylic acid
In: Dog B354
Man B354
→ 5-Methoxycarbonyl-1,4-dihydro-2-hydroxymethyl-6-methyl-4-(2-nitrophenyl) pyridine-3-carboxylic acid
Probably in: Dog B354
Man B354

Nifenazone (N-(2,3-dihydro-1,5-dimethyl-3-oxo-2-phenyl-1*H*-pyrazol-4-yl)-3-pyridinecarboxamide)
→ 4′-Hydroxynifenazone
In: Man D783
→ Nifenazone-N-oxide
In: Man D783
→ Nornifenazone
In: Man D783

Nifenazone-N-oxide
Formed from: Nifenazone

Niflumic acid (2-(*m*-trifluoromethylanilino) nicotinic acid)
→ 4′-Hydroxyniflumic acid
In: Dog A973
Man A973, A2684
Rat A1883
→ 5-Hydroxyniflumic acid
In: Dog A973
Man A973, A2684
Rat A1883
→ Niflumic acid-β-D-glucuronide
In: Dog A973
Man A973

Niflumic acid-β-D-glucuronide
Formed from: Niflumic acid

Nilvadipine (5-isopropyl 3-methyl 2-cyano-1,4-dihydro-6-methyl-4-(*m*- nitrophenyl) pyridine-3,5-dicarboxylate)
→ Dehydronilvadipine
In: Dog F17
Rat E847, F17, F484
See also: E818
→ Deisopropylnilvadipine
In: Dog E818
Rat E818, F484

Nimesulide (N-(4-nitro-2-phenoxyphenyl) methanesulphonamide)
→ *p*-Hydroxynimesulide
In: Man G435

Nimetazepam (1,3-dihydro-1-methyl-7-nitro-5-phenyl-2*H*-1,4-benzodiazepin-2-one)
Formed from: 2-(5-Cyano-N-methyl-3,5-dimethyl-3-azapentanamido)-5-nitrobenzophenone
→ 7-Amino-1,3-dihydro-1-methyl-5-phenyl-2*H*-1,4-benzodiazepin-2-one
In: Dog A1763
Rat A352
→ 1,3-Dihydro-7-nitro-5-phenyl-2*H*-1,4-benzodiazepin-2-one
In: Dog A1763
See also: A2016, A2101, A2102

Nimodipine (1,4-dihydro-2,6-dimethyl-4-(3-nitrophenyl)-3,5-pyridinedicarboxylic acid 2-methoxyethyl 1-methylethyl ester)
→ 1,4-Dihydro-3-(2-hydroxyethoxy)carbonyl-5-isopropoxycarbonyl-2,6-dimethyl-4-(*m*-nitrophenyl)pyridine
In: Rat G189
→ 1,4-Dihydro-5-isopropoxycarbonyl-2,6-dimethyl-4-(*m*-nitrophenyl)pyridine-3-carboxylic acid
In: Rat G189

→ 5-Isopropoxycarbonyl-3-(2-methoxyethoxy)carbonyl-2,6-dimethyl-4-(*m*-nitrophenyl)pyridine
In: Rat G189

Nipradilol (3,4-dihydro-8-(2-hydroxy-3-isopropylaminopropoxy)-3-nitrooxy-2*H*-1-benzopyran)
→ 8-(3-Amino-2-hydroxypropoxy)-3,4-dihydro-3-nitrooxy-2*H*-1-benzopyran
In: Rat D882
→ 3,4-Dihydro-3-hydroxy-8-(2-hydroxy-3-isopropylaminopropoxy)-2*H*-1-benzopyran
In: Rat D882
→ 4-Hydroxynipradilol
Probably in: Rat D882

Nirvanol
See: 5-Ethyl-5-phenylhydantoin

Nitecapone
See: 3-(3,4-Dihydroxy-5-nitrobenzylidene)-2,4-pentanedione

Nitracrine (N,N-dimethyl-N′-(1-nitro-9-acridinyl)-1,3-propanediamine)
→ 9-((3-Dimethylaminopropyl)amino)-1-(S-glutathionyl)acridine-N(10)-oxide
In: Rat H704
→ 2-Hydroxynitracrine
In: Rat H704

Nitracrine-N(10)-oxide
→ 9-((3-Dimethylaminopropyl)amino)-1-(S-glutathionyl)acridine-N(10)-oxide
In: Rat H704

Nitralin (4-methylsulphonyl-2,6-dinitro-N,N-dipropylaniline)
→ 7-Amino-2-ethyl-5-methylsulphonyl-1-propylbenzimidazole
In: Rat C750
→ N-Carboxyethyl-4-methylsulphonyl-2,6-dinitro-N-propylaniline
Probably in: Rat C750
→ 2-Methyl-7-methylsulphonyl-5-nitroquinoxaline
Probably in: Rat C750
→ 2-Methyl-5-methylsulphonyl-7-nitro-1-propylbenzimidazole
Probably in: Rat C750

Nitrazepam
See: 1,3-Dihydro-7-nitro-5-phenyl-2*H*-1,4-benzodiazepin-2-one

Nitrendipine (1,4-dihydro-2,6-dimethyl-4-(*m*-nitrophenyl)pyridine-3,5-dicarboxylic acid 3-ethyl ester 5-methyl ester)
→ 1,4-Dehydronitrendipine
In: Dog G398
Mouse G398
Rat G398
→ 1,4-Dihydro-2,6-dimethyl-4-(*m*-nitrophenyl)pyridine-3,5-dicarboxylic acid 3-ethyl ester
In: Dog G398
Mouse G398
Rat G398
→ 1,4-Dihydro-2,6-dimethyl-4-(*m*-nitrophenyl)pyridine-3,5-dicarboxylic acid 3-methyl ester
In: Dog G398
Mouse G398
Rat G398
→ 1,4-Dihydro-2-hydroxymethyl-6-methyl-4-(*m*-nitrophenyl)pyridine-3,5-dicarboxylic acid 3-ethyl ester 5-methyl ester
In: Dog G398
Mouse G398
Rat G398

***o*-Nitroacetanilide**
→ *o*-Acetamidoaniline
In: Saccharomyces F201
→ *o*-Nitroaniline
In: Sheep A2177

***m*-Nitroacetanilide**
Formed from: *m*-Nitroaniline
→ *m*-Acetamidoaniline
In: Saccharomyces F201
→ 2,4-Diacetamidophenol
Probably in: Hamster F764
Rat F764

***p*-Nitroacetanilide**
Formed from: *p*-Nitroaniline
→ *p*-Acetamidoaniline
In: Saccharomyces F201
→ *p*-Nitroaniline
In: Nocardia G343
Pseudomonas C691

o-Nitroacetophenone
→ 1-(*o*- Nitrophenyl)ethanol
In: Hamster F153

***p*-Nitroacetophenone**
→ *p*-Aminoacetophenone
In: Beef G401
→ 1-(*p*- Nitrophenyl)ethanol
In: Japanese monkey F241
Man A15, A353
Mouse E821, H230
Rabbit A3950, B386, B431, D125
Rat J744

***o*-Nitroaniline**
Formed from: *o*-Dinitrobenzene
o-Nitroacetanilide
→ 2-Amino-3-nitrophenol
In: Tomato F851
See also: E888
→ 4-Amino-3-nitrophenol
In: Tomato F851
See also: E888
→ *o*-Nitroaniline-N-β-D-glucoside
In: Tomato F851
→ *o*-Nitroaniline-N-β-D-glucuronide
Probably in: Rat E888
→ *o*-Phenylenediamine
In: Haloanaerobium G369
Sporohalobacter G369
Veillonella A2624

***m*-Nitroaniline**
Formed from: *m*-Dinitrobenzene
→ *p*-Acetamidophenol
Probably in: Hamster F764
Rat E888, F764, H815
→ *m*-Nitroacetanilide
In: Hamster F764
Rat E440, F764
→ *m*-Nitroaniline-N-β-D-glucuronide
In: Rat E888
→ Nitrobenzene
Probably in: Pseudomonas H288
→ *m*-Phenylenediamine
In: Haloanaerobium G369
Rat E888
Sporohalobacter G369
Veillonella A2624

***p*-Nitroaniline**
Formed from: L-Alanyl-*p*-nitroanilide
L-Arginyl-*p*-nitroanilide
α-N-Benzoyl-L-arginyl-*p*-nitroanilide
S-Benzyl-L-cysteinyl-*p*-nitroanilide
N-Benzyl-*p*-nitroaniline
p-Dinitrobenzene
L-Glutaminyl-*p*-nitroanilide
γ-L-Glutamyl-*p*-nitroanilide
Glycyl-*p*-nitroanilide
L-Leucyl-*p*-nitroanilide
L-Lysyl-*p*-nitroanilide
L-Methionyl-*p*-nitroanilide
p-Nitroacetanilide
p-Nitrophenyl azide
N-(*p*-Nitrophenyl)benzamide
N-(*p*-Nitrophenyl)phthalamidic acid
L-Phenylalanylglycyl-L-leucyl-*p*-nitroanilide
L-Phenylalanyl-*p*-nitroanilide
L-Prolyl-*p*-nitroanilide
Succinyl-L-alanyl-L-alanyl-*p*-nitroanilide
L-Tyrosyl-*p*-nitroanilide
L-Valyl-*p*-nitroanilide
→ 2-Amino-5-nitrophenol
In: Rat D401
See also: E888
→ N-Methyl-*p*-nitroaniline
In: Rabbit C333
→ *p*-Nitroacetanilide
In: Arthrobacter E276
Bacillus E276
Cordyceps E276
Hen E384
Nocardia G343
Pigeon E384
Rabbit E384, E399
Sheep A2788
→ *p*-Nitrophenylhydroxylamine
In: Rat D401
→ *p*-Phenylenediamine
In: Haloanaerobium G369
Sporohalobacter G369
Veillonella A2624
See also: E888

***o*-Nitroaniline-N-β-D-glucoside**
Formed from: *o*-Nitroaniline
***o*-Nitroaniline-N-β-D-glucuronide**
Formed from: *o*-Nitroaniline
***m*-Nitroaniline-N-β-D-glucuronide**
Formed from: *m*-Nitroaniline
***o*-Nitroanisole**
→ *o*-Methoxyaniline
In: Ascaris A304
Moniezia A1785
Rat D831
Saccharomyces F201
→ *o*-Nitrophenol
In: Rat D831
***m*-Nitroanisole**
→ *m*-Methoxyaniline
In: Ascaris A304
Moniezia A1785
Saccharomyces F201
***p*-Nitroanisole**
→ *p*-Methoxyaniline
In: Ascaris A304, A324
Moniezia A324
Sea urchin B703, B895
→ *p*-Nitrophenol
In: Baboon A3692
Crayfish E816
Cunninghamella A859
Man D135, J283
Methylosinus D513
Rhodococcus H809
Mouse B701
Prawn E816
Rabbit A2545, A3173, G17
Rat A46, A119, A127, A1071, A1953, A1954, A2565, A3048, B477, B820, C676, D98
Sea urchin B703
Starfish B703
***o*-Nitrobenzaldehyde**
→ *o*-Aminobenzaldehyde
In: Ascaris A304
Moniezia A1785
→ 2-Amino-4-hydroxy-4-(*o*-nitrophenyl) butyric acid
In: Pseudomonas D449
→ *o*-Nitrobenzoic acid
In: Man J24
Serratia G197
→ *o*-Nitrobenzyl alcohol
In: Dunaliella G459
Euglena G422
Hamster F153
Japanese monkey F241
Mung bean J910
Rabbit K569
Rat H759
Saccharomyces H425
***m*-Nitrobenzaldehyde**
Formed from: *m*-Nitrobenzoic acid
m-Nitrobenzyl alcohol
→ *m*-Aminobenzaldehyde
In: Ascaris A304
Moniezia A1785
→ *m*-Nitrobenzoic acid
In: Pseudomonas H624
Rat A1688, H761
→ *m*-Nitrobenzyl alcohol
In: Beef A614
Dunaliella G459
Euglena G422
Mouse E821
Monkey F241, F388
Mung bean J910
Pig F389
Rabbit K569
Saccharomyces H425
***p*-Nitrobenzaldehyde**
Formed from: *p*-Nitrobenzoic acid
p-Nitrobenzyl alcohol
N-(*p*-Nitrobenzyl)aniline
Probably formed from: Chloramphenicol
→ *p*-Aminobenzaldehyde
In: Ascaris A304
Beef G401
Moniezia A1785
→ 2-Amino-4-hydroxy-4-(*p*-nitrophenyl) butyric acid
In: Pseudomonas D449
→ *p*-Nitrobenzoic acid
In: Beef C508
Man J24

Pleurotus G668
Pseudomonas G1000
Rat B659, H761
See also: A2286, A2395, G980
→ *p*-Nitrobenzyl alcohol
In: Cochliobolus K378
Dunaliella G459
Euglena G422
Hamster F153
Man B569, K273
Mouse E821, F377, H230
Monkey F241, F388, F391
Mung bean J910
Pig A1679, F389
Rabbit A3500, A3950, B386, B431, D125, K569
Rat A2007, B659, C31, H759 J744, K546
Saccharomyces H425
→ (1*R*,2*S*)-1-(*p*-Nitrophenyl)propane-1,2-diol
In: Saccharomyces E487

o-Nitrobenzaldoxime
→ *o*-Nitrobenzaldoxime-β-D-(6-sulphato) glucopyanoside
In: Brassica F657
→ *o*-Nitrobenzyl-β-D-sulphatoglucoside
In: Brassica F657
→ *o*-Nitrophenylglucosinolate
In: Brassica F657

m-Nitrobenzaldoxime
→ *m*-Nitrobenzyl-β-D-sulphatoglucoside
In: Brassica F657

o-Nitrobenzaldoxime-β-D-(6-sulphato) glucopyanoside
Formed from: *o*-Nitrobenzaldoxime

o-Nitrobenzamide
Formed from: *o*-Nitrobenzonitrile
→ *o*-Nitrobenzoic acid
Probably in: Rhodococcus H952

m-Nitrobenzamidine
→ N-Hydroxy-*m*-nitrobenzamidine
In: Rabbit C573

7-Nitrobenzanthracene
→ 3,4-Dihydro-3,4-dihydroxy-7-nitrobenzanthracene
In: Rat C382
→ 8,9-Dihydro-8,9-dihydroxy-7-nitrobenzanthracene
In: Rat C382

Nitrobenzene
Formed from: Aniline
→ Aniline
In: Ascaris A304, A324
Boophilus A1024
Haloanaerobium G369, G921
Methanococcus H328
Microorganism A1359, J622
Moniezia A304, A1785
Rat A1359
Spider A1024
Sporohalobacter G369, G921
Veillonella A2624
See also: K520
→ Catechol
In: Pseudomonas J524
→ 1,2-Dihydro-1,2-dihydroxy-3-nitrobenzene
In: Pseudomonas H757
→ *o*-Nitrophenol
In: Rat A2160, D122
→ *m*-Nitrophenol
In: Mouse C530
Nitrosomonas H631
Rat A2160, C32, C530, D122
→ *p*-Nitrophenol
In: Mouse C530
Nitrosomonas H631
Rat A2160, C32, C530, D122
→ Nitrosobenzene
In: Microorganism C32
Pseudomonas K552
→ Phenylhydroxylamine
In: Spinach J501

o-Nitrobenzenesulphonic acid
→ 2,3-Dihydroxybenzenesulphonic acid
Probably in: Alcaligenes K293

m-Nitrobenzenesulphonic acid
→ *m*-Aminobenzenesulphonic acid
In: Veillonella A2624

o-Nitrobenzoic acid
Formed from: *o*-Nitrobenzaldehyde
o-Nitrobenzamide
o-Nitrobenzonitrile

o-Nitrobenzyl alcohol
2-Nitrobiphenyl
→ Anthranilic acid
In: Arthrobacter K407
Ascaris A304
Moniezia A1785
Veillonella A2624
→ *o*-Hydroxylaminobenzoic acid
In: Arthrobacter K407
→ *o*-Nitrobenzoyl-β-D-glucuronic acid
In: Rabbit H519

m-Nitrobenzoic acid
Formed from: *m*-Nitrobenzaldehyde
m-Nitrobenzonitrile
m-Nitrobenzyl alcohol
3-Nitrobiphenyl
→ *m*-Aminobenzoic acid
In: Ascaris A304
Moniezia A1785
Rat A3480
Veillonella A2624
See also: D383
→ *m*-Nitrobenzaldehyde
In: Nocardia G207
→ *m*-Nitrobenzoyl-β-D-glucuronic acid
In: Rabbit H519
→ *m*-Nitrobenzyl alcohol
In: Coriolus G821
→ *m*-Nitrohippuric acid
In: Rat D383

p-Nitrobenzoic acid
Formed from: *p*-Nitrobenzaldehyde
p-Nitrobenzonitrile
p-Nitrobenzoyl-L-alanine
N-(*p*-Nitrobenzoyl)aniline
p-Nitrohippuric acid
→ *p*-Aminobenzamide
In: Streptomyces E279
→ *p*-Aminobenzoic acid
In: Aerobacter A18, A1741, A1856
Ascaris A304, A324, A2037
Bacteroides A124
Bat A3867
Beef G401
Clostridiun A124
Cunninghamella A859
Escherichia A18, A124, A1856
Lactobacillus A124
Man A114, A621
Marmoset B802
Microorganism C435
Moniezia A324, A1785, A2037
Mouse A2037
Peptostreptococcus A124
Proteus A124, A1856
Rat A18, A114, A124, A3048, A3480
Salmonella A1856
Sheep A2037
Streptococcus A124, A1856
Veillonella A2624
See also: A653, A3049, A3849, D383
→ 4-Hydroxylaminobenzoic acid
In: Pseudomonas G980
→ *p*-Nitrobenzaldehyde
In: Nocardia G207
→ *p*-Nitrobenzoyl CoA
In: Pseudomonas G729, K796
→ *p*-Nitrobenzoyl-β-D-glucuronic acid
In: Bat A3867
Ferret A394, A3046
Rabbit H519
→ *p*-Nitrohippuric acid
In: Ferret A394, A3046
Marmoset A2802
Rat D383, F693

o-Nitrobenzonitrile
→ *o*-Aminobenzonitrile
In: Saccharomyces F201
→ *o*-Nitrobenzamide
In: Rhodococcus H952
→ *o*-Nitrobenzoic acid
In: Fusarium A3364
Rhodococcus H952

m-Nitrobenzonitrile
→ *m*-Aminobenzonitrile
In: Saccharomyces F201
→ *m*-Nitrobenzoic acid
In: Fusarium A3364
Nocardia A3420, D725
Rhodococcus H952

p-Nitrobenzonitrile
→ *m*-Aminobenzonitrile

In: Rat H305
Saccharomyces F201
→ *p*-Azobenzonitrile
In: Rat H305
→ *p*-Azoxybenzonitrile
In: Rat H305
→ *p*-Nitrobenzoic acid
In: Fusarium A3364
Nocardia D725
Rhodococcus H952

3-Nitrobenzophenone
→ 3-Aminobenzophenone
In: Saccharomyces F201

4-Nitrobenzophenone
→ 4-Aminobenzophenone
In: Saccharomyces F201

***p*-Nitrobenzoyl-L-alanine**
→ *p*-Nitrobenzoic acid
In: Corynebacterium E145

N-(*p*-Nitrobenzoyl)aniline
→ Aniline
In: Rabbit K435
→ *p*-Nitrobenzoic acid
In: Rabbit K435

***p*-Nitrobenzoyl CoA**
Formed from: *p*-Nitrobenzoic acid

***o*-Nitrobenzoyl-β-D-glucuronic acid**
Formed from: *o*-Nitrobenzoic acid

***m*-Nitrobenzoyl-β-D-glucuronic acid**
Formed from: *m*-Nitrobenzoic acid

***p*-Nitrobenzoyl-β-D-glucuronic acid**
Formed from: *p*-Nitrobenzoic acid

1-Nitro-3,4-benzpyrene
→ 1-Amino-3,4-benzpyrene
In: Rat E414
→ 7,8-Dihydro-7,8-dihydroxy-1-nitro-3,4-benzpyrene
In: Rat C577, D207
→ 9,10-Dihydro-9,10-dihydroxy-1-nitro-3,4-benzpyrene
In: Rat C577, D207

6-Nitro-3,4-benzpyrene
→ 6-Amino-3,4-benzpyrene
In: Microorganism D278
Mouse E971
→ 3,4-Benzpyrene
In: Mouse E971
Rat D207, E252, E467
→ *trans*-7,8-Dihydro-7,8-dihydroxy-6-nitro-3,4-benzpyrene
In: Rat D207
→ *trans*-9,10-Dihydro-9,10-dihydroxy-6-nitro-3,4-benzpyrene
In: Rat D207
→ 6-Hydroxy-3,4-benzpyrene
In: Mouse E971
Rat E252, E467
→ 1-Hydroxy-6-nitro-3,4-benzpyrene
In: Cunninghamella E298
Rat C231, C540, E467
→ 3-Hydroxy-6-nitro-3,4-benzpyrene
In: Cunninghamella E298
Rat C231, C383, C540, E467
→ 6-Nitro-3,4-benzpyrene quinol
In: Rat E467

6-Nitro-3,4-benzpyrene quinol
Formed from: 6-Nitro-3,4-benzpyrene

6-Nitro-3,4-benzpyren-1-yl-β-D-glucuronide
Formed from: 1-Hydroxy-6-nitro-3,4-benzpyrene

6-Nitro-3,4-benzpyren-3-yl-β-D-glucuronide
Formed from: 3-Hydroxy-6-nitro-3,4-benzpyrene

***o*-Nitrobenzyl alcohol**
Formed from: *o*-Nitrobenzaldehyde
o-Nitrotoluene
→ *o*-Aminobenzyl alcohol
In: Ascaris A304
Moniezia A1785
→ *o*-Nitrobenzoic acid
In: Rat C650, D427
See also: D383
→ *o*-Nitrobenzyl-β-D-glucuronide
In: Rat C650, D383, D427, D827
→ *o*-Nitrobenzyl sulphate
In: Rat D427

***m*-Nitrobenzyl alcohol**
Formed from: *m*-Nitrobenzaldehyde
m-Nitrobenzoic acid
m-Nitrotoluene
→ *m*-Aminobenzyl alcohol
In: Ascaris A304
Moniezia A1785

→ *m*-Nitrobenzaldehyde
In: Pseudomonas H624
Rhodopseudomonas D197
→ *m*-Nitrobenzoic acid
In: Rat C650, D427
See also: D383
→ *m*-Nitrobenzyl-β-D-glucuronide
In: Rat C650, D427, D827
→ *m*-Nitrobenzyl sulphate
In: Rat D427

p-Nitrobenzyl alcohol
Formed from: Chloramphenicol
p-Nitrobenzaldehyde
p-Nitrotoluene
→ *p*-Aminobenzyl alcohol
In: Ascaris A304
Moniezia A1785
→ *p*-(S-Glutathionyl)benzyl alcohol
Probably in: Rat D427
→ *p*-Nitrobenzaldehyde
In: Pleurotus G668
Pseudomonas G1000
Rhodopseudomonas D197
See also: A2286, A2395, C650, D427, G980
→ *p*-Nitrobenzyl-β-D-glucuronide
In: Rat C650, D427, D827
→ *p*-Nitrobenzyl sulphate
In: Rat C650, C765, D427, E650

N-(p-Nitrobenzyl)aniline
→ Aniline
In: Hamster G951
→ *p*-Nitrobenzaldehyde
In: Hamster G951

p-Nitrobenzyl chloride
→ S-(*p*-Nitrobenzyl)glutathione
In: Rat J583
See also: C878, C901

S-(p-Nitrobenzyl)-L-cysteine
Formed from: N-Acetyl-S-(*p*-nitrobenzyl)-L-cysteine

o-Nitrobenzyl-β-D-glucuronide
Formed from: *o*-Nitrobenzyl alcohol

m-Nitrobenzyl-β-D-glucuronide
Formed from: *m*-Nitrobenzyl alcohol

p-Nitrobenzyl-β-D-glucuronide
Formed from: *p*-Nitrobenzyl alcohol

S-(p-Nitrobenzyl)glutathione
Formed from: *p*-Nitrobenzyl chloride
p-Nitrobenzyl sulphate

o-Nitrobenzyl sulphate
Formed from: *o*-Nitrobenzyl alcohol

m-Nitrobenzyl sulphate
Formed from: *m*-Nitrobenzyl alcohol

p-Nitrobenzyl sulphate
Formed from: *p*-Nitrobenzyl alcohol
→ S-(*p*-Nitrobenzyl)glutathione
In: Rat C765

o-Nitrobenzyl-β-D-sulphatoglucoside
Formed from: *o*-Nitrobenzaldoxime

m-Nitrobenzyl-β-D-sulphatoglucoside
Formed from: *m*-Nitrobenzaldoxime

6-(p-Nitrobenzyl)thioinosine
Formed from: 6-(*p*-Nitrobenzyl) thioinosine-5′-phosphate

6-(p-Nitrobenzyl)thioinosine-5′-phosphate
→ 6-(*p*-Nitrobenzyl)thioinosine
In: HeLa cell D300

2-Nitrobiphenyl
→ *o*-Nitrobenzoic acid
In: Pseudomonas E41

3-Nitrobiphenyl
→ *m*-Nitrobenzoic acid
In: Pseudomonas E41

4-Nitrobiphenyl
→ 4-Aminobiphenyl
In: Rabbit E4
Rat A75
→ 2,3-Dihydro-2,3-dihydroxy-4′-nitrobiphenyl
In: Bacterium D911
→ 3,4-Dihydro-3,4-dihydroxy-4′-nitrobiphenyl
In: Bacterium D911
→ 4-Hydroxylaminobiphenyl
In: Rabbit E4
→ 2-Hydroxy-4′-nitrobiphenyl
In: Bacterium D911
→ 4-Hydroxy-4′-nitrobiphenyl
In: Bacterium D911
→ *p*-Nitrocinnamic acid
In: Bacterium D911
→ 3-(*p*-Nitrophenyl)muconolactone
In: Aspergillus H23

→ 4-Nitrosobiphenyl
In: Dog A880
Monkey A880

3-Nitrocatechol
Formed from: *m*-Nitrophenol
→ 3-Nitrocatechol-β-D-glucuronide
In: Rat J546

4-Nitrocatechol
Formed from: 2-Hydroxy-5-nitrophenyl sulphate
3,4-Methylenedioxynitrobenzene
m-Nitrophenol
p-Nitrophenol
4-Nitro-*o*-quinone
→ 2-Hydroxy-5-nitrophenyl-β-D-glucuronide
Probably in: Rat A127, D832, J546, K595
→ 2-Hydroxy-5-nitrophenyl sulphate
Probably in: Carpophyllum G819
Cystoceira G819
Frog D857, E841
Fucus G819
Himanthalia G819
Landsburgia G819
Rat A127, D832
Sargassum G819
Xenopus E841
→ Hydroxyquinol
In: Arthrobacter H270
Bacillus J679
Burkholderia K553
Sphingomonas K73
→ 2-Hydroxy-*p*-quinone
In: Phanerochaete G436
Pseudomonas J249
→ 2-Methoxy-4-nitrophenol
In: Rat A486, A2817
Streptomyces K652
With catechol O-methyltransferase A597
→ 2-Methoxy-5-nitrophenol
In: Rat A486
With catechol O-methyltransferase A597
→ 4-Nitroveratrole
In: Phanerochaete G436

3-Nitrocatechol-β-D-glucuronide
Formed from: 3-Nitrocatechol

Nitrocatechol sulphate
See: 2-Hydroxy-5-nitrophenyl sulphate

Nitrochlorphentermine
→ N-Hydroxychlorphentermine
In: Rat A3951

Nitrochlorphentermine
See: 2-(*p*-Chlorophenyl)-1,1-dimethylnitroethane

6-Nitrochrysene
→ 6-Aminochrysene
In: Man E833, F167
Rat E833
→ *trans*-1,2-Dihydro-1,2-dihydroxy-6-nitrochrysene
In: Man E833, F167
Rat E833
See also: E658
→ *trans*-9,10-Dihydro-9,10-dihydroxy-6-nitrochrysene
In: Man E833, F167
Rat E833

***o*-Nitrocinnamaldehyde**
→ *o*-Nitrocinnamic acid
In: Beef C508
Man J24

***p*-Nitrocinnamaldehyde**
→ *p*-Nitrocinnamic acid
In: Beef C508
Man J24

***o*-Nitrocinnamic acid**
Formed from: *o*-Nitrocinnamaldehyde

***m*-Nitrocinnamic acid**
→ *m*-Nitrostyrene
In: Catharanthus K244

***p*-Nitrocinnamic acid**
Formed from: 4-Nitrobiphenyl
p-Nitrocinnamaldehyde

Nitrodan
See: 3-Methyl-5-(*p*-nitrophenylazo)rhodanine

***p*-Nitrodibenzylamine**
Probably formed from: *p*-Nitrotribenzylamine

Nitrofen (2,4-dichlorophenyl 4-nitrophenyl ether)
→ 4-Aminophenyl 2,4-dichlorophenyl ether
In: Rat C543
Sheep A2535
→ 2-Chlorophenyl 4-nitrophenyl ether

In: Sheep A2535
→ 2,4-Dichloro-5-hydroxyphenyl 4-nitrophenyl ether
In: Rat C543
Sheep A2535
→ 2,4-Dichlorophenol
In: Sheep A2535

1-Nitrofluoranthene
→ 1-Aminofluoranthene
In: Rat F510

2-Nitrofluoranthene
→ 2-Aminofluoranthene
In: Rat G887

3-Nitrofluoranthene
→ 3-Aminofluoranthene
In: Rat F510, G887
→ 8-Hydroxy-3-nitrofluoranthene
Probably in: Rat G887
→ 3-Nitrosofluoranthene
In: Rabbit G361

7-Nitrofluoranthene
→ 7-Aminofluoranthene
In: Rat F510

8-Nitrofluoranthene
→ 8-Aminofluoranthene
In: Rat F510, G887

2-Nitrofluorene
Formed from: 2-Nitrosofluorene
→ 2-Aminofluorene
In: Rabbit E4
Rat C702
See also: E481, F953
→ 2-Hydroxylaminofluorene
In: Rabbit E4
See also: F953
→ 9-Hydroxy-2-nitrofluorene
In: Rat E675, E688

***m*-Nitrohippuric acid**
Formed from: *m*-Nitrobenzoic acid

***p*-Nitrohippuric acid**
Formed from: *p*-Nitrobenzoic acid
→ *p*-Aminohippuric acid
In: Ascaris A304
Moniezia A1785
→ *p*-Nitrobenzoic acid
In: Corynebacterium E145

Nitromethaqualone (3-(2-methoxy-4-nitrophenyl)-2-methyl-4-(3*H*)-quinazolinone)
→ Aminomethaqualone
In: Man C437
Rat C437
→ 2-Hydroxymethyl-3-(2-methoxy-4-nitrophenyl)-4(3*H*)-quinazolinone
In: Rat C437
→ 2-Methoxy-4-nitroaniline
In: Man C437

Nitromide
See: 3,5-Dinitrobenzamide

5-Nitro-α-naphthaldehyde
→ 5-Nitro-α-naphthoic acid
In: Man J24

1-Nitronaphthalene
→ 2-(S-Glutathionyl)-1,2-dihydro-1-hydroxy-5-nitronaphthalene
In: Mouse K12
→ 2-(S-Glutathionyl)-1,2-dihydro-1-hydroxy-8-nitronaphthalene
In: Mouse K12
Rat K12
→ α-Naphthylamine
In: Rabbit E4
Rat A75, J791
→ 1-Naphthylhydroxylamine
In: Rabbit A2173, E4

2-Nitronaphthalene
→ β-Naphthylamine
In: Rat A75
→ 2-Nitrosonaphthalene
In: Monkey A880
See also: B773

5-Nitro-α-naphthoic acid
Formed from: 5-Nitro-α-naphthaldehyde

2-Nitro-4-nitrosotoluene
Formed from: 2,4-Dinitrotoluene
→ 4-Amino-2-nitrotoluene
Probably in: Microorganism C263

2-Nitro-6-nitrosotoluene
Formed from: 2,6-Dinitrotoluene
→ 2-Hydroxylamino-6-nitrotoluene
Probably in: Microorganism G542
Salmonella G235

4-Nitro-2-nitrosotoluene
Formed from: 2,4-Dinitrotoluene

→ 2-Amino-4-nitrotoluene
Probably in: Microorganism C263

2′-Nitronorreticuline
→ 2′-Nitropapaverine
In: Papaver E777

2-Nitro-17β-oestradiol
→ 2-Nitro-17β-oestradiol-3-sulphate
In: Beef A1211

4-Nitro-17β-oestradiol
→ 4-Nitro-17β-oestradiol-3-sulphate
In: Beef A1211

2-Nitro-17β-oestradiol-3-sulphate
Formed from: 2-Nitro-17β-oestradiol

4-Nitro-17β-oestradiol-3-sulphate
Formed from: 4-Nitro-17β-oestradiol

2-Nitrooestrone
Formed from: 3-Acetyl-2-nitrooestrone

2-Nitrooxypropyl 3-carboxy-2,6-dimethyl-4-(*m*-nitrophenyl)pyridine-5-carboxylate
Probably formed from: 2-Nitrooxypropyl 3-nitrooxypropyl 2,6-dimethyl-4-(*m*-nitrophenyl)pyridine-5-carboxylate
→ 2-Nitrooxypropyl 3-carboxy-2,6-dimethyl-4-phenylpyridine-5-carboxylate
Probably in: Dog G593
Rat G593

3-Nitrooxypropyl 3-carboxy-2,6-dimethyl-4-(*m*-nitrophenyl)pyridine-5-carboxylate
Probably formed from: 2-Nitrooxypropyl 3-nitrooxypropyl 2,6-dimethyl-4-(*m*-nitrophenyl)pyridine-5-carboxylate
→ 3-Nitrooxypropyl 3-carboxy-2,6-dimethyl-4-phenylpyridine-5-carboxylate
Probably in: Dog G593
Rat G593

2-Nitrooxypropyl 3-carboxy-2,6-dimethyl-4-phenylpyridine-5-carboxylate
Probably formed from: 2-Nitrooxypropyl 3-carboxy-2,6-dimethyl-4-(*m*-nitrophenyl) pyridine-5-carboxylate

3-Nitrooxypropyl 3-carboxy-2,6-dimethyl-4-phenylpyridine-5-carboxylate
Probably formed from: 3-Nitrooxypropyl 3-carboxy-2,6-dimethyl-4-(*m*-nitrophenyl) pyridine-5-carboxylate

2-Nitrooxypropyl 3-carboxy-2-hydroxymethyl-6-methyl-4-(*m*-nitrophenyl)pyridine-5-carboxylate
Probably formed from: 2-Hydroxymethyl-5-(2-nitrooxypropoxycarbonyl)-3-(3-nitrooxypropoxycarbonyl)-6-methyl-4-(*m*-nitrophenyl)pyridine
→ 2-Hydroxypropyl 3-carboxy-2-hydroxymethyl-6-methyl-4-(*m*-nitrophenyl)pyridine-5-carboxylate
Probably in: Dog G593
Rat G593

3-Nitrooxypropyl 3-carboxy-2-hydroxymethyl-6-methyl-4-(*m*-nitrophenyl)pyridine-5-carboxylate
Probably formed from: 2-Hydroxymethyl-3-(2-nitrooxypropoxycarbonyl)-5-(3-nitrooxypropoxycarbonyl)-6-methyl-4-(*m*-nitrophenyl)pyridine
→ 3-Hydroxypropyl 3-carboxy-2-hydroxymethyl-6-methyl-4-(*m*-nitrophenyl)pyridine-5-carboxylate
Probably in: Dog G593
Rat G593

2-Nitrooxypropyl 3-nitrooxypropyl 1,4-dihydro-2,6-dimethyl-4-(*m*-nitrophenyl) pyridine-3,5-dicarboxylate
→ 2-Hydroxypropyl 3-nitrooxypropyl 1,4-dihydro-2,6-dimethyl-4-(*m*-nitrophenyl) pyridine-3,5-dicarboxylate
In: Dog G593
Rabbit H163
Rat G593
→ 3-Hydroxypropyl 2-nitrooxypropyl 1,4-dihydro-2,6-dimethyl-4-(*m*-nitrophenyl) pyridine-3,5-dicarboxylate
In: Dog G593
Rabbit H163
Rat G593
→ 2-Nitrooxypropyl 3-nitrooxypropyl 2,6-dimethyl-4-(*m*-nitrophenyl)pyridine-3,5-dicarboxylate
In: Dog G593
Rabbit H163
Rat G593

2-Nitrooxypropyl 3-nitrooxypropyl 2,6-dimethyl-4-(*m*-nitrophenyl)pyridine-3,5-dicarboxylate
Formed from: 2-Nitrooxypropyl 3-nitrooxypropyl 1,4-dihydro-2,6-dimethyl-4-(*m*-nitrophenyl)pyridine-3,5-dicarboxylate
→ 3,5-Dicarboxy-2,6-dimethyl-4-(*m*-nitrophenyl)pyridine
Probably in: Dog G593
Rat G593
→ 2-Hydroxymethyl-3-(2-nitrooxypropoxycarbonyl)-5-(3-nitrooxypropoxycarbonyl)-6-methyl-4-(*m*-nitrophenyl)pyridine
Probably in: Dog G593
Rat G593
→ 2-Hydroxymethyl-5-(2-nitrooxypropoxycarbonyl)-3-(3-nitrooxypropoxycarbonyl)-6-methyl-4-(*m*-nitrophenyl)pyridine
Probably in: Dog G593
Rat G593
→ 2-Hydroxy-3-nitrooxypropyl 3-carboxy-2,6-dimethyl-4-(*m*-nitrophenyl)pyridine-5-carboxylate
Probably in: Dog G593
Rat G593
→ 2-Hydroxypropyl 3-nitrooxypropyl 2,6-dimethyl-4-(*m*-nitrophenyl)pyridine-3,5-dicarboxylate
Probably in: Dog G593
Rabbit H163
Rat G593
→ 3-Hydroxypropyl 2-nitrooxypropyl 2,6-dimethyl-4-(*m*-nitrophenyl)pyridine-3,5-dicarboxylate
Probably in: Dog G593
Rabbit H163
Rat G593
→ 2-Nitrooxypropyl 3-carboxy-2,6-dimethyl-4-(*m*-nitrophenyl)pyridine-5-carboxylate
Probably in: Dog G593
Rat G593
→ 3-Nitrooxypropyl 3-carboxy-2,6-dimethyl-4-(*m*-nitrophenyl)pyridine-5-carboxylate
Probably in: Dog G593
Rat G593

2′-Nitropapaverine
Formed from: 2′-Nitronorreticuline

***p*-Nitrophenetole**
Probably formed from: *p*-Nitrosophenetole
→ *p*-Nitrophenol
In: Mouse B701
Rabbit A2545
Rat A119, B330

***p*-Nitrophenethyl bromide**
→ S-(*p*-2-Bromoethyl)phenyl)glutathione
Probably in: Pig C901

***o*-Nitrophenol**
Formed from: *o*-Aminophenol
o-Nitroanisole
Nitrobenzene
o-Nitrophenyl acetate
o-Nitrophenyl butyrate
o-Nitrophenyl-α-D-galactoside
o-Nitrophenyl-β-D-galactoside
o-Nitrophenyl-β-D-glucoside
o-Nitrophenyl-β-D-xyloside
→ *o*-Aminophenol
In: Ascaris A304
Haloanaerobium G369, G921
Moniezia A1785
Sporohalobacter G369, G921
Veillonella A2624
→ Catechol
In: Pseudomonas D44, E607, E726, H65
→ *o*-Nitrophenyl-β-D-glucoside
In: Gardenia E338
→ *o*-Nitrophenyl-β-D-glucuronide
In: Rat D831
→ *o*-Nitrophenyl sulphate
In: Rat D831

***m*-Nitrophenol**
Formed from: N-Acetyl-*m*-nitrophenyl-β-D-glucosaminide
Nitrobenzene
m-Nitrotoluene
→ *m*-Aminophenol
In: Ascaris A304
Haloanaerobium G369, G921
Moniezia A1785
Rana D857
Saccharomyces F201
Sporohalobacter G369, G921
Veillonella A2624

→ Hydroxyquinol
In: Pseudomonas H823
→ 3-Nitrocatechol
In: Pseudomonas E731
→ 4-Nitrocatechol
In: Penicillium H82
Rana D857
→ *m*-Nitrophenyl-β-D-glucoside
In: Gardenia E338
→ *m*-Nitrophenyl-β-D-glucuronide
In: Rana D857
Rat C530
→ *m*-Nitrophenyl sulphate
In: Man E575
Rana D857
Rat B941, C530, E575
→ Nitroquinol
In: Rana D857

p-Nitrophenol

Formed from: N-Acetyl-*p*-nitrophenyl-β-D-galactosaminide
N-Acetyl-*p*-nitrophenyl-β-D-glucosaminide
p-Aminophenol
Aniline
Bis(*p*-nitrophenyl) phosphate
Butyl *p*-nitrophenyl ether
Diethyl 4-nitrophenyl phosphate
Diethyl 4-nitrophenyl phosphorothioate
O,O-Dimethyl O-(4-nitrophenyl) phosphorothioate
Dimethyl 4-nitrophenyl phenylphosphorothioate
p-Ethoxynitrobenzene
Ethyl *p*-nitrophenyl phenylphosphate
Ethyl *p*-nitrophenyl phenylphosphorothioate
Ethyl *p*-nitrophenyl phosphorothioate
γ-Glutamyl-*p*-nitroanilide
Isopropyl *p*-nitrophenyl ether
p-Nitroanisole
Nitrobenzene
p-Nitrophenetole
p-Nitrophenoxyacetic acid
3-(*p*-Nitrophenoxy)-1-propene oxide
p-Nitrophenyl acetate
p-Nitrophenyl-α-L-arabinoside
p-Nitrophenyl butyrate
p-Nitrophenyl caprylate
p-Nitrophenyl-α-D-2-deoxygalactoside
p-Nitrophenyl-α-D-6-deoxygalactoside
p-Nitrophenyl (1*R*)-*cis*-3-(2,2-dibromovinyl)-2,2-dimethylcyclopropanecarboxylate
p-Nitrophenyl-α-L-fucoside
p-Nitrophenyl-β-D-fucoside
p-Nitrophenyl-α-D-galactoside
p-Nitrophenyl-β-D-galactoside
p-Nitrophenyl-α-D-glucoside
p-Nitrophenyl-β-D-glucoside
p-Nitrophenyl-β-D-glucuronide
p-Nitrophenyl-α-D-mannoside
p-Nitrophenyl-β-D-mannoside
p-Nitrophenyl 2-nitro-4-trifluoromethylphenyl ether
p-Nitrophenyl phosphate
p-Nitrophenyl propionate
p-Nitrophenyl sulphate
p-Nitrophenyl trimethylacetate
p-Nitrophenyl-β-D-xyloside
→ *p*-Aminophenol
In: Ascaris A304, A324
Haloanaerobium G369, G921
Microorganism C435
Moniezia A324, A1785
Rat A46
Sporohalobacter G369, G921
Veillonella A2624
See also: A2160, A3100, C530, E841
→ 4-Nitrocatechol
In: Bacillus J679
Frog E50, E841
Nocardia D182
Penicillium H82
Pseudomonas A3638
Rabbit E84
Rat A46, D832, G969
Rhodobacter J506
Sheep G131
Xenopus E841
With glucuronyltransferase C6
See also: A127, A1953, H270, H809

→ *p*-Nitrophenyl-β-D-glucoside
In: Agrostemma J709
Blowfly C494
Carrot H118
Cockroach C494
Cricket C494
Earwig C494
Gardenia E338
Glycine H660, J709
Housefly C494
Mealworm C494
Porina C494
Stick insect C494
Triticum J709
Wax moth C494
See also: A2025
→ *p*-Nitrophenyl-D-glucoside-6-phosphate
In: Blowfly C494
Cockroach C494
Earwig C494
Housefly A1604, C494
Mealworm C494
Wax moth C494
→ *p*-Nitrophenyl glucoside sulphate
Probably in: Tenebrio C494
→ *p*-Nitrophenyl-β-D-glucuronide
In: Blenny F248
Cat C562
Chicken A860, C12, C629, H889
Dog C562, H786
Flounder F248
Frog E50, E841
Guinea pig A862, A865, C562
Man B800, C654, F309, G968, J183, J416, J490, J783
Monkey A1997
Mouse C80, C266, C562, D315, D804, D858, H454
Perch F248
Pig A1997
Quail C562
Rabbit A43, A1996, A2084, B983, C562, F309, G17, J107
Rat A85, A127, A1071, A1668, A1862, A1997, A2565, A3827, B820, B962, C37, C530, C562, C672, D66, D103, D723, E219, F309, F799, G748, J107, J546, J550
Sculpin F248
Sheep J65, K463
Tree shrew A1997
Trout C562, F248
Xenopus E841
Zebra fish E813
See also: A1961
→ *p*-Nitrophenyl phosphate
In: Blowfly C494
Cockroach C494
Earwig C494
Housefly C494
Mealworm C494
Porina C494
Stick insect C494
Wax moth C494
→ *p*-Nitrophenyl sulphate
In: Beef D607, F196
Blowfly C494
Channel catfish K497
Chicken A860, C12, C629, H889
Crayfish E816
Cystoseira G819
Dog G712, J292
Frog E50, E841
Guinea pig G712
Housefly C494
Man C528, D233, D607, D977, D979, E194, E211, F116, F464, F692, F770, H276, J91, J324, J535, J571, J644, K257, K344, K405, K588
Mealworm C494
Mouse C530, D804, D858
Prawn E816
Rabbit G17, G712
Rat A85, A127, A1071, A2565, B820, B941, C441, C530, D10, D92, D103, D807, D933, E219, E575, F631, F799, G317, G675, K120, K735
Rhesus monkey H628
Sheep J65, K463
Starfish B703
Stick insect C494
Wax moth C494

Xenopus E841
Zebra fish E813
With glucuronyltransferase C6
→ Nitroquinol
In: Frog E50
See also: H809
→ Quinol
In: Moraxella G153
Penicillium E94
Pseudomonas H386
Rat J81
Sphingomonas K73
See also: A2321

p-Nitrophenoxyacetic acid
→ *p*-Nitrophenol
In: Escherichia H898

p-Nitrophenoxybutyric acid
Formed from: 4-Hydroxybutyl *p*-nitrophenyl ether

3-(p-Nitrophenoxy)lactic acid
Formed from: 3-(*p*-Nitrophenoxy)-1-propene oxide

2-(p-Nitrophenoxy)oxirane
Formed from: *p*-Nitrophenyl vinyl ether

N-(4-Nitro-2-phenoxyphenyl)methanesulphonamide
See: Nimesulide

3-(p-Nitrophenoxy)-1-propene oxide
→ 1-(S-Glutathionyl)-2-hydroxy-3-(*p*-nitrophenoxy)propane
In: Rat J583
See also: A2282, A3042, A3043, C562, C878
→ *p*-Nitrophenol
In: Rabbit A3042
Rat A3042
→ 3-(*p*-Nitrophenoxy)lactic acid
In: Rabbit A3042
Rat A3042

o-Nitrophenyl acetate
→ *o*-Nitrophenol
In: Rat F753

p-Nitrophenyl acetate
→ 4-(S-Glutathionyl)phenyl acetate
Probably in: Blue crab E434
→ *p*-Nitrophenol
In: Acinetobacter A962
Aspergillus J389
Bee F887, G551
Chromobacterium A1808
Dog A868
Hamster G744
Man A798, G251, H278, H465
Pig A333
Rat F753, G735

p-Nitrophenylacetic acid
Formed from: *p*-Nitrophenylacetylene
→ *p*-Nitrophenylacetylglycine
In: Ferret A394, A3046
Hen A395
Pigeon A395
→ *p*-Nitrophenylacetylornithine
In: Hen A395
→ *p*-Nitrophenylacetyltaurine
In: Ferret A394, A3046
Hen A395
Pigeon A395

p-Nitrophenylacetylene
→ *p*-Nitrophenylacetic acid
In: Rat E461

p-Nitrophenyl 6-O-acetyl-β-D-galactoside
Formed from: *p*-Nitrophenyl-β-D-galactoside

p-Nitrophenylacetylglycine
Formed from: *p*-Nitrophenylacetic acid

p-Nitrophenylacetylornithine
Formed from: *p*-Nitrophenylacetic acid

p-Nitrophenylacetyltaurine
Formed from: *p*-Nitrophenylacetic acid

p-Nitro-L-phenylalanine
→ *p*-Nitrophenylpyruvic acid
In: Hansenula D978

p-Nitrophenyl-α-L-arabinoside
→ *p*-Nitrophenol
In: Penicillium A3240

p-Nitrophenyl azide
→ *p*-Nitroaniline
In: Mouse G8

N-(p-Nitrophenyl)benzamide
→ N-(*p*-Aminophenyl)benzamide
In: Ascaris A2037
Moniezia A2037
Mouse A2037
Sheep A2037
→ Benzoic acid

In: Ascaris A2037
Moniezia A2037
Mouse A2037
Sheep A2037
→ *p*-Nitroaniline
In: Ascaris A2037
Moniezia A2037
Mouse A2037
Sheep A2037

o-Nitrophenyl butyrate
→ *o*-Nitrophenol
In: Rat F753

p-Nitrophenyl butyrate
→ *p*-Nitrophenol
In: Acinetobacter A962
Aspergillus J389
Dog A868
Man H465
Rabbit E26
Rat F753

p-Nitrophenyl caprylate
→ *p*-Nitrophenol
In: Acinetobacter A962
Contrast (in rat) F753

S-(o-Nitrophenyl)-L-cysteine
→ *o*-Nitrothiophenol
In: Escherichia E101
Salmonella E101
See also: F38

p-Nitrophenyl-α-D-2-deoxygalactoside
→ *p*-Nitrophenol
In: Aspergillus K437
Coffea K437
Mortierella K437

p-Nitrophenyl-α-D-6-deoxygalactoside
→ *p*-Nitrophenol
In: Coffea K437
Mortierella K437

p-Nitrophenyl (1R)-cis-3-(2,2-dibromovinyl)-2,2-dimethylcyclopropanecarboxylate
→ *p*-Nitrophenol
In: Mouse F289
Pig F289
Triatoma F289

2-(p-Nitrophenyl)-1,3-dithiolane
→ 2-(*p*-Nitrophenyl)-1,3-dithiolane-1-oxide
In: Rabbit F355

2-(p-Nitrophenyl)-1,3-dithiolane-1,3-dioxide
Formed from: 2-(*p*-Nitrophenyl)-1,3-dithiolane-1-oxide

2-(p-Nitrophenyl)-1,3-dithiolane-1-oxide
Formed from: 2-(*p*-Nitrophenyl)-1,3-dithiolane
→ 2-(*p*-Nitrophenyl)-1,3-dithiolane-1,3-dioxide
In: Rabbit F355

2-Nitro-p-phenylenediamine
Probably formed from: 2,4-Dinitroaniline
→ N^4-Acetyl-2-nitro-*p*-phenylenediamine
In: Rat C537, F828
→ 2,5-Diacetamidianiline
In: Rat C537

1-Nitro-2-phenylethane
Formed from: Phenylacetaldoxime

1-(o-Nitrophenyl)ethanol
Formed from: *o*-Nitroacetophenone

1-(p-Nitrophenyl)ethanol
Formed from: 1-Ethyl-4-nitrobenzene
p-Nitroacetophenone

p-Nitrophenyl-α-L-fucoside
→ *p*-Nitrophenol
In: Man A642
Rabbit A1221

p-Nitrophenyl-β-D-fucoside
→ *p*-Nitrophenol
In: Man A642
Secale K512

1-(((5-(4-Nitrophenyl)-2-furanyl)methylene)amino)-2,4-imidazolidinedione
See: Dantrolene

o-Nitrophenyl-α-D-galactoside
→ *o*-Nitrophenol
In: Man A642

o-Nitrophenyl-β-D-galactoside
→ *o*-Nitrophenol
In: Cicer K832
Escherichia A344
Macrophomina A1758
Picea K833
Scallop A376
Sclerotium A1759
Sugar cane B488

→ *o*-Nitrophenyl-β-D-galactosyl(1→3)-β-D-galactoside
In: Thermus J918
→ *o*-Nitrophenyl-β-D-galactosyl(1→6)-β-D-galactoside
In: Thermus J918

p-Nitrophenyl-α-D-galactoside
→ *p*-Nitrophenol
In: Bacillus B544
Bifidobacterium K291
Castanea B759
Coffea J595
Man A642
Pea A3137
Poterioochromonas B486
→ *p*-Nitrophenyl-α-D-galactosyl(1→3)-α-D-galactoside
In: Penicillium K283

p-Nitrophenyl-β-D-galactoside
→ *p*-Nitrophenol
In: Aspergillus H11
Cicer K832
Escherichia J399
Man A642, A789, C7
Pea A3137
Picea K833
Triglochin A3756
→ *p*-Nitrophenyl 6-O-acetyl-β-D-galactoside
In: Escherichia A731
Macrophomina A1758
Sclerotium A1759
Secale K512

p-Nitrophenyl-β-D-galactosyl(1→3)-α-D-N-acetylgalactosaminide
→ *p*-Nitrophenyl-β-D-galactosyl(1→3)-β-D-N-acetylglucosaminyl(1→6)-α-D-N-acetylgalactosaminide
In: Nocardia J698

p-Nitrophenyl-β-D-galactosyl(1→3)-β-D-N-acetylglucosaminyl(1→6)-α-D-N-acetylgalactosaminide
Formed from: *p*-Nitrophenyl-β-D-galactosyl(1→3)-α-D-N-acetylgalactosaminide

o-Nitrophenyl-β-D-galactosyl(1→3)-β-D-galactoside
Formed from: *o*-Nitrophenyl-β-D-galactoside

o-Nitrophenyl-β-D-galactosyl(1→6)-β-D-galactoside
Formed from: *o*-Nitrophenyl-β-D-galactoside

p-Nitrophenyl-α-D-galactosyl(1→3)-α-D-galactoside
Formed from: *p*-Nitrophenyl-α-D-galactoside

p-Nitrophenyl-β-gentiobioside
Probably formed from: *p*-Nitrophenyl-β-D-glucoside

p-Nitrophenyl-α-D-glucoside
→ *p*-Nitrophenol
In: Man A642

o-Nitrophenyl-β-D-glucoside
Formed from: *o*-Nitrophenol
→ *o*-Nitrophenol
In: Cherry K910
Cicer K832
Picea K833
Triglochin A3756
→ *o*-Nitrophenyl-β-galactosyl(1→6)-β-glucoside
In: Thermus J918
→ *o*-Nitrophenyl-β-glucosyl(1→3)-β-glucoside
In: Thermus J918

m-Nitrophenyl-β-D-glucoside
Formed from: *m*-Nitrophenol

p-Nitrophenyl-β-D-glucoside
Formed from: *p*-Nitrophenol
→ O-(β-D-Apiofuranosyl(1→2)-β-D-glucosyl)-*p*-nitrophenol
In: Petroselinum K725
→ 6′-O-Malonyl-*p*-nitrophenyl-β-D-glucoside
Probably in: Daucus H118
Glycine H660
→ *p*-Nitrophenol
In: Almond H660
Aspergillus F74
Cherry K910
Cicer K832
Man A642
Picea K833
Pig K309
Secale K512
Triglochin A3756
Zea K149
→ *p*-Nitrophenyl-β-gentiobioside

Probably in: Avena J7609
Wheat J709
→ *p*-Nitrophenyl-6″-O-malonyl-β-D-glucoside
In: Petroselinum K725

***p*-Nitrophenyl-D-glucoside-6-phosphate**
Formed from: *p*-Nitrophenol

***p*-Nitrophenyl glucoside sulphate**
Formed from: *p*-Nitrophenol

***o*-Nitrophenylglucosinolate**
Formed from: *o*-Nitrobenzaldoxime

***o*-Nitrophenyl-β-glucosyl(1→3)-β-glucoside**
Formed from: *o*-Nitrophenyl-β-D-glucoside

***o*-Nitrophenyl-β-glucosyl(1→6)-β-glucoside**
Formed from: *o*-Nitrophenyl-β-D-glucoside

***o*-Nitrophenyl-β-D-glucuronide**
Formed from: *o*-Nitrophenol

***m*-Nitrophenyl-β-D-glucuronide**
Formed from: *m*-Nitrophenol

***p*-Nitrophenyl-β-D-glucuronide**
Formed from: *p*-Nitrophenol
→ *p*-Nitrophenol
In: Alcaligees E323
Beef A1029, A1583
Escherichia A1029
Man A642
Rabbit A1221

S-(*o*-Nitrophenyl)glutathione
Formed from: *o*-Chloronitrobenzene
o-Dinitrobenzene

S-(*p*-Nitrophenyl)glutathione
Formed from: *p*-Chloronitrobenzene
p-Dinitrobenzene
O,O-Dimethyl O-(4-nitrophenyl) phosphorothioate
Ethyl *p*-nitrophenyl phenylphosphate
Ethyl *p*-nitrophenyl phenylphosphorothioate
→ N-Acetyl-S-(*p*-nitrophenyl)-L-cysteine
Probably in: Rat G871

***p*-Nitrophenylglycol**
Formed from: *p*-Nitrostyrene oxide

***p*-Nitrophenylhydroxylamine**
Formed from: *p*-Nitroaniline

***p*-Nitrophenyl laurate**
→ *p*-Nitrophenol
In: Man A798

***p*-Nitrophenyl-6″-O-malonyl-β-D-glucoside**
Formed from: *p*-Nitrophenyl-β-D-glucoside

***p*-Nitrophenyl-α-D-mannoside**
→ *p*-Nitrophenol
In: Dictyostelium A1774
Man A642

***p*-Nitrophenyl-β-D-mannoside**
→ *p*-Nitrophenol
In: Man A642

2-Nitro-N-(phenylmethyl)-1*H*-imidazole-1-acetamide
See: Benznidazole

3-(*p*-Nitrophenyl)mucocolactone
Formed from: 4-Nitrobiphenyl

***p*-Nitrophenyl 2-nitro-4-trifluoromethylphenyl ether** (fluorodifen)
→ *p*-Aminophenyl 2-nitro-4-trifluoromethylphenyl ether
In: Cucumber A483
→ 2-Amino-4-trifluoromethylphenyl *p*-nitrophenyl ether
In: Cucumber A483
→ *p*-Nitrophenol
In: Cucumber A483
Microorganism A2321
Nicotiana A524
Pea A1148
→ S-(2-Nitro-4-trifluoromethylphenyl) glutathione
Probably in: Pea A1148
Rat A2405

2-(*p*-Nitrophenyl)-1,3-oxathiolane
→ 2-(*p*-Nitrophenyl)-1,3-oxathiolane sulphoxide
In: Pig F176
Rat F176

2-(*p*-Nitrophenyl)-1,3-oxathiolane sulphone
Probably formed from: 2-(*p*-Nitrophenyl)-1,3-oxathiolane sulphoxide

2-(*p*-Nitrophenyl)-1,3-oxathiolane sulphoxide
Formed from: 2-(*p*-Nitrophenyl)-1,3-oxathiolane
→ 2-(*p*-Nitrophenyl)-1,3-oxathiolane sulphone
Probably in: Pig F176
Rat F176

p-Nitrophenyl 3-oxobutyl ether
Formed from: 3-Hydroxybutyl *p*-nitrophenyl ether
p-Nitrophenyl pent-1-yl ether
→ 2-Hydroxypent-1-yl *p*-nitrophenyl ether
Probably in: Rat D667
→ 4-Hydroxypent-1-yl *p*-nitrophenyl ether
In: Rat D667
p-Nitrophenyl phenylphosphoric acid
Formed from: Ethyl *p*-nitrophenyl phenylphosphorothioate
p-Nitrophenyl phosphate
Formed from: *p*-Nitrophenol
→ *p*-Nitrophenol
In: Ascaris A324
Escherichia A251
Man A642, A2543
Moniezia A324
Papaver A222
Rabbit A1221
Saccharomyces K864, K865
Serratia A251
Spinach A2623
N-(p-Nitrophenyl)phthalamidic acid
→ *p*-Nitroaniline
In: Xanthobacter J253
→ Phthalic acid
In: Xanthobacter J253
2-Nitro-1-phenylpropane
Formed from: Benzyl methyl ketoxime
α-Methyl-β-phenethylhydroxylamine
→ Amphetamine
In: Clostridium F897
Eubacterium F897
Mitsuokella F897
Peptostreptococcus F897
→ Phenylacetone
In: Rabbit A3215, B649
(1*R*,2*S*)-1-(p-Nitrophenyl)propane-1,2-diol
Formed from: *p*-Nitrobenzaldehyde
p-Nitrophenyl propionate
→ *p*-Nitrophenol
In: Acinetobacter A962
Aspergillus J389
Dog A868
Man H465

p-Nitrophenylpyruvic acid
Formed from: *p*-Nitro-L-phenylalanine
N-(p-Nitrophenyl)salicylamide
→ N-(*p*-Aminophenyl)salicylamide
In: Ascaris A2037
Moniezia A2037
Mouse A2037
Sheep A2037
→ *p*-Nitroaniline
In: Mouse A2037
→ Salicylic acid
In: Mouse A2037
β-(p-Nitrophenyl)serine
→ β-Hydroxy-*p*-nitrophenylpyruvic acid
In: Flavobacterium C206
p-Nitrophenylserinol
→ 1,3-Dihydroxy-1-(*p*-nitrophenylacetone)
In: Flavobacterium C206
o-Nitrophenyl sulphate
Formed from: *o*-Nitrophenol
m-Nitrophenyl sulphate
Formed from: *m*-Nitrophenol
p-Nitrophenyl sulphate
Formed from: *p*-Nitrophenol
→ *p*-Nitrophenol
In: Ascaris A324
Aspergillus B317
Cyanobacterium E179
Eubacterium E191, E432, F513, H354
Haemophilus H756
Klebsiella G756
Man A551, A2457, E398, K46
Moniezia A324
Peptococcus E449
Rabbit B481
Rat A1050
Streptomyces H603, H627, H588
p-Nitrophenyl 2,4,6-trichlorophenyl ether
→ *p*-Aminophenyl 2,4,6-trichlorophenyl ether
Probably in: Carp F703
p-Nitrophenyl trimethylacetate
→ *p*-Nitrophenol
In: Yeast A337
p-Nitrophenyl vinyl ether
→ 2-(*p*-Nitrophenoxy)oxirane
In: Rat E768

o-Nitrophenyl-β-D-xyloside
→ *o*-Nitrophenol
In: Picea K833

p-Nitrophenyl-β-D-xyloside
→ *p*-Nitrophenol
In: Bacillus A2061
Penicillium A3240
Secale K512

5-Nitro-m-phthalic acid
→ 5-Amino-*m*-phthalic acid
In: Veillonella A2624

1-Nitropyrene
Formed from: 1-Nitrosopyrene
→ 1-Aminopyrene
In: Bacteroides C791
Escherichia D217
Goldfish H877
Microorganism D278, D333, F841
Mouse E721, F787
Mycobacterium H336
Rabbit C702, D194, E4, G361
Rat D228, D271, D281, D571, J915
→ *cis*-4,5-Dihydro-4,5-dihydroxy-1-nitropyrene
In: Mycobacterium G329, H336
→ *trans*-4,5-Dihydro-4,5-dihydroxy-1-nitropyrene
In: Man F809, K79
Mouse D481, F721
Rabbit E753
Rat D333, D571, F710, F787
See also: D194, E689
→ *cis*-9,10-Dihydro-9,10-dihydroxy-1-nitropyrene
In: Mycobacterium G329, H336
→ 4-(S-Glutathionyl)-4,5-dihydro-5-hydroxy-1-nitropyrene
In: Rat F710
→ 5-(S-Glutathionyl)-4,5-dihydro-4-hydroxy-1-nitropyrene
In: Rat F710
→ 9-(S-Glutathionyl)-9,10-dihydro-10-hydroxy-1-nitropyrene
In: Rat F710
→ 10-(S-Glutathionyl)-9,10-dihydro-9-hydroxy-1-nitropyrene
In: Rat F710
→ 3-Hydroxy-1-nitropyrene
In: Man F809, K79
Mouse D481, E721, F787
Rabbit D194, E753
Rat D38, D271, D281, D333, D571, F710
→ 6-Hydroxy-1-nitropyrene
In: Cunninghamella D624
Man F809, K79
Mouse D481, E721, F787
Rabbit E753
Rat D38, D271, D333, D571
See also: D281
→ 8-Hydroxy-1-nitropyrene
In: Cunninghamella D624
Man F809, K79
Mouse D481, E721, F787
Rabbit E753
Rat D38, D271, D333, D571
→ 10-Hydroxy-1-nitropyrene
In: Rabbit D194
Rat D271

2-Nitropyrene
→ *trans*-4,5-Dihydro-4,5-dihydroxy-2-nitropyrene
In: Man K79
→ 6-Hydroxy-2-nitropyrene
In: Man K79

4-Nitropyrene
→ 4-Aminopyrene
In: Man K79
→ *trans*-9,10-Dihydro-9,10-dihydroxy-4-nitropyrene
In: Man K79
→ 6-Hydroxy-2-nitropyrene
In: Man K79
→ 9-Hydroxy-2-nitropyrene
In: Man K79

1-Nitropyrene-4,5-dione
Formed from: 4,5-Dihydro-4,5-dihydroxy-1-nitropyrene
→ 1-Aminopyrene-4,5-dione
In: Rat E689

1-Nitropyrene-4,5-oxide
→ 4,5-Dihydro-4,5-dihydroxy-1-nitropyrene
In: Rat E611

→ 4-(S-Glutathionyl)-4,5-dihydro-5-hydroxy-1-nitropyrene
In: Rat E611
→ 5-(S-Glutathionyl)-4,5-dihydro-4-hydroxy-1-nitropyrene
In: Rat E611

1-Nitropyrene-9,10-oxide
→ 9,10-Dihydro-9,10-dihydroxy-1-nitropyrene
In: Rat E611
→ 9-(S-Glutathionyl)-9,10-dihydro-10-hydroxy-1-nitropyrene
In: Rat E611
→ 10-(S-Glutathionyl)-9,10-dihydro-9-hydroxy-1-nitropyrene
In: Rat E611

1-Nitropyren-3-yl-β-D-glucuronide
Formed from: 3-Hydroxy-1-nitropyrene

1-Nitropyren-6-yl-β-D-glucuronide
Formed from: 6-Hydroxy-1-nitropyrene

1-Nitropyren-8-yl-β-D-glucuronide
Formed from: 8-Hydroxy-1-nitropyrene

Nitroquinol
Formed from: 4-Hydroxy-3-nitrobenzoic acid
m-Nitrophenol
p-Nitrophenol
→ Nitroquinol sulphate
Probably in: Frog D857

4-Nitroquinoline-N-oxide
→ 4-(S-Glutathionyl)quinoline-N-oxide
In: Japanese monkey F139
Mouse G428
→ 4-Hydroxylaminoquinoline-N-oxide
In: Ehrlich tumour A3214
Hamster A3214
Mouse G428, G995
Rat A75

Nitroquinol sulphate
Probably formed from: Nitroquinol

4-Nitro-*o*-quinone
Formed from: 2-Amino-4-nitrotoluene
→ 4-Nitrocatechol
In: Phanerochaete G436

2-Nitroresorcinol
→ 2-Aminoresorcinol
In: Veillonella A2624

4-Nitroresorcinol
→ Hydroxyquinol
In: Arthrobacter H270

Nitrosobenzene
Formed from: Aniline
Nitrobenzene
Probably formed from:
Phenylhydroxylamine
→ N-Phenylacetohydroxamic acid
In: Chlorella A3531
Rat J429
Yeast A2344
See also: K552
→ N-Phenylglycolhydroxamic acid
In: Chlorella A3531
Rat J429
→ Phenylhydroxylamine
In: Rat B174
See also: C32
→ N-Phenylisobutyrohydroxamic acid
In: Rat J429
→ N-Phenylisovalerohydroxamic acid
In: Rat J429

4-Nitrosobiphenyl
Formed from: 4-Aminobiphenyl
N-Hydroxy-4-acetamidobiphenyl
4-Hydroxylaminobiphenyl
4-Nitrobiphenyl
→ 4-Hydroxylaminobiphenyl
Probably in: Dog A880
Monkey A880

Nitrosochlorphentermine
Formed from: Chlorphentermine

N-Nitrosodibenzylamine
→ Bibenzyl
In: Rabbit A3868

N-Nitrosodiphenylamine
See: Diphenylnitrosamine

3-Nitrosofluoranthene
Formed from: 3-Nitrofluoranthene
→ 3-Hydroxylaminofluoranthene
Probably in: Rabbit G361

2-Nitrosofluorene
Formed from: 2-Aminofluorene
N-Hydroxy-2-acetamidofluorene
→ 2-Hydroxylaminofluorene
In: Horse H291
→ 2-Nitrofluorene

In: Ram C754
See also: F890
3-Nitrosofluorene
Formed from: N-Hydroxy-3-acetamidofluorene
1-Nitrosonaphthalene
→ 1-Naphthylhydroxylamine
In: Rabbit A2173
2-Nitrosonaphthalene
Formed from: 2-Nitronaphthalene
β-Naphthylamine
→ 2-Naphthylhydroxylamine
Probably in: Dog A880
Monkey A880
Rat B773
1-Nitroso-2-naphthol
→ 1,2-Dihydro-1-iminoxyl-2-oxonaphthalene radical
In: Horseradish E81
→ 1-Nitroso-2-naphthol-3,5-disulphonic acid
In: Horseradish E81
2-Nitroso-1-naphthol
→ 1,2-Dihydro-2-iminoxyl-1-oxonaphthalene radical
In: Horseradish E81
→ 2-Nitroso-1-naphthol-4-sulphonic acid
In: Horseradish E81
1-Nitroso-2-naphthol-3,5-disulphonic acid
Formed from: 1-Nitroso-2-naphthol
2-Nitroso-1-naphthol-4-sulphonic acid
Formed from: 2-Nitroso-1-naphthol
***o*-Nitrosonitrobenzene**
Formed from: *o*-Dinitrobenzene
***p*-Nitrosophenetole**
Formed from: *p*-Ethoxy-N-hydroxyacetanilide
→ *p*-Nitrosophenetole
Probably in: Rat B178
Nitrosophentermine (α,α-dimethyl-α-nitroso-β-phenylethane)
Formed from: Phentermine
1-Nitrosopyrene
Formed from: 1-Nitropyrene
→ 1-Aminopyrene
In: Mouse E721
→ 1-Nitropyrene
In: Mouse E721
***o*-Nitrosotoluene**
Formed from: *o*-Toluidine
→ N-Hydroxy-*o*-toluidine
In: Rat C857
***m*-Nitrostyrene**
Formed from: *m*-Nitrocinnamic acid
***m*-Nitrostyrene oxide**
→ 1-(S-Glutathionyl)-2-(*m*-nitrophenyl) ethanol
Probably in: Sheep A2282
***p*-Nitrostyrene oxide**
→ 1-(S-Glutathionyl)-2-(*p*-nitrophenyl) ethanol
Probably in: Sheep A2282
→ *p*-Nitrophenylglycol
In: Aspergillus H770, H922
Dab H812
Man C386
Rat C386
***p*-Nitrosulphinylanisole**
Formed from: *p*-Nitrothioanisole
→ *p*-Nitrosulphonylanisole
Probably in: Mortierella D759
***p*-Nitrosulphonylanisole**
Probably formed from: *p*-Nitrosulphinylanisole
2-Nitro-2-(4-(α,α,β,β-tetrafluorophenethyl)phenyl)propane
Formed from: α,α-Dimethyl-4-(α,α,β,β-tetrafluorophenethyl)benzylamine
***o*-Nitrothioanisole**
Formed from: *o*-Nitrothiophenol
***p*-Nitrothioanisole**
Formed from: *p*-Nitrothiophenol
→ *p*-Nitrosulphinylanisole
In: Horseradish E403
Mortierella D759
***o*-Nitrothiophenol**
Formed from: S-(*o*-Nitrophenyl)-L-cysteine
→ *o*-Nitrothioanisole
In: Tetrahymena E96
***p*-Nitrothiophenol**
→ *p*-Nitrothioanisole
In: Man D926
Rat C121
Tetrahymena E96

***o*-Nitrotoluene**
→ N-Acetyl-S-(*o*-nitrobenzyl)-L-cysteine
In: Rat D383
→ 3-Methylcatechol
In: Pseudomonas H344, J524
→ *o*-Nitrobenzyl alcohol
In: Pseudomonas G661, J219, J524
Rat C650, D827
See also: D383
→ *o*-Toluidine
In: Pseudomonas J196
Veillonella A2624

***m*-Nitrotoluene**
→ 3-Methylcatechol
In: Pseudomonas J524
→ 4-Methylcatechol
In: Pseudomonas J524
→ *m*-Nitrobenzyl alcohol
In: Pseudomonas G661, H624, J219
Rat C650, D827
See also: D383
→ *m*-Nitrophenol
In: Pseudomonas H624
→ *m*-Toluidine
In: Pseudomonas H624
Saccharomyces F201
Veillonella A2624

***p*-Nitrotoluene**
Formed from: *p*-Toluidine
→ 4-Methylcatechol
In: Pseudomonas J524
→ *p*-Nitrobenzyl alcohol
In: Eel A2286, A2395
Pseudomonas J219, J524
Rat A650, A2286, A2395, C650, C765, D827
See also: D383, G980
→ 2-Methyl-5-nitrophenol
In: Pseudomonas G661
→ *p*-Toluidine
In: Pseudomonas J196
Saccharomyces F201
Sea urchin B703
Veillonella A2624

4-Nitrotoluene-2-sulphonic acid
→ 4-Aminotoluene-2-sulphonic acid
In: Veillonella A2624

2-Nitro-*p*-toluidine
→ 3,4-Diaminotoluene
In: Veillonella A2624

3-Nitro-*o*-toluidine
→ 2,6-Diaminotoluene
In: Veillonella A2624

3-Nitro-*p*-toluidine
→ 3,4-Diaminotoluene
In: Veillonella A2624

4-Nitro-*o*-toluidine
→ 2,5-Diaminotoluene
In: Veillonella A2624

5-Nitro-*o*-toluidine
→ 2,4-Diaminotoluene
In: Veillonella A2624

***p*-Nitrotribenzylamine**
→ *p*-Nitrodibenzylamine
Probably in: Rat J471

2-Nitro-4-trifluoromethylphenol
→ 4-Trifluoromethylcatechol
In: Pseudomonas H65

4-Nitro-3-trifluoromethylphenol
→ 4-Amino-3-trifluoromethylphenol
In: Trout A544
→ 4-Nitro-3-trifluoromethylphenyl-β-D-glucuronide
In: Trout A544

3-Nitro-5-trifluoromethyl-*o*-phenylenediamine
Formed from: 2,6-Dinitro-4-trifluoromethylaniline
→ 2-Acetamido-6-nitro-4-trifluoromethylaniline
In: Streptomyces B13
→ N-((2-Amino-3-nitro-5-trifluoromethyl)-phenyl)methanesulphinamide
Probably in: Streptomyces B13
→ Bis(2-amino-3-nitro-5-trifluoromethylphenyl)diazene
Probably in: Streptomyces B13

4-Nitro-3-trifluoromethylphenyl-β-D-glucuronide
Formed from: 4-Nitro-3-trifluoromethylphenol

S-(2-Nitro-4-trifluoromethylphenyl)glutathione
Probably formed from: *p*-Nitrophenyl 2-nitro-4-trifluoromethylphenyl ether

5-Nitro-7-trifluoromethylquinoxaline
Formed from: Fluchloralin
3-Nitro-L-tyrosine
Probably formed from: L-Tyrosine
→ 4-Hydroxy-3-nitrophenylpyruvic acid
In: Hansenula D978
4-Nitro-4′-ureidodiphenylsulphone
Formed from: 4-Amino-4′-ureidodiphenylsulphone
4-Nitroveratrole (1,2-dimethoxy-4-nitrobenzene, 3,4-dimethoxy-1-nitrobenzene)
Formed from: 4-Nitrocatechol
Veratryl alcohol
→ Methoxy-*p*-quinone
In: Phanerochaete G436
4′-Nitrowarfarin
→ Dihydro-4′-nitrowarfarin
In: Rabbit G276
Nitroxynil
See: 4-Hydroxy-3-iodo-5-nitrobenzonitrile
Nogalamycin
→ 7-Deoxynogalamycin
In: Aeromonas A2670
Nomifensine (8-amino-1,2,3,4-tetrahydro-2-methyl-4-phenylisoquinoline)
→ 4′-Hydroxynomifensine
In: Man A3348, B373, E37
Monkey A3348
→ Nomifensine-β-D-glucuronide
In: Man B373
Nomifensine-β-D-glucuronide
Formed from: Nomifensine
3-Nonaprenyltoluquinone
Formed from: Homogentisic acid
Nonivamide (N-((4-hydroxy-3-methoxyphenyl) methyl)nonanamide)
→ ω-Hydroxynonivamide
In: Rat H410
Nonylbenzene
→ Benzoic acid
In: Beauveria D936
Paecilomyces D936
Penicillium D936
Verticillium D936
***o-n*-Nonylbenzoic acid**
→ (*S*)-3-*n*-Octylphthalide
In: Aspergillus J486
Pseudomonas J486
***p-n*-Nonylphenol**
→ 1-(*p*-Hydroxyphenyl)nonan-1-ol
In: Pseudomonas F381
→ 9-(*p*-Hydroxyphenyl)nonan-1-ol
In: Trout J412
See also: J379
→ 9-(*p*-Hydroxyphenyl)nonan-2-ol
In: Trout J412
→ *p-n*-Nonylphenyl-β-D-glucuronide
In: Trout J379
See also: K681
→ *p-n*-Nonylphenyl sulphate
In: Man K344
→ Phloretic acid
In: Trout K236
***p-n*-Nonylphenyl-β-D-glucuronide**
Formed from: *p-n*-Nonylphenol
***p-n*-Nonylphenyl sulphate**
Formed from: *p-n*-Nonylphenol
D-Noradrenaline (norepinephrine, D-(−)-3,4-dihydroxyphenylethanolamine)
Formed from: *erythro*-(+)-3,4-Dihydroxyphenylserine
threo-(+)-3,4-Dihydroxyphenylserine
3,4-Dihydroxyphenethylamine
4-Hydroxyphenylethanolamin-3-yl sulphate
D-Noradrenaline-3-sulphate
D-Normetanephrine
Octopamine
→ N-Acetyl-D-noradrenaline
In: Cockroach H835
→ D-Adrenaline
In: Beef A1463
Man A498
Rabbit A498, C182, C190
Rat A80, A935, C182, C190, E867, F397, F622
→ 3,4-Dihydroxymandelaldehyde
In: Bacteria B225
Dog A2512
Hymenolepis A3815
Rat A965, A1053, A1324, A2279
Salmo A1137
Scyliorhinus A1137

See also: A1909, A2411, A3079, B384, B943, C119, F2, F212, F778, G810, H115, H746
→ 3,4-Dihydroxyphenethylamine
Probably in: Man A1657
→ 1-(3,4-Dihydroxyphenyl)-2-iminoethanol
In: Arthrobacter B360
→ Feruloylnoradrenaline
In: Nicotiana D558, F72
→ Noradrenaline quinone
In: Mushroom D302
See also: A63, A624, A2556, B897, F944, G620
→ D-Noradrenaline-4-sulphate
Probably in: Eubacterium G602
Guinea pig E456, E695
Man A3814, C322, C523, D583, D607, D741
Rat A153, A1343, C185, C441, E695
→ Noradrenalone
In: Penicillium J16, K761
→ D-Normetanephrine
In: Cat C119
Coryphantha B807
Dog A849, A2512
Man A630, A3831, B118, E390
Pig C346, D652
Rabbit C119
Rat A486, A1257, A1337, D65, D704, H115, H746
Salmo A1137
Scyliorhinus A1137
Streptomyces K652
Trichostrongylus F2
See also: A2279, A2411
→ D-Norparanephrine
In: Rat A486
→ Octopamine
In: Rat A1257, A1337
→ Protocatechuic acid
In: Cockroach A2244
→ 1,2,3,4-Tetrahydro-4,6,7-trihydroxy-1-methylisoquinoline
Probably in: Rat A87

L-Noradrenaline
Formed from: *erythro*-3,4-Dihydroxy-D-phenylserine
erythro-3,4-Dihydroxy-L-phenylserine
→ L-Normetanephrine
In: Rat A486
→ L-Norparanephrine
In: Rat A486

Noradrenaline quinone
Formed from: D-Noradrenaline

D-Noradrenaline-3-sulphate
→ D-Noradrenaline
In: Man E177

D-Noradrenaline-4-sulphate
Formed from: D-Noradrenaline

Noradrenalone (3,4-dihydroxy-β-oxophenethylamine)
Formed from: Noradrenaline

Norbelladine (4-(((2-(4-hydroxyphenyl)ethyl)amino)methyl)-1,2-benzenediol)
→ Mesembrenol
In: Sceletium A210
→ Mesembrenone
In: Sceletium A210
→ Mesembrine
In: Sceletium A210

Norbenzphetamine
See: N-Benzylamphetamine

2,2′-Norberbamunine
Formed from: N-Methylcoclaurine

Norboldine ((6a*S*)-5,6,6a,7-tetrahydro-1,10-dimethoxy-4*H*-dibenzo[*de*,*g*]quinoline-2,9-diol)
→ Boldine
In: Litsea A3089

Norbuprenorphine
→ Norbuprenorphine-β-D-glucuronide
In: Man J416
Monkey H861
Rat J97a

Norbuprenorphine-β-D-glucuronide
Formed from: Norbuprenorphine

5″-Norcannabidiol-4″-oic acid
Formed from: 5″-Hydroxycannabidiol
→ 4″,5″-Bisnorcannabidiol-3″-oic acid
Probably in: Guinea pig B458
Man F503, G113
→ 6-Hydroxy-5″-norcannabidiol-4″-oic acid
Probably in: Guinea pig B458
Man F503, G113

Mouse A2085
→ 7-Hydroxy-5″-norcannabidiol-4″-oic acid
Probably in: Guinea pig B458
Man F503, G113
Mouse A2085

Norchlorcyclizine
Formed from: N-*tert*-butylnorchlorcyclizine
Chlorcyclizine

Norchlorphenoxamine
Formed from: Chlorphenoxamine
→ Bisnorchlorphenoxamine
In: Man E495

Norciramadol
Formed from: Ciramadol
→ Norciramadol-β-D-glucuronide
Probably in: Rat B990
Rhesus monkey B990

Norciramadol-β-D-glucuronide
Probably formed from: Norciramadol

N-Norclentiazem
Formed from: Clentiazem
→ (+)-(2*S*,3*S*)-3-Acetoxy-8-chloro-3,4-dihydro-4-(4-hydroxy-3-methoxyphenyl)-5-(2-(N-methylamino)ethyl)-4-oxo-1,5-(2*H*)-benzothiazepine
Probably in: Rat H76
→ N,N-Bisnorclentiazem
In: Rat H76
→ N,O-Bisnorclentiazem
Probably in: Rat H76

O-Norclentiazem
Formed from: Clentiazem

Norcocaine
Formed from: Cocaine
→ Benzoylnorecgonine
In: Rat A3721
→ N-Hydroxynorcocaine
In: Mouse C658
See also: D437, E201, E809, G38

Norcoclaurine (1,2,3,4-tetrahydro-6,7-dihydroxy-1-(*p*-hydroxybenzyl)isoquinoline)
Probably formed from: Didehydronorcoclaurine
3,4-Dihydroxyphenethylamine
Norcoclaurine-1-carboxylic acid
L-Tyrosine
→ Berberine
In: Coptis G749
→ Bisnorargemonine
In: Argemone K767
→ Coclaurine
In: Annona B137
Argemone H629, K783
Berberis H629, K933
Chonodendron H629
Coptis K785
Dicentra H629
Eschscholtzia K933
Fumaria H629
Papaver H629
Peumus K933
Thalictrum H242, H578
Tinospora H629
See also: A185
→ Columbamine
In: Berberis K767
→ Dihydromacarpine
In: Eschscholtzia K767
→ Jatrorrhizine
In: Berberis K767
→ N-Methylnorcoclaurine
In: Argemone H629
Berberis H629, K780
Chonodendron H629
Fumaria H629
Papaver H629
Tinospora H629
→ Norboldine
In: Peumus K767
→ Sanguinarine
In: Macleaya K767

Norcoclaurine-1-carboxylic acid
→ Norcoclaurine
Probably in: Annona B137

Norcodeine
Formed from: Codeine
→ Norcodeine-β-D-glucuronide
Probably in: Man H933
Rat A270
→ Normorphine
In: Rat E233
See also: A696

Norcodeine-β-D-glucuronide
Probably formed from: Norcodeine

Nordifloxacin
Formed from: Difloxacin

Nordimetacrine
Formed from: Dimetacrine

D-Norephedrine (α-hydroxy-D-amphetamine, α-(1-aminoethyl)benzenemethanol; this isomer assumed unless otherwise stated)
Formed from: D-Amphetamine
2-Aminopropiophenone
Cathinone
Oxyfedrine
→ Benzoic acid
In: Man A1226
Rabbit A996, A1226
→ N-Hydroxynorephedrine
In: Rabbit A2140
→ *p*-Hydroxynorephedrine
In: Man A1226
Rat A1226, A2553
See also: A528, A1473, A2232, A2348, A2485
→ 2-Oxo-1-phenyl-1-propanol
In: Rabbit A1226, A2140
→ 1-Hydroxy-2-oxo-1-phenylpropan oxime
In: Rabbit A2140

L-Norephedrine
Formed from: L-Amphetamine
L-Ephedrine
→ *p*-Hydroxynorephedrine
Probably in: Rat F481

Norephedrin-4-yl-β-D-glucuronide
Formed from: *p*-Hydroxynorephedrine

Norepinephrine
See: Noradrenaline

Norfemoxetine
Formed from: Femoxetine

Norfenfluramine (N-ethyl-3-trifluoromethylphenethylamine)
Formed from: N-(2-Benzoyloxyethyl) norfenfluramine
→ 1-(*m*-Trifluoromethylphenyl)acetone
Probably in: Guinea pig A1152

Norfentanyl
See: 4-(N-Propionylanilino)piperidine

Norfloxacin (1-ethyl-6-fluoro-1,4-dihydro-4-oxo-7-(1-piperazinyl)-3-quinolinecarboxylic acid)
Formed from: Pefloxacin

Norfluperlamine
Formed from: Fluperlamine

Norgallopamil
Formed from: Gallopamil
→ 3-Cyano-2-methyl-3-(3,4,5-trimethoxyphenyl)-6-hexanol
Probably in: Man F527
Rat F527
→ 3,4-Dimethoxyphenethylamine
Probably in: Rat F527
→ Norgallopamil-N-β-D-glucuronide
In: Rat F474

Norgallopamil-N-β-D-glucuronide
Formed from: Norgallopamil

Norglaucine (5,6,6a,7-tetrahydro-1,2,9,10-tetramethoxy-4*H*-dibenzo[*de*,*g*]quinoline)
Formed from: Glaucine

Norharman (9*H*-pyrido[3,4-*b*]indole)
Formed from: 1,2,3,4-Tetrahydro-β-carboline
L-Tryptophan
→ 2-Methyl-β-carbolinium
In: Man K631
→ 9-Methyl-β-carbolinium
In: Man K631

Norhydrocodone
Formed from: Hydrocodone

Norhyoscyamine
Formed from: Hyoscyamine
→ 6β-Hydroxynorhyoscyamine
In: Hyoscyamus E350

D-Norisolysergamide
Formed from: D-Isolysergamide

D-Norisolysergic diethylamide
Formed from: D-Isolysergic diethylamide

D-Norisolysergic dimethylamide
Formed from: D-Isolysergic dimethylamide

Norisomethergine
Formed from: Isomethergine

Norketotifen
Formed from: Ketotifen

Norlaudanidine (2-methoxy-5-(((1*R*)-1,2,3,4-tetrahydro-6,7-dimethoxy-1-isoquinolinyl)methyl)phenol)
Probably formed from: Norreticuline

→ Norlaudanosine
Probably in: Papaver A1637

Norlaudanosine ((1*S*)-1-((3,4-dihydroxyphenyl)methyl)-1,2,3,4-tetrahydro-6,7-dimethoxyisoquinoline)
Formed from: Laudanosine
Probably formed from: Norlaudanidine
→ Papaverine
In: Papaver A1637

Norlaudanosoline
See: Tetrahydropapaveroline

Norlergotrile
Formed from: Lergotrile

Norleucyl-β-naphthylamide
→ β-Naphthylamine
In: Beef A526

Norloprazolam
Formed from: Loprazolam

D-Norlysergamide
Formed from: D-Lysergamide

Norlysergic diethylamide
Formed from: Lysergic diethylamide

D-Norlysergic dimethylamide
Formed from: D-Lysergic dimethylamide

Normacromerine
See: 3,4-Dimethoxy-N-methylphenylethanolamine

Normacusine B (sarpagan-17-ol)
See: 10-Deoxysarpagine

Normeperidine
Formed from: Meperidine

D-Normetanephrine (4-hydroxy-3-methoxyphenylethanolamine)
Formed from: D-Noradrenaline
→ 3,4-Dimethoxy-N-methylphenylethanolamine
In: Coryphantha B807
→ 4-Hydroxy-3-methoxymandelaldehyde
In: Rat E339
See also: E390
→ D-Metanephrine
In: Beef A1463
Man A498
Rabbit A498
Rat A80
Sheep A3063
→ D-Noradrenaline
In: Bear A3063
Beef A3063
Cat A3063
Dog A3063
Guinea pig A3063
Man A3063, A3333
Mouse A3063
Pigeon A3063
Rabbit A3063
Rat A3063, A3333
→ D-Normetanephrine-4-β-D-glucuronide
In: Rat D289
→ D-Normetanephrine-4-sulphate
In: Rat D289, D583
→ Normetanephrone
In: Penicillium J16, K761

L-Normetanephrine
Formed from: L-Noradrenaline

D-Normetanephrine-4-β-D-glucuronide
Formed from: D-Normetanephrine

D-Normetanephrine-4-sulphate
Formed from: D-Normetanephrine

Normetanephrone (4-hydroxy-3-methoxy-α-oxophenethylamine)
Formed from: D-Normetanephrine

5-Normetapramine
Formed from: Metapramine

10-Normetapramine
Formed from: Metapramine

Normethadone
Formed from: 2-Dimethylamino-4,4-diphenyl-5-heptanone
Methadone
(Note. This spontaneously cyclises to 2-ethylidene-1,5-dimethyl-3,3-diphenyl-1-pyrrolidine, a recognised metabolite of methadone)

Normethergine
Formed from: Methergine

Normethotrimeprazine
→ Bisnormethotrimeprazine
In: Man A2904

Normethotrimeprazine sulphoxide
Probably formed from: Methotrimeprazine sulphoxide

Normorphine ((5α,6α)-7,8-didehydro-4,5-epoxymorphinan-3,6-diol)
Formed from: Morphine
Nalorphine
Norcodeine
→ 2,2′-Binormorphine
In: Horseradish A1357
→ Dihydronormorphine
In: Dog A3591
→ Morphine
Probably in: Rat A2475
→ Normorphine-3-β-D-glucuronide
In: Dog A3591
See also: A270, A2907, A3296, G909, H933
→ Normorphine-6-β-D-glucuronide
Probably in: Man A3296
→ Normorphinone
In: Dog A3591
Hamster K443
Rabbit H920

Normorphine-3-β-D-glucuronide
Formed from: Normorphine

Normorphine-6-β-D-glucuronide
Probably formed from: Normorphine

Normorphinone
Formed from: Normorphine

Nornaftifine
Formed from: Naftifine
→ *p*-Hydroxynornaftifine
Probably in: Man E32

Nornalmefene
Formed from: Nalmefene
→ Nornalmefene-β-D-glucuronide
Probably in: Rat H719

Nornalmefene-β-D-glucuronide
Probably formed from: Nornalmefene

Nornifenazone
Formed from: Nifenazone
→ Nornifenazone-β-D-glucuronide
Probably in: Man D283
→ Nornifenazone-N-oxide
Probably in: Man D283

Nornifenazone-β-D-glucuronide
Probably formed from: Nornifenazone

Nornifenazone-N-oxide
Probably formed from: Nornifenazone

O-Nornuciferine ((6a*R*)-5,6,6a,7-tetrahydro-1-methoxy-6-methyl-4*H*-dibenzo[*de*,*g*]quinolin-2-ol)
Formed from: N-Methylcrotsparine
N-Methylcrotsparinol

Nororientaline ((1*R*)-1,2,3,4-tetrahydro-1-((4-hydroxy-3-methoxyphenyl)methyl)-6-methoxy-7-isoquinolinol)
Formed from: 1-(3,4-Dihydroxybenzyl)-1,2,3,4-tetrahydro-7-hydroxy-6-methylisoquinoline
1,2,3,4-Tetrahydro-6,7-dihydroxy-1-(4-hydroxy-3-methoxybenzyl)isoquinoline
→ N-Methylnororientaline
In: Berberis K780

Noroxetorone
Formed from: Oxetorone

Noroxycodone
Formed from: Oxycodone

Noroxymorphone
Formed from: Naloxone
Naltrexone

D-Norparanephrine (3-hydroxy-4-methoxyphenylethanolamine)
Formed from: D-Noradrenaline

L-Norparanephrine
Formed from: L-Noradrenaline

Norpentazocine
Formed from: Pentazocine
→ Norpentazocine-β-D-glucuronide
Probably in: Rat A2134

Norpentazocine-β-D-glucuronide
Probably formed from: Norpentazocine

Norpethidine (4-phenyl-4-piperidinecarboxylic acid ethyl ester)
Formed from: Phenoperidine

Norphendimetrazine
Formed from: Phendimetrazine

Norprochlorperazine sulphoxide
Formed from: Perphenazine sulphoxide
Prochlorperazine sulphoxide

Norpromethazine
Formed from: Promethazine
→ Bisnorpromethazine
In: Rabbit B717
→ Hydroxynorpromethazine
In: Rabbit B717

→ N-Hydroxynorpromethazine
In: Pig G525
→ Norpromethazine sulphoxide
In: Pig G525
→ 10-(2′-Propanoneoxime)phenothiazine
In: Pig G525
Rabbit B717

Norpromethazine sulphone
Probably formed from: Promethazine sulphoxide
→ Bisnorpromethazine sulphone
Probably in: Rat B422

Norpromethazine sulphoxide
Formed from: Norpromethazine

Norprothanon
Formed from: Prothanon
→ Bisnorprothanon
In: Rat A3301
→ Hydroxynorprothanon
Probably in: Rat A3301

Norprotosinomenine (4′,7-di-O-methyltetrahydropapaveroline)
Formed from: 1,2,3,4-Tetrahydro-6,7-dihydroxy-1-(3-hydroxy-4-methoxybenzyl) isoquinoline
Probably formed from: Tetrahydropapaveroline
→ Cocculidine
In: Erythrina A3588

Norpseudocodamine
Probably formed from: Pseudocodamine

D-Norpseudoephedrine ((*αS*)-α-((1*S*)1-aminoethyl)benzenemethanol)
Formed from: Cathinone
→ 2-Imino-1-phenylpropan-1-ol
In: Arthrobacter B360

L-Norpseudoephedrine
Formed from: Cathinone
L-Pseudoephedrine

Norpseudolaudanine
Probably formed from: Pseudolaudanine

Norreticuline ((1*S*)-1,2,3,4-tetrahydro-1-((3-hydroxy-4-methoxyphenyl)methyl)-6-methoxy-7-isoquinolinol)
→ Norlaudanidine
Probably in: Papaver A1637
→ Reticuline
In: Berberis K780
With norreticuline N-methyltransferase E978
See also: B726

Norsalsolinol (1,2,3,4-tetrahydro-6,7-dihydroxyisoquinoline)
Formed from: 3,4-Dihydroxyphenethylamine
→ Norsalsolinol-N-sulphate
In: Rat E51
→ 1,2,3,4-Tetrahydro-6-hydroxy-7-methoxyisoquinoline
In: Rat A486
→ 1,2,3,4-Tetrahydro-7-hydroxy-6-methoxyisoquinoline
In: Rat A486

Norsalsolinol-N-sulphate
Formed from: Norsalsolinol

Norscopolamine
Formed from: Scopolamine
→ N-Hydroxynorscopolamine
In: Guinea pig A2816

Norsolorinic acid (1,3,6,8-tetrahydroxy-2-(1-oxohexyl)-9,10-anthracenedione)
Formed from hexanoic acid in Aspergillus H675

19-Nortestosterone
→ 17β-Oestradiol
In: Horse G395
→ Oestrone
In: Horse G395

5″-Nor-Δ^1-tetrahydrocannabinol-4″-oic acid
Formed from: Δ^1-Tetrahydrocannabinol
→ 5″-Nor-Δ^1-tetrahydro-6α-hydroxycannabinol-4″-oic acid
Probably in: Man C294

5″-Nor-Δ^6-tetrahydrocannabinol-4″-oic acid
Formed from: Δ^6-Tetrahydrocannabinol

5″-Nor-Δ^1-tetrahydro-6α-hydroxycannabinol-4″-oic acid
Probably formed from: 5″-Nor-Δ^1-tetrahydrocannabinol-4″-oic acid
→ 5″-Nor-Δ^1, $\Delta^{2''}$-tetrahydro-6α-hydroxycannabinol-4″-oic acid
Probably in: Man C294

5″-Nor-Δ^1, $\Delta^{2''}$-tetrahydro-6α-hydroxycannabinol-4″-oic acid
Probably formed from: 5″-Nor-Δ^1-tetrahydro-6α-hydroxycannabinol-4″-oic acid
N-(2)-Nor-D-tetrandrine
Formed from: D-Tetrandrine
O-Northebaine-6-ethyl ether
→ Codeinone
Probably in: Papaver C354
Northiaden (demethyldothiepin)
Formed from: Dosulepin
Dothiepin
Prothiaden
→ Bisdemethyldothiepin
In: Man E141
→ Northiaden sulphoxide
Probably in: Man A1500, A2797
Northiaden sulphoxide
Probably formed from: Northiaden
Northioridazine-5-sulphoxide
Probably formed from: Thioridazine-5-sulphoxide
Nortiflorex sulphone
Probably formed from: Tiflorex sulphone
Nortiflorex sulphoxide
Probably formed from: Tiflorex sulphoxide
Nortilidine
Formed from: Tilidine
→ Bisnortilidine
In: Dog A1440
Man A1440
Rabbit A1440
Rat A1440
Nortrimipramine
Formed from: Trimipramine
→ Bisnortrimipramine
In: Man F55
→ 10-Hydroxynortrimipramine
Probably in: Rat G269
Nortriptyline (10,11-dihydro-N-methyl-5*H*-dibenzo[*a*,*d*]cycloheptene-$\Delta^{5,\gamma}$-propylamine)
Formed from: Amitriptyline
→ Demethylnortriptyline
In: Man A1489, A3913, C26, E894, J162
Rat H887
See also: E892
→ 3-(10,11-Dihydro-5*H*-dibenzo[*a*,*d*]cyclohepten-5-ylidene)propionic acid
In: Rabbit C476
→ 1,2-Dihydro-1,2- dihydroxynortriptyline
Probably in: Man E894
→ *cis*-10,11-Dihydroxynortriptyline
In: Man E894
Rat E892, E931
→ *trans*-10,11-Dihydroxynortriptyline
In: Man E894
Rat E931
→ 2-Hydroxynortriptyline
In: Man E894, G241
Rat E892, E931
→ 10-Hydroxynortriptyline (*E* and *Z* isomers)
In: Man A1489, A3034, C26, C591, E894, E933, G241, J70, J162
Rat E892, E931, H887
See also: A1695, A2075, A3913, D739, G472
→ N-Hydroxynortriptyline
In: Guinea pig A888
Rabbit A888
Nortropisetron
Formed from: Tropisetron
Noruxepam
Formed from: Uxepam
L-Norvalyl-β-naphthylamide
→ β-Naphthylamine
In: Beef A526
Norwogonin (5,7,8-trihydroxyflavone)
Formed from: Chrysin
→ 3-(4-Carboxy-1,3-dioxobutyl)-4-hydroxy-6-phenyl-2-pyrone
In: Pseudomonas A1328
→ Wogonin
In: Orange A3987
Norzatosetron
Formed from: Zatosetron
Norzimetidine
Formed from: Zimetidine
→ Bisnorzimetidine
In: Dog B791
Noscapine ((3*S*)-6,7-dimethoxy-3-((5*R*)-5,6,7,8-tetrahydro-4-methoxy-6-methyl-1,3-

dioxolo[4,5-*g*]isoquinolin-5-yl)-1(3*H*)-isobenzofuranone)
→ Cotarnine
In: Man A3719
Rabbit A3719
Rat A3719
→ Hydroxycotarnine
In: Man A3719
Rabbit A3719
Rat A3719
→ Meconine
In: Man A3719
Rabbit A3719
Rat A3719
See also: B397

Novenamine (3-amino-7-((3-O-(aminocarbonyl)-6-deoxy-5-C-methyl-5-O-methyl-α-L-lyxo-hexopyranosyl)oxy)-4-hydroxy-8-methyl-2*H*-1-benzopyran-2-one)
Formed from: Novobiocin

Novobiocin
Formed from: 3-Aminoumbelliferone
threo-p-Hydroxyphenylserine
Methyl α-acetamido-*p*-coumarate
L-Tyrosine
→ Novenamine
Species unclear A540

Novocaine
See: 2-Diethylaminoethyl *p*-aminobenzoate

Nufenoxole (2-(3-(5-methyl-1,3,4-oxadiazol-2-yl)-3,3-diphenylpropyl)-2-azabicyclo[2.2.2]octane)
→ 2-(3-(5-Hydroxymethyl-1,3,4-oxadiazol-2-yl)-3,3-diphenylpropyl)-2-azabicyclo[2.2.1]octane
In: Monkey F585
Monkey F585
→ 5-Hydroxy-2-(3-(5-methyl-1,3,4-oxadiazol-2-yl)-3,3-diphenylpropyl)-2-azabicyclo[2.2.1]octane
In: Monkey F585
Monkey F585

O

Ochotensimine ((7′*S*)-3′,4′,6,8-tetrahydro-6′,7′-dimethoxy-2′-methyl-6-methylenespiro(7*H*-indeno[4,5-*d*]-1,3-dioxole-7,1′(2′*H*)-isoquinoline))
Formed from: 3,4-Dihydroxy-L-phenylalanine

Ochratoxin A (N-(((3*R*)-5-chloro-3,4-dihydro-8-hydroxy-3-methyl-1-oxo-1*H*-2-benzopyran-7-yl)carbonyl)-L-phenylalanine)
Formed from: L-Phenylalanine
→ 2′,3′-Dihydro-2′,3′-dihydroxyochratoxin A
In: Phenylbacterium C697
→ (4*R*)-4-Hydroxyochratoxin A
In: Rabbit C688
See also: B451, F332
→ (4*S*)-4-Hydroxyochratoxin A
In: Rabbit C688
→ 10-Hydroxyochratoxin A
In: Rabbit C688
→ *m*-Hydroxyochratoxin A
In: Phenylbacterium C697
→ Ochratoxin α
In: Microorganism D192
Phenylbacterium C697
Rat A558, A3924
With carboxypeptidase A A558
→ L-Phenylalanine
In: Microorganism D192
Rat A558
With carboxypeptidase A A558
See also: C697

Ochratoxin B (N-(((3*R*)-3,4-dihydro-8-hydroxy-3-methyl-1-oxo-1*H*-2-benzopyran-7-yl)carbonyl)-L-phenylalanine)
→ 7-Carboxy-3,4-dihydro-8-hydroxy-3-methylisocoumarin
In: Rat A558
→ L-Phenylalanine
In: Rat A558
With carboxypeptidase A A558

Ochratoxin α ((3*R*)-5-chloro-3,4-dihydro-8-hydroxy-3-methyl-1-oxo-1*H*-2-benzopyran-7-carboxylic acid)
Probably formed from: Ochratoxin A

Octachlorodibenzo-*p*-dioxin
Formed from: Pentachlorophenol
→ 1,2,3,4,6,7,8-Heptachlorodibenzo-*p*-dioxin
In: Microorganism H874
→ 1,2,3,4,6,7,9-Heptachlorodibenzo-*p*-dioxin
In: Microorganism H874
→ Tetrachlorocatechol
In: Phanerochaete H911

Octachlorostyrene
→ Heptachlorophenylacetic acid
In: Rat C635
→ Heptachlorostyrene
In: Rat C635

(5a*R*,6*S*,12*S*,14a*S*)-1,2,3,5a,6,11,12,14a-Octahydro-5a,6-dihydroxy-12-(2-hydroxy-2-methylpropyl)-9-methoxy-5*H*,14*H*-pyrrolo[1″,2″:4′,5′]pyrazino[1′,2′:1,6]pyrido[3,4-*b*]indole-5,14-dione
Formed from: Verruculogen

1,2,3,4,4a,5,6,10b-Octahydro-8,9-dihydroxy-4-*n*-propylbenzo[*f*]quinoline
→ 1,2,3,4,4a,5,6,10b-Octahydro-8-hydroxy-9-methoxy-4-*n*-propylbenzo[*f*]quinoline
In: Pig F633

((2α,6α,8α,9α,β)-Octahydro-3-hydroxy-2,6-methano-2*H*-quinolizin-8-yl) 1*H*-5-hydroxyindole-3-carboxylate
Formed from: 3-Dihydrodolasetron

((2α,6α,8α,9α,β)-Octahydro-3-hydroxy-2,6-methano-2*H*-quinolizin-8-yl) 1*H*-6-hydroxyindole-3-carboxylate
Formed from: 3-Dihydrodolasetron
((2α,6α,8α,9α,β)-Octahydro-3-hydroxy-2,6-methano-2*H*-quinolizin-8-yl) 1*H*-7-hydroxyindole-3-carboxylate
Formed from: 3-Dihydrodolasetron
1,2,3,4,4a,5,6,10b-Octahydro-8-hydroxy-9-methoxy-4-*n*-propylbenzo[*f*]quinoline
Formed from: 1,2,3,4,4a,5,6,10b-Octahydro-8,9-dihydroxy-4-*n*-propylbenzo[*f*]quinoline
***m*-(*cis*-Octahydro-3-methyl-2*H*-1,3-benzazocin-4-yl)phenol**
Formed from: Ciramadol
(2α,6α,9aβ)-Octahydro-3-oxo-2,6-methano-2*H*-quinolizin-8-yl 1*H*-indol-3-carboxylate
See: Dolasetron
(3*S*,7*R*,11*E*)-3,4,5,6,7,8,9,10-Octahydro-7,14,16-trihydroxy-3-methyl-1*H*-2-benzoxacyclotetradecin-1-one
See: α-Zearalenol
(3*S*,7*S*,11*E*)-3,4,5,6,7,8,9,10-Octahydro-7,14,16-trihydroxy-3-methyl-1*H*-2-benzoxacyclotetradecin-1-one
See: β-Zearalenol
2-Octaprenylphenol
→ Ubiquinone 40
In: Escherichia A3811
3-Octaprenyltoluquinol
Formed from: Homogentisic acid
Octoclothepine (1-(8-chloro-10,11-dihydrodibenzo[*b*,*f*]thiepin-10-yl)-4-methylpiperazine)
Formed from: Octoclothepine-N-oxide
Octoclothepine-N-oxide
→ Octoclothepine
In: Rat B935
***o*-Octopamine** (*o*-hydroxyphenethylamine)
Formed from: *o*-Tyramine
→ *o*-Hydroxymandelaldehyde
Probably in: Rat C823
***m*-Octopamine** (*m*-hydroxyphenethylamine)
Formed from: *m*-Tyramine
→ *m*-Hydroxymandelaldehyde
Probably in: Rat J287
Octopamine (*p*-hydroxyphenethylamine)
Formed from: D-Noradrenaline
Tyramine
→ N-Acetyloctopamine
In: Ascaridia G997
Boophilus G780
Cockroach H835
Periplaneta E922, G103, F393
Rhizoglyphus F669
See also: B765, B826
→ Feruloyloctopamine
In: Nocotiana D558, F72
→ *p*-Hydroxymandelaldehyde
In: Arthrobacter H912
Bacteria B225
Hymenolepis A3815
Man A1372
Periplaneta A2336
Rabbit A373
Rat A373, B389, E339
Rhizopus F669
See also: A3079, E386
→ *p*-Hydroxyphenylacetaldehyde
In: Pseudomonas K709
→ D-Noradrenaline
In: Mucuna E769
Trichostrongylus F2
See also: B897
→ Octopamine sulphate
In: Eubacterium G602
Lobster A3533
Man D583
→ *p*-Sympatol
In: Beef A1463
Man A2823
Monkey A2823
Rat A1311, A2823
Trichostrongylus F2
With phenylethanolamine N-methyltransferase A1721
See also: A1257, A1337
→ Tyramine
Probably in: Rabbit A373
Rat A373
Octopamine sulphate
Formed from: Octopamine

Octylbenzene
→ Benzoic acid
In: Paecilomyces D936
Penicillium D936
Verticillium D936
Octyl 1-(*m*-bromophenyl)ethylamine oxalamate
→ 1-(*m*-Bromophenyl)ethylamine
In: Candida J54
Octyl 1-(*o*-fluorophenyl)ethylamine oxalamate
→ 1-(*o*-Fluorophenyl)ethylamine
In: Candida J54
***n*-Octyl gallate**
→ *n*-Octyl gallate-β-D-glucuronide
In: Man G524, J617
***n*-Octyl gallate-β-D-glucuronide**
Formed from: *n*-Octyl gallate
Octyl L-γ-glutamyl dopamide
→ 3,4-Dihydroxyphenethylamine
In: Dog A3945
Octyl 1-(*o*-methoxyphenyl)ethylamine oxalamate
→ 1-(*o*-Methoxyphenyl)ethylamine
In: Candida J54
Octyl 1-(*m*-methoxyphenyl)ethylamine oxalamate
→ 1-(*m*-Methoxyphenyl)ethylamine
In: Candida J54
Octyl 1-(*p*-methoxyphenyl)ethylamine oxalamate
→ 1-(*p*-Methoxyphenyl)ethylamine
In: Candida J54
***p*-Octylphenol**
→ *p*-Octylphenyl sulphate
In: Man K344
4-*tert*-Octylphenol (4-(2,2,4,4,-tetramethylbut-4-yl)phenol)
→ 4-(2-Hydroxymethyl-2,4,4,-trimethylbut-4-yl)phenol
In: Rainbow trout K126
→ 4-(3-Hydroxy-2,2,4,4,-tetramethylbut-4-yl)catechol
In: Rainbow trout K126
→ 4-*tert*-Octylphenyl-β-D-glucuronide
In: Rainbow trout K126
→ 4-(2,2,4,4,-Tetramethylbut-4-yl)catechol
In: Rainbow trout K126
2-Octyl phenyl carbonate
→ Phenol
In: Acinetobacter F812
Octyl 1-phenylethylamine oxalamate
→ 1-Phenylethylamine
In: Candida J54
***n*-Octyl gallate-β-D-glucuronide**
Formed from: *n*-Octyl gallate
***p*-Octylphenyl sulphate**
Formed from: *p*-Octylphenol
(*S*)-3-*n*-Octylphthalide
Formed from: *o*-*n*-Nonylbenzoic acid
Octyl salicylate
→ Salicylic acid
In: Guinea pig D144
Odapipam
→ Odapipam-β-D-glucuronide
In: Dog H198
Man H198
Mouse H198
Rat H198
Odapipam-β-D-glucuronide
Formed from: Odapipam
Oduline ((5a*S*,7*R*,12b*R*,12c*R*)-1,2,3,5,5a,7,12b,12c-octahydro-1-methyl[1,3]dioxolo[6,7][2]benzopyrano[3,4-*g*]indol-7-ol)
Formed from: O-Methylbelladine
Oestetrol (15α-hydroxyoestriol)
Formed from: 15α-Hydroxy-17β-oestradiol
Probably formed from: Oestriol
→ Oestetrol N-acetylhexaminide
In: Man A501
→ Oestetrol-3-β-D-glucuronide
In: Guinea pig A467
Man A501
→ Oestetrol-17-β-D-glucuronide
Probably in: Man A501, A591, A1151
→ Oestetrol-3-sulphate
In: Man A1151
Oestetrol N-acetylhexaminide
Formed from: Oestetrol
Oestetrol-3-β-D-glucuronide
Formed from: Oestetrol
→ Oestetrol-3-β-D-glucuronide sulphate
Probably in: Man A501
Oestetrol-17-β-D-glucuronide
Formed from: Oestetrol

Oestetrol-3-β-D-glucuronide sulphate
Probably formed from: Oestetrol-3-β-D-glucuronide
Oestetrol-3-sulphate
Formed from: Oestetrol
17α-Oestradiol
Formed from: 17α-Oestradiol-17-N-acetylglucosaminide
17α-Oestradiol-3-β-D-glucoside
17α-Oestradiol-17-β-D-glucoside
Oestrone
Probably formed from: 17α-Oestradiol-3-sulphate
→ 2-Hydroxy-17α-oestradiol
In: Rabbit C395
→ 4-Hydroxy-17α-oestradiol
In: Rabbit C395
→ 17α-Oestradiol-3-β-D-glucoside
In: Rabbit A3784
→ 17α-Oestradiol-17-β-D-glucoside
In: Chimpanzee A2501
→ 17α-Oestradiol-3-β-D-glucuronide
Probably in: Beef A2451, A3995
Hen A401, A1040, A2208
Guinea pig A401
Rabbit A399, A401
→ 17α-Oestradiol-17-β-D-glucuronide
In: Hen A241, A3652
See also: A509, A737, A3995
→ 17α-Oestradiol-3-sulphate
In: Beef A1211
See also: A472, A2208, A3652, D486
→ Oestrone
In: Rabbit A157
17β-Oestradiol ((17β)-oestra-1,3,5(10)-triene-3,17-diol)
Formed from: Estramustine
17α-Ethynyl-17β-oestradiol
19-Nortestosterone
17β-Oestradiol-17-acetate
17β-Oestradiol-3-benzoate
17β-Oestradiol-17-β-cyclopentylpropionate
17β-Oestradiol-3-β-D-glucoside
17β-Oestradiol-17-β-D-glucoside
17β-Oestradiol-3-β-D-glucuronide
17β-Oestradiol-17-β-D-glucuronide
17β-Oestradiol-17-β-propionate
17β-Oestradiol-17-stearate
17β-Oestradiol-3-sulphate
17β-Oestradiol-17-valerate
Oestrone
3-O-Propyl-17β-oestradiol
Testosterone
→ 16-Epioestriol
In: Guinea pig A2583, A2868
Man A2309, A2465
Monkey A491, A3294
See also: A454, A951, A3652
→ 2-Hydroxy-17β-oestradiol
In: African catfish E966
Aspergillus D568
Caldariomyces D568
Hamster D514, H706, J171, J335
Man A2465, A3749, F627, G733, K621
Microorganism G489
Rabbit C395, E144
Rat A605, A1487, A2408, A2467, A3295, A3344, A3944, B681, C341, C833, D39, D514, D680, E40, E525 E538, F42, F627, F837, H342, J498, K603
Rhizopus D568
Streptomyces D568
See also: A375, A1575, A1844, A2969, A3749, D712
→ 4-Hydroxy-17β-oestradiol
In: Aspergillus D568
Hamster D514, H706, J171, J335
Man A 3749, G733, K621
Penicillium D568
Pig E78
Polyporus D568
Rabbit C395, E144
Rat A3344, D514, D680, E40, E342, E525, F42, F837, H342
Rhizopus D568
Streptomyces D568
→ 6α-Hydroxy-17β-oestradiol
In: Man A2465, K621
Rat A1487, A2192, E525, H342
→ 6β-Hydroxy-17β-oestradiol
In: Man A2465

Rat A1487, A2192, E525, H342
→ 7α-Hydroxy-17β-oestradiol
In: Man A2465
Rat A1487, A2192, A2408
→ 7β-Hydroxy-17β-oestradiol
In: Man A2465
→ 11β-Hydroxy-17β-oestradiol
In: Neurospora A944
→ 14α-Hydroxy-17β-oestradiol
In: Rat H342
→ 15α-Hydroxy-17β-oestradiol
In: Baboon A545
Man K621
Pig F511
Rat E525
See also: C833, J498
→ 3-O-Methyl-17β-oestradiol
In: Beef A3339
→ 17β-Oestradiol-17-arachidonate
In: Beef E431
→ 17β-Oestradiol-3-β-D-glucoside
In: Rabbit A3784
→ 17β-Oestradiol-3-β-D-glucuronide
In: Dog B689
Guinea pig A1936, A2825, A2870
Man A156, A250, A1196, K33, K134
Pig A1273
Rabbit F272
Rat C603, J498
Rhesus monkey A2897
See also: A399, A401, A454, A545, A1040, A1487, A1564, A2092, A2208, A2732, A2868, A3224, A3754, B258, B740, B800, F309, H209
→ 17β-Oestradiol-17-β-D-glucuronide
In: Dog B689
Hen A241, A3652
Man A156, A250, A1196
Mouse E899
Rabbit C603
Rat C603
Rhesus monkey A2897
→ 17β-Oestradiol-17-linoleate
In: Beef E431, E968
→ 17β-Oestradiol-17-linolenate
In: Beef E431, E968
→ 17β-Oestradiol-17-oleate
In: Beef E431, E968
→ 17β-Oestradiol-17-palmitate
In: Beef E431, E968
→ 17β-Oestradiol-17-palmitoleate
In: Beef E968
→ 17β-Oestradiol-3-sulphate
In: Beef A1211
Channel catfish K497
Guinea pig A1936, A2825, A2870
Hen D486
Man A236, G87
See also: A399, A472, A605, A1487, A2086, A2208, A3224, A3652, D500, E432, E840, E899, K120, K526
→ Oestriol
In: Baboon A545
Guinea pig A2583, E840
Hamster G625, H706, J171
Man A1196, A1457, A2309, A2465, E132
Monkey A491, A3294
Rabbit C384
Mouse E899
Rat A605, A936, A1487, A2192, A2408, A3281, C833, E525, E538, E840, H342
Streptomyces A2225
See also: A242, A375, A822, A951, A1280, A1936, A2569, A3224, A3241, A3754, A3786
→ Oestrone
In: Alcaligenes B454
Aspergillus D568
Baboon A545
Beef A3266, E19
Caldariomyces D568
Dog B689, K526
Elephant G599
Gliocladium D568
Guinea pig A1624, A2143, A2825, A2870
Hamster H706, J171
Hen B748, D486
Man A250, A1196, A1214, A1272, A1457, A2309, A2328 A2465, A3086, C701
Mouse A245, A2159, E821, E899
Neurospora A944
Peptococcus E499

Phanerochaete D568
Phaseolus B136
Pig C896, H209, E78
Pseudomonas A472, B454
Rabbit A157, A686
Rat A605, A1487, A2408, A3344, C341, C833, H342, J498
Rhesus monkey A491, A2897
Rhizopus D568
Staphylococcus B454
Streptococcus B454
See also: A236, A399, A454, A1575, A2569, A2583, A2868, A3224, A3786, B258, C280

17β-Oestradiol-17-acetate
Formed from: 17β-Oestradiol diacetate
→ 17β-Oestradiol
In: Man E439
Mouse J358
Rabbit J358
Rat J358
→ 17β-Oestradiol-17-acetate-3-sulphate
In: Beef A1211

17β-Oestradiol-17-acetate-3-sulphate
Formed from: 17β-Oestradiol-17-acetate

17α-Oestradiol-17-N-acetylglucosaminide
→ 17α-Oestradiol
In: Beef A3192

17α-Oestradiol-17-N-acetylglucosaminide-3-β-D-glucuronide
Probably formed from: 17α-Oestradiol-3-β-D-glucuronide

17β-Oestradiol-17-arachidonate
Formed from: 17β-Oestradiol

17β-Oestradiol-3-benzoate
→ 17β-Oestradiol
In: Beef A3266
Pig A102
See also: A2143, A2569, A3995, B258

17β-Oestradiol-17-β-cyclopentylpropionate
→ 17β-Oestradiol
In: Pig A102

17β-Oestradiol diacetate
→ 17β-Oestradiol-17-acetate
In: Mouse J358
Rabbit J358
Rat J358

17β-Oestradiol disulphate
Probably formed from: 17β-Oestradiol-3-sulphate
17β-Oestradiol-17-sulphate

17α-Oestradiol-3-β-D-glucoside
Formed from: 17α-Oestradiol
→ 17α-Oestradiol
In: Beef A3192

17β-Oestradiol-3-β-D-glucoside
→ 17β-Oestradiol
In: Beef A3192

17α-Oestradiol-17-β-D-glucoside
Formed from: 17α-Oestradiol
→ 17α-Oestradiol
In: Beef A3192
See also: A737
→ 17α-Oestradiol-17-β-D-glucoside-3-β-D-glucuronide
Probably in: Man A737

17β-Oestradiol-17-β-D-glucoside
→ 17β-Oestradiol
Probably in: Man A454

17α-Oestradiol-17-β-D-glucoside-3-β-D-glucuronide
Probably formed from: 17α-Oestradiol-17-β-D-glucoside
17α-Oestradiol-3-β-D-glucuronide

17α-Oestradiol-3-β-D-glucuronide
Formed from: 17α-Oestradiol
Oestrone-3-β-D-glucuronide
→ 17α-Oestradiol
In: Beef A3192
→ 17α-Oestradiol-17-N-acetylglucosaminide-3-β-D-glucuronide
Probably in: Rabbit A161, A246, A399, A509
→ 17α-Oestradiol-17-β-D-glucoside-3-β-D-glucuronide
Probably in: Bull A2451
Chimpanzee A2501
Rabbit A509
→ Oestrone-3-β-D-glucuronide
In: Rabbit A157

17β-Oestradiol-3-β-D-glucuronide
Formed from: 17β-Oestradiol
Oestrone-3-β-D-glucuronide
→ 17β-Oestradiol

In: Alcaligenes E323
Microorganism A3736
Pig A1273
See also: A2175
→ Oestrone-3-β-D-glucuronide
In: Man A1214, B827
Rabbit A157
Rat E699

17α-Oestradiol-17-β-D-glucuronide
Formed from: 17α-Oestradiol

17β-Oestradiol-17-β-D-glucuronide
Formed from: 17β-Oestradiol
Probably formed from: 17β-Oestradiol-17-β-D-glucuronide-3-sulphate
→ 2-Hydroxy-17β-oestradiol-17-β-D-glucuronide
In: Rat D63
→ 17β-Oestradiol
In: Alcaligenes E323
Pig C280
See also: A671
→ 17β-Oestradiol-17-β-D-glucuronide-3-sulphate
In: Guinea pig E840
Man A671
Rat E840

17β-Oestradiol-17-β-D-glucuronide-3-sulphate
Formed from: 17β-Oestradiol-17-β-D-glucuronide
→ 17β-Oestradiol-17-β-D-glucuronide
Probably in: Man A671

17β-Oestradiol-17-linoleate
Formed from: 17β-Oestradiol

17β-Oestradiol-17-linolenate
Formed from: 17β-Oestradiol

Oestradiol methyl ester
→ 16α-Hydroxyoestradiol methyl ester
In: Streptomyces A2225

17β-Oestradiol-17-oleate
Formed from: 17β-Oestradiol

17β-Oestradiol-6-one
→ 17β-Oestradiol-6-one-3-sulphate
In: Beef A1211

17β-Oestradiol-6-one-3-sulphate
Formed from: 17β-Oestradiol-6-one

17β-Oestradiol-17-palmitate
Formed from: 17β-Oestradiol

17β-Oestradiol-17-palmitoleate
Formed from: 17β-Oestradiol

17β-Oestradiol-17-propionate
→ 17β-Oestradiol
In: Pig A102

17β-Oestradiol-17-stearate
→ 17β-Oestradiol
In: Man E439

17α-Oestradiol-3-sulphate
Formed from: 17α-Oestradiol
Oestrone-3-sulphate
→ 17α-Oestradiol
Probably in: Mouse A245
→ 17α-Oestradiol disulphate
Probably in: Hen A3652

17β-Oestradiol-3-sulphate
Formed from: 17β-Oestradiol
Oestrone-3-sulphate
→ 2-Hydroxy-17β-oestradiol-3-sulphate
Probably in: Rat A1844
→ 15α-Hydroxy-17β-oestradiol-3-sulphate
In: Man A1588, D670
→ 17β-Oestradiol
In: Peptococcus E449
See also: A156, A245
→ 17β-Oestradiol-di-sulphate
In: Guinea pig A1936
See also: A399, A3652, E899
→ Oestriol-3-sulphate
In: Man A1588, E132
→ Oestrone-3-sulphate
In: Guinea pig A1624, A1936
Man A1214, B827
Mouse A245

17α-Oestradiol-17-sulphate
Formed from: 17α-Oestradiol

17β-Oestradiol-17-sulphate
Formed from: 3-Deoxy-17β-oestradiol-17-sulphate
→ 2-Hydroxy-17β-oestradiol-17-sulphate
In: Man D777
Rat C341, D254, E465
→ 4-Hydroxy-17β-oestradiol-17-sulphate
In: Man D777
→ 6β-Hydroxy-17β-oestradiol-17-sulphate
In: Man D777

→ 7α-Hydroxy-17β-oestradiol-17-sulphate
In: Man D777
→ 7β-Hydroxy-17β-oestradiol-17-sulphate
In: Man D777
→ 17β-Oestradiol disulphate
Probably in: Cat A399

17β-Oestradiol-17-valerate
→ 17β-Oestradiol
In: Man E439
Pig A102

17β-Oestradiol-17-valerinate
→ 17β-Oestradiol-17-valerinate-3-sulphate
In: Beef A1211

17β-Oestradiol-17-valerinate-3-sulphate
Formed from: 17β-Oestradiol-17-valerinate

Oestra-1,3,5(10),9-tetraene-12,17-dion-3-yl sulphate
Formed from: 3-Hydroxyoestra-1,3,5(10),9-tetraene-12,17-dione

1,3,5(10),16-Oestratetraen-3-ol
→ 1,3,5(10),16-Oestratetraen-3-ol-16α,17α-oxide
In: Rat B156
→ 1,3,5(10),16-Oestratetraen-3-ol-16β,17β-oxide
In: Rat B156

1,3,5(10),16-Oestratetraen-3-ol-16α,17α-oxide
Formed from: 1,3,5(10),16-Oestratetraen-3-ol
→ 16-Epioestriol
In: Rat B156

1,3,5(10),16-Oestratetraen-3-ol-16β,17β-oxide
Formed from: 1,3,5(10),16-Oestratetraen-3-ol
→ Epioestriol
In: Rat B156
→ Oestriol
In: Rat B156

Oestra-1,3,5(10),6-tetraene-17-on-3-yl sulphate
Formed from: 3-Hydroxyoestra-1,3,5(10),6-tetraene-17-one

Oestra-1,3,5(10),9-tetraene-17-on-3-yl sulphate
Formed from: 3-Hydroxyoestra-1,3,5(10),9-tetraene-17-one

Oestra-1,3,5(10)-triene-3,17-diol
See: 17β-Oestradiol

Oestra-1,3,5(10)-trien-16-on-3-yl sulphate
Formed from: 3-Hydroxyoestra-1,3,5(10)-trien-16-one

Oestra-1,3,5(10)-trien-3-yl sulphate
Formed from: 3-Hydroxyoestra-1,3,5(10)-triene

Oestriol (16α-hydroxy-17β-oestradiol)
Formed from: 16α-Hydroxyoestrone
16α-Hydroxy-19-oxotestosterone
16-Hydroxytestosterone
19-Nortestosterone
17β-Oestradiol
1,3,5(10),16-Oestratetraen-3-ol-16β,17β-oxide
Oestriol-3-β-D-glucuronide
Oestriol-17-β-D-glucuronide
Probably formed from: 16-Oxo-17β-oestradiol
→ 2-Hydroxyoestriol
In: Rat A1255, A2467, A3102, C242, C261
See also: A605, A951, A1280, A1487, A2408
→ 6ξ-Hydroxyoestrone
In: Rat C242
→ 16α-Hydroxyoestrone
In: Rat C242
See also: A545
→ Oestetrol
Probably in: Man A3241
→ Oestriol-3-β-D-glucuronide
In: Baboon A1111
Dog A1302, A2628
Man D133
Pig A1273
Rabbit A987
Rat A3102
Sheep A398
See also: A605, A951, A1564, A3091, A3224, F309, J617, K134
→ Oestriol-16-β-D-glucuronide
In: Baboon A1111
Dog A1302, A2628
Man A156, A822, D133
Monkey A2897
Rat A3102, C242
See also: A545, A987, A1196, E899, G968
→ Oestriol-17-β-D-glucuronide
In: Rat A3102
See also: E899
→ Oestriol-3-sulphate
In: Baboon A1111

Beef A1211
Dog A1302, A2628
Man D133
Sheep A398
See also: A951, A1936, A3224, A3754, C242, E840, E899
→ Oestrone
In: Rat A2549, A3644
→ 16-Oxo-17β-oestradiol
In: Rat A1255, A2549, A3644
See also: A545, A2309, A2583, A3102, F810

Oestriol-3,16-disulphate
Probably formed from: Oestriol-3-sulphate

Oestriol-3-β-D-glucuronide
Formed from: Oestriol
→ Oestriol
In: Alcaligenes E323
Guinea pig B519
Pig A1273
Rabbit B519
→ Oestriol-3-β-D-glucuronide sulphate
Probably in: Man D133
Rat A605

Oestriol-16-β-D-glucuronide
Formed from: Oestriol
→ Oestriol-16-β-D-glucuronide-3-sulphate
In: Alcaligenes E323
Baboon A1111
Man A822
See also: A2628

Oestriol-17-β-D-glucuronide
Formed from: Oestriol
→ Oestriol
In: Alcaligenes E323

Oestriol-3-β-D-glucuronide sulphate
Probably formed from: Oestriol-3-β-D-glucuronide

Oestriol-16-β-D-glucuronide-3-sulphate
Formed from: Oestriol-16-β-D-glucuronide

Oestriol-3-sulphate
Formed from: Oestriol
Oestriol-3-sulphate
→ Oestriol-3,16-disulphate
Probably in: Guinea pig A1936, A3754, B683

Oestrone (3-hydroxyoestra-1,3,5(10)-triene-17-one)
Formed from: 4-Androstene-3,17-dione
3-Deoxyoestrone
3-O-Methyloestrone
19-Nortestosterone
17α-Oestradiol
17β-Oestradiol
Oestriol
Oestrone acetate
Oestrone-3-β-D-glucuronide
Oestrone-3-sulphate
Probably formed from: 19-Hydroxy-4-androstene-3,17-dione
Oestrone benzoate
→ 9,11-Dehydrooestrone
In: Hamster G625
Rat G625
→ 1-(S-Glutathionyl)-2-hydroxyoestrone
In: Rat A1396
→ 4-(S-Glutathionyl)-2-hydroxyoestrone
In: Rat A1396
→ 2-Hydroxyoestrone
In: Hamster G625, J310
Rat A1280, A1396, A1625, A2175, A2295, A2408, A3295, G625
See also: A454, A490, A545, A605, A951, A1457, A1487, A3749, C341, D568, H342
→ 4-Hydroxyoestrone
In: Hamster D514, G625, J310
Rat J514
See also: A1457, A3749, D568
→ 6α-Hydroxyoestrone
In: Hamster G625
Pig C896
Rat A2408, G625
See also: A1487, A2465
→ 6β-Hydroxyoestrone
In: Hamster G625
Rat G625
See also: A1487, A2465
→ 7α-Hydroxyoestrone
In: Hamster G625
Pig C896
Rat A2408, G625
See also: A1487, A2465

→ 7β-Hydroxyoestrone
Probably in: Man A2465
→ 15α-Hydroxyoestrone
In: Man A1588
→ 15β-Hydroxyoestrone
In: Hamster G625
Rat G625
→ 16α-Hydroxyoestrone
In: Bacteroides B454
Guinea pig A2868
Man A1588, E132, J647
Rat A1280, A2408
Streptococcus B454
Streptomyces A2225, A3367, F810
See also: A242, A1487, A2309, H342
→ 16β-Hydroxyoestrone
In: Guinea pig A2868
See also: A1936
→ 17α-Oestradiol
In: Beef A2451, E19
Hen A241, A1496, A3652
Sheep A472, A1630, D876
See also: A399, A401, A509, A1040, A2208, A2569, A2583, A3266, A3995, D486
→ 17β-Oestradiol
In: Alcaligenes B454
Bacteroides B454
Cadariomyces D568
Chimpanzee A784
Cochliobolus K378
Curvularia D568
Guinea pig A1624, E840
Hanster G625
Hen A241, A401, A1496, A3652, B748
Man A518, A1272, C701, C736
Monkey A491, A2897
Mouse A2159
Pea A1921
Phaseolus B136
Polyporus D568
Pseudomonas B454
Rabbit A686, G714
Rat A375, A518, A1280, A2408, A3281, E840, G625
Sheep A472, A1243, A1630, D876
Streptomyces B454
See also: A529, A857, A951, A984, A1040, A1844, A1936, A2208, A2464, A2868, A3754, B910, C242, D49, E449, E464
→ Oestrone-3-D-glucoside
In: Chimpanzee A2501
Rabbit A3784
→ Oestrone-3-β-D-glucuronide
In: Cat C562
Dog C562
Guinea pig A242, A401, A1936, A2092, A3754, C562
Hen A1040, A3652
Man J783, K134
Monkey A951, A2897
Mouse D315
Pig A1273
Quail C562
Rabbit B983, C562, C603, F309
Rat A1280, A1368, A1564, A2732, B740, B910, C242, C562, F309
Trout C562
See also: A156, A250, A454, A2208, A2451, A2825, A2870, A3224, B689, C672, E899, H209, J498
→ Oestrone-3-sulphate
In: Beef A1211, D520
Guinea pig A1936, E840
Hen A3652
Man G87, K344
Rat B910, D500, E840
Sheep A472
See also: A236, A250, A605, A1196, A2208, A2825, A2870, A3224, A3754, B689, E456, E899

Oestrone acetate
→ Oestrone
In: Penicillium A933
Pig A102
With glutamate dehydrogenase A2004

Oestrone benzoate
→ Oestrone
Probably in: Rabbit A161

Oestrone-3-D-glucoside
Formed from: Oestrone

Oestrone-3-β-D-glucuronide
Formed from: 17α-Oestradiol-3-β-D-glucuronide

17β-Oestradiol-3-β-D-glucuronide
Oestrone
→ 17α-Oestradiol-3-β-D-glucuronide
In: Dog A1617
→ 17β-Oestradiol-3-β-D-glucuronide
In: Dog A1617
→ Oestrone
In: Alcaligenes E323
Dog A1617
Pig A1273

Oestrone-3-sulphate
Formed from: 17β-Oestradiol-3-sulphate
Oestrone
→ 2-Hydroxyoestrone-3-sulphate
In: Man A1022
Monkey A951
See also: A1844, A2175
→ 15α-Hydroxyoestrone-3-sulphate
In: Man A1588
→ 16α-Hydroxyoestrone-3-sulphate
In: Guinea pig A2868, B683
Man A1588, D670
See also: A3754
→ 16β-Hydroxyoestrone-3-sulphate
In: Guinea pig B683
Man A1588, D670
See also: A1936, A3754
→ 18-Hydroxyoestrone-3-sulphate
In: Man A1588
→ 17α-Oestradiol-3-sulphate
In: Hen A1040
Mouse A245
Sheep B121, D876
→ 17β-Oestradiol-3-sulphate
In: Guinea pig A242, A1624, A1936, A2868
Hen A1040
Man A518, A1624
Monkey A951
Mouse A245, E821
Rat A518, A1844, A3281, B910
Sheep B121, D876
See also: B683
→ Oestrone
In: Beef D520
Guinea pig A242, A2868, A3754
Man D49, D96, D120, D577, D808, E398, E464, J647
Microorganism A3736
Monkey A951
Mouse D502
Peptococcus E449
Phaseolus B136
Pig A984
Rat A375, A1844, A3281, B910, H392
See also: A1921, A2467

Ofloxacin (9-fluoro-2,3-dihydro-3-methyl-10-(4-methyl-1-piperazinyl)-7-oxo-7*H*-pyrido[1,2,3-*d*,*e*][1,4]benzoxazine-6-carboxylic acid)
→ Demethylofloxacin
In: Dog E47
Man F611
Monkey E47
Rat E47, F339
→ Ofloxacin-β-D-glucuronide
In: Dog E47
Man H764
Monkey E47
Rat E47, F339, F730
→ Ofloxacin-N-oxide
In: Dog E47
Man F611
Monkey E47
Rat E47, F339

Ofloxacin-β-D-glucuronide
Formed from: Ofloxacin

Ofloxacin-N-oxide
Formed from: Ofloxacin

Olanexidine
See: 1-(3,4-Dichlorobenzyl)-5-octylbiguanide

Olanzapine (2-methyl-4-(4-methyl-1-piperazinyl)-10*H*-thieno[2,3-*b*][1,5]benzodiazepine)
→ 4-Demethylolanzapine
In: Dog J150
Man J21, J96, J150
Mouse J150
Rat J150
→ 7-Hydroxyolanzapine
In: Dog J150
Mouse J150
Rat J150

→ 2-Hydroxymethyl-4-(4-methyl-1-piperazinyl)-10*H*-thieno[2,3-*b*][1,5]benzodiazepine
In: Dog J150
Man J21, J96, J150
Mouse J150
Rhesus monkey J150
Rat J150
→ Olanzapine-4′-β-D-glucuronide
In: Man J96
→ Olanzapine-10-β-D-glucuronide
In: Man J78
→ Olanzapine-4′-N-oxide
In: Dog J150
Man J21, J96
Mouse J150
Rat J150

Olanzapine-4′-β-D-glucuronide
Formed from: Olanzapine

Olanzapine-10-β-D-glucuronide
Formed from: Olanzapine

Olanzapine-4′-N-oxide
Formed from: Olanzapine
→ 7-Hydroxyolanzapine-4′-N-oxide
Probably in: Dog J150

Olivetol (5-pentylresorcinol)
→ 3-(3,5-Dihydroxyphenyl)propanal
In: Syncephalastrum D450
→ 5-(4-Hydroxypentyl)resorcinol
In: Syncephalastrum D450

Omeprazole (*R* and *S* isomers; 4-methoxy-2-(((4-methoxy-3,5-dimethyl-2-pyridinyl)methyl)sulphinyl)-1*H*-benzimidazole)
→ N-Acetyl-S-(5-methoxybenzimidazol-2-yl)-L-cysteine
In: Rat G300
→ 2-(((4-Hydroxy-3,5-dimethyl-2-pyridinyl)methyl)sulphinyl)-5-methoxy-1*H*-benzimidazole
In: Dog D844
Rat D844
→ 5-Hydroxy-2-(((4-methoxy-3,5-dimethyl-2-pyridinyl)methyl)sulphinyl)-1*H*-benzimidazole
In: Dog D844
Man K65, K488, K647
Rat D844
→ 6-Hydroxy-5-methoxy-2-(((4-methoxy-3,5-dimethyl-2-pyridinyl)methyl)sulphinyl)-1*H*-benzimidazole
In: Dog D844
Rat D844
→ 2-(2-(3-Hydroxymethyl-4-methoxy-5-methyl)pyridinyl)methylsulphinyl-5-methoxybenzimidazole
In: Man F19
→ 2-(2-(5-Hydroxymethyl-4-methoxy-3-methyl)pyridinyl)methylsulphinyl-5-methoxybenzimidazole
In: Dog D844
Man H596, J51, K65
Rat D844
→ Omeprazole sulphone
In: Man K65, K488

Omeprazole sulphone
Formed from: Omeprazole

Ommochrome (probably a mixture)
Formed from: 3-Hydroxy-L-kynurenine

Onapristone ((11β,13α,17α)-11-(4-(dimethylamino)phenyl)-17-hydroxy-17-(3-hydroxypropyl)oestra-4,9-dien-3-one)
→ Demethylonapristone
In: Man J288

Ondansetron (1,2,3,9-tetrahydro-9-methyl-3-((2-methyl-1*H*-imidazol-1-yl)methyl)-4*H*-carbazol-4-one)
→ Demethylondansetron
In: Man H476
Rat H732
→ 6-Hydroxyondansetron
In: Man H476
→ 7-Hydroxyondansetron
In: Man H476
Rat H732
See also: G870
→ 8-Hydroxyondansetron
In: Man H476
Rat H732

Opipramol (4-(3-(5*H*-dibenz[*b*,*f*]azepin-5-yl)propyl)-1-piperazineethanol)
→ Opipramol-10,11-oxide
In: Rat A2758

Opipramol-10,11-oxide
Formed from: Opipramol
Orange RN
See: Acid Orange 12
Orbifloxacin (*rel*-1-cyclopropyl-7-((3*R*,5*S*)-3,5-dimethyl-1-piperazinyl)-5,6,8-trifluoro-1,4-dihydro-4-oxo-3-quinolinecarboxylic acid)
→ N-Hydroxyorbifloxacin
In: Cat J898
→ Orbifloxacin-β-D-glucuronide
In: Beef K360
Cat J898
Pig K360
Orbifloxacin-β-D-glucuronide
Formed from: Orbifloxacin
Orchinol (9,10-dihydro-5,7-dimethoxy-2-phenanthrenol)
Formed from: 3,3′,5-Trihydroxybibenzyl
Orcinol (3,5-dihydroxytoluene)
→ 2,3,5-Trihydroxytoluene
In: Aspergillus E139
Pseudomonas A639, A2172, A2391, A2696
See also: E204
Orciprenaline
See: 3,5-Dihydroxy-N-isopropylphenylethanolamine
Orcylalanine
See: 2,4-Dihydroxy-6-methyl-L-phenylalanine
Org GB94
See: Mianserin
Orientaline (1,2,3,4-tetrahydro-7-hydroxy-1-(4-hydroxy-3-methoxybenzyl)-6-methoxy-2-methylisoquinoline)
→ 1,2,3,4-Tetrahydro-7-hydroxy-6-methoxy-2-methylisoquinoline
In: Nelumbo D566
→ Vanillyl alcohol
In: Nelumbo D566
Orientin (8-β-D-glucopyranosyl-3′,4′,5,7-tetrahydroxyflavone)
Formed from: Cinnamic acid
Naringenin
Probably formed from: Luteolin
Oripavine (3-O-demethylthebaine)
Formed from: Thebaine
→ Morphinone
In: Papaver D577
Oripavine 3-ethyl ester
Formed from: Thebaine
→ 3-Ethylmorphine
In: Papaver C400
Ornithuric acid (N^2,N^5-dibenzoyl-L-ornithine)
Formed from: Benzoic acid
S-(2-Ornithylamino-4-methoxy-5-(1-methyl-4-piperidinylamino)carboxyphenyl)glutathione
Formed from: Clebopride
L-Ornithyl-β-naphthylamide
→ β-Naphthylamine
In: Glycine A1596
Orobol-7-glucoside (oroboside, 3′,4′,5,7,-tetrahydroxyisoflavone-7-glucoside)
→ Orobol-7-glucoside-6″-malonate
In: Cicer D445
Orobol-7-glucoside-6″-malonate
Formed from: Orobol-7-glucoside
Oroboside
See: Orobol-7-glucoside
Orsellinic acid (2,4-dihydroxy-6-methylbenzoic acid)
→ 2,4-Dihydroxy-6-methyl-L-phenylalanine
In: Corn cockle A641
Orthanilic acid (*o*-aminobenzenesulphonic acid)
→ 2,3-Dihydroxybenzenesulphonic acid
In: Alcaligenes H239, H297, K293, K775
γ-Oryzanol (mixture of ferulate esters)
→ Ferulic acid
In: Rabbit B719
Rat C710, C849
→ 4′-Hydroxy-3′-methoxyacetophenone
In: Rabbit B719
Oryzemate (3-allyl-1,2-benzisothiazole-1,1-dioxide)
→ Allyl *o*-sulphamoylbenzoate
In: Rice A941
→ Saccharin
In: Rice A941
Oxacillin ((5-methyl-3-phenyl-4-isoxazolyl)penicillin)
→ (5-Hydroxymethyl-3-phenyl-4-isoxazolyl)penicillin
In: Man B356

Oxaflumazine (10-(3-(4-(2-(1,3-dioxan-2-yl)ethyl)-1-piperazinyl)propyl)-2-(trifluoromethyl)-10*H*-phenothiazine)
→ Oxaflumazine sulphoxide
In: Rodent A872
Oxaflumazine sulphoxide
Formed from: Oxaflumazine
Oxametacin (1-(*p*-chlorobenzoyl)-2-methyl-5-methoxy-3-indolylacetylhydroxamic acid)
→ Demethylindometacinamide-β-D-glucuronide
In: Man B867
→ Oxametacin-β-D-glucuronide
In: Man B867
Oxametacin-β-D-glucuronide
Formed from: Oxametacin
Oxamniquine (1,2,3,4-tetrahydro-6-hydroxymethyl-2-isopropylaminomethyl-7-nitroquinoline)
Formed from: Desoxymansil
→ 2-Aminomethyl-1,2,3,4-tetrahydro-6-hydroxymethyl-7-nitroquinoline
In: Mouse A3711
→ 1,2,3,4-Tetrahydro-6-hydroxymethyl-7-nitroquinoline-2-carboxylic acid
In: Dog A2793
Hamster A2793
Man A2793
Monkey A2793
Mouse A2793
Rat A2793
→ 1,2,3,4-Tetrahydro-2-isopropylaminomethyl-7-nitroquinoline-6-carboxylic acid
In: Dog A2793, A3939
Hamster A2793
Man A2793
Monkey A2793
Mouse A2793
Rat A2793
See also: A3711
Oxaniloyl-*p*-trifluoromethylaniline
Formed from: *p*-Trifluoromethylacetanilide
Oxapadol (1,4-epoxy-4,5-dihydro-1-phenyl-1*H*-[1,4]oxazepino[4,3-*a*]benzimidazole)
→ 2-(2-Benzimidazolyl)-4-hydroxymethyl-2-phenyl-1,3-dioxolone
Probably in: Rat B710
→ 7-Hydroxyoxapadol
Probably in: Dog B710
Man B710
Rat B710
→ 8-Hydroxyoxapadol
Probably in: Dog B710
Man B710
Rat B710
→ 9-Hydroxyoxapadol
Probably in: Dog B710
Man B710
Rat B710
Oxaphenamide (2-hydroxy-N-(4-hydroxyphenyl)benzamide)
Formed from: Etofesalamide
→ Oxaphenamide sulphate
Probably in: Rabbit K478
Oxaphenamide sulphate
Formed from: Oxaphenamide
Oxaprotiline (α-((dimethylamino)methyl)-9,10-ethanoanthracene-9(10*H*)-ethanol)
→ Demethyloxaprotiline
In: Dog E48
Man C864, C865, D547
→ 9,10-Ethanoanthracene-9(10*H*)-2′-hydroxypropionaldehyde
Probably in: Dog E48
→ 2-Hydroxyoxaprotiline
In: Dog E48
Rat E48
→ 3-Hydroxyoxaprotiline
In: Dog E48
Man C865
Rat E48
→ Oxaprotiline-β-D-glucuronide
In: Dog E48
Man C865, D547
Oxaprotiline-β-D-glucuronide
Formed from: Oxaprotiline
Oxaprozin (4,5-diphenyl-2-oxazolepropionic acid)
→ 4-(*p*-Hydroxyphenyl)-5-phenyl-2-oxazolepropionic acid
In: Dog A3450
Man B375

Monkey A3450
→ Oxaprozin-β-D-glucuronide
In: Dog A3450
Man B375
Monkey A3450

Oxaprozin-β-D-glucuronide
Formed from: Oxaprozin

Oxarbazole (9-benzoyl-1,2,3,4-tetrahydro-6-methoxycarbazole-3-carboxylic acid)
→ 9-Benzoyl-1,2,3,4-tetrahydro-6-hydroxycarbazole-3-carboxylic acid
In: Dog A2873
Guinea pig A2873
Man A2873
Monkey A2873
Rat A2873
→ 1,2,3,4-Tetrahydro-6-methoxycarbazole-3-carboxylic acid
In: Dog A2873
Monkey A2873

Oxatomide (1-(3-(4-(diphenylmethyl)-1-piperazinyl)propyl)-1,3-dihydro-2*H*-benzimidazol-2-one)
→ 1,3-Dihydro-5-hydroxy-2*H*-benzimidazol-2-one
In: Dog C870
Man C870
Rat C870
→ 2,3-Dihydro-2-oxo-1*H*-benzimidazole-1-propionic acid
In: Dog C870
Man C870
Rat C870
→ 1,3-Dihydro-1-(3-(1-piperazinyl)propyl)-2*H*-benzimidazol-2-one
In: Dog C870
Man C870
Rat C870
→ 1-(3-(4-(Diphenylmethyl)-1-piperazinyl)propyl)-1,3-dihydro-5-hydroxy-2*H*-benzimidazol-2-one
In: Dog C870
Man C870
Rat C870
→ 1-(3-(4-(Diphenylmethyl)-1-piperazinyl)propyl)-1,3-dihydro-6-hydroxy-2*H*-benzimidazol-2-one
In: Dog C870
Man C870
Rat C870

Oxazapetine (6,7-dihydro-6-(2-propenyl)-5*H*-dibenz[*c*,*e*]azepin-5-one)
Formed from: Dehydrazapetine

Oxazepam
See: 7-Chloro-1,3-dihydro-3-hydroxy-5-phenyl-2*H*-1,4-benzodiazepin-2-one

Oxazepam acetate
→ 7-Chloro-1,3-dihydro-3-hydroxy-5-phenyl-2*H*-1,4-benzodiazepin-2-one
In: Mouse A3728, B151

Oxazepam *n*-butyratate
→ 7-Chloro-1,3-dihydro-3-hydroxy-5-phenyl-2*H*-1,4-benzodiazepin-2-one
In: Mouse B151

Oxazepam diethylacetate
→ 7-Chloro-1,3-dihydro-3-hydroxy-5-phenyl-2*H*-1,4-benzodiazepin-2-one
In: Mouse B151

Oxazepam isobutyrate
→ 7-Chloro-1,3-dihydro-3-hydroxy-5-phenyl-2*H*-1,4-benzodiazepin-2-one
In: Mouse B151

Oxazepam α-methyl-β-phenylpropionate
→ 7-Chloro-1,3-dihydro-3-hydroxy-5-phenyl-2*H*-1,4-benzodiazepin-2-one
In: Mouse B151
→ α-Methyl-β-phenylpropionic acid
In: Mouse B151

Oxazepam γ-phenylbutyrate
→ 7-Chloro-1,3-dihydro-3-hydroxy-5-phenyl-2*H*-1,4-benzodiazepin-2-one
In: Mouse B151
→ 4-Phenylbutyric acid
In: Mouse B151

Oxazepam β-phenylpropionate
→ 7-Chloro-1,3-dihydro-3-hydroxy-5-phenyl-2*H*-1,4-benzodiazepin-2-one
In: Mouse B151
→ β-Phenylpropionic acid
In: Mouse B151

Oxazepam propionate
→ 7-Chloro-1,3-dihydro-3-hydroxy-5-phenyl-2*H*-1,4-benzodiazepin-2-one
In: Mouse B151

Oxazepam succinate half ester
→ 7-Chloro-1,3-dihydro-3-hydroxy-5-phenyl-2*H*-1,4-benzodiazepin-2-one
In: Mouse A1186
Rat A1186
Oxazepam succinate methyl ester
→ 7-Chloro-1,3-dihydro-3-hydroxy-5-phenyl-2*H*-1,4-benzodiazepin-2-one
In: Mouse B151
6*H*-Oxazolo[3,2-*a*:4,5-*b′*]diindole
Formed from: Indole
Oxcarbazepine (10-oxocarbamazepine)
→ 10,11-Dihydroxycarbamazepine
In: Man E49
→ 10-Hydroxycarbamazepine (*R* and *S* isomers)
In: Man E49, K102
→ *enol*-Oxcarbazepine-O-β-D-glucuronide
In: Man E49
→ *enol*-Oxcarbazepine-O-sulphate
In: Man E49
***enol*-Oxcarbazepine-O-β-D-glucuronide**
Formed from: Oxcarbazepine
***enol*-Oxcarbazepine-O-sulphate**
Formed from: Oxcarbazepine
Ozepinac (6,11-dihydro-1-oxodibenz[*b*,*e*]oxepin-3-acetic acid)
→ 6,11-Dihydro-1-oxodibenz[*b*,*e*]oxepin-3-acetyltaurine
In: Dog A3552
Mouse A3552
Rat A3552
→ Hydroxyozepinac
In: Dog A3552
Mouse A3552
Rat A3552
→ Ozepinac-β-D-glucuronide
In: Dog A3552
Mouse A3552
Rat A3552
Ozepinac-β-D-glucuronide
Formed from: Ozepinac
Oxetorone (6-(3-dimethylaminopropylidene)[2,3-*e*]benzofuro-12*H*-benzo[*b*]oxepine)
→ Noroxetorone
In: Rat A3557
→ Oxetorone-N-oxide
In: Rat A3557
Oxetorone-N-oxide
Formed from: Oxetorone
Oxfendazole (fenbendazole sulphoxide, methyl 5(6)-phenylsulphinyl-2-benzimidazolecarbamate)
Formed from: Fenbendazole
→ Fenbendazole
In: Ascaris A3849
Beef A1702
Moniezia A3849
Mouse A3849
→ Fenbendazole sulphone
In: Beef E637
Chicken E637
Duck E637
Goat E637
Rabbit E637
Sheep E637
Turkey E637
2-(4-(3-Oximinocyclohexyl)phenyl)propionic acid
See: Ximoprofen
2-Oximino-4-phenylbutyric acid
Probably formed from: 2-Amino-4-phenylbutyric acid
5-Oximino-5-(*p*-trifluoromethylphenyl)valeric acid
Probably formed from: 5-(2-Aminoethyl)oximino-5-(*p*-trifluoromethylphenyl)valeric acid
Oxindole (indolin-2-one)
Formed from: 3-Acetonyl-3-hydroxyoxindole
Indole
→ Dioxindole
Probably in: Microorganism F15
→ 5-Hydroxyoxindole
Probably in: Rat D158
→ Isatin
In: Microorganism F15, G330
See also: J314
Oxindole-3-acetic acid
Formed from: Indole-3-acetic acid
→ 2,3-Dihydro-7-hydroxy-2-oxo-1*H*-indole-3-acetic acid
In: Zea D543, D619
Oxindole-3-acetylaspartic acid
Formed from: Indole-3-acetylaspartic acid

Oxindol-5-yl sulphate
Probably formed from: 5-Hydroxyoxindole
Oxmetidine (1,2-(2-(5-methyl-4-imidazolylmethylthio)ethylamino)-5-(3,4-methylenedioxybenzyl)-4-pyrimidone)
→ 5-(3,4-Dihydroxybenzyl)-2-((5-methyl-4-imidazolylmethylthio)ethylamino)-4-pyrimidone
In: Dog C875
→ Oxmetidine-N-β-D-glucuronide
In: Rat C875
→ Oxmetidine-O-β-D-glucuronide
In: Dog C875
Rat C875
→ Oxmetidine sulphoxide
In: Rat C875
Oxmetidine-N-β-D-glucuronide
Formed from: Oxmetidine
Oxmetidine-O-β-D-glucuronide
Formed from: Oxmetidine
Oxmetidine sulphoxide
Formed from: Oxmetidine
→ Oxmetidine sulphoxide-β-D-glucuronide
Probably in: Rat C875
Oxmetidine sulphoxide-β-D-glucuronide
Probably formed from: Oxmetidine sulphoxide
10-Oxoamitriptyline
Probably formed from: 10-Hydroxyamitriptyline
3-Oxobenfuracarb
Probably formed from: 3-Hydroxybenfuracarb
1′-Oxobenzbromarone
Probably formed from: 1′-Hydroxybenzbromarone
6-Oxo-3,4-benzpyrene radical
Formed from: 3,4-Benzpyrene
6-Hydroxy-3,4-benzpyrene
6-Oxo-4″,5″-bisnorcannabidiol-3″-oic acid
Probably formed from: 6-Hydroxy-4″,5″-bisnorcannabidiol-3″-oic acid
4′-Oxobrimodinine
Formed from: Brimodinine
6-Oxobrotizolam
Probably formed from: 6-Hydroxybrotizolam
1′-Oxobufuralol
Formed from: 1′-Hydroxybufuralol
2-(3-Oxobutyl)-3-((2-(1*H*-tetrazol-5-yl)(1,1′-biphenyl)-4-yl)methyl)-1,3-diaza*spiro*[4,4]non-1-en-4-one
Probably formed from: 2-(3-Hydroxybutyl)-3-((2-(1*H*-tetrazol-5-yl)(1,1′-biphenyl)-4-yl)methyl)-1,3-diaza*spiro*[4,4]non-1-en-4-one
6-Oxocannabidiol-5″-oic acid
Probably formed from: 6-Hydroxycannabidiol-5″-oic acid
6-Oxocannabinol
Probably formed from: 6β-Hydroxycannabinol
7-Oxocannabinol
Probably formed from: 7-Hydroxycannabinol
→ Cannabinol-7-oic acid
In: Mouse H67
See also: A3405, A3436, B458
6-Oxocannabinol-5″-oic acid
Probably formed from: 6-Hydroxycannabinol-5″-oic acid
10-Oxocarbamazepine
See: Oxcarbazepine
3-Oxocarbofuran
Probably formed from: 3-Hydroxycarbofuran
→ 2,3-Dihydro-7-hydroxy-2,2-dimethyl-3-oxobenzofuran
Probably in: Aspergillus J641
Fusarium J641
Nicotiana A402
3-Oxochroman (3-chromanone)
→ (+)-3-Chromanol
In: Streptomyces E462
2′-Oxociglitazone
Probably formed from: 2′-Hydroxyciglitazone
3′-Oxociglitazone
Probably formed from: *cis*-3′-Hydroxyciglitazone
4′-Oxociglitazone
Probably formed from: *cis*-4′-Hydroxyciglitazone
4-Oxocilostamide
Formed from: Cilostamide
2′-Oxociprofloxacin
Formed from: Ciprofloxacin
→ 7-(2-Aminoethylamino)-1-cyclopropyl-6-fluoro-1,4-dihydro-4-oxoquinoline-3-carboxylic acid

Probably in: Beef F955
Man E230
Rhesus monkey E251

3′-Oxoclozapine
Formed from: Clozapine

Oxocotarnine
Formed from: Narcotine

2-(4-(2-Oxocyclohexylidenemethyl)phenyl) propionic acid
→ 2-(4-(2-Hydroxycyclohexylidene-methyl)phenyl)propionyltaurine
In: Dog H774

2-(4-(2-Oxocyclohexylmethyl)phenyl)propionic acid
→ 2-(4-(2-Hydroxycyclohexylmethyl) phenyl)propionic acid
In: Rabbit D125

2-(4-(3-Oxocyclohexyl)phenyl)propionic acid
Formed from: Ximoprofen
→ 2-(4-(3-Hydroxycyclohexyl)phenyl) propionic acid
Probably in: Baboon F498
Man F498
Rat F498

1-(4-Oxocyclohexyl)-1-phenyl-3-(1-pyrrolidinyl)-1-propanol
Probably formed from: 1-(*cis*-4-Hydroxycyclohexyl)-1-phenyl-3-(1-pyrrolidinyl)-1-propanol

2-(4-(2-Oxocyclopent-3-en-1-ylmethyl)phenyl)propionic acid
→ 2-(4-(2-Hydroxycyclopent-3-en-1-yl-methyl)phenyl)propionic acid
In: Rabbit D125

2-(*p*-((2-Oxocyclopentyl)methyl)phenyl)propionic acid
See: Loxoprofen

2-Oxocyproheptadine
Probably formed from: Cyproheptadine
→ Hydroxy-2-oxocyproheptadine
Probably in: Man A1879

7-Oxodehydroabietic acid
Probably formed from: 7-Hydroxydehydroabietic acid
→ 11,12-Dihydroxy-7-oxodehydroabietic acid
Probably in: Alcaligenes A932
Pseudomonas A932

10-Oxodemethylcyproheptadine
Probably formed from: Demethyl-cyproheptadine

6-Oxo-N-demethyldextromethorphan
Probably formed from: 6-Hydroxy-N-demethyldextromethorphan

6-Oxodextromethorphan
Probably formed from: 6-Hydroxydextromethorphan

11-Oxodibenzo[*c*,*f*][1,2]diazepine
Probably formed from: 11-Hydroxydibenzo[*c*,*f*][1,2]diazepine

α-Oxo-2-dibenzofuranbutyric acid
See: Furobufen

5-(2-(11-Oxododecyl)phenyl)-4,6-dithianonanedioic acid
Probably formed from: 5-(2-(11-Hydroxydodecyl)phenyl)-4,6-dithianonanedioic acid

2-Oxodoxapram
Formed from: Doxapram
→ 1-Ethyl-4-(N-formyl-N-(2-hydroxyethyl) aminoethyl)-3,3-diphenyl-2-pyrrolidinone
Probably in: Man G25

3-Oxodoxapram
Formed from: Doxapram
→ 4-(2-Aminoethyl)-1-ethyl-3,3-diphenyl-2-pyrrolidinone
Probably in: Man G25
→ 1-Ethyl-4-(2-(2-formylmethoxyethylamino)ethyl)-3,3-diphenyl-2-pyrrolidinone
Probably in: Man G25
→ 1-Ethyl-4-(2-(2-hydroxyethylamino)ethyl)-3,3-diphenyl-2-pyrrolidinone
Probably in: Man G25

4-Oxoestazolam
Probably formed from: 4-Hydroxyestazolam

9-Oxofluorene
See: Fluorenone

Oxoiprindole
Probably formed from: Hydroxyiprindole
Iprindole

α-(4-(1-Oxo-2-isoindolinyl)phenyl)propionic acid (indoprofen)
→ α-(4-(5-Hydroxy-1-oxo-2-isoindolinyl)phenyl)propionic acid

In: Hamster B686
See also: A1501
→ α-(4-(6-Hydroxy-1-oxo-2-isoindolinyl)phenyl)propionic acid
In: Hamster B686
See also: A1501

11-Oxo-11*H*-isoindolo[2,1-*a*]benzimidazole
Formed from: 3-(2-(2-(2-Benzimidazolyl)benzyloxy)ethyl)quinazolin-4-one
→ Hydroxy-11-oxo-11*H*-isoindolo[2,1-*a*]benzimidazole
Probably in: Animal E202

2-Oxoisolysergic diethylamide
Formed from: Isolysergic diethylamide

2-Oxolisuride
See: 2-Hydroxylisuride

Oxolortalamine
Formed from: Lortalamine

2-Oxolysergic diethylamide
Formed from: Lysergic diethylamide

5-Oxomefruside
Formed from: Mefruside

6′-Oxomesoridazine
Formed from: Mesoridazine

6′-Oxomesoridazine sulphoxide
Formed from: Mesoridazine sulphoxide

3-O-(2-(3-Oxomorpholino)ethyl)morphine
Probably formed from: 3-O-(2-(3-Hydroxymorpholino)ethyl)morphine
→ 3-O-(2-(3-Oxomorpholino)ethyl) normorphine
Probably in: Man F682

3-O-(2-(3-Oxomorpholino)ethyl)normorphine
Probably formed from: 3-O-(2-(3-Oxomorpholino)ethyl)morphine

***trans*-4-(5-Oxo-2-morpholinylmethoxy)-1,2-indandiol**
Formed from: 6-((Inden-7-yloxy)methyl)-3-morpholinone
trans-4-(2-Morpholinylmethoxy)-1,2-indandiol
→ *trans*-4-(5-Oxo-2-morpholinylmethoxy)-1,2-indandiol-β-D-glucuronide
In: Rat E837

***trans*-4-(5-Oxo-2-morpholinylmethoxy)-1,2-indandiol-β-D-glucuronide**
Formed from: *trans*-4-(5-Oxo-2-morpholinylmethoxy)-1,2-indandiol

6′-Oxonabilone
Formed from: 6′-Hydroxynabilone

7′-Oxonabilone
Probably formed from: 7′-Hydroxynabilone
→ 3-(4-Carboxy-1,1-dimethylbutyl)-6,6α,7,8,10,10α-hexahydro-1-hydroxy-6,6-dimethyl-9*H*-dibenzo[*b*,*d*]pyran-9-one
Probably in: Nocardia B37

(1-Oxo-2-naphthyl)semicarbazone
See: Naftazone

6-Oxoniaprazine
Formed from: Niaprazine

Oxonorfloxacin
Probably formed from: Oxopefloxacin

10-Oxonortriptyline (*E* and *Z* isomers)
Formed from: 10-Hydroxynortriptyline
→ 10-Hydroxynortriptyline
In: Man H470, K341, K358
Rat H470

6-Oxo-17β-oestradiol
Probably formed from: 6α-Hydroxy-17β-oestradiol

16-Oxo-17β-oestradiol
Formed from: Oestriol
→ 16-Epioestriol
In: Microorganism B454
See also: A2549, A3102, A3644, C242
→ 2-Hydroxy-16-oxo-17β-oestradiol
In: Rat A1255
→ Oestriol
Probably in: Man A1188
→ 16-Oxo-17β-oestradiol-β-D-glucuronide
In: Rat C242
See also: A545, A3102, A3652
→ 16-Oxo-17β-oestradiol sulphate
Probably in: Guinea pig A3754
Hen A3652

16-Oxo-17β-oestradiol-β-D-glucuronide
Probably formed from: 16-Oxo-17β-oestradiol

16-Oxo-17β-oestradiol sulphate
Probably formed from: 16-Oxo-17β-oestradiol

6-Oxooestrone
Probably formed from: 6β-Hydroxyoestrone

Oxopefloxacin
Formed from: Pefloxacin
→ Oxonorpefloxacin
Probably in: Dog D864
Man D864
Mouse D864
Rabbit D864
Rat D864

5-Oxophenmetrazine
Formed from: Phenmetrazine

N-(1-Oxo-6-phenoxyindan-5-yl)methanesulphonamide
Probably formed from: N-(1-Hydroxy-6-phenoxyindan-5-yl)methanesulphonamide

N-(3-Oxo-6-phenoxyindan-5-yl)methanesulphonamide
Probably formed from: N-(3-Hydroxy-6-phenoxyindan-5-yl)methanesulphonamide

β-Oxo-D-phenylalanine
Formed from: D-Phenylserine

3-Oxo-3-(phenylamino)propanoic acid
See: Malonanilic acid

4-Oxo-2-phenyl-4*H*-1-benzopyran-8-acetic acid
See: Flavoneacetic acid

2-(4-Oxo-2-phenyl-4*H*-benzopyran-7-oxy)acetic acid
Probably formed from: 3-(4-Oxo-2-phenyl-4*H*-benzopyran-7-oxy)lactic acid

3-(4-Oxo-2-phenyl-4*H*-benzopyran-7-oxy)lactic acid
Formed from: Flavodilol
→ 2-(4-Oxo-2-phenyl-4*H*-benzopyran-7-oxy)acetic acid
Probably in: Rat F26

Oxophenylbutazone
Formed from: γ-Hydroxyphenylbutazone

2-Oxo-3-phenylbutyric acid
→ L-2-Amino-3-phenylbutyric acid
In: Bacillus G168

2-Oxo-4-phenylbutyric acid
→ L-2-Amino-4-phenylbutyric acid
In: Bacillus G168

3-Oxo-1-phenyl-1-cyclohexene
Formed from: 3-Hydroxy-1-phenyl-1-cyclohexene

6-Oxo-1-phenyl-1-cyclohexene
Formed from: 6-Hydroxy-1-phenyl-1-cyclohexene

4-Oxo-1-(1-phenylcyclohexyl)piperidine
Probably formed from: 4-Hydroxy-1-(1-phenylcyclohexyl)piperidine
→ 4-(4-Oxopiperidin-1-yl)-4-phenylcyclohexanol
Probably in: Rabbit G29

2-Oxo-2-phenylethanol
Formed from: Acetophenone
Phenylglycol
Phenyramidol
→ Phenylglycol
Probably in: Rat A1745
→ Phenylglyoxal
Probably in: Nitrosomonas H219
Rat A1745, D223, D365

2-Oxo-2-phenylethylamine
→ Mandelic acid
In: Rat J287

1-((2-Oxo-2-phenyl)ethyl)imidazole
→ 1-((2-Hydroxy-2-phenyl)ethyl)imidazole
In: Man J61

N-((2-Oxo-2-phenyl)ethyl)-N-methylnitrosamine
Formed from: N-((2-Hydroxy-2-phenyl)ethyl)-N-methylnitrosamine

6-Oxo-α-phenyl-2-piperidineacetic acid
Probably formed from: Methyl 6-oxo-α-phenyl-2-piperidineacetate
α-Phenyl-2-piperidineacetic acid
→ 5-Hydroxy-6-oxo-α-phenyl-2-piperidineacetic acid
Probably in: Dog C14
Rat C14
→ 6-Oxo-α-phenyl-2-piperidineacetyl-β-D-glucuronic acid
Probably in: Dog C14

6-Oxo-α-phenyl-2-piperidineacetyl-β-D-glucuronic acid
Probably formed from: 6-Oxo-α-phenyl-2-piperidineacetic acid

1-Oxo-1-phenyl-2-propanol
Formed from: Benzaldehyde
l-Ephedrine

2-Oxo-1-phenyl-1-propanol (2-hydroxy-1-phenyl-1-propanone)
Formed from: D-Norephedrine
Propiophenone
→ Benzaldehyde
Probably in: Rabbit A1552
→ 1-Phenylpropane-1,2-diol
Probably in: Mouse A717
Rabbit A1552, A2135, A2140, B755
Rat B755

2-Oxo-3-phenyl-1-propanol
Formed from: Benzaldehyde
L-Phenylalaninol

3-Oxo-3-phenylpropionic acid (benzoylacetic acid)
Formed from: Benzoic acid
Probably formed from: β-Phenylhydracrylic acid
→ Benzoic acid
Probably in: Man A126
→ β-Phenylhydracrylic acid
Probably in: Horse C183

2-Oxo-1-phenylpyrrolidine
Formed from: RWJ-34103

2-Oxo-3-phenylvaleric acid
→ L-2-Amino-3-phenylvaleric acid
In: Bacillus G168

2-Oxo-5-phenylvaleric acid
→ L-2-Amino-5-phenylvaleric acid
In: Bacillus G168

4-Oxo-1-phthalazineacetic acid
Formed from: Zopolrestat

6-(3-Oxo-1-piperazinyl)-2-quinolone
Probably formed from: 6-(1-Piperazinyl)-2-quinolone

N-(3-(*m*-(2-Oxo-1-piperidinylmethyl)phenoxy)propyl)acetamide
Probably formed from: Roxatidine

4-Oxopiperidin-1-yl)-4-phenylcyclohexanol
Probably formed from: 4-(4-Oxo-1-(1-phenylcyclohexyl)piperidine

2-Oxopropiono-2′,6′-xylidide
→ 2-Hydroxypropiono-2′,6′-xylidide
In: Rabbit C9
Rat C9

2-Oxoproterguride
Formed from: Proterguride
→ 3-Hydroxy-2-oxoproterguride
Probably in: Rat G553

5-Oxoprotizinic acid
Formed from: Protizinic acid

1′-Oxoprotriptyline
Formed from: Protriptyline

1-(N-(2-Oxopyrrolidino)carbonylmethyl)-4-(3,4,5-trimethoxycinnamoyl)piperazine
Formed from: Cinepazide

2-(2-Oxopyrrolidino)-1-phenylpentane
Formed from: 1-Phenyl-2-pyrrolidinopentane

1*H*-4-Oxoquinaldine
Formed from: Quinaldine
→ N-Acetylisatinic acid
Probably in: Arthrobacter G274

2-Oxoroxatidine acetate
Formed from: Roxatidine acetate

1′-Oxosafrole
→ 3′-(S-Glutathionyl)-2′,3′-dihydro-1′-oxosafrole
In: Mouse D242
Rat D242

5′-Oxosameride
Probably formed from: 5′-Hydroxysameride

4-Oxosertrindole
Formed from: Sertrindole

5-Oxosertrindole
Formed from: 5-Hydroxysertrindole

4′-Oxosirdalud
Probably formed from: 4′-Hydroxysirdalud

18-Oxostrychnine
Formed from: Strychnine

1-Oxosulofenur
Probably formed from: 1-Hydroxysulofenur

3-Oxosulofenur
Probably formed from: 3-Hydroxysulofenur

6′-Oxosulphoridazine
Formed from: Sulphoridazine
→ 6′-Oxosulphoridazine sulphoxide
Probably in: Dog G538, G539
Man G538, G539
Rat G538, G69

2′-Oxosulphoridazine sulphoxide
Formed from: Sulphoridazine sulphoxide

6′-Oxosulphoridazine sulphoxide
Probably formed from: 6′-Oxosulphoridazine

6-Oxotargetrin
Probably formed from: 6-Hydroxytargetrin

7-Oxotargetrin
Probably formed from: 7-Hydroxytargetrin
→ 4-(1-(5,6,7,8-Tetrahydro-3,5,5,8,8-pentamethyl-7-oxo-2-naphthalenyl)-1-ethenyl)benzoyltaurine
Probably in: Man J290

4-Oxo-α-tetralone
Formed from: *o*-Succinylbenzoic acid

9-Oxotetranorcloprostenol
Probably formed from: Tetranorcloprostenol

6′-Oxothioridazine
Formed from: Thioridazine

2-Oxo-4-(*p*-tolyl)-3-butenoic acid
→ *p*-Methylcinnamaldehyde
In: Yeast F279

3-Oxotriazolophthalazine-4-β-D-glucuronide
Formed from: Apiracohl

5-Oxo-5-(*p*-trifluoromethylphenyl)valeric acid
Probably formed from: 5-(2-Aminoethyl)oximino-5-(*p*-trifluoromethylphenyl)valeric acid

3-Oxo-1-(2,3,4-trimethoxybenzyl)piperazine
Formed from: Trimetazidine

2-Oxo-2-(3,4,5-trimethoxyphenyl)ethylamine
→ 3,4,5-Trimethoxymandelic acid
In: Rat J287

10-Oxotrimipramine
Formed from: 10-Hydroxytrimipramine

β-Oxo-L-tryptophanyl-L-leucine
Probably formed from: *threo*-β-Hydroxy-L-tryptophanyl-L-leucine

β-Oxotryptophol (1-(3-indolyl)-1-oxoethanol)
Probably formed from: Indole-3-glyoxal

5-Oxoviloxazine
Formed from: Viloxazine
→ 5′-Hydroxy-5-oxoviloxazine
Probably in: Dog A355

19′-Oxovinblastine
Probably formed from: Vinblastine

6-Oxovincamine
Probably formed from: 6α-Hydroxyvincamine

5′-Oxoxilobam
Probably formed from: 5′-Hydroxyxilobam
→ N-(2,6-Dimethylphenyl)-N′-(N-methylsuccinamido)urea
Probably in: Dog G938
Mouse G938
Rat G938
→ 1-(2-Hydroxymethyl-6-methylphenyl)-3-(1-methyl-5-oxopyrrolidinylidene)urea
Probably in: Mouse G938
→ 4-Hydroxy-5′-oxoxilobam
Probably in: Mouse G938

4-Oxoxylazine
Formed from: Xylazine
→ N-(2,6-Dimethylphenyl)thiourea
Probably in: Horse G504

5-Oxozaleplon
Formed from: Zaleplon

Oxprenolol (2-(2-hydroxy-3-isopropylamino-1-propoxy)-1-(2-propen-1-oxy)benzene)
→ 2-(3-Amino-2-hydroxy-1-propoxy)-1-(2-propen-1-oxy)benzene
In: Man D664
See also: A1181
→ 2-(2-Carboxy-2-hydroxyethoxy)-1-(2-propen-1-oxy)benzene
In: Man A288, D664
Rat D561
→ 2-(Carboxymethoxy)-1-(2-propen-1-oxy)benzene
In: Man D664
Rat D561
→ 2-(2,3-Dihydroxypropoxy)-1-(2-propen-1-oxy)benzene
In: Man D664
Rat D561
→ 3-(2-Hydroxy-3-isopropylamino-1-propoxy)-4-(2-propen-1-oxy)phenol
In: Man A3741, B78, D664
Rat A3741, B78, B139, B505
→ 4-(2-Hydroxy-3-isopropylamino-1-propoxy)-3-(2-propen-1-oxy)phenol
In: Man A3741, B78, D664
Rat A3741, B78, B139, B505
→ Oxprenolol-β-D-glucuronide
In: Man A288, C426, D664
→ 2-(2-Propen-1-oxy)phenol
In: Man A288, D664

Oxprenolol-β-D-glucuronide
Formed from: Oxprenolol

Oxy(bis(bis(*p*-fluorophenyl)(methyl)(silane)))
Formed from: Flusilazole

Oxybutynin (4-diethylaminobutynyl α-cyclohexylmandelate, ditropan)
→ α-Cyclohexylmandelic acid
In: Rat E294
→ Desethyloxybutynin
In: Man G313, J620, J871
Rat E294

6α-Oxycodol
Formed from: Oxycodone
→ 6α-Oxycodol-N-oxide
Probably in: Rabbit C355

6β-Oxycodol
Formed from: Oxycodone
→ 7β-Hydroxy-6β-oxycodol
Probably in: Rabbit C162

6α-Oxycodol-N-oxide
Probably formed from: 6α-Oxycodol

Oxycodone ((5α)-4,5-epoxy-14-hydroxy-3-methoxy-17-methylmorphinan-6-one)
Formed from: 14β-Hydroxycodeinone
→ 14-Hydroxydihydrocodeine
In: Rabbit A3677
→ 14-Hydroxydihydroisocodeine
In: Rabbit A3677
→ 14-Hydroxydihydromorphine
In: Rabbit A3677
→ Noroxycodone
In: Rabbit A3677, C355
→ 6α-Oxycodol
In: Rabbit C355
→ 6β-Oxycodol
In: Rabbit C355
See also: J744
→ Oxycodone-N-oxide
In: Rabbit A3677, C355
→ Oxymorphone
In: Rabbit C355

Oxycodone-N-oxide
Formed from: Oxycodone

Oxyfedrine ((−)-ω-(1-hydroxy-1-phenylisopropylamino)-3′-methoxypropiophenone)
→ Norephedrine
In: Man A2204
Rat A2232

Oxyfluorofen (2-chloro-1-(3-ethoxy-4-nitrophenoxy)-4-trifluoromethylbenzene)
→ 4-(2-Chloro-4-trifluoromethylphenoxy)-2-ethoxyaniline
In: Rat A2748
→ 5-(2-Chloro-4-trifluoromethylphenoxy)-2-nitrophenol
In: Rat A2748

Oxyluciferin (Renilla)
Formed from: Luciferin (Renilla)

Oxyluciferin (Watasenia)
Formed from: Luciferin (Watasenia)

6α-Oxymorphol
Formed from: Oxymorphone

6β-Oxymorphol
Formed from: Oxymorphone

Oxymorphone ((5α)-4,5-epoxy-3,14-dihydroxy-17-methylmorphinan-6-one)
Formed from: Oxycodone
→ 6α-Oxymorphol
In: Dog C604
Guinea pig C604
Man C604
Rabbit C604
Rat C604
→ 6β-Oxymorphol
In: Guinea pig C604
Man C604
Rabbit C604
→ Oxymorphone-β-D-glucuronide
In: Man H933
Monkey H861
Rat J97a

Oxymorphone-β-D-glucuronide
Formed from: Oxymorphone

Oxypertine
See: 1-(2-(5,6-Dimethyoxy-2-methyl-3-indolyl)ethyl)-4-phenylpiperazine

Oxyphenbutazone
See: Oxyphenylbutazone

Oxyphenylbutazone (oxyphenbutazone)
Formed from Phenylbutazone
→ *p*,α-Dihydroxyphenylbutazone
Probably in: Man C287
Rat A296, D353
→ 4-Hydroxyoxyphenbutazone
In: Sheep A2201

Oxyprothepin (4-(10,11-dihydro-8-(methylthio)dibenzo[*b*,*f*]thiepin-10-yl)-1-piperazinepropanol)
Formed from: Oxyprothepin-8-sulphoxide
→ Oxyprothepin-8-sulphoxide
In: Rat C466

Oxyprothepin-5,8-dioxide
Probably formed from: Oxyprothepin-8-sulphoxide
→ Oxyprothepin-5-sulphoxide-8-sulphone
Probably in: Rat C466

Oxyprothepin-8-sulphoxide
Formed from: Oxyprothepin
→ Oxyprothepin-8-sulphoxide
In: Rat C466
→ Oxyprothepin
In: Guinea pig H259
Mouse H259
Rabbit H259
Rat H259
→ Oxyprothepin-5,8-dioxide
Probably in: Rat C466

Oxyprothepin-5-sulphoxide-8-sulphone
Probably formed from: Oxyprothepin-5,8-dioxide

β-Oxysydnocarb
Formed from: Sydnocarb

Oxytetracycline ((4*S*,4a*R*,5*S*,5a*R*,6*S*,12a*S*)-4-(dimethylamino)-1,4,4a,5,5a,6,11,12a-octahydro-3,5,6,10,12,12a-hexahydroxy-6-methyl-1,11-dioxo-2-naphthacenecarboxamide)
Formed from acetate in Streptomyces C848, D694, D778
Formed from malonate in Streptomyces C815
→ N-Demethyloxytetracycline
In: Hen K669
→ 4-Epioxytetracycline
In: Hen K669

P

PABA
See: *p*-Aminobenzoic acid
Paclitaxel
See: Taxol
Paeonidin (3′,4,5,7-tetrahydroxy-3′-methoxyflavylium)
Formed from: Cyanidin
→ Paeonidin-3-glucoside
In: Cabbage A1733
See also: A2739
Paeonidin-3-*p*-coumaroylglucoside
Probably formed from: Paeonidin-3-glucoside
Paeonidin-3-glucoside
Formed from: Chrysanthemin
Paeonidin
→ Paeonidin-3-*p*-coumaroylglucoside
Probably in: Vitis H886
Paeoniflorin ((1a*R*-(1aα,2β,3aα,5α,5aα,5bα)-5b-((benzoyloxy)methyl)tetrahydro-5-hydroxy-2-methyl-2,5-methano-1*H*-3,4-dioxacyclobuta[*c*,*d*]penyalen-1α(2*H*)-yl-β-D-glucopyranoside)
→ Paeonimetaboline 1
In: Bacteroides E573
Bifidobacterium E573
Clostridium E573
Lactobacillus E573
Streptococcus E573
Paeonimetaboline 1 ((2*S*-(2α,3aβ,5α,7aβ,8*R*))-tetrahydro-3α-hydroxy-7α,8-dimethyl-2,5-methano-1,3-benzodioxol-6(3a*H*)-one; mixture of 7*R* and 7*S*)
Formed from: Paeoniflorin
Paeonol
See: 2′-Hydroxy-4′-methoxyacetophenone
Pafenolol (N-(2-(4-(2-hydroxy-3-((1-methylethyl)amino)propoxy)phenyl)ethyl)-N′-(1-methylethyl)urea)
→ α-Hydroxypafenolol
In: Man G574
Rat G574
Palaudine (3′-demethylpapaverine)
Formed from: Papaverine
Palicoside ((2*S*,3*R*,4*S*)-3-ethenyl-2-(β-D-glucopyranosyloxy)-3,4-dihydro-4-(((*S*)-2,3,4,9-tetrahydro-2-methyl-1*H*-pyrido[3,4-*b*]indol-1-yl)methyl)-2*H*-pyran-5-carboxylic acid)
→ Palicoside aglycone
In: Catharanthus K428
Strychnos K428
Palicoside aglycone
Formed from: Palicoside
Pallidine ((9α,13α)-5,6,8,14-tetradehydro-2-hydroxy-3,6-dimethoxy-17-methylmorphinan-7-one)
Formed from: Reticuline
Palmatine (5,6-dihydro-2,3,9,10-tetramethoxydibenzo[*a*,*g*]quinolizinium)
Formed from: Columbamine
Reticuline
Tetrahydropalmatine
→ Corydaline
In: Corydalis H585
→ Corydalubine
In: Corydalis A3684
→ 13-Methylpalmatine
In: Corydalis H585
→ Tetrahydropalmatine
Probably in: Corydalis A3684, G136, H585
Palmitoyl-(+)-catechin
→ (+)-Catechin

Probably in: Rat C312

2-Palmitoyl-1-(*m*-phenoxybenzoyl)glycerol
Probably formed from: *m*-Phenoxybenzoyl CoA

1-Palmityl-1-phenyl-2-(*p*-toluoyl)hydrazine
Probably formed from: Toluic acid phenylhydrazide

Pamatolol (methyl 2-(4-(2-hydroxy-3-isopropylaminopropoxy)phenyl) ethylcarbamate)
→ *p*-(2-Hydroxy-3-isopropylaminopropoxy)benzoic acid
In: Rat B342
→ *p*-(2-Hydroxy-3-isopropylaminopropoxy)phenylacetic acid
In: Dog B342
Man B342
Rat B342
→ Methyl 2-(4-(2-carboxy-2-hydroxyethoxy)phenyl)ethylcarbamate
In: Dog B342
Man B342
Rat B342
→ Methyl 2-hydroxy-2-(4-(2-hydroxy-3-isopropylaminopropoxy)phenyl) ethylcarbamate
In: Rat B342

Panomifene ((*E*)-3,3,3-trifluoro-1-(4-((2-(2-hydroxyethyl)amino)ethoxy)phenyl)-1,2-diphenylpropene)
→ 1-(4-(2-Aminoethoxy)phenyl)-3,3,3-trifluoro-1,2-diphenylpropene
In: Dog H858
Man H858
Rat H858
→ 4-Hydroxypanomifene
In: Dog H858
→ 3,3,3-Trifluoro-1-(*p*-(2-hydroxyethoxy) phenyl)-1,2-diphenylpropene
In: Mouse H858
Rat H858
→ 3,3,3-Trifluoro-1-(*p*-hydroxyphenyl)-1,2-diphenylpropene
In: Dog H858
Man H858
Mouse H858
Rat H858

Papaveraldine ((6,7-dimethoxy-1-isoquinolinyl)(3,4-dimethoxyphenyl) methanone)
Formed from: Papaverine

Papaverine (6,7-dimethoxy-1-veratrylisoquinoline)
Formed from: Norlaudanosine
Tetrahydropapaveroline
→ 4′-Demethylpapaverine
In: Cat A309
Cunninghamella A2027
Dog A309
Guinea pig A309, A2027
Man A3051
Rat A305, A308, A309, A2027
Silene D599
See also: A1276, E736
→ 6-Demethylpapaverine
In: Aspergillus A2027
Cat A309
Dog A309
Guinea pig A309, A2027
Man A3051
Rabbit A309
Rat A305, A308, A309, A2027
Sepedonium A2027
Silene D599
Streptomyces A2027
See also: A1276
→ 7-Demethylpapaverine
In: Cat A309
Dog A309
Guinea pig A309
Man A3051
Rabbit A309
Rat A305, A308, A309, A2027
Sepedonium A2027
Streptomyces A2027
See also: A1276
→ 4-(3,4-Dimethoxybenzyl)-6,7-dimethoxyisocoumarin
In: Nocardia C268
→ 3-(3,4-Dimethoxybenzyl)-2-hydroxy-5,6-dimethoxy-1-indenone
In: Nocardia C268
→ 3,4-Dimethoxyphenylacetic acid

In: Nocardia C268
→ 2-Hydroxy-4,5-dimethoxyphenylacetaldehyde
Probably in: Nocardia C268
→ Palaudine
In: Man A3051
→ Papaveraldine
In: Cinchona E765
Glycyrrhiza E736
Parthenociccus E765
Phaseolus E765
Ptelia E765
→ Papaverine-N-oxide
In: Ochrosia E765
Rabbit B38
→ Papaverinol
In: Glycyrrhiza E736
Thevetia E765

Papaverine-N-oxide
Formed from: Papaverine

Papaverinol (α-(3,4-dimethoxyphenyl)-6,7-dimethoxy-1-isoquinolinemethanol)
Formed from: Papaverine

Paracetamol
See: *p*-Acetamidophenol

D-Paranephrine (D-3-hydroxy-4-methoxy-N-methylphenethylamine)
Formed from: D-Adrenaline

L-Paranephrine (L-3-hydroxy-4-methoxy-N-methylphenethylamine)
Formed from: L-Adrenaline

Paraoxon
See: Diethyl 4-nitrophenyl phosphate

Parathion
See: Diethyl 4-nitrophenyl phosphorothioate

Parbendazole
See: Methyl 5(6*H*)-butyl-2-benzimidazolecarbamate

Pargyline (N-benzyl-N-methylpropargylamine)
→ Benzaldehyde
Probably in: Rat D320, D669
→ Demethylpargyline
In: Cunninghamella C68
Rat A3005, D320, D669, E643
→ N-Methylbenzylamine
In: Cunninghamella C68
Rat A3005, D320, D669, E643
See also: A809
→ Pargyline-N-oxide
In: Guinea pig H282
Mouse H282
Rabbit H282
Rat D320, D669, E645, H282

Pargyline-N-oxide
Formed from: Pargyline

Paspalic acid (8,9-didehydro-6-methylergoline-8-carboxylic acid)
Formed from: Elymoclavine

Patuletin (3,3′,4′,5,7-pentahydroxy-6-methoxyflavone)
Formed from: Quercetagetin
→ Patuletin-3-sulphate
In: Flaveria F243, G769

Patuletin-3,3′-disulphate
Formed from: Patuletin-3-sulphate

Patuletin-3,4′-disulphate
Formed from: Patuletin-3-sulphate

Patuletin-3-sulphate
Formed from: Patuletin
→ Patuletin-3,3′-disulphate
In: Flaveria F243
→ Patuletin-3,4′-disulphate
In: Flaveria F243

Patulin (4-hydroxy-4*H*-furo[3,2-*c*]pyran-2(6*H*)-one)
Formed from: Gentisyl alcohol

Paxilline
Formed from: L-Tryptophan

Pazinadone (2-(7-chloro-1,8-naphthyridin-2-yl)-3-((1,4-dioxa-8-aza*spiro*[4,5]dec-8-yl)carbonylmethyl)isoindolin-1-one)
→ 2-(7-Chloro-1,8-naphthyridin-2-yl)-3-((4-hydroxypiperidin-1-yl)carbonylmethyl)isoindolin-1-one
In: Dog G695

Pedunculagin (cyclic 2,3:4,6-bis((1*S*)-4,4′,5,5′,6,6′-hexahydroxy(1,1′-biphenyl)-2,2′-dicarboxylate)-D-glucose ester)
→ Pedunculagin-O-sulphate
In: Eubacterium G475

Pedunculagin-O-sulphate
Formed from: Pedunculagin

Pefloxacin (1-ethyl-6-fluoro-1,4-dihydro-4-oxo-7-(4-methyl-1-piperazinyl)quinoline-3-carboxylic acid)
→ Norfloxacin
In: Dog D116, D864
Man D116, D864, E90
Monkey D116
Mouse D864
Rabbit D864
Rat D116, D864
→ Oxopefloxacin
In: Dog D864
Man D864
Mouse D864
Rabbit D864
Rat D864
→ Pefloxacin-β-D-glucuronide
In: Dog D116, D864
Man D864
Mouse D864
Rabbit D864
Rat D116, D864
→ Pefloxacin-N-oxide
In: Dog D116
Man D864, E90
Mouse D864
Rabbit D864
Rat D116

Pefloxacin-β-D-glucuronide
Formed from: Pefloxacin

Pefloxacin-N-oxide
Formed from: Pefloxacin

Pelargonidin (3,4′,5,7-tetrahydroxyflavylium)
Probably formed from: Aromadendrin
3,4-*cis*-Leucopelargonidin
→ Pelargonidin-3-glucoside
In: Carrot E280
Red cabbage A1733
Silene B404, B407
Zea E120
See also: D575

Pelargonidin-3-acetylglucoside
Formed from: Pelargonidin-3-glucoside

Pelargonidin-3-(4-caffeoylrhamnosyl(1 → 6))glucoside-5-glucoside
Formed from: Pelargonidin-3-rhamnosylglucoside

Pelargonidin-3-glucoside
Formed from: Pelargonidin
→ Pelargonidin-3-acetylglucoside
In: Zinnia G996
→ Pelargonidin-3-rhamnosylglucoside
In: Silene B409

Pelargonidin-5-glucoside-3-rhamnosylglucoside
Formed from: Pelargonidin-3-rhamnosylglucoside

Pelargonidin-3-rhamnosylglucoside
Formed from: Pelargonidin-3-glucoside
→ Pelargonidin-3-(4-caffeoylrhamnosyl (1 → 6))glucoside-5-glucoside
In: Silene B408
→ Pelargonidin-5-glucoside-3-rhamnosylglucoside
In: Silene B408

α-Peltatin (5,8,8a,9-tetrahydro-10-hydroxy-5-(4-hydroxy-3,5-dimethoxyphenyl) furo[3′,4′:6,7]naphtho[2,3-*d*]-1,3-dioxol-6(5a*H*)-one)
Formed from: 4′-Demethyldeoxypodophyllotoxin
→ β-Peltatin
Probably in: Podophyllum G201

β-Peltatin (5,8,8a,9-tetrahydro-10-hydroxy-5-(3,4,5-trimethoxyphenyl) furo[3′,4′:6,7]naphtho[2,3-*d*]-1,3-dioxol-6(5a*H*)-one)
Formed from: Deoxypodophyllotoxin
Probably formed from: α-Peltatin

Pencycuron (1-(*p*-chlorobenzyl)-1-cyclopentyl-3-phenylurea)
→ 1-(*p*-Chlorobenzyl)-1-cyclopentyl-3-(3,4-dihydroxyphenyl)urea
Probably in: Rabbit C413
→ 1-(*p*-Chlorobenzyl)-1-cyclopentyl-3-(*p*-hydroxyphenyl)urea
In: Rabbit C413
→ 1-(*p*-Chlorobenzyl)-1-(3-hydroxycyclopentyl)-3-phenylurea (*cis* and *trans* isomers)
In: Rabbit C413

Penfluridol (1-(4,4-bis(4-fluorophenyl)butyl)-4-(4-chloro-3-(trifluoromethyl)phenyl)-4-piperidinol)
→ 4,4-Bis(*p*-fluorophenyl)butanoic acid

In: Dog A3915
Man A3915
Rabbit A3915
Rat A3915
See also: A3322
→ 4-(4-Chloro-3-trifluoromethylphenyl)-4-hydroxypiperidine
In: Dog A3926
Man A3926
Rabbit A3926
Rat A3926
See also: A3322, A3915
→ Hydroxypenfluridol
In: Rat A3322
→ Penfluridol-N-oxide
In: Rat A3322

Penfluridol-N-oxide
Formed from: Penfluridol

Penicillin G
See: Benzylpenicillin

Penicillin V
See: Phenoxymethylpenicillin

Peniophorin (5-hydroxy-3,4-bis(4-hydroxyphenyl)-7-phenyl-2,6-benzofurandione)
Formed from: 4-Hydroxypulvinic acid
L-Phenylalanine
D-Tyrosine

Peniophoronin (9-hydroxy-1-(4-hydroxyphenyl)-4-phenyl-5*H*-benzofuro[5,4-*c*][1]benzopyran-2,5-(6*H*)-dione)
Formed from: 4-Hydroxypulvinic acid
L-Phenylalanine
D-Tyrosine

Penitrem A
Formed from: L-Tryptophan

Pentabromophenol
→ Tetrabromoquinol
Probably in: Mycobacterium G676

Pentachloroaniline
Formed from: Pentachloronitrobenzene
Probably formed from: Pentachloronitrosobenzene
→ Pentachloroaniline sulphamate
Probably in: Rat C794
→ 2,3,4,5-Tetrachloroaniline
Probably in: Allium B44
Rhesus monkey A3894, A3984
Goat C749
Rat C789, C794
Sheep C749
→ 2,3,4,6-Tetrachloroaniline
Probably in: Rat C789
→ 2,3,5,6-Tetrachloroaniline
Probably in: Rat C789

Pentachloroaniline sulphamate
Formed from: Pentachloroaniline

Pentachloroanisole
Formed from: Pentachlorophenol
→ Pentachlorophenol
In: Rainbow trout A3569
→ 2,3,4,5-Tetrachloroanisole
Probably in: Aeromonas A3805
Corynebacteria A3805
Cytophaga A3805
→ 2,3,4,6-Tetrachloroanisole
Probably in: Aeromonas A3805
Corynebacteria A3805
→ 2,3,5,6-Tetrachloroanisole
Probably in: Aeromonas A3805
Cytophaga A3805

Pentachlorobenzene
Formed from: N-Acetyl-S-(pentachlorophenyl)-L-cysteine
N-Acetyl-S-(pentachlorophenyl)-L-cysteine sulphoxide
Hexachlorobenzene
Lindane
Pentachloronitrobenzene
Pentachlorothiophenol
→ Pentachlorophenol
In: Monkey B32
Rabbit A1627
Rat A2626, A2872, H205
See also: A1563, A1950, A2444, A3894, B176
→ Pentachlorothiophenol
In: Rat H205
→ 1,2,3,4-Tetrachlorobenzene
In: Monkey B32
Rat A2626
See also: C881
→ 1,2,3,5-Tetrachlorobenzene
Probably in: Rat C881

→ 2,3,4,5-Tetrachlorophenol
In: Monkey B32
Rabbit A1627
Rat A2872
→ 2,3,5,6-Tetrachlorophenol
In: Monkey B32
→ 2,4,6-Trichlorophenol
In: Rat A2626

2,2′,3′,4,5-Pentachlorobiphenyl
→ 2,4,5-Trichlorobenzoic acid
In: Acinetobacter B125
Alcaligenes B125

2,2′,3,5′,6-Pentachlorobiphenyl
→ 2,2′,3′,5,6′-Pentachloro-3-hydroxybiphenyl
Probably in: Mouse A2456, A2529
Rat A2456, A2529
Quail A2456, A2529
→ 2,2′,3,5′,6-Pentachloro-4-hydroxybiphenyl
In: Quail A2456, A2529
→ 2,2′,3,5′,6-Pentachloro-5-hydroxybiphenyl
In: Mouse A2456, A2529
Rat A2456, A2529

2,2′,4,5,5′-Pentachlorobiphenyl
→ 2,2′,4′,5,5′-Pentachloro-3,4-dihydro-3,4-dihydroxybiphenyl
In: Rat A2112
→ 2,2′,4′,5,5′-Pentachloro-3-hydroxybiphenyl
In: Mouse A2166
Rat A2112, J278
See also: A2578
→ 2,2′,4′,5,5′-Pentachloro-4-hydroxybiphenyl
In: Rat J278
→ 2,2′,4′,5,5′-Pentachloro-3-methylthiobiphenyl
In: Rat J278
→ 2,2′,4′,5,5′-Pentachloro-4-methylthiobiphenyl
In: Rat J278

2,3,3′,4,4′-Pentachlorobiphenyl
→ 2,3,3′,4,4′-Pentachloro-2-hydroxybiphenyl
In: Mink H86
Mouse H86
→ 2,3,3′,4,4′-Pentachloro-5-hydroxybiphenyl
In: Mink H86
Mouse H86
→ 2′,3,3′,4,4′-Pentachloro-5-hydroxybiphenyl
In: Mink H86
Mouse H86
→ 2,3,3′,4′,5-Pentachloro-4-hydroxybiphenyl
In: Mink H86
Mouse H86
→ 2′,3,3′,4′,5-Pentachloro-4-hydroxybiphenyl
In: Mink H86
Mouse H86

2,3,4,5,6-Pentachlorobiphenyl
→ 2′,3′,4′,5′,6′-Pentachloro-3-hydroxybiphenyl
In: Rat A1444
→ 2′,3′,4′,5′,6′-Pentachloro-4-hydroxybiphenyl
In: Rat A1444

3,3′,4,4′,5-Pentachlorobiphenyl
→ 3,3′,4,4′,5′-Pentachloro-4-hydroxybiphenyl
In: Rat K200
→ 3,3′,4,4′,5′-Pentachloro-5-hydroxybiphenyl
In: Rat K200
→ 3,3′,4,4′,5′-Pentachloro-5-methylthiobiphenyl
In: Rat K200
→ 3,3′,4,4′,5′-Pentachloro-6-methylthiobiphenyl
In: Rat K200
→ 2,3′,4′5-Tetrachloro-3-hydroxybiphenyl
In: Guinea pig H942
Hamster H942
→ 3,3′,4′,5-Tetrachloro-4-hydroxybiphenyl
In: Hamster H942
→ 3,3′,4′,5′-Tetrachloro-4-hydroxybiphenyl
In: Rat K200

Pentachlorocyclohex-1-ene
→ *m*-Dichlorobenzene
In: Pea A1479, A1578
Zea A1479, A1578
→ Pentachlorophenol
Rat A2626
→ 1,2,3,4-Tetrachlorobenzene
In: Zea A1578
→ 1,2,3,5-Tetrachlorobenzene
In: Rat A3774
→ 1,2,4,5-Tetrachlorobenzene
In: Pea A1479, A1578
Rat A3774

Zea A1479, A1578
→ 1,2,3-Trichlorobenzene
In: Pea A1578
→ 1,2,4-Trichlorobenzene
In: Pea A1479, A1578
Zea A1479, A1578
→ 2,3,5-Trichlorophenol
In: Pea A1479
Zea A1479, A1578
→ 2,4,5-Trichlorophenol
In: Pea A1479, A1578
Zea A1479, A1578
→ 2,4,6-Trichlorophenol
In: Pea A1578
Rat A2626

1,2,3,7,8-Pentachlorodibenzo-*p*-dioxin
Formed from: 2,4,5-Trichlorophenol

1,2,4,7,8-Pentachlorodibenzo-*p*-dioxin
Formed from: 2,4,5-Trichlorophenol

2,2′,4′,5,5′-Pentachloro-3,4-dihydro-3,4-dihydroxybiphenyl
Formed from: 2,2′,4,5,5′-Pentachlorobiphenyl
→ 2,2′,4′,5,5′-Pentachloro-3,4-dihydroxybiphenyl
Probably in: Rat A2112

2,2′,3,5′,6-Pentachloro-3′,4-dihydroxybiphenyl
Probably formed from: 2,2′,3′,5,6′-Pentachloro-3-hydroxybiphenyl

2,2′,3,5′,6-Pentachloro-3′,5-dihydroxybiphenyl
Probably formed from: 2,2′,3,5′,6-Pentachloro-5-hydroxybiphenyl

2,2′,3,5′,6-Pentachloro-4,4′-dihydroxybiphenyl
Probably formed from: 2,2′,3′,5,6′-Pentachloro-4-hydroxybiphenyl

2,2′,3,5′,6-Pentachloro-4′,5-dihydroxybiphenyl
Probably formed from: 2,2′,3′,5,6′-Pentachloro-4-hydroxybiphenyl

2,2′,4′,5,5′-Pentachloro-3,4-dihydroxybiphenyl
Formed from: 2,2′,4′,5,5′-Pentachloro-3,4-dihydro-3,4-dihydroxybiphenyl

2′,3′,4′,5′,6′-Pentachloro-3,4-dihydroxybiphenyl
Probably formed from: 2′,3′,4′,5′,6′-Pentachloro-4-hydroxybiphenyl

2,2′,3′,5,5′-Pentachloro-4-hydroxybiphenyl
Formed from: 2,2′,3,3′5,5′-Hexachlorobiphenyl

2,2′,3′,5,6′-Pentachloro-3-hydroxybiphenyl
Probably formed from: 2,2′,3′,5,6′-Pentachlorobiphenyl
→ 2,2′,3,5′,6-Pentachloro-3′,4-dihydroxybiphenyl
Probably in: Quail A2456, A2529

2,2′,3′,5,6′-Pentachloro-4-hydroxybiphenyl
Formed from: 2,2′,3′,5,6′-Pentachlorobiphenyl
→ 2,2′,3,5′,6-Pentachloro-4,4′-dihydroxybiphenyl
Probably in: Quail A2456, A2529
→ 2,2′,3,5′,6-Pentachloro-4′,5-dihydroxybiphenyl
Probably in: Quail A2456, A2529

2,2′,3′,5,6′-Pentachloro-5-hydroxybiphenyl
Formed from: 2,2′,3′,5,6′-Pentachlorobiphenyl
→ 2,2′,3,5′,6-Pentachloro-3′,5-dihydroxybiphenyl
Probably in: Mouse A2456, A2529
Rat A2456, A2529

2,2′,4′,5,5′-Pentachloro-3-hydroxybiphenyl
Formed from: 2,2′,4,4′,5,5′-Hexachlorobiphenyl
2,2′,4,5,5′-Pentachlorobiphenyl

2,2′,4′,5,5′-Pentachloro-4-hydroxybiphenyl
Formed from: 2,2′,4,4′,5,5′-Hexachlorobiphenyl
2,2′,4,5,5′-Pentachlorobiphenyl

2′,3,3′,4,4′-Pentachloro-2-hydroxybiphenyl
Formed from: 2′,3,3′,4,4′-Pentachlorobiphenyl

2,3,3′,4,4′-Pentachloro-5-hydroxybiphenyl
Formed from: 2,3,3′,4,4′-Pentachlorobiphenyl

2′,3,3′,4,4′-Pentachloro-5-hydroxybiphenyl
Formed from: 2′,3,3′,4,4′-Pentachlorobiphenyl

2,3,3′,4′,5-Pentachloro-4-hydroxybiphenyl
Formed from: 2,3,3′,4,4′-Pentachlorobiphenyl

2′,3,3′,4′,5-Pentachloro-4-hydroxybiphenyl
Formed from: 2,3,3′,4,4′-Pentachlorobiphenyl

2′,3′,4′,5′,6′-Pentachloro-3-hydroxybiphenyl
Formed from: 2,3,,4,5,6-Pentachlorobiphenyl

2′,3′,4′,5′,6′-Pentachloro-4-hydroxybiphenyl
Formed from: 2,3,,4,5,6-Pentachlorobiphenyl
→ 2′,3′,4′,5′,6′-Pentachloro-3,4-dihydroxybiphenyl
Probably in: Rat A1444

2′,4,4′,5,5′-Pentachloro-2-hydroxybiphenyl
Formed from: 2,2′,4,4′,5,5′-Hexachlorobiphenyl
Probably formed from: 2′,3,4,4′,5,5′-Hexachloro-2-hydroxybiphenyl

3,3′,4,4′,5′-Pentachloro-5-hydroxybiphenyl
Formed from: 3,3′,4,4′,5-Pentachlorobiphenyl

3,3′,4′,5,5′-Pentachloro-4-hydroxybiphenyl
Formed from: 3,3′,4,4′,5-Pentachlorobiphenyl

2,2′,4′,5,5′-Pentachloro-3-methylsulphonylbiphenyl
Probably formed from: 2,2′,4′,5,5′-Pentachloro-3-methylthiobiphenyl

2,2′,4′,5,5′-Pentachloro-4-methylsulphonylbiphenyl
Probably formed from: 2,2′,4′,5,5′-Pentachloro-4-methylthiobiphenyl

3,3′,4′,5,5′-Pentachloro-5-methylsulphonylbiphenyl
Probably formed from: 3,3′,4′,5,5′-Pentachloro-5-methylthiobiphenyl

2,2′,4′,5,5′-Pentachloro-3-methylthiobiphenyl
Formed from: 2,2′,4′,5,5′-Pentachlorobiphenyl
→ 2,2′,4′,5,5′-Pentachloro-3-methylsulphonylbiphenyl
Probably in: Rat J278

2,2′,4′,5,5′-Pentachloro-4-methylthiobiphenyl
Formed from: 2,2′,4′,5,5′-Pentachlorobiphenyl
→ 2,2′,4′,5,5′-Pentachloro-4-methylsulphonylbiphenyl
Probably in: Rat J278

2,2′,4′,5,5′-Pentachloro-5-methylthiobiphenyl
Formed from: 2,2′,4′,5,5′-Pentachlorobiphenyl
→ 2,2′,4′,5,5′-Pentachloro-5-methylsulphonylbiphenyl
Probably in: Rat J278

2,2′,4′,5,5′-Pentachloro-6-methylthiobiphenyl
Formed from: 2,2′,4′,5,5′-Pentachlorobiphenyl

Pentachloronitrobenzene (terraclor)
→ S-(Aminotetrachlorophenyl)glutathione
In: Rat J445
→ Pentachloroaniline
In: Cress C777
Goat C794
Golden orfe B795
Monkey A3894, A3984
Onion B44
Rabbit A3822
Rat B176, J445
Sheep C784
Tetrahymena C136
→ Pentachlorobenzene
In: Monkey A3894, A3984
Onion B44
Rat B176
→ Pentachlorophenol
In: Golden orfe B795
Onion B44
→ S-(Pentachlorophenyl)glutathione
In: Arachis B816, B894
Rat J445
See also: 3822, B176
→ Pentachlorothiophenol
In: Golden orfe B795
Monkey A3984
Onion B44
Rat B176, J445
See also: A3894, C136, C777
→ 2,3,4,5-Tetrachloronitrobenzene
In: Rat B176
→ S-(Tetrachloronitrophenyl)glutathione
In: Arachis B816, B894
→ 2,3,4,5-Tetrachlorothiophenol
In: Cress C777
→ Trichlorobis(methylthio)aniline
In: Rat J445

Pentachloronitrosobenzene
→ Pentachloroaniline
Probably in: Rat C789

Pentachlorophenol
Formed from: Hexachlorobenzene
Pentachloroanisole
Pentachlorobenzene
γ-Pentachlorocyclohexene
Pentachloronitrobenzene
Veratrylglycerol β-(pentachlorophenyl) ether
→ 4-(2,6-Dimethoxy-4-(2,6-dimethoxy-4-(2,6-dimethoxy-4-(pentachlorophenoxy) phenoxy)phenoxy)phenoxy)-2,6-dimethoxyphenol
In: Rhizoctonia D690

→ 2,6-Dimethoxy-4-(pentachlorophenoxy)phenol
In: Rhizoctonia D690
→ 4-(2,6-Dimethoxy-4-(pentachlorophenoxy)phenoxy)-2,6-dimethoxyphenol
In: Rhizoctonia D690
→ 4-(2,6-Dimethoxy-4-(2,6-dimethoxy-4-(pentachlorophenoxy)phenoxy)phenoxy)-2,6-dimethoxyphenol
In: Rhizoctonia D690
→ Octachlorodibenzo-*p*-dioxin
In: Horseradish H837
→ Pentachloroanisole
In: Abalone G237
Aeromonas A3805
Arthrobacter C527
Bacillus A407
Brevibacterium A3805
Corynebacterium A3805
Cytophaga A3805
Minnow C856
Mycobacterium A3330, C547, D137
Onion B44
See also: B176, B795
→ Pentachlorophenyl acetate
In: Alcaligenes A3805
Arthrobacter A3805
Azotobacter A3805
Corynebacterium A3805
Cytophaga A3805
Pseudomonas A3805
→ Pentachlorophenyl-β-D-glucoside
In: Abalone H27
Glycine G247, H402
→ Pentachlorophenyl-β-D-glucuronide
In: Abalone G237
Bass H24
Goldfish A2960, A2963, B231
Man D316, F840
Minnow C856
Rainbow trout A3569, J582
Rat A2626, A3182, A3568, D316
Sea urchin G950
Topsmelt H93
Zebra fish E813
→ Pentachlorophenyl sulphate
In: Abalone G237, H27
Bass H24
Goldfish A2962, A2963, B231
Guinea pig E456
Oyster H27
Rainbow trout J582
Sea urchin G950
Topsmelt H93
Zebra fish E813
→ Tetrachlorocatechol
In: Mycobacterium C547, D137
Pseudomonas A3108
Rat F891
See also: A3805
→ Tetrachloro-4-(methylthio)phenol
Probably in: Rat H205
→ 2,3,4,5-Tetrachlorophenol
In: Alcaligenes A3805
Azotobacter A3805
Cytophaga A3805
Desulfitobacterium J232, K247
Microorganism G331, G667
Rat A2626, F891
See also: B176, E168, H205
→ 2,3,4,6-Tetrachlorophenol
In: Rat F891
→ 2,3,5,6-Tetrachlorophenol
In: Aeromonas A3805
Alcaligenes A3805
Azotobacter A3805
Cytophaga A3805
Flavobacterium A3805
Rat F891
See also: A3984, G471, G676
→ Tetrachloroquinol
In: Abalone G237, H27
Arthrobacter K756
Bass H24
Flavobacterium E265, G641, G746
Man D901
Mycobacterium C547, D137
Pseudomonas A3108

Rat A3182, A3568, A3660, C44, D901, E54, F891
Rhodococcus E324
See also: A407, A416, A2784, A2872, D552, H205, H769
→ Tetrachlororesorcinol
In: Rat F891

Pentachlorophenyl acetate
Formed from: Pentachlorophenol

S-(Pentachlorophenyl)-L-cysteine
Probably formed from: S-(Pentachlorophenyl)-L-glutamyl-L-cysteine
→ N-Acetyl-S-(pentachlorophenyl)-L-cysteine
Probably in: Rat J445
→ N-Malonyl-S-(pentachlorophenyl)-L-cysteine
Probably in: Arachis B816, B894

Pentachlorophenyl-β-D-glucoside
Formed from: Pentachlorophenol
→ Pentachlorophenyl-6″-O-malonyl-β-D-glucoside
In: Glycine G247

Pentachlorophenyl-β-D-glucuronide
Formed from: Pentachlorophenol

S-(Pentachlorophenyl)-L-glutamyl-L-cysteine
Probably formed from: S-(Pentachlorophenyl)glutathione
→ S-(Pentachlorophenyl)-L-cysteine
Probably in: Arachis B816
Rat J445

S-(Pentachlorophenyl)glutathione
Formed from: Pentachloronitrobenzene
Probably formed from: Pentachlorosulphinylanisole
→ S-(Pentachlorophenyl)-L-glutamyl-L-cysteine
Probably in: Arachis B816
Rat J445
→ S,S′-(Tetrachlorophenylene)diglutathione
Probably in: Arachis B816
Glycine B894

Pentachlorophenyl-6″-O-malonyl-β-D-glucoside
Formed from: Pentachlorophenyl-β-D-glucoside

Pentachlorophenyl sulphate
Formed from: Pentachlorophenol

Pentachlorosulphinylanisole
Formed from: Pentachlorothioanisole
→ S-(Pentachlorophenyl)glutathione
Probably in: Chicken G71
→ Pentachlorosulphonylanisole
Probably in: Allium B44
Rat G688

Pentachlorosulphonylanisole
Probably formed from: Pentachlorosulphinylanisole
→ 2,4,5,6-Tetrachloro-1,3-bis(S-glutathionyl)benzene
Probably in: Monkey G71

Pentachlorothioanisole
Formed from: Pentachlorothiophenol
→ N-Acetyl-S-(tetrachloro-4-methylthiophenyl)-L-cysteine
In: Rat B581, B936
→ Pentachlorosulphinylanisole
In: Rat G688
See also: B44
→ Pentachlorothiophenol
In: Rat C881
→ 2,3,5,6-Tetrachloro-1,4-bis(methylthio) benzene
In: Rat B936, F923
See also: A3984, B176, B581, C881, G688, J445
→ S-(Tetrachloromethylthiophenyl) glutathione
In: Rat C786, G688
See also: F894
→ Tetrachloro-4-(methylthio)thiophenol
Probably in: Rat G688
→ 2,3,5,6-Tetrachlorothioanisole
In: Rat C881
See also: G688

Pentachlorothiophenol
Formed from: N-Acetyl-S-(pentachlorophenyl)-L-cysteine
Hexachlorobenzene
Pentachlorobenzene
Pentachlorothioanisole
Probably formed from: N-Acetyl-S-(pentachlorophenyl)-L-cysteine sulphoxide
→ Pentachlorobenzene
In: Rat C881

→ Pentachlorothioanisole
In: Mycobacterium A3330
Rat C881
Tetrahymena E96
See also: A3894, A3984, B42, B44, B795, C787
→ Pentachlorothiophenyl-β-D-glucuronide
In: Rat H205, J445
→ 1,2,4,5-Tetrachlorobenzene
In: Rat C881
→ Tetrachloro(methylthio)phenol
Probably in: Rat H205
→ 2,3,4,5-Tetrachlorothioanisole
In: Rat C881
→ 2,3,4,6-Tetrachlorothioanisole
In: Rat C881
→ 2,3,4,5-Tetrachlorothiophenol
Probably in: Rat A2784, B176
→ 2,3,5,6-Tetrachlorothiophenol
In: Rat C881

Pentachlorothiophenyl-β-D-glucuronide
Formed from: Pentachlorothiophenol

Pentadecylbenzene
→ Benzoic acid
In: Candida A1206
→ Cinnamic acid
In: Candida A1206

Pentafluorobenzaldehyde
→ Pentafluoromandelonitrile
In: Almond K6

Pentafluoromandelonitrile
Formed from: Pentafluorobenzaldehyde

Pentafluorophenol
Formed from: Hexafluorobenzene
→ Pentafluorophenyl sulphate
In: Rat H769
→ Tetrafluorocatechol
In: Trichosporon H420
→ Tetrafluoroquinol
Probably in: Mycobacterium G676

Pentafluorophenyl sulphate
Formed from: Pentafluorophenol

1,2,3,4,6-Pentagalloylglucose
Formed from: 1,2,3,6-Tetragalloylglucose

Pentahydroxybenzene
Formed from: 1,2,3,5-Tetrahydroxybenzene
1,2,4,5-Tetrahydroxybenzene
→ 1,2,3,5-Tetrahydroxybenzene
Probably in: Pelobacter G601

2′,3′,4,4′,6′-Pentahydroxychalcone
Formed from: Caffeoyl CoA
→ Aureusidin
In: Antirrhinum K659
→ Eriodictyol
Probably in: Silene B404

***cis*-3,3′,4′,5,7-Pentahydroxyflavan**
See: (+)-Catechin

***trans*-3,3′,4′,5,7-Pentahydroxyflavan**
See: (+)-Epicatechin
(−)-Epicatechin

3′,4,4′,5,7-Pentahydroxyflavan
See: Luteoforol

3′,4′,5,7-Pentahydroxyflavanone
See: Dihydroquercetin

3,3′,4′,5′,7-Pentahydroxyflavanone
See: Dihydrorobinetin

***cis*-3,4,4′,5,7-Pentahydroxyflavan**
See: 3,4-*cis*-Leucopelargonidin

3′,4′,5,5′,7-Pentahydroxyflavanone
Formed from: Eriodictyol
→ Dihydromyricetin
In: Petunia D772

2′,3,4′,5,7-Pentahydroxyflavone
See: Morin

3,3′,4′,5,7-Pentahydroxyflavone
See: Quercetin

3,4′,5,7,8-Pentahydroxyflavone
See: 8-Hydroxykaempferol

3,3′,4′,5,7-Pentahydroxyflavylium
See: Cyanidin

3,3′,5,5′,7-Pentahydroxy-4′-methoxyflavan
Formed from: (−)-Epigallocatechin

3,4,5,5′,7-Pentahydroxy-3′-methoxyflavone
Probably formed from: Myricetin

3,3′,4′,5,7-Pentahydroxy-6-methoxyflavone
See: Patuletin

3,3′,4′,5,7-Pentahydroxy-8-methoxyflavone
See: 8-Methoxyquerctin

3,4′,5,5′,7-Pentahydroxy-3′-methoxyflavone
Probably formed from: Myricetin

3′,4′,5,5′,7-Pentahydroxy-3-methoxyflavone
Probably formed from: Myricetin

3′,4′,5,7,8-Pentahydroxy-3-methoxyflavone
See: 8-Hydroxy-3-O-methylquercetin
4′,5,6,7,8-Pentamethoxyflavone
See: Tangeretin
Pentamethylacetophenone imine
→ Pentamethylacetophenone oxime
In: Rabbit D955
Pentamethylacetophenone oxime
Formed from: Pentamethylacetophenone imine
Pentamethylpararosaniline
Formed from: Gentian violet
Pentamidine (1,5-bis(4-amidinophenoxy)pentane)
→ *p*-Amidinophenol
In: Rat G682
→ 1-(*p*-Amidinophenoxy)-5-(*p*-(hydroxyamidino)phenoxy)pentane
In: Rat G682
See also: J410
→ 5-(*p*-Amidinophenoxy)pentanol
In: Rat G682
→ 1,5-Bis(4-amidinophenoxy)-2-pentanol
In: Rat G682, J410
→ 1,5-Bis(4-amidinophenoxy)-3-pentanol
In: Rat G682, J410
→ N-Hydroxypentamidine
In: Rat F706, J410
Pentazocine ((2*R*,6*R*,11*R*)-*rel*-1,2,3,4,5,6-hexahydro-6,11-dimethyl-3-(3-methyl-2-butenyl-2,6-methano-3-benzazocin-8-ol)
→ 1,2,3,4,5,6-Hexahydro-8-hydroxy-α,6(eq),11(ax)-trimethyl-2,6-methano-3-benzazocine-3-(*trans*-2-buten-4-ol)
In: Mouse A1603
Rat B30
See also: A2134
→ Hydroxypentazocine
Probably in: Rat A2134, B30
→ Norpentazocine
In: Mouse A1603
Rat A2134, B30
→ Pentazocine-β-D-glucuronide
In: Rat A2134
Pentazocine-β-D-glucuronide
Formed from: Pentazocine
***p*-(Pent-1-en-1-yl)phenol**
Formed from: *p*-*n*-Pentylphenol
(1*E*)-1-Penten-4-yne-1,5-diylbis(2-hydroxy-4,1-phenylene)bis-β-D-glucopyranoside
See: Hypoxoside
Pentopril (1-(4-ethoxycarbonyl-2-methyl-2(*R*),4(*R*)-pentanoyl)-2,3-dihydro-2(*S*)-indole-2-carboxylic acid)
→ Desethylpentopril
In: Man E1
***o*-*n*-Pentylbenzoic acid**
→ (*S*)-3-*n*-Butylphthalide
In: Aspergillus J486
Pseudomonas J486
***p*-*n*-Pentylbenzoic acid**
→ *p*-*n*-Pentylbenzoyl CoA
In: Beef H740
***p*-*n*-Pentylbenzoyl CoA**
Formed from: *p*-*n*-Pentylbenzoic acid
7-Pentyloxyphenoxazone
See: Resorufin pentyl ether
***p*-*n*-Pentylphenol** (*p*-*n*-amylphenol)
→ 1-(*p*-Hydroxyphenyl)-1-pentanol
In: Pseudomonas F381
→ *p*-(Pent-1-en-1-yl)phenol
In: Penicillium J836
→ *p*-*n*-Pentylphenyl sulphate
In: Man K405
***p*-*n*-Pentylphenyl sulphate**
Formed from: *p*-*n*-Pentylphenol
5-Pentylresorcinol
See: Olivetol
Pentyl salicylate
→ Salicylic acid
In: Guinea pig D144
Perazine (10-(3-(4-methylpiperazinyl)propyl) phenothiazine)
→ Bisdemethylpromazine
In: Mouse A2265
Rabbit A2265
Rat A69, A2265
→ Demethylperazine
In: Rat A2183
→ 3-Hydroxyperazine
In: Rat A2183
→ Perazine sulphoxide
In: Rat A2183
→ N-(γ-Phenothiazinyl(10)propyl) ethylenediamine

In: Mouse A2265
Rabbit A2265
Rat A2265

Perazine sulphoxide
Formed from: Perazine

Perazin-3-yl-β-D-glucuronide
Formed from: 3-Hydroxyperazine

Perfluidone (1,1,1-trifluoro-N-(2-methyl-4-(phenylsulphonyl)phenyl) methanesulphonamide)
→ 1,1,1-Trifluoro-N-(4-((3-hydroxyphenyl)sulphonyl)-2-methylphenyl)methanesulphonamide
In: Arachis A1949
Cow A2052
→ 1,1,1-Trifluoro-N-(4-((4-hydroxyphenyl)sulphonyl)-2-methylphenyl)methanesulphonamide
In: Cow A2052

Perisoxal
See: 3-(1-Hydroxy-2-piperidinoethyl)-5-phenylisoxazole

Perisoxal-β-D-glucuronide
Formed from: 3-(1-Hydroxy-2-piperidinoethyl)-5-phenylisoxazole

[1*R*,*cis*]-Permethrin (3-phenoxybenzyl [1*R*,*cis*]-3-(2,2-dichlorovinyl)-2,2-dimethylcyclopropanecarboxylate)
→ 2′-Hydroxy-[1*R*,*cis*]-permethrin
In: Mouse A3804
Rat A1934, A3195, A3804
→ 4′-Hydroxy-*cis*-permethrin
In: Cabbage looper A3804
Carp A3853
Cotton A3195
Cow A3194
Housefly A3804
Mouse A3804
Rat A1934, A3804
Trout A3853
→ 6-Hydroxy-*cis*-permethrin
In: Mouse A3804
→ *c*-Hydroxy-*cis*-permethrin
In: Housefly A3804
Mouse A3804
Rat A3804
→ *t*-Hydroxy-*cis*-permethrin
In: Cabbage looper A3804
Carp A3853
Cotton A3195
Cow A3194
Housefly A3804
Mouse A3804
Rat A3804
Trout A3853
→ *m*-Phenoxybenzyl alcohol
In: Bean A3195
Bollworm A3515
Cabbage looper A3804
Carp A3853
Cotton A3195
Cow A3194
Housefly A3804
Mouse A3804
Rat A1823, A1934, A3804
Tobacco budworm A3515
Trout A3853
See also: A3511, K511

[1*R*,*trans*]-Permethrin (3-phenoxybenzyl [1*R*,*trans*]-3-(2,2-dichlorovinyl)-2,2-dimethylcyclopropanecarboxylate)
→ 2′-Hydroxy-*trans*-permethrin
In: Cotton A3195
→ 4′-Hydroxy-*trans*-permethrin
In: Carp A3853
Cotton A3195
Cow A3194
Hen A3511
Mouse A3804
Rat A3804
Trout A3853
→ 6-Hydroxy-*trans*-permethrin
In: Mouse A3804
→ *c*-Hydroxy-*cis*-permethrin
In: Carp A3853
Cow A3194
Mouse A3804
Rat A3804
Trout A3853
→ *t*-Hydroxy-*cis*-permethrin
In: Carp A3853
Cotton A3195
Cow A3194

Hen A3511
Mouse A3804
Rat A3804
Trout A3853
→ *m*-Phenoxybenzyl alcohol
In: Bean A3195
Bollworm A3515
Carp A3853
Cotton A3195
Cow A3194
Mouse A3804
Rat A1823, A1934, A3804
Tobacco budworm A3515
Trout A3853
See also: A3511, G679
→ 3-Phenoxybenzyl 2-carboxy-[1*R*,*trans*]-3-(2,2-dichlorovinyl)-2-methylcyclopropanecarboxylate
In: Mouse A3804

[1*RS*,*cis*]-Permethrin
→ 2′-Hydroxy-[1*RS*,*cis*]-permethrin
In: Rat A1934
→ 4′-Hydroxy-[1*RS*,*cis*]-permethrin
In: Rat A1934
→ *m*-Phenoxybenzyl alcohol
In: Rat A1934

[1*RS*,*trans*]-Permethrin
→ *m*-Phenoxybenzyl alcohol
In: Rat A1934

Peroxyphenylacetic acid
→ Benzyl alcohol
In: Rabbit C891

Perphenazine (4-(3-(2-chloro-10*H*-phenothiazin-10-yl)propyl)-1-piperazineethanol)
→ Bisdemethylpromazine
Probably in: Rat A67
→ N-(γ-(2-Chloro-10-phenothiazinyl)propyl)ethylenediamine
In: Rat A576, A1865
→ N-(γ-(2-Chloro-10-phenothiazinyl)propyl)-N′-hydroxymethylethylenediamine
In: Rat A1865
→ Perphenazine sulphoxide
In: Rat A67

Perphenazine sulphoxide
Formed from: Perphenazine
→ Norprochlorperazine sulphoxide
Probably in: Rat A67

Pethidine
See: Meperidine

Petunidin (2-(3,4-dihydroxy-5-methoxyphenyl)-3,5,7-trihydroxy-1-benzopyrilium)
Probably formed from: Delphinidin
→ Malvidin
Probably in: Petunia A2740

Petunidin-3-(*p*-coumaroyl)rutinoside
→ Petunidin-3-(*p*-coumaroyl)rutinoside-5-O-glucoside
In: Petunia D198

Petunidin-3-(*p*-coumaroyl)rutinoside-5-O-glucoside
Formed from: Petunidin-3-(*p*-coumaroyl)rutinoside

Petunidin-3-rhamnoside-5-glucoside
Probably formed from: Malvidin-3-rhamnoside-5-glucoside
→ Delphinidin-3-rhamnoside-5-glucoside
Probably in: Pea A194

Phaeomelanin
Formed from: 5-(S-(L-Cysteinyl))-3,4-dihydroxyphenethylamine
5-(S-(L-Cysteinyl))-3,4-dihydroxy-L-phenylalanine

Phaseollidin ((6a*R*,11a*R*)-6a,11a-dihydro-10-(3-methyl-2-butenyl)-6*H*-benzofuro[3,2-*c*][1]benzopyran-3,9-diol)
→ Phaseollidin hydrate
In: Fusarium B204, C58, C74

Phaseollidin hydrate
Formed from: Phaseollidin

Phaseollin ((6b*R*,12b*R*)-6b,12b-dihydro-3,3-dimethyl-3*H*,7*H*-furo[3,2-*c*:5,4-*f*′]bis[1]benzopyran-10-ol)
→ 6a,7-Dihydroxyphaseollin
In: Colletotrichum A190
→ 6a-Hydroxyphaseollin
In: Colletotrichum A190
→ 1a-Hydroxyphaseollone
In: Fusarium A209
→ Phaseollinisoflavan
In: Stemphylium A2283

Phaseollinisoflavan ((3*R*)-3,4-dihydro-2′,2′-dimethyl-(3,6′-bi-2*H*-1-benzopyran)-5′,7-diol)
Formed from: Phaseollin
Phellamurin (3,4′,5,7-tetrahydroxy-8-isoprenylflavanone)
→ 3-Carboxy-2,4,6-trihydroxyphenylacetic acid
In: Aspergillus A3306
→ *p*-Hydroxymandelic acid
In: Aspergillus A3306
→ Neophellamuretin
In: Aspergillus A3306
Phenacetin
See: *p*-Ethoxyacetanilide
Phenacetin glutathione
Formed from: *p*-Ethoxyacetanilide
Phenacyl chloride
→ 2-Chloro-1-phenylethanol
In: Corynebacterium J893
N³-Phenacyluracil
Formed from: N³-Phenacyluridine
N³-Phenacyluridine
→ N³-(α-Hydroxy-β-phenethyl)uridine
In: Mouse K121
Rabbit K569
→ N³-Phenacyluracil
In: Mouse K121
Phenanthrene
Formed from: Phenanthrene-9,10-oxide
→ 5,6-Benzocoumarin
In: Sphingomonas K610
→ 7,8-Benzocoumarin
In: Sphingomonas K610
→ Coumarin
In: Sphingomonas K610
→ (+)-*cis*-1,2-Dihydro-1,2-dihydroxyphenanthrene
In: Beijerinckia A2680
Pseudomonas A2680, K613
→ *trans*-1,2-Dihydro-1,2-dihydroxyphenanthrene
In: Coalfish B332
Cunninghamella D107
Dogfish C110
Flounder C110
Guinea pig A2968, G737
Lobster C110
Mouse A2968
Rat A2968, B772, G737
Trout C110
→ (+)-*cis*-3,4-Dihydro-3,4-dihydroxyphenanthrene
In: Beijerinckia A2680
Pseudomonas A2680, K613
→ *trans*-3,4-Dihydro-3,4-dihydroxyphenanthrene
In: Cunninghamella D107
Dogfish C110
Flounder C110
Guinea pig A2968, G737
Lobster C110
Mouse A2968
Phanerochaete H495
Rat A2968, B772, G737
Trout C110
→ *cis*-9,10-Dihydro-9,10-dihydroxyphenanthrene
In: Mycobacterium J315
Pseudomonas K613
→ Diphenic acid
In: Mycobacterium J315
Phanerochaete G622
Pleurotus J193
→ 1-Hydroxy-2-naphthoic acid
In: Bacillus K520
Brevibacterium K372
Burkholderia K284
Microorganism G680
Nocardioides K519
Pseudomonas H221, K284, K372
Sphingomonas K413, K610
→ 2-Hydroxy-1-naphthoic acid
In: Burkholderia K284
Pseudomonas K284
→ 1-Hydroxyphenanthrene
In: Coalfish B332
Cunninghamella F198
Dogfish C110
Flounder C110
Guinea pig G737
Lobster C110
Microorganism G680
Rat G737

Trout C110
See also: G474
→ 2-Hydroxyphenanthrene
In: Coalfish B332
Dogfish C110
Flounder C110
Guinea pig G737
Lobster C110
Rat G737
Trout C110
→ 3-Hydroxyphenanthrene
In: Coalfish B332
Dogfish C110
Flounder C110
Guinea pig G737
Lobster C110
Phanerochaete G406, H495
Rat G737
Trout C110
→ 4-Hydroxyphenanthrene
In: Coalfish B332
Dogfish C110
Flounder C110
Guinea pig A2968, G737
Lobster C110
Mouse A2968
Phanerochaete G406, H495
Rat A2968, G737
Trout C110
→ 9-Hydroxyphenanthrene
In: Coalfish B332
Dogfish C110
Flounder C110
Lobster C110
Phanerochaete G406, H495
Streptomyces F808
Trout C110
→ Phenanthrene-4-carboxylic acid
Probably in: Microorganism K253
→ Phenanthrene-9,10-oxide
In: Guinea pig A2968
Mouse A2968
Rat A2968, A3626
See also: B332, B772, C110, C630, F808, G406, G474, G737, H495, J193, J347, J637
→ Phenanthrene-9,10-quinone
In: Trametes J540
Wheat J637
→ Protocatechuic acid
In: Arthrobacter K372
Micrococcus K372
→ Salicylic acid
In: Pseudomonas H221

Phenanthrene-4-carboxylic acid
Probably formed from: Phenanthrene-4,5-dicarboxylic acid
→ *cis*-3,4-Dihydro-3,4-dihydroxyphenanthrene-4-carboxylic acid
Probably in: Mycobacterium J646

Phenanthrene-9-carboxylic acid
Formed from: 9-Formylphenanthrene

Phenanthrene-4,5-dicarboxylic acid
Probably formed from: *cis*-4,5-Dihydro-4,5-dihydroxypyrene
trans-4,5-Dihydro-4,5-dihydroxypyrene
→ Phenanthrene-4-carboxylic acid
Probably in: Mycobacterium H779, H884, J646

Phenanthrene-1,2-oxide
→ *trans*-1,2-Dihydro-1,2-dihydroxyphenanthrene
In: Rat A2680
→ S-(1,2-Dihydro-2-hydroxyphenanthr-1-yl)glutathione
Probably in: Sheep A2282

Phenanthrene-3,4-oxide
→ *trans*-3,4-Dihydro-3,4-dihydroxyphenanthrene
In: Rat A2680
→ S-(3,4-Dihydro-3-hydroxyphenanthr-4-yl)glutathione
Probably in: Sheep A2282

Phenanthrene-9,10-oxide
Formed from: Phenanthrene
→ *trans*-9,10-Dihydro-9,10-dihydroxyphenanthrene
In: Man A2486, A3217
Mouse A2486
Pleurotus J347
Rat A2486, A2680, A3031, A3166, B142
See also: A2968
→ S-(9,10-Dihydro-9-hydroxyphenanthr-10-yl)glutathione

Probably in: Sheep A2282, C472, C630, D235
→ Phenanthrene
In: Rat B558

Phenanthrene-9,10-quinone
Formed from: Phenanthrene
→ 9,10-Dihydroxyphenanthrene
In: Channel catfish H210
Cochliobolus K378
Guinea pig K773
Man K273
Rabbit H920
Rat H759, J744, K546
→ Diphenic acid
In: Phanerochaete G622
Pseudomonas J317
→ 9-Hydroxy-10-oxophenanthrene
In: Magnaporthe K704
→ Phthalic acid
In: Pseudomonas J317

Phenanthridine (benzo[*c*]quinoline)
→ 1,2-Dihydro-1,2-dihydroxyphenanthridine
In: Rat D338
→ 9,10-Dihydro-9,10-dihydroxyphenanthridine
In: Rat D338
→ 2-Hydroxyphenanthridine
In: Rat D338
→ Phenanthridine-N-oxide
In: Rat D338, J547
→ 6(5*H*)-Phenanthridinone
In: Chironomus K77
Rat D338
Streptomyces J648

Phenanthridine-N-oxide
Formed from: Phenanthridine

6(5*H*)-Phenanthridinone (phenanthridone)
Formed from: Florenone oxime
Phenanthridine

1,10-Phenanthroline-5,6-dione
→ 5,6-Dihydroxy-1,10-phenanthroline
In: Rabbit H920

Phenazepam
See: 7-Bromo-5-(*o*-chlorophenyl)-1,3-dihydro-2*H*-1,4-benzodiazepin-2-one

Phenazine-1-carboxylic acid
Formed from: Anthranilic acid
6-Hydroxyphenazine-1-carboxylic acid
Shikimic acid
Probably formed from: Phenazine-1,6-dicarboxylic acid

Phenazine-1,6-dicarboxylic acid
Probably formed from: Dimethyl phenazine-1,6-dicarboxylate
→ Phenazine-1-carboxylic acid
Probably in: Pseudomonas A2805, A3284

Phenazone
See: Antipyrine

Phenazopyridine (2,6-diamino-3-(phenylazo)pyridine)
→ *p*-Acetamidophenol
In: Rat G541
→ N-Acetyl-2,6-diamino-3-(phenylazo)pyridine
In: Guinea pig F621
Man F621
Mouse F621
Rat F621
→ 2,6-Diamino-3-(2-hydroxyphenylazo)pyridine
In: Mouse F621
Rat G541
→ 2,6-Diamino-3-(4-hydroxyphenylazo)pyridine
In: Guinea pig F621
Man F621
Mouse F621
Rat F621, G541
→ 2,6-Diamino-5-hydroxy-3-(phenylazo)pyridine
In: Guinea pig F621
Man F621
Mouse F621
Rat C621, F621, G541

Phencyclidine (1-(1-phenylcyclohexyl)piperidine)
→ 2-Cyano-1-(1-phenylcyclohexyl)piperidine
In: Rabbit C640
→ 3-Hydroxy-1-(1-phenylcyclohexyl)piperidine
In: Mouse E925
→ 4-Hydroxy-1-(1-phenylcyclohexyl)piperidine

In: Cat C873
Dog B882
Man F119, J149
Monkey C873
Mouse A3901, D504, F714
Rabbit B635, C873, G29
Rat C873
→ 1-(*m*-Hydroxyphenyl)-1-(1-piperidinyl) cyclohexane
In: Rat E910
→ Phenol
In: Rat A2635
→ 1-Phenylcyclohexylamine
In: Rabbit C598
Rat A2635, F892
→ N-(1-Phenylcyclohexyl)-5-aminopentan-1-al
In: Rabbit C598
See also: B635, B842, C873, D504
→ 3-Phenyl-3-piperidinocyclohexanol
In: Man F119
Mouse E925, F714
→ 4-Phenyl-4-piperidinocyclohexanol
In: Cat C873
Man F119, J149
Monkey C873
Mouse A3901, D504, F714
Rabbit B635, C873, F714
Rat C873
→ 1,2,3,4-Tetrahydro-N-(1-phenylcyclohexyl)pyridine
In: Rabbit C598, G29

Phencyclidine-N-oxide
→ 1-(4-Hydroxy-1-phenylcyclohexyl) piperidine-N-oxide
In: Rabbit C820

Phendimetrazine (2*S*,3*S*)-3,4-dimethyl-2-phenylmorpholine)
→ Norphendimetrazine
In: Rat C669

Phenelzine
See: β-Phenethylhydrazine

Phenethyl acetate
→ 2,3-Dihydroxyphenethyl acetate
In: Pseudomonas F283

Phenethylamine (2-phenylethylamine)
Formed from: N-Acetylphenethylamine
N-Ethoxycarbonylphenethylamine
Phenethylhydrazine
D-Phenylalanine
L-Phenylalanine
N-*n*-Propylphenethylamine
Tyramine
→ N-Acetylphenethylamine
In: Cunninghamella F345
Hen F694
Man K490
Prawn G824
Rat A807
Rhizoglyphus F669
→ Arborine
In: Glycosmis A538
→ Feruloylphenethylamine
In: Nicotiana F72
→ N-Hydroxyphenethylamine
In: Rabbit A1166
→ N-Methylphenethylamine
In: Phalaris A3891
Rat A1307, A1311, E6
→ Phenylacetaldehyde
In: Arthrobacter F626, H912
Escherichia E327
Hedgehog A434
Man A1901, A2010, A2967, A3957, B794, E55, F754
Mouse A1895
Octopus J88
Pig B206
Pseudomonas C517
Rabbit A72, A94
Rat A244, A1227, A1324, A1348, A2972, A3863, B660
Rhizoglyphus F669
Saccharomyces A3433
See also: A1715, A1904, A3079, C688, D909, J539
→ Phenylethanolamine
In: Rat A244, A368
Contrast (in man) A3258
See also: A1715, A2261
→ 1,2,3,4-Tetrahydroisoquinoline

In: Rat A2285
→ 1,2,3,4-Tetrahydro-1-methylisoquinoline
In: Mouse G800
Rat J360
→ *o*-Tyramine
In: Rat A370
→ *m*-Tyramine
In: Man B794
Rat A370
Contrast (in rat) C465
→ Tyramine
In: Man B794, E313
Rabbit A1717
Rat A244, A370
Contrast (in rat) C465

Phenethylamin-3-yl sulphate
Formed from: *m*-Tyramine

1-(Phenethyl)biguanide
See: Phenformin

1-Phenethylhydrazine
→ 1-Phenylethylamine
In: Microorganism A3616

β-Phenethylhydrazine (nardil, phenelzine)
→ N-Acetyl-β-phenethylhydrazine
In: Man A3857
Rabbit B978
Rat A3857, F10
→ N′-Methyl-N-(β-phenethyl)hydrazine
In: Beef G34, G249
→ Phenethylamine
In: Rat D318

Phenethylhydroxylamine
See: N-Hydroxyphenethylamine

N-(1-(2-Phenethyl-4-piperidinyl))aniline
Formed from: Fentanyl

N-(1-(2-Phenethyl-4-piperidinyl))lactanilide
Formed from: Fentanyl
→ 4-(N-Lactanilido)piperidine
In: Guinea pig D906
Rat D906

N-(1-(2-Phenethyl-4-piperidinyl))malonanilic acid
Formed from: Fentanyl

N-(1-(2-Phenethyl-4-piperidinyl))propionanilide
See: Fentanyl

***p*-Phenetidine**
See: *p*-Ethoxyaniline

Phenetole (ethoxybenzene)
→ *cis*-2,3-Dihydro-2,3-dihydroxyphenetole
In: Pseudomonas H251, J656
→ Ethenyloxybenzene
In: Beijerinckia H251
Pseudomonas H251
→ *o*-Ethoxyphenol
In: Pseudomonas H251
Rat G500
→ *m*-Ethoxyphenol
In: Pseudomonas H251
→ *p*-Ethoxyphenol
In: Pseudomonas H251
Rat G500
→ Phenol
In: Beijerinckia H251
Pseudomonas H251
Rat G500

Phenetsal (2-hydroxybenzoic acid 4-(acetylamino)phenyl ester)
Formed from: Benorylate
→ *p*-Acetamidophenol
In: Man F517
→ Salicylic acid
In: Man F517

Phenformin (1-(phenethyl)biguanide)
→ 4-Hydroxyphenformin
In: Dog A3854
Guinea pig A3851, A3854
Man C296
Rat A3851, A3854

Pheniramine (N,N-dimethyl-γ-phenyl-2-pyridinepropanamine)
→ Demethylpheniramine
In: Cunninghamella H456
→ Pheniramine-β-D-glucuronide
In: Man G18, H541
→ Pheniramine-N-oxide
In: Cunninghamella H456

Pheniramine-β-D-glucuronide
Formed from: Pheniramine

Pheniramine-N-oxide
Formed from: Pheniramine

Phenmetrazine (*threo*-3-methyl-2-phenylmorpholine)
Formed from: Tarugan

→ 4′-Hydroxyphenmetrazine
In: Man A2080
Rat A2080
Tamarin A2080
See also: A16
→ 5-Oxophenmetrazine
In: Man A2080
Rat A2080
Tamarin A2080
See also: A16

Phenobarbital (phenobarbitone, 5-ethyl-5-phenylbarbituric acid)
Formed from: N-Methoxymethylphenobarbital
Probably formed from: 2-Deoxyphenobarbital
1,5-Diethyl-5-phenylbarbituric acid
2-Hydroxyprimidone
Mephobarbital
→ 5-(3,4-Dihydro-3,4-dihydroxyphenyl)-5-ethylbarbituric acid
In: Guinea pig A470
Man A470
Rat A470
→ 5-Ethyl-5-(*p*-hydroxyphenyl)barbituric acid
In: Guinea pig A470
Man A470, A1706, A2136, A3914, K342
Rat A470, A1959, A3156, F286
See also: A1233, B642, E474
→ N-Hydroxyphenobarbital
In: Man A1968
Contrast: in man (wrong identity) A3914
→ Phenobarbital-N-glucoside
In: Man A3914, H173

Phenobarbital-N-glucoside
Formed from: Phenobarbital

Phenobarbital-4′-yl-β-D-glucuronide
Formed from: 5-Ethyl-5-(*p*-hydroxyphenyl)barbituric acid

Phenobarbital-4′-yl sulphate
Formed from: 5-Ethyl-5-(*p*-hydroxyphenyl)barbituric acid

Phenobarbitone
See: Phenobarbital

Phenobenzuron (N-(3,4-dichlorophenyl)-N-((dimethylamino)carbonyl)benzamide)
→ Diuron
In: Fusarium A990

Phenol
Formed from: Anisole
Acetophenone
Benzene
Benzenesulphonic acid
o-Bromophenol
m-Bromophenol
p-Bromophenol
p-tert-Butylphenyl diphenyl phosphate
Carfecillin
o-Chlorophenol
Diphenyl ether
Ferulic acid
Fomocaine
o-Fomocaine
m-Hydroxybenzoic acid
p-Hydroxybenzoic acid
4-Hydroxyphenyl phenyl ether
Isopropoxybenzene
N-Methyl-N-nitrosoaniline
2-Octyl phenyl carbonate
Phencyclidine
Phenetole
Phenoxyacetic acid
m-Phenoxybenzaldehyde
m-Phenoxybenzoic acid
p-Phenoxybenzoic acid
Phenyl acetate
Phenyl-N-acetyl-β-D-glucosaminide
Phenyl benzoate
Phenyl-β-D-galactoside
Phenyl-β-D-glucoside
Phenyl-β-D-glucuronide
Phenyl-β-maltoside
Phenyl nicotinate
Phenyl phosphate
Phenyl 2-(3-(1-piperidino)propyl)benzyl ether
Phenyl thioacetate
2,3,4,6-Tetrachlorophenol
D-Tyrosine
L-Tyrosine
Probably formed from: *p*-Aminophenol
2,4-Diaminophenol
Phenyl-6′-O-malonyl-β-D-glucoside

2,4,5-Trichlorophenol
→ Anisole
In: Guinea pig A115
Man A2956
Phellinus G208
Rabbit A115
Rat A115
→ Benzoic acid
In: Microorganism F177, F515, G167, G281, G424, G917, G923, H842, J12, K243
See also: F468
→ *o*-Bromophenol
In: Ulva K153
See also: E597
→ *p*-Bromophenol
In: Ulva K153
→ Catechol
In: Aspergillus H570
Bacillus A2397, F93
Candida A943, E204, K653
Escherichia G817
Frog D363
Mucuna E769
Penicillium H82
Pseudomonas E731, K433
Rana E841
Rat C542, F568, G295, F686, H818
Rhodotorula A1085
Streptomyces C510
Trichosporon A902, G196
Xenopus E841
See also: A768, A2100, A2160, A2260, A3190, A3795, B120, E565, G639, K362, K403, K885
Contrast: (in Mycobacterium) H824
(in rabbit) F238
→ Cyclohexanone
In: Pseudomonas B678
→ 2,2′-Dihydroxybiphenyl
In: Horseradish C468, D433
→ 4,4′-Dihydroxybiphenyl
In: Horseradish C468, D433, E297
Man E297
→ 3-(S-Glutathionyl)-4,4′dihydroxybiphenyl
In: Horseradish E297
→ *p*-Hydroxybenzoic acid
In: Microorganism G632, G923, H842
Pseudomonas G291
→ Phenol 3-hydroxybutyrate-4-hydroxybutyrate copopymer
In: Burkholderia K486
→ Phenylacetic acid
In: Microorganism J12
→ Phenyl-β-D-glucoside
In: Cepaea G60
Gardenia E338
Lemna H663
Manduca G816, G897
Millipede A2254
Trilobium G816
→ Phenyl-β-D-glucuronide
In: Antechinus B729
Bream B580
Bushbaby A3061
Capuchin A510, A3061
Cat A2387
Chimpanzee G639
Cynomolgus monkey A3061, G639
Dasyuroides B729
Dasyurus B729
Dog A510, A2387, B531
Elephant A2387
Ferret A510, A2387
Frog D363
Fruit bat A510
Golden orfe C424
Goldfish A2895, C404, C424
Hedgehog A510
Hen A14, A510, A860
Hydromys B729
Macropus B729
Man A3396, B531, F533, F903
Melomys B729
Mouse F533, H453, H574, K172
Notomys A3886, B729
Patas monkey A3061
Perch B580
Petaurus B729
Phascolarctos B729
Pig A510, A3717
Possum B729
Potorous B729

Pseudemys B729
Rabbit A510
Rainbow trout C424
Rat A3575, A3717, B164, B165, B531, B650, B729, B962, E14, E219, F312, F568, G295, H314, H453, H468, H522, H818
Rhesus monkey A510, A3061
Roach B580
Rudd B580
Sheep A3717
Sminthopsis B729
Squirrel monkey A510
Tamarin A3061
Tree shrew A3061
Zebra fish E813
→ Phenyl phosphate
In: Cat A2
Sheep A3717
→ Phenyl sulphate
In: Antechinus B729
Beef E800, D607
Bream B580
Bushbaby A3061
Capuchin A510, A3061
Carp A2442
Cat A510, A2387, A2686
Chimpanzee G639
Civet A1312, A2387
Clam A2442
Cynomolgus monkey A3061, G639
Dasyuroides B729
Dasyurus B729
Dog A510, A2387, B531
Elephant A2387
Eubacterium E432, G602, H439
Ferret A510, A2387
Frog D363, E841
Fruit bat A510
Genet A1312, A2387
Golden orfe C424
Goldfish A2385, A2442, B580, C424
Guppy B580
Haemophilus H439, H756
Hedgehog A510
Hen A14, A510, A860
Hyaena A2387
Hydromys B729
Klebsiella G756, H439
Lion A1312
Lobster A2442
Macropus B729
Man A3396, A3814, B531, B834, C523, D607, D613, D977, D979, E194, E198, E575, F533, F692, F770, F903, K405
Marmoset F226
Melomys B729
Minnow B580
Mouse B729, F533, H574, K172
Notomys A3886, B729
Patas monkey A3061
Perch B580
Petarus B729
Phascolarctos B729
Pig A3061
Possum B729
Potorous B729
Pseudomys B729
Rabbit A510
Rainbow trout C424
Rat A153, A510, A2442, A3717, B61, B164, B165, B531, B650, B729, B941, D748, E14, E219, E575, E630, E962, F312, F533, F568, G295, H314, H468, H522, H818, K120, K735
Rhesus monkey A510, A3061
Roach B580
Rudd B580
Sheep A3061
Sminthopsis B729
Squirrel monkey A510
Tamarin A3061
Tench B580
Tree shrew A3061
Xenopus E841
Zebra fish E813
→ Polyphenol
In: Horseradish K351
→ Quinol
Antechinus B729
Aspergillus H570
Capuchin A510
Cat A510, A2686

Dasyuroides B729
Dasyurus B729
Dog A510, B531
Ferret A510
Frog D363, E841
Hedgehog A510
Hydromys B729
Macropus B729
Man F533
Melomys B729
Millipede A2254
Mouse B729, F533, H574, J386, K172
Mycobacterium H824
Nitrosomomas H219, H631
Notomys B729
Petaurus B729
Phascolarctos B729
Pig A3717
Possum B729
Potorous B729
Pseudomys B729
Rabbit A510, F148, F238
Rat A3717, B531, B729, C542, F533, F568, G295
Sheep A3717
Sminthopsis B729
Squirrel monkey A510
Xenopus E841
Zebra fish E813
See also: A1218, A2100, A2387, A3886, B120, C424, D50, D358, D921, E220, F600, G60, G130, G639, H453, H818, J321
→ Resorcinol
In: Frog D363
Xenopus E841
Contrast (in Mycobacterium) H824
→ L-Tyrosine
In: Citrobacter B693, F38
Erwinia A708, B292, D868
Leptoglossus A2260
Microorganism A1164
Millipede A2254
Periplaneta D184

Phenol 3-hydroxybutyrate-4-hydroxybutyrate copopymer
Formed from: Phenol

Phenolphthalein (3,3′-bis(4-hydroxyphenyl)-1(3*H*)-isobenzofuranone)
Formed from: Phenolphthalein-β-D-glucuronide
Probably formed from: Phenolphthalein disulphate
→ 3-(3,4-Dihydroxyphenyl)-3-(*p*-hydroxyphenyl)-1(3*H*)-isobenzofuranone
In: Mouse J671
Rat J671
→ Phenolphthalein-β-D-glucuronide
In: Cat C562
Dog C562
Guinea pig D960, C562
Man B800, G524, G968, J783
Mouse C562, D315, J671
Quail C562
Rabbit A43, C562
Rat A3464, B740, C37, C562, C672, F688, J671
Trout C562
→ Phenolphthalein phenoxyl radical
In: Horseradish J145
→ Phenolphthalein sulphate
In: Beef A1211
Guinea pig D960
Mouse J671
Rat A3464, J671

Phenolphthalein-di-β-D-glucuronide
Formed from: Phenolphthalein-β-D-glucuronide

Phenolphthalein disulphate
→ Phenolphthalein
Probably in: Peptococcus E449

Phenolphthalein-β-D-glucuronide
Formed from: Phenolphthalein
→ Phenolphthalein
In: Alcaligenes E323
Beef A1029
Escherichia A1029
Man A789
Microorganism G567
→ Phenolphthalein-di-β-D-glucuronide
In: Mouse J671

Phenolphthalein phenoxyl radical
Formed from: Phenolphthalein

Phenolphthalein sulphate
Formed from: Phenolphthalein
Phenolsulphonephthalein
See: Phenol red
Phenol red (phenolsulphonephthalein, 4,4′-(1,1′-dioxido-3*H*-2,1-benzoxathiol-3-ylidene) bisphenol)
→ Phenol red-β-D-glucuronide
In: Dogfish A1743
Man G524
Shark flounder B436
Phenol red-β-D-glucuronide
Formed from: Phenol red
Phenoperidine (4-piperidinecarboxylic acid 1-(3-hydroxy-3-phenylpropyl)-4-phenylethyl ester)
→ Norpethidine
In: Man C326
Phenothiazine (thiodiphenylamine)
Formed from: Phenothiazine sulphoxide
→ 3-Hydroxyphenothiazine
In: Gerbil B735
Guinea pig A3923
Hamster B735
Mouse B735
Rat B735
→ N-Hydroxyphenothiazine
In: Guinea pig A322
Rabbit G322
Contrast (wrong identity) A2923, A3488
→ Phenothiazine-β-D-glucuronide
In: Gerbil B735
Guinea pig A3923
Hamster B735
Mouse B735
Rat B735
→ Phenothiazine-N-hydroperoxide
In: Guinea pig A322
Rabbit G322
Contrast (wrong identity) A2923, A3488
→ Phenothiazine-N-oxide
In: Guinea pig A322
Rabbit G322
→ Phenothiazine sulphoxide
In: Ascaris A3849
Gerbil B735
Guinea pig A3923
Hamster B735
Moniezia A3849
Mouse A3849, B735
Rat B735
→ Phenothiazin-10-yl-L-glutamyl-L-glutamic acid
In: Beef D410
→ Thionol
In: Gerbil B735
Mouse B735
Rat B735
Phenothiazine-β-D-glucuronide
Formed from: Phenothiazine
Phenothiazine-N-hydroperoxide
Formed from: Phenothiazine (challenged)
Phenothiazine-N-oxide
Formed from: Phenothiazine
Phenothiazine sulphoxide
Formed from: Phenothiazine
→ Phenothiazine
In: Ascaris A3849
Moniezia A3849
Mouse A3849
Phenothiazin-10-yl-L-glutamyl-L-glutamic acid
Formed from: Phenothiazine
N-(γ-Phenothiazin(10)propyl)ethylenediamine
Formed from: Perazine
γ-(Phenothiazinyl(10))propylamine
See: Bisdemethylpromazine
Phenothiazone (3*H*-phenothiazin-3-one)
Probably formed from: 3-Hydroxphenothiazine
Phenothrin (3-phenoxybenzyl-*d*-*trans*-chrysanthemumate)
→ *m*-Phenoxybenzyl alcohol
In: Dog A1521
Guinea pig A1521
Mouse A1521
Rabbit A1521
Rat A1521
See also: C63
***cis*-Phenothrin**
→ 3-(2-Hydroxymethyl-1-propenyl)-2,2-dimethylcyclopropanecarboxylic acid (3-phenoxyphenyl)methyl ester
In: Rat C63
→ *m*-Phenoxybenzaldehyde

Probably in: Rat C63

2-Phenoxyacetaldehyde

Formed from: 2-Phenoxyethanol

→ 2-Hydroxy-3-phenoxypropionitrile

In: Manihot K252 ((*S*)-isomer)

Prunus K252 ((*R*)-isomer)

Phenoxyacetic acid

Formed from: *o*-Methoxyphenoxyacetic acid

Phenoxymethylpenicillin

Veratrylglycerol β-phenyl ether

Probably formed from: 3-Phenoxylactic acid

→ *o*-Hydroxyphenoxyacetic acid

In: Aspergillus A2849

→ *p*-Hydroxyphenoxyacetic acid

In: Aspergillus A2849

Oat A3650

Pea A3650

→ Phenol

In: Microorganism H104

See also: D920

→ Phenoxyacetyl CoA

In: Penicillium A2291, C483

→ 2-Phenoxyethanol

In: Nocardia F651

Phenoxyacetyl CoA

Formed from: Phenoxyacetic acid

→ Phenoxymethylpenicillin

In: Penicillium D12

S-(Phenoxyacetyl)glutathione

→ Phenoxymethylpenicillin

In: Penicillium F711

***m*-Phenoxybenzaldehyde**

Formed from: Fenvalerate

Phenothrin

m-Phenoxybenzyl alcohol

m-Phenoxymandelonitrile

Probably formed from: Deltamethrin

cis-Phenothrin

→ *m*-Phenoxybenzoic acid

In: Hen F362

See also: A1934, A3199, A3800, A3804, C63, C127, C766, D443, D565

→ *m*-Hydroxybenaldehyde

Probably in: Pseudomonas G206

→ Phenol

In: Pseudomonas G206

→ *m*-Phenoxybenzoic acid

In: Rat H761

→ *m*-Phenoxybenzyl alcohol

Probably in: Cabbage D565

Cabbage looper A3804

Dog D443

Housefly A3804

Mouse A3800, A3804

→ *m*-Phenoxymandelonitrile

In: Prunus E353

***p*-Phenoxybenzaldehyde**

→ *p*-Phenoxybenzoic acid

In: Rat H761

→ *p*-Phenoxymandelonitrile (both isomers)

In: Almond J194, J302

Apple J302

Sorghum J194

Phenoxybenzamine (N-(2-chloroethyl)-N-(1-methyl-2-phenoxyethyl)benzenemethanamine)

→ N-Benzyl-N-phenoxyisopropylamine

In: Dog A1993

Rat A1993

Phenoxybenzene

See: Diphenyl ether

***m*-Phenoxybenzoic acid**

Formed from: *cis*-Cypermethrin

trans-Cypermethrin

Deltamethrin

1,2-Dipalmitoyl-3-(*m*-phenoxybenzoyl) glycerol

Fenvalerate

Fluvalinate

m-Phenoxybenzaldehyde

m-Phenoxybenzoylglucoside

Tralocythrin

Tralomethrin

Probably formed from: 1,3-Dipalmitoyl-2-(*m*-phenoxybenzoyl)glycerol

→ α-N-Acetyl-δ-N-(*m*-phenoxybenzoyl)-L-ornithine

In: Hen C40, E941

Quail F838

→ N^2-Benzoyl-N^5-*m*-phenoxybenzoyl-ornithine

In: Hen E941

→ N^5-Benzoyl-N^2-*m*-phenoxybenzoylornithine
In: Hen E941
→ *m*-Hydroxybenzoic acid
In: Hen G83, G979, H207
→ 2-Hydroxy-5-phenoxybenzoic acid
Probably in: Rat B199
→ 3-Hydroxy-5-phenoxybenzoic acid
Probably in: Mouse A3800, D697
→ 4-Hydroxy-3-phenoxybenzoic acid
In: Bacillus G679
→ *m*-(*o*-Hydroxyphenoxy)benzoic acid
Probably in: Cabbage D565
Hen D697
Mouse C127
Rat A1934, A3199, C127, C412
→ *m*-(*p*-Hydroxyphenoxy)benzoic acid
In: Cat B638
Duck B638
Gerbil B638
Guinea pig B638
Hamster B638
Hen G83, G979
Marmoset B638
Mouse B638
Rainbow trout E582
Rat B638, B640, C167, C836, D671, J877
Sheep B638
See also: A1521, A1823, A1934, A3194, A3199, B199, C63, C127, C132, C133, C412, C766, D224, D443, D565, D671, D697, E844, F838, G377
→ Phenol
In: Pseudomonas F640, F696, G206, H630
→ *m*-Phenoxybenzoyl CoA
In: Ferret B716
Gerbil B716
Hamster B716
→ *m*-Phenoxybenzoylglucoside
Probably in: Rat D671
→ *m*-Phenoxybenzoyl-β-D-glucuronic acid
In: Cow A3194
Duck B638
Ferret B638
Gerbil B638
Guinea pig B638
Hamster B638
Marmoset B638
Mouse B638, C127, C133
Rabbit B638
Rainbow trout E582, E844
Rat A1823, A1934, A3199, B199, B233, B638, B640, C127, D671, J877
See also: D443
→ *m*-Phenoxybenzoylglutamic acid
In: Cow A3194
Quail F838
→ *m*-Phenoxybenzoylglycylvaline
In: Duck B622, B638
Quail F838
→ *m*-Phenoxybenzoylserine
In: Quail F838
→ *m*-Phenoxybenzoyltaurine
In: Cat B638
Ferret B638
Gerbil B638
Hamster B638
Marmoset B638
Mouse A3250, B638, C127, C133, F838
Quail F838
Rainbow trout E582
→ *m*-Phenoxyhippuric acid
In: Cat B638
Cow A3194
Duck B638
Ferret B638
Guinea pig B638
Hamster B638
Marmoset B638
Mouse C127, B638
Quail F838
Rabbit B638
Rat A1521, A1823, A1934, A3199, B233, B638, C127, C167, C836, D671, J877
Sheep B638
See also: B199, D224, D443
→ Protocatechuic acid
In: Pseudomonas F640, H630

p-Phenoxybenzoic acid
Formed from: *p*-Phenoxybenzaldehyde
p-Phenoxybenzonitrile
→ Phenol

In: Pseudomonas F640, G206, H630
→ Protocatechuic acid
In: Pseudomonas F640, H630

_m_-Phenoxybenzonitrile
Formed from: Fastac

_p_-Phenoxybenzonitrile
→ *p*-Phenoxybenzoic acid
In: Rhodococcus H952

_m_-Phenoxybenzoyl CoA
Formed from: *m*-Phenoxybenzoic acid
→ 1,2-Dipalmitoyl-3-(*m*-phenoxybenzoyl) glycerol
In: Rat F749
→ 2-Hexadecyl-1-(*m*-phenoxybenzoyl) glycerol
Rat D416, F749
→ 2-Palmitoyl-3-(*m*-phenoxybenzoyl)glycerol
Probably in: Rat D416
→ *m*-Phenoxyhippuric acid
In: Ferret B716
Gerbil B716
Hamster B716

_m_-Phenoxybenzoylglucoside
Probably formed from: *m*-Phenoxybenzoic acid
→ *m*-Phenoxybenzoic acid
In: Rat B199

_m_-Phenoxybenzoyl-β-D-glucuronic acid
Formed from: *m*-Phenoxybenzoic acid

_m_-Phenoxybenzoylglutamic acid
Formed from: *m*-Phenoxybenzoic acid

_m_-Phenoxybenzoylglycylvaline
Formed from: *m*-Phenoxybenzoic acid

_m_-Phenoxybenzoylserine
Formed from: *m*-Phenoxybenzoic acid

_m_-Phenoxybenzoyltaurine
Formed from: *m*-Phenoxybenzoic acid

_m_-Phenoxybenzyl alcohol
Formed from: *trans*-Cypermethrin
Deltamethrin
Fluvalinate
[1*R*,*cis*]-Permethrin
[1*R*,*trans*]-Permethrin
[1*RS*,*cis*]-Permethrin
[1*RS*,*trans*]-Permethrin
Phenothrin
m-Phenoxybenzaldehyde
m-Phenoxybenzylgentiobioside
m-Phenoxybenzylgentiotrioside
m-Phenoxybenzyl-β-D-glucoside
Tralocythrin
Tralomethrin
→ *m*-(Hydroxyphenoxy)benzyl alcohol
In: Rat D671
See also: A3511, A3515, A3800, D443, D697
→ *m*-Phenoxybenzaldehyde
In: Pleurotus G668
See also: A1521, A1823, A1934, A3194, A3195, A3804, A3853, D671, J854
→ *m*-Phenoxybenzyl-β-D-glucuronide
In: Cow A3194
Mouse A3800
Rat D671

3-Phenoxybenzyl 2-carboxy-[1*R*,*trans*]-3-(2,2, dichlorovinyl)-2-methylcyclopropanecarboxylate
Formed from: *trans*-Permethrin

_m_-Phenoxybenzylgentiobioside
→ *m*-Phenoxybenzyl alcohol
In: Rat D671

_m_-Phenoxybenzylgentiotrioside
→ *m*-Phenoxybenzyl alcohol
In: Rat D671

_m_-Phenoxybenzyl-β-D-glucoside
→ *m*-Phenoxybenzyl alcohol
In: Rat D671

_m_-Phenoxybenzyl-β-D-glucuronide
Formed from: *m*-Phenoxybenzyl alcohol

α-Phenoxycarbonylbenzylpenicillin
See: Carfecillin

2-Phenoxyethanol
Formed from: Phenoxyacetic acid
2-Phenoxyethyl nicotinate
→ 2-Phenoxyacetaldehyde
In: Methylomonas A2712
Rhodopseudomonas D197

2-Phenoxyethyl nicotinate
→ 2-Phenoxyethanol
In: Man J141

_m_-Phenoxyhippuric acid
Formed from: *m*-Phenoxybenzoic acid
m-Phenoxybenzoyl CoA

N-(6-Phenoxyindan-5-yl)methanesulphonamide
→ N-(1-Hydroxy-6-phenoxyindan-5-yl)methanesulphonamide
In: Rat C485
→ N-(3-Hydroxy-6-phenoxyindan-5-yl)methanesulphonamide
In: Rat C485

Phenoxyisopropylamine
Probably formed from: N-Benzyl-N-phenoxyisopropylamine

3-Phenoxylactic acid
Formed from: Doxaminol
1,2-Epoxy-3-phenoxypropane
→ Phenoxyacetic acid
Probably in: Dog F550

***m*-Phenoxymandelonitrile**
Formed from: L-Cyhalothrin
cis-Cypermethrin
trans-Cypermethrin
Decamethrin
Deltamethrin
Fastac
Fenpyrithrin
Fenvalerate
Flucythrinate
Fluvalinate
m-Phenoxybenzaldehyde
Tralocythrin
Tralomethrin
→ *m*-Phenoxybenzaldehyde
With hydroxynitrile lyase H944
See also: A3804

***p*-Phenoxymandelonitrile**
Formed from: *p*-Phenoxybenzaldehyde

Phenoxymethylpenicillin (penicillin V)
Formed from: Phenoxyacetyl CoA
S-(Phenoxyacetyl)glutathione
→ Phenoxyacetic acid
In: Microorganism B456
Penicillium F646

2-Phenoxymuconic acid
Formed from: Diphenyl ether

***p*-Phenoxyphenol methanesulphonate**
→ *p*-(4-Hydroxyphenoxy)phenol methanesulphonate
In: Man A1120
Rat A1120

***p*-Phenoxyphenylhydroxylamine**
Formed from: N-Hydroxy-*p*-phenoxyacetanilide

2-(3-Phenoxyphenyl)propionic acid
See: (*R*)-Fenoprofen

***m*-Phenoxypropiophenone**
→ 2-Hydroxy-*m*-phenoxypropiophenone
In: Pseudomonas K524

Phenprobamate
See: γ-Phenylpropyl carbamate

Phenprocoumon (3-(α-ethylbenzyl)-4-hydroxycoumarin)
→ 4′-Hydroxyphenprocoumon
In: Man D801, D957
Rat A2365, B662
→ 6-Hydroxyphenprocoumon
In: Man D801, D957
Rat A2365, B662
→ 7-Hydroxyphenprocoumon
In: Man D801, D957
Rat A2365, B662
→ 8-Hydroxyphenprocoumon
In: Rat A2365, B662
→ 1-(*o*-Hydroxyphenyl)-3-phenylpentane-1,2-dione
In: Aspergillus F315
→ Phenprocoumon-β-D-glucuronide
In: Man A2606

Phenprocoumon-β-D-glucuronide
Formed from: Phenprocoumon

Phensuximide
See: N-Methyl-α-phenylsuccinimide

Phentermine (2-amino-2-methyl-1-phenylpropane, α,α-dimethylphenethylamine)
Formed from: N-Hydroxyphentermine
N-Isopropylphentermine
Mephentermine
N-Propylphentermine
→ *p*-Hydroxyphentermine
In: Cunninghamella F345
Man A1178
Rat A315, A1178, F519, G79, G570
See also: A2438, F335, G21
→ N-Hydroxyphentermine
In: Guinea pig C5
Man A1258

Rabbit A665, A1166, A1855, A2245, A2721, A2787, C5, F839
Rat A1966, A2417, C5, G79, G570
Species unclear A264
See also: A2008, F335, G21
→ Nitrosophentermine
In: Rabbit A2245, A2787

Phenthoate (S-(α-(carboethoxy)benzyl) O,O-dimethyl phosphorodithioate)
→ S-(α-Carboxybenzyl) O,O-dimethyl phosphorodithioate
In: Mouse A2661
→ S-(α-(Carboxyethoxy)benzyl) O,O-dimethyl phosphorothioate
In: Housefly A2661
Mouse D441
Rat D441
→ S-(α-(Carboxyethoxy)benzyl) O,O-dimethyl phosphorodithioate
In: Housefly A2661
Mouse A2661
→ Ethyl mandelate
In: Mouse D441
Rat D441
→ Mandelic acid
In: Housefly A2661
→ α-Methylmercaptphenylacetic acid
In: Housefly A2661

Phenylacetaldehyde
Formed from: N-Methylphenethylamine
Phenethylamine
2-Phenylethanol
α-Phenylhydracrylic acid
β-Phenylhydracrylic acid
Phenylpyruvic acid
Styrene oxide
Tropic acid
→ 3-Hydroxy-4-phenylbutan-2-one
In: Zygosaccharomyces K611
→ Phenylacetic acid
In: Achromobacter K869
Dunaliella G459
Escherichia E327
Flavobacterium E679, F780
Man J24
Pseudomonas H829
Rat A2973, H761
Thauera J397
Xanthobacter F440
See also: A923, H334
→ 2-Phenylethanol
In: Candida A2483
Chinese cabbage F931
Corynebacterium H881
Cucumber A2432
Euglena G422
Mung bean J207
Phanerochaete F866
Saccharomyces A3433
See also: A923
→ Phenyllactonitrile
In: Almond K320
Prunus F353

Phenylacetaldoxime
Probably formed from: N-Hydroxyphenethylamine
N-Hydroxy-L-phenylalanine
→ 2-Hydroxy-2-phenylacetaldoxime
Probably in: Millipede A2262
→ 1-Nitro-2-phenylethane
In: Rabbit A3315
→ Phenylacetonitrile
In: Bacillus J508, K411
See also: A2262, A3389

Phenylacetamide
Formed from: N-Ethylphenylacetamide
L-Phenylalanine
→ Phenylacetic acid
In: Absidia K514
Agrobacterium K514
Arthrobacter K514
Aspergillus K514
Aureobacterium K514
Aureobasidium K514
Bacillus J508, K514
Brevibacterium K514
Cellulomonas K514
Coprinus K514
Corynebacterium K514
Cunninghamella K514
Flammulina K514
Flavobacterium K514

Fusarium K514
Gibberella K514
Keratinomyces K514
Kocuria K514
Micrococcus K514
Mortierella K514
Mucor K514
Neosartorya K514
Nocardia G343, K514
Proteus K514
Pseudomonas K514
Pycnoporus K514
Rhizopus K514
Rhodococcus K514
Schizophyllum K514
Serratia K514
Stenotrophomonas K514
Talaromyces K514
Xanthomonas K514

2-(N-Phenylacetamidoxy)ethanol
Formed from: 1,2′-Bis(N-phenylacetamidoxy)ethoxyethane

2′-(N-Phenylacetamidoxy)ethoxyethanol
Formed from: 1,2′-Bis(N-phenylacetamidoxy)ethoxyethane

Phenyl acetate
Formed from: Acetophenone
→ Phenol
In: Acinetobacter A962
Arthrobacter A2380
Badger K863
Bee eater K863
Beef A2388, K863
Canada geese K863
Capybara K863
Cat K863
Chicken K863
Cormorant K863
Goat K863
Great tit K863
Guillemot K863
Japanese quail K863
Man F717, G251, K327
Mouse K863
Mute swan K863
Nocardia G343
Pig K863
Pigeon K863
Puffin K863
Rabbit K863
Rat F753, J198, K194, K863
Razorbill K863
Shag K863
Sheep K863
Starling K863
Tree sparrow K863
Trout K863

Phenylacetic acid
Formed from: Apigenin
Cinnamic acid
Daidzein
N,N-Diethylphenylacetamide
Diphenylmethane
n-Dodecylbenzene
Fentanyl
Ferulic acid
Fluoranthene
‘Fructose-phenylalanine’
Homophthalic acid
Luteolin
Naringenin
Phenol
Phenylacetaldehyde
Phenylacetamide
Phenylacetonitrile
Phenylacetylene
Phenylacetylglycine
trans-4-Phenyl-3-buten-2-one
4-Phenylbutyric acid
Phenylmalonic acid
Styrene
Probably formed from: *p*-Hydroxyphenylacetic acid
→ Arborine
In: Glycosmis A538
→ Benzoic acid
In: Microorganism H960
→ N^2,N^5-Bis(phenylacetyl)ornithine
Probably in: Hen A131, A196, A395
→ Cycloheptanecarboxylic acid
In: Alicyclobacillus H832
→ Cyclohexylacetic acid

In: Microorganism A3189
→ Cyclohexylideneacetic acid
In: Microorganism A3189
→ *cis*-2,3-Dihydro-2,3-dihydroxyphenylacetic acid
In: Bacterium B871, B950
→ 2,6-Dihydroxyphenylacetic acid
In: Aspergillus D685
Fusarium D685
Pellicularia D685
Penicillium D685
Phellus D685
Rhizopus D685
→ 3,4-Dihydroxyphenylacetic acid
In: Aspergillus D685
Penicillium D685
→ Homogentisic acid
In: Aspergillus D685
Fusarium D685
Gibberella D685
Mucor D685
Pellicularia D685
Phelinus D685
Rhizopus D685
Penicillium D685
→ *p*-Hydroxybenzoic acid
In: Poria A892
→ *o*-Hydroxyphenylacetic acid
In: Aspergillus A329, A707, A819, A2345, A2849, A3210, A3787, B133, B727, C484
Bacteria B950
Brevibacterium D623
Nocardia A1329
Pseudomonas C532
Trichosporon E332
See also: A1466, E746, F780, F870
→ *m*-Hydroxyphenylacetic acid
In: Aspergillus A707, A3210, A3787, C484
Bacteria B950
Brevibacterium D623
Rhizoctonia A203
Trichosporon B368, E332
See also: A1466, E746, F780
→ *p*-Hydroxyphenylacetic acid
In: Aspergillus A2691, A2849
Bacteria B950
Brevibacterium D623
Microorganism G159
Trichosporon E332
See also: A1011, F780, H49
→ Mandelic acid
In: Pseudomomas G929
→ Phenylacetyl-L-carnitine
In: Man K521
Rat K525
→ Phenylacetyl CoA
In: Man K220
Penicillium A2291, C483
Pseudomonas F65, G225, G890, G929, H340, K743
Rat A3578, B119
Thauera J397
→ 6-O-Phenylacetyl-β-D-glucose
In: Glycyrrhiza F445
See also: A3679
→ Phenylacetyl-β-D-glucuronic acid
In: Dog A196
Ferret A196
Fruit bat A1451
Mouse A196
Pigeon A395
Contrast (in ferret) A394
→ Phenylacetylglutamine
In: Baboon A196
Capuchin A196
Drill A196
Man A 128, A132, A196, H550
Marmoset A196
Monkey A196
Rat C308
→ Phenylacetylglycine
In: Bat A196, A1451, A3356
Bushbaby A196
Cat A196
Chicken A196, A395
Dog A196
Elephant A2387
Ferret A196, A394, A2058, A3046
Guinea pig A196
Hamster A196
Horse A129
Monkey A196

Mouse A196
Pigeon A196, A395
Rat A196, A384, A2144, A2958, A3578, C308
Sheep A130
Slow loris A196
→ Phenylacetyltaurine
In: Bushbaby A196
Cat A196
Dog A196
Ferret A196, A394, A2058, A3046
Hamster A196
Hen A395
Lobster C38
Man A196
Monkey A196
Mouse A196
Pigeon A196, A395
Rabbit A196
Rat A196
Slow loris A196
Contrast (in rat) C308
→ L-Phenylalanine
In: Microorganism A1164, H960
Ruminococcus C793
→ 4-Phenylbutyric acid
In: Rat B119
→ 2-Phenylethanol
In: Nocardia F651
See also: F445
→ 3-Phenylpropionic acid
In: Bacteria H960
→ Toluene
In: Clostridium D485

N-Phenylacetohydroxamic acid
Formed from: Nitrosobenzene

Phenylacetone (1-phenylpropan-2-one)
Formed from: D-Amphetamine
L-Amphetamine
N-Benzyl-N,α-dimethyl-β-phenethylamine
Benzyl methyl ketoxime
(*R*)-2-Hydroxy-2-phenylpropionitrile
N-Isopropylamphetamine
N-Methylamphetamine
α-Methyl-β-phenethylhydroxylamine
2-Nitro-1-phenylpropane
1-Phenyl-2-propanol
→ Benzyl alcohol
Probably in: Horse A915
Man B147
Marmoset B147
Mycobacterium B380
Pampas monkey B147
Rabbit A3621, B649
Rat A1982, B147
Rhesus monkey B147
Vervet B147
→ Phenylacetone cyanhydrin
In: Almond J334
Apple J334
→ 1-Phenyl-2-propanol
In: Guinea pig A3738
Rabbit A1398, A1692, A3451, A3621, A3738, B755, C25
Rat A3738, B755
See also: A915, A3013, A3315, B380
→ 1-Phenyl-2-propanon-1-ol
In: Rabbit A3451, A3621, B649
Rat B755

Phenylacetone cyanhydrin
Formed from: Phenylacetone

Phenylacetonitrile (benzyl cyanide)
Formed from: Benzyl glucosinolate
→ Mandelonitrile
In: Millipede A3389
See also: A2262, F485
→ Phenylacetic acid
In: Bacillus J508
Rhodococcus G172
See also: K514

ω-Phenylacetophenone (desoxybenzoin)
→ 1,2-Diphenylethanol
In: Helminthosporium G427

Phenylacetyl-L-aspartic acid
Probably formed from: 6-O-Phenylacetyl-β-D-glucose

Phenylacetyl-L-carnitine
Formed from: Phenylacetic acid

Phenylacetyl CoA
Formed from: Phenylacetic acid
Phenylpyruvic acid
→ Benzylpenicillin

In; Aspergillus F646
Penicillium D12, F646
→ Phenylglyoxylic acid
In: Thauera J397, J665, K173

Phenylacetylene
→ Phenylacetic acid
In: Rat E461

Phenyl-N-acetyl-β-D-glucosaminide
→ Phenol
In: Beef A1940
→ Phenyl-N-acetyl-β-D-glucosaminyl-β-(1→6)-N-acetyl-β-D-glucosaminide
In: Aspergillus A725
→ Phenyl-N,N-diacetyl-β-chitobioside
In: Aspergillus A725

Phenyl-N-acetyl-β-D-glucosaminyl-β-(1→6)-N-acetyl-β-D-glucosaminide
Formed from: Phenyl-N-acetyl-β-D-glucosaminide

6-O-Phenylacetyl-β-D-glucose
Formed from: Phenylacetic acid
→ Phenylacetyl-L-aspartic acid
Probably in: Glycine A3679

Phenylacetyl-β-D-glucuronic acid
Formed from: Phenylacetic acid

Phenylacetylglutamine
Formed from: Phenylacetic acid

S-(Phenylacetyl)glutathione
→ Benzylpenicillin
In: Penicillium F111

Phenylacetylglycine
Formed from: Phenylacetic acid
→ Benzylpenicillin
In: Escherichia D501
→ Phenylacetic acid
In: Escherichia D501

Phenylacetyltaurine
Formed from: Phenylacetic acid

Phenylacetyltropine
→ 6β-Hydroxyphenylacetyltropine
In: Hyoscyamus E350

L-Phenylalaninamide
→ L-Phenylalanine
In: Aspergillus A3607
Pseudomonas H435
→ Phenylpyruvic amide
In: Sporosarcina E445

D-Phenylalanine
Formed from: N-Acetyl-D-phenylalanine
5-Benzylhydantoin
L-Phenylalanine
→ Phenethylamine
In: Mouse A3612
→ L-Phenylalanine
In: Pseudomonas F110
→ Phenylpyruvic acid
In: Aspergillus A2345
Candida E97
Chicken C114
Frog C114
Man C114
Monkey C114
Mouse C114, C219
Nocardia E606, F394
Pig C114
Rat C114
Saccharomyces A3433
See also: A2691, B777, C134

L-Phenylalanine
Formed from: N-Acetyl-L-phenylalanine
N-Alanyl-L-phenylalanine
α-Aminocinnamic acid
L-Aspartyl-L-phenylalanine
5-Benzylhydantoin
Carbamoyl-L-phenylalanine
Carbobenzoxy-L-lysyl-L-phenylalanine
α-Chloracetamidocinnamic acid
Cinnamic acid
Glycyl-L-phenylalanine
N-Histidyl-L-phenylalanine
N-Leucyl-L-phenylalanine
Ochratoxin A
Ochratoxin B
Phenylacetic acid
L-Phenylalaninamide
D-Phenylalanine
L-Phenylalanylglycine
L-Phenylalanyl-β-naphthylamide
L-Phenylalanyl-*p*-nitroanilide
L-Phenylalanyl-L-phenylalanine
Phenylpyruvic acid
Pretyrosine
N-Seryl-L-phenylalanine

L-Tyrosine
Probably formed from: Aspartame
N-Formylcarbamoylaspartame
→ N-Adenylylphenylalanine
In: Penicillium K906
→ Alloimperatorin methyl ester
In: Thamnosma A1010
→ Apohyoscine
In: Datura A1638
→ Alloimperatorin methyl ether
In: Thamnosma A1010
→ 4-Allylphenol
Ocimum A1494
→ Arborine
In: Glycosmis A538
→ Benzoic acid
In: Anacystis A2792
Bacillus A1103
Penicillium E376
Rat A2144
→ N-Benzoylphenylalanine
In: Penicillium B808
See also: E376
→ Benzylaldoxime
In: Carica H947
→ N-(1-(*R*)-(Carboxy)ethyl)-L-phenylalanine
In: Arthrobacter J212
→ Cinnamic acid
In: Eubacterium H385
Lunaria J502
Nicotiana J819
Potato A606, J691
Penicillium E376
Peptostretococcus J305
Pimpinella H667
Rhizoctonia A1466
Rhodotorula H372, J34
Sweet potato F105
Ustilago J231
Zea A607
See also: A206, A2517, A2571, C646, C734
→ Curcumin
In: Curcuma A1147
→ Cyclopenine
In: Penicillium A2683
→ Cyclopeptine
In: Penicillium G535
→ Dechlorochloroflavonin
In: Aspergillus A154
→ Deoxyshikonin
In: Lithospermum H556
→ *cis*-2,3-Dihydro-2,3-dihydroxy-L-phenylalanine
In: Bacterium B871, B950
→ Ergotamine
In: Claviceps D946
→ Estragole
Ocimum A1494
→ Eucomin
In: Eucomis A228
→ Eugenol
Cinnamomum A2517
Ocimum A1494
→ Hydrangenol
In: Conocephalum A1924
→ N-Hydroxy-L-phenylalanine
In: Harpaphe A2262, A3389
→ Isoimperatorin
In: Thamnosma A1010
→ Isopimpinellin
In: Thamnosma A1010
→ Kievitone
In: Cowpea A225
→ Littorine
In: Datura A1638
→ Maclurin
Probably in: Anemarrhena B563
→ Mandelonitrile
Probably in: Millipede A503
→ 5-Methoxypodophyllotoxin
In: Linum F796
→ *p*-Methoxypropenylbenzene
In: Pimpinella A1253
→ 5-O-Methylxylerythrin
In: Peniophora A2737, A2738
→ Naringin
In: Grapefruit F227
→ Ochratoxin A
In: Aspergillus A608
→ Peniophorin
Peniophora A2737, A2738
→ Peniophoronin

Peniophora A2737, A2738
→ Phenethylamine
In: Bacillus K383
Bean E444
Man B566
Micrococcus A336, C62, G918
Mouse A3612
Papaver H427
Rabbit A1717
Rat A1715, B566, D6, E735
See also: A2848, C465, C819
→ Phenylacetamide
In: Pseudomonas C444, C546
→ D-Phenylalanine
In: Pseudomonas F110
→ α-D-L-Phenylalanine (D = ^{2}H)
In: Pseudomonas J883
→ β-Phenylalanine
Probably in: Taxus G738
→ β-Phenylhydracrylic acid
In: Cucumber K679
→ Phenylpyruvic acid
In: Acinetobacter C817
Anchusa E436
Arthrobacter A3208
Aspergillus A2345, C484
Azospirillum J22
Bacillus A1103, E358
Bacterium A2005, B950
Brevibacterium B266, D592, D724, E642
Bushbean A566, E444, E500
Candida A2483, E97
Cerastes G417
Chromobacterium E20
Clostridium A2571
Coprinus K563
Corynebacterium B266
Escherichia A3548, K715
Hansenula D978
Klebsiella B732
Man B677
Monkey A3570, K717
Morganella H307
Mung bean H826
Nectria A2156
Nocardia E606
Pseudomonas B672, C444, C546, D246
Pyrococcus H437
Rat A1715, A2144, A2268, A2339, A2342, A2958, A3416, A3570, B379, H608, K717, K718, K719
Rhizobium F268
Rhizoctonia A1466
Rhodococcus E451, K558
Saccharomyces A3433, C533
Sporosarcina E445
Streptomyces A2307
Thauera J397
Thermoactinomyces G222
Thermococcus H116
Trimeresurus E770
Trypanosoma A1619, G894
See also: A329, A364, A923, A2691, B444, C646
→ Podophyllotoxin
Juniperus J715
→ Precocene I
In: Ageratum F667
→ Precocene II
In: Ageratum F667
→ Psilotin
In: Psilotum C352
→ Resveratrol
In: Grape A2361
→ Sporopollenin
In: Tulip H29
→ Tenellin
In: Beauveria A2580
→ 2′4,4′-Trihydroxychalcone
In: Pueraria A2373
See also: A2742
→ (*S*)-Tropic acid
In: Datura A539, E420
→ L-*o*-Tyrosine
In: Bacteria B29, B123
Beef D524
Rat B484, C885, D701, E69, E574
With myeloperoxidase G265
→ L-*m*-Tyrosine
In: Bacteria B29, B123
Beef D524, F623
Rat B484, C885, D701, E69, E574

With myeloperoxidase G265
See also: A2141, C465, D6
→ L-Tyrosine
In: Beef A383, A1341, D542, F623
Flavobacterium B455
Man A1900, B777, C134, D277, D703, F736
Microorganism K390
Portulaca B930
Pseudomonas A757, C644
Rat A964, A1900, B484, C802, C885, D701, E574, G290
Rhodobacter J874
Streptomyces K317
Tobacco hornworm B574
With myeloperoxidase G265
See also: A1015, C465, D6, H385, J305, K169
→ Vignafuran
In: Vigna A3765
→ Xylerythrin
In: Peniophora A2737, A2738

α-D-L-Phenylalanine (Deuterium labelled)
Formed from: L-Phenylalanine

β-Phenylalanine
Probably formed from: L-Phenylalanine
→ Phenylisoserine
Probably in: Taxus G738

L-Phenylalanine methyl ester
→ Methyl phenylpyruvate
In: Flavobacterium C206
Sporosarcina E445

L-Phenylalanine mustard (4-(bis(2-chloroethyl) amino)-L-phenylalanine)
→ 4-(Bis(2-hydroxyethyl)amino)-L-phenylalanine
In: Mouse C188

L-Phenylalaninol
→ N-(1-(*R*)-(Carboxy)ethyl)-L-phenylalaninol
In: Arthrobacter J212
→ 2-Oxo-3-phenyl-1-propanol
In: Sporosarcina E445

L-Phenylalanyl-3,4-dihydroxyphenethylamine
L-Phenylalanyl-L-phenylalanyl-3,4-dihydroxyphenethylamine
L-Prolyl-L-phenylalanyl-3,4-dihydroxyphenethylamine
→ 3,4-Dihydroxyphenethylamine
In: Rat H432

L-Phenylalanylglycine
→ L-Phenylalanine
In: Hen G654

L-Phenylalanylglycyl-L-leucyl-*p*-nitroanilide
→ *p*-Nitroaniline
In: Escherichia E715

L-Phenylalanyl-β-naphthylamide
→ β-Naphthylamine
In: Agave A2745
Beef A526
Euonymus D216
Man A1219, A1397
Mytilus A1514
Rabbit A1589, B334
Rat A523
→ L-Phenylalanine
In: Agave A2745
Beef A526
Euonymus D216
Man A1219, A1397
Mytilus A1514
Rabbit A1589, B334
Rat A523

L-Phenylalanyl-*p*-nitroanilide
→ *p*-Nitroaniline
In: Euonymus C907
Nocardia G343
Rat C173
→ L-Phenylalanine
In: Euonymus C907
Nocardia G343
Rat C173

L-Phenylalanyl-L-phenylalanine
→ L-Phenylalanine
In: Brevibacterium E587

L-Phenylalanyl-L-phenylalanyl-3,4-dihydroxyphenethylamine
→ 3,4-Dihydroxyphenethylamine
In: Rat H432
→ L-Phenylalanyl-3,4-dihydroxyphenethylamine
In: Rat H432

L-Phenylalanyl-L-prolyl-3,4-dihydroxyphenethylamine
→ 3,4-Dihydroxyphenethylamine

In: Rat H432
→ L-Prolyl-3,4-dihydroxyphenethylamine
In: Rat H432

2-(Phenylamino)ethyl 5-(5,5-dimethyl-2-oxo-1,3,2-dioxaphosphorinan-2-yl)-1,4-dihydro-2,6-dimethyl-4-(*m*-nitrophenyl)-3-pyridinecarboxylate
Formed from: Efonidipine
→ 2-Aminoethyl 5-(5,5-dimethyl-2-oxo-1,3,2-dioxaphosphorinan-2-yl)-1,4-dihydro-2,6-dimethyl-4-(*m*-nitrophenyl)-3-pyridinecarboxylate
Probably in: Man H523
Rat H625

3-((4-(Phenylamino)phenyl)azo)benzenesulphonic acid
See: Acid Yellow 36

β-Phenylamphetamine
Formed from: Remacemide

Phenyl azide
→ *p*-Hydroxyphenyl azide
In: Mouse G8

1-Phenylazo-2-naphthol (sudan 1)
→ Benzyldiazonium
In: Tulipa K539
→ 1-(3,4-Dihydroxyphenylazo)-2-hydroxynaphthalene
Probably in: Rat H362
→ 2,6-Dihydroxy-1-phenylazonaphthalene
In: Tulipa K539
→ 2-Hydroxy-1-(4-hydroxyphenylazo)naphthalene
In: Tulipa K539

ω-Phenylazotoluene
→ ω-Phenylazoxytoluene
In: Guinea pig G258, G259
Hamster G258, G259
Mouse G258, G259
Rabbit G258, G259
Rat G258, G259

ω-Phenylazoxytoluene
Formed from: ω-Phenylazotoluene

α-Phenylbenzeneacetic acid 2-(diethylamino)ethyl ester
See: Adiphenine

α-Phenylbenzeneacetic acid 1-ethyl-3-piperidinyl ester
See: Piperidolate

2-Phenyl-1*H*-benzimidazole
See: Phenzidole

Phenyl benzoate
→ Benzoic acid
In: Pseudomonas K206
→ Phenol
In: Pseudomonas K206

2-Phenyl-1,2-benzoisoselenazol-3(2*H*)-one
See: Ebselen

3-Phenyl-2*H*-1-benzopyran
See: Isoflav-3-ene

2-Phenyl-4*H*-1-benzopyran-4α-ol
See: Flavan-4α-ol

2-Phenyl-4*H*-1-benzopyran-4β-ol
See: Flavan-4β-ol

2-Phenyl-4*H*-1-benzopyran-4-one
See: Flavone

3-Phenyl-4*H*-1-benzopyran-4-one
See: Isoflavone

2-Phenylbenzyloxyresorufin
→ Resorufin
In: Alligator J855
Rat J855

***endo*-2-Phenylbicyclo[2.2.1]heptane-2-carboxylic acid 3-diethylaminopropyl ester**
→ *endo*-5-Hydroxy-*endo*-2-phenylbicyclo[2.2.1]heptane-2-carboxylic acid 3-diethylaminopropyl ester
In: Rat F5
→ *exo*-5-Hydroxy-*endo*-2-phenylbicyclo[2.2.1]heptane-2-carboxylic acid 3-diethylaminopropyl ester
In: Rat F5
→ *exo*-6-Hydroxy-*endo*-2-phenylbicyclo[2.2.1]heptane-2-carboxylic acid 3-diethylaminopropyl ester
In: Rat F5

1-Phenylbutane-1,2-diol
Formed from: *trans*-1-Phenylbut-1-ene oxide

1-Phenylbutane-1,3-diol
Probably formed from: Phenyl-1-butanol

2-Phenylbutane-2,3-diol
Formed from: *cis*-2-Phenylbutene-2,3-oxide

trans-2-Phenylbutene-2,3-oxide

1-Phenyl-1-butanol
Formed from: *n*-Butylbenzene
→ Butyrophenone
Probably in: Rat H689
→ 1-Phenyl-1-butane-1,3-diol
Probably in: Rat H689

1-Phenyl-2-butanol
Formed from: 1-Phenyl-2-butanone

1-Phenyl-3-butanol
Formed from: *n*-Butylbenzene
1-Phenyl-3-butanone
1-Phenyl-3-butyl acetate
→ 1-Phenyl-3-butanone
Probably in: Rat H689

2-Phenyl-1-butanol
Formed from: 2-Phenylbutyric acid

2-Phenyl-2-butanol
Probably formed from: *sec*-Butylbenzene

3-Phenyl-1-butanol
Formed from: 3-Phenylbutyric acid
Probably formed from: *sec*-Butylbenzene

4-Phenyl-1-butanol
Formed from: 4-Phenylbutyric acid

1-Phenyl-1-butanone
See: Butyrophenone

1-Phenyl-2-butanone (benzyl ethyl ketone)
→ 1-Phenyl-2-butanol
In: Rabbit A3451
→ 1-Phenyl-2-butanon-1-ol
In: Rabbit A3451

1-Phenyl-3-butanone
Formed from: *trans*-4-Phenyl-3-buten-2-one
Probably formed from: 1-Phenyl-3-butanol
→ 1-Phenyl-3-butanol
In: Corynebacterium J893
Lactobacillus G148
With alcohol dehydrogenase G721

1-Phenyl-2-butanon-1-ol
Formed from: 1-Phenyl-2-butanone

Phenylbutazone (butazolidine, 4-butyl-1,2-diphenyl-3,5-pyrazolidinedione)
→ 4-Hydroxyphenylbutazone
In: Rat D353
→ γ-Hydroxyphenylbutazone
In: Goat G901
Horse D310
Man A3158
Rat A296
See also: B322, B419, C287
→ Oxyphenylbutazone
In: Horse D310
Man A1303, A3158
Rat A296, D353
See also: B322, C287
→ Phenylbutazone-4-β-D-glucuronide
Species unclear A3243
See also: C287
→ Phenylbutazone-O-β-D-glucuronide
Probably in: Man C287

Phenylbutazone-4-β-D-glucuronide
Formed from: Phenylbutazone

Phenylbutazone-O-β-D-glucuronide
Probably formed from: Phenylbutazone

***trans*-1-Phenylbut-1-ene oxide**
→ 1-Phenylbutane-1,2-diol
In: Man C386
Rat C386

***cis*-2-Phenyl-2-butene-2,3-oxide**
→ 2-(S-Glutathionyl)-2-phenyl-3-butanol
Probably in: Sheep A2282
→ 2-Phenylbutane-2,3-diol
In: Man E969

***trans*-2-Phenyl-2-butene-2,3-oxide**
→ 2-(S-Glutathionyl)-2-phenyl-3-butanol
Probably in: Sheep A2282
→ 2-Phenylbutane-2,3-diol
In: Man E969

4-Phenylbut-2-enoic acid
Formed from: 1-Phenyldodecane

4-Phenylbut-3-enoic acid
Formed from: 1-Phenyldodecane
→ Cinnamic acid
In: Nocardia A1329
→ 4-(*o*-Hydroxyphenyl)but-3-enoic acid
Probably in: Nocardia A1329

4-Phenyl-3-buten-2-ol
Formed from: *trans*-4-Phenyl-3-buten-2-one

1-Phenyl-1-buten-3-one
See: Benzalacetone

***trans*-4-Phenyl-3-buten-2-one** (methyl styryl ketone)
→ Benzaldehyde

Probably in: Mouse J293
Rat J160
→ 4-(S-Glutathionyl)-4-phenylbutan-2-one
Probably in: Armyworm E389
Cat C562
Dog C562
Guinea pig C562
Mouse C562, J293
Rabbit C562
Rat C562, J583
Quail C562
Trout C562
→ 4-Hydroxy-4-phenylbutan-2-ol
In: Mouse J293
→ Phenylacetic acid
In: Mouse J293
Rat J160
→ 1-Phenyl-3-butanone
In: Dog J731
Rat J731
→ 4-Phenyl-3-buten-2-ol
In: Dog J731
Mouse J293
Rat J160, J731

2-Phenyl-2-butenoylurea
Probably formed from: 3-Hydroxy-2-phenyl-2-butyrylurea

1-Phenyl-3-butyl acetate
→ 1-Phenyl-3-butanol
In: Pseudomonas K790

1-Phenyl-2-butylamine
→ 1-Phenyl-2-butylhydroxylamine
In: Rabbit D91

4-Phenylbutylamine
→ 4-Phenylbutyraldehyde
In: Arthrobacter H912
Pig A72

1-Phenyl-2-butylhydroxylamine
Formed from: 1-Phenyl-2-butylamine

4-Phenylbutyraldehyde
Formed from: 4-Phenylbutylamine
4-Phenylbutyric acid

4-Phenylbutyraldoxime
→ 4-Phenylbutyronitrile
In: Bacillus K411

2-Phenylbutyric acid
Formed from: 2-Phenylbutyrylurea
→ 2-Phenyl-1-butanol
In: Nocardia F651

3-Phenylbutyric acid
Formed from: 2-Phenyldodecane
→ 3-Phenyl-1-butanol
In: Nocardia F651

4-Phenylbutyric acid
Formed from: Oxazepam γ-phenylbutyrate
Phenylacetic acid
1-Phenyldodecane
→ Cinnamic acid
In: Nocardia A1329
See also: D746
→ *cis*-2,3-Dihydro-2,3-dihydroxyphenylbutyric acid
In: Bacterium B871, B950
→ 4-(*o*-Hydroxyphenyl)butyric acid
Probably in: Nocardia A1329
→ Phenylacetic acid
In: Man H550
See also: A1011, D746
→ 4-Phenyl-1-butanol
In: Nocardia F651
→ 4-Phenylbutyraldehyde
In: Nocardia G207

4-Phenylbutyronitrile
Formed from: 4-Phenylbutyraldoxime

2-Phenylbutyrylurea (pheneturide)
→ 3-Hydroxy-2-phenylbutyrylurea
In: Rat E320
→ 2-(*p*-Hydroxyphenyl)butyrylurea
In: Man E320
Rat E320
→ 2-Phenylbutyric acid
In: Man E320
Rat E320

5-(N-Phenylcarboxamido)barbituric acid
Formed from: Merbarone
→ 5-(N-(*p*-Hydroxyphenyl)carboxamido) barbituric acid
Probably in: Man F724

5-(N-Phenylcarboxamido)-2-thiobarbituric acid
See: Merbarone

2-Phenylchroman
→ 2-(*p*-Hydroxyphenyl)chroman
In: Absidia G14
3-Phenylchroman
→ 3-(*p*-Hydroxyphenyl)chroman
In: Absidia G14
Phenylcyclohexane
→ *cis*-3-Phenylcyclohexanol
In: Pseudomonas J55
→ *cis*-4-Phenylcyclohexanol
In: Pseudomonas J55
→ *trans*-4-Phenylcyclohexanol
In: Pseudomonas J55
***cis*-3-Phenylcyclohexanol**
Formed from: Phenylcyclohexane
***cis*-4-Phenylcyclohexanol**
Formed from: Phenylcyclohexane
***trans*-4-Phenylcyclohexanol**
Formed from: Phenylcyclohexane
2-Phenylcyclohexanone
→ *cis*-2-(2,3-Dihydro-2,3-dihydroxyphenyl) cyclohexanone
In: Escherichia K549
1-Phenyl-1-cyclohexene
→ *cis*-2-(2,3-Dihydro-2,3-dihydroxyphenyl) cyclohexene
In: Escherichia K549
→ 1,2-Dihydroxy-1-phenylcyclohexane
In: Mouse C641
Rat C829
→ 3-Hydroxy-1-phenyl-1-cyclohexene
In: Mouse C641
Rat C829
→ 6-Hydroxy-1-phenyl-1-cyclohexene
In: Mouse C641
1-Phenylcyclohexylamine
Formed from: N-Ethyl-1-phenylcyclohexylamine
1-(Hexahydroazepin-1-yl)-1-phenylcyclohexane
Phencyclidine
1-Phenyl-1-(pyrrolidin-1-yl)cyclohexane
N-(1-Phenylcyclohexyl)-5-aminopentan-1-al
Formed from: Phencyclidine
→ 5-(N-(1-Phenylcyclohexyl)amino)pentanoic acid
Probably in: Cat C873
Dog B842
Monkey C873
Mouse D504
Rabbit C598, C873
Rat C873
→ 5-(N-(1-Phenylcyclohexyl)amino)pentanol
Probably in: Cat C873
Monkey C873
Rabbit C873
Rat C873
5-(N-(1-Phenylcyclohexyl)amino)pentanoic acid
Probably formed from: N-(1-Phenylcyclohexyl)-5-aminopentan-1-al
5-(N-(1-Phenylcyclohexyl)amino)pentanol
Probably formed from: N-(1-Phenylcyclohexyl)-5-aminopentan-1-al
1-(1-Phenylcyclohexyl)piperidine
See: Phencyclidine
(1*R*,2*S*)-rel-2-phenylcyclopropanamine
See: Tranylcypromine
S-Phenyl-L-cysteine
Formed from: Thiophenol
Phenyl N,N′-diacetyl-β-chitobioside
Formed from: Phenyl N-acetyl-β-D-glucosaminide
→ Phenyl N,N′,N″-triacetyl-β-chitotrioside
Probably in: Aspergillus A725
2-Phenyl-1,3-di-(4-pyridyl)-2-propanol
→ 2-Phenyl-1,3-di-(4-pyridyl)-2-propanol-N-oxide
In: Dog A3469
Man A3469
Rat A3469
2-Phenyl-1,3-di-(4-pyridyl)-2-propanol-N-oxide
Formed from: 2-Phenyl-1,3-di-(4-pyridyl)-2-propanol
2-Phenyl-1,3-dithiolane
→ 2-Phenyl-1,3-dithiolane-1-oxide
In: Rabbit F355
2-Phenyl-1,3-dithiolane-1,3-dioxide
Formed from: 2-Phenyl-1,3-dithiolane-1-oxide
2-Phenyl-1,3-dithiolane-1-oxide
Formed from: 2-Phenyl-1,3-dithiolane
→ 2-Phenyl-1,3-dithiolane-1,3-dioxide
In: Rabbit F355
1-Phenyldodecane
→ 4-Phenylbut-2-enoic acid

In: Nocardia A1329
→ 4-Phenylbut-3-enoic acid
In: Nocardia A1329
→ 4-Phenylbutyric acid
In: Acinetobacter D746
Nocardia A1329, A2215

2-Phenyldodecane
→ 3-Phenylbutyric acid
In: Nocardia A2215

1-Phenyldodecane-*p*-sulphonic acid
→ 6-Phenylhexanoic acid
In: Cladosporium A1011

2-Phenyldodecane-*p*-sulphonic acid
→ 4-Phenylhexanoic acid
In: Pseudomonas A2215

3-Phenyldodecane-*p*-sulphonic acid
→ 3-Phenylpentanoic acid
In: Pseudomonas A2215

1,4-Phenylenebis(methylene)selenocyanate
→ Terephthalic acid
In: Rat J811
→ Tetraselenocyclophane
In: Rat J811

***o*-Phenylenediamine** (1,2-diaminobenzene)
Formed from: 2-Amino-4′-dimethylaminoazobenzene
o-Nitroaniline
Probably formed from: Methyl benzimidazol-2-ylcarbamate
→ 2,3-Diaminophenazine
With haemoglobin K510

***m*-Phenylenediamine** (1,3-diaminobenzene)
Formed from: *m*-Acetamidoaniline
3-Amino-4′-dimethylaminoazobenzene
m-Nitroaniline
→ *m*-Acetamidoaniline
Probably in: Rat E888

***p*-Phenylenediamine** (1,4-diaminobenzene)
Formed from: *p*-Acetamidoaniline
Acid red 151
4-Aminoazobenzene
4-Amino-4′-dimethylaminoazobenzene
4,4′-Diaminoazobenzene
p-Methyl red
Mordant orange 1
p-Nitroaniline
→ *p*-Acetamidoaniline
Probably in: Pseudomonas E395
→ N,N′-Diacetyl-*p*-phenylenediamine
In: Rat B308
→ 2,5-Diamino-N,N′-di-(4-aminophenyl)-2,5-cyclohexadiene-1,4-diimine
In: Horseradish K518
→ *p*-Quinone (identity assumed for lack of evidence)
Probably in: Botrytis A1930, A3974
Coriolus A3974, E369, K677
Nitrosomonas D564
Peach A3974
Spruce A3974
Trametes A3974
With caeruloplasmin A677

1-Phenylethanediol
See: Phenylglycol

Phenylethane-1,2-dione
See: Phenylglyoxal

Phenylethane-*p*-sulphonic acid
→ 4-Ethylcatechol
In: Alcaligenes A1323

1-Phenylethanol ((*R*) and (*S*) isomers)
Formed from: Acetophenone
N-((3,4-Dihydro-7-(piperazin-1-yl)-1(1*H*)-isoquinolinon-2-yl)acetyl)-3(*S*)-(ethynyl)-β-alanine phenethyl ester
Ethylbenzene
1-Phenylethyl acetate
Styrene
Probably formed from: Ethylbenzene hydroperoxide
→ Acetophenone
In: Azoarchus K241
Corynebacterium J893
Geotrichum H749
Lactobacillus G148
Nitrosomonas H219
Pseudomonas J219
See also: A287, A1745, B144, D223, D482, F870
Contrast (in Geotrichum) H749
→ (+)-*cis*-1-(2,3-Dihydro-2,3-dihydroxyphenyl)ethanol

In: Escherichia K549
Pseudomonas A914, K549
→ 1-(2,3-Dihydroxyphenyl)ethanol
In: Nocardia A3582
→ 1-Phenylethyl-β-D-glucoside
In: Almond H558
→ 1-Phenylethyl-β-D-glucuronide
In: Rat C670
→ Phenylglycol
Probably in: Man D482
Pseudomonas F870
Rat C670, D223

2-Phenylethanol
Formed from: Ethylbenzene
Phenylacetaldehyde
Phenylacetic acid
2-Phenylethyl acetate
Styrene
→ Phenylacetaldehyde
In: Bacteria K268
Rhodopseudomonas D197
Xanthobacter F440
See also: F870, H334
Contrast (in Corynebacterium) H881
→ 2-Phenylethyl-β-D-glucoside
In: Dioscoreophyllum F445
Mentha F786
Pyrus F786
Rosmarinus F786
Vitis F786

Phenylethanolamine (D and L isomers)
Formed from: Phenethylamine
→ Mandelaldehyde
In: Hedgehog A434
Rat A2950, A3258, A3664, B389
See also: A1715, A2261, A2415, G829, J287
→ N-Methylphenylethanolamine
In: Beef A368
Man A498, A2823
Monkey A2823
Rabbit A498
Rat A2823

1-Phenylethanone
See: Acetophenone

1-Phenylethyl acetate
→ 1-Phenylethanol
In: Helianthus J661
Solanum J661

2-Phenylethyl acetate
→ 2-Phenylethanol
In: Pseudomonas K790

1-Phenylethylamine ((*R*) and (*S*) isomers)
Formed from: (*S*)-N-Acetyl-1-phenylethylamine
3,3-Diphenyl-N-(1-phenylethyl) propylamine
Octyl 1-phenylethylamine oxalamate
1-Phenethylhydrazine
→ Acetophenone
With (*S*)-α-phenylethylamine pyruvate transaminase K449

2-Phenylethylamine
See: Phenethylamine

1-Phenylethyl-β-D-glucoside
Formed from: 1-Phenylethanol

2-Phenylethyl-β-D-glucoside
Formed from: 2-Phenylethanol

1-Phenylethyl-β-D-glucuronide
Probably formed from: 1-Phenylethanol

(+)-1-(1-Phenylethyl)-1*H*-imidazole
Probably formed from: (+)-1-(1-Phenylethyl)-1*H*-imidazole-5-carboxylic acid

(−)-1-(1-Phenylethyl)-1*H*-imidazole
Probably formed from: (−)-1-(1-Phenylethyl)-1*H*-imidazole-5-carboxylic acid

(+)-1-(1-Phenylethyl)-1*H*-imidazole-5-carboxylic acid
Formed from: (+)-Etomidate
→ (+)-1-(1-Phenylethyl)-1*H*-imidazole
Probably in: Rat A2633

(−)-1-(1-Phenylethyl)-1*H*-imidazole-5-carboxylic acid
→ (−)-1-(1-Phenylethyl)-1*H*-imidazole
Probably in: Rat A2633

N-(1-Phenylethyl)-2-propynylamine
Formed from: N-Methyl-N-(1-phenylethyl)-2-propynylamine

N-((2-Phenylethynyl)sulphonyl)morpholine
→ N-((2-Phenyloxo-2-ethane)sulphonyl) morpholine
In: Rat J73

Phenylformamidinesulphenic acid
Formed from: Phenylthiourea
→ Phenylformamidinesulphinic acid
Probably in: Pig B239
Phenylformamidinesulphinic acid
Probably formed from: Phenylformamidinesulphenic acid
→ Phenylformamidinesulphonic acid
Probably in: Pig B239
Phenylformamidinesulphonic acid
Probably formed from: Phenylformamidinesulphinic acid
Phenyl-β-D-galactoside
→ Phenol
In: Sugar cane B488
Sulfolobus G447
Phenyl-α-glucoside
Formed from: Phenyl-α-maltoside
Phenyl-β-D-glucoside
Formed from: Phenol
Phenyl-6′-O-malonyl-β-D-glucoside
→ Phenol
In: Bacillus A3029
Rat E14
Sulfolobus G447
Phenyl-β-D-glucuronide
Formed from: Phenol
→ Phenol
In: Goldfish C408
2-Phenylglutarimide
Formed from: Glutethimide
S-Phenylglutathione
Formed from: Benzene
→ N-Acetyl-S-phenyl-L-cysteine
Probably in: Rat A2100
1-O-Phenylglycerol
Formed from: Doxaminol
→ 1-Hydroxy-3-phenoxy-2-propyl sulphate
Probably in: Dog F550
→ 1-O-(*p*-Hydroxyphenyl)glycerol
Probably in: Dog F447, F550
D-Phenylglycine
Formed from: D-Phenylglycine methyl ester
Phenylglyoxylic acid
5-Phenylhydantoin
N-(*o*-Phthalyl)-D-phenylglycine
→ D-*p*-Hydroxyphenylglycine
In: Streptomyces K317
→ Phenylglyoxylic acid
In: Flavobacterium E355
Pseudomonas E42, J353
L-Phenylglycine
Formed from: N-Acetyl-L-phenylglycine
→ α-D-L-Phenylglycine (D = ^{2}H)
In: Pseudomonas J883
→ Phenylglyoxylic acid
In: Flavobacterium E355
Pseudomonas E42
α-D-L-Phenylglycine (D = ^{2}H)
Formed from: L-Phenylglycine
D-Phenylglycine methyl ester
→ Ampicillin
In: Escherichia D501
→ Cephalexin
In: Xanthomonas H64
→ D-Phenylglycine
In: Escherichia D501
Phenylglycol (1-phenylethanediol)
Formed from: Mandelaldehyde
D-Mandelic acid
L-Mandelic acid
Styrene oxide
Probably formed from: 2-Oxo-2-phenylethanol
1-Phenylethanol
Phenylglyoxal
→ Mandelaldehyde
Probably in: Bordetella G665
Man A688, D482, F330
Mouse H386
Nitrosomonas H219
Rat A569, A1713, A1745, A2093, A3833, D223, F496
→ 2-Oxo-2-phenylethanol
Probably in: Guinea pig C605
Nitrosomonas H219
Rabbit H920
Rat C605
→ Phenylglycol-2-acetate
Probably in: Rabbit A3489

Phenylglycol-2-acetate
Probably formed from: Phenylglycol
Phenylglyoxal (phenylethane-1,2-dione)
Probably formed from: 2-Oxo-2-phenylethanol
→ Mandelic acid
In: Rat J287
See also: D365
→ Phenylglyoxylic acid
Probably in: Rat A1745, D365
Phenylglyoxylic acid
Formed from: 3-Hydroxyflavone
D-Mandelic acid
L-Mandelic acid
Phenylacetyl CoA
Styrene
Probably formed from: Phenylglyoxal
→ Benzaldehyde
In: Acinetobacter D487
Flavobacterium E355
Pseudomonas E42, E714, G607
See also: A569, A3219, G333
→ Benzoyl CoA
In: Azoarcus K760
→ 2-Hydroxypropiophenone
In: Acinetobacter H403
Zygosaccharomyces K611
→ L-Mandelic acid
In: Micrococcus H668
Pseudomonas H668
Rat F496, J287
Streptococcus E271
→ D-Phenylglycine
In: Pseudomonas J353, K503
N^7-Phenylguanine
Formed from: Benzene
1-Phenylheptane
→ Cinnamic acid
In: Methylosinus B682
4-Phenylhexanoic acid
Formed from: 2-Phenyldodecane-*p*-sulphonic acid
6-Phenylhexanoic acid (phenylcaproic acid)
Formed from: 1-Phenyldecane-*p*-sulphonic acid
→ 4-Phenylbutyric acid
Probably in: Cladosporium A1011
6-Phenyl-5-hexen-2,4-dione
Probably formed from: 4-Hydroxy-6-phenyl-5-hexen-2-one
6-Phenyl-5-hexen-2-one
Probably formed from: 4-Hydroxy-6-phenyl-5-hexen-2-one
5-Phenylhydantoin
Formed from: 3-Ethyl-5-phenylhydantoin
→ D-Phenylglycine
In: Pseudomonas E391, E450
→ 2-Phenylhydantoic acid
Probably in: Beef A3462
Man B113
Rat A3462
β-Phenylhydracrylic acid (3-hydroxy-3-phenylpropionic acid)
Probably formed from: Cinnamic acid
3-Oxo-3-phenylpropionic acid
L-Phenylalanine
→ Benzoic acid
In: Nicotiana K679
→ 3-Oxo-3-phenylpropionic acid
Probably in: Man A126
→ Phenylacetaldehyde
In: Flavobacterium E679
Phenylhydrazine
→ Aniline
In: Microorganism A3616
Phenylhydroxylamine
Formed from: Aniline
Nitrobenzene
Nitrosobenzene
→ Aniline
In: Japanese quail B175
Rat B174
See also: C32, K520
→ Nitrosobenzene
In: Chlorella A3531
See also: B174, C489, C797
1-Phenylimidazole
→ 1-Phenylimidazole-3-β-D-glucuronide
In: Man K42
1-Phenylimidazole-3-β-D-glucuronide
Formed from: 1-Phenylimidazole-3-β-D-glucuronide

4-Phenylimidazolidin-2-one
Probably formed from: 1-Carboxymethyl-4-phenylimidazolidin-2-one
2-Phenylindole
→ 5-Hydroxy-2-phenylindole
In: Guinea pig G631
Hamster G631
Rabbit G631
Rat G631
→ 6-Hydroxy-2-phenylindole
In: Guinea pig G631
Hamster G631
Rabbit G631
Rat G631
→ 2-(*p*-Hydroxyphenyl)indole
In: Guinea pig G631
Hamster G631
Rabbit G631
Rat G631
N-Phenylisobutyrohydroxamic acid
Formed from: Nitrosobenzene
Phenylisoserine (3-amino-2-hydroxy-3-phenylpropionic acid)
Probably formed from: β-Phenylalanine
→ Cephalomannine
Probably in: Taxus G738
→ Taxol
In: Taxus G738
N-Phenylisovalerohydroxamic acid
Formed from: Nitrosobenzene
5-Phenyl-3-isoxazolylcarboxylic acid
Probably formed from: 5-Phenyl-3-isoxazolylglycollic acid
5-Phenylisoxazol-3-yl-β-D-glucuronide
Formed from: 3-Hydroxy-5-phenylisoxazole
5-Phenyl-3-isoxazolylglycollic acid
Formed from: 3-(1-Hydroxy-2-piperidinoethyl)-5-phenylisoxazole
→ 5-Phenyl-3-isoxazolylcarboxylic acid
Probably in: Rabbit A1809
5-Phenylisoxazol-3-yl sulphate
Formed from: 3-Hydroxy-5-phenylisoxazole
L-Phenyllactic acid (L-3-phenyllactic acid)
Formed from: Fructose-phenylalanine
Phenylpyruvic acid
→ Cinnamic acid
In: Clostridium K534
→ Phenylpyruvic acid
In: Candida D975
Lactobacillus E377
Neisseria F92
Pseudomonas K95
Rat B438
Rhodotorula G774
See also: A1387
→ (*S*)-Tropic acid
In: Datura A1638
2-Phenyllactic acid
Probably formed from: 2-Phenylpropionic acid
Phenyllactonitrile (2-hydroxy-3-phenylpropionitrile)
Formed from: Phenylacetaldehyde
N-Phenyllinoleamide
→ Aniline
In: Man G937
→ 9-Hydroxy-N-phenyllinoleamide
In: Man G937
→ 13-Hydroxy-N-phenyllinoleamide
In: Man G937
Phenyl-6′-O-malonyl-β-D-glucoside
→ Phenol
In: Rat E14
→ Phenyl-β-D-glucoside
In: Rat E14
Phenyl-α-maltoside
→ Phenyl-α-glucoside
In: Buckwheat A3873
See also: C847
Phenyl-β-maltoside
→ Phenol
In: Bacillus A3029
→ Phenyl-β-maltotrioside
In: Bacillus A3029
Phenyl-β-maltotrioside
Formed from: Phenyl-β-maltoside
Phenylmercapturic acid
See: N-Acetyl-S-phenyl-L-cysteine
Phenylmercuric acetate
→ Benzene
In: Microorganism A1003
Phenylmethaneselenic acid
Formed from: Benzyl selenocyanate

1-((2*E*)-(2-(Phenylmethylene)cyclohexyl)azetidine
See: Tazadolene

N-(2-(Phenylmethylene)cyclohexyl)-β-alanine
Formed from: Tazadolene

2-(((Phenylmethyl)imino)methyl)phenol
See: Saddamine

4-Phenylmuconolactone (2,5-dihydro-5-oxo-4-phenyl-2-furanacetic acid)
Probably formed from: 3,4-Dihydroxybiphenyl

2-Phenyl-4*H*-naphtho[1,2-*b*]pyran-4-one
See: 7,8-Benzoflavone

N-Phenyl-β-naphthylamine
→ N-(*p*-Hydroxyphenyl)-β-naphthylamine
In: Dog B901
Hamster B901
Man B901
Monkey B901
Mouse B901
Pig B901
Rat B901
→ 6-Hydroxy-N-phenyl-β-naphthylamine
In: Dog B901
Hamster B901
Man B901
Monkey B901
Mouse B901
Pig B901
Rat B901

Phenyl nicotinate
→ Phenol
In: Man J141

Phenyl nitroxide radical
Formed from: Aniline

1-Phenylnonane
→ Cinnamic acid
In: Nocardia A1329
→ Hydroxyphenylpentanoic acid
In: Nocardia A1329

5-(2-(8-Phenyloctyl)phenyl-4,6-dithianonanedioic acid
→ 5-(2-(8-Phenyloctyl)phenyl-4,6-dithianonanedioic acid-β-D-glucuronic acid
In: Guinea pig G384

5-(2-(8-Phenyloctyl)phenyl-4,6-dithianonanedioic acid-β-D-glucuronic acid
Formed from: 5-(2-(8-Phenyloctyl)phenyl-4,6-dithianonanedioic acid

S-((3-Phenyl-1,2,4-oxadiazol-5-yl)methyl) glutathione
Formed from: 5-Methyl-3-phenyl-1,2,4-oxadiazole
→ N-Acetyl-S-((3-phenyl-1,2,4-oxadiazol-5-yl)methyl)-L-cysteine
Probably in: Rat A1834

2-Phenyl-1,3-oxathiolane
→ 2-Phenyl-1,3-oxathiolane sulphoxide
In: Pig F176
Rat F176

2-Phenyl-1,3-oxathiolane sulphone
Probably formed from: 2-Phenyl-1,3-oxathiolane sulphoxide

2-Phenyl-1,3-oxathiolane sulphoxide
Formed from: 2-Phenyl-1,3-oxathiolane
→ 2-Phenyl-1,3-oxathiolane sulphone
Probably in: Pig F176
Rat F176

N-((2-Phenyloxo-2-ethane)sulphonyl)morpholine
Formed from: N-((2-Phenyloxo-2-ethynyl) sulphonyl)morpholine

3-Phenylpentanoic acid
Formed from: 3-Phenyldodecane-*p*-sulphonic acid

1-Phenylpentan-1-ol (1-hydroxy-1-phenyl-*n*-pentane)
→ Benzoic acid
In: Guinea pig A414
Mouse A414
→ 1-Phenylpent-1-yl-β-D-glucuronide
In: Guinea pig A414
Mouse A414

1-Phenyl-2-pentylamine
→ 1-Phenyl-2-pentylhydroxylamine
In: Rabbit D91

N-(1-Phenyl-2-pentyl)-γ-aminobutyric acid
Formed from: Prolintane

1-Phenylpent-1-yl-β-D-glucuronide
Formed from: 1-Phenylpentan-1-ol

1-Phenyl-2-pentylhydroxylamine
Formed from: 1-Phenyl-2-pentylamine

N-(1-Phenylpentan-2-yl)pyrrolidone
Formed from: Prolintane
→ 2-(3-Carboxypropylamino)-1-phenylpentane
Probably in: Man G63

Phenyl phosphate
Formed from: Phenol
→ Phenol
In: Man A789

Phenylphosphonic acid
Probably formed from: Ethyl phenylphosphonic acid
O-Methyl phenylphosphonic acid

S-Phenylphosphorothioate
Probably formed from: S,S-Diphenylphosphorodithioate

Phenylphosphorothioic acid
Probably formed from: O-Ethyl phenylphosphorothioic acid

6-Phenylpicolinic acid
Formed from: Biphenyl

1-Phenylpiperazine
Formed from: 2,3-Dihydro-2-oxo-3-(2-(4-phenylpiperazin-1-yl)ethyl)oxazolo[4,5-*b*] pyridine
1-(2-(5,6-Dimethoxy-2-methyl-3-indolyl)ethyl)-4-phenylpiperazine
→ 1-Phenylpiperazine-N-sulphate
In: Rat E51

1-Phenylpiperazine-N-sulphate
Formed from: 1-Phenylpiperazine

4-Phenylpiperidine
→ 4-Phenylpiperidine-N-sulphate
In: Rat E51

α-Phenyl-2-piperidineacetic acid (ritanilic acid)
Formed from: Methylphenidate
→ Ethylphenidate
Probably in: Man K38
→ α-(*p*-Hydroxyphenyl)-2-piperidineacetic acid
Probably in: Dog C14, A137
→ 6-Oxo-α-phenyl-2-piperidineacetic acid
Probably in: Dog C14, A137
Rat C14
→ α-Phenyl-2-piperidineacetic acid-1-carbamate
Probably in: Rat C14

α-Phenyl-2-piperidineacetic acid-1-carbamate
Probably formed from: α-Phenyl-2-piperidineacetic acid

α-Phenyl-2-piperidineacetic acid ethyl ester
See: Ethylphenidate

4-Phenylpiperidine-4-carboxylic acid
Probably formed from: N-Methyl-4-phenylpiperidine-4-carboxylic acid

4-Phenyl-4-piperidinecarboxylic acid ethyl ester
See: Norpethidine

4-Phenylpiperidine-N-sulphate
Formed from: 4-Phenylpiperidine

3-Phenyl-3-piperidinocyclohexanol
Formed from: Phencyclidine

4-Phenyl-4-piperidinocyclohexanol
Formed from: Phencyclidine
→ 4-(*p*-Hydroxyphenyl)-4-piperidinocyclohexanol
Probably in: Cat C873
Monkey C873
Rabbit C873
Rat C873
→ 4-(4-Hydroxypiperidino)-4-phenylcyclohexanol
Probably in: Cat C873
Monkey C873
Rabbit B635, C873
Rat C873

Phenyl 2-(3′-piperidinopropyl)benzyl ether
→ *p*-Hydroxyphenyl 2-(3′-piperidinopropyl)benzyl ether
In: Rat A2333
See also: A2383
→ Phenyl 2-(3′-piperidinopropyl)benzyl ether-N-oxide
In: Rat A2333, A2383

Phenyl 2-(3′-piperidinopropyl)benzyl ether-N-oxide
Formed from: Phenyl 2-(3′-piperidinopropyl)benzyl ether

Phenylpropane
→ 1-Phenyl-1-propanol
In: Rat J8

***erythro*-1-Phenylpropane-1,2-diol** (assumed unless otherwise stated)
Probably formed from: Benzaldehyde

2-Hydroxy-1-phenyl-1-propanone
2-Oxo-1-phenyl-1-propanol
1-Phenylpropane-1,2-dione
1-Phenyl-1-propanol
1-Phenyl-2-propanon-1-ol
→ 1-Phenyl-2-propanon-1-ol
Probably in: Rabbit B755
Rat B755, J8

***threo*-1-Phenylpropane-1,2-diol**
Formed from: Phenylprop-1-ene oxide
Probably formed from: 2-Oxo-1-phenyl-1-propanol

2-Phenylpropane-1,2-diol
Formed from: 2-Phenylpropene
2-Phenylpropene oxide
Probably formed from: 2-Phenyl-2-propanol

3-Phenylpropane-1,2-diol
Formed from: Allylbenzene-2,3-oxide

2-Phenylpropane-1,3-diol-1-carbamate
Formed from: Felbamate
→ 3-Carbamoyloxy-2-phenylpropionic acid
Probably in: Man G591, J454
→ 2-Phenylpropane-1,3-diol-1-carbamate-α-D-glucuronide
Probably in: Rat K56

2-Phenylpropane-1,3-diol-1-carbamate-α-D-glucuronide
Formed from: 2-Phenylpropane-1,3-diol-1-carbamate

2-Phenyl-1,3-propanediol dicarbamate
See: Felbamate

1-Phenylpropane-1,2-dione
→ 1-Phenylpropane-1,2-diol
Probably in: Japanese monkey F241

1-Phenylpropanesulphonic acid
→ 4-*n*-Propylcatechol
Probably in: Alcaligenes A1323

2-Phenylpropanesulphonic acid
→ 4-Isopropylcatechol
Probably in: Alcaligenes A1323

1-Phenyl-1-propanol
Formed from: Isopropylbenzene
1-Phenylpropane
Propiophenone
→ 1-Phenylpropane-1,2-diol
Probably in: Rat J8
→ Propiophenone
In: Corynebacterium J893
Rabbit B755
Rat B755

1-Phenyl-2-propanol
Formed from: α-Methyl-β-phenethylhydroxylamine
Phenylacetone
→ Phenylacetone
In: Rabbit B755
Rat B755

2-Phenyl-1-propanol
Formed from: Isopropylbenzene
2-Phenylpropionaldehyde
2-Phenylpropionic acid
→ 2-Phenylpropionic acid
Probably in: Rabbit G499
Rhodococcus H213

2-Phenyl-2-propanol
Formed from: α,α-Dimethylbenzyl hydroperoxide
→ Acetophenone
Probably in: Rat E382
→ 2-(*p*-Hydroxyphenyl)-2-propanol
Probably in: Rat E382

3-Phenylpropan-1-ol
Formed from: 3-Phenylpropionaldehyde
3-Phenylpropionic acid
3-Phenylpropyl acetate
Probably formed from: Cinnamyl alcohol

1-Phenyl-1-propanone
See: Propiophenone

1-Phenylpropan-2-one
See: Phenylacetone

1-Phenyl-2-propanone oxime
See: Benzyl methyl ketoxime

1-Phenyl-2-propanon-1-ol
Formed from: Phenylacetone
Probably formed from: 1-Phenylpropane-1,2-diol
→ *erythro*-1-Phenylpropane-1,2-diol
Probably in: Rabbit A3451, B649, B755
Rat B755
→ *threo*-1-Phenylpropane-1,2-diol
Probably in: Rabbit A3451

3-Phenyl-2-propenal
See: Cinnamaldehyde

Phenylprop-1-ene
→ Phenylprop-1-ene oxide
In: Coprinus K431
Phenylprop-2-ene
→ 2-Phenylpropane-1,2-diol
In: Pseudomonas K167
***cis*-Phenylprop-1-ene oxide**
→ 1-(S-Glutathiony)-1-phenyl-2-propanol
Probably in: Sheep A2282
→ 1-Phenylpropane-1,2-diol
In: Man E969
***trans*-Phenylprop-1-ene oxide**
→ 1-(S-Glutathiony)-1-phenyl-2-propanol
In: Rat D293
See also: A2282
→ 1-Phenylpropane-1,2-diol
In: Man E969
3-Phenyl-2-propenoic acid
See: Cinnamic acid
2-Phenylpropene oxide
→ 2-Phenylpropane-1,2-diol
In: Pig H753
3-Phenyl-2-propen-1-ol
Formed from: Cinnamyl alcohol
2-Phenylpropionaldehyde
→ 2-Hydroxy-3-methyl-3-phenylpropionitrile
In: Almond K320
→ 2-Phenylpropan-1-ol
In: Euglena G422
→ 2-Phenylpropionic acid
In: Euglena G422
3-Phenylpropionaldehyde
Formed from: β-Methylstyrene
3-Phenylpropionic acid
3-Phenylpropylamine
→ Colchicine
In: Colchicum D980
→ (*S*)-α-Hydroxy-4-phenylbutyronitrile
In: Hevea K231
→ 3-Phenylpropan-1-ol
In: Euglena G422
→ 3-Phenylpropionic acid
In: Euglena G422
Man J24
Mouse F432
Rat H761
Serratia G197
3-Phenylpropionaldoxime
Formed from: L-2-Amino-4-phenylbutyric acid
→ Gluconasturtiin
In: Nasturtium A162
→ 3-Phenylpropionitrile
In: Bacillus K411
2-(N-Phenyl-3-propionamidoxy)ethanol
Formed from: 1,2′-Bis(N-phenyl-3-propionamidoxy)ethoxyethane
2′-(N-Phenyl-3-propionamidoxy)ethoxyethanol
Formed from: 1,2′-Bis(N-phenyl-3-propionamidoxy)ethoxyethane
2-Phenylpropionic acid (hydratropic acid; both isomers)
Formed from: α-Methyl-α-phenylmalonic acid
2-Phenyl-1-propanol
2-Phenylpropionaldehyde
→ 2-Phenyllactic acid
Probably in: Rabbit G499
→ (*R*)-2-Phenyl-1-propanol
In: Gloeosporium D935
Glomerella D935
Nocardia F651
→ 2-Phenylpropionyl-β-D-glucuronic acid
In: Cat A2525
Ferret A3046
Man A2525
Monkey A2525
Mouse E66
Rabbit A2525
Rat A384, A2525, D838, E66, E937
→ 2-Phenylpropionylglycine
In: Cat A2525
Ferret A3046
Rat A2525
→ 2-Phenylpropionyltaurine
In: Cat A2525
Ferret A3046
→ *R*/*S* isomerization
In: Rat E66
3-Phenylpropionic acid (dihydrocinnamic acid)
Formed from: Cinnamic acid
Oxazepam β-phenylpropionate
3-Phenylpropionaldehyde

Phenylpyruvic acid
1-Phenyltridecane
Phenylvaleric acid
Phloretic acid
Propafenone
n-Propylbenzene
→ Cinnamic acid
In: Angiozanthus K225
Man A126
Microorganism A3189, C817, J12
→ *cis*-2,3-Dihydro-2,3-dihydroxyphenylpropionic acid
In: Acinetobacter C817
Bacteria B950
→ Hydroxyangiorufone
In: Angiozanthus K225
→ 3-Phenyl-1-propanol
In: Nocardia F651
→ 3-Phenylpropionaldehyde
In: Nocardia G207
See also: K222
→ 3-Phenylpropionyl-L-alanine
In: Man F674
→ 3-Phenylpropionylcarnitine
In: Man G266
→ 3-Phenylpropionyl gentiobioside
In: Aconitum F445
Coffea F445
→ Phloretic acid
In: Dioscoreophyllum F445
→ 3-Phenylpropionyl-β-D-glucuronic acid
In: Man F674
→ 2-Phenylpropionylglycine
In: Man F674

3-Phenylpropionitrile
Formed from: 3-Phenylpropionaldoxime

3-Phenylpropionyl-L-alanine
Formed from: 3-Phenylpropionic acid

3-Phenylpropionylcarnitine
Formed from: 3-Phenylpropionic acid

Phenylpropionyl CoA
→ 3,5-Dihydroxybibenzyl
In: Barlia C734
Epipactis G138
Pinus H607
See also: H582

3-Phenylpropionyl gentiobioside
Formed from: 3-Phenylpropionic acid

2-Phenylpropionyl-β-D-glucuronic acid
Formed from: 2-Phenylpropionic acid

3-Phenylpropionyl-β-D-glucuronic acid
Formed from: 3-Phenylpropionic acid

2-Phenylpropionylglycine
Formed from: 2-Phenylpropionic acid

3-Phenylpropionylglycine
Formed from: 3-Phenylpropionic acid

2-Phenylpropionyltaurine
Formed from: 2-Phenylpropionic acid

α-Phenyl-α-propoxybenzeneacetic acid 1-methyl-4-piperidinyl ester
See: Propiverine

3-Phenylpropyl acetate
→ 3-Phenylpropan-1-ol
In: Pseudomonas K790

3-Phenylpropylamine
Formed from: N-Acetyl-3-phenylpropylamine
→ 3-Phenylpropionaldehyde
In: Arthrobacter H912
Pig A72

γ-Phenylpropyl carbamate (phenprobamate)
→ γ-Hydroxy-γ-phenylpropyl carbamate
In: Dog A2249
Man A2249, A3226

N-2-(1-Phenylpropyl)-N′-*p*-tolylsulphonylurea
See: Tosifen

2-Phenylpyridine
→ N-Methyl-2-phenylpyridinium
In: Rabbit E824

3-Phenylpyridine
→ N-Methyl-3-phenylpyridinium
In: Rabbit E314, E824

4-Phenylpyridine
→ N-Methyl-4-phenylpyridinium
In: Man E178
Monkey E178
Mouse E178
Rabbit E178, E314, E824
Rat E178

1-Phenyl-1-(2-pyridyl)ethane
Probably formed from: 1-Phenyl-1-(2-pyridyl)ethylene

1-Phenyl-1-(2-pyridyl)ethanol
Formed from: Doxylamine

→ 1-Phenyl-1-(2-pyridyl)ethylene
Probably in: Man B909
Rat E11
Squirrel monkey E11

2-Phenyl-2-(2-pyridyl)ethanol
Probably formed from: 1-Phenyl-1-(2-pyridyl)ethylene

2-(1-Phenyl-1-pyridylethoxy)acetic acid
Probably formed from: 2-(1-Phenyl-1-pyridylethoxy)ethanol

2-(1-Phenyl-1-pyridylethoxy)ethanol
Formed from: Doxylamine
→ 2-(1-Phenyl-1-pyridylethoxy)acetic acid
Probably in: Rhesus monkey E92

1-Phenyl-1-(2-pyridyl)ethylene
Probably formed from: 1-Phenyl-1-(2-pyridyl)ethanol
→ 1-Phenyl-1-(2-pyridyl)ethane
Probably in: Man B909
Rat E11
Squirrel monkey E11
→ 1-Phenyl-1-(2-pyridyl)ethanol
Probably in: Rat E11
Squirrel monkey E11

2-(2-Phenyl-2-(2-pyridyl)ethyl)imidazole
Formed from: Midaglizole
→ 4-Hydroxy-2-(2-phenyl-2-(2-pyridyl)ethyl) imidazole
Probably in: Dog E842

3-Phenyl-3-(2-pyridyl)propioimidoylaminoacetic acid
Probably formed from: 4-Hydroxy-2-(2-phenyl-2-(2-pyridyl)ethyl)imidazole

3-Phenyl-3-(2-pyridyl)propionylaminoacetamide
Probably formed from: 4-Hydroxy-2-(2-phenyl-2-(2-pyridyl)ethyl)imidazole

1-Phenyl-2-pyrrolidinopentane
See: Prolintane

1-Phenyl-1-(pyrrolidin-1-yl)cyclohexane
→ 1-Phenylcyclohexylamine
In: Rat F892

Phenylpyruvic acid (3-phenylpyruvic acid)
Formed from: D-Phenylalanine
L-Phenylalanine
L-Phenyllactic acid
L-*threo*-Phenylserine
Probably formed from: Prephenic acid
→ Benzoic acid
In: Microorganism H960
→ 2-Benzylmalic acid
In: Chromobacterium E20
→ *cis*-2,3-Dihydro-2,3-dihydroxyphenylpyruvic acid
Probably in: Bacteria B950
→ *o*-Hydroxyphenylacetic acid
Probably in: Man A2141
Rat A1715
→ Mandelic acid
In: Rat G829
→ Phenylacetaldehyde
In: Candida A2483
Thauera J397
See also: A329, A364, A1103, A1466, A2156, A2345, A2571, A2691, A3433, B950, C484, C646, E97, H960, J305
→ Phenylacetyl CoA
In: Pyrococcus H306
→ L-Phenylalanine
In: Bacillus E445, G168
Bacteria H960
Brevibacterium B266, D90, E642
Claviceps E234
Corynebacterium B266
Escherichia A3548, K94
Hansenula D978
Klebsiella B732
Lactobacillus H954
Methanophilus H114
Microtetraspora F222
Mouse C345
Mung bean B290
Neisseria F92
Nocardia F394
Oak A1238
Pig F950
Protozoa H960
Pseudomonas B672
Rat A1015, A3026, B438, D196, K758
Rhodococcus E377, E451, K558
Sporosarcina E423
Streptomyces C695
See also: A1387, J12
→ Phenyllactic acid (both isomers)

In: Bacteria H960
Candida A2483, D975
Lactobacillus E587
Streptococcus E271
See also: A1103, A1715, A2571, A3433, C646, E97, J305
→ 3-Phenylpropionic acid
Bacteria H960
→ *enol*-Phenylpyruvic acid
In: Man J475
Pig G916

***enol*-Phenylpyruvic acid**
Formed from: Phenylpyruvic acid

Phenylpyruvic amide
Formed from: L-Phenylalaninamide

N-Phenylquinolinium
→ N-Phenyl-2-quinolone
In: Guinea pig D147
Rabbit D147
→ N-Phenyl-4-quinolone
In: Guinea pig D147
Rabbit D147

N-Phenyl-2-quinolone
Formed from: N-Phenylquinolinium

N-Phenyl-4-quinolone
Formed from: N-Phenylquinolinium

N-Phenylquinoneimine
Probably formed from: *p*-Hydroxydiphenylamine

D-Phenylserine (D-3-phenylserine)
Formed from: Benzaldehyde
→ β-Oxo-D-phenylalanine
In: Pseudomonas H91

L-*threo*-Phenylserine
→ Phenylpyruvic acid
In: Pseudomonas H569

L-Phenylserine-3,4-quinone
Probably formed from: *erythro*-3,4-Dihydroxy-L-phenylserine

1-Phenyl-1-stearyl-2-(*p*-toluoyl)hydrazine
Probably formed from: Toluic acid phenylhydrazide

***p*-Phenylstyrene**
→ 4-(1,2-Dihydroxyethyl)biphenyl
In: Rat A2859

***p*-Phenylstyrene oxide**
→ 2-(4-Biphenylyl)-1-(S-glutathionyl)ethanol
Probably in: Sheep A2282

2-Phenylsuccinic acid monoamide
Probably formed from: α-Phenylsuccinimide

α-Phenylsuccinimide
→ 2-Phenylsuccinic acid monoamide
Probably in: Beef A3462
Rat A3462

Phenyl sulphate
Formed from: Phenol

Phenylsulphinylcarbanilide
Formed from: Phenylthiocarbanilide

5-(+)-(*Z*)-7-((1*R*,2*S*,3*S*,4*S*)-3-Phenylsulphonylaminobicyclo[2.2.1]hept-2-yl)heptenoic acid
→ 7-(5-Hydroxy-3-phenylsulphonylaminobicyclo[2.2.1]hept-2-yl)heptenoic acid
In: Man J79
Rat J79
→ 7-(6-Hydroxy-3-phenylsulphonylaminobicyclo[2.2.1]hept-2-yl)heptenoic acid
In: Man J79
Rat J79
→ 5-((1*R*,2*S*,3*S*,4*S*)-3-Phenylsulphonylaminobicyclo[2.2.1]hept-2-yl)pentenoic acid
In: Rat J79
See also: H474

5-((1*R*,2*S*,3*S*,4*S*)-3-Phenylsulphonylaminobicyclo[2.2.1]hept-2-yl)pentanoic acid
Probably formed from: 5-((1*R*,2*S*,3*S*,4*S*)-3-Phenylsulphonylaminobicyclo[2.2.1]hept-2-yl)pentenoic acid
→ 5-(5-Hydroxy-3-phenylsulphonylaminobicyclo[2.2.1]hept-2-yl)pentanoic acid
In: Rat J79
→ 5-(6-Hydroxy-3-phenylsulphonylaminobicyclo[2.2.1]hept-2-yl)pentanoic acid
In: Rat J79

5-((1*R*,2*S*,3*S*,4*S*)-3-Phenylsulphonylaminobicyclo[2.2.1]hept-2-yl)pentenoic acid
Formed from: 5-(+)-(*Z*)-7-((1*R*,2*S*,3*S*,4*S*)-3-Phenylsulphonylaminobicyclo[2.2.1]hept-2-yl)heptenoic acid
→ 5-(5-Hydroxy-3-phenylsulphonylaminobicyclo[2.2.1]hept-2-yl)pentenoic acid
In: Man J79
Rat J79
→ 5-(6-Hydroxy-3-phenylsulphonylaminobicyclo[2.2.1]hept-2-yl)pentenoic acid
In: Man J79
Rat J79
→ 5-((1*R*,2*S*,3*S*,4*S*)-3-Phenylsulphonylaminobicyclo[2.2.1]hept-2-yl)pentanoic acid
Probably in: Man J79
Rat J79
→ 3-((1*R*,2*S*,3*S*,4*S*)-3-Phenylsulphonylaminobicyclo[2.2.1]hept-2-yl)propionic acid
Probably in: Man
J79
Rat H474, J79

3-((1*R*,2*S*,3*S*,4*S*)-3-Phenylsulphonylaminobicyclo[2.2.1]hept-2-yl)propionic acid
Probably formed from: 5-((1*R*,2*S*,3*S*,4*S*)-3-Phenylsulphonylaminobicyclo[2.2.1]hept-2-yl)pentenoic acid
→ 3-(5-Hydroxy-3-phenylsulphonylaminobicyclo[2.2.1]hept-2-yl)propionic acid
In: Man J79
Rat J79
→ 3-(6-Hydroxy-3-phenylsulphonylaminobicyclo[2.2.1]hept-2-yl)propionic acid
In: Man J79
Rat J79

2-Phenyl-1-tetralone
→ 2-(*p*-Hydroxyphenyl)-1-tetralone
In: Absidia G14

3-Phenyl-1-tetralone
→ 3-(*p*-Hydroxyphenyl)-1-tetralone
In: Absidia G14

N-Phenyl-N′-(1,2,3-thiadiazol-5-yl)urea
See: Thidiazuron

Phenyl thioacetate
→ Phenol
In: Pig F289

5-(Phenylthio)-1*H*-benzimidazol-2-ylcarbamic acid methyl ester
See: Fenbendazole

Phenylthiocarbanilide
→ Phenylsulphinylcarbanilide
In: Mouse D446
Pig D446

Phenylthiourea
→ Phenylformamidinesulphenic acid
In: Pig B239
See also: A2549

Phenyltin
Formed from: Diphenyltin

N-(α-Phenyl-β-(*p*-tolyl)ethyl)glutaric acid monoamide
Formed from: N-(α-Phenyl-β-(*p*-tolyl)ethyl)linoleamide

N-(α-Phenyl-β-(*p*-tolyl)ethyl)linoleamide
→ N-(α-Phenyl-β-(*p*-tolyl)ethyl)glutaric acid monoamide
In: Rat A1770
→ N-(α-Phenyl-β-(*p*-tolyl)ethyl)succinic acid monoamide
In: Rat A1770

N-(α-Phenyl-β-(*p*-tolyl)ethyl)succinic acid monoamide
Formed from: N-(α-Phenyl-β-(*p*-tolyl)ethyl)linoleamide

5-Phenyl-5-(*p*-tolyl)hydantoin
→ 5-(*p*-Hydroxymethylphenyl)-5-phenylhydantoin
In: Rat A2402

Phenyl N,N′,N″-triacetyl-β-chitotrioside
Probably formed from: Phenyl N,N′-diacetyl-β-chitobioside

1-Phenyltridecane
→ Cinnamic acid
In: Acinetobacter D746

→ 3-Phenylpropionic acid
In: Acinetobacter D746

3-Phenyl-3-(*p*-trifluoromethylphenoxy) propylamine
Formed from: N-Methyl-3-phenyl-3-(*p*-trifluoromethylphenoxy)propylamine

1-Phenyl-1-(*m*-trifluoromethylphenyl)propane-1,2-diol
Formed from: 1-Phenyl-1-(*m*-trifluoromethylphenyl)propanol

1-Phenyl-1-(*m*-trifluoromethylphenyl)propane-1,2-diol
Formed from: 1-Phenyl-1-(*m*-trifluoromethylphenyl)propanol

1-Phenyl-1-(*m*-trifluoromethylphenyl)propane-1,3-diol
Formed from: 1-Phenyl-1-(*m*-trifluoromethylphenyl)propanol
→ 1-(*p*-Hydroxyphenyl)-1-(*m*-trifluoromethylphenyl)propane-1,3-diol
Probably in: Man E814

1-Phenyl-1-(*m*-trifluoromethylphenyl)propanol (flumecinol)
→ 1-(*p*-Hydroxyphenyl)-1-(*m*-trifluoromethylphenyl)propanol
In: Man E814
Rat G526
→ 1-Phenyl-1-(*m*-trifluoromethylphenyl) propane-1,2-diol
In: Man E814
→ 1-Phenyl-1-(*m*-trifluoromethylphenyl) propane-1,3-diol
In: Man E814
→ 1-Phenyl-1-(*m*-trifluoromethylphenyl)-1-propyl-β-D-glucuronide
In: Man E814
→ 1-Phenyl-1-(*m*-trifluoromethylphenyl)-1-propyl sulphate
In: Man E814

1-Phenyl-1-(*m*-trifluoromethylphenyl)-1-propyl-β-D-glucuronide
Formed from: 1-Phenyl-1-(*m*-trifluoromethylphenyl)propanol

1-Phenyl-1-(*m*-trifluoromethylphenyl)-1-propyl sulphate
Formed from: 1-Phenyl-1-(*m*-trifluoromethylphenyl)propanol

α-Phenyl-N-(2,4,6-trimethylphenyl)nitrone
Formed from: N-Benzyl-2,4,6-trimethylaniline

1-Phenylundecane-*p*-sulphonic acid
→ Phenylvaleric acid
In: Cladosporium A1011

Phenylurea
Formed from: Thidiazuron

Phenylvaleric acid
Formed from: 1-Phenylundecane-*p*-sulphonic acid
→ 3-Phenylpropionic acid
Probably in: Cladosporium A1011

Phenyl vinyl ketone
Formed from: (1-Hydroxyallyl)benzene

Phenyramidol (N-(2-pyridyl)phenylethanolamine)
→ Hydroxyphenyramidol
In: Rat A1852
→ Mandelic acid
In: Man A3285
→ 2-Oxo-2-phenylethanol
In: Rat A3285
→ Phenyramidol-β-D-glucuronide
In: Rat A1852

Phenytoin
See: 5,5-Diphenylhydantoin

Phenzidole (2-phenyl-1*H*-benzimidazole)
→ 5-O-β-D-Glucuronyl-2-phenylbenzimidazole
In: Sheep A2205

Phillygenin (4((1*S*,3a*R*,4*R*,6a*R*)-4-(3,4-dimethoxyphenyl)tetrahydro-1*H*,3*H*-furo[3,4-*c*]furan-1-yl)-2-methoxyphenol)
Formed from: Coniferyl alcohol

Phloretic acid (*p*-hydroxyphenylpropionic acid)
Formed from: Caffeic acid
Cetraxate
p-Coumaric acid
Dihydro-2′,4,4′-trihydroxychalcone
Dihydro-2′,4,6′-trihydroxychalcone
1-(2,6-Dihydroxy-3,4-dimethoxyphenyl)-3-(*p*-hydroxyphenyl)propan-1-one
9-(*p*-Hydroxyphenyl)nonylamine
p-Hydroxyphenylpropylamine
p-(*n*-Nonyl)phenol
Phloretin
Proanthocyanidin B_3
4-(2,4,6-Trihydroxyphenyl)flavan-3-ol

Probably formed from: 4-(2-Carboxyethyl)phenyl *trans*-4-aminomethylcyclohexanecaboxylate
p-Hydroxyphenylpentanoic acid
→ *p*-Coumaric acid
Probably in: Man A126
Rat D159
Sceletium A3136
→ Dihydrocaffeic acid
In: Mucuna E769, F112
Potato D964
See also: D650, K140
→ 2,5-Dihydroxyphenylpropionic acid
In: Bacterium A489
Pseudomonas A1480
→ 3-(*p*-Hydroxyphenyl)propanol
Probably in: Alnus A3144
Betula A3144
→ *m*-Hydroxyphenylpropionic acid
In: Microorganism A2767, A3990
→ Marchantin A
In: Marchantia J902
→ 3-Phenylpropionic acid
In: Microorganism A2767, K574
See also: D654

Phloretin (2′,4′,6′-trihydroxy-3-(*p*-hydroxyphenyl)propiophenone)
Formed from: Phloridzin
→ Phloretic acid
In: Microorganism D159
Rat D159
→ Phloroglucinol
In: Microorganism D159
Rat D159

Phloretin-2′-β-D-glucoside
See: Phloridzin

Phloridzin (phloretin-2′-β-D-glucoside)
→ Phloretin
Probably in: Microorganism D159
Rat D159

Phloroglucinol (1,3,5-trihydroxybenzene)
Formed from: Apigenin
Biochanin A
Dihydrophloroglucinol
Eriodictyol
Genistein
Naringenin
Phloretin
Proanthocyanidin B_3
Pyrogallol
1,2,3,5-Tetrahydroxybenzene
2,4,6-Trihydroxybenzoic acid
4-(2,4,6-Trihydroxyphenyl)flavan-3-ol
Probably formed from:
Protocatechuoylphloroglucinolcarboxylic acid
→ Dihydrophloroglucinol
In: Coprococcus C99
Eubacterium E388, F75
Pelobacter E189, G795
Penicillium F639
Rhodopseudomonas A1370
See also: G772
→ 3,5-Dihydroxyphenyl-α-D-glucoside
In: Bacillus H234
→ 3,5-Dihydroxyphenyl sulphate
In: Man E575
Rat E575
→ Hydroxyquinol
In: Penicillium F639
See also: G292
→ Pyrogallol
In: Fusarium C707
→ Resorcinol
In: Bradyrhizobium J484
Penicillium F639
See also: E742, G292
→ 1,2,3,5-Tetrahydroxybenzene
In: Candida E204
Pseudomonas A2391
Rhodococcus E565, G820, H121
See also: A1930, H878
Does not yield rutin in buckwheat A179

Phloroglucinolcarboxylic acid
See: 2,4,6-Trihydroxybenzoic acid

Phloroisobutyrophenone (2′,4′,6′-trihydroxyisobutyrophenone)
Formed from isobutyryl CoA in Humulus J887, K175

Phloroisovalerophenone (2′,4′,6′-trihydroxyisovalerophenone)
Formed from isovaleryl CoA in Humulus J887, K175

Pholcodine ((5α,6α)-7,8-didehydro-4,5-epoxy-17-methyl-3-(2-(4-morpholinyl)ethoxy) morphinan-6-ol)
→ 3-O-(2-Hydroxyethyl)normorphine
In: Rat F682
→ 3-O-(2-(3-Hydroxymorpholino)ethyl) morphine
Probably in: Man F682
→ Morphine
In: Man F682, F801, H496, H497
→ 3-O-(2-Morpholinoethyl)normorphine
In: Rat F682
→ Pholcodine-β-D-glucuronide
In: Man F801, H496

Pholcodine-β-D-glucuronide
Formed from: Pholcodine

Phomazarine (7-butyl-5,10-dihydro-3,4,9-trihydroxy-8-methoxy-5,10-dioxobenzo[*g*] quinoline-2-carboxylic acid)
Formed from acetate and malonate in Pyrenochaeta B65

Phosphanilic acid (4-aminophenyl)phosphonic acid)
→ N-Acetylphosphanilic acid
In: Man B690
Rat B690

***o*-Phthalaldehyde** (*o*-formylbenzaldehyde)
→ Phthalide
In: Euglena G422

Phthaldehydic acid
See: *o*-Formylbenzoic acid

Phthalazine
Formed from: Hydralazine
→ 1(2*H*)-Phthalazinone
In: Arthrobacter J33

1(2*H*)-Phthalazinone
Formed from: Budralazine
Hydralazine
Phthalazine
→ 6-Hydroxy-1(2*H*)-phthalazinone
Probably in: Rat A1503, A1816

2-91-Phthalazinyl)hydrazinecarboxylic acid ethyl ester
See: Apiracohl

Phthalic acid (benzene-1,2-dicarboxylic acid)
Formed from: Anthraquinone
Benzalphthalide
N-Benzylphthalamidic acid
N-(1-(1*H*)-(1-Carboxy-1-(*n*-heptyl)) imidazol-4-yl)phthalamidic acid
5-(*o*-Carboxy-N-benzoyl)amino-2-nitrobenzoic acid
N-Carboxymethylphthalamidic acid
N-(*o*-Chlorophenyl)phthamidic acid
1,4-Dihydroxyanthraquinone
Fluorene
o-Hydroxymethylbenzoic acid
N-Isopropylphthalamidic acid
N-(*p*-Nitrophenyl)phthalamidic acid
Phenanthrene-9,10-quinone
Phthalic acid mono-*n*-butyl ester
Phthalic acid monoethyl ester
Phthalic acid mono-(2-ethylhexyl) ester
Phthalic acid monomethyl ester
N-(*o*-Phthalyl)-L-aspartyl-L-phenylalanine methyl ester
N-(*o*-Phthalyl)-D-phenylalanyl-L-alanine
N-(*o*-Phthalyl)-L-phenylalanyl-L-alanine
N-(*o*-Phthalyl)-D-phenylglycine methyl ester
o-Phthalylcarbacephen
N-(*o*-Phthalyl)-L-leucine
N-(*o*-Phthalyl)-L-methionine
N-(*o*-Phthalyl)-D-phenylglycine
Pyrene
Probably formed from: *o*-Formylbenzoic acid
o-Hydroxymethylbenzoic acid
Phthalic acid mono-5-hexenyl ester
→ Benzoic acid
In: Microorganism D31
→ 3,4-Dihydro-3,4-dihydroxyphthalic acid
In: Rhodococcus H89, J211
See also: H789
→ 3,4-Dihydro-3,4-dihydroxyphthalic acid
In: Pseudomonas K922
→ 4,5-Dihydroxyphthalic acid
In: Moraxella J27

Pseudomonas A3096
See also: A2653, A3528, B2
→ 3-Hydroxyphthalic acid
In: Corynebacterium A2999
→ Protocatechuic acid
In: Earthworm H9
Moraxella J27
Rhodococcus J211
See also: K622

Phthalic acid monobenzyl ester
Formed from: Benzyl *n*-butyl phthalate

Phthalic acid mono-*n*-butyl ester
Formed from: Benzyl *n*-butyl phthalate
Di-*n*-butyl phthalate
→ Phthalic acid
In: Beef J718
Microorganism K248
Nocardia A3528
Rat A2376, A3866
See also: K261
→ Phthalic acid mono-(3-hydroxybutyl) ester
Probably in: Beef J718
Rat A2376
→ Phthalic acid mono-(4-hydroxybutyl) ester
Probably in: Rat A2376
→ Phthalic acid mono-*n*-butyl ester-β-D-glucuronic acid
In: Beef J718
Hamster C600
Rat C600

Phthalic acid mono-*n*-butyl ester-β-D-glucuronic acid
Formed from: Phthalic acid mono-*n*-butyl ester

Phthalic acid mono-(4-carboxybutyl) ester
Probably formed from: Phthalic acid mono-(5-carboxypentyl) ester
→ Phthalic acid mono-(3-carboxypropyl) ester
Probably in: Rat C869

Phthalic acid mono-(2-(2-carboxyethyl)butyl) ester
Probably formed from: Phthalic acid mono-(2-ethylhexyl) ester

Phthalic acid mono-(4-carboxy-2-ethylbutyl) ester
Probably formed from: Phthalic acid mono-(2-ethylhexyl) ester

Phthalic acid mono-(2-carboxyethyl) ester
Probably formed from: Phthalic acid mono-(3-carboxypropyl) ester
→ Phthalic acid mono-(carboxymethyl) ester
Probably in: Rat C869

Phthalic acid mono-(5-carboxy-2-ethyl-4-hydroxypentyl) ester
Probably formed from: Phthalic acid mono-(2-ethyl-4-hydroxyhexyl) ester
→ Phthalic acid mono-(5-carboxy-2-ethyl-4-oxopentyl) ester
Probably in: Rat C887

Phthalic acid mono-(5-carboxy-2-ethyl-4-oxopentyl) ester
Probably formed from: Phthalic acid mono-(5-carboxy-2-ethyl-4-hydroxypentyl) ester

Phthalic acid mono-(5-carboxy-2-ethylpentyl) ester
Probably formed from: Phthalic acid mono-(5-carboxy-2-ethylhexyl) ester
Phthalic acid mono-(2-ethylhexyl) ester
Phthalic acid mono-(2-ethyl-6-hydroxyhexyl) ester
→ Phthalic acid mono-(3-carboxy-2-ethylpropyl) ester
Probably in: Rat A842

Phthalic acid mono-(3-carboxy-2-ethylpropyl) ester
Probably formed from: Phthalic acid mono-(5-carboxy-2-ethylpentyl) ester
Phthalic acid mono-(2-ethylhexyl) ester
Phthalic acid mono-(2-ethyl-5-oxohexyl) ester

Phthalic acid mono-(2-carboxyhexyl) ester
Probably formed from: Phthalic acid mono-(2-carboxymethylhexyl) ester

Phthalic acid mono-(2-(carboxymethyl)butyl) ester
Probably formed from: Phthalic acid mono-(2-ethylhexyl) ester

Phthalic acid mono-(carboxymethyl) ester
Probably formed from: Phthalic acid mono-(2-carboxyethyl) ester

Phthalic acid mono-(2-carboxymethylhexyl) ester
Probably formed from: Phthalic acid mono-(2-hydroxyethylhexyl) ester
→ Phthalic acid mono-(2-carboxyhexyl) ester

Probably in: African green monkey B579
Man D360
Rat C887
→ Phthalic acid mono-(2-carboxymethyl-5-hydroxyhexyl) ester
Probably in: Rat C887
→ Phthalic acid mono-(2-carboxymethyl-4-oxohexyl) ester
Probably in: Rat C887

Phthalic acid mono-(2-carboxymethyl-5-hydroxyhexyl) ester
Probably formed from: Phthalic acid mono-(2-carboxymethylhexyl) ester

Phthalic acid mono-(2-carboxymethyl-4-oxohexyl) ester
Probably formed from: Phthalic acid mono-(2-carboxymethylhexyl) ester

Phthalic acid mono-(8-carboxyoctyl) ester
Probably formed from: Phthalic acid mono-(9-decenyl) ester

Phthalic acid mono-(5-carboxypentyl) ester
Probably formed from: Phthalic acid mono-(9-decenyl) ester
Phthalic acid mono-(5-hexenyl) ester
→ Phthalic acid mono-(4-carboxybutyl) ester
Probably in: Rat C869

Phthalic acid mono-(3-carboxypropyl) ester
Probably formed from: Phthalic acid mono-(4-carboxybutyl) ester
→ Phthalic acid mono-(2-carboxyethyl) ester
Probably in: Rat C869

Phthalic acid monocyclohexyl ester
Formed from: Phthalic acid dicyclohexyl ester

Phthalic acid mono-(9-decenyl) ester
Formed from: Di-(9-decenyl) phthalate
→ Phthalic acid mono-(8-carboxyoctyl) ester
Probably in: Rat C869
→ Phthalic acid mono-(5-carboxypentyl) ester
Probably in: Rat C869

Phthalic acid mono-(5,6-dihydroxyhexyl) ester
Probably formed from: Phthalic acid mono-(5,6-epoxyhexyl) ester

Phthalic acid mono-(5,6-epoxyhexyl) ester
Probably formed from: Phthalic acid mono-5-hexenyl ester
→ Phthalic acid mono-(5,6-dihydroxyhexyl) ester
Probably in: Rat C869

Phthalic acid monoethyl ester
Formed from: Diethyl phthalate
→ Phthalic acid
In: Micrococcus D686

Phthalic acid mono-(2-ethylhexyl) ester
Formed from: Di-(2-ethylhexyl) phthalate
→ Phthalic acid
In: Catfish A2239
Earthworm H9
Rat A2883
Trout A1990
See also: G876
→ Phthalic acid mono-(2-(2-carboxyethyl) butyl) ester
Probably in: Man D360
→ Phthalic acid mono-(4-carboxy-2-ethylbutyl) ester
Probably in: African green monkey B579
Rat C887
→ Phthalic acid mono-(5-carboxy-2-ethylpentyl) ester
Probably in: Rat C887
→ Phthalic acid mono-(3-carboxy-2-ethylpropyl) ester
Probably in: African green monkey B579
Rat C887
→ Phthalic acid mono-(2-(carboxymethyl) butyl) ester
Probably in: Man D360
→ Phthalic acid mono-(2-ethylhexyl) ester-β-D-glucuronic acid
In: Guinea pig F743
Trout A1990
See also: B579
→ Phthalic acid mono-(2-ethyl-4-hydroxyhexyl) ester
In: Rat D377
See also: B579, C887
→ Phthalic acid mono-(2-ethyl-5-hydroxyhexyl) ester
In: Rat A2883, D377
See also: A842, B579, C869, D360, G91
→ Phthalic acid mono-(2-ethyl-6-hydroxyhexyl) ester

In: Rat D377
See also: B579, G91
→ Phthalic acid mono-(2-(1-hydroxyethyl)hexyl) ester
In: Rat D377
See also: D887
→ Phthalic acid mono-(2-(2-hydroxyethyl)hexyl) ester
Probably in: African green monkey B579
Man D360
Rat C887

Phthalic acid mono-(2-ethylhexyl) ester-β-D-glucuronic acid
Formed from: Phthalic acid mono-(2-ethylhexyl) ester

Phthalic acid mono-(2-ethyl-4-hydroxyhexyl) ester
Formed from: Phthalic acid mono-(2-ethylhexyl) ester
→ Phthalic acid mono-(5-carboxy-2-ethyl-4-hydroxypentyl) ester
Probably in: Rat C887
→ Phthalic acid mono-(2-ethyl-4-hydroxyhexyl) ester-β-D-glucuronide
Probably in: African green monkey B579
→ Phthalic acid mono-(2-ethyl-4-oxohexyl) ester
Probably in: Rat C887

Phthalic acid mono-(2-ethyl-5-hydroxyhexyl) ester
Formed from: Phthalic acid mono-(2-ethylhexyl) ester
→ Phthalic acid mono-(2-ethyl-5-hydroxyhexyl) ester-β-D-glucuronde
Probably in: African green monkey B579
→ Phthalic acid mono-(2-ethyl-5-oxohexyl) ester
Probably in: African green monkey B579
Man D360
Mouse G91
Rat A842, A2883, C887
→ Phthalic acid mono-(5-hydroxy-2-(1-hydroxyethyl)hexyl) ester
Probably in: Rat C887

Phthalic acid mono-(2-ethyl-6-hydroxyhexyl) ester
Formed from: Phthalic acid mono-(2-ethylhexyl) ester
→ Phthalic acid mono-(5-carboxy-2-ethylpentyl) ester
Probably in: Mouse G91
→ Phthalic acid mono-(2-ethyl-6-hydroxyhexyl) ester-β-D-glucuronide
Probably in: African green monkey B579

Phthalic acid mono-(2-ethyl-4-hydroxyhexyl) ester-β-D-glucuronide
Probably formed from: Phthalic acid mono-(2-ethyl-4-hydroxyhexyl) ester

Phthalic acid mono-(2-ethyl-5-hydroxyhexyl) ester-β-D-glucuronide
Probably formed from: Phthalic acid mono-(2-ethyl-5-hydroxyhexyl) ester

Phthalic acid mono-(2-ethyl-6-hydroxyhexyl) ester-β-D-glucuronide
Probably formed from: Phthalic acid mono-(2-ethyl-6-hydroxyhexyl) ester

Phthalic acid mono-(2-ethyl-4-oxohexyl) ester
Probably formed from: Phthalic acid mono-(2-ethyl-4-hydroxyhexyl) ester

Phthalic acid mono-(2-ethyl-5-oxohexyl) ester
Probably formed from: Phthalic acid mono-(2-ethyl-5-hydroxyhexyl) ester
→ Phthalic acid mono-(5-carboxy-2-ethylpentyl) ester
Probably in: African green monkey B579
Rat A2883
→ Phthalic acid mono-(2-ethyl-5-oxohexyl) ester-β-D-glucuronic acid
Probably in: African green monkey B579

Phthalic acid mono-(2-ethyl-5-oxohexyl) ester-β-D-glucuronic acid
Probably formed from: Phthalic acid mono-(2-ethyl-5-oxohexyl) ester

Phthalic acid mono-(5-hexenyl) ester
Formed from: Di-(5-hexenyl) phthalate
→ Phthalic acid
Probably in: Rat C869
→ Phthalic acid mono-(5,6-epoxyhexyl) ester
Probably in: Rat C869

Phthalic acid mono-(3-hydroxybutyl) ester
Phthalic acid mono-*n*-butyl ester

Phthalic acid mono-(4-hydroxybutyl) ester
Phthalic acid mono-*n*-butyl ester

Phthalic acid mono-(2-(1-hydroxyethyl)hexyl) ester
Formed from: Phthalic acid mono-(2-ethylhexyl) ester

Phthalic acid mono-(2-(2-hydroxyethyl)hexyl) ester
Probably formed from: Phthalic acid mono-(2-ethylhexyl) ester
→ Phthalic acid mono-(2-carboxymethylhexyl) ester
Probably in: African green monkey B579
Man D360
Rat C887, D377
→ Phthalic acid mono-(2-(2-hydroxyethyl)hexyl) ester-β-D-glucuronide
Probably in: African green monkey B579

Phthalic acid mono-(2-(2-hydroxyethyl)hexyl) ester-β-D-glucuronide
Probably formed from: Phthalic acid mono-(2-(2-hydroxyethyl)hexyl) ester

Phthalic acid mono-(5-hydroxy-2-(1-hydroxyethyl)hexyl) ester
Probably formed from: Phthalic acid mono-(2-ethyl-5-hydroxyhexyl) ester

Phthalic acid monomethyl ester
Formed from: Dimethyl phthalate
→ Phthalic acid
In: Aspergillus K622
Microorganism K282
Xanthobacter J253

Phthalic acid mono-*n*-octyl ester
Formed from: Di-*n*-octyl phthalate

Phthalide (1(3*H*)-isobenzofuranone)
Formed from: *o*-Formylbenzoic acid
o-Formylbenzonitrile
o-Phthalaldehyde

Phthalidyl theophylline
→ *o*-Formylbenzoic acid
In: Rat E344

***o*-(N-Phthalimido)acetophenone**
→ N-(*o*-Acetophenone)phthalamic acid
In: Rat E914
→ *o*-(N-Phthalimido)benzoic acid
In: Rat E914

***o*-(N-Phthalimido)benzoic acid**
Formed from: *o*-(N-Phthalimido) acetophenone

α-Phthalimidoglutarimide
See: Thalidomide

Phthalonitrile (1,2-benzodinitrile)
→ *o*-Cyanobenzoic acid
Probably in: Fusarium A3364

N-(*o*-Phthalyl)-L-aspartyl-L-phenylalanine methyl ester
→ Phthalic acid
In: Xanthobacter J253

***o*-Phthalyl carbacephen**
→ Phthalic acid
In: Xanthobacter J253

N-(*o*-Phthalyl)-L-leucine
→ Phthalic acid
In: Xanthobacter J253, K791

N-(*o*-Phthalyl)-L-methionine
→ Phthalic acid
In: Xanthobacter J253

N-(*o*-Phthalyl)-D-phenylalanyl-L-alanine
→ Phthalic acid
In: Xanthobacter K793

N-(*o*-Phthalyl)-L-phenylalanyl-L-alanine
→ Phthalic acid
In: Xanthobacter K793

N-(*o*-Phthalyl)-D-phenylglycine
→ D-Phenylglycine
In: Xanthobacter J253
→ Phthalic acid
In: Xanthobacter J253

N-(*o*-Phthalyl)-D-phenylglycine methyl ester
→ Phthalic acid
In: Xanthobacter K793

Phyllodulcin ((3*R*)-3,4-dihydro-8-hydroxy-3-(3-hydroxy-4-methoxyphenyl)-1*H*-2-benzopyran-1-one)
Probably formed from: Hydrangenol

***trans*-Phylloquinone**
See: Vitamin K_1

Physcion (1,8-dihydroxy-3-methoxy-6-methyl-9,10-anthracenedione)
Formed from: Emodin
→ 3-Carboxy-1,8-dihydroxy-6-methoxyanthraquinone

In: Mouse E273
Rat E273
→ Emodin
In: Mouse E273
Rat E273

Phytylplastoquinol
See: 2,3-Dimethyl-5-phytylquinol

Phytyltoluquinol (2-methyl-6-phytylquinol)
Formed from: Homogentisic acid
→ 2,3-dimethyl-5-phytylquinol
In: Spinach A3972, B575

Piceid (3-hydroxy-5-((1*E*)-2-(4-hydroxyphenyl)ethenyl)phenyl-β-D-glucopyranoside)
→ 3,5-Dihydroxybenzaldehyde
In: Pseudomonas F370
→ *p*-Hydroxybenzaldehyde
In: Pseudomonas F370

Picein (*p*-hydroxyacetophenone-β-D-glucoside)
→ *p*-Hydroxyacetophenone
In: Spruce K833

Picenadol (rel-3-((3*R*,4*S*)-1,3-dimethyl-4-propyl-4-piperidinyl)phenol)
→ Demethylpicenadol
In: Man F601
→ Picenadol-β-D-glucuronide
In: Man F601
→ Picenadol-O-sulphate
In: Man F601

Picenadol-β-D-glucuronide
Formed from: Picenadol

Picenadol-O-sulphate
Formed from: Picenadol

Picene
→ *trans*-1,2-Dihydro-1,2-dihydroxypicene
In: Rat E757, F360
→ *trans*-3,4-Dihydro-3,4-dihydroxypicene
In: Rat E757, F360
→ *trans*-5,6-Dihydro-5,6-dihydroxypicene
In: Rat E757, F360
→ 2-Hydroxypicene
In: Rat E757, F360
→ 4-Hydroxypicene
In: Rat E757, F360

Picobenzide (3,5-dimethyl-N-((4-pyridyl)methyl)benzamide)
→ Picobenzide-N-oxide
In: Man E107

Picobenzide-N-oxide
Formed from: Picobenzide

Picric acid (2,4,6-trinitrophenol)
→ 2-Amino-4,6-dinitrophenol
In: Bacteria A1467
See also: A2624

Picumast (7-(4-(*p*-chlorobenzyl)piperazin-1-yl)propoxy-3,4-dimethylcoumarin)
→ 7-(4-(*p*-Chlorobenzyl)piperazin-1-yl) propoxy-3-hydroxymethyl-4-methylcoumarin
In: Man F413

Picumeterol ((α*R*)-4-amino-3,5-dichloro-α-(((6-(2-(2-pyridinyl)ethoxy)hexyl) amino)methyl)benzenemethanol)
→ 4-Amino-N-(5-carboxypentyl)-3,5-dichlorophenylethanolamine
In: Dog H323
Rat H323

Pigment IP/II
Formed from: Aklanonic acid

Pigment Violet 12
See: 1,4-Dihydroxyanthraquinone

Pimobendan (4,5-dihydro-6-(2-(4-methoxyphenyl)-1*H*-benzimidazol-5-yl)-5-methyl-3-(2*H*)-pyridazinone)
→ O-Demethylpimobendan
In: Man H311, J801
→ Pimobendan-N-β-D-glucuronide
In: Man H311

Pimobendan-N-β-D-glucuronide
Formed from: Pimobendan

Pinaverium (4-((2-bromo-4,5-dimethoxyphenyl)methyl)-4-(2-(2-(6,6-dimethylbicyclo[3.1.1]hept-2-yl) ethoxy)ethyl)morpholinium)
→ Hydroxypinaverium
In: Dog C150
Man C150
Rat C150

Pinazepam (7-chloro-1,3-dihydro-5-phenyl-1-(2-propynyl)-2*H*-1,4-benzodiazepin-2-one)
→ 7-Chloro-1,3-dihydro-5-phenyl-2*H*-1,4-benzodiazepin-2-one
In: Dog A3006
Guinea pig A3006

Man C259
Rat A3006, B933

Pindolol (4-(2-hydroxy-3-isopropylaminopropoxy)indole)
→ 4-(2-Hydroxy-3-isopropylaminopropoxy)anthranilic acid
In: Man E835
Rat A1738
→ 4-(2-Hydroxy-3-isopropylaminopropoxy) indolin-2-one
In: Cat A1738
→ 3-Hydroxypindolol
In: Monkey B398
Rat A1738
→ 5-Hydroxypindolol
In: Man E835
Monkey A1738, B398
Rat A1738
→ 6-Hydroxypindolol
In: Man E835
Rat A1738
→ 3-(4-Indolyloxy)lactic acid
In: Monkey B398
Rabbit A1738
→ Pindolol-β-D-glucuronide
In: Man E835
Monkey A1738, B398

Pindolol-β-D-glucuronide
Formed from: Pindolol

Pinocembrin ((2*S*)-2,3-dihydro-5,7-dihydroxy-2-phenyl-4*H*-1-benzopyran-4-one)
Formed from: 2′,4′,6′-Trihydroxychalcone

(+)-Pinoresinol (4,4′-((1*S*,3a*R*,4*R*,6a*S*)-tetrahydro-1*H*,3*H*-furo[3,4-*c*]furan-1,4-diyl)bis(2-methoxyphenol))
Formed from: Coniferyl alcohol
→ Lariciresinol
In: Forsythia G998, H883
→ Piperitol
In: Sesamum J516
→ (+)-Sesamin
In: Sesamum J516
→ (+)-Sesamolin
In: Sesamum J516

Pinosylvine (3,5-dihydroxystilbene)
Formed from: Cinnamoyl CoA
p-Coumaroyl CoA

Pioglitazone (5-((4-(2-(5-ethyl-2-pyridinyl)ethoxy)phenyl)methyl)-2,4-thiazolidinedione)
→ 5-((4-(2-(5-(2-Hydroxyethyl)-2-pyridinyl)ethoxy)phenyl)methyl)-2,4-thiazolidinedione
In: Rat G946
→ 5-((4-(2-Hydroxy-2-(5-ethyl-2-pyridinyl)ethoxy)phenyl)methyl)-2,4-thiazolidinedione
In: Rat G946

Pipamperone (1-(3-(*p*-fluorobenzoyl)propyl)-1,4′-bipiperidine-4-carboxamide)
→ Dihydropipamperone
In: Man F236
→ 1-(3-(*p*-Fluorobenzoyl)propyl)-hydroxy-1,4′-bipiperidine-4′-carboxamide
In: Man F236

Pipecolo-2′,6′-xylidide
Formed from: Bupivacaine
Ropivacaine
→ 3′-Hydroxypipecolo-2′,6′-xylidide
Probably in: Man H548
→ 4′-Hydroxypipecolo-2′,6′-xylidide
Probably in: Monkey A255

2-Piperazino-3*H*-2-quinazolone
See: Centpiperalone

3-((Piperazin-1-yl)ethoxy)-*p*-cymene
→ 5-Isopropyl-2-methyl-4-(piperazin-1-yl)ethoxyphenol
In: Cunninghamella J132
→ 4-Isopropyl-3-(2-(piperazin-1-yl) ethoxy)benzyl alcohol
In: Cunninghamella J132

3-((1-Piperazinyl)methyl)benzoic acid
Probably formed from: 1-(3-((1-Piperazinyl)methyl)benzoyl)piperidine
→ 3-((1-Piperazinyl)methyl)benzoyl-β-D-glucuronic acid
Probably in: Dog J581

1-(3-((1-Piperazinyl)methyl)benzoyl)piperidine
Formed from: Mazapertine
→ N-(4-Formylbutyl)-3-((1-piperazinyl)methyl)benzamide
Probably in: Dog J581

→ 3-((1-Piperazinyl)methyl)benzoic acid
Probably in: Dog J581

3-((1-Piperazinyl)methyl)benzoyl-β-D-glucuronic acid
Probably formed from: 3-((1-Piperazinyl)methyl)benzoic acid

3-(3-(Piperazin-1-yl)propyl)-5-(1,2,4-triazol-4-yl)-1*H*-indole
Formed from: 3-(3-(4-(2-(*m*-Fluorophenyl)ethyl)piperazin-1-yl)propyl)-5-(1,2,4-triazol-4-yl)-1*H*-indole

6-(1-Piperazinyl)-2-quinolone
Formed from: 6-(4-(3,4-Dimethoxybenzoyl)-1-piperazinyl)-3,4-dihydro-2(1*H*)-quinolone
→ 6-(3-Oxo-1-piperazinyl)-2-quinolone
Probably in: Man F318
Monkey F318
Mouse F318
Rat F318

4-Piperidinecarboxylic acid 1-(3-hydroxy-3-phenylpropyl)-4-phenylethyl ester
See: Phenoperidine

2-Piperidinoethyl 3-methylflavone-8-carboxylate
See: Flavoxate

(α*S*)-*rel*-α-(2*R*)-2-Piperidinyl-2,8-bis (trifluoromethyl)-4-quinolinemethanol
See: Mefloquine

***erythro*-Piperidin-2-yl-3,4-dihydroxyphenylmethanol**
See: Rimiterol

N-(2-Piperidinylmethyl)-2,5-bis(2,2,2-trifluoroethoxy)benzamide
See: Flecainide

***m*-(Piperidin-1-ylmethyl)phenol**
Formed from: Roxatidine
Roxatidine acetate

3-(*m*-(1-Piperidinylmethyl)phenoxy)propionic acid
Formed from: 2-(2-Hydroxyethyl-1-thio)-N-(3-(3-(piperidinylmethyl)phenoxy)propyl) acetamide
Roxatidine
Loxtidine
Probably formed from: 3-(*m*-(1-Piperidinylmethyl)phenoxy)propylamine

3-(*m*-(1-Piperidinylmethyl)phenoxy)propylamine
Formed from: Roxatidine acetate
→ 3-(*m*-(1-Piperidinylmethyl)phenoxy) propionic acid
Probably in: Dog E904
Rat E904

N-(1-(2-Piperidinyl))propionanilide
Formed from: Fentanyl

Piperidolate (α-phenylbenzeneacetic acid 1-ethyl-3-piperidinyl ester)
→ Diphenylacetic acid
In: Rat A1504

Piperitenone (3-methyl-6-(1-methylethylidene)-2-cyclohexen-1-one)
→ *p*-Cresol
In: Rat J560

Piperitol (4-((1*R*,3a*S*,4*R*,6a*S*)-4-(1,3-benzodioxol-5-yl)tetrahydro-1*H*,3*H*-furo[3,4-*c*]furan-1-yl)-2-methoxyphenol)
Formed from: (+)-Pinoresinol

Piperonal (3,4-methylenedioxybenzaldehyde)
Probably formed from: 3,4-Methylenedioxybenzylamine
Piperonyl alcohol
→ 3,4-Methylenedioxymandelonitrile
In: Almond J194
Sorghum J194
→ Piperonylic acid
In: Hyphomicrobium C123
Rat D132
See also: G865
→ Piperonyl alcohol
In: Man F87

Piperonyl acetate
→ Piperonyl alcohol
Probably in: Rabbit B188

Piperonylacetone
See: 3,4-Methylenedioxyphenylacetone

Piperonyl alcohol (3,4-methylenedioxybenzyl alcohol)
Formed from: Piperonal
Probably formed from: Piperonyl acetate
Piperonyl isobutyrate
→ Piperonal
In: Man F87
See also: B188, D132

Piperonyl butoxide
→ 4-(((Butoxy)ethoxy)ethoxy)methyl-5-propylcatechol

In: Rat A3947

Piperonylglycine
Formed from: Piperonylic acid

Piperonylic acid (3,4-methylenedioxybenzoic acid)
Formed from: 3,4-Methylenedioxybenzonitrile
Piperonal
Safrole
Probably formed from: 3,4-Methylenedioxycinnamic acid
3,4-Methylenedioxyphenylacetone
3,4-Methylenedioxyphenylglyoxylic acid
3,4-Methylenedioxyphenylhydracrylic acid
→ Piperonylglycine
In: Guinea pig A1544
Man H736
Rat A1544, C164, D132
→ Protocatechuic acid
In: Pseudomonas A2449
See also: D132, G865

Piperonylideneacetone
→ 3,4-Methylenedioxyphenylacetone
Probably in: Rabbit B188

Piperonyl isobutyrate
→ Piperonyl alcohol
Probably in: Rabbit B188

Piribedil (1-(3,4-methylenedioxybenzyl)-4-(2-pyrimidinyl)piperazine)
Formed from: Piribedil methane sulphonate
→ 1-(3,4-Dihydroxybenzyl)-4-(2-pyrimidinyl)piperazine
In: Man A1069, A1582
Rat A318, A1057, A1582
→ 4-(5-Hydroxy-2-pyrimidinyl)-1-(3,4-methylenedioxybenzyl)piperazine
In: Man A1069, A1582
Rat A1057, A1582
→ Piribedil-N-oxide
In: Rat A1057, A1582
See also: A1069

Piribedil methane sulphonate
→ Piribedil
In: Man A1069

Piribedil-N-oxide
Formed from: Piribedil

Piritrexim (6-(2,5-dimethoxyphenyl)methyl)-5-methylpyrido[2,3-*d*]pyrimidine-2,4-diamine)
→ 2′-Demethylpiritrexim
In: Dog G44
Rat F757
→ 5′-Demethylpiritrexim
In: Dog G44
Rat F757
→ 3′-Hydroxypiritrexim
In: Dog G44
Rat F757

Piroglíride (N-(1-methyl-2-pyrrolidinylidene)-N′-phenyl-1-pyrrolidinecarboximidamide)
→ N-(5-Hydroxy-1-methyl-2-pyrrolidinylidene)-N′-phenyl-1-pyrrolidinecarboximidamide
In: Dog J693
Man J693
Monkey J693
Rat J693
→ N′-(*p*-Hydroxyphenyl)-N-(1-methyl-2-pyrrolidinylidene)-1-pyrrolidinecarboximidamide
In: Dog J693
Man J693
Monkey J693
Rat J693

Piroheptine (3-(10,11-dihydro-5*H*-dibenzo[*a*,*d*]cyclohepten-5-ylidene)-1-ethyl-2-methylpyrrolidine)
→ 2-Hydroxypiroheptine
In: Rabbit A1535

Piroxicam (4-hydroxy-2-methyl-N-(2-pyridyl)-2*H*-1,2-benzothiazine-3-carboxamide-1,1-dioxide)
Formed from: Ampiroxicam
→ 3-Carboxy-2-methyl-4-oxobenzothiazine-1,1-dioxide
In: Monkey B618
→ 5′-Hydroxypiroxicam
In: Dog B618, B793
Man H744
Monkey B618
Rat B618, B793
→ 6-Methyl-6*H*-7-oxopyrido[1,2-*a*]pyrimidino[5,4-*c*]benzothiazine-5,5-dioxide
In: Dog B618

→ N-Methylsaccharin
In: Dog B618, B793
Rat B793
→ Saccharin
In: Dog B618
Monkey B618
Rat B618

Pirprofen (2-(3-chloro-4-(3-pyrrolinyl) phenyl)propionic acid)
→ 4-Amino-3-chloro-α-methylphenylacetic acid
In: Man C41
Mouse C41
Rat C41
→ 2-(3-Chloro-4-(1-pyrrolyl)phenyl)propionic acid
In: Man C41, C422
Monkey C41, C422
Mouse C41, C422
Rat C41, C422
→ Pirprofen-β-D-glucuronide
In: Man C41
Monkey C41
Mouse C41
Rat F564
→ Pirprofen oxide
In: Man C41, C422
Monkey C41, C422
Mouse C41, C422

Pirprofen-β-D-glucuronide
Formed from: Pirprofen

Pirprofen oxide
Formed from: Pirprofen
→ Dihydrodihydroxypirprofen
Probably in: Man C41
Mouse C41
Rat C41
→ Pirprofen oxide-β-D-glucuronide
Probably in: Man C41
Mouse C41
Rat C41

Pirprofen oxide-β-D-glucuronide
Formed from: Pirprofen oxide

Pisatin ((6a*R*,12a*R*)-3-methoxy-6*H*-[1,3] dioxolo[5,6]benzofuro[3,2-*c*][1] benzopyran-6a(12a*H*)-ol)
Formed from: Formononetin
6a-Hydroxymaackiain
→ 3,6a-Dihydroxy-8,9-methylenedioxypterocarpan
In: Fungus K430
Fusarium A3414, B382

Piscerythrone (3′-(3,3-dimethylallyl)-2′,4′,5,7-tetrahydroxy-5′-methoxyisoflavone)
→ 3-(2,3-Dihydro-3,5-dihydroxy-2,2-dimethyl-1-benzopyran-6-yl)-5,7-dihydroxy-4*H*-1-benzopyran-4-one
In: Aspergillus G264
Botrytis G264
→ 3-(2,3-Dihydro-3,5-dihydroxy-2,2-dimethyl-1-benzopyran-8-yl)-5,7-dihydroxy-4*H*-1-benzopyran-4-one
In: Aspergillus G264
Botrytis G264
→ 3-(2,3-Dihydro-4-hydroxy-2-(2-hydroxy-2-propyl)benzofuran-5-yl)-5,7-dihydroxy-4*H*-1-benzopyran-4-one
In: Aspergillus G264
Botrytis G264
→ 3-(2,3-Dihydro-4-hydroxy-2-(2-hydroxy-2-propyl)benzofuran-7-yl)-5,7-dihydroxy-4*H*-1-benzopyran-4-one
In: Aspergillus G264
Botrytis G264

Piscidone (6′-(3,3-dimethylallyl)-4′,5,5′,7-tetrahydroxy-2′-methoxyisoflavone)
→ 3-(2,3-Dihydro-3,8-dihydroxy-7-methoxy-2,2-dimethyl-1-benzopyran-8-yl)-5,7-dihydroxy-4*H*-1-benzopyran-4-one
In: Botrytis G264

3-Pivaloylmorphine
→ Morphine
In: Man K517

Pivhydrazine (N-benzyl-N′-(2,2-dimethylpropionyl)hydrazine)
→ Benzoic acid
In: Rat A3922

Plastoquinone (2,3-dimethyl-5-((2*E*,6*E*,10*E*,14*E*,18*E*,22*E*,26*E*,30*E*)-3,7,11,15,19,23,27,31,35-nonamethyl-2,6,10,14,18,22,26,30,34-hexatriacontanonaenyl)-2,5-cyclohexadiene-1,4-dione)
Formed from: Homogentisic acid

Platydesmine ((2*S*)-2,3-dihydro-4-methoxy-α,α-dimethylfuro[2,3-*b*]quinoline-2-methanol)
→ Dictamnine
In: Skimmia A1676

Platydesminium (N-methylplatydesmine)
Formed from: 3-Dimethylallyl-2,4-dihydroxyquinoline

Plumbagin (5-hydroxy-2-methyl-1,4-naphthalenedione)
Formed from acetate, malonate and propionate in Drosophyllum A181
→ 3-Hydroxyplumbagin
In: Pseudomonas F46

Podophyllotoxin ((5*R*,5a*R*,8a*R*,9*R*)-5,8,8a,9-tetrahydro-9-hydroxy-5-(3,4,5-trimethoxyphenyl)furo[3′,4′:6,7]naphtho[2,3-*d*]-1,3-dioxol-6(5a*H*)-one)
Formed from: Deoxypodophyllotoxin
Ferulic acid
Podophyllotoxone
L-Phenylalanine
Yatein
Probably formed from: 4′-Demethylpodophyllotoxin
→ Podophyllotoxone
In: Podophyllum D274

Podophyllotoxone ((5a*R*,8a*R*,9*R*)-5a,6,8a,9-tetrahydro-9-(3,4,5-trimethoxyphenyl) furo[3′,4′:6,7]naphtho[2,3-*d*]-1,3-dioxole-5,8-dione)
Formed from: Podophyllotoxin
→ Podophyllotoxin
In: Podophyllum D274

Poly(*o*-aminophenol)
Formed from: *o*-Aminophenol

Polycatechol
Formed from: Catechol

Poly(*o*-cresol)
Formed from: *o*-Cresol

Poly(2,5-diaminobenzenesulphonic acid)
Formed from: 2,5-Diaminobenzenesulphonic acid

Poly(2,6-dimethoxyphenol)
Formed from: 2,6-Dimethoxyphenol

Polyguaiacol
Formed from: Guaiacol

Poly(4-hydroxy-3,5-dimethoxybenzoic acid)
Formed from: 4-Hydroxy-3,5-dimethoxybenzoic acid

Poly(4-hydroxy-3-methoxyacetophenone)
Formed from: 4-Hydroxy-3-methoxyacetophenone

Poly(4-hydroxystyrene)
→ Poly(4-hydroxystyrene)quinone
In: Mushroom K285

Poly(4-hydroxystyrene)quinone
Formed from: Poly(4-hydroxystyrene)

Poly(*o*-methoxyaniline)
Formed from: *o*-Methoxyaniline

Polyphenol
Formed from: Phenol

Polyvanillin
Formed from: Vanillin

Ponceau 3R
See: 2-Hydroxy-1-(2,4,5-trimethylphenylazo)naphthalene-3,6-disulphonic acid

Ponceau SX
See: 1-Hydroxy-2-(5-sulpho-2,4-xylylazo)naphthalene-4-sulphonic acid

Poncirenin
Formed from: Poncirin
→ Ponciretin
In: Streptococcus H218

Ponciretin
Formed from: Poncirenin
Poncirin

Poncirin ((2*S*)-7-((2-O-(6-deoxy-α-L-mannopyranosyl)-β-D-glucopyranosyl)oxy)-2,3-dihydro-5-hydroxy-2-(4-methoxyphenyl)-4*H*-1-benzopyran-4-one)
→ Isosakuranetin
In: Microorganism J525
→ Poncirenin
In: Streptococcus H218
→ Ponciretin

In: Streptococcus H218

Practolol (eraldin, 3-(4-acetamidophenoxy)-2-hydroxy-N-isopropyl)propylamine)
→ 1-(4-Acetamido-3-hydroxyphenoxy)-3-isopropylaminopropan-2-ol
In: Hamster A3705, D954
Rat A3705, K122
→ Deacetylpractolol
In: Marmoset A3705
Rat K122
→ Practolol-β-D-glucuronide
Probably in: Man A2752

Practolol-β-D-glucuronide
Probably formed from: Practolol

Pramiverine (N-isopropyl-4,4-diphenylcyclohexylamine)
→ 4,4-Diphenylcyclohexanol
In: Dog A2577
Man A2577
Monkey A2577
Rat A2577
→ 4,4-Diphenylcyclohexylamine
In: Dog A2577
Man A2577
Monkey A2577
Rat A2577

Pranidipine (1,4-dihydro-2,6-dimethyl-4-(3-nitrophenyl)-3,5-pyridinedicarboxylic acid methyl (2*E*)-3-phenyl-2-propenyl ester)
→ 1,4-Dehydropranidipine
In: Man J606
→ 1,4-Dihydro-2,6-dimethyl-4-(*m*-nitrophenyl)pyridine-3,5-dicarboxylic acid 3-methyl ester
In: Man J606

Pranolium (2-hydroxy-N,N-dimethyl-N-(1-methylethyl)-3-(1-naphthalenyloxy)-1-propanaminium)
→ α-Naphthol
In: Baboon C96
Dog C96
Rhesus monkey C96

Pranoprofen (2-(5*H*-[1]benzopyrano[2,3-*b*]pyridin-7-yl)propionic acid)
→ Pranoprofen glucoside
In: Mouse G176, G178
→ Pranoprofen-β-D-glucuronide
In: Mouse G176, G178

Pranoprofen glucoside
Formed from: Pranoprofen

Pranoprofen-β-D-glucuronide
Formed from: Pranoprofen

Pratensein (3′,5,7-trihydroxy-4′-methoxyisoflavone)
→ Dihydropratensein
In: Fusarium E724

Pratensein-7-glucoside
→ Pratensein-7-glucoside-6″-malonate
In: Cicer D445

Pratensein-7-glucoside-6″-malonate
Formed from: Pratensein-7-glucoside

Prazepam
See: 7-Chloro-1-cyclopropylmethyl-1,3-dihydro-5-phenyl-2*H*-1,4-benzodiazepin-2-one

Praziquantel (2-(cyclohexylcarbonyl)-1,2,3,6,7,11b-hexahydro-4-pyrazino[2,1-*a*]isoquinolin-4-one)
→ 4-*cis*-Hydroxypraziquantel
In: Man K638, K835
→ 4-*trans*-Hydroxypraziquantel
In: Man K638, K835
→ 6-Hydroxypraziquantel
In: Man K638

Prazocin (4-amino-2-(4-(2-furoyl)piperazin-1-yl)-6,7-dimethoxyquinazoline)
→ 4-Amino–6,7-dimethoxy-2-(4-(1-piperazinyl)quinazoline
In: Dog A2030, A2785
Rat A2030, A2785
→ 6-O-Demethylprazocin
In: Dog A2030, A2785
Rat A2030, A2785
→ 7-O-Demethylprazocin
In: Dog A2030, A2785
Rat A2030, A2785
→ 2,4-Diamino-6,7-dimethoxyquinazoline
In: Dog A2030, A2785
Rat A2030, A2785

Precarthamin
→ Carthamin
In: Carthamus K557, K579

Precocene I (7-methoxy-2,2-dimethyl-2*H*-benzo[*b*]pyran)
Formed from: L-Phenylalanine
Formed from acetate and mevalonate in
Ageratum B821
Contrast (in Ageratum) F667
→ 3,4-Dihydroxy-7-methoxy-2,2-dimethylchromane
In: Locust B447

Precocene II (6,7-dimethoxy-2,2-dimethyl-2*H*-benzo[*b*]pyran)
Formed from: L-Phenylalanine
Formed from acetate and mevalonate in
Ageratum B821
Contrast (in Ageratum) F667
→ (+)-3-Chromanol
In: Aspergillus E462
Brevilegnia E462
Rhodotorula E462
Streptomyces E462
Stysanus E462
Syncephalastrum E462
→ *cis*-3,4-Dihydro-3,4-dihydroxy-6,7-dimethoxy-2,2-dimethyl-2*H*-benzo[*b*]pyran
In: Aspergillus E462
Brevilegnia E462
Cabbage looper B210
Rhodotorula E462
Streptomyces E462
Stysanus E462
Syncephalastrum E462
→ *trans*-3,4-Dihydro-3,4-dihydroxy-6,7-dimethoxy-2,2-dimethyl-2*H*-benzo[*b*]pyran
In: Aspergillus E462
Brevilegnia E462
Cabbage looper B210
Rhodotorula E462
Streptomyces E462
Stysanus E462
Syncephalastrum E462
See also: E442
→ 3,4-Dihydro-3-hydroxy-6,7-dimethoxy-2,2-dimethyl-2*H*-benzo[*b*]pyran
In: Cabbage looper B210
→ 6-Hydroxy-7-methoxy-2,2-dimethyl-2*H*-benzo[*b*]pyran
In: Cabbage looper B210
Gastrimargus B439
Melanopus B439
See also: E442, E491
→ 7-Hydroxy-6-methoxy-2,2-dimethyl-2*H*-benzo[*b*]pyran
In: Cabbage looper B210
Gastrimargus B439
Melanopus B439
→ 2-Hydroxymethyl-6,7-dimethoxy-2-methyl-2*H*-benzo[*b*]pyran
In: Melanopus E491

Prefix
See: 2,6-Dichlorothiobenzamide

Preforan
See: *p*-Nitrophenyl 2-nitro-4-trifluoromethylphenyl ether

Premazepam (3,7-dihydro-6,7-dimethoxy-5-phenylpyrrolo[3,4-*e*][1,4]diazepin-2(1*H*)-one)
→ 4-Amino-3-benzoyl-2,5-dihydro-1,2-dimethyl-5-oxopyrrole
In: Dog C834
Rat C834
→ 3-Benzoyl-4-(N-(N-glycollyl)glycyl)amino-1,2-dimethylpyrrole
In: Rat C834
→ 3-Benzoyl-4-(N-(N-methyl)glycyl)amino-1,2-dimethylpyrrole
In: Dog C834
Rat C834
→ 3,7-Dihydro-6-hydroxymethyl-7-methyl-5-phenylpyrrolo[3,4-*e*][1,4]diazepin-2(1*H*)-one
In: Dog C834
Rat C834
→ 3,7-Dihydro-6-methyl-5-phenylpyrrolo[3,4-*e*][1,4]diazepin-2(1*H*)-one
In: Dog C834
Rat C834

Prenalterol (2-hydroxy-3-(4-hydroxyphenoxy)-N-(isopropyl)propylamine)
→ β-(*p*-Hydroxyphenoxy)lactic acid
In: Dog B992

→ Prenalterol-β-D-glucuronide
In: Dog B992
Man B992
Mouse B992
Rat B992
→ Prenalterol-O-sulphate
In: Dog B992
Man B992
Rat B992

Prenalterol-β-D-glucuronide
Formed from: Prenalterol

Prenalterol-O-sulphate
Formed from: Prenalterol

Prenazone
See: Feprazone

Prenylamine (N-(3,3-diphenylpropyl)-α-methylphenethylamine)
→ Amphetamine
In: Man A528

Prephenic acid (1-carboxy-4-hydroxy-α-oxo-2,5-cyclohexadiene-1-propanoic acid)
→ *p*-Hydroxyphenylpyruvic acid
Probably in: Reseda A1601
→ Phenylpyruvic acid
Probably in: Reseda A1601

Pretyrosine (arogenate, β-(1-carboxy-4-hydroxy-2,5-cyclohexadien-1-yl)alanine)
→ L-Tyrosine
In: Brevibacterium B47
Claviceps E234
Corynebacterium B47
Flavobacterium C115
Microtetraspora F222
Streptomyces C695

Primaclone
See: 2-Deoxyphenobarbital

Primaquine (8-((1-aminopent-4-yl)amino)-6-methoxyquinoline)
→ 8-(4-Amino-1-methylbutylamino)-6-hydroxyquinoline
In: Dog C66
Rat H136
→ 8,8′-Bis((4-amino-1-methylbutyl)amino)-5,5′-biquinoline
In: Horseradish G544
→ 8-(3-Formyl-1-methylpropylamino)-6-methoxyquinoline
In: Beef G230
See also: C279, D117, D131, E150, E705, F321, F437
→ 5-Hydroxyprimaquine
In: Dog C66
Rat H136
→ 4-((6-Methoxyquinolin-8-yl)amino)-*n*-pentanol
In: Hamster F321

Primidone
See: 2-Deoxyphenobarbital

Prizidilol (6-(2-(3-((1,1-dimethylethyl)amino)-2-hydroxypropoxy)phenyl)-3-(2*H*)-pyridazinone hydrazone)
→ N-Acetylprizidilol
In: Rat E207

Proanthocyanidin A_2 ((2*R*,3*R*,8*S*,14*R*,15*R*)-2,8-bis(3,4-dihydroxyphenyl)-3,4-dihydro-8,14-methano-2*H*,14*H*-1-benzopyrano[7,8-*d*][1,3]benzodioxocin-3,5,11,13,15-pentol)
Formed from: Cinnamic acid

Proanthocyanidin B_3 ((2*R*,2′*R*,3*S*,3′*S*,4*S*)-2,2′-bis(3,4-dihydroxyphenyl)-3,3′,4,4′-tetrahydro-(4,8′-bi-2*H*-1-benzopyran)-3,3′,5,5′,7,7′-hexol)
→ *p*-Hydroxyphenylacetic acid
In: Microorganism D654
→ *p*-Hydroxyphenyllactic acid
In: Microorganism D654
→ *m*-Hydroxyphenylpropionic acid
In: Microorganism D654
→ *m*-Hydroxyphenyl)-3-(2,4,6-trihydroxyphenyl)propan-2-ol
In: Microorganism D654
→ Melilotic acid
In: Microorganism D654
→ Phloretic acid
In: Microorganism D654
→ Phloroglucinol
In: Microorganism D654

Probenecid
See: *p*-Dipropylsulphamoylbenzoic acid

Procainamide
See: *p*-Amino-N-(2-diethylaminoethyl) benzamide

Procaine
See: 2-Diethylaminoethyl *p*-aminobenzoate

Procarbazine (N-isopropyl-α-(2-methylhydrazino)-*p*-toluamide)
→ *p*-Aminomethyl-N-isopropylbenzamide
Probably in: Microorganism A3616
→ N-Isopropylbenzamide radical
In: Horseradish D292
Ram D292

Procaterol (8-hydroxy-5-(1-hydroxy-2-isopropylaminobutyl)carbostyril)
→ 5-(2-Amino-1-hydroxybutyl)-8-hydroxycarbostyril
In: Dog A3483, A3856
Man A3483
Rat A3054, A3483
→ 5-Formyl-8-hydroxycarbostyril
In: Dog A3483, A3856
Man A3483
Rat A3054, A3483
→ Procaterol-β-D-glucuronide
In: Dog A3483, A3856
Man A3483
Rat A3054, A3483
→ Procaterol-O-sulphate
In: Rat A3054

Procaterol-β-D-glucuronide
Formed from: Procaterol

Procaterol-O-sulphate
Formed from: Procaterol

Procetofene
See: Fenofibrate

Prochloraz (N-propyl-N-(2-(2,4,6-trichlorophenoxy)ethyl)-1*H*-imidazole-1-carboxamide)
→ 2,4,6-Trichlorophenol
In: Rat G26, G717
→ 2,4,6-Trichlorophenoxyacetaldehyde
Probably in: Rat G26, G717
→ N-(2,4,6-Trichlorophenoxyethyl)-N-propylurea
Probably in: Rat G26

Prochlorperazine (2-chloro-10-(3-(4-methyl-1-piperazinyl)propyl)-10*H*-phenothiazine)
→ Bisdemethylprochlorperazine
In: Rat A69
→ N-(γ-(2-Chloro-10-phenothiazinyl) propyl)ethylenediamine
In: Rat A576
→ N-(γ-(2-Chloro-10-phenothiazinyl)propyl)-N′-methylethylenediamine
In: Rat A1865
→ Prochlorperazine-N-oxide
In: Rat A67
→ Prochlorperazine sulphoxide
In: Rat A67

Prochlorperazine-N-oxide
Formed from: Prochlorperazine

Prochlorperazine sulphoxide
Formed from: Prochlorperazine

Procyanidin polymer
Probably formed from: (+)-Catechin

Procyanidin B$_2$ ((2*R*,2′*R*,3*R*,3′*R*,4*R*)-2,2′-bis(3,4-dihydroxyphenyl)-3,3′,4,4′-tetrahydro-(4,8′-bi-2*H*-1-benzopyran)-3,3′,5,5′,7,7′-hexol)
Formed from: Cinnamic acid

Procyanidin B$_3$ ((2*R*,2′*R*,3*S*,3′*S*,4*S*)-2,2′-bis(3,4-dihydroxyphenyl)-3,3′,4,4′-tetrahydro-(4,8′-bi-2*H*-1-benzopyran)-3,3′,5,5′,7,7′-hexol)
Formed from: Cinnamic acid

Procyanidin B$_4$ ((2*R*,2′*R*,3*S*,3′*R*,4*S*)-2,2′-bis(3,4-dihydroxyphenyl)-3,3′,4,4′-tetrahydro-(4,8′-bi-2*H*-1-benzopyran)-3,3′,5,5′,7,7′-hexol)
Formed from: Cinnamic acid

Procyclidine (α-cyclohexyl-α-phenyl-1-pyrrolidinepropanol)
→ 1-(3-Hydroxycyclohexyl)-1-phenyl-3-(1-pyrrolidinyl)-1-propanol (*cis* and *trans* isomers)
In: Man A3907
Rat E788
→ 1-(*cis*-4-Hydroxycyclohexyl)-1-phenyl-3-(1-pyrrolidinyl)-1-propanol
In: Rat D240, E788
→ 1-(*trans*-4-Hydroxycyclohexyl)-1-phenyl-3-(1-pyrrolidinyl)-1-propanol
In: Rat D240, E788

Procymidone (N-(3,5-dichlorophenyl)-1,2-dimethylcyclopropane-1,2-dicarboximide)
→ 3-Hydroxyprocymidone
In: Rabbit J488

Prodolic acid (1,3,4,9-tetrahydro-1-*n*-propylpyrano[3,4-*b*]indole-1-acetic acid)
→ Hydroxyprodolic acid
In: Rat A2477

Proflavine (3,6-diaminoacridine)
→ N-Acetylproflavine
In: Channel catfish J407
Rainbow trout J142
→ Proflavine-β-D-glucuronide
In: Channel catfish J407
Rainbow trout J142
Proflavine-β-D-glucuronide
Formed from: Proflavine
Profluralin (N-(cyclopropylmethyl)-2,6-dinitro-N-bromo-4-(trifluoromethyl)benzenamine)
→ N-Cyclopropylmethyl-2,6-dinitro-4-trifluoromethylaniline
In: Rat A2912
→ 2,6-Dinitro-N-propyl-4-trifluoromethylaniline
In: Rat A2912
Proglumetacin
→ 1-(*p*-Chlorobenzoyl)-5-methoxy-2-methylindole-3-acetic acid
In: Dog E410
→ 2-(4-(3-Hydroxypropyl)piperazin-1-yl)ethyl (1-(*p*-chlorobenzoyl)-5-methoxy-2-methylindol-3-yl)acetate
In: Dog E410
→ Proglumide
In: Dog E410
Proglumide (4-(benzoylamino)-5-(dipropylamino)-5-oxopentanoic acid)
Formed from: Proglumetacin
Prolan (1,1-bis(*p*-chlorophenyl)-2-nitropropane)
→ 2-Amino-1,1-bis(*p*-chlorophenyl) propane
In: Insect A2190
Mouse A2190
Prolintane (1-phenyl-2-pyrrolidinopentane)
→ N-(4-Hydroxy-1-phenylpent-2-yl) pyrrolidine
In: Man G63
Rabbit A1274
→ N-(5-Hydroxy-1-phenylpent-2-yl) pyrrolidine
In: Man G63
→ *p*-Hydroxyprolintane
In: Man G63
Rabbit A1274
Rat A265
→ N-(1-Phenyl-2-pentyl)-γ-aminobutyric acid
In: Rabbit A1274
Rat A265
→ N-(1-Phenylpent-2-yl)pyrrolidone
In: Man G63
Rabbit A1274
Rat A265
L-Prolyl-3,4-dihydroxyphenethylamine
Formed from: L-Phenylalanyl-L-prolyl-3,4-dihydroxyphenethylamine
→ 3,4-Dihydroxyphenethylamine
In: Rat H432
L-Prolyl-N-formyl-L-kynurenine
Formed from: L-Prolyl-L-tryptophan
L-Prolyl-L-glutaminyl-L-glutaminyl-L-prolyl-L-tyrosine
→ Di-(L-prolyl-L-glutaminyl-L-glutaminyl-L-prolyl-L-tyrosine)
In: Horseradish J327
→ Tetra-(L-prolyl-L-glutaminyl-L-glutaminyl-L-prolyl-L-tyrosine)
In: Horseradish J327
→ Tri-(L-prolyl-L-glutaminyl-L-glutaminyl-L-prolyl-L-tyrosine)
In: Horseradish J327
L-Prolyl-L-glutaminyl-L-glutaminyl-L-prolyl-L-tyrosyl-L-prolyl-L-glutaminyl-L-glutaminyl-L-prolyl-L-alanine
→ Di-(L-prolyl-L-glutaminyl-L-glutaminyl-L-prolyl-L-tyrosyl-L-prolyl-L-glutaminyl-L-glutaminyl-L-prolyl-L-alanine)
In: Horseradish J327
L-Prolyl-β-naphthylamide
→ β-Naphthylamide
In: Agave A2745
Arabidopsis K224
Beef A526
Pseudomonas G643
L-Prolyl-*p*-nitroanilide
Formed from: Succinyl-L-alanyl-L-alanyl-L-prolyl-*p*-nitroanilide
→ *p*-Nitroaniline
In: Nocardia G343
L-Prolyl-L-phenylalanyl-3,4-dihydroxyphenethylamine
→ 3,4-Dihydroxyphenethylamine

In: Rat H432
→ L-Phenylalanyl-3,4-dihydroxyphenethylamine
In: Rat H432
L-Prolyl-L-prolyl-3,4-dihydroxyphenethylamine
→ 3,4-Dihydroxyphenethylamine
In: Rat H432
L-Prolyl-L-tryptophan
→ L-Prolyl-N-formyl-L-kynurenine
In: Wheat E24
Promazine (10-(3-dimethylaminopropyl) phenothiazine)
Formed from: Chlorpromazine
→ 10-(2-Carboxyethyl)phenothiazine
In: Rabbit E822
→ Demethylpromazine
In: Horse B609
See also: A69
→ 2-Hydroxypromazine
Probably in: Man B753
→ 3-Hydroxypromazine
In: Horse B609
See also: B753
→ Promazine-N-oxide
In: Horse B609
→ Promazine radical
Species unclear K386
→ Promazine sulphoxide
Probably in: Man B753
Promazine-N-oxide
Formed from: Promazine
→ Promazine-N-oxide sulphoxide
Probably in: Horse B609
Promazine-N-oxide sulphoxide
Probably formed from: Promazine-N-oxide
Promazine radical
Formed from: Promazine
Promazine sulphoxide
Probably formed from: Promazine
Promazin-2-yl-β-D-glucuronide
Formed from: 2-Hydroxypromazine
Promazin-3-yl-β-D-glucuronide
Formed from: 3-Hydroxypromazine
Promethazine (N,N,α-trimethyl-10*H*-phenothiazine-10-ethanamine)
→ Hydroxypromethazine
In: Man J869
Rabbit B717
→ Norpromethazine
In: Rabbit B717
→ Promethazine-N-oxide
In: Pig G365
Rabbit B717
→ Promethazine sulphoxide
In: Dog J869
See also: B422
Promethazine-N-oxide
Formed from: Promethazine
Promethazine sulphone
Formed from: Promethazine sulphoxide
→ Norpromethazine sulphone
Probably in: Rat B422
Promethazine sulphoxide
Formed from: Promethazine
→ Promethazine sulphone
Probably in: Rat B422
Prominal
See: Mephobarbital
Pronethalol
See: 2-Isopropylamino-1-(2-naphthyl)ethanol
Propachlor (2-chloro-N-isopropylacetamide)
→ *p*-Acetamidophenol
In: Rat B93
→ 2-(S-Glutathionyl)-N-isopropyl-N-phenylacetamide
In: Rat B669, B707, C421
See also: A1682, A2405, B91, B92, B93, C407, C936, D932, E790, F588, F900, G552
→ N-(*p*-Hydroxyphenyl)-N-isopropyl-2-methylsulphonylacetamide
In: Rat A1682, B93, C421
Sheep A1682, B91
→ N-(*p*-Hydroxyphenyl)-2-methylsulphonylacetamide
In: Microorganism C421
Pig E790
Rat B93, C421
→ N-(1-Hydroxyprop-2-yl)-2-methylsulphonyl-N-phenylacetamide
In: Microorganism C421
Pig E790
Rat A1682, B93, C421
Sheep A1682, B91

→ N-Isopropylacetanilide
In: Chicken D932
Rat C421
→ N-Isopropyl-2-mercaptoacetanilide
In: Bacteroides C407
Eubacterium C407
Megasphaera C407
Rat B707
→ N-Isopropyl-2-methylsulphonylacetanilide
In: Rat B936
→ N-Isopropyloxanilic acid
In: Chicken D932
→ 2-Methylsulphonyl-N-phenylacetamide
In: Microorganism C421
Rat A1682, B93, C421
Sheep A1682

Propafenone (2′-(2-hydroxy-3-propylaminopropoxy)-3-phenylpropiophenone)
→ Depropylpropafenone
In: Dog D439
Man K636
Rat J538, K343
→ *o*-(2-Hydroxy-3-propionaldoxy)-*ω*-phenylpropiophenone
In: Dog D439
Man D200
→ *o*-Hydroxy-ω-phenylpropiophenone
In: Dog D439
Man D200
→ 4′-Hydroxypropafenone
In: Dog E18
Rat J538
See also: D439
→ 5-Hydroxypropafenone
In: Dog D439, E18
Man D200, K636
Mouse H227
Rat J538, K343
See also: C913
→ 3-Phenylpropionic acid
In: Dog D439
→ Propafenone-β-D-glucuronide
Probably in: Dog D439, E18
Man D200

Propafenone-β-D-glucuronide
Formed from: Propafenone

Propanidid (propyl 4-(N,N-diethylcarbamoylmethoxy)-3-methoxyphenylacetate)
→ 4-((N,N-Diethyl)acetamid-2-oxy)-3-methoxyphenylacetic acid
In: Pig A2809
Rat C677

Propanil
See: 3′,4′-Dichloropropionanilide

Propanilide
See: Propionanilide
Formed from: Sufentanil

10-(2′-Propanoneoxime)phenothiazine
Formed from: N-Hydroxynorpromethazine
Norpromethazine
Promethazine

Propantheline (N-methyl-N-(1-methylethyl)-N-(2-((9*H*-xanthen-9-ylcarbonyl)oxy)ethyl)-2-propanaminium)
→ Hydroxypropantheline
Probably in: Man A3490
→ Xanthanoic acid
In: Man A3490

N-Propargylbenzylamine
See: Demethylpargyline

2-(2-Propen-1-oxy)phenol
Formed from: Oxprenolol
→ 1-Acetoxy-2-(2-propen-1-oxy)benzene
Probably in: Man A288

1-Propenylbenzene
→ 3-(1-Propenyl)catechol
In: Pseudomonas F283

***p*-(Propen-2-yl)benzoic acid**
Probably formed from: *p*-(2-Hydroxy-2-propyl)benzoic acid
→ *p*-(Propen-2-yl)benzoylglycine
Probably in: Guinea pig C574
Rat C574

***p*-(Propen-2-yl)benzoylglycine**
Probably formed from: *p*-(Propen-2-yl)benzoic acid

3-(1-Propenyl)catechol
Formed from: 1-Propenylbenzene

4-(1-Propenyl)catechol
Formed from: Isosafrole

p-(1-Propenyl)phenol (anol)
Formed from: *trans-p*-Methoxypropenylbenzene
p-(*n*-Propyl)phenol
→ *p*-Coumaric acid
Probably in: Mouse H725
Rat H725
→ Epoxypseudoisoeugenol 2-methylbutyrate
In: Pimpinella J188
→ *p*-(1-Propenyl)phenyl sulphate
In: Mouse H725
Rat H725
→ *p*-Tyrosol
Probably in: Mouse H725
Rat H725

p-(1-Propen-2-yl)phenol
Formed from: *p*-Isopropylphenol

p-(1-Propenyl)phenyl sulphate
Formed from: *p*-(1-Propenyl)phenol

Propham
See: Isopropyl carbanilate

p-Propionamidophenol
Formed from: Alfentanil
Propionanilide
Sufentanil
→ *p*-Aminophenol
Probably in: Rat E895

Propionanilide (propanilide)
Formed from: Alfentanil
Sufentanil
→ Aniline
In: Aspergillus E322
Red rice A3773
→ *p*-Propionamidophenol
Probably in: Dog F567
Man F567, H176
Rat F567

4-(N-Propionylanilino)piperidine (norfentanyl)
Formed from: Fentanyl
N-(1-(2-(*p*-Hydroxy)phenyl)ethyl-4-piperidinyl)propionanilide
→ 4-(N-Anilino)piperidine
Probably in: Man J266

3-Propionylchloramphenicol
Formed from: Chloramphenicol

o-Propionylmercaptobenzoic acid
→ *o*-Mercaptobenzoic acid
In: Guinea pig H214

6-Propionylmorphine
Formed from: Dipropionylmorphine

Propionylsalicylic acid
→ Salicylic acid
In: Acinetobacter A962

Propiophenone (1-phenyl-1-propanone)
Formed from: 1-Phenyl-1-propanol
→ 2-Oxo-1-phenyl-1-propanol
In: Rabbit B755
Rat B755, J8
→ 1-Phenyl-1-propanol
In: Corynebacterium J893
Lactobacillus G148
Rabbit B755
Rat B755, J8

Propiverine (α-phenyl-α-propoxybenzeneacetic acid 1-methyl-4-piperidinyl ester)
→ Demethylpropiverine
In: Rat H552
→ 1-Methyl-4-piperidinyl α-(2-carboxyethoxy)-α,α-diphenylacetate
In: Rat H552
→ N-Methylpiperidinyl-4-benzilate
In: Rat E628, H552
→ Propiverine-N-oxide
In: Man E629
Rat H552
→ O-(*n*-Propyl)benzilic acid
In: Rat E628

Propiverine-N-oxide
Formed from: Propiverine

Propofol
See: 2,6-Diisopropylphenol

Propoxur (2-(1-methylethoxy)phenol methylcarbamate)
→ 2-Isopropylphenol
In: Rat B730

p-Propoxyacetanilide
→ *p*-Acetamidophenol
In: Cunninghamella F590
→ *p*-(2-Hydroxypropoxy)acetanilide
In: Cunninghamella F590
→ *p*-Propoxyaniline
In: Mouse C609

***p*-Propoxyamphetamine**
→ *p*-Hydroxyamphetamine
In: Cunninghamella F893
***p*-Propoxyaniline**
Formed from: *p*-Propoxyacetanilide
***p*-Propoxybenzaldehyde**
→ *p*-Propoxybenzoic acid
In: Beef C508
***p*-Propoxybenzoic acid**
Formed from: *p*-Propoxybenzaldehyde
4-Propoxycatechol
Formed from: *p*-Propoxyphenol
→ Propoxy-*o*-quinone
In: Mushroom H907
7-Propoxycoumarin
→ Umbelliferone
In: Dog C160
Guinea pig C160
Monkey C160
Mouse C160
Rabbit C160
Rat C160
1-(1-Propoxy)naphthalene
→ 1-(2-Hydroxypropoxy)naphthalene
In: Rat B582
→ α-Naphthol
In: Rat B582
1-(2-Propoxy)naphthalene
→ 1-(1-Hydroxy-2-propoxy)naphthalene
In: Rat B582
→ α-Naphthol
In: Rat B582
Propoxyphene
See: 3-Dimethylamino-1,2-diphenyl-2-butyl propionate
***p*-Propoxyphenol**
→ 4-Propoxycatechol
In: Mushroom H907
Propoxy-*o*-quinone
Formed from: 4-Propoxycatechol
Propranolol (inderal; *R* and *S* isomers; 1-((1-methylethyl)amino)-3-(1-naphthalenyloxy)-2-propanol))
Formed from: O-Acetylpropranolol
Butyryl propranolol
Cyclohexanoyl propranolol
→ 2-Hydroxy-3-(α-naphthyloxy) propionaldehyde
In: Rat B980
See also: A620, A628, A1142, A1414, B55, C581, C620, C625, C746, C840, D521, F312, F346, H144
→ 2-Hydroxy-3-(α-naphthyloxy)propylamine
In: Cunninghamella F346, G493
Dog B55, C581, C620
Guinea pig E869
Man D343, G868, H54, K504
Rat A620, B55, C234, C840, D98, F312, G836, H409, K504
Sheep H144
→ 2-Hydroxypropranolol
In: Man B987
Rat B987, E915
→ 3-Hydroxypropranolol
In: Rat B55
See also: E917
→ 4-Hydroxypropranolol
In: Cunninghamella F346, G493
Dog B55, C581, C620, C625, D376, D521
Guinea pig E869
Hamster C620
Man A3918, B323, B987, C625, C746, D343, D376, D527, G868, H54, K504, K578
Rat A620, A3918, B55, B987, C19, C620, C843, D98, D527, E915, F312, G836, K504
Sheep H144
See also: D859, E916, E917
→ 5-Hydroxypropranolol
In: Guinea pig E869
Hamster C620
Man B987, G868, H54, K504
Rat B987, C620, C843, E915, G836, K504
Sheep H144
→ 6-Hydroxypropranolol
In: Rat E915
See also: E917
→ 7-Hydroxypropranolol
In: Dog B987
Hamster C620
Man B987, H54

Rat B987, C620, C843, G836, K504
→ 8-Hydroxypropranolol
In: Cunninghamella G493
→ 1-(1-Hydroxy-2-propylamino)-3-(1-naphthyloxy)-2-propanol
In: Rat E149
→ α-Naphthol
In: Dog B55
Rat A620, B55, C840, F312
→ Propranolol dihydrodiol
In: Horse F387
→ Propranolol-O-β-D-glucuronide
In: Dog C581, C827, D398, D521
Man A1712, A2440, B323, C257, C746, D343, D527, H933
Monkey H861
Rat C620

Propranolol dihydrodiol
Formed from: Propranolol

Propranolol-O-β-D-glucuronide
Formed from: Propranolol

Propyl acitretinate
Formed from: Acitretinoyl CoA

N-Propylalanylaniline
→ Aniline
In: Mouse A830

N-(*n*-Propyl)alanyl-*o*-toluidine
→ *o*-Toluidine
In: Mouse A830

Propyl *p*-aminobenzoate
→ Propyl *p*-sulphamidobenzoate
In: Eubacterium H439
Haemophilus H439
Klebsiella H439

4-(2-*n*-Propylaminoethyl)-2(3*H*)-indolone
Formed from: 4-(2-Di-*n*-propylaminoethyl)-2(3*H*)-indolone

2-(*n*-Propylamino)-8-hydroxytetralin
Formed from: 2-(Di-*n*-propylamino)-8-hydroxytetralin
→ 2-(*n*-Propylamino)-8-hydroxytetralin-O-β-D-glucuronide
Probably in: Rat H132

2-(*n*-Propylamino)-8-hydroxytetralin-O-β-D-glucuronide
Formed from: 2-(*n*-Propylamino)-8-hydroxytetralin

N-Propylamphetamine
Formed from: N-Methyl-N-propylamphetamine
→ Amphetamine
In: Cunninghamella A3878
Mycobacterium B380
Rat A2850
→ N-Hydroxy-N-propylamphetamine
In: Cunninghamella A3878, F345
Rat A2850
→ *p*-Hydroxy-N-propylamphetamine
In: Cunninghamella A3878, F345
Rat A1998
→ N-(2-Hydroxypropyl)amphetamine
In: Rat A2850
→ N-((1-Methyl-2-phenyl)ethyl)-1-propanimine-N-oxide
In: Cunninghamella F345
Rat A2850

N-Propylaniline
Formed from: N,N-Dipropylaniline
N-Methyl-N-propylaniline

***p*-Propylanisole**
See: *p*-Methoxypropylbenzene

***n*-Propylbenzene**
Formed from: Ferulic acid
→ 3-*n*-Propylcatechol
In: Pseudomonas B267
→ 3-Phenylpropionic acid
In: Pseudomonas B267

O-(*n*-Propyl)benzilic acid
Formed from: Propiverine
→ Benzilic acid
Probably in: Rat E628
→ 5-Methyl-2,2-diphenyl-1,4-dioxan-3-one
Probably in: Rat E628

***p-n*-Propylbenzoic acid**
→ *p*-*n*-Propylbenzoyl-β-D-glucuronic acid
In: Rat J549

1-Propyl-1*H*-benzotriazole
→ 1-(2-Hydroxypropyl)-1*H*-benzotriazole
In: Rat C240

2-Propylbenzotriazole
→ 4,5-Dihydro-4,5-dihydroxy-2-propylbenzotriazole
In: Rat C419

→ 5-Hydroxy-2-methylbenzotriazole
In: Rat C419
→ 4-Hydroxy-2-propylbenzotriazole
In: Rat C419

p-n-Propylbenzoyl-β-D-glucuronic acid
Formed from: *p-n*-Propylbenzoic acid

Propylcannabinol
Formed from: Cannabinol

Propyl β-carboline-3-carboxylate
→ β-Carboline-3-carboxylic acid
In: Rat C220

3-n-Propylcatechol
Formed from: *n*-Propylbenzene
o-Propylphenol
→ 2-Hydroxy-6-oxonona-2,4-dienoic acid
Probably in: Pseudomonas B267

4-n-Propylcatechol
Formed from: Dihydrosafrole
Probably formed from: 4-Allylcatechol
1-Phenylpropanesulphonic acid

6-n-Propylchromone-2-carboxylic acid
→ 6-*n*-(2-Carboxyethyl)chromone-2-carboxylic acid
In: Rat G986
→ 6-*n*-(1-Hydroxypropyl)chromone-2-carboxylic acid
In: Rat G986
→ 6-*n*-(2-Hydroxypropyl)chromone-2-carboxylic acid
In: Rat G986

Propyl 4-(N,N-diethylcarbamoylmethoxy)-3-methoxyphenylacetate
See: Propanidid

Propyl gallate
→ Gallic acid
In: Klebsiella E113
→ Propyl gallate-α-D-glucoside
Probably in: Bacillus H234
→ Propyl gallate-β-D-glucuronide
In: Man G524
→ Propyl gallate-O-sulphate
In: Eubacterium G475

Propyl gallate-α-D-glucoside
Formed from: Propyl gallate

Propyl gallate-β-D-glucuronide
Formed from: Propyl gallate

Propyl gallate-O-sulphate
Formed from: Propyl gallate

n-Propyl-L-γ-glutamyl dopamide
→ 3,4-Dihydroxyphenethylamine
In: Dog A3945

N-(n-Propyl)glycyl-2,6-dimethylanilide
→ 2,6-Dimethylanilide
In: Mouse A830

n-Propyl p-hydroxybenzoate (propylparaben)
→ *p*-Hydroxybenzoic acid
In: Cat A3387
→ *n*-Propyl *p*-hydroxybenzoate-β-D-glucuronide
In: Man G524

n-Propyl p-hydroxybenzoate-β-D-glucuronide
Formed from: *n*-Propyl *p*-hydroxybenzoate

n-Propylnoradrenaline
→ 3-Hydroxy-4-methoxy-N-propylphenylethanolamine
In: Rat A486
→ *n*-Propylnormetanephrine
In: Rat A486

n-Propylnormetanephrine
Formed from: *n*-Propylnoradrenaline

3-O-Propyl-17β-oestradiol
→ 17β-Oestradiol
In: Rat A3786
→ 3-O-Propyl-17β-oestriol
In: Rat A3786
→ 3-O-Propyloestrone
In: Rat A3786

3-O-Propyloestriol
Formed from: 3-O-Propyl-17β-oestradiol

3-O-Propyloestrone
Formed from: 3-O-Propyl-17β-oestradiol

Propylparaben
See: *n*-Propyl *p*-hydroxybenzoate

N-n-Propylphenethylamine
Probably formed from: N,N-Di-*n*-propylphenethylamine
→ Phenethylamine
In: Cunninghamella F345

o-n-Propylphenol
→ 3-*n*-Propylcatechol
In: Pseudomonas E958, J392

p-Propylphenol
→ 3-(*p*-Hydroxyphenyl)propanol

In: Pseudomonas F88
→ *p*-(1-Hydroxy-*n*-propyl)phenol (both isomers)
In: Byssochlamys K164
Penicillium J664, J680, J836
Pseudomonas F381, K177
See also: D696
→ *p*-Methoxypropylbenzene
In: Pimpinella J188
→ *p*-(1-Propenyl)phenol
In: Penicillium J664, J836
→ *p*-Propylphenyl-β-D-glucuronide
In: Man G524
→ *p*-Propylphenyl sulphate
In: Man E575, K405
Rat E575

N-Propylphentermine
→ *p*-Hydroxy-N-propylphentermine
In: Cunninghamella F345
→ Phentermine
In: Cunninghamella F345

***p*-Propylphenyl-β-D-glucuronide**
Formed from: *p*-Propylphenol

***p*-Propylphenyl sulphate**
Formed from: *p*-Propylphenol

N-Propyl-2′,6′-pipecoloxylidide
See: Ropivacaine

N-(*n*-Propyl)-N-(4-pyridinyl)-1*H*-indolamine
See: Besipirdine

Propyl salicylate
→ Salicylic acid
In: Guinea pig D144

4-(Propylselenyl)catechol
Formed from: *p*-(Propylselenyl)phenol
→ 4-(Propylselenyl)-*o*-quinone
In: Mushroom H907

***p*-(Propylselenyl)phenol**
→ 4-(Propylselenyl)catechol
In: Mushroom H907

4-(Propylselenyl)-*o*-quinone
Formed from: 4-(Propylselenyl)catechol

O-Propylsterigmatocystin
→ Aflatoxin B_1
In: Aspergillus K249

β-*n*-Propylstyrene oxide
→ (*R*,*R*)-*threo*-1,2-dihydroxy-1-phenylpentane
In: Rabbit J17

Propyl *p*-sulphamidobenzoate
Formed from: Propyl *p*-aminobenzoate

***p*-Propylsulphamoylbenzoic acid**
Formed from: *p*-Dipropylsulphamoylbenzoic acid

Propylsulphinylbenzene
Formed from: Propylthiobenzene

2-(N-Propyl-N-(2-thienyl)ethylamino)tetralin-5-yl-β-D-glucuronide
Formed from: 5-Hydroxy-2-(N-propyl-N-(2-thienyl)ethylamino)tetralin

Propylthiobenzene
→ Propylsulphinylbenzene
In: Mortierella D759
Pseudomonas H165

(5-(Propylthio)-1*H*-benzimidazol-2-yl)carbamic acid methyl ester
See: Albendazole

Propyphenazone (1,2-dihydro-1,5-dimethyl-4-(1-methylethyl)-2-phenyl-3*H*-pyrazol-3-one)
→ Demethylpropyphenazone
In: Man B10

Proterguride (N,N-diethyl-N′-((8α)-6-propylergolin-8-yl)urea)
→ 2-Amino-1,2,3,4,4a,5,6,10b-octahydro-6-oxo-4-propylbenzo[*f*]quinoline
In: Rat G553
→ 3,4-Dehydroproterguride
In: Rat G553
→ Desethylproterguride
In: Rat G553
→ 2-Oxoproterguride
In: Rat G553
→ Proterguride-6-N-oxide
In: Rat G553

Proterguride-6-N-oxide
Formed from: Proterguride

Prothanon (9,9-dioxopromethazine)
→ 1-(2,3-Dihydroxypropyl)-9,9-dioxophenothiazine
In: Rat A3301
→ 9,9-Dioxophenothiazine
In: Man A682
Rat A3301
→ Hydroxyprothanon
In: Rat A3301

→ Norprothanon
In: Man A682
Rat A3301
→ Prothanon-N-oxide
In: Rat A3301

Prothanon-N-oxide
Formed from: Prothanon

Prothiaden
See: Dosulepin

Protizinic acid (7-methoxy-α,10*H*-phenothiazine-2-acetic acid)
→ 7-Demethylprotizinic acid
In: Rat A3385
→ 5-Oxoprotizinic acid
In: Rat A3385

Protoaphin aglycone ((1*R*-(1a,3b,4a,5(1′*R**,3′*R**,4′*S**)))-3,3′,4,4′-tetrahydro-4,4′,7,7′,9,9′,10-pentahydroxy-1,1′,3,3′-tetramethyl-(5,6′-bi-1*H*-naphtho[2,3-*c*]pyran)-5′,10′-dione)
→ Xanthoaphin
In: Aphid A3112

Protoberberine (5,6-dihydrodibenzo[*a*,*g*] quinolizinium)
Formed from: Tetrahydroprotoberberine

Protocatechualdehyde (3,4-dihydroxybenzaldehyde)
Formed from: Astringin
3,4-Dihydroxybenzyl alcohol
3,4-Dihydroxybenzylamine
3,4-Dihydroxymandelic acid
3,4-Dihydroxy-L-phenylalanine
3,4-Dihydroxyphenylglycine
erythro-3,4-Dihydroxy-L-phenylserine
threo-3,4-Dihydroxy-L-phenylserine
m-Hydroxybenzaldehyde
p-Hydroxybenzaldehyde
4-Hydroxy-3-nitrobenzaldehyde
Protocatechuic acid
Vanillin
Probably formed from: 3,4-Dihydroxybenzyl alcohol
3,4-Dihydroxyphenylglyoxylic acid
m-Hydroxybenzaldehyde
Isovanillin
→ 3,4-Dihydroxybenzyl alcohol
In: Phanerochaete F866
→ Isovanillin
In: Nicotiana B490
Rat A486
With catechol O-methyltransferase A597
→ Protocatechuic acid
In: Acinetobacter F366
Aspergillus A596
Desulphovibrio F666
Streptomyces C284
See also: G865
→ Vanillin
In: Nicotiana A3778, B490
Rat A486, A2817
With catechol O-methyltransferase A597
See also: A2315, G782

Protocatechuamide
Formed from: Protocatechuic acid

Protocatechuic acid (3,4-dihydroxybenzoic acid)
Formed from: *p*-Carboxybenzenesulphonic acid
3-Carboxybiphenyl ether
(+)-Catechin
Catechol
Conidendrin
Cyclohexanecarboxylic acid
Dehydrpdivanillyl alcohol
3,4-Diacetoxybenzoic acid
1,2-Dihydro-1,2-dihydroxyterephthalic acid
3,4-Dihydroxyphthalic acid
4,5-Dihydroxyphthalic acid
Gallic acid
m-Hydroxybenzoic acid
p-Hydroxybenzoic acid
4-Hydroxyisophthalic acid
p-Hydroxylaminobenzoic acid
Isophthalic acid
Isovanillic acid
D-Noradrenaline
L-Noradrenaline
Phenanthrene
m-Phenoxybenzoic acid
p-Phenoxybenzoic acid
Phthalic acid
Piperonylic acid
Protocatechualdehyde

Protocatechuic acid-3-O-sulphate
Protocatechuic acid-4-O-sulphate
Protocatechuoylphloroglucinolcarboxylic acid
Terephthalic acid
Vanillic acid
Probably formed from: Ethyl protocatechuate
4-Fluoro-3-hydroxybenzoic acid
3-Hydroxyphthalic acid
Methyl protocatechuate
→ δ-Carboxy-α-hydroxy-*cis,cis*-muconic semialdehyde
In: Bacillus A2827, A3887
Pseudomonas A1090, D771
See also: B226, J653
→ β-Carboxy-*cis,cis*-muconic acid
In: Bradyrhizobium J484
Hydrogenophaga J218
Nocardia A960
Pseudomonas A563, A2399, D771
Streptomyces B877
See also: A1464
→ γ-Carboxy-*cis,cis*-muconic semialdehyde
In: Micrococcus K753
See also: A698
→ 4-Carboxy-*o*-quinone
Probably in: Botrytis A1930
→ Catechol
In: Bacterium C396
Clostridium F402, H576, J184
Microorganism G632
See also: A2837, B2, D771, F266, J669
→ 3,4-Dihydroxybenzoyl-D-glucose
In: Cucumis A611
→ 3,4-Dihydroxyhippuric acid
Probably in: Quercus K737
Rabbit A1991
Rat A1991
→ 2,3,6,7,9,10-Hexahydro-2,3,7,8-tetraoxoanthracene
In: Klebsiella H748
→ *m*-Hydroxybenzoic acid
In: Microorganism B321
See also: A647
→ 2-Hydroxy-*cis,cis*-muconic semialdehyde
In: Bacillus A1525
→ Hydroxyquinol
In: Candida J461
Sporotrichum B73, C57
Trichosporon B368, E288
See also: B34
→ Isovanillic acid
In: Nicotiana B490
Pampas grass A187
Rat A486, A678, A2817, C169, D297
See also: E374, F266
→ Protocatechualdehyde
In: Neurospora A750
→ Protocatechuamide
In: Streptomyces E279
→ Protocatechuic acid-α-D-glucoside
Probably in: Bacillus H234
→ Protocatechuic acid-3-β-D-glucoside
In: Lobelia H686
→ Protocatechuic acid-4-β-D-glucoside
In: Manduca G897
Nicotiana B442
→ Protocatechuic acid-3-O-sulphate
In: Rat B724
→ Protocatechuic acid-4-O-sulphate
In: Rat B724
→ 2,4,5-Trihydroxybenzoic acid
In: Pseudomonas A326
→ Vanillic acid
In: Nicotiana A3778, B490
Pampas grass A187
Phanerochaete J339
Rat A486, A678, A2817, C169, D279
See also: A647, A1241, A2315, A2449, A3757, B854, F266

Protocatechuic acid-α-D-glucoside
Formed from: Protocatechuic acid

Protocatechuic acid-3-β-D-glucoside
Formed from: Protocatechuic acid

Protocatechuic acid-4-β-D-glucoside
Formed from: Protocatechuic acid

Protocatechuic acid-3-O-sulphate
Formed from: Protocatechuic acid
→ Protocatechuic acid

In: Rat B724, B854

Protocatechuic acid-4-O-sulphate
Formed from: Protocatechuic acid
→ Protocatechuic acid
In: Rat B724, B854

Protocatechuoylphloroglucinolcarboxylic acid
Formed from: Quercetin
→ Phloroglucinol
Probably in: Bradyrhizobium G185
Rhizobium G185
→ Protocatechuic acid
Probably in: Bradyrhizobium G185
Pseudomonas A1328
Rhizobium G185, H280

Protokylol (4-(2-((2-(1,3-benzodioxol-5-yl)-1-methylethyl)amino)-1-hydroxyethyl)-1,2-benzenediol)
→ Protokylol quinone
Probably in: Beef A624

Protokylol quinone
Probably formed from: Protokylol

Protopine (4,6,7,14-tetrahydro-5-methylbis[1,3]benzodioxolo[4,5-*c*:5′,6′-*g*] azecin-13(5*H*)-one)
Formed from: 3,4-Dihydroxy-L-phenylalanine
N-Methylstylopine
cis-7,8,13,13a-Tetrahydro-N-methylcoptisinium
→ Sanguinarine
In: Macleaya C696
See also: F699

Protosinomenine ((1*S*)-1,2,3,4-tetrahydro-1-((3-hydroxy-4-methoxyphenyl)methyl)-7-methoxy-2-methyl-6-isoquinolinol)
→ Tetrahydro-3,9-dihydroxy-2,10-dimethoxyprotoberberine
In: Berberis D615, D878
Eschscholtzia H680

Protriptyline (N-methyl-5*H*-dibenzo[*a*,*d*] cycloheptene-5-propylamine)
→ N-Acetoxyprotriptyline
In: Candida J726
Cunninghamella J726
Fusarium J726
→ N-Acetylprotriptyline
In: Fusarium J726
→ 5-Aminopropyl-5*H*-dibenzo[*a*,*d*] cycloheptene
In: Fusarium J726
Rat A1880
→ 5-(2-Carboxyethyl)-5*H*-dibenzo[*a*,*d*] cycloheptene
In: Fusarium J726
→ 2-Hydroxyprotriptyline
In: Cunninghamella J726
→ 10-Hydroxyprotriptyline
In: Rat A2926
→ 1′-Oxoprotriptyline
In: Fusarium J726
→ Protriptyline-10,11-oxide
In: Rat A1880, A2926

Protriptyline-10,11-oxide
Formed from: Protriptyline
→ 10,11-Dihydro-10,11-dihydroxyprotriptyline
Probably in: Rat A2926

Prulifloxacin
See: 6-Fluoro-1-methyl-7-(4-((5-methyl-2-oxo-1,3-dioxol-4-yl)methyl)-1-piperazinyl)-4-oxo-1*H*,4*H*-[1,3]thiazeto[3,2-*a*]quinoline-3-carboxylic acid

Prunasin (mandelonitrile-β-D-glucoside)
Formed from: Amygdalin
Mandelonitrile
→ Mandelonitrile
In: Cherry H695, K910
Davallia K910
Prunus H343

Prunetin (4′,5-dihydroxy-7-methoxyisoflavone)
Formed from: Genistein

Prunin (naringenin-7-β-D-glucoside)
Formed from: Naringenin
Naringin
→ 7-O-(β-D-Apiofuranosyl(1→2)-β-D-glucosyl)naringenin
In: Parsley K827
→ Malonylprunin
In: Parsley D684
→ Naringenin
In: Man J821
Matthiola A3890
Pig K309

→ Neohesperidin
In: Grapefruit F419
Pumello F419

Pseudobaptigenin (7-hydroxy-3′,4′-methylenedioxyisoflavone)
→ Maackiain
In: Clover A3474

Pseudobenzoylecgonine
Formed from: Pseudococaine
→ Pseudobenzoylnorecgonine
Probably in: Rat A2424

Pseudobenzoylnorecgonine
Probably formed from: Pseudobenzoylecgonine

Pseudococaine ((1*R*,2*S*,3*S*,5*S*)-3-(benzoyloxy)-8-methyl-8-azabicyclo[3.2.1]octane-2-carboxylic acid methyl ester)
→ Pseudobenzoylecgonine
In: Rat A2424
→ Pseudonorcocaine
In: Rat A2424

Pseudocodamine (2-methoxy-4-((1,2,3,4-tetrahydro-6,7-dimethoxy-2-methyl-1-isoquinolinyl)methyl)phenol)
Formed from: Laudanosine
→ Norpseudocodamine
Probably in: Dog E855
Man E855
Rabbit E855

Pseudocumene
See: 1,2,4-Trimethylbenzene

Pseudocumidine
See: 2,4,5-Trimethylaniline

D-Pseudoephedrine ((α*S*)-α-((1*S*)-1-(methylamino)ethyl)benzenemethanol)
→ Benzoic acid
In: Rabbit A996
Rat A1279
→ *p*-Hydroxy-D-pseudoephedrine
In: Rat A1279
→ 2-Methylimino-1-phenylpropan-1-ol
In: Arthrobacter B360
→ D-Norpseudoephedrine
In: Rabbit A996

L-Pseudoephedrine
Formed from: Deprenyl
→ Benzoic acid
In: Rabbit A996
→ L-Norpseudoephedrine
In: Rabbit A996

Pseudoisoeugenol-α,β-oxide 2-methylbutyrate
Formed from: L-Phenyllactic acid

Pseudolaudanine (1-((3,4-dimethoxyphenyl)methyl)-1,2,3,4-tetrahydro-7-methoxy-2-methyl-6-isoquinolinol)
Formed from: Laudanosine
→ Norpseudolaudanine
Probably in: Dog E855
Man E855
Rabbit E855

Pseudomorphine
Formed from: Morphine

Pseudonorcocaine ((1*R*, 2*S*,3*S*,5*S*)-3-(benzoyloxy)-8-azabicyclo[3.2.1]octane-2-carboxylic acid methyl ester)
Formed from: Pseudococaine

D-Pseudonoroephedrine
Formed from: Cathinone

L-Pseudonoroephedrine
Formed from: Cathinone

Psilotin ((6*S*)-6-((4-(β-D-glucopyranosyloxy)phenyl)-5,6-dihydro-2*H*-pyran-2-one)
Formed from: L-Phenylalanine
→ Psilotin epoxide
Probably in: Psilotum C352

Psilotin epoxide
Probably formed from: Psilotin

Psoralen (7*H*-furo[3,2-*g*][1]benzopyran-7-one)
Formed from: Marmesin
→ Bergapten
Probably in: Ficus A548

Psoralen quinone (4*H*-furo[3,2-*g*][1]benzopyran-4,7,9-trione)
Formed from: 8-Methoxypsoralen

2,4(1*H*,3*H*)-Pteridinedione
See: Lumazine

Pteroic acid (4-(((2-amino-1,4-dihydro-4-oxo-6-pteridinyl)methyl)amino)benzoic acid)
→ *p*-Aminobenzoic acid
In: Pseudomonas A971

***trans*-Pterostilbene** (4-hydroxy-3′,5′-dimethoxystilbene)
→ 1,3-Di-O-methyl-ε-viniferin
In: Botrytis J823

→ Pterostilbene-*trans*-dehydrodimer
In: Botrytis K147

Pterostilbene-*trans*-dehydrodimer (3-(3,5-dimethoxyphenyl)-5-(2-(3,5-dimethoxyphenyl)ethenyl)-2-(*p*-hydroxyphenyl)benzofuran)
Formed from: *trans*-Pterostilbene

Pteroyldiglutamic acid
Formed from: Folic acid
Pteroylheptaglutamic acid
Pteroylpentaglutamic acid
→ Folic acid
In: Chicken A1340
→ Pteroyltriglutamic acid
In: Pig K455

Pteroylglutamic acid
See: Folic acid

Pteroylheptaglutamic acid
Formed from: Pteroylhexaglutamic acid
Probably formed from: Folic acid
→ Pteroyldiglutamic acid
In: Chicken A1339, A1340
→ Pteroyloctaglutamic acid
In: Pig K455

Pteroylhexaglutamic acid
Formed from: Folic acid
Pteroylpentaglutamic acid
→ Pteroylheptaglutamic acid
In: Pig K455

Pteroyloctaglutamic acid
Formed from: Pteroylheptaglutamic acid

Pteroylpentaglutamic acid
Formed from: Folic acid
Pteroyltetraglutamic acid
→ Pteroyldiglutamic acid
In: Chicken A1339
→ Pteroylhexaglutamic acid
In: Pig K455

Pteroyltetraglutamic acid
Formed from: Folic acid
Pteroyltriglutamic acid
→ Pteroylpentaglutamic acid
In: Pig K455

Pteroyltriglutamic acid
Formed from: Folic acid
Pteroyldiglutamic acid
→ Pteroyltetraglutamic acid
In: Pig K455

Puberulin (6,8-dimethoxy-7-((3-methyl-2-butenyl)oxy)-2*H*-1-benzopyran-2-one)
Formed from: Scopoletin

Puerarin (8β-D-glucopyranosyl-7-hydroxy-3-(4-hydroxyphenyl)-4*H*-1-benzopyran-4-one)
Formed from: 2′,4,4′-Trihydroxychalcone
→ Daidzein
In: Microorganism J525
Rat H408
→ Puerarin-7-β-D-glucuronide
In: Rat H408
→ Puerarin-4′-sulphate
In: Rat H408

Puerarin-7-β-D-glucuronide
Formed from: Puerarin

Puerarin-4′-sulphate
Formed from: Puerarin

Pulvinic acid ((α*E*)-α-(3-hydroxy-5-oxo-4-phenyl-2(5*H*)-furanylidene)benzeneacetic acid)
→ 4-Hydroxypulvinic acid
Probably in: Peniophora A2738

Pummerer's ketone (4a,9b-dihydro-8,9b-dimethyl-3(4*H*)-dibenzofuranone)
Formed from: *p*-Cresol

Puromycin (3′-(((2*S*)-2-amino-3-(4-methoxyphenyl)-1-oxopropyl)amino)-3′-deoxy-N,N-dimethyladenosine)
Formed from: O-Demethylpuromycin
→ N-Acetylpuromycin
In: Streptomyces D879

Purpurogallin (2,3,4,6-tetrahydroxy-5*H*-benzocyclohepten-5-one)
Formed from: Pyrogallol

Pyramidone
See: Aminopyrine

Pyrene
→ *cis*-4,5-Dihydro-4,5-dihydroxypyrene
In: Mycobacterium J646
→ 1-Hydroxypyrene
In: Aspergillus H388
Crinipellis H879
Cunninghamella E181
Digitalis H876
Dog C318
Hamster F449

Man F449, H393
Marasmiellus H879
Marasmius H879
Pig C499
Procellio H870, K260
Rainbow trout G874
Rat B903, J848
Wheat H876
→ 6-Hydroxypyrene
Crinipellis H879
Marasmiellus H879
Marasmius H879
→ Phthalic acid
In: Mycobacterium H844
→ Pyrene-4,5-oxide
In: Rat A739
See also: B903, E913, H779, H844, H879

Pyrene-4,5-oxide
Formed from: Pyrene
→ *trans*-4,5-Dihydro-4,5-dihydroxypyrene
In: Rat A739
See also: B903, E913, H779
→ S-(4,5-Dihydro-4-hydroxypyren-5-yl) glutathione
In: Man E928
→ S-(4,5-Dihydro-5-hydroxypyren-4-yl) glutathione
In: Rat C472
See also: D235, D842

Pyrene-1,6-quinone
Probably formed from: 1,6-Dihydroxypyrene

Pyrene-1,8-quinone
Probably formed from: 1,8-Dihydroxypyrene

Pyren-1-yl glucoside
Probably formed from: 1-Hydroxypyrene

Pyren-1-yl-β-D-glucuronide
Probably formed from: 1-Hydroxypyrene

Pyren-1-yl sulphate
Probably formed from: 1-Hydroxypyrene

Pyren-6-yl sulphate
Probably formed from: 6-Hydroxypyrene

Pyribenzamine
See: Tripelennamine

3-Pyridinecarboxylic acid 2-(2-(4-chlorophenoxy)-2-methyl-1-oxopropoxy)-1,3-propanediyl ester
See: Binifibrate

5-((4-(2-(2-Pyridinylamino)ethoxy)phenyl)methyl)-2,4-thiazolidinedione
Formed from: 5-((4-(2-(methyl-2-pyridinylamino)ethoxy)phenyl)methyl)-2,4-thiazolidinedione
→ 5-((4-(2-(3-Hydroxy-2-pyridinylamino)ethoxy)phenyl)methyl)-2,4-thiazolidinedione
Probably in: Dog H715
Rat H715
→ 5-((4-(2-(5-Hydroxy-2-pyridinylamino)ethoxy)phenyl)methyl)-2,4-thiazolidinedione
Probably in: Rat H715
→ 5-((4-(2-(2-Pyridinylamino)ethoxy) phenyl)methyl)-2,4-thiazolidinedione-β-D-glucuronide
Probably in: Rat H715

5-((4-(2-(2-Pyridinylamino)ethoxy)phenyl)methyl)-2,4-thiazolidinedione-β-D-glucuronide
Probably formed from: 5-((4-(2-(2-Pyridinylamino)ethoxy)phenyl)methyl)-2,4-thiazolidinedione

2-((Pyridin-4-yl)ethoxyphen-4-yl)carbonyl-2(*S*)-phenylsulphonylamino-β-alanine
Formed from: 2-((Pyridin-4-yl)ethoxyphen-4-yl)carbonyl-2(*S*)-phenylsulphonylamino-β-alanine ethyl ester

2-((Pyridin-4-yl)ethoxyphen-4-yl)carbonyl-2(*S*)-phenylsulphonylamino-β-alanine ethyl ester
→ 2-((Pyridin-4-yl)ethoxyphen-4-yl)carbonyl-2(*S*)-phenylsulphonylamino-β-alanine
In: Dog J71
Monkey J71

6-(3-Pyridinyl)-2(1*H*)-quinolinone
→ 6-(3-Pyridinyl)-2(1*H*)-quinolinone-N-oxide
In: Dog G410

6-(3-Pyridinyl)-2(1*H*)-quinolinone-N-oxide
Formed from: 6-(3-Pyridinyl)-2(1*H*)-quinolinone

9*H*-Pyrido[3,4-*b*]indole
See: Norharman

N-(2-Pyridyl)phenylethanolamine
See: Phenyramidol

5-(*p*-(2-Pyridylsulphamoyl)phenylazo)salicylic acid
See: Sulphasalazine
4-((3-(2-Pyridyl)-3-(*p*-tolyl)-2-propen-1-yl) amino)butyric acid
Formed from: Triprolidine
Pyrilamine ((2*Z*)-N-((4-methoxyphenyl)methyl)-N′,N′-dimethyl-N-2-pyridinyl-1,2-ethanediamine)
→ Anisaldehyde
Probably in: Rat E907
→ N-Demethylpyrilamine
In: Aspergillus E882
Candida E882
Cunninghamella E882
Debaryomyces E882
Monkey H100
Mouse H100
Mucor E882
Penicillium E882
Rat H100
Saccharomyces E882
Syncephalastrum E882
→ O-Demethylpyrilamine
In: Aspergillus E882
Candida E882
Cunninghamella E882
Debaryomyces E882
Man G933
Monkey H100
Mouse H100
Mucor E882
Penicillium E882
Rat E905, E907, E930, F577, H100
Saccharomyces E882
Syncephalastrum E882
→ Hydroxypyrilamine
In: Man G933
Monkey H100
Mouse H100
Rat H100
→ Pyrilamine-β-D-glucuronide
In: Man G18, H841
See also: G933
→ Pyrilamine-N-oxide
In: Aspergillus E882
Candida E882
Cunninghamella E882
Debaryomyces E882
Mucor E882
Penicillium E882
Rat E905
Saccharomyces E882
Syncephalastrum E882
Pyrilamine-β-D-glucuronide
Formed from: Pyrilamine
Pyrilamine-N-oxide
Formed from: Pyrilamine
Pyrocatechualdehyde
See: 2,3-Dihydroxybenzaldehyde
Pyrocatechuic acid (2,3-dihydroxybenzoic acid)
Formed from: Anthranilic acid
o-Chlorobenzoic acid
2,3-Diacetoxybenzoic acid
2,3-Dihydroxy-L-phenylalanine
2,3-Dimethoxybenzoic acid
N,N-Dimethylanthranilic acid
m-Hydroxybenzoic acid
N-Methylanthranilic acid
Salicylic acid
Probably formed from: 2,3-Dihydroxybenzaldehyde
→ 3-Carboxy-2-hydroxy-*cis,cis*-muconic semialdehyde
Probably in: Bacillus G766
→ 3-Carboxy-2-hydroxyphenyl-D-glucoside
In: Cucumis A611
See also: B442
→ α-Carboxy-*cis,cis*-muconic acid
In: Alcaligenes K775
Tecoma A202, A2241
→ Catechol
In: Aspergillus B83, J206
Bacterium C396
Trichosporon B368, B790
→ 2,3-Dihydroxybenzoyladenylic acid
In: Escherichia K830
Vibrio K666
→ Pyrogallol
In: Trichosporon D232
→ 2,3,4-Trihydroxybenzoic acid
In: Aspergillus A1130

Comamonas J519
Pseudomonas A327, A1229
→ 2,3,6-Trihydroxybenzoic acid
In: Pseudomonas A326
→ *o*-Vanillic acid
In: Phanerochaete J339
See also: A3757
→ Vibriobactin
In: Vibrio K666

Pyrocatechol
See: Catechol

Pyrogallol (1,2,3-trihydroxybenzene)
Formed from: Catechol
2,6-Dihydroxybenzoic acid
Gallic acid
3-Methoxycatechol
Phloroglucinol
Pyrocatechuic acid
Resorcinol
2,3,4-Trihydroxybenzoic acid
1,2,3-Trimethoxybenzene
→ 1,3-Dihydroxy-2-methoxybenzene
Probably in: Rat K575
→ 2,3-Dihydroxy-*cis,cis*-muconic semialdehyde
Probably in: Pseudomonas G355, J649
→ 2,3-Dihydroxyphenyl sulphate
In: Rat A1240
→ 2,6-Dihydroxyphenyl sulphate
In: Rat A1240
→ α-Hydroxy-*cis,cis*-muconic acid
In: Pseudomonas B547
Trichosporon C84
→ 3-Methoxycatechol
Probably in: Beef D422
Nicotiana A3778
→ Phloroglucinol
In: Eubacterium E338, E507, G772
Pelobacter E189, F434, G795
→ Purpurogallin
Probably in: Amasya apple H567
Banana A218
Coriolus E369
Dimocarpus K96
Monocillium C519
Mushroom B897
Potato A3190
Tea A972
→ Pyrogallol-β-D-glucoside
In: Manduca G897
→ Pyrogallol-β-D-glucuronide
Probably in: Rat K575
→ 2-Pyrone-6-carboxylic acid
In: Pseudomonas B547
→ Resorcinol
In: Escherichia + Streptococcus E113
→ 1,2,3,4-Tetrahydroxybenzene
Probably in: Rhodococcus E565
→ 1,2,3,5-Tetrahydroxybenzene
In: Pelobacter G621
→ 2,3,4-Trihydroxy-L-phenylalanine
In: Escherichia A342
See also: A708

Pyrogallol-β-D-glucoside
Formed from: Pyrogallol

Pyrogallol-β-D-glucuronide
Probably formed from: Pyrogallol

1-(N-Pyrrolidinocarbonylmethyl)-4-(3,4,5-trihydroxycinnamoyl)piperazine
Probably formed from: 4-(4-Hydroxy-3,5-dimethoxycinnamoyl)-1-(N-pyrrolidinocarbonylmethyl)piperazine

1-(N-Pyrrolidinocarbonylmethyl)-4-(3,4,5-trimethoxycinnamoyl)piperazine
See: Cinepazide

3′-Pyrrolidinyl 3-carboxy-2,6-dimethyl-4-(*m*-nitrophenyl)pyridine-5-carboxylate
Probably formed from: Methyl 3′-pyrrolidinyl 2,6-dimethyl-4-(*m*-nitrophenyl)pyridine-3,5-dicarboxylate

2-(3-Pyrrolidinyl)-2,2-diphenylacetamide
Formed from: Darifenacin

N-((2-Pyrrolidinyl)methyl)-5-sulphamoyl-2-anisidide
Formed from: Sulpiride

L-Pyrrolidinyl-β-naphthylamide
→ β-Naphthylamine
In: Pseudomonas G643

Pyrroloquinoline quinol
Probably formed from: Pyrroloquinoline quinone

Pyrroloquinoline quinone (4,5-dihydro-4,5-dioxo-1*H*-pyrrolo[2,3-*f*]quinoline-2,7,9-tricarboxylic acid)
→ Pyrroloquinoline quinol
Probably in: Pseudomonas E73

ε-Pyrromycinone (rutilantinone, (1*R*,2*R*,4*S*)-2-ethyl-1,2,3,4,6,11-hexahydro-2,4,5,7,10-pentahydroxy-6,11-dioxo-1-naphthacenecarboxylic acid methyl ester)
→ Cinerubin A
In: Streptomyces D170, D171
→ Cinerubin B
In: Streptomyces D170, D171
→ ζ-Pyrromycinone
In: Streptomyces D170

ζ-Pyrromycinone ((1*R*,2*R*)-2-ethyl-1,2,3,4,6,11-hexahydro-2,5,7,10-tetrahydroxy-6,11-dioxo-1-naphthacenecarboxylic acid methyl ester)
Formed from: ε-Pyrromycinone

Q

Quazepam (7-chloro-5-(*o*-fluorophenyl)-1,3-dihydro-1-(2,2,2-trifluoroethyl)-2*H*-1,4-benzodiazepine-2-thione)
→ 7-Chloro-5-(*o*-fluorophenyl)-1,3-dihydro-1-(2,2,2-trifluoroethyl)-2*H*-1,4-benzodiazepin-2-one
In: Hamster D392
Man D336, D492
Mouse D392, D843

Quazodine (4-ethyl-6,7-dimethoxyquinazoline)
→ 4-Ethyl-7-hydroxy-6-methoxyquinazoline
In: Dog A3212
Man A3212
Rat A3212

Quercetagetin (3,3′,4′,5,6,7-hexahydroxyflavone)
→ Patuletin
In: Calamondin orange A3987
→ 4′,5,6-Trihydroxy-3,3′,7-trimethoxyflavone
In: Calamondin orange A3987

Quercetagetin-7-glucoside
→ Quercetagetin-7-glucoside-6″-malonate
In: Cicer D445

Quercetagetin-7-glucoside-6″-malonate
Formed from: Quercetagetin-7-glucoside

Quercetin (3,3′,4′,5,7-pentahydroxyflavone)
Formed from: Caffeic acid
Dihydroquercetin
Kaempferol
Quercetin-3-β-D-glucoside
Quercetin-4′-β-D-glucoside
Quercitrin
Rutin
Probably formed from: Quercetin-4′-sulphate
→ 3,4-Dihydroxyphenylacetic acid
In: Clostridium F203, G319
Microorganism K205
→ 2,3,3,3′,4′,5,7-Heptahydroxyflavan-4-one
Probably in: Clover F825
→ *m*-Hydroxyphenylacetic acid
In: Microorganism A283
→ Isorhamnetin
In: Citrus A3435, A3987
Lotus C905
Nicotiana A3757
Pig K505
Rat J690
Spinach C94
Trillium K250
Zea F173, G223
See also: A3778, F367, G124, G137
→ Kaempferol
In: Pig K505
→ Protocatechuoylphloroglucinolcarboxylic acid
In: Aspergillus K214
See also: A1328, G185, H280
→ Quercetin-3-O-(2″-O-β-D-glucopyranosyl)-β-D-galactopyranoside
In: Petunia H658
→ Quercetin-3-β-D-glucoside
In: Alnus B680
Cabbage A1733
Carrot E280
Crocus G124
Hippeastrum E636
Ipomoea G124
Narcissus B680
Norway spruce G215
Pisum C90
Quercus B680
Secale B680
Tulipa B680
Vitis G137

Zea E120
See also: E977, G816, G897, H402, K889
→ Quercetin-4′-β-D-glucoside
In: Onion G248
See also: K809
→ Quercetin-7-β-D-glucoside
In: Chrysanthemum J320
Parsley K739
Petunia K303
See also: K556
→ Quercetin-β-D-glucuronide
In: Man G524, J416, J783, K467
Rat J690
→ Quercetin quinone
Probably in: Horseradish A1183
Vicia J858
→ Quercetin quinone methide
In: Horseradish K535
→ Quercetin-3-sulphate
In: Flaveria F210, F243, G769
See also: F513, H354
→ Quercetin-3-D-xyloside
In: Euonymus G444
→ Rhamnetin
In: Citrus A3435, A3987
Prunus G982
→ Tamarixetin
In: Pig K505
→ 3,3′,4′,7-Tetrahydroxy-5-methoxyflavone
In: Calamondin orange A3987
→ 3′,4′,5,7-Tetrahydroxy-3-methoxyflavone
In: Apple D258
Calamondin orange A3987, B830
Chrysosplenium B830, B832, C579, C792, D584, D589
Lotus C905
Nicotiana A3757
→ 3,3′,5,7-Tetrahydroxy-4′-methoxyflavone
In: Calamondin orange A3987
Nicotiana A3757
See also: C100

Quercetin-3-diglucoside
Formed from: Quercetin-3-β-D-glucoside
→ Quercetin-3-triglucoside
In: Pisum C90

Quercetin-3,4′-di-β-D-glucoside
Formed from: Quercetin-3-β-D-glucoside

Quercetin-3,7-di-β-D-glucoside
Formed from: Quercetin-3-β-D-glucoside

Quercetin dihydrodiol
See: 2,3,3,3′,4′,5,7-Heptahydroxyflavan-4-one

Quercetin-3,3′-disulphate
→ Quercetin-3,3′,7-trisulphate
In: Flaveria F210
See also: F513

Quercetin-3,4′-disulphate
Formed from: Quercetin-3-sulphate
Quercetin-3,4′,7-trisulphate
→ Quercetin-3-sulphate
Probably in: Helix F102
→ Quercetin-3,4′,7-trisulphate
In: Flaveria F210

Quercetin-3,7-disulphate
Formed from: Quercetin-3,4′,7-trisulphate

Quercetin-4′,7-disulphate
→ Quercetin-4′-sulphate
In: Helix F102
→ Quercetin-7-sulphate
In: Helix F102

Quercetin-3-O-(2″-O-β-D-glucopyranosyl)-β-D-galactopyranoside
Formed from: Quercetin

Quercetin-3-β-D-glucoside (isoquercitrin)
Formed from: Quercetin
→ Quercetin-3-diglucoside
In: Pisum C90
→ Quercetin-3-malonylglucoside
In: Parsley D684

Quercetin-4′-β-D-glucoside
Formed from: Quercetin
→ Quercetin
In: Man J821
Pig K309

Quercetin-7-β-D-glucoside
Formed from: Quercetin
→ Quercetin-7-β-D-glucoside-6″-malonate
In: Cicer D445

Quercetin-7-β-D-glucoside-6″-malonate
Formed from: Quercetin-7-β-D-glucoside

Quercetin-β-D-glucuronide
Formed from: Quercetin

→ Quercetin-β-D-glucuronide sulphate
Probably in: Rat J690

Quercetin-β-D-glucuronide sulphate
Probably formed from: Quercetin-β-D-glucuronide

Quercetin-3-malonylglucoside
Formed from: Quercetin-3-glucoside

Quercetin quinone
Formed from: Quercetin

Quercetin quinone methide
Formed from: Quercetin

Quercetin-3-rhamnoside
See: Quercitrin

Quercetin-3-rutinoside
See: Rutin

Quercetin-3-sulphate
Formed from: Quercetin
Probably formed from: Quercetin-3,4′-disulphate
→ Quercetin-3,3′-disulphate
In: Flaveria F210, F243
→ Quercetin-3,4′-disulphate
In: Flaveria F210, F243

Quercetin-4′-sulphate
Formed from: Quercetin-4′,7-disulphate
→ Quercetin
Probably in: Helix F102

Quercetin-7-sulphate
Formed from: Quercetin-4′,7-disulphate

Quercetin-3-triglucoside
Formed from: Quercetin-3-diglucoside

Quercetin-3,3′,7-trisulphate
Formed from: Quercetin-3,3′-disulphate

Quercetin-3,4′,7-trisulphate
Formed from: Quercetin-3,4′-disulphate
→ Quercetin-3,4′-disulphate
In: Helix F102
→ Quercetin-3,7-disulphate
In: Helix F102

Quercetin-3-D-xyloside
Formed from: Quercetin

Quercitrin (quercetin-3-rhamnoside)
→ O-Methylquercitrin
In: Nicotiana A3757
→ Quercetin
In: Bacteroides E515, E977
Fusobacterium K205

Questin (1,6-dihydroxy-8-methoxy-3-methyl-9,10-anthracenedione)
Formed from: Emodin
→ Sulochrin
In: Aspergillus K701

Quinaldic acid (quinoline-2-carboxylic acid)
→ 4-Hydroxyquinoline-2-carboxylic acid
In: Serratia K770
→ Kynurenic acid
In: Alcaligenes J32
Pseudomonas K768

Quinaldine (2-methylquinoline)
Formed from: Quinaldine-N-oxide
→ 1-*H*-4-Oxoquinaldine
In: Arthrobacter G274, G803, J33

Quinaldine-N-oxide
→ Quinaldine
In: Saccharomyces H894

Quinalizarin (1,2,5,8-tetrahydroxyanthraquinone)
→ Quinalizarin-β-D-glucoside
In: Cinchona K889
→ Quinalizarin-β-D-glucuronide
In: Man G968, J416, J783

Quinalizarin-β-D-glucoside
Formed from: Quinalizarin

Quinalizarin-β-D-glucuronide
Formed from: Quinalizarin

Quinazoline
→ *cis*-5,6-Dihydro-5,6-dihydroxyquinazoline
In: Pseudomonas E541, H50
→ 4-Hydroxyquinazoline
In: Arthrobacter J33
Pseudomonas E541, H50, K765
Rabbit A3695
Rat A3695

2,4(1*H*,3*H*)-Quinazolinedione
Probably formed from: 4-Hydroxyquinazoline

Quinfamide (1-dichloroacetyl-6-(2-furoyloxy)-1,2,3,4-tetrahydroquinoline)
→ 1-Acetyl-1,2,3,4-tetrahydro-6-hydroxyquinoline
In: Rat C327
→ 1,2,3,4-Tetrahydro-6-hydroxy-1-hydroxyacetylquinoline
In: Rat C327
→ 1-Dichloroacetyl-1,2,3,4-tetrahydro-6-hydroxyquinoline

In: Rat C327
Quinic acid (1,3,4,5-tetrahydroxycyclohexanecarboxylic acid)
→ Gallic acid
In: Enterobacter A718
Pseudomonas A718
Quinidine (β-quinine, 6-methoxy-α-(5-vinyl-2-quinuclidinyl)-4-quinolinemethanol)
Formed from: Quinidinone
Quininone
→ 6-Demethylquinidine
In: Rabbit C289
Rat C627
→ 10,11-Dihydro-10,11-dihydroxyquinidine
In: Rabbit C289
→ 2′-Hydroxyquinidine
In: Guinea pig G508
Rabbit C289, G508
Man C290
→ 3-Hydroxyquinidine
In: Rabbit C289
Man C290, E942, K47
Rat C627
→ 5′-Hydroxyquinidine
In: Man A2615
→ Quinidine-1-oxide
In: Man C271, K47
Quinidine-1-oxide
Formed from: Quinidine
Quinidinone (6′-methoxycinchonan-9-one)
→ Quinidine
In: Cinchona E357
→ Quinine
In: Cinchona E357
Quinifuryl (N-(4-(diethylamino)-1-methylbutyl)-2-(2-(5-nitro-2-furanyl)ethenyl)-4-quinolinecarboxamide)
→ *o*-Aminobenzyl alcohol
Probably in: Mouse K87
Quinine ((8α,9*R*)-6′-methoxycinchonan-9-ol)
Formed from: Quinidinone
→ Dihydroquinine
In: Man B399
→ 6′-Hydroxycinchonidine
In: Man B399, J826
→ 2′-Hydroxyquinine
In: Guinea pig G508
Man J826
Rabbit A543, G508
See also: J6, A1365
→ 3-Hydroxyquinine
In: Adelie penguin K300
Man B399, F913, J7, J826, K300
See also: J6
→ Quinine-β-D-glucuronide
Probably in: Man J6
→ Quinine-10,11-oxide
In: Man B399
→ Quinine-N-oxide
In: Man A2782
→ Quininone
In: Cinchona E357
Man J826
β-Quinine
See: Quinidine
Quinine-β-D-glucuronide
Probably formed from: Quinine
Quinine-10,11-oxide
Formed from: Quinine
→ 10,11-Dihydro-10,11-dihydroxyquinine
Probably in: Man B399, J826
Quinine-N-oxide
Formed from: Quinine
Quininone (2′-quinone)
Formed from: Quinine
→ Quinidine
In: Hansenula G134
Quinizarin
See: 1,4-Dihydroxyanthraquinone
Quinizarin-4-β-D-glucoside
See: 4-β-D-Glucopyranosyloxy-1-hydroxyanthraquinone
Quinol (1,4-dihydroxybenzene, hydroquinone)
Formed from: *p*-Acetamidophenol
N-Acetyl-*p*-hydroxyphenyl-β-D-glucosaminide
Aniline
Arbutin
Bis(4-hydroxyphenyl) ether
p-Bromophenol
p-Chlorophenol
1,4-Dimethoxybenzene

p-Fluorophenol
p-Hydroxyacetophenone
p-Hydroxybenzoic acid
p-Hydroxybenzonitrile
p-Hydroxybenzophenone
p-Hydroxybenzyl alcohol
p-Hydroxyphenoxyacetic acid
p-Hydroxyphenyl acetate
p-Hydroxyphenyl-β-D-glucuronide
p-Hydroxyphenyl sulphate
Hydroxyquinol
p-Methoxyphenol
p-Nitrophenol
Phenol
Quinol diacetate
p-Quinone

Probably formed from: *p*-Hydroxybenzaldehyde
m-(*p*-Hydroxyphenoxy)benzoic acid

→ Arbutin
In: Antennaria F873
Datura A1725, H300
Gardenia E338
Juglans A3472
Manduca G816, G897
Millipede A2254
Periplaneta G816
Rauwolfia G726, K346

→ Benzoic acid
In: Bacterium H110

→ Gentisic acid
In: Desulfococcus H293

See also: H110

→ α-Hydroxymuconic semialdehyde
Probably in: Penicillium E94

→ *p*-Hydroxyphenyl-α-D-glucoside
In: Bacillus H296

→ *p*-Hydroxyphenyl-β-D-glucuronide
In: Capuchin A510
Dog B531
Man J364
Rat B531, B962, D462, J575
Squirrel monkey A510

See also: A3717, B729

→ *p*-Hydroxyphenyl phosphate
Probably in: Pig A3717
Rat A3717
Sheep A3717

→ *p*-Hydroxyphenyl sulphate
In: Cat A510
Dog A510, B531
Ferret A510
Hedgehog A510
Man J364
Rabbit A510
Rana E841
Rat B531, D462, J575
Xenopus E841

See also: A2686, A3717, B729, H818

→ Hydroxyquinol
In: Aspergillus H236, H570
Candida E204, K653
Pseudomonas H386
Rabbit F238
Rhodococcus E565
Trichosporon A902

See also: A2100, H818

→ Quinol-2-S-glutathione
In: Mouse J321
Rat J321

See also: D50, F600

→ *p*-Quinone
In: Coriolus A3974
Man F125, G130
Mouse F416
Peach A3974
Rhus A3974
Trametes A3974

See also: A702, A1930, A9138, B491, B897, C519, E369, E407, F944

Quinol-2,5-bis(S-glutathione)
Probably formed from: Quinol-2-S-glutathione
→ Quinol-2,3,5-tris(S-glutathione)
Probably in: Mouse J321
Rat J321

Quinol-2,6-bis(S-glutathione)
Probably formed from: Quinol-2-S-glutathione

Quinol diacetate
→ Quinol
In: Acinetobacter A962

Quinol diglucoside
Probably formed from: Arbutin
Quinol-2-S-glutathione
Probably formed from: Quinol
→ S-(2,5-Dihydroxyphenyl)-L-cysteine
Species unclear D50
See also: F600
→ Quinol-2,5-bis(S-glutathione)
Probably in: Mouse J321
Rat J321
→ Quinol-2,6-bis(S-glutathione)
Probably in: Mouse J321
Rat J321
Quinoline
Formed from: Quinoline-N-oxide
→ *o*-Aminophenylhydracrylic acid
In: Pseudomonas B466
→ Anthranilic acid
Pseudomonas H50
→ *cis*-5,6-Dihydro-5,6-dihydroxyquinoline
Pseudomonas E541, H50
→ *trans*-5,6-Dihydro-5,6-dihydroxyquinoline
Rat C458
→ 2,3-Dihydroxyphenylpropionic acid
Pseudomonas F454
→ 2-Hydroxyquinoline
In: Comamonas K769
Desulfobacterium H917
Microorganism J367
Nocardia E694
Pseudomonas D923, F454, G151, K766
Rhodococcus K7687
→ 3-Hydroxyquinoline
In: Pseudomonas E541, F454
Rat C458
→ 4-Hydroxyquinoline
In: Arthrobacter J33
→ 6-Hydroxyquinoline
Pseudomonas H50
→ 8-Hydroxyquinoline
In: Nocardia E694
Pseudomonas E541, H50
→ N-Methylquinolinium
In: Rabbit E824
→ Quinoline-N-oxide
In: Cunninghamella H332
Rat C458
Quinoline-2-carboxylic acid
See: Quinaldic acid
Quinoline-3-carboxylic acid
Formed from: 3-Formylquinoline
Quinoline-4-carboxylic acid
Formed from: 4-Formylquinoline
Quinoline-N-oxide
Formed from: Quinoline
→ Quinoline
In: Saccharomyces H894
3-(Quinolin-2(1*H*)-on-6-oxy)lactic acid
Formed from: Toborinone
Quinolin-8-yl glucoside-6-phosphate
Probably formed from: 8-Hydroxyquinoline
Quinolin-6-yl-β-D-glucuronide
Probably formed from: 6-Hydroxyquinoline
Quinolin-8-yl-β-D-glucuronide
Probably formed from: 8-Hydroxyquinoline
Quinolin-8-yl sulphate
Probably formed from: 8-Hydroxyquinoline
2-(3-(2-Quinolinylmethoxy)phenylamino)benzoic acid
→ 2-(3-(2-Quinolinylmethoxy) phenylamino)benzoyl-β-D-glucuronic acid
In: Dog F508
Guinea pig F508
Man F508
Rat F508
2-(3-(2-Quinolinylmethoxy)phenylamino) benzoyl-β-D-glucuronic acid
Formed from: 2-(3-(2-Quinolinylmethoxy) phenylamino)benzoic acid
Quinol monoacetate
See: *p*-Hydroxyphenyl acetate
2(1*H*)-Quinolone
See: 2-Hydroxyquinoline
4(1*H*)-Quinolone
See: 4-Hydroxyquinoline
γ-Quinolone
See: 4-Hydroxyquinoline
Quinol-2,3,5-tris(S-glutathione)
Probably formed from: Quinol-2,5-bis (S-glutathione)
***o*-Quinone** (3,5-cyclohexadiene-1,2-dione)
Formed from: Catechol

***p*-Quinone** (benzoquinone)
  Formed from: N,N-Dimethyl-*p*-anisidine
    p-Phenylenediamine
    Quinol
  → Luciferin (firefly)
    In: Firefly A2437
  → Quinol
    In: Beef G401
    Cochliobolus K378
    Guinea pig K773
    Sporotrichum B301
    Thauera K173
    Zea A217

2-(*o*-Quinon-4-yl)glycine
  Formed from: 3,4-Dihydroxyphenylglycine

Quinoxaline
  → *cis*-5,6-Dihydro-5,6-dihydroxyquinoxaline
    In: Pseudomonas E541, H50
  → 2-Hydroxyquinoxaline
    In: Pseudomonas E541, H50
    Rabbit A3695
    Rat A3695
    Streptomyces J648
  → 5-Hydroxyquinoxaline
    In: Pseudomonas E541, H50
  → N-Methylquinoxalinium
    In: Rabbit E842

2,6-Quinoxalinedimethanol-1,4-dioxide
  See: Dioxidine

Quinoxalinedimethanol-1,4-dioxide diacetate
  See: Quinoxidine

2(1*H*)-Quinoxalinone
  See: 2-Hydroxyquinoxaline

Quinoxidine (2,3-quinoxalinedimethanol-1,4-dioxide diacetate)
  → Dioxidine
    In: Man A1292
    Rat A1292

10-(3-Quinuclidinylmethyl)phenothiazine
  → 10-(3-Quinuclidinylmethyl)phenothiazine-N-oxide
    In: Dog A1747
    Rat A1747
  → 10-(3-Quinuclidinylmethyl)phenothiazine sulphoxide
    In: Dog A1747
    Rat A1747

10-(3-Quinuclidinylmethyl)phenothiazine-N-oxide
  Formed from: 10-(3-Quinuclidinylmethyl)phenothiazine
  → 10-(3-Quinuclidinylmethyl)phenothiazine-N-oxide sulphoxide
    Dog A1747
    Rat A1747

10-(3-Quinuclidinylmethyl)phenothiazine-N-oxide sulphoxide
  Probably formed from: 10-(3-Quinuclidinylmethyl)phenothiazine-N-oxide

10-(3-Quinuclidinylmethyl)phenothiazine sulphone
  Formed from: 10-(3-Quinuclidinylmethyl)phenothiazine sulphoxide

10-(3-Quinuclidinylmethyl)phenothiazine sulphoxide
  Formed from: 10-(3-Quinuclidinylmethyl)phenothiazine
  → 10-(3-Quinuclidinylmethyl)phenothiazine sulphone
    In: Dog A1747
    Rat A1747

R

Ractopamine (4-hydroxy-α-(((3-(4-hydroxyphenyl)-1-methylpropyl) amino)methyl)benzenemethanol)
→ Ractopamine-10-β-D-glucuronide
In: Rat H171
Turkey G586, K28
→ Ractopamine-10′-β-D-glucuronide
In: Rat H171
Turkey G586, K28
→ Ractopamine-O-sulphate
In: Rat H171

Ractopamine-10-β-D-glucuronide
Formed from: Ractopamine
→ Ractopamine-10-β-D-glucuronide-10′-sulphate
Probably in: Rat H171

Ractopamine-10′-β-D-glucuronide
Formed from: Ractopamine

Ractopamine-10-β-D-glucuronide-10′-sulphate
Probably formed from: Ractopamine-10-β-D-glucuronide

Ractopamine-O-sulphate
Formed from: Ractopamine

Raspberry ketone
See: 4-(*p*-Hydroxyphenyl)butan-2-one

Raucaffricine ((17*R*,19*E*,21α)-17-(acetyloxy)-1,2,19,20-tetradehydro-1-demethylajmalan-21-yl-β-D-glucopyranoside)
→ Vomilenine
In: Rauwolfia K939

Rebeccamycin (1,11-dichloro-12,13-dihydro-12-(4-O-methyl-β-D-glucopyranosyl)-5*H*-indole [2,3-*a*]pyrrolo[3,4-*c*]carbazole-5,7(6*H*)-dione)
Formed from: L-Tryptophan

Reboxetine (2-(α-(2-ethoxyphenoxy)benzyl) morpholine; *R*,*R* and *S*,*S* isomers)
→ Desethylreboxetine
In: Man J777
See also: G462
→ 2-(α-(2-Ethoxy-4-hydroxyphenoxy) benzyl)morpholine
In: Man J777
See also: G462
→ 2-(α-(2-Ethoxy-5-hydroxyphenoxy) benzyl)morpholine
In: Man J777
See also: G462
→ 2-(α-(2-Ethoxyphenoxy)benzyl)-3-oxomorpholine
Probably in: Dog G462
Man G462
Monkey G462
Rat G462
→ 2-(α-(2-Ethoxyphenoxy)benzyl)-5-oxomorpholine
Probably in: Dog G462
Man G462
Monkey G462
Rat G462
→ Reboxetine-N-oxide
Probably in: Dog G462
Man G462
Monkey G462
Rat G462

Reboxetine-N-oxide
Probably formed from: Reboxetine

Recainam (N-(2,6-dimethylphenyl)-N′-(3-((1-methylethyl)amino)phenyl)urea)
→ N-(2-Carboxyethyl)-N′-(2,6-dimethylphenyl)urea
In: Rat F571
→ Deisopropylrecainam
In: Dog F571
Man F571

Monkey F571
Mouse F571
→ N-(2,6-Dimethylphenyl)urea
In: Rat F571
→ *m*-Hydroxyrecainam
In: Dog F571
Monkey F571
Rat F571
→ *p*-Hydroxyrecainam
In: Dog F571
Man F571
Monkey F571
Mouse F571
Rat F571

Recipavrin
See: 3-Dimethylamino-1,1-diphenylbutane

Reduced Michler's ketone
See: 4,4′-Methylenebis((N,N-dimethyl)aniline)

Remacemide (2-amino-N-(1-methyl-1,2-diphenylethyl)acetamide)
→ β-Phenylamphetamine
In: Rat H59

Remikiren ((1*S*-(1*R**(*R**(*R**))2*S**,3*R**))-N-(1-(cyclohexylmethyl)-3-cyclopropyl-2,3-dihydroxypropyl)-α-((2-(((1,1-dimethylethyl) sulphonyl)methyl)-1-oxo-3-phenylpropyl)amino)-1*H*-imidazole-4-propanamide)
→ 2-(*tert*-Butylsulphonylmethyl)phenylpropionic acid
In: Rat H524
→ (1*S*-(1*R**(*R**(*R**))2*S**,3*R**))-N-(3-Cyclopropyl-2,3-dihydroxypropyl)-1-((4-hydroxycyclohexyl)methyl)-α-((2-(((1,1-dimethylethyl)sulphonyl)methyl)-1-oxo-3-phenylpropyl)amino)-1*H*-imidazole-4-propanamide
In: Rat H524

Reproterol (7-(3-(2-(3,5-dihydroxyphenyl)-2-hydroxyethylamino)propyl)theophylline)
→ 1,2,3,4-Tetrahydro-4,6,8-trihydroxy-2-(3-theophyllin-7-ylpropyl)isoquinoline
In: Rat A3565, A3663, C97

Resacetophenone
See: 2′,4′-Dihydroxyacetophenone

Reserpine (3,4,5-trimethoxybenzoyl methyl reserpate)
→ Syringomethyl reserpate
Probably in: Dog C118
Rabbit C118
Rat C118
→ 3,4,5-Trimethoxybenzoic acid
In: Rat A1169

Reserpiline (4a,5,7,8,13,13b,14,14a-octahydro-10,11-dimethoxy-4-methyl-4*H*-indolo[2,3-*a*]pyrano[3,4-*g*]quinolizine-1-carboxylic acid methyl ester)
Formed from: Strictosidine

***cis*-Resmethrin** (5-benzyl-3-furylmethyl *cis*-chrysanthemate)
→ 5-Benzyl-3-furoic acid
In: Mouse A2289
Rat A2044, A2289
→ 5-Benzyl-3-furylmethanol
In: Mouse A2289
Rat A2289

***trans*-Resmethrin**
→ 5-Benzyl-3-furoic acid
In: Mouse A2289
Rat A2044, A2289
→ 5-Benzyl-3-furylmethanol
In: Mouse A2289
Rat A2289

Resorcinol (1,3-dihydroxybenzene)
Formed from: Daidzein
Dihydro-2′,4,4′-trihydroxychalcone
Dihydro-2′,4,6′-trihydroxychalcone
2,4-Dihydroxybenzoic acid
2,6-Dihydroxybenzoic acid
4′,7-Dihydroxyflavone
1,3-Dimethoxybenzene
Hydroxyquinol
m-Methoxyphenol
Metoxadiazone
Phenol
Phloroglucinol
Pyrogallol
→ 7-Amino-3,6-bis(2,4-dihydroxyphenyl)-2-phenoxazone
In: Piricularia B143
→ 3,6-Bis(2,4-dihydroxyphenyl)-7-hydroxy-2-phenoxazone
In: Piricularia B143
→ 1,3-Dihydroxy-3,5-cyclohexadiene

In: Microorganism F844
→ 2,4-Dihydroxy-L-phenylalanine
In: Erwinia A708
Escherichia A2242, B292
→ *m*-Hydroxyphenyl-β-D-glucoside
In: Datura A1725
Gardenia E338
→ *m*-Hydroxyphenyl-β-D-glucuronide
In: Rat B962, C479, E532
→ *m*-Hydroxyphenyl sulphate
In: Haemophilus H756
Man D977
Rana E841
Rat E532
Xenopus E841
→ Hydroxyquinol
In: Aspergillus E139
Bradyrhizobium J484
Candida K653
Pelobacter F434, G621
Pseudomonas A2172, A2391, A2696
Trichosporon A902, G196
See also: C475, E204, E565, G817, H6
→ 5-Oxo-2-hexenoic acid
In: Microorganism H13
→ Pyrogallol
In: Azotobacter B862
→ Resorcinol polymer
In: Algae J720

Resorcinol polymer
Formed from: Resorcinol

α-Resorcylic acid
See: 3,5-Dihydroxybenzoic acid

β-Resorcylic acid
See: 2,4-Dihydroxybenzoic acid

γ-Resorcylic acid
See: 2,6-Dihydroxybenzoic acid

Resorufin (7-hydroxyphenoxazone)
7-Ethoxyresorufin
7-Methoxyphenoxazone
Resorufin benzyl ether
Resorufin pentyl ether
2-Phenylbenzyloxyresorufin

Resorufin benzyl ether (7-benzyloxyphenoxazone)
→ Resorufin
In: Alligator J855
Catfish E209
Dog H486
Man J63
Rat D421, J63, J855, J904

Resorufin ethyl ether
See: 7-Ethoxyresorufin

Resorufin pentyl ether (7-pentyloxyphenoxazone)
→ Resorufin
In: Alligator J855
Man J63, J245
Rat D421, J63, J855, J891

Respilene
See: Zipeprol

Resveratrol (3,4′,5-trihydroxystilbene)
Formed from: *p*-Coumaroyl CoA
L-Phenylalanine
→ Batatasin I
In: Dioscorea D136
→ 3,4′-Dihydroxy-5-methoxystilbene
Probably in: Picea D213
→ 7-Hydroxy-2,4,6-trimethoxyphenanthrene
Probably in: Dioscorea D136
→ Resveratrol-β-D-glucuronide
In: Man K354
→ Resveratrol sulphate
In: Man K118, K124
→ ε-Viniferin
In: Botrytis J823

Resveratrol-β-D-glucuronide
Formed from: Resveratrol

Resveratrol sulphate
Formed from: Resveratrol

Reticuline ((1*S*)-1,2,3,4-tetrahydro-1-((3-hydroxy-4-methoxyphenyl)methyl)-6-methoxy-2-methyl-7-isoquinolinol)
Formed from: 1,2-Dehydroreticulinium
1-(3,4-Dihydroxybenzyl)-1,2,3,4-tetrahydro-7-hydroxy-6-methoxy-2-methylisoquinoline
3′-Hydroxy-N-methyl-(*S*)-coclaurine
Norreticuline
1,2,3,4-Tetrahydro-6,7-dihydroxy-1-(3-hydroxy-4-methoxybenzyl)isoquinoline
1,2,3,4-Tetrahydro-7-hydroxy-1-(3-hydroxy-4-methoxybenzyl)-6-methoxyisoquinoline
→ Adiantifoline

In: Thalictrum E15
→ Berberine
Probably in: Corydalis A3684
→ Boldine
In: Litsea A3089
→ Bulbocapnine
In: Corydalis A565
→ Cheilanthifoline
In: Papaver A3141
→ Columbamine
In: Berberis B573, E418
→ Coreximine
In: Rat A2187, A2937, A3183, B478
→ 1,2-Dehydroreticulinium
Probably in: Papaver A1620, A2516, A3345
→ Dehydroscoulerine
In: Berberis E418
→ 4-Hydroxyreticuline
In: Nelumbo D566
→ (+)-Isoboldine
In: Litsea A1547
Rat A2937, A3183, B478
Thalictrum F229
→ Jatrorrhizine
In: Berberis E482
→ Pallidine
In: Rat A2937, A3183, B478
→ Palmatine
In: Berberis B573
Cocculus B726
→ Salutaridine
In: Papaver K759
Pig H365
Rat E562
→ Scoulerine
In: Berberis D615, D874
Eschscholtzia H680, K705
Macleaya A1565
Papaver A3141
Rat A2937, A3183, B478
See also: E978
→ Tetrahydropalmatine
In: Cocculus B726

Retigabine
See: Ethyl N-(2-amino-4-(4-fluorobenzylamino)phenyl)carbamate

Rhamnazin (3,4′,5-trihydroxy-3′-methoxyflavone)
Formed from: Isorhamnetin
Rhamnetin

Rhamnetin (3,3′,4′,5-tetrahydroxyflavone)
Formed from: Quercetin
→ Rhamnazin
In: Citrus A3435
→ Rhamnetin-3-β-D-glucoside
In: Norway spruce G215
→ Rhamnetin-4′-O-glucoside
In: Onion G248
→ Rhamnetin-3-sulphate
In: Flaveria F243, G769
→ 3,4′,5-Trihydroxy-3′,7-dimethoxyflavone
Probably in: Lotus C905
→ 3′,4′,5′-Trihydroxy-3,7-dimethoxyflavone
In: Apple D258
Chrysosplenium C792

Rhamnetin-3,4′-disulphate
→ Rhamnetin-3-sulphate
In: Flaveria F210
Helix F102

Rhamnetin-3-β-D-glucoside
Formed from: Rhamnetin

Rhamnetin-4′-O-glucoside
Formed from: Rhamnetin

Rhamnetin-3-sulphate
Formed from: Rhamnetin
Rhamnetin-3,4′-disulphate

Rhamnocitrin (3,4′,5-trihydroxy-7-methoxyflavone)
Formed from: Kaempferol
→ Rhamnocitrin-3-O-glucosylgalactoside
In: Petunia H658

Rhamnocitrin-3-O-glucosylgalactoside
Formed from: 3,4′,5-Trihydroxy-7-methoxyflavone

Rhaponticin (3-hydroxy-5-((1*E*)-2-(3-hydroxy-4-methoxyphenyl)ethenyl)phenyl-β-D-glucuronide)
→ Rhapontigenin
In: Man K527

Rhapontigenin (1,3-dihydroxy-5-((1*E*)-2-(3-hydroxy-4-methoxyphenyl)ethenyl)benzene)
Formed from: Rhaponticin

Rhein (9,10-dihydro-4,5-dihydroxy-9,10-dioxo-2-anthracenecarboxylic acid)
Formed from: Rhein-1-β-D-glucoside
Probably formed from: 8-Glucosylrhein
Sennidin A
Sennidin B
→ 8-Hydroxyrhein
Probably in: Dog J143
→ Rhein-1-β-D-glucoside
In: Dog J143
See also: E681
→ Rhein-β-D-glucuronide
In: Dog J143
→ Rhein-O-sulphate
In: Dog J143

Rhein-9-anthrone (9,10-dihydro-4,5-dihydroxy-10-oxo-2-anthracenecarboxylic acid)
Formed from: Sennidin A
Sennidin B
Sennoside A

Rhein-1-β-D-glucoside
Formed from: Rhein
→ Rhein
In: Microorganism C103

Rhein-β-D-glucuronide
Formed from: Rhein

Rhein-O-sulphate
Formed from: Rhein

Rhodamine 110 (3,6-diamino-9-(2-carboxyphenyl)xanthylium)
Formed from: Rhodamine 123

Rhodamine 123 (3,6-diamino-9-(2-methoxycarbonylphenyl)xanthylium)
→ Rhodamine 110
In: Rat G165

Rhododendrin ((1*R*)-3-(4-hydroxyphenyl)-1-methylpropyl-β-D-glucopyranoside)
Formed from: *p*-Coumaryl alcohol
3-(*p*-Hydroxyphenyl)butan-2-ol
4-(*p*-Hydroxyphenyl)propanol

Rhododendrol
See: 4-(*p*-Hydroxyphenyl)butan-2-ol

ζ-Rhodomycinone ((1*R*,2*R*)-2-ethyl-1,2,3,4,6,11-hexahydro-2,5,7,12-tetrahydroxy-6,11-dioxo-1-naphthacenecarboxylic acid methyl ester)
Formed from: ε-Rhodomycinone

ε-Rhodomycinone ((1*R*,2*R*,4*S*)-2-ethyl-1,2,3,4,6,11-hexahydro-2,4,5,7,12-pentahydroxy-6,11-dioxo-1-naphthacenecarboxylic acid methyl ester)
Formed from: Aklanonic acid
Aklaviketone
Aklavinone
→ Carminomycinone
Probably in: Streptomyces B571
→ Daunorubicin aglycone
Probably in: Streptomyces B598
→ ζ-Rhodomycinone
In: Streptomyces B443

Riboflavinal
See: Vitamin B_2 aldehyde

Riboflavine
Formed from: 6,7-Dimethyl-8-ribityllumazine
Flavine adenine dinucleotide
Lumazine
Riboflavine-5′-phosphate
Vitamin B_2 aldehyde
Formed from guanosine in Salmonella A2604
→ 7-Carboxylumichrome
In: Rat E406
→ 8-Carboxylumichrome
In: Rat E406
→ 10-Carboxymethyl-7,8-dimethylisoalloxazine
In: Microorganism A660
→ Dihydroriboflavine
In: Eubacterium B426
Man A3008
→ 5′-(α-Glucosyl)riboflavine
In: Barley C771, C846
Pig C847
→ β-Glucosylriboflavine
In: Barley C771
→ 10-(2-Hydroxyethyl)isoalloxazine
In: Rat E406
→ 7α-Hydroxyriboflavine
In: Rat C703, E406
→ 8α-Hydroxyriboflavine
In: Rat C703, E406

→ Lumiflavine
In: Rat E406
→ Riboflavine-5′-phosphate
In: Rat A919, C145, E406
See also: A3523, B428
→ Vitamin B_2 aldehyde
In: Schizophyllum C731, C850, J393

Riboflavine-4′5′-cyclic phosphate
→ Riboflavine-5′-phosphate
In: Rat J630

Riboflavine-5′-phosphate (flavine mononucleotide)
Formed from: Flavine adenine dinucleotide
Riboflavine
Riboflavine-4′5′-cyclic phosphate
→ Dihydroriboflavine-5′-phosphate
In: Beneckea A3095, A3209
Escherichia A1228
Eubacterium B426
Man A3008
Photobacterium A2243
→ 5,6-Dimethylbenzimidazole
In: Propionibacterium B428
→ Flavine adenine dinucleotide
In: Mouse A3780
Rat A919
See also: C145
→ Riboflavine
In: Mouse A3780

Riboflavinoic acid
See: Vitamin B_2 acid

Rifabutin
→ 25-O-Deacetylrifabutin
In: Man F453, J273, J282, J309
Rat J273
→ 27-O-Demethylrifabutin
In: Man J273, J309
Rat J273
→ 20-Hydroxyrifabutin
In: Man J273, J309
Rat J273
→ 31-Hydroxyrifabutin
In: Man F453, J273, J309
Rat J273
→ 32-Hydroxyrifabutin
In: Man F453, J273, J309
Rat J273

Rifalazil
→ 25-Deacetylrifalazil
In: Man J750, K117
Rat J750
→ 32-Hydroxyrifalazil
In: Man J750, K117
Rat J750

Rifampicin (rifampin, 3-(((4-methyl-1-piperazinyl)imino)methyl)rifamycin)
Formed from: 3-Amino-5-hydroxybenzoic acid
→ 25-Deacetylrifampicin
In: Guinea pig A1080
Man A1080, A2452, A3157, J282
→ N-Demethylrifampicin
In: Man A2452
→ 3-Formylrifampicin
In: Man A2452
→ Rifampicin quinone
In: Man A3399

Rifampicin quinone
Formed from: Rifampicin

Rifampin
See: Rifampicin

Rifamycin B
Formed from: 3-Amino-5-hydroxybenzoic acid
Rifamycin S
→ Rifamycin O
In: Monocillium C519, C862
→ Rifamycin S
In: Monocillium C862

Rifamycin L (4-(hydroxyacetate)rifamycin)
Formed from: Rifamycin S

Rifamycin LM427
→ 25-O-Deacetylrifamycin LM427
In: Man D924
→ 30-Hydroxyrifamycin LM427
Probably in: Man D924
→ 31-Hydroxyrifamycin LM427
In: Man D924

Rifamycin O (4-O-(carboxymethyl)-1-deoxy-1,4-dihydro-4-hydroxy-1-oxorifamycin γ-lactone)
Formed from: 3-Amino-5-hydroxybenzoic acid
Rifamycin B

Rifamycin S (1,4-dideoxy-1,4-dihydro-1,4-dioxorifamycin)
Formed from: 3-Amino-5-hydroxybenzoic acid
Rifamycin B
→ Rifamycin B
In: Nocardia C101
→ Rifamycin L
In: Nocardia C101

Rifapentine (3-(((4-cyclopentyl-1-piperazinyl)imino)methyl)rifamycin)
→ 25-Deacetylrifapentine
In: Cynomolgus monkey J431
Man J430
Mouse J431
Rat J431
→ 3-Formylrifapentine
In: Cynomolgus monkey J431
Man J430
Mouse J431
Rat J431

Rilmazafone
See: 5-((2-Aminoacetamido)methyl)-1-(4-chloro-2-(*o*-chlorobenzoyl)phenyl)-N,N-dimethyl-1*H*-1,2,4-triazole-3-carboxamide

Riluzole (6-(trifluoromethoxy)-2-benzothiazolamine)
→ 2-Amino-6-hydroxybenzothiazole
In: Man J391
→ 4-Hydroxyriluzole
In: Man J391
→ 5-Hydroxyriluzole
In: Man J391
→ 7-Hydroxyriluzole
In: Man J391
→ N-Hydroxyriluzole
In: Man J391

Rimiterol (*erythro*-piperidin-2-yl-3,4-dihydroxyphenylmethanol)
→ 3-O-Methylrimiterol
In: Man A279
See also: D652
→ Rimiterol-β-D-glucuronide
In: Rat A284
→ Rimiterol-O-sulphate
In: Man A279, A284

Rimiterol-β-D-glucuronide
Formed from: Rimiterol

Rimiterol-O-sulphate
Formed from: Rimiterol

Risperidone (3-(2-(4-(6-fluoro-1,2-benzisoxazol-3-yl)-1-piperidinyl)ethyl)-6,7,8,9-tetrahydro-2-methyl-4*H*-pyrido[1,2-*a*]pyrimidin-4-one)
→ 7-Hydroxyrisperidone
In: Dog G860
Man G845
Rat G860
→ 9-Hydroxyrisperidone
In: Dog G860
Man G845, J371, K1, K399
Rat G860

Ritanilic acid
See: α-Phenyl-2-piperidineacetic acid

Ritodrine
See: *p*-Hydroxy-N-(2-(*p*-hydroxyphenyl)ethyl)norephedrine

Ritonavir ([5*S*-(5*R**,8*R**,10*R**,11*R**)]-10-hydroxy-2-methyl-5-(1-methylethyl)-1-(2-(1-methylethyl)-4-thiazolyl)-3,6-dioxo-8,11-bis(phenylmethyl)-2,4,7,12-tetraazatridecan-13-oic acid 5-thiazolylmethyl ester)
→ De(isopropylthiazolylmethyl)ritonavir
In: Dog J146
Man J146
Rat J146
→ N-Demethylritonavir
In: Man J420
See also: J146
→ De(thiazolylmethylcarbonyl)ritonavir
In: Dog J146
Man J146
Rat J146
→ [5*S*-(5*R**,8*R**,10*R**,11*R**)]-10-Hydroxy-1-(2-(1-hydroxy-1-methylethyl)-4-thiazolyl)-2-methyl-5-(1-methylethyl)3,6-dioxo-8,11-bis(phenylmethyl)-2,4,7,12-tetraazatridecan-13-oic acid 5-thiazolylmethyl ester
In: Dog J146
Man J146, J420
Rat J146
→ [5*S*-(5*R**,8*R**,10*R**,11*R**)]-10-Hydroxy-2-hydroxymethyl-5-(1-methylethyl)-1-(2-(1-methylethyl)-4-thiazolyl)-3,6-dioxo-8,11-

bis(phenylmethyl)-2,4,7,12-tetraazatride-can-13-oic acid 5-thiazolylmethyl ester
In: Dog J146
→ Ritonavir-β-D-glucuronide
In: Dog J146

Ritonavir-β-D-glucuronide
Formed from: Ritonavir

Rizatriptan (N,N-dimethyl-5-(1*H*-1,2,4-triazol-1-ylmethyl)-1*H*-indole-3-ethanamine)
→ Demethylrizatriptan
In: Man J802
→ 6-Hydroxyrizatriptan
In: Man J802
→ Rizatriptan-N^{10}-oxide
In: Man J802
→ 5-((2-Triazolo)methyl)indole-3-acetic acid
In: Man J802

Rizatriptan-N^{10}-oxide
Formed from: Rizatriptan

Robaxin
See: Methocarbamol

Robendine (1,3-bis(*p*-chlorobenzylideneamino) guanidine)
→ 3-Amino-4-(*p*-chlorobenzylideneamino)-5-*p*-chlorophenyl)-4*H*-triazole
In: Chicken A2253
→ *p*-Chlorobenzoic acid
In: Chicken A2253

Robinetinidin (3,7-dihydroxy-2-(3,4,5-trihydroxyphenyl)-1-benzopyrilium)
Formed from: Dihydrorobinetin

Robinin (kaempferol-7-rhamnoside-3-robinoside)
→ Kaempferol
In: Bacteroides E515, E977

Rofecoxib (4-(4-(methylsulphonyl)phenyl)-3-phenyl-2(5*H*)-furanone)
→ 3′4′-Dihydro-3′,4′-dihydroxyrofecoxib
In: Dog K139
Rat K139
→ *cis*-3,4-Dihydrorofecoxib
In: Dog K139
Rat K139
→ *trans*-3,4-Dihydrorofecoxib
In: Dog K139
Rat K139
→ 4′-Hydroxyrofecoxib
In: Dog K139
Rat K139
→ 5-Hydroxyrofecoxib
In: Dog K139
Rat K139

Rolipram (4-(3-cyclopentyloxy-4-methoxyphenyl)-2-pyrrolidinone)
→ O-Demethylrolipram
In: Cynomolgus monkey G710
Man G590, G710
Rat G590, G710
→ 4-(3-Hydroxy-4-methoxyphenyl)-2-pyrrolidone
In: Cynomolgus monkey G710
Man G590, G710
Rabbit G710
Rhesus monkey G710
Rat G590, G710
→ 2′-Hydroxyrolipram
In: Cynomolgus monkey G710
Man G590, G710
Rat G590, G710
→ 3′-Hydroxyrolipram
In: Cynomolgus monkey G710
Man G590, G710
Rabbit G710
Rhesus monkey G710
Rat G590, G710
→ 5″-Hydroxyrolipram
In: Man G590, G710
Rat G590

Ropinirole (4-(2-(dipropylamino)ethyl)-1,3-dihydro-2*H*-indol-2-one)
→ 4-Carboxymethyloxindole
In: Cynomolgus monkey J561
Man J561
Mouse J561
→ Depropylropinrole
In: Cynomolgus monkey J561
Man J276, J561
Mouse J561
→ 7-Hydroxyropinrole
In: Cynomolgus monkey J561
Man J276, J561
Mouse J561
Rat J561

Ropivacaine (*S* isomer assumed; N-propyl-2′,6′-pipecoloxylidide)
→ 2′-Hydroxymethyl-6′-methyl-N-propylpipecoloxylidide
In: Man J44
See also: J45
→ 3-Hydroxyropivacaine
In: Man J44, J45, J889
Rat K129
→ 4-Hydroxyropivacaine
In: Man J44, J45
Rat K129
→ (*S*)-Pipecolo-2′,6′-xylidide
In: Man H548, J45

Roquefortine ((3*E*,5a*S*,10b*R*,11a*S*)-10b-(1,1-dimethyl-2-propenyl)-6,10b,11,11a-tetrahydro-3-(1*H*-imidazol-4-ylmethylene-2*H*-pyrazino[1′,2′:1,5]pyrrolo[2,3-*b*]indole-1,4(3*H*,5a*H*)-dione)
Formed from: L-Tryptophan

Roquinimex (linomide, 1,2-dihydro-4-hydroxy-1,N-dimethyl-N-phenyl-2-oxoquinoline-3-carboxamide)
→ 1,2-Dihydro-4-hydroxy-N-(*p*-hydroxyphenyl)-1,N-dimethyl-2-oxoquinoline-3-carboxamide
In: Man K24
Mouse K24
Rat K24
→ 1,2-Dihydro-4-hydroxy-1-methyl-N-phenyl-2-oxoquinoline-3-carboxamide
In: Mouse K24
Rat K24
→ 1,2-Dihydro-4-hydroxy-N-methyl-N-phenyl-2-oxoquinoline-3-carboxamide
In: Mouse K24
Rat K24
→ 6-Hydroxyroquinimex
In: Man K128
Mouse K24
Rat K24
→ 7-Hydroxyroquinimex
In: Man K128
Mouse K24
Rat K24
→ 8-Hydroxyroquinimex
In: Man K128
Mouse K24
Rat K24

Rosiglitazone
See: 5-((4-(2-(Methyl-2-pyridinylamino)ethoxy)phenyl)methyl)-2,4-thiazolidinedione

Rosmarinic acid ((α*R*)-α-(((2*E*)-3-(3,4-dihydroxyphenyl)-1-oxo-2-propenyl)oxy)-3,4-dihydroxybenzenepropanoic acid)
Formed from: Caffeic acid
Caffeoyl CoA
Caffeoyl *p*-hydroxyphenyllactic acid ester
p-Coumaroyl 3,4-dihydroxyphenyllactic acid ester
3,4-Dihydroxyphenyllactic acid
Probably formed from: 3,4-Dihydroxy-L-phenylalanine
→ Caffeic acid
In: Rat J686
→ Caffeoyl CoA
In: Coleus G420
→ 3,4-Dihydroxyphenyllactic acid
In: Coleus G420
→ O-Methylrosmarinic acid
In: Parsley F160
→ Rosmarinic acid-O-sulphate
In: Eubacteroum G475

Rosmarinic acid-O-sulphate
Formed from: Rosmarinic acid

Rotenone ((2*R*,6a*S*,12a*S*)-1,2,12,12a-tetrahydro-8,9-dimethoxy-2-(1-methylethenyl)[1]benzopyrano[3,4-*b*]furo[2,3-*h*][1]benzopyran-6-(6a*H*)-one)
→ 1′,2′-Dihydro-1′,2′-dihydroxyrotenone
In: Cunninghamella C416
→ 2′-Hydroxyrotenone
In: Cunninghamella C416

Roxatidine (2-hydroxy-N-(3-(*m*-(1-piperidinylmethyl)phenoxy)propyl)acetamide)
Formed from: 2-Acetoxy-N-(3-(*m*-(1-piperidinylmethyl)phenoxy)propyl)acetamide
Formed from: Roxatidine acetate
→ N-(3-(*m*-(2-Oxo-1-piperidinylmethyl)phenoxy)propyl)acetamide
Probably in: Dog E904
Rat E904

Roxatidine acetate
→ 3-Hydroxyroxatidine acetate
In: Dog E452
Rat E452
→ 2-Oxoroxatidine acetate
In: Rat E452
→ *m*-(1-Piperidinylmethyl)phenol
In: Dog E904
Rat E904
→ 3-(*m*-(1-Piperidinylmethyl)phenoxy) propylamine
In: Dog E904
Rat E904
→ Roxatidine
In: Dog E904
Rat E904

Roxifiban (methyl N^2-(1-butyloxycarbonyl)-N^3-(2-(3-(formamidinophenyl)isoxazolin-5(*R*)-yl)acetyl)-2,3-diaminopropionate)
→ Demethylroxifiban
In: Man K356
Microorganism K356
Mouse K356
Rat K356
→ Methyl N^2-(1-butyloxycarbonyl)-N^3-(2-(3-(formamidinophenyl)isoxazolin-5(*S*)-yl)acetyl)-2,3-diaminopropionate
In: Man K356
Microorganism K356
Mouse K356
Rat K356

Rubazoic acid (rubazonic acid; 4-((1,5-dihydro-3-methyl-5-oxo-1-phenyl-4*H*-pyrazol-4-ylidene)amino)-2,4-dihydro-5-methyl-2-phenyl-3*H*-pyrazol-3-one)
Probably formed from: 4-Aminoantipyrine
4-Methylrubazoic acid

Rubazonic acid
See: Rubazoic acid

Rubeomycin A
→ Rubeomycin B
In: Rhodosporidium D19
Rhodotorula D19
Torulopsis D19

Rubeomycin B
Formed from: Rubeomycin A

Rubidomycin
See: Daunorubicin

Rubidazone
→ Daunorubicin
In: Mouse B217

Rubomycin C
See: Daunorubicin

Rubrofusarin (5,6-dihydroxy-8-methoxy-2-methyl-4*H*-naphtho[2,3-*b*]pyran-4-one)
Formed from acetic acid in Fusarium C448

Rubroskyrin ((8*S*,8a*S*,9*S*,17*S*,17a*R*,19*R*)-7,8,8a,9-tetrahydro-1,4,6,8,10,15,19-heptahydroxy-2,12-dimethyl-9,17-methanonaphtho[2′,3′:5,6]cyclohept[1,2-*d*]anthracene-5,11,14,16,18(17*H*)-pentone)
Formed from: Catenarin
Emodinanthrone

Rutaretin ((2*S*)-2,3-dihydro-9-hydroxy-2-(1-hydroxy-1-methylethyl)-7*H*-furo[3,2-*g*][1]benzopyran-7-one)
Formed from: Marmesin
→ Marmesin
In: Ruta A547

Rutilantinone
See: ε-Pyrromycinone

Rutin (quercetin-3-rutoside)
→ 3,4-Dihydroxyphenylacetic acid
In: Man B918
Rat C824
→ 7-O-(β-Hydroxyethyl)rutin
In: Rat A271
→ 4-Hydroxy-3-methoxyphenylacetic acid
In: Guinea pig A1375
Man B918
Rat C824
→ *m*-Hydroxyphenylhydracrylic acid
In: Man B918
→ *m*-Hydroxyphenylpropionic acid
In: Rat C824
→ 4-Methylcatechol
In: Man B918
Rat C824
→ O-Methylrutin
In: Nicotiana A3757
→ Quercetin
In Bacteroides E515, E977

Microorganism A283, J525
→ Rutin-α-D-glucoside
In: Pig C847
→ Rutin quinone
Probably in: Horseradish A1183
→ Rutin-O-sulphate
In: Eubacterium H354

Rutin-α-D-glucoside
Formed from: Rutin

Rutin quinone
Formed from: Rutin

Rutin-O-sulphate
Formed from: Rutin

RWJ 34130
→ 2-Oxo-1-phenylpyrrolidine
In: Rat K183
→ RWJ 34130 sulphoxide
In: Rat K183

RWJ 34130 sulphoxide
Formed from: RWJ 34130

Ryosidine (4-(*o*-difluoromethoxyphenyl)-1,4-dihydro-3,5-di-(methoxycarbonyl)-2,6-dimethylpyridine)
→ 3-Carboxy-4-(*o*-difluoromethoxyphenyl)-2-(α-hydroxy)methyl-5-methoxycarbonyl-6-methylpyridine
In: Man D635
→ 3-Carboxy-4-(*o*-hydroxyphenyl)-5-methoxycarbonyl-2,6-dimethylpyridine
In: Man D635

S

Saccharin (1,2-benzoisothiazol-3(2*H*)-one 1,1-dioxide)
Formed from: Benz[*d*]isothiazoline-1,1-dioxide
Oryzemate
Piroxicam
o-Toluenesulphonamide
Probably formed from: 2,3-Dihydro-3-hydroxybenz[*d*]isothiazoline-1,1-dioxide
N-Methylsaccharin
→ *o*-Carboxybenzenesulphonamide
In: Rat A500, A1568
Contrast (in guinea pig) A559
(man) A2039
(rat) A22, A559, A2039
→ Saccharin-N-D-glucoside
Probably in: Oryza A941
→ *o*-Sulphobenzoic acid
In: Rat A500

Saccharin-N-D-glucoside
Formed from: Saccharin

Saddamine (2-(((phenylmethyl)imino)methyl) phenol)
→ Benzoic acid
In: Man B155

Safrole (4-allyl-1,2-methylenedioxybenzene)
→ 4-Allylcatechol
In: Guinea pig A1544
Rat A1544, B3109, B3567, B3947, C564
See also: B528
→ *p*-Cresol
In: Rat A3567
→ Epoxyhydroxysafrole
In: Rat A3567
→ 4-(1-Hydroxyallyl)-1,2-methylenedioxybenzene
In: Guinea pig A833, A1544
Hamster A833
Rat A833, A1544, B3109, C564
→ 4-(3-Hydroxyallyl)-1,2-methylenedioxybenzene
In: Rat A3109
→ Isosafrole
Probably in: Rat B528
→ Piperonylic acid
In: Guinea pig A1544
Rat A1544
→ Safrole-2,3-oxide
In: Rat A3109, B528
See also: A1544

Safrole glycol (4-(2,3-dihydroxypropyl)-1,2-methylenedioxybenzene)
Formed from: Safrole-2,3-oxide
→ 4-(2,3-Dihydroxypropyl)catechol
Probably in: Rat A1544
→ 1-(3,4-Methylenedioxyphenyl)glycerol
Probably in: Rat A1544
→ 3,4-Methylenedioxyphenyllactic acid
Probably in: Rat A1544, C564

Safrole-2,3-oxide
Formed from: Safrole
→ Safrole glycol
In: Guinea pig A1544, G86
Rabbit A100
Rat A1544, C564, G86
See also: A3109, B528

Sakuranetin (4′,5-dihydroxy-7-methoxyflavanone)
Formed from: Naringenin

Salazopyrin
See: Sulphasalazine

Salbutamol
See: N-*tert*-Butyl-4-hydroxy-3-hydroxymethylphenylethanolamine

Salbutamol-β-D-glucuronide
Formed from: N-*tert*-Butyl-4-hydroxy-3-hydroxymethylphenylethanolamine
Salbutamol sulphate
Formed from: N-*tert*-butyl-4-hydroxy-3-hydroxymethylphenylethanolamine
Salicin (*o*-hydroxymethylphenyl-β-D-glucuronide)
Formed from: Salicyl alcohol
Probably formed from: Leiocarposide
→ Salicin-β-D-glucoside
In: Man A499
→ Salicyl alcohol
In: Man A499
Picea K833
Triglochin A3756
See also: C910, F263
Salicin-β-D-glucoside
Formed from: Salicin
Salicyl alcohol (*o*-hydroxybenzyl alcohol, saligenin)
Formed from: Ferulic acid
o-Methoxybenzyl alcohol
Salicin
Salicylaldehyde
Salithion
Probably formed from: Di-*o*-cresyl *o*-hydroxyphenyl phosphate
→ *o*-Bromophenol
In: Ulva K153
See also: E579
→ *o*-Hydroxybenzyl-β-D-glucoside
In: Datura A1725
See also: G897
→ *o*-Hydroxymethylphenyl-α-D-glucoside
Probably in: Bacillus H234
→ *o*-Hydroxymethylphenyl sulphate
Probably in: Man E575
Rat E575
→ Salicin
In: Almond H558
Datura A1725
Gardenia D71, D715
Trilobium G816
See also: E681
→ Salicylaldehyde
In: Acinetobacter E663
Pseudomonas A3837
See also: C910, D325, G917
Salicylaldehyde (*o*-hydroxybenzaldehyde)
Formed from: *o*-Hydroxybenzylidenepyruvic acid
o-Hydroxyphenylglyoxylic acid
Salicyl alcohol
Salicylaldoxime
Salicylic acid
Probably formed from: *o*-Hydroxybenzylamine
Salicylidenebenzylamine
→ Catechol
In: Pig H482
→ Salicyl alcohol
In: Dunaliella G459
Euglena G422
Man F87
Mung bean J910
See also: C411
→ Salicylaldehyde-β-D-glucoside
In: Manduca G816, G897
Melanopus G816
Trilobium G816
→ Salicylic acid
In: Acinetobacter E663, F336
Arthrobacter J33
Helix G804
Man J24
Mouse G804
Pseudomonas K495
Serratia G197
Streptomyces C284
See also: C411, G356, G865
Salicylaldehyde-β-D-glucoside (helicin)
Formed from: Salicylaldehyde
Salicylaldoxime
→ Salicylaldehyde
In: Rabbit E358
Salicylamide (*o*-hydroxybenzamide)
Formed from: *o*-Ethoxybenzamide
Ethyl 2-carbamoyloxybenzoate
o-Hydroxybenzonitrile
Salicylhydroxamic acid
N-Salicyl-α-hydroxyglycine

Salicylic acid
→ Gentisamide
In: Man A1199
Mouse E683
Rat F84, F671, H738
→ Salicylamide-2-β-D-glucuronide
In: Dog E285
Man A1199, A2246, A2605, A2840, D8
Mouse E683
Rabbit B778, E285
Rat B778, F84, H738
→ Salicylamide-2-sulphate
In: Beef D607
Dog D612
Eubacterium E432, G602
Man A1199, A2246, A2840, C523, D8, D607, D977, D979
Mouse E683
Rabbit B778
Rat B778, C441, D807, F84, H738

Salicylamide-2-β-D-glucuronide
Formed from: Salicylamide

Salicylamide-2-sulphate
Formed from: Salicylamide

Salicylanilide (2-hydroxy-N-phenylbenzamide)
Formed from: Benzanilide
→ Aniline
In: Mouse A2037
Sheep A2037
→ 2,4′-Dihydroxybenzanilide
In: Rat B970
→ 2,5-Dihydroxybenzanilide
In: Rat B970
→ Salicylic acid
In: Mouse A2037
Sheep A2037

Salicylhydroxamic acid
→ Salicylamide
In: Guinea pig C897
Rabbit C897
Rat H16

N-Salicyl-α-hydroxyglycine
Formed from: *o*-Hydroxyhippuric acid
→ Salicylamide
In: Man K660

Salicylic acid (*o*-hydroxybenzoic acid)
Formed from: Acetylsalicylic acid
2-(*p*-Aminobenzoyloxy)benzoic acid
5-Amino-2-hydroxybenzoic acid
2-Aminonaphthalene-1-sulphonic acid
Benorylate
Benzoic acid
Butyl salicylate
n-Butyrylsalicylic acid
N-(2-Chloro-4-nitrophenyl)salicylamide
N-(*o*-Chlorophenyl)salicylamide
5-Chlorosalicylic acid
4-Chloro-2,2′,3-trihydroxybiphenyl
o-Cresol
Decyl salicylate
4,5-Dihydro-1,3-dimethyl-1*H*-pyrazol[3,4-*b*][1,4]benzoxazepine
Dibenzofuran
Eterylate
Ethylbenzene
Ethyl 2-carbamoyloxybenzoate
Ethyl salicylate
Hexyl salicylate
o-Hydroxybenzonitrile
3-Hydroxyflavone
o-Hydroxyhippuric acid
2-Hydroxy-6-(*o*-hydroxyphenyl)-6-oxohexa-2,4-dienoic acid
2-Hydroxynaphthalene-1-sulphonic acid
2-Hydroxy-4-(3-oxo-3*H*-benzofuran-2-yliden)but-2-enoic acid
2-Hydroxy-1,4-naphthoquinone
Isobutyrylsalicylic acid
o-Methoxybenzoic acid
Methyl salicylate
Naphthalene-1-sulphonic acid
Naphthalene-2-sulphonic acid
N-(*p*-Nitrophenyl)salicylamide
Octyl salicylate
Pentyl salicylate
Phenanthrene
Phenetsal
Propionylsalicylic acid
Propyl salicylate
Salicylaldehyde
Salicylanilide

Salicylic ethylene glycol ester
Salsalate
Salicyloylglutamic acid
Styrene
Probably formed from: 2-Hydroxyisophthalic acid
Maltol salicylate
→ *o*-Carboxyphenyl-β-D-glucoside
In: Catharanthus E681
Mallotus E681
Nicotiana H257, H833, J694
Perilla E681
Potato J691
Vicia H31
See also: A1725, G897, H593
→ *o*-Carboxyphenyl-β-D-glucuronide
In: Horse B801
Man A412, G203
Rabbit A2355
Rat C237, F580
See also: A495, A978, A1590, C374, D665, D668, F588, H287
→ *o*-Carboxyphenyl sulphate
In: Rat C237
→ Catechol
In: Amycolatopsis F615
Pseudomonas A609, B751, D918, G434, H506
Streptomyces F615
Trichosporon B368, D232
See also: C475, K432
→ 2,3-Dihydro-2,3,6-trihydroxybenzoic acid
Probably in: Pseudomonas B465
→ 2,4-Dihydroxybenzoic acid
In: Aspergillus A2292
→ Gentisic acid
In: Cucumis A611
Dog A978
Fagopyrum H31
Horse B801
Lignobacter B503
Man A412, F134, J567
Micrococcus D56
Nocardia B2
Pig A978
Pseudomonas G280
Rat F580
Rhizobium C500
Rhodococcus H341, H438
See also: A499, A876, B201, C374, D479, D668, E819, F588, G203, J917, K432, K696
→ *o*-Hydroxyhippuric acid
In: Dog A495, A978
Goat A495
Horse A495, B801
Man A412, A980, A3098, C374, D665
Monkey A555
Mouse A3361
Pig A495, A978
Rabbit A2355, H519
Rat B56, F580, H287, J549
See also: A499, A876, A1590, B201, E819, F588, G203
→ Isophthalic acid
In: Microorganism G917
→ Methyl salicylate
In: Clarkia J521, K239
→ Pyrocatechuic acid
In: Man F134, J567
Rat E272, F580. H817, J754
See also: A2217, E819
→ Salicylaldehyde
In: Neurospora A750
→ Salicylamide
In: Streptomyces E279
→ Salicyloyl CoA
In: Bacterium J213
Man K220
→ Salicyloyl-β-D-glucoside
In: Glycine A3585
Mung bean A3585
Nicotiana H257, J694
See also: A1725
→ Salicyloyl-β-D-glucuronic acid
In: Horse B801
Man A412
Rabbit A2355
Rat F850, J549
See also: F588, C374, D668, G203

Salicylic ethylene glycol ester
→ Salicylic acid

In: Mouse A3361

Salicylidenebenzylamine
→ Salicylaldehyde
Probably in: Dog C411
Rat C411

Salicyloyl CoA
Formed from: Salicylic acid
→ Benzoyl CoA
In: Bacterium J213

Salicyloyl-β-D-glucoside
Formed from: Salicylic acid

Salicyloyl-β-D-glucuronic acid
Formed from: Salicylic acid

Salicyloylglutamic acid
→ Salicylic acid
In: Rabbit G731

Salicyluric acid
See: *o*-Hydroxyhippuric acid

Salidroside (2-(*p*-hydroxyphenyl)ethyl-β-D-glucoside)
Formed from: *p*-Tyrosol

Saligenin
See: Salicyl alcohol

Saligenin cyclic-*o*-cresyl phosphate
Probably formed from: *o*-Cresyl *o*-hydroxyphenyl phosphate

Salioxon (2-methoxy-4*H*-1,3,2-benzodioxaphosphorin-2-oxide)
Formed from: Salithion
→ Demethylsalioxon
Probably in: Man A1562
Rat A1562

Salithion (2-methoxy-4*H*-1,3,2-benzodioxaphosphorin-2-sulphide)
→ Demethylsalithion
In: Bean A1562
Rat A1562
→ O-(*o*-Hydroxy)benzyl phosphorothioate
In: Rat A1562
→ Salicyl alcohol
In: Bean A1562
Rat A1562
Rice A1562
→ Salioxon
In: Rat A1562

Salmeterol (4-hydroxy-α′-(((2-(4-methoxyphenyl)-1-methylethyl) amino)methyl)-1,3-benzenedimethanol)
→ 4-Hydroxy-α-(((6-hydroxyhexyl)-amino)methyl)-1,3-benzenedimethanol
In: Man G838
See also: J868
→ 4-Hydroxy-α-(((6-(4-hydroxy-4-phenylbutoxy)hexyl)amino)methyl)-1,3-benzenedimethanol
In: Man G838, H705
Rat G838
→ Salmeterol-β-D-glucuronide
In: Beef J580
Mouse G838
Rabbit G838
Rat G838

Salmeterol-β-D-glucuronide
Formed from: Salmeterol

Salsalate (2-hydroxybenzoic acid 2-carboxyphenyl ester)
→ Salicylic acid
In: Man D479

Salsolidine ((1*S*)-1,2,3,4-tetrahydro-6,7-dimethoxy-1-methylisoquinoline)
→ N-Methylsalsolidine
In: Beef J899

Salsoline
See: 1,2,3,4-Tetrahydro-6-hydroxy-7-methoxy-1-methylisoquinoline

Salsolinol (1,2,3,4-tetahydro-6,7-dihydroxy-1-methylisoquinoline; both isomers)
Formed from: 3,4-Dihydroxyphenethylamine
→ N-Methyl-(*R*)-salsolinol
In: Man J472
→ 1,2,3,4-Tetahydro-6-hydroxy-7-methoxy-1-methylisoquinoline
In: Corydalis G425
Pig J363
Rat D27
Contrast (in Corydalis) G470
→ 1,2,3,4-Tetahydro-7-hydroxy-6-methoxy-1-methylisoquinoline
In: Corydalis G425, G470
Pig J363
Rat D27

Salsolinol-1-carboxylic acid (1-carboxy-1,2,3,4-tetrahydro-6,7-dihydroxy-1-methylisoquinoline)
→ 1,2,3,4-Tetahydro-6-hydroxy-7-methoxy-1-methylisoquinoline-1-carboxylic acid
In: Rat D27

Salsolinol-3-carboxylic acid (3-carboxy-1,2,3,4-tetrahydro-6,7-dihydroxy-1-methylisoquinoline)
→ 1,2,3,4-Tetahydro-6-hydroxy-7-methoxy-1-methylisoquinoline-3-carboxylic acid
In: Rat D27
→ 1,2,3,4-Tetahydro-7-hydroxy-6-methoxy-1-methylisoquinoline-3-carboxylic acid
In: Rat D27

Saluamine (4-chloro-5-sulphamoylanthranilic acid)
Formed from: Furosemide

Salutaridine (5,6,8,14-tetradehydro-4-hydroxy-3,6-dimethoxy-17-methylmorphinan-7-one)
Formed from: Reticuline
→ Salutaridinol
In: Papaver K758

Salutaridinol (5,6,8,14-tetradehydro-3,6-dimethoxy-17-methylmorphinan-4,7-diol)
Formed from: Reticuline
Salutaridine
→ Salutaridinol-7-O-acetate
In: Papaver J1, K784, K786
→ Thebaine (spontaneously from salutaridinol-7-O-acetate)
In: Papaver H255
See also: A1620

Salutaridinol-7-O-acetate
Formed from: Salutaridinol

Sameridine (N-ethyl-1-hexyl-N-methyl-4-phenyl-4-piperidinecarboxamide)
→ 5′-Hydroxysameridine
In: Rat K43
→ 6′-Hydroxysameridine
In: Rat K43

Sampangine (7H-naphtho[1,2,3-*ij*][2,7]naphthyridin-7-one)
→ *enol*-Sampangine-β-D-glucuronide
In: Rat K466

***enol*-Sampangine-β-D-glucuronide**
Formed from: Sampangine

Sanguinarine (13-methyl[1,3]benzodioxolo[5,6-*c*]-1,3-dioxolo[4,5-*i*]phenanthridinium)
Formed from: Dihydrosanguinarine
Norcoclaurine
Protopine
→ Chelirubine
In: Macleaya C696
→ Dihydrosanguinarine
In: Verticillium A676
See also: F699

Saquinavir ((1*S*,2*R*)-3-(3*S*,4α*S*,8α*S*))-N[1]-(((1,1-dimethylethyl)amino) carbonyl)octahydro-2(1*H*)-isoquinolinyl)-2-hydroxy-1-(phenylmethyl)propyl)-2-((2-quinolinylcarbonyl)amino)butanediamide)
→ N[1]-(((1,1-Dimethylethyl)amino)carbonyl)octahydrohydroxy-2(1*H*)-isoquinolinyl)-2-hydroxy-1-(phenylmethyl)propyl)-2-((2-quinolinylcarbonyl)amino)butanediamide
In: Man J97b
→ ((1*S*,2*R*)-3-(3*S*,4α*S*,8α*S*))-N[1]-(2-Hydroxy-3-(3-(((1-hydroxymethyl-1-methylethyl)amino)carbonyl)octahydro-2(1*H*)-isoquinolinyl)-1-(phenylmethyl) propyl)-2-((2-quinolinylcarbonyl)amino) butanediamide
In: Man J97b

Sativan ((3*R*)-3-2,4-dimethoxyphenyl)-3,4-dihydro-2*H*-1-benzopyran-7-ol)
Formed from: Medicarpin
Vestitol

Sceletinone ((3a*R*,7a*S*)-1,2,3,3a,7,7a-hexahydro-3a-(4-hydroxyphenyl)-1-methyl-6*H*-indol-6-one)
→ Mesembrenol
In: Sceletium A3136

Schizandrin
→ 7,8-Dihydroschizandrin
In: Rat H87

Scoparone
See: 6,7-Dimethoxycoumarin

Scopolamine (hyoscine, (α*S*)-α-hydroxymethylbenzeneacetic acid (1α,2β,4β,5α,7β)-9-methyl-3-oxa-9-azatricyclo[3.3.1.0^{2,4}]non-7-yl ester)
Formed from: 6,7-Dehydrohyoscyamine
Probably formed from: Hyoscyamine
→ Aposcopolamine

In: Guinea pig G19, G957
Mouse G19
Rabbit G19
Rat G19
Solandra A1621
→ *m*-Hydroxyscopolamine
In: Guinea pig G19
Rat G19
→ *p*-Hydroxyscopolamine
In: Rat G19
→ Norscopolamine
In: Guinea pig A2816, G19
Mouse G19
Rabbit G19
Rat G19
Solandra A1621
→ (*S*)-Tropic acid
In: Guinea pig G19
Mouse G19
Rabbit A2507, G19

Scopoletin (7-hydroxy-6-methoxycoumarin)
Formed from: Cinnamic acid
6,7-Dihydroxycoumarin
6,7-Dimethoxycoumarin
→ 6,7-Dihydroxycoumarin
Probably in: Mouse A3623
→ Puberulin
In: Agathosma D73
→ Scopolin
In: Duboisia H621
Forsythia A3365
Manduca G816
Nicotiana B442, B819, H621, K393, K576
Solanum H621, K576
→ Scopoletin-β-D-glucuronide
In: Man G968, J416

Scopoletin-7-glucoside
See: Scopolin

Scopoletin-β-D-glucuronide
Formed from: Scopoletin

Scopolin (scopoletin-7-glucoside)
Formed from: 7-Hydroxy-6-methoxycoumarin
Scopoletin

Scoulerine ((13a*S*)-5,8,13,13a-tetrahydro-3,10-dimethoxy-6*H*-dibenzo[*a*,*g*]quinolizine-2,9-diol)
Formed from: Dehydroscoulerin
Reticuline
→ (*S*)-Cheilanthifoline
In: Eschscholtzia K914
→ (*S*)-Tetrahydrocolumbamine
In: Argemone D77
Berberis D581

Scutellarein (4′,5,6,7-tetrahydroxyflavone)
→ Scutellarein-7-β-D-glucuronide
In: Scutellaria K402
→ 4′,5,7-Trihydroxy-6-methoxyflavone
In: Calamondin orange A3987
→ 5,6,7-Trihydroxy-4′-methoxyflavone
In: Calamondin orange A3987

Scutellarein-7-β-D-glucuronide
Formed from: Scutellarein

Scytalone
See: 3,6,8-Trihydroxy-1-tetralone

Seclazone (7-chloro-3,3a-dihydro-2*H*,9*H*-isoxazolo[3,2-*b*][1,3]benzoxazin-9-one)
→ 5-Chlorosalicylic acid
In: Man A1392
Rat A34, A1884

7′,8′-Seco-8′-carboxybromocryptine
Formed from: Bromocryptine

7′,8′-Seco-8′-carboxyisobromocryptine
Formed from: Bromocryptine

(−)-Secoisolariciresinol ((*R*-(*R**,*R**))-2,3-bis((4-hydroxy-3-methoxyphenyl)methyl)-1,4-butanediol)
Formed from: (+)-Lariciresinol

Selegiline
See: Deprenyl

Selenoanisole
→ Selenoanisole selenoxide
In: Mortierella D78

Selenoanisole selenoxide
Formed from: Selenoanisole

2-Selenobenzanilide
Formed from: Ebselen
→ 2,2′-Diselenobisbenzanilide
Probably in: Rat E808, F127
→ 2-(Methylseleno)benzanilide
Probably in: Man F235, F311

Pig F235, F311
Rat E808, F127, F235, F311
→ 2-Selenobenzanilide-β-D-glucuronide
Probably in: Man F235
Pig F235
Rat E808, F127, F235

2-Selenobenzanilide-β-D-glucuronide
Probably formed from: 2-Selenobenzanilide

Selenosinigrin
Formed from selenate in horseradish A159

Semotiadil ((+)-(*R*)-2-(5-methoxy-2-((N-methyl-N-(2-((3,4-methylenedioxy) phenoxy)ethyl)amino)propoxy)phenyl)-4-methyl-2*H*-1,4-benzothiazin-3(4*H*)-one)
→ 5′-Demethylsemotiadil
In: Rat H313, J130
→ (+)-(*R*)-2-(2-((N-(2-((3,4-Dihydroxy) phenoxy)ethyl)-N-methylamino)propoxy)-5-methoxyphenyl)-4-methyl-2*H*-1,4-benzothiazin-3(4*H*)-one
In: Rat J130
See also: H313
→ Hydroxysemotiadil
In: Rat H313

Sennidin A ((9*R*,9′*R*)-rel-(+)-9,9′,10,10′-tetrahydro-4,4′,5,5′-tetrahydroxy-10,10′-dioxo-(9,9′-bianthracene))-2,2′-dicarboxylic acid)
Formed from: Sennidin A-8-glucoside
Sennoside A
→ Rhein
Probably in: Rat B333
→ Rhein-9-anthrone
In: Bifidobacterium C399
Clostridium C399
Eubacterium C399
Lactobacillus C399
Microorganism C103
Peptostreptococcus C399
Proteus C399
See also: D905, B333

Sennidin B ((9*R*,9′*S*)-rel-(+)-9,9′,10,10′-tetrahydro-4,4′,5,5′-tetrahydroxy-10,10′-dioxo-(9,9′-bianthracene))-2,2′-dicarboxylic acid)
Formed from: Sennidin B-8-glucoside
Sennoside B
→ Rhein
Probably in: Rat B333
→ Rhein-9-anthrone
In: Microorganism C103
See also: B333

Sennidin A-8-glucoside
Formed from: Sennoside A
→ Sennidin A
In: Bifidobacterium H105
Microorganism C399

Sennidin B-8-glucoside
Formed from: Sennoside B
→ Sennidin B
In: Bifidobacterium H68, H105
Microorganism C399

Sennoside A ((*R**,*R**)-5,5′-bis(β-D-glucopyranosyloxy)-9,9′,10,10′-tetrahydro-4,4′-dihydroxy-10,10′-dioxo-(9,9′-bianthracene))-2,2′-dicarboxylic acid)
→ 8-Glucosylrhein
In: Bifidobacterium B696
Clostridium B696
Peptostreptococcus B696
→ 8-Glucosylrheinanthrone
In: Lactobacillus B696
→ Sennidin A
In: Bifidobacterium B696
Clostridium B696
Microorganism C103
→ Sennidin A-8-glucoside
In: Bifidobacterium H105
Microorganism C399
→ Sennoside B
In: Bifidobacterium B696
Eubacterium B696
Peptostreptococcus B696
→ Sennosidin A
In: Rat B333

Sennoside B ((9*R*,9′*S*)-5,5′-bis(β-D-glucopyranosyloxy)-9,9′,10,10′-tetrahydro-4,4′-dihydroxy-10,10′-dioxo-(9,9′-bianthracene))-2,2′-dicarboxylic acid)
Formed from: Sennoside A
→ Sennidin B-8-glucoside
In: Bifidobacterium H68, H105
Microorganism C399

See also: C103
→ Sennosidin B
In: Rat B333
Seratrodast (ζ-(2,4,5-trimethyl-3,6-dioxo-1,4-cyclohexadien-1-yl)-benzeneheptanoic acid)
→ 7-(5-Hydroxymethyl-3,6-dimethyl-*p*-quinon-2-yl)-7-phenylheptanoic acid
In: Man J80
→ 4′-Hydroxyseratrodast
In: Man J80
Serotonin
See: 5-Hydroxytryptamine
Serpentine ((19α)-3,4,5,6,16,17-hexadehydro-16-(methoxycarbonyl)-19-methyloxayohimbanium)
Formed from: Strictosidine
Sertindole (1-(2-(4-(5-chloro-1-(4-fluorophenyl)-1*H*-indol-3-yl)-1-piperidinyl)ethyl)-2-imidazolidinone)
→ 5-Chloro-1-(*p*-fluorophenyl)-3-(piperidin-4-yl)indole
In: Dog H471
Rat H471
→ 4,5-Dehydrosertindole
In: Dog H471
Rat H471
→ 5-Hydroxysertindole
In: Dog H471
Rat H471
→ 4-Oxosertindole
In: Dog H471
Rat H471
Sertraline ((1*S*,4*S*)-4-(3,4-dichlorophenyl)-1,2,3,4-tetrahydro-N-methyl-α-naphthylamine)
→ Demethylsertraline
In: Glycine K361
Man H899, J729
→ 4-(3,4-Dichlorophenyl)-2-hydroxytetralone
In: Dog F82
Rat F82
→ N-Hydroxysertraline
In: Dog F82
Rat F82
→ Sertraline carbamoyl-O-β-D-glucuronide
In: Dog F16, F82
Rat F82
Sertraline carbamoyl-O-β-D-glucuronide
Formed from: Sertraline
L-Seryl-β-naphthylamide
→ β-Naphthylamine
In: Agave A2745
Glycine A1596
Rabbit B334
L-Seryl-L-phenylalanine
→ L-Phenylalanine
In: Brevibacterium E587
1-Seryl-2-(2,3,4-trihydroxybenzyl)hydrazine
See: Benserazide
L-Seryl-L-tyrosine
Formed from: L-Seryl-L-tyrosyl-β-naphthylamide
L-Seryl-L-tyrosyl-β-naphthylamide
→ β-Naphthylamine
In: Man A2149
Rat A523
→ L-Seryl-L-tyrosine
In: Man A2149
Rat A523
(+)-Sesamin ((1*S*,3a*R*,4*S*,6a*R*)-5,5′-(tetrahydro-1*H*,3*H*-furo[3,4-*c*]furan-1,4-diyl)bis(1,3-benzodioxole))
Formed from: (+)-Pinoresinol
L-Tyrosine
Sesamol
See: 3,4-Methylenedioxyphenol
(+)-Sesamolin ((1*S*,3a*R*,4*R*,6a*R*)-5-(4-(1,3-benzodioxol-5-yloxy)tetrahydro-1*H*,3*H*-furo[3,4-*c*]furan-1-yl)-1,3-benzodioxole)
Formed from: (+)-Pinoresinol
Setoclavine ((8α)-9,10-didehydro-6,8-dimethylergolin-8-ol)
Formed from: Agroclavine
Sevin
See: 1-Naphthyl methylcarbamate
Shikimic acid (3,4,5-trihydroxy-1-cyclohexene-1-carboxylic acid)
→ *p*-Amino-L-phenylalanine
Probably in: Streptomyces A385
→ Benzoic acid
In: Rat A2706
→ 3-Carboxy-L-phenylalanine

In: Iris A1601
Reseda A1601
→ 3-Carboxy-L-tyrosine
In: Iris A1601
Reseda A1601
→ Catechol
In: Rat A2706
→ 1,6-Dihydroxyphenazine-5,10-dioxide
In: Brevibacterium A2941
Chromobacterium A958
→ γ-Glutaminyl-4-hydroxybenzene
In: Agaricus B810
→ *p*-Hydroxybenzoic acid
In: Lithospermum A3763
→ Menaquinone MK-9 (II-H_2)
In: Mycobacterium A691
→ Phenazine-1-carboxylic acid
In: Pseudomonas A958
→ *o*-Succinylbenzoic acid
Probably in: Zea A175

Shikonin (5,8-dihydroxy-2-((1*R*)-1-hydroxy-4-methyl-3-pentenyl)-1,4-naphthalenedione)
Formed from: Geranylhydroquinone
Probably formed from: Deoxyshikonin
→ O-Acetylshikonin
Probably in: Lithospermum H556
→ O-β,β-Dimethylacrylylshikonin
Probably in: Lithospermum H556
→ O-β-Hydroisovalerylshikonin
Probably in: Lithospermum H556
→ 1′-Hydroxyshikonin
In: Rat K178
→ 6-Hydroxyshikonin
In: Rat K178
→ 7-Hydroxyshikonin
In: Rat K178
→ O-Isobutyrylshikonin
Probably in: Lithospermum H556
→ O-Isovalerylshikonin
Probably in: Lithospermum H556
→ O-α-Methyl-*n*-butyrylshikonin
Probably in: Lithospermum H556

Siduron
See: 1-(2-Methylcyclohexyl)-3-phenylurea

Sildenafil (viagra, 1-(4-ethoxy-3-(6,7-dihydro-1-methyl-7-oxo-3-propyl-1*H*-pyrazolo[4,3-*d*]pyrimidin-5-yl)phenylsulphonyl)-4-methylpiperazine)
→ 1-(4-Ethoxy-3-(6,7-dihydro-3-(hydroxypropyl)-1-methyl-7-oxo-1*H*-pyrazolo[4,3-*d*]pyrimidin-5-yl)phenylsulphonyl)-4-methylpiperazine
In: Man A557
Mouse A557
Rabbit A557
Rat A557
→ 1-(4-Ethoxy-3-(6,7-dihydro-1-methyl-7-oxo-3-propyl-1*H*-pyrazolo[4,3-*d*]pyrimidin-5-yl)phenylsulphonyl)-hydroxy-4-methylpiperazine
In: Dog A557
Rat A557
→ 1-(4-Ethoxy-3-(6,7-dihydro-1-methyl-7-oxo-3-propyl-1*H*-pyrazolo[4,3-*d*]pyrimidin-5-yl)phenylsulphonyl)piperazine
In: Dog A557
Man A557
Mouse A557
Rabbit A557
Rat A557
→ 1-(4-Ethoxy-3-(6,7-dihydro-7-oxo-3-propyl-1*H*-pyrazolo[4,3-*d*]pyrimidin-5-yl)phenylsulphonyl)-4-methylpiperazine
In: Dog A557
Rat A557

Silvex (2-(2,4,5-trichlorophenoxy)propionic acid)
→ 2,4,5-Trichlorophenol
In: Microorganism A2749

Silybin (isomer unclear)
Formed from: Coniferyl alcohol
Dihydroquercetin
→ Dehydrosilybin
Probably in: Rat A2276
→ Silybin-β-D-glucuronide
In: Rat A2276
→ Silybin sulphate
In: Rat A2276

Silybin A (2-((2*R*,3*R*)-2,3-dihydro-3-(4-hydroxy-3-methoxyphenyl)-2-(hydroxymethyl)-1,4-benzodioxin-6-yl)-2,3-dihydro-3,5,7-trihydroxy-4*H*-1-benzopyran-4-one)
→ Silybin A-5-β-D-glucuronide
In: Sheep K337

→ Silybin A-7-β-D-glucuronide
In: Sheep K337
→ Silybin A-20-β-D-glucuronide
In: Man K337
Sheep K337

Silybin-β-D-glucuronide
Formed from: Silybin

Silybin A-5-β-D-glucuronide
Formed from: Silybin A

Silybin A-7-β-D-glucuronide
Formed from: Silybin A

Silybin A-20-β-D-glucuronide
Formed from: Silybin A

Silybin sulphate
Formed from: Silybin

Simaomycin α
Formed from acetate in Actinomadura F245

Sinactine (6,6a,11,14-tetrahydro-8,9-dinethoxy-12*H*-benzo[*a*]-1,3-benzodioxolo[4,5-*g*]quinolizine)
Formed from: Dehydrosinactine

Sinapaldehyde (4-hydroxy-3,5-dimethylcinnamaldehyde)
Formed from: 3,4-Dihydroxy-4-methylcinnamaldehyde
Sinapoyl CoA
Sinapyl alcohol
Probably formed from: Sinapic acid
→ Sinapic acid
In: Pseudomonas J707
See also: C447
→ Sinapyl alcohol
In: Aralia G815
Glycine A2325, B922
Mung bean J910
Populus D54
Spruce B922
Wheat H510
See also: A2744

Sinapic acid (4-hydroxy-3,5-dimethylcinnamic acid)
Formed from: 5-Methoxycaffeic acid
Methyl sinapate
Sinapaldehyde
Sinapine
Sinapoyl-β-D-glucoside
Probably formed from: 3,4,5-Trimethoxy-cinnamic acid
→ Dihydrosinapic acid
In: Microorganism A744, D894
→ 4-Hydroxy-3,5-dimethylphenylhydracrylic acid
Probably in: Rhodotorula E262
Schizophyllum C447
Trametes A2744
→ 4-Hydroxy-3,5-dimethoxystyrene
In: Brettanomyces E263
Microorganism F390
→ 5-Methoxycaffeic acid
In: Microorganism A261
→ Sinapaldehyde
Probably in: Trametes A2744
→ Sinapic acid-4-β-D-glucoside
In: Brassica J712
Nicotiana B442
Rose K809
Tomato B846
→ Sinapoyl CoA
In: Glycine A2148, A2514
Leek A2514
Petunia A2845
Phaseolus A2514
→ Sinapoyl-β-D-glucose
In: Chenopodium G255
Quercus K737
Raphanus B440, B599, K786
Tomato B846
→ 3,4,5-Trihydroxycinnamic acid
Probably in: Clostridium G367

Sinapic acid choline ester
See: Sinapine

Sinapic acid-β-D-glucoside
Formed from: Sinapic acid

Sinapine (Sinapic acid choline ester)
Formed from: Sinapoyl-β-D-glucose
→ Sinapic acid
In: Raphanus B263

Sinapinic acid (misprint for sinapic acid?)
→ Radical
In: Coriolus K677

Sinapoyl CoA
Formed from: Sinapic acid

5-Methoxycaffeoyl CoA
→ 5-Methoxyconiferaldehyde
In: Glycine A2576
→ Sinapaldehyde
In: Eucalyptus H330
Glycine B922
Populus D54
Spruce B922
→ Sinapoylgalactarate
In: Secale K722
→ Sinapoylglucarate
In: Secale K722
→ Sinapoylglucarolactone
In: Secale K722
→ Sinapoylquinic acid
In: Secale K722
→ Sinapoylshikimic acid
In: Secale K722
→ Sinapoyltyramine
In: Nicotiana D558

Sinapoylgalactarate
Formed from: Sinapoyl CoA

Sinapoylglucarate
Formed from: Sinapoyl CoA

Sinapoylglucarolactone
Formed from: Sinapoyl CoA

Sinapoyl-β-D-glucoside
Formed from: Sinapic acid
→ 1,2-Di-O-sinapoyl-β-D-glucoside
In: Raphanus K724
→ 5‴-Methoxycelosianin II
In: Chenopodium G255
→ Sinapic acid
In: Raphanus B599, B881, K724, K787
→ Sinapine
In: Raphanus K720
Sinapis K720

Sinapoylquinic acid
Formed from: Sinapoyl CoA

Sinapoylshikimic acid
Formed from: Sinapoyl CoA

Sinapoyltyramine
Formed from: Sinapoyl CoA
Tyramine

Sinapyl alcohol (4-hydroxy-3,5-dimethylcinnamyl alcohol)
Formed from: Sinapaldehyde
→ Sinapaldehyde
In: Aralia G815
Glycine A2325
See also: C447
→ Sinapyl alcohol-4-β-D-glucoside
In: Forsythia A3365
Rose K809

Sinapyl alcohol-4-β-D-glucoside
Formed from: Sinapyl alcohol

Sirdalud (5-chloro-4-(2-imidazolin-2-ylamino)-2,1,3-benzothiadiazole)
→ 7-(N-Acetyl-S-(L-cysteinyl))sirdalud
In: Dog F487
Man F487
Mouse F487
Rabbit F487
Rat F487
→ 5-Chloro-4-guanidino-2,1,3-benzothiadiazole
In: Dog F487
Man F487
Mouse F487
Rabbit F487
Rat F487
→ 4′,5′-Dehydrosirdalud
In: Dog F487
Man F487
Mouse F487
Rabbit F487
Rat F487
→ 1,3-Dihydrosirdalud-2,2-dioxide
In: Dog F487
Man F487
Mouse F487
Rabbit F487
Rat F487
→ 4′-Hydroxysirdalud
In: Dog F487
Man F487
Mouse F487
Rabbit F487
Rat F487
→ 7-Hydroxysirdalud
In: Dog F487
Man F487
Mouse F487

Rabbit F487
Rat F487

Skatole (3-methylindole)
Formed from: L-Tryptophan
→ 3-(N-Acetyl-S-(L-cysteinyl)methyl)indole
In: Goat G702
Mouse G702
→ *o*-Formamidiacetophenone
In: Wheat C559
See also: J780
→ 3-(S-Glutathionyl)methylindole
In: Goat D829
→ 3-Hydroxy-3-methylindolenine
In: Man J730
See also: K434
→ 4-Hydroxyskatole
Probably in: Goat G702
→ 5-Hydroxyskatole
In: Pig J780
See also: G702
→ 6-Hydroxyskatole
In: Pig J780
→ Indole-3-methanol
In: Pig J780
Pseudomonas A3734
See also: A3009, G702
→ 3-Methyleneindolenine
In: Man J730
See also: E656
→ 3-Methyloxindole
In: Microorganism G330
Pig J780
See also: G702

SKF 525A
See: 2-Diethylaminoethyl 2,2-diphenylvalerate

Skimmianine (4,7,8-trimethoxyfuro[2,3-*b*]quinoline, β-fagarine)
Formed from: Dictamnine
4-Methoxy-3-prenylquinol-2-one

Skimmin (7-(β-D-glucopyranosyloxy)-2*H*-1-benzopyran-2-one)
Formed from: Umbelliferone

Skyrin (2,2′,4,4′,5,5′-hexahydroxy-7,7′-dimethyl-1,1′-bianthraquinone)
Formed from: 2-Acetyl-1,8-dihydroxy-3-(2-oxopropyl)naphthalene
Catenarin
Emodin
Emodinanthrone
Iridoskyrin

Solan (*m*-chloro-2-methyl-*p*-valerotoluidide)
→ 3-Chloro-4-methylaniline
Bacillus A1006
Fusarium B237

Sophoricoside (3-(4-(β-D-glucopyranosyloxy)phenyl)-5,7-dihydroxy-4*H*-1-benzopyran-4-one)
→ 7-O-Methylsophoricoside
In: Prunus G982

Sophorol ((+)-2,3-dihydro-7-hydroxy-3-(6-hydroxy-1,3-benzodioxol-5-yl)-4*H*-1-benzopyran-1-one)
→ Maackiain
Probably in: Pisum F860

Sparfloxacin (*cis*-5-amino-1-cyclopropyl-7-(3,5-dimethyl-1-piperazinyl)-6,8-difluoro-1,4-dihydro-4-oxo-3-quinolinecarboxylic acid)
→ Sparfloxacin-β-D-glucuronide
In: Man H764
Rat G334

Sparfloxacin-β-D-glucuronide
Formed from: Sparfloxacin

Sparsiflorine ((6a*S*)-5,6,6a,7-tetrahydro-2-methoxy-4*H*-dibenzo[*de*,*g*]quinoline-1,10-diol)
Formed from: Coclaurine

Speciociliatine ((α*E*,2*S*,3*S*,12b*R*)-3-ethyl-1,2,3,4,6,7,12,12b-octahydro-8-methoxy-α-(methoxymethylene)indolo[2,3-*a*]quinolizine-2-acetic acid)
Formed from: Strictosidine

***trans*-Sphagnum acid** ((2*E*)-3-(4-hydroxyphenyl)-2-pentenedioic acid)
Formed from: *p*-Coumaric acid

***Spiro*(9*H*-fluorene-3′,9-piperidine)-2′,6′-dione**
Probably formed from: 6′-Hydroxy*spiro*(9*H*-fluorene-3′,9-piperidine)-2′-one

Sporopollenin
Formed from: L-Phenylalanine

Staurosporine ((9*S*,10*R*,11*R*,13*R*)-2,3,10,11,12,13-hexahydro-10-methoxy-9-methyl-11-(methylamino)-9,13-epoxy-1*H*,9*H*-

diindolo[1,2,3-*gh*:3′,2′,1′-*lm*]pyrrolo[3,4-*j*][1,7]benzodiazonin-1-one)
Formed from: L-Tryptophan

Stearyl gallate
→ Stearyl gallate-O-sulphate
In: Eubacterium G475

Stearyl gallate-O-sulphate
Formed from: Stearyl gallate

Steffimycin (4-((6-deoxy-2-O-methyl-α-L-mannopyranosyl)oxy)-3,4-dihydro-2,5,7-trihydroxy-3,9-dimethoxy-2-methyl-1,6,11(2*H*)-naphthacenetrione)
→ 7-Deoxysteffimycinone
In: Aeromonas A2670, A2754, A2756
Citrobacter A2670, A2756
Escherichia A2670, A2756

Steffimycin B (4-((6-deoxy-2,4-di-O-methyl-α-L-mannopyranosyl)oxy)-3,4-dihydro-2,5,7-trihydroxy-3,9-dimethoxy-2-methyl-1,6,11(2*H*)-naphthacenetrione)
→ 7-Deoxysteffimycinone B
In: Aeromonas A2670

Steffimycinol (7,8,9,10-tetrahydro-1,7,8,10,11-pentahydroxy-3,9-dimethoxy-8-methyl-5,12-naphthacenedione)
Formed from: Steffimycinone
→ 7-Deoxysteffimycinol
In: Aeromonas A2759

Steffimycinone (3,4-dihydro-2,4,5,7-tetrahydroxy-3,9-dimethoxy-2-methyl-1,6,11(2*H*)-naphthacenetrione)
→ 7-Deoxysteffimycinone
In: Aeromonas A2756
Citrobacter A2670
→ Steffimycinol
In: Streptomyces A2759, B571

Sterigmatocystin ((3a*R*,12c*S*)-3a,12c-dihydro-8-hydroxy-6-methoxy-7*H*-furo[3′,2′:4,5]furo[3,2-*b*]xanthen-7-one)
Formed from: Demethylsterigmatocystin
→ Aflatoxin B_1
In: Aspergillus A917, A2594, C888, E544, E691, G154, F398, K249
→ Aflatoxin B_2
In: Aspergillus D745
→ Aflatoxin G_1
In: Aspergillus E544, F398
→ O-Methylsterigmatocystin
In: Aspergillus C888, D745, E373, E419, E512, E763, E964, F79, F281, F398, F782
See also: E767
→ Sterigmatocystin-β-D-glucuronide
In: Monkey A796, A799

Sterigmatocystin-β-D-glucuronide
Formed from: Sterigmatocystin

Sterosan
See: 5,7-Dichloro-8-hydroxy-2-methylquinoline

***cis*-Stilbene** (1,1′-(1,2-ethenediyl)bisbenzene)
Formed from: *cis*-Stilbene oxide
→ *meso*-Hydrobenzoin
In: Rabbit A78
→ *cis*-Stilbene oxide
In: Rabbit A78, A86

***trans*-Stilbene**
Formed from: *trans*-Stilbene oxide
→ 4-Hydroxy-*trans*-stilbene
In: Aspergillus A1218
Cunninghamella A1218
Rabbit A1991
Rat A1421, A1991
Rhizopus A1218
Streptomyces A1218

***cis*-Stilbene oxide**
Formed from: *cis*-Stilbene
→ 2-(S-Glutathionyl)-1-hydroxydibenzyl
Probably in: Sheep A2282
→ D-*threo*-Hydrobenzoin
In: Man G195
Rabbit A78, A86, F333
→ *cis*-Stilbene
In: Bifidobacterium K25
Clostridium K25
Escherichia K25
Pseudomonas K25
Salmonella K25
Streptococcus K25

***trans*-Stilbene oxide**
→ *meso*-Hydrobenzoin
In: Man C386, G195
Mouse B648, C861, D643, E56, E826
Rat C386, E759, G195
See also: D609

→ *trans*-Stilbene
In: Bifidobacterium K25
Clostridium K25
Escherichia K25
Pseudomonas K25
Salmonella K25
Streptococcus K25

4-Stilbenyl nitroxide radical
Formed from: 4-Aminostilbene

Stilboestrol (α,α′-diethylstilbenediol, diethylstilboestrol, 4,4′-((1*E*)-1,2-diethyl-1,2-ethenediyl)bisphenol)
→ Dienoestrol (several isomers)
In: Chimpanzee A3307
Hamster B194, E646, E801
Horseradish A3419, D166
Mouse A2592, A3419, D166
Rhesus monkey A3307
→ 1-Hydroxydienoestrol
In: Chimpanzee A3307
Hamster B194
Mouse A2592
Rhesus monkey A3307
→ 3-(*p*-Hydroxyphenyl)hex-2-en-4-one
In: Horseradish D166
Mouse D166
→ *p*-Hydroxypropiophenone
In: Horseradish D166
Mouse D166
→ *cis*-ω-Hydroxypseudostilboestrol
In: Mouse A2592
→ *trans*-ω-Hydroxypseudostilboestrol
In: Mouse A2592
See also: B194
→ 1-Hydroxystilboestrol
In: Chimpanzee A3307
Hamster B194, E646, E801
Mouse A2592
Rhesus monkey A3307
→ 3-Hydroxystilboestrol
In: Hamster B194, E801, E646
Rat A3295, B324
See also: B6, D514
→ Indenoestrol A
In: Hamster B194
→ 4′-O-Methylstilboestrol
In: Mouse C725
→ Stilboestrol-4-β-D-glucuronide
In: Cat C562
Dog C562
Guinea pig C562
Hen A399
Man J783
Mouse A1534, C562
Quail C562
Rabbit C562
Rat A1977, A3479, C562
Sheep A455
Trout C562
→ Stilboestrol quinone
In: Hamster H69
Horseradish D892
Rat H69
→ Stilboestrol-4-sulphate
In: Beef A1211
Hen A399
Man A2086, K344
Monkey A2086
Rat A2086

Stilboestrol-4,4′-disulphate
→ Stilboestrol-4-sulphate
In: Rat A3519

Stilboestrol epoxide
→ *p*-Hydroxypropiophenone
In: Beef D100
Rat B100

Stilboestrol-4-β-D-glucuronide
Formed from: Stilboestrol
→ Stilboestrol
In: Microorganism A866

Stilboestrol-4-β-D-glucuronide-4′-sulphate
Formed from: Stilboestrol-4-sulphate

Stilboestrol quinone
Formed from: Stilboestrol

Stilboestrol-4-sulphate
Formed from: Stilboestrol
Stilboestrol-4,4′-disulphate
→ Stilboestrol
In: Rat A3519
→ Stilboestrol-4-β-D-glucuronide-4′-sulphate

In: Rat A3519
See also: A399

Stiripentol (3-hydroxy-4,4-dimethyl-1-(3,4-methylenedioxyphenyl)-1-pentene; assumed to be *R*)
→ 3,5-Dihydroxy-4,4-dimethyl-1-(3,4-methylenedioxyphenyl)-1-pentene
In: Rat F574
→ 1-(3,4-Dihydroxyphenyl)-3-hydroxy-4,4-dimethyl-1-pentene
In: Man D264, E852
Rat F574, G831
→ (*S*)-Stiripentol
In: Rat G940, G942

(*S*)-Stiripentol
Formed from: (*R*)-Stiripentol
Does not yield (*R*)-stiripentol in rat G940, G942

Stizolobic acid
Formed from: 3,4-Dihydroxy-L-phenylalanine

Stobadine ((−)-*cis*-2,8-dimethyl-2,3,4,4a,5,9b-hexahydro-1*H*-pyrido[4,3-*b*]indole)
→ Demethylstobadine
In: Rat E802, F348
→ Stobadine-N-oxide
In: Rat E802, F326, F348

Stobadine-N-oxide
Formed from: Stobadine

Strictosamide (isovincoside lactam, (15β,16α,17β)-19,20-didehydro-16-ethenyl-17-(β-D-glucopyranosyloxy)oxayohimban-21-one)
Probably formed from: Strictosidine
→ Camptothecin
In: Camptotheca B86

Strictosidine (isovincoside)
Formed from: Tryptamine
L-Tryptophan
→ Ajmalicine
In: Catharanthus A3142, A3153, A3257, A3861, B270
See also: D20
→ Ajmaline
In: Rauwolfia B270
→ Aricine
In: Rauwolfia B270
→ Camptothecin
In: Camptotheca B87
→ Catharanthine
In: Catharanthus A3142, A3257, A3861, B270
→ Cathenamine
Probably in: Catharanthus A3762
→ 4,21-Dehydrocorynantheine
In: Catharanthus A3151
→ 10-Deoxysarpagine
In: Rauwolfia H676
→ Isoreserpiline
In: Rauwolfia B270
→ Isoreserpinine
In: Rauwolfia B270
→ Mitragynine
In: Rauwolfia A3153, B270
→ Reserpiline
In: Rauwolfia A3153, B270
→ Serpentine
In: Catharanthus A3257
→ Speciociliatine
In: Mitragyna A3153, B270
→ Strictosamide
Probably in: Camptotheca B86
→ Strictosidine aglycone
In: Amsonia K868
Catharanthus K428, K868
Strychnos K428
Tabernaemontana H571
→ Tetrahydroalstonine
In: Catharanthus A3142
→ Vindoline
In: Catharanthus A3142, A3257, A3861, B270
→ α-Yohimbine
In: Rauwolfia A3153, B270

Strictosidine aglycone
Formed from: Strictosidine

Strychnine (strychnidin-10-one)
Formed from: Strychnine-N-oxide
→ 2-Hydroxystrychnine
In: Dog G191
Guinea pig G191
Mouse G191
Rabbit D830, G191
Rat F329, G191
Rodent F475
→ 16-Hydroxystrychnine

In: Rat F329, G191, F475
See also: G927
→ 22-Hydroxystrychnine
In: Guinea pig F911
Rat G191
→ 18-Oxostrychnine
In: Rodent F475
→ Strychnine-N-oxide
In: Rabbit D830
Rat F329, G191
Rodent F475
→ Strychnine-21,22-oxide
In: Rabbit D830
Rat F329, G191
Rodent F475

Strychnine-21,22-oxide
Formed from: Strychnine
→ 21,22-Dihydro-21α,22α-dihydroxystrychnine
Probably in: Rat F329
→ 21,22-Dihydro-21α,22β-dihydroxystrychnine
Probably in: Rat F329

Strychnine-N-oxide
Formed from: Strychnine
→ Strychnine
In: Microorganism G927

Stylopine ((12b*S*)-6,7,12b,13-tetrahydro-4*H*-bis[1,3]benzodioxolo[5,6-*a*:4′,5′-*g*]quinolizine)
Formed from: (*S*)-Cheilanthifoline
→ N-Methylstylopine
In: Berberis E299
Corydalis E299
Dicentra E299
Eschscholtzia K782
Fumaria E299
Papaver E299

Styrene (ethenylbenzene)
Formed from: Cinnamic acid
Styrene oxide
Probably formed from: Ethylbenzene
→ Benzaldehyde
In: Coprinus K431
Horseradish E308
Man K431
See also: K612
→ 2-Bromo-1-phenylethanol
In: Corallina E597
→ *cis*-2,3-Dihydro-2,3-dihydroxystyrene
In: Pseudomonas J656
See also: H253
→ (*R*)-S-(2-Hydroxy-1-phenylethyl)glutathione
In: Horseradish E80
→ (*S*)-S-(2-Hydroxy-1-phenylethyl)glutathione
In: Horseradish E80
→ (*R*)-S-(2-Hydroxy-2-phenylethyl)glutathione
In: Horseradish E80
See also: E205
→ (*S*)-S-(2-Hydroxy-2-phenylethyl)glutathione
In: Horseradish E80
See also: E205
→ *p*-Hydroxystyrene
In: Rat A3833, D295
→ (*R*)-Mandelic acid
In: Rat F930
→ (*S*)-Mandelic acid
In: Man K612
Rat F930
→ Phenylacetic acid
In: Pseudomonas C532
Rat B196
→ 1-Phenylethanol
In: Pseudomonas F870
→ 2-Phenylethanol
In: Pseudomonas F870
→ Phenylglyoxylic acid
In: Man K612
Rat F930
→ Salicylic acid
In: Pseudomonas F870
See also: K612
→ Styrene oxide
In: Coprinus K431
Corynebacterium H881
Flounder D387
Horseradish E308
Man D69, G735, K431, K489, K612
Methylococcus F278

Microorganism H659
Mouse G765
Nitrosomonas H219
Pseudomonas H369, K167
Rat A1796, A3624, G765
See also: A569, A688, A2693, A3059, A3822, A3833, B282, B714, D48, D499, F330, G638, H825, J164, J219, K368
→ 2-Vinylmuconic acid
In: Rhodococcus H253

Styrene episulphide
→ 2-(S-Glutathionyl)-2-phenylethanethiol
Probably in: Sheep A2282

Styrene oxide
Formed from: Styrene
→ 2-(S-Glutathionyl)-1-phenylethanol
In: Man E928
Rat C194, D842, F133
See also: A3059, B714, C605, D387, G636, K638
→ 2-(S-Glutathionyl)-2-phenylethanol
In: Man E928
Rabbit A1713
Rat A1713, A2093, C194, D293, D842, F133
See also: A143, A1694, A1964, A2132, A2282, A2509, A2974, A3059, A3461, A3790, B714, C4, C21, C605, D387, K368
→ Phenylacetaldehyde
In: Corynebacterium H881, J517
Microorganism H659
Pseudomonas H369
Xanthobacter F440
See also: K368
→ Phenylglycol
In: Bass B69
Black drum B69
Cat C562
Clam B69
Crab B69
Dog C562
Dogfish B69
Eel B69
Flounder B69
Guinea pig C562, C605
Jack crevalle B69
King of Norway B69
Lobster B69
Man A79, A3217, B63, B220, B942, C4, C386, E969, K662
Mangrove snapper B69
Mouse A2006, C562, E826, G765, K368, K592
Mullet B69
Nitrosomonas H219
Pseudomonas G429
Quail C562
Rabbit A1694, A2132, A2509, C562
Rat A1713, A1796, A1894, A1964, A2093, A3031, A3421, B63, B142, B192, B220, C194, C314, C386, C562, C605, E759, G765
Ray B69
Redfish B69
Shark B69
Sheepshead B69
Skate B69
Streptomyces K394
Teleost A1694
Trout C562
See also: A569, A3489, A3624, A3833, D69
→ Styrene
In: Bifidobacterium K25
Clostridium K25
Escherichia K25
Lactobacillus K25
Pseudomonas K25
Salmonella K25
Streptomyces K25
→ *o*-Tyrosol
In: Guinea pig C605
Rat C605
→ *p*-Tyrosol
In: Guinea pig C605
Mouse K368
Rat C605

2-Succinamidofluorene
→ 2-Aminofluorene
In: Guinea pig A808
Mouse A808
Rabbit A808
Rat A808

Succinyl-L-alanyl-L-alanyl-L-alanyl-*p*-nitroanilide
→ L-Alanyl-*p*-nitroanilide
In: Rat C693

Succinyl-β-alanyl-L-alanyl-L-alanyl-*p*-nitroanilide
→ L-Alanyl-*p*-nitroanilide
In: Rat C693
→ *p*-Nitroaniline
In: Guinea pig D230
Man B946, C178, C213, C391

Succinyl-L-alanyl-L-alanyl-L-isoleucyl-*p*-nitroanilide
→ L-Isoleucyl-*p*-nitroanilide
In: Rat C693

Succinyl-L-alanyl-L-alanyl-L-leucyl-*p*-nitroanilide
→ L-Leucyl-*p*-nitroanilide
In: Rat C693

Succinyl-L-alanyl-L-alanyl-L-prolyl-*p*-nitroanilide
→ L-Prolyl-*p*-nitroanilide
In: Rat C693

Succinyl-L-alanyl-L-alanyl-L-valyl-*p*-nitroanilide
→ L-Valyl-*p*-nitroanilide
In: Rat C693

Succinyl-L-alanyl-glycyl-L-alanyl-*p*-nitroanilide
→ L-Alanyl-*p*-nitroanilide
In: Rat C693

Succinyl-L-alanyl-L-leucyl-L-alanyl-*p*-nitroanilide
→ L-Alanyl-*p*-nitroanilide
In: Rat C693

Succinyl-L-asparaginyl-L-alanyl-L-alanyl-L-alanyl-*p*-nitroanilide
→ L-Alanyl-*p*-nitroanilide
In: Rat C693

***o*-Succinylbenzoic acid** (4-(2′-carboxyphenyl)-4-oxobutyric acid)
Formed from: Isochorismic acid
Probably formed from: 2-Hydroxy-6-succinyl-2,4-cyclohexadiene-1-carboxylic acid
Shikimic acid
→ 2-Carboxy-4-oxo-2-prenyl-1-tetralone
In: Catalpa B518
→ 2-(3-Carboxypropionyl)benzoyl CoA
In: Mycobacterium E594
→ Catalpalactone
In: Catalpa B48
→ Catalponone
In: Catalpa B48, B518
→ 1,4-Dihydroxy-2-naphthic acid
In: Escherichia A2774
→ 3-Hydroxydehydro-iso-α-lapachone
In: Catalpa B48
→ 2-Hydroxy-1,4-naphthoquinone
In: Impatiens A1639
See also: A174, A3032
→ 5-Hydroxy-1,4-naphthoquinone
In: Juglans A1639
→ α-Lapachone
In: Catalpa B48
→ 1,4-Naphthoquinone
In: Juglans A3470
→ 4-Oxo-α-tetralone
In: Juglans A3470
→ *o*-Succinylbenzoyl CoA
In: Escherichia E594
Galium E594
Mycobacterium B916, C157, C337, E594, K744
→ Vitamin K_1
In: Zea A175, A3973

***o*-Succinylbenzoyl CoA**
Formed from: *o*-Succinylbenzoic acid
→ 1,4-Dihydroxy-2-naphthoic acid
In: Escherichia C337, E594
Galium E594
Mycobacterium B916, C337, E594, K744
Contrast (wrong identity) D710
→ *o*-Succinylbenzoyl di-CoA
In: Escherichia E594
Galium E594
Mycobacterium E594

***o*-Succinylbenzoyl di-CoA**
Probably formed from: *o*-Succinylbenzoyl CoA

Succinyl-glycyl-L-alanyl-L-alanyl-*p*-nitroanilide
→ L-Alanyl-*p*-nitroanilide
In: Rat C693

Succinyl-L-leucyl-L-alanyl-L-alanyl-*p*-nitroanilide
→ L-Alanyl-*p*-nitroanilide
In: Rat C693

N-Succinylmofegiline
Formed from: Mofegiline

Sudan 1
See: 1-Phenylazo-2-naphthol

Sudoxicam (4-hydroxy-2-methyl-N-(2-thiazolyl)-2*H*-1,2-benzothiazine-3-carboxamide-1,1-dioxide)
→ 3-(N′-Carboxymethyl)formylureido-4-hydroxy-2-methyl-2*H*-1,2-benzothiazine-1,1-dioxide
In: Dog A2137
Monkey A2137
Rat A2137

Sufentanil (N-(4-methoxymethyl)-(1-(2-(2-thienyl)ethyl)-4-piperidinyl)-N-phenylpropanamide)
→ Demethylsufentanil
In: Dog F567
Man F567
Rat E895, F567
→ N-(4-(Methoxymethyl)-4-piperidinyl)-N-phenylpropanamide
In: Dog E895, F567
Man F567
Rat E895, F567
→ Propionanilide
In: Dog F567
Man F567
Rat F567
→ *p*-Propionamidophenol
In: Dog E895
Rat E895

Sulfurmycin A
Formed from: Sulfurmycin B
Sulfurmycinone

Sulfurmycin B
Formed from: Sulfurmycinone
→ Sulfurmycin A
In: Streptomyces D17

Sulfurmycinone ((1*R*,2*S*,4*S*)-1,2,3,4,6,11-hexahydro-2,4,5,7-tetrahydroxy-6,11-dioxo-2-(2-oxopropyl)-1-naphthacenecarboxylic acid methyl ester)
→ Sulfurmycin A
In: Streptomyces D170
→ Sulfurmycin B
In: Streptomyces D170

Sulindac
See: (1*Z*)-5-Fluoro-2-methyl-1-((4-(methylsulphinyl)phenyl)methylene)-1*H*-indene-3-acetic acid

Sulindac sulphide
→ *cis*-5-Fluoro-2-methyl-1-((*p*-methylsulphinyl)benzylidenyl)indene-3-acetic acid
In: Man K508

Sulmet
See: Sulphamethazine

Sulochrin (2-(2,6-dihydroxy-6-methylbenzoyl)-5-hydroxy-3-methoxybenzoic acid methyl ester)
Formed from: Questin
→ Bisdechlorogeodin
In: Aspergillus C556

Suloctidil (1-(4-isopropylthiophenyl)-2-N-octylaminopropanol)
→ 1-(4-Isopropylsulphonylphenyl)-2-aminopropanol
In: Man A2910
Contrast (artifact) B255

Sulofenur (N-(((4-chlorophenyl)amino)carbonyl)-2,3-dihydro-1*H*-indene-5-sulphonamide)
→ *p*-Chloroaniline
In: Monkey F731
Mouse F731
Rat F731
→ N-(*p*-Chlorophenyl)-N′-(indene-5-sulphonyl)urea
In: Man G514
Monkey G514
Rat G514
→ N-(*p*-Chlorophenyl)-N′-(indene-6-sulphonyl)urea
In: Man G514
Monkey G514
Rat G514
→ 1-Hydroxysulofenur
In: Man F731
Monkey F731
Mouse F731
Rat F731, G898
→ 3-Hydroxysulofenur
In: Mouse F731
Rat F731

Sulphadiazine (2-sulphanilamidopyrimidine)
→ N^4-Acetylsulphadiazine
In: Beef E734, G913
Goat A43, A362
Horse F928

Pig D873, F918
Rat A3672
See also: A1988, B110, B201
→ 4-Amino-N-(4-hydroxy-2-pyrimidinyl)benzenesulphonamide
In: Monkey H637
Pig F918
→ 4-Hydroxy-2-sulphanilamidopyrimidine
In: Beef E734
Goat A362
Horse F928
Pig D873
Rat A3672
→ 5-Hydroxy-2-sulphanilamidopyrimidine
In: Horse F928
Rat A3672
Rhesus monkey H637
→ Sulphadiazine-N^1-β-D-glucuronide
In: Marmoset B201
Rat B201
→ Sulphadiazine-N^4-β-D-glucuronide
In: Marmoset B201
Rabbit B201
Rat A3672, B201
→ Sulphadiazine-N^4-sulphate
In: Marmoset B201
Rabbit B201
Rat B201

Sulphadiazine-N^1-β-D-glucuronide
Formed from: Sulphadiazine

Sulphadiazine-N^4-β-D-glucuronide
Formed from: Sulphadiazine

Sulphadiazine-N^4-sulphate
Formed from: Sulphadiazine

Sulphadimethoxine (N^1-(2,6-dimethoxy-4-pyrimidinyl)sulphanilamide)
→ N^4-Acetylsulphadimethoxine
In: Cat A2387
Cepaea F354
Chicken H352
Civet A1312, A2387
Ferret A2387
Genet A1312, A2387
Lion A1312
Lobster E708
Man F344
Pig G734
Turtle F541
→ N^1-(2-Hydroxy-6-methoxy-4-pyrimidinyl)sulphanilamide
In: Cepaea F354
Hen H352
Turtle F541
→ N^1-(6-Hydroxy-2-methoxy-4-pyrimidinyl)sulphanilamide
In: Cepaea F354
Hen H352, H356
Turtle F541
→ 3-Hydroxysulphadimethoxine
In: Pig G734
→ Sulphadimethoxine-N-β-D-glucoside
In: Pig G734
→ Sulphadimethoxine-N^1-β-D-glucuronide
In: Dog A2387
Hen H352
Man F344
Hyaena A2387
→ Sulphadimethoxine-N^4-β-D-glucuronide
In: Guinea pig A1418
Hyaena A2387
See also: H734

Sulphadimethoxine-N-β-D-glucoside
Formed from: Sulphadimethoxine

Sulphadimethoxine-N^1-β-D-glucuronide
Formed from: Sulphadimethoxine

Sulphadimethoxine-N^4-β-D-glucuronide
Formed from: Sulphadimethoxine

Sulphadimidine
See: Sulphamerazine

Sulphadoxine (5,6-dimethoxy-4-sulphanilamidopyrimidine)
→ N^4-Acetylsulphadoxine
In: Goat A1119

Sulphamerazine (sulphadimidine, sulphamezathine, N^1-(4-methyl-2-pyrimidinyl)sulphanilamide)
→ N^4-Acetylsulphamerazine
In: Beef E734
Horse F928
Pig F918
→ 4-Amino-N-(4-hydroxymethyl-2-pyrimidinyl)benzenesulphonamide
In: Beef E734
Horse F928

Pig F918
→ 4-Amino-N-(4-hydroxy-6-methyl-2-pyrimidinyl)benzenesulphonamide
In: Pig F918
→ 5-Hydroxysulphamerazine
In: Horse F928
→ Sulphamerazine-N^1-β-D-glucoside
In: Man F6

Sulphamerazine-N^1-β-D-glucoside
Formed from: Sulphamerazine

Sulphamethazine (sulphadimidine, 4-amino-N-(4,6-dimethyl-2-pyrimidinyl)benzenesulphonamide)
→ N^4-Acetylsulphamethazine
In: Beef D869, E734
Goat A1119
Horse D786, F928
Man F70, F954, J 507
Mouse J405
Pig B620, D846, F918
Rabbit A1037
Rat D284, F506, G575
Sheep J348
See also: A869, A2562, A2909, A3119, A3253, A3857, B110, B700, C302
→ 4-Amino-N-(6-hydroxymethyl-4-methyl-2-pyrimidinyl)benzenesulphonamide
In: Beef E734
Horse F928
→ 4-Amino-N-(2-(4-hydroxymethyl-6-methyl)pyrimidinyl)benzenesulphonamide
In: Horse D786
→ N-(4,6-Dimethyl-2-pyrimidinyl)benzenesulphonamide
In: Pig B620, B718, D846
Rat E243, F506
→ N-(4-Hydroxymethyl-6-methyl-2-pyrimidinyl)sulphanilamide
In: Beef D869
Rat G575
→ 4-Hydroxysulphamethazine
In: Rhesus monkey H637
→ 5-Hydroxysulphamethazine
In: Beef D869
Horse D786, F928
Rat G575
Rhesus monkey H637
→ Sulphamethazine-N^1-β-D-glucoside
In: Man F6
→ Sulphamethazine-N^4-β-D-glucoside
In: Pig B620, D846
Rat D284, F506
→ Sulphamethazine-N^4-lactoside
In: Beef G373

Sulphamethazine-N^1-β-D-glucoside
Formed from: Sulphamethazine

Sulphamethazine-N^4-β-D-glucoside
Formed from: Sulphamethazine

Sulphamethazine-N^4-lactoside
Formed from: Sulphamethazine

Sulphamethomidine (N^1-(6-methoxy-2-methyl-4-pyrimidinyl)sulphanilamide)
→ N^4-Acetylsulphamethomidine
In: Man A601
Pseudemys F553
Rabbit A601
→ O-Demethylsulphamethomidine
In: Pseudemys F553
→ Sulphamethomidine-N^1-β-D-glucuronide
In: Man A601
Rabbit A601
→ Sulphanilamide
In: Pseudemys F553

Sulphamethomidine-N^1-β-D-glucuronide
Formed from: Sulphamethomidine

Sulphamethoxazole (sulphisomezole, sulphamethoxizole, N^1-(5-methyl-3-isoxazolyl)sulphanilamide)
Formed from: Sulphamethoxazole hydroxylamine
→ N^4-Acetylsulphamethoxazole
In: Cepaea D106, E552
Goat A1119
Man A600, H405, H502
Rat A926
Turtle D23, E553
→ N-Chlorosulphamethoxazole
With myeloperoxidase G694
→ N^1-(5-Hydroxymethyl-3-isoxazolyl)sulphanilamide
In: Cepaea E552
Man A600, H502

Turtle E553
→ N-Hydroxysulphamethoxazole
In: Man G606, H405
Rat H184
→ Sulphamethoxazole-N^1-β-D-glucoside
In: Man F6
→ Sulphamethoxazole-N^1-β-D-glucuronide
In: Man A600, H405, H502
→ Sulphamethoxazole-2′-β-D-glucuronide
In: Man A600

Sulphamethoxazole-N^1-β-D-glucoside
Formed from: Sulphamethoxazole

Sulphamethoxazole-N^1-β-D-glucuronide
Formed from: Sulphamethoxazole

Sulphamethoxazole-2′-β-D-glucuronide
Formed from: Sulphamethoxazole

Sulphamethoxazole hydroxylamine
→ Sulphamethoxazole
In: Dog K338
Man K338

Sulphamethoxizole
See: Sulphamethoxazole

Sulphamezathine
See: Sulphamerazine

Sulphamonomethoxine (4-methoxy-6-sulphanilamidopyrimidine)
→ N^4-Acetylsulphamonomethoxine
In: Cepaea F356
Man F593
→ 6-Demethylsulphamonomethoxine
In: Man F593
Pseudemys F551
→ N^1-(2-Hydroxy-6-methoxy-4-pyrimidinyl)sulphanilamide
In: Cepaea F356
Man F593
Pseudemys F551
→ Sulphamonomethoxine-N^1-β-D-glucuronide
In: Man A601, F593

Sulphamonomethoxine-N^1-β-D-glucuronide
Formed from: Sulphamonomethoxine

3-(Sulphamoylmethyl)-1,2-benzisoxazole
See: Zonisamide

***o*-Sulphamoylbenzoic acid**
See: *o*-Carboxybenzenesulphonamide

***p*-Sulphamoylbenzoic acid**
See: *p*-Carboxybenzenesulphonamide

***o*-Sulphamoylbenzyl alcohol**
Formed from: Benz[*d*]isothiazoline-1,1-dioxide
o-Toluenesulphonamide
Probably formed from: 2,3-Dihydro-3-hydroxybenz[*d*]isothiazole-1,1-dioxide

***o*-Sulphamoylmethylcarbonylphenol**
Formed from: Zonisamide
→ *o*-Sulphamoylmethylcarbonylphenyl-β-D-glucuronide
Probably in: Rat F594

***o*-Sulphamoylmethylcarbonylphenyl-β-D-glucuronide**
Probably formed from: *o*-Sulphamoylmethylcarbonylphenol

Sulphanilamide (*p*-aminobenzenesulphonamide)
Formed from: 7-Acetamido-1-hydroxy-2-(*p*-sulphamoylphenylazo)naphthalene-3,6-disulphonic acid
Sulphamethomidine
Probably formed from: Asulam
→ *p*-Acetamidobenzenesulphonamide
In: Goat A1119
Rabbit A1037
Padmelon C403
Possum C403
Rat C403
Tasmanian devil C403
See also: B110
→ N^1-Acetylsulphanilamide
In: Padmelon C403
Possum C403
Rat C403
Tasmanian devil C403

2-Sulphanilamidopyridine
See: Sulphapyridine

2-Sulphanilamidopyrimidine
See: Sulphadiazine

Sulphanilic acid (*p*-aminobenzenesulphonic acid)
Formed from: Acid Orange 6
Acid Orange 7
Acid Red 151
4-Amino-3-carboxy-5-hydroxy-1-(*p*-sulphophenyl)pyrazole

3-Carboxy-5-hydroxy-1-(*p*-sulphophenyl)-4-*p*-sulphophenylazo)pyrazole
p-(2-Hydroxy-1-naphthylazo)benzenesulphonic acid
2-Hydroxy-1-(*p*-sulphophenylazo)naphthalene-6-sulphonic acid
Methyl orange
Mordant Yellow 10
1-(*p*-Sulphophenylazo)-4-anthranol
1-(*p*-Sulphophenylazo)-4-naphthol
→ *p*-Acetamidobenzenesulphonic acid
In: Rabbit E399
→ 3,4-Dihydroxybenzenesulphonic acid
In: Hydrogenophaga G785
Microorganism E775, H808
See also: K293

Sulphapyridine (2-sulphanilamidopyridine)
Formed from: Sulphasalazine
Salicylazosulphapyridine
→ N^4-Acetylsulphapyridine
In: Microorganism E648, E807
See also: A516, G842
→ 4′-Hydroxysulphapyridine
In: Pseudemys G455
→ 5′-Hydroxysulphapyridine
In: Pseudemys G455
See also: A1023

Sulphasalazine (salazopyrin, 5-(*p*-(2-pyridylsulphamoyl)phenylazo)salicylic acid)
→ 5-Amino-2-hydroxybenzoic acid
In: Man A714
Microorganism A516, C128, E648, E807
Rat A714
See also: A1023
→ Sulphapyridine
In: Microorganism A516, C128, E648, E807
Man A714, B421
Mouse G842
Rat A714
See also: A1023

N-Sulphatoxy-4-acetamidobiphenyl
Formed from: N-Hydroxy-4-acetamidobiphenyl

4-(N-Sulphatoxyamino)biphenyl
Formed from: 4-Hydroxylaminobiphenyl

***m*-(*o*-Sulphatoxyphenoxy)benzoic acid**
Formed from: *m*-(*o*-Hydroxyphenoxy)benzoic acid

***m*-(*p*-Sulphatoxyphenoxy)benzoic acid**
Formed from: *m*-(*p*-Hydroxyphenoxy)benzoic acid

***m*-Sulphatoxyphenyl-β-D-glucuronide**
Probably formed from: *m*-Hydroxyphenyl-β-D-glucuronide

Sulphatroxazole (4-amino-N-(4,5-dimethyl-3-isoxazolyl)benzenesulphonamide)
→ N^4-Acetylsulphatroxazole
In: Beef F929
→ 3-((*p*-Aminophenyl)sulphon)amido)-5-hydroxymethyl-4-methylisoxazole
In: Beef F929

2-Sulphinylbenzimidazole
Formed from: 2-Mercaptobenzimidazole

Sulphinpyrazole (sulphinpyrazone, 1,2-diphenyl-4-(2-(phenylsulphinyl)ethyl-3,5-pyrazolidinedione)
Formed from: 1,2-Diphenyl-4-(2-(phenylthio)ethyl)-3,5-pyrazolidinedione
→ 3,5-Dioxo-1,2-diphenyl-4-(2′-phenylsulphonylethyl)pyrazolidine
In: Dog B711
Guinea pig B711
Monkey B711
Pig B711
Rabbit B711
Rat B711
→ 3,5-Dioxo-1,2-diphenyl-4-(2′-phenylthioethyl)pyrazolidine
In: Dog B711
Escherichia H560
Guinea pig B711
Man B510, B856, C561
Microorganism E241
Monkey B711
Pig B711
Rabbit B711, B856, D114, E152, H561
Rat B711, H561
→ 4-Hydroxysulphinpyrazole
In: Dog B711
Guinea pig B711

Man B510
Pig B711
Rat B711
→ *p*-Hydroxysulphinpyrazole
In: Dog B711
Guinea pig B711
Monkey B711
Pig B711
Rat B711
→ Sulphinpyrazole-4-β-D-glucuronide
In: Animal A3243
Rat A772

Sulphinpyrazole-4-β-D-glucuronide
Formed from: Sulphinpyrazole

Sulphinylmethyl-6-thiomethylbenzonitrile
Formed from: 2,6-Dichlorobenzonitrile

Sulphisomezole
See: Sulphamethoxazole

Sulphisomidine (4-amino-N-(2,6-dimethyl-4-pyrimidinyl)benzenesulphonamide)
→ N-Acetylsulphisomidine
In: Goat B77
Pseudemys F552

Sulphisoxazole (N^1-(3,4-dimethyl-5-isoxazolyl)sulphanilamide)
→ N^4-Acetylsulphisoxazole
In: Man A453
Rat D385

***o*-Sulphobenzoic acid**
Formed from: Saccharin

Sulphobromophthalein (bromosulphophthalein, BSP)
→ S-(3,3-Bis(4-hydroxy-3-sulphophenyl)-4,5,7-tribromophthalidyl)glutathione
In: Cat C562
Dog C562
Guinea pig C562
Mouse C562
Quail C562
Rabbit C562
Rat C562
Trout C562
See also: F804, G843

3-Sulphocatechol
See: 2,3-Dihydroxybenzenesulphonic acid

4-Sulphocatechol
See: 3,4-Dihydroxybenzenesulphonic acid

1-(N-(Sulphooxy)benzenepropanimidate) 1-thio-β-D-glucopyranose
See: Gluconasturtiin

1-(N-(Sulphooxy)-1*H*-indole-3-ethanimidate) 1-thio-β-D-glucopyranose
See: Glucobrassicin

1-(*p*-Sulphophenylazo)-4-anthranol
→ 4-Amino-1-hydroxyanthracene
In: Pseudomonas D180
→ Sulphanilic acid
In: Pseudomonas D180

1-(*p*-Sulphophenylazo)-4-naphthol
→ 1-Amino-4-naphthol
In: Pseudomonas D180
→ Sulphanilic acid
In: Pseudomonas D180

***p*-Sulphophenylhydrazine**
Formed from: 3-Carboxy-5-hydroxy-1-(*p*-sulphophenyl)-4-(*p*-sulphophenylazo)pyrazole

Sulphoridazine (thioridazine-2-sulphone)
Formed from: Mesoridazine
Thioridazine
→ Demethylsulphoridazine
In: Man E342
→ 7-Hydroxysulphoridazine
In: Dog G539
Man E342, G539
→ 6′-Oxosulphoridazine
In: Dog G539
Man G539
Rat G69
→ Sulphoridazine-N-oxide
In: Dog G539
Man G539
→ Sulphoridazine sulphoxide
In: Dog G539
Man G539, E342
Rat G69
See also: G538, G705, D366

Sulphoridazine-N-oxide
Formed from: Sulphoridazine

Sulphoridazine sulphoxide
Formed from: Sulphoridazine
→ Demethylsulphoridazine sulphoxide
Probably in: Dog G705
Man G705

Rat G69, G705
→ 2′-Oxosulphoridazine sulphoxide
Probably in: Man E342

5-Sulphosalicylic acid (2-hydroxy-5-sulphobenzoic acid)
Formed from: Naphthalene-1,6-disulphonic acid
Naphthalene-2,6-disulphonic acid
→ 3,4-Dihydroxybenzenesulphonic acid
Probably in: Pseudomonas A3066
→ Gentisic acid
In: Pseudomonas E980

Sulpiride (N-((1-ethyl-2-pyrrolidinyl)methyl)-5-sulphamoyl-2-anisamide)
→ N-((1-Ethyl-5-oxo-2-pyrrolidinyl)methyl)-5-sulphamoyl-2-anisamide
In: Monkey A1976, A3590, C463
Rat A3564
→ N-((1-Ethyl-2-pyrrolidinyl)methyl)-2-hydroxy-5-sulphamoylbenzamide
In: Rat A3564
→ N-((2-Pyrrolidinyl)methyl)-5-sulphamoyl-2-anisidide
In: Rat A3564

Sulprostone (7-(3-hydroxy-2-(3-hydroxy-4-phenoxy-1-butenyl)-5-oxocyclopentyl)-N-(methylsulphonyl)-5-heptenamide)
→ 7-(3-Hydroxy-2-(3-hydroxy-4-phenoxy-1-butenyl)-5-oxocyclopentyl)-5-heptenamide
In: Guinea pig G50
Man G50
→ 7-(2-(3-Hydroxy-4-phenoxy-1-butenyl)-5-oxocyclopentyl)-N-(methylsulphonyl)-5-heptenamide
In: Guinea pig G50
Man G50

Sulpyrine
See: Dipyrone

Sultopride (N-((1-ethyl-2-pyrrolidyl)methyl)-2-methoxy-5-sulphonylethylbenzamide)
→ Demethylsultopride
In: Dog D412
Rabbit D412
Rat D174, D412
→ N-((1-Ethyl-5-oxo-2-pyrrolidyl)methyl)-2-methoxy-5-sulphonylethylbenzamide
In: Dog D412
Rabbit D412
Rat D174, D412
→ N-((1-Ethyl-2-pyrrolidyl)methyl)-2-methoxy-5-sulphobenzamide
In: Rat D174
→ 5-Ethylsulphonyl-2-methoxybenzoic acid
In: Dog D412
Man D412
Rabbit D412
Rat D174, D412
→ N-(2-Pyrrolidylmethyl)-5-sulphonylethylbenzamide
In: Dog D412
Rabbit D412
Rat D412

Sumatriptan (3-(2-(dimethylamino)ethyl-N-methyl-1*H*-indole-5-methanesulphonamide)
→ 5-(Methanesulphonamido)indole-3-acetic acid
In: Man H245
Rat J62

Sumicidin
See: Fenvalerate

Sumithion
See: Fenitrothion

Sunset Yellow
See: 2-Hydroxy-1-(4-sulphophenylazo)naphthalene-6-sulphonic acid

Sunset Yellow-O-sulphate
Formed from: 2-Hydroxy-1-(4-sulphophenylazo)naphthalene-6-sulphonic acid

Supidimide (3-(2,3-dihydro-1,1-dioxido-3-oxo-1,2-benzisothiazol-2-yl)-2-oxopiperidine)
→ 2-Carboxy-N-(2-oxopiperidin-3-yl)benzenesulphonamide
In: Rat C388
→ 3-(2,3-Dihydro-1,1-dioxido-3-oxo-1,2-benzisothiazol-2-yl)-6-hydroxy-2-oxopiperidine
In: Rat C388

Suprofen (2-(4-(2-thienylcarbonyl)phenyl)propionic acid)
→ 2-(*p*-Carboxyphenyl)propionic acid
In: Mouse D138, D346
Rat D346, D378
→ 3-Hydroxy-2-(4-(2-thienylcarbonyl)phenyl)propionic acid

In: Man H229
→ 2-(4-(5-Hydroxy-2-thienylcarbonyl)phenyl)propionic acid
In: Monkey D346
Mouse D346
Rat D346, D378, F130
→ Suprofen-β-D-glucuronide
In: Man G562
→ 2-(4-(2-Thienylcarbonyl)phenyl)propionyl-taurine
In: Rat D346
→ 2-(4-((2-Thienyl)hydroxymethyl)phenyl)propionic acid
In: Dog D346
Guinea pig D346
Rat D346

Suprofen-β-D-glucuronide
Formed from: Suprofen

Sustiva
See: Efavirenz

Suxibuzone (4-butyl-4-(β-carboxypropionyloxy-methyl)-1,2-diphenyl-3,5-pyrazolidinedione)
→ 4-Hydroxymethyloxyphenylbutazone-β-D-glucuronide
In: Man C287
→ 4-Hydroxymethylphenylbutazone-β-D-glucuronide
In: Dog B322
Man B419, C287
Monkey B322
→ Phenylbutazone
In: Dog B322
Man B419, C287
Monkey B322
Rabbit B322
Rat B322
→ Suxibuzone-β-D-glucuronide
In: Dog B322
Man B419, C287
Monkey B322

Suxibuzone-β-D-glucuronide
Formed from: Suxibuzone

Sydnocarb (3-(1-methyl-2-phenylethyl)-5-(((phenylamino)carbonyl)amino)-1,2,3-oxazolium inner salt)
→ β-Oxysydnocarb
In: Man B31

***m*-Sympatol** (*m*-hydroxy-N-methylphenylethanolamine, neosynephrine, phenylephrine; both isomers)
→ L-Adrenaline
In: Horseradish B925
See also: C212
→ *m*-Hydroxymandelic acid
In: Man B171, C107, C894
Rat C894, J287
→ *m*-Hydroxyphenylglycol
In: Man C894
Rat C894
→ *m*-Sympatol-β-D-glucuronide
In: Man B171, C894
Rat C894
→ *m*-Sympatol-O-sulphate
In: Man C894
Rat C894

***p*-Sympatol** (synephrine, *p*-hydroxy-N-methylphenylethanolamine)
Formed from: Octopamine
→ Adrenaline
In: Mushroom E460
See also: B897
→ *p*-Hydroxymandelaldehyde
In: Man A3791
Nocardia E615
→ *p*-Hydroxyphenylacetaldehyde
In: Arthrobacter A2209, A2915
Nocardia E615
See also: B782
→ 1-(*p*-Hydroxyphenyl)-2-methyliminoethanol
In: Arthrobacter B360
→ N-Methyl-*p*-sympatol
In: Beef A1463

***m*-Sympatol-β-D-glucuronide**
Formed from: *m*-Sympatol

***m*-Sympatol-O-sulphate**
Formed from: *m*-Sympatol

Synephrine
See: *p*-Sympatol

Syringaldazine (4-hydroxy-3,5-dimethoxybenzaldehyde ((4-hydroxy-3,5-dimethoxyphenyl)methylene)hydrazone)
→ Radical
In: Coriolus K677

Syringaldehyde
See: 4-Hydroxy-3,5-dimethoxybenzaldehyde
Syringic acid
See: 4-Hydroxy-3,5-dimethoxybenzoic acid
Syringic acid-4-β-D-glucoside
Formed from: 4-Hydroxy-3,5-dimethoxybenzoic acid
Syringin (4-hydroxy-3,5-dimethoxybenzyl alcohol-4-β-D-glucoside)
→ 4-Hydroxy-3,5-dimethoxybenzyl alcohol
In: Spruce K833
Syringomethyl reserpate
Probably formed from: Reserpine
Syrosingopine
→ Methyl reserpate
Probably in: Dog C118
Rabbit C118
Rat C118
Syringylglycerol
Formed from: 1,3-Dihydroxy-1,2-bis(4-hydroxy-3,5-dimethoxyphenyl)propane
→ ω-Hydroxyacetosyringone
Probably in: Coriolus E547
Syringylglycerol β-guaiacyl ether
→ Guaiacol
In: Coriolus E572
→ 4-Hydroxy-3,5-dimethoxybenzaldehyde
In: Coriolus E572
→ 3-Hydroxy-2-(*o*-methoxyphenoxy)propionic acid
In: Coriolus E572
→ 2-(*o*-Methoxyphenoxy)ethanol
In: Coriolus E572
Syringylglycerol β-vanillic acid ether
Formed from: Syringylglycerol β-vanillin ether
→ Glycerol 2-vanillic acid ether
Probably in: Fusarium C117
Syringylglycerol β-vanillin ether
→ 3,5-Dimethoxyquinone
In: Fusarium C117
→ Syringylglycerol β-vanillic acid ether
In: Fusarium C117
Syrosingopine ((3β,16β,17α,18β,20α)-18-((4-((ethoxycarbonyl)oxy)-3,5-dimethoxybenzoyl)oxy)-11,17-dimethoxyyohimban-16-carboxylic acid methyl ester)
→ Syringomethyl reserpate
Probably in: Dog C118
Rabbit C118
Rat C118

T

T_1
See: 3-Iodo-L-thyronine
rT_1
See: 3′-Iodo-L-thyronine
T_3
See: 3,3′,5-Triiodo-L-thyronine
T_4
See: L-Thyroxine
rT_3
See: 3,3′,5′-Triiodo-L-thyronine
2,4,5-T
See: 2,4,5-Trichlorophenoxyacetic acid
Tabersonine ((5α,12*R*,19α)-2,3,6,7-tetradehydroaspidospermidine-3-carboxylic acid methyl ester)
→ Vindoline
In: Catharanthus F843
Tacrine (9-amino-1,2,3,4-tetrahydroacridine)
→ 1-Hydroxytacrine
In: Dog H133, J151
Hamster H133
Man G697, G961, H133, J151
Mouse H133
Rat F575, G697, G961, H61, H133, H718, J151, J195
→ 2-Hydroxytacrine
In: Dog H133, J151
Hamster H133
Man G697, G961, H133, J151
Mouse H133
Rat F575, H61, H133, H718, J151
→ 3-Hydroxytacrine
In: Man J195
Rat J195
→ 4-Hydroxytacrine
In: Dog H133, J151
Hamster H133
Man G697, H133, J151
Mouse H133
Rat F575, H61, H133, H718, J151
→ 7-Hydroxytacrine
In: Dog H133
Hamster H133
Man G697, H133
Mouse H133
Rat H133
→ 1,2,3,4-Tetrahydro-9-hydroxylaminoacridine
In: Rat F80
Talampicillin ((2*S*,5*R*,6*R*)-6-(((2*R*)-aminophenylacetyl)amino)-3,3-dimethyl-7-oxo-4-thia-1-azabicyclo[3.2.0]heptane-2-carboxylic acid 1,3-dihydro-3-oxo-1-isobenzofuranyl ester)
→ Ampicillin
In: Dog A3058
Rat A3058
→ *o*-Formylbenzoic acid
In: Dog A3058
Man A3058
Rat A3058
Taleranol
See: β-Zearalenol
Tamarixetin (3,3′,5,7-tetrahydroxy-4′-methoxyflavone)
Formed from: Quercetin
Tamarixetin-3,3′-disulphate
Formed from: Tamarixetin-3-sulphate
Tamarixetin-3,4′-disulphate
Formed from: Tamarixetin-3-sulphate
Tamarixetin-3,7-disulphate
→ Tamarixetin-3-sulphate
In: Helix F102

Tamarixetin-3-sulphate
Formed from: Tamarixetin-3,7-disulphate
→ Tamarixetin-3,3′-disulphate
In: Flaveria F210, F243
→ Tamarixetin-3,4′-disulphate
In: Flaveria F210

Tamoxifen (*trans*-1-(*p*-β-dimethylaminoethoxyphenyl)-1,2-diphenylbut-1-ene)
→ Demethyltamoxifen
In: Man A3235, C316, F712, G844, H182, K465
Mouse F712
Rat B966, E213, F712, G587, G589, H572
See also: C575
→ 1-(*p*-(2-Dimethylaminoethoxy)phenyl)-1,2-diphenyl-1-buten-3-ol-N-oxide
In: Rat D608
→ 1-(*p*-β-Dimethylaminoethoxyphenyl)-1-hydroxy-1,2-diphenylbutane
In: Monkey A253
→ 1-(*p*-Dimethylaminoethoxyphenyl)-1-(*p*-hydroxyphenyl)-2-phenylbut-1-ene
In: Chicken D326
Dog A253
Man A252, F712, J900, K465
Monkey A253
Mouse D326, F712, J900
Rat D326, D837, E213, F131, F712, F772, H572, G587, J589, J900
See also: C575, J589
→ 4-(1,2-Diphenyl-1-butenyl)phenoxyacetic acid
In: Rat F560, H450
→ *trans*-1-(*p*-(β-Hydroxyethoxy)phenyl)-1,2-diphenylbut-1-ene
In: Man C191, C316
Rat D837
→ 1-(*p*-Hydroxphenyl)-1,2-diphenylbut-1-ene
In: Dog A253
See also: B966
→ α-Hydroxytamoxifen
In: Man H182, J900
Mouse J900
Rat H572, J900
→ Tamoxifen epoxide
In: Rat D608
→ Tamoxifen-N-oxide
In: Man G844
Rat A3943, B966, D837, E213, G587, G589, H572

Tamoxifen epoxide
Formed from: Tamoxifen

Tamoxifen-N-oxide
Formed from: Tamoxifen
→ 1-(*p*-(2-Dimethylaminoethoxy)phenyl)-1,2-diphenyl-1-buten-3-ol-N-oxide
In: Rat D608
See also: E213
→ 1-(*p*-(2-Dimethylaminoethoxy)phenyl)-1-(*p*-hydroxyphenyl)-2-phenylbut-1-ene-N-oxide
Probably in: Rat E213

Tamsulosin ((−)-(*R*)-5-(2-((2-(*o*-ethoxyphenoxy)ethyl)amino)propyl)-2-methoxybenzenesulphonamide)
→ Demethyltamsulosin
In: Dog H528
Man H714, J442
Rat H528
→ Desethyltamsulosin
In: Dog H528
Man H714, J442
Rat H528
→ (−)-(*R*)-5-(2-((2-(2-Ethoxy-4-hydroxyphenoxy)ethyl)amino)propyl)-2-methoxybenzenesulphonamide
In: Dog H528
Man H714, J442
Rat H528
→ (−)-(*R*)-5-(2-((2-(2-Ethoxy-5-hydroxyphenoxy)ethyl)amino)propyl)-2-methoxybenzenesulphonamide
In: Dog H528
Man H714, J442
Rat H528
→ *o*-Ethoxyphenoxyacetic acid
In: Dog H528
Man H714, J442
Rat H528

Tangeretin (4′,5,6,7,8-pentamethoxyflavone)
→ 4′-Hydroxy-,5,6,7,8-tetramethoxyflavone
In: Man G549

Rat G549, J300, K582
→ 6-Hydroxy-4′,5,7,8-tetramethoxyflavone
In: Man K582

Tannin (gallotannin, tannic acid)
→ Caffeic acid
In: Mouse A647
→ *m*-Digallic acid
In: Mouse A647
→ (−)-Epigallocatechin gallate
In: Mouse A647
→ Gallic acid
In: Citrobacter K146
Mouse A647
See also: K911
→ (−)-Gallocatechin gallate
In: Mouse A647

Targetrin (4-(1-(5,6,7,8-tetrahydro-3,5,5,8,8-pentamethyl-2-naphthalenyl)-1-ethenyl) benzoic acid)
→ 6-Hydroxytargetrin
In: Man J290
→ 7-Hydroxytargetrin
In: Man J290

Tartrazine
See: 3-Carboxy-5-hydroxy-1-(*p*-sulphophenyl)-4-(*p*-sulphophenylazo)pyrazole

Tarugan (2,3-dimethyl-4-(3′-methyl-2′-phenylmorpholinomethyl)-1-phenylpyrazolone)
→ Phenmetrazine
In: Man A1019

Taxifolin
See: Dihydroquercetin

Taxiphyllin ((αS)-α-(β-D-glucopyranosyloxy)-4-hydroxybenzeneacetonitrile)
→ *p*-Hydroxymandelonitrile
In: Triglochin A3756

Taxol (paclitaxel)
Formed from: Phenylisoserine
Formed from taxa-4(20),11(12)-dien-5α-ol in Taxus K104
→ Baccatin III
In: Eucalyptus K597
Rat F595, H325, H504
→ 7-Epitaxol
In: Man H505
Marchantia K641
→ 8-Hydroxymethyl-8-demethytaxol
In: Rat H504
→ 2-(*m*-Hydroxy)taxol
In: Rat F595, G992, H479, H504, J481
→ 3′-(*p*-Hydroxy)taxol
In: Man H504, H654, J612, J772, J863
Rat F595, G992, H479, H504, J481
→ 6α-Hydroxytaxol
In: Man G863, H260, H177, H504, H505, H622, H654, J612, J772, J863
Rat H479
See also: H503
Contrast (wrong identity) J481

Taxotere
See: Docetaxel

Tazadolene (1-((2*E*)-2-(phenylmethylene) cyclohexyl)azetidine)
→ *m*-Hydroxytazadolene
In: Dog G375
→ *p*-Hydroxytazadolene
In: Dog G375
→ N-(2-(Phenylmethylene)cyclohexyl)-β-alanine
In: Dog G375

Tazarotene (6-((3,4-dihydro-4,4-dimethyl-2*H*-1-benzothiopyran-6-yl)ethynyl)-3-pyridinecarboxylic acid ethyl ester)
→ Tazarotenic acid
In: Man J331
Rat J331

Tazarotenic acid (6-((3,4-dihydro-4,4-dimethyl-2*H*-1-benzothiopyran-6-yl)ethynyl)-3-pyridinecarboxylic acid)
Formed from: Tazarotene

Tazofelone ((*R* and *S* isomers; 5-((3,5-bis(1,1-dimethylethyl)-4-hydroxyphenyl)methyl)-4-thiazolidinone)
→ 3,5-Di-*tert*-butyl-1-hydroxy-4-oxo-1-((thiazolidin-4-on-5-yl)methyl)cyclohexa-2,5-dione
In: Man H830, K3
→ Tazofelone sulphoxide
In: Man H830, K3

Tazofelone sulphoxide
Formed from: Tazofelone

Tecnazene
See: 2,3,5,6-Tetrachloronitrobenzene

Teflubenzuron (N′-(3,5-dichloro-2,4-difluorobenzoyl)-N-(2,6-difluorobenzoyl)urea
→ 2,6-Difluorobenzamide
In: Rat J260
→ 2,6-Difluorobenzoic acid
In: Rat J260

Temazepam
See: 7-Chloro-1,3-dihydro-3-hydroxy-1-methyl-5-phenyl-2*H*-1,4-benzodiazepin-2-one

Temazepam-β-D-glucuronide
Probably formed from: 7-Chloro-1,3-dihydro-3-hydroxy-1-methyl-5-phenyl-2*H*-1,4-benzodiazepin-2-one

Tenellin (3-((2*E*,4*E*)-4,6-dimethyl-1-oxo-2,4-octadienyl)-1,4-dihydroxy-5-(4-hydroxyphenyl)-2(1*H*)-pyridinone)
Formed from: L-Phenylalanine

Tenidap (5-chloro-2,3-dihydro-3-(hydroxy-(2-thienyl)methylene)-2-oxo-1*H*-indol-1-ylcarboxamide)
→ 3′-Hydroxytenidap
Probably in: Rat J92
→ 4-Hydroxytenidap
Probably in: Rat J92
→ 5′-Hydroxytenidap
In: Rat J92
→ 6-Hydroxytenidap
In: Rat J92
→ Tenidap-β-D-glucuronide
In: Rat J92

Tenidap-β-D-glucuronide
Formed from: Tenidap

Teniposide
→ Free radical
In: Horseradish E74
Sheep E74

Terazocin (1-(4-amino-6,7-dimethoxy-2-quinazolinyl)-4-(tetrahydro-2-furanyl)carbonylpiperazine)
→ 4-Amino-2-(4-(4-carboxy-2-hydroxybutyryl)piperazin-1-yl)-6,7-dimethoxyquinazoline
In: Rat E477
→ 4-Amino-6,7-dimethoxy-2-(1-piperazinyl)quinazoline
In: Rat E477
→ 6-Demethylterazocin
In: Rat E477
→ 7-Demethylterazocin
In: Rat E477

Terbinafine (N-((2*E*)-6,6-dimethyl-2-hepten-4-ynyl)-N-methyl-1-naphthalenemethanamine)
→ Demethylterbinafine
In: Man J763
→ 1-(N-((6-Hydroxymethyl)-6-methylhept-2-en-4-yn-1-yl)-N-((methylamino)methyl) naphthalene
In: Man J763
→ 1-((Methylamino)methyl)naphthalene
In: Man J763
→ 1-Naphthaldehyde
In: Man J763
→ Terbinafine dihydrodiol
In: Man J763

Terbinafine dihydrodiol
Formed from: Terbinafine
→ Demethylterbinafine dihydrodiol
Probably in: Man J763

Terbutaline (1-(3,5-dihydroxyphenyl)-2-*tert*-butylaminoethanol)
Formed from: Ibuterol
Probably formed from: N-*tert*-butyl-3-hydroxy-5-dimethylcarbamoyloxyphenyl ethanolamine)
→ Terbutaline-β-D-glucuronide
In: Rat A256, A266, D344
→ Terbutaline sulphate
In: Dog A2564
Man A2564, G307, H85

Terbutaline-β-D-glucuronide
Formed from: Terbutaline

Terbutaline sulphate
Formed from: Terbutaline

Terephthalaldehyde (4-formylbenzaldehyde)
→ *p*-Formylbenzoic acid
In: Euglena G422
See also: A2094
→ *p*-Hydroxmethylbenzaldehyde
In: Dunaliella G459
Euglena G422
→ (*S*)-4-(2-Hydroxypropionyl)benzaldehyde
With benzoylformate decarboxylase K657
→ Terephthalic acid
In: Streptomyces C284

Terephthalaldehydic acid
See: *p*-Formylbenzoic acid
Terephthalic acid (benzene-1,4-dicarboxylic acid)
Formed from: Di-(2-ethylhexyl) terephthalate
p-Formylbenzoic acid
Monomethyl terephthalate
1,4-Phenylenebis(methylene)selenocyanate
Terephthalaldehyde
Tetraselenocyclophane
→ 1,2-Dihydro-1*R*,2*S*-dihydroxyterephthalic acid
In: Comamonas H396
→ Protocatechuic acid
In: Comamonas H396
Nocardia B2
Pseudomonas C91
Terephthalonitrile (1,4-benzodinitrile)
→ *p*-Cyanobenzoic acid
Probably in: Fusarium A3364
Nocardia D725
Terfenadine (α-(4-(1,1-dimethylethyl)phenyl)-4-(hydroxy(diphenyl)methyl)-1-piperidinebutanol)
→ 1-(4-*tert*-Butylphenyl)-4-(4-(hydroxy(diphenyl)methyl)piperidin-1-yl)-1-butanone
In: Rat H53
→ Diphenyl 4-piperidinylmethanol
In: Man G573, H53, H196
Rat H53
→ 4-(4-(Hydroxy(diphenyl)methyl)piperidin-1-yl)-1-(4-(1-hydroxymethyl)-1-methylethyl)phenyl)-1-butanol
In: Man G573, H53, H196
Rat H53
See also: H324, J616
Terfenadine acid
See: 1-(4-(1-Carboxy-1-methylethyl)phenyl)-4-(4-(hydroxy(diphenyl)methyl)piperidin-1-yl)-1-butanol
Terguride (N,N-diethyl-N′-((8α)-6-methylergolin-8-yl)urea)
→ Desethylterguride
In: Dog F337
Guinea pig F337
Monkey F337
Rat F337
Terodiline (N-(*tert*-butyl)-4,4-diphenyl-2-butylamine)
→ N-(*tert*-Butyl)-4-hydroxy-4,4-diphenyl-2-butylamine
In: Rat E839
→ N-(*tert*-Butyl)-4-(*p*-hydroxyphenyl)-4-phenyl-2-butylamine
In: Dog D784, F677
Man F33, F412
Rat D774, E839
→ N-(2-Hydroxymethyl-2-propyl)-4,4-diphenyl-2-butylamine
In: Man F33, F412
Rat D784
***m*-Terphenyl**
→ 1,3-Bis(muconolacton-3-yl)benzene
In: Aspergillus H23
→ *m*-(*p*-Hydroxyphenyl)phenylmuconolactone
In: Aspergillus H23
→ 3,4,4″-Trihydroxy-*m*-terphenyl
In: Neurospora H23
Terrachlor
See: Pentachloronitrobenzene
Territrem A
→ 4-(methyl)-Hydroxyterritrem A
In: Rat F740
Territrem B
→ 4′-O-Demethylterritrem B
In: Rat F740
→ 4-(methyl)-Hydroxyterritrem B
In: Rat F740
Tersavid (N^1-benzyl-N^2-pivaloylhydrazine)
→ Benzoic acid
In: Rat A2104
→ Tersavid-β-D-glucuronide
Probably in: Rat A2104
Tersavid-β-D-glucuronide
Formed from: Tersavid
Tertatolol (1-((3,4-dihydro-2*H*-1-benzothiopyran-8-yl)oxy)-3-((1,1-dimethylethyl)amino)-2-propanol)
→ 4-Hydroxytertatolol
In: Man H549
→ Tertatolol-β-D-glucuronide
In: Man H549

Tertatolol-β-D-glucuronide
Formed from: Tertalol
Testosterone
→ 17β-Oestradiol
In: Horse E631
5,5′,5″,5‴-Tetraacetamido-2,2′,2″,2‴-tetrahydroxyquaterphenyl
Probably formed from: 5,5′,5″-Triacetamido-2,2′,2″-trihydroxyterphenyl
1,1′,8,8′-Tetraacetoxy-10,10′-bisanthrone
Formed from: 1,8,9-Trihydroxyanthracene-1,8-diacetate
2,3,4,6-Tetra-O-acetyl-1-((N-acetyl-N-phenylamino)oxy)-1-deoxy-β-D-glucopyranoside
→ 1-((N-Acetyl-N-phenylamino)oxy)-1-deoxy-β-D-glucopyranoside
In: Aspergillus J389
Tetra-(N-acetyl-L-tyrosine)
Formed from: N-Acetyl-L-tyrosine
Probably formed from: Tri-(N-acetyl-L-tyrosine)
Tetrabenazine ((3*R*,11b*R*)-rel-1,3,4,6,7,11b-hexahydro-9,10-dimethoxy-3-(2-methylpropyl)-2*H*-benzo[*a*]quinolizin-2-one)
→ Benzophenone diphenylhydrazone
In: Man E875
Rat E875
Tetrabromobisphenol A (4,4′-isopropylidene-bis(2,6-dibromophenol))
→ 2,2-Bis(*p*-hydroxyphenyl)propane
In: Microorganism K529
→ Tetrabromobisphenol A-β-D-glucuronide
In: Rat J841, K127
Tetrabromobisphenol A-di-β-D-glucuronide
Formed from: Tetrabromobisphenol A-β-D-glucuronide
Tetrabromobisphenol A-β-D-glucuronide
Formed from: Tetrabromobisphenol A
→ Tetrabromobisphenol A-di-β-D-glucuronide
In: Rat J841, K127
→ Tetrabromobisphenol A-β-D-glucuronide sulphate
Probably in: Rat J841, K127
Tetrabromobisphenol A-β-D-glucuronide sulphate
Probably formed from: Tetrabromobisphenol A-β-D-glucuronide
2,3,7,8-Tetrabromodibenzodioxin
→ 4,4′,5,5′-Tetrabromo-2,2′-dihydroxydiphenyl ether
Probably in: Rat G567
→ 2,3,7-Tribromodibenzodioxin
In: Rat G567
4,4′,5,5′-Tetrabromo-2,2′-dihydroxydiphenyl ether
Probably formed from: 2,3,7,8-Tetrabromodibenzodioxin
2,2′,4,4′-Tetrabromodiphenyl ether
→ Tetrabromo-hydroxydiphenyl ether
In: Mouse H929
Rat H929
Tetrabromo-hydroxydiphenyl ether
Formed from: 2,2′,4,4′-Tetrabromodiphenyl ether
Tetrabromoquinol
Probably formed from: Pentabromophenol
→ Hydroxyquinol
In: Mycobacterium G676
Tetracaine (amethocaine, 4-(butylamino)benzoic acid 2-(dimethylamino)ethyl ester)
→ Tetracaine-N-oxide
In: Horse A2708
Mouse A2708
Rabbit A2708
Rat A2708
Tetracaine-N-oxide
Formed from: Tetracaine
2,3,4,5-Tetrachloroaniline
Probably formed from: Pentachloroaniline
2,3,4,5-Tetrachloronitrobenzene
2,3,4,6-Tetrachloroaniline
Probably formed from: Pentachloroaniline
2,3,5,6-Tetrachloroaniline
Formed from: 2,3,5,6-Tetrachloronitrobenzene
Probably formed from: Pentachloroaniline
→ 4-Amino-2,3,5,6-tetrachlorophenol
Probably in: Rat H512
2,3,4,5-Tetrachloroanisole
Probably formed from: Pentachloroanisole
2,3,4,5-Tetrachlorophenol

2,3,4,6-Tetrachloroanisole
Formed from: 2,3,4,6-Tetrachlorophenol
Probably formed from: Pentachloroanisole
2,3,5,6-Tetrachloroanisole
Probably formed from: Pentachloroanisole
2,3,5,6-Tetrachlorophenol
2,2′,3,3′-Tetrachloroazobenzene
Formed from: 2,3-Dichloroaniline
2,2′,4,4′-Tetrachloroazobenzene
Formed from: 2,4-Dichloroaniline
2,2′,5,5′-Tetrachloroazobenzene
Formed from: 2,5-Dichloroaniline
3,3′,4,4′-Tetrachloroazobenzene
Formed from: 3,4-Dichloroaniline
→ N,N′-Bis(3,4-Dichlorophenyl)-N-hydroxyhydrazine
In: Rat H535
→ 3,4-Dichloroaniline
In: Rat H535
3,3′,5,5′-Tetrachloroazobenzene
Formed from: 3,5-Dichloroaniline
3,3′,4,4′-Tetrachloroazoxybenzene
Formed from: 3,4-Dichloroaniline
1,2-Dichloro-4-nitrobenzene
1,2,3,4-Tetrachlorobenzene
Formed from: Hexachlorobenzene
Lindane
Pentachlorobenzene
γ-Pentachlorocyclohex-1-ene
→ N-Acetyl-S-(2,3,4,5-tetrachlorophenyl)-L-cysteine
In: Monkey E535, E635
→ 2,3,4,5-Tetrachlorophenol
In: Monkey E635
Rabbit A1627
See also: B923, J702
→ 2,3,4,6-Tetrachlorophenol
In: Rabbit A1627
See also: B923
→ 2,3,4,5-Tetrachlorophenylsulphinic acid
In: Monkey E535
→ 2,3,4,5-Tetrachlorothioanisole
In: Monkey E535
1,2,3,5-Tetrachlorobenzene
Formed from: Lindane
γ-Pentachlorocyclohex-1-ene
Probably formed from: Pentachlorobenzene
→ 2,3,4,5-Tetrachlorophenol
In: Monkey E535
Rabbit A1627
See also: A3101
→ 2,3,4,6-Tetrachlorophenol
In: Monkey E535
See also: A2444
→ 2,3,5,6-Tetrachlorophenol
In: Monkey E535
Rabbit A1627
→ 2,3,5,6-Tetrachlorophenylsulphinic acid
In: Monkey E535
→ 1,3,5-Trichlorobenzene
Probably in: Rat A3774
→ 3,4,6-Trichlorocatechol
In: Pseudomonas G227
1,2,4,5-Tetrachlorobenzene
Formed from: N-Acetyl-S-(2,3,5,6-tetrachlorophenyl)-L-cysteine
Hexachlorobenzene
Lindane
γ-Pentachlorocyclohex-1-ene
Pentachlorothiophenol
S,S′-(Tetrachloro-*p*-phenylene)di-L-cysteine
2,3,5,6-Tetrachloronitrobenzene
→ 2,3,5,6-Tetrachlorophenol
In: Rabbit A1627
See also: A3822
2,3,4,5-Tetrachlorobenzoic acid
Formed from: 2,3,4,5-Tetrachlorobiphenyl
2,2′,3,3′-Tetrachlorobiphenyl
→ 2,3-Dichlorobiphenyl
In: Acinetobacter B125
Alcaligenes A1822, B125
2,2′,4,4′-Tetrachlorobiphenyl
→ 2,4-Dichlorobenzoic acid
In: Acinetobacter B125
2,2′,5,5′-Tetrachlorobiphenyl
→ 2,2′,5,5′-Tetrachlorobiphenyl-3,4-oxide
Probably in: Rat C328
→ 2,2′,5,5′-Tetrachloro-3,4-dihydro-3,4-dihydroxybiphenyl
In: Monkey A2334
Rabbit A328
Rat B108, C738
See also: A2300, A2638

→ 2,2′,5,5′-Tetrachloro-3-hydroxybiphenyl
In: Hamster H583
Monkey A2300
Rabbit A328
Rat A2169, A2404, A2638, B108, C738, G218, G419, H545
→ 2,2′,5,5′-Tetrachloro-4-hydroxybiphenyl
In: Rabbit A328
Rainbow trout A2443
Rat A2404, C738, G419
See also: A701, A2334, A2578
→ 2,2′,5,5′-Tetrachloro-3-methylthiobiphenyl
In: Mouse A1810
See also: B898
→ 2,2′,5,5′-Tetrachloro-4-methylthiobiphenyl
In: Mouse A1810
See also: B898

2,3′,4,4′-Tetrachlorobiphenyl
→ 2,3′,4,4′-Tetrachloro-3-hydroxybiphenyl
In: Rat A1446, H545
→ 2,3′,4,4′-Tetrachloro-5-hydroxybiphenyl
In: Rat A1446, H545

2,3,4,5-Tetrachlorobiphenyl
→ 2,3,4,5-Tetrachlorobenzoic acid
In: Acinetobacter B125
Alcaligenes B125

2,3′,4′,5-Tetrachlorobiphenyl
→ 2,3′,4′,5-Tetrachloro-3-hydroxybiphenyl
In: Guinea pig H942
Hamster H942
Rat H545, H942, H943, J278
→ 2,3′,4′,5-Tetrachloro-3-hydroxybiphenyl
In: Hamster H942, K265
Rat H545, H942, H943, J278
→ 2,3′,4,5-Tetrachloro-3-methylthiobiphenyl
In: Rat J278
See also: H943
→ 2,3′,4,5-Tetrachloro-4-methylthiobiphenyl
In: Rat J278
See also: H943

2,3′,5,5′-Tetrachlorobiphenyl
→ 2,3′,5,5′-Tetrachloro-4-hydroxybiphenyl
In: Hamster K265

2,3,4′,6-Tetrachlorobiphenyl
→ 2,3,4′,6-Tetrachloro-4-methylsulphonylbiphenyl
In: Mouse E246

2,3,5,6-Tetrachlorobiphenyl
→ 2′,3′,5′,6′-Tetrachloro-3-hydroxybiphenyl
In: Rat A1444
→ 2′,3′,5′,6′-Tetrachloro-4-hydroxybiphenyl
In: Rat A1444

3,3′,4,4′-Tetrachlorobiphenyl
→ 3,4-Dichlorobenzoic acid
In: Acinetobacter B125
→ 3,3′,4,4′-Tetrachloro-2-hydroxybiphenyl
In: Mouse D856, F29
Rose K509
Scup J148
Tomato K509
→ 3,3′,4,4′-Tetrachloro-5-hydroxybiphenyl
In: Beluga whale K570
Lettuce K509
Mouse F29
Rat E779, F488, G218, H160, H545, J263
Scup J148
See also: A953
→ 3,3′,4,4′-Tetrachloro-6-hydroxybiphenyl
In: Lettuce K509
Rose K509
Sunflower K509
Tomato K509
→ 3,3′,4′,5-Tetrachloro-4-hydroxybiphenyl
In: Beluga whale K570
Mouse F29
Rat E779, F488, G218, H160, H545, J263
Scup J148
→ 4,4′,5,5′-Tetrachloro-2-hydroxybiphenyl
In: Rat H160
Scup J148
→ 3,3′,4,4′-Tetrachloro-5-methylthiobiphenyl
In: Rat J263
→ 3,3′,4,4′-Tetrachloro-6-methylthiobiphenyl
In: Rat J263
→ 3,3′,4′-Trichloro-4-hydroxybiphenyl
In: Rat 488
See also: F29
→ 3′,4,4′-Trichloro-6-methylsulphonyl-3-methylthiobiphenyl
In: Rat J263

3,3′,5,5′-Tetrachlorobiphenyl
→ 3,3′,5,5′-Tetrachloro-4-hydroxybiphenyl

In: Rat G952

2,2′,5,5′-Tetrachlorobiphenyl-3,4-oxide
Probably formed from: 2,2′,5,5′-Tetrachlorobiphenyl
→ 2,2′,5,5′-Tetrachloro-3-hydroxybiphenyl
In: Rat C738
→ 2,2′,5,5′-Tetrachloro-3,4-dihydro-3,4-dihydroxybiphenyl
In: Rat C738

1,1,1,2-Tetrachloro-2,2-bis(*p*-chlorophenyl)ethane
→ 1,1-Dichloro-2,2-bis(*p*-chlorophenyl) ethylene
In: Mouse E845
Rat E845

2,4,5,6-Tetrachloro-1,3-bis(S-glutathionyl)benzene
Probably formed from: Pentachlorosulphonylanisole

2,2′,5,5′-Tetrachloro-4,4′-bis(methylsulphonyl) biphenyl
Formed from: 2,2′,5,5′-Tetrachloro-4-methylsulphonylbiphenyl

2,3,5,6-Tetrachloro-1,4-bis(methylthio)benzene
Formed from: Pentachlorothioanisole
S,S′-(Tetrachloro-*p*-phenylene)-di-L-cysteine

Tetrachlorocatechol
Formed from: Octachlorodibenzo-*p*-dioxin
Pentachlorophenol
Probably formed from: 2,3,4,5-Tetrachlorophenol
→ Tetrachlorocatechol diacetate
Probably in: Alcaligenes A3805
→ Tetrachloro-2-methoxyphenol
In: Mycobacterium A3330, D137
→ 3,4,6-Trichlorocatechol
In: Bacteria F883

Tetrachlorocatechol diacetate
Probably formed from: Tetrachlorocatechol

1,2,3,4-Tetrachlorodibenzo-*p*-dioxin
→ 1,2,3,4-Tetrachloro-hydroxydibenzo-*p*-dioxin
In: Rat A3809

1,3,7,8-Tetrachlorodibenzo-*p*-dioxin
→ 1,3,7,8-Tetrachloro-2-hydroxydibenzo-*p*-dioxin
In: Dog C700

1,4,7,8-Tetrachlorodibenzo-*p*-dioxin
→ 4,5-Dichlorocatechol
In: Rat H940
→ 1,4,7,8-Tetrachloro-2-hydroxydibenzo-*p*-dioxin
In: Rat H940
→ 4,7,8-Trichloro-1-hydroxydibenzo-*p*-dioxin
In: Rat H940

2,3,7,8-Tetrachlorodibenzo-*p*-dioxin
Probably formed from: 1,2,3,4,7,8-Hexachlorodibenzo-*p*-dioxin
1,2,3,6,7,8-Hexachlorodibenzo-*p*-dioxin
→ 4,5-Dichlorocatechol
In: Phanerochaete H911
→ 1,2-Dichloro-4,5-dimethoxybenzene
In: Dog C853
→ 2,3,7,8-Tetrachloro-1-hydroxydibenzo-*p*-dioxin
In: Rat C165
→ 1,3,7,8-Tetrachloro-2-methoxydibenzo-*p*-dioxin
Probably in: Dog C853
→ 2,3,7-Trichlorodibenzo-*p*-dioxin
In: Microorganism H874
→ 2,3,7-Trichloro-8-hydroxydibenzo-*p*-dioxin
In: Microorganism H874
Rat C165
See also: C853

2,3,7,8-Tetrachlorodibenzofuran
→ 2,3,7,8-Tetrachloro-4-hydroxydibenzofuran
In: Rat F864, H78
Rainbow trout G775
→ 2,7,8-Trichlorodibenzofuran
In: Rat F864

2,2′,5,5′-Tetrachloro-3,4-dihydro-3,4-dihydroxybiphenyl
Formed from: 2,2′,5,5′-Tetrachlorobiphenyl
2,2′,5,5′-Tetrachlorobiphenyl-3,4-oxide
→ 2,2′,5,5′-Tetrachloro-3,4-dihydro-3,4,x′-trihydroxybiphenyl
Probably in: Rhesus monkey A2334
→ 2,2′,5,5′-Tetrachloro-3,4-dihydroxybiphenyl
In: Burkholderia J860
Comamonas K196

***cis*-2′,4,5,5′-Tetrachloro-2,3-dihydro-2,3-dihydroxybiphenyl**
Probably formed from: 2,3′,4′,5-Tetrachlorobiphenyl

2,2′,5,5′-Tetrachloro-3,4-dihydro-3,4,x′-trihydroxybiphenyl
Probably formed from: 2,2′,5,5′-Tetrachloro-3,4-dihydro-3,4-dihydroxybiphenyl

2,3,5,6-Tetrachloro-1,4-dihydroxybenzene
See: Tetrachloroquinol

2,2′,5,5′-Tetrachloro-3,3′-dihydroxybiphenyl
Probably formed from: 2,2′,5,5′-Tetrachloro-3-hydroxybiphenyl

2,2′,5,5′-Tetrachloro-3,4-dihydroxybiphenyl
Formed from: 2,2′,5,5′-Tetrachloro-3,4-dihydro-3,4-dihydroxybiphenyl

2,2′,5,5′-Tetrachloro-3-hydroxybiphenyl
Probably formed from: 2,2′,5,5′-Tetrachloro-4-hydroxybiphenyl

2,3′,5′,6′-Tetrachloro-3,4-dihydroxybiphenyl
Probably formed from: 2,3′,5′,6′-Tetrachloro-4-hydroxybiphenyl

3,3′,4,4′-Tetrachloro-2,5-dihydroxybiphenyl
Probably formed from: 3,3′,4,4′-Tetrachloro-5-hydroxybiphenyl

3,3′,4,4′-Tetrachloro-5,6-dihydroxybiphenyl
Probably formed from: 3,3′,4,4′-Tetrachloro-5-hydroxybiphenyl

3,3′,5,5′-Tetrachloro-4,4′-dihydroxybiphenyl
Probably formed from: 3,3′,4′,5-Tetrachloro-4-hydroxybiphenyl

Tetrachloro-1,2-dimethoxybenzene
Formed from: Tetrachloro-2-methoxyphenol

Tetrachloro-1,4-dimethoxybenzene
Formed from: Tetrachloro-4-methoxyphenol

2,2′,6,6′-Tetrachloro-4,4′-dimethylazoxybenzene
Formed from: 2,6-Dichloro-4-methylaniline

2,4,4′,5-Tetrachlorodiphenyl ether
→ *p*-Chlorophenol
In: Rat A3503
→ 2′,4,4′,5′-Tetrachloro-2-hydroxydiphenyl ether
Probably in: Rat A3503
→ 2,4,5-Trichlorophenol
In: Rat A3503

3,3′,4,4′-Tetrachlorodiphenyl ether
→ 3,4-Dichlorophenol
In: Rat A3503
→ 3′,4,4′,5-Tetrachloro-2-hydroxydiphenyl ether
Probably in: Rat A3503
→ 3′,4,4′,5-Tetrachloro-3-hydroxydiphenyl ether
In: Rat A3503

Tetrachloroguaiacol
See: Tetrachloro-2-methoxyphenol

2,2′,5,5′-Tetrachloro-3-hydroxybiphenyl
Formed from: 2,2′,5,5′-Tetrachlorobiphenyl
2,2′,5,5′-Tetrachlorobiphenyl-3,4-oxide
→ 2,2′,5,5′-Tetrachloro-3,3′-dihydroxybiphenyl
Probably in: Rat C738
→ 2,2′,5,5′-Tetrachloro-3,4-dihydroxybiphenyl
In: Rat C738

2,2′,5,5′-Tetrachloro-3-hydroxybiphenyl
Formed from: 2,2′,5,5′-Tetrachlorobiphenyl

2,2′,5,5′-Tetrachloro-4-hydroxybiphenyl
Formed from: 2,2′,5,5′-Tetrachlorobiphenyl
→ 2,2′,5,5′-Tetrachloro-3,4-dihydroxybiphenyl
Probably in: Rhesus monkey A2334

2,3′,4,4′-Tetrachloro-3-hydroxybiphenyl
Formed from: 2,3′,4,4′-Tetrachlorobiphenyl

2,3′,4,4′-Tetrachloro-5-hydroxybiphenyl
Formed from: 2,3′,4,4′-Tetrachlorobiphenyl

2,3′,4′,5-Tetrachloro-3-hydroxybiphenyl
Formed from: 3,3′,4,4′,5-Pentachlorobiphenyl
2,3′,4′,5-Tetrachlorobiphenyl

2,3′,4′,5-Tetrachloro-4-hydroxybiphenyl
Formed from: 2,3′,4′,5-Tetrachlorobiphenyl

2,3′,5,5′-Tetrachloro-4-hydroxybiphenyl
Formed from: 2,3′,5,5′-Tetrachlorobiphenyl

2′,3′,5′,6′-Tetrachloro-3-hydroxybiphenyl
Formed from: 2,3,5,6-Tetrachlorobiphenyl

2′,3′,5′,6′-Tetrachloro-4-hydroxybiphenyl
Formed from: 2,3,5,6-Tetrachlorobiphenyl
→ 2′,3′,5′,6′-Tetrachloro-3,4-dihydroxy-biphenyl
Probably in: Rat A1444

3,3′,4,4′-Tetrachloro-2-hydroxybiphenyl
Formed from: 3,3′,4,4′-Tetrachlorobiphenyl

3,3′,4,4′-Tetrachloro-5-hydroxybiphenyl
Formed from: 3,3′,4,4′-Tetrachlorobiphenyl
→ 3,3′,4,4′-Tetrachloro-2,5-dihydroxybiphenyl
Probably in: Rat F488
→ 3,3′,4,4′-Tetrachloro-5,6-dihydroxybiphenyl
Probably in: Rat F488

3,3′,4,4′-Tetrachloro-6-hydroxybiphenyl
Probably formed from: 3,3′,4,4′-Tetrachlorobiphenyl

3,3′,4′5-Tetrachloro-4-hydroxybiphenyl
Formed from: 3,3′,4,4′,5-Pentachlorobiphenyl
3,3′,4,4′-Tetrachlorobiphenyl
→ 3,3′,5,5′-Tetrachloro-4,4′-dihydroxybiphenyl
Probably in: Rat F488
→ 3,3′,4′,5-Tetrachloro-4-hydroxybiphenyl-4′,5′-oxide
Probably in: Rat F488

3,3′,4′5′-Tetrachloro-4-hydroxybiphenyl
Formed from: 3,3′,4,4′,5-Pentachlorobiphenyl

3,3′,5,5′-Tetrachloro-4-hydroxybiphenyl
Formed from: 3,3′,5,5′-Tetrachlorobiphenyl

4,4′,5,5′-Tetrachloro-2-hydroxybiphenyl
Formed from: 3,3′,4,4′-Tetrachlorobiphenyl

3,3′,4′,5-Tetrachloro-4-hydroxybiphenyl-4′,5′-oxide
Probably formed from: 3,3′,4′,5-Tetrachloro-4-hydroxybiphenyl

1,2,3,4-Tetrachloro-hydroxydibenzo-*p*-dioxin
Formed from: 1,2,3,4-Tetrachlorodibenzo-*p*-dioxin

1,3,7,8-Tetrachloro-2-hydroxydibenzo-*p*-dioxin
Formed from: 1,3,7,8-Tetrachlorodibenzo-*p*-dioxin

1,4,7,8-Tetrachloro-2-hydroxydibenzo-*p*-dioxin
Formed from: 1,4,7,8-Tetrachlorodibenzo-*p*-dioxin
→ 1,4,7,8-Tetrachloro-2-hydroxydibenzo-*p*-dioxin-β-D-glucuronide
In: Rat H940

2,3,7,8-Tetrachloro-1-hydroxydibenzo-*p*-dioxin
Formed from: 2,3,7,8-Tetrachlorodibenzo-*p*-dioxin

1,4,7,8-Tetrachloro-2-hydroxydibenzo-*p*-dioxin-β-D-glucuronide
Formed from: 1,4,7,8-Tetrachloro-2-hydroxydibenzo-*p*-dioxin

2,3,7,8-Tetrachloro-4-hydroxydibenzofuran
Formed from: 2,3,7,8-Tetrachlorodibenzofuran
→ 2,3,7,8-Tetrachloro-4-hydroxydibenzofuran-β-D-glucuronide
In: Rainbow trout G775

2,3,7,8-Tetrachloro-4-hydroxydibenzofuran-β-D-glucuronide
Formed from: 2,3,7,8-Tetrachloro-4-hydroxydibenzofuran

2,4,4′,5-Tetrachloro-2-hydroxydiphenyl ether
Probably formed from: 2,4,4′,5-Tetrachlorodiphenyl ether

3′,4,4′,5-Tetrachloro-2-hydroxydiphenyl ether
Probably formed from: 3,3′,4,4′-Tetrachlorodiphenyl ether

3′,4,4′,5-Tetrachloro-3-hydroxydiphenyl ether
Probably formed from: 3,3′,4,4′-Tetrachlorodiphenyl ether

3,4,5,6-Tetrachloro-2-hydroxythiophenol
Formed from: 2,3,4,5-Tetrachlorophenol

2,4,5,6-Tetrachloroisophthalonitrile
See: Chlorothalonil

Tetrachloro mercaptothioanisole
Formed from: Hexachlorobenzene

Tetrachloro-2-methoxyphenol
Formed from: Tetrachlorocatechol
→ Tetrachloro-1,2-dimethoxybenzene
In: Mycobacterium A3330, D137

Tetrachloro-4-methoxyphenol
Formed from: Tetrachloroquinol
→ Tetrachloro-1,4-dimethoxybenzene
In: Mycobacterium A3330, D137
See also: A416, C547

S-(Tetrachloromethylsulphinylphenyl)-L-cysteine
Probably formed from: S-(Tetrachloromethylsulphinylphenyl)-L-cysteinylglycine
→ N-Acetyl-S-(tetrachloromethylsulphinylphenyl)-L-cysteine
Probably in: Rat G688
→ 3-(Tetrachloromethylsulphinylphenylthio) lactic acid
Probably in: Rat G688

S-(Tetrachloromethylsulphinylphenyl)-L-cysteinylglycine
Probably formed from: S-(Tetrachloromethylsulphinylphenyl)glutathione
→ S-(Tetrachloromethylsulphinylphenyl)-L-cysteine
Probably in: Rat F894, G688

S-(Tetrachloromethylsulphinylphenyl)glutathione
Probably formed from: S-(Tetrachloromethylthiophenyl)glutathione
→ S-(Tetrachloromethylsulphinylphenyl)-L-cysteinylglycine
Probably in: Rat F894, G688
→ S-(Tetrachloromethylsulphonylphenyl) glutathione
Probably in: Rat F894

3-(Tetrachloromethylsulphinylphenylthio)lactic acid
Probably formed from: S-(Tetrachloromethylsulphinylphenyl)-L-cysteine

2,2′,5,5′-Tetrachloro-3-methylsulphonylbiphenyl
Probably formed from: 2,2′,5,5′-Tetrachloro-3-methylthiobiphenyl

2,2′,5,5′-Tetrachloro-4-methylsulphonylbiphenyl
Probably formed from: 2,2′,5,5′-Tetrachloro-4-methylthiobiphenyl
→ 2,2′,5,5′-Tetrachloro-4,4′-bis (methylsulphonyl)biphenyl
Probably in: Mouse B898

2,3′,4′,5-Tetrachloro-3-methylsulphonylbiphenyl
Probably formed from: 2,3′,4′,5-Tetrachloro-3-methylthiobiphenyl

2,3′,4′,5-Tetrachloro-4-methylsulphonylbiphenyl
Probably formed from: 2,3′,4′,5-Tetrachloro-4-methylthiobiphenyl

2,3,4′,6-Tetrachloro-4-methylsulphonylbiphenyl
Probably formed from: 2,3,4′,6-Tetrachlorobiphenyl

S-(Tetrachloromethylsulphonylphenyl)glutathione
Probably formed from: S-(Tetrachloromethylsulphinylphenyl)glutathione

2,2′,5,5′-Tetrachloro-3-methylthiobiphenyl
Formed from: 2,2′,5,5′-Tetrachlorobiphenyl
→ 2,2′,5,5′-Tetrachloro-3-methylsulphonylbiphenyl
Probably in: Mouse A1810
Rat B898

2,2′,5,5′-Tetrachloro-4-methylthiobiphenyl
Formed from: 2,2′,5,5′-Tetrachlorobiphenyl
→ 2,3′,5,5′-Tetrachloro-4-methylsulphonylbiphenyl
Probably in: Mouse A1810
Rat B898

2,3′,4′,5-Tetrachloro-3-methylthiobiphenyl
Formed from: 2,3′,4′,5-Tetrachlorobiphenyl
→ 2,3′,4′,5-Tetrachloro-3-methylsulphonylbiphenyl
Probably in: Rat H943, J278

2,3′,4′,5-Tetrachloro-4-methylthiobiphenyl
Formed from: 2,3′,4′,5-Tetrachlorobiphenyl
→ 2,3′,4′,5-Tetrachloro-4-methylsulphonylbiphenyl
Probably in: Rat H943, J278

3,3′,4,4′-Tetrachloro-5-methylthiobiphenyl
Formed from: 3,3′,4,4′-Tetrachlorobiphenyl

3,3′,4,4′-Tetrachloro-6-methylthiobiphenyl
Formed from: 3,3′,4,4′-Tetrachlorobiphenyl

Tetrachloro-4-(methylthio)phenol
Probably formed from: Pentachlorothiophenol
→ Trichloro-bis(methylthio)phenol
Probably in: Rat H205

S-(Tetrachloro(methylthio)phenyl)-L-cysteinylglycine
Probably formed from: S-(Tetrachloromethylthiophenyl)glutathione

S-(Tetrachloro(methylthio)phenyl)glutathione
Formed from: Pentachlorothioanisole
→ S-(Tetrachloromethylsulphinylphenyl) glutathione
Probably in: Rat F894, G688
→ S-(Tetrachloromethylthiophenyl)-L-cysteinylglycine
Probably in: Rat F894, G688

Tetrachloro-4-(methylthio)thiophenol
Probably formed from: Pentachlorothioanisole

2,3,5,6-Tetrachlorothioanisole
→ 2,3,5,6-Tetrachloro-1,4-bis(methylthio) benzene
In: Rat C881
→ Tetrachloro-4-(methylthio)thiophenyl-β-D-glucuronide
Probably in: Rat C881
→ 2,3,5,6-Tetrachlorothioanisole
In: Rat C881

Tetrachloro-4-(methylthio)thiophenyl-β-D-glucuronide
Probably formed from: Tetrachloro-4-(methylthio)thiophenol
1,2,3,4-Tetrachloronaphthalene
→ 5,6,7,8-Tetrachloro-α-naphthol
In: Pig A1829
→ 5,6,7,8-Tetrachloro-β-naphthol
In: Pig A1829
5,6,7,8-Tetrachloro-α-naphthol
Formed from: 1,2,3,4-Tetrachloronaphthalene
5,6,7,8-Tetrachloro-β-naphthol
Formed from: 1,2,3,4-Tetrachloronaphthalene
2,3,4,5-Tetrachloronitrobenzene
Formed from: Pentachloronitrobenzene
→ 2,3,4,5-Tetrachloroaniline
Probably in: Rat B176
2,3,4,6-Tetrachloronitrobenzene (tecnazene)
→ 2,3,5,6-Tetrachloroaniline
In: Rat H512
→ 1,2,4,5-Tetrachlorobenzene
In: Rat H512
→ S-(2,3,5,6-Tetrachlorophenyl)-L-cysteinylglycine
In: Rat H512
S-(Tetrachloronitrophenyl)glutathione
Formed from: Pentachloronitrobenzene
→ N-Malonyl-S-(tetrachloronitrophenyl)-L-cysteine
Probably in: Arachis B894
2,3,4,5-Tetrachlorophenol
Formed from: β-Lindane
Pentachlorobenzene
Pentachlorophenol
1,2,3,4-Tetrachlorobenzene
1,2,3,5-Tetrachlorobenzene
→ 2,3,4,5-Tetrachloroanisole
Probably in: Rat B176
→ Tetrachlorocatechol
Probably in: Pseudomonas J702
→ 3,4,5,6-Tetrachloro-2-hydroxythiophenol
In: Rat A2872
→ 2,3,5-Trichlorophenol
Probably in: Rat E121
→ 2,4,5-Trichlorophenol
In: Microorganism G331
See also: Rat E121
→ 3,4,5-Trichlorophenol
In: Desulfitobacterium J232
See also: E168, G667
→ 2,3,4,5-Tetrachlorophenyl-β-D-glucuronide
Probably in: Rat A2626
→ Trichloroquinol
In: Arthrobacter K756
Azobacter H884
Rat A3661
Rhodococcus E324
2,3,4,6-Tetrachlorophenol
Formed from: β-Lindane
Pentachlorophenol
1,2,3,4-Tetrachlorobenzene
1,2,3,5-Tetrachlorobenzene
→ Phenol
Probably in: Pseudomonas C893
→ 2,3,4,6-Tetrachloroanisole
In: Aspergillus A1510
Paecilomyces A1510
Penicillium A1510
Scopulariopsis A1510
→ 2,3,4-Trichlorophenol
In: Desulfitobacterium J232
See also: E121
→ 2,4,6-Trichlorophenol
In: Desulfomonile G471
See also: E121
→ Trichloroquinol
In: Azotobacter H884
Rat A3661
Rhodococcus E324
2,3,5,6-Tetrachlorophenol
Formed from: Pentachlorophenol
1,2,3,5-Tetrachlorobenzene
1,2,4,5-Tetrachlorobenzene
→ 2,6-Dimethoxy-4-(2,3,5,6-tetrachlorophenoxy)phenol
In: Rhizoctonia D690
→ 4-(2,6-Dimethoxy-4-(2,3,5,6-tetrachlorophenoxy)phenoxy)-2,6-dimethoxyphenol
In: Rhizoctonia D690
→ 4-(4-(2,6-Dimethoxy-4-(2,3,5,6-tetrachlorophenoxy)phenoxy)-2,6-dimethoxyphenoxy)-2,6-dimethoxyphenol

In: Rhizoctonia D690
→ 4-(4-(4-(2,6-Dimethoxy-4-(2,3,5,6-tetrachlorophenoxy)phenoxy)-2,6-dimethoxyphenoxy)-2,6-dimethoxyphenoxy)-2,6-dimethoxyphenol
In: Rhizoctonia D690
→ 2,3,5,6-Tetrachloroanisole
Probably in: Aeromonas A3805
Cytophaga A3805
→ Tetrachloroquinol
Flavobacterium G641, G746
Rat A3661
Rhodococcus E324
→ 2,3,5-Trichlorophenol
In: Desulfitobacterium J232
→ 2,3,6-Trichlorophenol
In: Desulfomonile G471
→ 3,4,6-Trichlorocatechol
In: Burkholderia J177

S-(2,3,5,6-Tetrachlorophenyl)-L-cysteinylglycine
Formed from: 2,3,5,6-Tetrachloronitrobenzene
→ N-Acetyl-S-(L-cysteinyl)-2,3,5,6-tetrachlorobenzene
Probably in: Rat H512
→ 2,3,5,6-Tetrachlorothiophenol
Probably in: Rat H512

S,S′-(Tetrachloro-*p*-phenylene)di-L-cysteine
Probably formed from: S,S′-(Tetrachlorophenylene)diglutathione
→ 2,3,5,6-Tetrachlorothioanisole
In: Rat C881
→ 2,3,5,6-Tetrachloro-1,4-bis(methylthio)benzene
In: Rat C881

S,S′-(Tetrachlorophenylene)diglutathione
Probably formed from: S-(Pentachlorophenyl)glutathione
→ S,S′-(Tetrachloro-*p*-phenylene)di-L-cysteine
Probably in: Arachis B816

2,3,4,5-Tetrachlorophenyl-β-D-glucuronide
Probably formed from: 2,3,4,5-Tetrachlorophenol

2,3,4,5-Tetrachlorophenylsulphinic acid
Formed from: 1,2,3,4-Tetrachlorobenzene

2,3,5,5-Tetrachlorophenylsulphinic acid
Formed from: 1,2,3,5-Tetrachlorobenzene

Tetrachloroquinol (2,3,5,6-tetrachloro-1,4-dihydroxybenzene)
Formed from: Pentachlorophenol
2,3,5,6-Tetrachlorophenol
Hydroxyquinol
In: Mycobacterium G676
→ Tetrachloro-4-methoxyphenol
In: Mycobacterium A3330, D137
See also: A416, C547
→ Tetrachloroquinol-β-D-glucuronide
Probably in: Rat A3568
→ Tetrachloro-*p*-quinone
Probably in: Rat F891
→ Trichloro-S-glutathionylquinol
In: Phanerochaete K138
→ Trichloroquinol
In: Phanerochaete K138
See also: E265, E509, F891

Tetrachloroquinol-β-D-glucuronide
Probably formed from: Tetrachloroquinol

Tetrachloro-*p*-quinone
Probably formed from: Tetrachloroquinol

Tetrachlororesorcinol
Formed from: Pentachlorophenol

α,α′,4,4′-Tetrachlorostilbene
Formed from: *p*-Chlorobenzotrichloride

2,3,5,6-Tetrachlorosulphinylanisole
Probably formed from: 2,3,5,6-Tetrachlorothioanisole
→ 2,3,5,6-Tetrachlorosulphonylanisole
Probably in: Rat H512

2,3,5,6-Tetrachlorosulphonylanisole
Probably formed from: 2,3,5,6-Tetrachlorosulphinylanisole

2,3,4,5-Tetrachlorothioanisole
Formed from: Pentachloronitrobenzene
Pentachlorothiophenol
1,2,3,4-Tetrachlorobenzene
→ 2,3,4-Trichlorothioanisole
Probably in: Squirrel monkey E535

2,3,4,6-Tetrachlorothioanisole
Formed from: Pentachlorothiophenol

2,3,5,6-Tetrachlorothioanisole
Formed from: Pentachlorothioanisole
Tetrachloro-4-(methylthio)phenol
S,S′-(Tetrachloro-*p*-phenylene)di-L-cysteine

Probably formed from: 2,3,5,6-Tetrachlorothiophenol
→ Tetrachloro-4-(methylthio)thiophenol
Probably in: Rat H512
→ 2,3,5,6-Tetrachlorosulphinylanisole
Probably in: Rat H512

2,3,5,6-Tetrachlorothiophenol
Probably formed from: N-Acetyl-S-(2,3,5,6-tetrachlorophenyl)-L-cysteine
Pentachlorothiophenol
S-(2,3,5,6-Tetrachlorophenyl)-L-cysteinylglycine
→ Bis(2,3,5,6-tetrachlorophenyl)disulphide
Probably in: Rat H512
→ Methyl 2,3,5,6-tetrachlorophenyl disulphide
Probably in: Rat H512
→ 2,3,5,6-Tetrachlorothioanisole
Probably in: Rat C881, H512

Tetrachlorvinphos
See: Gardona

Tetracycline ((4*S*,4a*S*,5a*S*,6*S*,12a*S*)-4-(dimethylamino)-1,4,4a,5,5a,6,11,12a-octahydro-3,6,10,12,12a-pentahydroxy-6-methyl-1,11-dioxo-2-naphthacenecarboxamide)
→ N-Demethyltetracycline
In: Hen K669
→ 4-Epitetracycline
In: Hen K669

Tetraethyl 2-((3-bisphosphono)propyl)-5-hydroxy-1-methyl-4-phenylpyrimidin-6-one
Formed from: Tetraethyl 2-((3-bisphosphono)propyl)-1-methyl-4-phenylpyrimidin-6-one
→ Tetraethyl 2-((3-bisphosphono)propyl)-5-hydroxy-1-methyl-4-phenylpyrimidin-6-one-β-D-glucuronide
In: Rat H491

Tetraethyl 2-((3-bisphosphono)propyl)-5-hydroxy-1-methyl-4-phenylpyrimidin-6-one-β-D-glucuronide
Formed from: Tetraethyl 2-((3-bisphosphono)propyl)-5-hydroxy-1-methyl-4-phenylpyrimidin-6-one

Tetraethyl 2-((3-bisphosphono)propyl)-1-methyl-4-phenylpyrimidin-6-one
→ Tetraethyl 2-((3-bisphosphono)propyl)-5-hydroxy-1-methyl-4-phenylpyrimidin-6-one
In: Rat H491
→ Triethyl 2-((3-bisphosphono)propyl)-1-methyl-4-phenylpyrimidin-6-one
In: Rat H491

2,3,5,6-Tetrafluoroaniline
→ 4-Amino-2,3,5,6-tetrafluorophenol
In: Rat H273

2,3,4,6-Tetrafluorobenzaldehyde
→ 2,3,4,6-Tetrafluorobenzoic acid
In: Mouse G98

2,3,4,6-Tetrafluorobenzoic acid
Formed from: 2,3,4,6-Tetrafluorobenzaldehyde

Tetrafluorocatechol
Formed from: Pentafluorophenol

2,3,5,6-Tetrafluoro-4-hydroxybenzoic acid
→ Tetrafluoroquinol
In: Candida J461

4′-(α,α,β,β-Tetrafluorophenethyl)acetophenone
Formed from: α,α-Dimethyl-4-(α,α,β,β-tetrafluorophenethyl)benzylamine

2-(4-(α,α,β,β-Tetrafluorophenethyl)phenyl)propan-1,2-diol
Probably formed from: 2-(4-(α,α,β,β-Tetrafluorophenethyl)phenyl)-2-propanol

2-(4-(α,α,β,β-Tetrafluorophenethyl)phenyl)-2-propanol
Formed from: α,α-Dimethyl-4-(α,α,β,β-tetrafluorophenethyl)benzylamine
→ 2-(4-(α,α,β,β-Tetrafluorophenethyl)benzyl methyl ether
In: Baboon A2116, A3021
Dog A2116, A3021
Man A2116, A3021
Monkey A2116, A3021
Rat A2116
→ 2-(4-(α,α,β,β-Tetrafluorophenethyl)phenyl)propane-1,2-diol
In: Baboon A2116, A3021
Dog A2116, A3021
Man A2116, A3021
Monkey A2116, A3021
Rat A2116
→ 2-(4-(α,α,β,β-Tetrafluorophenethyl)phenyl)propene
In: Baboon A2116
Dog A2116

Man A2116
Monkey A2116
Rat A2116
→ 2-(4-(α,α,β,β-Tetrafluorophenethyl)phenyl)prop-2-yl-β-D-glucuronide
In: Baboon A2116
Dog A2116
Man A2116
Monkey A2116
Rat A2116

2-(4-(α,α,β,β-Tetrafluorophenethyl)phenyl)propene
Probably formed from: 2-(4-(α,α,β,β-Tetrafluorophenethyl)phenyl)-2-propanol

2-(4-(α,α,β,β-Tetrafluorophenethyl)phenyl)prop-2-yl-β-D-glucuronide
Probably formed from: 2-(4-(α,α,β,β-Tetrafluorophenethyl)phenyl)-2-propanol

2,3,5,6-Tetrafluorophenol
→ 3,4,6-Trifluorocatechol
In: Trichosporon H420

Tetrafluoroquinol
Formed from: 2,3,5,6-Tetrafluoro-4-hydroxybenzoic acid
Probably formed from: Pentafluorophenol
→ Hydroxyquinol
In: Mycobacterium G676

3′,3″,5′,5″-Tetrafluorostilboestrol
→ 3′,3″,5′,5″-Tetrafluorostilboestrol quinone
In: Horseradish D892

3′,3″,5′,5″-Tetrafluorostilboestrol quinone
Formed from: 3′,3″,5′,5″-Tetrafluorostilboestrol

1,2,3,6-Tetragalloylglucose
→ 1,2,3,4,6-Pentagalloylglucose
In: Quercus K726

Tetrahydroalstonine (16,17-didehydro-19-methyloxayohimban-16-carboxylic acid methyl ester)
Formed from: Cathenamine
4,21-Dehydrogeissoschizine
Strictosidine
Tryptamine

Tetrahydroberberine (canadine, 5,8,13,13a-tetrahydro-9,10-dimethoxy-6*H*-benzo[*g*]-1,3-benzodioxolo[5,6-*a*]quinolizine)
Formed from: Berberine
7,8-Dihydroberberine
Tetrahydrocolumbamine
→ Allocryptopine
In: Corydalis H585
Dicentra J496
Macleaya C696
→ Berberine
In: Berberis E732, K947
Coptis E678, E732
Corydalis G136, H585
Coscinium H267
Dicentra J496
Thalictrum 242, H401
→ Chelerythrine
In: Dicentra J496
Macleaya C696
→ Dehydrocanadine
Probably in: Berberis D145
→ 7,8-Dihydroberberine
In: Macleaya C696
See also D470, D526
→ N-Methylcanadine
In: Berberis E299
Corydalis E299, H585
Dicentra E299
Eschscholtzia K782
Fumaria E299
Papaver E299
→ Thalictricavine
In: Corydalis H585

Tetrahydroberbinium
See: N-Methylcanadine

Δ[1]-Tetrahydrocannabinol
(Δ[9]-tetrahydrocannabinol)
→ 4″,5″-Bisnor-Δ[1]-*trans*-tetrahydrocannabinol-3″,7-dioic acid
In: Man D456
Rabbit A1486, B89
→ 4″,5″-Bisnor-Δ[1]-tetrahydrocannabinol-3″-oic acid
In: Guinea pig A2846
Man C294
Rabbit B89
→ Cannabinol
In: Rat A2211
See also: A675
→ Cannabinol-7-oic acid

In: Rhesus monkey A1454
→ 4′-Depentyl-Δ¹-tetrahydrocannabinol-4′-acrylic-7-oic acid
In: Rabbit B89
→ *cis*-Hexahydrocannabinol-7-oic acid
In: Mouse A3238
→ *trans*-Hexahydrocannabinol-7-oic acid
In: Mouse A3238
→ 5″-Nor-Δ¹-tetrahydrocannabinol-4″-oic acid
In: Guinea pig A2846
→ Δ¹-Tetrahydrocannabinol-2′-β-D-glucuronide
In: Rabbit B102
See also: C805
→ Δ¹-Tetrahydrocannabinol palmitate
Probably in: Rat A2613
→ Δ¹-Tetrahydro-1α,2α-dihydroxycannabinol
In: Man D139
Rabbit A108
Rat B530, G95
Squirrel monkey A758
→ Δ¹-Tetrahydro-4″,7-dihydroxycannabinol
In: Guinea pig B458
Man C2
Mouse A3155
See also: B102
→ Δ¹-Tetrahydro-1″-hydroxycannabinol
In: Guinea pig B458
Thamnidium B742
→ Δ¹-Tetrahydro-2″-hydroxycannabinol
In: Fusarium B742
Gibberella B742
Guinea pig B458, F819
Man C2
Thamnidium B742
→ Δ¹-Tetrahydro-3″-hydroxycannabinol
In: Chaetomium A3628
Fusarium B742
Gibberella B742
Guinea pig B458, F819
Monkey H418
Mouse E7, E411, E598
Rat F599
Thamnidium B742
→ Δ¹-Tetrahydro-4″-hydroxycannabinol
In: Fusarium B742
Gibberella B742
Guinea pig B458, F7
Mouse E7, F7
Rabbit F7
Rat F7
Syncephalastrum A2895
Thamnidium B742
→ Δ¹-Tetrahydro-5″-hydroxycannabinol
In: Guinea pig F7
Mouse F7
Rabbit A3449, F7
See also: B89
→ Δ¹-Tetrahydro-6α-hydroxycannabinol
In: Guinea pig B458, F819
Man C2
Monkey H418
Mouse A1376, A1391, E7, E411, H67
Rabbit A108
Thamnidium B742
See also: A680, A758, E598, F599
→ Δ¹-Tetrahydro-6β-hydroxycannabinol
In: Guinea pig B458
Man C2
Monkey H418
Thamnidium B742
See also: A680, B89
→ Δ¹-Tetrahydro-7-hydroxycannabinol
In: Chaetomium A3628
Guinea pig B458
Man A680, C2, G298
Monkey A438, H418
Mouse A879, A1376, E7, E411, E598, H67
Rabbit A108, A389, A3449, B89
Rat A116, A389, A393, A767, A949, B68, F599, G95
Thamnidium B742
See also: A440, A1267, A2563, B89, B102, B458, C294, D456
→ Δ¹-Tetrahydro-2″-hydroxycannabinol-7-oic acid
In: Rabbit B102
→ 2″,3″,4″,5″-Tetranor-Δ¹-tetrahydrocannabinol-1″-oic acid
Probably in: Guinea pig A2846
Rabbit B89

→ 3″,4″,5″-Trisnor-Δ^{1}-tetrahydrocannabinol-2″-oic acid
In: Guinea pig A2846
→ 3″,4″,5″-Trisnor-Δ^{1}-tetrahydro-2″-hydroxycannabinol
Probably in: Rabbit B89

$\Delta^{1(7)}$-Tetrahydrocannabinol
→ *cis*-Hexahydrocannabinol
In: Mouse B604
→ *trans*-Hexahydrocannabinol
In: Mouse B604
→ $\Delta^{1(7)}$-Tetrahydrocannabinol-1β,7-oxide
In: Rat B367
→ $\Delta^{1(7)}$-Tetrahydro-2α-hydroxycannabinol
In: Rat B367
→ $\Delta^{1(7)}$-Tetrahydro-2β-hydroxycannabinol
In: Rat B367
→ $\Delta^{1(7)}$-Tetrahydro-2″-hydroxycannabinol
In: Rat B367
→ $\Delta^{1(7)}$-Tetrahydro-3″-hydroxycannabinol
In: Rat B367
→ $\Delta^{1(7)}$-Tetrahydro-4″-hydroxycannabinol
In: Rat B367
→ $\Delta^{1(7)}$-Tetrahydro-5″-hydroxycannabinol
In: Rat B367
→ $\Delta^{1(7)}$-Tetrahydro-6α-hydroxycannabinol
In: Rat B367
→ $\Delta^{1(7)}$-Tetrahydro-6β-hydroxycannabinol
In: Rat B367

$\Delta^{3,4}$-Tetrahydrocannabinol
($\Delta^{6\alpha,10\alpha}$-tetrahydrocannabinol)
→ Cannabinol
In: Mycobacterium A3010, A3071
→ Δ^{2}-Tetrahydro-4-hydroxycannabinol
In: Bacillus A3071
Mycobacterium A3010
→ $\Delta^{3,4}$-Tetrahydro-4″-hydroxycannabinol
In: Bacillus A3010, A3071
→ $\Delta^{3,4}$-Tetrahydro-5-oxocannabinol
In: Streptomyces A3010, A3071

$\Delta^{6\alpha,10\alpha}$-Tetrahydrocannabinol
See: $\Delta^{3,4}$-Tetrahydrocannabinol

Δ^{6}-Tetrahydrocannabinol
(Δ^{8}-tetrahydrocannabinol)
→ 4″,5″-Bisnor-Δ^{6}-tetrahydrocannabinol-3″-oic acid
In: Guinea pig A3676, B583
Rat A3676
→ Cannabinol
In: Rat A2211
→ *cis*-Hexahydrocannabinol-7-oic acid
In: Mouse A3238
→ *trans*-Hexahydrocannabinol-7-oic acid
In: Mouse A3238
→ Hexahydro-1,6-dihydroxycannabinol
In: Guinea pig B583
→ 5″-Nor-Δ^{6}-tetrahydrocannabinol-4″-oic acid
In: Guinea pig A3676
→ Δ^{6}-Tetrahydrocannabinol-4-β-D-glucuronide
In: Mouse A3724
See also: B666
→ Δ^{6}-Tetrahydrocannabinol-1,6-oxide
In: Man D139
Mouse A3425, C278, C375, C455, H67
See also: C749
→ Δ^{6}-Tetrahydrocannabinol palmitate
Probably in: Rat A2613
→ Δ^{6}-Tetrahydro-1″-hydroxycannabinol
In: Guinea pig B583, E897
Monkey B79
→ Δ^{6}-Tetrahydro-2″-hydroxycannabinol
In: Guinea pig B583, E897
Monkey B79
→ Δ^{6}-Tetrahydro-3″-hydroxycannabinol
In: Guinea pig B583
Monkey B79
Mouse E897
→ Δ^{6}-Tetrahydro-4″-hydroxycannabinol
In: Guinea pig B583, E897, F7
Monkey B79
Mouse F7
Rabbit F7
Rat F7
Syncephalastrum A2595
→ Δ^{6}-Tetrahydro-5″-hydroxycannabinol
In: Guinea pig B583
Mouse E897, F7
Rabbit F7
→ Δ^{6}-Tetrahydro-5α-hydroxycannabinol
In: Guinea pig B583, E897, E945

Monkey A758, B79
Mouse B50
→ Δ^6-Tetrahydro-5β-hydroxycannabinol
In: Guinea pig B583, E897, E945, F7
Monkey A758, B79
Mouse B50
Rat F7
→ Δ^6-Tetrahydro-7-hydroxycannabinol
In: Guinea pig B583, E897, E945, F7
Man C749
Monkey A2210, B79
Mouse B50, H67
Rabbit A1106, F7
Rat A389, A949, F7
See also: A2563, A3676, B752
→ Δ^6-Tetrahydro-1″-hydroxycannabinol-7-oic acied
In: Guinea pig B583
See also: A2210
→ 2″,3″,4″,5″-Tetranor-Δ^6-tetrahydrocannabinol-1″-oic acid
In: Guinea pig B583
Rat A3676
→ 3″,4″,5″-Trisnor-Δ^6-tetrahydrocannabinol-2″-oic acid
In: Guinea pig A3676
Mouse A3676

Δ^7-Tetrahydrocannabinol
→ α-Hexahydrocannabinol
In: Mouse B604
→ β-Hexahydrocannabinol
In: Mouse B604

Δ^8-Tetrahydrocannabinol
See: Δ^6-Tetrahydrocannabinol

Δ^9-Tetrahydrocannabinol
See: Δ^1-Tetrahydrocannabinol

Δ^1-Tetrahydrocannabinol-4″,7-dioic acid
Probably formed from: Δ^1-Tetrahydrocannabinol-7-oic acid

Δ^1-Tetrahydrocannabinol-2′-β-D-glucuronide
Formed from: Δ^1-Tetrahydrocannabinol

Δ^6-Tetrahydrocannabinol-4′-β-D-glucuronide
Formed from: Δ^6-Tetrahydrocannabinol

Δ^1-Tetrahydrocannabinol-5″-oic acid
Formed from: Δ^1-Tetrahydro-5″-hydroxycannabinol
→ 4″,5″-Bisnor-Δ^1-tetrahydrocannabinol-3″-oic acid
Probably in: Mouse D345

Δ^1-Tetrahydrocannabinol-7-oic acid
Formed from: Δ^1-Tetrahydro-7-oxocannabinol
→ Δ^1-Tetrahydrocannabinol-4″,7-dioic acid
Probably in: Rabbit B102
→ Δ^1-Tetrahydro-1″-hydroxycannabinol-7-oic acid
Probably in: Guinea pig B458
Man C294
→ Δ^1-Tetrahydro-2″-hydroxycannabinol-7-oic acid
Probably in: Guinea pig B458
Man C294
→ Δ^1-Tetrahydro-3″-hydroxycannabinol-7-oic acid
Probably in: Guinea pig B458
Man C294
→ Δ^1-Tetrahydro-4″-hydroxycannabinol-7-oic acid
Probably in: Guinea pig B458
Man C294, D456
→ Δ^1-Tetrahydro-6α-hydroxycannabinol-7-oic acid
Probably in: Guinea pig B458

Δ^6-Tetrahydrocannabinol-5″-oic acid
→ 4″,5″-Bisnor-Δ^6-tetrahydrocannabinol-3″-oic acid
In: Mouse D345

Δ^6-Tetrahydrocannabinol-7-oic acid
Formed from: Δ^6-Tetrahydro-7-hydroxycannabinol
Δ^6-Tetrahydro-7-oxocannabinol
→ Δ^6-Tetrahydro-2″-hydroxycannabinol-7-oic acid
Probably in: Guinea pig B583
Mouse B50
→ Δ^6-Tetrahydro-3″-hydroxycannabinol-7-oic acid
Probably in: Guinea pig B583
Mouse B50
→ Δ^6-Tetrahydro-4″-hydroxycannabinol-7-oic acid
Probably in: Guinea pig B583
Mouse B50

→ Δ^6-Tetrahydro-5α-hydroxycannabinol-7-oic acid
Probably in: Mouse B50
→ Δ^6-Tetrahydro-5β-hydroxycannabinol-7-oic acid
Probably in: Mouse B50

Δ^1-Tetrahydrocannabinol-1α,2α-oxide
Formed from: Δ^1-Tetrahydrocannabinol
→ Hexahydro-1α,2β-dihydroxycannabinol
In: Rat B416
See also: B530
→ Hexahydro-1β,2α-dihydroxycannabinol
In: Mouse D359
See also: B530
→ Δ^1-Tetrahydro-2″-hydroxycannabinol-1α,2α-oxide
In: Rat B416
→ Δ^1-Tetrahydro-3″-hydroxycannabinol-1α,2α-oxide
In: Rat B416
→ Δ^1-Tetrahydro-4″-hydroxycannabinol-1α,2α-oxide
In: Rat B416
→ Δ^1-Tetrahydro-6β-hydroxycannabinol-1α,2α-oxide
In: Rat B416
→ Δ^1-Tetrahydro-7-hydroxycannabinol-1α,2α-oxide
In: Rat B416

Δ^1-Tetrahydrocannabinol-1β,2β-oxide
→ Hexahydro-1β,2α-dihydroxycannabinol
In: Mouse D359

$\Delta^{1(7)}$-Tetrahydrocannabinol-1β,7-oxide
Formed from: $\Delta^{1(7)}$-Tetrahydrocannabinol
→ Hexahydro-1α,7-dihydroxycannabinol
Probably in: Rat B367
→ Hexahydro-1β,7-dihydroxycannabinol
Probably in: Rat B367

Δ^6-Tetrahydrocannabinol-1,6-oxide
(both isomers)
Formed from: Δ^6-Tetrahydrocannabinol
→ Hexahydro-1α,6β-dihydroxycannabinol
In: Man D139
Mouse D438
See also: A3425
→ Hexahydro-1β,6α-dihydroxycannabinol
In: Mouse D438

Δ^1-Tetrahydrocannabinol palmitate
Probably formed from: Δ^1-Tetrahydrocannabinol

Δ^6-Tetrahydrocannabinol palmitate
Probably formed from: Δ^6-Tetrahydrocannabinol

Δ^1-Tetrahydrocannabinol-5″-yl-β-D-glucuronide
Formed from: Δ^1-Tetrahydro-5″-hydroxycannabinol

Δ^1-Tetrahydrocannabinol-7-yl-β-D-glucuronide
Formed from: Δ^1-Tetrahydro-7-hydroxycannabinol

1,2,3,4-Tetrahydro-β-carboline
(tetrahydronorharman)
Formed from: 1,2,3,4-Tetrahydro-β-carboline-1-carboxylic acid
Tryptamine
→ 1,2-Dihydro-β-carbolin-1-one
In: Rat C344, D129
→ Norharman
In: Rat C344, D129
→ 1,2,3,4-Tetrahydro-6-hydroxy-β-carboline
In: Rat C344, D129
→ 1,2,3,4-Tetrahydro-7-hydroxy-β-carboline
In: Rat C344, D129

1,2,3,4-Tetrahydro-β-carboline-1-carboxylic acid
→ 1,2,3,4-Tetrahydro-β-carboline
In: Mouse D874

Tetrahydrocolumbamine ((13a*S*)-5,8,13,13a-tetrahydro-3,9,10-trimethoxy-6*H*-dibenzo[*a*,*g*]quinolizin-2-ol)
Formed from: Columbamine
Scoulerine
→ Canadine
In: Coptis E732
Thalictrum H301
→ Columbamine
Probably in: Berberis D470
→ Tetrahydropalmatine
In: Berberis D470

Tetrahydrocoptisine
Formed from: Coptisine

Tetrahydrocurcumin
Formed from: Curcumin
→ Hexahydrocurcumin
Probably in: Rat A3492
→ Tetrahydrocurcumin-β-D-glucuronide

Probably in: Mouse J577
Rat A3492

Tetrahydrocurcumin-β-D-glucuronide
Probably formed from: Tetrahydrocurcumin

5,8,13,13a,Tetrahydro-6*H*-dibenzo[*a*,*g*]quinolizine
See: Tetrahydroprotoberberine

(+)-N-(2,3,4,5-Tetrahydro-1-((3,5-dihydroxyadamantan-1-yl)methyl)-2,4-dioxo-5-phenyl-1,5-benzodiazepin-3-yl)-N′-phenylurea
Probably formed from: (+)-N-(2,3,4,5-Tetrahydro-1-((3-hydroxyadamantan-1-yl)methyl)-2,4-dioxo-5-phenyl-1,5-benzodiazepin-3-yl)-N′-phenylurea

***cis*-1,2,9,10-Tetrahydro-1(*R*),2(*S*)-dihydroxyanthracene**
Formed from: 9,10-Dihydroanthracene

1,2,3,4-Tetrahydro-3β,4α-dihydroxybenzo[*c*] phenanthrene-1β,2β-oxide
→ 1α-(S-Glutathionyl)-1,2,3,4-tetrahydro-2β,3β,4α-trihydroxybenzo[*c*]phenanthrene
In: Rat H395

***cis*-7,8,9,10-Tetrahydro-9,10-dihydroxy-3,4-benzpyrene**
Formed from: 7,8-Dihydro-3,4-benzpyrene

***trans*-7,8,9,10-Tetrahydro-9,10-dihydroxy-3,4-benzpyrene**
Formed from: 7,8-Dihydro-3,4-benzpyrene

7,8,9,10-Tetrahydro-7α,8β-dihydroxy-3,4-benzpyrene-9β,10β-oxide
Formed from: *trans*-7,8-Dihydro-7,8-dihydroxy-3,4-benzpyrene

7,8,9,10-Tetrahydro-7β,8α-dihydroxy-3,4-benzpyrene-9α,10α-oxide
Formed from: *trans*-7,8-Dihydro-7,8-dihydroxy-3,4-benzpyrene
→ 10β-(S-Glutathionyl)-7,8,9,10-tetrahydro-7β,8α,9α-trihydroxy-3,4-benzpyrene
In: Rat J395
→ 7,8,9,10-Tetrahydro-7β,8α,9α,10α-tetrahydroxy-3,4-benzpyrene
Probably in: Mouse F150
Rat A3093
→ 7,8,9,10-Tetrahydro-7β,8α,9α,10β-tetrahydroxy-3,4-benzpyrene
Probably in: Cunninghamella B452
Mouse F150
Rat A3093

7,8,9,10-Tetrahydro-7β,8α-dihydroxy-3,4-benzpyrene-9β,10β-oxide
Formed from: *trans*-7,8-Dihydro-7,8-dihydroxy-3,4-benzpyrene
→ 7,8,9,10-Tetrahydro-7β,8α,9β,10α-tetrahydroxy-3,4-benzpyrene
Probably in: Rat A3093
→ 7,8,9,10-Tetrahydro-7β,8α,9β,10β-tetrahydroxy-3,4-benzpyrene
Probably in: Rat A3093

1,2,3,4-Tetrahydro-1-(3,4-dihydroxybenzyl) isoquinoline
→ 1,2,3,4-Tetrahydro-1-(3-hydroxy-4-methoxybenzyl)isoquinoline
Probably in: Mouse K664

Δ^1-Tetrahydro-1″,2″-dihydroxycannabinol
Probably formed from: Δ^1-Tetrahydro-1″-hydroxycannabinol

Δ^1-Tetrahydro-1″,3″-dihydroxycannabinol
Probably formed from: Δ^1-Tetrahydro-1″-hydroxycannabinol

Δ^1-Tetrahydro-1″,4″-dihydroxycannabinol
Probably formed from: Δ^1-Tetrahydro-1″-hydroxycannabinol

Δ^1-Tetrahydro-1″,6β-dihydroxycannabinol
Probably formed from: Δ^1-Tetrahydro-1″-hydroxycannabinol

Δ^1-Tetrahydro-1″,7-dihydroxycannabinol
Probably formed from: Δ^1-Tetrahydro-7-hydroxycannabinol
→ Δ^1-Tetrahydro-1″-hydroxycannabinol-7-oic acid
Probably in: Mouse E897

Δ^1-Tetrahydro-2″,6α-dihydroxycannabinol
Probably formed from: Δ^1-Tetrahydro-2″-hydroxycannabinol

Δ^1-Tetrahydro-2″,6β-dihydroxycannabinol
Probably formed from: Δ^1-Tetrahydro-2″-hydroxycannabinol

Δ^1-Tetrahydro-2′,7-dihydroxycannabinol
Probably formed from: Δ^1-Tetrahydro-7-hydroxycannabinol

Δ^1-Tetrahydro-2″,7-dihydroxycannabinol
Formed from: Δ^1-Tetrahydro-2″-hydroxycannabinol
Probably formed from: Δ^1-Tetrahydro-7-hydroxycannabinol

→ Δ^1-Tetrahydro-2″-hydroxycannabinol-7-oic acid
Probably in: Mouse F492

Δ^1-Tetrahydro-3″,6α-dihydroxycannabinol
Probably formed from: Δ^1-Tetrahydro-3″-hydroxycannabinol

Δ^1-Tetrahydro-3″,6β-dihydroxycannabinol
Probably formed from: Δ^1-Tetrahydro-3″-hydroxycannabinol
→ Δ^1-Tetrahydro-3″-hydroxy-6-oxocannabinol
Probably in: Guinea pig B458

Δ^1-Tetrahydro-3″,7-dihydroxycannabinol
Formed from: Δ^1-Tetrahydro-3″-hydroxycannabinol
Probably formed from: Δ^1-Tetrahydro-7-hydroxycannabinol
→ Δ^1-Tetrahydro-3″-hydroxycannabinol-7-oic acid
Probably in: Mouse E897, F492
Rabbit B102
→ Δ^1-Tetrahydro-3″,6α,7-trihydroxycannabinol
Probably in: Mouse F492

Δ^1-Tetrahydro-4″,6α-dihydroxycannabinol
Probably formed from: Δ^1-Tetrahydro-4″-hydroxycannabinol

Δ^1-Tetrahydro-4″,6β-dihydroxycannabinol
Probably formed from: Δ^1-Tetrahydro-4″-hydroxycannabinol
→ Δ^1-Tetrahydro-4″-hydroxy-6-oxocannabinol
Probably in: Guinea pig B458
Mouse A3155

Δ^1-Tetrahydro-4″,7-dihydroxycannabinol
Formed from: Δ^1-Tetrahydrocannabinol
Δ^1-Tetrahydro-4″-hydroxycannabinol
Δ^1-Tetrahydro-7-hydroxycannabinol
→ Δ^1-Tetrahydro-4″-hydroxycannabinol-7-oic acid
Probably in: Mouse A3155, E897, F526
Rabbit B102
→ Δ^1-Tetrahydro-4″,6α,7-trihydroxycannabinol
Probably in: Mouse A3155, F526

Δ^1-Tetrahydro-6α,7-dihydroxycannabinol
Formed from: Δ^1-Tetrahydro-6α-hydroxycannabinol
Probably formed from: Δ^1-Tetrahydro-7-hydroxycannabinol
→ Δ^1-Tetrahydro-7-hydroxy-6-oxocannabinol
Probably in: Guinea pig B458
→ Δ^1-Tetrahydro-1″,6α,7-trihydroxycannabinol
Probably in: Guinea pig B458
→ Δ^1-Tetrahydro-2″,6α,7-trihydroxycannabinol
Probably in: Guinea pig B458
→ Δ^1-Tetrahydro-3″,6α,7-trihydroxycannabinol
Probably in: Guinea pig B458
→ Δ^1-Tetrahydro-4″,6α,7-trihydroxycannabinol
Probably in: Guinea pig B458

Δ^1-Tetrahydro-6β,7-dihydroxycannabinol
Probably formed from: Δ^1-Tetrahydro-7-hydroxycannabinol
→ Δ^1-Tetrahydro-1″,6β,7-trihydroxycannabinol
Probably in: Guinea pig B458
→ Δ^1-Tetrahydro-2″,6β,7-trihydroxycannabinol
Probably in: Guinea pig B458
→ Δ^1-Tetrahydro-3″,6β,7-trihydroxycannabinol
Probably in: Guinea pig B458
→ Δ^1-Tetrahydro-4″,6β,7-trihydroxycannabinol
Probably in: Guinea pig B458

$\Delta^{1(7)}$-Tetrahydro-2β,4″-dihydroxycannabinol
Probably formed from: $\Delta^{1(7)}$-Tetrahydro-2β-hydroxycannabinol

$\Delta^{1(7)}$-Tetrahydro-2″,6α-dihydroxycannabinol
Probably formed from: $\Delta^{1(7)}$-Tetrahydro-6α-hydroxycannabinol

$\Delta^{1(7)}$-Tetrahydro-2″,6β-dihydroxycannabinol
Probably formed from: $\Delta^{1(7)}$-Tetrahydro-6β-hydroxycannabinol

$\Delta^{1(7)}$-Tetrahydro-3″,6α-dihydroxycannabinol
Probably formed from: $\Delta^{1(7)}$-Tetrahydro-6α-hydroxycannabinol

$\Delta^{1(7)}$-Tetrahydro-4″,6α-dihydroxycannabinol
Probably formed from: $\Delta^{1(7)}$-Tetrahydro-6α-hydroxycannabinol

$\Delta^{1(7)}$-Tetrahydro-4″,6β-dihydroxycannabinol
Probably formed from: $\Delta^{1(7)}$-Tetrahydro-6β-hydroxycannabinol

Δ^{6}-Tetrahydro-1″,3″-dihydroxycannabinol
Probably formed from: Δ^{6}-Tetrahydro-1″-hydroxycannabinol

Δ^{6}-Tetrahydro-1″,4″-dihydroxycannabinol
Probably formed from: Δ^{6}-Tetrahydro-1″-hydroxycannabinol

Δ^{6}-Tetrahydro-1″,7-dihydroxycannabinol
Probably formed from: Δ^{6}-Tetrahydro-1″-hydroxycannabinol
Δ^{6}-Tetrahydro-7-hydroxycannabinol
→ Δ^{6}-Tetrahydro-1″-hydroxycannabinol-7-oic acid
Probably in: Guinea pig B583

Δ^{6}-Tetrahydro-2″,7-dihydroxycannabinol
Probably formed from: Δ^{6}-Tetrahydro-2″-hydroxycannabinol
Δ^{6}-Tetrahydro-7-hydroxycannabinol
→ Δ^{6}-Tetrahydro-2″,5α,7-trihydroxycannabinol
Probably in: Mouse B50
→ Δ^{6}-Tetrahydro-2″,5β,7-trihydroxycannabinol
Probably in: Mouse B50
→ Δ^{6}-Tetrahydro-2″-hydroxycannabinol-7-oic acid
Probably in: Guinea pig B583

Δ^{6}-Tetrahydro-3″,7-dihydroxycannabinol
Probably formed from: Δ^{6}-Tetrahydro-3″-hydroxycannabinol
Δ^{6}-Tetrahydro-7-hydroxycannabinol
→ Δ^{6}-Tetrahydro-3″,5α,7-trihydroxycannabinol
Probably in: Mouse B50
→ Δ^{6}-Tetrahydro-3″,5β,7-trihydroxycannabinol
Probably in: Mouse B50
→ Δ^{6}-Tetrahydro-3″-hydroxycannabinol-7-oic acid
Probably in: Guinea pig B583

Δ^{6}-Tetrahydro-4″,5-dihydroxycannabinol
Probably formed from: Δ^{6}-Tetrahydro-4″-hydroxycannabinol

Δ^{6}-Tetrahydro-4″,7-dihydroxycannabinol
Probably formed from: Δ^{6}-Tetrahydro-4″-hydroxycannabinol
Δ^{6}-Tetrahydro-7-hydroxycannabinol
→ Δ^{6}-Tetrahydro-4″,5α,7-trihydroxycannabinol
Probably in: Mouse B50
→ Δ^{6}-Tetrahydro-4″,5β,7-trihydroxycannabinol
Probably in: Mouse B50
→ Δ^{6}-Tetrahydro-4″-hydroxycannabinol-7-oic acid
Probably in: Guinea pig B583

Δ^{6}-Tetrahydro-5α,7-dihydroxycannabinol
Probably formed from: Δ^{6}-Tetrahydro-5α-hydroxycannabinol
Δ^{6}-Tetrahydro-7-hydroxycannabinol
→ Δ^{6}-Tetrahydro-7-hydroxy-5-oxocannabinol
Probably in: Mouse B50

Δ^{6}-Tetrahydro-5β,7-dihydroxycannabinol
Probably formed from: Δ^{6}-Tetrahydro-5β-hydroxycannabinol
Δ^{6}-Tetrahydro-7-hydroxycannabinol

Δ^{1}-Tetrahydro-4″,6α-dihydroxycannabinol-7-oic acid
Probably formed from: Δ^{1}-Tetrahydro-,4″-hydroxycannabinol-7-oic acid
Δ^{1}-Tetrahydro-4″,6α,7-trihydroxycannabinol

Δ^{1}-Tetrahydro-4″,6α-dihydroxycannabinol-7-oic acid
Probably formed from: Δ^{1}-Tetrahydro-,4″-hydroxycannabinol-7-oic acid

Δ^{6}-Tetrahydro-2″,5α-dihydroxycannabinol-7-oic acid
Probably formed from: Δ^{6}-Tetrahydro-5α-hydroxycannabinol-7-oic acid

Δ^{6}-Tetrahydro-2″,5β-dihydroxycannabinol-7-oic acid
Probably formed from: Δ^{6}-Tetrahydro-5β-hydroxycannabinol-7-oic acid

Δ^{6}-Tetrahydro-3″,5α-dihydroxycannabinol-7-oic acid
Probably formed from: Δ^{6}-Tetrahydro-5α-hydroxycannabinol-7-oic acid

Δ^6-Tetrahydro-3″,5β-dihydroxycannabinol-7-oic acid
Probably formed from: Δ^6-Tetrahydro-5β-hydroxycannabinol-7-oic acid

Δ^6-Tetrahydro-4″,5α-dihydroxycannabinol-7-oic acid
Probably formed from: Δ^6-Tetrahydro-5α-hydroxycannabinol-7-oic acid

Δ^6-Tetrahydro-4″,5β-dihydroxycannabinol-7-oic acid
Probably formed from: Δ^6-Tetrahydro-5β-hydroxycannabinol-7-oic acid

1,2,3,4-Tetrahydro-r-1,t-2-dihydroxychrysene-t-3,4-oxide
→ 4-(S-Glutathionyl)-1,2,3,4-tetrahydro-r-1,t-2,3-trihydroxychrysene
In: Rat H395

1,2,3,4-Tetrahydro-c-3,t-4-dihydroxydibenz[*a*,*h*]anthracene-r-1,2-oxide
→ 1-(S-Glutathionyl)-1,2,3,4-tetrahydro-2,c-3,t-4-trihydroxydibenz[*a*,*h*]anthracene
In: Rat H395

Tetrahydro-1,9-dihydroxy-2,10-dimethoxyprotoberberine
Formed from: Crassifoline

Tetrahydro-2,3-dihydroxy-9,10-dimethoxyprotoberberine
→ Tetrahydrojatrorrhizine
In: Argemone D77

Tetrahydro-3,9-dihydroxy-2,10-dimethoxyprotoberberine
Formed from: Protosinomenine

1,2,3,4-Tetrahydro-2,3-dihydroxy-1,4-di-(methylthio)naphthalene
Probably formed from: S-(1,2-Dihydro-2-hydroxynaphth-1-yl)glutathione

1,2,3,4-Tetrahydro-2,4-dihydroxy-1,3-di-(methylthio)naphthalene
Probably formed from: 1,2-Dihydro-1-hydroxy-2-methylthionaphthalene

Δ^1-Tetrahydro-2″,7-dihydroxy-*n*-heptylcannabinol
Probably formed from: Δ^1-Tetrahydro-7-hydroxy-*n*-heptylcannabinol
→ Δ^1-Tetrahydro-2″,3″,7-trihydroxy-*n*-heptylcannabinol
Probably in: Mouse D357

Δ^1-Tetrahydro-3″,7-dihydroxy-*n*-heptylcannabinol
Probably formed from: Δ^1-Tetrahydro-7-hydroxy-*n*-heptylcannabinol

Δ^1-Tetrahydro-6α,7-dihydroxy-*n*-heptylcannabinol
Probably formed from: Δ^1-Tetrahydro-7-hydroxy-*n*-heptylcannabinol

Δ^1-Tetrahydro-6β,7-dihydroxy-*n*-heptylcannabinol
Probably formed from: Δ^1-Tetrahydro-7-hydroxy-*n*-heptylcannabinol

Δ^1-Tetrahydro-2″,3″-dihydroxy-*n*-heptylcannabinol-7-oic acid
Probably formed from: Δ^1-Tetrahydro-2″-hydroxy-*n*-heptylcannabinol-7-oic acid

1,2,3,4-Tetrahydro-1-(3,4-dihydroxybenzyl)-6,7-dihydroxyisoquinoline
See: Tetrahydropapaveroline

1,2,3,4-Tetrahydro-6,7-dihydroxy-1-(*p*-hydroxybenzyl)isoquinoline
See: Norcoclaurine

1,2,3,4-Tetrahydro-6,7-dihydroxy-1-(3-hydroxy-4-methoxybenzyl)isoquinoline
Formed from: Tetrahydropapaveroline
→ Norprotosinomenine
In: Argemone D77
→ Reticuline
In: Litsea A3222

1,2,3,4-Tetrahydro-6,7-dihydroxy-1-(4-hydroxy-3-methoxybenzyl)isoquinoline
Formed from: Tetrahydropapaveroline
→ Nororientaline
In: Argemone K783

1,2,3,4-Tetrahydro-6,7-dihydroxy-1-(3-hydroxy-4-methoxybenzyl)-2-methylisoquinoline
Formed from: Laudanosoline

1,2,3,4-Tetrahydro-6,7-dihydroxyisoquinoline
See: Norsalsolinol

1,2,3,4-Tetrahydro-6,7-dihydroxy-8-methoxy-1,N-dimethylisoquinoline
→ 1,2,3,4-Tetrahydro-6-hydroxy-7,8-dimethoxy-1,N-dimethylisoquinoline
In: Lophophora A3600
→ 1,2,3,4-Tetrahydro-7-hydroxy-6,8-dimethoxy-1,N-dimethylisoquinoline
In: Lophophora A3600

1,2,3,4-Tetrahydro-6,8-dihydroxy-7-methoxy-1,N-dimethylisoquinoline
Formed from: 1,2,3,4-Tetrahydro-6,7,8-trihydroxy-1,N-dimethylisoquinoline

1,2,3,4-Tetrahydro-7,8-dihydroxy-6-methoxy-1,N-dimethylisoquinoline
→ 1,2,3,4-Tetrahydro-7-hydroxy-6,8-dimethoxy-1,N-dimethylisoquinoline
In: Lophophora A3600
→ 1,2,3,4-Tetrahydro-8-hydroxy-6,7-dimethoxy-1,N-dimethylisoquinoline
In: Lophophora A3600

1,2,3,4-Tetrahydro-4,6-dihydroxy-7-methoxyisoquinoline
Formed from: 4-Hydroxynorsalsolinol

1,2,3,4-Tetrahydro-4,7-dihydroxy-6-methoxyisoquinoline
Formed from: 4-Hydroxynorsalsolinol

1,2,3,4-Tetrahydro-7,8-dihydroxy-6-methoxyisoquinoline
→ 1,2,3,4-Tetrahydro-7-hydroxy-6,8-dimethoxyisoquinoline
Probably in: Lophophora A3600
→ 1,2,3,4-Tetrahydro-8-hydroxy-6,7-dimethoxyisoquinoline
Probably in: Lophophora A3600

1,2,3,4-Tetrahydro-7,8-dihydroxy-6-methoxy-1-methylisoquinoline
→ 1,2,3,4-Tetrahydro-7-hydroxy-6,8-dimethoxy-1-methylisoquinoline
Probably in: Lophophora A3600
→ 1,2,3,4-Tetrahydro-8-hydroxy-6,7-dimethoxy-1-methylisoquinoline
Probably in: Lophophora A3600

Tetrahydro-1′,7-dihydroxy-*abn*-methyl-Δ^6-cannabinol
Probably formed from: Tetrahydro-7-hydroxy-*abn*-methyl-Δ^6-cannabinol

Tetrahydro-5α,7-dihydroxy-methyl-Δ^6-cannabinol
Probably formed from: Tetrahydro-7-hydroxy-methyl-Δ^6-cannabinol

1,2,3,4-Tetrahydro-6,7-dihydroxy-1-methylisoquinoline
See: Salsolinol

1,2,3,4-Tetrahydro-6,7-dihydroxy-2-methylisoquinoline (N-methylnorsalsolinol)
Formed from: 3,4-Dihydroxy-N-methylphenethylamine
→ 1,2,3,4-Tetrahydro-6-hydroxy-7-methoxy-2-methylisoquinoline
In: Rat A486
→ 1,2,3,4-Tetrahydro-7-hydroxy-6-methoxy-2-methylisoquinoline
In: Rat A486

1,2,3,4-Tetrahydro-6,7-dihydroxy-1-methylisoquinoline-1-carboxylic acid
See: Salsolinol-1-carboxylic acid

1,2,3,4-Tetrahydro-6,7-dihydroxy-1-methylisoquinoline-3-carboxylic acid
See: Salsolinol-3-carboxylic acid

(+)-1,2,3,4-Tetrahydro-*cis*-(1*R*,2*S*)-dihydroxynaphthalene
Formed from: 1,2-Dihydronaphthalene
→ 3,4-Dihydro-1,2-dihydroxynaphthalene
In: Pseudomonas K348
Sphingomonas H880

(−)-1,2,3,4-Tetrahydro-*cis*-(1*R*,2*S*)-dihydroxynaphthalene
Formed from: 1,2-Dihydronaphthalene
→ 3,4-Dihydro-1,2-dihydroxynaphthalene
In: Pseudomonas K348

(+)-1,2,3,4-Tetrahydro-*cis*-(1*S*,2*R*)-dihydroxynaphthalene
Formed from: 1,2-Dihydronaphthalene

5,6,7,8-Tetrahydro-2,5-dihydroxynaphthalene
Probably formed from: 5,6,7,8-Tetrahydro-2-naphthol
→ 1,2,3,4-Tetrahydro-6-hydroxy-1-oxo-naphthalene
Probably in: Penicillium J836

Δ^1-Tetrahydro-3″,7-dihydroxy-6-oxocannabinol
Probably formed from: Δ^1-Tetrahydro-3′,6,7-trihydroxycannabinol

Δ^1-Tetrahydro-4″,7-dihydroxy-6-oxocannabinol
Probably formed from: Δ^1-Tetrahydro-4″-hydroxy-6-oxocannabinol

1,2,3,4-Tetrahydro-3,4-dihydroxy-2-oxodibenzanthracene
Probably formed from: *trans*-1,2,3,4-Tetrahydro-1,2,3,4-tetrahydroxydibenzanthracene

1,2,3,4-Tetrahydro-3,4-dihydroxy-1-oxonaphthalene
Formed from: α-Naphthol

4-(1-(5,6,7,8-Tetrahydro-6,7-dihydroxy-3,5,5,8,8-pentamethyl-2-naphthalenyl)-1-ethenyl) benzoyltaurine
Probably formed from: 6,7-Dihydroxytargetrin

***cis*-3,4,9,10-Tetrahydro-(3*S*),(4*R*)-dihydroxyphenanthrene**
Formed from: 9,10-Dihydrophenanthrene

1,3,4,5-Tetrahydro-2,7-dihydroxy-5-phenyl-2*H*-1,4-benzodiazepine
Probably formed from: 1,3,4,5-Tetrahydro-7-hydroxy-5-phenyl-2*H*-1,4-benzodiazepin-2-one

1,2,3,4-Tetrahydro-3,4-dihydroxy-1-oxonaphthalene
Formed from: α-Naphthol

(*S*)-(−)-1,2,3,4-Tetrahydro-5,6-dihydroxy-N-(N-((2-phenyl)ethyl)-6-hexylamino)-N-propyl-2-naphthylamine
→ (*S*)-(−)-1,2,3,4-Tetrahydro-5,6-dihydroxy-N-(N-((2-phenyl)ethyl)-6-hexylamino)-N-propyl-2-naphthylamine-β-D-glucuronide
In: Rat K352
→ (*S*)-(−)-1,2,3,4-Tetrahydro-6-hydroxy-5-methoxy-N-(N-((2-phenyl)ethyl)-6-hexylamino)-N-propyl-2-naphthylamine
In: Rat K352

(*S*)-(−)-1,2,3,4-Tetrahydro-5,6-dihydroxy-N-(N-((2-phenyl)ethyl)-6-hexylamino)-N-propyl-2-naphthylamine-β-D-glucuronide
Formed from: (*S*)-(−)-1,2,3,4-Tetrahydro-5,6-dihydroxy-N-(N-((2-phenyl)ethyl)-6-hexylamino)-N-propyl-2-naphthylamine

7-(3-(Tetrahydro-2,4-dihydroxy-2*H*-pyran-4-yl) phenyl)methoxy-4-phenylnaphtho[2,3-*c*]furan-1(3*H*)-one
Formed from: 7-(3-(Tetrahydro-4-hydroxy-2*H*-pyran-4-yl)phenyl)methoxy-4-phenylnaphtho[2,3-*c*]furan-1(3*H*)-one

1,2,3,4-Tetrahydro-6,7-dihydroxy-1-(3,4,5-trimethoxybenzyl)isoquinoline
See: Trimetoquinol

6a,7,10,10a-Tetrahydro-1,4′-dihydroxy-6,6,9-trimethyl-3-*n*-pentyldibenzo[*b*,*d*]pyran
Formed from: 6a,7,10,10a-Tetrahydro-1-hydroxy-6,6,9-trimethyl-3-*n*-pentyldibenzo[*b*,*d*]pyran

5,8,13,13a-Tetrahydro-9,10-dimethoxy-6*H*-benzo[*g*]-1,3-benzodioxolo[5,6-*a*] quinolizine
See: Tetrahydroberberine

1,2,3,4-Tetrahydro-6,7-dimethoxy-1-methylisoquinoline
Probably formed from: 1,2,3,4-Tetrahydro-7-hydroxy-6-methoxy-1-methylisoquinoline

1,2,3,4-Tetrahydro-6,7-dimethoxy-N-methylisoquinoline
Formed from: Armepavine

1,2,3,4-Tetrahydro-6,7-dimethoxy-3-(2-(2-oxo-1-piperidinyl)ethyl-5*H*-1-benzopyrano[3,4-*c*] pyridin-5-one
Formed from: 1,2,3,4-Tetrahydro-6,7-dimethoxy-3-(2-(1-piperidinyl)ethyl-5*H*-1-benzopyrano[3,4-*c*]pyridin-5-one

1,2,3,4-Tetrahydro-8,9-dimethoxy-3-(2-(1-piperidinyl)ethyl-5*H*-1-benzopyrano[3,4-*c*] pyridin-5-one
→ 1,2,3,4-Tetrahydro-8,9-dimethoxy-3-(2-(2-oxo-1-piperidinyl)ethyl-5*H*-1-benzopyrano[3,4-*c*]pyridin-5-one
In: Dog A1960
→ 1,2,3,4-Tetrahydro-8-hydroxy-9-methoxy-3-(2-(1-piperidinyl)ethyl-5*H*-1-benzopyrano[3,4-*c*]pyridin-5-one
Probably in: Dog A1960

2,3,4,9-Tetrahydro-N,N-dimethyl-1*H*-carbazol-3-amine
See: Cyclindole

1,2,3,4-Tetrahydro-1,2-dimethylisoquinoline
Formed from: 1,2,3,4-Tetrahydro-1-methylisoquinoline

2,3,5,6-Tetrahydro-5,6-dioxo-1*H*-indole-2-carboxylic acid
See: Dopachrome

Tetrahydrofolic acid (N-(4-(((2-amino-1,4,5,6,7,8-hexahydro-4-oxo-6-pteridinyl)methyl)amino) benzoyl)-L-glutamic acid)
Formed from: Dihydrofolic acid
10-Formyltetrahydrofolic acid

5,10-Methylenetetrahydrofolic acid
5-Methyltetrahydrofolic acid
→ Dihydrofolic acid
In: Man A2946
→ 10-Formyltetrahydrofolic acid
In: Clostridium E360
→ γ-Glutamyltetrahydrofolic acid
In: Escherichia A1362
Pig E277
→ Tetrahydropteroyldiglutamic acid
In: Pig K455

Δ^1-Tetrahydro-*n*-heptylcannabinol
→ Δ^1-Tetrahydro-6β-hydroxy-*n*-heptylcannabinol
Probably in: Mouse D357
→ Δ^1-Tetrahydro-7-hydroxy-*n*-heptylcannabinol
Probably in: Mouse D357

Δ^1-Tetrahydro-*n*-heptylcannabinol-7-oic acid
Formed from: Δ^1-Tetrahydro-7-hydroxy-*n*-heptylcannabinol
→ Δ^1-Tetrahydro-6α-hydroxy-*n*-heptylcannabinol-7-oic acid
Probably in: Mouse D357
→ Δ^1-Tetrahydro-2″-hydroxy-*n*-heptylcannabinol-7-oic acid
Probably in: Mouse D357
→ Δ^1-Tetrahydro-3″-hydroxy-*n*-heptylcannabinol-7-oic acid
Probably in: Mouse D357

Tetrahydrohomofolic acid
Formed from: Dihydrohomofolic acid

(+)-N-(2,3,4,5-Tetrahydro-1-((3-hydroxyadamantan-1-yl)methyl)-2,4-dioxo-5-phenyl-1,5-benzodiazepin-3-yl)-N′-(*p*-hydroxyphenyl)urea
Formed from: (+)-N-(2,3,4,5-Tetrahydro-1-((3-hydroxyadamantan-1-yl)methyl)-2,4-dioxo-5-phenyl-1,5-benzodiazepin-3-yl)-N′-phenylurea

(+)-N-(2,3,4,5-Tetrahydro-1-((3-hydroxyadamantan-1-yl)methyl)-2,4-dioxo-5-phenyl-1,5-benzodiazepin-3-yl)-N′-phenylurea
Formed from: (+)-N-(2,3,4,5-Tetrahydro-1-((adamantan-1-yl)methyl)-2,4-dioxo-5-phenyl-1,5-benzodiazepin-3-yl)-N′-phenylurea
→ (+)-N-(2,3,4,5-Tetrahydro-1-((3,5-dihydroxyadamantan-1-yl)methyl)-2,4-dioxo-5-phenyl-1,5-benzodiazepin-3-yl)-N′-phenylurea
Probably in: Dog H926, J414
Man H926
Rat H926, J414
→ (+)-N-(2,3,4,5-Tetrahydro-1-((3-hydroxyadamantan-1-yl)methyl)-2,4-dioxo-5-phenyl-1,5-benzodiazepin-3-yl)-N′-(*p*-hydroxyphenyl)urea
Probably in: Dog H926, J414
Man H926
Rat H926, J414

(+)-N-(2,3,4,5-Tetrahydro-1-((4-hydroxyadamantan-1-yl)methyl)-2,4-dioxo-5-phenyl-1,5-benzodiazepin-3-yl)-N′-phenylurea
Formed from: (+)-N-(2,3,4,5-Tetrahydro-1-((adamantan-1-yl)methyl)-2,4-dioxo-5-phenyl-1,5-benzodiazepin-3-yl)-N′-phenylurea

1,2,3,4-Tetrahydro-1-(*p*-hydroxybenzyl)-6,7-dimethoxy-2-methylisoquinoline
See: Armepavine

Tetrahydro-4-(*p*-hydroxybenzyl)-1-methylpyridine
→ 4-(*p*-Hydroxybenzyl)-1-methylpyridinium
Probably in: Rat F816

Δ^6-Tetrahydro-(2-hydroxy-*n*-butyl)cannabinol-7-oic acid
Probably formed from: *n*-Butyl-Δ^6-tetrahydrocannabinol-7-oic acid

Δ^6-Tetrahydro-(3-hydroxy-*n*-butyl)cannabinol-7-oic acid
Probably formed from: *n*-Butyl-Δ^6-tetrahydrocannabinol-7-oic acid

Δ^1-Tetrahydro-1″-hydroxycannabinol
Formed from: Δ^1-Tetrahydrocannabinol
→ Δ^1-Tetrahydro-1″,6β-dihydroxycannabinol
Probably in: Guinea pig B458

Δ^1-Tetrahydro-2″-hydroxycannabinol
Formed from: Δ^1-Tetrahydrocannabinol
→ Δ^1-Tetrahydro-2″,6α-dihydroxycannabinol
In: Mouse F492
→ Δ^1-Tetrahydro-2″,6β-dihydroxycannabinol
In: Mouse F492
→ Δ^1-Tetrahydro-2″,7-dihydroxycannabinol
In: Mouse F492

Δ^1-Tetrahydro-3″-hydroxycannabinol
Formed from: Δ^1-Tetrahydrocannabinol
→ Δ^1-Tetrahydro-3″,6α-dihydroxycannabinol
In: Mouse F492
→ Δ^1-Tetrahydro-3″,6β-dihydroxycannabinol
Probably in: Guinea pig B458
→ Δ^1-Tetrahydro-3″,7-dihydroxycannabinol
In: Mouse F492
→ 3″,4″,5″-Trisnor-2″-formyl-Δ^1-tetrahydro-7-hydroxycannabinol
In: Mouse F492
→ 3″,4″,5″-Trisnor-Δ^1-tetrahydro-3″-hydroxycannabinol
In: Mouse F492
→ 3″,4″,5″-Trisnor-Δ^1-tetrahydrocannabinol-2″-oic acid
In: Mouse F492

Δ^1-Tetrahydro-4″-hydroxycannabinol
Formed from: Δ^1-Tetrahydrocannabinol
→ 4″,5″-Bisnor-Δ^1-tetrahydrocannabinol-3″-oic acid
In: Mouse F526
→ Δ^1-Tetrahydro-4″,6α-dihydroxycannabinol
In: Mouse F526
→ Δ^1-Tetrahydro-4″,6β-dihydroxycannabinol
Probably in: Guinea pig B458
→ Δ^1-Tetrahydro-4″,7-dihydroxycannabinol
In: Mouse F526
→ Δ^1-Tetrahydro-4″-oxocannabinol
In: Mouse F526

Δ^1-Tetrahydro-5″-hydroxycannabinol
Formed from: Δ^1-Tetrahydrocannabinol
→ Δ^1-Tetrahydrocannabinol-5″-oic acid
In: Mouse D345
See also: B89
→ Δ^1-Tetrahydrocannabinol-3′-yl-β-D-glucuronide
In: Rabbit A3449
→ Δ^1-Tetrahydro-5″-hydroxycannabinol-3′-β-D-glucuronide
In: Rabbit A3449

Δ^1-Tetrahydro-6α-hydroxycannabinol
Formed from: Δ^1-Tetrahydrocannabinol
→ Δ^1-Tetrahydro-6α,7-dihydroxycannabinol
In: Mouse B614
→ Δ^1-Tetrahydro-6-oxocannabinol
In: Mouse B614
See also: A758, A1376, C2

Δ^1-Tetrahydro-6β-hydroxycannabinol
Formed from: Δ^1-Tetrahydrocannabinol
→ 4″,5″-Bisnor-Δ^1-tetrahydro-6β-hydroxycannabinol-3″-oic acid
Probably in: Rabbit B102
→ Δ^1-Tetrahydro-6-oxocannabinol
In: Mouse B614
See also: G298
→ 3″,4″,5″-Trisnor-Δ^1-tetrahydro-6β-hydroxycannabinol-2″-oic acid
Probably in: Rabbit B102

Δ^1-Tetrahydro-7-hydroxycannabinol
(Δ^9-tetrahydro-11-hydroxycannabinol)
Formed from: Δ^1-Tetrahydrocannabinol
Δ^1-Tetrahydro-7-hydroxycannabinol palmitate
→ 7-Acetoxy-Δ^1-tetrahydrocannabinol
Probably in: Rat A389
→ Δ^1-Tetrahydrocannabinol-7-yl-β-D-glucuronide
In: Rabbit A3449
→ Δ^1-Tetrahydro-1″,7-dihydroxycannabinol
Probably in: Guinea pig B458
Mouse E598
→ Δ^1-Tetrahydro-2′,7-dihydroxycannabinol
In: Mouse B614
→ Δ^1-Tetrahydro-2″,7-dihydroxycannabinol
Probably in: Guinea pig B458
Man C2
Mouse E598
→ Δ^1-Tetrahydro-3″,7-dihydroxycannabinol
Probably in: Chaetomium A3628
Guinea pig B458
Man C2
Mouse E598
→ Δ^1-Tetrahydro-4″,7-dihydroxycannabinol
Probably in: Mouse E598
→ Δ^1-Tetrahydro-6α,7-dihydroxycannabinol
Probably in: Guinea pig B458
Man A680, C2
Mouse E411, E598
Rabbit A108
Rat A116, A389
→ Δ^1-Tetrahydro-6β,7-dihydroxycannabinol

Probably in: Guinea pig B458
Man C2
→ Δ^{1}-Tetrahydro-7-hydroxycannabinol-3′-β-D-glucuronide
In: Rabbit A3449
→ Δ^{1}-Tetrahydro-7-hydroxycannabinol-1,2-oxide
Probably in: Man C2
→ Δ^{1}-Tetrahydro-7-hydroxycannabinol palmitate
In: Rat B68
See also: A2613
→ Δ^{1}-Tetrahydro-7-oxocannabinol
In: Rat H67
See also: A1267, E598

$\Delta^{1(7)}$-Tetrahydro-2α-hydroxycannabinol
Formed from: $\Delta^{1(7)}$-Tetrahydrocannabinol

$\Delta^{1(7)}$-Tetrahydro-2β-hydroxycannabinol
Formed from: $\Delta^{1(7)}$-Tetrahydrocannabinol
→ $\Delta^{1(7)}$-Tetrahydro-2β,4″-dihydroxycannabinol
Probably in: Rat B367

$\Delta^{1(7)}$-Tetrahydro-2″-hydroxycannabinol
Formed from: $\Delta^{1(7)}$-Tetrahydrocannabinol

$\Delta^{1(7)}$-Tetrahydro-3″-hydroxycannabinol
Formed from: $\Delta^{1(7)}$-Tetrahydrocannabinol

$\Delta^{1(7)}$-Tetrahydro-4″-hydroxycannabinol
Formed from: $\Delta^{1(7)}$-Tetrahydrocannabinol

$\Delta^{1(7)}$-Tetrahydro-5″-hydroxycannabinol
Formed from: $\Delta^{1(7)}$-Tetrahydrocannabinol

$\Delta^{1(7)}$-Tetrahydro-6α-hydroxycannabinol
Formed from: $\Delta^{1(7)}$-Tetrahydrocannabinol
→ $\Delta^{1(7)}$-Tetrahydro-2″,6α-dihydroxycannabinol
Probably in: Rat B367
→ $\Delta^{1(7)}$-Tetrahydro-3″,6α-dihydroxycannabinol
Probably in: Rat B367
→ $\Delta^{1(7)}$-Tetrahydro-4″,6α-dihydroxycannabinol
In: Rat B367

$\Delta^{1(7)}$-Tetrahydro-6β-hydroxycannabinol
Formed from: $\Delta^{1(7)}$-Tetrahydrocannabinol
→ $\Delta^{1(7)}$-Tetrahydro-2″,6β-dihydroxycannabinol
Probably in: Rat B367
→ $\Delta^{1(7)}$-Tetrahydro-4″,6β-dihydroxycannabinol
Probably in: Rat B367

$\Delta^{3,4}$-Tetrahydro-4″-hydroxycannabinol
Formed from: $\Delta^{3,4}$-Tetrahydrocannabinol

Δ^{6}-Tetrahydro-1″-hydroxycannabinol
Formed from: Δ^{6}-Tetrahydrocannabinol
→ Δ^{6}-Tetrahydro-1″,3″-dihydroxycannabinol
Probably in: Guinea pig B583
→ Δ^{6}-Tetrahydro-1″,4″-dihydroxycannabinol
Probably in: Guinea pig B583
→ Δ^{6}-Tetrahydro-1″,7-dihydroxycannabinol
Probably in: Guinea pig B583

Δ^{6}-Tetrahydro-2″-hydroxycannabinol
Formed from: Δ^{6}-Tetrahydrocannabinol
→ Δ^{6}-Tetrahydro-2″,7-dihydroxycannabinol
Probably in: Mouse E897
Guinea pig B583

Δ^{6}-Tetrahydro-3″-hydroxycannabinol
Formed from: Δ^{6}-Tetrahydrocannabinol
→ Δ^{6}-Tetrahydro-3″,7-dihydroxycannabinol
Probably in: Mouse E897
Guinea pig B583
→ Δ^{6}-Tetrahydro-1,3″,6-trihydroxycannabinol
Probably in: Guinea pig B583

Δ^{6}-Tetrahydro-4″-hydroxycannabinol
Formed from: Δ^{6}-Tetrahydrocannabinol
→ Δ^{6}-Tetrahydro-4″,5-dihydroxycannabinol
Probably in: Rhesus monkey B79
→ Δ^{6}-Tetrahydro-4″,7-dihydroxycannabinol
Probably in: Mouse E897
Guinea pig B583
→ Δ^{6}-Tetrahydro-1,4″,6-trihydroxycannabinol
Probably in: Guinea pig B583

Δ^{6}-Tetrahydro-5″-hydroxycannabinol
Formed from: Δ^{6}-Tetrahydrocannabinol

Δ^{6}-Tetrahydro-5α-hydroxycannabinol
Formed from: Δ^{6}-Tetrahydrocannabinol
→ Δ^{6}-Tetrahydro-5α,7-dihydroxycannabinol
Probably in: Man E897
Guinea pig B583
→ Δ^{6}-Tetrahydro-5-oxocannabinol
In: Guinea pig E945
Man K322
Rat K397
See also: A758

Δ^6-Tetrahydro-5β-hydroxycannabinol
Formed from: Δ^6-Tetrahydrocannabinol
→ Δ^6-Tetrahydro-5β,7-dihydroxycannabinol
Probably in: Man E897
Guinea pig B583
→ Δ^6-Tetrahydro-5-oxocannabinol
In: Guinea pig E945
Man K322
Rat K397

Δ^6-Tetrahydro-7-hydroxycannabinol
Formed from: Δ^6-Tetrahydrocannabinol
→ Δ^6-Tetrahydrocannabinol-7-oic acid
In: Guinea pig B583
Rabbit A1106
→ Δ^6-Tetrahydro-1″,7-dihydroxycannabinol
Probably in: Guinea pig B583
Rhesus monkey B79
→ Δ^6-Tetrahydro-2″,7-dihydroxycannabinol
Probably in: Guinea pig B583
Mouse B50
Rhesus monkey B79
→ Δ^6-Tetrahydro-3″,7-dihydroxycannabinol
Probably in: Guinea pig B583
Mouse B50
→ Δ^6-Tetrahydro-4″,7-dihydroxycannabinol
Probably in: Guinea pig B583
Mouse B50
Rhesus monkey B79
→ Δ^6-Tetrahydro-5α,7-dihydroxycannabinol
Probably in: Guinea pig B583
Mouse B50
Rat A389, A1122
→ Δ^6-Tetrahydro-5β,7-dihydroxycannabinol
Probably in: Guinea pig B583
Mouse B50
Rat A389
Rhesus monkey B79
→ Δ^1-Tetrahydro-7-hydroxycannabinol acetate
Probably in: Mouse E897
→ Δ^1-Tetrahydro-7-hydroxycannabinol palmitate
Probably in: Rat A2613
→ Δ^6-Tetrahydro-7-hydroxy-3′-O-methylcannabinol
In: Rat A1122
→ Δ^6-Tetrahydro-7-oxocannabinol
In: Mouse H67, J451
Rabbit B59

Δ^9-Tetrahydro-11-hydroxycannabinol
See: Δ^1-Tetrahydro-7-hydroxycannabinol

Δ^6-Tetrahydro-7-hydroxycannabinol acetate
Probably formed from: Δ^6-Tetrahydro-7-hydroxycannabinol

Δ^1-Tetrahydro-5″-hydroxycannabinol-3′-β-D-glucuronide
Formed from: Δ^1-Tetrahydro-5″-hydroxycannabinol

Δ^1-Tetrahydro-7-hydroxycannabinol-3′-β-D-glucuronide
Formed from: Δ^1-Tetrahydro-7-hydroxycannabinol

Δ^1-Tetrahydro-1″-hydroxycannabinol-7-oic acid
Probably formed from: Δ^1-Tetrahydrocannabinol-7-oic acid

Δ^1-Tetrahydro-2″-hydroxycannabinol-7-oic acid
Formed from: Δ^1-Tetrahydrocannabinol
Probably formed from: Δ^1-Tetrahydrocannabinol-7-oic acid
Δ^1-Tetrahydro-2″,7-dihydroxycannabinol

Δ^1-Tetrahydro-3″-hydroxycannabinol-7-oic acid
Probably formed from: Δ^1-Tetrahydrocannabinol-7-oic acid
Δ^1-Tetrahydro-3″,7-dihydroxycannabinol

Δ^1-Tetrahydro-4″-hydroxycannabinol-7-oic acid
Probably formed from: Δ^1-Tetrahydrocannabinol-7-oic acid
Δ^1-Tetrahydro-4″,7-dihydroxycannabinol
→ Δ^1-Tetrahydro-4″,6α-dihydroxycannabinol-7-oic acid
Probably in: Mouse A3155

Δ^1-Tetrahydro-6α-hydroxycannabinol-7-oic acid
Probably formed from: Δ^1-Tetrahydrocannabinol-7-oic acid

Δ^6-Tetrahydro-1″-hydroxycannabinol-7-oic acid
Formed from: Δ^6-Tetrahydro-1″,7-dihydroxycannabinol
Probably formed from: Δ^6-Tetrahydrocannabinol
→ Δ^6-Tetrahydro-1″-oxocannabinol-7-oic acid
Probably in: Monkey A2210

Δ^6-Tetrahydro-2″-hydroxycannabinol-7-oic acid
Formed from: Δ^6-Tetrahydro-2″,7-dihydroxycannabinol
Δ^6-Tetrahydrocannabinol-7-oic acid

Δ^6-Tetrahydro-3″-hydroxycannabinol-7-oic acid
Formed from: Δ^6-Tetrahydro-3″,7-dihydroxycannabinol
Δ^6-Tetrahydrocannabinol-7-oic acid

Δ^6-Tetrahydro-4″-hydroxycannabinol-7-oic acid
Formed from: Δ^6-Tetrahydro-4″,7-dihydroxycannabinol
Δ^6-Tetrahydrocannabinol-7-oic acid
→ Δ^6-Tetrahydro-4″-oxocannabinol-7-oic acid
Probably in: Guinea pig B583

Δ^6-Tetrahydro-5α-hydroxycannabinol-7-oic acid
Probably formed from: Δ^6-Tetrahydrocannabinol-7-oic acid
→ Δ^6-Tetrahydro-2″,5α-dihydroxycannabinol
Probably in: Mouse B50
→ Δ^6-Tetrahydro-3″,5α-dihydroxycannabinol
Probably in: Mouse B50
→ Δ^6-Tetrahydro-4″,5α-dihydroxycannabinol
Probably in: Mouse B50
→ Δ^6-Tetrahydro-5-oxocannabinol-7-oic acid
Probably in: Mouse B50

Δ^6-Tetrahydro-5β-hydroxycannabinol-7-oic acid
Probably formed from: Δ^6-Tetrahydrocannabinol-7-oic acid
→ Δ^6-Tetrahydro-2″,5β-dihydroxycannabinol-7-oic acid
Probably in: Mouse B50
→ Δ^6-Tetrahydro-3″,5β-dihydroxycannabinol-7-oic acid
Probably in: Mouse B50
→ Δ^6-Tetrahydro-4″,5β-dihydroxycannabinol-7-oic acid
Probably in: Mouse B50

Δ^1-Tetrahydro-2″-hydroxycannabinol-1α,2α-oxide
Formed from: Δ^1-Tetrahydrocannabinol-1α,2α-oxide

Δ^1-Tetrahydro-3″-hydroxycannabinol-1α,2α-oxide
Formed from: Δ^1-Tetrahydrocannabinol-1α,2α-oxide

Δ^1-Tetrahydro-4″-hydroxycannabinol-1α,2α-oxide
Formed from: Δ^1-Tetrahydrocannabinol-1α,2α-oxide

Δ^1-Tetrahydro-6β-hydroxycannabinol-1α,2α-oxide
Formed from: Δ^1-Tetrahydrocannabinol-1α,2α-oxide

Δ^1-Tetrahydro-7-hydroxycannabinol-1α,2α-oxide
Formed from: Δ^1-Tetrahydrocannabinol-1α,2α-oxide
Probably formed from: Δ^1-Tetrahydro-7-hydroxycannabinol

Δ^1-Tetrahydro-7-hydroxycannabinol palmitate
Probably formed from: Δ^1-Tetrahydro-7-hydroxycannabinol
→ Δ^1-Tetrahydro-7-hydroxycannabinol
With cholesterol esterase B68
Lipase B68

Δ^6-Tetrahydro-7-hydroxycannabinol palmitate
Probably formed from: Δ^6-Tetrahydro-7-hydroxycannabinol

1,2,3,4-Tetrahydro-6-hydroxy-β-carboline
Formed from: 5-Hydroxytryptamine
1,2,3,4-Tetrahydro-β-carboline
1,2,3,4-Tetrahydro-6-hydroxy-β-carboline-1-carboxylic acid

1,2,3,4-Tetrahydro-7-hydroxy-β-carboline
Formed from: 1,2,3,4-Tetrahydro-β-carboline

1,2,3,4-Tetrahydro-6-hydroxy-β-carboline-1-carboxylic acid
→ 1,2,3,4-Tetrahydro-6-hydroxy-β-carboline
In: Rat E171

Tetrahydro-4-(4-hydroxy-3,5-dimethoxybenzoyl)-1,4-oxazine
Probably formed from: Tetrahydro-4-(3,4,5-trimethoxybenzoyl)-1,4-oxazine

1,2,3,4-Tetrahydro-6-hydroxy-7,8-dimethoxy-1,N-dimethylisoquinoline
Formed from: 1,2,3,4-Tetrahydro-6,7-dihydroxy-8-methoxy-1,N-dimethylisoquinoline

1,2,3,4-Tetrahydro-7-hydroxy-6,8-dimethoxy-1,N-dimethylisoquinoline
Formed from: 1,2,3,4-Tetrahydro-6,7-dihydroxy-8-methoxy-1,N-dimethylisoquinoline
1,2,3,4-Tetrahydro-7,8-dihydroxy-6-methoxy-1,N-dimethylisoquinoline
→ 1,2,3,4-Tetrahydro-6,7,8-trimethoxy-1,N-dimethylisoquinoline
In: Lophophora A3600

1,2,3,4-Tetrahydro-8-hydroxy-6,7-dimethoxy-1,N-dimethylisoquinoline
Formed from: 1,2,3,4-Tetrahydro-7,8-dihydroxy-6-methoxy-1,N-dimethylisoquinoline

1,2,3,4-Tetrahydro-7-hydroxy-6,8-dimethoxyisoquinoline
Probably formed from: 1,2,3,4-Tetrahydro-7,8-dihydroxy-6-methoxyisoquinoline
→ 1,2,3,4-Tetrahydro-6,7,8-trimethoxyisoquinoline
In: Lophophora A3600

1,2,3,4-Tetrahydro-8-hydroxy-6,7-dimethoxyisoquinoline
Probably formed from: 1,2,3,4-Tetrahydro-7,8-dihydroxy-6-methoxyisoquinoline

1,2,3,4-Tetrahydro-7-hydroxy-6,8-dimethoxy-1-methylisoquinoline
Probably formed from: 1,2,3,4-Tetrahydro-7,8-dihydroxy-6-methoxy-1-methylisoquinoline
→ 1,2,3,4-Tetrahydro-6,7,8-trimethoxy-1-methylisoquinoline
In: Lophophora A3600

1,2,3,4-Tetrahydro-7-hydroxy-6,8-dimethoxy-N-methylisoquinoline
→ 1,2,3,4-Tetrahydro-6,7,8-trimethoxy-1-methylisoquinoline
In: Lophophora A3600

1,2,3,4-Tetrahydro-8-hydroxy-6,7-dimethoxy-1-methylisoquinoline
Probably formed from: 1,2,3,4-Tetrahydro-7,8-dihydroxy-6-methoxy-1-methylisoquinoline

(1*S*)-1,2,3,4-Tetrahydro-1-(2-(3-hydroxy-4,5-dimethoxyphenyl)ethyl)-6-methoxy-2-methyl-7-isoquinolinol
See: Autumnaline

Tetrahydro-4-(4-hydroxy-3,5-dimethoxysulphinylbenzoyl)-1,4-oxazine
Probably formed from: Tetrahydro-4-(4-hydroxy-3,5-dimethoxythiobenzoyl)-1,4-oxazine

Tetrahydro-4-(4-hydroxy-3,5-dimethoxythiobenzoyl)-1,4-oxazine
Formed from: Trithiozine
→ Tetrahydro-4-(4-hydroxy-3,5-dimethoxysulphinylbenzoyl)-1,4-oxazine
Probably in: at 3260

1,2,3,4-Tetrahydro-N-(*p*-(1-hydroxyethyl)benzoyl)isoquinoline
Formed from: N-(*p*-Ethylbenzoyl)-1,2,3,4-tetrahydroisoquinoline

Δ^1-Tetrahydro-6β-hydroxy-*n*-heptylcannabinol
Probably formed from: Δ^1-Tetrahydro-*n*-heptylcannabinol

Δ^1-Tetrahydro-7-hydroxy-*n*-heptylcannabinol
Probably formed from: Δ^1-Tetrahydro-*n*-heptylcannabinol
→ Δ^1-Tetrahydro-2″,7-dihydroxy-*n*-heptylcannabinol
Probably in: Mouse D357
→ Δ^1-Tetrahydro-3″,7-dihydroxy-*n*-heptylcannabinol
Probably in: Mouse D357
→ Δ^1-Tetrahydro-6α,7-dihydroxy-*n*-heptylcannabinol
Probably in: Mouse D357
→ Δ^1-Tetrahydro-6β,7-dihydroxy-*n*-heptylcannabinol
Probably in: Mouse D357
→ Δ^1-Tetrahydro-*n*-heptylcannabinol-7-oic acid
Probably in: Mouse D357

Δ^1-Tetrahydro-2″-hydroxy-*n*-heptylcannabinol-7-oic acid
Probably formed from: Δ^1-Tetrahydro-*n*-heptylcannabinol-7-oic acid
→ Δ^1-Tetrahydro-2″,3″-dihydroxy-*n*-heptylcannabinol-7-oic acid
Probably in: Mouse D357

Δ^1-Tetrahydro-3″-hydroxy-*n*-heptylcannabinol-7-oic acid
Probably formed from: Δ^1-Tetrahydro-*n*-heptylcannabinol-7-oic acid

Δ^1-Tetrahydro-6α-hydroxy-*n*-heptylcannabinol-7-oic acid
Probably formed from: Δ^1-Tetrahydro-*n*-heptylcannabinol-7-oic acid

1,2,3,4-Tetrahydro-6-hydroxy-1-hydroxyacetylquinoline
Formed from: Quinfamide

1,2,3,4-Tetrahydro-7-hydroxy-1-(*p*-hydroxybenzyl)-6-methoxyisoquinoline
See: Coclaurine

1,2,3,4-Tetrahydro-7-hydroxy-1-(3-hydroxy-4-methoxybenzyl)-6-methoxyisoquinoline
→ Reticuline
In: Litsea A3222

1,2,3,4-Tetrahydro-7-hydroxy-1-(4-hydroxy-3-methoxybenzyl)-6-methoxy-2-methylisoquinoline
See: Orientaline

2,3,4,5-Tetrahydro-4-hydroxy-1-(2-hydroxy-2-(β-naphthyl)ethyl)-2-oxoimidazole
Probably formed from: 1-(2-Hydroxy-2-(β-naphthyl)ethyl)hydantoin

1,2,3,4-Tetrahydro-4-hydroxyisoquinoline
Formed from: 1,2,3,4-Tetrahydroisoquinoline

1,2,3,4-Tetrahydro-9-hydroxylaminoacridine
Formed from: Tacrine

(*E*)-3,4,5,6-Tetrahydro-3-(2-hydroxy-4-methoxybenzylidene)-2,3′-bipyridine
Formed from: (*E*)-3-(2,4-Dimethoxybenzylidene)-3,4,5,6-tetrahydro-2,3′-bipyridine
→ (*E*)-3,4,5,6-Tetrahydro-3-(2-hydroxy-4-methoxybenzylidene)-2,3′-bipyridine-β-D-glucuronide
Probably in: Rat J728

(*E*)-3,4,5,6-Tetrahydro-3-(4-hydroxy-2-methoxybenzylidene)-2,3′-bipyridine
Formed from: (*E*)-3-(2,4-Dimethoxybenzylidene)-3,4,5,6-tetrahydro-2,3′-bipyridine
→ (*E*)-3,4,5,6-Tetrahydro-3-(4-hydroxy-2-methoxybenzylidene)-2,3′-bipyridine-β-D-glucuronide
Probably in: Rat J728

(*E*)-3,4,5,6-Tetrahydro-3-(2-hydroxy-4-methoxybenzylidene)-2,3′-bipyridine-β-D-glucuronide
Probably formed from: (*E*)-3,4,5,6-Tetrahydro-3-(2-hydroxy-4-methoxybenzylidene)-2,3′-bipyridine

(*E*)-3,4,5,6-Tetrahydro-3-(4-hydroxy-2-methoxybenzylidene)-2,3′-bipyridine-β-D-glucuronide
Probably formed from: (*E*)-3,4,5,6-Tetrahydro-3-(4-hydroxy-2-methoxybenzylidene)-2,3′-bipyridine

1,2,3,4-Tetrahydro-1-(3-hydroxy-4-methoxybenzyl)isoquinoline
Probably formed from: 1,2,3,4-Tetrahydro-1-(3,4-dihydroxybenzyl)isoquinoline

Tetrahydro-7-hydroxy-6-methoxy-β-carboline
Formed from: Tetrahydro-6-methoxy-β-carboline

1,2,3,4-Tetrahydro-6-hydroxy-7-methoxyisoquinoline
Formed from: 1,2,3,4-Tetrahydro-6,7-dihydroxyisoquinoline

1,2,3,4-Tetrahydro-7-hydroxy-6-methoxyisoquinoline
Formed from: 1,2,3,4-Tetrahydro-6,7-dihydroxyisoquinoline

1,2,3,4-Tetrahydro-6-hydroxy-7-methoxy-1-methylisoquinoline (salsoline)
Formed from: Salsolinol

1,2,3,4-Tetrahydro-7-hydroxy-6-methoxy-1-methylisoquinoline (isosalsoline)
Formed from: Salsolinol
→ N-Methylisosalsoline
In: Beef J899
Corydalis G425
→ 1,2,3,4-Tetrahydro-6,7-dimethoxy-1-methylisoquinoline
Probably in: Tenuis G470

1,2,3,4-Tetrahydro-6-hydroxy-7-methoxy-2-methylisoquinoline
Formed from: 1,2,3,4-Tetrahydro-6,7-dihydroxy-2-methylisoquinoline

1,2,3,4-Tetrahydro-7-hydroxy-6-methoxy-2-methylisoquinoline (coripalline)
Formed from: N-Methylcoclaurine
Orientaline
1,2,3,4-Tetrahydro-6,7-dihydroxy-2-methylisoquinoline
→ 1,2,3,4-Tetrahydro-7-hydroxy-6-methoxy-2-methylisoquinolinium
Probably in: Nelumbo D566

1,2,3,4-Tetrahydro-6-hydroxy-7-methoxy-1-methylisoquinoline-1-carboxylic acid
Formed from: Salsolinol-1-carboxylic acid

1,2,3,4-Tetrahydro-6-hydroxy-7-methoxy-1-methylisoquinoline-3-carboxylic acid
Formed from: Salsolinol-3-carboxylic acid

1,2,3,4-Tetrahydro-7-hydroxy-6-methoxy-1-methylisoquinoline-3-carboxylic acid
Formed from: Salsolinol-3-carboxylic acid

1α,3,6α,11α-Tetrahydro-1α-hydroxy-9-methoxy-3-oxo-6*H*-[1,3]dioxolo[5,6]benzofuro[3,2-*c*][1]benzopyran
Formed from: Maackiain

1,2,3,4-Tetrahydro-1-(4-hydroxy-3-methoxyphenethyl)-2-methylisoquinoline
Probably formed from: Cinnamic acid

(*S*)-(−)-1,2,3,4-Tetrahydro-6-hydroxy-5-methoxy-N-(N-((2-phenyl)ethyl)-6-hexylamino)-N-propyl-2-naphthylamine
Formed from: (*S*)-(−)-1,2,3,4-Tetrahydro-5,6-dihydroxy-N-(N-((2-phenyl)ethyl)-6-hexylamino)-N-propyl-2-naphthylamine

(1*S*)-1,2,3,4-Tetrahydro-1-((3-hydroxy-4-methoxyphenyl)methyl)-6-methoxy-7-isoquinolinol
See: Norreticuline

(1*R*)-1,2,3,4-Tetrahydro-1-((4-hydroxy-3-methoxyphenyl)methyl)-6-methoxy-7-isoquinolinol
See: Nororientaline

(1*S*)-1,2,3,4-Tetrahydro-1-((3-hydroxy-4-methoxyphenyl)methyl)-7-methoxy-2-methyl-6-isoquinolinol
See: Protosinomenine

(1*S*)-1,2,3,4-Tetrahydro-1-((3-hydroxy-4-methoxyphenyl)methyl)-7-methoxy-2-methyl-8-isoquinolinol
See: Crassifoline

1,2,3,4-Tetrahydro-8-hydroxy-9-methoxy-3-(2-(1-piperidinyl)ethyl)-5*H*-1-benzopyrano[3,4-*c*]pyridin-5-one
Probably formed from: 1,2,3,4-Tetrahydro-8,9-dimethoxy-3-(2-(1-piperidinyl)ethyl)-5*H*-1-benzopyrano[3,4-*c*]pyridin-5-one

Tetrahydro-2-hydroxy-2-methylbenzopyran
Formed from: 3-Acetylcoumarin

1,2,3,4-Tetrahydro-N-(*p*-hydroxymethylbenzoyl)isoquinoline
Formed from: 1,2,3,4-Tetrahydro-N-(*p*-toluoyl)isoquinoline

1,2,3,4-Tetrahydro-N-(*p*-hydroxymethylbenzoyl)-2-methylquinoline
Formed from: 1,2,3,4-Tetrahydro-2-methyl-N-(*p*-toluoyl)quinoline

1,2,3,4-Tetrahydro-N-(*p*-hydroxymethylbenzoyl)-4-methylquinoline
Formed from: 1,2,3,4-Tetrahydro-4-methyl-N-(*p*-toluoyl)quinoline

Δ^6-Tetrahydro-1′-hydroxy-methylcannabinol
Formed from: Δ^6-Tetrahydro-methylcannabinol

Δ^1-Tetrahydro-6α-hydroxy-methylcannabinol
Formed from: Δ^1-Tetrahydro-methylcannabinol

Δ^1-Tetrahydro-7-hydroxy-methylcannabinol
Formed from: Δ^1-Tetrahydro-methylcannabinol
→ Δ^1-Tetrahydro-methylcannabinol-7-oic acid
Probably in: Mouse E623

Δ^6-Tetrahydro-7-hydroxy-3′-O-methylcannabinol
Formed from: Δ^6-Tetrahydro-7-hydroxycannabinol

Δ^6-Tetrahydro-7-hydroxy-methylcannabinol
Formed from: Δ^6-Tetrahydro-methylcannabinol
→ Δ^6-Tetrahydro-5α,7-dihydroxy-methylcannabinol
Probably in: Mouse E623
→ Δ^6-Tetrahydro-methylcannabinol-7-oic acid
Probably in: Mouse E623

Δ^6-Tetrahydro-1-hydroxy-*abn*-methylcannabinol
Formed from: Δ^6-Tetrahydro-*abn*-methylcannabinol

Δ^6-Tetrahydro-1′-hydroxy-*abn*-methylcannabinol
Formed from: Δ^6-Tetrahydro-*abn*-methylcannabinol

Δ^6-Tetrahydro-7-hydroxy-*abn*-methylcannabinol
Formed from: Δ^6-Tetrahydro-*abn*-methylcannabinol
→ Δ^6-Tetrahydro-1″,7-dihydroxy-*abn*-methylcannabinol
Probably in: Mouse E623
→ Δ^6-Tetrahydro-*abn*-methylcannabinol-7-oic acid
Probably in: Mouse E623

Δ^1-Tetrahydro-1′-hydroxy-methylcannabinol-7-oic acid
Probably formed from: Δ^1-Tetrahydro-methylcannabinol-7-oic acid

Δ^6-Tetrahydro-1′-hydroxy-methylcannabinol-7-oic acid
Probably formed from: Δ^6-Tetrahydro-methylcannabinol-7-oic acid

Δ^6-Tetrahydro-5α-hydroxy-methylcannabinol-7-oic acid
Probably formed from: Δ^6-Tetrahydro-methylcannabinol-7-oic acid

Δ^6-Tetrahydro-5β-hydroxy-methylcannabinol-7-oic acid
Probably formed from: Δ^6-Tetrahydro-methylcannabinol-7-oic acid

1,2,3,4-Tetrahydro-5-hydroxy-1-methyl-β-carboline
Formed from: 1,2,3,4-Tetrahydro-1-methyl-β-carboline

1,2,3,4-Tetrahydro-6-hydroxy-1-methyl-β-carboline
Formed from: 5-Hydroxytryptamine
1,2,3,4-Tetrahydro-1-methyl-β-carboline
→ 1,2,3,4-Tetrahydro-6-hydroxy-1-methyl-β-carboline-N-oxide
Probably in: Rat A908
→ 1,2,3,4-Tetrahydro-6-methoxy-1-methyl-β-carboline
Probably in: Rat A908

1,2,3,4-Tetrahydro-6-hydroxy-2-methyl-β-carboline
Formed from: 5-Hydroxy-N-methyltryptamine

1,2,3,4-Tetrahydro-7-hydroxy-1-methyl-β-carboline
Formed from: 1,2,3,4-Tetrahydro-1-methyl-β-carboline

1,2,3,4-Tetrahydro-6-hydroxy-1-methyl-β-carboline-N-oxide
Probably formed from: 1,2,3,4-Tetrahydro-6-hydroxy-1-methyl-β-carboline

1,2,3,4-Tetrahydro-6-hydroxymethyl-2-isopropylaminomethyl-7-nitroquinoline
See: Oxamniquine

1,2,3,4-Tetrahydro-4-hydroxy-1-methylisoquinoline
Formed from: 1,2,3,4-Tetrahydro-1-methylisoquinoline

1,2,3,4-Tetrahydro-2-hydroxymethyl-6-methyl-7-nitroquinoline
Probably formed from: Desoxymansil

1,2,3,4-Tetrahydro-6-hydroxymethyl-7-nitroquinoline-2-carboxylic acid
Formed from: Oxamniquine

1,2,3,4-Tetrahydro-N-(*p*-hydroxymethylphenylacetyl)isoquinoline
Formed from: 1,2,3,4-Tetrahydro-N-(*p*-tolylacetyl)isoquinoline

1,2,3,4-Tetrahydro-4-hydroxy-2-methyl-N-(*p*-toluoyl)quinoline
Formed from: 1,2,3,4-Tetrahydro-2-methyl-N-(*p*-toluoyl)quinoline

1,2,3,4-Tetrahydro-4-hydroxy-4-methyl-N-(*p*-toluoyl)quinoline
Formed from: 1,2,3,4-Tetrahydro-4-methyl-N-(*p*-toluoyl)quinoline

1,2,3,4-Tetrahydro-1-hydroxynaphthalene
Formed from: α-Tetralone

Tetrahydro-5-hydroxynorharman
Formed from: L-Tryptophan

Tetrahydro-6-hydroxynorharman
Formed from: L-Tryptophan

Δ^1-Tetrahydro-3″-hydroxy-6-oxocannabinol
Probably formed from: Δ^1-Tetrahydro-3″,6β-dihydroxycannabinol

Δ^1-Tetrahydro-4″-hydroxy-6-oxocannabinol
Probably formed from: Δ^1-Tetrahydro-4″,6β-dihydroxycannabinol
→ Δ^1-Tetrahydro-4″,7-dihydroxy-6-oxocannabinol
Probably in: Mouse A3155

Δ^1-Tetrahydro-7-hydroxy-6-oxocannabinol
Probably formed from: Δ^1-Tetrahydro-6α,7-dihydroxycannabinol

Δ^6-Tetrahydro-7-hydroxy-5-oxocannabinol
Probably formed from: Δ^6-Tetrahydro-5α,7-dihydroxycannabinol

1,2,3,4-Tetrahydro-5-hydroxy-1-oxonaphthalene
Formed from: 5-(3-*tert*-Butylamino-2-hydroxypropoxy)-1,2,3,4-tetrahydro-1-oxonaphthalene

1,2,3,4-Tetrahydro-6-hydroxy-1-oxonaphthalene
Formed from: 5,6,7,8-Tetrahydro-2,5-dihydroxynaphthalene

4-(1-(5,6,7,8-Tetrahydro-6-hydroxy-3,5,5,8,8-pentamethyl-2-naphthalenyl)-1-ethenyl) benzoyltaurine
Probably formed from: 6-Hydroxytargetrin

4-(1-(5,6,7,8-Tetrahydro-7-hydroxy-3,5,5,8,8-pentamethyl-2-naphthalenyl)-1-ethenyl) benzoyltaurine
Probably formed from: 7-Hydroxytargetrin

2,3,5,6-Tetrahydro-2-(2-hydroxyphenoxymethyl)-1,4-oxazine
Formed from: Viloxazine
→ 2,3,5,6-Tetrahydro-2-(2-hydroxyphenoxymethyl)-1,4-oxazine-O-sulphate
In: Rat A355

2,3,5,6-Tetrahydro-2-(2-hydroxyphenoxymethyl)-1,4-oxazine-N-hippurate-O-sulphate
Probably formed from: 2,3,5,6-Tetrahydro-2-(2-hydroxyphenoxymethyl)-1,4-oxazine-O-sulphate

2,3,5,6-Tetrahydro-2-(2-hydroxyphenoxymethyl)-1,4-oxazine-O-sulphate
Formed from: 2,3,5,6-Tetrahydro-2-(2-hydroxyphenoxymethyl)-1,4-oxazine
→ 2,3,5,6-Tetrahydro-2-(2-hydroxyphenoxymethyl)-1,4-oxazine-N-hippurate-O-sulphate
Probably in: Rat A355

1,3,4,5-Tetrahydro-7-hydroxy-5-phenyl-2*H*-1,4-benzodiazepin-2-one
→ 1,3,4,5-Tetrahydro-2,7-dihydroxy-5-phenyl-2*H*-1,4-benzodiazepin-2-one
Probably in: Rat A2858

(*R*)-1,2,3,6-Tetrahydro-1-((3-(*p*-hydroxyphenyl)-3-cyclohexen-1-yl)methyl)-4-phenylpyridine
Formed from: (*R*)-1,2,3,6-Tetrahydro-4-phenyl-1-((3-phenyl-3-cyclohexen-1-yl)methyl)-pyridine
→ (*R*)-1,2,3,6-Tetrahydro-1-((3-(*p*-hydroxyphenyl)-3-cyclohexen-1-yl)methyl)-4-phenylpyridine-O-sulphate
In: Cynomolgus monkey J448
Man J448
→ (*R*)-1,2,3,6-Tetrahydro-4-(*p*-hydroxyphenyl)-1-((3-(*p*-hydroxyphenyl)-3-cyclohexen-1-yl)methyl)pyridine
Probably in: Cynomolgus monkey J448
Man J448

(*R*)-1,2,3,6-Tetrahydro-1-((3-(*p*-hydroxyphenyl)-3-cyclohexen-1-yl)methyl)-4-phenylpyridine-O-sulphate
Formed from: (*R*)-1,2,3,6-Tetrahydro-1-((3-(*p*-hydroxyphenyl)-3-cyclohexen-1-yl)methyl)-4-phenylpyridine

(*R*)-1,2,3,6-Tetrahydro-1-((*p*-hydroxyphenyl)-3-cyclohexen-1-yl)methyl)-4-(*p*-sulphatoxyphenyl)pyridine
Probably formed from: (*R*)-1,2,3,6-Tetrahydro-4-(*p*-hydroxyphenyl)-1-((3-(*p*-hydroxyphenyl)-3-cyclohexen-1-yl)methyl) pyridine

(*R*)-1,2,3,6-Tetrahydro-4-(*p*-hydroxyphenyl)-1-((3-(*p*-hydroxyphenyl)-3-cyclohexen-1-yl)methyl)pyridine
Probably formed from: (*R*)-1,2,3,6-Tetrahydro-1-((3-(*p*-hydroxyphenyl)-3-cyclohexen-1-yl)methyl)-4-phenylpyridine
(*R*)-1,2,3,6-Tetrahydro-4-(*p*-hydroxyphenyl)-1-((3-phenyl-3-cyclohexen-1-yl)methyl) pyridine
→ (*R*)-1,2,3,6-Tetrahydro-1-((3-(*p*-hydroxy-phenyl)-3-cyclohexen-1-yl)methyl)-4-(*p*-sulphatoxyphenyl)pyridine
Probably in: Cynomolgus monkey J448
Man J448
→ (*R*)-1,2,3,6-Tetrahydro-4-(*p*-hydroxyphenyl)-1-((3-(*p*-sulphatoxyphenyl)-3-cyclohexen-1-yl)methyl)pyridine
Probably in: Cynomolgus monkey J448

(*R*)-1,2,3,6-Tetrahydro-4-(*p*-hydroxyphenyl)-1-((3-(*p*-hydroxyphenyl)-3-cyclohexen-1-yl)methyl)pyridine-O,O′-disulphate
Probably formed from: (*R*)-1,2,3,6-Tetrahydro-4-(*p*-hydroxyphenyl)-1-((3-(*p*-sulphatoxyphenyl)-3-cyclohexen-1-yl)methyl)pyridine

1,2,3,4-Tetrahydro-1-((4-hydroxyphenyl)methyl)-6,7-isoquinolinediol
See: Higenamine

(1*R*)-1,2,3,4-Tetrahydro-1-((4-hydroxyphenyl)methyl)-7-methoxy-6-isoquinolinol
See: Isococlaurine

1,2,3,4-Tetrahydro-1-((4-hydroxyphenyl)methyl)-6-methoxy-2-methyl-7-isoquinolinol
See: N-Methylcoclaurine

(*R*)-1,2,3,6-Tetrahydro-4-(*p*-hydroxyphenyl)-1-((3-phenyl-3-cyclohexen-1-yl)methyl)pyridine
Formed from: (*R*)-1,2,3,6-Tetrahydro-4-phenyl-1-((3-phenyl-3-cyclohexen-1-yl)methyl)pyridine
→ (*R*)-1,2,3,6-Tetrahydro-4-(*p*-hydroxyphenyl)-1-((3-phenyl-3-cyclohexen-1-yl)methyl)pyridine-O-sulphate
In: Cynomolgus monkey J448
→ (*R*)-1,2,3,6-Tetrahydro-4-(*p*-hydroxyphenyl)-1-((3-(*p*-hydroxyphenyl)-3-cyclohexen-1-yl)methyl)pyridine
Probably in: Cynomolgus monkey H752

(*R*)-1,2,3,6-Tetrahydro-4-(*p*-hydroxyphenyl)-1-((3-phenyl-3-cyclohexen-1-yl)methyl)pyridine-O-sulphate
Formed from: (*R*)-1,2,3,6-Tetrahydro-4-(*p*-hydroxyphenyl)-1-((3-phenyl-3-cyclohexen-1-yl)methyl)pyridine

1,2,3,6-Tetrahydro-4-(*p*-hydroxyphenyl)pyridine
Formed from: 1-Cyclopropyl-1,2,3,6-tetrahydro-4-phenylpyridine

(*R*)-1,2,3,6-Tetrahydro-4-(*p*-hydroxyphenyl)-1-((3-(*p*-sulphatoxyphenyl)-3-cyclohexen-1-yl)methyl)pyridine
Probably formed from: (*R*)-1,2,3,6-Tetrahydro-4-(*p*-hydroxyphenyl)-1-((3-(*p*-hydroxyphenyl)-3-cyclohexen-1-yl)methyl)pyridine
→ (*R*)-1,2,3,6-Tetrahydro-4-(*p*-hydroxyphenyl)-1-((3-(*p*-hydroxyphenyl)-3-cyclohexen-1-yl)methyl)pyridine-O,O′-disulphate
Probably in: Cynomolgus monkey J448

2,3,4,5-Tetrahydro-7-(4-hydroxypiperidino)imidazo[2,1-*b*]quinazolin-2(1*H*)-one
Formed from: 2,3,4,5-Tetrahydro-7-piperidinoimidazo[2,1-*b*]quinazolin-2(1*H*)-one

Δ^1-Tetrahydro-2″-hydroxypropylcannabinol
Formed from: Δ^1-Tetrahydropropylcannabinol

Δ^1-Tetrahydro-7-hydroxypropylcannabinol
Formed from: Δ^1-Tetrahydropropylcannabinol
→ Δ^1-7-Carboxytetrahydropropylcannabinol
Probably in: Mouse E625

Δ^6-Tetrahydro-2″-hydroxypropylcannabinol
Formed from: Δ^6-Tetrahydropropylcannabinol

Δ^6-Tetrahydro-7-hydroxypropylcannabinol
Formed from: Δ^6-Tetrahydropropylcannabinol
→ Δ^6-Tetrahydropropylcannabinol-7-oic acid
Probably in: Mouse E625

1,2,3,4-Tetrahydro-5-hydroxy-2-(N-propyl-N-2-thienylethylamino)naphthalene
→ 1,2,3,4-Tetrahydro-5-hydroxy-2-(N-propyl-N-2-thienylethylamino)naphthalene-β-D-glucuronide
In: Monkey F596
→ 1,2,3,4-Tetrahydro-5-hydroxy-2-(N-propyl-N-2-thienylethylamino)naphthalene-O-sulphate
In: Monkey F596

1,2,3,4-Tetrahydro-5-hydroxy-2-(N-propyl-N-2-thienylethylamino)naphthalene-β-D-glucuronide
Formed from: 1,2,3,4-Tetrahydro-5-hydroxy-2-(N-propyl-N-2-thienylethylamino)naphthalene

1,2,3,4-Tetrahydro-5-hydroxy-2-(N-propyl-N-2-thienylethylamino)naphthalene-O-sulphate
Formed from: 1,2,3,4-Tetrahydro-5-hydroxy-2-(N-propyl-N-2-thienylethylamino)naphthalene

7-(3-(Tetrahydro-4-hydroxy-2*H*-pyran-4-yl)phenyl)methoxy-3-hydroxy-4-phenylnaphtho[2,3-*c*]furan-1(3*H*)-one
Formed from: 7-(3-(Tetrahydro-4-hydroxy-2*H*-pyran-4-yl)phenyl)methoxy-4-phenylnaphtho[2,3-*c*]furan-1(3*H*)-one

7-(3-(Tetrahydro-4-hydroxy-2*H*-pyran-4-yl)phenyl)methoxy-4-phenylnaphtho[2,3-*c*]furan-1(3*H*)-one
→ 7-(3-(1-Carboxymethyl-1,3-dihydroxypropyl)phenyl)methoxy-4-phenylnaphtho[2,3-*c*]furan-1(3*H*)-one
In: Monkey H137

Rat H137
→ 7-(3-(Tetrahydro-2,4-dihydroxy-2*H*-pyran-4-yl)phenyl)methoxy-4-phenyl-naphtho[2,3-*c*]furan-1(3*H*)-one
In: Man H137
Monkey H137
Rat H137
→ 7-(3-(Tetrahydro-4-hydroxy-2*H*-pyran-4-yl)phenyl)methoxy-3-hydroxy-4-phenyl-naphtho[2,3-*c*]furan-1(3*H*)-one
In: Monkey H137
Rat H137

1-((5,6,7,8-Tetrahydro-6-hydroxy-5,5,8,8-tetramethyl-2-naphthyl)carbamoyl)benzoic acid
Formed from: 1-((5,6,7,8-Tetrahydro-5,5,8,8-tetramethyl-2-naphthyl)carbamoyl)benzoic acid
→ 1-((5,6,7,8-Tetrahydro-6-hydroxy-5,5,8,8-tetramethyl-2-naphthylcarbamoyl)benzoyltaurine
Probably in: Rat H128, H863
→ 1-((5,6,7,8-Tetrahydro-5,5,8,8-tetramethyl-6-oxo-2-naphthyl)carbamoyl)benzoic acid
Probably in: Rat H128, H863

1-((5,6,7,8-Tetrahydro-7-hydroxy-5,5,8,8-tetramethyl-2-naphthylcarbamoyl)benzoic acid
Formed from: 1-((5,6,7,8-Tetrahydro-5,5,8,8-tetramethyl-2-naphthylcarbamoyl)benzoic acid
→ 1-((5,6,7,8-Tetrahydro-7-hydroxy-5,5,8,8-tetramethyl-2-naphthylcarbamoyl)benzoyltaurine
Probably in: Rat H128, H863

1-((5,6,7,8-Tetrahydro-6-hydroxy-5,5,8,8-tetramethyl-2-naphthylcarbamoyl)benzoyltaurine
Probably formed from: 1-((5,6,7,8-Tetrahydro-6-hydroxy-5,5,8,8-tetramethyl-2-naphthylcarbamoyl)benzoic acid

1-((5,6,7,8-Tetrahydro-7-hydroxy-5,5,8,8-tetramethyl-2-naphthylcarbamoyl)benzoyltaurine
Probably formed from: 1-((5,6,7,8-Tetrahydro-7-hydroxy-5,5,8,8-tetramethyl-2-naphthylcarbamoyl)benzoic acid

2,3,4,5-Tetrahydro-5-hydroxy-N-(*p*-toluoyl)-1*H*-benzazepine
Formed from: 2,3,4,5-Tetrahydro-N-(*p*-toluoyl)-1*H*-benzazepine

Tetrahydro-2-hydroxy-4-(3,4,5-trimethoxybenzoyl)-1,4-oxazine
Probably formed from: Tetrahydro-4-(3,4,5-trimethoxybenzoyl)-1,4-oxazine

6a,7,10,10a-Tetrahydro-1-hydroxy-6,6,9-trimethyl-3-*n*-pentyldibenzo[*b*,*d*]pyran
→ 6a,7,10,10a-Tetrahydro-1,4′-dihydroxy-6,6,9-trimethyl-3-*n*-pentyldibenzo[*b*,*d*]pyran
In: Cunninghamella A2194

5,6,7,8-Tetrahydro-6-(1-imidazolyl-methyl)naphthalene-2-carboxylic acid
→ 5,6,7,8-Tetrahydro-6-(1-imidazolyl-methyl)naphthalene-2-carboxyl-β-D-glucuronic acid
In: Man F566

5,6,7,8-Tetrahydro-6-(1-imidazolyl-methyl)naphthalene-2-carboxyl-β-D-glucuronic acid
Formed from: 5,6,7,8-Tetrahydro-6-(1-imidazolylmethyl)naphthalene-2-carboxylic acid

Tetrahydro-4-(*p*-iodobenzyl)-1-methylpyridine
→ 4-(*p*-Iodobenzyl)-1-methylpyridinium
Probably in: Rat F816

1,2,3,4-Tetrahydro-2-isopropylaminomethyl-7-nitroquinoline-6-carboxylic acid
Formed from: Oxamniquine

1,2,3,4-Tetrahydroisoquinoline
Formed from: Phenethylamine
→ Isoquinoline
In: Rat F725
→ 1,2,3,4-Tetrahydro-4-hydroxyisoquinoline
In: Rat F725, G453
→ 1,2,3,4-Tetrahydroisoquinoline-N-sulphate
In: Guinea pig E456
Man H276, J324
→ 1,2,3,4-Tetrahydro-N-methylisoquinoline
In: Man F228
Pig K923
Rabbit E6
Rat F725

1,2,3,4-Tetrahydroisoquinoline-N-sulphate
Formed from: 1,2,3,4-Tetrahydroisoquinoline

1,2,3,4-Tetrahydroisoquinoline-7-sulphonamide
→ 1,2,3,4-Tetrahydroisoquinoline-7-sulphonamide-2-β-D-glucuronide
In: Dog D648

1,2,3,4-Tetrahydroisoquinoline-7-sulphonamide-2-β-D-glucuronide
Formed from: 1,2,3,4-Tetrahydroisoquinoline-7-sulphonamide

1,2,3,6-Tetrahydro-4-(*o*-isopropylphenyl)-1-methylpyridine
→ 4-(*o*-Isopropylphenyl)-1-methylpyridinium
Probably in: Beef F395
Man F395

Tetrahydrojatrorrhizine (5,8,13,13a-tetrahydro-2,9,10-trimethoxy-6*H*-dibenzo[*a*,*g*]quinolizin-3-ol)
Formed from: Jatrorrhizine
Tetrahydro-2,3-dihydroxy-1,10-dimethoxyprotoberberine
→ Jatrorrhizine
Probably in: Berberis D145

Tetrahydromagnolol
Formed from: Dihydromagnolol
→ 2,2′-Dihydroxy-5-(1-propen-1(*E*))-yl)-5′-propylbiphenyl
Probably in: Rat D549, D625

Tetrahydro-4-(*o*-methoxybenzyl)-1-methylpyridine
→ 4-(*o*-Methoxybenzyl)-1-methylpyridinium
Probably in: Rat F816

Tetrahydro-4-(*p*-methoxybenzyl)-1-methylpyridine
→ 4-(*p*-Methoxybenzyl)-1-methylpyridinium
Probably in: Rat F816

1,2,3,4-Tetrahydro-6-methoxycarbazole-3-carboxylic acid
Formed from: Oxarbazole

1,2,3,4-Tetrahydro-6-methoxy-β-carboline
Formed from: 5-Methoxytryptamine
→ 1,2,3,4-Tetrahydro-6-hydroxy-β-carboline
In: Rat A742
→ 1,2,3,4-Tetrahydro-7-hydroxy-6-methoxy-β-carboline
In: Rat A742

1,2,3,4-Tetrahydro-6-methoxy-1-methyl-β-carboline
Probably formed from: 1,2,3,4-Tetrahydro-6-hydroxy-1-methyl-β-carboline

(6a*R*)-5,6,6a,7-Tetrahydro-1-methoxy-6-methyl-4*H*-dibenzo[*de*,*g*]quinolin-2-ol
See: O-Nornuciferine

5,6,7,8-Tetrahydro-4-methoxy-6-methyl-1,3-dioxolo[4,5-*g*]isoquinoline
See: Hydrocotarnine

(1*S*)-1,2,3,4-Tetrahydro-6-methoxy-1-methyl-7-isoquinolinol
See: Isosalsoline

2-(1,2,3,4-Tetrahydro-7-methoxynaphth-1-yl) ethylamine
→ N-Acetyl-2-(1,2,3,4-tetrahydro-7-methoxynaphth-1-yl)ethylamine
In: Man K490

1,2,3,6-Tetrahydro-4-(*o*-methoxyphenyl)-1-methylpyridine
→ 4-(*o*-Methoxyphenyl)-1-methylpyridinium
Probably in: Beef F395
Man F395

1,2,3,6-Tetrahydro-4-(*m*-methoxyphenyl)-1-methylpyridine
→ 4-(*m*-Methoxyphenyl)-1-methylpyridinium
Probably in: Beef F395

Tetrahydro-4-(*o*-methylbenzyl)-1-methylpyridine
→ 4-(*o*-Methylbenzyl)-1-methylpyridinium
Probably in: Rat F816

Tetrahydro-4-(*m*-methylbenzyl)-1-methylpyridine
→ 4-(*m*-Methylbenzyl)-1-methylpyridinium
Probably in: Rat F816

Tetrahydro-4-(*p*-methylbenzyl)-1-methylpyridine
→ 4-(*p*-Methylbenzyl)-1-methylpyridinium
Probably in: Rat F816

Tetrahydro-4-(α-methylbenzyl)-1-methylpyridine
→ 4-(α-Methylbenzyl)-1-methylpyridinium
Probably in: Rat F816

***cis*-7,8,13,13a-Tetrahydro-N-methylberbinium**
→ Allocryptopine
In: Macleaya C696
→ Chelerythrine
In: Macleaya C696

Δ^1-Tetrahydro-methylcannabinol
→ Δ^1-Tetrahydro-6α-hydroxy-methylcannabinol
In: Mouse E623
→ Δ^1-Tetrahydro-7-hydroxy-methylcannabinol
In: Mouse E623

Δ^6-Tetrahydro-methylcannabinol
→ Δ^6-Tetrahydro-1′-hydroxy-methylcannabinol
In: Mouse E623
→ Δ^6-Tetrahydro-7-hydroxy-methylcannabinol
In: Mouse E623

Δ^6-Tetrahydro-*abn*-methylcannabinol
→ Δ^6-Tetrahydro-1-hydroxy-*abn*-methylcannabinol
In: Mouse E623
→ Δ^6-Tetrahydro-1′-hydroxy-*abn*-methylcannabinol
In: Mouse E623
→ Δ^6-Tetrahydro-7-hydroxy-*abn*-methylcannabinol
In: Mouse E623

Δ^1-Tetrahydro-methylcannabinol-7-oic acid
Probably formed from: Δ^1-Tetrahydro-7-hydroxy-methylcannabinol
→ Δ^1-Tetrahydro-1′-hydroxy-methyl-cannabinol-7-oic acid
Probably in: Mouse E623

Δ^6-Tetrahydro-methylcannabinol-7-oic acid
Probably formed from: Δ^6-Tetrahydro-7-hydroxy-methylcannabinol
→ Δ^6-Tetrahydro-1′-hydroxy-methylcannabinol-7-oic acid
Probably in: Mouse E623
→ Δ^6-Tetrahydro-5α-hydroxy-methylcannabinol-7-oic acid
Probably in: Mouse E623
→ Δ^6-Tetrahydro-5β-hydroxy-methylcannabinol-7-oic acid
Probably in: Mouse E623

Δ^6-Tetrahydro-*abn*-methylcannabinol-7-oic acid
Probably formed from: Δ^6-Tetrahydro-7-hydroxy-*abn*-methylcannabinol

1,2,3,4-Tetrahydro-1-methyl-β-carboline
(methtryptoline)
Formed from: 1,2,3,4-Tetrahydro-1-methyl-β-carboline-1-carboxylic acid
→ 1,2,3,4-Tetrahydro-5-hydroxy-1-methyl-β-carboline
In: Rat E626
→ 1,2,3,4-Tetrahydro-6-hydroxy-1-methyl-β-carboline
In: Rat E171, E626
→ 1,2,3,4-Tetrahydro-7-hydroxy-1-methyl-β-carboline
In: Rat E171, E626

1,2,3,4-Tetrahydro-2-methyl-β-carboline
Formed from: N-Methyltryptamine

1,2,3,4-Tetrahydro-1-methyl-β-carboline-1-carboxylic acid
Formed from: Tryptamine
→ Eleagnine
In: Eleagnus C353
→ Harmalan
In: Rat E505, E739
→ Harman
In: Passiflora C353, E505
Rat E739
→ 1,2,3,4-Tetrahydro-1-methyl-β-carboline
In: Mouse D874

1,2,3,4-Tetrahydro-1-methyl-β-carboline-3-carboxylic acid
Formed from: 1,1′-Ethylidenebis(tryptophan)

(6a*R*)-5,6,6a,7-Tetrahydro-6-methyl-4*H*-dibenzo[*de*,*g*]quinoline-10,11-diol
See: Apomorphine

***cis*-7,8,13,13a-Tetrahydro-N-methylcoptisinium**
→ Protopine
In: Macleaya C696

1,2,3,4-Tetrahydro-1-methylisoquinoline (both isomers)
Formed from: 1,2,3,4-Tetrahydroisoquinoline

Phenethylamine
→ 1,2-Dihydro-1-methylisoquinoline
In: Rat F725
→ 1,2,3,4-Tetrahydro-1,2-dimethylisoquinoline
In: Rat F725
→ 1,2,3,4-Tetrahydro-4-hydroxy-1-methylisoquinoline

In: Rat F725

1,2,3,4-Tetrahydro-N-methylisoquinoline
Formed from: 1,2,3,4-Tetrahydroisoquinoline

1,2,3,4-Tetrahydro-6-methyl-2-(((1-methylethyl)amino)methyl)-7-nitroquinoline
See: Desoxymansil

1,2,3,9-Tetrahydro-9-methyl-3-((2-methyl-1*H*-imidazol-1-yl)methyl)-4*H*-carbazol-4-one
See: Ondansetron

1,2,3,6-Tetrahydro-1-methyl-4-(*o*-methylphenyl)pyridine
→ 1-Methyl-4-(*o*-methylphenyl)pyridinium
Probably in: Beef 395
Man F395

1,2,3,6-Tetrahydro-1-methyl-4-(*m*-methylphenyl)pyridine
→ 1-Methyl-4-(*m*-methylphenyl)pyridinium
Probably in: Beef 395
Man F395

1,2,3,6-Tetrahydro-1-methyl-4-(*p*-methylphenyl)pyridine
→ 1-Methyl-4-(*p*-methylphenyl)pyridinium
Probably in: Beef 395
Man F395

Tetrahydro-1-methyl-4-(1-naphthylmethyl)pyridine
→ 1-Methyl-4-(1-naphthylmethyl)pyridinium
Probably in: Rat F816

Tetrahydro-α-N-methylpalmatine
Formed from: Tetrahydropalmatine

3,4,5,6-Tetrahydro-5-methyl-1-phenyl-1*H*-2,5-benzoxazocine
See: Nefopam

1,2,3,6-Tetrahydro-1-methyl-4-(2-phenylethyl)pyridine
→ 1-Methyl-4-(2-phenylethyl)pyridinium
Probably in: Rat F816

1,2,3,6-Tetrahydro-1-methyl-4-phenylpyridine (MPTP)
Formed from: 1,2,3,6-Tetrahydro-4-phenylpyridine
→ 2,3-Dihydro-1-methyl-4-phenylpyridinium
In: Beef F458
Man D987, E43
Mouse E684, G326
Rat D821, E919, F678
See also: F253
→ 5,6-Dihydro-1-methyl-4-phenyl-2(1*H*)-pyridone
In: Rat E711, K32
See also: E720
→ 1-Methyl-4-phenyl-2-piperidinone
In: Rat E711
→ 1-Methyl-4-phenylpyridinium
In: Beef F458
Man E3
Monkey D222, D509, E720
Mouse E3, E684, E720, E970, G326
Rat D141, E3, E711, E919, F678, K32
See also: F395, F816
Contrast (rat and pig) D509
→ 1,2,3,6-Tetrahydro-1-methyl-4-phenylpyridine-N-oxide
In: Man D987
Mouse D987, E684, E970, E979, G326
Pig G392, H361
Rat E919, G392, K32
→ 1,2,3,6-Tetrahydro-4-phenylpyridine
In: Man J75, J172
Mouse D987, E684, E720, E979
Rat E919, K32

1,2,3,6-Tetrahydro-1-methyl-4-phenylpyridine-N-oxide
Formed from: 1,2,3,6-Tetrahydro-1-methyl-4-phenylpyridine

1,2,3,6-Tetrahydro-1-methyl-4-(*o*-propylphenyl)pyridine
→ 1-Methyl-4-(*o*-propylphenyl)pyridinium
Probably in: Beef F395
Man F395

Tetrahydro-5-methylpteroylpentaglutamic acid
Formed from: Folic acid

2,3,4,9-Tetrahydro-1-methyl-1*H*-pyrido[3,4-*b*]indole
See: Eleagnine

1,2,3,4-Tetrahydro-2-methyl-4-quinazolinone
Formed from: Anthranilic acid

Tetrahydro-N-methylquinoline
Formed from: Tetrahydroquinoline

Tetrahydro-2-methyl-N-(*p*-toluoyl)quinoline
→ Tetrahydro-N-(*p*-hydroxymethylbenzoyl)-2-methylquinoline
In: Cunninghamella D810

→ Tetrahydro-4-hydroxy-2-methyl-N-(*p*-toluoyl)quinoline
In: Cunninghamella D810
Tetrahydro-4-methyl-N-(*p*-toluoyl)quinoline
→ Tetrahydro-N-(*p*-hydroxymethylbenzoyl)-4-methylquinoline
In: Cunninghamella D810
→ Tetrahydro-4-hydroxy-4-methyl-N-(*p*-toluoyl)quinoline
In: Cunninghamella D810
1,2,3,4-Tetrahydronaphthalene
See: Tetralin
1,2,3,4-Tetrahydro-2-naphthoic acid
Probably formed from: 2-Naphthoic acid
5,6,7,8-Tetrahydro-2-naphthoic acid
Probably formed from: 2-Naphthoic acid
1,2,3,4-Tetrahydro-1-naphthol
Formed from: Tetralin
α-Tetralone
→ 1,2,3,4-Tetrahydronaphth-1-yl-β-D-glucuronide
In: Man A1891
Mouse C80
Rat B962
→ α-Tetralone
In: Guinea pig E67
Japanese monkey F241
Mouse E821
Rabbit A2032
1,2,3,4-Tetrahydro-2-naphthol
Formed from: Tetralin
β-Tetralone
→ 1,2,3,4-Tetrahydronaphth-2-yl-β-D-glucuronide
In: Man A1891
5,6,7,8-Tetrahydro-1-naphthol
→ 5,6,7,8-Tetrahydronaphth-1-yl-β-D-glucuronide
In: Beef A1211
→ 5,6,7,8-Tetrahydronaphth-1-yl sulphate
In: Rat B962
5,6,7,8-Tetrahydro-2-naphthol
→ 7,8-Dihydro-2-naphthol
Probably in: Penicillium J836
→ 5,6,7,8-Tetrahydro-2,5-dihydroxynaphthalene
Probably in: Penicillium J836
→ 5,6,7,8-Tetrahydronaphth-2-yl sulphate
In: Beef A1211
1,2,3,4-Tetrahydronaphth-1-yl-β-D-glucuronide
Formed from: 1,2,3,4-Tetrahydro-1-naphthol
1,2,3,4-Tetrahydronaphth-2-yl-β-D-glucuronide
Formed from: 1,2,3,4-Tetrahydro-2-naphthol
5,6,7,8-Tetrahydronaphth-1-yl-β-D-glucuronide
Formed from: 5,6,7,8-Tetrahydro-1-naphthol
Tetrahydro-3-(1-naphthylmethylfuran-5-ylpropionic acid
Formed from: Nafronyl
Tetrahydro-β-(1-naphthylmethyl)-2-furanpropionic acid
Formed from: Nafronyl
5,6,7,8-Tetrahydronaphth-1-yl sulphate
Formed from: 5,6,7,8-Tetrahydro-1-naphthol
5,6,7,8-Tetrahydronaphth-2-yl sulphate
Formed from: 5,6,7,8-Tetrahydro-2-naphthol
Tetrahydronorharmine
See: Tetrahydro-β-carboline
3-(Tetrahydrooxazol-2-yl)indole
Formed from: Camalexine
Δ^1-Tetrahydro-4″-oxocannabinol
Formed from: Δ^1-Tetrahydro-4″-hydroxycannabinol
Δ^1-Tetrahydro-6-oxocannabinol
Formed from: Δ^1-Tetrahydro-6α-hydroxycannabinol
Δ^1-Tetrahydro-6β-hydroxycannabinol
Δ^1-Tetrahydro-7-oxocannabinol
Formed from: Δ^1-Tetrahydro-7-hydroxycannabinol
→ Δ^1-Tetrahydrocannabinol-7-oic acid
In: Mouse H67
See also: B458
$\Delta^{3,4}$-Tetrahydro-5-oxocannabinol
Formed from: $\Delta^{3,4}$-Tetrahydrocannabinol
Δ^6-Tetrahydro-5-oxocannabinol
Formed from: Δ^6-Tetrahydro-5α-hydroxycannabinol
Δ^6-Tetrahydro-5β-hydroxycannabinol
Δ^6-Tetrahydro-7-oxocannabinol
Formed from: Δ^6-Tetrahydro-7-hydroxycannabinol
→ Δ^6-Tetrahydrocannabinol-7-oic acid
In: Mouse F723, H67, K307

Δ^6-Tetrahydro-1″-oxocannabinol-7-oic acid
Formed from: Δ^6-Tetrahydro-1″-hydroxycannabinol-7-oic acid
Δ^6-Tetrahydro-4″-oxocannabinol-7-oic acid
Formed from: Δ^6-Tetrahydro-4″-hydroxycannabinol-7-oic acid
Δ^6-Tetrahydro-5-oxocannabinol-7-oic acid
Formed from: Δ^6-Tetrahydro-5α-hydroxycannabinol-7-oic acid
(2*S*,4*E*)-1,2,3,4-tetrahydro-4-(oxoethylidene)-2,6-pyridinedicarboxylic acid
See: Betalamic acid
1,2,3,4-Tetrahydro-1-oxonaphthalene
See: α-Tetralone
1,2,3,4-Tetrahydro-2-oxonaphthalene
See: β-Tetralone
1,2,3,4-Tetrahydro-1-oxonaphth-5-yloxyacetic acid
Formed from: 5-(3-*tert*-Butylamino-2-hydroxypropoxy)-1,2,3,4-tetrahydro-1-oxonaphthalene
β-(1,2,3,4-Tetrahydro-1-oxonaphth-5-yloxy)lactic acid
Formed from: 5-(3-*tert*-Butylamino-2-hydroxypropoxy)-1,2,3,4-tetrahydro-1-oxonaphthalene
2,3,4,5-Tetrahydro-7-(2-oxopiperidino) imidazo[2,1-*b*]quinazolin-2(1*H*)-one
Formed from: 2,3,4,5-Tetrahydro-7-piperidinoimidazo[2,1-*b*]quinazolin-2(1*H*)-one
→ 7-(4-Carboxybutyl)amino-2,3,4,5-tetrahydroimidazo[2,1-*b*]quinazolin-2(1*H*)-one
Probably in: Rat H463
1,2,3,4-Tetrahydro-4-oxo-N-(*p*-toluoyl)quinoline
Formed from: 1,2,3,4-Tetrahydro-N-(*p*-toluoyl)quinoline
Tetrahydropalmatine ((13a*S*)-5,8,13,13a-tetrahydro-2,3,9,10-tetramethoxy-6*H*-dibenzo[*a*,*g*]qinolizine)
Formed from: Palmatine
Reticuline
Tetrahydrocolumbamine
→ Corydaline
In: Corydalis A3684
→ Muramine
In: Corydalis H585
See also: A1923
→ Palmatine
In: Berberis K947
Cocculus B726
Corydalis H585
→ Tetrahydro-α-N-methylpalmatine
In: Corydalis H585
Tetrahydropapaverine (1-((3,4-dimethoxyphenyl)methyl)-1,2,3,4-tetrahydro-6,7-dimethoxyisoquinoline)
→ Laudanosine
In: Berberis B573, K780
Tetrahydropapaveroline (norlaudanosoline, 1-(3,4-dihydroxybenzyl)-1,2,3,4-tetrahydro-6,7-dihydroxyisoquinoline)
Formed from: 3,4-Dihydroxyphenethylamine
3,4-Dihydroxyphenylacetaldehyde
Tetrahydropapaveroline-1-carboxylic acid
Probably formed from: Dihydropapaveroline
→ Dihydropapaveroline
Probably in: Papaver A1513
→ 1-(3,4-Dihydroxybenzyl)-1,2,3,4-tetrahydro-6-hydroxy-7-methoxyisoquinoline
In: Adlumia D77
Argemone D77
Berberis K783
Chelidonium K783
Cissampelos D77
Corydalis D77
Eschscholtzia K783
Fumaria D77
Papaver K783
Rat A3497, C661
Thalictrum D77
→ 1-(3,4-Dihydroxybenzyl)-1,2,3,4-tetrahydro-7-hydroxy-6-methoxyisoquinoline
In: Argemone D77
Rat A3497, C661
→ Hydrastine
Probably in: Hydrastis A3684
→ Laudanosoline
Probably in: Rat A53
→ Norprotosinomenine

Probably in: Erythrina A3588
→ Papaveroline
In: Papaver A1620
→ Reticuline
In: Litsea A3222
Papaver A1620
→ 1,2,3,4-Tetrahydro-6,7-dihydroxy-1-(3-hydroxy-4-methoxybenzyl)isoquinoline
In: Coptis K572
→ 1,2,3,4-Tetrahydro-6,7-dihydroxy-1-(4-hydroxy-3-methoxybenzyl)isoquinoline
In: Rat C661
→ 2,3,9,10-Tetrahydroxyberberine
In: Rat A1445
→ 2,3,10,11-Tetrahydroxyberberine
In: Rat A1445

Tetrahydropapaveroline-1-carboxylic acid
Formed from: 3,4-Dihydroxyphenethylamine
3,4-Dihydroxy-L-phenylalanine
Probably formed from: 3,4-Dihydroxyphenylpyruvic acid
→ O-Methyltetrahydropapaveroline-1-carboxylic acid
In: Argemone D77
→ Tetrahydropapaveroline
In: Papaver A1513
See also: A3222

7,8,9,10-Tetrahydro-1,7,8,9,10-pentahydroxy-3,4-benzpyrene
Probably formed from: 7,8-Dihydro-7β,8α-dihydroxy-3,4-benzpyrene-9α,10α-oxide

7,8,9,10-Tetrahydro-3,7,8,9,10-pentahydroxy-3,4-benzpyrene
Probably formed from: 7,8-Dihydro-7β,8α-dihydroxy-3,4-benzpyrene-9α,10α-oxide

1,2,3,4-Tetrahydro-1,2,3,5,9-pentahydroxychrysene
Probably formed from: *trans*-1,2-Dihydro-1,2,9-trihydroxychrysene

7,8,9,10-Tetrahydro-1,7,8,10,11-pentahydroxy-3,9-dimethoxy-8-methyl-5,12-anthracenedione
See: Steffimycinol

(1*R*,2*S*,3*R*,4*S*)-1,2,3,4-Tetrahydro-1,2,3,4,5-pentahydroxy-7-methoxy-2-methyl-9,10-anthracenedione
See: Altersolanol A

4-(1-(5,6,7,8-Tetrahydro-3,5,5,8,8-pentamethyl-2-naphthalenyl)-1-ethenyl)benzoic acid
See: Targetrin

4-(1-(5,6,7,8-Tetrahydro-3,5,5,8,8-pentamethyl-7-oxo-2-naphthalenyl)-1-ethenyl)benzoyltaurine
Probably formed from: 7-Oxotargetrin

(*R*)-1,2,3,6-Tetrahydro-4-phenyl-1-((3-phenyl-3-cyclohexen-1-yl)methyl)pyridine
→ (*R*)-1,2,3,6-Tetrahydro-1-((3-(*p*-hydroxyphenyl)-3-cyclohexen-1-yl)methyl)-4-phenylpyridine
In: Cynomolgus monkey H752, J448
Dog H752
Man H752
Rat H752
→ (*R*)-1,2,3,6-Tetrahydro-4-(*p*-hydroxyphenyl)-1-((3-phenyl-3-cyclohexen-1-yl)methyl)pyridine
In: Cynomolgus monkey H752, J448

1,2,3,4-Tetrahydro-N-(1-phenylcyclohexyl)pyridine
Formed from: Phencyclidine
→ 4-(1,2,3,4-Tetrahydropyridin-1-yl)-4-phenylcyclohexanol
Probably in: Rabbit G29

2,3,4,6-Tetrahydro-1-(1-phenylcyclohexyl)pyridinium
→ 5-Formyl-2,3,4,6-tetrahydro-1-(1-phenylcyclohexyl)pyridinium
In: Rat G209

1,2,3,6-Tetrahydro-4-phenylpyridine
Formed from: 1-Cyclopropyl-1,2,3,6-tetrahydro-4-phenylpyridine
1,2,3,6-Tetrahydro-N-methyl-4-phenylpyridine
→ 1,2,3,6-Tetrahydro-N-methyl-4-phenylpyridine
In: Man E178
Monkey E178
Mouse E178
Rabbit E178
Rat E6, E178
See also: K923
→ 1,2,3,6-Tetrahydro-4-phenylpyridine-N-sulphate
In: Guinea pig H455
Man J571

Rabbit H455, J914
Rat E45, E51

1,2,3,6-Tetrahydro-4-phenylpyridine-N-sulphate
Formed from: 1,2,3,6-Tetrahydro-4-phenylpyridine

2,3,4,5-Tetrahydro-7-piperidinoimidazo[2,1-*b*]quinazolin-2(*H*)-one
→ 2-Amino-3-carboxymethyl-3,4-dihydro-6-piperidinoquinazoline
In: Rat H463
→ 2,3,4,5-Tetrahydro-7-(4-hydroxypiperidino)imidazo[2,1-*b*]quinazolin-2(*H*)-one
In: Rat H463
→ 2,3,4,5-Tetrahydro-7-(2-oxopiperidino)imidazo[2,1-*b*]quinazolin-2(*H*)-one
In: Rat H463

(−)-1,3,4,5-Tetrahydro-4-propylaminobenz[*c*,*d*]indole-6-carboxamide
Formed from: (−)-4-Dipropylamino-1,3,4,5-tetrahydrobenz[*c*,*d*]indole-6-carboxamide

Δ^1-Tetrahydropropylcannabinol (4″,5″-bisnor-Δ^1-tetrahydrocannabinol)
→ Δ^1-Tetrahydro-2″-hydroxypropylcannabinol
In: Mouse E625
→ Δ^1-Tetrahydro-7-hydroxypropylcannabinol
In: Mouse E625

Δ^6-Tetrahydropropylcannabinol (4″,5″-bisnor-Δ^6-tetrahydrocannabinol)
→ Δ^6-Tetrahydro-2″-hydroxypropylcannabinol
In: Mouse E625
→ Δ^6-Tetrahydro-7-hydroxypropylcannabinol
In: Mouse E625

Δ^1-Tetrahydropropylcannabinol-7-oic acid
Probably formed from: Δ^1-Tetrahydro-7-hydroxypropylcannabinol

Δ^6-Tetrahydropropylcannabinol-7-oic acid
Probably formed from: Δ^6-Tetrahydro-7-hydroxypropylcannabinol

1,3,4,9-Tetrahydro-1-*n*-propylpyrano[3,4-*b*]indole-1-acetic acid
See: Prodolic acid

Tetrahydroprotoberberine (5,8,13,13a-tetrahydro-6*H*-dibenzo[*a*,*g*]quinolizine)
→ Protoberberine
In: Berberis D145

Tetrahydropteroyldiglutamic acid
Formed from: Tetrahydrofolic acid
→ Tetrahydropteroyltriglutamic acid
In: Pig K455

Tetrahydropteroylheptaglutamic acid
Formed from: 5-Methyltetrahydropteroylheptaglutamic acid
Tetrahydropteroylhexaglutamic acid
→ Dihydropteroylheptaglutamic acid
In: Man A2946
→ Tetrahydropteroyloctaglutamic acid
In: Pig K455

Tetrahydropteroylhexaglutamic acid
Formed from: Tetrahydropteroylpentaglutamic acid
→ Tetrahydropteroylheptaglutamic acid
Probably in: Pig E277

Tetrahydropteroyloctaglutamic acid
Formed from: Tetrahydropteroylheptaglutamic acid

Tetrahydropteroylheptaglutamic acid
Formed from: Folic acid
5-Methyltetrahydropteroylpentaglutamic acid
Probably formed from: Tetrahydropteroyltetraglutamic acid
→ Dihydropteroylpentaglutamic acid
In: Man A2946
→ Tetrahydropteroylhexaglutamic acid
In: Pig K455
See also: E277

Tetrahydropteroyltetraglutamic acid
Formed from: Tetrahydropteroyltriglutamic acid
→ Tetrahydropteroylpentaglutamic acid
Probably in: Pig E277

Tetrahydropteroyltriglutamic acid
Formed from: Tetrahydropteroyldiglutamic acid
5-Methyltetrahydropteroyltriglutamic acid
→ Dihydropteroyltriglutamic acid
In: Man A2946
→ Tetrahydropteroyltetraglutamic acid
In: Pig K455
See also: E277

4′-O-Tetrahydropyranyladriamycin
→ Adriamycin
In: Mouse D785
→ 7-Deoxyadriamycin
In: Mouse D785
→ 13-Dihydro-4′-O-tetrahydropyranyladriamycin
In: Mouse D785

4′-O-Tetrahydropyranyldaunomycin
→ 4′-O-Dihydropyranyldaunomycin
In: Streptomyces B600

4-(1,2,3,4-Tetrahydropyridin-1-yl)-4-phenylcyclohexanol
Probably formed from: 1,2,3,4-Tetrahydro-N-(1-phenylcyclohexyl)pyridine

Tetrahydroquinoline
→ Tetrahydro-N-methylquinoline
In: Rabbit E6
Rat K923
→ Tetrahydroquinoline-N-sulphate
In: Guinea pig E456
Rat E51

Tetrahydroquinoline-N-sulphate
Formed from: Tetrahydroquinoline

1,2,19,20-Tetrahydroraucaffricine
→ 1,2,19,20-Tetrahydrovomilenine
In: Rauwolfia K939

1,2,9,10-Tetrahydro-1,2,9,10-tetrahydroxybenz[*j*]aceanthrylene
Probably formed from: *trans*-1,2-Dihydro-1,2-dihydroxybenz[*j*]aceanthrylene

1,2,3,4-Tetrahydro-1,2,3,4-tetrahydroxybenzanthracene
Formed from: *trans*-1,2-Dihydro-1,2-dihydroxybenzanthracene
trans-3,4-Dihydro-3,4-dihydroxybenzanthracene

8,9,10,11-Tetrahydro-8,9,10,11-tetrahydroxybenzanthracene
Formed from: *trans*-10,11-Dihydro-10,11-dihydroxybenzanthracene
Probably formed from: 8,9-Dihydro-8,9-dihydroxybenzanthracene

(+)-(1*R*,2*S*,7*R*,8*S*)-*cis*,*cis*-1,2,7,8-Tetrahydro-1,2,7,8-tetrahydroxybenzo[*b*]naphtho[2,1-*d*]thiophene
Formed from: *cis*-1,2-Dihydro-1,2-dihydroxybenzo[*b*]naphtho[2,1-*d*]thiophene

7,8,9,10-Tetrahydro-7,8,9,10-tetrahydroxy-3,4-benzpyrene (stereochemistry unclear)
trans-7,8-Dihydro-7,8-dihydroxy-3,4-benzpyrene
9,10-Dihydro-9,10-dihydroxy-3,4-benzpyrene
→ 7,8,9,10-Tetrahydro-7,8,9,10-tetrahydroxy-3,4-benzpyrene-β-D-glucuronide
Probably in: Man A387

7,8,9,10-Tetrahydro-7α,8β,9α,10α-tetrahydroxy-3,4-benzpyrene
Formed from: *trans*-7,8-Dihydro-7,8-dihydroxy-3,4-benzpyrene

7,8,9,10-Tetrahydro-7α,8β,9α,10β-tetrahydroxy-3,4-benzpyrene
Formed from: *trans*-7,8-Dihydro-7,8-dihydroxy-3,4-benzpyrene

7,8,9,10-Tetrahydro-7α,8β,9β,10α-tetrahydroxy-3,4-benzpyrene
Formed from: *trans*-7,8-Dihydro-7,8-dihydroxy-3,4-benzpyrene

7,8,9,10-Tetrahydro-7α,8β,9β,10α-tetrahydroxy-3,4-benzpyrene
Probably formed from: 7,8,9,10-Tetrahydro-7β,8α-dihydroxy-3,4-benzpyrene-9α,10α-oxide

7,8,9,10-Tetrahydro-7β,8β,9α,10β-tetrahydroxy-3,4-benzpyrene
Probably formed from: 7,8,9,10-Tetrahydro-7β,8α-dihydroxy-3,4-benzpyrene-9α,10α-oxide

7,8,9,10-Tetrahydro-7βϖ,8α,9β,10α-tetrahydroxy-3,4-benzpyrene
Probably formed from: 7,8,9,10-Tetrahydro-7β,8α-dihydroxy-3,4-benzpyrene-9β,10β-oxide

7,8,9,10-Tetrahydro-7βϖ,8α,9β,10β-tetrahydroxy-3,4-benzpyrene
Probably formed from: 7,8,9,10-Tetrahydro-7β,8α-dihydroxy-3,4-benzpyrene-9β,10β-oxide

4,5,9,10-Tetrahydro-4,5,9,10-tetrahydroxybenz[*e*]pyrene
Formed from: *trans*-9,10-Dihydro-9,10-dihydroxybenz[*e*]pyrene

7,8,9,10-Tetrahydro-7,8,9,10-tetrahydroxy-3,4-benzpyrene-β-D-glucuronide
Probably formed from: 7,8,9,10-Tetrahydro-7,8,9,10-tetrahydroxy-3,4-benzpyrene

(+)-(3*S*,4*R*,9*S*,10*R*)-*cis*,*cis*-3,4,9,10-Tetrahydro-3,4,9,10-tetrahydroxychrysene
Formed from: (+)-(3*S*,4*R*)-*cis*-3,4-Dihydro-3,4-dihydroxychrysene

1,2,3,4-Tetrahydro-1,2,3,4-tetrahydroxydibenz[*c*,*h*]acridine
Formed from: *trans*-3,4-Dihydro-3,4-dihydroxydibenz[*c*,*h*]acridine

***trans*-1,2,3,4-Tetrahydro-1,2,3,4-tetrahydroxydibenzanthracene**
Probably formed from: 3,4-Dihydro-3,4-dihydroxydibenzanthracene
→ 1,2,3,4,8,9-Hexahydro-1,2,3,4,8,9-hexahydroxydibenzanthracene
Probably in: Rat G442
→ 1,2,3,4-Tetrahydro-3,4-dihydroxy-2-oxodibenzanthracene
Probably in: Rat E556

3,4,10,11-Tetrahydro-3,4,10,11-tetrahydroxydibenzanthracene
Probably formed from: 3,4-Dihydro-3,4-dihydroxydibenzanthracene
→ 1,2,3,4,10,11-Hexahydro-1,2,3,4,10,11-hexahydroxydibenzanthracene
Probably in: Rat G442

3,4,12,13-Tetrahydro-3,4,12,13-tetrahydroxydibenzanthracene
Probably formed from: 3,4-Dihydro-3,4-dihydroxydibenzanthracene

***trans*-3,4,8,9-Tetrahydro-3,4,8,9-tetrahydroxydibenz[*a*,*j*]anthracene**
Formed from: Dibenz[*a*,*j*]anthracene

Δ^1-Tetrahydro-2″,3″,6α,7-tetrahydroxy-*n*-heptylcannabinol
Probably formed from: Δ^1-Tetrahydro-2″,3″,7-trihydroxy-*n*-heptylcannabinol

(8*S*,10*S*)-7,8,9,10-Tetrahydro-6,8,10,11-tetrahydroxy-8-hydroxyacetyl-1-methoxy-5,12 naphthacenedione
Probably formed from: 7-O-((2,6-Dideoxy-2-fluoro-α-L-talopyranosyl) adriamycinone

(5*S*,7*S*)-5,7,9,10-Tetrahydro-5,7,9,10-tetrahydroxy-2-methoxy-7-methyl-1,4-anthracenedione
See: (1*S*,3*S*)-Austrocortirubin

8,9,10,11-Tetrahydro-8,9,10,11-tetrahydroxy-7-methylbenzanthracene
Probably formed from: 8,9-Dihydro-8,9-dihydroxy-7-methylbenzanthracene

1,2,3,4-Tetrahydro-1β,2α,3α,4β-tetrahydroxynaphthalene
Probably formed from: 1,2-Dihydro-1β,2α-dihydroxynaphthalene-3α,4α-oxide

1,2,3,4-Tetrahydro-1β,2α,3β,4α-tetrahydroxynaphthalene
Probably formed from: 1,2-Dihydro-1β,2α-dihydroxynaphthalene-3α,4α-oxide

7,8,9,10-Tetrahydro-7,8,9,10-tetrahydroxy-1-nitro-3,4-benzpyrene
Formed from: *trans*-7,8-Dihydro-7,8-dihydroxy-1-nitro-3,4-benzpyrene
trans-9,10-Dihydro-9,10-dihydroxy-1-nitro-3,4-benzpyrene

4,5,9,10-Tetrahydro-4,5,9,10-tetrahydroxy-1-nitropyrene
Formed from: 4,5-Dihydro-4,5-dihydroxy-1-nitropyrene

Tetrahydro-2,3,9,10-tetrahydroxyprotoberberine
Formed from: Laudanosoline

1-((5,6,7,8-Tetrahydro-5,5,8,8-tetramethyl-2-naphthyl)carbamoyl)benzoic acid
→ 1-((5,6,7,8-Tetrahydro-6-hydroxy-5,5,8,8-tetramethyl-2-naphthyl)carbamoyl)benzoic acid
In: Rat H128, H863
→ 1-((5,6,7,8-Tetrahydro-7-hydroxy-5,5,8,8-tetramethyl-2-naphthyl)carbamoyl)benzoic acid
In: Rat H128, H863
→ 1-((5,6,7,8-Tetrahydro-5,5,8,8-tetramethyl-2-naphthyl)carbamoyl)benzoyltaurine

In: Rat H128, H863

1-((5,6,7,8-Tetrahydro-5,5,8,8-tetramethyl-2-naphthyl)carbamoyl)benzoyltaurine
Formed from: 1-((5,6,7,8-Tetrahydro-5,5,8,8-tetramethyl-2-naphthyl)carbamoyl)benzoic acid

1-((5,6,7,8-Tetrahydro-5,5,8,8-tetramethyl-6-oxo-2-naphthyl)carbamoyl)benzoic acid
Probably formed from: 1-((5,6,7,8-Tetrahydro-6-hydroxy-5,5,8,8-tetramethyl-2-naphthyl)carbamoyl)benzoic acid

1,2,3,4-Tetrahydro-1-thianaphthalene
See: Thiochroman

1,2,3,4-Tetrahydro-1-thianaphthalene-1,1-dioxide
Probably formed from: 1,2,3,4-Tetrahydro-1-thianaphthalene-1-oxide

1,2,3,4-Tetrahydro-1-thianaphthalene-1-oxide
Probably formed from: Thiochroman
→ 1,2,3,4-Tetrahydro-1-thianaphthalene-1,1-dioxide
Probably in: Helminthosporium G427
Mortierella G427

2,3,4,5-Tetrahydro-N-(*p*-toluoyl)-1*H*-1-benzazepine
→ 2,3,4,5-Tetrahydro-5-hydroxy-N-(*p*-toluoyl)-1*H*-1-benzazepine
In: Cunninghamella D810

1,2,3,4-Tetrahydro-N-(*p*-toluoyl)isoquinoline
→ 1,2,3,4-Tetrahydro-N-(*p*-hydroxybenzoyl) isoquinoline
In: Cunninghamella D810

1,2,3,4-Tetrahydro-N-(*p*-toluoyl)quinoline
→ 1,2,3,4-Tetrahydro-4-oxo-N-(*p*-toluoyl) quinoline
In: Cunninghamella D810

1,2,3,4-Tetrahydro-N-(*p*-tolylacetyl)isoquinoline
→ 1,2,3,4-Tetrahydro-N-(*p*-hydroxymethyl-phenylacetyl)isoquinoline
In: Cunninghamella D810

Tetrahydro-2-(3-trifluoromethylphenyl)-1,4-oxazine
→ 2-Amino-1-(*m*-trifluoromethylphenylethanol
In: Dog A3852
Rat A3852
→ *m*-Trifluoromethyl benzoic acid
In: Dog A3852
Rat A3852
→ *m*-Trifluoromethylmandelic acid
In: Dog A3852
Rat A3852

S-(5,6,8,9-Tetrahydro-5,6,9-trihydroxybenzanthr-8-yl)glutathione
Probably formed from: 5,6-Dihydro-5,6-dihydroxybenzanthracene

S-(5,6,8,9-Tetrahydro-6,8,9-trihydroxybenzanthr-5-yl)glutathione
Probably formed from: 5,6-Dihydro-5,6-dihydroxybenzanthracene

S-(8,9,10,11-Tetrahydro-8,9,10-trihydroxybenzanthr-11-yl)glutathione
Probably formed from: 8,9-Dihydro-8,9-dihydroxybenzanthracene

S-(7,8,9,10-Tetrahydro-7,8,9-trihydroxy-3,4-benzpyren-10-yl)glutathione
Probably formed from: 7,8-Dihydro-7,8-dihydroxy-3,4-benzpyrene

S-(7,8,9,10-Tetrahydro-8,9,10-trihydroxy-3,4-benzpyren-7-yl)glutathione
Probably formed from: *trans*-9,10-Dihydro-9,10-dihydroxy-3,4-benzpyrene

Δ^1-Tetrahydro-1″,6α,7-trihydroxycannabinol
Probably formed from: Δ^1-Tetrahydro-6α,7-dihydroxycannabinol

Δ^1-Tetrahydro-1″,6β,7-trihydroxycannabinol
Probably formed from: Δ^1-Tetrahydro-6β,7-dihydroxycannabinol

Δ^1-Tetrahydro-2″,6α,7-trihydroxycannabinol
Probably formed from: Δ^1-Tetrahydro-6α,7-dihydroxycannabinol

Δ^1-Tetrahydro-2″,6β,7-trihydroxycannabinol
Probably formed from: Δ^1-Tetrahydro-6β,7-dihydroxycannabinol

Δ^1-Tetrahydro-3″,6α,7-trihydroxycannabinol
Probably formed from: Δ^1-Tetrahydro-3″,7-dihydroxycannabinol
Δ^1-Tetrahydro-6α,7-dihydroxycannabinol
→ Δ^1-Tetrahydro-3″,7-dihydroxy-6-oxocannabinol
Probably in: Mouse F492

Δ^1-Tetrahydro-3″,6β,7-trihydroxycannabinol
Probably formed from: Δ^1-Tetrahydro-6β,7-dihydroxycannabinol

Δ^1-Tetrahydro-4″,6α,7-trihydroxycannabinol
Probably formed from: Δ^1-Tetrahydro-4″,7-dihydroxycannabinol
Δ^1-Tetrahydro-6α,7-dihydroxycannabinol

Δ^1-Tetrahydro-4″,6β,7-trihydroxycannabinol
Probably formed from: Δ^1-Tetrahydro-6β,7-dihydroxycannabinol

Δ^6-Tetrahydro-2″,5α,7-trihydroxycannabinol
Probably formed from: Δ^6-Tetrahydro-2″,7-dihydroxycannabinol

Δ^6-Tetrahydro-2″,5β,7-trihydroxycannabinol
Probably formed from: Δ^6-Tetrahydro-2″,7-dihydroxycannabinol

Δ^6-Tetrahydro-3″,5α,7-trihydroxycannabinol
Probably formed from: Δ^6-Tetrahydro-3″,7-dihydroxycannabinol

Δ^6-Tetrahydro-3″,5β,7-trihydroxycannabinol
Probably formed from: Δ^6-Tetrahydro-3″,7-dihydroxycannabinol

Δ^6-Tetrahydro-4″,5α,7-trihydroxycannabinol
Probably formed from: Δ^6-Tetrahydro-4″,7-dihydroxycannabinol

Δ^6-Tetrahydro-4″,5β,7-trihydroxycannabinol
Probably formed from: Δ^6-Tetrahydro-4″,7-dihydroxycannabinol

1,2,3,4-Tetrahydro-6,7,8-trihydroxy-1,N-dimethylisoquinoline
→ 1,2,3,4-Tetrahydro-6,8-dihydroxy-7-methoxy-1,N-dimethylisoquinoline
In: Lophophora A3600

Δ^1-Tetrahydro-2″,3″,7-trihydroxy-*n*-heptylcannabinol
Probably formed from: Δ^1-Tetrahydro-2″,7-dihydroxy-*n*-heptylcannabinol
→ Δ^1-Tetrahydro-2″,3″,6α,7-tetrahydroxy-*n*-heptylcannabinol
Probably in: Mouse D357

7,8,9,10-Tetrahydro-6,8,11-trihydroxy-8-hydroxyacetyl-1-methoxy-5,12-naphthacenedione (7-deoxyadriamycin aglycone, 7-deoxyadriamycinone)
Formed from: Adriamycin
→ 7-Deoxyadriamycinol aglycone
In: Streptomyces A2755
See also: A1989

7,8,9,10-Tetrahydro-6,8,11-trihydroxy-8-(1-hydroxyethyl)-1-methoxy-5,12-naphthacenedione
See: 7-Deoxydaunorubicinol aglycone

1,2,3,4-Tetrahydro-4,6,7-trihydroxyisoquinoline
See: 4-Hydroxynorsalsolinol

(1*S*,3*S*)-1,2,3,4-Tetrahydro-1,3,8-trihydroxy-6-methoxy-3-methyl-9,10-anthracenedione
See: (1*S*,3*S*)-Austrocortilutein

5,6,6a,7-Tetrahydro-1,10,11-trihydroxy-2-methoxy-6-methyl-1*H*-dibenzo[*de*,*g*]quinoline
Formed from: N-Methylcoclaurine

Tetrahydro-2,9,10-trihydroxy-3-methoxyprotoberberine
Formed from: 6-O-Methyllaudanosoline

1,2,3,4-Tetrahydro-4,6,7-trihydroxy-1-methylisoquinoline
Probably formed from: D-Noradrenaline

5,6,7,8-Tetrahydro-3,6,9-trihydroxy-1-methyl-8-oxo-2-anthracenecarboxylic acid methyl ester
See: Aloesaponol 1

1,2,3,4-Tetrahydro-1,2,3-trihydroxy-4-methylthionaphthalene
Probably formed from: S-(1,2-Dihydro-2-hydroxynaphth-1-yl)glutathione

4,5,7,8-Tetrahydro-4,7,8-trihydroxy-5-oxo-3,4-benzpyrene
Probably formed from: 4,5-Dihydro-4-hydroxy-5-oxo-3,4-benzpyrene-7,8-oxide

1,2,3,4-Tetrahydro-4,6,8-trihydroxy-2-(3-theophyllin-7-ylpropyl)isoquinoline
Formed from: Reproterol
→ 1,2,3,4-Tetrahydro-4,6,8-trihydroxy-2-(3-theophyllin-7-ylpropyl)isoquinoline-β-D-glucuronide
In: Rat C97

1,2,3,4-Tetrahydro-4,6,8-trihydroxy-2-(3-theophyllin-7-ylpropyl)isoquinoline-β-D-glucuronide
Formed from: 1,2,3,4-Tetrahydro-4,6,8-trihydroxy-2-(3-theophyllin-7-ylpropyl) isoquinoline

Tetrahydro-4-(3,4,5-trimethoxybenzoyl)-1,4-oxazine
Formed from: Trithiozine

→ N,N-Bis(2-hydroxyethyl)-3,4,5-trimethoxybenzamide
Probably in: Man C304
→ Tetrahydro-4-(4-hydroxy-3,5-dimethoxybenzoyl)-1,4-oxazine
Probably in: Rat A3260
→ Tetrahydro-2-hydroxy-4-(3,4,5-trimethoxybenzoyl)-1,4-oxazine
Probably in: Man C304
→ 3,4,5-Trimethoxybenzoic acid
Probably in: Rat A3260

1,2,3,4-Tetrahydro-6,7,8-trimethoxy-1,N-dimethylisoquinoline
Formed from: 1,2,3,4-Tetrahydro-7-hydroxy-6,8-dimethoxy-1,N-dimethylisoquinoline

1,2,3,4-Tetrahydro-6,7,8-trimethoxyisoquinoline
Formed from: 1,2,3,4-Tetrahydro-7-hydroxy-6,8-dimethoxyisoquinoline

1,2,3,4-Tetrahydro-6,7,8-trimethoxy-1-methylisoquinoline
Formed from: 1,2,3,4-Tetrahydro-7-hydroxy-6,8-dimethoxy-1-methylisoquinoline

1,2,3,4-Tetrahydro-6,7,8-trimethoxy-N-methylisoquinoline
Formed from: 1,2,3,4-Tetrahydro-7-hydroxy-6,8-dimethoxy-N-methylisoquinoline

Tetrahydro-4-(3,4,5-trimethoxythiobenzoyl)-1,4-oxazine
See: Trithiozine

Tetrahydrovomilenine
Formed from: 1,2,19,20-Tetrahydroraucaffricine

1,2,5,8-Tetrahydroxyanthraquinone
See: Quinalizarin

3′,4,4′,6-Tetrahydroxyaurone
See: Aureusidin

1,2,3,4-Tetrahydroxybenzene
Formed from: 2,3,4-Trihydroxybenzoic acid
Probably formed from: Pyrogallol

1,2,3,5-Tetrahydroxybenzene
Formed from: 3,5-Dimethoxyquinone
Gallic acid
Phloroglucinol
Pyrogallol
2,4,6-Trihydroxybenzoic acid
Probably formed from: Pentahydroxybenzene
Phloroglucinolcarboxylic acid
→ Hydroxyquinol
In: Pelobacter F434
→ Pentahydroxybenzene
Probably in: Pelobacter F434, G621
→ Phloroglucinol
In: Pelobacter F434, G621

1,2,4,5-Tetrahydroxybenzene
Formed from: 2,5-Dihydroxyquinone
Hydroxyquinol
→ Hydroxyquinol
In: Pelobacter G621
→ Pentahydroxybenzene
Probably in: Pelobacter G621

2,3,4,6-Tetrahydroxybenzoic acid
Probably formed from: 2,4,6-Trihydroxybenzoic acid

2,3′,4,6-Tetrahydroxybenzophenone
Formed from: *m*-Hydroxybenzoyl CoA
2,4,6-Trihydroxybenzophenone
→ Mangostin
In: Garcinia E738

2,3,9,10-Tetrahydroxyberbine
Formed from: Tetrahydropapaveroline

2,3,10,11-Tetrahydroxyberbine
Formed from: Tetrahydropapaveroline
→ Coreximine
Probably in: Rat A1445
→ 2,3,10-Trihydroxy-11-methoxyberbine
In: Rat C661
→ 2,3,11-Trihydroxy-10-methoxyberbine
In: Rat C661
→ 2,10,11-Trihydroxy-3-methoxyberbine
In: Rat C661
→ 3,10,11-Trihydroxy-2-methoxyberbine
In: Rat C661

2,2′,3,3′-Tetrahydroxybibenzyl
Probably formed from: 2,3-Dihydroxybibenzyl

3,3′,4,4′-Tetrahydroxybibenzyl
Probably formed from: 3,4,4′-Trihydroxybibenzyl

2,2′,3,3′-Tetrahydroxybiphenyl
Probably formed from: 2,2′,3-Trihydroxybiphenyl
→ 6-(2,3-Dihydroxyphenyl)-2-hydroxy-6-oxomuconic acid
In: Pseudomonas H22

1,1′,8,8′-Tetrahydroxybisanthrone
Probably formed from: 1,8-Dihydroxy-9-anthrone

4,4′,5,5′-Tetrahydroxy-2,2′-bis(hydroxymethyl)-(9,9′-bianthracene)-10,10′(9*H*,9′*H*)-dione
See: Aloe-emodin bianthrone

2′,3,4,4′-Tetrahydroxychalcone (butein)
Formed from: Caffeoyl CoA
2′,3,4′-Trihydroxy-4-methoxychalcone
2′,4,4′-Trihydroxychalcone
2′,4,4′-Trihydroxy-3-methoxychalcone
→ 2′,4,4′-Trihydroxy-3-methoxychalcone
In: Rat C665

2′,4,4′,6′-Tetrahydroxychalcone (naringenin-chalcone)
Formed from: *p*-Coumaroyl CoA
→ Aureusidin
In: Antirrhinum K659
→ Naringenin
In: Dianthus 392
Tulipa B377
See also: B404

2′,4,4′,6′-Tetrahydroxychalcone-4′-neohesperoside
→ 4′,5,7-Trihydroxyflavanone-4′-neohesperoside
Probably in: Grapefruit A2518

3′,4′,5,6-Tetrahydroxy-3,7-dimethoxyflavone
→ 4′,5,6-Trihydroxy-3,3′,7-trimethoxyflavone
In: Chrysosplenium B832, C579
→ 3′4′,5-Trihydroxy-3,6,7-trimethoxyflavone
In: Chrysosplenium B832, C579, C792

3′,4′,5,7-Tetra-O-(β-hydroxyethyl)rutin
Probably formed from: 3′,4′,7-Tri-O-(β-hydroxyethyl)rutin
→ 3′,4′,5.7-Tetra-O-(β-hydroxyethyl)rutin aglycone
In: Microorganism A283
→ 3′,4′,5,7-Tetra-O-(β-hydroxyethyl)rutin-β-D-glucuronide
In: Rat A271

3′,4′,5,7-Tetra-O-(β-hydroxyethyl)rutin aglycone
Formed from: 3′,4′,5,7-Tetra-O-(β-hydroxyethyl)rutin

3′,4′,5,7-Tetra-O-(β-hydroxyethyl)rutin-β-D-glucuronide
Formed from: 3′,4′,5,7-Tetra-O-(β-hydroxyethyl)rutin

3′,4′,5,7-Tetra-O-(β-hydroxyethyl)rutoside
Unchanged in rat A733

3′,4′,5,7-Tetrahydroxyflavanone-7-O-glucoside
→ 5,7-Dihydroxychromone-7-O-glucoside
In: Mentha A2642
→ Hydroxyquinol
In: Mentha A2642

2,4′,5,7-Tetrahydroxyflavone
→ 2,4′,5,7-Tetrahydroxyflavone-6-C-glucoside
Probably in: Fagopyrum E614

4′,5,7,8-Tetrahydroxyflavone
→ 4′,5,7-Trihydroxy-8-methoxyflavone
In: Calamondin orange A3987
→ 5.7,8-Trihydroxy-4′-methoxyflavone
In: Calamondin orange A3987

2,4′,5,7-Tetrahydroxyflavone-6-C-glucoside
Probably formed from: 2,4′,5,7-Tetrahydroxyflavone

2,4′,5,7-Tetrahydroxyisoflavanone
Formed from: Naringenin

2′,4′,5,7-Tetrahydroxyisoflavanone
Formed from: 2′,4′,5,7-Tetrahydroxyisoflavone

2′,4′,5,7-Tetrahydroxyisoflavone (2′-hydroxygenistein)
→ 2′,4′,5,7-Tetrahydroxyisoflavanone
In: Glycine K702
→ 2′,4′,7-Trihydroxy-5-methoxyisoflavone
In: Lupin E756, F182

4′,5,6,7-Tetrahydroxyisoflavone
Formed from: Genistein

4′,5,7,8-Tetrahydroxyisoflavone
Formed from: Genistein

3′,4′,5,7-Tetrahydroxyisoflavone-7-glucoside
See: Orobol-7-glucoside

5-(3′,5′,8′,9′-Tetrahydroxy-4′-(*p*-mentha-1-en-3-yl)phenyl)propionic acid
Probably formed from: 5-(3′,5′-Dihydroxy-4′-(*p*-mentha-1,8(9)-dien-3-yl)phenyl)propionic acid

2′,4,4′,6′-Tetrahydroxy-3-methoxychalcone
Formed from: Feruloyl CoA

3,3′,5,7-Tetrahydroxy-4′-methoxyflavan
Formed from: (+)-Catechin
(+)-Epicatechin
(−)-Epicatechin

3,4′,5,7-Tetrahydroxy-3′-methoxyflavan
Formed from: (+)-Catechin
(−)-Epicatechin
Probably formed from: (+)-Epicatechin
→ 3,4′,5,7-Tetrahydroxy-3′-methoxyflavan-5-β-D-glucuronide
In: Rat J608
See also: B534, C311, C486, E245
→ 3,4′,5,7-Tetrahydroxy-3′-methoxyflavan-O-sulphate
Probably in: Man C486
Marmoset C249
Rat C249, C311, E245

3′,4′,5,7-Tetrahydroxy-3-methoxyflavan
→ 3′,4′,5,7-Tetrahydroxy-3-methoxyflavan-5-β-D-glucuronide
In: Rat C587
See also: D629
→ *cis*-3′,4′-5,7-Tetrahydroxy-3-methoxyflavan sulphate
In: Man D629
→ 4′,5,7-Trihydroxy-3,3′-dimethoxyflavan
In: Man C722, D629
See also: C587

3,4′,5,7-Tetrahydroxy-3′-methoxyflavan-5-β-D-glucuronide
Probably formed from: 3,4′,5,7-Tetrahydroxy-3′-methoxyflavan
→ 3,4′,5,7-Tetrahydroxy-3′-methoxyflavan-5-β-D-glucuronide-O-sulphate
Probably in: Man C486
Rat E245

3′,4′,5,7-Tetrahydroxy-3-methoxyflavan-5-β-D-glucuronide
Probably formed from: 3′,4′,5,7-Tetrahydroxy-3-methoxyflavan

3,4′,5,7-Tetrahydroxy-3′-methoxyflavan-5-β-D-glucuronide-O-sulphate
Probably formed from: 3,4′,5,7-Tetrahydroxy-3′-methoxyflavan-5-β-D-glucuronide

3′,4′,5,7-Tetrahydroxy-3-methoxyflavan sulphate
Formed from: 3′,4′,5,7-Tetrahydroxy-3-methoxyflavan

***cis*-3,4′,5,7-Tetrahydroxy-3′-methoxyflavan-O-sulphate**
Probably formed from: *cis*-3,4′,5,7-Tetrahydroxy-3′-methoxyflavan

3,3′,4′,7-Tetrahydroxy-5-methoxyflavone
Formed from: Quercetin

3,3′,5,7-Tetrahydroxy-4′-methoxyflavone
Formed from: Quercetin

3′,4′,5,7-Tetrahydroxy-3-methoxyflavone
Formed from: Quercetin
→ 5-Hydroxy-3,3′,4′,7-tetramethoxyflavone
In: Calamondin orange A3987
→ 3′,4′,5-Trihydroxy-3,7-dimethoxyflavone
In: Apple D258
Calamondin orange A3987
Chrysosplenium C579, C792, D589, E764, F184
See also: B830

1,2,3,8-Tetrahydroxy-6-methylanthraquinone
Formed from: Emodin

1,2,3,4-Tetrahydroxynaphthalene
Probably formed from: 1,2,4-Trihydroxynaphthalene

1,3,6,8-Tetrahydroxynaphthalene
→ 3,6,8-Trihydroxy-1-tetralone
In: Magnaporthe K663
Pyricularia K703
Verticillium K593

1,4,5,8-Tetrahydroxynaphthalene
Formed from: 5,8-Dihydroxy-1,4-naphthoquinone

1,3,6,8-Tetrahydroxy-2-(1-oxohexyl)-9,10-anthracenedione
See: Norsolorinic acid

2,3,4,5-Tetrahydroxyphenylpropionic acid
Probably formed from: 2,3-Dihydro-2,3,4,5-tetrahydroxyphenylpropionic acid

2′,4′,5,7-Tetrahydroxy-3′-prenylisoflavone
→ 2′,4′,7-Trihydroxy-5-methoxy-3′-prenylisoflavone
In: Lupin F182

3,3″,4,4″-Tetrahydroxy-*m*-terphenyl
Probably formed from: 3,4,4″-Trihydroxy-*m*-terphenyl

3,3′,5,5′-Tetraiodothyroacetic acid
→ 3,3′,5,5′-Tetraiodothyroacetic acid-4′-sulphate
In: Monkey B770
Rat B770

3,3′,5,5′-Tetraiodothyroacetic acid-4′-sulphate
Formed from: 3,3′,5,5′-Tetraiodothyroacetic acid

3,3′,5,5′-Tetraiodothyropropionic acid
→ Thyropropionic acid
Probably in: Rat A1095

Tetralin (1,2,3,4-tetrahydronaphthalene)
→ 1,2,3,4-Tetrahydro-1-naphthol
In: Man A1891
→ 1,2,3,4-Tetrahydro-2-naphthol
In: Man A1891

α-Tetralone (1,2,3,4-tetrahydro-1-oxonaphthalene)
Formed from: 1-Aminotetralin
1,2-Dihydronaphthalene
1,2,3,4-Tetrahydro-1-naphthol
→ 1,2,3,4-Tetrahydro-1-naphthol
In: Rabbit D125

β-Tetralone (1,2,3,4-tetrahydro-2-oxonaphthalene)
Formed from: 1-Aminotetralin
→ 1,2,3,4-Tetrahydro-2-naphthol
In: Rabbit D125

3,3′,5,5′-Tetramethoxydiphenoquinone
Formed from: 2,6-Dimethoxyphenol

2′,3′,5′,6′-Tetramethylacetophenone imine
→ 2′,3′,5′,6′-Tetramethylacetophenone oxime
In: Rabbit D955

2′,3′,5′,6′-Tetramethylacetophenone oxime
Formed from: 2′,3′,5′,6′-Tetramethylacetophenone imine

N,N,9,9-Tetramethyl-10(9*H*)-acridinepropanamine
See: Dimetacrine

N,N,N′,N′-Tetramethylbenzidine
→ N,N,N′-Trimethylbenzidine
Probably in: Glycine K419
Man K419

3,3′,5,5′-Tetramethylbiphenyl-4,4′-bisdiglycidyl ether
→ 3,3′,5,5′-Tetramethylbiphenyl-4-(2,3-dihydroxypropyl) ether-4′-glycidyl ether
In: Man K112, K113
Mouse K112, K113
Rat K112, K113

3,3′,5,5′-Tetramethylbiphenyl-4,4′-bis(2,3-dihydroxypropyl) ether
Formed from: 3,3′,5,5′-Tetramethylbiphenyl-4-(2,3-dihydroxypropyl) ether-4′-glycidyl ether

3,3′,5,5′-Tetramethylbiphenyl-4-(2,3-dihydroxypropyl) ether-4′-glycidyl ether
Formed from: 3,3′,5,5′-Tetramethylbiphenyl-4,4′-bisdiglycidyl ether
→ 3,3′,5,5′-Tetramethylbiphenyl-4,4′-bis(2,3-dihydroxypropyl) ether
In: Man K112, K113
Mouse K112, K113
Rat K112, K113

4-(2,2,4,4-Tetramethylbut-4-yl)catechol
Formed from: 4-*tert*-Octylphenol
→ 4-(2,2,4,4-Tetramethylbut-4-yl)catechol-β-D-glucuronide
Probably in: Rainbow trout K126

4-(2,2,4,4-Tetramethylbut-4-yl)catechol-β-D-glucuronide
Probably formed from: 4-(2,2,4,4-Tetramethylbut-4-yl)catechol

4-(2,2,4,4,-Tetramethylbut-4-yl)phenol
See: 4-*tert*-Octylphenol

N,N,N′,N′-Tetramethyl-*p*-phenylenediamine
→ Pigment
Probably in: Neisseria A2689
Pseudomonas C72
→ N,N,N′-Trimethyl-*p*-phenylenediamine
In: Rat D578

Tetramethylquinol (duroquinol)
Formed from: Tetramethylquinone
→ Tetramethylquinone
In: Arum A3740
See also: D706

Tetramethylquinone
Formed from: Tetramethylquinol
→ Tetramethylquinol
In: Beef G401
Thauera K173

***d*-Tetrandrine** (d assumed if not stated)
→ O-Demethyltetrandrine
In: Rat C759
→ N-(2)-Nor-*d*-tetrandrine
In: Cunninghamella A2768
Man C464

Rat C464
Streptomyces A2608
→ Tetrandrine-2′-oxide
In: Man C464
Rat C464, C759

Tetrandrine-2′-oxide
Formed from: Tetrandrine

2,2′,6,6′-Tetranitro-4,4′-azoxytoluene
Formed from: 2,4,6-Trinitrotoluene

Tetranorcloprostenol
Formed from: Cloprostenol
→ 9-Oxotetranorcloprostenol
Probably in: Marmoset A3845
Rat A3845

2″,3″,4″,5″-Tetranorcannabidiol-1″-oic acid
Probably formed from: 3″,4″,5″-Trisnorcannabidiol-2″-oic acid
→ 2″,3″,4″,5″-Tetranor-6-hydroxycannabidiol-1″-oic acid
Probably in: Dog G113
Man G113
→ 2″,3″,4″,5″-Tetranor-7-hydroxycannabidiol-1″-oic acid
Probably in: Man G113
Rat G113

2″,3″,4″,5″-Tetranor-6,7-dihydroxycannabidiol-1″-oic acid
Probably formed from: 4″,5″-Bisnor-6-hydroxycannabidiol-3″-oic acid
2″,3″,4″,5″-Tetranor-6-hydroxycannabidiol-1″-oic acid
2″,3″,4″,5″-Tetranor-7-hydroxycannabidiol-1″-oic acid
→ 2″,3″,4″,5″-Tetranor-6,7-dihydroxycannabidiol-1″-oic acid
Probably in: Dog F557

2″,3″,4″,5″-Tetranor-6-hydroxycannabidiol-1″-oic acid
Probably formed from: 2″,3″,4″,5″-Tetranorcannabidiol-1″-oic acid
→ 2″,3″,4″,5″-Tetranor-6,7-dihydroxycannabidiol-1″-oic acid
Probably in: Dog F557

2″,3″,4″,5″-Tetranor-7-hydroxycannabidiol-1″-oic acid
Probably formed from: 2″,3″,4″,5″-Tetranorcannabidiol-1″-oic acid
→ 2″,3″,4″,5″-Tetranor-6,7-dihydroxycannabidiol-1″-oic acid
Probably in: Dog G113

2″,3″,4″,5″-Tetranor-Δ^1-tetrahydrocannabinol-1″,7-dioic acid
Probably formed from: 2″,3″,4″,5″-Tetranor-Δ^1-tetrahydro-7-hydroxycannabinol-1″-oic acid
→ 2″,3″,4″,5″-Tetranor-Δ^1-tetrahydro-6β-cannabinol-1″,7-dioic acid
Probably in: Rabbit B89

2″,3″,4″,5″-Tetranor-Δ^6-tetrahydrocannabinol-1″,7-dioic acid
Probably formed from: 2″,3″,4″,5″-Tetranor-Δ^6-tetrahydrocannabinol-1″-oic acid

2″,3″,4″,5″-Tetranor-Δ^1-tetrahydrocannabinol-1″-oic acid
Probably formed from: 4″,5″-Bisnor-Δ^1-tetrahydrocannabinol-3″-oic acid
Δ^1-Tetrahydrocannabinol
→ 2″,3″,4″,5″-Tetranor-Δ^1-tetrahydro-6α-hydroxycannabinol-1″-oic acid
Probably in: Rabbit B89
→ 2″,3″,4″,5″-Tetranor-Δ^1-tetrahydro-6β-hydroxycannabinol-1″-oic acid
Probably in: Rabbit B89
→ 2″,3″,4″,5″-Tetranor-Δ^1-tetrahydro-7-hydroxycannabinol-1″-oic acid
Probably in: Rabbit B89

2″,3″,4″,5″-Tetranor-Δ^6-tetrahydrocannabinol-1″-oic acid
Formed from: Δ^6-Tetrahydrocannabinol
Probably formed from: 4″,5″-Bisnor-Δ^6-tetrahydrocannabinol-3″-oic acid
→ 2″,3″,4″,5″-Tetranor-Δ^6-tetrahydrocannabinol-1″,7-dioic acid
Probably in: Rat A3676

2″,3″,4″,5″-Tetranor-Δ^1-tetrahydro-6β-cannabinol-1″,7-dioic acid
Probably formed from: 2″,3″,4″,5″-Tetranor-Δ^1-tetrahydrocannabinol-1″,7-dioic acid

2″,3″,4″,5″-Tetranor-Δ^1-tetrahydro-6α-hydroxycannabinol-1″-oic acid
Probably formed from: 2″,3″,4″,5″-Tetranor-Δ^1-tetrahydrocannabinol-1″-oic acid

2″,3″,4″,5″-Tetranor-Δ¹-tetrahydro-6β-hydroxycannabinol-1″-oic acid
Probably formed from: 2″,3″,4″,5″-Tetranor-Δ¹-tetrahydrocannabinol-1″-oic acid

2″,3″,4″,5″-Tetranor-Δ¹-tetrahydro-7-hydroxycannabinol-1″-oic acid
Probably formed from: 2″,3″,4″,5″-Tetranor-Δ¹-tetrahydrocannabinol-1″-oic acid
→ 2″,3″,4″,5″-Tetranor-Δ¹-tetrahydrocannabinol-1″,7-dioic acid
Probably in: Rabbit B89

1,1,1′,1′-Tetraphenyldimethyl ether
Formed from: Diphenylmethane
→ Benzhydrol
In: Pseudomonas A2831

Tetra-(L-prolyl-L-glutaminyl-L-glutaminyl-L-prolyl-L-tyrosine)
Formed from: L-Prolyl-L-glutaminyl-L-glutaminyl-L-prolyl-L-tyrosine
Probably formed from: Tri-(L-prolyl-L-glutaminyl-L-glutaminyl-L-prolyl-L-tyrosine)

Tetraselenocyclophane
Formed from: 1,4-Phenylenebis(methylene) selenocyanate

Thalicarpine ((6a*S*)-9-(4,5-dimethoxy-2-(((1*S*)-1,2,3,4-tetrahydro-6,7-dimethoxy-2-methyl-1-isoquinolinyl)methyl)phenoxy)-5,6,6a,7-tetrahydroxy-1,2,10-trimethoxy-6-methyl-4*H*-dibenzo[*de*,*g*]quinoline)
Formed from: Isoboldine
Thalmelatine
→ Adiantifoline
In: Thalictrum E15
See also: F229
→ Hernandalinol
In: Streptomyces A3004
→ Thalmelatidine
In: Thalictrum E15
See also: F229

Thalictricavine ((13*S*,13a*R*)-5,8,13,13a-tetrahydro-9,10-dimethoxy-13-methyl-6*H*-benzo[*g*]-1,3-benzodioxolo[5,6-*a*]quinolizine)
Formed from: Berberine
Mesothalictricavine
13-Methylberberine
Tetrahydroberberine
→ 13-Methylberberine
In: Corydalis H585
Dicentra J496
→ α-N-Methylthalictricavine
In: Dicentra J496
See also: H585

Thalidomide (N-(2,6-dioxo-3-piperidyl) phthalimide, α-phthalimidiglutarimide; isomer unclear)
→ Thalidomide oxide
Probably in: Rat B859

(*R*)-Thalidomide
→ 5′-Hydroxy-(*R*)-thalidomide
Probably in: Rat H778

(*S*)-Thalidomide
→ 5-Hydroxy-(*S*)-thalidomide
In: Rat H778

Thalidomide oxide
Probably formed from: Thalidomide

Thalmelatidine
Formed from: Thalicarpine

Thalmelatine
→ Thalicarpine
In: Thalictrum E15

Thebaine ((5α)-6,7,8,14-tetradehydro-4,5-epoxy-3,6-dimethoxy-17-methylmorphinan)
Formed from: Salutaridinol
→ Codeinone
Probably in: Papaver A1620
Rat A270
→ 14β-Hydroxycodeinone
In: Rat A270
→ Oripavine
In: Rat A270
→ Thebaine-N-oxide
In: Horseradish C414

Thebaine-N-oxide
Formed from: Thebaine

Thiabendazole (2-(4′-thiazolyl)benzimidazole)
→ 2-Acetylbenzimidazole
In: Mouse G260
→ Benzimidazol-2-ylglyoxal
In: Mouse G964
→ 4-Hydroxythiabendazole
In: Mouse G260
→ 5-Hydroxythiabendazole
In: Beef J542

Lepomis H822
Mouse G260
Pig J542
Rabbit H717, J542
Rat A3706, J542
Sheep J542

Thiacetazone
See: *p*-Acetamidobenzaldehyde thiosemicarbazone

Thiamphenicol (2,2-dichloro-N-((1*R*,2*R*)-2-hydroxy-1-(hydroxymethyl)-2-(4-(methylsulphonyl)phenyl)ethyl)acetamide)
Formed from: Thiamphenicol palmitate
→ Thiamphenicol aldehyde
In: Rat B601

Thiamphenicol aldehyde (1-formyl-N-((1*R*,2*R*)-2-hydroxy-1-(hydroxymethyl)-2-(4-(methylsulphonyl)phenyl)ethyl)formamide)
Formed from: Thiamphenicol
→ Thiamphenicol oxamic acid
In: Rat B601

Thiamphenicol oxamic acid (1-carboxy-N-((1*R*,2*R*)-2-hydroxy-1-(hydroxymethyl)-2-(4-(methylsulphonyl)phenyl)ethyl)formamide)
Formed from: Thiamphenicol aldehyde

Thiamphenicol palmitate
→ Thiamphenicol
In: Man A1261
Mouse A1261

Thianthrene (diphenyl disulphide)
Formed from: Thiophenol
→ *cis*-1,2-Dihydro-1,2-dihydroxythianthrene
In: Pseudomonas J332
→ 1-Hydroxythianthrene
Probably in: Rat B281, H869
→ 2-Hydroxythianthrene
Probably in: Rat B281, H869
→ Thianthrene sulphoxide
In: Rat H869

Thianthrene disulphoxide
Formed from: Thianthrene sulphoxide

Thianthrene sulphoxide
Formed from: Thianthrene
→ Thianthrene disulphoxide
In: Rat H869

Thiapride ((N-(2-(diethylamino)ethyl)-2-methoxy-5-(methylsulphonyl)benzamide)
→ Desethylthiapride
In: Rat A3643
→ Thiapride-N-oxide
In: Rat A3643

Thiapride-N-oxide
Formed from: Thiapride

3-(2-Thiazolyl)-1*H*-indole
See: Camalexine

Thidiazuron (N-phenyl-N′-(1,2,3-thiadiazol-5-yl)urea)
→ *p*-Hydroxythidiazuron
In: Chicken A3196
Goat A3196
→ Phenylurea
In: Goat A3196

2-(4-(2-Thienylcarbonyl)phenyl)malonic acid
Probably formed from: 3-Hydroxy-2-(4-(2-thienylcarbonyl)phenyl)propionic acid

2-(4-(2-Thienylcarbonyl)phenyl)propionic acid
Formed from: Suprofen
→ 2-(4-((2-Thienyl)hydroxymethyl)phenyl) propionyltaurine
Probably in: Dog D346

2-(4-(2-Thienylcarbonyl)phenyl)propionyltaurine
Formed from: Suprofen

2-(4-((2-Thienyl)hydroxymethyl)phenyl) propionyltaurine
Probably formed from: 2-(4-((2-Thienyl) hydroxymethyl)phenyl)propionic acid

***p*-(Thioacetamido)phenol**
Formed from: Thioacetanilide
→ *p*-Acetamidophenol
Probably in: Rat C572

Thioacetanilide
→ *p*-(Thioacetamido)phenol
In: Rat C572

***p*-Thioanisidine**
Formed from: *p*-Thioanisidine sulphoxide
→ N-Acetyl-*p*-thioanisidine
In: Guinea pig A511
Mouse A511
Rabbit A511
Rat A511
→ *p*-Thioanisidine sulphoxide
In: Guinea pig A511

Mouse A511
Rabbit A511
Rat A511

p-Thioanisidine sulphoxide
Formed from: *p*-Thioanisidine
→ N-Acetyl-*p*-thioanisidine sulphoxide
Probably in: Guinea pig A511
Mouse A511
Rabbit A511
Rat A511
→ *p*-Thioanisidine
In: Aspergillus A511
→ *p*-Thioanisidine sulphone
In: Guinea pig A511
Mouse A511
Rabbit A511
Rat A511

p-Thioanisidine sulphone
Formed from: *p*-Thioanisidine sulphoxide
→ N-Acetyl-*p*-thioanisidine sulphone
Probably in: Guinea pig A511
Mouse A511
Rabbit A511
Rat A511

Thioanisole
Formed from: Thiophenol
→ *cis*-2,3-Dihydro-2,3-dihydroxythioanisole
In: Pseudomonas J656
→ Thioanisole sulphoxide
In: Caldariomyces J892
Horseradish E403
Mortierella D78, D759
Mouse D446
Pig D446
Pseudomonas H165

Thioanisole sulphoxide
Formed from: O-Ethyl S-phenyl ethylphosphonodithioate
Thioanisole
→ Methyl phenyl sulphone
Probably in: Housefly A3613
Microorganism A1462
Mouse A3613
Rhizopus A2539

Thiobenzamide
→ Thiobenzamide sulphoxide
In: Mouse C877
Pig C877
Rabbit C877
Rat C612, C877

Thiobenzamide sulphoxide
Formed from: Thiobenzamide
→ Benzamide
In: Rat C612

4,4′-Thiobis(benzenethiol)
→ S-Methyl-4,4′-thiobis(benzenethiol)
In: Brassica K492, K594

4,4′-Thiobis(6-*tert*-butyl-*m*-cresol)
→ 4,4′-Thiobis(6-*tert*-butyl-*m*-cresol)-β-D-glucuronide
In: Rat C580, D720

4,4′-Thiobis(6-*tert*-butyl-*m*-cresol)-β-D-glucuronide
Formed from: 4,4′-Thiobis(6-*tert*-butyl-*m*-cresol)

2,2′-Thiobis(4,6-dichlorophenol)
See: Bithionol

Thiochroman (1,2,3,4-tetrahydro-1-thianaphthalene)
→ 1,2,3,4-Tetrahydro-1-thianaphthalene-1-oxide
In: Helminthosporium G427
Mortierella G427

Thiochroman-4-ol
→ Thiochroman-4-one
In: Hamster K443
Rabbit A2032

Thiochroman-4-one
Formed from: Thiochroman-4-ol
→ Thiochroman-4-one sulphoxide
In: Cunninghamella E46

Thiochroman-4-one sulphoxide
Formed from: Thiochroman-4-one

2-(Thiocyanomethylthio)benzothiazole
→ 2-Mercaptobenzothiazole
In: Rat J210
→ 2-(Mercaptomethylthio)benzothiazole
In: Rat J210

Thiodiphenylamine
See: Phenothiazine

Thionol (7-hydroxy-3*H*-phenothiazin-3-one)
Formed from: Phenothiazine
→ Thionol-β-D-glucuronide

Probably in: Gerbil B735
Mouse B735
Rat B735
Thionol-β-D-glucuronide
Formed from: Thionol
Thiophenol (benzenethiol)
Formed from: O-Ethyl S-phenyl ethylphosphonodithioate
Thiophenyl acetate
Thiophenyl butyrate
Thiophenyl propionate
→ S-Phenyl-L-cysteine
In: Escherichia J485
→ Thianthrene
In: Horseradish J593
Mushroom J593
→ Thioanisole
In: Brassica K492, K594
Man C917, D926
Thiophenyl acetate
→ Thiophenol
In: Rabbit E26
Pig A1871
Thiophenyl butyrate
→ Thiophenol
In: Rabbit E26
***p*-(S-Thiophenyl)phenylhydroxylamine**
Formed from: N-Hydroxy-*p*-(S-thiophenyl) acetanilide
Thiophenyl propionate
→ Thiophenol
In: Rabbit E26
Thioproperazine (N,N-dimethyl-10-(3-(4-methyl-1-piperazinyl)propyl)-10*H*-phenothiazine-2-sulphonamide)
→ Thioproperazine-N-oxide
In: Pig A2861
Thioproperazine-N-oxide
Formed from: Thioproperazine
Thioridazine (10-(2-(1-methyl-2-piperidinyl) ethyl)-2-(methylthio)-10*H*-phenothiazine)
Probably formed from: Mesoridazine
→ N-Demethylthioridazine
In: Rat K618
→ S-Demethylthioridazine
In: Man D366
See also: E124
→ 3-Hydroxythioridazine
In: Man D366
→ 7-Hydroxythioridazine
In: Man D366
→ Mesoridazine
In: Beef A3445
Dog G705
Man A1573, A3121, A3188, D366, F96, F436, G705
Mouse H166
Rat G705, K618
→ 6′-Oxothioridazine
In: Dog J208
Rat J208
→ Sulphoridazine
In: Dog G705
Man F96, F436, G705
Rat G705, K618
→ Thioridazine-N-oxide
In: Mouse H166
Rat E124
→ Thioridazine-5-sulphoxide
In: Beef A3445
Dog G705, J208
Man A3188, D366, G705
Mouse H166
Rat E124, G705, J208, K618
Thioridazine-2,5-disulphoxide
Formed from: Thioridazine-5-sulphoxide
Thioridazine-N-oxide
Formed from: Thioridazine
→ Thioridazine-N-oxide-2-sulphoxide
Probably in: Mouse H166
Thioridazine-N-oxide-2-sulphoxide
Probably formed from: Thioridazine-N-oxide
Thioridazine-5-sulphoxide
Formed from: Thioridazine
→ Thioridazine-2,5-disulphoxide
In: Beef A3445
→ Northioridazine-5-sulphoxide
Probably in: Dog G705
Man G705
Rat G705
Thiosalicylic acid
See: *o*-Mercaptobenzoic acid

Thioxanthene
→ 9-Hydroxythioxanthene
In: Pseudomonas H672
→ Thioxanthene sulphoxide
In: Pseudomonas H672
Thioxanthene sulphone
Probably formed from: Thioxanthene sulphoxide
→ 9-Hydroxythioxanthene sulphone
Probably in: Pseudomonas H672
Thioxanthene sulphoxide
Formed from: Thioxanthene
→ Thioxanthene sulphone
Probably in: Pseudomonas H672
Thioxanthone (9*H*-thioxanthen-9-one)
Probably formed from: 9-Hydroxythioxanthene
→ Thioxanthone sulphoxide
In: Cunninghamella E46
Mortierella E46
Thioxanthone sulphone
Formed from: Thioxanthone sulphoxide
Probably formed from: 9-Hydroxythioxanthene sulphone
Thioxanthone sulphoxide
Formed from: Thioxanthone
→ Thioxanthone sulphone
In: Cunninghamella E46
Mortierella E46
Thioxyloside (4-cyanophenyl-1,5-dithio-β-D-xylopyranoside)
→ Thioxyloside-2-β-D-glucuronide
In: Man J584
Rat J584
→ Thioxyloside-3-β-D-glucuronide
In: Rat J584
→ Thioxyloside-4-β-D-glucuronide
In: Man J584
Thioxyloside-2-β-D-glucuronide
Formed from: Thioxyloside
Thioxyloside-3-β-D-glucuronide
Formed from: Thioxyloside
Thioxyloside-4-β-D-glucuronide
Formed from: Thioxyloside
Thonzylamine (N-((4-methoxyphenyl)methyl)-N′,N′-dimethyl-N-2-pyrimidinyl-1,2-ethanediamine)
→ Thonzylamine-N-β-D-glucuronide
In: Man H541
Thonzylamine-N-β-D-glucuronide
Formed from: Thonzylamine
L-Threonyl-β-naphthylamide
→ β-Naphthylamine
In: Beef A526
Thymohydroquinone glucoside
Formed from: 2-Isopropyl-5-methylquinol
Thymol (2-isopropyl-5-methylphenol)
→ 2-Isopropyl-5-methylquinol
In: Man B494
Mucor A832
→ Thymol-β-D-glucuronide
In: Man A277, G524, J416
Rabbit B494
→ Thymol-O-sulphate
In: Mouse H959
Thymol-β-D-glucuronide
Formed from: Thymol
Thymol-O-sulphate
Formed from: Thymol
Thymoquinone (2-methyl-5-(1-methylethyl)-2,5-cyclohexadiene-1,4-dione)
Formed from: 4-Iodothymol
Thymoxamine (4-(2-dimethylaminoethoxy)-5-isopropyl-2-methylphenyl acetate (ester))
→ Deacetylthymoxamine
In: Absidia J133
Acremonium J133
Aspergillus J133
Cat D788
Cunninghamella J133
Curvularia J133
Cylindrocarpon J133
Dog D788, E136
Fusarium J133
Geotrichum J133
Gliocladium J133
Glomerella J133
Man D788, D789, F301
Mortierella J133
Mucor J133
Rabbit E136
Rat B418, D787, D788, E136, F301, J493
Rhizopus J133
Sporotrichum J133

Syncephalastrum J133
Thamnostylum J133
Zygodesmus J133
→ Demethylthymoxamine
In: Rat J493

Thyroacetic acid (4-(*p*-hydroxyphenoxy) phenylacetic acid)
Probably formed from: L-Thyronine

D-Thyronine (4-(4-hydroxyphenoxy)-D-phenylalanine)
Probably formed from: D-Thyroxine

L-Thyronine (4-(4-hydroxyphenoxy)-L-phenylalanine)
Probably formed from: 3,3′,5-Triiodo-L-thyronine
→ Thyroacetic acid
Probably in: Man A468
→ L-Thyronine-4′-sulphate
In: Monkey B770
Rat B770

L-Thyronine-4′-sulphate
Formed from: L-Thyronine

Thyropropionic acid
Probably formed from: 3,3′,5,5′-Tetraiodo-thyropropionic acid

D-Thyroxine
→ D-Thyronine
Probably in: Rat A1095

L-Thyroxine (T_4, 3,3′,5,5′-tetraiodo-L-thyronine, *p*-(4-hydroxy-3,5-diiodophenoxy)-3,5-L-phenylalanine)
Formed from: L-Thyroxine-4′-β-D-glucuronide
L-Thyroxine-O-sulphate
3,3′,5-Triiodo-L-thyronine
→ 3,5-Diiodo-L-tyrosine
In: Man D553
Rat C809, D186
→ L-Thyronine
In: Man A468
See also: A1095
→ L-Thyroxine-4′-β-D-glucuronide
In: Rainbow trout J38, J319
Rat B721, B771, G741, J798
→ L-Thyroxine-O-sulphate
In: Man K90, K620
Rat K649
→ 3,3′,5-Triiodo-L-thyronine
In: Beef D461, E188
Dog C854, H882
Man A782, E193, E601, H882
Oreochromis H923
Quail E196
Rainbow trout J38
Rat A1548, A2682, A3117, A3312, A3400, A3549, A3752, B126, B504, B721, C809, D79, D930, E184, E185, F821, G337, H587
See also: K242
→ 3,3′,5′-Triiodo-L-thyronine
In: Dog C854
Man C760, E193
Rabbit F373
Rat A3117, A3312, A3549, A3752, B126, B504, B956, C109, D930, G337, K69

L-Thyroxine-4′-β-D-glucuronide
→ L-Thyroxine
In: Rat G463

L-Thyroxine sulphamate
→ 3,3′,5-Triiodo-L-thyronine sulphamate
In: Rat G337
→ 3,3′,5′-Triiodo-L-thyronine sulphamate
In: Rat G337

L-Thyroxine-O-sulphate
Formed from: L-Thyroxine
→ L-Thyroxine
In: Microorganism D594
Trout K242
→ 3,3′,5-Triiodo-L-thyronine sulphate
In: Rat G337

Tianeptine (7-((3-chloro-6,11-dihydro-6-methyl-5,5-dioxidodibenzo[*c*,*f*][1,2]thiazepin-11-yl) amino)heptanoic acid)
→ 5-((3-Chloro-6,11-dihydro-6-methyl-5,5-dioxodibenzo[*c*,*f*]-1,2-thiazepin-11-yl) amino)pentanoic acid
In: Man F576
→ 3-Chloro-6,11-dihydro-6-methyl-5,5,11-trioxodibenzo[*c*,*f*]-1,2-thiazepine
In: Man F576
→ 11,N-Dehydrotianeptine

In: Man F576

Tiaramide (4-((5-chloro-2-oxo-3(2*H*)-benzothiazolyl)acetyl)-1-piperazineethanol)
→ 1-((5-Chloro-2-oxo-3(2*H*)-benzothiazolyl) acetyl)piperazine
In: Dog A2099, C298
Guinea pig C298
Man B981, C301
Monkey A2099, C298
Mouse A2099, C298
Rat A2097, A2099, C298, C301, C487
→ 4-((5-Chloro-2-oxo-3(2*H*)-benzothiazolyl) acetyl)-1-piperazineacetic acid
In: Dog A2099, C298
Man B981, C298
Monkey A2099, C298, C318
Mouse A2099, C298
Rat A2097, A2099, C298, C318
→ Tiaramide-β-D-glucuronide
In: Dog A2099, C298
Man A2843, B981, C298
Monkey A2843, A2099, C298
Rat A2843
→ Tiaramide-N-oxide
In: Dog A2099
Guinea pig C298
Man B981, C298, C301
Monkey C298
Mouse A2099, C298
Rat A2097, A2099, C298, C301
→ Tiaramide sulphate
In: Guinea pig H455
Man H455
Mouse C571
Rat C298, C571, E45

Tiaramide-β-D-glucuronide
Formed from: Tiaramide

Tiaramide-N-oxide
Formed from: Tiaramide
→ Tiaramide
In: Rat A1528, A1847, A1913

Tiaramide sulphate
Formed from: Tiaramide

Tiazuril (2-(3,5-dimethyl-4-(*p*-chlorophenylthio) phenyl)-*as*-triazine-3,5-(2*H*,4*H*)-dione)
→ Hydroxytiazuril
In: Chicken A2750
Rat A2750
→ Tiazuril sulphoxide
In: Chicken A2750
Dog A2750
Rat A2750

Tiazuril sulphone
Formed from: Tiazuril sulphoxide
→ Hydroxytiazuril sulphone
Probably in: Chicken A2750
Rat A2750

Tiazuril sulphoxide
Formed from: Tiazuril
→ Hydroxytiazuril sulphoxide
Probably in: Chicken A2750
Rat A2750
→ Tiazuril sulphone
In: Dog A2750

Ticlopidine (5-((2-chlorophenyl)methyl)-4,5,6,7-tetrahydrothieno[3,2-*c*]pyridine)
→ *o*-Chlorobenzoic acid
In: Rat C52
→ Ticlopidine-N-oxide
In: Rat C52

Ticlopidine-N-oxide
Formed from: Ticlopidine

Tienilic acid (2,3-dichloro-4-(2-thienylcarbonyl) phenoxy)acetic acid)
→ 4-Carboxy-2,3-dichlorophenoxyacetic acid
In: Dog A3554
Mouse A3554
Pig A3554
Rat A3554
→ Dihydrotienilic acid
In: Dog A3554
Mouse A3554
Pig A3554
Rat A3554
→ 5-Hydroxytienilic acid
In: Man D435
Rat D435, F130, F132

Tienocarbine (7,8,9,10-tetrahydro-1,9-dimethyl-6*H*-pyrido[4,3-*b*]thieno[3,2-*e*]indole)
→ Demethyltienocarbine
In: Rat D25

Tiflorex (2-ethylamino-1-(*m*-trifluoronethylthiophenyl)propane)
→ Tiflorex sulphoxide
In: Rat B252, B623

Tiflorex sulphone
Probably formed from: Tiflorex sulphoxide
→ Nortiflorex sulphone
Probably in: Rat B252, B623

Tiflorex sulphoxide
Formed from: Tiflorex
→ Nortiflorex sulphoxide
Probably in: Rat B252, B623
→ Tiflorex sulphone
Probably in: Rat B252, B623

Tilidine ((1*R*,2*S*)-rel-2-(dimethylamino)-1-phenyl-3-cyclohexane-1-carboxylic acid ethyl ester)
→ Nortilidine
In: Dog A1440
Man A1440
Rabbit A1440
Rat A1440, A2763

Timiperone (4′-fluoro-4-(2-thioxo-1-benzimidazolinyl)piperidino)butyrophenone)
→ 2,3-Dihydro-1-(4-piperidinyl)-2-thioxo-1-benzimidazole
In: Dog C16
Monkey C16
Rat B804, C16
→ 4-(4-(2,3-Dihydro-2-thioxo-1*H*-benzimidazol-1-yl)piperidinyl)-1-fluorophenyl)-1-butanol
In: Dog C16
Man J846
Monkey C16
Rat B804, C16
→ *p*-Fluorohippuric acid
In: Dog C16
Monkey C16
Rat C16

Tiospirone (8-(4-(4-(1,2-benzothiazol-3-yl)-1-piperazinyl)butyl)-8-aza*spiro*(4,5)-decane-7,9-dione)
→ 1-(Benzothiazol-3-yl)piperazine sulphone
In: Man F732
→ 3′-Hydroxytiospirone
In: Man F732
→ 6′-Hydroxytiospirone
In: Man F732

Tirapazamine (3-amino-1,2,4-benzotriazine-1,4-dioxide)
→ 3-Amino-1,2,4-benzotriazine-1-oxide
In: Hamster E972
Mouse F747, H32
→ Free radical
In: Rat G351

Tiropramide (N-benzoyl-O-(2-diethylaminoethyl)-L-tyrosyl-dipropylamide)
→ N-Benzoyl-O-(2-diethylaminoethyl)-L-tyrosine
In: Rat F97
→ N-Benzoyl-O-(2-diethylaminoethyl)-L-tyrosyl-N-(3-hydroxypropyl)propylamide
In: Rat F32, F97
→ N-Benzoyl-L-tyrosyl dipropylamide
In: Rat F97
→ Depropyltiropramide
In: Rat F32, F97
→ Desethyltiropramide
In: Rat F32, F97

TNT
See: 2,4,6-Trinitrotoluene

Toborinone (6-(3-(3,4-dimethoxybenzylamino)-2-hydroxypropoxy)-2(1*H*)-quinolinone)
→ 6-(3-Amino-2-hydroxypropoxy)-2(1*H*)-quinolinone
In: Dog J154
Rat J154
→ 4-(S-(L-Cysteinyl))-6-(3-(3,4-dimethoxybenzylamino)-2-hydroxypropoxy)-3,4-dihydro-3-hydroxy-2(1*H*)-quinolinone
In: Dog J154
Rat J154
→ 6-(2,3-Dihydroxypropoxy)-2(1*H*)-quinolinone
In: Dog J154
Rat J154
→ 6-(3-(3,4-Dimethoxybenzylamino)-2-hydroxypropoxy)-3-(S-glutathionyl)-3,4-dihydro-4-hydroxy-2(1*H*)-quinolinone)
In: Dog J154
Rat J154

→ 6-(3-(3,4-Dimethoxybenzylamino)-2-hydroxypropoxy)-8-hydroxy-2(1*H*)-quinolinone
In: Dog J154
Rat J154
→ 6-(3-(3-Hydroxy-4-methoxybenzylamino)-2-hydroxypropoxy)-2(1*H*)-quinolinone
In: Dog J154
Rat J154
→ 6-(3-(4-Hydroxy-3-methoxybenzylamino)-2-hydroxypropoxy)-2(1*H*)-quinolinone
In: Dog J154
Rat J154
→ 3-(Quinolin-2(1*H*)-on-6-oxy)lactic acid
In: Dog J154
Rat J154
→ Toborinone-β-D-glucuronide
In: Dog J154
Rat J154

Toborinone-β-D-glucuronide
Formed from: Toborinone

Tocainide (tocainamide, 2-amino-2′,6′-propionoxyxylidide)
→ Tocainide carbamoyl O-β-D-glucuronide
In: Dog C371
Guinea pig C371
Monkey C371
Rabbit C371
See also: B331

Tocainide carbamoyl O-β-D-glucuronide
Formed from: Tocainide

α-Tocopherol (vitamin E, (2*R*)-3,4-dihydro-2,5,7,8-tetramethyl-2-((4*R*,8*R*)-4,8,12-trimethyltridecyl)-2*H*-1-benzopyran-6-ol)
Formed from: Homogentisic acid
β-Tocopherol
γ-Tocopherol
α-Tocopheryl acetate
→ 6-(3′-Carboxybutyl)-2,3,5-trimethyl-1,4-benzoquinone
Probably in: Rabbit A1198
→ 6-(2-Carboxyethyl)-6-hydroxy-2,5,7,8-tetramethylchroman
In: Man H117, K604
→ 6-(5′-Carboxy-3′-hydroxy-3′-methylpentyl)-2,3,5-trimethyl-1,4-benzoquinone
In: Rat G293
See also: A1198
→ α-Tocopheronolactone
in: Man K604

β-Tocopherol ((2*R*)-3,4-dihydro-2,5,8-trimethyl-2-((4*R*,8*R*)-4,8,12-trimethyltridecyl)-2*H*-1-benzopyran-6-ol)
Probably formed from: δ-Tocopherol
→ α-Tocopherol
In: Euglena G754
Spinach G612

γ-Tocopherol ((2*R*)-3,4-dihydro-2,7,8-trimethyl-2-(4,8,12-trimethyltridecyl)-2*H*-1-benzopyran-6-ol)
Formed from: 2,3-Dimethyl-5-phytylquinol
Homogentisic acid
→ α-Tocopherol
In: Capsicum D580, D699
Euglena G754
Spinach A3972, B575, G612
See also: D902

δ-Tocopherol ((2*R*)-3,4-dihydro-2,8-dimethyl-2-((4*R*,8*R*)-4,8,12-trimethyltridecyl)-2*H*-1-benzopyran-6-ol)
→ β-Tocopherol
Probably in: Spinach G612

α-Tocopherolquinol (2-((3*R*,7*R*,11*R*)-3-hydroxy-3,7,11,15-tetramethylhexadecyl)-3,5,6-trimethyl-1,4-benzenediol)
Formed from: α-Tocopherolquinone

α-Tocopherolquinone (2-((3*R*,7*R*,11*R*)-3-hydroxy-3,7,11,15-tetramethylhexadecyl)-3,5,6-trimethyl-2,5-cyclohexadiene-1,4-dione)
→ α-Tocopherolquinol
In: Rat G681

α-Tocopheronolactone (2,3,5-trimethyl-6-(2-(tetrahydro-2-methyl-5-oxo-2-furanyl)ethyl)-2,5-cyclohexadiene-1,4-dione)
Formed from: α-Tocopherol

α-Tocopheryl acetate
→ α-Tocopherol
In: Chicken B132
See also: A1198

Tofizopam (1-(3,4-dimethoxyphenyl)-5-ethyl-7,8-dimethoxy-4-methyl-5*H*-2,3-benzodiazepine)
→ 3′-Demethyltofizopam
In: Monkey C274, D548
Rabbit C274, D548

→ 4′-Demethyltofizopam
In: Dog C274, D548
Man C274, D548
Monkey C274, D548
Rabbit C274, D548
Rat C274, D548
→ 7-Demethyltofizopam
In: Dog C274, D548
Man C274, D548
Monkey C274, D548
Rabbit C274, D548
Rat C274, D548
→ 8-Demethyltofizopam
In: Monkey D548
Rabbit C274

Tolamolol (1-(2-(4-carbamoylphenoxy) ethylamino)-3-(2-methylphenoxy) propan-2-ol)
→ 1-(2-(4-Carbamoylphenoxy)ethylamino)-3-(-4-hydroxy-2-methylphenoxy)propan-2-ol
In: Guinea pig A363
Mouse A363
Rabbit A363
Rat A363
→ 1-(2-(*p*-Carboxyphenoxy)ethylamino)-3-(2-methylphenoxy)propan-2-ol
In: Guinea pig A363
Mouse A363
Rabbit A363
Rat A363

Tolazamide (1-(hexahydroazepin-1-yl)-3-(*p*-tolyl)sulphonylurea)
→ 1-(Hexahydroazepin-1-yl)-3-(*p*-hydroxymethylphenyl)sulphonylurea
In: Man A3563
Rat A3563
→ 1-(Hexahydro-4-hydroxyazepin-1-yl)-3-(*p*-tolyl)sulphonylurea
In: Man A3563
See also: A1416
→ *p*-Toluenesulphonamide
In: Man A3563
Rat A3563

Tolbutamide (1-butyl-3-(*p*-tolyl)sulphonylurea)
→ 1-Butyl-3-(*p*-hydroxymethylphenyl) sulphonylurea
In: Dog C826
Guinea pig G146
Hamster G146
Man H925, J299, J619
Monkey C826
Mouse G146, K161
Rabbit C826
Rat C826, G146
See also: G882, K22
→ *p*-Tolylsulphonylurea
In: Dog C826
Monkey C826
Rabbit C826

Tolcapone (3,4-dihydroxy-4′-methyl-5-nitrobenzophenone)
→ 3-O-Methyltolcapone
In: Man H592, K298
→ Tolcapone-3-β-D-glucuronide
In: Man K298, K331
Rat J546, J886

Tolcapone-3-β-D-glucuronide
Formed from: Tolcapone

***o*-Tolidine** (3,3′-dimethylbenzidine)
Formed from: Acid Red 114
Direct Blue 14
Direct Blue 25
Direct Blue 53
Direct Orange 6
Direct Red 2
Direct Red 39
→ N-Acetyl-*o*-tolidine
In: Rat C471
See also: C541

Tolmesoxide (4,5-dimethoxy-2-methylphenyl methyl sulphoxide)
→ 5-Hydroxy-4-methoxy-2-methylphenyl methyl sulphone
In: Dog B400
Man B400
Rat B400
→ 4-Hydroxy-5-methoxy-2-methylphenyl methyl sulphoxide
In: Dog B400
Man B400
Rat B400

→ 2-Hydroxymethyl-4,5-dimethoxyphenyl methyl sulphone
In: Dog B400
Man B400
Rat B400

Tolmetin (2-carboxymethyl-1-methyl-5-(*p*-methylbenzoyl)pyrrole)
→ 5-(*p*-Hydroxymethylphenyl)-1-methylpyrrole-2-acetic acid
In: Rat A3666
See also: A32, A1875, A2121, A2986, A3666, B105
→ Tolmetin-β-D-glucuronide
In: Man B105

Tolmetin-β-D-glucuronide
Formed from: Tolmetin

Toloxatone (5-hydroxymethyl-3-(3-methylphenyl)-2-oxazolidinone)
→ 3-Carboxyphenyl-5-hydroxymethyl-2-oxazolidinone
In: Man A3709
→ Hydroxytoloxatone
In: Man A3709
→ Toloxatone-β-D-glucuronide
In: Man A3709

Toloxatone-β-D-glucuronide
Formed from: Toloxatone

Tolterodine (N,N-diisopropyl-3-(2-hydroxy-5-methylphenyl)phenylpropanamine)
→ Deisopropyltolterodine
In: Dog J419
Man J375, J419
Mouse J419
Rat J419
→ 3-(2-Hydroxy-5-hydroxymethylphenyl)-N,N-diisopropyl-3-phenylpropylamine
In: Dog J419
Man J375, J419, J480
Mouse J419
Rat J419
→ 3-(2-Hydroxy-5-methylphenyl)-3-(*p*-hydroxyphenyl)-N,N-diisopropylpropylamine
See also: Rat J419
→ Tolterodine-β-D-glucuronide
In: Dog J419
Mouse J419

Tolterodine-β-D-glucuronide
Formed from: Tolterodine

Toltrazuril (1-methyl-3-(3-methyl-4-((trifluoromethyl)thio)phenoxy)phenyl)-1,3,5-triazine-2,4,6(1*H*,3*H*,5*H*)-trione)
→ Toltrazuril sulphoxide
In: Rat G507

Toltrazuril sulphone
Formed from: Toltrazuril sulphoxide

Toltrazuril sulphoxide
Formed from: Toltrazuril
→ Toltrazuril sulphone
In: Rat H79

***o*-Tolualdehyde** (*o*-methylbenzaldehyde)
Formed from: *o*-Methylbenzyl alcohol
→ *o*-Methylbenzyl alcohol
In: Chlorella K176
Dunaliella G459, K176
Euglena G422
Isochrysis K176
Mung bean J910
Nannochloris K176
Porphyridium K176
Saccharomyces H425
→ (1*R*,2*S*)-1-(*o*-Methylphenyl)propane-1,2-diol
In: Saccharomyces E487
→ *o*-Toluic acid
In: Acinetobacter F366
Man J24
Mouse H150

***m*-Tolualdehyde** (*m*-methylbenzaldehyde)
Formed from: *m*-Methylbenzyl alcohol
m-Methylmandelonitrile
m-Toluic acid
→ *m*-Methylbenzyl alcohol
In: Chlorella K176
Dunaliella G459, K176
Euglena G422
Isochrysis K176
Mung bean J910
Nannochloris K176
Porphyridium K176
Saccharomyces H425
→ *m*-Toluic acid
In: Acinetobacter F366

Euglena G422
Man J24
Mouse H150
Pseudomonas K495
Rat H761

***p*-Tolualdehyde** (*p*-methylbenzaldehyde)
Formed from: *p*-Methylbenzyl alcohol
p-Methylbenzylamine
p-Methylmandelonitrile
p-Methylphenylglyoxylic acid
p-Toluic acid
→ 4,4′-Dimethylbenzoin
In: Pseudomonas K424
→ *p*-Methylbenzyl alcohol
In: Dunaliella G459
Euglena G422
Mung bean J910
Phanerochaete F866
Saccharomyces H425
→ *p*-Methylmandelonitrile
In: Almond J194
Sorghum J194
With (*S*)-oxynitrilase F680
→ (1*R*,2*S*)-1-(*p*-Methylphenyl)propane-1,2-diol
In: Saccharomyces E487
→ *p*-Toluic acid
In: Acinetobacter F366
Beef C508
Euglena G422
Man J24
Mouse H150
Pseudomonas K495
Rat H761
See also: G865, H616

***m*-Toluamide** (*m*-methylbenzamide)
Formed from: N-Ethyl-*m*-toluamide
→ *m*-Hydroxymethylbenzamide
In: Rat J572

Toluene
Formed from: Benzene
Dibenzyl disulphide
Dibenzyl sulphoxide
Ferulic acid
Phenylacetic acid
Probably formed from: *p*-Toluenesulphonic acid
2,4,6-Triaminotoluene
→ N-Acetyl-S-benzyl-L-cysteine
In: Man J479
Rat C98
→ N-Acetyl-S-(L-cysteinyl)toluene
In: Man J497
→ Benzyl alcohol
In: Aspergillus A2849
Dog H460
Escherichia K491
Man F598, J164
Mortierella F823
Nitrosomonas H219, H631
Pseudomonas A1009, G652, J219
Rat D122, F170, F358, F598
Salmon E710
See also: B24, D718, J497, K523
→ (*R*)-(+)-Benzylsuccinic acid
In: Azoarcus K238
Desulfobacula J847
Thauera J191, J651, J817
→ *cis*-2,3-Dihydro-2,3-dihydroxytoluene
In: Pseudomonas A3184, B291, E354, G661, J656, K699
Rhodococcus K654
See also: G653
→ *o*-Cresol
In: Aspergillus A1218, A2849
Bacterium F163
Cunninghamella A1218
Dog H460
Helicostylum A1218
Man B18, D718, F598, J164, J497
Penicillium A1218
Rat D122, D718, F170, F358
Rhizopus A1218
Streptomyces A1218
See also: F857, J344
→ *m*-Cresol
In: Rat D122, D718, F170, F358
→ *p*-Cresol
In: Aspergillus A2849
Cunninghamella A1218
Dog H460
Man D718, F358, F598, J164
Nitrosomonas H631

Penicillium A1218
Pseudomonas G212, K203
Rat D122, D718, F170, F790
Rhizopus A1218
See also: F163
→ *o*-Xylene
In: Man F853
→ *m*-Xylene
In: Man F853
→ *p*-Xylene
In: Man F853

Toluene-1,2-oxide
→ S-(1,2-Dihydro-2-hydroxy-2-methylphenyl)glutathione
Probably in: Sheep A2282

***o*-Toluenesulphonacetamide**
Formed from: *o*-Toluenesulphonamide

***o*-Toluenesulphonamide**
→ *o*-Carboxybenzenesulphonamide
In: Man A3242
Rat A559, A2107, A3242
→ Saccharin
In: Rat A2107
→ *o*-Sulphamoylbenzyl alcohol
In: Man A3242
Rat A3242
→ *o*-Toluenesulphonacetamide
In: Man A3242
Rat A2107, A3242

***p*-Toluenesulphonamide**
Formed from: Tolazamide
→ N-Acetyl-*p*-toluenesulphonamide
In: Rat A3041
→ *p*-Hydroxymethylbenzenesulphonamide
In: Rat A3041
See also: A559, A2021

***p*-Toluenesulphonic acid**
→ *p*-Hydroxymethylbenzenesulphonic acid
In: Pseudomonas F188
→ 4-Methylcatechol
In: Alcaligenes H297, K775
Pseudomonas J657
→ Toluene
Probably in: Rhodococcus H618

Toluene-α-sulphonic acid
Formed from: S-Benzyl O-ethyl phenylphosphonothioate

Toluene-*p*-sulphonyl-L-arginine
Formed from: Toluene-*p*-sulphonyl-L-arginine methyl ester

Toluene-*p*-sulphonyl-L-arginine methyl ester
→ Toluene-*p*-sulphonyl-L-arginine
In: Rice D743

***p*-Toluenesulphonyl-4-hydroxypiperidine**
Formed from: *p*-Toluenesulphonylpiperidine

Toluene-*p*-sulphonyl-L-lysine
Formed from: Toluene-*p*-sulphonyl-L-lysine methyl ester

Toluene-*p*-sulphonyl-L-lysine methyl ester
→ Toluene-*p*-sulphonyl-L-lysine
In: Rice D743

***p*-Toluenesulphonylpiperidine**
→ *p*-Hydroxymethylbenzenesulphonyl piperidine
In: Rat C23
→ *p*-Toluenesulphonyl-4-hydroxypiperidine
In: Rat C23

***o*-Toluic acid** (*o*-methylbenzoic acid)
Formed from: 5-Chloro-2-methylbenzoic acid
o-Methylbenzamide
2-Methylbenzanilide
o-Tolualdehyde
o-Tolunitrile
o-Xylene
→ *o*-Cresol
In: Pseudomonas G434, G448
→ 1,2-Dihydro-1,2-dihydroxy-6-methylbenzoic acid
Probably in: Pseudomonas B465
→ *o*-Hydroxymethylbenzoic acid
In: Pseudomonas A1048
→ *o*-Methylbenzoyl-β-D-glucuronic acid
In: Rabbit H519
→ *o*-Methylbenzyl alcohol
In: Rhizopus A1048
→ *o*-Methylbenzoyl CoA
In: Pseudomonas F65
→ *o*-Methylhippuric acid
In: Man F206
Rabbit H519

***m*-Toluic acid** (*m*-methylbenzoic acid)
Formed from: *o*-Cresol
(3-Methylbenzyl)succinic acid
3-Methylcyclohexanecarboxylic acid

m-Tolualdehyde
m-Tolunitrile
m-Xylene
→ 1,2-Dihydro-1,2-dihydroxy-5-methylbenzoic acid
Probably in: Pseudomonas B465
→ 2-Hydroxy-3-methylbenzoic acid
Probably in: Pseudomonas A1301
→ *m*-Methylbenzoyl-β-D-glucuronic acid
In: Rabbit H519
→ 3-Methylcatechol
Probably in: Pseudomonas A2765
→ 4-Methylcatechol
Probably in: Pseudomonas A2765
→ *m*-Methylhippuric acid
In: Man B309, D482
Rabbit H519
Rat B24, F455, F899
→ *m*-Tolualdehyde
In: Clostridium G416
Nocardia G207

***p*-Toluic acid** (*p*-methylbenzoic acid)
Formed from: Bitolterol
p-Methylbenzoylthiamine pyrophosphoric acid
4-Methylcyclohexanecarboxylic acid
p-Tolualdehyde
p-Tolunitrile
p-Toluoyl chloride phenylhydrazide
N-(*p*-Toluoyl)indoline
p-Xylene
→ *p*-Cresol
Probably in: Pseudomonas A1301
→ 1,2-Dihydro-1,2-dihydroxy-4-methylbenzoic acid
Probably in: Pseudomonas B465
→ *p*-Hydroxymethylbenzoic acid
In: Comamonas G405
→ *p*-Methylbenzoyl CoA
In: Beef H740
Pseudomonas G729
→ *p*-Methylbenzoyl-β-D-glucuronic acid
In: Rabbit H519
Rat J549
→ *p*-Methylhippuric acid
In: Rabbit H519
Rat C153, C681, J549
→ *p*-Tolualdehyde
In: Clostridium G416
Nocardia G207
→ *p*-Toluoyl-L-threonine
In: Streptomyces H269

***p*-Toluic acid phenylhydrazide**
Formed from: *p*-Toluoyl chloride phenylhydrazide
→ Benzene
In: Rat A2542
→ 1-Lauryl-1-phenyl-2-(*p*-toluoyl)hydrazine
Probably in: Sheep B15

***o*-Toluidine** (*o*-methylaniline)
Formed from: Aptocaine
N-(*n*-Butylalanyl)-*o*-toluidide
N,N-Diethylalanyl-*o*-toluidide
N,N-Diethylglycyl-*o*-toluidide
N-Ethylalanyl-*o*-toluidide
N-Isobutylalanyl-*o*-toluidide
o-Methylacetanilide
N-Methylglycyl-*o*-toluidide
o-Nitrotoluene
N-(*n*-Propylalanyl)-*o*-toluidide
→ *o*-Aminobenzyl alcohol
Probably in: Rat B170
→ 2-Amino-*m*-cresol
In: Rat B170
→ 4-Amino-*m*-cresol
In: Man D135
Rat B170
→ *o*,*o*′-Azoxytoluene
In: Rat B170
→ N-Hydroxy-*o*-toluidine
In: Rat C857
→ *o*-Methylacetanilide
In: Pseudomonas J196
Rabbit E399
Rat B170
→ 3-Methylcatechol
In: Rhodococcus D263, G321
→ *o*-Nitrosotoluene
In: Rat B170

***m*-Toluidine** (*m*-methylaniline)
Formed from: 4-Dimethylamino-3′-methylazobenzene

m-Methylacetanilide
3′-Methyl-4-methylaminoazobenzene
m-Nitrotoluene
→ *m*-Aminobenzyl alcohol
Probably in: Rat A3783, A3963, B27
→ 4-Amino-*o*-cresol
In: Man D135
See also: A3783, B27
→ *m*-Methylacetanilide
In: Rabbit E399
See also: A673, A3835, B27
→ 4-Methylcatechol
In: Pseudomonas E164

***p*-Toluidine** (*p*-methylaniline)
Formed from: N-Benzyl-*p*-toluidine
p-Methylacetanilide
p-Nitrotoluene
→ N-(*p*-Aminobenzyl)-*p*-toluidine
In: Rat G580
→ *p*-Methylacetanilide
In: Guinea pig F109
Pseudomonas J196
Rabbit E399, F109
Rat F109
Sea urchin B703
→ *p*-Methylformanilide
In: Guinea pig F109
Rabbit F109
Rat F109
→ *p*-Methylnitrosotoluene
In: Pea C797
Rat G580
With chloroperoxidase A3579
→ *m*-Methyl-*p*-toluidine
In: Rabbit E6
→ *p*-Nitrotoluene
In: Serratia K143
→ *p*-Tolylhydroxylamine
In: Rabbit F839

***o*-Tolunitrile** (*o*-methylbenzonitrile)
→ *o*-Hydroxymethylbenzonitrile
In: Rat H305
→ *o*-Methylbenzamide
In: Rhodococcus G172, H952
→ *o*-Toluic acid
In: Fusarium A3364
Rhodococcus G172

***m*-Tolunitrile** (*m*-methylbenzonitrile)
→ *m*-Toluic acid
In: Fusarium A3364
Nocardia A3420
Rhodococcus G172

***p*-Tolunitrile** (*p*-methylbenzonitrile)
→ *p*-Hydroxymethylbenzonitrile
In: Mortierella F823
→ *p*-Toluic acid
In: Fusarium A3364
Nocardia A3420
Rhodococcus G172, H952

***p*-Toluoyl chloride phenylhydrazide**
→ Aniline
In: Sheep B15
→ *p*-Toluic acid
In: Sheep B15
→ *p*-Toluic acid phenylhydrazide
In: Sheep B15

***p*-Toluoyl-4-hydroxypiperidine**
Formed from: *p*-Toluoylpiperidine

N-(*p*-Toluoyl)indoline
→ Indoline
In: Cunninghamella D810
→ *p*-Toluic acid
In: Cunninghamella D810

***p*-Toluoylpiperidine**
→ *p*-Hydroxymethylbenzoylpiperidine
In: Rat C23
→ *p*-Toluoyl-4-hydroxypiperidine
In: Rat C23

***p*-Toluoyl-L-threonine**
Formed from: *p*-Toluic acid
→ *p*-Toluoyl-L-threonyl-L-alloisoleucine
Probably in: Streptomyces H269
→ *p*-Toluoyl-L-threonyl-L-isoleucine
Probably in: Streptomyces H269
→ *p*-Toluoyl-L-threonyl-D-valine
Probably in: Streptomyces H269
→ *p*-Toluoyl-L-threonyl-L-valine
Probably in: Streptomyces H269

***p*-Toluoyl-L-threonyl-L-alloisoleucine**
Probably formed from: *p*-Toluoyl-L-threonine

***p*-Toluoyl-L-threonyl-L-isoleucine**
Probably formed from: *p*-Toluoyl-L-threonine

p-Toluoyl-L-threonyl-D-valine
Probably formed from: *p*-Toluoyl-L-threonine
p-Toluoyl-L-threonyl-L-valine
Probably formed from: *p*-Toluoyl-L-threonine
Toluquinone (methylbenzoquinone)
Probably formed from: Methylquinol
→ Methylquinol
In: Beef G401
Guinea pig K773
Sporotrichum B301
S-(p-Tolyl)glutathione
Formed from: N,N,N-Trimethyl-*p*-toluidinium
p-Tolylhydroxylamine
Formed from: *p*-Toluidine
2-(o-Tolyloxy)acetic acid
Formed from: Mephenesin
3-(o-Tolyloxy)lactic acid
Formed from: Mephenesin
m-(p-Tolyloxy)benzaldehyde
→ *m*-(*p*-Tolyloxy)benzoic acid
In: Rat H761
p-(p-Tolyloxy)benzaldehyde
→ *p*-(*p*-Tolyloxy)benzoic acid
In: Rat H761
m-(p-Tolyloxy)benzoic acid
Formed from: *m*-(*p*-Tolyloxy)benzaldehyde
p-(p-Tolyloxy)benzoic acid
Formed from: *p*-(*p*-Tolyloxy)benzaldehyde
cyclic-Tolyl phosphate
Probably formed from: *o*-Cresyl phosphate
p-Tolylsulphonamide
Probably formed from: *p*-Tolylsulphonylurea
p-Tolylsulphonylurea
Formed from: Tolbutamide
→ *p*-Tolylsulphonamide
Probably in: Dog C826
Tomoxiprole (2-(4-methoxyphenyl)-3-(1-methylethyl)-3*H*-naphth[1,2-*d*]imidazole)
→ Demethyltomoxiprole
In: Rat D913
Topazolin (4′,5,7-trihydroxy-3-methoxy-6-(3,3-dimethylallyl)flavone)
→ 6-(2,3-Dihydroxy-3-methylbutyl)-4′,5,7-trihydroxy-3-methoxyflavone
In: Botrytis G264
→ 2,3-Dihydro-4-hydroxy-7-(*p*-hydroxyphenyl)-2-(2-hydroxy-2-propyl)-6-methoxy-5*H*-[3,2-*g*][1]furobenzopyran-5-one
In: Botrytis G264
→ 6,7-Dihydro-5,7-dihydroxy-2-(*p*-hydroxyphenyl)-3-methoxy-8,8-dimethyl-4*H*,8*H*-[1,2-*b*:5,4-*b*′]benzodipyran-4-one
In: Botrytis G264
Topotecan (10-((dimethylamino)methyl)-4-ethyl-4,9-dihydroxy-1*H*-pyrano[3′,4:6,7]indolizino[1,2-*b*]quinoline-3,14(4*H*, 12*H*)-dione)
→ Demethyltopotecan
In: Man J844
Rat J704
→ Topotecan-β-glucuronide
In: Man J844
Rat J704
Topotecan-β-glucuronide
Formed from: Topotecan
(S)-Torosachrysone gentiobioside ((3*S*)-3,4-dihydro-3,8,9-trihydroxy-6-methoxy-3-methyl-1(2*H*)-anthracenone gentiobioside)
→ (1*S*,3*S*)-Austrocortilutein
In: Dermocybe K606
→ (1*S*,3*S*)-Austrocortirubin
In: Dermocybe K606
Tosifen (N-2-(1-phenylpropyl)-N′-*p*-tolylsulphonylurea)
→ N-(*p*-Hydroxymethylphenylsulphonyl)-N′-2-(1-phenylpropyl)urea
In: Dog A2877
Rat A2877
Trachelogenin ((3*S*,4*S*)-4-((3,4-dimethoxyphenyl)methyl)dihydro-3-hydroxy-3-((4-hydroxy-3-methoxyphenyl)methyl)-2(3*H*)-furanone)
Formed from: O-Demethyltrachelogenin
Tracheloside
→ O-Demethyltrachelogenin
Probably in: Rat H48
Tracheloside ((3*S*,4*S*)-4-((3,4-dimethoxyphenyl)methyl)-3-((4-(β-D-glucopyranosyloxy) 3-methoxyphenyl)methyl)dihydro-3-hydroxy-2(3*H*)-furanone)
→ Trachelogenin
In: Rat H48

Tralocythrin (3-(1,2-dibromo-2,2-dichloroethyl)-2,2-dimethylcyclopropanecarboxylic acid cyano(3-phenoxyphenyl)methyl ester)
→ Cypermethrin
In: Rat C412
→ *m*-Phenoxybenzoic acid
In: Rat C412
→ *m*-Phenoxybenzyl alcohol
In: Rat C412
→ *m*-Phenoxymandelonitrile
In: Bacillus G891

Tralomethrin (2,2-dimethyl-3-(1,2,2,2-tetrabromoethyl)cyclopropanecarboxylic acid cyano(3-phenoxyphenyl)methyl ester)
→ Deltamethrin
In: Rat C412
→ *m*-Phenoxybenzoic acid
In: Rat C412
→ *m*-Phenoxybenzyl alcohol
In: Bacillus G891
Rat C412
→ *m*-Phenoxymandelonitrile
In: Bacillus G891

Tramadol (1-(3-methoxyphenyl)-2-dimethylaminomethylcyclohexanol)
→ O-Demethyltramadol
In: Man K634

Tranylcypromine ((1*R*,2*S*)-rel-2-phenylcyclopropanamine)
→ *p*-Hydroxytranylcypromine
Rat E148

Trazodone (2-(3-(4-(3-chlorophenyl)-1-piperazinyl)propyl)-*s*-triazolo[4,3-*a*]pyridin-3(2*H*)-one)
→ 1-(*m*-Chlorophenyl)piperazine
In: Man J423
Rat B341, C50
→ *p*-Hydroxytrazodone
In: Man A2162, A2808
Rabbit A1438, A1776
Rat A285, A1438
See also: A294
→ β-(3-oxo-*s*-triazolo[4,3-*a*]pyridin-2-yl) propionic acid
In: Rabbit A1776
→ Trazodone dihydrodiol
In: Man A2162, A2808
→ Trazodone-N-β-D-glucuronide
In: Man H162
→ Trazodone-N-oxide
In: Man A2162, A2808

Trazodone dihydrodiol
Formed from: Trazodone
→ Trazodone dihydrodiol-N-oxide
In: Man A2808

Trazodone dihydrodiol-N-oxide
Probably formed from: Trazodone dihydrodiol

Trazodone-N-β-D-glucuronide
Formed from: Trazodone

Trazodone-N-oxide
Formed from: Trazodone

Tremetone (1-((2*R*)-2,3-dihydro-2-(1-methylethenyl)-5-benzofuranyl)ethanone)
Probably formed from acetate in Eupatorium A191
→ Dehydrotremetone
In: Eupatorium A191, A193

5,5′,5″-Triacetamido-2,2′,2″-trihydroxyterphenyl
Probably formed from: 5,5′-Diacetamido-2,2′-dihydroxybiphenyl
→ 5,5′,5″,5‴-Tetraacetamido-2,2′,2″,2‴-tetrahydroxyquaterphenyl
Probably with xanthine oxidase F171

1,8,9-Triacetoxyanthracene
→ 1,8-Diacetoxy-9-anthrone
In: Animal B150

Tri-(N-acetyl-L-tyrosine)
Formed from: N-Acetyl-L-tyrosine
Di-(N-acetyl-L-tyrosine)
→ Tetra-(N-acetyl-L-tyrosine)
In: Horseradish J327

1,2,4-Triaminobenzene
Formed from: 2,4-Diaminoazobenzene
2,4-Dinitroaniline
2-Nitro-*p*-phenylenediamine
Probably formed from: 1,2-Diamino-4-(*p*-fluorobenzyl)aminobenzene
→ 3,4-Diaminoacetanilide
Probably in: Pseudomonas E393
Rat J586

2,4,5-Triamino-6-benzyloxypyrimidine
Formed from: 2,4-Diamino-6-benzyloxy-5-nitrosopyrimidine

2,4,6-Triaminotoluene
Probably formed from: 2,4-Diamino-6-mitrotoluene
→ Toluene
Probably in: Desulphovibrio G719, G806

Triamterene (2,4,7-triamino-6-phenylpteridine)
→ *p*-Hydroxytriamterene
In: Rat A1871

Triazolam (8-chloro-6-(*o*-chlorophenyl)-1-methyl-4*H*-*s*-triazolo[4,3-*a*][1,4] benzodiazepine)
→ 8-Chloro-6-(2-chloro-hydroxyphenyl)-1-methyl-4*H*-*s*-triazolo[4,3-*a*][1,4] benzodiazepine
In: Dog A3702
Rat A3702
→ 8-Chloro-6-(*o*-chlorophenyl)-1-hydroxymethyl-4*H*-*s*-triazolo[4,3-*a*][1,4] benzodiazepine
In: Dog A2422, A3702
Man J20, K389
Mouse K389
Rat A3702
→ 2′,5-Dichloro-2-(3-methyl-4*H*-1,2,4-triazol-4-yl)benzophenone
In: Dog A3702
Rat A3702
→ 4-Hydroxytriazolam
In: Dog A2422, A3702
Man J20, K389
Mouse K389
Rat A3702
→ α-Hydroxytriazolam
In: Man J20

5-((2-Triazolo)methyl)indole-3-acetic acid
Formed from: Rizatriptan

***s*-Triazolo[3,4-*a*]phthalazine**
Formed from: Budralazine
Hydralazine
→ 8-Hydroxy-*s*-triazolo[3,4-*a*]phthalazine
Probably in: Rat A1503, A1816

Tribenoside (ethyl 3,5,6-tris-O-(phenylmethyl)-D-glucofuranoside)
→ Benzoic acid
Probably in: Man C469

Tribenzylamine
→ Benzaldehyde
Probably in: Rat K296
→ Dibenzylamine
In: Rat K296
→ Tribenzylamine-N-oxide
In: Rat K296

Tribenzylamine-N-oxide
Formed from: Tribenzylamine

2,4,6-Tribromoaniline
Formed from: N-Benzyl-2,4,6-tribromoaniline

1,2,4-Tribromobenzene
→ 2,4,5-Tribromophenol
In: Rabbit A1825
→ 2,4,6-Tribromophenol
In: Rabbit A1825

1,3,5-Tribromobenzene
→ 2,3,5-Tribromophenol
In: Rabbit A1825
→ 2,4,6-Tribromophenol
Probably in: Rabbit A1825

2,3,7-Tribromodibenzodioxin
Formed from: 2,3,7,8-Tetrabromodibenzodioxin
→ 2,3,7-Tribromo-hydroxydibenzodioxin
Probably in: Rat G537

2,3,7-Tribromo-hydroxydibenzodioxin
Probably formed from: 2,3,7-Tribromodibenzodioxin

2,3,5-Tribromophenol
Probably formed from: 1,3,5-Tribromobenzene

2,4,5-Tribromophenol
Formed from: 1,2,4-Tribromobenzene

2,4,6-Tribromophenol
Formed from: 1,2,4-Tribromobenzene
1,3,5-Tribromobenzene
Probably formed from: 2,4-Dibromobenzene
2,6-Dibromobenzene
→ 2,6-Dibromoquinol
In: Azotobacter H884
Flavobacterium G641

3,4′,5-Tribromosalicylanilide
→ 3,4′-Dibromo-5-hydroxysalicylanilide
In: Rat B709
→ 3,5-Dibromo-4′-hydroxysalicylanilide
In: Rat B709

2,4,6-Tri-*tert*-butylphenyl
→ 2,4,6-Tri-*tert*-butylphenoxy radical

Probably in: Rat C490

2,4,6-Tri-*tert*-butylphenoxy radical
Probably formed from: 2,4,6-Tri-*tert*-butylphenyl

1,2,4-Tricarboxybenzene
Formed from: Fluorene-2-carboxylic acid

2,3,4-Trichloroacetanilide
See: 2,3,4-Trichloroaniline

2,4,5-Trichloroacetanilide
See: 2,4,5-Trichloroaniline

3,4,5-Trichloroacetanilide
See: 3,4,5-Trichloroaniline

2′,4′,5′-Trichloroacetophenone
Formed from: 2-Chloro-1-(2,4,5-trichlorophenyl)vinyl methyl phosphate
Gardona
→ 1-(2,4,5-Trichlorophenyl)ethanol
Probably in: Beef A3807
Pig A3807
Sheep A3807

2,3,4-Trichloroaniline
→ 2,2′,3,3′,4,4′-Hexachloroazobenzene
In: Horseradish A661
→ 2,3,4-Trichloroacetanilide
In: Rainbow trout G703
Swordtail G703

2,4,5-Trichloroaniline
→ 2,4,5-Trichloroacetanilide
In: Rainbow trout G703

2,4,6-Trichloroaniline
Formed from: Banamate
N-Benzyl-2,4,6-trichloroaniline

3,4,5-Trichloroaniline
→ 3,4,5-Trichloroacetanilide
In: Rainbow trout G703
Swordtail G703

2,3,4-Trichloroanisole
Formed from: 2,3,4-Trichlorophenol

2,3,5-Trichloroanisole
Formed from: 2,3,5-Trichlorophenol

2,3,6-Trichloroanisole
Formed from: 2,3,6-Trichlorophenol

2,4,5-Trichloroanisole
Formed from: 2,4,5-Trichlorophenol

2,4,6-Trichloroanisole
Formed from: 2,4,6-Trichlorophenol

3,4,5-Trichloroanisole
Formed from: 3,4,5-Trichlorophenol

1,2,3-Trichlorobenzene
Formed from: Lindane
γ-Pentachlorocyclohex-1-ene
→ *m*-Dichlorobenzene
In: Microorganism E686
→ 2,3,4-Trichlorophenol
In: Rabbit A1627
→ 2,3,5-Trichlorophenol
In: Rabbit A1627

1,2,4-Trichlorobenzene
Formed from: Lindane
γ-Pentachlorocyclohex-1-ene
→ N-Acetyl-S-(2,3,5-trichlorophenyl)-L-cysteine
In: Rat B988
See also: G64
→ N-Acetyl-S-(2,4,5-trichlorophenyl)-L-cysteine
In: Rat B988
→ *p*-Dichlorobenzene
In: Microorganism E686
→ 1,2-Dihydro-1,2-dihydroxy-3,4,6-trichlorobenzene
In: Rhesus monkey B988
→ S-(Trichlorodihydrohydroxyphenyl) glutathione
In: Rat G64
→ S-(Trichlorophenyl)glutathione
In: Rat G64
→ 2,3,5-Trichlorophenol
In: Rabbit A1627
Rat B988
Rhesus monkey B988
See also: G64
→ 2,4,5-Trichlorophenol
In: Rabbit A1627
Rat B988
Rhesus monkey B988
See also: G227
→ 2,3,5-Trichlorothiophenol
In: Rat B988
→ 2,4,5-Trichlorothiophenol
In: Rat B988

1,3,5-Trichlorobenzene
Formed from: Lindane
Hexachlorobenzene
Probably formed from: 1,2,3,5-Tetrachlorobenzene
→ *m*-Dichlorobenzene
In: Microorganism E686
→ 2,3,5-Trichlorophenol
In: Rabbit A1627
→ 2,4,6-Trichlorophenol
In: Pseudomonas C64
Rabbit A1627
See also: A2789, A3101

2,3,4-Trichlorobenzoic acid
Formed from: 2,3,4-Trichlorobiphenyl

2,4,5-Trichlorobenzoic acid
Formed from: 2,2′,3′,4,5-Pentachlorobiphenyl
2,4,5-Trichlorobiphenyl

2,4,6-Trichlorobenzoic acid
Formed from: 2,4,6-Trichlorobiphenyl

2,2′,5-Trichlorobiphenyl
→ 2,5-Dichlorobenzoic acid
In: Acinetobacter B125
Alcaligenes A1822, B125
→ 2,2′,5-Trichloro-4-hydroxybiphenyl
In: Rat A2607
See also: A2578
→ 2,2′,5′-Trichloro-6-hydroxybiphenyl
In: Rat A2607

2,3,4-Trichlorobiphenyl
→ 2,3,4-Trichlorobenzoic acid
In: Acinetobacter B125
Alcaligenes B125

2′,3,4-Trichlorobiphenyl
→ 3,4-Dichlorobiphenyl
In: Acinetobacter B125
Alcaligenes B125

2,3′,5-Trichlorobiphenyl
→ 2,5-Dichlorobenzoic acid
In: Acinetobacter B125
Alcaligenes A1822, B125
→ *cis*-2′,5,5′-Trichloro-2,3-dihydro-2,3-dihydroxybiphenyl
Probably in: Burkholderia K547

2,4,4′-Trichlorobiphenyl
→ 2,4-Dichlorobenzoic acid
In: Acinetobacter B125
Alcaligenes A1822

2,4,5-Trichlorobiphenyl
→ 2,4,5-Trichlorobenzoic acid
In: Acinetobacter B125
Alcaligenes A1822, B125

2,4′,5-Trichlorobiphenyl
→ 2,5-Dichlorobenzoic acid
In: Acinetobacter B125
Alcaligenes A1822, B125
→ *cis*-2′,4,5′-Trichloro-2,3-dihydro-2,3-dihydroxybiphenyl
Probably in: Burkholderia K547
→ 2′,4,5′-Trichloro-3-hydroxybiphenyl
Probably in: Nereis A3291
Rhesus monkey A3275
→ 2,4′,5-Trichloro-3-hydroxybiphenyl
In: Rat J278
→ 2,4′,5-Trichloro-4-hydroxybiphenyl
In: Rat J278
→ 2,4′,5-Trichloro-3-methylthiobiphenyl
In: Rat J278
→ 2,4′,5-Trichloro-4-methylthiobiphenyl
In: Rat J278

2,4,6-Trichlorobiphenyl
→ 2,4,6-Trichlorobenzoic acid
In: Alcaligenes B125
→ 2′,4′,6′-Trichloro-4-hydroxybiphenyl
In: Rat A1423, A1444, A2470

3,4,5-Trichlorobiphenyl
→ *cis*-3′,4′,5′-Trichloro-2,3-dihydro-2,3-dihydroxybiphenyl
Probably in: Burkholderia K547

1,1,1-Trichloro-2,2-bis(*p*-chlorophenyl)ethane
(DDT, dichlorodiphenyltrichloroethane)
→ 2,2-Bis(*p*-chlorophenyl)acetic acid
In: Hamster A1380, A1742
Mouse A1380, A1742, C439
→ 1,1-Dichloro-2,2-bis(*p*-chlorophenyl)ethane
In: Acanthamoeba A3521
Hare A812
Hamster A1380
Hydrogenomonas A531
Mouse A1380, C439
Sole A3418

Trout A2396, A3397
Vivipera A3518
Yeast A1508
→ 1,1-Dichloro-2,2-bis(*p*-chlorophenyl) ethylene
In: Acanthamoeba A3521
Amphidinium A786
Culex A1425
Cyclotella A786
Housefly F435
Isochrysis A786
Locust A913
Mouse A1380, A1742, C439
Olisthodiscus A786
Phagocata A1395
Rat A928, E428
Skeletonema A786
Sole A3418
Spodoptera A1076
Tetraselmis A786
Trout A2396, A3397
Vivipera A3518
→ 2,2,2-Trichloro-1,1-bis(*p*-chlorophenyl) ethanol
Probably in: Mouse A1742
Trout A237
→ 1,1,1-Trichloro-2-(4-chloro-2,3-dihydro-2,3-dihydroxyphenyl)-2-(*p*-chlorophenyl)ethane
In: Alcaligenes J895
→ 1,1,1-Trichloro-2-(*o*-chlorophenyl)-2-(*p*-chlorophenyl)ethane
In: Turkey A480

2,2,2-Trichloro-1,1-bis(*p*-chlorophenyl)ethanol
(dicofol, kelthane)
Probably formed from: 1,1,1-Trichloro-2,2-bis(*p*-chlorophenyl)ethane
→ 4,4′-Dichlorobenzhydrol
Probably in: Mouse B326
→ 2,2-Dichloro-1,1-bis(*p*-chlorophenyl) ethanol
In: Mouse E845
Rat E845

1,1,1-Trichloro-2,2-bis(*p*-hydroxyphenyl)ethane
Formed from: 1,1,1-Trichloro-2-(*p*-hydroxyphenyl)-2-(*p*-methoxyphenyl)ethane
→ 1,1,1-Trichloro-2,2-bis(*p*-hydroxyphenyl) ethane-β-D-glucuronide
Probably in: Goat C198

1,1,1-Trichloro-2,2-bis(*p*-hydroxyphenyl)ethane-β-D-glucuronide
Probably formed from: 1,1,1-Trichloro-2,2-bis(*p*-hydroxyphenyl)ethane

1,1,1-Trichloro-2,2-bis(*p*-methoxyphenyl)ethane
(methoxychlor)
→ 1,1-Dichloro-2,2-bis(*p*-methoxyphenyl) ethane
In: Chicken D178
Goat D178
→ 1,1-Dichloro-2,2-bis(*p*-methoxyphenyl) ethylene
In: Chicken D178
Goat D178
Pine A891
See also: C175, C198
→ 1,1,1-Trichloro-2-(3-hydroxy-4-methoxyphenyl)-2-(*p*-methoxyphenyl) ethane
In: Man H57
Rat H57
→ 1,1,1-Trichloro-2-(*p*-hydroxyphenyl)-2-(*p*-methoxyphenyl)ethane
In: Chicken D178
Goat C198, D178
Man H57, J439
Rat H57
See also: C175

Trichlorobis(methylthio)aniline
Formed from: Pentachloronitrobenzene

Trichlorobis(methylthio)phenol
Probably formed from: Tetrachloro (methylthio)phenol

1,1,1-Trichloro-2,2-bis(*p*-nitrophenyl)ethane
→ 1,1-Dichloro-2,2-bis(*p*-nitrophenyl) ethylene
In: Housefly F435

3,4,4′-Trichlorocarbanilide
→ 3,4,4′-Trichlorocarbanilide-N-β-D-glucuronide
In: Man A2884, A2885
Monkey A2884
→ 3,4,4′-Trichlorocarbanilide-N′-β-D-glucuronide

In: Man A2884, A2885
Monkey A2884
→ 3′,4,4′-Trichloro-2-hydroxycarbanilide
In: Man A2884, A2885
Monkey A2884
Rat A2076, A2884
See also: A3797
→ 3′,4,4′-Trichloro-3-hydroxycarbanilide
In: Man A2884
Monkey A2884, A3797
Rat A2076, A2884
→ 4,4′,5-Trichloro-2-hydroxycarbanilide
In: Man A2884, A2885
Monkey A2884
Rat A2076, A2884

3,4,4′-Trichlorocarbanilide-N-β-D-glucuronide
Formed from: 3,4,4′-Trichlorocarbanilide

3,4,4′-Trichlorocarbanilide-N′-β-D-glucuronide
Formed from: 3,4,4′-Trichlorocarbanilide

3′,4,4′-Trichlorocarbanilid-2-yl-β-D-glucuronide
Formed from: 3′,4,4′-Trichloro-2-hydroxycarbanilide

3′,4,4′-Trichlorocarbanilid-3-yl-β-D-glucuronide
Formed from: 3′,4,4′-Trichloro-3-hydroxycarbanilide

4,4′,5-Trichlorocarbanilid-2-yl-β-D-glucuronide
Formed from: 4,4′,5-Trichloro-2-hydroxycarbanilide

3′,4,4′-Trichlorocarbanilid-2-yl sulphate
Formed from: 3′,4,4′-Trichloro-2-hydroxycarbanilide
→ 3′,4,4′-Trichloro-2-hydroxycarbanilide
In: Helix F124

3′,4,4′-Trichlorocarbanilid-3-yl sulphate
Formed from: 3′,4,4′-Trichloro-3-hydroxycarbanilide

4,4′,5-Trichlorocarbanilid-2-yl sulphate
Formed from: 4,4′,5-Trichloro-2-hydroxycarbanilide

3,4,5-Trichlorocatechol
Formed from: 2,3,4-Trichlorophenol

3,4,5-Trichlorophenol
→ 3,4-Dichlorocatechol
Probably in: Microorganism F883
→ 3,5-Dichlorocatechol
In: Bacteria F883
→ 3,4,5-Trichloro-2-methoxyphenol
Probably in: Rhodococcus F383

3,4,6-Trichlorocatechol
Formed from: 1,2,3,5-Tetrachlorobenzene
Tetrachlorocatechol
1,2,4,5-Tetrachlorophenol
2,3,5-Trichlorophenol
2,3,6-Trichlorophenol
2,4,5-Trichlorophenol
Probably formed from: 2,3,5,6-Tetrachlorophenol
→ 3,4,6-Trichloro-2-methoxyphenol
Probably in: Rhodococcus F383
→ 2,3,5-Trichloro-*cis*,*cis*-muconic acid
In: Pseudomonas J702

1,1,1-Trichloro-2-(4-chloro-2,3-dihydro-2,3-dihydroxyphenyl)-2-(*p*-chlorophenyl)ethane
Formed from: 1,1,1-Trichloro-2,2-bis(*p*-chlorophenyl)ethane

1,1,1-Trichloro-2-(2-chloro-3,4-dihydroxyphenyl)-2-(*p*-chlorophenyl)ethane
Probably formed from: 1,1,1-Trichloro-2-(2-chloro-3-hydroxyphenyl)-2-(*p*-chlorophenyl) ethane
→ 1,1,1-Trichloro-2-(2-chloro-4-hydroxy-3-methoxyphenyl)-2-(*p*-chlorophenyl)ethane
Probably in: Chicken A2368
Rat A991

1,1,1-Trichloro-2-(2-chloro-4-hydroxy-3-methoxyphenyl)-2-(*p*-chlorophenyl)ethane
Probably formed from: 1,1,1-Trichloro-2-(2-chloro-3,4-dihydroxyphenyl)-2-(*p*-chlorophenyl)ethane
→ 1,1-Dichloro-2-(2-chloro-4-hydroxy-3-methoxyphenyl)-2-(*p*-chlorophenyl)ethane
Probably in: Rat A991

1,1,1-Trichloro-2-(2-chloro-3-hydroxyphenyl)-2-(*p*-chlorophenyl)ethane
Formed from: 1,1,1-Trichloro-2-(*o*-chlorophenyl)-2-(*p*-chlorophenyl)ethane
→ 1,1,1-Trichloro-2-(2-chloro-3,4-dihydroxyphenyl)-2-(*p*-chlorophenyl) ethane
Probably in: Chicken A2368
Rat A991

1,1,1-Trichloro-2-(*o*-chlorophenyl)-2-(*p*-chlorophenyl)ethane
Formed from: 1,1,1-Trichloro-2,2-bis(*p*-chlorophenyl)ethane
→ 2-(*o*-Chlorophenyl)-2-(*p*-chlorophenyl) acetic acid
In: Chicken A2368
Rat A991
→ 1,1-Dichloro-2-(*o*-chlorophenyl)-2-(*p*-chlorophenyl)ethane
In: Chicken A2368
Rat A991
→ 1,1-Dichloro-2-(*o*-chlorophenyl)-2-(*p*-chlorophenyl)ethylene
In: Chicken A2368
→ 1,1,1-Trichloro-2-(2-chloro-3-hydroxyphenyl)-2-(*p*-chlorophenyl)ethane
In: Chicken A2368
Rat A991
Does not yield 1,1,1-trichloro-2,2-bis(*p*-chlorophenyl)ethane in monkey or rat A441

1,2,4-Trichlorodibenzo-*p*-dioxin
→ 1,2,4-Trichloro-hydroxydibenzo-*p*-dioxin
In: Rat A3809

2,3,7-Trichlorodibenzo-*p*-dioxin
Formed from: 2,3,7,8-Tetrachlorodibenzo-*p*-dioxin
→ 2-Chlorodibenzo-*p*-dioxin
Probably in: Microorganism H874

***cis*-2′,4,5′-Trichloro-2,3-dihydro-2,3-dihydroxybiphenyl**
Probably formed from: 2,4′,5-Trichlorobiphenyl

***cis*-2′,5,5′-Trichloro-2,3-dihydro-2,3-dihydroxybiphenyl**
Probably formed from: 2,3′,5-Trichlorobiphenyl

***cis*-3′,4′,5′-Trichloro-2,3-dihydro-2,3-dihydroxybiphenyl**
Probably formed from: 3,4,5-Trichlorobiphenyl

S-(Trichlorodihydrohydroxyphenyl)glutathione
Formed from: 1,2,4-Trichlorobenzene

2′,4′,6′-Trichloro-3,4-dihydroxybiphenyl
Probably formed from: 2′,4′,6′-Trichloro-4-hydroxybiphenyl
→ 2′,4′,6′-Trichloro-4-hydroxy-3-methoxybiphenyl
Probably in: Rat A1423, A2470

4′,4′,5-Trichloro-2,2′-dihydroxycarbanilide
Probably formed from: 3′,4′,4′-Trichloro-2-hydroxycarbanilide
→ 4′,4′,5-Trichloro-2,2′-dihydroxycarbanilide-β-D-glucuronide
Probably in: Man A2884
Rat A2076, A2884
Rhesus monkey A2884
→ 4′,4′,5-Trichloro-2,2′-dihydroxycarbanilide-O-sulphate
Probably in: Rat A2076

4′,4′,5-Trichloro-2,3′-dihydroxycarbanilide
Probably formed from: 3′,4,4′-Trichloro-3-hydroxycarbanilide
→ 4′,4′,5-Trichloro-2,3′-dihydroxycarbanilide-β-D-glucuronide
Probably in: Rat A2076
→ 4′,4′,5-Trichloro-2,3′-dihydroxycarbanilide-O-sulphate
Probably in: Rat A2076

4′,4′,5-Trichloro-2,2′-dihydroxycarbanilide-β-D-glucuronide
Probably formed from: 4′,4′,5-Trichloro-2,2′-dihydroxycarbanilide

4′,4′,5-Trichloro-2,3′-dihydroxycarbanilide-β-D-glucuronide
Probably formed from: 4′,4′,5-Trichloro-2,3′-dihydroxycarbanilide

4′,4′,5-Trichloro-2,2′-dihydroxycarbanilide-O-sulphate
Probably formed from: 4′,4′,5-Trichloro-2,2′-dihydroxycarbanilide

4′,4′,5-Trichloro-2,3′-dihydroxycarbanilide-O-sulphate
Probably formed from: 4′,4′,5-Trichloro-2,3′-dihydroxycarbanilide

2′,4,4′-Trichloro-2,3-dihydroxydiphenyl ether
Probably formed from: 5-Chloro-2-(2,4-dichlorophenoxy)phenol

2′,4,4′-Trichloro-2,5-dihydroxydiphenyl ether
Probably formed from: 5-Chloro-2-(2,4-dichlorophenoxy)phenol

2′,4,4′-Trichloro-2,6-dihydroxydiphenyl ether
Probably formed from: 5-Chloro-2-(2,4-dichlorophenoxy)phenol
4,4′,6-Trichloro-2,2′-dihydroxydiphenyl ether
Probably formed from: 5-Chloro-2-(2,4-dichlorophenoxy)phenol
4,4′,6-Trichloro-2′,3-dihydroxydiphenyl ether
Probably formed from: 5-Chloro-2-(2,4-dichlorophenoxy)phenol
1,1,1-Trichloro-2-(3,4-dihydroxyphenyl)-2-(*p*-methoxyphenyl)ethane
Probably formed from: 1,1,1-Trichloro-2-(3-hydroxy-4-methoxyphenyl)-2-(*p*-methoxyphenyl)ethane
3,4,5-Trichloro-2,6-dimethoxyphenol
Probably formed from: 3,4,5-Trichloro-2-methoxyphenol
2,2′,4-Trichlorodiphenyl ether
→ *o*-Chlorophenol
In: Rat A3503
→ 3,4-Dichlorophenol
In: Rat A3503
→ 2′,4′,6-Trichloro-2-hydroxydiphenyl ether
Probably in: Rat A3503
Trichloro-S-glutathionylisophthalonitrile
(3 isomers)
Probably formed from: Chlorothalonil
Trichloro-S-glutathionylquinol
Formed from: Tetrachloroquinol
→ Trichloroquinol
In: Phanerochaete K138
2,2′,5′-Trichloro-4-hydroxybiphenyl
Formed from: 2,2′,5′-Trichlorobiphenyl
2,2′,5′-Trichloro-6-hydroxybiphenyl
Formed from: 2,2′,5′-Trichlorobiphenyl
2,4′,5-Trichloro-3-hydroxybiphenyl
Formed from: 2,4′,5-Trichlorobiphenyl
2′,4,5′-Trichloro-3-hydroxybiphenyl
Formed from: 2,4′,5-Trichlorobiphenyl
2,4′,5-Trichloro-4-hydroxybiphenyl
Formed from: 2,4′,5-Trichlorobiphenyl
2′4′,6′-Trichloro-4-hydroxybiphenyl
Formed from: 2,4,6-Trichlorobiphenyl
→ 2′4′,6′-Trichloro-3,4-dihydroxybiphenyl
Probably in: Rat A1423, A2470
3,3′,4′-Trichloro-4-hydroxybiphenyl
Formed from: 3,3′,4,4′-Tetrachlorobiphenyl
3′,4,4′-Trichloro-2-hydroxycarbanilide
Formed from: 3,4,4′-Trichlorocarbanilide
3′,4,4′-Trichlorocarbanilid-2-yl sulphate
→ 3′,4,4′-Trichlorocarbanilid-2-yl-β-D-glucuronide
Probably in: Man A2884
Monkey A2884
Rat A2076, A2884
→ 3′,4,4′-Trichlorocarbanilid-2-yl sulphate
In: Man A2884, A2885
Monkey A2884
Rat A2076, A2884
→ 4,4′,5-Trichloro-2,2′-dihydroxycarbanilide
In: Man A2884
Monkey A2884
Rat A2076, A2884
3′,4,4′-Trichloro-3-hydroxycarbanilide
Formed from: 3,4,4′-Trichlorocarbanilide
→ 3′,4,4′-Trichlorocarbanilid-3-yl-β-D-glucuronide
In: Man A2884
Monkey A2884, A3797
Rat A2076, A2884
→ 3′,4,4′-Trichlorocarbanilid-2-yl sulphate
In: Man A2884
Monkey A2884, A3979
Rat A2076, A2884
→ 4,4′,5-Trichloro-2,3′-dihydroxycarbanilide
Probably in: Rat A2076
4,4′,5-Trichloro-2-hydroxycarbanilide
Formed from: 3,4,4′-Trichlorocarbanilide
→ 4,4′,5-Trichlorocarbanilid-2-yl-β-D-glucuronide
Probably in: Man A2884
Monkey A2884
Rat A2076, A2884
→ 4,4′,5-Trichlorocarbanilid-2-yl sulphate
In: Man A2884, A2885
Monkey A2884
Rat A2076, A2884
1,2,4-Trichloro-hydroxydibenzo-*p*-dioxin
Formed from: 1,2,4-Trichlorodibenzo-*p*-dioxin

2,3,7-Trichloro-8-hydroxydibenzo-*p*-dioxin
Formed from: 2,3,7,8-Tetrachlorodibenzo-*p*-dioxin

4,7,8-Trichloro-1-hydroxydibenzo-*p*-dioxin
Formed from: 1,4,7,8-Tetrachlorodibenzo-*p*-dioxin
→ 4,7,8-Trichloro-1-hydroxydibenzo-*p*-dioxin-β-D-glucuronide
In: Rat H940

4,7,8-Trichloro-1-hydroxydibenzo-*p*-dioxin-β-D-glucuronide
Formed from: 4,7,8-Trichloro-1-hydroxydibenzo-*p*-dioxin

2,7,8-Trichloro-3-hydroxydibenzofuran
Formed from: 2,3,7,8-Tetrachlorodibenzofuran

2′,4′,6-Trichloro-2-hydroxydiphenyl ether
Probably formed from: 2,2′,4′-Trichlorodiphenyl ether

2′,4′,6′-Trichloro-4-hydroxy-3-methoxybiphenyl
Probably formed from: 2′,4′,6′-Trichloro-3,4-dihydroxybiphenyl

1,1,1-Trichloro-2-(3-hydroxy-4-methoxyphenyl)-2-(*p*-methoxyphenyl)ethane
Formed from: 1,1,1-Trichloro-2,2-bis(*p*-methoxyphenyl)ethane
→ 1,1,1-Trichloro-2-(3,4-dihydroxyphenyl)-2-(*p*-methoxyphenyl)ethane
Probably in: Man H57
Rat H57
→ 1,1,1-Trichloro-2-(*p*-hydroxyphenyl)-2-(3-hydroxy-4-methoxyphenyl)ethane
Probably in: Man H57
Rat H57

2,4,6-Trichloro-3-hydroxyphenoxyacetic acid
Probably formed from: 2,4,6-Trichlorophenoxyacetic acid

2,4,6-Trichloro-3-hydroxyphenoxyethanol
Probably formed from: 2,4,6-Trichlorophenoxyethanol

N-(2,4,6-Trichloro-3-hydroxyphenoxyethyl)-N-propylurea
Probably formed from: N-(2,4,6-Trichlorophenoxyethyl)-N-propylurea

1,1,1-Trichloro-2-(*p*-hydroxyphenyl)-2-(3-hydroxy-4-methoxyphenyl)ethane
Probably formed from: 1,1,1-Trichloro-2-(3-hydroxy-4-methoxyphenyl)-2-(*p*-methoxyphenyl)ethane

1,1,1-Trichloro-2-(*p*-hydroxyphenyl)-2-(*p*-methoxyphenyl)ethane
Probably formed from: 1,1,1-Trichloro-2,2-bis(*p*-methoxyphenyl)ethane
→ 4-Hydroxy-4′-methoxybenzophenone
Probably in: Chicken D178
→ 1,1,1-Trichloro-2,2-bis(*p*-hydroxyphenyl) ethane
In: Man H57
Rat H57
See also: C175, C198, D178
→ 1,1,1-Trichloro-2-(*p*-hydroxyphenyl)-2-(*p*-methoxyphenyl)ethane-β-D-glucuronide
Probably in: Chicken D178

1,1,1-Trichloro-2-(*p*-hydroxyphenyl)-2-(*p*-methoxyphenyl)ethane-β-D-glucuronide
Probably formed from: 1,1,1-Trichloro-2-(*p*-hydroxyphenyl)-2-(*p*-methoxyphenyl)ethane

2,3,5-Trichloro-4-methoxyphenol
Formed from: Trichloroquinol

3,4,5-Trichloro-2-methoxyphenol
Probably formed from: 3,4,5-Trichlorocatechol
→ 3,4,5-Trichloro-2,6-dimethoxyphenol
Probably in: Arthrobacter C527
→ 3,4,5-Trichloroveratrole
In: Arthrobacter C527
See also: F383

3,4,6-Trichloro-2-methoxyphenol
Probably formed from: 3,4,6-Trichlorocatechol
→ 3,4,6-Trichloroveratrole
Probably in: Rhodococcus F383

4,5,6-Trichloro-2-methoxyphenol
Probably formed from: 3,4,6-Trichlorocatechol
→ 4,5,6-Trichloro-2-methoxyphenol-β-D-glucuronide
In: Blenny F248
Flounder F248
Perch F248
Sculpin F248
Trout F248
→ 3,4,5-Trichloroveratrole
In: Acinetobacter D535

Pseudomonas D535

4,5,6-Trichloro-2-methoxyphenol-β-D-glucuronide
Formed from: 4,5,6-Trichloro-2-methoxyphenol

2,4′,5-Trichloro-3-methylsulphinylbiphenyl
Probably formed from: 2,4′,5-Trichloro-3-methylthiobiphenyl
→ 2,4′,5-Trichloro-3-methylsulphonylbiphenyl
Probably in: Rat J278

2,4′,5-Trichloro-4-methylsulphinylbiphenyl
Probably formed from: 2,4′,5-Trichloro-4-methylthiobiphenyl
→ 2,4′,5-Trichloro-4-methylsulphonylbiphenyl
Probably in: Rat J278

2,4′,5-Trichloro-3-methylsulphonylbiphenyl
Probably formed from: 2,4′,5-Trichloro-3-methylsulphinylbiphenyl

2,4′,5-Trichloro-4-methylsulphonylbiphenyl
Formed from: 4-(N-Acetyl-S-(L-cysteinyl))-2,4′,5-trichlorobiphenyl
Probably formed from: 2,4′,5-Trichloro-4-methylsulphinylbiphenyl

3′,4,4′-Trichloro-6-methylsulphonyl-3-methylthiobiphenyl
Formed from: 3,3′,4,4′-Tetrachlorobiphenyl

2,4′,5-Trichloro-3-methylthiobiphenyl
Formed from: 2,4′,5-Trichlorobiphenyl
→ 2,4′,5-Trichloro-3-methylsulphinylbiphenyl
Probably in: Rat J278

2,4′,5-Trichloro-4-methylthiobiphenyl
Formed from: 2,4′,5-Trichlorobiphenyl
→ 2,4′,5-Trichloro-4-methylsulphinylbiphenyl
Probably in: Rat J278

2,5,6-Trichloro-4-methylthioisophthalonitrile
Formed from: Chlorthalonil
→ 2,5-Dichloro-4,6-bis(methylthio) isophthalonitrile
Probably in: Rat J453

2,4,5-Trichlorophenacyl chloride
Formed from: Gardona
2-Chloro-1-(2,4,5-trichlorophenyl)vinyl methyl phosphate
→ 2-Chloro-1-(2,4,5-trichlorophenyl)ethanol
In: Chicken A3198
See also: A3807
→ 1-(2,4,5-Trichlorophenyl)ethanol
In: Chicken A3198
→ S-(2,4,5-Trichlorophenacyl)glutathione
In: Beef A3807
Pig A3807
Sheep A3807

S-(2,4,5-Trichlorophenacyl)glutathione
Probably formed from: 2,4,5-Trichlorophenacyl chloride

2,3,4-Trichlorophenol
Formed from: Hexachlorobenzene
2,3,4,6-Tetrachlorophenol
1,2,3-Trichlorobenzene
→ 2,4-Dichlorophenol
In: Desulfomonile G471
→ 3,4-Dichlorophenol
In: Desulfitobacterium J232
Microorganism G331
→ 2,3-Dichloroquinol
In: Arthrobacter K756
→ 2,3,4-Trichloroanisole
In: Mycobacterium A3330
Rhodococcus F383
→ 3,4,5-Trichlorocatechol
In: Rhodococcus F383

2,3,5-Trichlorophenol
Formed from: γ-Pentachlorocyclohex-1-ene
2,3,5,6-Tetrachlorophenol
1,2,3-Trichlorobenzene
1,2,4-Trichlorobenzene
1,3,5-Trichlorobenzene
Probably formed from: 2,3,4,5-Tetrachlorophenol
→ 2,3-Dichlorophenol
Desulfomonile G471
→ 3,5-Dichlorophenol
In: Desulfitobacterium J232
Microorganism G331
→ 2,3,5-Trichloroanisole
In: Mycobacterium A3330
→ 3,4,6-Trichlorocatechol
In: Rhodococcus F383
→ 2,3,5-Trichlorophenyl-β-D-glucuronide
In: Monkey B988

2,3,6-Trichlorophenol
Formed from: 2,3,5,6-Tetrachlorophenol
→ 2,3-Dichlorophenol
In: Microorganism H509

→ 2,5-Dichlorophenol
In: Microorganism G331, H509
→ 2,6-Dichlorophenol
In: Desulfomonile G471
→ 2,6-Dimethoxy-4-(2,3,6-trichlorophenoxy) phenol
In: Rhizoctonia D690
→ 4-(2,6-Dimethoxy-4-(2,3,6-trichlorophenoxy)phenoxy)-2,6-dimethoxyphenol
In: Rhizoctonia D690
→ 4-(4-(2,6-Dimethoxy-4-(2,3,6-trichlorophenoxy)phenoxy)-2,6-dimethoxyphenoxy)-2,6-dimethoxyphenol
In: Rhizoctonia D690
→ 4-(4-(4-(2,6-Dimethoxy-4-(2,3,6-trichlorophenoxy)phenoxy)-2,6-dimethoxyphenoxy)-2,6-dimethoxyphenoxy)-2,6-dimethoxyphenol
In: Rhizoctonia D690
→ 3-Methoxy-5-(2,3,6-trichlorophenoxy)-*o*-quinone
In: Rhizoctonia D690
→ 2,3,6-Trichloroanisole
In: Mycobacterium A3330

2,4,5-Trichlorophenol
Formed from: Haloprogin
Hexachlorobenzene
γ-Pentachlorocyclohex-1-ene
Silvex
2,4,4′,5-Tetrachlorodiphenyl ether
2,3,4,5-Tetrachlorophenol
2,4,5-Trichlorophenoxyacetic acid
1,2,4-Trichlorobenzene
→ 3,5-Dichlorocatechol
Probably in: Pseudomonas B16
→ 2,4-Dichlorophenol
In: Desulfomonile G471
→ 3,4-Dichlorophenol
In: Microorganism G331, G667
→ 2,5-Dichloroquinol
In: Arthrobacter K750
Azotobacter H884
Pseudomonas J177
→ 4-(2,6-Dimethoxy-4-(2,4,5-trichlorophenoxy)phenoxy)-2,6-dimethoxyphenol
In: Rhizoctonia D690
→ 4-(4-(2,6-Dimethoxy-4-(2,4,5-trichlorophenoxy)phenoxy)-2,6-dimethoxyphenoxy)-2,6-dimethoxyphenol
In: Rhizoctonia D690
→ 1,2,3,4,6,7-Hexachlorodibenzo-*p*-dioxin
In: Horseradish F378
→ 1,2,3,6,7,9-Hexachlorodibenzo-*p*-dioxin
In: Horseradish F378
→ 1,2,3,6,8,9-Hexachlorodibenzo-*p*-dioxin
In: Horseradish F378
→ 1,2,4,6,7,9-Hexachlorodibenzo-*p*-dioxin
In: Horseradish F378
→ 1,2,4,6,8,9-Hexachlorodibenzo-*p*-dioxin
In: Horseradish F378
→ 3-Methoxy-5-(2,4,5-trichlorophenoxy)-*o*-quinone
In: Rhizoctonia D690
→ 1,2,3,7,8-Pentachlorodibenzo-*p*-dioxin
In: Horseradish F378
→ 1,2,4,7,8-Pentachlorodibenzo-*p*-dioxin
In: Horseradish F378
→ Phenol
Probably in: Pseudomonas C893
→ 2,4,5-Trichloroanisole
In: Mycobacterium A3330
Rhodococcus F383
Tetrahymena E96
→ 3,4,6-Trichlorocatechol
In: Rhodococcus F383
→ 2,4,5-Trichlorophenyl-β-D-glucuronide
In: Monkey B988
→ 2,4,5-Trichlorophenyl sulphate
In: Dog A477
Mouse A477
Rabbit A477
Rat A477

2,4,6-Trichlorophenol
Formed from: α-Lindane
ε-Lindane
Pentachlorobenzene
γ-Pentachlorocyclohex-1-ene
Prochloraz
2,3,4,6-Tetrachlorophenol
1,3,5-Trichlorobenzene
→ 2,4-Dichlorophenol

In: Desulfitobacterium J232
Flavobacterium G641
Microorganism G331, G667
→ 2,6-Dichloroquinol
In: Arthrobacter K750
Azotobacter H884
Pseudomonas J177
→ 2,4,6-Trichloroanisole
In: Arthrobacter C527
Mycobacterium A3330

3,4,5-Trichlorophenol
Formed from: 2,3,4,5-Tetrachlorophenol
→ 3,5-Dichlorophenol
In: Microorganism G331
See also: E168
→ 2,6-Dichloroquinol
In: Azotobacter H884
→ 3,4,5-Trichloroanisole
In: Arthrobacter C527
Mycobacterium A3330
→ 3,4,5-Trichlorocatechol
In: Rhodococcus F383

2,4,6-Trichlorophenoxyacetaldehyde
Probably formed from: Prochloraz
→ 2,4,6-Trichlorophenoxyacetic acid
Probably in: Rat G26
→ 2,4,6-Trichlorophenoxyethanol
Probably in: Rat G26

2,4,5-Trichlorophenoxyacetic acid (2,4,5-T)
→ 2,4,5-Trichlorophenol
In: Microorganism H104
Mouse A1405
Pseudomonas B16, C893, H644
Rat A1405
Sheep A2366
→ 2,4,5-Trichlorophenoxyacetylaspartic acid
In: Glycine A3202
→ 2,4,5-Trichlorophenoxyacetyl-β-D-glucuronic acid
Probably in: Rat E534
→ 2,4,5-Trichlorophenoxyacetylglutamic acid
In: Glycine A3202
→ 2,4,5-Trichlorophenoxyacetylglycine
In: Mouse A1405
Rat A1405, E534
→ 2,4,5-Trichlorophenoxyacetyltaurine
In: Dogfish A2359
Lobster C38
Mouse A1405
Rat A1405, E534

2,4,6-Trichlorophenoxyacetic acid
Probably formed from: 2,4,6-Trichlorophenoxyacetaldehyde
→ 2,4,6-Trichloro-3-hydroxyphenoxyacetic acid
Probably in: Rat G717

2,4,5-Trichlorophenoxyacetylaspartic acid
Formed from: 2,4,5-Trichlorophenoxyacetic acid

2,4,5-Trichlorophenoxyacetyl-β-D-glucuronic acid
Formed from: 2,4,5-Trichlorophenoxyacetic acid

2,4,5-Trichlorophenoxyacetylglutamic acid
Formed from: 2,4,5-Trichlorophenoxyacetic acid

2,4,5-Trichlorophenoxyacetylglycine
Formed from: 2,4,5-Trichlorophenoxyacetic acid

2,4,5-Trichlorophenoxyacetyltaurine
Formed from: 2,4,5-Trichlorophenoxyacetic acid

2,4,6-Trichlorophenoxyethanol
Probably formed from: 2,4,6-Trichlorophenoxyacetaldehyde
→ 2,4,6-Trichloro-3-hydroxyphenoxyethanol
Probably in: Rat G717
→ 2,4,6-Trichlorophenoxyethyl-β-D-glucuronide
In: Rat G26, G717

2,4,6-Trichlorophenoxyethyl-β-D-glucuronide
Formed from: 2,4,6-Trichlorophenoxyethanol

N-(2,4,6-Trichlorophenoxyethyl)-N-propylurea
Formed from: Prochloraz
→ N-(2,6-Dichloro-4-hydroxyphenoxyethyl)-N-propylurea
Probably in: Rat G26
→ N-(2,4,6-Trichloro-3-hydroxyphenoxyethyl)-N-propylurea
Probably in: Rat G26

2,3,4-Trichloro-L-phenylalanine
→ 2,3,4-Trichlorophenylpyruvic acid
In: Phaseolus E444

2,4,5-Trichloro-L-phenylalanine
→ 2,4,5-Trichlorophenylpyruvic acid
In: Phaseolus E444

S-(Trichlorophenyl)-L-cysteinylglycine
Probably formed from: S-(Trichlorophenyl)glutathione

2,4,5-Trichlorophenyl)ethanol
Formed from: Gardona
2,4,5-Trichlorophenacyl chloride
Probably formed from: 2′,4′,5′-Trichloroacetophenone

2,3,5-Trichlorophenyl-β-D-glucuronide
Formed from: 2,3,5-Trichlorophenol

2,4,5-Trichlorophenyl-β-D-glucuronide
Formed from: 2,4,5-Trichlorophenol

S-(Trichlorophenyl)glutathione
Formed from: 1,2,4-Trichlorobenzene
→ L-Glutamyl-S-(trichlorophenyl)-L-cysteine
Probably in: Rat G64
→ S-(Trichlorophenyl)-L-cysteinylglycine
Probably in: Rat G64

2,4,5-Trichlorophenylglycol
Formed from: Gardona

2,4,6-Trichlorophenylhydrazine
Formed from: Banamite

2,3,4-Trichlorophenylpyruvic acid
Formed from: 2,3,4-Trichloro-L-phenylalanine

2,4,5-Trichlorophenylpyruvic acid
Formed from: 2,4,5-Trichloro-L-phenylalanine

2,4,5-Trichlorophenyl sulphate
Formed from: 2,4,5-Trichlorophenol

Trichloroquinol
Formed from: 2,3,4,5-Tetrachlorophenol
2,3,4,6-Tetrachlorophenol
Tetrachloroquinol
2,3,6-Trichlorophenol
Trichloro-S-glutathionylquinol
→ 3,5-Dichloro-2-(S-glutathionyl)quinol
In: Phanerochaete K138
→ 2,3-Dichloroquinol
In: Rhodococcus E509
→ 2,5-Dichloroquinol
In: Rhodococcus E509
→ 2,6-Dichloroquinol
In: Phanerochaete K138
See also: E265
→ 2,3,5-Trichloro-4-methoxyphenol
In: Rhodococcus E509
→ Trichloro-*p*-quinone
Probably in: Rat F891

Trichloro-*p*-quinone
Probably formed from: Trichloroquinol

2,3,4-Trichlorothioanisole
Probably formed from: 2,3,4,5-Tetrachlorothioanisole

2,4,5-Trichlorothioanisole
Formed from: 2,4,5-Trichlorothiophenol

2,3,5-Trichlorothiophenol
Formed from: 1,2,4-Trichlorobenzene

2,4,5-Trichlorothiophenol
Formed from: 1,2,4-Trichlorobenzene
→ 2,4,5-Trichlorothioanisole
In: Mycobacterium A3330

3,4,5-Trichloroveratrole
Formed from: 3,4,5-Trichloro-2-methoxyphenol
4,5,6-Trichloro-2-methoxyphenol

3,4,6-Trichloroveratrole
Probably formed from: 3,4,6-Trichloro-2-methoxyphenol

Tri-*o*-cresyl phosphate
→ *o*-Cresol
In: Cat D375
Rat G129
→ Di-*o*-cresyl phosphate
In: Cat D375
Rat G129

2,4,6-Tri-(3′,5′-di-*tert*-butylcyclohexa-2′,5′-dien-4′-on-1′-ylidenemethyl)mesitylene
Formed from: 2,4,6-Tri-(3′,5′-di-*tert*-butyl-4′-hydroxybenzyl)mesitylene

2,4,6-Tri-(3′,5′-di-*tert*-butyl-4′-hydroxybenzyl) mesitylene
→ 2,4,6-Tri-(3′,5′-di-*tert*-butylcyclohexa-2′,5′-dien-4′-on-1′-ylidenemethyl)mesitylene
In: Rat A515

3-(2′,4′,5′-Triethoxybenzoyl)propionic acid
→ 3-(2′,4′-Diethoxy-5′-hydroxybenzoyl) propionic acid
In: Dog A3056
Man A3491
Rat A3056, A3251
→ 3-(2′,5′-Diethoxy-4′-hydroxybenzoyl) propionic acid

In: Dog A3056
Man A3491
Rat A3056, A3251
→ 3-(2′,4′-Diethoxy-5′-(2-hydroxyethoxy) benzoyl)propionic acid
In: Dog A3056
Rat A3056, A3251
→ 3-(2′,5′-Triethoxybenzoyl)propionyl-β-D-glucuronic acid
In: Man A3491

3-(2′,5′-Triethoxybenzoyl)propionyl-β-D-glucuronic acid
Formed from: 3-(2′,4′,5′-Triethoxybenzoyl)propionic acid

Triethyl 2-((3-bisphosphono)propyl)-1-methyl-4-phenylpyrimidin-6-one
Formed from: Tetraethyl 2-((3-bisphosphono) propyl)-1-methyl-4-phenylpyrimidin-6-one

Triflubazam (1-methyl-5-phenyl-7-trifluoromethyl-1*H*-1,5-benzodiazepine-2,4(3*H*,5*H*)-dione)
→ Demethyltriflubazam
In: Dog A1864
Man A1873
Monkey A1864
→ 3′,4′-Dihydro-3′,4′-dihydroxytriflubazam
In: Dog A1864
→ 4′-Hydroxytriflubazam
In: Dog A1864
Man A1873
Monkey A1864

Trifluoperazine-N-β-D-glucuronide
Formed from: 10-(3-(4-methylpiperazin-1-yl)propyl)-2-trifluoromethyl-10*H*-phenothiazine

Trifluoperazine-1″-oxide
Formed from: 10-(3-(4-methylpiperazin-1-yl)propyl)-2-trifluoromethyl-10*H*-phenothiazine

Trifluoperazine-4″-oxide
Formed from: 10-(3-(4-methylpiperazin-1-yl)propyl)-2-trifluoromethyl-10*H*-phenothiazine

2,2,2-Trifluoroacetophenone
→ 2,2,2-Trifluoro-1-phenylethanol
In: Bacillus G826
Candida G826
Deleya G826
Micrococcus C25
Rabbit C25
Rhodotorula G826
Saccharomyces G826
Torulopsis G826
Xanthomonas G826

N-Trifluoroacetyladriamycin
Formed from: N-Trifluoroacetyladriamycin-14-valerate
→ N-Trifluoroacetyladriamycinol
Probably in: Mouse A2710
Rat A3517

N-Trifluoroacetyladriamycinol
Formed from: N-Trifluoroacetyladriamycin

N-Trifluoroacetyladriamycin-14-valerate
→ N-Trifluoroacetyladriamycin
In: Mouse A2710
Rat A3517

N-Trifluoroacetyldoxorubicin
Formed from: N-Trifluoroacetyldoxorubicin-14-valerate
→ 13-Dihydro-N-trifluoroacetyldoxorubicin
Probably in: Rat D172
→ Doxorubicin
Probably in: Rat D172

N-Trifluoroacetyldoxorubicin-14-valerate
→ N-Trifluoroacetyldoxorubicin
In: Rat D172

N-Trifluoroacetylepinine
→ N-Trifluoroacetylepinine quinone
With tyrosinase K691

N-Trifluoroacetylepinine quinone
Formed from: N-Trifluoroacetylepinine

2,3,6-Trifluoroaniline
→ 4-Amino-2,3,5-trifluorophenol
In: Rat H273

2,4,6-Trifluoroaniline
Formed from: N-Benzyl-2,4,6-trifluoroaniline

N-(2,4,6-Trifluorobenzyl)aniline
→ Aniline
In: Hamster G951

3,4,5-Trifluorocatechol
Formed from: 2,3,4-Trifluorophenol
3,4,5-Trifluorophenol

3,4,6-Trifluorocatechol
Formed from: 2,3,5,6-Tetrafluorophenol

2,3,5-Trifluorophenol

3,3,3-Trifluoro-1-(*p*-(2-hydroxyethoxy)phenyl)-1,2-diphenylpropene
Formed from: Panomifene

3,3,3-Trifluoro-1-(*p*-hydroxyphenyl)-1,2-diphenylpropene
Formed from: Panomifene

1,1,1-Trifluoro-N-(4-((3-hydroxyphenyl)sulphonyl)-2-methylphenyl)methanesulphonamide
Formed from: Perfluidone
→ 1,1,1-Trifluoro-N-(4-((3-hydroxyphenyl) sulphonyl)-2-methylphenyl) methanesulphonamide-β-D-glucuronide
In: Arachis A1949

1,1,1-Trifluoro-N-(4-((4-hydroxyphenyl)sulphonyl)-2-methylphenyl)methanesulphonamide
Formed from: Perfluidone

1,1,1-Trifluoro-N-(4-((3-hydroxyphenyl)sulphonyl)-2-methylphenyl)methanesulphonamide-β-D-glucuronide
Formed from: 1,1,1-Trifluoro-N-(4-((3-hydroxyphenyl)sulphonyl)-2-methylphenyl) methanesulphonamide

6-(Trifluoromethoxy)-2-benzothiazolamine
See: Riluzole

***p*-Trifluoromethylacetanilide**
Formed from: *p*-Trifluoromethylaniline
→ N-Oxaniloyl-*p*-trifluoromethylaniline
In: Dog D321
Rat D321

***p*-Trifluoromethylaniline**
→ 4-Amino-2-trifluoromethylphenol
In: Dog D321
→ *p*-Trifluoromethylacetanilide
In: Rabbit E399
See also: D321

***m*-Trifluoromethylanisole**
→ *m*-Trifluoromethylphenol
In: Sporomusa J199

Trifluoromethylbenzene
→ 1,2-Dihydro-1,2-dihydroxy-3-trifluoromethylbenzene
In: Pseudomonas J656
See also: F283

***o*-Trifluoromethylbenzaldehyde**
→ *o*-Trifluoromethylbenzyl alcohol
In: Saccharomyces H425

***m*-Trifluoromethylbenzaldehyde**
→ *m*-Trifluoromethylbenzyl alcohol
In: Saccharomyces H425

***p*-Trifluoromethylbenzaldehyde**
→ *p*-Trifluoromethylbenzyl alcohol
In: Saccharomyces H425

***o*-Trifluoromethylbenzoic acid**
→ *cis*-1,2-Dihydro-1,2-dihydroxy-2-trifluoromethylbenzoic acid
In: Pseudomonas H649
→ *o*-Trifluoromethylbenzoyl-β-D-glucuronic acid
In: Rat J549

***m*-Trifluoromethylbenzoic acid**
Formed from: N-Ethyl-α-methyl-3-trifluoromethylphenethylamine
Tetrahydro-2-(3-trifluoromethylphenyl)-1,4-oxazine
Probably formed from: *m*-Trifluoromethylmandelic acid
→ 1,2-Dihydro-2-hydroxy-3-trifluoromethylbenzoic acid
In: Pseudomonas E698
→ 2-Hydroxy-6-oxo-6-trifluoromethylhexa-2,4-dienoic acid
In: Microorganism H42
→ *m*-Trifluoromethylbenzoyl-β-D-glucuronic acid
In: Rat J549
→ *m*-Trifluoromethylhippuric acid
In: Dog A631, A3852
Monkey A631
Rat A3123, A3730, A3852, J549

***p*-Trifluoromethylbenzoic acid**
Formed from: Fluvoxamine
→ *cis*-1,2-Dihydro-1,2-dihydroxy-4-trifluoromethylbenzoic acid
Probably in: Pseudomonas H649
→ *p*-Trifluoromethylbenzoyl-β-D-glucuronic acid
In: Rat F818, J549
→ *p*-Trifluoromethylhippuric acid
In: Dog C652
Hamster C652
Mouse C652
Rat C652
Contrast (in rat) J549

5-Trifluoromethyl-2-benzothiazole
Probably formed from: 5-Trifluoromethyl-2-benzothiazole-2-carboxylic acid

5-Trifluoromethyl-2-benzothiazole-2-carboxylic acid
Formed from: Zopolrestat
→ 5-Trifluoromethyl-2-benzothiazole
Probably in: Rat J457

***o*-Trifluoromethylbenzoyl-β-D-glucuronic acid**
Formed from: *o*-Trifluoromethylbenzoic acid

***m*-Trifluoromethylbenzoyl-β-D-glucuronic acid**
Formed from: *m*-Trifluoromethylbenzoic acid

***p*-Trifluoromethylbenzoyl-β-D-glucuronic acid**
Formed from: *p*-Trifluoromethylbenzoic acid

***o*-Trifluoromethylbenzyl alcohol**
Formed from: *o*-Trifluoromethylbenzaldehyde

***m*-Trifluoromethylbenzyl alcohol**
Formed from: *m*-Trifluoromethylbenzaldehyde

***p*-Trifluoromethylbenzyl alcohol**
Formed from: *p*-Trifluoromethylbenzaldehyde
p-Trifluoromethyltoluene

3-Trifluoromethylcatechol
Formed from: *m*-Fluorotrifluoromethylbenzene
Probably formed from: *cis*-1,2-Dihydro-1,2-dihydroxy-3-trifluoromethylbenzene

4-Trifluoromethylcatechol
Formed from: 2-Nitro-4-trifluoromethylphenol
Probably formed from: *m*-Trifluoromethylphenol
→ 2-Methoxy-4-trifluoromethylphenol
In: Rat A486
With catechol O-methyltransferase A597
→ 2-Methoxy-5-trifluoromethylphenol
In: Rat A486
With catechol O-methyltransferase A597

***m*-Trifluoromethylhippuric acid**
Formed from: *m*-Trifluoromethylbenzoic acid

***p*-Trifluoromethylhippuric acid**
Formed from: *p*-Trifluoromethylbenzoic acid

***m*-Trifluoromethylmandelic acid**
Formed from: Tetrahydro-2-(3-trifluoromethylphenyl)-1,4-oxazine
Probably formed from: α-Methoxy-*m*-trifluoromethylphenylacetic acid
→ *m*-Trifluoromethylbenzoic acid
Probably in: Dog A631
Monkey A631

***o*-Trifluoromethylphenol**
→ *o*-Trifluoromethylphenyl-β-D-glucuronide
In: Rat H468, H522
→ *o*-Trifluoromethylphenyl sulphate
In: Man E575
Rat E575, H468, H522

***m*-Trifluoromethylphenol**
Formed from: 3-(*m*-Trifluoromethylphenoxy)-1,2-propanediol 1-carbamate
m-Trifluoromethylanisole
→ *m*-Trifluoromethylcatechol
Probably in: Pseudomonas A2391
→ *m*-Trifluoromethylphenyl-β-D-glucuronide
In: Rat H468, H522
→ *m*-Trifluoromethylphenyl sulphate
In: Rat H468, H522

***p*-Trifluoromethylphenol**
→ *p*-Trifluoromethylphenyl-β-D-glucuronide
In: Rat H468, H522
→ *p*-Trifluoromethylphenyl sulphate
In: Man E575
Rat E575, H468, H522

2-Trifluoromethylphenothiazine
Formed from: Flurazicine

N-(γ-(2-Trifluoromethyl-10-phenothiazinyl)propyl) ethylenediamine
Formed from: Fluphenazine
1-(3-(4-Methylpiperazin-1-yl)propyl) trifluoromethylphenothiazine

2-(*m*-Trifluoromethylphenoxy)acetic acid
Probably formed from: 3-(*m*-Trifluoromethylphenoxy)lactic acid

3-(*m*-Trifluoromethylphenoxy)lactic acid
Formed from: 3-(*m*-Trifluoromethylphenoxy)-1,2-propanediol 1-carbamate
→ 2-(*m*-Trifluoromethylphenoxy)acetic acid
Probably in: Man A386

3-(*m*-Trifluoromethylphenoxy)-1,2-propanediol 1-carbamate
→ 3-(*m*-Trifluoromethylphenoxy)lactic acid
In: Man A386
→ 3-(*m*-Trifluoromethylphenoxy)-1,2-propanediol 1-carbamate-β-D-glucuronide
In: Man A386

3-(*m*-Trifluoromethylphenoxy)-1,2-propanediol 1-carbamate-β-D-glucuronide
Formed from: 3-(*m*-Trifluoromethylphenoxy)-1,2-propanediol 1-carbamate

1-(*m*-Trifluoromethylphenyl)acetone
Formed from: Dexfenfluramine
Probably formed from: N-Ethyl-α-methyl-3-trifluoromethylphenethylamine
Norfenfluramine
→ 1-(*m*-Trifluoromethylphenyl)-2-propanol
Probably in: Guinea pig A1152
Man C410
Rat J325

2-((3-(Trifluoromethyl)phenyl)amino)benzoic acid
See: Flufenamic acid

2-((3-(Trifluoromethyl)phenyl)amino)benzoic acid 2-(2-hydroxyethoxy)ethyl ester
See: Etofenamate

***o*-Trifluoromethylphenyl-β-D-glucuronide**
Formed from: *o*-Trifluoromethylphenol

***m*-Trifluoromethylphenyl-β-D-glucuronide**
Formed from: *m*-Trifluoromethylphenol

***p*-Trifluoromethylphenyl-β-D-glucuronide**
Formed from: *p*-Trifluoromethylphenol

***p*-Trifluoromethylphenylhydroxylamine**
Formed from: *p*-Trifluoromethylaniline

1-(*m*-Trifluoromethylphenyl)piperazine
Formed from: Antrafenine

2-(4-(*m*-Trifluoromethylphenyl)piperazinyl) ethanol
Formed from: Antrafenine

2-(4-(3-Trifluoromethylphenyl)piperazinyl)ethyl 2-(7-trifluoromethyl-4-quinolinylamino) benzoate
See: Antrafenine

1-(*m*-Trifluoromethylphenyl)-1,2-propanediol
Probably formed from: 1-(*m*-Trifluoromethylphenyl)-2-propanol
→ 1-(*m*-Trifluoromethylphenyl)-1,2-propanediol-β-D-glucuronide
Probably in: Man G494

1-(*m*-Trifluoromethylphenyl)-1,2-propanediol-β-D-glucuronide
Probably formed from: 1-(*m*-Trifluoromethylphenyl)-1,2-propanediol

1-(*m*-Trifluoromethylphenyl)-2-propanol
Probably formed from: 1-(*m*-Trifluoromethylphenyl)acetone
→ 1-(*m*-Trifluoromethylphenyl)-1,2-propanediol
Probably in: Man C410, G126, G494
Rat J325

***o*-Trifluoromethylphenyl sulphate**
Formed from: *o*-Trifluoromethylphenol

***m*-Trifluoromethylphenyl sulphate**
Formed from: *m*-Trifluoromethylphenol

***p*-Trifluoromethylphenyl sulphate**
Formed from: *p*-Trifluoromethylphenol

N-(7-Trifluoromethyl-4-quinolinyl)anthranilic acid
Formed from: Antrafenine

N-(8-Trifluoromethyl-4-quinolinyl)anthranilic acid
Formed from: Floctafenin

4-(3-(2-(Trifluoromethyl)-9*H*-thioxanthen-9-ylidene)propyl)-1-piperazineethanol
See: Flupentixol

***p*-Trifluoromethyltoluene**
→ *p*-Trifluoromethylbenzyl alcohol
In: Mortierella F823

4-Trifluoromethylveratrole (1,2-dimethoxy-4-(trifluoromethyl)benzene)
→ 2-Methoxy-5-trifluoromethylphenol
Probably in: Sporomusa J199

2,3,4-Trifluorophenol
→ 3,4,5-Trifluorocatechol
In: Trichosporon H420

2,3,5-Trifluorophenol
→ 3,5-Difluorocatechol
In: Trichosporon H420
→ 3,4,6-Trifluorocatechol
In: Trichosporon H420

2,3,6-Trifluorophenol
→ 3,4-Difluorocatechol
In: Trichosporon H420
→ 3,6-Difluorocatechol
In: Trichosporon H420

3,4,5-Trifluorophenol
→ 3,4,5-Trifluorocatechol
In: Trichosporon H420

2,2,2-Trifluoro-1-phenylethanol
Formed from: 2,2,2-Trifluoroacetophenone

Triflupromazine
See: 10-(3-Dimethylaminopropyl)-2-trifluoromethylphenothiazine
Triflupromazine-N-β-D-glucuronide
Formed from: 10-(3-Dimethylaminopropyl)-2-trifluoromethylphenothiazine
1,2,6-Tri-O-galloyl-β-D-glucose
Formed from: 1,6-Di-O-galloyl-β-D-glucose
Galloyl-β-D-glucose
Triglochinin ((2*Z*,4*E*)-4-(cyano(β-D-glucopyranosyloxy)methylene)-2-hexenedioic acid)
Formed from: 3,4-Dihydroxymandelonitrile
p-Hydroxyphenylacetonitrile
ω,3′,4′-Trihydroxyacetophenone
Formed from: 4-(1,2-Dihydroxyethyl)-*o*-quinone
3,4-Dihydroxyphenylglycol
2′,4′,5′-Trihydroxyacetophenone
Probably formed from: 2′,4′-Dihydroxyacetophenone
→ 2′,4′,5′-Trihydroxyacetophenone-5′-sulphate
Probably in: Rat K209
2′,4′,5′-Trihydroxyacetophenone-5′-sulphate
Probably formed from: 2′,4′,5′-Trihydroxyacetophenone
2,10,11-Trihydroxyamitriptyline
Probably formed from: 2,10-Dihydroxyamitriptyline
2,4,5-Trihydroxyamphetamine
Probably formed from: 2,4,5-Trihydroxy-N-methylamphetamine
1,8,9-Trihydroxyanthracene-1,8-diacetate
→ Chrysazin diacetate
In: Mouse D286
→ 1,1′,8,8′-Tetraacetoxy-10,10′-bisanthrone
In: Mouse D286
1,2,7-Trihydroxyanthraquinone
See: Anthrapurpurin
2,4,5-Trihydroxybenzaldehyde
Probably formed from: 2,4-Dihydroxybenzaldehyde
3,4,5-Trihydroxybenzaldehyde
→ Gallic acid
In: Burkholderia K107
1,2,3-Trihydroxybenzene
See: Pyrogallol
1,2,4-Trihydroxybenzene
See: Hydroxyquinol
1,3,5-Trihydroxybenzene
See: Phloroglucinol
2,3,4-Trihydroxybenzoic acid
Formed from: 2,4-Dihydroxybenzoic acid
Pyrocatechuic acid
→ 3-Carboxy-2-hydroxy-*cis*,*cis*-muconic acid
Probably in: Pseudomonas A2694
→ Pyrogallol
In: Pelobacter E189
→ 1,2,3,4-Tetrahydroxybenzene
In: Sporotrichum B73, C57
→ 2,3,4-Trihydroxybenzoyladenylic acid
In: Escherichia K830
2,3,5-Trihydroxybenzoic acid
Formed from: 3,5-Dihydroxybenzoic acid
→ ε-Carboxy-α,γ-dihydroxy-*cis*,*cis*-muconic semialdehyde
Probably in: Bacillus A3735
→ Hydroxyquinol
In: Trichosporon B790
2,3,6-Trihydroxybenzoic acid
Formed from: Pyrocatechuic acid
→ Hydroxyquinol
In: Trichosporon B790
2,4,5-Trihydroxybenzoic acid
Formed from: 2,4-Dihydroxybenzoic acid
Gentisic acid
Protocatechuic acid
2,4,6-Trihydroxybenzoic acid (phloroglucinolcarboxylic acid)
Formed from: (+)-Catechin
Genistein
Kaempferol
Luteolin
Probably formed from: 3-Carboxy-2,4,6-trihydroxyphenylacetic acid
2,4,6-Trihydroxybenzoylformic acid
→ 4-Carboxy-3,5-dihydroxyphenyl sulphate
Probably in: Eubacterium H439
Haemophilus H439
Klebsiella H439
→ Phloroglucinol

In: Bradyrhizobium J484
Pelobacter E189
See also: A3306, A3777, E742, G292, H280
→ 1,2,3,5-Tetrahydroxybenzene
In: Sporotrichum C57
See also: A3777
→ 2,3,4,6-Tetrahydroxybenzoic acid
Probably in: Brassica H878

3,4,5-Trihydroxybenzoic acid
See: Gallic acid

3,4,5-Trihydroxybenzonitrile
Probably formed from: 3-Bromo-4,5-dihydroxybenzonitrile

2,2′,4-Trihydroxybenzophenone
Probably formed from: 2,2′-Dihydroxy-4-methoxybenzophenone

2,3′,4-Trihydroxybenzophenone
Probably formed from: 2,4-Dihydroxybenzophenone

2,4,6-Trihydroxybenzophenone
Formed from: Benzoyl CoA
→ 2,3,4,6-Tetrahydroxybenzophenone
In: Hypericum J509

2,3,4-Trihydroxybenzoyladenylic acid
Formed from: 2,3,4-Trihydroxybenzoic acid

2,4,6-Trihydroxybenzoylformic acid
Formed from: Naringenin
→ 2,4,6-Trihydroxybenzoic acid
Probably in: Horseradish A3777

1-(2,3,4-Trihydroxybenzyl)hydrazine
→ 1-(2,4-Dihydroxy-3-methoxybenzyl)hydrazine
Probably in: Pig B117

3,3′,5-Trihydroxybibenzyl
Formed from: *m*-Hydroxyphenylpropionyl CoA
→ Batatasin III
In: Dioscorea D136
→ Orchinol
Orchis C734

3,4,4′-Trihydroxybibenzyl
Probably formed from: 3,4,4′-Trihydroxy-*trans*-stilbene
→ 3,3′,4,4′-Tetrahydroxybibenzyl
Probably in: Rabbit A1991
Rat A1991

3,4′,5-Trihydroxybibenzyl
Formed from: *p*-Hydroxyphenylpropionyl CoA

2,2′,3-Trihydroxybiphenyl
Formed from: Dibenzofuran
2,2′-Dihydroxybiphenyl
→ 2-Hydroxy-6-(*o*-hydroxyphenyl)-6-oxomuconic acid
In: Pseudomonas H22
→ 2,2′,3,3′-Tetrahydroxybiphenyl
In: Pseudomonas H22
See also: E960

2,3,4′-Trihydroxybiphenyl
Formed from: 3,4′-Dihydroxybiphenyl

2,3,5-Trihydroxybiphenyl
Formed from: 2,5-Dihydroxybiphenyl

3,4,4′-Trihydroxybiphenyl
Probably formed from: 3,4-Dihydroxybiphenyl
3,4′-Dihydroxybiphenyl
→ 3,4′-Dihydroxy-4-methoxybiphenyl
Probably in: Rabbit A2911
Rat A3808
→ 4,4′-Dihydroxy-3-methoxybiphenyl
Probably in: Mouse B615
Rabbit A2911
Rat A3808, B615

2″,6,7-Trihydroxycannabidiol
Probably formed from: 6,7-Dihydroxycannabidiol

3″,6,7-Trihydroxycannabidiol
Probably formed from: 3″,7-Dihydroxycannabidiol

4″,6,7-Trihydroxycannabidiol
Probably formed from: 4″,7-Dihydroxycannabidiol
6,7-Dihydroxycannabidiol

5″,6,7-Trihydroxycannabidiol
Probably formed from: 5″,7-Dihydroxycannabidiol
6,7-Dihydroxycannabidiol
→ 6,7-Dihydroxycannabidiol-5″-oic acid
Probably in: Dog F557

2′,3,4-Trihydroxychalcone
→ 2′,4-Dihydroxy-3-methoxychalcone
In: Rat C665

2′,4,4-Trihydroxychalcone
Formed from: Morin

2′,4,4′-Trihydroxychalcone (isoliquiritigenin)
Formed from: *p*-Coumaroyl CoA
Daidzein
L-Phenylalanine
→ 4′,6-Dihydroxy-2-(α-hydroxybenzyl) coumaranone
In: Cicer A1729
→ 4,4′-Dihydroxy-2′-methoxychalcone
In: Alfalfa G610, H908
Glycyrrhiza K886
→ Formononetin
Probably in: Medicago A3725
Onobrychis A2742
→ Garbanzol
In: Cicer A1729
→ Puerarin
Pueraria A2373
→ 2′,4,4′-Trihydroxychalcone-1,2-dioxetane
In: Cicer A1728, A1729
Horseradish A1728, A1729

2′,4′,6′-Trihydroxychalcone
Formed from: Cinnamoyl CoA
→ Butein
In: Dahlia J626
→ Pinocembrin
In: Cephalocereus H633

2′,4,4′-Trihydroxychalcone-1,2-dioxetane
Formed from: 2′,4,4′-Trihydroxychalcone

2′,4,4′-Trihydroxychalcone-4′-glucoside
→ 2′,4,4′-Trihydroxychalcone-4′-glucoside-6″-malonate
In: Cicer D445

2′,4,4′-Trihydroxychalcone-4′-glucoside-6″-malonate
Formed from: 2′,4,4′-Trihydroxychalcone-4′-glucoside

α,6,8-Trihydroxycinmethylin
Probably formed from: α,8-Dihydroxycinmethylin
→ 2-((2-Carboxyphenyl)methoxy)-6-hydroxy-4-(1-hydroxy-1-methylethyl)-1-methyl-7-oxabicyclo[2.2.1]heptane
Probably in: Goat F98

α,6,9-Trihydroxycinmethylin
Probably formed from: α,9-Dihydroxycinmethylin

Trihydroxycinmetropium
Probably formed from: Dihydroxycinmetropium

3,4,5-Trihydroxycinnamic acid
Probably formed from: Sinapic acid
→ 5-Methoxycaffeic acid
In: Nicotiana A3757
See also: F367

4,6,7-Trihydroxy-5-coumaranoyl-β-acrylic acid
Probably formed from: 8-Hydroxypsoralen

4′,5,7-Trihydroxy-3,3′-dimethoxyflavan (3,3′-di-O-methylcatechin)
Formed from: 3′,4′,5,7-Tetrahydroxy-3-methoxyflavan
→ 4′,5,7-Trihydroxy-3,3′-dimethoxyflavan-β-D-glucuronide
Probably in: Man D629
Rat C587
→ 4′,5,7-Trihydroxy-3,3′-dimethoxyflavan sulphate
Probably in: Rat C587

4′,5,7-Trihydroxy-3,3′-dimethoxyflavan-β-D-glucuronide
Probably formed from: 4′,5,7-Trihydroxy-3,3′-dimethoxyflavan

4′,5,7-Trihydroxy-3,3′-dimethoxyflavan sulphate
Probably formed from: 4′,5,7-Trihydroxy-3,3′-dimethoxyflavan

3,5,5′-Trihydroxy-4′,7-dimethoxyflavone
→ 5-Hydroxy-3,3′,4′,7-tetramethoxyflavone
In: Calamondin orange A3987

3,4′,5-Trihydroxy-3′,7-dimethoxyflavone
Probably formed from: Rhamnetin
→ 3,4′,5-Trihydroxy-3′,7-dimethoxyflavone-3-O-(2″-O-β-D-glucopyranosyl)-β-D-galactopyranoside
In: Petunia H658

3′,4′,5-Trihydroxy-3,7-dimethoxyflavone
Probably formed from: Rhamnetin
3′,4′,5,7-Tetrahydroxy-3-methoxyflavone
→ 3′,5-Dihydroxy-3,4′,7-trimethoxyflavone
In: Chrysosplenium C792, D584, D589
See also: D258
→ 4′,5-Dihydroxy-3,3′,7-trimethoxyflavone
Probably in: Chrysosplenium C579

3′,5,5′-Trihydroxy-4′,7-dimethoxyflavone
→ 3′,5,5′-Trihydroxy-4′,7-dimethoxyflavone-5′-glucoside
In: Chrysosplenium C755

3,5,7-Trihydroxy-3′,4′-dimethoxyflavone
→ 3,5-Dihydroxy-3′,4′,7-trimethoxyflavone
In: Calamondin orange A3987
→ 5,7-Dihydroxy-3,3′,4′-trimethoxyflavone
In: Calamondin orange A3987

4,6,7-Trihydroxy-5,8-dimethoxyflavone
Formed from: 4,6-Dihydroxy-5,7,8-trimethoxyflavone

3,4′,5-Trihydroxy-3′,7-dimethoxyflavone-3-O-(2″-O-β-D-glucopyranosyl)-β-D-galactopyranoside
Formed from: 3,4′,5-Trihydroxy-3′,7-dimethoxyflavone

3′,5,5′-Trihydroxy-4′,7-dimethoxyflavone-5′-glucoside
Formed from: 3′,5,5′-Trihydroxy-4′,7-dimethoxyflavone

2,2′,3-Trihydroxydiphenyl ether
Formed from: Dibenzo-*p*-dioxin

3′,4′,7-Tri-O-(β-hydroxyethyl)quercetin
Formed from: 3′,4′,7-Tri-O-(β-hydroxyethyl)rutin

3′,4′,7-Tri-O-(β-hydroxyethyl)rutin
Probably formed from: 4′,7-Di-O-(β-hydroxyethyl)rutin
→ 3′,4′,5,7-Tetra-O-(β-hydroxyethyl)rutin
In: Rat A271
See also: A271
→ 3′,4′,7-Tri-O-(β-hydroxyethyl)quercetin
In: Mouse A2358
Rat A271
→ 3′,4′,7-Tri-O-(β-hydroxyethyl)rutin-β-D-glucuronide
Probably in: Rat A271

3′,4′,7-Tri-O-(β-hydroxyethyl)rutin-β-D-glucuronide
Formed from: 3′,4′,7-Tri-O-(β-hydroxyethyl)rutin

2,4′,7-Trihydroxyflavanone
→ Formononetin
In: Glycyrrhiza K630

2,5,7-Trihydroxyflavanone
→ 2,5,7-Trihydroxyflavanone-6-glucoside
Probably in: Fagopyrum E614

4′,5,7-Trihydroxyflavanone
See: Naringenin

2,5,7-Trihydroxyflavanone-6-glucoside
Probably formed from: 2,5,7-Trihydroxyflavanone

4′,5,7-Trihydroxyflavanone-7-neohesperidoside
Probably formed from: 2′,4,4′,6′-Trihydroxychalcone-4′-neohesperidoside

3′,4′,5-Trihydroxyflavone
Probably formed from: 4′,5-Dihydroxyflavone

3,4′,7-Trihydroxyflavone
→ 3,4′,7-Trihydroxyflavone-7-β-D-glucoside
In: Parsley K739

4′,5,7-Trihydroxyflavone
See: Apigenin

3,4′,7-Trihydroxyflavone-7-β-D-glucoside
Formed from: 3,4′,7-Trihydroxyflavone

2,4,5-Trihydroxy-6-(S-glutathionyl)phenethylamine-*p*-quinone
Formed from: 2,4,5-Trihydroxyphenethylamine

1,6,8-Trihydroxy-3-hydroxymethylanthraquinone
Formed from: Emodin
→ 3-Formyl-1,6,8-trihydroxyanthraquinone
Probably in: Mouse E151, E273
Rat E273

2′,4′,6′-Trihydroxy-3-(*p*-hydroxyphenyl)propiophenone
See: Phloretin

2,4′,7-Trihydroxyisoflavanone
Formed from: Liquiritigenin
→ Daidzein
In: Pueraria F845

2′,4′,7-Trihydroxyisoflavanone
Formed from: 2′,4′,7-Trihydroxyisoflavone

4′,5,7-Trihydroxyisoflavanone
Formed from: Genistein
→ 5,6-Dihydro-3-(*p*-hydroxyphenyl)-4,6-dioxo-4*H*-pyran
In: Fusarium C474
→ 4′,5,7-Trihydroxyisoflavanone-β-D-glucuronide
Probably in: Rat K363

4′,5,7-Trihydroxyisoflavanone-β-D-glucuronide
Probably formed from: 4′,5,7-Trihydroxyisoflavanone

2′,4′,7-Trihydroxyisoflavone
→ 2′,4′,7-Trihydroxyisoflavanone
In: Glycine K702

3′,4′,7-Trihydroxyisoflavone
Probably formed from: Daidzein

4′,6,7-Trihydroxyisoflavone
Formed from: 4′,7-Dihydroxy-6-methoxyisoflavone
→ 4′,6-Dihydroxy-7-methoxyisoflavone
In: Alfalfa G322
→ 4′,7-Dihydroxy-6-methoxyisoflavone
In: Calamondin orange A3987
→ 6,7-Dihydroxy-4′-methoxyisoflavone
Probably in: Alfalfa H875
Calamondin orange A3987

4′,7,8-Trihydroxyisoflavone
Probably formed from: Daidzein

1,3,5-Trihydroxy-2-methoxybenzene
Formed from: 3,5-Dihydroxy-4-methoxybenzoic acid

2,3,10-Trihydroxy-11-methoxyberbine
Formed from: 2,3,10,11-Tetrahydroxyberbine

2,3,11-Trihydroxy-10-methoxyberbine
Formed from: 2,3,10,11-Tetrahydroxyberbine

2,10,11-Trihydroxy-3-methoxyberbine
Formed from: 2,3,10,11-Tetrahydroxyberbine

3,10,11-Trihydroxy-2-methoxyberbine
Formed from: 2,3,10,11-Tetrahydroxyberbine

2,3,4′-Trihydroxy-3′-methoxybiphenyl
Formed from: Catechol
Guaiacol

1,3,6-Trihydroxy-7-methoxy-2,8-bis(3-methyl-2-butenyl)-9*H*-xanthen-9-one
See: Mangostin

2′,6,6′-Trihydroxy-2-methoxycarbonyl-4′-methylbenzophenone
See: Moniliphenone

2′,3,4′-Trihydroxy-4-methoxychalcone
→ 2′,3,4,4′-Tetrahydroxychalcone
In: Microorganism C665
Rat C665

2′,4,4′-Trihydroxy-3-methoxychalcone
Formed from: 2′,3,4,4′-Tetrahydroxychalcone
→ 2′,3,4,4′-Tetrahydroxychalcone
In: Microorganism C665
Rat C665

2′,4,4′-Trihydroxy-6′-methoxychalcone
→ Eucomin
In: Eucomis A228

2′,4,4′-Trihydroxy-6′-methoxychalcone-4′-neohesperidoside
→ 4′,7-Dihydroxy-5-methoxyflavanone-7-neohesperidoside
Probably in: Grapefruit A2518

4′,5,7-Trihydroxy-3-methoxy-6-(3,3-dimethylallyl)flavone
See: Topazolin

3′,4′,7-Trihydroxy-5-methoxyflavanone
Formed from: 4′,7-Dihydroxy-5-methoxyflavanone

3′,5,7-Trihydroxy-4′-methoxyflavanone
See: Hesperetin

4′,5,7-Trihydroxy-3′-methoxyflavanone
See: Homoeriodictyol

3,4′,5-Trihydroxy-3′-methoxyflavone
See: Rhamnazin

3,4′,5-Trihydroxy-7-methoxyflavone
See: Rhamnocitrin

3′,4′,5-Trihydroxy-7-methoxyflavone
Formed from: Luteolin

3,4′,7-Trihydroxy-3′-methoxyflavone
Formed from: Fisetin

3,5,7-Trihydroxy-4′-methoxyflavone
See: Kaempferide

3′,5,7-Trihydroxy-4′-methoxyflavone
See: Diosmetin

4′,5,7-Trihydroxy-3-methoxyflavone
Formed from: Kaempferol

4′,5,7-Trihydroxy-3′-methoxyflavone
See: Chrysoeriol

4′,5,7-Trihydroxy-6-methoxyflavone
Formed from: Scutellarein

4′,5,7-Trihydroxy-8-methoxyflavone
Formed from: 4′,5,7,8-Tetrahydroxyflavone

5,6,7-Trihydroxy-4′-methoxyflavone
Formed from: Scutellarein

5,7,8-Trihydroxy-4′-methoxyflavone
Formed from: 4′,5,7,8-Tetrahydroxyflavone
→ 5,7-Dihydroxy-4′,8-dimethoxyflavone
Probably in: Calamondin orange A3987
→ 5,8-Dihydroxy-4′,7-dimethoxyflavone
Probably in: Calamondin orange A3987

3,4′,5-Trihydroxy-3′-methoxyflavone disulphate
See: Veronicafolin-3,4′-disulphate

4′,5,7-Trihydroxy-3′-methoxyflavone-7-O-glucoside
Formed from: Luteolin-7-O-glucoside

2′,5′,7-Trihydroxy-4′-methoxyisoflavanone
See: Vestitone

3′,5,7-Trihydroxy-4′-methoxyisoflavanone
See: Dihydropratensein

2′,4′,7-Trihydroxy-5-methoxyisoflavone
Formed from: 2′,4′,5,7-Tetrahydroxyisoflavone

2′,5,7-Trihydroxy-4′-methoxyisoflavone
Formed from: Biochanin A

3′,5,7-Trihydroxy-4′-methoxyisoflavone
See: Pratensein

5,6,7-Trihydroxy-4′-methoxyisoflavone
Formed from: Biochanin A

3,5,7-Trihydroxy-2-(4-methoxyphenyl)-8-(3-methyl-2-butenyl)-4*H*-1-benzopyran-4-one
See: Icaritin

2′,4′,7-Trihydroxy-5-methoxy-3′-prenylisoflavone
Formed from: 2′,4′,5,7-Trihydroxy-3′-prenylisoflavone

2,4,5-Trihydroxy-N-methylamphetamine
Probably formed from: 3,4-Dihydroxy-N-methylamphetamine
→ 2,4,5-Trihydroxyamphetamine
Probably in: Rat G354

1,4,5-Trihydroxy-2-methyl-9,10-anthracenedione
See: Islandicin

1,3,8-Trihydroxy-6-methyl-9(10*H*)-anthracenone
See: Emodin anthrone

1,3,8-Trihydroxy-6-methylanthraquinone
See: Emodin

3,4′,5-Trihydroxy-6′-methyldibenzo-α-pyrone
See: Alternariol

3,4,5-Trihydroxy-N-methylphenethylamine
→ 3,5-Dihydroxy-4-methoxy-N-methylphenethylamine
In: Lophophora A3600

1,2,4-Trihydroxynaphthalene
Probably formed from: 1,2-Dihydroxynaphthalene
→ 1,2,3,4-Tetrahydroxynaphthalene
Probably in: Rat B602

1,3,8-Trihydroxynaphthalene
→ Vermelone
In: Magnaporthe K704
Pyricularia K703
Verticillium K593

1,4,5-Trihydroxynaphthalene
Formed from: 5-Hydroxy-1,4-naphthoquinone
→ 4,8-Dihydroxy-1-naphthalenyl-β-D-glucoside
Probably in: Juglans A3470

2,5,7-Trihydroxy-1,4-naphthalenedione
See: Flaviolin

2,5,8-Trihydroxy-1,4-naphthoquinone
Formed from: 5,8-Dihydroxy-1,4-naphthoquinone

2,4,5-Trihydroxyphenethylamine
→ 2,4-Dihydroxy-5-methoxyphenethylamine
With catechol O-methyltransferase A2070
→ 4-Hydroxy-5-(2-aminoethyl)-*o*-quinone
Probably in: Man A63
→ 2,4,5-Trihydroxy-6-(S-glutathionyl) phenethylamine-*p*-quinone
In: Mouse A2898
Rat A2898
→ 2,4,5-Trihydroxyphenethylamine-β-D-glucuronide
In: Rat K595
→ 2,4,5-Trihydroxyphenethylamine-O-sulphate
In: Man D583

3,4,5-Trihydroxyphenethylamine
→ 3-Hydroxy-5-(2-aminoethyl)-*o*-quinone
Probably in: Man A63

2,4,5-Trihydroxyphenethylamine-β-D-glucuronide
Formed from: 2,4,5-Trihydroxyphenethylamine

2,4,5-Trihydroxyphenethylamine-O-sulphate
Formed from: 2,4,5-Trihydroxyphenethylamine

2,3,5-Trihydroxyphenylacetic acid
Formed from: 6-(Carboxypentyl)-2,4-dihydroxybenzoic acid

2,4,5-Trihydroxyphenylacetic acid
Formed from: 3,4-Dihydroxyphenylacetic acid
3,4-Dihydroxyphenylpyruvic acid
Probably formed from: Homogentisic acid

3,4,5-Trihydroxyphenylacetic acid
Probably formed from: 3,4-Dihydroxyphenyl-acetic acid
2,3,4-Trihydroxy-L-phenylalanine
Formed from: Pyrogallol
2,4,5-Trihydroxy-L-phenylalanine
(6-hydroxydopa)
Probably formed from: 2,4-Dihydroxy-L-phenylalanine
3,4-Dihydroxy-L-phenylalanine
3,4,5-Trihydroxy-L-phenylalanine
Formed from: 3,4-Dihydroxy-L-phenylalanine
4-(2,4,6-Trihydroxyphenyl)flavan-3-ol
→ *p*-Hydroxyphenylacetic acid
In: Microorganism D654
→ *p*-Hydroxyphenyllactic acid
In: Microorganism D654
→ *m*-Hydroxyphenylpropionic acid
In: Microorganism D654
→ 1-(*m*-Hydroxyphenyl)-3-(2,4,6-trihydroxyphenyl)propan-2-ol
In: Microorganism D654
→ Melilotic acid
In: Microorganism D654
→ Phloretic acid
In: Microorganism D654
→ Phloroglucinol
In: Microorganism D654
3,4,5-Trihydroxyphenylpropionic acid
Formed from: 3,4-Dihydroxy-5-methoxyphenylpropionic acid
3,5-Dihydroxy-4-methoxyphenylpropionic acid
→ 3,4-Dihydroxy-5-methoxyphenylpropionic acid
In: Rat A745
→ 3,5-Dihydroxy-4-methoxyphenylpropionic acid
In: Rat A745
→ 3,5-Dihydroxyphenylpropionic acid
In: Microorganism A744
Rat A745
3,6α,9-Trihydroxypterocarpan
Formed from: 3,9-Dihydroxypterocarpan
→ 2-Dimethylallyl-3,6α,9-trihydroxypterocarpan
In: Glycine K905
→ 4-Dimethylallyl-3,6α,9-trihydroxypterocarpan
In: Glycine K904, K905
1,4,5-Trihydroxypyrene
Probably formed from: 1-Hydroxypyrene
3,3′,5-Trihydroxystilbene
Formed from: *m*-Coumaroyl CoA
3,4,4′-Trihydroxy-*trans*-stilbene
Probably formed from: 3,4-Dihydroxy-*trans*-stilbene
4,4′-Dihydroxy-*trans*-stilbene
→ 3,4,4′-Trihydroxybibenzyl
Probably in: Rabbit A1991
Rat A1991
3,4′,5-Trihydroxystilbene
See: Resveratrol
3,4,4″-Trihydroxy-*m*-terphenyl
Formed from: *m*-Terphenyl
→ 3,4,4″-Trihydroxy-*m*-terphenyl
Probably in: Aspergillus H23
2,4,8-Trihydroxy-1-tetralone
Formed from: 3,5-Dihydroxy-1,4-naphthoquinone
→ 4,8-Dihydroxy-1-tetralone
Probably in: Wangiella D760
3,4,8-Trihydroxy-1-tetralone
Formed from: 2,5-Dihydroxy-1,4-naphthoquinone
3,6,8-Trihydroxy-1-tetralone (scytalone, (3*R*)-3,4-dihydro-3,6,8-trihydroxy-1(2*H*)-naphthalenone)
Formed from: 1,3,6,8-Tetrahydroxynaphthalene
Formed from acetate in Phialophora A1399
2′,5,5′-Trihydroxy-3,4′,6,7-tetramethoxyflavone
→ 2′,5,5′-Trihydroxy-3,4′,6,7-tetramethoxyflavone-2′-glucoside
In: Chrysosplenium C755
→ 2′,5,5′-Trihydroxy-3,4′,6,7-tetramethoxyflavone-5′-glucoside
In: Chrysosplenium C755
2′,5,5′-Trihydroxy-3,4′,6,7-tetramethoxyflavone-2′-glucoside
Formed from: 2′,5,5′-Trihydroxy-3,4′,6,7-tetramethoxyflavone

2′,5,5′-Trihydroxy-3,4′,6,7-tetramethoxyflavone-5′-glucoside
Formed from: 2′,5,5′-Trihydroxy-3,4′,6,7-tetramethoxyflavone

2,3,5-Trihydroxytoluene
Formed from: 3,4-Dihydroxy-5-methylbenzoic acid
Orcinol

2,3,6-Trihydroxytoluene
Formed from: 2-Methylresorcinol

2,4,5-Trihydroxytoluene
Formed from: 2-Hydroxy-5-methylquinone
→ 2,4-Dihydroxy-5-methyl-6-oxo-2,4-hexadienoic acid
In: Burkholderia J903
See also: H238, K445

2′,5,5′-Trihydroxy-3,4′,7-trimethoxyflavone
→ 2′,5,5′-Trihydroxy-3,4′,7-trimethoxyflavone-2′-glucoside
In: Chrysosplenium C755
→ 2′,5,5′-Trihydroxy-3,4′,7-trimethoxyflavone-5′-glucoside
In: Chrysosplenium C755

3′,4′,5-Trihydroxy-3,6,7-trimethoxyflavone
Formed from: 3′,4′,5,6-Tetrahydroxy-3,7-dimethoxyflavone

3′,5,6-Trihydroxy-3,4′,7-trimethoxyflavone
Formed from: 3′,5-Dihydroxy-3,4′,7-trimethoxyflavone

4′,5,6-Trihydroxy-3,3′,7-trimethoxyflavone
Formed from: Quercetagetin
3′,4′,5,6-Tetrahydroxy-3,7-dimethoxyflavone
→ 4′,5-Dihydroxy-3,3′,6,7-tetramethoxyflavone
In: Chrysosplenium B832, D584, D589

5,5′,6-Trihydroxy-3,4′,7-trimethoxyflavone
→ 5,5′,6-Trihydroxy-3,4′,7-trimethoxyflavone-5′-glucoside
In: Chrysosplenium C755

2′,5,5′-Trihydroxy-3,4′,7-trimethoxyflavone-2′-glucoside
Formed from: 2′,5,5′-Trihydroxy-3,4′,7-trimethoxyflavone

2′,5,5′-Trihydroxy-3,4′,7-trimethoxyflavone-5′-glucoside
Formed from: 2′,5,5′-Trihydroxy-3,4′,7-trimethoxyflavone

5,5′,6-Trihydroxy-3,4′,7-trimethoxyflavone-5′-glucoside
Formed from: 5,5′,6-Trihydroxy-3,4′,7-trimethoxyflavone

2,4,6-Triiodophenol
→ 2,6-Diiodoquinol
In: Flavobacterium G641

3,3′,5-Triiodothyroacetic acid
→ 3,3′-Diiodothyroacetic acid
In: Rat F218, G685
→ 3,3′,5-Triiodothyroacetic acid-4′-sulphate
In: Monkey B770
Rat B770

3,3′,5-Triiodothyroacetic acid-4′-sulphate
Formed from: 3,3′,5-Triiodothyroacetic acid
→ 3,3′-Diiodothyroacetic acid-4′-sulphate
In: Rat F218

3,3′,5-Triiodo-L-thyronine (T_3)
Formed from: 3,5-Diiodo-L-thyronine
L-Thyroxine
3,3′,5-Triiodo-L-thyronine-β-D-glucuronide
→ 3,3′-Diiodo-L-thyronine
In: Dog H882
Guinea pig D910
Man H882
Oreochromis H923
Rabbit F373
Rat A3549, A3752, B126, B721, C109, D930, F365, G337, G614, J197
See also: F807, K242
→ 3,5-Diiodo-L-thyronine
In: Man K105
Rat E185, F821, G614
→ L-Thyronine
Probably in: Rat A1095
→ L-Thyroxine
Probably in: Rat C163
→ 3,3′,5-Triiodo-L-thyronine-β-D-glucuronide
In: Rainbow trout J38, J319
Rat B721, C724, F365, G741
→ 3,3′,5-Triiodo-L-thyronine-O-sulphate
In: Eubacterium E432, G602
Man E672, K90, K620
Monkey B770

Rat B721, B770, C724, E827, J495, K649
→ 3,3′,5-Triiodothyropyruvic acid
In: Rabbit K713

3,3′,5′-Triiodo-L-thyronine (rT_3)
Formed from: L-Thyroxine
→ 3,3′-Diiodo-L-thyronine
In: Dog H882, K639
Man H882
Oreochromis H923
Rat A3549, A3752, B126, B553, B721, B833, B914, B956, C48, C724, D79, D531, D735, E184, E363, G337, K639
Ring dove G170
Sceloporus K395
See also: A3889, K242
→ 3,3′,5′-Triiodo-L-thyronine-β-D-glucuronide
In: Man G524
Rat B721, B770, G741
→ 3,3′,5′-Triiodo-L-thyronine-O-sulphate
In: Man K90, K620
Monkey B771
Rat B771, K649

3,3′,5-Triiodo-L-thyronine-β-D-glucuronide
Formed from: 3,3′,5-Triiodo-L-thyronine
→ 3,3′,5′-Triiodo-L-thyronine
In: Rat G463

3,3′,5′-Triiodo-L-thyronine-β-D-glucuronide
Formed from: 3,3′,5′-Triiodo-L-thyronine

3,3′,5-Triiodo-L-thyronine sulphamate
Formed from: L-Thyroxine sulphamate
→ 3,3′-Diiodo-L-thyronine sulphamate
In: Rat G337

3,3′,5′-Triiodo-L-thyronine sulphamate
Formed from: L-Thyroxine sulphamate
→ 3,3′-Diiodo-L-thyronine sulphamate
In: Rat G337

3,3′,5-Triiodo-L-thyronine-O-sulphate
Formed from: L-Thyroxine-O-sulphate
3,3′,5-Triiodo-L-thyronine
→ 3,3′-Diiodo-L-thyronine-O-sulphate
In: Dog H882
Man H882
Rat C717, F365, G337
→ 3,3′,5-Triiodo-L-thyronine
In: Man E600
Microorganism D594
Rainbow trout K242
Rat E600, G614

3,3′,5′-Triiodo-L-thyronine-O-sulphate
Formed from: L-Thyroxine-O-sulphate
→ 3,3′-Diiodo-L-thyronine-O-sulphate
In: Rat G337

3,3′,5-Triiodothyropropionic acid
→ 3,3′,5-Triiodothyropropionic acid-4′-sulphate
In: Monkey B770
Rat B770

3,3′,5-Triiodothyropropionic acid-4′-sulphate
Formed from: 3,3′,5-Triiodothyropropionic acid

3,3′,5-Triiodothyropyruvic acid
Formed from: 3,3′,5-Triiodo-L-thyronine

Trimebutine (3,4,5-trimethylbenzoic acid 2-(dimethylamino)-2-phenylbutyl ester)
→ 2-Dimethylamino-2-phenylbutan-2-ol
In: Dog F28
Rat F28

Trimetazidine (1-(2,3,4-trimethoxybenzyl) piperazine)
Formed from: 1-(Bis(*p*-Fluorophenyl)methyl)-4-(2,3,4-trimethoxybenzyl)piperazine
→ N-Formyltrimetazidine
In: Man H521
→ 3-Hydroxy-2,4-dimethoxybenzylpiperazine
In: Mammalia A448
See also: H521
→ N-Methyltrimetazidine
In: Man H521
→ 3-Oxo-1-(2,3,4-trimethoxybenzyl) piperazine
In: Man H521
→ 1-(2,3,4-Trimethoxybenzyl)-N^4-acetylpiperazine
In: Mammalia A448
Man H521

Trimetazon (4-(4,4-dimethyl-3-oxopentyl)-3,5-dioxo-1,2-diphenylpyrazolidine)
→ 4-(4,4-Dimethyl-3-oxopentyl)-3,5-dioxo-1-(*p*-hydroxyphenyl)-2-phenylpyrazolidine
In: Rat A2284

→ 4-(4,4-Dimethyl-3-oxopentyl)-4-hydroxy-3,5-dioxo-1,2-diphenylpyrazolidine
In: Rat A2284
→ 4-(3-Hydroxy-4,4-dimethylpentyl)-3,5-dioxo-1,2-diphenylpyrazolidine
In: Rat A2284
→ 4-(4-Hydroxy-4-methyl-3-oxopentyl)-3,5-dioxo-1,2-diphenylpyrazolidine
In: Rat A2284
→ 4-(3-Hydroxy-4-methylpentyl)-3,5-dioxo-1,2-diphenylpyrazolidine
In: Rat A2284

Trimethoprim (2,4-diamino-5-(3,4,5-trimethoxybenzyl)pyrimidine)
→ 3′-Demethyltrimethoprim
In: Beef A2296
Goat A3478
Man A1165
Pig A3478
Contrast (in Pseudemys) F254
→ 4′-Demethyltrimethoprim
In: Man A1165
→ α-Hydroxytrimethoprim
In: Man A1165
→ Trimethoprim-1-oxide
In: Guinea pig E624
Hamster E624
Man A1165
Mouse E624
Rat A1017, E624
→ Trimethoprim-3-oxide
In: Guinea pig E624
Hamster E624
Man A1165
Mouse E624
Rat A1017, E624
Contrast (in Pseudemys) F181, F254
→ Trimethoprim iminoquinone methide
In: Man K235

Trimethoprim iminoquinone methide
Formed from: Trimethoprim

Trimethoprim-1-oxide
Formed from: Trimethoprim

Trimethoprim-3-oxide
Formed from: Trimethoprim

3′,4′,5′-Trimethoxyacetophenone
Formed from: 4′-Hydroxy-3′,5′-dimethoxyacetophenone

3,4,5-Trimethoxyallylbenzene
See: Elemicin

2,3,4-Trimethoxybenzaldehyde
Probably formed from: 1-(Bis(*p*-fluorophenyl) methyl)-4-(2,3,4-trimethoxybenzyl)piperazine

3,4,5-Trimethoxybenzaldehyde
Formed from: 4-Hydroxy-3,5-dimethoxybenzaldehyde
3,4,5-Trimethoxybenzoic acid
Probably formed from: 3,4,5-Trimethoxybenzylamine
→ 3,4,5-Trimethoxybenzoic acid
In: Desulphovibrio F666
Serratia G197
See also: G865
→ 3,4,5-Trimethoxybenzyl alcohol
In: Phanerochaete F866

3,4,5-Trimethoxybenzamide
Probably formed from: N-(3,4,5-Trimethoxybenzoyl)-2-aminoethoxyacetic acid

3-(3,4,5-Trimethoxybenzamido)piperidine
→ 3-(3-Hydroxy-4,5-dimethoxybenzamido) piperidine
In: Dog B357
Guinea pig B357
Mouse B357
Rat B357
→ 3-(4-Hydroxy-3,5-dimethoxybenzamido) piperidine
In: Dog B357
Guinea pig B357
Mouse B357
Rat B357
→ 5-(3,4,5-Trimethoxybenzamido)-2-piperidone
In: Dog B357
Guinea pig B357
Man B357
Mouse B357
Rat B357
→ 3,4,5-Trimethoxybenzoic acid
In: Dog B357
Guinea pig B357
Mouse B357

Rat B357

5-(3,4,5-Trimethoxybenzamido)-2-piperidone
Formed from: 3-(3,4,5-Trimethoxybenzamido) piperidine

1,2,3-Trimethoxybenzene
→ Pyrogallol
In: Clostridium G367

2,4,5-Trimethoxybenzoic acid
→ 4-Methoxygentisic acid
In: Arthrobacter A2773

3,4,5-Trimethoxybenzoic acid
Formed from: Dilazap
4-Hydroxy-3,5-dimethoxybenzoic acid
1-(3-Hydroxypropyl)-4-(3-(3,4,5-trimethoxybenzoyl)propyl)piperazine
Mescaline
Reserpine
3,4,5-Trimethoxybenzaldehyde
3-(3,4,5-Trimethoxybenzamido)piperidine
Trimetozine
Probably formed from: Tetrahydro-4-(3,4,5-trimethoxybenzoyl)-1,4-oxazine
3,4,5-Trimethoxycinnamic acid
→ 4-Hydroxy-3,5-dimethoxybenzoic acid
In: Pseudomonas B501
Vanilla F824
Wheat A709
See also: D16, F655, G931
→ 3,4,5-Trimethoxybenzaldehyde
In: Nocardia G207
With NAD(P)-dependent benzaldehyde oxidoreductase E248
→ 3,4,5-Trimethoxybenzoyl-β-D-glucose
In: Quercus K737
→ 3,4,5-Trimethoxybenzyl alcohol
In: Coriolus G821
See also: D453

N-(3,4,5-Trimethoxybenzoyl)-2-aminoethoxyacetic acid
Formed from: Trimetozine
→ 3,4,5-Trimethoxybenzamide
Probably in: Dog A3302

N-(3,4,5-Trimethoxybenzoyl)diethanolamine
Formed from: Trimetozine
→ N-(2-Hydroxyethyl)-N-(3,4,5-trimethoxybenzoyl)glycine
Probably in: Dog A3302
→ N-(3,4,5-Trimethoxybenzoyl) diethanolamine-β-D-glucuronide
In: Dog A3302
→ N-(3,4,5-Trimethoxybenzoyl)ethanolamine
Probably in: Dog A3302

N-(3,4,5-Trimethoxybenzoyl)diethanolamine-β-D-glucuronide
Formed from: N-(3,4,5-Trimethoxybenzoyl)diethanolamine

N-(3,4,5-Trimethoxybenzoyl)ethanolamine
Probably formed from: N-(3,4,5-Trimethoxybenzoyl)diethanolamine

3,4,5-Trimethoxybenzoyl-β-D-glucose
Formed from: 3,4,5-Trimethoxybenzoic acid

3,4,5-Trimethoxybenzoyl methyl reserpate
See: Reserpine

N-(3,4,5-Trimethoxybenzoyl)morpholine
See: Trimetozine

1-(2,3,4-Trimethoxybenzyl)-N^4-acetylpiperazine
Formed from: Trimetazidine

3,4,5-Trimethoxybenzyl alcohol
Formed from: 3,4,5-Trimethoxybenzaldehyde
3,4,5-Trimethoxybenzoic acid

3,4,5-Trimethoxybenzylamine
→ 3,4,5-Trimethoxybenzaldehyde
Probably in: Rat G865

2′,4,4′-Trimethoxychalcone
→ 2′,4′-Dimethoxyacetophenone
In: Rat A1171
→ 4-Hydroxy-2′,4′-dimethoxychalcone
In: Rat A1171

3,4,5-Trimethoxycinnamic acid
Formed from: 3,4,5-Trimethoxyphenylpropionic acid
Probably formed from: 3-(3,4,5-Trimethoxyphenyl)-2-propen-1-ol
→ 3-Hydroxy-4,5-dimethoxycinnamic acid
In: Rat A745
See also: B198, D16
→ Sinapic acid
Probably in: Microorganism A261
→ 3,4,5-Trimethoxybenzoic acid
Probably in: Vanilla F824
→ 3,4,5-Trimethoxycinnamoyl CoA
In: Glycine A2148
→ 3,4,5-Trimethoxyphenylpropionic acid
In: Microorganism A744

Rat A745
See also: B198
3,4,5-Trimethoxycinnamoyl CoA
Formed from: 3,4,5-Trimethoxycinnamic acid
N-(3,4,5-Trimethoxycinnamoyl)piperazine
Formed from: 1-(N-Isopropylcarbamoylmethyl)-4-(3,4,5-trimethoxycinnamoyl) piperazine
2-(4-(3,4,5-Trimethoxycinnamoyl)-1-piperazinyl) acetamide
Formed from: Cinpropazide
4-(3,4,5-Trimethoxycinnamoyl)-1-piperazinylacetic acid
Formed from: Ethyl 4-(3,4,5-Trimethoxycinnamoyl)-1-piperazinylacetate
4,7,8-Trimethoxyfuro[2,3-*b*]quinoline
See: Skimmianine
2,3,4-Trimethoxymandelic acid
Formed from: N,N-Dimethyl-2-oxo-2-(2,3,4-trimethoxyphenyl)ethylamine
2,3,4-Trimethoxyphenylethanolamine
3,4,5-Trimethoxymandelic acid
Formed from: N,N-Dimethyl-2-oxo-2-(3,4,5-trimethoxyphenyl)ethylamine
2-Oxo-2-(3,4,5-trimethoxyphenyl) ethylamine
3,4,5-Trimethoxy-N-methylphenethylamine
Formed from: Mescaline
4-Hydroxy-3,5-dimethoxy-N-methylphenethylamine
2,4,5-Trimethoxyphenethylamine
→ 2,4,5-Trimethoxyphenylacetaldehyde
Probably in: Rat G865
2,4,6-Trimethoxyphenethylamine
Does not yield 2,4,5-trimethoxyphenylacetic acid in rat G865
3,4,5-Trimethoxyphenol
Probably formed from: Enciprazine
2,4,5-Trimethoxyphenylacetaldehyde
Probably formed from: 2,4,5-Trimethoxyphenylethylamine
→ 2,4,5-Trimethoxyphenylacetic acid
Probably in: Rat G865
3,4,5-Trimethoxyphenylacetaldehyde
Formed from: Mescaline
→ 3,4,5-Trimethoxyphenylacetic acid
In: Mouse A117, H634
Rat A925
→ 3,4,5-Trimethoxyphenylethanol
In: Rat A925
2,4,5-Trimethoxyphenylacetic acid
Probably formed from: 2,4,5-Trimethoxyphenylacetaldehyde
3,4,5-Trimethoxyphenylacetic acid
Formed from: 3,4,5-Trimethoxyphenylacetaldehyde
→ 3,4-Dihydroxy-5-methoxyphenylacetic acid
Probably in: Acetobacterium F655
Arthrobacter B931
3,4,5-Trimethoxyphenylethanol
Formed from: 3,4,5-Trimethoxyphenylacetaldehyde
2,3,4-Trimethoxyphenylethanolamine
→ 2,3,4-Trimethoxymandelic acid
In: Rat J287
1-(3,4,5-Trimethoxyphenyl)glycerol
Probably formed from: 3-(3,4,5-Trimethoxyphenyl)-2-propen-1-ol
3,4,5-Trimethoxyphenyllactic acid
Probably formed from: 3-(3,4,5-Trimethoxyphenyl)propan-1,2-diol
1-(3,4,5-Trimethoxyphenyl)propane-1,2-diol
Formed from: Isoelemicin
3-(3,4,5-Trimethoxyphenyl)propane-1,2-diol
Probably formed from: Isoelemicin-2,3-oxide
→ 3-(3-Hydroxy-4,5-dimethoxyphenyl) propane-1,2-diol
Probably in: Rat B198
→ 3,4,5-Trimethoxyphenyllactic acid
Probably in: Rat B198
1-(3,4,5-Trimethoxyphenyl)-2-propen-1-ol
Formed from: Elemicin
→ 1-(3,4,5-Trimethoxyphenyl)glycerol
Probably in: Rat B198
3-(3,4,5-Trimethoxyphenyl)-2-propen-1-ol
Formed from: Elemicin
Isoelemicin
→ 3,4,5-Trimethoxycinnamic acid
Probably in: Rat B198
3,4,5-Trimethoxyphenylpropionic acid
Formed from: 3,4,5-Trimethoxycinnamic acid
→ Dihydrosinapic acid
In: Microorganism A261, A744

→ 3-Hydroxy-4,5-dimethoxyphenylpropionic acid
In: Microorganism A744
→ 3,4,5-Trimethoxycinnamic acid
In: Rat A745
→ 3,4,5-Trimethoxyphenylpropionylglycine
Probably in: Rat B198

3,4,5-Trimethoxyphenylpropionylglycine
Probably formed from: 3,4,5-Trimethoxyphenylpropionic acid

3,4,5-Trimethoxypropenylbenzene
See: Isoelemicin

2′,4′,6′-Trimethylacetophenone imine
→ 2′,4′,6′-Trimethylacetophenone oxime
In: Rabbit D955

2′,4′,6′-Trimethylacetophenone oxime
Formed from: 2′,4′,6′-Trimethylacetophenone imine

4-(2-Trimethylammonioethyl)-*o*-quinone
Probably formed from: 3,4-Dihydroxyphenethyl trimethyl ammonium

2,4,5-Trimethylaniline (pseudocumidine)
Formed from: 2-Hydroxy-1-(2,4,5-trimethylphenylazo)naphthalene-3,6-disulphonic acid
→ 2,2′,4,4′,5,5′-Hexamethylazobenzene
In: Geotrichum A661
Horseradish A661

2,4,6-Trimethylaniline (mesidine)
Formed from: N-Benzyl-2,4,6-trimethylaniline
→ 2,2′,4,4′,6,6′-Hexamethylazobenzene
In: Horseradish A661

2,4,6-Trimethylbenzaldehyde
Formed from: N-(2,4,6-Trimethylbenzyl) aniline

1,2,3-Trimethylbenzene (hemimellitene)
→ 2,3-Dimethylbenzoic acid
Probably in: Rat A1561

1,2,4-Trimethylbenzene (pseudocumene)
→ 2,4-Dimethoxybenzyl alcohol
In: Rat F328
See also: A1561
→ 2,5-Dimethoxybenzyl alcohol
In: Rat F328
See also: A1561
→ 3,4-Dimethoxybenzyl alcohol
In: Pseudomonas D236
Rat F328
See also: A1561
→ 2,3,5-Trimethylphenol
In: Rat F328
→ 2,3,6-Trimethylphenol
In: Rat F328
→ 2,4,5-Trimethylphenol
In: Rat F328

1,3,5-Trimethylbenzene
See: Mesitylene

N,α,α-Trimethylbenzeneethanamide
See: Mephentermine

N,N,N′-Trimethylbenzidine
Formed from: N,N,N′,N′-Tetramethylbenzidine

3,4,5-Trimethylbenzoic acid 2-(dimethylamino)-2-phenylbutyl ester
See: Trimebutine

7,8,10-Trimethylbenzo[*g*]pteridine-2,4(3*H*,10*H*)-dione
See: Lumiflavin

N-(2,4,6-Trimethylbenzyl)aniline
→ Aniline
In: Hamster G951
→ 2,4,6-Trimethylbenzaldehyde
In: Hamster G951

3,3,5-Trimethylcyclohexyl mandelate (cyclandelate)
→ Mandelic acid
In: Rat F479

ζ-(2,4,5-Trimethyl-3,6-dioxo-1,4-cyclohexadien-1-ylbenzeneheptanoic acid
See: Seratrodast

6,6,9-Trimethyl-3-pentyl-6*H*-dibenzo[*b*,*d*]pyran-1-ol
See: Cannabinol

2,3,5-Trimethylphenol
Formed from: 1,2,4-Trimethylbenzene
→ 2,2′-Dihydroxy-3,3′,4,4′,6,6′-hexamethylbiphenyl
Probably in: Horseradish J592

2,3,6-Trimethylphenol
Formed from: 1,2,4-Trimethylbenzene

2,4,5-Trimethylphenol
Formed from: 1,2,4-Trimethylbenzene

2,4,6-Trimethylphenol
Formed from: Mesitylene

N,N,α-Trimethyl-10*H*-phenothiazine-10-ethanamine
See: Promethazine

N,N,2-Trimethyl-*p*-phenylenediamine
Formed from: 4-Dimethylamino-2-methylazobenzene

N,N,N′-Trimethyl-*p*-phenylenediamine
Formed from: N,N,N′,N′-Tetramethyl-*p*-phenylenediamine

4,5′,8-Trimethylpsoralen
→ 5′-Carboxy-4,8-dimethylpsoralen
In: Man A2669
Mouse A2669, A3233
→ 5′-Hydroxymethyl-4,8-dimethylpsoralen
Probably in: Mouse A3233

1,2,5-Trimethyl-6*H*-pyrido[4,3-*b*]carbazole
See: 9-Hydroxy-2-methylolivacinium

2,2,4-Trimethyl-6-quinolone
Formed from: 1,2-Dihydro-6-hydroxy-2,2,4-trimethylquinoline

Trimethyl 3-(*o*-quinon-4-yl)propyl ammonium
Formed from: 3-(3,4-Dihydroxyphenyl)propyl trimethyl ammonium

4,5,6-Trimethylresorcinol
Formed from: 4,6-Dihydroxy-2,3-dimethylbenzyl alcohol

9-(Trimethylsilyl)methylanthracene
→ 9-Hydroxymethylanthracene
In: Horseradish H796
With cytochrome P450 H796

N,N,N-Trimethyl-*p*-toluidinium
→ S-(*p*-Tolyl)glutathione
In: Cynomolgus monkey E874

1,N,N-Trimethyltryptamine
Formed from: 1,N-Dimethyltryptamine
N,N-Dimethyltryptamine

N,N,N-Trimethyl-L-tryptophan
See: Hypaphorine

Trimetoquinol (1,2,3,4-tetrahydro-6,7-dihydroxy-1-(3,4,5-trimethoxybenzyl) isoquinoline)
→ 2,3-Dihydroxy-9,10,11-trimethoxytetrahydroprotoberberine
In: Rat C325

Trimetozine (N-(3,4,5-trimethoxybenzoyl) morpholine)
→ *m*-Demethyltrimetozine
In: Dog A3302
Rat A3304
→ *p*-Demethyltrimetozine
In: Dog A3302
Rat A3304
→ 2-Hydroxy-N-(3,4,5-trimethoxybenzoyl)morpholine
In: Dog A3302
Rat A3304
→ 3,4,5-Trimethoxybenzoic acid
In: Dog A3302
Rat A3304
→ N-(3,4,5-Trimethoxybenzoyl)-2-aminoethoxyacetic acid
In: Dog A3302
→ N-(3,4,5-Trimethoxybenzoyl) diethanolamine
In: Dog A3302

Trimetrexate (5-methyl-6-(((3,4,5-trimethoxyphenyl)amino)methyl)-2,4-quinazolinediamine)
→ 2,4-Diamino-6-((4-hydroxy-3,5-dimethoxyphenyl)amino)methyl-5-methylquinazoline
Probably in: Dog F603
Rat F603
→ 2,4-Diamino-5-methyl-6-quinazolinecarboxylic acid
In: Dog F603
Rat F603

Trimipramine (2′-methylimipramine)
→ 2-Hydroxytrimipramine
In: Man F55, G991
Rat G269
→ 10-Hydroxytrimipramine
In: Rat G269
→ Iminodibenzyl
In: Man F55
→ Nortrimipramine
In: Glycine K359
Man F55
Rat G269
→ Trimipramine-N-β-D-glucuronide
In: Man H162

Trimipramine-N-β-D-glucuronide
Formed from: Trimipramine

Trimophonium
See: *m*-Hydroxyphenyl trimethylammonium
2,4,6-Trinitrobenzaldehyde
Probably formed from: 2,4,6-Trinitrotoluene
→ 2-Acetamido-4,6-dinitrobenzaldehyde
Probably in: Myriophyllum K416
1,3,5-Trinitrobenzene
→ 3,5-Dinitroaniline
In: Pseudomonas H288
See also: A2624, H328
2,4,6-Trinitrobenzoic acid
→ 4-Amino-2,6-dinitrobenzoic acid
Probably in: Veillonella A2624
2,4,6-Trinitrophenol
See: Picric acid
2,4,6-Trinitrotoluene (TNT)
→ 2-Amino-4,6-dinitrotoluene
In: Bjerkandera K370
Croseus K364
Desulphovibrio G806
Earthworm K511
Man E349
Microorganism K396
Phanerochaete K202, K543
Pseudomonas J181
Rabbit E349
Rat D614, E349
→ 4-Amino-2,6-dinitrotoluene
In: Bjerkandera K370
Croseus K364
Desulphovibrio G806
Earthworm K511
Escherichia A2624
Man E349
Microorganism K396
Phanerochaete K202, K543
Pseudomonas A2624, H957, J181
Rabbit E349
Rat D614, E349
Veillonella A2624
See also: H215, H810
→ 2,4-Dinitrotoluene
In: Pseudomonas J181
→ 6-Hydroxy-2,4-dinitrotoluene
Probably in: Myriophyllum K416
→ 2-Hydroxylamino-4,6-dinitrotoluene
In: Phanerochaete K202
→ 4-Hydroxylamino-2,6-dinitrotoluene
In: Clostridium A2624
Escherichia A2624
Phanerochaete K202
Pseudomonas H957, J181
Veillonella A2624
→ 2,2′,6,6′-Tetranitro-4,4′-azoxytoluene
In: Clostridium A2624
Escherichia A2624
Pseudomonas A2624
Veillonella A2624
→ 2,4,6-Trinitrobenzaldehyde
Probably in: Myriophyllum K416
Tripamide (N-(4-aza-indo-tricyclo[5.2.1.0$^{2.6}$]decan-4-yl)-4-chloro-3-sulphamoylbenzamide)
→ 4-Chloro-3-sulphamoylbenzamide
In: Rat B585, B805
→ 4-Chloro-3-sulphamoylbenzoic acid
In: Rabbit C452
Rat B585, B805, C452
→ 4-Chloro-3-sulphamoylbenzoyl-(N′-acetyl)hydrazide
In: Rat B585
→ 3-Hydroxytripamide
In: Rat B585, B805
→ 8-Hydroxytripamide
In: Rat B805
→ 9-Hydroxytripamide
In: Rat B585
Tripelennamine (pyribenzamine, N-benzyl-N′,N′-dimethyl-N-(2-pyridyl)ethylenediamine)
→ Benzaldehyde
Probably in: Rat E907
→ Benzyl alcohol
In: Rat F297
→ 2-Benzylaminopyridine
In: Rat F297, F577
→ N-Benzyl-N-(4-hydroxy-2-pyridyl)-N′,N′-dimethylethylenediamine
In: Rat F297
→ Demethyltripelennamine
In: Cunninghamella E851
Rat A379, F297
→ N-(2-Dimethylaminoethyl)benzylamine
In: Rat F577

→ *p*-Hydroxytripelennamine
In: Rat A379
→ α-Hydroxytripelennamine
In: Man G409
Rat E907, E930, G409
→ Tripelennamine-N-β-D-glucuronide
In: Man A2115, G18, G509
Rabbit C30
→ Tripelennamine-N-oxide
In: Cunninghamella E851
Man A2115

Tripelennamine-N-β-D-glucuronide
Formed from: Tripelennamine

Tripelennamine-N-oxide
Formed from: Tripelennamine

Triphenylene
→ *trans*-1,2-Dihydro-1,2-dihydroxytriphenylene
In liver microsomes E540
→ 1-Hydroxytriphenylene
In liver microsomes E540
→ 2-Hydroxytriphenylene
In liver microsomes E540

Triphenylene-1,2-oxide
→ *trans*-1,2-Dihydro-1,2-dihydroxytriphenylene
In liver microsomes E540

Triphenyleno[4,5-*bcd*]thiophene
→ Triphenyleno[4,5-*bcd*]thiophene sulphone
In: Rat E166

Triphenyleno[4,5-*bcd*]thiophene sulphone
Formed from: Triphenyleno[4,5-*bcd*]thiophene

Triphenyllead
→ Benzene
In: Rat A3810

Triphenyl phosphate
→ Diphenyl phosphate
In: Rat D440

Triphenyltin
→ Diphenyltin
In: Rat A1932

Triprolidine (2-((*E*)-(1-(4-methylphenyl)-3-(1-pyrrolidinyl)-1-propenyl)pyridine)
→ (*E*)-2-(1-(*p*-Carboxyphenyl)-3-(1-pyrrolidinyl)propenyl)pyridine
In: Dog G512
Mouse G510
→ (*E*)-4-Hydroxy-2-(3-(1-pyrrolidinyl)-1-(*p*-tolyl)propenyl))pyridine
In: Mouse G510
→ 4-((3-(2-Pyridyl)-3-(*p*-tolyl)-2-propen-1-yl)amino)butyric acid
In: Dog G512
Mouse G510

Tri-(L-prolyl-L-glutaminyl-L-glutaminyl-L-prolyl-L-tyrosine)
Formed from: L-Prolyl-L-glutaminyl-L-glutaminyl-L-prolyl-L-tyrosine
Probably formed from: Di-(L-prolyl-L-glutaminyl-L-glutaminyl-L-prolyl-L-tyrosine)
→ Tetra-(L-prolyl-L-glutaminyl-L-glutaminyl-L-prolyl-L-tyrosine)
Probably in: Horseradish J327

2,3,5-Tris(CoA-S-yl)-4-dimethylaminophenol
Formed from: *p*-Dimethylaminophenol

2,5,6-Tris(S-(L-cysteinyl))-3,4-dihydroxy-L-phenylalanine
Probably formed from: 2,5-Bis(S-(L-cysteinyl))-3,4-dihydroxy-L-phenylalanine

3″,4″,5″-Trisnorcannabidiol-2″-oic acid
Probably formed from: 4″,5″-Trisnorcannabidiol-3″-oic acid
→ 2″,3″,4″,5″-Tetranorcannabidiol-1″-oic acid
Probably in: Man F503, G113

N,N,O-Trisnorclentiazem
Probably formed from: N,O-Bisnorclentiazem

3″,4″,5″-Trisnor-2″,6-dihydroxycannabidiol
Probably formed from: 4″,5″-Bisnorcannabidiol-3″-oic acid
3″,4″,5″-Trisnor-2″-hydroxycannabidiol

3″,4″,5″-Trisnor-2″,7-dihydroxycannabidiol
Probably formed from: 3″,4″,5″-Trisnor-2″-hydroxycannabidiol
→ 3″,4″,5″-Trisnor-2″-hydroxycannabidiol-7-oic acid
Probably in: Man F503, G113
Rat G113

3″,4″,5″-Trisnor-2″-formyl-Δ[1]-tetrahydro-7-hydroxycannabidiol
Formed from: Δ[1]-Tetrahydro-3″-hydroxycannabidiol

→ 3″,4″,5″-Trisnor-Δ1-tetrahydro-7-hydroxycannabidiol-2″-oic acid
Probably in: Mouse F492

3″,4″,5″-Trisnor-2″-hydroxycannabidiol
Probably formed from: Cannabidiol
→ 3″,4″,5″-Trisnor-2″,6-dihydroxycannabidiol
Probably in: Man F503
→ 3″,4″,5″-Trisnor-2″,7-dihydroxycannabidiol
Probably in: Man F503

3″,4″,5″-Trisnor-Δ1-tetrahydrocannabidiol-2″,7-dioic acid
Probably formed from: 3″,4″,5″-Trisnor-Δ1-tetrahydrocannabidiol-2″-oic acid

3″,4″,5″-Trisnor-2″-hydroxycannabidiol-7-oic acid
Probably formed from: 3″,4″,5″-Trisnor-2″,7-dihydroxycannabidiol

3″,4″,5″-Trisnor-Δ1-tetrahydrocannabidiol-2″-oic acid
Formed from: 4″,5″-Bisnor-Δ1-tetrahydrocannabidiol-3″-oic acid
Δ1-Tetrahydrocannabidiol
Δ1-Tetrahydro-3″-hydroxycannabidiol
→ 3″,4″,5″-Trisnor-Δ1-tetrahydrocannabidiol-2″,7-dioic acid
Probably in:Man C294
Mouse F526
→ 3″,4″,5″-Trisnor-Δ1-tetrahydro-6β-hydroxycannabidiol-2″-oic acid
Probably in:Man C294
Mouse F526
→ 3″,4″,5″-Trisnor-Δ1-tetrahydro-7-hydroxycannabinol-2″-oic acid
Probably in: Man C294
Mouse F526

3″,4″,5″-Trisnor-Δ6-tetrahydrocannabinol-2″-oic acid
Formed from: Δ6-Tetrahydrocannabinol
Probably formed from: 4″,5″-Bisnor-Δ1-tetrahydrocannabidiol-3″-oic acid

3″,4″,5″-Trisnor-Δ1-tetrahydro-2″,6β-dihydroxycannabinol
Probably formed from: 3″,4″,5″-Trisnor-Δ1-tetrahydro-2″-hydroxycannabinol
→ 3″,4″,5″-Trisnor-Δ1-tetrahydro-2″,6β,7-trihydroxycannabinol
Probably in: Mouse F492

3″,4″,5″-Trisnor-Δ1-tetrahydro-2″,7-dihydroxycannabinol
Probably formed from: 3″,4″,5″-Trisnor-Δ1-tetrahydro-2″-hydroxycannabinol
→ 3″,4″,5″-Trisnor-Δ1-tetrahydro-2″-hydroxycannabinol-7-oic acid
Probably in: Mouse F492

3″,4″,5″-Trisnor-Δ1-tetrahydro-6β,7-dihydroxycannabinol-2″-oic acid
Probably formed from: 3″,4″,5″-Trisnor-Δ1-tetrahydro-6β-hydroxycannabinol-2″-oic acid

3″,4″,5″-Trisnor-Δ1-tetrahydro-2″-hydroxycannabinol
Formed from: Δ1-Tetrahydro-3″-hydroxycannabinol
Probably formed from: Δ1-Tetrahydrocannabinol
→ 3″,4″,5″-Trisnor-Δ1-tetrahydro-2″,6β-dihydroxycannabinol
Probably in: Mouse F492
Rabbit B89
→ 3″,4″,5″-Trisnor-Δ1-tetrahydro-2″,7-dihydroxycannabinol
Probably in: Mouse F492
→ 3″,4″,5″-Trisnor-Δ1-tetrahydro-2″-hydroxycannabidiol-7-oic acid
Probably in: Man C294
Rabbit B89

3″,4″,5″-Trisnor-Δ1-tetrahydro-2″-hydroxycannabidiol-7-oic acid
Probably formed from: 3″,4″,5″-Trisnor-Δ1-tetrahydro-2″,7-dihydroxycannabinol
3″,4″,5″-Trisnor-Δ1-tetrahydro-2″-hydroxycannabinol

3″,4″,5″-Trisnor-Δ1-tetrahydro-6β-hydroxycannabidiol-2″-oic acid
Probably formed from: 3″,4″,5″-Trisnor-Δ1-tetrahydrocannabidiol-2″-oic acid
3″,4″,5″-Trisnor-Δ1-tetrahydro-6β-hydroxycannabinol

3″,4″,5″-Trisnor-Δ1-tetrahydro-7-hydroxycannabidiol-2″-oic acid
Probably formed from: 3″,4″,5″-Trisnor-Δ1-tetrahydrocannabidiol-2″-oic acid

3″,4″,5″-Trisnor-Δ¹-tetrahydro-6β-hydroxycannabinol-2″-oic acid
Probably formed from: 3″,4″,5″-Trisnor-Δ¹-tetrahydrocannabinol
Δ¹-Tetrahydro-6β-hydroxycannabinol
→ 3″,4″,5″-Trisnor-Δ¹-tetrahydro-6β,7-dihydroxycannabinol-2″-oic acid
Probably in: Mouse F526

3″,4″,5″-Trisnor-Δ¹-tetrahydro-7-hydroxycannabinol-2″-oic acid
Probably formed from: 3″,4″,5″-Trisnor-2″-formyl-Δ¹-tetrahydro-7-hydroxycannabinol
3″,4″,5″-Trisnor-Δ¹-tetrahydrocannabinol-2″-oic acid

3″,4″,5″-Trisnor-Δ¹-tetrahydro-2″,6β,7-trihydroxycannabinol
Probably formed from: 3″,4″,5″-Trisnor-Δ¹-tetrahydro-2,6β-dihydroxycannabinol

N,N,O-Trisnorvenlafaxine
Probably formed from: N,N-Bisnorvenlafaxine
→ N,N,O-Trisnorvenlafaxine-β-D-glucuronide
Probably in: Man G564
Rat G564
Rhesus monkey G564

N,N,O-Trisnorvenlafaxine-β-D-glucuronide
Probably formed from: N,N,O-Trisnorvenlafaxine

Trithion (S-(((4-chlorophenyl)thio)methyl) O,O-diethyl phosphorodithioate)
→ Trithion sulphoxide
Probably in: Rat A2639

Trithion sulphoxide
Probably formed from: Trithion
→ *p*-Chlorothiophenol
Probably in: Rat A2639

Trithiozine (tetrahydro-4-(3,4,5-trimethoxythiobenzoyl)-1,4-oxazine)
→ Tetrahydro-4-(4-hydroxy-3,5-diimethoxythiobenzoyl)-1,4-oxazine
In: Man C304
Rat A3260
→ Tetrahydro-4-(3,4,5-trimethoxybenzoyl)-1,4-oxazine
In: Dog A3023
Man C304
Rat A3023, A3260
→ Trithiozine sulphoxide
In: Rat A3260

Trithiozine sulphoxide
Formed from: Trithiozine

Tritoqualine (7-amino-4,5,6-triethoxy-3-(5,6,7,8-tetrahydro-4-methoxy-6-methyl-1,3-dioxolo[4,5-*g*]isoquinolin-5-yl)-1(3*H*)-isobenzofuranone)
→ 7-Amino-4,5,6-triethoxyphthalide
In: Dog E105
Mouse E105
Rat E105
→ Cotarnine
In: Dog E105
Mouse E105
Rat E105
→ 3,4-Dihydro-6-hydroxy-7,8-dimethoxy-2-methylisoquinolinium
In: Dog E105
Mouse E105
Rat E105

Troglitazone (5-((4-((3,4-dihydro-6-hydroxy-2,5,7,8-tetramethyl-2*H*-1-benzopyran-2-yl)methoxy)phenyl)methyl)-2,4-thiazolidinedione)
Formed from: Troglitazone-O-sulphate
→ 5-((4-((3,4-Dihydro-6-hydroxy-5-hydroxymethyl-2,7,8-trimethyl-2*H*-1-benzopyran-2-yl)methoxy)phenyl)methyl)-2,4-thiazolidinedione
In: Man J770
→ 5-((4-(2,5-Dihydroxy-3,4,6-trimethylphenyl)-2-hydroxy-2-methylbutoxy)benzyl)-2,4-thiazolidinedione
In: Cynomolgus monkey K601
Man J770, K135
Pig K135
→ 5-(4-(2-Hydroxy-2-methyl-4-(3,5,6-trimethyl-1,4-benzoquinon-2-yl)butoxy) benzyl)-2,4-thiazolidinedione
In: Cynomolgus monkey K601
→ Troglitazone-β-D-glucuronide
In: Cynomolgus monkey K601
Man J770

Pig K135
→ Troglitazone-O-sulphate
In: Cynomolgus monkey K601
Man J770, K135
Pig K135

Troglitazone-β-D-glucuronide
Formed from: Troglitazone

Troglitazone-O-sulphate
Formed from: Troglitazone
→ Troglitazone
In: Rat K120

Tropacocaine (benzoylpseudotropeine, (3-exo)-8-methyl-8-azabicyclo[3.2.1]octan-3-ol benzoate)
→ Benzoic acid
In: Rabbit A2507

Tropesin (1-(4-chlorobenzoyl)-5-methoxy-2-methyl-1*H*-indole-3-acetic acid 2-carboxy-2-phenylethyl ester)
→ 1-(*p*-Chlorobenzoyl)-5-methoxy-2-methylindole-3-acetic acid
In: Man C223

(*S*)-Tropic acid (3-hydroxy-2-phenylpropionic acid)
Formed from: Methylscopolammoniun
→ Hyoscyamine
Probably in: Datura A1638
→ Phenylacetaldehyde
In: Flavobacterium F780
Pseudomonas H829
→ Tropic acid-β-D-glucoside
In: Panax F255

Tropic acid-β-D-glucoside
Formed from: Tropic acid

Tropine 5-hydroxyindole-3-carboxylate
Formed from: Tropine indole-3-carboxylate
→ Tropine 5-hydroxyindole-3-carboxylate-β-D-glucuronide
In: Man F363
Rat F363
→ Tropine 5-hydroxyindole-3-carboxylate-O-sulphate
In: Man F363

Tropine 6-hydroxyindole-3-carboxylate
Formed from: Tropine indole-3-carboxylate
→ Tropine 6-hydroxyindole-3-carboxylate-β-D-glucuronide
In: Rat F363
→ Tropine 6-hydroxyindole-3-carboxylate-O-sulphate
In: Man F363

Tropine 7-hydroxyindole-3-carboxylate
Formed from: Tropine indole-3-carboxylate
→ Tropine 7-hydroxyindole-3-carboxylate-β-D-glucuronide
In: Rat F363
→ Tropine 7-hydroxyindole-3-carboxylate-O-sulphate
In: Man F363

Tropine 5-hydroxyindole-3-carboxylate-β-D-glucuronide
Formed from: Tropine 5-hydroxyindole-3-carboxylate

Tropine 6-hydroxyindole-3-carboxylate-β-D-glucuronide
Formed from: Tropine 6-hydroxyindole-3-carboxylate

Tropine 7-hydroxyindole-3-carboxylate-β-D-glucuronide
Formed from: Tropine 7-hydroxyindole-3-carboxylate

Tropine 5-hydroxyindole-3-carboxylate-O-sulphate
Formed from: Tropine 5-hydroxyindole-3-carboxylate

Tropine 6-hydroxyindole-3-carboxylate-O-sulphate
Formed from: Tropine 6-hydroxyindole-3-carboxylate

Tropine 7-hydroxyindole-3-carboxylate-O-sulphate
Formed from: Tropine 7-hydroxyindole-3-carboxylate

Tropine indole-3-carboxylate
→ Demethyltropine indole-3-carboxylate
In: Man F363
Rat F363
→ Tropine 5-hydroxyindole-3-carboxylate
In: Man F363
Rat F363
→ Tropine 6-hydroxyindole-3-carboxylate
In: Man F363
Rat F363
→ Tropine 7-hydroxyindole-3-carboxylate

In: Man F363
Rat F363
→ Tropine-N-oxide indole-3-carboxylate
In: Man F363
Rat F363

Tropine-N-oxide indole-3-carboxylate
Formed from: Tropine indole-3-carboxylate

Tropisetron (1*H*-indole-3-carboxylic acid (3-*endo*)-8-methyl-8-azabicyclo[3.2.1]oct-3-yl ester)
→ 5-Hydroxytropisetron
In: Dog J214
Man G388, H584, J214
Rat J214
→ 6-Hydroxytropisetron
In: Dog J214
Man G388, G870, H584, J214
Rat J214
→ 7-Hydroxytropisetron
In: Man G388, G870
→ Nortropisetron
In: Man G388, G870
→ Tropisetron-N-oxide
In: Dog J214
Man G388, G870, H584, J214
Rat J214

Tropisetron-N-oxide
Formed from: Tropisetron

Trovafloxacin ((1α,5α,6α)-7-(6-amino-3-azabicyclo[3.1.0]hex-3-yl)-1-(2,4-difluorophenyl)-6-fluoro-1,4-dihydro-4-oxo-1,8-naphthyridine-3-carboxylic acid)
→ N-Acetyltrovafloxacin
In: Dog J69
Man J140
Rat J69
→ Trovafloxacin-β-D-glucuronide
In: Dog J69
Rat J69
→ Trovafloxacin sulphamate
In: Man J140

Trovafloxacin-β-D-glucuronide
Formed from: Trovafloxacin

Trovafloxacin sulphamate
Formed from: Trovafloxacin

Tryptamine
Formed from: L-Tryptophan
Probably formed from: N-Methyltryptamine
→ N-Acetyltryptamine
In: Cockroach H835
Drosophila H758
Man K490
Periplaneta J632, J633
Prawn G824
Quail H685
Rat A807, A1350, D282, F874
Rhizoglyphus F669
Sheep D282, J850
Zygosaccharomyces A923
→ Ajmalicine
In: Catharanthus A2741, A3069
See also: B576, C248
→ 19-Epiajmalicine
In: Catharanthus A2741, A3069
→ Geissoschizine
In: Catharanthus B46
→ 5-Hydroxytryptamine
In: Peganum A1330, F9, G906
→ β-Hydroxytryptamine
In: Beef B469
Pseudomonas A3009, B84
Rat B469
→ Indole-3-acetaldehyde
In: Arthrobacter F626, H912
Barley A446
Beef A1285
Crown gall A1527
Hedgehog A434
Hymenopelis A3815
Lampetra H581
Man A2010, E55
Pea A755
Pig B206
Pseudomonas C517
Rat A968, A1053, A1282, A1285, B653
Rhizoglyphus F669
Sheep A535
Tetrahymena A3131
Tomato A446
See also: A993, A1922, C181, D221, E518, E578, J539, K541, K550
→ Indole-3-glyoxal
Probably in: Pseudomonas A3009

→ 1-Methyltryptamine
In: Rabbit C333
→ N-Methyltryptamine
In: Phalaris A3891
Rabbit A111, B741, C111, C156, D646, E6, J884
Rat A80, A111, A968, A1307, A1311, A1667
See also: A1310, E974
→ Strictosidine
In: Catharanthus A3762, B71, B265
Rauwolfia F63
Tabernaemontana H571
See also B86, C248
→ Tetrahydroalstonine
In: Catharanthus A3069, B46
→ 1,2,3,4-Tetrahydro-β-carboline
In: Beef B35
Chick A1428
Rabbit A111
Rat A104, A111, A1428, A2285
See also: A3946, C679
→ 1,2,3,4-Tetrahydro-1-methyl-β-carboline-1-carboxylic acid
In: Rat E311, E505
→ Vindoline
In: Catharanthus A3737
See also: B362, B576

Tryptamin-5-yl-β-D-glucuronide
Formed from: 5-Hydroxytryptamine

Tryptamin-5-yl sulphate
Formed from: 5-Hydroxytryptamine

D-Tryptophan
Formed from: 5-Indolylmethylhydantoin
Probably formed from: Indole-3-pyruvic acid
→ 3-Formylindole
In: Orobanche C181
→ N-Formyl-D-tryptophan
In: Rabbit A845
Wheat E24
→ β-Hydroxy-D-tryptophan
Probably in: Pseudomonas A3007
→ Indole
With tryptophanase J250
→ Indole-3-acetic acid
In: Hordeum F672
→ Indole-3-glyoxal
In: Pseudomonas A3009
→ Indole-3-pyruvic acid
In: Alaska pea F405
Man C114
Monkey C114
Mouse C114, C219
Pig C114
Pseudomonas B855
Rat C114
Wheat H841
Zea G888
Contrast (in carp, catfish, crucian carp, dace, electric ray, rainbow trout) C219

L-Tryptophan
Formed from: N-Acetyl-L-tryptophan
L-Alanyl-L-tryptophan
N-Carbamyl-L-tryptophan
N-Chloroacetyl-L-tryptophan
Indole
Indole-3-acetic acid
Indole-3-pyruvic acid
5-Indolylmethylhydantoin
Isatin
L-Leucyl-L-tryptophan
L-Tryptophanylglycine
L-Tryptophanyl-β-naphthylamide
Probably formed from: N-Malonyl-L-tryptophan
→ N-Acetyl-L-tryptophan
In: Claviceps A3655
Cordyceps A993
Xanthomonas E435
→ Akuammicine
In: Catharanthus A3976, B576
→ 3-Aminomethylindole
Probably in: Phalaris A1928
→ Anthramycin
In: Streptomyces A1677
→ β-Carboline-1-propionic acid
In: Ailanthus E304
→ Chaetoglobosin A
In: Chaetomium C732
→ Chaetoglobosin B
In: Chaetomium C732
→ Cryptoechinuline A

In: Aspergillus A2338
→ α,β-Dehydrotryptophan
In: Chromobacterium H248, H664
→ 11-Demethyltomaymycin
In: Streptomyces A2154
→ 4-Dimethylallyl-L-tryptophan
In: Claviceps A3599
→ Echinuline
In: Aspergillus A2338
See also: C816
→ N-Formyl-L-kynurenine
In: Escherichia A433
Mouse E30
Wheat C599, E24
See also: A674, A845, A1195, A3534, A3535, A3551, B790, B855, F169, G787
→ Glucobrassicin
In: Horseradish C745
→ Harman
In: Phaseolus A2158
→ 3-Hydroxy-3-methylbutenylpaspalinine
In: Claviceps F280
→ β-Hydroxy-L-tryptophan
Probably in: Pseudomonas A3007
→ 5-Hydroxy-L-tryptophan
In: Bacillus A1244
Beef A1341
Helix A1051
Hymenolepis C524, D105
Man G393
Rabbit E269
Rat A3611, D64, E269
Sedum D41
See also: A1938, A3534, A3535, C898, E967, K548, K550
→ Hypaphorine
In: Lens A234
See also: A1618
→ Indole
In: Clostridium A2571
Escherichia A430, A433, E101
Microorganism D477, G485
Salmonella E101
→ Indole-3-acetaldehyde (direct)
In: Arabidopsis K605
Pseudomonas G396
→ Indole-3-acetaldoxime
In: Cabbage F659
→ Indole-3-acetamide
In: Azospirillum G776
Bradyrhizobium E754, F788
Coprinus K563
Poncirus G760
Pseudomonas E435, H384
Streptomyces E578
Xanthomonas E435
See also: C444
→ Indole-3-acetic acid (not via indole-3-acetaldehyde)
In: Ustilago H902
→ Indole-3-acrylic acid
Probably in: Clostridium A2571
→ Indole-3-glycolaldehyde
In: Pseudomonas B98
→ Indole-3-glyoxal
In: Pseudomonas A3009, B98
→ Indole-3-pyruvic acid
In: Azospirillum F49, J22
Bacillus A2791, E385, E445
Barley A446
Brevibacterium D724
Bushbean A566
Candida E97
Coprinus K563
Cryptococcus B483
Enterbacter H348
Escherichia A3548, K715
Flavobacterium C206
Klebsiella B732
Monkey A3570, K717
Mung bean H826
Pea G483
Phaseolus A463
Pseudomonas B672, B855, G396
Pyrococcus H437
Rat A2339, A2342, A2847, A3026, A3570, B379, D196, K717
Rhizobium F268
Rhodococcus E451
Saccharomyces C533, G320
Sporosarcina E445
Streptomyces A2307, D148

Taphrina A405
Thermococcus H116
Tomato A446, G483
Trypanosoma B863
Wheat H841
Zea G483, G888
See also: A923, A2571, A3551, C444, E405, E435, E578, G598, J189, K259
→ 2-Methyl-L-tryptophan
In: Streptomyces K752
→ β-Methyl-L-tryptophan
In: Streptomyces D148, D183
→ Norharman
In: Phaseolus A2158
→ Paxilline
In: Penicillium F280
→ Penitrem A
In: Penicillium F280
→ Rebeccamycin
In: Saccharothrix E772, F784
→ Roquefortine
In: Penicillium F280
→ Skatole
In: Microorganism D477, G485
→ Staurosporine
In: Streptomyces E670
→ Strictosidine
In: Catharanthus A3205
See also: B576
→ Tetrahydro-5-hydroxynorharman
In: Peganum G906
→ Tetrahydro-6-hydroxynorharman
In: Peganum G906
→ Tryptamine
In: Bacillus D555
Barley A446
Catharanthus D982
Cordyceps A993
Cucumber A1172
Hypericum K550
Lactobacillus K259
Man C566
Micrococcus A336, C62, D539, G918
Mycobacterium F189
Phalaris E974
Pig A1906
Rat C566, D64, E735
Streptomyces E578
Tabernaemontana H571
Tomato A446
Walnut K548
See also: A433, A1922, B98, D590, D794, F371
→ L-Tryptophan-1-β-D-glucoside
In: Pear K577
→ Violacein
In: Chromobacterium K481

L-Tryptophanamide
→ α,β-Dehydrotryptophanamide
Probably in: Chromobacterium H248, H644
→ β-Hydroxy-L-tryptophanamide
Probably in: Pseudomonas A3007

L-Tryptophan-1-β-D-glucoside
Formed from: L-Tryptophan

L-Tryptophan methyl ester
→ N-Formyl-L-tryptophan methyl ester
In: Wheat C559, E24
→ β-Hydroxy-L-tryptophan methyl ester
In: Pseudomonas B84
→ 1-Methyl-L-tryptophan methyl ester
In: Rabbit C333
→ N-Methyl-L-tryptophan methyl ester
In: Rabbit C156

L-Tryptophanyl-L-alanine
→ N-Formyl-L-kynureninyl-L-alanine
In: Wheat E24

L-Tryptophanylglycine
→ N-Formyl-L-kynureninylglycine
In: Wheat C559, E24
→ L-Tryptophan
In: Hen G654

L-Tryptophanyl-β-naphthylamide
→ β-Naphthylamime
In: Agave A2745
Arabidopsis K224
Bacteria C146
Beef A526
Mytilus A1514
→ L-Tryptophan
In: Agave A2745
Arabidopsis K224

Bacteria C146
Beef A526
Mytilus A1514

L-Tryptophanyl-L-leucine
→ N-Formyl-L-kynureninyl-L-leucine
In: Wheat C559, E24
→ *erythro*-β-Hydroxy-L-tryptophanyl-L-leucine
In: Pseudomonas B917
→ *threo*-β-Hydroxy-L-tryptophanyl-L-leucine
In: Pseudomonas B917

L-Tryptophanyl-L-phenylalanine
→ N-Formyl-L-kynureninyl-L-phenylalanine
In: Wheat C559
→ β-Hydroxy-L-tryptophanyl-L-phenylalanine
Probably in: Pseudomonas A3007

L-Tryptophanyl-L-tryptophan
→ β-Hydroxy-L-tryptophanyl-L-tryptophan
Probably in: Pseudomonas A3007

L-Tryptophanyl-L-tyrosine
→ N-Formyl-L-kynureninyl-L-tyrosine
In: Wheat E24

Tryptophol (2-(3-indolyl)ethanol)
Formed from: Indole-3-acetaldehyde
→ O-Acetyltryptophol
In: Ceratocystis A1922
See also: A3551
→ β-Hydroxytryptophol
Probably in: Pseudomonas A3007
→ Indole-3-acetaldehyde
In: Cucumber A835, A957
Phycomyces E443
Rhizobium E405
See also: E729
→ Indole-3-glyoxal
Probably in: Pseudomonas A3009
→ Tryptophol glucoside
In: Orobanche C181
→ Tryptophol-β-D-glucuronide
In: Pea A755

Tryptophol glucoside
Formed from: Tryptophol

Tryptophol-β-D-glucuronide
Formed from: Tryptophol

Tsugaresinol
See: Conidendrin

Tuberin (N-(2-(4-methoxyphenyl)ethenyl) formamide)
Formed from glycine in Streptomyces D541

Tubulosine ((1*R*)-1-(((2*S*,3*R*,11b*S*)-3-ethyl-1,3,4,6,7,11b-hexahydro-9,10-dimethoxy-2*H*-benzo[*a*]quinolizin-2-yl)methyl)-2,3,4,9-tetrahydro-1*H*-pyrido[3,4-*b*]indol-6-ol)
Formed from: Deoxytubulosine

Tulobuterol (N-(*tert*-butyl)-*o*-hydroxyphenylethanolamine)
→ 3-Hydroxytulobuterol
In: Dog C43
Guinea pig C43
Rabbit C43
Rat C43
→ 4-Hydroxytulobuterol
In: Dog C43
Guinea pig C43
Rabbit C43
Rat C43
→ 5-Hydroxytulobuterol
In: Dog C43
Guinea pig C43
Rabbit C43
Rat C43

Tylophorine ((13a*S*)-9,11,12,13,13a,14-hexahydro-2,3,6,7-tetramethoxydibenzo[*f*,*h*] pyrrolo[1,2-*b*]isoquinoline)
Formed from: Cinnamic acid
L-Tyrosine

***o*-Tyramine** (*o*-hydroxyphenethylamine)
Formed from: Phenethylamine
L-*o*-Tyrosine
→ *o*-Hydroxyphenylacetaldehyde
In: Rat A3667
See also: G865, J411
→ *o*-Octopamine
In: Rat C823

***m*-Tyramine** (*m*-hydroxyphenethylamine)
Formed from: 3,4-Dihydroxyphenethylamine
3-Hydroxy-4-methoxyphenethylamine
3-Methoxytyramine
Phenethylamine
L-*m*-Tyrosine
→ 3,4-Dihydroxyphenethylamine

Probably in: Man B273
Contrast (in rat) A2961
→ *m*-Hydroxyphenylacetaldehyde
In: Rat A2961, A3667, B660
See also: A3493, B162, B273, B794, G865
→ *m*-Octopamine
Probably in: Locust B907
Rat B273
→ Phenethylamin-3-yl sulphate
In: Man C523, D741
Does not yield phenethylamine in rat C465

Tyramine (*p*-tyramine)
Formed from: N-Acetyltyramine
3,4-Dihydroxyphenethylamine
N-Hydroxytyramine
N-Methyltyramine
Phenethylamine
L-Tyrosine
→ N-Acetyltyramine
In: Candida A2822
Cockroach H835
Drosophila A643
Enterobacteria A2889
Fungus A705
Rat A411
Musca G914
Schistocerca A2804, B826
See also: A410
→ β-Alanyltyramine
In: Manduca E783
→ Cinnamoyltyramine
In: Nicotiana D558, E531
→ *p*-Coumaroyltyramine
In: Nicotiana D558, E531, F72
→ Dauricine
In: Menispermum H220
→ 3,4-Dihydroxyphenethylamine
In: Avocado A3975
Berberis F855
Locust A2804, B765
Mucuna E769, F112
Mushroom H907
Nematode A2572
Thalictrum H242
See also: A2617, A3788, B826, B897
→ Feruloyltyramine
In: Nicotiana D558, E531, F72
→ *p*-Hydroxybenzoic acid
In: Fungus A705
→ *p*-Hydroxyphenylacetaldehyde
In: Alcaligenes H367
Arthrobacter F626, H912
Aspergillus A2888, A3377
Bacteria B225
Cat C135
Epidermophyton A2888
Guinea pig A3958
Hedgehog A434
Hymenolepis A3815
Klebsiella A2888
Man F754
Micrococcus K420
Microsporon A2888
Mucor A2888
Penicillium A2888
Pig A2587, B206
Proteus A2888
Pseudomonas C517
Rabbit A94
Rat A1331, A1348, A1613, A3667, B660, B661, F271
Staphylococcus A2888
Streptococcus A2888
Trichophyton A2888
See also: A410, A411, A561, A705, A1157, A1909, A2617, A2888, A2889, A3788, B794, F380, G865, J539
→ *p*-Hydroxyphenylacetaldoxime
In: Man J336
Pig J336
→ N-Methyl-*p*-tyramine
In: Cat C135
Opuntia A3977
Rabbit A1311, E6
Rat A1311
See also: B577
→ Octopamine
In: Beef C387, K154, K157
Cat C135
Hymenolepis C502
Locust B765
Man E386, K459

Rat A1331
With dopamine β-hydroxylase A1721
See also: A469, A1337, A2336, A2617, A3788, B826
→ Phenethylamine
In: Rabbit A1717, A2179
Contrast (in rat) C465
→ Sinapoyltyramine
In: Nicotiana D558
→ Tyramine-β-D-glucuronide
In: Guinea pig A3203
Rat A411, A3203
→ Tyramine-4-sulphate
In: Cat C135
Dog J292
Eubacterium G475, G602, H345
Guinea pig A2066
Man A1, A365, C322, J239, K120
Marmoset F226
Mouse A2066
Rat A2066, K120
Macaca A2066

Tyramine-β-D-glucuronide
Formed from: Tyramine

Tyramine-4-sulphate
Formed from: Tyramine

L-*o*-Tyrosine
Formed from: L-Phenylalanine
→ 2,3-Dihydroxy-L-phenylalanine
Probably in: Bacteria B29, B123
→ *o*-Hydroxyphenylacetamide
Probably in: Pseudomonas D246
→ *o*-Tyramine
In: Barley A1210
Man C566
Micrococcus C62
Rat C566, J411

D-*m*-Tyrosine
Formed from: 5-(*m*-Hydroxybenzyl)hydantoin
→ *m*-Hydroxyphenylpyruvic acid
In: Rat A3493
→ D-*m*-Tyrosine-O-sulphate
In: Man H888

L-*m*-Tyrosine
Formed from: L-Phenylalanine
Probably formed from: *m*-Hydroxyphenylpyruvic acid
3-Methoxytyrosine
→ *m*-Coumaric acid
In: Parsley H647
→ 3,4-Dihydroxy-L-phenylalanine
In: Beef F623
Rat A2584, A2715
See also: A2961
→ *m*-Hydroxymandelic acid
In: Rat A3493
→ *m*-Hydroxyphenylacetaldehyde
In: Pig A2425
→ *m*-Hydroxyphenylacetamide
Probably in: Pseudomonas D246
→ *m*-Hydroxyphenylpyruvic acid
In: Rat A2961, A3493
→ *m*-Tyramine
In: Barley A1210
Man C566
Micrococcus C62
Mouse C514
Periplaneta E448
Rat A2961, A3493, C566
Streptococcus A537
See also: B273, C465
→ L-*m*-Tyrosine-O-sulphate
In: Man H888

D-Tyrosine
Formed from: 5-(*p*-Hydroxybenzyl)hydantoin
D-Tyrosine-4-sulphate
→ 3,4-Dihydroxy-D-phenylalanine
In: Avocado A3975
Banana A541
Portulaca K93
→ *p*-Hydroxyphenylpyruvic acid
In: Candida E97
→ 5-O-Methylxylerythrin
In: Peniophora A2737, A2738
→ Peniophorin
In: Peniophora A2737, A2738
→ Peniophoronin
In: Peniophora A2737, A2738
→ Phenol
In: Symbiobacterium G603
See also: A708
→ Xylerythrin
In: Peniophora A2737, A2738

→ D-Tyrosine-4-sulphate
In: Man H888, J644

L-Tyrosine
Formed from: N-Acetyl-L-tyrosine
L-Alanyl-L-tyrosine
N-Carbamyl-L-tyrosine
L-Glutamyl-L-tyrosine
Glycyl-L-tyrosine
5-(*p*-Hydroxybenzyl)hydantoin
p-Hydroxyphenylacetic acid
p-Hydroxyphenylpyruvic acid
3-Iodo-L-tyrosine
L-Leucyl-L-tyrosine
Phenol
L-Phenylalanine
Pretyrosine
L-Tyrosine methyl ester
L-Tyrosine-4-phosphate
L-Tyrosine-4-sulphate
L-Tyrosyl-L-glutamic acid
L-Tyrosyl-L-glutamine
L-Tyrosyl-β-naphthylamide
L-Tyrosyl-*p*-nitroanilide
Probably formed from: L-Tyrosine ethyl ester
→ N-Acetyl-L-tyrosine
Probably in: Claviceps A3655
→ Anthramycin
In: Streptomyces A1677
→ Aristolochic acid I
In: Aristolochia E773
→ Berberine
In: Coptis G749
→ 3-Bromo-L-tyrosine
In: Aplysina H696
→ 3-Chloro-L-tyrosine
In: Man J226
→ *p*-Coumaric acid
In: Enterobacteria A2889
Fungus A2889
Mycobacteria A2889
Trichosporon B34
Zea A607
See also: A2336, A2571, A2767, B502
→ *p*-Cresol
In: Microorganism H842
Rat A603
→ 2,3-Dihydroxy-L-phenylalanine
Probably in: Aspergillus A3377
→ 2,4-Dihydroxy-L-phenylalanine
In: Aeromonas D677
→ 3,4-Dihydroxy-L-phenylalanine
In: Agaricus A3135
Aspergillus A1131, F192
Avocado A3975
Banana A541
Beef A383
Berberis F855
Gerbil H867
Horseradish A2458, B925
Locust A2804, B765
Mucuna A3988, C775, E769, F112, F958
Mushroom C442, H907
Nematode A2572
Planorbis A2055
Portulaca K93
Potato A3190, D964
Rana A1208
Rat A1173, A2570
Neuroblastoma A850
PC12 cells K169
See also: A218, A1364, A1900, A1930, A2568, B162, B574, B897, C784, F251, F623, G449, J313, J642
→ L-Dityrosine
In: Horseradish J327, J401
Rat G616
→ Flexirubin
In: Flexibacter A3760, A3761
→ Hydrangenol
In: Conocephalum A1924
→ *p*-Hydroxybenzaldehyde
In: Odonthalia B300
Sea anemone D245
Sorghum A3175, A3331
→ *p*-Hydroxybenzoic acid
In: Anacystis A2792
Trichosporon B34
→ *p*-Hydroxybenzylglucosinolate
In: Sinapis J689
→ *p*-Hydroxyphenylacetaldoxime
In: Sorghum A3175, A3331, F547, H423, K711

Triglochin K216
→ *p*-Hydroxyphenylacetamide
In: Pseudomonas D246
See also: C444
→ *p*-Hydroxyphenylpyruvic acid
In: Anchusa E436
Arthrobacter A3208
Azospirillum J22
Bacillus A2791, E385, E445
Bacterium A2005
Brevibacterium B266, D592, D724, E642
Bushbean A566
Candida E97
Cat A1197
Corynebacterium B266
Drosophyllum A181
Escherichia A3548, K715
Flavobacterium C206
Hansenula D978
Man A3368, C210
Monkey A3570, K717
Mung bean H826
Naja J684
Nocardia E606
Odonthalia B300
Periplaneta A2336
Pseudomonas D246
Pyrococcus H437
Rat A1383, A2268, A2339, A2342, A2847, A3570, B379, K717, K718
Rhizobium F268
Rhodobacter J874
Rhodococcus E451
Saccharomyces C533
Sporosarcina E445
Streptomyces A2307, H275
Thalictrum H242
Thermococcus H116
Trimeresurus E770
Trypanosoma A1619, G894
See also: A923, A2651, A2767, A2889, A3377, B455, C444, C484
→ *p*-Hydroxyphenylpyruvic acid oxime
In: Sorghum A3175, A3331
Contrast (wrong identity) F547
→ N-Hydroxy-L-tyrosine
In: Sorghum A3175, A3335, G228, H331
→ 3-Iodo-L-tyrosine
In: Goat A997
→ Kyotorphin
In: Rat K857
→ 3-Nitro-L-tyrosine
In: Mouse K385
→ Norcoclaurine
Probably in: Annona B137
Croton A185
→ Novobiocin
In: Streptpmyces A945
→ Phenol
In: Citrobacter F196
Erwinia A708, B292
Leptoglossus A2260
Microorganism H842
Millipede A2254
Symbiobacterium G603
Rat A603
→ L-Phenylalanine
In: Hen A418
See also: D184
→ Sesamin
In: Sesamum A1159
→ Tylophorine
Probably in: Tylophora A1549
→ Tyramine
In: Acinetobacter H302
Bacillus H302
Barley A1210
Candida A2822, H302
Citrobacter H302
Cryptococcus H302
Enterobacteria A2889
Enterococcus H302
Escherichia H302
Fungus A2889
Hafnia H302
Lactobacillus H302, J342, K277
Locust A2804
Micrococcus A336, C62, G918
Mycobacteria A2889, F189
Papaver B464, H427, H557
Penicillium A2822
Periplaneta A2336, E448

Pig A1906, B464
Proteus H302
Providencia B760
Pseudomonas H302
Rat C465, D6, E735, F863
Saccharomyces H302
Salmonella H302
Schistocerca B826
Serratia H302
Shigella H302
Staphylococcus H302
Streptococcus A537, B464, D868, F197
Thalictrum H242
Torulopsis H302
Yersinia H302
See also: A564, A3377, B765, B826, B836, C465, C819
→ β-Tyrosine
In: Bacillus B316, K842
→ L-Tyrosine-O-β-D-glucoside
In: Manduca B574, J40
→ L-Tyrosine-4-sulphate
In: Dog H371
Man H371, H888

β-Tyrosine (β-amino-4-hydroxybenzenepropanoic acid)
Formed from: L-Tyrosine

L-Tyrosine ethyl ester
→ L-Tyrosine
Probably in: Rat B836

L-Tyrosine-O-β-D-glucoside
Formed from: L-Tyrosine

L-Tyrosine methyl ester
→ 3,4-Dihydroxy-L-phenylalanine methyl ester
In: Mucuna E769
Mushroom H907
→ L-Tyrosine
In: Rat F362
→ L-Tyrosine methyl ester-4-sulphate
In: Guinea pig E456
Rat B941

L-Tyrosine methyl ester-4-sulphate
Formed from: L-Tyrosine methyl ester

L-Tyrosine-4-phosphate
→ L-Tyrosine
In: Rat F362

D-*m*-Tyrosine-O-sulphate
Formed from: D-*m*-Tyrosine

L-*m*-Tyrosine-O-sulphate
Formed from: L-*m*-Tyrosine

D-Tyrosine-4-sulphate
Formed from: D-Tyrosine
→ D-Tyrosine
In: Man A3839

L-Tyrosine-4-sulphate
Formed from: L-Tyrosine
→ L-Tyrosine
In: Comamonas B159
Man A3839
Rabbit B481

***o*-Tyrosol** (2-(*o*-hydroxyphenyl)ethanol)
Formed from: Coumarin
o-Hydroxyphenylacetaldehyde
Styrene oxide
→ *o*-Hydroxyphenylacetic acid
In: Rat C863

***m*-Tyrosol** (2-(*m*-hydroxyphenyl)ethanol)
Formed from: *m*-Hydroxyphenylacetaldehyde
→ *m*-Hydroxyphenylacetaldehyde
In: Rat A2961

***p*-Tyrosol** (2-(*p*-hydroxyphenyl)ethanol)
Formed from: *p*-Hydroxyphenylacetaldehyde
Styrene oxide
Probably formed from: *p*-(1-Propenyl)phenol
→ 2-(3,4-Dihydroxyphenyl)ethanol
In: Mucuna E769
Mushroom H907
→ Salidroside
In: Rhodiola J645
→ *p*-Tyrosol-β-D-glucuronide
Probably in: Rat A411
→ *p*-Tyrosol sulphate
Probably in: Rat A411

***p*-Tyrosol-β-D-glucuronide**
Probably formed from: *p*-Tyrosol

***p*-Tyrosol sulphate**
Probably formed from: *p*-Tyrosol

L-Tyrosyl-L-arginine
See: Kyotorphin

L-Tyrosyl-L-glutamic acid
→ L-Tyrosine
In: Aspergillus A3658

L-Tyrosyl-L-glutamine
→ L-Tyrosine
In: Aspergillus A3658

L-Tyrosyl-3-(2-naphthyl)-D-alanyl-L-leucine
Probably formed from: L-Tyrosyl-3-(2-naphthyl)-D-alanyl-L-leucyl-L-arginyl-L-prolylglycinamide
→ 3-(2-Naphthyl)-D-alanyl-L-leucine
Probably in: Rhesus monkey D835

L-Tyrosyl-3-(2-naphthyl)-D-alanyl-L-leucyl-L-arginyl-L-prolylglycinamide
→ 3-(2-Naphthyl)-D-alanyl-L-leucyl-L-arginyl-L-prolylglycinamide
Probably in: Rhesus monkey D835
→ L-Tyrosyl-3-(2-naphthyl)-D-alanyl-L-leucine
Probably in: Rhesus monkey D835

L-Tyrosyl-β-naphthylamide
→ β-Naphthylamine
In: Agave A2745
Arabidopsis K224
Bacteria C146
Beef A526
Euonymus D216
Glycine A1596
Mytilus A1514
Rabbit B334
Rat A523
Streptococcus C277
→ L-Tyrosine
In: Agave A2745
Arabidopsis K224
Bacteria C146
Beef A526
Euonymus D216
Glycine A1596
Mytilus A1514
Rabbit B334
Rat A523
Streptococcus C277

L-Tyrosyl-*p*-nitroanilide
→ *p*-Nitroaniline
In: Rat C173
→ L-Tyrosine
In: Rat C173

U

Ubiquinol 40
Formed from: Ubiquinone 40
Ubiquinone-0
See: 2,3-Dimethoxy-5-methyl-*p*-quinone
Ubiquinone-5 (ubiquinone-1)
→ 5,6-Dimethoxy-3-methyl-2-prenylquinol
In: Beef G401
Ubiquinone-30 (coenzyme Q_6)
Formed from: 4-Carboxy-2-hexaprenyl-6-methylphenol
→ 2,3-Dimethoxy-5-methyl-*p*-quinone
Probably formed from: 5-Demethyl-ubiquinone 30
Ubiquinone-40 (coenzyme Q_8)
Formed from: 2-Octaprenylphenol
→ Ubiquinol 40
Probably in: Pig K814
→ Ubiquinone 40 semiquinone radical
In: Escherichia K661
Ubiquinone-45 (coenzyme Q_9)
Formed from: 2,3-Dimethoxy-5-methyl-*p*-quinone
Dihydroubiquinone 45
Ubiquinone-50 (coenzyme Q_{10})
Formed from: 2,3-Dimethoxy-5-methyl-*p*-quinone
Probably formed from: 5-Demethyl-ubiquinone 30
Ubiquinone 40 semiquinone radical
Formed from: Ubiquinone 40
Umbellic acid
See: 2,4-Dihydroxycinnamic acid
Umbelliferone (7-hydroxycoumarin)
Formed from: 7-Butoxycoumarin
p-Coumaric acid
Coumarin
7-Ethoxycoumarin
7-Methoxycoumarin
7-Propoxycoumarin
7-Vinyloxycoumarin
→ *p*-Coumaric acid
In: Agathosma D73
→ 7-Demethylsuberosin
In: Ruta A163
→ 3,7-Dihydroxycoumarin
In: Mouse B190
→ 6,7-Dihydroxycoumarin
In: Cichorium D687
Man G954
Mouse B190
See also: C774, D73, G605
→ 7,8-Dihydroxycoumarin
In: Mouse B190
→ Marmesin
In: Ruta A547
→ Skimmin
In: Cichorium D687
Nicotiana B442, B819
→ Umbelliferone-β-D-glucoside
In: Bupleurum E681
Catharanthus E681
Datura E681
Gardenia E681
Lithospermum E681
Manduca G816
Perilla E681
Periplaneta G816
Trilobium G816
→ Umbelliferone-β-D-glucuronide
Chicken B468
Dog H786
Guinea pig A3970, A3971, B303
Man A3553, G968, J416, J490
Rainbow trout C610, H487

Rat A3970, B165, C160, C195, C563, D128, H697, K129
→ Umbelliferone-O-sulphate
In: Guinea pig A3970, A3971, B303
Rat A3970, C563, D155, D817, H697, K129

Umbelliferone-β-D-glucoside
Formed from: Umbelliferone

Umbelliferone-β-D-glucuronide
Formed from: Umbelliferone

Umbelliferone-O-sulphate
Formed from: Umbelliferone

Umbelliprenin (7-(((2*E*,6*E*)-3,7,11-trimethyl-2,6,10-dodecatrienyl)oxy)-2*H*-1-benzopyran-2-one)
Formed from glycine in Thamnosma A1012

Ungeremine (4,5-dihydro-2-hydroxy[1,3]dioxolo[4,5-*j*]pyrrolo[3,2,1-*de*]phenanthridinium)
Formed from: Lycorine

Urapidil (6-(3-(4-(*o*-methoxyphenyl)piperazinyl)propylamino)-1,3-dimethyluracil)
→ 6-(3-(4-(4-Hydroxy-2-methoxyphenyl)piperazinyl)propylamino)-1,3-dimethyluracil
In: Dog D202
Man D202, E155
Mouse D202
Rat D202
→ 6-(3-(4-(*o*-Hydroxyphenyl)piperazinyl)propylamino)-1,3-dimethyluracil
In: Dog D202
Man D202
Mouse D202
Rat D202
→ 6-(3-(4-(*o*-Methoxyphenyl)piperazinyl)propylamino)-1-methyluracil
In: Dog D202
Man D202
Mouse D202
Rat D202

4-Ureidoetodolac
Formed from: Etodolac

Usnic acid (both isomers; 2,6-diacetyl-7,9-dihydroxy-8,9b-dimethyl-1,3-(2*H*,9b*H*))
→ 3′-(3-Acetyl-4-hydroxy-1-methyl-2-oxocyclohexa-3,5-dienyl)-2′,4′,6′-trihydroxy-5′-methylacetophenone
In: Evernia B774
See also: E126
→ Deacetylusnic acid
In: Mortierella A2521

Uxepam (4-carbamoyl-7-chloro-1,3,4,5-tetrahydro-1-methyl-5-phenyl-2*H*-1,4-benzodiazepin-2-one, 7-chloro-1,2,3,5-tetrahydro-1-methyl-2-oxo-5-phenyl-4*H*-1,4-benzodiazepine-4-carboxamide)
→ 4-Carbamoyl-7-chloro-1,3,4,5-tetrahydro-5-phenyl-2*H*-1,4-benzodiazepin-2-one
In: Dog E158
Man E158
Rat E158
→ 7-Chloro-1,3-dihydro-1-methyl-5-phenyl-2*H*-1,4-benzodiazepin-2-one
In: Man E158
Rat D763, E158
→ Noruxepam
In: Rat D763

V

Valeranilide
→ Aniline
In: Oryza A3773
(*S*)-N-Valeryl-N-(2′-(1*H*-tetrazol-5-yl)(1,1′-biphenyl-4-yl)methyl)valine
See: Valsartan
Valsartan ((*S*)-N-Valeryl-N-((2′-(1*H*-tetrazol-5-yl)(1,1′-biphenyl)-4-yl)methyl)valine)
→ N-(4-Hydroxypentanoyl)-N-(2′-(1*H*-tetrazol-5-yl)(1,1′-biphenyl-4-yl)methyl)valine
In: Man J72
L-Valyl-β-naphthylamide
→ β-Naphthylamine
In: Agave 2745
Aspergillus A3607
Beef A526
Glycine A1596
Rabbit B334
Rat A523
L-Valyl-*p*-nitroanilide
Formed from: Succinyl-L-alanyl-L-alanyl-L-valyl-*p*-nitroanilide
→ *p*-Nitroaniline
In: Nocardia G343
Vanillactic acid (4-hydroxy-3-methoxyphenyllactic acid)
Formed from: 3,4-Dihydroxyphenyllactic acid
Vanilpyruvic acid
Probably formed from: 2,3-Dihydro-2,3-dihydroxyeugenol
→ 3,4-Dihydroxyphenyllactic acid
In: Rat A2961
→ Vanilpyruvic acid
In: Rat A2961
***o*-Vanillic acid** (2-hydroxy-3-methoxybenzoic acid)
Formed from: 2-hydroxy-3-methoxybenzaldehyde
Pyrocatechuic acid
Vanillic acid (4-hydroxy-3-methoxybenzoic acid)
Formed from: ω-(4-carboxy-2-methoxyphenoxy)acetovanillone
5-Carboxyvanillic acid
5-Chloro-4-hydroxy-3-methoxybenzoic acid
Dehydrodiconiferyl alcohol
Dehydrodiisoeugenol
4-Ethoxy-3-methoxybenzoic acid
4-Ethoxy-3-methoxyphenylglycol
β-vanillic acid ether
Protocatechuic acid
Vanillin
Veratric acid
Probably formed from: 5-Carboxyvanillic acid
4-Hydroxy-3-methoxy-β-oxophenylpropionic acid
→ 2-(5-Carboxy-2-hydroxy-3-methoxyphenyl)-6-methoxy-*p*-quinone
In: Rhizoctonia C276
→ 2-(4-Carboxy-2-methoxyphenoxy)-6-methoxy-*p*-quinone
In: Rhizoctonia C276
→ 5,5′-Dicarboxy-2,2′-dihydroxy-3,3′-dimethoxybiphenyl
In: Rhizoctonia C276
→ 3,4-Dihydroxy-5-methoxybenzoic acid
Probably in : Botrytis A1930
→ Guaiacol
In: Nocardia E374, K406
Paecilomyces F199
Streptomyces B763, B877

See also: A361, D181, 367
→ Methoxyquinol
In: Candida J461
Fungus C285
Horseradish A3892
Paecilomyces F199
Phanerochaete B312
Polyporus A961
Sporotrichum B73, B381, C57
→ Protocatechuic acid
In: Acetobacterium F655, J677
Arthrobacter A672
Bacterium J361
Desulfomicrobium J491
Fusarium E115
Nocardia A960, B2, E374, F266
Pseudomonas D16, E733, E957
Sphingomonas K166
Sporomusa J199
Streptomyces B763
See also: A3068, B445, D779, D929, F3, F402, G275, G865, G931
→ Vanillic acid-β-D-glucoside
In: Nicotiana B442
→ Vanillic acid-4-sulphate
In: Rat B854
→ Vanillin
In: Neurospora A750, J723
Nocardia F266, K406
Pycnoporus D569, K72
Sporotrichum B381
With NAD(P) dependent benzaldehyde oxidoreductase E248
→ Vanilloyl-D-glucose
In: Cucumis A611
Glycine A3585
Mung bean A3585
Quercus K737
Raphanus B440
Sweet potato D70
See also: G897
→ Vanilloylglycine
In: Rat A361
→ Veratric acid
In: Nocardia E374, F266
Phanerochaete G782

Vanillic acid-β-D-glucoside
Formed from: Vanillic acid
Vanillic acid-4-sulphate
Formed from: Vanillic acid
***o*-Vanillin**
See: 2-Hydroxy-3-methoxybenzaldehyde
Vanillin (4-hydroxy-3-methoxybenzaldehyde)
Formed from: 1,2-Bis(4-hydroxy-3-methoxyphenyl)ethylene
1-(4-Ethoxy-3-methoxyphenyl)-2-(4-hydroxymethyl-2-methoxyphenoxy) propane-1,3-diol
Feruloyl CoA
4-Formyl-2-methoxyphenoxyacetic acid
4-Hydroxy-3-methoxybenzylamine
4-Hydroxy-3-methoxystyrene
Isoeugenol
Isorhapontin
Protocatecualdehyde
Vanillic acid
Veratrylglycerol β-guaiacyl ether
Vanillyl alcohol
Probably formed from: 4-Hydroxy-3-methoxyphenylglyoxylic acid
→ 2-Amino-4-hydroxy-4-(4-hydroxy-3-methoxyphenyl)butyric acid
In: Pseudomonas D449
→ Catechol
In: Clostridium G367
→ Cryptostyline I
In: Cryptostylis A1515
→ Polyvanillin
In: Bjerkandera K565
→ Protocatechualdehyde
In: Microorganism A261
See also: A361
→ Vanillic acid
In: Acinetobacter F366
Arthrobacter J33
Aureobasidium E402
Brettanomyces H356
Burkholderia K655
Clostridium H102, J864
Desophovibrio F666
Fusarium E115
Glucuronobacter G161

Horseradish A3892
Lithospermum G229
Pseudomonas F3, F244, J650, K107
Rat A361
Rhodopseudomonas C733, G246
Serratia G197
Streptomyces C284, C699, C818, F225
Wheat K668
See also: A2744, A3068, D574, D920, D929, E957, G865, H102, H309
→ Vanillin-β-D-glucoside
In: Agrotis G816
Coffea G930
Gardenia G930
Melanopus G816
Manduca G816, G897
Nicotiana G930
Periplaneta G816
Prunus G930
Trilobium G816
→ Vanillin-β-D-glucuronide
In: Man G524
→ Vanillin-O-sulphate
In: Dog J292
Rat C441
→ Vanillyl alcohol
In: Aureobasidium E402
Brettanomyces H356
Dunaliella G459
Euglena G422
Gluconobacter G161
Phanerochaete F866
Pseudomonas F3
Rat A361
Sporotrichum B381
Streptomyces E279
Wheat K668
See also: A2744, D569, D574, F266, H309, K406
→ Veratraldehyde
In: Phanerochaete G782
Pimpinella J188
See also: D569, H309

Vanillin acid-β-D-glucoside
Formed from: Vanillin

Vanillin-β-D-glucuronide
Formed from: Vanillin

Vanillin-O-sulphate
Formed from: Vanillin

Vanilloyl-D-glucose
Formed from: Vanillic acid

Vanilloylglycine
Formed from: Vanillic acid

Vanillyl alcohol (4-hydroxy-3-methoxybenzyl alcohol)
Formed from: 3,4-Dihydroxybenzyl alcohol
Orientaline
Vanillin
Probably formed from: 4-Ethoxy-3-methoxybenzyl alcohol
→ 3,4-Dihydroxybenzyl alcohol
Probably in : Rat A361
→ Guaiacol
In: Rat A361
→ 2-Methoxy-4-methylphenol
Probably in : Rat A361
→ Vanillin
In: Acinetobacter E663
Byssochlamys K164
Horseradish A3892
Methylosinus F404
Penicillium G662, J354, K761
Phanerochaete H613
Pleurotus F867
Rat A361
Rhodopseudomonas C733, D197
See also: E402
→ Veratryl alcohol
In: Phanerochaete G782

Vanilmandelic acid
See: 4-Hydroxy-3-methoxymandelic acid

Vanilpyruvic acid
Formed from: 3-Methoxy-L-tyrosine
Vanillactic acid
→ 4-Hydroxy-3-methoxyphenylacetic acid
In: Rat A3229
→ Vanillactic acid
In: Coleus G289
Rat A3229

Variabilin (6a-hydroxy-3,9-dimethoxypterocarpan)
Formed from: Medicarpin

Velnacrine (9-amino-1,2,3,4-tetrahydro-1-acridinol)
→ *cis*-3-Hydroxyvelnacrine
In: Dog G840
Man G840
Rat G840
→ *cis*-4-Hydroxyvelnacrine
In: Dog G840
Man G840
Rat G840
→ *trans*-4-Hydroxyvelnacrine
In: Dog G840
Man G840
Rat G840
→ 7-Hydroxyvelnacrine
In: Dog G840
Man G840
Rat G840

Venlafaxine (1-(2-dimethylamino-1-(*p*-methoxyphenyl)ethyl)cyclohexanol)
→ N-Demethylvenlafaxine
In: Dog G564
Man G564, J295, K103
Mouse G564
Rat G564
→ O-Demethylvenlafaxine
In: Dog G564
Man G564, J294, J295, K103
Mouse G564
Rat G564
→ 1-(2-Dimethylamino-1-(*p*-methoxyphenyl)ethyl)cyclohexae-1,4-diol
In: Mouse G564
Rat G564
Rhesus monkey G564
→ 5-(*p*-Methoxyphenyl)-3-methyl-1-oxa-3-aza*spiro*[5.5]undecan-9-ol
In: Rat G564
Rhesus monkey G564

Verapamil
See: 5-(N-(3,4-Dimethoxyphenethyl) methylamino)-2-(3,4-dimethoxyphenyl)-2-isopropylvaleronitrile

Veratraldehyde (3,4-dimethoxybenzaldehyde)
Formed from: α-Benzylveratryl alcohol
1,2-Bis(3,4-dimethoxyphenyl)-1,3-propanediol
3,4-Dihydro-2-methylpapaverinium
3,4-Dimethoxyallylbenzene
3,4-Dimethoxyphenylacetic acid
3,4-Dimethoxyphenylglycol
1-(3,4-Dimethoxyphenyl)-1-oxopropan-2,3-diol
2-Methylpapaverinium
Vanillin
Veratric acid
Veratryl alcohol
Veratrylamine
Veratrylglycerol β-guaiacyl ether
Veratrylglycerol β-phenyl ether
→ 3,4-Dimethoxymandelonitrile (both isomers)
In: Almond J194, J302
Apple J302
Cherry J302
Sorghum J194
→ Veratric acid
In: Alcaligenes H8
Arion G804
Aureobasidium E402
Limax G804
Methylosinus F404
Mouse G804, F432
Rhodopseudomonas G246
Serratia G197
Streptomyces C284
See also: A1765, B38, B548, D335, G637, G865
→ Veratryl alcohol
In: Alcaligenes H8
Aureobasidium E402
Dunaliella G459
Euglena G422
Phanerochaete F866
See also: D569, F698, G637

Veratric acid (3,4-dimethoxybenzaldehyde)
Formed from: 5-Carboxy-2-(3,4-dimethoxyphenyl)-2,3-dihydro-3-hydroxymethyl-7-methoxybenzofuran
Isovanillic acid
Vanillic acid

Veratraldehyde
Veratroyl-β-D-glucuronic acid
Probably formed from: 3,4-Dimethoxy-cinnamic acid
3,4-Dimethoxyphenylhydracrylic acid
3,4-Dimethoxyphenylpropionic acid
→ 3,4-Dimethoxyphenol
In: Sporotrichum B73
→ Isovanillic acid
In: Nocardia A960, D779, F266
Pseudomonas E957
Rat A1241
See also: B445
→ Vanillic acid
In: Glycine A3585
Mung bean A3585
Nocardia A960, B445, D779, F266
Polyporus A1117
Pycnoporus D569
Pseudomonas E957
Sporomusa G931
Streptomyces B763
Wheat A709
See also: A1765, D16, J199
→ Veratraldehyde
In: Neurospora A750
Nocardia G207
Phlebia F698
With NAD(P)-dependent benzaldehyde oxidoreductase E248
See also: F824, H8, K896
→ Veratroyl-β-D-glucose
In: Quercus K737
→ Veratroylglycine
In: Rat A1765
→ Veratryl chloride
Probably in : Bjerkandera J482

Veratrole (1,2-dimethoxybenzene)
Formed from: Guaiacol
→ Catechol
In: Clostridium G367
See also: G931
→ Guaiacol
In: Rat J201

Veratroyl-β-D-glucose
Formed from: Veratric acid

Veratroyl-β-D-glucuronic acid
→ Veratric acid
In: Beef A1029
Escherichia A1029

Veratroylglycine
Formed from: Veratric acid

Veratryl alcohol (3,4-dimethoxybenzyl alcohol)
Formed from: Vanillyl alcohol
Veratraldehyde
Veratryl-β-D-glucuronide
Veratrylglycerol β-(pentachlorophenyl) ether
→ 2-Bromo-4,5-dimethoxybenzyl alcohol
In: Glycine K439
→ 3,4-Dimethoxybenzyl-β-D-glucoside
In: Verbesina K98
→ 4,5-Dimethoxy-2-nitrobenzaldehyde
In: Phanerochaete J624
→ 4-Nitroveratrole
In: Phanerochaete J624
→ Veratraldehyde
In: Acinetobacter F859, E663
Arion G804
Aureobasidiun E402
Bjerkandera F610
Helix G804
Limax G804
Methylosinus F404
Phanerochaete D758, F257, F610, H613, J621
Pleurotus F867, K373
See also: E392, E653

Veratrylamine (3,4-dimethoxybenzylamine)
→ 4-Hydroxy-3-methoxybenzylamine
In: Rat G865
→ Veratraldehyde
Probably in : Rat G865

Veratryl chloride
Probably formed from: Veratric acid

Veratryl-β-D-glucuronide
→ Veratryl alcohol
In: Beef A1029
Escherichia A1029

Veratrylglycerol
Formed from: 3,4-Dimethoxycinnamyl alcohol
1-(3,4-Dimethoxyphenyl)-1-oxopropan-2,3-diol
→ 1-(3,4-Dimethoxyphenyl)-1-oxopropan-2,3-diol
In: Coriolus D335

Veratrylglycerol β-coniferyl ether
→ Veratrylglycerol β-(4-formyl-2-methoxyphenoxy) ether
In: Phanerochaete D18
→ Veratrylglycerol β-(4-(1-glyceryl)-2-methoxyphenoxy) ether
In: Phanerochaete D18

Veratrylglycerol β-(4-formyl-2-methoxyphenoxy) ether
Formed from: Veratrylglycerol β-coniferyl ether

Veratrylglycerol β-(4-(1-glyceryl)-2-methoxyphenoxy) ether
Formed from: Veratrylglycerol β-coniferyl ether

Veratrylglycerol β-guaiacyl ether
→ Guaiacol
In: Acinetobacter G637
→ 2-(*o*-Methoxyphenoxy)ethanol
In: Acinetobacter F859
Microorganism D920
→ Vanillin
In: Microorganism D920
See also: G637
→ Veratraldehyde
In: Acinetobacter G637

Veratrylglycerol β-(pentachlorophenyl) ether
→ Pentachlorophenol
In: Phanerochaete E392
→ Veratryl alcohol
In: Phanerochaete E392

Veratrylglycerol β-phenyl ether
→ 3-(3,4-Dimethoxyphenyl)-3-hydroxy-2-phenoxypropionic acid
In: Corynebacterium B548
→ Phenoxyacetic acid
In: Corynebacterium B548
→ Veratraldehyde
In: Corynebacterium B548

Verlukast ((*S*)-(3-(((3-(2-(7-chloroquinolin-2-yl)-(*E*)-ethenyl)phenyl)-3-dimethylamino-3-oxopropylthio)methyl)thio)propionic acid)
→ (*S*)-(3-(((3-(2-(7-chloroquinolin-2-yl)-(*E*)-ethenyl)phenyl)-3-(N-hydroxymethyl-N-methylamino)-3-oxopropylthio) methyl)thio)propionic acid
In: Rat G84
→ Demethylverlukast
In: Rat G84
→ 12-(S-Glutathionyl)-11,12-dihydroverlukast
In: Rat H464
→ 6″-Hydroxyverlukast
In: rat G696
→ Verlukast-5″,6″-oxide
In: Rat G839
See also: G696
→ Verlukast-S-oxide
In: Rat G84
→ Verlukast-S′-oxide
In: Rat G84

Verlukast-5″,6″-oxide
Formed from: Verlukast
→ 5″,6″-Dihydro-5″,6″-dihydroxyverlukast
Probably in : Rat G696, G839

Verlukast-S-oxide
Formed from: Verlukast

Verlukast-S′-oxide
Formed from: Verlukast

Veronicafolin-3,4′-disulphate (veronicafolin: 3,3′,5-trihydroxy-4′-methoxyflavone (Chemical Abstracts) or 3,4′,5-trihydroxy-3′-methoxyflavone [F102])
→ Veronicafolin-3,-sulphate
In: Helix F102

Vermelone (3,8-dihydroxy-1-tetralone)
Formed from: 1,3,8-Trihydroxynaphthalene

Veronicafolin-3-sulphate
Formed from: Veronicafolin-3,4′-disulphate

Verruculogen
→ Iso-TR-2
In: Sheep E782
→ (5a*R*,6*S*,12*S*,14a*S*)-1,2,3,5a,6,11,12,14a-Octahydro-5a,6-dihydroxy-12-(2-hydroxy-2-methylpropyl)-9-methoxy-5*H*,14*H*-pyrrolo[1″,2″:4′,5′]pyrazino[1′,2′:1,6]pyrido[3,4-*b*]indole-5,14-dione

In: Sheep E782

Versicolorin A ((3a*S*,12a*S*)-3a,12a-dihydro-4,6,8-trihydroxyanthra[2,3-*b*]furo[3,2-*d*]furan-5,10-dione)
Formed from: Versicolorin B
Versicolorin A hemiacetal
Formed from acetate in Aspergillus A1824
Probably formed from: Versiconal acetate
→ Aflatoxin B_1
In: Aspergillus A1824, F260
See also: D57, E744

Versicolorin B ((3a*S*,12a*S*)-2,3,3a,12a-tetrahydro-4,6,8-trihydroxyanthra[2,3-*b*]furo[3,2-*d*]furan-5,10-dione)
Formed from: Versicolorin C
Probably formed from: Averufin
→ Aflatoxin B_2
In: Aspergillus F260
→ Versicolorin A
In: Aspergillus G344
See also: F260

Versicolorin C ((3a*R*,12a*S*)-rel-2,3,3a,12a-tetrahydro-4,6,8-trihydroxyanthra[2,3-*b*]furo[3,2-*d*]furan-5,10-dione)
→ Versicolorin B
In: Aspergillus G344

Versicolorin A hemiacetal
→ Versicolorin A
In: Aspergillus G120
See also: F260

Versiconal acetate ((α*S*)-α-(2-(acetyloxy)ethyl)-9,10-dihydro-1,3,6,8-tetrahydroxy-9,10-dioxo-2-anthraceneacetaldehyde)
Formed from: Averufin
→ Versicolorin A
Probably in: Aspergillus D57

Versiconal hemiacetal acetate
Formed from acetate in Aspergillus A2536
→ Versiconal hemiacetal alcohol
Probably in : Aspergillus F461

Versiconal hemiacetal alcohol
Probably formed from: Versiconal hemiacetal acetate

Vestitol ((3*R*)-3,4-dihydro-3-(2-hydroxy-4-methoxyphenyl)-2*H*-1-benzopyran-7-ol)
Formed from: Medicarpin
→ Medicarpin
In: Clover A3723
Medicago A3723
→ Sativan
Medicago A3723

Vestitone (2′,5′,7-trihydroxy-4′-methoxyisoflavanone)
Formed from: Biochanin A
→ Medicarpin
In: Cicer K897

Viagra
See: Sildenafil

Vicenin (6,8-bis(C-glucosyl)-4′,5,7-trihydroxyflavanone)
Formed from: Cinnamic acid

Vicianin ((α*R*)-α-((6-O-α-L-arabinopyranosyl-β-D-glucopyranosyl)oxy)benzeneacetonitrile)

Vignafuran (3-methoxy-4-(6-methoxy-2-benzofuranyl)phenol)
Formed from: L-Phenylalanine

Viloxazine (vivalan, 2-(2-ethoxyphenoxymethyl)-2,3,5,6-tetrahydro-1,4-oxazine)
→ N-Acetylviloxazine
In: Dog A355
→ 4′-Hydroxyviloxazine
In: Man A354
See also: A355
→ 5′-Hydroxyviloxazine
In: Man A354
→ N-Methylviloxazine
In: Dog A355
Man A354
→ 5-Oxoviloxazine
In: Dog A355
→ 2,3,5,6-Tetrahydro-2-(2-hydroxyphenoxymethyl)-1,4-oxazine
In: Rat A355

Vinamidine ((2*R*,3*E*,7a*R*,12b*S*,13*S*)-3-ethylidene-1,3,4,6,7,12b-hexahydro-2*H*-2,7a-methanoindolo[2,3-*a*]quinolizine-13-carboxylic acid methyl ester)
Formed from: 3′,4′-Anhydrovinblastine

Vinblastine (vincaleukoblastine)
Formed from: Anhydrovinblastine
Catharanthine
Vindoline

Formed from loganin in Catharanthus A1384
→ 3′,4′-Anhydrovinblastine-N-oxide
Probably in : Dog F662
→ Catharinine
In: Man F231
→ Deacetylvinblastine
In: Beef D890
Man A3124
See also: F662
→ 16-Deoxyvinblastine-(15*H*)-15,16-oxide
In: Man F231
→ 10′-Hydroxyvinblastine
In: Streptomyces A1431
→ 19′-Oxovinblastine
Probably in : Dog F662
→ Vinblastine-7′,20′-ether
In: Streptomyces A1431

Vinblastine-7′,20′-ether
Formed from: Vinblastine

Vincaleukoblastine
See: Vinblastine

Vincamine (3α,14β,16α)-14,15-dihydro-14-hydroxyeburnamenine-14-carboxylic acid methyl ester)
→ 6α-Hydroxyvincamine
In: Rat B733
See also: B4
→ 6β-Hydroxyvincamine
In: Rat B733
→ Vincamone
In: Rat B733

Vincamone (3α,16α)-eburnamenin-14-one)
Formed from: Vincamine

Vincoside
Does not yield ajmalicine, catharanthine, serpentine or vindoline in Catharanthus A3257

Vincristine (22-oxovincaleukoblastine)
→ 4-Deacetylvincristine
In: Dog F668
→ 4′-Deoxy-3′-hydroxyvincristine
Probably in : Dog F668
→ 3′,4′-Epoxyvincristine-N-oxide
Probably in : Dog F668
→ N-Formylcatharinine
In: Horseradish J385
With caeruloplasmin J385

Vindoline ((2β,3β,4β,5α,12*R*,19α)-4-acetyloxy-6,7-didehydro-3-hydroxy-16-methoxy-1-methylaspidospermine-3-carboxylic acid methyl ester)
Formed from: Deacetylvindoline
Strictosidine
Tabersonine
Tryptamine
→ 3′,4′-Anhydrovinblastine
In: Catharanthus E745
Horseradish E722
→ Deacetylvindoline
In: Catharanthus A3204
→ Dihydrovindoline
In: Catharanthus A3204
→ Dihydrovindoline ether
In: Streptomyces A2984
→ Vinblastine
In: Catharanthus A3608
→ Vindoline dimer
In: Streptomyces A2984

Vindoline dimer
Formed from: Vindoline

Vineomycin A_1
Formed from acetate in Streptomyces C282

Vineomycin B_2
Formed from acetate in Streptomyces C282

ε-Viniferin (5-((2*R*,3*R*)-2,3-dihydro-6-hydroxy-2-(4-hydroxyphenyl)-4-((1*E*)-2-(4-hydroxyphenyl)ethenyl)-3-benzofuranyl)-1,3-benzenediol)
Formed from: Resveratrol

Vinorine (17*R*,19*E*)-1,2,19,20-tetradehydro-1-demethylajmalan-17-ol acetate)
→ Vomilenine
In: Rauwolfia H414

Vinorelbine
→ Deacetylvinorelbine
In: Man G713
→ 15,16-Epoxyvinorelbine
In: Rat H941
→ 10′-Hydroxyvinorelbine
In: Rat H941
→ 11′-Hydroxyvinorelbine
In: Rat H941
→ 19′-Hydroxyvinorelbine

In: Rat H941
Vintriptol
→ Vintriptol acid
In: Man G439
Vintriptol acid
Formed from: Vintriptol
***p*-Vinylbenzoic acid**
Formed from: *p*-Methylstyrene
→ *p*-Vinylhippuric acid
In: Rat C153, C681
3-Vinylcatechol
Formed from: *cis*-2,3-Dihydro-2,3-dihydroxy-styrene
Styrene
→ 2-Hydroxy-6-oxoocta-2,4,7-trienoic acid
In: Rhodococcus H253
4-Vinylguaiacol
See: 4-Hydroxy-3-methoxystyrene
***p*-Vinylhippuric acid**
Formed from: *p*-Vinylbenzoic acid
7-Vinyloxycoumarin
→ Umbelliferone
In: Rat F105
→ 7-Vinyloxycoumarin oxide
In: Rat F105
7-Vinyloxycoumarin oxide
Formed from: 7-Vinyloxycoumarin
4-Vinylphenol
See: *p*-Hydroxystyrene
***p*-Vinyl-L-phenylalanine**
Formed from: *p*-Vinylphenypyruvic acid
***p*-Vinylphenyl-β-D-glucoside**
→ O-(β-D-Apiofuranosyl(1 → 2)-β-D-glucosyl)-*p*-vinylphenol
In: Parsley K827
***p*-Vinylphenylpyruvic acid**
→ *p*-Vinyl-L-phenylalanine
Bacillus G168
4-Vinylsyringol
See: 4-Hydroxy-3,5-dimethoxystyrene
Violacein ((3*E*)-3-(1,2-dihydro-5-(5-hydroxy-1*H*-indol-3-yl)-2-oxo-3*H*-pyrrol-3-ylidene)-1,3-dihydro-2*H*-indol-2-one)
Vinpocetine
See: Ethyl apovincaminate
2-Vinylphenol
See: *o*-Hydroxystyrene
3-Vinylphenol
See: *m*-Hydroxystyrene
Vitamin B_2 acid (riboflavinoic acid)
Probably formed from: Vitamin B_2 aldehyde
Vitamin B_2 aldehyde (riboflavinal, 10-(4-formyl-2,3,4-trihydroxybutyl)-7,8-dimethylisoalloxazine)
Formed from: Riboflavine
→ Vitamin B_2 acid
Probably in : Schizophyllum C731, J393
Vitamin B_{12}
Formed from: 5-Hydroxybenzimidazole
→ Adenosylcobalamin
In Spirometra A3168
→ Hydroxycobalamin
In Spirometra A3168
Vitamin E
See: α-Tocopherol
Vitamin K_1 (2-methyl-3-phytyl-1,4-naphthoquinone, *trans*-phylloquinone; assumed unless otherwise stated)
Formed from: *o*-Succinylbenzoic acid
Vitamin K_1 epoxide
→ 3-(5′-Carboxy-3′-methyl-2′-pentenyl)-2-methyl-1,4-naphthoquinone
In: Rabbit A1198
See also: A793
→ 3-(5′-Carboxybutyl)-2-methyl-1,4-naphthoquinone
In: Rabbit A1198
See also: A793
→ 2,3-Dihydro-2-hydroxy-2-methyl-3-phytylnaphthoquinone
In: Beef D683
→ 2,3-Dihydro-3-hydroxy-2-methyl-3-phytylnaphthoquinone
In: Beef D683
→ 1,4-Dihydroxy-2-methyl-3-phytylnaphthoquinone
Probably in : Rat D84
→ Vitamin K_1 epoxide
In: Rat A51, A1356, A3441
Vitamin $K_{2(10)}$
See: Menaquinone 2

Vitamin $K_{2(15)}$
See: Menaquinone 3
Vitamin $K_{2(20)}$ (menaquinone 4)
Formed from: Menadiol diphosphate
Vitamin $K_{2(45)}$
Formed from: Menadiol diphosphate
Vitamin $K_{2(50)}$
Formed from: Menadiol diphosphate
Vitamin K_1 epoxide
Formed from: Vitamin K_1
→ Vitamin K_1
In: Rat B338, D84, D324
Vitavax
See: 3-Carboxanilido-5,6-dihydro-2-methyl-1,4-oxathiin
Vitexin (8-β-D-glucopyranosyl-5,7-dihydroxy-2-(4-hydroxyphenyl)-4*H*-1-benzopyran-4-one)
Formed from: Naringenin
Probably formed from: Apigenin
→ Vitexin-2″-D-glucoside
In: Silene B203
Vitexin-2″-D-glucoside
Formed from: Vitexin
Vivalan
See: Viloxazine
VMA
See: 4-Hydroxy-3-methoxymandelic acid
Volazocine ((2*R*,6*R*,11*R*)-rel-3-(cyclopropylmethyl)-1,2,3,4,5,6-hexahydro-6,11-dimethyl-2,6-methano-3-benzazocine)
→ Hydroxyvolazocine
In: Rat A2134
Vomilenine (17*R*,19*E*,21α)-1,2,19,20-tetradehydro-1-demethylajmalan-17,21-diol 17-acetate)
Formed from: Raucaffricine
Vinorine

W

Warfarin (1-(4-hydroxy-3-coumarinyl)-1-phenyl-2-butanone; stereochemistry unclear)
→ Dihydrowarfarin
In: Catharanthus G880
Rabbit A3980
→ 1-(*o*-Hydroxyphenyl)-3-phenylhexan-1,2,5-trione
In: Aspergillus F315
→ 3′-Hydroxywarfarin
In: Cunninghamella G204
→ 4′-Hydroxywarfarin
In: Beauveria G645, G706, J82
Cunninghamella F144
Man F117
Rabbit A3980
Rat A2853, F907
Streptomyces J82
See also: A2643
→ 6-Hydroxywarfarin
In: Catharanthus G880
Cow A2024
Coypu A2024
Deer A2024
Jackdaw A2024
Man F117
Mouse A2024
Owl A2024
Pig A2024
Pigeon A2024
Quail A2024
Rabbit A2024, A3980
Rat A496, A2024, A2853, F907, J48
Squirrel A2024
→ 7-Hydroxywarfarin
In: Cow A2024
Coypu A2024
Deer A2024
Jackdaw A2024
Man F117, H925
Mouse A2024
Owl A2024
Pig A2024
Pigeon A2024
Quail A2024
Rabbit A2024, A3980
Rat A2024, A2853, F907, J48
Squirrel A2024
Streptomyces J82
→ 8-Hydroxywarfarin
In: Cow A2024
Coypu A2024
Deer A2024
Jackdaw A2024
Man F117, H711
Mouse A2024
Owl A2024
Pig A2024
Pigeon A2024
Quail A2024
Rabbit A2024
Rat A2024, A2853, F907, J48
Squirrel A2024
→ 10-Hydroxywarfarin
In: Catharanthus G880
→ 12-Hydroxywarfarin
Streptomyces J82
→ Warfarin-α-diketone
In: Beauveria G645, G706

(*R*)-Warfarin
→ Dehydrowarfarin
In: Man C798
Mouse D143
Rat B166, B311
→ (*R*,*R*)-Dihydrowarfarin

In: Man A1675
Rat A1659
→ (*R*,*S*)-Dihydrowarfarin
In: Man A1470, A1675
Rabbit G276
Rat A1659
See also: A140, A1669
→ 1-Hydroxy-1-(4′-hydroxy-3′-coumarinyl)-1-phenyl-3-butanone
In: Rat A1801
See also: B166
→ 4′-Hydroxy-(*R*)-warfarin
In: Man C798
Mouse D143
Rabbit A3731
Rat A1659, A1801, B166, B311, B662, D716
→ 6-Hydroxy-(*R*)-warfarin
In: Man A1470, C798, H481, H697a, H698
Mouse D143
Rabbit A3731
Rat A1659, A1801, A2762, B166, B311, B662, D716
→ 7-Hydroxy-(*R*)-warfarin
In: Man A1675, C798, H697a
Mouse D143
Rabbit A3731
Rat A1659, A1801, A2762, B166, B311, B662, D716
→ 8-Hydroxy-(*R*)-warfarin
In: Man H481, H697a, H698
Mouse D143
Rabbit A3731
Rat A1659, A1801, A2762, B166, B311, B662, D716
→ 10-Hydroxy-(*R*)-warfarin
In: Man C798, H697a, H698
Mouse D143

(*S*)-Warfarin
→ Dehydrowarfarin
In: Man C798
Mouse D143
Rat B166, B311
→ (*S*,*R*)-Dihydrowarfarin
In: Man A1675
Rat A1659
→ (*S*,*S*)-Dihydrowarfarin
In: Man A1470, A1675
Rabbit G276
Rat A1659
See also: A140, A1669
→ 1-Hydroxy-1-(4′-hydroxy-3′-coumarinyl)-1-phenyl-3-butanone
In: Rat A1801
See also: B166
→ 4′-Hydroxy-(*S*)-warfarin
In: Mouse D143
Rabbit A3731
Rat A1659, A1801, A2762, B166, B311, B662, D716
→ 6-Hydroxy-(*S*)-warfarin
In: Man A1470, C798, H697a, H698
Mouse D143
Rabbit A3731
Rat A1659, A1801, B166, B311, B662, D716
→ 7-Hydroxy-(*S*)-warfarin
In: Man A1470, A1675, C798, H202, H697a, H698
Mouse D143
Rabbit A3731
Rat A1659, A1801, B166, B311, B662, D716
→ 8-Hydroxy-(*S*)-warfarin
In: Mouse D143
Rabbit A3731
Rat A1659, A1801, A2762, B166, B311, B662, D716
→ 10-Hydroxy-(*S*)-warfarin
In: Man C798
Mouse D143
→ 11-Hydroxy-(*S*)-warfarin
In: Rat G596, J224

Warfarin-α-diketone
Formed from: Warfarin

Wellbutrin
See: Bupropion

Wighteone (5,7-dihydroxy-3-(4-hydroxyphenyl)-6-(3-methyl-2-butenyl)-4-*H*-1-benzopyran-4-one)
→ 2′,3′-Dihydro-2′,3′-dihydroxywighteone
In: Aspergillus E8
Botrytis E8

→ Dihydrofuranisoflavone
In: Aspergillus E8
Botrytis E8
→ Dihydropyranisoflavone
In: Aspergillus E8
Botrytis E8

Wogonin (5,7-dihydroxy-8-methoxyflavone)
Formed from: Norwogonin
Wogonin-β-D-glucuronide
→ 5-Hydroxy-7,8-dimethoxyflavone
Probably in: Citrus A3987
→ Wogonin-7-β-D-glucuronide
In: Scutellaria K402

Wogonin-β-D-glucuronide (position unclear)
→ Wogonin
In: Scutellaria H747

Wogonin-7-β-D-glucuronide
Formed from: Wogonin

X

Xamoterol (N-(2-((2-hydroxy-3-(4-hydroxyphenoxy)propyl)amino)ethyl)-4-morpholinecarboxamide)
→ Xamoterol-*p*-β-D-glucuronide
In: Dog F285
Rat D386, E224, F312
→ Xamoterol-*p*-O-sulphate
In: Dog F285
Man D386
Rat E224, F312

Xamoterol-*p*-β-D-glucuronide
Formed from: Xamoterol

Xamoterol-*p*-O-sulphate
Formed from: Xamoterol

Xanthanoic acid (9*H*-xanthene-9-carboxylic acid)
Formed from: Propantheline
→ Hydroxyxanthanoic acid
Probably in: Man A3490
→ Xanthanoyl-*p*-β-D-glucuronic acid
Probably in: Man A3490

Xanthanoyl-*p*-β-D-glucuronic acid
Probably formed from: Xanthanoic acid

Xanthene
→ Xanthone
In: Pseudomonas H254

9*H*-Xanthene-9-carboxylic acid
See: Xanthanoic acid

9*H*-Xanthenone
See: Xanthone

Xanthoaphin (1-(4-hydroxyphenyl)-4-(4-methoxyphenyl)buta-1,3-diene-2,3-diisonitrile)
Formed from: Protoaphin aglycone

Xanthocillin (1-(4-hydroxyphenyl-2,3-diisocyano-4-(4-methoxyphenyl)-1,3-butadiene)
Formed from glycine and serine in Streptomyces D541

Xanthommatin (3-carboxy-1-hydroxy-5*H*-pyrido[3,2-*a*]phenoxazin-5-one)
Formed from: 3-Hydroxy-L-kynurenine
→ Dihydroxanthommatin
In: Drosophila E192, E387

Xanthone (9*H*-xanthen-9-one)
Formed from: Xanthene
→ 2-Hydroxyxanthone
In: Rat A290
→ 3-Hydroxyxanthone
In: Rat A290
→ 4-Hydroxyxanthone
In: Rat A290

Xanthone-1-acetic acid
→ 1-Carboxyxanthone
In: Mouse G936
→ 3-Hydroxyxanthone-1-acetic acid
In: Mouse G936
→ Xanthone-1-acetyl-β-D-glucuronic acid
In: Mouse G936

Xanthone-1-acetyl-β-D-glucuronic acid
Formed from: Xanthone-1-acetic acid

Xanthotoxol
See: 8-Hydroxypsoralen

Xanthurenic acid (4,8-dihydroxy-2-carboxylic acid)
Formed from: 3-Hydroxy-L-kynurenine
8-Methoxykynurenic acid
→ 8-Methoxykynurenic acid
In: Man A1550
Rat A1506
→ Xanthurenic acid-4,8-diglucoside
Probably in: Hornet A1195
→ Xanthurenic acid-8-O-β-D-glucoside
In: Drosophila F636

Xanthurenic acid-4,8-diglucoside
Formed from: Xanthurenic acid

Xanthurenic acid-8-O-β-D-glucoside
Formed from: Xanthurenic acid
Xilobam (1-(2,6-dimethoxyphenyl)-3-(1-methyl-2-pyrrolidinylidene)urea)
→ N-(2,6-Dimethylphenyl)urea
In: Mouse G938
Rat G938
→ 5′-Hydroxyxilobam
In: Dog G938
Man G938
Mouse G938
Ximoprofen (2-(4-(3-oximinocyclohexyl) phenyl)propionic acid)
→ 2-(4-(3-Oxocyclohexyl)phenyl)propionic acid
In: Baboon F498
Man F498
Rat F498
Xylazine (2-(2,6-dimethylphenylamino)-5,6-dihydro-4*H*-1,3-thiazine)
→ 3′-Hydroxyxylazine
In: Horse G504
Rat G504
→ 4′-Hydroxyxylazine
In: Horse G504
Rat G504
→ 4-Oxoxylazine
In: Horse G504
Rat G504
***o*-Xylene** (1,2-dimethylbenzene)
Formed from: Toluene
→ N-Acetyl-S-(*o*-methylbenzyl)-L-cysteine
In: Man F206
Rat C98
→ 2,3-Dimethylphenol
In: Aspergillus A2775
Gliocladium A2775
Pseudomonas E463, J357
Rat A2160
Streptomyces A2775
→ 3,4-Dimethylphenol
In: Aspergillus A2775
Pseudomonas J357
Rat A2160
→ *o*-Methylbenzyl alcohol
In: Pseudomonas J219
→ *o*-Toluic acid
In: Man F206
***m*-Xylene** (1,3-dimethylbenzene)
Formed from: Toluene
→ 4,5-Dihydro-4,5-dihydroxy-*m*-xylene
Probably in: Pseudomonas A1313
→ 2,4-Dimethylphenol
In: Aspergillus A2775
Gliocladium A2775
Man B309, D482
Pseudomonas J357
Rat E93, F899
Streptomyces A2775
→ 2,6-Dimethylphenol
In: Aspergillus A2775
→ *m*-Methylbenzyl alcohol
In: Pseudomonas J219, J357
Rat E93
→ (3-Methylbenzyl)succinic acid
In: Azoarcus K271
→ *m*-Toluic acid
In: Man B309, D482
Rat B24, F455, F899
Thauera H616
See also: A1301
***p*-Xylene** (1,4-dimethylbenzene)
Formed from: Toluene
→ 2,3-Dihydro-2,3-dihydroxy-*p*-xylene
In: Pseudomonas A1313
→ 2,5-Dimethylphenol
In: Aspergillus A2775
Gliocladium A2775
Pseudomonas J357
Rat A59, A2160
→ *p*-Methylbenzyl alcohol
In: Mortierella F823
Nitrosomonas H219
Pseudomonas A1009, J219, J357
Rat A59
See also: A1301, A1996, A2160, A3448, H616
→ *p*-Toluic acid
In: Rabbit A280
Xylerythrin (5-hydroxy-3-(4-hydroxyphenyl)-4,7-diphenyl-2,6-benzofurandione)
Formed from: 4-Hydroxypulvinic acid
L-Phenylalanine
D-Tyrosine

Xylobiosylcatechol
Formed from: Catechol

O-β-D-Xylopyranosyl(1 → 1)-O-(5-O-((*E*)-*p*-coumaroyl)-α-L-arabinofuranosyl(1 → 3))-O-β-D-xylopyranosyl(1 → 4)-D-xylopyranose
→ *p*-Coumaric acid
In: Neocallomastix G327

O-β-D-Xylopyranosyl(1 → 4)-O-(5-O-feruloyl-α-L-arabinofuranosyl(1 → 3)-O-β-D-xylopyranosyl(1 → 4)-D-xylopyranose
→ Ferulic acid
In: Aspergillus H381

β-Xylosylcatechol
Formed from: Catechol

3,4-Xylyl N-hydroxymethylcarbamate
Formed from: 3,4-Xylyl N-methylcarbamate

3,4-Xylyl N-methylcarbamate (meobal)
→ 4-Hydroxymethyl-3-methylphenyl N-methylcarbamate
In: Bean A1209
→ 3,4-Xylyl N-hydroxymethylcarbamate
In: Bean A1209

Y

Yangonin (4-methoxy-6-((1*E*)-2-(4-methoxyphenyl)ethenyl)-2*H*-pyran-2-one)
→ *p*-Hydroxykawain
Probably in: Rat A3485

Yatein ((3*R*,4*R*)-4-(1,3-benzodioxol-5-ylmethyl)dihydro-3-((3,4,5-trimethoxyphenyl)methyl)-2(3*H*)-furanone)
→ Podophyllotoxin
In: Podophyllum E237

Yohimbine ((16α,17α)-17-hydroxyyohimban-16-carboxylic acid methyl ester)
→ 10-Hydroxyyohimbine
In: Man K264
→ 11-Hydroxyyohimbine
In: Man K264

α-Yohimbine (isoyohimbine, (16β,17α,20α)-17-hydroxyyohimban-16-carboxylic acid methyl ester)
Formed from: Strictosidine

Z

Zafirlukast (5-(cyclopentoxycarbonylamino)-3-((2-methoxy-4-((2-methylphenyl)sulphonyl)aminocarbonyl)phenylmethyl)-1-methylindole)
→ 5-Amino-3-((2-methoxy-4-((2-methylphenyl)sulphonyl)aminocarbonyl)phenylmethyl)-1-methylindole
In: Dog J455
Mouse J455
Rat J455
→ 3-(4-Carboxy-2-methoxyphenylmethyl)-5-(cyclopentoxycarbonylamino)-1-methylindole
In: Dog J455
Mouse J455
Rat J455
→ 5-(Cyclopentoxycarbonylamino)-1-hydroxymethyl-3-((2-methoxy-4-((2-methylphenyl)sulphonyl)aminocarbonyl)phenylmethyl)indole
In: Dog J455
Mouse J455
Rat J455
→ 5-(Cyclopentoxycarbonylamino)-3-((4-((hydroxy-2-methylphenyl)sulphonyl)aminocarbonyl)-2-methoxyphenylmethyl)-1-methylindole
In: Dog J455
Mouse J455
Rat J455
→ 5-((2-Hydroxycyclopentoxy)carbonylamino)-3-((2-methoxy-4-((2-methylphenyl)sulphonyl)aminocarbonyl)phenylmethyl)-1-methylindole
In: Dog J455
Mouse J455
Rat J455
→ 5-((3-Hydroxycyclopentoxy)carbonylamino)-3-((2-methoxy-4-((2-methylphenyl)sulphonyl)aminocarbonyl)phenylmethyl)-1-methylindole
In: Dog J455
Mouse J455
Rat J455

Zaleplon (N-(3-(3-cyanopyrazolo[1,5-*a*]pyrimidin-7-yl)phenyl)-N-ethylacetamide)
→ Desethylzaleplon
In: Cynomolgus monkey J570
Man J299
Rat J570
→ 5-Oxozaleplon
In: Cynomolgus monkey J570
Rat J570

Zamifenacin ((3*R*)-(+)-diphenylmethoxy-1-(2-(3,4-methylenedioxyphenyl)ethyl)piperidine))
→ 1-(2-(3,4-Dihydroxyphenyl)ethyl)-3-diphenylmethoxypiperidine
In: Man H547
→ Hydroxyzamifenacin
In: Mouse H547
Rat H547

Zatosetron (5-chloro-2,3-dihydro-2,2-dimethyl-N-((3-*endo*)-8-methyl-8-azabicyclo[3.2.1]oct-3-yl)-7-benzofurancarboxamide)
→ 3-Hydroxyzatosetron
In: Man G556
→ Norzatosetron
In: Man G556
→ Zatosetron-N-oxide
In: Man G556

Zatosetron-N-oxide
Formed from: Zatosetron

α-Zearalenol (zeranol, (3*S*,7*R*,11*E*)-3,4,5,6,7,8,9,10-octahydro-7,14,16-trihydroxy-3-methyl-1*H*-2-benzoxacyclotetradecin-1-one)
Formed from: Zearalenone
→ α-Zearalenol-β-D-glucoside
Probably in: Zea G261
→ α-Zearalenol-β-D-glucuronide
In: Pig F716
Rat F716
See also: E264, H51
→ α-Zearalenol-O-sulphate
In: Pig F716
Rat F716
See also: E264
→ Zearalenone
In: Man C82

β-Zearalenol (taleranol, (3*S*,7*S*,11*E*)-3,4,5,6,7,8,9,10-octahydro-7,14,16-trihydroxy-3-methyl-1*H*-2-benzoxacyclotetradecin-1-one)
Formed from: Zearalenone
→ β-Zearalenol-β-D-glucoside
Probably in: Zea G261
→ β-Zearalenol-β-D-glucuronide
In: Pig F716
Rat F716
See also: H51
→ β-Zearalenol-O-sulphate
In: Pig F716
Rat F716

α-Zearalenol-β-D-glucoside
Probably formed from: α-Zearalenol

β-Zearalenol-β-D-glucoside
Probably formed from: β-Zearalenol

α-Zearalenol-β-D-glucuronide
Formed from: α-Zearalenol

β-Zearalenol-β-D-glucuronide
Formed from: β-Zearalenol

α-Zearalenol-O-sulphate
Formed from: α-Zearalenol

β-Zearalenol-O-sulphate
Formed from: β-Zearalenol

Zearalenone (2,4-dihydroxy-6-(10-hydroxy-6-oxo-*trans*-1-undecenyl)benzoic acid μ-lactone)
Formed from: α-Zearalenol
Formed from acetate and malonate in Fusarium A1191
→ 6-Carboxypentyl-2,4-dihydroxybenzoic acid
In: Fusarium A1817
→ 3′-Hydroxyzearalenone (2 isomers)
In: Fusarium B17
→ 8′-Hydroxyzearalenone
In: Fusarium A1817
→ α-Zearalenol
In: Guinea pig C858
Man C82
Microorganism D192
Mouse C858
Pig C858, H51
Rabbit C858
Rat A3825, C719, C858
Sheep E718
Turkey E264
Zea G261
See also: B511
→ β-Zearalenol
In: Guinea pig C858
Microorganism D192
Mouse C858
Pig C858
Rabbit C858
Rat C858
Sheep E718
Zea G261
See also: C430
→ Zearalenone-β-D-glucoside
In: Zea G261
→ Zearalenone-β-D-glucuronide
In: Pig F716
Rat A3825, F716
Turkey E264
→ Zearalenone-O-sulphate
In: Pig F716, H51
Rat F716
Turkey E264

Zearalenone-β-D-glucoside
Formed from: Zearalenone

Zearalenone-β-D-glucuronide
Formed from: Zearalenone

Zearalenone-O-sulphate
Formed from: Zearalenone

Zectran
See: 4-Dimethylamino-3,5-dimethylphenyl methylcarbamate
Zenarestat (3-(4-bromo-2-fluorobenzyl-7-chloro-2,4-dioxo-1,2,3,4-tetrahydroquinazolin-1-yl)acetic acid)
→ Hydroxyzenarestat
In: Rat G58
See also: G372
→ Zenarestat-β-D-glucuronide
In: Rat G11
Zenarestat-β-D-glucuronide
Formed from: Zenarestat
Zeranol
See: α-Zearalenol
Zetidoline (1-(3-chlorophenyl)-3-(2-(3,3-dimethyl-1-azetidinyl)ethyl)imidazolin-2-one)
→ 1-(3-Chloro-4-hydroxyphenyl)-3-(2-(3,3-dimethyl-1-azetidinyl)ethyl)imidazolin-2-one
In: Dog D384
Rat D384, D645
→ 1-(5-Chloro-2-hydroxyphenyl)-3-(2-(3,3-dimethyl-1-azetidinyl)ethyl)imidazolin-2-one
In: Rat D645
→ 1-(*m*-Chlorophenyl)-3-(2-(3,3-dimethyl-1-azetidinyl)ethyl)-5-hydroxyimidazolin-2-one
In: Rat D645
→ 1-(*m*-Chlorophenyl)-3-(2-hydroxyethyl) imidazolidin-2-one
In: Rat D645
→ Zetidoline-N-oxide
In: Rat D645
Zetidoline-N-oxide
Formed from: Zetidoline
Zileuton (N-(1-benzo[*b*]thien-2-ylethyl)-N-hydroxyurea; both *R* and *S*)
→ N-(1-Benzo[*b*]thien-2-ylethylurea
In: Man H469
→ Hydroxyzileuton
In: Man H469
→ Zileuton -β-D-glucuronide
In: Man G305, H142
→ Zileuton sulphoxide
In: Man H469
Zileuton-β-D-glucuronide
Formed from: Zileuton
Zileuton sulphoxide
Formed from: Zileuton
Zimeldine
See: Zimelidine
(*E*)-Zimelidine
→ (*E*)-Zimelidine-N-oxide
In: Pig E936
Rat E936
Zimelidine (zimeldine, (*Z*)-3-(4-bromophenyl)-N,N-dimethyl-3-(3-pyridyl)allylamine)
→ 3-(*p*-Bromophenyl)-3-(3-pyridyl) acraldehyde
In: Dog B791
Rat B791
→ Norzimelidine
In: Dog B791
Man B828
Rat B791
→ Zimelidine-N-oxide
In: Dog B791
Pig E936
Rat B791, E936
Zimelidine-N,N′-dioxide
Formed from: Zimelidine-N-oxide
(*E*)-Zimelidine-N-oxide
Formed from: (*E*)-Zimelidine
Zimelidine-N-oxide
Formed from: Zimelidine
→ Zimelidine-N,N′-dioxide
In: Dog B791
Rat B791
Zingerol (4-(4-hydroxy-3-methylphenyl)butan-2-ol)
Formed from: Zingerone
Zingerone (4-(4-hydroxy-3-methylphenyl)butan-2-one)
→ 4-(3,4-Dihydroxyphenyl)butan-2-one
In: Rat A1786
→ 2-(3,4-Dihydroxyphenyl)ethanol
In: Rat A1786
→ 1-Hydroxy-4-(4-hydroxy-3-methylphenyl) butan-2-one
In: Rat A1786
→ 4-Hydroxy-4-(4-hydroxy-3-methylphenyl) butan-2-one

In: Rat A1786
→ 4-(4-Hydroxy-3-methylphenyl)butan-1,2-diol
In: Rat A1786
→ 4-(4-Hydroxy-3-methylphenyl)butan-2,3-diol
In: Rat A1786
→ Zingerol
In: Rat A1786

Zipeprol (1-(2-hydroxy-3-methoxy-3-phenylpropyl)-4-(2-methoxy-2-phenylethyl)piperazine, respilene)
→ 1-(2-Hydroxy-3-methoxy-3-phenylpropyl) piperazine
In: Man A2936, A3298, B766
Rabbit C161
→ Mandelic acid
In: Man B766
→ 1-(2-Methoxy-3-phenylethyl)piperazine
In: Man A2936, A3298, B766
Rabbit C161
See also: A3300

Ziprasidone (5-(2-(4-(1,2-benzisothiazol-3-yl)-1-piperazin-1-yl)ethyl)-6-chloro-1,3-dihydro-2*H*-indol-2-one)
→ 1-(1,2-Benzisothiazol-3-yl)piperazine
In: Man J279
Rat J97
→ 6-Chloro-5-formylmethyloxindole
In: Man J279
See also: J97, K470
→ 6-Chloro-5-((piperazin-1-yl)ethyl)oxindole
In: Man J279
→ Dihydro-S-methylziprasidone
In: Man J279
Rat J281
→ Ziprasidone-N-oxide
In: Rat J97
→ Ziprasidone sulphoxide
In: Man J279
Rat J97

Ziprasidone-N-oxide
Formed from: Ziprasidone

Ziprasidone sulphone
Probably formed from: Ziprasidone sulphoxide

Ziprasidone sulphoxide
Formed from: Ziprasidone
→ 1-(4-Amino-5-carboxymethyl-2-chlorophenethyl)-4-(1,2-benzisothiazol-3-yl)piperazine sulphoxide
Probably in: Rat J97
→ 1-(2-(6-Chlorooxindol-5-yl)ethyl)-4-(2-sulphonamido)benzoyl)piperazine
Probably in: Rat J97
→ Ziprasidone sulphone
Probably in: Man J279, K470
Rat J97

Zolazepam (4-(*o*-fluorophenyl)-6,8-dihydro-1,3,8-trimethylpyrazolo[3,4-*e*][1,4]diazepin-7(1*H*)-one)
→ 1-Demethylzolazepam
In: Dog A1572
Monkey A1572
Rat A1572
→ 8-Demethylzolazepam
In: Dog A1572
Monkey A1572
Rat A1572
→ 6-Hydroxyzolazepam
In: Dog A1572
Monkey A1572
Rat A1572

Zolimidine (2-(4′-methylsulphonylphenyl) imidazo[1,2-*a*]pyridine)
→ 5,6-Dihydro-5,6-dihydroxy-2-(4′-methylsulphonylphenyl)imidazo[1,2-*a*] pyridine
In: Man A1557
→ 7,8-Dihydro-7,8-dihydroxy-2-(4′-methylsulphonylphenyl)imidazo[1,2-*a*] pyridine
In: Man A1557

Zolmitriptan ((*S*)-4-((3-(2-dimethylaminoethyl)-1*H*-indol-5-yl)methyl)-2-oxazolidine)
→ Demethylzolmitriptan
In: Man J741
→ (*S*)-4-((3-Formylmethyl-1*H*-indol-5-yl)methyl)-2-oxazolidine
See also: Man J741
→ Zolmitriptan-N-oxide
In: Man J741

Zolmitriptan-N-oxide
Formed from: Zolmitriptan
Zolpiden (N,N,6-trimethyl-2-(4-methylphenyl) imidazo[1,2-*a*]pyridine-3-acetamide)
→ 6-Hydroxymethyl-N,N-dimethyl-2-(4-methylphenyl)imidazo[1,2-*a*]pyridine-3-acetamide
In: Man H478
→ N-Hydroxymethyl-N,6-dimethyl-2-(4-methylphenyl)imidazo[1,2-*a*]pyridine-3-acetamide
In: Man H478
→ 2-((*p*-Hydroxymethyl)phenyl)-N,N,6-trimethylimidazo[1,2-*a*]pyridine-3-acetamide
In: Man H478
Zomepirac (5-(*p*-chlorobenzoyl)-1,4-dimethyl-1*H*-pyrrole-2-acetic acid)
→ *p*-Chlorobenzoic acid
In: Hamster B75
Man B107
Monkey B107
Mouse B75, B107
Rat B75, B107
→ 5-(*p*-Chlorobenzoyl)-4-hydroxymethyl-1-methyl-1*H*-pyrrole-2-acetic acid
In: Hamster B75
Man B75, B107, D255
Mouse B75, B107
Rabbit B75
Rat B75, B107
→ Zomepirac-β-D-glucuronide
In: Man B75, B107, C34, D255
Monkey B75, B107
Mouse B75, B107
Rabbit B75
Rat B107
Zomepirac-β-D-glucuronide
Formed from: Zomepirac
Zonisamide (3-(sulphamoylmethyl)-1,2-benzisoxazole)
→ N-Acetylzonisamide
In: Rat F594
→ *o*-(1-Amino-2-sulphamoylethyl)phenol
In: Rat F594
→ 3-Carboxy-1,2-benzisoxazole
In: Rat F594
→ *o*-Sulphamoylmethylcarbonylphenol
In: Rabbit H532
Rat F594, G59, G690, H703
→ Zonisamide-β-D-glucuronide
In: Rat F594
Zonisamide-β-D-glucuronide
Formed from: Zonisamide
Zopiclone (4-methyl-1-piperazinecarboxylic acid 6-(5-chloro-2-pyridinyl)-6,7-dihydro-7-oxo-5*H*-pyrrolo[3,4-*b*]pyrazin-5-yl ester)
→ Demethylzopiclone
In: Man J767
→ Zopiclone-N-oxide
In: Man J767
Zopiclone-N-oxide
Formed from: Zopiclone
Zopolrestat (3,4-dihydro-4-oxo-3-((5-trifluoromethyl-2-benzothiazolyl)methyl)-1-phthalazineacetic acid)
→ 3,4-Dihydro-5-hydroxy-4-oxo-3-((5-trifluoromethyl-2-benzothiazolyl)methyl)-1-phthalazineacetic acid
Probably in: Dog J459
Rat J457
→ 3,4-Dihydro-6-hydroxy-4-oxo-3-((5-trifluoromethyl-2-benzothiazolyl)methyl)-1-phthalazineacetic acid
Probably in: Dog J459
Rat J457
→ 3,4-Dihydro-7-hydroxy-4-oxo-3-((5-trifluoromethyl-2-benzothiazolyl)methyl)-1-phthalazineacetic acid
Probably in: Dog J459
Rat J457
→ 3,4-Dihydro-3-((4-hydroxy-5-trifluoromethyl-2-benzothiazolyl) methyl)-4-oxo-1-phthalazineacetic acid
Probably in: Dog J459
Rat J457
→ 3-((3-(S-Glutathionyl)-3,4-dihydro-4-hydroxy-5-trifluoromethyl-2-benzothiazolyl)methyl)-3,4-dihydro-4-oxo-1-phthalazineacetic acid
Probably in: Dog J459
Rat J457
→ 3-((4-(S-Glutathionyl)-3,4-dihydro-3-hydroxy-5-trifluoromethyl-

2-benzothiazolyl)methyl)-3,4-dihydro-4-oxo-1-phthalazineacetic acid
Probably in: Rat J457
→ 4-Oxo-1-phthalazineacetic acid
In: Rat J457
→ 5-Trifluoromethyl-2-benzothiazole-2-carboxylic acid
In: Rat J457
→ Zopolrestat-β-D-glucuronide
In: Dog J459
Rat J457

Zopolrestat-β-D-glucuronide
Formed from: Zopolrestat

Zotepine (2-chloro-11-(2-dimethylaminoethoxy) dibenzo[*b*,*f*]thiepine)
→ Demethylzotepine
In: Man J555
→ 2-Hydroxyzotepine
In: Man J555
→ 3-Hydroxyzotepine
In: Man J555
→ Zotepine-N-oxide
In: Man J555
→ Zotepine sulphoxide
In: Man J555

Zotepine-N-oxide
Formed from: Zotepine

Zotepine sulphoxide
Formed from: Zotepine

Zoxazolamine
See: 2-Amino-5-chlorobenzoxazole

Zyvox
See: Linezolid

Part 2
Enzymes and Reactions

Prior to 1972 many of the key enzyme systems described here were extensively studied, and were reviewed in my previous volume. Many of these systems have been further studied since 1972, and the new material is reviewed here. However, in some cases the newly published material is no more than a tidying-up of loose ends; no attempt has been made to re-introduce material published prior to 1972, except for enzymes given an E.C. number that would not otherwise be listed; these are listed with minimal data.

The policy for reviewing reactions and enzymes has been, not to write exhaustive reviews on each reaction type, but to give a representative set of data for those that have received extensive experimental study, and to concentrate relatively more effort on those that have not been extensively studied or are novel. Where extensive studies have been carried out in several species, the results for each species (and in some cases, each specific organs in that species) have been grouped together in a single paragraph; occasionally results are grouped according to substrate.

The molecular weights quoted are in Daltons, and temperature in degrees Celsius; the units are not mentioned in the text.

P450 Enzymes

Many microsomal reactions are catalyzed by enzymes of the P450 group (E.C. 1.14.14.1), particularly where xenobiotics are involved. Although most P450 studies have been carried out with liver microsomes, P450-containing microsomes are obtained from many tissues, and also from plants. The name P450 refers to Pigment with maximal optical absorption at 450 nm, observed with the carbon monoxide complex formed from the reduced enzyme; this binding inhibits the enzyme. The haem moiety binds molecular oxygen, which hydroxylates the substrate with the additional formation of water. Electron transport into the system is usually mediated by the flavin moiety of NADPH: cytochrome P450 reductase.

In this review no attempt has been made to rationalize isozyme names; over the years numerous isozymes have been detected and improved methods have been developed for distinguishing between isozymes, resulting in improved systems of naming (see below).

Some compounds have been found to be specific substrates for individual isozymes, and these have been used as probes for their identification. Studies for isozyme identification have been so repetitive that in general they have not been reported in the compound section of this book.

Some commonly used P450 probes

Isozyme	*Probe*	*Reaction*
CYP1A2	Phenacetin	de-ethylation
	7-Ethoxyresorufin	de-ethylation
CYP1A/12	7-Ethoxyresorufin	de-ethylation

Table (*Continued*)

Isozyme	*Probe*	*Reaction*
CYP2A6	Coumarin	7-hydroxylation
CYP2B6	Bupropion	*t*-butyl hydroxylation
	(*S*)-Mephenytoin	N-demethylation
	7-Ethoxy-4-trifluoromethylcoumarin	de-ethylation
CYP2C8	Taxol	6α-hydroxylation
CYP2C9	Tolbutamide	methyl hydroxylation
	Diclofenac	4′-hydroxylation
CYP2C18	2-(2,3-Dichloro-4-(4-hydroxybutoxy)benzoyl)thiophene	5-hydroxylation
CYP2C19	(*S*)-Mephenytoin	4′-hydroxylation
CYP2D6	Dextromethorphan	demethylation
	Bufuralol	1′-hydroxylation
CYP2E1	*p*-Nitrophenol	*o*-hydroxylation
	Chlorzoxazone	6-hydroxylation
CYP3A4	Testosterone	6β-hydroxylation
	Medazolam	α-hydroxylation
CYP3A5	Testosterone	6β-hydroxylation
CYP4A9	Lauric acid	12-hydroxylation
CYP 4A11	Lauric acid	12-hydroxylation

Monoclonal antibody studies are also used for the identification of several isozymes, for instance, CYP2C8, CYP3A5 and CYP4A11.

The major isozymes are 1A2, 2C9, 2C19, 2D6 and 3A4. Structural variants have been detected in these groups, such as 2C9 with a difference involving an arginyl/cysteinyl residue and 2D6 with a methionyl/valyl difference.

The reactions carried out by P450 enzymes are listed under the relevant reactions in the Reactions section of this book. However, the following publications are illustrative, particularly of earlier naming systems.

A naming system for rabbit liver microsomal isozymes was introduced with LM designations, followed by a subscript numbering scheme. LM_1, molecular weight 47 000 and LM_7, molecular weight 60 000, were not induced by phenobarbital or by β-naphthoflavone. LM_2, molecular weight 50 000, was induced by phenobarbital, and LM_4 by β-naphthoflavone. The substrates studied showed relatively poor discrimination between isozymes [A2147]. The isozymes show quantitatively different reaction rates for the formation of 4 tetrahydrotetraols formed from *trans*-benzpyrene-7,8-dihydrodiol, via the two corresponding 9, 10-oxides. The P450 types were then (1978) designated LM_1, LM_2, LM_{3b}, $LM_{4(BNF)}$, $LM_{4(PB)}$ and LM_7 [A3605]. This redundant system was also used with monohydroxylation of benzpyrene [A2327].

Rat microsomal enzyme is induced by phenobarbital and 3-methylcholanthrene; the induced enzymes differ slightly in the absorption maxima of the carbon monoxide adducts (450 and 448 nm respectively) [A734]. Multiple forms with molecular weights of about 50 000 were separated [A3028]. In one study (1978) with solubilized enzyme from male rat, four fractions were obtained by an initial separation, each of which could be separated into eight haem-containing bands [A3447]. Another study detected 21 fractions. Benzpyrene hydroxylase was associated with P448 isozymes, pI 6.83, 6.55 and 6.36; the second of these fractions also exhibited 7-ethoxycoumarin de-ethylase activity [D740]. One publication designated two of these as $P\text{-}448_{MC}$ and $P\text{-}448_{HCB}$; they catalyse hydroxylation at different positions on 2-acetamidofluorene [D269]. Isozymes designated as A, B and D N-hydroxylate 2-acetamidofluorene [D142]. The primary structures of male rat isozymes f, g and h, molecular weights

51 000, 50 000 and 51 000 respectively differ from one another and from isozymes a-e, and this is confirmed by antibody studies. Isozymes f and g showed a low activity towards all the substrates tested; h N-demethylated benzphetamine and hydroxylated zoxazolamine and 17β-oestradiol [D39]. These naming systems are also redundant.

Liver microsomes from phenobarbital-treated mice yield 4 P450 fractions with marginally different absorption spectra [A1699].

A monkey liver P450, molecular weight 50 000, catalyzes oxidations on a series of substrates, in the presence of a hydrogen transport system [D510].

Guinea pig isozymes, designated as the IIB family, molecular weight 52 000, hydroxylate Δ^9-tetrahydrocannabinol and aniline, and demethylate *p*-nitroanisole [F819].

A mathematical analysis based on potential energy profiles during the reaction, based on arene oxide ring opening, gave a close approximation to the extent that the NIH shift (the migration of a substituent to a position *ortho* to its original position at the hydroxylation site) was observed for *para* hydroxylation, in agreement with the hypothesis that there is a *p*-quinoid intermediate. The data also support the hypothesis that there may be a pathway involving initial formation of a tetrahedral intermediate in the absence of epoxide formation [E284].

A full assessment of P450 enzymes is found in David Lewis "Guide to Cytochromes P450", Taylor & Francis, London (2001).

1. Reactions involving the formation and degradation of the aromatic nucleus

1.1 *Aromatization of 6-membered carbon and heterocyclic rings.*

Prephenate dehydrogenase (E.C. 1.3.1.12)

Prephenate → *p*-hydroxyphenylpyruvate

The amino acid composition of A. aerogenes enzyme, molecular weight 76 000, has been determined; it is involved in the biosynthesis of tyrosine [K873].

Prephenate dehydrogenase (NADP$^+$) (E.C. 1.3.1.13)

The mung bean (Vigna radiata) enzyme, molecular weight 52 000, requires NADPH; it may be identical with pretyrosine dehydrogenase [B290]

Prephenate dehydratase (E.C. 4.2.1.51)

Salmonella typhimurium enzyme, optimum pH 8.5, is part of a complex with chorismate mutase (E.C. 5.4.99.5). The reaction product, phenylpyruvate, is further converted into phenylalanine [K859].

Arogenate (pretyrosine) dehydrogenase (E.C. 1.3.1.43)

Arogenate + NAD(P) →
L-tyrosine + NAD(P)H

Vigna radiata (mung bean) enzyme, molecular weight 52 000, may be identical with prephenate dehydrogenase [B290].

Nicotiana silvestris enzyme is inhibited by L-tyrosine and substituted tyrosines [C501].

Corynebacterium glutamicum and Brevibacterium flavum enzymes have molecular weights of 158 000, and B. ammoniagenes 68 000. They are not inhibited by tyrosine, and *p*-chloromercuribenzoate inhibition is reversed by thiols. Contrary to some earlier reports, these organisms lack the *p*-hydroxyphenylpyruvate pathway for the formation of tyrosine [B47].

Actinoplanes missouriensis enzyme, molecular weight 68 000, has an optimum at pH 9.5 [F277].

Streptomyces phaeochromogenes enzyme is a dimer with subunit molecular weight 28 100 and pI 4.45 [D880].

This reaction has also been observed in Microtetraspora glauca [F222], Claviceps [E234], Streptomycetes [C695] and Flavobacterium devorans [C115].

Phenylalanine formation from arogenate

This reaction has been detected in Claviceps [E234].

Anthranilate synthase; (E.C. 4.1.3.27)

Chorismate + glutamine or NH_3 →
anthranilate

Saccharomyces cerevisiae enzyme is dimeric, molecular weight 130 000, and subunit molecular weight 64 000 [D512]. Enterobacter liquefaciens and Erwinia carotavora enzymes have molecular weights of 140 000, and Aeromonas formicans

220 000 [A1481]. Claviceps enzyme has an optimum pH of 7.8 or 8.6 using glutamine as co-substrate. It requires Mg^{2+}, partly replaceable by Mn^{2+} or Co^{2+}, is stimulated by K^+, Na^+ or NH_3^+ and is inhibited by tryptophan, chanoclavine, elymoclavine, indoleacrylate and prephenate [A2552].

Shikimate forms phenazine-1-carboxylate in Pseudomonas aureofaciens [A958]; earlier publications suggest that the pathway is via chorismate and anthranilate.

p-Aminobenzoate formation from chorismate

Streptomyces coelicolor catalyzes this reaction, but isochorismate is not a substrate; however, both compounds are precursors of *p*-aminobenzoate in S. aminophilus and Enterobacter aerogenes; isochorismate may initially be converted into chorismate [F111]. S. griseus enzyme, molecular weight 50 000 uses ammonia or glutamine as the source of the amino group [D734].

p-Aminophenylalanine formation from chorismate

In Streptomyces this reaction is an initial step in the formation of chloamphenicol. Its formation is considered to be a four-stage biosynthesis including a transamination, and the reaction is stimulated by an aminotransferase from the crude extract, as well as by several other enzymes present [A2461, B158].

Benzoate formation from shikimate

This reaction has been observed in rat [A2706].

p-Hydroxybenzoate formation from shikimate

This reaction occurs in Lithospermum erythrorhizon as an early step in the formation of the naphthoquinone shikonin and γ-glutaminyl-4-hydroxybenzene [A3763, B810].

Phenylalanine formation from shikimate

Phenylalanine and tyrosine are formed from shikimate in Reseda lutea, R. odorata and Iris. Previous publications indicate an involvement of prephenate [A1601].

3-Carboxyphenylalanines formed from shikimate

3-Carboxyphenylalanine and 3-carboxytyrosine are formed from shikimate in Reseda lutea, R. odorata and Iris, as well as phenylalanine and tyrosine. The carboxyl group arises from shikimate; it is not considered that carboxylation of the parent amino acid is involved [A1601].

Phenazine-1-carboxylate and iodinin formation from shikimate

These reactions involve the formation of a phenazine ring system, apparently incorporating two shikimate molecules, probably with anthranilate as an early intermediate.

The formation of phenazine-1-carboxylate has been observed in Pseudomonas aureofaciens [A958].

The formation of iodinin has been observed in Brevibacterium/Chromobacterium iodinum [A958, A2941].

Menaquinone MK-9(II-H_2) formation from shikimate

This reaction, which involves the formation of a naphthoquinone, has been observed in Mycobacterium phlei [A691].

Catechol formation from shikimate and dehydroshikimate

The first reaction has been observed in rat, [A2706] and the second in E. coli, where it is a major pathway for the anaerobic utilization of glucose [G366].

o-Succinylbenzoate formation from isochorismate

E. coli and Aerobacter aerogenes enzymes catalyze this reaction, with α-oxoglutarate as co-substrate. It appears that earlier claims that chorismate is the substrate were incorrect. The enzyme, optimum pH 8.3, requires thiamine pyrophosphate as coenzyme, and it may be a Mn^{2+}-containing enzyme [D682, E430, E457]. The reaction product is the starting point for the formation of a range of substituted naphthalene natural products.

3-Amino-5-hydroxybenzoate formation from 5-deoxy-5-amino-3-dehydroshikimate

This reaction has been detected in Amycolatopsis mediterranei. The enzyme, which contains bound pyridoxal phosphate, catalyzes both an α,β-dehydration and a stereospecific 1,4-enolisation during the reaction sequence. The product is a key intermediate in the formation of rifamycin [J643].

cis-Dihydrobenzene-1,2-diol dehydrogenase; (E.C. 1.3.1.19)

cis-1, 2-Dihydroxycyclohexa-3, 5-diene → catechol

Mouse kidney contains four soluble (three minor) isozymes. Two minor ones are immunologically identical with aldehyde reductase (E.C. 1.2.1 class) and 3α-hydroxysteroid dehydrogenase (E.C. 1.1.1.213). The other two, molecular weight 39 000, require $NAD(P)^+$ as coenzyme. Both *cis* and *trans* isomers are substrates, as well as other dihydrodiols; *p*-nitrobenzaldehyde and quinones are also reduced [F377].

An enzyme in Bacterium acts only on *cis*-isomers [B871]. Pseudomonas putida enzyme, a homotetramer, molecular weight 102 000, acts on a range of aromatic hydrocarbon *cis*-dihydrodiols except those substituted on K-regions [A1298, A1591].

Pseudomonas putida *cis*-naphthalene-1,2-dihydrodiol dehydrogenase (E.C. 1.3.1.29) and *cis*-biphenyl-2,3-dihydrodiol dehydrogenase both oxidize *cis*-naphthalene-1,2-dihydrodiol and *cis*-biphenyl-2,3-dihydrodiol as well as *cis*-2,2′,5,5′-tetrachlorobiphenyl-3,4-dihydrodiol. They both require NAD^+ [K196].

Pseudomonas *cis*-chlorobenzene dihydrodiol dehydrogenase acts on *cis*-(1*R*,2*S*)-indan-1,2-diol, but not on *cis*-(1*S*,2*R*)-indan-1,2-diol. Some enantiomeric selectivity is observed with *p*-halotoluene-2,3-dihydrodiols, *cis*-1,2-dihydroxytetralin and *cis*-naphthalene-1,2-dihydrodiol [K348].

Xanthobacter flavus enzyme is a homotetramer, monomeric molecular weight 26 500 and pI 5.4. It requires $NADP^+$ (NAD^+ is less effective) and acts on *cis*-3,6-dichlorobenzene-1,2-dihydrodiol and benzene-1,2-dihydrodiol [J186].

Bacillus enzyme is a homohexamer, monomeric molecular weight 29 500, pI 6.4 and optimum pH 9.8. It oxidizes *cis*-toluene-2,3-dihydrodiol, and is stable up to 80° [E368].

trans-Dihydrobenzene-1,2-diol dehydrogenase; (E.C. 1.3.1.20)

trans-1, 2-Dihydroxycyclohexa-3, 5-diene → catechol

This and similar enzymes aromatize *trans*-dihydrodiols formed from benzene and a range of analogues formed from polynuclear hydrocarbons, including those that are carcinogenic. This reaction is one step in the potential degradation of these hydrocarbons; the catechols so formed could undergo ring fission, at least in microorganisms.

An enzyme in pig lens oxidizes this and other similar substrates, utilising $NADP^+$. Quinones and nitrobenzaldehydes are also substrates for the reverse reaction [F389].

Monkey liver cytosol contains four isozymes, molecular weights in the range 36 000–39 000, each with a similar range of specificities, and optima at pH 5.8, 6.2, 7.9 and 8.7. One isozyme was formerly known as indanol dehydrogenase (E.C. 1.1.1.112) [F388]. One at least is a dimer that is inactive towards *cis*- isomers [F391].

Hamster liver cytosol contains five isozymes with a broad specificity, molecular weight about 35 000 [F153].

Guinea pig liver contains four major and four minor isozymes, molecular weights about 34 000, except for two minor ones, molecular weights 26 500 and 14 500. The specificity is broad [E67].

Rat liver cytosolic enzyme requires NADP as co-substrate, and acts on *trans*-naphthalene-1,2-dihydrodiol and *trans*-benzpyrene-7,8-dihydrodiol [E583].

Mouse liver enzyme has been separated into four isozymes that require NADP. Two are monomers, molecular weights 30 000 and 34 000, and two dimers, molecular weights 64 000 and 65 000, pI 8.1, 6.2, 5.5 and 5.4, respectively [C712]. Another study found two cytosolic forms that were identified as 17β-hydroxysteroid dehydrogenase (major; E.C. 1.1.1.63 and 1.1.1.64) and aldehyde reductase. The major one also oxidizes a range of alcohols and reduces a range of aldehydes and ketones [E821].

Beef liver cytosolic enzyme contains three enzymes that are active towards *trans*-benzene dihydrodiol. One is 3α-hydroxysteroid dehydrogenase and a second is a high K_m aldehyde reductase. The third also acts on other dihydrodiols including *trans*-naphthalene dihydrodiol; it is a distinctly different activity [G750].

1,6-Dihydroxycyclohexa-2,4-diene-1-carboxylate dehydrogenase (E.C. 1.3.1.25)

Alcaligenes eutropus enzyme, molecular weight 95 000 and optimum pH 8.0 may be a homotetramer. It forms catechol and CO_2, and requires NAD^+; $NADP^+$ is ineffective, and no other cofactors are required [K944].

***cis*-1,2-dihydroxycyclohexa-3,5-diene-1-carboxylate dehydrogenase** (E.C. 1.3.1.55)

Pseudomonas putida and Acinetobacter calcoaceticus enzymes, molecular weight 28 000 form catechol from the substrate [K763].

2,3-Dihydro-2,3-dihydroxybenzoate dehydrogenase; (E.C. 1.3.1.28)

Aerobacter aerogenes and E. coli enzymes require NAD^+ as co-substrate [K876].

***o*-Succinylbenzoate synthase**

Amycolaptosis enzyme acts on 2-hydroxy-6-succcinyl-2,5-cyclohexadienecarboxylate as substrate [K68].

Kynurenate-7.8-dihydrodiol dehydrogenase (E.C. 1.3.1.18)

Pseudomonas fluorescens enzyme requires NAD^+ for the formation of 7,8-dihydroxykynurenate [K874].

***trans*-Acenaphthene-1,2-dihydrodiol dehydrogenase** (E.C. 1.10.1.1)

Rat liver cytosolic enzyme acts on the (–), but not on the (+) isomer (an inhibitor), with $NADP^+$ as co-substrate, whereas NAD^+ is inactive. A further substrate is (–)-1-phenylethanol. Activity is also found in mouse, guinea pig, rabbit, hamster, dog, cat and pig liver. Some of these can accept NAD^+ as co-substrate; it has been suggested that more than one enzyme may be involved [K817].

Aromatization of dihydropyridines

Nilvadipine, a 1,4-dihydro-4-phenylpyridine is aromatized in rat liver microsomes, and utilises NADPH. The reaction is inhibited by P450 inhibitors [E847]. Many metabolic studies on calcium channel blockers with similar structures to nilvadipine have demonstrated that a major proportion of each drug is similarly aromatized.

A large number of studies with the neurotoxin 1,2,3,6-tetrahydro-1-methyl-4-phenylpyridine (MPTP), which causes a drug-induced parkinsonism, have demonstrated that the proximal toxin is 1-methyl-4-phenylpyridinium (MPP^+). Metabolic studies have detected the formation of MPP^+ in mouse, monkey, rat, man and beef (e.g. [E3, D222, D141, E3, F458] respectively). The formation of 2,3-dihydro-1-methyl-4-phenylpyridinium has also been observed, and this in turn is converted into MPP^+, at least in mouse [E684]. A major study with human placental MAOA and beef liver MAOB found that a large range of MPTP analogues, particularly those with substituents on the aryl moiety are substrates. Although the products were not identified, it was assumed that they were MPP^+ analogues [F395].

Tetrahydroprotoberberine aromatization

Berberis wilsonae enzyme is a flavoprotein, molecular weight 100 000 and optimum pH 8.9. Tetrahydroprotoberberine yields protoberberine, presumably via the 7,14-dehydroberbinium analogue; canadine and tetrahydrojatrorrhizine are also substrates. The reaction requires oxygen and forms peroxide as the second product. The enzyme is specific for (*S*)- isomers [D145].

Corydalis cava bulb protoberberine reductase, optimum pH 7.5, requires NADH for the reduction of protoberberines to (14*R*)-tetrahydroberberines, apparently in two stages. For instance, both berberine and 7,8-dihydroberberine are reduced to canadine. The reaction, which dearomatizes the heterocyclic ring, is reversible; both 7,8-dihydroberberine and (*R*)-canadine are oxidized to berberine, with $NADP^+$ as coenzyme. Palmatine, dehydrosinactine, coptisine, columbamine, jatrorrhizine and dehydroscoulerine are also substrates [G136].

Tetrahydroberberine (canadine) aromatization

Thalictrum minus tetrahydroberberine oxidase is composed of 3 isozymes, two of which are specific. All have an optimum at pH 9.0. The molecular weight of one of the specific enzymes is more than 200 000, and the other, a trimer, is 145 000 [H401].

Coptis japonica (*S*)-tetrahydroberberine oxidase (E.C. 1.3.3.8) is a dimer, molecular weight 58 000, and requires oxygen and Fe; it is highly specific. It aromatizes one of the heterocyclic rings [E678, E732].

Berberis aggregata aromatizes tetrahydroberberine, tetrahydrocolumbamine and tetrahydropalmatine, with the formation of a quaternary amino moiety [K947].

Aromatization of dihydromacarpine

Eschscholtzia californica enzyme, molecular weight 56 000, pI 8.8 and optimum pH 7.0, requires oxygen to oxidize dihydromacarpine to macarpine. This completes the aromatization of the ring system, including the formation of a quaternary nitrogen [E604].

Dihydrobenzophenanthridine oxidase
(E.C. 1.5.3.12)

Sanguinaria canadensis enzyme is composed of three isozymes, molecular weights 77 000, 67 000 and 59 000 and optimum pH 7.0. It converts dihydrosanguinarine into sanguinarine and dihydrochelerythrine into chelerythrine, aromatizing the heterocyclic ring [G723]. Eschscholtzia californica enzyme, molecular weight 56 000, pI 8.8 and optimum pH 7.0, requires oxygen to convert dihydromacarpine into macarpine [E604]. In all of these a quaternary base is formed.

Many plant alkaloids contain an aromatic heterocyclic ring moiety. There may be an aromatizing step similar to this one, but in most cases the enzymology has not been studied.

D-Dopachrome tautomerase

Human enzyme is found in erythrocytes and other blood cells, but not in plasma, with 5,6-dihydroxyindole as the product [H872].

The enzyme that enolizes indole-3-pyruvate and *p*-hydroxyphenylpyruvate converts

D-dopachrome into 5,6-dihydroxyindole. The enzyme source is not clear from Chemical Abstracts; it is claimed to be a rat enzyme, but it may be the same as that described in another publication, found in human lymphocytes. The latter is a macrophage migration inhibition factor in addition to its tautomerase activity [J475, J514].

L-Dopachrome tautomerase (E.C. 5.3.3.12)

Mouse melanoma enzyme forms 5,6-dihydroxyindole-2-carboxylate as the initial product. The apoenzyme is activated by Zn^{2+}, but not by Fe^{2+} or Cu^{2+} [H338].

Locusta migratoria enzyme, dopachrome conversion factor, (dopachrome Δ-isomerase, E.C. 5.3.3.12), molecular weight 85 000, forms 5,6-dihydroxyindole from L-dopachrome; it also acts on L-dopachrome methyl ester and methyldopachrome, but not on their D-isomers or dopaminechrome [J842].

Bombyx mori dopa quinone imine conversion factor, optimum pH 7.5–9, forms 5,6-dihydroxyindole from L- (but not D-) dopachrome [H429].

Aromatization of lindane

Lindane (the isomer used is not clear, but is presumed to be the active γ-isomer) is converted into 1,2,4-trichlorobenzene in bean, and into 1,2,3- and 1,2,4-trichlorobenzene in Zea mays [A766].

1.2 *Formation of carbon ring systems*

Naphthalene derivatives formed from *o*-succinylbenzoate

a. Naphthoate synthetase (E.C. 4.1.3.36)

E. coli enzyme, molecular weight 45 000, requires acetyl CoA, ATP and Mg^{2+} to form 1,4-dihydroxy-2-naphthoate. In crude extracts, the addition of farnesyl pyrophosphate results in the formation of menaquinone-3 at the expense of 1,4-dihydroxy-2-naphthoate [A2774]. Studies with Mycobacterium phlei, E. coli and Galium mollugo have found that *o*-succinylbenzoyl CoA is formed as an intermediate [C337, E594]. M. phlei enzyme has a molecular weight of 44 000 and optimum pH 6.9. The succinyl carboxyl is retained as the carboxyl group in 1,4-dihydroxy-2-naphthoate [A2943].

b. Phylloquinone formation

In Zea mays phylloquinone is formed from *o*-succinylbenzoate, retaining its structural integrity [A175, A3973].

c. 1,4-Naphthoquinone formation

Juglans regia catalyzes the formation of 1,4-naphthoquinone and juglone (5-hydroxy-1,4-naphthoquinone) [A3470]. 1,4-Dihydroxy-2-naphthoate is an intermediate in juglone formation as well as in lawsone (2-hydroxy-1,4-naphthoquinone) formation [A1639].

Cannabidiolate synthetase

Cannabis sativa enzyme, molecular weight 74 000 and pI 6.1, catalyzes the ring closure of cannabigerolate and cannabinerolate into cannabidiolate. The enzyme is not an oxygenase or peroxidase; presumably it is a dehydrogenase, which forms a cyclohexene ring system [J205].

Pinosylvin (3,5-dihydroxystilbene) synthase (E.C. 2.3.1.146)

Pine (Pinus sylvestris) enzyme catalyzes the condensation of cinnamoyl CoA with malonyl CoA to form a second aromatic ring. With *p*-coumaroyl CoA the hydroxylated analogue, resveratrol is formed [B313].

Dioscorea shows similar metabolic reactions with cinnamoyl CoA, *m*-hydroxyphenylpropionyl

CoA and analogues; the products, such as resveratrol and pinosylvin are intermediates in the formation of hircinol and batatasins [D136].

This reaction occurs poorly in Barlia longibracteata with cinnamoyl CoA, *m*-coumaroyl CoA and *p*-coumaroyl CoA as substrates. The corresponding dihydro substrates form the analogous bibenzyls, but more effectively [C734].

Epipactis palustris enzyme, molecular weight 85 000, shows a similar specificity and co-substrate requirement to those for the above enzymes [G138].

Bibenzyl synthase

Barlia longibracteata enzyme acts on CoA conjugates of 3-phenylpropionic acid; the *m*- and *p*-hydroxy analogues also form the corresponding bibenzyls [C734].

Bletilla striata enzyme is a dimer, monomeric molecular weight 46 000, which condenses *m*-hydroxyphenylpropionyl CoA with malonyl CoA to form 3,3′,5-trihydroxybibenzyl [H101].

Epipactis palustris enzyme, molecular weight 85 000 acts on *m*-hydroxyphenylpropionyl CoA; a good additional substrate is phenylpropionyl CoA, which yields dihydropinosylvin, but the corresponding cinnamoyl CoAs are poor substrates [G138]. This distinguishes the enzyme from stilbene synthase, which it closely resembles.

Chalcone synthase (E.C. 2.3.1.74)

This reaction is the first step in the reaction sequence that leads to the formation of polycyclic flavonoid plant pigments from cinnamates, forming a second aromatic ring.

Buckwheat enzyme is a homodimer, molecular weight 83 000, pI 5.2 and optimum pH 8.0. It condenses malonyl CoA with *p*-coumaroyl CoA, feruloyl CoA and caffeoyl CoA [E180].

Cephalocereus enzyme converts cinnamoyl CoA into 2′,4′,6′-trihydroxychalcone [H633].

Daucus carota enzyme exhibits optima at pH 7.9 and 6.8 with *p*-coumaroyl CoA and caffeoyl CoA, respectively [D698].

Dianthus caryophyllus enzyme exhibits pH optima of 8.0 and 7.0 with *p*-coumaroyl CoA and caffeoyl CoA, respectively [C392].

Glycine max enzyme, molecular weight 75 000, is composed of three isozymes, pI 5.45 (main), 5.35 and 5.5, and pH optima of 7.5 and 6.5 with *p*-coumaroyl CoA and caffeoyl CoA, respectively [E579]. A reductase is involved in the reaction sequence [E661].

Glycyrriza echinata enzyme acts on *p*-coumaroyl CoA, malonyl CoA and NADPH [E619].

Parsley enzyme, molecular weight 77 000 appears to be a dimer (monomeric molecular weight 42 000) with *p*-coumaroyl CoA and malonyl CoA as substrates [B352, K808].

Phaseolus vulgaris enzyme, molecular weight 77 000 and optimum pH 8.0, acts on malonyl CoA and *p*-coumaroyl CoA [C813].

Rye enzyme requires malonyl CoA as co-substrate, and acts on *p*-coumaroyl CoA (optimum pH 8) and caffeoyl CoA (optimum pH 6.5) to form 2′,4,4′,6-tetrahydroxy- and 2′,3,4,4′,6′-pentahydroxychalcones (precursors of naringenin and eriodictyol), respectively [E662].

Spinach enzyme is composed of two isozymes, pH optimum 7.5–8, with *p*-coumaroyl CoA, feruloyl CoA and caffeoyl CoA as substrates [D596].

A tulip anther enzyme, molecular weight 55 000 and optimum pH 8.0 acts on *p*-coumaroyl CoA, feruloyl CoA and caffeoyl CoA to form naringenin, homoeriodictyol and eriodictyol respectively; it is inhibited by CoA, flavanones and thiols. The preparation was claimed to be free from chalcone-flavanone isomerase activity; the expected chalcone intermediates were not detected [A3792]. A similar enzyme found in Haplopappus gracilis, optimum pH about 8 for *p*-coumaroyl CoA, and 6.5–7 for caffeoyl CoA was called flavanone synthase. The reaction was not stoichiometric, with small amounts of by-products such as benzalacetones being formed [A3362]; a similar series of reactions was found in Petroselinum crispum [A2618]. These publications all come from early studies on the enzyme system, and despite the claims that chalcone-flavanone isomerase activity was absent, there must be a suspicion that the preparations were all contaminated with this enzyme.

Benzophenone synthase (E.C. 2.3.1.151)

Centaurium erythraea enzyme, optimum pH 7.5, acts on *m*-hydroxybenzoyl CoA and malonyl CoA to form 2,3′,4,6-tetrahydroxybenzophenone [J35].

Norsolorinate synthase

Aspergillus parasiticus converts hexanoate or pentanoate into norsolorinate, a polyphenolic anthraquinone, which is a postulated precursor of aflatoxin B_1. With pentanoate, an additional reaction product with a 5-oxopentane side-chain, instead of a 6-oxohexane side-chain, has been detected; if 6-fluorhexanoate is used, 6′-fluoronorsolorinate is formed [H675].

Phloroisovalerophenone synthase

Humulus lupulus enzyme, found in cone glandular hairs, is a homodimer, monomeric molecular weight 45 000 and pI 6.1; the amino acid sequence has been determined. It utilizes one mol of isovaleryl CoA, and three mol of malonyl CoA to form the aromatic nucleus. Replacement of the former with isobutyryl CoA yields the corresponding isobutyrophenone [K175].

6-Methylsalicylate synthase

Penicillium patulum enzyme is a tetramer, molecular weight 750 000. It condenses acetyl CoA, and requires NADPH [G751].

Benzoates from pyrones

Macrophoma commelinae converts a series of 2-pyrones into the corresponding benzoates, usually in good yield. The carbonyl group becomes the benzoate carboxyl, and the 3, 4, 5 and 6-substituents on the pyrone become the 2, 3, 4 and 5-substituents on the benzoate [E766].

Purpurogallin formation

Pyrogallol → purpurogallin

This reaction is catalyzed by peroxidases (E.C. 1.11.1.7) from peanut, with peroxide as co-substrate. Four isozymes are found with pH optima at 6, 6.4, 8 and 8 [A2519].

Salutaridine synthase (E.C. 1.1.3.35)

Papaver somniferum enzyme, probably a P450 that requires oxygen and NADPH, is found in root, shoot and capsules, but not in latex. The enzyme acts on (*R*)- (but not (*S*)-) reticuline, and the reaction involves the formation of a 6-membered carbon ring by linkage between the two aromatic nuclei, one of which is converted into a cyclohexadienone system [K759].

4,5-Methylenechrysene formation

This reaction has been detected in rat liver cytosol, with 5-methychrysene as substrate [G118].

1.3 *Formation of heterocyclic ring systems*

Indole-3-glycerol-phosphate synthase; (E.C. 4.1.1.48)

1-(*o*-Carboxyphenylamino)-1-deoxyribulose-5-phosphate → indole-3-glycerolphosphate

The Bacillus subtilis enzyme, molecular weight 23 500, differs from N-(5′-phosphoribosyl) anthranilate isomerase (E.C. 5.3.1.24) [B97].

Hordenine cyclization

Mushroom tyrosinase, with peroxide, acts on hordenine with an optimum at pH 6.7 to form a compound whose spectra indicates that the product is N,N-dimethylindoliumolate, presumably via a quinone [K374].

Chalcone-flavanone isomerase (chalcone isomerase; E.C. 5.5.1.6)

Grapefruit enzyme acts on chalcone-4′-neohesperosides with a free, unhindered 4-hydroxyl group. Other structural features required for substrate activity are the presence of either 2,6-dihydroxy or 2-hydroxy-4-methoxy groups. It is reversibly inhibited by cyanide but not by azide, EDTA, Hg^{2+} or *p*-chloromercuribenzoate [A2518]. Tulipa petal enzyme is cytosolic [A2523].

There are many publications on chalcone synthase in which this activity is part of the reaction system, and its requirement is implicit in the formation of all flavonoids.

Riboflavin formation

This involves two enzymes in the later part of the reaction sequence, 6,7-dimethyl-8-ribityllumazine synthase and riboflavin synthase (E.C. 2.5.1.9):

5-Amino-6-ribitylamino-2,4(1*H*,3*H*)-pyrimidinedione + (3*S*)-3,4-dihydroxy-2-butanone-4-phosphate → 6,7-dimethyl-8-ribityllumazine

6,7-Dimethyl-8-ribityllumazine → riboflavin + 5-amino-6-ribitylamino-2,4(1*H*,3*H*)-pyrimidinedione

Two synthases in Bacillus subtilis have molecular weights of 70 000 and 1 000 000. The smaller (more active) molecule appears to be a homotrimer; the larger molecule is composed of one of the above molecules as well as about 60, possibly identical, polypeptide chains of a different type [B315]. Further studies on the larger molecule have shown that it is a complex of an icosahedral capsid of 60 β units surrounding a core of three α units. The β units catalyze the first reaction, and the α units the second reaction. In the first reaction the natural (*S*)-butanone can be replaced by the (*R*)-isomer; the latter reacts at 1/6 of the rate for the (*R*)-isomer. The second reaction involves a dismutation, which forms both riboflavin and the substrate for the first reaction [H398].

Formyltetrafolate cyclo-ligase (E.C. 6.3.3.2)

Sheep liver enzyme, optimum pH about 4.8, forms $N^{5,10}$-methenyltetrahydrofolate with hydrolysis of the ATP co-substrate to ADP. It is inhibited by thiol-binding reagents [K826].

5,10-Methylenetetrahydrofolate reductase (E.C. 1.7.99.5; formerly 1.1.1.68)

E. coli enzyme, optimum pH 6.3–6.4, requires $FADH_2$ to form 5-methyltetrahydrofolate [K851].

Methylenetetrafolate dehydrogenase ($NADP^+$) (E.C. 1.5.1.5)

Calf thymus enzyme, optimum pH 6.5 is very unstable, and it is protected by thiols and glycerol. The reaction is reversible, forming $N^{5,10}$-methenyltetrahydrofolate preferentially [K935].

Methylenetetrafolate dehydrogenase (NAD^+) (E.C. 1.5.1.15)

Clostridium formicoaceticum enzyme is a homodimer, molecular weight 60 000 and Stokes radius 3.32nm. The reaction forms 5,10-methenyltetrahodrofolate, activation energy 8.5 kcal/mol; $NADP^+$ is not an alternative co-substrate [K928].

Formiminotetrafolate cyclodeaminase (E.C. 4.3.1.4)

Pig liver enzyme forms 5,10-methenyltetrahydrofolate and ammonia as reaction products [K937].

Tetrahydroisoquinoline and tetrahydro-β-carboline formation

A large number of plant alkaloids contain a ring system based on the formation of 1,2,3,4-tetrahydroisoquinolines (THIQ). This reaction can occur spontaneously (Pictet-Spengler reaction); for instance, under physiological conditions dopamine condenses with acetaldehyde to form salsolinol (6,7-dihydroxy) and small amounts of its isomer isosalsolinol (7,8-dihydroxy). In principle, any aldehyde or ketone may react with any phenethylamine analogue activated in the *meta* position. The small amounts of salsolinol found in man are probably formed mainly spontaneously.

In rat brain, a mitochondrial enzyme catalyzes the formation of 1-methylTHIQ from phenethylamine and pyruvate [J360]. Tryptamine forms 1,2,3,4-tetrahydro-β-carboline [A2285]. The enzyme requires 5-methyltetrahydrofolate as methyl donor; it also acts on N-methyltryptamine to form 1,2,3,4-tetrahydro-2-methyl-β-carboline. It is inhibited by dopac and by other catecholamine and indoleamine metabolites [A104]. In brain stem, serotonin and acetaldehyde form 1-methyl-1,2,3,4-tetrahydro-6-hydroxy-β-carboline [A908].

Rat brain and rabbit lung form 1,2,3,4-tetrahydro-β-carboline from tryptamine, 1,2,3,4-tetrahydro-2-methyl-β-carboline from N-methyltryptamine, 1,2,3,4-tetrahydro-6-hydroxy-β-carboline from serotonin and 1,2,3,4-tetrahydro-6-hydroxy-2-methyl-β-carboline from N-methylserotonin [A111].

In pig liver, 5,10-methylenetetrahydrofolate reductase oxidizes N^5-methyltetrahydrofolate to 5,10-methylenetetrahydrofolate, which (presumably non-enzymatically) forms norsalsolinol from dopamine with formaldehyde from 5,10-methylenetetrahydrofolate [A2346].

Eschscholtzia tenuifolia (*S*)-norlaudanosoline synthase ((*S*)-norcoclaurine synthase, E.C. 4.2.1.78), molecular weight 15 500 and optimum pH 7.5 or 7.8, is composed of four isozymes, with dopamine and 3,4-dihydroxyphenylacetaldehyde as substrates; the latter can be replaced by other phenylacetaldehydes, but not by pyruvates. The product is the L-(−) compound. Berberis regeliana, Corydalis sempervireus (misprint for sempervirens?), Dicentra spectabilis, E. pulchella, Fumaria officinalis, Glaucium rubrum, Meconopsis cambrica, Papaver somniferum, Ranunculus flammula and Thalictrum tuberosum also show this activity [B919, C859].

Berberis stolonifera also contains (*S*)-norlaudanosoline (norcoclaurine) synthase (E.C. 4.2.1.78) [K932, K933].

Strictosidine synthase (E.C. 4.3.3.2)

Rauwolfia serpentina enzyme, which is monomeric, molecular weight 30 000, pI 4.5 and optimum pH 6.5, contains 5.3% carbohydrate. It catalyzes the ring closure of tryptamine with the aldehyde group of secologanin, to form the tetrahydro-β-carboline strictosidine as the product [F63].

Catharanthus roseus enzyme, molecular weight 38 000, pI 4.6 and optimum pH 5.0–7.5 (or 6.8) forms strictosidine, but none of its epimer vincoside [B71, B265].

Chanoclavine 1 cyclase

Claviceps enzyme, which forms agroclavine, requires ATP and $NAD(P)^+$, but not FAD or oxygen. An aldehyde appears to be an intermediate. The enzyme develops rapidly in the transition from trophophase to idiophase, and decreases sharply after 9–10 days fermentation [A893].

Berberine bridge formation ((S)-Reticuline oxidoreductase E.C. 1.5.3.9)

Berberis beaniana enzyme, molecular weight 49 000 or 52 000, pI 4.9 and optimum pH 8.9 converts (*S*)-reticuline into (*S*)-scoulerine, involving ring closure with the N-methyl group. The reaction requires oxygen, and peroxide is the second product. (*S*)-Protosinomenine and (*S*)-laudanosoline also undergo a similar reaction. It is inhibited by *o*-phenanthroline and by reducing agents [D615, D878].

Eschscholtzia californica enzyme contains one mol of FAD, but no metal. The substrates are 1-benzyl-N-methyltetrahydroisoquinolines, with closure of the *ortho* position in the benzyl moiety to the N-methyl ring; several compounds with this structure are substrates [H680]. The putative molecular weight, based on gene sequence is 57 352 (excluding carbohydrate) [K705].

Macleaya microcarpa enzyme has an optimum pH of 7.5–8.2 [A1565].

This reaction has been observed in rat liver microsomes [B478] and in Papaver somniferum [A3141].

Cinnabarinate synthesis (phenoxazinone synthase, formerly E.C. 1.10.3.4)

$$\text{3-Hydroxyanthranilate} + O_2 \rightarrow \text{cinnabarinate}$$

In mouse, one enzyme appears to be catalase (E.C. 1.11.1.6), based on results from animals with an inborn metabolic error in which catalase is absent. The catalase and haemoglobin-catalysed reactions both require Cu^{2+} or Mn^{2+} [A3587]. Baboon and beef liver synthases require Mn^{2+}, these also appear to be catalase [A3083].

Goat bladder cinnabarinate synthase, molecular weight about 55 000 and optimum pH 7.2, is inhibited by divalent cations (except Mn^{2+}) and by chelating agents. The activation energy at 20° is 12.36 kcal/mol [K279].

Two Streptomyces antibioticus enzymes, molecular weights 900 000 and 200 000, appear to be monomeric and dimeric. They are not interconverted; the monomers are considered to be different [B924]. S. antibioticus enzyme acts on 3-hydroxy-4-methylanthranilate to form actinocin; this is a key reaction in the formation of actinomycin [E676]. A Drosophila melanogaster phenoxazine synthase appears to act by preventing the formation of an inhibitor rather than by directly catalyzing the formation of cinnabarinate (or xanthommatin from 3-hydroxykynurenine). It requires Mn^{2+} [A2574].

Formation of esculetin

Mushroom tyrosinase oxidizes *cis*-caffeic acid to esculetin [D598].

Dehydrodicinnamic acid dilactone formation

Inonotus (a mushroom) enzyme, molecular weight 39 000, optimum pH about 6 and stable up to about 60°, forms dehydrodicaffeic acid dilactone from caffeate. The reaction involves hydroxylation of caffeate at the β-position and dimer formation (with loss of the double bond) at the α-position and lactonization between the side chains. It is stimulated by a range of divalent cations, and inhibited or inactivated by cyanide, diethyldithiocarbamate and several physiological reducing agents. The reaction is also catalysed by peroxidase and peroxide, or by *o*-diphenol oxidase (E.C. 1.10.3.1) [A3016, A3019].

Sorghum bicolor enzyme, called ferulic dimerase, acts on ferulate [A1641].

Protoaphin dehydratase (cyclizing) (E.C. 4.2.1.73)

Woolly aphid (Eriosoma lanigerum) protoaphin dehydratase (E.C. 4.2.1.73), molecular weight 120 000 is a glycoprotein, which acts on protoaphin aglycone to form xanthoaphin. The enzyme is inhibited by naphthoresorcinol [A3112].

Pummerer's ketone formation

Pisum sativum ascorbate peroxidase (E.C. 1.11.1.11) acts on *p*-cresol to form

Pummerer's ketone, a tetrahydrodibenzofuran [K404].

Dihydrovindoline ether formation

Streptomyces griseus catalyzes the formation of an internal ether bond in vindoline between an existing hydroxyl group and a double bonded carbon, with loss of the double bond [A2984].

Methylenedioxy ring formation

Berberis stolonifera berberine synthase (columbamine oxidase, E.C. 1.1.3.26), molecular weight 32 000 and optimum pH 8.9, is a Fe^{2+}-containing enzyme which forms berberine from columbamine, by ring closure between adjacent hydroxyl and methoxy groups. Tetrahydrocolumbamine is not a substrate [D536].

Thalictrum tuberosum microsomal (*S*)-canadine synthase (E.C. 1.1.3.36, a P450), optimum pH 8.5 requires NADPH, forming a methylenedioxy bridge in (*S*)-tetrahydrocolumbamine to form (*S*)-canadine. The enzyme shows high selectivity [H301].

Eschscholtzia californica microsomes contain (*S*)-cheilanthifoline synthase (E.C. 1.1.3.33), optimum pH 7.9 and activation energy 54 kj/mol, and (*S*)-stylopine synthase (E.C. 1.1.3.32), optimum pH 8.0 and activation energy 25 kj/mol. Both are inducible P450 enzymes requiring oxygen and NADPH (NADH is inactive), but addition of NADH activates the reaction. Both are highly specific; the former acts on (*S*)-scoulerine and the latter on (*S*)-cheilanthifoline [K914].

6β-Hydroxyhyoscyamine epoxidase (E.C. 1.14.11.14)

Hyoscyamus niger enzyme requires α-oxoglutarate, ascorbate and Fe^{2+} in the formation of scopolamine [K887].

Pterocarpin synthase (E.C. 1.1.1.246)

Cicer arietinum culture enzyme, optimum pH 6.0, which requires NADPH (NADH is poor) acts on vestitone, forming a cyclic ether by reduction [K879].

Dioxin formation

Horseradish peroxidase catalyzes the oxidation of pentachlorophenol to form octachloro-*p*-dioxin [H837].

Dihydrogeodin oxidase

Aspergillus terreus enzyme, an intensely blue protein, molecular weight 153 000, is a homodimer containing eight Cu/subunit. It is inhibited by azide and ethylxanthate, but not by cyanide or diethyldithiocarbamate. It forms geodin, with the formation of an internal ether bond, and the oxidation of one of the aryl rings to a 4-oxo-2,5-diene ring to form a *spiro*-compound [E300]. Although it is highly specific, it has some activity towards sulochrin and griseophenone B [C556].

Catechol oxidase (dimerising) (E.C. 1.1.3.14)

This spinach leaf enzyme, optimum pH 7.4, forms diphenylenedioxide-2,3-quinone with oxygen (1.5 mol) as co-substrate. It is activated by FAD, and inhibited by cyanide; it is not a copper-containing polyphenol oxidase [K880].

Rifamycin B oxidase (E.C. 1.10.3.6)

Monocillium oxidizes rifamycin B to rifamycin O. The reaction involves the oxidation of a phenolic ring to a *gem*-diol monoether quinone analogue, followed by ring closure of the acetoxy group to form a *gem*-diol ester. Other substrates are rifamycin SV and simple catechols and quinols. In contrast, Pleurotus laccase scarcely acts on rifamycin B [C519]. The same reaction is observed in Humicola [D922].

Phenanthridinone formation

Rat liver converts fluorenone oxime into phenanthridinone. The enzyme appears to be both mitochondrial and microsomal. It requires NADPH, but not oxygen; it is not inhibited by carbon monoxide. The reaction involves ring expansion, with the oxime moiety being converted into a heterocyclic nitrogen, apparently with migration of the hydroxyl group to the carbon adjacent to the nitrogen [A3230].

Stilbene oxidase

Botrytis cinerea enzyme is laccase-like (E.C. 1.10.3.2), molecular weight 32 000, with two isozymes, pI 4.3 and 4.35. Pterostilbene, 4,4′-dihydroxystilbene and resveratrol are substrates, the latter forming ε-viniferin. The reaction involves dimerization, with the formation of a benzofuran ring system [J823].

Strictosidine conversion into 10-deoxysarpagine

Rauwolfia serpentina P450 catalyzes this reaction; it utilizes NADPH and oxygen. The reaction involves a C–C ring closure [H676].

Phthalide formation

Pseudomonas putida and Aspergillus niger oxidize 2-ethylbenzoate to (*S*)-3-methylphthalide. 2-*n*-Pentyl- and 2-*n*-nonylbenzoates similarly form the corresponding butyl and octyl phthalides [J486].

Geotrichum candidum converts methyl *o*-acetylbenzoate into 3-methylphthalide; this reaction is also observed in Mucor javanicus, M. heimalis, Endomyces magnusii, E. resii and Saccharomyces [J486].

Aflatoxin B_1 formation

Aspergillus flavus enzyme, molecular weight 64 000–70 000 (by two methods) and optimum pH 8 forms aflatoxin B_1 from sterigmatocystin; the enzyme(s) is activated by Zn^{2+}, Co^{2+} and Mn^{2+}. The reaction involves the conversion of two 6-membered fused rings, one aromatic, into two fused 5-membered rings, one of which is a lactone. It has been suggested that O-methylsterigmatocystin is an intermediate; presumably a number of other compounds are also intermediates. Dihydrosterigmatocystin yields dihydroaflatoxin B_1 [E691, G154].

Aspergillus parasiticus enzyme has an optimum at pH 7.5–7.8 [A2594].

1.4 *Polymerization reactions*

Formation of biphenyl compounds and oligomers

L-Tyrosine, α-methyl-L-tyrosine and morphine are dimerized by rat gut peroxidase [G617]. Horseradish peroxidase dimerizes tyrosine [J401] and a range of tyrosine-containing peptides. With N-acetyltyrosine and some peptides the products include trimers and tetramers [J327]. The site of reaction appears to be *ortho* to the hydroxyl group, forming a carbon-carbon bond.

Lactoperoxidase oxidizes guaiacol to 4,4′-dihydroxy-3,3′-dimethoxybiphenyl and 3,3′-dimethoxy-4,4′-biphenylquinone, and a mixture of guaiacol and catechol forms 2′,3′,4-trihydroxy-3-methoxybiphenyl [J352].

Human neutrophil myeloperoxidase and horseradish peroxidase convert phenol into 4,4′-dihydroxybiphenyl [E297]. Horseradish peroxidase also converts phenol into 2,2′-dihydroxybiphenyl and polymers [C468, D433].

Silkmoth larva and locust convert a mixture of 4-substituted catechols and resorcinols into biphenyls. For instance, N-acetyldopamine and 4-ethylresorcinol form 6′-acetamidoethyl-3-ethyl-3′,4,4′,6-tetrahydroxybiphenyl [G346].

Pisum sativum ascorbate peroxidase forms a biphenyl from *p*-cresol [K404].

A laccase-like copper-containing phenol oxidase in Rhizoctonia praticola, optimum pH 6.7–6.9 and molecular weight 78 000, converts 2,6-dimethoxyphenol into 3,3′,5,5′-tetramethoxybiphenoquinone [A3836].

Formation of higher polyphenols

Bjerkandera adusta manganese peroxidase (E.C. 1.11.1.13) converts guaiacol into a mixture of polymers, molecular weights up to about 3200. A range of other phenols, syringic acid and *o*-anisidine are also substrates [K565].

Trimerisation of indole

Oak, horse chestnut and Rhus oxidize indole to 2,2′-bis(3-indolyl)indoxyl [A3125].

Pterostilbene dimerisation

Botrytis cinerea laccase-like stilbene oxidase converts pterostilbene into a dimer (3-(3,5-dimethoxyphenyl)-5-(2-(3,5-dimethoxyphenyl) ethenyl)-2-(*p*-hydroxyphenyl)benzofuran) [K147].

Coniferyl alcohol dehydrogenase (E.C. 1.1.1.194)

Prunus strobus, Abies balsama, Laryx laricina, Picea rubens and Pinus banksiana enzymes act on coniferyl alcohol to form dehydrodiconiferyl alcohol and pinoresinol. The reactions involve condensation of the side chain to the aryl nucleus to form dehydrodiconiferyl alcohol, and between two side chains of coniferyl alcohol to form pinoresinol. Several isozymes are involved, which are glycoproteins [G745].

Sitka spruce (Picea sitchensis) xylem contains two glycoprotein isozymes, molecular weights 62 000 or 80 000 (depending on method; the higher value may be a methodological artifact); it is not a peroxidase [J820].

Berbabamunine synthase (E.C. 1.1.3.34)

Berberis stolonifera enzyme is a P450, molecular weight 46 000, pI 6.05 and optimum pH 8–8.5 or 7.2–7.5, depending on conditions. Substrates include (*R*)- and (*S*)-N-methylcoclaurine, and (*S*)-coclaurine, which are dimerized by stereospecific oxidative phenol coupling, without transfer of activated oxygen to the substrate molecules [G736].

Vindoline dimerisation

Streptomyces griseus catalyzes the formation of a complex dimer from vindoline, probably via dihydrovindoline ether that involves a head-to-head C-C bond formation adjacent to the non-indole nitrogen [A2984].

1.5 *Reactions involving carbon ring fission*

Catechol 1,2-dioxygenase (pyrocatechase; E.C. 1.13.11.1)

Catechol →*cis*, *cis*-muconate

Rhodococcus rhodochrous enzyme, a homodimer with optimum pH 9, acts on catechol, 3- and 4-methylcatechol. The molecular weight of the monomer, based on mass spectrometry and genetic coding is 31 558 or 31 539, respectively [J703]. R. erythropolis enzyme, monomeric molecular weight 36 000–37 000 contains 1.3 mol Fe/mol. Pyrogallol, as well as catechol, 3- and 4-methylcatechol are substrates, but protocatechuate is not. It is inactivated at 50°. Its amino acid composition has been determined [D247].

Acinetobacter radioresistens enzyme is a homodimer, molecular weight 78 000. It is unusual in that it dissociates into an active monomer in 0.5M sodium sulphate. It contains 0.96 mol Fe^{3+}/subunit. It is more highly specific than Rhodococcus rhodochrous enzyme; 3- and 4-methylcatechol are poor substrates [J390].

A. calcoaceticus enzyme, molecular weight 81 000–85 000 (using different measuring techniques), is composed of two monomers and contains two mol Fe/mol. Additional substrates (poor) are 3- and 4-methylcatechol and 3-isopropylcatechol, but a number of other catechols are not substrates. It has a broad optimum at pH 7–9, which coincides with its stability range. Its amino acid composition has been determined; methionine is the amino terminal residue [A2640].

Rhizobium leguminosarum enzyme is a homodimer, molecular weight 70 000 and optimum pH 9–9.5, which contains one mol Fe/mol [F647]. R. trifolii enzyme is also a dimer, molecular weight 107 000, containing 1 mol Fe^{3+}/mol monomer [D672].

Candida tropicalis enzyme, optimum pH 7.6–8.0, acts on catechol, 4-methylcatechol, 3-and 4-chlorocatechol but not on other catechols [E204].

Trichosporon cutaneum enzyme has molecular weight 105 000 and 35 000 for holoenzyme and monomer respectively. Its specificity is broad, acting on catechol, 4-methylcatechol, pyrogallol and hydroxyquinol [C84].

Pseudomonas pyrocatechase II acts on catechols substituted at positions 3 and 4 with methyl, chloro or fluoro groups [B754]. P. arvilla enzyme converts pyrogallol into both α-hydroxymuconic acid and 2-pyrone-6-carboxylic acid; the latter is formed by ring closure of the ring fission product, possibly without prior release from the enzyme [B547].

Brevibacterium enzyme contains Fe^{3+}, apparently sulphur-bound [A1231].

Enzyme from an unspecified bacterium has an optimum between pH 7 and 10 [B525].

Catechol 2,3-dioxygenase (metapyrocatechase; E.C. 1.13.11.2)

Catechol → 2-hydroxymuconic semialdehyde

A study carried out on extradiol dioxygenases from a series of 7 Pseudomonas strains demonstrated that each enzyme has its own quantitative specificity towards catechol, 3- and 4-methylcatechol, 4-chlorocatechol and 2,3-dihydroxybiphenyl, ranging from good activity for most of these substrates to good activity for only 2,3-dihydroxybiphenyl [H272]. One Pseudomonas enzyme is a homotetramer, with apparent molecular weight for monomer and tetramer of 33 000 and 110 000, respectively. Its optimum pH is 8–8.5 and is stable up to 70°. It acts on catechol, 3- and 4-methylcatechol; 3-fluorocatechol, and 4-chlorocatechol are poor substrates [J649]. Another study by the same research team confirmed many of these results, but gave a tetrameric molecular weight of 120 000 [G355]. P. putida enzyme is probably a homotetramer with subunit molecular weight 34 000 [K204].

P. arvilla enzyme oxidizes pyrogallol to α-hydroxymuconic acid [B547] and is inhibited competitively by *o*-nitrophenol or *m*-phenanthroline relative to catechol, but non-competitively relative to oxygen [A1031]. It is not inhibited by superoxide dismutase (E.C. 1.15.1.1), hence, superoxide is not the oxidizing species, nor by compounds capable of trapping singlet oxygen [A1047]. P. aeruginosa enzyme is inhibited by ATP and Mg^{2+} [A2832].

Bacillus thermoleovorans enzyme is a homotetramer, subunit molecular weight 34 700, pI 4.8, optimum pH 7.2, and contains one Fe/monomer. It is inactivated rapidly at 70° [K228].

Enzyme from a thermophilic Bacillus is inactivated by high concentrations of oxygen [J834].

Hydroxyquinol 1,2-dioxygenase (E.C. 1.13.11.37)

Burkholderia cepacia enzyme is a dimer, molecular weight 68 000. It is highly specific, forming maleylacetate [J251].

Azotobacter enzyme is a dimer, monomeric molecular weight 34 000, which is activated by Fe^{2+}. An additional substrate is 6-chloro-1,2,4-trihydroxybenzene [H602].

Phanerochaete chrysosporium enzyme is a dimer, molecular weight 90 000. It incorporates molecular oxygen into the product. It is highly specific, but also oxidizes catechol [H284].

This reaction has also been observed in Pseudomonas [A2697], Sporotrichum [B124], Trichosporon [B368, D232], Arthrobacter [H270], Aspergillus [H236], Bradyrhizobium [J484] and Streptomyces [H347].

2,4,5-Trihydroxytoluene oxygenase

Burkholderia enzyme is apparently a dimer; the molecular weight was found to be 67 000 or 78 000 by different methods. It contains two Fe/mol and is stimulated by Fe^{2+} and ascorbate. The product is 2,4-dihydroxy-5-methyl-*cis, cis*-muconic semialdehyde. Other catechols are also substrates [J903].

3-Methylcatechol 1,2-dioxygenase

3-Methylcatechol → 2-methyl-*cis*, *cis*-muconate

This reaction has been observed with protocatechuate 3,4-dioxygenase (E.C. 1.13.11.3) from Hydrogenophaga palleroni [J218], catechol 1,2-dioxygenases (E.C. 1.13.11.1) from Pseudomonas, Alcaligenes eutrophus, Trichosporon cutaneum, Rhodococcus rhodochrous and R. erythropolis, as well as by activated sludge microorganisms [A3769, B754, C84, D247, E31, J703].

3-Methylcatechol 2,3-dioxygenase

3-Methylcatechol → 2-hydroxy-6-methyl-*cis*, *cis*-muconic semialdehyde

This reaction has been observed in several subspecies of Pseudomonas putida, in Burkholderia cepacia, Sphingomonas and Bacillus stearothermophilus, in activated sludge microorganisms [A2397, E31, J654, K5], as well as being catalysed by chlorocatechol 2,3-dioxygenase and 1,2-dihydroxynaphthalene oxygenase (see below).

4-Methylcatechol 2,3-dioxygenase

4-Methylcatechol → 2-hydroxy-5-methyl-*cis*, *cis*-muconic semialdehyde

Bacillus stearothermophilus catalyses this reaction [A2397], and many reports suggest that this reaction may be common in other microorganisms.

4-Methylcatechol 3,4-dioxygenase

4-Methylcatechol → 3-methyl-*cis*, *cis*-muconate

This reaction has been observed with protocatechuate 3,4-dioxygenase from Hydrogenophaga palleronii [J218], and catechol 1,2-dioxygenases from Pseudomonas, Alcaligenes eutrophus, Trichosporon cutaneum, Rhizobium trifolii, Aspergillus niger, Candida tropicalis, Rhodococcus rhodochrous and R. erythropolis [A3769, B754, C84, C342, D247, D672, E204, G647, J703].

4-Methylcatechol 4,5-dioxygenase

4-Methylcatechol → 2-hydroxy-4-methyl-*cis*, *cis*-muconic semialdehyde

Candida tropicalis catalyzes this reaction [A943], and many reports suggest that this reaction may be common in microorganisms.

2,3-Dihydroxybiphenyl 1,2-dioxygenase (E.C. 1.13.11.39)

Pseudomonas enzyme is an octamer, monomeric molecular weight 33 200. The product is 2-hydroxy-6-oxo-6-phenylhexa-2,4-dienoate. Catechol, 3- and 4-methylcatechol are poor substrates [G771]. P. pseudoalcaligenes and P. aeruginosa enzymes, molecular weights 250 000 and monomeric molecular weights 31 000, contain Fe^{2+}. They are specific for the 1,2-position of 2,3-dihydroxybiphenyl [E329].

Pyrocatechuate 2,3-dioxygenase (E.C. 1.13.11.28)

Tecoma stans enzyme is found both in the lamellar structure of chloroplasts and in the cytosol. The properties of both enzymes are similar, except that the particulate enzyme is somewhat more stable. Copper-chelating reagents are inhibitory, especially diethyldithiocarbamate. A range of heavy metal ions at mM concentration are inhibitory, with some quantitive differences between chloroplast and cytosolic enzymes; in particular, the former is totally inhibited by Cd^{2+}, whereas the latter is unaffected [A202]. It is very labile. Additional substrates include 2,3-dihydroxy-4-isopropylbenzoate and 2,3-dihydroxy-4-methylbenzoate, but not a range of other catechols. Thiols and thiol-binding reagents are inhibitory. Pre-incubation with substrate does not prevent this inhibition, although it protects against thermal inactivation [A2241].

2,3-Dihydroxybenzoate 3,4-dioxygenase (E.C. 1.13.11.14)

Pseudomonas fluorescens enzyme requires oxygen, and forms α-hydroxymuconic semialdehyde and CO_2; other catechols, including catechol, are not substrates [K945].

Protocatechuate 2,3-dioxygenase

Bacillus macerans acts on protocatechuate to form 5-carboxy-2-hydroxymuconic semialdehyde [A3887].

Protocatechuate 3,4-dioxygenase; (E.C. 1.13.11.3)

Protocatechuate → β-carboxymethyl-*cis*, *cis*-muconate

Tecoma stans enzyme, molecular weight 150 000 and optimum pH 5.2, requires Fe^{2+}. It is highly specific, and is not a monophenol monooxygenase. Thiol-binding reagents are inhibitory, and this is partially reversed by thiols [B223].

Acinetobacter calcoaceticus enzyme (electrophoretically homogeneous) is inhibited by many substrate analogues that do not undergo oxidation. Its optimum pH is 8.5–9, and is inactivated at 60°. Amino acid analyses demonstrate a quite different composition from Pseudomonas enzyme [A2407].

Azotobacter vinelandii enzyme, molecular weight variously estimated as 480 000 and 510 000, is composed of monomers, molecular weight 23 300 and 25 250. The native enzyme contains 10 mol of iron [B262].

Brevibacterium fuscum enzyme, molecular weight 315 000 may be a pentamer, and is composed of subunits, molecular weights 40 000 and 22 500. It contains essential Fe^{3+}. Amino acid and spectral analyses have been carried out [D86].

Hydrogenophaga palleronii enzyme not only acts on protocatechuate, but also on dopac, caffeate and some neutral catechols as well as 4-sulphocatechol. Its molecular weight is 97 000 and is composed of monomers, molecular weights 23 000 and 28 500. The specificity of Agrobacterium radiobacter is similar, except that 4-sulphocatechol is not a substrate [J218].

Moraxella isozymes have molecular weights of 220 000 and 158 000. These both appear to be polymers of structure $(\alpha\beta)_3$ and $(\alpha\beta)_4$ [F62].

Pseudomonas fluorescens enzyme is wine red in colour, molecular weight 409 000. Its amino acid composition has been determined [C729]. It is also found in P. aeruginosa [B547]. 3-Fluoro-4-hydroxybenzoate is strongly inhibitory [A1958]. The latter enzyme is thought to utilize superoxide as oxidant [B299]. It contains non-haem Fe^{3+}, removable by reduction and chelation [A152]. Protocatechualdehyde inhibits competitively relative to protocatechuate, and non-competitively relative to oxygen [A1031].

Rhizobium leguminosarum enzyme is induced by protocatechuate [F224].

Protocatechuate 4,5-dioxygenase; (E.C. 1.13.11.8)

Protocatechuate → γ-carboxy-α-hydroxymuconic semialdehyde

Rhizobium leguminosarum enzyme is a dimer, molecular weight 120 000 and optimum pH 9.5 [H127].

Gentisate 1,2-dioxygenase; (E.C. 1.13. 11.4)

Gentisate → 6-carboxy-4-hydroxy-*cis, cis*-muconic semialdehyde (maleylpyruvate)

Pseudomonas acidovorans and P. testosteroni enzymes differ in molecular weight (164 000 and 158 000, respectively) and amino acid composition. They are similar in requiring Fe^{2+}, and show a broad specificity toward substituted gentisates; 1,4-dihydroxy-2-naphthoate, an analogue of gentisate with a fused ring system, is also a substrate for the P. acidovorans enzyme [F649]. P. alcaligenes enzyme is a tetramer, molecular weight 154 000, optimum pH 8 and pI 4.8–5.0. P. putida enzyme is a dimer, molecular weight 82 000, optimum pH 8 and pI 4.6–4.8. Enzymes from both species are activated by Fe^{2+} [K4].

Klebsiella pneumoniae enzyme is a tetramer, molecular weight 159 000, optimum pH 8–9 and pI 4.7 [J246]. Bacillus stearothermophilis enzyme appears to be a hexamer, monomeric molecular weight 40 000 that requires Fe^{2+} [H893]. Rhodococcus erythropolis enzyme, a homooctamer, molecular weight 328 000, is activated by Fe^{2+}. Other substrates are gentisates substituted with methyl or halogen [H441].

2,3-Dihydroxy-*p*-cumate 3,4-dioxygenase

Bacillus enzyme has an optimum pH of 8.2. The reaction product is 3-carboxy-2-hydroxy-7-methyl-6-oxoocta-2,4-dienoate. Other substrates include pyrocatechuate, 3-methylcatechol, 3-isopropylcatechol and 2,3-dihydroxy-*p*-toluate, but not catechol or 4-substituted catechols. This distinguishes the enzyme from catechol 2,3-dioxygenase [G766, G768].

Homogentisicase (homogentisate 1,2-dioxygenase; E.C. 1.13.11.5)

Homogentisate + O_2 → maleylacetoacetate

Rat liver enzyme shows a diurnal rhythm, with a sharp decrease in activity at 1800–2100, immediately after light is withdrawn. This coincides with maximal activity in the diurnal rhythm for tyrosine transaminase (E.C. 2.6.1.5) [A1060].

Mouse liver enzyme appears to be a homotrimer, molecular weight 149 000, optimum pH 6.1 and pI about 8. It has an absolute requirement for Fe^{2+}; ascorbate is required, probably to keep Fe in the reduced form [H436].

Rabbit liver enzyme is a heterodimer monomeric molecular weight 230 000 and 200 000, and an optimum at about pH 5. It is also found in human embryo [H768].

A homogentisicase has been found in Drosophyllum lusitanicum [A1149].

Aspergillus enzyme requires Fe^{2+}; other cations such as Fe^{3+}, Cu^{2+}, Co^{2+} and Mn^{2+} cannot substitute for Fe^{2+}. This enzyme is very labile in the presence of oxygen [B848]. Another study found that A. nidulans enzyme is 'rather stable' and highly specific [H900].

3,4-Dihydroxyphenylacetate 2,3-dioxygenase (E.C. 1.13.11.15)

3,4-Dihydroxyphenylacetate → δ-carboxymethyl-α-hydroxy-*cis, cis*-muconic semialdehyde

Bacillus brevis enzyme is a tetramer, molecular weight 140 000, with monomeric molecular weight 36 000. It is not activated by Fe^{2+} or inhibited by CN^- (atypical for this type of enzyme, which is usually activated by Fe^{2+}), peroxide or diethyldithiocarbamate. It contains two mol of Mn/mol, probably as Mn^{2+}; it is claimed to be the first Mn-containing oxygen-activating enzyme described [B947]. It may utilize superoxide as co-substrate [B299].

Bacillus stearothermophilus enzyme is a globular protein, molecular weight 106 000, composed of monomers, molecular weight 33 000–35 000, and is stable up to 55°. It acts on 3,4-dihydroxyphenylacetate, 3,4-dihydroxymandelate and dihydrocaffeate with a similar V_{max}; protocatechuate is a poorer substrate. Catechol, 4-methylcatechol, L-dopa and dopamine are not substrates, indicating a narrower specificity than for enzyme from Pseudomonas ovalis. It shows a broad optimum about pH 8.4–8.7. Its amino acid composition has been determined [A2827].

Pseudomonas ovalis enzyme (colourless), is apparently a homopolymer, molecular weight 140 000 and monomeric molecular weight 35 000. It contains 4–5 mol of non-haem iron (Fe^{2+}). Its amino acid composition has been determined, and the carbonyl terminal amino acid is leucine. Many catechols are substrates, but often with a very low oxidation rate. It is inactivated by oxygen or by other oxidizing agents; reactivation is brought about by Fe^{2+} or reducing agents, suggesting that inactivation involves oxidation to Fe^{3+} [A1090, A1230].

Klebsiella pneumoniae enzyme, a homotetramer, molecular weight 102 000, requires four mol of Mg^{2+}/mol [H232].

Arthrobacter synephrinum enzyme, molecular weight 282 000, additionally acts on 3,4-dihydroxymandelate and dihydrocaffeate [A3279].

3-Carboxyethylcatechol 2,3-dioxygenase (E.C. 1.13.11.16)

Achromobacter enzyme requires Fe^{2+} [K819].

Caffeate 3,4-dioxygenase (E.C. 1.13.11.22)

This reaction in Hydrogenophaga palleronii is catalyzed by protocatechuate 3,4-dioxygenase (see above) [J218].

L-Dopa 4,5-dioxygenase

Amanita muscaria enzyme is a heteropolymer, monomeric molecular weight 22 000. The muconic semialdehyde formed is an intermediate in the formation of betalamic acid. Other catechols are also substrates [G119].

Questin monooxygenase (E.C. 1.14.13.43)

Aspergillus terreus enzyme is composed of two proteins, optimum pH 7.5, requires oxygen and NADPH and is very unstable. The reaction involves opening the quinone ring to form a carboxylic acid [K701].

3,4-Dihydroxyphenylalanine: oxygen 2,3-oxidoreductase (recyclizing) (stizolobinate synthase, E.C. 1.3.11.30)

Stizolobium hassjoo seedling enzyme, optimum pH 7.6 is highly specific. It requires Zn^{2+}, is unstable especially in air and is inhibited by thiol-binding compounds [K706, K708].

3,4-Dihydroxyphenylalanine: oxygen 4,5-oxidoreductase (recyclizing) (stizolobate synthase, E.C. 1.3.11.29)

Stizolobium hassjoo seedling enzyme, optimum pH 7.4 is highly specific. It requires Zn^{2+}, is unstable especially in air and is inhibited by thiol-binding compounds [K706, K708].

1,2-Dihydroxynaphthalene oxygenase

Pseudomonas putida enzyme, molecular weight in excess of 275 000, and subunit molecular weight 19 000, yields 2-hydroxychromene-2-carboxylate, presumably a secondary product from *cis*-4-(*o*-hydroxyphenyl)-2-oxobut-3-enoate, the fission product. It requires Fe^{2+}; iron-binding compounds are inhibitory, and this is reversed by Fe^{2+}. 3-Methylcatechol is also a substrate, but most other catechols are not [B527].

1-Hydroxy-2-naphthoate 1,2-dioxygenase (E.C. 1.13.11.38)

Nocardioides enzyme, monomeric molecular weight 45 000 and native molecular weight 270 000, contains one mol Fe^{2+}/mol, which is essential for activity. The product is 4-(*o*-carboxyphenyl)-2-oxobutyrate [J598].

Coccus enzyme requires oxygen and is activated by Fe^{2+} [E730].

Diphenyl ether fission

Rat liver converts thyroxine into diiodotyrosine; the reaction requires oxygen [C809]. It is also catalyzed by human thyroid peroxidase [D553].

Pseudomonas cruciviae degrades diphenyl ether to phenol, probably with 2-phenoxymuconate as an intermediate [E41].

***o*-Aminophenol 1,6-dioxygenase**

Pseudomonas arvilla pyrocatechase oxidizes the substrate to 2-aminomuconic semialdehyde [A3636]. P. pseudoalcaligenes enzyme, molecular weight 140 000 and monomeric molecular weights 35 000 and 39 000, requires oxygen , with an apparent requirement for Fe^{2+}. It also acts on catechol, 6-amino-3-methylphenol, 2-amino-3-methylphenol and 2-amino-4-chlorophenol, but other catechols and quinols are not substrates [J255, J329].

A bacterial enzyme, molecular weight 140 000 and optimum pH 7.5 is stable between pH 7.5 and 9 [H915].

3-Hydroxyanthranilate 3,4-dioxygenase (E.C. 1.13.11.6)

$$\text{3-Hydroxyanthranilate} + O_2 \rightarrow \alpha\text{-amino-}\beta\text{-carboxymuconic semialdehyde}$$

The rat brain and liver enzymes, molecular weight 37 000–38 000, appear to be identical. It has a broad optimum at about pH 6–5 [E472]. Baboon liver enzyme has a similar molecular weight, optimum pH 7.4–7.6. It requires Fe^{2+} for activity [A2280] as do the vervet monkey and rat liver enzymes [A912, E114]. The latter enzyme exists as several isozymes. Beef kidney enzyme has a molecular weight of 34 000, and readily polymerizes to an inactive form [A2430].

4-Sulphocatechol 3,4-dioxygenase

Hydrogenophaga palleronii protocatechuate 3,4-dioxygenase acts on 4-sulphocatechol as well as on protocatechuate, dopac, caffeate and some neutral catechols. Its molecular weight is 97 000 and is composed of monomers, molecular weights 23 000 and 28 500. The product is 3-sulphomuconate [J218].

Chlorocatechol 2,3-dioxygenase (catechol 1,6-dioxygenase)

$$\text{3-Chlorocatechol} \rightarrow \text{2-hydroxymuconic acid}$$

Pseudomonas putida enzyme is a tetramer, monomeric molecular weight 33 400 and pI 7.1. Other substrates include catechol, and catechols substituted with methyl or halogen. Although the product formed from 3-chlorocatechol is 2-hydroxymuconic acid indicating the formation of the acid chloride by proximal fission, the product formed from 3,5-dichlorocatechol indicates distal (1,6) fission. This enzyme is atypical for extradiol dioxygenases in that it is rapidly inactivated above 40° [J515].

Chlordiazon catechol dioxygenase (E.C. 1.13.11.36)

Phenylobacterium immobilis enzyme, pI 4.5, contains one Fe/subunit, and has been crystallized. The product is a hydroxymuconic semialdehyde [C332].

Cyclopropane ring fission

Rat liver microsomes act on 1,1-dichloro-*cis*-diphenylcyclopropane to form several products such as 2-chloro-3-oxo-1,3-diphenyl-1-propene. This involves an unusual type of ring fission which is considered to form 2,3-dichloro-1,3-diphenyl-1-propene as the initial product [F573].

Methylococcus capsulatus soluble methane monooxygenase (E.C. 1.14.13.25) converts cyclopropylbenzene into benzyl alcohol. 3-Phenylprop-2-en-1-ol is also formed, and this may be an intermediate [F278].

1.6 *Reactions involving heterocyclic ring fission*

D-Tryptophan 2,3-dioxygenase (D-tryptophan pyrrolase)

D-Tryptophan + O_2 → D-kynurenine

The properties of rat liver enzyme are very similar to those of L-tryptophan 2,3-dioxygenase, and it is readily induced by cortisone, D- or L-tryptophan. An intestinal mucosal enzyme that is less readily induced acts on both isomers of tryptophan [A598].

Wheat germ enzyme acts on both D- and L-tryptophan, and substituted analogues. The enzyme is found as 3 molecular forms [E24].

L-Tryptophan 2,3-dioxygenase (L-tryptophan pyrrolase; E.C. 1.13.11.11)

L-Tryptophan + O_2 → L-formylkynurenine

This enzyme in vertebrates is the first in the sequence for the oxidation of dietary tryptophan ingested in excess of requirements for protein synthesis and hormone formation.

The enzyme is found (in decreasing order of activity) in mammalian liver, lung and brain from rat, gerbil and mouse. In rabbit, lung shows the highest activity [J830].

Rats receiving an I.P. dose of ethanol four hours prior to death showed a twofold increase in both enzyme quantity and total enzyme activity. This is caused by an increase in the half-life of the enzyme from 2.3 to 3.9 h. [A2059]. In contrast, treatment with sucrose or ethanol for two days resulted in no change in the amount of 'holoenzyme', but a 30 per cent decrease in 'total enzyme'. The effect of cortisol in conjunction with sucrose appears to result entirely from decreased 'holoenzyme', but with no effect on 'total enzyme'. On its own, cortisol markedly increases both enzyme forms and eliminates the effects of ethanol. In this study it appears that the two forms were distinguished by measuring activity with and without added haematin [A2062]. Rat cerebral activity is increased by administration of L-tryptophan [A1289].

Drosophila melanogaster eye enzyme is mainly cytosolic, but a small proportion is insoluble [A1194].

Wheat germ enzyme acts on both D- and L-tryptophan, and on a range of tryptophan containing peptides, as well as on tryptophan analogues substituted on both ionic groups or the 5- position, all with similar activity. The enzyme is found as three molecular forms [C559, E24].

Bacillus brevis (a thermophile) enzyme is a haemoprotein, which appears to be a homotetramer, molecular weight 110 000. The enzyme is stabilized by L-tryptophan [F737]. Pseudomonas acidovorans enzyme is a tetramer, containing two moles both of haem and copper (Cu^+ is the active form) per mole. Tryptophan interacts with both haem and copper, and molecular oxygen binds to the enzyme-tryptophan complex, possibly to the copper. Studies with chelating agents demonstrate that although Cu^{2+}-binding agents do not inhibit, Cu^+-binding agents are inhibitory. After oxidation diethyldithiocarbamate prevents reactivation by ascorbate [A792, A1426]. P. fluorescens enzyme does not contain copper; inhibition by copper-binding reagents appears to be by other mechanisms [A1096]. Streptomyces parvulis enzyme (optimum pH 8.3) requires ascorbate, and can utilize 5-methyl-, 5-fluoro- and α-methyl-L-tryptophan as additional substrates; D- isomers are not substrates. It is probably a haemoprotein, molecular weight 88 000 [E588, E589].

Peptide tryptophan 2,3-dioxygenase (E.C. 1.13.11.26)

Wheat germ (optimum pH about 8) and rat liver enzymes act on tryptophan residues in polypeptides such as pepsin and trypsin, converting them into kynurenine residues [K823].

Indole 2,3-dioxygenase; (E.C. 1.13.11.17)

Indole + O_2 → anthranilate

In rat, indole is converted into N-formylanthranilate, which is the expected initial product of indole ring fission, as well as anthranilate [D158].

Jasminum grandiflorum leaf enzyme, optimum pH 4.8, forms anthranilate from the substrate. Other substrates are 5-hydroxy- and 5-bromoindole. FAD is co-substrate and Cu^{2+} is also required; Cu^{2+}-chelating reagents are inhibitory [B70].

Tecoma stans leaf enzyme, optimum pH 5.2, utilizes two molecules of oxygen. Other substrates are 5-hydroxyindole, 5-bromoindole and 5-methylindole. It is not inhibited by thiols or by thiol-binding reagents, copper or non-haem iron chelators, nor by atebrin, which suggests that it is not a flavoprotein. It is inactivated by dialysis, but the cofactor has not been identified [C170].

Zea mays leaf enzyme, molecular weight 98 000, exhibits an optimum at pH 5.0. The initial product appears to be 2-formamidobenzaldehyde, although the products detected are anthranil and anthranilate; their formation requires four oxygen atoms. Copper-binding reagents are inhibitors, reversible by Cu^{2+}, which also activates dialyzed enzyme [D334].

Aspergillus niger forms both N-formylanthranilate and anthranilate, and enzymes for both reactions were found [F417].

A similar reaction occurs in wheat seedling, converting skatole into *o*-formamidoacetophenone. The enzyme, which is composed of two isozymes, is different from tryptophan 2,3-dioxygenase [C559].

Indoleamine 2,3-dioxygenase (E.C. 1.13.11.42)

Mouse liver indoleamine 2,3-dioxygenase is totally different from rat liver tryptophan 2,3-dioxygenase, although tryptophan is the substrate. It is induced by lipopolysaccharide (which suppresses tryptophan 2,3-dioxygenase) and is found in almost all tissues examined, except trachea, bladder and spleen. In most tissues the activity is low, but good activity is found in lung, colon, liver, caecum, seminal vesicle and especially in epididymis [E30].

2,3-Dihydroxyindole 2,3-dioxygenase (E.C. 1.13.11.23)

A microorganism enzyme, optimum pH 8.0 and stable at pH 7–9, requires oxygen. Analogues of the substrate are not oxidized [K821].

Quinaldate 4-oxidoreductase (E.C. 1.3.99.18)

Alcaligenes enzyme, molecular weight 155 000, requires oxygen and NADH, and is activated by Fe^{2+}. The reaction product is kynurenate [J32].

Pseudomonas enzyme, optimum pH 8.0, pI 4.6 and molecular weight 300 000, is composed of subunits, molecular weights 90 000, 34 000 and 20 000 (consistent with a $\alpha_2\beta_2\gamma_2$ structure), contains molybdenum, iron, acid labile sulphur and FAD. It forms kynurenic acid from quinaldate. Heavy metals are inhibitory [K768].

Serratia marcescens enzyme (inducible), molecular weight 95 000–100 000 is a heterodimer, monomeric molecular weights 75 000–80 000 and 18 000–19 000; it contains Fe and Mo [K770].

7,8-Dihydroxykynurenate 8,8a-dioxygenase (E.C. 1.13.11.10)

This reaction forms 5-(3-carboxy-3-oxopropenyl)-4,6-dihydroxypyridine-2-carboxylate. The enzyme (source unstated) requires oxygen and NAD(P)H [K853].

Coumarin 3,4-oxide fission

This compound rapidly degrades to *o*-hydroxyphenylacetaldehyde in aqueous solution; it is postulated that this is an integral step in the route by which coumarin is degraded by mouse liver microsomes [H807].

Quercetin 2,3-dioxygenase (E.C. 1.13.11.24)

Aspergillus flavus enzyme contains copper but not iron. The products are 2-protocatechuoylphloroglucinolcarboxylic acid and carbon monoxide [K820].

Catechin oxygenase

Chaetomium cupreum enzyme, molecular weight 40 000, optimum pH 7.0 and pI 9.1, is a glycoprotein and is specific for catechin. The reaction products are catechol, protocatechuate and phloroglucinolcarboxylate [G814].

2-Hydroxychromene-2-carboxylate isomerase

Pseudomonas enzyme, molecular weight 25 000–27 000 and optimum pH about 8, is stable at pH 3–10. Glutathione appears to be both cofactor and stabiliser (at 50°). The reaction product is *o*-hydroxybenzylidenepyruvate, and lies on the degradation pathway of Tobias acid to salicylate [J711].

Maackiain fission

Stemphylium botryosum reductively opens the furan ring to form 2′,7-dihydroxy-4′,5′-methylenedioxyisoflavan [A2283].

Flavonol 2,4-dioxygenase

Aspergillus niger enzyme grown on rutin is a trimer, molecular weight 130 000–170 000 (clearly a mixture), monomeric molecular weights about 65 000, 55 000 and 33 000. It is a copper-containing glycoprotein composed of about 50 per cent N-linked oligomannose-type glycan chains. The products from quercetin are carbon monoxide and protocatechuoylphloroglucinolcarboxylate. It is potently inhibited by ethylxanthate, which specifically reduces Cu^{2+}. It is inhibited by diethyldithiocarbamate without loss of copper [K214].

(+)-Pinoresinol/(+)-lariciresinol reductase

Forsythia intermedia enzyme is composed of two isozymes, molecular weight 35 000, separable by anion-exchange chromatography. The reaction involves the successive fission of both ether rings in pinoresinol to larisiresinol and (−)-secoisolarisiresinol [H883].

Usnic acid dehydrogenases ((*S*)-usnate reductase, E.C. 1.1.1.199)

Evernia prunastri enzyme, molecular weight 450 000, requires NADH for the reduction of L-usnic acid. D-Usnic acid dehydrogenase is composed of two isozymes. The reaction results in the opening of the ether ring [B774, E126].

Phosphoribosyl isomerisation

Phosphoribosylanthranilate isomerase cleaves the ribosyl ether bond to form *o*-carboxyphenylaminodeoxyribulose-5-phosphate. The molecular weight of the enzymes from Enterobacter hafniae, Aeromonas formicans and Proteus is 67 000. This is a key reaction step in the formation of tryptophan; the next step is the formation of indole-3-glycerol phosphate, which either forms tryptophan directly, or forms indole as an intermediate [A1481].

(–)-Epigallocatechin ether fission

Musa accuminata forms 2-hydroxy-3-(2,4,6-trihydroxyphenyl)-1-(3,4,5-trihydroxyphenyl)-1-propanone from (–)-epigallocatechin; this presumably is an oxidative fission [K423].

3,4-Dihydroxyquinoline 2,4-dioxygenase
(E.C. 1.13.99.5)

Arthrobacter and Pseudomonas putida enzymes form N-acetyl- and N-formylanthranilic acids, respectively, from 1*H*-3,4-dihydroxyquinaldine and 1*H*-3,4-dihydroxyquinoline, respectively, with release of carbon monoxide. Two oxygen atoms are incorporated, indicating oxygenolytic attack at C-2 and C-4 of both substrates [K774].

5,10-Methylenetetrahydrofolate dehydrogenase
(E.C. 1.5.1.5 and 1.5.1.15; $NADP^+$ and NAD^+-requiring, respectively)

5, 10-Methylenetetrahydrofolate → 10-formyltetrahydrofolate

Pig liver enzyme appears to be a complex with 5,10-methenyltetrahydrofolate cyclohydrolase and 10-formyltetrahydrofolate synthetase (E.C. 6.3.4.3) [A2844].

In dihydrofolate-deficient immature domestic poultry chicks, activity is reduced by 25 per cent by folate or oestradiol, and these effects are additive [A1680].

Pisum sativum enzyme appears to be identical with 5,10-methenyltetrahydrofolate cyclohydrolase, but is separable from 10-formyltetrahydrofolate synthetase. Its molecular weight is 38 500, and it requires NADP [H619], with optimum pH 7.8 [A1429].

Saccharomyces cerevisiae enzyme is composed of two cytosolic isozymes; one requires NADPH and the other NADH [H794].

Peptostreptococcus productus 5,10-methylenetetrahydrofolate dehydrogenase, molecular weight 66 000 is a dimer. The product is said to be 5,10-methenyltetrahydrofolate, which makes it a different enzyme from others listed under this heading [F869].

5,10-Methylenetetrahydrofolate reductase
(E.C. 1.1.1.68)

5, 10-Methylenetetrahydrofolate ↔ 5-methyltetrahydrofolate.

Pig liver enzyme, molecular weight 77 300, contains one mol of FAD [C85].

Rat liver and brain enzymes are cytosolic, optimum pH 6.6 [A1586, A2502]. The reverse reaction requires FAD as hydrogen acceptor. An additional substrate for the reverse reaction is 5-methyltetrahydropteroylpentaglutamate [A2502].

Human platelet enzyme releases formaldehyde from 5-methyltetrahydrofolate [A2446].

Clostridium formicoaceticum enzyme is an octomer composed of two different monomers, molecular weights 26 000 and 35 000, and contains iron and zinc, acid-labile sulphur and FAD. It is inactivated by oxygen. For the reverse reaction, methylene blue, menadione, benzyl viologen or FAD can be used as oxidant [D260].

Peptostreptococcus productus enzyme is a dimer, molecular weight 66 000 [F869].

5,10-Methenyltetrafolate cyclohydrolase
(E.C. 3.5.4.9)

Pisum sativum enzyme, molecular weight 38 500, which requires NADP, could not be separated from 5,10-methylenetetrafolate dehydrogenase (E.C. 1.5.1.5 and 1.5.1.15). It is inhibited by dihydrofolates [H619].

Pig liver enzyme is associated with a complex of enzymes that are involved with folate reactions [A2844].

Hydantoin hydrolysis (c.f. E.C. 3.5.2.2)

Flavobacterium reversibly hydrolyzes a range of 5-aralkylhydantoins to N-carbamoylamino acids (the corresponding L-amino acids are released by a further enzymatic hydrolysis). Both D- and L-N-carbamoylamino acids are formed with most substrates (DL mixtures), but the L-isomer is preferred. The optimum pH is about 8.6 [F938, F940].

Pseudomonas is also effective in catalyzing the reaction with 5-(*p*-hydroxyphenyl)hydantoin, both isomers of which yield D-*p*-hydroxyphenylglycine, but Achromobacter delicatulus is much less effective than Pseudomonas. The optimum pH for the Pseudomonas enzyme is 8.0 for a range of aryl-substituted hydantoins. A racemization step accompanied by asymmetric hydrolysis has been suggested [E391, E450, F943].

Methyl tetrahydroprotoberberine 1,4-monooxygenase (E.C. 1.14.13.37)

Corydalis vaginans enzyme fissions the heterocyclic ring system of (*S*)-N-methylcanadine and N-methylstylopine to form an oxo-azadecane ring system with both aromatic moieties unchanged [E577].

Morpholine and piperazine ring fission

Many studies have shown that these ring systems are subject to fission reactions. For instance, in rat the morpholino ring of doxapram is oxidized to bis(2-hydroxyethyl)amino and 2-hydroxyethylamino groups [A1389]. In rat trifluoperazine, fluphenazine, prochlorperazine, perazine and chlorcyclizine are oxidized to phenylenediamines [A576].

Lactone hydrolysis

Human serum Type Q and Type R hydrolases act on the lactones 2-coumaranone, dihydrocoumarin and homogentisic lactone as well as on many aliphatic lactones to release the free acids. The hydrolysis rate is not as great as for phenyl acetate, but it is much greater than for paraoxon [K327].

Human serum arylesterase (see paraoxonase) hydrolyzes the lactone analogue 4-((5-methyl-2-oxo-1,3-dioxol-4-yl)methylthio)benzenesulphonate [G892].

Acinetobacter calcoaceticus lactonohydrolase, molecular weight 30 000 and optimum pH 7, is specific for 3,4-dihydrocoumarin; its amino acid sequence has been determined. It also shows bromoperoxidase activity towards monochlorodimedon with peroxide in the presence of an organic acid; this is due to the formation of a low-molecular weight brominating species [K429].

Dioxin fission

Phanerochaete sordida acts on 2,3,7,8-tetrachlorodibenzo-*p*-dioxin and octachlorodibenzo-*p*-dioxin to form 4,5-dichlorocatechol and tetrachlorocatechol respectively [H911].

Methylenedioxy ring fission

Rat liver microsomes oxidize 3,4-methylenedioxymethamphetamine to 3,4-dihydroxymethamphetamine; the (+)-isomer is the preferred substrate. 3,4-Methylenedioxyamphetamine is also a substrate. It appears to be mediated by a NADH-dependent P450 enzyme [A1570, F805]. A range of other methylenedioxybenzenes are substrates, with the methylenedioxy carbonyl being oxidized to carbon monoxide, which then inhibits the reaction; the enzyme is a phenobarbital-induced P450. The optimum pH is 7.4 [A3947].

Rabbit liver P450IIB4 and probably other P450 isozymes oxidize methylenedioxybenzene, methylenedioxyamphetamine and methylenedioxymethamphetamine. It requires NADPH, and is inhibited by carbon monoxide [F761].

Pseudomonas converts piperonylate into vanillate and protocatechuate. The authors suggested (probably erroneously) that vanillate is the initial product [A2449].

Dibenzothiophene sulphone monooxygenase

Rhodococcus erythropolis enzyme, which has been crystallized, is a homodimer, molecular weight 97 000, monomeric molecular weight 50 000 and optimum pH about 7.5. It converts the above compound into 2′-hydroxybiphenyl-2-sulphonate; dibenz[*c*,*e*][1,2]oxathiin-6,6-dioxide is a second substrate. The terminal amino acid sequence has been determined. 8-Hydroxyquinoline, α,α-bipyridyl, Mn^{2+} and Ni^{2+} are inhibitory. This enzyme is the second in a sequence that desulphurises thiophenes; further steps desulphinate the reaction product [K315].

1.7 *Dearomatization*

Benzene 1,2-dioxygenase; (E.C. 1.14.12.3)

Benzene → *cis*-1, 2-dihydroxycyclohexa-3, 5-diene (*cis*-1, 2-dihydro-1, 2-dihydroxybenzene)

Mammalia and other vertebrates differ from microorganisms in that the latter form *cis*-dihydrodiols instead of *trans*- analogues (which arise from aryl epoxides) from benzene and polynuclear hydrocarbons.

Pseudomonas putida enzyme has a broad specificity for monosubstituted benzenes, with oxidation adjacent to the substituent group [J656]. It is composed of a dioxygenase, molecular weight 215 300, and an electron transport protein, molecular weight 12 300. Both are iron-sulphur proteins, the first with two 2Fe-2S clusters and the second with one such cluster [B45]. A further publication claims that the molecular weights are 186 000 and 21 000 respectively, with a third component, a flavoprotein, molecular weight 60 000. The system requires Fe^{2+} and NADH [A1611].

The reaction has also been observed in Moraxella [A638] and Rhodococcus [H873].

Toluene 2,3-dioxygenase (E.C. 1.14.12.11)

Pseudomonas putida enzyme, molecular weight 151 000, is composed of subunits, molecular weights 52 500 and 20 800. It requires NADH, Fe^{2+}, NADH cytochrome c reductase (E.C. 1.6.99.3) and a Fe-S enzyme (ferredoxin $_{TOL}$). The product is toluene-2,3-dihydrodiol [B291]. This enzyme (initially the product was mis-identified as toluene-1,2-dihydrodiol) can utilise Fe^{3+}, Mg^{2+}, Ca^{2+} and Cu^{2+} (in decreasing order of effectiveness) in place of Fe^{2+} [A3184].

Naphthalene 1,2-dioxygenase (E.C. 1.14.12.12)

Naphthalene → *cis*-1, 2-dihydro-1, 2-dihydroxynaphthalene

Pseudomonas enzyme, which yields the (+)-dihydrodiol, is a three-protein system, and utilizes NAD(P)H and oxygen stoichiometrically [C144]. P. putida has been identified as one species exhibiting this activity [A1591].

Corynebacterium renale enzyme, a heterodimer, molecular weight 99 000, monomeric molecular weight 43 000 and 56 000, is not a P450. The reaction is stoichiometric, with an optimum at pH 6.5. The enzyme contains one FAD and one Fe^{2+}/mol, and requires NADH, which can be replaced by peroxide. Catalase, superoxide dismutase and hydroxyl radical scavengers are inhibitory. Hydroxyl radical is considered to be the oxidizing species, forming a dihydrohydroxynaphthalene radical. A hydroxyl ion then adds, to form a *cis*-1,2-dihydrodiol [B949, C714].

Biphenyl dioxygenase

Burkholderia enzyme forms *cis*-biphenyl-2,3-dihydrodiol from biphenyl. A number of polychlorinated biphenols are also substrates [K547].

Benzoate 1,2-dioxygenase (E.C. 1.14.12.10)

Pseudomonas arvilla enzyme is composed of NADH-cytochrome c reductase and the oxygenase. The latter, molecular weight 270 000–280 000 (depending on method), pI 4.5 and Stokes radius 5.6 nm, contains 10 mol of Fe, and about eight of labile sulphur/mol, but no haem or flavin. The product is benzoate-1,2-dihydrodiol. A range of benzoates monosubstituted with halide, methyl, methoxy, amino or hydroxyl groups in *o*-, *m*- and *p*-positions are substrates, although some of these are essentially inactive [B465].

Phthalate oxygenase

Rhodococcus erythropolis enzyme, optimum pH 6.5, is a tetramer, monomeric molecular weight 56 000. The reaction product is phthalate-3,4-dihydrodiol. Oxygen and NADH are required [H89].

Phthalate 4,5-dioxygenase (E.C. 1.14.12.7)

Pseudomonas cepacia enzyme is a two-component system, composed of a Fe-S-protein, molecular weight 34 000, with NADH-dependent oxidoreductase activity; its function is to keep the oxygenase (the second component) in a reduced form. The oxygenase is a nonhaem iron protein, molecular weight 217 000, monomeric molecular weight 48 000. The reaction is stoichiometric, requiring oxygen and NADH. 4-Chlorophthalate is also a substrate [K922].

Terephthalate 1,2-dioxygenase (E.C. 1.14.2.15)

Comamonas testosteroni enzyme, molecular weight 126 000, appears to be a tetramer with monomers, molecular weight 49 000 and 18 000; the N-terminals have been identified. The product is (1*R*,2*S*)-dihydroxy-3,5-cyclohexadiene-1,4-dicarboxylic acid. The enzyme requires a second protein fraction, oxygen and NADH, and its activity is enhanced by Fe^{2+}; it contains 2Fe-2S units. The specificity is narrow, but 1,4-dicarboxynaphthalene is a second substrate [H396].

Kynurenate 7,8-hydroxylase (E.C. 1.14.99.2)

This Pseudomonas fluorescens enzyme (incorrectly named kynurenine 7,8-hydroxylase in the E.C. list) forms kynurenate-7,8-dihydrodiol [K874].

Benzoyl CoA reductase (E.C. 1.3.99.15)

Thauera aromatica enzyme, molecular weight 160 000–170 000; is a heterotetramer, which contains FAD and ferredoxin. The product is 3,4-dihydrobenzoyl CoA [J527]. The subunit molecular weights are 48 000, 45 000, 38 000 and 32 000, and it contains 11 mol of Fe and acid-labile sulphur. It is greenish-brown with an absorption band typical of Fe-S. The reaction requires Mg^{2+} and ATP; Mn^{2+}, Fe^{2+} and (less effective) Co^{2+} can replace Mg^{2+}. It requires a strong reducing agent such as Ti(III) for activity, and oxygen inactivates. Several analogues of the substrate, with a single substituent on the nucleus are also reduced [J175].

Anthraniloyl CoA monooxygenase (E.C. 1.14.13.40)

Pseudomonas enzyme, a homodimer, molecular weight 170 000 and pI 5.3, requires NADH and oxygen. Three products are formed, 5-hydroxyanthraniloyl CoA, 2-amino-5-hydroxycyclohex-1-enecarboxyl CoA and 2-amino-5-oxocyclohex-1-enecarboxyl CoA, the latter being the main product. The aromatic and reduced ring systems are considered to be formed by separate routes [F463, F465, K949].

Azoarcus evansii 2-aminobenzoyl CoA monooxygenase/reductase is a flavoprotein that requires oxygen. It is postulated that 5-hydroxylation with migration of the hydrogen to position 6 is followed by a NADH-dependent reduction at the same catalytic locus to form 2-amino-5-oxocyclohex-1-enecarboxyl CoA [K215].

Hydroxyquinol reductase

Desulphovibrio inopinatus enzyme forms, probably, 1,2,4-trihydroxycyclohexa-1,3-diene from hydroxyquinol [K483].

Phloroglucinol reductase

Eubacterium oxidoreducens enzyme, molecular weight 78 000, monomeric molecular weight 33 000, and optimum pH 7.8, requires NADPH; NADH is ineffective. It is not a metalloenzyme or a flavoprotein. The reaction, which is reversible, forms dihydrophloroglucinol [F75].

Penicillium simplissimum forms dihydrophloroglucinol; the enzyme requires NADPH, but NADH is inactive [F639].

Coprococcus enzyme, molecular weight 130 000 and optimum pH 7.4, requires NADPH. The reaction is reversible, but only slowly. It is inactivated at 53° [C99].

Tetrahydroxynaphthalene reductase (E.C. 1.1.1.252)

Pyricularia oryzae enzyme reduces 1,3,6,8-tetrahydroxynaphthalene to scytalone (3,6,8-trihydroxy-1-tetralone) and 1,3,8-trihydroxynaphthalene to vermelone [K703].

Magnaporthe grisea enzyme (vermelone formation) requires NADPH as co-substrate, and is inhibited by tricyclazole. It is an essential step in the formation of a melanin that is required for the initiation of blast disease in rice [K704].

Anhydrotetracycline oxygenase (E.C. 1.14.13.38)

Streptomyces aureofaciens enzyme, monomeric molecular weight 57 500 is a dimer, which hydroxylates one of the aromatic rings of anhydrotetracycline at the 6 position (already substituted with a methyl group); in effect a hydration of a double bond. It requires NADPH and oxygen to form 12-dehydrotetracycline [K900, K938].

Other reactions that involve dearomatization are found in sections 1.8, 2.2 and 2.9.2

1.8 *Ring expansion*

Alicyclobacillus converts phenylacetate into cycloheptanecarboxylate. There is a similar reaction with L-phenylalanine [H832].

2. Oxidations and reductions involving the aromatic nucleus and non-organic substituents

2.1 *Hydroxylation of the aromatic nucleus*

2.1.1 Hydroxylations associated with physiologically important amino acids and hormones

Phenylalanine hydroxylase (phenylalanine 4-monooxygenase; E.C. 1.14.16.1)

L-Phenylalanine →L-tyrosine

This vertebrate enzyme degrades dietary phenylalanine in excess of requirements for protein and hormone formation. In a group of closely associated inherited diseases in man (e.g. phenylketonuria, hyperphenylalaninaemia) that result in moderate to severe mental retardation, this enzyme is either inactive or has a very low activity, or, rarely, the enzyme for the formation of the tetrahydropteridine cofactor is inactive. The mental retardation is caused by grossly elevated concentrations of phenylalanine in body fluids, especially during brain development in the first few years of life.

One study on human liver enzyme demonstrated a molecular weight of 54 000 and pI 5.0–5.2 [F826]. Another study indicated a molecular weight of 165 000, apparently a trimer. It is not a phosphorylated enzyme, nor is it activated by phosphorylation; attempted phosphorylation does not incorporate phosphate into the enzyme molecule [D14]. Its activity increases twofold from foetus to adult [A3020].

Rat liver enzyme, which is activated by lysolecithin, acts on 4-methylphenylalanine to yield 3-methyltyrosine and 4-hydroxymethylphenylalanine. Isotope studies demonstrate that the hydroxyl groups originate from molecular oxygen and not from water. Kinetic studies with substrate labelled with different hydrogen isotopes suggest that the two products are formed by different mechanisms. In consequence, the authors question the generally accepted hypothesis that the formation of epoxide intermediates is obligatory in all aromatic hydroxylations [G150]. Purified enzyme is found as 2 forms, molecular weights 240 000 and 51 000; the former, pI 5.6, may be a tetramer. Analyses found 0.6 mol of Fe and 0.3 mol of phosphate per subunit. The amino acid composition has been determined [B19]. Another study indicated the presence of three isozymes in liver, pI 5.2, 5.3 and 5.6, one kidney isozyme, pI 5.35 and a hepatoma enzyme, pI 5.2 [A1736].

Chromobacterium violaceum enzyme is a monomer, molecular weight 32 000 and pI 4.5. The amino acid composition has been determined. It does not contain Fe, but (in line with other enzymes of this type) requires a tetrahydropteridine cofactor [A3884].

Studies with Pseudomonas enzyme and phenylalanine labelled with D or T at the *p*- position showed a 10-fold isotope effect between H and D labelled substrate and 2.8-fold between D and T labelled substrate, with migration of the substituent to the *meta* position. This 'NIH Shift' has been interpreted as indicative of an epoxide intermediate (but see above) [A757]. Enzyme found in a Pseudomonas species has a molecular weight less than 30 000 [A55].

Phenylalanine hydroxylation by xanthine oxidase (E.C. 1.1.3.22)

The product of this reaction, using enzyme from cow's milk with hypoxanthine as co-substrate, is a roughly equimolar mixture of *o*-, *m*- and *p*-tyrosines. The reaction is prevented by superoxide dismutase (E.C. 1.15.1.1) and catalase (E.C. 1.11.1.6); hydroxyl radical scavengers also prevent the reaction. Hence, hydroxyl radicals are considered to be the oxidizing species [D524].

Tyrosine 3-monooxygenase (tyrosine hydroxylase, E.C. 1.14.16.2)

L-Tyrosine+O_2→
3, 4-dihydroxy-L-phenylalanine

This enzyme catalyzes a key reaction in the formation of the neurotransmitters dopamine, noradrenaline and adrenaline.

Human medulla enzyme contains four isozymes [G190]. Brain enzyme is found mainly in the caudate and the substantia nigra, with smaller but significant amounts in the pons, mid-brain, mammillary body, amygdala, hypothalamus and nucleus accumbens [A3609] as well as in gastric mucosa and duodenum [J464].

As anticipated from the known disease process, post-mortem parkinsonian brain activity is decreased about five fold in caudate nucleus, putamen and substantia nigra, but not in other brain areas [A381].

Phosphorylation of human neuroblastoma enzyme activates by reducing the K_m for cofactor [A3687]. Human phaeochromocytoma enzyme has an optimum at pH 7.8 [A3442].

Rat striatal enzyme has a molecular weight of 61 300. Enzymatic phosphorylation using ATP and cAMP-dependent protein kinase incorporates one mol of phosphate/mol of hydroxylase. This form has a lower K_m for biopterin cofactor than the non-phosphorylated form. It is unstable, but inactivation does not involve phosphate removal [D733]. The molecular weight of caudate nucleus enzyme is 65 000, 130 000 in sympathetic ganglia and more than 200 000 in locus coeruleus, hypothalamus and adrenal, and both the small and large forms are found in substantia nigra. These forms, which may be polymeric with nucleotides being involved in the polymerisation, respond differently to tyrosine hydroxylase antibodies [A45]. Another study on rat enzyme claims that phosphorylation increases V_{max} without affecting K_m [A3687]. Striatal enzyme has an optimum at pH 6.0 [A3442]. Administration of the phenylalanine hydroxylase inhibitor *p*-chlorophenylalanine to rats increases the formation of the subsidiary products *o*- and *m*-tyrosine, whereas α-methyltyrosine, an inhibitor of tyrosine hydroxylase decreases the formation of these isomers, suggesting that they are formed by tyrosine hydroxylase [E69]. Brain synaptosomal enzyme is activated by lysolecithin and phosphatidylserine [A1698]. Apomorphine administration to rats leads to a 100 per cent increase in the activity of the enzyme in adrenal over a period of three days. This appears to be due to an increase in the absolute amount of enzyme [A1614]. It is not present in rat stomach tissues [H620].

Cultured rat phaeochromocytoma enzyme requires tetrahydrobiopterin as co-substrate. Peroxide, rather than being a substrate is inhibitory. The enzyme appears to be identical with that from other tissues, and contains one Fe^{2+}/ subunit. The cofactor is converted into a carbinolamine during hydroxylation; the authors in a different study suggest that a peroxytetrahydropterin is the hydroxylating species, which can be replaced by peroxide [G290, G290a].

As well as L-tyrosine, beef and adrenal chromaffin enzymes hydroxylate L-phenylalanine, forming a little *m*-tyrosine as well as L-tyrosine, both of which are converted into 3,4-dihydroxy-L-phenylalanine [F623]. Cytosolic adrenal medulla enzyme, molecular weight 280 000, appears to be a homotetramer, pI 6.0 and optimum pH 6.8. Its amino acid composition and terminal amino acids have been determined [A3442, C799]. Particulate adrenal enzyme can be solubilised with α-chymotrypsin to yield an enzyme, molecular weight 34 000 that contains Fe. Only the unsolubilized enzyme can be activated by phosphorylation or by phospholipids in a manner similar to beef brain enzyme (see below) [A3017].

Beef corpus striatum enzyme (molecular weight about 60 000) is activated by incubation with ATP and cyclic AMP-dependent protein kinase, which yields a phosphorylated hydroxylase containing one mol of phosphate [B413]. Caudate nucleus enzyme is activated by phosphatidylserine and other polyanions, which reduce K_m for the pterin cofactor. Phosphatidylserine raises the optimum pH (from about 6) by 1 unit, whereas the effect of heparin is marginal. The enzyme is strongly inhibited by tyrosine at concentrations higher than 0.05 mM. Phenylalanine as a substrate has the same V_{max} as tyrosine, but high reaction rates only occur at high concentrations relative to physiological levels [A383].

In rabbit the enzyme is uniformly distributed between the white and grey matter in spinal cord [A2462].

Enzyme in neonatal mouse superior ganglia cultured cells is induced 30–40 per cent by actinomycin D or 80 per cent by K^+; their effects are additive. It is postulated that the effect of actinomycin D is to inhibit the formation of a substance that represses transcription of mRNA [A3].

Sulphate or heparin increase V_{max} for dog hypothalamus enzyme; heparin also decreases K_m for cofactor [A2064].

Guinea pig atrial enzyme is inhibited by 3,4-dihydroxyphenylglycol and by noradrenaline at 10^{-5}–10^{-6} M. Both compounds compete with cofactor, but 4-hydroxy-3-methoxymandelate, 3,4-dihydroxymandelate and 4-hydroxy-3-methoxyphenylglycol are poor inhibitors [A2060]. Vas deferens enzyme is activated by 10^{-6}M Ca^{2+}; it decreases K_m for both substrate and cofactor [A44].

Other studies on mammalian enzymes (including some in which the species is not stated) confirm activation by phosphorylation [A3442, A3686].

Chicken embryo brain enzyme is not detected at 10 days incubation, but is found in all regions at 14 days, and increases rapidly over the next 4 days [A1413].

Kinetic studies indicate a sequential reaction, and not a ping-pong mechanism for the reaction [A1904].

Helix pomatia enzyme is mainly soluble. Its optimum pH is 6.5, requires a tetrahydropteridine cofactor and is stimulated by Fe^{2+} and catalase. Inhibitors include dopamine, 6-hydroxydopamine, serotonin, noradrenaline and dodecyl sulphate [A1349]. Planorbis (water snail) enzyme shows very similar properties [A2055].

Hydroxylation by monophenol monooxygenase (tyrosinase; E.C. 1.10.3.1)

This type of reaction usually forms catechols from phenols. Its physiological importance in vertebrates is the formation of L-dopa from L-tyrosine by tyrosinase, usually in skin; L-dopa is the precursor of melanin. The presence of unusually high urinary concentrations of intermediate metabolites associated with melanin formation is observed (in man) both after exposure to sunlight, and in patients with melanoma without exposure to the sun. These metabolites can be used as a marker in the diagnosis of melanoma. In mouse pups devoid of tyrosine hydroxylase tyrosinase appears to be responsible for the appearance of significant amounts of neural catecholamines [K89].

Sea anemone tentacle enzyme is particulate, and oxidizes tyrosine and dopa; the latter reacts further to form 5-hydroxydopa. It is inhibited by diethyldithiocarbamate [D245].

Avocado enzyme hydroxylates D- and L-tyrosine and tyramine as well as *p*-cresol, but much more slowly than the oxidation of D- and L-dopa and other catechols to quinones. The initial time-lag in the reaction is eliminated by ascorbate or L-dopa [A3975]. (The time-lag reported in early studies on tyrosinase was caused by the time required for the build-up in dopa concentrations; dopa, as well as other catechols, acts as co-substrate which is oxidized to a quinone).

Banana pulp enzyme is found in both soluble and particulate fractions. It oxidizes both D- and

L-tyrosine, and also L-dopa. Ascorbate activates, and diethyldithiocarbamate is inhibitory [A541].

Berberis stolonifera phenoloxidase, molecular weight 60 000 and optimum pH 6.0, oxidizes both L-tyrosine and tyramine, as well as some phenolic tetrahydroisoquinoline natural products. Both ascorbate and oxygen are required for activity [F855].

Horseradish peroxidase and peroxide act on mixtures of dopa and phenols; the phenols accelerate the oxidation of dopa, although when the phenol used is tyrosine there is no indication of tyrosine hydroxylation. Peroxidase forms dopa from tyrosine in the presence of oxygen and dihydrofumarate, however [A2458].

Enzyme from Mucuna pruriens, in which one of the highest naturally-occurring concentrations of L-dopa is found (rendering its fruit toxic) acts on a range of phenols including L-tyrosine, generating catechols [E769]. This is the basis of a synthetic process for making L-dopa.

Papaver somniferum tyrosinase acts on tyrosine [A171].

Of the 17 phenoloxidase isozymes detected in potato, five act on tyrosine as well as L-dopa [A3190].

Portulaca grandiflora tyrosinase, molecular weight 53 000 and optimum pH 5.7, acts on L-tyrosine, but D-tyrosine is a poor although significant substrate. The dopa formed is further oxidized via dopaquinone to *cyclo*-dopa, an intermediate in the formation of betanidin [K93].

In wheat the enzyme that hydroxylates tyrosine can be separated electrophoretically from the enzyme that forms dopaquinone from L-dopa [A176], suggesting that it is a different class of enzyme from most tyrosinases, which catalyse these reactions sequentially with a single enzyme.

Agoricus (presumably a misprint for Agaricus) bisporus tyrosinase oxidizes N-acetyltyramine with a lag period (typical for phenols), but the lag for hordenine is indefinitely long. The lag is eliminated by the addition of a trace of a catechol. The normal lag period is terminated by the autocatalytic oxidation of phenol to catechol. N,N-Dipropyldopamine oxidation yields as final product N,N-dipropylindoliumolate [J460]. Mushroom (presumably A. bisporus) tyrosinase, molecular weight 122 500, has a Stokes radius stated to be 42.75×10^{-8} cm^2 sec^{-1} (units meant to be cm?) [B489]. It is not affected by superoxide dismutase or by superoxide scavengers [D2].

Aspergillus oryzae enzyme is activated at pH 2–5, and the resultant tyrosinase activity has an optimum at pH 6.0 [A1131].

Vibrio tyrosinaticus is composed of two tyrosinases, molecular weights 41 000 and 38 500; it does not cross-react with antiserum to hamster melanoma tyrosinase. It acts on L-tyrosine and slightly on the D- isomer and *m*-tyrosine, but catechol and L-phenylalanine are not substrates. It is inactivated by diethyldithiocarbamate, and this is reversed by Cu^{2+}, Mn^{2+}, Cd^{2+} or Fe^{2+} [A141].

Tyramine 3-hydroxylase

This reaction has been detected in nematode, locust, Thalictrum, avocado, Mucuna and mushroom [A2572, A2804, A3975, B765, E769, F112, F855, H242, H907]. At least in some instances, the reaction is probably catalysed by the same enzyme that forms L-dopa from L-tyrosine.

p-Hydroxyphenylpyruvate oxidase;
(4-hydroxyphenylpyruvate dioxygenase, E.C. 1.13.11.27)

p-Hydroxyphenylpyruvate + O_2 →
CO_2 + homogentisate

This is a key vertebrate enzyme, and is involved the catabolism of tyrosine. Tyrosine is a major dietary amino acid, and it is also formed from dietary phenylalanine, quantitatively greatly in excess of amounts required for the formation of protein and hormones, and a mechanism is required for its disposal. A deficiency of activity is found in premature infants and in the inherited diseases classified as Tyrosinosis. In man mutations in the gene controlling this enzyme lead to tyrosinaemia type III and hawkinsinuria [K665].

Mechanistic studies using beef liver enzyme and substrate specifically labelled with deuterium in the methylene group indicate that side-chain migration occurs with retention of configuration at the methylene group [B904]. Studies using an enzyme with unspecified origin, but presumably mammalian, have shown that the phenolic hydroxyl group is not exchanged during the reaction. This indicates that, although still possible, an intermediate with a peroxide 1,4-bridge is most unlikely [A844]. 1-Carboxymethyl-1-hydroxy-4-oxocyclohexa-2,5-diene is not an intermediate [A346].

Human enzyme is a dimer, molecular weight 87 000. It acts on the *keto* isomer with optima at pH 4.5 and 7.8. It is activated by reducing agents such as ascorbate and is very sensitive to inactivation by peroxide. Iron- and copper-chelating reagents are inhibitory, and reactivation by dialysis indicates that the chelators do not remove the metal from the enzyme molecule. It exists in three forms with different pI between 6.5 and 7.5. These appear to be dimers of two monomeric forms [A3128, A3129].

A genetic defect in mice leads to a deficiency of this enzyme [G216].

Pig enzyme is inhibited by pentafluorophenylpyruvate and thiophenyl oxalate. It is inactivated by tetrafluoro-4-hydroxyphenylpyruvate and by 2- and 3-thienylpyruvate [H382]. The enzyme is stable only after purification. During storage it polymerizes, and this is reversed by thiols. The molecular weight is 89 000 by ultracentrifugation, and 52 000 and 44 000, respectively, by gel filtration and electrophoresis. It contains Cu and Fe (not stoichiometric). The temperature coefficient was found to be unusually high, about five (even higher than that reported by Goodwin, B.L. (1972) Tyrosine Catabolism, Clarendon Press, Oxford). The activity is stimulated by some hydrophilic solvents, including alcohols and ethers [C208].

Rat liver enzyme, molecular weight 63 000 and pI 5.85, is inactivated by dialysis and other processes that remove small molecules, and is reactivated by Fe^{2+} and dichlorophenolindophenol [A2702]. At birth, about 25 per cent of the enzyme is in an active form, and this increases to 90 per cent in the adult [A771]. The molecular weight is 45 085 by mass spectrometry (theoretical value 45 082). It also shows α-oxoisocaproate dioxygenase activity. At the C-terminal a 14 amino acid sequence is essential for activity; a mutant with a deletion of these amino acids produces an inactive protein [J235].

Bamboo enzyme is highly specific, molecular weight about 10 000. It is stoichiometric, and no intermediate has been detected [B238]. Zea mays enzyme, molecular weight 43 000 and optimum pH 7.3, requires a reducing system or ascorbate and catalase [J247]. Its presence has also been reported in Drosophyllum lusitanicum [A1149].

Enzyme from Pseudomonas strain PJ874 is blue and contains both iron and zinc. It catalyzes a stoichiometric reaction involving the *keto* isomer of the substrate, with an optimum and maximal stability at pH 7.9. It appears to act by a mono-iso-ordered bi-bi mechanism, in which Fe^{3+} is reduced [C365, C728]. Another publication claims that enzyme from strain PJ874 is a tetramer, monomeric molecular weight 36 000, optimum pH 7 and pI 4.8, and contains both Fe and Cu. The Fe content parallels activity during purification and the ratio of activity towards *p*-hydroxyphenylpyruvate and phenylpyruvate remains constant [A3106].

Tryptophan 5-hydroxylase; (tryptophan 5-monooxygenase, E.C. 1.14.16.4)

L-Tryptophan
$+ O_2 \rightarrow$ 5-hydroxy-L-tryptophan

This reaction is a key step in the formation of the neurotransmitter serotonin, as well as the hormone melatonin, which responds to light and dark periods and is associated with the biological clock in vertebrates.

Activity in brain (presumably human) median and dorsal raphe nuclei is 50–100 times greater than in caudate nucleus and hippocampus [A1690].

Mouse gut mucosal enzyme is probably of enterochromaffin origin [E503]. Brain enzyme is activated by phosphorylation and inactivated by phosphatase [A3813]. Mouse mastocytoma enzyme, pI 6.0, has a molecular weight of 270 000, and despite a monomeric molecular weight of 53 000 it is claimed to be a tetramer. L-Phenylalanine and (marginally) L-tyrosine are also substrates [C216].

Rat brain stem enzyme, which is stimulated by Fe^{2+}, is composed of two isozymes, one of which has a molecular weight of 300 000. Despite a monomeric molecular weight of 59 000 it is also claimed to be a tetramer. L-Phenylalanine is a substrate, but not L-tyrosine [C113]. Influences *in vivo* that reduce brain tryptophan concentration cause the enzyme activity to increase, apparently by an increase in V_{max} [A1920], an effect that should sustain brain serotonin levels in face of alterations in tryptophan availability. Liver enzyme (that also acts on phenylalanine) is stimulated by 5-fluorotryptophan and 7-azatryptophan by up to 20-fold, and at higher cofactor concentrations by phenylalanine and thienylalanine [A791]. Pineal enzyme shows a diurnal rhythm with raised activities at night, and this is eliminated in constant light. Cycloheximide but not actinomycin D causes a rapid loss in activity, which suggests that enzyme activity is controlled at the translation step. Sulphydryl compounds protect the enzyme from inactivation at 0° but not at higher temperatures, and dithiothreitol reactivates the inactivated enzyme. Rapid *in vivo* inactivation is caused by *p*-chlorophenylalanine, and this is reversed within 24h [A3237]. Presumably this diurnal rhythm is associated with the diurnal rhythm of melatonin concentrations.

Pig brain stem enzyme, molecular weight 55 000–60 000, is not stimulated by Fe^{2+} or by chelating agents and is unstable during storage [A1306].

The activity of chick brain enzyme, optimum pH 7.8–8.0 [A3236], increases about 10-fold just before and after hatching [A3130].

Enzyme from yellowfin tuna resembles mammalian enzyme. It is a trimer, molecular weight 280 000 [H910]. Skipjack liver enzyme is a homotrimer, molecular weight 288 000 and optimum pH 8.0 [H394].

Sedum morgaianum enzyme has an optimum at pH 7.5 [D41].

Chromobacterium violaceum enzyme, optimum pH 7, is inducible by tryptophan and (better) phenylalanine. D-Tryptophan is also a substrate, but phenylalanine is not. It requires oxygen, reduced pteridine cofactor and a thiol. Both phenylalanine and *p*-chlorophenylalanine are inhibitory [A1597].

Tryptamine 5-hydroxylation

Peganum harmala enzyme acts on tryptamine, α-methyl-, N-methyl- and 6-fluorotryptamine [A1330, F9, G906].

Kynurenine 3-hydroxylase; (kynurenine 3-monooxygenase; E.C. 1.14.13.9)

L-Kynurenine → 3-hydroxy-L-kynurenine

Studies on rabbit, rat, gerbil and mouse demonstrate that activity is found in liver, lung and brain, in decreasing order [J830].

Rat liver enzyme is associated with the mitochondrial outer membrane. When solubilized it can be separated into two fractions by chromatography; the major one has a molecular weight of 200 000, optimum pH 8 and pI 5.4. It contains dissociable FAD (one mol/mol) that cannot be replaced by FMN or riboflavin. It is activated by FAD, dithiothreitol and phosphatidylcholine, and maximal activity occurs in 10 mM KCl. Inhibition is brought about by *p*-chloromercuribenzoate and bathocuproinsulphonate [A2270, A2293, A2482].

Honeybee eye enzyme, which requires NADPH, has an optimum at pH 7.25 [A1064]. Drosophila melanogaster eye enzyme appears to be mitochondrial [A1194].

Anthranilate 3-hydroxylase; (anthranilate 3-monooxygenase; E.C. 1.14.16.3)

Anthranilate → 3-hydroxyanthranilate

Aspergillus niger enzyme, molecular weight 43 000–45 000, optimum pH 8.2 and pI 5.36, requires FAD. Purified enzyme also catalyzes the formation of pyrocatechuate, which suggests that it also shows anthranilate 2,3-hydroxylase (deaminating) activity (E.C. 1.14.13.35); it is claimed that both activities reside on the same molecule. Cu^{2+} and *p*-chloromercuribenzoate are inhibitory [D231].

This reaction is also observed in rat liver and brain [A3653, F676].

Anthranilate 5-hydroxylase

This reaction is observed in rat liver [A3653, D296].

Indole hydroxylases

Indole 4-hydroxylase is found in pumpkin [A3530]. A similar reaction occurs in the mycological formation of psilocine, but the reaction does not appear to have been studied at an enzyme level.

Indole 5-hydroxylase is also present in pumpkin [A3530]. Both 5- and 6-hydroxylases are present in Tradescantia, Zebrina and Seterasia [A3538], and in rat [D158].

Oestradiol hydroxylases (and analogues)

a. 2- And 4-hydroxylase

Rat anterior pituitary enzyme catalyzes these reactions, in particular 4-hydroxylation. The latter is considered to be NADPH-dependent, whereas 2-hydroxylation may be peroxidase-dependent [F42]. Rat liver microsomes catalyze these reactions by 3 P450s, $P450_{VT\text{-}A}$, $P450_{PCN\text{-}E}$ and $P450_{ISF\text{-}G}$. The first two are constitutive, male-specific and are induced by testosterone [E40]. A report on 2-hydroxylation suggested that P450 is not involved, at least in the brain of male rats. This activity is greater in male rats than in female, and is found (in decreasing order of activity) in liver, brain, kidney, testis, adrenal and lung, with activity in the lung about 1 per cent of that in liver. Other organs show negligible activity. Activity is reduced by castration, and in hyperthyroid and euthyroid rats the activity is lower than in control animals [A3944]. A further study on brain enzyme suggested that it is probably P450, and that the enzyme also acts on 17α-ethynyloestradiol, stilboestrol and oestrone to yield catechols [A3295]. 17β-Oestradiol-17-sulphate is a substrate, undergoing both reactions, with lower activity for 4-hydroxylation. Again, highest activity is observed in liver, and much lower in kidney, brain, heart, lung, testis, ovary and uterus. At least in liver, there is no constant ratio between the two activities suggesting enzyme heterogeneity [E465]. Kidney enzyme is microsomal and lung enzyme is mitochondtial [B681].

Pig ovary enzyme catalyses 2-hydroxylation in particular, optimum pH 7.8; the enzyme requires NAD(P)H [E78].

Rabbit hypothalamus enzyme appears to be a soluble peroxidase with an optimum at pH 6 and pI about 7.7. It is stimulated by cumene hydroperoxide and yields similar amounts of 2- and 4-hydroxy-17β-oestradiol [E144].

Horse oestrogen 2-hydroxylase, a P450, is associated with testosterone aromatase [H891].

b. 3-Hydroxylase

Oestradiol is formed from 2-hydroxy-3-deoxyoestradiol by a P450 enzyme found in liver microsomes. This is possibly the same enzyme that 2-hydroxylates 17β-oestradiol [B1].

c. 6α-Hydroxylase

A human enzyme hydroxylates 17β-oestradiol at the 6α-position [A2465, K621]. The same reaction is observed in rat [A1487, A2192, E525]. Oestrone is a substrate in rat and hamster [A2408, C896, G625].

d. 6β-Hydroxylase (E.C. 1.14.99.11)

Rat liver and brain micromes hydroxylate 17β-oestradiol at the 6β-position [A1487, A2192, H342, E525]. The same reaction is observed in man [A2465]. Oestrone is a substrate in rat and hamster [A2408, C896, G625].

e. 7α-Hydroxylase

A human enzyme hydroxylates 17β-oestradiol at the 7α-position [A2465]. The same reaction is observed in rat [A1487, A2192]. Oestrone is a substrate in rat and hamster [A2408, C896, G625].

f. 7β-Hydroxylase

A human enzyme hydroxylates 17β-oestradiol at the 7β-position [A2465].

g. 11β-Hydroxylase

This activity is found in Neurospora, with 17β-oestradiol as substrate [A944].

h. 14α-Hydroxylase

Rat liver micromes hydroxylate 17β-oestradiol and oestriol at the 14α-position [H342].

i. 15α-Hydroxylase

A human enzyme hydroxylates 17β-oestradiol at the 15α-position [K621]. The same reaction is observed in baboon, rat and pig [A545, E525, F511]. Oestrone is a substrate in man [A1588].

j. 15β-Hydroxylase

This activity has been detected in kidney and liver microsomes from hamster and rat, with oestrone as substrate [G625].

k. 16α-Hydroxylase

Human foetal liver enzyme hydroxylates 17β-oestradiol (to oestriol) and oestrone at the 16α-position, but these are poor substrates, whereas their 3-sulphate esters are readily hydroxylated. The optimum pH is 7 [E132]. The same reaction is observed in rat liver microsomes [H342] as well as in a number of other mammalian species.

l. 16β-Hydroxylase

In man 17β-oestradiol is converted into 16-epioestriol by hydroxylation at the 16β-position [A2309]. The same reaction is observed in guinea pig [A2868], monkey [A491, A3294] and rat [A2583].

2.1.2 Hydroxylations of natural products and miscellaneous compounds associated with normal animal physiology

Phenol 2-monooxygenase (E.C. 1.14.13.7)

Phenol → catechol

Trichosporon cutaneum enzyme acts on resorcinol and *m*-cresol to yield hydroxyquinol and 4-methylcatechol respectively, as well as on phenol [G196]. Its molecular weight is 148 000, it contains one mol/mol of FAD and requires NADPH. The optimum pH is 7.2–7.6. Bleaching by dithionite, which inactivates the enzyme is readily reversible. It is also inactivated by heavy metals and *p*-chloromercuribenzoate, the latter being reversed by dithiothreitol. The specificity is broad, with substrates not being limited to monophenols; catechol, for instance, yields pyrogallol. Activity towards hydroxylated benzyl alcohols, aldehydes and benzoic acids is negligible [A902].

Bacillus stearothermophilus enzyme requires NADH for activity [F47].

Comamonas testosteroni contains a phenol hydroxylase [K362].

Hydroquinone hydroxylase

Candida parapsilosus enzyme is a homodimer, monomeric molecular weight 76 000 containing dissociable FAD. It catalyses *ortho* hydroxylation of a range of phenols [K653].

Rhodococcus enzyme has an optimum at pH 7.9 and requires NADPH. Its specificity for monophenols is broad, but catechol, nitrophenols and protocatechuate are not substrates [E565].

Phloroglucinol 2-monooxygenase

Rhodococcus enzyme, molecular weight 155 000 and optimum pH forms 1,2,3,5-tetrahydroxybenzene from phloroglucinol [H121].

Orcinol 2-monooxygenase; (E.C. 1.14.13.6)

Orcinol + $O_2 \rightarrow$ 2, 3, 5-trihydroxytoluene

Pseudomonas putida enzyme, molecular weight 63 000–68 000 or 70 000, contains 1 mol of dissociable FAD. It utilizes one mol of oxygen and NAD(P)H; the latter can be replaced partially by FMN. Without orcinol, O_2 reacts with NADH and water to yield hydrogen peroxide, and at 60° the reaction with orcinol uncouples by about 50 per cent with resultant peroxide formation. Other substrates include resorcinol, 4-bromo- and 4-methylresorcinol, whereas non-phenolic substrate analogues accelerate peroxide formation. Resorcinol hydroxylase is very similar to this enzyme, but is distinguished by, for instance, absorption spectra, amino acid composition, stability to oxygen, specificity and reaction with antisera. For instance, *m*-hydroxybenzyl alcohol and a range of phenols are substrates for orcinol hydroxylase but not for resorcinol hydroxylase, whereas resorcinol hydroxylase acts on *m*-ethylphenol, but orcinol hydroxylase does not [A2172, A2391, A2696].

Resorcinol hydroxylase

Resorcinol + $O_2 \rightarrow$ hydroxyquinol

This Pseudomonas putida activity is very similar to that of orcinol hydroxylase (see above), with some distinctive differences. Its molecular weight is 68 000–70 000; it could not be distinguished from orcinol hydroxylase on the basis of many parameters, however [A2391, A2696]. Rhodococcus enzyme has an optimum at pH 7.0 [G893].

Benzoate 2-hydroxylase

Benzoate → salicylate

Benzoate is hydroxylated in all three positions by human blood granulocytes. It is postulated that the reaction is mediated by hydroxyl radical [E65].

Nicotiana tabacum enzyme is a soluble P450, molecular weight 160 000. It is specific for the formation of salicylate [H681].

This reaction has been observed in potato leaf [J691], rice [H593], Bacillus [A2217], Aspergillus niger and Cuninghamella bainieri [A1218].

Benzoate 3-hydroxylase

This reaction has been observed in man [E65], Flavobacterium [E355] and Pseudomonas [D771].

Benzoate 4-hydroxylase (E.C. 1.14.13.12)

Benzoate + $O_2 \rightarrow$ *p*-hydroxybenzoate

Rhodotorula graminis enzyme is membrane-bound with optimum pH of 7.6. It requires NADPH and is stimulated by FAD. It differs from enzyme found in filamentous fungi, which requires a pteridine cofactor [D620].

Pseudomonas enzyme, molecular weight 120 000 and optimum pH 7.2, requires a tetrahydropteridine and oxygen, and is activated by Fe^{2+}. The reaction is not stoichiometric; a slight excess of oxygen is utilized [A2799].

Aspergillus niger enzyme, optimum pH 6.2, is specific for benzoate, requires a tetrahydropteridine and Fe^{2+}, and uses equimolar amounts of oxygen and NADPH. Thiol groups in the enzyme are required for activity [A1606].

Benzoyl CoA 3-monooxygenase (E.C. 1.14.13.58)

Pseudomonas enzyme, a monomeric flavoprotein, molecular weight 65 000, requires FAD or FMN, and NADPH [H397].

m-Hydroxybenzyl alcohol 6-hydroxylase

Penicilliun patulum microsomal enzyme, optimum pH 7.5, requires NADPH in the formation of gentisyl alcohol. It is suggested that the enzyme is m-cresol 2-hydroxylase [A2314].

Salicylate 3-hydroxylase

Salicylate → 2, 3-dihydroxybenzoate (pyrocatechuate)

Rat liver enzyme is considered to act via a free radical that non-specifically causes hydroxylation of salicylate rather than by the action of a hydroxylase [J754].

Salicylate 5-hydroxylase

Salicylate → gentisate

Rat liver enzyme is considered to act via a free radical that non-specifically causes hydroxylation of salicylate rather than by the action of a hydroxylase [J754].

Rhodococcus erythropolis enzyme, a homotetramer containing FAD, molecular weight 205 000, optimum pH 7.9 and pI 6.3, requires NADH, but NADPH is ineffective. Other substrates are 2,3- and 2,4-dihydroxybenzoates [H438].

Lignobacter enzyme requires NAD(P)H. Other (poor) substrates include p-hydroxybenzoate, 2,3-, 2,4-, 2,6- and 3,4-dihydroxybenzoates [B503].

3-Hydroxybenzoate 2-monooxygenase
(E.C. 1.14.99.23)

An enzyme in a Pseudomonas testosteroni mutant is induced by benzoate [K748].

m-Hydroxybenzoate 4-hydroxylase;
(3-hydroxybenzoate 4-monooxygenase, E.C. 1.14.13.23, formerly E.C. 1.14.99.13)

m-Hydroxybenzoate + O_2 → protocatechuate

Aspergillus niger enzyme is a flavoprotein, optimum pH 7.2, requires NADPH and is inhibited by superoxide dismutase. Other substrates include 3-hydroxyanthranilate, pyrocatechuate, 3,5-dihydroxybenzoate and 3-hydroxy-5-methoxybenzoate. It is also inhibited by heavy metals, o-phenanthroline, salicylaldoxime, m-aminobenzoate, diethyldithiocarbamate and sulphydryl-binding reagents [A1130, A3526].

Comamonas (Pseudomonas) testosteroni enzyme, molecular weight 71 000, optimum pH in the range 6–7.5 (buffer-dependent), contains FAD and is activated by FAD and NADPH; the latter cannot be replaced by NADH. Other substrates include gentisate, pyrocatechuate and several other acids with a m-hydroxyl group, but not other analogues, which may be inhibitory [A1229, J519]. There is up to 70 per cent uncoupling of the oxygenation, when peroxide is formed, depending on substrate [A327].

3-Hydroxybenzoate 6-hydroxylase
(E.C. 1.14.13.24)

m-Hydroxybenzoate → gentisate

Klebsiella pneumoniae enzyme is monomeric, molecular weight 42 000. It contains FAD and requires NAD(P)H. Inactivation is rapid at 45° [G199, H399].

Pseudomonas aeruginosa enzyme requires oxygen and NAD(P)H, with a range of 3-hydroxybenzoates as substrates; there is 0–30 per cent decoupling of hydroxylation, depending on substrate [A236].

Pseudomonas cepacia enzyme, molecular weight 44 000 and optimum pH 8, contains one mol/mol of FAD with a requirement for NAD(P)H, and is inducible [E287]. The reaction mechanism has been suggested to involve random binding of substrate and NADH to the oxidized enzyme followed by reduction of the enzyme and release of NAD^+. Oxygen then binds, followed by release of water and gentisate from the now oxidised holoenzyme [E289].

Rhodococcus erythropolis enzyme is a homotetramer, molecular weight 196 000, optimum pH 8.6 and pI 6.7. It contains FAD and requires NADH, but NADPH is ineffective. A second substrate is 2,3-dihydroxybenzoate [H438].

p-Hydroxybenzoate 1-hydroxylase

p-Hydroxybenzoate → gentisate

Klebsiella pneumoniae enzyme requires NAD(P)H; the activity is rapidly destroyed at 45° [G199]. It is also found in Bacillus and Bacterium [A489, B161].

p-Hydroxybenzoate 3-hydroxylases
(4-hydroxybenzoate 3-monooxygenase, E.C. 1.14.13.2 and 1.14.13.33)

p-Hydroxybenzoate + O_2 → protocatechuate

Corynebacterium cyclohexanicum enzyme is a monomer, molecular weight 47 000, and contains FAD. It requires Mg^{2+} and NADH or NADPH, and is specific for p-hydroxybenzoate [D554].

Moraxella enzyme is a dimer, molecular weight 85 000 and pI 6.55. It is specific for p-hydroxybenzoate [G793].

Pseudomonas fluorescens enzyme is mainly a dimer, molecular weight 70 000–75 000 and pI 5.8; it also exists as higher polymers. An amino acid analysis has been performed [B249]; Ser 212 is an important binding site [K109]. The reaction mechanism is complex, where the initial step appears to be oxidation of the FAD moiety to peroxyflavin. Other substrates include p-aminobenzoate and 2,4-dihydroxybenzoate; the latter yields 2,3,4- and 2,4,5-trihydroxybenzoates [A2554, K109]. Anions, such as Cl^-, Br^-, I^-, F^- and CNS^- compete with NADPH cofactor, quenching the fluorescence [A1082]. P. cephacia enzyme is inducible [E287].

Rhizobium leguminosarum enzyme is induced by 4-hydroxybenzoate [F224]

Rhodococcus erythropolis enzyme is a homotetramer, molecular weight 196 000, optimum pH 8.4 and pI 6.7, and contains FAD. It requires NADH, but NADPH is ineffective. A second substrate is 2,4-dihydroxybenzoate [H438].

4-Hydroxy-3-methylbenzoate hydroxylase

4-Hydroxy-3-methylbenzoate →
4-hydroxy-3-hydroxymethylbenzoate

Pseudomonas putida enzyme, which is composed of two proteins, requires oxygen and NAD(P)H. The proteins involved appear to be a hydroxylase and an electron-transferring protein acting on NADH; this makes it an unusual type of mixed-function oxidase [F223, F624].

3-Hydroxybenzoyl CoA 6-hydroxylase

Pseudomonas enzyme requires NAD(P)H to form gentisoyl CoA [H397].

Anthraniloyl CoA monooxygenase
(E.C. 1.14.13.40)

This activity is described in Section 1.7.

Phenylacetate 2-hydroxylase

Phenylacetate → o-hydroxyphenylacetate

Aspergillus niger microsomal enzyme, optimum pH 7.8, requires NADPH and oxygen. It is stimulated by thiols, and inhibited by thiol-binding reagents and by CO; it therefore appears to be a P450. It is highly specific [B133].

This activity has been demonstrated in bacteria [B950], Aspergillus fumigatus [C484], A. sojae [A2345], Trichosporon cutaneum [E332], Brevibacterium linens [D623], Pseudomonas fluorescens [C532] and Nocardia salmonicolor [A1329].

Phenylacetate 3-hydroxylase

Phenylacetate → m-hydroxyphenylacetate

Rhizoctonia solani enzyme, optimum pH 5–6, requires NAD(P)H and tetrahydrofolate. Benzoate, cinnamate, phenylpropionate and aryloxyacetates are not substrates.

Diethyldithiocarbamate and α,α-dipyridyl are inhibitors [A203].

This activity has been demonstrated in bacteria [B950], Aspergillus niger [A3787], A. fumigatus [C484], Brevibacterium linens [D623] and Trichosporon cutaneum [E332].

Phenylacetate 4-hydroxylase

Phenylacetate → *p*-hydroxyphenylacetate

This activity has been demonstrated in bacteria [B950], Brevibacterium linens [D623], Trichosporon cutaneum [E332] and Aspergillus niger [A329].

Mandelate 4-monooxygenase (E.C. 1.14.16.6)

L-Mandelate → *p*-hydroxymandelate

Pseudomonas convexa enzyme, molecular weight 91 000 and optimum pH 5.4, is inducible. It requires a tetrahydropteridine, NADPH, Fe^{2+} and oxygen, and is highly specific for mandelate. The activity is inhibited by thiol-binding reagents [A2662].

***o*-Hydroxyphenylacetate 5-hydroxylase**

o-Hydroxyphenylacetate → homogentisate

Rhodococcus erythropolis enzyme, molecular weight 45 000, contains FAD and requires NADH for activity [J23].

***m*-Hydroxyphenylacetate 4-hydroxylase** (E.C. 1.14.13.23)

m-Hydroxyphenylacetate → 3, 4-dihydroxyphenylacetate

Klebsiella pneumoniae requires NADH or NADPH [G214]. The reaction has also been observed in Escherichia [B517], Trichosporon [E332], and possibly other microorganisms.

3-Hydroxyphenylacetate 6-hydroxylase (E.C. 1.14.13.63)

m-Hydroxyphenylacetate → homogentisate

Flavobacterium enzyme, which is mainly dimeric with one FAD/subunit and optimum pH 8.3, requires NAD(P)H as co-substrate; the amino acid composition and N-terminal sequence have been determined. It is stable between pH 5 and 9. Inactivation by thiol-binding reagents is reversed by dithiothreitol. Other substrates (less good) are 3,4-dihydroxyphenylacetate and *p*-hydroxyphenylacetate, with hydroxylation exclusively at position 6 [K779].

Rhodococcus erythropolis enzyme, molecular weight 45 000, contains FAD and requires NADH for activity [J23]. Flavobacterium enzyme, optimum pH 8.3, requires NAD(P)H. Poorer substrates include 3,4-dihydroxyphenylacetate and *p*-hydroxyphenylacetate, whereas homogentisate is not a substrate [G404].

***p*-Hydroxyphenylacetate 1-hydroxylase** (E.C. 1.14.13.18)

p-Hydroxyphenylacetate → homogentisate

In this reaction the side chain migrates during the hydroxylation ('NIH shift').

Bacterium enzyme, optimum pH about 7.5, is stoichiometric for oxygen and NAD(P)H, and is slightly stimulated by Mg^{2+}; its stability range is pH 6–9. Other substrates include *p*-hydroxyphenylpyruvate, 3,4-dihydroxyphenylacetate, *p*-hydroxyphenylpropionate and (poor) *p*-hydroxybenzoate. It is inhibited by iodoacetate, iodoacetanilide, *p*-chloromercuribenzoate, iodosobenzoate, N-ethylmaleimide, Hg^{2+} and EDTA [A489].

Pseudomonas acidovorans enzyme requires FAD, Mg^{2+} and NAD(P)H, and is inhibited by KCl. It is unstable when purified. Substrate substituted with deuterium in the *ortho* position shows 50 per cent label retention, consistent with no NIH shift. Other substrates include dopac, homovanillate, *p*-hydroxyphenoxyacetate, *p*-hydroxyphenylacetate, *p*-hydroxyphenylpropionate, 4-hydroxy-2-methylphenylacetate and *p*-hydroxyhydratropate [A1480].

Rhodococcus erythropolis enzyme, molecular weight 45 000, contains FAD and requires NADH for activity. This enzyme is not identical with *o*-hydroxyphenylacetate 5-hydroxylase or *m*-hydroxyphenylacetate 6-hydroxylase; these enzymes are all different [J23].

p-Hydroxyphenylacetate 2-hydroxylase

This activity is associated with Flavobacterium 3-hydroxyphenylacetate 6-hydroxylase [K779].

p-Hydroxyphenylacetate 3-hydroxylase
(E.C. 1.14.13.2 and 1.14.13.3)

p-Hydroxyphenylacetate → 3, 4-dihydroxyphenylacetate

Studies on the reaction mechanism of Pseudomonas putida enzyme demonstrate the formation of three flavin-oxygen intermediates [H206]. Klebsiella pneumoniae enzyme requires NADH but not NADPH, which indicates that the enzyme is different from *m*-hydroxyphenylacetate 4-hydroxylase [G214].

trans-Cinnamate 2-monooxygenase
(E.C. 1.14.13.14)

Melilotus alba enzyme, optimum pH 7.0, is found in chloroplasts, attached largely to lamellar membranes. It is activated by glucose-6-phosphate or by light [A145].

trans-Cinnamate 3-monooxygenase

Cinnamate + O_2 → *m*-coumarate

This reaction may be involved in orchinol formation from L-phenylalanine in Orchis species [C734].

trans-Cinnamate 4-monooxygenase
(E.C. 1.14.13.11)

Cinnamate + O_2 → *p*-coumarate

Capsicum annuum enzyme, optimum pH 8, requires oxygen and NADPH. It is microsomal, possibly a P450 [F689].

Gherkin enzyme, optimum pH 7.5, is found in endoplasmic reticulum and requires NADPH; NADH is inactive. Inhibition by 2-mercaptoethanol and carbon monoxide is consistent with it being a P450 [A3146].

Petroselinum crispum enzyme is specific for *trans*-cinnamate; *cis*- is inhibitory. Reduction of the double bond prevents hydroxylation [A2820].

Enzyme from roots of Quercus pedunculata is microsomal [A622].

Sorghum etiolated seedling enzyme, a microsomal P450, requires oxygen and NADPH [K916].

Swede enzyme is also microsomal, requiring NADPH. It is inhibited by FMN, FAD, KCl and slightly by caffeate. Carbon monoxide inhibition is partially reversed by light [A198].

Melilotate hydroxylase (melilotate 3-monooxygenase, E.C. 1.14.13.4)

Melilotate → 2, 3-dihydroxyphenylpropionate

Enzyme from a Bacterium (probably a Pseudomonad) is a tetramer, molecular weight 250 000 and contains four mol/mol of FAD. Its amino acid composition has been determined [A861]. A reaction mechanism has been proposed which includes binding of NADH [A863].

o-Coumarate hydroxylase

o-Coumarate + O_2 → 2, 3-dihydroxycinnamate

This activity has been found in Pseudomonas [A934].

m-Coumarate hydroxylase

m-Coumarate + O_2 → caffeate

Mouse liver enzyme requires oxygen and NADPH [A2871].

***p*-Coumarate 3-monooxygenase** (E.C. 1.14.18.1)

p-Coumarate + O_2 → caffeate

Mytilus edulis hydroxyindole oxidase (which forms pigments from indoles) contains copper and haem iron (1:1 ratio) and catalyzes the above reaction. It appears to require peroxide, but not a reduced pyridine nucleotide [B314].

In Helianthus annuus, label is incorporated from $^{18}O_2$ into the reaction product [A178].

Ipomoea batatas root enzyme is monomeric, molecular weight 33 000, optimum pH 7.0 and pI 8.3. It hydroxylates *p*-coumarate, *p*-coumaroyl-β-D-glucose and *p*-cresol, but not a range of other phenols. Dopa and caffeate are also oxidized, presumably to quinones [F842].

Mung bean enzyme, optimum pH 5.0, (not a polyphenol oxidase) is specific for *p*-coumarate. It requires oxygen and NAD(P)H, ascorbate or 5,6,7,8-tetrahydro-6,7-dimethylpterin [F45].

Potato tuber enzyme requires FAD (or FMN) and NAD(P)H [D964].

Sorghum bicolor leaf enzyme, optimum pH about 6, requires ascorbate or a reduced pyridine nucleotide as electron donor. Activity is far higher for catechols than for *p*-coumarate, but shows no activity towards tyrosine. It exists as many interconvertible polymers, molecular weights 60 000–1 500 000 [A2433].

The reaction is also carried out by mushroom tyrosinase [H907], Mucuna pruriens phenoloxidase [E769] and Alnus rubra [A3343].

Ferulate 5-hydroxylase

Poplar stem enzyme is a P450 enzyme that requires NADPH; NADH is ineffective. It is different from cinnamate 4-hydroxylase (E.C. 1.14.13.11) [D179].

5-O-(4-Coumaroyl)-D-quinate 3′-monooxygenase (E.C. 1.14.13.36)

Daucus carota enzyme, which requires oxygen, ascorbate and NADPH but not NADH, is highly specific. The product is chlorogenate [E536].

***p*-Coumaroyl CoA 4-hydroxylase**

Parsley enzyme is cytosolic, with a sharp optimum at pH 6.5. It requires ascorbate (a specific requirement) and Zn^{2+} or Ca^{2+}. The product is caffeoyl CoA [F66].

Coumarin 3-hydroxylase

This activity has been found in mammalia, including beef, gerbil, hamster, man, mouse rabbit and rat [A5, G73, G604, H154, J913].

Coumarin 7-hydroxylase

Coumarin → umbelliferone

Cumene hydroperoxide can act as the oxidant for rat and rabbit microsomal enzyme (considered to be P450) [A335]. Another study indicated that the enzyme is present in liver from man (very active), mouse, rabbit and guinea pig (less active), but absent from rat liver [D559]. In mouse liver it is found in microsomes and mitochondria [F8], the microsomal enzyme being $IIP450_{15\alpha}$ [F100].

Isoflavone 2′-hydroxylase

Cicer arietinum microsomal enzyme acts on biochanin A and formononetin with oxygen and NADPH as co-substrates; NADH is poorly active [K751].

Flavanone 3′-hydroxylase (flavonoid 3′-monooxygenase, E.C. 1.14.13.21)

Horseradish peroxidase oxidizes naringenin to many products including eriodictyol [A3777].

Matthiola incana enzyme is microsomal and requires NADPH. It oxidizes naringenin and dihydrokaempferol [B679].

Petunia hybrida enzyme (flavonoid 3′,5′-hydroxylase) is found in flower microsomes, and requires oxygen and NADPH, but not NADH. It oxidizes naringenin to eriodictyol, and eriodictyol to 3′,4′,5,5′,7-pentahydroxyflavanone [H310].

Zea mays enzyme, optimum pH 8.5, is found in seedling microsomes. It requires oxygen and NADPH, and oxidizes kaempferol, naringenin and apigenin [D947].

This reaction has also been observed in Sinningia cardinalis [E570], Verbena hybrida [D471], parsley [C772], Dianthus caryophyllus [C392] and in rat liver microsomes [J300].

Isoflavone 3′-hydroxylase (E.C. 1.14.13.52)

Cicer arietinum microsomal enzyme acts on biochanin A and formononetin to form calycosin and pratensein respectively, with oxygen and NADPH as co-substrates; NADH is poorly active. Genistein and daidzein are extremely poor substrates [K751].

Flavanone 4′-hydroxylase

This activity has been observed in rat, Absidia, Gibberella and Streptomyces [A1512, B222, F673, G14].

Flavonol 6-hydroxylase

Chrysosplenium americanum enzyme, molecular weight 42 000–45 000, requires Fe^{2+} and α-oxoglutarate. Substrates are 3,4′,7-tri-O-methylquercetin and 3,4′,7-tri-O-methylquercetagetin [K658].

Isoflavone 6- and 8-hydroxylases

Aspergillus saitoi converts genistein into 8-hydroxygenistein and daidzein into 8-hydroxydaidzein, probably with hydroxylation of the aglycones [K155].

A partly characterized bacterium converts genistein into 4′,5,6,7- and 4′5,7,8-tetrahydroxyisoflavones, and biochanin A into 5,6,7-trihydroxy-4′-methoxyisoflavone [J627].

Taxifolin 8-monooxygenase (E.C. 1.14.13.19)

Pseudomonas enzyme, which forms dihydrogossypetin is a flavoprotein requiring oxygen and NAD(P)H [K917].

Juglone 3-monooxygenase (E.C. 1.14.99.27)

Pseudomonas putida enzyme is composed of two dimeric isozymes, which require oxygen, molecular weights 56 000 and 59 000, one of which is induced by juglone. Further substrates include naphthazarin, 1,4-naphthoquinone and 2-chloro-1,4-naphthoquinone [K897].

Cyclopenine 3′-hydroxylase

Penicillium cyclopium enzyme requires oxygen and a hydrogen donor, e.g. NAD(P)H, ascorbate or tetrahydropteridine to form cyclopenol. It is inhibited by cyanide or thiocyanate, but not by carbon monoxide. It also hydroxylates a number of analogues [A1633].

Dihydrosanguinarine 10-monooxygenase (E.C. 1.14.13.56)

Eschscholtzia californica enzyme is a microsomal P450, specific for position 10. It requires oxygen, not replaceable by peroxide or superoxide [K777].

Dihydrochelirubine 12-monooxygenase (E.C. 1.14.13.57)

Thalictrum bulgaricum enzyme, optimum pH 8.5, is a microsomal P450, specific for position 12. It requires oxygen and NAD(P)H [K776].

2.1.3 Hydroxylations of xenobiotics

Benzene hydroxylation

Rat liver mitochondrial enzyme, which requires NADPH, has a molecular weight of 52 000 [F686].

Methylococcus capsulatus methane monooxygenase (E.C. 1.14.13.25) hydroxylates benzene and halobenzenes by a NIH shift mechanism (products not stated) [F278].

Many other studies have also detected this reaction.

m-Cresol hydroxylase

Penicillium patulum enzyme acts on *m*-cresol to form methylquinol, as well as to form *m*-hydroxybenzyl alcohol. The enzyme, optimum pH 7.5, is a P450, which requires oxygen and NADPH, and is inhibited by carbon monoxide and cytochrome c. *m*-Cresol is a precursor of the antibiotic patulin, the formation of which requires both ring hydoxylation and side chain oxidation as well as ring fission and lactonisation [A1468].

Aniline hydroxylations

Aniline → *o*-aminophenol

This reaction has been detected in beef, mouse, pig, rat, sheep, rainbow trout, Boophilus and Nephila [A1795, A2788, A3460, D532, E60, J582].

Aniline → *m*-aminophenol

This reaction has been detected in Apocynum, Catharanthus and Conium [A2619].

Aniline → *p*-aminophenol

Species in which activity has been found in liver microsomes include bandicoot, beef, bettong, kangaroo, man, monkey, mouse, pig, possum, quokka, rat, shrew, tree shrew, rainbow trout, Boophilus, Cunninghamella and Nephila [A1218, A1795, A1997, A2420, A3460, B145, D532, E60, J639]. In mammalia the reaction requires three factors for maximal activity: cytochrome P450 or P448, a reductase and a lipid component [A627, A735]. Cumene hydroperoxide can act as the oxidant [A335]. The activity of aniline-4-hydroxylase is not detectable in rat before birth. After birth, activity increases 30-fold to a maximum at weaning, and then declines until at six months it has decreased by 90 per cent [A71]. At birth, the activity is five times greater in female than in male rats, with higher activity at 10 than at five weeks age. The development pattern only follows the activity of P450 in broad outline [A736].

Besides P450, catalase, haemoglobin and myoglobin can catalyse this reaction [B940, D43, H614]. Human and sheep erythrocytes oxidize aniline to *p*-aminophenol, with NADPH as co-substrate; the enzyme appears to be oxyhaemoglobin [B64]. Human foetal haemoglobin is more active than adult haemoglobin [B145]. Other substrates hydroxylated are *o*- and *m*-toluidine and N-methylaniline, and associated reactions involve N- and O-demethylation [D135]. Ferrihaemoglobin α-chains are less active than β-chains [D283]. Methaemoglobin is also active [D43]. In haemoglobin and methaemoglobin hydroxylation of aniline, ascorbate and dihydrofumarate can replace NADH as electron donor [D557].

Chlorobenzene hydroxylation

Rat liver forms all three chlorophenols from chlorobenzene. Pre-treatment with 3-methylcholanthrene induces *o*-hydroxylation, whereas phenobarbital induces the formation of all three phenols [A2320].

Methylococcus capsulatus soluble methane monooxygenase (E.C. 1.14.13.25) hydroxylates chlorobenzene and a number of analogues by a NIH shift mechanism [F278].

Chlorophenol 4-monoxygenase

Burkholderia cepacia chlorophenol 4-monooxygenase not only catalyzes the formation of benzoates from benzaldehydes, but also the formation of *o*-hydroxybenzaldehydes, for instance with syringaldehyde as substrate. The reaction involves a NIH shift of the formyl group [K107].

2,4-Dichlorophenol hydroxylase (E.C. 1.14.13.20)

Acinetobacter enzyme is a homotetramer, molecular weight 240 000 and optimum pH 7.6; it requires FAD and NADPH, but NADH is not so effective; FMN and riboflavin are inactive. It also acts on *p*-chlorophenol and 4-chloro-2-methylphenol, but some non-substrate chlorophenols induce the reduction of oxygen to peroxide [C232].

Proteobacteria enzyme, molecular weight 256 000, subunit molecular weight 65 000, pI 5.2 and optimum pH 8.0, requires NAD(P)H; the product is 3,5-dichlorocatechol. It has a broad specificity, but not for compounds in which both *ortho* positions are blocked [J466].

Pseudomonas cepacia enzyme is a homotetrameric flavoprotein, monomeric molecular weight 69 000 that requires FAD and NADPH; NADH is not so effective [G221].

4-Nitrophenol 2-monooxygenase (E.C. 1.14.13.29)

Nocardia enzyme, optimum pH 7.3, is soluble, requires oxygen, NAD(P)H and FAD. It is inducible, inactivated on dialysis and is somewhat unstable, even when frozen [D182].

5-Hydroxyisophthalate 4-hydroxylase

Bacterium enzyme, which is highly specific, is a flavoprotein containing FAD that cannot be replaced by FMN; it requires NAD(P)H [A2985].

Hydroxylation of phenoxyacetate

Aspergillus niger hydroxylates phenoxyacetate in *o*- and *p*-positions; oat and pea hydroxylate in the *p*-position. Arene oxides are postulated intermediates [A2849].

Biphenyl hydroxylations

Biphenyl → 2-, 3- and 4-hydroxybiphenyl, and dihydroxybiphenyls

Hamster liver enzyme is a mixture of P450 and P448; 2-hydroxylation is induced preferentially by phenobarbital, and 3-methylcholanthrene induces 2- and 4- monohydroxylations [A323]. Carbon monoxide, SKF 525A and NADH inhibit 4-hydroxylation [A11].

Rat liver 2-hydroxylation (but not 3- and 4-hydroxylation) is induced by corticosteroids, especially by betamethasone. 4-Hydroxylation is, quantitatively, most important, and 3-hydroxylation the least [A3680]. All three monohydroxylations are induced by 3-methylcholanthrene [B195, A3680], and 3- and 4-hydroxylation by phenobarbital [A2371, A3680]. The reaction system involves P450, NADPH and NADPH-cytochrome c reductase; NADPH and reductase can be replaced by cumene hydroperoxide [A2371].

Rabbit intestinal and liver microsomal enzyme activity increases two- to four-fold from nine days after birth to the adult level [A1980].

Avocado mesocarp 2-hydroxylase is found in both microsomal and cytosolic fractions; microsomal activity is increased by pre-incubation with safrole or 3,4-benzpyrene. Activity in both fractions is only slightly inhibited by classical P450 inhibitors such as carbon monoxide or SKF 525A [A11].

Benzpyrene hydroxylases

The intense interest in this and related enzyme activities is consequential to the central role that 3,4-benzpyrene plays in the aetiology of human lung cancer.

Several publications have briefly reported on benzpyrene hydroxylase without specifying the site of hydroxylation. Many publications (see 3,4-Benzpyrene) report on 3-hydroxylation, but a significant number additionally report hydroxylation at positions 1 (rat, man and sole, e.g [A458, C473, J94]), 4 (rat [C204]), 7 (rat, man, mouse and scup, e.g. [A458, B122, D965, G445]), 8 (rainbow trout [A1424]) and 9 (rat, monkey, hamster, rabbit, mouse, man and Saccharomyces, e.g. [A458, A2327, A2379, A2814, A3326, A3412, A3681]). 3-Hydroxylation (E.C. 1.14.14.2) occurs in a large range of species, including man, rat [A13], monkey [A1869], tree shrew, pig [A1997], rabbit [A2729], camel [H103], quokka, kangaroo, bandicoot, [A2420], mouse [A3681], guinea pig [B82], pigeon, crow, kite, egret [C307], trout [A2145], goldfish, bullhead [D465], bluegill [E480], scup [D965], sole [G374], killifish [B259], mullet [C141], barnacle [B744], Saccharomyces [A3326] and Candida [B775].

Studies on monohydroxylation of benzpyrene and other polynuclear hydrocarbons is complicated by the ready dehydration of dihydrodiols under acid conditions to monophenols; the former are metabolic products formed by hydration of epoxides generated in metabolism of polynuclear hydrocarbons. In rat, it has been suggested that at least some 3-hydroxybenzpyrene is formed by a spontaneous rearrangement of benzpyrene-2,3-oxide [A3277].

In man, 3-hydroxylation has been described in bronchus [A13], placenta, [A114], lung [A1644], liver [A2379], blood [B122], hair follicle [B350], bladder [C138], hepatoma [E480], melanocyte [G360] and P450 isozymes [H10]. Placental enzyme is largely microsomal [A985].

In man and rat, liver enzyme is primarily microsomal, and is activated by low molecular weight cofactors. In man, the activity is twice as high in smokers compared with non-smokers. In rat, activity is three times higher in males than in females, but there are only small sex differences in man, rabbit and guinea pig [A313]. Rat liver microsomal enzyme activity is reduced by extraction with organic solvents [A331].

In rabbit, the liver enzyme activity remains low up to 16 days after birth, and then increases to or above the adult level at about 30 days [A1980].

2-Hydroxybiphenyl 3-monooxygenase (E.C. 1.14.13.44)

Pseudomonas azelaica enzyme, pI 6.3, is a homotetramer, molecular weight 256 000 and monomeric molecular weight 60 000, each monomer containing one mol of unbound FAD. It requires NADH and oxygen; the latter is incorporated into the substrate. Other 2-hydroxybiphenyls and substituted phenols are also substrates. Uncoupling of the reaction results in the formation of peroxide [E960, J392].

D-Amphetamine hydroxylation

Rat liver activity is mainly microsomal, but some is mitochondrial. Microsomal activity requires NADPH, with a fairly sharp optimum at pH 7.0 [A77].

Indole-3-butyrate 4-hydroxylation

Bupleurum and Phytolacca enzymes, molecular weights 10 000–12 000 (both native and denatured) and optimum pH 5–6, hydroxylate the indole nucleus [J838].

Carbostyril formation from quinolines

Rat and guinea pig liver aldehyde oxidase (E.C. 1.2.3.1) oxidize several N-alkylquinoliniums and analogues. N-Methyl- and N-phenylquinolinium both form the corresponding carbostyrils and 4-quinolones, and a similar reaction has been observed with N-methylphenanthridinium and N-methyl-5,6-benzoquinolinium [D147].

Comamonas testosteroni quinoline 2-oxidoreductase, molecular weight 360 000, is composed of subunits, molecular weights 87 000, 32 000 and 22 000, and contains molybdenum, iron, acid labile sulphur, FAD and molybdopterin cytosine dinucleotide [K769].

Pseudomonas putida quinoline oxidoreductase, molecular weight 300 000 is composed of subunits, molecular weights 85 000, 30 000 and 20 000. It contains eight Fe and two FAD/mol as well as molybdopterin cytosine dinucleotide. Other substrates are 5-, 6-, 7- and 8-hydroxyquinoline, and 8-chloroquinoline; the product from quinoline is carbostyril [F856, G151, K766].

Rhodococcus quinoline 2-oxidoreductase (E.C. 1.3.99.17), optimum pH 9.5, molecular weight 300 000, monomeric molecular weights 82 000, 32 000 and 18 000, contains molybdenum, iron, acid labile sulphur, FAD and molybdopterin cytosine dinucleotide. Further substrates include quinolines substituted with hydroxyl, methyl and chloro groups [K767].

This reaction has also been detected in Nocardia [E694] and Desulfobacterium [H917].

Quinaldine 4-oxidoreductase

Rat and guinea pig liver aldehyde oxidase forms 4-quinolones (see Carbostyril formation, above) [D147].

Arthrobacter enzyme, molecular weight 340 000 and monomeric molecular weights 82 000, 35 000 and 22 000, contains molybdenum, iron and FAD. Molybdenum is present as molybdopterin cytosine dinucleotide. It forms 4(1*H*)-quinolones from several quinolines, and isoquinolines and analogues form the corresponding 1(2*H*)-oxo compounds. Aldehydes are also substrates [G803, J33].

Quinoline-4-carboxylate 2-oxidoreductase (E.C. 1.3.99.19)

An enzyme in Agrobacterium species, molecular weight 320 000, is composed of subunits, molecular weights 85 000, 35 000 and 21 000. It contains eight mol of Fe, two Mo, and two FAD; molybdopterin cytosine dinucleotide is required for activity. Carbostyrils are formed from quinoline, 4-carboxyquinoline, 4-chloroquinoline and 4-methylquinoline [G778].

4-Hydroxyquinoline 3-monooxygenase (E.C. 1.14.13.62)

Pseudomonas putida enzyme is a trimer, molecular weight 126 000. It requires oxygen and NADH, and it is specific for 1*H*-4-oxoquinoline [G812].

2-Hydroxyquinoline 8-monooxygenase (E.C. 1.14.13.61)

Pseudomonas putida enzyme is a highly specific two-component system. One is a yellow reductase, molecular weight 38 000 which contains FAD and [2 Fe-2 S] units. It transfers electrons to an oxygenase, a homohexamer, monomeric molecular weight 55 000, and contains about six [2 Fe-2 S] units and additional Fe. The oxygenase requires the reductase, oxygen and NADH for activity, and activity is enhanced by polyethylene glycol and Fe^{2+} [K778].

Isoquinoline 1-oxidoreductase (E.C. 1.3.99.16)

Pseudomonas diminuta enzyme is dimeric, monomeric molecular weight 16 000 and 80 000, and pI in the range 6.2–6.8. It contains molybdenum, iron, acid labile sulphur, phosphate and cytosine monophosphate (probably as molybdopterin cytosine dinucleotide), but no FAD. It requires an electron acceptor, but not oxygen or NAD, and acts on isoquinoline and quinazoline (to form 1- and 4-oxo compounds, respectively) as well as analogues; quinolines are not substrates [K765]. DNA sequencing indicates monomeric molecular weights of 16 399 and 84 249 for the monomers [K764].

Debrisoquine 4-hydroxylase

Human enzyme is microsomal, requiring NADPH. It is not found in all human subjects [C324].

Rat liver microsomal enzyme has been purified as a specific P450. It also N- and O-dealkylates other compounds [D98].

Hydroxylation with internal hydroxyl transfer

Horse liver alcohol dehydrogenase (E.C. 1.1.1.1) acts on 2-hydroxylaminofluorene to form 2-amino-1- and 3-hydroxyfluorene [H291].

Comamonas enzyme reduces 1-chloro-4-nitrobenzene anaerobically to 2-amino-5-chlorophenol, suggesting that there is an initial reduction to the hydroxylamine, followed by a Bamberger rearrangement, with hydroxyl migration to form the final product [K80].

2.1.4 Hydroxylation with elimination of substituent

4-Hydroxybenzoate 1-hydroxylase (decarboxylating) (E.C. 1.14.13.64)

Candida parapsilosis enzyme is a monomer, molecular weight 50 000 and optimum pH 8 containing FAD, which requires oxygen and NAD(P)H. Quinol is formed, with molecular oxygen incorporated into the new hydroxyl group. It also acts on a range of ring-substituted 4-hydroxybenzoates [J461].

This reaction has also been found in Pycnoporus cinnabarinus [D569].

Gentisate 1-hydroxylase (decarboxylating)

Trichosporon cutaneum acts on gentisate to form hydroxyquinol [B368].

3,4-Dihydroxybenzoate 1-hydroxylase (decarboxylating)

Trichosporon cutaneum enzyme is highly specific. It requires oxygen and NADH, and forms hydroxyquinol with the release of carbon dioxide [E288]

Salicylate 1-monooxygenase (E.C. 1.14.13.1)

e.g. Salicylate → catechol + CO_2

Pseudomonas putida enzyme is a monomer, molecular weight 54 000. The amino acid composition and the terminal amino acids have been determined [A1216]. Another study using a different strain found a molecular weight of 45 000, with a different amino acid composition [F658]. The reaction requires oxygen and NADH, with carbon dioxide as the second product. Salicylaldehyde is also a substrate, and this releases formate [B751]. By using specifically ring-labelled salicylate it has been found that decarboxylation and hydroxylation with P. cepacia enzyme occur at the same carbon atom [B883]. It is an inducible enzyme [E287]. A Pseudomonas enzyme also acts on 3- and 5-chlorosalicylate and 3,5-dichlorosalicylate [H506].

Trichosporon cutaneum enzyme, molecular weight 45 300 and optimum pH 7.5, contains FAD. It also acts on salicylates substituted with a hydroxyl group at positions 3, 4, 5 and 6, an amino or chloro group at 4 or 5, a methyl at 4, or a methoxy or fluoro at 5 [D232].

This reaction has been found to occur non-enzymatically in rat, catalysed by free radicals formed by the parkinsonism-inducing ion MPP^+ [J681]. In rat liver the formation of catechol and other products from salicylate is catalysed by the action of hydroxyl free radicals rather than by the direct action of a decarboxylating hydroxylase [J754].

Vanillate 1-hydroxylase (decarboxylating)

Sporotrichum pulverulentum enzyme acts on vanillate, protocatechuate, gallate, 2,4-dihydroxybenzoate and *p*-hydroxybenzoate. Homovanillate and 2,3,4-trihydroxybenzoate are less effective as substrates, whereas gentisate, syringate, ferulate, veratrate and *p*-methoxybenzoate are poor substrates, and benzoate and *m*-methoxybenzoate are not substrates. The product from vanillate is methoxyquinol [B73]. The enzyme has a molecular weight of 65 000 and requires NADPH and FAD for maximal activity. Tiron, Cu^{2+}, Ag^{+}, Hg^{2+} and *p*-chloromercuribenzoate are inhibitory, whereas EDTA, diethyldithiocarbamate and Fe^{3+} are not. Further substrates are 3,4-dihydroxy-5-methoxybenzoate and 2,4,6-trihydroxybenzoate [C57].

Phanerochaete chrysosporium enzyme, optimum pH 7.5–8.5, requires NAD(P)H and oxygen, and is cytosolic [B312].

4-Hydroxyisophthalate hydroxylase

Pseudomonas enzyme, which is different from *p*-hydroxybenzoate hydroxylase, is a homodimer, molecular weight 103 000, containing one mol of FAD, with protocatechuate as the reaction product. It also acts on 5-sulphosalicylate at a much lower rate; other compounds are not substrates. It is inhibited by substrate analogues and thiol-binding compounds, and is stabilized by thiols [A3066].

Aniline oxidation to catechol

Nocardia carries out this reaction, with the incorporation of molecular oxygen; a cyclic peroxide intermediate has been suggested [A1520].

Anthranilate 3-monooxygenase (deaminating) (E.C. 1.14.13.35; formerly E.C. 1.14.12.2)

Anthranilate → pyrocatechuate

Trichosporon cutaneum enzyme, molecular weight 95 000 and monomeric molecular weight 50 000, contains two mol of FAD and exhibits a sharp optimum pH at 7.7. The oxygen in position 2 comes from water, and in position 3 from molecular oxygen. Other substrates are N-methyl- and N,N-dimethylanthranilate. Under some conditions 3-hydroxyanthranilate is also formed; this requires a reducing agent that acts on imines [C393, E270].

Aspergillus niger, molecular weight 44 000, pI 5.36 and optimum pH 8.2, requires FAD and Fe^{2+}. It is inhibited by Cu^{2+}, *o*-phenanthroline and *p*-chloromercuribenzoate. Another activity associated with this enzyme is anthranilate 3-hydroxylation [A778, B327, D231].

This reaction has also been observed in Aspergillus soyae [A1299].

Anthranilate hydroxylase (with deamination and decarboxylation; E.C. 1.14.12.1)

Pseudomonas cepacia 2-halobenzoate 1,2-dioxygenase acts on anthranilate and other benzoates with a considerable range of substituents [G434].

Trichosporon cutaneum enzyme, molecular weight 94 000 and optimum pH 7.7, appears to be a dimer. Substrates are anthranilate and N-methylanthranilate; some benzoates are non-substrate effectors, with the formation of peroxide from oxygen [C393].

This reaction has been observed in Micrococcus, Aspergillus soyae and Pseudomonas pyrrocinia [A1299, C120, D56].

4-Aminobenzoate hydroxylase (E.C. 1.14.13.27)

4-Aminobenzoate → *p*-aminophenol + CO_2

Agaricus bisporus enzyme, molecular weight 49 000, pI 6.0 and optimum pH 6–8 (partly dependent on cofactor), contains about one mol of FAD. It requires oxygen and NAD(P)H. Other substrates include aminobenzoates and *p*-hydroxybenzoate. Unlike 4-aminobenzoate, these substrates also form peroxide non-stoichiometrically, whereas peroxide is formed stoichiometrically relative to cofactor oxidation in the presence of other benzoates that are not substrates. Neither FMN nor riboflavin can replace FAD [D770, E162].

2-Aminobenzenesulphonate 2,3-dioxygenase (E.C. 1.14.12.14)

Alcaligenes enzyme is monomeric, molecular weight 42 000, and requires two oxygen molecules to form 2,3-dihydroxybenzenesulphonate. Maximal activity is found near the end of the exponential growth phase [H239, K775].

Alcaligenes enzyme, molecular weight 134 000 according to another study, is composed of two pairs of monomers, monomeric molecular weights 45 000 and 16 000, with one [2Fe-2S] centre associated with each of the larger chains. Inhibition by *o*-phenanthroline indicates the presence of another Fe-binding site. The N-terminal sequences have been determined. The product formed from 2-aminobenzenesulphonate is 2,3-dihydroxybenzenesulphonamide. Other substrates are benzenesulphonate, and benzene-sulphonates substituted with nitro, amino, chloro and hydroxyl groups; the reaction products from these compounds were not identified [K293].

4-Sulphobenzoate 3,4-dioxygenase (E.C. 1.14.12.8)

Comamonas testosteroni contains two isozymes. One is a red dimer, monomeric molecular weight 50 000, and the other a yellow monomer, molecular weight 36 000, with NADH and Fe^{2+} as cofactors. The enzyme is highly specific, forming sulphite and protocatechuate. The reaction does not appear to involve two steps or to involve an intermediate dihydrodiol [G244].

Pseudomonas putida oxidizes benzenesulphonate to catechol and *p*-toluenesulphonate to 4-methylcatechol [J657]. The same reaction is observed in Alcaligenes [H297].

Salicylaldehyde hydroxylation

Pig liver microsomal flavin-containing monooxygenase-1 forms catechol and formate in equimolecular amounts from salicylaldehyde [H482].

Dechlorination with concomitant hydroxylation by microorganisms

a. 4-Chlorobenzoate dehalogenase (E.C. 3.8.1.6)

p-Halobenzoate → *p*-hydroxybenzoate

Studies have found that this reaction occurs in three stages: conjugation with CoA (E.C. 6.2.1.33) followed by dehalogenation (E.C. 3.8.1.7) and hydrolysis (E.C. 6.2.1.33) [K191]; detailed information is found under these headings.

An Arthrobacter enzyme, optimum pH 6.8, is activated by Mn^{2+} and inhibited by peroxide. Other substrates are *p*-fluoro- and *p*-bromobenzoates [D544]. Another study claims that although *p*-iodobenzoate is a substrate *p*-fluorobenzoate is not, nor are *p*-chlorophenylacetate or *p*-chlorocinnamate. The molecular weight is about 45 000 and the optimum pH 7–7.5. Unlike all other similar dehydrogenases reported at the time it is not inhibited by EDTA or activated by Mn^{2+} [E752].

A Pseudomonas dehalogenase requires ATP, CoA and Mg^{2+} [G205]. The incorporated oxygen comes entirely from water, indicating that molecular oxygen is not involved [D473] (but see *b.* below). This enzyme, optimum pH 7–7.5, is activated by Mn^{2+} or Co^{2+}, and is inhibited by EDTA. It acts on chloro-, bromo- and iodobenzoates, but not on *p*-fluorobenzoate [E359].

b. p-Chlorobenzoyl CoA dehalogenase (E.C. 3.8.1.7)

p-Chlorobenzoyl CoA →
p-hydroxybenzoyl CoA

This is the second step in the reaction sequence leading to the oxidative dehalogenation of *p*-chlorobenzoate.

Pseudomonas 4-chlorobenzoyl CoA dehalogenase also acts on the bromo- and iodo- analogues, but not on the fluoro- analogue [G920, H645]. About 75 per cent of the incorporated oxygen comes from molecular oxygen, and the remainder from water [H216]. It is a homotetramer with high temperature stability, molecular weight 120 000, pI 6.7 and optimum pH 10 [H645, K191].

Acinetobacter enzyme also acts on bromo-analogues [G896].

Arthrobacter enzyme is a tetramer, monomeric molecular weight 33 000, pI 6.1 and optimum pH 8, with a stability range pH 6.5–10. It also acts on the fluoro-, bromo- and iodo- analogues, but not on *o*- and *m*- chloro analogues [H279].

c. Chlorophenol 4-monooxygenase

Burkholderia enzyme is a two-component system. One is a reductase, molecular weight 22 000 that contains FAD, and the other has a molecular weight of 58 000. The reaction requires oxygen and NADH. 2,4,5-Trichlorophenol is oxidized to 2,5-dichloroquinol and then to 5-chloro-2-hydroxyquinol. Other substrates are 2,3,5,6-tetrachlorophenol, 2,4,6-trichlorophenol and 2,5-dichlorophenol [J177].

Dechlorination with catechol formation

a. 2-Chlorobenzoate 1,2-dioxygenase
(E.C. 1.14.12.13)

Pseudomonas cepacia forms pyrocatechuate from 2-chlorobenzoate [F401].

b. 2-Halobenzoate 1,2-dioxygenase

o-Halobenzoate → catechol

Pseudomonas cepacia enzyme is composed of two protein fractions, and acts on a range of *o*-halobenzoates as well as other *o*-substituted benzoates. In many instances the products have not been identified, but benzoate, *o*-toluate, *m*-hydroxybenzoate and *p*-hydroxybenzoate form benzoate-1,2-dihydrodiol, *o*-cresol, gentisate and quinol respectively; in none of these compounds is there a readily displaceable *ortho* substituent, which drives the reaction in a different direction [G434].

c. 4-Chlorophenylacetate 3,4-dioxygenase
(E.C. 1.14.12.9)

p-Chlorophenylacetate →
3, 4-dihydroxyphenylacetate

This Pseudomonas activity involves a two-component, intensely red-brown enzyme system. One of the enzymes is a homotrimer, molecular weight 140 000 and pI 5.0. In the reaction two oxygen atoms are incorporated from molecular oxygen [E163].

d. 2,4,6-Trichlorophenol 4-monooxygenase

Azotobacter enzyme is a homotetramer, molecular weight 240 000, which forms 2,6-dichloroquinol from 2,4,6-trichlorophenol. It requires FAD, and utilizes one mol of oxygen and two mol of NADH. Other substrates include a range of *para*-chlorinated or brominated phenols; *o*-chlorophenol is a poor substrate [H884].

Pentachlorophenol monooxygenase
(E.C. 1.14.13.50)

Arthrobacter enzyme, optimum pH 7.5, is stimulated by EDTA and requires NADPH and oxygen, with formation of tetrachloroquinol. Other substrates include 2,3,4- 2,4,5-, and 2,4,6-trichlorophenol, and 2,3,4,5-tetrachlorophenol, but other analogues are not substrates [K750].

A Flavobacterium enzyme, molecular weight 63 000, pI 4.3 and optimum pH 7.5–8.5, can exist as polymeric forms. The monomer is a flavoprotein containing one mol of FAD, and the reaction requires two mol of NADPH. It hydroxylates pentachlorophenol to tetrachloroquinol [G272].

Sphingomonas pentachlorophenol 4-monooxygenase, which requires NADPH, is a flavoprotein, molecular weight predicted to be 59 993 with 538 amino acid residues by DNA studies. Other substrates include 4-nitrocatechol and *p*-nitrophenol [K73].

Quinol formation from *para* substituted phenols

Rat liver microsomes form quinol from phenols substituted in the *para* position with nitro, cyano, hydroxymethyl, acetyl, benzoyl or halide groups. Results from experiments using $^{18}O_2$ suggest that the initial reaction is the formation of a 4-hydroxy-1-oxo-2,5-diene, involving oxygen transport by the Fe in the enzyme. This intermediate may then either form quinol directly, by elimination of the substituent (as a positively charged ion) from the 4-position, or the formation of *p*-quinone, with the substituent forming a negatively charged ion. Trapping experiments suggest that both mechanisms may occur, depending on the identity of the substituent. In the case of *p*-cresol the side chain is not eliminated; instead there is a NIH shift to form toluquinol [J81].

Defluorination with catechol formation

Pseudomonas converts 3-fluorinated compounds into the corresponding 2,3-dihydroxy analogues. Although fluorobenzene is not a substrate, toluene, anisole, fluorobenzene, chlorobenzene, bromobenzene, iodobenzene, benzonitrile and benzyl alcohol fluorinated at position 3 are substrates [F76].

2-Nitrotoluene 2,3-dioxygenase

2-Nitrotoluene + O_2 → 3-methylcatechol + nitrite

Pseudomonas enzyme is a three-component system which forms 3-methylcatechol as well as *o*-nitrobenzyl alcohol from 2-nitrotoluene (the best substrate). A series of other nitrotoluenes and nitrobenzene are also catechol-forming substrates. Both oxygen atoms are incorporated from molecular oxygen [H344, J524].

2-Nitrophenol 2-monooxygenase (E.C. 1.14.13.31)

o-Nitrophenol + O_2 → catechol + nitrite

Pseudomonas forms catechol and nitrite from *o*-nitrophenol [E726]. P. putida enzyme, molecular weight 58 000–65 000, optimal activity and optimum stability at pH 7.5–8.0, requires NADPH and oxygen. It is activated by Mg^{2+} or Mn^{2+}, but not by flavins. It is very sensitive to heat inactivation, but is stabilized by substrate. It also acts on a range of other *o*-nitrophenols, but not all *o*-nitrophenols are substrates [E607, H65, K898].

2,4-Dinitrotoluene 3,4-dioxygenase

2,4-Dinitrotoluene + O_2 →
4-methyl-5-nitrocatechol + nitrite

A Pseudomonas enzyme catalyzes this reaction [G407, J524].

***p*-Nitrophenol hydroxylation**

p-Nitrophenol + O_2 → quinol + nitrite

Penicillium chrysogenum enzyme is found in a membrane fraction. It requires oxygen and NAD(P)H, and is stimulated by FAD; the oxygen atom is incorporated from molecular oxygen [E94].

Moraxella forms quinol and nitrite from *p*-nitrophenol [G153].

4-Nitrocatechol hydroxylation

Sphingomonas pentachlorophenol 4-monooxygenase (see above) oxidizes 4-nitrocatechol to hydroxyquinol, and *p*-nitrophenol (poorly) to quinol, with the release of nitrite [K73].

Bacillus sphaericus and Arthrobacter enzymes form hydroxyquinol with the release of nitrite [H270, J679].

6-Nitrobenzo[*a*]pyrene hydroxylation

Rat lung and liver catalyse the conversion of 6-nitrobenzo[*a*]pyrene into 6-hydroxybenzo[*a*]pyrene, also with the formation of a trace of benzpyrene [E467]. The same reaction has been observed in mouse [E971].

2.2 *Formation of quinones and analogues from catechols, quinols and other precursors*

Monophenol monooxygenase (E.C. 1.14.18.1)

Tyrosinase (classically from mushroom or skin melanocytes), polyphenol oxidase (from potato) and laccase (E.C. 1.10.3.2; originally from the lacquer tree) are old classifications for enzymes that are now grouped together under the heading of Monophenol Monooxygenase, although these classifications are still commonly found in the literature. In the current section, this classification (except polyphenol oxidase) has been used because it is still in such common use. They are all copper-containing enzymes, but they differ in specificity; laccases preferentially oxidize *p*-phenylenediamine relative to oxidation of catechols to quinones, but they differ from tyrosinase in that they do not oxidize tyrosine to dopa. However, the distinctions between these classes of enzyme are not clear-cut.

One study with *o*- and *p*-diphenol oxidases (considered to be catecholases and laccases respectively) from apple, banana, Agaricus, Glomerella, Sclerotina (*o*-oxidases), peach, spruce, Botrytis, Coriolus, Trametes, Glomerella and Rhus (*p*-oxidases) distinguishes between the activites on the basis of the following parameters:

Some distinguishing features between catechol oxidase and laccase

Test	*Catechol oxidase*	*Laccase*
Catechols	Oxidized	Oxidized
Quinols	Zero or poor oxidation	Oxidized
p-Phenylenediamine	Zero or poor oxidation	Oxidized
p-Cresol	Orange-red pigment*	–
1-Naphthol	–	Purple colour*
Cinnamates	Inhibitory	No inhibition
PVP	Inhibitory	No inhibition
Cationic detergents	No inhibition	Inhibitory
Dodecyl sulphate	Activates	No action

* with whole microorganisms [A3974].

2.2.1. Tyrosinase and unspecified types.

a. Plant and fungal enzymes

Apple enzyme is composed of isozymes, molecular weights 46 000, 120 000, 460 000 and 800 000. They are considered to be polymers of a common monomer [A1133].

Amasya apple enzyme contains three isozymes, with optima at pH 6.6–9.0, dependent on the substrate; substrates are catechol, 4-methylcatechol, pyrogallol and L-dopa [H567].

Four Arachis peroxidase isozymes, which additionally exhibit polyphenol oxidase activity towards L-dopa, have optima at pH 7.6, 8.0, 8.0 and 8.0 [A2519].

Artichoke enzyme, molecular weight 116 000, pI 4.5, and broad optimum pH 5–8 is activated by Cu^{2+} and Fe^{3+}. Substrates include catechol, caffeate and 5-O-caffeoylquinate [F664].

Avocado cultivars show different isozyme patterns. Two cultivars each contain six major isozymes and probably some trace components, and a third cultivar contains a smaller number of isozymes. It would appear that some isozymes are cultivar-specific. They oxidize L-dopa and other catechols [A1636].

Banana fruit pulp contains a soluble enzyme that acts on catechol, but with rapid inactivation. Denaturation above 70° has an activation energy of 85 000 cal/mol; at lower temperatures a slower inactivation occurs, activation energy 18 000 cal/mol [A212]. More detailed studies have demonstrated nine isozymes separable by electrophoresis in the inner pulp, eight in the outer pulp and 10 in the skin. They all act on catechol, but each has its own pattern of specificity towards 4-methylcatechol, dopamine, pyrogallol, *d*-catechin, caffeate and DL-dopa. Two in the inner pulp, and one in the outer pulp act on L-tyrosine. Inhibitors include diethyldithiocarbamate, cyanide, metabisulphite, mercaptoethanol, propane-1,2-dithiol and mercaptoethanol [A218].

Cabbage phloroglucinol oxidase, molecular weight 39 000–40 000, which oxidizes phloroglucinol and phloroglucinolcarboxylate (the reaction products are unclear) additionally shows peroxidase properties towards guaiacol, with optimum pH 7.6 and 6.4 for these reactions respectively. Catechols are not substrates for the oxidase activity. It is moderately stable at 100° [H412].

Helianthus (sunflower) enzyme exhibits optima at pH 5.3, 5.8, 6.0 and 6.5, with best stability between pH 5.5 and 7 [A2152, A2153].

Ipomoea batatas (sweet potato) enzyme is composed of 12 isozymes, optimum pH 6–6.5 [F208].

Longan (Dimocarpus longan) fruit peel enzyme, optimum pH 6.5, acts on pyrogallol, catechol and 4-methylcatechol, but not on chlorogenate, *p*-cresol, resorcinol or tyrosine. It is inhibited by glutathione, cysteine, thiourea, Fe^{2+}, Sn^{2+}, and is activated by Mn^{2+} and Cu^{2+} [K96].

Green olive enzyme, molecular weight 42 000, has been separated into seven to nine bands of activity by isoelectric focussing, with pI 5, 5.6 and 7 for the main bands, and pI 4.8–5.6 for the remainder. They act on catechols including DL-dopa, dopamine and catechol, but monophenols are not substrates. Inhibition (in decreasing order, at 10^{-3}–10^{-5} M) is brought about by diethyldithiocarbamate, 2-mercaptoethanol, bisulphite, thiourea and 8-hydroxyquinoline. Halides inhibit at higher concentration, and chloride inhibition is strongly pH-dependent [A2360].

Pear enzyme (Bartlett) contains two isozymes that act on chlorogenate with an optimum at pH 4 [A1007].

Enzymes from two species of Phoenix exhibit a molecular weight of 17 000–17 500, optimum pH about 6 and maximal stability about 7. Substrates include catechol and other diphenols, but not monophenols or *meta*-diphenols [F151].

Potato enzyme is a complex mixture, with 17 isozymes separable by electrophoresis with L-dopa as substrate. Only five of these act on L-tyrosine, a pattern of specificities that is repeated by other catechols and monophenols. Isoelectric focussing shows multiple peaks in the range of pH 4.0–4.7 and 5.1–5.4. The monomeric molecular weight is 36 000, and monomer, dimer, tetramer, octamer and polymeric forms are found. Only high molecular weight forms exhibit monophenol oxidase activity [A2997, A3190].

A spinach phenoloxidase is a monomer, molecular weight 36 000. It dimerizes in the presence of 5 mM pyrocatechualdehyde without loss of activity; other aldehydes are ineffective. On freezing, it polymerizes [A1735]. Root enzyme develops after germination and differs from two chloroplast oxidases [A1730].

Sugar cane enzyme is composed of two fractions, molecular weight 32 000 and 130 000, both with broad optima between pH 4.5 and 7.5. It acts on chlorogenate, a better substrate than caffeate or L-dopa. Thiols, especially thioglycollate, inhibit non-competitively [A180].

Tea catechol oxidase, optimum pH 5.7 and pI 4.1, acts on catechol, pyrogallol and a series of catecholic flavonoids. Flavonoid gallate esters undergo oxidative degallation. A flavanol gallate oxidase exhibits pI 9.6 [A972] and contains Fe^{2+} [A1004]. In another study six isozymes were detected, optimum pH 6.2–6.6 [A1128].

Vitis vinifera (grape) enzyme is devoid of laccase activity. Substrates include 4-methylcatechol, (+)-catechin, caffeate, chlorogenate, and caffeoyltartrate. *p*-Coumarate and *p*-coumaroyltartrate, but not quinol are substrates [E458]. It exists as multiple forms, eight having been detected [A1038], and pI 4.7, 4.9 and 5.2 [A208]. Kinetic studies indicate random binding of 4-methylcatechol and oxygen followed by release of quinone, then a second catechol molecule binds and is oxidized to quinone [A1631].

Wheat contains an enzyme that acts on L-dopa [A176].

A study on *o*-diphenol oxidases has detected activity in apple, banana, Agaricus, Sclerotina and Glomerella. 4-Methylcatechol is substrate in all these species, and chlorogenate in some. Inhibition is caused by cinnamate, *p*-coumarate, ferulate, usually competitively, and by polyvinylpyrrolidine. They are activated by dodecyl sulphate [A3974].

Agaricus bisporus enzyme is composed of three dimeric isozymes, pI 5.12, 5.41 and 6.25, formed from combinations of two different subunits. L-Dopa and catechol are substrates; only one hydroxylates tyrosine effectively [A3135]. It is inhibited competitively by thiambutosine [A2185]. 4-*tert*-Butylcatechol is oxidized to 4-*tert*-butyl-*o*-quinone, with ethyl hydroperoxide as 2-electron acceptor [A1495].

Mushroom (Xerocomus) enzyme is inhibited by lamprene [A2185].

Portabella mushroom tyrosinase has a molecular weight of 48 000–50 000 and pI 5.1–5.3 [K101].

Mushroom tyrosinase acts on 3,4-dihydroxybenzyl cyanide to form the corresponding *o*-quinone. The time sequence indicates the initial formation of a semiquinone radical from a single electron oxidation, followed by dismutation. The quinone rapidly forms the corresponding quinone methide [K834]. It also oxidizes poly(*p*-hydroxystyrene) to the corresponding quinone, but only a small proportion of the phenolic groups are oxidized [K285].

Streptomyces glaucescens tyrosinase, molecular weight 29 100, optimum pH 6.8, pI 6.95 and maximal stability at pH 8.0, acts on both L-tyrosine methyl ester and L-dopa; it contains one mol of Cu^{+}. K_m increases as pH decreases below 8, due to a change of ionisation at the active centre; K_m is also strongly dependent on oxygen concentration for the oxidation of catechols but not monophenols [A749].

b. Animal enzymes

Beef uterus enzyme has been separated into two components, one with molecular weight 80 000. Substrates are catecholamines and analogues; for instance, oxidation of adrenaline is the first step in the formation of adrenochrome. No activity was observed with monophenols or *m*-diphenols [A624].

Calliphora erythrocepha (blowfly) enzyme is found as a pro-enzyme, molecular weight 115 000 [A900].

Corcyra cephalonica enzyme, which is found in larval and pupal haemolymph, oxidizes a series of catechols related to L-dopa and catecholamines, including N-acetyldopamine [A2556].

Fleshfly enzyme, obtained from the anterior end of the third instar, molecular weight 400 000, appears to be a lipoprotein. It acts on dopa, methylcatechol and *p*-cresol, but tyrosine is not a substrate [A7]. Probably at least five isozymes oxidize L-dopa [A27].

Studies on the oxidation of L-dopa by frog epidermis enzyme suggests a uni-uni-bi-bi ping-pong mechanism [D15]. It is present as a pro-enzyme whose amino acid composition has been determined, and is activated by peptidases. It appears to exist as monomeric (molecular weight 50 000) and tetrameric forms, each monomer containing one copper atom. Both tyrosine and dopa are substrates [A1208].

Gerbil eye enzyme, which is involved in melanogenesis utilizes both L-tyrosine and L-dopa; the latter is the cofactor in the oxidation of the former [H687].

The sea squirt Halocynthia roretzi polyphenol oxidase is a heterotetramer, molecular weight 170 000, monomeric molecular weights 55 000 and 30 000. It acts on L-tyrosine, DL-dopa and catechol with an optimum at pH 6.4 [J313].

Mytilus edulis byssus enzyme is polymeric, monomeric molecular weight 120 000 and optimum pH 8–8.5. 4-Methylcatechol is substrate, and it requires sodium chloride concentrations of 0.5–0.6 M [D726].

Squid tyrosinase, which oxidizes L-dopa, exhibits three precipitin lines [A883].

Caeruloplasmin oxidizes D- and L-dopa, dopamine and *p*-phenylenediamine; the catechol substrates inhibit each other, but not the diamine oxidation. This suggests two separate oxidation sites on the molecule [A677].

2.2.2. Enzymes classified as laccase and quinol oxidases

Sarcophaga bullata laccase, optimum pH 4.5, acts on quinol and a range of catechols including L-dopa and N-acetyldopamine. It is inhibited by azide [E407].

Silkworm laccase, optimum pH 5.5, acts on quinol and L-dopa [A702].

Arum maculatum quinol oxidase, a mitochondrial enzyme that oxidizes menadiol, is very unstable after solubilization [E87].

Lentinus edodes laccase, molecular weight 74 000, pI 3.42, and optimum 4–4.2, has a carbohydrate content of 7.5 per cent. Its absorption spectrum is indicative of a copper-containing protein. [H834].

Mangifera (mango) laccase oxidizes quinol as well as 4-methylcatechol, but *p*-cresol is not a substrate. A similar enzyme is found in Schinus, Pistacia and Pleiogynium [A3138].

Peach laccase, molecular weight 73 500, contains two Cu/mol; its amino acid composition has been determined. It is inhibited by diethyldithiocarbamate [A186].

Shinus molle enzyme, molecular weight 105 000 and optimum pH 6.2, acts on quinol. It contains three to four mol/mol copper and is about 50 per cent polysaccharide composed of a variety of saccharide residues. Its amino acid composition has been determined [B491].

Botrytis cinerea laccase, a blue enzyme, acts on quinol, a range of catechols, monophenols and phenylenediamines at a similar rate, whereas tyrosine is oxidized at a considerably slower rate [A1930, E458]. Optimum pH is substrate-dependent, between pH 3.5 and 5; it is stable down to pH 2.5, its pI, but is inactivated above pH 7. It is rapidly inactivated above 40°. Slight inhibition is caused by cyanide and EDTA [A1930].

Coriolus hirsutus laccase is a glycoprotein molecular weight 73 000, optimum pH 2.5–4.0 (substrate dependent) and pI 7.4, with three copper-containing centres. The N-terminal sequence shows low homology with other Basidiomycetes laccases. Oxidation of 2,2′-azino-bis(3-ethylbenzothiazoline-6-sulphonic acid) forms a cation radical. Many phenols and *p*-phenylenediamine are substrates [K677].

Trametes polyphenol oxidase is composed of two isozymes considered to be laccases, molecular weights 61 000 and 90 000, and pI 3.4 and 2.7 respectively, with optimum pH 4.5–5.0. The larger molecule is stable up to 50°, whereas the smaller is stable up to 60°, at neutral to acidic pH. They act on (+)-catechin, but not on tyrosine, and are inhibited by copper chelators, reducing agents and N-bromosuccinimide, but not by carbon monoxide [K106].

A study has detected *p*-diphenol oxidase in peach, spruce, Rhus, Botrytis, Coriolus, Trametes and Glomerella. Substrates include quinol, *p*-phenylenediamine and 4-methylcatechol in all species, whereas catechol and chlorogenate are substrates in some species. Inhibition is not brought about by cinnamate, dodecyl sulphate or polyvinylpyrrolidine [A3974].

4-Methyl-5-nitrocatechol oxygenase

Burkholderia enzyme, molecular weight 60 000–65 000, contains one mol of FAD, and requires oxygen and NAD(P)H. The product is 2-hydroxy-5-methyl-*p*-quinone. 4-Nitrocatechol is also a substrate, but other catechols are not [J249].

Chlorogenate oxidase

Ipomoea batatas enzyme is composed of two isozymes, molecular weights 39 000 and 40 000 [J822].

Tyrosylprotein oxidation

Arthrobacter globiformis oxidizes a tyrosyl residue in the active centre of phenylethylamine oxidase to a topaquinone residue. The reaction requires two molecules of oxygen, and one mol of peroxide is formed. The stoichiometry suggests that a trihydroxyphenylalanyl residue is an intermediate [J905].

Phloroglucinol oxidation

Brassica oleracea contains three isozymes, two of which have molecular weights 43 000 and 32 000, and optimum pH 8.0 and 7.4 respectively, which oxidize phloroglucinol and phloroglucinolcarboxylate but not other phenols; they are polyphenol oxidases which are activated by Mn^{2+}. The products are possibly quinones. At lower pH these enzymes show peroxidase activity, inhibited by Mn^{2+} [H878].

Naphthohydroquinone dehydrogenase

Human enzyme is found in an inactive form in saliva, and is activated by gastric juice. It is a metalloenzyme that requires oxygen, but no cofactors, oxidizing rifampicin to its quinone; quinol is not a substrate. It is inhibited by sulphide, cysteine, cyanide and citrate [A3399].

***trans*-Dihydrodiols as quinone precursors**

Rat liver cytosol 3-hydroxysteroid dihydiol dehydrogenase converts *trans*-dihydrodiols into *o*-quinones. Substrates are benzene dihydrodiol and 3,4-dihydrodiols of benzanthracene, 7-methylbenzanthracene, 12-methylbenzanthracene and 7,12-dimethylbenzanthracene [E740]. Enzyme from rat lens, optimum pH 9.0, oxidizes naphthalene-1,2-dihydrodiol to the corresponding 1,2-dione. It has been identified as aldol reductase [J552].

It is claimed that benzpyrene 3,6-quinone formation and reduction in rat liver microsomes are catalyzed by different enzymes [A3637].

Polynuclear hydrocarbon quinone formation

Because of their key role in carcinogenesis by xenobiotics, polynuclear hydrocarbons have been the subject of a large number of metabolic studies, which have demonstrated that they can be oxidized to quinones. Most of the studies have been carried out solely by examining the identity of metabolic end-products. Benzpyrene (benzo[*a*]pyrene) illustrates these findings; *trans*-dihydrodiols are probable intermediates (see above).

Studies with benzpyrene have shown that 1,6-, 3,6- and 6,12-quinones are formed, the oxo groups of which are conjugated through the aromatic nucleus. Most studies have been carried out in rats, but these reactions have also been observed in man, mouse, monkey, rabbit, trout, catfish, goldfish, bullhead, sea anemone, starfish, Petroselinum, Glycine max, Pycnoporus, Phanerochaete and Cunninghamella.

In man, the reaction is catalyzed by P450 [H10] and in melanocyte [G360].

In rat the reaction occurs in liver nucleus [A3164, A3782], liver microsomes [G445], and is also catalyzed by cytosolic lipoxygenase (E.C. 1.13.11.12) [G101]. With liver microsomes benzpyrene metabolism is slower with peroxide as oxidant than with oxygen, which has suggested a peroxide-dependent oxidation as a second pathway [B656]; it is possible that peroxide may have released oxygen spontaneously or by the action of traces of catalase, however. A liver microsomal P450 designated $P450_{MC}$ catalyzes these reactions [E226].

In mouse the reaction occurs in liver microsomes [G445].

In rabbit the reaction occurs in lung [A2394].

In trout and catfish the reaction occurs in hepatocytes [J311].

The activity in goldfish and bullhead liver is an order of magnitude lower than in rat [D465].

In sea anemone, the reaction is catalyzed by microsomes [J503].

Pycnoporus cinnabarinus enzyme is a laccase [J907].

Crinipellis, Marasmius and Marasmiellus (fungi) act on pyrene-1,6- and 1,8-diols to form the corresponding quinines [H879].

Phaseollin oxidation

Fusarium solani oxidizes phaseollin to 1α-hydroxyphaseollone, with the oxidation of one of the aryl rings to a 4-oxo-2,5-diene ring [A209].

Lipoxygenase as peroxidase

Glycine max lipoxygenase and peroxide oxidize 5-(S-cysteinyl)-L-dopa and 5-S-cysteinyldopamine. Both substrates form two types of phaeomelanin [J252].

Guaiacol peroxidase

Spruce needle chloroplast enzyme, a tetramer, monomeric molecular weight 41 000, pI 4.4 and optimum pH 5–6.8, may be a glycoprotein. The reaction it catalyzes is not stated in Chemical Abstracts, but it may yield *o*-quinone [J318].

Arachis peroxidase isozymes have optima at pH 4.4, 5.2, 5.6 and 6.4 [A2519].

Chloroperoxidase (E.C. 1.11.1.10) in Caldariomyces fumago is composed of two isozymes, molecular weights 46 000 and 40 000. They produce a brown pigment from guaiacol [A3764].

2-Amino-4-nitrotoluene oxidation

Phanerochaete chrysosporium manganese peroxidase (E.C. 1.11.1.13) oxidizes this compound to 4-nitro-*o*-quinone, with the release of methanol [G436].

Catechol formation with desulphonation

Both Alcaligenes and Pseudomonas putida desulphonate benzenesulphonate and toluene-*p*-sulphonate, forming catechol and 4-methylcatechol respectively [H297, J657].

Quinone methide formation

Sarcophaga bullata forms quinone methides from 4-alkylcatechols, (including 4-methylcatechol), 3,4-dihydroxyphenylethanol, 3,4-dihydroxyphenylacetic acid and N-acetyldopamine, with quinones as intermediates [C692, G184]. The quinone methides are unstable and further products are formed by attack at the α-carbon, to form α-hydroxy-3,4-catechols; in aqueous methanol some of the methoxy analogue is formed, suggesting that water, rather than a hydroxyl ion may be the other reactant [C692, F793]. In Sarcophaga bullata, Manduca sexta and Periplaneta americana N-acetylnoradrenaline is formed from N-acetyldopamine by this pathway [F68].

Sarcophaga bullata 4-alkyl-*o*-quinone: 2-hydroxy-*p*-quinonemethide isomerase, molecular weight 98 000 and optimum pH 6.0, is found in larval haemolymph. It acts on quinones of 3,4-dihydroxyphenylethanol, N-(β-alanyl)dopamine and N-acetyldopamine. A further attack on the quinones formed from these products form in turn the corresponding α-oxo products [F792, F793].

Mushroom tyrosinase converts 3,4-dihydroxymandelic acid into 3,4-dihydroxybenzaldehyde, possibly with a quinone methide intermediate [E963].

2.3 *Free radicals from phenols*

Hydroxybenzpyrenes

Rat liver oxidizes 1-, 2-, 4-, 5-, 6-, 7-, 8-, 9-, 10- and 11-hydroxybenzpyrene to free radicals. The formation of radicals from 1-, 4-, 9- and 11-hydroxybenzpyrene requires NADPH [B476].

2.4 *Oxidation of nuclear amino groups and analogues*

Formation of substituted hydroxylamines

$$R_1R_2NH + O_2 \rightarrow R_1R_2NOH$$

Dibenzylamine is substrate for liver microsomal enzyme in guinea pig, rabbit, hamster, mouse and chick, but it is virtually absent from cat liver. In rabbit the optimum pH is 6.9 with NADPH as cofactor, and 7.7 with NADH as cofactor. Inhibitors for one or both systems include cyanide, *p*-chloromercuribenzoate, FMN and to a lesser extent cysteine, N-ethylmaleimide, 1,10-phenanthroline, azide, imidazole and SKF 525-A [A1737].

Mouse and pig liver enzymes contain FAD, require NAD(P)H, with an optimum at about pH 9 [D446]. Substrates include N-methyl- and

N-ethylaniline and various secondary aliphatic amines, with oxygen as oxidizing agent. It is solubilized from microsomes with Triton X-45 and Triton X-102 [K160].

Aniline is substrate for a rabbit liver microsomal enzyme. Inhibition by acylation using diethyl pyrocarbonate is partially reversed by hydroxylamine, and is prevented by pyridine. Photo-inactivation occurs in the pH range 6–8 with rose bengal as photon acceptor [A1746].

Maize microsomes N-hydroxylate 2-hydroxy-1,4-benzoxazin-3-one, with optimum pH 7.5. The enzyme appears to be a NADPH-requiring P450 [G245].

There are many publications dealing with N-hydroxylation of primary amines of polynuclear hydrocarbons and their amides, in particular carcinogens, such as 1- and 2-aminonaphthalene, 4-biphenylamine and 2-aminofluorene. Few studies have been carried out at an enzyme level beyond determining that hydroxylation of 2-aminonaphthalene by pig liver microsomes requires oxygen and NADPH [A1791]. In addition, this reaction type has been observed in beef, dog, monkey, guinea pig, man, mouse and hamster; and bladder is a further organ in which it occurs [A880, B172, B179, C397, C727, D738].

N-Oxide formation (dimethylaniline monooxygenase (N-oxide-forming); E.C. 1.14.13.8)

$$R_1R_2R_3N \rightarrow R_1R_2R_3NO$$

Human foetal liver N,N-dimethylaniline-oxidizing activity requires NADPH. It is not inhibited by CO, which indicates that it is not a P450 enzyme [A1143].

N,N-Dimethylaniline is substrate for a rabbit liver microsomal enzyme. Inhibition using diethyl pyrocarbonate is partially reversed by hydroxylamine and is prevented by pyridine. Photo-inactivation occurs in the pH range 6–8 with rose bengal as photon acceptor [A 1746]. The enzyme system is composed of P448, NADPH-cytochrome c reductase and a lipid cofactor. The authors of this work postulate the existence of two N-oxide forming microsomal systems, one a haemoprotein and the other a flavoprotein oxidase. The flavoprotein is more stable at 37° and pH 7.4 than the P450, which, after solubilization is separable into two fractions, one of which is far less stable than the other. Both the P450 and the flavoprotein systems have similar activity [A3161, B230]. Lung activity, optimum pH 8.9, is three times greater than in liver. Lung enzyme is stimulated by Mg^{2+} and, surprisingly, by low concentrations of Hg^{2+}, whereas both these ions inhibit liver enzyme. Both are activated by 1 mM Ni^{2+} and inhibited by 10 mM Ni^{2+} [A1144].

Mouse liver dimethylaniline-oxidizing activity increases threefold just after birth, and then only slightly until 15 days post-partum. In males it is then near the adult level. In females it increases by a further 50% to the adult level by 25 days post-partum [A2498].

Mouse and pig liver microsomal enzymes, optimum pH about 9, contain FAD and require NAD(P)H. Substrates include N,N-dimethyl- and N,N-diethylaniline, and imipramine [D446]. Pig liver enzymes appears to be the same as alkylhydrazine N-oxidase [A1041].

Oxidation of *p*-phenylenediamines

This type of reaction has been extensively studied, but few have studied the reaction products. A free radical is likely to be the initial product, but in lieu of further information the product is recorded as a quinone in the compounds section for the sake of completeness, although this may be incorrect.

Beef lactoperoxidase acts on N,N-dimethyl-*p*-anisidine with ethyl hydroperoxide as oxidant in the presence of bromide. The product is *p*-quinone. It is considered that the peroxidase releases bromine, which is the oxidizing species [F500].

Nitrosomonas europaea enzyme, molecular weight 127 500, subunit molecular weight 40 100, and pI 4.63 contains copper. It utilizes oxygen in the oxidation of *p*-phenylenediamine, presumably to a free radical [D564]. Caeruloplasmin shows similar activity [A677]. Peach, spruce, Botrytis,

Trametes, and Coriolus versicolor laccases oxidize *p*-phenylenediamine [A1930, A3974, E369].

Oxidation of *o*-dianisidine

Peroxidases from Arachis oxidize *o*-dianisidine; the identity of the product is uncertain. Four isozymes are found with pH optima at 4, 4.4, 6, and 6.4 [A2519].

p-Benzoquinone imine formation

Human P450 oxidizes paracetamol to an active intermediate. Trapping with glutathione indicates that it is N-acetyl-*p*-benzoquinone imine [K330].

Oxidation of amino to nitroso groups

Human haemoglobin oxidizes *p*-chloroaniline and 3,4-dichloroaniline to nitrosobenzenes and nitrobenzene [C879].

Rat liver microsomes oxidize *p*-toluidine to 4-methylnitrosobenzene [G580].

Pea seed microsomes oxidize *p*-chloroaniline to *p*-chlorophenylhydroxylamine and *p*-chloronitrosobenzene, as well as to *p*-chloronitrobenzene; it was demonstrated that this represented a stepwise oxidation sequence. Nitrosobenzenes were also formed from aniline, *p*-toluidine, *p*-bromobenzene and 3,4-dichloroaniline [C797].

Chloroperoxidase, with peroxide, oxidizes *p*-chloroaniline to *p*-chloronitrosobenzene with optimum pH 4.4 [A3838]. A similar reaction is observed with a series of *para*-substituted anilines [A3579].

Oxidation of amino to nitro groups

$$R.NH_2 \rightarrow R.NO_2$$

Human oxyhaemoglobin oxidizes *p*-chloroaniline to *p*-chloronitrobenzene, apparently with *p*-chloronitrosobenzene as intermediate; the same reaction was observed with 3,4-dichloroaniline. The reaction is claimed to be a peroxidase type [C879].

Rat liver microsomes oxidize dapsone to 4-amino-4′-nitrodiphenylsulphone, and rat and chicken liver microsomes oxidize 4-amino-4′-ureidodiphenylsulphone to 4-nitro-4′-ureidodiphenylsulphone [A970].

Nitrosomonas oxidizes aniline to nitrobenzene [H219].

Serratia marcescens chloroperoxidase oxidizes 2- and 4-aminophenol, 2-, 3- and 4-chlorophenol, *p*-toluidine and *p*-aminobenzoate to the corresponding nitro compounds in high yield [K143].

Pseudomonas putida bromoperoxidase acts on aniline and peroxide to form azobenzene, azoxybenzene and nitrobenzene; it is suggested that this is the reaction sequence [H12].

Oxidation of hydroxylamines to nitro groups

Rat and rabbit liver oxidize N-hydroxyphentermine to 2-nitro-1-phenylpropane. It appears that superoxide is the oxidizing species; superoxide dismutase is inhibitory, and peroxide is not involved. It is thought that the reaction may be non-enzymatic [B654]. Rat liver microsomal enzyme is inhibited by haemoglobin, catalase or carbon monoxide [A3671].

Rabbit liver acts on N-hydroxyamphetamine [A3621].

Oxidation of hydroxylamino to nitroso groups

Rat haemoglobin oxidizes 4-hydroxylaminobiphenyl to 4-nitrosobiphenyl [G5].

Chloroperoxidase, with peroxide, oxidizes *p*-chlorophenylhydroxylamine to *p*-chloronitrosobenzene; the reaction is more rapid than with *p*-chloroaniline as substrate [A3838].

Oxidation of hydroxylamino to nitro groups

Ram seminal vesicle oxidizes 2-hydroxylaminofluorene to 2-nitrofluorene [C754].

Oxidation of nitroso to nitro groups

Ram seminal vesicle oxidizes 2-nitrosofluorene to 2-nitrofluorene [C754].

Mouse oxidizes 1-nitrosopyrene to 1-nitropyrene [E721].

Aniline conversion into azobenzenes

Horseradish peroxidase and Geotrichum candidum aniline oxidase both oxidize aniline and a range of substituted anilines, particularly those substituted with halogens. These are converted into azobenzenes [A661]. Geotrichum candidum aniline oxidase and peroxidase both oxidize aniline to azobenzene, optimum 5.0 (with oxygen) and pH 4.5 (with peroxide) respectively [A690].

Pseudomonas putida bromoperoxidase acts on aniline and peroxide to form azobenzene [H12].

Human placenta and Glycine max lipoxygenases act on 4-aminobiphenyl and linoleate to form 4,4′-azobis(biphenyl). A radical reaction appears to be involved [J95].

Ram seminal vesicle and horseradish peroxidase both oxidize 2-aminofluorene to 2,2′-azobisfluorene [C754].

Oxidation of azo to azoxy compounds

Azobenzene N-oxidase is found in (in reducing order) hamster, guinea pig, mouse, rabbit and rat. The reaction has been observed in liver, lung, heart and kidney and the activity in hamster liver is mainly microsomal. Substrates (in reducing order) are ω,ω′-azotoluene, ω-(phenylazo)toluene and azobenzene [G258, G259].

Nitrone formation

Hamster liver microsomes catalyze the formation of nitrones from some but not all benzanilides. These include benzanilide, 4-chlorobenzanilide, 4-methylbenzanilide, 2,4,6-trimethylbenzanilide, 2′,4′-dichlorobenzanilide and 2′,6′-dichlorobenzanilide [G951].

Nitroxide radical formation

$$R.NH_2 \rightarrow R.NH\text{–}O$$

Rabbit liver microsomes oxidize a range of carcinogenic amines including 2-aminofluorene, 2-aminonaphthalene, 2-aminoanthracene, 4-aminostilbene, 4-aminoazobenzene and benzidine as well as aniline, N-methylaniline and phenacetin to the corresponding nitroxide radicals. The reaction requires the participation of NADPH [B184].

2.5 *Oxidation of heterocyclic amines*

Hydroxyindole oxidase

An enzyme in Mytilus edulis oxidizes indoles such as serotonin, 5-hydroxytryptophan, psilocin and 5-hydroxyindole, presumably to pigments. Mono- and diphenols, such as adrenaline, L-dopa, tyrosine and *p*-coumarate are also substrates. The enzyme contains Cu and haem in a 1:1 ratio, and appears to require peroxide, but not a pyridine nucleotide. It is inhibited by copper-chelators [B314].

2.6 *Reduction of nitro groups and analogues*

Reduction of nitro compounds to anilines

$$R.NO_2 \rightarrow R.NH_2$$

This reaction often, if not always, involves the formation of intermediate nitroxides and hydroxylamines, although in many studies these intermediates were not detectable.

Human placenta reduces *p*-nitrobenzoate to *p*-aminobenzoate. Fractionation indicates that the active enzymes are haemoglobin or methaemoglobin. Studies with pure proteins found that the reaction can be catalyzed by haem-containing proteins such as methaemoglobin, metmyoglobin and cytochrome c, with NADPH as reducing agent. Placental reduction and the haem-model system (haematin, NADPH, FMN and *p*-nitrobenzoate) are inhibited by carbon monoxide [A621].

Formation of *p*-aminobenzoate from *p*-nitrobenzoate is observed in conventional rats, but is, relatively, very low in germ-free animals. Studies with microorganisms have found reduction by the gut flora Lactobacillus plantarum, Bacteroides fragilis, Peptostreptococcus productus, Clostridium, Proteus mirabilis, S. faecalis and E. coli [A124]. The reduction rate in rat gut wall and contents is reduced by treatment with oral antibiotics, whereas liver activity is unaffected [A3480].

Rat liver microsomal reduction of *p*-nitrobenzoate is inhibited by carbon monoxide and oxygen; presumably the enzyme is a P450. The reaction is enhanced four-fold by haemoglobin (boiled or native) and haemin. In contrast, azoreductase acting on neoprontosil is not activated in this manner [A3049].

Rabbit liver microsomal P450 reduces 1-nitropyrene and 2-nitrofluorene; the reaction requires NADPH (NADH is a poor cofactor). A cytosolic reductase has been identified as aldehyde oxidase, with electron-donating co-substrates. N′-Methylnicotinamide is best, but methotrexate, 2-hydroxypyrimidine and a series of aldehydes can also be used, but not NAD(P)H [C702]. Liver aldehyde oxidase (E.C. 1.2.3.1) also reduces 1-nitropyrene and 3-nitrofluoranthrene. FMN and FAD are alternative cofactors; oxygen is inhibitory. The intermediate nitroso and hydroxylamino analogues have been detected [G361].

p-Nitrobenzoate is reduced in the nematode Ascaris lumbricoides. Activity is found in the intestinal brush border but not in other tissues. The enzyme is cytosolic, it has a stability range of pH 4–7 (it rapidly loses activity at 0° outside this range), and has a sharp optimum at pH 6.5. It requires NADH, a low molecular weight thiol such as glutathione and a divalent cation such as Mn^{2+} or Cu^{2+}; NADPH is not a cofactor [A349]. Its molecular weight is about 130 000. A large range of nitrobenzenes and analogues are reduced to the corresponding anilines, and no intermediates are detected. This activity could not be separated from azoreductase by ammonium sulphate fractionation [A304].

Mouse and sheep liver, Ascaris suum and Moniezia expansa enzymes require NADH and GSH for the reduction of disophenol, nitroxynil and nitrodan [A3694]. The same species (except A. lumbricoides) reduce a range of nitrobenzanilides, all at about 10 per cent of the rate for *p*-nitrobenzoate [A2037].

Moniezia expansa enzyme acts on nitrobenzene and a range of substituted nitrobenzenes, with optimum pH 6.4–6.7; the preparation was free from xanthine and aldehyde oxidase activities. It was stimulated by hypoxanthine, and inhibited by FAD, FMN, riboflavin, allopurinol, dicoumarol, 5-nitro-2-furaldehyde, azide and cyanide [A1785]. The enzyme, molecular weight 125 000, is cytosolic, requires NADH (NADPH is not so good), and glutathione or cysteine. It is not inhibited by oxygen, carbon monoxide, EDTA or by azide [A1741]; the properties are so similar to those of azoreductase that they may be the same enzyme. M. expansa (pI 4.50) and Ascaris lumbricoides (pI 4.75) azo and nitroreductases could not be separated by isoelectric focussing [A1792].

Bacteroides fragilis contains four isozymes that reduce 1-nitropyrene to 1-aminopyrene. Isozyme I, molecular weight 52 000 and optimum pH 7.5, requires NADH for activity. The corresponding parameters, respectively, for isozyme II are 320 000, 6.3 and NADPH, for isozyme III 180 000, 7–8 and NADPH, and for isozyme IV 680 000, 5–9 and NADPH [C791].

E. coli contains four isozymes that reduce 1-nitropyrene to 1-aminopyrene, with NAD(P)H as co-substrate. Additionally, they appear to reduce FAD, and three of the isozymes reduce nitrofurazone [D217].

Reduction of nitro compounds to hydroxylamines

Spinach leaf ferredoxin: NADP oxidoreductase (E.C. 1.18.1.2) reduces nitrobenzene stoichiometrically to phenylhydroxylamine [J501].

Rat nitroquinoline-N-oxide reductase (E.C. 1.6.6.10), optimum pH 6.4, is highly specific. Supernatant enzyme (found in many tissues) requires NAD(P)H, and the minor, microsomal enzyme requires NADH, and utilizes two mol of co-substrate [K903].

Dog and human liver microsomes reduce N-hydroxysulphamethoxazole to sulphamethoxazole with NADH as co-substrate; NADPH is not so good. The enzymes from both species appear to have both low and high affinity components. They are not inhibited by oxygen, carbon monoxide, azide, or by other P450 inhibitors [K338].

Phenylhydroxylamine is reduced to aniline in rat and quail [B174, B175].

Reduction of nitroso compounds to hydroxylamines and amines

Human erythrocytes condense nitrosobenzene with glutathione to form glutathione sulphinanilide, a reaction that is probably non-enzymatic. This compound is reduced by two mol of NADPH to form aniline, as well as glutathione and glutathione sulphinate [B177, B228].

Rat liver reduces nitrosobenzene to phenylhydroxylamine and aniline [B174] and 1-nitrosonaphthalene to 1-naphthylhydroxylamine [A2173].

Horse liver alcohol dehydrogenase reduces 2-nitrosofluorene to 2-hydroxylaminofluorene and 2-aminofluorene, and *p*-nitrosophenol to *p*-aminophenol [H291].

Mouse reduces 1-nitrosopyrene to 1-aminopyrene [E721].

Reduction of hydroxylamines to amines

Mouse liver enzyme reduces 4-hydroxylamino-quinoline-1-oxide to 4-aminoquinoline-1-oxide. The activity is increased by administering molybdenum salts, and reduced by tungsten salts. It is suggested that, as with the other molybdenum-containing enzymes, xanthine oxidase and sulphite oxidase, tungsten slowly reduces the body load of molybdenum, thereby reducing the availability of molybdenum for the enzyme [A3227].

Reduction of hydroxylamides to amides

Rat liver microsomal reductase, which acts on N-hydroxy-2-acetamidofluorene to form 2-acetamidofluorene, requires NADPH. It is inhibited by oxygen and carbon monoxide; inhibition and induction studies suggest that it belongs to the P450 group, probably a P448. Activity is also found in lung, kidney and small intestine, and in decreasing order in hamster, guinea pig, rabbit, rat and mouse [A3965].

Horseradish peroxidase and horse cytochrome c catalyze a complex series of reactions with N-hydroxy-2- and 3-acetamidofluorenes as substrates. It is postulated that a free radical (apparently a nitroxyl radical, with the acetyl moiety still *in situ*) formed by the action of peroxide undergoes dismutation to form the corresponding nitroso, acetamido and N-acetoxyacetamidofluorenes [D32]. A similar series of complex reactions is found in rat and rabbit with N-hydroxy-2-acetamidofluorene, N-hydroxy-4-acetamidobiphenyl, N-hydroxyphenacetin and N-hydroxy-4-chloroacetanilide as substrates [B178].

Rabbit blood reduces both N-hydroxy-2-acetamidofluorene and N-hydroxyphenacetin to the parent amines, using NAD(P)H and FAD, and the reaction is inhibited by oxygen and carbon monoxide. Haemoglobin and catalase (native or boiled) catalyse the reaction; it is postulated that the reduction is a direct action of the haem moiety [K111].

N-Oxide reduction

Rat liver cytosolic 4-(N,N-dimethylamino)azobenzene-N-oxide reductase is a tetramer, molecular weight 370 000 and requires NADPH for oxygen removal; other N-oxides tested were not substrates. Lower activities are found in other tissues [E187].

A rat liver mitochondrial enzyme, bound to the inner membrane, requires NAD(P)H and reduces dimethylaniline-N-oxide as well as aliphatic N-oxides. The activity is also found in microsomes, but the relative reaction rates for the substrates are different from those in mitochondria. It is inhibited by oxygen, and partially by carbon monoxide [A1913].

Rat liver microsomal enzyme, apparently a P450, reduces dimethylaniline-N-oxide. It requires NADH, and is inhibited by oxygen or carbon monoxide. It also reduces imipramine-N-oxide [A1847].

A cytochrome c reduces dimethylaniline-N-oxide [B185].

Reduction of N-nitroso compounds

Rabbit liver converts N-nitrosodibenzylamine into bibenzyl, which is thought to be a breakdown product of an unstable 1-hydroxy-2, 2-dibenzylhydrazine [A3868].

Azobenzene reductase (E.C. 1.6.6.7)

$$Ar{-}N{=}N{-}Ar' \rightarrow ArNH_2 + Ar'NH_2$$

Rat microsomal enzyme reduces a range of substituted azobenzenes, including methyl red, *p*-methyl red and methyl orange [G511]. A liver enzyme that acts on methyl red but not on analogues lacking an *o*-carboxyl group is probably dimeric, monomeric molecular weight 30 000, and contains two mol FAD/mol [B28]. A reductase has been detected with an optimum at pH 6.2 [A2351]. Liver DT diaphorase (E.C. 1.8.1.4), molecular weight 30 000, reduces methyl red but not butter yellow [A2730].

Rabbit liver cytosol aldehyde oxidase (E.C. 1.2.3.1) acts as an azoreductase towards methyl red, amaranth, methyl orange and *p*-dimethylaminoazobenzene, with an electron donor such as acetaldehyde as second substrate [C884].

In perfused rat liver neoprontosil is reduced to sulphanilamide. It requires a pO_2 less than 200 mm for maximal activity, and above 400 mm the enzyme is inactive. This may explain previously reported low activities in liver [A3048].

Methyl red azoreductase is present in liver cytosol from rat, guinea pig, rabbit and hamster, but the activity in mouse and sheep liver is much lower. Microsomal enzyme activity is similar to that in cytosol, but is less variable between species. Particularly in rat, but less so in mouse, 3-methylcholanthrene induces both cytosolic and microsomal enzymes [B187].

Moniezia expansa enzyme acts on azobenzene and a range of substituted azobenzenes, with optimum pH 6.4–6.7; the preparation was free from xanthine and aldehyde oxidase activities. It is stimulated by hypoxanthine, and inhibited by FAD, FMN, riboflavin, allopurinol, dicoumarol, 5-nitro-2-furaldehyde, azide and cyanide [A1785]. The enzyme, molecular weight 125 000, is cytosolic, requires NADH (NADPH is not so good), and glutathione or cysteine. It is not inhibited by oxygen, carbon monoxide, EDTA or azide [A1741]; the properties are so similar to those of nitroreductase that they may be the same enzyme. M. expansa (pI 4.50) and Ascaris lumbricoides (pI 4.75) azo and nitroreductases could not be separated by isoelectric focussing [A304, A1792].

A Pseudomonas enzyme, molecular weight in the range 20 000–22 000 and optimum pH 6.2–6.8 depending on the substrate, requires NAD(P)H. Studies with 1-naphthols substituted at the 4-position with the azo moiety found that absence of the 1-hydroxyl or the presence of a hydroxyl in another position may reduce the activity to zero. Phenolic analogues are also inactive [D180].

Cautobacter subvibrioides enzyme has a molecular weight of about 30 000 [K314].

Shigella dysenteriae contains two azoreductases. One is a dimer, molecular weight 28 000, and the other a monomer, molecular weight 11 000. Both require NAD(P)H and are flavoproteins containing one mol FMN/mol. Substrates are Ponceau SX, Tartrazine, Amaranth and Orange II [G767].

Azide reduction

In man, the azo group in azidomorphine is reduced to an amino group [A3050].

Mouse reduces *m*-azidopyrimethamine to the *m*-amino analogue [F319].

Substituted hydroxamates from nitroso compounds

In this reaction nitrosobenzenes are converted into N-substituted hydroxamates; the oxo group is reduced to a hydroxyl, with N-conjugation.

Rat liver enzyme is associated with an organelle larger than microsomes (mitochondria?), and requires glucose-6-phosphate as the energy source. 4-Chloronitrosobenzene forms N-(4-chlorophenyl)glycollhydroxamate [B130]. A similar reaction occurs with aryl-substituted nitrosobenzenes and several 2-oxoacids such as pyruvate, with elimination of the carboxyl group and the formation of N-aryl-N-hydroxyacylamides. For instance, 2-oxoisocaproate and nitrosobenzene form N-hydroxy-N-phenylisobutyramide [J429].

In Chlorella both nitrosobenzene and phenylhydroxylamine are substrates, but it appears that phenylhydroxylamine conversion to nitrosobenzene occurs first. Both N-phenylacetohydroxamate and N-phenylglycollhydroxamate are formed. It is postulated that 'active acetaldehyde' and 'active glycollaldehyde' (presumably bound to thiamine) are involved as co-substrates [A3531].

Yeast enzyme appears to be a pyruvate decarboxylase (E.C. 1.2.4.1), with α-hydroxyethylthiamine as coenzyme [A2344].

2.7 *Oxidation and reduction of sulphur atoms*

Few of the reactions in this section have been studied at an enzyme level. The examples quoted for each reaction are purely illustrative.

Oxidation of thioethers to sulphoxides

$$\mathrm{R{-}S{-}R'} \rightarrow \mathrm{R{-}\overset{\overset{\displaystyle O}{\|}}{S}{-}R'}$$

A soluble enzyme that oxidizes chlorpromazine is found in guinea pig liver and serum, but only in small amounts in kidney, spleen and pancreas; it is not a P450. The molecular weight is 85 000, with an optimum at pH 7.4. It is rapidly inactivated at 50° [A1353].

In mouse liver, Ascaris suum and Moniezia expansa fenbendazole, albendazole, bithionol and phenothiazine are oxidized to sulphoxides, optimum pH 7.0–7.2 [A3849].

Mouse and pig liver microsomal FAD-containing monooxygenases, optimum pH about 9, require NAD(P)H, and oxidize nitrogen and sulphur atoms. Substrates include thioanisole, benzyl methyl sulphide, diphenyl sulphide, benzyl disulphide, benzyl mercaptan, 2-mercaptobenzimidazole, phenylthiocarbamide and thiocarbanilide [D446].

Rabbit (the enzyme source may be liver) flavin-containing monooxygenase acts on *p*-tolyl and 2-naphthyl thioethers with methyl or ethyl thio substituents. The reaction rate rapidly falls as the length of the alkyl group increases [H484].

Rat liver microsomal albendazole monooxygenase (E.C. 1.14.13.32) contains FAD; it is not a P450 [K749]

An enzyme (source unclear) converts promazine and chlorpromazine into cation radicals which disproportionate, with sulphoxides as one product [K386].

Horseradish peroxidase forms sulphoxides from thioanisole and *para* substituted thioanisoles, with peroxide as co-substrate. The substituents were methyl, chloro, acetamido, acetoxy, nitrile and nitro groups [E403].

Oxidation of sulphoxides to sulphones

Many sulphur-containing drugs are oxidized to the corresponding sulphones, and the intermediate sulphoxides are sometimes detected.

In rat and dog liver butaperazine is oxidized to the sulphone, a sulphoxide is the presumed intermediate [A295].

In mouse liver, Ascaris suum and Moniezia expansa, albendazole sulphoxide is further oxidized to the sulphone [A3849].

Dimerization of thiophenol (thiol oxidase; E.C. 1.8.3.2)

ArSH → ArSSAr

Horseradish peroxidase and mushroom tyrosinase oxidize thiophenol to diphenyl disulphide [J593].

Reduction of sulphoxides

Many sulphinyl analogues of drugs and xenobiotics, unlike sulphones, are reduced to the parent thio compound, for instance N-methylsulphinylbenzamide in rat [D705], and sulphinylpyrazole in mammalia and microorganisms [B711, E241, H560].

In mouse liver, Ascaris suum and Moniezia expansa the sulphoxides of fenbendazole, albendazole, bithionol and phenothiazine are reduced to the parent compounds, optimum pH 7.2–7.4 [A3849].

Reduction of disulphide bond

Rat liver mitochondria reduce 5,5′-dithiobis(2-nitrobenzoate) to 5-mercapto-2-nitrobenzoate [D611].

Methylthio incorporation

Rat liver converts N-hydroxy-2-acetamidofluorene into 2-amino-3-methylthiofluorene, with N-acetylmethionine as co-substrate. The first reaction appears to be:

$$\mathrm{R{-}\underset{\displaystyle OH}{\underset{|}{N}}{-}COCH_3 \rightarrow R{-}NH{-}O{-}COCH_3}$$

The product then reacts with N-acetylmethionine to form a sulphonium condensate, with the sulphur atom bound to position 3 of the fluorene moiety, which then, by fission of the S–C bond in the methionine moiety forms 2-amino-3-methylthiofluorene. A similar reaction is observed with N-hydroxy-4-acetamidobiphenyl as substrate [A729].

2.8 *Reduction of quinones and analogues*

Quinone reductases (NAD(P)H: quinone oxidoreductase, E.C. 1.6.99.2, 1.6.99.5 and 1.6.99.6)

Quinone ↔ quinol

Human erythrocyte NAD(P)H dehydrogenases reduce *p*-quinone, but rather poorly [B260].

Rat liver mitochondrial reductase forms the quinol from 5-demethylubiquinone 9, with NADH as co-substrate [A3084].

Guinea pig lens $NADPH_2$: quinone reductase (E.C. 1.5.5.5) is ζ-crystallin, which requires oxygen and NADPH, with peroxide and superoxide as products. A range of quinones and 2,6-dichlorophenolindophenol are substrates [K773].

Ubiquinol cytochrome-c reductase (E.C. 1.10.2.2) has been reviewed [K816]. Beef heart enzyme is composed of many isozymes, seven of which have molecular weights in the range 4000–29 000 [K930].

Pig liver mitochondrial electron-transferring flavoprotein dehydrogenase (E.C. 1.5.5.1), optimum pH 5.2 and molecular weight 69 000–73 000 (dependent on method) acts on ubiquinone. Its amino acid composition has been determined [K814].

Sporotrichum pulverolentum enzyme reduces a series of quinones including *p*-quinone, toluquinone, methoxyquinone, 1,4-naphthoquinone, menadione and 4,5-dimethoxy-*o*-quinone [B301].

Phanerochaete chrysosporium enzyme is soluble, possibly a homodimer, molecular weight 44 000, optimum pH 5.0–6.5 and pI 4.3, which contains flavin mononucleotide and requires NAD(P)H. Substrates are 2-methoxy- and 2,6-dimethoxy-*p*-quinone, and menadione [H611].

Hydrogen: quinone oxidoreductase
(E.C. 1.12.99.3)

Wolinella succinogenes enzyme utilizes molecular hydrogen to reduce 2,3-dimethyl-1,4-naphthoquinone. This cytoplasmic membrane enzyme is composed of three proteins including cytochrome b [K772].

Benzpyrene-3,6-quinone reduction

Rat liver microsomes reduce this quinone to benzpyrene-3,6-diol, with NADPH as co-substrate [A3637].

Rabbit liver morphine 6-dehydrogenase reduces polynuclear quinones with NAD(P)H as coenzyme, but does not reduce *p*-quinones [H920].

Dichlorophenolindophenol (quinone imine) reduction (DT diaphorase; NAD(P)H dehydrogenase (quinone), E.C. 1.6.99.2)

$$\text{2, 6-Dichloroindophenol} \rightarrow \text{N-(3, 5-dichloro-4-hydroxy)-4-hydroxyaniline}$$

Two human erythrocyte NAD(P)H dehydrogenases, molecular weight about 18 000, reduce a range of substrate types including 2,6-dichlorophenolindophenol [B260]. The rat liver enzyme exhibits complex kinetics, which possibly involves two separate active sites and the interconversion of heat- and cold-stable forms [A2223].

Rat liver cytosolic quinone reductase, which requires NAD(P)H, reduces N-acetyl-*p*-benzoquinone imine, 2-amino-1,4-naphthoquinone imine, N,N-dimethylindoaniline and 2-acetamido-N,N-dimethylindoaniline [E214].

Bacillus stearothermophilus NADH: dichlorophenolindophenol oxidoreductase is a flavoprotein, molecular weight 43 000, activated by FMN or cyanide. It is inhibited by EDTA or *p*-chloromercuribenzoate, but it is stable up to 70° [B725].

Phanerochaete chrysosporium benzoquinone reductase also exhibits this activity [H611]. This reaction has been detected in Escherichia [A1228], beef [A1899], Diphyllobothrium [A679], Pseudomonas [A2952], Saccharomyces cerevisiae [K229], guinea pig [K773] and Eubacterium [B426].

Xanthommatin reductase (E.C. 1.3.1.41)

Drosophila melanogaster enzyme, which utilizes NADH, forms 5,12-dihydroxanthommatin by reduction of the quinoneimine moiety [E192].

2.9 *Halogens and the aromatic nucleus*

Halogenation

The thyroid hormone thyroxine is formed by the incorporation of iodide into tyrosyl residues of the protein thyroglobulin; this is explored in an extensive early literature.

Human neutrophil myeloperoxidase acts on ticlopidine both to dechlorinate and to introduce a further chlorine in the 2 position [K53].

Iodinating enzyme present in goat submaxillary gland appears to be iodide peroxidase (E.C. 1.11.1.8), optimum pH 4. It iodinates L-tyrosine and L-3-iodotyrosine, as well as exchanging labelled iodine in 3,5-diiodotyrosine with inorganic iodine [A997].

Glycine max seed coat peroxidase acts on veratryl alcohol, with bromide and peroxide, to form 2-bromo-4,5-dimethoxybenzyl alcohol quantitatively, with optimum pH below 2.5.

Pyrazole and indole are also brominated; iodide but not chloride can substitute for bromide [K439].

Bromoperoxidase (non-haem type) from the red alga Corallina pilulifera catalyses a variety of reactions. Anisole is brominated at all the nuclear positions, 1-methoxynaphthalene is 4-brominated, phenol and salicyl alcohol yield 2,4,6-tribromophenol. In addition styrene and analogues are brominated on the side-chain [E597].

Ulva lactuca (a green marine alga) bromoperoxidase forms 2,4,6-tribromophenol from phenol, and for a series of phenols with carbon side chains at the *ortho* and *para* positions (listed below) the side chains are replaced by bromine. Phenol forms *o*- and *p*-bromophenol, 2,4- and 2,6-dibromophenol as intermediates. In contrast, *o*-hydroxybenzyl alcohol forms *o*-, but little *p*-bromophenol. *p*-Hydroxybenzaldehyde forms *p*-bromophenol. *p*-Hydroxyphenylacetate forms a little *p*-bromophenol, a trace of 2,4-dibromophenol, but no 2,6-dibromophenol. *p*-Hydroxybenzoate and *p*-hydroxybenzyl alcohol form *p*-bromophenol and 2,4-dibromophenol, but no *o*-bromophenol or 2,6-dibromophenol [K153].

Pseudomonas putida bromoperoxidase acts on aniline with bromide and peroxide to form *o*- and *p*-bromophenol; other anilines are not substrates. In the absence of bromide azobenzene, azoxybenzene and nitrobenzene are formed [H12].

Streptomyces venezuelae bromoperoxidase-catalase (a homogeneous enzyme), molecular weight about 130 000, appears to be a dimer, pI 4.5 and optimum pH 6.9. It brominates 2-(3,5-dibromo-2-methoxyphenyl)-1-methylpyrrole to form one pentabromo and two tetrabromo analogues. The position of the bromo substituents is not known, but some appear to be on the aromatic nucleus [F276].

Chlorination by molecular chlorine

Myeloperoxidase in human neutrophils generates molecular chlorine with peroxide and chloride; the chlorine reacts with L-tyrosine to yield 3-chloro-L-tyrosine [J226]. The potential scope of this reaction for synthetic chemistry is enormous, in which free halogens can react with a large range of compounds under controlled conditions.

Deiodination of thyroxine and analogues (without hydroxylation; E.C. 1.11.1.8 and 3.8.1.4)

Rat liver type I enzyme deiodinates thyroxine and a series of partially deiodinated analogues and their O-sulphate and N-sulphonate conjugates [G337]. Deiodination of the outer ring of thyroxine and analogues is optimal at pH 6–6.5; inner ring deiodination is optimal at above pH 8.5, whereas inner ring deiodination of T_3 ocurs at pH 8 [A3752]; a second apparent optimum for rT_3 of 4.5 with no apparent formation of halogenated products at physiological pH is an experimental artifact arising from the rapid disappearance of rT_3 at physiological pH under the conditions used [A3117, A3549]. A monomeric 5′-deiodinase, molecular weight 56 000 appears to be a metalloenzyme, possibly containing iron [F821]. The observed similarities between 5′-deiodination of rT_3 and rT_2 in rat liver suggest that a single enzyme is involved [B833]. No cofactors were detected in one study, nor was oxygen involved. Inhibition has been observed with both particulate and cytosolic enzymes using Zn^{2+} or propylthiouracil, but not with α-methyltyrosine [A3312, A3549, A3889].

Solubilized liver microsomal 5′-deiodinase has a molecular weight of 55 000 and pI 5.7 [F802]. Another study claims that both 5- and 5′-deiodinase activities are entirely microsomal [B504]. Parenchymal cell microsomal enzymes convert thyroxine into T_3 and rT_3, both of which are further converted into 3,3′-T_2, the deiodination of rT_3 being by far the fastest [B126]. Another study reports that the molecular weight of solubilized microsomal enzyme is 65 000 or 200 000, depending on the method; the reason for the differencies between these results is obscure [B427]. A further study on microsomal enzyme found that it is a lipoprotein that requires Fe^{2+}, but it is not a P450 [A1095].

A rat enzyme that deiodinates T_2 to T_1 is found mainly in liver. Some is microsomal, but most of the activity is associated with the plasma membrane. It is stimulated by dithiothreitol and inhibited by oxygen. It is claimed that its properties closely resemble those of 5′-deiodinase [C108].

Liver, thyroid and kidney enzymes that 5′-deiodinate iodothyronines contain one mol of selenium/mol, presumably present as selenocysteine [F882]. Astrocyte type III iodothyronine deiodinase, which removes an iodine from the inner ring of triiodothyronine is also a selenoprotein [J197].

Glial cell Type I deiodinase is a dimer, with monomeric molecular weight of 27 000, and Type II a hexamer, monomeric molecular weight 29 000. Both are membrane-bound [G121].

5′-Deiodinase from rat cerebral cortex appears to involve two activities, one with low affinity and the other with high affinity [B914]. Its optimum pH is 8.0, and it requires dithiothreitol. Its distribution through the brain is different from that of thyroxine and T_3 5-deiodinases [C48].

Rat pineal 5′-deiodinase has an optimum pH of 6.5 [E104].

Rat 5′-deiodinase (type 3) activity is found in the pituitary; the enzyme acting on thyroxine may be different from the one that deiodinates rT_3 [D79]. In pregnant rat uterus, Type 3 deiodinase at day nine of pregnancy is found in mesometrial and antimesometrial decidual tissue. At days 12 and 13 it is localised to epithelial cells lining the uterine lumen [K69].

Human kidney microsomal thyroxine 5′-deiodinase has an optimum at pH 6.5. Another enzyme, a 5-deiodinase, forms rT_3 [E193].

In beef cattle the conversion of T_4 into T_3 is most active in liver, with lesser activity in kidney. Trace activity is found in muscle [D461].

In monkey hepatocarcinoma deiodination of the outer ring of thyroxine and analogues is optimal at pH 6.3; inner ring deiodination is optimal at pH 7.9; activity is enhanced by dithiothreitol and inhibited by Zn^{2+} [A3889].

Reductive dehalogenation

3-Chloro-4-hydroxybenzoate → *p*-hydroxybenzoate

Desulfitobacterium chlororespirans dehalogenase, optimum pH 6.5, fulfils the entire energy requirements for the growth of this organism. The best co-substrate reductant is reduced methylviologen. It dechlorinates a series of chlorinated benzoates, phenylacetates and phenols *ortho* to the hydroxyl group [J232]. D. dehalogenans enzyme, optimum pH 8.2, also requires reduced methyl viologen as electron donor. It acts on a range of chlorinated phenols, and converts 3-chloro-4-hydroxyphenyl acetate into 4-hydroxyphenyl acetate [K247].

Desulfomonile tiedje enzyme acts on *m*-chlorobenzoate to yield benzoate, and on analogues with various substituents in the 5- and 6-positions with reduced methylviologen as reductant. It is a heterodimer, subunit molecular weights 64 000 and 37 000 and optimum pH 7.2. It may be a haem protein [G152, H642].

Tetrachloroquinol reductive dehalogenase

In Phanerochaete chrysosporium dehalogenation takes place in two steps. The first is the replacement of a chlorine atom with glutathione, and the second, catalysed by glutathione conjugate reductase, removes the glutathione moiety, with glutathione or another thiol as co-substrate. The first step is catalyzed by a membrane-bound enzyme, optimum pH 6.3, and the second enzyme is cytosolic, optimum pH 8.3. Other substrates are trichloroquinol and 2,6-dichloroquinol, which gives a pathway for serial dechlorination [K138].

Dehalogenation of other xenobiotics

Human neutrophil myeloperoxidase dechlorinates ticlopidine [K53].

2.10 *Epoxide formation and reduction*

Epoxide formation

Epoxides are key intermediates in the metabolism of aromatic polynuclear hydrocarbons in animals, and in the formation of DNA adducts that lead to carcinogenesis. They are very reactive and usually cannot be detected, except as their degradation products, such as *trans*-dihydrodiols and glutathione conjugates. Most of the published information on the pathways can be found under dihydrodiol formation or the *See also* entries under the oxide formation from the parent hydrocarbons in Part 1.

Rat liver cytochrome P450c catalyses the formation of naphthalene and anthracene 1,2-oxides. Trapping experiments indicate that the (+)-(1*R*,2*S*) enantiomers are the predominent products [D261]. However, cytochrome P450b forms the (1*S*,2*R*) oxides predominently from these compounds [D792]. P450c forms 5,6- and 8,9-oxides from benz[a]anthracene. Liver microsomes form principally (+)-(3*S*,4*R*)- and (+)-(5*S*,6*R*)-oxides from benzo[*c*]phenanthrene [E748]. Pyrene forms pyrene-4,5-oxide [A739].

Rat lung and liver form the 4,5-oxide from benzo[*a*]pyrene and 5,6-oxide from benz[*a*]anthracene and 7-methylbenz[*a*]anthracene [A70].

Rabbit lung and guinea pig liver microsomes form 4,5-oxide from benzo[*a*]pyrene [B286, B812].

Benzene yields benzene oxide in rat; it is found in blood [J351].

Carbamazepine forms a 10,11-epoxide in man [A3122].

The role of epoxides as mandatory intermediates in the formation of phenols in xenobiotics, including polynuclear hydrocarbons, has not been unambiguously demonstrated. Epoxides have been suggested as intermediates in the hydroxylation of bromobenzene by rat liver microsomes, involving a non-enzymatic step. Dihydrodiols formed from epoxides are dehydrated to phenols under acidic conditions, often used in extraction of metabolites from urine [B825]. Acetanilide hydroxylation has been considered to involve an epoxide [A3626]. On the other hand, a range of hydroxylations in Nitrosomonas europaea by ammonia monooxygenase involve a NIH shift mechanism, but the reaction is considered not to involve epoxide formation [H631].

6β-Hydroxyhyoscyamine epoxidase (E.C. 1.14.11.14) is described in section 1.3.

Vitamin K epoxidase

Human liver enzyme shows high activity at 10–30 weeks gestation, and then declines to the adult level at birth [K414].

Epoxide reduction

Rat liver microsomes reduce benzpyrene-4,5- and 7,8-oxides to the parent hydrocarbon, with NADP as co-substrate. Activity is enhanced in 3-methylcholanthrene-treated rats, and further increased by riboflavin. Milk xanthine oxidase also catalyzes the reaction, with enhancement by riboflavin. Other oxides, including styrene oxide, show little or no reduction [B558]. Rat liver reductase system is composed of cytochrome P448, NADPH, a reductase and a lipid fraction. Methylcholanthrene-induced P448 can be replaced less effectively by P450, either from methylcholanthrene- or phenobarbital-treated animals. The purified reductase used was free from P450 [A3527]. Activity is stimulated by FMN or methylviologen and inhibited by oxygen, dimethylaniline-N-oxide, cumene hydroperoxide or carbon monoxide. It is absent at birth, but the activity develops by 28 days post partum to about 50 per cent of that found at 49 days [A2601]. Other substrates include the 5,6-oxides of benzanthracene, 7-methylbenzanthracene, 7-hydroxymethylbenzanthracene and 7,12-dimethylbenzanthracene, benzanthracene-8,9-oxide and 3-methylcholanthrene-11,12-oxide. A dihydromonohydroxy compound is not an intermediate [A350].

Vitamin K epoxide reductase
(E.C. 1.1.4.1 and 1.1.4.2)

Human liver enzyme activity is lower than adult at 10–30 weeks gestation, it is then very variable until birth, after which it settles to the adult level [K414].

2.11 *Deamination of arylamines without hydroxylation*

Desulphovibrio converts aniline into benzene; the amino group is released as ammonia. 2,4-Dinitrophenol forms phenol with the release of ammonia but not nitrite; reduction to 2,4-diaminophenol appears to be the initial step [G827].

2.12 *Dehydroxylation of phenols*

a. Dehydroxylation

Many studies have detected the removal of the *para* hydroxyl group in catechols, usually by gut flora. Little information is available about the enzymes involved. Dehydroxylation of dihydrocaffeate occurs, for instance, in conventional rats, but not in germ-free animals [A557]. Rats, fed with L-dopa and dopamine form small amounts of *m*-hydroxy analogues and metabolites [A2961]; the excretion of *m*-hydroxy metabolites formed from dietary plant catecholic compounds is dramatically reduced in germ-free rats compared with conventional rats, indicating that gut flora are involved [H73]. However, studies with rat brain striatum have shown that trace amounts of dopamine are converted into *m*- and *p*-tyramine [C465].

A denitrifying bacterium converts salicylate into benzoate. Enzymes have been detected that catalyze salicylate CoA ligation and reductive dehydroxylation of salicyloyl CoA as part of the reaction sequence [J213].

A mixture of E. coli and S. faecalis converts pyrogallol into resorcinol; pure cultures did not catalyze the reaction. The activity is inducible [E113].

Human faeces dehydroxylate phloretate to phenylpropionate [A2767].

b. Apparent intermolecular hydroxyl transfer

Pelobacter acidigallici forms phloroglucinol from pyrogallol in the presence of 1,2,3,5-tetrahydroxybenzene stoichiometrically. It is considered that the reaction sequence is:

$$\text{Pyrogallol} + 1,2,3,5\text{-tetrahydroxybenzene} \rightarrow 1,2,3,5\text{-tetrahydroxybenzene} + \text{phloroglucinol}.$$

A similar reaction occurs with other phenols; some of the listed products have not been fully characterized:

$$1,2,3,5\text{-Tetrahydroxybenzene} \rightarrow \text{phloroglucinol} + \text{pentahydroxybenzene}$$

$$1,2,3,5\text{-Tetrahydroxybenzene} + \text{hydroxyquinol} \rightarrow \text{phloroglucinol} + 1,2,4,5\text{-tetrahydroxybenzene}$$

$$1,2,3,5\text{-Tetrahydroxybenzene} + \text{resorcinol} \rightarrow \text{phloroglucinol} + \text{hydroxyquinol}$$

$$\text{Pyrogallol} + \text{hydroxyquinol} \rightarrow \text{resorcinol} + 1,2,4,5\text{-tetrahydroxybenzene}$$

$$\text{Hydroxyquinol} \rightarrow \text{resorcinol} + 1,2,4,5\text{-tetrahydroxybenzene}\ [\text{F434}].$$

Dihydrophloroglucinol formation appears to be the first reaction step with phloroglucinol as substrate; the enzyme for this step, optimum pH 7.2, requires NADPH [E189].

Eubacterium oxidoreducens pyrogallol-phloroglucinol isomerase (E.C. 1.97.1.2) converts pyrogallol into phloroglucinol [E388]; the reaction requires 1,2,3,5-tetrahydroxybenzene [G772]. The reaction is apparently anaerobic, optimum pH 7.3, and requires dimethyl sulphoxide, which is converted into methyl sulphide. Hydroxyquinol forms resorcinol and 2,6-dihydroxy-*p*-quinone. The reaction sequence proposed is:

Pyrogallol → 1,3-dihydroxy-2-oxocyclohexa-3,5-diene → 3-hydroxy-*o*-quinone → 3,5-dihydroxy-1,2-dioxocyclohex-3-ene → 1,2,3,5-tetrahydroxybenzene → 2,3,5-trihydroxy-1-oxocyclohexa-3,5-diene → 1,2-dihydroxy-3,5-dioxocyclohexane → phloroglucinol

This mechanism may be different from that for Pelobacter [E507].

Penicillium simplissimum converts phloroglucinol into a mixture of hydroxyquinol and resorcinol. It is proposed that dihydrophloroglucinol is an intermediate [F639].

2.13 *Nitro group addition and removal*

Nitro substitution

Phanerochaete chrysosporium lipid peroxidase with peroxide and tetranitromethane acts on veratryl alcohol to form several products, which include 3,4-dimethoxynitrobenzene and 4,5-dimethoxy-2-nitrobenzaldehyde. The reaction appears to be the result of the formation of nitro radicals, which then nitrate with or without displacement of the side chain [J624].

Mouse brain forms nitrotyrosine from tyrosine; nitric oxide synthase (E.C. 1.14.13.39) appears to be responsible [K385].

Removal of nitro groups

In rat 2,3,5,6-tetrachloronitrobenzene forms 1,2,4,5-tetrachlorobenzene. The mechanism is unclear, but conjugation with glutathione may be involved [H512]. 6-Nitrobenzpyrene yields benzpyrene; this reaction occurs in liver [D207, E252] and is also observed in mouse [E971].

3. Oxidations and reductions of substituent side chains and non-aromatic ring systems (without altering chain length)

3.1 *Hydroxylation of the carbon side-chain*

Dopamine-β-hydroxylase (dopamine-β-monooxygenase; DBH; E.C. 1.14.17.1)

$$Ar.CH_2.CH_2NH_2 + \text{ascorbate} + O_2 \rightarrow Ar.CHOH.CH_2NH_2 + \text{dehydroascorbate}$$

Substrates for this enzyme are phenethylamines, including tyramine and dopamine. The physiologically important substrate dopamine yields the neurotransmitter noradrenaline.

Human serum enzyme requires ascorbate and is stimulated by fumarate, N-ethylmaleimide and by low concentrations of copper. Two isozymes are found, molecular weights 368 000 and 188 000, both with optima at pH 5.0. The K_m for the smaller isozyme is 10 times greater than for the larger isozyme [A3375].

Human umbilical cord plasma activity is about 2 per cent of the adult level [A3403]. Plasma DBH activity shows a familial correlation [A1336]. A diurnal rhythm has been observed for the serum enzyme, in which bed rest is partially responsible for a drop in activity. On a normal regime, a rise of 10 per cent after waking is followed by a steady maximal activity during the afternoon and a decline in the evening and night [A1245]. Measurements of plasma enzyme in different human subjects by immunoassay shows a three-fold range, whereas measurement by activity shows a 150-fold range, suggesting that much of the enzyme is present in an inactive form [A1158]. Dialysis increases activity, suggesting the presence of a low-molecular weight inhibitor. Pure beef enzyme is inhibited by boiled human plasma, suggesting that this inhibitor is heat stable [A1956].

Human adrenal enzyme does not act on dopamine 3-or 4-sulphate [K459].

Beef adrenal enzyme, molecular weight 290 000, exhibits an optimum at pH 5.0 [C387]. However, a further study claims a molecular weight of 290 000 in the pH range 4–11, and that urea can cause it to dissociate into three inactive subunits [A885]. According to another study the molecular weight is 130 000, and is probably a homotetramer composed of two pairs of monomers that are bound by disulphide bridges [A1093]. The dimer and tetramer are reversibly interconverted, polymerisation (from molecular weight 145 000 to 290 000, with an increase in Stokes radius from 5.8 to 6.9 nm) is enhanced by increasing the pH from 5.0 to 5.7. With tyramine as substrate, changes in activity correlate with polymerization of the enzyme [K157]. Since the pH of chromaffin granules is 5.6, this polymerisation is probably physiologically important. Dissociation of the tetramer into the dimer is caused by firm binding to ADP; this is independent of pH between pH 5 and 7. At pH 5.5 the K_m for both ascorbate and tyramine is lower than at neutrality without affecting V_{max} [K154]. The enzyme contains two Cu/subunit, but it is not stimulated by Ni^{2+}, Co^{2+}, Mn^{2+}, Fe^{3+} or Zn^{2+}; the amino acid composition has been measured [A885, K156]. Medullary enzyme has an optimum pH 4.8–5.5, and this is sharpened, moved and skewed to a lower pH as the concentration of tyramine is raised from 0.2–10 mM [A1008].

All four subunit chains from chromaffin granule enzyme, solubilized with sodium dodecyl sulphate, contain about 4.8 per cent of carbohydrate, most of which is mannose. In consequence, it is able to complex with concanavalin A from which it can be displaced by α-methyl-D-mannoside; this property has led to a purification procedure [A332, K193].

The enzyme has been detected in cerebrospinal fluid [A2838] and in lymph [A1136].

Beef enzyme is released from adrenal by treatment with acetylcholine [A3532]; it is present in the cytoplasm of all medullary cells [A1409], where it is partially (1/3) membrane-bound, with the remainder soluble. These forms are immunologically identical. The soluble enzyme appears to be homogeneous, and the insoluble to be represented by two forms. Partial proteolysis of the insoluble form releases enzyme that has the same properties as the soluble form [A1907, A3580].

Beef adrenal enzyme is inhibited by the anti-thyroid agents methimazole and propylthiouracil, by thiouracil and 2-mercaptoimidazole [A1918]. Thioureas inhibit by several mechanisms; N-phenyl-N′-3-(4*H*-1,2,4-triazolyl)thiourea and N-(*n*-butyl)-N′-3-(4*H*-1,2,4-triazolyl)thiourea both inhibit noncompetitively relative to substrate, but the former is a mixed type inhibitor with respect to ascorbate, and Cu^{2+} does not reverse the inhibition. Inhibition by N-(*n*-butyl)-N′-3-(4*H*-1,2,4-triazolyl)thiourea is noncompetitive relative to ascorbate, and inhibition is reversed by Cu^{2+} [A797]. Catalase protects against denaturation, and diethylpyrocarbamate, a histidine-binding compound, inactivates it. It is inhibited by diethyldithiocarbamate [A1097]. Incubation with cysteine inactivates it irreversibly. Although the mechanism has not been identified [A800], reduction of disulphide bridges possibly followed by bonding with cysteine may occur. It is not stimulated by N-ethylmaleimide [A3375].

Beef adrenal enzyme acts on D-amphetamine, forming *l*-norephedrine, thus removing the *pro-R* hydrogen [A1101], which is replaced by a hydroxyl group while retaining the configuration, i.e., there is no Walden-type inversion [A1410]. Isotope effects indicate that β-C–H cleavage is the rate-limiting step, with a secondary, possibly steric effect from the α-CH [A3689]. The enzyme acts on 2-indanamine, which is a cyclic analogue of phenethylamine in which rotation at the hydroxylation site is prevented. The product is almost stereochemically pure *trans*-(1*S*, 2*S*)-2-amino-1-indanol [J666]. The reaction rate is reduced by 50 per cent in D_2O, possibly by D exchange in a histidyl residue involved in H transfer [A330]. Mechanistic studies suggest that a ping-pong mechanism is not involved; it is suggested that a ter bi sequential type is more likely, with electron donor, oxygen and substrate adding in that sequence [A1465]. Adrenal enzyme activity is enhanced by a superoxide-generating system, and is inhibited by superoxide dismutase [A1163]. Activation by fumarate or sodium chloride shifts the optimum pH from about 6.2 to 7, with increased activity at the optimum. This is thought to be due to anion binding at a basic group adjacent to an ionizable group at the active site [A1093]. A study in the absence of fumarate suggested that ascorbate and tyramine add sequentially; the apparent ping-pong mechanism only occurs at saturating concentrations of fumarate (acetate or chloride show a similar effect) [K189].

Pure beef enzyme, free from sulphatase activity, is claimed to act on dopamine-3-sulphate to generate unbound noradrenaline. The reaction is inhibited by fusaric acid, a DBH inhibitor [A3334].

Beef brain enzyme has a molecular weight of 400 000, optimum pH 5.0 [C387].

Rat brain enzyme exhibits maximal activity in the presence of N-ethylmaleimide (50 mM) and Cu^{2+}; it may activate by interaction with an endogenous inhibitor [A1723]. The enzyme exists as multiple forms, which may be interconverted, with molecular weights 73 000 and 77 000 [K460]. It is inhibited by some methimazole analogues including 1-cyclohexyl-2-mercaptoimidazole [A1697].

Rat serum enzyme activity remains fairly constant from birth until weaning, followed by a 50 per cent drop over five days, and then a slow decline to the adult level. In heart the activity increases fourfold between birth and weaning, and then remains constant [A1700]. The enzyme has also been detected in rat stomach, heart, stellate ganglion, superior cervical ganglion and salivary gland as well as adrenal [A368, H620].

In dog, hypotension produced by haemorrhage raises plasma DBH activity by 40 per cent [A3576].

Hen adrenal enzyme appears to be a homotetramer, molecular weight 320 000 and optimum pH 5–6, with subunits linked by disulphide bridges. It requires ascorbate (ferrocyanide can act as substitute) and fumarate (acetate is a poor substitute). The product from substrate β-labelled with tritium retains an excess of label. Kinetic studies have suggested a ping-pong mechanism [C56].

Opuntia converts N-methyltyramine into normacromerine (N-methyl-3,4-dimethoxyphenylethanolamine) [A3977].

Tryptamine and tryptophan β-hydroxylation

An enzyme in beef adrenal medulla and in most rat tissues acts on tryptamine; it is unclear whether this has any physiological role [B469].

Pseudomonas indole-3-alkane α-hydroxylase (tryptophan 2′-dioxygenase, E.C. 1.13.99.3) acts on tryptamine to form (*R*)-β-hydroxytryptamine, removing the *pro-S* hydrogen from C-2 of the side chain. It also acts on tryptophan methyl ester and DL-homotryptophan. A mixture of (2*S*, 3*R*)- and (2*R*, 3*S*)-β-methyltryptophan is also hydroxylated, but a mixture of (2*R*, 3*R*)- and (2*S*, 3*S*)-β-methyltryptophan is inactive [B84]. The molecular weight of the enzyme is 250 000, pI 4.8 and optimum pH about 4. Its amino acid composition has been determined, and it contains 0.8 mol Fe/mol. Cyanide and hydroxylamine are inhibitory. A range of tryptophans and tryptamines as well as indoles with 3-alkyl moieties are also substrates. L-Tryptophan is hydroxylated and decarboxylated to β-hydroxytryptamine, with decarboxylation apparently as the second reaction step. Decarboxylation does not occur with, for instance, indole-3-propionate or peptide-linked tryptophan [A3007]. During the reaction an oxygen atom is introduced from water, but not from molecular oxygen; oxygen can be replaced by other electron acceptors. The reaction produces an olefinic intermediate that is then hydrated, possibly spontaneously [A3148].

NADPH-dependent L-phenylalanine monooxygenase

Carica papaya enzyme is found only in leaves. It is highly specific; other phenylalanines are not substrates. The reaction involves oxidative decarboxylation to form benzylaldoxime [H947].

Mandelic acid formation from phenylacetic acids

$$R.CH_2.COOH \rightarrow R.CHOH.COOH$$

Substrates include phenylacetate and *p*-hydroxyphenylacetate. Species in which the reaction has been detected include plants (barley, wheat, parsley and Glycine max) and microorganisms (Aspergillus and Pseudomonas) [A329, A2691, A3594, G929].

Hydroxyphenylacetonitrile 2-monooxygenase
(E.C. 1.14.13.42)

Etiolated sorghum seedling enzyme requires oxygen and NADP to form the corresponding mandelonitrile [K711].

Hydroxylation of nuclear methyl groups

$$R.CH_3 \rightarrow R.CH_2OH$$

a. Polynuclear hydrocarbons

Rat liver enzyme oxidizes 7-methyl- and 7,12-dimethylbenzanthracene, incorporating oxygen from molecular oxygen, but not from water [A339].

b. Phenols

A Pseudomonas putida enzyme, which acts on 4-hydroxy-3-methylbenzoate is a 2-protein system, molecular weight 115 000, that appears to be composed of a flavin-containing hydroxylase and an electron-transfer system that uses NADH. It is claimed to be an unusual type of mixed-function oxidase [F223]. The electron transfer system appears to be a flavocytochrome c [A3870].

p-Cresol is converted into *p*-hydroxybenzyl alcohol by Pseudomonas 4-cresol dehydrogenase (hydroxylating) (E.C. 1.17.99.1), a dimer, molecular weight 115 000. One component is a c-type cytochrome and the other a flavoprotein [A3276]; the oxygen is incorporated from water [A3829, K410]. Studies on *p*-cresol oxidation with 3 Pseudomonas strains including P. putida found that each has a different molecular weight and K_m. The enzyme is anaerobic, with 8α-(O-tyrosyl)FAD as cofactor [B853, K841]. It further oxidizes the substrate to the corresponding aldehyde [A3829].

Pseudomonas *p*-cresol methylhydroxylase is a cytochrome c flavoprotein that catalyzes the incorporation of oxygen from water, and oxidizes the product further to the aldehyde [K203]. The enzyme is a tetramer with two pairs of identical polypeptide chains, one with molecular weight 119 000, and the other a cytochrome, molecular weight 9300; each of the latter binds a flavin molecule. In the wild-type enzyme this is covalently bound, but in enzyme expressed in E. coli it is not covalently bound [K289]. Growth on *p*-cresol induces the *p*-cresol oxidizing enzyme, which is a dimer, molecular weight 100 000, composed of a flavoprotein and cytochrome c, with different molecular weights and K_m (not quoted) for the enzyme from three different strains [A3870, B853]. *p*-Cresol and *p*-ethylphenol are oxidized adjacent to the aromatic nucleus. A ping-pong mechanism has been suggested for the oxidation [E361].

3,5-Xylenol is oxidized in Pseudomonas putida by an enzyme that does not oxidize *p*-cresol. It requires NADH, and is inhibited by cyanide but not by carbon monoxide [A3870].

Penicillium simplicissimum vanillyl alcohol oxidase (E.C. 1.1.3.38) is a flavoprotein that acts on *p*-cresol [G662]; it also forms *p*-hydroxbenzaldehyde very slowly [J680].

P. patulum enzyme is microsomal, optimum pH 7.5, and requires molecular oxygen and NADPH for oxidation of *p*-cresol. Carbon monoxide and cytochrome c are inhibitory [A1468].

Achromobacter enzyme, molecular weight 130 000, is composed of subunits, molecular weights 54 000 and 12 500. The substrate is *p*-cresol [F868].

c. Toluenes and xylenes

Pseudomonas aeruginosa enzyme is composed of three proteins, and requires FAD and NADH [A1009].

E. coli xylene monooxygenase oxidizes toluene and pseudocumene to the corresponding alcohols and aldehydes [K491].

Ethylbenzene dehydrogenases

Ethylbenzene → 1-phenylethanol

Azoarcus enzyme, which is membrane-bound, requires quinone as an electron acceptor; the product is pure (*S*)-1-phenylethanol. Not all analogues are substrates, but propylbenzene and *p*-ethylfluorobenzene are hydroxylated [K241].

Byssochlamys fulva vanillyl alcohol oxidase is a homodimer, monomeric molecular weight 58 000, which oxidizes *p*-hydroxyphenylethane and *p*-hydroxyphenylpropane to the corresponding (*S*)-1-(*p*-hydroxyphenyl)alcohols [K164].

E.coli enzyme forms mainly the (*R*)-isomer [G740]

Mortierella isabellina enzyme forms both (*R*)- and (*S*)-phenylethanols, and the identity of the *para* substituent determines which predominates. In no case was one isomer found to be the exclusive product. With bromo and nitro substitution the enantiomeric enrichment was only slight, whereas with methoxy and chloro (*S*) predominates (between 20 per cent and 40 per cent enantiomeric excess), but with cyano, methyl, ethyl and fluoro (*R*) predominates [D969].

A Penicillium simplicissimum vanillyl alcohol oxidase acts on 4-allyl- and 4-alkylphenols (1–3 C chain). Further oxidation forms the corresponding acetophenone and propiophenone. *p*-Propylphenol also forms a small amount of *p*-coumaryl alcohol, and ethylphenol and propylphenol form vinylphenol and propenylphenol respectively. A quinone methide intermediate is probable for the formation of (*R*)-benzylic alcohols, which are formed in about 94 per cent enantiomeric purity. The hydroxyl group comes from water [J16, J664, J680].

The (*S*)-isomer is formed by Pseudomonas from ethylbenzene [J219]. P. fluorescens eugenol dehydrogenase forms (*S*)-1-(*p*-hydroxyphenyl)ethanol and (*S*)-1-(*p*-hydroxyphenyl)propanol from the corresponding 4-alkylphenols [K177].

Pseudomonas putida 4-ethylphenol methylenehydroxylase is a flavocytochrome c composed of two pairs of subunits, molecular weight about 120 000. One subunit, molecular weight 50 000 is a flavoprotein and the other, molecular weight 10 000 is a cytochrome c. The mechanism involves dehydrogenation followed by hydration to form 1-(*p*-hydroxyphenyl)ethanol. It acts on a range of 4-alkylphenols with up to nine carbon atoms in the side chain and on 5-indanol (a cyclic analogue of 4-ethylphenol). *p*-Cresol and 2,4-xylenol are substrates; and 1-(*p*-hydroxyphenyl)ethanol is further oxidized to *p*-hydroxyacetophenone. It is considered that the reaction proceeds via a quinone methide [F381]. In another study the enzyme was called 4-ethylphenol methylenehydroxylase [E396].

This type of reaction has also been observed in man [D482], rat [D365], rabbit [A287], chinook salmon [E710], Desulfobacula toluolica [J847], Nocardia tartaricans [B144] and in Nitrosomonas [H219]; in N. europaea ammonia monooxygenase catalyzes the reaction [H631]; in many studies, particularly the early ones the chirality of the product was not established.

4-Allylphenol ω-hydroxylation

Penicillium simplicissimum vanillyl alcohol oxidase hydroxylates eugenol and chavicol to the corresponding cinnamyl alcohols [H389]. The enzyme also carries out a range of other activities, such as oxidizing secondary vanillyl alcohols, including phenylethanolamines, to the corresponding ketones [J16].

Byssochlamys fulva enzyme is a homodimer, monomeric molecular weight 58 000, which additionally oxidizes vanillyl alcohol to vanillin, and (*p*-hydroxyphenyl)alkanes to (*S*)-1-(*p*-hydroxyphenyl)alcohols [K164].

Pseudomonas fluorescens eugenol dehydrogenase, a dimer, monomeric molecular weights 10 000 and 58 000, is a flavoprotein that requires an oxidizing agent, such as ferricyanide. It forms coniferyl alcohol from eugenol, as well as oxidizing *p*-hydroxybenzyl alcohols [J890].

Acetophenone ω-hydroxylation

Solanum khasianum enzyme, which hydroxylates the terminal carbon of acetovanillone, requires oxygen and NADPH. Its inhibition properties are like those of P450 [H906]. A similar reaction probably occurs in Pseudomonas with acetophenone [J219].

In rat a similar reaction occurs with paeonol [E849].

Cannabinoid side chain hydroxylases

Extensive studies with cannabinoids have demonstrated that hydroxylations occur at various positions in the molecule; however, little appears to have been done at an enzyme level. The literature on this subject is extensive, and the references given (for Δ^1-tetrahydrocannabinol) are illustrative.

Cannabinoid 1″-hydroxylation

This reaction has been detected in guinea pig and Thamnidium [B458, B742].

Cannabinoid 2″-hydroxylation

This reaction has been detected in man, guinea pig, Fusarium, Gibberella and Thamnidium [B458, B742, C2, F819].

Cannabinoid 3″-hydroxylation

This reaction has been detected in mouse, rat, guinea pig, monkey, Chaetomium, Fusarium, Gibberella and Thamnidium [A3628, B458, B742, E598, F599, F819, H418].

Cannabinoid 4″-hydroxylation

This reaction has been detected in guinea pig, rat, rabbit, Fusarium, Gibberella, Syncephalastrum and Thamnidium [A2595, B458, B742, E7, F7]

Cannabinoid 5″-hydroxylation

This reaction has been detected in mouse, rabbit and guinea pig [A3449, F7].

Cannabinoid 7-hydroxylation

This reaction has been detected in rat, rabbit, monkey, man, mouse and Chaetomium [A108, A116, A438, A680, A879, A3628].

Agroclavine hydroxylation

Claviceps microsomal enzyme, a P450 that requires NADPH, hydroxylates agroclavine to elymoclavine [B809].

3.2 *Alkyl oxidation to ketone*

Phenylacetyl CoA: acceptor oxidoreductase

$$\text{Phenylacetyl CoA} + 2\ \text{quinone} + 2\ H_2O \rightarrow \text{phenylglyoxylate} + 2\ \text{quinol} + \text{CoASH}$$

Thauera aromatica enzyme is a membrane-bound trimer, subunit molecular weights 93 000, 27 000 and 26 000, with an assumed native molecular weight of 280 000. Analyses indicate that this contains 0.66 Mo, 30 Fe and 25 acid-labile S. The reaction is anaerobic, with duroquinone, menadione or (best) dichlorophenolindophenol as electron acceptors. Phenylglyoxylyl CoA is an intermediate, but mandelyl CoA is not [K173].

3.3 *Oxidations and reductions of alcohols, aldehydes and ketones*

Alcohol dehydrogenases (aryl-alcohol dehydrogenase; E.C. 1.1.1.90, aryl-alcohol dehydrogenase (NADPH); E.C. 1.1.1.91)

$$R.CH_2OH \rightarrow R.CHO$$

Human brain enzyme is composed of at least five isozymes, pI 5.3, 6.0, 6.3, 7.0 and 7.9. They all act on *p*-nitrobenzaldehyde and indole-3-acetaldehyde, but are distinguished by their relative activities towards menadione, daunorubicin, *p*-hydroxyphenylacetaldehyde and *p*-hydroxymandelaldehyde as potential substrates. All require NADPH; one that can also utilize NADH appears to be succinic semialdehyde reductase (E.C. 1.2.1.16/ E.C. 1.2.1.24). The pI 7.9 enzyme is strongly inhibited by quercetin and quercitrin [B569].

Human brain aflatoxin B_1 aldehyde reductase is identical with succinic semialdehyde reductase, which also reduces phenanthrene-9,10-quinone, phenylglyoxal and *p*-nitrobenzaldehyde [K273].

Horse liver enzyme oxidizes benzyl alcohol reversibly; the enzyme has two active sites [A2174].

A Glycine max (soyabean) cinnamyl alcohol dehydrogenase isozyme (E.C1.1.1.195), molecular weight 43 000 and optimum pH 9.2, oxidizes coniferyl alcohol. A second isozyme in addition reversibly oxidizes several substituted cinnamyl alcohols with optimum pH 8.8, and optimum pH 6.6 for reduction [A2325].

Rye aromatic alcohol dehydrogenase is composed of 3 NADP-dependent isozymes, one

NAD-dependent isozyme and one without nucleotide specifity [D45].

Acinetobacter calcoaceticus benzyl alcohol dehydrogenase is a monomer, molecular weight 38 923 with 370 amino acid residues, based on nucleotide sequencing. It requires NADH and zinc; it is a member of a family of zinc-dependent long-chain alcohol dehydrogenases (E.C. 1.1.1.192) [J621]. Other substrates include coniferyl alcohol, cinnamyl alcohol and other (unspecified) aromatic alcohols, but few aliphatic alcohols are substrates [E473]. Another study claims that the enzyme is a tetramer with a monomeric molecular weight consistent with the above value. The optimum pH for oxidation is 9.2, and for reduction 8.9 [E596].

Azoarcus 1-phenylethanol dehydrogenase, which is inducible, only acts on the (*S*)-isomer; it requires NAD^+ [K241].

Geotrichum candidum converts (*S*)-arylethanols into (*R*)-arylethanols via acetophenones; (*R*)-arylethanols are not substrates. A range of phenylethanols substituted on the aromatic nucleus with chloro, methyl and methoxy groups (but not in the *ortho* position) are also substrates [H749].

A methanol dehydrogenase (E.C. 1.1.1.244) in Methylomonas methanica, molecular weight 60 000, oxidizes 2-phenoxyethanol and a range of aliphatic alcohols at similar rates, but does not act on benzyl alcohol or secondary alcohols. It requires NH_4^+, with optimum pH 9.5 [A2712].

Mycobacterium tuberculosis enzyme has an optimum of 6.5–8, depending on the nature of the buffer and other parameters. It oxidizes benzyl alcohols as well as reducing benzaldehyde. It is inhibited by *p*-chloromercuribenzoate, benzoate and *o*-phenanthroline [A150].

Penicillium urticae 3-hydroxybenzyl-alcohol dehydrogenase (E.C. 1.1.1.97), molecular weight 120 000 and optimum pH 7.6, requires NADP. It is a key enzyme in the formation of patulin from 6-methylsalicylate [K946].

Penicillium simplicissimum vanillyl alcohol oxidase (E.C. 1.1.3.38) is an octomer, monomeric molecular weight 65 000. It is a flavoprotein (1 mol/mol of monomer), with covalently bound 8a-(N^3-histidyl)FAD. It is highly specific for *p*-hydroxybenzyl alcohols [G662]. It forms the corresponding acetophenone and propiophenone from *p*-ethylphenol and *p*-propylphenol; there is good evidence that the corresponding alcohols are intermediates [J664, J680].

Phanerochaete chrysosporium enzyme, molecular weight 78 000, has a FAD prosthetic group. It is specific for benzyl alcohols [H613].

Pleurotus eryngii alcohol oxidase (E.C. 1.1.3.13), molecular weight 72 600 and pI 3.9, contains 15 per cent carbohydrate. It oxidizes a series of benzyl and cinnamyl alcohols [G668]. P. pulmonarius enzyme which contains 14 per cent carbohydrate, molecular weight 70 500 and pI 3.95, has been crystallized. DNA studies indicate that it is composed of 593 amino acid residues, including a signal peptide of 27 amino acid residues [K373].

Pleurotus ostreatus veratryl alcohol oxidase is a glycoprotein containing FAD, optimum pH 6.5 (broad), which forms veratraldehyde, with oxygen forming peroxide. A range of benzyl alcohols and cinnamyl alcohols are oxidized [F867].

Polystictus versicolor aryl alcohol oxidase is found in the media surrounding the mycelia. It acts on several benzyl alcohols and 2-hydroxymethylnaphthalene, but other alcohols are only marginally active [K878].

Pseudomonas fluorescens enzyme, a dimer, monomeric molecular weights 10 000 and 58 000, is a flavoprotein requiring an oxidizing agent, such as ferricyanide. It oxidizes *p*-hydroxybenzyl alcohols as well as forming coniferol alcohol from eugenol [J890].

Pseudomonas putida grown on 3,5-xylenol contains two NAD^+-dependent alcohol dehydrogenases, molecular weights 122 000 and 145 000. When grown on *p*-cresol a single NAD^+-dependent alcohol dehydrogenase develops, molecular weight 75 000. They all have an optimum pH of 9.5 or higher, and oxidize a range of substituted benzyl alcohols with minor differences in specificity. The xylenol-induced enzymes that oxidize benzyl alcohol, *m*- and *p*-hydroxybenzyl alcohols undergo spontaneous inactivation, and are protected by dithiothreitol. Inactivation by *p*-chloromercuribenzoate is partly prevented by substrate [A1366, A3837].

Pseudomonas putida 4-ethylphenol methylenehydroxylase (see Ethylbenzene dehydrogenases, above) oxidizes 1-(*p*-hydroxyphenyl)ethanol to *p*-hydroxyacetophenone [F381].

Pseudomonas syringae D-phenylserine dehydrogenase, molecular weight 31 000 and optimum pH 10.4, requires $NADP^+$ [H91].

Rhodopseudomonas acidophila *p*-hydroxybenzyl alcohol dehydrogenase (E.C. 1.1.1.90), molecular weight 27 000 and pI 7.4, has an optimum pH 6.5 for oxidation and pH 9 (broad) for the reverse reaction. Substrates include cinnamyl alcohol, phenylethanol and a series of benzyl alcohol analogues substituted in the *m*- and *p*-positions, but not *o*-analogues [D197].

Thauera benzyl alcohol dehydrogenase, which is a homotetramer, molecular weight 160 000, requires NAD^+ [H610].

A benzyl alcohol dehydrogenase has been detected in a Bacterium [A730].

Aromatic aldehyde and ketone reductases (aryl alcohol oxidase, E.C. 1.1.3.7; c.f. aryl-alcohol dehydrogenase ($NADP^+$); E.C. 1.1.1.91)

R′.CO.R → R′.CHOH.R, where R is H or an alkyl group and R′ an aryl or an aralkyl group.

Four or possibly 5 isozymes of aldehyde reductase in human brain reduce indole-3-acetaldehyde and *p*-nitrobenzaldehyde, with pI 5.3 (this isozyme appears to be E.C. 1.1.1.2), 6.0, 6.3, and the fourth fraction shows two values at 7.0 (minor) and 7.9. They require NADPH, although one isozyme can utilize NADH. Menadione, daunorubicin, *p*-hydroxyphenylacetaldehyde and *p*-hydroxymandelaldehyde are substrates for at least 2 of the isozymes. [B569].

Human erythrocyte 4-nitroacetophenone reductase, optimum pH about 7, is not cytochrome c. It requires NADPH, but not NADH. It is unstable at 50°, and is moderately unstable at 4° and is inhibited by *p*-nitrobenzaldehyde, *p*-chloromercuribenzoate and N-ethylmaleimide, and slightly by methanol. It is claimed to differ from other side chain keto reductases. A similar enzyme is found in rat [A15].

Beef enzyme is widely distributed throughout the brain [A1908]. It requires NADH; NADPH is a poor co-substrate. The optimum pH for reduction is 6.8, and 10 for the reverse reaction. Some aliphatics as well as *p*-nitrobenzaldehyde are substrates [A614].

Pig brain contains two isozymes, one low- and the other high-affinity. Both reduce 3,4-dihydroxy- and 4-hydroxy-3-methoxyphenylacetaldehydes as well as *l*-*p*-hydroxyphenylglycolaldehyde and D-3,4-dihydroxyphenylglycolaldehyde; the high affinity isozyme also reduces 5-hydroxyindole-3-acetaldehyde [A1287]. Another study found the molecular weight of the cytosolic enzyme to be 29 000 and pI 5.8. It utilizes NADPH in the reduction of benzaldehydes, *p*-hydroxymandelaldehyde, indole-3-acetaldehyde and some aliphatic aldehydes (especially lactaldehyde and glyceraldehyde; acetaldehyde is very poor), but it is ineffective in reducing ketones [A1679].

Monkey (apparently rhesus) brain enzymes have a similar specificity to that of pig brain enzyme. Both isozymes require NADPH; one is low affinity and high activity and the other high affinity and low activity [A1908].

Rat liver aldehyde reductase isozymes, pI 6.5 and 6.9, reduce 3,5-dihydroxyphenylacetaldehyde and *p*-nitrobenzaldehyde [B659]. An enzyme tentatively identified as E.C. 1.1.1.2 has an optimum pH of 6–8 for the forward reaction, whereas the optimum for the reverse reaction (on daunorubicinol) is above pH 10. Both aliphatic and aromatic aldehydes are reduced, the best being 4-carboxybenzaldehyde; adriamycin is reduced, but only poorly. Barbital, warfarin and phenobarbital are inhibitors, but is activated by NaCl, with peak activity enhanced by 50 per cent above the base-line activity at ionic strength 0.4. With 50 mM sodium sulphate there is 40 per cent activation, which declines to zero at ionic strength 0.3–0.4, with inhibition at higher concentrations [A2007]. A rat liver aldehyde reductase, optimum pH 6.5, reduces many benzaldehydes, as well as quinones and phenylglyoxal. It differs in specificity from other aldehyde reductases in that it reduces *o*-quinones; it is the only one found in this study that reduces aflatoxin B_1-dialdehyde, a very poor substrate [H759].

Rat liver cytosol contains a highly specific 2-carboxybenzaldehyde reductase, molecular weight 64 000; the 3- and 4-carboxy analogues are not substrates. It requires NAD(P)H and a thiol for activity [G239].

Rat aflatoxin B_1 aldehyde reductase 2 is found as multiple forms, with molecular weights in the range 36 800 and 38 000. They also act on other aldehydes and quinones [K546].

A rabbit liver cytosolic enzyme, molecular weight 33 000 and optimum pH 6.2, reduces the oxo group of loxoprofen and requires NADPH. It also reduces benzaldehydes, acetophenones and various other oxo-compounds. A similar activity that reduces loxoprofen is found in guinea pig [D125]. Another study with rabbit liver cytosol found four isozymes, designated F_1, F_2, F_3 and F_4. F_2 is an aldehyde reductase, whereas the others are aromatic aldehyde/ketone reductases. F_1 and F_3 are monomeric with molecular weights 38 000 and 29 000 respectively. F_4 appears to be trimeric, molecular weight 78 000, and monomeric molecular weights 24 000 and 26 000. Substrates include 3- and 4-benzoylpyridines and *p*-nitroacetophenone [B431]. Flavonoids are inhibitory [K560]. A further publication reports molecular weights that differ appreciably from these values; it also describes a benzaldehyde-reducing enzyme that does not reduce acetophenones, whereas two acetophenone-reducing enzymes also reduce some benzaldehydes [B386]. Seven isozymes were detected in one study. Two are aldehyde reductases, the major having pI 6.0–6.3 and the minor 7.8. A major enzyme, pI 4.9, is active towards *p*-nitrobenzaldehyde and *p*-nitroacetophenone, but not naloxone or naltrexone. Four others, including one major enzyme, which reduce naloxone and naltrexone as well as other carbonyl compounds, were named dihydromorphinone reductases, pIs 5.4–5.5, 6.5, 6.6 and 6.9–7.2 [A3950]. Rabbit and chicken dihydromorphinone reductases require NADPH, and are found mainly in liver cytosol, although some activity is present in kidney and lung [A2290].

An enzyme in rabbit, rat and guinea pig liver that reduces substituted benzoylpyridines is found in both microsomes and cytosol (the latter not in rat), and requires NADPH. Cytosolic enzyme is inhibited by heavy metals, *o*-phenanthroline and azide among other compounds, whereas the microsomal enzyme is inhibited by Hg^{2+}, *p*-chloromercuribenzoate and N-ethylmaleimide [A2834].

Sheep heart enzyme reduces a range of phenylglycolaldehydes and oxidizes *p*-hydroxybenzaldehyde, and requires NADP as co-substrate [A57].

Cucumis sativus (cucumber) cytoplasmic indole-3-acetaldehyde reductase (E.C. 1.1.1.191) is highly specific and requires NADPH [F462]. Another study identified 3 indole-3-acetaldehyde reductase isozymes. Two, molecular weights 17 000 and 52 000 and optimum pH 5.2, require NADPH. A NADH-specific enzyme (E.C. 1.1.1.190) has a molecular weight of 332 000 and optimum pH 7.0. They all reduce phenylacetaldehyde and (poorly) *trans*-cinnamaldehyde. Some aliphatics are poor substrates, but benzaldehydes are not reduced [A2432].

A mung bean monomeric enzyme, molecular weight 36 000 and optimum pH 6.2–7.5, reduces eutypine irreversibly, and a series of benzaldehydes and cinnamaldehyde with NADPH as cofactor; it appears that no tests were made on the reversibility with these other substrates [J910].

Populus euramericana stem cinnamyl alcohol dehydrogenase, molecular weight 36 000 and pI 5.6, requires NADPH for the reduction of coniferaldehyde, *p*-coumaraldehyde and sinapaldehyde, and is inhibited by 1,10-phenanthroline and sulphydryl-binding compounds [D54].

Both spruce and Glycine max contain cinnamyl alcohol: $NADP^+$ dehydrogenases (E.C. 1.1.1.195) that reduce coniferaldehyde, *p*-coumaraldehyde and sinapaldehyde. Both are dimers, monomeric molecular weight about 35 000 [B922]. Another study on the Glycine enzyme found a molecular weight of 40 000 for a zinc-containing isozyme that reduces coniferaldehyde. It requires NADPH, and is inhibited by thiol-binding reagents [B95].

Swede enzyme, which is composed of three isozymes, is part of the system for forming coniferyl alcohol from ferulate. It requires NADPH (NADH is inactive) for the reduction of coniferaldehyde, and is reversible [A216]. Forsythia enzyme is very similar [A848].

Candida guilliermondii contains a phenylacetaldehyde reductase that requires NAD(P)H [A2483].

Corynebacterium phenylacetaldehyde reductase requires NADH. It also reduces nuclear substituted phenylacetaldehydes, phenacyl chloride, acetophenone, propiophenone and 4-phenylbutan-2-one stereoselectively to the (*S*)-alcohol (phenacyl chloride yields the (*R*)-isomer with the same spatial geometry). In the reverse reaction the (*S*)-alcohols are substrates [J893].

Geotrichum candidum reduces acetophenones [H749].

Lactobacillus kefir acetophenone reductase, optimum pH 7.0, which is protected by Mg^{2+}, requires NADPH, but NADH is inactive. The optimum pH for reduction is 7.0, and for oxidation 8.0. Propiophenone, a range of acetophenones, and some (but not all) other analogues tested, as well as benzaldehyde, are substrates. The product from acetophenone is (*R*)-(+)-1-phenylethanol, which is a substrate for the reverse reaction, although (*S*)-1-phenylethanol is not [Gl48].

Phycomyces blakesleeanus indole-3-acetaldehyde reductase is a tetramer, monomeric molecular weight 38 000, pI 5.4 and optimum pH 6–8. It requires NAD(P)H as cofactor [F846].

A Pseudomonas guaiacylglycerol β-guaiacyl ether dehydrogenase reduces the α-oxo analogue of this compound [G357].

Tryptophol oxidase

Tryptophol → indole-3-acetaldehyde

Cucumber enzyme is inhibited by the reaction product, and this inhibition is reversed by oxygen but not by the substrate. Several auxins are inhibitory [A835, A957]; this undoubtedly prevents over-production of IAA.

Phaseolus vulgaris, molecular weight 56 000, requires oxygen, and forms peroxide as a second product [H554].

Phycomyces blakesleeanus enzyme, molecular weight 56 000 and optimum pH 6–8, forms indole-3-acetaldehyde and possibly peroxide. It is activated by FAD and inhibited by Hg^{2+}, iodoacetate, and by 4 μM indole-3-acetaldehyde [E443].

Mandelate dehydrogenases and oxidases

R.CHOH.COOH → R.CO.COOH

Rhodotorula graminis L-(+)-mandelate dehydrogenase is a tetramer, monomeric molecular weight 59 100, that contains one mol each of haem and FMN per subunit. The optimum pH is 7.9, and pI 4.4. It acts on mandelate and a series of nuclear-substituted analogues [H90] and is stereospecific [G904]. Both D- and L-dehydrogenases are inducible. L-Dehydrogenase, optimum pH 7.0, requires dichlorophenolindophenol as reductant, and may be membrane-bound; activity is enhanced by phenazine methosulphate. D-Dehydrogenase, optimum pH 9.0, is not membrane-bound, and requires NAD^+ [D475].

Acinetobacter calcoaceticus enzymes, specific for each stereoisomer are found in cytoplasmic membranes. After solubilization, the D-dehydrogenase has a molecular weight of 59 700, optimum pH 8.0 and pI 5.5 [D870] and are very similar to the corresponding D- and L-lactate dehydrogenases (E.C. 1.1.2.4 and 1.1.2.3 respectively) [D674]. A novel enzyme found in a mutant strain but not in the wild-type organism, acts solely on the D-isomer. It is membrane-bound, and its pH and temperature dependence are similar to those of the L-dehydrogenase (the activity found in most strains). Inhibitors include L-mandelate [A1649, C801].

Aspergillus niger D-mandelate oxidase (not a dehydrogenase), optimum pH 7.6, is particulate and very unstable. It is specific for the D-isomer of several mandelates. Neither NAD nor $NADP^+$ are involved, and metal ions are not activators. It appears to require cytochrome c and molecular oxygen, which cannot be replaced by other oxidizing agents, and peroxide is not formed as a second product. Heavy metal ions are inhibitors [A1204].

Pseudomonas putida (*S*)-mandelate dehydrogenase oxidizes mandelate and indole-3-glycollate, and acts very slowly on plenyllactate, indole-3-lactate and some aliphatic analogues [K95]. It is membrane-bound, with a binding segment of about 39 residues [K230].

Rhizobium leguminosarum enzyme is induced by 4-hydroxymandelate [F224].

A Bacterium (unidentified) enzyme, optimum pH 9.5, oxidizes D-VMA to the corresponding benzoylformate, but D-mandelate is not a substrate [J256]. Another study has identified a L-dehydrogenase in a Bacterium [A730].

Indole-3-acetaldehyde reductase (E.C. 1.1.1.190 and 1.1.1.191)

Indole-3-acetaldehyde → tryptophol

Human brain enzyme acts on indole-3-acetaldehyde [B569]. Rat liver enzyme [A2007] and pig brain enzyme [A1679] also reduce other aldehydes (see above).

Brassica campestris (Chinese cabbage), molecular weight 32 000 and optimum pH 6–7 requires NADPH. Other substrates are benzaldehyde and phenylacetaldehyde, but not 3-formylindole. It has also been found in B. napus, B. oleracea, Arabidopsis thahana and Sinapis alba [F931].

Cucumis sativa (cucumber) enzyme, which requires NADPH, is cytoplasmic and is specific for indole-3-acetaldehyde [F462].

Mung bean enzyme, a dimer, monomeric molecular weight 39 000, requires NADPH. It also reduces benzaldehyde and phenylacetaldehyde [J207].

Phycomyces blakesleeanus enzyme, molecular weight 38 000, pI 5.4 and optimum pH 6–8 requires NAD(P)H [F846].

This reduction has also been detected in Orobanche gracilis, O. lutea, O. ramosa [C181], Zygosaccharomyces [A923] and pea [A755].

Benzoylformate reductase

Benzoylformate → mandelate

This activity has been found in rats [J287], with a preponderance (10:1) of the (*R*)-isomer formed [F496].

Pseudomonas polycolor and Micrococcus freundii catalyse this reaction; this enables racemic mandelate to be converted into the (*R*)-isomer, because these organisms contain an enzyme that converts (*S*)-mandelate into benzoylformate, but is inactive towards (*R*)-mandelate [H668].

S. faecalis enzyme is dimeric, molecular weight 72 000, pI 4.9 and optimum pH 4.5, also reduces phenylpyruvate and some aliphatics, but not *p*-hydroxyphenylpyruvate or *p*-hydroxybenzoylformate. The optimum pH for the reverse reaction is 9.2 [E271].

Hydroxyphenylpyruvate reductase (E.C. 1.1.1.237)

Coleus blumei enzyme, which requires NADH reduces *p*-hydroxyphenylpyruvate and 3,4-dihydroxyphenylpyruvate to the corresponding lactates [E660].

Aromatic α-ketoacid reductase ((*R*)-aromatic lactate dehydrogenase; 1.1.1.222, diiodophenylpyruvate reductase; E.C. 1.1.1.96)

1. Reduction

Dog heart enzyme is a cytosolic dimer, monomeric molecular weight 40 000, pI 5.4, which requires NADH. Activity is also found in brain, kidney and liver, and is considered to be associated with an isozyme of malate dehydrogenase. The best substrate is 3,5-diiodophenylpyruvate, with good activity towards phenylpyruvate and indole-3-pyruvate [A2917].

In vivo studies have demonstrated this activity in rat [A2961, A3327]; reduction is catalyzed by lactate dehydrogenase (E.C. 1.1.2.3) and aromatic alpha-keto acid reductase [A3327]. Highest activity is found in heart, and (in reducing order) in kidney, muscle and liver. 3,4-Dihydroxyphenylpyruvate is 10 times as active as 3-methoxy-4-hydroxyphenylpyruvate. Oxamate (a lactate dehydrogenase inhibitor) does not inhibit liver mitochondrial enzyme [A2983].

In a range of animals cytoplasmic malate dehydrogenase (E.C. 1.1.1.37) has been found to be identical with aromatic alpha-keto acid reductase (with *p*-hydroxyphenylpyruvate as substrate), with lactate dehydrogenase accounting for a minimal proportion of the total activity found in these species. The studies were carried out on flight muscle of Falco, Milvago, Herpetotheres, Phalcobanes, Spiziapteryx and Polyhierax, Palaemonedes (a marine invertebrate), and frog liver and muscle. The activity in Fundulus grandis (a marine fish) is identified as lactate dehydrogenase [E561].

Coleus blumei enzyme has an optimum pH between 6.5 and 7.0, requires NAD(P)H, and reduces the physiologically important pyruvates *m*- and *p*-hydroxy-, 3,4-dihydroxy- and 4-hydroxy-3-methoxyphenylpyruvates [E660, G289].

Candida guilliermondi enzyme requires NAD(P)H [A2483].

Candida maltosa enzyme is a tetramer, molecular weight 250 000–280 000, monomeric molecular weight 68 000. It requires Mn^{2+} and NAD(P)H for reduction; the reaction is reversible, with optimum pH 6.5 for reduction, and 9.5 for oxidation. Substrates studied are phenylpyruvate, *p*-hydroxyphenylpyruvate, indole-3-pyruvate and the corresponding lactates [D975].

Lactobacillus casei D-hydroxyisocaproate dehydrogenase reduces phenylpyruvate to D-phenyllactate [E587].

2. Oxidation (indolelactate dehydrogenase; E.C. 1.1.1.110)

Formation of phenylpyruvate from phenyllactate has been recorded in rat, Candida, Lactobacillus, Neisseria, Pseudomonas and Rhodotorula [B438, D975, E377, F92, G774, K95]. A similar reaction has been detected in rat for *m*-hydroxyphenyllactate and vanillactate [A2961]. In addition, *p*-hydroxyphenyllactate is oxidized by Neisseria gonorrhoeae [F92], and indole-3-lactate by Candida maltosa (see above) [D975].

Cinnamyl alcohol dehydrogenase (E.C. 1.1.1.195)

Aralia cordata enzyme is a heterodimer, molecular weight 72 000. The reaction is reversible, the reverse requiring NADPH. Substrates are coniferaldehyde, sinapaldehyde, coniferyl alcohol and sinapyl alcohol [G815].

Eucalyptus gunnii contains two isozymes; one is monomeric, molecular weight 38 000, and the other, molecular weight 83 000, is a heterodimer [G660].

Loblolly pine enzyme is a dimer, monomeric molecular weight 44 000. It reduces coniferaldehyde and sinapaldehyde [G484].

Nicotiana enzyme, which is composed of two isozymes, molecular weights 42 500 and 44 000, requires NADP [G458].

Spruce (Picea abies) enzyme is a dimer, molecular weight 42 000. Substrates are coniferaldehyde, *p*-coumaraldehyde, coniferyl alcohol and *p*-coumaryl alcohol with NADP(H) as co-substrate [G781].

Etiolated wheat seedlings contain three isozymes; the molecular weights of two of these are 40 000 and 45 000. Substrates are coniferaldehyde, *p*-coumaraldehyde and sinapaldehyde [H510].

Among those species studied, the enzyme is not found in Pteridophyta or monocotyledonous angiosperms (except Zea). It is mostly found in gymnosperms and dicotyledonous angiosperms. It is usually a single enzyme, except in a range of Salix species (three to eight isozymes, usually four). It is not the same as alcohol dehydrogenase; it requires NADP, whereas alcohol dehydrogenase requires NAD [A1732].

Benzyl 2-methyl-hydroxybutyrate dehydrogenase (E.C. 1.1.1.217)

Reduction of benzyl 3-oxo-2-methybutyrate to (2*R*,3*S*)- and (2*S*,3*S*)-benzyl 3-hydroxy-2-methylbutyrate in Candida albicans, Endomycopsis fibligera, Hansenula anomala, Lipomyces starkeyi, Pichia farinosa, P. membranaefaciens, Rhodotorula glutinis, Saccharomyces cerevisia and S. acidifaciens has been detected [K940].

Methyl 2-oxo-3-phenylbutyrate reduction

This reaction has been found in Candida albicans, Endomycopsis fibligera, Hansenula anomala, Kloeckera saturnus, Lipomyces starkeyi, Pichia farinosa, P. membranaefaciens, Rhodotorula glutinis, Saccharomyces cerevisiae, S. acidifaciens, S. delbruechii and S. fermentati. Different organisms form different ratios of stereoisomers [K883].

2-Hydroxy-6-oxo-6-phenylhexa-2,4-dienoate reductase (E.C. 1.3.1.40)

Pseudomonas cruciviae enzyme is composed of three isozymes. One, molecular weight 170 000, requires NADPH, and also reduces the methyl ester of the above compound (which forms 2,6-dioxo-6-phenylhexanoate) [E156].

Indanol dehydrogenase

(*S*)-l-Indanol ↔ 1-indanone

Human placenta oxidizes 1-indanol to 1-indanone, with $NAD(P)^+$ as cofactor. Most of the activity is present in microsomes, with some in mitochondria but little in the cytoplasm [D415].

Japanese monkey liver cytosolic enzyme, molecular weight 36 000 and pI 8.7, requires $NAD(P)^+$ for oxidation and NADPH for reduction; the amino acid composition has been determined. The specificity is broad for cyclic alcohols such as (*S*)-1-indanol, benzene-1,2-dihydrodiol and 1-hydroxytetralin, and for aldehydes and ketones in which the oxo-group is conjugated with the aromatic nucleus, such as benzaldehydes and acetophenones. The activity with (*R*)-1-indanol is much lower than with (*S*)-1-indanol [F241]. Four isozymes have been found, two major and two minor; classical indanol dehydrogenase is one of the major isozymes. Quantitative studies show that (*S*)-l-indanol and 1-acenaphthenol are the best substrates. One minor isozyme has a similar specificity, and differs mainly in that the pI is 7.9. The other major isozyme, molecular weight 38 000 and pI 6.2, shows a similar activity for each substrate studied. The product from benzene-1,2-dihydrodiol is catechol. The other minor isozyme that was studied does not oxidize (*S*)-1-indanol, but does oxidize dihydrodiols to catechols. The main activity for these enzymes is the reduction of nitrobenzaldehydes [F388].

Rabbit liver cytosolic enzyme is not separable from 3-hydroxyhexobarbital dehydrogenase, and like the monkey enzyme the specificity is broad [A2032].

Oestradiol 17α-dehydrogenase (E.C. 1.1.1.48)

Rabbit liver enzyme (see oestradiol 17β-dehydrogenase) oxidizes 17α-oestradiol and its 3-glucuronide [A157]. Chicken liver enzyme is marginally active towards 17α-oestradiol [B748].

Oestradiol 17β-dehydrogenase (E.C. 1.1.1.62)

17β-Oestradiol ↔ oestrone

Human ovary enzyme, optimum pH 8.1 and 6.9 for the forward and reverse reactions respectively, is cytosolic, with NADP(H), or less effectively NAD(H) as cofactors for reduction; 3-methoxyoestrone and 3-methoxy-17β-oestradiol are better substrates than the parent compounds [A3086].

Human endometrium enzyme utilizes $NAD(P)^+$; reduction is not stimulated by

NADPH [A732]. Its activity increases at the end of the proliferation phase of the oestrus cycle, reaches its maximum value by the mid-secretory phase and falls towards its original value at the end of this phase. It then remains constant at about 5 per cent of the maximum value throughout the proliferative phase [A1104].

Human enzymes, both foetal and maternal, have an optimum pH of about 9, and are very unstable at $-20°$ [A 1551]. The reverse reaction, which is catalyzed by placental enzyme, is inhibited by ATP, especially with NADPH as cofactor. In contrast, ATP inhibition is more marked with NADH when 16α-hydroxyoestrone is the substrate [A1775]. Placental enzyme activity is not affected by prostaglandins [A374]. Kidney enzyme oxidizes 17β-oestradiol and its 3-sulphate and glucuronide conjugates [A1214]. Both human and rat erythrocyte enzymes reduce oestrone and its sulphate conjugate [A518].

Rabbit liver enzyme oxidizes 17α- and 17β-oestradiol and their 3-glucuronides. Soluble enzymes have been separated into three fractions. One, that oxidizes both 17α-compounds, has been further separated into five sub-fractions by isoelectric focussing, each of which exhibits different kinetics. The second fraction oxidizes 17β-oestradiol, and the third 17β-oestradiol-*β*-D-glucuronide [A157].

Sheep ovary 17β-hydroxysteroid: $NAD(P)^+$ dehydrogenase has a molecular weight of 70 000 and optimum pH 9.2 [A2389].

Rat liver microsomal enzyme reduces 16α-chlorooestrone; oestrone is inhibitory [A1749].

Chicken liver enzyme is composed of three isozymes, pI 6.0, 6.8 and 6.9, and optimum pH 9.9 for the forward reaction. Two isozymes have molecular weights of 43 000 and 97 000. The reaction requires $NADP^+$; the reverse reaction utilizes NADPH. 17α-Oestradiol is marginally active, but oestriol is not a substrate. *p*-Chloromercuribenzoate is inhibitory [B748].

Cochliobolus lunatus enzyme acts on oestrone and alkyl steroids as well as quinones, aldehydes and ketones [K378].

Flavanone reduction (E.C. 1.1.1.234)

Flavanone → hydroxyflavan

Cryptomeria japonica enzyme is cytosolic, molecular weight 133 000 and optimum pH 7. It requires NADPH with (+)-aromadendrin and (+)-dihydroquercetin as substrates [E758].

Matthiola incana flower enzyme, optimum pH about 6, requires NADPH; NADH is not so good. It reduces (+)-aromadendrin to 3,4-*cis*-3,4,4′,5,7-pentahydroxyflavan; it also reduces (+)-dihydroquercetin and (+)-dihydromyricetin [D782].

Dihydrokaempferol 4-reductase (E.C. 1.1.1.219)

Matthiola incana flower enzyme, optimum pH about 6, forms *cis*-3,4-leucopelargonidin from dihydrokaempferol. Other substrates are (+)-dihydroquercetin and (+)-dihydromyricetin [D782].

Codeinone reductase (E.C. 1.1.1.247)

Papaver somniferum enzyme, a monomer molecular weight 35 000 requires NADPH for reduction of (−)-codeinone and a range of morphinan ketones, but does not act on other aldehydes and ketones [K756, K757].

Naloxone reductase (E.C. 1.1.1.128)

Rat liver enzyme, molecular weight 34 000 and pI 5.9, which reduces naloxone to 6β-naloxol, is identical to 3α-steroid dehydrogenase. It also reduces benzaldehydes, acetophenones, quinones and non-aryl ketones [J744]. In guinea pig the 6α-isomer is formed [F579].

Daunorubicin reductase

Human liver contains four isozymes, three with optimum pH 6.0, and the fourth, possibly an aldehyde reductase, optimum pH 8.5. Their molecular weights are in the range of 30 000–40 000 [C531]. Two isozymes of human brain aldehyde reductase reduce daunorubicin, pI 5.3 and 7.9 [B569].

Rabbit liver contains two reductases with optima at pH 6.0 and 8.5. Their properties indicate that they are ketone reductase and aldehyde reductase respectively [B402]. One, tentatively identified as E.C. 1.1.1.2, reduces daunorubicin less well than other substrates. Sodium chloride activates, with maximal activity at an ionic strength of 0.4, but it causes less activation at higher concentration. Sodium sulphate inhibits at ionic strength 1, but it activates at low concentration [A2007]. In one study only two of seven carbonyl reductases (E.C. 1.1.1.184) examined, pI 4.9 and 6.0–6.3, reduced daunorubicin [A3950].

Rat liver enzyme, molecular weight 39 000, pI 6.3 and optimum pH 8.5–9.0, is probably monomeric and requires NADPH. It also reduces some sugar aldehydes and straight chain aldehydes. The amino acid composition has been determined [A50, A1327].

Two classes of reductase found in both human and rabbit liver have optima at pH 6.0 and 8.5, whereas in mouse and rat liver the optimum pH is 8.5. These activities can be further separated by isoelectric focussing: Rabbit 'pH 6.0' enzyme: molecular weight 32 300, 3 isozymes, pI 4.8, 5.3 and 6.3; 'pH 8.5' enzyme: molecular weight 36 000, 2 isozymes, pI 5.9 and 6.3, with numerous minor forms in both classes. Human 'pH 6.0' enzyme: molecular weight 34 500; 'pH 8.5' enzyme, molecular weight 38 700; these are less clearly separated into isozymes than rabbit enzyme. Relative to rabbit, the activities in mouse, rat and man are very low. Adriamycin is also a substrate for human and rabbit enzymes, molecular weights 34 500 (man) and 32 800 (rabbit). In man, the isozymes are similar to daunorubicin reductases with four isozymes, all with pI 5.4, as well as a number of minor isozymes [B169].

Steffimycinone reductase

This reaction, which occurs in Streptomyces nogalacter, forms steffimycinol with an optimum at pH 7. It requires NADPH; NADH is inactive [A2759].

Salutaridine reductase (E.C. 1.1.1.248)

Papaver somniferum enzyme, molecular weight 52 000, pI 4.4 and optimum pH 6.0–6.5 (reverse reaction pH 9.0–9.5), forms (7*S*)-salutaridinol, a precursor of morphine. It is highly specific [K758].

Other ketone reductases

Rabbit liver enzyme, which is cytosolic, molecular weight 33 000 and optimum pH 6.2, requires NADPH. It reduces aryl ketones as well as alkyl ketones such as 2-(4-(2-oxocyclopentyl-methyl)phenyl)propionate [D125].

Aldehyde dehydrogenases (E.C. 1.2.1.3, 1.2.1.4, 1.2.1.5, 1.2.1.7, 1.2.1.28, 1.2.1.29, 1.2.1.39)

$$R.CHO \rightarrow R.COOH$$

Pig brain contains an enzyme that oxidizes the physiologically important aldehydes 5-hydroxyindole-3-acetaldehyde, 4-hydroxy-3-methoxy- and 3,4-dihydroxyphenylacetaldehydes, D-3,4-dihydroxyphenylglycollaldehyde as well as *d*- and *l*-*p*-hydroxyphenylglycollaldehyde [A1287].

Rat liver enzyme is found in mitochondria (three isozymes, pI 5.4, 5.6 and 6.9) and cytosol (five isozymes, pI 5.8, 6.05, 6.15, 6.6 and 7.4). All of them oxidize *p*-nitrobenzaldehyde and 3,4-dihydroxyphenylacetaldehyde [B659]. Cytosolic enzyme oxidizes phenylacetaldehyde, and is induced by phenobarbital, DDT, polychlorobiphenyls and other xenobiotics [A2973]. Another study found three peaks of activity towards *m*-nitrobenzaldehyde. One isozyme required NADPH and a second required NADH. Both were identified as E.C. 1.1.1.2, the second being a 3α-hydroxysteroid dehydrogenase. The third required NADH, and was identified as alcohol dehydrogenase, E.C. 1.1.1.1 [A1688].

Cucumber enzyme oxidizes indole-3-acetaldehyde to indole-3-acetate. It is a metalloflavoprotein that does not require cofactors [A3378].

Achromobacter euridice phenylacetaldehyde dehydrogenase (E.C. 1.2.1.39), optimum pH 8.9, requires NAD^+ (NADH does not catalyze the reverse reaction) and a monovalent cation (K^+ is best). The reaction is irreversible and is highly specific; there is some action with indole-3-acetaldehyde, but other aldehydes show poor activity. It is unstable, but it is protected by substrates or 10 per cent acetone [K869].

Acinetobacter calcoaceticus benzaldehyde dehydrogenase II has a molecular weight of 51 654, with 484 amino acid residues, based on nucleotide sequencing [J621]. It is a tetramer, optimum pH 9.5 [E596]. The reaction is not reversible [E473].

Dehydrogenase I, which is involved in mandelate metabolism, is a tetramer, subunit molecular weight 56 000, optimum pH 9.5 and pI 5.5. It requires NAD^+; $NADP^+$ is less effective. It oxidizes a large range of benzaldehydes. There is an associated esterase activity with *p*-nitrophenyl acetate as substrate [F366].

A Flavobacterium constitutive benzaldehyde dehydrogenase requires NAD^+, whereas a dehydrogenase induced by phenylglycine requires phenazinemethosulphate [E355].

Pleurotus eryngii aryl alcohol dehydrogenase oxidizes benzaldehydes and cinnamaldehydes, but mostly at a slow rate [G668].

Pseudomonas coniferaldehyde dehydrogenase, apparently a homodimer, molecular weight about 86 000 and optimum pH 8.8, requires NAD^+. Other substrates are *trans*-cinnamaldehyde, sinapaldehyde and benzaldehyde, but not vanillin [J707].

Pseudomonas putida produces two different aldehyde dehydrogenases depending on whether it is grown on *p*-cresol or 3,5-xylenol. The former enzyme is stable at 4°, but the latter enzyme is somewhat unstable at 4°; stability is improved in 10 per cent ethanol [A30]. Rhodopseudomonas acidophila contains two aldehyde dehydrogenases, one of which preferentially acts on aliphatic, and the other on aromatic aldehydes. The latter is a dimer, molecular weight about 70 000, optimum pH 9.0 and pI 4.74. It oxidizes cinnamaldehyde and a series of benzaldehydes [G246].

Streptomyces aldehyde oxidase, molecular weight about 80 000, oxidizes vanillin [F225].

Benzaldehyde dehydrogenase I has been found in a Bacterium [A730].

Aryl-aldehyde oxidase (E.C. 1.2.3.9)

Streptomyces viridosporus enzyme oxidizes vanillin and a range of benzaldehydes, but is inactive towards phthalaldehyde and aliphatics. It requires oxygen; peroxide is a second product [C284].

3-Formylindole/indole-3-acetaldehyde oxidases (E.C. 1.2.3.7)

Citrus enzyme, molecular weight 200 000 and optimum pH 7.5, forms peroxide as well as indole-3-carboxylate; it is highly specific [H80].

Pea indole-3-acetaldehyde oxidase is composed of two isozymes that are not activated by pyridine nucleotides. The main has optimum pH 4.5, and the other optimum pH 7.0, with 3-formylindole as another substrate. It is not a dismutase [G446, G520]. Another report states that the optimum pH is 8.0; it requires oxygen [B861].

3.4 *Reductions of acids*

Aryl-aldehyde dehydrogenase (E.C. 1.2.1.30)

a. Ar.COOH → Ar.CHO

Clostridium formicoaceticum reductase, a dimer, monomeric molecular weight 67 000, contains iron and tungsten. It reduces a range of benzoates [G416].

Neurospora crassa enzyme has an optimum pH of 7. Benzoate initially yields benzoyl AMP, which then produces benzaldehyde. In the presence of hydroxylamine, benzoate yields benzoylhydroxylamate. Many benzoates and cinnamates are substrates, but a large proportion of these are at best very poor substrates.

The reaction scheme suggested is:

$$\text{Aryl acid} + Mg^{2+} + \text{ATP} \leftrightarrow \text{pyrophosphate} + \text{enzyme.acyl AMP}$$

$$\text{Enzyme.acyl AMP} + \text{NADPH} \leftrightarrow \text{aldehyde} + \text{AMP} + \text{NADP}^{+}$$

[A750].

Nocardia enzyme, molecular weight 140 000 or 163 000 (depending on the measurement method), reduces benzoate, and requires ATP and NADPH [H955]. N. asteroides enzyme, a monomer, molecular weight 152 000, requires ATP, NADPH and Mn^{2+}; benzoyl AMP is an intermediate in the reduction of benzoate. The enzyme acts on substituted benzoates, preferentially on *meta*-substituted acids, but usually not on those with *ortho* substituents. Some longer chain acids are also substrates [G207].

b. Ar.COOH→ Ar.CH$_2$OH

Rat brain enzyme reduces dopac to 3,4-dihydroxyphenylethanol, optimum pH 7.5; it is unclear whether an alcohol dehydrogenase is part of the system. It is activated by Zn^{2+}, Mn^{2+}, Co^{2+} and Cu^{2+}, and EDTA is inhibitory. 3,4-Dihydroxymandelate, 4-hydroxy-3-methoxymandelate and 4-hydroxy-3-methoxyphenylacetate are not substrates [H798].

Cinnamoyl CoA reductase (E.C. 1.2.1.44)

$$\text{Cinnamoyl CoA} \rightarrow \text{cinnamaldehyde}$$

Eucalyptus gunnii enzyme (cinnamoyl CoA: $NADP^{+}$ oxidoreductase), molecular weight 38 000 and pI 7, acts on CoA conjugates of ferulic, sinapic and *o*-coumaric acids [H330].

Glycine max enzyme, molecular weight 38 000 and optimum pH 6.1, requires NADPH. Substrates include CoA conjugates of ferulic, sinapic, *p*-coumaric, 5-hydroxyferulic, caffeic and cinnamic acids. It is inhibited by thiol-binding reagents [A2576]. Spruce and Glycine enzymes are similar; in another study they were found to be dimeric, with the molecular weight of the native enzymes about 70 000. The reaction is reversible [B922].

Populus euramericana enzyme, a homodimer, monomeric molecular weight 40 000, pI 7.5, is found in the stem. Substrates include CoA conjugates of ferulic, sinapic and *p*-coumaric acids, with NADPH as co-substrate [D54].

3.5 *Deamination*

D-Phenylglycine dehydrogenase

Flavobacterium enzyme, which requires phenazinemethosulphate as cofactor, acts on a number of D-amino acids, but not on the L-isomers. The product formed from D-phenylglycine is phenylglyoxylate [E355].

D-Amino-acid oxidase (E.C. 1.4.3.3)

$$R.CHNH_2.COOH + O_2 + H_2O \rightarrow R.CO.COOH + NH_3 + H_2O_2$$

Mouse enzyme acts on phenylalanine and tryptophan [C219]. In one mouse strain some animals are devoid of activity; the genetics demonstrate the involvement of an inactive allele [C219a].

Pig kidney enzyme has a monomeric molecular weight of 38 000 or 39 600 using different methods [A151].

Rat kidney enzyme acts on D-dopa, and is inhibited by D-alanine and benzoate. The appearance of 3,4-dihydroxyphenylpyruvate predominates over formation of dopamine in kidney (presumably formed by reverse transamination of 3,4-dihydroxyphenylpyruvate followed by decarboxylation), but only at high substrate concentrations [A1393].

Studies on human, African green monkey, rat, pig and mouse kidney found activity with phenylalanine and tryptophan, but in chicken and frog the tryptophan activity was missing [C114].

Fish enzymes (electric eel, rainbow trout, carp, crucian carp and catfish) act on phenylalanine but not on tryptophan [C219].

L-Amino-acid oxidase (E.C. 1.4.3.2)

$$R.CHNH_2.COOH + O_2 + H_2O \rightarrow R.CO.COOH + NH_3 + H_2O_2$$

These enzymes are an important component of snake venoms, and their concentration is sufficiently high in some to enable the raw venom to be used as a reagent for measuring specific amino acids.

Cerastes vipera oxidase is a dimer containing FAD, molecular weight 122 000, monomeric molecular weights 61 000 and 64 000 and optimum pH 7.5; phenylalanine is a substrate. It is activated by Cu^{2+} and Mn^{2+}, and inhibited by EDTA [G417].

Trimeresurus (Taiwan habu snake) venom enzyme is dimeric, molecular weight 140 000 and pI 5.4, and contains two mol FMN/mol. Substrates are phenylalanine and tyrosine [E770]. Another study found pI 8.4 for habu snake enzyme. During purification the specificity changes, suggesting that there are several isozymes [A636].

Viper palaestinae enzyme is composed of three isozymes, molecular weight 130 000 and optimum pH 8.8 with kynurenine as substrate [A1044].

Neurospora crassa enzyme, molecular weight 300 000, and optimum pH 9.5 probably contains four FAD/mol; phenylalanine is a substrate [A404].

L-Phenylalanine dehydrogenase (L-phenylalanine: NAD^+ oxidoreductase, deaminating, E.C. 1.4.1.20)

$$\text{L-Phenylalanine} + NAD^+ \leftrightarrow \text{phenylpyruvate} + NH_4^+ + NADH$$

Sporosarcina ureae and Bacillus sphaericus enzymes have been crystallized. They are both probably homooctamers, molecular weights 305 000 and 340 000, pI 5.3 and 4.3, optimum pH 10.5 and 11.3 (oxidation), and 9.0 and 10.3 (reduction) respectively. S. ureae enzyme acts on phenylalanine (best) and on tyrosine, tryptophan, phenylalaninamide, phenylalanine methyl ester and L-phenylalaninol. B. sphaericus acts on phenylalanine and tyrosine, but poorly if at all on the other foregoing compounds [E445]. S. ureae enzyme aminates pyruvates [E423]; its molecular weight has also been reported to be 290 000 [D919].

Microbacterium enzyme, a homooctomer, monomeric molecular weight 41 000 and pI 5.8, requires NAD^+ [J534].

Thermoactinomyces intermedius enzyme is a hexamer, monomeric molecular weight 41 000, that acts reversibly on L-phenylalanine [G222].

Nocardia enzyme is monomeric, molecular weight 42 000 and optimum pH 10 with phenylpyruvate, *p*-hydroxyphenylpyruvate and indole-3-pyruvate as substrates; other pyruvates are not substrates. The enzyme does not act on D-phenylalanine or other amino acids [F394].

Rhodococcus enzyme, optimum pH 10.1 for the forward reaction and 9.25 for the reverse, acts on phenylalanine (best), tyrosine and tryptophan and the corresponding pyruvates [E451]. It requires NAD^+ or analogues; NADPH and analogues are inactive. The reaction sequence is binding of NAD^+ prior to phenylalanine; the products are released in the order ammonia, phenylpyruvate and NADH [K558].

Bacillus sphaericus enzyme acts on several phenylpyruvates with *para* substituents, and phenylpyruvate analogues with different side chain lengths. The products are the corresponding amino acids, which are formed quantitively [G168].

Activity has only been found in one out of many Brevibacterium strains tested [D90].

L-Tryptophan dehydrogenase and 2′,3′-dioxygenase (E.C. 1.4.1.19)

Pea (optimum pH 8.5), maize and tomato enzymes require NAD(P) for the reversible formation of indole-3-pyruvate [G483]. It is found additionally in Prosopis juliflora (mesquite) and wheat, but not in Brassica [D968].

Chromobacterium violaceum L-tryptophan 2′,3′-dioxygenase forms 2′,3′-dehydrotryptophan as the initial product, which can then isomerize to the imine, from which indole-3-pyruvate is formed by hydrolysis. The reaction, which is specific for the L-isomer, requires an unmodified indole nucleus and a carboxyl group. It also acts on some peptide hormones that contain tryptophan residues as well as other tryptophan analogues [H664]. The enzyme is polymeric, molecular weight 68 0000, with monomeric molecular weights 74 000 and 14 000, pI 4 and a broad optimum pH 3–8. It is a haem-containing mixed function oxidase, which forms peroxide from oxygen. N-substituted tryptophans (which cannot isomerize) form 2′,3′-dehydro analogues of the substrates. Tryptamine, N-substituted tyrosine and phenylalanine are not substrates [H248]. Reaction with N-benzyloxycarbonyltryptophan involves a *syn* elimination [A2770].

In Pseudomonas N-acetyl-L-tryptophanamide forms N-acetyl-2′,3′-dehydrotryptophanamide in two steps, the first of which is enzymatic, and forms 5-(3-indolyl)-2-methyl-2-oxazoline-4-carboxamide. This then forms 2′,3′-dehydrotryptophanamide non-enzymatically with a time lag; at a lower pH, β-hydroxytryptophanamide is formed. The latter compound is further oxidized to β-oxotryptophanamide [A3617, C511]. Other products are indole-3-glycolaldehyde and probably indole-3-glyoxal [A3009].

Monoamine oxidases (MAO) (E.C. 1.4.3.4)

$$R.CH_2.NH_2 + O_2 + H_2O \rightarrow R.CHO + NH_3 + H_2O_2$$

These enzymes play an important part in the physiological control system for the catecholamine neurotransmitters in animals. Studies on these enzymes have led to the development of drugs that affect the amount of catecholamine available for neurotransmission, and, in consequence, valuable treatments for neurologically-induced illness have been developed as a result of the enormously extensive studies on this enzyme system. Of the many thousands of studies on monoamine oxidases it is possible to include in this review only a few that address basic enzyme properties.

a. Mitochondrial MAO.

MAO activity, which is associated with the inner mitochondrial membrane, is separated into two forms, distinguishable by inactivation by low concentrations of clorgyline at 10^{-8} M (MAOA) and deprenyl at 10^{-6} M (MAOB); these are ‘suicide’ inhibitors for MAO. In general, serotonin is oxidized by MAOA, and phenethylamine by MAOB [A1613, A1909, A2727, A2588]. Neither inhibitor is cleanly associated with MAOA or B; both inhibitions overlap in a concentration-dependent manner. Both clorgyline and pargyline (also an A inhibitor) and deprenyl initially inhibit reversibly, and this is followed by an irreversible phase. The conversion rate of clorgyline-inhibited MAOB (rat liver) into irreversibly inactivated enzyme is much slower than for MAOA. Deprenyl-inhibited MAOB is slowly converted into an irreversibly inactivated form, whereas there is no irreversible inactivation of MAOA by deprenyl [A1670, C46].

After solubilising human frontal lobe MAO with octylglucoside, MAOA and B were separated by chromatography on DEAE-Sepharose CL-6B [C682].

Studies using dopamine specifically labelled with deuterium in the α-position have demonstrated that the *R*- and not the *S*-deuterium is removed [D606]. α,α-D_2-Phenethylamines (phenethylamine, tyramine and *m*-tyramine) are oxidized by rat liver microsomal enzyme at 25% of the rate for unlabelled compounds; the β,β-D_2-analogues show a slight enhancement in activity [B660].

Kinetic studies suggest that the substrate amino group hydrogen bonds to an amino group in the enzyme. Introduction of a β-or *p*-hydroxyl group leads to a sharp drop in binding free energy, accompanied by a shift from B to A type substrates, possibly with a shift in orientation of the bound substrate [A1957].

Human MAOB acts on phenethylamine, dopamine and tyramine [F754]. By using clorgyline to inhibit MAOA and deprenyl to inhibit MAOB, studies on liver, kidney and cerebral enzyme have found that serotonin and 3-methoxytyramine are substrates for MAOA and phenethylamine for MAOB, whereas dopamine, tyramine and tryptamine are substrates for both MAOA and B [A2967, E55].

Phenethylamine is oxidized by human lung MAOA and B, whereas serotonin only by MAOA [A3957].

Dopamine is oxidized mainly by MAOB in dopamine-rich human brain areas, whereas MAOA contributes significantly to dopamine oxidation in most other tissues [A3688].

Monoamine oxidase activity is almost unaltered from normal in a range of areas in postmortem human parkinsonian brain [A381].

Human placental MAOA oxidizes phenethylamine and serotonin [A3957], as well as kynuramine, MPTP and a large range of MPTP analogues substituted in the phenyl ring [F395].

Activity of human platelet enzyme, which is almost exclusively MAOB [A1901] fluctuates during the oestrous cycle, with a maximum about the time of ovulation, and then drops by about 23 per cent to a minimum 5–11 days later [A367]. Activity is evenly distributed throughout human heart [A756]. A particulate enzyme in blood, thought to be mitochondrial, has an optimum pH between 6 and 8.8 depending on substrate, apparently without correlation with A and B status [A2010].

Rat liver enzyme can be distinguished into two distinct activities by the action of clorgyline and aryl nitriles [A367]. It is inhibited by the thiol-binding compounds nitroprusside and 5,5′-dithiobis(2-nitrobenzoate). Neither arsenite nor thiols are inhibitory [A238]. Hydroxylamine rapidly inhibits activity towards tyramine and benzylamine within one minute, whereas after one hour the activity towards tyramine, but not benzylamine is increased [A 1760]. It migrates electrophoretically as several bands, but after treatment with perchlorate, which releases lipid without change in activity, only one band is found [A 1039]. Using tyramine as substrate K_m for MAOA and B are different [B661]. Both MAOA and B oxidize tryptamine, N-methyl-tryptamine, serotonin and 5-methoxytryptamine, whereas N-methylserotonin, bufotenine and N,N-dimethylserotonin are selective for MAOA, and N,N-dimethyltryptamine is selective for MAOB [B653].

Another study found that rat brain MAOA is heat stable, whereas B is labile. A and B are partially separated by sucrose density gradient centrifugation [A1227, A1309, Al841]; more MAOB is found in high-density mitochondria [A1909]. Both A and B are fairly evenly distributed throughout brain [A2386]. Brain enzyme is inhibited reversibly by β-propiolactone and β-nitropropionate [A1800], by apomorphine [A2069] and atropine (also heart enzyme) [A1898]. The specificity of A and B are similar to human enzyme (above) [A1324]. Brain mitochondrial enzyme has an optimum pH of 7.5 for dopamine and tyramine, 8.2–8.5 for serotonin and tryptamine, and 9.1 for kynuramine. Similar results were obtained with beef brain mitochondrial enzyme [A1285]. Brain MAOA oxidizes dopamine better than MAOB [A2484]; *m*- and *p*-tyramine are substrates for both isozymes, whereas *o*-tyramine is substrate only for MAOB [A3667]. The ratio of B to A in synaptosomal mitochondria is about 1/2 that for extrasynaptosomal mitochondria from whole brain; the ratio is lower in striatal mitochondria than in cerebellar, and in cortical mitochondria it is lower still [A2487]. One study, however, suggested that both isozymes are probably extraneuronal [A3619]. Rat skeletal muscle contains both MAOA and B; A is more heat labile than B [A1798].

Rat brain enzyme activity increases sixfold to the adult level in the first 10 days following parturition [A2490].

Spinal cord transection in rat has no effect on cord MAOA or B activity [A562].

3-(N-Cyclopropyl)-5-phenylethylamino-1,2,4-oxadiazole is specific inhibitor for rat liver MAOA. Substitution with methyl or chloro in the *p*-position of the phenyl ring does not alter selectivity. Other substrates substituted on the *m*-position partially or completely lose their selectivity for MAOA [A794].

Rat A and B arterial isozymes are partially soluble. After treatment with 6-hydroxydopamine 70 per cent of MAOA activity was lost, whereas B was unaffected [A1045]. Rat lung shows both MAOA and B activities [A2161, A3863]. Another study found MAOB in pancreas as well as in lung, liver, kidney and brain [A2835].

Rat placental MAO activity increases threefold from day 15–20 of gestation, and then declines by 50 per cent at term [A2469].

Rat heart enzyme is mainly MAOB at three weeks post partum, a mixture of A and B at eight weeks and only A in the adult [A3665].

Studies on partially lysed MAOA and B from rat hepatoma indicate that the peptide chains are different in these isozymes [A3683].

In rat, a series of oxo analogues of phenylethanolamines are converted into mandelic acids, with phenylglyoxals as the initial products. The reaction is much faster than with the corresponding phenylethanolamines [J287]. Further studies (Goodwin, B.L., Ruthven, C.R.J., unpublished) with human placental MAO showed that MAOA oxidizes these β-oxophenethylamines.

Beef liver MAOB acts on benzylamine, MPTP and a large number of MPTP analogues substituted on the phenyl ring [F395]. These results (and the results for human MAOA) suggest that MAO is involved in MPTP-induced parkinsonism, and if parkinsonism is caused by the action of an environmental amine, the involvement of MAOB might explain the claimed beneficial effects of deprenyl in parkinsonism; MPTP does not appear to be the enviromental toxin [K948].

A study on beef brain cortical enzyme, using clorgyline, pargyline, harmaline and deprenyl as inhibitors suggested that the inhibition pattern is too complex to be simply due to MAOA and B. A two-stage inhibition is postulated, involving a rapid reversible step at two different binding sites, followed by a slow irreversible phase [A2181]. Treatment of this enzyme with cyanoacetyl hydroxyethylhydrazide induced the ability to oxidize histamine, cadaverine and analogues; this was prevented by pre-treatment with clorgyline, but not with deprenyl [A1916].

Guinea pig brain activity increases 2.5-fold in the first 35 days following parturition, to 50 per cent of the adult level [A2490].

In hamster the ratio of MAOB to A activity varies from tissue to tissue. The ratio in brain is about 0.03, and in heart, lung, liver and spleen it is higher, but less than 1. In rat, the ratio is also consistently less than 1 in these tissues, but the ratio between tissues is different from that found in hamster. In rabbit the ratio for the above tissues is close to 1 [A2588].

Japanese monkey platelet enzyme is mostly B, with three components, molecular weights about 60 000, pI 5.5, 6.5 and 7.0, whereas liver enzyme has pI 6.5. They can be solubilized with Triton X-100 [F758].

Mouse lung mitochondria, after fractionation by sucrose density gradient centrifugation, are separated into fractions with more MAOB than A in the higher density mitochondria [A1895]. Enzyme activity in the eye decreased after day one post-partum, increased to a peak at day five and then fell to the adult level at 14 days [A946]. In brain, it increased from day one to the adult level at two weeks except in cerebellum, where a slight increase continued until after week six [A1105]. Using tryptamine as substrate, brain and liver enzymes are inhibited by a range of β-carbolines, especially harmine and harmaline [A2178].

Pig brain enzyme shows a time-dependent inhibition by several N-substituted propargylamines, although this study found that not all compounds with this structure are inhibitory [A74]. A pig liver mitochondrial enzyme that acts on benzylamine (a benzylamine oxidase?) is a polymer with molecular weight about 1 200 000 containing about eight Cu and subunit molecular weight 146 000 [A1033]. Pig dental pulp membrane-bound enzyme (presumably mitochondrial) oxidizes a range of amines, particularly phenethylamine and tryptamine [B206]. Pig heart enzyme oxidizes serotonin, tyramine and benzylamine, it is inhibited by clorgyline, and is considered to be heterogeneous [A2587]. The development pattern differs from organ to organ. Liver enzyme activity falls by 40 per cent at birth, and then increases threefold to the adult level at one month. Kidney and spleen activities rise steadily from 10 days pre-partum until 70 days post-partum. Heart activity reaches a maximum near birth, and then declines by 80 per cent over two months. Brain activity declines slightly near birth, and then increases slightly to the adult level; adrenal enzyme shows a similar pattern [A1569].

Rabbit lung and brain enzymes are inhibited reversibly by imipramine, DMI and DDMI [A94]. The ratio of MAOA to B in heart, kidney, lung, liver, spleen and brain is about 1 [A2588], and both A and B are found in platelets [A2326].

Terbutaline and orciprenaline are not substrates for rat and human liver MAO [A724].

The ratio of chick CNS MAOA and B depends on site, using phenethylamine and serotonin as markers. The highest ratio is found in spinal cord and lowest in cerebrum [A2386]. Brain enzyme activity is detectable at 14 days incubation, and increases steadily until 2 days post-hatching [A1413].

Squid brain enzyme can be differentiated into A and B [E147].

b. Benzylamine oxidase (E.C. 1.4.3.6).

In man, the enzyme is found in all vascular tissues, localized in smooth muscle [B537]. Serum activity is reduced in burns and cancer patients [A2965]. It appears to be present in lung and placenta; however, benzylamine oxidation is considered to be catalyzed by MAOB in liver [A3957].

Human plasma and rat lung enzymes are not inhibited by deprenyl at 10^{-4} M (both MAOA and B are inactivated at this concentration) or by clorgyline, but semicarbazide, procarbazine and carbidopa are inhibitory. It is found (high activity) in aorta and lung, with slightly less in colon, ileum stomach, portal vein and duodenum, with low activity in some other tissues, including serum. Rat enzyme also oxidizes dopamine and phenethylamine, but these are not substrates in man [A2835, A3690].

Beef aorta enzyme is separable into two fractions; the major is particulate (probably not mitochondrial), and the minor cytosolic. N-Methylbenzylamine is a substrate only for the particulate enzyme. Carbonyl-binding reagents, especially pargyline are inhibitory [A574]. It is proposed that the mechanism for beef liver enzyme is ping-pong, with an imine intermediate being hydrolyzed to ammonia and aldehyde, with reduced flavoenzyme reacting with oxygen to form peroxide [A967].

Pig plasma enzyme is a dimer, monomeric molecular weight 95 000. Water of hydration bound to enzyme Cu^{2+} rapidly exchanges [A1025]. Heart enzyme has a molecular weight of 97 000, the same as plasma enzyme. It, also, is inhibited by carbonyl-binding reagents, and contains two Cu/mol [J833].

Pig dental pulp enzyme (soluble) oxidizes benzylamine, tryptamine and (less well) tyramine, but serotonin and phenethylamine are very poor substrates. This distinguishes it from MAOA and B [B206].

Rat brain contains a soluble enzyme that oxidizes kynuramine, but is not inhibited by deprenyl or clorgyline; this may be benzylamine oxidase [C186].

Rabbit enzyme that oxidizes mescaline appears to be similar to benzylamine oxidase [A1666].

c. Microorganism MAO.

Arthrobacter globiformis contains a soluble, Cu^{2+}-dependent enzyme (E.C. 1.4.3.6) that oxidizes phenethylamine. The peptide chain contains a 2,4,5-trihydroxyphenylalanyl residue as its quinone; this acts as cofactor in the reaction. It is formed spontaneously from a specific tyrosyl residue by the action of Cu^{2+} [H298]. The enzyme is a homodimer, molecular weight 141 000, with optimal activity and stability at pH 6.5. The products are presumably aldehydes, with peroxide as a second product. Other substrates include 3-phenylpropylamine, tyramine, dopamine, octopamine, tryptamine and 4-phenylbutylamine, but not benzylamine or aliphatic amines; carbonyl-binding reagents are inhibitors [F626, H912].

E. coli contains a phenethylamine oxidase with the above quinone cofactor, which is formed spontaneously by incubation of the enzyme at 30° and pH 6, especially at low enzyme concentration [H422].

Micrococcus luteus enzyme is a homodimer containing FAD, monomeric molecular weight 49 000 by standard methods; DNA studies indicate a molecular weight of 49 100, with 443 amino acid residues. Substrates include tyramine, adrenaline, dopamine and noradrenaline. It is inhibited by reagents specific for MAOA and B [K420].

MAO is membrane bound in some bacterial species, but is not present in others. It has been detected in Klebsiella, Escherichia, Salmonella, Pseudomonas, Brevibacterium and Micrococcus, and is induced by tyramine. Substrates include tyramine, dopamine, and (except in Micrococcus) octopamine and noradrenaline [B225].

d. Unspecified enzymes.

Helix pomatia enzyme is found in crop and nervous system, and a little in the heart [A657].

Hymenolepis diminuta enzyme is membrane-bound. It oxidizes (in decreasing order) dopamine, adrenaline, noradrenaline, tryptamine, tyramine and octopamine, but not serotonin or benzylamine. Inhibitors include cupferron, α,α-dipyridyl, iodoacetamide, pargyline, nialamide and iproniazid, but not azide, hydroxylamine or semicarbazide [A3815].

Tetrahymena pyriformis MAO oxidizes tryptamine, dopamine and serotonin. The activity increases in the pH range 6.5–7.8 [A3131].

Spermine: oxygen oxidoreductase (deaminating; E.C. 1.4.3.3)

Sheep plasma enzyme acts on benzylamine and nuclear-substituted benzylamines, tyramine, tryptamine and serotonin [A535].

Monoamine dehydrogenases

Rat brain enzyme, which has also been called ephedrine-neotetrazolium chloride reductase (neotetrazolium chloride is the hydrogen-receptor co-substrate), is mainly mitochondrial, but some is cytosolic, with a dialysable heat-labile cofactor; $NADP^+$ activates it. Substrates include ephedrine, adrenaline, tryptamine, serotonin, tyramine and noradrenaline, optimum pH 7.5. It is inhibited irreversibly by cyanide, but not by MAO inhibitors such as iproniazid or pargyline; iron chelators, such as *o*-phenanthroline and α,α-dipyridyl, are inhibitory. The nature of the reaction is not specified in Chemical Abstracts, but it appears to act on the amino groups, probably to form carbonyl compounds [A1053, A2489, A2490, A2802].

An Alcaligenes faecalis enzyme called aromatic amine dehydrogenase acts on tyramine and a range of phenethylamines to form aldehyde and ammonia, but not peroxide. The protein chain contains a crosslinked tryptophan tryptophanylquinone group, which is part of the redox system [H367, H421].

Pseudomonas aromatic amine dehydrogenase (aralkylamine dehydrogenase, E.C. 1.4.99.4) is an inducible tetramer composed of two pairs of monomers, molecular weights 46 000 and 8000 and optimum pH 7.5–8.0, depending on substrate (tyramine, serotonin, tryptamine, phenethylamine, benzylamine and, best, dopamine); the products are ammonia and aldehydes. Traces of iron and copper are found, but these may be artifacts. The amino acid composition has been determined [C517, K709].

Octopamine hydro-lyase (E.C. 4.2.1.87)

Pseudomonas aeruginosa enzyme is soluble and forms *p*-hydroxyphenylacetaldehyde [K709].

Synephrinase (E.C. 4.2.1.88)

Arthrobacter synephrinum enzyme, optimum pH 8.0 is cytosolic. It converts *p*-sympatol into *p*-hydroxyphenylacetaldehyde and methylamine, apparently anaerobically. It requires Mg^{2+} or Ca^{2+}, and is further activated by thiols. Tyramine, N-methyltyramine and hordenine are not substrates [A2915].

Nocardia also demonstrates this activity [E615].

Diamine oxidase (E.C. 1.4.3.6)

p-Dimethylaminomethylbenzylamine is oxidized by human kidney enzyme. A study on inhibitors revealed that a series of dimethylsulphonium and trimethylammonium compounds with a hydrocarbon chain consisting of up to 12 methylene units, trimethylsulphonium, tetramethylammonium and ammonium compounds as well as phenelzine are competitive inhibitors, whereas pargyline is an uncompetitive inhibitor. A series of isothiuronium and guanidinium compounds with a chain consisting of up to 12 methylene units, 5-methylisothiuronium, guanidine, methylguanidine, hydralazine, iproniazid, nialamide and other compounds are noncompetitive inhibitors [A1686].

An enzyme in pig kidney that is claimed to be diamine oxidase acts on *p*-substituted benzylamines and many aliphatic amines [A3693]; it may be identical with benzylamine oxidase (R. Lewinsohn, personal communication).

3.6 *Oxidative removal of substituents on amino groups*

N-Demethylation

$$R_1R_2NR_3 \rightarrow R_1R_2NH$$

Aminopyrine N-demethylase activity is lower in red-winged blackbird than in rat; this difference parallels lower levels of P450 in bird microsomes [A358]. In male rat, activity increases 20-fold in the first 10 weeks post-partum, most of which occurs in the first four weeks. In female rat, activity is three times higher at birth than in the male, and it increases a further threefold in the following five weeks [A736]. The results of another study on rat liver differed considerably from this in detail [A71]. In rabbit, the activity increases sharply from a low level at about 20 days post-partum, and then remains level or declines slightly to the adult level [A1980].

Aminopyrine is demethylated by catalase and cumene hydroperoxide by a free-radical mechanism. It is suggested that the active site for this reaction differs from the one involving hydrogen peroxide [B378].

d-Benzphetamine is demethylated as the preferred substrate by all six P450 fractions separated electrophoretically in one study on human liver [C798].

p-Chloro-N-methylaniline demethylase is found in chick embryo liver and human foetal liver, adrenal, brain and lung [B815].

N,N-Dimethylaniline is demethylated by rat P450, horseradish peroxidase, beef liver catalase, human placental and Glycine max lipoxygenases, lactoperoxidase and haemoglobin, whale myoglobin and chloroperoxidase. Using substrate labelled with deuterium on one methyl group an isotope effect is observed with all these enzymes, the extent of which is specific for each enzyme. High isotope effects found with most of the haemoproteins tested are considered to indicate a rate-limiting step of hydrogen atom abstraction from the α-carbon (in this case defined as the methyl group), whereas deprotonization of the α-carbon is postulated for a low isotope effect, as observed with P450 and chloroperoxidase [C814]. Lipoxygenases also act on a range of other N-methylated anilines, with the formation of formaldehyde, but without any evidence for the formation of N-oxides as intermediates. Glycine enzyme has an optimum pH of 6.5 [K419].

N-Ethyl-N-methylaniline is dealkylated to both monodealkylated products. NADH enhances dealkylation, but not N-oxidation; dealkylation of the corresponding N-oxide is very feeble. These results, combined with the effects of selective inhibitors indicate that most of the dealkylation occurs directly, rather than via the N-oxide [A310].

Ethylmorphine N-demethylase is found in quokka, kangaroo, bettong, bandicoot, and possum (marsupials), as well as in rat [A2420]. Activity of rat and mouse liver enzymes is increased by EDTA and decreased by Fe^{2+}; this probably correlates with inactivation by lipid peroxides [A89].

Imipramine demethylase activity found in rat hepatocyte microsomes is suppressed by SKF 525-A, a P450 inhibitor [K168]. Glycine lipoxygenase catalyzes demethylation, with peroxide as oxidant; apparently a free radical is an intermediate, and formaldehyde is released. A number of imipramine analogues are also demethylated [K359].

Human haemoglobin can dealkylate N-methylaniline and benzphetamine [D135].

N-Methylephedrine is demethylated by rat liver microsomes. The enzyme is inhibited by CO or SKF 525-A, but not by methimazole or by pre-incubation [F501].

Rat liver P450 N-demethylates 15 different compounds, utilizing either O_2 and NADPH, or hydrogen peroxide. There is no correlation between the reaction rates corresponding to these mechanisms for the different compounds used; it is therefore considered that these are independent reactions [C669].

In rat, guinea pig and mouse N-demethylase activity is correlated with P450 activity [A627].

With some tertiary amines demethylation occurs in two stages, with the amine N-oxide as intermediate. This is described under the enzymes involved.

Demethylation of N-methylated amides

Rat liver microsomes demethylate N,N-dimethylbenzamide to form N-methylbenzamide and formaldehyde, with the corresponding N-hydroxymethyl compound as an intermediate. Several *para*-substituted N,N-dimethylbenzamides are also substrates [G683].

Amides from hippurates

Human bifunctional peptidylglycine α-amidating monooxygenase requires ascorbate, copper and oxygen, with hippurates as substrates. Salicyluric acid initially forms N-salicyl-α-hydroxyglycine, which then forms salicylamide and glyoxylate [K660].

Lipoxygenase (E.C. 1.13.11.12)

Glycine max enzyme, optimum pH 6.5, dealkylates aminopyrine by a free radical reaction, forming formaldehyde and peroxide. This is claimed to be a novel reaction [J526].

Formyltetrahydrofolate dehydrogenase (E.C. 1.5.1.6)

Pig liver enzyme, optimum pH 7.5, requires NADP; NADPH is ineffective. Tetrahydrofolate is formed, with the release of carbon dioxide from the formyl moiety [K877].

Pteridine/folate fission

Pseudsomonas acidovorans enzyme, named pteridine 6-methylaminohydrolase, acts on pteroate or folate (it is unclear which compounds were tested) to cleave the C–N bond linking the pteridine and aminobenzoate moieties. Structural considerations indicate that this is an oxidative reaction, with formation of *p*-aminobenzoate [A971].

Dealkylation of quaternary bases

N-Methylnaltrexone is demethylated by rat, dog, man and mouse [F486].

N-Debutylation

Man and monkey debutylate bupivacaine, but the reaction was not further studied [A255, J881, K14].

N-Oxide dealkylation

$$R_1R_2R_3NO \rightarrow R_1R_2NH$$

a. Aryl N-oxides (dimethylaniline-N-oxide aldolase; E.C. 4.1.2.24)

N,N-Dimethylaniline and *p*-cyano-N,N-dimethylaniline N-oxides are demethylated by bacterial $P450_{CAM}$ and rabbit liver $P450_{LM2}$ [D40]. Cytochrome c is also able to catalyse this reaction [B185].

b. Alkyl N-oxides

This reaction is observed *in vivo* in mammalia with N-oxides of tricyclic drugs such as imipramine [A3656], but is not well documented. It is complicated in that reduction to the parent drug followed by demethylation without involvement of the N-oxide may contribute to the observed results, although in man the urinary excretion of DMI is very much greater after dosing with imipramine-N-oxide than after imipramine (personal unpublished observation).

3.7 *Oxidations and reductions involving nitrogen atoms*

Tyrosine N-monooxygenase (E.C. 1.14.13.41)

Sorghum bicolor enzyme, which catalyzes this reaction is a P450. This is a key step in the formation of dhurrin [A3175, A3335, G228, H331, K711].

Oxidation of amino to nitro groups

$$R.NH_2 \rightarrow R.NO_2$$

Chlorphentermine is oxidized to 1-(*p*-chlorophenyl)-2-methyl-2-nitropropane in rabbit and man; it appears that the corresponding hydroxylamine is an intermediate [A177].

Oxidation of hydroxylamines to oximes

Rat liver microsomes oxidize N-hydroxyamphetamine to phenylacetone oxime. The enzyme is probably not P450, and oxygen is probably not the oxidizing species [B664].

Amidoxime formation from amidine

Rabbit liver microsomes convert benzamidine into benzamidoxime [E238, F230, G798]; the reaction has also been observed in rat [G990].

Conversion of imines into oximes

$$R.\ C = NH.\ CH_3 \rightarrow R.\ C = NOH.\ CH_3$$

Rabbit liver microsomes act on some acetophenone imines, including 2,6-di-, 2,4,6-tri-, 2,3,5,6-tetra- and 2,3,4,5,6-pentamethylacetophenone imine as well as 2,6-dichloroacetophenone imine. The reaction is also observed in hamster, rat, mouse, guinea pig and ferret. The product is mainly the *anti*-oxime and usually a small proportion (<10 per cent) of the *syn*-isomer [D955].

Rabbit enzyme, which acts on 2,6-dimethyl-acetophenone imine and 2,6-dichloroaceto-phenone imine, is found in liver, with lesser amounts in lung [C243].

Oxime reduction to hydroxylamines

Rat liver microsomal (not P450) and cytosolic oxime reductases which act on acetophenone oxime require NADPH [A1452].

Oxime oxidation to nitro compounds

Both rat and rabbit liver oxidize 1-phenyl-2-propanone oxime to 2-nitro-1-phenylpropane [A1982, A3315].

Oxo formation from oxime

Rabbit liver aldehyde oxidase acts on the oximes of acetophenone, salicylaldehyde and benzamidine to form the corresponding oxo compounds [E358].

Aldoxime and nitrile formation from tyrosine

Triglochin maritima and Sorghum bicolor P450 both oxidize L-tyrosine to *p*-hydroxyphenyl-acetaldoxime. It is postulated that the reaction sequence is:

L-Tyrosine → N-hydroxy-L-tyrosine → N, N-dihydroxy-L-tyrosine → 3-(*p*-hydroxyphenyl)-2-nitrosopropionate;

the latter then decarboxylates and rearranges to the oxime. Further reaction steps lead to the formation of *p*-hydroxyphenylacetonitrile and *p*-hydroxymandelonitrile [H423, K216, K711]; the first of these steps has been confirmed directly in Sorghum, and the formation of dhurrin and *p*-hydroxymandelonitrile from the aldoxime confirms the side-chain hydroxylation reaction [A2914, J504]. Other potential precursors of *p*-hydroxymandelonitrile, such as *p*-hydroxy-phenylacetamide, 1-nitro-2-(*p*-hydroxyphenyl)-ethane, *p*-hydroxyphenylacetaldoxime, tyramine and N-hydroxytyramine are ineffective in Sorghum [A3335].

Salicylhydroxamate reductase

Rat liver enzyme is a heterodimer, molecular weight 140 000–150 000 and pI 5.4. The product is salicylamide. Another substrate (poor) is N-hydroxy-2-acetamidofluorene [H16]. Both guinea pig and rabbit liver aldehyde oxidases catalyze this reaction, with an electron donor. Anthranilhydroxamate is another substrate [C897].

Aldehyde formation from N-nitrosoamine

Mouse liver microsomes form benzaldehyde from N-methyl-N-nitrosobenzylamine, with NADPH. The enzyme has also been found in mouse forestomach, as well as oesophagus, forestomach and liver of rat (in all cases, female) [B131].

N-Oxide formation with tertiary alkylamines

$$R_1R_2R_3N \rightarrow R_1R_2R_3NO$$

Rat liver and brain imipramine N-oxidase is a microsomal flavoprotein. After solubilization 2 isozymes were found, molecular weights 57 000 and 61 000, with optimum pH 8.5; the activity was rapidly lost at 45° [J174]. It is postulated that it is different from the dimethylaniline-oxidizing enzyme [C196]. N-Methylephedrine is oxidized by a rat liver flavoprotein, and inhibited by methimazole or by pre-incubation, but not by CO or SKF 525-A [F501]. Further studies on the liver enzyme found that two systems were involved, and its activity was lost rapidly by pre-incubation at pH 8.5, with little loss at pH 7.5. A high-affinity enzyme was inhibited by methimazole, an inhibitor of the flavin-containing monooxygenase, whereas a low-affinity enzyme was inhibited by SKF 525-A, suggesting that it is a P450 [K168].

Mouse and pig liver microsomal enzymes, which have an optimum pH of about 9, contain FAD and require NAD(P)H. Substrates include imipramine, as well as N,N-dimethyl- and N,N-diethylaniline [D446].

Methylphenyltetrahydropyridine N-monooxygenase (E.C. 1.13.12.11)

Mouse microsomal enzyme acts on MPTP to form the N-oxide [E979].

N-Oxide reductase

Rat liver reduction of the N-oxides of imipramine, cyclobenzaprine and brucine is observed in cytosol, mitochondria and microsomes, and requires NAD(P)H and menadione (or any of a range of quinones). It has been suggested that the cytosolic enzyme is DT-diaphorase plus a haem protein. The first reaction is considered to be the reduction of a quinone by a quinone reductase, followed by a non-enzymatic reduction of the N-oxide by the quinol, mediated by the haem group of haemoproteins [J563].

A rat liver mitochondrial enzyme, bound to the inner membrane, requires NAD(P)H to reduce imipramine-N-oxide as well as aromatic N-oxides. The activity is also found in microsomes, but the relative reaction rates for the substrates are different from those in mitochondria. It is inhibited by oxygen, and mitochondrial enzyme is partially inhibited by carbon monoxide [A1913].

Rat liver microsomes reduce octoclothepine-N-oxide to octoclothepine anaerobically (oxygen inhibits), with NADPH as co-substrate. It is also inhibited by carbon monoxide [B935]. Imipramine and tiaramide N-oxides are substrates, and the reaction requires NADH. It is inhibited by oxygen or carbon monoxide; the enzyme appears to be a P450 [A1847]

Rat liver xanthine oxidase or crude liver extract reduce benzydamine-N-oxide, optimum pH 9 for the xanthine-dependent reduction or pH 7 for the crude extract; there are at least two enzymes in the crude extract. There is a requirement for NAD(P)H or FMN [B307].

Rat blood reduces the N-oxides of imipramine, brucine and cyclobenzaprine, with menadione or other quinols as electron acceptor, and NAD(P)H as coenzyme. The enzyme appears to be haemoglobin [J876].

Rat, mouse, hamster, pig and rabbit liver cytosolic aldehyde oxidase as well as rabbit liver microsomes reduce the N-oxides of imipramine and cyclobenzaprine. Rabbit microsomal enzyme requires NAD(P)H. Cytosolic enzyme is not activated by NAD(P)H, but several compounds, including aldehydes, N′-methylnicotinamide and especially 2-hydroxypyrimidine are effective; with the latter compound the reaction rate is much greater than with microsomes [D218].

Reflux of N-oxides at pH 14 can remove the oxygen moiety [H894].

3.8 *O-Dealkylation*

e.g. 4-Methoxybenzoate monooxygenase (O-demethylating) (E.C. 1.14.99.15)

$$\mathrm{Ar.O.CH_2.R \rightarrow ArOH + RCHO}$$

Human haemoglobin can demethylate *p*-anisidine, anisole, *p*-methoxyphenol and *p*-nitroanisole [D135].

Rat liver microsomes demethylate *o*-, *m*- and *p*-methoxyphenols and *o*-, *m*- and *p*-anisoles. Studies with $^{18}O_2$ show a high oxygen incorporation with *o*- and *p*-methoxyphenols, and low incorporation with *m*-methoxyphenol. This suggests that demethylation of *o*- and *p*-methoxyphenols involves a quinone-type intermediate, with elimination of the entire methoxy moiety [J201]. 7-Ethoxycoumarin deethylase is composed of two isozymes, one low and the other high affinity [D666]. An activity confined to microsomes that is inhibited by carbon monoxide demethylates 2-methoxyoestrone and 2-methoxyoestradiol; it has a requirement for NADPH, suggesting a P450 involvement. Each compound appears to be demethylated by a separate isozyme [B668]. Hepatocyte microsomal enzyme demethylates 2,6-dichloro-4-nitroanisole with a broad optimum between pH 6.0–7.5 [A2855]. 1,3-Dichloro-2-methoxy-5-nitrobenzene demethylase, which is inhibited by vinyl chloride, may be composed of two isozymes. It is not induced by pre-treatment with 3-methylcholanthrene, i.e., it is not a P448 [A3495].

Rabbit liver microsomes demethylate *p*-nitroanisole and *p*-nitrophenetole. With NADPH as co-substrate (optimum pH 7.4) carbon monoxide inhibits, but with NADH as co-substrate (optimum pH 6.0) carbon monoxide is not inhibitory; cyanide inhibits neither demethylation, demonstrating that cytochromes are not involved in the hydrogen transport system [A2545].

Rabbit 7-ethoxycoumarin deethylase activity begins to increase from a low level at day nine post partum in liver and at day 16 in small intestine. At 30 days the activity is at least as great as that in the adult [A1980].

7-Alkoxycoumarin dealkylation has been detected in mouse, guinea pig, rabbit, dog and monkey, with methyl, ethyl and propyl as the alkyl groups; it exhibits a requirement for NADPH and oxygen. 7-Butoxycoumarin is also dealkylated in rat; induction studies indicate that this involves a different enzyme [C160].

Activity of guinea pig, rat and mouse liver microsomal enzyme activity is correlated with P450 activity [A627].

Erythrocyte lysates from bear, beef, cat, dog, guinea pig, man, mouse, rabbit, pigeon, rat and sheep demethylate normetanephrine to noradrenaline, and at least in rat dopa, dopamine and adrenaline are formed from the corresponding 3-O-methyl compounds [A3063].

Red-winged blackbird liver microsomal enzyme acts on *p*-nitroanisole; the measured activity is lower than in rat, and follows a lower P450 content [A358].

Bacillus oxidizes *p*-hydroxyphenoxyacetate to quinol with the incorporation of oxygen from molecular oxygen [A3639].

Chaetomium piluliferum and Xerocomus badius (fungi) enzymes, optimum pH 5 and 7 respectively act on 3- and 4-methoxybenzoate, veratrate and 3,4-dimethoxycinnamate, with oxygen and NADH as co-substrates [K896].

Pseudomonas fluorescens *meta*-demethylating system is composed of two enzymes, one, a non-haem Fe-containing monooxygenase, is a dimer, molecular weight 118 000, and the other a NADH-dependent reductase, molecular weight 80 000 [A3627].

Pseudomonas putida demethylating system is composed of two enzymes, one a Fe-containing monooxygenase, molecular weight 120 000, and the other a Fe–S-containing NADH-dependent reductase, molecular weight 42 000 and optimum pH 8.0. Both are highly unstable in the presence of oxygen. The system demethylates *m*- and *p*-methoxybenzoate; use of uncoupling substrate analogues results in formation of peroxide [A903, A2952, K800]. P. putida Fe–S-containing reductase, pI 4.72, contains about 30 per cent carbohydrate; its amino acid composition has been determined. The Fe–S moiety is plant ferredoxin type, i.e., it contains S^{2-}. The enzyme has been given the trivial name Putidamonooxin I [A2954]; the two components have been called putidamonooxin and NADH-putidamonooxin oxido-reductase [K800]. The monooxygenase contains protein, iron and labile sulphur in the ratio 1:3.3:3.3 (the expected value is 1:4:4). Inhibition is brought about by Cu^{2+}, Hg^{2+}, *p*-chloromercuribenzoate and bathophenanthrolinedisulphonic acid [K801]. The optical absorption is altered by substrate. It has been postulated that this is a previously unreported enzyme type [A2955]. A Pseudomonas enzyme has been described that acts on *p*-methoxybenzoate and is specific for the *para* isomer [A563].

7-Ethoxyresorufin deethylase

Human liver enzyme is a P450 [J299]; many studies have found that 7-ethoxyresorufin can be used as a specific probe for CYP1A2. Activity is also found in lymphocytes [J245].

Rat liver microsomal enzyme dealkylates ethyl-, benzyl- and pentylresorufins [D421]. Liver, kidney and lung activity is associated with just one P450, at that time designated as P448. It is induced by the carcinogens benzo[*a*]pyrene and 1,2,5,6-dibenzanthracene, but not by benzo[*e*]pyrene and anthracene [C748]. Enzyme is also found in lung [J904] and intestine [A2127].

Alligator activity is much lower than in rat [J855].

Channel catfish liver microsomes dealkylate alkylresorufins with methyl, ethyl and benzyl, but not pentyl alkyl groups. After treating animals with 3-methylcholanthrene, all these compounds are dealkylated [E209].

This activity has been found in many other species, including dog, cat, rabbit, guinea pig, hamster, mouse, quail and trout [A2129, C562].

Carbon monoxide-dependent O-demethylase

Clostridium thermoaceticum enzyme demethylates methoxybenzenes and methoxybenzoates. Nitrogen or hydrogen cannot replace carbon monoxide, which is postulated to act on CoA to form acetyl CoA with the released methyl group [K457].

2,4,5-Trichlorophenoxyacetate oxygenase

Pseudomonas cepacia enzyme is a two-component system, one of which is a red protein and the other is a tetramer composed of two pairs of identical subunits, molecular weights 49 000 and 24 000. The reaction requires oxygen and NADH to form 2,4,5-trichlorophenol and glyoxylic acid [H644].

3.9 *Oxidations and reductions of non-aromatic ring systems*

Indane hydroxylation

Pseudomonas putida toluene dioxygenase (E.C. 1.14.12.11) hydroxylates a range of substituted indanes in the 1-position. About 70 per cent of the oxygen comes from water [E567, K207]. Naphthalene dioxygenase (E.C. 1.14.12.12) yields (+)-(1*S*)-indanol and (+)-*cis*-(1*R*,2*S*)indanediol as well as further oxidation products [H580].

Rhodococcus strains form *cis*-(1*S*,2*R*)-indanediol, with one exception, which forms the *trans*-(1*R*,2*R*)-isomer [K305].

In rat 1- and 2-hydroxylation occurs as well as the formation of *cis*- and *trans*-indane-1,2-diol [E533, F617].

1*H*-4-Oxoquinoline monooxygenase

Pseudomonas putida enzyme is a trimer, molecular weight 126 000, which requires oxygen and NADH. It is specific for 1*H*-4-oxoquinoline, which it hydroxylates at position 3 [G812].

Flavanone 2-hydroxylase

Licorice flavone synthase II acts on naringenin to form 2-hydroxynaringenin as a first step in flavanone formation [K290].

Flavanone 3-hydroxylase (E.C. 1.14.11.9)

Petunia hybrida enzyme, optimum pH 6.5–7.0, requires Fe^{2+}, ascorbate and α-oxoglutarate. It oxidizes naringenin to dihydrokaempferol, eriodictyol to dihydroquercetin and 3′,4′,5,5′,7-pentahydroxyflavanone to dihydromyricetin [D772]. In another study the enzyme was found to have a molecular weight of 74 000 and optimum pH 8.5. It oxidizes (2*S*)-naringenin to (2*R*,3*R*)-dihydrokaempferol and (2*S*)-eriodictyol to (2*R*,3*R*)-dihydroquercetin, but (2*R*)-naringenin is not a substrate [E72].

Tulipa anther enzyme oxidizes naringenin to dihydrokaempferol [F162].

This reaction has also been observed in Haplopappus gracilis [A200] and Petroselinum crispum [B957].

Matthiola incana requires oxygen, Fe^{2+}, ascorbate and α-oxoglutarate, forming dihydrokaempferol, succinate and carbon dioxide; eriodictyol is a second substrate [B679].

Isoliquiritigenin 3-hydroxylase

Dahlia variabilis petal enzyme, optimum pH 7.5, appears to be a cytochrome P450. The product is butein [J626].

Dihydrofolate reductase (tetrahydrofolate dehydrogenase; E.C. 1.5.1.3)

Calf and rat brain enzyme activities, which are low relative to rat liver, are inhibited by methotrexate [A3366].

Pig enzyme, molecular weight 20 000 and pI 8.6, is inhibited by methotrexate [A2480].

Calf and rat brain enzyme activities, which are low relative to rat liver, are inhibited by methotrexate [A3366].

Rabbit brain enzyme is found in cortex, stem, cerebellum and striatum, with highest activity in cerebellum, and a three-fold activity range between tissues [A3401]. Liver and brain enzymes, optimum pH 4.8, require NADPH. Brain activity is 15 per cent of that for liver. Both activities are abolished by methotrexate [A2011].

In dihydrofolate-deficient immature domestic poultry chicks, activity is doubled by folate or oestradiol, but these effects are not additive [A1680].

Drosophila menanogaster enzyme, molecular weight in the range 17 000–22 000, is monomeric, with optima at pH 4.8 and 8.5 [K458].

Daucus carota enzyme is cytosolic and appears to be a homotrimer, molecular weight 183 000, monomeric molecular weight 58 400 and optimum pH 5.9; it requires NADPH. It does not cross-react with antibodies against the human enzyme [K199].

Glycine max enzyme requires NADPH for activity; NADH is poor. It is inhibited by phosphate, but Cu^{2+} with phosphate activates. It is inhibited by N-ethylmaleimide, glutathione or aminopterin, and several iron-chelating compounds inhibit in the presence of phosphate, but diethyldithiocarbamate activates [A2343].

Helianthus enzyme, optimum pH 5.9–7.2 (buffer dependent) requires $NADPH_2$; the activity increases with increasing callus age [K461].

E. coli contains two isozymes, molecular weights 18 500. The activity of one is nearly constant in the pH range 4–9, whereas the other shows high activity at pH 4, but decreases almost to zero at pH 9 [A3113].

Lactobacillus leichmannii enzyme, molecular weight 20 000, Stokes radius 0.34 nm and optimum pH 7.4 is composed of 168 amino acid residues; the terminal amino acids have been determined and the presence of one active thiol has been detected. It is unstable above 35°. There are complex relationships between KCl and urea concentrations and the kinetics of the reaction. The reaction rate is greater at pH 6.5 than at 8.5, and the reaction rate is nearly linear with time. 1M KCl inhibits especially at pH 6.5, with non-linear reaction rates. Urea (4M) inhibits strongly, especially at pH 8.5. Urea and KCl together inhibit less at pH 6.5 and more at pH 8.5 than urea alone. Methotrexate (40 nM) is inhibitory in all situations, and the presence of urea and KCl together dramatically increases its effect [A6, K484].

The complete amino acid sequence of a Streptococcus faecium mutant has been determined [A2222].

Flavin mononucleotide reductases (NAD(P)H: flavin oxidoreductase; E.C. 1.6.8.1)

Human erythrocyte NADPH dehydrogenase reduces FMN, FAD, riboflavin, as well as methylene blue and dichlorophenolindophenol [A3008].

Benecka harveyi, a luminous bacterium, contains two isozymes; one (molecular weight 30 000) requires NADH, and the other (molecular weight 40 000) NADPH [A2305, A3095]. Another study found that the former isozyme is monomeric, molecular weight 24 000, and that NADPH is a poor co-substrate. FAD is another substrate [A3209].

Eubacterium enzyme, molecular weight 260 000 and optimum pH 6.8, has an absolute requirement for NADH. It reduces FMN, riboflavin, as well as methylene blue, menadione and dichlorophenolindophenol [B426].

Acetylindoxyl oxidase (E.C. 1.7.3.2)

Zea enzyme, optimum pH above 9, is found in leaf of etiolated seedlings and shoot sap (only after dialysis), and requires oxygen [K902].

Cannabinoid ring hydroxylases

Extensive studies with cannabinoids have demonstrated that hydroxylations occur at various positions in the molecule; however, few studies appear to have been carried out at an enzyme level. The literature on this subject is very extensive, and the references given are purely illustrative.

Cannabinoid 6α-hydroxylation

This reaction has been detected in rabbit, guinea pig, man, mouse and Thamnidium [A108, A1391, B742, C2, E7, F819].

Cannabinoid 6β-hydroxylation

This reaction has been detected in man, monkey, guinea pig and Thamnidium [B458, B742, C2, H418].

Hydroxylations of the side chains are described elsewhere.

Codeine 14β-hydroxylase

Pseudomonas putida carries out this reaction; codeinone is also a substrate [K311].

Desacetoxyvindoline hydroxylation

Catharanthus roseus enzyme, molecular weight 45 000 is probably monomeric, and is composed of three isozymes, pI 4.6, 4.7 and 4.8. Hydroxylation occurs at the 4-position, to form deacetylvindoline [H5].

3,9-Dihydroxypterocarpan 6a-hydroxylase

Glycine max P450 D6aH, molecular weight 55 000, catalyzes the reaction; it requires NADPH, and dilauroylphosphatidylcholine is the best activator found in a study to identify its lipid requirement [F242].

Geissoschizine dehydrogenase (E.C. 1.3.1.36)

Catharanthus roseus enzyme is highly specific and requires $NADP^+$; other oxidoreductive cofactors are inactive. It catalyzes the removal of the 21α hydrogen, to form a quaternary base [K700].

2′-Hydroxydaidzein reductase (E.C. 1.3.1.51)

Glycine max enzyme, optimum pH 7.0, is monomeric, molecular weight 34 700; it reduces the 2,3-bond. Other substrates include 2′-hydroxyformononetin and 2′-hydroxygenistein [K702].

Hyoscyamine (6*S*)-dioxygenase (E.C. 1.14.11.11)

Hyoscyamus enzyme, molecular weight 38 000–40 000, is found exclusively in roots. It is thought to be involved in the pathway for scopolamine formation [G135]. H. niger enzyme, molecular weight 41 000 and optimum pH 7.8, requires α-oxoglutarate and is activated by Fe^{2+}, catalase or a reducing agent, such as ascorbate. Other substrates include a series of tropine esters [E350].

Hyoscyamus niger root enzyme requires Fe^{2+}, oxygen and α-oxoglutarate (which yields carbon dioxide and succinate), and is stimulated by ascorbate. It is also found in Datura fastuosa (but not other Datura species), Atropa belladonna and Duboisia leichhardtii [K885].

Morphine 6-dehydrogenase (E.C. 1.1.1.218)

Morphine → morphinone

Hamster liver cytosolic enzyme, molecular weight 38 000, optimum pH 9.3 and pI 5.6, requires NAD^+; $NADP^+$ is a poor co-substrate. Substrates include morphine, codeine, normorphine, 1-indanol, 1-acenaphthenol, 4-chromanol and thiochroman-4-ol [K443].

Rabbit liver cytosoic enzyme, molecular weight 36 000, pI 6.4 and optimum pH 9.4, requires

NAD(P)$^+$, with NAD$^+$ preferred. Codeine, ethylmorphine, normorphine, 1-indanol and phenylglycol are also substrates. The enzyme also reduces polynuclear quinones with NAD(P)H as coenzyme, but not *p*-quinones [H920].

Pseudomonas putida enzyme, molecular weight 32 000 and optimum pH 9.5 oxidizes morphine and codeine, and reduces codeinone with NADPH as coenzyme [G181].

Morphinone reductase

Codeinone → hydrocodone

Pseudomonas putida enzyme, molecular weight 80 000 contains one mol of FMN. A second substrate is morphinone [H892]. It requires NADH for the reduction of the double bond conjugated with the oxo group. The reaction sequence is:

Formation of a charge-transfer intermediate with NADH;
Reduction of the flavin moiety;
Formation of reduced enzyme-codeinone complex;
Flavin re-oxidation;
Hydrocodone release [J635].

Protopine 6-monooxygenase (E.C. 1.14.13.55)

Eschscholtzia californica forms dihydrosanguinarine from protopine; 6-hydroxylation is considered to be the first reaction step [F699].

Pterocarpan hydroxylase

Ascochyta rabei (a fungus) enzyme, molecular weight 58 000, requires FAD, NAD(P)H and oxygen. It oxidizes medicarpin and maackiain at the 1α carbon to form the corresponding 1α-hydroxy-1,4-dien-3-ones [G213].

Vinorine hydroxylation

Rauwolfia serpentina enzyme is a P450 enzyme, optimum pH 8.3, which hydroxylates vinorine to vomilenine with oxygen and NADPH as co-substrates. This is a key step in the formation of ajmaline [H414].

Vitamin K_1-2,3-epoxide reductase (tertiary alcohol-forming)

Beef liver microsomal enzyme, molecular weight 25 000 appears to be a homodimer which requires dithiothreitol and 3-((3-cholamidopropyl)dimethylammonio)-1-propanesulphonic acid and is inhibited by high concentrations of glycerol. The reaction products are 2,3-dihydro-2- and 3-hydroxy-2-methyl-3-phytylnaphthoquinone [D683].

Zearalenone reductase

Liver enzyme (microsomal), molecular weight 230 000, optimum pH 4.5 and 7.4 has been detected in (in reducing order) cow, mouse, pig, rat, rabbit and guinea pig; the figures shown are for rat. Hamster optima, pH 5.5 and 8.0 with NADH, are raised by 0.5 pH unit by substitution with NADPH. The products are both 6α- and 6β-zearalenol. Except with guinea pig enzyme, where neither product predominates at any pH, the β-isomer predominates at neutral pH, whereas the α-isomer predominates at acidic pH [C719, C858].

3.10 *Sulphur replacement*

$$\mathrm{R{-}\overset{\overset{\displaystyle O}{\|}}{P}{=}S \rightarrow R{-}\overset{\overset{\displaystyle O}{\|}}{P}{=}O}$$

Parathion, for instance, is a substrate for this class of reaction, with paraoxon as product. Little is known about the reaction.

The reaction has been observed in mouse liver [D858], rabbit [A1995], rat liver [A881] and microorganism [A2435], but not in lobster hepatocytes [A2567].

3.11 *Dehydroxylation*

Rat kidney dehydroxylates oestriol, with oestrone as the final product [A2549, A3644].

Rat and rabbit convert octopamine into *p*-hydroxyphenylacetate; antibiotics do not affect the reaction, unlike aryl dehydroxylation [A373].

3.12 *Side chain halogenation*

Bjerkandera acts on benzoate and *p*-hydroxybenzoate to form veratryl chloride. This is claimed to be the first report of a benzylic halide natural product [J482].

3.13 *Oxidative rearrangement*

Acetophenone oxygenase

A Pseudomonas putida arylketone monooxygenase, molecular weight 70 000, catalyzes a Baeyer-Villiger type oxidation, converting acetophenones into phenyl acetates. It contains one mol FAD and requires oxygen and NADPH; NADH is inactive. Its N-terminal sequence has been determined. Substrates include acetophenone, *p*-hydroxyacetophenone and *p*-hydroxypropiophenone; benzophenone and cyclohexylacetone are not substrates [K698]. Arthrobacter and Alcaligenes also exhibit the same reaction; monochloroanisoles are additional substrates [A2380, G107].

A similar reaction has been reported with *p*-hydroxypropiophenone in Pseudomonas [K698].

Alcaligenes enzyme acts on *o*-, *m*- and *p*-chloroacetophenone to form the corresponding phenyl acetates [G107].

Arthrobacter enzyme requires one mol each of NADPH and molecular oxygen for the reaction [A2380].

4. Formation and degradation of side-chains

4.1 *Side-chain formation*

Nuclear C-methyl incorporation

1. Benzene and toluene

Human bone marrow supplemented with S-adenosylmethionine converts benzene into toluene, which in turn yields *o*-, *m*- and *p*-xylenes [F853]. This may account for the apparent hydroxymethylation of benzene (Sloane, N.H., *Biochim. Biophys. Acta* (1965), **107**, 599; Sloane, N.H., Heinemann, M., Biochim. Biophys. Acta (1970), **201**, 384).

2. Polunuclear hydrocarbons

Rat is able to incorporate a methyl group at positions 7 and 12 of benzanthracene and possibly its metabolites. Incorporation was not observed with a range of non-carcinogenic analogues [F534]. The enzyme has been located in rat lung cytosol. Incorporation of a second methyl group occurs with both monomethylated benzanthracenes to yield 7,12-dimethylbenzanthracene, with S-adenosylmethionine as the methyl donor [F742, G491]. With rat liver cytosol chrysene forms 6-methylchrysene [G118].

3. Natural products.

Euglena gracilis γ-tocopherol methyltransferase (c.f E.C. 2.1.1.64), molecular weight 150 000 and optimum pH 7.5, incorporates a methyl group into both γ-tocopherol and β-tocopherol, in both cases forming α-tocopherol [G754].

In Streptomyces antibioticus 3-hydroxyanthranilate 4-C-methyltransferase (E.C. 2.1.1.97), a key enzyme in the formation of actinomycin, yields 3-hydroxy-4-methylanthranilate. The enzyme, molecular weight 36 000, which is stimulated by EDTA or mercaptoethanol, has negligible activity towards a range of substrate analogues [E725, F934]. The co-substrate is S-adenosylmethionine [E676]. The optimum pH lies between 6 and 8, depending on the buffer [E542].

C-Formylation

Rat brain and liver mitochondria incorporate a formyl group at the 5-position of 2,3,4,5-tetrahydro-1-(1-phenylcyclohexyl)pyridinium, with N^5-formyltetrahydrofolate as co-substrate [G209].

Incorporation of carboxyl group into phenols without hydroxyl removal

Phenol is a substrate for this reaction type [G923]; bacteria form *p*-hydroxybenzoate reversibly from phenol and carbon dioxide (E.C. 4.1.1.61), optimum pH 6.5–7.0. A two-stage mechanism has been suggested [F468].

Desulfococcus forms gentisate from quinol and carbon dioxide [H293].

Incorporation of carboxyl group into aniline

Rhodococcus erythropolis cells incorporate carbon dioxide into aniline (optimum at pH 7–7.5) to form anthranilate [D273]. Fratearia also catalyses the reaction [D595].

Incorporation of carboxyl group into polynuclear hydrocarbons

Microorganisms act on naphthalene to form 2-naphthoate with incorporation of carbon dioxide. Phenanthrene is also carboxylated, but the position is unclear [K253].

Incorporation of carboxyl group with hydroxyl removal

Phenol is converted into benzoate by microorganisms; the reaction sequence has not been clarified [F177, G281]. Substituted phenols with an *ortho* substituent yield *m*-substituted benzoates, with elimination of a hydroxyl group. Substrates include catechol, as well as phenols *ortho* substituted with halides, amino or carboxyl groups [G917].

A bacterial consortium acts on phenol to form benzoate, apparently involving *para* carboxylation as well as a dehydroxylation step [G424].

Bacterium forms benzoate from quinol and bicarbonate; it was postulated that gentisate was an intermediate [H110]. Another study found that the reaction occurs in species that dehydroxylate *p*-hydroxybenzoate [G167].

Other microorganisms, including those found in pig manure form benzoate from phenol, apparently without *p*-hydroxybenzoate as an intermediate [F515, H842].

The reverse reaction is described under ***p*-Hydroxy- and 3,4-dihydroxybenzoate decarboxylases** (below).

Tyrosine phenol-lyase (E.C. 4.1.99.2)

e.g, L-Serine + phenol ↔ L-tyrosine

Citrobacter freundii enzyme has a broad optimum from pH 6.8 to above 9 and acts on phenol to generate L-tyrosine, with S-(*o*-nitrophenyl)-L-cysteine as co-substrate [F38]. C. intermedius enzyme forms tyrosine from phenol, glycine and formaldehyde, but only in the presence of an aldolase (E.C. 4.1.2.13) [B693].

Erwinia herbicola enzyme has an optimum pH of about 8 for amino acid synthesis, and is denatured by high concentrations of phenol or catechol [A851]. The enzyme also acts on L-dopa, and the reverse reaction has been observed with phenol, catechol, resorcinol and pyrogallol. In intact cells D-tyrosine is converted into L-tyrosine, presumably by side-chain removal and reconstruction [A708].

Escherichia intermedia enzyme acts on resorcinol to form 2,4-dihydroxy-L-phenylalanine, with S-methyl-L-cysteine as co-substrate, which can be replaced by L-tyrosine; presumably the enzyme releases L-serine from L-tyrosine, which then condenses with resorcinol. The configuration with both D- and L-tyrosine is retained at C-3 [A2242, E401]. Phenols that are substrates include phenol, *o*- and *m*-cresol, *o*- and *m*-chlorophenol, catechol, resorcinol, pyrogallol and hydroxyquinol [B292]. The enzyme requires pyridoxal phosphate as cofactor; its N-oxide and 2′-hydroxypyridoxal phosphate are not so good as cofactors [A155].

Leptoglossus phyllopus enzyme, which is found in the ventral abdominal gland interconverts phenol and L-tyrosine [A2260].

Symbiobacterium thermophilum enzyme, optimum pH 7 and pI 4.8 is heat stable. It is activated by K^+ and NH_4^+, but not by Na^+ or Mg^{2+}, with both D- and L-tyrosine as substrates. It converts catechol into L-dopa, with pyruvate and ammonia as co-substrates [G603].

Activity has been found in Pseudomonas, Xanthomonas, Alcaligenes, Achromobacter, Escherichia, Aerobacter, Erwinia, Proteus, Salmonella and Bacillus, but generally only in a small proportion of the strains tested. In this study it was not found in numerous other bacterial genera, fungi, Actinomycetes or yeasts [A692].

Catechol is a substrate for enzyme found in Erwinia, Symbiobacterium and Citrobacter [e.g. E959, G603, J512], and leads to a potentially viable method for the commercial production of L-dopa.

Tryptophan synthase; (E.C. 4.2.1.20)

Indole + L-serine →L-tryptophan

Enzymes from Daucus carota and Nicotiana tabacum utilize as additional substrates 4-, 5- and 6-fluoroindole, 5-hydroxy-, 5-methoxy- and 5-methylindole. Several other substituted indoles are at best poor substrates [B874].

In Juglans regia the enzyme is associated with a particulate fraction that is not mitochondrial, with optimum pH 7–8 [A818].

The reaction has also been studied in Zea mays [H126].

In Enterobacteriaceae genera the activity (using pyruvate and ammonia instead of serine) has been found only in some strains and species of Escherichia, Kluyvera, Enterobacter, Erwinia and Proteus. Other genera tested and found inactive include Pseudomonas, Azotobacter, Aeromonas, Rhizobium, Alcaligenes, Salmonella and Clostridium. Between four and 475 strains of each genus were tested [A781].

Vibrio enzyme requires pyridoxal phosphate, and can use S-methylcysteine in place of cysteine, optimum pH 8.0 [B85].

E. coli enzyme, optimum pH 9, yields 5-hydroxytryptophan from 5-hydroxyindole [A430].

Salmonella typhimurium enzyme, which requires pyridoxal phosphate for activity, is a complex of two pairs of different subunits [K275]. This and E. coli enzymes catalyze the reverse reaction, but only slowly [E101].

The reverse reaction has been described under Tryptophanase.

L-Tryptophan 2-C-methyltransferase
(E.C. 2.1.1.106)

Streptomyces laurentii enzyme, optimum pH 7.8 (sharp) transfers the methyl group from L-methionine to the nucleus of L-tryptophan to form L-2-methyltryptophan with retention of configuration. Indole-3-pyruvate and D-tryptophan (poor) are also substrates, but not indole [K752].

L-Tryptophan 4-dimethylallyltransferase
(E.C. 2.5.1.34)

This compound, which is a key intermediate in the formation of ergot alkaloids, is formed from L-tryptophan in Claviceps paspali [A3599].

Aspulvinone dimethylallyltransferase
(E.C. 2.5.1.35)

Aspergillus terreus enzyme, molecular weight 240 000–270 000 (monomer 45 000) and optimum pH 7.0 acts on aspulvinone E or aspulvinone G and dimethylallyl pyrophosphate to form a series of dimethylallyl-substituted analogues [K829].

Trihydroxypterocarpan dimethylallyltransferase
(E.C. 2.5.1.36)

Glycine enzyme acts on 3,6a,9-trihydroxypterocarpan and dimethylallyl pyrophosphate to form 2- and 4-dimethylallyl-3,6a,9-trihydroxypterocarpan [K905]. The enzyme is found after elicitor treatment (Phytophthora megasperma) [K904].

Formation of ubiquinone, tocopherol and analogues by prenylation

In man 2,3-dimethoxy-5-methyl-*p*-quinone (ubiquinone 0) is precursor to ubiquinones 30, 45 and 50, and 'menadione diphosphate' is precursor for vitamins $K_{2(20)}$, $K_{2(45)}$ and $K_{2(50)}$ [A774].

Rat liver and brain mitochondria contain 4-hydroxybenzoate: polyprenyl transferase (c.f. E.C. 2.5.1.31), with polyprenyl pyrophosphates containing 8–10 isoprene units as the second substrate. *p*-Aminobenzoate is also a substrate. Inhibitors (not very effective) include serotonin, dopamine, noradrenaline, aspirin and other salicylates. Maximum activity is found in heart, and it is also found in kidney and spleen [A147].

Homogentisate is the substrate from which many prenylated compounds are formed.

3-Octa- and 3-nonaprenyltoluquinols are formed in Euglena gracilis and sugar beet, with octaprenyl- and nonaprenyl pyrophosphates as co-substrates respectively; γ-tocopherol is another product [A201, A1584, G754]. Plastoquinones, α- and γ-tocopherol are formed in spinach and lettuce chloroplasts, with phytyltoluquinone as a possible intermediate [A2522, D902].

Saccharomyces (baker's yeast) contains a mitochondrial 4-hydroxybenzoate: polyprenyl transferase with an optimum at pH 7 that synthesizes the corresponding 2-polyprenylphenols with two, three, four, five, eight, nine and 10 isoprene units. Formation of the triprenyl product requires *trans, trans*- farnesyl pyrophosphate. The synthesis is stimulated by Mg^{2+} and inhibited by phosphate [A147].

An E. coli enzyme system acts on *p*-hydroxybenzoate and polyprenylpyrophosphates to form 3-octaprenyl- and 3-nonaprenyl-4-hydroxybenzoates [A1524].

Benzaldehyde side chain carboligation

Pseudomonas putida benzoylformate decarboxylase (which requires thiamine pyrophosphate) exhibits a second reaction in which benzaldehydes form (*R*)-benzoins [K424].

4.2 *Decarboxylation reactions of phenolic groups without hydroxylation*

p-Hydroxy- and 3,4-dihydroxybenzoate decarboxylases (E.C. 4.1.1.61 and E.C. 4.1.1.63 respectively)

Clostridium hydroxybenzoicum enzyme, a homohexamer, subunit molecular weight 57 000, pI 5.1 and optimum pH 5.6–6.2, reversibly decarboxylates *p*-hydroxybenzoate and protocatechuate to form phenol and catechol respectively [H576].

a. p-Hydroxybenzoate decarboxylase

A microorganism enzyme, molecular weight 420 000 (subunit molecular weight 119 000) and pI 5.6 acts on *p*-hydroxybenzoate to yield phenol [K573].

A bacterial enzyme, optimum pH 6.5–7.0 reversibly decarboxylates *p*-hydroxybenzoate specifically; it requires Mn^{2+} and phosphate, and is rapidly inactivated by oxygen [F468].

A mixture of microorganisms decarboxylates several benzoates reversibly [G632].

b. 3,4-Dihydroxybenzoate decarboxylase

Clostridium hydroxybenzoicum enzyme, molecular weight 270 000 and optimum pH 7.0 appears to be a homotetramer, and is absolutely specific for protocatechuate. The reaction is reversible, in favour of the acid. No coenzyme is required. A second enzyme was separated in the same study; the N-terminal sequences of these enzymes were different [J184].

6-Methylsalicylate decarboxylase (E.C. 4.1.1.24)

This activity, which forms *m*-cresol is found in Valsa friesii [A1049].

Orsellinate decarboxylase (E.C. 4.1.1.58)

Gliocladium roseum enzyme catalyzes the formation of orcinol and carbon dioxide; the enzyme is activated by azide [K908].

Pyrocatechuate decarboxylase (E.C. 4.1.1.46)

$$\text{Pyrocatechuate} \rightarrow \text{catechol} + CO_2$$

An enzyme in Aspergillus niger is a homotetramer, monomeric molecular weight 28 000 and optimum pH 5.2 [E417]. Another study found the native molecular weight to be 150 000, but with the same optimum pH. It is inhibited by cyanide and borohydride [B83]. A. oryzae enzyme, which does not require cofactors, appears to have a histidyl residue at the active centre [J206].

Bacterium decarboxylates pyrocatechuate as well as gallate, protocatechuate and *m*-hydroxybenzoate [C396].

Trichosporon cutaneum is a dimer, molecular weight 66 000. Other substrates include 2,3,5- and 2,3,6-trihydroxybenzoates [B790].

Gentisate decarboxylase (E.C. 4.1.1.62)

Klebsiella aerogenes enzyme, optimum pH 5.9, which is soluble, does not require oxygen and releases carbon dioxide [K943].

Gallate decarboxylase (E.C. 4.1.1.59)

Pantoea (formerly Enterobacter) agglomerans enzyme, a homohexamer, molecular weight 320 000, requires Fe (which makes it unique among similar decarboxylases), and is inhibited by Fe^{2+}-binding reagents. It is highly specific, forming pyrogallol and carbon dioxide [J859].

A bacterium has been described that decarboxylates gallate and several other benzoates [C396].

3,4-Dihydroxyphthalate decarboxylase (E.C. 4.1.1.69)

Protocatechuate is formed from 3,4-dihydroxyphthalate in Micrococcus; no enzymology has been described [K753].

4,5-Dihydroxyphthalate carboxylyase (E.C. 4.1.1.55)

4, 5-Dihydroxyphthalate → protocatechuate + CO_2

Pseudomonas fluorescens enzyme is probably a hexamer, molecular weight 420 000, monomeric molecular weight 66 000 and optimum pH 6.8. It is inhibited by *p*-chloromercuribenzoate [D538]. P. testosteroni enzyme, which is induced by phthalate, appears to be a tetramer, molecular weight 150 000 and monomeric molecular weight 38 000. It also forms *m*-hydroxybenzoate from 4-hydroxyphthalate [A3332].

***p*-Cresol formation from *p*-hydroxyphenylacetate**

Clostridium difficile enzyme, which is unstable, requires amino acids (serine or threonine) or the corresponding pyruvates as cofactor, or dithionite [D917].

4.3 *Decarboxylation reactions of side-chains*

Auxin oxidases

Zea mays enzyme is composed of two isozymes, molecular weights 32 500 (main) and 54 500, with a requirement for Mn^{2+} and *p*-coumarate. The product with IAA is indole-3-methanol [F798]. This product is also found in wheat [D150] and Pinus sylvestris [D597]. Lupinus alba forms, in addition, 3-hydroxymethyloxindole, 3-methyleneoxindole and 3,3′-bisindolylmethane [F847]. Horseradish peroxidase also forms indole-3-methanol, 3-formylindole, 3-methyleneoxindole and 3-hydroxymethyloxindole from IAA [C525, D234, F961].

3-Methyleneoxindole formation (indoleacetate oxidase)

Indole-3-acetate → 3-methyleneoxindole

Arachis contains four peroxidase isozymes with indoleacetate oxidase and polyphenol oxidase activities. Each isozyme has a different optimum pH for each type of substrate; the optima for indoleacetate oxidase are 7.2–7.6 [A2519].

Bean (Phaseolus vulgaris) etiolated seedling root oxidase activity is increased by treating the plants with naphthenate (identity unspecified), cyclohexanecarboxylate and cyclopentylacetate; there is no *in vitro* effect with these compounds [A220].

Three commercial sources of horseradish peroxidase contain a total of 42 isozymes by isoelectric focussing, and the oxidase-peroxidase ratio is essentially identical for all these isozymes [A3293]. Other studies have failed to separate peroxidase and indole oxidase activities in peroxidase preparations from horseradish and Betula (yellow birch) leaves (20 and 13 isozymes respectively) [A3428, A3429].

Studies on Nicotiana indicate that the reaction is entirely mediated by peroxidases. At least four isozymes are found by electrophoresis, and indoleacetate oxidase is separated from monophenol monooxygenases [A221].

Oat coleoptile peroxidase is separable by electrophoresis into eight isozymes, six of which exhibit indole-3-acetate oxidase activity [A1135].

Peach seed enzyme, optimum pH 4.5–5.0, requires Mn^{2+} and 2,4-dichlorophenol [A655].

Wheat enzyme is a peroxidase, which requires Mn^{2+} and a phenolic cofactor without addition of peroxide. With peroxide, ferulate and *p*-coumarate are also oxidized. A coloured free radical appears to be the first product [A219].

4-Hydroxymandelate oxidase (E.C. 1.1.3.19)

Pseudomonas convexa enzyme, molecular weight 155 000 requires Mn^{2+}, oxygen and FAD or FMN in oxidatively decarboxylating 4-hydroxy-mandelate to form *p*-hydroxybenzaldehyde. It is inhibited by thiols, EDTA, cyanide and 8-hydroxyquinoline [A2681].

3,4-Dihydroxymandelate decarboxylation with tyrosinase

Mushroom enzyme forms protocatechualdehyde from 3,4-dihydroxymandelate. It has been postulated that the mechanism involves quinone formation, followed by tautomerization to a quinone methide, with oxidative decarboxylation [D865, E963].

Benzoylformate (phenylglyoxylate) decarboxylase (E.C. 4.1.1.7)

R.CO.COOH → R.CHO

Pseudomonas putida decarboxylates *p*-hydroxyphenylglyoxylate to form *p*-hydroxybenzaldehyde [E669]. The enzyme requires thiamine pyrophosphate with phenylglyoxylate, *p*-methylphenylglyoxylate, *p*-chloromethylphenylglyoxylate and *p*-fluoromethylphenylglyoxylate as substrates. *p*-Bromomethylphenylglyoxylate is inhibitory; it reacts with the cofactor to yield bromide and (*p*-methylbenzoyl)thiamine pyrophosphate which in turn forms *p*-toluate. A small proportion of *p*-chloromethylphenylglyoxylate exhibits the same side-reaction [E714]. Other substrates are phenylglyoxylate and *m*-hydroxyphenylglyoxylate [E771].

Acinetobacter calcoaceticus, a tetramer, monomeric molecular weight 58 000 and optimum pH 5.9, requires thiamine pyrophosphate, which is firmly bound. It appears to be highly specific [D809].

Moraxella enzyme acts on *o*-hydroxyphenyl-glyoxylate to form salicylaldehyde [G356].

This activity has also been observed in Acinetobacter, Bacterium and Flavobacterium [A730, D487, E355].

L-Arylamino acid decarboxylases (E.C. 4.1.1.28)

$R.CHNH_2.COOH \rightarrow R.CH_2NH_2$

1. Phenylalanines and tryptophans

Animal enzymes have developed genetically with specificities that enable the non-protein amino acids L-dopa, the precursor of the catecholamine neurotransmitters and 5-hydroxy-tryptophan, the precursor of the neurotransmitter serotonin, to be readily decarboxylated. These enzymes have little activity towards the amino acids tyrosine, tryptophan and phenylalanine, which are ingested in large amounts and are utilized extensively for protein synthesis. The ready decarboxylation of these amino acids would swamp the organism with amines that would interfere with neurotransmission, as well as possibly compromising the availability of amino acids for protein synthesis.

Human phaeochromocytoma enzyme appears to be a dimer, monomeric molecular weight 50 000 and pI 5.7. Substrates are L-dopa, 5-hydroxy-L-tryptophan and L-*threo*-3,4-dihydroxyphenylserine [D793, G452]. Human kidney enzyme is a dimer, molecular weight 100 000. Phenylalanine, tyrosine and tryptophan are not substrates [G104].

The presence of the enzyme in human brain has been a matter of controversy. For instance in one study only a few P.M. brains showed any activity, and these were considered to show activity only because the enzyme had been carried there by blood as a result of trauma [A3753]. This was clearly erroneous, since the brain needs to synthesize these neurotransmitters. In P.M. parkinsonian brain dopa decarboxylase activity is reduced about 10-fold in those areas that normally show high activity, especially caudate nucleus, putamen, substantia nigra and putamen. Only a small decrease has been found in other brain regions, and an increase has been detected in cerebellar cortex [A381]. In aromatic L-amino acid decarboxylase deficiency, a rare autosomal recessive, a catecholamine deficiency is observed, with associated symptoms [K380].

Rat kidney enzyme is a homodimer, monomeric molecular weight 48 000, and pI 6.7. The amino acid composition has been determined. It is immunologically identical with enzyme found in striatum, adrenal medulla, pineal and liver. The optimum pH for L-dopa is 7.0, 5-hydroxytryptophan 7.7, tyrosine 8.2, tryptophan 8.6 and phenylalanine 8.2. The V_{max} for the last three substrates is low, and K_m is much higher than physiological concentrations [E735]. In agreement with this, there is little evidence for tyrosine decarboxylation in rat or mouse liver under more moderate conditions [A2559].

Rat brain enzyme, molecular weight 115 000 and pH 6.4–6.5, contains firmly bound pyridoxal phosphate. Substrates are L-dopa and 5-hydroxytryptophan, but not tryptophan or tyrosine. The reaction products serotonin and dopamine are inhibitory, as are N-ethylmaleimide, dodecyl sulphate, Cu^{2+}, Fe^{2+} and Ni^{2+}. In dilute solution of either pyridoxal phosphate or the amino acid substrate, an excess of the second substrate is inhibitory [A2378]. Liver enzyme is inhibited by serotonin and dopamine (competitive), and by α-methyl-*m*-tyrosine (competitive for dopa but non-competitive for 5-hydroxytryptophan). Dopamine binds better at pH 6.8 than at 7.8, serotonin binds with equal firmness under these conditions, whereas α-methyl-*m*-tyrosine and α-methyldopa bind less firmly at pH 6.8 [A2105].

A tyrosine decarboxylase, optimum pH 8.0, well distributed through rat brain and kidney as well as human kidney; is thought to be 'ordinary' L-amino acid decarboxylase, based on the ratio of activity with 5-hydroxytryptophan as substrate and on antibody tests. It is stimulated by a series of cyclic organic solvents such as benzene, toluene, cyclohexane and cyclopentane, as well as by phenol. A series of other amino acids are decarboxylated, but information about relative reaction rates was not given [C566]. Rat kidney activity increases fivefold between two days post-partum and six-eight weeks, whereas brain activity remains constant until about five weeks [A2195].

In vivo studies with rats show that although *o*- and *m*-tyrosine and 3-hydroxy-4-methoxy-phenylalanine readily yield decarboxylation products, a series of methoxyphenylalanines are not significantly decarboxylated [A2961, J411].

Pig kidney enzyme has a molecular weight of 86 000 and pI 5.0 according to one study [F197]. In another study, the molecular weight was found to be 112 000, with subunit molecular weights 57 000, 40 000 and 21 000 with dopa, 5-hydroxytryptophan, tryptophan and tyrosine as substrates [A1906]. It converts L-dopa substituted with deuterium on the α-carbon into *S*-[α-D]-dopamine whereas L-dopa in D_2O yields *R*-[α-D]-dopamine [D606]. About 0.02 per cent of L-dopa and *m*-tyrosine, and 2 per cent of α-methyl-*m*-tyrosine and α-methyldopa undergo a subsidiary reaction at pH 6.5. For L-dopa the subsidiary products are 3,4-dihydroxyphenyl-acetaldehyde and pyridoxamine phosphate, which, in the absence of further pyridoxal phosphate cofactor, inactivates the enzyme. The proportion of substrate undergoing this subsidiary reaction relative to decarboxylation is independent of pH in the range 6.5–9 [A2425]. Another study on this alternative reaction found that α-methyldopa uses molecular oxygen and yields phenylacetone and ammonia as products, and 5-hydroxytryptophan forms 5-hydroxyindole-3-acetaldehyde. The amount of these subsidiary products formed greatly exceeded the amount of pyridoxal phosphate present, indicating that it is not simply a decarboxylation-dependent transamination with pyridoxal phosphate as amino acceptor. Nor is it an oxidative deamination [B780].

Beef adrenal medulla enzyme is present in the cytoplasm of all cells [A1409], molecular weight 50 000, with at least five components with pI in the range 4.9–5.3 [G452].

Mouse enzyme has an optimum pH of 6.6 [A1297]. Both D- and L-phenylalanine are decarboxylated, but not by prior conversion of the D-to the L- isomer. Decarboxylation of D-tyrosine is not inhibited by L-tyrosine decarboxylase inhibitors [A3612]. In eye the activity of the enzyme acting on 5-hydroxytryptophan rises rapidly in the first three days after birth, and then declines to the adult level at day seven [A946]. In brain, the activity reaches the adult level in cerebellum, medulla and pons two weeks after birth, in the mesencephalon and diencephalon at four weeks and in the cerebral hemispheres after six weeks [A1067].

The enzyme is found in islet cells from hamster, rabbit and guinea pig [A1911].

Embryo chick brain enzyme is detected at 14 days incubation, but with very low activity in cerebellum. Activity increases 10-fold just before hatching [A1413]. In pineal, 5-hydroxytryptophan decarboxylase is inactive at 12 days incubation, and at hatching the activity is 50 per cent of the activity found 16 days post-hatching. In cerebrum the pattern of development is similar, except that it appears two days earlier [A160].

Skipjack tuna liver enzyme, which acts on 5-hydroxytryptophan, has a molecular weight of 110 000 [H683].

In leech, the enzyme is present in nerves [A659].

Mytilus edulis and Elliptio complanata dopa decarboxylases are inhibited by α-methyldopa, N^1-(DL-seryl)-N^2-(2,3,4-trihydroxybenzyl) hydrazine and hydrazinomethyl-3,4-dihydroxyphenylalanine [A1084].

Aedes aegypti (mosquito) enzyme, molecular weight 112 000 [B738] is induced by β-ecdysone, and this effect is enhanced by β-ecdysone administered together with dibutyryl cyclic AMP, but cyclic AMP on its own has no effect [A1027].

Calliphora vicuna larval dopa decarboxylase, molecular weight 90 000–96 000 is composed of two subunits, molecular weights 40 000 and 46 000. 5-Hydroxytryptophan, tryptophan, tyrosine and phenylalanine are not substrates. Al^{3+} and Mn^{2+} are activators, whereas Cu^{2+} and Hg^{2+} are inhibitory. N-Acetyldopamine, the product that is used in the formation of polymers, is also inhibitory [A2151].

Ceratitis capitata enzyme was found to have very similar properties to the human enzyme (see above) [G104].

Periplaneta americana decarboxylates L-dopa, L-tyrosine, 5-hydroxytryptophan and *m*-tyrosine. Two isozymes are found; one, molecular weight 120 000 acts on dopa and the other, molecular weight 100 000, on tyrosine. Both have an optimum pH of 7.5 [E448].

Barley (Hordeum vulgare) enzyme, optimum pH 7.3, is specific for L-isomers. Substrates include dopa, *o*-, *m*- and *p*-tyrosine. Competitive inhibition is observed between the substrates, and caffeate and *p*-coumarate are also competitive inhibitors. Other inhibitors include hydroxylamine, semicarbazide, Fe^{2+}, Cu^{2+} and Hg^{2+}. The enzyme is also found in Triticum aestivum, Zea mays and Cytisus scoparius [A1210].

Papaver somniferum contains two decarboxylases which require pyridoxal phosphate (like other decarboxylases). They act on L-tyrosine or, better, L-dopa, with a broad optimum, pH 7.5–8.5 [H557]. Another study found the optimum pH to be 7. Substrate inhibition is observed with pyridoxal phosphate and L-dopa at 1 mM, and by cyanide, semicarbazide, hydroxylamine and *p*-chloromercuribenzoate [A3143].

Tomato shoot enzyme decarboxylates L-tryptophan and 5-hydroxytryptophan; D-tryptophan, L-phenylalanine and L-tyrosine are not substrates. It requires pyridoxal phosphate and has an optimum pH of 8.0. Inhibition is brought about by pre-incubation with 2 mM Fe^{2+}, Fe^{3+}, Ca^{2+}, Co^{2+} and Zn^{2+}, but not by Mg^{2+} or Mn^{2+} [A536].

Bacillus tryptophan decarboxylase, molecular weight 150 000 and optimum pH 7, requires pyridoxal phosphate [D555].

Lactobacillus brevis tyrosine decarboxylase, optimum pH 5.0, requires pyridoxal phosphate; L-dopa is a less good substrate. It is stabilized by substrate and coenzyme, and is inhibited by glycerol, mercaptoethanol and tyramine [K277].

Purified Micrococcus percitreus enzyme, a dimer, molecular weight 101 000, subunit molecular weight 48 000, and optimum pH 9.0, requires pyridoxal phosphate. Compounds found to be substrates are tryptophan, 5-hydroxytryptophan, phenylalanine, *o*- and *p*-tyrosine (*m*-tyrosine is a poor substrate), dopa and its 2,4-dihydroxy analogue, 2- and 3-methyltyrosine, 2- and 3-chlorotyrosine, 3,5-dibromotyrosine, 5-fluorotryptophan, 5-methyltryptophan and 3-hydroxykynurenine. Substitution with a methyl group on the α-carbon almost or totally eliminates activity [A336, C62].

S. faecalis enzyme requires pyridoxal phosphate [G332]. Substrates include tyrosine and dopa, and after pre-treatment of cells with toluene *m*-tyrosine is a substrate [A537]. The enzyme is a dimer, molecular weight 143 000 and pI 4.4; a second component, pI 4.5, is eliminated by removal of pyridoxal phosphate and other low molecular weight compounds [F197]. Another study found pI 3.2 and 4.5 for isozymes [A3573].

Tyrosine decarboxylase (E.C. 4.1.1.25) is found in one or more strains of Acinetobacter, Bacillus, Candida, Citrobacter, Cryptococcus, Enterococcus, E. coli, Hafnia, Lactobacillus, Pseudomonas, Proteus, Saccharomyces, Serratia, Shigella, Staphylococcus aureus, Torulopsis and Yersinia [H302].

2. Phenylserines

$$R.CHOH.CHNH_2.COOH \rightarrow R.CHOH.CH_2NH_2$$

Pig kidney enzyme, optimum pH 8.5, decarboxylates L-*threo*-3,4-dihydroxyphenylserine to yield noradrenaline. It is specific for the L-isomer, requires pyridoxal phosphate and is inhibited competitively by L-dopa, L-5-hydroxytryptophan, and to some extent by D-dopa. D-*threo*-3,4-Dihydroxyphenylserine inhibits competitively at low concentration, but non-competitively at high concentration. It is inactivated above 40° [A2593].

Rat heart decarboxylates L-*threo*-3,4-dihydroxyphenylserine; the enzyme is specific for the L-isomer, and requires low concentrations of pyridoxal phosphate; high concentrations are inhibitory. The optimum pH is 8.5 [A3882].

Rat and human enzymes appear to be the same as the normal amino acid decarboxylase [C566].

A crude decarboxylase preparation from a species that Chemical Abstracts does not specify has an optimum pH of 7.8–8.2. L-*threo*-3,4-Dihydroxyphenylserine and (a poorer substrate) D-*threo*-3,4-dihydroxyphenylserine are substrates [A3177].

L-Phenylalanine oxidase (deaminating and decarboxylating) (E.C. 1.13.12.9)

L-Phenylalanine → phenylpyruvate

L-Phenylalanine → phenylacetamide

Pseudomonas enzyme is a dimer, molecular weight about 140 000, which contains two mol of FAD. The amino acid composition has been determined. It catalyzes two reactions; one is the formation of phenylpyruvate with a sharp optimum at pH 11 and a broad plateau between pH 4 and 9 with elimination of ammonia and water. The other is the formation of phenylacetamide with elimination of carbon dioxide, which shows a broad optimum between pH 5 and 9. Molecular oxygen is incorporated into the carbonyl of the amide. The enzyme is specific for L-isomers, and tyrosine, *o*-, and *m*-tyrosine, *p*-fluorophenylalanine and tryptophan are additional substrates. *m*-Tyrosine is the only substrate where deamination is the major pathway; with some others the reaction is almost exclusively amide formation. It is inhibited by Hg^{2+}, but milder inhibitors such as Fe^{2+} and Cu^{2+} markedly reduce the proportion of substrate undergoing the deamination reaction; otherwise the proportion of phenylalanine deaminated is about 20 per cent of the total [C444, C546, D246].

Tryptophan 2-monooxygenase (E.C. 1.13.12.3)

Tryptophan → indole-3-acetamide

Coprinus enzyme, molecular weight 420 000 and optimum pH 9.0, appears to be a hexamer, each monomer binding one FAD non-covalently. This reaction is a minor pathway with L-tryptophan oxidase (see below) [K563].

The reaction has been observed in Poncirus trifoliata, Azospirillum brasiliense, Xanthomonas campestris, Streptomyces, Bradyrhizobium, Pseudomonas savastanoi and P. campestris [E435, E578, F788, G760, G776, H384].

Tryptophan oxidases

Arabidopsis thaliana enzyme formation of indole-3-acetate involves a single enzyme complex, molecular weight 160 000–180 000. It requires oxygen, but this is not incorporated into the product; indole-3-acetonitrile may be the intermediate [K605].

Coprinus enzyme, molecular weight 420 000, optimum pH 7.0 and stability range pH 6.0–10.5, appears to be a hexamer, each monomer binding one FAD non-covalently. The main reaction forms indole-3-pyruvate and indole-3-acetate; a smaller amount (10 per cent) of indole-3-acetamide is formed. Slight activity is found with L-phenylalanine and L-tryptophan [K563].

Tryptophan 2′-dioxygenase (E.C. 1.13.99.3)

Pseudomonas enzyme, molecular weight 280 000 is a haemoprotein; it is not a catalase or peroxidase. L-Tryptophan forms indole-3-glycolaldehyde (or possibly indole-3-glyoxal); a number of indoles are also substrates, but decarboxylation is not a pre-requisite for activity [A3009]. Another study on an enzyme, molecular weight 250 000, optimum pH about 4 and pI 4.8 with an apparently different specificity contained iron; its amino acid composition was determined. There are indications that α-hydroxylation is the first reaction step in a sequence that may include decarboxylation [A3007].

Phenylpyruvate decarboxylase (E.C. 4.1.1.43)

R.CO.COOH → R.CHO

Acinetobacter calcoaceticus enzyme, a tetramer, monomeric molecular weight 56 800 and optimum pH 7.0, requires thiamine pyrophosphate. The product is phenylacetaldehyde; the enzyme appears to be highly specific [D809].

Candida guilliermondii enzyme requires thiamine pyrophosphate and Mg^{2+} [A2483].

Thauera aromatica also catalyzes this reaction [J397].

***p*-Hydroxyphenylpyruvate decarboxylase**

R.CO.COOH → R.COOH

Arthrobacter enzyme forms *p*-hydroxyphenyl-acetate from *p*-hydroxyphenylpyruvate without formation of the aldehyde. The enzyme, optimum pH about 7.5, requires thiamine pyrophosphate, FAD, glutathione and Mg^{2+} or Mn^{2+} [A3208].

Indole-3-pyruvate decarboxylase (E.C. 4.1.1.74)

R.CO.COOH → R.CHO

The reaction has been found in Bragyrhizobium elkanii [J189], E. coli [H348] and Azospirillum lipoferum [F49]. Enterobacter cloacae enzyme has been reviewed [K792].

Indole-3-pyruvate ferredoxin oxidoreductase

R.CO.COOH → R.CO.CoA

Pyrococcus furiosus enzyme is a tetramer composed of two pairs of subunits, molecular weights 23 000 and 66 000, optimum pH 8.5–10.5, with one thiamine, a cluster of 4 $[4Fe\text{-}4S]^{2+,\,1+}$ and one $[3Fe\text{-}4S]^{0,\,1+}$ units, and a requirement for CoA. The products from indole-3-pyruvate, phenylpyruvate and *p*-hydroxyphenylpyruvate are the corresponding acetyl CoA compounds. It is only active at high temperature (90°), and is inhibited by oxygen [H306].

Mandelates formed from phenylpyruvates

$R.CH_2.CO.COOH \rightarrow R.CHOH.COOH$

This reaction has been detected in rat with phenylpyruvate as substrate [G829].

Amycolatopsis orientalis converts *p*-hydroxyphenylpyruvate into *p*-hydroxymandelate; molecular oxygen is incorporated into both CO_2 and the benzylic hydroxyl. This appears to be one of the steps in the formation of vancomycin [K499].

Decarboxylation of cinnamic acids

$R.CH{=}CHCOOH \rightarrow R.CH{=}CH_2$

Bacillus subtilis enzyme is a dimer, molecular weight 45 000, with a broad optimum at about pH 5. Substrates include ferulate, *p*-coumarate and caffeate [J623]. B. pumilis enzyme is a dimer, molecular weight 45 000 and optimum pH 5.5. Substrates are *p*-coumarate and ferulate [H346].

Pseudomonas fluorescens enzyme is a dimer, molecular weight 40 000 and optimum pH 7.3. Substrates are *p*-coumarate and ferulate, but not *o*- and *m*-coumarate [H364].

Polyporus circinata enzyme has an optimum at pH 6.1. *p*-Coumarate and caffeate are substrates, but not cinnamate or some other aromatic acids. The products are the corresponding styrenes [A2690].

Cladosporium phlei enzyme acts on *p*-coumarate to yield *p*-hydroxystyrene; caffeate and ferulate are also substrates, with a slight preference for the *cis*-isomers. The acrylate side-chain and the *p*-hydroxyl group are essential for activity. The enzyme is stable at −20°, but is rapidly inactivated at 35°; the temperature coefficient for inactivation appears to be unusually large. Thiols protect the enzyme from inactivation by iodoacetate and *p*-chloromercuribenzoate. Maleate and some cinnamates are inhibitory, but chelators, fumarate and acrylate are not [A2675].

Saccharomyces cerevisiae decarboxylates 3,4-dimethoxycinnamate to yield 3,4-dimethoxystyrene. Substrate labelled with deuterium at position 2 on the side-chain totally retains the label conformation [A2157].

Klebsiella oxytoca enzyme acts on *p*-coumarate, caffeate, ferulate and 2,4-dihydroxycinnamate, but only on the (*E*)- (i.e., *trans*) isomers [J853].

Lactobacillus plantarum enzyme is a homotetramer, molecular weight 93 000 and optimum pH 5.5–6.0; it does not require metal ions for activity. Substrates are *p*-coumarate and caffeate, but not ferulate [H897].

Candida lambrica enzyme has an optimum at pH 6.5 [K70].

This reaction has been observed in a large range of microorganisms. A major study on the decarboxylation of ferulate, isoferulate, *o*- *m*- and *p*-methoxycinnamate, *o*- *m*- and *p*-coumarate, caffeate, 5-methoxyferulate, *p*-methylcinnamate, *o*- *m*- and *p*-chlorocinnamate and 2,6-dichlorocinnamate was carried out by a series of unspecified microorganisms. Ferulate decarboxylation was used as a marker to identify the reaction in Aspergllus carneus, A. ochraeus and A. terreus, Bacillus pumilus, Candida intermedia, Corynespora cassiicola, Curvularia affinis, C. clavata and C. lunata, Fusarium coerulum, F. dimerum, F. eumartii, F. moniliforme, F. oxysporum, F. roseum, F. solani and F. tritinctum, Hansenula anomala, H. beckii, H. capsulata, H. henricii and H. minuta, and Saccharomyces cerevisiae [F390], Aspergillus niger, Bacillus megaterium and B. subtilis, Mycobacterium, Nocardia, Penicillium frequentans, Pseudomonas putida, Rhizopus arrhizus, Rhodotorula rubra and Streptomyces rimosus [G895]. Other active species include Brettanomyces anomalus, Cladosporium herbarum, Erwinia uredovora, Klebsiella oxytoca, Rhizoctonia solani, Rhodotorula rubra and R. minuta [G797, H356, K70].

In some instances the reaction product is an ethylbenzene analogue [e.g. E263, H52, H356]; the corresponding styrene is a probable intermediate.

Arylmalonate decarboxylase (E.C. 4.1.1.76)

Alcaligenes bronchisepticus enzyme, molecular weight 24 000 and optimum pH 8.5, decarboxylates α-aryl-α-methylmalonates, where the aryl group may be phenyl, phenyl substituted with halide, methyl or methoxy groups, or by naphthyl groups [G716]. The enzyme contains 240 amino acid residues, with a corresponding molecular weight of 24 737 [K794].

E. coli enzyme, which is stereospecific, forms (*R*)-α-fluorophenylacetate from α-fluorophenylmalonate [K232, K576].

Phaseolus hydroxycinnamate side chain oxidase

P. mungo enzyme, molecular weight about 30 000–40 000 and optimum pH 7.5 apparently decarboxylates *p*-coumarate; the identity of the product is not stated in Chemical Abstracts. The enzyme requires cysteine, and is inhibited by iron and copper chelators [D591].

Tropate dehydrogenase

Pseudomonas enzyme, which forms phenylacetaldehyde from tropate, has an optimum pH 9.5 and requires NAD^+ [H829].

Precarthamin decarboxylase

Carthamus tinctorius enzyme, molecular weight 33 000, activation energy 19.7 kcal/mol and optimum pH 5.0, forms carthamin; it is found in immature flowers. It is inhibited by divalent cations and reducing agents [K557, K579].

Decarboxylation with bromination

Glycine max seed coat peroxidase, with bromide and peroxide as co-substrates, converts 3,4-dimethoxycinnamate into *trans*-ω-bromo-3,4-dimethoxystyrene [K439].

Ethanolamine incorporation with decarboxylation

Debaryomyces polymorphus replaces the carboxyl group in maesanin (a quinone) with an ethanolamino group to form a substituted aniline [K388].

4.4 *Other reactions involving side-chain shortening and removal*

Polynuclear hydrocarbon demethylation

Rat lung cytosol demethylates 7-methyl- and 7,12-dimethylbenzanthracene [G491]. Liver cytosol demethylates 5- and 6-methylchrysene [G118].

Tryptophanase and tryptophan indole-lyase (E.C. 1.13.11.11 and 4.1.99.1)

$$\text{L-Tryptophan} \rightarrow \text{pyruvate} + NH_3 + \text{indole}$$

In E. coli the reaction, optimum pH 8.5 is

reversed by high concentrations of pyruvate and ammonium ion [A430]. D-Tryptophan is also a substrate, but only in high concentrations of $(NH_4)_2HPO_4$ [J250].

Salmonella typhimurium and E. coli tryptophan synthase (E.C. 4.2.1.20), free from any tryptophanase (the latter catalyses the reaction rapidly) slowly catalyzes the reaction. It is not inhibited by (3*R*)-2,3-dihydro-L-tryptophan, a tryptophanase inhibitor, but is inhibited by (3*S*)-2,3-dihydro-L-tryptophan, a tryptophan synthase inhibitor. Another substrate is S-(*o*-nitrophenyl)-L-cysteine, which yields *o*-nitrothiophenol, but the reaction rate is slower than with tryptophanase [E101].

Proteus rettgeri enzyme synthesises tryptophans from indole, 5-methylindole, 5-hydroxyindole and 5-aminoindole [B292].

Gut flora, protozoa and Clostridium catalyze the reaction [A2571, D477, G485].

Debenzylation of tetrahydroisoquinolines

Nelumbo nucifera peroxidase releases vanillyl alcohol from orientaline, and *p*-hydroxybenzyl alcohol from N-methylcoclaurine and armepavine [D566].

(+)-Usnate deacetylation

Mortierella isabellina enzyme, molecular weight 76 000 and optimum pH 7, forms (+)-2-deacetylusnate and acetate from usnate; it is claimed to be a hydrolytic reaction. It is activated by divalent cations (Co, Ni, Mn, Mg, and Zn), but a large number of potential inhibitors do not act on this enzyme. It is specific for the (+)-isomer, and it does not act on several analogues [A2521].

Kynureninase (kynurenine hydrolase, hydroxykynureninase; E.C. 3.7.1.3)

L-Kynurenine → anthranilate

Human liver enzyme, which requires pyridoxal phosphate, is both cytosolic and mitochondrial. Both forms have similar properties, with molecular weight 130 000 and pI 5.9. They act on kynurenine, or, better, 3-hydroxykynurenine [D272].

Mouse liver enzyme is inhibited by Zn^{2+} and activated by Mn^{2+} [A97]. The affinity for 3-hydroxykynurenine is much greater than for kynurenine. 3-Hydroxyanthranilate is inhibitory [A854].

Activity is found in liver, lung and brain (in decreasing order of activity) from rabbit, rat, gerbil and mouse [J830].

Rat liver enzyme does not act on 5-hydroxykynurenine; 3-hydroxykynurenine is a better substrate than kynurenine. Several substituted hydrazines are inhibitory, and interaction with the pyridoxal phosphate coenzyme is thought to be the mechanism [A2546, A2664].

Suncus murinus liver enzyme, a dimer, molecular weight 110 000, optimum pH 8.5 and pI 6.4, requires pyridoxal phosphate, with kynurenine and 3-hydroxykynurenine as substrates. Traces are found in other organs [F175].

Neurospora crassa enzyme kynureninase I, which has been crystallized, is inducible [K227], molecular weight 105 000. It is inhibited by hydroxylamine, phenylhydrazine, semicarbazide and borohydride; all but the latter inactivation are reversed by pyridoxal phosphate. EDTA and divalent ions are not inhibitory. L-3-Hydroxykynurenine is a better substrate, whereas N-formyl-L-kynurenine is less effective; this contrasts with enzyme from Pseudomonas marginalis for which both compounds are less effective substrates than kynurenine [A2766]. The inducible enzyme has an optimum pH 8.5 and pI 4.90 [B26]; the latter is a homodimer, and contains one mol of pyridoxal phosphate [C245]. Kynureninase II is constitutive [K227], molecular weight 110 000, optimum pH 8.5 and pI 4.75. L-3-Hydroxykynurenine is a better substrate than kynurenine [B26].

Both constitutive and inducible enzymes are found in Mucor ambiguus and M. javanicis, Gibberella fujikuroi, Neurospora sitophila and N. tetrasperma, Aspergillus niger, A. oryzae and

A. wentii, Fusarium oxysporum, Penicillium notatum, P. purpurogenum and P. urticae. Only a constitutive enzyme is found in Rhizopus oryzae [B26].

Rhizopus stolonifer enzyme acts on 3-hydroxykynurenine and kynurenine. Enzyme in Penicillium roqueforti, Pseudomonas fluorescens and A. niger are induced by tryptophan [A3659].

A Streptomyces parvulus enzyme, molecular weight 82 000, acts on kynurenine, and second enzyme, molecular weight 56 000, acts on 3-hydroxykynurenine [B335].

Studies on induction in Penicillium roqueforti, A. niger, Pseudomonas fluorescens and Rhizopus stolonifer indicate that different enzymes act on kynurenine and 3-hydroxykynurenine [A2150].

Pseudomonas marginalis enzyme acts on kynurenine in the presence of pyridoxal phosphate and benzaldehyde to form 2-amino-4-hydroxy-4-phenylbutyrate. This is considered to arise from the trapping of an amino acid β-carbanion that would normally form alanine in the absence of benzaldehyde. Benzaldehyde can be replaced by *o*- and *p*-nitrobenzaldehyde, 3- and 4-formylpyridine and poorly by vanillin, but not by aliphatic aldehydes as trapping agents [D449]. P. fluorescens enzyme acts on β-methyl-L-kynurenine [K233].

Benzaldehyde lyase (benzoin aldolase) (E.C. 4.1.2.38)

Benzoin → benzaldehyde

Pseudomomas fluorescens enzyme, molecular weight 80 000 (monomer 53 000) and optimum pH 7.5–8.5, requires thiamine pyrophosphate and a divalent cation to form benzaldehyde, irreversibly. Anisoin is the only substrate analogue tested with activity [F221].

Hydroxynitrilase (E.C. 4.1.2.39)

Hevea brasiliensis (rubber) enzyme acts on acetone cyanhydrin and mandelonitrile to form cyanide and the corresponding oxo-compound; a similar reaction is observed with cassava [K795].

This reverse reaction with this enzyme is described in section 4.5.

Aldehyde formation from benzonitrile

Sorghum forms *p*-hydroxybenzaldehyde from *p*-hydroxyphenylacetonitrile; *p*-hydroxymandelonitrile is considered to be the enzymatic product of a peroxidase-type reaction on *p*-hydroxyphenylacetonitrile; it then breaks down spontaneously to the aldehyde. The reaction requires oxygen, and is stimulated by peroxide and Mn^{2+}, but cyanide is inhibitory, presumably reversing the spontaneous breakdown step. The same reaction was observed with horseradish peroxidase [A3471].

Phenylserine aldolase (E.C. 4.1.2.26)

A human brain enzyme converts L-*threo*-3,4-dihydroxyphenylserine into protocatechualdehyde and glycine. The enzyme, optimum pH 8.4, is mainly cytosolic and requires pyridoxal phosphate. The *erythro*-isomer is less active, and the D-isomers are not substrates [E371]. The activity is also found in rat [H231].

Benzoates from phenethylamines

A mouse enzyme that requires oxygen and NAD(P)H to convert mescaline into 3,4,5-trimethoxybenzoate, shows highest activity in brain. The reaction is not inhibited by monoamine, or by diamine oxidase inhibitors [A117].

β-Oxidation

$$R.CH_2.\ CH_2.\ COOH \rightarrow R.\ COOH$$

A large number of studies have detected metabolic products that imply the degradation of side-chain carboxylic acids, with the side chain reduced in length by a multiple of two carbon atoms.

In Trichosporon cutaneum an intermediate in this reaction, *p*-hydroxyphenylhydracrylate is converted into *p*-hydroxybenzaldehyde; ATP and CoA are required [B34]. Flavobacterium converts phenylhydracrylate into phenylacetaldehyde [E679].

A number of studies with cannabinoids have demonstrated degradation of the alkyl side chain by Nocardia, involving removal of two carbon units, presumably with initial oxidation of the terminal methyl group [e.g A3010].

Pseudomonas fluorescens 4-hydroxycinnamoyl CoA hydratase/lyase is a homodimer, molecular weight 63 000, (monomeric molecular weight calculated to be 31 010 from the gene sequence), pI 5.2 (calculated as 5.63 from gene sequence) and optimum pH 8.5–9.5. The product is a benzaldehyde; the reaction sequence is postulated to involve the formation of phenylhydracryloyl CoA, with acetyl CoA elimination. Substrates are the CoA conjugates of ferulate, caffeate and *p*-coumarate, but not cinnamate, sinapate or *o*-coumarate. It is inactivated by iodoacetamide [K92].

Skatole formation

Gut flora convert L-tryptophan into skatole [G485]. With rumen organisms up to twice as much tryptophan is converted into skatole as into indole [D477].

Double bond fission by lipase

Lipase and peroxide convert isoeugenol into vanillin [K626].

Lignin degrading peroxidase (diarylpropane oxygenase)

Phanerochaete chrysosporium enzyme, molecular weight 42 000 requires peroxide. It acts on 1,2-bis(3,4-dimethoxyphenyl)propanediol to form veratraldehyde and 1-(3,4-dimethoxyphenyl) ethanediol; the latter is further degraded to veratraldehyde and 3,4-dimethoxy-ω-hydroxyacetophenone. Other studies found a molecular weight of 41 000 for a major isozyme, and 39 000 and 43 000 for two minor isozymes. It contains Fe, probably as haem, and the main isozyme contains 6 per cent carbohydrate. A number of analogues of the above compounds are substrates [C401, D467, D655]. This enzyme is essential for the organism's ability to degrade wood.

Lignostilbene α, β-dioxygenase (E.C. 1.13.11.43)

Pseudomonas enzyme, optimum pH 8.5 and molecular weight 94 000, is a homodimer containing Fe. It requires oxygen for the inter-phenyl double bond fission of 1,2-bis(4-hydroxy-3-methoxyphenyl)ethylene to form vanillin. *p*-Hydroxystyrene is another substrate; other analogues may have marginal activity [K754].

Noscapine fission

In rat, rabbit and man noscapine is converted into cotarnine, hydrocotarnine and meconine. The reaction involves a novel C–C bond fission between two ring systems; the products demonstrate, in one instance only, hydroxylation of the heterocyclic ring at the site of fission [A3719].

C-Glucoside removal

Gut flora remove the 8-C-β-D-glucopyranosyl group from puerarin [J525]. This reaction is also observed in rat [H408].

Human gut flora remove the C-glucoside group of aloesin; this was claimed to be the first report of this reaction type [G226].

Gut flora remove the C-glucoside group present in barbaloin [F44].

Oxidative de-esterification

Rat and dog remove the isopropyl group from barnon. It was suggested that the reaction was oxidative, with formation of acetone [A3057].

In rat dimethyl 2,6-dimethyl-4-(*o*-nitrophenyl)-3,5-pyridinedicarboxylate is mono-demethylated; there is no hydrolytic cleavage, and the reaction appears to oxidative [F540].

Tea catechol oxidase, optimum pH 5.7, oxidatively degallates (−)-epigallocatechin gallate and (−)-epicatechin gallate to release gallate; it is not an esterase [A972].

Anthracycline glycoside reductase

Rat liver P450, optimum pH 7.4 removes the glycoside moiety of aclacinomycin A reductively, utilizing NAD(P)H as co-substrate, with 7-deoxyaklavinone and 7-deoxyaklavinone dimer as products; the reaction removes the oxygen at the 7-position. Two other compounds, designated MA 144M1 and M144N1 are claimed to be substrates; their identities are not disclosed in Chemical Abstracts [B229].

Aeromonas enzyme, molecular weight 35 000, requires NADH and is inhibited by oxygen but not by cyanide or EDTA. It cleaves steffimycin to 7-deoxysteffimycinone [A2754]. The same reaction is found in Aeromonas hydrophila, E. coli and Citrobacter freundii [A2756, A2759].

Streptomyces steffisburgensis enzyme, which requires NADH, cleaves daunomycin, daunomycinol and adriamycin to the corresponding 7-deoxy aglycones [A2755, A2757].

Many studies in animals with daunorubicin and analogues have demonstrated the formation of 7-deoxyaglycones; presumably a similar reaction occurs.

Daunorubicin/adriamycin oxidative deglycosylation

Cows' milk xanthine oxidase forms semiquinone free radicals from these compounds, which then yield the corresponding 7-deoxyaglycones [B370].

2,4′-Dihydroxyacetophenone dioxygenase (E.C. 1.13.11.41)

Alcaligenes enzyme acts on the above compound to form formate and *p*-hydroxybenzoate; molecular oxygen is incorporated into both products [E173].

Phloretin hydrolase (E.C. 3.7.1.4)

Phloretin is converted into phloretate and phloroglucinol in rat and microorganisms [D159].

2-Hydroxy-6-oxo-6-phenylhexa-2,4-dienoate hydrolase (E.C. 3.7.1.8)

Burkholderia cepacia enzyme, a homotetramer, monomeric molecular weight 32 000, forms benzoate and 2-hydroxypenta-2,4-dienoate from 2-hydroxy-6-oxo-6-phenylhexa-2,4-dienoate. A similar reaction is observed with compounds in which the phenyl group is replaced by an alkyl group or the phenyl group is in position 5, and with a range of analogues halogenated on the nucleus and the side chain. The reaction is stated to be a hydrolysis, and is involved in the degradation of polychlorinated biphenyls [J710, K522].

Pseudomonas cruciviae enzyme (grown on biphenyl – the substrate is a biphenyl metabolite), molecular weight 160 000 and monomeric molecular weight 29 000, has an optimum at pH 4.7 [E85].

Transmethylations

a. Methylenetetrahydrofolate-tRNA (uracil-5-) methyltransferase (E.C. 2.1.1.74)

S. faecalis enzyme transfers the methyl group from methylenetetrahydrofolate to form tetrahydrofolate; the reaction requires $FADH_2$ [K837].

b. 5-Methyltetrahydropteroyltriglutamate-homocysteine methyltransferase (E.C. 2.1.1.14)

E. coli enzyme, molecular weight 84 000 is polymeric and requires Mg^{2+} (Mn^{2+} is less effective); it forms methionine and tetrahydropteroyltriglutamate. 5-Methyltetrahydrofolate is not a substrate [K807, K842].

c. 5-Methyltetrahydrofolate-homocysteine methyltransferase (E.C. 2.1.1.13)

E. coli enzyme, molecular weight 150 000 for urea-resolved apoenzyme (native molecular weight appears to be 205 000), contains bound cobalamin and requires Mg^{2+}, ATP, FAD and NADH; it forms methionine and tetrahydrofolate. The Stokes radius is 5.42 nm [K804, K807].

Enzyme from green string bean, barley sprout and spinach acts additionally on 5-methyltetrahydropteroyltriglutamate [K802].

d. Tetrahydropteroylglutamate methyltransferase

Rat liver enzyme, optimum pH 6.7, acts on 5-methyltetrahydropteroylglutamate and 5-methyltetrahydropteroylpentaglutamate with L-homocysteine as co-substrate to form methionine and the demethylated glutamates [A2502].

Phaseolus vulgaris enzyme, molecular weight 40 000 and optimum pH 6.5, catalyzes the reaction anaerobically [A3085].

Tyrosine phenol-lyase is described in 4.1.

4.5 *Chain lengthening reactions*

Hydroxynitrile lyases (oxynitrilase; mandelonitrile lyase and hydroxymandelonitrile lyase, E.C. 4.1.2.10 and 4.1.2.11 respectively)

e.g. Mandelonitrile $\leftrightarrow$ benzaldehyde + HCN

Sorghum bicolor and almond catalyze the formation of hydroxynitriles from benzaldehydes to form (S) and (R) isomers respectively. A range of benzaldehydes, substituted in the p-position, are substrates [H208, J194]. The molecular weight of Sorghum enzyme is 95 000 [G263].

Ximenia americana enzyme is composed of two isozymes, one of which is minor. The other, molecular weight 38 000, optimum pH 5.5 and pI 3.9, is a glycoprotein, and unlike Prunus enzyme it does not appear to be a flavoprotein. Only one isomer of mandelonitrile, (presumably (R)) is substrate [F265].

(R)-Oxynitrilase is found in apple, apricot, cherry, plum, and almond kernel. Substrates studied were benzaldehyde, 3,4-isopropylenedioxybenzaldehyde, 3,4-dimethoxybenzaldehyde, 4-acetoxy-3-acetoxymethylbenzaldehyde, and benzaldehydes substituted in the p-position with acetoxy, *tert*-butylcarbonyloxy, phenoxy, isopropyl or hydroxyl groups. In most cases the enantiomeric purity of the product is in excess of 90 per cent. Almond enzyme acts on all these substrates, and apple on all but p-acetoxybenzaldehyde. Only benzaldehyde is substrate for all species studied; of the remaining species only cherry and apple enzymes act on substituted benzaldehydes [J302].

Almond enzyme acts on cinnamaldehyde, 1-(formylmethoxy)naphthalene, acetophenone, 2-acetyl-6-methoxynaphthalene and pentachlorobenzaldehyde. Almond, peach and loquat all act on cinnamaldehyde and some benzaldehydes [K6, K320]. Prunus amygdalus enzyme, used in conjunction with an organic solvent yields oxynitriles with high enantomeric purity (presumably (R)) from benzaldehyde, m-phenoxybenzaldehyde and phenylacetaldehyde as well as from some aliphatic aldehydes [E353].

Prunus serotina leaf and stem enzyme hydrolyzes (R)-mandelonitrile; Prunus enzyme acts on benzaldehyde [H343, H781].

Phlebodium aureum (fern) enzyme exists as three polymeric forms, molecular weights about 170 000, all with monomeric molecular weight 20 000 and optimum pH 6.5. They act on benzaldehyde and are (R)-specific. Unlike many analogous enzymes from other species it is not a flavoprotein. Iodoacetamide, diethylpyrocatechuate and Ag^{+} are inhibitors;

they are more susceptible to inhibition than P. amygdalus enzyme [H781].

An (*S*)-oxynitrilase (source uncertain) differs from (*R*)-oxynitrilase in that it acts on aromatic aldehydes, whereas (*R*)-oxynitrilase acts on both aromatic and aliphatic aldehydes. Its substrates are benzaldehydes, substituted in the *para* position with H, methyl, chloro or hydroxyl groups, or in the *meta* position with chloro, bromo, hydroxyl, methoxy or phenoxy groups. In most cases the product is enantiomerically at least 90 per cent pure, except for *p*-chlorobenzaldehyde which yields the racemic mixture [F680].

D-Phenylserine formation

Arthrobacter acts on glycine and benzaldehyde to form D-phenylserine. The reaction requires mercaptoethanol, pyridoxal phosphate and Mn^{2+} [G140].

Propiophenone formation

Acinetobacter calcoaceticus acts on benzoylformate and acetaldehyde to form (*S*)-2-hydroxypropiophenone [H403].

2-Hydroxypropiophenone formation

Pseudomonas putida benzoylformate decarboxylase acts on aryl aldehydes and acetaldehyde with thiamine diphosphate as coenzyme, to form 2-hydroxypropiophenones. A large number of benzaldehydes are substrates, with low activity for *o*-substituted benzaldehydes [K524].

Acetophenone formation from benzoate

Streptomyces levoris converts *p*-aminobenzoate into *p*-aminoacetophenone [C350].

Benzalacetone synthase

Raspberry converts *p*-coumaroyl CoA and malonyl CoA into *p*-hydroxyphenylbutan-2-one, via *p*-hydroxyphenylbut-3-en-2-one [J26].

Benzylsuccinate synthase

Thauera aromatica enzyme, molecular weight 220 000 is composed of four subunits, molecular weights 94 000, 90 000, 12 000 and 10 000; the substrates are toluene and succinate, and more than 95 per cent of the product is the (+)-isomer [J817]. The enzyme is very sensitive to oxygen [J651]. Another report states that the co-substrate is fumarate [J191].

Azoarcus enzyme is anaerobic, and uses fumarate as co-substrate. Besides toluene, it acts on xylenes, monofluorotoluenes and benzaldehyde. Studies with several toluene analogues found that the hydrogen abstracted from the toluene methyl group is retained in the succinyl moiety [K238].

The reaction is also found in Desulfobacula toluolica [J847].

L-Phenylacetylcarbinol formation

Candida utilis forms this compound from benzaldehyde and pyruvate by the action of pyruvate decarboxylase. The product is used in ephedrine synthesis [H771, J192].

Zygosaccharomyces bisporus enzyme acts on pyruvate and aldehydes. Benzaldehyde forms 1-hydroxy-1-phenyl-2-propanone; phenylacetaldehyde is another substrate [K611]. Candida and Saccharomyces exhibit the same reaction [J192].

Side-chain C-methyltransferases

Streptomyces flocculus forms (2*S*,3*R*)-β-methyltryptophan from L-tryptophan and *S*-adenosylmethionine [D183].

Streptomyces griseus indolepyruvate C-methyltransferase (E.C. 1.1.1.47), molecular weight 55 000–59 000 (by different methods) and optimum pH 7.5–8.5 is stable at 0°, but activity is lost on freezing or heating; it does not require cofactors. It introduces a methyl group into the side chain to form (*S*)-3-methylindolepyruvate, with S-adenosylmethionine as co-substrate. It is inhibited by thiol-binding reagents and iron-zinc chelators, as well as by indolmycin, the product of the reaction sequence. Phenylpyruvate and *p*-hydroxyphenylpyruvate are also substrates [A2281, A2307, K847]. Evidence has been presented that the same enzyme is responsible for the methylation of tryptophan and indolepyruvate. Streptonigrin is reckoned to be the final product of this reaction sequence [D148].

C-Glucoside formation

Fagopyrum esculentum UDP-glucose: 2-hydroxyflavanone-6 (or 8)-glucosyltransferase, molecular weight 41 000 and optimum pH 9.8, acts on 2,4′,5,7-tetrahydroxyflavone and 2,5,7-trihydroxyflavanone, but not on analogues lacking the 7-hydroxyl group [E614].

C-Glucuronide formation

In mouse Δ^6-tetrahydrocannabinol forms the corresponding C-4′-glucuronide [A3724], in man feprazone forms a C-4-glucuronide [A3916], and sulphinpyrazole (species unclear) forms C-4-glucuronide [A3243].

L-Phenylalanine from phenylacetate, and allied reactions

This reaction has been detected in Ruminococcus albus [C793], in rumen bacteria and protozoa [H960]. Another study with rumen microorganisms found in addition that *p*-hydroxyphenylacetate forms L-tyrosine and IAA forms L-tryptophan [A1164].

Indole-3-butyrate formation

Zea mays enzyme, molecular weight 31 000 and optimum pH 4.8, catalyzes this reaction, with IAA as substrate. It requires acetyl CoA and ATP as cofactors [G758, H694].

5. Conjugation and substitution reactions

5.1 *Ester formation* A. Carboxylate esters

Acetyl CoA: benzyl alcohol acetyltransferase

Clarkia breweri flower enzyme acts on benzyl alcohol to form benzyl acetate [J521, K240].

Mandelonitrile ester formation

Pseudomonas forms O-acetylmandelonitriles from *m*-methoxy- and 3,4-dimethoxymandelonitriles, with vinyl acetate as co-substrate [H303].

O-Acetylmandelic acid formation

Commercial lipase forms this ester from mandelic acid and vinyl acetate [K310].

Chloramphenicol O-acetyltransferase (E.C. 2.3.1.28)

This enzyme has been purified by affinity chromatography from a chloramphenicol-resistant E. coli strain [A164]; acetyl CoA is the co-substrate.

In Klebsiella the 3-acetate is the major product, whereas some 1-acetate is formed [A460].

Streptomyces griseus forms both the 1- and 3-acetates as well as the propionate, isobutyrate, butyrate and isovalerate at the 3-position [A2476].

The activity has been detected in duck [G945].

Oestradiol esters

Beef placental acyl CoA: oestradiol-17β acyltransferase, optimum pH 5.0 forms oleoyl, palmitoyl, palmitoleoyl, linoleoyl, linolenoyl and possibly arachidonoyl 17β esters with oestradiol [E431, E968].

Acetyl CoA deacetylvindoline 4-O-acetyltransferase

A plant enzyme, molecular weight 50 000, forms vindoline [J701].

Acetyl CoA deacetylvindoline 17-O-acetyltransferase (E.C. 2.3.1.107)

Catharanthus roseus enzyme is found in vindoline-containing plant parts [K931].

Acetyl CoA: 10-hydroxytaxane O-acetyltransferase

Taxus chinensis enzyme is monomeric, molecular weight 71 000, pI 5.6 and optimum pH 9.0 and is specific for the 10β-isomer. Other substrates are 10-deacetylbaccatin III as well as other analogues that lack an aryl moiety [K91].

Indoleacetylglucose inositol O-acetyltransferase (E.C. 2.3.1.72)

Zea mays acts on indole-3-acetylglucose and *myo*-inositol to form indole-3-acetyl-*myo*-inositol and glucose [B433, K746].

N-Hydroxyarylamine O-acetyltransferase

Hamster enzyme acetylates the hydroxyl group of N-hydroxy-2-acetamidofluorene [G757].

The enzyme is cytosolic, molecular weight 33 000, and requires acetyl CoA. It also shows arylhydroxamate N,O-acetyltransferase (E.C. 2.3.1.118) and arylamine N-acetyltransferase activities [E110].

The activity is also found in rat liver [C70, G811]; all the three above activities are found associated with the rat enzyme, molecular weight 32 000 [H566]

Aromatic hydroxylamine O-acetyltransferase
(E.C. 2.3.1.56)

N-Hydroxy-4-acetamidobiphenyl
+ N-hydroxybiphenyl →
N-hydroxy-4-aminobiphenyl
+ N-acetoxy-4-aminobiphenyl

Rat enzyme, optimum pH 7.5, is found in liver with lesser amounts in kidney, small intestinal mucosa, spleen and mammary tissue. The reaction involves the transfer of an acetyl group from N-hydroxy-2-acetamidofluorene or N-hydroxy-4-acetamidobiphenyl to 4-hydroxylaminobiphenyl or 2-hydroxylaminofluorene, forming the corresponding acethydroxamic acids. It is activated by NAD(P)H or cysteine [K747].

Esterification by transacetylation and transformylation

Hamster liver N,O-acetyltransferase converts N-hydroxy-2-acetamidofluorene into N-acetoxyaminofluorene. Other substrates are N-hydroxy-2-acetamidofluorene substituted at the 7 position with bromo, acetyl, ethyl or ethoxy groups [H106]. It is a glycoprotein, molecular weight 60 000 and pI 5.4. It also deacetylates the product as well as hydrolyzing *p*-nitrophenyl acetate [G744].

Dog liver N,O-acyltransferase is a glycoprotein trimer, monomeric molecular weight about 60 000, and about 20 000 of this is glycoside. The pI of the aglycone enzyme is 5.6, and 5.4–5.6 for the native enzyme, which appears to be heterogeneous. The substrate is N-hydroxy-2-formamidofluorene [H247].

Rat liver enzyme acts on N-hydroxy-2-formamidofluorene and N-hydroxy-2-acetamidofluorene. Four isozymes have been found, molecular weights 60 000, 61 000, 180 000 and 60 000, with pI 5.0, 5.5, 6.0 and 6.5 respectively. The largest isomer is a trimer. These enzymes also catalyze deacetylation reactions [G735].

Rat liver arylhydroxamate N,O-acetyltransferase (E.C. 2.3.1.118), molecular weight 38 500, pI 4.5 and optimum pH 7.0, is oxygen-labile. This enzyme activates the substrates to bind with tRNA [D447].

Acetyl CoA: salutaridinol-7-O-acetyltransferase
(E.C. 2.3.1.150)

Papaver somniferum enzyme, molecular weight 50 000, pI 4.8 and optimum pH 6–9, acts on salutaridinol with acetyl CoA as co-substrate. This is a key reaction in the formation of opium alkaloids; the reaction product spontaneously eliminates the acetate moiety with the formation of an epoxide ring to position 5, forming thebaine [J1, K784, K786].

Glycoside esterification

Chenopodium rubrum hydroxycinnamoyl-transferase acts on both (*R*)- and (*S*)-amaranthin, with sinapoylglucose, feruloylglucose, caffeoylglucose and *p*-coumaroylglucose as acyl donors (celosianin II is a typical product); the glycoside is esterified at the 2″ position [G255].

Gentiana triflora anthocyanin 5-aromatic acyltransferase (E.C. 2.3.1.153), molecular weight 52 000 and pI 4.6 is monomeric. Substrates are delphinidin- and cyanidin-3,5-diglucosides which form the corresponding 5-O-glucoside-6″-O-hydroxycinnamates; *p*-coumaroyl CoA and caffeoyl CoA are co-substrates [K788].

Silene dioica petal hydroxycinnamoyl CoA: anthocyanidin 3-rhamnosyl(1 → 6)glucoside

4‴-hydroxycinnamoyl transferase, molecular weight 56 000 and optimum pH 7.6–7.8, condenses *p*-coumaroyl or caffeoyl CoA with cyanidin or pelargonidin 3-rhamnosyl (1→6)glucosides and cyanidin 3-rhamnosyl (1→6)glucoside-5-glucoside [B411].

Zinnia elegans enzyme, which requires acetyl CoA, acts on cyanidin and pelargonidin 3-glucosides to form acetylglucosides [G996].

E. coli acetyl CoA: galactoside 6-O-acetyltransferase (E.C. 2.3.1.18) acts on *p*-nitrophenyl-β-D-galactoside, to form the 6-O-acetylgalactoside [A731].

O-Malonyltransferases

Malonyl CoA + flavonoid-O-β-glucoside → CoA + O-malonylflavonoid-O-β-glucoside

Petroselinum crispum isozymes, molecular weight 50 000 and optimum pH about 8, require malonyl CoA. One isozyme (E.C. 2.3.1.115) acts effectively on 7-glucosides of apigenin, luteolin, diosmetin, naringenin and on apiin, but 3-glucosides are poor substrates, with malonyl CoA as co-substrate. A number of non-flavonoid glucosides are also substrates. The other isozyme (E.C. 2.3.1.116) acts on 3-glucosides of kaempferol, quercetin and isorhamnetin, but poorly on 7-glucosides; they are immunologically distinct [D684, K725, K727].

Chickpea (Cicer) malonyl CoA: isoflavone 7-O-glucoside-6″-O-malonyltransferase (E.C. 2.3.1.115), molecular weight 112 000, pI 5.3 and optimum pH 8.0, acts on 7-glucosides of biochanin A, genistein, pratensein, orobol, formononetin, apigenin, luteolin, kaempferol, quercetin and quercetagetin, maackiain 3-glucoside, 2′,4,4′-trihydroxychalcone-4′-glucoside and indoxyl-β-D-glucoside, and slowly on a number of other glucosides, including many non-flavonoid compounds [D445].

Glycine max enzyme, molecular weight 48 000 acts on pentachlorophenyl-β-D-glucoside and malonyl CoA to form the 6-O-malonylglucoside [G247]. Another substrate is *p*-nitrophenyl-β-D-glucoside [H660]. Daucus carota acts on β-glucosides of *p*-nitrophenol and 3,4-dichloroaniline [H118]. A similar reaction is found in wheat [J709].

Rosmarinic acid synthetase (E.C. 2.3.1.140)

Coleus blumei enzyme, optimum pH 7–7.5, acts on caffeoyl CoA and (*R*)-3,4-dihydroxyphenyllactate to form rosmarinic acid by esterification of the alcoholic hydroxyl of 3,4-dihydroxyphenyllactate with caffeate; it is specific for the (*R*)-isomer, and the reaction is reversible. Analogues can be formed by using combinations of substrates, which include *p*-coumaroyl CoA and *p*-hydroxyphenyllactate [G420].

Glycerol ester formation

Human hepatocytes incorporate the substituted benzoic acid lifibrol into triglycerides, with palmitic, stearic, oleic, linoleic, linolenic and arachidonic acid as other components of the triglyceride [H533].

Rat liver monoacylglycerol acyltransferase acts on the CoA derivatives of benzoate, 3-phenoxybenzoate and 1-naphthylacetate to incorporate the acyl moieties into 2-hexadecylglycerol. They are also incorporated into 1,2-dipalmitoylglycerol by the action of diacylglycerol acyltransferase (E.C. 2.3.1.20) [F749]. Rat hepatocytes form triglycerides in which ibuprofen and 3-phenoxybenzoate are incorporated. The product structures are uncertain. Incorporation is not observed with more polar acids, such as 3-phenylbutyrate [F745].

Adipocytes form fenbufenoylglycerol from fenbufen, which then condenses with palmitoyl CoA to form a diester with liver monoacylglycerol acyltransferase as catalyst [H795].

Chlorogenic acid: chlorogenate caffeoyl transferase

2 Chlorogenate (3-O-caffeoylquinate) → isochlorogenate (3, 5-di-O-caffeoyllquinate) + quinate

Ipomoea batatas (sweet potato) tuber enzyme, molecular weight 25 000, pI 4.6 and optimum pH 5.0 is strictly specific; it requires no cofactors [F179].

Chlorogenate-glucarate O-hydroxycinnamoyltransferase (E.C. 2.3.1.98)

Tomato cotyledon enzyme, optimum pH 5.7, pI 5.75 and molecular weight 40 000 is monomeric; the reaction forms quinic acid and 2-O-caffeoylglucarate from chlorogenate and glucarate; galactarate is also an acceptor. The activation energy is 57 kj/mol [K721, K723].

Quinate O-hydroxycinnamoyltransferase (E.C. 2.3.1.99)

Feruloyl CoA + quinate →
O-feruloylquinate + CoA

This enzyme is found in Secale cereale, with CoA esters of ferulate, caffeate, sinapate and *p*-coumarate as substrates [K722]. It is also found in tomato [K723].

Potato tuber enzyme consists of three isozymes, molecular weight about 41 500, two of which are specific for quinate, and the third additionally shows slight activity towards shikimate. The substrates are CoA derivatives of *p*-coumarate, ferulate and caffeate, with chlorogenate as the product from caffeate; the free acids are inhibitors [A3755]. Another study found significant activity with shikimate. In the presence of CoA, the reverse reaction is found with chlorogenate and 5′-(*p*-coumaroyl)quinate. The activity in tubers is modulated by the temperature conditions of storage [A3145].

Ipomoea batatas enzyme, molecular weight 25 000, pI 8.6 and optimum pH 6.0, acts on quinate, and only poorly on shikimate. Substrates are D-glucosides of cinnamate, *p*-coumarate and caffeate [F143].

Studies with tomato have shown that conjugation is strictly restricted to the 5′-position in quinate [A3149].

The activity is found in all of 30 Angiosperms studied, suggesting that it is found in most if not all Angiosperms. However no activity was found in Pinus pinea (Gymnosperm) or Ceratopteris thalictroides (Pteridophyta) [A3758].

Galactarate O-hydroxycinnamoyltransferase (E.C. 2.3.1.130)

Feruloyl CoA + galatarate →
O-feruloylgalactarate + CoA

This enzyme is found in Secale cereale, with CoA esters of ferulate, caffeate, sinapate and *p*-coumarate as substrates [K722].

Shikimate O-hydroxycinnamoyltransferase (E.C. 2.3.1.133)

Feruloyl CoA + shikimate →
O-feruloylshikimate + CoA

This enzyme is found in Secale cereale, with CoA esters of ferulate, caffeate, sinapate and *p*-coumarate as substrates [K722].

***p*-Hydroxycinnamoyl CoA: shikimate *p*-hydroxycinnamoyl transferase**

Cichorium endiva enzyme, molecular weight 58 000 and optimum pH 6.5, carries out this reaction reversibly, with the formation of *p*-coumaroylshikimate and CoA. With the exception of Ceratopteris, this activity was found in every plant genus examined: Agrostemma, Betula, Capsicum, Caragana, Catalpa, Catharanthus, Coffea, Coleus, Datura, Drosophyllum, Forsythia, Galium, Geum, Juglans, Linum, Lonicera, Lycopersicon, Malus, Nicotiana, Padus, Petroselinum, Phaseolus, Pimpinella, Pinus, Rheum, Rauwolfia, Rubia, Salix, Sedum, Solanum and Stevia [B202].

Glucarate O-hydroxycinnamoyltransferase

Feruloyl CoA + glucarate →
O-feruloylglucarate + CoA

This enzyme is found in Secale cereale, with CoA esters of ferulate, caffeate, sinapate and *p*-coumarate as substrates [K722].

Glucarolactone O-hydroxycinnamoyltransferase

Feruloyl CoA + glucaralactone → O-feruloylglucaralactone + CoA

This enzyme is found in Secale cereale, with CoA esters of ferulate, caffeate, sinapate and *p*-coumarate as substrates [K722].

Alcohol O-cinnamoyltransferase (E.C. 2.3.1.152)

The existence of this enzyme is based on the presence of 1-O-*trans*-cinnamoyl-β-D-glucopyranosyl(1→6)-β-glucopyranose and analogues in Physalis peruvianis, Pisidium guajava and Vaccinium vitis-idaea [K789].

β-Glucogallin O-galloyltransferase (E.C. 2.3.1.90)

Quercus robur leaf enzyme, optimum pH 6.0–6.5, catalyzes dismutation of 1-galloyl-β-D-glucose, with the formation of glucose and digalloylglucose as products [K894].

Rhus typhina leaf enzyme specifically acts on 1,6-di-O-galloyl-β-D-glucose and 1-galloyl-β-D-glucose to form 1,2,6-tri-O-galloyl-β-D-glucose [K888].

Glucogallin-tetrakisgalloylglucose O-galloyltransferase (E.C. 2.3.1.143)

Quercus robur enzyme, molecular weight 260 000 and optimum pH 6.3 is found in young leaves, and is stable between pH 5.0 and 6.5. It forms 1,2,3,5,6-pentagalloylglucose from 1,2,3,6-tetragalloylglucose with 1-galloyl-β-D-glucose as acyl donor [K726].

Sinapoylglucose-choline O-sinapoyltransferase (E.C. 2.3.1.91)

Sinapis alba and Raphanus enzymes, molecular weights 60 000 and optimum pH 7.2 (Raphanus) and 7.6 (Sinapis) act on choline and 1-O-sinapoyl-β-D-glucose to form sinapine; 1-feruloyl- and 1-(*p*-coumaroyl)glucose are also substrates. There is no requirement for cations or thiols; the activation energy is 53 kj/mol [K720].

Phosphatidylcholine conjugation

Adipocytes form fenbufenoyl phosphatidylcholine from fenbufen [H795].

Phenylacetylcarnitine formation

This compound is formed in rat; it is also found in patients with phenylketonuria, where phenylacetic acid levels are raised [K521].

Benzoylmalate formation

Glycine max forms this compound from benzoate; the time course of the reaction indicates that benzoylglucose is an intermediate [A3679].

Benzyl alcohol forms O-benzoyl-L-malate in barley, presumably via benzoate [A1726].

SAM: benzoic acid carboxyl methyltransferase

Antirrhinum majus enzyme, optimum pH 7.5 and molecular weight 100 000 is dimeric. It is activated by K^+ or NH_4^+, and is highly specific for the formation of methyl benzoate [K650].

SAM: salicylic acid carboxyl methyltransferase

Clarkia breweri enzyme is a dimer, molecular weight 40 300; the amino acid sequence has been determined from the corresponding DNA sequence. The product is methyl salicylate [J521, K239].

B. Sulphate esters

Phenolsulphotransferases (E.C. 2.8.2.1)

$$ArOH + PAPS \rightarrow ArOSO_3H + PAP$$

Studies in the period covered by this review detected two forms of this enzyme, one of which acts on simple phenols (PST-P), and the other on catecholamines (PST-M). This section is in consequence divided into three parts.

a. General studies.

The enzyme requires sulphate and sulphate-activating enzymes [A2065] or added phosphoadenosine phosphosulphate (PAPS) [B72, D979].

Human enzymes acting on dopamine and *p*-nitrophenol are found in many brain areas; one is thermolabile and the other thermostable [D233]. Placental enzyme acts on catecholamines and simple phenols [C322]. Human enzyme from different tissues shows different ratios of activity toward phenol and dopamine. Using these substrates differences in thermal stabilities and inhibition properties were detected, and this has been interpreted to indicate the presence of PST-P and PST-M [B834]. A study with the three isozymes SULT1A1, SULT1A3 and SULT1B2, prepared by recombinant techniques (PSTP-M and thyroid hormone sulphotransfeases respectively), found that bisphenol A, 4-nonylphenol, stilboestrol and 17α-ethynyloestradiol are primarily substrates for PSTP. Dopamine is sulphated mainly by PSTM, although it is a significant substrate for PSTP. 4-Octylphenol and *p*-nitrophenol are good substrates for both these isozymes. None of the above compounds are good substrates for SULT1B2, although most of them show some activity [K344].

Human erythrocyte enzyme with HMPG as substrate has an optimum pH of 7.5 and uses PAPS as co-substrate. Endogenous inhibitors have been detected [B72]. It is cytosolic, and also acts on *p*-acetamidophenol, serotonin, phenol, noradrenaline, tyramine and dopamine, with increasing activity [A3814].

Human lung enzyme, molecular weight 35 000–38 000, which appears not to be PST-P or M, requires PAPS, with a broad specificity for simple phenols and catecholamines; dopamine is the best substrate [D979].

Rat brain enzyme acts on HMPG, DHPG, HMPE and pyrogallol, the latter yielding three products; neither serotonin nor normetanephrine are substrates [A1240]. Another study found an optimum pH in the range 6.8–8.6; other substrates are HVA, VMA, dopac, dopamine and noradrenaline. Neither Mg^{2+} nor mercaptoethanol activate, but ATP is inhibitory [A153]. A further study found a molecular weight of 69 000 and an optimum pH 5.5–6.4. Substrate studies found that simple phenols, VMA, HMPG, HVA, dopamine, adrenaline, noradrenaline and dopamine are substrates, whereas DOPAC, DOMA, HVA, DHPG, metadrenaline, normetadrenaline, serotonin, vanillyl alcohol and *p*-hydroxybenzaldehyde are not [C441]. Rat cerebral enzyme, which acts on hydroxyl groups in tyrosyl peptides, requires PAPS and is found primarily in microsomes [E341].

Rat liver cytosolic enzyme has been separated chromatographically into five fractions, all with subunit molecular weights in the range 28 000–36 000. Three of the fractions show high activity towards *p*-nitrophenol, and the remaining two are much less active [D933]. Two isozymes of a dimeric rat liver enzyme, molecular weights about 65 000, act on simple phenols and catecholamines, optimum pH 5.5 and 6.2 for β-naphthol and 8 for tyrosine methyl ester. They were found to have almost identical amino acid compositions [B61, B941]. Protocatechuate is sulphated in the 3- and 4- positions in the ratio 1:4.8, with an optimum pH 7.4 [B724]. N-Hydroxy-2-acetamidofluorene is a poor substrate [G317].

A rat stomach enzyme, molecular weight 32 000, has a double optimum at pH 5.4 and 6.6 and acts on *p*-nitrophenol, α- and β-naphthols and salicylamide as well as HMPE and HMPG [D807].

Rat platelet enzyme acts on *p*-nitrophenol, and the activity parallels that found in liver [F631].

This activity has been detected liver from rat, rabbit, mouse, cat and guinea pig, mouse kidney, rabbit adrenal and rabbit brain, but not in human liver, rat kidney and intestine or mouse brain [A2067].

Using isoprenaline as substrate, activity has been found in liver from monkey, dog, mouse, rat, rabbit and guinea pig. Activity was also found in kidney, small intestine and lung from only some of these species [B665].

Rat and rabbit enzymes show optimal activity at pH 7.8 [A2065].

Chick embryo enzyme that acts on 4-methylumbelliferone is detectable at five days incubation, with an increase until 20 days. Based on specific activity, a small peak in activity at seven days is followed by a larger, broader peak at 15 days [A3328].

Channel catfish (Ictalurus) enzyme acts on benzpyrene phenols and other phenols. One isozyme, molecular weight 41 000,which is found in liver and gut is immunologically similar to human enzyme, but a liver isozyme, molecular weight 31 000 is not [K497].

b. PST-M.

Human hepatoma enzyme requires Mn^{2+}. Dopamine is a good substrate, and dopa, tyrosine and *m*-tyrosine are lesser substrates; the D-isomers are better substrates than L-[H888]. Small intestine enzyme is a dimer, molecular weight 69 000, and is thermolabile. Dopamine and *p*-nitrophenol are substrates at high concentration [F116].

Human liver enzyme appears to act on L-T_3, but only poorly [E672].

Human ileum enzyme, molecular weight 69 000, which acts on dopamine is thermolabile. It appears that phenol is also a substrate, but only at high concentration [F770].

Human platelet enzyme acts on *m*-tyramine, noradrenaline, adrenaline, serotonin, *p*-hydroxyamphetamine, isoprenaline, salbutamol, α-naphthol, paracetamol, and at high concentration (1 mM) salicylamide [C523]. Dopamine is sulphated primarily (88 per cent) in the 3-position [E186]. The enzyme activity is not altered in either phaeochromocytoma or uraemic patients [E195].

Human brain enzyme, molecular weight 250 000 and optimum pH 7.0, acts on a large number of phenolic analogues of dopamine and noradrenaline, as well as on serotonin. It is inhibited by sodium chloride, but this is slight at physiological concentrations. The kinetics are apparently complex [D583]. It is inhibited by about 45 per cent by 0.3 M NaCl [E198]. Another study on brain cortex enzyme found a molecular weight of 62 000 and an optimum pH 7.8–8.0. Catecholamine analogues are much better substrates than serotonin [D741].

Human and rat liver enzymes act on 4-hydroxypropranolol with high (3 μM) affinity. Although the K_m is the same for both isomers of 4-hydroxypropranolol the V_{max} is different, which leads to stereospecificity [F735].

Rat brain enzyme, which requires PAPS as co-substrate, acts on N-acetyldopamine with optimum pH 9.0. It was not determined which sulphotransferase is responsible for another optimum found at pH 6.6 [E493]. PST-M appears to be absent from rat kidney [E100].

The enzyme is absent from dog platelets [E186].

In marmoset the enzyme acts on tyramine. It is found in lung, liver, small intestine, with small amounts in stomach and kidney, and only traces in brain [F226].

Adrenaline is sulphated in dog, monkey, rat, guinea pig and rabbit, with activity in liver and small intestine, but not in brain or heart [A3444].

Tyramine is sulphated in macaque liver with optimum pH 9.2–9.4. Activity is found (in decreasing order) in macaque, rat, mouse and guinea pig. In macaque the enzyme is more active in intestine than in liver, but in the other species studied the intestinal activity is very low [A2066].

c. PST-P.

Human lung enzyme is a 'mixed' type that acts on adrenaline and noradrenaline as well as simple phenols [D977]. Small intestine enzyme acts on *p*-acetamidophenol [F116]. In liver the sulphation of L-T_3 correlates with PST-P activity [E672].

Human platelet enzyme acts on paracetamol, salicylamide and phenol at low (5 μM) concentration [C523]. Enzyme designated SULT1A1, which acts on 2-naphthol, is assayed by the reverse reaction; with PAP and *p*-nitrophenyl sulphate as substrates; *p*-nitrophenol is released [K46].

Human liver enzyme, which exists as two isozymes that act on *p*-nitrophenol, is thermostable. It also acts on dopamine for which it has a low affinity, optimum pH 7.15 [E211]. Purified enzyme is dimeric, molecular weight 68 000, and acts on phenol and *p*-nitrophenol. Antibody studies suggest that it is closely related to PST-M [F692].

Human brain enzyme has been separated into two fractions; P_1, which acts on phenol, optimum pH 8.5, is inhibited 80 per cent by 0.3 M sodium chloride, whereas the P_{11} form is much less affected by ions. P_1 is inhibited by phosphoadenosine phosphate competitively relative to PAPS, and by ATP non-competitively relative to phenol; both enzymes are inactivated by 2,6-dichloro-4-nitrophenol [E198].

Human ileum enzyme, molecular weight 69 000, is thermostable [F770].

Two alleles of the human enzyme are found. In one ^{213}Arg is replaced by His, resulting in an enzyme with low activity [K257].

Human and rat liver enzymes show a low affinity (500 μM) for 4-hydroxypropranolol, and unlike PST-M they exhibit no stereospecificity [F735].

Rat liver cytosol contains two isozymes, one of which is thermolabile, optimum pH 9, and the other isozyme, optimum pH 6.4, is more stable; they both act on paracetamol [D460]. The former must be considered to be PST-M despite the recorded specificity. Kidney enzyme acts on *p*-nitrophenol, α- and β-naphthols and salicylamide, but not on catecholamines or catecholamine metabolites [D807].

Dog liver enzyme, molecular weight 32 000–34 000, reacts to antibodies like a PST-P. It acts on α-naphthol, vanillin, *p*-nitrophenol, tyramine, serotonin and dopamine. The V_{max} for dopamine is low although the K_m is similar to that for α-naphthol. The activity towards serotonin and tyramine is low with a low V_{max} and K_m greater than 100 μM (1 μM for α-naphthol) [J292].

Beef adrenal medulla enzyme acts on HMPG, but there appears to be little activity towards catecholamines [D187].

Marmoset enzyme, which is found in lung, liver, small intestine, with small amounts in stomach and kidney, acts on phenol. It is almost absent from brain [F226].

Thyroid hormone sulphotransferases

Human PST-P and PST-M (designated SULT1A1 and SULT1A3 respectively) both require PAPS as co-substrate and conjugate 3,3′-T_2, rT_3, T_3 and T_4 with decreasing activity; these substrates are mutually inhibitory. 3,5-T_2, 3-T_1 and T_0 are not substrates. Inhibition studies suggest that there are other similar sulphating enzymes in liver and kidney [K90]. Human liver T_3-conjugating activity is thermostable and correlates well with PST-P, the main liver sulphoconjugase, but not with PST-M activity. However, purified PST-M does conjugate T_3 [E672].

Rat liver iodothyronine sulphotransferase, optimum pH 8.0, conjugates 3,3′-T_2 and T_3 [J495]. Arylsulphotransferases I and IV from rat liver and monkey hepatoma conjugate 3,3′-T_2, rT_3, T_3, 3-T_1 3′-T_1 and T_0, 3,3′,5-triiodothyropropionate, 3,3′,5-triiodothyroacetate and tetraiodothyroacetate (only sulphotransferase I), but not T_4, 3,5-T_2 or tetraiodothyropropionate [B770].

Rat uterus iodothyronine sulphotransferase, optimum pH 6.0, conjugates 3,3′-T_2, T_3 and T_4 [K649].

Eubacterium trans-sulphatase (see below) sulphates T_3 [E432, G602].

Tyrosine-O-sulphate formation

Studies with tyrosine have usually failed to detect sulphate conjugation. However, studies with Caco-2 human gut epithelial cells, human Chang liver cells and canine Madin-Derly liver cells have detected this reaction [H371]. Human sulphotransferase M acts on D-tyrosine [J644].

Tyrosine ester sulphotransferase (E.C. 2.8.2.9)

In male rat liver this activity appears to be identical with sulphotransferase IV (E.C. 2.8.2.1) [K735], optimum pH 8 [B941].

Tyrosylpeptide sulphotransferase (E.C. 2.8.2.20)

Human liver enzyme is microsomal, optimum pH 6.4, and requires Mn^{2+} [F752]. Human tyrosylprotein sulphotransferase, molecular weight about 50 000–54 000, which is found in the Golgi network of practically every tissue, especially liver, requires PAPS [J724].

Rat cerebral enzyme, which acts on the hydroxyl groups in tyrosyl peptides, requires PAPS and is found primarily in microsomes [E341]. Rat liver microsomal enzyme also requires PAPS [J614], molecular weight 50 000–54 000 and optimum pH 5.5 [J532].

Oestrogen sulphotransferases (E.C. 2.8.2.4)

In human liver both oestrone and 17β-oestradiol are sulphated at the 3 position. Both PST-P and dehydroepiandrosterone sulphotransferase catalyse the reaction, but not PST-M [G87].

A study on beef adrenal enzyme, optimum pH 7–8, demonstrated that it acts on a large range of oestrones and analogues. At least some of them are sulphated at the 3-position, but not at the 2- or 4-position. Some simple phenols are very poorly sulphated, but many hydroxylated polynuclear hydrocarbons are substrates. For optimal reaction a lipophilic group *para* to the hydroxyl group and an oxygen atom on the D ring for hydrogen bonding to the enzyme less than 0.372 nm above the ring are required [A1211].

In beef placenta oestrone and oestrone-3-sulphate are interconverted, but each reaction occurs in separate parts of the placenta [D520].

In guinea pig oestrone is sulphated, and both oestrone-3-sulphate and 17β-oestradiol-3-sulphate are converted into17β-oestradiol disulphate without exchange of sulphate [A1936].

Rat liver cytosol enzyme, optimum pH 5.5–6.0, acts on oestradiol and oestrone [D500].

Oestrone is sulphated in sheep [A472].

Activity in gilt uterus appears at day three of the oestrus cycle, rising sharply at day six to a maximum at day nine [A1450].

In hen oestrone, 16-epioestriol, 16-oxo-17β-oestradiol, 17α-oestradiol and 17β-oestradiol are sulphated at the 3-position, and both oestradiols are disulphated [A3652].

Flavonoid sulphotransferases

Three sulphotransferases, molecular weight 35 000, have been isolated from Flaveria chloraefolia shoot tips. A transferase specific for the 3 position (E.C. 2.8.2.25), pI 5.4, shows optima at pH 6.5 and 8.5, and acts (in decreasing order) on rhamnetin, isorhamnetin, quercetin, patuletin and kaempferol, but not on quercetagetin, gossypetin or myricetin. The 3′-specific enzyme (E.C. 2.8.2.26), pI 6.0 and optimum pH 7.5, acts (in decreasing order) on 3-sulphate esters of quercetin, patuletin and tamarixetin but not kaempferol or isorhamnetin, nor on aglycones. The 4′-specific enzyme (E.C. 2.8.2.27), pI 5.1 and optimum pH 7.5, acts (in decreasing order) on 3-sulphates of quercetin, kaempferol, isorhamnetin and patuletin, but not tamarixetin, nor on aglycones [F243, G769]. A 7-specific enzyme acts on isorhamnetin-3-sulphate, the 3,3′- and 3,4′-disulphates of quercetin, but not on quercetin-3- or 3′-sulphate, nor on aglycones (E.C. 2.8.2.28) [F210].

N-Hydroxyarylamine sulphotransferase

Human adrenal and liver enzymes act on 4-hydroxylaminobiphenyl [H304, H434];

presumably substitution occurs on the hydroxyl group.

N-Hydroxylaminofluorene is O-sulphated in mouse [D756].

N-Hydroxy-2-acetamidofluorene is sulphated in man, monkey, mouse, and rat liver [A1840, A2086, A2206, D756, K34]. This compound is a poor although significant substrate for rat liver cytosolic phenolsulphotransferase; two N-hydroxyarylamine sulphotransferase isozymes have been found for which N-hydroxy-2-acetamidofluorene and *p*-nitrophenol are both good substrates [G317]. Although in many cases the group conjugated has not been determined, the amido nitrogen is improbable. This reaction is part of the carcinogenic activation process of N-hydroxy-2-acetamidofluorene [G675].

Alcohol sulphotransferase (E.C. 2.8.2.2)

The benz[*a*]anthracene metabolites 12-hydroxymethyl-7-methylbenz[*a*]anthracene, 7-hydroxymethyl-12-methylbenz[*a*]anthracene and 7,12-dihydroxymethylbenz[*a*]anthracene are sulphated by dehydroepiandrosterone-steroid sulphotransferase in human liver, which differs from PST-P and M [J683].

Sulphatoglucoside formation

Brassica forms *o*-nitrobenzyl-β-D-sulphatoglucopyranoside and *E*- and *Z*-*o*-nitrobenzaldoxime-β-D-(6-sulphato)glucopyranoside as well as *o*-nitrophenylglucosinolate from *o*-nitrobenzaldoxime. Other substrates are benzaldoxime, nitro and halogen-substituted benzaldoximes, as well as some aliphatic analogues. It would appear that hydrolysis and reductive steps precede conjugation [F657].

Renilla-luciferin sulphotransferase (E.C. 2.8.2.10)

Renilla reniformis enzyme is specific for Renilla-luciferin [K831].

Trans-sulphatase (arylsulphate sulphotransferase, E.C. 2.8.2.22)

These enzymes transfer the sulphate group from, for instance, *p*-nitrophenol to another phenolic group.

Human gut flora (Eubacterium) enzyme uses 4-methylumbelliferyl and *p*-nitrophenyl sulphates as sulphate donors, and tyramine, rutin, quercetin, esculetin (at position 6), baicalin and, by far the best, baicalein as sulphate receivers [H354]. It is a homotetramer, pI 3.9, molecular weight 315 000 and optimum pH 8–9 using tyramine as substrate. Another sulphate donor is *p*-acetylphenyl sulphate. Acceptors include α- and β-naphthol, oestradiol, phenol, tyrosine methyl ester, tyrosine-containing peptides, tyramine, triiodothyronine, ethyl *p*-hydroxybenzoate, 4-methylumbelliferone, salicylamide, *p*-acetamidophenol and the catecholamines adrenaline and dopamine, and dopa, specifically at the 4-position [E191, E432]. In another study with this species quercetin formed the 3,3′-disulphate and the 3,3′,7-trisulphate, but a monosulphate was not detected [F513].

Haemophilus enzyme is a tetramer, molecular weight 290 000, with *p*-nitrophenyl sulphate, 4-methylumbelliferyl sulphate, α- and β-naphthyl sulphates as sulphate donors. Sulphate acceptors include phenol, resorcinol and α-naphthol [H756].

Klebsiella enzyme, which sulphates phenol and requires Mg^{2+}, is a dimer, molecular weight 160 000. The pIs for the subunits are 5.3 and 10–10.5 [G756].

5.2 *Sulphamate formation*

Amine sulphotransferase (E.C. 2.8.2.3)

$$R.NH_2 + PAPS \rightarrow R.NH.SO_2.OH + PAP$$

Guinea pig amine N-sulphotransferase requires PAPS as co-substrate. The substrate requires an unprotonated amino group for activity. Substrates include aniline, *p*-chloroaniline, β-naphthylamine, DMI, tetrahydroquinoline, tetrahydroisoquinoline, and some aliphatic amines. The optimum pH is specific for each substrate, and lies in the range 6–10. Many phenols are also substrates, including oestrone and tyrosine [E456].

Human liver cytosolic hydroxysteroid sulphotransferase (E.C. 2.8.2.15) N-sulphates aniline, MPTP and and 1-((5-chloro-2-oxo-3(2*H*)-benzothiazolyl)acetyl)piperazine, and differs from the enzyme that sulphates phenols [J571].

Rabbit liver cytosol contains two enzymes, one of which acts on DMI and male steroid hormones, but not on phenols. The other, molecular weight 34 000, is almost inactive towards phenols and alcohols, but sulphates aniline, DMI, MPTP and 1-((5-chloro-2-oxo-3(2*H*)-benzothiazolyl)acetyl)piperazine [J914].

In rat and mouse liver the enzyme that sulphates 1-((5-chloro-2-oxo-3(2*H*)-benzothiazolyl)acetyl)piperazine (dealkyltiaramide) is different from the one that sulphates the primary alcohol group of tiaramide. The pH optima are modified by buffers, but the values for dealkyltiaramide is about 9–10, and for tiaramide 8–9. Rat enzyme is markedly inhibited by 0.5 M NaCl and KCl, whereas mouse enzyme is only slightly affected by these salts. Activity is much lower in female rat than in male [C571].

Pig liver enzyme acts on many amines with PAPS as co-substrate; the optimum pH varies from 6 to 10, depending on substrate. Aniline, *p*-chloroaniline, demethylimipramine, β-naphthylamine, tetrahydro-4-phenylpyridine, tetrahydroisoquinoline and tetrahydroquinoline are substrates [K923].

5.3 *Glycoside formation*

Flavone apiosyltransferase (E.C. 2.4.2.25)

Parsley enzyme, optimum pH 7.0, requires UDPapiose, but no coenzymes. It acts on flavone-7-O-β-D-glucosides; for instance 7-O-β-D-glucosylapigenin yields apiin [K827].

Arabinoside formation

Silene dioica enzyme, optimum pH 7.2–7.5 acts on isovitexin and UDParabinose to form isovitexin-2″-O-arabinoside. Isovitexin-7-O-xyloside is also a substrate, but isovitexin-7-O-glucoside is not, nor can UDPglucose or UDPrhamnose substitute for the co-substrate. The reaction is stimulated by Mg^{2+} or by Mn^{2+} [A3759].

Zea mays indolylacetylinositol arabinosyltransferase (E.C. 2.4.2.34) activity is found in immature kernels, to form indol-3-ylacetyl-*myo*-inositol arabinoside [K734].

Fructoside formation

Claviceps purpurea β-D-fructofuranoside, optimum pH 5.7 or 6.5, depending on substrate, forms elymoclavine-O-β-D-fructofuranoside and elymoclavine-1-O-β-D-fructofuranosyl(2 → 1)-O-β-D-fructofuranoside from elymoclavine. Other substrates are chanoclavine, lysergol and dihydrolysergol [F516].

Galactoside formation

Zea indolylacetyl-*myo*-inositol galactosyltransferase (E.C. 2.4.1.156) requires UDPgalactose to form indolylacetyl-*myo*-inositol galactoside, a major Zea indole. Activity is found in immature kernels [K811].

Aspergillus oryzae β-galactosidase transfers the galactoside moiety of *p*-nitrophenyl-β-D-galactoside to elymoclavine, chanoclavine, lysergol, 9,10-dihydrolysergol and ergometrine [H11].

E. coli β-D-galactosidase acts on 4-methylumbelliferyl-β-D-xyloside and *p*-nitrophenyl-β-D-galactoside to form galactosyl(β1 → 4)xylyl(β)-4-methylumbelliferone and a small proportion of the (β1 → 3) analogue [J399].

α-Glucoside formation

Bacillus subtilis enzyme, molecular weight 54 000 or 65 000 (different methods), with an optimum pH for starch as co-substrate of 6–7; it acts on quinol [H296].

Leuconostoc mesenterioides sucrose phosphorylase (E.C. 2.4.1.7) transfers α-D-glucose units from sucrose to the 4′- and 4″- positions of (-)-epigallocatechin gallate, forming 4′-mono- and 4′,4″-diglucosides [J5].

Formation of ester β-D-glucosides

Although it is generally assumed that these plant natural products contain a β-linked glucopyranose moiety, these data have often not been demonstrated.

Chenopodium rubrum enzyme requires UDPglucose and acts on ferulate, sinapate, *p*-coumarate and cinnamate [G255].

In Glycine max, benzoate is converted into benzoylglucose transiently. Other substrates are α-naphthylacetate, α-naphthoate and probably phenylacetate [A3679].

Sweet potato (Ipomoea batatas) tuber enzyme, molecular weight of 45 000, is specific for carboxyl-containing compounds. Substrates include cinnamate, *o*- and *p*-coumarate, caffeate, ferulate, benzoate and vanillate (E.C. 2.4.1.177) [D70, F145].

Quercus robur leaf gallate 1-β-D-glucosyltransferase (E.C. 2.4.1.136), optimum pH 6.5–7.0 and molecular weight 68 000, forms 1-galloyl-β-D-glucoside; it is specific for UDPglucose. A number of benzoic acids are substrates (except salicylic acid); cinnamic acids are less good [K736, K737].

Raphanus sativus enzyme, optimum pH 5.8–6.0 or about 7, is found in dry seed and during germination. Co-substrate is UDP- or TDPglucose; other analogues are inactive. The reaction is enhanced by thiols and is reversibly inactivated by thiol-binding compounds.
A reversible reaction is found with sinapate (best), benzoate, vanillate, anthranilate, *p*-hydroxybenzoate, cinnamate, ferulate, caffeate, 3,5-dimethoxycinnamate and coumarate(s), but 3,5-dihydroxybenzoate is a poor substrate [B440, B599].

This reaction is found widely, including Coleus, Pilea, Cistus, Cestrum, Glycyrrhiza echinata, Prunus serotina, Glycine max, mung bean and ripe tomato [A3475, A3585, B493, C756, F445]. Benzoate forms benzoylgentiobioside as well as benzoylglucose in Aconitum japonicum, Coffea arabica, Dioscoreophyllum cumminsii and Nicotiana tabacum; it is unclear whether this involves a stepwise glycoside addition [F445].

Cucumis sativus forms glucose esters with a series of mono- and dihydroxybenzoates [A611].

Brassica napus and UDPglucose: sinapic acid glucosyltransferase (E.C. 2.4.1.120), molecular weight 42 000 and pI 5, conjugates sinapate reversibly [J712]. Raphanus sativus enzyme also forms 1-O-sinapoyl-β-glucose reversibly [K787].

Ester glucoside at C-6

Glycyrrhiza echinata, Dioscoreophyllum cumminsii and Aconitum japonicum conjugate phenylacetate with the 6-hydroxyl of glucose [F445].

Formation of ether D-glucosides

A. Animals

In man, both 3,6-dihydroxy-3,4-benzpyrene and 1-naphthol form glucosides with added UDPglucose, the former yielding the 6-glucoside. With added UDPgalacturonate the corresponding galacturonides are formed [H575].

Pig liver, riboflavin-conjugating enzyme, pI 3.7, optimum pH 6.0, transfers the α-glucosyl group from phenyl-α-maltoside or other maltosides to form riboflavin glucoside. The enzyme is stable at pH 3.5–9, and at 55°. Other substrates are esculetin and rutin [C847].

Coho salmon forms a glucoside from 1-naphthol [A3879].

Housefly microsomal enzyme, optimum pH 8.5, requires UDPglucose as well as Mg^{2+}; other divalent cations are not so good. It conjugates 1-naphthol [A2238].

Enzymes in Melanoplus sanguanipes (grasshopper), Periplaneta americana, Manduca sexta, Agrotis ipsilon (cutworm), Tenebrio molitor and Trilobium confusum act on a range of phenols (E.C. 2.4.1.35). Guaiacol is a substrate in all these species, but phenol, vanillin, salicylaldehyde, umbelliferone and a number of other phenols are substrates in only a few of these species, except that M. sexta has a broad specificity for simple phenols and natural products. UDPglucose or (less good) GTPglucose and TDPglucose are co-substrates [G816, G897].

p-Nitrophenol forms a glucoside in cockroach, stick insect, cricket, mealworm, Porina and wax moth larvae, housefly, blowfly and earwig [C494].

Drosophila xanthurenic acid: UDPglucosyltransferase, optimum pH 7.1, which conjugates the 8-hydroxyl group is both microsomal and cytosolic and requires Mg^{2+} or Mn^{2+} [F636].

B. Plants and microorganisms

a. Broad specificity; xenobiotic substrates.

Many studies on plants do not differentiate between α- and β-glycosides. Most glycosides are, in fact, β-glycosides, and where the stereospecificity of synthesis is not stated, β- can be assumed.

Soyabean (Glycine max) enzyme, molecular weight 47 000 and pI 4.8, acts on pentachlorophenol [G247]. Other substrates are quercetin and 2,2-bis(*p*-chlorophenyl)acetate. It appears to be a monomer, molecular weight about 50 000 and pI 4.9 [H402]. G. max, Triticum aestivum (wheat) Avena fatua and Agrostemma githago form *p*-nitrophenyl-β-D-glucoside [H660, J709].

Daucus carota (carrot) enzyme acts on *p*-nitrophenol [H118].

Gardenia jasminoides forms β-D-glucosides from phenol, catechol, resorcinol, quinol, *o*-, *m*- and *p*-nitrophenol [E338].

b. Flavonoid 3-glucosyltransferases (UDPglucose: flavonol 3-O-D-glucosyltransferase, E.C. 2.4.1.91; UDPglucose: cyanidin 3-O-D-glucosyltransferase, E.C. 2.4.1.115).

Daucus carota UDPG: cyanidin 3-O-glucosyltransferase, optimum pH 8.0, requires ascorbate. Other substrates include delphinidin, pelargonidin, quercetin and kaempferol, but not naringenin, eriodictyol or dihydroquercetin [E280].

Hippeastrum petal enzyme is a homodimer, molecular weight 49 000, pI 5.6 and optimum pH 5, and requires UDPglucose as co-substrate. Good substrates include kaempferol, dihydrokaempferol and quercetin, whereas naringenin, isosakuranetin, pelargonidin cyanidin and malvidin are poor substrates [E636]. UDPG: anthocyanidin 3-O-glucosyltransferase in Hippeastrum and Tulipa petal and leaf is cytosolic [A2523].

Norway spruce needle flavonol 3-O-β-D-glycosyltransferase, molecular weight 40 000, optimum pH 8.4 and pI 5.0, acts on kaempferol reversibly. Other substrates are isorhamnetin, quercetin, rhamnetin and myricetin [G215].

Petunia hybrida enzyme, optimum pH about 8.2, acts on cyanidin and delphinidin. Substrates are mutually inhibitory [A3191].

Pisum sativum enzyme, optimum pH about 8, acts on kaempferol, quercetin and myricetin [C90].

Red cabbage seedling UDPG: cyanidin 3-O-glucosyltransferase, optimum pH about 8, acts additionally on pelargonidin, paeonidin, malvidin, kaempferol, quercetin, isorhamnetin, myricetin and fisetin [A1733].

Senecio hybridus UDPglucose: cyanidin 3-O-glucosyltransferase, exhibits molecular weight 52 000 and optimum pH 7.5 [J722].

Silene dioica UDPG: cyanidin 3-O-glucosyltransferase, optimum pH 7.5, appears to have a molecular weight of 60 000 and can exist as a dimer. It also acts on pelargonidin and delphinidin, but not on flavonols nor on the 5-position of anthocyanins [B407].

Pollen from Alnus, Quercus, Narcissus, Tulipa and Secale contains flavonol 3-O-glycosyltransferase, possibly associated with the pollen wall. Substrates are quercetin and kaempferol [B680].

The reaction has also been observed in Crocus sativum, Ipomoea batatas, Vitis and Zea mays [E120, G124, G137].

A. niger enzyme, molecular weight about 370 000 and pI 4 may act on the 3 position of cyanidin [C900].

c. Flavonoid 4′- glucosyltransferase

An Allium cepa enzyme, molecular weight 48 000, optimum pH 7.7–8.0 and pI 6.0 is specific for the 4′-position of catechol flavonoids including quercetin, myricetin, rhamnetin, kaempferol, luteolin and apigenin (in decreasing order). It does not require Mn^{2+}, Mg^{2+} or Ca^{2+}, and it is unaffected by EDTA, but Cu^{2+} is inhibitory. It is inhibited by *p*-chloromercuribenzoate, reversible by mercaptoethanol; it requires thiols for stability [G248].

d. Flavonoid 2′- and 5′- glucosyltransferase

Chrysosplenium americanum enzyme, molecular weight 42 000, pI 5.1 and optimum pH 7.5–8 requires UDPglucose. It acts on a range of polyhydroxylated polymethoxyflavonols, and substrates with both 2′- and 5′- hydroxyls yield both monoglucosides [C755].

e. Anthocyanin 5-O-glucosyltransferase

Petunia hybrida pollen enzyme, optimum pH 8.3, acts on the 3-(*p*-coumaroyl)rutinosides of delphinidin and petunidin, but not on the corresponding rutinosides. The enzyme does not have a cation requirement [D198].

f. UDPglucose: anthocyanidin-3-rhamnosylglucoside 5-O-glucosyltransferase (E.C. 2.4.1.116)

Silene dioica enzyme, molecular weight 55 000 and optimum pH 7.4 for anthocyanidin-3-rhamnosylglucoside, has an optimum at pH 6.5 with anthocyanidin-3-glucoside as substrate [B410]; it requires Ca^{2+}, and EDTA is inhibitory. Pelargonidin-3-rhamnosylglucoside is also a substrate, but cyanidin-3-glucoside is only marginally active at pH 6.5 [B408].

g (1). Flavonoid 7- glucosyltransferase (E.C. 2.4.1.81)

Chrysanthemum segetum UDPG: 3′,4′,5,7-tetrahydroxyflavone 7-O-glucosyltransferase has an optimum pH of 6.85–8.25, depending on substrate; substrates include gossypetin, quercetin, kaempferol, myricetin, naringenin, eriodictyol, apigenin and luteolin [J320].

Petroselinum crispum (hortense) UDPG: luteolin β-D-glucosyltransferase (E.C. 2.4.1.81), molecular weight 50 000, optimum pH 7.5 and stability range pH 6.5–9.5 acts on a range of flavonoids, with luteolin as the best substrate. TDPglucose is an alternative co-substrate [K739].

Petunia hybrida leaf contains three isozymes, molecular weight about 54 000 and optimum pH 7.5. Substrates include naringenin, hesperetin, kaempferol, quercetin, apigenin and luteolin [K303].

g (2). UDPG: isoflavone 7-O-β-D-glucosyltransferase (E.C. 2.4.1.170)

Cicer orietinum enzyme, molecular weight 50 000, optimum pH 8.5–9.0 and pI 5.4 acts on biochanin and formononetin; many other flavonoids are not substrates [K738].

g (3). Flavanone 7-O-β-D-glucosyltransferase (E.C. 2.4.1.185)

Citrus paradisi (grapefruit) enzyme, optimum pH 6.5–7.5 and activation energy 7.8 kcal/mol is found in leaves. Different substrates demonstrate the presence of several isozymes; hesperetin isozymes pI 4.0, 4.4 and 4.5; naringenin pI 4.4 and 3.9. It differs from the enzyme that acts on naringenin chalcone [K891]. Another study, with naringenin and hesperetin as substrates reported molecular weight 55 000, optimum pH 7.5–8.0 and pI 4.3, with no action on other flavones or flavanones [K810].

h. UDPglucose: salicyl alcohol glucosyltransferases.

The reaction is catalysed by almond β-glucosidase with D-glucose as co-substrate. 1-Phenylethanol is another substrate [H558].

In Datura innoxa only a small proportion of salicyl alcohol is converted into salicin, whereas most is converted into *o*-hydroxybenzyl-β-D-glucose [A1725].

Gardenia jasminoides enzyme, molecular weight 51 000 and optimum pH 9.0–9.5, is specific for salicyl alcohol, yielding only salicin [D71, D715].

Trilobium confusum phenol β-glucosyltransferase acts on salicyl alcohol, but similar enzymes from other insect species studied are apparently inactive towards this substrate. UDPG is co-substrate, and GDPG and TDGP are somewhat less active [G816].

i. UDPglucose: p-hydroxybenzoate glucosyltransferase (E.C. 2.4.1.194)

Lithospermum erythrorhizon, molecular weight 47 500, pI 5.0 and optimum pH 7.8, conjugates the phenolic group. It is highly specific, with *p*-nitrophenol only marginally active. This reaction is an early step in the formation of shikonin [G341]. Another publication claims that the molecular weight is 51 000 [J365].

Pinus densiflora pollen enzyme, molecular weight 33 000 and optimum pH 7.5, requires UDPG and is activated by Ca^{2+} and EDTA. It acts on the phenolic group of *p*-hydroxybenzoate [F108].

j. Quinol

Datura innoxa, D. meteloides, Antennaria microphylla and Gardenia jasminoides catalyze this reaction [A1725, E338, F873, H300].

Juglans regia leaf and J. major callus enzymes, optimum pH 7–7.5, convert quinol into arbutin. It is specific for UDPglucose, and requires mercaptoethanol [A3472].

Rauwolfia serpentina forms arbutin and, apparently, quinol diglucoside [G726]. The enzyme molecular weight is 52 000 [K346].

Bacillus subtilis enzyme, molecular weight 67 000, pI 5.1 and optimum pH 6, is stable between pH 5 and 8. It is not activated by Ca^{2+}. Malto-oligosaccharides are much better glucose donors that maltose. As well as quinol many other phenols are substrates [H234].

k. Vanillin

Coffea arabica, Prunus amygdalus, Gardenia jasminoides and Nicotiana tabacum enzymes act on vanillin [G930].

l. Coniferyl alcohol (E.C. 2.4.1.111)

Forsythia ovata enzyme(s) convert coniferyl alcohol into coniferin (4-glucoside) and isoconiferin (side-chain glucoside). The optimum pH is buffer-dependent, in the range 7.5–8.0. It requires a thiol and is inactivated by thiol-binding reagents, but divalent cations are not required. It also acts on sinapyl alcohol, ferulate, scopoletin, hydrangetin and isorhamnetin, but not on several other analogues. A study carried out on a large range of species found this activity in all of them, including bryophytes, pteridophytes, gymnosperms, dicotyledons and monocotyledons. Highest activity was found in Ginkgo biloba, Pinus pinea, Picea glauca and Forsythia ovata [A3365].

Paul's scarlet rose enzyme, molecular weight 52 000 and optimum pH 7.5, requires UDPglucose and a thiol. It is specific for coniferyl and sinapyl alcohols, with lesser activity on cinnamic acids and some flavonoids. The crude enzyme is unstable, even frozen; stability is improved by mercaptoethanol and ethylene glycol [K809].

m. UDPG: o-dihydroxycoumarin 7-O-glucosyltransferase (E.C. 2.4.1.104)

Nicotiana tabacum culture enzyme, molecular weight 49 000, pI 5.0 and optimum pH 7.5, acts on esculetin, scopoletin, daphnetin, hydrangetin and umbelliferone to form cichoriin, scopolin, daphnin, hydrangin and skimmin respectively; there is no cation requirement, but divalent cations and thiol-binding reagents are inhibitors. Caffeate (yields 4-glucoside), protocatechuate, vanillate and syringate are also substrates, whereas ferulate, 5-hydroxyferulate, sinapate and pyrocatechuate are poor substrates [B442, B819, K556].

n. Alizarin 2-β-glucosyltransferase
(E.C. 2.4.1.103)

Streptomyces aureofaciens enzyme, optimum pH 7.1, requires UDPglucose and acts on alizarin, 2-hydroxy-, 1,3-dihydroxy-, 2,6-dihydroxy- and 3-hydroxy-1-methoxyanthraquinones [B462, B463].

o. Hydroxyanthracenequinone glucosyltransferase (E.C. 2.4.1.181)

Cinchona succirubra enzyme, molecular weight 50 000 and optimum pH 7, is composed of five isozymes pI 4.1, 4.3, 4.5, 4.8 and 5.3. A number of hydroxyanthraquinones including alizarin, chrysophanol and emodin are substrates, as well as hydroxycinnamates and flavones [K889].

p. Mandelonitriles (E.C. 2.4.1.85 and 2.4.1.178)

Prunus serotina (black cherry) enzyme, optimum pH 7–8, requires UDPglucose, and acts on the side-chain hydroxyl group of mandelonitrile to form prunasin. Benzyl alcohol, mandelate and benzoate are also conjugated, but prunasin is not a substrate [C756].

Sorghum bicolor conjugates the (*S*)-isomer of *p*-hydroxymandelonitrile to form dhurrin. The amino acid composition has been determined. Other substrates are mandelonitrile and (poor) benzyl alcohol and benzoate [K313].

Triglochin maritima forms side-chain glucosides from *p*-hydroxymandelonitrile and 3,4-dihydroxymandelonitrile [D58].

q. Oligoglycosides

Petunia pollen flavonol 3-O-glycosyltransferase is composed of two isozymes, which act on kaempferol and quercetin specifically at the 3 position. Other substrates include myricetin, the 3-galactoside, 3-glucoside and 7-rhamnoglucoside of kaempferol, 4′-O-methylkaempferol, 7-O-methyl- and 3′,7-di-O-methylquercetin. The products are the corresponding 3-O-(2″-O-β-D-glucopyranosyl)-β-D-galactopyranosides and, apparently, where appropriate the corresponding 3-diglucoside [H658].

An enzyme in Pisum sativum converts kaempferol-, quercetin- and myricetin-3-glucosides into (2→1)β-diglucosides. A second enzyme acts on these diglucosides to form the corresponding (2→1)β-triglucosides. Both have optimum pH about 8 [C90].

Triticum aestivum, Glycine max, Avena fatua and Agrostemma githago form β-D-glucoside and β-D-gentiobioside conjugates from *p*-nitrophenol; the former may be an intermediate in formation of the latter conjugate [J709].

r. UDPglucose: 2-coumarate O-β-D-glucosyltransferase (E.C. 2.4.1.114)

Melilotus alba enzyme is cytosolic [B369].

s. UDPglucose: cis-p-coumaric acid β-D-glucosyltransferase

Sphagnum fallax enzyme, molecular weight 56 000 and optimum pH 9.3 acts on *cis*- but not *trans-p*-coumarate, and only poorly on *cis*-caffeate. Conjugation occurs at the *para* position [J384].

t. UDPglucose: betanidin β-D-glucosyltransferases

Dorotheanthus bellidiformis enzymes, molecular weight 55 000 and optimum pH 7.5 act on the 5- position to form betanin, and the 6- position to form gomophrenin I. The first enzyme is composed of three isozymes, whereas the second is a single form [J185].

u. Vitexin and isovitexin (E.C. 2.4.1.105 and 2.4.1.106 respectively)

In Silene alba, 2″-O-glucosylating enzymes require Mg^{2+}, Mn^{2+}, Co^{2+} or Ca^{2+}. One acts on vitexin, optimum pH 7.5, and the other on isovitexin, optimum pH 8.5 [B203]. The latter enzyme requires UDPglucose, and also acts on isoorientin [B385].

v. Tyrosine β-glucosyltransferase

Manduca sexta larval fat body enzyme, optimum pH 7.5–9, is specific for tyrosine and requires UDPglucose; TPDglucose is not as good. It is activated by Mg^{2+} [J40].

w. Indoxyl: UDPglucose glucosyltransferase

This reaction is of historical interest, since it forms indoxyl glucoside (indican) which is stored in Isatis tinctoria (woad), and on hydrolysis the released indoxyl is spontaneously oxidized to indigo.

Baphicacanthus cusia enzyme, molecular weight 60 000, optimum pH 8.5 and pI 6.5, forms indican from indoxyl. The substrates include 4-, 5-, 6- and 7-hydroxyindole, but not oxindole [K347].

Polyganum tinctorium indican synthase is composed of two isozymes, one of which has a molecular weight of 53 000 or 48 000 (different methods) and optimum pH about 10 [K375].

x. UDPG: scopoletin β-D-glucosyltransferase (E.C. 2.4.1.128)

Nicotiana enzyme, which is stimulated by 2,4-D, forms scopolin [K741].

y. 2,4-Dihydroxy-7-methoxy-2H-1,4- -3(4H)-one 2-D-glucosyltransferase (E.C. 2.4.1.202)

Zea mays enzyme, molecular weight 50 000 and optimum pH 8.5, is composed of two isozymes, activated by Ca^{2+} or Mg^{2+} and stabilized by reducing agents [K892].

N-Glucoside formation

5-Aminosalicylate forms an N-β-D-glucoside that is unstable at acid pH. It can also be formed spontaneously at neutral pH [F456].

In Daucus carota 3,4-dichloroaniline forms the β-D-glucoside [H118]. The corresponding reaction in Glycine max is catalyzed by an enzyme, molecular weight 48 000 [G247].

Glucosinolate formation

Brassica converts *o*-nitrobenzaldoxime into *o*-nitrophenylglucosinolate [F657].

Glucuronides (E.C. 2.4.1.17)

All glucuronyl transferases are at present classified under this E.C. number.

A number of glucuronyl transferases have been classified in one publication, although this classification is undoubtedly incomplete. For instance, an enzyme designated as UGT 1A1 acts on bilirubin and phenols, UGT 1A3 conjugates primary, secondary and tertiary amines, UGT 1A4 acts on tertiary amines and alcohols, UGT 1A6 on planar phenols and UGT 1A9 on bulky phenols [J416].

Ester glucuronide synthesis

Many carboxylic acids are conjugated with glucuronic acid, but the enzymology is not as well understood as the formation of ether glucuronides. The following studies are illustrative.

In ferret, benzoate, *p*-nitrobenzoate, diphenylacetate, α-naphthylacetate and (less effective) β-naphthylacetate and hydratropate are substrates, whereas phenylacetate, *p*-chlorophenylacetate, *p*-nitrophenylacetate and indole-3-acetate are not substrates [A394, A3046].

Benzoate and phenylacetate are conjugated in fruit bat [A884, A1451].

A guinea pig liver enzyme acts on anthranilate, with UDPglucuronate as co-substrate [A865].

In human patients with an inborn metabolic error that eliminates bilirubin glucuronidation the conjugation of the acyl group of diflunisal was unimpaired [J183].

In rabbit, all substituted benzoates studied were conjugated. Substituents included nitro, halogens, methyl, amino and hydroxyl in all the possible positions on the aromatic nucleus [H519].

The rat liver enzyme that conjugates (*R*)- and (*S*)-naproxen is composed of at least two active fractions, pI 7.8 and 8.7. One of the fractions preferentially acts on the (*R*)-isomer, and the other on the (*S*)-isomer [F726].

Benzoate and *p*-aminobenzoate are conjugated in rat, mouse, hamster, dog and marmoset [A651, A3771, B201, C381, D914].

Benzoate is conjugated in rat, bandicoot, Tasmanian devil, potoroos, padmelon, brushtail possum, sugar glider and echidna [D914].

In pigeon diphenylacetate is well conjugated, whereas substituted phenylacetates and indole-3-acetate are at best poorly conjugated. Hen did not conjugate any of the phenylacetates studied [A395]. Diphenylacetate is conjugated in a range of mammalian species [A2527].

Ether β-D-glucuronides

The co-substrate for this reaction is UDPglucuronate.

a. Phenolic and bilirubin conjugation

Clinically, bilirubin conjugation is important in human neonatal jaundice because high concentrations of unconjugated bilirubin cause kernicterus and brain damage.

In patients with an inborn metabolic error that eliminates bilirubin glucuronidation the conjugation of phenolic substrates was unimpaired [J183].

Purified rat liver enzyme, molecular weight 57 000, is induced by phenobarbital and acts on phenols, but not on bilirubin [A3827]. In different strains of rat the activity towards chloramphenicol does not correlate with activity towards bilirubin [A2511]. Bilirubin conjugating enzyme shows different properties from that which acts on 1-naphthol; for instance, galactosamine causes a reduction in 1-naphthol conjugation, whereas bilirubin is unaffected [A372].

b. Phenolic and alicyclic glucuronides

A number of studies have shown that, with or without induction with phenobarbital or 3-methylcholanthrene, solubilized enzyme can be crudely separated into two fractions with different properties [e.g. A1919, A1961, B983].

Human brain isozyme UGT1A6 acts on serotonin [K108]. Gut and lymphocyte enzymes have a broad specificity, with a large number of substrates [B800, J783].

Solubilized mouse liver microsomal enzyme has been fractionated into two types, pI 6.7 and 8.5. The first is induced by 3-methylcholanthrene and acts on the 'type 2' substrates, 4-hydroxybiphenyl, morphine, naphthol, 2-, 6-, 8- and 9-hydroxy-3,4-benzpyrene. The second is induced by 3-methylcholanthrene and phenobarbital, and acts on the 'type 1' substrates, 3-, 10-, 11- and 12-hydroxy-3,4-benzpyrene, phenolphthalein and oestrone. Both fractions act on the 'type 3' substrates *p*-nitrophenol, 4-methylumbelliferone, 1-, 4-, 5- and 7-hydroxy-3,4-benzpyrene, and 2-hydroxybiphenyl. All the 'pI 8.5' activities are induced by 3-methylcholanthrene [C459, D315].

Mouse brain activity is less than 1 per cent of that found in liver. It acts on *o*- and *p*-nitrophenol, and 4-hydroxybiphenyl [C266].

The activity of rat liver enzyme that acts on morphine increases 20-fold between days one and 20 post partum; the increase in gut enzyme activity is much less [C221]. The crude enzyme has a broad specificity [B962]; its molecular weight is 57 000 [A3827].

Rat liver glucuronidation of *p*-nitrophenol and 4-methylumbelliferone is inhibited by 5-(*p*-hydroxyphenyl)-5-phenylhydantoin [A1668].

Rat liver microsomal enzyme acts on a range of nitrocatechols, but compounds in which the aromatic moiety is conjugated with unsaturated groups react much less effectively [J546].

In rat *p*-hydroxyamphetamine is conjugated in liver, but not in lung, brain, heart, kidney, intestine and spleen [A4].

Rat liver nuclei contain some conjugating enzyme; most of this is associated with the nuclear envelope [A2002].

Rat conjugates 2-phenyl-1,3-propanediol monocarbamate. The claim that the product is an α-glucuronide is presumably a misprint; the claim goes without comment [K56].

Purified rabbit liver microsomal enzyme, solubilized with the non-ionic detergent Emulgen 911, conjugates *p*-nitrophenol, but not oestrone; an oestrone-conjugating enzyme is also found (see below) [B983].

Guinea pig liver microsomal enzyme, optimum pH 7.5, conjugates 7-hydroxychlorpromazine. This and rat enzyme also conjugate several hydroxylated metabolites of chlorpromazine, promazine and imipramine; the guinea pig enzyme also acts on oestrone and oestradiol [A2092]. Studies on the kinetics showed an atypical pattern at low co-substrate concentration, and it is claimed that this cannot be explained by multiple enzyme forms [A865].

HMPE is conjugated in rat, rabbit, guinea pig and human liver, but not in mouse and cat liver, rat kidney or gut [A2067]

Dog liver microsomal enzyme, molecular weight 50 000, acts on phenol and on the 3-position of morphine [H786].

Rainbow trout liver microsomal enzyme acts on T_3 and T_4, with optimum pH greater than 8.5 and 6.8–7.8 respectively [J319].

c. Oestrogens

1. 17β-Oestradiol

In rat liver microsomes both 3- and 17-glucuronidation occurs [J498] in the ratio 1:2.5, apparently by three different enzymes, each with different activities at the two sites; in rabbit the ratio is 1:20, apparently catalyzed by a single enzyme [C603]. In rhesus monkey 17-glucuronidation is preponderant [A2897].

3-Glucuronidation is found in guinea pig, man and pig, and rabbit blastocyte [A1273, A1936, F272, K33].

3- And 17-glucuronidation is found in man [A1196]; 17-glucuronidation is found in human kidney [A250], but in gut mucosa 3-glucuronidation is the sole reaction [K134].

2. Oestriol-3-glucuronide

This reaction has been observed in dog [A1302, A2628], man [D133], pig liver, kidney and gut [A1273] rabbit [A987], rat [A3102] and sheep [A398].

3. Oestriol-16α-glucuronide

In baboon oestriol is conjugated at the 16 position in liver and kidney, whereas 3-conjugation occurs elsewhere in the body, but not in kidney [A1111].

The reaction is observed in rat, where liver is one site of reaction [A3102, C242], rhesus monkey liver [A2897], dog [A1302, A2628] and man, where kidney is one site of reaction [A156, A822, D133].

4. Oestriol-17-glucuronide

The reaction has been observed in rat [A3102].

5. Oestrone-3-glucuronide

The reaction has been observed in cat, dog, guinea pig, hen, human gut, mouse liver microsomes, pig liver, kidney and gut, quail, rabbit, rat, rhesus monkey and trout [A242, A401, A951, A1040, A1273, A1936, A2897, C562, C603, D315, F272, J498, J783]. Rat liver enzyme is induced by 3-methylcholanthrene (but not by phenobarbital), with a change in kinetic properties, which suggests that the induced enzyme is different from non-induced enzyme [A1368].

Rabbit liver oestrone glucuronyltransferase also conjugates (less effectively) 17β-oestradiol, oestriol, some phenols and (ring) substituted anilines [F309]. Another study found that the purified microsomal enzyme, solubilized with the non-ionic detergent Emulgen 911, does not conjugate *p*-nitrophenol [B983].

d. Miscellaneous reactions

UDPglucuronate: baicalein 7-O-glucuronosyltransferase

Scutellaria baicalensis enzyme is dimeric, molecular weight 110 000, monomeric molecular weight 52 000 and pI 4.8. It is specific for baicalein and flavones with a substituent *ortho* to the 7-hydroxyl group [K402].

Luteolin glucuronosyltransferases

Rye primary leaves contain the enzymesf luteolin 7-O-glucuronosyltransferase (E.C. 2.4.1.189), molecular weight 34 000, pI 4.80, optimum pH 6.5 and 8.5 and activation energy 23 kj/mol which forms luteolin-7-O-β-D-glucuronide, luteolin-7-O-β-D-glucuronide 7-O-glucuronosyltransferase (E.C. 2.4.1.190), molecular weight 37 000, pI 4.75, optimum pH 6.5 and activation energy 50 kj/mol which forms luteolin-7-O-β-D-diglucuronide, and luteolin-7-O-β-D-diglucuronide 4′-O-glucuronosyltransferase (E.C. 2.4.1.191), molecular weight 29 000, pI 4.75, optimum pH 7 and activation energy 38 kj/mol which forms luteolin-7-O-(β-D-glucuronosyl(1→2)-β-D-glucuronide)-4′-O-β-D-glucuronide [K890].

Sugar glucuronides

Chenopodium rubrum enzyme acts on feruloylglucose and *p*-coumaroylglucose, the latter forming feruloylglucuronosyglucose, with UDPglucuronate as co-substrate [G255].

Hydroxylamine glucuronides

Human and rat liver act on N-hydroxy-2-acetamidofluorene [E219, H92]; rabbit liver conjugates N-hydroxyphenacetin and N-hydroxyacetanilide [C588]. In rat, the reaction is found in liver and mammary microsomes [D273]. *p*-Chloro-*N*-hydroxyacetanilide is another substrate for rat liver enzyme [A1840].

N-Glucuronide formation

Human embryo kidney UDPglucuronyltransferase UGT 1A3 and UGT 1A4 glucuronidate primary amines including 2-, 3- and 4-aminobiphenyl, benzidine, *p*-phenetidine, aniline, dapsone, 7-amino-4-methylcoumarin, α- and β-naphthylamine and 2-aminofluorene, but not 2- and 4-aminophenol or secondary amines including diphenylamine and demethylclozapine. Demethylimipramine, nortriptyline, tertiary amines, including amitriptyline, trifluoperazine, diphenhydramine, pyrilamine, triflupromazine, loxapine, meperidine, imipramine, and clozapine (both the secondary and tertiary amino groups) are not substrates [H541, J416].

In man (where studied, liver microsomes contain the enzyme activity) quaternary glucuronides are formed by tertiary amine: UDPglucuronosytransferase with amitriptyline, chlorpheniramine, chlorpromazine, clomipramine, clozapine, cyproheptadine, diphenhydramine, doxepin, doxylamine, imipramine, lamotrigine, loxapine, pheniramine, pyrilamine, trazodone, trimipramine, and tripelennamine as substrates [C614, E932, F718, F769, G18, G492, H162, J137]. Antibody studies show that it differs from phenol UDPglucuronosytransferase [G509].

Rabbit liver oestrone UDPglucuronyltransferase acts on aniline, 4-aminobiphenyl, α- and β-naphthylamine [D288, F309]. A human liver enzyme, pI 6.2 acts on 4-aminobiphenyl and α-naphthylamine, but another, pI 7.4 acts on α-naphthylamine but not on 4-aminobiphenyl [F309]. In rat the reaction on α- and β-naphthylamine has been found in hepatocytes [G743].

Although it has been suggested that quaternary glucuronides are formed only in primates, guinea pig glucuronidates lamotrigine on a heterocyclic nitrogen [F763], and rat liver glucuronidates amitriptyline, imipramine and chlorpromazine [C614, G492].

Lactoside formation

Cow (milk) forms the N^4-lactoside of sulphamethazine from the parent compound [G373].

Rhamnoside formation (E.C. 2.4.1.159)

Silene dioica (red campion) petal enzyme, molecular weight 45 000 and optimum pH 8.1, acts on cyanidin-3-glucoside with UDP-L-rhamnose as co-substrate to form cyanidin-3-(6″-rhamnosyl)glucoside. It is stimulated by Mg^{2+}, Mn^{2+} and Co^{2+}. Other substrates are the 3-O-glucosides of pelargonidin and delphinidin; cyanidin-3,5-diglucoside is a poorer substrate [B405, B409].

Xyloside formation

Euonymus alatus leaf UDP-D-xylose: flavonol 3-O-xylosyltransferase (E.C. 2.4.2.35), molecular weight 48 000, pI 6.1 and optimum pH 7.0, acts on kaempferol, quercetin and fisetin, but not on dihydroflavonols [G444].

Penicillium acts on catechol and xylan to form β-xylosylcatechol and xylobiosylcatechol [J487].

Sinapoylglucose-sinapoylglucose O-sinapyltransferase (E.C. 2.3.1.103)

Raphanus sativus, molecular weight 55 000 and optimum pH 8.0 does not require cations or thiols. It catalyzes dismutation of the substrate with activation energy 62 kj/mol to form 1,2-di-O-sinapoyl-β-D-glucose; substrates are 1-sinapoylglucose and 1-feruloylglucose [K724].

Galloyl exchange of glucosides

A Quercus robur enzyme, optimum pH 6.0–6.5, catalyzes a rapid exchange between glucose and 1-galloyl-β-D-glucose [K894].

5.4 *Conjugation of thiols*

CoA conjugation

$$\text{RCOOH} + \text{CoA} + \text{ATP} \rightarrow \text{AcylCoA} + \text{AMP} + \text{pyrophosphate}$$

Beef liver mitochondrial enzyme, molecular weight 66 000, acts on a series of substituted benzoates, but not on those substituted in the *o*-position, nor on naphthylacetates. This is the first step in the formation of hippurates [H740].

Human liver contains two isozymes, both of which conjugate benzoate, *p*-hydroxybenzoate and phenylacetate, one conjugates salicylate, but only poorly, and the other conjugates a naphthylacetate [K220].

Rat liver microsomal (*R*)-ibuprofenoyl CoA synthetase is identical with long chain acyl CoA synthetase (E.C. 6.2.1.3), molecular weight 72 000, which requires ATP and Mg^{2+}; (*S*)-ibuprofen is inhibitory. The CoA ester is involved in chiral inversion of the substrate [J234].

Cactus (Cephalocereus senilis) (hydroxy)cinnamate: CoA ligase acts on cinnamate, *p*-coumarate, caffeate, *p*-methoxycinnamate and ferulate, but not on sinapate [H633].

Forsythia ferulate: CoA ligase requires ATP and CoA; Mg^{2+} and a thiol are also required for maximal activity. It is an enzyme in the sequence that converts ferulate into coniferyl alcohol [A848].

In Glycine max two *p*-coumarate: CoA ligase isozymes (E.C. 6.2.1.12) have been detected [A2271], optimum pH 7.8 and 8.5. Both act on cinnamate, *o*-, *m*- and *p*-coumarate, caffeate, ferulate and isoferulate, whereas the pH 7.8 enzyme acts on sinapate,

3,4-dimethoxycinnamate and 3,4,5-trimethoxy-cinnamate. *Cis-p*-coumarate is only 1 per cent as effective as the *trans*-isomer. The reaction requires ATP and a divalent cation (Mg^{2+}, Mn^{2+}, Co^{2+}, or Ni^{2+}; Ca^{2+} and Zn^{2+} are relatively ineffective). Both are susceptible to substrate inhibition, and it is activated by tris buffer (at less than 100 mM for the 'pH 7.8' isozyme and 300–400 mM for the 'pH 8.5' isozyme) [A2148].

Pea seedling contains one major isozyme, molecular weight 75 000, and two minor isozymes, molecular weight less than 20 000. There is some evidence for the interconversion of these forms. Substrates include *p*-coumarate, ferulate, caffeate, isoferulate and cinnamate, but sinapate, 3,4-dimethoxycinnamate and *p*-methoxycinnamate are not substrates. A minor isozyme acts on all these compounds except cinnamate. Studies with runner bean, Glycine max shoot, aged cucumber, aged marrow, aubergine, red and green pepper and leek show that each species has its own specificity. *p*-Coumarate and isoferulate are substrates in all these species, ferulate in all but green pepper, caffeate in all but aubergine and red pepper, whereas sinapate is a substrate in runner bean, G. max and leek, *p*-methoxycinnamate in runner bean and leek, but cinnamate is substrate only in G. max and marginally in cucumber [A2514].

Three isozymes are found in Petunia, all with optimum pH 7.5–7.8. One acts on *p*-coumarate and caffeate, a second on *p*-coumarate and sinapate, whereas the third acts on cinnamate, *p*-coumarate, ferulate and (poorly) caffeate [A2845].

Poplar stem hydroxycinnamoyl: CoA ligase, which is composed of three isozymes, pI 4.4, 5 and 6.35, acts on *p*-coumarate, with different distribution in xylem sclerenchyma and parenchyma; there is a correlation between the different forms and the monomeric composition of the lignins in these tissues [C740].

In potato *p*-coumarate: CoA ligase (E.C. 6.2.1.12) is found only on ageing after slicing the tuber, presumably as a response to injury. Substrates are *p*-coumarate, ferulate, caffeate, *m*- and *p*-methoxycinnamate [A204].

In swede *p*-coumarate: CoA ligase is only found in aged slices, and only one form is found. It is one of the enzymes in the sequence that converts ferulate into coniferyl alcohol [A216].

Tomato *p*-coumarate: CoA ligase acts on *o*-, *m*- and *p*-coumarate, caffeate, ferulate and isoferulate, and poorly on *m*- and *p*-methoxy-cinnamates; other cinnamates without a hydroxyl group are not substrates. This enzyme is part of the sequence that forms chlorogenate and 5′-(*p*-coumaroyl)quinate [A3149].

Willow stem enzyme, molecular weight 57 000 and optimum pH 7, acts on *p*-coumarate, ferulate and (best) caffeate; cinnamate and sinapate are not substrates [D466].

Desulfitomaculum enzyme acts on several benzoates [K210].

Mycobacterium phlei *o*-succinylbenzoate-CoA ligase (E.C. 6.2.1.26) requires ATP and CoA, with the formation of AMP. It is stable at pH 1 [K744].

An enzyme in Penicilliun chrysogenum, optimum pH 6.8, acts on phenylacetate, phenoxyacetate, *o*- and *p*-methylphenoxyacetate, but not phenylpropionate. This enzyme is in the sequence that leads to the formation of penicillins [A2291, C483].

In Pseudomonas *p*-chlorobenzoate: CoA ligase (E.C. 6.2.1.33) is a homodimer, subunit molecular weight 57 000, pI 5.3 and optimum pH 8.4. It requires CoA, ATP and a divalent cation (Mg^{2+}, Co^{2+} or Mn^{2+}). Substrates include benzoate, *p*-halobenzoates, with lower activity on some other *para*-disubstituted benzoates. This is the first reaction in the sequence that converts *p*-chlorobenzoate into *p*-hydroxybenzoate [G729, H645, K191], forming *p*-chlorobenzoyl CoA [H216]. A Pseudomonas aromatic acid: CoA ligase contains three isozymes. One is a dimer, molecular weight 120 000 and optimum pH 8.5–9.2, another is a monomer, molecular weight 60 000 and optimum pH 9.3, and the third, molecular weight 65 000 and optimum pH 8.5. They act on benzoate, anthranilate, monofluorobenzoates, and the two monomeric isozymes also act on *o*-toluate [G342].

A study in Pseudomonas with phenylacetate as substrate (phenylacetate: CoA ligase, E.C. 6.2.1.30) found a molecular weight of 52 000 and optimum pH 8.5 for the enzyme which is insensitive to oxygen, confirmed the requirement for Mg^{2+} or Mn^{2+}, and found that ATP is degraded to AMP and pyrophosphate. The enzyme is unstable, and is stabilized by glycerol and/or phenylacetate. Analysis of the N-terminal sequence showed little congruence with similar ligases. It is found in cells grown anaerobically in the presence of phenylacetate; analogues are ineffective. Aerobic growth produces low enzyme activity [G890]. Another study in P. putida found a molecular weight of 48 000 (denatured) and pI 5.4–4.9, with a requirement for ATP, CoA and Mg^{2+} or Mn^{2+}. Other acids are not substrates [K743].

Pseudomonas putida cinnamoyl: CoA ligase has an optimum pH of 6.5 and molecular weight of 53 000. Good substrates include *p*-coumarate, ferulate and caffeate, whereas *m*-coumarate, 3,4-methylenedioxycinnamate, *p*-methoxycinnamate and 3,4,5-trihydroxycinnamate are poor substrates. A series of other cinnamates are not substrates [B298].

Two isozymes are found in a denitrifying Pseudomonas. Both act on anthranilate and *o*-fluorobenzoate; only one isozyme acts on *o*-toluate, *m*- and *p*-hydroxybenzoates. Phenylacetate is substrate for at least one isozyme [F65]. A study on Pseudomonas has detected anthranilate: CoA ligase (E.C. 6.2.1.32) [K742].

Rhodopseudomonas palustris benzoate: CoA ligase (E.C. 6.2.1.25) requires Mg^{2+} and ATP, with benzoate, *o*- and *p*-fluorobenzoates and *o*-chlorobenzoate as substrates, but other analogues are inactive [E605]. *p*-Hydroxybenzoate: CoA ligase (E.C. 6.2.1.27) requires Mg^{2+} and ATP; acetyl CoA is not an acetyl donor [F48].

An anerobic syntrophic benzoate-degrading culture benzoyl: CoA ligase, molecular weight about 420 000, subunit molecular weight 58 000 and optimum pH 8, requires acetyl CoA and Mg^{2+}. It acts on benzoate, monofluorobenzoates and some heterocyclic acids [H500].

Succinyl CoA: benzylsuccinate CoA transferase

Thauera aromatica acts reversibly on (*R*)-(+)-benzylsuccinate, with succinyl CoA (a common CoA donor) as co-substrate, forming (*R*)-(+)-benzylsuccinyl CoA and succinate. This is an early step in the reaction sequence that leads from toluene to benzoyl CoA [K302].

Cinnamoyl CoA: phenyllactate CoA transferase

Clostridium sporogenes enzyme, molecular weight 46 000 forms cinnamate and (*R*)-phenyllactyl CoA from cinnamoyl CoA and (*R*)-phenyllactate [K534].

Glutathione S-aryltransferase (E.C. 2.5.1.18)

$$ArR + GSH \rightarrow GSAr + RH;$$

R is usually a halide or nitro group.

Human platelet cytosolic enzyme, molecular weight 60 000 and pI 4.65, acts on 1-chloro-2,4-dinitrobenzene. It is present in chorial villi, and a little is found in amnion [A3953].

Japanese and rhesus monkey enzymes act on nitro compounds with the release of nitrite. Substrates (in decreasing order) include *o*-dinitrobenzene, 4-nitroquinoline-N-oxide, 3,4-dinitrobenzoate and *p*-dinitrobenzene, with slight or no activity towards other compounds. Results suggest non-reactivity when the nitro group is *o*- or *p*- to a carboxyl, hydroxyl or ether group, or *m*- to a second nitro group [F139].

In female mouse, enzyme acting on 1-chloro-2,4-dinitrobenzene is found in liver, lung and intestine (in reducing order). Activity is low one month after birth and rises to a peak at nine months and then falls rapidly to a stable level at 12–13 months, similar to the one month level [B875].

Pig brain enzyme is composed of four isozymes that act on a number of nitrobenzenes. The main one is dimeric, molecular weight 43 000 and optimum pH 6.5–7.5; its amino acid composition has been determined. Substrates include *o*- and *p*-dinitrobenzene, 1-chloro-2,4-dinitrobenzene,

1,2-dichloro-4-nitrobenzene, 2,4-, 2,5- and 3,4-dinitrobenzoate, *p*-nitrobenzyl chloride, 2,5-dinitrophenol and *p*-nitrophenethyl bromide [C901].

Rabbit enzyme is found in liver, gut and lung. Activity is low at birth and increases slowly to the adult level by four months. In lung the corresponding activity increase is twofold. In kidney the activity falls 50 per cent at birth, and then increases threefold to the adult level at weaning [A2132].

Rat liver cytosol contains five isozymes that act on 1-chloro-2,4-dinitrobenzene [D303]; based on fractionation and induction studies, activity on this compound and on 1,2-dichloro-4-nitrobenzene appears to be due to different enzymes. Activity is found in microsomes, mitochondria and cytosol. Phenobarbital induces microsomal and cytosolic enzymes, but 3-methylchloanthrene and 2,3,7,8-tetrachlorodibenzo-*p*-dioxin induce only cytosolic enzyme [B702]. Another study detected about eight isozymes [A3509]. Rat liver enzyme is inhibited by 10^{-6} M trialkyl- and triaryltin and triethyllead, but less effectively by dialkyl analogues or by triethylgermanium [A1664].

Enzyme acting on 1-chloro-2,4-dinitrobenzene is found in rabbit, guinea pig, hamster, rat, dog and human bladder [C685].

Enzyme that acts on 1-chloro-2,4-dinitrobenzene and/or 1,2-dichloro-4-nitrobenzene is found in liver and kidney from sheepshead, pinfish, jack crevalle, southern flounder, winter flounder, Atlantic stingray, bluntnose ray, small skate, nurse shark, spiny lobster and blue crab [B69].

Callinectus sapidus (blue crab) hepatopancreas contains two isozymes, pI 5.7 and 5.9. Substrates include 1-chloro-2,4-dinitrobenzene, *p*-nitrophenyl acetate, *p*-nitrobenzyl chloride, bromosulphophthalein and benzpyrene-4,5-oxide. The amino acid composition of both isozymes has been determined [E434].

Three enzymes from Melanogaster act on 1-chloro-2,4-dinitrobenzene, but not on styrene oxide or on 2-(*p*-nitrophenoxy)propene oxide [D303].

Ascaris suum and Moniezia expansa enzymes, molecular weight 37 000 with a broad optimum at about pH 8.3, act on 1-chloro-2,4-dinitrobenzene. Activity is stimulated slightly by Co^{2+}, and inhibited by Cu^{2+}, Fe^{3+} and Hg^{2+}. The product is S-(2,4- (not 2,3- as stated) dinitrophenyl)glutathione [A3038].

Glutathione S-epoxidetransferases

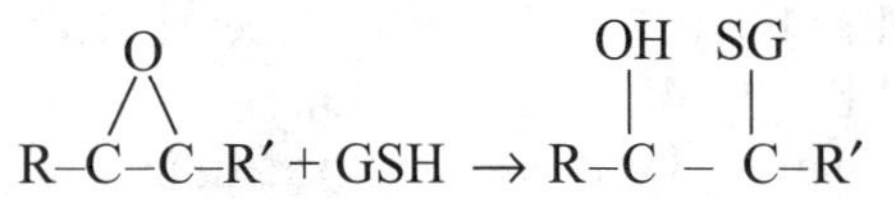

Human enzyme has been fractionated into isozymes designated π (acidic, pI < 5.5, in placenta), α–ε (basic, pI > 7.5, in liver) and μ (neutral, in liver). Substrates studied were benzpyrene-4,5-oxide, styrene oxide and pyrene-4,5-oxide. They are distinguished by their activity on styrene oxide. π-Enzyme substitutes the substrate with glutathione proximal to the aromatic nucleus, α–ε distal to the nucleus and μ forms both products [E928].

Human placental enzyme (cytosolic) acts on styrene oxide with a broad optimum (pH 8.5–9). The activity decreases between 25 weeks gestation and term [C21].

In rat liver two isozymes, designated A_2 and C_2, have been separated. The C_2 enzyme is specific for oxides with (*R*)-configuration, and is high-affinity. (5*S*, 6*R*)- and (5*R*, 6*S*)-benzanthracene-5,6-oxides, and (4*S*, 5*R*)- and (4*R*, 5*S*)-benzpyrene-4,5-oxides, phenanthrene-9,10-oxide and pyrene-4,5-oxide all form products solely with the (*S*, *S*)-configuration. Analogues with at least one aryl-heterocyclic (aza) ring form a lower proportion of the (*S*, *S*) product. The A_2 enzyme forms a lower proportion of the (*S*, *S*) product for all substrates, and is low-affinity. A further isozyme designated B has a stereospecificity similar to that of A [C472, D235]. Another study found that isozymes A and E act on epoxides [J583]. This activity is optimal at pH 10 [D313].

In rat lung six isozymes are found, with pI 4.8, 5.3, 6.0, 6.8, 7.2 and 8.8. They are all dimers, with common monomers [D912].

Rat cytosolic enzyme has been found in liver, lung, testis and heart. These (unfractionated) act preferentially on (*R*)-configured oxides (benzanthracene-5,6-oxide, benzpyrene-4,5-oxide, styrene oxide and pyrene-4,5-oxide); in particular, the heart enzyme shows a 93 per cent selectivity. With (*R*,*S*)- substrates there is little stereoselectivity for the products [D842]. Liver and kidney activity is about 5 per cent of the adult level at birth, and increases linearly to the adult level by 40–45 days [A3043].

Besides liver and kidney, most rat tissues show some activity, including ovary and adrenal. Two major and two or three minor components have been separated, with pI in the range 6.5–7.5 [A143, A1964, A3440, A3678].

Rat ovary activity increases threefold from birth to near the adult level at 30 days post partum. In adrenal, activity remains nearly constant for the first 60 days, then declines by about 70 per cent to the adult level. These development patterns differ somewhat depending on whether benzpyrene-4,5-oxide or styrene oxide are used as substrate, suggesting the presence of more than one enzyme [A3461].

Rabbit enzyme (cytosolic) acts on styrene oxide with highest activity in liver (optimum pH 7), and with lesser activity in kidney, lung (optimum pH 7.5) and intestine [A2509]. In liver there is little activity at birth; it increases to the adult level at weaning. In kidney it increases twofold to the adult level between birth and weaning. Intestinal activity falls at birth, then increases twofold to the adult level at weaning. In lung, the increase to the adult level is only small [A2132].

Mouse enzymes are separable into three α-class isozymes, one π-class isozyme and one μ-class isozyme (based on differing specificities), all of which conjugate (−)-*anti*- and (+)-*syn*-11,12-dihydro-11,12-dihydroxybenzo[*g*]chrysene-13,14-oxide, especially an α-class isozyme designated GSTA1-1 [K270].

Sheep liver enzyme, molecular weight about 40 000, requires no cofactors [A143]. It has a broad specificity for arene and alkene oxides, and also acts on styrene sulphide [A2282].

Activity is much less in rabbit and man than in guinea pig [A3678].

Callinectus sapidus hepatopancreas glutathione S-aryltransferase isozymes also act on benzpyrene-4,5-oxide [E434].

Enzyme that acts on styrene oxide and benzpyrene-4,5-oxide is found in liver and kidney from Atlantic stingray, black drum, bluntnose ray, croaker, dogfish, eel, jack crevalle, king of Norway, large skate, lobster, mangrove snapper, mullet, nurse shark, pigfish, redfish, sea bass, sheepshead, small skate, southern flounder, thorny skate and winter flounder [B69]. Activity in liver from teleost species is similar to that in rat liver, and usually higher than in elasmobranch species [A1694].

Oxyhaemoglobin as glutathionyltransferase

In man, 4-dimethylaminophenol forms a series of products, triply substituted with glutathione or thiol groups, apparently at positions 2, 3 and 5, with up to three of either group in any one product [F138]. Presumably, the thiol products are formed by breakdown of glutathione derivatives.

Glutathione S-aralkyltransferase (E.C. 2.5.1.18)

In rat, (1-bromoethyl)benzene is converted into N-acetyl-S-(1-phenylethyl)-L-cysteine and (2-bromoethyl)benzene into N-acetyl-S-(2-phenylethyl)-L-cysteine, which are final products from glutathione conjugates. The pattern of mercapturate formation is different from that with styrene, suggesting that styrene formation is not involved [A3708].

Glutathione conjugation with peroxidase

A glutathione thiyl radical formed by the action of peroxidase has been detected, which is believed to conjugate with the terminal carbon of styrene, followed by oxidation adjacent to the aromatic nucleus, to yield the corresponding dihydrohydroxy compound. In the absence of oxygen several conjugates are formed, including (*R*)- and (*S*)-S-((2-hydroxy-1- and 2-phenyl)ethyl)glutathione [E80, E205]. N-Acetylbenzidine is a substrate, forming N-acetyl-N′-(glutathion-S-yl)benzidine-S-oxide [K412].

Glutathione S-aralkylsulphate transferase?

A cytosolic preparation from in rat liver converts 7-hydroxymethyl-12-methylbenzanthracene sulphate into S-((12-methylbenzanthracenyl)-7-methyl)glutathione [B965].

Glutathione conjugation, with esters as substrates

Five out of eight rat liver enzymes that act on 1-chloro-2,4-dinitrobenzene also de-esterify methylparathion with glutathione as acceptor, forming S-(*p*-nitrophenyl)glutathione [A3509]. Other products are O-methyl-O-(*p*-nitrophenyl)phosphorothioate, O-(*p*-nitrophenyl)phosphorothioate and S-methylglutathione [E225]. Similar reactions have been detected in human placenta [E64].

Chicken liver enzyme demethylates tetrachlorvinphos with glutathione as methyl acceptor [A2532].

Glutathione conjugation, with ethers as substrates

Pea seedling cytosolic enzyme, optimum pH 9.3–9.5, acts on 2,4′-dinitro-4-trifluoromethyldiphenyl ether and glutathione to form *p*-nitrophenol and, probably, S-(2-nitro-4-trifluoromethyl)glutathione, but several other diphenyl ethers are not substrates. It is inhibited by several substituted ureas, aniline derivatives, sulphobromophthalein and by substituted diphenyl ethers [A1148].

Glutathione conjugation, with cysteine conjugates as substrates

Mouse acts on the cysteine conjugate of paracetamol to form the corresponding glutathione conjugate in place of cysteine [D341].

Glutathione conjugation, with glucuronide as substrates

Rat liver glutathione S-transferases act on 1-O-clofibrylglucuronide to form S-(*p*-chlorophenoxy-2-methylpropanoyl)glutathione. The reaction also occurs spontaneously [H141].

Glutathione conjugation with tertiary amine elimination

Cynomolgus monkey acts on melphalan (*p*-(bis(2-chloroethyl)amino)phenylalanine) to form *p*-(S-glutathionyl)phenylalanine. It is proposed that the reaction involves the formation of a cyclic aziridinium (a quaternary) ion. The authors, in support of this mechanism, quote that N,N-dimethyl-*p*-toluidine is not a substrate, but trimethyl-*p*-toluidinium does yield the corresponding glutathione adduct [E874].

Cysteine conjugation with thiophenol

E. coli enzyme acts on thiophenol, pyridoxal phosphate, ammonium ion and serine to form S-phenyl-L-cysteine [J485].

Thiol S-methyltransferases (e.g. E.C. 2.1.1.9)

Human kidney thiopurine methyltransferase (E.C. 2.1.1.67) acts on thiophenol as well as on amino-, carboxyl-, methoxy-, methyl-, halogen-, nitro- and acetamido-substituted thiophenols [C917, D926].

Rat liver enzyme acts on thiophenols with S-adenosylmethionine as co-substrate [C121].

Brassica oleracea thiol methyltransferase is composed of five isozymes, molecular weights 26 000–31 000 with optima at pH 5, 6 and 8. Substrates include 4,4′-thiobisbenzenethiol, thiosalicylic acid and thiophenol; thiobenzoic acid is a poor substrate, but phenols are not substrates [K594].

Tetrahymena thermophila enzyme is cytoplasmic, molecular weight 41 000 and optimum pH 7.5. Substrates are thiophenols substituted with nitro or chloro groups, as well as thiosalicylate and thiobenzoate, but phenols and anilines are not substrates; S-adenosylmethionine is the co-substrate [E96].

S-Adenosylmethionine: thioether S-methyltransferase

Mouse lung enzyme, molecular weight 28 000, pI 5.3 and optimum pH 6.3, is mostly cytosolic, with highest activity in lung, but with some in liver. It acts mainly on aliphatics, including selenium and tellurium analogues, as well as on thioethers. Benzyl methyl thioether is a substrate, but the product has not been identified; by analogy with dimethylselenide, which forms trimethylselenonium ion, it would appear to be benzyl dimethyl sulphide. The reaction is inhibited by sinefungin and by S-adenosylhomocysteine [E677].

5.5 *Formation of amides and substituted amides*

Amide formation

In Streptomyces violaceoniger benzoic acid and a series of substituted benzoates form the corresponding benzamides [E279].

Coprinus L-tryptophan oxidase forms indole-3-acetamide as a subsidiary reaction [K563].

Formylation of arylamines

$$RNH_2 + \text{N-formyl-L-kynurenine} \rightarrow RNHCHO + \text{L-kynurenine}$$

2-Aminoanthraquinone is formylated in rat [A3818].

4-Aminobiphenyl, α- and β-naphthylamine, 2-aminofluorene and 1-aminopyrene are formylated in rabbit, guinea pig and rat, with N-formyl-L-kynurenine as formyl donor. In rabbit liver the rate of formylation is greater than acetylation. Rat also formylates *p*-aminophenol, *p*-chloroaniline, *p*-toluidine and *p*-anisidine [F109]; hepatocytes are active towards *p*-aminophenol [K60].

2-Aminofluorene is formylated by Pacific oyster [F890].

1-Aminopyrene is formylated both by gut flora [F841] and in goldfish [H877].

5-Aminosalicylate is formylated in man and pig. Studies with rat liver have shown a requirement for N-formyl-L-kynurenine [F924].

2-Formamido-4,6-dinitrotoluene and 4-formamido-2,6-dinitrotoluene are formed from TNT by Phanerochaete chrysosporium [K202].

Formylation of alkylamines

A number of studies in man, rabbit, guinea pig and rat have demonstrated that 4-formamidoantipyrine is formed as a major product from aminopyrine [A1941, A1942, A2362, A2372, B10, H167]. Some authors have claimed that it is not clear whether the formyl group arises by formylation of 4-aminoantipyrine or by oxidation of a methyl group [A1939]; others claim that methyl oxidation is definitely involved [A1941]. Presumably 4-methylaminoantipyrine, which some studies have detected as a metabolite, is the proximal substrate for this putative oxidase.

Formate dihydrofolate ligase (E.C. 6.3.4.17)

Human breast cancer and Lactobacillus casei enzymes, optimum pH 8.5, require ATP, formate and Mg^{2+} to form 10-formyldihydrofolate; it is slightly activated by K^+ [K745].

Formyltetrahydrofolate synthetase (formate tetrahydrofolate ligase, E.C. 6.3.4.3)

$$\text{THF} + \text{ATP} + \text{formate} \rightarrow \text{10-formylTHF} + \text{ADP} + \text{P}_i$$

Pig liver enzyme is part of a complex composed of methylenetetrahydrofolate dehydrogenase (E.C. 1.5.1.5), 10-formyltetrahydrofolate synthetase (E.C. 6.3.4.3) and methenyltetrahydrofolate cyclohydrolase (E.C. 3.5.4.9). The molecular weight is 150 000 (gel filtration) or 100 000 (gel electrophoresis) [A2844].

In immature chick, 10-formyltetrahydrofolate synthetase activity is decreased by about 40 per cent by folate or oestradiol; these effects are additive [A1680].

In Clostridium cylindrosporum, an enzyme that requires Mg^{2+} acts on tetrahydrofolate reversibly, with ADP and formyl phosphate as co-substrates, to yield N^{10}-formyltetrahydrofolate and ATP [E360].

Arylamine acetyltransferase (E.C. 2.3.1.5)

$$\text{Ar.NH}_2 + \text{acetyl CoA} \rightarrow \text{Ar.NHCOCH}_3$$

Procainamide and sulphadimidine are acetylated in man; about 50 per cent of subjects are fast acetylators, and the remainder are slow [A2562].

Rat and rabbit liver parenchymal cell enzyme acetylates sulphamethazine, sulphanilamide and *p*-aminobenzoate; in addition sulphadiazine is acetylated in rabbit cells. Non-parenchymal cells show no activity [B110].

2,4-Toluenediamine forms 4-acetamido-2-aminotoluene and a trace of 2-acetamido-4-aminotoluene in liver from hamster, guinea pig, rabbit, mouse and rat, with only marginal activity in man and none in dog. In hamster, guinea pig, rabbit and mouse the diacetyl derivative is also formed. The enzyme is cytosolic and requires acetyl CoA as co-substrate. Mouse enzyme shows a broad optimum at pH 7.5; in rat the activity decreases steadily from pH 6 to 9, and in rabbit there is a broad optimum at pH 6–7.5. In hamster and rabbit the activity is also found in kidney, intestinal mucosa and lung, but in some cases these tissues form diacetyl derivatives only poorly [A311].

In rabbit at least two enzymes are found. One, 'hepatic', found mainly in liver and intestine acts on sulphamethazine, optimum pH 5.5–7, and an 'extrahepatic' enzyme, optimum pH 6–7 which acts on *p*-aminobenzoate, is found in all tissues studied especially liver and gut. 'Hepatic' enzyme is stable at 4°. The study was carried out on both fast and slow sulphamethazine acetylators, based on liver activity. The enzyme in spleen and kidney does not correlate with the liver activity and may represent another enzyme; gut enzyme appears to be a mixture of both enzymes. 'Extrahepatic' enzyme is unstable at 4° [A869]. Liver enzyme has a molecular weight of 33 500, pI 5.2 and a broad optimum pH 5.9–8.6. A large number of anilines are substrates including some substituted in the *ortho* position [E399].

Acetylation of sulphonamides and other amines in rabbits appears to be controlled by an autosomal gene, with heterozygotes acetylating rapidly, and affected homozygotes (about 25 per cent of the population) acetylating slowly. There are virtually no animals that acetylate at an intermediate rate. Sulphamethazine forms N^4-acetylsulphamethazine [A1037].

Mouse enzyme, molecular weight 31 000, with a broad optimum near neutrality, acts on *p*-aminobenzoate; it is identical with arylhydroxamic acid: N,O-acetyltransferase (E.C. 2.3.1.118) [F174]. The same identity has been demonstrated with a rat liver cytosolic enzyme, molecular weight 32 000 [H566].

Hamster liver cytosol contains two isozymes, with *p*-aminobenzoate, 2-aminofluorene and 4-aminobiphenyl as substrates [F114].

Rat and sheep pineal enzymes, which act on aniline and *p*-phenetidine, require acetyl CoA as co-substrate. Two isozymes appear to be present, one of which acts predominantly on arylamines and the other on aralkylamines. The activity of the former is unaffected by pre-treatment with isoproterenol or cycloheximide [D282].

In sheep, acetylation of aniline is observed in rumen, duodenum, jejunum, ileum and colon. Other substrates include *p*-aminophenol, *p*-aminobenzoate, *p*-anisidine and *p*-nitroaniline [A2788].

Chick pineal enzyme acts on aniline, *p*-phenetidine and 2-aminofluorene, but with much lower activity towards phenethylamine and 5-methyoxytryptamine. Activity is also found in kidney [F694]. Liver enzyme, molecular weight about 34 000 and optimum pH about 8.6, acts on *p*-nitroaniline. Its activity is lower than in pigeon liver, but higher than in rabbit liver [E384].

Aniline acetylating enzyme requiring acetyl CoA is found in Bacillus cereus and Cordyceps militaris, whereas Arthrobacter, Nocardia, Pseudomonas, Serratia and a range of other Bacillus species are poor sources of the activity. The optimum pH of B. cereus enzyme is 7.0; it is not stable above 30°. Other substrates include *p*-nitroaniline, 3,5-dimethyl-4-nitroaniline, *p*-nitrobenzalhydrazone, 4-amino-4′-nitroazobenzene, but not a range of other nitroanilines and phenylhydrazones. All 41 strains tested in this study showed some activity [E276].

Pseudomonas aeruginosa enzyme, which acts on 2-aminofluorene, has an optimum pH of 8.5 [J652].

Both Enterobacter aerogenes and Aeromonas hydrophila enzymes, molecular weights 44 900 and optimum pH 7.5 and 7.0 respectively, act on 2-aminofluorene [J677, J816].

Lactobacillus acidophilus enzyme, molecular weight 44 900 and optimum pH 7.0, acts on 2-aminofluorene. It is inhibited by Zn^{2+}, Ca^{2+}, Mg^{2+}, Fe^{2+} and Cu^{2+}. Pre-incubation with acetyl CoA protects the enzyme against inactivation by iodoacetamide [K208].

Enterobius vermicularis (nematode) enzyme, molecular weight 44 900 and optimum pH 7.5, acts on 2-aminofluorene and is inhibited by Zn^{2+}, Ca^{2+} and Fe^{2+} [K174].

Acetyl derivatives of primary alkylamines

$$R.NH_2 + \text{acetyl CoA} \rightarrow R.NHCOCH_3$$

Rat brain enzyme, which acts on phenethylamine and tryptamine, requires acetyl CoA. Activity is evenly distributed throughout all the brain areas examined, with the highest activity in cerebellum [A807].

Rat pineal enzyme, which is involved in melatonin synthesis, is so unstable that it has proved extremely difficult to isolate [H758]. It exists in two forms, with molecular weights 10 000 and 95 000, and acts on serotonin [E497].

Rat liver enzyme, which acts on tryptamine and serotonin, is inhibited competitively by melatonin at 10^{-6} M, whereas the pineal enzyme is not inhibited [A1350].

Rat and sheep pineal enzyme, which acts on tryptamine, serotonin, 6-fluorotryptamine and 5-methoxytryptamine (forming melatonin), requires acetyl CoA. Two enzymes appear to be present, one of which acts predominantly on arylamines and the other on aralkylamines. The latter is induced 100-fold by treatment with isoproterenol in rat, and cycloheximide treatment of sheep reduces nocturnal activity by 90 per cent [D282].

Chick pineal acetylates serotonin. At 16–20 days age the enzyme shows a diurnal rhythm in the presence of diurnal lighting. This rhythm is much decreased in constant dark, and is eliminated in constant light. Its activity increases before hatching and then remains fairly constant, irrespective of lighting conditions. In contrast, cerebral and cerebellar activity increases only marginally between 11 days incubation and adult [A1626].

Quail duodenum enzyme, optimum pH 5.8, acts on tryptamine, serotonin and 5,6-dihydroxytryptamine [H685].

Macrobrachium rosenberghii (prawn) optic lobe enzyme, optimum pH 6.5, acts on tryptamine, 5-methoxytryptamine and phenethylamine, but it has no action on aniline or *p*-phenetidine [G824].

Shrimp acetyl CoA: arylamine N-acetyltransferase acts on 2-aminofluorene and *p*-aminobenzoate [K485].

Drosophila melanogaster nervous system enzyme, molecular weight 29 500, acts on tyramine, serotonin and dopamine. Activity is much greater than for monoamine oxidase or catecholamine O-methyltransferase. This, in conjunction with high dopa decarboxylase activity, directs dopa to the formation of N-acetyldopamine [A643, H758].

Periplaneta americana indoleamine N-acetyltransferase, optimum pH 6.0 and 9–9.5, is found only in accessory female reproductive glands and in head ganglia. It acts on tryptamine and serotonin, with acetyl CoA as co-substrate [J632, J633].

Musca domestica enzyme, which acts on tyramine, is a monomer, molecular weight 27 600, optimum pH 7.2 and pI 5.8 [G914].

Cockroach testicular enzyme, molecular weight 28 000–30 000 (depending on method) and optimum pH 6.0, requires acetyl CoA. Substrates are tryptamine, serotonin, dopamine, octopamine, noradrenaline, tyramine and 5(?)-methoxytryptamine [H835].

An enzyme in Boophilus microplus acts on octopamine [G780].

D-Tryptophan N-acetyltransferase (E.C. 2.3.1.34)

Saccharomyces cerevisiae enzyme is specific for the D-isomer, and probably requires acetyl CoA [K893].

L-Aminoacid acetyltransferase (E.C. 2.3.1.1)

$$R.CHNH_2.COOH + \text{acetyl CoA} \rightarrow R.CHNHCOCH_3.COOH + CoA$$

Xanthomonas campestris acetylates L-tryptophan, but the reaction is not found in Pseudomonas syringae [E435].

In Claviceps purpurea L-tryptophan and probably L-tyrosine are acetylated [A3655].

Two strains of Cordyceps militaris acetylate L-tryptophan [A993].

Mercapturates (acetyl CoA: S-substituted cysteine N-acetyltransferase, E.C. 2.3.1.80)

Many xenobiotics undergo conjugation with glutathione, and these conjugates can be hydrolyzed to the corresponding S-(L-cysteine) conjugates. They are further metabolized by acetylation to mercapturate (N-acetyl-L-cysteine) conjugates. Few studies on the acetylation step have been carried out at an enzyme level.

Activity is found towards S-benzyl-L-cysteine in rat liver and kidney microsomes. In males the activity at five days after birth is 1–2 per cent of the adult level, but it is somewhat higher in females. The activity then increases slowly, followed by an exponential phase to reach the adult level in liver at 40 days, whereas in kidney it exceeds the adult level at 40 days [A3043].

6-Aminopenicillanic acid – phenylacetyl CoA transferase

The pure enzyme from Penicillium chrysogenum is identical with 6-aminopenicillanic acid phenoxyacetyl CoA transferase. Its optimum pH is 7.0, it requires Mg^{2+}, is stimulated by thiols and is inhibited by thiol-binding reagents. Both aromatic substrates are equally good, but *p*-methylphenoxyacetyl Co A is a less effective substrate [D12]. The reaction is an essential step in the formation of penicillins. The enzyme is also found in Aspergillus nidulans [F646].

Puromycin N-acetyltransferase

Streptomyces alboniger enzyme, molecular weight 23 000, acetylates the tyrosyl residue of puromycin. It also acts on O-demethylpuromycin and its analogue chryscandin [D879].

Conjugation of carboxylic acids

a. Alanine

Alanine is conjugated with indole-3-acetate in crown gall [A2843] and with 2,4-dichlorophenoxyacetate in Glycine and Zea [A3200].

b. Arginine

Rat brain synaptosome tyrosine-arginine ligase (E.C. 6.3.2.24), molecular weight 240 000–245 000, optimum pH 7.5–9.0 and pI 6.1–6.2, requires ATP and Mg^{2+} to form L-tyrosyl-L-arginine [K857].

c. Aspartate

N-Benzoylaspartate is formed in barley from benzyl alcohol, presumably with benzoate as intermediate [A1726].

Oxindole-3-acetylaspartate and indole-3-acetylaspartate are formed from indole-3-acetate in a Populus hybrid [H349]. Indole-3-acetylaspartate is formed in Dalbergia dolichopetala seeds [E729], pea, Teucrium canadense, sycamore, crown gall and Arabidopsis [A635, A2843, A3220, E307, J713]. This reaction in Vicia faba is the first step of a reaction sequence in which the indole nucleus is oxidized [E437].

Glycine max conjugates α-naphthylacetate and phenylacetate with aspartate. The corresponding acylglucosides appear to be transitory intermediates [A3679].

Aspartate is conjugated with 2,4-dichlorophenoxyacetate in Glycine max and corn [A3200].

d. Benzoyladenylate formation

This is formed from benzoate in N. crassa [A750].

e. Benzoylhydroxylamate formation

This is formed from benzoate and hydroxylamine in N. crassa [A750].

f. Glutamate

In cow and quail, *m*-phenoxybenzoate is conjugated with glutamate [A3194, F838].

Indole 3-acetate is conjugated with glutamate in Teucrium canadense, Arabidopsis and crown gall [A2843, E307, J713]. This conjugate is formed with 2,4-dichlorophenoxyacetate in Glycine max and corn [A3200].

Dihydrofolate synthase (E.C. 6.3.2.12) acts on dihydropteroate with glutamate and ATP as co-substrates, to form dihydrofolate and ADP. Serratia indica enzyme requires a univalent cation (K^+, NH_4^+, Rb^+) and a divalent cation (Mg^{2+}, Mn^{2+}, Fe^{2+}) [A2836]. Pea enzyme is mainly mitochondrial, from which it is easily released. Activity is maximal six to eight days after gemination. Spinach leaf enzyme is also mitochondrial [A2473]. Pea seedling enzyme, molecular weight 56 000 and optimum pH 8.8, also requires a univalent cation (K^+, NH_4^+, Rb^+) and a divalent cation (Mg^{2+}, Mn^{2+}) and is specific for dihydropteroate; pteroate and tetrahydropteroate are not substrates. It is inhibited by ADP and Ag^+; *p*-chloromercuribenzoate inhibition is reversed by 2-mercaptoethanol [A1363, A2240]. E. coli enzyme acts on dihydropteroate and requires a univalent cation (K^+, NH_4^+, Rb^+) and a divalent cation (Mg^{2+}, Mn^{2+}). It is specific for dihydropteroate; pteroate, tetrahydropteroate and polyglutamates are not substrates, and dihydrofolate is inhibitory [A1360, A1362].

g. Glutamine

In man, phenylacetylglutamine is formed from phenylacetate, and this conjugate is essentially the only one formed in man with this acid [A128, A132, H550]; it is a major urinary component.

In addition to man, phenylacetylglutamine is formed from phenylacetate in rhesus, cynomolgus, green, red bellied, mona and squirrel monkeys, mangabey, drill, baboon, capuchin and marmoset [A196].

Indole 3-acetate is conjugated with glutamine in man, Macaca, Cercopithecus, Papio, Saimiri Aotus and Cebus, but not in a range of other mammals [A274].

Diphenylacetate is conjugated in rat [C313].

h. Glycine and hippurate conjugates; relationship with other conjugation reactions

Many acids are conjugated with glycine, in particular benzoates; only a small proportion of those studied are listed here.

The first step in the formation of hippurates is formation of CoA conjugates, an ATP-driven reaction [H740]. Both these reactions occur in the soluble fraction of liver mitochondria [A1369].

Lipoate reduces the formation of hippurate; it appears to deplete hepatic CoA [H722].

The formation of hippurates is just one of several possible conjugation routes. A study on conjugation of phenylacetate in different animal species found that in man and both old and new world monkeys there is a little (none in man) conjugation with glycine, most conjugation is with glutamine and a little with taurine. In all other mammals and birds studied, conjugation was observed with glycine, but none with glutamine, and in most cases, some conjugation with taurine. In dog, ferret and mouse, some acyl glucuronide is formed [A196]. Other studies in rat found a small proportion diverted to the glutamine and glucuronide pathways [C308, C381]. Glycine conjugation is found in elephant and hyaena [A2387].

In rat, a linear relationship has been demonstrated between the polarity of *p*-substituted benzoates and the proportion converted into hippurates [F693].

Glycine conjugates are formed in ferret from benzoate, *p*-nitrobenzoate, phenylacetate, *p*-chlorophenylacetate, *p*-nitrophenylacetate, 1- and 2-naphthylacetate and indole-3-acetate, whereas taurine conjugates are formed from phenylacetate, *p*-chlorophenylacetate, *p*-nitrophenylacetate, 1- and 2-naphthylacetate and indole-3-acetate. Glucuronides are formed from benzoate, *p*-nitrobenzoate, 1- and 2-naphthylacetate and diphenylacetate, but not from indole-3-acetate [A394].

m-Phenoxybenzoate is conjugated with glycine in a range of mammals and in mallard, and all of them in addition form the glucuronide, and several conjugate it with taurine [B638].

In rabbit, both hippurates and glucuronides are formed from benzoate, monochlorobenzoates, monofluorobenzoates, 3- and 4-methylbenzoates, *p*-hydroxy-, *p*-cyano- and *p*-methoxybenzoates, whereas only glycine conjugates are formed with nitrobenzoates, aminobenzoates, *o*-methyl-, *o*- and *m*-hydroxy-, and 4-acetamidobenzoates [H519].

Indole-3-acetate is conjugated with glycine in a range of mammals including new world monkeys, but not by old world monkeys or man. In man it is mainly conjugated as acyl glucuronide (but not in any of the other species studied) and with glutamine. Taurine conjugates are found in monkeys, Mustela and Columbia [A274].

In man and rat, kidney and liver are sites of glycine conjugate formation [G834, J908].

In rat brain, both benzoate and *m*-trifluoromethylbenzoate form hippurates [A3730].

In rat and a number of marsupials benzoylglucuronide is formed as well as hippurate [D914].

Salicylic acid is conjugated with glycine in rat [B56], and is an important metabolite of aspirin [B201]; this is also a finding in many species including man.

In marmoset the proportion of benzoate conjugated with glycine decreases with increasing dose [B201].

Benzoate is mainly converted by horse into hippurate, with a trace as glucuronide. Salicylate is similarly conjugated. 2-Naphthylacetate is primarily conjugated with glycine, with lesser amounts of glucuronide and taurine conjugate [B801].

Lion, civet and genet form hippurate from benzoate, and in the same species 1-naphthylacetate is conjugated with glycine, whereas taurine and glucuronide conjugates are formed only in civet and genet [A1312].

In pipistrelle bat both benzoate and phenylacetate are conjugated with glycine [A3356].

i. Glycylvaline

m-Phenoxybenzoate forms this conjugate in mallard, but a large number of mammalian species do not exhibit this reaction [B622, B638]. It has also been observed in quail [F838].

j. Leucine

This conjugate is formed with 2,4-dichlorophenoxyacetate in Glycine max and Zea mays [A3200].

k. Lysine

Benzoate and *p*-chlorobenzoate form N^2-benzoyl-N^6-*p*-chlorobenzoyl- and N^6-benzoyl-N^2-*p*-chlorobenzoyllysines in chicken [A2253].

Pseudomonas syringae, subspecies savastanoi, forms indoleacetyl-lysine synthetase (E.C. 6.3.2.20) [K921].

l. Ornithurate and analogues

Ornithurate is formed from benzoate in chicken, pheasant and Coturnix [A1156], quail [E516] and Tuatara [B449].

Symmetrical ornithine conjugates are formed from phenylacetate, *p*-nitrophenylacetate, *p*-chlorophenylacetate and 1-naphthylacetate in hen, but not in pigeon [A131, A395]. A similar conjugate is formed with *p*-chlorobenzoate in chicken [A2253].

m. Mixed ornithine conjugates

In quail and chicken, *m*-phenoxybenzoate is conjugated with N^2-acetylornithine at the 5 position [C40, F838]. Additionally, in chicken benzoate, *m*-hydroxybenzoate and *m*-phenoxybenzoate form small amounts of other ornithine conjugates; N^2-benzoyl-N^5-*m*-phenoxybenzoyl, N^5-benzoyl-N^2-*m*-phenoxybenzoyl, N^2-benzoyl-N^5-*m*-hydroxybenzoyl and N^5-benzoyl-N^2-*m*-hydroxybenzoyl (main conjugate) [E941].

N^2-Benzoyl-N^5-*p*-chlorobenzoyl-, N^5-benzoyl-N^2-*p*-chlorobenzoyl-, N^2-benzoyl-N^5-*p*-hydroxybenzoyl- and N^5-benzoyl-N^2-*p*-hydroxybenzoylornithine conjugates are formed with a mixture of *p*-chlorobenzoate, benzoate and *p*-hydroxybenzoate in chicken [A2253].

n. Phenylalanine

This conjugate is formed with 2,4-dichlorophenoxyacetate in Glycine max and Zea mays [A3200].

E. coli phenylalanine N-acetyltransferase (E.C. 2.3.1.53), optimum pH 8, requires acetyl CoA and acts on L-phenylalanine, L-phenylalanyl-L-tyrosine and some aliphatic amino acids [K924]. A similar reaction is involved in human phenylketonuria, where N-acetyl-L-phenylalanine is a major metabolite.

o. Pteroylpolyglutamates

Pig liver folylpoly-γ-glutamate synthetase (E.C. 6.3.2.17) is composed of two isozymes, both with monomeric molecular weight 62 000 and optimum pH 9.5. The reaction requires a reducing agent, Mg^{2+} and a monovalent cation; K^+ is best, with folate, dihydrofolate and tetrahydrofolate as substrates. The reaction sequence is, for instance:

$$\text{THF} + \text{ATP} \rightarrow \text{ADP} + P_i + \text{tetrahydrofolylglutamates}$$

Depending on substrate, the products contain two to seven glutamyl residues, and it appears that the glutamate residues are added sequentially [E277].

In E. coli, dihydropteroate is conjugated with glutamate to yield dihydrofolate; pteroate and tetrahydropteroate are not substrates, nor are oligoglutamyl analogues. It requires ATP, Mg^{2+} or Mn^{2+}, and NH_4^+, K^+ or Rb^{2+}. Tetrahydrofolate yields its mono- and then the diglutamyl analogue [A1362].

Folylpolyglutamate synthetase (E.C. 6.3.2.17) from pig liver acts on pteroates, dihydropteroates and tetrahydropteroates with up to seven glutamate residues, building up the chain one unit at a time. The reaction rate decreases sharply with increasing chain length [K455].

p. Serine

E. coli 2,3-dihydroxybenzoate-serine ligase (E.C. 6.3.2.14) acts on L-serine and pyrocatechuate. Its appearance is prevented by the presence of Fe^{2+} [K855].

q. Taurine

In rainbow trout, both benzoate and *m*-phenoxybenzoate form taurine conjugates; benzoate also undergoes this reaction in bluegill, perch and minnow [E582, E911], channel catfish [F554] and in southern flounder [E900].

m-Phenoxybenzoate is conjugated with taurine in mouse and quail [A3250, C127, C133, F838]. A large study found that marmoset, hamster, ferret, cat, gerbil and mouse form the taurine conjugate, but this conjugate was not detected in rat, guinea pig, rabbit, sheep or mallard [B638].

An extensive study with phenylacetate found taurine conjugates in man, old and new world monkeys, bushbaby, slow loris, dog, cat, ferret, rabbit, mouse, rat, hamster and pigeon, although in most cases this was a minor conjugation route. The reaction was not detected in guinea pig, vampire bat or chicken [A196].

Taurine conjugates are formed in ferret from phenylacetate, *p*-chlorophenylacetate, *p*-nitrophenylacetate, 1- and 2-naphthylacetate and indole-3-acetate [A394].

A study with indole-3-acetate found conjugation with taurine in old and new world monkeys (Macaca, Cercopithecus, Papio, Saimiri, Aotus and Cebus), Mustela and Columbia, but not in a range of other mammals [A274].

In pigeon and hen, taurine conjugates are formed with phenylacetate, *p*-chlorophenylacetate, *p*-nitrophenylacetate and 1-naphthylacetate; indole-3-acetate is conjugated in pigeon [A395].

r. L-Tryptophan

This conjugate is formed with 2,4-dichlorophenoxyacetate in Glycine max and Zea mays [A3200].

s. Tyramine

Nicotiana tabacum, infected with TMV, contains cytosolic feruloyl CoA: tyramine N-feruloyl CoA transferase (E.C. 2.3.1.110), molecular weight 45 000. It has a broad specificity, with octopamine, dopamine, noradrenaline, 3-methoxytyramine, tyramine, phenethylamine and (poor) *p*-sympatol as acyl receptors, and the CoA conjugates of ferulate, cinnamate, *p*-coumarate, sinapate, dephosphoferuloyl CoA and feruloyl-4′-phosphopantetheine as donors [D558, F72].

t. Valine

Indole-3-acetate is conjugated with valine in crown gall [A2843].

N-Malonyltransferases

Arachis seedling contains D-tryptophan N-malonyltransferase (E.C. 2.3.1.112), anthranilate N-malonyltransferase (E.C. 2.3.1.113) and 3,4-dichloroaniline N-malonyltransferase (E.C. 2.3.1.114), molecular weights 38 000, 50 000 and 45 000 respectively [K727].

Daucus carota catalyzes this reaction with 3,4-dichloroaniline to form N-malonyl-3,4-dichloroaniline [H118].

Peptide formation with heterocyclic amines

A calf enzyme acts on phenothiazine to form N-conjugated short chain peptides of phenothiazine and 3-hydroxyphenothiazine; the peptide chains appear to have little similarity in structure [D410].

Amidation by transacetylation and transformylation

Liver from several mammalian species transfers the formyl group from 4-formamidobiphenyl to α- and β-naphthylamine, from α-formamidonaphthalene to 4-aminobiphenyl, 2-aminofluorene and β-naphthylamine, from β-formamidonaphthalene to 4-aminobiphenyl, 2-aminofluorene and α-naphthylamine, and from 1-formamidopyrine to 4-aminobiphenyl [F109].

5-Aminosalicylate is formylated in rat and human liver. In rat the reaction proceeds when the reactants are fortified with N-formyl-L-kynurenine, which suggests that the reaction is a transformylation [F924].

E. coli amino acid transacetylase transfers an acetyl group from a range of N-acetyl-L-amino acids to L-phenylalanine [K924].

N-Carbamoylation

Dog metabolizes methylphenidate to form the corresponding carbamide [C14].

5.6 *Phosphorylation reactions*

Phosphate ester formation

Although O-phosphorylation by protein kinases is a well-established mechanism for activation of enzymes, few studies have detected the formation of phosphate esters in phase 2 metabolism of xenobiotics; this may be a consequence of technical problems in handling phosphate ester metabolites.

Man, monkey and rat form phosphate esters from N-hydroxy-2-acetamidofluorene [A2086].

In sheep, phenol is converted into phenol and quinol phosphates [A3717]. In cat phenol is phosphorylated, but not naphthols [A2].

In housefly, and probably in grass grub and blowfly, both 1-naphthol and *p*-nitrophenol yield phosphate and glucoside-6-phosphate esters [A1604]. These reactions have also been detected with *p*-nitrophenol in cockroach, stick insect, mealworm, Porina and wax moth larvae, housefly, blowfly and earwig, except that the glucoside ester has not been detected with mealworm and Porina [C494].

Phenylalanine adenylyltransferase (E.C. 2.7.7.54)

Penicillium cyclopium acts on ATP and phenylalanine to form N-adenylylphenylalanine and pyrophosphate; it appears to be an intermediate in the formation of cyclo-(anthaniloyl-phenylalanyl) [K906].

Anthranilate adenylyltransferase (E.C. 2.7.7.55)

Penicillium cyclopium acts on ATP and anthranilate to form N-adenylylanthranilate and pyrophosphate; it appears to be an intermediate in the formation of cyclo-(anthaniloyl-phenylalanyl) [K906].

Flavokinase (E.C. 2.7.2.26)

$$\text{Riboflavin} \rightarrow \text{FMN}$$

This reaction has been detected in rat [A919, C145, E406].

FMN adenylyltransferase (E.C. 2.7.7.2)

$$\text{FMN} + \text{ATP} \rightarrow \text{FAD} + \text{pyrophosphate}$$

This activity is found in rat and mouse [A919, A3780]. In plants it is activated by Mg^{2+} and slightly by Zn^{2+}; it is found in Arachis hypogea, Cajanus indicus, Cannavalia gladiata, Cicer arietinum, Dolichos lablab, D. biflorus, Ipomoea batatas, Papaver sativum, Phaseolus radiatus, Ricinus communis, R. mungo, Sesamum indicum and Solanum tuberosum [K828].

(2,3-Dihydroxybenzoyl)adenylate synthase (E.C. 2.7.7.58)

E. coli enzyme, molecular weight 59 000 requires ATP, releasing pyrophosphate. It also acts on several *o*-hydroxybenzoates, but not on other analogues [K830].

Actinomycin synthetase I

Streptomyces antibioticus enzyme acts on ATP and 3-hydroxy-4-methylanthranilate to form 3-hydroxy-4-methylanthraniloyladenylate, with the release of pyrophosphate. This reaction is in the sequence that leads to the formation of actinomycin D [G791].

Protein-tyrosine kinase (E.C. 2.7.1.112)

This enzyme is found in chicken and man; it is enhanced by infection with Rous sarcoma virus [K918]. Rat membrane vesicle enzyme requires ATP as phosphorylating agent, and is enhanced by low Zn^{2+} concentrations; it phosphorylates the tyrosine residues of several proteins [K918, K919].

Phenylalanine tRNA ligase (E.C. 6.1.1.20)

This activity has been detected in rat liver and Saccharomyces cerevisiae [K941].

Tyrosine tRNA ligase (E.C. 6.1.1.1)

Pea root contains two isozymes, molecular weights about 50 000 and 70 000, which require ATP and Mg^{2+}. It acts on tRNA from plants and beef, but not E. coli [K825].

Tryptophan tRNA ligase (E.C. 6.1.1.2)

Beef pancreas enzyme, molecular weight 110 000 appears to be a tetramer composed of 2 pairs of similar chains [K854].

5.7 *Ether formation*

Catechol O-methyltransferase (COMT, E.C. 2.1.1.6)

$$Ar.(OH)_2 + SAM \rightarrow Ar.OH.OMe$$

a. Animals

This reaction is one of the major pathways by which the catecholamine hormones are inactivated. It is of particular importance in Parkinson's disease, in which the synthesis of dopamine is seriously compromised by destruction of dopaminergic neurones. Inhibition of COMT would potentially reduce the losses of dopamine, with possible alleviation of symptoms; some nitrocatechol substrate analogues are strongly inhibitory.

Human brain enzyme, which is 55–80 per cent microsomal, acts on noradrenaline and dopamine [B118]. Its activity is scarcely altered in parkinsonism [A381].

Human erythrocyte enzyme is inhibited by S-adenosylhomocysteine [A2956].

Human liver and placental enzymes act on noradrenaline, adrenaline, dopamine and isoproterenol in the *meta* position. The K_m for the placental enzyme is lower than for the liver enzyme [A3831]. Placental enzyme, which is activated by cysteine, is monomeric, molecular weight 23 000 [G171]. Another study found the molecular weight to be 25 000, with pI 5.3 [F795]. A further study on placenta isolated an enzyme, molecular weight 49 000 and pI 5.0 [A2905].

Human liver, brain and placental enzymes are composed of two isozymes, similar to the corresponding rat enzyme (molecular weights 24 000 and 47 500, with pI 4.9 and 4.8, and Stokes radii 2.01 and 2.87 nm respectively). These forms are not inter-converted, and are considered to be genetically different [A3793].

In man and rat, the small intestinal enzyme is found in mucous membrane and muscle layers, in fractions, both particulate and soluble [E693].

In pig liver, methylation of dopac is inhibited competitively by 2,3-dihydroxypyridine, which is not a substrate [C211]. Substrates include D-adrenaline, benserazide, carbidopa, 5-S-(L-cysteinyl)-L-dopa, α-difluoromethyldopa, D- and L-dopa, dopac, α-fluoromethyldopa and α-methyldopa [B813]. In liver and brain, the enzyme is cytosolic only, and is evenly distributed throughout brain [A2977]. Antibody and other studies suggest that the brain and liver enzymes are identical [A3550].

In pig kidney, the activity rises steadily from 10 days before birth to 70 days post partum. In heart, maximal activity is found around birth, after which it declines slightly. In spleen it peaks 15 days post partum, and then declines by 40 per cent. In brain the activity fluctuates around birth, and then remains steady from two weeks post partum. In adrenal, activity doubles at birth and then slowly declines [A1569].

In mouse liver, 70 per cent of the activity is associated with the plasma membrane [A345].

Rat cerebral microvessels show activity, considered to be extraneuronal [A3619]; it is both soluble and membrane-bound [C477]. Two forms of liver enzyme are found, molecular weight 25 000 and pI 5.1; they differ in internal disulphide bonding [F795, F797]. The ratio of activity in liver, brain and eythrocytes is 100:10:1, with optimum pH 9, 8.2 and 9 respectively. The ratio of *meta* to *para* methylation is similar in brain and erythrocyte, and different from that in liver. A number of substituted catechols are substrates [A2817].

Rat liver enzyme, optimum pH 7.3–8.2, requires Mg^{2+} and is inhibited by Ca^{2+}. Protocatechuate is both *meta* and *para*-methylated [A56, C169]; this is also observed with caffeate [A187]. After purification by affinity chromatography, its molecular weight is 23 000, and the pure enzyme is very unstable [A340]. A minor isozyme has molecular weight 47 500. The major and minor isozymes have pI 4.9 and 4.8, and Stokes radii 2.01 and 2.87 nm respectively [A3793]. Another study found two isozymes, pI 4.66 (main) and 4.54, both optimum pH 7.8–8.2 [A1408]. An insoluble isozyme with optimum pH 8.5, molecular weight 21 000 and a requirement for Mg^{2+}, is similar to cytosolic enzyme after solubilization [A2586]. It is inhibited by a series of 3-nitrocatechols with IC_{50} about 10–500 nM and, less effectively, by 4-nitrocatechols and 3-cyanocatechols with an electron-withdrawing group *meta* to the nitro group or other substituent. For 3-nitrocatechols, best inhibition occurs when the substituent is *n*-pentoyl or conjugated heterocyclic groups of ester and amide types. Many of these compounds are relatively non-toxic, which makes them potentially useful as drugs for sparing dopamine in the treatment of parkinsonism [E923]. A particularly good inhibitor is RO-4-4602, which is effective at 10^{-5} M [A2585].

Rat lung, kidney, liver and brain enzymes are both membrane-bound and cytosolic. Using dopamine and protocatechuate as substrates the *meta*/*para* ratio for methylation is four to six with cytosolic enzyme, but higher with membrane enzyme (8–60). Dopamine shows higher ratio values than other substrates. The ratio increases as substrate concentration is reduced, particularly for dopamine [D297].

Rat erythrocyte enzyme is both particulate and cytosolic, optimum pH about 8.1 [A457]. Two cytosolic isozymes are found, with pI 4.70 for the particulate enzyme [A1408], they exhibit a similar specificity, requirement for Mg^{2+} and optimum pH, but they differ in thermal stability and K_m [A1382].

A major study with liver enzymes examined *meta* and *para* methylation. For neutral and acidic substrates, the *m*/*p* ratio is independent of pH, but for amine substrates the ratio decreases with increasing pH. In this study the ratio was independent of enzyme source (three rat strains, guinea pig, rabbit, mouse and monkey), and was the same for particulate and cytosolic enzymes (in contrast with above results), but it depended on the concentration of divalent ions. Polar side-chain substituents inhibit *p*-substitution [A486]. Resorcinols are not substrates [A724].

Rabbit, rat and guinea pig lung enzymes, which require Mg^{2+}, act on *l*-proterenol, optima pH 7.2 for microsomal enzyme and broad (7.5–9) for soluble enzyme [A115].

Tetrahymena pyriformis enzyme is cytosolic [A3131].

A theoretical analysis of the action of COMT has been undertaken [B545].

b. Plants.

Alfalfa root nodule enzyme, molecular weight about 103 000, acts on caffeate and 5-hydroxyferulate [B487].

Apple fruit enzyme, molecular weight 78 000, pI 5.25 and optimum pH 6.8, acts on a range of catechols, including cinnamates and flavonoids. The report implied that Mg^{2+} is not required for activity [D258].

Aspen xylem enzyme, molecular weight 40 000, acts on caffeate and 5-hydroxyferulate [G624].

Bambusa contains caffeate O-methyltransferase (E.C. 2.1.1.68), pI 4.61 [A169].

Brassica oleracea enzyme contains two isozymes, both with molecular weight 42 000 and optimum pH 7.6, but with different pI. They are not activated by cations. Substrates include caffeate, 5-hydroxyferulate, quercetin, 3,4,5-trihydroxycinnamate and esculetin, but not chlorogenate or protocatechuate [F367].

Chrysosplenium americanum enzyme, molecular weight 65 000, pI 5.4 and optimum pH 7.5–8.5, requires Mg^{2+} and methylates the 3′ position of 3,7-di-O-methylquercetagetin [B832, C579].

Cortaderia selloana (pampas grass) enzyme acts on caffeate to yield only ferulate. Protocatechuate is methylated in both positions; *meta*-methylation activity is largely destroyed by heating the enzyme preparation at 45°, whereas *para*-methylation is retained [A187].

Daucus carota S-adenosylmethionine: caffeoyl CoA 3-O-methyltransferase (E.C. 2.1.1.104) requires Mg^{2+}; other catechols are inactive, but caffeoyl-3′-dephosphoCoA is a substrate. It is inhibited by S-adenosyl-L-homocysteine [F94, K731].

Glycine max culture yields two isozymes of caffeate O-methyltransferase (E.C. 2.1.1.68); one is unstable at 4°. The stable one, which acts on caffeate (forming ferulate) and 5-hydroxyferulate, is considered to be involved in lignin formation, and the other, which acts on luteolin and quercetin, to be involved in flavonoid synthesis [A2434, K728].

Lotus corniculatus flower bud enzyme, pI 5.1 and optimum pH 7.7, acts on position 3′ of flavonols [C905].

Nicotiana leaf contains three isozymes, molecular weights 90 000, 93 000 and 100 000, with pI 4.80, 5.21 and 4.74 respectively, is stable at 0°. Most substrates are methylated at the *meta* position, but with some *p*-substitution for protocatechualdehyde, esculetin and protocatechuate; in the latter case *p*-substitution predominates with the 'pI 5.21' enzyme [B490]. Many catechols are substrates [A3778]. Cell cultures also have a broad specificity; isozymes, molecular weights 70 000 and 74 000, and optimum pH 8.3 and 7.3, respectively, have been detected. In this study caffeate was methylated primarily in the *m*-position, for esculetin the ratio was near unity, and for quercetin *p*-substitution predominated. In each case the 'pH 8.3' enzyme *p*-methylated to a greater extent than the 'pH 7.3' enzyme [A3437, A3757].

Parsley S-adenosyl-L-methionyl: caffeoyl CoA 3-O-methyltransferase (E.C. 2.1.1.104), molecular weight 48 000, pI 5.7 and optimum pH 7.5, requires Mg^{2+}, and is activated by fairly high concentrations of phosphate or by tris and sodium chloride; EDTA, Mn^{2+} and Ca^{2+} are inhibitory. Methyl caffeate, chlorogenate, rosmarinate and *trans*-5-O-caffeoylshikimate are also substrates, but not caffeate [F160].

Parsley luteolin O-methyltransferase (E.C. 2.1.1.42), molecular weight about 48 000 and optimum pH about 9.7 requires Mg^{2+}. Luteolin and its 7-glucoside are the best substrates; other catechols have a much lower affinity [K805].

Petunia hybrida petal anthocyanin methyltransferase is cytosolic, converting cyanidin into paeonidin [C743].

A poplar hybrid (tremula/alba) caffeoyl CoA O-methyltransferase is found in all lignifying cells [K625]. Populus trichocarpa enzyme is composed of two isozymes [K516].

Spinach beet caffeate O-methyltransferase (E.C. 2.1.1.68), optimum pH 6.5 does not require Mg^{2+}; it has a broad specificity [A2315].

Trillium apetalon leaf enzyme is dimeric, molecular weight 78 000, optimum pH 7 and pI 5.3. It is stimulated by EDTA and dithiothreitol, and is inhibited by *p*-chloromercuribenzoate, iodoacetate and by several divalent heavy metal ions [K250].

Vitis vinifera S-adenosylmethionine: cyanidin-3-glucoside 3′-O-methyltransferase, optimum pH 7.7–9.8, requires Mg^{2+}. It is less active towards cyanidin than its glucoside [H886].

Zea mays S-adenosylmethionine: quercetin-3′-O-methyltransferase, optimum pH 8.5, requires dithioerythritol, and Mg^{2+} or Mn^{2+}. Eriodictyol, quercetin and luteolin are substrates [F173, G223].

c. Microorganism

Streptomyces griseus enzyme, molecular weight 36 000, pI 4.4 and optimum pH 7.5 requires Mg^{2+}. Substrates include simple catechols and esculetin [K652].

Phenol O-methyltransferases (E.C. 2.1.1.25)

a. Animals

Human erythrocyte stroma activity (with *p*-acetamidophenol as substrate) correlates highly with aliphatic S-methyltransferase (E.C. 2.1.1.9), suggesting that they are the same enzyme. The optimum pH is 9.0, and it is stable at room temperature. It does not require Mg^{2+}, and is not inhibited by Ca^{2+}, but tropolone and

S-adenosylhomocysteine are inhibitory [A2956, C741].

Rabbit, rat and guinea pig lung enzymes act on phenol and phenols substituted with one or two methyl groups (all possible structures were tested), or with ethyl and bromo groups [A115].

Rat liver mitochondrial 5-demethylubiquinone-9 methyltransferase which requires S-adenosylmethionine is found on the inner membrane [A3084].

b. Plants

Alfalfa (Medicago) and licorice (Glycyrrhiza) chalcone 2′-O-methyltransferases require S-adenosylmethionine and act on isoliquiritigenin and licodione respectively (H908].

Alternaria alternata alternariol O-methyltransferase, molecular weight 43 000, requires S-adenosylmethionine [G473]. A. tenuis enzyme, molecular weight 110 000 and optimum pH 8.0 requires Mg^{2+} [E301].

Apple fruit enzyme (quercetin O-methyltransferase), which has a requirement for Mg^{2+}, acts on the 3 position of quercetin, and the 7 position of 3-O-methylquercetin. It has a structural requirement for 3′,4′-dihydroxy groups for substrate activity. The molecular weight is 47 000, shows a double optimum pH at 7.3 and 8.3 and pI 4.9 [D258].

Argemone platyceras S-adenosylmethionine: (*R*)- and (*S*)-norlaudanosoline 6-O-methyltransferase (E.C. 2.1.1.128), molecular weight 47 000 and optimum pH 7.5, acts on the 6- and to a small extent, the 7-position of norlaudanosoline. It does not act on other classes of phenols. This activity is also found in Adlumia fungosa, Argemone intermedia, Berberis henryana, B. wilsoniae, B. stolonifera, Chelidonium majus, Cissampelos mucronata, Corydalis sempervirens, C. pallida, Eschscholtzia tenuifolia, Fumaria officinalis, Glaucium flavum, Papaver somniferum, Thalictrum tuberosum and T. sparsiflorum [D77]. A. platyceras S-adenosylmethionine: 6-O-methylnorlaudanosoline 5′-O-methyltransferase (E.C. 2.1.1.121; it should be called 3′-methyltransferase) forms nororientaline, with methylation on the benzyl moiety. It is highly specific; it does not act on (*R*)- or (*S*)-norlaudanosoline, laudanosoline, simple phenols, flavanoids or coumarins, amongst others [D76].

Berberis koetineana S-adenosyl: 3′-hydroxy-N-methyl-(*S*)-coclaurine-4′-O-methyltransferase (E.C. 2.1.1.116), molecular weight 40 000 and optimum pH 8.5, is highly specific for the (*S*)-isomer. It acts on other analogues, such as norlaudanosoline and (*S*)-3′-hydroxycoclaurine. The enzyme is not separable from 6-O-methyltransferase activity [F854].

Berberis wilsoniae and B. aggregata S-adenosyl-L-methionine: columbamine O-methyltransferase (E.C. 2.1.1.89), molecular weight 52 000 and optimum pH 8.9 are vesicle-bound enzymes that methylate columbamine at the 2 position to form palmatine, but they are inactive towards tetrahydrocolumbamine [E268]. This activity is highly specific [E256].

Berberis (*S*)-scoulerine 9-O-methyltransferase (E.C. 2.1.1.117), molecular weight 63 000 and optimum pH 8.9, requires S-adenosylmethionine to form (*S*)-tetrahydrocolumbamine [D581].

Calamondin orange (Citrus mitis) has been demonstrated to contain, besides catechol O-methyltransferase, enzymes that O-methylate a range of flavones and isoflavones at the 4′, 5, 6, 7 and 8 positions, and possibly at the 3′ position, including substrates with methoxy substituents [A3987].

Catharanthus roseus 11-O-demethyl-17-O-deacetylvindoline O-methyltransferase (E.C. 2.1.1.94) is highly specific, and is involved in the formation of vindoline [K729].

Chrysosplenium americanum flavonol 3-O-methyltransferase (E.C. 2.1.1.76), molecular weight 65 000, pI 4.8 and optimum pH 7.5–8.5, requires S-adenosylmethionine and Mg^{2+}. It acts only on the 3-position of quercetin. The activity is also found in Calamondin orange and Nicotiana [B830, B832, C579].

Chrysosplenium americanum flavonol 6-O-methyltransferase, molecular weight 65 000, pI 5.7 and optimum pH 7.5–8.5, requires S-adenosylmethionine and Mg^{2+}. It acts on 3,7-di-O-methyl- and 3,3′,7-tri-O-methylquercetagetin. Similar activity on quercetin is also found in orange and tobacco [B830, B832, C579].

Cicer arietinum culture isoflavone 4′-methyltransferase (E.C. 2.1.1.46), molecular weight 110 000 and optimum pH 9, is specific for 4′-hydroxyisoflavones such as daidzein. It is inhibited by heavy metals, *p*-chloromercuribenzoate and S-adenosylhomocysteine. It is considered to be an ordered bi bi reaction with S-adenosylmethionine and S-adenosylhomocysteine as leading reaction partners [A1493].

Citrus aurantium S-adenosylmethionine: eriodictyol 4′-O-methyltransferase, molecular weight 52 000 and optimum pH about 7.5 is activated by Mg^{2+} or by EDTA [H591].

Coptis japonica S-adenosylmethionine: 3′-hydroxy-N-methylcoclaurine-4′-O-methyltransferase, molecular weight 80 000 and optimum pH 8.0 is dimeric and forms reticuline by an ordered bi bi reaction. It also acts on (*R*,*S*)-laudanosoline and (*R*,*S*)-norlaudanosoline, and is inhibited by several divalent cations [K572]. (*R*,*S*)-Norcoclaurine 6-O-methyltransferase (E.C. 2.1.1.128), molecular weight 95 000 and pI 4.7 is dimeric, and requires S-adenosylmethionine to form coclaurine by a bi-bi ping-pong mechanism. It is inhibited by divalent cations [D77, K785].

Daucus carota root 6-hydroxymellein O-methyltransferase (E.C. 2.1.1.107), molecular weight 76 000, pI 5.7 and optimum pH 7.5–8.0, requires S-adenosylmethionine for methylation at the 6-position, and is inhibited by both reaction products. It is not found in fresh root, but is induced by the action of, for instance, 2-chloroethylphosphonic acid [J13, K730].

Eschscholtzia californica 10-hydroxydihydrosanguinarine 10-O-methyltransferase (E.C. 2.1.1.119), molecular weight 49 000 and optimum pH 8.5 is highly specific, forming dihydrochelirubine [K777].

Glycyrrhiza echinata licodione 2′-O-methyltransferase (E.C. 2.1.1.65) is specific for the *o*-hydroxyl group; another substrate is isoliquiritigenin, but other compounds are poor substrates [K886].

Lentinus lepideus enzyme O-methylates the methyl esters of *p*-coumaric and sinapic acids, but is inactive towards the acids themselves and the corresponding methyl benzoates [A214].

Lotus corniculatus flower bud enzyme, pI 5.5 and optimum pH 7.9 or 8.1, acts on the 8-position of 8-hydroxykaempferol and 8-hydroxyquercetin (E.C. 2.1.1.88). The enzyme contains a labile thiol group and requires S-adenosylmethionine, Mg^{2+}; EDTA is inhibitory [C905, D904].

Yellow lupin root isoflavone 5-O-methyltransferase, molecular weight 55 000, pI 5.2 and optimum pH 7, requires S-adenosylmethionine, but not Mg^{2+}. Substrates include genistein, derrone, 2′-hydroxygenistein, the 8-prenyl analogues 2,3-dehydrokievitone (best), lupiwighteone, and exhibit some activity towards caffeate and 2′-hydroxy-3′-prenylgenistein [F182].

Medicago sativa isoflavone 7-O-methyltransferase acts on daidzein, genistein and other analogues [G322].

Ocimum basilicum (basil) SAM: chavicol O-methyltransferase (SAM: allylphenyl O-methyltransferase), optimum pH 7.5, acts on chavicol but not eugenol, with high activity almost exclusively in young leaves. Another cultivar shows activity towards both chavicol and eugenol [K692].

Parsley culture S-adenosymethionine: xanthotoxol and bergaptol methyltransferases (E.C. 2.1.1.70 and 2.1.1.69 respectively) form xanthotoxin and bergapten respectively [D983].

Pimpinella anisum S-adenosylmethionine: anol O-methyltransferase acts on anol, an obligatory intermediate in the formation of epoxypseudoisoeugenol 2-methylbutyrate. Other substrates include eugenol, chavicol, dihydroanol and vanillin, but *p*-methoxycinnamyl alcohol, *p*-coumarate and *p*-coumaryl alcohol are not substrates [J188].

Pisum S-adenosylmethionine: 6a-hydroxymaackiain 3-O-methyltransferase is composed of two isozymes, molecular weights 43 000 and 66 000 and pI 4.9 and 5.2, respectively, with optimum pH 7.9 [F396].

Prunus flavonoid 7-O-methyltransferase, molecular weight 36 000, pI 4.1 and optimum pH

7.5, acts on sophoricoside, genistein and quercetin [G982].

Rice (Oryza sativa) flavone 7-O-methyltransferase acts on naringenin, apigenin, luteolin and kaempferol as well as caffeate; isoflavones are not substrates. The enzyme is only found in UV-irradiated plants [J190].

Robinia pseudoacacia seedlings and other plant parts contain apigenin 4′-O-methyltransferase (E.C. 2.1.1.75), optimum pH 9.0; it forms acacetin [C100].

Ruta graveolens enzyme, molecular weight 85 000–110 000 (more than one enzyme?) is specific for the 5 and 8 positions of bergaptol and analogues. The substrates studied were bergaptol, xanthotoxol, 5-hydroxyxanthotoxin (optimum pH 7.3–8) and 8-hydroxybergapten (optimum pH 8.5–9) [A3572].

Silene pratensis isoorientin 3′-O-methyltransferase (E.C. 2.1.1.78) forms isoscoparin [C351].

Thalictrum bulgaricum 12-hydroxydihydrochelirubine 12-methyltransferase (E.C. 2.1.1.120) is specific, forming dihydromacarpine; it is probably cytosolic. A 10-methyltransferase is also found [K776].

Thalictrum minus S-adenosylmethionine: norcoclaurine 6-O-methyltransferase is a dimer, molecular weight 72 000, pI 4.3 and optimum pH 9.0, and is inhibited by S-adenosylhomocysteine [H578].

Tinospora cordifolia and a number of other plant genuses contain (*R*,*S*)-norcoclaurine 6-O-methyltransferase (E.C. 2.1.1.128) [H629].

c. Microorganisms

Aspergillus parasiticus demethylsterigmatocystin O-methylase (E.C. 2.1.1.109), molecular weight 150 000, monomeric molecular weight 43 000, pI 4.4 and a broad optimum pH 6.5–9, appears to methylate dihydrodemethylsterigmatocystin as a second substrate [J511]. The enzyme is composed of two types of subunit and requires S-adenosylmethionine [E964] to catalyze a key reaction in the formation of aflatoxins. It is inhibited by thiol-binding reagents and heavy metals [K266]. Sterigmatocystin 6-O-methylase (E.C. 2.1.1.110) is also found [K732].

Aspergillus terreus emodin O-methyltransferase is probably a hexamer, molecular weight 332 000, pI 4.4 and optimum pH 7–8 that requires S-adenosylmethionine to form questin. It is highly specific [G633], but the close analogues, catenarin, ω-hydroxyemodin and 2-chloroemodin are also substrates, all with S-adenosylmethionine as co-substrate [C334].

Mycobacterium O-methylates pentachlorophenol, tetrachloroquinol and tetrachlorocatechol [A3330, D137].

A Phanerochaete chrysosporium O-methyltransferase, a homodimer, molecular weight 71 000, requires S-adenosylmethionine. It acts on isovanillate, protocatechuate, *m*-hydroxybenzoate, gallate and several other hydroxybenzoates, specifically methylating at the *meta* position [J339].

A Phanerochaete chrysosporium enzyme is monomeric, molecular weight 53 000 with optimum pH 7–9. A series of potential simple phenolic substrates has demonstrated a structural requirement for an *ortho* or *para* substituent for significant activity, but not all compounds with this structure are substrates [G782].

Rhodococcus chlorophenolicus O-methylates several halogenated quinols [E761].

Saccharomyces cerevisiae hexaprenyl dihydroxybenzoate methyltransferase (E.C. 2.1.1.114) acts on 3-hexaprenyl-4,5-dihydroxybenzoate and S-adenosylmethionine; the nucleotide sequence has been determined [K781].

Iodophenol O-methyltransferase (E.C. 2.1.1.26)

The reference provided supporting this E.C. numbered reaction is incorrect; it has not been possible to find alternative bibliography.

Hydroxyindole O-methyltransferase (E.C. 2.1.1.4)

This enzyme catalyzes an essential step in the formation of the pineal hormone melatonin from N-acetylserotonin, which, among other activities, is involved in controlling the biological clock in vertebrates.

Both beef and chicken pineal enzymes are dimeric, monomeric molecular weight 39 000. They differ in specificity, electrophoretic properties and antibody reactions. Serotonin and bufotonin are poor substrates; the enzyme is subject to substrate inhibition by N-acetylserotonin (mM range) [C522]. Another study with beef enzyme indicated that the native enzyme is a mixture of polymeric forms, optimum pH 8.1 [A3834].

Beef and rat pineal enzymes are inhibited by pyridoxal phosphate, apparently by forming an inactive Schiff base with L-S-adenosylmethionine that interacts with the enzyme. The inhibition is competitive with S-adenosylmethionine and non-competitive with N-acetylserotonin. Inhibition by pyridoxal phosphate is abolished by 10^{-4} M noradrenaline [A803, A2493]. The enzyme also methylates 17β-oestradiol at position 3 [A3339].

Chick pineal enzyme at 15 days age showed an activity rise with age faster in illuminated animals than in those kept under dark conditions, but those exposed to normal diurnal lighting showed an intermediate rate of increase, but without a diurnal rhythm in enzyme activity [A249].

Trout retina enzyme has optima at pH 7.6 and 8.4 [D30]. In steelhead trout the activity increased during the dark period (pre-midnight – 4 a.m.), and then declined steadily by 50 per cent during the day [A1485].

Tocopherol O-methyltransferase (E.C. 2.1.1.95)

Capsicum chromoplast membranes convert γ-tocopherol into α-tocopherol; the reaction requires S-adenosylmethionine [D580].

O-Demethylpuromycin O-methyltransferase (E.C. 2.1.1.38)

Streptomycin alboniger enzyme, optimum pH 8, is highly specific and requires S-adenosylmethionine [K846].

Tetrahydrocolumbamine 2-O-methyltransferase (E.C. 2.1.1.89)

Berberis aggregata enzyme requires S-adenosylmethionine to form tetrahydropalmatine; it is highly specific [K947].

(*S*)-Scoulerine 9-O-methyltransferase (E.C. 2.1.1.117)

Berberis enzyme, molecular weight 63 000 and optimum pH 8.9, acts on (*S*)-scoulerine to form (*S*)-tetrahydrocolumbamine, by methylation at position 9 [D581].

Methylation of tertiary alcohol

Dog, monkey, baboon and man O-methylate 1,1-dimethyl-2-(4-(α,α,β,β-tetrafluoroethyl) phenyl)ethanol. This is claimed to be the first report of tertiary alcohol methylation [A3021].

Methylation of *gem*-diol

Rat forms 9-hydroxy-9-methoxycicloprofen from cicloprofen [A2353].

Methylation with methyl chloride

Phanerochaete chrysosporium acts on isovanillate to form veratryl alcohol, probably with veratrate as the intermediate. S-Adenosylmethionine is not the methyl donor; the best donor is methyl chloride, which may be formed from methionine [J237].

Phellinus pomaceus O-methylates phenol, with chloromethane as co-substrate [G208].

Formation of polyphenoxyphenols

Rhizoctonia praticola extracellular laccase catalyzes a series of complex reactions between syringate and pentachlorophenol or any of a number of polychlorinated phenols, as substrates. Although the structures of the products were not fully characterized, some were *para*-linked polyphenoxyphenols and others appeared to be *o*-quinones, although the mass spectral data were equivocal for some quinones. The products with syringate and pentachlorophenol were 2,6-dimethoxy-4-pentachlorophenoxyphenol and 4-pentachlorophenoxypoly(2,6-dimethoxyphenoxy)-2,6-dimethoxyphenol, where 'poly' is one, two or three units, depending on the substrate mixture. Syringate and 2,4,5-trichlorophenol form 5-(2,4,5-trichlorophenoxy)-3-methoxy-*o*-quinol [D690].

5.8 *N-Alkylation*

N-Methyltransferases

a. Phenylethanolamine N-methyltransferase (PNMT; E.C. 2.1.1.28)

This enzyme is essential for the formation of the catecholamine hormone adrenaline.

Human and monkey enzymes are found in most brain areas. In monkey brain there was a range of one order of magnitude in activity between different areas, with little in the frontal cortex and cerebellum. In human brain the activity was similar to that in monkey brain areas, with little in the frontal cortex or mamillary bodies [A2823]. Human and rabbit adrenal enzymes act on noradrenaline, normetanephrine, phenylethanolamine, 3,4-dichlorophenylethanolamine and 3,4-dichlorophenylethylenediamine. They are inhibited by α-methylphenethylamines with a range of small substituents on the aromatic nucleus. The affinity is almost identical for enzymes from both species [A498].

Beef adrenal medulla enzyme, molecular weight 30 000, is composed of four isozymes, pI 5.1, 5.2, 5.3 and 5.4 [E397]. In contrast to rat adrenal, beef adrenal enzyme is strongly activated by inorganic phosphate and, to a lesser extent by chloride and sulphate at 200 mM [D294]; in contrast another study claimed that it is inhibited 60 per cent by 75 mM NaCl [A1685], as well as by a series of substrate analogues, especially phenylethanolamines, but not competitively [A1643]. It is present in only some cells [A1409].

Beef brain isozymes, molecular weights 32 000 and 65 000, are possibly a monomer and dimer, with optimum pH 8–9, depending on substrate. Substrates include normetanephrine, octopamine, metanephrine, *p*-sympatol, noradrenaline and adrenaline, with the reaction rate reduced by about 80 per cent for the monomethyl substrates. In some cases there is substrate inhibition [A1463].

A dog liver enzyme, not found in monkey or rat liver, requires S-adenosylmethionine as co-substrate, but not 5-methyltetrahydrofolate. It acts on 7,8-dichlorotetrahydroisoquinoline (best), with optimum pH 8.0. Other substrates include epinine, dopamine, adrenaline and (poor) noradrenaline, phenethylamine and phenylethanolamine, but not indolamines [C657].

Rabbit adrenal enzyme has many properties in common with human enzyme (see above) [A498]. It is inhibited by 2-aminotetralins, and by 2,3,4,5-tetrahydro-1*H*-2-benzazepines substituted with one or more chloro groups on the aromatic moiety [A1851, A3443].

Rat brain activity is low at birth, and then starts to increase at three days until at two weeks post partum it reaches 10 times the adult level [A2665]. It is found primarily in the medulla oblongata with lesser amounts in locus ceruleus, basal hypothalamus and some other areas, but not in cerebellar cortex, hippocampus or olfactory bulb [A2823]. Brain and adrenal enzymes are inhibited by 1,2,3,4-tetrahydroquinoline-7-sulphonamide at 10^{-6} M [A3439].

Rat stomach exhibits this activity [H620], as do rat retinal neurones [E253]. Adrenal enzyme forms a transient ternary complex with S-adenosylmethionine during the methylation step [A623]. Adrenal enzyme is composed of two isozymes, molecular weight 37 000; they are not interconvertible [A144].

b. Arylamine/2-substituted ethylamine N-methyltransferases

Human brain cytosolic enzyme, optimum pH 8.25, methylates 1,2,3,4-tetrahydroisoquinoline with S-adenosylmethionine as co-substrate [F228].

Rabbit liver isozymes (E.C. 2.1.1.49), molecular weight 30 000, pI 4.9 and 5.1 show slightly different kinetics towards a wide range of substrates, including tryptamine and substituted tryptamines, phenethylamine and amphetamine, anilines, benzylamine, heterocyclic amines, including tetrahydroquinoline and tetrahydroisoquinoline, and DMI [A1311, C156, E6]. Another study found the enzyme to be monomeric, molecular weight 27 000, pI 4.8 and optimum pH 7.5. Its amino acid composition has been determined. It acts on a range of anilines and tryptamines with S-adenosylmethionine as co-substrate, but tyramine, indole and benzylamine are not substrates. The mechanism appears to be rapid equilibrium random bi-bi [C333].

Rabbit lung indoleamine N-methyltransferase is monomeric, molecular weight 31 500 and optimum pH 7.9, with tryptamine as substrate. It is activated by dithiothreitol [C111].

Rabbit and human lung indoleamine N-methyltransferases (E.C. 2.1.1.4) act on N-methyltryptamine. Rabbit enzyme in particular is strongly inhibited by 2,3,4,6,7,8-hexahydro-pyrrolo[1,2-*a*]pyrimidine [A1662]. Rabbit lung enzyme, which acts on tryptamines and N-methyltryptamines, is inhibited by several N,N-dimethyltryptamines and by S-adenosylhomocysteine [A106]. This lung enzyme is not found in guinea pig or rat [A115].

Rabbit lung and rat lung enzymes N-methylate tryptamine, N-methyltryptamine, serotonin and N-methylserotonin, with S-adenosylmethionine or 5-methyltetrahydrofolate as co-substrate [A111].

Rat brain enzyme requires N-methyltetrahydrofolate as co-substrate. Indoleamines, phenethylamines, octopamine and other phenylethanolamines are substrates; one product is N,N-dimethyltryptamine [A1307, A1311]. The optimum pH is about 6.4, and may be modified by metabisulphite. The reaction rate is highest for primary amines and least for tertiary amines, if they react at all. The molecular weight is much higher than for COMT; catechols are not substrates [A80, A1310].

Rat liver benzyltetrahydroisoquinoline methyltransferase, molecular weight 27 500 and optimum pH 7.7–8.0, acts on tetrahydropapaveroline and requires S-adenosylmethionine as co-substrate. The context suggests that it is an 'ordinary' N-methyltransferase. Inhibition is brought about by pyrogallol, N-ethylmaleimide and *p*-chloromercuribenzoate [A53].

Rat skin enzyme, which methylates dopamine and noradrenaline, is different from PNMT although noradrenaline is a substrate [F622]. It also occurs in some organs after denervation, which should eliminate PNMT [E867].

Tinospora cordifolia enzyme, a monomer, molecular weight 85 000 and optimum pH 8.6, N-methylates (*S*)-norcoclaurine and (*S*)-coclaurine, but not the (*R*)- isomers. Dicentra spectabilis enzyme shows the same stereospecificity, but Fumaria capreolata, Chonodendron tomentosum, Argemone platyceras, Papaver somniferum, Berberis stolonifera and B. juliana act on both (*R*)- and (*S*)-isomers [H629].

Tyramine N-methyltransferase (E.C. 2.1.1.27)

This activity is found in Opuntia [A3977]. The formation of hordenine in barley implies that the same reaction occur in that species.

β-Carboline 2-N-methyltransferase

Beef brain cytosolic enzyme, optimum pH 8.5–9, is activated by Fe or Mn salts. The substrates are 9-methylnorharman and S-adenosylmethionine [H949].

N-(1(*R*)-(carboxyl)ethyl)-(*S*)-norvaline: NAD$^+$ oxidoreductase (L-norvaline forming)

Arthrobacter enzyme forms secondary amines with pyruvate as amine acceptor. As well as phenylalanine and phenylalaninol a range of aliphatic amino acids are substrates to form the corresponding N-(1-(*R*)-(carboxyl)ethyl)-(*S*)-

amino acids. Primary amines and amino acid esters are not substrates [J212].

Methyleneimine formation

Rat brain converts phenethylamine into methylene-β-phenylethylimine, with 5-methyltetrahydrofolate as methylene donor; the reaction is considered to be enzymatic [A2285], although the product is the Schiff base that is spontaneously formed with formaldehyde.

Formimino transfer

An enzyme found in a range of mammalia and organs acts on tetrahydrofolate and formimino-L-glutamic acid to form 5-formiminotetrahydrofolate [K937]

Quaternary amine formation

Rabbit liver S-adenosylmethionine N-methyltransferase acts on a range of pyridines, including 2-, 3- and 4-phenylpyridines, 3-benzylpyridine, 7-azaindole, quinoline, isoquinoline and quinoxaline [E824].

S-Adenosylmethionine: (*S*)-tetrahydroberberine N-methyltransferase is found in Corydalis vaginans, Berberis stenophylla, Dicentra spectabilis, Fumaria officinalis and Papaver somniferum. C. vaginans enzyme, molecular weight 72 000 and optimum pH 8, acts on (*S*)-stylopine and canadine; it is specific for (*S*)-isomers [E299].

Tetrahydropteroylglutamate methyltransferase

Rat liver enzyme, optimum pH 6.7, acts on 5-methyltetrahydropteroylglutamate and 5-methyltetrahydropteroylpentaglutamate with L-homocysteine as co-substrate to form methionine and the demethylated glutamates [A2502].

Phaseolus vulgaris enzyme, molecular weight 40 000 and optimum pH 6.5, catalyzes the reaction anaerobically [A3085].

N,N-Dimethyladrenaline formation?

A study on retinal catecholamines tentatively identified the presence of N-methyladrenaline [A8]; further tests showed that the chromatographic standard used was in fact N,N-dimethyladrenaline, raising the possibility that quaternary catecholamines may be physiologically important in animals.

N-Ethylation

Man, rat and rabbit have been claimed to form N-ethyl-N-demethylmianserin from mianserin; the data presented are equivocal [B610].

N-(β-N-Acetylglucosaminide) formation

This reaction has been found in monkey, with delaviridine as substrate [J275].

5.9 *Silane anhydrides formation*

Goat acts on bis(*p*-fluorophenyl)methyl(1*H*-1,2,4-triazole-1-ylmethyl)silane to form oxy(bis(bis(*p*-fluorophenyl)(methyl)(silane))) [K144].

6. Elimination of substituents

6.1 *Ester hydrolysis*

Arylesterase (E.C. 3.1.1.2)

$$ArOOCR + H_2O \rightarrow ArOH + RCOOH$$

A human serum enzyme, molecular weight about 43 000 and pI 5.1, is a glycoprotein, with phenyl acetate and paraoxon as substrates [F717]. Liver aspirin esterase (E.C. 3.1.1.55) has an optimum at pH 5–6.5 [C779]. Erythrocyte aspirin esterase is an intracellular cytosolic enzyme, molecular weight 95 000. It appears to be different from 'nonspecific' esterases [C705]. Lens isozymes (molecular weights 200 000 and 30 000) hydrolyze 4-methylumbelliferyl palmitate [J233]. Liver microsomes hydrolyze *p*-nitrophenyl acetate, propionate and butyrate [H465]. Plasma heroin hydrolase is an arylesterase, optimum pH about 7.5, which also acts on *p*-nitrophenyl acetate and *p*-nitrophenyl laurate, but not on *p*-nitrophenyl phosphate [A798, G251].

Rat liver aspirin hydrolase is equally divided between cytosol and particulate fractions. Two soluble isozymes, molecular weight 35 000 (serine esterases) and three high molecular weight enzymes (molecular weight about 220 000) with different immunological properties from the low molecular weight esterases have been detected. The serine esterases both act on 4-methylumbelliferyl acetate and substituted phenyl and naphthyl acetates. Butyrate esters are also hydrolyzed, but caprylate esters only poorly. 4-Acetoxybenzoate is also hydrolyzed [F753]. pH optima are at 5.5 and 7.4 [C779]. Apomorphine is formed from its dipropionyl, diisobutyryl and dipivaloyl esters. Activity is found in most tissues, especially plasma, where activity is 10 times greater than in human plasma [A2013]. Rat liver hydrolyzes phenyl acetate; two enzyme fractions were identified, both with optimum pH 8.5 and a requirement for Ca^{2+}; they exhibit slight stability differences at 35–40°. One was identified as paraoxonase and the other as arylesterase [K194].

In both rat and mouse aspirin esterases I and II are found in all tissues, especially liver and kidney [H921].

Mouse liver aspirin esterase has an optimum at pH 7.4 [C779]. Two isozymes are found in liver and kidney [K148].

Beef plasma phenyl acetate hydrolase (E.C. 3.1.1.2; molecular weight 440 000) activity is markedly reduced by solvent extraction; it may be a lipoprotein [A2388].

Cat liver aspirin esterase has a broad optimum at pH 8 [C779]. Serum hydrolyzes 1-naphthyl acetate, *p*-nitrophenyl acetate, propionate and butyrate. Some of the isozymes do not hydrolyze 1-naphthyl acetate. The major isozyme has an optimum at pH 7.8 [A1598].

Dog liver and kidney enzyme has been separated by electrophoresis into nine bands of activity that hydrolyze 1-naphthyl acetate, and *p*-nitrophenyl acetate, propionate and butyrate. The major bands (optimum pH 7.8) are non-specific carboxylesterases [A868].

Guinea pig liver microsomal carboxylesterase, monomeric molecular weight 55 000 and optimum pH 8, is specific for salicylates with acyl chain length two (aspirin), three, four, five, six, eight and 10. Substrate inhibition is observed with increasing chain length. Thioaspirin and 1-naphthyl acetate are also substrates. It is very sensitive to inhibition by the carboxylesterase inhibitor bis(*p*-nitrophenyl)phosphate [D144]. Cerebral cortex enzyme is composed of two isozymes, optimum pH 7.6, molecular weights 78 000 and 180 000, and pI 5.1 and 5.8 respectively. They are specific in requirement for Mn^{2+}; other divalent cations are inhibitory [K861].

Hamster liver microsomal N,O-acetyltransferase hydrolyzes *p*-nitrophenyl acetate [G744].

Pig pancreas *p*-nitrophenyl acetate hydrolase (a lipase) forms an acetyl lipase as an intermediate [A333]. Liver aspirin esterase, molecular weight 160 000, has an optimum at pH 7.4 [E259].

Rabbit serum arylester hydrolase, molecular weight 150 000–200 000, monomeric molecular weights 40 000–45 000 and 47 000–54 000, which also acts on paraoxon, requires Ca^{2+} [E25]. It also hydrolyzes *p*-nitrophenyl acetate and butyrate [E26]. Rabbit gastric mucosal enzyme is cytosolic, molecular weight 66 000 or 59 000 (depending on method) and optimum pH 8.6 [A3910].

Chicken liver microsomal enzyme, optimum pH 8.6, which is highly specific for α-tocopheryl acetate, requires bile salts [B132].

A general study on birds and mammals detected phenyl acetate-hydrolyzing arylesterase in serum from badger, capybara, cat, goat, mouse, ox, pig, rabbit, rat, rabbit, sheep, bee eater, Canada geese, chicken, cormorant, great tit, guillemot, Japanese quail, mute swan, pigeon, puffin, razorbill, shag, starling, tree sparrow, and in trout. All the mammals (except capybara) also showed paraoxonase activity [K863].

Moniezia expansa and Ascaris lumbricoides aspirin hydrolases, molecular weight 87 000, have optima at pH 7.0. Activity is enhanced by Ca^{2+} and by low molecular weight thiols, but Cd^{2+}, Cu^{2+}, Hg^{2+}, La^{3+}, Zn^{2+}, *p*-chloromercuribenzoate, EDTA, F^{-}, N-ethylmaleimide, and paraoxon are inhibitory. 2-Naphthyl acetate and 4-methylumbelliferyl acetate are hydrolyzed by enzyme, molecular weights 30 000–300 000; this range suggests that it is a mixture [A3040].

Festuca pratensis leaf naphthyl acetate esterase is composed of five isozymes, molecular weight 55 000. During senescence two additional esterases appear, with the same molecular weight. They do not act on a series of aliphatic esters [A2515].

Petroselinum crispum leaf malonyl esterases, monomeric molecular weight 35 000, exhibit pI 3.8, 3.9, 4.0 and 4.05; all are dimers, except the 'pI 3.8' enzyme, which is monomeric. It acts on 1-naphthyl acetate and propionate; it also forms apigenin-7-O-(6-O-malonylglucoside) from apigenin-7-O-glucoside [C721].

Acinetobacter lwoffii hydrolyzes a large range of phenyl esters including salicylates, *p*-nitrophenyl acetate and higher esters, acetoxybenzoates, diacetoxybenzoates and naphthyl acetates [A962].

Aspergillus oryzae enzyme, molecular weight 35 000, hydrolyzes *p*-nitrophenyl acetate, propionate and butyrate, with the reaction rate decreasing with increasing molecular weight [J389].

Saccharomyces cerevisiae enzyme, a dimer, molecular weight 84 000 and optimum pH 8.0 is moderately stable at 70°. Substrates are 1- and 2-naphthyl esters of fatty acids with 2–10 carbon atom chain [E258].

Yeast (species not stated) carboxypeptidase hydrolyzes *p*-nitrophenyl trimethylacetate [A337].

Glutamate dehydrogenase (from an unspecified source) hydrolyzes 3-acetyloestrone and 3-acetyl-2-nitrooestrone, with a proportion of the acetyl group being incorporated into the enzyme, inactivating it. It is inhibited by ADP, GTP, NADH and especially oestrone, suggesting that hydrolysis takes place at the oestrogen-binding site [A2004].

Megachile rotundata (solitary bee) [F887], bee [G551] and Chromobacterium [A1808] hydrolyse *p*-nitrophenyl acetate.

Arylesterase (carboxylesterase type; E.C. 3.1.1.1)

$$ArCOOR + H_2O \rightarrow ArCOOH + ROH$$

Rabbit serum cocaine esterase (E.C. 3.1.1.1) has an optimum pH of 8.9 [A2507].

Raphanus sativus (radish) sinapine esterase (E.C. 3.1.1.49), which is specific for sinapine, has an optimum at pH 8.5. The products are sinapate and choline. It is not affected by normal esterase inhibitors, such as eserine [B263].

Orsellinate-depside hydrolase (E.C. 3.1.1.40)

Lasallia pustulata (lichen) enzyme, molecular weight 42 000 is composed of four isozymes, stable at 57°. Substrates include lecanoric acid, evernic acid, gyrophoric acid, methyl lecanorate and erythrin [K862].

Aralkyl esters

Rabbit serum atropine esterase, optimum pH 8.9, additionally acts on scopolamine. It is different from the enzyme that hydrolyzes cocaine [A2507].

Streptomyces ferulate esterase hydrolyzes methyl ferulate. There are two components, molecular weight 29 000, optimum pH 5.5, pI 7.9 and 8.5 [G363].

Microorganisms can hydrolyze phenyl-substituted *trans*-2-(alkoxycarbonylethyl)lactams, with the removal of the alkyl group [K110].

Feruloyl esterase

A. niger contains two isozymes, molecular weights 132 000 and 29 000, pI 3.0 and 3.6 respectively, which hydrolyze methyl ferulate and methyl *p*-coumarate. Methyl sinapate is hydrolysed only by the second isozyme and methyl caffeate only by the first isozyme [G915].

Penicillium expansum enzyme, molecular weight 65 000 and optimum pH 5.6, acts on methyl *p*-coumarate, methyl ferulate, and *p*-coumarate and ferulate esters of glycosides [J499]. P. pinophilum enzyme has molecular weight 57 000, pI 4.6 and optimum pH 6.0 [G805].

Neocallimastix enzyme is a dimer, molecular weight 11 000, monomeric molecular weight 5800, pI 4.7 and optimum pH 7.2, which hydrolyses a number of glycoside *p*-coumaroyl esters [G327].

Aspergillus awamori enzyme acts on α-naphthyl esters of acetic, propionic and butyric acids, but activity with the valerate ester is marginal [J831].

Tannase (E.C. 3.1.1.20)

Aspergillus niger enzyme hydrolyzes tannin, chebulinic acid, *m*-digallic acid, methyl gallate and methyl protocatechuate; hydrolysis requires at least two hydroxyls (other than *ortho*) in the acid component of the substrate. It is found only in plants, especially in plant galls rich in tannin, and in bacteria and fungi grown on tannin [K710].

Methylumbelliferyl acetate deacetylase (E.C. 3.1.1.56)

Human erythrocyte enzyme (called esterase D) additionally hydrolyzes the butyrate analogue. It is composed of three electrophoretically separated types determined by two autosomal alleles (one uncommon). It is relatively stable, and was found in all tissues examined, including liver, spleen and fibroblast [K712].

Heroin esterase

Human serum hydrolyzes heroin to 6-acetylmorphine; the only serum esterase that acts on heroin is cholinesterase [B213]. Heroin is hydrolyzed by a plasma arylesterase, optimum pH about 7.5, which also acts on *p*-nitrophenyl acetate and *p*-nitrophenyl laurate, but not on *p*-nitrophenyl phosphate [A798].

Sugar acetates

The Aspergillus oryzae enzyme (Taka diastase) that hydrolyzes *p*-nitrophenyl acetate also hydrolyzes the acetyl groups in polyacetylated oligosaccharide phenolic conjugates [J389].

Chlorogenate hydratase (E.C. 3.1.1.42)

Aspergillus niger enzyme is a tetramer, molecular weight 240 000, stability range pH 3.0–8.5, activation energy 6.0 kcal/mol and a broad optimum at pH 6.5, is composed of several isozymes, pI 4.0–4.5. It is highly specific; there is slight activity on isochlorogenate. It has been crystallized and the amino acid composition determined [K909, K911].

Meperidine esterase

Rat liver microsomal enzyme, optimum pH 8–9.5 has a heat of activation of 4.6×10^6/mol, with the formation of meperidinic acid [C8].

Bis(2-ethylhexyl)phthalate esterase (E.C. 3.1.1.60) and related reactions

Many studies have demonstrated in animal species, although not at an enzyme level, that phthalate diesters are hydrolyzed stepwise. These include di-(*n*-butyl) phthalate, di-(cyclohexyl) phthalate, di-(9-decenyl) phthalate, di-(2-ethylhexyl) phthalate, di-(5-hexenyl) phthalate and dimethyl phthalate. These reactions have been observed in man [D360], rat, baboon, ferret [A1652, C869, G876], monkey [B579], beef [J718], rainbow trout [A1990], catfish [A2239], minnow, [A2798], earthworm [H9] and microorganisms [K248].

Rainbow trout dealkylates di-(2-ethylhexyl) phthalate. The reaction is inhibited by some methylenedioxy compounds at mM concentrations [A3710].

Wheat contains 12 esterases, only one of which, molecular weight 38 000 (subunit 22 000) acts on the above substrate, to remove one alkyl group. A far better substrate is *p*-nitrophenyl octanoate [D259].

Permethrinase

Bacillus cereus enzyme, molecular weight 61 000 and optimum pH 7.5 hydrolyzes permethrin and a series of α-cyano pyrethroids, fenvalerate, fluvalinate, fastac, deltamethrin, cyfluthrin, flucythrinate, tralocythrin, fenpyrithrin, tralomethrin, cyhalothrin and flumethrin, with release of *m*-phenoxybenzoate or 4-fluoro-3-phenoxybenzoate [G891].

Thiophenyl ester hydrolysis

Rabbit serum arylester hydrolase hydrolyzes phenylthioacetate, propionate and butyrate [E26]. Guinea pig microsomal carboxylesterase also catalyzes this reaction [D144].

CoA thioesterase

p-Hydroxybenzoyl CoA thioesterase (E.C. 3.1.2.23) is found in Corynebacterium sepedonicum [J178] and Pseudomonas [H645]. The latter is a homotetramer, monomeric molecular weight 16 000. It also hydrolyzes benzoyl CoA [K191]. This is the third step in the reaction sequence that converts *p*-chlorobenzoate into *p*-hydroxybenzoate [K798].

Arylsulphatases (E.C. 3.1.6.1)

$$Ar.O.SO_3^- + H_2O \rightarrow Ar.OH + SO_4^{2-} + H^+$$

Beef liver sulphatase A is a glycoprotein, molecular weight 107 000 containing eight galactose, 14 mannose, 18 glucosamine and eight sialic acid residues. Neuraminidase removes the sialic acid residues, presumably terminal groups, with little effect on activity [A148]. Brain sulphatase B, a globular protein, molecular weight 60 000, has been separated into 7 fractions by ion-exchange chromatography [A921].

Beef retinal arylsulphatase A, molecular weight 100 000, exhibits optima at pH 4.6 and 5.7, and arylsulphatase B, molecular weight 40 000 has an optimum at pH 5.2 [B247].

Human liver arylsulphatase A is a glycoprotein, molecular weight 104 500 at pH 8.1, which is composed of two different subunits. At pH 5.5 a tetramer is formed [A2801]. Another study found a molecular weight of 134 000, and pI 4.7 [A2918] or 4.3 [A2457]. The heterodimer is composed of monomers, molecular weights variously assessed as 69 000 and 57 000, 66 000 and 53 000, or 59 000 and 49 000, but apparently they are not found in stoichiometric amounts. It is possible that the larger monomer is convertible into the smaller monomer [A2801, B221].

Human placenta arylsulphatase B is composed of three isozymes, optimum pH 5.8, and molecular weights 48 000, 60 000 and 71 500 [A595]. Brain enzyme, solubilized with lysolecithin, has an optimum at pH 6.8, and molecular weight 103 000–105 000, with subunits, molecular weights 25 000 and 47 000. It contains two mol of sialic acid, which is not essential for activity; 4-methylumbelliferyl sulphate is substrate [A942]. Liver arylsulphatase B, molecular weight 50 000, optimum pH 6.1 and pI 7.5, is inactivated by freezing. It hydrolyzes 4-nitrocatechol sulphate and *p*-nitrophenyl sulphate, and is only slightly inhibited by divalent cations [A2457].

Human placental arylsulphatase C is a homotrimer, molecular weight 238 000. It acts on 4-nitrocatechol sulphate and oestrone sulphate [E398].

Human arylsulphatase hydrolyzes both D- and L-tyrosine-4-sulphate at 5 per cent of the rate for nitrocatechol sulphate, with optimum pH 5.3–5.5 [A3839]. Enzyme found in amniotic fluid cells has an optimum pH of 5.0 [A551]. Parotid and submandibular saliva, tears and sweat contain arylsulphatase A, as does rat saliva [A2839].

Rabbit testis sulphatase A is a dimer at pH 7.1, molecular weight 110 000, and a tetramer at pH 5.0. The optimum pH is time dependent; it is 5.2 after five minutes incubation and 4.9 after 90 min. It acts on 4-nitrocatechol sulphate, but not on *p*-nitrophenyl sulphate [A2945]. Sperm enzyme may contain two components, the main one being composed of A and B, but not C, with optima at pH 4.8, 5.6 and 6.0. It acts on 4-nitrocatechol sulphate and *p*-nitrophenyl sulphate [A1145]. Liver sulphatase A, monomeric molecular weight 70 000, is a dimer at pH 7.4 and a tetramer at pH 4.8 [A1585].

Rabbit liver arylsulphatase A is a dimeric glycoprotein, molecular weight 140 000. The amino acid composition has been determined; it also contains 4.6 per cent carbohydrate (mannose, N-acetylglucosamine and sialic acid). 4-Nitrocatechol sulphate is a good substrate, with lower activity towards other substrates [B481].

Rabbit brain enzyme is found in all the subcellular fractions tested. Its activity remains nearly constant at 1/3 of the adult level from 12–30 days post partum [A1221].

Rat liver arylsulphatase IV (tyrosine ester type) has a molecular weight of 30 000 [H483]. Liver arylsulphatase A has an optimum pH of 4.9–5.9, depending on the assay conditions, molecular weight 130 000 at pH 7.5 and 400 000 at pH 5.0. Arylsulphatase B, molecular weight 34 000–66 000 and optimum pH 5.9, appears to be quite different from beef liver arylsulphatase B [A966]. Two rat isozymes, pI 4.0 (minor) and pI 6.4 are active towards 4-nitrocatechol sulphate; they are not heparin sulphamidases [A575]. Arylsulphatase A is found mainly in parenchymal cells, and arylsulphatase B in non-parenchymal cells. Both activities are also found in adrenal, brain, testis, spleen and kidney [A969]. Rat liver enzyme hydrolyzes both protocatechuate mono-O-sulphates [B724].

Rat mast cell enzyme has a molecular weight of 150 000 and optimum pH 5.0 [A2815].

Rat brain enzyme, which acts on 4-methylumbelliferyl sulphate, is found in neurones and glial cells [A512]. It has an optimum at pH 6.8 for 4-methylumbelliferyl sulphate, and pH 7.2 for *p*-nitrophenyl sulphate [A1050].

Sheep brain arylsulphatase A is a glycoprotein containing 25 per cent mannose and glucose, with 0.5 per cent sialic acid [A2374]. Arylsulphatase B is also a glycoprotein [A2068].

Chicken embryo brain activity increases from day 14 to day 18, and then remains steady until hatching [A836].

Helix pomatia enzyme acts on a range of polysulphated flavonoids. Sulphates at positions 4′ and 7 are far more susceptible to hydrolysis than 3-sulphates, which enables the 3-sulphates to be prepared by differential hydrolysis. Quercetin-3,4′,7-tri-O-sulphate yields quercetin-3,4′-di-O-sulphate and quercetin-3,7-di-O-sulphate as intermediates [F102].

Cystoseira tamariscifolia (seaweed) enzyme, molecular weight 166 000 and optimum pH 6.1, hydrolyzes the sulphates of *p*-nitrophenol,

4-nitrocatechol and phloroglucinol. Activity is also found in Fucus, Carpophyllum, Himanthalia, Landsburgia and Sargassum species [G819].

Aspergillus oryzae arylsulphatase II is a homodimer, molecular weight about 95 000 and optimum pH 5.5 [A146]. A. sojae alkaline arylsulphatase III has an optimum at pH 8.5. It is stable at 20° between pH 8 and 10, acts on *p*-nitrophenyl sulphate, and is inhibited by Ag^{+}, Hg^{2+}, Zn^{2+}, borate and fluoride. A. awamori enzyme shows similar properties [B317].

Cyanobacterium plectonema enzyme, which hydrolyzes nitrocatechol sulphate is cytosolic, with optimum pH about 10. It does not require cations for activity [E179].

Comamonas terrigena tyrosine sulphate sulphohydrolase, which exists as two isozymes, is separated electrophoretically from arylsulphatase. One is inducible by tyrosine, and the other is probably constitutive [B159].

Klebsiella enzymes have molecular weights 47 000 and 45 000, with optimum pH 7.5 [A2776].

Streptomyces griseorubiginosus enzyme, molecular weight 45 000, pI 4.95 and optimum pH 8.5, requires Ca^{2+} for the hydrolysis of etoposide-4′-sulphate and *p*-nitrophenyl sulphate [H603].

Glutamate dehydrogenase (E.C. 1.4.1.3 and 1.4.1.4)

This enzyme hydrolyses 3-acetoxy-2-nitrooestrone and 3-acetoxyoestrone, the acetyl group being incorporated into the enzyme molecule, with inactivation. It is inhibited by ADP, GTP, NADH and, especially, oestradiol [A2004].

Nitrate ester hydrolysis

Dog and rat hydrolyze side-chain nitrate esters of a potential calcium-channel blocker [G593].

6.2 *Hydrolysis of glycosides*

β-N-Acetylgalactosaminidase

Human amniotic fluid cell enzyme, which acts on 4-methylumbelliferyl-β-N-acetylgalactosaminide has an optimum pH of 3.6 [A551].

Rat brain enzyme, which hydrolyzes the same substrate, is found in neurones and glial cells [A512].

Rabbit brain enzyme activity increases about threefold from six days gestation to birth, and is then near the adult level [A1221].

Sheep liver enzyme, optimum pH between 5 and 7, is unstable above 50° [B500].

Physarium polycephalum contains two isozymes that act on both 4-methylumbelliferyl-β-N-acetylgalactosaminide and 4-methylumbelliferyl-β-N-acetylglucosaminide [A23].

β-D-N-Acetylglucosamidase

Beef liver enzyme hydrolyzes phenolic substrates substituted with nitro, chloro, hydroxyl, methyl and methoxy groups on the aromatic nucleus [A1940]. It also hydrolyzes 17α-oestradiol-17-β-D-N-acetylglucosamide [A3192].

Human amniotic fluid cell enzyme, optimum pH 4.6, acts on 4-methylumbelliferyl-β-D-N-acetylglucosamide [A551]. Epidermis enzyme, optimum pH 3.7–3.9, hydrolyzes 4-methylumbelliferyl-β-D-N-acetylglucosamide and *p*-nitrophenyl-β-D-N-acetylglucosamide, [A642]. Activity is found in leucocytes [A789] and in urine [A1845].

Rabbit brain enzyme activity rises about threefold from six days gestation to birth, to near the adult level [A1221].

Rat brain enzyme is found in neurones and glial cells, and hydrolyzes 4-methylumbelliferyl-β-D-N-acetylglucosamide [A512].

Sheep liver enzyme, optimum pH between 5 and 7 is unstable above 50° [B500].

Physarium polycephalum contains two isozymes that hydrolyze both 4-methylumbelliferyl-β-N-acetylgalactosaminide and 4-methylumbelliferyl-β-N-acetylglucosaminide [A23].

α-L-Arabinosidase

Penicillium wortmannii extracellular xylosidase (see below) hydrolyzes *p*-nitrophenyl-α-L-arabinopyanoside [A3240].

α-L-Fucosidase

Human epidermis enzyme acts on *p*-nitrophenyl-α-L-fucoside [A642]. The enzyme is also found in urine [A1845].

Rabbit brain enzyme activity increases marginally during gestation, to about the adult level [A1221].

β-D-Fucosidase

Human epidermis enzyme hydrolyzes *p*-nitrophenyl-β-D-fucoside [A642].

Secale cereale β-glucosidase hydrolyzes the same compound [K512].

α-D-Galactosidase (c.f. E.C. 3.2.1.22)

Human amniotic fluid cell enzyme, optimum pH 4.2, hydrolyzes 4-methylumbelliferyl-α-D-galactoside [A551]. Epidermis enzyme, optimum pH 3.7, hydrolyzes 4-methylumbelliferyl-α-D-galactoside and *p*-nitrophenyl-α-D-galactoside [A642]. Activity is also found in leucocytes [A789].

Rat brain enzyme is found in neurones and glial cells, and hydrolyzes 4-methylumbelliferyl-α-D-galactoside [A512].

Castanea sativa (sweet chestnut) seed contains two isozymes, molecular weights 215 000 and 53 000, pH optima 4.5 and 6, respectively, which hydrolyze *p*-nitrophenyl-α-D-galactoside much more effectively than oligosaccharides. The smaller isozyme is the main one in fresh seeds, but the larger one is the major one in 14 week old dry seeds [B759].

Activity in pea embryo increases during germination, but little change is found in cotyledon. Pisum elatus enzyme has a broad optimum pH at 3–5.5, and P. sativum at pH 3–7 [A3137].

Poteriochromonas malhamensis enzyme, molecular weight 360 000 and optimum pH 7, has maximal stability at pH 8. Substrates are *p*-nitrophenyl-α-D-galactoside and 4-methylumbelliferyl-α-D-galactoside [B486].

Secale cereale β-glucosidase hydrolyzes *p*-nitrophenyl-β-galactoside [K512].

Bacillus stearothermophilus contains two isozymes, molecular weights 280 000 and 325 000. Both appear to be homotetramers, and hydrolyze *p*-nitrophenyl-α-D-galactoside [B544].

2- And 6-deoxy-*p*-nitrophenyl-α-galactoside, but not 3- or 4-deoxy-*p*-nitrophenyl-α-galactoside, are hydrolyzed by enzyme from Coffea, Mortierella vinacea and A. niger [K437].

β-D-Galactosidase (c.f. E.C. 3.2.1.23)

Human amniotic fluid cell enzyme, which hydrolyzes 4-methylumbelliferyl-β-D-galactoside has an optimum pH of 3.6 [A551]. Epidermis enzyme hydrolyzes 4-methylumbelliferyl-β-D-galactoside and *p*-nitrophenyl-β-D-galactoside, optimum pH 4.5 [A642]. Activity is found in leucocytes [A789].

Rabbit brain enzyme activity remains fairly constant prior to birth, at about twice the adult level [A1221].

Rat brain enzyme, with 4-methylumbelliferyl-β-D-galactoside as substrate, is found in neurones and glial cells [A512].

Queen scallop contains an isozyme, molecular weight 148 000 and optimum pH near 6, and a second isozyme, which is active at pH 3 [A376].

Activity in pea embryo increases during germination, but little change in activity is found in cotyledon. Pisum elatus and P. sativum enzymes have an optimum pH 4–4.5 [A3137].

Petunia hybrida enzyme, optimum pH 4.3, is composed of five isozymes, pI 5.1, 5.65, 5.9, 6.1 and 6.5. It hydrolyzes 5-bromo-4-chloro-3-indolyl-β-D-galactoside [B495].

Sugar cane enzyme, optimum pH 4.25, which is found primarily in cell walls hydrolyses phenyl- and *o*-nitrophenyl-β-D-galactosides [B488].

Macrophomina phaseoli enzyme is a glycoprotein, containing about 12 per cent of carbohydrate. Its optimum pH is 5.0, and is stable up to 55° and between pH 4 and 8. It hydrolyses *o*- and *p*-nitrophenyl-β-D-galactosides, but not N-acetyl-β-D-glucosaminides, α-D-galactosides, α-fucosides or β-xylosides. It is inhibited slightly by Hg^{2+}, but not by other heavy metals [A1758].

Sclerotium tuliparum contains two isozymes, with optimum pH 2.0. It is stable up to 50° and between pH 3 and 6. It acts on *o*-nitrophenyl-β-D-galactoside but only poorly on *p*-nitrophenyl-β-D-galactoside. It is inactivated by N-bromosuccinimide but not by heavy metals [A1759].

α-D-Glucosidase (E.C. 3.2.1.20)

Human amniotic fluid cell enzyme hydrolyzes 4-methylumbelliferyl-α-D-glucoside [A551]. Epidermis enzyme optimum pH 3.7, hydrolyzes 4-methylumbelliferyl-α-D-glucoside and *p*-nitrophenyl-α-D-glucoside [A642].

Rat brain enzyme is found in neurones and glial cells, with 4-methylumbelliferyl-α-D-glucoside as substrate [A512].

Buckwheat enzyme hydrolyzes phenyl-α-maltoside to phenyl-α-glucoside [A3873].

β-D-Glucosidase (c.f. E.C. 3.2.1.21)

Beef liver enzymes hydrolyze 3-glucosides of 17α- and 17β-oestradiol (95 per cent cytosolic) and 17α-oestradiol-17-β-D-glucoside (95 per cent microsomal); the 3-glucosidase is much more active than the 17-glucosidase [A3192].

Human epidermis enzyme, optimum pH 4.8, hydrolyzes 4-methylumbelliferyl-β-D-glucoside and *p*-nitrophenyl-β-D-glucoside [A642].

Pig liver enzyme, molecular weight 55 000, acts on 7-glucosides of several flavonoids, but some other flavonoid glucosides are not substrates, nor are 7-rhamnoglucosides [K309].

Rat brain enzyme, which hydrolyzes 4-methylumbelliferyl-β-D-glucoside, is found in neurones and glial cells [A512].

Pinus banksiana lignifying xylem contains two glycoprotein isozymes, molecular weights 110 000 and 90 000, both pI 3.8, which hydrolyze *E*-coniferin to coniferyl alcohol [H308].

Secale cereale (rye) enzyme, pI 4.9–5.1, molecular weight about 300 000 and monomeric molecular weight 60 000, hydrolyzes glucosides as well as fucosides, galactosides and xylosides [K512].

Triglochin maritima contains two isozymes, both of which hydrolyze β-D-glucosides of *o*- and *p*-nitrophenol, 4-methylumbelliferone and salicin, as well as *p*-nitrophenyl-β-D-galactoside. One, molecular weight 125 000 and optimum pH 5.2, hydrolyzes the aliphatic natural product triglochinin, and the other, molecular weight about 250 000 and optimum pH 5.0 in addition hydrolyzes taxiphyllin to *p*-hydroxymandelonitrile [A3756].

Amygdalin-β-glucosidase (E.C. 3.2.1.117)

Prunus serotina (cherry) enzyme, optimum pH 5.5 forms prunasin, but prunasin is not a substrate. It acts on *o*- and *p*-nitrophenyl-β-glucosides, but not α-glucosides [K910].

Coniferin-β-glucosidase (E.C. 3.2.1.126)

Cicer arietinum enzyme, optimum pH 5, isozymes pI 8.5–10, which all appear to be heterodimers, molecular weight 110 000 and monomeric molecular weights 63 000 and 43 000. Other β-glucosides (poorer substrates) and β-galactosides are also hydrolyzed [K832].

Spruce (Picea abies) enzyme, optimum pH 4.5–5.5, is composed of several isozymes, one of which is monomeric, molecular weight 58 000, found in cell wall particulates from root and seed.

Several other glycosides as well as some galactosides are also substrates [K833].

Prunasin-β-glucosidase (E.C. 3.2.1.118)

Prunus serotina (cherry) enzyme, optimum pH 6.5 hydrolyses prunasin. It acts on *o*- and *p*-nitrophenyl-β-glucosides, but not α-glucosides [K910].

Strictosidine-β-glucosidase (E.C. 3.2.1.105)

Catharanthus roseus cell culture enzyme, optimum pH 6.2, forms strictosidine aglycone. It is also found in C. pusilus, C. trichophyllus and Amsonia salicifolia [K868].

Vicianin-β-glucosidase (E.C. 3.2.1.119)

Davallia trichomanoides enzyme, optimum pH 6.0, hydrolyzes vicianin to mandelonitrile and vicianoside. It also acts slowly on (*R*)-amygdalin and (*R*)-prunasin to release mandelonitrile [K910].

β-D-Glucuronidase (E.C. 3.2.1.31)

Beef liver enzyme has an optimum pH of 4–5, according to one study. It hydrolyzes phenolic substrates, benzyl-β-D-glucuronide and acyl glucuronides of benzoate, veratrate and indole-3-acetate [A1029]. In another study, the optimum pH was found to be 4.4 and molecular weight 290 000, and composed of two isozymes, pI 5.1 and 5.9. It is a glycoprotein, containing mannose, galactose, glucose, glucosamine and sialic acid [A1583]. It also hydrolyzes 17α-oestradiol-3-β-D-glucuronide [A3192].

Human amniotic fluid cell enzyme, optimum pH 3.6, hydrolyzes 4-methylumbelliferyl-β-D-glucuronide [A551]. Epidermis enzyme hydrolyzes 4-methylumbelliferyl-β-D-glucuronide and *p*-nitrophenyl-β-D-glucuronide, optimum pH 4.5 [A642]. Activity is also found in leucocytes [A789] and platelets [A810].

Rabbit brain enzyme activity declines by about 50 per cent to the adult level between 12 days gestation and birth [A1221].

Rat brain enzyme is found in neurones and glial cells; 4-methylumbelliferyl-β-D-glucuronide is substrate [A512]. Five liver isozymes are heterotetramers composed of 3 different subunits, molecular weights 58 700, 60 200 and 62 900, which appear to be glycoproteins. The isozymes differ in inactivation rates in 3M guanidinium solution. 4M urea inhibits reversibly, and sodium dodecyl sulphate inactivates irreversibly [A1629].

Juglans regia enzyme, molecular weight 64 000 and pI 8.9, hydrolyzes hydrojuglone-β-D-glucuronide [J695].

Scutellaria baicalensis enzyme (baicalinase), molecular weight 55 000, pI 5.4 and optimum pH 4.7, hydrolyzes wogonin-β-D-glucuronide, and baicalin to baicalein [H747].

Alcaligenes isozymes I and II, molecular weights 75 000 and 300 000 and optimum pH 7.5 and 6.0, respectively, are both inhibited by saccharo-1,4-lactone, the classical β-glucuronidase inhibitor. Besides phenolic glucuronides, oestrone, oestriol and oestradiol glucuronides, with the substituent at the 3-, 16α- and 17β- positions, are hydrolyzed by one or both isozymes [E323].

E. coli enzyme hydrolyses phenolic substrates (optimum pH 5.5–6.5), benzyl-β-D-glucuronide and acyl glucuronides of benzoate, veratrate and indole-3-acetate (optimum pH 5.5) [A1029].

N-Glucuronide hydrolysis

The quaternary glucuronide of croconazole is hydrolyzed by β-glucuronidase [F490].

α-D-Mannosidase (c.f. E.C. 3.2.1.24)

Human amniotic fluid cell enzyme, optimum pH 4.0, hydrolyses 4-methylumbelliferyl-α-D-mannoside [A551]. Epidermis enzyme, optimum pH 4.3, hydrolyzes 4-methylumbelliferyl-α-D-mannoside and *p*-nitrophenyl-α-D-mannoside

[A642]. The enzyme is also found in urine [A1845].

Rabbit brain enzyme activity increases about sevenfold between 12 and 18 days gestation, and then remains constant near the adult level [A1221].

Rat brain enzyme is found in neurones and glial cells; 4-methylumbelliferyl-α-D-mannoside is substrate [A512]. Gastrocnemius muscle activity is found between pH 5 and 7. The activity above pH 6 is mostly destroyed above 55°, and the main residual enzyme is active below pH 6. Some is residual activity found in heated enzyme preparations above pH 6; this is activated by Co^{2+} and Cd^{2+}, but these ions do not activate the (thermally stable) activity observed below pH 6 [A1842].

Dictyostelium (slime mould) enzyme, which hydrolyzes *p*-nitrophenyl- and 4-methylumbelliferyl-α-D-mannosides, has been separated into two isozymes that are active below pH 4.5. Both lose activity rapidly at 55°, but a residual activity is much more thermally stable [A1774].

β-D-Mannosidase (E.C. 3.2.1.25)

Human epidermis enzyme hydrolyzes *p*-nitrophenyl-β-D-mannoside [A642].

Rat brain enzyme, with 4-methylumbelliferyl-β-D-mannoside as substrate, is found in neurones and glial cells [A512].

Quercitrinase (E.C. 3.2.1.66)

Aspergillus flavus enzyme, optimum pH 6.4, acts on quercitrin to release quercetin and L-rhamnose. It is fairly specific, but myricitrin and robinin are also substrates [K866].

Tetra-N-acetylchitotetraosidase

Vibrio alginolyticus chitinase, optimum pH 5.5, hydrolyzes 3,4-dinitrophenyl tetra-N-acetylchitotetraoside to 3,4-dinitrophenol. It is inhibited by α-chytin or colloidal chitin, is stable at 40°, but is inactivated at 60° [A2722].

β-D-Xylosidase

Mouse fibroblasts probably contain this enzyme [C77].

Secale cereale β-glucosidase hydrolyzes *p*-nitrophenyl-β-xyloside [K512].

Penicillium wortmannii extracellular enzyme, molecular weight 100 000, pI 5.0 and optimum pH 3.3–4.0, hydrolyzes *o*- and *p*-nitrophenyl-β-D-xylosides. Its amino acid composition has been determined. It is inactivated by N-bromosuccinimide [A3240].

Bacillus pumilus enzyme is inducible (not stated how) and hydrolyzes *p*-nitro- and *p*-fluorophenyl-β-D-xylosides [A2061].

6.3 *Hydrolysis of amides*

Primary amides

Few studies appear to have been carried out on this reaction at an enzyme level.

Pseudomonas chlororaphis enzyme, a dimer, molecular weight 105 000 and optimum pH 7.0–8.6, hydrolyzes benzamide and phenylalaninamide, but not N-methylbenzamide [H435].

Benzamide is hydrolyzed by mouse and sheep liver [A2037], Rhodococcus rhodochrous [J857], and Nocardia globerula [G343]. Benzamide and several mono- and dichlorobenzamides are hydrolyzed in Aspergillus flavus and A. niger [A640].

Formylase (kynurenine formamidase; E.C. 3.5.1.9)

Mouse enzyme, molecular weight 60 000–70 000, is probably a dimer. Eserine is inhibitory [B251].

Pig enzyme has been separated into three bands of activity [A2893].

Rat liver enzyme, molecular weight about 35 000 and pI 4.9, hydrolyses formyl-L-kynurenine [A2893]. Brain enzyme, optimum pH about 7.6, hydrolyzes formyl-L-kynurenine and 5-hydroxyformyl-L-kynurenine; it is unclear whether 5-hydroxyformyl-D-kynurenine is a substrate [A1284].

Chicken yolk sac membrane contains two isozymes, molecular weights 73 000 (major) and 78 000 (minor); only the latter is inhibited by eserine. The major isozyme disappears from liver after hatching and is replaced by greatly increased levels of the other isozyme [B251].

Drosophila melanogaster contains two isozymes, one with molecular weight 60 000 and a broad optimum, pH 6.7–7.8, and the other, molecular weight 31 000 and optimum pH 6.5–8.0. They are not interconvertible [A2269]. In eye the enzyme is mainly cytosolic [A1194].

A study on molecular weights obtained values of 60 000 and 36 500 for Drosophila virilis isozymes. The larger form in beef liver has molecular weight 59 000, with 58 500 in yeast (species not stated), 59 000 in chicken liver and 60 000 in Musca domestica. The smaller form has molecular weight 30 500 in Rana pipiens liver and 36 000 in mouse liver [A2269].

Microorganisms contain two classes of isozyme, molecular weights 54 000–59 000 and 25 000–30 000. Both are found in Candida lipolytica, C. guilliermondii, Hansenula henricii, Pichia guilliermondii, Rhodotorula rubra, Rhodosporidium toruloides and R. sphaerocarpum. The larger isozyme is found in H. fabiani, C. utilis and Saccharomyces cerevisiae [A3880]. H. henricii formamidase I, molecular weight 56 000 and optimum pH 7–8 may be a dimer of formamidase II, optimum pH 5.5–7.0, with N-formylanthranilate and N-formylkynurenine as substrates [A3776]. In Streptomyces parvulus the molecular weights are 42 000 and 24 000 [B788].

A study on 90 strains of Mycobacterium found activity only in M. fortuitum and other rapidly growing strains [A584].

Formyltetrahydrofolate deformylase;
(E.C. 3.5.1.10)

An E. coli enzyme, which converts this compound into tetrahydrofolate and formate is a hexamer, monomeric molecular weight 32 000. It is activated by methionine [H380].

An enzyme, source unstated, but possibly from microorganisms, expresses two different mechanisms. One is hydrolysis, and the other is a dehydrogenase reaction with NAD^+ as co-substrate. Mutation studies have shown that the full molecular length is required for the dehydrogenase activity, but the hydrolysis reaction requires only part of the molecule [K316].

Acetyl aminoacid hydrolysis

Rat liver mercapturic acid deacylase (which differs from other classes of acylase) is a tetramer, molecular weight about 145 000. It deacetylates *p*-nitrobenzyl-, benzyl- and *p*-bromo-phenylmercapturates (i.e., S-substituted N-acetyl-L-cysteines), as well as N-acetyl-L-tyrosine and N-chloroacetyl-L-tyrosine. Acylase III, molecular weight 55 000, also hydrolyzes all these compounds [C29].

Rat kidney enzyme, which is not acylase I, hydrolyzes N-acetyl-L-phenylalanine, N-acetyl-L-tryptophan and N-acetyl-L-tyrosine [A3520].

Brevibacterium enzyme, which may be a homodimer, molecular weight 50 000 and optimum pH 7.5, hydrolyzes α-acetamidocinnamate (α,β-dehydro-N-acetyl-phenylalanine). It is inhibited by Hg^{2+}, Cd^{2+} and *p*-chloromercuribenzoate. Slight activation is found with dithiothreitol, which appears to be entirely due to protection from inactivation. The product is an iminoacid that can be hydrolyzed to phenylpyruvate or reduced enzymatically to L-phenylalanine. Other substrates include a range of dipeptides [E587].

Hippurate hydrolase (E.C. 3.5.1.32)

Streptococcus enzyme has a molecular weight in excess of 70 000 and optimum pH 7.1–9.0 [A824].

Corynebacterium equi N-benzoyl-L-alanine amidohydrolase (E.C. 3.4.17.6) hydrolyzes hippurate and a number of *para*-substituted hippurates (see above) [E145]. Only some strains of Streptomyces hydrolyze hippurate [A435].

Hydrolysis of L-aminoacid β-naphthylamides

Human lens enzyme hydrolyzes alanyl, arginyl, glycyl, leucyl, lysyl, methionyl and phenylalanyl-β-naphthylamides (βNA), with optimum pH 6–8, depending on substrate. It is activated by Co^{2+} and Mn^{2+}, and (less) by Ca^{2+}. Thiols stimulate, dithiothreitol stabilizes it, and *p*-chloromercuribenzoate inactivates [A1397]. Synovial membrane and fluid contain two isozymes, activated slightly by Co^{2+} and Mn^{2+}, and inhibited by puromycin and Cu^{2+} [A751]. Two muscle isozymes have been shown to hydrolyze some β-naphthylamides; a third (main) hydrolyzes aspartyl, glycyl, lysyl and phenylalanyl-βNA as well as some *p*-nitroanilides. The enzyme is inactivated by EDTA, and activity is restored by several divalent ions [A1219].

Human and pig epidermal enzymes, which are both soluble and membrane-bound, hydrolyse leucyl-βNA. The particulate enzymes, optimum pH 7.0 (man) and 6.6 (pig), are inhibited by puromycin or (partially) by EDTA. They do not require heavy metals or thiols for activity. Pig enzyme is moderately stable at 70°. The soluble enzymes, which are inhibited by puromycin and thiol-binding reagents and activated by dithiothreitol, are inactivated at 60° [A3067].

Rabbit small intestinal enzyme, which hydrolyzes α-L-glutamyl-βNA is found mainly in brush border. It is a different enzyme from L-leucyl-βNA arylamidase [A2347]. Its molecular weight is 225 000, optimum pH about 7 and hydrolyzes alanyl, arginyl, α-glutamyl, leucyl, lysyl, phenylalanyl, alanyl-alanyl, alanyl-leucyl, leucyl-alanyl-βNAs [A1589]. Lung enzyme, which is apparently an aminoendopeptidase, hydrolyzes alanyl, arginyl, α-glutamyl, histidyl, leucyl, lysyl, methionyl, phenylalanyl, seryl and valyl-βNAs, but not prolyl nor isoleucyl-βNAs [B334]. Skeletal muscle enzyme, molecular weight 340 000, is composed of three subunits, molecular weights 51 000, 72 000 and 92 000, and optimum pH 7.5–8, is also an aminoendopeptidase designated Hydrolase H; it requires 40 mM mercaptoethanol for maximal activity. It hydrolyses α-N-benzoylarginyl-βNA and leucyl-βNA, but other α-N-benzoylaminoacid-βNas are not hydrolyzed [B524, B787].

Rat brush border enzyme hydrolyzes alanyl, α-glutamyl, γ-glutamyl, histidyl, isoleucyl, leucyl, lysyl, phenylalanyl, tyrosyl, valyl, alanyl-alanyl and seryl-tyrosyl-βNAs. Except for γ-glutamyl-βNA, hydrolysis is inhibited by puromycin or EDTA [A523].

Intestinal enzyme activity is maximal in jejunum from Pottos monkey and tamarin, and maximal in ileum from tree shrew, slow loris, capuchin, owl monkey and macaque. It is evenly spread through the gut of squirrel monkey, one species of tamarin and baboon. [A977].

Chick skeletal muscle enzyme aminopeptidase H, molecular weight 400 000 with subunit molecular weight 55 000 and optimum pH 8.0 is activated by reducing agents. It hydrolyzes α-N-benzoylarginyl-βNA and leucyl-βNA as well as many small peptides, such as L-tryptophanylglycine and L-phenylalanylglycine [G654].

Penaeus setiferus (shrimp) muscle lysyl-βNA arylamidase, optimum pH 6.5, is activated by Co^{2+} and Mn^{2+}, and inhibited by puromycin [A1286].

Mytilus edulis L-leucyl-βNA arylamidase, optimum pH 7.2, is found in hepatopancreas, mantle, gills, foot, siphuncles and adductor muscle. Phenylalanyl, tyrosyl and tryptophanyl-βNA are hydrolyzed, apparently by the same enzyme. 8M urea, *p*-chloromercuribenzoate, iodoacetamide and mercaptoethanol are inhibitory [A1514].

Nemertean worm enzymes have optimum pH in the range of 6.9–10.5, depending on species, and are activated by Ca^{2+} [A1622].

Agave americana enzyme, optimum pH 7–8, hydrolyzes alanyl, glycyl, isoleucyl, lysyl, phenylalanyl, prolyl, seryl, tryptophanyl, tyrosyl and valyl-βNA. It is inhibited by low concentrations of ascorbate, and by diethyl pyrocarbonate [A2745].

Euonymus enzyme, molecular weight 62 000 and optimum pH 7.6, hydrolyzes leucyl, phenylalanyl and tyrosyl-βNA [D216].

Glycine max root enzyme, molecular weight 63 000 and optimum pH 8, acts on N^{α}- alanyl, arginyl, benzoylarginyl, histidyl, isoleucyl, lysyl, methionyl, ornithyl, seryl, tyrosyl and valyl-βNA (in decreasing order of activity) as well as on leucyl-phenylalanyl and leucyl-tyrosyl-βNA [A1596].

Pisum sativum germinating seeds contain two isozymes that hydrolyze leucyl-βNA, only one of which is inhibited by *o*-phenanthroline. The activity of both increases to a plateau after 24 h germination, and then from six days it decays to zero at 20–24 days [A3213].

E. coli enzyme, molecular weight 80 000, pI 5.7 and optimum pH 7.5, hydrolyzes leucyl-βNA [E586].

Flavobacterium meningosepticum enzyme, molecular weight 500 000 is composed of subunits, molecular weight 62 000. The optimum pH is 7.5 for leucyl-βNA and 6.2 for cysteine di-βNA [C73].

Streptococcus durans β-naphthylamidase hydrolyzes leucyl, alanyl, tyrosyl and aspartyl-βNA. Two forms are found, an arylamidase, molecular weight 80 000 and an aminopeptidase, molecular weight 300 000 [C277].

Hydrolysis of L-aminoacid *p*-nitroanilides

Guinea pig kidney enzyme, molecular weight 120 000 and optimum pH 8.0, which hydrolyses succinyl-trialanyl-*p*-nitroanilide(pNA) but only acts poorly on other analogues, requires Ca^{2+} for activity. Liver enzyme is composed of 2 isozymes, optimum pH 7.0 and 8.0 [C178, D230].

A major human muscle enzyme hydrolyzes lysyl, leucyl and glutamate-pNA as well as some amino acid β-naphthylamides. It is inactivated by EDTA, and activity is restored by several divalent ions [A1219]. The optimum pH is 7.3 [A2450].

Human uterus enzyme, molecular weight 71 000 and optimum pH 7.4–7.5, hydrolyzes succinyl-trialanyl-pNA, but not a series of other peptides; it is claimed to be different from other enzymes of this type [C391]. A kidney enzyme of this type, pI 4.9 and optimum pH 8.0, is activated by Ca^{2+}, and inhibited by EDTA or *o*-phenanthroline [B946]. The analogous serum enzyme, molecular weight 200 000–10 000 000, is apparently a series of polymers of the 200 000 monomer. It is inhibited by EDTA, and reactivated by Ca^{2+}; a second enzyme is not inhibited by EDTA [C213].

Rat cardiac muscle contains two isozymes, molecular weights 257 000 and 105 000, optimum pH 7.0 and 7–8 respectively, and they hydrolyze leucyl-pNA. They are inhibited by puromycin, *p*-hydroxymercuribenzoate, *o*-phenanthroline and bivalent cations [A3126]. Skeletal enzyme is a mixture of a monomer, molecular weight 122 000, a dimer and a polymer, molecular weight greater than 1 000 000. The dimer pI is 5.2, and the monomer pI a triplet at about 4.9; the optimum pH is 7.3. No cations are required. Substrates are alanyl, leucyl, lysyl, phenylalanyl and tyrosyl-pNA [C173].

Glycine max root enzyme, optimum pH 8 and molecular weight 63 000, acts on N^{α}-benzoylarginyl, glycyl, leucyl and lysyl-pNA [A1596].

Vigna unguiculata enzyme, molecular weight 60 000, is stable at pH 7–9 and 40°. It hydrolyses leucyl and N^{α}-benzoylarginyl-pNA [C171].

Streptococcus durans β-naphthylamidase (see above) also hydrolyzes leucyl, alanyl, glycyl, and lysyl-pNA [C277].

Nocardia globerula aryl acylamidase is a homodimer, molecular weight 126 000 and optimum pH about 9.5–10.5. It hydrolyses alanyl, arginyl, glycyl, leucyl, lysyl, methionyl, phenylalanyl, prolyl and valyl-pNA, and also a range of aryl-substituted acetanilides, amides and esters [G343].

Bacillus sphaericus enzyme, molecular weight 75 000, optimum pH 7.0–8.5, is stable at −15°, but it is rapidly inactivated at 40°. It is inhibited by mM heavy metal ions. It hydrolyses L-alanyl-pNA as well as many other anilides [A1006].

Aminopeptidases (E.C. 3.4.11 group)

Aspergillus oryzae aminopeptidase IV, molecular weight 130 000 and optimum pH 7.0, hydrolyses a number of tyrosine-containing dipeptides, but not L-amino acid β-naphthylamides [A3658].

Brevibacterium acylase (see Acetyl amino acid hydrolysis, below) hydrolyzes a range of phenylalanine, tyrosine and tryptophan-containing dipeptides [E587].

Glutathione conjugate hydrolysis

Many xenobiotics are converted into mercapturates prior to excretion. It is considered that conjugation with glutathione is a prerequisite, and these conjugates are then hydrolyzed to the corresponding S-cysteinyl conjugates. These reactions have received little study at the enzyme level.

A compound or mixture described as (−)-*anti*-benzo[a]pyrene-(7*R*, 8*S*)-diol 9*S*, 10*R* epoxide glutathione conjugate is hydrolyzed by a commercial γ-glutamyltranspeptidase to form the corresponding cysteinylglycine conjugate [K286].

Pteroyl-γ-oligoglutamyl endopeptidase (E.C. 3.4.19.8)

Chicken intestine enzyme, optimum pH 80 000 and pI 4.8, yields pteroylglutamyl-γ-glutamate from the corresponding oligoglutamates with 5 or 7 glutamyl residues, but not with three residues [A1339].

N-Benzoyl-L-alanine amidohydrolase (E.C. 3.4.17.6)

Corynebacterium equi enzyme is a homohexamer, molecular weight 230 000, pI 4.6 and optimum pH 8.0. It hydrolyzes N-benzoyl-L-alanine and several *para*-substituted analogues as well as hippurates and N-benzoyl-L-aminobutyrate [E145].

Indole-3-acetyl-L-aspartate hydrolase

Enterobacter enzyme, molecular weight 45 000 and optimum pH 8–8.5, is specific for this peptide [J254].

Non-aminoacid arylamidases

Small intestine enzyme hydrolyzes 2-acetamidofluorene as well as butyric, benzoic and succinic analogues, and 4-acetamidobiphenyl. Activity is found in rat, man, mouse and guinea pig. Acetanilides are not substrates [A808].

Mouse liver enzyme, which hydrolyzes 2-acetamidofluorene and 2,4-diacetamidotoluene is mainly microsomal, with some cytosolic [A3949].

Rat liver microsomal enzyme is solubilized with phospholipase A, and precipitated without inactivation by a rabbit antibody [A738].

Sheep brain enzyme hydrolyzes *o*-nitroacetanilide. The optimum is somewhere between pH 6.5 and 8, probably about 7.5; the data are imprecise. It is inhibited by 1 mM choline, succinylcholine, benzoylcholine, acetylcholine and (best) butyrylcholine, and by serotonin and 5-hydroxytryptophan [A2177].

Ascaris lumbricoides enzyme, molecular weight 90 000 and optimum pH 7.4, hydrolyzes acetanilides substituted on the aromatic nucleus. The enzyme is cytosolic, with good activity in intestinal epithelium, with less in reproductive organs. Best substrates, hydrolyzed at a similar rate to acetanilide are *meta*-substituted; *ortho*-substituted substrates are least hydrolyzed. It is inhibited by Cu^{2+}, Zn^{2+}, N-ethylmaleimide, *p*-chloromercuribenzoate, and especially by anthelmintic organophosphates [A2036].

Moniezia expansa (tapeworm) cytosolic enzyme, molecular weight 95 000 and optimum pH 7.4, hydrolyzes acetanilide and a series of *o*-, *m*- and *p*-substituted acetanilides. It is inhibited by Cu^{2+}, Zn^{2+}, N-ethylmaleimide, *p*-chloromercuribenzoate, and anthelmintic organophosphates [A1755].

Moniezia expansa, Ascaris lumbricoides, mouse and sheep liver enzymes hydrolyze a range of substituted benzanilides. Reaction rates are similar for all compounds tested, except that an *ortho* hydroxyl group markedly reduces the reaction rate, and the rate is increased 10-fold in *p*-nitrobenzoyl analogues [A2037].

Red rice enzyme, optimum pH about 8.3, hydrolyzes propionanilides substituted with one or two chloro groups on the aromatic nucleus. The enzyme hydrolyzes propionanilides, acetanilides, valeranilides and butyranilides in reducing order, but is not active towards 2-methacrylanilides. It is inactivated at above 40° [A3773].

Taraxacum officinale (dandelion) enzyme, optimum pH 7.4–7.8, hydrolyzes propionanilides substituted with one or two chloro groups on the aromatic nucleus. The enzyme hydrolyzes 3,4-dichloropropionanilide and its acetanilide, valeranilide, butyranilide and 2-methacrylanilide analogues in reducing order. It is inhibited by mM Hg^{2+}, Cu^{2+}, catechol, *p*-chloromercuribenzoate, *p*-quinone, *o*-iodosobenzoate and N-ethylmaleimide, and is inactivated at above 50° [A230].

Bacillus sphaericus enzyme, molecular weight 75 000 and optimum pH 7.0–8.5, is stable at −15°, but it is rapidly inactivated at 40°. It is inhibited by mM heavy metal ions. It hydrolyzes linuron, propanil, karsil, dicryl, solan and prophan, most of which are anilides, as well as L-alanyl-pNA [A1006].

Nocardia globerula aryl acylamidase not only hydrolyzes L-aminoacid *p*-nitroanilides (see above), but also a range of aryl-substituted acetanilides [G343].

Pseudomonas fluorescens enzyme is monomeric, molecular weight 52 500 and optimum pH 8.6; it hydrolyzes acetanilides. It is unstable at 60° [C691].

Rat gut flora hydrolyze a range of acetanilides [A261].

Arylalkyl acylamidase (E.C. 3.5.1.76)

Pseudomonas putida enzyme, optimum pH 8 and molecular weight 150 000, appears to be a homotetramer. It shows strict specificity; it hydrolyzes N-acetyl arylalkylamines such as N-acetyl-2-phenylethylamine, but is inactive towards acetanilides. It is activated by Zn^{2+}, Mg^{2+} and Mn^{2+}; other heavy metals are inhibitory. The amino acid composition has been determined [K790].

Phthalyl amidase (3.5.1.79)

Xanthobacter agilis enzyme, whose activity is enhanced by growth on phthalate, is monomeric, molecular weight 49 900 (49 724 calculated from DNA sequence) and pI 5.5. It acts on a wide range of *o*-phthalylamides as well as numerous other amides. This reaction has been suggested as a synthetic method for completion of hydrolysis of the phthalimido protecting group used in the Gabriel synthesis of primary amines. The phthalimido group is notoriously difficult to remove chemically, whereas hydrolysis to phthalylamide by classical chemical techniques is easier [J253, K793].

β-Lactamase (E.C. 3.5.2.6)

Klebsiella oxytoca enzyme, molecular weight 27 000, pI 5.34 and optimum pH 7.0, acts on benzylpenicillin [G193].

Dihydropyrimidinase (E.C. 3.5.2.2)

Rat and calf liver enzymes, molecular weight 252 000 (subunit 59 000) and 266 000 (subunit 62 000), respectively, hydrolyze 5-phenylhydantoin and α-phenylsuccinimide [A3462].

γ-Glutamyl transpeptidase (E.C. 2.3.2.2)

Rat kidney enzyme acts on γ-glutamyl-*p*-nitroanilide and glycylglycine to form *p*-nitroaniline; the other product is presumably a tripeptide [A3043].

Indomethacin hydrolysis

Pig liver hydrolyzes the benzoyl moiety on the indole nitrogen. The purified enzyme is inactive toward esters [J248].

6.4 *Hydrolysis of phosphates*

Acid phosphatase (E.C. 3.1.3.2)

The enzyme is found in human platelets [A810], leucocytes (phenyl phosphate as substrate) [A789] and in epidermis with optimum pH 5–6, the value depending on substrate (*p*-nitrophenyl and 1-naphthyl phosphates) [A642].

Rabbit brain enzyme activity increases about twofold from 6–14 days gestation, and then remains constant at the adult level [A1221].

Spinach leaf enzyme is a dimer, molecular weight 92 000. It contains Mn and a polysaccharide component, and hydrolyzes *p*-nitrophenyl phosphate [A2623].

Polysphondylium pallidum (slime mould) acid phosphatase I, molecular weight 150 000 and optimum pH 3.5, is stable at 4°. It hydrolyzes *p*-nitrophenyl phosphate and, better, sugar phosphates [A158].

E. coli enzymes, pI 4.9, 5.3 and 6.4, hydrolyze *p*-nitrophenyl phosphate [A1511].

Alkaline phosphatase (E.C. 3.1.3.1)

Enzyme is found in human leucocytes, with phenyl phosphate as substrate [A789]. Kidney enzyme, molecular weight 180 000, pI 4.7 and optimum pH 11.4, is activated by Mg^{2+} and Ni^{2+}, whereas most other divalent cations are slightly inhibitory. Monovalent ions activate at low concentration. It is inhibited by N-bromosuccinimide, *p*-chloromercuribenzoate, iodoacetate, EDTA, L-cysteine and *o*-phenanthroline, and is inactivated at 37° [A1807]. Placental and intestinal enzymes, which hydrolyze phenyl phosphate, is inhibited by inorganic phosphate at 10–50 mM, particularly the placental enzyme. It is activated, with *p*-nitrophenyl phosphate as substrate, by Mg^{2+} and inactivated by Zn^{2+}, whereas biliary, bone, liver and kidney enzymes are activated by both ions [A1943, A2543].

Rat intestinal enzyme is composed of isozymes, molecular weights 64 000, 79 000 and 92 000, but another measurement method gives a value of about 160 000, suggesting that it may be dimeric. It contains about 20 per cent polysaccharide that contains glucose, galactose and mannose residues [A1650]. Rat placental enzyme, a dimer, molecular weight 135 000, hydrolyzes *p*-nitrophenyl phosphate. It is activated by Mg^{2+}, but only in the presence of Zn^{2+}; excess Zn^{2+}, sufficient to block the Mg^{2+}-binding sites inactivates it [A240].

Intestinal enzyme activity is greatest in duodenum or proximal jejunum in tree shrew, capuchin, squirrel monkey, Pottos monkey and tamarin. It is evenly spread through jejunum and proximal ileum in slow loris, macaque and baboon. Peak activity is found in proximal ileum of owl monkey. Activity is partially lost in 10 minutes at 56° [A977].

Egg yolk enzyme which hydrolyzes *p*-nitrophenyl phosphate with optimum pH 9.2–9.6, is composed of three fractions, pI 3.65, 3.95 and 4.25, with maximal stability at pH 5–8.5. It is activated by Mg^{2+}, but less so by other divalent cations [A2063].

E. coli enzymes, pI 4.8 and 5.3, hydrolyse *p*-nitrophenyl phosphate [A1511]. In another study three isozymes were detected, with a broad optimum at pH 8–11. It is inactivated irreversibly at pH 2.3, but is stable at 80° [A251].

Serratia marcescens enzyme has a sharp optimum at pH 9. It is inactivated reversibly at pH 2.3, and is unstable at 80° [A251].

4-Nitrophenylphosphatase (E.C. 3.1.3.41)

Saccharomyces cerevisiae enzyme molecular weight 60 000 and optimum pH 8.0–8.2, is highly specific; it requires Mg^{2+} and is stimulated by some other divalent cations; others antagonise Mg^{2+} activation [K864, K865].

FMN phosphatase

This activity, which produces riboflavine from FMN, has been detected in mouse [A3780].

FAD pyrophosphatase (E.C. 3.6.1.18)

$$FAD + H_2O \rightarrow AMP + FMN$$

This activity has been detected in mouse [A3780].

Protein-tyrosine phosphatase (E.C. 3.1.3.48)

Rat muscle membrane vesicle and liver enzyme acts on phosphorylated IgG heavy chain [K867, K919].

Phosphodiesterase (E.C. 3.1.4.1)

E. coli enzymes, pI 5.0, 5.3, 5.6 and 6.3, hydrolyze bis(*p*-nitrophenyl) phosphate [A1511]. A similar reaction has been observed in rat [A2591].

Paraoxonase (E.C. 3.1.8.1)

$$\text{Aryl dialkyl phosphate} + H_2O \rightarrow \text{dialkyl phosphate} + \text{hydroxyaryl}$$

Human serum arylesterase that acts on paraoxon has a molecular weight of 380 000, with monomeric molecular weight 46 000 [F717, G892].

A general study on birds and mammals detected paraoxonase activity in badger, capybara, cat, goat, mouse, ox, pig, rabbit, rat and sheep, but not in capybara, trout or birds [K863].

Activity has been detected in cow, dog, guinea pig, horse, human, rabbit, rat and sheep serum. A phosphorylphosphatase, molecular weight about 500 000 catalyzes the reaction. A study on a large number of rabbits found that the serum activity is monophasic, whereas a similar study in humans showed a biphasic distribution [A1296, J553]. The rat liver enzyme is microsomal, optimum pH 7.5–7.8, and probably requires Ca^{2+} [D72]. Activity is also found in mouse liver [D804].

Cat serum enzyme appears to be carboxylesterase [A1598].

E. coli enzymes have pI 5.3, 5.6 and 6.2 [A1511].

6.5 *Miscellaneous reactions*

Nitrilase (E.C. 3.5.5.1)

Brassica campestris enzyme, optimum pH 7.5, hydrolyzes indole-3-acetonitrile [A3986].

Agrobacterium tumefaciens enzyme is a homotetramer, molecular weight 102 000, which acts on indole-3-acetonitrile to form indole-3-acetamide. A Rhizobium specoes shows similar activity [H366].

Fusariun solani enzyme, molecular weight 620 000, subunit molecular weight 76 000, pI 4.2 and optimum pH 7.8–9.1, hydrolyzes benzonitrile to benzoate with an activation energy of 48.4 kJ/mol. A range of halogen, nitro, hydroxy and methylbenzonitriles and other nitriles are substrates. The rates are greatest for *para*-substituted benzonitriles and least for *ortho*-substituted compounds, except for tolunitriles where *m*-tolunitrile is most rapidly hydrolysed. The enzyme is inhibited by thiol-binding reagents as well as by some heavy metals [A3364].

Nocardia enzyme, molecular weight 560 000, is composed of monomers, molecular weight 45 000. It forms benzoates and ammonia, apparently without the intermediate formation of amides. Subunits form an active dodecamer with substrate as catalyst; this polymerization is highly dependent on temperature, pH and enzyme concentration, with maximal rate at pH 7.2. The optimum pH for the enzyme is 8.0 or 7–9.5, with activation energy for the reaction of 51.8 kJ/mol above 20°; below 20° it increases to 119 kJ/mol, the increase possibly being caused by enzyme association. Thiol-binding reagents are inhibitory. Substrates are benzonitriles substituted with halogens, nitro, methyl and hydroxyl groups, although *o*-substituted compounds are usually not substrates; the nitrile group must be substituted directly onto the aromatic nucleus. The rates for *m*- and *p*-substituted compounds are usually similar, except for nitro-substituted compounds, where the *p*-substituent eliminates the reaction. With benzonitrile, benzamide is not an intermediate. The reaction shows a lag period. The enzyme is unstable at 0° or on freezing [A1367, A3420, D725].

Rhodococcus rhodochrous amidase hydrolyzes benzonitrile quantitatively, but much more slowly than benzamide [J857].

Cysteine conjugate β-lyase (E.C. 4.4.1.13)

Gut and other microflora, especially Eubacterium limosum act on 2-(S-cysteinyl)-N-isopropylacetamide and S-(2-benzothiazolyl)cysteine to form the corresponding mercapto analogues. An enzyme is found in rat liver and gut, pig liver and kidney, onion root and corn root that acts on S-(2-benzothiazolyl)cysteine, but not on 2-(S-cysteinyl)-N-isopropylacetamide; there appears to be an inverse relationship between species that contain the enzyme and the presence of glutathione-S-transferase [D356]. Although this reaction type may be involved in the degradation of many glutathione conjugates, the reaction is poorly documented.

Glutathione elimination

Rat liver acts on S-(2,4-dichlorophenacyl)glutathione to form 2,4-dichloroacetophenone and 1-(2,4-dichlorophenyl)ethanol [A1752].

7. Transfer reactions

7.1 *Transamination*

D-Phenylglycine aminotransferase

Pseudomonas stutzeri enzyme is a homodimer, molecular weight 92 000, pI 5.0 and optimum pH 9–10. The co-substrate is α-oxoglutarate; the reverse reaction requires L-glutamate. Another substrate is *p*-hydroxy-D-phenylglycine; otherwise the enzyme is specific [J353, K503].

D-*p*-Hydroxyphenylglycine aminotransferase
(E.C. 2.6.1.72)

Pseudomonas putida acts on D-*m*- and *p*-hydroxyphenylglycine to form the corresponding phenylglyoxylates; the reaction appears to be reversible [K716].

L-Tyrosine α-oxoglutarate transaminase
(E.C. 2.6.1.5)

This enzyme is the first in the sequence that leads to the oxidation of tyrosine to CO_2 and water, and was one of the first biochemical systems studied in mammalia. Most key studies were carried out prior to the period covered here. Many enzymes of this class show a broad specificity.

Rat small intestinal enzyme is a dimer, molecular weight 90 000 and optimum pH 7.5–8.0. The pure enzyme acts on tyrosine, phenylalanine, tryptophan and aspartate [A1580]. Another publication claims that the enzyme is a homodimer, molecular weight about 100 000, and each monomer has a pyridoxal phosphate-binding site [B12]. In a further study two major components (TATI and TATII) were found to be immunologically identical. Both are dimers, molecular weights about 90 000; acrylamide electrophoresis detected one component, molecular weight 44 000 in TATI, and two components, molecular weight 44 000 and 46 000 in TATII [A2916]. Another study found the molecular weight for the soluble dimer to be 104 000, and the monomer 52 000, pI 4.3, and with no saccharide content [A2930]. Four isozymes have been separated from rat, one of which is found in all tissues [A816].

Rat brain enzyme acts on tyrosine as well as other amino acids [A894]. It is mainly mitochondrial, molecular weight 63 000, acting on tyrosine, phenylalanine, tryptophan, 5-hydroxytryptophan, kynurenine and 3-hydroxykynurenine, but much less rapidly than for aspartate; it is suggested that it is L-aspartate α-oxoglutarate transaminase (E.C. 2.6.1.1) [A2342]. Another study found that it is associated with the inner mitochondrial membrane. It is a tetramer, molecular weight 100 000 and monomeric molecular weight 25 000, containing two mol of pyridoxal phosphate. Both serine and alanine were found to be N-terminal amino acids, suggesting that it contains two moles of each monomer. Tyrosine and 3-iodotyrosine were the best substrates, which are transaminated by a ping-pong bi-bi mechanism. Homogentisate, fumarate, maleate and noradrenaline are inhibitory [A1383].

Soluble rat liver enzyme, molecular weight 50 000, has been separated into three isozymes, and occasionally a fourth has been found [A3388]. Multiple peaks of activity were found in one study, and these were altered considerably by the preparative procedures, suggesting that the enzyme may be particularly susceptible to artifact formation [A382]. One of at least two rat liver cytosolic isozymes is a small molecule, molecular weight about 23 000, inducible by cortisol or by a high dose of tyrosine and methionine [A462]. It is inhibited by MK486, which also inhibits transamination of L-dopa and 3-methoxytyrosine; both liver and kidney mitochondrial enzymes are unaffected by MK486 [A456].

Rat liver enzyme shows its highest activity at birth [A771]. Activity rises slightly during gestation to the adult level at birth, it then decreases rapidly by 60 per cent, followed by an increase to the adult level at weaning. There may be a further decrease before adulthood [A71]. Liver enzyme shows a diurnal rhythm, with low daytime levels, commencing when the pups eyes open at 11–15 days after birth. During weaning high activity is found throughout the whole cycle, and after weaning a high amplitude diurnal rhythm is established [A527]; the activity increases 300 per cent at night [D196]. Rat liver enzyme is induced by the action of corticosteroids [A462, D196]; it is induced by hydrocortisone or quinolinate during the phase in the diurnal rhythm in which activity is normally rising. Tryptophan also induces, poorly, possibly functioning through the formation and action of quinaldic acid. The diurnal activity is eliminated by cycloheximide [A1060].

Trypanosoma bruceri enzyme requires pyridoxal phosphate for activity; phenylalanine is not a substrate [A1619]. T. cruzi enzyme is dimeric, monomeric molecular weight 45 000, pI 4.6–4.8 and optimum pH 7.0. It has a broad specificity [G894].

Anchusa officinalis enzyme, optimum pH about 9, transaminates tyrosine and phenylalanine, but not tryptophan or L-dopa. Oxaloacetate can replace α-oxoglutarate as co-substrate [E436].

Bushbean (Phaseolus vulgaris) seedling enzyme, molecular weight 128 000 and optimum pH 8.3, transaminates tyrosine, phenylalanine and tryptophan; aspartate appears to be the best substrate. Glyoxylate cannot substitute for α-oxoglutarate as co-substrate, and no requirement for pyridoxal phosphate was found. The Stokes radius is 4.3 nm. Highest activity is found 8–14 days after germination, depending on the organ and light conditions [A566]. This appears to be the same enzyme described under phenylalanine α-oxoglutarate transaminase.

Activity is found in Drosophyllum lusitanicum [A1149].

Bacillus brevis enzyme is dimeric, molecular weight 71 000 and optimum pH 6.5–8.6. It contains one pyridoxal phosphate/monomer. Phenylalanine, tyrosine and tryptophan are substrates, but it is much less effective with aliphatic amino acids [E385]. B. subtilis contains two aromatic acid transaminases, molecular weights 64 000 and 73 000 and optimum pH about 7.2, which act on tyrosine and phenylalanine, and are stable at 55°. Histidinol phosphate transaminase (E.C. 2.6.1.9), molecular weight 54 000 and optimum pH also about 7.2, acts on both these substrates with pyridoxal phosphate as coenzyme, with histidinol phosphate or (poorer) α-oxoglutarate as amino acceptor. It is rapidly inactivated at 55° [A2791, A3546].

Brevibacterium linens enzyme acts on phenylalanine and tryptophan as well as on tyrosine; pyruvate cannot substitute for α-oxoglutarate. Two isozymes are found, optimum pH 9.0; one is L-aryl amino acid aminotransferase, molecular weight 126 000, and the other is L-aspartate aminotransferase, molecular weight 81 000 [D724].

Brevibacterium flavum and Corynebacterium glutamicum both contain two isozymes, molecular weights 155 000 and 260 000, with a pyridoxal phosphate content. They act on phenylpyruvate and *p*-hydroxyphenylpyruvate, with glutamate, tyrosine or phenylalanine as co-substrate, and with leucine also active with the larger enzyme [B266].

E. coli aromatic amino acid transaminase (E.C. 2.6.1.57), molecular weight 90 000 and optimum pH 7.2–7.6 or 8.0, acts on tyrosine and phenylalanine, and at far higher concentrations tryptophan and aliphatic amino acids are also substrates, all with a similar V_{max}, with pyridoxal phosphate as co-enzyme and oxaloacetate (best) or α-oxoglutarate as amino acceptors. The reverse reaction, with glutamate as co-substrate, transaminates phenylpyruvate and *p*-hydroxyphenylpyruvate at 50 μM concentration, and on oxaloacetate and 4-methylpentanoate at 3 mM. Aspartate aminotransferase (E.C. 2.6.1.1), molecular weight 84 000 and optimum pH 6.6–7.8 or 8–9, has a similar specificity, but it is active towards all the substrates at about 1 mM concentration. The amino acid compositions of both enzymes are similar, and both are dimers [A3548, K715].

Thermococcus litoralis enzyme is composed of two isozymes, molecular weights 45 000 and 47 000 that require pyridoxal phosphate (which forms a pyridoxamine intermediate in the transamination) and α-oxoglutarate. Tyrosine, phenylalanine and tryptophan, but not aliphatic amino acids are substrates. The enzymes are inactive at 30°, but active at 95–100° [H116].

Pyrococcus furiosus enzyme, molecular weight 44 000 acts on tyrosine, phenylalanine and tryptophan, with pyridoxal phosphate and α-oxoglutarate as co-substrates. It loses 50 per cent of its activity at 95° in 16 h [H437].

Phenylalanine α-oxoglutarate transaminase
(E.C. 2.6.1.64)

Rat small intestine mucosal enzyme is a homodimer, molecular weight 100 000, pI 8.5 and optimum pH about 8, is induced in vitamin B_6 deficiency. It has no aspartate transaminase activity; aliphatic amino acids are not substrates, and tryptophan, tyrosine and 5-hydroxytryptophan are relatively poor substrates. It is specific for α-oxoglutarate. This activity is not found in other tissues [B379]. Liver enzyme (cytosolic) is separable from phenylalanine pyruvate transaminase (E.C. 2.6.1.58), is more heat labile, and is probably not inducible by glucagon [A478, A2304]. Small intestinal tyrosine α-oxoglutarate transaminase acts on phenylalanine, tryptophan and aspartate [A1580]. Activity has also been found in rat brain [A894].

Bushbean transaminases I and II, molecular weights 100 000 and 110 000 respectively and optimum pH 8.5, can substitute oxaloacetate for α-oxoglutarate as co-substrate. Substrates include phenylalanine, mono- and dichlorophenylalanines, monofluorophenylalanines and other *p*-substituted phenylalanines, but because aspartate is the best substrate these enzymes are classified as L-aspartate transaminase. The co-substrate is α-oxoglutarate; oxaloacetate can substitute for it, but only with a few substrates [E500, E640].

Bacterium enzyme, molecular weight 90 000 and pI 4.4, transaminates only phenylalanine and tyrosine. It is not activated by pyridoxal phosphate, and the enzyme shows no indication of a pyridoxal phosphate content or requirement. Oxaloacetate, but not pyruvate can substitute for α-oxoglutarate [A2005].

Candida guilliermondii enzyme requires pyridoxal phosphate [A2483].

Glutamine-phenylpyruvate transaminase
(E.C. 2.6.1.64)

Rat kidney enzyme appears to be a homodimer, monomeric molecular weight 48 000, and is found both in the cytosol and mitochondria. It contains firmly bound pyridoxal phosphate and is identical with L-phenylalanine-α-keto-γ-methiolbutyrate transaminase (the best substrate) [K718, K719].

L-Dopa transamination (E.C. 2.6.1.49)

Rat liver transaminase, part of which is particulate, converts 3,4-dihydroxyphenylpyruvate into L-dopa, with glutamate, aspartate, tyrosine, phenylalanine or tryptophan (best) as co-substrate [A1393]. Activity on L-dopa is found in brain, with α-oxoglutarate or pyruvate as co-substrate [A894].

Alcaligenes faecalis transaminase converts 3,4-dihydroxyphenylpyruvate into L-dopa, with glutamate, aspartate or phenylalanine as co-substrate, with optimum pH about 10, 9.5 and 7.5 respectively, possibly involving different enzymes. The stability range is pH 7–9.5 [A956].

Enterobacter cloacae enzyme converts 3,4-dihydroxyphenylpyruvate into L-dopa, with glutamate or glycine as co-substrate; other amino acids are poor co-substrates. The optimum pH is 7.5–8.5, depending on co-substrate. Activity is rapidly lost at 45° outside the pH range 7–9 [A1616].

L-Kynurenine (3-hydroxykynurenine) aminotransferases

Human brain activity has been separated from other transaminases, with pyruvate, oxaloacetate or α-oxoglutarate (E.C. 2.6.1.7) as co-substrates [A894]. Two isozymes have been separated. One, optimum pH 9.6 requires pyruvate, and the other, optimum pH 7.4, requires pyruvate or α-oxoglutarate as co-substrates [G141].

Human liver kynurenine: glyoxylate aminotransferase (E.C. 2.6.1.63), molecular weight 90 000–100 000, also acts on phenylalanine and many aliphatics; many compounds are amino acceptors. It is identical with serine pyruvate transaminase (E.C. 2.6.1.51) and alanine transaminase (E.C. 2.6.1.2) [B677].

Rat liver enzymes are found mainly in mitochondria with smaller amounts in nuclei, lysozomes and cytosol, but not in microsomes [A2892]. They can utilize pyruvate (optimum pH 8.0–8.5) and α-oxoglutarate (optimum pH 6.0–6.5) [A2275, A2892]. Another study found the latter activity in liver, brain and small intestine; enzymes in these tissues appear to be identical. Other substrates, tyrosine, phenylalanine, tryptophan and 5-hydroxytryptophan, are appreciably better than kynurenine; pyruvate is not a co-substrate. The molecular weight is 88 000 with optimum pH 8–8.5. This activity may be identical with mitochondrial tyrosine transaminase and aspartate α-oxoglutarate transaminase. A second liver isozyme, molecular weight 100 000 and optimum pH 6–6.5, which is specific for kynurenine and α-oxoglutarate, is not found in brain or small intestine [A2339]. Identity of the mitochondrial activity (which requires pyridoxal phosphate) with α-aminoadipate transaminase (E.C. 2.6.1.39) has also been suggested [C557], and in another study, with 3,5-diiodotyrosine transaminase (E.C. 2.6.1.24). Both kidney and liver enzymes are inactivated at 70°. They are inhibited by straight chain dicarboxylic acids; a study on the relationship between inhibitory potential and chain length of these acids found that acids with three or fewer methylene units were almost inactive, a maximum in potential was found with four methylene units, a minimum for six, and a further maximum for eight, but longer chain compounds were not studied [A3398]. The same researchers conclude that this activity is the same as α-aminoadipate transaminase, based on subcellular distribution and differences between the sexes and inhibition by dicarboxylates. Highly purified enzyme is active towards α-aminoadipate, which is also a competitive inhibitor [A338].

Rat kidney enzymes are both mitochondrial and cytosolic. One, which requires α-oxoglutarate as co-substrate has an optimum at pH 6–6.5, and the other, requiring pyruvate as co-substrate has an optimum at pH 8–8.5 [A2273]. A cytosolic enzyme is considered to be identical with α-aminoadipate aminotransferase. It appears to be a homodimer, molecular weight 85 000 and pI 6.56, and is inactivated irreversibly by L-seryl-O-sulphate [A3094]. The pyruvate-requiring isozyme appears to be identical with the corresponding brain enzyme, which is found in astrocyte-like cells. It may be a homodimer with monomeric molecular weight 48 000. Phenylpyruvate, oxaloacetate and a range of other 2-oxo-fatty acids can also act as co-substrates [F800, G423].

Rat brain enzyme has been separated into two isozymes. The preferred co-substrate for one, optimum pH 9.1, is pyruvate and is inhibited by glutamine. The preferred substrate for the other is 3-hydroxykynurenine, optimum pH about 7, and shows a broad specificity for oxo co-substrates [J470]. Kynurenine/glutamine aminotransferase K is both cytosolic and mitochondrial, and requires pyruvate as co-substrate [H608].

Rat intestinal enzyme appears to be identical with tyrosine α-oxoglutarate transaminase (see above) [A1580]. In another study other substrates were 3- and 5-hydroxykynurenine, optimum pH 8.0–8.5; but α-oxoglutarate was a poor co-substrate [A1005].

A study on rat and monkey liver enzymes (E.C. 2.6.1.63), both molecular weight 80 000, pI 8.0 and optimum pH 8–8.5 are mitochondrial, with glyoxylate as coenzyme; this can be replaced by α-oxoglutarate only in rat. It is induced by glucagon, at least in rat. Aromatic amino acids are substrates including kynurenine, especially phenylalanine [K717].

Aedes aegypti (mosquito) enzyme transaminates 3-hydroxykynurenine and pyruvate [J668].

Activity is found in several species of Saccharomyces, and in Pichia, Debaryomyces and Hansenula, with optimum pH 8 for Hansenula. Co-substrates are glyoxylate, pyruvate or α-oxoglutarate, but oxaloacetate is poor. It is inhibited by 1 mM D-cycloserine and hydroxylamine, partially reversible by pyridoxal phosphate, which is probably the coenzyme [A2337].

Hansenula schneggii enzyme, optimum pH 8.0, is a homodimer, monomeric molecular weight 52 000, and each monomer contains one mol of pyridoxal phosphate. The amino acid composition has been determined. The specificity is broad; other (poorer) substrates include phenylalanine and tyrosine and their nitro-substituted analogues, 3-hydroxykynurenine, dopa, benzoylalanine and many aliphatic amino acids, with α-oxoglutarate, α-oxoadipate, phenylpyruvate, *p*-hydroxyphenylpyruvate or indole-3-pyruvate as co-substrates [D978].

L-Tryptophan transaminases

Rat brain enzymes use pyruvate and α-oxoglutarate as co-substrates [A894]. In midbrain phenylpyruvate is 150 times better than α-oxoglutarate as co-substrate; *p*-hydroxyphenylpyruvate is less effective [A3026]. A brain enzyme found in cytosol and synaptosomes, but not in mitochondria, molecular weight about 55 000, pI 6.2 and optimum pH 8.0, transaminates tryptophan, 5-hydroxytryptophan and tyrosine, but not phenylalanine or histidine. The activity with aspartate is much greater than for these amino acids; it is considered to be identical with L-aspartate: 2-oxoglutarate aminotransferase (E.C. 2.6.1.1). Oxaloacetate is co-substrate; α-oxoglutarate is less good [A2847].

Rat liver cytosol contains three isozymes, one of which is specific for tryptophan, a second also acts on aspartate and the third on tyrosine [A644].

Enzyme is found in most plant families including Amaranthaceae, Balsaminaceae, Compositae, Cruciferae, Labiatae, Leguminosae, Moraceae, Papaveraceae, Solanaceae and Umbelliferae. It is present mainly in stem tips and leaves, but less in stem and root. It is not found in Fucus spiralis (an algal seaweed) [A753].

Phaseolus aureus enzyme requires α-oxoglutarate as co-substrate [A463].

Zea mays contains two isozymes, molecular weights 45 000 and 80 000 and optimum pH 8–9, which require α-oxoglutarate as co-substrate [G888].

Agrobacterium tumefaciens tryptophan phenylpyruvate aminotransferase (E.C. 2.6.1.28) has an optimum at pH 9.6 [B568].

Klebsiella aerogenes aminotransferase I is probably a dimer, monomeric molecular weight 42 000, pI 6.64 and a broad optimum at pH 8. It requires pyridoxal phosphate, and it is considered to be a flavoprotein. Substrates are tryptophan, phenylalanine and phenylpyruvate, with oxaloacetate and α-oxoglutarate as co-substrate for the forward reaction, and glutamate for the reverse. For tryptophan, oxaloacetate is the better co-substrate. Aminotransferase IV, molecular weight 100 000 and optimum pH 9, is probably aspartate aminotransferase. Substrates and co-substrates are the same as for aminotransferase I. It is stable at 55° [B732].

Pseudomonas L-tryptophan α-oxocaproate aminotransferase (E.C. 2.6.1.27) and molecular weight 90 000–110 000, additionally transaminates tyrosine and 5-hydroxytryptophan. Several long-chain pyruvates as well as phenylpyruvate and indole-3-pyruvate are also co-substrates, but pyruvate and α-oxoglutarate are inactive [B672].

Rhizobacterium enzyme, optimum pH 8.0, requires α-oxoglutarate and pyridoxal phosphate [E982].

Saccharomyces cerevisiae contains two transaminases, one of which acts on tryptophan, phenylalanine and tyrosine with pyruvate, phenylpyruvate or α-oxoglutarate as co-substrate. The other acts on tryptophan, with pyruvate as co-substrate [C533].

Streptomyces griseus enzyme, which is involved in indolmycin biosynthesis, requires α-oxoglutarate and pyridoxal phosphate. Other substrates are β-methyltryptophan, phenylalanine and tyrosine [A2307].

L-Phenylalanine pyruvate aminotransferases

Rat liver enzyme is found in peroxisomes and mitochondria, and probably represents several isozymes [A3416]. It is induced by cAMP [A1248]. Induction by tetraiodoglucagon is blocked by cycloheximide and actinomycin D, demonstrating that induction involves protein synthesis [A981]. Glucagon increases activity fourfold in 24 h or eightfold in eight days, and this is enhanced by theophylline, which is an inhibitor of cAMP hydrolysis [A478]. In addition to glucagon induction it is activated twofold by heating at 40°–70° (it is inactivated at higher temperature) [A2459]. It is claimed to be asparagine-pyruvate aminotransferase (E.C. 2.6.1.14) [E331]. Another study which claims that it is probably identical with histidine-pyruvate aminotransferase, found two isozymes, molecular weight about 74 000. One is a homodimer, pI 7.8, monomeric molecular weight 42 000, and the other is a heterodimer, pI 7.3, subunit molecular weights 42 000 and 44 000. Both are immunologically identical, but with different amino acid composition. Several aliphatic amino acids are better substrates than histidine, but not as good as phenylalanine [A3606]. It is separable from phenylalanine-α-oxoglutarate aminotransferase (see above) [A2304].

It is also found in brain [A369, A894]. Rat brain kynurenine/glutamine aminotransferase K (above) also acts on phenylalanine [H608].

Rat liver activity increases about fivefold just before birth, followed by a decrease of about 30 per cent at five days post partum, and then it slowly doubles to the adult level. A similar pattern is observed in kidney, but with smaller fluctuations. Brain enzyme shows minimal changes in activity, with a small fluctuation at birth. In human adult liver the activity is 10 times greater than in foetal liver [A3020].

Monkey and rat liver enzymes, molecular weight 80 000, pI 8.0 and optimum pH 8–8.5, act on phenylalanine, tyrosine, tryptophan, 5-hydroxytryptophan and kynurenine (in decreasing order of activity in rat) with glyoxylate, pyruvate, hydroxypyruvate, 2-oxo-4-methylthiobutyrate and (in rat) 2-oxoglutarate as co-substrates. It is different from kynurenine transaminase [A3570].

Tyrosine-oxaloacetate aminotransferase

E. coli enzyme contains two isozymes, molecular weights 82 000 and 88 000; the pI, about 4.5, differs by 0.05 unit between isozymes. Growth on tyrosine repressed one isozyme, but not the other [A1224].

Aromatic amino acid: oxaloacetate aminotransferase (E.C. 2.6.1.60)

Monkey and rat liver enzyme, molecular weight 80 000, pI 8.0 and optimum pH 8.0–8.5, is induced by glucagon; it is different from kynurenine transaminase. Co-substrates are glyoxylate, pyruvate, hydroxypyruvate and 2-oxo-4-methylthiobutyrate; α-oxoglutarate is active only in rat [K714, K907].

L-Tyrosine: phenylpyruvate aminotransferase

Rat liver enzyme is not induced by cortisol or by other agents that activate tyrosine-α-oxoglutarate aminotransferase in vivo. Its activity increases by 40 per cent at night [D196].

Diiodotyrosine transaminase (E.C. 2.6.1.24)

Rat kidney mitochondrial enzyme, optimum pH 6.4 requires pyridoxal phosphate and α-oxoglutarate, and is specific for L-acids. Substrates include diiodotyrosine, dibromotyrosine, dichlorotyrosine; monoiodotyrosine (optimum pH 7.2), phenylalanine, tryptophan and *p*-fluorophenylalanine are poor substrates [K925].

Thyroid hormone transaminase (E.C. 2.6.1.26)

Rabbit enzyme, molecular weight 95 000 appears to be a homodimer, and is found mainly in liver and kidney, with smaller amounts in heart, brain, lung and skeletal muscle, but none in serum. It acts on T_3, 3-iodotyrosine, 3,5-diiodotyrosine and 3,5-dinitrotyrosine, with α-oxoglutarate as co-substrate; oxaloacetate and pyruvate are alternative co-substrates, except for 3,5-dinitrotyrosine, which is specific for α-oxoglutarate [K713].

D-Tryptophan aminotransferase

Alaska pea contains a D-tryptophan aminotransferase with pyruvate or α-oxoglutarate as co-substrate. It is specific for the D-isomer [F405].

A D-tryptophan aminotransferase, molecular weight 55 000, which requires pyruvate as co-substrate, is found in Zea mays, optimum pH 8–9 [G888].

5-Hydroxytryptophan aminotransferases

Rat brain contains this activity, with pyruvate and α-oxoglutarate as co-substrates [A894, A2339, A2342, A2847]. Activity has also been found in Hansenula schneggii [D978], monkey and rat liver [A3570] and rat small intestine [B379]. This activity is associated with non-specific aminotransferases.

(*S*)-α-Phenethylamine: pyruvate transaminase

A patent has described the formation of acetophenone from (*S*)-α-phenethylamine, apparently with pyruvate forming optically-active alanine as the second product. The source of the enzyme was not stated in Chemical Abstracts [K449].

Phenylacetone transamination

p-Methyoxyphenylacetone is transaminated to (*S*)-(+)-*p*-methoxyamphetamine, optimum pH about 6. The most effective amino donor is L-alanine. Activity is found in Brevibacterium, Chromobacterium, Flavobacterium, Mycobacterium, Pseudomonas and Sarcina. B. linens enzyme also acts on 3,4-dimethyoxyphenylacetone and 4-(*p*-methoxyphenyl)-2-butanone, again to form (*S*)-amphetamine analogues [F813].

Amination with L-hydroxyisocaproate dehydrogenase

Lactobacillus confusus L-hydroxyisocaproate dehydrogenase forms L-phenylalanine from phenylpyruvate, NADH and NH_4^+ [H954].

7.2 *Isomerization*

Many stereoisomerisation reactions are carried out with the formation of well-defined intermediates, such as oxo compounds for amino acids and mandelic acids. These are described elsewhere.

Phenylpyruvate tautomerase (E.C. 5.3.2.1)

R. CH_2–CO.CO_2H ↔ R.CH=CHOH.CO_2H

Since its discovery the tautomerase has been an enzyme looking for a role, since no physiological or biochemical role has been discovered for this tautomerization. It may be an activity not associated with its primary function.

Human lymphocyte enzyme, molecular weight 12 000, acts on phenylpyruvate and *p*-hydroxyphenylpyruvate. It functions as a macrophage migration inhibitory factor [J475].

The enzyme that converts D-dopachrome into 5,6-dihydroxyindole enolizes indole-3-pyruvate and *p*-hydroxyphenylpyruvate; it is suggested that these two activities are associated with the same protein [J514].

Mouse enzyme (additionally acts as macrophage migration inhibition factor) is inhibited competitively by (*E*)-2-fluoro-*p*-hydroxycinnamate; its activity is profoundly affected by mutation at Pro-1 [K274, K559].

Mandelates from phenylglyoxals

R.CO.CHO → R.CHOH.CO_2H

Administration of phenylglyoxals to rats results in the formation of the corresponding mandelates. Potential intermediates, e.g. phenylglyoxylates were not detected, nor did their administration result in the formation of equivalent quantities of mandelates. This reaction involves an internal oxidation-reduction, apparently in a single step [J287].

Styrene oxide isomerase (E.C. 5.3.99.7)

Xanthobacter forms phenylacetaldehyde from styrene oxide [F440]. The same reaction is found in Corynebacterium [H881] and Pseudomonas putida [H369]. The Corynebacterium enzyme, optimum pH 7, is highly specific and acts preferentially on the (*S*)-isomer. The reaction is irreversible [J517].

Quinone methide isomerase

Sarcophaga bullata larval haemolymph 4-alkyl-*o*-quinone/2-hydroxy-*p*-quinone methide isomerase is a dimer, molecular weight 98 000 and optimum pH 6.0, which acts on quinones of N-acetyldopamine, N-β-alanyldopamine and 3,4-dihydroxyphenylethanol to form quinone methides. These then spontaneously hydrate to form, for instance, N-acetylnoradrenaline. The quinones formed from these products can react further, finally to form, for instance, N-acetyl-β-oxophenethylamine. Dihydrocaffeoyl methylamide quinone forms caffeoyl methylamide and a little 3,4-dihydroxyphenylhydracryloyl methylamide [F792, F793]. The reaction sequence for N-acetyldopamine involves, firstly quinone formation by the action of phenoloxidase, followed by quinone methide formation, which then either hydrates to form N-acetylnoradrenaline or, by the action of a tautomerase forms β-acetamido-3,4-dihydroxystyrene [F386].

This reaction has also been observed in Manduca sexta and Periplaneta americana [F68].

***Cis-trans* isomerisation**

Cis- and *trans*-cinnamates are isomerized in plants in which phenylalanine ammonia lyase is active, by the action of UV radiation in sunlight; this may be important in the formation of cinnamates and their metabolic products, such as flavonoids [H615].

(−)-Epigallocatechin epimerase

Musa accuminata fruit converts this compound into (−)-gallocatechin; the reaction involves *R*/*S*-isomerisation of the phenyl moiety [K423].

Cicloprofen and analogue isomerization

Rat converts (*R*)- ibuprofen to (*S*)-ibuprofen, with quantitative loss of one hydrogen atom at

C-2 without isotope effect, as one step in the sequence. It is postulated that the mechanism involves CoA conjugation, followed by reversible enolization and hydrolysis. All the reaction steps except, perhaps, the final one are considered to be reversible [F733, G304].

Rat, dog, man and monkey convert (−)-cicloprofen into (+)-cicloprofen. The reverse reaction is, at best, slow [A1696, A1767, A2111].

Rat and sheep convert (*R*)-(−)-fenoprofen to (*S*)-(+)-fenoprofen [H597].

7.3 *Migrations*

Formation of (*S*)-tropic acid

Datura innoxa forms tropate from cinnamate. Labelling experiments show that C-2 of the side chain is converted into the hydroxymethyl moiety [A2557]. Starting with L-phenylalanine, isotope studies show that the carboxyl group migrates, with retention of the configuration [E420].

Isoflavone synthase

The reaction is important in the formation of this major class of plant natural products from flavones.

Glycine max isoflavone synthase is microsomal, and requires NADPH and oxygen. The reaction involves the migration of the benzene ring from C-2 to C-3 of the coumarin ring, with the formation of a double bond at position 2–3; a quinonoid intermediate linked to C-2 and C-3 is postulated. Substrates include naringenin and liquiritigenin, which form genistein and daidzein respectively [D468].

Glycyrrhiza echinata 2-hydroxyisoflavanone synthase is a P450 enzyme, with liquiritigenin and genistein as substrates; the reaction appears to be a 'NIH'-type of reaction, involving hydroxylation and migration of the substituent. Usually daidzein and genistein are found as the products; it is unclear whether these are artifacts in all experiments, or whether they are always formed enzymatically, since they can be formed from the 2-hydroxy intermediates by treatment with acid [K308].

Pueraria lobata microsomes convert liquiritigenin into 2,4′,7-trihydroxyisoflavanone, with NADPH as cofactor [F845].

Tyrosine 2,3-aminomutase (E.C. 5.4.3.6)

Bacillus brevis enzyme, optimum pH 8.5, pI 4.6 and molecular weight 75 000, converts L-tyrosine into β-tyrosine, with ATP as co-substrate. It is suggested that there may be a Schiff base intermediate [B316, K824].

A similar reaction may occur with phenylalanine in the formation of the phenylisoserine moiety found in taxol [G738].

Side chain migration without hydroxylation

Human anaerobic faecal organisms convert *p*-hydroxyphenylpropionate into *m*-hydroxyphenylpropionate [A2767, A3990].

7.4 *Racemization*

Amino acid racemase (E.C. 5.1.1.10)

Pseudomonas putida enzyme, molecular weight 62 000–65 000, pI 7.6 and optimum pH about 8, acts on D- and L-phenylalanine as well as on aliphatic amino acids [F110]. The enzyme does not racemize phenylglycine, but it catalyzes α-D labelling in the presence of D_2O, with configuration retention [J883].

Phenylalanine racemase (E.C. 5.1.1.11)

Bacillus brevis enzyme, molecular weight 100 000, requires ATP, its activity is enhanced by dithiothreitol and is reversible. The optimum pH for the reaction on the L-isomer is 8.5, and for the reverse 7.9 [K860].

Mandelate racemase (E.C. 5.1.2.2)

Pseudomonas enzyme acts on the (*R*)-isomer [K310]. It was originally detected in P. fluorescens; both isomers were converted into the DL mixture [K950]. In view of other research it must be considered that this reaction is the consequence of reversible formation of phenylglyoxylic acid.

8. Formation and reactions of non-aryl double and triple bonds

8.1 *Double and triple bond formation*

L-Phenylalanine ammonia lyase (E.C. 4.3.1.5)

L-Phenylalanine ↔ *trans*-cinnamate + NH_3

Aesculus hippocastanum enzyme is composed of two isozymes [A1632].

Bamboo (Bambusa oldhami) shoot enzyme, pI 4.7 and molecular weight 360 000, monomeric molecular weight 86 000, is probably tetrameric. After harvesting, there is a sharp increase in activity that is only partly due to *de novo* synthesis. Two minor fractions have a molecular weight of at least 600 000 [E983].

Brassica juncea enzyme, optimum pH 9.0, is a tetramer, monomeric molecular weight 40 000 [J513].

Cucumis sativus enzyme is inhibited by D-phenylalanine, cinnamate, *o*-coumarate, gallate, quercetin and kaempferol [A173].

In gherkin an inhibitor is present that reversibly binds to the enzyme. The inhibitor is a small molecule associated with a larger molecule (molecular weight greater than 12 000); the association breaks down during storage [A224].

Hordeum vulgare enzyme is found in chloroplasts. In etiolated seedlings four minutes exposure to red light results in increased activity some hours later [A223].

Melilotus alba enzyme, optimum pH 8.7, is found in apical buds and axillary leaves. It is inhibited by cinnamate and *o*-coumarate, but not by *p*-coumarate, coumarin, melilotate or *o*-coumarate glucoside [B369].

Nicotiana tabacum enzyme is inhibited by 0.1–0.25 M D-mannitol, D-sorbitol, D-glucose, D-mannose and D-sorbose [A1925].

Osimum basilicum enzyme is a homotetramer, molecular weight 152 000 [J243].

Parsley enzyme contains a dehydroalanine group that is essential for activity; it is formed after protein synthesis. *m*-Tyrosine, unlike tyrosine, is also a substrate [H647].

Pisum sativum seedling enzyme is composed of two isozymes, one of which is formed in the dark, and the other in the light [A847].

Quercus pedunculata roots contain two isozymes, one of which is microsomal, and the other is associated with mitochondria and microbodies [A622].

Radish enzyme, pI 4.9 and molecular weight 280 000, is found only when grown in light. In the absence of light an inactive protein is found, with identical properties including immunoelectrophoresis, pI and sedimentation velocity [A1083].

Wheat enzyme, molecular weight 325 000 is composed of monomers, molecular weights 75 000 and 85 000. The amino acid composition has been determined [A464].

Activity has been found in barnyard grass (Echinochloa crusgalli), bracken fern (Ptreridium spp.), cocklebur (Xanthium pensylvanicum), corn (Zea mays), cotton (Gossypium hirsutum), crabgrass (Digitaria sanguinalis), green and yellow foxtail (Setaria viridis and S. glauca respectively), johnsongrass (Sorghum halepeuse), morning glory (Ipomoea purpurea), yellow and purple nutsedge (Cyperus esculentus and C. rotundus), pigweed (Amaranthus retroflexus) soya (Glycine max) and wheat (Triticum vulgare) [A189].

Alternaria cultures produce two different isozymes. One, designated 68 (the age of the culture in hours), molecular weight 560 000 and optimum pH 8.2, is inhibited by benzoate, whereas the other, designated 116, molecular weight 380 000 and optimum pH 8.8, is inhibited by cinnamate, *p*-coumarate, benzoate and *p*-hydroxybenzoate [A20].

In Rhizoctonia solani the enzyme only develops in the presence of phenylalanine [A1466].

Rhodotorula glutinis enzyme is a tetramer, monomeric molecular weight 83 000. It is inactivated by nitromethane, apparently by bonding with the dehydroalanine imine prostheic group [A1642].

Ustilago maydis (fungus), molecular weight 320 000 (monomer 80 000) has an optimum at pH 8.8–9.2. Phenylalanine, but not tyrosine is the substrate [J231].

Enzyme activity is modulated by *cis-trans* isomerisation of the product by UV radiation [H615].

L-Tyrosine ammonia-lyase

L-Tyrosine ↔ *p*-coumarate + NH_3

Activity has been found in barnyard grass, bracken fern, cocklebur, corn, cotton, crabgrass, green and yellow foxtail, johnsongrass, yellow and purple nutsedge, pigweed, soya bean and wheat. The ratio relative to phenylalanine ammonia-lyase activity is different for each species. Green foxtail enzyme is not separated chromatographically from phenylalanine ammonia-lyase [A189].

Zea mays enzyme (which also acts on phenylalanine) eliminates the 3-*pro-S* hydrogen [A607].

Cinnamates

Although some studies implying the formation of cinnamates from phenylpropionates have been carried out, the enzymes involved remain obscure.

Phenyllactate dehydratase

Clostridium sporogenes enzyme, which is a trimer, monomeric molecular weights 40 000, 43 000 and 46 000, interconverts (*R*)-phenyllactate and *trans*-cinnamate. The 46 000 component is a cinnamoyl CoA: phenyllactate CoA transferase, and the rest of the complex, which retains dehydratase activity, requires FAD, Mg^{2+}, NADH and cinnamoyl CoA [K534].

Phenylserine dehydratase

Pseudomonas pickettii enzyme forms ammonia and phenylpyruvate from L-*threo*-phenylserine [H569].

Substituted acetaldoxime dehydratase

(indoleacetaldoxine dehydratase, E.C. 4.2.1.29)

e.g. Indole-3-acetaldoxime →
indole-3-acetonitrile

Chinese cabbage, Isatis tinctoria, Helianthus annuus and Zea mays catalyze this reaction with indole-3-acetaldoxime as substrate [D896].

Bacillus acts on (*Z*)-3-phenylpropionaldoxime [J508]. Further studies found that the enzyme, phenylacetaldoxime dehydratase, is monomeric, molecular weight about 40 000 and optimum pH 7.0, which is stable at neutral pH up to 45°. It contains loosely bound protohaem IX and requires FMN. Activity is also enhanced by Fe^{2+}, Sn^{2+}, SO_3^{2-} or azide and by the absence of oxygen. It is inhibited by chelators, carbonyl-binding reagents, electron donors, ferrocyanide and ferricyanide. The gene codes for 351 amino acid residues, corresponding with a molecular weight 40 018. The (*Z*)-isomer is the preferred substrate; other substrates include phenylacetaldoxime, 4-phenylbutyraldoxime, *p*-chlorophenylacetaldoxime, *p*-methoxyphenylacetaldoxime and indole-3-acetaldoxime, as well as an unspecified naphthoaldoxime and aliphatic analogues [K411].

This activity has been detected in banana and Aspergillus niger [K882].

Sorghum bicolor cytochrome P450 acts on (*Z*)-*p*-hydroxyphenylacetaldoxime [A2914, J504].

Flavone formation from flavanone (flavonol synthase)

Antirrhinum majus enzyme, optimum pH about 7, is found in microsomes from flowers; it requires NADPH. It converts naringenin into apigenin and eriodictyol into luteolin, with the formation of a double bond in the heterocyclic ring. The enzyme is also found in Verbena hybrida and Taraxacum officinale, plants that form flavones; the enzyme has not been found in species that do not contain flavones [C65].

Petunia hybrida flower bud enzyme, optimum pH 6.5, requires α-oxoglutarate, ascorbate and Fe^{2+}. Substrates are dihydrokaempferol and dihydroquercetin, but not dihydromyricetin; kaempferol and quercetin are the products [D950].

Dianthus caryophyllus enzyme, optimum pH 7.4, requires α-oxoglutarate, ascorbate and Fe^{2+}. Substrates are dihydrokaempferol and dihydroquercetin [G761].

DDT dehydrochlorinase (E.C. 4.5.1.1)

$$(p\text{-}ClC_6H_4)_2CHCCl_3 \rightarrow (p\text{-}ClC_6H_4)_2C{=}CCl_2 + H^+ + Cl^-$$

Two dimeric isozymes are found in housefly, both with pI 7.1. The subunits in the main isozyme have molecular weights 23 000 and 25 000, and in the minor isozyme both have a molecular weight of 25 000. They exhibit marked glutathione S-transferase activity towards chloronitrobenzenes, but another glutathione S-transferase (E.C. 2.5.1.18) has no DDT dehydrochlorinase activity [D332].

Schistocerca gregaria (desert locust) enzyme activity is highest in cuticle, moderate in fat body and low in gut and flight muscle [A3185].

A DDT dehydrochlorinase contained phosphatidylcholine, phosphatidylserine and their lyso- analogues. Both saturated and unsaturated C_{12}–C_{18} (especially C_{16} and C_{18}) fatty acid components were found [A2258].

8.2 *Reactions of double bonds*

Double bond reductases

a. Cinnamate

Cinnamate → phenylpropionate

Although studies on this reaction have demonstrated that *m*- and *p*-coumarates are reduced to the corresponding phenylpropionates in rat [Al334], Lactobacillus, Proteus and Clostridium [A653], little has been discovered relating to the enzymes involved. Cinnamate is reduced in Streptomyces setonii [C818] and in mixed microorganisms [J12], and indole-3-acrylate may be reduced in Clostridium [A2571].

Arthrobacter coumarate reductase (E.C. 1.3.1.11) requires NADH and is specific for *o*-coumarate, with very low activity for some analogues. The reaction is not reversible [K870].

b. Muconate (E.C. 1.3.1.40)

Pseudomonas cruciviae enzyme is composed of three isozymes. One, molecular weight 170 000 requires NADPH. It reduces one of the double bonds in 2-hydroxy-6-oxo-6-phenylhexa-2,4-dienoate (its methyl ester is also a substrate) to form 2,6-dioxo-6-phenylhexanoate [E156].

c. 3-Methyleneoxindole reductase (E.C. 1.3.1.17)

Pea enzyme, which requires NADP, is composed of four or five isozymes; it forms 3-methyloxindole [K875].

d. Biochanin-A reductase (E.C. 1.3.1.46)

Fusarium javanicum enzyme, a homodimer, molecular weight 87 000, monomeric molecular weight 43 000 and optimum pH 7.5, requires NADPH. It converts biochanin A into 5,7-dihydroxy-4′-methoxyisoflavanone by reducing the double bond in the heterocyclic ring. It is highly specific for biochanin A, although it will act on pratensein and genistein [E724].

Styrene monooxygenase

Styrene → styrene oxide

Rat enzyme is found in liver with some in lung, but little in heart, spleen or brain. It requires NAD(P)H, and is inhibited by metyrapone or SKF 525-A [A1796].

Human leucocyte myeloperoxidase and Coprinus cinereus peroxidase convert styrene and peroxide into styrene oxide and benzaldehyde. Coprinus enzyme forms benzaldehyde as the main product, and a significant amount of benzaldehyde is formed by myeloperoxidase. Substituted styrenes are also substrates [K431].

Horseradish peroxidase forms styrene oxide from styrene in the presence of glutathione, oxygen and peroxide; the incorporated oxygen comes from molecular oxygen. It is suggested that a glutathione thiyl radical is involved [E308].

Pseudomonas fluorescens forms epoxides from styrene, methyl *trans*-cinnamate, *trans*-cinnamyl alcohol and its acetate ester [K167].

Methylococcus capsulatus soluble methane monooxygenase (E.C. 1.14.13.25) converts styrene into styrene oxide [F278].

Dehydrocyclopeptine epoxidase

Penicillium cyclopium enzyme, molecular weight in excess of 480 000, forms an epoxide with the side chain double bond. The enzyme is a mixed function oxidase that may contain flavin, and requires oxygen. NAD(P)H, ascorbate or 6-methyltetrahydropteridine are cofactors. It is inhibited by cyanide, thiocyanate, heavy metals and iron-chelating reagents, but is stimulated by Fe^{2+} [A2734].

Dioxetane formation

Garbanzo and horseradish peroxidase act on 2′,4,4′-trihydroxychalcone to form a dioxetane (peroxide) ring with the double bond. It uses oxygen stoichiometrically, with peroxide as catalyst. It is inhibited by Mn^{2+} and a range of reducing agents, especially *p*-phenylenediamine [A1729].

Hydration of double bonds in unsaturated fatty acids

This type of reaction, which is analogous to one of the steps in the β-oxidation of fatty acids, is implicit in the degradation of cinnamates and longer chain phenolic fatty acids and hydrocarbons. Little information at an enzyme level is available.

Kievitone hydratase

Fusarium solani kievitone hydratase is composed of two isozymes, both of which are dimeric. It hydrates the prenyl double bond to form a tertiary alcohol. Although there have been technical problems in assessing their molecular weights, the molecular weights of the monomers appear to be 47 000 and 49 000. They are glycoproteins, pI 5.1 and optimum pH 5.5. Fusarium also hydrates phaseollidin, but a different enzyme appears to be involved [C58, C713, F687].

Phaseollidin hydratase (E.C. 4.2.1.97)

Fusarium solani enzyme forms phaseollidin hydrate by addition of water to the side-chain double bond. It is separable from kievitone hydratase, which catalyzes a similar reaction [K797].

1,2-Dehydroreticulinium reductase (E.C. 1.5.1.27)

Papaver somniferum enzyme, molecular weight 30 000 and optimum pH 8.5, requires NADPH and is highly specific, reducing the carbon-nitrogen double bond to form (*R*)-reticuline. Its activity is highest four days after germination, and then declines; it is present only where morphinan alkaloids are found [K771].

Halogenation

Bromoperoxidase (non-haem type) from the red alga Corallina pilulifera catalyzes a series of reactions: styrene and *trans*-cinnamyl alcohol add HBr to form 1-phenyl-2-bromoethanol and 2-bromo-1-phenylpropane-1,3-diol respectively, and cinnamate yields a mixture of 2-bromophenylhydracrylate and ω-bromostyrene. In addition anisole and other compounds are halogenated on the aromatic nucleus. It is postulated that the reaction proceeds via a halonium cation intermediate [E597].

Sulphonation

In rat N-(4-chloro-2-fluoro-5-((1-methyl-2-propynyl)oxy)phenyl)-3,4,5,6-tetrahydrophthalimide forms 4-chloro-2-fluoro-5-hydroxyphenyl-1-sulpho-1,2-cyclohexanedicarboximide and several other metabolites containing a sulpho group, in which a C-sulphonic acid is formed by addition at the double bond. It is postulated that a glutathione adduct may be involved [G568].

Amination of hydrocarbon double bonds

Rat liver has been claimed to aminate myristicin to form an amphetamine [A699, A975]. Another study failed to confirm this reaction type (Goodwin, B.L., Ruthven, C.R.J., unpublished).

9. Hydration of epoxides

Epoxide hydratase; styrene oxides (E.C. 3.3.2.3)

$$\mathrm{R{-}\overset{\;O}{\overbrace{HC{-}CH}}{-}R' + H_2O \rightarrow R{-}CHOH - CHOH{-}R'}$$

This reaction converts styrene oxide and analogues into side-chain diols.

Solubilized horse liver microsomal enzyme has a molecular weight of 50 000 [H905].

Human liver cytosolic enzyme exists in two forms. One, monomeric molecular weight 49 000 and pI 9.2, hydrolyzes *cis*-stilbene oxide. The other, a dimer, monomeric molecular weight 61 000, hydrolyzes the *trans*-isomer [G195]. Another study with liver enzyme that hydrolyzes the *trans*-isomer found a molecular weight of 58 000 for the soluble enzyme and 50 000 for microsomal hydroxylase. The soluble enzyme was very active towards stilbene oxide and naphthalene-1,2-oxide, but inactive towards benzpyrene-4,5-oxide, and almost inactive towards *p*-nitrostyrene oxide. Another enzyme, which hydrolyzed *p*-nitrostyrene oxide, was less active towards *trans*-stilbene oxide. The specificity of microsomal enzyme was similar to that from rat liver (below), with good activity towards styrene oxide, *p*-nitrostyrene oxide and *trans*-stilbene oxide, but naphthalene-1,2-oxide was the best substrate studied and with significant activity towards benzpyrene-4,5-oxide. All the enzymes studied showed low activity towards *trans*-β-ethylstyrene oxide, and *cis*-β-ethylstyrene oxide was scarcely hydrolyzed [C386].

Mouse liver nuclear envelope enzyme hydrolyzes *trans*-stilbene oxide, optimum pH 6.5. Induction and inhibition studies demonstrate that this enzyme is different from the one hydrolyzing benzpyrene-4,5-oxide [D609, E56]; antibodies to it do not cross-react with other epoxide hydrolases [C861]. In kidney and liver the activity is both cytosolic and particulate, the latter being found in peroxisomes [D643]; another study found activity in both cytosol and the mitochondrial fraction. The mitochondrial enzyme was found mainly in the matrix and the intermembrane space fraction. Both mitochondrial and soluble enzymes have a molecular weight in the range 120 000–140 000 [B648]. An enzyme that acts on 1-(4′-ethylphenoxy)-3,7-dimethyl-6,7-epoxy-*trans*-2-octene is found in liver, especially in the soluble and mitochondrial fractions. The soluble enzyme is also found in kidney, lung, testes and spleen (in order of reducing activity). The level differs between male and female animals, and increases about four-fold between five and 10 weeks post partum. The molecular weight is 130 000 [A3955]. Microsomal enzyme, molecular weight 50 000, cross-reacts with antiserum prepared with rat liver enzyme [A2006].

Rabbit enzyme activity is highest in liver (optimum pH 7), slightly lower in kidney but with good activity in lung (optimum pH 7.5) and intestine [A2509]. Liver activity is low at birth, and increases steadily after weaning to the adult level. Gut and kidney enzyme activities follow a similar time course, but lung activity increases twofold to the adult level at weaning, decreases slightly for two to four months, and then increases [A2132]. *cis*-Stilbene oxide hydrolysis is inhibited competitively by *trans*-stilbeneimine [A86].

Rat enzyme is found in liver, kidney, spleen (only in males), but little or none is found in lung, heart or brain [A1796]. It acts on *cis*-β-ethyl-, *n*-propyl-, *n*-butyl- and *n*-hexylstyrene oxides to form the analogous (*R*, *R*)-*threo*-diols [J17]. *trans*-Stilbene oxide is also a substrate for liver cytosolic enzyme, which differs from the microsomal enzyme; it is a homodimer, molecular weight 120 000, and optimum pH 7.4 [E759]. Another publication claims that the molecular weight of the microsomal enzyme is 53 000–54 000, and is polymerizable. The amino acid composition has been determined [A3421]. The specificity is essentially the same as for the human enzyme (above).

Activity towards styrene oxide has been found in liver and kidney from Atlantic stingray, black drum, bluntnose ray, dogfish, eel, southern flounder, winter flounder, jack crevalle, king of Norway, mangrove snapper, mullet, nurse shark, redfish, sea bass, sheepshead, large skate, small skate, thorny skate, lobster, spiny lobster, blue crab, rock crab and clam [B69].

Teleost species show activity in liver at a similar level to that found in rabbit liver [A1694].

Aspergillus enzyme acts only on the (*R*)-isomer of *p*-nitrostyrene [H922].

Pseudomonas enzyme, pI 5.5, hydrolyzes styrene oxide and a series of aliphatic epoxides [G429].

Streptomonas antibioticus enzyme hydrolyzes (*S*)-styrene oxide to (*S*)-phenylglycol [K394].

Epoxide hydratase; polynuclear hydrocarbons

This activity is a key step in the metabolism of polynuclear hydrocarbons in animals. *In vivo* studies usually fail to detect the epoxide metabolites because of their instability, and the final products, *trans*-dihydrodiols and glutathione conjugates are those usually detected; phenolic metabolites are often assumed to be products arising from epoxide intermediates.

Human liver enzyme exhibits an optimum at pH 8.6–8.9. Substrates (in decreasing order of activity) include phenanthrene-9,10-oxide, benzpyrene-4,5-, 7,8- and 9,10-oxides, benzanthracene-5,6-oxide, naphthalene-1,2-oxide, 3-methylcholanthrene-11,12-oxide, dibenzo[*a*,*h*]anthracene-5,6-oxide and benzpyrene-11,12-oxide; the latter is only marginally active. Styrene oxide is also a substrate [A95, A3217]. Human enzyme is found in liver (best), kidney, lung, spleen and gut (least) [C507]. Its molecular weight is 49 000; it is immunologically distinct from rat liver enzyme, although the specificity is identical [A3031, B63].

Mouse liver enzyme, with benzpyrene-4,5-oxide as substrate, is found in microsomes, with negligible activity in the soluble and mitochondrial fractions [B648].

Studies on rat liver enzyme have found that the addition of water to naphthalene- and anthracene-1,2-oxides occurs exclusively at the 2 position. When (−)-(1*R*,2*S*)-naphthalene-1,2-oxide is used as substrate, both (−)-(1*R*,2*R*)- and (+)-(1*S*,2*S*)-naphthalene-1,2-dihydrodiols are formed in nearly equal amounts [D261]. Specificity for both nuclear and mitochondrial enzymes is similar to that for human enzyme (above), but with much lower activity in the nucleus. Although both these enzymes have molecular weights of 49 000, their properties, especially induction, indicate that the enzymes are different [B142]. Studies on microsomal enzyme, both in rat and man (including activity towards styrene oxide) indicate the presence of multiple forms [B220]. Studies in rats found that (−)-benzpyrene-7,8-dihydrodiol yields tetrahydro-7β,8α,9β,10β-, 7β,8α,9α,10β-, 7β,8α,9β,10α- and 7β,8α,9α,10α-tetrahydroxy-benzpyrenes, in reducing amounts [A3093]. Benzpyrene-4,5-oxide (racemic) yields the 4,5-dihydrodiol highly enriched in (−) isomer, with oxygen from water incorporated almost equally at positions 4 and 5, whereas hydrolysis of 7,8- and 9,10- oxides shows almost no stereospecificity, with water incorporated almost exclusively at position 8 with 7,8-oxide as substrate [A3166]. Liver microsomal enzyme has optimum pH values between 7.4 and 8.9, depending on the substrate used [A3031]. Highest activity is found in liver, with lower but significant activity in testis, kidney and ovary. Low or trace activity is found in lung, adrenal, fat, bladder, prostate, trachea, tongue, oesophagus, membraneous and glandular stomach, small intestine, caecum, colon, epidermis, cutis, subcutis, submaxillay gland, spleen, thymus, brain, heart and triceps. It is absent from blood [A1894].

Rat adrenal enzyme activity remains approximately constant after about eight days post-partum. Ovarian enzyme activity is low before weaning; at weaning the activity increases eightfold to the adult level [A3461]. Immunological studies indicate that liver microsomal activity towards benzpyrene-4,5-oxide and styrene oxide reside on the same molecule [A2445]. Inhibition with styrene oxide allows benzpyrene-4,5-oxide to accumulate as a metabolite of benzpyrene [A739].

Skin microsomal enzyme, which has been found in man, rat and mouse, acts on all of a range of epoxides tested. Rat enzyme has a broad optimum with maximal activity at pH 9, whereas human enzyme has a constant activity between pH 7 and 10. 1,1,1-Trichloropropene and cyclohexene oxides are inhibitory [A2486].

Activity towards benzpyrene-4,5-oxide has been found in liver and kidney from Atlantic stingray, dogfish, eel, sheepshead, small skate, winter flounder, lobster, spiny lobster, blue crab, rock crab and clam [B69].

The quinoline verlukast is epoxidized at positions 5 and 6 in rat *in vivo*, and the dihydrodiol formed from this is found in rhesus monkey [G839].

10. Light-forming reactions

Cypridina-luciferin 2-monooxygenase
(E.C. 1.13.12.6)

Cypridina hilgendorfii enzyme, molecular weight 68 000 with monomeric molecular weight 10 000–13 700 (method-dependent) requires oxygen, and produces a bluish luminescence. The amino acid composition has been determined [K803, K815].

Firefly luciferase (Photinus-luciferin 4-monooxygenase (ATP-hydrolyzing); E.C. 1.13.12.7)

The enzyme, molecular weight 50 000 with one luciferin-binding site, readily dimerizes. The dimer binds one ATP and one $ATPMg_2$ [A905].

Both D- and L-luciferin (a benzthiazole) produce light; with the latter substrate L-luciferyladenylate is an intermediate; presumably D-luciferyladenylate is an intermediate with the oxidative decarboxylation of D-luciferin. The reaction is stimulated by pyrophosphate and pyrophosphatase [J204, K850]. Three luminescent components have been separated chromatographically; one is associated with the enzyme, a second behaves like luciferyl adenylate, but the third was not characterized. ATP is required for their formation [K447]. The visible spectrum is dependent on the environment [K849].

Treatment with 2-benzothiazolesulphonate, a luciferin analogue, changes the emission to red light after a short exposure, whereas longer exposure totally inhibits light emission. L-1-Chloro-4-phenyl-3-(*p*-tolylsulphonamido)-2-butanone inhibits without changing the emission wavelength [A1522].

Renilla-luciferin 2-monooxygenase
(E.C. 1.13.12.5)

This reaction has been reviewed; the products are carbon dioxide and oxyluciferin [K812, K813, K926]. It is found in Renilla, Aequorea, Carvernularia, Leioptilus, and probably Pelagia and other phyla such as cephalopod molluscs [K884].

Watasenia-luciferin 2-monooxygenase
(E.C. 1.13.12.8)

This squid enzyme opens an amide bond with decarboxylation [K899].

Bibliography

A1. Smith, I, Mitchell, P.D., *Biochem. Soc. Trans.* (1974) **2**, 620.

A2. Capel, I.D., Millburn, P., Williams, R.T., *Biochem. Soc. Trans.* (1974) **2**, 305.

A3. Mackay, A.V.P., Iverson, L.L., *Biochem. Soc. Trans.* (1974) **2**, 669.

A4. Dingell, J.V., Dring, L.G., Smith, R.L., Williams, R.T., *Biochem. Soc. Trans.* (1974) **2**, 306.

A5. Gangolli, S.D., Shilling, W.H., Grasso, P., Gaunt, I.F., *Biochem. Soc. Trans.* (1974) **2**, 310.

A6. Fennell, F.A.M., Harding, N.G.L., *Biochem. Soc. Trans.* (1974) **2**, 340.

A7. Hughes, L., Price, G.M., *Biochem. Soc. Trans.* (1974) **2**, 336.

A8. Plummer, C.K., Kilpatrick, I.C., James, M.W., Goodwin, B., Phillipson, O.T., *Eur. J. Pharmacol.* (1998) **148**, 481–482.

A9. Wiseman, A., Gondal, J.A., Sims, P., *Biochem. Soc. Trans.* (1975) **3**, 278.

A10. Siddik, Z.H., Barnes, R.D., Dring, L.G., Smith, R.L., Williams, R.T., *Biochem. Soc. Trans.* (1975) **3**, 290.

A11. McPherson, F.J., Markham, A., Bridges, J.W., Hartman, G.C., Parke, D.V., *Biochem. Soc. Trans.* (1975) **3**, 281.

A13. Pal, K., Grover, P.L., Sims, P., *Biochem. Soc. Trans.* (1975) **3**, 174.

A14. Capel, I.D., Millburn, P., Williams, R.T., *Biochem. Soc. Trans.* (1975) **3**, 875.

A15. Cohen, G.M., Flockhart, I.R., *Biochem. Soc. Trans.* (1975) **3**, 305.

A16. Franklin, R.B., Dring, L.G., Williams, R.T., *Biochem. Soc. Trans.* (1974) **2**, 877.

A17. Dixon, P.A.F., Uwaifo, A.O., Caldwell, J., Smith, R.L., *Biochem. Soc. Trans.* (1974) **2**, 879.

A18. Zachariah, P., *Fed. Proc.* (1974) **33**, 471.

A19. Pantuck, E.J., Hsiao, K.-C., Kaplan, S.A., Kuntzman, R., Conney, A.H., *Fed. Proc.* (1974) **33**, 471.

A20. Pridham, J.B., Woodhead, S., *Biochem. Soc. Trans.* (1974) **2**, 1070.

A21. Köster, U., Caldwell, J., Smith, R.L., *Biochem. Soc. Trans.* (1974) **2**, 881.

A22. Ball, L.M., Renwick, A.G., Williams, R.T., *Biochem. Soc. Trans.* (1974) **2**, 1084.

A23. Kilpatrick, D.C., Stirling, J.L., *Biochem. Soc. Trans.* (1975) **3**, 246.

A27. Hughes, L., Price, G.M., *Biochem. Soc. Trans.* (1975) **3**, 72.

A30. Keat, M.J., Hopper, D.J., *Biochem. Soc. Trans.* (1975) **3**, 385.

A31. Thomas, B.H., Wong, L.T., Hynie, I., Zeitz, W., *Biochem. Soc. Trans.* (1975) **3**, 687.

A32. Sumner, D., Dayton, P., Cucinell, S., Plostnieks, J., *Fed. Proc.* (1974) **33**, 471.

A33. Dixon, W.R., Brooks, M.A., Postma, E., Hackman, M.R., Schwartz, M.A.,

A34. Edelson, J., Schuster, E.B., *Fed. Proc.* (1974) **33**, 472.

A35. Hes, J., Sternson, L.A., *Fed. Proc.* (1974) **33**, 472.

A36. Murphy, P.J., Williams, T.L., Bellamy, G.M., *Fed. Proc.* (1974) **33**, 472.

A37. Takanashi, S., Bachur, N.R., *Fed. Proc.* (1974) **33**, 499.

A38. Kadar, D., Conn, A.W., Tang, H.Y., *Fed. Proc.* (1974) **33**, 499.

A39. Rubin, R.J., Miller, J., *Fed. Proc.* (1974) **33**, 499.

A40. Lan, S.J., El-Hawey, A.M., Dean, A.V., Poutsiaka, J.W., Schreiber, E.C., *Fed. Proc.* (1974) **33**, 499.

A41. Waterbury, L.D., *Fed. Proc.* (1974) **33**, 500.

A42. Ikeda, G.J., Estep, C.B., Wiemeler, L.H., *Fed. Proc.* (1974) **33**, 534.

A43. Litterst, C.L., Mimnaugh, E.G., Gram, T.E., *Fed. Proc.* (1974) **33**, 533.

A44. Morganroth, V.H. III, Broadle-Biber, M.C., Roth, R.H., *Fed. Proc.* (1974) **33**, 535.

A45. Tong, H.J., Reis, D.J., *Fed. Proc.* (1974) **33**, 535.

A46. Chrastil, J., Wilson, J.T., *Fed. Proc.* (1974) **33**, 573.

A47. Parli, C.J., Hicks, J., *Fed. Proc.* (1974) **33**, 560.

A48. Hinson, J.A., Mitchell, J.R., Jollow, D.J., *Fed. Proc.* (1974) **33**, 573.

A49. Wolf, D.E., VandenHeuvel, W.J.A., Walker, R.W., Koniuszy, F.R., Jacob, T.A., Wolf, F.J., *Fed. Proc.* (1974) **33**, 574.

A50. Felsted, R.L., Bachur, N.R., *Fed. Proc.* (1974) **33**, 1382.

A51. Knauer, T.E., Willingham, A.K., Matschiner, J.T., *Fed. Proc.* (1974) **33**, 1314.

A52. Crawford, R.L., Hareland, W.L., Chapman, P.J., *Fed. Proc.* (1974) **33**, 1370.

A53. Meyerson, L.R., Davis, V.E., *Fed. Proc.* (1975) **34**, 508.

A54. Tu, M-h, Chen, C., *Fed. Proc.* (1975) **34**, 625.

A55. Letendre, C.H., Kato, T., Dickens, G., Guroff, G., *Fed. Proc.* (1975) **34**, 624.

A56. Weinshilboum, R.M., Raymond, F.A., *Fed. Proc.* (1975) **34**, 778.

A57. Smolen, A., Anderson, A.D., *Fed. Proc.* (1975) **34**, 778.

A58. Lin, C-c, Chang, R., Zampaglione, N., *Fed. Proc.* (1975) **34**, 785.

A59. Harper, C., *Fed. Proc.* (1975) **34**, 785.

A60. Sanders-Bush, E., Oates, J.A., Bush, M.T., *Fed. Proc.* (1975) **34**, 809.

A61. Butler, T.C., Dudley, K.H., *Fed. Proc.* (1975) **34**, 738.

A62. Parli, C.J., Schmidt, B., *Fed. Proc.* (1975) **34**, 813.

A63. Barrass, B.C., Coult, D.B., Rich, P., Tutt, K.J., *Biochem. Pharmacol.* (1974) **23**, 47–56.

A64. Huffman, D.H., Shoeman, D.W., Azarnoff, D.L., *Biochem. Pharmacol.* (1974) **23**, 197–202.

A65. Alvin, J., Dixit, B.N., *Biochem. Pharmacol.* (1974) **23**, 139–152.

A66. Seiler, N., Demisch, L., *Biochem. Pharmacol.* (1974) **23**, 259–272.

A67. Gaertner, H.J., Breyer, U., Liomin, G., *Biochem. Pharmacol.* (1974) **23**, 303–312.

A68. Grover, P.L., Hewer, A., Sims, P., *Biochem. Pharmacol.* (1974) **23**, 323–332.

A69. Breyer, U., Gaertner, H.J., Prox, A., *Biochem. Pharmacol.* (1974) **23**, 313–322.

A70. Grover, P.L., *Biochem. Pharmacol.* (1974) **23**, 333–343.

A71. Mukhtar, H., Sahib, M.K., Kidwai, J.R., Grover, P.L., Hewer, A., Sims, P., *Biochem. Pharmacol.* (1974) **23**, 345–349.

A72. Williams, C.H., Grover, P.L., Hewer, A., Sims, P., *Biochem. Pharmacol.* (1974) **23**, 615–628.

A73. Sipes, I.G., Gigon, P.L., Krishna, G., *Biochem. Pharmacol.* (1974) **23**, 451–455.

A74. Williams, C.H., Lawson, J., *Biochem. Pharmacol.* (1974) **23**, 629–636.

A75. Poirer, L.A., Weisburger, J.H., Williams, C.H., Grover, P.L., Hewer, A., Sims, P., *Biochem. Pharmacol.* (1974) **23**, 661–669.

A76. Niwaguchi, T., Inoue, T., Nakahara, Y, *Biochem. Pharmacol.* (1974) **23**, 1073–1078.

A77. Rommelspacher, H., Honecker, H., Schulze, G., Strauss, S.M., *Biochem. Pharmacol.* (1974) **23**, 1065–1071.

A78. Watabe, T., Akamatsu, K., *Biochem. Pharmacol.* (1974) **23**, 1079–1085.

A79. Oesch, F., Thoenen, H., Fahrlaender, H., *Biochem. Pharmacol.* (1974) **23**, 1307–1317.

A80. Laduron, P.M., Gommeren, W.R., Leysen, J.E., *Biochem. Pharmacol.* (1974) **23**, 1599–1608.

A81. Baselt, R.C., Bickel, M.H., *Biochem. Pharmacol.* (1974) **23**, 1405–1411.

A82. Norman, B.H., Poore, R.E., Neal, R.E., *Biochem. Pharmacol.* (1974) **23**, 1733–1744.

A83. Baetke, K.P., Gough, B.J., Shellenberger, T.E., *Biochem. Pharmacol.* (1974) **23**, 1745–1752.

A84. Mishra, L.C., Parmar, A.S., Mead, J.A.R., *Biochem. Pharmacol.* (1974) **23**, 1827–1834.

A85. Gessner, T., *Biochem. Pharmacol.* (1974) **23**, 1809–1816.

A86. Watabe, T., Akamatsu, K., *Biochem. Pharmacol.* (1974) **23**, 1845–1851.

A87. Davis, V.E., Cashaw, J.L., McLaughlan, B.R., Hamlin, T.A., *Biochem. Pharmacol.* (1974) **23**, 1877–1889.

A88. Ohira, H., Kaseda, H., Kido, R., *Biochem. Pharmacol.* (1974) **23**, 1918–1920.

A89. Kamataki, T., Kitagawa, H., *Biochem. Pharmacol.* (1974) **23**, 1915–1918.

A90. Mulder, G.J., Hagedoorn, A.H., *Biochem. Pharmacol.* (1974) **23**, 2101–2109.

A91. Aitio, A., *Biochem. Pharmacol.* (1974) **23**, 2203–2205.

A92. Autrup, H., Thurlow, B.J., Warwick, G.P., *Biochem. Pharmacol.* (1974) **23**, 2341–2345.

A93. Fuller, R.W., Rousch, B.W., *Biochem. Pharmacol.* (1974) **23**, 2208–2209.

A94. Roth, J.A., Gillis, C.N., *Biochem. Pharmacol.* (1974) **23**, 2537–2545.

A95. Booth, J., Sims, P., *Biochem. Pharmacol.* (1974) **23**, 2547–2555.

A96. Varga, F., Fischer, E., Szily, T.S., *Biochem. Pharmacol.* (1974) **23**, 2617–2623.

A97. El-Sewedy, S.M., Abdel-Tawab, G.A., El-Zoghby, S.M., Zeitoun, R., Mostafa, M.H., Shalaby, Sh.M., *Biochem. Pharmacol.* (1974) **23**, 2557–2565.

A98. Niwaguchi, T., Inoue, T., Sakai, T., *Biochem. Pharmacol.* (1974) **23**, 3063–3066.

A99. Ball, C.R., Double, J.A., *Biochem. Pharmacol.* (1974) **23**, 3173–3177.

A100. Watabe, T., Akamatsu, K., *Biochem. Pharmacol.* (1974) **23**, 2839–2844.

A101. Jonsson. J.A., *Biochem. Pharmacol.* (1974) **23**, 3191–3197.

A102. Schöttler, C., Krisch, K., *Biochem. Pharmacol.* (1974) **23**, 2867–2875.

A103. Beckett, A.H., Mihailova, D., *Biochem. Pharmacol.* (1974) **23**, 3347–3352.

A104. Waldmeier, P.C., Feldtrauer, J.J., Maitre, L., *Biochem. Pharmacol.* (1975) **24**, 1209–1213.

A105. Landsberg, L., Berardino, M.B., Silva, P., *Biochem. Pharmacol.* (1975) **24**, 1167–1174.

A106. Lin, R.-L., Narasimhachari, N., *Biochem. Pharmacol.* (1975) **24**, 1239–1240.

A107. Mulder, G.J., Bleeker, V., *Biochem. Pharmacol.* (1975) **24**, 1481–1484.

A108. Ben-Zvi, Z., Burstein, S., *Biochem. Pharmacol.* (1975) **24**, 1130–1131.

A109. Borchardt, R.T., *Biochem. Pharmacol.* (1975) **24**, 1542–1544.

A111. Rosengarten, H., Meller, E., Friedhoff, A.J., *Biochem. Pharmacol.* (1975) **24**, 1759–1764.

A112. Giraldi, T., Nisi, C., Sava, G., *Biochem. Pharmacol.* (1975) **24**, 1793–1797.

A113. Assandri, A., Zerial, A., Strolin-Benedetti, M., Perazzi, A., *Biochem. Pharmacol.* (1975) **24**, 1767–1770.

A114. Juchau, M.R., Zachariah, P.K., *Biochem. Pharmacol.* (1975) **24**, 227–233.

A115. Law, F.C.P., Drach, J.C., Sinsheimer, J.E., *Biochem. Pharmacol.* (1975) **24**, 385–390.

A116. Siemens, A.J., Kalant, H., *Biochem. Pharmacol.* (1975) **24**, 755–762.

A117. Demisch, L., Seiler, N., *Biochem. Pharmacol.* (1975) **24**, 575–580.

A118. Bock, K.W., Winne, D., *Biochem. Pharmacol.* (1975) **24**, 859–862.

A119. Al-Gailany, K.A.S., Bridges, J.W., Netter, K.J., *Biochem. Pharmacol.* (1975) **24**, 867–870.

A120. Benkert, K., Fries, W., Kiese, M., Lenk, W., *Biochem. Pharmacol.* (1975) **24**, 1375–1380.

A121. Laduron, P., Leysen, J., *Biochem. Pharmacol.* (1975) **24**, 929–932.

A122. Pantuck, E.J., Hsiao, K.-C., Kaplan, S.A., Kuntzman, R., Conney, A.H., *J. Pharmacol. Exp. Ther.* (1974) **191**, 45–52.

A123. Alvin, J., Goh, E., Bush, M.T., *J. Pharmacol. Exp. Ther.* (1975) **194**, 117–125.

A124. Wheeler, L.A., Soderberg, F.P., Goldman, P., *J. Pharmacol. Exp. Ther.* (1975) **194**, 135–144.

A125. Bartholini, G., Constantinidis, J., Puig, M., Tissot, R., Pletscher, A., *J. Pharmacol. Exp. Ther.* (1975) **193**, 523–532.

A126. Schotten, C., *Z. Physiol. Chem.* (1882–1883) **7**, 2239.

A127. Chrastil, J., Wilson, J.T., *J. Pharmacol. Exp. Ther.* (1975) **193**, 631–638.

A128. Thierfelder, H., Sherwin, C.P., *Ber. Chem. Ges.* (1914) **47**, 2630–2634; *Z. Physiol. Chem.* (1915) **94**, 1–9.

A129. Salkowski, E., *Z. Physiol. Chem.* (1885) **9**, 229–237.

A130. Vasiliu, H., *Mitt. Landw. Inst. Univ. Breslau* (1909) **4**, 703.

A131. Totani, G., *Z. Physiol. Chem.* (1910) **68**, 75–82.

A132. Sherwin, C.P., Wolf, M., Wolf, W., *J. Biol. Chem.* (1919) **37**, 113–119.

A133. Misra, A.L., Bloch, R., Vadlamani, N.L., Mulé, S.J., *J. Pharmacol. Exp. Ther.* (1974) **188**, 34–44.

A134. Abrams, L.S., Elliott, H.W., *J. Pharmacol. Exp. Ther.* (1974) **189**, 285–292.

A135. Cohen, M.L., Trevor, A.J., *J. Pharmacol. Exp. Ther.* (1974) **189**, 351–358.

A136. Klaassen, C.D., Fitzgerald, T.J., *J. Pharmacol. Exp. Ther.* (1974) **191**, 548–556.

A137. Faraj, B.A., Israili, Z.H., Perel, J.M., Jenkins, M.L., Holtzman, S.G., Cucinell, S.A., Dayton, P.G., *J. Pharmacol. Exp. Ther.* (1974) **191**, 535–547.

A138. Takanashi, S., Bachur, N.R., *J. Pharmacol. Exp. Ther.* (1974) **195**, 41–49.

A139. Bachur, N.R., Hildebrand, R.C., Jaenke, R.S., *J. Pharmacol. Exp. Ther.* (1974) **191**, 331–340.

A140. Moreland, T.A., Hewick, D.S., *Biochem. Pharmacol.* (1975) **24**, 1953–1957.

A141. Pomerantz, S.H., Murthy, V.V., *Arch. Biochem. Biophys.* (1974) **160**, 73–82.

A142. Sloane, N.H., Davis, T.K., *Arch. Biochem. Biophys.* (1974) **163**, 46–52.

A143. Hayakawa, T., Lemahieu, R.A., Udenfriend, S., *Arch. Biochem. Biophys.* (1974) **162**, 223–230.

A144. Lee, H.-S., Schulz, A.R., Fuller, R.W., *Arch. Biochem. Biophys.* (1974) **163**, 735–741.

A145. Gestetner, B., Conn, E.E., *Arch. Biochem. Biophys.* (1974) **163**, 617–624.

A146. Sampson, E.J., Vergara, E.V., Fedor, J.M., Funk, M.O., Benkovic, S.J., *Arch. Biochem. Biophys.* (1975) **169**, 372–383.

A147. Alam, S.S., Nambudiri, A.M.D., Rudney, H., *Arch. Biochem. Biophys.* (1975) **171**, 183–190.

A148. Graham, E.R.B., Roy, A.B., *Biochim. Biophys. Acta* (1973) **329**, 88–92.

A149. Kreuzaler, F., Hahlbrock, K., *Arch. Biochem. Biophys.* (1975) **169**, 84–90.

A150. Sloane, N.H., *Biochim. Biophys. Acta* (1973) **327**, 11–19.

A151. Curti, B., Ronchi, S., Branzoli, U., Ferri, G., Williams, C.H., Jr., *Biochim. Biophys. Acta* (1973) **327**, 266–273.

A152. Fujiwara, M., Nozaki, M., *Biochim. Biophys. Acta* (1973) **327**, 306–312.

A153. Foldes, A., Meek, J.L., *Biochim. Biophys. Acta* (1973) **327**, 365–374.

A154. Marchelli, R., Vining, L.C., *Can. J. Biochem.* (1973) **51**, 1624–1629.

A155. Kumagai, H., Yamada, H., Masugi, F., Fukui, S., *Biochim. Biophys. Acta* (1973) **327**, 510–514.

A156. Hobkirk, R., Green, R.N., Nilsen, M., Jennings, B.A., *Can. J. Biochem.* (1974) **52**, 9–14.

A157. Hasnaim, S., Williamson, D.G., *Can. J. Biochem.* (1974) **52**, 120–125.

A158. Horgen, I.A., Horgen, P.A., O'Day, D.H., *Can. J. Biochem.* (1974) **52**, 126–136.

A159. Stewart, J.M., Nigam, S.N., McConnell, W.B., *Can. J. Biochem.* (1974) **52**, 144–145.

A160. Wainwright, S.D., *Can. J. Biochem.* (1974) **52**, 149–154.

A161. Gwilliam C., Paquet, A., Williamson, D.G., Layne, D.S., *Can. J. Biochem.* (1974) **52**, 452–455.

A162. Dörnemann, Löffelhardt, W., Kindl, H., *Can. J. Biochem.* (1974) **52**, 916–921.

A163. Ellis, B.E., Brown, S.A., *Can. J. Biochem.* (1974) **52**, 734–738.

A164. Guitard, M., Daigneault, R., *Can. J. Biochem.* (1974) **52**, 1087–1090.

A168. Yamasaki, K., Tamaki, T., Uzawa, S., Sankawa, U., Shibata, S., *Phytochemistry* (1973) **12**, 2877–2882.

A169. Shimada, M., Kuroda, H., Higuchi, T., *Phytochemistry* (1973) **12**, 2873–2875.

A170. Frear, D.S., Swanson, H.R., *Phytochemistry* (1974) **13**, 357–360.

A171. Roberts, M.F., *Phytochemistry* (1973) **12**, 119–123.

A172. Pont, V., Jarczyk, H.J., Collet, G.F., Thomas, R., *Phytochemistry* (1974) **13**, 785–792.

A173. Iredale, S.E., Smith, H., *Phytochemistry* (1974) **13**, 575–583.

A174. Grotzinger, E., Campbell, I.M., *Phytochemistry* (1974) **13**, 923–926.

A175. Thomas, G., Threlfall, D.R., *Phytochemistry* (1974) **13**, 807–813.

A176. Taneja, S.R., Sachar, R.C., *Phytochemistry* (1974) **13**, 1367–1371.

A177. Caldwell, J., Köster, U., Smith, R.L., Williams, R.T., *Biochem. Pharmacol.* (1975) **24**, 2225–2232.

A178. Fritz, G.J., King, R.W., Dougherty R.C., *Phytochemistry* (1974) **13**, 1473–1475.

A179. Ali, M.A., Kagan, J., *Phytochemistry* (1974) **13**, 1479–1482.

A180. Coombs, J., Baldry, C., Bucke, C., Long, S.P., *Phytochemistry* (1974) **13**, 1203–1208.

A181. Durand, R., Zenk, M.H., *Phytochemistry* (1974) **13**, 1483–1492.

A182. Nagels, L., Parmentier, F., *Phytochemistry* (1974) **13**, 2759–2762.

A183. van Brederode, J., van Nigtevecht, G., *Phytochemistry* (1974) **13**, 2763–2766.

A184. Chang, C., Kimler, L., Mabry, T.J., *Phytochemistry* (1974) **13**, 2771–2775.

A185. Bhakuni, D.S., Satish, S., Uprety, H., Kapil, R.S., *Phytochemistry* (1974) **13**, 2767–2769.

A186. Lehman, E., Harel, E., Mayer, A.M., *Phytochemistry* (1974) **13**, 1713–1717.

A187. Finkle, B.J., Kelly, S.H., *Phytochemistry* (1974) **13**, 1719–1725.

A188. Barz, W., Mohr, F., Teufel., *Phytochemistry* (1974) **13**, 1785–1787.

A189. Jangaard, N.O., *Phytochemistry* (1974) **13**, 1765–1768.

A190. Burden, R.S., Bailey, J.A., Vincent, G.G., *Phytochemistry* (1974) **13**, 1789–1791.

A191. Lin, T.-J., Ramstad, E., Heinstein, P., *Phytochemistry* (1974) **13**, 1809–1815.

A192. Thomas, G., Threlfall, D.R., *Phytochemistry* (1974) **13**, 1825–1833.

A193. Lin, T.-J., Ramstad, E., Heinstein, P., *Phytochemistry* (1974) **13**, 1817–1823.

A194. Statham, C.M., Crowden, R.K., *Phytochemistry* (1974) **13**, 1835–1840.

A195. Rosenberg, H., Stohs, S.J., *Phytochemistry* (1974) **13**, 1861–1863.

A196. James, M.O., Smith, R.L., Williams, R.T., Reidenberg, M., *Proc. Roy. Soc. Lond. B* (1972) **182**, 25–35.

A197. Milborrow, B.V., Harthoorn, P.A., *Phytochemistry* (1975) **14**, 2371–2374.
A198. Hill, A.C., Rhodes, J.C., *Phytochemistry* (1975) **14**, 2387–2391.
A199. Vaughan, P.F.T., Eason, R., Paton, J.Y., Ritchie, G.A., *Phytochemistry* (1975) **14**, 2383–2386.
A200. Fritsch, H., Grisebach, H., *Phytochemistry* (1975) **14**, 2437–2442.
A201. Thomas, G., Threlfall, D.R., *Phytochemistry* (1975) **14**, 2607–2615.
A202. Sharma, H.K., Vaidyanathan, C.S., *Phytochemistry* (1975) **14**, 2135–2139.
A203. Kohmoto, K., Nishimura, S., *Phytochemistry* (1975) **14**, 2131–2133.
A204. Rhodes, M.J.C., Wooltorton, L.S.C., *Phytochemistry* (1975) **14**, 2161–2164.
A205. Shomaya, Y., Yagi, M., Nishioka, I., Yamauchi, T., *Phytochemistry* (1975) **14**, 2189–2192.
A206. Majak, W., Towers, G.H.N., *Phytochemistry* (1973) **12**, 2189–2195.
A207. Melin, D., *Phytochemistry* (1975) **14**, 2193–2197.
A208. Dubernet, M., Ribéreau-Gayon, P., *Phytochemistry* (1974) **13**, 1085–1087.
A209. Van den Heuvel, J., Van Etten, H.D., Serum, J.W., Coffen, D.L., Williams, T.H., *Phytochemistry* (1974) **13**, 1129–1131.
A210. Jeffs, P.W., Campbell, H.F., Farrier, D.S., Ganguli, G., Martin, N.H., Molina, G., *Phytochemistry* (1974) **13**, 933–945.
A212. Padrón, M.P., Lozano, J.A., González, G., *Phytochemistry* (1975) **14**, 1959–1963.
A213. Terashima, N., Mori, I., Kanda, T., *Phytochemistry* (1975) **14**, 1991–1992.
A214. Wat, C.-K., Towers, G.H.N., *Phytochemistry* (1975) **14**, 663–666.
A215. Sterier, A.M., *Phytochemistry* (1975) **14**, 1993–1996.
A216. Rhodes, M.J.C., Wooltorton, L.S.C., *Phytochemistry* (1975) **14**, 1235–1240.
A217. Hirasawa, E., Suzuki, Y., *Phytochemistry* (1975) **14**, 99–101.
A218. Montgomery, M.W., Sgarbieri, V.C., *Phytochemistry* (1975) **14**, 1245–1249.
A219. Macháčková, I., Gančeva, K., Zmrhal, Z., *Phytochemistry* (1975) **14**, 1251–1254.
A220. Loh, J.W.C., Severson, J.G., Jr., *Phytochemistry* (1975) **14**, 1265–1267.
A221. Shinsi, H., Noguchi, M., *Phytochemistry* (1975) **14**, 1255–1258.
A222. Antoun, M.D., Roberts, M.F., *Phytochemistry* (1975) **14**, 1275–1278.
A223. Saunders, J.A., McClure, J.W., *Phytochemistry* (1975) **14**, 1285–1289.
A224. French, C.J., Smith, H., *Phytochemistry* (1975) **14**, 963–966.
A225. Munn, C.B., Drysdale, R.B., *Phytochemistry* (1975) **14**, 1303–1307.
A226. Vance, C.P., Nambudiri, A.M.D., Wat, C.-K., Towers, G.H.N., *Phytochemistry* (1975) **14**, 967–969.
A227. Deyrup, C.L., Allen, C.M., Jr., *Phytochemistry* (1975) **14**, 971–973.
A228. Dewick, P.M., *Phytochemistry* (1975) **14**, 983–988.
A229. Dewick, P.M., *Phytochemistry* (1975) **14**, 979–982.
A230. Hoagland, R.E., *Phytochemistry* (1975) **14**, 383–386.
A231. Hösel, W., Frey, G., Barz, W., *Phytochemistry* (1975) **14**, 417–422.
A232. Hofinger, M., Monseur, X., Pais, M., Jarreau, F.X., *Phytochemistry* (1975) **14**, 475–477.
A233. Leete, E., *Phytochemistry* (1975) **14**, 471–474.
A234. Kutáček, M., Hofinger, M., Králová, M., *Arch. Int. Physiol. Biochim.* (1970) **79**, 699–709.
A235. De Nicola, M.G., Amico, V., Sciuto, S., Piattelli, M., *Phytochemistry* (1975) **14**, 479–481.
A236. Hobkirk, R., Mellor, J.D., Nilsen, M., *Can. J. Biochem.* (1975) **53**, 903–906.
A237. Zinck, M.E., Addison, R.F., *Can. J. Biochem.* (1975) **53**, 636–639.
A238. Symes, A.L., Sourkes, T.L., *Can. J. Biochem.* (1975) **53**, 910–913.

A239. Duncan, R.J.S., *Can. J. Biochem.* (1975) **53**, 920–922.

A240. PetitClerc, C., Delisle, M., Martel, M., Fecteau, C., Brière, N., *Can. J. Biochem.* (1975) **53**, 1089–1100.

A241. Robinson, A.R., Baker, S.J., Common, R.H., *Can. J. Biochem.* (1975) **53**, 1085–1088.

A242. Hobkirk, R., Nilsen, M., Jennings, B., *Can. J. Biochem.* (1975) **53**, 1133–1135.

A243. Moore, G.J., Benoiton, N.L., *Can. J. Biochem.* (1975) **53**, 1145–1149.

A244. Wu, P.H., Boulton, A.A., *Can. J. Biochem.* (1975) **53**, 42–50.

A245. Hobkirk, R., Nilsen, M., *Can. J. Biochem.* (1975) **53**, 1232–1235.

A246. Paquet, A., Layne, D.S., *Can. J. Biochem.* (1975) **53**, 293–297.

A247. Safe, S., Hutzinger, O., Ecobichon, D.J., Grey, A.A., *Can. J. Biochem.* (1975) **53**, 415–420.

A248. Santimone, M., *Can. J. Biochem.* (1975) **53**, 649–657.

A249. Wainwright, S.D., *Can. J. Biochem.* (1975) **53**, 438–443.

A250. Mellor, J.D., Hobkirk, R., *Can. J. Biochem.* (1975) **53**, 779–783.

A251. Bhatti, A.R., *Can. J. Biochem.* (1975) **53**, 819–822.

A252. Fromson, J.M., Pearson, S., Bramah, S., *Xenobiotica* (1973) **3**, 711–714.

A253. Fromson, J.M., Pearson, S., Bramah, S., *Xenobiotica* (1973) **3**, 693–709.

A254. Chapman, D.I., Marcroft, J., *Xenobiotica* (1973) **3**, 745–751.

A255. Goehl, T.J., Davenport, J.B., Stanley, M.J., *Xenobiotica* (1973) **3**, 761–772.

A256. Conway, W.D., Singhvi, M., Gibaldi, M., Boyes, R.N., *Xenobiotica* (1973) **3**, 813–821.

A257. Bolt, H.M., Kappus, H., Remmer, H., *Xenobiotica* (1973) **3**, 773–785.

A258. Kanai, Y., *Xenobiotica* (1974) **4**, 441–456.

A259. Chasseaud, L.F., Hawkins, D.R., Fry, B.J., Lewis, J.D., Saggers, V.H., Sword, I.P., *Xenobiotica* (1974) **4**, 393–407.

A260. Douch, P.G.C., *Xenobiotica* (1974) **4**, 457–475.

A261. Smith, G.E., Griffiths, L.A., *Xenobiotica* (1974) **4**, 477–487.

A262. Smith, R.L., Timbrell, J.A., *Xenobiotica* (1974) **4**, 503–508.

A263. Smith, R.L., Timbrell, J.A., *Xenobiotica* (1974) **4**, 489–501.

A264. Beckett, A.H., Bélanger, P.M., *Xenobiotica* (1974) **4**, 509–519.

A265. Yoshihara, S., Yoshimura, H., *Xenobiotica* (1974) **4**, 529–535.

A266. Singhvi, S.M., Conway, W.D., Gibaldi, M., Boyes, R.N., *Xenobiotica* (1974) **4**, 563–570.

A267. Gentil, A., Lasne, C., Chouroulinkov, I., *Xenobiotica* (1974) **4**, 537–548.

A268. Andoh, Y.A., Renwick, A.G., Williams R.T., *Xenobiotica* (1974) **4**, 571–583.

A269. Axelson, J.E., Gibaldi, M., Conway, W.D., *Xenobiotica* (1974) **4**, 585–593.

A270. Misra, A.L., Pontani, R.B., Mulé, S.J., *Xenobiotica* (1974) **4**, 17–32.

A271. Barrow, A., Griffiths, L.A., *Xenobiotica* (1974) **4**, 1–16.

A272. Allan, R.J., Roxon, J.J., *Xenobiotica* (1974) **4**, 637–643.

A273. Capel, I.D., Millburn, P., Williams, R.T., *Xenobiotica* (1974) **4**, 601–615.

A274. Bridges, J.W., Evans, M.E., Idle, J.R., Millburn, P., Osiyemi, F.O., Smith, R.L., Williams, R.T., *Xenobiotica* (1974) **4**, 645–652.

A275. Ikeda, G.J., Stanaszek, R.S., Mueller, S.L., *Xenobiotica* (1974) **4**, 653–663.

A277. Noall, M.W., Knight, V., Hargrove, W.W., Elledge, B.W., *Anal. Biochem.* (1975) **69**, 10–15.

A278. Barrow, A., *Xenobiotica* (1974) **4**, 669–680.

A279. Evans, M.E., Shenfield, G.M., Thomas, N., Walker, S.R., Paterson, J.W., *Xenobiotica* (1974) **4**, 681–692.

A280. Carlone, M.F., Fouts, J.R., *Xenobiotica* (1974) **4**, 705–715.

A281. Gandolfi, A.J., Buhler, D.R., *Xenobiotica* (1974) **4**, 693–704.

A282. Schwandt, H.-J., Sadeé, W., Beyer, K.-H., *Xenobiotica* (1974) **4**, 733–741.
A283. Barrow, A., Griffiths, L.A., *Xenobiotica* (1974) **4**, 743–754.
A284. Griffin, J.P., Williams, J.R.B., Maughan, E., *Xenobiotica* (1974) **4**, 755–764.
A285. Yamato, C., Takahashi, T., Fujita, T., Kuriyama, S., Hirose, N., *Xenobiotica* (1974) **4**, 765–777.
A286. Williams, F.M., Briant, R.H., Dollery, C.T., Davies, D.S., *Xenobiotica* (1974) **4**, 345–353.
A287. Kiese, M., Lenk, W., *Xenobiotica* (1974) **4**, 337–343.
A288. Riess, W., Huerzeler, H., Raschdorf, F., *Xenobiotica* (1974) **4**, 365–373.
A289. Bolt, A.G., Graham, G., Wilson, P., *Xenobiotica* (1974) **4**, 355–363.
A290. Griffiths, L.A., *Xenobiotica* (1974) **4**, 375–382.
A291. Smyth, R.D., Polk, A., Diamond, J., Burns, B.J., Herczeg, T., *Xenobiotica* (1974) **4**, 383–391.
A292. Beckett, A.H., Midha, K.K., *Xenobiotica* (1974) **4**, 297–311.
A293. Miller, J.J., Powell, G.M., Olavesen, A.H., Curtis, C.G., *Xenobiotica* (1974) **4**, 285–289.
A294. Yamato, C., Takahashi, T., Fujita, T., *Xenobiotica* (1974) **4**, 313–326.
A295. Bruce, R.B., Turnbull, L.B., Newman, J.H., Kinzie, J.M., Morris, P.H., Pinchbeck, F.M., *Xenobiotica* (1974) **4**, 197–207.
A296. Bakke, O.M., Draffan, G.H., Davies, D.S., *Xenobiotica* (1974) **4**, 237–254.
A297. Tanayama, S., Momose, S., Kanai, Y., Shirakawa, Y., *Xenobiotica* (1974) **4**, 219–227.
A298. Kohli, J.D., Khanna, R.N., Gupta, B.N., Dhar, M.M., Tandon, J.S., Sircar, K.P., *Xenobiotica* (1974) **4**, 97–100.
A299. Grantham, P.H., Mohan, L.C., Weisburger, E.K., Fales, H.M., *Xenobiotica* (1974) **4**, 69–76.
A300. Henderson, P.T., Vree, T.B., van Ginneken, C.A.M., van Rossum, J.M., *Xenobiotica* (1974) **4**, 121–130.
A301. Tanayama, S., Shirakawa, Y., Kanai, Y., Suzuoki, Z., *Xenobiotica* (1974) **4**, 33–47.
A302. Tanayama, S., Momose, S., Takagaki, E., *Xenobiotica* (1974) **4**, 57–65.
A303. Tanayama, S., Kanai, Y., *Xenobiotica* (1974) **4**, 49–56.
A304. Douch, P.G.C., *Xenobiotica* (1975) **5**, 401–406.
A305. Belpaire, F.M., Bogaert, M.G., Rosseel, M.T., Anteunis, M., *Xenobiotica* (1975) **5**, 413–420.
A306. Keysell, G.R., Booth, J., Sims, P., *Xenobiotica* (1975) **5**, 439–448.
A307. Gorrod, J.W., Temple, D.J., Beckett, A.H., *Xenobiotica* (1975) **5**, 453–463.
A308. Belpaire, F.M., Bogaert, M.G., *Xenobiotica* (1975) **5**, 431–438.
A309. Belpaire, F.M., Bogaert, M.G., *Xenobiotica* (1975) **5**, 421–429.
A310. Gorrod, J.W., Temple, D.J., Beckett, A.H., *Xenobiotica* (1975) **5**, 465–474.
A311. Glinsukon, T., Benjamin, T., Grantham, P.H., Weisburger, E.K., Roller, P.P., *Xenobiotica* (1975) **5**, 475–483.
A312. Wiśniewska-Knypl, J.M., Jablońska, J.K., *Xenobiotica* (1975) **5**, 511–519.
A313. Pelkonen, O., Kaltiala, E.H., Kärki, N.T., Jalonen, K., Pyörälä, K., *Xenobiotica* (1975) **5**, 501–509.
A314. Ryrfeldt, Å, Bodin, N.-O., *Xenobiotica* (1975) **5**, 521–529.
A315. Cho, A.K., Hodshon, B.J., Lindeke, B., Jonsson, J., *Xenobiotica* (1975) **5**, 531–538.
A316. Gigon, P.L., *Xenobiotica* (1975) **5**, 585–594.
A317. Dubin, P., Wright, K.L., *Xenobiotica* (1975) **5**, 563–571.
A318. Fanelli, R., Frigerio, A., Garattini, S., *Xenobiotica* (1975) **5**, 595–600.
A319. Kodama, R., Sonoda, T., Yano, T., Furukawa, K., Amano, H., Noda, K., Ide, H., *Xenobiotica* (1975) **5**, 601–609.

A321. Cheng, L.K., Fung, H.-L., *Xenobiotica* (1975) **5**, 611–624.
A322. Beckett, A.H., Al-Sarraj, S., Essien, E.E., *Xenobiotica* (1975) **5**, 325–355.
A323. Burke, M.D., Bridges, J.W., *Xenobiotica* (1975) **5**, 357–376.
A324. Douch, P.G.C., Blair, S.S.B., *Xenobiotica* (1975) **5**, 279–292.
A325. Tatsumi, K., Kitamura, S., Yoshimura, H., Tanaka, S., Hashimoto, K., Igarashi, T., *Xenobiotica* (1975) **5**, 377–388.
A326. Groseclose, E.E., Ribbons, D.W., *Biochem. Biophys. Res. Commun.* (1973) **55**, 897–903.
A327. Michalover, J.L., Ribbons, D.W., *Biochem. Biophys. Res. Commun.* (1973) **55**, 888–896.
A328. Gardner, A.M., Chen, J.T., Roach, J.A.G., Ragelis, E.P., *Biochem. Biophys. Res. Commun.* (1973) **55**, 1377–1384.
A329. Kishore, G., Sugumaran, M., Vaidyanathan, C.S., *Biochem. Biophys. Res. Commun.* (1974) **56**, 851–859.
A330. Aunis, D., Miras-Portugal, M.T., Mandel, P., *Biochem. Biophys. Res. Commun.* (1974) **57**, 1192–1199.
A331. Vore, M., Hamilton, J.G., Lu, A.Y.H., *Biochem. Biophys. Res. Commun.* (1974) **56**, 1038–1044.
A332. Rush, R.A., Thomas, P.E., Kindler, S.H., Udenfriend, S., *Biochem. Biophys. Res. Commun.* (1974) **57**, 1301–1305.
A333. Sémériva, M., Chapus, C., Bovier-Lapierre, C., Desnouelle, P., *Biochem. Biophys. Res. Commun.* (1974) **58**, 808–813.
A334. Swenson, D.H., Miller, E.C., Miller, J.A., *Biochem. Biophys. Res. Commun.* (1974) **60**, 1036–1043.
A335. Rahimtula, A.D., O'Brien, P.J., *Biochem. Biophys. Res. Commun.* (1974) **60**, 440–447.
A336. Nakazawa, H., Kumazai, H., Yamada, H., *Biochem. Biophys. Res. Commun.* (1974) **61**, 75–82.
A337. Nakagawa, Y., Kaiser, E.T., *Biochem. Biophys. Res. Commun.* (1974) **61**, 730–734.
A338. Tobes, M.C., Mason, M., *Biochem. Biophys. Res. Commun.* (1975) **62**, 309–397.
A339. Grandjean, C., Cavalieri, E., *Biochem. Biophys. Res. Commun.* (1974) **61**, 912–919.
A340. Borchardt, R.T., Cheng, C.-F., Thakker, D.R., *Biochem. Biophys. Res. Commun.* (1975) **63**, 69–77
A341. Nishikimi, M., *Biochem. Biophys. Res. Commun.* (1975) **63**, 92–98.
A342. Rapp, P., Kumagai, H., Yamada, H., Ueno, T., Fukami, H., *Biochem. Biophys. Res. Commun.* (1975) **64**, 241–247.
A343. Schreiber, W., *Biochem. Biophys. Res. Commun.* (1975) **63**, 509–514.
A344. Fink, A.L., Angelides, K.J., *Biochem. Biophys. Res. Commun.* (1975) **64**, 701–708.
A345. Aprille, J.R., Malamud, D.F., *Biochem. Biophys. Res. Commun.* (1975) **64**, 1293–1302.
A346. Nakai, C., Nozaki, M., Hayashi, O., Saito, I., Matsuura, T., *Biochem. Biophys. Res. Commun.* (1975) **67**, 590–595.
A347. Hsiung, K.P., Kuan, S.S., Guilbault, G.G., *Biochem. Biophys. Res. Commun.* (1975) **66**, 1225–1230.
A348. Oelschäger, H.A.H., Temple, D.J., Temple, C.F., *Xenobiotica,* (1975) **5**, 309–323.
A349. Douch, P.G.C., *Xenobiotica,* (1975) **5**, 293–302.
A350. Booth, J., Hewer, A., Keysell, G.R., Sims, P., *Xenobiotica,* (1975) **5**, 197–203.
A351. Matsubara, T., Nakamura, Y., Tochino, Y., *Xenobiotica,* (1975) **5**, 205–212.
A352. Yanagi, Y., Haga, F., Endo, M., Kitagawa, S., *Xenobiotica,* (1975) **5**, 245–257.
A353. Cohen, G.M., Flockhart, I.R., *Xenobiotica,* (1975) **5**, 213–222.

A354. Case, D.E., Reeves, P.R., *Xenobiotica,* (1975) **5**, 113–129.

A355. Case, D.E., Illston, H., Reeves, P.R., Shuker, B., Simons, P., *Xenobiotica,* (1975) **5**, 83–111.

A356. Forshell, G.P., Case, D.E., Reeves, P.R., *Xenobiotica,* (1975) **5**, 73–82.

A357. Bobik, A., Holder, G.M., Ryan, A.J., Wiebe, L.I., *Xenobiotica,* (1975) **5**, 65–72.

A358. Pan, H.P., Hook, G.E.R., Fouts, J.R., *Xenobiotica,* (1975) **5**, 17–24.

A359. Kodama, R., Yano, T., Furukawa, K., Noda, K., Ide, H., *Xenobiotica,* (1975) **5**, 39–48.

A360. Thompson, R.M., Gerber, N., Seibert, R.A., *Xenobiotica,* (1975) **5**, 145–153.

A361. Strand, L.P., Scheline, R.R., *Xenobiotica,* (1975) **5**, 49–63.

A362. Atef, M., Nielsen, P., *Xenobiotica,* (1975) **5**, 167–172.

A363. Wood, B.A., Stopher, D.A., Monro, A.M., *Xenobiotica,* (1975) **5**, 183–195.

A364. Narayanan, T.K., Rao, G.R., *Biochem. Biophys. Res. Commun.* (1974) **58**, 728–735.

A365. Smith, I., Mitchell, P.D., *Biochem. J.* (1974) **142**, 189–191.

A366. Belmaker, R.H., Murphy, D.L., Wyatt, R.J., Loriaux, L., *Arch. Gen. Psychiat.* (1974) **31**, 553–556.

A367. Houslay, M.D., Tipton, K.F., *Biochem. J.* (1974) **139**, 645–652.

A368. Molinoff, P.B., Weinshilboum, R., Axelrod, J., *J. Pharmacol. Exp. Ther.* (1971) **178**, 425–431.

A369. Brand, L.M., Harper, A.E., *Proc. Soc. Exp. Biol. Med.* (1974) **147**, 211–215.

A370. Boulton, A.A., Dyck, L.E., Durden, D.A., *Life Sci.* (1974) **15**, 1673–1683.

A371. Bronaugh, R.L., McMurty, J., Hoehn, M.M., Rutledge, C.O., *Biochem. Pharmacol.* (1975) **24**, 1317–1320.

A372. Abou-El-Makarem, M.M., Otani, G., Bock, K.W., *Biochem. Soc. Trans.* (1975) **3**, 881–883.

A373. Honma, Kuriowa, Y., Hamada, A., *Chem. Pharm. Bull.* (1975) **23**, 1745–1751.

A374. Shaw, M.A., Jeffery, J., *Biochem. Soc. Trans.* (1975) **3**, 889–890.

A375. Holler, M., Napp, M., Breuer, H., *Biochem. Soc. Trans.* (1975) **3**, 890–892.

A376. Donnelly, W.J., *Biochem. Soc. Trans.* (1975) **3**, 1208–1210.

A377. Lenk, W., *Biochem. Soc. Trans.* (1975) **3**, 961–962.

A378. Bronaugh, R.L., Hattox, S.E., Hoehn, M.M, Murphy, R.C., Rutledge, C.O., *J. Pharmacol. Exp. Ther.* (1975) **195**, 441–452.

A379. Subba Rao, G., Krishna, G., Gillette, J.R., *J. Pharmacol. Exp. Ther.* (1975) **195**, 433–440.

A380. Francis, T.J.R., Bick, R.J., Callaghan, P., Hopkins, R.P., *Biochem. Soc. Trans.* (1975) **3**, 1244–1246.

A381. Lloyd, K.G., Davidson, L., Hornykiewicz, O., *J. Pharmacol. Exp. Ther.* (1975) **195**, 453–464.

A382. Johnson, R.W., Grossman, A., *Biochem. Biophys. Res. Commun.* (1974) **59**, 520–526.

A383. Lloyd, T., Kaufman, S., *Biochem. Biophys. Res. Commun.* (1974) **59**, 1262–1269.

A384. Dixon, P.A.F., Caldwell, J., Woods, C.J., Smith, R.L., *Biochem. Soc. Trans.* (1976) **4**, 143–145.

A385. O'Neill, W.P., Nystrom, R.F., Rinehart, K.L., Jr., Gottlieb, D., *Biochemistry* (1973) **12**, 4775–4784.

A386. Forist, A.A., Buhler, D.R., Harpootlian, H., Judy, R.W., *J. Med. Chem.* (1971) **14**, 969–971.

A387. Horning, M.G., Stratton, C., Wilson, A., Horning, E.C., Hill, R.M., *Anal. Lett.* (1971) **4**, 537–545.

A388. Beckett, A.H., Al-Sarraj, S., *J. Pharm. Pharmacol.* (1972) **24**, 174–176.

A389. Wall, M.E., *Ann. N. Y. Acad. Sci.* (1971) **191**, 23–39.

A390. Imomura, H., Matsui, E., Kato, Y., Furuta, T., *Yakugaku Zasshi* (1971) **91**, 811–817.

A391. Meshi, T., Sugihara, J., Sato, Y., *Chem. Pharm. Bull.* (1971) **19**, 1546–1556.

A392. Medenwald, H., Schlossman, K., Wuenschke, C., *Arzneim.-Forsch.* (1972) **22**, 53–56.

A393. Sumner, B., *Ann. N. Y. Acad. Sci.* (1971) **191**, 61–67.

A394. Idle, J.R., Millburn, P., Williams, R.T., *Biochem. Soc. Trans.* (1976) **4**, 139–141.

A395. Idle, J.R., Millburn, P., Williams, R.T., *Biochem. Soc. Trans.* (1976) **4**, 141–143.

A396. Valerino, D.M., Johns, D.G., Zuharko, D.S., Oliviero, V.T., *Biochem. Pharmacol.* (1972) **21**, 821–831.

A397. Queisnerova, M., Metys, J., Svatek, E., *Cesk. Farm.* (1971) **20**, 379–382.

A398. Miyazaki, T., Peric-Golia, L., Slaunwhite, W.R., Jr., Sandberg, A.A., *Endocrinology* (1972) **90**, 516–524.

A399. Quamme, G.A., Layne, D.S., Williamson, D.G., *Comp. Biochem. Physiol. A* (1971) **40**, 257–263.

A401. Quamme, G.A., Layne, D.S., Williamson, D.G., *Can. J. Physiol. Pharmacol.* (1972) **50**, 45–47.

A402. Ashworth, R.J., Sheets, T.J., *J. Agric. Food Chem.* (1972) **20**, 407–12.

A403. Tucker, B.V., Pack, D.E., *J. Agric. Food Chem.* (1972) **20**, 412–416.

A404. Aurich, H., Luppa, D., Schuecker, G., *Acta Biol. Med. Ger.* (1972) **28**, 209–220.

A405. LeHir, M., Schneider, A., *C. R. Acad. Sci., Ser. D* (1972) **274**, 2034–2037.

A406. Huffman, D.H., Bachur, N.R., *Cancer Res.* (1972) **32**, 600–605.

A407. Suzuki, T., Nose, K., *Noyaku Seisan Gijutsu* (1971) **26**, 21–24.

A408. Beckett, A.H., Shenoy, E.V.B., Salmon, J.A., *J. Pharm. Pharmacol.* (1972) **24**, 194–202.

A409. Beckett, A.H., Vaughan, D.P., Essien, E.E., *J. Pharm. Pharmacol.* (1972) **24**, 244.

A410. Tacker, M., Creaven, P.J., McIsaac, W.M., *J. Pharm. Pharmacol.* (1972) **24**, 247–249.

A411. Tacker, M., McIsaac, W.M., Creaven, P.J., *J. Pharm. Pharmacol.* (1972) **24**, 245–246.

A412. Levy, G., Tsuchiya, T., Amsel, L.P., *Clin. Pharmacol. Ther.* (1972) **13**, 258–268.

A413. Goenecha, S., Hecke-Seibicke, E., *Z. Klin. Chem. Klin. Biochem.* (1972) **10**, 112–113.

A414. Ito, K., Murata, T., *Oyo Yakuri,* (1971) **5**, 851–856.

A415. DiFazio, C.A., Brown, R.E., *Anesthesiology* (1972) **36**, 238–243.

A416. Jakobson, I., Yllner, S., *Acta Pharmacol. Toxicol.* (1971) **29**, 513–534.

A417. Anderson, J.R., Horsgood, R.K., *Soil Biol. Biochem.* (1971) **3**, 271–280.

A418. Ishibashi, T., *Agric. Biol. Chem.* (1972) **36**, 596–603.

A419. Herbert, R.B., Holliman, F.G., Ibberson, P.N., *J. Chem. Soc., Chem. Commun.* (1972) 355–356.

A420. Knuppen, R., Ball, P., Haupt, O., Breuer, H., *Hoppe-Seyler's Z. Physiol. Chem.* (1972) **353**, 565–568.

A421. DiSanto, A.R., Wagner, J.G., *J. Pharm. Sci.* (1972) **61**, 598–602.

A422. Glasson, B., Benakis, A., Thomasset, M., *Arzneim-Forsch.* (1971) **21**, 1985–1992.

A423. Chang, T., Glazko, A.J., *Anesthesiology* (1972) **36**, 401–404.

A424. Takahashi, Y., Araki, K., Tsubaki, T., *Igaku No Ayumi* (1971) **79**, 175–176.

A425. Arcamone, F., Glaesser, A.H., Minghetti, A., Nicolella, V., *Boll. Chim. Farm.* (1971) **110**, 704–711.

A426. Back, D.J., Calvey, T.N., *Br. J. Pharmacol.* (1973) **44**, 534–543.

A427. Beckett, A.H., Shenoy, E.V.B., Brookes, L.G., *J. Pharm. Pharmacol.* (1972) **24**, 281–288.

A428. Dubnick, B., Towne, C.A., *Toxicol. Appl. Pharmacol.* (1972) **22**, 82–92.

A429. Fujita, T., Kawasaki, N., *Yakugaku Zasshi* (1972) **92**, 356–358.

A430. Watanabe, T., Snell, E.E., *Proc. Nat. Acad. Sci. U.S.A.* (1972) **69**, 1086–1090.

A431. Uesugi, Y., Tomizawa, C., *Agr. Biol. Chem.* (1972) **36**, 313–317.

A432. Bhattacharya, S., *Enzymologia* (1972) **42**, 107–114.

A433. Ola, F.N., Ordonez, C.R., *Rev. Latinoamer. Microbiol.* (1971) **13**, 285–289.

A434. Blaschko, H., *Acta Physiol. Scand.* (1972) **84**, 524–527.

A435. Ziegler, P., Kutzner, H.J., *Naturwissenschaften* (1972) **59**, 123.

A436. Cohen, A.I., Dreyfuss, J., Fales, H.M., *J. Med. Chem.* (1972) **15**, 542–545.

A438. Ho, B.T., Estevez, V., Englert, L.F., *J. Pharm. Pharmacol.* (1972) **24**, 414–416.

A439. Mixich, G., *Helv. Chim. Acta* (1972) **55**, 1031–1038.

A440. Burstein, S., Rosenfeld, J., Wittstruck, T., *Science* (1972) **176**, 422–423.

A441. Cranmer, M.F., *Bull. Environ. Contam. Toxicol.* (1972) **7**, 121–124.

A442. Sherman, M., Beck, J., Herrick, R.B., *J. Agric. Food Chem.* (1972) **20**, 617–624.

A443. Andrawes, N.R., Chancey, E.L., Crabtree, R.J., Herrett, R.A., Weiden, M.H.J., *J. Agric. Food Chem.* (1972) **20**, 608–617.

A444. Takahashi, G., Yasuhira, K., *Cancer Res.* (1972) **32**, 710–715.

A445. Chkanikov, D.I., Makeev, A.M., Pavlova, N.N., Dubovoi, V.P., *Fiziol. Rast* (1972) **19**, 436–442.

A446. Gibson, R.A., Schneider, E.A., Wightman, F., *J. Exp. Bot.* (1972) **23**, 381–399.

A447. Ball, P., Knuppen, R., Haupt, M., Breuer, H., *Eur. J. Biochem.* (1972) **26**, 560–569.

A448. Naito, S., Osumi, S., Sekishiro, K., Hirose, M., *Chem. Pharm. Bull.* (1972) **20**, 682–688.

A449. Irino, O., Tateishi, M., Miura, C., Fukawa, K., *Chem. Pharm. Bull.* (1972) **20**, 734–740.

A450. Dreyfuss, J., Bigger, J.T., Jr., Cohen, A.I., Schreiber, E.C., *Clin. Pharmacol. Ther.* (1972) **13**, 366–371.

A451. Imai, K., Sugiura, M., Kubo, H., Tamura, Z., Ohya, K., Tsunakawa, N., Hirayama, K., Narabayashi, H., *Chem. Pharm. Bull.* (1972) **20**, 759–764.

A452. Karim, A., Ranney, R.E., Evensen, K.L., Clark, M.L., *Clin. Pharmacol. Ther.* (1972) **13**, 407–419.

A453. Kaplan, S.A., Weinfeld, R.E., Abruzzo, C.W., Lewis, M., *J. Pharm. Sci.* (1972) **61**, 773–778.

A454. Hobkirk, R., Nilsen, M., Williamson, D.G., Layne, D.S., *J. Clin. Endocrinol. Metab.* (1972) **34**, 690–694.

A455. Huber, T.L., Horn, G.W., Beadle, R.E., *J. Anim. Sci.* (1972) **34**, 786–790.

A456. Fellman, J.K., Roth, E.S., Heriza, E.L., Fujita, T.S., *Biochem. Pharmacol.* (1976) **25**, 222–223.

A457. Roffman, M., Reigle, T.G., Orsulak, P.J., Schildkraut, J.J., *Biochem. Pharmacol.* (1976) **25**, 208–209.

A458. Croy, R.G., Selkirk, J.K., Harvey, R.G., Engel, J.F., Gelboin, H.V., *Biochem. Pharmacol.* (1976) **25**, 227–230.

A459. Norman, B.J., Neal, R.A., *Biochem. Pharmacol.* (1976) **25**, 37–45.

A460. Marca, G., Codazza, D., *Arch. Vet. Ital.* (1971) **22**, 137–143.

A461. Glinsukon, T., Benjamin, T., Grantham, P.H., Lewis, N.L., *Biochem. Pharmacol.* (1976) **25**, 95–97.

A462. Litwack, G., Lichtask, E., *Nature New Biol.* (1972) **237**, 149–150.

A463. Truelsen, T.A., *Physiol. Plant.* (1972) **26**, 289–295.

A464. Nari, J., Mouttet, C., Pinna, M.H., Ricard, J., *FEBS Lett.* (1972) **23**, 220–224.

A466. Minamikawa, T., Yoshida, S., Hasegawa, M., Komagata, K., Kato, K., *Agric. Biol. Chem.* (1972) **36**, 773–778.

A467. Jirku, H., Kadner, S., Levitz, M., *Steroids* (1972) **19**, 535–547.

A468. Pittman, C.S., Buck, M.W., Chambers, J.B., Jr., *J. Clin. Invest.* (1972) **51**, 1759–1766.

A469. Brandau, K., Axelrod, J., *Naunyn-Schmiedeberg's Arch. Pharmacol.* (1972) **273**, 123–133.

A470. Harvey, D.J., Glazener, L., Stratton, C., Nowlin, J., Hill, R.M., Horning, M.G., *Res. Commun. Pathol. Pharmacol.* (1972) **3**, 557–565.

A471. Mussini, E., Marcucci, F., Fanelli, R., Garattini, S., *Chem.-Biol. Interact.* (1972) **5**, 73–76.

A472. Ainsworth, L., *Steroids* (1972) **19**, 595–603.

A474. Nicholls, P.J., Orton, T.C., *Br. J. Pharmacol.* (1972) **45**, 48–59.

A475. Reidenberg, M.M., *Proc. Soc. Exp. Biol. Med.* (1972) **140**, 1059–1061.

A476. Brown, L.W., Forist, A.A., *J. Pharm. Sci.* (1972) **61**, 858–860.

A477. Weikel, J.H., Jr., Bartek, M.J., *Toxicol. Appl. Pharmacol.* (1972) **22**, 375–386.

A478. Fuller, R.W., Snoddy, H.D., Bromer, W.W., *Mol. Pharmacol.* (1972) **8**, 345–352.

A479. Mehendale, H.M., Dorough, H.W., *Pestic. Biochem. Physiol.* (1971) **1**, 307–318.

A480. Simpson, C.F., Thompson, N.P., Neilson, J.T., *Bull. Environ. Contam. Toxicol.* (1972) **7**, 277–283.

A481. McBain, J.B., Hoffman, L.J., Menn, J.J., Casida, J.E., *Pestic. Biochem. Physiol.* (1971) **1**, 356–365.

A482. Dann, R.C., Mitscher, L.A., Couri, D., *Res. Commun. Pathol. Pharmacol.* (1972) **3**, 667–675.

A483. Eastin, E.F., *Weed Res.* (1972) **20**, 255–260.

A485. Wallnoefer, P.R., Safe, S., Hutzinger, O., *Pestic. Biochem. Physiol.* (1971) **1**, 458–463.

A486. Creveling, C.R., Morris, N., Shimizu, H., Ong, H.H., Daly, J., *Mol. Pharmacol.* (1972) **8**, 398–409.

A487. Kaufman, D.D., Plimmer, J.R., Iwan, J., Klingebiel, U.I., *J. Agric. Food Chem.* (1972) **20**, 916–919.

A488. Johnston, H.W., Briggs, G.G., Alexander, M., *Soil Biol. Biochem.* (1972) **4**, 187–190.

A489. Blakley, E.R., *Can. J. Microbiol.* (1972) **18**, 1247–1255.

A490. Honma, S., Nambara, T., *Chem. Pharm. Bull.* (1972) **20**, 1343–1344.

A491. Breckwoldt, M., Flickinger, G.L., Murawec, T., Touchstone, J.C., *Endocrinology* (1972) **91**, 1–5

A492. Sisenwine, S.F., Tio, C.O., Shrader, S.R., Ruelius, H.W., *Arzneim.-Forsch.* (1972) **22**, 682–687.

A493. Miyake, T., Suzuki, Y., *Vitamins* (1972) **45**, 340–343.

A494. Chang, T., Okerholm, R.A., Glazko, A.J., *Anal. Lett.* (1972) **5**, 195–202.

A495. Davis, L.E., Westfall, B.A., *Amer. J. Vet. Res.* (1972) **33**, 1253–1262.

A496. Losito, R., Rousseau, M.A., *Thromb. Diath. Haemorrh.* (1972) **27**, 318–326.

A498. Fuller, R.W., Roush, B.W., *Int. J. Biochem.* (1972) **3**, 225–228.

A499. Steinegger, E., Hoevel, H., *Pharm. Acta Helv.* (1972) **47**, 222–234.

A500. Kennedy, G., Fancher, O.E., Calandra, J.C., Keller, R.E., *Food Cosmet. Toxicol.* (1972) **10**, 143–149.

A501. Jirku, H., Kadner, S., Levitz, M., *Steroids* (1972) **19**, 519–534.

A502. Siegel, M.R., Zabbia, A.J., Jr., *Phytopathology* (1972) **62**, 630–634.

A503. Towers, G.H.N., Duffey, S.S., Siegel, S.M., *Can. J. Zool.* (1972) **50**, 1047–1050.

A504. Vree, T.B., Muskens, A.T.J.M., Van Rossum, J.M., *Arch. Int. Pharmacodyn. Ther.* (1972) **197**, 392–395.

A505. Movrin, M., *Bull. Sci. Cons. Acad. Sci. Arts R. S. F. Yougoslavie, Sect. A* (1972) **17**, 2–3.

A506. Borga, O., Garle, M., Gutova, M., *Pharmacology* (1972) **7**, 129–137.

A507. Baumel, I.P., Gallagher, B.B., Mattson, R.H., *Arch. Neurol.* (1972) **27**, 34–41.

A508. Paulson, G.D., Docktor, M.M., Jacobsen, A.M., Zaylskie, R.G., *J. Agric. Food Chem.* (1972) **20**, 867–876.

A509. Layne, D.S., Quamme, G.A., Labow, R.S., Mellor, J.D., Polakova, A., Williamson, D.G., *Horm. Steroids, Proc. Int. Congr., 3rd* (1970) 317–322.

A510. Capel, I.D., French, M.R., Millburn, P., Smith, R.L., Williams, R.T., *Biochem. J.,* (1972) **127**, 25P-26P.

A511. Kexel, H., Schmidt, H.-L., *Biochem. Pharmacol.* (1972) **21**, 1009–1017.

A512. Raggavan, S.S., Rhoads, D.B., Kanfer, J.N., *Biochim. Biophys. Acta* (1972) **268**, 755–760.

A513. Meek, J.L., Neff, N.H., *J. Pharmacol. Exp. Ther.* (1972) **181**, 457–462.

A515. Wright, A.S., Crowne, R.S., Potter, D., *Xenobiotica* (1972) **2**, 7–23.

A516. Peppercorn, M.A., Goldman, P., *J. Pharmacol. Exp. Ther.* (1972) **181**, 555–562.

A517. Bartsch, H., Miller, J.A., Miller, E.C., *Biochim. Biophys. Acta* (1972) **273**, 40–51.

A518. Mulder, E., Lamers-Stahlhofen, G.J.M., van der Molen, H.J., *Biochim. Biophys. Acta* (1972) **260**, 290–297.

A519. Burns, D.J.W., Engel, L.L., Bethune, J.L., *Biochemistry* (1972) **11**, 2699–2703.

A520. Bullock, F.J., Bruni, R.J., Asbell, M.A., *J. Pharmacol. Exp. Ther.* (1972) **182**, 70–76.

A521. Sonntag, A.C., Stenger, V.G., Vogh, B.P., Maren, T.H., *J. Pharmacol. Exp. Ther.* (1972) **182**, 34–47.

A522. Jacobson, M.M., Tsai, I., Poland, A., Kuntzman, R., *J. Pharmacol. Exp. Ther.* (1972) **182**, 312–319.

A523. Auricchio, S., Pierro, M., Andria, G., de Ritis, G., *Biochim. Biophys. Acta* (1972) **274**, 420–425.

A524. Locke, R.K., Baron, R.L., *J. Agric. Food Chem.* (1972) **20**, 861–867.

A525. Feung, C.-S., Hamilton, R.H., Witham, F.H., Mumma, R.O., *Plant Physiol.* (1972) **50**, 80–86.

A526. Oya, H., Kato, T., Nagatsu, I., Nagatsu, T., *Arch. Oral Biol.* (1972) **17**, 1245–1248.

A527. Osawa, K., Abe, T., Muto, Y., Hosoya, N., *Life Sci* (1972) **11**, Pt 1, 693–700.

A528. Slechtova, Chundela, B., *Activ. Nerv. Super.* (1972) **14**, 119–120.

A529. Nambara, T., Numazawa, M., Ishioka, S., *Chem. Pharm. Bull.* (1972) **20**, 1145–1149.

A530. DiSanto, A.R., Wagner, J.C., *J. Pharm. Sci.* (1972) **61**, 1086–1090.

A531. Pfaender, F.K., Alexander, M., *J. Agric. Food Chem.* (1972) **20**, 842–846.

A532. Raha, C.R., *Indian J. Biochem. Biophys.* (1972) **9**, 105–110.

A533. Schröder, H., Campbell, D.E.S., *Clin. Pharmacol. Ther.* (1972) **13**, 539–551.

A534. Sethunathan, N., Yoshida, T., *Inst. Environ. Sci., Tech. Meet., Proc.* (1972) **18**, 355–257.

A535. Rucker, R.B., Goettlich-Riemann, W., *Enzymologia* (1972) **43**, 33–44.

A536. R.A., Barrett, G., Wightman, F., *J. Exp. Bot.* (1972) **23**, 775–786.

A537. Blaschko, H., Longlands, M.G., *Naunyn-Schmiedeberg's Arch. Pharmacol.* (1972) **273**, 430–433.

A538. O'Donovan, D.G., Horan, H., *Biochem. Physiol. Alkaloide, Int. Symp. 4th* (1969) 505–506.

A539. Liebisch, H.W., Bhavsar, G.C., Schaller, H.J., *Biochem. Physiol. Alkaloide, Int. Symp. 4th* (1969) 233–237.

A540. Sebek, O.K., Hoeksema, H., *J. Antibiot.* (1972) **25**, 434–436.

A541. Nagatsu, I., Sudo, Y., Nagatsu, T., *Enzymologia* (1972) **43**, 25–31.

A542. Frigerio, A., Fanelli, R., Biandrate, P., Passerini, G., Morselli, P.L., Garrattini, S., *J. Pharm. Sci.* (1972) **61**, 1144–1147.

A543. Watabe, T., Kiyonaga, K., *J. Pharm. Pharmacol.* (1972) **24**, 625–630.

A544. Lech, J.J., Costrini, N.U., *Comp. Gen. Pharmacol.* (1972) **3**, 160–166.

A545. Leung, K., Merkatz, I., Solomon, S., *Endocrinology* (1972) **91**, 523–528.

A546. Murray, K., Duggleby, C.J., Sala-Trepat, J.M., Williams, P.A., *Eur. J. Biochem.* (1972) **28**, 301–310.

A547. Dall'Acqua, F., Capozzi, A., Marciani, S., Caporale, G., *Z. Naturforsch. B* (1972) **27**, 813–817.

A548. Caporale, G., Dall'Acqua, F., Marciani, S., *Z. Naturforsch. B* (1972) **27**, 871–872.

A549. Sugiyama, H., Shigematsu, H., Kanki, T., *Appl. Entomol. Zool.* (1972) **7**, 61–70.

A551. Den Tandt, W.R., *Clin. Chim. Acta* (1972) **40**, 199–204.

A552. Wright, S.J.L., Forey, A., *Soil Biol. Biochem.* (1972) **4** 207–213.

A553. Wan, S.H., Riegelman, S., *J. Pharm. Sci.* (1972) **61**, 1278–1284.

A554. Wan, S.H., Von Lehmann, B., Riegelman, S., *J. Pharm. Sci.* (1972) **61**, 1288–1292.

A555. Wan, S.H., Riegelman, S., *J. Pharm. Sci.* (1972) **61**, 1284–1287.

A556. Marchal, P., Rigaud, J., *Arch. Mikrobiol.* (1972) **85**, 1–5.

A557. Peppercorn, M.A., Goldman, P., *Proc. Nat. Acad. Sci. U.S.A.* (1972) **69**, 1413–1415.

A558. Doster, R.C., Sinnhuber, R.O., *Food Cosmet. Toxicol.* (1972) **10**, 389–394.

A559. Minegishi, K., Asahina, M., Yamaha, T., *Chem. Pharm. Bull.* (1972) **20**, 1351–1356.

A560. Bull, D.L., Whitten, C.J., *J. Econ. Entomol.* (1972) **65**, 973–976.

A561. Smith, I., March, S.E., Gordon, A.J., *Clin. Chim. Acta* (1972) **40**, 415–419.

A562. Goridis, C., Meek, J.L., Neff, N.H., *Life Sci.* (1972) **11** (Pt. 1) 861–866.

A563. Buswell, J.A., Mahmood, A., *Arch. Mikrobiol.* (1972) **84**, 275–286.

A564. Tocci, P.M., Phillips, J., Sager, R., *Clin. Chim. Acta* (1972) **40**, 449–453.

A565. Blaschke, G., *Biochem. Physiol. Alkaloide, Int. Symp., 4th* (1969) , 283–286.

A566. Forest, J.C., Wightman, F., *Can. J. Biochem.* (1972) **50**, 813–829.

A567. Chang, T., Okerholm, R.A., Glazko, A.J., *Res. Commun. Chem. Pathol. Pharmacol.* (1972) **4**, 13–23.

A568. Ikeda, G.J., *Biochem. Pharmacol.* (1972) **21**, 2150–2153.

A569. Ikeda, M., Ohtsuji, H., Imamura, T., *Xenobiotica* (1972) **2**, 101–106.

A570. Hook, G.E.R., Bend, J.R., Hoel, D., Fouts, J.R., Gram, T.E., *J. Pharmacol. Exp. Ther.* (1972) **182**, 474–490.

A571. Labow, R.S., Layne, D.S., *Biochem. J.* (1972) **128**, 491–497.

A572. Ackermann, E., *Biochem. Pharmacol.* (1972) **21**, 2169–2180.

A573. Kamm, J.J., Szuna, A., Kuntzman, R., *J. Pharmacol. Exp. Ther.* (1972) **182**, 507–514.

A574. Harris, E.D., O'Dell, B.L., *Biochem. Biophys. Res. Commun.* (1972) **48**, 1173–1178.

A575. Friedman, Y., Arsenis, C., *Biochem. Biophys. Res. Commun.* (1972) **48**, 1133–1139.

A576. Gaertner, H.J., Breyer, U., *Arzneim.-Forsch.* (1972) **22**, 1084–1085.

A578. Thomas, R.C., Judy, R.W., *J. Med. Chem.* (1972) **15**, 964–968.

A579. Garcia, E.E., Benjamin, L.E., Fryer, R.I., Sternbach, L.H., *J. Med. Chem.* (1972) **15**, 986–987.

A580. Pekas, J.C., *Toxicol. Appl. Pharmacol.* (1972) **23**, 62–70.

A581. Porter, C.C., Torchiana, M.L., Stone, C.A., *Life Sci.* (1972) **11** (Pt. 1) 787–795.

A582. Patterson, D.S.P., Roberts, B.A., *Food Cosmet. Toxicol.* (1972) **10**, 501–512.

A583. Moriyama, H., Sugiyama, H., Shigematsu, H., *Pestic. Biochem. Physiol.* (1972) **2**, 1–7.

A584. Magalhaes, M., Campos, G., Almeida, T., *Rev. Microbiol.* (1972) **3**, 45–46.

A585. Rosengarten, H., Meller, E., Friedhoff, A.J., *Res. Commun. Chem. Pathol. Pharmacol.* (1972) **4**, 457–465.

A586. Kazano, H., Kearney, P.C., Kaufman, D.D., *J. Agric. Food Chem.* (1972) **20**, 975–979.

A587. Amsel, L.P., Davison, C., *J. Pharm. Sci.* (1972) **61**, 474–475.

A588. Davies, D.S., *Bull. Physio-Pathol. Resp.* (1972) **8**, 679–683.

A589. Marsel, J., Doering, G., Remberg, G., Spiteller, G., *Z. Rechtsmed.* (1972) **70**, 245–250.

A590. Taylor, J.A., *Clin. Pharmacol. Ther.* (1972) **13**, 710–718.

A591. Fishman, J., Schut, H., Solomon, S., *J. Clin. Endocrinol. Metab.* (1972) **35**, 339–344.

A592. Remberg, G., Marsel, J., Doering, G., Spiteller, G., *Arch. Toxikol.* (1972) **29**, 153–157.

A593. Harvey, G., Pierrepoint, C.G., Anderson, A.B.M., Turnbull, A.C., Griffiths, K., *Biol. Reprod.* (1972) **7**, 15–24.

A594. Tate, S.S., Orlando, J., Meister, A., *Proc. Nat. Acad. Sci. U.S.A.* (1972) **69**, 2505–2508.

A595. Gniot-Szulzycka, J., *Acta Biochim. Pol.* (1972) **19**, 181–190.

A596. Bizen, T., Hara, S., Ohba, T., Murakami, H., Sugama, S., *Hakko Kogaku Zasshi* (1972) **50**, 710–715.

A597. Katz, R., Jacobson, A.E., *Mol. Pharmacol.* (1972) **8**, 594–599.

A598. Loh, H.H., Berg, C.P., *J. Nutr.* (1972) **102**, 1331–1339.

A600. Ueda, M., Orita, K., Koizumi, T., *Chem. Pharm. Bull.* (1972) **20**, 2047–2050.

A601. Ueda, M., Tsurui, Y., Koizumi, T., *Chem. Pharm. Bull.* (1972) **20**, 2042–2046.

A602. Sullivan, L.J., Eldridge, J.M., Knaak, J.B., Tallant, M.J., *J. Agric. Food Chem.* (1972) **20**, 980–985.

A603. Aarbakke, J., Bakke, O.M., *Scand. J. Gastroenterol.* (1972) **7**, 417–421.

A604. Gál. E.M., *Brain Res.* (1972) **44**, 309–312.

A605. Wortmann, W., Cooper, D.Y., Touchstone, J.C., *Steroids* (1972) **20**, 321–330.

A606. Wightman, R.H., Staunton, J., Battersby, A.R., Hanson, K.R., *J. Chem. Soc., Perkin Trans. 1* (1972) 2355–2364.

A607. Strange, P.C., Staunton, J., Wiltshire, H.R., Battersby, A.R., Hanson, K.R., Havir, E.A., *J. Chem. Soc., Perkin Trans. 1* (1972) 2364–2372.

A608. Maebayashi, Y., Miyaki, K., Yamazaki, M., *Chem. Pharm. Bull.* (1972) **20**, 2172–2175.

A609. Chakrabarty, A.M., *J. Bacteriol.* (1972) **112**, 815–823.

A610. Thomas, J., Meffin, P., *J. Med. Chem.* (1972) **15**, 1046–1049.

A611. Cooper-Driver, G., Corner-Zamodits, J.J., Swain, T., *Z. Naturforsch. B* (1972) **27**, 943–946.

A613. Raig, P., Ammon, E.R., *Arzneim.-Forsch.* (1972) **22**, 1399–1404.

A614. Erwin, V.G., Heston, W.D.W., Tabakoff, B., *J. Neurochem.* (1972) **19**, 2260–2278.

A615. Motoyama, N., Dauterman, W.C., *Pestic. Biochem. Physiol.* (1972) **2**, 170–177.

A616. Somani, S.M., Roberts, J.B., Thomas, B.H., Wilson, A., *Eur. J. Pharmacol.* (1970) **12**, 114–119.

A617. Mather, L.E., Thomas, J., *Life Sci.* (1972) **11** (Pt. 1) , 915–919.

A618. Lin. C., Chang, R., Magat, J., Symchowicz, S., *J. Pharm. Pharmacol.* (1972) **24**, 911–913.

A619. Kaiser, C., Swagzdis, J.E., Flanagan, T.L., Lester, B.M., Burghard, G.L., Green, H., Zirkle, C.L., *J. Med. Chem.* (1972) **15**, 1146–1150.

A620. Tindell, G.L., Walle, T., Gaffney, T.E., *Life Sci.* (1972) **11** (Pt. 2) , 1029–1036.

A621. Symms, K.G., Juchau, M.R., *Biochem. Pharmacol.* (1972) **21**, 2519–2527.

A622. Alibert, G., Ranjeva, R., Boudet, A., *Biochim. Biophys. Acta* (1972) **279**, 282–289.
A623. Pohorecky, L.A., Baliga, B.S., *Biochem. Pharmacol.* (1972) **21**, 2859–2866.
A624. Leonardi, R.G., Inchiosa, M.A., Jr., *Biochem. Pharmacol.* (1972) **21**, 2804–2811.
A625. Goddard, P., Hill, M.J., *Biochim. Biophys. Acta* (1972) **280**, 336–343.
A626. Gabler, W.L., Hubbard, G.L., *Biochem. Pharmacol.* (1972) **21**, 3071–3073.
A627. Flynn, E.J., Lynch, M., Zannoni, V.G., *Biochem. Pharmacol.* (1972) **21**, 2577–2590.
A628. Walle, T., Ishizaki, T., Gaffney, T.E., *J. Pharmacol. Exp. Ther.* (1972) **183**, 508–512.
A629. Gerber, N., Lynn, R., Oates, J., *Ann. Intern. Med.* (1972) **77**, 765–771.
A630. Danon, A., Sapira, J.D., *Clin. Pharmacol. Ther.* (1972) **13**, 916–922.
A631. Intoccia, A.P., Hwang, B., Joseph, G., Konicki, G., Walkenstein., S.S., *Neuropharmacology* (1972) **11**, 761–775.
A632. Guibbert, D., Duperray, B., Pacheco, H., Tomatis, O., Turusov, V., *Therapie* (1972) **27**, 907–918.
A633. Knowles, C.O., Sonawane, B.R., *Bull. Environ. Contamin. Toxicol.* (1972) **8**, 73–76.
A634. Janko, Z., Stefaniak, H., Czerwinska, E., *Pr. Inst. Przem. Org.* (1970) **2**, 241–257.
A635. Davis, P.J., *Physiol. Plant.* (1972) **27**, 262–270.
A636. Nakano, K., Inamasu, Y., Hagihara, S., Obo, F., *Acta Med. Kagoshima* (1972) **14**, 229–239.
A637. Morgenstern, A.P., Nauta, W.T., *Arzneim.-Forsch.* (1972) **22**, 1766–1768.
A638. Högen, T., Jaenicke, L., *Eur. J. Biochem.* (1972) **30**, 369–375.
A639. Ribbons, D.W., Ohta, Y., Higgins, I.J., *Mol. Basis Electron Transp., Proc. Miami Winter Symp.* (1972) 251–274.
A640. Fournier, J.C., Catroux, G., C. R. *Acad. Sci., Ser. D* (1972) **275**, 1723–1726.
A641. Schuette, H.R., Mueller, P., *Biochem. Physiol. Pflanz.* (1972) **163**, 518–523.
A642. Ohkawara, A., Halprin, K.M., Taylor, J.R., Levine, V., *Br. J. Dermatol.* (1972) **87**, 450–459.
A643. Dewhurst, S.A., Croker, S.G., Ikeda, K., McCaman, R.E., *Comp. Biochem. Physiol. B* (1972) **43**, 975–981.
A644. Wada, F., Sugitachi, A., Usami, M., Taguchi, K., Ikeda, C., Sakamoto, Y., *J. Vitaminol.* (1974) **18**, 119–124.
A645. Ryan, K.J., Naftolin, F., Reddy, V., Flores, F., Petro, Z., *Am. J. Obstet. Gynecol.* (1972) **114**, 454–460.
A646. Schaumloeffel, E., Clausnitzer, M., Heck, R., Jaeger, R., Klonk, G., Muelverstedt, F., Kreyling, H., *Arzneim.-Forsch.* (1972) **22**, 1643–1650.
A647. Milic, B.L., Stojanović, S., *J. Sci. Food Agric.* (1972) **23**, 1163–1167.
A648. Banci, F., Cartoni, G.P., Cavalli, A., Monai, A., *Arzneim.-Forsch.* (1972) **22**, 1724–1726.
A649. Besserer, K., *Arzneim.-Forsch.* (1972) **22**, 1737–1738.
A650. Ikeda, M., Ohtsuji, H., Imamura, T., *Jap. J. Pharmacol.* (1972) **22**, 479–491.
A651. Kato, Y., *Yakugaku Zasshi* (1972) **92**, 1152–1156.
A652. Borud, O., Midtvedt, T., Gjessing, L.R., *Acta Pharmacol. Toxicol.* (1972) **31**, 540–549.
A653. Soleim, H., Scheline, R.R., *Acta Pharmacol. Toxicol.* (1972) **31**, 471–480.
A654. Bordeleau, L.M., Bartha, R., *Can. J. Microbiol.* (1972) **18**, 1857–1864.
A655. Ritzert, R.W., Hemphill, D.D., Sell, H.M., Bukovac, M.J., *J. Am. Soc. Hort. Sci.* (1972) **97**, 48–51.
A656. Kreuzaler, F., Hahlbrock, K., *FEBS Lett.* (1972) **28**, 69–72.
A657. Cardot, J., *C. R. Soc Biol.* (1972) **166**, 336–339.
A658. Juorio, A.V., *Br. J. Pharmacol.* (1972) **45**, 166p-167p.

A659. Coggeshall, R.E., Dewhurst, S.A., Weinreich, D., McCaman, R.E., *J. Neurobiol.* (1972) **3**, 259–265.

A660. West, D.W., Owen, E.C., *Br. J. Nutr.* (1973) **29**, 33–41.

A661. Bordeleau, L.M., Bartha, R., *Can. J. Microbiol.* (1972) **18**, 1873–1882.

A662. Thompson, R.M., Gerber, N., Seibert, R.A., Desiderio, D.M., *Res. Commun. Chem. Pathol. Pharmacol.* (1972) **4**, 543–552.

A663. Stoll, R.G., Holmes, P.D., Roth, S.B., Wagner, J.G., *Res. Commun. Chem. Pathol. Pharmacol.* (1972) **4**, 327–338.

A664. Chasseaud, L.F., Hawkins, D.R., Moore, D.H., Hathaway, D.E., *Arzneim.-Forsch.* (1972) **22**, 2003–2005.

A665. Cho, A.K., Lindeke, B., Hodshon, B.J., *Res. Commun. Chem. Pathol. Pharmacol.* (1972) **4**, 519–528.

A666. Douglas, J.S., Nicholls, P.J., *J. Pharm. Pharmacol.* (1972) **24** (Suppl) , 150P.

A667. Valerino, D.M., *Res. Commun. Chem. Pathol. Pharmacol.* (1972) **4**, 529–542.

A668. VandenHeuvel, W.J.A., Smith, J.L., Silber, R.H., *J. Pharm. Sci.* (1972) **61**, 1997–1998.

A669. Osumi, Y., Wada, I., Fujiwara, M., *Jap. J. Pharmacol.* (1972) **22**, 723–726.

A670. Jirku, H., Kadner, S., Levitz, M., *J. Clin. Endocrinol. Metab.* (1972) **35**, 522–534.

A671. Musey, P.I., Green, R.N., Hobkirk, R., *J. Clin. Endocrinol. Metab.* (1972) **35**, 448–457.

A672. Raymond, D.D., Alexander, M., *Pestic. Biochem. Physiol.* (1972) **2**, 270–277.

A673. Baba, S., Mori, Y., Toyoshi, K., *Yakugaku Zasshi* (1972) **92**, 1364–1369.

A674. Watanabe, M., Minegishi, K., Tsutsui, Y., *Cancer Res.* (1972) **32**, 2049–2053.

A675. Hollister, L.E., Kanter, S.L., Moore, F., Green, D.E., *Clin. Pharmacol. Ther.* (1972) **13**, 849–855.

A676. Howell, C.R., Stipanovic, R.D., Bell, A.A., *Pestic. Biochem. Physiol.* (1972) **2**, 364–370.

A677. Løvstad, R.A., *Acta Chem. Scand.* (1972) **26**, 2832–2836.

A678. Thomas, H., Mueller-Enoch, D., Roth, S., *Hoppe-Seyler's Z. Physiol. Chem.* (1972) **353**, 1894–1898.

A679. Salminen, K., *Comp. Biochem. Physiol. B.* (1973) **44**, 283–289.

A680. Wall, M.E., Brine, D.R., Pitt, C.G., Perez-Reyes, M., *J. Am. Chem. Soc.* (1972) **94**, 8579–8581.

A681. Yamamoto, M., Nishida, H., Inui, T., Ozaki, A., *Hakko Kogaku Zasshi* (1972) **50**, 884–892.

A682. Bornschein, I., Pfeifer, S., *Pharmazie* (1972) **27**, 561–562.

A683. Bornschein, I., Scheffner, J., Pfeifer, S., Pharmazie, (1972) **27**, 616–617.

A684. Conolly, M.E., Davies, D.S., Dollery, C.T., Morgan, C.D., Paterson, J.W., Sandler, M., *Br. J. Pharmacol.* (1972) **46**, 458–472.

A685. Shimeno, H., Yoshimura, H., *Yakugaku Zasshi* (1972) **92**, 1376–1379.

A686. Crocker, J.M., Thomas, G.H., *J. Endocrinol.* (1973) **56**, 213–218.

A687. Baba, S., Kawai, K., Horie, M., Shida, Y., *Yakugaku Zasshi* (1972) **92**, 1569–1571.

A688. Horiguchi, S., Teramoto, K., *Sangyo Igaku* (1972) **14**, 288–289.

A689. Al-Rubae, A.Y., Knowles, C.O., *J. Econ. Entomol.* (1972) **65**, 1600–1603.

A690. Bordeleau, L.M., Bartha, R., *Can. J. Microbiol.* (1972) **18**, 1865–1871.

A691. Baldwin, R.M., Snyder, C.D., Rapoport, H., *J. Am. Chem. Soc.* (1973) **95**, 276–278.

A692. Enei, H., Matsui, H., Yamashita, K., Okumura, S., Yamada, H., *Agric. Biol. Chem.* (1972) **36**, 1861–1868.

A693. Turano, P., Turner, W.J., Manian, A.A., *J. Chromatogr.* (1973) **75**, 277–293.

A694. Brown, L.W., Forist, A.A., *J. Pharm. Sci.* (1973) **62**, 145–146.

A695. Ramsay, C.H., Bodin, N.O., Hansson, E., *Arzneim.-Forsch.* (1972) **22**, 1962–1970.

A696. Ebbinghausen, W.O.R., Mowat, J., Vestergaard, P., *J. Pharm. Sci.* (1973) **62**, 146–148.

A697. Baba, S., Kawai, K., *Yakugaku Zasshi* (1972) **92**, 1534–1539.

A698. Dennis, D.A., Chapman, P.J., Dagley, S., *J. Bacteriol.* (1973) **113**, 521–523.

A699. Braun, U., Kalbhen, D.A., *Deut. Med. Wochenscr.* (1972) **97**, 1614–1615.

A700. Kaufman, D.D., Plimmer, J.R., Klingebiel, U.I., *J. Agric. Food Chem.* (1973) **21**, 127–132.

A701. Hutzinger, O., Nash, D.M., Safe, S., De Freitas, A.S.W., Norstrom, R.W., Wildish, D.J., Zitko, V., *Science* (1972) **178**, 312–314.

A702. Yamazaki, H.I., *Insect Biochem.* (1972) **2**, 431–444.

A703. Miyata, K., Noguchi, Y., Enomoto, M., *Jap. J. Exp. Med.* (1972) **42**, 483–489.

A704. Cardillo, R., Fuganti, C., Ghiringhelli, D., Giangrasso, D., Grasselli, P., *Tetrahedron Lett.* (1972) 4875–4878.

A705. Tanaka, T., *Bull. Pharm. Res. Inst. Osaka* (1972) **98/99**, 1–9.

A706. Ahmed, M., Focht, D.D., *Can. J. Microbiol.* (1973) **19**, 47–52.

A707. Sugumaran, M., Ramanarayanan, M., Vaidyanathan, C.S., *FEBS Lett.* (1973) **29**, 69–72.

A708. Enei, H., Matsui, H., Okumura, S., Yamada, H., *Agric. Biol. Chem.* (1972) **36**, 1869–1876.

A709. Harms, H., Priess, I., *Planta* (1973) **109**, 307–315.

A710. Queisnerova, M., Grimova, J., *Cesk. Farm.* (1972) **21**, 325–328.

A711. Thenen, S.W., Shin, Y.S., Stokstad, E.L.R., *Proc. Soc. Exp. Biol. Med.* (1973) **142**, 638–641.

A712. De Leenheer, A., Heyndrickx, A., *J. Pharm. Sci.* (1973) **62**, 31–36.

A713. Mitchard, M., Kendall, M.J., Chan, K., *J. Pharm. Pharmacol.* (1972) **24**, 915.

A714. Peppercorn, M.A., Goldman, P., *Gastroenterology* (1973) **64**, 240–245.

A715. Parker, S.E., Weinstock, J., *J. Med. Chem.* (1973) **16**, 34–37.

A716. Sawada, H., Yano, H., Kido, A., *Yakugaku Zasshi* (1972) **92**, 1237–1241.

A717. Baba, S., Enogaki, K., Matsuda, A., Nagase, Y., *Yakugaku Zasshi* (1972) **92**, 1270–1274.

A718. Korth, H., *Arch. Mikrobiol.* (1973) **89**, 67–72.

A719. Zingales, I.A., *J. Chromatogr.* (1973) **75**, 55–78.

A721. Laham, S., *Ann. Occup. Hyg.* (1972) **15**, 203–209.

A723. Imamaliev, A.I., Semykina, E.E., *Dokl. Akad. Nauk Uzb. SSR* (1972) **29**, 59–60.

A724. Persson, K., Persson, K., *Xenobiotica* (1972) **2**, 375–382.

A725. Mega, T., Ikenaka, T., Matsushima, Y., *J. Biochem.* (1972) **72**, 1197–1203.

A726. Rubin, A., Warrick, P., Wolen, R.L., Chernish, S.M., Ridolfo, A.S., Gruber, C.M., Jr., *J. Pharmacol. Exp. Ther.* (1972) **183**, 449–457.

A727. BeMiller, J.N., Colilla, W., *Phytochemistry* (1972) **11**, 3393–3402.

A728. Katchen, B., Buxbaum, S., Meyer, J., Ning, J., *J. Pharmacol. Exp. Ther.* (1973) **184**, 453–464.

A729. Bartsch, H., Dworkin, M., Miller, J.A., Miller, E.C., *Biochim. Biophys. Acta* (1972) **286**, 272–298.

A730. Livingstone, A., Fewson, C.A., *Biochem. J.,* (1972) **130**, 937–946.

A731. Musso, R.E., Zabin, I., *Biochemistry* (1973) **12**, 553–557.

A732. Sweat, M.L., Young, R.B., *Biochim. Biophys. Acta* (1973) **296**, 189–195.

A733. Barrow, A., Griffiths, L.A., *Xenobiotica* (1972) **2**, 575–586.

A734. West, S.B., Lu, A.Y.H., *Arch. Biochem. Biophys.* (1972) **153**, 298–303.

A735. Lu, A.Y.H., Jacobson, M., Levin, W., West, S.B., Kuntzman, R., *Arch. Biochem. Biophys.* (1972) **153**, 294–297.

A736. Mac Leod, S.M., Renton, K.W., Eade, N.R., *J. Pharmacol. Exp. Ther.* (1972) **183**, 489–498.

A737. Williamson, D.G., Layne, D.S., Nilsen, M., Hobkirk, R., *Can. J. Biochem.* (1972) **50**, 958–962.

A738. Akao, T., Omura, T., *J. Biochem.* (1972) **72**, 1245–1266.

A739. Grover, P.L., Hewer, A., Sims, P., *Biochem. Pharmacol.* (1972) **21**, 2713–2726.

A740. Mehta, S.U., Mattoo, A.K., Modi, V.V., *Biochem. J.,* (1972) **130**, 159–166.

A741. Pütter, J., Schlossmann, K., *Biochim. Biophys. Acta.* (1972) **286**, 186–188.

A742. Ho, B.T., Taylor, D., Walker, K.E., McIsaac, W., *Xenobiotica* (1972) **2**, 349–362.

A743. Robertson, A., Glynn, J.P., Watson, A.K., *Xenobiotica* (1972) **2**, 339–347.

A744. Meyer, T., Scheline, R.R., *Xenobiotica* (1972) **2**, 383–390.

A745. Meyer, T., Scheline, R.R., *Xenobiotica* (1972) **2**, 391–398.

A746. Buckland, J.L., Collins, R.F., Pullin, E.M., *Pestic. Sci.* (1973) **4**, 148–160.

A749. Lerch, K., Ettlinger, L., *Eur. J. Biochem.* (1972) **31**, 427–437.

A750. Gross, G.G., *Eur. J. Biochem.* (1972) **31**, 585–592.

A751. Lemperg, R., Lundgren, E., *Clin. Chim. Acta* (1973) **43**, 321–325.

A752. Abou-Chaar, C.I., Guenther, H.F., Manuel, M.F., Robbers, J.E., Floss, H.G., *Lloydia* (1972) **35**, 272–279.

A753. Truelsen, T.A., *Physiol. Plant.* (1973) **28**, 67–70.

A754. Sutamihardja, T.M., Ishikura, A., Nagamura, Y., Ishiguro, I., *Experientia* (1973) **29**, 26–28.

A755. Magnus, V., Iskric, S., Kveder, S., *Planta* (1973) **110**, 57–62.

A756. Gennser, G., Rosengren, E., Von Studnitz, W., *Experientia* (1973) **29**, 20–22.

A757. Bowman, W.R., Gretton, W.R., Kirby, G.W., *J. Chem. Soc., Perkin Trans. 1* (1973) 218–220.

A758. Gurny, O., Maynard, D.E., Potcher, R.G., Kierstead, R.W., *J. Am. Chem. Soc.* (1972) **94**, 7928–7929.

A759. Bigler, F., Quitt, P., Vecchi, M., Vetter, W., *Arzneim.-Forsch.* (1972) **22**, 2191–2198.

A760. Henderson, G.L., Wilson, B.K., *Res. Commun. Chem. Pathol. Pharmacol.* (1973) **5**, 1–8.

A761. Baty, J.D., Evans, D.A.P., *J. Pharm. Pharmacol.* (1973) **25**, 83–84.

A762. Campbell, D.B., *Vie. Med. Can. Fr.* (1973) **2**, 34–40.

A763. Kopitar, Z., Jauch, R., Hankwitz, R., Pelzer, H., Maass, D., Willim, K.-D., *Eur. J. Pharmacol.* (1973) **21**, 6–10.

A764. Huffman, D.H., Benjamin, R.F., Bachur, N.R., *Clin. Pharmacol. Ther.* (1972) **13**, 895–905.

A765. Clasing, D., Bohn, G., *Sportarzt. Sportmed.* (1972) **23**, 313–316.

A766. Mueller, W., Korte, F., *Proc. Int. Gas Chromatogr. Mass Spectrom.* (1972) 131–161.

A767. Ho, B.T., Estevez, V.S., *Res. Commun. Chem. Pathol. Pharmacol.* (1973) **5**, 215–218.

A768. Bayly, R.C., Wigmore, G.J., *J. Bacteriol.* (1973) **113**, 1112–1120.

A770. Lehrmann, C., Ullrich, V., Rummel, W., *Naunyn.-Schmiedeberg's Arch. Pharmacol.* (1973) **276**, 89–98.

A771. Goswami, M.N.D., Rosenberg, A.J., Meury, F., *Develop. Biol.* (1973) **30**, 129–136.

A772. Aldridge, A., Johnson, G.E., *Can. J. Pharmacol.* (1972) **50**, 1061–1065.

A773. Tatematsu, A., Nadai, T., Yoshizumi, H., Sakurai, H., Furukawa, H., Hayashi, M., *Shitsuryo Bunseki* (1972) **20**, 339–346.

A774. Ritzl, F., *Ber. Kernforschungsanlage Juelich, Juel. 888-ME* (1972)

A775. Inoue, K., *Vitamins* (1972) **46**, 239–243.

A776. Marhold, J., Rambousek, V., Pípalová, J., Kroupa, J., Matrka, M., *Neoplasma* (1973) **20**, 27–30.

A777. Tomlin, A.D., Forgash, A.J., *J. Econ. Entomol.* (1973) **66**, 6–9.

A778. Kumar, R.P., Sreeleela, N.S., Subba Rao, P.V., Vaidyanathan, C.S., *J. Bacteriol.* (1973) **113**, 1213–1216.

A779. Knowles, C.O., Shrivastava, S.P., *J. Econ. Entomol.* (1973) **66**, 75–79.

A780. Shetty, A.S., Gaertner, F.H., *J. Bacteriol.* (1973) **113**, 1127–1133.

A781. Nakazawa. H., Enei, H., Okumura, S., Yamada, H., *Agric. Biol. Chem.* (1972) **36**, 2523–2528.

A782. Surks, M.I., Schadlow, A.R., Stock, J.M., Oppenheimer, J.H., *J. Clin. Invest.* (1973) **52**, 805–811.

A783. Phillips, J.P., Forrest, H.S., Kulkarni, A.D., *Genetics* (1973) **73**, 45–56.

A784. Shinada, T., Ryan, K.J., *Steroids* (1973) **21**, 233–244.

A785. Kimura, Y., Kume, M., Kageyama, K., *Oyo Yakuri* (1972) **6**, 955–971.

A786. Rice, C.P., Sikka, H.C., *J. Agric. Food Chem.* (1973) **21**, 148–152.

A787. Terayama, H., Naruke, T., Kasai, S., Numata, M., Nakayama, H., Saito, T., Irikura, T., *Chem. Pharm. Bull.* (1973) **21**, 12–20.

A788. Stenersen, J., Gilman, A., Verdanis, A., *J. Agric. Food Chem.* (1973) **21**, 166–171.

A789. Avila, J.L., Convit, J., *Biochim. Biophys. Acta* (1973) **293**, 397–408.

A791. Sullivan, P.T., Kester, M.V., Norton, S.J., *Biochim. Biophys. Acta* (1973) **293**, 343–350.

A792. Brady, F.O., Monaco, M.E., Forman, H.J., Schutz, G., Feigelson, P., *J. Biol. Chem.* (1972) **247**, 7915–7922.

A793. Shearer, M.J., Barkham, P., *Biochim. Biophys. Acta* (1973) **297**, 300–312.

A794. Long, R.F., Mantle, T.J., Wilson, K., *Biochem. Pharmacol.* (1976) **25**, 247–252.

A795. Sandler, M., Bonham Carter, S., Hunter, K.R., Stern, G.M., *Nature* (1973) **241**, 439–443.

A796. Steyn, M., Thiel, P.G., *Biochem. Pharmacol.* (1976) **25**, 265–266.

A797. Oyama, H., Izumi, H., Ozawa, H., *Biochem. Pharmacol.* (1976) **25**, 277–280.

A798. Smith, D.A., Cole, W.J., *Biochem. Pharmacol.* (1976) **25**, 367–370.

A799. Thiel, P.G., Steyn, M., *Biochem. Pharmacol.* (1976) **25**, 3267–3273.

A800. Izumi, H., Oyama, H., Hayakari, M., Ozawa, H., *Biochem. Pharmacol.* (1976) **25**, 488–489.

A801. Mathieu, P., Greffe, J., Deruaz, D., Guilluy, R., Gjessing, L., *Biochem. Pharmacol.* (1976) **25**, 497–500.

A802. Hinson, J.A., Mitchell, J.R., Jollow, D.J., *Biochem. Pharmacol.* (1976) **25**, 599–601.

A803. Nir, I., Hirschmann, N., Sulman, F.G., *Biochem. Pharmacol.* (1976) **25**, 581–583.

A804. Josting, D., Winne, D., Bock, K.W., *Biochem. Pharmacol.* (1976) **25**, 613–616.

A805. Baumel, I.P., Gallagher, B.B., DiMicco, J., Dionne, R., *J. Pharmacol. Exp. Ther.* (1976) **196**, 180–187.

A806. Merits, I., *Biochem. Pharmacol.* (1976) **25**, 829–833.

A807. Hsu, L.L., Geyer, M.A., Mandell, A.J., *Biochem. Pharmacol.* (1976) **25**, 815–819.

A808. Nagasawa, H.T., Kuo, T.H., Shirota, F.N., Alexander, C.S., *Biochem. Pharmacol.* (1976) **25**, 855–858.

A809. Durden, D.A., Philips, S.R., Boulton, A.A., *Biochem. Pharmacol.* (1976) **25**, 858–859.

A810. Day, H.J., Sohim, N.O., *Scand. J. Haematol.* (1973) **10**, 136–143.

A811. Kodama, K., Umehara, K., Shimizu, K., Nakatani, S., Minoda, Y., Yamada, K., *Agric. Biol. Chem.* (1973) **37**, 45–50.

A812. Ortivein, L., *Anz. Schaedlingsk Pflanzenschutz* (1972) **45**, 182–187.

A813. Hoffman, L.J., Ross, J.H., Menn, J.J., *J. Agric. Food Chem.* (1973) **21**, 156–163.

A814. Bedford, C.T., Knittel, P., Money, T., Phillips, G.T., Salisbury, P., *Can. J. Chem.* (1973) **51**, 694–697.

A815. Flesher, J.W., Sydnor, K.L., *Int. J. Cancer* (1973) **11**, 433–437.

A816. Pitot, H.C., Iwasaki, Y., Inoue, H., Kasper, C., Mohrenweiser, H., *Isozymes Enzyme Regul. Cancer, Proc. Conf.* (1971) 191–204.

A817. Batkai, L., Nogradi, M., Farkas, L., Feuer, L., Horvath, I., *Arch. Mikrobiol.* (1973) **90**, 165–166.

A818. Grosse, W., Witzsche, H., *Planta* (1973) **111**, 65–71.

A819. Ueno. T., Yoshizako, F., Nishimura, A., *Can. J. Microbiol.* (1973) **19**, 393–395.

A820. Roncucci, R., Simon, M.J., Gillet, C., Lambelin, G., Kaisin, M., *Proc. Int. Symp. Gas Chromatogr. Mass Spectrom.* (1972) 431–449.

A821. Borthakur, P.C., *Indian J. Physiol. Pharmacol.* (1972) **16**, 345–348.

A822. Støa, K.F., Skulstad, P.A., *Steroids Lipids Res.* (1972) **3**, 299–314.

A823. Roerig, S., Fujimoto, J.M., Wang, R.I.H., *Proc. Soc Exp. Biol. Med.* (1973) **143**, 230–233.

A824. Ferrieri, P., Wannamaker, L.W., Nelson, J., *Infec. Immunity* (1973) **7**, 747–752.

A825. Saarivirta, M., *Farm. Aikak.* (1973) **82**, 11–15.

A826. Kataoka, S., Taira, K., Ariyoshi, T., Takabatake E., *Chem. Pharm. Bull.* (1973) **21**, 358–365.

A828. Murata, H., Kougo, K., Yasumura, A., Nakajima, E., Shindo, H., *Chem. Pharm. Bull.* (1973) **21**, 404–414.

A829. Vikhlyaev, Y.I., Zherdev, V.P., *Farmakol. Toksikol.* (1973) **36**, 167–172.

A830. Åkerman, S.B.A., Ross, S.B., Telč, A., *Acta Pharmacol. Toxicol.* (1973) **32**, 88–96.

A831. Viau, J.-P., Epps, J.E., Di Carlo, F.J., *J. Pharm. Sci.* (1973) **62**, 641–645.

A832. Herber, R., Lepage, M., Pierfitte, M., Villoutreix, J., *C. R. Soc. Biol.* (1972) **166**, 1087–1090.

A833. Borchert, P., Wislocki, P.G., Miller, J.A., Miller, E.C., *Cancer Res.* (1973) **33**, 575–589.

A835. Percival, F.W., Purves, W.K., Vickery, L.E., *Plant Physiol.* (1973) **51**, 739–743.

A836. Fedecka-Bruner, B., Marin, L., Dameron, F., *C. R. Acad. Sci., Ser. D* (1973) **276**, 2325–2328.

A837. Kleiner, H., Brouet-Yager, M., *Clin. Chim. Acta* (1973) **45**, 109–112.

A838. Eschenhof, E., *Arzneim.-Forsch.* (1973) **23**, 390–400.

A839. Bod, P., Szarka, E., Gyimesi, J., Horváth, G., Vajna-Méhesfalvi, Z., Horváth, I., *J. Antibiot.* (1973) **26**, 101–103.

A840. Walkenstein, S.S., Intoccia, A.P., Flanagan, T.L., Hwang, B., Flint, B., Weinstock, J., Villani, A.J., Blackburn, D., Green, H., *J. Pharm. Sci.* (1973) **62**, 580–584.

A841. Sigel, C.W., Brent, D.A., *J. Pharm. Sci.* (1973) **62**, 694–695.

A842. Albro, P.W., Thomas, R., Fishbein, L., *J. Chromatogr.* (1973) **76**, 321–330.

A844. Widman, A.S., Soloway, A.H., Stern, R.L., Bursey, M.M., *Bioorg. Chem.* (1973) **2**, 176–178.

A845. Loh, H.H., Berg, C.P., *J. Nutr.* (1973) **103**, 397–406.

A846. Steck, W., Gamborg, O.L., Bailey, B.K., *Lloydia* (1973) **36**, 93–95.

A847. Attridge, T.H., Smith, H., *Plant Sci. Lett.* (1973) **1**, 247–252.

A848. Gross, G.G., Stöckigt, J., Mansell, R.L., Zenk, M.H., *FEBS Lett.* (1973) **31**, 283–286.

A849. Hempel, K., Lange, H.W., Kayser, E.F., Röger, L., Hennemann, H., Heidland, A., *Naunyn-Schmiedeberg's Arch. Pharmacol.* (1973) **277**, 373–386.

A850. Imashuku, S., Inui, A., Nakamura, T., Tanaka, J., Miyake, S., *J. Clin. Endocrinol. Metab.* (1973) **36**, 831–936.

A851. Enei, H., Matsui, H., Nakazawa, H., Okumura, S., Yamada, H., *Agric. Biol. Chem.* (1973) **37**, 493–499.

A853. Arndts, D., Pollmann, W., *Naunyn-Schmiedeberg's Arch. Pharmacol.* (1973) **277**, 349–362.

A854. McDermott, C.E., Casciano, D.A., Gaertner, F.H., *Biochem. Biophys. Res. Commun.* (1973) **51**, 813–818.

A855. Blount, J.F., Mohacsi, E., Vane, F.M., Mannering, G.J., *J. Med. Chem.* (1973) **16**, 352–355.

A856. Holtzman, J.L., Seligman, A.M., *Arch. Biochem. Biophys.* (1973) **155**, 237–246.

A857. Payne, A.H., Lawrence, C.C., Foster, D.L., Jaffe, R.B., *J. Biol. Chem.* (1973) **248**, 1598–1602.

A858. Smith, D.S., Peterson, R.E., Fujimoto, J.M., *Biochem. Pharmacol.* (1973) **22**, 485–492.

A859. Ferris, J.P., Fasco, M.J., Stylianopoulou, F.L., Jerina, D.M., Daly, J.W., Jeffrey, A.M., *Arch. Biochem. Biophys.* (1973) **156**, 97–103.

A860. Quebbeman, A.J., Anders, M.W., *J. Pharmacol. Exp. Ther.* (1973) **184**, 695–708.

A861. Strickland, S., Massey, V., *J. Biol. Chem.* (1973) **248**, 2944–2952.

A862. Zakim, D., Goldenberg, J., Vessey, D.A., *Biochim. Biophys. Acta* (1973) **309**, 67–74.

A863. Strickland, S., Massey, V., *J. Biol. Chem.* (1973) **248**, 2953–2962.

A865. Vessey, D.A., Goldenberg, J., Zakim, D., *Biochim. Biophys. Acta* (1973) **309**, 58–66.

A866. Fischer, L.J., Kent, T.H., Weissinger, J.L., *J. Pharmacol. Exp. Ther.* (1973) **185**, 163–170.

A867. Zakrewski, S.F., Grzelakowska-Sztabert, B., *J. Biol. Chem.* (1973) **248**, 2684–2690.

A868. Ecobichon, D.J., *Can. J. Biochem.* (1973) **51**, 506–513.

A869. Hearse, D.J., Weber, W.W., *Biochem. J.,* (1973) **132**, 519–526.

A870. Rance, M.J., Shillingford, J.S., *Biochem. Pharmacol.* (1976) **25**, 735–741.

A871. Mendoza, C.E., Augustinsson, K.-B., Axenfors, B., *Biochem. Pharmacol.* (1976) **25**, 701–705.

A872. Boissier, J.R., Gerardin, A., Dumont, C., *Arch. Pharm. Fr.* (1972) **30**, 851–860.

A873. Nishikawa, T., Mineura, K., Takahira, H., *Yakugaku Zasshi* (1973) **93**, 232–238.

A874. Murata, T., Ito, K., Minowa, T., Kirioka, K., *Oyo Yakuri* (1972) **6**, 1507–1514.

A875. Cradock, J.C., Egorin, M.J., Bachur, N.R., *Arch. Int. Pharmacodyn. Ther.* (1973) **202**, 48–61.

A876. Kimura, E.T., Young, P.R., Dodge, P.E., Swett, L.R., *Arch. Int. Pharmacodyn. Ther.* (1973) **202**, 119–134.

A877. Taska, R.J., Schoolar, J.C., *Arch. Int. Pharmacodyn. Ther.* (1973) **202**, 66–78.

A878. Chen, C.-T., Samejima, K., Tamura, Z., *Igaku No Ayumi* (1973) **84**, 195–196.

A879. Ryrfeldt, A., Ramsay, C.H., Nilsson, I.M., Widman, M., Agurell, S., *Acta Pharm. Suec.* (1973) **10**, 13–28.

A880. Radimski, J.L., Conzelman, G.M., Jr., Rey, A.A., Brill, E., *J. Nat. Cancer Inst.* (1973) **50**, 989–995.

A881. Lichtenstein, E.P., Fuhremann, T.W., Hochberg, A.A., Zahlten, R.N., Stratman, F.W., *J. Agric. Food Chem.* (1973) **21**, 416–424.

A882. Viallard, J.L., Motta, C., Dastugue, G., *Pathol.-Biol.* (1973) **21**, 329–333.

A883. Del Vecchio, V.G., Lenox, R., Burnett, J.B., *Experientia* (1973) **29**, 344–345.

A884. Bababunmi, E.A., Smith, R.L., Williams, R.T., *Life Sci.* (1973) **12** (Pt 2) , 317–326.

A885. Foldes, A., Jeffrey, P.L., Preston, B.N., Austin, L., *J. Neurochem.* (1973) **20**, 1431–1442.

A886. Lemoyne, C., Jerome, H., Kamoun, P., *Biochimie* (1973) **55**, 233–235.

A887. Lin, C.-C.L., Blair, G.E., Cassady, J.M., Gröger. D., Maier, W., Floss, H.G., *J. Org. Chem.* (1973) **38**, 2249–2251.

A888. Beckett, A.H., Al-Sarraj, S., *J. Pharm. Pharmacol.* (1973) **25**, 335–336.

A889. Ishikawa, K., Akiyama, M., Nohara, Y., Tsuchiya, H., Negi, S., *Nichidai Igaku Zasshi* (1973) **32**, 499–507.

A890. Csetenyi, J., Baker, K.M., Frigerio, A., Morselli, P.L., *J. Pharm. Pharmacol.* (1973) **25**, 340–341.

A891. Sundaram, K.M.S., Smith, G.G., O'Brien, W., Bonnett, D., *Chem. Contr. Res. Inst., Ottawa, Inform. Rep. CC-X* (1972) 31.

A892. Li, C.Y., Lu, K.C., Trappe, J.M., Bollen, W.B., *Can. J. Bot.* (1973) **25**, 827–828.

A893. Erge, D., Maier, W., Groeger, D., *Biochem. Physiol. Pflanz.* (1973) **164**, 234–247.

A894. Minatogawa, Y., Noguchi, T., Kido, R., *J. Neurochem.* (1973) **20**, 1479–1481.

A896. Weinges, K., *Acta Phys. Chim. Debrecina* (1971) **17**, 265–272.

A897. Flores, F., Naftolin, F., Ryan, K.J., White, R.J., *Science* (1973) **180**, 1074–1075.

A898. Goodall, McC., *Advan. Neurol.* (1972) **1**, 517–524.

A899. Kato, Y., *Yakugaku Zasshi* (1973) **93**, 397–400.

A900. Munn, E.A., Bufton, S.F., *Eur. J. Biochem.* (1973) **35**, 3–10.

A901. Brookes, L.G., Forrest, I.S., *Res. Commun. Chem. Pathol. Pharmacol.* (1973) **5**, 741–758.

A902. Neujahr, H.Y., Gaal, A., *Eur. J. Biochem.* (1973) **35**, 386–400.

A903. Bernhardt, F.H., Erdin, N., Staudinger, H., Ullrich, V., *Eur. J. Biochem.* (1973) **35**, 126–134.

A905. McElroy, W.D., DeLuca, M., *Chemiluminescence Biolumin., Pap. Int. Conf.* (1972) 285–311.

A906. Uehleke, H., Tabarelli, S., *Naunyn-Schmiedeberg's Arch. Pharmacol.* (1973) **278**, 55–68.

A907. Caldwell, J., Hawksworth. G.M., *J. Pharm. Pharmacol.* (1973) **25**, 422–424.

A908. Dajani, R.M., Saheb, S.E., *Ann. N. Y. Acad. Sci.* (1973) **215**, 120–123.

A909. Burdfield, P.A., Calvey, T.N., Roberts, J.B., *J. Pharm. Pharmacol.* (1973) **25**, 428–429.

A910. Gradnik, R., Fleischmann, L., *Pharm. Acta Helv.* (1973) **48**, 181–191.

A911. Fleischmann, L., *Pharm. Acta Helv.* (1973) **48**, 192–199.

A912. Savage, N., Soldin, S.J., Levy, P.R., *Recl. Trav. Chim. Pays-Bas* (1973) **92**, 723–730.

A913. Naqvi, S.N.H., Muzaffar, S.A., Ashrafi, S.H., *Z. Pflanzenkr. Pflanzenschutz.* (1970) **77**, 577–581.

A914. Gibson, D.T., Gschwendt, B., Yeh, W.K., Kobal, V.M., *Biochemistry* (1973) **12**, 1520–1527.

A915. Chapman, D.I., Marcroft, J., *Xenobiotica* (1973) **3**, 49–61.

A917. Hsieh, D.P.H., Lin, M.T., Yao, R.C., *Biochem. Biophys. Res. Commun.* (1973) **52**, 992–997.

A918. Baugh, C.M., Krumdieck, C.L., Nair, M.G., *Biochem. Biophys. Res. Commun.* (1973) **52**, 27–34.

A919. Fazekas, A.G., Sandor, T., *Can. J. Biochem.* (1973) **51**, 772–782.

A920. Chang, R.L., *Biochem. Pharmacol.* (1973) **22**, 1389–1392.

A921. Bleszynski, W.S., Roy, A.B., *Biochim. Biophys. Acta* (1973) **317**, 164–171.

A922. Inaba, T., Brien, J.F., *J. Chromatogr.* (1973) **80**, 161–165.

A923. Rosazza, J.P., Juhl, R., Davis, P., *Appl. Microbiol.* (1973) **26**, 98–105.

A924. Sawada, H., Fukumoto, M., Yano, H., Hara, A., Kido, A., *Gifu Daigaku* (1972) **20**, 619–632.

A925. Ho, B.T., Pong, S.F., Browne, R.G., Walker, K.E., *Experientia* (1973) **29**, 275–277.

A926. Kitakaze, T., Ito, K., Ogawa, Y., *Chemotherapy* (1973) **21**, 224–228.

A927. Chang, T.T.L., Kuhlman, C.F., Schillings, R.T., Sisenwine, S.F., Tio, C.O., Ruelius, H.W., *Experientia* (1973) **29**, 653–655.

A928. Datta, P.R., *Pestic. Symp., Collect. Pap. Inter-Amer. Conf. Toxicol. Occup. Med. 6th, 7th* (1968–1970) 41–45.

A929. Mabee, M.S., Chipley, J.R., *Appl. Microbiol.* (1973) **25**, 763–769.

A931. Dünges, W., *Nature New Biol.* (1973) **243**, 60–61.

A932. Biellman, J.F., Branlant, G., Gero-Robert, M., Poiret, M., *Tetrahedron* (1973) **29**, 1237–1241.

A933. Carlström, K., Krook, K., *Acta Chem. Scand.* (1973) **27**, 1240–1248.

A934. Nakayama, Y., Nonomura, S., Tatsumi, C., *Agric. Biol. Chem.* (1973) **37**, 1423–1437.

A935. Pendleton, R.G., Snow, I.B., Gessner, G., Green, H., *Arch. Int. Pharmacodyn. Ther.* (1973) **203**, 368–375.

A936. Presl, J., Herzmann, J., Roehling, S., Horsky, J., *Endocrinol. Exp.* (1973) **7**, 119–123.

A938. Wiese, M.V., Vargas, J.M., Jr., *Pestic. Biochem. Physiol.* (1973) **3**, 214–222.

A939. Thorn, G.D., *Pestic. Biochem. Physiol.* (1973) **3**, 137–140.

A940. Lee, S.S., Griffin, D.A., Fang, S.C., *Weed Res.* (1973) **13**, 234–235.

A941. Uchiyama, M., Abe, H., Sato, R., Shimura, M., Watanabe, T., *Agric. Biol. Chem.* (1973) **37**, 737–745.

A942. Perumal, A.S., Robins, E., *J. Neurochem.* (1973) **21**, 459–471.

A943. Hashimoto, K., *J. Gen. Appl. Microbiol.* (1973) **19**, 171–187.

A944. Maugra, M., Savigny, P., Lematre, J., C. R. *Acad. Sci., Ser. D* (1973) **276**, 3221–3223.

A945. Calvert, R.T., Spring, M.S., Stoker, J.R., *J. Pharm. Pharmacol.* (1973) **25**, 511–512.

A946. Smith, M.D., *Comp. Gen. Pharmacol.* (1973) **4**, 175–178.

A947. Gabler, W.L., Hubbard, G.L., *Arch. Int. Pharmacodyn. Ther.* (1973) **203**, 72–91.

A948. Kondo, M., Maeda, T., Tanji, S., *Oyo Yakuri* (1973) **7**, 15–24.

A949. Ho, B.T., Estevez, V.S., Englert, L.F., *J. Pharm. Pharmacol.* (1973) **25**, 488–490.

A950. Nambara, T., Honma, S., Asama, M., Akiyama, S., Nokubo, M., *Chem. Pharm. Bull.* (1973) **21**, 914–915.

A951. Wortmann, W., Johnston, D.E., Wortmann, B., Touchstone, J.C., *J. Steroid Biochem.* (1973) **4**, 271–276.

A952. Turner, B.C., Caro, J.H., *J. Environ. Qual.* (1973) **2**, 245–247.

A953. Yoshimura, H., Yamamoto, H., *Chem. Pharm. Bull.* (1973) **21**, 1168–1169.

A954. Koes, M.T., Forrester, L.J., Brown, H.D., *Food Cosmet. Toxicol.* (1973) **11**, 463–466.

A955. Kinoshita, N., Shears, B., Gelboin, H.V., *Cancer Res.* (1973) **33**, 1937–1944.

A956. Nagasaki, T., Sugita, M., Fukawa, H., *Agric. Biol. Chem.* (1973) **37**, 1701–1706.

A957. Vickery, L.E., Sherwin, J.E., Purves, W.K., *Plant Growth Subst., Proc. Int. Conf. 7th* (1970) 91–95.

A958. Hollstein, U., McCamey, D.A., *J. Org. Chem.* (1973) **38**, 3415–3417.

A959. Wallnoefer, P.R., Engelhardt, G., Safe, S., Hutzinger, O., *Chemosphere* (1973) **2**, 69–72.

A960. Crawford, R.L., McCoy, E., Harkin, J.M., Kirk, T.K., Obst, J.K., *Appl. Microbiol.* (1973) **26**, 176–184.

A961. Kirk, T.K., Lorenz, L.F., *Appl. Microbiol.* (1973) **26**, 173–175.

A962. Grant, D.J.W., *J. Appl. Bacteriol.* (1973) **36**, 47–59.

A963. Ohmori, T., Ikai, T., Minoda, Y., Yamada, K., *Agric. Biol. Chem.* (1973) **37**, 1599–1605.

A964. Bagchi, S.P., Zarycki, E.P., *Biochem. Pharmacol.* (1973) **22**, 1353–1368.

A965. Graefe, K.H., Stefano, F.J.E., Langer, S.Z., *Biochem. Pharmacol.* (1973) **22**, 1147–1160.

A966. Worwood, M., Dodgson, K.S., Hook, G.E.R., Rose, F.A., *Biochem. J.* (1973) **134**, 183–190.

A967. Oi, S., Yasunobu, K.T., *Biochem. Biophys. Res. Commun.* (1973) **53**, 631–637.

A968. Wu, P.H., Boulton, A.A., *Can. J. Biochem.* (1973) **51**, 1104–1112.

A969. Hook, G.E.R., Dodgson, K.S., Rose, F.A., Worwood, M., *Biochem. J.* (1973) **134**, 191–195.

A970. Tyler, T.R., Buhs, R.P., VandenHeuvel, W.J.A., *Biochem. Pharmacol.* (1973) **22**, 1383–1385.

A971. Bergius, L.M.M. Court, G.M., Dow, J., Harding, N.G.L., *Biochem.Soc. Trans.* (1973) **1**, 408–409.

A972. Coggon, P., Moss, G.A., Sanderson, G.W., *Phytochemistry* (1973) **12**, 1947–1955.

A973. Lan, S.J., Chando, T.J., Weliky, I., Schreiber, E.C., *J. Pharmacol. Exp. Ther.* (1973) **186**, 323–330.

A974. Siegel, M.R., *Phytopathology* (1973) **63**, 890–896.

A975. Braun, U., Kalbhen, D.A., *Pharmacology* (1973) **9**, 312–316.

A976. Grover, P.L., Hewer, A., Sims, P., *FEBS Lett.* (1973) **34**, 63–68.

A977. Welsh, J.D., Russell, L.C., *Int. J. Biochem.* (1973) **4**, 379–384.

A978. Davis, L.E., Westfall, B.A., Short, C.R., *Am. J. Vet. Res.* (1973) **34**, 1105–1108.

A979. Bickel, M.H., Minder, R., Di Francesco, C., *Experientia* (1973) **29**, 960–961.

A980. Von Lehmann, B., Wan, S.H., Riegelman, S., Becker, C., *J. Pharm. Sci.* (1973) **62**, 1493–1486.

A981. Fuller, R.W., Baker, J.C., Bromer, W.C., *Endocrinology* (1973) **93**, 740–742.

A982. Walker, N., *Soil Biol. Biochem.* (1973) **5**, 525–530.

A983. Dalezios, J.I., Hsieh, D.P.H., Wogan, G.N., *Food Cosmet. Toxicol.* (1973) **11**, 605–616.

A984. Perry, J.S., Heap, R.B., Amoroso, E.C., *Nature* (1973) **245**, 45–47.

A985. Juchau, M.R., Smuckler, E.A., *Toxicol. Appl. Pharmacol.* (1973) **26**, 163–179.

A986. Baker, K.M., Csetenyi, J., Frigerio, A., Morselli, P.L., Parravicini, F., Pifferi, G., *J. Med. Chem.* (1973) **16**, 703–705.

A987. Kirdani, R., Musey, P., Varkarakio, M.J., Sandberg, A.A., *Steroids* (1973) **22**, 1–18.

A989. Le Bourhis, B., *Parfums, Cosmet., Savons Fr.* (1973) **3**, 450–456.

A990. Fournier, J.C., Soulas, G., Catroux, G., *C. R. Acad. Sci., Ser. D* (1973) **276**, 2993–2996.

A991. Feil, V.J., Lamoureux, C.J.H., Styrvoky, E., Zaylskie, R.G., Thacker, E.J., Holman, G.M., *J. Agric. Food Chem.* (1973) **21**, 1072–1078.

A992. Battersby, A.R., Kelsey, J.E., Staunton, J., Suckling, K.E., *J. Chem. Soc., Perkin Trans. 1* (1973) 1609–1615.

A993. Hoeregott, H., *Biochem. Physiol. Pflanz.* (1973) **164**, 500–508.

A994. Ikeda, G.J., *J. Med. Chem.* (1973) **16**, 1157–1161.

A995. Nilsson, I., Agurell, S., Nilsson, J.L.G., Widman, M., Leander, K., *J. Pharm. Pharmacol.* (1973) **25**, 486–487.

A996. Feller, D.R., Basu, P., Mellon, W., Curott, J., Malspeis, L., *Arch. Int. Pharmacodyn. Ther.* (1973) **203**, 187–199.

A997. Mahajani, U., Haldar, I., Datta, A.G., *Eur. J. Biochem.* (1973) **37**, 541–552.

A998. Cone, E.J., *Tetrahedron Lett* (1973) 2607–2610.

A999. Saarivirta, M., *Planta Med.* (1973) **24**, 112–119.

A1000. Sullivan, H.R., Due, S.L., *J. Med. Chem.* (1973) **16**, 909–913.

A1001. Christiansen, J., Gram, L.F., *J. Pharm. Pharmacol.* (1973) **25**, 604–608.

A1002. Laanio, T.L., Kearney, P.C., Kaufman, D.D., *Pestic. Biochem. Physiol.* (1973) **3**, 271–277.

A1003. Nelson, J.D., Blair, W., Brinckman, F.E., Colwell, R.R., Iverson, W.P., *Appl. Microbiol.* (1973) **26**, 321–326.

A1004. Singh, D.V., Mukherjee, P., *Curr. Sci.* (1973) **42**, 391.

A1005. Noguchi, T., Nakamura, J., Kido, R., *Life Sci.* (1973) **13**, 1001–1010.

A1006. Engelhardt, G., Wallnöfer, P.R., Plapp, R., *Appl. Microbiol.* (1973) **26**, 709–719.

A1007. Rivas, N. de J., Whitaker, J.R., *Plant Physiol.* (1973) **52**, 501–507.

A1008. Miras-Portugal, M.T., Aunis, D., Mandel, P., *FEBS Lett.* (1973) **34**, 140–142.

A1009. Nozaka, J., *Osaka Daigaku Igaku Zasshi* (1971) **23**, 269–276.

A1010. Kutney, J.P., Verma, A.K., Young, R.N., *Tetrahedron* (1973) **29**, 2645–2660.

A1011. Willetts, A.J., *Antonie van Leeuwenhoek, J. Microbiol. Serol.* (1973) **39**, 585–597.

A1012. Kutney, J.P., Verma, A.K., Young, R.N., *Tetrahedron* (1973) **29**, 2661–2671.

A1013. Kutney, J.P., Salisbury, P.J., Verma, A.K., *Tetrahedron* (1973) **29**, 2673–2681.

A1014. Volz, M., *Arzneim.-Forsch.* (1973) **23**, 1024–1027.

A1015. Walser, M., Lund, P., Ruderman, N.B., Coulter, A.W., *J. Clin. Invest.* (1973) **52**, 2865–2877.

A1016. Zimmer, H., Kokosa, J., Garteiz, D.A., *Arzneim.-Forsch.* (1973) **23**, 1028–1029.

A1017. Brooks, M.A., De Silva, J.A.F., D'Arconte, L., *J. Pharm. Sci.* (1973) **62**, 1395–1397.

A1018. Gerber, N., Seibert, R.A., Thompson, R.M., *Res. Commun. Chem. Pathol. Pharmacol.* (1973) **6**, 499–511.

A1019. Cartoni, G.P., Cavalli, A., Giarusso, A., Rosati, F., *J. Chromatogr.* (1973) **84**, 419–422.

A1020. Horning, M.G., Butler, C., Harvey, D.J., Hill, R.M., Zion, T.E., *Res. Commun. Chem. Pathol. Pharmacol.* (1973) **6**, 565–578.

A1021. Stillwell, W.G., Stafford, M., Horning, M.G., *Res. Commun. Chem. Pathol. Pharmacol.* (1973) **6**, 579–590.

A1022. Fishman, J., Yoshizawa, I., Hellman, L., *Steroids* (1973) **22**, 401–411.

A1023. Schroder, H., Schroder, B.M., *Acta Pharm. Suec.* (1973) **10**, 263–268.

A1024. Holder, G.M., Willcox, S., *Life Sci.* (1973) **13**, 391–400.

A1025. Boden, N., Charlton, S.C., Holmes, M.C., Knowles, P.F., *Biochem. Soc. Trans.* (1973) **1**, 1008.

A1026. Shindo, H., Komai, T., Kawai, K., *Chem. Pharm. Bull.* (1973) **21**, 2031–2038.

A1027. Fuchs, S., Schlaeger, D.A., *Biochem. Biophys. Res. Commun.* (1973) **54**, 784–789.

A1028. Swenson, D.H., Miller, J.A., Miller, E.C., *Biochem. Biophys. Res. Commun.* (1973) **53**, 1260–1267.

A1029. Tomašić, J., Keglević, D., *Biochem. J.* (1973) **133**, 789–795.

A1030. Wallace, J.W. Grisebach, H., *Biochim. Biophys. Acta* (1973) **304**, 837–841.

A1031. Hori, K., Hashimoto, T., Nozaki, M., *J. Biochem.* (1973) **74**, 375–384.

A1032. Bacher, A., Mailänder, B., *J. Biol. Chem.* (1973) **248**, 6227–6231.

A1033. Carper, W.R., Stoddard, D.D., Martin, D.F., *Biochem. Biophys. Res. Commun.* (1973) **54**, 721–725.

A1034. Booth, J., Keysell, G.R., Sims, P., *Biochem. Pharmacol.* (1973) **22**, 1781–1791.

A1035. Borud, O., Midtvedt, T., Gjessing, L.R., *Acta Pharmacol. Toxicol.* (1973) **33**, 308–316.

A1036. Horng, A.J., Yang, S.F., *Biochim. Biophys. Acta* (1973) **321**, 456–460.

A1037. Gordon, G.R., Shafizadeh, A.G., Peters, J.H., *Xenobiotica* (1973) **3**, 133–150.

A1038. Harel, E., Mayer, A.M., Lehman, E., *Phytochemistry* (1973) **12**, 2649–2654.

A1039. Houslay, M.D., Tipton, K.F., *Biochem. J.* (1973) **135**, 173–186.

A1040. Robinson, A.R., Henneberry, G.O., Common, R.H., *Biochim. Biophys. Acta* (1973) **326**, 93–102.

A1041. Prough, R.A., *Arch. Biochem. Biophys.* (1973) **158**, 442–444.

A1042. Booth, G.M., Yu, C.-C., Hansen, D.J., *J. Environ. Qual.* (1973) **2**, 408–411.

A1043. Klug, E., *Deut. Apoth.-Ztg.* (1973) **113**, 989–990.
A1044. Shalam, N., Bdolah, A., *Comp. Biochem. Physiol. B* (1973) **46**, 691–698.
A1045. Coquil, J.F., Goridis, C., Mack, J., Neff, N.H., *Brit. J. Pharmacol.* (1973) **48**, 590–599.
A1046. Sankawa, U., Ebizuka, Y., Shibata, S., *Tetrahedron Lett.* (1973) 2125–2128.
A1047. Sternson, L.A., Sternson, A.W., *Chem.-Biol. Interact.* (1973) **7**, 143–150.
A1048. Engelhardt, G., Wallnoefer, P., *Arch. Mikrobiol.* (1973) **93**, 229–237.
A1049. Slater, G.P., Child, J.J., Hogge, L.R., *Can. J. Microbiol.* (1973) **19**, 1169–1171.
A1050. Perumal, A.S., Robins, E., *Brain Res.* (1973) **59**, 349–358.
A1051. Osborne, N.N., *Brit. J. Pharmacol.* (1973) **48**, 546–549.
A1052. Cohen, M.L., Chan, S.-L., Way, W.L., Trevor, A.J., *Anesthesiology* (1973) **39**, 370–376.
A1053. Sen, R., Parmar, S.S., Guha, S.R., *J. Pharm. Pharmacol.* (1973) **25**, 835–837.
A1054. Ciaranello, R.D., Barchas, J.D., *Progr. Brain Res.* (1973) **39**, 217–223.
A1055. Bacher, A., Gilch, B., Rappold, H., Lingens, F., *Z. Naturforsch, C* (1973) **28**, 614–617.
A1056. Marchetti, E., Mattalia, G., Bergesi, G., *Arzneim.-Forsch.* (1973) **23**, 1291–1295.
A1057. Jenner, P., Taylor, A.R., Campbell, D.B., *J. Pharm. Pharmacol.* (1973) **25**, 749–750.
A1058. Litwin, A., Adams, L.E., Hess, E.V., McManus, J., Zimmer, H., *Arthrhitis Rheum.* (1973) **16**, 217–220.
A1059. Kelsey, M.I., Moscatelli, E.A., *J. Chromatogr.* (1973) **85**, 65–74.
A1060. Hardeland, R., *Int. J. Biochem.* (1973) **4**, 357–364.
A1061. Soini, J., Majasaari, K., *Acta Chem. Scand.* (1973) **27**, 2115–2120.
A1062. Stoeckigt, J., Mansell, R.L., Gross, G.G., Zenk, M.H., *Z. Pflanzenphysiol.* (1973) **70**, 305–307.
A1063. Holmstead, R.L., Fukuto, T.R., March, R.B., *Arch. Environ. Contam. Toxicol.* (1973) **1**, 133–147.
A1064. Dustman, J.H., *Hoppe-Seyler's Z. Physiol. Chem.* (1973) **354**, 1068–1072.
A1065. Eiseman, J.L., Gehring, P.J., Gibson, J.E., *Toxicol. Appl. Pharmacol.* (1974) **27**, 140–144.
A1066. Dunn, G.L., Gallagher, G., Jr., Davis, L.D., Hoover, J.R.E., Stedman, R.J., *J. Med. Chem.* (1973) **16**, 996–1002.
A1067. Baker, P.C., Hoff, K.M., Smith, M.D., *Brain Res.* (1973) **58**, 147–155.
A1068. Yoshida, H., Niihara, H., Kametani, K., *Yakuzaigaku* (1973) **33**, 71–78.
A1069. Campbell, D.B., Jenner, P., Taylor, A.R., *Advan. Neurol.* (1973) **3**, 199–213.
A1070. Allison, M.J., Robinson, I.M., Baetz, A.L., *J. Bacteriol.* (1974) **117**, 175–180.
A1071. Minck, K., Schupp, R.R., Illing, H.P.A., Kahl, G.F., Netter, K.J., *Naunyn-Schmiedeberg's Arch. Pharmacol.* (1973) **279**, 347–360.
A1072. Thomas, H., Mueller-Enoch, D., Lax, E.R., *Hoppe-Seyler's Z. Physiol. Chem.* (1973) **354**, 1097–1102.
A1073. Sellberg, A., Ando, G., Moller, J.K., Hansen, T., *Farm. Revy* (1973) **72**, 373–376.
A1074. Kagiwada, K., Shoji, S., Kamasuka, T., Yoshioka, M., Koizumi, T., Hase, T., *Oyo Yakuri* (1973) **7**, 889–901.
A1075. Takada, S., Tanji, S., Isgiyama, N., *Oyo Yakuri* (1973) **7**, 881–887.
A1076. Balba, M.H., Abdallah, M.D., Zazou, M.H., Rahal, A., *Toxicology* (1973) **1**, 171–174.
A1077. Briggs, G.G., Walker, N., *Soil Biol. Biochem.* (1973) **5**, 695–697.
A1078. Takaji, K., Suzuki, T., Saitoh, Y., Nishihara, K., *Yakuzaigaku* (1971) **31**, 213–219.
A1079. Daniel. J.W., Green, T., Phillips, P.J., *Food Cosmet. Toxicol.* (1973) **11**, 781–792.
A1080. Yokosawa, A., Arai, H., Sato, H., Motomiya, M., Oka, S., *Sci. Rep. Res.*

Inst., Tohoku Univ., Ser. C (1972) **19**, 177–182.

A1081. Waterbury, L.D., Wendel, O.T., Pearce, L.A., *Res. Commun. Chem. Pathol. Pharmacol.* (1973) **6**, 855–865.

A1082. Steennis, P.J., Cordes, M.M., Hilkens, J.G.H., Müller, F., *FEBS Lett.* (1973) **36**, 177–180.

A1083. Blondel, J.-D., Huault, C., Faye, L., Rollin, P., Cohen, P., *FEBS Lett.* (1973) **36**, 239–244.

A1084. Paparo, A.A., Tate, S.S., *Comp. Gen. Pharmacol.* (1973) **4**, 383–391.

A1085. Nei, N., Tanaka, Y., Takata, N., *Hakko Kogaku Zasshi* (1974) **52**, 28–34.

A1086. von Bahr, C., Rane, A., Orrenius, S., Sjoqvist, F., *Acta Pharmacol. Toxicol.* (1974) **34**, 58–64.

A1087. de Silva, J.A.F., Bekersky, I., Puglisi, C.V., *J. Chromatogr. Sci.* (1973) **11**, 547–554.

A1088. Smith, R.V., Cook, M.R., *J. Pharm. Sci.* (1974) **63**, 161–162.

A1089. Merrill, A.H., Campbell, T.C., *Toxicol. Appl. Pharmacol.* (1974) **27**, 210–213.

A1090. Ono-Kamimoto, M., *J. Biochem.* (1973) **74**, 1049–1059.

A1091. Archakov, A.I., Karuzina, I.I., Kokareva, I.S., Bachmanova, G.I., *Biochem. J.* (1973) **136**, 371–379.

A1092. Landsberg, L., Taubin, H.L., *Biochem. Pharmacol.* (1973) **22**, 2789–2800.

A1093. Craine, J.E., Daniels, G.H., Kaufman, S., *J. Biol. Chem.* (1973) **248**, 7838–7844.

A1094. Lin. C.-C., Magat, J., Chang, R., McGlotten, J., Symchowicz, S., *J. Pharmacol. Exp. Ther.* (1973) **187**, 415–422.

A1095. Ushijima, Y., Suwa, K., Nakano, M., *Biochim. Biophys. Acta* (1973) **320**, 284–294.

A1096. Ishimura, Y., Hayaishi, O., *J. Biol. Chem.* (1973) **248**, 8610–8612.

A1097. Aunis, D., Miras-Portugal, M.-T., Mandel, P., *Biochim. Biophys. Acta* (1973) **327**, 313–327.

A1098. Davidson, G.E., Brown, J.P., Scott, J.M., *Biochem. Soc. Trans.* (1973) **1**, 1182–1184.

A1099. Wada, G.H., Fellman, J.H., *Biochemistry* (1973) **12**, 5212–5217.

A1100. Schill, L., Grisebach, H., *Hoppe-Seyler's Z. Physiol. Chem.* (1973) **354**, 1555–1562.

A1101. Taylor, K.B., *J. Biol. Chem.* (1974) **249**, 454–458.

A1102. Palmer, F.B.St.C., *Comp. Biochem. Physiol. B.* (1974) **47**, 515–519.

A1103. Suemitsu, R., Boku, S., Matsuoka, T., *Doshisha Daigaku Rikogaku Kenkyu Hokoku* (1973) **14**, 119–127.

A1104. Tseng, L., Gurpide, E., *Endocrinology* (1974) **94**, 419–423.

A1105. Baker, P.C., Hoff, K.M., Smith, M.D., *Brain Res.* (1974) **65**, 255–264.

A1106. Mechoulam, R., Ben Zvi, Z., Agurell, S., Nilsson, I.M., Nilsson, J.L.G., Edery, H., Grunfeld, Y., *Experientia* (1973) **29**, 1193–1195.

A1107. Granerus, A.K., Jagenburg, R., Svanborg, A., *Naunyn-Schmiedeberg's Arch. Pharmacol.* (1973) **280**, 429–439.

A1108. Turnbull, L.B., Johnson, C.P. III, Chen, Y.H., Sancilio, L.F., Bruce, R.B., *J. Med. Chem.* (1974) **17**, 45–48.

A1109. Zimmer, A., Stahl, E., Bücheler, A., *Arzneim.-Forsch.* (1973) **23**, 1798–1801.

A1110. Banerjee, N.C., Mishra, S.S., *Indian Vet. J.* (1973) **50**, 450–452.

A1111. Musey, P.I., Kirdani, R.Y., Bhanalaph, T., Sandberg, A.A., *Steroids* (1973) **22**, 795–817.

A1112. Purchase, I.F.H., Steyn, M., Gilfillan, T.C., *Chem-Biol. Interact.* (1973) **7**, 283–287.

A1113. Regan, L.A., Frear, D.S., Swanson, H.R., *Weed Sci.* (1973) **21**, 542–549.

A1114. Mori, Y., Toyoshi, K., Baba, S., *Chem. Pharm. Bull.* (1973) **21**, 2577–2584.

A1115. Corbella, A., Gariboldi, P., Jomni, G., Forgione, A., Marcucci, F., Martelli, P., Mussini, E., Mauri, F., *J. Chem. Soc., Chem. Commun.* (1973) 721–722.

A1116. Starovoitov, I.I., Pshirkov, S.Y., Nefedova, M.Y., Yakovlev, G.I., Adanin, V.M., Skyrabin, G.K., *Dokl. Akad. Nauk. SSR* (1973) **213**, 1210–1212.

A1117. Kirk, T.K., Lorenz, L.F., *Appl. Microbiol.* (1974) **27**, 360–367.

A1118. Chyczewski, T., *Pol. J. Pharmacol. Pharm.* (1973) **25**, 373–378.

A1119. Nielsen, P., *Acta Vet. Scand.* (1973) **14**, 647–649.

A1120. Sun, F.F., *J. Pharm. Sci.* (1973) **62**, 1780–1784.

A1121. Schwartz, M.A., Postma, E., Kolis, S.J., Leon, A.S., *J. Pharm. Sci.* (1973) **62**, 1776–1779.

A1122. Estevez, V.Z., Englert, L.F., Ho, B.T., *Res. Commun. Chem. Pathol. Pharmacol.* (1973) **6**, 821–827.

A1123. Sullivan, H.R., Due, S.L., McMahon, R.E., *Res. Commun. Chem. Pathol. Pharmacol.* (1973) **6**, 1072–1078.

A1124. Awazu, S., Imaoka, K., *Yakugaku Zasshi* (1973) **93**, 1603–1610.

A1125. Patel, J.M., Pawar, S.S., *Indian J. Med. Res.* (1973) **61**, 1666–1671.

A1126. Inoue, K., Itokawa, Y., *Biochem. Med.* (1973) **8**, 450–463.

A1128. Perera, K.P.W.C., Wickremasinghe, R.L., *Tea Quart.* (1972) **43**, 153–163.

A1129. Dutkiewicz, T., Szymanska, J., *Bromatol. Chem. Toksykol.* (1973) **6**, 323–327.

A1130. Kumar, R.P., Subba Rao, P.V., Vaidyanathan, C.S., *Indian J. Biochem Biophys.* (1973) **10**, 184–190.

A1131. Hanedo, K., Takeda, I., *Hakko Kogaku Zasshi* (1974) **52**, 106–113.

A1132. Floss, H.G., Tcheng-Lin, M., Chang, C.-j., Naidoo, B., Blair, G.E., Abou-Chaar, C.I., Cassady, J.M., *J. Am. Chem. Soc.* (1974) **96**, 1898–1909.

A1133. Samarodova-Bianki, G.B., *Tr. Prikl. Bot. Genet. Selek.* (1972) **48**, 195–201.

A1134. Thakur, M.L., Ibrahim, R.K., *Z. Pflanzenphysiol.* (1974) **71**, 391–397.

A1135. Gordon, W.R., Henderson, J.H.M., *Can. J. Bot.* (1973) **51**, 2047–2052.

A1136. Åberg, H.E., Hansson, H.E., Wetterberg, L., Ross, S.B., Frödén, Ö., *Life Sci.* (1974) **14**, 65–71.

A1137. Mazeaud, M.M., Mazeaud, F., *Comp. Gen. Pharmacol.* (1973) **4**, 209–217.

A1138. Ishii, A., Deguchi, T., Takahira, H., *Yakugaku Zasshi* (1973) **93**, 1383–1390.

A1139. Fleischmann, L., Gradnik, B., *Arzneim.-Forsch.* (1973) **23**, 1455–1457.

A1140. Nogami, H., Hanano, M., Awazu, S., Imaoka, K., *Yakugaku Zasshi* (1973) **93**, 1585–1592.

A1141. Nogami, H., Hanano, M., Awazu, S., Imaoka, K., *Yakugaku Zasshi* (1973) **93**, 1593–1602.

A1142. Bakke, O.M., Davies, D.S., Davies, L., Dollery, C.T., *Life Sci.* (1973) **13**, 1665–1675.

A1143. Rane, A., *Clin. Pharmacol. Ther.* (1974) **15**, 32–38.

A1144. Devereux, T.R., Fouts, J.R., *Chem.-Biol. Interact.* (1974) **8**, 91–105.

A1145. Yang, C.H., Srivastava, P.N., *Proc. Soc. Exp. Biol. Med.* (1974) **145**, 721–725.

A1146. Irikura, T., Naruke, T., Kasai, S., Akiyama, Y., Hashimoto, M., *Oyo Yakuri* (1973) **7**, 985–990.

A1147. Roughley, P.J., Whiting, D.A., *J. Chem. Soc., Perkin Trans. 1* (1973) 2379–2388.

A1148. Frear, D.S., Swanson, H.R., *Pestic. Biochem. Physiol.* (1973) **3**, 473–482.

A1149. Durand, R., Zenk, M.H., *FEBS Lett.* (1974) **39**, 218–220.

A1150. Reidenberg, M.M., Drayer, D., De Marco, A.L., Bello, C.T., *Clin. Pharmacol. Ther.* (1973) **14**, 970–977.

A1151. Nagatomi, K., Osawa, Y., Kirdani, R.Y., Sandberg, A.A., *J. Clin. Endocrinol. Metab.* (1973) **37**, 887–900.

A1152. Beckett, A.H., Ogunbona, F.A., *J. Pharm. Pharmacol.* (1973) **25** (Suppl.) 170P.

A1153. Jacquot, C., Rapin, J.R., Wepierre, J., Cohen, Y., *Arch. Int. Pharmacodyn. Ther.* (1974) **207**, 298–304.

A1154. Lefournier, C., Cohadon, F., Jullien, I., Giraud, J.G., Pinhas, H., *C. R. Acad. Sci., Ser. C* (1973) **277**, 1675–1677.
A1155. Bartholini, G., Lloyd, K.G., Pletscher, A., *Life Sci.* (1974) **14**, 323–328.
A1156. Bunden, K., Kawakami, S., Chiaravanond, P., Kratzer, F.H., *Poult. Sci.* (1973) **52**, 1984–1985.
A1157. Fuganti, C., Ghiringhelli, D., Grasselli, P., Santopietro-Amisano, A., *J. Chem. Soc., Chem. Commun.* (1973) 862–863.
A1158. Rush, R.A., Thomas, P.E., Nagatsu, T., Udenfriend, S., *Proc. Nat. Acad. Sci. U. S. A.* (1974) **71**, 872–874.
A1159. Jain, S.C., Khanna, P., *Indian J. Exp. Biol.* (1973) **11**, 578.
A1160. Yeh, S.Y., *Experientia* (1974) **30**, 264–266.
A1161. Weiss, R., Brauer, J., Goertz, U., Petry, R., *Arzneim.-Forsch.* (1973) **24**, 345–348.
A1162. Das, N.P., *Planta Med.* (1974) **25**, 274–280.
A1163. Liu, T.Z., Sehn, J.-T., Ganong, W.F., *Proc. Soc. Exp. Biol. Med.* (1974) **146**, 37–40.
A1164. Kristensen, S., *Brit. J. Nutr.* (1974) **31**, 357–365.
A1165. Sigel, C.W., Grace, M.E., Nichol, C.A., *J. Infect. Dis.* (1973) **128** (Suppl.) S580–S583.
A1166. Lindeke, B., Cho, A.K., Thomas, T.L., Michelson, L., *Acta Pharm. Suec.* (1973) **10**, 493–506.
A1167. Testa, B., *Acta Pharm. Suec.* (1973) **10**, 441–454.
A1168. Verebly, K., Inturrisi. C.E., *Clin. Pharmacol. Ther.* (1974) **15**, 302–309.
A1169. Stawarz, R.J., Stitzel, R.E., *Pharmacology* (1974) **11**, 178–191.
A1170. Grundon, M.F., Harrison, D.M., Spyropoulos, C.G., *J. Chem. Soc., Chem. Commun.* (1974) 51–52.
A1171. Chabannes, B., Grandgeorge, M., Duperray, B., Pacheco, H., *Chim. Ther.* (1973) **8**, 621–625.
A1172. Elkinawy, M., Raa, J., *Physiol. Plant.* (1974) **30**, 172–175.
A1173. Javoy, F., Agid, Y., Bouvet, D., Glowinski, J., *J. Pharm. Pharmacol.* (1974) **26**, 179–185.
A1174. Bachur, N.R., Egorin, M.J., Hildebrand, R.C., *Biochem. Med.* (1973) **8**, 352–361.
A1175. Gupta, R.N., Chittim, B.G., Keane, P.M., *J. Chromatogr. Sci.* (1974) **12**, 67–70.
A1178. Cho, A.K., *Res. Commun. Chem. Pathol. Pharmacol.* (1974) **7**, 67–78.
A1180. Tamilski, T.S., Morreal, C.E., Dao, T.L., *Cancer Res.* (1973) **33**, 3117–3122.
A1181. Leeson, G.A., Garteiz, D.A., Knapp, W.C., Wright, G.J., *Drug Metab. Dispos.* (1973) **1**, 565–568.
A1182. Miyakawa, M., Harada, T., Yoshida, O., *Igaku To Seibutsugaku* (1973) **86**, 355–360.
A1183. Schreiber, W., *FEBS Lett.* (1974) **41**, 50–52.
A1184. Verkhovtseva, M.I., Suvorov, N.N., Semenov, L.F., Chernov, G.A., *Farmakol. Toksikol.* (1974) **37**, 39–41.
A1185. Lindström, A., Olsson, B., Pettersson, G., *Eur. J. Biochem.* (1974) **42**, 377–381.
A1186. Salmona, M., Saronio, C., Bianchi, R., Marcucci, F., Mussini, E., *J. Pharm. Sci.* (1974) **63**, 222–225.
A1187. Zak, S.B., Gilleran, T.G., Karliner, J., Lukas, G., *J. Med. Chem.* (1974) **17**, 381–382.
A1188. De la Torre, B., Wiqvist, N., Diczfalusy, E., *Acta Endocrinol.* (1974) **75**, 159–172.
A1189. Algeri, S., Turner, A.J., Frigerio, A., Baker, K.M., *Riv. Farmacol. Ter.* (1973) **4**, 121a-131a.
A1190. Jigami, Y., Omori, T., Minoda, Y., Yamada, K., *Agric. Biol. Chem.* (1974) **38**, 467–469.
A1191. Steele, J.A., Lieberman, J.R., Mirocha, C.J., *Can. J. Microbiol.* (1974) **20**, 531–534.
A1192. Spokes, J.R., Walker, N., *Arch. Microbiol.* (1974) **96**, 125–134.
A1193. Focht, D.D., Joseph, H., *Can. J. Microbiol.* (1974) **20**, 631–635.

A1194. Sullivan, D.T., Grillo, S.L., Kitos, R.J., *J. Exp. Zool.* (1974) **188**, 225–233.
A1195. Ishiguro, I., Ikeno, T., *Seikagaku* (1974) **46**, 67–73.
A1196. Hobkirk, R., Nilsen, M., *J. Steroid Biochem.* (1974) **5**, 15–20.
A1197. Guldberg, A.M.R., Guldberg, H.C., *Neuropharmacology* (1973) **12**, 1135–1144.
A1198. Watanabe, M., Toyoda, M., Imada, I., Morimoto, H., *Chem. Pharm. Bull.* (1974) **22**, 176–182.
A1199. Song, C.S., Bonkowsky, H.L., Tschudy, D.P., *Clin. Pharmacol. Ther.* (1974) **15**, 431–435.
A1201. Sano, M., Hakusui, H., *Chem. Pharm. Bull.* (1974) **22**, 696–706.
A1202. Spirtes, M.A., *Advan. Biochem. Psychopharmacol.* (1974) **9**, 399–404.
A1203. Lauriere, J., Jemmali, M., Frayssinet, C., *Ann. Nutr. Aliment.* (1973) **27**, 335–343.
A1204. Ramanarayanan, M., Vaidyanathan, C.S., *Indian J. Biochem. Biophys.* (1973) **10**, 254–256.
A1205. Balinova, A., *Khim. Sel. Khoz.* (1974) **12**, 382–385.
A1206. Jigami, Y., Omori, T., Minoda, Y., Yamada, K., *Agric. Biol. Chem* (1974) **38**, 401–408.
A1207. Benezet, H.J., Matsumura, F., *J. Agric. Food Chem.* (1974) **22**, 427–430.
A1208. McGuire, J., Newman, D., Barisas, G., *Yale J. Biol. Med.* (1973) **46**, 572–582.
A1209. Ohkawa, H., Yoshihara, R., Kohara, T., Miyamoto, J., *Agric. Biol. Chem.* (1974) **38**, 1035–1044.
A1210. Hosoi, K., *Plant Cell Physiol.* (1974) **15**, 429–440.
A1211. Rohzin, J., Soderstrom, R.L., Brooks, S.C., *J. Biol. Chem.* (1974) **249**, 2079–2087.
A1212. Hirata, F., Hayaishi, O., Tokuyama, T., Senoh, S., *J. Biol. Chem.* (1974) **249**, 1311–1313.
A1213. Birkenmeier, E., Alpers, D.H., *Biochim. Biophys. Acta* (1974) **350**, 100–112.
A1214. Hobkirk, R., Green, R.N., Nilsen, M., Jennings, B.A., *Can. J. Biochem.* (1974) **52**, 15–20.
A1215. Findlay, J.K., Siekmann, L., Breuer, H., *Biochem. J.* (1974) **137**, 263–272.
A1216. Takemori, S., Hon-Nami, K., Kawahara, F., Katagiri, M., *Biochim. Biophys. Acta* (1974) **342**, 137–144.
A1217. Brown, J.P., Davidson, G.E., Scott, J.M., *Biochim. Biophys. Acta* (1974) **343**, 78–88.
A1218. Smith, R.V., Rosazza, J.P., *Arch. Biochem. Biophys.* (1974) **161**, 551–558.
A1219. Bury, A.F., Pennington, R.J., *Biochem. Soc. Trans.* (1974) **1**, 1306–1307.
A1220. Tanayama, S., Tsuchida, E., Suzuoki, Z., *Xenobiotica* (1973) **3**, 643–656.
A1221. Clarke, J.T.R., *Can. J. Biochem.* (1974) **52**, 294–303.
A1222. Briant, R.H., Blackwell, E.W., Williams, F.M., Davies, D.S., Dollery, C.T., *Xenobiotica* (1973) **3**, 787–799.
A1223. Kampffmeyer, H.G., *Biochem. Pharmacol.* (1974) **23**, 713–724.
A1224. Mavrides, C., Orr, W., *Biochim. Biophys. Acta* (1974) **336**, 70–78.
A1226. Sinsheimer, J.E., Dring, L.G., Williams, R.T., *Biochem. J.* (1973) **136**, 763–771.
A1227. Yang, H.-Y.T., Neff, N.H., *J. Pharmacol. Exp. Ther.* (1973) **187**, 365–371.
A1228. Siegel, L.M., Davis, P.S., Kamin, H., *J. Biol. Chem.* (1974) **249**, 1572–1586.
A1229. Michalover, J.L., Ribbons, D.W., *Biochem. Biophys. Res. Commun.* (1974) **56**, 1102–1110.
A1230. Ono-Kamimoto, M., Senoh, S., *J. Biochem.* (1974) **75**, 329–333.
A1231. Nagami, K., Senoh, S., *J. Biochem.* (1974) **75**, 389–398.
A1232. Pruitt, A.W., Faraj, B.A., Dayton, P.G., *J. Pharmacol. Exp. Ther.* (1974) **188**, 248–256.
A1233. Alvin, J., Bush, M.T., *J. Pharmacol. Exp. Ther.* (1974) **188**, 8–14.
A1234. Kanai, Y., Kobayashi, T., Tanayama, S., *Xenobiotica* (1973) **3**, 657–670.

A1235. Tanayama, S., *Xenobiotica* (1973) **3**, 671–680.

A1236. Cardillo, R., Fuganti, C., Ghiringhelli, D., Giangrasso, D., Grasselli, P., Santopietro-Amisano, A., *Tetrahedron* (1974) **30**, 459–461.

A1237. Coutinho, C.B., Cheripko, J.A., Carbone, J.J., Manning, J.E., Boff, E., *Xenobiotica* (1973) **3**, 681–690.

A1238. Gadal, P., Boudet, A., Robreau, G., *Plant Sci. Lett.* (1974) **3**, 55–60.

A1239. Bodnaryk, R.P., Brunet, P.C.J., Koeppe, J.K., *J. Insect Physiol.* (1974) **20**, 911–923.

A1240. Eccleston, D., Ritchie, I.M., *J. Neurochem.* (1973) **21**, 635–646.

A1241. Thomas, H., Mueller-Enoch, D., *Naturwissenschaften* (1974) **61**, 222.

A1243. Rossier, G., Pierrepoint, C.G., *J. Reprod. Fert.* (1974) **37**, 43–49.

A1244. Daum, J., Kieslich, K., *Naturwissenschaften* (1974) **61**, 167–168.

A1245. Okada, T., Fujita, T., Ohta, T., Kato, T., Ikuta, K., Nagatsu, T., *Experientia* (1974) **30**, 605–607.

A1246. Hackman, M.R., Brooks, M.A., de Silva, J.A.F., Ma, T.S., *Anal. Chem.* (1974) **46**, 1075–1082.

A1247. Daenens, P., Van Boven, M., *Arzneim.-Forsch.* (1974) **24**, 195–202.

A1248. Fuller, R.W., Baker, J.C., Snoddy, H.D., *Biochem. Med.* (1974) **9**, 301–308.

A1249. Leighty, E.G., Fentiman, A.F., Jr., *Res. Commun. Chem. Pathol. Pharmacol.* (1974) **8**, 65–74.

A1250. Kiese, M., Renner, G., *Naunyn-Schmiedeberg's Arch. Pharmacol.* (1974) **283**, 143–150.

A1251. Smith, A.E., Cullimore, D.R., *Can. J. Microbiol.* (1974) **20**, 773–776.

A1252. Fuganti, C., Ghiringhelli, D., Grasselli, P., *J. Chem. Soc., Chem. Commun.* (1974) 350–351.

A1253. Manitto, P., Monti, G., Gramatica, P., *Tetrahedron Lett.* (1974) 1567–1568.

A1254. Erge, D., Groeger, D., *Biochem. Physiol. Pflanz.* (1974) **165**, 479–485.

A1255. Nambara, T., Ishiguro, J., Kawarada, Y., Tajima, H., *Chem. Pharm. Bull.* (1974) **22**, 889–893.

A1256. Wilk, S., Watson, E., *Front. Catecholamine Res., Proc. Int. Catecholamine Symp., 3rd* (1973) 1067–1069.

A1257. Brandau, K., Axelrod, J., *Front. Catecholamine Res., Proc. Int. Catecholamine Symp., 3rd* (1973) 129–131.

A1258. Beckett, A.H., Bélanger, P.M., *J. Pharm. Pharmacol.* (1974) **26**, 205–206.

A1259. Bodin, N.-O., *Life Sci.* (1974) **14**, 685–692.

A1260. Kitagawa, H., Nakata, R., Yano, H., *Oyo Yakuri* (1973) **7**, 1231–1240.

A1261. Della Bella, D., Veronese, M., Marca, G., Franceschinis, R., *Arzneim.-Forsch.* (1974) **24**, 836–839.

A1262. Takahashi, T., Sato, Y., *Radioisotopes* (1973) **22**, 351–359.

A1263. Sun, F.F., Stafford, J.E., *J. Pharm. Sci.* (1974) **63**, 539–544.

A1264. Drayer, D.E., Reidenberg, M.M., Sevy, R.W., *Proc. Soc. Exp. Biol. Med.* (1974) **146**, 358–363.

A1265. Koizumi, T., Ueda, M., Takada, S., *Chem. Pharm. Bull.* (1974) **22**, 894–906.

A1266. Rendić, S., Šunjić, V., Kajfež, F., Klasinc, L., Mildner, P., *Chimia* (1974) **28**, 232–234.

A1267. Ben-Zvi, Z., Burstein, S., *Res. Commun. Chem. Pathol. Pharmacol.* (1974) **8**, 223–229.

A1268. Steen, R.C., Schultz, I.R., Zimmerman, D.C., Fleeker, J.R., *Weed Res.* (1974) **14**, 23–28.

A1269. Dierickx, P.J., Vendrig, J.C., *Z. Pflanzenphysiol.* (1974) **72**, 11–22.

A1270. Fisher, J., Walsh, C., *J. Am. Chem. Soc.* (1974) **96**, 4345–4346.

A1271. Bond, P.A., Howlett, D.R., *Biochem. Med.* (1974) **10**, 219–228.

A1272. Gabb, R.G., Stone, G.M., *J. Endocrinol.* (1974) **62**, 109–123.

A1273. Rao, G.S., Rao, M.L., Haueter, G., Breuer, H., *Hoppe-Seyler's Z. Physiol. Chem.* (1970) **355**, 881–890.
A1274. Yoshihara, S., Yoshimura, H., *Chem. Pharm. Bull.* (1974) **22**, 714–717.
A1275. Tyce, G.M., Sharpless, N.S., Owen, C.A., Jr., *Advan. Neurol.* (1974) **5**, 327–329.
A1276. Belpaire, F.M., Bogaert, M.G., *Arch. Int. Pharmacodyn. Ther.* (1974) **208**, 362.
A1277. Murray, J.F., Jr., Gordon, G.R., Peters, J.H., *Proc. West. Pharmacol. Soc.* (1974) **17**, 150–154.
A1278. Forist, A.A., Pulliam, A.L., Kaiser, D.G., *Res. Commun. Chem. Pathol. Pharmacol.* (1974) **8**, 385–388.
A1279. Jacquot, C., Bralet, J., Cohen, Y., *C. R. Soc. Biol.* (1973) **167**, 1789–1794.
A1280. Honma, S., Nambara, T., *Chem. Pharm. Bull.* (1974) **22**, 687–695.
A1281. Numazawa, M., Tanaka, Y., Momono, Y., Nambara, T., *Chem. Pharm. Bull.* (1974) **22**, 663–668.
A1282. Asaad, M.M., Barry, H., III, Clarke, D.E., Dixit, B.N., *Brit. J. Pharmacol.* (1974) **50**, 277–282.
A1284. Tsuda, H., Noguchi, T., Kido, R., *J. Neurochem.* (1974) **22**, 679–683.
A1285. Achee, F.M., Togulga, G., Gabay, S., *J. Neurochem.* (1974) **22**, 651–661.
A1286. Bauer, B.A., Eitenmiller, R.R., *J. Food Sci.* (1974) **39**, 10–14.
A1287. Duncan, R.J.S., Sourkes, T.L., *J. Neurochem.* (1974) **22**, 663–669.
A1288. Rivera-Calimlim, L., *Brit. J. Pharmacol.* (1974) **50**, 259–263.
A1289. Gál, E.M., *J. Neurochem.* (1974) **22**, 861–863.
A1290. Grover, T.A., Piette, L.H., Manian, A.A., *Advan. Biochem. Psychopharmacol.* (1974) **9**, 561–569.
A1291. Taylor, J.A., Jr., *Drug Metab. Dispos.* (1974) **2**, 221–227.
A1292. Rudzit, E.A., Bespalov, A.M., Tagirov, R.F., *Farmakol. Toksikol.* (1974) **37**, 472–474.
A1293. Harrison, S.D., Jr., Bosin, T.R., Maickel, R.P., *Drug Metab. Dispos.* (1974) **2**, 228–236.
A1294. Hsieh, D.P.H., Dalezios, J.I., Krieger, R.I., Masri, M.S., Haddon, W.F., *J. Agric. Food Chem.* (1974) **22**, 515–517.
A1295. Masri, M.S., Haddon, W.F., Lundin, R.E., Hsieh, D.P.H., *J. Agric. Food Chem.* (1974) **22**, 512–515.
A1296. Zech, R., Zürcher, K., *Comp. Biochem. Physiol. B.* (1974) **48**, 427–433.
A1297. Malacrida, A., Cima, L., Gasperi, G., Grigolo, A., *Ital. J. Biochem.* (1974) **23**, 12–18.
A1298. Patel, T.R., Gibson, D.T., *J. Bacteriol.* (1974) **119**, 879–888.
A1299. Yuasa, K., Ishizuka, K., Kaburaki, S., Sakasai, T., *Nippon Nogei Kagaku Kaishi* (1974) **48**, 313–318.
A1301. Davey, J.F., Gibson, D.T., *J. Bacteriol.* (1974) **119**, 923–929.
A1302. Kirdany, R.Y., Sandberg, A.A., *Steroids* (1974) **23**, 667–686.
A1303. Midha, K.K., McGilveray, I.J., Charette, C., *J. Pharm. Sci.* (1974) **63**, 1234–1239.
A1304. Chin, B.H., Eldridge, J.M., Sullivan, L.J., *Clin. Toxicol.* (1974) **7**, 37–56.
A1305. Kanto, J., Iisalo, E., Lehtinen, V., Salminen, J., *Psychopharmacologia* (1974) **36**, 123–131.
A1306. Youdim, M.B.H., Hamon, M., Bourgoin, S., *Adv. Biochem. Psychopharmacol.* (1974) **11**, 13–17.
A1307. Hsu, L.L., Mandell, A.J., *Adv. Biochem. Psychopharmacol.* (1974) **11**, 75–84.
A1308. Ōya, H., Harada, M., Nagatsu, T., *Arch. Oral Biol.* (1974) **19**, 489–491.
A1309. Youdim, M.B.H., *Adv. Biochem. Psychopharmacol.* (1974) **11**, 59–63.
A1310. Leysen, J., Laduron, P., *Adv. Biochem. Psychopharmacol.* (1974) **11**, 65–74.
A1311. Banerjee, S.P., Snyder, S.H., *Adv. Biochem. Psychopharmacol.* (1974) **11**, 85–93.
A1312. French, M.R., Babalunmi, E.A., Golding, R.R., Bassir, O., Caldwell, J.,

Smith, R.L., Williams, R.T., *FEBS Lett.* (1974) **46**, 134–137.

A1313. Gibson, D.T., Mahadevan, V., Davey, J.F., *J. Bacteriol.* (1974) **119**, 930–936.

A1314. Dorough, H.W., McManus, J.P., Kumar, S.S., Cardona, R.A., *J. Agric. Food Chem.* (1974) **22**, 642–645.

A1315. Pree, D.J., Saunders, J.L., *J. Agric. Food Chem.* (1974) **22**, 620–625.

A1316. Büchi, G.H., Müller, P.M., Roebuck, B.D., Wogan G.N., *Res. Commun. Chem. Pathol. Pharmacol.* (1974) **8**, 585–592.

A1317. Ivie, G.W., Clarke, D.E., Rushing, D.D., *J. Agric. Food Chem.* (1974) **22**, 632–634.

A1318. Polan, C.E., Hayes, J.R., Campbell, T.C., *J. Agric. Food Chem.* (1974) **22**, 635–638.

A1319. Beynon. K.I., Roberts, T.R., Wright, A.N., *Pestic. Sci.* (1974) **5**, 429–442.

A1320. Lin, R.-L., Narasimhachari, N., *Res. Commun. Chem. Pathol. Pharmacol.* (1974) **8**, 535–542.

A1321. Akatsuka, T., Kasakura, N., Soejima, M., *Ibaraki Daigaku Nogakubu Gakujutsu Hokoku* (1973) 43–48.

A1322. Fuganti, C., *Gazz. Chim. Ital.* (1973) **103**, 1255–1258.

A1323. Bird, J.A., Cain, R.B., *Biochem. J.* (1974) **140**, 121–134.

A1324. Yang, H.-Y.T., Neff, N.H., *J. Pharmacol. Exp. Ther.* (1974) **189**, 733–740.

A1325. Sciuto, S., Oriente, G., Piattelli, M., Impellizzeri, G., Amico, V., *Phytochemistry* (1974) **13**, 947–951.

A1327. Felsted, R.L., Gee, M., Bachur, N.R., *J. Biol. Chem.* (1974) **249**, 3672–3679.

A1328. Schultz, E., Engle, F.E., Wood, J.M., *Biochemistry* (1974) **13**, 1768–1776.

A1329. Sariaslani, F.S., Harper, D.B., Higgins, I.J., *Biochem. J.* (1974) **140**, 31–45.

A1330. Nettleship, L., Slaytor, M., *Phytochemistry* (1974) **13**, 735–742.

A1331. Wu, P.H., Boulton, A.A., *Can. J. Biochem.* (1974) **52**, 374–381.

A1332. Nambudiri, A.M.D., Vance, C.P., Towers, G.H.N., *Biochim. Biophys. Acta* (1974) **343**, 148–155.

A1333. Leinweber, F.-J., Di Carlo, F.J., *J. Pharmacol. Exp. Ther.* (1974) **189**, 271–277.

A1334. Ranganathan, S., Ramasarma, T., *Biochem. J.* (1974) **140**, 517–522.

A1336. Weinshilboum, R.M., Raymond, F.A., Elveback, L.R., Weidman W.H., *Biochem. Pharmacol.* (1974) Suppl. Pt. 2, 930–936.

A1337. Brandau, K., Axelrod, J., *Biochem. Pharmacol.* (1974) Suppl. Pt. 1, 83–85.

A1339. Saini, P.K., Rosenberg, I.H., *J. Biol. Chem.* (1974) **249**, 5131–5134.

A1340. Rosenberg, I.H., Neumann, H., *J. Biol. Chem.* (1974) **249**, 5126–5130.

A1341. Nukiwa, T., Tohyama, C., Okita, C., *Biochem. Biophys. Res. Commun.* (1974) **60**, 1029–1035.

A1342. Buswell, J.A., *Biochem. Biophys. Res. Commun.* (1974) **60**, 934–941.

A1343. Meek, J.L., Foldes, A., *Biochem. Pharmacol.* (1974) Suppl. Pt. 1, 116–120.

A1344. Nayak, P.K., Misra, A.L., Mulé, S.J., *J. Pharmacol. Exp. Ther.* (1976) **196**, 556–569.

A1345. Marietta, M.P., White, P.F., Pudwill, C.R., Way, W.L., Trevor, A.J., *J. Pharmacol. Exp. Ther.* (1976) **196**, 536–544.

A1346. Lowenthal, D.T., Øie, S., van Stone, J.C., Briggs, W.A., Levy, G., *J. Pharmacol. Exp. Ther.* (1976) **196**, 570–578.

A1347. Yeh, S.Y., Gorodetzky, C.W., McQueen, R.L., *J. Pharmacol. Exp. Ther.* (1976) **196**, 249–256.

A1348. Grosso, D.S., Gawienowski, A.M., *Biochem. Pharmacol.* (1976) **25**, 957–961.

A1349. Osborne, N.N., Guthrie, P.B., Neuhoff, V., *Biochem. Pharmacol.* (1976) **25**, 925–931.

A1350. Howd, R.A., Seo, K.S., Wurtman, R.J., *Biochem. Pharmacol.* (1976) **25**, 977–978.

A1351. Misra, A.L., Pontani, R.B., Vadlamani, N.L., Mulé, S.J., *J. Pharmacol. Exp. Ther.* (1976) **196**, 257–268.

A1352. Booth, J., Sims, P., *Biochem. Pharmacol.* (1976) **25**, 979–980.

A1353. Prema, K., Gopinathan, K.P., *Biochem. Pharmacol.* (1976) **25**, 1299–1303.

A1354. Cardona, R.A., King, C.M., *Biochem. Pharmacol.* (1976) **25**, 1051–1056.

A1355. Meyerson, L.R., McMurtrey, K.D., Davis, V.E., *Biochem. Pharmacol.* (1976) **25**, 1013–1020.

A1356. Willingham, A.K., Laliberte, R.E., Bell, R.G., Matschiner, J.T., *Biochem. Pharmacol.* (1976) **25**, 1063–1066.

A1357. Roerig, D.L., Reak, C.J., Wang, R.I.H., *Biochem. Pharmacol.* (1976) **25**, 1075–1080.

A1358. Jordan, B.J., Rance, M.J., *J. Pharm. Pharmacol.* (1974) **26**, 359–361.

A1359. Reddy, B.G., Pohl, L.R., Krishna, G., *Biochem. Pharmacol.* (1976) **25**, 1119–1122.

A1360. Webb, S.R., Ferone, R., *Biochim. Biophys. Acta* (1976) **422**, 419–426.

A1361. Clarke, K.F., Callely, A.G., Livingstone, A., Fewson, C.A., *Biochim. Biophys. Acta* (1975) **404**, 169–179.

A1362. Griffin, M.J., Brown, G.M., *J. Biol. Chem.* (1964) **239**, 310–316.

A1363. Ikeda, M., Iwai, K., *Plant Cell Physiol.* (1970) **11**, 639–656.

A1364. Lunt, D.O., Evans, W.C., *Biochem. Soc. Trans.* (1976) **4**, 491–492.

A1365. Hayes, J.S., Brendel, K., *Biochem. Pharmacol.* (1976) **25**, 1495–1500.

A1366. Keat, M.J., Hopper, D.J., *Biochem. Soc. Trans.* (1976) **4**, 494–495.

A1367. Harper, D.B., *Biochem. Soc. Trans.* (1976) **4**, 502–504.

A1368. Gessner, T., *Biochem. Soc. Trans.* (1976) **4**, 522–524.

A1369. Gatley, S.J., Sherratt, H.S.A., *Biochem. Soc. Trans.* (1976) **4**, 525–526.

A1370. Whittle, P.J., Lunt, D.O., Evans, W.C., *Biochem. Soc. Trans.* (1976) **4**, 490–491.

A1371. McPherson, F.J., Bridges, J.W., Parke, D.V., *Biochem. Pharmacol.* (1976) **25**, 1345–1350.

A1372. Hengstmann, J.H., Konen, W., Konen, C., Eichelbaum, M., Dengler, H.J., *Naunyn-Schmiedeberg's Arch. Pharmacol.* (1974) **283**, 93–106.

A1374. Del Villar, E., Sanchez, E., Tephly, T.R., *Drug Metab. Dispos.* (1974) **2**, 370–374.

A1375. Gazave, J.M., Parrot, J.L., *Journ. Angeiol. Lang. Fr.* (1973) 272–281.

A1376. Jones, G., Widman, M., Agurell, S., Lindgren, J.E., *Acta Pharm. Suec.* (1974) **11**, 283–294.

A1377. Rane, A., *J. Pediatr.* (1974) **85**, 543–545.

A1378. Das, N.P., *Drug Metab. Dispos.* (1974) **2**, 209–213.

A1379. Willis, D.E., Addison, R.F., *Comp. Gen. Pharmacol.* (1974) **5**, 77–81.

A1380. Wallcave, L., Bronczyk, S., Gingell, R., *J. Agric. Food Chem.* (1974) **22**, 904–908.

A1382. Roffman, M., Reigle, T.G., Orsulak, P.J., Schildkraut, J.J., *Adv. Biochem. Psychopharmacol.* (1974) **12**, 189–194.

A1383. Mandel, P., Aunis, D., *Ciba Found. Symp.* (1974) **22**, 67–83.

A1384. Daddona, P.E., Hutchinson, C.R., *J. Am. Chem. Soc.* (1974) **96**, 6806–6807.

A1385. Gross, D., Lehmann, H., Schuette, H.R., *Biochem. Physiol. Pflanz.* (1974) **166**, 281–287.

A1386. Emson, P.C., Fonnum, F., *J. Neurochem.* (1974) **22**, 1079–1088.

A1387. Wada, H., *Nippon Nogei Kagaku Kaishi* (1974) **48**, 351–357.

A1388. Stenger, V.G., Maren, T.H., *Comp. Gen. Pharmacol.* (1974) **5**, 23–35.

A1389. Takada, T., *Shinshu Igaku Zasshi* (1973) **21**, 549–560.

A1390. Murphy, P.J., Bernstein, J.R., McMahon, R.E., *Mol. Pharmacol.* (1974) **10**, 634–639.

A1391. Ben-Zvi, Z., Burnstein, S., Zikopoulos, J., *J. Pharm. Sci.* (1974) **63**, 1173–1174.

A1392. Edelson, J., Schuster, E., Shahinian, S., Douglas, J.F., *Arch. Int. Pharmacodyn. Ther.* (1974) **209**, 66–74.

A1393. Shindo, H., Maeda, T., *Chem. Pharm. Bull.* (1974) **22**, 1721–1730.

A1394. Skutches, C.L., Smith, F.H., *J. Am. Oil Chem. Soc.* (1974) **51**, 413–415.

A1395. Phillips, J., Wells, M., Chandler, C., *Bull. Environ. Contam. Toxicol.* (1974) **12**, 355–358.

A1396. Hoppen, H.O., Sieckmann, L., Breuer, H., *Hoppe-Seyler's Z. Physiol. Chem.* (1974) **355**, 1305–1315.

A1397. Swanson, A.A., Truesdale, A.W., *Ophthalmic Res.* (1974) **6**, 235–244.

A1398. Hucker, H.B., *Drug Metab. Dispos.* (1973) **1**, 332–336.

A1399. Aldridge, D.C., Davies, A.B., Jackson, M.R., Turner, W.B., *J. Chem. Soc., Perkin Trans. 1* (1974) 1540–1541.

A1400. Rabiant, J., Oliver, C., Ancher, J.F., Sergant, M., Bouvet, P., *Ann. Pharm. Fr.* (1974) **32**, 183–192.

A1401. Matin, S.B., Callery, P.S., Zweig, J.S., O'Brien, A., Rapoport, R., Castagnoli, N., Jr., *J. Med. Chem.* (1974) **17**, 877–882.

A1402. Mrochek, J.E., Katz, S., Christie, W.H., Dinsmore, S.R., *Clin. Chem.* (1974) **20**, 1086–1096.

A1403. Irikura, T., Nagatsu, Y., Baba, S., Sone, H., *Yakugaku Zasshi* (1974) **94**, 802–810.

A1404. Blackwell, E.W., Briant, R.H., Conolly, M.E., Davies, D.S., Dollery, C.T., *Br. J. Pharmacol.* (1974) **50**, 587–591.

A1405. Grunow, W., Boehme, C., *Arch. Toxicol.* (1974) **32**, 217–225.

A1407. Knowles, C.O., Aziz, S.A., *J. Econ. Entomol.* (1974) **67**, 574–576.

A1408. Bohoun, C., Assicot, M., *Biochem. Pharmacol.* (1974) Suppl. Pt. 1, 63–68.

A1409. Goldstein, M., Anagnoste, B., Freedman, L.S., Roffman, M., Ebstein, R.P., Park, D.H., Fuxe, K., Hökfelt, T., *Biochem. Pharmacol.* (1974) Suppl. Pt. 1, 27–36.

A1410. Battersby, A.R., Sheldrake, P.W., Staunton, J., Williams, D.C., *J. Chem. Soc., Chem. Commun.* (1974) 566–567.

A1411. Ishiguro, I., Naito, J., Ishikura, A., Shiotani, T., *Yakugaku Zasshi* (1974) **94**, 1224–1231.

A1413. Waymire, J.C., Vernadakis, A., Weiner, N., *Adv. Behav. Biol.* (1974) **8**, 149–170.

A1414. Walle, T., Morrison, J.I., Tindell, G.L., *Res. Commun. Chem. Pathol. Pharmacol.* (1974) **9**, 1–9.

A1415. Trenner, N.R., Carlin, J.R., Arison, B.H., VandenHeuvel, W.J.A., *J. Pharm. Sci.* (1974) **63**, 1312–1313.

A1416. Forist, A.A., Judy, R.W., *J. Pharm. Pharmacol.* (1974) **26**, 565.

A1418. Miyazaki, Y., Ogino, O., Kamiya, A., Hori, R., Arita, T., *Yakugaku Zasshi* (1974) **94**, 1090–1094.

A1419. Gros, P.M., Davi, H.J., Chasseaud, L.F., Hawkins, D.R., *Arzneim.-Forsch.* (1974) **24**, 1385–1390.

A1420. Frigerio, A., Baker, K.M., Morselli, P.L., *Mass Spectrom. Biochem. Med., Symp.* (1973) 65–82.

A1421. Scheline, R.R., *Experientia* (1974) **30**, 880–881.

A1422. Booth, J., Sims, P., *FEBS Lett.* (1974) **47**, 30–33.

A1423. Goto, M., Sugiura, K., Hattori, M., Miyagawa, T., Okamura, M., *Chemosphere* (1974) **3**, 227–232.

A1424. Pedersen, M.G., Hershberger, W.K., Juchau, M.R., *Bull. Environ. Contam. Toxicol.* (1974) **12**, 481–486.

A1425. Kalra, R.L., *J. Med. Entomol.* (1974) **11**, 437–441.

A1426. Feigelson, P., Brady, F.O., *Bioelectrochem. Bioenergy* (1974) **1**, 67–72.

A1427. Ohira, H., *Wakayama Igaku* (1974) **25**, 131–138.

A1428. Mandel, L.R., Rosegay, A., Walker, R.W., VandenHeuvel, W.J.A., Rokach, J., *Science* (1974) **186**, 741–743.

A1429. Suzuki, N., Iwai, K., *J. Nutr. Sci. Vitaminol.* (1974) **20**, 89–96.

A1430. Bueckmann, D., *Zool. Jahrb., Abt. Allg. Zool. Physiol. Tiere* (1974) **78**, 257–278.
A1431. Neuss, N., Mallett, G.E., Brannon, D.R., Mabe, J.A., Horton, H.R., Huckstep, L.L., *Helv. Chim. Acta* (1974) **57**, 1886–1890.
A1432. Stafford, M., Kellermann, G., Stillwell, R.N., Horning, M.G., *Res. Commun. Chem. Pathol. Pharmacol.* (1974) **8**, 593–606.
A1433. Shibata, Y., Tsurumi, Y., Seino, A., Fukimbara, T., *Hakko Kogaku Zasshi* (1974) **52**, 720–729.
A1434. Dudley, K.H., Bius, D.L., Waldrop, C.D., *Drug Metab. Dispos.* (1974) **2**, 113–122.
A1435. Ambre, J.J., Fischer, L.J., *Drug Metab. Dispos.* (1974) **2**, 151–158.
A1436. Vickers, S., Stuart, E.K., Bianchine, J.R., Hucker, H.B., Jaffe, M.E., Rhodes, R.E., Vanden Heuvel, W.J.A., *Drug Metab. Dispos.* (1974) **2**, 6–22.
A1437. Tetsuo, M., Kuhara, T., Yoshino, M., Kumitomo, J., Juiichi, M., Matsumoto, I., *IRCS Libr. Compend.* (1974) **2**, 1456.
A1438. Fujita, R., Yamato, C., Takahashi, T., Ohtake, S., *Mod. Probl. Pharmacopsychiatry* (1974) **9**, 56–64.
A1439. Sawada, H., Hara, A., Yano, H., Kido, A., *Gifu Daigaku Igakubu Kiya* (1973) **21**, 248–259.
A1440. Vollmer, K.O., Achenbach, H., *Arzneim.-Forsch.* (1974) **24**, 1237–1242.
A1441. Conway, W.D., Melethil, S., *J. Pharm. Sci.* (1974) **63**, 1551–1554.
A1442. Santagostino, G., Maffei, F.R., Pirillo, D., *J. Pharm. Sci.* (1974) **63**, 1690–1692.
A1443. Mitchell, J.R., Thorgeirsson, S.S., Potter, W.Z., Jollow, D.J., Keiser, H., *Clin. Pharmacol. Ther.* (1974) **16**, 676–684.
A1444. Goto, M., Sugiura, K., Hattori, M., Miyagawa, T., Okamura, M., *Chemosphere* (1974) **3**, 233–238.
A1445. Cashaw, J.L., McMurtrey, K.D., Brown, H., Davis, V.E., *J. Chromatogr.* (1974) **99**, 567–573.
A1446. Yoshimura, H., Yamamoto, H., Kinoshita, H., *New Methods Environ. Chem. Toxicol., Collect. Pap. Res. Conf. New Methodol. Ecol. Chem.* (1973) 291–297.
A1447. Van Boven, M., Janssen, G., Daenens, P., *Mikrochim. Acta* (1974) 603–610.
A1448. Perez-Silva, M.G., *Biol. R. Soc. Esp. Hist. Nat., Secc. Biol.* (1973) **77**, 177–180.
A1449. Sparnins, V.L., Chapman, P.J., Dagley, S., *J. Bacteriol.* (1974) **120**, 159–167.
A1450. Pack, B.A., Brooks, S.C., *Endocrinology* (1974) **95**, 1680–1690.
A1451. Ette, S.I., French, M.R., Smith, R.L., Williams, R.T., *FEBS Lett.* (1974) **49**, 134–136.
A1452. Hes, J., Sternson, L.A., *Drug Metab. Dispos.* (1974) **2**, 345–350.
A1453. Boerner, U., Roe, R.L., Becker, C.E., *J. Pharm. Pharmacol.* (1974) **26**, 393–398.
A1454. Ben-Zvi, Z., Bergen, J.R., Burstein, S., *Res. Commun. Chem. Pathol. Pharmacol.* (1974) **9**, 201–204.
A1455. Negishi, T., Mineura, K., Marumo, H., *Oyo Yakuri* (1974) **8**, 1059–1066.
A1456. Pitre, D., Felder, E., Farmaco, Ed. Sci. (1974) **29**, 793–801.
A1457. Williams, J.G., Lingcope, C., Williams, K.I.H., *Steroids* (1974) **24**, 687–701.
A1458. Goto, M., Sugiura, K., Hattori, M., Miyagawa, T., Okamura. M., *New Methods Environ. Chem. Toxicol., Collect. Pap. Res. Conf., New Methodol. Ecol. Chem.* (1973) 299–302.
A1460. Juchau, M.R., Namkung, M.J., *Drug Metab. Dispos.* (1974) **2**, 380–385.
A1461. Quiram, D.R., Weinshilboum, R.M., *Biochem. Pharmacol.* (1976) **25**, 1727–1732.
A1462. Flashinski, S.J., Lichtenstein, E.P., *Can. J. Microbiol.* (1974) **20**, 399–411.
A1463. Eagles, P.A.M., Iqbal, M., *Brain Res.* (1974) **80**, 177–180.
A1464. Willetts, A.J., *Antonie van Leeuwenhoek* (1974) **40**, 547–549.

A1465. Ljones, T., Flatmark, T., *FEBS Lett.* (1974) **49**, 49–52.

A1466. Kalghatgi, K.K., Nambudiri, A.M.D., Bhat, J.V., Subba Rao, P.V., *Indian J. Biochem. Biophys.* (1974) **11**, 116–118.

A1467. Gvozdyak, P.I., Mogilevich, N.F., Rotmistrov, M.N., *Mikrobiol. Zh.* (1974) **36**, 424–428.

A1468. Murphy, G., Vogel, G., Krippahl, G., Lynen, F., *Eur. J. Biochem.* (1974) **49**, 443–455.

A1469. Tsuda, H., *Wakayama Igaku* (1974) **25**, 7–12.

A1470. Lewis, R.J., Trager, W.F., Chan, K.K., Breckenridge, A., Orme, M., Roland, M., Schary, W., *J. Clin. Invest.* (1974) **53**, 1607–1617.

A1471. Uhlig, W., Bönisch, H., Trendelenberg, U., *Naunyn-Schmiedeberg's Arch Pharmacol.* (1974) **283**, 245–261.

A1472. Kurihara, T., Igarashi, H., *Tokyo Yakka Daigaku Kenkyu Nempo* (1973) 27–33.

A1473. Sever, P.S., Caldwell, J., Dring, L.G., Williams, R.T., *Eur.J. Clin. Pharmacol.* (1973) **6**, 177–180.

A1474. Mishra, L.C., Parmar, A.S., Mead, J.A.R., *Cancer Res.* (1974) **34**, 3187–3191.

A1475. Bornschein, I., Borchert, H.H., Pfeifer, S., *Pharmazie* (1974) **29**, 730–731.

A1476. Balba, M., Saha, J.G., *Environ. Lett.* (1974) **7**, 181–194.

A1477. Syoyama, M., Miyachi, Y., *Eisei Kagaku* (1974) **20**, 213–216.

A1478. Rouchaud, J.P., Decallonne, J.R., Meyer, J.A., *Phytopathology* (1974) **64**, 1513–1517.

A1479. Moza, P., Mostafa, I., Klein, W., *Chemosphere* (1974) **3**, 255–258.

A1480. Hareland, W.A., Crawford, R.L., Chapman, P.J., Dagley, S., *J. Bacteriol.* (1975) **121**, 272–285.

A1481. Largen, M., Belser, W.L., *J. Bacteriol.* (1975) **121**, 239–249.

A1482. Starovoitov, P.I., Pskirkov, S.Y., Nefedova, M.Y., Yakovlev, G.I., Zyakan, A.M., Adanin, V.M., *Izv. Akad. Nauk SSSR, Ser. Khim.* (1974) 2338–2345.

A1483. Mandel, L.R., Walker, R.W., *Life Sci.* (1974) **15**, 1457–1463.

A1484. Grundon, M.F., Harrison, D.M., Sypropoulos, C.G., *J. Chem. Soc., Perkin Trans. 1* (1974) 2181–2184.

A1485. Smith, J.R., Weber, L.J., *Proc. Soc. Exp. Biol. Med.* (1974) **147**, 441–443.

A1486. Nordqvist, M., Agurell, S., Binder, M., Nilsson, I.M., *J. Pharm. Pharmacol.* (1974) **26**, 471–473.

A1487. Ball, P., Hoppen, O., Knuppen, R., *Hoppe-Seyler's Z. Physiol. Chem.* (1974) **355**, 1451–1462.

A1488. Chatterjie, N., Fujimoto, J.M., Inturrisi, C.E., Roerig, S., Wang, R.I.H., Bowen, D.V., Field, F.H., Clarke, D.D., *Drug Metab. Dispos.* (1974) **2**, 401–405.

A1489. Alexanderson, B., Borgå, O., *Eur. J. Clin. Pharmacol.* (1973) **5**, 174–180.

A1490. Midha, K.K., *J. Chromatogr.* (1974) **101**, 10–14.

A1491. Billings, R.E., McMahon, R.E., Blake, D.A., *Life Sci.* (1974) **14**, 1437–1446.

A1492. Freeman, J.J., Sulser, F., *Neuropharmacology* (1974) **13**, 1187–1190.

A1493. Wengenmayer, H., Ebel, J., Grisebach, H., *Eur. J. Biochem.* (1974) **50**, 135–143.

A1494. Manitto, P., Monti, D., Gramatica, P., *J. Chem. Soc., Perkin Trans. 1* (1974) 1727–1431.

A1495. Strothkamp, K.G., Mason, H.S., *Biochem. Biophys. Res. Commun.* (1974) **61**, 827–832.

A1496. Chan, A.H.-H., Common, R.H., *Comp. Biochem. Physiol. B* (1974) **49**, 105–111.

A1497. Mateju, J., Cudlin, J., Hovorkova, N., Blumauerova, M., Vanek, Z., *Folia Microbiol.* (1974) **19**, 307–316.

A1498. Yoshizumi, H., Nadai, T., Tatematsu, A., *Shitsuryo Bunseki* (1974) **22**, 151–162.

A1499. Krugers Dagneau, P.G.L.C., Klein Elhorst, J.T., *Pharm. Weekbl.* (1975) **110**, 1–9.

A1500. Queisnerova, M., Svatek, E., Mansfeld, V., *Act. Nerv. Super.* (1974) **16**, 179.

A1501. Goldaniga, G.C., Pianezzola, E., Valzelli, G., *Arzneim.-Forsch.* (1974) **24**, 1603–1606.

A1502. Vickers, S., Stuart, E.K., Hucker, H.B., VandenHeuvel, W.J.A., *J. Med. Chem.* (1975) **18**, 134–138.

A1503. Sano, M., Moroi, R., Ono, K., *Chem. Pharm. Bull.* (1974) **22**, 3006–3009.

A1504. Yamamoto, K., Kinoshita, Y., Arisada, T., *Oyo Yakuri* (1976) **8**, 1115–1122.

A1505. Louis, W.J., Doyle, A.E., Sampson, R.G., *Clin. Exp. Pharmacol. Physiol.* (1974) **1**, 341–346.

A1506. Ohira, H., Kaseda, H., Kido, R., *Wakayama Med. Rep.* (1973) **16**, 21–23.

A1507. Jollow, D.J., Thorgeirsson, S.S., Potter, W.Z., Hashimoto, M., Mitchell, J.R., *Pharmacology* (1974) **12**, 251–271.

A1508. Kokke, R., Kujala, T., *Comp. Stud. Food Environ. Contam., Proc. Symp.* (1973) 551–558.

A1509. Schuette, H.R., Goehler, K.D., *Isotopenpraxis* (1974) **10**, 353–356.

A1510. Gee, J.M., Peel, J.L., *J. Gen. Microbiol.* (1974) **85**, 237–243.

A1511. Zech, R, R., Wigand, K.D., *Experientia* (1975) **31**, 157–158.

A1512. Udupa, S.R., Chadha, M.S., India, A.E.C., *Bhabha At. Res. Cent.* (1974) **BARC-764**, 33–37.

A1513. Wilson, M.L., Coscia, C.J., *J. Am. Chem. Soc.* (1975) **97**, 431–432.

A1514. Graff, G.L.A., Brouet-Yager, M., Kleiner, H., *Comp. Biochem. Physiol. B* (1974) **49**, 381–386.

A1515. Agurell, S., Granelli, I., Leander, K., Rosenblom, J., *Acta Chem. Scand., Ser B* (1974) **B28**, 1175–1179.

A1516. Ishiguro, I., Naito, J., Shinohara, R., Ishikura, A., *Yakugaku Zasshi* (1974) **94**, 1232–1239.

A1517. Koseki, Y., Itoh, T., Nakaya, S., *Nippon Yakurigaku Zasshi* (1973) **69**, 869–880.

A1518. Drayer, D.E., Strong, J.M., Jones, B., Sandler, A., Reidenberg, M.M., *Drug Metab. Dispos.* (1974) **2**, 499–505.

A1519. Cone, E.J., Gorodetzky, C.W., Yeh, S.Y., *Drug Metab. Dispos.* (1974) **2**, 506–512.

A1520. Bachofer, R., Lingens, F., Schaefer, W., *FEBS Lett.* (1974) **50**, 288–290.

A1521. Miyamoto, J., Suzuki, T., Nakae, C., *Pestic. Biochem. Physiol.* (1974) **4**, 438–450.

A1522. White, E.H., Branchini, B.R., *J. Am. Chem. Soc.* (1975) **97**, 1243–1245.

A1523. Masri, M.S., Booth, A.N., Hsieh, D.P.H., *Life Sci.* (1974) **15**, 203–212.

A1524. El Hachimi, Z., Samuel, O., Azerad, R., *Biochimie* (1974) **56**, 1239–1247.

A1525. Crawford, R.L., *J. Bacteriol.* (1975) **121**, 531–536.

A1526. Reif, V.D., Sinsheimer, J.E., Ward, J.C., Schteingart, D.E., *J. Pharm. Sci.* (1974) **63**, 1730–1736.

A1527. Davies, F.S., Mau, S.-L.C., Nooden, L.D., *Physiol. Plant.* (1975) **33**, 39–41.

A1528. Sugiura, M., Iwasaki, K., Noguchi, H., Kato, R., *Life Sci.* (1974) **15**, 1433–1442.

A1529. Bachur, N.R., *J. Pharmacol. Exp. Ther.* (1971) **177**, 573–578.

A1530. Morreal, C.E., Alks, V., Spiess, A.J., *Biochem. Pharmacol.* (1976) **25**, 1927–1930.

A1531. Aitio, A., Hartiala, J., Uotila, P., *Biochem. Pharmacol.* (1976) **25**, 1919–1920.

A1532. Pendleton, R.G., Kaiser, C., Gessner, G., *J. Pharmacol. Exp. Ther.* (1976) **197**, 623–632.

A1533. Warsh, J.J., Stancer, H.C., *J. Pharmacol. Exp. Ther.* (1976) **197**, 545–555.

A1534. Shah, H.C., McLachlan, J.A., *J. Pharmacol. Exp. Ther.* (1976) **197**, 687–696.

A1535. Tokuma, Y., Koda, S., Tsubouchi, S., Morimoto, Y., *Bull. Chem. Soc. Jpn.* (1975) **48**, 294–297.

A1536. Stillwell, W.G., Horning, M.G., *Res. Commun. Chem. Pathol. Pharmacol.* (1974) **9**, 601–619.

A1537. Peters, J.H., Gordon, G.R., Biggs, J.T., Jr., Levy, L., *Proc. Soc. Exp. Biol. Med.* (1975) **148**, 251–255.

A1538. Tanaka, S., Shimizu, H., Nakayama, H., Takenoshita, M., Ichishita, H., *Nippon Yakurigaku Zasshi* (1973) **69**, 419–428.

A1539. Belleville, M., Lechevin, J.C., Meynaud, A., Grand, M., Fontaine, L., *Eur. J. Med. Chem.-Chim. Ther.* (1974) **9**, 543–547.

A1540. Schoenharting, M., Mende, G., Seibert, G., *Hoppe-Seyler's Z. Physiol. Chem.* (1974) **355**, 1391–1399.

A1541. Kupferberg, H.J., Yonekawa, W., *Drug Metab. Dispos.* (1975) **3**, 26–29.

A1542. Bollag, J.-M., Czaplicki, E.J., Minard, R.S., *J. Agric. Food Chem.* (1975) **23**, 85–90.

A1543. Corner, E.D.S., Kilvington, C.C., O'Hara, S.C.M., *J. Mar. Biol. Assoc. UK* (1973) **53**, 819–832.

A1544. Stillwell, W.G., Carman, M.J., Bell, L., Horning, M.G., *Drug Metab. Dispos.* (1974) **2**, 489–498.

A1545. Wheeler, L., Strother, A., *Drug Metab. Dispos.* (1974) **2**, 533–538.

A1546. Suzuki, Y., Nishikawa, Y., Mitsuda, H., *J. Nutr. Sci. Vitaminol.* (1974) **20**, 301–316.

A1547. Tewari, S., Bhakuni, D.S., Kapil, R.S., *J. Chem Soc., Chem. Commun.* (1974) 940–941.

A1548. Hesch, R.D., Brunner, G., Söling, H.D., *Clin. Chim. Acta* (1975) **59**, 209–213.

A1549. Mulchandani, N.B., Iver, S.S., Badheka, L.P., *India, A.E.C., Bhabha At. Res. Cent.,* (1974) **BARC-764**, 18–26.

A1550. Leklem, J.E., Brown, R.R., Hankes, L.V., Jansen, C.R., *Biochem. Med.* (1975) **12**, 32–44.

A1551. Plotti, G., Menini, E., Bompiani, A., *J. Endocrinol.* (1975) **64**, 103–109.

A1552. Kawai, T., Baba, S., *Chem. Pharm. Bull.* (1975) **23**, 289–293.

A1553. Bajwa, R.S., Anderson, J.A., *J. Pharm. Sci.* (1975) **64**, 343–344.

A1554. Chang, T., Okerholm, R.A., Glazko, A.J., *Res. Commun. Chem. Pathol. Pharmacol.* (1974) **9**, 391–404.

A1555. Albert, K.S., Hallmark, M.R., Sakmar, E., Weidler, D.J., Wagner, J.G., *Res. Commun. Chem. Pathol. Pharmacol.* (1974) **9**, 463–469.

A1556. Naito, S.-I., Yamamoto, T., *J. Pharm. Sci.* (1975) **64**, 253–258.

A1557. Almirante, L., Danieli, B., Frigerio, A., Mugnaini, A., Picco, S., *Farmaco, Ed. Sci.* (1974) **29**, 941–956.

A1558. Helia, O., Vargova, E., Olschlager, H., Jindra, A., *Cesk. Farm.* (1974) **23**, 308–311.

A1559. Kampffmeyer, H.G., *Naunyn-Schmiedeberg's Arch. Pharmacol.* (1974) **286**, 221–225.

A1560. Tiedge, J.M., Hagedorn, M.L., *J. Agric. Food Chem.* (1975) **23**, 77–81.

A1561. Mikulski, P.I., Wiglusz, R., *Toxicol. Appl. Pharmacol.* (1975) **31**, 21–31.

A1562. Mihara, K., Miyamoto, J., *Agric. Biol. Chem.* (1974) **38**, 1913–1924.

A1563. Mehendale, H.M., Fields, M., Matthews, H.B., *J. Agric. Food Chem.* (1975) **23**, 261–265.

A1564. Lucier, G.W., *J. Steroid Biochem.* (1974) **5**, 681–685.

A1565. Rink, E., Böhm, H., *FEBS Lett.* (1975) **49**, 396–399.

A1566. Lafont, P., Lafont, J., *Ann Microbiol.* (1974) **125B**, 451–457.

A1567. Narasimhachari, N., Lin, R.-L., *Brain Res.* (1975) **87**, 126–129.

A1568. Lethco, E.J., Wallace, W.C., *Toxicology* (1975) **3**, 287–300.

A1569. Stanton, H.C., Cornejo, R.A., Marsmann, H.J., Brown, L.J., Mueller, R.L., *Arch. Int. Pharmacodyn. Ther.* (1975) **213**, 128–144.

A1570. Marquardt, G.M., Di Stefano, V., *Life Sci.* (1974) **15**, 1603–1610.

A1571. Hart, A., Orr, D.L.J., *J. Pharm. Pharmacol.* (1974) **26**, Suppl., 70P-71P.

A1572. Baukema, J., Okerholm, R.A., Glazko, A.J., *Res. Commun. Chem. Pathol. Pharmacol.* (1975) **10**, 227–238.

A1573. Gruenke, L.D., Craig, J.C., Dinovo, E.C., Gottschalk, L.A., *Res. Commun. Chem. Pathol. Pharmacol.* (1975) **10**, 221–225.

A1574. Hintze, K.L., Wold, J.S., Fischer, L.J., *Drug Metab. Dispos.* (1975) **3**, 1–9.

A1575. Williams, M.C., Helton, E.D., Goldzieher, J.W., *Steroids* (1975) **25**, 229–246.

A1576. Safe, S., Platonow, N., Hutzinger, O., *J. Agric. Food Chem.* (1975) **23**, 259–261.

A1577. Williams, J.G., Longcope, C., Williams, K.I.H., *Steroids* (1975) **25**, 343–354.

A1578. Mostafa, I.Y., Moza, P.N., *Egypt. J. Chem.* (1973) 235–242.

A1579. Laham, S., Dulos, E., Chang, W., *Biomedicine* (1974) **21**, 299–302.

A1580. Okuno, E., Minatogawa, Y., Noguchi, T., Kido, R., *Life Sci.* (1975) **16**, 883–890.

A1581. Steyn, P.S., Vleggaar, R., Pitout, M.J., Steyn, M., Thiel, P.G., *J. Chem. Soc., Perkin Trans. 1* (1974) 2551–2552.

A1582. Millard, B.J., Campbell, D.B., Jenner, P., Taylor, A.R., *Mass Spectrom. Biochem. Med., Symp.* (1973) 1–8.

A1583. Himeno, M., Hashiguchi, Y., Kato, K., *J. Biochem.* (1974) **76**, 1243–1252.

A1584. Thomas, G., Threlfall, D.R., *Biochem. J.* (1974) **142**, 437–440.

A1585. Lee, G.D., Van Etten, R.L., *Arch. Biochem. Biophys.* (1975) **166**, 280–294.

A1586. Burton, E.G., Sallach, H.J., *Arch. Biochem. Biophys.* (1975) **166**, 483–494.

A1587. Nakatani, M., Morimoto, M., Noguchi, T., Kido, R., *Biochem. J.* (1974) **143**, 303–310.

A1588. Ingelman-Sundberg, M., Rane, A., Gustafsson, J.-Å., *Biochemistry* (1975) **14**, 429–437.

A1589. Takesue, Y., *J. Biochem.* (1975) **77**, 103–115.

A1590. Pütter, J., Daniels, R., *Arzneim.-Forsch.* (1974) **24**, 1833–1834.

A1591. Jeffrey, A.M., Yeh, H.J.C., Jerina, D.M., Patel, T.R., Davey, J.F., Gibson, D.T., *Biochemistry* (1975) **14**, 575–583.

A1592. van der Hoeven, T.A., Haugen, D.A., Coon, M.J., *Biochem. Biophys. Res. Commun.* (1974) **60**, 569–575.

A1593. Johnsson, J.A., *Biochem. Pharmacol.* (1974) **23**, 3191–3197.

A1594. Shin, Y.S., Buehring, K.U., Stokstad, E.L.R., *J. Biol. Chem.* (1974) **249**, 5772–5777.

A1596. Hoagland, R.E., Graf, G., *Can. J. Biochem.* (1974) **53**, 903–910.

A1597. Letendre, C.H., Dickens, G., Guroff, G., *J. Biol. Chem.* (1974) **249**, 7186–7191.

A1598. Ecobichon, D.J., *Can. J. Biochem.* (1974) **52**, 1073–1078.

A1599. Catelani, D., Colombi, A., *Biochem. J.* (1974) **143**, 431–434.

A1600. Larsen, P.O., Wieczorkowska, E., *Biochim. Biophys. Acta* (1975) **381**, 409–415.

A1601. Larsen, P.O., Onderka, D.K., *Biochim. Biophys. Acta* (1975) **381**, 397–408.

A1602. Chen, C., Shaw, Y.-S.,, *Biochem. J.* (1974) **144**, 479–501.

A1603. James, S.P., Pugh, G.C., Waring, R.H., *Xenobiotica* (1974) **4**, 521–528.

A1604. Heenan, M.P., Smith, J.V., *Biochem. J.* (1974) **144**, 303–310.

A1606. Reddy, C.C., Vaidyanathan, C.S., *Biochim. Biophys. Acta* (1975) **384**, 46–57.

A1607. Chang, H.-H.S., Robinson, A.R., Common, R.H., *Can. J. Biochem.* (1975) **53**, 223–230.

A1608. Mathieu, P., Charvet, J.-C., Greffe, J., *Biochem. Pharmacol.* (1975) **24**, 43–48.

A1609. Fellman, J.H., Wada, G.H., Roth, E.S., *Biochim. Biophys. Acta* (1975) **381**, 9–16.

A1610. Mulder, G.J., Hayen-Keulemans, K., Sluiter, N.E., *Biochem. Pharmacol.* (1975) **24**, 103–107.

A1611. Axcell, B.C., Geary, P.J., *Biochem. J.* (1975) **146**, 173–183.

A1612. Workman, P., Ball, C.R., Double, J.A., *Biochem. Pharmacol.* (1976) **25**, 1139–1144.
A1613. Ekstedt, B., *Biochem. Pharmacol.* (1976) **25**, 1133–1138.
A1614. Quick, M., Sourkes, T.L., *Biochem. Pharmacol.* (1976) **25**, 1157–1166.
A1615. Nemoto, N., Gelboin, H.V., *Biochem. Pharmacol.* (1976) **25**, 1221–1226.
A1616. Nagasaki, T., Sugita, M., Fukawa, H., Lin, H.-T., *Agric. Biol. Chem.* (1975) **39**, 363–369.
A1617. Collins, D.C., Preedy, J.R.K., *Steroids* (1975) **25**, 427–439.
A1618. Clark, A.G., *Microbios* (1974) **11A**, 29–35.
A1619. Stibbs, H.H., Seed, J.R., *Int. J. Biochem.* (1975) **6**, 197–203.
A1620. Antoun, M.D., Roberts, M.F., *J. Pharm. Pharmacol.* (1974) **26**, Suppl., 114P-115P.
A1621. Evans, W.C., Ghani, A., Treagust, P.G., *J. Pharm. Pharmacol.* (1974) **26**, Suppl., 112P.
A1622. Slinger, I., *Comp. Biochem. Physiol. C.* (1975) **50**, 1–4.
A1624. Hobkirk, R., Nilsen, M., Jennings, B., *Can. J. Biochem.* (1975) **53**, 1333–1336.
A1625. Fishman, J., Norton, B., *Endocrinology* (1975) **96**, 1054–1059.
A1626. Fraser, I.H., Wainwright, S.D., *Can. J. Biochem.* (1976) **54**, 103–109.
A1627. Kohli, J. Jones, D., Safe, S., *Can. J. Biochem.* (1976) **54**, 203–208.
A1628. Lin, K.-T., Crawhall, J.C., *Can. J. Biochem.* (1976) **54**, 423–431.
A1629. Potier, M., Gianetto, R., *Can. J. Biochem.* (1976) **54**, 321–326.
A1630. Tsang, C.P.W., *Can. J. Biochem.* (1976) **54**, 666–669.
A1631. Lerner, H.R., Mayer, A.M., *Phytochemistry* (1976) **15**, 57–60.
A1632. Charriere-Ladreix, Y., *Phytochemistry* (1975) **14**, 1727–1731.
A1633. Richter, I., Luckner, M., *Phytochemistry* (1976) **15**, 67–70.
A1634. Kuroda, H., Shimada, M., Higuchi, T., *Phytochemistry* (1975) **14**, 1759–1763.
A1635. Wallace, J.W., *Phytochemistry* (1975) **14**, 1765–1768.
A1636. Kahn, V., *Phytochemistry* (1976) **15**, 267–272.
A1637. Uprety, H., Bhakuni, D.S., Kapil, R.S., *Phytochemistry* (1975) **14**, 1535–1537.
A1638. Evans, W.C., Woolley, J.G., *Phytochemistry* (1976) **15**, 287–289.
A1639. Müller, W.-U., Leistner, E., *Phytochemistry* (1976) **15**, 407–410.
A1640. Ellis, B.E., *Phytochemistry* (1976) **15**, 489–491.
A1641. Stafford, H.A., Brown, M.A., *Phytochemistry* (1976) **15**, 465–469.
A1642. Havir, E.A., Hanson, K.R., *Biochemistry* (1975) **14**, 1620–1626.
A1643. Hoffman, A.R., Ciaranello, R.D., Axelrod, J., *Biochem. Pharmacol.* (1975) **24**, 544–546.
A1644. Cohen, G.M., Haws, S.M., Moore, B.P., Bridges, J.W., *Biochem. Pharmacol.* (1976) **25**, 2561–2570.
A1645. Duhm, B., Maul, W., Medenwald, H., Patzschke, K., Wegner, K., *Arzneim.-Forsch.* (1967) **17**, 672–687.
A1646. Sullivan, H.R., Billings, R.E., Occolowitz, J.L., Boaz, H.E., Marshall, F.J., McMahon, R.E., *J. Med. Chem.* (1970) **13**, 904–909.
A1647. Kodama, K., Nakatani, S., Umehara, K., Shimizu, K., Minoda, Y., Yamada, K., *Agric. Biol. Chem.* (1970) **34**, 1320–1324.
A1648. Omori, T., Jigami, Y., Minoda, Y., *Agric. Biol. Chem.* (1974) **38**, 409–415.
A1649. Fewson, C.A., Beggs, J.D., Seenan, E., Ahlquist, E.F., *Biochem. Soc. Trans.* (1976) **4**, 614–615.
A1650. Malik, N., Butterworth, P.J., *Biochem. Soc. Trans.* (1976) **4**, 646–649.
A1651. Elliott, G.-R., Odam, E.M., Townsend, M.G., *Biochem. Soc. Trans.* (1976) **4**, 615–617.
A1652. Lake, B.G., Phillips, J.C., Hodgson, R.A., Severn, B.J., Gangolli, S.D., Lloyd,

A.G., *Biochem. Soc. Trans.* (1976) **4**, 654–655.
A1653. Angelo, M., Dring, L.G., Lancaster, R., Smith, R.L., *Biochem. Soc. Trans.* (1976) **4**, 704–706.
A1655. Hopkins, R., Martin, L., Bland, R., *Biochem. Soc. Trans.* (1976) **4**, 726–729.
A1657. Sandler, M., Ruthven, C.R.J., Wood, C., *Int. T. Neuropharmacol.* (1964) **3**, 123–128.
A1658. Burke, M.D., Prough, R.A., *Biochem. Pharmacol.* (1976) **25**, 2187–2195.
A1659. Pohl, L.R., Nelson, S.D., Porter, W.R., Trager, W.F., Fasco, M.J., Baker, F.D., Fenton, J.W. II, *Biochem. Pharmacol.* (1976) **25**, 2153–2162.
A1660. Harvey, D.J., Martin, B.R., Paton, W.D.M., *Biochem. Pharmacol.* (1976) **25**, 2217–2219.
A1661. Yang, S.K., Gelboin, H.V., *Biochem. Pharmacol.* (1976) **25**, 2221–2225.
A1662. Mandel, L.R., *Biochem. Pharmacol.* (1976) **25**, 2251–2256.
A1663. Bock, K.W., Van Ackeren, G., Lorch, F., Birke, F.W., *Biochem. Pharmacol.* (1976) **25**, 2351–2356.
A1664. Henry, R.A., Byington, K.H., *Biochem. Pharmacol.* (1976) **25**, 2291–2295.
A1665. McLean, A., Newell, D., Baker, G., *Biochem. Pharmacol.* (1976) **25**, 2331–2335.
A1666. Riceberg, L.J., Simon, M., Van Vunakis, H., Abeles, R.H., *Biochem. Pharmacol.* (1975) **24**, 119–125.
A1667. Boarder, M.R., Oon, M.C.H., Rodnight, R., *Biochem. Pharmacol.* (1976) **25**, 2109–2112.
A1668. Batt, A.-M., Ziegler, J.-M., Siest, G., *Biochem. Pharmacol.* (1975) **24**, 152–154.
A1669. Hewick, D.S., McEwen, J., *J. Pharm. Pharmacol.* (1973) **25**, 458–465.
A1670. Egashira, T., Ekstedt, B., Oreland, L., *Biochem. Pharmacol.* (1976) **25**, 2583–2586.
A1671. Tyce, G.M., Sharpless, N.S., Owen, C.A., Jr., *Biochem. Pharmacol.* (1976) **25**, 2635–2641.
A1672. Lee, P.W., Allahyari, R., Fukuto, T.R., *Biochem. Pharmacol.* (1976) **25**, 2671–2674.
A1673. Smith, R.V., Erhardt, P.W., Neumeyer, J.L., Borgman, R.J., *Biochem. Pharmacol.* (1976) **25**, 2106–2107.
A1674. Collins, J.F., Grundon, M.F., *J. Chem. Soc., Chem. Commun.* (1969) 621–622.
A1675. Chan, K.K., Lewis, R.J., Trager, W.F., *J. Med. Chem.* (1972) **15**, 1265–1270.
A1676. Grundon, M.F., James, K.J., *J. Chem. Soc., Chem. Commun.* (1971) 1311–1312.
A1677. Hurley, L.H., Zmijewski, M., *J. Chem. Soc., Chem. Commun.* (1974) 337–338.
A1678. Carroll, J., Armstrong, L.M., *Biochem. Soc. Trans.* (1976) **4**, 871–873.
A1679. Turner, A.J., Tipton, K.F., *Eur. J. Biochem.* (1972) **30**, 361–368.
A1680. Burns, R.A., Jackson, N., *Biochem. Soc. Trans.* (1976) **4**, 914–917.
A1681. Whitehead, V.M., Stein, H.A., *Biochem. Soc. Trans.* (1976) **4**, 918–920.
A1682. Bakke, J.E., Feil, V.J., Price, C.E., *Biomed. Mass Spectrom.* (1976) **3**, 226–229.
A1683. Due, S.L., Sullivan, H.R., McMahon, R.E., *Biomed. Mass Spectrom.* (1976) **3**, 217–225.
A1684. Jarman, M., Melzack, D.H., Ross, W.C.H., *Biochem. Pharmacol.* (1976) **25**, 2475–2478.
A1685. Cubeddo X, L., Gimenez, M., Goldstein, D.J., *Biochem. Pharmacol.* (1976) **25**, 2537–2538.
A1686. Shindler, J.S., Bardsley, W.G., *Biochem. Pharmacol.* (1976) **25**, 2689–2694.
A1687. Mansell, R.L., Stoeckigt, J., Zenk, M.H., *Z. Pflanzenphysiol.* (1972) **68**, 286–288.
A1688. Pietruszko, R., Chen, F.-F., *Biochem. Pharmacol.* (1976) **25**, 2721–2725.
A1689. Chin, W.-T., Stone, G.M., Smith, A.E., *J. Agric. Food Chem.* (1970) **18**, 709–712.

A1690. Meek, J.L., Lofstrandh, S., *Fed. Proc.* (1976) **35**, 241.
A1691. Paulson, G.D., Jacobson, A.M., *J. Agric. Food Chem.* (1974) **22**, 629–631.
A1692. Wright, J., Cho, A.K., Gal, J., *Fed. Proc.* (1976) **35**, 243.
A1693. Lynn, R.K., Leger, R.M., Gordon, W.P., Olsen, G.D., Smith, R., Gerber, N., Wendel, H.A., *Fed. Proc.* (1976) **35**, 243.
A1694. James, M.O., Van Stee, E.W., Bend, J.R., *Fed. Proc.* (1976) **35**, 244.
A1695. Hucker, H.B., Balletto, A.J., Demetriades, J., Arison, B.H., Zacchei, A.G., *Fed. Proc.* (1976) **35**, 244.
A1696. Lan, S.-j., Kripalani, K.J., Dean, A.V., Di Fazio, L.T., Schreiber, E.C., *Fed. Proc.* (1976) **35**, 244.
A1697. Fuller, R.W., Ho, P.P.K., Snoddy, H.D., *Fed. Proc.* (1976) **35**, 267.
A1698. Raese, J.D., Patrick, R.L., Barchas, J.D., *Fed. Proc.* (1976) **35**, 485.
A1699. Huang, M.-T., West, S.B., Lu, A.Y.H., *Fed. Proc.* (1976) **35**, 281.
A1700. Olukotun, A., Weinshilboum, R., *Fed. Proc.* (1976) **35**, 485.
A1701. Honohan, H., Enderlin, F.E., Ryerson, B.A., *Fed. Proc.* (1976) **35**, 328.
A1702. Bell, J.P., Tomlinson, R.V., *Fed. Proc.* (1976) **35**, 487.
A1703. Hwang, S.W., Hodgson, J.R., Short, R.D., Lee, C.C., *Fed. Proc.* (1976) **35**, 487.
A1704. Fujimoto, J.M., Roerig, S.C., Wang, R.I.H., *Fed. Proc.* (1976) **35**, 487.
A1705. Horning, M.G., Thenot, J.-P., Haegele, K., Nowlin, J., Stafford, M., Sommer, K.R., *Fed. Proc.* (1976) **35**, 488.
A1706. White, M.P., *Fed. Proc.* (1976) **35**, 565.
A1707. Gerber, N., Olsen, G.D., Swanson, B.N., Leger, R.M., Smith, R., Lynn, R.K., *Fed. Proc.* (1976) **35**, 565.
A1708. Lemberger, L., Rubin, A., Warrick, P., Crabtree, R., Rowe, H., *Fed. Proc.* (1976) **35**, 566.
A1709. Stillwell, W.G., Butler, C., Gregory, P., Horning, M.G., Hill, R.M., *Fed. Proc.* (1976) **35**, 566.
A1710. Kennedy, K.A., Fischer, L.J., *Fed. Proc.* (1976) **35**, 665.
A1711. Di Carlo, F.J., Leinweber, F.-J., Szpiech, J.M., Burt, R.L., Davidson, I.W.F., *Fed. Proc.* (1976) **35**, 566.
A1712. Walle, T., Conradi, E.C., Walle, U.K., Gaffney, T.E., *Fed. Proc.* (1976) **35**, 665.
A1713. Bend, J.R., Ben-Zvi, Z., Law, F.C.P., Ryan, A.J., *Fed. Proc.* (1976) **35**, 666.
A1714. Cohen, G.M., Moore, B.P., *Biochem. Pharmacol.* (1976) **25**, 1623–1629.
A1715. Blau, K., Goodwin, B., Woods, H.F., Youdim, M.B.H., *Br. J. Pharmacol.* (1976) **58**, 474P.
A1716. Bronaugh, R.L., Wenger, G.R., Garver, D.L., Rutledge, C.O., *Biochem. Pharmacol.* (1976) **25**, 1679–1681.
A1717. Silkaitis, R.P., Mosnaim, A.D., *Brain Res.* (1976) **114**, 105–115.
A1718. Roerig, D.L., Hasegawa, A.T., Wang, R.I.H., *J. Pharmacol. Exp. Ther.* (1976) **199**, 93–102.
A1719. Butler, T.C., Dudley, K.H., Johnson, D., Roberts, S.B., *J. Pharmacol. Exp. Ther.* (1976) **199**, 82–92.
A1720. Kaul, B., Millian, S.J., Davidow, B., *J. Pharmacol. Exp. Ther.* (1976) **199**, 171–178.
A1721. Tallman, J.F., Saavedra, J.M., Axelrod, J., *J. Pharmacol. Exp. Ther.* (1976) **199**, 216–221.
A1722. Sonntag, A.C., Stenger, V.G., Maren, T.H., *J. Pharmacol. Exp. Ther.* (1972) **182**, 48–55.
A1723. Yu, P.H., Boulton, A.A., Wu, P.H., *Can. J. Biochem.* (1976) **54**, 988–991.
A1724. Bickel, H., Schultz, G., *Phytochemistry* (1976) **15**, 1253–1255.
A1725. Tabata, M., Ikeda, F., Hiraoka, N., *Phytochemistry* (1976) **15**, 1225–1229.
A1726. Juneja, P.S., Gholson, R.K., *Phytochemistry* (1976) **15**, 647–649.
A1727. Phillipson, J.D., Handa, S.S., *Phytochemistry* (1976) **15**, 605–608.
A1728. Wong, E., Wilson, J.M., *Phytochemistry* (1976) **15**, 1325–1332.

A1729. Wilson, J.M., Wong, E., *Phytochemistry* (1976) **15**, 1333–1341.
A1730. Satö, M., *Phytochemistry* (1976) **15**, 1845–1847.
A1731. Saini, M.S., Cheng, M., Anderson, J.A., *Phytochemistry* (1976) **15**, 1497–1500.
A1732. Mansell, R.L., Babbel, G.R., Zenk, M.H., *Phytochemistry* (1976) **15**, 1849–1853.
A1733. Saleh, N.A.M., Poulton, J.E., Grisebach, H., *Phytochemistry* (1976) **15**, 1865–1868.
A1734. Mulchandani, N.B., Iyer, S.S., Badheka, L.P., *Phytochemistry* (1976) **15**, 1697–1699.
A1735. Satö, M., *Phytochemistry* (1976) **15**, 1665–1667.
A1736. Tourian, A., *Biochem. Biophys. Res. Commun.* (1976) **68**, 51–55.
A1737. Beckett, A.H., Gibson, G.G., *Xenobiotica* (1975) **5**, 677–686.
A1738. Kiechel, J.R., Niklaus, P., Schreier, E., Wagner, H., *Xenobiotica* (1975) **5**, 741–754.
A1739. Toft, P., Dring, L.G., Hirom, P.C., Williams, R.T., Donetti, A., Midgley, J.M., *Xenobiotica* (1975) **5**, 729–739.
A1740. Belvedere, G., Rovei, V., Pantarotto, C., Frigerio, A., *Xenobiotica* (1975) **5**, 765–771.
A1741. Douch, P.G.C., *Xenobiotica* (1975) **5**, 773–780.
A1742. Gingell, R., *Xenobiotica* (1976) **6**, 15–20.
A1743. Guarino, A.M., Anderson, J.B., *Xenobiotica* (1976) **6**, 1–13.
A1744. Thomas, J., Morgan, D., Vine, J., *Xenobiotica* (1976) **6**, 39–48.
A1745. Sullivan, H.R., Miller, W.M., McMahon, R.E., *Xenobiotica* (1976) **6**, 49–54.
A1746. Hlavica, P., Kehl, M., *Xenobiotica* (1976) **6**, 679–689.
A1747. Uzan, A., Gueremy, C., LeFur, G., *Xenobiotica* (1976) **6**, 649–665.
A1748. Mori, Y., Toyoshi, K., Baba, S., *Chem. Pharm. Bull.* (1976) **24**, 500–506.
A1749. Nambara, T., Nokubo, M., *Chem. Pharm. Bull.* (1976) **24**, 162–165.
A1750. Arfwidsson, A., Borg, K.O., Hoffmann, K.-J., Skånberg, I., *Xenobiotica* (1976) **6**, 691–711.
A1751. Page, J.G., Sullivan, H.R., Due, S.L., *Xenobiotica* (1976) **6**, 713–723.
A1752. Crawford, M.J., Hutson, D.H., King, P.A., *Xenobiotica* (1976) **6**, 745–762.
A1753. Wiebkin, P., Fry, J.R., Jones, C.A., Lowing, R., Bridges, J.W., *Xenobiotica* (1976) **6**, 725–743.
A1754. Mitchell, S.C., Waring, R.H., *Xenobiotica* (1976) **6**, 763–768.
A1755. Douch, P.G.C., Gahagan, H.M., *Xenobiotica* (1976) **6**, 769–773.
A1756. Mori, Y., Toyoshi, K., Baba, S., *Chem. Pharm. Bull.* (1976) **24**, 865–870.
A1757. Maeda, T., Shindo, H., *Chem. Pharm. Bull.* (1976) **24**, 1104–1106.
A1758. Sugiura, M., Suzuki, M., Sasaki, M., Shimomura, T., *Chem. Pharm. Bull.* (1976) **24**, 794–799.
A1759. Sugiura, M., Suzuki, M., Sasaki, M., Shimomura, T., *Chem. Pharm. Bull.* (1976) **24**, 788–793.
A1760. Kamijo, K., Watanabe, T., *Chem. Pharm. Bull.* (1976) **24**, 698–704.
A1761. Crew, M.C., Melgar, M.D., George, S., Greenough, R.C., Szpiech, J.M., Di Carlo, F.J., *Xenobiotica* (1976) **6**, 89–100.
A1762. Reynolds, C.N., Wilson, K., Burnett, D., *Xenobiotica* (1976) **6**, 113–124.
A1763. Yanagi, Y., Haga, F., Endo, M., Kitagawa, S., *Xenobiotica* (1976) **6**, 101–112.
A1764. Burnett, D., Reynolds, C.N., Wilson, K., Francis, J.R., *Xenobiotica* (1976) **6**, 125–134.
A1765. Solheim, E., Scheline, R.R., *Xenobiotica* (1976) **6**, 137–150.
A1767. Kripalani, K.J., El-Abdin, A.Z., Dean, A.V., Schreiber, E.C., *Xenobiotica* (1976) **6**, 159–169.
A1768. Hirohashi, A., Nagata, A., Miyawaki, H., Nakatani, H., Toki, K., *Xenobiotica* (1976) **6**, 329–337.

A1769. Lan, S.J., Dean, A.V., Walker, B.D., Schreiber, E.C., *Xenobiotica* (1976) **6**, 171–183.

A1770. Nagata, A., *Xenobiotica* (1976) **6**, 339–344.

A1771. Sever, P.S., Dring, L.G., Williams, R.T., *Xenobiotica* (1976) **6**, 345–353.

A1772. Caldwell, J., Moffatt, J.R. Smith, R.L., *Xenobiotica* (1976) **6**, 275–280.

A1773. Gorrod, J.W., Temple, D.J., *Xenobiotica* (1976) **6**, 265–274.

A1774. Kilpatrick, D.C., *Biochem. Soc. Trans.* (1976) **4**, 1083–1085.

A1775. Shaw, M.A., Jeffery, J., *Biochem. Soc. Trans.* (1976) **4**, 1097–1098.

A1776. Yamato, C., Takahashi, T., Fujita, T., *Xenobiotica* (1976) **6**, 295–306.

A1777. Burke, M.D., Upshall, D.G., *Xenobiotica* (1976) **6**, 321–328.

A1778. Tateishi, M., Shimizu, H., *Xenobiotica* (1976) **6**, 207–215.

A1779. Russo, R., Bartošek, I., Villa, S., Guaitani, A., Garattini, S., *Xenobiotica* (1976) **6**, 201–215.

A1780. Smith, G.E., Griffiths, L.A., *Xenobiotica* (1976) **6**, 217–236.

A1781. Cheng, L.K., Fung, H.-L., *Xenobiotica* (1976) **6**, 237–248.

A1782. Waring, R.H., Pheasant, A.E., *Xenobiotica* (1976) **6**, 257–262.

A1783. Davis, M., Harrison, N.G., Ideo, G., Portmann, B., Labadarios, D., Williams, R., *Xenobiotica* (1976) **6**, 249–255.

A1784. James, M.O., Bend, J.R., *Xenobiotica* (1976) **6**, 393–398.

A1785. Douch, P.G.C., *Xenobiotica* (1976) **6**, 399–404.

A1786. Monge, P., Scheline, R., Solheim, E., *Xenobiotica* (1976) **6**, 411–423.

A1787. Allen, J.G., Brown, A.N., Marten, T.R., *Xenobiotica* (1976) **6**, 405–409.

A1788. Tateishi, M., Shimizu, H., *Xenobiotica* (1976) **6**, 431–439.

A1789. Cameron, B.D., Chasseaud, L.F., Hawkins, D.R., Taylor, T., *Xenobiotica* (1976) **6**, 441–455.

A1790. Franklin, R.A., Aldridge, A., *Xenobiotica* (1976) **6**, 499–508.

A1791. Poulsen, L.L., Masters, B.S.S., Ziegler, D.M., *Xenobiotica* (1976) **6**, 481–498.

A1792. Douch, P.G.C., *Xenobiotica* (1976) **6**, 531–536.

A1793. Misra, A.L., Patel, M.N., Alluri, V.R., Mulé, S.J., Nayak, P.K., *Xenobiotica* (1976) **6**, 537–552.

A1794. Wong, L.T., Solomonraj, G., Thomas, B.H., *Xenobiotica* (1976) **6**, 575–584.

A1795. Willcox, S., Weatherby, R.P., Holder, G.M., *Xenobiotica* (1976) **6**, 565–573.

A1796. Salmona, M., Pachecka, J., Cantoni, L., Belvedere, G., Mussini, E., Garattini, S., *Xenobiotica* (1976) **6**, 585–591.

A1797. Pachecka, J., Salmona, M., Cantoni, L., Mussini, E., Pantarotto, C., Frigerio, A., Belvedere, G., *Xenobiotica* (1976) **6**, 593–598.

A1798. Arora, R.C., Meltzer, H.Y., *Biochem. Pharmacol.* (1977) **26**, 45–49.

A1799. Cowan, D.A., Huizing, G., Beckett, A.H., *Xenobiotica* (1976) **6**, 605–616.

A1800. Mohammed, Y.S., Mahfouz, M.M., Osman, M.Y., *Biochem. Pharmacol.* (1977) **26**, 62–63.

A1801. Pohl, L.R., Porter, W.R., Trager, W.F., Fasco, M.J., Fenton, J.W., *Biochem. Pharmacol.* (1977) **26**, 109–114.

A1807. Sugiura, M., Hirano, K., Iino, S., Suzuki, H., Oda, T., *Chem. Pharm. Bull.* (1976) **24**, 1698–1703.

A1808. Sugiura, M., Isobe, M., *Chem. Pharm. Bull.* (1976) **24**, 1822–1828.

A1809. Hashimoto, S., Shizu, M., Takahashi, S., *Chem. Pharm. Bull.* (1976) **24**, 1757–1764.

A1810. Mio, T., Sumino, K., Mizutani, T., *Chem. Pharm. Bull.* (1976) **24**, 1958–1960.

A1811. Suzuki, T., Takeda, M., *Chem. Pharm. Bull.* (1976) **24**, 1967–1975.

A1812. Suzuki, T., Takeda, M., *Chem. Pharm. Bull.* (1976) **24**, 1976–1982.

A1813. Suzuki, T., Takeda, M., *Chem. Pharm. Bull.* (1976) **24**, 1983–1987.

A1814. Kawai, K., Baba, S., *Chem. Pharm. Bull.* (1976) **24**, 2728–2732.
A1815. Suzuki, T., Takeda, M., *Chem. Pharm. Bull.* (1976) **24**, 1988–1991.
A1816. Moroi, R., Ono, K., Saito, T., Sano, M., Akimoto T., *Chem. Pharm. Bull.* (1976) **24**, 2850–2858.
A1817. Steele, J.A., Mirocha, C.J., Pathre, S.V., *J. Agric. Food Chem.* (1976) **24**, 89–97.
A1818. Ivie, G.W., Bull, D.L., Witzel, D.A., *J. Agric. Food Chem.* (1976) **24**, 147–151.
A1819. Bull, D.L., Ivie, G.W., *J. Agric. Food Chem.* (1976) **24**, 143–146.
A1820. Riggin, R.M., McCarthy, M.J., Kissinger, P.T., *J. Agric. Food Chem.* (1976) **24**, 189–191.
A1821. Benezet, H.J., Knowles, C.O., *J. Agric. Food Chem.* (1976) **24**, 152–154.
A1822. Furukawa, K., Matsumura, F., *J. Agric. Food Chem.* (1976) **24**, 251–256.
A1823. Elliott, M., Janes, N.F., Pulman, D.A., Gaughan, L.C., Unai, T., Casida, J.E., *J. Agric. Food Chem.* (1976) **24**, 270–276.
A1824. Lee, L.S., Bennett, J.W., Cucullu, A.F., Ory, R.L., *J. Agric. Food Chem.* (1976) **24**, 1167–1170.
A1825. Ruzo, L.O., Safe, S., Hutzinger, O., *J. Agric. Food Chem.* (1976) **24**, 291–293.
A1826. Hsieh, D.P.H., Lin, M.T., Yao, R.C., Singh, R., *J. Agric. Food Chem.* (1976) **24**, 1170–1174.
A1827. Santi, R., Gozzo, F., *J. Agric. Food Chem.* (1976) **24**, 1229–1235.
A1828. Smith, A.E., Phillips, D.V., *J. Agric. Food Chem.* (1976) **24**, 294–296.
A1829. Ruzo, L., Jones, D., Safe, S., Hutzinger, O., *J. Agric. Food Chem.* (1976) **24**, 581–583.
A1830. Nye, D.E., Hurst, H.E., Dorough, H.W., *J. Agric. Food Chem.* (1976) **24**, 371–377.
A1831. Iverson, F., *J. Agric. Food Chem.* (1976) **24**, 1238–1241.
A1832. Bull, D.L., Whitten, C.J., Ivie, G.W., *J. Agric. Food Chem.* (1976) **24**, 601–605.
A1833. Wiedmann, J.L., Ecke, G.G., Still, G.G., *J. Agric. Food Chem.* (1976) **24**, 588–592.
A1834. Schickedantz, P.D., Skladanowski, M.A., Zaletel, J., Marmor, R.S., Minnemeyer, H.J., *J. Agric. Food Chem.* (1976) **24**, 876–881.
A1835. Honohan, T., Hale, R.L., Brown, J.P., Wingard, R.E., Jr., *J. Agric. Food Chem.* (1976) **24**, 906–911.
A1836. Feung, C.-s., Hamilton, R.H., Mumma, R.O., *J. Agric. Food Chem.* (1976) **24**, 1013–1015.
A1837. Atallah, Y.H., Whitacre, D.M., Dorough, H.W., *J. Agric. Food Chem.* (1976) **24**, 1007–1012.
A1838. Crecelius, C.S., Knowles, C.O., *J. Agric. Food Chem.* (1976) **24**, 1018–1023.
A1839. Morgan, C.D., Sandler, M., Davies, D.S., Conolly, M., Paterson, J.W., Dollery, C.T., *Biochem. J.* (1969) **114**, 8P.
A1840. Mulder, G.J., Hinson, J.A., Gillette, J.R., *Biochem. Pharmacol.* (1977) **26**, 189–196.
A1841. Owen, F., Bourne, R.C., Lai, J.C.K., Williams, R., *Biochem. Pharmacol.* (1977) **26**, 289–292.
A1842. Wallace, R.R., Lewis, M.H.R., *Biochem. Soc. Trans.* (1977) **5**, 231–233.
A1843. Noda, K., Noguchi, H., Okui, M., Tada, K., Morimoto, Y., Kozatani, J., *Arzneim.-Forsch.* (1972) **22**, 732–743.
A1844. Höller, M., Dengler, K., Breuer, H., *Biochem. Soc. Trans.* (1977) **5**, 243–245.
A1845. Price, R.G., Thorpe, R., Tucker, S.M., Phillips, N.C., *Biochem. Soc. Trans.* (1977) **5**, 248–250.
A1846. Davis, D.R., Yeary, R.A., *Biochem. Pharmacol.* (1977) **26**, 535–539.
A1847. Sugiura, M., Iwasaki, K., Kato, R., *Biochem. Pharmacol.* (1977) **26**, 489–495.
A1848. Cohen, G.M., Moore, B.P., Bridges, J.W., *Biochem. Pharmacol.* (1977) **26**, 551–553.

A1849. Arnett, C.D., Callery, P.S., Zenker, N., *Biochem. Pharmacol.* (1977) **26**, 377–380.

A1850. Murphy, P.J., Williams, T.L., Kau, D.L.K., *J. Pharmacol. Exp. Ther.* (1976) **199**, 423–431.

A1851. Fuller, R.W., Molloy, B.B., *Biochem. Pharmacol.* (1977) **26**, 446–447.

A1852. Gerber, N., Seibert, R.A., Desiderio, D.M., Thompson, R.M., *Drug. Metab. Dispos.* (1974) **2**, 140–147.

A1853. Dorough, H.W., Cardona, R.A., Lolmaugh, S., *Drug. Metab. Dispos.* (1974) **2**, 129–139.

A1855. Cho, A.K., Lindeke, B., Sum, C.Y., *Drug. Metab. Dispos.* (1974) **2**, 1–8.

A1856. Zachariah, P.K., Juchau, M.R., *Drug. Metab. Dispos.* (1974) **2**, 74–78.

A1857. Klutch, A., *Drug. Metab. Dispos.* (1974) **2**, 23–30.

A1858. Hucker, H.B., Balletto, A.J., Stauffer, S.C., Zacchei, A.G., Arison, B.H., *Drug. Metab. Dispos.* (1974) **2**, 406–415.

A1859. Thomson, J.A., Holtzman, J.L., *Drug. Metab. Dispos.* (1974) **2**, 577–582.

A1861. Carter, J.E., Roll, D.B., Petersen, R.V., *Drug. Metab. Dispos.* (1974) **2**, 341–344.

A1862. Sanchez, E., Tephly, T.R., *Drug. Metab. Dispos.* (1974) **2**, 247–253.

A1863. de Sagher, R.M., De Leenheer, A.P., Cruyl, A.A., Claeys, A.E., *Drug. Metab. Dispos.* (1975) **3**, 423–429.

A1864. Alton, K.B., Patrick, J.E., Shaw, C., McGuire, J.L., *Drug. Metab. Dispos.* (1975) **3**, 445–452.

A1865. Gaertner, H.J., Liomin, G., Villumsen, D., Bertele, R., Breyer, U., *Drug. Metab. Dispos.* (1975) **3**, 437–444.

A1866. Tocco, D.J., Breault, G.O., Zacchei, A.G., Steelman, S.L., Perrier, C.V., *Drug. Metab. Dispos.* (1975) **3**, 453–466.

A1867. Chang, R.L., Zampaglione, N., Lin, C., *Drug. Metab. Dispos.* (1975) **3**, 487–493.

A1868. Davison, C., Benziger, D., Fritz, A., Edelson, J., *Drug. Metab. Dispos.* (1975) **3**, 520–524.

A1869. Juchau, M.R., Namkung, M.J., Berry, D.L., Zachariah, P.K., *Drug. Metab. Dispos.* (1975) **3**, 494–501.

A1870. Allen, J.G., East, P.B., Francis, R.J., Haigh, J.L., *Drug. Metab. Dispos.* (1975) **3**, 332–337.

A1871. Kau, S.T., Sastry, B.V.R., Alvin, J.D., Bush, M.T., *Drug. Metab. Dispos.* (1975) **3**, 345–351.

A1872. Harper, C., Drew, R.T., Fouts, J.R., *Drug. Metab. Dispos.* (1975) **3**, 381–388.

A1873. Alton, K.B., Grimes, R.M., Shaw, C., Patrick, J.E., McGuire, J.L., *Drug. Metab. Dispos.* (1975) **3**, 352–360.

A1874. Statham, C.N., Pepple, S.K., Lech, J.J., *Drug. Metab. Dispos.* (1975) **3**, 400–406.

A1875. Sumner, D.D., Dayton, P.G., Cucinell, S.A., Plostnieks, J., *Drug. Metab. Dispos.* (1975) **3**, 283–286.

A1876. Lucek, R.W., Coutinho, C.B., Cheripko, J.A., Ryan, J., Schwartz, M.A., *Drug. Metab. Dispos.* (1975) **3**, 297–302.

A1877. Stillwell, W.G., Gregory, P.A., Horning, M.G., *Drug. Metab. Dispos.* (1975) **3**, 287–296.

A1878. Lan, S.J., El-Hawey, A.M., Dean, A.V., Schreiber, E.C., *Drug. Metab. Dispos.* (1975) **3**, 171–179.

A1879. Porter, C.C., Arison, B.H., Gruber, V.F., Titus, D.C., VandenHeuvel, W.J.A., *Drug. Metab. Dispos.* (1975) **3**, 189–197.

A1880. Hucker, H.B., Balletto, A.J., Demetriades, J., Arison, B.H., Zacchei, A.G., *Drug. Metab. Dispos.* (1975) **3**, 80–84.

A1881. Boobis, A.R., Powis, G., *Drug. Metab. Dispos.* (1975) **3**, 63–68.

A1882. Schillings, R.T., Sisenwine, S.F., Schwartz, M.H., Ruelius, H.W., *Drug. Metab. Dispos.* (1975) **3**, 85–88.

A1883. Lan, S.-J., Chando, T.J., Schreiber, E.C., *Drug. Metab. Dispos.* (1975) **3**, 96–103.

A1884. Edelson, J., Douglas, J.F., Ludwig, B.J., *Drug. Metab. Dispos.* (1973) **1**, 737–741.

A1885. Hucker, H.B., Stauffer, S.C., White, S.D., Rhodes, R.E., Arison, B.H., *Drug. Metab. Dispos.* (1973) **1**, 721–736.

A1886. Dayton, P.G., Perel, J.M., Cunningham, R.F., Israili, Z.H., Weiner, I.M., *Drug. Metab. Dispos.* (1973) **1**, 742–751.
A1887. Merits, I., Anderson, D.J., Sanders, R.C., *Drug. Metab. Dispos.* (1973) **1**, 691–697.
A1888. Lan, S.-J., Chando, T.J., Cohen, A.I., Weliky, I., Schreiber, E.C., *Drug. Metab. Dispos.* (1973) **1**, 619–627.
A1889. Lin, C., Chang, R., Casmer, C., Symchowicz, S., *Drug. Metab. Dispos.* (1973) **1**, 611–618.
A1890. Parli, C.J., Lee, N.W., McMahon, R.E., *Drug. Metab. Dispos.* (1973) **1**, 628–633.
A1891. Drayer, D.E., Reidenberg, M.M., *Drug. Metab. Dispos.* (1973) **1**, 577–579.
A1892. Sisenwine, S.F., Kimmel, H.B., Liu, A.L., Segaloff, A., Ruelius, H.W., *Drug. Metab. Dispos.* (1973) **1**, 537–542.
A1893. Scott, K.N., Couch, M.W., Wilder, B.J., Williams, C.M., *Drug. Metab. Dispos.* (1973) **1**, 506–515.
A1894. Oesch, F., Glatt, H., Schmassen, H., *Biochem. Pharmacol.* (1977) **26**, 603–607.
A1895. Gallagher, B.M., *Biochem. Pharmacol.* (1977) **26**, 935–938.
A1896. Kimura, T., Kodama, M., Nagata, C., *Biochem. Pharmacol.* (1977) **26**, 671–674.
A1897. Nemoto, N., Takayama, S., *Biochem. Pharmacol.* (1977) **26**, 679–684.
A1898. Mohammed, Y.S., Mahfouz, M.M., *Biochem. Pharmacol.* (1977) **26**, 871–874.
A1899. Verboom, C.N., de Vries, J., van Bree, L., Nauta, W.T., *Biochem. Pharmacol.* (1977) **26**, 685–689.
A1900. Bagchi, S.P., Smith, T.M., *Biochem. Pharmacol.* (1977) **26**, 900–902.
A1901. Donnelly, C.H., Murphy, D.L., *Biochem. Pharmacol.* (1977) **26**, 853–858.
A1903. Cayen, M.N., Ferdinandi, E.S., Greselin, E., *J. Pharmacol. Exp. Ther.* (1977) **200**, 33–43.
A1904. Musacchio, J.M., Craviso, G.L., *Biochem. Pharmacol.* (1974) Suppl. Pt. 1, 6–11.
A1906. Dairman, W., Christenson, J., Udenfriend, S., *Biochem. Pharmacol.* (1974) Suppl. Pt. 1, 20–26.
A1907. Thomas, J.A., Van Orden, L.S. III, Redick, J.A., Kopin, I.J., *Biochem. Pharmacol.* (1974) Suppl. Pt. 1, 37–39.
A1908. Erwin, V.G., *Biochem. Pharmacol.* (1974) Suppl. Pt. 1, 110–115.
A1909. Neff, N.H., Yang, H.-Y.T., Goridis, C., *Biochem. Pharmacol.* (1974) Suppl. Pt. 1, 86–90.
A1911. Lebovitz, H.E., Downs, R.W., Jr., *Biochem. Pharmacol.* (1974) Suppl. Pt. 2, 688–690.
A1912. Gerber, M., Leger, R.M., Gordon, P., Smith, R.G., Bauer, J., Lynn, R.K., *J. Pharmacol. Exp. Ther.* (1977) **200**, 487–495.
A1913. Sugiura, M., Kato, R., *J. Pharmacol. Exp. Ther.* (1977) **200**, 25–32.
A1914. Friedman, P.A., Smith, M.W., *Biochem. Pharmacol.* (1977) **26**, 804–805.
A1915. Numazawa, M., Soeda, N., Moro, S., Nambara, T., *Biochem. Pharmacol.* (1977) **26**, 769–773.
A1916. Baumanis, E.A., Kalnina, I.E., Moskvitina, T.A., Gorkin, V.Z., *Biochem. Pharmacol.* (1977) **26**, 1059–1063.
A1917. Coutts, R.T., Kovach, S.H., *Biochem. Pharmacol.* (1977) **26**, 1043–1049.
A1918. Hidaka, H., Nagasaka, A., *Biochem. Pharmacol.* (1977) **26**, 1092–1094.
A1919. Bock, K.W., von Clausbruch, U.C., Josting, D., Ottenwälder, H., *Biochem. Pharmacol.* (1977) **26**, 1097–1100.
A1920. Neckers, L.M., Biggio, G., Moja, E., Meek, J.L., *J. Pharmacol. Exp. Ther.* (1977) **201**, 110–116.
A1921. Young, I.J., Knights, B.A., Hillman, J.R., *Nature* (1977) **267**, 429.
A1922. Fenn, P., Durbin, R.D., Kuntz, J.E., *Phytochemistry* (1977) **16**, 899–901.
A1923. Rönsch, H., *Phytochemistry* (1977) **16**, 691–698.
A1924. Gorham, J., *Phytochemistry* (1977) **16**, 915–918.

A1925. Kopp, M., Fritig, B., Hirth, L., *Phytochemistry* (1977) **16**, 895–898.

A1926. Kamel, M.Y., Saleh, N.A., Ghazy, A.M., *Phytochemistry* (1977) **16**, 521–524.

A1927. Yagi, A., Ogata, Y., Yamauchi, T., Nishioka, I., *Phytochemistry* (1977) **16**, 1098–1100.

A1928. Leete, E., Minich, M.L., *Phytochemistry* (1977) **16**, 149–150.

A1929. Dewick, P.M., *Phytochemistry* (1977) **16**, 93–97.

A1930. Dubernet, M., Ribereau-Gayon, P., Lerner, H.R., Harel, E., Mayer, A.M., *Phytochemistry* (1977) **16**, 191–193.

A1931. Simpson, T.J., Holker, J.S.E., *Phytochemistry* (1977) **16**, 229–233.

A1932. Kimmel, E.C., Fish, R.H., Casida, J.E., *J. Agric. Food Chem.* (1977) **25**, 1–9.

A1933. Kuhn, P.J., Smith, D.A., Ewing, D.F., *Phytochemistry* (1977) **16**, 296–297.

A1934. Gaughan, L.C., Unai, T., Casida, J.E., *J. Agric. Food Chem.* (1977) **25**, 9–17.

A1935. Sundström. G., *J. Agric. Food Chem.* (1977) **25**, 18–21.

A1936. Hobkirk, R., Freeman, D.J., Harvey, P.R.C., Nilsen, M., Jennings, B., *Can. J. Biochem.* (1977) **55**, 390–397.

A1937. Gandolfi, A.J., Buhler, D.R., *J. Agric. Food Chem.* (1977) **25**, 21–25.

A1938. Wainwright, S.D., *Can. J. Biochem.* (1977) **55**, 415–423.

A1939. Goromaru, T., Noda, A., Matsuyama, K., Iguchi, S., *Chem. Pharm. Bull.* (1976) **24**, 1376–1383.

A1940. Tanaka, M., Kyosaka, S., Sekiguchi, Y., *Chem. Pharm. Bull.* (1976) **24**, 3144–3148.

A1941. Noda, A., Goromaru, T., Tsubone, N., Matsuyama, K., Iguchi, S., *Chem. Pharm. Bull.* (1976) **24**, 1502–1505.

A1942. Noda, A., Tsubone, N., Mihara, M., Goromaru, T., Iguchi, S., *Chem. Pharm. Bull.* (1976) **24**, 3229–3231.

A1943. Sugiura, M., Hirano, K., Iino, S., Suzuki, H., Oda, T., *Chem. Pharm. Bull.* (1977) **25**, 352–354.

A1944. Chang, K.-M., Knowles, C.O., *J. Agric. Food Chem.* (1977) **25**, 493–501.

A1945. Murakoshi, I., Ogawa, M., Toriizuka, Haginiwa, J., *Chem. Pharm. Bull.* (1977) **25**, 527–528.

A1946. Miaullis, J.B., Hoffman, L.J., De Baun, J.R., Menn, J.J., *J. Agric. Food Chem.* (1977) **25**, 501–506.

A1947. Gorbach, S.G., Kuenzler, K., Asshauer, J., *J. Agric. Food Chem.* (1977) **25**, 507–511.

A1948. Archer, T.E., Stokes, J.D., Bringhurst, R.S., *J. Agric. Food Chem.* (1977) **25**, 536–541.

A1949. Lamoureux, G.L., Stafford, L.E., *J. Agric. Food Chem.* (1977) **25**, 512–517.

A1950. Samborn, J.R., Childers, W.F., Hansen, L.G., *J. Agric. Food Chem.* (1977) **25**, 551–553.

A1951. Olson, L.E., Allen, J.L., Hogan, J.W., *J. Agric. Food Chem.* (1977) **25**, 554–556.

A1952. Davis, J.E., Mende, T.J., *J. Pharmacol. Exp. Ther.* (1977) **201**, 490–497.

A1953. Burgschat, H., Netter, K.J., *J. Pharmacol. Exp. Ther.* (1977) **201**, 482–489.

A1954. Thurman, R.G., Marazzo, D.P., Jones, L.S., Kauffman, F.C., *J. Pharmacol. Exp. Ther.* (1977) **201**, 498–506.

A1955. Ruo, T.-I., Loo, S.-S., Chen, C., *Fed. Proc.* (1977) **36**, 663.

A1956. Hogue-Angeletti, R.A., Lysiak, W., *Fed. Proc.* (1977) **36**, 802.

A1957. Zeller, E.A., Huprikar, S.V., Wilson, W.E., *Fed. Proc.* (1977) **36**, 721.

A1958. Phillips, R.S., May, S.W., *Fed. Proc.* (1977) **36**, 832.

A1959. Touchstone, J.C., Levin, S.S., Novack, B., Cooper, D.Y., *Fed. Proc.* (1977) **36**, 843.

A1960. Leinweber, F.-J., Di Carlo, F.J., *Fed. Proc.* (1977) **36**, 938.

A1961. Del Villar, E., Sanchez, E., Tephly, T.R., *Fed. Proc.* (1977) **36**, 844.

A1962. Perel, J.M., Manian, A.A., *Fed. Proc.* (1977) **36**, 939.

A1963. Findlay, J.W.A., Butz, R.F., Koster, R., Welch, R.M., *Fed. Proc.* (1977) **36**, 939.

A1964. Bend, J.R., Mukhtar, H., Foureman, G.L., Lee, I.P., *Fed. Proc.* (1977) **36**, 961.

A1965. Greenlee, W.F., Poland, A.P., *Fed. Proc.* (1977) **36**, 939.

A1966. Sum, C.Y., Jonsson, J., *Fed. Proc.* (1977) **36**, 961.

A1967. Lynn, R.K., Bowers, J.L., Gerber, N., Smith, R.G., Thompson, R.M., Griffin, D., Deinzer, M., *Fed. Proc.* (1977) **36**, 961.

A1968. Tang, B.K., Inaba, T., Kalow, W., *Fed. Proc.* (1977) **36**, 966.

A1969. Baldwin, J.R., Witiak, D.T., Feller, D.R., *Fed. Proc.* (1977) **36**, 962.

A1970. Horning, M.G., Lertratanangkoon, K., *Fed. Proc.* (1977) **36**, 966.

A1971. Callaghan, J.T., *Fed. Proc.* (1977) **36**, 967.

A1972. Radomski, J.L., Moreno, H.R., Hearn, W.L., Radomski, T., *Fed. Proc.* (1977) **36**, 999.

A1973. Swanson, B.N., Lynn, R.K., *Fed. Proc.* (1977) **36**, 971.

A1974. Smith, B.R., Bend, J.R., *Fed. Proc.* (1977) **36**, 999.

A1975. Levine, W.G., Finkelstein, T., *Fed. Proc.* (1977) **36**, 1030.

A1976. Imondi, A.R., Alam, A., Brennan, J.J., Hagerman, L.M., *Fed. Proc.* (1977) **36**, 1031.

A1977. Lasker, J., Rickert, D.E., *Fed. Proc.* (1977) **36**, 1031.

A1978. Robinson, W.T., Bruderlein, H., Kraml, M., Rochefort, J.G., Zarowny, D., Dvornik, D., *Fed. Proc.* (1977) **36**, 1032.

A1979. Marhold, J., Rambousek, V., Pípalová, J., Matrka, M., *Neoplasma* (1969) **16**, 191–194.

A1980. Tredger, J.M., Chhabra, R.S., Fouts, J.R., *Drug Metab. Dispos.* (1976) **4**, 17–24.

A1981. Marhold, J., Rambousek, V., Pípalová, J., Matrka, M., *Neoplasma* (1969) **16**, 53–56.

A1982. Coutts, R.T., Dawe, R., Dawson, G.W., Kovach, S.H., *Drug Metab. Dispos.* (1976) **4**, 35–39.

A1983. Spaulding, T.C., Kotake, A.N., Takemori, A.E., *Drug Metab. Dispos.* (1976) **4**, 45–52.

A1984. Rash, J.J., Lynch, M.J., *Drug Metab. Dispos.* (1976) **4**, 59–64.

A1985. Roerig, S., Fujimoto, J.M., Wang, R.I.H., Pollock, S.H., Lange, D., *Drug Metab. Dispos.* (1976) **4**, 53–58.

A1986. Shargel, L., Dorrbecker, S.A., Levitt, M., *Drug Metab. Dispos.* (1976) **4**, 65–71.

A1987. Shargel, L., Dorrbecker, S.A., *Drug Metab. Dispos.* (1976) **4**, 72–78.

A1988. Weber, W.W., Miceli, J.N., Hearse, D.J., Drummond, G.S., *Drug Metab. Dispos.* (1976) **4**, 94–101.

A1989. Takanashi, S., Bachur, N.R., *Drug Metab. Dispos.* (1976) **4**, 79–87.

A1990. Melancon, M.J., Jr., Lech, J.J., *Drug Metab. Dispos.* (1976) **4**, 112–118.

A1991. Tay, L.K., Sinsheimer, J.E., *Drug Metab. Dispos.* (1976) **4**, 154–163.

A1992. Kolis, S.J., Williams, T.H., Schwartz, M.A., *Drug Metab. Dispos.* (1976) **4**, 169–176.

A1993. Knapp, D.R., Holcombe, N.H., Krueger, S.A., Privitera, P.J., *Drug Metab. Dispos.* (1976) **4**, 164–168.

A1994. Tyler, T.R., Lee, J.T., Flynn, H., VandenHeuvel, J.A., *Drug Metab. Dispos.* (1976) **4**, 177–179.

A1995. Kamataki, T., Lee Lin, M.C.M., Belcher, D.H., Neal, R.H., *Drug Metab. Dispos.* (1976) **4**, 180–189.

A1996. Chhabra, R.S., Fouts, J.R., *Drug Metab. Dispos.* (1976) **4**, 208–214.

A1997. Litterst, C.L., Gram, T.E., Mimnaugh, E.G., Leber, P., Emmerling, D., Freudenthal, R.A., *Drug Metab. Dispos.* (1976) **4**, 203–207.

A1998. Coutts, R.T., Dawson, G.W., Kazakoff, C.W., Wong, J.Y., *Drug Metab. Dispos.* (1976) **4**, 256–261.

A1999. Dugger, H.A., Coombs, R.A., Schwarz, H.J., Migdalof, B.H., Orwig, B.A., *Drug Metab. Dispos.* (1976) **4**, 262–268.

A2000. Misra, A.L., Bloch, R., Vardy, J., Mulé, S.J., Verebely, K., *Drug Metab. Dispos.* (1976) **4**, 276–280.

A2001. Turnbull, L.B., Teng, L., Newman, J., Chremos, A.N., Bruce, R.B., *Drug Metab. Dispos.* (1976) **4**, 269–275.

A2002. Wishart, G.J., Fry, D.T., *Biochem. Soc. Trans.* (1977) **5**, 705–706.

A2003. Karoum, F., Garrison, C.K., Neff, N.H., Wyatt, R.J., *J. Pharmacol. Exp. Ther.* (1977) **201**, 654–661.

A2004. Pons, M., Michel, F., Julliard, J., Descomps, B., Castes de Paulet, A., *Biochem. Soc. Trans.* (1977) **5**, 724–725.

A2005. Jack, G.W., *Biochem. Soc. Trans.* (1977) **5**, 726–728.

A2006. Knowles, R.G., Burchell, B., *Biochem. Soc. Trans.* (1977) **5**, 731.

A2007. Felsted, R.L., Richter, D.R., Bachur, N.R., *Biochem. Pharmacol.* (1977) **26**, 1117–1124.

A2008. Beckett, A.H., Belanger, P.M., *Biochem. Pharmacol.* (1977) **26**, 1455–1462.

A2009. Cohen, G.M., Moore, B.P., *Biochem. Pharmacol.* (1977) **26**, 1481–1487.

A2010. Nyholm-Poulsen, L., Squires, R.F., *Biochem. Pharmacol.* (1977) **26**, 1444–1445.

A2011. Spector, R., Levy, P., Abelson, H.T., *Biochem. Pharmacol.* (1977) **26**, 1507–1511.

A2012. Ames, M.M., Melmon, K.L., Casagnoli, N., Jr., *Biochem. Pharmacol.* (1977) **26**, 1757–1762.

A2013. Baldessarini, R.J., Kula, N.S., Walton, K.G., Borgman, R.J., *Biochem. Pharmacol.* (1977) **26**, 1749–1756.

A2014. Nemoto, N., Takayama, S., Gelboin, H.V., *Biochem. Pharmacol.* (1977) **26**, 1825–1829.

A2015. Jonsson, J., *Xenobiotica* (1977) **7**, 93–94.

A2016. Illing, H.P.A., Johnson, P., Rising, P.A., Yeomans, M.A., *Xenobiotica* (1977) **7**, 99.

A2017. Horning, M.G., Butler, C., Gregory, P., Lertratanangkoon, L., Stillwell, W.G., *Xenobiotica* (1977) **7**, 94.

A2018. Van Der Veen, F., De Jongh, G.D., *Xenobiotica* (1977) **7**, 99–100.

A2019. Kolar, G.F., *Xenobiotica* (1977) **7**, 100–101.

A2020. Ribas, B., Sanchez, M.I., *Xenobiotica* (1977) **7**, 101.

A2021. Ball, L.M., Renwick, A.G., Williams, R.T., *Xenobiotica* (1977) **7**, 101.

A2022. Noordhoek, J., Van Der Berg, A.P., *Xenobiotica* (1977) **7**, 103.

A2023. Hutson, D.H., Holmes, D.S., Crawford, M.J., *Xenobiotica* (1977) **7**, 107.

A2024. Townsend, M.G., Tarrant, K.A., *Xenobiotica* (1977) **7**, 114–115.

A2025. Elmamlouk, T.H., Gessner, T., *Xenobiotica* (1977) **7**, 111–112.

A2026. Metzler, M., Neumann, H.-G., *Xenobiotica* (1977) **7**, 117–132.

A2027. Rosazza, J.P., Kammer, M., Youel, L., Smith, R.V., Erhardt, P.W., Truong, D.H., Leslie, S.W., *Xenobiotica* (1977) **7**, 133–143.

A2028. Mitoma, C., Saito, T., Howd, R.A., *Xenobiotica* (1977) **7**, 165–179.

A2029. Forshell, G.P., *Xenobiotica* (1977) **7**, 153–163.

A2030. Taylor, J.A., Twomey, T.M., Von Wittenau, M.S., *Xenobiotica* (1977) **7**, 357–364.

A2031. Morgan, D.J., Smyth, M.P., Thomas, J., Vine, J., *Xenobiotica* (1977) **7**, 365–375.

A2032. Takenoshita, R., Toki, S., *Xenobiotica* (1977) **7**, 383–391.

A2033. McMahon, R.E., Sullivan, H.R., *Xenobiotica* (1977) **7**, 377–382.

A2034. Wright, J., Cho, A.K., Gal, J., *Xenobiotica* (1977) **7**, 257–266.

A2035. Hutson, D.H., Crayford, J.V., Hoadley, E.C., *Xenobiotica* (1977) **7**, 279–300.

A2036. Douch, P.G.C., Gahagan, H.M., *Xenobiotica* (1977) **7**, 309–314.

A2037. Douch, P.G.C., Gahagan, H.M., *Xenobiotica* (1977) **7**, 301–307.

A2038. Hawkins, D.R., Moore, D.H., Chasseaud, L.F., *Xenobiotica* (1977) **7**, 315–325.
A2039. Ball, L.M., Renwick, A.G., Williams, R.T., *Xenobiotica* (1977) **7**, 189–203.
A2040. Gentil, A., Lasne, C., Chouroulinkov, I., *Xenobiotica* (1977) **7**, 221–233.
A2041. Mitoma, C., Onodera, T., Takegoshi, T., Thomas, D.W., *Xenobiotica* (1977) **7**, 205–220.
A2044. Ueda, K., Gaughan, L.C., Casida, J.E., *J. Agric. Food Chem.* (1975) **23**, 106–115.
A2045. Lin, T.H., North, H.H., Menzer, R.E., *J. Agric. Food Chem.* (1975) **23**, 253–256.
A2046. Safe, S., Platonow, N., Hutzinger, O., *J. Agric. Food Chem.* (1975) **23**, 259–261.
A2047. Lin, T.H., North, H.H., Menzer, R.E., *J. Agric. Food Chem.* (1975) **23**, 257–258.
A2048. Guirguis, G.N., Brindley, W.A., *J. Agric. Food Chem.* (1975) **23**, 274–279.
A2050. Suzuki, T., Uchiyama, M., *J. Agric. Food Chem.* (1975) **23**, 281–286.
A2051. Jordan, L.S., Zurqiyah, A.A., De Muir, A.R., Clerx, W.A., *J. Agric. Food Chem.* (1975) **23**, 286–289.
A2052. Ivie, G.W., *J. Agric. Food Chem.* (1975) **23**, 869–872.
A2053. Safe, S., Hutzinger, O., Jones, D., *J. Agric. Food Chem.* (1975) **23**, 851–853.
A2054. Reif, V.D., Littleton, B.C., Sinsheimer, J.E., *J. Agric. Food Chem.* (1975) **23**, 996–999.
A2055. Osborne, N.N., *Biochem. Soc. Trans.* (1977) **5**, 858–863.
A2056. Emudianughe, T.S., Caldwell, J., Smith, R.L., *Biochem. Soc. Trans.* (1977) **5**, 1006–1008.
A2057. Michel-Lipowsky, M., Caldwell, J., Smith, R.L., *Biochem. Soc. Trans.* (1977) **5**, 1005–1006.
A2058. Hirom, P.C., Idle, J.R., Millburn, P., Williams, R.T., *Biochem. Soc. Trans.* (1977) **5**, 1033–1035.
A2059. Badawy, A.A.-B., Evans, M., *Biochem. Soc. Trans.* (1977) **5**, 1035–1036.
A2060. Rubio, M.C., Langer, S.Z., *Naunyn-Schmiedeberg's Arch. Pharmacol.* (1973) **280**, 315–330.
A2061. Van Doorslaer, E., Kersters-Hilderson, H., De Bruyne, C.K., *Biochem. Soc. Trans.* (1977) **5**, 1113–1114.
A2062. Evans, M., Badawy, A.A.-B., *Biochem. Soc. Trans.* (1977) **5**, 1037–1038.
A2063. Debruyne, I., Stockx, J., *Biochem. Soc. Trans.* (1977) **5**, 1109–1111.
A2064. Kuczenski, R.T., Mandell, A.J., *J. Neurochem.* (1972) **19**, 131–137.
A2065. Wong, K.P., *J. Neurochem.* (1975) **24**, 1059–1063.
A2066. Wong, K.P., *Biochem. J.* (1976) **160**, 491–493.
A2067. Wong, K.P., *Biochem. J.* (1976) **158**, 33–37.
A2068. Balasubramanian, K.A., Bachhawat, B.K., *J. Neurochem.* (1976) **27**, 485–492.
A2069. Di Chiara, G., Balakleevsky, A., Porceddu, M.L., Tagliamonte, A., Gessa, G.L., *J. Neurochem.* (1974) **23**, 1105–1108.
A2070. Daly, J., Horner, L., Witkop, B., *J. Am. Chem. Soc.* (1961) **83**, 4787–4792.
A2071. Kawai, K., Baba, S., *Chem. Pharm. Bull.* (1975) **23**, 920–922.
A2072. Schalzing, G., *Drug Metab. Dispos.* (1977) **5**, 104–115.
A2073. Breyer, U., Schmalzing, G., *Drug Metab. Dispos.* (1977) **5**, 97–103.
A2074. De Long, A.F., Smyth, R.D., Polk, A., Nayak, R.K., Martin, G., Douglas, G.H., Reavey-Cantwell, N.H., *Drug Metab. Dispos.* (1977) **5**, 122–131.
A2075. Hucker, H.B., Balletto, A.J., Demetriades, J., Arison, B.H., Zacchei, A.G., *Drug Metab. Dispos.* (1977) **5**, 132–142.
A2076. Jeffcoat, A.R., Handy, R.W., Francis, M.T., Willis, S., Wall, M.E., Birch, C.G., Hiles, R.A., *Drug Metab. Dispos.* (1977) **5**, 157–166.
A2077. Midha, K.K., Cooper, J.K., By, A., Ethier, J.-C., *Drug Metab. Dispos.* (1977) **5**, 143–148.

A2078. Rane, A., Shand, D.G., Wilkinson, G.R., *Drug Metab. Dispos.* (1977) **5**, 179–184.
A2079. Edelson, J., McMullen, J.P., *Drug Metab. Dispos.* (1977) **5**, 185–190.
A2080. Franklin, R.B., Dring, L.G., Williams, R.T., *Drug Metab. Dispos.* (1977) **5**, 223–233.
A2081. Cunningham, R.F., Perel, J.M., Israili, Z.H., Dayton, P.G., *Drug Metab. Dispos.* (1977) **5**, 205–210.
A2082. Lin, C., Li, Y., McGlotten, J., Morton, J.B., Symchowicz, S., *Drug Metab. Dispos.* (1977) **5**, 234–238.
A2083. Benziger, D.P., Balfour, K., Clemans, S., Edelson, J., *Drug Metab. Dispos.* (1977) **5**, 239–245.
A2084. Del Villar, E., Sanchez, E., Tephly, T.R., *Drug Metab. Dispos.* (1977) **5**, 273–278.
A2085. Martin, B.R., Harvey, D.J., Paton, W.D.M., *Drug Metab. Dispos.* (1977) **5**, 259–267.
A2086. Namkung, M.J., Zachariah, P.K., Juchau, M.R., *Drug Metab. Dispos.* (1977) **5**, 288–294.
A2087. Di Giovanni, J., Slaga, T.J., Berry, D.L., Juchau, M.R., *Drug Metab. Dispos.* (1977) **5**, 295–301.
A2088. Henderson, G.L., North-Root, H., Kuttab, S.H., *Drug Metab. Dispos.* (1977) **5**, 321–328.
A2089. Gammans, R.E., Sehon, R.D., Anders, M.W., Hanna, P.E., *Drug Metab. Dispos.* (1977) **5**, 310–316.
A2090. Hubbard, J.W., Midha, K.K., Cooper, J.K., *Drug Metab. Dispos.* (1977) **5**, 329–334.
A2091. Yeh, S.Y., McQuinn, R.I., Gorodetzky, C.W., *Drug Metab. Dispos.* (1977) **5**, 335–342.
A2092. Dingell, J.V., Sossi, N., *Drug Metab. Dispos.* (1977) **5**, 397–404.
A2093. Ryan, A.J., Bend, J.R., *Drug Metab. Dispos.* (1977) **5**, 363–367.
A2094. Shiobara, Y., *Xenobiotica* (1977) **7**, 457–468.
A2096. Higuchi, S., Sasaki, H., Shiobara, Y., Sado, T., *Xenobiotica* (1977) **7**, 469–479.
A2097. Noguchi, H., Okui, M., Noda, K., Kato, R., *Xenobiotica* (1977) **7**, 491–503.
A2098. Salhab, A.S., Abramson, F.P., Geelhoed, G.W., Edwards, G.S., *Xenobiotica* (1977) **7**, 401–408.
A2099. Noguchi, H., Okui, M., Noda, K., Sugiura, M., Tanaka, Y., Kato, R., *Xenobiotica* (1977) **7**, 505–516.
A2100. Timbrell, J.A., Mitchell, J.R., *Xenobiotica* (1977) **7**, 415–423.
A2101. Rising, P.A., Illing, H.P.A., Johnson, P., Yeomans, M.A., *Xenobiotica* (1977) **7**, 425–437.
A2102. Illing, H.P.A., Yeomans, M.A., Rising, P.A., Johnson, P., *Xenobiotica* (1977) **7**, 439–454.
A2103. Barford, P.A., Blair, J.A., Staff, F.J., Malghani, M.A.K., *Biochem. Soc. Trans.* (1977) **5**, 1316–1318.
A2104. Bolton, G.C., Griffiths, L.A., *Biochem. Soc. Trans.* (1977) **5**, 1358–1361.
A2105. Bender, D.A., Coulson, W.F., *Biochem. Soc. Trans.* (1977) **5**, 1353–1356.
A2106. Baty, J.D., Robinson, P.R., *Biochem. Soc. Trans.* (1977) **5**, 1361–1362.
A2107. Renwick, A.G., Ball, L.M., *Biochem. Soc. Trans.* (1977) **5**, 1363–1365.
A2108. Tong, S., Ioannides, C., Parke, D.V., *Biochem. Soc. Trans.* (1977) **5**, 1374–1377.
A2109. Burke, M.D., Benford, D.J., Bridges, J.W., Parke, D.V., *Biochem. Soc. Trans.* (1977) **5**, 1370–1372.
A2110. Thompson, R.M., Beghin, J., Fife, W.K., Gerber, N., *Drug Metab. Dispos.* (1976) **4**, 349–356.
A2111. Lan, S.J., Kripalani, K.J., Dean, A.V., Egli, P., DiFazio, L.T., Schreiber, E.C., *Drug Metab. Dispos.* (1976) **4**, 330–339.
A2112. Chen, P.R., McKinney, J.D., Matthews, H.B., *Drug Metab. Dispos.* (1976) **4**, 362–367.
A2113. Di Carlo, F.J., Herzig, D.J., Kusner, E.J., Schumann, P.R., Melgar, M.D., George, S., Crew, M.C., *Drug Metab. Dispos.* (1976) **4**, 368–371.

A2114. Turnbull, L.B., Teng, L., Newman, J., Bruce, R.B., Maynard, W.R., *Drug Metab. Dispos.* (1976) **4**, 379–386.

A2115. Chaudhuri, N.K., Servando, O.A., Manniello, M.J., Luders, R.C., Chao, D.K., Bartlett, M.F., *Drug Metab. Dispos.* (1976) **4**, 372–378.

A2116. Zacchei, A.G., Weidner, L.L., Besselaar, G.H., Raftery, E.B., *Drug Metab. Dispos.* (1976) **4**, 387–401.

A2117. Dvorchic, B.H., Stenger, V.G., Quattropani, S.L., *Drug Metab. Dispos.* (1976) **4**, 423–429.

A2118. Sum, C.Y., Cho, A.K., *Drug Metab. Dispos.* (1976) **4**, 436–441.

A2119. Hinson, J.A., Mitchell, J.R., *Drug Metab. Dispos.* (1976) **4**, 430–435.

A2120. Pohl, R.J., Philpot, R.M., Fouts, J.R., *Drug Metab. Dispos.* (1976) **4**, 442–450.

A2121. Migdalof, B.H., Yorgey, K.A., Popkave, C.G., *Drug Metab. Dispos.* (1976) **4**, 468–473.

A2122. Zacchei, A.G., Wishousky, T.I., Arison, B.H., Fanelli, G.M., Jr., *Drug Metab. Dispos.* (1976) **4**, 479–489.

A2123. Dayton, H.E., Inturrisi, C.E., *Drug Metab. Dispos.* (1976) **4**, 474–478.

A2124. Zacchei, A.G., Wishousky, T.I., *Drug Metab. Dispos.* (1976) **4**, 490–498.

A2125. Lynn, R.K., Olsen, G.D., Leger, R.M., Gordon, W.P., Smith, R.G., Gerber, N., *Drug Metab. Dispos.* (1976) **4**, 504–509.

A2126. Fusaro, G.A., Garteiz, D.A., Chan, K.Y., Leeson, G.A., *Drug Metab. Dispos.* (1976) **4**, 562–567.

A2127. Stohs, S.J., Grafström, R.C., Burke, M.D., Orrenius, S., *Drug Metab. Dispos.* (1976) **4**, 517–521.

A2128. Midha, K.K., Cooper, J.K., McGilveray, I.J., Coutts, R.T., Dawe, R., *Drug Metab. Dispos.* (1976) **4**, 568–576.

A2129. Burke, M.D., Prough, R.A., Mayer, R.T., *Drug Metab. Dispos.* (1977) **5**, 1–8.

A2130. Somani, S.M., Anderson, J.H., *Drug Metab. Dispos.* (1977) **5**, 15–18.

A2131. Thompson, J.A., Holtzman, J.L., *Drug Metab. Dispos.* (1977) **5**, 9–14.

A2132. James, M.O., Foureman, G.L., Law, F.C., Bend, J.R., *Drug Metab. Dispos.* (1977) **5**, 19–28.

A2133. Melancon, M.J., Jr., Lech, J.J., *Drug Metab. Dispos.* (1977) **5**, 29–36.

A2134. Lynn, R.K., Smith, R.G., Leger, R.M., Deinzer, M.L., Griffin, D., Gerber, N., *Drug Metab. Dispos.* (1977) **5**, 47–55.

A2135. Feller, D.R., Malspeis, L., *Drug Metab. Dispos.* (1977) **5**, 37–46.

A2136. Whyte, M.P., Dekaban, A.S., *Drug Metab. Dispos.* (1977) **5**, 63–70.

A2137. Hobbs, D.C., Twomey, T.M., *Drug Metab. Dispos.* (1977) **5**, 75–81.

A2138. Maeda, T., Shindo, H., *Chem. Pharm. Bull.* (1977) **25**, 1992–2000.

A2139. Maeda, T., Miyakoshi, N., Shindo, H., *Chem. Pharm. Bull.* (1977) **25**, 2001–2012.

A2140. Beckett, A.H., Jones, G.R., Al-Sarraj, S., *J. Pharm. Pharmacol.* (1974) **26**, 945–951.

A2141. Hoskins, J.A., *Biochem. Biophys. Res. Commun.* (1977) **77**, 50–56.

A2142. Wechter, W.J., Loughhead, D.G., Reischer, R.J., Van Giessen, G.J., Kaiser, D.G., *Biochem. Biophys. Res. Commun.* (1974) **61**, 833–837.

A2143. Eaton, G.G. Goy, R.W., Resko, J.A., *Horm. Behav.* (1975) **6**, 81–97.

A2144. Rowe, V.D., Fales, H.M., Pisano, J.J., Andersen, A.E., Guroff, G., *Biochem. Med.* (1975) **12**, 123–136.

A2145. Ahokas, J.T., Pelkonen, O., Karki, N.T., *Biochem. Biophys. Res. Commun.* (1975) **63**, 635–641.

A2146. Abou-Donia, M.B., Dieckert, J.W., *Toxicol. Appl. Pharmacol.* (1975) **31**, 32–46.

A2147. Haugen, D.A., Van der Hoeven, T.A., Coon, M.J., *J. Biol. Chem.* (1975) **250**, 3567–3570.

A2148. Knobloch, K.-H., Hahlbrock, K., *Eur. J. Biochem.* (1975) **52**, 311–320.

A2149. Bury, A.F., Pennington, R.J., *Biochem. J.* (1975) **145**, 413–416.

A2150. Shetty, A.S., Gaertner, F.H., *J. Bacteriol.* (1975) **122**, 235–244.
A2151. Fragoulis, E.G., Sekeris, C.E., *Arch. Biochem. Biophys.* (1975) **168**, 15–25.
A2152. Judel, G.-K., *Biochem. Physiol. Pflanz.* (1975) **167**, 243–251.
A2153. Judel, G.-K., *Biochem. Physiol. Pflanz.* (1975) **167**, 233–241.
A2154. Hurley, L.H., Gairola, C., Zmijewski, M.J., Jr., *J. Chem. Soc., Chem. Commun.* (1975) 120–121.
A2155. Hopper, D.J., Taylor, D.G., *J. Bacteriol.* (1975) **122**, 1–6.
A2156. Brown, A.E., Swinburne, T.R., *Phytopathol. Z.* (1974) **81**, 254–257.
A2157. Manitto, P., Gramatica, P., Ranzi, B.M., Monti, D., *Bull. Liazon, Groupe Polyphenols* (1974) 5, 7, 11.
A2158. Veliky, I.A., Barber, K.M., *Lloydia* (1975) **38**, 125–130.
A2159. Thorsen, T., J, *Steroid Biochem.* (1975) **6**, 35–41.
A2160. Tomaszewski, J.E., Jerina, D.M., Daly, J.W., *Biochemistry* (1975) **14**, 2024–2031.
A2161. Bahkle, Y.S., Youdim, M.B.H., *J. Physiol.* (1975) **248**, 23P-25P.
A2162. Baiocchi, L., Frigerio, A., Giannangeli, M., Palazzo, G., *Arzneim.-Forsch.* (1974) **24**, 1699–1706.
A2163. Pfeifer, S., Pohlmann, H., *Acta Pharm. Suec.* (1974) **11**, 645–646.
A2164. Cone, E.J., Gorodetzky, C.W., Yeh, S.Y., *J. Pharm. Sci.* (1975) **64**, 618–621.
A2165. Gigon, P.L., *Arch. Int. Pharmacodyn. Ther.* (1975) **213**, 211–221.
A2166. Sundström, G., Wachtmeister, C.A., *Chemosphere* (1975) **4**, 7–11.
A2167. Lui, H., Sweeney, G.D., *FEBS Lett.* (1975) **51**, 225–226.
A2168. Safe, S., Hutzinger, O., Ecobichon, D., *Experientia* (1974) **30**, 720–721.
A2169. Van Miller, J.P., Hsu, I.C., Allen, J.R., *Proc. Soc. Exp. Biol. Med.* (1975) **148**, 682–687.
A2170. Hosokawa, S., Miyamoyo, J., *Bochu-Kagaku* (1974) **39**, 49–53.
A2171. Statham, C.N., Lech, J.J., *J. Fish. Res. Board Can.* (1975) **32**, 515–522.
A2172. Ohta, Y., Higgins, I.J., Ribbons, D.W., Groseclose, E.E., *J. Biol. Chem.* (1975) **250**, 3814–3825.
A2173. Sternson, L.A., *Experientia* (1975) **31**, 268–270.
A2174. Hadorn, M., John, V.A., Meier, F.K., Dutter, H., *Eur. J. Biochem.* (1975) **54**, 65–73.
A2175. Honma, S., Nambara, T., *Chem. Pharm. Bull.* (1975) **23**, 787–793.
A2176. Lesca, P., Lecointe, P., Paoletti, C., Mansuy, D., *Biochem. Pharmcol.* (1977) **26**, 2169–2173.
A2177. Oommen, A., Balasubramanian, A.S., *Biochem. Pharmcol.* (1977) **26**, 2163–2167.
A2178. Buckholtz, N.S., Boggan, W.O., *Biochem. Pharmcol.* (1977) **26**, 1991–1996.
A2179. Mosnaim, A.D., Edstrand, D.L., Wolf, M.E., *Biochem. Pharmcol.* (1977) **26**, 1725–1728.
A2180. Mohn, G., Niderlaender, P., Kessler, C., *Biochem. Pharmcol.* (1977) **26**, 1885–1892.
A2181. Achee, F.M., Gabay, S., *Biochem. Pharmcol.* (1977) **26**, 1637–1644.
A2182. Patsalos, P.N., Lascelles, P.T., *Biochem. Pharmcol.* (1977) **26**, 1929–1933.
A2183. Breuer, U., Winne, D., *Biochem. Pharmcol.* (1977) **26**, 1275–1280.
A2184. Batt, A.M., Ziegler, J.M., Loppinet, V., Siest, G., *Biochem. Pharmcol.* (1977) **26**, 1354–1355.
A2185. Lavrijsen, K., Lavrijsen, P., Lontie, R., *Biochem. Pharmcol.* (1977) **26**, 1345–1347.
A2186. Parli, C.J., Schmidt, B., *Res. Commun. Chem. Pathol. Pharmacol.* (1975) **10**, 601–604.
A2187. Kametani, T., Takemura, M., Takahashi, K., Ihara, M., Fukumoto, K., *Heterocycles* (1975) **3**, 139–142.
A2188. Bonnichsen, R., Dimberg, R., Mårde, Y., Ryhage, R., *Clin. Chim. Acta* (1975) **60**, 67–75.

A2189. Funaki, H., Matsuda, T., Shirasawa, E., Fujii, J., Shindo, T., *Kyoto Furitsu Ika Daigaku Zasshi* (1975) **84**, 39–46.

A2190. Hirwe, A.S., Metcalf, R.L., Lu, P.-Y., Chio, L.-C., *Pestic. Biochem. Physiol.* (1975) **5**, 65–72.

A2191. Hoagland, R.E., Graf, G., Handel, E.D., *Weed Res.* (1974) **14**, 371–374.

A2192. Gustafsson, J.Å., Ingelman-Sundberg, M., *J. Biol. Chem.* (1975) **250**, 3451–3458.

A2193. Williams, R.J., Evans, W.C., *Biochem. J.* (1975) **148**, 1–10.

A2194. Fager, E.E., Wideburg, N.E., US 3,864,492 (4–2–1975) .

A2195. Awapara, J., Saine, S., *J. Neurochem.* (1975) **24**, 817–818.

A2197. Pong, S.F., Huang, C.L., *J. Pharm. Sci.* (1974) **63**, 1527–1532.

A2198. Belleville, M., Meynaud, A., Lechevin, J.C., Grand, M., Fontaine, L., *Therapie* (1974) **29**, 829–840.

A2199. Ariga, T., Tanaka, K., Hattori, K., Hioki, M., Shindo, H., *Sankyo Kenkyusho Nempo* (1974) **26**, 94–105.

A2200. Andresen, B.D., Hammer, R.H., Templeton, J.L., Moldowan, M.J., Panzik, H.L., *Res. Commun. Chem. Pathol. Pharmacol.* (1975) **10**, 443–453.

A2201. Portoghese, P.S., Svanborg, K., Samuelsson, B., *Biochem. Biophys. Res. Commun.* (1975) **63**, 748–755.

A2202. Frigerio, A., *Cron. Chim.* (1974) **44**, 12–18.

A2203. Funaki, H., Matsuda, T., *Kyoto Furitsu Ika Daigaku Zasshi* (1975) **84**, 1–9.

A2204. Appel, E., Planz, G., Palm, D., Grobecker, H., *Eur. J. Clin. Pharmacol.* (1975) **8**, 161–166.

A2205. Epel'dimov, L.S., *Sb. Nauch. Rab. Sib. Nauch.-Issled. Vet. Inst.* (1970) **17**, 249–252.

A2206. King, C.M., Olive, C.W., *Cancer Res.* (1975) **35**, 906–912.

A2207. Mashev, N., *Fiziol. Rast.* (*Sofia*) (1974) **1**, 47–54.

A2208. Chan, A.H.H., Robinson, A.R., Common, R.H., *Steroids* (1975) **25**, 677–695.

A2209. Devi, N.A., Kutty, R.K., Vasantharajan, V.N., Subba Rao, P.V., *J. Bacteriol.* (1975) **122**, 866–873.

A2210. Gau, W., Bieniek, D., Coulston, F., Korte, F., *Chemosphere* (1974) **3**, 71–76.

A2211. McCallum, N.K., Yagen, B., Levy, S., Mechoulam, R., *Experientia* (1975) **31**, 520–521.

A2212. Hendel, J., *Acta Pharmacol. Toxicol.* (1975) **37**, 17–22.

A2213. Sawada, H., Hara, A., *Yakugaku Zasshi* (1975) **95**, 430–438.

A2214. Engelhardt, G., Wallnöfer, P.R., Hutzinger, O., *Bull. Envron. Contam. Toxicol.* (1975) **13**, 342–347.

A2215. Baggi, G., Catelani, D., Colombi, A., Galli, E., Treccani, V., *Ann. Microbiol. Enzimol.* (1974) **24**, 317–323.

A2216. Kitagawa, H., Miyazu, K., Tsuruta, M., Satoh, H., Yokoshima, T., Nanbo, T., Takemoto, M., Ushioda, K., Hasegawa, M., *Jap. J. Pharmacol.* (1975) **25**, 1–14.

A2217. Spokes, J.R., Walker, N., *Ann. Microbiol. Enzimol.* (1974) **24**, 307–315.

A2218. Reidenberg, M.M., Drayer, D.E., Levy, M., Warner, H., *Clin. Pharmacol. Ther.* (1975) **17**, 722–730.

A2219. Mandelli, M., Morselli, P.L., Nordio, S., Pardi, G., Principi, N., Sereni, F., Tognoni, G., *Clin. Pharmacol. Ther.* (1975) **17**, 564–572.

A2221. Levy, G., Khanna, N.N., Soda, D.M., Tsuzuki, O., Stern, L., *Pedriatics* (1975) **55**, 818–825.

A2222. Peterson, D.L., Gleisner, J.M., Blakley, R.L., *J. Biol. Chem.* (1975) **250**, 4945–4954.

A2223. Hollander, P.M., Bartfai, T., Gatt, S., *Arch. Biochem. Biophys.* (1975) **169**, 568–576.

A2224. Kozlovskii, A.G., Koshcheenko, K.A., Arinbasarov, M.U., Baklashova, T.G., Adanin, V.M., Yakovlev, G.I., *Izv. Akad.*

Nauk SSSR, Ser. Khim. (1975) 1155–1158.

A2225. Schubert, K., Le Dinh Phai, Kaufmann, G., Hoerhold, C., *Acta Biol. Med. Ger.* (1975) **34**, 173–180.

A2226. Hino, M., Nagatsu, T., Kakumu, S., Okuyama, S., Yoshii, Y., Nagatsu, I., *Clin. Chim. Acta* (1975) **62**, 5–11.

A2227. Gibson, D.T., Mahadevan, V., Jerina, D.M., Yagi, H., Yeh, H.J.C., *Science* (1975) **189**, 295–297.

A2229. Yoshino, H., Okumura, M., *J. Pharm. Sci.* (1975) **64**, 1071–1073.

A2230. Langone, J.J., Van Vunakis, H., Bachur, N.R., *Biochem. Med.* (1975) **12**, 283–289.

A2231. Chan, K., Kendall, M.J., Wells, W.D.E., Mitchard, M., *J. Pharm. Pharmacol.* (1975) **27**, 235–241.

A2232. Appel, E., Planz, G., Grobecker, H., Palm, D., *Wirkungsweise Oxyfedrin, Int. Arbeitstag* (1971) 127–133.

A2233. Tatematsu, A., Nadai, T., Kume, M., Hoshika, S., *Oyo Yakuri* (1975) **9**, 195–203.

A2234. Sugihara, J., Sato, Y., *Radioisotopes* (1975) **24**, 167–173.

A2235. Pfeifer, S., Pohlmann, H.P., Bornschein, I., Kraft, R., *Pharmazie* (1975) **30**, 290–296.

A2236. Ida, S., Oguri, K., Yoshimura, H., *Yakugaku Zasshi* (1975) **95**, 564–569.

A2237. Horvath, R.S., Dotzlaf, J.E., Kreger, R., *Bull. Envron. Contam. Toxicol.* (1975) **13**, 357–361.

A2238. Kumar, S.S., Dorough, H.W., *Insect Biochem.* (1975) **5**, 265–273.

A2239. Stalling, D.L., Hogan, J.W., Johnson, J.L., *Environ. Health Perspect.* (1973) 159–173.

A2240. Iwai, K., Ikeda, M., *J. Nutr. Sci. Vitaminol.* (1975) **21**, 7–18.

A2241. Sharma, H.K., Vaidyanathan, C.S., *Eur. J. Biochem.* (1975) **56**, 163–171.

A2242. Sawada, S., Kumagai, H., Yamada, H., Hill, R.K., *J. Am. Chem. Soc.* (1975) **97**, 4334–4337.

A2243. Charlier, J., Gerlo, E., *Arch. Int. Physiol. Biochim.* (1975) **83**, 354–356.

A2244. Lake, C.R., Mills, R.R., Koeppe, J.K., *Insect Biochem.* (1975) **5**, 223–229

A2245. Beckett, A.H., Bélanger, P.M., *J. Pharm. Pharmacol.* (1974) **26**, 558–560.

A2246. Houston, J.B., Levy, G., *Nature* (1975) **255**, 78–79.

A2247. Frigerio, A., Sossi, N., Belvedere, G., Pantarotto, C., Garattini, S., *J. Pharm. Sci.* (1974) **63**, 1536–1540.

A2249. Farrier, D.S., *Arzneim.-Forsch.* (1975) **25**, 813–817.

A2250. Honma, S., Kanbekawa, A., *Yakugaku Zasshi* (1975) **95**, 680–689.

A2251. Gal, J., Gruenke, L.D., Castagnoli, N., Jr., *J. Med. Chem.* (1975) **18**, 683–688.

A2252. Tatsumi, K., Nakano, M., Kitamura, S., Yoshimura, H., *Yakugaku Zasshi* (1975) **95**, 690–697.

A2253. Zulalian, J., Champagne, D.A., Wayne, R.S., Blinn, R.C., *J. Agric. Food Chem.* (1975) **23**, 724–730.

A2254. Duffey, S.S., Blum, M.S., *Insect Biochem.* (1977) **7**, 57–65.

A2255. Borg, K.O., Carlsson, E., Hoffman, K.-J., Jonsson, T.E., Thorin, H., Wallin, B., *Acta Pharmacol Toxicol.* (1975) **36** (Suppl. V) , 125–135.

A2256. Burstein, S., Varanelli, C., *Res. Commun. Chem. Pathol. Pharmacol.* (1975) **11**, 343–354.

A2257. Conway, W.D., Lee, F.H., Neufeld, L., *J. Pharm. Sci.* (1975) **64**, 1158–1162.

A2258. Dinamarca, M.L., Ramirez, A., Ballester, E., Del Villar, E., Capdevila, J., Maccioni, R., *Int. J. Biochem.* (1975) **6**, 405–412.

A2259. Siiteri, P.K., Thompson, E.A., *J. Steroid Biochem.* (1975) **6**, 317–322.

A2260. Duffey, S.S., Aldrich, J.R., Blum, M.S., *Comp. Biochem. Physiol.* (1977) **56B**, 101–102.

A2261. Hoag, G.N., Hill, A., Zaleski, W., *Clin. Biochem.* (1977) **10**, 181–182.

A2262. Duffey, S.S., Underhill, E.W., Towers, G.H.N., *Comp. Biochem. Physiol.* (1974) **47B**, 753–766.
A2263. Flockhart, I.R., Smith, R.L., Williams, R.T., *Proc. Eur. Soc. Study Drug Toxic.* (1973) **14**, 36–43.
A2264. Sellman, R., Kanto, J., Pekkarinen, J., *Acta Pharmacol. Toxicol.* (1975) **37**, 242–249.
A2265. Breyer, U., Gaertner, H.J., *Proc. Eur. Soc. Study Drug Toxic.* (1973) **14**, 59–66.
A2266. Ruzo, L.O., Safe, S., Hutzinger, O., Platonow, N., Jones, D., *Chemosphere* (1975) **4**, 121–123.
A2267. Selander, H.G., Jerina, D.M., Piccolo, D.E., Berchtold, G.A., *J. Am. Chem. Soc.* (1975) **97**, 4428–4430.
A2268. Okuno, E., Minatogawa, Y., Noguchi, T., Kido, R., *Life Sci.* (1975) **17**, 211–218.
A2269. Moore, G.P., Sullivan, D.T., *Biochim. Biophys. Acta* (1975) **397**, 468–477.
A2270. Nishimoto, Y., *Aichi Ika Daigaku Igakukai Zasshi* (1975) **3**, 69–77.
A2271. Knobloch, K.H., Hahlbrock, K., *Planta Med.* (1975) Suppl. 102–106.
A2272. Wurm, R., Weyhenmeyer, R., Renz, P., *Eur. J. Biochem.* (1975) **56**, 427–432.
A2273. Noguchi, T., Nakatani, M., Minatogawa, Y., Morimoto, M., Kido, R., *Hoppe-Seyler's Z. Physiol. Chem.* (1975) **356**, 1245–1250.
A2275. Nakatani, M., *Wakayama Igaku* (1974) **25**, 359–368.
A2276. Buelles, H., Buelles, J., Krumbiegel, G., Mennicke, W.H., Nitz, D., *Arzneim.-Forsch.* (1975) **25**, 902–905.
A2278. Squires, R.F., *J. Neurochem.* (1975) **24**, 47–50.
A2279. Braestrup, C., Nielsen, M., *J. Pharm. Pharmacol.* (1975) **27**, 413–419.
A2280. Savage, N., Levy, P.R., *Int. J. Biochem.* (1975) **6**, 459–466.
A2281. Speedie, M.K., Hornemann, U., Floss, H.G., *Methods Enzymol.* (1975) **43**, 498–502.
A2282. Hayakawa, T., Udenfriend, S., Yagi, H., Jerina, D.M., *Arch. Biochem. Biophys.* (1975) **170**, 438–451.
A2283. Higgins, V.J., *Physiol. Plant Pathol.* (1975) **6**, 5–18.
A2284. Peterkova, M., Grimova, J., *Cesk. Farm.* (1975) **24**, 128–132.
A2285. Hsu,L.L., Mandell, A.J., *Life Sci.* (1975) **17**, 387–396.
A2286. Omori, S., Ikeda, S., Ogata, M., *Igaku To Seibutsugaku* (1975) **90**, 129–131.
A2287. Ando, M., Nakagawa, M., Nakamura, T., Tomita, K., *Agric. Biol. Chem.* (1975) **39**, 803–809.
A2288. Kuhr, R.J., Davis, A.C., *Pestic. Biochem. Physiol.* (1975) **5**, 330–337.
A2289. Ueda, K., Gaughan, L.C., Casida, J.E., *Pestic. Biochem. Physiol.* (1975) **5**, 280–294.
A2290. Pollock, S.H., *Life Sci.* (1975) **17**, 465–475.
A2291. Brunner, R., Roehr, M., *Methods Enzymol.* (1975) **43**, 482–487.
A2292. Ramanarayanan, M., Vaidyanathan, C.S., *Indian J. Exp. Biol.* (1975) **13**, 393–396.
A2293. Nishimoto, Y., Takeuchi, F., Shibata, Y., *J. Biochem.* (1975) **78**, 573–581.
A2294. Belvedere, G., Pantarotto, C., Frigerio, A., *Res. Commun. Chem. Pathol. Pharmacol.* (1975) **11**, 221–232.
A2295. Fishman, J., Norton, B., *Adv. Biosci.* (1975) **15**, 123–131.
A2296. White, G., *Chem. Ind.* (*London*) (1975) 589–590.
A2297. Sugiyama, M., Miyamoto, T., Yonezawa, H., Sato, E., Matsuoka, Y., *Oyo Yakuri* (1975) **9**, 733–742.
A2298. Lesko, S., Caspary, W., Lorentzen, R., Ts'o, P.O.P., *Biochemistry* (1975) **14**, 3978–3984.
A2299. Zimmer, H., Glaser, R., Kokosa, J., Garteiz, D.A., Hess, E.V., Litwin, A., *J. Med. Chem.* (1975) **18**, 1031–1033.
A2300. Hsu, I.C., Van Miller, J.P., Allen, J.R., *Bull. Environ. Contam. Toxicol.* (1975) **14**, 233–240.

A2301. Sundström, G., Jansson, B., Jensen, S., *Nature* (1975) **255**, 627–628.

A2302. Safe, S., Ruzo, L.O., Jones, D., Platonow, N.S., Hutzinger, O., *Can. J. Physiol. Pharmacol.* (1975) **53**, 392–396.

A2304. Shih, J.C., *Life Sci.* (1975) **17**, 627–632.

A2305. Gerlo, E., Charlier, J., *Eur. J. Biochem.* (1975) **57**, 461–467.

A2307. Speedie, M.K., Hornemann, U., Floss, H.G., *J. Biol. Chem.* (1975) **250**, 7819–7825.

A2308. Hext, P.M., Rose, F.A., *Biochem. J.* (1975) **150**, 175–181.

A2309. Lisboa, B.P., Sauer, H.-D., *J. Steroid Biochem.* (1975) **6**, 1131–1136.

A2310. Malspeis, L., Bathala, M.S., Ludden, T.M., Bhat, H.B., Frank, S.G., Sokoloski, T.D., Morrison, B.E., Reuning, R.H., *Res. Commun. Chem. Pathol. Pharmacol.* (1975) **12**, 43–65.

A2311. Ojima, N., Ogura, K., Seto, S., *J. Chem. Soc., Chem. Commun.* (1975) 717–718.

A2312. Sanchez, E., Tephly, T.R., *Mol. Pharmacol.* (1975) **11**, 613–620.

A2313. Stillwell, W.G., *Res. Commun. Chem. Pathol. Pharmacol.* (1975) **12**, 25–41.

A2314. Murphy, G., Lynen, F., *Eur. J. Biochem.* (1975) **58**, 467–475.

A2315. Poulton, J.E., Butt, V.S., *Biochim. Biophys. Acta* (1975) **403**, 301–314.

A2316. Crawford, R.L., *Appl. Microbiol.* (1975) **30**, 439–444.

A2317. Hasegawa, M., Takai, H., Matsubara, I., *J. Pharm. Sci.* (1975) **64**, 1732–1733,

A2318. Urakubo, G., Kido, Y., Ikebuchi, H., Nagamatsu, K., *Radioisotopes* (1975) **24**, 473–478.

A2319. Verebely, K., Chedekel, M.A., Mulé, S.J., Rosenthal, D., *Res. Commun. Chem. Pathol. Pharmacol.* (1975) **12**, 67–84.

A2320. Selander, H.G., Jerina, D.M., Daly, J.W., *Arch. Biochem. Biophys.* (1975) **168**, 309–321.

A2321. Tewfik, M.S., Hamdi, Y.A., *Soil Biol. Biochem.* (1975) **7**, 79–82.

A2322. Ueyama, I., Takase, I., *Agric. Biol. Chem.* (1975) **39**, 1719–1727.

A2324. Yang, S.K., Dower, W.V., *Proc. Nat. Acad. Sci. U.S.A.* (1975) **72**, 2601–2605.

A2325. Wyrambik, D., Grisebach, H., *Eur. J. Biochem.* (1975) **59**, 9–15.

A2326. Edwards, D.J., Chang, S.-S., *Life Sci.* (1975) **17**, 1127–1134.

A2327. Wiebel, F.J., Selkirk, J.K., Gelboin, H.V., Haugen, D.A., Van der Hoeven, T.A., Coon, M.J., *Proc. Nat. Acad. Sci. U.S.A.* (1975) **72**, 3917–3920.

A2328. Krishnan, A.R., Bajaj, B.K., Hingorani, V., Laumas, K.R., *Acta Endocrinol. (Copenhagen)* (1975) **80**, 719–731.

A2329. Ferrari, V., Della Bella, D., *Postgrad. Med. J.* (1975) **50**, Suppl 5, 17–22.

A2330. Iribe, K., Ueno, M., Nakamura, N., *Nippon Daigaku Yakugaku Kenkyu Hokoku* (1974) **14**, 15–26.

A2331. Beckett, A.H., Coutts, R.T., Gibson, G.G., *J. Pharm. Pharmacol.* (1975) **27**, 659–665.

A2332. Takanashi, S., Tohira, Y., Suzuki, H., Ogiya, S., Sakuma, A., Kimura, K., *Yakugaku Zasshi* (1975) **95**, 897–900.

A2333. Kamenikova, L., *Zentralbl. Pharm., Pharmakother. Laboratoriumsdiagn.* (1975) **114**, 709–711.

A2334. Hsu, I.C., Van Miller, J.P., Seymour, J.L., Allen, J.R., *Proc. Soc. Exp. Biol. Med.* (1975) **150**, 185–188.

A2335. Halladay, S.C., Carter, D.E., Sipes, I.G., Brodie, B.B., Bressler, R., *Life Sci.* (1975) **17**, 1003–1009.

A2336. Lake, C.R., Mills, R.R., *Insect Biochem.* (1975) **5**, 659–669.

A2337. Yonaha, K., Moriguchi, M., Hirasawa, T., Yamamoto, T., Soda, K., *Bull. Inst. Chem. Res., Kyoto Univ.* (1975) **53**, 315–318.

A2338. Cardillo, R., Fuganti, C., Ghiringelli, D., Grasselli, P., Gatti, G., *J. Chem. Soc., Chem. Commun.* (1975) 778–779.

A2339. Noguchi, T., Minatogawa, Y., Okuno, E., Nakatani, M., Morimoto, M., Kido, R., *Biochem. J.* (1975) **151**, 399–406.

A2341. Hart, A., Orr, D.L.J., *Antonie van Leeuwenhoek* (1975) **41**, 239–247.

A2342. Noguchi, T., Nakatani, M., Minatogawa, Y., Okuno, E., Kido, R., *J. Neurochem.* (1975) **25**, 579–582.

A2343. Reddy, V., Rao, N.A., *Indian J. Biochem. Biophys.* (1975) **12**, 75–80.

A2344. Corbett, M.D., Cahoy, J.E., Chipko, B.R., *J. Nat. Cancer Inst.* (1975) **55**, 1247–1258.

A2345. Yuasa, K., Ishizuka, K., Kaburaki, S., Sakasai, T., *Agric. Biol. Chem.* (1975) **39**, 2199–2206.

A2346. Pearson, A.G.M., Turner, A.J., *Nature* (Lond.) (1975) **258**, 173–174.

A2347. Andria, G., Morzi, A., Auricchio, S., *Biochim. Biophys. Acta* (1976) **419**, 42–50.

A2348. Honecker, H., Coper, H., *Naunyn-Schmiedeberg's Arch. Pharmacol.* (1975) **291**, 111–121.

A2349. Liss, E., Robertson, A., *Arzneim.-Forsch.* (1975) **25**, 1792–1793.

A2350. Rance, M.J., Shillingford, J.S., *Xenobiotica* (1977) **7**, 529–536.

A2351. Autrup, H., Warwick, G.P., *Chem.-Biol. Interact.* (1975) **11**, 329–342.

A2352. Beckett, A.H., Rahman, N.N., *Xenobiotica* (1977) **7**, 537–547.

A2353. Dean, A.V., Lan, S.J., Kripalani, K.J., Di Fazio, L.T., Schreiber, E.C., *Xenobiotica* (1977) **7**, 549–560.

A2354. Brewster, D., Jones, R.S., Parke, D.V., *Xenobiotica* (1977) **7**, 601–609.

A2355. Davison, C., Dorrbecker, B.R., Edelson, J., *Xenobiotica* (1977) **7**, 561–571.

A2356. Mc Elhatton, P.R., Sullivan, F.M., Toseland, P.A., *Xenobiotica* (1977) **7**, 611–615.

A2357. Mc Elhatton, P.R., Sullivan, F.M., Toseland, P.A., *Xenobiotica* (1977) **7**, 617–622.

A2358. Hackett, A.M., Griffiths, L.A., *Xenobiotica* (1977) **7**, 641–651.

A2359. Guarino, A.M., James, M.O., Bend, J.R., *Xenobiotica* (1977) **7**, 623–631.

A2360. Ben-Shalom, N., Kahn, V., Harel, E., Mayer, A.M., *Phytochemistry* (1977) **16**, 1153–1158.

A2361. Langcake, P., Pryce, R.J., *Phytochemistry* (1977) **16**, 1193–1196.

A2362. Iguchi, S., Goromaru, T., Noda, A., *Chem. Pharm. Bull.* (1975) **23**, 932–934.

A2363. Watabe, T., Akamatsu, K., *Biochem. Pharmacol.* (1975) **24**, 442–444.

A2364. Honma, S., Nambara, T., *Chem. Pharm. Bull.* (1975) **23**, 787–793.

A2365. Haddock, R.E., Trager, W.F., Pohl, L.R., *J. Med. Chem.* (1975) **18**, 519–523.

A2366. Clark, D.E., Palmer, J.S., Radeleff, R.D., Crookshank, H.R., Farr, F.M., *J. Agric. Food Chem.* (1975) **23**, 573–578.

A2367. Miyazaki, S., Sikka, H.C., Lynch, R.S., *J. Agric. Food Chem.* (1975) **23**, 365–368.

A2368. Feil, V.J., Lamoureux, C.H., Zaylskie, R.G., *J. Agric. Food Chem.* (1975) **23**, 382–388.

A2369. Honma, S., Kambekawa, A., *Chem. Pharm. Bull.* (1975) **23**, 1045–1056.

A2370. Yeh, S.Y., *J. Pharmacol. Exp. Ther.* (1975) **192**, 201–210.

A2371. Burke, M.T., Mayer, R.T., *Drug Metab. Dispos.* (1975) **3**, 245–253.

A2372. Iguchi, S., Goromaru, T., Noda, A., Tsubone, N., *Chem. Pharm. Bull.* (1975) **23**, 1889–1890.

A2373. Inoue, T., Fujita, M., *Chem. Pharm. Bull.* (1974) **22**, 1422–1423.

A2374. Balasubramanian, K.A., Bachhawat, B.K., *Biochim. Biophys. Acta* (1975) **403**, 113–121.

A2376. Williams, D.T., Blanchfield, B.J., *J. Agric. Food Chem.* (1975) **23**, 854–857.

A2377. Rozman, K., Mueller, W., Iatropoulos, M., Coulston, F., Korte, F., *Chemosphere* (1975) **4**, 289–298.

A2378. Corgier, M., Pacheco, H., *Biochimie* (1975) **57**, 1005–1017.

A2379. Selkirk, J.K., Croy, R.G., Whitlock, J.P., Jr., Gelboin, H.V., *Cancer Res.* (1975) **35**, 3651–3655.

A2380. Cripps, R.E., *Biochem. J.* (1975) **152**, 233–241.

A2382. Goeber, B., Pankow, K., Pfeifer, S., Kraft, R., *Pharmazie* (1975) **30**, 754–755.

A2383. Kamenikova, L., Jindra, A., *Pharmazie* (1975) **30**, 730–732.

A2384. Sullivan, H.R., Due, S.L., McMahon, R.E., *J. Pharm. Pharmacol.* (1975) **27**, 728–732.

A2385. Kobayashi, K., Akitake, H., Matsuda, C., Kimura, S., *Nippon Suisan Gakkaishi* (1975) **41**, 1277–1282.

A2386. Suzuki, O., Yagi, K., *Experientia* (1976) **32**, 13–14.

A2387. Caldwell, J., French, M.R., Idle, J.R., Renwick, A.G., Bassir, O., Williams, R.T., *FEBS Lett.* (1975) **60**, 391–395.

A2388. Don, M.M., Masters, C.J., Winzor, D.J., *Biochem. J.* (1975) **151**, 625–630.

A2389. Michel, F., Nicolas, J.C., Crastes de Paulet, A., *Biochimie* (1975) **57**, 1131–1140.

A2390. Baumann, P., *Psychopharmacologia* (1975) **45**, 39–45.

A2391. Ohta, Y., Ribbons, D.W., *Eur. J. Biochem.* (1976) **61**, 259–269.

A2392. Andresen, B.D., Templeton, J.L., Hammer, R.H., Panzik, H.L., Moldowan, M.J., *Res. Commun. Chem. Pathol. Pharmacol.* (1975) **12**, 627–634.

A2393. Somani, S.M., *Eur. J. Pharmacol.* (1975) **30**, 336–343.

A2394. Bingham, E., Niemeier, R., Dalbey, W., *Fed. Proc.* (1976) **35**, 81–84.

A2395. Ohmori, S., Ikeda, M., Ogata, M., *Physiol. Chem. Phys.* (1975) **7**, 477–480.

A2396. Atchison, G.J., Johnson, H.E., *Trans. Am. Fish Soc.* (1975) **104**, 782–784.

A2397. Buswell, J.A., *J. Bacteriol.* (1975) **124**, 1077–1083.

A2399. Sparnins, V.L., Dagley, S., *J. Bacteriol.* (1975) **124**, 1374–1381.

A2400. Beckett, A.H., Huizing, G., *J. Pharm. Pharmacol.* (1975) **27**, Suppl., 42P.

A2401. Baba, S., Kuroki, M., Suzuki, K., Kazuya, Y., *Radioisotopes* (1975) **24**, 614–618.

A2402. Thompson, R.M., Gerber, N., *Res. Commun. Chem. Pathol. Pharmacol.* (1976) **13**, 41–54.

A2403. Galeazzi, R.L., Sheiner, L.B., Lockwood, T., Benet, L.Z., *Clin. Pharmacol. Ther.* (1976) **19**, 55–62.

A2404. Yoshimura, H., Yamamoto, H., Yonezawa, K., *Fukuoka Igaku Zasshi* (1975) **66**, 555–562.

A2405. Lamoureux, G.L., Davison, K.L., *Pestic. Biochem. Physiol.* (1975) **5**, 497–506.

A2406. Fox, C.H., Selkirk, J.K., Price, F.M., Croy, R.G., Sanford, K.K., Cottler-Fox, M., *Cancer Res.* (1975) **35**, 3551–3557.

A2407. Hou, C.T., Lillard, M.O., Schwartz, R.D., *Biochemistry* (1976) **15**, 582–588.

A2408. Hoppen, H.O., Bohlig, H.J., Werner, M., Knuppen, R., *Hoppe-Seyler's Z. Physiol. Chem.* (1976) **357**, 67–74.

A2409. Wang, K.C., Yeh, T., *T'ai-wan Yao Hsueh Tsa Chih* (1973) **25**, 45–48.

A2410. Stubenrauch, G., Gelbke, H.P., Knuppen, R., *Hoppe-Seyler's Z. Physiol. Chem.* (1976) **357**, 75–80.

A2411. Inue, A., *Nippon Shonika Gakkai Zasshi* (1976) **80**, 48–59.

A2412. Barrow, K.D., Quigley, F.R., *Tetrahedron Lett.* (1975) 4269–4270.

A2413. Akhtar, M.N., Boyd, D.R., Thompson, N.J., Koreeda, M., Gibson, D.T., Mahadevan, V., Jerina, D.M., *J. Chem. Soc., Perkin Trans. 1* (1975) 2506–2511.

A2414. Schillings, R.T., Sisenwine, S.F., Ruelius, H.W., *Drug Metab. Dispos.* (1977) **5**, 425–435.

A2415. Williams, C.H., *Biochem. Soc. Trans.* (1977) **5**, 1770–1771.

A2416. Roerig, S.C., Fujimoto, J.M., Wang, R.I.H., *Drug Metab. Dispos.* (1977) **5**, 454–463.

A2417. Sum, C.Y., Cho, A.K., *Drug Metab. Dispos.* (1977) **5**, 464–468.

A2418. Ericsson, O., Danielsson, B., *Drug Metab. Dispos.* (1977) **5**, 497–502.

A2419. Marty, F., Straw, J.A., *Drug Metab. Dispos.* (1977) **5**, 482–486.

A2420. McManus, M.E., Ilett, K.F., *Drug Metab. Dispos.* (1977) **5**, 503–510.
A2421. Billings, R.E., McMahon, R.E., Ashmore, J., Wagle, S.R., *Drug Metab. Dispos.* (1977) **5**, 518–526.
A2422. Eberts, F.S., Jr., *Drug Metab. Dispos.* (1977) **5**, 547–555.
A2423. Harvey, D.J., Glazener, L., Johnson, D.B., Butler, C.M., Horning, M.G., *Drug Metab. Dispos.* (1977) **5**, 527–546.
A2424. Misra, A.L., Pontani, R.B., *Drug Metab. Dispos.* (1977) **5**, 556–563.
A2425. O'Leary, M.H., Baughn, R.L., *J. Biol. Chem.* (1977) **252**, 7168–7173.
A2426. Jacobs, S.A., Stoller, R.G., Chabner, B.A., Johns, D.G., *J. Clin. Invest.* (1976) **57**, 534–538.
A2427. Zweig, J.S., Castagnoli, N., Jr., *Psychopharmacol. Commun.* (1975) **1**, 359–371.
A2428. Kaiser, D.G., Vangiessen, G.J., Reischer, R.J., Wechter, W.J., *J. Pharm. Sci.* (1976) **65**, 269–273.
A2429. Hengstmann, J.H., Weyand, U., Dengler, H.J., *Eur. J. Clin. Pharmacol.* (1975) **9**, 179–187.
A2430. Koontz, W.A., Shiman, R., *J. Biol. Chem.* (1976) **251**, 368–377.
A2431. Sever, P.S., Dring, L.G., Williams, R.T., *Eur. J. Clin. Pharmacol.* (1975) **9**, 193–198.
A2432. Brown, H.M., Purves, W.K., *J. Biol. Chem.* (1976) **251**, 907–913.
A2433. Stafford, H.A., *Plant Physiol.* (1976) **57**, 320–324.
A2434. Poulton, J., Grisebach, H., Ebel, J., Schaller-Hekeler, B., Hahlbrock, K., *Arch. Biochem. Biophys.* (1976) **173**, 301–305.
A2435. Munnecke, D.M., Hsieh, D.P.H., *Appl. Environ. Microbiol.* (1976) **31**, 63–69.
A2437. Okada, K., Iio, H., Goto, T., *J. Chem. Soc., Chem. Commun.* (1976) 32.
A2438. Beckett, A.H., Bélanger, P.M., *J. Pharm. Pharmacol.* (1975) **27**, 928–936.
A2439. Ball, P., Stubenrauch, G., Knuppen, R., *Hoppe-Seyler's Z. Physiol. Chem.* (1976) **357**, 81–88.
A2440. Ehrsson, H., *J. Pharm. Pharmacol.* (1975) **27**, 971–973.
A2441. Shigematsu, H., Yamano, S., Yoshimura, H., *Arch. Biochem. Biophys.* (1976) **173**, 178–186.
A2442. Kobayashi, S., Kimura, S., Akitake, H., *Nippon Suisan Gakkaishi* (1976) **42**, 171–177.
A2443. Melancon, M.J., Jr., Lech, J.J., *Bull. Environ. Contam. Toxicol.* (1976) **15**, 181–188.
A2444. Kohli, J., Weisgerber, I., Klein, W., *Pestic. Biochem. Physiol.* (1976) **6**, 91–97.
A2445. Oesch, F., Bentley, P., *Nature* (1976) **259**, 53–55.
A2446. Stebbins, R.D., Meller, E., Rosengarten, H., Friedhoff, A., Silber, R., *Arch. Biochem. Biophys.* (1976) **173**, 673–679.
A2448. Sato, Y., Machida, T., Oda, T., *Tetrahedron Lett.* (1975) 4571–4574.
A2449. Vasavada, P.C., Forney, F.W., *Antonie van Leeuwenhoek* (1975) **41**, 479–491.
A2450. Kar, N.C., Pearson, C.M., *Proc. Soc. Exp. Biol. Med.* (1976) **151**, 583–586.
A2451. Leung, B.S., Pearson, J.R., Martin, R.P., *J. Steroid Biochem.* (1975) **6**, 1477–1481.
A2452. Winsel, K., Iwainsky, H., Werner, E., Eule, H., *Pharmazie* (1976) **31**, 95–99.
A2453. Reiter, S., Salvesen, B., *Medd. Nor. Farm. Selsk.* (1974) **36**, 23–29.
A2454. Selkirk, J.K., Croy, R.G., Gelboin, H.V., *Cancer Res.* (1976) **36**, 922–926.
A2455. Giardina, E.-G.V., Dreyfuss, J., Bigger, J.T., Jr., Shaw, J.M., Schreiber, E.C., *Clin. Pharm. Ther.* (1976) **19**, 339–351.
A2456. Sundstrom, G., Jansson, B., *Chemosphere* (1975) **4**, 361–370.
A2457. Agogbua, S.I.O., Wynn, C.H. *Biochem. J.* (1976) **153**, 415–421.
A2458. Smith, P.I., Swan, G.A., *Biochem. J.* (1976) **153**, 403–408.

A2459. Shih, J.C., Chiu, R.H., Chan, Y.L., *Biochem. Biophys. Res. Commun.* (1975) **68**, 1348–1355.
A2460. Walker, N., Janes, N.F., Spokes, J.R., Van Berkum, P., *J. Appl. Bacteriol.* (1975) **39**, 281–286.
A2461. Jones, A., Vining, L.C., *Can. J. Microbiol.* (1976) **22**, 237–244.
A2462. Zivin, J.A., Reid, J.L., Tappaz, M.L., Kopin, I.J., *Brain Res.* (1976) **105**, 151–156.
A2463. Crawford, R.L., *Can. J. Microbiol.* (1976) **22**, 276–280.
A2464. Jenkins, G., Heap, R.B., *Nature* (1976) **259**, 330–331.
A2465. Ball, P., Farthmann, E., Knuppen, R., *J. Steroid Biochem.* (1976) **7**, 139–143.
A2466. Karim, A.K.M.B., Evans, D.A.P., *J. Med. Genet.* (1976) **13**, 17–19.
A2467. Elce, J.S., Lai, P.P.C., *Steroids* (1976) **27**, 335–352.
A2468. Rauws, A.G., Olling, M., Freudenthal, J., ten Ham, M., *Toxicol. Appl. Pharmacol.* (1976) **35**, 333–339.
A2469. Chen, C.H., Klein, D.C., Robinson, J.C., *J. Reprod. Fert.* (1976) **46**, 477–479.
A2470. Goto, M., Sugiura, K., Hattori, M., Miyagawa, T., Okamura, M., *Chemosphere* (1974) **3**, 227–232.
A2472. Goto, M., Sugiura, K., Hattori, M., Miyagawa, T., Okamura, M., *New Methods in Environmental Chemistry and Toxicology*, F. Coulston, F. Korte, Eds, *International Academic Printing Co., Totsuka, Tokyo* (1973) 299–302.
A2473. Ikeda, M., Iwai, K., *J. Nutr. Sci. Vitaminol.* (1975) **21**, 1–6.
A2474. Küpfer, A., Bircher, J., Preisig, R., *J. Pharmacol. Exp. Ther.* (1977) **203**, 493–499.
A2475. Iwamoto, K., Klaassen, C.D., *J. Pharmacol. Exp. Ther.* (1977) **203**, 365–376.
A2476. El-Kersh, T.A., Plourde, J.R., *J. Antibiot.* (1976) **29**, 292–302.
A2477. Robinson, W.T., Martel, R.R., Ferdinandi, E., Kraml, M., *J. Pharmacol. Exp. Ther.* (1977) **203**, 729–735.
A2478. Maass, W.S.G., Hutzinger, O., Safe, S., *Arch. Environ. Contam. Toxicol.* (1975) **3**, 470–478.
A2479. Buchenauer, H., *Z. Pflanzenkr. Pflanzenschutz* (1975) **82**, 604–613.
A2480. Poe, M., Bennett, C.D., Donoghue, D., Hirshfield, J.M., Williams, M.N., Hoogsteen, K., *Chem. Biol. Pteridines, Proc. Int. Symp., 5th* (1975) 51–59.
A2481. Kadlubar, F.F., Miller, J.A., Miller, E.C., *Cancer Res.* (1976) **36**, 1196–1206.
A2482. Nishimoto, Y., *Aichi Ika Daigaku Igakukai Zasshi* (1976) **4**, 6–15.
A2483. Gopalakrishna, Y., Narayanan, T.K., Rao, G.R., *Biochem. Biophys. Res. Commun.* (1976) **69**, 417–422.
A2484. Waldmeier, P.C., Delini-Stula, A., Maitre, L., *Naunyn-Schmiedeberg's Arch. Pharmacol.* (1976) **292**, 9–14.
A2485. Tan, K.H., *Comp. Biochem. Physiol. C* (1975) **52**, 127–131.
A2486. Oesch, F., Schmassmann, H., Bentley, P., *Biochem. Pharmacol.* (1978) **27**, 17–20.
A2487. Student, A.K., Edwards, D.J., *Biochem. Pharmacol.* (1977) **26**, 2337–2342.
A2488. Malejka-Giganti, D., McIver, R.C., Glasebrook, A.L., Gutmann, H.R., *Biochem. Pharmacol.* (1978) **27**, 61–69.
A2489. Ghosh, S.K., Guha, S.R., *Biochem. Pharmacol.* (1978) **27**, 112–114.
A2490. Ghosh, S.K., Guha, S.R., *J. Neurochem.* (1972) **19**, 229–231.
A2491. Guha, S.R., Ghosh, S.K., *Biochem. Pharmacol.* (1970) **19**, 2929–2932.
A2492. Workman, P., Double, J.A., *Biochem. Pharmacol.* (1978) **27**, 199–202.
A2493. Borchardt, R.T., Wu, Y.S., Wu, B.S., *Biochem. Pharmacol.* (1978) **27**, 120–122.
A2494. Ryrfeldt, Å., Nilsson, E., *Biochem. Pharmacol.* (1978) **27**, 301–305.
A2495. Kuhn, C.M., Schanberg, S.M., Breese, G.R., *Biochem. Pharmacol.* (1978) **27**, 343–351.

A2496. Pohl, L.R., Nelson, S.D., Krishna, G., *Biochem. Pharmacol.* (1978) **27**, 491–496.

A2497. Shiverick, K.T., Solomon, S., *Biochem. Pharmacol.* (1978) **27**, 451–455.

A2498. Wirth, P.J., Thorgeirsson, S.S., *Biochem. Pharmacol.* (1978) **27**, 601–603.

A2499. Hsu, L.L., *Res. Commun. Chem. Pathol. Pharmacol.* (1975) **12**, 355–362.

A2501. Labow, R.S., Williamson, D.G., Layne, D.S., Collins, D.C., *Can. J. Biochem.* (1975) **53**, 1028–1030.

A2502. Cheng, F.W., Shane, B., Stokstad, E.L.R., *Can. J. Biochem.* (1975) **53**, 1020–1027.

A2503. Connors, T.A., Hickman, J.H., Jarman, M., Melzack, D.H., Ross, W.C., *Biochem. Pharmacol.* (1975) **24**, 1665–1670.

A2504. Feung, C.-S., Hamilton, R.H., Mumma, R.O., *J. Agric. Food Chem.* (1975) **23**, 373–376.

A2507. Cauthen, S.E., Ellis, R.D., Larrison, S.B., Kidd, M.R., *Biochem. Pharmacol.* (1976) **25**, 181–185.

A2509. James, M.O., Fouts, J.R., Bend, J.R., *Biochem. Pharmacol.* (1976) **25**, 187–193.

A2510. Schneck, D.W., Sprouse, J.S., Hayes, A.H., Jr., Shiroff, R.A., *J. Pharmacol. Exp. Ther.* (1978) **204**, 212–218.

A2511. Young, W.S., III, Lietman, P.S., *J. Pharmacol. Exp. Ther.* (1978) **204**, 203–211.

A2512. Brandãu, F., *J. Pharmacol. Exp. Ther.* (1977) **203**, 23–29.

A2513. Freeman, D.S., Gjika, H.B., Van Vunakis, H., *J. Pharmacol. Exp. Ther.* (1977) **203**, 203–212.

A2514. Wallis, P.J., Rhodes, M.J.C., *Phytochemistry* (1977) **16**, 1891–1894.

A2515. Thomas, H., Bingham, M.J., *Phytochemistry* (1977) **16**, 1887–1889.

A2516. Hodges, C.C., Horn, J.S., Rapoport, H., *Phytochemistry* (1977) **16**, 1939–1942.

A2517. Senanyake, U.M., Wills, R.B.H., Lee, T.H., *Phytochemistry* (1977) **16**, 2032–2033.

A2518. Raymond, W.R., Maier, V.P., *Phytochemistry* (1977) **16**, 1535–1539.

A2519. Srivastava, O.P., van Huystee, R.B., *Phytochemistry* (1977) **16**, 1527–1530.

A2520. Bailey, J.A., Burden, R.S., Mynett, A., Brown, C., *Phytochemistry* (1977) **16**, 1541–1544.

A2521. Kutney, J.P., Ebizuka, Y., Salisbury, P.J., Wat, C.-K., Towers, G.H.N., *Phytochemistry* (1978) **17**, 49–52.

A2522. Bickel, H., Palme, L., Schultz, G., *Phytochemistry* (1978) **17**, 119–124.

A2523. Hrazdina, G., Wagner, G.J., Siegelman, H.W., *Phytochemistry* (1978) **17**, 53–56.

A2524. Dixon, P.A.F., Caldwell, J., Smith, R.L., *Xenobiotica* (1977) **7**, 695–706.

A2525. Dixon, P.A.F., Caldwell, J., Smith, R.L., *Xenobiotica* (1977) **7**, 707–715.

A2526. Kirkpatrick, D., Weston, K.T., Hawkins, D.R., Chasseaud, L.F., Elsom, L.F., *Xenobiotica* (1977) **7**, 747–755.

A2527. Dixon, P.A.F., Caldwell, J., Smith, R.L., *Xenobiotica* (1977) **7**, 717–725.

A2528. Honohan, T., Enderlin, F.E., Ryerson, B.A., Parkinson, T.M., *Xenobiotica* (1977) **7**, 765–774.

A2529. Sundström, G., Jansson, B., *Chemosphere* (1975) **4**, 361–370.

A2530. Tulp, M.T.M., Sundström, G., Hutzinger, O., *Chemosphere* (1976) **5**, 425–432.

A2531. Tulp, M.T.M., Sundström, G., de Graaf, J.C., Hutzinger, O., *Chemosphere* (1977) **6**, 109–116.

A2532. Akhtar, M.H., Foster, T.S., *J. Agric. Food Chem.* (1977) **25**, 1017–1022.

A2533. Sundström, G., Hutzinger, O., Safe, S., *Chemosphere* (1976) **5**, 249–253.

A2534. Knowles, C.O., Benezet, H.J., *J. Agric. Food Chem.* (1977) **25**, 1022–1026.

A2535. Hunt, L.M., Chamberlain, W.F., Gilbert, B.N., Hopkins, D.E., Gingrich, A.R., *J. Agric. Food Chem.* (1977) **25**, 1062–1065.

A2536. Fitzell, D.L., Singh, R., Hsieh, D.P.H., Motell, *J. Agric. Food Chem.* (1977) **25**, 1193–1197.

A2537. Hayes, J.R., Polan, C.E., Campbell, T.C., *J. Agric. Food Chem.* (1977) **25**, 1189–1193.

A2538. Monard, R.D., Russel, S., Bollag, J.-M., *J. Agric. Food Chem.* (1977) **25**, 841–844.

A2539. Lichtenstein, E.P., Parlar, H., Korte, F., Suss, A., *J. Agric. Food Chem.* (1977) **25**, 845–851.

A2540. Chu, I., Villeneuve, D.C., Secours, V., Viau, A., *J. Agric. Food Chem.* (1977) **25**, 881–883.

A2541. Subba-Rao, R.V., Alexander, M., *J. Agric. Food Chem.* (1977) **25**, 855–858.

A2542. Jaglan, P.S., Nappier, J.L., Hornish, R.E., Friedman, A.R., *J. Agric. Food Chem.* (1977) **25**, 963–964.

A2543. Sugiura, M., Hirano, K., Iino, S., Suzuki, H., Oda, T., *Chem. Pharm. Bull.* (1977) **25**, 653–656.

A2544. Moroi, R., Ono, K., Saito, T., Akimoto, T., Sano, M., *Chem. Pharm. Bull.* (1977) **25**, 830–835.

A2545. Shigematsu, H., Yamano, S., Yoshimura, H., *Chem. Pharm. Bull.* (1977) **25**, 825–830.

A2546. Bender, D.A., Smith, W.R.D., *Biochem. Soc. Trans.* (1978) **6**, 120–122.

A2547. Ziegler, D.M., *Biochem. Soc. Trans.* (1978) **6**, 94–96.

A2548. Franklin, R.A., Aldridge, A., White, C de B., *Br. J. Clin. Pharmacol.* (1975) **2**, 472.

A2549. Höller, M., Lisboa, B.P., Dengler, K., Breuer, H., *Biochem. Soc. Trans.* (1978) **6**, 136–138.

A2550. Dahl, S.G., *Clin. Pharmacol. Ther.* (1976) **19**, 435–442.

A2551. Chundela, B., Šlechtová, R., *J. Chromatogr.* (1976) **119**, 609–612.

A2552. Schmauder, H.P., Groeger, D., *Biochem. Physiol. Pflanz.* (1976) **169**, 471–486.

A2553. Thiercelin, J.F., Jacquot, C., Rapin, J.R., Cohen, Y., *Arch. Int. Pharmacodyn. Ther.* (1976) **220**, 153–163.

A2554. Entsch, B., Ballou, D.P., Massey, V., *J. Biol. Chem.* (1976) **251**, 2550–2563.

A2555. Davis, P.J., Gustafson, M.E., Rosazza, J.P., *J. Bacteriol.* (1976) **126**, 544–546.

A2556. Raghavan, K.G., Nadkarni, G.B., *Insect Biochem.* (1976) **6**, 201–205.

A2557. Prabhu, B.V., Gibson, C.A., Schramm, L.C., *Lloydia* (1976) **39**, 79–81.

A2558. Coupland, R.E., Kobayashi, S., Kent, C., *J. Endocrinol.* (1976) **69**, 139–148.

A2559. Fellman, J.H., Roth, E.S., Fujita, T.S., *Arch. Biochem. Biophys.* (1976) **174**, 560–565.

A2560. Bhaduri, A.P., Garg, N.K., Mukherjee, S.K., *Indian J. Biochem. Biophys.* (1975) **12**, 413–414.

A2561. Allen, J.G., Burden, T., Haigh, J.L., Parkes, M.W., Fothergill, G.A., *Arzneim.-Forsch.* (1976) **26**, 568–572.

A2562. Frislid, K., Berg, M., Hansteen, V., Lunde, P.K.M., *Eur. J. Clin. Pharmacol.* (1976) **9**, 433–438.

A2563. Erdmann, G., Just, W.W., Thel, S., Werner, G., Wiechmann, M., *Psychopharmacology* (1976) **47**, 53–58.

A2564. Davies, D.S., George, C.F., Blackwell, E., Conolly, M.E., Dollery, C.T., *Br. J. Clin. Pharmacol.* (1974) **1**, 129–136.

A2565. Moldeus, P., Vadi, H., Berggren, M., *Acta Pharmacol. Toxicol.* (1976) **26**, 17–32.

A2566. Safe, S., Jones, D., Hutzinger, O., *J. Chem. Soc., Perkin Trans. 1* (1976) 357–359.

A2567. Elmamlouk, T.H., Gessner, T., *Comp. Biochem. Physiol. C* (1976) **53**, 19–24.

A2568. De Belleroche, J.S., Bradford, H.F., Jones, D.G., *J. Neurochem.* (1976) **26**, 561–571.

A2569. Kaltenbach, C.C., Dunn, T.G., Koritnik, D.R., Tucker, W.F., Batson, D.B., Staigmiller, R.B., Niswender, G.D., *J. Toxicol. Environ. Health* (1976) **1**, 607–616.

A2570. Bagchi, S.P., Smith, T.M., *Brain Res.* (1976) **107**, 591–601.

A2571. Elsden, S.R., Hilton, M.G., Waller, J.M., *Arch. Microbiol.* (1976) **107**, 283–288.

A2572. Kisiel, M.J., Deubert, K.H., Zuckerman, B.M., *Exp. Aging Res.* (1976) **2**, 37–44.

A2573. French, D.J., Vance, C.P., Towers, G.H.N., *Phytochemistry* (1976) **15**, 564–566.

A2574. Ryall, R.L., Ryall, R.G., Howells, A.J., *Insect Biochem.* (1976) **6**, 135–142.

A2575. Trost, B.N., Müller, J., *Acta Endocrinol (Copenhagen)* (1976) **82**, 353–365.

A2576. Wengenmayer, H., Ebel, J., Grisebach, H., *Eur. J. Biochem.* (1976) **65**, 529–536.

A2577. Diekmann, H.W., Bühring, K.U., Garbe, A., Nowak, H., *Arzneim.-Forsch.* (1976) **26**, 722–726.

A2578. Ghiasuddin, S.M., Menzer, R.E., Nelson, J.O., *Toxicol. Appl. Pharmacol.* (1976) **36**, 187–194.

A2579. Kapoor, S.K., Kalra, R.L., *J. Food Sci. Technol.* (1975) **12**, 227–230.

A2580. Wright, J.L.C., Vining, L.C., McInnes, A.G., Smith, D.G., Walter, J.A., *Can. J. Biochem.* (1977) **55**, 678–685.

A2582. Brown, S.A., Sampathkumar, S., *Can. J. Biochem.* (1977) **55**, 686–692.

A2583. Hobkirk, R., Nilsen, M., *Can. J. Biochem.* (1977) **55**, 769–771.

A2584. Tong, J.H., Smyth, R.G., Benoiton, N.L., D'Iorio, A., *Can. J. Biochem.* (1977) **55**, 1103–1107.

A2585. Bu, N.T., Kuchel, O., Parent, M.T., *Can. J. Biochem.* (1977) **55**, 771–773.

A2586. Tong, J.H., D'Iorio, A., *Can. J. Biochem.* (1977) **55**, 1108–1113.

A2587. Lyles, G.A., Greenawalt, J.W., *Biochem. Pharmacol.* (1978) **27**, 923–935.

A2588. Edwards, D.J., Malsbury, C.W., *Biochem. Pharmacol.* (1978) **27**, 959–963.

A2589. Jones, C.A., Moore, B.P., Cohen, G.M., Fry, J.R., Bridges, J.W., *Biochem. Pharmacol.* (1978) **27**, 693–702.

A2590. Tateishi, M., Suzuki, S., Shimizu, H., *Biochem. Pharmacol.* (1978) **27**, 809–810.

A2591. Brandt, E., Heymann, E., *Biochem. Pharmacol.* (1978) **27**, 733–737.

A2592. Metzler, M., McLachan, J.A., *Biochem. Pharmacol.* (1978) **27**, 1087–1094.

A2593. Inagaki, C., Tanaka, C., *Biochem. Pharmacol.* (1978) **27**, 1081–1086.

A2594. Singh, R., Hsieh, D.P.H., *Appl. Environ. Microbiol.* (1976) **31**, 743–745.

A2595. Robertson, L.W., Lyle, M.A., Billets, S., *Biomed. Mass Spectrom.* (1975) **2**, 266–271.

A2597. Fadda, F., Medda, L., Liguori, G., *Boll. Soc. Ital. Biol. Sper.* (1976) **52**, 7–9.

A2600. Yadgiri, B., Tulpule, P.G., *Indian J. Dairy Sci.* (1975) **28**, 108–112.

A2601. Kato, R., Iwasaki, K., Shiraga, T., Noguchi, H., *Biochem. Biophys. Res. Commun.* (1976) **70**, 681–687.

A2602. De Frank, J.J., Ribbons, D.W., *Biochem. Biophys. Res. Commun.* (1976) **70**, 1129–1135.

A2603. Yeh, E.-L., Landis, D.H., Maas, J.W., *Biochem. Med.* (1975) **14**, 411–416.

A2604. Mailänder, B., Bacher, A., *J. Biol. Chem.* (1976) **251**, 3623–3628.

A2605. Adachi, Y., Wakisaka, G., Yamamoto, T., *Gastroenterol. Jpn.* (1975) **10**, 120–131.

A2606. Heni, N., Glogner, P., *Naunyn-Schmiedeberg's Arch. Pharmacol.* (1976) **293**, 183–186.

A2607. Jequidt, J., Erb, F., Pommery, J., Colein, P., *Eur. J. Toxicol. Environ. Hyg.* (1976) **9**, 23–29.

A2608. Davis, P.J., Rosazza, J.P., *J. Org. Chem.* (1976) **41**, 2548–2551.

A2609. Rosival, L., Vargová, M., Szokolayová, J., Čerey, K., Hladká, A., Bátora, V., Kovačičová, J., Truchlik, Š., *Pestic. Biochem. Physiol.* (1976) **6**, 280–286.

A2610. Gyarmati, L., Csontos, A., Racz, I., Satory, E., Harsanyi, K., *Acta Pharm. Hung.* (1976) **46**, 64–72.

A2611. Martin, B., Nordqvist, M., Agurell, S., Lindgren, J.E., Leander, K., Binder, M.,

J. Pharm. Pharmacol. (1976) **28**, 275–279.

A2612. Brodie, R.R., Chasseaud, L.F., Elsom, F.F., Franklin, E.R., Taylor, T., *Arzneim.-Forsch.* (1976) **26**, 896–901.

A2613. Leighty, E.G., Fentiman, A.F., Jr., Foltz, R.L., *Res. Commun. Chem. Pathol. Pharmacol.* (1976) **14**, 13–28.

A2614. Belvedere, G., Pantarotto, C., Rovei, V., Frigerio, A., *J. Pharm. Sci.* (1976) **65**, 815–821.

A2615. Beermann, B., Leander, K., Lindstrom, B., *Acta Chem. Scand. B* (1976) **30**, 465.

A2617. Jones, R.A.D., Pollitt, R.J., *J. Pharm. Pharmacol.* (1976) **28**, 461–462.

A2618. Hrazdina, G., Kreuzaler, F., Hahlbrock, K., Grisebach, H., *Arch. Biochem. Biophys.* (1976) **175**, 392–399.

A2619. Carew, D.P., Bainbridge, T., *Lloydia* (1976) **39**, 147–149.

A2620. Westerinck, B.H.C., Korf, J., *Eur. J. Pharmacol.* (1976) **37**, 249–255.

A2621. Inagaki, C., Fujiwara, H., Tanaka, C., *Jpn. J. Pharmacol.* (1976) **26**, 380–382.

A2622. Castagnoli, N., Zweig, J.S., Weinkam, R.J., *Adv. Mass Spectrom. Biochem. Med.* (1976) **1**, 119–127.

A2623. Fujimoto, S., Nakagawa, T., Ishimitsu, S., Ohara, A., *Chem. Pharm. Bull.* (1977) **25**, 1459–1462.

A2624. McCormick, N.G., Feeherry, F.E., Levinson, H.S., *Appl. Environ. Microbiol.* (1976) **31**, 949–958.

A2625. Malspeis, L., Ludden, T.M., Bathala, M.S., Morrison, B.E., Feller, D.R., Reuning, R.H., *Res. Commun. Chem. Pathol. Pharmacol.* (1976) **14**, 393–406.

A2626. Engst, R., Macholz, R.M., Kujawa, M., Lewerenz, H.J., Plas, R., *J. Environ. Sci. Health, Part B* (1976) **B11**, 95–117.

A2628. Collins, D.C., Musey, P.I., Preedy, J.R.K., *Steroids* (1976) **28**, 67–80.

A2630. Kieslich, K., Vidic, H.J., Petzoldt, K., Hoyer, G.A., *Chem. Ber.* (1976) **19**, 2259–2265.

A2632. Bertoli, M., Conti, M., Cova, A., Setnikar, I., *Pharmacol. Res. Commun.* (1976) **8**, 417–428.

A2633. Meuldermans, W.E.G., Lauwers, W.F.J., Heykants, J.J.P., *Arch. Int. Pharmacodyn. Ther.* (1976) **221**, 140–149.

A2634. Meuldermans, W.E.G., Hurkmans, R.M.A., Lauwers, W.F.J., Heykants, J.J.P., *Eur. J. Drug Metab. Pharmacokinet.* (1976) **1**, 35–40.

A2635. Wong, L.K., Biemann, K., *Clin. Toxicol.* (1976) **9**, 583–591.

A2636. Elsom, L.F., Hawkins, D.R., Chasseaud, L.F., *J. Chromatogr.* (1976) **123**, 463–467.

A2637. Franklin, R.A., *Br. J. Clin. Pharmacol.* (1976) **3**, 513–514.

A2638. Norback, D.H., Seymour, J.L., Knierien, K.M., Peterson, R.E., Allen, J.R., *Res. Commun. Chem. Pathol. Pharmacol.* (1976) **14**, 527–533.

A2639. Menn, J.J., De Baun, J.R., Hoffman, L.J., Ross, J.H., *Environ. Qual. Saf.* (1975) Suppl. 3 (Pesticides) 382–388.

A2640. Patel, R.N., Hou, C.T., Felix, A., Lillard, M.O., *J. Bacteriol.* (1976) **127**, 536–544.

A2641. Tank, A.W., Weiner, H., Thurman, J.A., *Ann. N.Y. Acad. Sci.* (1976) **273**, 219–226.

A2642. Janystyn, B., Stocker, M., *Z. Naturforsch. C: Biosci.* (1976) **31C**, 408–410.

A2643. Pohl, L.R., Nelson, S.D., Garland, W.A., Trager, W.F., *Biomed. Mass Spectrom.* (1975) **2**, 23–30.

A2644. Gyarmati, L., Csontos, A., Racz, I., Satori, E., Harsanyi, K., *Pharmazie* (1976) **31**, 246–247.

A2645. Misra, A.L., Vardy, J., Bloch, R., Mulé, S.J., Deneau, G.A., *J. Pharm. Pharmacol.* (1976) **28**, 316–318.

A2647. Sawada, H., Hara, A., *Experientia* (1976) **32**, 987–988.

A2648. Kolar, G.F., Schlesiger, J., *Chem.-Biol. Interact.* (1976) **14**, 301–311.

A2649. Bauer, J.E., Gerber, N., Lynn, R.K., Smith, R.G., Thompson, R.M., *Experientia* (1976) **32**, 1032–1033.

A2650. Meyer, T., Larsen. J.C., Hansen, E.V., Scheline, R.R., *Acta Pharmacol. Toxicol.* (1976) **39**, 433–441.

A2651. Sparnins, V.L., Chapman, P.J., *J. Bacteriol.* (1976) **127**, 362–366.

A2653. Engelhardt, G., Wallnöfer, P.R., Rast, H.G., Fiedler, F., *Arch. Microbiol.* (1976) **109**, 109–114.

A2654. Johnson, R.D., Ruthven, C.R.J., Goodwin, B.L., Sandler, M., *J. Neural Transm.* (1976) **38**, 181–191.

A2655. Alm, P., Petterson, G.B., *Acta Pharm. Suec.* (1976) **13**, 167–174.

A2656. Rehm, K., Hempel, K., Maurer-Schulze, B., Schlemmer, W., *Arzneim.-Forsch.* (1976) **26**, 813–820.

A2657. Kato, Y., Arima, N., Nishimine, H., *Yakugaku Zasshi* (1976) **96**, 819–826.

A2658. Delong, A.F., Smyth, R.D., Polk, A., Nayak, R.K., Reavey-Cantwell, N.H., *Arch. Int. Pharmacodyn. Ther.* (1976) **222**, 322–331.

A2659. Stillwell, W.G., Myran, C.S., Stewart, J.T., *Res. Commun. Chem. Pathol. Pharmacol.* (1976) **14**, 605–619.

A2660. Meyer, T., Scheline, R.R., *Acta Pharmacol. Toxicol.* (1976) **39**, 419–432.

A2661. Takade, D.Y., Allsup, T., Khasawinah, A., Kao, T.S., Fukuto, T.R., *Pestic. Biochem. Physiol.* (1976) **6**, 367–376.

A2662. Bhat, S.G., Vaidyanathan, C.S., *Arch. Biochem. Biophys.* (1976) **176**, 314–323.

A2663. Bachofer, R., *Zentralbl., Bakteriol., Parasitenkl., Infektionskr. Hyg., Abt. 1: Orig., Reihe B* (1976) **162**, 153–156.

A2664. De Antoni, A., Costa, C., Allegri, G., *Acta Vitaminol. Enzymol.* (1975) **29**, 339–343.

A2665. Kouvelas, E.D., Savakis, C.E., Tzebelikos, E.T., Bonatsos, G., Mitrossilis, S., *Experientia* (1976) **32**, 1136–1138.

A2666. Yasui, M., Sakakibara, J., Syoyama, M., Miyauchi, Y., *Kagaku Keisatsu Kenkyusho Hokaku* (1974) **27**, 28–33.

A2667. Lo, C.M., Kwok, M.-L., Wurtman, R.J., *Neuropharmacology* (1976) **15**, 395–402.

A2668. Lee, F.H., Conway, W.D., *Pharmacol. Res. Commun.* (1976) **8**, 469–475.

A2669. Mandula, B.B., Pathak, M.A., Dudek, G., *Science* (1976) **193**, 1131–1134.

A2670. Marshall, V.P., Reisender, E.A., Reineke, L.M., Johnson, J.H., Wiley, P.F., *Biochemistry* (1976) **15**, 4139–4145.

A2671. Andrews, R.S., Bond, C.C., Burnett, J., Saunders, A., Watson, K., *J. Int. Med. Res.* (1976) **4**, Suppl. 4, 34–39.

A2672. Blase, B.W., Loomis, T.A., *Toxicol. Appl. Pharmacol.* (1976) **37**, 481–490.

A2673. Dahms, R., Gurtoo, H.L., *Res. Commun. Chem. Pathol. Pharmacol.* (1976) **15**, 11–20.

A2674. Bhat, S.G., Vaidyanathan, C.S., *J. Bacteriol.* (1976) **127**, 1108–1118.

A2675. Harada, T., Mino, Y., *Can. J. Microbiol.* (1976) **22**, 1258–1262.

A2676. Kato, Y., Arima, N., Nishimine, H., *Yakugaku Zasshi* (1976) **96**, 941–944.

A2677. Verebey, K., Volavka, J., Mulé, S.J., Resnick, R.B., *Clin. Pharmacol. Ther.* (1976) **20**, 315–328.

A2678. Bajaj, K.L., Miller, I.R., Bhatia, I.S., *Indian J. Exp. Biol.* (1976) **14**, 329–331.

A2679. Greenblatt, D.J., Schillings, R.T., Kyriakopoulos, A.A., Shader, R.I., Sisenwine, S.F., Knowles, J.A., Ruelius, H.W., *Clin. Pharmacol. Ther.* (1976) **20**, 329–341.

A2680. Jerina, J.M., Selander, H., Yagi, H., Wells, M.C., Davey, J.F., Mahadevan, V., Gibson, D.T., *J. Am. Chem. Soc.* (1976) **98**, 5988–5996.

A2681. Bhat, S.G., Vaidyanathan, C.S., *Eur. J. Biochem.* (1976) **68**, 323–331.

A2682. Visser, T.J., Van der Does-Tobé, I., Docter, R., Hennemann, G., *Biochem. J.* (1976) **157**, 479–482.

A2683. Kirby, G.W., Narayanaswami, S., *J. Chem. Soc., Perkin Trans. 1* (1976) 1564–1567.

A2684. Cohen, A.I., Weliky, I., Lan, S.J., Levine, S.D., *Biomed. Mass Spectrom.* (1974) **1**, 1–4.

A2685. Borchert, H.H., Bornschein, I., Pfeifer, S., *Pharmazie* (1976) **31**, 579.

A2686. Miller, J.J., Powell, G.M., Olavesen, A.H., Curtis, C.G., *Toxicol. Appl. Pharmacol.* (1976) **38**, 47–57.

A2687. Jindrova, N., *Pharmazie* (1976) **31**, 580–581.

A2688. Robocker, W.C., Zamora, B.A., *Weed Sci.* (1976) **24**, 435–438.

A2689. Devoe, I.W., Gilchrist, J.E., *J. Bacteriol.* (1976) **128**, 144–148.

A2690. Bayne, H.G., Finkle, B.J., Lundin, R.E., *J. Gen. Microbiol.* (1976) **95**, 188–190.

A2691. Kishore, G., Sugumaran, M., Vaidyanathan, C.S., *J. Bacteriol.* (1976) **128**, 182–191.

A2692. De Frank, J.J., Ribbons, D.W., *J. Bacteriol.* (1977) **129**, 1356–1364.

A2693. Engström, K., Härkönen, H., Kalliokoski, P., Rantanen, J., *Scand. J. Work, Environ. Health* (1976) **2**, 21–26.

A2694. De Frank, J.J., Ribbons, D.W., *J. Bacteriol.* (1977) **129**, 1365–1374.

A2695. Chapman, P.J., Ribbons, D.W., *J. Bacteriol.* (1976) **125**, 975–984.

A2696. Ohta, Y., Ribbons, D.W., *Flavin & Flavoproteins, ed. T.P. Singer, Elsevier, Amsterdam* (1976) 134–144.

A2697. Chapman, P.J., Ribbons, D.W., *J. Bacteriol.* (1976) **125**, 985–998.

A2698. Franklin, R.A., Aldridge, A., *Br. J. Clin. Pharmacol.* (1976) **3**, 497–502.

A2699. Meuldermans, W., Lauwers, W., Knaeps, A., Heykants, J., *Arzneim.-Forsch.* (1977) **27**, 828–832.

A2700. Sandler, M., Johnson, R.D., Ruthven, C.R.J., Reid, J.L., Calne, D.B., *Nature* (1974) **247**, 364–366.

A2701. Engelhardt, G., Rast, H.G., Wallnöfer, P.R., *Arch. Microbiol.* (1977) **114**, 25–33.

A2702. Lin, K.-T., Crawhall, J.C., *Can. J. Biochem.* (1976) **54**, 423–431.

A2704. Sütfeld, R., Wiermann, R., *Z. Pflanzenphysiol.* (1976) **79**, 467–472.

A2706. Brewster, D., Jones, R.S., Parke, D.V., *Biochem. Soc. Trans.* (1976) **4**, 518–521.

A2707. Murillo, E., Plumpton, S.E., Gaunt, J.K., *Biochem. Soc. Trans.* (1976) **4**, 486–487.

A2708. Momose, A., Fukuda, J., *Chem. Pharm. Bull.* (1976) **24**, 1637–1640.

A2709. Ackermann, E., Richter, K., Sage, S., *Biochem. Pharmacol.* (1976) **25**, 1557–1560.

A2710. Israel, M., Pegg, W.J., Wilkinson, P.M., *J. Pharmacol. Exp. Ther.* (1978) **204**, 696–701.

A2711. Naito, S., Shioda, K., Sawada, M., Niki, S., Awataguchi, M., Mizutani, M., *Chem. Pharm. Bull* (1976) **24**, 1943–1947.

A2712. Patel, R.N., Hou, C.T., Felix, A., *J. Bacteriol.* (1978) **133**, 641–649.

A2713. Kolar, G.F., Schlesiger, J., *Cancer Lett.* (1975) **1**, 43–47.

A2714. Kolar, G.F., Schlesiger, J., *Chemotherapy* (1976) **8**, 91–96.

A2715. Hollunger, G., Persson, S.-Å., *Acta Pharmacol. Toxicol.* (1974) **34**, 391–398.

A2716. McCormick, N.G., Cornell, J.H., Kaplan, A.M., *Appl. Environ. Microbiol.* (1978) **35**, 945–948.

A2717. Beckett, A.H., Gibson, G.G., *Xenobiotica* (1978) **8**, 73–83.

A2718. Lan, S.J., Dean, A.V., Kripalani, K.J., Cohen, A.I., *Xenobiotica* (1978) **8**, 121–131.

A2719. Beckett, A.H., Haya, K., *Xenobiotica* (1978) **8**, 85–96.

A2720. Schmalzing, G., Breyer, U., *Xenobiotica* (1978) **8**, 45–54.

A2721. Beckett, A.H., Belanger, P.M., *Xenobiotica* (1978) **8**, 55–60.

A2722. Aribisala, O.A., Gooday, G.W., *Biochem. Soc. Trans.* (1978) **6**, 568–569.

A2723. Forrest, I.S., Green, D.E., Usdin, E., *Fed. Proc.* (1978) **37**, 271.

A2724. Halpaap, K., Horning, E.C., *Fed. Proc.* (1978) **37**, 465.

A2725. Koop, D.R., Novak, R.F., Hollenberg, P.F., *Fed. Proc.* (1978) **37**, 304.

A2726. Jordan, R.A., Franklin, M.R., *Fed. Proc.* (1978) **37**, 561.

A2727. Edwards, D.J., *Fed. Proc.* (1978) **37**, 576.

A2728. Smith, B.R., Ball, L.M., Bend, J.R., *Fed. Proc.* (1978) **37**, 643.

A2729. Wolf, C.R., Ball, L., Bend, J.R., Philpot, R.M., *Fed. Proc.* (1978) **37**, 643.

A2730. Huang, M.T., Miwa, G.T., Lu, A.Y.H., *Fed. Proc.* (1978) **37**, 769.

A2731. Collier, T.K., Thomas, L.C., Malins, D.C., *Fed. Proc.* (1978) **37**, 693.

A2732. Vore, M., *Fed. Proc.* (1978) **37**, 767.

A2733. Jacob, T.A., Buhs, R.P., Rosegay, A., Carlin, J., VandenHeuvel, W.J.A., Wolf, F.J., *Fed. Proc.* (1978) **37**, 813.

A2734. Voigt, S., Luckner, M., *Phytochemistry* (1977) **16**, 1651–1655.

A2735. Wolf, F.J., Steffens, J.J., Alvaro, R.F., Jacob, T.A., *Fed. Proc.* (1978) **37**, 814.

A2736. Inoue, K., Ueda, S., Shiobara, Y., Inouye, H., *Phytochemistry* (1977) **16**, 1689–1694.

A2737. von Massow, F., *Phytochemistry* (1977) **16**, 1695–1698.

A2738. von Massow, F., Noppel, H.E., *Phytochemistry* (1977) **16**, 1699–1701.

A2739. Steiner, A.M., *Phytochemistry* (1977) **16**, 1703–1704.

A2740. Kho, K.F.F., *Phytochemistry* (1978) **17**, 245–248.

A2741. Treimer, J.F., Zenk, M.H., *Phytochemistry* (1978) **17**, 227–231.

A2742. Dewick, P.M., *Phytochemistry* (1978) **17**, 249–250.

A2743. Fairbairn, J.W., Handa, S.S., Gürkan, E., Phillipson, J.D., *Phytochemistry* (1978) **17**, 261–262.

A2744. Nishida, A., Fukuzumi, T., *Phytochemistry* (1978) **17**, 417–419.

A2745. du Toit, P.J., Schabort, J.C., *Phytochemistry* (1978) **17**, 371–375.

A2746. Bristol, D.W., Ghanuni, A.M., Oleson, A.E., *J. Agric. Food Chem.* (1977) **25**, 1308–1314.

A2747. Hass, J.R., Jao, L.T., Wilson, N.K., Matthews, H.B., *J. Agric. Food Chem.* (1977) **25**, 1330–1333.

A2748. Adler, I.L., Jones, B.M., Wargo, J.P., Jr., *J. Agric. Food Chem.* (1977) **25**, 1339–1341.

A2749. Ou, L.T., Sikka, H.C., *J. Agric. Food Chem.* (1977) **25**, 1336–1339.

A2750. Lynch, M.J., Figdor, S.K., *J. Agric. Food Chem.* (1977) **25**, 1344–1353.

A2751. Ivie, G.W., *J. Agric. Food Chem.* (1978) **26**, 81–89.

A2752. Reeves, P.R., Case, D.E., Jepson, H.T., McCormick, D.J., Nicholls, J.T., Felix, R.H., Holt, P.J.L., Zacharias, F.J., Fluke, R.W., *J. Pharmacol Exp. Ther.* (1978) **205**, 489–498.

A2753. Ivie, G.W., Wright, J.E., *J. Agric. Food Chem.* (1978) **26**, 90–94.

A2754. McCarville, M., Marshall, V., *Biochem. Biophys. Res. Commun.* (1977) **74**, 331–335.

A2755. Marshall, V.P., McGovren, J.P., Richard, F.A., Richard, R.E., Wiley, P.F., *J. Antibiot.* (1978) **23**, 336–342.

A2756. Wiley, P.F., Marshall, V.P., *J. Antibiot.* (1975) **28**, 838–840.

A2757. Marshall, V.P., Reisender, E.A., Wiley, P.F., *J. Antibiot.* (1976) **29**, 966–968.

A2758. Frigerio, A., Pantarotto, C., *J. Pharm. Pharmacol.* (1976) **28**, 665–666.

A2759. Wiley, P.F., Koert, J.M., Elrod, D.W., Reisender, E.A., Marshall, V.P., *J. Antibiot.* (1977) **30**, 649–654.

A2760. Butler, T.C., Dudley, K.H., Johnson, J., Roberts, S.B., *J. Pharmacol Exp. Ther.* (1976) **199**, 82–92.

A2761. Yakatan, G.J., Maness, D.D., Scholler, J., Novick, W.J., Jr., Doluisio, J.T., *J. Pharm. Sci.* (1976) **65**, 1456–1460.

A2762. Pohl, L.R., Bales, R., Trager, W.F., *Res. Commun. Chem. Pathol. Pharmacol.* (1976) **15**, 233–256.

A2763. Vollmer, K.O., Poisson, A., *Arzneim.-Forsch.* (1976) **26**, 1827–1836.

A2765. Knackmuss, H.J., Hellwig, M., Lackner, H., Otting, W., *Eur. J. Appl. Microbiol.* (1976) **2**, 267–276.

A2766. Tanizawa, K., Yamamoto, T., Soda, K., *FEBS Lett.* (1976) **70**, 235–238.

A2767. Curtius, H.C., Mettler, M., Ettlinger, L., *J. Chromatogr.* (1976) **126**, 569–580.

A2768. Davis, P.J., Wiese, D.R., Rosazza, J.P., *Lloydia* (1977) **40**, 239–246.

A2769. Rosazza, J.P., Stocklinski, A.W., Gustafson, M.A., Adrian, J., Smith, R.V., *J. Med. Chem.* (1975) **18**, 791–794.

A2770. Gustafson, M.E., Miller, D., Davis, P.J., Rosazza, J., Chang, C., Floss, H.G., *J. Chem. Soc., Chem. Commun.* (1977) 842–843.

A2771. Rosazza, J.P., Foss, P., Lemberger, M., Sih, C.J., *J. Pharm. Sci.* (1974) **63**, 544–547.

A2772. Otten, S., Rosazza, J.P., *Appl. Environ. Microbiol.* (1978) **35**, 554–557.

A2773. Lee, Y.-L.T., Sparnins, V.L., Dagley, S., *Appl. Environ. Microbiol.* (1978) **35**, 817–819.

A2774. Bryant, R.W., Jr., Bentley, R., *Biochemistry* (1976) **15**, 4792–4796.

A2775. Smith, R.V., Humphrey, D.W., Engel, K.O., Rosazza, J.P., *Appl. Environ. Microbiol.* (1976) **31**, 448–449.

A2776. Okamura, H., Yamada, T., Murooka, Y., Harada, T., *Agric. Biol. Chem.* (1976) **40**, 2071–2076.

A2777. Martin, B., Agurell, S., Nordqvist, M., Lindgren, J.E., *J. Pharm. Pharmacol.* (1976) **28**, 603–608.

A2778. Pitre, D., *Farmaco, Ed. Prat.* (1976) **31**, 516–528.

A2779. Lynn, R.K., Leger, R.M., Gerber, N., Smith, R.G., *Res. Commun. Chem. Pathol. Pharmacol.* (1976) **15**, 1–10.

A2780. Pitre, D., Fumagalli, L., *Farmaco, Ed. Prat.* (1976) **31**, 529–539.

A2781. Pitre, D., Felder, E., *Farmaco, Ed. Prat.* (1976) **31**, 540–546.

A2782. Jovanovic, J., Remberg, G., Ende, M., Spiteller, G., *Arch. Toxicol.* (1976) **35**, 137–139.

A2783. Pitre, D., Maffei-Facino, F., *Farmaco, Ed. Sci.* (1976) **31**, 755–762.

A2784. Koss, G., Koransky, W., Steinbach, K., *Arch. Toxicol.* (1976) **35**, 107–114.

A2785. Althuis, T.H., Hess, H.-J., *J. Med. Chem.* (1976) **20**, 146–149.

A2786. Sanders-Bush, E., Oates, J.A., Bush, M.T., *Life Sci.* (1976) **19**, 1407–1411.

A2787. Beckett, A.H., Bélanger, P.M., *J. Pharm. Pharmacol.* (1976) **28**, 692.

A2788. Turner, J.C., Green, R.S., Kelly, W.J., *Gen. Pharmacol.* (1976) **7**, 289–291.

A2789. Engst, R., Macholz, R.M., Kujawa, M., *Bull. Environ. Contam. Toxicol.* (1976) **16**, 248–252.

A2790. Ruisinger, S., Klages, U., Lingens, F., *Arch. Microbiol.* (1976) **110**, 253–256.

A2791. Weigent, D.A., Nester, E.W., *J. Biol. Chem.* (1976) **251**, 6974–6980.

A2792. Löffelhardt, W., Kindl, H., *Z. Naturforsch. C, Biosci.* (1976) **31C**, 693–699.

A2793. Kaye, B., Woolhouse, N.M., *Ann. Trop. Med. Parasitol.* (1976) **70**, 323–328.

A2794. Bending, M.R., Bennett, P.N., Rowland, M., Steiner, J., *Br. J. Clin. Pharmacol.* (1976) **3**, 956P.

A2795. Gamble, J.A.S., Dundee, J.W., Gray, R.C., *Br. J. Anaesth.* (1976) **48**, 1087–1090.

A2797. Queisnerova, M., Svatek, E., Metys, J., *Cesk. Farm.* (1976) **25**, 232–235.

A2798. Mayer, F.L., *J. Fish Res. Board Can.* (1976) **33**, 2610–2613.

A2799. Channa Reddy, C., Vaidyanathan, C.S., *Arch Biochem. Biophys.* (1976) **177**, 488–498.

A2801. Draper, R.K., Fiskum, G.M., Edmond, J., *Arch Biochem. Biophys.* (1976) **177**, 525–538.

A2802. Nandi, N., Guha, S.R., *Indian J. Exp. Biol.* (1976) **14**, 663–665.

A2803. Verebely, K., Volavka, J., Mulé, S.J., Resnick, R., *Clin. Pharmacol. Ther.* (1975) **18**, 180–190.

A2804. Vaughan, P.F.T., Neuhoff, V., *Brain Res.* (1976) **117**, 175–180.

A2805. Hollstein, U., Krisov, G.E., Mock, D.L., *Tetrahedron Lett.* (1976) 3267–3270.

A2806. Francis, R.J., Allen, J.G., East, P.B., Ruane, R.J., *Eur. J. Drug Metab. Pharmacokinet.* (1976) **1**, 113–124.

A2808. Jauch, R., Kopitar, Z., Prox, A., Zimmer, A., Karlsson, M., Beschke, K., Bücheler, A., Schmaus, H., Wachsmuth, H., *Arzneim.-Forsch.* (1976) **26**, 2084–2089.

A2809. Porsius, A.J., Van Zwieten, P.A., *Arzneim.-Forsch.* (1976) **26**, 1999–2001.

A2810. Vereczkey, L., Szporny, L., *Arzneim.-Forsch.* (1976) **26**, 1933–1938.

A2811. Smith, G.E., Griffiths, L.A., *Experientia* (1976) **32**, 1556–1557.

A2812. Thakker, D.R., Yagi, H., Lu, A.Y.H., Levin, W., Conney, A.H., Jerina, D.M., *Proc. Nat. Acad. Sci. U.S.A.* (1976) **73**, 3381–3385.

A2813. Karnetova, J., Mateju, J., Sedmera, P., Vokoun, J., Vanek, Z., *J. Antibiot.* (1976) **29**, 1199–1202.

A2814. Hundley, S.G., Freudenthal, R.I., *Cancer Res.* (1977) **37**, 244–249.

A2815. Orange, R.P., Moore, E.G., *J. Immunol.* (1976) **117**, 2191–2196.

A2816. Phillipson, J.D., Handa, S.S., Gorrod, J.W., *J. Pharm. Pharmacol.* (1976) **28**, 687–691.

A2817. Bade, P., Christ, W., Rakow, D., Coper, D., *Life Sci.* (1976) **19**, 1833–1844.

A2818. Martin, L.E., Hopkins, R., Bland, R., *Br. J. Clin. Pharmacol.* (1976) **3**, Suppl. 3, 695–710.

A2819. Verbeeck, R., Tjandramaya, T.B., Verberckmoes, R., De Schepper, P.J., *Br. J. Clin. Pharmacol.* (1976) **3**, 1033–1039.

A2820. Pfändler, R., Scheel, D., Sandermann, H., Jr., Grisebach, H., *Arch. Biochem. Biophys.* (1977) **178**, 315–316.

A2821. Larsen, J.C., Tarding, F., *Acta Pharmacol. Toxicol.* (1976) **39**, 525–535.

A2822. Ito, K., Tanaka, T., *Euphoria Cacophoria* (1976) **3**, 1–8.

A2823. Lew, J.Y., Matsumoto, Y., Pearson, J., Goldstein, M., Hokfelt, T., Fuxe, K., *Brain Res.* (1977) **119**, 199–210.

A2824. Bollag, J.-M., Russel, S., *Microb. Ecol.* (1976) **3**, 65–73.

A2825. Freeman, D.J., Hobkirk, R., *Steroids* (1976) **28**, 613–619.

A2826. Cowan, D.A., *Methodol. Dev. Biochem.* (1976) **5**, 193–201.

A2827. Jamaluddin, M.P., *J. Bacteriol.* (1978) **129**, 690–697.

A2828. Breimer, D.D., Baars, A.J., *Pharmacology* (1976) **14**, Suppl. 1, 30–47.

A2829. Zweig, J.S., Castagnoli, N., Jr., *J. Med. Chem.* (1977) **20**, 414–421.

A2830. Takimoto, Y., Miyamoto, J., *Nippon Noyaku Gakkaishi* (1976) **1**, 261–271.

A2831. Subba-Rao, R.V., Alexander, M., *Appl. Environ. Microbiol.* (1977) **33**, 101–108.

A2832. Kachhy, A.N., Modi, V.V., *Indian J. Biochem. Biophys.* (1976) **13**, 234–236.

A2833. Lin, J.-K., Proc. *Nat. Sci Council, Part 2* (*Taiwan*) (1975) **8**, 75–95.

A2834. Sawada, H., Hara, A., *Drug Metab. Dispos.* (1978) **6**, 205–212.

A2835. Lewinsohn, R., Böhm, K.-H., Glover, V., Sandler, M., *Biochem. Pharmacol.* (1978) **27**, 1857–1863.

A2836. Ikeda, M., Iwai, K., *J. Nutr. Sci. Vitaminol.* (1976) **22**, 365–373.

A2837. Beggs, J.D., Cook, A.M., Fewson, C.A., *J. Gen. Microbiol.* (1976) **96**, 365–374.

A2838. Goldstein, D., Cubeddo, X.L., *Rev. Fac. Farm. Univ. Cent. Venez.* (1976) **33**, 53–68.

A2839. Watson, R.V., Landman, A.D., *Experientia* (1977) **33**, 49–51.

A2840. Alam, S.N., Roberts, R.J., Fischer, L.J., *J. Pedriatr.* (*St Louis*) (1977) **90**, 130–135.

A2842. Strolin Benedetti, M., Larue, D.A., Perrier, C.V., Sitavanc, L., *Arzneim.-Forsch.* (1977) **27**, 158–161.

A2843. Feung, C.S., Hamilton, R.H., Mumma, R.O., *Plant Physiol.* (1976) **58**, 666–669.

A2844. Tan, L.U.L., Drury, E.J., MacKenzie, R.E., *J. Biol. Chem.* (1977) **252**, 1117–1122.

A2845. Ranjeva, R., Boudet, A.M., Faggion, R., *Biochimie* (1976) **58**, 1255–1262.

A2846. Martin, B.R., Harvey, D.J., Paton, W.D.M., *J. Pharm. Pharmacol.* (1976) **28**, 773–774.

A2847. Minatogawa, Y., Noguchi, T., Kido, R., *J. Neurochem.* (1976) **27**, 1097–1101.

A2849. Boyd, D.R., Campbell, R.M., Craig, H.C., Watson, C.G., Daly, J.W., Jerina, D.M., *J. Chem. Soc., Perkin Trans. 1* (1976) 2438–2443.

A2850. Coutts, R.T., Dawson, G.W., Beckett, A.H., *J. Pharm. Pharmacol.* (1976) **28**, 815–821.

A2851. Pantarotto, C., Cappellini, L., Negrini, P., Frigerio, A., *J. Chromatogr.* (1977) **131**, 430–436.

A2853. Powell, M.L., Pope, B., Elmer, G.W., Trager, W.F., *Life Sci.* (1977) **20**, 171–177.

A2854. Inaba, T., Umeda, T., Endrenyi, L., Mahon, W.A., Kalow, W., *Biochem. Pharmacol.* (1978) **27**, 1151–1158.

A2855. Hultman, D., Sundh, K., Wachtmeister, C.A., Arrhenius, E., *Biochem. Pharmacol.* (1978) **27**, 1129–1134.

A2856. Landsberg, L., Berardino, M.B., Stoff, J., Young, J.B., *Biochem. Pharmacol.* (1978) **27**, 1365–1371.

A2857. Parli, C.J., Schmidt, B., Shaar, C.J., *Biochem. Pharmacol.* (1978) **27**, 1405–1408.

A2858. Ötvös, L., Tegyey, Z., Vereczkey, L., Ledniczky, M., Tamás, J., Pálosi, E., Szporny, L., *Drug Metab. Dispos.* (1978) **6**, 213–217.

A2859. Hanzlik, R.P., Shearer, G.O., Hamburg, A., Gillesse, T., *Biochem. Pharmacol.* (1978) **27**, 1435–1439.

A2860. Gammans, R.E., Sehon, R.D., Anders, M.W., Hanna, P.E., *Drug Metab. Dispos.* (1978) **6**, 226–231.

A2861. Sofer, S.S., Ziegler, D.M., *Drug Metab. Dispos.* (1978) **6**, 232–239.

A2862. Hignite, C.E., Tschung, C., Huffman, D.H., Azarnoff, D.L., *Drug Metab. Dispos.* (1978) **6**, 288–295.

A2863. Juchau, M.R., Namkung, M.R., Jones, A.H., Di Giovanni, J., *Drug Metab. Dispos.* (1978) **6**, 273–281.

A2864. Wiebe, L.I., Mercer, J.R., Ryan, A.J., *Drug Metab. Dispos.* (1978) **6**, 296–302.

A2865. Zacchei, A.G., Wishousky, T.I., Arison, B.H., Hitzenberger, G., *Drug Metab. Dispos.* (1978) **6**, 303–312.

A2866. Thakker, D.R., Yagi, H., Lehr, R.E., Levin, W., Buening, M., Lu, A.Y.H., Chang, R.L., Wood, A.W., Conney, A.H., Jerina, D.M., *Mol. Pharmacol.* (1978) **14**, 502–513.

A2867. Ludden, T.M., Bathala, M.S., Malspeis, L., Drum, M.A., Reuning, R.H., *Drug Metab. Dispos.* (1978) **6**, 321–328.

A2868. Harvey, P.R.C., Hobkirk, R., *Steroids* (1977) **30**, 115–128.

A2869. Bhatia, I.S., Bajaj, K.L., Chakravarti, P., *Indian J. Exp. Biol.* (1977) **15**, 118–120.

A2870. Freeman, D.J., Hobkirk, R., *Biochim. Biophys. Acta* (1978) **540**, 478–483.

A2871. Bajaj, K.L., Singh, J., Chakravarti, P., *Indian J. Exp. Biol.* (1977) **15**, 381–383.

A2872. Koss, G., Koransky, W., *Pentachlorophenol (ed. K. Ranga Rao), Plenum, NY* (1978) 131–137.

A2873. Benziger, D., Dorrbecker, S., Goehl, T., Edelson, J., *Drug Metab. Dispos.* (1978) **6**, 8–15.

A2874. Bosin. T.R., Bickers, R.G., Dinner, A., *Drug Metab. Dispos.* (1978) **6**, 31–37.

A2875. Szabadi, R.R., McQueen, C.A., Drummond, G.S., Weber, W.W., *Drug Metab. Dispos.* (1978) **6**, 16–20.

A2876. Warren, J.T., Allen, R., Carter, D.E., *Drug Metab. Dispos.* (1978) **6**, 38–44.

A2877. Lin, C., Foester, R., Lim, J., Hsu, T., Morton, J.B., Symchowicz, S., *Drug Metab. Dispos.* (1978) **6**, 50–58.

A2878. Hunt, R.J., Miller, K.W., *Drug Metab. Dispos.* (1978) **6**, 75–81.

A2879. El-Hawari, A.M., Plaa, G.L., *Drug Metab. Dispos.* (1978) **6**, 59–69.

A2880. Dromgoole, S.H., Nyman, K.E., Furst, D.E., Kolman, W.A., Paulus, H.E., *Drug Metab. Dispos.* (1978) **6**, 102–104.

A2881. Breyer-Pfaff, U., Kreft, H., Rassner, H., Prox, A., *Drug Metab. Dispos.* (1978) **6**, 114–119.

A2882. Nerland, D.E., Mannering, G.J., *Drug Metab. Dispos.* (1978) **6**, 150–153.

A2883. Villeneuve, D.C., Secours, V., Franklin, C., Rock, G., Viau, A., *Drug Metab. Dispos.* (1978) **6**, 146–149.

A2884. Birch, C.G., Hiles, R.A., Eichhold, T.H., Jeffcoat, A.R., Handy, R.W., Hill, J.M., Willis, S.L., Hess, T.R., Wall, T.R., *Drug Metab. Dispos.* (1978) **6**, 169–176.

A2885. Hiles, R.A., Birch, C.G., *Drug Metab. Dispos.* (1978) **6**, 177–183.

A2886. Kerdel, F.A., Bick, R.J., Hopkins, R.P., Callaghan, P., *Biochem. Soc. Trans.* (1978) **6**, 785–787.

A2887. Hucker, H.B., Balletto, A.J. Arison, B.H., Zacchei, A.G., *Drug Metab. Dispos.* (1978) **6**, 184–192.

A2888. Ito, K., Tanaka, T., *Bull. Pharm. Res. Inst.* (1973) **100**, 1–11.

A2889. Ito, K., Tanaka, T., *Euphoria Cacophoria* (1974) **1**, 30–36.

A2890. Yamada, H., Yoshida, H., Nakazawa, H., Kumagai, H., *Acta Vitaminol. Enzymol.* (1975) **29**, 248–251.

A2891. Ishiguro, I., Naito, J., Shinohara, R., *Acta Vitaminol. Enzymol.* (1975) **29**, 278–282.

A2892. Kido, R., Noguchi, T., *Acta Vitaminol. Enzymol.* (1975) **29**, 307–312.

A2893. Krisch, K., Arndt, R., Junge, W., Menge, U., Michelssen, K., *Acta Vitaminol. Enzymol.* (1975) **29**, 302–306.

A2894. Kido, R., Noguchi, T., Minatogawa, Y., *Acta Vitaminol. Enzymol.* (1975) **29**, 344–345.

A2895. Kobayashi, K., Kimura, S., Shimizu, E., *Nippon Suisan Gakkaishi* (1976) **42**, 1365–1372.

A2896. Thonnart, N., Vanhaelen, M., Vanhaelen-Fastré, R., *J. Med. Chem.* (1977) **20**, 604–606.

A2897. Musey, P.I., Collins, D.C., Preedy, J.R., *Steroids* (1977) **29**, 93–104.

A2898. Liang, Y.O., Plotsky, P.M., Adams, R.N., *J. Med. Chem.* (1977) **20**, 581–583.

A2899. Ackermann, E., Richter, K., *Eur. J. Clin. Pharmacol.* (1977) **11**, 43–49.

A2900. Kochlar, M.M., Davidson, P.P., Bhushan, R.S., *Res. Commun. Chem. Pathol. Pharmacol.* (1977) **16**, 251–263.

A2901. Wright, J., Cho, A.K., Gal, J., *Life Sci.* (1977) **20**, 467–473.

A2902. Bidard, J.N., Cronenberger, L., *C. R. Hebd. Seances Acad. Sci., Ser. D* (1977) **284**, 389–392.

A2903. Whitacre, D.M., Badie, M., Schwemmer, B.A., Diaz, L.I., *Bull. Environ. Contam. Toxicol.* (1976) **16**, 689–696.

A2904. Dahl, S.G., Garle, M., *J. Pharm. Sci.* (1977) **66**, 190–193.

A2905. Darmenton, P., Cronenberger, L., Pacheco, H., *Biochimie* (1976) **58**, 1401–1403.

A2906. Draffan, G.H., Clare, R.A., Murray, S., Bellward, G.D., Davies, D.S., Dollery, C.T., *Adv. Mass Spectrom. Biochem. Med.* (1977) **2**, 389–394.

A2907. Yeh, S.Y., McQuinn, R.L., Gorodetzky, C.W., *J. Pharm. Sci.* (1977) **66**, 201–204.

A2908. Greenblatt, D.J., Joyce, T.H., Comer, W.H., Knowles, J.A., Shader, R.I., Kyriakopoulos, A.A., MacLauchlin D.S., Ruelius, H.W., *Clin. Pharmacol. Ther.* (1977) **21**, 222–230.

A2909. Talseth, T., Landmark, K.H., *Eur. J. Clin. Pharmacol.* (1977) **11**, 33–36.

A2910. Roncucci, R., Simon, M.J., Debast, K., Vandriessche, R., Gillet, C., Lambelin, G., *Adv. Mass Spectrom. Biochem. Med.* (1977) **2**, 37–51.

A2911. Meyer, T., *Acta Pharmacol. Toxicol.* (1977) **40**, 193–200.

A2912. Nelson, J.O., Kearney, P.C., Plimmer, J.R., Menzer, R.E., *Pestic. Biochem. Physiol.* (1977) **7**, 73–82.

A2913. Meyer, T., Bakke, T., *Acta Pharmacol. Toxicol.* (1977) **40**, 201–208.

A2914. Shimada, M., Conn, E.E., *Arch. Biochem. Biophys.* (1977) **180**, 199–207.

A2915. Veeraswamy, M., Abitha Devi, N., Krishnan Kutty, R.K., Subba Rao, P.V., *Biochem. J.* (1977) **159**, 807–809.

A2916. Roewekamp, W., Sekeris, C.E., Staerk, J., *FEBS Lett.* (1977) **73**, 225–228.

A2917. Takada, Y., Noguchi, T., Kido, R., *Life Sci.* (1977) **20**, 609–616.

A2918. Collins, J., Yamada, W., Worth, W., Austin, J., *Adv. Exp. Med. Biol.* (1976) **68**, 225–232.

A2919. Davis, P.J., Wiese, D., Rosazza, J.P., *J. Chem. Soc., Perkin Trans. 1* (1977) 1–6.

A2920. De Ridder, J.J., *Adv. Mass Spectrom. Biochem. Med.* (1977) **2**, 365–373.

A2922. Gescher, A., Stevens, M.F.G., *J. Pharm. Pharmacol.* (1976) **28**, Suppl. 17P.

A2923. Beckett, A.H., Navas, G.E., *J. Pharm. Pharmacol.* (1976) **28**, Suppl. 16P.

A2924. Kuhara, T., Tetsuo, M., Yoshino, M., Matsumoto, I., *Adv. Mass Spectrom. Biochem. Med.* (1977) **2**, 53–63.

A2925. McGilveray, I.J., Midha, K.K., Cooper, J.K., Coutts, R.T., Dawe, R., *J. Pharm. Pharmacol.* (1976) **28**, Suppl., 55P.

A2926. Belvedere, G., Frigerio, A., Pantarotto, C., Rovei, V., *Adv. Mass Spectrom. Biochem. Med.* (1977) **2**, 91–100.

A2927. Belvedere, G., Frigerio, A., Pantarotto, C., Rovei, V., *Adv. Mass Spectrom. Biochem. Med.* (1977) **2**, 101–118.

A2928. Cox, P.J., Farmer, P.B., Foster, A.B., Jarman, M., *Adv. Mass Spectrom. Biochem. Med.* (1977) **2**, 149–160.

A2929. Trebbi, A., Gervasi, G.B., Comi, V., Frigerio, A., *Adv. Mass Spectrom. Biochem. Med.* (1977) **2**, 133–148.

A2930. Belarbi, A., Bollack, C., Befort, N., Beck, J.P., Beck, G., *FEBS Lett.* (1977) **75**, 221–225.

A2932. Koshcheenko, K.A., Baklashova, T.G., Kozlovskii, A.G., Arinbasarov, M.U., Skryabin, G.K., *Prikl. Biokhim. Mikrobiol.* (1977) **13**, 248–254.

A2934. Beckett, A.H., Vutthikongsirigool, W., *J. Pharm. Pharmacol.* (1976) **28**, Suppl., 54P.

A2936. Beckett, A.H., Achari, R., *J. Pharm. Pharmacol.* (1976) **28**, Suppl., 57P.

A2937. Kametani, T., Ohta, Y., Takemura, M., Ihara, M., Fukumoto, K., *Heterocycles* (1977) **6**, 415–421.

A2938. Russel, S., Bollag, J.M., *Acta Microbiol. Pol.* (1977) **26**, 59–64.

A2939. Ranney, R.E., Oppermann, J.A., Muldoon, E., McMahon, F.G., *J. Toxicol. Environ. Health* (1976) **2**, 441–451.

A2941. Hollstein, U., Mock, D.L., Sibbitt, R.R., Roisch, U., Lingens, F., *Tetrahedron Lett.* (1978) 2987–2990.

A2943. McGovern, E.P., Bentley, R., *Arch. Biochem. Biophys.* (1978) **188**, 56–63.

A2945. Yang, C.-H., Srivastava, P.N., *Biochem. J.* (1976) **159**, 133–142.

A2946. Bertino, J.R., Coward, J.K., Cashmore, A., Chello, P., Panichajakul, S., Horvath, C.G., Stout, R.W., *Biochem. Soc. Trans.* (1976) **4**, 853–856.

A2947. Hayashi, M., Fuwa, T., Awazu, S., Hanano, M., *Chem. Pharm. Bull.* (1976) **24**, 2589–2596.

A2950. Edwards, D.J., *Life Sci.* (1978) **23**, 1201–1208.

A2951. Zacchei, A.G., Wishousky, T.I., Watson, L.S., *Drug Metab. Dispos.* (1978) **6**, 313–320.

A2952. Bernhardt, F.-H., Pachowsky, H., Staudinger, H., *Eur. J. Biochem.* (1975) **57**, 241–256.

A2953. Bernhardt, F.-H., Nastainczyk, W., Seydewitz, V., *Eur. J. Biochem.* (1977) **72**, 107–115.

A2954. Bernhardt, F.-H., Heymann, E., *IRCS Med. Sci.* (1975) **3**, 463.

A2955. Bernhardt, F.-H., Nastainczyk, W., *IRCS Med. Sci.* (1974) **2**, 1268.

A2956. Pazmiño, P.A., Weinshilboum, R.M., *Clin. Chim. Acta* (1978) **89**, 317–329.

A2957. Stillwell, W.G., Bouwsma, O.J., Thenot, J.-P., Horning, M.G., Griffin, G.W.,

Ishikawa, K., Takaku, M., *Res. Commun. Chem. Pathol. Pharmacol.* (1978) **20**, 509–530.

A2958. Haley, C.J., Harper, A.E., *Arch. Biochem. Biophys.* (1978) **189**, 524–530.

A2959. Midha, K.K., Coutts, R.T., *J. Chromatogr.* (1977) **144**, 280–283.

A2960. Kobayashi, K., Kimura, S., Shimizu, E., *Nippon Suisan Gakkaishi* (1977) **43**, 601–607.

A2961. Goodwin, B.L., Ruthven, C.R.J., King, G.S., Sandler, M., Leask, B.G.S., *Xenobiotica* (1978) **8**, 629–651.

A2962. Akitake, H., Kobayashi, K. *Bull. Jap. Soc. Sci. Fisheries* (1975) **41**, 321–327.

A2963. Kobayashi, K., *Environ. Sci. Res.* (1978) **12**, 89–105.

A2964. Marquardt, G.M., Di Stefano, V., Ling, L.L., *Biochem. Pharmacol.* (1978) **27**, 1503–1505.

A2965. Lewinsohn, R., *Clin. Chim. Acta* (1977) **81**, 247–256.

A2966. Kaplan, E., Gutmann, H.R., Emory, T.H., *Biochem. Pharmacol.* (1978) **27**, 1581–1589.

A2967. Roth, J.A., Feor, K., *Biochem. Pharmacol.* (1978) **27**, 1606–1608.

A2968. Chaturapit, S., Holder, G.M., *Biochem. Pharmacol.* (1978) **27**, 1865–1871.

A2969. Numazawa, M., Shirao, R., Soeda, N., Nambara, T., *Biochem. Pharmacol.* (1978) **27**, 1833–1838.

A2970. Eyer, P., Gaber, H., *Biochem. Pharmacol.* (1978) **27**, 2215–2221.

A2971. Eyer, P., Kampffmeyer, H.G., *Biochem. Pharmacol.* (1978) **27**, 2223–2228.

A2972. Dial, E.J., Clarke, D.E., *Biochem. Pharmacol.* (1978) **27**, 2374–2375.

A2973. Deitrich, R.A., Blundeau, P., Roper, M., Schmuck, J., *Biochem. Pharmacol.* (1978) **27**, 2343–2347.

A2974. Baars, A.J., Jansen, M., Breimer, D.D., *Biochem. Pharmacol.* (1978) **27**, 2487–2494.

A2975. Billings, R.E., Murphy, P.J., McMahon, R.E., Ashmore, J., *Biochem. Pharmacol.* (1978) **27**, 2525–2529.

A2976. Vore, M., Bauer, J., Pascucci, V., *J. Pharmacol. Exp. Ther.* (1978) **206**, 439–447.

A2977. Goldberg, R., Tipton, K.F., *Biochem. Pharmacol.* (1978) **27**, 2623–2629.

A2978. Yamato, C., Kinoshita, K., *J. Pharmacol. Exp. Ther.* (1978) **206**, 468–474.

A2979. Misra, A.L., Mulé, S.J., Bloch, R., Bates, T.R., *J. Pharmacol. Exp. Ther.* (1978) **206**, 475–491.

A2980. Bolanowska, W., Gessner, T., *J. Pharmacol. Exp. Ther.* (1978) **206**, 233–238.

A2981. Wilk, S., Mizoguchi, H., Orlowski, M., *J. Pharmacol. Exp. Ther.* (1978) **206**, 227–232.

A2982. Greenlee, W.F., Poland, A., *J. Pharmacol. Exp. Ther.* (1978) **205**, 596–605.

A2983. Maeda, T., Shindo, H., *Chem. Pharm. Bull.* (1978) **26**, 2054–2057.

A2984. Nabih, T., Youel, L., Rosazza, J.P., *J. Chem. Soc., Perkin Trans. 1* (1978) 757–762.

A2985. Elmorsi, E.A., Hopper, D.J., *Biochem. Soc. Trans.* (1978) **6**, 958–959.

A2986. Ayres, J.W., Sakmar, E., Hallmark, M.R., Wagner, J.G., *Res. Commun. Chem. Pathol. Pharmacol.* (1977) **16**, 475–483.

A2987. Landi, M., Rimorini, N., Zerilli, L.F., Gallo, G.G., Rovei, V., Frigerio, A., *Adv. Mass Spectrom. Biochem. Med.* (1977) **2**, 395–403.

A2988. Reinhold, V.N., Bruni, R.J., *Biomed. Mass Spectrom.* (1976) **3**, 335–339.

A2989. Harvey, D.J., Martin, B.R., Paton, W.D.M., *Res. Commun. Chem. Pathol. Pharmacol.* (1977) **16**, 265–279.

A2990. Pantarotto, C., Cappellini, L., De Pascale, A., Frigerio, A., *J. Chromatogr.* (1977) **134**, 307–314.

A2991. Frigerio, A., Lanzoni, J., Pantarotto, C., Rossi, E., Rovei, V., Zanol, M., *J. Chromatogr.* (1977) **134**, 299–305.

A2992. Knoll, R., Christ, W., Müller-Oerlinghausen, B., Coper, H., *Naunyn-*

Schmiedeberg's Arch. Pharmacol. (1977) **297**, 195–200.

A2993. Rozman, K., Mueller, W., Coulston, F., Korte, F., *Chemosphere* (1977) **6**, 81–84.

A2994. Fonseka, K., Widman, M., *J. Pharm. Pharmacol.* (1977) **29**, 12–14.

A2995. Bakke, J.E., Feil, V.J., Price, C.E., Zaylskie, R.G., *Biomed. Mass Spectrom.* (1976) **3**, 299–315.

A2996. Parisi, G., D'Amora, D., Franco, A.R., *Bull. Soc. Ital. Biol. Sper.* (1976) **52**, 1072–1077.

A2997. Mathies, G., Belitz, H.-D., *Z. Lebensm.-Unters. Forsch.* (1977) **163**, 279–282.

A2998. Peele, J.D., Jr., Oswald, E.O., *Biochim. Biophys. Acta.* (1977) **497**, 598–607.

A2999. Harada, T., Koiwa, S., *J. Ferment. Technol.* (1977) **55**, 97–102.

A3000. Mitsugi, K., Sano, K., Yokozeki, K., Yamada, K., Noda, I., Kagawa, T., Eguchi, C., Yasuda, N., Tamura, F., Togo, K., (5–4–1977) *U.S.* 4, 016, 037.

A3001. Bakke, J.E., Price, C.E., Zaylskie, R.G., *Biomed. Mass. Spectrom.* (1976) **3**, 329–334.

A3002. Sherman, A.D., Hsiao, W.-C., Gál, E.M., *Neuropharmacology* (1977) **16**, 17–24.

A3003. Farrier, D.S., *Arzneim.-Forsch.* (1977) **27**, 575–583.

A3004. Nabih, T., Davis, P.J., Caputo, J.F., Rosazza, J.P., *J. Med. Chem.* (1977) **20**, 914–917.

A3005. Diehl, E., Najm, S., Wolff, R.E., Eberhardt, J.D., *J. Pharmacol.* (1976) **7**, 563–575.

A3006. Comi, V., Fossati, A., Gervasi, G.B., *Farmaco, Ed. Sci.* (1977) **32**, 278–285.

A3007. Roberts, J., Rosenfeld, H.J., *J. Biol. Chem.* (1977) **252**, 2640–2647.

A3008. Yabisui, T., Matsuki, T., Tanishima, K., Takeshita, M., Yoneyama, Y., *Biochem. Biophys. Res. Commun.* (1977) **76**, 174–182.

A3009. Takai, K., Ushiro, H., Noda, Y., Narumiya, S., Tokuyama, T., Hayaishi, O., *J. Biol. Chem.* (1977) **252**, 2648–2656.

A3010. Abbott, B.J., Fukuda, D.S., Archer, R.A., *Experientia* (1977) **33**, 718–720.

A3011. Dziadkowiec, I., Czarnik, Z., Rembiesa, R., *Endocrinol. Exp.* (1977) **11**, 19–22.

A3012. Bešić, J., Grims, M., Fuks, Ž, Beus, I., Rulnjević, J., *Acta Pharm. Jugosl.* (1977) **27**, 13–16.

A3013. Beckett, A.H., Haya, K., *J. Pharm. Pharmacol.* (1977) **29**, 89–95.

A3015. Gurtoo, H.L., Motycka, L., *Cancer Res.* (1976) **36**, 4663–4671.

A3016. Kumada, Y., Takeuchi, T., Umezawa, H., *Agric. Biol. Chem.* (1977) **41**, 869–876.

A3017. Hoeldtke, R., Kaufman, S., *J. Biol. Chem.* (1977) **252**, 3160–3169.

A3018. Hjelmeland, L.M., Aronow, L., Trudell, J.R., *Biochem. Biophys. Res. Commun.* (1977) **76**, 541–549.

A3019. Kumada, Y., Takeuchi, T., Umezawa, H., *Agric. Biol. Chem.* (1977) **41**, 877–885.

A3020. Sanchez-Urretia, L., Greengard, O., *Biochim. Biophys. Acta* (1977) **497**, 682–689.

A3021. Zacchei, A.G., Rhodes, R.E., Christy, M.E., *Biomed. Mass Spectrom.* (1976) **3**, 286–294.

A3022. Beckett, A.H., Bélanger, P.M., *Br. J. Clin. Pharmacol.* (1977) **4**, 193–200.

A3023. Pacifici, G.M., Bianchetti, G., Frigerio, A., Gomeni, R., Morselli, P.L., Garattini, S., Pifferi, G., Carpi, C., *Eur. J. Drug Metab. Pharmacokinet.* (1976) **1**, 141–147.

A3024. Dittmann, B., Renner, G., *Naunyn-Schmiedeberg's Arch. Pharmacol.* (1977) **296**, 87–89.

A3026. Lees, G.J., *Life Sci.* (1977) **20**, 1749–1762.

A3027. Roebuck, B.D., Wogan, G.N., *Cancer Res.* (1977) **37**, 1649–1656.

A3028. Guengerich, F.P., *J. Biol. Chem.* (1977) **252**, 3970–3979.

A3029. Ishikura, K., Nitta, Y., Watanabe, T., *J. Biochem.* (1977) **81**, 1187–1192.

A3030. Fukuzumi, T., Katayama, Y., *Mokuzai Gakkaishi* (1977) **23**, 214–215.

A3031. Lu, A.Y.H., Jerina, D.M., Levin, W., *J. Biol. Chem.* (1977) **252**, 3715–3723.

A3032. Guest, J.R., *J. Bacteriol.* (1977) **130**, 1038–1046.

A3033. Okerholm, R.A., Keeley, F.J., Peterson, F.E., Glazko, A.J., *Proc. R. Soc. Med.* (1976) **69** (Suppl. 2) , 11–14.

A3034. Alván, G., Borgå, O., Lind, M., Palmér, L., Siwers, B., *Eur. J. Clin. Pharmacol.* (1977) **11**, 219–224.

A3035. Benjamin, R.S., Riggs, C.E., Jr., Bachur, N.R., *Cancer Res.* (1977) **37**, 1416–1420.

A3036. Bogatskii, A.V., Golovenko, N.Y., Andronati, S.A., Kolomeichenko, G.Y., Zhilina, Z.I., *Dokl. Akad. Nauk SSSR* (1977) **234**, 215–218.

A3037. Chatfield, D.H., Green, J.N., *Xenobiotica* (1978) **8**, 133–144.

A3038. Douch, P.G.C., Buchanan, L.L., *Xenobiotica* (1978) **8**, 171–176.

A3039. Nation, R.L., Triggs, E.J., Vine, J., *Xenobiotica* (1978) **8**, 165–169.

A3040. Douch, P.G.C., *Xenobiotica* (1978) **8**, 177–182.

A3041. Ball, L.M., Williams, R.T., Renwick, A.G., *Xenobiotica* (1978) **8**, 183–190.

A3042. James, S.P., Pheasant, A.E., Solheim, E., *Xenobiotica* (1978) **8**, 219–228.

A3043. James, S.P., Pheasant, A.E., *Xenobiotica* (1978) **8**, 207–217.

A3044. Hakusui, H., Tachizawa, H., Sano, M., *Xenobiotica* (1978) **8**, 229–238.

A3045. Leinweber, F.-J., Greenough, R.C., Di Carlo, F.J., *Xenobiotica* (1978) **8**, 239–243.

A3046. Idle, J.R., Millburn, P., Williams, R.T., *Xenobiotica* (1978) **8**, 253–264.

A3047. Tsuji, H., Yoshimura, H., Tsukamoto, H., *Xenobiotica* (1978) **8**, 245–251.

A3048. Jonen-Kern, R., Jonen, H.G., Schupp, R.R., Minck, K., Kahl, G.F., Netter, K.J., *Xenobiotica* (1978) **8**, 271–280.

A3049. Jonen, H.G., Bonow, J.H., Kugler, M., Kahl, R., Kahl, G.F., *Xenobiotica* (1978) **8**, 281–288.

A3050. Cone, E.J., *Xenobiotica* (1978) **8**, 301–304.

A3051. Belpaire, F.M., Rosseel, M.T., Bogaert, M.G., *Xenobiotica* (1978) **8**, 297–300.

A3052. Reeves, P.R., Barnfield, D.J., Longshaw, S., McIntosh, D.A.D., Winrow, M.J., *Xenobiotica* (1978) **8**, 305–311.

A3053. Reeves, P.R., McAinsh, J., McIntosh, D.A.D., Winrow, M.J., *Xenobiotica* (1978) **8**, 313–320.

A3054. Shimizu, T., Mori, H., Tabusa, E., Morita, S., Miyamoto, G., Yasuda, Y., Nakagawa, K., *Xenobiotica* (1978) **8**, 349–358.

A3055. Sullivan, H.R., McMahon, R.E., Roffey, P., Hoffman, D.G., Marshall, F.J., Billings, R.E., Benslay, D.N., Marshall, W.S., *Xenobiotica* (1978) **8**, 333–340.

A3056. Tanayama, S., Kobayashi, T., Kanai, Y., *Xenobiotica* (1978) **8**, 365–375.

A3057. Bedford, C.T., Crayford, J.V., Hutson, D.H., Wiggins, D.E., *Xenobiotica* (1978) **8**, 383–395.

A3058. Jeffery, D.J., Jones, H., Langley, P.F., *Xenobiotica* (1978) **8**, 419–427.

A3059. Seutter-Berlage, F., Delbressine, L.P.C., Smeets, F.L.M., Ketelaars, H.C.J., *Xenobiotica* (1978) **8**, 413–418.

A3060. Pottier, J., Busigny, M., Raynaud, J.P., *Xenobiotica* (1978) **8**, 429–437.

A3061. Mehta, R., Hirom, P.C., Millburn, P., *Xenobiotica* (1978) **8**, 445–452.

A3062. Chkanikov, D.I., Pavlova, N.N., Makeev, A.M., Nazarova, T.A., Makoveichuk, A.Y., *Fiziol. Rast.* (*Moscow*) (1977) **24**, 566–573.

A3063. Tyce, G.M., *Res. Commun. Chem. Pathol. Pharmacol.* (1977) **16**, 669–685.

A3064. Rouchaud, J.P., Lhoest, G.J., Mercier, M.J., Meyer, J.A., *Pestic. Sci.* (1977) **8**, 23–30.

A3065. Rouchaud, J.P., Decallonne, J.R., Meyer, J.A., *Pestic. Sci.* (1977) **8**, 31–34.

A3066. Elmorsi, E.A., Hopper, D.J., *Eur. J. Biochem.* (1977) **76**, 197–208.

A3067. Gray, G.M., Tabiowo, A., Trotter, M.D., *Biochem. J.* (1977) **161**, 667–675.

A3068. Tadasa, K., *Agric. Biol. Chem.* (1977) **41**, 925–929.

A3069. Stöckigt, J., Husson, H.P., Kan-Fan, C., Zenk, M.H., *J. Chem. Soc., Chem. Commun.* (1977) 164–166.

A3070. Arinbasarov, M.U., Baklashova, T.G., Kozlovskii, A.G., *Prikl. Biokhim. Mikrobiol.* (1977) **13**, 427–430.

A3071. Fukuda, D., Archer, R.A., Abbott, B.J., *Appl. Environ. Microbiol.* (1977) **33**, 1134–1140.

A3073. Little, G.J., Dale, A.D., Ord, D.A., Marten, T.R., *Anal. Chem.* (1977) **49**, 1311–1313.

A3074. Edelson, J., Douglas, J.F., Ludwig, B.J., *Arzneim.-Forsch.* (1977) **27**, 789–793.

A3075. Burt, A.J., Chasseaud, L.F., Franklin, E.R., Hawkins, D.R., Taylor, T., *Eur. J. Drug. Metab. Pharmacokinet.* (1976) **1**, 167–172.

A3076. Frigerio, A., Sossi, N., *Cron. Chim.* (1977) **49**, 16–21.

A3077. Kiseleva, V.N., Gilev, A.P., Parinov, V.Y., Satcs, V., Germane, S., *Byull. Eksp. Biol. Med.* (1977) **83**, 702–704.

A3078. Tulp, M.T.M., Tillmanns, G.M., Hutzinger, O., *Chemosphere* (1977) **6**, 223–230.

A3079. Gillis, C.N., Roth, J.A., *Br. J. Pharmacol.* (1977) **59**, 585–590.

A3080. Engelhardt, G., Wallnöfer, P., Fuchsbichler, G., Baumeister, W., *Chemosphere* (1977) **6**, 85–92.

A3081. Kandasamy, D., Chendrayan, K., Rajukkannu, K., Balasubramanian, M., *Curr. Sci.* (1977) **46**, 280–281.

A3083. Savage, N., Prinz, W., *Biochem. J.* (1977) **161**, 551–554.

A3084. Houser, R.M., Olson, R.E., *J. Biol. Chem.* (1977) **252**, 4017–4021.

A3085. Grabau, W., Ruediger, H., *Hoppe-Seyler's Z. Physiol. Chem.* (1977) **358**, 695–698.

A3086. Pittaway, D.E., Andersen, R.N., Givens, J.R., *Acta Endocrinol. (Copenhagen)* (1977) **85**, 624–635.

A3087. Noda, Y., Takai, K., Tokuyama, T., Narumiya, S., Ushiro, H., Hayaishi, O., *J. Biol. Chem.* (1977) **252**, 4413–4415.

A3088. Kho, K.F.F., Bolsman-Louwen, A.C., Vuik, J.C., Bennink, G.J.H., *Planta* (1977) **135**, 109–118.

A3089. Bhakuni, D.S., Tewari, S., Kapil, R.S., *J. Chem. Soc., Perkin Trans 1* (1977) 706–709.

A3090. Forsman, A., Fölsch, G., Larsson, M., Öhman, R., *Curr. Ther. Res. Clin. Exp.* (1977) **21**, 606–617.

A3091. Kirdani, R.Y., Barua, N.R., Sandberg, A.A., *J. Clin. Endocrinol. Metab.* (1977) **44**, 1121–1129.

A3092. Estevez, V.S., Ho, B.T., Englert, L.F., *Res. Commun. Chem. Pathol. Pharmacol.* (1977) **17**, 179–182.

A3093. Thakker, D.R., Yagi, H., Akagi, H., Koreeda, M., Lu, A.Y.H., Levin, W., Wood, A.W., Conney, A.H., Jerina, D.M., *Chem.-Biol. Interact.* (1977) **16**, 281–300.

A3094. Tobes, M.C., Mason, M., *J. Biol. Chem.* (1977) **252**, 4591–4599.

A3095. Jablonski, E., De Luca, M., *Biochemistry* (1977) **16**, 2932–2936.

A3096. Nakazawa, T., Hayashi, E., *J. Bacteriol.* (1977) **131**, 42–48.

A3097. Goewert, R.R., Sippel, C.J., Olson, R.E., *Biochem. Biophys. Res. Commun.* (1977) **77**, 599–605.

A3098. Furst, D.E., Gupta, N., Paulus, H.E., *J. Clin. Invest.* (1977) **60**, 32–42.

A3099. Jacobs, S.A., Stoller, R.G., Chabner, B.A., Johns, D.G., *Cancer Treat. Rep.* (1976) **61**, 651–656.

A3100. Gibson, W.L., Brown, L.R., *Dev. Ind. Microbiol.* (1975) **16**, 77–87.

A3101. Stein, K., Portig, J., Koransky, W., *Naunyn-Schmiedeberg's Arch. Pharmacol.* (1977) **298**, 115–128.

A3102. Nambara, T., Kawarada, Y., *Chem. Pharm. Bull.* (1977) **25**, 942–948.

A3103. Forsman, A., Larsson, M., *Curr. Ther. Res.* (1978) **24**, 567–568.

A3105. Balba, M.T., Evans, W.C., *Biochem. Soc. Trans.* (1977) **5**, 302–304.

A3106. Lindstedt, S., Odelhog, B., Rundgren, M., *Biochemistry* (1977) **16**, 3369–3377.

A3107. Bullock, P., Spanner, S., Ansell, G.B., *Biochem. Soc. Trans.* (1977) **5**, 166–168.

A3108. Suzuki, T., *J. Environ. Sci. Health, Pt. B* (1977) **B12**, 113–127.

A3109. Doumas, J., Maume, B.F., *C. R. Seances Soc. Biol. Ses. Fil.* (1977) **171**, 108–114.

A3112. Cameron, D.W., Sawyer, W.H., Trikojus, V.M., *Aust. J. Biol. Sci.* (1977) **30**, 173–181.

A3113. Baccanari, D.P., Averett, D., Briggs, C., Burchall, J., *Biochemistry* (1977) **16**, 3566–3572.

A3114. Joshi, P.R., Shiralkar, N.D., *J. Food Sci. Technol.* (1977) **14**, 77–79.

A3117. Hüfner, M., Grussendorf, M., Ntokalou, M., *Clin. Chim. Acta* (1977) **78**, 251–259.

A3118. Mori, H., Kido, M., Murakami, N., Morita, S., Kohri, H., Nakagawa, K., Uno, T., Ishihara, T., *Yakugaku Zasshi* (1977) **97**, 350–358.

A3119. Bevill, R.F., Sharma, R.M., Meachum, S.H., Wozniak, S.C., Bourne, D.W.A., Dittert, L.W., *Am. J. Vet. Res.* (1977) **38**, 973–977.

A3120. Wretlind, M., Pilbrant, A., Sundwall, A., Vessman, J., *Acta Pharmacol. Toxicol.* (1977) **40**, Suppl. 1, 28–39.

A3121. Axelsson, R., Mårtensson, E., *Curr. Ther. Res. Clin. Exp.* (1977) **21**, 561–586.

A3122. Pynnönen, S., Yrjänä, T., *Int. J. Clin. Pharmacol. Biopharm.* (1977) **15**, 222–226.

A3123. Sherman, A.D., Gál, E.M., *Neuropharmacology* (1977) **16**, 309–315.

A3124. Owellen, R.J., Hartke, C.A., Hains, F.O., *Cancer Res.* (1977) **37**, 2597–2602.

A3125. Medvedev, V.A., Korshikov, I.I., Bashkatov, V.G., Tarabrin, V.P., *Fiziol. Rast.* (Moscow) (1977) **24**, 858–860.

A3126. Bury, A.F., Coolbear, T., Savery, C.R., *Biochem. J.* (1977) **163**, 565–570.

A3127. Yang, S.K., Roller, P.P., Gelboin, H.V., *Biochemistry* (1977) **16**, 3680–3687.

A3128. Lindblad, B., Lindstedt, G., Lindstedt, S., Rundgren, M., *J. Biol. Chem.* (1977) **252**, 5073–5084.

A3129. Rundgren, M., *J. Biol. Chem.* (1977) **252**, 5085–5093.

A3130. Suzuki, O., Noguchi, E., Yagi, K., *Brain Res.* (1977) **131**, 379–382.

A3131. Feldman, J.M., Roche, J.M., Blum, J.J., *J. Protozool.* (1977) **24**, 459–462.

A3132. Coldwell, B.B., Thomas, B.H., Whitehouse, L.W., Wong, L.T., Hynie, I., *Proc. Eur. Soc. Toxicol.* (1976) **17**, 269–276.

A3133. Brüsewitz, G., Cameron, B.D., Chasseaud, L.F., Görler, K., Hawkins, D.R., Koch, H., Mennicke, W.H., *Biochem. J.* (1977) **162**, 99–107.

A3135. Goodenough, P.W., *Phytochemistry* (1978) **17**, 633–636.

A3136. Jeffs, P.W., Karle, J.M., Martin, N.H., *Phytochemistry* (1978) **17**, 719–728.

A3137. Marbach, I., Mayer, A.M., Maron, R., *Phytochemistry* (1978) **17**, 655–657.

A3138. Joel, D.M., Marbach, I., Mayer, A.M., *Phytochemistry* (1978) **17**, 796–797.

A3139. Saini, M.S., Anderson, J.A., *Phytochemistry* (1978) **17**, 799–800.

A3140. Yagi, A., Yamanouchi, M., Nishioka, I., *Phytochemistry* (1978) **17**, 895–897.

A3141. Furuya, T., Nakano, M., Yoshikawa, T., *Phytochemistry* (1978) **17**, 891–893.

A3142. Brown, R.T., Leonard, J., Sleigh, S.K., *Phytochemistry* (1978) **17**, 899–900.

A3143. Roberts, M.F., Antoun, M.D., *Phytochemistry* (1978) **17**, 1083–1087.

A3144. Klischies, M., Zenk, M.H., *Phytochemistry* (1978) **17**, 1281–1284.

A3145. Rhodes, M.J.C., Wooltorton, L.S.C., *Phytochemistry* (1978) **17**, 1225–1229.

A3146. Billett, E.E., Smith, H., *Phytochemistry* (1978) **17**, 1511–1516.

A3147. Stöckigt, J., Zenk, M.H., *FEBS Lett.* (1974) **42**, 131–134.

A3148. Rosenfeld, H.G., Watanabe, K.A., Roberts, J., *J. Biol. Chem.* (1977) **252**, 6970–6973.

A3149. Rhodes, M.J.C., Wooltorton, L.S.C., *Phytochemistry* (1978) **17**, 947–951.

A3150. Besic, J., Grims, M., Fuks, Z., Beus, I., Rulnjevic, J., Gjuris, V., Vrhovac, B., *Acta Pharm. Jugosl.* (1976) **26**, 107–114.

A3151. Stöckigt, J., Rueffer, M., Zenk, M.H., Hoyer, G.-A., *Planta Med.* (1978) **33**, 188–192.

A3152. Klischies, M., Stöckigt, J., Zenk, M.H., *Phytochemistry* (1978) **17**, 1523–1525.

A3153. Rueffer, M., Nagakura, N., Zenk, M.H., *Tetrahedron Lett.* (1978) 1593–1596.

A3154. Nagakura, N., Höfle, G., Coggiola, D., Zenk, M.H., *Planta Med.* (1978) **34**, 381–389.

A3155. Harvey, D.J., Paton, W.D.M., *Res. Commun. Chem. Pathol. Pharmacol.* (1978) **21**, 435–446.

A3156. Caldwell, J., Croft, J.E., Smith, R.L., Snedden, W., *Br. J. Pharmacol.* (1977) **60**, 295P-296P.

A3157. Komatsu, T., *Chemotherapy (Tokyo)* (1977) **25**, 565–574.

A3158. Aarbakke, J., Bakke, O.M., Milde, E.J., Davies, D.S., *Eur. J. Clin. Pharmacol.* (1977) **11**, 359–366.

A3159. Dieterle, W., Faigle, J.W., Montigel, C., Sulc, M., Theobald, W., *Eur. J. Clin. Pharmacol.* (1977) **11**, 367–375.

A3160. Danielson, T.J., Davis, B.A., Boulton, A.A., *Can. J. Physiol. Pharmacol.* (1977) **55**, 439–443.

A3161. Hlavica, P., Kehl, M., *Biochem. J.* (1977) **164**, 487–496.

A3162. Cone, E.J., Gorodetzky, C.W., *Probl. Drug Depend.* (1975) 338–347.

A3163. Kampffmeyer, H.G., *Eur. J. Drug. Metab. Pharmacokinet.* (1976) **1**, 182–187.

A3164. Bresnick, E., Stoming, T.A., Vaught, J.B., Thakker, D.R., Jerina, D.M., *Arch. Biochem. Biophys.* (1977) **183**, 31–37.

A3165. Phillips, J.C., Hardy, K., Richards, R., Cottrell, R.C., Gangolli, S.D., *Toxicology* (1977) **7**, 257–263.

A3166. Thakker, D.R., Yagi, H., Levin, W., Lu, A.Y.H., Conney, A.H., Jerina, D.M., *J. Biol. Chem.* (1977) **252**, 6328–6334.

A3167. Ito, S., Prota, G., *Experientia* (1977) **33**, 1118–1119.

A3168. Tkachuck, R.D., Weinstein, P.P., Mueller, J.F., *J. Parasitol.* (1977) **63**, 694–700.

A3169. Sauber, K., Müller, R., Keller, E., Eberspächer, J., Lingens, F., *Z. Naturforsch. C: Biosci.* (1977) **32C**, 557–562.

A3171. North-Root, H., Kuttab, S.H., Henderson, G.L., *Probl. Drug Depend.* (1975) 645–657.

A3172. Schaller, K.H., Schütz, H.W., Geldmacher v. Mallinckrodt, M., Gossler, K., Bost, H.P., *Acta Pharmacol. Toxicol.* (1977) **41**, Suppl., 230–238.

A3173. Itakura, N., Fisher, A.B., Thurman, R.G., *J. Appl. Physiol.* (1977) **43**, 238–245.

A3174. Roubal, W.T., Collier, T.K., Malins, D.C., *Arch. Environ. Contam. Toxicol.* (1977) **5**, 513–529.

A3175. Møller, B.L., Conn, E.E., *Proc. West. Pharmacol. Soc.* (1977) **20**, 103–107.

A3177. Fujiwara, H., Inagaki, C., Ikeda, Y., Tanaka, C., *Nippon Yakurigaku Zasshi* (1976) **72**, 891–898.

A3178. Nagy, A., *J. Pharm. Pharmacol.* (1977) **29**, 104–107.

A3179. Sherman, A.D., Gal, E.M., *Psychopharmacol. Commun.* (1976) **2**, 421–427.

A3180. Skrumsager, B.K., Henderson, G.L., *Proc. West. Pharmacol. Soc.* (1977) **20**, 473–476.

A3181. North-Root, H., Kuttab, S.H., Henderson, G.L., *Probl. Drug Depend.* (1975) 1012–1017.

A3182. Braun, W.H., Young, J.D., Blau, G.E., Gehring, P.J., *Toxicol. Appl. Pharmacol.* (1977) **41**, 395–406.

A3183. Kametani, T., Ohta, Y., Takemura, M., Ihara, M., Fukumoto, K., *Bioorg. Chem.* (1977) **6**, 249–256.

A3184. Yeh, W.K., Gibson, D.T., Liu, T.-N., *Biochem. Biophys. Res. Commun.* (1977) **78**, 401–410.

A3185. Nath, A., Mehrotra, K.N., *Indian J. Exp. Biol.* (1977) **15**, 665–666.
A3186. Mårtensson, E., Nyberg, G., Axelsson, R., Serck-Hansen, K., *Curr. Ther. Res.* (1975) **18**, 687–699.
A3187. Niederwieser, A., Matasovic, A., Tippett, P., Danks, D.M., *Clin. Chim. Acta* (1977) **76**, 345–356.
A3188. Axelsson, R., *Curr. Ther. Res.* (1977) **21**, 587–605.
A3189. Balba, M.T., Evans, W.C., *Biochem. Soc. Trans.* (1979) **7**, 4–3–405.
A3190. Matheis, G., Belitz, H.-D., *Z. Lebensm. Unters.-Forsch.* (1975) **157**, 221–227.
A3191. Kho, K.K.F., Kamsteeg, J., Van Brederode, J., *Zeitschr. Pflanzenphysiol.* (1978) **88**, 449–464.
A3192. Williamson, D.G., Gwilliam, C., Layne, D.S., *Can. J. Biochem.* (1977) **55**, 41–42.
A3194. Gaughan, L.C., Ackerman, M.E., Unai, T., Casida, J.E., *J. Agric. Food Chem.* (1978) **26**, 613–618.
A3195. Gaughan, L.C., Casida, J.E., *J. Agric. Food Chem.* (1978) **26**, 525–528.
A3196. Benezet, H.J., Knowles, C.O., Campbell, J.R., Savage, J.E., *J. Agric. Food Chem.* (1978) **26**, 622–627.
A3197. Goettfert, J.K., Parlar, H., Korte, F., *J. Agric. Food Chem.* (1978) **26**, 628–632.
A3198. Akhtar, M.H., *J. Agric. Food Chem.* (1978) **26**, 932–935.
A3199. Ruzo, L.O., Unai, T., Casida, J.E., *J. Agric. Food Chem.* (1978) **26**, 918–925.
A3200. Feung, C.S., Loerch, R., Hamilton, R.H., Mumma, R.O., *J. Agric. Food Chem.* (1978) **26**, 1064–1067.
A3201. Yang, R.S.H., Pittman, K.A., Rourke, D.R., Stein, V.B., *J. Agric. Food Chem.* (1978) **26**, 1076–1083.
A3202. Arjmand, M., Hamilton, R.H., Mumma, R.O., *J. Agric. Food Chem.* (1978) **26**, 1125–1128.
A3203. Wong, K.P., *Biochem. Soc.* (1977) **164**, 529–531.
A3204. Carew, D.P., Krueger, R.J., *Phytochemistry* (1977) **16**, 1461–1462.
A3205. Stöckigt, J., Zenk, M.H., *FEBS Lett.* (1977) **79**, 233–237.
A3206. Kiwada, H., Hayashi, M., Fuwa, T., Awazu, S., Hanano, M., *Chem. Pharm. Bull.* (1977) **25**, 1566–1573.
A3208. Blackley, E.R., *Can. J. Microbiol.* (1977) **23**, 1128–1139.
A3209. Michaliszyn, G.A., Wing, S.S., Meighen, E.A., *J. Biol. Chem.* (1977) **252**, 7495–7497.
A3210. Yoshizako, F., Chubachi, M., Nishimura, A., Ueno, T., *Can. J. Microbiol.* (1977) **23**, 1140–1144.
A3212. Bartek, M.J., La Budde, J.A., *Toxicol. Appl. Pharmacol.* (1976) **36**, 307–322.
A3213. Collier, M.D., Murray, D.R., *Aust. J. Plant Physiol.* (1977) **4**, 571–582.
A3214. Biaglow, J.E., Jacobson, B.E., Nygaard, O.F., *Cancer Res.* (1977) **37**, 3306–3313.
A3215. Jonsson, J., Kammerer, R.C., Cho, A.K., *Res. Commun. Chem. Pathol. Pharmacol.* (1977) **18**, 75–82.
A3217. Kapitulnik, J., Levin, W., Lu, A.Y.H., Morecki, R., Dansette, P.M., Jerina, D.M., Conney, A.H., *Clin. Pharmacol. Ther.* (1977) **21**, 158–165.
A3218. Cerniglia, C.E., Gibson, D.T., *Appl. Environ. Microbiol.* (1977) **34**, 363–370.
A3219. Rao, D.N.R., Vaidyanathan, C.S., *Can. J. Microbiol.* (1977) **23**, 1496–1499.
A3220. Maillard, F., Zryd, J.P., *Can. J. Bot.* (1977) **55**, 2530–2534.
A3221. Schrull, R., Becker, H., *Planta Med.* (1977) **32**, 27–32.
A3222. Bhakuni, D.S., Singh, A.N., Tewari, S., Kapil, R.S., *J. Chem. Soc., Perkin Trans. 1* (1977) 1662–1666.
A3223. Jacques, D., Opie, C.T., Porter, L.J., Haslam, E., *J. Chem. Soc., Perkin Trans. 1* (1977) 1637–1643.
A3224. Bottoms, G.D., Coppoc, G.L., Monk, E., Moore, A.B. Roesel, O.F., Regnier, F.E., *J. Anim. Sci.* (1977) **45**, 674–685.
A3225. Ball, P., Stubenrauch, G., Knuppen, R., *J. Steroid Biochem.* (1977) **8**, 989–993.
A3226. Horie, M., Baba, S., *Chem. Pharm. Bull.* (1978) **26**, 1039–1043.

A3227. Yamane, Y., Sakai, K., Morita, I., *Chem. Pharm. Bull.* (1978) **26**, 479–503.
A3228. Nagamatsu, K., Kido, Y., Urakubo, G., *Chem. Pharm. Bull.* (1978) **26**, 1267–1273.
A3229. Maeda, T., Tanaka, M., Tanaka, K., Shindo, H., *J. Pharm. Dyn.* (1978) **1**, 288–300.
A3230. Sternson, L.A., Hincal, F., *Biochem. Pharmacol.* (1978) **27**, 1919–1924.
A3231. Wiebkin, P., Fry, J.R., Jones, C.A., Lowing, R.K., Bridges, J.W., *Biochem. Pharmacol.* (1978) **27**, 1899–1907.
A3232. Moldéus, P., *Biochem. Pharmacol.* (1978) **27**, 2859–2863.
A3233. Mandula, B.B., Pathak, M.A., *Biochem. Pharmacol.* (1979) **28**, 127–132.
A3234. Jennette, K.W., Bornstein, W., Chuang, A.H.L., Bresnick, E., *Biochem. Pharmacol.* (1979) **28**, 338–339.
A3235. Adam, H.K., Douglas, E.J., Kemp, J.V. *Biochem. Pharmacol.* (1979) **28**, 145–147.
A3236. McLennan, I.S., Lees, G.J., *J. Neurochem.* (1978) **30**, 429–436.
A3237. Sitaram, B.R., Lees, G.J. *J. Neurochem.* (1978) **31** 1021–1026.
A3238. Harvey, D.J., Martin, B.R., Paton, W.D.M., *J. Pharm. Pharmacol.* (1977) **29**, 495–497.
A3239. Yoshida, A., Yoshioka, M., Sakai, T., Tamura, Z., *Chem. Pharm. Bull.* (1978) **26**, 1177–1781.
A3240. Deleyn, F.C., Claeyssens, M., Van Beeumen, J., De Bruyne, C.K., *Can. J. Biochem.* (1978) **56**, 43–50.
A3241. Schut, H.A.J., Bowman, J.M., Solomon, S., *Can. J. Biochem.* (1978) **56**, 101–106.
A3242. Renwick, A.G., Ball, L.M., Corina, D.L., Williams, R.T., *Xenobiotica* (1978) **8**, 461–474.
A3243. Richter, W.J., Alt, K.O., Dieterle, W., Faigle, J.W., Kriemler, H.-P., Mory, H., Winkler, T., *Helv. Chim. Acta* (1975) **58**, 2512–2517.
A3244. Renwick, A.G., Williams, R.T., *Xenobiotica* (1978) **8**, 475–486.
A3245. Renwick, A.G., *Xenobiotica* (1978) **8**, 487–495.
A3246. Noordhoek, J., Van Den Berg, A.P., Savenije-Chapel, E.M., Koopman-Kool, E., *Xenobiotica* (1978) **8**, 515–522.
A3247. Vine. J., Morgan, D., Thomas, J., *Xenobiotica* (1978) **8**, 509–513.
A3248. Godeneche, D., Madelmont, J.C., Moreau, M.F., Plagne, R., *Xenobiotica* (1978) **8**, 597–603.
A3249. Emudianughe, T.S., Caldwell, J., Dixon, P.A.F., Smith, R.L., *Xenobiotica* (1978) **8**, 525–534.
A3250. Hutson, D.H., Casida, J.E., *Xenobiotica* (1978) **8**, 565–571.
A3251. Kobayashi, T., Tanayama, S., Kanai, Y., *Xenobiotica* (1978) **8**, 535–545.
A3252. Jori, A., De Ponte, P., Caccia, S., *Xenobiotica* (1978) **8**, 583–588.
A3253. Glowinski, I.B., Radtke, H.E., Weber, W.W., *Mol. Pharmacol* (1978) **14**, 940–949.
A3254. Zacchei, A.G., Wishousky, T.I., Dziewanowska, Z.E., De Schepper, P.J., Hitzenberger, G., *Eur. J. Drug Metab. Pharmacokinet.* (1977) **1**, 37–45.
A3255. Katayama, Y., Fukuzumi, T., *Mokuzai Gakkaishi* (1978) **24**, 643–649.
A3256. Stöckigt, J.M., *J. Chem. Soc., Chem. Commun.* (1978) 1097–1099.
A3257. Stöckigt, J.M., Zenk, M.H., *J. Chem. Soc., Chem. Commun.* (1978) 646–648.
A3258. Edwards, D.J., Rizk, M., *Clin. Chim. Acta* (1979) **95**, 1–10.
A3259. Niederwieser, A., Wadman, S.K., Danks, D.M., *Clin. Chim. Acta* (1978) **90**, 195–200.
A3260. Ventura, P., Cornelli, U., Pifferi, G., Frigerio, A., *Farmaco, Ed. Sci.* (1977) **32**, 640–649.
A3261. Goewert, R.R., Sippel, C.J., Grim, M.F., Olson, R.E., *FEBS Lett.* (1978) **87**, 219–221.
A3262. Baird, W.M., Chern, C.J., Diamond, L., *Cancer Res.* (1977) **37**, 3190–3197.

A3263. Wallnöfer, P.R., Tillmans, G., Engelhardt, G., *Pestic. Biochem. Physiol.* (1977) **7**, 481–485.
A3264. Jones, R.S., Mendis, D., Parke, D.V., *Biochim. Biophys. Acta* (1977) **500**, 124–131.
A3266. Dunn, T.G., Kaltenbach, C.C., Koritnik, D.R., Turner, D.L., Niswender, G.D., *J. Anim. Sci.* (1977) **45**, 659–673.
A3267. Nakai, Y., Kanai, Y., Tanayama, S., *Takeda Kenkyusho Hô* (1977) **36**, 46–52.
A3268. Hulhoven, R., Desager, J.P., Sokal, G., Harvengt, C., *Eur. J. Cancer* (1977) **13**, 1065–1069.
A3269. Mussini, E., Marcucci, F., Airoldi, L., Facchinetti, T., Garattini, S., *J. Pharm. Sci.* (1977) **66**, 1482–1483.
A3270. Nagy, A., Johansson, R., *Psychopharmacology (Berlin)* (1977) **54**, 125–131.
A3272. Jonsson, J., *Res. Commun. Chem. Pathol. Pharmacol.* (1977) **18**, 189–199.
A3273. Yamamoto, H., Kuki, M., Ozaki, M., *Oyo Yakuri* (1977) **14**, 97–108, 109–114.
A3274. Nagy, A., Treiber, L., *J. Pharm. Pharmacol.* (1973) **25**, 599–603.
A3275. Felt, G.R., Mueller, W.F., Iatropoulos, M.J., Coulston, F., Korte, F., *Toxicol. Appl. Pharmacol.* (1977) **41**, 619–627.
A3276. Hopper, D.J., Taylor, D.G., *Biochem. J.* (1977) **167**, 155–162.
A3277. Yang, S.K., Roller, P.P., Fu, P.P., Harvey, R.G., Gelboin, H.V., *Biochem. Biophys. Res. Commun.* (1977) **77**, 1176–1182.
A3278. Ohtsuki, K., Sakurai, Y., Shiota, H., *J. Ferment. Technol.* (1977) **55**, 476–484.
A3279. Krishnan Kutty, R., Abitha Devi, N., Veeraswamy, M., Ramesh, S., Subba Rao, P.V., *Biochem. J.* (1977) **167**, 163–170.
A3280. Nagy, A., Johansson, R., *Naunyn-Schmiedeberg's Arch. Pharmacol.* (1975) **290**, 145–150.
A3281. Höller, M., Grochtmann, W., Napp, M., Breyer, H., *Biochem. J.* (1977) **166**, 363–371.
A3283. Pantarotto, C., Belvedere, G., Frigerio, A., *Farmaco, Ed. Prat.* (1976) **31**, 509–515.
A3284. Messenger, A.J.M., Turner, J.M., *Biochem. Soc. Trans.* (1979) **6**, 1326–1328.
A3285. Gönechea, S., Eckhardt, G., Göbel, K.J., *J. Clin. Chem. Clin. Biochem.* (1977) **15**, 489–498.
A3286. Stewart, D.J., Inaba, T., Kalow, W., *J. Pharmacol. Exp. Ther.* (1978) **207**, 171–177.
A3287. Pang, K.S., Gillette, J.R., *J. Pharmacol. Exp. Ther.* (1978) **207**, 178–194.
A3288. Nozu, T., Aoyagi, T., Setoyama, K., Sawa, T., Tanaka, I., *Oyo Yakuri* (1977) **14**, 1–14.
A3289. Kuhn, C.M., Schanberg, S.M., *J. Pharmacol. Exp. Ther.* (1978) **207**, 544–554.
A3290. Helton, E.D., Williams, M.C., Goldzieher, J.W., *Steroids* (1977) **30**, 71–83.
A3291. Ernst, W., Goerke, H., Weber, K., *Chemosphere* (1977) **6**, 559–568.
A3292. Bates, G.W., Edman, C.D., Porter, J.C., MacDonald, P.C., *J. Clin Endocrinol. Metab.* (1977) **45**, 1120–1123.
A3293. Hoyle, M.C., *Plant Physiol.* (1977) **60**, 787–793.
A3294. Breckwoldt, M., Zahradnik, H.P., Lee, Z.Y., *Proc. Serono Symp.* (1976) **8** (Ovul. Hum.) 71–78.
A3295. Paul, S.M., Axelrod, J., Dilberto, E.J., Jr., *Endocrinology (Philadelphia)* (1977) **101**, 1604–1610.
A3296. Yeh, S.Y., Gorodetzky, C.W., Krebs, H.A., *J. Pharm. Sci.* (1977) **66**, 1288–1293.
A3297. Lau, S.S., Domino, E.F., *Biomed. Mass Spectrom.* (1977) **4**, 317–321.
A3298. Beckett, A.H., Achari, R., *J. Pharm. Pharmacol.* (1977) **29**, 589–592.
A3299. Davis, C.M., Fenimore, D.C., *J. Pharm. Pharmacol.* (1977) **29**, 573–574.
A3300. Beckett, A.H., Achari, R., *J. Pharm. Pharmacol.* (1977) **29**, 645.

A3301. Borchert, H.H., Bornschein, I., Pfeifer, S., Kraft, R., *Pharmazie* (1977) **32**, 489–496.
A3302. Tohno, M., Doi, Y., Kajikawa, N., Ofuji, T., Nadai, T., *Oyo Yakuri* (1977) **14**, 289–302.
A3303. Jindra, A., Vargova, E., Helia, O., *Pharmazie* (1977) **32**, 496–498.
A3304. Sakai, Y., Kajikawa, N., Tohno, M., Ofuji, T., *Oyo Yakuri* (1977) **14**, 303–308.
A3305. Midha, K.K., Hindmarsh, K.W., McGilveray, I.J., Cooper, J.K., *J. Pharm. Sci.* (1977) **66**, 1596–1602.
A3306. Sakai, S., *Appl. Environ. Microbiol.* (1977) **34**, 500–505.
A3307. Metzler, M., Müller, W., Hobson, W.C., *J. Toxicol. Environ. Health* (1977) **3**, 439–450.
A3308. Hussein, Y.A., *Phyton* (*Buenos Aires*) (1977) **35**, 79–84.
A3309. Leinweber, F.J., Greenough, R.C., Di Carlo, F.J., *J. Pharm. Sci.* (1977) **66**, 1570–1577.
A3310. Du Suich, P., Erill, S., *Clin. Pharmacol. Ther.* (1977) **22**, 588–595.
A3311. Kazyak, L., Kelley, J.A., Cella, J.A., Droege, R.E., Hilpert, L.R., Permisohn, R.C., *Clin. Chem.* (1977) **23**, 2001–2006.
A3312. Cavalieri, R.R., Gavin, L.A., Bui, F., McMahon, F., Hammond, M., *Biochem. Biophys. Res. Commun.* (1977) **79**, 897–902.
A3313. Razzaque, A., Ellis, B.E., *Planta* (1977) **137**, 287–291.
A3315. Beckett, A.H., Jones, G.R., *J. Pharm. Pharmacol.* (1977) **29**, 416–421.
A3316. Goeber, B., Brandt, K.P., Pfeifer, S., Otto, A., *Pharmazie* (1977) **32**, 543–544.
A3318. Shader, R.I., Greenblatt, D.J., Harmatz, J.S., Franke, K., Koch-Weser, J., *J. Clin. Pharmacol.* (1977) **17**, 709–718.
A3319. Takada, M., *J. Toxicol. Sci.* (1976) **1**, 26–31.
A3321. Prough, R.A., Sipal, Z., Jakobsson, S.W., *Life Sci.* (1977) **21**, 1629–1635.
A3322. Nakamura, S., Sugihara, J., Takahashi, T., *Oyo Yakuri* (1977) **14**, 515–522.
A3323. Billings, R.E., McMahon, R.E., *Mol. Pharmacol.* (1978) **14**, 145–154.
A3324. Berlin-Wahlén, A., Sandberg, R., *Acta Pharm. Suec.* (1977) **14**, 321–334.
A3325. Kehr, W., Zimmermann, R., Thiede, M., *Naunyn-Schmiedeberg's Arch. Pharmacol.* (1977) **300**, 19–23.
A3326. Woods, L.F.J., Wiseman, A., *Proc. Chem. Soc.* (1979) **7**, 124–127.
A3328. Collett, R.A., Ungkitchanukit, A., *Biochem. Soc. Trans.* (1979) **7**, 132–134.
A3329. Reynolds, G.P., Riederer, P., Sandler, M., *Biochem. Soc. Trans.* (1979) **7**, 143–145.
A3330. Suzuki, T., *J. Pestic. Sci.* (1978) **3**, 441–443.
A3331. Møller, B.L., *Analyt. Biochem.* (1977) **81**, 292–304.
A3332. Nakazawa, T., Hayashi, E., *Appl. Environ. Microbiol.* (1978) **36**, 264–269.
A3333. Tyce, G.M., Sharpless, N.S., Owen, C.A., Jr., *Am. J. Physiol.* (1978) **235**, E150-E157.
A3334. Buu, N.T., Kuchel, O., *Life Sci.* **24**, 783–790 (1979) and *Adv. Biosci.* (1979) **20**, 95–100.
A3335. Møller, B.L., Conn, E.E., *J. Biol. Chem.* (1979) **254**, 8575–8583.
A3336. Yokoyama, T., Hayase, Y., Aikawa, H., Odaka, Y., Miyamoto, S., Mory, Y., *Oyo Yakuri* (1977) **13**, 457–464.
A3337. Dencker, H., Dencker, S.J., Green, A., Nagy, A., *Clin. Pharmacol. Ther.* (1976) **19**, 584–586.
A3338. Bakke, J.E., Price, C.E., *J. Environ. Sci. Health, Pt. B* (1977) **B12**, 251–260.
A3339. Weisz, J., O'Brien, L.V., Lloyd, T., *Endocrinol.* (*Philadelphia*) (1978) **102**, 330–333.
A3340. Laborde, A.L., Gibson, D.T., *Appl. Environ. Microbiol.* (1977) **34**, 783–790.
A3341. Mitsuda, H., Nakajima, K., Yamada, Y., *J. Nutr. Sci. Vitaminol.* (1977) **23**, 413–422.

A3342. Horitsu, H., Takada, M., Idaka, E., Tomoyeda, M., Ogawa, T., *Eur. J. Appl. Microbiol.* (1977) **4**, 217–224.
A3343. Li, C.Y., *Lloydia* (1977) **40**, 298–300.
A3344. Ball, P., Haupt, M., Knuppen, R., *Acta Endocrinol. (Copenhagen)* (1978) **87**, 1–11.
A3345. Borkowski, P.R., Horn, J.S., Rapoport, H., *J. Am. Chem. Soc.* (1978) **100**, 276–281.
A3346. Teng, L., Bruce, R.B., Dunning, L.K., *J. Pharm. Sci.* (1977) **66**, 1615–1618.
A3347. Nozu, T., Suwa, T., Setoyama, K., Kurachi, M., Tanaka, I., *Yakugaku Zasshi* (1977) **97**, 1189–1194.
A3348. Kellner, H.-M., Baeder, C., Christ, O., Heptner, W., Hornke, I., Ings, R.M.J., *Br. J. Clin. Pharmacol.* (1977) **4** (Suppl 2) , 109S-116S.
A3349. Butler, T.C., Dudley, K.H., Mitchell, G.N., Johnson, D., *Arch. Int. Pharmacodyn. Ther.* (1977) **228**, 4–9.
A3350. Sato, Y., Meshi, T., Nakamura, S., Kawanishi, M., *Oyo Yakuri* (1977) **13**, 649–657.
A3351. Pekas, J.C., Bakke, J.E., Giles. J.L., Price, C.E., *J. Environ. Sci. Health, Pt B.* (1977) **B12**, 261–268.
A3352. Cerniglia, C.E., Gibson, D.T., *Arch. Biochem. Biophys.* (1978) **186**, 121–127.
A3353. Casey, J., Threlfall, D.R., *FEBS Lett.* (1978) **85**, 249–253.
A3356. Angelo, M.M., Idle, J.R., *Comp. Biochem. Physiol. C* (1977) **58**, 57–59.
A3358. Pfeifer, S., Bornschein, I., Gehrke, A., *Pharmazie* (1977) **32**, 587–592.
A3359. Cone, E.J., Phelps, B.A., Gorodetzky, C.W., *J. Pharm. Sci.* (1977) **66**, 1709–1713.
A3360. Hindmarsh, K.W., Hamon, N.W., Le Gatt, D.F., *Can. J. Pharm. Sci.* (1977) **12**, 103–105.
A3361. Yokoshima, T., Nanpo, T., Takemoto, M., Mitsugi, K., Itoyama, H., Uchimura, K., *Iyakuhin Kenkyu* (1976) **7**, 392–405.
A3362. Saleh, N.A.M., Fritsch, H., Kreuzaler, F., Grisebach, H., *Phytochemistry* (1978) **17**, 183–186.
A3363. Braier, L., *Rev. Bras. Pesqui. Med. Biol.* (1977) **10**, 319–323.
A3364. Harper, D.B., *Biochem. J.* (1977) **167**, 685–692.
A3365. Ibrahim, R.K., *Z. Pflanzenphysiol.* (1977) **85**, 253–262.
A3366. Lynn, R., Rueter, M.E., Guynn, R.W., *J. Neurochem.* (1977) **29**, 1147–1149.
A3367. Iida, M., Iizuki, H., *Z. Allg. Mikrobiol.* (1977) **17**, 507–512.
A3368. Fell, V., Hoskins, J.A., Pollitt, R.J., *Clin. Chim. Acta* (1978) **83**, 259–269.
A3369. Leinweber, F.S., Szpiech, J.M., Di Carlo, F.J., *Pharmacology* (1978) **16**, 70–77.
A3370. Sellman, R., Hurme, M., Kanto, J., *Eur. J. Clin. Pharmacol.* (1977) **12**, 209–212.
A3371. Wagner, J., Faigle, J.W., Imhof, P., Liehr, G., *Arzneim.-Forsch.* (1977) **27**, 2388–2395.
A3372. Nakagawa, M., Ando, M., *Agric. Biol. Chem.* (1977) **41**, 1975–1984.
A3374. Jacobson, K.B., *Arch. Biochem. Biophys.* (1978) **186**, 84–88.
A3375. Ikeno, T., Hashimoto, S., Kuzuya, H., Nagatsu, T., *Mol. Cell Biochem.* (1977) **18**, 117–123.
A3376. Shane, B., Stokstad, E.L.R., *J. Gen. Microbiol.* (1977) **103**, 261–270.
A3377. Yuasa, K., Ishizuka, K., Sakasai, T., *Agric. Biol. Chem.* (1978) **42**, 167–169.
A3378. Bower, P.J., Brown, H.M., Purves, W.K., *Plant Physiol.* (1978) **61**, 107–110.
A3379. Heckendorf, A.H., Hutchinson, C.R., *Tetrahedron Lett.* (1977) 4153–4154.
A3380. Fujita, M., Inoue, T., *Tetrahedron Lett.* (1977) 4503–4506.
A3381. Raffaeli, E., Maffei Fancino, R., Salmona, M., Pitré, D., *Pharmacol. Res. Commun.* (1977) **9**, 833–846.
A3382. Heykants, J.J.P., Meuldermans, W.E.G., Knaeps, A.G., Michiels, L.J.M., *Drug Metab. Pharmacokinet.* (1977) **2**, 81–91.
A3383. Constantin, M., Pognat, J.F., *Arzneim.-Forsch.* (1978) **28**, 64–72.

A3384. Eckhardt, G., Gönechea, S., Tondorf, G., *J. Clin. Chem. Clin. Biochem.* (1977) **15**, 687–690.

A3385. Ohnishi, H., Tanaka, S., Nagata, T., Takahashi, M., Ishiyama, Y., Hayashi, Y., Kamiyama, S., Kosuzume, H., *Iyakuhin Kenkyu* (1975) **6**, 426–433.

A3386. Chang, S.C., *J. Econ. Entomol.* (1978) **71**, 31–39.

A3387. Phillips, J.C., Topp, C.S., Gangolli, S.D., *Toxicology Lett.* (1978) **2**, 237–242.

A3388. Donner, P., Wagner, H., Kröger, H., *Biochem. Biophys. Res. Commun.* (1978) **80**, 766–772.

A3389. Duffey, S.S., Towers, G.H.N., *Can. J. Zool.* (1978) **56**, 7–16.

A3390. Tulp, M.T.M., Olie, K., Hutzinger, O., *Biomed. Mass Spectrom.* (1977) **4**, 310–316.

A3391. Pollock, R.J., Kaufman, S., *J. Neurochem.* (1978) **30**, 253–256.

A3392. Kennedy, K.A., Halmi, K.A., Fisher, L.J., *Life Sci.* (1977) **21**, 1813–1819.

A3393. Ishii, Y., Sakai, Y., Matsuda, T., Nozu, K., Inoue, M., Tono, M., Ofuji, T., *Oyo Yakuri* (1975) **10**, 645–652.

A3394. Schutz, H.W., *Beitr. Gerichtl. Med.* (1977) **35**, 85–87.

A3395. Inoue, S., Sugiyama, M., Tatewaki, N., Ando, T., Kono, T., Fujioka, M., *Iyakuhin Kenkyu* (1975) **6**, 267–276.

A3396. Graber, H., Benesch, L., Arr, M., Magyar, T., *Int. J. Clin. Pharmacol. Biopharm.* (1978) **16**, 59–62.

A3397. Adddison, R.F., Willis, D.E., *Toxicol. Appl. Pharmacol.* (1978) **43**, 303–315.

A3398. Tobes, M.C., Mason, M., *Life Sci.* (1978) **22**, 793–801.

A3399. Nakagawa, H., Machita, M., Sawara, S., *Kokuritsu Kekkaku Ryoyosho Kyoda Kenju Nempo* (1975) **8**, 49–55.

A3400. Kaplan, M.M., Utiger, R.D., *J. Clin. Invest.* (1978) **61**, 459–471.

A3401. Spector, R., Levy, P., Abelson, H.T., *J. Neurochem.* (1977) **29**, 919–921.

A3402. Smith, R.G., Doyle, D.G., Jr., Lynn, R.K., Gerber, N., *Biomed. Mass Spectrom.* (1977) **4**, 275–279.

A3403. Weinshilboum, R.M., Dunnette, J., Raymond, F., Kleinberg, F., *Experientia* (1978) **34**, 310–311.

A3404. Leinweber, F.J., Szpiech, J.M., Di Carlo, F.J., *J. Pharm. Sci.* (1978) **67**, 129–131.

A3405. Harvey, D.J., Martin, B.R., Paton, W.D.M., *Biomed. Mass Spectrom.* (1977) **4**, 364–370.

A3406. Hayashi, R., Okada, M., Nakazawa, T., Tanaka, Y., Nobuhiro, T., Miyakoshi, N., Sato, Y., *Oyo Yakuri* (1975) **10**, 761–771.

A3407. Ono, K., Okazaki, O., Saito, T., Sano, M., Akimoto, T., *Oyo Yakuri* (1976) **12**, 269–277.

A3408. Nakagawa, H., Matsubara, Y., Yamano, S., Hiraoka, K., Suga, T., *Oyo Yakuri* (1975) **10**, 841–848.

A3409. Kôsaka, K., Nakagawa, H., Matsubara, Y., Matsumoto, Y., Suga, T., *Oyo Yakuri* (1975) **10**, 459–464.

A3410. Wyndham, C., Safe, S., *Biochemistry* (1978) **17**, 208–215.

A3411. Silas, J.H., Lennard, M.S., Tucker, G.T., Smith, A.J., Malcolm, S.L., Marten, T.R., *Br. J. Clin. Pharmacol.* (1978) **5**, 27–34.

A3412. Selkirk, J.K., *Nature* (1977) **270**, 604–607.

A3413. Tulp. M.T.M., Schmitz, R., Hutzinger, O., *Chemosphere* (1978) **7**, 103–108.

A3414. Lappe, U., Barz, W., *Z. Naturforsch. C: Biosci.* (1978) **33C**, 301–302.

A3415. Kay, M.H., Radford, C., Alworth, W.L., *J. Chem. Soc., Chem. Commun.* (1978) 22–23.

A3416. Noguchi, T., Minatogawa, Y., Takada, Y., Okuno, E., Kido, R., *Biochem. J.* (1978) **170**, 173–175.

A3417. Ernst, W., *Veroeff. Inst. Meeresforsch. Bremerhaven.* (1979) **17**, 233–240.

A3418. Ernst, W., Goerke, H., *Marine Biol.* (1974) **24**, 287–304.

A3419. Metzler, M., McLachlan, J.A., *Biochem. Biophys. Res. Commun.* (1978) **85**, 874–884.
A3420. Harper, D.B., *Biochem. J.* (1977) **165**, 309–319.
A3421. Lu, A.Y.H., *J. Biol. Chem.* (1975) **250**, 8283–8288.
A3422. Moore, B.P., Cohen, G.M., Bridges, J.W., Parke, D.V., *Biochem. Soc. Trans.* (1977) **5**, 1368–1370.
A3423. Yesair, D.W., McNitt, S., Bittman, L., *Microsomes Drug. Oxid., Proc. Int. Symp., 3rd* (1976) 688–697.
A3425. Harvey, D.J., Paton, W.D.M., *J. Pharm. Pharmacol.* (1977) **29**, 498–500.
A3426. Cerniglia, C.E., Morgan, J.C., Gibson, D.T., *Biochem. J.* (1979) **180**, 175–185.
A3427. Dodge, R.H., Cerniglia, C.E., Gibson, D.T., *Biochem. J.* (1979) **178**, 223–230.
A3428. Hoyle, M.C., *Plant Physiol.* (1972) **50**, 15–18.
A3429. Gove, J.P., Hoyle, M.C., *Plant Physiol.* (1975) **56**, 684–687.
A3430. Cerniglia, C.E., Hebert, R.L., Szaniszlo, P.J., Gibson, D.T., *Arch. Microbiol.* (1978) **117**, 135–143.
A3431. Hindmarsh, K.W., Seifert, B.W., Watrous, W.M., Johnson, D.D., Crawford, R.D., *Can. J. Pharmaceut. Sci.* (1978) **13**, 59–60.
A3432. Chang, S.C., Stokes, J.B., *J. Econ. Entomol.* (1979) **72**, 15–19.
A3433. Yuasa, K., Ishizuka, K., Kaburaki, S., Sakasai, T., *Agric. Biol. Chem.* (1976) **40**, 1679–1685.
A3434. Cone, E.J., Darwin, W.D., Gorodetzky, C.W., Tan, T., *Drug Metab. Dispos.* (1978) **6**, 488–493.
A3435. Brunet, G., Saleh, A.M., Ibrahim, R.K., *Z. Naturforsch.* (1978) **33C**, 786–788.
A3436. Yisak, W.-A., Widman, M., Lindgren, J.-E., Agurell, S., *J. Pharm. Pharmacol.* (1977) **29**, 487–490.
A3437. Tsang, Y.-F., Ibrahim, R.K., *Z. Naturforsch.* (1979) **34C**, 46–50.
A3438. Andresen, B.D., Davis, F.T., Long, M.D., *J. Pharm. Sci.* (1979) **68**, 283–288.
A3439. Pendleton, R.G., Gessner, G., Weiner, G., Jenkins, B., Sawyer, J., Bondinell, W., Intoccia, A., *J. Pharmacol. Exp. Ther.* (1979) **208**, 24–30.
A3440. Van Cantfort, J., Manil, L., Gielen, J.E., Glatt, H.R., Oesch, F., *Biochem. Pharmacol.* (1979) **28**, 455–460.
A3441. Friedman, P.A., Smith, M.W., *Biochem. Pharmacol.* (1979) **28**, 937–938.
A3442. Hegstrand, L.R., Simon, J.R., Roth, R.H., *Biochem. Pharmacol.* (1979) **28**, 519–523.
A3443. Fuller, R.W., Molloy, B.B., Hemrick, S.K., *Biochem. Pharmacol.* (1979) **28**, 528–530.
A3444. Wong, K.P., *Biochem. J.* (1978) **174**, 777–782.
A3445. Traficante, L.J., Siekierski, J., Sakalis, G., Gershon, S., *Biochem. Pharmacol.* (1979) **28**, 621–626.
A3446. Nelson, S.D., Garland, W.A., Mitchell, J.R., Vaishnav, Y., Statham, C.N., Buckpitt, A.R., *Drug Metab. Dispos.* (1978) **6**, 363–367.
A3447. Warner, M., La Marca, M., Neims, A.H., *Drug Metab. Dispos.* (1978) **6**, 353–362.
A3448. Patel, J.M., Harper, C., Drew, R.T., *Drug Metab. Dispos.* (1978) **6**, 368–374.
A3449. Pallante, S., Lyle, M.A., Fenselau, C., *Drug Metab. Dispos.* (1978) **6**, 389–395.
A3450. Janssen, F.W., Kirkman, S.K., Knowles, J.A., Ruelius, H.W., *Drug Metab. Dispos.* (1978) **6**, 465–475.
A3451. Kammerer, R.C., Cho, A.K., Jonsson, J., *Drug Metab. Dispos.* (1978) **6**, 396–402.
A3452. Kadar, D., Lovrics, A.J., Endrenyi, L., *Drug Metab. Dispos.* (1978) **6**, 476–480.
A3453. Lynn, R.K., Smith, R.G., Thompson, R.M., Deinzer, M.L., Griffin, D., Gerber, N., *Drug Metab. Dispos.* (1978) **6**, 494–501.
A3454. Demisch, L., Kaczmarczyk, R., Seiler, N., *Drug Metab. Dispos.* (1978) **6**, 507–509.
A3455. Piafsky, K.M., Rane, A., *Drug Metab. Dispos.* (1978) **6**, 502–503.

A3456. Illing, H.P.A., Fromson, J.M., *Drug Metab. Dispos.* (1978) **6**, 510–517.
A3457. Mayo, B.C., Brodie, R.R., Chasseaud, L.F., Hawkins, D.R., *Drug Metab. Dispos.* (1978) **6**, 518–527.
A3458. Branfman, A.R., Bruni, R.J., Reinhold, V.N., Silveira, D.M., Chadwick, M., Yesair, D.W., *Drug Metab. Dispos.* (1978) **6**, 542–548.
A3459. Chadwick, M., Silveira, D.M., Platz, B.B., Hayes, D., *Drug Metab. Dispos.* (1978) **6**, 528–541.
A3460. Kao, J., Faulkner, J., Bridges, J.W., *Drug Metab. Dispos.* (1978) **6**, 549–555.
A3461. Mukhtar, H., Philpot, R.M., Bend, J.R., *Drug Metab. Dispos.* (1978) **6**, 577–583.
A3462. Maguire, J.H., Dudley, K.H., *Drug Metab. Dispos.* (1978) **6**, 601–605.
A3463. Lucier, G.W., McDaniel, O.S., Schiller, C.M., Matthews, H.B., *Drug Metab. Dispos.* (1978) **6**, 584–590.
A3464. Andersson, B., Berggren, M., Moldeús, P., *Drug Metab. Dispos.* (1978) **6**, 611–616.
A3465. Midha, K.K., Hubbard, J.W., Bailey, K., Cooper, J.K., *Drug Metab. Dispos.* (1978) **6**, 623–630.
A3466. Ruenitz, P.C., *Drug Metab. Dispos.* (1978) **6**, 631–636.
A3467. Vickers, S., Duncan, C.A., White, S.D., Breault, G.O., Royds, R.B., De Schepper, P.J., Tempero, K.F., *Drug Metab. Dispos.* (1978) **6**, 640–646.
A3468. Hucker, H.B., Stauffer, S.C., Balletter, A.J., White, S.D., Zacchei, A.G., Arison, B.H., *Drug Metab. Dispos.* (1978) **6**, 659–672.
A3469. Schwartz, M.A., Williams, T.H., Kolis, S.J., Postma, E., Sasso, G.J., *Drug Metab. Dispos.* (1978) **6**, 647–653.
A3470. Müller, W.-U., Leistner, E., *Phytochemistry* (1978) **17**, 1735–1738.
A3471. Liljegren, D.R., *Phytochemistry* (1978) **17**, 1695–1699.
A3472. Müller, W.-U., Leistner, E., *Phytochemistry* (1978) **17**, 1739–1742.
A3473. Shaw, G.J., Chick, M., Hodges, R., *Phytochemistry* (1978) **17**, 1743–1745.
A3474. Dewick, P.M., Ward, D., *Phytochemistry* (1978) **17**, 1751–1754.
A3475. Molderez, M., Nagels, L., Parmentier, F., *Phytochemistry* (1978) **17**, 1747–1750.
A3476. Shaw, G.J., Milne, G.W.A., Minghetti, A., *Phytochemistry* (1979) **18**, 178–179.
A3477. Charrière-Ladreix, Y., *Phytochemistry* (1979) **18**, 43–45.
A3478. Nielsen, P., Dalgaard, L., *Xenobiotica* (1978) **8**, 657–664.
A3479. Lasker, J., Rickert, D.E., *Xenobiotica* (1978) **8**, 665–672.
A3480. Gardner, D.M., Renwick, A.G., *Xenobiotica* (1978) **8**, 679–690.
A3481. Dawes, C.P., James, S.P., Majer, J.R., *Xenobiotica* (1978) **8**, 673–678.
A3482. Risdall, P.C., Adams, S.S., Crampton, E.L., Marchant, B., *Xenobiotica* (1978) **8**, 691–704.
A3483. Shimizu, T., Mori, H., Tabusa, E., Nakahawa, K., *Xenobiotica* (1978) **8**, 705–710.
A3484. Gorrod, J.W., Temple, D.J., Beckett, A.H., *Xenobiotica* (1979) **9**, 17–25.
A3485. Rasmussen, A.K., Scheline, R.R., Solheim, E., Hänsel, R., *Xenobiotica* (1979) **9**, 1–16.
A3486. Kanai, Y., Nakai, Y., Nakajima, N., Shigeharu, T., *Xenobiotica* (1979) **9**, 33–50.
A3487. Mahgoub, A., Idle, J.R., Smith, R.L., *Xenobiotica* (1979) **9**, 51–56.
A3488. Beckett, A.H., Navas, G.E., *Xenobiotica* (1978) **8**, 721–736.
A3489. Battistini, C., Bellucci, G., Mastrorilli, E., *Xenobiotica* (1979) **9**, 57–61.
A3490. Vose, C.W., Prout, M., Haskins, N.T., Ford, C., Palmer, R.F., Tidd, M.J., *Xenobiotica* (1978) **8**, 745–752.
A3491. Kobayashi, T., Kanai, Y., Tanayama, S., *Xenobiotica* (1978) **8**, 737–744.
A3492. Holder, G.M., Plummer, J.L., Ryan, A.J., *Xenobiotica* (1978) **8**, 761–768.

A3493. Tong, J.H., Smyth, R.G., D'Iorio, A., *Biochem. Pharmacol.* (1979) **28**, 1029–1036.
A3494. Kaplan, E., Gutmann, H.R., *Biochem. Pharmacol.* (1979) **28**, 1609–1614.
A3495. Hultmark, D., Sundh, K., Johansson, L., Arrhenius, E., *Biochem. Pharmacol.* (1979) **28**, 1587–1590.
A3496. Stegeman, J.J., Binder, R.L., *Biochem. Pharmacol.* (1979) **28**, 1686–1688.
A3497. Meyerson, L.R., Cashaw, J.L., McMurtrey, K.D., Davis, V.E., *Biochem. Pharmacol.* (1979) **28**, 1745–1752.
A3498. Kazyak, L., Anthony, R.M., Sunshine, I., *J. Anal. Tox.* (1979) **3**, 67–71.
A3499. Caldwell, J., Emudianughe, T.S., *Biochem. Soc. Trans.* (1979) **7**, 521–522.
A3500. Ahmed, N.K., Felsted, R.L., Bachur, N.R., *J. Pharmacol. Exp. Ther.* (1979) **209**, 12–19.
A3501. Emudianughe, T.S., Caldwell, J., Smith, R.L., *Biochem. Soc. Trans.* (1979) **7**, 522–523.
A3502. Kapetanović, I.M., Strong, J.M., Mieyal, J.J., *J. Pharmacol. Exp. Ther.* (1979) **209**, 20–24.
A3503. Tulp, M.T.M., Sundström, G., Martron, L.B.J.M., Hutzinger, O., *Xenobiotica* (1979) **9**, 65–77.
A3504. Hoffman, K.-J., Arfwidsson, A., Borg, K.O., Skånberg, I., *Xenobiotica* (1979) **9**, 93–106.
A3505. Hoffman, K.-J., Skånberg, I., Borg, K.O., *Xenobiotica* (1979) **9**, 79–91.
A3506. Kimura, K., Nagata, A., Miyawaki, H., *Xenobiotica* (1979) **9**, 119–127.
A3507. Nicholls, P.J., Ross, M.S.F., Luscombe, D.K., *Xenobiotica* (1979) **9**, 129–140.
A3508. Frear, D.S., Swanson, H.R., Mansager, E.R., Wien, R.G., *J. Agric. Food Chem.* (1978) **26**, 1347–1351.
A3509. Motoyama, N., Dauterman, W.C., *J. Agric. Food Chem.* (1978) **26**, 1296–1301.
A3510. VandenHeuvel, W.J.A., Wolf, D.E., Arison, B.H., Buhs, R.P., Carlin, J.R., Ellsworth, R.L., Jacob, T.A., Koniuszy, F.R., Smith, J.L., Trenner, N.R., Walker, R.W., Wolf, F.J., *J. Agric. Food Chem.* (1978) **26**, 1357–1364.
A3511. Gaughan, L.C., Robinson, R.A., Casida, J.E., *J. Agric. Food Chem.* (1978) **26**, 1374–1380.
A3515. Bigley, W.S., Plapp, F.W., Jr., *J. Agric. Food Chem.* (1978) **26**, 1128–1134.
A3517. Israel, M., Wilkinson, P.M., Pegg, W.J., Frei, E. III, *Cancer Res.* (1978) **38**, 365–370.
A3518. Yadav, D.V., Agarwal, H.C., Pillai, M.K.K., *Bull. Environ. Contam. Tox.* (1978) **19**, 300–306.
A3519. Barford, P.A., Olavesen, A.H., Curtis, C.G., Powell, G.M., *Biochem. J.* (1977) **164**, 423–430.
A3520. Endo, Y., *Biochim. Biophys. Acta* (1978) **523**, 207–214.
A3521. Pollero, R., de Pollero, S.C., *Bull. Environ. Contam. Tox.* (1978) **19**, 345–350.
A3522. Di Carlo, F.J., Leinweber, F.J., Szpiech, J.M., Davidson, I.W.F., *Clin. Pharmacol. Ther.* (1977) **22**, 858–863.
A3523. Harzer, G., Rokos, H., Otto, M.K., Bacher, A., Ghisla, S., *Biochim. Biophys. Acta* (1978) **540**, 48–54.
A3524. Sugiyama, M., Okuyama, Y., Yoshimoto, Y., *Yakugaku Zasshi* (1978) **98**, 31–40.
A3525. Fahl, W.E., Jefcoate, C.R., Kasper, C.B., *J. Biol. Chem.* (1978) **253**, 3106–3113.
A3526. Kumar, A.A., Vaidyanathan, C.S., Rao, N.A., *Indian J. Biochem. Biophys.* (1978) **15**, 5–13.
A3527. Yamazoe, Y., Sugiura, M., Kamataki, T., Kato, R., *FEBS Lett.* (1978) **88**, 337–340.
A3528. Engelhardt, G., Wallnöfer, P.R., *Appl. Environ. Microbiol.* (1978) **35**, 243–246.
A3529. Yu, P.H., *Anal. Biochem.* (1978) **86**, 498–504.
A3530. Horvath, M.M., *Bot. Kozl.* (1977) **64**, 117–119.

A3531. Corbett, M.D., Chipko, B.R., Paul, J.H., *J. Environ. Pathol. Toxicol.* (1978) **1**, 259–266.

A3532. Aunis, D., Serck-Hanssen, G., Helle, K.B., *Gen. Pharmacol.* (1978) **9**, 37–43.

A3533. Kennedy, M.B., *J. Neurochem.* (1978) **30**, 315–320.

A3534. Davison, C., Scrime, M., Edelson, J., *Arch. Int. Pharmacodyn. Ther.* (1977) **230**, 4–18.

A3535. Gál, E.M., Sherman, A.D., *J. Neurochem.* (1978) **30**, 607–613.

A3537. Marshall, V.P., McGovren, J.P., Richard, F.A., Richard, R.E., Wiley, P.F., *J. Antibiot.* (1978) **31**, 336–342.

A3538. Horvath, M.M., *Acta Biol. Szeged.* (1977) **23**, 69–71.

A3540. Auvinen, J., Marvola, M., *Acta Pharm. Fenn.* (1978) **87**, 37–42.

A3541. Greenblatt, D.J., Shader, R.I., *Lancet i* (1978) 720.

A3543. Mercier, M.J., Razzouk, C., Riberfroid, M., *Proc. Eur. Soc. Toxicol.* (1977) **18**, 329–331.

A3545. Croteau, R., *J. Food Biochem.* (1977) **1**, 317–326.

A3546. Mavrides, C., Comerton, M., *Biochim. Biophys. Acta* (1978) **524**, 60–67.

A3547. Inaba, T., Stewart, D.J., Karlow, W., *Clin. Pharmacol. Ther.* (1978) **23**, 547–552.

A3548. Powell, J.T., Morrison, J.F., *Eur. J. Biochem.* (1978) **87**, 391–400.

A3549. Hüfner, M., Grussendorf, M., *Clin. Chim. Acta* (1978) **85**, 243–251.

A3550. Gulliver, P.A., Tipton, K.F., *J. Neurochem.* (1979) **32**, 1525–1529.

A3551. Fenn, P., Durbin, R.D., Kuntz, J.E., *Physiol. Plant Pathol.* (1978) **12**, 297–309.

A3552. Hakusui, H., Suzuki, W., Takegoshi, T., Saito, T., Sano, M., *Arzneim.-Forsch.* (1978) **28**, 456–461.

A3553. Ritschel, W.A., Brady, M.E., Tan, H.S.I., Hoffman, K.A., Yiu, I.M., Grummich, K.W., *Eur. J. Clin. Pharmacol.* (1977) **12**, 457–461.

A3554. Levron, J.C., Le Fur, J.M., Adnot, P., Dormard, Y., Cucumel, N., Enjoubault, G., *Eur. J. Drug Metab. Pharmacokinet.* (1977) **2**, 107–120.

A3555. Sugiyama, M., Tatewaki, N., Sugimoto, T., Okuyama, Y., Nakagawa, T., *Yakugaku Zasshi* (1978) **98**, 302–311.

A3556. Ledniczky, M., Szinai, I., Ujszászi, K., Holly, S., Kemeny, V., Mády, G., Ötvös, L., *Arzneim.-Forsch.* (1978) **28**, 673–677.

A3557. Rossi, E., De Pascale, A., Negrini, P., Zanol, M., Frigerio, A., *J. Chromatogr.* (1978) **152**, 228–233.

A3558. Lallemand, J.Y., Lemaitre, P., Beeley, L., Lesca, P., Mansuy, D., *Tetrahedron Lett.* (1978) 1261–1264.

A3559. Andreasen, F., Hansen, H.E., Mikkelsen, E., *Eur. J. Clin. Pharmacol.* (1978) **13**, 41–48.

A3560. Pettersen, J.E., Ulsaker, G.A., Jellum, E., *J. Chromatogr.* (1978) **145**, 413–420.

A3561. Glazko, A.J., Chang, T., Borondy, P.E., Dill, W.A., Young, R., Croskey, L., *Curr. Ther. Res.* (1978) **23**, 22–41.

A3562. Mandell, A.I., Stentz, F, Kitabchi, A.E., *Ophthalmology (Rochester, Minn.)* (1978) **85**, 268–275.

A3563. Thomas, R.C., Duchamp, D.J., Judy, R.W., Ikeda, G.J., *J. Med. Chem.* (1978) **21**, 725–732.

A3564. Dross, K., *Arzneim.-Forsch.* (1978) **28**, 824–827.

A3565. Niebch, Von G., Klingler, K.H., Eikelmann, G., Kucharczyk, N., *Arzneim.-Forsch.* (1978) **28**, 765–767.

A3566. Hashimoto, K., Sasaki, Y., Ohara, Y., Matsuzawa, T., *Nippon Suisan Gakkaishi* (1978) **44**, 623–629.

A3567. Levi, P., Janiaud, P., Delaforge, M., Morizot, J.P., Maume, B.F., Padieu, P., *C. R. Seances Soc. Biol. Ses. Fill.* (1977) **171**, 1034–1040.

A3568. Ahlborg, U.G., *Environ. Sci. Res.* (1978) **12**, 115–130.

A3569. Lech, J.J., Glickman, A.H., Statham, C.N., *Environ. Sci. Res.* (1978) **12**, 107–113.

A3570. Harada, I., Noguchi, T., Kido, R., *Hoppe-Seyler's Z. Physiol. Chem.* (1978) **359**, 481–488.

A3571. Wallnöfer, P., Söhlemann, F., Öhlmann, L., *Pflanzenschutz.-Nachr.* (1976) **29**, 236–253.

A3572. Thompson, H.J., Sharma, S.K., Brown, S.A., *Arch. Biochem. Biophys.* (1978) **188**, 271–281.

A3573. Allenmark, S., Servenius, B., *J. Chromatogr.* (1978) **153**, 239–245.

A3574. Cubbedu, L.X., Hoffmann, I.S., Ferrari, G.B., *J. Pharmacol. Exp Ther.* (1979) **209**, 165–175.

A3575. Illing, H.P.A., House, E.S.A., *Biochem. Soc. Trans.* (1979) **7**, 652–654.

A3576. Pinardi, G., Talmaciu, R.K., Santiago, E., Cubeddo, L.X., *J. Pharmacol. Exp Ther.* (1979) **209**, 176–184.

A3577. Küpfer, A., Bircher, J., *J. Pharmacol. Exp. Ther.* (1979) **209**, 190–195.

A3578. Loo, Y.H., Fulton, T., Miller, K.A., Wisniewski, H.M., *J. Neurochem.* (1979) **32**, 1699–1705.

A3579. Corbett, M.D., Baden, D.G., Chipko, B.R., *Bioorg. Chem.* (1979) **8**, 91–95.

A3580. Bjerrum, O.J., Helle, K.B., Bock, E., *Biochem. J.* (1979) **181**, 231–237.

A3581. Balemans, M.G.M., Noordegraaf, E.M., Bary, F.A.M., Van Berlo, M.F., *Experientia* (1978) **34**, 887–888.

A3582. Cripps, R.E., Trudgill, P.W., Whateley, J.G., *Eur. J. Biochem.* (1978) **86**, 175–186.

A3584. Roebuck, B.D., Siegel, W.G., Wogan, G.N., *Cancer Res.* (1978) **38**, 999–1002.

A3585. Barz, W., Schlepphorst, R., Wilhelm, P., Kratzl. K., Tengler, E., *Z. Naturforsch. C, Biosci.* (1978) **33C**, 363–367.

A3587. Feinstein, R.N., *Proc. Soc. Exp. Biol. Med.* (1978) **158**, 398–401.

A3588. Bhakuni, D.S., Singh, A.N., *J. Chem. Soc., Perkin Trans 1* (1978) 618–622.

A3589. Morishita, S., Baba, S., Nagase, Y., *J. Pharm. Sci.* (1978) **67**, 757–761.

A3590. Imondi, A.R., Alam, A.S., Brennan, J.J., Hagerman, L.M., *Arch. Int. Pharmacodyn. Ther.* (1978) **232**, 79–91.

A3591. Yeh, S.Y., McQuinn, R.L., Krebs, H.A., Gorodetzky, C.W., *J. Pharm. Sci.* (1978) **67**, 878–880.

A3592. Haegele, K.D., McLean, A.J., Du Suich, P., Barron, K., Laquer, J., McNay, J.L., Carrier, O., *Br. J. Clin. Pharmacol.* (1978) **5**, 489–494.

A3593. Van Boven, M., *Toxicol. Lett.* (1978) **2**, 17–21.

A3594. Meyer, E., Barz, W., *Planta Med.* (1978) **33**, 336–344.

A3595. Gioia, B., Frigerio, A., Vigevani, A., *Adv. Mass Spectrom.* (1978) **7B**, 1622–1627.

A3596. Harrison, J.M., Clarke, R.S., Inch, T.D., Upshall, D.G., *Experientia* (1978) **34**, 698–699.

A3598. Pezzuto, J.M., Yang, C.S., Yang, S.K., McCourt, D.W., Gelboin, H.V., *Cancer Res.* (1978) **38**, 1241–1245.

A3599. Petroski, R.J., Kelleher, W.J., *Lloydia* (1978) **41**, 332–341.

A3600. Basmadjian, G.P., Hussain, S.F., Paul, A.G., *Lloydia* (1978) **41**, 375–380.

A3601. Larsen, P.O., Wieczorkowska, E., *Biochim. Biophys. Acta* (1978) **542**, 253–262.

A3602. Wilson, J.T., Howell, R.L., Holladay, M.W., Brilis, G.M., Chrastil, J., Watson, J.T., Taber, D.F., *Clin. Pharmacol. Ther.* (1978) **23**, 635–643.

A3603. Schweitzer, J.W., Schwartz, R., Friedhoff, A.J., *Arch. Int. Pharmacodyn. Ther.* (1978) **231**, 21–29.

A3604. Ebel, S., Schütz, H., *Z. Rechtsmed.* (1978) **81**, 107–117.

A3605. Deutsch, J., Leutz, J.C., Yang, S.K., Gelboin, H.V., Chiang, Y.L., Vatsis, K.P., Coon, M.J., *Proc. Nat. Acad. Sci. U.S.A.* (1978) **75**, 3123–3127.

A3606. Shih, J.C., Chan, Y.-L., *Arch. Biochem. Biophys.* (1978) **189**, 454–464.

A3607. Nakadai, T., Nasuno, S., *Agric. Biol. Chem.* (1978) **42**, 1291–1292.

A3608. Hassam, S.B., Hutchinson, C.R., *Tetrahedron Lett.* (1978) 1681–1684.

A3609. Mackay, A.V.P., Davies, P., Dewar, A.J., Yates, C.M., *J. Neurochem.* (1978) **30**, 827–839.

A3611. Lane, J.D., Aprison, M.H., *J. Neurochem.* (1978) **30**, 671–678.

A3612. Borison, R.L., Maple, P.J., Havdala, H.S., Diamond, B.I., *Res. Commun. Chem. Pathol. Pharmacol.* (1978) **21**, 363–366.

A3613. Lee, P.W., Allahyari, R., Fukuto, T.R., *Pestic. Biochem. Physiol.* (1978) **9**, 23–32.

A3614. Gulliver, P.A., Tipton, K.F., *Eur. J. Biochem.* (1978) **88**, 439–444.

A3615. Shetty, A.S., Gaertner, F.H., *FEMS Microbiol. Lett.* (1978) **3**, 259–263.

A3616. Bolton, G.C., Griffiths, L.A., *Experientia* (1978) **34**, 1484–1486.

A3617. Noda, Y., Takai, K., Tokuyama, T., Narumiya, S., Ushiro, H., Hayaishi, O., *J. Biol. Chem.* (1978) **253**, 4819–4822.

A3618. Nelson S.D., Nelson, W.L., Trager, W.F., *J. Med. Chem.* (1978) **21**, 721–725.

A3619. Lai, F.M., Spector, S., *Arch. Int. Pharmacodyn. Ther.* (1978) **233**, 227–234.

A3620. Duboff, G.S., Zarafonetis, C.J.D., *Res. Commun. Chem. Pathol. Pharmacol.* (1978) **20**, 585.

A3621. Kammerer, R.C., Jonsson, J., Gal, J., Cho, A.K., *Life Sci.* (1978) **23**, 283–290.

A3622. Lind, C., Vadi, H., Ernster, L., *Arch. Biochem. Biophys.* (1978) **190**, 97–108.

A3623. Zutshi, U., Rao, P.G., Soni, A., Atal, C.K., *Indian J. Exp. Biol.* (1978) **16**, 836–838.

A3624. Watabe, T., Isobe, M., Yoshikawa, K., Takabatake, E., *J. Pharmacobio.-Dyn.* (1978) **1**, 98–104.

A3625. Moore, B.P., Cohen, G.M., *Cancer Res.* (1978) **38**, 3066–3075.

A3626. Rahimtula, A.D., O'Brien, P.J., Seifried, H.E., Jerina, D.M., *Eur. J. Biochem.* (1978) **89**, 133–141.

A3627. Pietrowski, R.A., Cartwright, N.J., *Microbios* (1977) **19**, 89–102.

A3628. Christie, R.M., Richards, R.W., Watson, W.P., *Aust. J. Chem.* (1978) **31**, 1799–1807.

A3629. Viukari, M., Linnoila, M., *Ann. Clin. Res.* (1977) **9**, 284–286.

A3631. Martin, L.E., Parke, D.V., Symons, A.M., Wade, A., *Br. J. Pharmacol.* (1978) **63**, 352P.

A3632. Reynolds, C.N., Wilson, K., Burnett, D., *Eur. J. Clin. Pharmacol.* (1978) **13**, 285–289.

A3633. Takada, M., *J. Toxicol. Sci.* (1978) **3**, 51–56.

A3634. Wilson, K., Reynolds, C.N., Burnett, D., *Eur. J. Clin. Pharmacol.* (1978) **13**, 291–297.

A3635. Pal, S.P., Mitra, R., *Calcutta Med. J.* (1977) **74**, 95–100.

A3636. Que, L., Jr., *Biochem. Biophys. Res. Commun.* (1978) **84**, 123–129.

A3637. Capdevila, J., Estabrook, R.W., Prough, R.A., *Biochem. Biophys. Res. Commun.* (1978) **83**, 1291–1298.

A3638. Sundhakar-Barik, Siddaramappa, R., Wahid, P.A., Sethunathan, N., *Antonie van Leeuwenhoek*, (1978) **44**, 171–176.

A3639. Crawford, R.L., *FEMS Microbiol. Lett.* (1978) **4**, 233–234.

A3640. Martin, M., Dewick, P.M., *Tetrahedron Lett.* (1978) 2341–2344.

A3641. Gupta, R.K., Haslam, E., *J. Chem. Soc., Perkin Trans 1* (1978) 892–896.

A3642. Ehrsson, H., Eksborg, S., Wallin, I., *Eur. J. Drug Metab. Pharmacokinet.* (1978) **3**, 125–128.

A3643. Strolin-Benedetti, M., Donath, A., Frigerio, A., Morgan, K.T., Laville, C., Malnoe, A., *Ann. Pharm. Fr.* (1978) **36**, 279–288.

A3644. Höller, M., Lisboa, B.P., Dengler, K., Breuer, H., *Biochem. Soc. Trans.* (1978) **6**, 136–138.

A3645. Aarbakke, J., *Acta Pharmacol. Toxicol.* (1978) **43**, 64–68.

A3646. Yasuhara, M., Kobayashi, H., Kimura, T., Muranishi, S., Sezaki, H., *J. Pharmacobio-Dyn.* (1978) **1**, 122–131.

A3647. Yasuhara, M., Kobayashi, H., Muranishi, S., Sezaki, H., Kimura, T., *J. Pharmacobio-Dyn.* (1978) **1**, 114–121.

A3648. Kujawa, M., Engst, R., Macholz, R., *Environ. Pollut. Hum. Health, Proc. Int. Symp., 1st* (1975) 661–672.

A3649. Bresnick, E., *Adv. Enzyme Regul.* (1978) **16**, 347–361.

A3650. Hutber, G.N., Lord, E.I., Loughman, B.C., *J. Exp. Bot.* (1978) **29**, 619–629.

A3652. Baker, S.J., Common, R.H., *Steroids* (1978) **32**, 95–107.

A3653. Ueda, T., Otsuka, H., Goda, K., Ishiguro, I., Naito, J., Kotake, Y., *J. Biochem.* (1978) **84**, 687–696.

A3654. Sawada, Y., Hayashi, M., Awazu, S., Hanano, M., *Chem. Pharm. Bull.* (1978) **26**, 1357–1363.

A3655. Liang, H.-J.L., Anderson, J.A., *Phytochemistry* (1978) **17**, 597–598.

A3656. Nagy, A., *Acta Pharmacol. Toxicol.* (1978) **42**, 68–72.

A3657. Cone, E.J., Gorodetzky, C.W., Darwin, W.D., Carroll, F.I., Brine, G.A., Welch, C.D., *Res. Commun. Chem. Path. Pharmacol.* (1978) **20**, 413–433.

A3658. Nakadai, T., Nasuno, S., *Agric. Biol. Chem.* (1977) **41**, 1657–1666.

A3659. Shetty, A.S., Gaertner, F.H., *J. Bacteriol.* (1975) **122**, 235–244.

A3660. Ahlborg, U.G., Larsson, K., Thunberg, T., *Arch. Toxicol.* (1978) **40**, 45–53.

A3661. Ahlborg, U.G., Larsson, K., Thunberg, T., *Arch. Toxicol.* (1978) **40**, 63–74.

A3662. Mansour, B., Ullrich, V., Pfleger, K., *Biochem. Pharmacol.* (1979) **28**, 2321–2326.

A3663. Kucharczyk, N., Yang, J., Sofia, R.D., *Biochem. Pharmacol.* (1979) **28**, 2219–2222.

A3664. Suzuki, O., Katsumata, Y., Oya, M., Matsumoto, T., *Biochem. Pharmacol.* (1979) **28**, 2327–2332.

A3665. Edwards, D.J., Pak, K.Y., Venetti, M.C., *Biochem. Pharmacol.* (1979) **28**, 2337–2343.

A3666. Hashimoto, M., Miyazaki, H., Nambu, K., Fujii, T., Arakawa, S., Yoshida, K., Utsui, Y., *Drug Metab. Dispos.* (1979) **7**, 20–23.

A3667. Suzuki, O., Oya, M., Katsumata, Y., *Biochem. Pharmacol.* (1979) **28**, 2682–2684.

A3668. Beaubien, A.R., Pakuts, A.P., *Drug Metab. Dispos.* (1979) **7**, 34–39.

A3669. Ruelius, H.W., Tio, C.O., Knowles, J.A., McHugh, S.L., Schillings, R.T., Sisenwine, S.F., *Drug Metab. Dispos.* (1979) **7**, 40–43.

A3670. Sullivan, H.R., Roffey, P., McMahon, R.E., *Drug Metab. Dispos.* (1979) **7**, 76–80.

A3671. Sum, C.Y., Cho, A.K., *Drug Metab. Dispos.* (1979) **7**, 65–69.

A3672. Woolley, J.L., Jr., Sigel, C.W., *Drug Metab. Dispos.* (1979) **7**, 94–99.

A3673. Dugger, H.A., Madrid, V.O., Talbot, K.C., Coombs, R.A., Orwig, B.A., *Drug Metab. Dispos.* (1979) **7**, 132–137.

A3674. Eichelbaum, M., Ende, M., Remberg, G., Schomerus, M., Dengler, H.J., *Drug Metab. Dispos.* (1979) **7**, 145–148.

A3675. Lynn, R.K., Bauer, J.E., Gordon, W.P., Smith, R.G., Griffin, D., Thompson, R.M., Jenkins, R., Gerber, N., *Drug Metab. Dispos.* (1979) **7**, 138–144.

A3676. Agurell, S., Edward, C., Halldin, M., Leander, K., Levy, S., Lindgren, J.-E., Mechoulam, R., Nordqvist, M., Ohlsson, A., *Drug Metab. Dispos.* (1979) **7**, 155–161.

A3677. Ishida, T., Oguri, K., Yoshimura, H., *Drug Metab. Dispos.* (1979) **7**, 162–165.

A3678. Pacifici, G.M., Boobis, A.R., Brodie, M.J., Davies, D.S., *Biochem. Soc. Trans.* (1979) **7**, 1060–1061.

A3679. Schlepphorst, R., Barz, W., *Planta Med.* (1979) **36**, 333–342.

A3680. Bridges, J.W., McKillop, D., Cox, P.J., Farmer, P.B., Foster, A.B., Jarman, M.,

Biochem. Soc. Trans. (1979) **7**, 1073–1075.

A3681. Mattison, D.R., West, D.R., Menard, R.H., *Biochem. Pharmacol.* (1979) **28**, 2101–2104.

A3682. Ritschel, W.A., Brady, M.E., Tan, H.S.I., *Int. J. Clin. Pharmacol. Biopharm.* (1979) **17**, 99–103.

A3683. Cawthon, R.M., Breakefield, X.O., *Nature* (1979) **281**, 692–694.

A3684. Holland, H.L., Jeffs, P.W., Capps, T.M., Maclean, D.B., *Can. J. Chem.* (1979) **57**, 1588–1597.

A3685. Gaertner, F.H., Shetty, A.S., *Biochim. Biophys. Acta* (1977) **482**, 453–460.

A3686. Raese, J.D., Edelman, A.M., Lazar, M.A., Barchas, J.D., *Catecholamines: Basic Clin. Front., Proc. Inst. Catecholamine Symp., 4th*, (1978) 46–48.

A3687. Joh, T.H., Park, D.H., Reis, D.J., *Catecholamines: Basic Clin. Front., Proc. Inst. Catecholamine Symp., 4th*, (1978) 43–45.

A3688. Glover, V., Elsworth, J.D., Sandler, M., *Catecholamines: Basic Clin. Front., Proc. Inst. Catecholamine Symp., 4th*, (1978) 201–203.

A3689. Kato, T., Nagatsu, T., Hashimoto, Y., Miyazaki, H., Levitt, M., Perel, J.M., *Catecholamines: Basic Clin. Front., Proc. Inst. Catecholamine Symp., 4th*, (1978) 144–146.

A3690. Lewinsohn, R., Glover, V., Böhm, K.-H., Sandler, M., *Catecholamines: Basic Clin. Front., Proc. Inst. Catecholamine Symp., 4th*, (1978) 207–209.

A3691. Boobis, A.R., Murray, S., Jones, H., Reid, J.L., Davies, D.S., *Catecholamines: Basic Clin. Front., Proc. Inst. Catecholamine Symp., 4th*, (1978) 213–215.

A3692. Gee, S.J., Krieger, R.I., *Xenobiotica* (1979) **9**, 459–465.

A3693. Bardsley, W.G., Ashford, J.S., Hill, C.M., *Biochem. J.* (1971) **122**, 557–567.

A3694. Douch, P.G.C., Buchanan, L.L., *Xenobiotica* (1979) **9**, 467–474.

A3695. Stubley, C., Stell, J.G.P., Mathieson, D.W., *Xenobiotica* (1979) **9**, 475–484.

A3696. Hackett, A.M., Marsh, I., Barrow, A., Griffiths, L.A., *Xenobiotica* (1979) **9**, 491–502.

A3697. Midha, K.K., Bailey, K., Cooper, J.K., Hubbard, J.W., *Xenobiotica* (1979) **9**, 485–490.

A3698. Ekonomov, A.L., Rodionov, A.P., Zherdev, V.P., Vikhlyaev, Y.I., *Xenobiotica* (1979) **9**, 503–510.

A3699. Polgár, M., Vereczkey, L., Szporny, L., Czira, G., Tamás, J., Gács-Baitz, E., Holly, S., *Xenobiotica* (1979) **9**, 511–520.

A3700. Cooper, S.F., Dugal, R., Bertrand, M.J., *Xenobiotica* (1979) **9**, 405–414.

A3701. Kitchen, I., Tremblay, J., André, J., Dring, L.G., Idle, J.R., Smith, R.L., Williams, R.T., *Xenobiotica* (1979) **9**, 397–404.

A3702. Kitagawa, H., Esumi, Y., Kurosawa, S., Sekine, S., Yokoshima, T., *Xenobiotica* (1979) **9**, 429–439.

A3703. Pan, H.P., Fouts, J.R., Devereux, T.R., *Xenobiotica* (1979) **9**, 441–446.

A3704. Grantham, P.H., Benjamin, T., Tahan, L.C., Roller, P.P., Miller, J.R., Weisburger, E.K., *Xenobiotica* (1979) **9**, 333–341.

A3705. Reeves, P.R., McCormick, D.J., Jepson, H.T., *Xenobiotica* (1979) **9**, 453–458.

A3706. Wilson, C.G., Parke, D.V., Green, J., Cawthorne, M.A., *Xenobiotica* (1979) **9**, 343–351.

A3707. Portig, J., Kraus, P., Stein, K., Koransky, W., Noack, G., Gross, B., Sodomann, S., *Xenobiotica* (1979) **9**, 353–378.

A3708. Seutter-Berlage, F., Delbressine, L.P.C., Smeets, F.L.M., Wagenaars-Zegers, M.A.P., *Xenobiotica* (1979) **9**, 311–316.

A3709. Malnoë, A., Strolin-Benedetti, M., *Xenobiotica* (1979) **9**, 281–288.

A3710. Melancon, M.J., Jr., Lech, J.J., *Xenobiotica* (1979) **9**, 317–322.

A3711. Woolhouse, N.M., Kaye, B., Parke, D.V., *Xenobiotica* (1979) **9**, 323–332.

A3712. Sisenwine, S.F., Tio, C.O., Ruelius, H.W., *Xenobiotica* (1979) **9**, 237–246.
A3713. Ono, K., Saito, T., Sasano, H., Moroi, R., Sano, M., Akimoto, T., *Xenobiotica* (1979) **9**, 227–236.
A3714. Jorritsma, J., Meerman, J.H.N., Vonk, R.J., Mulder, G.J., *Xenobiotica* (1979) **9**, 247–252.
A3715. Douch, P.G.C., *Xenobiotica* (1979) **9**, 263–268.
A3716. Shiobara, Y., Ogiso, T., *Xenobiotica* (1979) **9**, 157–164.
A3717. Kao, J., Bridges, J.W., Faulkner, J.K., *Xenobiotica* (1979) **9**, 141–147.
A3718. Aimoto, T., Ito, O., Kimura, R., Murata, T., Ito, K., *Xenobiotica* (1979) **9**, 173–180.
A3719. Tsunoda, N., Yoshimura, H., *Xenobiotica* (1979) **9**, 181–187.
A3720. Siddik, Z.H., Barnes, R.D., Dring, L.G., Smith, R.L., Williams, R.T., *Biochem. Pharmacol.* (1979) **28**, 3081–3091.
A3721. Misra, A.L., Pontani, R.B., Vadlamani, N.L., *Xenobiotica* (1979) **9**, 189–199.
A3722. Tank, A.W., Weiner, H., *Biochem. Pharmacol.* (1979) **28**, 3139–3147.
A3723. Dewick, P.M., Martin, M., *Phytochemistry* (1979) **18**, 591–596.
A3724. Levy, S., Yagen, B., Mechoulam, R., *Science* (1978) **200**, 1391–1392.
A3725. Dewick, P.M., Martin, M., *Phytochemistry* (1979) **18**, 597–602.
A3726. Murphy, P.J., Williams, T.L., Smallwood, J.K., Bellamy, G., Molloy, B.B., *Life Sci.* (1978) **23**, 301–311.
A3727. Essien, E.E., *Niger. J. Pharm.* (1978) **9**, 63–69.
A3728. Maksay, G., Tegyey, Z., Ötvös, L., *J. Pharm. Sci.* (1978) **67**, 1208–1210.
A3729. Phillipson, J.D., El-Dabbas, S.W., Gorrod, J.W., *Eur. J. Drug Metab. Pharmacokinet.* (1978) **3**, 119–124.
A3730. Gal, E.M., Sherman, A.D., *Commun. Psychopharmacol.* (1977) **1**, 353–361.
A3731. Fasco, M.J., Vatsis, K.P., Kaminsky, L.S., Coon, M.J., *J. Biol. Chem.* (1978) **253**, 7813–7820.
A3732. Nemoto, N., Takayama, S., Gelboin, H.V., *Chem.-Biol. Interact.* (1978) **23**, 19–30.
A3733. Degen, G.H., Neumann, H.-G., *Chem.-Biol. Interact.* (1978) **22**, 239–255.
A3734. Ushiro, H., Takai, K., Noda, Y., Narumiya, S., Tokuyama, T., Hayaishi, O., *J. Biol. Chem.* (1978) **253**, 9002–9008.
A3735. Crawford, R.L., Perkins, P.E., *FEMS Microbiol. Lett.* (1978) **4**, 161–162.
A3736. Lombardi, P., Goldin, B., Boutin, E., Gorbach, S.L., *J. Steroid Biochem.* (1978) **9**, 795–801.
A3737. Stuart, K.L., Kutney, J.P., Honda, T., Worth, B.R., *Heterocycles* (1978) **9**, 1419–1427.
A3738. Coutts, R.T., Jones, G.R., Townsend, R.E., *J. Pharm. Pharmacol.* (1978) **30**, 415–418.
A3739. Klutch, A., Levin, W., Chang, R.L., Vane, F., Conney, A.H., *Clin. Pharmacol. Ther.* (1978) **24**, 287–293.
A3740. Huq, S., Palmer, J.M., *FEBS Lett.* (1978) **95**, 217–220.
A3741. Nelson, W.L., Burke, T.R., Jr., *J. Med. Chem.* (1978) **21**, 1185–1186.
A3742. Renner, G., Richter, E., Schuster, K.P., *Chemosphere* (1978) **7**, 663–668.
A3743. Grunow, W., Altmann, H.J., Böhme, C., *Food Cosmet. Toxicol.* (1978) **16**, 349–353.
A3744. Vaught, J.B., Gurtoo, H.L., Paigen, B., Minowada, J., Sartori, P., *Cancer Lett. (Shannon)* (1978) **5**, 261–168.
A3745. Baird, W.M., Chemerys, R., Chern, C.J., Diamond, L., *Cancer Res.* (1978) **38**, 3432–3437.
A3746. Husain, M.M., Dehnen, W., Monch, W., *Toxicol. Eur. Res.* (1978) **1**, 75–83.
A3747. Chauhan, Y.S., Rathore, V.S., Garg, G.K., Bhargava, A., *Biochem. Biophys. Res. Commun.* (1978) **83**, 1237–1245.
A3748. Kabuto, S., Tamaki, T., Koide, T., Ishihama, H., *Yakugaku Zasshi* (1978) **98**, 944–949.

A3749. Ball, P., Knuppen, R., *J. Clin. Endokrinol. Metab.* (1978) **47**, 732–737.

A3750. Dixon, P.A.F., Caldwell, J., *Eur. J. Metab. Pharmacokinet.* (1978) **3**, 95–98.

A3752. Visser, T.J., Fekkes, D., Docter, R., Hennemann, G., *Biochem. J.* (1978) **174**, 221–229.

A3753. Sacks, W., Shane, L.L., Sacks, S., *Res. Commun. Psychol. Psychiatry Behav.* (1978) **3**, 287–312.

A3754. Hobkirk, R., Nilsen, M., Mori, J., *Endocrinology (Philadelphia)* (1978) **103**, 1227–1233.

A3755. Rhodes, M.J.C., Wooltorton, L.S.C., Lourenço, E.J., *Phytochemistry* (1979) **18**, 1125–1129.

A3756. Nahrstedt, A., Hösel, W., Walther, A., *Phytochemistry* (1979) **18**, 1137–1141.

A3757. Tsang, Y.-F., Ibrahim, R.K., *Phytochemistry* (1979) **18**, 1131–1136.

A3758. Ulbrich, B., Zenk, M.H., *Phytochemistry* (1979) **18**, 929–933.

A3759. Heinsbroek, R., van Brederode, J., van Nigtevecht, G., Kamsteeg, J., *Phytochemistry* (1979) **18**, 935–937.

A3760. Achenbach, H., Böttger, A., Kohl, W., Fautz, E., Reichenbach, H., *Phytochemistry* (1979) **18**, 961–963.

A3761. Fautz, E., Reichenbach, H., *Phytochemistry* (1979) **18**, 957–959.

A3762. Stöckigt, J., *Phytochemistry* (1979) **18**, 965–971.

A3763. Inouye, H., Ueda, S., Inoue, K., Matsumura, H., *Phytochemistry* (1979) **18**, 1301–1308.

A3764. Sae, A.S.W., Cunningham, B.A., *Phytochemistry* (1979) **18**, 1785–1787.

A3765. Martin, M., Dewick, P.M., *Phytochemistry* (1979) **18**, 1309–1317.

A3766. Ziv, G., Dekker, A., Hageman, R., *Refu. Vet.* (1978) **35**, 41–57.

A3767. Yisak, W., Agurell, S., Lindgren, J.E., Widman, M., *J. Pharm. Pharmacol.* (1978) **30**, 462–463.

A3768. Midgley, I., Hawkins, D.R., Chasseaud, L.F., *Arzneim.-Forsch.* (1978) **28**, 1911–1916.

A3769. Dorn, E., Knackmuss, H.J., *Biochem. J.* (1978) **174**, 85–94.

A3770. Mass, M.J., Kaufman, D.G., *Cancer Res.* (1978) **38**, 3861–3866.

A3771. Kao, J., Jones, C.A., Fry, J.R., Bridges, J.W., *Life Sci.* (1978) **23**, 1221–1228.

A3772. Davis, D.G., Hodgson, R.H., Dusbabek, K.E., Hoffer, B.L., *Physiol. Plant.* (1978) **44**, 87–91.

A3773. Hoagland, R.E., *Plant Cell Physiol.* (1978) **19**, 1019–1027.

A3774. Engst, R., Macholz, R.M., Kujawa, M., *J. Environ. Sci. Health B* (1978) **B13**, 425–427.

A3775. Nau, H., Brendel, K., Liddiard, C., *Drug Metab. Rev.* (1979) **9**, 127–142.

A3776. Bode, R., Birnbaum, D., *Biochem. Physiol. Pflanz.* (1979) **174**, 26–38.

A3777. Patzlaff, M., Barz, W., *Z. Naturforsch., C: Biosci.* (1978) **33C**, 675–684.

A3778. Legrand, M., Fritig, B., Hirth, L., *Planta* (1978) **144**, 101–108.

A3779. Shimazono, H., Kinjo, K., Nishimura, H., Kawachi, S., *Mokuzai Gakkaishi* (1978) **24**, 837–840.

A3780. Okuda, J., Nagamine, J., Okumura, M., Yagi, K., *J. Nutr. Sci. Vitaminol.* (1978) **24**, 505–510.

A3781. Niwaguchi, T., Inoue, T., *J. Chromatogr.* (1978) **161**, 223–229.

A3782. Bresnick, E., Chuang, A.H.L., Bornstein, W.A., *Chem.-Biol. Interact.* (1979) **24**, 111–115.

A3783. Mori, Y., Hori, T., Toyoshi, K., *Gann* (1978) **69**, 757–762.

A3784. Comerton, A.M., Layne, D.S., Williamson, D.G., *Can. J. Biochem.* (1978) **56**, 1082–1085.

A3785. Bernhardt, F.H., Heymann, E., Taylor, P.S., *Eur. J. Biochem.* (1978) **92**, 209–223.

A3786. Tresca, J.P., Ponsard, G., Degrelle, H., Jayle, M.-F., *Can. J. Biochem.* (1978) **56**, 985–992.

A3787. Sugumaran, M., Vaidyanathan, C.S., *J. Indian Inst. Sci.* (1978) **60**, 125–141.

A3788. Jones, R.A.D., Lee, C.R., Pollitt, R.J., *Stable Isot., Proc. Int. Symp.* (1977) 247–252.

A3789. Hoskins, J.A., Pollitt, R.J., *Stable Isot., Proc. Int. Symp.* (1977) 253–260.

A3790. Smith, B.R., Maguire, J.H., Ball, L.M., Bend, J.R., *Fed. Proc.* (1978) **37**, 2480–2484.

A3791. Hengstmann, J.H., Aulepp, H., *Arzneim.-Forsch.* (1978) **28**, 2326–2331.

A3792. Sütfeld, R., Kehrel, B., Wiermann, R., *Z. Naturforsch., C: Biosci.* (1978) **33C**, 841–846.

A3793. Huh, M. M.-O., Friedhoff, A.J., *J. Biol. Chem.* (1979) **254**, 299–308.

A3794. Lee, J.K., Kim, K.C., *Han'guk Nonghwa Hakhoechi* (1978) **21**, 197–203.

A3795. Frederick, J.R., Pye, E.K., *Enzyme Eng.* (1975) **3**, 263–271.

A3796. Kellermann, G.H., Luyten-Kellermann, M., *Life Sci.* (1978) **23**, 2485–2490.

A3797. Hiles, R.A., Caudill, D., Birch, C.G., Eichholt, T., *Toxicol. Appl. Pharmacol.* (1978) **46**, 593–608.

A3798. Lynn, R.K., Holm, R., Swanson, B.N., Smith, R., Gerber, N., *J. Clin. Pharmacol.* (1979) **19**, 20–30.

A3799. Rast, H.G., Engelhardt, G., Wallnöfer, P.R., Oehlmann, L., Wagner, K., *J. Agric. Food Chem.* (1979) **27**, 699–702.

A3800. Ruzo, L.O., Engel, J.L., Casida, J.E., *J. Agric. Food Chem.* (1979) **27**, 725–731.

A3801. Chin, B.H., Eldridge, J.M., Anderson, J.H., Jr., Sullivan, L.J., *J. Agric. Food Chem.* (1979) **27**, 716–720.

A3802. Wright, F.C., Riner, J.C., *J. Agric. Food Chem.* (1979) **27**, 576–577.

A3803. Shimabukuro, R.H., Walsh, W.C., Hoerauf, R.A., *J. Agric. Food Chem.* (1979) **27**, 615–623.

A3804. Shono, T., Ohsawa, K., Casida, J.E., *J. Agric. Food Chem.* (1979) **27**, 316–325.

A3805. Rott, B., Nitz, S., Korte, F., *J. Agric. Food Chem.* (1979) **27**, 306–310.

A3806. Zenser, T.V., Mattammal, M.B., Davis, B.B., *J. Pharmacol. Exp. Ther.* (1979) **208**, 418–421.

A3807. Akhtar, M.H., Foster, T.S., *J. Agric. Food Chem.* (1979) **27**, 113–116.

A3808. Halpaap, K., Horning, M.G., Horning, E.C., *J. Chromatogr.* (1978) **166**, 479–490.

A3809. Tulp, M.T.M., Hutzinger, O., *Chemosphere* (1978) **7**, 761–768.

A3810. Williams, B., Dring, L.G., Williams, R.T., *Toxicol. Appl. Pharmacol.* (1978) **46**, 567–578.

A3811. Knoell, H.E., *FEBS Lett.* (1979) **97**, 155–158.

A3812. Pettersen, J.E., *Biomed. Mass Spectrom.* (1978) **5**, 488–490.

A3813. Lysz, T.W., Sze, P.Y., *J. Neurosci. Res.* (1978) **3**, 411–418.

A3814. Hart, R.F., Renskers, K.J., Nelson, E.B., Roth, J.A., *Life Sci.* (1979) **24**, 125–130.

A3815. Sanchez, M.S., Barrett, J., *Parasitology* (1979) **78**, 1–5.

A3816. Magyar, K., Lengyel, M., Knoll, J., *Acta Physiol. Acad. Sci. Hung.* (1978) **51**, 401–411.

A3817. Vane, F.M., Ellis, D.H., Rao, J., Petrin, M., Blount, J.F., Mohacsi, E., Szuna, A., Kamm, J.J., *Biomed. Mass. Spectrom.* (1978) **5**, 498–507.

A3818. Gothoskar, S.V., Benjamin, T., Roller, P.R., Weisburger, E.K., *Xenobiotica* (1979) **9**, 533–537.

A3819. Neye, J., Fenner, H., *Arch. Pharm. (Weinheim, Ger.)* (1978) **311**, 1042–1050.

A3820. Ramsey, J.C., Young, J.D., *Scand. J. Work, Environ. Health* (1978) **4** (Suppl. 2) , 84–91.

A3821. Fuchs, A., De Vries, F.W., *Antonie van Leeuwenhoek* (1978) **44**, 283–292.

A3822. Renner, G., Schuster, K.P., Kuhnhen, H.B., Stachel, H.D., *Chemosphere* (1978) **7**, 943–947.

A3823. Malins, D.C., Collier, T.K., Thomas, L.C., Roubal, W.T., *Int. J. Environ. Anal. Chem.* (1979) **6**, 55–66.

A3824. Wallnöfer, P.R., Ziegler, W., Engelhardt, G., Rothmeier, H., *Chemosphere* (1978) **7**, 967–972.

A3825. Kiessling, K.-H., Pettersson, H., *Acta Pharmacol. Toxicol.* (1978) **43**, 285–290.

A3826. Gaal, A., Neujahr, H.Y., *J. Bacteriol.* (1979) **137**, 13–21.

A3827. Burchell, B., *Biochem. J.* (1978) **173**, 749–757.

A3828. Stockigt, J., *J. Chem. Soc., Chem. Commun.* (1978) 1097–1099.

A3829. Hopper, D.J., *Biochem. J.* (1978) **175**, 345–347.

A3830. Brine, G.A., Brine, D.R., Welch, C.D., Bondeson, V.C., Carroll, F.I., *Res. Commun. Chem. Pathol. Pharmacol.* (1978) **22**, 455–464.

A3831. Burba, J.V., *Can. J. Physiol. Pharmacol.* (1979) **57**, 213–216.

A3832. Neal, G.E., Colley, P.J., *Biochem. J.* (1978) **174**, 839–851.

A3833. Pantarotto, C., Fanelli, R., Bidoli, F., Morazzoni, P., Salmona, M., Szczawinska, K., *Scand. J. Work Environ. Health* (1978) **4** (Suppl. 2) , 67–77.

A3834. Kuwano, R., Yoshida, Y., Takahashi, Y., *J. Neurochem.* (1978) **31**, 815–824.

A3835. Mori, Y., Hori, T., Toyoshi, K., Horie, M., *J. Pharmaco.-Bio. Dyn.* (1978) **1**, 192–201.

A3836. Bollag, J.M., Sjoblad, R.D., Liu, S.-Y., *Can. J. Microbiol.* (1979) **25**, 229–233.

A3837. Keat, M.J., Hopper, D.J., *Biochem. J.* (1978) **175**, 659–667.

A3838. Corbett, M.D., Chipko, B.R., Baden, D.G., *Biochem. J.* (1978) **175**, 353–360.

A3839. Fluharty, A.L., Stevens, R.L., Goldstein, E.B., Kihara, H., *Biochim. Biophys. Acta* (1979) **566**, 321–326.

A3840. Stierlin, H., Faigle, J.W., Sallmann, A., Küng, W., Richter, W.J., Kriemler, H.-P., Alt, K.O., Winkler, T., *Xenobiotica* (1979) **9**, 601–610.

A3841. Haefelfinger, P., *HRC & CC, J. High Resolut. Chromatogr. Chromatogr. Commun.* (1979) **2**, 39–42.

A3842. Swaisland, A.J., Franklin, R.A., White, A.C., *Br. J. Clin. Pharmacol.* (1979) **7**, 120–123.

A3843. Balogh, E., Ajtay, M.K., Szoes. I., *Rev. Med. (Tirgu-Mures, Rom)* (1978) **24**, 32–35.

A3844. Stierlin, H., Faigle, J.W., *Xenobiotica* (1979) **9**, 611–621.

A3845. Bourne, G.R., Moss, S.R., Phillips, P.J., Shuker, B., *Xenobiotica* (1979) **9**, 623–631.

A3846. Skaare, J.U., Solheim, E., *Xenobiotica* (1979) **9**, 649–657.

A3847. Šípal, Z., Ahlenius, T., Bergstrand, A., Rodriquez, L., Jakobsson, S.W., *Xenobiotica* (1979) **9**, 633–645.

A3848. Skaare, J.U., *Xenobiotica* (1979) **9**, 659–668.

A3849. Douch, P.G.C., Buchanan, L.L., *Xenobiotica* (1979) **9**, 675–679.

A3850. Danhof, M., Krom, D.P., Breimer, D.D., *Xenobiotica* (1979) **9**, 695–702.

A3851. Guest, D., King, L.J., Parke, D.V., *Xenobiotica* (1979) **9**, 681–693.

A3852. Cucharczyk, N., Yang, J.T., Valia, K.H., Stiefel, F.J., Sofia, R.D., *Xenobiotica* (1979) **9**, 703–711.

A3853. Glickman, A.H., Shono, T., Casida, J.E., Lech, J.J., *J. Agric. Food Chem.* (1979) **27**, 1038–1041.

A3854. Alkalay, D., Volk, J., Roth, W., Klemani, L., Bartlett, M.F., *J. Pharm. Sci.* (1979) **68**, 156–160.

A3855. Fletcher, C.L., Kaufman, D.D., *J. Agric. Food Chem.* (1979) **27**, 1127–1130.

A3856. Yasuda, Y., Fujisawa, N., Morita, S., Kohri, H., *Arzneim.-Forsch.* (1979) **29**, 261–265.

A3857. Tilstone, W.J., Margot, P., Johnstone, E.C., *Psychopharmacology (Berlin)* (1979) **60**, 261–263.

A3861. Battersby, A.R., Lewis, N.G., Tippett, J.M., *Tetrahedron Lett.* (1978) 4849–4852.

A3862. Crooks, P.A., Breakefield, X.O., Sulens, C.H., Castiglione, C.M., Coward, J.K., *Biochem. J.* (1978) **176**, 187–196.

A3863. Kung, H.C., Wilson, A.G.E., *Life Sci.* (1979) **24**, 425–431.

A3864. Williams, T.H., Sasso, G.J., Ryan, J.J., Schwartz, M.A., *J. Med. Chem.* (1979) **22**, 436–440.

A3865. Kimura, M., Masubayashi, K., Hakusui, H., Sano, M., Akimoto, T., *Rinsho Yakuri* (1970) **7**, 161–170.

A3866. Kaneshima, H., Yamaguchi, T., Itoh, K., *Bull. Environ. Contam. Toxicol.* (1978) **20**, 725–728.

A3867. Emerole, G.O., French, M.R., Williams, R.T., *Biochem. Exp. Biol.* (1977) **13**, 385–390.

A3868. Gal, J., Estin, C.D., Moon, B.J., *Biochem. Biophys. Res. Commun.* (1978) **85**, 1466–1467.

A3869. Gross, D., Laanio, T., Dupuis, G., Esser, H.O., *Pestic. Biochem. Physiol.* (1979) **10**, 49–59.

A3870. Keat, M.J., Hopper, D.J., *Biochem. J.* (1978) **175**, 649–658.

A3871. Umetsu, N., Fahmy, M.A.H., Fukuto, T.R., *Pestic. Biochem. Physiol.* (1979) **10**, 104–119.

A3872. Tanizawa, K., Soda, K., *J. Biochem. (Tokyo)* (1979) **85**, 901–906.

A3873. Chiba, S., Kanaya, K., Hiromi, K., Shimomura, T., *Agric. Biol. Chem.* (1979) **43**, 237–242.

A3875. Fujiwara, M., Shibata, M., Nomiyama, Y., Sugimoto, T., Hirata, F., Tokuyama, T., Senoh, S., Hayaishi, O., *Proc. Natl. Acad. Sci. U.S.A.* (1979) **76**, 1145–1149.

A3876. Sawada, H., Hara, A., *Yakugaku Zasshi* (1979) **99**, 30–37.

A3878. Coutts, R.T., Foster, B.C., Jones, G.R., Myers, G.E., *Appl. Environ. Microbiol.* (1979) **37**, 429–432.

A3879. Collier, T.K., Thomas, L.C., Malins, D.C., *Comp. Biochem. Physiol. C* (1978) **61C**, 23–28.

A3880. Bode, R., Birnbaum, D., *Z. Allge. Mikrobiol.* (1979) **19**, 221–222.

A3881. Loveland, P.M., Nixon, J.E., Pawlowski, N.E., Eisele, T.A., Libbey, L.M., Sinnhuber, R.O., *J. Environ. Pathol. Toxicol.* (1979) **2**, 707–718.

A3882. Seuferer, S.L., Braymer, H.D., Dunn, J.J., *Pestic. Biochem. Physiol.* (1979) **10**, 174–180.

A3883. Kennedy, K.A., Fischer, L.J., *Drug Metab. Dispos.* (1979) **7**, 319–324.

A3884. Nakata, H., Yamauchi, T., Fujisawa, H., *J. Biol. Chem.* (1979) **254**, 1829–1833.

A3885. Summers, M.C., Markovic, R., Klinman, J.P., *Biochemistry* (1979) **18**, 1969–1979.

A3886. Wheldrake, J.F., Baudinette, R.V., Hewitt, S., *Comp. Biochem. Physiol. C* (1978) **61C**, 103–107.

A3887. Crawford, R.L., Bromley, J.W., Perkins-Olson, P.E., *Appl. Environ. Microbiol.* (1979) **37**, 614–618.

A3888. Ohmura, I., Inagaki, C., Araki, H., Tanaka, C., *Jpn. J. Pharmacol.* (1978) **28**, 747–753.

A3889. Sorimachi, K., Robbins, J., *Biochim. Biophys. Acta* (1979) **583**, 443–453.

A3890. Forkmann, G., *Phytochemistry* (1979) **18**, 1973–1975.

A3891. Mack, J.P.G., Slaytor, M., *Phytochemistry* (1979) **18**, 1921–1925.

A3892. Krisnangkura, K., Gold, M.H., *Phytochemistry* (1979) **18**, 2019–2021.

A3893. Jones, D.H., Kohli, J., Safe, S., *Xenobiotica* (1979) **9**, 733–736.

A3894. Kögel, W., Müller, W.F., Coulston, F., Korte, F., *J. Agric. Food Chem.* (1979) **27**, 1181–1185.

A3895. Marquis, L.Y., Shimabukuro, R.H., Stolzenberg, G.E., Feil, V.J., Zaylskie, R.G., *J. Agric. Food Chem.* (1979) **27**, 1148–1156.

A3896. Chin, B.H., Sullivan, L.J., Eldridge, J.E., *J. Agric. Food Chem.* (1979) **27**, 1395–1398.

A3897. Chin, B.H., Sullivan, L.J., *J. Agric. Food Chem.* (1979) **27**, 1419–1420.

A3898. Levine, W.G., Finkelstein, T.T., *J. Pharmacol. Exp. Ther.* (1979) **208**, 399–405.

A3899. Corke, C.T., Bunce, N.J., Beaumont, A.-L., Merrick, R.L., *J. Agric. Food Chem.* (1979) **27**, 644–646.

A3900. Man, M., Alozie, S.O., Woods, L.A., Martin, B.R., Harris, L.S., *Drug Metab. Dispos.* (1980) **8**, 55–59.
A3901. Martin, B.R., Vincek, W.C., Balster, R.L., *Drug Metab. Dispos.* (1980) **8**, 49–54.
A3902. Plummer, J.L., Smith, R.L., Ball, L.M., Bend, J.R., *Drug Metab. Dispos.* (1980) **8**, 68–72.
A3903. Talseth, T., Haegle, K.D., McNay, J.L., *Drug Metab. Dispos.* (1980) **8**, 73–76.
A3904. Segura, J., Bakke, O.M., Huizing, G., Beckett, A.H., *Drug Metab. Dispos.* (1980) **8**, 87–92.
A3905. Slack, J.A., Ford-Hutchinson, A.W., *Drug Metab. Dispos.* (1980) **8**, 84–86.
A3906. Martin, J.R., George, J.W., Pohl, L.R., *Drug Metab. Dispos.* (1980) **8**, 93–97.
A3907. Paeme, G., Sonck, W., Tourwé, D., Grimée, R., Vercruysse, A., *Drug Metab. Dispos.* (1980) **8**, 115–116.
A3908. Chien, M.M., Rosazza, J.P., *Drug Metab. Dispos.* (1979) **7**, 211–214.
A3909. Ruenitz, P.C., McLennon, T.P., Sternson, L.A., *Drug Metab. Dispos.* (1979) **7**, 204–207.
A3910. Spenney, J.G., Nowell, R.M., *Drug Metab. Dispos.* (1979) **7**, 215–219.
A3911. Kolis, S.J., Williams, T.H., Postma, E.J., Sasso, J., Confalone, P.N., Schwartz, M.A., *Drug Metab. Dispos.* (1979) **7**, 220–225.
A3912. Andrews, L.S., Pohl, L.R., Hinson, J.A., Fisk, C.L., Gillette, J.R., *Drug Metab. Dispos.* (1979) **7**, 296–300.
A3913. Biggs, S.R., Chasseaud, L.F., Hawkins, D.R., Midgley, I., *Drug Metab. Dispos.* (1979) **7**, 233–236.
A3914. Tang, B.K., Kalow, W., Grey, A.A., *Drug Metab. Dispos.* (1979) **7**, 315–318.
A3915. Grindel, J.M., Migdalof, B.H., Cressman, W.A., *Drug Metab. Dispos.*, (1979) **7**, 325–329.
A3916. Yamaguchi, H., Kubo, J., Sekine, K., Naruchi, T., Hashimoto, Y., Kato, R., *Drug Metab. Dispos.* (1979) **7**, 340–344.
A3917. Tocco, D.J., De Luna, F.A., Duncan, A.E.W., Walker, R.W., Arison, B.H., VandenHeuvel, W.J.A., *Drug Metab. Dispos.* (1979) **7**, 330–334.
A3918. Nelson, W.L., Powell, M.L., *Drug Metab. Dispos.* (1979) **7**, 351–355.
A3919. Bopp, R.J., Nash, J.F., Ridolfo, A.S., Shepard, E.R., *Drug Metab. Dispos.* (1979) **7**, 356–359.
A3920. Perez, J., Sitar, D.S., Ogilvie, R.I., *Drug Metab. Dispos.* (1979) **7**, 383–387.
A3921. Andresen, B.D., Davis, F.T., *Drug Metab. Dispos.* (1979) **7**, 360–365.
A3922. Bolton, G.C., Griffiths, L.A., *Drug Metab. Dispos.* (1979) **7**, 388–392.
A3923. Mitchell, S.C., Waring, R.H., *Drug Metab. Dispos.* (1979) **7**, 399–403.
A3924. Galtier, P., Charpenteau, J.-L., Alvinerie, M., Labouche, C., *Drug Metab. Dispos.* (1979) **7**, 429–434.
A3925. Kreft, H., Breyer-Pfaff, U., *Drug Metab. Dispos.* (1979) **7**, 404–410.
A3926. Grindel, J.M., Migdalof, B.H., Hills, J.F., *Drug Metab. Dispos.* (1979) **7**, 448.
A3927. Hashimoto, M., Sekine, Y., Muto, J., Yamazuchi, T., Maeda, T., Utsui, Y., Yoshida, K., Arakawa, S., Shimokawa, N., Shimakawa, K., *Drug Metab. Dispos.* (1979) **7**, 435–441.
A3928. Thonnart, N., Vanhaelen, M., Vanhaelen-Fastré, R., *Drug Metab. Dispos.* (1979) **7**, 449–450.
A3929. Küpfer, A., Brilis, G.M., Watson, J.T., Harris, T.M., *Drug Metab. Dispos.* (1980) **8**, 1–4.
A3930. Wong, L.K., Wang, C.-L.A., Daniel, F.B., *Drug Metab. Dispos.* (1980) **8**, 28–33.
A3931. Osterloh, J.D., Karakaya, A., Carter, D.E., *Drug Metab. Dispos.* (1980) **8**, 12–15.
A3932. Al-Turk, W.A., Stohs, S.J., Roche, E.B., *Drug Metab. Dispos.* (1980) **8**, 44–45.
A3934. Southwell, I.A., Flynn, T.M., Degabrielle, R., *Xenobiotica* (1980) **10**, 17–23.

A3935. Koschier, F.J., Pritchard, J.B., *Xenobiotica* (1980) **10**, 1–6.

A3936. Billings, R.E., Whitaker, G.N., McMahon, R.E., *Xenobiotica* (1980) **10**, 33–36.

A3937. Pal, K., Grover, P.L., Sims, P., *Xenobiotica* (1980) **10**, 25–31.

A3938. Hawkins, D.R., Chasseaud, L.F., Woodhouse, R.N., Cresswell, D.G., *Xenobiotica* (1980) **10**, 81–95.

A3939. Kaye, B., Roberts, D.W., *Xenobiotica* (1980) **10**, 97–101.

A3940. Sato, T., Konishi, M., Ago, T., Nakagawa, Y., Iwatani, K., Mori, Y., *Xenobiotica* (1980) **10**, 125–133.

A3941. Willems, A.G.M., Overmars, H., Scherpenisse, P., de Lange, N., Post, L.C., *Xenobiotica* (1980) **10**, 103–112.

A3942. Baldwin, M.K., Hutson, D.H., *Xenobiotica* (1980) **10**, 135–144.

A3943. Foster, A.B., Griggs, L.J., Jarman, M., van Maanen, J.M.S., Schulten, H.-R., *Biochem. Pharmacol.* (1980) **29**, 1977–1979.

A3944. Hoffman, A.R., Paul, S.M., Axelrod, J., *Biochem. Pharmacol.* (1980) **29**, 83–87.

A3945. Minard, F.N., Grant, D.S., Cain, J.C., Jones, P.H., Kyncl, J., *Biochem. Pharmacol.* (1980) **29**, 69–75.

A3946. Barker, S.A., Monti, J.A., Christian, S.T., *Biochem. Pharmacol.* (1980) **29**, 1049–1057.

A3947. Yu, L.-S., Wilkinson, C.F., Anders, M.W., *Biochem. Pharmacol.* (1980) **29**, 1113–1122.

A3948. Mitra, C., Guha, S.R., *Biochem. Pharmacol.* (1980) **29**, 1213–1216.

A3949. Glinsukon, T., Lookyee, A., Toskulkao, C., Shank, R.C., *Biochem. Pharmacol.* (1980) **29**, 1189–1191.

A3950. Felsted, R.L., Richter, D.R., Jones, D.M., Bachur, N.R., *Biochem. Pharmacol.* (1980) **29**, 1503–1516.

A3951. Caldwell, J., Michel-Lipowsky, M., Smith, R.L., *Biochem. Pharmacol.* (1980) **29**, 1325–1330.

A3952. Chasseaud, L.F., Down, W.H., Grover, P.L., Sacharin, R.M., Sims, P., *Biochem. Pharmacol.* (1980) **29**, 1589–1590.

A3953. Polidoro, G., Di Ilio, C., Del Boccio, G., Zulli, P., Federici, G., *Biochem. Pharmacol.* (1980) **29**, 1677–1680.

A3954. Kennedy, M.W., Carpenter, N.K., Dymerski, P.P., Adams, S.M., Kaminsky, L.S., *Biochem. Pharmacol.* (1980) **29**, 727–736.

A3955. Gill, S.S., Hammock, B.D., *Biochem. Pharmacol.* (1980) **29**, 389–395.

A3956. Sgaragli, G., Della Corte, L., Puliti, R., De Sarlo, F., Francalanci, R., Guarna, A., *Biochem. Pharmacol.* (1980) **29**, 763–769.

A3957. Lewinsohn, R., Glover, V., Sandler, M., *Biochem. Pharmacol.* (1980) **29**, 777–781.

A3958. Das, P.K., Guha, S.R., *Biochem. Pharmacol.* (1980) **29**, 2049–2053.

A3959. McLean, A., Newell, D., Baker, G., Connors, T., *Biochem. Pharmacol.* (1980) **29**, 2039–2047.

A3960. Sugawara, Y., Fujihara, M., Miura, Y., Hayashida, K., Takahashi, T., *Chem. Pharm. Bull.* (1978) **26**, 3312–3321.

A3961. Reynolds, G.P., Elsworth, J.D., Blau, K., Sandler, M., Lees, A.J., Stern, G.M., *Br. J. Clin. Pharmacol.* (1978) **6**, 542–544.

A3962. Taylor, S.M., Rodgers, R.M., Lynn, R.K., Gerber, N., *J. Pharmacol. Exp. Ther.* (1980) **213**, 289–299.

A3963. Mori, Y., Hori, T., Toyoshi, K., *Chem. Pharm. Bull.* (1979) **27**, 235–237.

A3964. Carpenter, H.M., Mudge, G.H., *J. Pharmacol. Exp. Ther.* (1980) **213**, 350–354.

A3965. Yamazoe, Y., Ishii, K., Yamaguchi, N., Kamataki, T., Kato, R., *Biochem. Pharmacol.* (1980) **29**, 2183–2188.

A3966. Tyce, G.M., Owen, C.A., Jr., *Biochem. Pharmacol.* (1979) **28**, 3271–3278.

A3967. Grafström, R., Ormstad, K., Moldéus, P., Orrenius, S., *Biochem. Pharmacol.* (1979) **28**, 3573–3579.

A3968. Miner, D.J., Kissinger, P.T., *Biochem. Pharmacol.* (1979) **28**, 3285–3290.
A3969. Tyce, G.M., Owen, C.A., Jr., *Biochem. Pharmacol.* (1979) **28**, 3279–3284.
A3970. Dawson, J.R., Bridges, J.W., *Biochem. Pharmacol.* (1979) **28**, 3291–3297.
A3971. Dawson, J.R., Bridges, J.W., *Biochem. Pharmacol.* (1979) **28**, 3299–3305.
A3972. Soll, J., Schultz, G., *Phytochemistry* (1980) **19**, 215–218.
A3973. Hutson, K.G., Threlfall, D.R., *Phytochemistry* (1980) **19**, 535–537.
A3974. Walker, J.R.L., McCallion, R.F., *Phytochemistry* (1980) **19**, 373–377.
A3975. Kahn, V., Pomerantz, S.H., *Phytochemistry* (1980) **19**, 379–385.
A3976. Scott, A.I., Mizukami, H., Hirata, T., Lee, S.-L., *Phytochemistry* (1980) **19**, 488–489.
A3977. Keller, W.J., *Phytochemistry* (1980) **19**, 413–414.
A3978. Suzuki, O., Oya, M., Katsumata, Y., *Biochem. Pharmacol.* (1980) **29**, 2663–2667.
A3979. Tateishi, K., Koitabashi, C., Ichihara, S., *Biochem. Pharmacol.* (1980) **29**, 2705–2708.
A3980. Wong, L.T., Solomonraj, G., *Xenobiotica* (1980) **10**, 201–210.
A3981. Ribéreau-Gayon, G., Palfreyman, M.G., Zraïka, M., Wagner, J., Jung, M.J., *Biochem. Pharmacol.* (1980) **29**, 2465–2469.
A3982. Huizing, G., Beckett, A.H., Segura, J., Bakke, O.M., *Xenobiotica* (1980) **10**, 211–218.
A3983. Barrow, A., Brownsill, R.D., Spalton, P.N., Walls, C.M., Gunn, Y., Haskins, N.J., Rose, D.A., Palmer, R.F., *Xenobiotica* (1980) **10**, 219–228.
A3984. Kögel, W., Müller, W.F., Coulston, F., Korte, F., *J. Agric. Food Chem.* (1979) **27**, 1181–1185.
A3985. Marquis, L.Y., Shimabukuro, R.H., Stolzenberg, G.E., Feil, V.J., Zaylskie, R.G., *J. Agric. Food Chem.* (1979) **27**, 1148–1156.
A3986. Rausch, T., Hilgenberg, W., *Phytochemistry* (1980) **19**, 747–750.
A3987. Brunet, G., Ibrahim, R.K., *Phytochemistry* (1980) **19**, 741–746.
A3988. Remmen, S.F.A., Ellis, B.E., *Phytochemistry* (1980) **19**, 1421–1423.
A3989. Manley, S., Chapman, D.J., *Phytochemistry* (1980) **19**, 1453–1457.
A3990. Fuchs-Mettler, M., Curtius, H.-C., Baerlocher, K., Ettlinger, L., *Eur. J. Biochem.* (1980) **108**, 527–534.
A3991. Andresen, B.D., Davis, F.T., Long, M.D., *J. Pharm. Sci.* (1979) **68**, 283–288.
A3992. Yoshizawa, H., Nagafusa, H., Migita, Y., Kondoh, H., Nozu, T., Suwa, T., Tanaka, I., *Oyo Yakuri* (1978) **16**, 1087–1093.
A3993. Yoshitake, A., Kawahara, K., Shono, F., Izawa, A., Komatsu, T., *Chemotherapy (Tokyo)* (1978) **26** (Suppl. 4) , 96–100.
A3993a. Yoshitake, A., Kawahara, K., Shono, F., Izawa, A., Komatsu, T., *Chemotherapy (Tokyo)* (1978) **26** (Suppl. 4) , 91–95.
A3994. Kanazawa, J., Tomizawa, C., *Arch. Environ. Contam. Toxicol.* (1978) **7**, 397–407.
A3995. Rao, P.N., Purdy, R.H., Williams, M.C., Moore, P.H., Jr., Goldzieher, J.W., Layne, D.S., *J. Steroid Biochem.* (1979) **10**, 179–185.
A3996. Nomeir, A.A., Dauterman, W.C., *Pestic. Biochem. Physiol.* (1979) **10**, 190–196.
B1. Numazawa, M., Soeda, N., Kiyono, Y., Nambara, T., *J. Steroid Biochem.* (1979) **10**, 227–233.
B2. Engelhardt, G., Rast, H.G., Wallnöfer, P.R., *FEMS Microbiol. Lett.* (1979) **5**, 245–251.
B3. Cerniglia, C.E., Gibson, D.T., Van Baalen, C., *Biochem. Biophys. Res. Commun.* (1979) **88**, 50–58.
B4. Vigano, V., Paracchini, S., Pesce, E., Piacenza, G., *Quant. Mass Spectrom. Life Sci.* (1978) **2**, 357–366.
B5. Lai, A.A., Min, B.H., Garland, W.A., Levy, R.H., *J. Pharmacokinet. Biopharm.* (1979) **7**, 87–95.

B6. Weidenfeld, J., Carter, P., Reinhold, V.N., Tanner, S.B.IV, Engel, L.L., *Biomed. Mass Spectrom.* (1978) **5**, 587–590.

B7. Coutts, R.T., Dawson, G.W., Jones, G.R., *Res. Commun. Chem. Pathol. Pharmacol.* (1978) **22**, 589–592.

B8. Essac, E.G., Matsumura, F., *Pestic. Biochem. Physiol.* (1979) **10**, 67–78.

B9. Mori, Y., Hori, T., Toyoshi, K., Horie, M., *Chem. Pharm. Bull.* (1979) **27**, 554–557.

B10. Volz, M., Kellner, H.-M., *Br. J. Clin. Pharmacol.* (1980) **10**, 299S-308S.

B11. Hartimann, J., Reineke, W., Knackmuss, H.J., *Appl. Environ. Microbiol.* (1979) **37**, 421–428.

B12. Lee, K.-L., Roberson, L.E., Kenney, F.T., *Anal. Biochem* (1979) **95**, 188–193.

B13. Lusby, W.R., Oliver, J.E., Kearney, P.C., *J. Agric. Food Chem.* (1980) **28**, 641–644.

B14. Ohsawa, K., Casida, J.E., *J. Agric. Food Chem.* (1980) **28**, 250–255.

B15. Jaglan, P.S., *J. Agric. Food Chem.* (1980) **28**, 682–684.

B16. Rosenberg, A., Alexander, M., *J. Agric. Food Chem.* (1980) **28**, 297–302.

B17. Pathre, S.V., Fenton, S.W., Mirocha, C.J., *J. Agric. Food Chem.* (1980) **28**, 421–424.

B18. Angerer, J., *Int. Arch. Occup. Environ. Health* (1979) **43**, 63–67.

B19. Nakata, H., Fujisawa, H., *Biochim. Biophys. Acta* (1980) **614**, 313–327.

B20. Hinson, J.A., Nelson, S.D., Gillette, J.R., *Mol. Pharmacol.* (1979) **15**, 419–427.

B21. Jindra, A., Jindrova, N., Helia, O., *Eur. J. Drug Metab. Pharmacokinet.* (1978) **4**, 211–216.

B22. Volz, M., Christ, O., Kellner, H.M., Kuch, H., Fehlhaber, H.W., Gantz, D., Hadju, P., Cavagna, F., *Br. J. Clin. Pharmacol.* (1979) **7** (Suppl. 1), 41–50.

B24. Ogata, M., Fujii, T., *Int. Arch. Occup. Environ. Health* (1979) **43**, 45–51.

B25. Andrews, P.A., Brenner, D.E., Chou, F.-T.E., Kubo, H., Bachur, N.R., *Drug Metab. Dispos.* (1980) **8**, 152–156.

B26. Tanizawa, K., Soda, K., *J. Biochem.* (1979) **85**, 1367–1375.

B27. Mori, Y., Yamamoto, T., Toyoshi, K., *Chem. Pharm. Bull.* (1979) **27**, 379–385.

B28. Huang, M.-T., Miwa, G.T., Lu, A.Y.H., *J. Biol. Chem.* (1979) **254**, 3930–3934.

B29. Buck, R., Eberspächer, J., Lingens, F., *Liebigs Ann. Chem.* (1979) 564–571.

B30. Fukawa, K., Irino, O., Maruyama, K., Ohsawa, Y., Saitoh, K., *Chem. Pharm. Bull.* (1979) **27**, 578–585.

B31. Kholodov, L.E., Lil'in, E.T., *Mater. Resp. Rasshir. Konf. Farmakol. Gruz., 2nd* (1977) 84–85.

B32. Rozman, K., Williams, J., Müller, W.F., Coulston, F., Korte, F., *Bull. Environ. Contam. Toxicol.* (1979) **22**, 190–195.

B33. Smith, B.R., Bend, J.R., *Cancer Res.* (1979) **39**, 2051–2056.

B34. Sparnins, V.L., Burbee, D.G., Dagley, S., *J. Bacteriol.* (1979) **138**, 425–430.

B35. Turner, A.J., Pearson, A.G.M., Mason, R.J., *Dev. Biochem.* (1978) **4**, 501–506.

B36. Howie, M.B., Bourke, E., *Clin. Sci.* (1979) **56**, 9–14.

B37. Archer, R.A., Fukuda, D.S., Kossoy, A.D., Abbott, B.J., *Appl. Environ. Microbiol.* (1979) **37**, 965–971.

B38. Walterova, D., Nemeckova, A., Preininger, V., Simanek, V., Dolejs, L., *Heterocycles* (1979) **12**, 247–251.

B39. Idle, J.R., Mahgoub, A., Angelo, M.M., Dring, L.G., Lancaster, R., Smith, R.L., *Br. J. Clin. Pharmacol.* (1979) **7**, 257–266.

B40. Stocklinski, A.W., De Loach, G.D., Williams, H., *Res. Commun. Chem. Pathol. Pharmacol.* (1979) **24**, 397–400.

B41. Dodge, R.H., Cerniglia, C.E., Gibson, D.T., *Biochem. J.* (1979) **178**, 223–230.

B42. Koss, G., Koransky, W., Steinbach, K., *Arch. Toxicol.* (1979) **42**, 19–31.

B44. Begum, S., Scheunert, I., Haque, A., Klein, W., Korte, F., *Pestic. Biochem. Physiol.* (1979) **11**, 189–200.
B45. Crutcher, S.E., Geary, P.J., *Biochem. J.* (1979) **177**, 393–400.
B46. Lee, S.L., Hirata, T., Scott, A.I., *Tetrahedron Lett.* (1979) 691–694.
B47. Fazel, A.M., Jensen, R.A., *J. Bacteriol.* (1979) **138**, 805–815.
B48. Inouye, H., Ueda, S., Inoue, K., Shiobara, Y., Wada, I., *Tetrahedron Lett.* (1978) 4551–4554.
B49. Wolf, D.E., VandenHeuvel, W.J.A., Tyler, R.W., Walker, R.W., Koniuszy, F.R., Gruber, V., Arison, B.H., Rosegay, A., Jacob, T.A., Wolf, F.J., *Drug Metab. Dispos.* (1980) **8**, 131–136.
B50. Harvey, D.J., Paton, W.D.M., *Drug Metab. Dispos.* (1980) **8**, 178–186.
B51. Karlaganis, G., Küpfer, A., Bircher, J., Schlunegger, U.P., Gfeller, H., Bigler, P., *Drug Metab. Dispos.* (1980) **8**, 173–277.
B52. Roerig, S.C., Fujimoto, J.M., Wang, R.I.H., *Drug Metab. Dispos.* (1980) **8**, 295–299.
B53. Hinson, J.A., Pohl, L.R., Monks, T.J., Gillette, J.R., Guengerich, F.P., *Drug Metab. Dispos.* (1980) **8**, 289–294.
B54. Ohmiya, Y., Mehendale, H.M., *Drug Metab. Dispos.* (1980) **8**, 313–318.
B55. Vu, V.T., Abramson, F.P., *Drug Metab. Dispos.* (1980) **8**, 300–304.
B56. Bekersky, I., Colburn, W.A., Fishman, L., Kaplan, S.A., *Drug Metab. Dispos.* (1980) **8**, 319–324.
B57. Maguire, J.H., Butler, T.C., Dudley, K.H., *Drug Metab. Dispos.* (1980) **8**, 325–331.
B58. Yamamoto, K., Tajima, K., Mizutani, T., *J. Pharmaco.-Bio. Dyn.* (1979) **2**, 164–168.
B59. Watanabe, K., Yamamoto, I., Oguri, K., Yoshimura, H., *Biochem. Biophys. Res. Commun.* (1979) **88**, 178–182.
B60. Chou, M.W., Easton, G.D., Yang, S.K., *Biochem. Biophys. Res. Commun.* (1979) **88**, 1085–1091.
B61. Sekura, R.D., Jakoby, W.B., *J. Biol. Chem.* (1979) **254**, 5658–5663.
B62. Better, J., Gatenbeck, S., *Acta Chem. Scand., Ser. B* (1979) **B33**, 235–238.
B63. Lu, A.Y.H., Thomas, P.E., Ryan, D., Jerina, D.M., Levin, W., *J. Biol. Chem.* (1979) **254**, 5878–5881.
B64. Blisard, K.S., Mieyal, J.J., *J. Biol. Chem.* (1979) **254**, 5104–5110.
B65. Birch, A.J., Simpson, T.J., *J. Chem. Soc., Perkin Trans. 1* (1979) 816–822.
B66. Huizing, G., Beckett, A.H., Segura, J., *Pharm. Weekbl., Sci. Ed.* (1979) **1**, 436–443.
B67. Angmor, J.D., Dewick, P.M., Evans, W.C., *Planta Med.* (1979) **35**, 342–347.
B68. Leighty, E.G., *Res. Commun. Chem. Pathol. Pharmacol.* (1979) **24**, 393–396.
B69. James, M.O., Bowen, E.R., Dansette, P.M., Bend, J.R., *Chem.-Biol. Interact.* (1979) **25**, 321–344.
B70. Divakar, N.G., Subramanian, V., Sugumaran, M., Vaidyanathan, C.S., *Plant Sci. Lett.* (1979) **15**, 177–181.
B71. Mizukami, H., Nordlov, H., Lee, S.-L., Scott, A.I., *Biochemistry* (1979) **18**, 3760–3763.
B72. Anderson, R.J., Weinshilboum, R.M., *J. Lab. Clin. Med.* (1979) **94**, 158–171.
B73. Buswell, J.A., Ander, P., Pettersson, B., Eriksson, K.E., *FEBS Lett.* (1979) **103**, 98–101.
B74. Connor, M.J., Pheasant, A.E., Blair, J.A., *Biochem. J.* (1979) **178**, 795–797.
B75. Grindel, J.M., O'Neill, P.J., Yorgey, K.A., Schwartz, M.H., McKown, L.A., Migdalof, B.H., Wu, W.-N., *Drug Metab. Dispos.* (1980) **8**, 343–348.
B77. Jha, H.N., Banerjee, N.C., *Indian Vet. J.* (1979) **56**, 19–22.
B78. Nelson, W.L., Burke, T.R., *J. Med. Chem.* (1979) **22**, 1082–1088.
B79. Halldin, M., Widman, M., Martin, B., *Acta Pharm. Suec.* (1979) **16**, 34–40.
B80. Poupaert, J.H., Adline, J., Claesen, M.H., De Laey, P., Dumont, P.A., *J. Med. Chem.* (1979) **22**, 1140–1142.

B81. Vereczkey, L., Szentirmay, Z., Szporny, L., *Arzneim.-Forsch.* (1979) **29**, 953–956.

B82. Hill, D.L., Shih, T.-W., Struck, R.F., *Cancer Res.* (1979) **39**, 2528–2531.

B83. Ramachandran, A., Subramanian, V., Sugumaran, M., Vaidyanathan, C.S., *FEMS Microbiol. Lett.* (1979) **5**, 421–425.

B84. Tsai, M.-D., Floss, H.G., Rosenfeld, H.J., Roberts, J., *J. Biol. Chem.* (1979) **254**, 6437–6443.

B85. Whitt, D.D., Klug, M.J., De Moss, R.D., *Arch. Microbiol.* (1979) **122**, 169–175.

B86. Hutchinson, C.R., Heckendorf, A.H., Straughn, J.L., Daddona, P.E., Cane, D.E., *J. Am. Chem. Soc.* (1979) **101**, 3358–3369.

B87. Donehower, R.C., Hande, K.R., Drake, J.C., Chabner, B.A., *Clin. Pharmacol. Ther.* (1979) **26**, 63–72.

B88. Frigerio, A., Negrini, P., Cappelini, L., Rotilio, D., De Pascale, A., Rossi, E., *Eur. Drug Metab. Pharmacokinet.* (1979) **4**, 35–41.

B89. Nordqvist, M., Lindgren, J.E., Agurell, S., *J. Pharm. Pharmacol.* (1979) **31**, 231–237.

B90. Andrews, L.S., Sasame, H.A., Gillette, J.R., *Life Sci.* (1979) **25**, 567–572.

B91. Bakke, J.E., Price, C.E., *J. Environ. Sci. Health, Pt. B* (1979) **B14**, 291–304.

B92. Pekas, J.C., Larsen, G.L., Feil, V.J., *J. Toxicol. Environ. Health* (1979) **5**, 653–662.

B93. Bakke, J.E., Price, C.E., *J. Environ. Sci. Health, Pt. B* (1979) **B14**, 427–441.

B94. Dybing, E., Søderlund, E., Haug, L.T., Thorgeirsson, S.S., *Cancer Res.* (1979) **39**, 3268–3275.

B95. Wyrambik, D., Grisebach, H., *Eur. J. Biochem.* (1979) **97**, 503–509.

B97. Hoch, S.O., *J. Bacteriol.* (1979) **139**, 362–368.

B98. Narumiya, S., Takai, K., Tokuyama, T., Noda, Y., Ushiro, H., Hayaishi, O., *J. Biol. Chem.* (1979) **254**, 7007–7015.

B99. Baxter, R.L., Dorschel, C.A., Lee, S.-L., Scott, A.I., *J. Chem. Soc., Chem. Commun.* (1979) 257–259.

B100. Rogawski, M.A., Roth, R.H., Aghajanian, G.K., *J. Neurochem.* (1979) **32**, 1219–1226.

B102. Nordqvist, M., Agurell, S., Rydberg, M., Falk, L., Ryman, T., *J. Pharm. Pharmacol.* (1979) **31**, 238–243.

B103. Yamaguchi, H., Sekine, K., Kubo, J., Naruchi, T., Hashimoto, Y., Kato, R., *Iyakuhin Kenkyu* (1979) **10**, 491–498.

B104. Kangas, L., *Acta Pharmacol. Toxicol.* (1979) **45**, 16–19.

B105. Grindel, J.M., Migdalof, B.H., Plostnieks, J., *Clin. Pharmacol. Ther.* (1979) **26**, 122–128.

B106. Ekonomov, A.L., Zherdev, V.P., Rodionov, A.P., Vikhlyaev, Y.I., *Khim.-Farm. Zh.* (1979) **13**, 11–16.

B107. Wu, W.N., Weaner, L.E., Kalbron, J., O'Neill, P.J., Grindel, J.M., *Drug Metab. Dispos.* (1980) **8**, 349–352.

B108. Preston, B.D., Allen, J.R., *Drug Metab. Dispos.* (1980) **8**, 197–204.

B109. Egorin, M.J., Clawson, R.E., Ross, L.A., Chou, F.-T.E., Andrews, P.A., Bachur, N.R., *Drug Metab. Dispos.* (1980) **8**, 353–362.

B110. Suolinna, E.-M., *Drug Metab. Dispos.* (1980) **8**, 205–207.

B111. Levine, W.G., *Drug Metab. Dispos.* (1980) **8**, 212–217.

B112. Gaver, R.C., Vasiljev, M., Wong, H., Monkovic, I., Swigor, J.E., van Harken, D.R., Smyth, R.D., *Drug Metab. Dispos.* (1980) **8**, 230–235.

B113. Bius, D.L., Yonekawa, W.D., Kupferberg, H.J., Cantor, F., Dudley, K.H., *Drug Metab. Dispos.* (1980) **8**, 223–239.

B114. Duggan, D.E., Hook, K.F., Hwang, S.S., *Drug Metab. Dispos.* (1980) **8**, 241–246.

B115. Konishi, M., Agoh, T., Sato, T., Konaka, R., Mori, Y., *Drug Metab. Dispos.* (1980) **8**, 253–259.

B116. Burchell, B., Bock, K.W., *Biochem. Pharmacol.* (1980) **29**, 3204–3207.

B117. Hagan, R.M., Raxworthy, M.J., Gulliver, P.A., *Biochem. Pharmacol.* (1980) **29**, 3123–3126.

B118. Roth, J.A., *Biochem. Pharmacol.* (1980) **29**, 3119–3122.

B119. Loo, Y.H., Miller, K.A., Nowlin, J., Horning, M.G., *Life Sci.* (1980) **26**, 657–663.

B120. Rickert, D.E., Baker, T.S., Bus, J.S., Barrow, C.S., Irons, R.D., *Toxicol. Appl. Pharmacol.* (1979) **49**, 417–423.

B121. Tsang, C.P.W., Hackett, A.J., *Theriogenology* (1979) **11**, 429–439.

B122. Okano, P., Miller, H.N., Robinson, R.C., Gelboin, H.V., *Cancer Res.* (1979) **39**, 3184–3193.

B123. Buck, R., Eberspächer, J., Lingens, F., *Hoppe-Seyler's. Z. Physiol. Chem.* (1979) **360**, 957–969.

B124. Buswell, J.A., Eriksson, K.-E., *FEBS Lett.* (1979) **104**, 258–260.

B125. Furukawa, K., Tomizuka, N., Kamibayashi, A., *Appl. Environ. Microbiol.* (1979) **38**, 301–310.

B126. Fekkes, D., Van Overmeeren-Kaptein, E., Docter, R., Hennemann, G., Visser, T.J., *Biochim. Biophys. Acta* (1979) **587**, 12–19.

B128. Macnicoll, A.D., Burden, P.M., Robeiro, O., Hewer, A., Grover, P.L., Sims, P., *Chem.-Biol. Interact.* (1979) **26**, 121–132.

B129. Morton, K.C., King, C.M., Baetcke, K.P., *Cancer Res.* (1979) **39**, 3107–3113.

B130. Corbett, M.D., Baden, D.G., Chipko, B.R., *Bioorg. Chem.* (1979) **8**, 227–235.

B131. Schweinsberg, F., Kouros, M., *Cancer Lett. (Shannon, Irel.)* (1979) **7**, 115–120.

B132. Kartha, V.N.R., Bai, N.J., George, T., Krishnamurthy, S., *J. Biosci.* (1979) **1**, 159–168.

B133. Sugumaran, M., Vaidyanathan, C.S., *FEMS Microbiol. Lett.* (1979) **5**, 427–430.

B134. MacLeod, M.C., Cohen, G.M., Selkirk, J.K., *Cancer Res.* (1979) **39**, 3463–3470.

B135. Yang, S.K., Chou, M.W., Weems, H.B., Fu, P.P., *Biochem. Biophys. Res. Commun.* (1979) **90**, 1136–1141.

B136. Yang, I.J., Knights, B.A., Hillman, J.R., *Z. Pflanzenphysiol.* (1979) **94**, 307–316.

B137. Prakash, O., Bhakuni, D.S., Kapil, R.S., *J. Chem. Soc., Perkin Trans. 1* (1979) 1515–1518.

B138. Cone, E.J., Darwin, W.D., Gorodetzky, C.W., *J. Pharm. Pharmacol.* (1979) **31**, 314–317.

B139. Nelson, W.L., Burke, T.R., Jr., *J. Med. Chem.* (1979) **22**, 1088–1092.

B141. Gewirtz, D.A., White, J.C., Randolph, J.K., Goldmann, I.D., *Cancer Res.* (1979) **39**, 2914–2918.

B142. Bornstein, W.A., Levin, W., Thomas, P.E., Ryan, D.E., Bresnick, E., *Arch. Biochem. Biophys.* (1979) **197**, 436–446.

B143. Mythirayee, C., Krishnamurty, V., Madhavakrishna, W., *Leather Sci. (Madras)* (1979) **26**, 213–217.

B144. Cox, D.P., Goldsmith, C.D., *Appl. Environ. Microbiol.* (1979) **38**, 514–520.

B145. Blisard, K.S., Mieyal, J.J., *Biochem. Biophys. Res. Commun.* (1980) **96**, 1261–1266.

B147. Caldwell, J., O'Gorman, J., Smith, R.L., Bassir, O., Fafunso, M.A., French, M.R., *Br. J. Pharmacol.* (1979) **66**, 156P-157P.

B148. Caldwell, J., Emudianughe, T.S., Smith, R.L., *Br. J. Pharmacol.* (1979) **66**, 421P-422P.

B149. Chheda, G.B., Hong, C.I., Gurwara, S.K., Mittelman, A., *J. Pharm. Sci.* (1979) **68**, 1054–1056.

B150. Wiegerbe, W., Gerber, A., Kappler, J., Bayerl, C., *Arzneim.-Forsch.* (1979) **29**, 1083–1088.

B151. Maksay, G., Tegyey, Z., Kemény, V., Lukovits, I., Ötvös, L., Palosi, E., *J. Med. Chem.* (1979) **22**, 1436–1443.

B152. Shingu, H., Ohmizo, H., Katano, T., Naito, S., Goto, K., Fujiwara, H.,

Tatematsu, A., Yoshizumi, H., Nadai, T., *Iyakuhin Kenkyu* (1979) **10**, 623–636.
B153. Shingu, H., Uesaka, E., Fujiwara, H., *Iyakuhin Kenkyu* (1979) **10**, 615–622.
B154. Komiyama, T., Oki, T., Inui, T., Takeuchi, T., Umezawa, H., *Gann* (1979) **70**, 403–410.
B155. Al-Ani, M.R., Wood, S.G., Lawson, A., *Eur. J. Drug Metab. Pharmacokinet.* (1979) **4**, 117–119.
B156. Watabe, T., Ichihara, S., Sawahata, T., *J. Biol. Chem.* (1979) **254**, 10720–10727.
B157. Klečka, G.M., Gibson, D.T., *Biochem. J.* (1979) **180**, 639–645.
B158. Francis, M.M., Westlake, D.W.S., *Can. J. Microbiol.* (1979) **25**, 1408–1415.
B159. Fitzgerald, J.W., Maca, H.W., Rose, F.A., *J. Gen. Microbiol.* (1979) **111**, 407–415.
B161. Buswell, J.A., Clark, J.S., *J. Gen. Microbiol.* (1979) **96**, 209–213.
B162. Narasimhachari, N., *Stable Isot., Proc. Int. Conf. 3rd* (1978) 393–402.
B163. Mandelbaum-Shavit, F., Blondheim, S.H., *Biochem. Pharmacol.* (1981) **30**, 65–69.
B164. Weitering, J.G., Krijgsheld, K.R., Mulder, G.J., *Biochem. Pharmacol.* (1979) **28**, 757–762.
B165. Shirkey, R.J., Kao, J., Fry, J.R., Bridges, J.W., *Biochem. Pharmacol.* (1979) **28**, 1461–1466.
B166. Fasco, M.J., Piper, L.J., Kaminsky, L.S., *Biochem. Pharmacol.* (1979) **28**, 97–103.
B167. Siliciano, R.F., Margolis, S., Lietman, P.S., *Biochem. Pharmacol.* (1978) **27**, 2759–2762.
B168. Szelfer, S.J., Acara, M., *J. Pharmacol. Exp. Ther.* (1979) **210**, 295–300.
B169. Ahmed, N.K., Felsted, R.L., Bachur, N.R., *Biochem. Pharmacol.* (1978) **27**, 2713–2719.
B170. Son, O.S., Everett, D.W., Fiala, E.S., *Xenobiotica* (1980) **10**, 457–468.
B171. Crowley, J.P., Couch, M.W., Williams, C.M., Threatte, R.M., Fregly, M.J., *Clin. Chim. Acta* (1981) **109**, 125–131.
B172. McMahon, R.E., Turner, J.C., Whitaker, G.W., *Xenobiotica* (1980) **10**, 469–481.
B173. Yamazoe, Y., Yamaguchi, N., Kamataki, T., Kato, R., *Xenobiotica* (1980) **10**, 483–494.
B174. Eyer, P., Kampffmeyer, H., Maister, H., Rösch-Oehme, E., *Xenobiotica* (1980) **10**, 499–516.
B175. Blaauboer, B.J., van Holsteijn, C.W.M., Wit, J.G., *Xenobiotica* (1980) **10**, 495–498.
B176. Renner, G., *Xenobiotica* (1980) **10**, 537–550.
B177. Eyer, P., Lierheimer, E., *Xenobiotica* (1980) **10**, 517–526.
B178. Lenk, W., Scharmer, U., *Xenobiotica* (1980) **10**, 573–591.
B179. Razzouk, C., Mercier, M., Roberfroid, M., *Xenobiotica* (1980) **10**, 565–571.
B180. Huizing, G., Beckett, A.H., *Xenobiotica* (1980) **10**, 593–602.
B181. Gorrod, J.W., Patterson, L.H., *Xenobiotica* (1980) **10**, 603–610.
B182. Szinicz, L.L., Weger, N., *Xenobiotica* (1980) **10**, 611–620.
B183. Elbers, R., Kampffmeyer, H.G., Rabes, H., *Xenobiotica* (1980) **10**, 621–632.
B184. Stier, A., Clauss, R., Lücke, A., Reitz, I., *Xenobiotica* (1980) **10**, 661–673.
B185. Renner, G., Gaber, H., *Xenobiotica* (1980) **10**, 633–644.
B186. Ito, S., Teradaira, R., Fujita, K., *Biochem. Pharmacol.* (1980) **29**, 3277–3280.
B187. Renton, K.W., *Xenobiotica* (1980) **10**, 243–246.
B188. Wright, D.J., Holder, G.M., *Xenobiotica* (1980) **10**, 265–270.
B189. Doyle, B., Griffiths, L.A., *Xenobiotica* (1980) **10**, 247–256.
B190. Legrum, W., Netter, K.J., *Xenobiotica* (1980) **10**, 271–279.
B191. Kutzman, R.S., Meyer, G.-J., Wolf, A.P., *Xenobiotica* (1980) **10**, 281–288.
B192. Parkki, M.G., *Xenobiotica* (1980) **10**, 307–310.

B193. Facchini, V., Griffiths, L.A., *Xenobiotica* (1980) **10**, 289–297.

B194. Gottschlich, R., Metzler, M., *Xenobiotica* (1980) **10**, 317–327.

B195. Benford, D.J., Bridges, J.W., Parke, D.V., *Xenobiotica* (1980) **10**, 329–336.

B196. Delbressine, L.P.C., Ketelaars, H.C.J., Seutter-Berlage, F., Smeets, F.L.M., *Xenobiotica* (1980) **10**, 337–342.

B197. Inoue, T., Niwaguchi, T., Murata, T., *Xenobiotica* (1980) **10**, 343–348.

B198. Solheim, E., Scheline, R.R., *Xenobiotica* (1980) **10**, 371–380.

B199. Crayford, J.V., Hutson, D.H., *Xenobiotica* (1980) **10**, 355–364.

B200. Mayo, B.C., Hawkins, D.R., Hümpl, M., Chasseaud, L.F., Girkin, R., *Xenobiotica* (1980) **10**, 413–420.

B201. Hall, B.E., James, S.P., *Xenobiotica* (1980) **10**, 421–434.

B202. Ulbrich, B., Zenk, M.H., *Phytochemistry* (1980) **19**, 1625–1629.

B203. Heinsbroek, R., Van Brederode, J., Van Nigtevecht, G., Maas, J., Kamsteeg, J., Besson, E., Chopin, J., *Phytochemistry* (1980) **19**, 1935–1937.

B204. Smith, D.A., Kuhn, P.J., Bailey, J.A., Burden, R.S., *Phytochemistry* (1980) **19**, 1673–1675.

B205. Hodges, C.C., Rapoport, H., *Phytochemistry* (1980) **19**, 1681–1684.

B206. Norqvist, A., Fowler, C.J., Oreland, L., *Biochem. Pharmacol.* (1981) **30**, 403–409.

B207. Corbett, M.D., Chipko, B.R., *Bioorg. Chem.* (1980) **9**, 273–287.

B208. Akhtar, M.H., Foster, T.S., *J. Agric. Food Chem.* (1980) **28**, 698–704.

B209. Oehler, D.D., Ivie, G.W., *J. Agric. Food Chem.* (1980) **28**, 686–689.

B210. Soderlund, D.M., Messeguer, A., Bowers, W.S., *J. Agric. Food Chem.* (1980) **28**, 724–731.

B211. Gomes, U.C.R., McCarthy, B.W., Shanley, B.C., *Biochem. Pharmacol.* (1981) **30**, 571–575.

B212. Lau, S.S., Abrams, G.D., Zannoni, V.G., *J. Pharmacol. Exp. Ther.* (1980) **214**, 703–708.

B213. Lockridge, O., Mottershaw-Jackson, N., Eckerson, H.W., La Du, B.N., *J. Pharmacol. Exp. Ther.* (1980) **215**, 1–8.

B214. Davis, D.S., Hughes, H., *Br. J. Pharmacol.* (1979) **66**, 425P.

B215. Kucharczyk, N., Ballard, F., Yang, J., Myers, G., Sofia, R.D., *Arch. Int. Pharmacodyn. Ther.* (1979) **239**, 180–194.

B216. Pottier, J., Busigny, M., Raynaud, J.P., *Eur. J. Drug Metab. Pharmacokinet.* (1979) **4**, 109–115.

B217. Baurain, R., Deprez-De Campeneere, D., Trouet, A., *Cancer Chemother. Pharmacol.* (1979) **2**, 37–41.

B218. Russel, S., *Zesz. Nauk. Szk. Gl. Gospod. Wiejsk.-Akad. Roln. Warszawie, Rozpr. Nauk.* (1978) **101** (33pp).

B219. Zia, H., O'Donnell, J.P., Ma, J.K.H., *J. Pharm. Sci.* (1979) **68**, 1335–1336.

B220. Guengerich, F.P., Wang, P., Mitchell, M.B., Mason, P.S., *J. Biol. Chem.* (1979) **254**, 12248–12254.

B221. James, G.T., Austin, J.H., *Clin. Chim. Acta* (1979) **98**, 103–111.

B222. Buset, H., Scheline, R.R., *Biomed. Mass Spectrom.* (1979) **6**, 212–220.

B223. Mohan, V.P., Kishore, G., Sugumaran, M., Vaidyanathan, C.S., *Plant Sci. Lett.* (1979) **16**, 267–272.

B224. Crawford, R.L., Olson, P.E., Frick, T.D., *Appl. Environ. Microbiol.* (1979) **38**, 379–384.

B225. Murooka, Y., Doi, N., Harada, T., *Appl. Environ. Microbiol.* (1979) **38**, 565–569.

B226. Klages, U., Lingens, F., *FEMS Microbiol. Lett.* (1979) **6**, 201–203.

B227. Hufford, C.D., Collins, C.C., Clark, A.M., *J. Pharm. Sci.* (1979) **68**, 1239–1243.

B228. Dölle, B., Töpner, W., Neumann, H.-G., *Xenobiotica* (1980) **10**, 527–536.

B229. Komiyama, T., Oki, T., Inui, T., Takeuchi, T., Umezawa, H., *Gann* (1979) **70**, 395–401.

B230. Hlavica, P., Hülsmann, S., *Biochem. J.* (1979) **182**, 109–116.

B231. Kobayashi, K., Nakamura, N., *Nippon Suisan Gakkaishi* (1979) **45**, 1085–1088.

B232. Sundaram, K.M.S., Szeto, S.Y., *J. Environ. Sci. Health Part B* (1979) **B14**, 589–602.

B233. Ohkawa, H., Kaneko, H., Tsuji, H., Miyamoto, J., *Nippon Noyaku Gakkaishi* (1979) **4**, 143–155.

B234. Hecht, S.S., La Voie, E., Mazzarese, R., Hirota, N., Ohmori, T., Hoffmann, D., *JNCI, J. Nat. Cancer Inst.* (1979) **63**, 855–861.

B235. Cerniglia, C.E., Gibson, D.T., *J. Biol. Chem.* (1979) **254**, 12174–12180.

B236. Grossman, J.C., Khan, M.A.Q., *Comp. Biochem. Physiol. C* (1979) **63C**, 251–260.

B237. Engelhardt, G., Wallnöfer, P.R., Ziegler, W., *Chemosphere* (1979) **8**, 725–730.

B238. Maruyama, E., Moteki, I., Asano, Y., Sueyasu, K., Mineyuki, K., Hashimoto, K., Hasegawa, C., *Kaseigaku Zasshi* (1979) **30**, 603–607.

B239. Poulsen, L.L., Hyslop. R.M., Ziegler, D.M., *Arch. Biochem. Biophys.* (1979) **198**, 78–88.

B240. Pachecka, J., Krzywicka, K., *Pol. J. Pharmacol. Pharm.* (1979) **31**, 187–191.

B241. Kolar, G.F., Carubelli, R., *Cancer Lett. (Shannon, Irel.)* (1979) **7**, 209–214.

B242. Kucharczyk, N., Edelson, J., Sofia, R.D., Ludwig, B.J., Shahinian, S., Schuster, E., Ballard, F.H., Yang, J., Myers, G., *Arzneim.-Forsch.* (1979) **29**, 1550–1556.

B243. Moretó, M., Goñalons, E., Mylonakis, N., Giráldez, A., Torralba, A., *Arzneim.-Forsch.* (1979) **29**, 1561–1564.

B244. Hackett, A.M., Griffiths, L.A., *Biochem. Soc. Trans.* (1979) **7**, 645–646.

B245. Noda, K., Suzuki, A., Okui, M., Noguchi, H., Nishiura, M., Nishiura, N., *Arzneim.-Forsch.* (1979) **29**, 1595–1600.

B247. Hara, S., Hayasaka, S., Mizuno, K., *Exp. Eye Res.* (1979) **29**, 277–289.

B248. Tanizawa, K., Soda, K., *J. Biochem. (Tokyo)* (1979) **86**, 1199–1209.

B249. Müller, F., Voordouw, G., Van Berkel, W.J.H., Steenis, P.J., Visser, S., Van Rooijen, P.J., *Eur. J. Biochem.* (1979) **101**, 235–244.

B250. Otsuka, H., Anderson, J.A., Floss, H.G., *J. Chem. Soc., Chem. Commun.* (1979) 660–662.

B251. Seifert, J., Casida, J.E., *Comp. Biochem. Physiol. C* (1979) **63C**, 123–127.

B252. André, J.L., Dring, L.G., Gillet, G., Mas-Chamberlin, C., *Br. J. Pharmacol.* (1979) **66**, 506P.

B253. Caldwell, J., Emudianughe, T.S., *Biochem. Soc. Trans.* (1979) **7**, 521–522.

B255. Cautreels, W., Martens, M., Roncucci, R., Gillet, C., Debast, K., Lambelin, G., *Rec. Dev. Mass Spectrom. Biochem. Med.* (1979) **2**, 85–96.

B256. Gaetani, M., Yamaguchi, H., Vidi, A., Hashimoto, Y., Donetti, A., *Pharmacol. Res. Commun.* (1979) **11**, 719–730.

B258. Yamamoto, T., Honjo, H., Yamamoto, T., Oshima, K., Otsubo, K., Okada, H., Shibata, K., *J. Steroid Biochem.* (1979) **11**, 1287–1294.

B259. Von Hofe, E.H., Duncan, K.L., Beal, M.L., Mondal, S., Puffer, H.W., *Proc. West. Pharmacol. Soc.* (1979) **22**, 281–284.

B260. Ueberschär, K.H., Kible, S., Laule, G., Maurer, P., Wallenfels, K., *Hoppe-Seyler's Z. Physiol. Chem.* (1979) **360**, 1409–1419.

B261. Batt, A.M., MacKenzie, P., Hanninen, O., Vainio, H., *Med. Biol.* (1979) **57**, 281–286.

B262. Durham, D.R., Stirling, L.A., Ornston, L.N., *Biochemistry* (1980) **19**, 149–155.

B263. Nurmann, G., Strack, D., *Z. Naturforsch. C, Biosci.* (1979) **34C**, 715–720.

B265. Treimer, J.F., Zenk, M.H., *Eur. J. Biochem.* (1979) **101**, 225–233.

B266. Fazel, A.M., Jensen, R.A., *J. Bacteriol.* (1979) **140**, 580–587.

B267. Jigami, Y., Kawasaki, Y., Omori, T., Minoda, Y., *Appl. Environ. Microbiol.* (1979) **38**, 783–788.

B268. Davis, P.J., Glade, J.C., Clark, A.M., Smith, R.V., *Appl. Environ. Microbiol.* (1979) **38**, 891–893.

B269. Johnson, D.B., Swanson, M.J., Barker, C.W., Fanska, C.B., Murrill, E.E., *Prep. Biochem.* (1979) **9**, 391–406.

B270. Nagakura, N., Rüffer, M., Zenk, M.H., *J. Chem. Soc., Perkin Trans. 1* (1979) 2308–2312.

B273. Fell, V., Greenway, A.M., Hoskins, J.A., *Biochem. Med.* (1979) **22**, 246–255.

B275. Yoshida, K., Nambu, K., Arakawa, S., Miyazaki, H., Hashimoto, M., *Biomed. Mass Spectrom.* (1979) **6**, 253–259.

B276. Casalini, C., Cesarano, G., Mascellani, G., Tamagnone, G., Giumanini, A.G., *Recent Dev. Mass Spectrom. Biochem. Med.* (1979) **2**, 107–119.

B277. Cerna, J., Zidek, Z., Janku, I., *J. Pharm. Pharmacol.* (1979) **31**, 711–712.

B278. Suzuki, T., Rikihisa, T., *Arzneim.-Forsch.* (1979) **29**, 1707–1715.

B279. Egyházi, E., Shugar, D., *FEBS Lett.* (1979) **107**, 431–435.

B280. Cappellini, L., De Pascale, A., Negrini, P., Frigerio, A., Rossi, E., Testoni, G., *Boll. Chim. Farm.* (1979) **118**, 316–331.

B281. Mitchell, S.C., Waring, R.H., *IRCS Med. Sci.: Libr. Compend.* (1979) **7**, 582.

B282. Guillemin, M.P., Bauer, D., *Int. Arch. Occup. Environ. Health* (1979) **44**, 249–263.

B283. Lin, S.-N., Chen, C.-Y., Murphy, S.D., Caprioli, R.M., *J. Agric. Food Chem.* (1980) **28**, 85–88.

B284. Eastman, A., Bresnick, E., *Cancer Res.* (1979) **39**, 4316–4321.

B285. MacNicoll, A.D., Burden, P.M., Rattle, H., Grover, P.L., Sims, P., *Chem.-Biol. Interact.* (1979) **27**, 365–379.

B286. Ball. L.M., Plummer, J.L., Smith, B.R., Bend, J.R., *Med. Biol.* (1979) **57**, 298–305.

B287. Grafström, R., Moldéus, P., Andersson, B., Orrenius. S., *Med. Biol.* (1979) **57**, 287–293.

B288. Son, O.S., Fowble, J.W., Miller, D.D., Feller, D.R., *Toxicol. Appl. Pharmacol.* (1979) **51**, 367–377.

B289. Anderson, N.G., Lieberman, S., *Endocrinology (Baltimore)* (1980) **106**, 13–18.

B290. Rubin, J.L., Jensen, R.A., *Plant Physiol.* (1979) **64**, 727–734.

B291. Subramanian, V., Liu, T.-N., Yeh, W.K., Gibson, D.T., *Biochem. Biophys. Res. Commun.* (1979) **91**, 1131–1139.

B292. Yamada, H., Kumagai, H., *Pure Appl. Chem.* (1978) **50**, 1117–1127.

B293. Ruddick, J.A., Craig, J., Stavric, B., Willes, R.F., Collins, B., *Food Cosmet. Toxicol.* (1979) **17**, 435–442.

B294. Varanasi, U., Gmur, D.J., Treseler, P.A., *Arch. Environ. Contam. Toxicol.* (1979) **8**, 673–692.

B295. Griffiths, L.A., Facchini, V., *Rec. Dev. Mass Spectrom. Biochem. Med.* (1979) **2**, 121–126.

B297. Short, R.D., Dacre, J.C., Lee, C.C., *Experientia* (1979) **35**, 1625–1627.

B298. Zenk, M.H., Ulbrich, B., Busse, J., Stöckigt, J., *Anal. Biochem.* (1979) **101**, 182–187.

B299. Mayer, R., Widom, J., Que, L., Jr., *Biochem. Biophys. Res. Commun.* (1980) **92**, 285–291.

B300. Manley, S.L., Chapman, D.J., *Plant Physiol.* (1979) **64**, 1032–1038.

B301. Buswell, J.A., Hamp, S., Eriksson, K.E., *FEBS Lett.* (1979) **108**, 229–232.

B303. Dawson, J.R., Bridges, J.W., *Biochem. Soc. Trans.* (1979) **7**, 1106–1107.

B304. Kiwada, H., Awazu, S., Hanano, M., *J. Pharmacobio.-Dyn.* (1979) **2**, 356–364.

B305. Bässman, H., Böttcher, J., Schüppel, R., *Naunyn-Schmiedeberg's Arch. Pharmacol.* (1979) **309**, 203–205.

B306. Chien, M.M., Rosazza, J.P., *J. Nat. Prod.* (1979) **42**, 643–647.
B307. Kataoka, S., Naito, T., *Chem. Pharm. Bull.* (1979) **27**, 2913–2920.
B308. Nakao, M., Takeda, Y., *Yakugaku Zasshi* (1979) **99**, 1149–1153.
B309. Riihimaki, V., Pfaffli, P., Savolainen, K., Pekari, K., *Scand. J. Work Environ. Health* (1979) **5**, 217–231.
B310. Mehta, R., Meredith-Brown, M., Cohen, G.M., *Chem.-Biol. Interact.* (1979) **28**, 345–358.
B311. Kaminsky, L.S., Fasco, M.J., Guengerich, F.P., *J. Biol. Chem.* (1980) **255**, 85–91.
B312. Yajima, Y., Enoki, A., Mayfield, M.B., Gold, M.H., *Arch. Microbiol.* (1979) **123**, 319–321.
B313. Schöppner, A., Kindl, H., *FEBS Lett.* (1979) **108**, 349–352.
B314. Kampa, L., Peisach, J., *J. Biol. Chem.* (1980) **255**, 595–601.
B315. Bacher, A., Baur, R., Eggers, U., Harders, H.D., Otto, M.K., Schnepple, H., *J. Biol. Chem.* (1980) **255**, 632–637.
B316. Parry, R.J., Kurylo-Borowska, Z., *J. Am. Chem. Soc.* (1980) **102**, 836–837.
B317. Sakurai, Y., Isobe, K., Shiota, H., *Agric. Biol. Chem.* (1980) **44**, 1–7.
B319. Nakagawa, A., Ohuchi, R., Yoshizawa, I., *J. Pharmacobio.-Dyn.* (1979) **2**, 365–373.
B320. Farmer, P.B., Foster, A.B., Jarman, M., Newell, D.R., Oddy, M.R., Kiburis, J.H., *Chem.-Biol. Interact.* (1979) **28**, 211–224.
B321. Balba, M.T., Clarke, N.A., Evans, W.C., *Biochem. Soc. Trans.* (1979) **7**, 1115–1116.
B322. Shindo, T., Yasuda, Y., Taira, K., Mitani, N., Kanda, A., Akazawa, A., *Yakugaku Zasshi* (1979) **99**, 1186–1200.
B323. Walle, T., Conradi, E.C., Walle, U.K., Fagan, T.C., Gaffney, T.E., *Clin. Pharmacol. Ther.* (1980) **27**, 23–31.
B324. Sossi, N., Strong, J.M., Dingell, J.V., *Pharmacology* (1980) **20**, 173–180.
B325. O'Dea, R.F., Mirkin, B.L., *Clin. Pharmacol. Ther.* (1980) **27**, 37–43.
B326. Tabata, K., Miyata, T., Saito, T., *Appl. Entomol. Zool.* (1979) **14**, 490–493.
B327. Subramanian, V., Sugumaran, M., Vaidyanathan, C.S., *Indian J. Biochem. Biophys.* (1979) **16**, 370–374.
B328. Murakami, N., Yoshikawa, N., Hisatsuka, K., *Jpn. Kokai Tokkyo Koho* (5–1–1980) 80 00,307
B329. Robbers, J.E., Otsuka, H., Floss, H.G., Arnold, E.V., Clardy, J., *J. Org. Chem.* (1980) **45**, 1117–1121.
B330. Shigematsu, H., Yoshida, K., *Chem. Pharm. Bull.* (1979) **27**, 3002–3008.
B331. Elvin, A.T., Keenaghan, J.B., Byrnes, E.W., Tenthorey, P.A., McMaster, P.D., Takman, B.H., Lalka, D., Manion, C.V., Baer, D.T., Wolshin, E.M., Meyer, M.B., Ronfeld, R.A., *J. Pharm. Sci.* (1980) **69**, 47–49.
B332. Solbakken, J.E., Palmork, K.H., Neppelberg, T., Scheline, R.R., *Acta Pharmacol. Toxicol.* (1980) **46**, 127–132.
B333. Lemli, J., Lemmens, L., *Pharmacology* (1980) **20** (Suppl. 1), 50–57.
B334. Singh, H., Kalnitsky, G., *J. Biol. Chem.* (1980) **255**, 369–374.
B335. Troost, T., Hitchcock, M.J.M., Katz, E., *Biochim. Biophys. Acta* (1980) **612**, 97–106.
B336. Schreiber, A., Hellwig, M., Dorn, E., Reineke, W., Knackmuss, H.J., *Appl. Environ. Microbiol.* (1980) **39**, 58–67.
B337. Cook, A.M., Daughton, C.G., Alexander, M., *Biochem. J.* (1979) **184**, 453–455.
B338. Carlisle, T.L., Suttie, J.W., *Biochemistry* (1980) **19**, 1161–1167.
B339. Engesser, K.H., Schmidt, E., Knackmuss, H.J., *Appl. Environ. Microbiol.* (1980) **39**, 68–73.
B341. Melzacka, M., Boksa, J., Maj, J., *J. Pharm. Pharmacol.* (1979) **31**, 855–856.
B342. Hoffmann, K.J., Skånberg, I., Borg, K.O., *Eur. J. Drug Metab. Pharmacokinet.* (1979) **4**, 163–173.

B343. Jauch, V.R., Griesser, E., Oesterhelt, G., *Arzneim.-Forsch.* (1980) **30**, 60–67.

B344. Baines, A.D., Chan, W., *Life Sci.* (1980) **26**, 253–259.

B345. Uba, K., Miyata, N., Watanabe, K., Hirobe, M., *Chem. Pharm. Bull.* (1980) **28**, 356–358.

B347. Stocklinski, A.W., Ware, O.B., Oberst, T.J., *Life Sci.* (1980) **26**, 365–369.

B348. Wong, L.T., Whitehouse, L.W., Solomonraj, G., Paul, C.J., Thomas, B.H., *J. Anal. Toxicol.* (1979) **3**, 260–263.

B350. Vermorken, A.J.M., Goos, C.M., Rölofs, H.M.G., Henderson, P.T., Bloemendal, H., *Toxicology* (1979) **14**, 109–116.

B351. Keller, E., Eberspächer, J., Lingens, F., *Z. Naturforsch. C: Biosci.* (1979) **34C**, 914–922.

B352. Heller, W., Hahlbrock, K., *Arch. Biochem. Biophys.* (1980) **200**, 617–619.

B353. Cerniglia, C.E., Dodge, R.H., Gibson, D.T., *Bot. Mar.* (1980) **23**, 121–124.

B354. Kondo, S., Kuchiki, A., Yamamoto, K., Akimoto, K., Takahashi, K., Awata, N., Sugimoto, I., *Chem. Pharm. Bull.* (1980) **28**, 1–7.

B355. Nozu, T., Suwa, T., Migita, Y., Tanaka, I., *Oyo Yakuri* (1979) **18**, 497–505.

B356. Murai, Y., Nakagawa, T., Uno, T., *Chem. Pharm. Bull.* (1980) **28**, 362–364.

B357. Irikura, T., Kasuga, K., Abe, Y., Kasai, S., Kinoshita, U., *Oyo Yakuri* (1979) **18**, 619–634.

B358. Klečka, G.M., Gibson, D.T., *Appl. Environ. Microbiol.* (1980) **39**, 288–296.

B359. Takahashi, O., Hiraga, K., *Food Cosmet. Toxicol.* (1979) **17**, 451–454.

B360. Klamann, E., Lingens, F., *Z. Naturforsch. C, Biosci.* (1980) **35C**, 80–87.

B361. Daumy, G.O., McColl, A.S., Andrews, G.C., *J. Bacteriol.* (1980) **141**, 293–296.

B362. Leete, E., *J. Nat. Prod.* (1980) **43**, 130–135.

B363. Healy, J.B., Jr., Young, L.Y., Reinhard, M., *Appl. Environ. Microbiol.* (1980) **39**, 436–444.

B365. Cox, D.P., Williams, A.L., *Appl. Environ. Microbiol.* (1980) **39**, 320–326.

B366. Kahn, G.C., Boobis, A.R., Blair, L., Brodie, M.J., Davies, D.S., *Br. J. Clin. Pharmacol.* (1980) **9**, 284P.

B367. Binder, M., Barlage, U., *Helv. Chim. Acta* (1980) **63**, 255–267.

B368. Anderson, J.J., Dagley, S., *J. Bacteriol.* (1980) **141**, 534–543.

B369. Poulton, J.E., McRee, D.E., Conn, E.E., *Plant Physiol.* (1980) **65**, 171–175.

B370. Pan, S.-S., Bachur, N.R., *Mol. Pharmacol.* (1980) **17**, 95–99.

B371. Rueffer, M., Kan-Fan, C., Husson, H.P., Stöckigt, J., Zenk, M.H., *J. Chem. Soc., Chem. Commun.* (1979) 1016–1018.

B372. Nozu, T., Suwa, T., Fukushima, K., Ohzeki, M., *Oyo Yakuri* (1979) **18**, 835–843.

B373. Hornke, I., Fehlhaber, H.W., Girg, M., Jantz, H., *Br. J. Clin. Pharmacol.* (1980) **9**, 255–264.

B374. Ofuji, T., Kawahara, K., Matsuda, T., Uda, K., Koutani, Y., *Oyo Yakuri* (1979) **18**, 1055–1062.

B375. Janssen, F.W., Jusko, W.J., Chiang, S.T., Kirkman, S.K., Southgate, P.J., Coleman, A.J., Ruelius, H.W., *Clin. Pharmacol. Ther.* (1980) **27**, 352–362.

B376. Macnicoll, A.D., Grover, P.L., Sims, P., *Chem.-Biol. Interact.* (1980) **29**, 169–188.

B377. Sütfeld, R., Wiermann, R., *Arch. Biochem. Biophys.* (1980) **201**, 64–72.

B378. Sayo, H., Hosokawa, M., *Chem. Pharm. Bull.* (1980) **28**, 683–685.

B379. Noguchi, T., Hayashi, S., *J. Biol. Chem.* (1980) **255**, 2267–2269.

B380. Coutts, R.T., Foster, B.C., *Can. J. Microbiol.* (1980) **26**, 343–349.

B381. Ander, P., Hatakka, A., Eriksson, K.E., *Arch. Microbiol.* (1980) **125**, 189–202.

B382. Fuchs, A., de Vries, F.W., Platero Sanz, M., *Physiol. Plant Pathol.* (1980) **16**, 119–133.

B383. Cerniglia, C.E., Van Baalen, C., Gibson, D.T., *Arch. Microbiol.* (1980) **125**, 203–207.

B384. Brandão, F., Paiva, M.Q., Guimarães, S., *Naunyn-Schmiedeberg's Arch. Pharmacol.* (1980) **311**, 1–7.

B385. van Brederode, J., Chopin, J., Kamsteeg, J., van Nigtevecht, G., Heinsbroek, R., *Phytochemistry* (1979) **18**, 655–656.

B386. Sawada, H., Hara, A., *Biochem. Pharmacol.* (1979) **28**, 1089–1094.

B387. Autrup, H., *Biochem. Pharmacol.* (1979) **28**, 1727–1730.

B388. Duquette, P.H., Holtzman, J.L., *J. Pharmacol. Exp. Ther.* (1979) **211**, 213–218.

B389. Suzuki, O., *Biochem. Pharmacol.* (1979) **28**, 2327–2332.

B390. Siddik, Z.H., Barnes, R.D., Dring, L.G., Smith, R.L., Williams, R.T., *Biochem. Pharmacol.* (1979) **28**, 3093–3101.

B391. Tierney, B., Bresnick, E., Sims, P., Grover, P.L., *Biochem. Pharmacol.* (1979) **28**, 2607–2610.

B392. Mehta, R., Cohen, G.M., *Biochem. Pharmacol.* (1979) **28**, 2479–2484.

B393. Willes, R.F., Stavric, B., Craig, J.C., Ruddick, J.A., *Toxicol. Appl. Pharmacol.* (1980) **54**, 276–284.

B394. Böttcher, J., Bässman, H., Schüppel, R., *Naunyn-Schmiedeberg's Arch. Pharmacol.* (1981) **316** (Suppl.), R5.

B395. Böttcher, J., Bässman, H., Schüppel, R., *Biochem., Biophys. Regul. Cytochrome P-450, Proc. Eur. Meet. Cytochrome P-450, 3rd.* (1980) 81–83.

B396. Sullivan, H.R., Miller, W.M., Stark, D.G., Wood, P.G., *Xenobiotica* (1981) **11**, 9–22.

B397. Tsunoda, N., Yoshimura, H., *Xenobiotica* (1981) **11**, 23–32.

B398. Maurer, G., Donatsch, P., Galliker, H., Kiechel, J.R., Meier, J., *Xenobiotica* (1981) **11**, 33–41.

B399. Liddle, C., Graham, G.G., Christopher, R.K., Bhuwapathanapun, S., Duffield, A.M., *Xenobiotica* (1981) **11**, 81–87.

B400. Greenslade, D., Havler, M.E., Humphrey, M.J., Jordan, B.J., Lewis, C.J., Rance, M.J., *Xenobiotica* (1981) **11**, 89–96.

B401. Michelot, J., Madelmont, J.C., Jordan, D., Mornex, R., Meyniel, G., *Xenobiotica* (1981) **11**, 123–130.

B402. Ahmed, N.K., Felsted, R.L., Bachur, N.R., *Xenobiotica* (1981) **11**, 131–136.

B403. Midha, K.K., Cooper, J.K., Bailey, K., Hubbard, J.W., *Xenobiotica* (1981) **11**, 137–146.

B404. Kamsteeg, J., van Brederode, J., van Nigtevecht, G., *Phytochemistry* (1980) **19**, 1459–1462.

B405. Kamsteeg, J., van Brederode, J., van Nigtevecht, G., *Phytochemistry* (1979) **18** 659–660.

B407. Kamsteeg, J., van Brederode, J., van Nigtevecht, G., *Biochem. Genet.* (1978) **16**, 1045–1058.

B408. Kamsteeg, J., van Brederode, J., van Nigtevecht, G., *Biochem. Genet.* (1978) **16**, 1059–1071.

B409. Kamsteeg, J., van Brederode, J., van Nigtevecht, G., *Z. Naturforsch.* (1980) **35C**, 249–257.

B410. Kamsteeg, J., van Brederode, J., van Nigtevecht, G., *Z. Pflanzenphysiol.* (1980) **96**, 87–93.

B411. Kamsteeg, J., van Brederode, J., Hommels, C.H., van Nigtevecht, G., *Biochem. Physiol. Pflanz* (1980) **175**, 403–411.

B412. Ruo, T.I., Morita, Y., Atkinson, A.J., Jr., Henthorn, T., Thenot, J.-P., *J. Pharmacol. Exp. Ther.* (1980) **216**, 357–362.

B413. Edelman, A.M., Raese, J.D., Lazar, M.A., Barchas, J.D., *J. Pharmacol. Exp. Ther.* (1980) **216**, 647–653.

B414. Varanasi, U., Gmur, D.J., Reichert, W.L., *Arch. Environ. Contam. Tox.* (1981) **10**, 203–214.

B415. Otsuka, H., Quigley, F.R., Gröger, D., Anderson, J.A., Floss, H.G., *Planta Med.* (1980) **40**, 109–119.

B416. Ohlsson, A., Emanuelson, I., *Acta Pharm. Suec.* (1979) **16**, 396–407.

B417. Kataoka, S., Nishimura, K., Naito, T., *Chem. Pharm. Bull.* (1979) **27**, 2890–2903.

B418. Feniou, C., Neau, B., Prat, G., Cheze, C., Fauran, F., Roquebert, J., *J. Pharm. Pharmacol.* (1980) **32**, 104–107.

B419. Kageyama, T., Oono, F., Yasuda, Y., Shindo, T., Mitani, N., Maruden, A., Akazawa, A., *Rinsho Yakuri* (1979) **10**, 525–533.

B420. Bornschein, I., Kraft, R., Pfeifer, S., Ullman, M., Langner, H., *Pharmazie* (1979) **34**, 732–735.

B421. Berlin, C.M., Jr., Yaffe, S.J., *Dev. Pharmacol. Ther.* (1980) **1**, 31–39.

B422. Bornschein, I., Pfeifer, S., *Pharmazie* (1979) **34**, 750.

B423. Grantham, P.H., Mohan, L., Benjamin, T., Roller, P.P., Miller, J.R., Weisburger, E.K., *J. Environ. Pathol. Toxicol.* (1979) **3**, 149–166.

B424. Salcher, O., Van Pee, K.H., *Z. Naturforsch. C, Biosci.* (1980) **35C**, 340–341.

B425. Novak, R.F., Koop, D.R., Hollenberg, P.F., *Mol. Pharmacol.* (1980) **17**, 128–136.

B426. Lipsky, R.H., Hylemon, P.B., *Biochim. Biophys. Acta* (1980) **612**, 328–336.

B427. Fekkes, D., Van Overmeeren, E., Hennemann, G., Visser, T.J., *Biochim. Biophys. Acta* (1980) **613**, 41–51.

B428. Hörig, J.A., Renz, P., *Eur. J. Biochem.* (1980) **105**, 587–592.

B429. Cerniglia, C.E., Van Baalen, C., Gibson, D.T., *J. Gen. Microbiol.* (1980) **116**, 485–494.

B430. Tsujimura, R., Akeda, S., *Mie. Med. J.*, (1979) **29**, 99–108.

B431. Sawada, H., Hara, A., Nakayama, T., Kato, F., *J. Biochem.* (1980) **87**, 1153–1165.

B432. Bhakuni, D.S., Jain, S., Chaturvedi, R., *Tetrahedron* (1979) **35**, 2323–2326.

B433. Michalczuk, L., Bandurski, R.S., *Biochem. Biophys. Res. Commun.* (1980) **93**, 588–592.

B434. Schwartz, R.D., Williams, A.L., Hutchinson, D.B., *Appl. Environ. Microbiol.* (1980) **39**, 702–708.

B435. Valcavi, U., Caponi, R., Martelli, P., Minoja, F., *Eur. J. Drug Metab. Pharmacokinet.* (1979) **4**, 231–235.

B436. Pritchard, J.B., Anderson, J.B., Rall, D.P., Guarino, A.M., *Comp. Biochem. Physiol. C* (1980) **65C**, 99–104.

B437. Hümpel, M., Illi, V., Milius, W., Wendt, H., Kurowski, M., *Eur. J. Drug Metab. Pharmacokinet.* (1979) **4**, 237–243.

B438. Collier, V.U., Butler, D.O., Mitch, W.E., *Amer. J. Physiol.* (1980) **238**, E450-E457.

B439. Bergot, B.J., Judy, K.J., Schooley, D.A., Tsai, L.W., *Pestic. Biochem. Physiol.* (1980) **13**, 95–104.

B440. Strack, D., *Z. Naturforsch. C, Biosci.* (1980) **35C**, 204–208.

B441. Nomeir, A.A., Hajjar, N.P., Hodgson, E., Dauterman, W.C., *Pestic. Biochem. Physiol.* (1980) **13**, 112–120.

B442. Ibrahim, R.K., Boulay, B., *Plant Sci. Lett.* (1980) **18**, 177–184.

B443. Blumauerova, M., Kralovcova, E., Mateju, J., Jizba, J., Vanek, Z., *Folia Microbiol. (Prague)* (1979) **24**, 117–127.

B444. Sugden P.H., *FEBS Lett.* (1980) **114**, 127–131.

B445. Eggeling, L., Sahm, H., *Arch. Microbiol.* (1980) **126**, 141–148.

B446. Rominger, K.L., Hitzenberger, G., *Int. J. Clin. Pharmacol. Ther. Toxicol.* (1980) **18**, 150–157.

B447. Pratt, G.E., Jennings, R.C., Hamnett, A.F., Brooks, G.T., *Nature* (1980) **284**, 320–323.

B448. Sparling, J., Fung, D., Safe, S., *Biomed. Mass Spectrom.* (1980) **7**, 13–19.

B449. Jordan, T.W., Smith, J.N., Vaughn, L., *Biochem. Syst. Ecol.* (1980) **8**, 101–103.

B450. Kato, S., McKinney, J.D., Matthews, H.B., *Toxicol. Appl. Pharmacol.* (1980) **53**, 389–398.

B451. Størmer, F.C., Pederson, J.I., *Appl. Environ. Microbiol.* (1980) **39**, 971–975.

B452. Cerniglia, C.E., Gibson, D.T., *J. Biol. Chem.* (1980) **255**, 5159–5163.

B453. Perin-Roussel, O., Croisy-Delcey, M., Mispelter, J., Saguem, S., Chalvet, O., Ekert, B., Fouquet, J., Jacquignon, P., Lhoste, J.M., Muel, B., Zajela, F., *Cancer Res.* (1980) **40**, 1742–1749.

B454. Järvenpää, P., Kosunen, T., Fotsis, T., Adlercrentz, H., *J. Steroid Biochem.* (1980) **13**, 345–349.

B455. Krauss, G., Süssmuth, R., Lingens, F., *Hoppe-Seyler's Z. Physiol. Chem.* (1980) **361**, 801–807.

B456. Diers, I.V., Emborg, C., *Br. U.K. Patent Appl.* (28–11–1979) 2, 021, 119.

B457. Bélanger, P.M., Lalande, M., *Life Sci.* (1980) **26**, 2037–2044.

B458. Harvey, D.J., Martin, B.R., Paton, W.D.M., *J. Pharm. Pharmacol.* (1980) **32**, 267–271.

B461. Hart, S., Calder, I., Ross, B., Tange, J., *Clin. Sci.* (1980) **58**, 379–384.

B462. Matějů, J., Cudlín, J., Steinerová, N., Blumauerová, M., Vaněk, Z., *Folia Microbiol. (Prague)* (1979) **24**, 205–210.

B463. Matějů, J., Cudlín, J., Steinerová, N., Blumauerová, M., Vaněk, Z., *Folia Microbiol. (Prague)* (1978) **23**, 337–340.

B464. Battersby, A.R., Chrystal, E.J.J., *J. Chem. Soc., Perkin Trans 1* (1980) 31–32.

B465. Yamaguchi, M., Fujisawa, H., *J. Biol. Chem.* (1980) **255**, 5058–5063.

B466. Kucher, R.V., Turovskii, A.A., Dzumedzei, N.V., Shevchenko, A.G., *Mikrobiol. Zh. (Kiev)* (1980) **42**, 284–287.

B467. Lamm, L., Hörig, J.A., Renz, P., Heckmann, G., *Eur. J. Biochem.* (1980) **109**, 115–118.

B468. Cacini, W., Ritschel, W.A., *Arch. Int. Pharmacodyn. Ther.* (1980) **243**, 197–208.

B469. Kveder, S., Medugorac, A., Plavšić, V., Iskrić, S., *Period. Biol.* (1978) **80**, 163–167.

B470. Fujise, H., *Nagoya Med. J.* (1977) **22**, 185–194.

B471. Lhoest, G., Poupaert, J.H., Claesen, M., *Eur. J. Mass Spectrom. Biochem., Med. Environ. Res.* (1980) **1**, 57–59.

B472. Hermansson, J., Glaumann, H., Karlen, B., Von Bahr, C., *Acta Pharmacol. Toxicol.* (1980) **47**, 49–52.

B473. Rovei, V., Sanjuan, M., Mitchard, M., *Ann. Chim. (Rome)* (1977) **67**, 733–743.

B475. Parli, C.K., Krieter, P., Schmidt. B., *Arch. Biochem. Biophys.* (1980) **203**, 161–166.

B476. Kimura, T., Kodama, M., Nagata, C., *Gann* (1980) **71**, 417–418.

B477. Alpert, A.J., Cavalieri, E.L., *J. Med. Chem.* (1980) **23**, 919–927.

B478. Kametani, T., Kanaya, N., Ohta, Y., Ihara, M., *Heterocycles* (1980) **14**, 963–970.

B479. Szwajcer, E., Szewczuk, A., Mordarski, M., *Acta Biochim. Pol.* (1980) **27**, 123–134.

B480. Nixon, P.F., Bertino, J.R., *Methods Enzymol.* (1980) **66**, 547–553.

B481. Waheed, A., Van Etten, R.L., *Biochim. Biophys. Acta* (1980) **614**, 92–101.

B482. Katayama, T., Nakatsubo, F., Higuchi, T., *Arch. Microbiol.* (1980) **126**, 127–132.

B483. Gunasekaran, M., *Mycologia* (1980) **72**, 578–585.

B484. Ishimitsu, S., Fujimoto, S., Ohara, A., *Chem. Pharm. Bull.* (1980) **28**, 1653–1655.

B485. Stöckigt, J., Rueffer, M., Kan-Fan, C., Husson, H.P., *Planta Med.* (1980) **39**, 73–76.

B486. Dey, P.M., Kauss, H., *Phytochemistry* (1981) **20**, 45–48.

B487. Vance, C.P., Bryan, J.W., *Phytochemistry* (1981) **20**, 41–43.

B488. Etcheberrigaray, J.L., Vattuone, M.A., Sampietro, A.R., *Phytochemistry* (1981) **20**, 49–51.

B489. Sharma, R.C., Ali, R., *Phytochemistry* (1981) **20**, 399–401.

B490. Collendavelloo, J., Legrand, M., Geoffroy, P., Barthelemy, J., Fritig, B., *Phytochemistry* (1981) **20**, 611–616.

B491. Bar-Nun, N., Mayer, A.M., Sharon, N., *Phytochemistry* (1981) **20**, 407–408.

B492. Jaroszewski, J.W., Ettlinger, M.G., *Phytochemistry* (1981) **20**, 819–821.

B493. Nagels, L., Molderez, M., Parmentier, F., *Phytochemistry* (1981) **20**, 965–967.

B494. Takada, M., Agata, I., Sakamoto, M., Yagi, N., Hayashi, N., *J. Toxicol. Sci.* (1979) **4**, 341–349.

B495. Komp, M., Hess, D., *Phytochemistry* (1981) **20**, 973–976.

B496. Tateishi, M., Shiozaki, S., Iida, Y., Koitabashi, C., *Oyo Yakuri* (1980) **19**, 191–199.

B497. Ohi, T., Shibamoto, N., Iguchi, H., Shirai, M., *Jpn. J. Antibiot.* (1980) **33**, 163–168.

B498. Lhoest, G., *Eur. J. Drug Metab. Pharmacokinet.* (1980) **5**, 41–43.

B499. Oketani, Y., Kobayashi, N., Mitsuzono, T., Uesugi, T., *Oyo Yakuri* (1980) **19**, 383–393.

B500. Donoso, L.A., Spikes, J.D., *Enzyme* (1980) **25**, 111–117.

B501. Donnelly, I.M., Dagley, S., *J. Bacteriol.* (1980) **142**, 916–924.

B502. Lambert, M.A., Moss, C.W., *J. Clin. Microbiol.* (1980) **12**, 291–293.

B503. Buswell, J.A., Paterson, A., Salkinoja-Salonen, M.S., *FEMS Microbiol. Lett.* (1980) **8**, 135–137.

B504. Auf dem Brinke, D., Koehrle, J., Koedding, R., Hesch, R.D., *J. Endocrinol. Invest.* (1980) **3**, 73–76.

B505. Burke, T.R., Jr., Howald, W.N., Nelson, W.L., *Res. Commun. Chem. Pathol. Pharmacol.* (1980) **28**, 399–412.

B506. Mimura, K., Baba, S., *Chem. Pharm. Bull.* (1980) **28**, 1704–1710.

B507. Chan, K.K., Nayar, M.S.B., Cohen, J.L., Chlebowski, R.T., Liebman, H., Stolinsky, D., Mitchell, M.S., Farquhar, D., *Res. Commun. Chem. Pathol. Pharmacol.* (1980) **28**, 551–561.

B508. Coutts, R.T., Foster, B.C., Pasutto, F.M., *Can. J. Microbiol.* (1980) **26**, 844–849.

B509. Brunström, A., Ingelman-Sundberg, M., *Biochem. Biophys. Res. Commun.* (1980) **95**, 431–439.

B510. Dieterle, W., Faigle, J.W., Moppert, J., *Arzneim.-Forsch.* (1980) **30**, 989–993.

B511. Palyusic, M., Hagler, W.M., Horvath, L., Mirocha, C.J., *Acta Vet. Acad. Sci. Hung.* (1980) **28**, 159–166.

B512. Chen, S.-C., Nakamura, H., Tamura, Z., *Chem. Pharm. Bull.* (1980) **28**, 1403–1407.

B513. Ishimitsu, T., Hirose, S., *Chem. Pharm. Bull.* (1980) **28**, 2272–2273.

B514. Cerniglia, C.E., Mahaffey, W., Gibson, D.T., *Biochem. Biophys. Res. Commun.* (1980) **94**, 226–232.

B516. Ishii, H., Niwaguchi, T., Nakahara, Y., Hayashi, M., *J. Chem. Soc., Perkin Trans. 1* (1980) 902–905.

B517. Cooper, R.A., Skinner, M.A., *J. Bacteriol.* (1980) **143**, 302–306.

B518. Inoue, K., Ueda, S., Shirobara, Y., Inoye, H., *Tetrahedron Lett.* (1980) **21**, 621–622.

B519. Miyazaki, T., Mizukoshi, H., Araki, Y., Shimizu, N., *Endocrinol. Jpn.* (1980) **27**, 175–182.

B521. Fumero, S., Mondino, A., Silvestri, S., Zanolo, G., De Marchi, G., Pedrazzini, S., *Arzneim.-Forsch.* (1980) **30**, 1253–1256.

B523. MacLeod, M.C., Levin, W., Conney, A.H., Lehr, R.E., Mansfield, B.K., Jerina, D.M., Selkirk, J.K., *Carcinogenesis (N.Y.)* (1980) **1**, 165–173.

B524. Okitani, A., Nishimura, T., Otsuka, Y., Matsukura, U., Kato, H., *Agric. Biol. Chem.* (1980) **44**, 1705–1708.

B525. Chung, Y.-B., Lee, H.-J., *Taehan Hwahakhoe Chi* (1980) **24**, 25–33.

B526. Parker, R.J., Hirom, P.C., Millburn, P., *Xenobiotica* (1980) **10**, 689–703.

B527. Patel, J.R., Barnsley, E.A., *J. Bacteriol.* (1980) **143**, 668–673.

B528. Delaforge, M., Janiaud, P., Levi, P., Morizot, J.P., *Xenobiotica* (1980) **10**, 737–744.

B529. Greenslade, D., Havler, M.E., Humphrey, M.J., Jordan, B.J., Rance, M.J., *Xenobiotica* (1980) **10**, 753–760.

B530. Ben-Zvi, Z., *Xenobiotica* (1980) **10**, 805–809.

B531. Filer, C.W., Humphrey, M.J., Jeffery, D.J., Jones, K.H., Langley, P.F., *Xenobiotica* (1980) **10**, 761–769.

B532. Mbanefo, C., Bababunmi, E.A., Mahgoub, A., Sloan, T.P., Idle, J.R., Smith, R.L., *Xenobiotica* (1980) **10**, 811–818.

B533. Islam, S.I., Idle, J.R., Smith, R.L., *Xenobiotica* (1980) **10**, 819–825.

B534. Shaw, I.C., Griffiths, L.A., *Xenobiotica* (1980) **10**, 905–911.

B535. Higuchi, S., Shiobara, Y., *Xenobiotica* (1980) **10**, 889–896.

B536. Inoue, T., Niwaguchi, T., Murata, T., *Xenobiotica* (1980) **10**, 913–920.

B537. Lewinsohn, R., *J. Pharm. Pharmacol.* (1981) **33**, 569–575.

B538. Vargay, Z., Simon, G., Winter, M., Szuts, T., *Eur. J. Drug Metab. Pharmacokinet.* (1980) **5**, 69–74.

B539. Wright, G.C., Herrett, R.J., Heotis, J.P., Butterfield, J.L., *J. Pharm. Sci.* (1980) **69**, 842–845.

B541. Tegyey, Z., Maksay, G., Kardos, J., Ötvös, L., *Experientia* (1980) **36**, 1031–1032.

B542. Moldeus, P., Rahimtula, A., *Biochem. Biophys. Res. Commun.* (1980) **96**, 469–475.

B544. Pederson, D.M., Goodman, R.E., *Can. J. Microbiol.* (1980) **26**, 978–984.

B545. Shinagawa, Y., Shinagawa, Y. *Int. J. Quantum Chem.* (1980) **18**, 521–526.

B546. Tittman, U., Lingens, F., *FEMS Microbiol. Lett.* (1980) **8**, 255–258.

B547. Saeki, Y., Nozaki, M., Senoh, S., *J. Biol. Chem.* (1980) **255**, 8465–8471.

B548. Rast, H.G., Engelhardt, G., Ziegler, W., Wallnöfer, P.R., *FEMS Microbiol. Lett.* (1980) **8**, 259–263.

B549. Smith, R.V., Davis, P.J., Clark, A.M., Glover-Milton, S., *J. Appl. Bacteriol.* (1980) **49**, 65–73.

B550. Franklin, R.A., Heatherington, K., Morrison, B., Ward, T., *Xenobiotica* (1981) **11**, 159–165.

B551. Clothier, R.H., Dewar, J.R., Santos, M.A., North, A.D., Foster, S., Balls, M., *Xenobiotica* (1981) **11**, 149–157.

B553. Eisenstein, Z., Balsam, A., Garber, J.R., Ingbar, S.H., *Endocrinology (Baltimore)* (1980) **107**, 530–537.

B554. Mizojiri, K., Yamamoto, E., Nakajima, Y., Okabe, H., Sugeno, K., Watanabe, J., *Chem. Pharm. Bull.* (1980) **28**, 2235–2239.

B555. Loiseau, G.P., Millischer, R.J., Lohier, P.Y., Mardiguian, J.S., Gilede, A.M.P., Ginoccio, A.V., *J. Pharm. Pharmacol.* (1980) **32**, 483–488.

B556. Gönechea, S., Eckhardt, G., Fahr, W., *Arzneim.-Forsch.* (1980) **30**, 1580–1584.

B557. Malvy, C., Paoletti, C., *Biochem. Biophys. Res. Commun.* (1980) **96**, 363–369.

B558. Sugiura, M., Yamazoe, Y., Kamatake, T., Kato, R., *Cancer Res.* (1980) **40**, 2910–2914.

B559. Namkung, M.J., Juchau, M.R., *Toxicol. Appl. Pharmacol.* (1980) **55**, 253–259.

B560. Barber, J., Staunton, J., *J. Chem. Soc., Chem. Commun.* (1980) 552–554.

B561. Shen, A.L., Fahl, W.E., Jefcoate, C.R., *Arch. Biochem. Biophys.* (1980) **204**, 511–523.

B562. Fujita, M., Inoue, T., *Chem. Pharm. Bull.* (1980) **28**, 2476–2481.
B563. Fujita, M., Inoue, T., *Chem. Pharm. Bull.* (1980) **28**, 2482–2486.
B564. Wal, J.-M., Peleran, J.-C., Bories, G.F., *FEBS Lett.* (1980) **119**, 38–42.
B565. Kaka, J.S., Hayton, W.L., *J. Pharmacokinet. Biochem.* (1980) **8**, 193–202.
B566. Williams, M.C., Goldzieher, J.W., *Steroids* (1980) **36**, 255–282.
B567. Levin, W., Buening, M.K., Wood, A., Chang, R.L., Kedzierski, B., Thakker, D.R., Boyd, D.R., Gadaginamath, G.S., Armstrong, R.N., Yagi, H., Karle, J.M., Slaga, T.J., Jerina, D.M., Conney, A.H., *J. Biol. Chem.* (1980) **255**, 9067–9076.
B568. Sukanya, N.K., Vaidyanathan, C.S., *J. Indian Inst. Sci.* (1979) **61**, 51–62.
B569. Hoffman, P.L., Wermuth, B., Von Wartburg, J.P., *Adv. Exp. Med. Biol.* (1980) **132**, 749–759.
B570. Cerniglia, C.E., Gibson, D.T., *Proc. Nat. Acad. Sci. U.S.A.* (1980) **77**, 4554–4558.
B571. Yoshimoto, A., Oki, T., Takeuchi, T., Umezawa, H., *J. Antibiot.* (1980) **33**, 1158–1166.
B572. Maier, W., Erge, D., Gröger, D., *Planta Med.* (1980) **40**, 104–108.
B573. Kelleher, W.J., Rother, A., Wellmann, E., Grisebach, H., *Planta Med.* (1980) **40**, 127–136.
B574. Kramer, K.J., Hopkins, T.L., Ahmed, R.F., Mueller, D., Lookhart, G., *Arch. Biochem. Biophys.* (1980) **205**, 146–155.
B575. Soll, J., Schultz, G., *Phytochemistry* (1980) **19**, 215–218.
B576. Scott, A.I., Mizukame, H., Hirata, T., Lee, S.-L., *Phytochemistry* (1980) **19**, 488–489.
B577. Keller, W.J., *Phytochemistry* (1980) **19**, 413–414.
B578. Rickert, D.E., Long, R.M., *Drug Metab. Dispos.* (1981) **9**, 226–232.
B579. Albro, P.W., Hass, J.R., Peck, C.C., Odam, D.G., Corbett, J.T., Bailey, F.J., Blatt, H.E., Barrett, B.B., *Drug Metab. Dispos.* (1981) **9**, 223–225.
B580. Layiwola, P.J., Linnecar, D.F.C., *Xenobiotica* (1981) **11**, 167–171.
B581. Bakke, J.E., Aschbacher, P.W., Feil, V.J., Gustaffson, B.E., *Xenobiotica* (1981) **11**, 173–178.
B582. Hunter, W.H., Wilson, P., *Xenobiotica* (1981) **11**, 179–188.
B583. Harvey, D.J., Paton, W.D.M., *Res. Commun. Subst. Abuse* (1981) **2**, 193–201.
B584. Brewster, D., Humphrey, M.J., McLeavy, M.A., *Xenobiotica* (1981) **11**, 189–196.
B585. Horie, T., Ohno, T., Kinoshita, K., *Xenobiotica* (1981) **11**, 197–206.
B586. Sloane, N., *Xenobiotica* (1981) **11**, 267–274.
B587. Bachmann, M., Schlatter, Ch., *Xenobiotica* (1981) **11**, 217–225.
B588. Minard, R.D., Liu, S.-Y., Bollag, J.-M., *J. Agric. Food Chem.* (1981) **29**, 250–253.
B589. Lusby, W.R., Oliver, J.E., Smith, R.H., Jr., Kearney, P.C., Finegold, H., *J. Agric. Food Chem.* (1981) **29**, 245–250.
B590. Chrzanowski, R.L., Jelinek, A.G., *J. Agric. Food Chem.* (1981) **29**, 580–587.
B591. Bull, D.L., Ivie, G.W., *J. Agric. Food Chem.* (1981) **29**, 121–125.
B592. Chiba, M., Veres, D.F., *J. Agric. Food Chem.* (1981) **29**, 588–590.
B593. Daumy, G.O., McColl, A.S., Andrews, G.C., *J. Chem. Soc., Chem. Commun.* (1979) 1073.
B594. Gil, V., MacLeod, A.J., *Phytochemistry* (1980) **19**, 1365–1368.
B595. Ishii, H., Hayashi, M., Niwaguchi, T., Nakahara, Y., *Chem. Pharm. Bull.* (1979) **27**, 3029–3038.
B596. Ishii, H., Hayashi, M., Niwaguchi, T., Nakahara, Y., *Chem. Pharm. Bull.* (1979) **27**, 1570–1575.
B597. Szwajcer, E., Szewczuk, A., Mordarski, M., *Biotechnol. Bioeng.* (1981) **23**, 1675–1681.

B598. Yoshimoto, A., Oki, T., Umezawa, H., *J. Antibiot.* (1980) **33**, 1199–1201.

B599. Nurmann, G., Strack, D., *Z. Pflanzenphysiol.* (1981) **102**, 11–17.

B600. Oki, T., Takatsuki, Y., Tobe, H., Yoshimoto, A., Takeuchi, T., Umezawa, H., *J. Antibiot.* (1981) **34**, 1229–1231.

B601. Martin, J.L., Gross, B.J., Morris, P., Pohl, L.R., *Drug Metab. Dispos.* (1980) **8**, 371–375.

B602. Horning, M.G., Stillwell, W.G., Griffin, G.W., Tsang, W.-S., *Drug Metab. Dispos.* (1980) **8**, 404–414.

B603. Rodgers, R.M., Taylor, S.M., Dickinson, R.G., Ilias, A.M., Lynn, R.K., Gerber, N., *Drug Metab. Dispos.* (1980) **8**, 390–396.

B604. Harvey, D.J., Gill, E.W., Slater, M., Paton, W.D.M., *Drug Metab. Dispos.* (1980) **8**, 439–445.

B605. Frincke, J.M., Henderson, G.L., *Drug Metab. Dispos.* (1980) **8**, 425–433.

B606. Kojima, S., Maruyama, K., Babasaki, T., *Drug Metab. Dispos.* (1980) **8**, 463–466.

B607. Bond, J.A., Rickert, D.E., *Drug Metab. Dispos.* (1981) **9**, 10–14.

B608. Sisenwine, S.F., Tio, C.O., *Drug Metab. Dispos.* (1981) **9**, 37–42.

B609. Dewey, E.A., Maylin, G.A., Ebel, J.G., Henion, J.D., *Drug Metab. Dispos.* (1981) **9**, 30–36.

B610. De Jongh, G.D., van den Wildenberg, H.M., Nieuwenhuyse, H., van der Veen, F., *Drug Metab. Dispos.* (1981) **9**, 48–53.

B611. Hackett, A.M., Griffiths, L.A., *Drug Metab. Dispos.* (1981) **9**, 54–59.

B612. Ratnayake, J.H., Hanna, P.E., Anders, M.W., Duggan, D.E., *Drug Metab. Dispos.* (1981) **9**, 85–87.

B613. Acara, M., Gessner, T., Greizerstein, H., Trudnowski, R., *Drug Metab. Dispos.* (1981) **9**, 75–79.

B614. Burstein, S., Shoupe, T.S., *Drug Metab. Dispos.* (1981) **9**, 94–96.

B615. Halpaap-Wood, K., Horning, E.C., Horning, M.G., *Drug Metab. Dispos.* (1981) **9**, 97–102.

B616. Toda, T., Oshino, N., *Drug Metab. Dispos.* (1981) **9**, 108–113.

B617. Halpaap-Wood, K., Horning, E.C., Horning, M.G., *Drug Metab. Dispos.* (1981) **9**, 103–107.

B618. Hobbs, D.C., Twomey, T.M., *Drug Metab. Dispos.* (1981) **9**, 114–118.

B619. Cayen, M.N., Kraml, M., Robinson, W.T., Dvornik, D., *Drug Metab. Dispos.* (1981) **9**, 119–124.

B620. Paulson, G.D., Giddings, J.M., Lamoureux, C.H., Mansager, E.R., Struble, C.B., *Drug Metab. Dispos.* (1981) **9**, 142–146.

B621. Raitano, L.A., Slikker, W., Jr., Hill, D.E., Hadd, H.E., Cairns, T., Helton, E.D., *Drug Metab. Dispos.* (1981) **9**, 129–134.

B622. Huckle, K.R., Climie, I.J.G., Hutson, D.H., Millburn, P., *Drug Metab. Dispos.* (1981) **9**, 147–149.

B623. Mas-Chamberlin, C., Gillet, G., André, J., Gomeni, R., Dring, L.G., Morselli, P.L., *Drug Metab. Dispos.* (1981) **9**, 150–155.

B624. Warrander, A., Metcalf, R., Fromson, J.M., *Drug Metab. Dispos.* (1981) **9**, 161–167.

B625. Fraire, C., Lecointe, P., Paoletti, C., *Drug Metab. Dispos.* (1981) **9**, 156–160.

B626. Procaccini, R.L., Klunk, L.J., *Drug Metab. Dispos.* (1981) **9**, 202–206.

B627. Uetrecht, J.P., Woosley, R.L., Freeman, R.W., Sweetman, B.J., Oates, J.A., *Drug Metab. Dispos.* (1981) **9**, 183–187.

B628. Haugen, D.A., *Drug Metab. Dispos.* (1981) **9**, 212–218.

B629. Belleville, M., Grand, M., Briet, P., *Drug Metab. Dispos.* (1981) **9**, 233–239.

B630. van Bladeren, P.J., Delbressine, L.P.C., Hoogeterp, J.J., Beaumont, H.G.M., Breimer, D.D., Seutter-Berlage, F., van der Gen, A., *Drug Metab. Dispos.* (1981) **9**, 246–249.

B631. Egorin, M.J., Clawson, R.E., Ross, L.A., Chou, F.-T.E., Andrews, P.A., Bachur, N.R., *Drug Metab. Dispos.* (1981) **9**, 240–245.

B632. Hubbard, J.W., Bailey, K., Midha, K.K., Cooper, J.K., *Drug Metab. Dispos.* (1981) **9**, 250–254.

B633. McLean, S., Davies, N.W., Watson, H., Favretto, W.A., Bignell, J.C., *Drug Metab. Dispos.* (1981) **9**, 255–260.

B634. Leinweber, F.-J., Szuna, A.J., Williams, T.H., Sasso, G.J., De Barbieri, B.A., *Drug Metab. Dispos.* (1981) **9**, 284–291.

B635. Kammerer, R.C., Schmitz, D.A., Di Stefano, E.W., Cho, A.K., *Drug Metab. Dispos.* (1981) **9**, 274–278.

B636. Yamazoe, Y., Ishii, K., Kamataki, T., Kato, R., *Drug Metab. Dispos.* (1981) **9**, 292–296.

B637. Anders, M.W., Ratnayake, J.H., Hanna, P.E., Fuchs, J.A., *Drug Metab. Dispos.* (1981) **9**, 307–310.

B638. Huckle, K.R., Hutson, D.H., Millburn, P., *Drug Metab. Dispos.* (1981) **9**, 352–359.

B639. Gemborys, M.W., Mudge, G.H., *Drug Metab. Dispos.* (1981) **9**, 340–351.

B640. Huckle, K.R., Chipman, J.K., Hutson, D.H., Millburn, P., *Drug Metab. Dispos.* (1981) **9**, 360–368.

B641. Wall, M.E., Brine, D.R., Peres-Reyes, M., *Drug Metab. Dispos.* (1981) **9**, 369–375.

B642. Hooper, W.D., Kunze, H.E., Eadie, M.J., *Drug Metab. Dispos.* (1981) **9**, 381–385.

B643. Bondesson, U., Hartvig, P., Danielsson, B., *Drug Metab. Dispos.* (1981) **9**, 376–380.

B644. Koster, H., Halsema, I., Scholtens, E., Knippers, M., *Biochem. Pharmacol.* (1981) **30**, 2569–2575.

B645. Ruchirawat, M., Grantham, P.H., Benjamin, T., Weisburger, E.K., *Biochem. Pharmacol.* (1981) **30**, 2715–2718.

B646. Boobis, A.R., Kahn, G.C., White, C., Brodie, M.J., Davies, D.S., *Biochem. Pharmacol.* (1981) **30**, 2451–2456.

B647. Ruenitz, P.C., Toledo, M.M., *Biochem. Pharmacol.* (1981) **30**, 2203–2207.

B648. Gill, S.S., Hammock, B.D., *Biochem. Pharmacol.* (1981) **30**, 2111–2120.

B649. Kammerer, R.C., Di Stefano, E., Jonsson, J., Cho, A.K., *Biochem. Pharmacol.* (1981) **30**, 2257–2263.

B650. Hogg, S.I., Curtis, C.G., Upshall, D.G., Powell, G.M., *Biochem. Pharmacol.* (1981) **30**, 1551–1555.

B651. Fry, J.R., *Biochem. Pharmacol.* (1981) **30**, 1915–1919.

B652. Chow, Y.M., Gutmann, H.R., Kaplan, E., *Biochem. Pharmacol.* (1981) **30**, 1253–1261.

B653. Suzuki, O., Katsumata, Y., Oya, M., *Biochem. Pharmacol.* (1981) **30**, 1353–1358.

B654. Maynard, M.S., Cho, A.K., *Biochem. Pharmacol.* (1981) **30**, 1115–1119.

B655. Vodicnik, M.J., Franklin, R.B., Elcombe, C.R., Lech, J.J., *Biochem. Pharmacol.* (1981) **30**, 1091–1097.

B656. Renneberg, R., Capdevila, J., Chacos, N., Estabrook, R.W., Prough, R.A., *Biochem. Pharmacol.* (1981) **30**, 843–848.

B657. Chipman, J.K., Hirom, P.C., Frost, G.S., Millburn, P., *Biochem. Pharmacol.* (1981) **30**, 937–944.

B658. Krzywicka, K.A., Pachecka, J., *Biochem. Pharmacol.* (1981) **30**, 803–805.

B659. Tank, A.W., Weiner, H., Thurman, J.A., *Biochem. Pharmacol.* (1981) **30**, 3265–3275.

B660. Yu, P.H., Barclay, S., Davis, B., Boulton, A.A., *Biochem. Pharmacol.* (1981) **30**, 3089–3094.

B661. Fowler, C.J., Tipton, K.F., *Biochem. Pharmacol.* (1981) **30**, 3329–3332.

B662. Porter, W.R., Wheeler, C., Trager, W.F., *Biochem. Pharmacol.* (1981) **30**, 3099–3104.

B663. Ohmiya, Y., Mehendale, H.M., *Biochem. Pharmacol.* (1982) **31**, 157–162.

B664. Matsumoto, R.M., Cho, A.K., *Biochem. Pharmacol.* (1982) **31**, 105–108.

B665. Wong, K.P., *Biochem. Pharmacol.* (1982) **31**, 59–62.

B666. Watanabe, K., Yamamoto, I., Oguri, K., Yoshimura, H., *J. Pharmacobio-Dyn.* (1980) **3**, 482–484.

B667. Kalow, W., Otton, S.V., Kadar, D., Endrenyi, L., Inaba, T., *Can. J. Physiol. Pharmacol.* (1980) **58**, 1142–1144.

B668. Hoffman, A.R., Paul, S.M., Axelrod, J., *Endocrinology (Baltimore)* (1980) **107**, 1192–1197.

B669. Bakke, J.E., Gustafsson, J.-A., Gustafsson, B.E., *Science* (1980) **210**, 433–435.

B670. Hietanen, E., *Pharmacology* (1980) **21**, 233–243.

B671. Marnett, L.J., Bienowski, M.J., *Biochem. Biophys. Res. Commun.* (1980) **96**, 639–647.

B672. Koide, Y., Honma, M., Shimomura, T., *Agric. Biol. Chem.* (1980) **44**, 2013–2019.

B673. Bhakuni, D.S., Singh, A.N., Jain, S., *Tetrahedron* (1980) **36**, 2149–2151.

B674. Capdevila, J., Orrenius, S., *FEBS Lett.* (1980) **119**, 33–37.

B675. Tohno, M., Kajikawa, N., Yamamoto, T., Ofuji, T., *Yakugaku Zasshi* (1980) **100**, 933–940.

B677. Okuno, E., Minatogawa, Y., Nakamura, M., Kamoda, N., Nakanishi, J., Makino, M., Kido, R., *Biochem. J.* (1980) **189**, 581–590.

B678. Balba, M.T., Evans, W.C., *Biochem. Soc. Trans.* (1980) **8**, 452–453.

B679. Forkmann, G., Heller, W., Grisebach, H., *Z. Naturforsch. C, Biosci.* (1980) **35C**, 691–695.

B680. Wiermann, R., Buth-Weber, M., *Protoplasma* (1980) **104**, 307–313.

B681. Numazawa, M., Kiyono, Y., Nambara, T., *J. Steroid Biochem.* (1980) **13**, 1101–1105.

B682. Higgins, I.J., *Ger. Offen* (9–10–1980) 3, 008, 665.

B683. Hobkirk, R., Mori, J., Nilsen, M., *J. Steroid Biochem.* (1980) **13**, 945–950.

B684. Jakubowski, Z., Angielski, S., *Pol. J. Pharmacol. Pharm.* (1980) **32**, 37–46.

B685. Caldwell, J., Davies, S., Smith, R.L., *Br. J. Pharmacol.* (1980) **70**, 112P-113P.

B686. Goldaniga, G.C., Montesanti, L., Moro, E., Pianezzola, E., Valzelli, G., *Arzneim.-Forsch.* (1980) **30**, 1654–1658.

B687. Chasseaud, L.F., Taylor, T., Brodie, R.R., *J. Pharm. Pharmacol.* (1980) **32**, 652–653.

B688. Nguyen Van Bac, Moisand, C., Gouyette, A., Muzard, G., Nguyen Dat Xuong, *Cancer Treat. Rep.* (1980) **64**, 879–887.

B689. Longcope, C., Yesair, D.W., Williams, K.I.H., Callahan, M.M., Bourget, C., Brown, S.K., Carraher, M.S., Flood, C., Rachwall, P.C., *J. Steroid Biochem.* (1980) **13**, 1047–1055.

B690. Lee, F.H., Van Harken, D.R., Vasiljev, M., Smyth, R.D., Hottendorf, G.H., *Antimicrob. Agents Chemother.* (1980) **18**, 746–752.

B691. Bock, K.W., Lilienblum, W., Pfeil, H., *FEBS Lett.* (1980) **121**, 269–272.

B692. Von der Trenk, T., Sandermann, H., Jr., *FEBS Lett.* (1980) **119**, 227–231.

B693. Faleev, N.G., Sadovnikova, M.S., Martinkova, N.S., Belikov, V.M., *Enzyme Microb. Technol.* (1980) **2**, 305–308.

B694. Barber, J., Staunton, J., *J. Chem. Soc., Chem. Commun.* (1980) 1163–1164.

B695. Klages, U., Lingens, F., *Zentralbl. Bakteriol. Abt. 1 Orig. C:* (1980) **1**, 215–223.

B696. Kobashi, K., Nishimura, T., Kusaka, M., Hattori, M., Namba, T., *Planta Med.* (1980) **40**, 225–236.

B698. Hwang, B.Y.-H., Kuo, G.Y., Miao, C., Intoccia, A.P., *Xenobiotica* (1981) **11**, 311–318.

B699. Schmid, J., Prox, A., Reuter, A., Zipp, H., Koss, F.W., *Eur. J. Drug Metab. Pharmacokinet.* (1980) **5**, 81–92.

B700. Jeyakumar, L.H., French, M.R., *Xenobiotica* (1981) **11**, 319–321.

B701. Pannatier, A., Testa, B., Etter, J.-C., *Xenobiotica* (1981) **11**, 345–350.

B702. Mukhtar, H., Baars, A.J., Breimer, D.D., *Xenobiotica* (1981) **11**, 367–371.

B703. Landrum, P.F., Crosby, D.G., *Xenobiotica* (1981) **11**, 351–361.

B704. Climie, I.J.G., Hutson, D.H., Stoydin, G., *Xenobiotica* (1981) **11**, 401–424.

B705. Woo, G.K., Williams, T.H., Kolis, S.J., Warinsky, D., Sasso, G.J., *Xenobiotica* (1981) **11**, 373–384.

B706. Stocklinski, A.W., *Xenobiotica* (1981) **11**, 425–432.

B707. Larsen, G.L., Bakke, J.E., *Xenobiotica* (1981) **11**, 473–480.

B708. Norbury, H.M., Waring, R.H., *Xenobiotica* (1981) **11**, 511–518.

B709. Norbury, H.M., Waring, R.H., *Xenobiotica* (1981) **11**, 501–508.

B710. Ancher, J.F., Donath, A., Malnoë, A., Morizur, J.P., Strolin Benedetti, M., *Xenobiotica* (1981) **11**, 519–530.

B711. Dieterle, W., Faigle, J.W., *Xenobiotica* (1981) **11**, 559–568.

B712. Emslie, K.R., Calder, I.C., Hart, S.J., Tange, J.D., *Xenobiotica* (1981) **11**, 579–587.

B713. Struck, R.F., Shih, T.W., Johnston, T.P., Kirk, M.C., Hill, D.L., *Xenobiotica* (1981) **11**, 569–577.

B714. Delbressine, L.P.C., Van Bladeren, P.J., Smeets, F.L.M., Seutter-Berlage, F., *Xenobiotica* (1981) **11**, 589–594.

B715. Midgley, I., Biggs, S.R., Hawkins, D.R., Chasseaud, L.F., Darragh, A., Brodie, R.R., Walmsley, L.M., *Xenobiotica* (1981) **11**, 595–608.

B716. Huckle, K.R., Tait, G.H., Millburn, P., Hutson, D.H., *Xenobiotica* (1981) **11**, 635–644.

B717. Clement, B.A., Beckett, A.H., *Xenobiotica* (1981) **11**, 609–618.

B718. Paulson, G.D., Struble, C., *Life Sci.* (1980) **27**, 1811–1817.

B719. Fujiwara, S., Sakurai, S., Noumi, K., Sugimoto, I., Awata, N., *Yakugaku Zasshi* (1980) **100**, 1011–1018.

B720. Omiya, Y., Mehendale, H.M., *Pharmacology* (1981) **22**, 172–182.

B721. Sorimachi, k., Niwa, A., Yasumura, Y., *Biochim. Biophys. Acta* (1980) **633**, 134–143.

B722. Ohkawa, H., Nambu, K., Miyamoto, J., *Nippon Noyako Gakkaishi* (1980) **5**, 215–223.

B723. Johnson, E.F., Levitt, D.S., Muller-Eberhard, U., Thorgerirsson, S.S., *Cancer Res.* (1980) **40**, 4456–4459.

B724. Pennings, E.J.M., Van Kempen, G.M.J., *Biochem. J.* (1980) **191**, 133–138.

B725. Mains, I., Power, D.M., Thomas, E.W., Buswell, J.A., *Biochem. J.* (1980) **191**, 457–465.

B726. Bhakuni, D.S., Dewan, S., Jain, S., Gupta, S., *Tetrahedron* (1980) **36**, 2491–2495.

B727. Yoshizako, F., Nishimura, A., Ueno, T., *Annu. Rep. Radiat. Cent. Osaka Prefect.* (1979) **20**, 115–118.

B728. Rubio, F., Seawall, S., Pocelinko, R., De Barbieri, B., Benz, W., Berger, L., Morgan, L., Pao, J., Williams, T.H., Koechlin, B., *J. Pharm. Sci.* (1980) **69**, 1245–1253.

B729. Baudinette, R.V., Wheldrake, J.F., Hewitt, S., Hawke, D., *Aust. J. Zool.* (1980) **28**, 511–520.

B730. Foss, W., Krechniak, J., *Arch. Toxicol., Suppl.* (1980) **4**, 346–349.

B731. Gill, S.S., *Bull. Environ. Contam. Toxicol.* (1980) **25**, 697–671.

B732. Paris, C.G., Masianik, B., *J. Bacteriol.* (1981) **145**, 266–271.

B733. Vereczkey, L., Tamas, J., Czira, G., Szporny, L., *Arzneim.-Forsch.* (1980) **30**, 1860–1865.

B735. Mitchell, S.C., *Comp. Biochem. Physiol. C* (1980) **67C**, 199–202.

B736. Hufford, C.D., Capiton, G.A., Clarke, A.M., Baker, J.K., *J. Pharm. Sci.* (1981) **70**, 151–155.

B737. La Voie, E.J., Hecht, S.S., Amin, S.S., Bedenko, V., Hoffmann, D., *Cancer Res.* (1980) **40**, 4528–4532.

B738. Kang, S.-H., Fuchs, M.S., Webb, P.M., *Insect Biochem.* (1980) **10**, 501–508.

B739. Lynn, R.K., Danielson, D.W., Ilias, A.M., Kennish, J.M., Wong, K., Matthews, H.B., *Toxicol. Appl. Pharmacol.* (1980) **56**, 248–258.

B740. Bansal. S.K., Gessner, T., *Anal. Biochem.* (1980) **109**, 321–329.

B741. Fuller, R.W., Hemrick, Lücke, S.K., *Res. Commun. Chem. Pathol. Pharmacol.* (1980) **30**, 401–408.

B742. Binder, M., Popp, A., *Helv. Chim. Acta* (1980) **63**, 2515–2518.

B743. Balba, M.T., Evans, W.C., *Biochem. Soc. Trans.* (1980) **8**, 624–625.

B744. Stegeman, J.J., Kaplan, H.B., *Comp. Biochem. Physiol. C* (1981) **68C**, 55–61.

B745. Bhavnani, B.R., Woolever, C.A., *Endocrinology (Baltimore)* (1981) **108**, 232–238.

B746. Bansal, S.K., Zaleski, J., Gessner, I., *Biochem. Biophys. Res. Commun.* (1981) **98**, 131–139.

B748. Renwick, A.G.C., Soon, C.Y., Chambers, S.M., Brown, C.R., *J. Biol. Chem.* (1981) **256**, 1881–1887.

B749. Gozukara, E.M., Belvedere, G., Robinson, R.C., Deutsch, J., Coon, M.J., Guengerich, F.P., Gelboin, H.V., *Mol. Pharmacol.* (1981) **19**, 153–161.

B750. Yokoyama, M.T., Carlson, J.R., *Appl. Environ. Microbiol.* (1981) **41**, 71–76.

B751. Suzuki, K., Katagiri, M., *Biochim. Biophys. Acta* (1981) **657**, 530–534.

B752. Watanabe, K., Yamamoto, I., Oguri, K., Yoshimura, H., *J. Pharmacobio.-Dyn.* (1980) **3**, 686–691.

B753. Forrest, I.S., Green, D.E., Serra, M.T., Usdin, E., *Dev. Neurosci.* (1980) **7**, 151–154.

B754. Schmidt, E., Remberg, G., Knackmuss, H.J., *Biochem. J.* (1981) **192**, 331–337.

B755. Coutts, R.T., Prelusky, D.B., Jones, G.R., *Can. J. Physiol. Pharmacol.* (1981) **59**, 195–201.

B756. Wong, K., Addison, R.F., Law, F.C.P., *Bull. Environ. Contam. Toxicol.* (1981) **26**, 243–247.

B757. Caporale, G., Innocenti, G., Guiotto, A., Rodighiero, P., Dall'Acqua, F., *Phytochemistry* (1981) **20**, 1283–1287.

B758. Enoki, A., Yajima, Y., Gold, M.H., *Phytochemistry* (1981) **20**, 1543–1546.

B759. Dey, P.M., *Phytochemistry* (1981) **20**, 1493–1496.

B760. Chen, K.C.S., Holmes, K.K., *Anal. Biochem.* (1981) **111**, 60–66.

B761. Hara, E., Kawajiri, K., Goto, O., Tagashira, Y., *Cancer Res.* (1981) **41**, 253–257.

B762. Ito, S., Fujita, K., *Biochim. Biophys. Acta* (1981) **672**, 151–157.

B763. Sutherland, J.B., Crawford, D.L., Pometto, A.L.III, *Appl. Environ. Microbiol.* (1981) **41**, 442–448.

B765. Mir, A.K., Vaughan, P.F.T., *J. Neurochem.* (1981) **36**, 441–446.

B766. Gielsdorf, W., Toffel-Nadolny, P., *J. Clin. Chem. Clin. Biochem.* (1981) **19**, 25–30.

B767. Hawkins, D.R., Chasseaud, L.F., Weston, K.T., *Eur. J. Drug Metab. Pharmacokinet.* (1980) **5**, 145–152.

B768. Maffei, F.R., Carini, M., Bertuletti, R., Tofanetti, O., *Pharmacol. Res. Commun.* (1981) **13**, 121–132.

B769. Simmonds, R.G., Woodage, T.J., Duff, S.M., Green, J.N., *Eur. J. Drug Metab. Pharmacokinet.* (1980) **5**, 169–172.

B770. Sekura, R.D., Sato, K., Cahnmann, H.J., Robbins, J., Jakoby, W.B., *Endocrinology (Baltimore)* (1981) **108**, 454–456.

B771. Sorimachi, K., Niwa, A., Yasumura, Y., *Endocrinol. Jpn.* (1980) **27**, 631–636.

B772. Nordqvist, M., Thakker, D.R., Vyas, K.P. Yagi, H., Levin, W., Ryan, D.E., Thomas, P.E., Conney, A.H., Jerina, D.M., *Mol. Pharmacol.* (1981) **19**, 168–178.

B773. Kadlubar, F.F., Unruh, L.E., Flammang, T.J., Sparks, D., Mitchum, R.K., Mulder, G.J., *Chem.-Biol. Interact.* (1981) **33**, 129–147.

B774. Estévez, M.P., Legaz, E., Olmeda, L., Pérez, F.J., Vincente, C., *Z. Naturforsch. C, Biosci.* (1981) **36C**, 35–39.

B775. Cerniglia, C.E., Crow, S.A., *Arch. Microbiol.* (1981) **129**, 9–13.

B776. Matsumura, K., Takahara, Y., Aratani, T., Nishide, K., Kato, H., Kiriyama, N., *Yakugaku Zasshi* (1981) **101**, 198–201.

B777. Tokuhisa, S., Saisu, K., Naruse, K., Yoshikawa, H., Baba, S., *Chem. Pharm. Bull.* (1981) **29**, 514–518.

B778. Shibasaki, J., Konishi, R., Koike, M., Imamura, A., Sueyasu, M., *J. Pharmacobio.-Dyn.* (1981) **4**, 91–100.

B779. Wu, J., Wong, L.K., *Appl. Environ. Microbiol.* (1981) **41**, 843–845.

B780. Barboni, E., Borri Voltatorini, C., D'Erme, M., Fiori, A., Minelli, A., Rosei, M.A., Turano, C., *Biochem. Biophys. Res. Commun.* (1981) **99**, 576–583.

B782. William, F., Mahadevan, A., *Z. Pflanzenkr. Pflanzenschutz.* (1980) **87**, 738–744.

B783. Maxwell, G.D., Moore, M.M., Hildebrand, J.G., *Insect Biochem.* (1980) **10**, 657–665.

B784. Chou, M.W., Yang, S.K., Sydor, W., Yang, C.S., *Cancer Res.* (1981) **41**, 1559–1564.

B785. Savides, M.C., Oehme, F.W., *Toxicology* (1980) **17**, 355–363.

B786. McMahon, R.E., Turner, J.C., Whitaker, G.W., Sullivan, H.R., *Biochem. Biophys. Res. Commun.* (1981) **99**, 662–667.

B787. Okitani, A., Nishimura, T., Kato, H., *Eur. J. Biochem.* (1981) **115**, 269–274.

B788. Katz, E., *Dev. Biochem.* (1980) **16**, 159–177.

B789. Davis, P.J., Rosazza, J.P., *Bioorg. Chem.* (1981) **10**, 97–103.

B790. Anderson, J.J., Dagley, S., *J. Bacteriol.* (1981) **146**, 291–297.

B791. Lundström. J., Högberg, T., Gosztonyi, T., De Paulis, T., *Arzneim.-Forsch.* (1981) **31**, 486–494.

B792. Tanabe, H., Kubo, J., Yamaguchi, H., Hashimoto, Y., *Iyakuhin Kenkyu* (1981) **12**, 252–262.

B793. Tanaka, T., Enogaki, K., Kano, H., Murakami, M., Noguchi, Y., *Yakuri to Chiryo* (1980) **8**, 4673–4682.

B794. Davis, B.A., Boulton, A.A., *Eur. J. Mass Spectrom. Biochem. Med. Environ. Res.* (1980) **1**, 149–153.

B795. Bahig, M.E., Kraus, A., Klein, W., Korte, F., *Chemosphere* (1981) **10**, 319–322.

B796. Nakatsubo, F., Kirk, T.K., Shimada, M., Higuchi, T., *Arch. Microbiol.* (1981) **128**, 416–420.

B797. Schwartz, R.D., *Enzyme Microb. Technol.* (1981) **3**, 158–159.

B798. Klages, U., Markus, A., Lingens, F., *J. Bacteriol.* (1981) **146**, 64–68.

B799. Corbett, M.D., Corbett, B.R., *Appl. Environ, Microbiol.* (1981) **41**, 942–949.

B800. Li, H.C., Porter, N., Holmes, G., Gessner, T., *Xenobiotica* (1981) **11**, 647–654.

B801. Marsh, M.V., Caldwell, J., Smith, R.L., Horner, M.W., Houghton, E., Moss, M.S., *Xenobiotica* (1981) **11**, 655–663.

B802. Kuzniar, E.J.A., James, S.P., *Xenobiotica* (1981) **11**, 675–683.

B803. Bradbury, A., Powell, G.M., Curtis, C.G., Rhodes, C., *Xenobiotica* (1981) **11**, 665–674.

B804. Sudo, K., Tachizawa, H., Sano, M., *Xenobiotica* (1981) **11**, 685–691.

B805. Horie, T., Ohno, T., Kinoshita, K., Kitagawa, H., Kitada, M., *Xenobiotica* (1981) **11**, 693–699.

B806. Fujita, M., Inoue, T., *Phytochemistry* (1981) **20**, 2183–2185.

B807. Keller, W.J., *Phytochemistry* (1981) **20**, 2165–2167.

B808. Doerfler, D.L., Bird, B.A., Campbell, I.M., *Phytochemistry* (1981) **20**, 2303–2304.

B809. Kim, I.-S., Kim, S.-U., Anderson, J.A., *Phytochemistry* (1981) **20**, 2311–2314.

B810. Stüssi, H., Rast, D.M., *Phytochemistry* (1981) **20**, 2347–2352.

B811. Bird, C.R., Smith, T.A., *Phytochemistry* (1981) **20**, 2345–2346.

B812. Colby, H.D., Johnson, P.B., Pope, M.R., Zulkoski, J.S., *Biochem. Pharmacol.* (1982) **31**, 639–646.

B813. Gordonsmith, R.H., Raxworthy, M.J., Gulliver, P.A., *Biochem. Pharmacol.* (1982) **31**, 433–437.

B814. Maus, T.P., Pearson, R.K., Anderson, R.J., Woodson, L.C., Reiter, C., Weinshilboum, R.M., *Biochem. Pharmacol.* (1982) **31**, 849–856.

B815. Rifkind, A.B., Petschke, T., *J. Pharmacol. Exp. Ther.* (1981) **217**, 572–578.

B816. Lamoureux, G.L., Rusness, D.G., *J. Agric. Food Chem.* (1980) **28**, 1057–1070.

B817. Barber, P.J., Wilson, B.J., *Biochem. Pharmacol.* (1980) **29**, 2577–2582.

B818. Tateishi, M., Koitabashi, C., Ichihara, S., *Biochem. Pharmacol.* (1980) **29**, 2705–2708.

B819. Ibrahim, R.K., *Phytochemistry* (1980) **19**, 2459–2460.

B820. Reinke, L.A., Belinsky, S.A., Evans, R.K., Kauffman, F.C., Thurman, R.G., *J. Pharmacol. Exp. Ther.* (1981) **217**, 863–870.

B821. Vyas, A.V., Mulchandani, N.B., *Phytochemistry* (1980) **19**, 2597–2598.

B822. Küpfer, A., Roberts, R.K., Schenker, S., Branch, R.A., *J. Pharmacol. Exp. Ther.* (1981) **218**, 193–199.

B823. Fischer, L.J., Green, M.D., Harman, A.W., *J. Pharmacol. Exp. Ther.* (1981) **219**, 281–286.

B824. Pang, K.S., Koster, H., Halsema, I.C.M., Scholtens, E., Mulder, G.J., *J. Pharmacol. Exp. Ther.* (1981) **219**, 134–140.

B825. Lau, S.S., Zannoni, V.G., *J. Pharmacol. Exp. Ther.* (1981) **219**, 563–572.

B826. Mir, A.K., Vaughan, P.F.T., *Insect Biochem.* (1981) **11**, 571–577.

B827. Hobkirk, R., Green, R.N., Nilsen. M., Jennings, B.A., *Can. J. Biochem.* (1974) **52**, 15–20.

B828. Brown, D., Scott, D.H.T., Scott, D.B., Meyer, M., Westerlund, D., Lundström, J., *Eur. J. Clin. Pharmacol.* (1980) **17**, 111–116.

B829. Ohmiya, Y., Mehendale, H.M., *Pharmacology* (1981) **22**, 172–182.

B830. De Luca, V., Brunet, G., Khouri, H., Ibrahim, R., *Z. Naturforsch.* (1982) **37C**, 134–135.

B831. Davis, P.J., Talaat, R.E., *Appl. Environ. Microbiol.* (1981) **41**, 1243–1247.

B832. Ibrahim, R.K., De Luca, V., Jay, M., Voirin, B., *Naturwissenschaften* (1982) **69**, 41–42.

B833. Fekkes, D., Hennemann, G., Visser, T.J., *Biochem. Pharmacol.* (1982) **31**, 1705–1709.

B834. Rein, G., Glover, V., Sandler, M., *Biochem. Pharmacol.* (1982) **31**, 1893–1897.

B835. Parkinson, A., Safe, S., *Biochem. Pharmacol.* (1982) **31**, 1849–1856.

B836. Edwards, D.J., *Life Sci.* (1982) **30**, 1427–1434.

B837. Cerniglia, C.E., Dodge, R.H., Gibson, D.T., *Chem.-Biol. Interact.* (1982) **38**, 161–173.

B838. Blecher, H., Blecher, R., Wegst, W., Eberspaecher, J., Lingens, F., *Xenobiotica* (1981) **11**, 749–754.

B839. Davi, H., Dupont, P., Jeanniot, J.P., Roncucci, R., Cautreels, W., *Xenobiotica* (1981) **11**, 735–747.

B840. Franklin, R.A., Pierce, D.M., *Xenobiotica* (1981) **11**, 755–762.

B841. Ingebrigtsen, K., Skare, J.U., Nafstad, I., Førde, M., *Xenobiotica* (1981) **11**, 795–800.

B842. Baker, J.K., Wohlford, J.G., Bradbury, B.J., Wirth, P.W., *J. Med. Chem.* (1981) **24**, 666–669.

B843. Stopher, D.A., Beresford, A.P., Kinns, M., Bowen, D.V., *Xenobiotica* (1981) **11**, 849–856.

B844. Kuninaka, T., Sugai, K., Saito, T., Mori, N., Kimura, R., Murata, T., *yakugaku Zasshi* (1981) **101**, 232–238.

B845. Thakker, D.R., Levin, W., Buening, M., Yagi, H., Lehr, R.E., Wood, A W., Conney, A.H., Jerina, D.M., *Cancer Res.* (1981) **41**, 1389–1396.

B847. Black, O., Jr., Murrill, E., Fanska, C., *Dig. Dis. Sci.* (1981) **26**, 358–363.

B848. Sonoda, M., Sakagishi, Y., *Appl. Spectrosc.* (1981) **35**, 328–331.

B849. Kunz, D.A., Chapman, J.J., *J. Bacteriol.* (1981) **146**, 179–191.

B850. Waller, A.R., Chasseaud, L.F., *Food Cosmet. Toxicol.* (1981) **19**, 1–6.

B851. Kärenlampi, S.O., Hynninen, P.H., *Chemosphere* (1981) **10**, 391–396.

B852. Gruger, E.H., Jr., Schnell, J.V., Fraser, P.S., Brown, D.W., Malins, D.C., *Aquat. Toxicol.* (1981) **1**, 37–48.

B853. McIntire, W., Edmondson, D.E., Hopper, D.J., Singer, T.P., *Biochemistry* (1981) **20**, 3068–3075.

B854. Pennings, E.J.M., Van Kempen, G.M.J., *Biochem. J.* (1981) **193**, 869–874.

B855. Salcher, O., Lingens, F., *J. Gen. Microbiol.* (1980) **121**, 465–471.

B856. Jakobsen, P., Pederson, A.K., *J. Pharm. Pharmacol.* (1981) **33**, 89–92.

B857. Caccia, S., Ballabio, M., Garattini, S., *J. Pharm. Pharmacol.* (1981) **33**, 185–187.

B858. Edwards, D.J., Rizk, M., *J. Neurochem.* (1981) **36**, 1641–1647.

B859. Gordon, G.B., Spielberg, S.P., Blake, D.A., Balasubramanian, V., *Proc. Nat. Acad. Sci. U.S.A.* (1981) **78**, 2545–2548.

B860. Salem, Z., Murray, T., Yunis, A.A., *J. Lab. Clin. Med.* (1981) **97**, 881–886.

B861. Miyata, S., Suzuki, Y., Kamisaka, S., Masuda, Y., *Physiol. Plant.* (1981) **51**, 402–406.

B862. Groseclose, E.E., Ribbons, D.W., *J. Bacteriol.* (1981) **146**, 460–466.

B863. Hall, J.E., Dahm, K.H., Seed, J.R., *Comp. Biochem. Physiol. B* (1981) **68B**, 521–526.

B864. Beck, O., Jonsson, G., *J. Neurochem.* (1981) **36**, 2013–2018.

B865. Negrini, P., Cappellini, L., De Pascale, A., Frigerio, A., Macciantelli, D., McCamish, M., Placucci, G., Rossi, E., Testoni, G., *Anal. Chem. Symp. Ser.* (1979) **4**, 253–265.

B866. Larsson, H., Lund, J., *Acta Pharmacol. Toxicol.* (1981) **48**, 424–432.

B867. Vergin, H., Ferber, H., Brunner, F., Kukovetz, W.R., *Arzneim.-Forsch.* (1981) **31**, 513–518.

B868. Smallridge, R.C., Burman, K.D., Ward, K.E., Wartofsky, L., Dimond, R.C., Wright, F.D., Latham, K.R., *Endocrinology (Baltimore)* (1981) **108**, 2336–2345.

B869. Armstrong, R.N., Kedzierski, B., Levin, W., Jerina, D.M., *J. Biol.Chem.* (1981) **256**, 4726–4733.

B870. Goewert, R.R., Sippel, C.J., Olson, R.E., *Biochemistry* (1981) **20**, 4217–4223.

B871. Wegst, W., Tittman, U., Eberspächer, J., Lingens, F., *Biochem. J.* (1981) **194**, 679–684.

B872. Enoki, A., Goldsby, G.P., Krisnangkura, K., Gold, M.H., *FEMS Microbiol. Lett.* (1981) **10**, 373–377.

B873. Lasker, J.M., Graham, D.G., Abou-Donia, M.B., *Biochem. Pharmacol.* (1982) **31**, 1961–1967.

B874. Widholm, J.M., *Plant Physiol.* (1981) **67**, 1101–1104.

B875. Stohs, S.J., Al-Turk, W.A., Angle, C.R., *Biochem. Pharmacol.* (1982) **31**, 2113–2116.

B876. Kutney, J.P., Aweryn, B., Choi, L.S.L., Kolodziejczyk, P., Kurz, W.G.W., Chatson, K.B., Constabel, F., *Heterocycles* (1981) **16**, 1169–1171.

B877. Pometto, A.L.III, Sutherland, J.B., Crawford, D.L., *Can. J. Microbiol.* (1981) **27**, 636–638.

B878. Mitchell, S.C., Waring, R.H., *Pestic. Sci.* (1981) **12**, 79–85.

B879. Ito, R., Tsutsui, S., Hashimoto, S., Nishizawa, M., Kuninaka, T., Sugai, K., Mori, N., *Toho Igakkai Zasshi* (1980) **27**, 620–625.

B880. Hallberg, E., Rydström, J., *Acta Chem. Scand. Ser. B* (1981) **B35**, 145–146.

B882. Carroll, F.I., Brine, G.A., Boldt, K.G., Cone, E.J., Yousefnejad, D., Vaupel, D.B., Buchwald, W.F., *J. Med. Chem.* (1981) **24**, 1047–1051.

B883. Hamzah, R.Y., Tu, S.-G., *J. Biol. Chem.* (1981) **256**, 6392–6384.

B884. Henderson, G.L., Frincke, J.M., Garber, R.J., Knight, H.J., *Proc. West. Pharmacol. Soc.* (1981) **24**, 137–140.

B885. Mori, M., Naruse, Y., Kozuka, H., *Chem. Pharm. Bull.* (1981) **29**, 1147–1150.

B886. Yano, K., Takahashi, M., *Lab. Pract.* (1981) **30**, 468.

B887. Franklin, E.J., Knowles, C.O., *Pestic. Sci.* (1981) **12**, 133–141.

B889. Kamimura, S., Okuda, M., Kitahara, N., Hamano, K., Totori, T., Kanno, S., Hayashi, R., *Oyo Yakuri* (1980) **20**, 745–751.

B890. Yeh, S.Y., Krebs, H.A., Changchit, A., *J. Pharm. Sci.* (1981) **70**, 867–870.

B891. Orwig, B.A., Dugger, H.A., Bhuta, S.I., Talbot, K.C., Schwartz, H.J., *Arzneim.-Forsch.* (1981) **31**, 904–911.

B892. Zumstein, A., Karduck, W., Starke, K., *Naunyn-Schmiedeberg's Arch. Pharmacol.* (1981) **316**, 205–217.

B894. Lamoureux, G.L., Gouot, J.M., Davis, D.G., Rusness, D.G., *J. Agric. Food Chem.* (1981) **29**, 996–1002.

B895. Landrum, P.F., Crosby, D.G., *Ecotoxicol. Environ. Saf.* (1981) **5**, 240–254.

B896. Nesnow, S., Bergman, H., *Cancer Res.* (1981) **41**, 2621–2626.

B897. Passi, S., Nazzaro-Porro, M., *Br. J. Dermatol.* (1981) **104**, 659–665.

B898. Bergman, A., Brandt, I., Darnerud, P.O., Wachtmeister, C.A., *Xenobiotica* (1982) **12**, 1–7.

B899. Nagasawa, T., Utagawa, T., Goto, J., Kim, C.-J., Tani, Y., Kumagai, H., Yamada, H., *Eur. J. Biochem.* (1981) **117**, 33–40.

B900. Assandri, A., Omodei-Salè, A., Ferrari, P., Tuan, G., Perazzi, A., Ripamonti, A., Martinelli, E., *Xenobiotica* (1982) **12**, 19–30.

B901. Anderson, M.M., Mitchum, R.K., Beland, F.A., *Xenobiotica* (1982) **12**, 31–43.

B903. Jacob, J., Grimmer, G., Raab, G., Schmoldt, A., *Xenobiotica* (1982) **12**, 45–53.

B904. Leinberger, R., Hull, W.E., Simon, H., Retey, J., *Eur. J. Biochem* (1981) **117**, 311–318.

B905. Brilon, C., Beckmann, W., Knackmuss, H. *J. Appl. Environ. Microbiol.* (1981) **42**, 44–55.

B907. David, J.-C., Coulon, J.-F., Lafon-Cazal, M., Vinson, D., *Experientia* (1981) **37**, 804–805.

B908. Schmid, K., Küng, W., Riess, W., Dollery, C.T., Harland, S.J., *Arzneim.-Forsch.* (1981) **31**, 1143–1147.

B909. Gielsdorf, W., Schubert, K., *J. Clin. Chem. Clin. Biochem.* (1981) **19**, 485–490.

B910. Schwenk, M., Frank. B., Bolt, H.M., Winne, D., *Arzneim.-Forsch.* (1981) **31**, 1254–1257.

B911. Inaba, T., Uchino, H., Kalow, W., *Res. Commun. Chem. Pathol. Pharmacol.* (1981) **33**, 3–8.

B913. Black, O., Jr., Murrill, E., Fanska, C., *Res. Commun. Chem. Pathol. Pharmacol.* (1981) **33**, 103–118.

B914. Visser, T.J., Leonard, J.L., Kaplan, M.M., Larsen, P.R., *Biochem. Biophys. Res. Commun.* (1981) **101**, 1297–1304.

B916. Heide, L., Leistner, E., *FEBS Lett.* (1981) **128**, 201–204.

B917. Ito, S., Takai, K., Tokuyama, T., Hayaishi, O., *J. Biol. Chem.* (1981) **256**, 7834–7843.

B918. Baba, S., Furuta, T., Horie, M., Nakagawa, H., *J. Pharm. Sci.* (1981) **70**, 780–782.

B919. Rueffer, M., El-Shagi, H., Nagakura, N., Zenk, M.H., *FEBS Lett.* (1981) **129**, 5–9.

B920. Mimura, K., Baba, S., *Chem. Pharm. Bull.* (1981) **29**, 2043–2050.

B921. Pope, D.J., Gilbert, A.P., Easter, D.J., Chan, R.P., Turner, J.C., Gottfried, S., Parke, D.V., *J. Pharm. Pharmacol.* (1981) **33**, 302–308.

B922. Luderitz, T., Grisebach, H., *Eur. J. Biochem.* (1981) **119**, 115–124.

B923. Richter, E., Renner, G., Bayerl, J., Wick, M., *Chemosphere* (1981) **10**, 779–785.

B924. Choy, H.A., Jones, G.H., *Arch. Biochem. Biophys.* (1981) **211**, 55–65.

B925. Klibanov, A.M., Berman, Z., Alberti, B.N., *J. Am. Chem. Soc.* (1981) **103**, 6263–6264.

B927. Iyayi, C.B., Dart, R.K., *Microbios Lett.* (1980) **15**, 127–133.

B929. Caspary, W.J., Carroll, C., Morton, R., Dinces, N., Ts'O, P.O.P., Harvey, R.G., *Chem.-Biol. Interact.* (1981) **36**, 311–317.

B930. Endress, R., *Plant Physiol.* (1981) **68**, 272–274.

B931. Donnelly, M.I., Chapman, P.J., Dagley, S., *J. Bacteriol.* (1981) **147**, 477–481.

B932. Schaumberg, J.P., Rosazza, J.P., *J. Nat. Prod.* (1981) **44**, 478–481.

B933. Marcucci, F., Airoldi, L., Zavattini, G., Mussini, E., *Eur. J. Drug Metab. Pharmacokinet.* (1981) **6**, 109–114.

B934. Wong, L.T.L., Whitehouse, L.W., Solomonraj, G., Paul, C.J., *Toxicol. Lett.* (1981) **9**, 145–151.

B935. Helia, O., Vargova, E., Jindra, A., *Eur. J. Drug Metab. Pharmacokinet.* (1981) **6**, 155–158.

B936. Gustafsson, J.A., Rafter, J.J., Bakke, J.E, Gustafsson, B.E., *Nutr. Cancer* (1981) **2**, 224–231.

B937. Goos, C.M.A.A., Hukkelhoven, M.W.A.C., Vermorken, A.J.M., Henderson, P.T., Bloemendal, H., *Exp. Eye Res.* (1981) **33**, 345–350.

B940. Akhrem, A.A., Kiseleva, S.N., Kiselev, P.A., *React. Kinet. Catal. Lett.* (1981) **16**, 259–263.

B941. Sekura, R.D., Jakoby, W.B., *Arch. Biochem. Biophys.* (1981) **211**, 352–359.

B942. O'Neill, V.A., Rawlins, M.D., Chapman, P.H., *Br. J. Clin. Pharmacol.* (1981) **12**, 517–521.

B943. Fernandez-Pardal, J., Gimeno, M.F., Gimeno, A.L., *Can. J. Physiol. Pharmacol.* (1981) **59**, 1245–1249.

B944. Prough, R.A., Saeki, Y., Capdevila, J., *Arch. Biochem. Biophys.* (1981) **212**, 136–146.

B945. Miller, M.E., Garland, W.A., Min, B.H., Ludwick, B.T., Ballard, R.H., Levy, R.H., *Clin. Pharmacol. Ther.* (1981) **30**, 343–347.

B946. Ishida, M., Ogawa, M., Kosaki, G., Mega, T., Ikenaka, T., *Biochem. Int.* (1981) **3**, 239–246.

B947. Que, L., Jr., Widom, J., Crawford, R.L., *J. Biol. Chem.* (1981) **256**, 10941–10944.

B948. Kurechi. T., Kikugawa, K., Aoshima, S., *Chem. Pharm. Bull.* (1981) **29**, 2351–2358.

B949. Dua, R.J., Meera, S., *Eur. J. Biochem.* (1981) **120**, 461–465.

B950. Wegst, W., Lingens, F., *Hoppe-Seyler's Z. Physiol. Chem.* (1981) **362**, 1219–1227.

B952. Gielsdorf, W., *Z. Rechtsmed.* (1981) **87**, 117–127.

B953. Narige, T., Mizumura, M., Okuizumi, N., Matsumoto, K., Furukawa, Y., Hondo, T., *Yakuri to Chiryo* (1981) **9**, 1885–1892.

B954. Norback, D.H., Mack, E., Reddy, G., Britt, J., Hsia, M.T., *Res Commun. Chem. Pathol. Pharmacol.* (1981) **32**, 71–85.

B955. Tahara, S., Hafsah, Z., Ono, A., Asaishi, E., Mizutani, J., *Agric. Biol. Chem.* (1981) **45**, 2253–2258.

B956. Tanaka, K., Inada, M., Ishii, H., Naito, K., Nishikawa, M., Mashio, Y., Imura, H., *Endocrinology (Baltimore)* (1981) **109**, 1619–1624.

B957. Britsch, L., Heller, W., Grisebach, H., *Z. Naturforsch. C, Biosci.* (1981) **36C**, 742–750.

B958. Higgins, V.J., Ingham, J.L., *Phytopathology* (1981) **71**, 800–803.

B960. Sonobe, H., Kamps, La V.R., Mazzola, E.P., Roach, J.A.G., *J. Agric. Food Chem.* (1981) **29**, 1125–1129.

B961. Zangouras, A., Caldwell, J., Hutt, A.J., Smith, R.L., *Biochem. Pharmacol.* (1981) **30**, 1383–1386.

B962. Okulicz-Kozaryn, I., Schaeffer, M., Batt, A.-M., Siest, G., Loppinet, V., *Biochem. Pharmacol.* (1981) **30**, 1457–1461.

B964. Marsh, M.J., Hutt, A.J., Caldwell, J., Smith, R.L., Horner, M.W., Houghton, E., Moss, M.S., *Biochem. Pharmacol.* (1981) **30**,1879–1882.

B965. Watabe, T., Ishizuka, T., Ozawa, N., Isobe, M., *Biochem. Pharmacol.* (1982) **31**, 2542–2544.

B966. Bates, D.J., Foster, A.B., Griggs, L.J., Jarman, M., Leclercq, G., Devleeschouwer, N., *Biochem. Pharmacol.* (1982) **31**, 2823–2827.

B967. Prosser, C.I., Roy, A.B., *Biochem. Pharmacol.* (1982) **31**, 2675–2676.

B968. Ito, S., Fujita, K., *Biochem. Pharmacol.* (1982) **31**, 2887–2889.

B969. Emslie, K.R., Calder, I.C., Hart, S.J., Tange, J.D., *Xenobiotica* (1982) **12**, 77–82.

B970. Mitchell, S.C., Norbury, H.M., Waring, R.H., Gadsden, P.M., Wood, P.B., *Xenobiotica* (1982) **12**, 93–99.

B971. Gunnarsson, P.O., Andersson, S.-B., Ellman, M., Olsson, A., *Xenobiotica* (1982) **12**, 83–91.

B972. Bolanowska, W., Gessner, T., *Xenobiotica* (1982) **12**, 125–127.

B973. Clark, B., Smith, D.A., Eason, C.T., Parke, D.V., *Xenobiotica* (1982) **12**, 147–153.

B974. Fitzloff, J.F., Portig, J., Stein, K., *Xenobiotica* (1982) **12**, 197–202.

B975. Wood, S.G., Kirkpatrick, D., Jackson, A.J.S., Hawkins, D.R., Down, W.H., Chasseaud, L.F., Biggs, S.R., Darragh, A., *Xenobiotica* (1982) **12**, 165–176.

B976. Neill, E.A.M., Chapple, D.J., *Xenobiotica* (1982) **12**, 203–210.

B977. Phillippe, M., Niloff, J.M., *Am. J. Obstet. Gynecol.* (1982) **143**, 782–787.

B978. Hein, D.W., Weber, W.W., *Drug Metab. Dispos.* (1982) **10**, 225–229.

B979. Yang, S.K., *Drug Metab. Dispos.* (1982) **10**, 205–211.

B980. Chen, C.-H., Nelson, W.L., *Drug Metab. Dispos.* (1982) **10**, 277–278.

B981. Klunk, L.J., Riska, P.S., Maynard, D.E., *Drug Metab. Dispos.* (1982) **10**, 241–245.

B982. Feng, P.C.C., Fenselau, C.C., Jacobson, R.R., *Drug Metab. Dispos.* (1982) **10**, 286–289.

B983. Tukey, R.H., Robinson, R., Holm, B., Falany, C.N., Tephly, T.R., *Drug Metab. Dispos.* (1982) **10**, 97–101.

B984. Cook, C.S., Karim, A., Sollman, P., *Drug Metab. Dispos.* (1982) **10**, 116–121.

B985. Levine, W.G., Lu, A.Y.H., *Drug Metab. Dispos.* (1982) **10**, 102–109.

B986. Melancon, M.J., Rickert, D.E., Lech, J.J., *Drug Metab. Dispos.* (1982) **10**, 128–133.

B987. Walle, T., Oatis, J.E., Jr., Walle, U.K., Knapp, D.R., *Drug Metab. Dispos.* (1982) **10**, 122–127.

B988. Lingg, R.D., Kaylor, W.H., Pyle, S.M., Kopfler, F.C., Smith, C.C., Wolfe, G.F., Cragg, S., *Drug Metab. Dispos.* (1982) **10**, 134–141.

B989. Nelis, H.J.C.F., de Leenheer, A.P., *Drug Metab. Dispos.* (1982) **10**, 142–146.

B990. Sisenwine, S.F., Tio, C.O., Ruelius, H.W., *Drug Metab. Dispos.* (1982) **10**, 161–167.

B991. Chow, S.A., Fischer, L.J., *Drug Metab. Dispos.* (1982) **10**, 156–160.

B992. Hoffman, K.-J., Arfwidsson, A., Borg, K.O., *Drug Metab. Dispos.* (1982) **10**, 173–179.

C1. Minnigh, M.B., Zemaitis, M.A., *Drug Metab. Dispos.* (1982) **10**, 183–188.

C2. Halldin, M.M., Widman, M., von Bahr, C., Lindgren, J.-E., Martin, B.R., *Drug Metab. Dispos.* (1982) **10**, 297–301.

C3. Lertratanangkoon, K., Horning, M.G., *Drug Metab. Dispos.* (1982) **10**, 1–10.

C4. Pacifici, G.M., Rane, A., *Drug Metab. Dispos.* (1982) **10**, 302–305.

C5. Florence, V.M., Di Stefano, E.W., Sum, C.Y., Cho, A.K., *Drug Metab. Dispos.* (1982) **10**, 312–315.

C6. Fenselau, C., Feng, P.C.C., Chen, T., Johnson, L.P., *Drug Metab. Dispos.* (1982) **10**, 316–318.

C7. Yuen, C.-T., Price, R.G., Chattagoon, L., Richardson, A.C., Praill, P.F.G., *Clin. Chim. Acta* (1982) **124**, 195–204.

C8. Yeh, S.Y., *Drug Metab. Dispos.* (1982) **10**, 319–325.

C9. Gal, J., French, T.A., Zysset, T., Haroldsen, P.E., *Drug Metab. Dispos.* (1982) **10**, 399–404.

C10. Egger, H., Bartlett, F., Itterly, W., Rodebaugh, R., Shimanskas, C., *Drug Metab. Dispos.* (1982) **10**, 405–412.

C11. Kojima, S., Honda, T., Nakagawa, M., Kiyozumi, M., Takadate, A., *Drug Metab. Dispos.* (1982) **10**, 429–433.

C12. Diamond, G.L., Quebbemann, A.J., *Drug Metab. Dispos.* (1981) **9**, 402–409.

C13. Maguire, J.H., Kraus, B.L., Butler, T.C., Dudley, K.H., *Drug Metab. Dispos.* (1981) **9**, 393–401.

C14. Egger, H., Bartlett, F., Dreyfuss, R., Karliner, J., *Drug Metab. Dispos.* (1981) **9**, 415–423.

C15. Benziger, D.P., Fritz, A.K., Clemans, S.D., Edelson, J., *Drug Metab. Dispos.* (1981) **9**, 424–427.

C16. Tachizawa, H., Sudo, K., Sasano, H., Sano, M., *Drug Metab. Dispos.* (1981) **9**, 442–448.

C17. Hucker, H.B., Balletto, A.J., White, S.D., Zacchei, A.G., *Drug Metab. Dispos.* (1981) **9**, 428–433.

C18. Ruenitz, P.C., *Drug Metab. Dispos.* (1981) **9**, 456–460.

C19. Thompson, J.A., Hull, J.E., Norris, K.J., *Drug Metab. Dispos.* (1981) **9**, 466–471.

C20. Gyurik, R.J., Chow, A.W., Zaber, B., Brunner, E.L., Miller, J.A., Villani, A.J., Petka, L.A., Parish, R.C., *Drug Metab. Dispos.* (1981) **9**, 503–508.

C21. Pacifici, G.M., Rane, A., *Drug Metab. Dispos.* (1981) **9**, 472–475.

C22. Feng, P.C.C., Fenselau, C.C., Jacobson, R.R., *Drug Metab. Dispos.* (1981) **9**, 521–524.

C23. Shono, T., Ohmizu, Y., Toda, T., Oshino, N., *Drug Metab. Dispos.* (1981) **9**, 476–480.

C24. Bakke, J.E., Rafter, J., Larsen, G.L., Gustaffson, J.Å., *Drug Metab. Dispos.* (1981) **9**, 525–528.

C25. Gal, J., De Vito, D., Harper, T.W., *Drug Metab. Dispos.* (1981) **9**, 557–560.

C26. Mellström, B., von Bahr, C., *Drug Metab. Dispos.* (1981) **9**, 565–568.

C27. Tybring, G., von Bahr, C., Bertilsson, L., Collste, H., Glaumann, H., Solbrand, M., *Drug Metab. Dispos.* (1981) **9**, 561–564.

C28. Pacifici, G.M., Rane, A., *Drug Metab. Dispos.* (1981) **9**, 569–572.

C29. Suzuki, S., Tateishi, M., *Drug Metab. Dispos.* (1981) **9**, 573–577.

C30. Lehmann, J.P., Fenselau, C., *Drug Metab. Dispos.* (1982) **10**, 446–449.

C31. Morris, P.L., Burke, T.R., Jr., George, J.W., Pohl, L.R., *Drug Metab. Dispos.* (1982) **10**, 439–445.

C32. Levin, A.A., Dent, J.G., *Drug Metab. Dispos.* (1982) **10**, 450–454.

C33. Long, R.M., Rickert, D.E., *Drug Metab. Dispos.* (1982) **10**, 455–458.

C34. Hasegawa, J., Smith, P.C., Benet, L.Z., *Drug Metab. Dispos.* (1982) **10**, 469–473.

C35. Cummings, S.W., Guengerich, F.P., Prough, R.A., *Drug Metab. Dispos.* (1982) **10**, 459–464.

C36. Parker, R.J., Priester, E.R., Sieber, S.M., *Drug Metab. Dispos.* (1982) **10**, 499–504.

C37. Fuhrman-Lane, C., Fujimoto, J.M., *Drug Metab. Dispos.* (1982) **10**, 505–509.

C38. James, M.O., *Drug Metab. Dispos.* (1982) **10**, 516–522.

C39. Johnson, S.W., *Drug Metab. Dispos.* (1982) **10**, 510–515.

C40. Huckle, K.R., Stodyn, G., Hutson, D.H., Millburn, P., *Drug Metab. Dispos.* (1982) **10**, 523–528.

C41. Egger, H., Bartlett, F., Yuan, H.-P., Karliner, J., *Drug Metab. Dispos.* (1982) **10**, 529–536.

C42. Goromaru, T., Matsuura, H., Furuta, T., Baba, S., Yoshimura, N., Miyawaki, T., Sameshima, T., *Drug Metab. Dispos.* (1982) **10**, 542–546.

C43. Matsumura, K., Kubo, O., Sakashita, T., Kato, H., Watanabe, K., Hirobe, M., *Drug Metab. Dispos.* (1982) **10**, 537–541.

C44. Rizzardini, M., Smith, A.G., *Biochem. Pharmacol.* (1982) **31**, 3543–3548.

C45. Shono, T., Ohmizu, Y., Toda, T., Oshino, N., *Drug Metab. Dispos.* (1982) **10**, 555–557.

C46. Fowler, C.J., Mantle, T.J., Tipton, K.F., *Biochem. Pharmacol.* (1982) **31**, 3555–3561.

C47. Ross, D., Farmer, P.B., Gescher, A., Hickman, J.A., Threadgill, M.D., *Biochem. Pharmacol.* (1982) **31**, 3621–3627.

C48. Tanaka, K., Inada, M., Mashio, Y., Ishii, H., Naito, K., Nishikawa, M., Takahashi, K., Imura, H., *Acta Endocrinol.* (1982) **101**, 371–376.

C49. Sava, G., Giraldi, T., Lassiani, L., Nisi, C., Farmer, P.B., *Biochem. Pharmacol.* (1982) **31**, 3629–3634.

C50. Caccia, S., Ballabio, M., Samanin, R., Zanini, M.G., Garattini, S., *J. Pharm. Pharmacol.* (1981) **33**, 477–478.

C51. Cayen, M.N., Kraml, M., Ferdinandi, E.S., Eckhardt, S., Greselin, E., Dvornik, D., *Drug Metab. Rev.* (1981) **12**, 339–362.

C52. Tuong, A., Bouyssou, A., Paret, J., Cuong, T.G., *Eur. J. Drug Metab. Pharmacokinet.* (1981) **6**, 91–98.

C53. Lehmann, K.A., Moeseler, G., Daub, D., *Anaesthetist* (1981) **30**, 461–466.

C54. Kaminsky, L.S., Kennedy, M.W., Adams, S.M., Guengerich, F.P., *Biochemistry* (1981) **20**, 7379– 7384.

C55. Lau, S.S., Zannoni, V.G., *Mol. Pharmacol.* (1981) **20**, 234–235.

C56. Long, R.A., Weppelman, R.M., Taylor, J.E., Tolman, R.L., Olson, G., *Biochemistry* (1981) **20**, 7423–7431.

C57. Buswell, J.A., Eriksson, K.E., Pettersson, B., *J. Chromatogr.* (1981) **215**, 99–108.

C58. Smith, D.A., Harver, J.M., Cleveland, T.E., *Phytopathology* (1981) **71**, 1212-1215.

C59. Otten, S.L., Rosazza, J.P., *J. Nat. Prod.* (1981) **44**, 562–568.

C60. Yoshizako, F., Nishimura, A., Ueno, T., *Annu. Rep. Radiat. Cent. Osaka Prefect.* (1980) **21**, 71–74.

C61. Green, T.R., Schaefer, R.E., *Biochemistry* (1981) **20**, 7483–7487.

C62. Nakazawa, H., Kumagai, H., Yamada, H., *Agric. Biol. Chem.* (1981) **45**, 2543–2552.

C63. Kaneko, H., Ohkawa, H., Miyamoto, J., *Nippon Noyaku Gakkaishi* (1981) **6**, 169–182.

C64. Ballschmiter, K., Scholz, C., *Angew. Chem.* (1981) **93**, 1026–1027.

C65. Stotz, G., Forkmann, G., *Z. Naturforsch. C: Biosci.* (1981) **36C**, 737–741.

C66. Strother, A., Fraser, I.M., Allahyari, R., Tilton, B.E., *Bull. W.H.O.* (1981) **59**, 413–425.

C67. Ferber, H., Vergin, H., Hitzenberger, G., *Eur. J. Clin. Pharmacol.* (1981) **19**, 431–435.

C68. Coutts, R.T., Foster, B.C., Pasutto, F.M., *Life Sci.* (1981) **29**, 1951–1958.

C69. Lhoest, G., *Eur. J. Mass Spectrom. Biochem., Med. Environ. Res.* (1980) **1**, 197–204.

C70. Reigh, D.L., Floyd, R.A., *Cancer Biochem. Biophys.* (1981) **5**, 213–217.

C71. Hecht, S.S., La Voie, E.J., Bedenko, V., Pingaro, L., Katayama, S., Hoffmann, D., Sardella, D.J., Boger, E., Lehr, R.E., *Cancer Res.* (1981) **41**, 4341–4345.

C72. Yang, T., *Eur. J. Biochem.* (1981) **121**, 335–341.

C73. Yoshimoto, T., Tsuru, D., *Agric. Biol. Chem.* (1981) **45**, 2449–2455.

C74. Kuhn, P.J., Smith, D.A., *Ann. Phytopathol.* (1980) **12**, 453–456.

C76. Dart, R.K., Iyayi, C.B., *Microbios Lett.* (1981) **16**, 27–33.

C77. Zaidi, K.F., Creme, S., Leaback, D.H., *FEBS Lett.* (1981) **134**, 91–95.

C78. Weber, W.W., King, C.M., *Methods. Enzymol.* (1981) **77**, 272–280.

C79. Lehmann, K.A., Hunger, L., Brandt, K., Daub, D., *Anaesthetist* (1983) **32**, 165–173.

C80. Batt, A.M., Martin, N., Siest, G., *Toxicol. Lett.* (1981) **9**, 355–360.

C81. Lehmann, K.A., Weski, C., Hunger, L., Heinrich, C., Daub, D., *Anaesthetist* (1982) **31**, 221–227.

C82. Thouvenot, D., Morfin, R., Di Stefano, S., Picart, D., *Eur. J. Biochem.* (1981) **121**, 139–145.

C83. Amin, S., Juchatz, A., Furuya, K., Hecht, S.S., *Carcinogenesis* (1981) **2**, 1027–1032.

C84. Itoh, M., *Agric. Biol. Chem.* (1981) **45**, 2787–2796.

C85. Doubner, S.C., Matthews, R.G., *J. Biol. Chem.* (1982) **257**, 140–145.

C86. Franck, B., Strange, A, *Liebig's Ann. Chem.* (1981) 2106–2116.

C87. Raza, H., Khanna, S.K., Singh, G.B., Murti, C.R.K., *Indian J. Exp. Biol.* (1982) **20**, 48–51.

C88. Koruna, I., Ryska, M., Roubal, Z., Kuchar, M., Grimova, J., Nemecek, O., *Cesk. Farm.* (1981) **30**, 257–262.

C89. Gorrod, J.W., Gooderham, N.J., *Eur. J. Drug Metab. Pharmacokinet.* (1981) **6**, 195–206.

C90. Jourdan, P.S., Mansell, R.L., *Arch. Biochem. Biophys.* (1982) **213**, 434–443.

C91. Elmorsi, E.A., Hopper, D.J., *Biochem. Soc. Trans.* (1981) **9**, 431.

C93. Hassam, S.B., Floss, H.G., *J. Nat. Prod.* (1981) **44**, 756–758.

C94. Charriere-Ladreix, Y., Douce, R., Joyard, J., *FEBS Lett.* (1981) **133**, 55–58.

C95. Pederson, S.B., Alhede, B., Buchardt, O., Moller, J., Bock, K., *Arzneim.-Forsch.* (1981) **31**, 1944–1948.

C96. Allan, L., Barrow, A., Haskins, N.J., Rogers, R.F., Rose, D.A., Palmer, R.F., *Eur. J. Drug Metab. Pharmacokinet.* (1981) **6**, 179–187.

C97. Kucharczyk, N., Yang, J.T., Sofia, R.D., *Arzneim.-Forsch.* (1981) **31**, 2085–2088.

C98. Van Doorn, R., Leijdekkers, C.M., Bos, R.P., Brouns, R.M.E., Henderson, P.T., *JAT, J. Appl. Toxicol.* (1981) **1**, 236–242.

C99. Patel, T.R., Jure, K.G., Jones, G.A., *Appl. Environ. Microbiol.* (1981) **42**, 1010–1017.

C100. Kuroki, G., Poulton, J.E., *Z. Naturforsch C: Biosci.* (1981) **36C**, 916–920.

C101. Ghisalba, O., Roos, R., Schupp, T., Nuesch, J., *J. Antibiot.* (1982) **35**, 74–80.

C102. Combie, J., Blake, J.W., Nugent, T.E., Tobin, T., *Clin. Chem.* (1982) **28**, 83–86.

C103. Dreessen, M., Eyssen, H., Lemli, J., *J. Pharm. Pharmacol.* (1981) **33**, 679–681.

C104. Nagatsu, Y., Endo, K., Irikura, T., *Chemotherapy (Tokyo)* (1981) **29** (Suppl 4), 119–127.

C105. Hirata, T., Sakano, S., Suga, T., *Experientia* (1981) **37**, 1252–1253.

C106. Ozaki, T., Uchida, H., Irikura, T., *Chemotherapy (Tokyo)* (1981) **29** (Suppl 4), 128–135.

C107. Hengstmann, J.H., Goronzy, J., *Eur. J. Clin. Pharmacol.* (1982) **21**, 335–341.

C108. Chopra, I.J., Teco, G.N.C., *Endocrinology (Baltimore)* (1982) **110**, 89–97.

C109. Roti, E., Fang, S.L., Braverman, L.E., Emerson, C.H., *Endocrinology* (Baltimore) (1982) **110**, 34–37.
C110. Solbakken, J.E., Palmork, K.H., *Comp. Biochem. Physiol. C* (1981) **70C**, 21–26.
C111. Irace, G., Colonna, G., Camardella, M., Pietra, G.D., Porta, R., *Biochemistry* (1982) **21**, 1464–1470.
C112. Sherman, P.A., Sander, E.G., *Biochem. Biophys. Res. Commun.* (1981) **103**, 997–1005.
C113. Nakata, H., Fujisawa, H., *Eur. J. Biochem.* (1982) **122**, 41–47.
C114. Konno, R., Yasumura, Y., *Dobutsugaku Zasshi* (1981) **90**, 368–373.
C115. Keller, B., Keller, E., Süssmuth, R., Lingens, F., *FEMS Microbiol. Lett* (1982) **13**, 3–4.
C116. Srivastava, L.P., Khanna, S.K., Singh, G.B., Murti, C.R.K., *Environ. Res.* (1982) **27**, 185–189.
C117. Katayama, T., Nakatsubo, F., Higuchi, T., *Arch. Microbiol.* (1981) **130**, 198–203.
C118. Suzuki, Y., *Toho Igakkai Zasshi* (1981) **28**, 260–274.
C119. Graefe, K.H., *Naunyn-Schmiedeberg's Arch. Pharmacol.* (1981) **318**, 71–82.
C120. Lübbe, C., Salcher, O., Lingens, F., *FEMS Microbiol. Lett.* (1982) **13**, 31–33.
C121. Weisiger, R.A., Jakoby, W.B., *Methods Enzymol.* (1981) **77**, 257–262.
C123. Köhler, J., Schwartz, A.C., *Can. J. Microbiol.* (1982) **28**, 65–72.
C124. Gardner, A.M., Alvarez, G.H., Ku, Y., *Bull. Environ. Contam. Toxicol.* (1982) **28**, 91–96.
C125. Perkins, E.G., Baker, D.H., Johnson, G.H., Makowski, E., *Prog. Food Nutr. Sci.* (1981) **5**, 229–242.
C126. Ohmiya, Y., Mehendale, H.M., *Res. Commun. Psychol. Psychiatry Behav.* (1981) **6**, 205–212.
C127. Kaneko, H., Ohkawa, H., Miyamoto, J., *Nippon Nogaku Gakkaishi* (1981) **6**, 317–326.
C128. Campbell, D., *Sven. Farm. Tidskr.* (1981) **85**, 6–9.
C129. Heizmann, P., Ziegler, W.H., *Arzneim.-Forsch.* (1981) **31**, 2220–2223.
C130. Tajima, K., Yamamoto, K., Mizutani, T., *Chem. Pharm. Bull.* (1981) **29**, 3738–3741.
C131. Decad, G.M., Graichen, M.E., Dent, J.G., *Toxicol. Appl. Pharmacol.* (1982) **62**, 325–334.
C132. Crawford, M.J., Croucher, A., Hutson, D.H., *Pestic. Sci.* (1981) **12**, 399–411.
C133. Hutson, D.H., Gaughan, L.C., Casida, J.E., *Pestic. Sci.* (1981) **12**, 385–398.
C134. Lehmann, W.D., Theobald, N., Fischer, R., Heinrich, H.C., *Clin. Chim. Acta* (1983) **128**, 181–198.
C135. Garcha, G., Imrie, P.R., Marley, E., Thomas, D.V., *J. Psychiat. Res.* (1983) **17**, 75–92.
C136. Murphy, S.E., Drotar, A., Fall, R., *Chemosphere* (1982) **11**, 33–39.
C137. Lehmann, K.A., Hunger, L., Brandt, K., Daub, D., *Anaesthetist* (1983) **32**, 165–173.
C138. Moore, B.P., Hicks, R.M., Knowles, M.A., Redgrave, S., *Cancer Res.* (1982) **42**, 642–648.
C139. Hinton, D.E., Klaunig, J.E., Jack, R.M., Lipsky, M.M., Trump, B.F., *ASTM Spec. Tech. Publ.* (1981) **737**, 215–225.
C141. Tan, B., Kilgore, M.V., Elam, D.L., Jr., Melius, P., Schoor, W.P., *ASTM Spec. Tech. Publ.* (1981) **737**, 239–246.
C142. Ketterer, B., Srai, S.K.S., Waynforth, B., Tullis, D.L., Evans, F.E., Kadlubar, F.F., *Chem.-Biol. Interact.* (1982) **38**, 287–302.
C143. Liu, S.-Y., Minard, R.D., Bollag, J.M., *Soil Sci. Soc. Am. J.* (1981) **45**, 1100–1105.
C144. Ensley, B.D., Gibson, D.T., Laborde, A.L., *J. Bacteriol.* (1982) **149**, 948–954.
C145. Ono, S., Hirano, H., Sato, Y., *Exp. Eye Res.* (1982) **34**, 297–301.

C146. Hwang, S.Y., Oishi, K., Aida, K., *Agric. Biol. Chem.* (1982) **46**, 285–287.
C147. Morris, C.M., Barnsley, E.A., *Can. J. Microbiol.* (1982) **28**, 73–79.
C148. Adams, J.D., Jr., Baillie, T.A., Trevor, A.J., Castagnoli, N., Jr., *Biomed. Mass Spectrom.* (1981) **8**, 527–538.
C150. Borchers, F., Achtert, G., *Anal. Chem. Symp. Ser.* (1981) **7**, 39–56.
C151. Sakamoto, M., Chiba, K., Agata, I., Higa, T., Kiyota, M., Hayashi, N., Yagi, N., Itoh, M., Sekikawa, H., Takada, M., *J. Toxicol. Sci.* (1981) **6**, 307–314.
C152. Yoshizawa, H., Uchimaru, R., Kamataki, T., Kato, R., Ueno, Y., *Cancer Res.* (1982) **42**, 1120–1124.
C153. Bergemalm-Rynell, K., Steen, G., *Toxicol. Appl. Pharmacol.* (1982) **62**, 19–31.
C154. Cerniglia, C.E., Freeman, J.P., Franklin, W., Pack, L.D., *Biochem. Biophys. Res. Commun.* (1982) **107**, 1224–1229.
C155. Nemoto, N., Takayama, S., *Toxicol. Lett.* (1982) **10**, 75–79.
C156. Lyon, E.S., Jakoby, W.B., *Methods Enzymol.* (1981) **77**, 263–266.
C157. Kolkmann, R., Knauel, G., Arendt, S., Leistner, E., *FEBS Lett.* (1982) **137**, 53–56.
C160. Matsubara, T., Yoshihara, E., Iwata, T., Tochino, Y., Hachino, Y., *Jpn. J. Pharmacol.* (1982) **32**, 9–21.
C161. Achari, R., Beckett, A.H., *J. Pharm. Pharmacol.* (1981) **33**, 747–748.
C162. Ishida, T., Oguri, K., Yoshimura, H., *J. Pharmacobio.-Dyn.* (1982) **5**, 134–136.
C163. Guha, M., Hati, R.N., Datta, A.G., *J. Biosci.* (1981) **3**, 239–248.
C164. Klungsoeyr, J., Scheline, R.R., *Acta Pharmacol. Toxicol.* (1981) **49**, 305–312.
C165. Sawahata, T., Olson, J.R., Neal, R.A., *Biochem. Biophys. Res. Commun.* (1982) **105**, 341–346.
C166. Kaka, J.S., Somani, S.M., Schaeffer, D.J., *EcoToxicol. Environ. Saf.* (1982) **6**, 35–40.
C167. Huckle, K.R., Hutson, D.H., Millburn, P., *Biochem. Soc. Trans.* (1981) **9**, 578–579.
C169. Borchardt, R.T., *Methods Enzymol.* (1981) **77**, 267–272.
C170. Kinapuli, S.P., Vaidyanathan, C.S., *Plant Sci. Lett.* (1982) **24**, 183–188.
C171. Ainouz, I.L., Benevides, N.M.B., *Biol. Plant.* (1982) **24**, 39–47.
C172. Boutin, J.A., Jacquier, A., Neiger, A., Batt, A.A., Marliere, P., Grosdidier, J., Gazaigne, J., Schmitt, J., Siest, G., *Ann. Med. Nancy Est.* (1981) **20**, 1199–1202.
C173. Jacobs-Sturm, A., Dahlmann, B., Reinauer, H., *Biochim. Biophys. Acta* (1982) **715**, 34–41.
C174. Chin, C.C., Pineda, J., Warren, J.C., *J. Biol. Chem.* (1982) **257**, 2225–2229.
C175. Baarschers, W.H., Bharath, A.I., Elvish, J., Davies, M., *Can. J. Microbiol.* (1982) **28**, 176–179.
C176. Challis, I.R., Adcock, J.W., *Pestic. Sci.* (1981) **12**, 638–644.
C177. Wilson, K., Burnett, D., Oram, M., Reynolds, C.T., *Eur. J. Drug Metab. Pharmacokinet.* (1981) **6**, 289–295.
C178. Sasaki, M., Sugiyama, Y., Kunimatsu, M., Katagiri, K., *J. Appl. Biochem.* (1981) **3**, 400–410.
C179. Light, D.R., Waxman, D.J., Walsh, C., *Biochemistry* (1982) **21**, 2490–2498.
C180. Koster, H., Halsema, I., Scholtens, E., Pang, K.S., Mulder, G.J., *Biochem. Pharmacol.* (1982) **31**, 3023–3028.
C181. Magnus, V., Šimaga, Š., Iskrić, S., Kveder, S., *Plant Physiol.* (1982) **69**, 853–858.
C182. Fuller, R.W., Hemrick-Luecke, S.K., *Biochem. Pharmacol.* (1982) **31**, 3128–3130.
C183. Marsh, M.V., Caldwell, J., Hutt, A.J., Smith, R.L., Horner, M.W., Houghton, E., Moss, M.S., *Biochem. Pharmacol.* (1982) **31**, 3225–3230.
C184. Brundin, A., Ratnayake, J.H., Sunram, J.M., Anders, M.W., *Biochem. Pharmacol.* (1982) **31**, 3885–3890.

C185. Wang, P.-C., Kuchel, O., Buu, N.T., Genest, J., *J. Neurochem.* (1983) **40**, 1435–1440.
C186. Shekhar, C., Mayanil, K., Baquer, N.Z., *Biochem. Pharmacol.* (1982) **31**, 3925–3927.
C187. Wong, K.P., Yeo, T., *Biochem. Pharmacol.* (1982) **31**, 4001–4003.
C188. Suzukake, K., Vistica, B.P., Vistica, D.T., *Biochem. Pharmacol.* (1983) **32**, 165–167.
C189. Lee, I.P., Mukhtar, H., Suzuki, K., Bend, J.R., *Biochem. Pharmacol.* (1983) **32**, 159–161.
C190. Fuller, R.W., Hemrick-Luecke, S.K., Perry, K.W., *Biochem. Pharmacol.* (1983) **32**, 215–220.
C191. Bain, R.R., Jordan, V.C., *Biochem. Pharmacol.* (1983) **32**, 373–375.
C192. Nohmi, T., Miyata, R., Yoshikawa, K., Nakadate, M., Ishidate, M., Jr., *Biochem. Pharmacol.* (1983) **32**, 735–738.
C193. Sasaki, K., Hosoya, R., Wang, Y.-M., Raulston, G.L., *Biochem. Pharmacol.* (1983) **32**, 503–507.
C194. Watabe, T., Ozawa, N., Hiratsuka, A., *Biochem. Pharmacol.* (1983) **32**, 777–785.
C195. Koster, A.S., Noordhoek, J., *Biochem. Pharmacol.* (1983) **32**, 895–900.
C196. Ohmiya, Y., Mehendale, H.M., *Biochem. Pharmacol.* (1983) **32**, 1281–1285.
C197. Mita, S., Yamazoe, Y., Kamataki, T., Kato, R., *Biochem. Pharmacol.* (1983) **32**, 1179–1182.
C198. Davison, K.L., Feil, V.J., Lamoureux, C.J.H., *J. Agric. Food Chem.* (1982) **30**, 130–137.
C200. Chrzanowski, R.L., Leitch, R.E., *J. Agric. Food Chem.* (1982) **30**, 155–160.
C201. You, I.S., Bartha, R., *J. Agric. Food Chem.* (1982) **30**, 274–277.
C202. Newton, J.F., Braselton, W.E., Jr., Kuo, C.H., Kluwe, W.M., Gemborys, M.W., Mudge, G.H., Hook, J.B., *J. Pharmacol. Exp. Ther.* (1982) **221**, 76–79.
C203. Gollamudi, R., Glassman, J.M., *J. Appl. Toxicol.* (1982) **2**, 21–26.
C204. Gozukara, E.M., Guengerich, F.P., Miller, H., Gelboin, H.V., *Carcinogenesis* (1982) **3**, 129–133.
C205. Amin, S., La Voie, E.J., Hecht, S.S., *Carcinogenesis* (1982) **3**, 171–174.
C206. Beschle, H.G., Süssmuth, R., Lingens, F., *Hoppe-Seyler's Z. Physiol. Chem.* (1982) **363**, 439–444.
C207. Weston, A., Grover, P.L., Sims, P., *Biochem. Biophys. Res. Commun.* (1982) **105**, 935–941.
C208. Roche, P.A., Moorehead, T.J., Hamilton, G.A., *Arch. Biochem. Biophys.* (1982) **216**, 62–73.
C210. Wadman, S.K., Duran, M., Ketting, D., Bruinvis, L., van Sprang, F.J., Berger, R., Smit, G.P.A., Steinmann, B., Leonard, J.V., Divry, P., Farriaux, J.P., Cartigny, B., *Clin. Chim. Acta* (1983) **130**, 231–238.
C211. Raxworthy, M.J., Youde, I.R., Gulliver, P.A., *Biochem. Pharmacol.* (1983) **32**, 1361–1364.
C212. Crowley, J.R., Williams, C.M., Fregley, M.J., *J. Pharm. Pharmacol.* (1983) **35**, 264–265.
C213. Sasaki, M., Yoshikane, K., Nobata, E., Katagiri, K., Takeuchi, T., *J. Biochem.* (1981) **89**, 609–614.
C215. Oram, M., Wilson, K., Burnett, D., Al-Dabbagh, S.G., Idle, J.R., Smith, R.L., *Eur. J. Clin. Pharmacol.* (1982) **23**, 147–150.
C216. Nakata, H., Fujisawa, H., *Eur. J. Biochem.* (1982) **124**, 595–601.
C217. Inaba, T., Uchino, H., Kadar, D., Kalow, W., *Res. Commun. Chem. Pathol. Pharmacol.* (1981) **32**, 236–244.
C218. Ohtsuki, K., Kabashima, J., Nakao, T., *Tokyo-toritsu Eisei Kenkyushu Kenkyu Nempo* (1981) 51–52.

C219. Konno, R., Uchiyama, S., Yasumura, Y., *Comp. Biochem. Physiol.* (1982) **71B**, 735–738.

C219a. Konno, R., Yoshiro, Y., *Genetics* (1983) **103**, 277–285.

C220. Simonsen, H., Nielsen, M., Braestrup, C., *Acta Pharmacol. Toxicol.* (1982) **50**, 89–92.

C221. Pacifici, G.M., Rane, A., *Dev. Pharmacol. Ther.* (1981) **3**, 160–167.

C222. Coutts, R.T., Jones, G.R., Liu, S.-F., *Proc. West. Pharmacol. Soc.* (1982) **25**, 119–123.

C223. Koruna, I., Ryska, M., Roubal, Z., *Cesk. Farm.* (1982) **31**, 67–70.

C224. Sakai, T., Nagao, S., *Nihon Daigaku No-Juigakbu Gakujutsu Kenkyu Hokoku* (1982) 188–194.

C225. Lhoest, G., *Anal. Chem., Symp. Ser.* (1981) **7**, 99–109.

C226. Yamazaki, H., Kumahashi, H., Masuda, N., Kishi, T., Kakiuchi, Y., Watabe, K., Kondo, M., *Chemosphere* (1982) **11**, 337–347.

C229. Cooper, C.S., Pal, K., Hewer, A., Grover, P.L., Sims, P., *Carcinogenesis* (1982) **3**, 203–210.

C230. Wong, P.K., Hampton, M.J., Floyd, R.A., *Prostaglandins Relat. Lipids* (1982) **2**, 167–179.

C231. Fu, P.P., Chou, M.W., Yang, S.K., Beland, F.A., Kadlubar, F.F., Casciano, D.A., Heflich, R.H., Evans, F.E., *Biochem. Biophys. Res. Commun.* (1982) **105**, 1037–1043.

C232. Beadle, C.A., Smith, A.R.W., *Eur. J. Biochem.* (1982) **123**, 323–332.

C233. Markus, A., Klages, U. Lingens, F., *Hoppe-Seyler's Z. Physiol. Chem.* (1982) **363**, 431–437.

C234. Fujita, S., Usui, T., Suzuki, M., Hirasawa, M., Mori, Y., Naganuma, H., Suzuki, T., *Biochem. Biophys. Res. Commun.* (1982) **105**, 1233–1239.

C235. Banks, S.W., Steele, M.J., Ward, D., Dewick, P.M., *J. Chem. Soc., Chem. Commun.* (1982) 157–158.

C236. Böttcher, J., Bässmann, H., Schüppel, R., *J. Pharm. Pharmacol.* (1982) **34**, 168–175.

C237. Iwamoto, K., Takei, M., Watanabe, J., *J. Pharm. Pharmacol.* (1982) **34**, 176–180.

C238. Dell, H.D., Fiedler, J., Kamp, R., Kurz, J., Wünschel, C., *Arch. Pharm. (Weinheim, Ger.)* (1982) **315**, 416–422.

C239. Essien, E.E., Afamefuna, G.C., *Clin. Chem. (Washington, DC)* (1982) **28**, 1148–1152.

C240. Hoffmann, H., Pooth, R., *Arch. Pharm. (Weinheim, Ger.)* (1982) **315**, 422–428.

C241. Notarianni, L.J., Oldham, H.G., Bennett, P.N., Southgate, C.C.B., Parfitt, R.T., *Adv. Exp. Med. Biol.* (1982) **136B**, 1077–1083.

C242. Höller, M., Weber, H., Breuer, H., *Acta Endocrinol. (Copenhagen)* (1982) **100**, 57–62.

C243. Gorrod, J.W., Christou, M., *Adv. Exp. Med. Biol.* (1982) **136B**, 1133–1142.

C244. Thakker, D.R., Levin, W., Yagi, H., Tada, M., Ryan, D.E., Thomas, P.E., Conney, A.H., Jerina, D.M., *J. Biol. Chem.* (1982) **257**, 5103–5110.

C245. Tanizawa, K., Yamamoto, T., Soda, K., *Bull. Inst. Chem. Res., Kyoto Univ.* (1981) **59**, 311–320.

C247. Eaton, R.W., Ribbons, D.W., *Arch. Biochem. Biophys.* (1982) **216**, 289–295.

C248. Balsevich, J., Constabel, F., Kurz, W.G.W., *Planta Med.* (1982) **44**, 231–233.

C249. Hackett, A.M., Shaw, I.C., Griffiths, L.A., *Experientia* (1982) **38**, 538–540.

C250. Imbs, J., Leibenguth, A., Koffel, J.C., Jung, L., *Arzneim.-Forsch.* (1982) **32**, 503–508.

C251. Horn, A.S., Griever-Kazemier, H., Dijkstra, D., *J. Med. Chem.* (1982) **25**, 993–996.

C252. Braverman, A.M., *Eur. J. Rheumatol. Inflammation* (1982) **5**, 304–307.

C253. Miyazaki, H., Nambu, K., Hashimoto, M., *Life Sci.* (1982) **30**, 2203–2206.

C254. Cerniglia, C.E., Freeman, J.P., Franklin, W., Pack, L.D., *Carcinogenesis* (1982) **3**, 1255–1260.

C255. Cerniglia, C.E., Fu, P.P., Yang, S.K., *Appl. Envron. Microbiol.* (1982) **44**, 682–689.

C256. Cerniglia, C.E., Althaus, J.R., Evans, F.E., Freeman, J.P., Mitchum, R.K., Yang, S.K., *Chem.-Biol. Interact.* (1983) **44**, 119–132.

C257. Silber, B., Holford, N.G.H., Riegelman, S., *J. Pharm. Sci.* (1982) **71**, 699–704.

C258. Lasker, J.M., Huang, M.T., Conney, A.H., *Science* (1982) **216**, 1419–1421.

C259. Pacifici, G.M., Placidi, G.F., Fornaro, P., Gomeni, R., *Eur. J. Clin. Pharmacol.* (1982) **22**, 225–228.

C260. Otsuka, M., Kurozumi, T., Usuki, S., Furuuchi, S., Miura, Y., Harigaya, S., *J. Pharmacobio.-Dyn.* (1982) **5**, S-59.

C261. Höller, M., Weber, H., Breuer, H., *Acta Endocrinol.* (1982) **100**, 274–278.

C262. Pue, M.A., Frost, G.S., Hirom, P.C., *Biochem. Soc. Trans.* (1982) **10**, 112.

C263. Guest, D., Schnell, S.R., Rickert, D.E., Dent, J.G., *Toxicol. Appl. Pharmacol.* (1982) **64**, 160–168.

C264. Boobis, A.R., Brodie, M.J., McManus, M.E., Staiano, N., Thorgeirsson, S.S., Davies, D.S., *Adv. Exp. Med. Biol.* (1982) **136B**, 1193–1201.

C265. Fahl, W.E., *Arch. Biochem. Biophys.* (1982) **216**, 581–592.

C266. Shigezane, J., Oguri, K., Mishima, M., Yoshimura, H., *J. Pharmacobio.-Dyn.* (1982) **5**, S-61.

C267. Fu, P.P., Chou, M.W., Yang, S.K., *Biochem. Biophys. Res. Commun.* (1982) **106**, 940–946.

C268. Hauer, B., Haase-Aschoff, K., Lingens, F., *Hoppe-Seyler's Z. Physiol. Chem.* (1982) **363**, 499–506.

C269. Babu, B.H., Vaidyanathan, C.S., *FEMS Microbiol. Lett.* (1982) **14**, 101–106.

C270. Dayer, P., balant, L., Courvoisier, F., Kupfer, A., Kubli, A., Gorgia, A., Fabre, J., *Eur. J. Drug Metab. Pharmacokinet.* (1982) **7**, 73–77.

C271. Guentert, T.W., Daly, J.J., Riegelman, S., *Eur. J. Drug Metab. Pharmacokinet.* (1982) **7**, 31–38.

C273. Cantavelli, G., Gentili, P., Manzardo, S., Ravenna, F., Riva, G., *Farmaco, Ed. Sci.* (1982) **37**, 310–318.

C274. Tomori, É., Horváth, G., Elekes, I., Láng, T., Kórösi, J., *J. Chromatogr.* (1982) **241**, 89–99.

C275. Ishida, T., Asakawa, Y., Takemoto, T., *Hiroshima Kogyo Daigaku Kenkyo Kiyo* (1982) **16**, 79–91.

C276. Bollag, J.M., Liu, S.-Y., Minard, R.D., *Soil Biol. Biochem.* (1982) **14**, 157.

C277. Machuga, E.J., *J. Bacteriol.* (1982) **150**, 747–754.

C278. Narimatsu, S., Watanabe, K., Yamamoto, I., Yoshimura, H., *J. Pharmacobio.-Dyn.* (1982) **5**, S-58.

C279. Baker, J.K., McChesney, J.D., Hufford, C.D., Clark, A.M., *J. Chromatogr.* (1982) **230**, 69–77.

C280. Pohland, R.C., Coppoc, G.L., Bottoms, G.D., Moore, A.B., *J. Anim. Sci.* (1982) **55**, 145–152.

C282. Imamura, N., Kakinuma, K., Ikekawa, N., Tanaka, H., Ohmura, S., *J. Antibiot.* (1982) **35**, 602–608.

C283. van Bladeren, P.J., Armstrong, R.N., Cobb, D., Thakker, D.R., Ryan, D.E., Thomas, P.E., Sharma, N.D., Boyd, D.R., Levin, W., Jerina, D.M., *Biochem. Biophys. Res. Commun.* (1982) **106**, 602–609.

C284. Crawford, D.L., Sutherland, J.B., Pometto, A.L.III, Miller, J.M., *Arch. Microbiol.* (1982) **131**, 351–355.

C285. Buswell, J.A., Eriksson, K.E., Gupta, J.K., Hamp, S.G., Nordh, I., *Arch. Microbiol.* (1982) **131**, 366–374.

C286. MacKinney, A.A., Jr., Shinnick, F.L., Gorder, C., *Res. Commun. Chem. Pathol. Pharmacol.* (1982) **36**, 395–411.

C287. Yasuda, Y., Shindo, T., Mitani, N., Ishida, N., Oono, F., Kageyama, T., *J. Pharm. Sci.* (1982) **71**, 565–572.
C288. Hoskins, J.A., Farmer, P.B., *Anal. Chem. Symp. Ser.* (1982) **11**, 223–228.
C289. Guentert, T.W., Huang, J.D., Øie, S., *J. Pharm. Sci.* (1982) **71**, 812–815.
C290. Russo, J., Jr., Russo, M.E., Smith, R.A., Pershing, L.K., *J. Clin. Pharmacol.* (1982) **22**, 264–270.
C291. Parkinson, A., Safe, S., *Adv. Exp. Med. Biol.* (1982) **136A**, 745–752.
C292. Dahl, S.G., Bratlid, T., Lingjaerde, O., *Ther. Drug Monit.* (1982) **4**, 81–87.
C293. Ruo, T.I., Thenot, J.P., Stec, G.P., Atkinson, A.J., Jr., *Ther. Drug Monit.* (1981) **3**, 231–237.
C294. Halldin, M.M., Carlsson, S., Kanter, S.L., Widman, M., Agurell, S., *Arzneim.-Forsch.* (1982) **32**, 764–768.
C295. Twele, R., Spiteller, G., *Arzneim.-Forsch.* (1982) **32**, 759–763.
C296. Oates, N.S., Shah, R.R., Idle, J.R., Smith, R.L., *Clin. Pharmacol. Ther.* (1982) **32**, 81–89.
C297. Facino, R.M., Pitre, D., Carini, M., *Farmaco, Ed. Sci.* (1982) **37**, 463–474.
C298. Noguchi, H., Tada, K., Iwasaki, K., *Xenobiotica* (1982) **12**, 211–220.
C299. Coutts, R.T., Jones, G.R., *Res. Commun. Chem. Pathol. Pharmacol.* (1982) **36**, 173–176.
C300. Sakai, T., Niwaguchi, T., Murata, T., *Xenobiotica* (1982) **12**, 233–239.
C301. Iwasaki, K., Noguchi, H., Kamataki, T., Kato, R., *Xenobiotica* (1982) **12**, 221–226.
C302. Islam, S.I., *Xenobiotica* (1982) **12**, 323–328.
C303. Sportsøl, S., Scheline, R.R., *Xenobiotica* (1982) **12**, 249–257.
C304. Renwick, A.G., Pettet, J.L., Gruchy, B., Corina, D.L., *Xenobiotica* (1982) **12**, 329–343.
C305. Słojkowska, Z., Krasuska, H.J., Pachecka, J., *Xenobiotica* (1982) **12**, 359–364.
C306. Hard, S.J., Healey, K., Smail, M.C., Calder, I.C., *Xenobiotica* (1982) **12**, 381–386.
C307. Husain, M.M., Kumar, A., Mukhtar, H., *Xenobiotica* (1982) **12**, 375–380.
C308. Jones, A.R., *Xenobiotica* (1982) **12**, 387–395.
C309. Chipman, J.K., Bhave, N.A., Hirom, P.C., Millburn, P., *Xenobiotica* (1982) **12**, 397–404.
C310. Fears, R., Baggaley, K.H., Walker, P., Hindley, R.M., *Xenobiotica* (1982) **12**, 427–433.
C311. Shaw, I.C., Hackett, A.M., Griffiths, L.A., *Xenobiotica* (1982) **12**, 405–416.
C312. Hackett, A.M., Griffiths, L.A., *Xenobiotica* (1982) **12**, 447–456.
C313. Michelot, J., Madelmont, J.C., Rousset, B., Labarre, P., Mornex, R., Meyniel, G., *Xenobiotica* (1982) **12**, 457–462.
C314. Kennedy, S.M.E., Burchell, B., *Biochem. Pharmacol.* (1983) **32**, 2029–2032.
C315. Christ, W., Hecker, W., Gindler, K., Stille, G., *Biochem. Pharmacol.* (1983) **32**, 1961–1964.
C316. Kemp, J.V., Adam, H.K., Wakeling, A.E., Slater, R., *Biochem. Pharmacol.* (1983) **32**, 2045–2052.
C317. Willenberg, P., Ullrich, V., Rummel, W., *Biochem. Pharmacol.* (1983) **32**, 2103–2107.
C318. Noguchi, H., Iwasaki, K., *Xenobiotica* (1982) **12**, 499–508.
C319. Ross, D., Farmer, P.B., Gescher, A., Hickman, J.A., *Biochem. Pharmacol.* (1983) **32**, 1773–1781.
C320. Meerman, J.H.N., Sterenborg, H.M.J., Mulder, G.J., *Biochem. Pharmacol.* (1983) **32**, 1587–1593.
C321. Blanck, A., Rane, A., Toftgård, R., Gustafsson, J.Å., *Biochem. Pharmacol.* (1983) **32**, 1547–1552.
C322. Sodha, R.J., Glover, V., Sandler, M., *Biochem. Pharmacol.* (1983) **32**, 1655–1657.

C324. Kahn, G.C., Boobis, A.R., Murray, S., Plummer, S., Brodie, M.J., Davies, D.S., *Br. J. Clin. Pharmacol.* (1982) **13**, 594P.

C325. Williams, D.A., Maher, T.J., Zaveri, D.C., *Biochem. Pharmacol.* (1983) **32**, 1447–1448.

C326. Isherwood, C.N., Murray, G.R., Chan, K., Calvey, T.N., Williams, N.E., *Br. J. Clin. Pharmacol.* (1982) **13**, 612P.

C327. Baker, J.F., O'Melia, P.E., Benziger, D.P., Clemans, S.D., Edelson, J., *Arch. Int. Pharmacodyn. Ther.* (1982) **258**, 29–38.

C328. Forgue, S.T., Allen, J.R., *Chem.-Biol. Interact.* (1982) **40**, 233–245.

C329. Shimada, K., Yumura, Y., Nambara, T., *J. Pharmacobio.-Dyn.* (1982) **5**, 448–450.

C330. Somani, S.M., Khalique, A., *J. Toxicol. Environ. Health* (1982) **9**, 889–897.

C331. Chiu, P.L., Fu, P.P., Yang, S.K., *Biochem. Biophys. Res. Commun.* (1982) **106**, 1405–1411.

C332. Müller, R., Schmitt, S., Lingens, F., *Eur. J. Biochem.* (1982) **125**, 579–584.

C333. Lyon, E.S., Jakoby, W.B., *J. Biol. Chem.* (1982) **257**, 7531–7535.

C334. Fujii, I., Ebizuka, Y., Sankawa, U., *Chem. Pharm. Bull.* (1982) **30**, 2283–2286.

C335. Canonica, L., Galliani, G., Lesma, G., Tollari, J., *Int. Conf. Chem. Biotechnol. Biol. Act. Nat. Prod., 1st* (1981) **2**, 522–527.

C336. Simpson, T.J., De Jesus, A.E., Steyn, P.S., Vleggaar, R., *J. Chem. Soc., Chem. Commun.* (1982) 631–632.

C337. Heide, L., Arendt, S., Leistner, E., *J. Biol. Chem.* (1982) **257**, 7396–7400.

C338. Hiraku, S., Muryobayashi, K., Ito, H., Inagawa, T., Tsuboshima, M., *Iyakuhin Kenkyu* (1982) **13**, 756–765.

C339. Tanaka, H., Sasaki, H., Komatsu, S., Yamagishi, O., Kojima, H., Shibuya, M., Aoyama, T., *Oyo Yakuri* (1982) **23**, 751–765.

C340. MacLeod, M.C., Mansfield, B.K., Selkirk, J.K., *Chem.-Biol. Interact.* (1982) **40**, 275–285.

C341. Watanabe, K., Yoshizawa, I., *J. Pharmacobio.-Dyn.* (1982) **5**, 340–347.

C342. Ninnekar, H.Z., Vaidyanathan, C.S., *J. Indian Inst. Sci.* (1981) **63**, 131–136.

C343. Hawkins, D.R., Biggs, S.R., Brodie, R.R., Chasseaud, L.F., Midgley, I., *J. Pharm. Pharmacol.* (1982) **34**, 299–303.

C344. Greiner, B., Rommelspacher, H., *Proc. Clin. Biol. Res.* (1982) **90**, 201–208.

C345. Mwaluko, G., Weinshilboum, R., *Br. J. Clin. Pharmacol.* (1982) **14**, 231–239.

C346. Raxworthy, M.J., Gulliver, P.A., *J. Steroid Biochem.* (1982) **17**, 17–21.

C347. Zmijewski, M.J., Jr., *Tetrahedron Lett.* (1982) **23**, 1775–1776.

C348. Gorst-Allman, C.P., Steyn, P.S., Vleggaar, R., *J. Chem. Soc., Chem. Commun.* (1982) 562–563.

C349. Eckenrode, F., Rosazza, J.P., *J. Nat. Prod.* (1982) **45**, 226–228.

C350. Aleksintseva, O.A., Tsyganov, V.A., Namestnikova, V.P., *Antibiotiki* (1982) **27**, 493–495.

C351. van Brederode, J., Kamps-Heinsbroek, R., Mastenbroek, O., *Z. Pflanzenphysiol.* (1982) **106**, 43–53.

C352. Leete, E., Muir, A. Towers, G.H.N., *Tetrahedron Lett.* (1982) **23**, 2635–2638.

C353. Herbert, R.B., Mann, J., *J. Chem. Soc., Perkin Trans. 1* (1982) 1523–1525.

C354. Brochmann-Hanssen, E., Cheng, C.Y., *J. Nat. Prod.* (1982) **45**, 434–436.

C355. Ishida, T., Oguri, K., Yoshimura, H., *J. Pharmacobio.-Dyn.* (1982) **5**, 521–525.

C356. Decouvelaere, B., Terlain, B., Bieder, A., *Therapie* (1982) **37**, 249–257.

C357. Ishida, T., Yamano, S., Toki, S., *Adv. Endog. Exog. Opioids, Proc. Int. Narc. Res. Conf.,* 12th (1981) 481–483.

C358. Tsuchiya, K., Uchida, M., Sugimoto, T., *Nippon Noyaku Gakkaishi* (1982) **7**, 187–193.

C359. Marsden, P.J., Kuwano, E., Fukuto, T.R., *Pestic. Biochem. Physiol.* (1982) **18**, 38–48.

C360. Kadlubar, F.F., Frederick, C.B., Weiss, C.C., Zenser, T.V., *Biochem. Biophys. Res. Commun.* (1982) **108**, 253–258.

C361. Reed, G.A., Marnett, L.J., *J. Biol. Chem.* (1982) **257**, 11368–11376.

C362. McCarthy, D.J., Struck, R.F., Shih, T.W., Suling, W.J., Hill, D.L., Enke, S.E., *Cancer Res.* (1982) **42**, 3475–3479.

C364. Weltring, K.M., Mackenbrock, K., Barz, W., *Z. Naturforsch. C: Biosci.* (1982) **37C**, 570–574.

C365. Lindstedt, S., Rundgren, M., *J. Biol. Chem.* (1982) **257**, 11922–11931.

C366. Bakke, J.E., Feil, V.J., Struble, C., *Biomed. Mass Spectrom.* (1982) **9**, 246–251.

C367. Netrval, J., Vodnansky, M., *Kvansky Prum.* (1982) **28**, 131–134.

C368. Vachta, J., Gold-Aubert, P., *Eur. J. Drug Metab. Pharmacokinet.* (1982) **7**, 147–156.

C369. Allan, R.J., Watson, T.R., *Eur. J. Drug Metab. Pharmacokinet.* (1982) **7**, 131–136.

C370. Uchino, H., Inaba, T., Kalow, W., *Xenobiotica* (1983) **13**, 155–162.

C371. Gipple, K.J., Chan, K.T., Elvin, A.T., Lalka, D., Axelson, J.E., *J. Pharm. Sci.* (1982) **71**, 1011–1014.

C372. Görler, K., Krumbiegel, G., Mennicke, W.H., Siehl, H.-U., *Xenobiotica* (1982) **12**, 535–542.

C373. Yamato, C., Takahashi, T., Kinoshita, K., Fujita, T., *Xenobiotica* (1982) **12**, 549–559.

C374. Hutt, A.J., Caldwell, J., Smith, R.L., *Xenobiotica* (1982) **12**, 601–610.

C375. Narimatsu, S., Watanabe, K., Yamamoto, I., Yoshimura, H., *Xenobiotica* (1982) **12**, 561–565.

C376. Niwaguchi, T., Inoue, T., Suzuki, S., *Xenobiotica* (1982) **12**, 617–625.

C377. Dow, J., Strolin-Benedetti, M., *Xenobiotica* (1982) **12**, 633–643.

C378. Maggs, J.L., Grabowski, P.S., Rose, M.E., Park, B.K., *Xenobiotica* (1982) **12**, 657–668.

C379. Hutt, A.J., Navas, G.E., Beckett, A.H., *Xenobiotica* (1982) **12**, 645–655.

C380. Sasaki, H., Kamimura, H., Shiobara, Y., Esumi, Y., Takaichi, M., Yokoshima, T., *Xenobiotica* (1982) **12**, 803–812.

C381. Thabrew, M.I., Olorunsogo, O.O., Olowookere, J.O., Babamunmi, E.A., *Xenobiotica* (1982) **12**, 849–853.

C382. Fu, P.P., Yang, S.K., *Biochem. Biophys. Res. Commun.* (1983) **115**, 123–129.

C383. Chou, M.W., Evans, F.E., Yang, S.K., Fu, P.P., *Carcinogenesis* (1983) **4**, 699–702.

C384. Padron, V.A., Lemon, H.M., Stohs, S.J., *Horm. Met. Res.* (1982) **14**, 323–326.

C385. Bhavnani, B.R., Woolever, C.A., *J. Steroid Biochem.* (1982) **17**, 217–223.

C386. Wang, P., Meijer, J., Guengerich, P., *Biochemistry* (1982) **21**, 5769–5776.

C387. Matsui, H., Yamamoto, C., Nagatsu, T., *J. Neurochem.* (1982) **39**, 1066–1071.

C388. Becker, R., Frankus, E., Graudums, I., Günzler, W.A., Helm, F.Ch., Flohé, L., *Arzneim.-Forsch.* (1982) **32**, 1101–1111.

C389. Benziger, D.P., Edelson, J., *Antimicrob. Agents Chemother.* (1982) **22**, 475–481.

C390. Kinoshita, T., Konieczny, M., Santella, R., Jeffrey, A.M., *Cancer Res.* (1982) **42**, 4032–4038.

C391. Ito, A., Honda, M., Mori, Y., *Biochem. Med.* (1982) **28**, 32–40.

C392. Spribille, R., Forkmann, G., *Planta* (1982) **155**, 176–182.

C393. Powlowski, J., Dagley, S., *Dev. Biochem.* (1982) **21**, 339–341.

C394. Opdycke, J.C., Miller, R.W., Menzer, R.E., *J. Agric. Food Chem.* (1982) **30**, 1223–1227.

C395. Hersey, R.M., Lloyd, T., Maclusky, N.J., Weisz, J., *Endocrinology (Baltimore)* (1982) **111**, 1734–1736.

C396. Yoshida, H., Tani, Y., Yamada, H., *Agric. Biol. Chem.* (1982) **46**, 2539–2546.

C397. Razzouk C., Roberfroid, M.B., *Chem.-Biol. Interact.* (1982) **41**, 251–264.
C398. Baxter, R.L., Hasan, M., Mackenzie, N.E., Scott, A.I., *J. Chem. Soc., Chem. Commun.* (1982) 791–793.
C399. Hattori, M., Kim, G., Motoike, S., Kobashi, K., Namba, T., *Chem. Pharm. Bull.* (1982) **30**, 1338–1346.
C400. Brochmann-Hanssen, E., Okamoto, Y., *J. Nat. Prod.* (1980) **43**, 731–735.
C401. Tien, M., Kirk, T.K., *Science* (1983) **221**, 661–663.
C403. McLean, S., Galloway, H., Butler, S., Whittle, D., Nicol, S.C., *Xenobiotica* (1983) **13**, 81–85.
C404. Nagel, R., Urich, K., *Xenobiotica* (1983) **13**, 97–100.
C405. Midgley, I., Fowkes, A.G., Chasseaud, L.F., Hawkins, D.R., Girkin, R., Kesselring, K., *Xenobiotica* (1983) **13**, 87–95.
C406. Layiwola, P.J., Linnecar, D.F.C., Knights, B., *Xenobiotica* (1983) **13**, 107–113.
C407. Larsen, G.L., Bakke, J.E., *Xenobiotica* (1983) **13**, 115–126.
C408. Layiwola, P.J., Linnecar, D.F.C., Knights, B., *Xenobiotica* (1983) **13**, 27–29.
C409. Marten, T.R., Ruane, R.J., East, P.B., Malcolm, S.L., *Xenobiotica* (1983) **13**, 1–18.
C410. Midha, K.K., Hawes, E.M., Cooper, J.K., Hubbard, J.W., *Xenobiotica* (1983) **13**, 31–38.
C411. Kirkpatrick, D., Hawkins, D.R., Chasseaud, L.F., Finn, C.M., Conway, B., Al-Ani, M.R., Al-Sayyab, A.F., *Xenobiotica* (1983) **13**, 53–64.
C412. Cole, L.M., Ruzo, L.O., Wood, E.J., Casida, J.E., *J. Agric. Food. Chem.* (1982) **30**, 631–636.
C413. Ueyama, I., Kurogochi, S., Kobori, I., Hoshnino, T., Ishii, Y., Takase, I., *J. Agric. Food. Chem.* (1982) **30**, 1061–1067.
C414. Vágújfalvi, D., Petz-Stifer, M., *Phytochemistry* (1982) **21**, 1533–1536.
C415. Böttcher, J., Bässmann, H., Schüppel, R., Lehmann, W.D., *Naunyn-Schmiedemerg's Arch. Pharmacol.* (1982) **321**, 226–233.
C416. Sariaslani, F.S., Rosazza, J.P., *Appl. Environ. Microbiol.* (1983) **45**, 616–621.
C417. Rosazza, J.P., Eckenrode, F., Gustafson, M.E., Nabih, T., Peczynska-Czoch, W., Sariaslani, F.S., Schaumberg, J.P., *Cancer Treat. Symp.* (1983) **1**, 51–58.
C418. Böttcher, J., Bässmann, H., Schüppel, R., *Drug Metab. Dispos.* (1982) **10**, 90–91.
C419. Hoffmann, H., Pooth, R., *Arch. Pharm. (Weinheim, Ger.)* (1983) **316**, 724–725.
C420. Emudianughe, T.S., Caldwell, J., Smith, R.L., *Xenobiotica* (1983) **13**, 133–138.
C421. Rafter, J.J., Gustafsson, J.Å., Bakke, J.E., Larsen, G.L., Norin, K.E., Gustafsson, B.E., *Xenobiotica* (1983) **13**, 171–178.
C422. Chaudhury, N.K., Ball, T.J., *J. Org. Chem.* (1982) **47**, 5196–5198.
C424. Nagel, R., *Xenobiotica* (1983) **13**, 101–106.
C425. Houston, J.B., Cassidy, M.K., *J. Pharm. Pharmacol.* (1982) **34**, 536–538.
C426. Dayer, P., Glasson, P., Balant, L., Striberni, R., Fabre, J., *Eur. J. Drug Metab. Pharmacokinet.* (1983) **8**, 181–188.
C427. Chiu, P.-L., Weems, H.B., Wong, T.K., Fu, P.P., Yang, S.K., *Chem.-Biol. Interact.* (1983) **44**, 155–168.
C428. Ueyama, I., Kurogochi, S., Kobori, I., Hoshino, Y., Takase, I., *Agric. Food Chem.* (1982) **30**, 1061–1067.
C430. Migdaloff, B.H., Dugger, H.A., Heider, J.G., Coombs, R.A., Terry, M.K., *Xenobiotica* (1983) **13**, 209–221.
C432. Marsh, M.V., Caldwell, J., Sloan, T.P., Smith, R.L., Horner, M., Moss, M.S., *Xenobiotica* (1983) **13**, 233–240.

C433. Inoue, T., Suzuki, S., Niwaguchi, T., *Xenobiotica* (1983) **13**, 241–249.

C434. McKay, G., Cooper, J.K., Hawes, E.M., Roy, S.D., Midha, M.M., *Xenobiotica* (1983) **13**, 257–264.

C435. Rowland, I.R., Mallett, A.K., Wise, A., Bailey, E., *Xenobiotica* (1983) **13**, 251–256.

C436. Vicario, G.P., Battaglia, R., Bernardi, L., Arcamone, F., *Farmaco, Ed. Sci.* (1982) **37**, 651–662.

C437. Van Boven, M., Daenens, P., *J. Pharm. Sci.* (1982) **71**, 1152–1157.

C438. Neidlein, R., Eder, T., *Arzneim.-Forsch.* (1982) **32**, 1292–1298.

C439. Gold, B., Brunk, G., *Chem.-Biol. Interact.* (1982) **41**, 327–339.

C440. Razzouk, C., Batardy-Gregoire, M., Roberfroid, M., *Cancer Res.* (1982) **42**, 4712–4718.

C441. Borchardt, R.T., Schasteen, C.S., *Biochim. Biophys. Acta* (1982) **708**, 272–279.

C442. Agrup, G., Rorsman, H., Rosengren, E., *Acta Derm.-Venereol.* (1982) **62**, 371–376.

C443. Brochmann-Hanssen, E., Cheng, C.Y., Chiang, H.C., *J. Nat. Prod.* (1982) **45**, 629–634.

C444. Koyama, H., *J. Biochem.* (*Tokyo*) (1982) **92**, 1235–1240.

C445. Nambu, K., Yoshida, K., Matsumoto, K., Morino, N., Fujii, T., Furukawa, H., Matsunaga, Y., Miyazaki, H., Arakawa, S., Hashimoto, M., *Yakuri to Chiryo* (1982) **10**, 4445–4458.

C446. Willeke, U., Barz, W., *Z. Naturforsch. C: Biosci.* (1982) **37C**, 861–869.

C447. Iyayi, C.B., Dart, R.K., *Microbios* (1982) **34**, 167–176.

C448. Abell, C., Garson, M.J., Leeper, F.J., Staunton, J., *J. Chem. Soc., Chem. Commun* (1982) 1011–1013.

C449. Kitamura, S., Tatsumi, K., *Jpn. J. Pharmacol.* (1982) **32**, 833–838.

C450. Crespi-Perellino, N., Grein, A., Merli, A., Minghetti, A., Spalla, C., *Experientia* (1982) **38**, 1455–1456.

C452. Horie, T., Ohno, T., Kinoshita, K., *Jpn. J. Pharmacol.* (1982) **32**, 1041–1049.

C453. Faed, E.M., Dobbs, B.R., Lee, D., *Proc. Univ. Otago Med. Sch.* (1982) **60**, 29–31.

C454. Harron, D.W.G., Arndts, D.E., Shanks, R.G., *Int. J. Pharm.* (1982) **12**, 277–283.

C455. Yamamoto, I., Narimatsu, S., Watanabe, K., Yoshimura, H., *Biochem. Biophys. Res. Commun.* (1982) **109**, 922–926.

C456. Zannoni, V.G., Marker, E.K., Lau, S.S., *Drug-Nutr. Interact.* (1982) **1**, 193–204.

C457. Hart, S.J., Calder, I.C., Tange, J.D., *Eur. J. Drug Metab. Pharmacokinet.* (1982) **7**, 203–222.

C458. Tada, M., Takahashi, K., Kawazoe, Y., *Chem. Pharm. Bull.* (1982) **30**, 3834–3837.

C459. Owens, I.S., Mackenzie, P.I., *Biochem. Biophys. Res. Commun.* (1982) **109**, 1075–1082.

C460. Blakley, E.R., Papish, B., *Can. J., Microbiol.* (1982) **28**, 1037–1046.

C461. Schuesler-Van Hees, M.T.I.W., Beijersbergen van Henegouwen, G.M.J., *Pharm. Weekbl., Sci. Ed.* (1982) **4**, 176–182.

C462. Wilking, N., Appelgren, L.E., Carlström, K., Norggren, T., Theve, T., *Rev. Endocr.-Relat. Cancer, Suppl. 9* (1982) 233–239.

C463. Brennan, J.J., Imondi, A.R., Westmoreland, D.G., Williamson, M.J., *J. Pharm. Sci.* (1982) **71**, 1199–1203.

C464. Lin, M., Zhang, W., Zhao, X., Lu, J., Wang, M., Chen, L., *Yaoxue Xuebao* (1982) **17**, 728–735.

C465. Dyck, L.E., Yang, C.R., Boulton, A.A., *J. Neurosci. Res.* (1983) **10**, 211–220.

C466. Franc, Z., Smolik, S., Selecka, A., Janda, J., *Cesk. Farm.* (1982) **31**, 337–342.

C467. Narasimhachari, N., Saady, J.J., Joseph, A., Ettigi, P., Friedel, R.O., *J. Clin. Psychopharmacol.* (1982) **2**, 413–416.

C468. Sawahata, T., Neal, R.A., *Biochem. Biophys. Res. Commun.* (1982) **109**, 988–994.
C469. Siufi, A., Pommier, F., *Eur. J. Drug Metab. Pharmacokinet.* (1982) **7**, 223–228.
C470. Leegwater, D.C., Van der Greef, J., Bos, K.D., *Meded. Fac. Landbouwwet, Rijksuniv. Gent.* (1982) **47**, 401–408.
C471. Tanaka, K., Mii, T., Marui, S., Matsubara, I., Igaki, H., *Ind. Health* (1982) **20**, 227–235.
C472. Cobb, D., Boehlert, C., Lewis, D., Armstrong, R.N., *Biochemistry* (1983) **22**, 805–812.
C473. Gmur, D.J., Varanasi, U., *Carcinogenesis* (1982) **3**, 1397–1403.
C474. Willeke, U., Barz, W., *Arch. Microbiol.* (1982) **132**, 266–269.
C475. Shailubhai, K., Somayaji, R., Rao, N.N., Modi, V.V., *Experientia* (1983) **39**, 70–72.
C476. Beckett, A.H., Hutt, A.J., *J. Pharm. Pharmacol.* (1982) **34**, 614.
C477. Rivett, A.J., Francis, A., Roth, J.A., *J. Neurochem.* (1983) **40**, 215–219.
C478. Flanagan, R.J., Storey, G.C.A., Holt, D.W., Farmer, P.B., *J. Pharm. Pharmacol.* (1982) **34**, 638–643.
C479. Merker, P.C., Yeung, D., Doughty, D., Nacht, S., *Res. Commun. Chem. Pathol. Pharmacol.* (1982) **38**, 367–388.
C480. Giannangeli, M., Interdonato, N., Baiocchi, L., *Boll. Chim. Farm.* (1982) **121**, 465–474.
C481. Böttcher, J., Bässmann, H., Schüppel. R., *J. Pharm. Pharmacol.* (1984) **36**, 391–398.
C482. Struble, C.B., Feil, V.J., Pekas, J.C., Gerst, J.W., *Pestic. Biochem. Physiol.* (1983) **19**, 85–94.
C483. Kogekar, R.C., Deshpande, V.N., *Indian J. Biochem. Biophys.* (1982) **19**, 257–261.
C484. Yoshizako, F., Nishimura, A., Ueno, T., *Nippon Nogei Kagaku Kaishi* (1982) **56**, 1135–1141.
C485. Krause, W., Kühne, G., Schillinger, E., *Xenobiotica* (1983) **13**, 265–272.
C486. Hackett, A.M., Griffiths, L.A., Broillet, A., Wermeille, M., *Xenobiotica* (1983) **13**, 279–286.
C487. Iwasaki, K., Shiraga, T., Noda, K., Tada, K., Noguchi, H., *Xenobiotica* (1983) **13**, 273–278.
C488. Leinweber, F.-J., Loh, A.C., Szuna, A.J., Carbone, J.J., Williams, T.H., Sasso, G.J., Tilley, J.W., Pace, D., Dahlen, P., Kovacs, J.L., Kleavans, L.R., Cazes, M., *Xenobiotica* (1983) **13**, 287–294.
C489. Blaauboer, B.J., Van Holsteijn, C.W.M., *Xenobiotica* (1983) **13**, 295–302.
C490. Takahashi, O., Hiraga, K., *Xenobiotica* (1983) **13**, 319–326.
C491. Maftouh, M., Meunier, G., Dugue, B., Monsarrat, B., Meunier, B., Paoletti, C., *Xenobiotica* (1983) **13**, 303–310.
C492. French, M.C., Harrison, J.M., Inch, T.D., Leadbeater, L., Newman, J., Upshall, D.G., Powell, G.M., *Xenobiotica* (1983) **13**, 345–359.
C493. Furnival, B., Harrison, J.M., Newman, J., Upshall, D.G., *Xenobiotica* (1983) **13**, 361–372.
C494. Ngah, W.Z.W., Smith, J.N., *Xenobiotica* (1983) **13**, 383–389.
C495. French, M.C., Harrison, J.M., Newman, J., Upshall, D.G., Powell, G.M., *Xenobiotica* (1983) **13**, 373–381.
C496. Beckett, A.H., Hutt, A.J., Navas, G.E., *Xenobiotica* (1983) **13**, 391–405.
C497. Galtier, P., Coche, Y., Alvinerie, M., *Xenobiotica* (1983) **13**, 407–413.
C498. Maggs, J.L., Grimmer, S.F.M., Orme, M.L'E., Breckenridge, A.M., Park, B.K., Gilmore, I.T., *Xenobiotica* (1983) **13**, 421–431.
C499. Keimig, S.D., Kirby, K.W., Morgan, D.P., Keiser, J.E., Hubert, T.D., *Xenobiotica* (1983) **13**, 415–420.
C500. Muthukumar, G., Arunakumari, A., Mahadevan, A., *Plant. Soil* (1982) **69**, 163–169.

C501. Gaines, C.G., Byng, G.S., Whitaker, R.J., Jensen, R.A., *Planta* (1982) **156**, 233–240.
C502. Ribeiro, P., Webb, R.A., *Mol. Biochem. Parasitol.* (1983) **7**, 53–62.
C503. Sabater, F., Acosta, M., Sanchez-Bravo, J., Cuello, J., Del Rio, J.A., *Physiol. Plant.* (1983) **57**, 75–78.
C504. Lindén, H.B., Niemi, S., *Acta Pharmacol. Toxicol.* (1982) **51**, 434–440.
C505. Fu, P.P., Yang, S.K., *Biochem. Biophys. Res. Commun.* (1982) **109**, 927–934.
C506. Kunapuli, S.P., Vaidyanathan, C.S., *Plant Physiol.* (1983) **71**, 19–23.
C507. Parkki, M.G., Ahotupa, M., Toikkanen, S., *Dev. Biochem.* (1982) **23**, 749–750.
C508. Pelsy, G., Klibanov, A.M., *Biochim. Biophys. Acta* (1983) **742**, 352–357.
C509. Fisher, J., Ramakrishnan, K., Becvar, J.E., *Biochemistry* (1983) **22**, 1347–1355.
C510. Antai, S.P., Crawford, D.L., *Can. J. Microbiol.* (1983) **29**, 142–143.
C511. Zavala, F., Takai, K., Hayaishi, O., *J. Biol. Chem.* (1983) **258**, 344–351.
C512. Francis, R.J., East, P.B., Larman, J., *J. Clin. Pharmacol.* (1982) **23**, 529–533.
C513. Mayo, B.C., Biggs, S.R., Hawkins, D.R., Chasseaud, L.F., Darragh, A., Baldock, G.A., Whitby, B.R., *J. Pharmacobio.-Dyn.* (1982) **5**, 951–964.
C514. Boulton, A.A., Juorio, A.V., *Experientia* (1983) **39**, 130–134.
C515. Nanbo, T., *J. Pharmacobio.-Dyn.* (1982) **5**, 988–993.
C516. Steiner, E., Von Bahr, C., Rane, A., *Dev. Pharmacol. Ther.* (1982) **5**, 14–20.
C517. Iwaki, M., Yagi, T., Horiike, K., Saeki, Y., Ushijima, T., Nozaki, M., *Arch. Biochem. Biophys.* (1983) **220**, 253–262.
C518. Dujovne, C.A., Pitterman, A., Vincek, W.C., Dobrinska, M.R., Davies, R.O., Duggan, D.E., *Clin. Pharmacol. Ther.* (1983) **33**, 172–177.
C519. Han, M.H., Seong, B.-L., Son, H.-J., Mheen, T.-I., *FEBS Lett.* (1983) **151**, 36–40.
C520. Mass, M.J., Kaufman, D.G., *Basic Life Sci.* (1983) **24**, 331–351.
C521. Wong, T.K., Chiu, P.L., Fu, P.P., Yang, S.K., *Nat. Conf. Ser.* 1 (1983) 5P 27–43.
C522. Nakane, M., Yokoyama, E., Deguchi, T., *J. Neurochem.* (1983) **40**, 790–796.
C523. Bonham Carter, S.M., Rein, G., Glover, V., Sandler, M., Caldwell, J., *Br. J. Clin. Pharmacol.* (1983) **15**, 323–330.
C524. Ribeiro, P., Webb, R.A., *Int. J. Parasitol.* (1983) **13**, 101–106.
C525. Grambow, H.J., Langenbek-Schwich, B., *Planta* (1983) **157**, 131–137.
C526. De Beer, J.O., Dierckx, P.J., De Bruyn, A., *Toxicol. Lett.* (1983) **15**, 309–316.
C527. Nielson, A.H., Allard, A.S., Hynning, P.A., Remberger, M., Landner, L., *Appl. Environ. Microbiol.* (1983) **45**, 774–783.
C528. Reiter, C., Werness, P.G., Van Loon, J., Smith, L.H., Weinshilboum, R.M., *Br. J. Clin. Pharmacol.* (1983) **15**, 211–219.
C529. Torii, H., Yoshida, K., Doi, T., Tsukamoto, T., Matsuoka, C., Tanayama, S., *Takeda Kenkyushoho* (1982) **41**, 180–191.
C530. Rickert, D.E., Bond, J.A., Long, R.M., Chism, J.P., *Toxicol. Appl. Pharmacol.* (1983) **67**, 206–214.
C531. Felsted, R.L., Bachur, N.R., *Prog. Clin. Biol. Res.* (1982) **114**, 291–305.
C532. Baggi, G., Boga, M.M., Catelani, D., Galli, E., Treccani, V., *Syst. Appl. Microbiol.* (1983) **4**, 141–147.
C533. Kradolfer, P., Niederberger, P., Hütter, R., *Arch. Microbiol.* (1982) **133**, 242–248.
C537. Nakao, M., Takeda, Y., *J. Toxicol. Environ. Health* (1983) **11**, 93–100.
C538. Arnold, T.H., Jr., Epps, J.M.III, Cook, H.R., Hamrick, M.E., *Res. Commun. Chem. Pathol. Pharmacol.* (1983) **39**, 381–398.
C539. Peters, J.H., Gordon, G.R., Kashiwase, D., Acton, E.M., Hunt, C.A., *Cancer Res.* (1983) **43**, 1477–1487.
C540. Fu, P.P., Chou, M.W., *Dev. Biochem.* (1982) **23**, 71–74.

C541. Bowman, M.C., Nony, C.R., Billedean, S.M., Martin, J.L., Thompson, H.C., Jr., Lowry, L.K., *J. Anal. Toxicol.* (1983) **7**, 55–60.

C542. Sawahata, T., Neal, R.A., *Mol. Pharmacol.* (1983) **23**, 453–460.

C543. Costlow, R.D., Manson, J.M., *Toxicology* (1983) **26**, 11–23.

C544. Josephy, P.D., Eling, T.E., Mason, R.P., *Mol. Pharmacol.* (1983) **23**, 461–466.

C545. Kozaki, A., Sakurai, S., Awata, N., *Yakugaku Zasshi* (1982) **102**, 1177–1183.

C546. Koyama, H., *J. Biochem.* (1983) **93**, 1313–1319.

C547. Suzuki, T., *J. Pestic. Sci.* (1983) **8**, 419–428.

C548. Saijo, R., *Agric. Biol. Chem.* (1983) **47**, 455–460.

C549. Ungell, A.-L., Nilsson, S., *Mol. Physiol.* (1983) **3**, 183–188.

C550. Klungsoeyr, J., Scheline, R.R., *Biomed. Mass Spectrom.* (1982) **9**, 323–329.

C553. Rietveld, E.C., Plate, R., Seutter-Berlage, F., *Arch. Toxicol.* (1983) **52**, 199–207.

C554. Panthananickal, A., Weller, P., Marnett, L.J., *J. Biol. Chem.* (1983) **258**, 4411–4418.

C555. Quistad, G.B., Mulholland, K.M., *J. Agric. Food Chem.* (1983) **31**, 585–589.

C556. Fujii, I., Ijima, H., Ebizuka, Y., Sankawa, U., *Chem. Pharm. Bull.* (1983) **31**, 337–340.

C557. Takeuchi, F., Otsuka, H., Shibata, Y., *Biochim. Biophys. Acta* (1983) **743**, 323–330.

C558. Melzacka, M., Danek, L., *Pharmacopsychiatria* (*Stuttgart*) (1983) **16**, 30–34.

C559. Sburlati, A.R., Frydman, R.B., *Plant Physiol.* (1983) **71**, 822–827.

C560. Kloss, M.W., Rosen, G.M., Rauckman, E.J., *Mol. Pharmacol.* (1983) **23**, 482–485.

C561. Mahony, C., Wolfram, K.M., Nash, P.V., Bjornsson, T.D., *Clin. Pharmacol. Ther.* (St. Louis) (1983) **33**, 491–497.

C562. Gregus, Z., Watkins, J.B., Thompson, T.N., Harvey, M.J., Rozman, K., Klaassen, C.D., *Toxicol. Appl. Pharmacol.* (1983) **67**, 430–441.

C563. Conway, J.G., Kauffman, F.C., Ji, S., Thurman, R.G., *Mol. Pharmacol.* (1982) **22**, 509–516.

C564. Klungsøyr, J., Scheline, R.R., *Acta Pharmacol. Toxicol.* (1983) **52**, 211–216.

C566. Bowsher, R.R., Henry, D.P., *J. Neurochem.* (1983) **40**, 992–1002.

C567. Chern, W.H., Dauterman, W.C., *Toxicol. Appl. Pharmacol.* (1983) **67**, 303–309.

C568. Illing, H.P.A., Watson, K.V., Chamberlain, J., *Xenobiotica* (1983) **13**, 531–538.

C569. Gorrod, J.W., Patterson, L.H., *Xenobiotica* (1983) **13**, 513–520.

C570. Illing, H.P.A., Bevan, C.D., Robinson, J.D., Chamberlain, J., Fromson, J.M., *Xenobiotica* (1983) **13**, 539–553.

C571. Iwasaki, K., Shiraga, T., Nodi, K., Tada, K., Noguchi, H., *Xenobiotica* (1983) **13**, 565–574.

C572. Trennery, P.N., Waring, R.H., *Xenobiotica* (1983) **13**, 475–482.

C573. Clement, B., *Xenobiotica* (1983) **13**, 467–473.

C574. Walde, A., Ve, B., Scheline, R.R., Monge, P., *Xenobiotica* (1983) **13**, 503–512.

C575. Wilking, N., Appelgren, L.-E., Carlström, K., Pousette, Å., Theve, N.O., *Acta Pharmacol. Toxicol.* (1982) **50**, 161–168.

C576. Fu, P.P., Yang, S.K., *Carcinogenesis* (1983) **4**, 979–984.

C577. Chou, M.W., Fu, P.P., *Biochem. Biophys. Res. Commun.* (1983) **117**, 541–548.

C578. Schmid, S.E., Au, W.Y.W., Hill, D.E., Kadlubar, F.F., Slikker, W., Jr., *Drug Metab. Dispos.* (1983) **11**, 531–536.

C579. De Luca, V., Ibrahim, R.K., *Phytochemistry* (1982) **21**, 1537–1540.

C580. Birnbaum, L.S., Eastin, W.C., Jr., Johnson, L., Matthews, H.B., *Drug Metab. Dispos.* (1983) **11**, 537–543.

C581. Walle, T., Wilson, M.T., Walle, U.K., Bai, S.A., *Drug Metab. Dispos.* (1983) **11**, 544–549.

C582. Moustafa, M.A.A., Claesen, M., Adline, J., Vandervorst, D., Poupaert, J.H., *Drug Metab. Dispos.* (1983) **11**, 574–580.

C583. MacGregor, T.R., Nastasi, L., Farina, P.R., Keirns, J.J., *Drug Metab. Dispos.* (1983) **11**, 568–573.

C584. Guarna, A., Della Corte, L., Giovannini, M.G., De Sarlo, F., Sgaragli, G., *Drug Metab. Dispos.* (1983) **11**, 581–584.

C585. Le Bigot, J.F., Cresteil, T., Kiechel, J.R., Beaune, P., *Drug Metab. Dispos.* (1983) **11**, 385–389.

C586. Hooper, W.D., Treston, A.M., Jacobsen, N.W., Dickinson, R.G., Eadie, M.J., *Drug Metab. Dispos.* (1983) **11**, 607–610.

C587. Hackett, A.M., Griffiths, L.A., *Drug Metab. Dispos.* (1983) **11**, 602–606.

C588. Feng, P.C.C., Fenselau, C., Colvin, M.E., Hinson, J.A., *Drug Metab. Dispos.* (1983) **11**, 103–108.

C589. Emudianughe, T.S., Caldwell, J., Sinclair, K.A., Smith, R.L., *Drug Metab. Dispos.* (1983) **11**, 97–102.

C590. Lynn, R.K., Garvie-Gould, C., Milam, D.F., Scott, K.F., Eastman, C.L., Rodgers, R.M., *Drug Metab. Dispos.* (1983) **11**, 109–114.

C591. Mellström, B., Bertilsson, L., Bergersson, C., Göransson, M., Von Bahr, C., *Drug Metab. Dispos.* (1983) **11**, 115–119.

C592. Teshima, R., Nagamatsu, K., Ikebuchi, H., Kido, Y., Terao, T., *Drug Metab. Dispos.* (1983) **11**, 152–157.

C593. Morris, P.L., Burke, T.R., Jr., Pohl, L.R., *Drug Metab. Dispos.* (1983) **11**, 126–130.

C594. Egorin, M.J., Andrews, P.A., Nakazawa, H., Bachur, N.R., *Drug Metab. Dispos.* (1983) **11**, 167–171.

C595. Duncan, J.D., Hallström, G., Paulsen-Sörman, Lindeke, B., Cho, A.K., *Drug Metab. Dispos.* (1983) **11**, 15–20.

C596. Abou-Donia, M.B., Kinnes, C.G., Abdo, K.M., Bjornsson, T.D., *Drug Metab. Dispos.* (1983) **11**, 31–36.

C597. Hansen, D.K., Hodes, M.E., *Drug Metab. Dispos.* (1983) **11**, 21–24.

C598. Hallström, G., Kammerer, R.C., Nguyen, C.H., Schmitz, D.A., Di Stefano, E.W., Cho, A.K., *Drug Metab. Dispos.* (1983) **11**, 47–53.

C599. Hucker, H.B., Hutt, J.E., White, S.D., Arison, B.H., Zacchei, A.G., *Drug Metab. Dispos.* (1983) **11**, 54–58.

C600. Foster, P.M.D., Cook, M.W., Thomas, L.V., Walters, D.G., Gangolli, S.D., *Drug Metab. Dispos.* (1983) **11**, 59–61.

C601. Garland, W.A., Miwa, B.J., Dairman, W., Kappell, B., Chiueh, M.C.C., Divoll, M., Greenblatt, D.J., *Drug Metab. Dispos.* (1983) **11**, 70–72.

C602. Sundheimer, D.W., Brendel, K., *Drug Metab. Dispos.* (1983) **11**, 433–440.

C603. Falany, C.N., Chowdhury, J.R., Chowdhury, N.R., Tephly, T.R., *Drug Metab. Dispos.* (1983) **11**, 426–432.

C604. Cone, E.J., Darwin, W.D., Buchwald, W.F., Gorodetzky, C.W., *Drug Metab. Dispos.* (1983) **11**, 446–450.

C605. Nakatsu, K., Hugenroth, S., Sheng, L.-S., Horning, E.C., Horning, M.G., *Drug Metab. Dispos.* (1983) **11**, 463–470.

C606. Sloan, T.P., Lancaster, R., Shah, R.R., Idle, J.R., Smith, R.L., *Br. J. Clin. Pharmacol.* (1983) **15**, 443–450.

C607. Maurer, G., Schreier, E., Delaborde, S., Loosli, H.R., Nufer, R., Shukla, A.P., *Eur. J. Drug Metab. Pharmacokinet.* (1982) **7**, 281–292.

C608. Duquette, P.H., Peterson, F.J., Crankshaw, D.L., Lindemann, N.J., Holtzman, J.L., *Drug Metab. Dispos.* (1983) **11**, 477–480.

C609. Estus, G.S., Mieyal, J.J., *Drug Metab. Dispos.* (1983) **11**, 471–476.

C610. Andersson, T., Förlin, L., Hansson, T., *Drug Metab. Dispos.* (1983) **11**, 494–498.
C611. Bekersky, I., Popick, A., *Drug Metab. Dispos.* (1983) **11**, 512–513.
C612. Hanzlik, R.P., Cashman, J.R., *Drug Metab. Dispos.* (1983) **11**, 201–205.
C613. Streeter, A.J., Timbrell, J.A., *Drug Metab. Dispos.* (1983) **11**, 184–189.
C614. Lehman, J.P., Fenselau, C., De Paulo, J.R., *Drug Metab. Dispos.* (1983) **11**, 221–225.
C615. Levine, W.G., Lee, S., *Drug Metab. Dispos.* (1983) **11**, 239–243.
C616. Breger, R.K., Novak, R.F., Franklin, R.B., Rickert, D., Lech, J.J., *Drug Metab. Dispos.* (1983) **11**, 319–323.
C617. Kolis, S.J., Postma, E.J., Williams, T.H., Sasso, G.J., Schwartz, M.A., *Drug Metab. Dispos.* (1983) **11**, 324–328.
C618. Bories, G.F., Peleran, J.C., Wal, J.M., Corpet, D.E., *Drug Metab. Dispos.* (1983) **11**, 249–254.
C619. Hartmann, F., Gruenke, L.D., Craig, J.C., Bissell, D.M., *Drug Metab. Dispos.* (1983) **11**, 244–248.
C620. Bargar, E.M., Walle, U.K., Bai, S.A., Walle, T., *Drug Metab. Dispos.* (1983) **11**, 266–272.
C621. Bailey, K., Thomas, B.H., Vézina, M., Whitehouse, L.W., Zeitz, W., Solomonraj, G., *Drug Metab. Dispos.* (1983) **11**, 277–278.
C622. Wong, F.A., Bateman, C.P., Shaw, C.J., Patrick, J.E., *Drug Metab. Dispos.* (1983) **11**, 301–307.
C623. Rodgers, R.M., Garvie-Gould, C., Scott, K.F., Milam, D.F., Lynn, R.K., *Drug Metab. Dispos.* (1983) **11**, 293–300.
C624. Furuuchi, S., Otsuka, M., Miura, Y., Harigaya, S., *Drug Metab. Dispos.* (1983) **11**, 371–376.
C625. Walle, T., Walle, U.K., Knapp, D.R., Conradi, E.C., Bargar, E.M., *Drug Metab. Dispos.* (1983) **11**, 344–349.
C626. Suzuki, T., Hashimura, Y., Takeyama, S., *Drug Metab. Dispos.* (1983) **11**, 377–386.
C627. Yu, V.C., De Lamirande, E., Horning, M.G., Pang, K.S., *Drug Metab. Dispos.* (1982) **10**, 568–572.
C628. Maguire, J.H., Butler, T.C., Dudley, K.H., *Drug Metab. Dispos.* (1982) **10**, 595–598.
C629. Diamond, G.L., Anders, M.W., Tremaine, L.M., Quebbemann, A.J., *Drug Metab. Dispos.* (1982) **10**, 573–578.
C630. Lertratanangkoon, K., Horning, M.G., Middleditch, B.S., Tsang, W.-S., Griffin, G.W., *Drug Metab. Dispos.* (1982) **10**, 614–623.
C631. Stogniew, M., Fenselau, C., *Drug Metab. Dispos.* (1982) **10**, 609–613.
C632. Janssen, F.W., Kirkman, S.K., Fenselau, C., Stogniew, M., Hofmann, B.R., Young, E.M., Ruelius, H.W., *Drug Metab. Dispos.* (1982) **10**, 599–604.
C633. Sisenwine, S.F., Tio, C.O., Hadley, F.V., Liu, A.L., Kimmel, H.B., Ruelius, H.W., *Drug Metab. Dispos.* (1982) **10**, 605–608.
C634. Stillwell, W.G., Horning, M.G., Griffin, G.W., Tsang, W.-S., *Drug Metab. Dispos.* (1982) **10**, 624–631.
C635. Chu, I., Villeneuve, D.C., Secours, V., Benoit, F.M., Viau, A., *Drug Metab. Dispos.* (1982) **10**, 632–635.
C636. Leeling, J.L., Evans, J.V., Helms, R.J., Ryerson, B.A., *Drug Metab. Dispos.* (1982) **10**, 649–653.
C637. Yu, C.C., Atallah, Y.H., Whitacre, D.M., *Drug Metab. Dispos.* (1982) **10**, 645–648.
C638. Claesen, M., Moustafa, M.A.A., Adline, J., Vandervorst, D., Poupaert, J.H., *Drug Metab. Dispos.* (1982) **10**, 667–671.
C639. Igari, Y., Sugiyama, Y., Sawada, Y., Iga, T., Hanano, M., *Drug Metab. Dispos.* (1982) **10**, 676–679.
C640. Ward, D.P., Trevor, A.J., Kalir, A., Adams, J.D., Baillie, T.A., Castagnoli, N., Jr., *Drug Metab. Dispos.* (1982) **10**, 690–695.
C641. Martin, B.R., Bailey, B.B., Awaya, H., May, E.L., Narasimhachari, N., *Drug Metab. Dispos.* (1982) **10**, 685–689.

C642. Chen, M.-L., Chiou, W.L., *Drug Metab. Dispos.* (1982) **10**, 706–707.

C643. Redalieu, E., Bartlett, M.F., Waldes, L.M., Darrow, W.R., Egger, H., Wagner, W.E., *Drug Metab. Dispos.* (1982) **10**, 708–709.

C644. Stocklein, W., Eisgruber, A., Schmidt, H.-L., *Biotechnol. Lett.* (1983) **5**, 703–708.

C645. Lenzen, S., Rustenbeck, I., Panten, U., *J. Biol. Chem.* (1984) **259**, 2043–2046.

C646. Bourgeau, G., Mayrand, D., *Can. J. Microbiol.* (1983) **29**, 1184–1189.

C647. Deesen, P.E., Leyland-Jones, B., *Drug Metab. Dispos.* (1984) **12**, 9–13.

C648. Strother, A., Allahyari, R., Buchholz, J., Fraser, I.M., Tilton, B.E., *Drug Metab. Dispos.* (1984) **12**, 35–44.

C649. Fong, K.-L.L., Hwang, B.Y.-H., *Drug Metab. Dispos.* (1984) **12**, 14–19.

C650. De Bethizy, J.D., Rickert, D.E., *Drug Metab. Dispos.* (1984) **12**, 45–50.

C651. Lindstrom, T.D., Whitaker, G.W., *Drug Metab. Dispos.* (1984) **12**, 72–76.

C652. Ruijten, H.M., de Bree, H., Borst, A.J.M., de Lange, N., Scherpenisse, P.M., Vincent, W.R., Post, L.C., *Drug Metab. Dispos.* (1984) **12**, 82–92.

C653. Uetrecht, J.P., Sweetman, B.J., Woosley, L., Oates, J.A., *Drug Metab. Dispos.* (1984) **12**, 77–81.

C654. Bock, K.W., Lilienblum, W., von Bahr, C., *Drug Metab. Dispos.* (1984) **12**, 93–97.

C655. Maftouh, M., Monsarrat, B., Rao, R., Meunier, B., Paoletti, C., *Drug Metab. Dispos.* (1984) **12**, 111–119.

C656. Ruenitz, P.C., Bagley, J.R., Mokler, C.M., *Biochem. Pharmacol.* (1983) **32**, 2941–2947.

C657. Fong, K.-L.L., Hwang, B.Y.-H., *Biochem. Pharmacol.* (1983) **32**, 2781–2786.

C658. Shuster, L., Casey, E., Welankiwar, S.S., *Biochem. Pharmacol.* (1983) **32**, 3045–3051.

C659. Leone, R.M., *Biochem. Pharmacol.* (1983) **32**, 3120–3122.

C660. Schnellman, R.G., Putnam, C.W., Sipes, I.G., *Biochem. Pharmacol.* (1983) **32**, 3233–3239.

C661. Cashaw, J.L., Ruchirawat, S., Nimit, Y., Davies, V.E., *Biochem. Pharmacol.* (1983) **32**, 3163–3169.

C662. Wood, S.G., John, B.A., Chasseaud, L.F., Johnstone, I., Biggs, S.R., Hawkins, D.R., Priego, J.G., Darragh, A., Lambe, R.F., *Xenobiotica* (1983) **13**, 731–742.

C663. Sakai, T., Niwaguchi, T., Kimura, R., Murata, T., *Xenobiotica* (1983) **13**, 715–724.

C664. Eyer, P., Hell, W., *Xenobiotica* (1983) **13**, 649–659.

C665. Brown, S., Griffiths, S.A., *Xenobiotica* (1983) **13**, 669–682.

C666. Bakke, J.E., Bergman, Å.L., Brandt, I., Darnerud, P., Struble, C., *Xenobiotica* (1983) **13**, 597–605.

C667. Siegers, C.-P., Rozman, K., Klaassen, C.D., *Xenobiotica* (1983) **13**, 591–596.

C668. Capdevila, J., Saeki, Y., Falck, J.R., *Xenobiotica* (1983) **13**, 105–118.

C669. Estabrook, R.W., Martin-Wixtrom, C., Saeki, Y., Renneberg, R., Hildebrandt, A., Werringloer, J., *Xenobiotica* (1984) **14**, 87–104.

C670. Climie, I.J.G., Hutson, D.H., Stoydin, G., *Xenobiotica* (1983) **13**, 611–618.

C671. Maggs, J.L., Gabowski, P.S., Park, B.K., *Xenobiotica* (1983) **13**, 619–626.

C672. Ullrich, D., Bock, K.W., *Biochem. Pharmacol.* (1984) **33**, 97–101.

C673. Law, F.C.P., Song, Y.Y., Chakrabarty, S., *Xenobiotica* (1983) **13**, 627–633.

C674. Anders, M.W., Sunram, J.M., Wilkinson, C.F., *Biochem. Pharmacol.* (1984) **33**, 577–580.

C675. Gold, B., Brunk, G., *Biochem. Pharmacol.* (1984) **33**, 979–982.

C676. Reinke, L.A., Harmon, T., Belinsky, S.A., Kauffman, F.C., Thurman, R.G., *Biochem. Pharmacol.* (1984) **33**, 1315–1321.

C677. Mentlein, R., Heymann, E., *Biochem. Pharmacol.* (1984) **33**, 1243–1248.
C679. Barker, S.A., Beaton, J.M., Christian, S.T., Monti, J.A., Morris, P.E., *Biochem. Pharmacol.* (1984) **33**, 1395–1400.
C680. Curt, G.A., Jolivet, J., Bailey, B.D., Carney, D.N., Chabner, B.A., *Biochem. Pharmacol.* (1984) **33**, 1681–1685.
C681. Heinonen, T.H., *Biochem. Pharmacol.* (1984) **33**, 1585–1593.
C682. Pearce, L.B., Roth, J.A., *Biochem. Pharmacol.* (1984) **33**, 1809–1811.
C683. Eacho, P.I., Johnson, P.B., Pope, M.R., Colby, H.D., *Biochem. Pharmacol.* (1984) **33**, 1811–1813.
C684. Watabe, T., Hiratsuka, A., Ishikawa, K., Isobe, M., Ozawa, N., *Biochem. Pharmacol.* (1984) **33**, 1839–1844.
C685. Whyte, W., Pong, R.S., Buckpitt, A.R., *Biochem. Pharmacol.* (1984) **33**, 1813–1816.
C686. Mosnaim, A.D., Wolf, M.E., Madubuike, U.P., *Biochem. Pharmacol.* (1984) **33**, 1993–1996.
C687. Cooke, W.J., Cooke, L.M., *J. Pharmacol. Exp.Ther.* (1983) **225**, 85–93.
C688. Størmer, F.C., Støren, O., Hansen, C.E., Pedersen, J.I., Aasen, A.J., *Appl. Environ. Microbiol.* (1983) **45**, 1183–1187.
C689. Lin, L.O., Yunis, A.A., *Pharmacology* (1983) **27**, 58–64.
C691. Hammond, P.M., Price, C.P., Scawen, M.D., *Eur. J. Biochem.* (1983) **132**, 651–655.
C692. Sugumaran, M., Lipke, H., *FEBS Lett.* (1983) **155**, 65–68.
C693. Kocna, P., Kasafirek, E., Fric, P., Slaby, J., *Experientia* (1983) **39**, 389–390.
C694. Robert, T., Williams, D.J., *J. Chem. Soc., Chem. Commun.* (1983) 128–130.
C695. Keller, B., Keller, E., Görisch, H., Lingens, F., *Hoppe-Seyler's Z. Physiol. Chem.* (1983) **364**, 455–459.
C696. Takao, N., Kamigauchi, M., Okada, M., *Helv. Chim. Acta* (1983) **66**, 473–484.
C697. Wegst, W., Lingens, F., *FEMS Microbiol. Lett.* (1983) **17**, 341–344.
C698. Wong, C.C., Chiu, K.W., Choy, Y.M., *Comp. Biochem. Physiol. B* (1983) **74B**, 739–742.
C699. Pometto, A.L.III, Crawford, D.L., *Appl. Environ. Microbiol.* (1983) **45**, 1582–1585.
C700. Poiger, H., Buser, H.R., *Environ. Sci. Res.* (1983) **26**, 483–492.
C701. Milewich, L., Hendricks, T.S., Romero, L.H., *J. Steroid Biochem.* (1982) **17**, 669–674.
C702. Kitamura, S., Narai, N., Hayashi, M., Tatsumi, K., *Chem. Pharm. Bull.* (1983) **31**, 776–779.
C703. Ohkawa, H., Ohishi, N., Yagi, K., *J. Biol. Chem.* (1983) **258**, 5629–5633.
C704. Leclaire, J., Dansette, P.M., Fortsmeyer, D., Mansuy, D., *Dev. Biochem.* (1982) **23**, 95–98.
C705. Costello, P.B., Green, F.A., *Arthritis Rheum.* (1983) **26**, 541–547.
C706. Lübbe, C., Van Pee, K.H., Salcher, O., Lingens, F., *Hoppe-Seyler's Z. Physiol. Chem.* (1983) **364**, 447–453.
C707. Walker, J.R.L., Taylor, B.G., *Arch. Microbiol.* (1983) **134**, 123–126.
C708. Harrison, M.P., *Methodol. Surv. Biochem. Anal.* (1983) **12**, 191–196.
C709. Maruyama, K., *Kyoyobu kenkyu Hokoku* (1982) **18**, 63–73.
C710. Shigeru, F., Sakurai, S., Sugimoto, I., Awata, N., *Chem. Pharm. Bull.* (1983) **31**, 645–652.
C712. Bolcsak, L.E., Nerland, D.E., *J. Biol. Chem.* (1983) **258**, 7252–7255.
C713. Cleveland, T.E., Smith, D.A., *Physiol. Plant Pathol.* (1983) **22**, 129–142.
C714. Rao, P.S., Srinivasan, M., *J. Biosci.* (1983) **5**, 7–19.
C715. Maurer, G., Schreier, E., Delaborde, S., Nufer, R., Shukla, A.P., *Eur. J. Drug Metab. Pharmacokinet.* (1983) **8**, 51–62.
C716. Cerniglia, C.E., Althaus, J.R., Evans, F.E., Freeman, J.P., Mitchum, R.K., yang, S.K., *Chem.-Biol. Interact.* (1983) **44**, 119–132.

C717. Visser, T.J., Mol, J.A., Otten, M.H., *Endocrinology (Baltimore)* (1983) **112**, 1547–1549.

C719. Tashiro, F., Shibata, A., Nishimura, N., Ueno, Y., *J. Biochem.* (1983) **93**, 1557–1566.

C720. Milewich, L., Garcia, R.L., Johnson, A.R., *J. Clin. Endocrinol. Metab.* (1983) **57**, 8–14.

C721. Matern, U., *Arch. Biochem. Biophys.* (1983) **224**, 261–271.

C722. Wermeille, M., Turin, E., Griffiths, L.A., *Eur. J. Drug Metab. Pharmacokinet.* (1983) **8**, 77–84.

C723. Franklin, R.A., Robson, P., Stevenson, D., *Eur. J. Clin. Pharmacol.* (1983) **24**, 629–634.

C724. Otten, M.H., Mol, J.A., Visser, T.J., *Science* (1983) **221**, 81–83.

C725. Maydl, R., Newbold, R.R., Metzler, M., McLachlan, J.A., *Endocrinology (Baltimore)* (1983) **113**, 146–151.

C727. Poupko, J.M., Radomski, T., Santella, R.M., Radomski, J.L., *JNCI, J. Natl. Cancer Inst.* (1983) **70**, 1077–1080.

C728. Rundgren, M., *Eur. J. Biochem.* (1983) **133**, 657–663.

C729. Pujar, B.G., Ribbons, D.W., *Indian J. Biochem. Biophys.* (1983) **20**, 112–114.

C730. Barnsley, E.A., *J. Bacteriol.* (1983) **154**, 113–137.

C731. Tachibana, S., Murakami, T., *Mol. Cell. Biochem.* (1983) **51**, 149–160.

C732. Sekita, S., Yoshihira, K., Natori, S., *Chem. Pharm. Bull.* (1983) **31**, 490–498.

C733. Yamanaka, K., Tsuyuki, Y., *Agric. Biol. Chem.* (1983) **47**, 1361–1362.

C734. Fritzemeier, K.H., Kindl, H., *Eur. J. Biochem.* (1983) **133**, 545–550.

C735. Christ, W., Hecker, W., Junge, H., Stille, G., *Naunyn-Schmiedeberg's Arch. Pharmacol.* (1983) **323**, 176–182.

C736. Vittek, J., Gordon, G.G., Munnangi, P.R., Rappaport, S.C., Southren, A.L., *J. Dent. Res.* (1983) **62**, 341–343.

C737. Lecointe, P., Puget, A., *C. R. Seances Acad. Sci., Ser. 3* (1983) **296**, 279–282.

C738. Preston, B.D., Miller, J.A., Miller, E.C., *J. Biol. Chem.* (1983) **258**, 8304–8311.

C739. Loeppky, R.N., Outram, J.R., *IARC Sci. Publ.* (1982) **41**, 459–472.

C740. Grand, C., Boudet, A., Boudet, A.M., *Planta* (1983) **158**, 225–229.

C741. Keith, R.A., Abraham, R.T., Pazmiño, P., Weinshilboum, R.M., *Clin. Chim. Acta* (1983) **131**, 257–272.

C742. Claus, G., Kutzner, H.J., *Syst. Appl. Microbiol.* (1983) **4**, 169–180.

C743. Jonsson, L.M.V., Donker-Koopman, W.E., Uitslager, P., Schram, A.W., *Plant Physiol.* (1983) **72**, 287–290.

C744. Raemsch, K.D., Sommer, J., *Hypertension (Dallas)* (1983) **5**, 18–24.

C745. Helmlinger, J., Rausch, T., Hilgenberg, W., *Physiol. Plant.* (1983) **58**, 302–310.

C746. Silber, B.M., Holford, N.H.G., Riegelman, S., *J. Pharm. Sci.* (1983) **72**, 725–732.

C747. Neidlein, R., Hofmann, G., *Arzneim.-Forsch.* (1983) **33**, 920–925.

C748. Ioannides, C., Lum, P.Y., Parke, D.V., *Xenobiotica* (1984) **14**, 119–137.

C749. Yamamoto, I., Narimatsu, S., Watanabe, K., Shimonishi, T., Yoshimura, H., Nagano, T., *Chem. Pharm. Bull.* (1983) **31**, 1784–1787.

C750. Crayford, J.V., Hutson, D.H., Stoydin, G., *Xenobiotica* (1984) **14**, 221–233.

C751. Hamill, S., Cooper, D.Y., *Xenobiotica* (1984) **14**, 139–149.

C752. Heijbroek, W.M.H., Muggleton, D.F., Parke, D.V., *Xenobiotica* (1984) **14**, 235–247.

C753. Yoshida, K., Torii, H., Sugiyama, Y., Fujita, T., Tanayama, S., *Xenobiotica* (1984) **14**, 249–257.

C754. Boyd, J.A., Harvan, D.J., Eling, T.E., *J. Biol. Chem.* (1983) **258**, 8246–8254.

C755. Bajaj, K.L., de Luca, V., Khouri, H., Ibrahim, R.K., *Plant Physiol.* (1983) **72**, 891–896.

C756. Poulton, J.E., Shin, S.-I., *Z. Naturforsch. C: Biosci.* (1983) **38C**, 369–374.

C758. Beart, P.M., Rowe, P.R., Louis, W.J., *J. Pharm. Pharmacol.* (1983) **35**, 519–520.

C759. He, S., Huang, Z., *Tianjin Yiyao* (1983) **11**, 231–233.

C760. Suzuki, M., Yoshida, K., Sakurada, T., Kitaoka, H., Kaise, K., Kaise, N., Fukazawa, H., Yamamoto, M., Saito, S., Yoshinaga, K., *Tohoku J. Exp. Med.* (1983) **140**, 311–318.

C763. Mårdh, G., *J. Neurochem.* (1983) **41**, 299–301.

C764. Walde, A., Scheline, R.R., *Acta Pharmacol. Toxicol.* (1983) **53**, 57–63.

C765. de Bethizy, J.D., Rickert, D.E., *Biochem. Biophys. Res. Commun.* (1983) **114**, 500–504.

C766. Akhtar, M.H., *J. Agric. Food Chem.* (1983) **31**, 1080–1083.

C767. Jacob, J., Schmoldt, A., Grimmer, G., *Carcinogenesis* (1983) **4**, 905–910.

C768. Bunce, N.J., Merrick, R.L., Corke, C.T., *J. Agric. Food Chem.* (1983) **31**, 1071–1075.

C769. Otten, M.H., de Herder, W.W., Hazenberg, M.P., van de Boom, M., Hennemann, G., *FEMS Microbiol. Lett.* (1983) **18**, 75–77.

C770. Murakoshi, I., Koide, C., Ikegami, F., Nasu, K., *Chem. Pharm. Bull.* (1983) **131**, 1777–1779.

C771. Suzuki, Y., Uchida, K., *Nogaku Kenkyu* (1983) **60**, 81–89.

C772. Hagmann, M.L., Heller, W., Grisebach, H., *Eur. J. Biochem.* (1983) **134**, 547–554.

C773. Meuldermans, W., Hencrickx, J., Hurkmans, R., Swysen, E., Woestenborghs, R., Lauwers, W., Heykants, J., *Arzneim.-Forsch.* (1983) **33**, 1142–1151.

C774. Sariaslani, F.S., Rosazza, J.P., *Appl. Environ. Microbiol.* (1983) **46**, 468–474.

C775. Wichers, H.J., Malingre, T.M., Huizing, H.J., *Planta* (1983) **158**, 482–486.

C776. Masunaga, S., Urushigawa, Y., Yonezawa, Y., *Chemospere* (1983) **12**, 1075–1082.

C777. Renner, G., Hopfer, C., *Chemospere* (1983) **12**, 967–970.

C778. Benthe, H.F., Thieme, R.A., *Biomed. Mass Spectrom.* (1983) **10**, 65–73.

C779. Ali, B., Kaur, S., *J. Pharm. Exp. Ther.* (1983) **226**, 589–594.

C780. Girkin, R., Brodie, R.R., Midgley, I., Hawkins, D.R., Chasseaud, L.F., *Eur. J. Drug Metab. Pharmacokinet.* (1983) **8**, 109–116.

C781. Miura, S., Hamada, T., Satomi, O., Midorikawa, T., Awata, N., *Oyo Yakuri* (1983) **25**, 1015–1024.

C782. Miyazaki, H., Fujii, T., Yoshida, K., Arakawa, S., Furukawa, H., Suzuki, H., Kagemoto, A., Hashimoto, M., Tamaki, N., *Eur. J. Drug Metab. Pharmacokinet.* (1983) **8**, 117–131.

C783. Priego, J.G., Maroto, M.L., Ortega, M.P., Armijo, M., *Arch. Farmacol. Toxicol.* (1983) **9**, 39–46.

C784. Boels, P.J., Verbeuren, T.J., Vanhoutte, P.M., *Naunyn-Schmiedeberg's Arch. Pharmacol.* (1983) **324**, 7–14.

C785. Sharman, D.F., Kelly, M.J., Loncar-Stevanovic, H., Comp. *Biochem. Physiol. C* (1983) **75C**, 217–222.

C786. Aschbacher, P.W., Mitchell, A.D., *Chemosphere* (1983) **12**, 961–965.

C787. Renner, G., *Life Sci.* (1983) **33**, 1427–1431.

C788. Chou, M.W., Evans, F.E., Yang, S.K., Fu, P.P., *Carcinogenesis* (1983) **4**, 699–702.

C789. Renner, G., *Chemospere* (1983) **12**, 971–973.

C791. Kinouchi, T., Ohnishi, Y., *Appl. Environ. Microbiol.* (1983) **46**, 596–604.

C792. Ibrahim, R.K., De Luca, V., Brunet, G., *Plant Tissue, Cult., Proc. Int. Congr. Plant Tissue Cell Cult., 5th* (1982) 369–370.

C793. Stack, R.J., Hungate, R.E., Opsahl, W.P., *Appl. Environ. Microbiol.* (1983) **46**, 539–544.

C793a. Otsuka, M., Kurozumi, T., Furuuchi, S., Usuki, S., Kotera, K., Hariyaga, S.,

Chem. Pharm. Bull. (1983) **31**, 2438–2450.
C793b. Farina, P., Benfenati, E., Reginato, R., Torti, L., D'Incalci, M., Threadgill, M.D., Gescher, A., *Biomed. Mass Spectrom.* (1983) **10**, 485–488.
C794. Aschbacher, P.W., Feil, V.J., *J. Agric. Food. Chem.* (1983) **31**, 1150–1158.
C796. Kim, S.U., Cho, Y.J., Floss, H.G., Anderson, J.A., *Planta Med.* (1983) **48**, 145–148.
C797. Corbett, M.D., Corbett, B.R., *J. Agric. Food. Chem.* (1983) **31**, 1276–1282.
C798. Wang, P.P., Beaune, P., Kaminsky, L.S., Dannan, G.A., Kadlubar, F.F., Larrey, D., Guengerich, F.P., *Biochemistry* (1983) **22**, 5375–5383.
C799. Oka, K., Kojima, K., Nagatsu, T., *Biochem. Int.* (1983) **7**, 387–393.
C800. McCorkindale, N.J., Hayes, D., Johnston, G.A., Clutterbuck, A.J., *Phytochemistry* (1983) **22**, 1026–1028.
C801. Hills, C.A., Fewson, C.A., *Biochem. J.* (1983) **209**, 379–386.
C802. Moldawer, L.L., Kawamura, I., Bistrian, B.R., Blackburn, G.L., *Biochem. J.* (1983) **210**, 811–817.
C803. Cozzi, P., Rampinelli, D., Tonani, R., Meinardi, G., Trane, F., *Farmaco, Ed. Prat.* (1983) **38**, 344–351.
C804. Otsuka, M., Furuuchi, S., Harigaya, S., *Chem. Pharm. Bull.* (1983) **31**, 2799–2809.
C805. Watanabe, K., Kita, M., Yamamoto, I., Oguri, K., Yoshimura, H., *J. Pharmacobio.-Dyn.* (1983) **6**, 581–587.
C806. Anderounmu, A.F., Fleckenstein, L., *J. Pharmacol. Exp. Ther.* (1983) **226**, 633–639.
C807. Eweiss, N., Nicholls, P.J., Askam, V., *IRCS Med. Sci., Lib. Compend.* (1983) **11**, 843–844.
C808. Martin, B.W., Ivie, G.W., Bailey, E.M., *Arch. Environ. Contam. Toxicol.* (1983) **12**, 499–507.
C809. Balsam, A., Sexton, F., Borges, M., Ingbar, S.H., *J. Clin. Invest.* (1983) **72**, 1234–1245.
C810. Moss, E.J., Judah, D.J., Przybylski, M., Neal, G.E., *Biochem. J.* (1983) **210**, 227–233.
C811. La Voie, E.J., Adams, E.A., Hoffmann, D., *Carcinogenesis* (1983) **4**, 1133–1138.
C812. Banks, R.B., Smith, T.J., Hanna, P.E., *J. Med. Chem.* (1983) **26**, 1780–1784.
C813. Whitehead, I.M., Dixon, R.A., *Biochim. Biophys. Acta* (1983) **747**, 298–303.
C814. Miwa, G.T., Walsh, J.S., Kedderis, G.L., Hollenberg, P.F., *J. Biol. Chem.* (1983) **258**, 14445–14449.
C815. Thomas, R., Williams, D.J., *J. Chem. Soc., Chem. Commun.* (1983) 677–679.
C816. Harrison, D.M., Quinn, P., *J. Chem. Soc., Chem. Commun.* (1983) 879–880.
C817. Keil, H., Tittmann, U., Lingens, F., *Syst. Appl. Microbiol.* (1983) **4**, 313–325.
C818. Sutherland, J.B., Crawford, D.L., Pometto, A.L.III, *Can. J. Microbiol.* (1983) **29**, 1253–1257.
C819. Camirand, A., Phipps, J., Wightman, F., *Can. J. Bot.* (1983) **61**, 2302–2308.
C820. Hallström, G., Nguyen, C.H., Cho, A.K., *Acta Pharm. Suec.* (1983) **20**, 181–192.
C822. Igarashi, K., Kasuya, F., Fukui, M., *J. Pharmacobio.-Dyn.* (1983) **6**, 538–550.
C823. James, M.I., Midgley, J.M., Williams, C.M., *J. Pharm. Pharmacol.* (1983) **35**, 559–565.
C824. Baba, S., Furuta, T., Fujioka, M., Goromaru, T., *J. Pharm. Sci.* (1983) **72**, 1155–1158.
C825. Lussiana, J.P., Jacquot, C., Plat, M., Viel, C., *Ann. Pharm. Fr.* (1983) **41**, 151–158.
C826. Gee, S.J., Green, C.E., Tyson, C.A., *Drug Metab. Dispos.* (1984) **12**, 174–178.
C827. Wilson, B.K., Thompson, J.A., *Drug Metab. Dispos.* (1984) **12**, 161–164.
C829. Cook, C.E., Brine, D.R., Tallent, C.R., *Drug Metab. Dispos.* (1984) **12**, 186–192.

C830. Streeter, D.G., Pfadenhauer, E.H., *Drug Metab. Dispos.* (1984) **12**, 199–203.
C831. Monks, T.J., Lau, S.S., Pohl, L.R., Gillette, J.R., *Drug Metab. Dispos.* (1984) **12**, 193–198.
C832. Dodion, P., Egorin, M.J., Tamburini, J.M., Riggs, C.E., Jr., Bachur, N.R. *Drug Metab. Dispos.* (1984) **12**, 209–216.
C833. Cheng, K.-C., Schenkman, J.B., *Drug Metab. Dispos.* (1984) **12**, 222–234.
C834. Assandri, A., Barone, D., Ferrari, P., Perazzi, A., Ripamonti, A., Tuan, G., Zerilli, L.F., *Drug Metab. Dispos.* (1984) **12**, 257–263.
C835. Vickers, S., Duncan, C.A.H., Ramjit, H.G., Dobrinska, M.R., Dollery, C.T., Gomez, H.J., Leidy, H.L., Vincek, W.C., *Drug Metab. Dispos.* (1984) **12**, 242–246.
C836. Huckle, K.R., Hutson, D.H., Millburn, P., *Drug Metab. Dispos.* (1984) **12**, 264–265.
C837. Lau, S.S., Monks, T.J., Gillette, J.R., *Drug Metab. Dispos.* (1984) **12**, 291–296.
C838. McDonald, J.J., Cerniglia, C.E., *Drug Metab. Dispos.* (1984) **12**, 330–336.
C839. Pang, K.S., Huang, J.C., Finkle, C., Kong, P., Cherry, W.F., Fayz, S., *Drug Metab. Dispos.* (1984) **12**, 314–322.
C840. Nelson, W.L., Bartels, M.J., *Drug Metab. Dispos.* (1984) **12**, 345–352.
C841. Dodion, P., Chang, B.K., Egorin, M.J., Olman, E.A., Engisch, K.L., Bachur, N.R., *Drug Metab. Dispos.* (1984) **12**, 365–370.
C843. Nelson, W.L., Bartels, M.J., *Drug Metab. Dispos.* (1984) **12**, 382–384.
C844. Kocna, P., Fric, P., Slabý, J., Kasafírek, E., *Hoppe-Seyler's Z. Physiol. Chem.* (1980) **361**, 1401–1412.
C845. Corbett, M.D., Chipko, B.R., Batchelor, A.O., *Biochem. J.* (1980) **187**, 893–903.
C846. Suzuki, Y., Uchida, K., *Agric. Biol. Chem.* (1983) **47**, 1145–1147.
C847. Suzuki, Y., Uchida, K., *Methods Enzymol.* (1980) **66**, 327–333.
C848. Thomas, R., Williams, D.J., *J. Chem. Soc., Chem. Commun.* (1983) 128–130.
C849. Fujiwara, S., Sakurai, S., Sugimoto, I., Awata, N., *Chem. Pharm. Bull.* (1983) **31**, 645–652.
C850. Tachibana, S., Oka, M., *J. Nutr. Sci. Vitaminol.* (1980) **26**, 419–426.
C851. Price, C.P., Hammond, P.M., Scawen, M.D., *Clin. Chem.* (1983) **29**, 358–361.
C852. Moss, E.J., Manson, M.M., Neal, G.E., *Carcinogenesis* (1984) **5**, 869–874.
C853. Poiger, H., Buser, H.-R., Weber, H., Zweifel, U., Schlatter, C., *Experientia* (1982) **38**, 484–486.
C854. Yoshida, K., Sakurada, T., Kitaoka, H., Fukazawa, H., Kaise, N., Kaise, K., Yamamoto, M., Saito, S., Yoshinaga, K., *Endocrinol. Jpn.* (1983) **30**, 211–217.
C855. Veerkamp, W., Pel, R., Hutzinger, O., *Chemosphere* (1983) **12**, 1337–1343.
C856. Huckins, J.N., Petty, J.D., *Arch. Environ. Contam. Toxicol.* (1983) **12**, 667–671.
C857. Kulkarni, B., Fiala, E.S., Weisburger, J.H., *Carcinogenesis* (1983) **4**, 1275–1279.
C858. Ueno, Y., Tashiro, F., Kobayashi, T., *Food Chem. Toxicol.* (1983) **21**, 167–173.
C859. Schumacher, H.-M., Rüffer, M., Nagakura, N., Zenk, M.H., *Planta Med.* (1983) **48**, 212–220.
C860. Golbeck, J.H., Albaugh, S.A., Radmer, R., *J. Bacteriol.* (1983) **156**, 49–57.
C861. Guenthner, T.M., Oesch, F., *J. Biol. Chem.* (1983) **258**, 15054–15061.
C862. Seong, B.L., Son, H.J., Mheen, T.I., Han, M.H., *J. Antibiot.* (1983) **36**, 1402–1404.
C863. Norman, R.L., Wood, A.W., *Drug Metab. Dispos.* (1984) **12**, 543–549.
C864. Dieterle, W., Faigle, J.W., Imhof, P., Sulc, M., Wagner, J., *Xenobiotica* (1984) **14**, 303–310.
C865. Dieterle, W., Faigle, J.W., Kriemler, H.-P., Winkler, T., *Xenobiotica* (1984) **14**, 311–319.

C866. Allen, J.G., Galloway, D.B., Ehsanullah, R.S.B., Ruane, R.J., Bird, H.A., *Xenobiotica* (1984) **14**, 321–326.

C867. Haddock, R.E., Jeffery, D.J., Lloyd, J.A., Thawley, A.R., *Xenobiotica* (1984) **14**, 327–337.

C868. Kühler, T., *Xenobiotica* (1984) **14**, 417–428.

C869. Albro, P.W., Corbett, J.T., Marbury, D., Parker, C., *Xenobiotica* (1984) **14**, 389–398.

C870. Meuldermans, W., Hendrickx, J., Knaeps, F., Lauwers, W., Heykants, J., Grindel, J.M., *Xenobiotica* (1984) **14**, 445–462.

C871. Assandri, A., Perazzi, A., Ferrari, P., Omodei-Salé, A., Ripamonti, A., Tuan, G., Galliani, G., Zerilli, L.F., *Xenobiotica* (1984) **14**, 429–443.

C872. Hackett, A.M., Shaw, I.C., Griffiths, L.A., *Xenobiotica* (1984) **14**, 491–499.

C873. Kammerer, R.C., Schmitz, D.A., Cho, A.K., *Xenobiotica* (1984) **14**, 475–482.

C874. Lindstrom, T.D., Whitaker, G.W., *Xenobiotica* (1984) **14**, 503–508.

C875. Cox, P.J., Hejbroek, W.M.H., Taylor, D.C., Wood, P.J., Cresswell, P.R., Pepper, E.S., Moore, P.J., Mitchell, R.C., Blake, T.J.A., Dawborne, J.S., Metcalf, R., Johnson, R., *Xenobiotica* (1984) **14**, 509–514.

C876. Lau, S.S., Monks, T.J., Greene, K.E., Gillette, J.R., *Xenobiotica* (1984) **14**, 539–543.

C877. Tynes, R.E., Hodgson, E., *Xenobiotica* (1984) **14**, 515–520.

C878. Gawai, K.R., Pawar, S.S., *Xenobiotica* (1984) **14**, 605–607.

C879. Lenk, W., Sterzl, H., *Xenobiotica* (1984) **14**, 581–588.

C880. Groenewoud, G., Hundt, H.K.L., *Xenobiotica* (1984) **14**, 711–717.

C881. Renner, G., Nguyen, P.-T., *Xenobiotica* (1984) **14**, 693–704.

C882. Bäärnhielm, C., Skånberg, I., Borg, K.O., *Xenobiotica* (1984) **14**, 719–726.

C883. Ishimitsu, T., Hirose, S., Asahara, K., Imaizumi, M., *Chem. Pharm. Bull.* (1984) **32**, 3320–3322.

C884. Kitamura, S., Tatsumi, K., *Chem. Pharm. Bull.* (1983) **31**, 3334–3337.

C885. Ishimitsu, S. Fujimoto, S., Ohara, A., *Chem. Pharm. Bull.* (1984) **32**, 2439–2441.

C886. Mellström, B., Bertilsson, L., Lou, Y.C., Saewe, J., Sjøqvist, F., *Clin. Pharmacol. Ther.* (1983) **34**, 516–520.

C887. Albro, P.W., Tondear, I., Marbury, D., Jordan, S., Schroeder, J., Corbett, J.T., *Biochim. Biophys. Acta* (1983) **760**, 283–292.

C888. Jeenah, M.S., Duttoh, M.F., *Biochem. Biophys. Res. Commun.* (1983) **116**, 1114–1118.

C889. Weston, A., Grover, P., Sims, P., *Carcinogenesis* (1983) **4**, 1307–1311.

C890. Lewis, D.A., Armstrong, R.N., *Biochemistry* (1983) **22**, 6297–6303.

C891. McCarthy, M.B., White, R.E., *J. Biol. Chem.* (1983) **258**, 11610–11616.

C893. Karns, J.S., Kilbane, J.J., Duttagupta, S., Chakrabarty, A.M., *Appl. Environ. Microbiol.* (1983) **46**, 1176–1181.

C894. Ibrahim, K.E., Midgley, J.M., Crowley, J.R., Williams, C.M., *J. Pharm. Pharmacol.* (1983) **35**, 144–147.

C895. Rollema, H., Grol, C.J., *Pharm. Weekbl., Sci. Ed.* (1983) **5**, 159–164.

C896. Maschler, I., Ball, P., Bayerköhler, G., Gaues, J., Knuppen, R., *Steroids* (1983) **41**, 597–607.

C897. Sugihara, K., Kitamura, S., Tatsumi, K., *Chem. Pharm. Bull.* (1983) **31**, 3366–3369.

C898. Legay, C., Faudon, M., Hery, F., Ternaux, J.P., *Neurochem. Int.* (1983) **5**, 721–727.

C899. Burka, L.T., Plucinski, T.M., Macdonald, T.L., *Proc. Nat. Acad. Sci. U.S.A.* (1983) **80**, 6680–6684.

C900. Blom, H., *Food Chem.* (1983) **12**, 197–204.

C901. Asaoka, K., Takahashi, K., *J. Biochem.* (1983) **94**, 1191–1199.

C902. Canonica, L., Galliani, G., Rindone, B., Tollari, S., Andreoni, V., Galli, E., *Experientia* (1983) **39**, 1273–1275.

C903. Medinsky, M.A., Dent, J.G., *Toxicol. Appl. Pharmacol.* (1983) **68**, 359–366.

C904. Clark, A.M., Clinton, R.T., Baker, J.K., Hufford, C.D., *J. Pharm. Sci.* (1983) **72**, 1288–1290.

C905. Jay, M., De Luca, V., Ibrahim, R., *Z. Naturforsch.* (1983) **38C**, 413–417.

C907. Tazaki, K., Ishikura, N., *Plant Cell Physiol.* (1983) **24**, 1263–1268.

C909. Bhakuni, D.S., Jain, S., Chaturvedi, R., *J. Chem. Soc., Perkin Trans. 1* (1983) 1949–1952.

C910. Pasteels, J.M., Rowell-Rahier, M., Braekman, J.C., Dupont, A., *Physiol. Entomol.* (1983) **8**, 307–314.

C911. Sum, C.Y., yacobi, A., Kartzinel, R., Stampfli, H., Davis, C.S., Lai, C.M., *Clin. Pharmacol. Ther.* (1983) **34**, 427–434.

C912. Otsuka, M., Furuuchi, S., Usuki, S., Nitta, S., Harigaya, S., *J. Pharmacobio.-Dyn.* (1983) **6**, 708–720.

C913. Marchesini, B., Boschi, S., Berti, C., *J. Chromatogr.* (1983) **278**, 173–178.

C914. Tanaka, Y., Nishikawa, Y., Hayashi, R., *Chem. Pharm. Bull.* (1983) **31**, 3656–3664.

C915. Mimura, K., Baba, S., *Chem. Pharm. Bull.* (1983) **31**, 3698–3706.

C916. Melikian, A.A., La Voie, E.J., Hecht, S.s., Hoffmann, D., *Carcinogenesis* (1983) **4**, 843–849.

C917. Woodson, L.C., Ames, M.M., Selassie, C.D., Hansch, C., Weinshilboum, R.M., *Mol. Pharmacol.* (1983) **24**, 471–478.

C919. Wong. L.K., Dru, J., Lin, L.-S., Knapp, J., *Appl. Environ. Microbiol.* (1983) **46**, 1239–1242.

C920. Kawamori, M., Hashimoto, Y., Katsumata, R., Okashi, R., Takayama, K., *Agric. Biol. Chem.* (1983) **47**, 2503–2509.

D1. Sasaki, H., Kamimura, H., Kaniwa, H., Kawamura, S., Takaichi, M., Yokoshima, T., *Xenobiotica* (1984) **14**, 621–631.

D2. Kahn, V., Golan-Goldhirsh, A., Whitaker, J.R., *Phytochemistry* (1983) **22**, 1875–1883.

D3. Tadasa, K., Kayahara, H., *Agric. Biol. Chem.* (1983) **47**, 2639–2640.

D4. Elliott, D.C., *Physiol. Plant.* (1983) **59**, 428–437.

D5. Weidolf, L., Borg, K.O., Hoffmann, K.-J., *Xenobiotica* (1984) **14**, 657–666.

D7. Tajima, K., Yamamoto, K., Mizutani, T., *Chem. Pharm. Bull.* (1983) **31**, 3671–3677.

D8. De Boer, A.G., Breimer, D.D., Gubbens-Stibbe, J.M., Bosma, A., *Biopharm. Drug Dispos.* (1983) **4**, 321–330.

D9. Gupte, S.M., Bartels, M.J., Kerr, B.M., Laganiere, S., Silber, B.M., Nelson, W.L., *Res. Commun. Chem. Pathol. Pharmacol.* (1983) **42**, 235–244.

D10. Mizuma, T., Hayashi, M., Awazu, S., *J. Pharmacobio.-Dyn.* (1983) **6**, 851–858.

D11. Kotaki, H., Yamamura, Y., Tanimura, Y., Saitoh, Y., Nakagawa, F., Tamura, Z., *J. Pharmacobio.-Dyn.* (1983) **6**, 881–887.

D12. Kogekar, R.G., Deshpande, V.N., *Indian J. Biochem. Biophys.* (1983) **20**, 208–212.

D13. Dunn, B.B., Stack, M.E., Park, D.L., Joshi, A., Friedman, L., King, R.L., *J. Toxicol. Environ. Health* (1983) **12**, 283–289.

D14. Abita, J.P., Blandin-Savoja, F., Rey, F., *Biochem. Int.* (1983) **7**, 727–737.

D15. Galindo, J.D., Pedreno, E., Garcia-Carmona, F., Garcia-Canovas, F., Solano Munoz, F., Lozano, J.A., *Int. J. Biochem.* (1983) **15**, 1455–1461.

D16. Taylor, B.F., *Appl. Environ. Microbiol.* (1983) **46**, 1286–1292.

D17. Hoshino, T., Sekine, Y., Fujiwara, A., *J. Antibiot.* (1983) **36**, 1458–1463.

D18. Umezawa, T., Nakatsubo, F., Higuchi, T., *Agric. Biol. Chem.* (1983) **47**, 2677–2681.

D19. Ogawa, Y., Mizukoshi, S., Mori, H., *J. Antibiot.* (1983) **36**, 1561–1563.

D20. Balsevich, J., Kurz, W.G.W., *Planta Med.* (1983) **49**, 79–84.

D21. Ghosh, D.K., Mishra, A.K., *Curr. Microbiol.* (1983) **9**, 219–224.

D22. El-Rashidy, R., Niazi, S., *Biopharm. Drug Dispos.* (1983) **4**, 389–396.

D23. Vree, T.B., Vree, J.B., *J. Vet. Pharmacol. Ther.* (1983) **6**, 237–240.

D24. Overmars, H., Scherpenisse, P.M., Post, L.G., *Eur. J. Drug Metab. Pharmacokinet.* (1983) **8**, 269–280.

D25. Dell, H.-D., Kamp, R., Brons, B., Schöllnhammar, G., *Arzneim.-Forsch.* (1983) **33**, 1633–1637.

D26. Vachta, J., Valter, K., Gold-Aubert, P., *Eur. J. Drug Metab. Pharmacokinet.* (1983) **8**, 297–308.

D27. Collins, M.A., Origitano, T.C., *J. Neurochem.* (1983) **41**, 1569–1575.

D28. Innocenti, G., Dall'Acqua, F., Caporale, G., *Phytochemistry* (1983) **22**, 2207–2209.

D29. Cerniglia, C.E., Freeman, J.P., Althaus, J.R., van Baalen, C., *Arch. Microbiol.* (1983) **136**, 177–183.

D30. Gern, W.A., Wechsler, E., Duvall, D., *Gen. Comp. Endocrinol.* (1984) **53**, 169–178.

D31. Taylor, B.F., Ribbons, D.W., *Appl. Environ. Microbiol.* (1983) **46**, 1276–1281.

D32. Ritter, C.L., Malejka-Giganti, D., Polnaszek, C.F., *Chem.-Biol. Interact.* (1983) **46**, 317–334.

D33. Cerniglia, C.E., Lambert, K.J., Miller, D.W., Freeman, J.P., *Appl. Environ. Microbiol.* (1984) **47**, 111–118.

D34. Boux, L.J., Duke, C.C., Holder, G.M., Ireland, C.M., Ryan, A.J., *Carcinogenesis* (1983) **4**, 1429–1435.

D35. Campbell, N.R.C., Dunnette, J.H., Mwaluko, G., Van Loon, J., Weinshilboum, R.M., *Clin. Pharmacol. Ther. (St. Louis)* (1984) **35**, 55–63.

D36. Hodgson, R.M., Weston, A., Grover, P.L., *Carcinogenesis* (1983) **4**, 1639–1643.

D37. Stier, C.T., Jr., McKendall, G., Itskovitz, H.D., *J. Pharmacol. Exp. Ther.* (1984) **228**, 53–56.

D38. Bond, J.A., *Mutat. Res.* (1983) **124**, 315–324.

D39. Ryan, D.E., Iida, S., Wood, A.W., Thomas, P.E., Lieber, C.S., Levin, W., *J. Biol. Chem.* (1984) **259**, 1239–1250.

D40. Heimbrook, D.C., Murray, R.I., Egeberg, K.D., Sligar, S.G., Nee, M.W., Bruice, T.C., *J. Am. Chem. Soc.* (1984) **106**, 1514–1515.

D41. Reynolds, J.D., Kimbrough, T.D., Weekley, L.B., *Pflanzenphysiol.* (1983) **112**, 465–479.

D42. Lowery, C., Orton, T.C., Garner, R.C., *Toxicology* (1983) **29**, 131–141.

D43. Cambou, B., Guillochon, D., Thomas, D., *Enzyme Microbiol. Technol.* (1984) **6**, 11–17.

D44. Zayer, J., Kearney, P.C., *J. Agric. Food Chem.* (1984) **32**, 238–242.

D45. Jaaska, V., Jaaska, V., *Biochem, Physiol. Pflanz.* (1984) **179**, 21–30.

D46. Jarman, M., Foster, A.B., Goss, P.E., Griggs, L.J., Howe, I., Coombes, R.C., *Biomed. Mass Spectrom.* (1983) **10**, 620–625.

D47. Böcker, R., Estler, C.J., *Drugs Exp. Clin. Res.* (1983) **9**, 813–819.

D48. Loef, A., Gullstrand, E., Nordqvist, M.B., *Scand. J. Work Environ. Health* (1983) **9**, 419–430.

D49. Carlström, K., Von Uexkuell, A.-K., Einhorn, N., Fredricsson, B., Lunell, N.-O., Sundelin, P., *Acta Obstet. Gynecol. Scand.* (1983) **62**, 519–524.

D50. Lunte, S.M., Kissinger, P.T., *Chem.-Biol. Interact.* (1983) **47**, 195–212.

D51. Skelly, M.F., Shertzer, H.G., *Exp. Lung Res.* (1983) **5**, 259–268.

D52. Larson, J.D., Lamoureux, G.L., *J. Agric. Food Chem.* (1984) **32**, 177–182.

D53. Swallow, W.H., Pal, K., Phillips, D.H., Grover, P.L., Sims, P., *Chem.-Biol. Interact.* (1983) **47**, 347–361.

D54. Sarni, F., Grand, C., Boudet, A.M., *Eur. J. Biochem.* (1984) **139**, 259–265.

D55. Jeffery, D.R., Roth, J.A., *J. Neurochem.* (1984) **42**, 826–832.

D56. Haribabu, B., Kamath, A.V., Vaidyanathan, C.S., *FEMS Microbiol. Lett.* (1984) **21**, 197–200.

D57. Townsend, C.A., Christensen, S.B., *Tetrahedron* (1983) **39**, 3575–3582.

D58. Hösel, W., Schiel, O., *Arch. Biochem. Biophys.* (1984) **229**, 177–186.

D59. Reineke, W., Knackmuss, H.-J., *Appl. Environ. Microbiol.* (1984) **47**, 395–402.

D60. Della Corte, L., Sgaragli, G.P., *Pharmacol. Res. Commun.* (1984) **16**, 207–213.

D61. Gervasi, P.G., Benetti, D., Citti, L., Del Monte, M., Turchi, G., *Experientia* (1984) **40**, 180–182.

D62. Hjelle, J.J., Klaassen, C.D., *J. Pharmacol. Exp. Ther.* (1984) **228**, 407–413.

D63. Watanabe, K., Yoshizawa, I., *Steroids* (1983) **42**, 163–170.

D64. Knowles, R.G., Pogson, C.I., *J. Neurochem.* (1984) **42**, 677–684.

D65. Köster, G., Göde, E., Breuer, H., *J. Neurochem.* (1984) **42**, 788–797.

D66. Rush, G.F., Newton, J.F., Hook, J.B., *J. Pharmacol. Exp. Ther.* (1983) **227**, 658–662.

D67. Kitani, H., Tada, M., Morita, T., Koshida, Y., *Chem.-Biol. Interact.* (1983) **47**, 123–132.

D68. Coles, B., Srai, S.K.S., Waynforth, H.B., Ketterer, B., *Chem.-Biol. Interact.* (1983) **47**, 307–323.

D69. Wigaeus, E., Löf, A., Bjurström, R., Byfaelt Nordqvist, M., *Scand. J. Work, Environ. Health* (1983) **9**, 479–488.

D70. Shimizu, T., Kojima, M., *J. Biochem.* (1984) **95**, 205–212.

D71. Terao, T., Ohashi, H., Mizukami, H., *Plant Sci. Lett.* (1984) **33**, 47–52.

D72. McIlvain, J.E., Timoszyk, J., Nakatsugawa, T., *Pestic. Biochem. Physiol.* (1984) **21**, 162–169.

D73. Brown, S.A., Rivett, D.E.A., Thompson, H.J., *Z. Naturforsch. C: Biosci.* (1984) **39C**, 31–37.

D74. Williams, D.E., Buhler, D.R., *Cancer Res.* (1983) **43**, 4752–4756.

D75. Reid, R.A., Skiera, L.A., *Biochem. Soc. Trans.* (1983) **11**, 787.

D76. Rueffer, M., Nagakura, N., Zenk, M.H., *Planta Med.* (1983) **49**, 196–198.

D77. Rueffer, M., Nagakura, N., Zenk, M.H., *Planta Med.* (1983) **49**, 131–137.

D78. Holland, H.L., Carter, I.M., *Bioorg. Chem.* (1983) **12**, 1–7.

D79. Maeda, M., Ingbar, S.H., *Endocrinology (Baltimore)* (1984) **114**, 747–752.

D80. Nagashima, H., Tanaka, Y., Watanabe, H., Hayashi, R., Kawada, K., *Chem. Pharm. Bull.* (1984) **32**, 251–257.

D81. Baty, J.D., Willis, R.G., Koh, Y.K., *Anal. Proc. (Lond.)* (1984) **21**, 14–16.

D82. Dahlin, D.C., Miwa, G., Lu, A.Y.H., Nelson, S.D., *Proc. Natl. Acad. Sci. U.S.A.* (1984) **81**, 1327–1331.

D83. Gunnarsson, P.O., Andersson, S.B., Johansson, S.A., *Eur. J. Drug Metab. Pharmacokinet.* (1993) **8**, 395–402.

D84. Lee, J.J., Fasco, M.J., *Biochemistry* (1984) **23**, 2246–2252.

D85. Enriquez, P.M., Divincenzo, G.D., *Toxicology* (1984) **29**, 337–343.

D86. Whittaker, J.W., Lipscomb, J.D., Kent, T.A., Münck, E., *J. Biol. Chem.* (1984) **259**, 4466–4475.

D87. Tabata, M., Umetani, Y., Shima, K., Tanaka, S., *Plant Cell, Tissue Organ. Cult.* (1984) **3**, 3–9.

D88. Peters, J.H., Gordon, G.R., Kashiwase, D., Acton, E.M., *Cancer Res.* (1984) **44**, 1453–1459.

D89. Wang, T., Riden, D.M., Wolfenden, H.T., Wooley, R.L., Wood, A.J.J., Wilkinson, G.R., *J. Pharmacol. Exp.Ther.* (1984) **228**, 605–611.

D90. Hammel, W., Weiss, N., Kula, M.R., *Arch. Microbiol.* (1984) **137**, 47–52.

D91. Duncan, J.D., Hallström, G., Florence, V.M., Cho, A.K., Lindeke, B.G.A., *Acta Pharm. Suec.* (1983) **20**, 331–340.

D92. Mizuma, T., Yamaguchi, H., Hayashi, M., Awazu, S., *Chem. Pharm. Bull.* (1983) **31**, 4565–4567.

D93. Rosen, G.M., Rauckman, E.J., Ellington, S.P., Dahlin, D.C., Christie, J.L., Nelson, S.D., *Mol. Pharmacol.* (1984) **25**, 151–157.

D94. Meuldermans, W., Hurkmans, R., Swysen, E., Hendrickx, J., Thijssen, J., Lauwers, W., Heykants, J., *Eur. J. Drug Metab. Pharmacokinet.* (1983) **8**, 335–349.

D96. Mitchell, B.F., Cross, J., Hobkirk, R., Challis, J.R.G., *J. Clin. Endocrinol. Metab.* (1984) **58**, 845–849.

D97. Allan, R.J., Watson, T.R., *Eur. J. Drug Metab. Pharmacokinet.* (1983) **8**, 373–381.

D98. Larrey, D., Distlerath, L.M., Dannan, G.A., Wilkinson, G.R., Guengerich, F.P., *Biochemistry* (1984) **23**, 2787–2795.

D99. Suzuki, T., Sakoda, S., Ueji, M., Kishimoto, S., *Life Sci.* (1985) **36**, 435–442.

D100. Jellinck, P.H., Løvsted, J., Newcombe, A.M., *Biochem. Pharmacol.* (1983) **32**, 3247–3249.

D102. Monsarrat, B., Maftouh, M., Meunier, G., Dugue, B., Bernadou, J., Armand, J.P., Picard-Fraire, C., Meunier, B., Paoletti, C., *Biochem. Pharmacol.* (1983) **32**, 3887–3890.

D103. Tremaine, L.M., Diamond, G.L., Quebbemann, A.J., *Biochem. Pharmacol.* (1984) **33**, 419–427.

D104. Vágújfalvi, D., Petz-Stifter, M., *Phytochemistry* (1982) **21**, 1533–1536.

D105. Ribeiro, P., Webb, R.A., *Comp. Biochem. Physiol.* (1984) **79C**, 159–164.

D106. Vree, T.B., Vree, M.L., *J. Vet. Pharmacol. Ther.* (1984) **7**, 239–241.

D107. Cerniglia, C.E., Yang, S.K., *Appl. Environ. Microbiol.* (1984) **47**, 119–124.

D108. Grimmer, G., Jacob, J., Schmoldt, A., Raab, G., Mohr, U., Emura, M., *Polynucl. Aromat. Hydrocarbons: Form. Metab. Meas. Int. Symp. 7th* (1982) 521–532.

D109. Cerniglia, C.E., Miller, D.W., Yang, S.K., Freeman, J.P., *Appl. Environ. Microbiol.* (1984) **48**, 294–300.

D110. Jacob, J., Schmoldt, A., Grimmer, G., Hamann, M., Raab, G., *Polynucl. Aromat. Hydrocarbons: Form. Metab. Meas. Int. Symp. 7th* (1982) 627–641.

D111. Andrews, S.M., Griffiths, L.A., *Xenobiotica* (1984) **14**, 751–760.

D112. van der Merwe, P.J., Hundt, H.K.L., *Xenobiotica* (1984) **14**, 795–802.

D113. Rosen, G.M., Rauckman, E.J., Wilson, R.L., Jr., *Xenobiotica* (1984) **14**, 785–794.

D114. Strong, H.A., Renwick, A.G., George, C.F., *Xenobiotica* (1984) **14**, 815–826.

D115. Bell, J.A., Bradbury, A., Jenner, W.N., Manchee, G.R., Martin, L.E., *Biochem. Soc. Trans.* (1983) **11**, 715–716.

D116. Montay, G., Goueffon, Y., Roquet, F., *Antimicrob. Agents Chemother.* (1984) **25**, 463–472.

D117. Clark, A.M., Baker, J.K., McChesney, J.D., *J. Pharm. Sci.* (1984) **73**, 502–506.

D118. Sudo, K., Hashimoto, K., Kurata, T., Okazaki, O., Tsumura, M., Tachizawa, H., *Chemotherapy (Tokyo)* (1984) **32** (Suppl 1), 1203–1210.

D119. Sgaragli, G.P., Della Corte, L., *Psychopharmacol. Bull.* (1984) **20**, 165–167.

D120. Jasonni, M., Bulletti, C., Franceschetti, F., Bonavia, M., Ciotti, P., Bolelli, G.F., Armani, A., Flamigni, C., *Maturitas* (1984) **5**, 251–257.

D122. Hanzlik, R.P., Hogberg, K., Judson, C.M., *Biochemistry* (1984) **23**, 3048–3055.

D124. Kamaya, Y., Higuchi, T., *FEMS Microbiol. Lett.* (1984) **22**, 89–92.

D125. Tanaka, Y., Nishikawa, Y., Matsuda, K., Yamazaki, M., Hayashi, R., *Chem. Pharm. Bull.* (1984) **32**, 1040–1048.

D126. Sugihara, J., Sugawara, Y., Ando, H., Harigaya, S., Etoh, A., Kohno, K., *J. Pharmacobio.-Dyn.* (1984) **7**, 24–32.

D127. Shoemaker, D.D., Cysyk, R.L., Gormeley, P.E., DeSouza, J.J.V., Malspeis, L., *Cancer Res.* (1984) **44**, 1939–1945.

D128. Conway, J.G., Kauffman, F.C., Tsukada, T., Thurman, R.G., *Mol. Pharmacol.* (1984) **25**, 487–493.

D129. Greiner, B., Rommelspacher, H., *Naunyn-Schmiedeberg's Arch. Pharmacol.* (1984) **325**, 349–355.

D130. Rane, A., Säwe, J., Lindberg, B., Svensson, J.-O., Garle, M., Erwald, R., Jorulf, H., *J. Pharmacol. Exp. Ther.* (1984) **229**, 571–576.

D131. Baker, J.K., Bedford, J.A., Clark, A.M., *Pharm. Res.* (1984) 98–100.

D132. Klungsøyr, J., Scheline, R.R., *Acta Pharm. Suec.* (1984) **21**, 67–72.

D133. Levitz, M., Kadner, S., Young, B.K., *J. Steriod Biochem.* (1984) **20**, 971–974.

D134. Knogge, W., Weissenböck, G., *Eur. J. Biochem.* (1984) **140**, 113–118.

D135. Starke, D.W., Blisard, K.S., Mieyal, J.J., *Mol. Pharmacol.* (1984) **25**, 467–475.

D136. Fritzemeier, K.H., Kindl, H., Schlösser, E., *Z. Naturforsch C: Biosci.* (1984) **39C**, 217–221.

D137. Suzuki, T., *Nippon Nokayu Gakkaishi* (1983) **8**, 419–428.

D138. Mori, Y., Yokoya, F., Sakai, Y., Toyoshi, K., Baba, S., *Chem. Pharm. Bull.* (1984) **32**, 1106–1112.

D139. Yamamoto, I., Norimatsu, S., Shimonishi, T., Watanabe, K., Yoshimura, H., *J. Pharmacobio.-Dyn.* (1984) **7**, 254–262.

D140. Del Tito, B.J., Jr., Mukhtar, H., Bickers, D.R., *J. Invest. Dermatol.* (1984) **82**, 378–380.

D141. Chiba, K., Trevor, A., Castagnoli, N., Jr., *Biochem. Biophys. Res. Commun.* (1984) **120**, 574–578.

D142. Lotlikar, P.D., Pandey, R.N., Clearfield, M.S., Paik, S.M., *Toxicol. Lett.* (1984) **21**, 111–118.

D143. Kaminsky, L.S., Dannan, G.A., Guengerich, F.P., *Eur. J. Biochem.* (1984) **141**, 141–148.

D144. White, K.N., Hope, D.B., *Biochim. Biophys. Acta* (1984) **785**, 138–147.

D145. Amann, M., Nagakura, N., Zenk, M.H., *Tetrahedron Lett.* (1984) **25**, 953–954.

D147. Taylor, S.M., Stubley-Beedham, C., Stell, J.G.P., *Biochem. J.* (1984) **220**, 67–74.

D148. Speedie, M.K., Hartley, D.L., *J. Antibiot.* (1984) **37**, 159–166.

D149. Ohta, Y., Ishiguro, I., Naito, J., Shinohara, R., *Biochem. Int.* (1984) **8**, 617–627.

D150. Langenbeck-Schwich, B., Granbow, H.J., *Physiol. Plant.* (1984) **61**, 125–129.

D151. Liu, D., Thomson, K., Anderson, A.C., *Appl. Environ. Microbiol.* (1984) **47**, 1295–1298.

D152. Elmer, G.W., Remmel, R.P., *Xenobiotica* (1984) **14**, 829–840.

D153. Moran, E., O'Kennedy, R., Thornes, R.D., *Biochem. Soc. Trans.* (1984) **12**, 459.

D154. Lindstrom, T.D., Whitaker, T.D., Whitaker, G.W., *Xenobiotica* (1984) **14**, 841–847.

D155. Paterson, P., Fry, J.R., Horner, S.A., *Xenobiotica* (1984) **14**, 849–859.

D156. Yamamoto, T., Takano, R., Egashira, T., Yamanaka, Y., *Xenobiotica* (1984) **14**, 867–875.

D157. Yamada, H., Baba, T., Hirata, Y., Oguri, K., Yoshimura, H., *Xenobiotica* (1984) **14**, 861–866.

D158. Acheson, R.M., Nwankwo, J.O., *Xenobiotica* (1984) **14**, 877–883.

D159. Monge, P., Solheim, E., Scheline, R.R., *Xenobiotica* (1984) **14**, 917–924.
D160. Hutson, D.H., Lakeman, S.K., Logan, C.J., *Xenobiotica* (1984) **14**, 925–934.
D161. Kaye, B., Offerman, J.L., Reid, J.L., Elliott, H.L., Hillis, W.S., *Xenobiotica* (1984) **14**, 935–945.
D162. Dowell, P.S., Pierce, D.M., Franklin, R.A., Norris, R., Harries, H., Whitehead, H., *Xenobiotica* (1984) **14**, 947–953.
D163. Boux, L.J., Holder, G.M., *Xenobiotica* (1985) **15**, 11–20.
D164. Sakai, T., Niwaguchi, T., Kimura, R., Murata, T., *Xenobiotica* (1985) **15**, 31–40.
D165. Shimizu, T., Mori, H., Yasuda, Y., Morita, S., Nakagawa, K., *Xenobiotica* (1985) **15**, 21–29.
D166. Metzler, M., Haaf, H., *Xenobiotica* (1985) **15**, 41–49.
D167. Breyer-Pfaff, U., Kroker, M., Winkler, T., Kriemler, P., *Xenobiotica* (1985) **15**, 57–66.
D168. Hoshino, T., Fujiwara, A., *J. Antibiot.* (1983) **36**, 1463–1467.
D169. Oida, T., Yoshida, K., Kagemoto, A., Sekine, Y., Higashijima, T., *Xenobiotica* (1985) **15**, 87–96.
D170. Hoshino, T., Setoguchi, Y., Fujiwara, A., *J. Antibiot.* (1984) **37**, 1469–1472.
D171. Hoshino, T., Fujiwara, A., *J. Antibiot.* (1984) **37**, 1473–1474.
D172. Zini, G., Vicario, G.P., Lazzati, M., Arcamone, F., *Cancer Chemother. Pharmacol.* (1984) **12**, 154–156.
D173. Owen, J.A., Nakatsu, K., *Can. J. Physiol. Pharmacol.* (1984) **62**, 452–456.
D174. Dross, K., Sillmann, P., Küster, J., *Eur. J. Drug Metab. Pharmacokinet.* (1984) **9**, 1–9.
D175. Green, M.D., Fischer, L.J., *Toxicol. Appl. Pharmacol.* (1984) **74**, 125–133.
D176. Marshall, T.C., Dorough, H.W., *J. Agric. Food Chem.* (1984) **32**, 882–886.
D177. Shimada, K., Yumura, Y., Terashima, E., Nambara, T., *J. Steroid Biochem.* (1984) **20**, 1163–1167.
D178. Davison, K.L., Lamoureux, C.-J.H., Feil, V.J., *J. Agric. Food Chem.* (1984) **32**, 900–908.
D179. Grand, C., *FEBS Lett.* (1984) **169**, 7–11.
D180. Zimmermann, T., Gasser, F., Kulla, H.G., Leisinger, T., *Arch. Microbiol.* (1984) **138**, 37–43.
D181. Kuwahara, M., Glenn, J.K., Morgan, M.A., Gold, M.H., *FEBS Lett.* (1984) **169**, 247–250.
D182. Mitra, D., Vaidyanathan, C.S., *Biochem. Int.* (1984) **8**, 609–615.
D183. Hartley, D.L., Speedie, M.K., *Biochem. J.* (1984) **220**, 309–313.
D184. Sin, F.Y.T., Brunet, P.C.J., Sin, I.L., *Comp. Biochem. Physiol. B: Comp. Biochem.* (1984) **77B**, 791–798.
D185. Sewell, G.J., Soper, C.J., Parfitt, R.T., *Appl. Microbiol. Biotechnol.* (1984) **19**, 247–251.
D186. Meinhold, H., Buchholz, R., *Nuklearmedizin, Suppl. (Stuttgart)* (1984) **20**, 786–789.
D187. Racz, K., Buu, N.T., Kuchel, O., *Can. J. Physiol. Pharmacol.* (1984) **62**, 622–626.
D188. Koepke, S.R., Tondear, Y., Farrelly, J.G., Stewart, M.L., Michejda, C.J., Kroeger-Koepke, M.B., *Biochem. Pharmacol.* (1984) **33**, 1509–1513.
D189. Wang, X., Xie, J., Song, Z., *Yaoxue Xuebao* (1984) **19**, 90–95.
D190. Murakami, N., Uchida, M., Yabutani, K., Okada, M., Aizawa, H., *Nippon Noyaku Gakkaishi* (1983) **8**, 483–491.
D191. Uchida, M., Ogawa, K., Sugimoto, T., Aizawa, H., *Nippon Noyaku Gakkaishi* (1983) **8**, 537–544.
D192. Kiessling, K.H., Pettersson, H., Sandholm, K., Olsen, M., *Appl. Environ. Microbiol.* (1984) **47**, 1070–1073.
D193. Isobe, N., Matsuo, M., Miyamoto, J., *Nippon Noyaku Gakkaishi* (1983) **8**, 551–559.

D194. King, L.C., Kohan, M.J., Ball, L.M., Lewtas, J., *Cancer Lett.* (*Shannon, Irel.*) (1984) **22**, 255–262.

D195. Hossaini, M.A., Manouchehri, A.V., Zareh, Z., *J. Radioanal. Nucl. Chem.* (1984) **83**, 371–377.

D196. Chia, Y.C., Smith, G.W., Lees, G.J., *Life Sci.* (1984) **34**, 2443–2452.

D197. Yamanaka, K., Minoshima, R., *Agric. Biol. Chem.* (1984) **48**, 1161–1171.

D198. Jonsson, L.M.V., Aarsman, M.E.G., van Diepen, J., de Vlaming, P., Smit, N., Schram, A.W., *Planta* (1984) **160**, 341–347.

D199. Masana, M., De Toranzo, E.G.D., Castro, J.A., *Dev. Toxicol. Environ. Sci.* (1983) **11**, 383–386.

D200. Hege, H.G., Hollmann, M., Kaumeier, S., Lietz, H., *Eur. J. Drug Metab. Pharmacokinet.* (1984) **9**, 41–55.

D201. Helia, O., Cizmarik, J., *Pharmazie* (1984) **39**, 189.

D202. Sturm, E., Zech, K., *Biomed. Mass. Spectrom.* (1984) **11**, 211–216.

D203. Tanabe, H., Kageyama, H., Kubo, J., Naruchi, T., *Iyakuhin Kenkyu* (1984) **15**, 467–476.

D204. Kvasničková, E., Nobilis, M., Hais, I.M., *J. Chromatogr.* (1984) **295**, 201–209.

D205. Hoskins, J.A., Holliday, S.B., Greenway, A.M., *Biomed. Mass. Spectrom.* (1984) **11**, 296–300.

D206. Matsuo, M., Mihara, K., Okuno, M., Ohkawa, H., Miyamoto, J., *Food Chem. Toxicol.* (1984) **22**, 345–354.

D207. Raha, C., Hines, R., Bresnick, E., *J. Chromatogr.* (1984) **291**, 231–239.

D208. Angelino, S.A.G.F., van Valkengold, B.H., Buurman, D.J., van der Plas, H.C., Müller, F., *J. Heterocycl. Chem.* (1984) **21**, 107–112.

D209. Yang, S.K., Fu, P.P., *Chem.-Biol. Interact.* (1984) **49**, 71–88.

D210. Wagner, C., Eckardt, K., Schumann, G., Ihn, W., Tresselt, D., *J. Antibiot.* (1984) **37**, 691–692.

D211. Collins, J., Hegeman, G., *Arch. Microbiol.* (1984) **138**, 153–160.

D212. Schwartz, H.S., Paul, B., *Cancer Res.* (1984) **44**, 2480–2484.

D213. Rolfs, C.H., Kindl, H., *Plant Physiol.* (1984) **75**, 489–492.

D214. Kojima, S., Honda, T., Babasaki, T., Kiyozumi, M., Nakagawa, M., *Eisei Kagaku* (1984) **30**, 91–95.

D215. Isobe, N., Matsuo, M., Miyamoto, J., *Nippon Noyaku Gakkaishi* (1983) **8**, 451–465.

D216. Tazaki, K., Ishikura, N., *Plant Cell Physiol.* (1984) **25**, 731–737.

D217. Narai, N., Kitamura, S., Tatsumi, K., *J. Pharmacobio.-Dyn.* (1984) **7**, 407–413.

D218. Kitamura, S., Tatsumi, K., *Biochem. Biophys. Res. Commun.* (1984) **121**, 749–754.

D219. Tahara, S., Nakahara, S., Mizutani, J., Ingham, J.L., *Agric. Biol. Chem.* (1984) **48**, 1471–1477.

D220. Bull, D.L., Ivie, G.W., Beier, R.P., Pryor, N.W., Oertli, E.H., *J. Chem. Ecol.* (1984) **10**, 893–911.

D221. Sandberg, G., *Planta* (1984) **161**, 398–403.

D222. Langston, J.W., Irwin, I., Langston, E.B., Forno, L.S., *Neurosci. Lett.* (1984) **48**, 87–92.

D223. Engström, K.M., *Scand. J. Work, Environ. Health* (1984) **10**, 83–87.

D224. Staiger, L.E., Quistad, G.B., *J. Agric. Food Chem.* (1984) **32**, 1130–1133.

D225. Jerina, D.M., van Bladeren, P.J., Yagi, H., Gibson, D.T., Mahadevan, V., Neese, A.S., Koreeda, M., Sharma, N.D., Boyd, D.R., *J. Org. Chem.* (1984) **49**, 3621–3628.

D226. Loveland, P.M., Nixon, J.E., Bailey, G.S., *Comp. Biochem. Physiol. C, Comp. Pharmacol. Toxicol.* (1984) **78C**, 13–19.

D227. McMurtrey, K.D., Knight, T.J., *Mutat. Res.* (1984) **140**, 7–11.

D228. Saito, K., Kamataki, T., Kato, R., *Cancer Res.* (1984) **44**, 3169–3173.

D229. Moore, B.P., Potter, P.M., Hicks, R.M., *Carcinogenesis* (1984) **5**, 949–954.

D230. Suzuki, T., Sasaki, M., Katagiri, K., *Nagoya Med. J.* (1984) **28**, 157–169.

D231. Schott, H.-H., Zinke, K., Langenbach-Schmidt, D., *Prog. Tryptophan Serotonin Res., Proc. Meet. Int. Study Group Tryptophan Res. ISTRY, 4th* (1983) 773–776.

D232. Sze, I.S.-Y., Dagley, S., *J. Bacteriol* (1984) **159**, 353–359.

D233. Young, W.F., Jr., Okazaki, H., Laws, E.R., Jr., Weinshilboum, R.M., *J. Neurochem.* (1984) **43**, 706–715.

D234. Kobayashi, S., Sugioka, K., Nakano, H., Nakano, M., Tero-Kubota, S., *Biochemistry* (1984) **23**, 4589–4597.

D235. Boehlert, C.C., Armstrong, R.N., *Biochem. Biophys. Res. Commun.* (1984) **121**, 980–986.

D236. Baggi, G., Catelani, D., Sorlini, C., Treccani, V., *Ann. Microbiol.* (1982) **32**, 37–45.

D237. Sieben, R, Philippi, U., Eich, E., *J. Nat. Prod.* (1984) **47**, 433–438.

D240. Paeme, G., Grimee, R., Vercruysse, A., *Eur. J. Drug Metab. Pharmacokinet.* (1984) **9**, 103–108.

D241. Sakamoto, M., Kobayashi, S., Nakano, M., Negishi, T., *Sci. Total Environ.* (1984) **37**, 247–251.

D242. Fennell, T.R., Miller, J.A., Miller, E.C., *Cancer Res.* (1984) **44**, 3231–3240.

D243. Reed, G.A., Grafstrom, R.C., Krauss, R.S., Autrup, H., Eling, T.E., *Carcinogenesis* (1984) **5**, 955–960.

D244. Tawata, S., *Ryukyu Daigaku Nogakubu Gakujutsu Hokoku* (1983) 217–226.

D245. Carlberg, M., Jergil, B., Lindbladh, C., Rosengren, E., *Gen. Pharmacol.* (1984) **15**, 301–307.

D246. Koyama, H., *J. Biochem.* (1984) **96**, 421–427.

D247. Aoki, K., Konohana, T., Shinke, R., Nishira, H., *Agric. Biol. Chem.* (1984) **48**, 2087–2095.

D248. Schocken, M.J., Gibson, D.T., *Appl. Environ. Microbiol.* (1984) **48**, 10–16.

D249. Marchiset, D., Aubert, C., Sumitapura, Y.C., Egre, A., Cano, J.P., *Eur. J. Drug Metab. Pharmacokinet.* (1984) **9**, 123–128.

D250. Doi, T., Tsukamoto, T., Torii, H., Yoshida, K., Tanayama, S., *Takeda Kenkyushoho* (1984) **43**, 19–32.

D252. Motohashi, M., Yoshida, K., Torii, H., Tsukamoto, T., Tanayama, S., *Takeda Kenkyushoho* (1984) **43**, 33–38.

D253. Alken, R.G., Belz, G.G., Haegele, K.D., Meinicke, T., Schechter, P.J., *Clin. Pharmacol. Ther.* (*St. Louis*) (1984) **36**, 209–216.

D254. Watanabe, K., Kimura, R., Yoshizawa, I., *Chem. Pharm. Bull.* (1984) **32**, 2745–2751.

D255. O'Neill. P.J., Yorgey, K.A., Renzi, N.L., Jr., Williams, R.L., Benet, L.Z., *J. Clin. Pharmacol.* (1982) **22**, 470–476.

D256. Suzuki, H., Nakane, H., Kawamura, M., Yoshizawa, M., Takeshita, E., Saruta, T., *Am. J. Physiol.* (1984) **247**, E285–E290.

D257. Elliott, B.M., *Carcinogenesis* (1984) **5**, 1051–1055.

D258. Machiex, J.J., Ibrahim, R.K., *Biochem. Physiol. Pflanz.* (1984) **179**, 659–664.

D259. Krell, H.W., Sanderman, H., Jr., *Eur. J. Biochem.* (1984) **143**, 57–62.

D260. Clark, J.E., Ljungdahl, L.G., *J. Biol. Chem.* (1984) **259**, 10845–10849.

D261. van Bladeren, P.J., Vyas, K.P., Sayer, J.M., Ryan, D.E., Thomas, P.E., Levin, W., Jerina, D.M., *J. Biol. Chem.* (1984) **259**, 8966–8973.

D262. Youde, I.R., Raxworthy, M.J., Gulliver, P.A., Dijkstra, D., Horn, A.S., *J. Pharm. Pharmacol.* (1984) **36**, 309–313.

D263. Appel, M., Raabe, T., Lingens, F., *FEMS Microbiol Lett.* (1984) **24**, 123–126.

D264. Levy, R.H., Lin, H.S., Blehaut, H.M., Tor, J.A., *J. Clin. Pharmacol.* (1983) **23**, 523–533.

D266. Correia, M.A., Krowech, G., Caldera-Munoz, P., Yee, S.L., Straub, K., Castagnoli, N., Jr., *Chem.-Biol. Interact.* (1984) **51**, 13–24.

D267. Fischer, V., Mason, R.P., *J. Biol. Chem.* (1984) **259**, 10284–10288.

D268. Bodor, N., Visor, G., *Exp. Eye Res.* (1984) **38**, 621–626.

D269. Goldstein, J.A., Weaver, R., Sundheimer, D.W., *Cancer Res.* (1984) **44**, 3768–3771.

D271. Bond, J.A., Mauderly, J.L., *Cancer Res.* (1984) **44**, 3924–3929.

D272. Inada, J., Okuno, E., Kimura, M., Kido, R., *Int. J. Biochem.* (1984) **16**, 623–628.

D273. Aoki, K., Shinke, R., Nishira, H., *Agric. Biol. Chem.* (1984) **48**, 2309–2313.

D274. Jackson, D.E., Dewick, P.M., *Phytochemistry* (1984) **23**, 1037–1042.

D275. Jackson, D.E., Dewick, P.M., *Phytochemistry* (1984) **23**, 1029–1035.

D276. Tomoda, A., Shirasawa, E., Nagao, S., Minami, M., Yoneyama, Y., *Biochem. J.* (1984) **222**, 755–760.

D277. Uebelhack, R., Franke, L., Kutter, D., Thoma, J., Seidel, K., *Biomed. Biochim. Acta* (1984) **43**, 1033–1035.

D278. Cerniglia, C.E., Howard, P.C., Fu, P.P., Franklin, W., *Biochem. Biophys. Res. Commun.* (1984) **123**, 262–270.

D279. Cerniglia, C.E., Freeman, J.P., Althaus, J.R., Van Baalen, C., *Drug Chem. Toxicol.* (1984) **1**, 381–394.

D281. El-Bayoumy, K., Hecht, S.S., *Cancer Res.* (1984) **44**, 4317–4322.

D282. Voisin, P., Namboordiri, M.A.A., Klein, D.C., *J. Biol. Chem.* (1984) **259**, 10913–10918.

D283. Ferraiolo, B.L., Onady, G.M., Mieyal, J.J., *Biochemistry* (1984) **23**, 5528–5534.

D284. Zulalian, J., Stout, S.J., Babcock, C.N., Lucas, L.M., Miller, P., Orloski, E.J., *J. Agric. Food Chem.* (1984) **32**, 1434–1440.

D285. Goromaru, T., Matsuura, H., Furuta, T., Baba, S., *Chem. Pharm. Bull.* (1984) **32**, 3179–3186.

D286. Plumier, E., Wappes, B., Wiegrebe, W., *Arch. Dermatol. Res.* (1984) **276**, 406–408.

D287. Alvan, G., Grind, M., Graffner, C., Sjöqvist, F., *Clin. Pharmacol. Ther.* (*St. Louis*) (1984) **36**, 515–519.

D288. Lilienblum, W., Bock, K.W., *Biochem. Pharmacol.* (1984) **33**, 2041–2046.

D289. Yoneda, S., Alexander, N., Vlachakis, N.D., *Biochem. Pharmacol.* (1984) **33**, 2029–2032.

D290. Logan, C.J., Cottee, F.H., Page, J.A., *Biochem. Pharmacol.* (1984) **33**, 2345–2346.

D291. Powis, G., Svingen, B.A., Dahlin, D.C., Nelson, S.D., *Biochem. Pharmacol.* (1984) **33**, 2367–2370.

D292. Sinha, B.K., *Biochem. Pharmacol.* (1984) **33**, 2777–2781.

D293. Watabe, T., Ozawa, N., Hiratsuka, A., Saito, Y., Tsurumori, T., *Biochem. Pharmacol.* (1984) **33**, 2687–2690.

D294. Park, D.H., Joh, T.H., *Biochem. Pharmacol.* (1984) **33**, 2911–2913.

D295. Watabe, T., Hiratsuka, A., Sone, T., Ishihama, T., Endoh, K., *Biochem. Pharmacol.* (1984) **33**, 3101–3103.

D296. Naito, J., Sasaki, E., Ohta, Y., Shinohara, R., Ishiguro, I., *Biochem. Pharmacol.* (1984) **33**, 3195–3200.

D297. Nissinen, E., *Biochem. Pharmacol.* (1984) **33**, 3105–3108.

D298. Docherty, M.d'A., Cohen, G.M., *Biochem. Pharmacol.* (1984) **33**, 3201–3208.

D299. Schnellmann, R.G., Volp, R.F., Putnam, C.W., Sipes, I.G., *Biochem. Pharmacol.* (1984) **33**, 3503–3509.

D300. Ogbunde, P.O.J., Gati, W.P., Paterson, A.R.P., *Biochem. Pharmacol.* (1984) **33**, 3561–3566.

D301. Eacho, P.I., O'Donnell, J.P., Colby, H.D., *Biochem. Pharmacol.* (1984) **33**, 3627–3632.

D302. Jimenez, M., Garcia-Canovas, F., Garcia-Carmona, F., Lozano, J.A.,

Iborra, J.L., *Biochem. Pharmacol.* (1984) **33**, 3689–3697.

D303. Jensen, M., Baars, A.J., Breimer, D.D., *Biochem. Pharmacol.* (1984) **33**, 3655–3659.

D304. Watabe, T., Hiratsuka, A., Tsurumori, T., *Biochem. Pharmacol.* (1984) **33**, 4051–4056.

D305. Nishimoto, M., Varanasi, U., *Biochem. Pharmacol.* (1985) **34**, 263–268.

D306. Blée, E., Casida, J.E., Durst, F., *Biochem. Pharmacol.* (1985) **34**, 389–390.

D307. Ross, D., Larsson, R., Andersson, B., Nilsson, U., Lindqvist, T., Lindeke, T., Moldeus, P., *Biochem. Pharmacol.* (1985) **34**, 343–351.

D308. Caccia, S., Fong, M.H., Garattini, S., Notarnicola, A., *Biochem. Pharmacol.* (1985) **34**, 393–394.

D309. Maftouh, M., Amiar, Y., Picard-Fraire, C., *Biochem. Pharmacol.* (1985) **34**, 427–428.

D310. Smith, P.B.W., Caldwell, J., Smith, R.L., Horner, M.W., Houghton, E., Moss, M.S., *Biochem. Pharmacol.* (1985) **34**, 459–460.

D311. Schmeiser, H.H., Pool, B.L., Wiessler, M., *Biochem. Pharmacol.* (1985) **34**, 455–456.

D312. Sutton, J.D., Sangster, S.A., Caldwell, J., *Biochem. Pharmacol.* (1985) **34**, 465–466.

D313. Hukkelhoven, M.W.A.C., van Pelt, F.N.A.M., Dijkstra, A.C., Vermorken, A.J.M., *Biochem. Pharmacol.* (1985) **34**, 703–705.

D314. Yang, C.Y., Zukowski, K., Lee, M.-S., *Biochem. Pharmacol.* (1985) **34**, 837–841.

D315. Mackenzie, P.I., Joffe, M.M., Munson, P.J., Owens, I.S., *Biochem. Pharmacol.* (1985) **34**, 737–746.

D316. Lilienblum, W., *Biochem. Pharmacol.* (1985) **34**, 893–894.

D317. Lind, C., *Biochem. Pharmacol.* (1985) **34**, 895–897.

D318. Dyck, L.E., Durden, D.A., Boulton, A.A., *Biochem. Pharmacol.* (1985) **34**, 1925–1929.

D319. Schut, H.A.J., Dixit, R., Loeb, T.R., Stoner, G.D., *Biochem. Pharmacol.* (1985) **34**, 969–976.

D320. Weli, A.M., Lindeke, B., *Biochem. Pharmacol.* (1985) **34**, 1993–1998.

D321. Wilson, I.D., Macdonald, C.M., Fromson, J.M., Troke, J.A., Hillbeck, D., *Biochem. Pharmacol.* (1985) **34**, 2025–2028.

D322. Chang, W.-M., Lin, J.-K., Wu, K.-C., Hsiung, K.-P., *Biochem. Pharmacol.* (1985) **34**, 2566–2569.

D323. Singh, P.D.A., Jackson, J.R., James, S.P., *Biochem. Pharmacol.* (1985) **34**, 2207–2209.

D324. Reuter, I.L., Dose, K., Guenther, T., Wörner, W., Oesch, F., *Biochem. Pharmacol.* (1985) **34**, 2617–2620.

D325. Chiueh, C.C., Ohata, M., Jonas, L.A., Greenblatt, D.J., Rapoport, S.I., *Drug Metab. Dispos.* (1985) **13**, 1–4.

D326. Lyman, S.D., Jordan, V.C., *Biochem. Pharmacol.* (1985) **34**, 2787–2794.

D327. Roy, S.D., McKay, G., Hawes, E.M., Midha, K.K., *J. Chromatogr.* (1984) **310**, 307–317.

D328. Iguchi, S., Tanaka, T., Kaneo, Y., Goromaru, T., *J. Pharmacobio.-Dyn.* (1984) **7**, 702–705.

D331. Thakker, D.R., Yagi, H., Sayer, J.M., Kapur, U., Levin, W., Chang, R.L., Wood, A.W., Conney, A.H., Jerina, D.M., *J. Biol. Chem.* (1984) **259**, 11249–11256.

D332. Clark, A.G., Shamaan, N.A., *Pestic. Biochem. Physiol.* (1984) **22**, 249–261.

D333. El-Bayoumy, K., Reddy, B., Hecht, S.S., *Carcinogenesis* (1984) **5**, 1371–1373.

D334. Pundir, C.S., Garg, G.K., Rathore, V.S., *Phytochemistry* (1984) **23**, 2423–2427.

D335. Kamaya, Y., Higuchi, T., *FEMS Microbiol. Lett.* (1984) **24**, 225–229.

D336. Zampaglione, N., Hilbert, J.M., Ning, J., Chung, M., Gural, R., Symchowicz, S., *Drug Metab. Dispos.* (1985) **13**, 25–29.

D337. Yost, G.S., Findley, B.L., *Drug Metab. Dispos.* (1985) **13**, 5–8.

D338. La Voie, E.J., Adams, E.A., Shigematsu, A., Hoffmann, D., *Drug Metab. Dispos.* (1985) **13**, 71–75.

D339. Oravec, C.T., Samuel, M.J., D'Ambrosio, S.M., *Drug Metab. Dispos.* (1985) **13**, 76–80.

D340. Hultin, T.A., Weber, W.W., *Drug Metab. Dispos.* (1985) **13**, 148–150.

D341. Fischer, L.J., Green, M.D., Harman, A.W., *Drug Metab. Dispos.* (1985) **13**, 121–126.

D342. Jurima, M., Inaba, T., Kalow, W., *Drug Metab. Dispos.* (1985) **13**, 151–155.

D343. Walle, T., Walle, U.K., Olanoff, L.S., *Drug Metab. Dispos.* (1985) **13**, 204–209.

D344. Koster, A.S., Frankhuijzen-Sierevogel, A.C., Noordhoek, J., *Drug Metab. Dispos.* (1985) **13**, 232–238.

D345. Harvey, D.J., Leuschner, J.T.A., *Drug Metab. Dispos.* (1985) **13**, 215–219.

D346. Mori, Y., Kuroda, N., Sakai, Y., Yokoya, F., Toyoshi, K., Baba, S., *Drug Metab. Dispos.* (1985) **13**, 239–245.

D347. Suzuki, T., Hashimura, Y., Takeyama, S., *Drug Metab. Dispos.* (1985) **13**, 246–254.

D348. Davi, H., Carayon, A., Marti, E., Sales, Y., Cautreels, W., *Xenobiotica* (1985) **15**, 97–106.

D349. Streeter, A.J., Timbrell, J.A., *Drug Metab. Dispos.* (1985) **13**, 255–259.

D350. West, P.R., Harman, L.S., Josephy, P.D., Mason, R.P., *Biochem. Pharmacol.* (1984) **33**, 2933–2936.

D351. Cravedi, J.-P., Heuillet, G., Peleran, J.-C., Wal, J.-M., *Xenobiotica* (1985) **15**, 115–121.

D352. Abbott, F.S., Slatter, J.G., Burton, R., Kong, G.I., *Xenobiotica* (1985) **15**, 129–140.

D353. Alexander, D.M., Mathew, G.E.A., Wilson, B.J., *Xenobiotica* (1985) **15**, 123–128.

D354. Mairizis, J.C., Madelmont, J.C., Parry, D., Meyniel, G., Buisson, J.P., Royer, R., *Xenobiotica* (1985) **15**, 141–148.

D355. Fischbach, T., Lenk, W., *Xenobiotica* (1985) **15**, 149–164.

D356. Larsen, G.L., *Xenobiotica* (1985) **15**, 199–209.

D357. Harvey, D.J., *Xenobiotica* (1985) **15**, 187–197.

D358. Gad-El Karim, M.M., Sadugopa Ramanujam, V.M., Legator, M.S., *Xenobiotica* (1985) **15**, 211–220.

D359. Narimatsu, S., Yamamoto, I., Yoshimura, H., *Xenobiotica* (1985) **15**, 227–235.

D360. Schmid, P., Schlatter, C., *Xenobiotica* (1985) **15**, 251–256.

D361. Kaye, B., Rance, D.J., Waring, L., *Xenobiotica* (1985) **15**, 237–242.

D362. Mancinelli, A., Guiso, G., Garattini, S., Urso, R., Caccia, S., *Xenobiotica* (1985) **15**, 257–265.

D363. Beyer, J., Frank, G., *Xenobiotica* (1985) **15**, 277–280.

D364. Pontani, R.B., Vadlamani, N.L., Misra, A.L., *Xenobiotica* (1985) **15**, 287–297.

D365. Engström, K., Elovaara, E., Aitio, A., *Xenobiotica* (1985) **15**, 281–286.

D366. Papadopoulos, A.S., Crammer, J.L., Cowan, D.A., *Xenobiotica* (1985) **15**, 309–316.

D367. Logan, C.J., Hutson, D.H., Hesk, D., *Xenobiotica* (1985) **15**, 391–397.

D368. Bieder, A., Decouvelaere, B., Gaillard, C., Gaillot, J., Depaire, H., Raynaud, L., Snozzi, C., *Xenobiotica* (1985) **15**, 421–435.

D369. Kamimura, H., Sasaki, H., Kawamura, S., *Xenobiotica* (1985) **15**, 413–420.

D370. Smith, D.A., Johnson, P.M., Wilkinson, D.J., *Xenobiotica* (1985) **15**, 437–444.

D371. Haley, C.S., Waring, R.H., Mitchell, S.C., Shah, R.R., Idle, J.R., Smith, R.L., *Xenobiotica* (1985) **15**, 445–450.

D372. Lindstrom, T.D., Lacefield, W.B., Whitaker, G.W., *Drug Metab. Dispos.* (1984) **12**, 691–697.

D373. Wall, M.E., Perez-Reyes, M., Brine, D.R., Cook, C.E., *Drug Metab. Dispos.* (1984) **12**, 677–682.

D374. Babiek, L.M., Cherry, W.F., Fayz, S., Pang, K.S., *Drug Metab. Dispos.* (1984) **12**, 698–704.

D375. Nomeir, A.A., Abou-Donia, M.B., *Drug Metab. Dispos.* (1984) **12**, 705–711.

D376. Bai, S.A., Walle, T., *Drug Metab. Dispos.* (1984) **12**, 749–754.

D377. Albro, P.W., Chae, K., Philpot, R., Corbett, J.T., Schroeder, J., Jordan, S., *Drug Metab. Dispos.* (1984) **12**, 742–748.

D378. Mori, Y., Sakai, Y., Kuroda, N., Yokoya, F., Toyoshi, K., Horie, M., Baba, S., *Drug Metab. Dispos.* (1984) **12**, 767–771.

D379. Meuldermans, W., Hendrickx, J., Lauwers, W., Swysen, E., Hurkmans, R., Knaeps, F., Woestenborghs, R., Heykants, J., *Drug Metab. Dispos.* (1984) **12**, 772–781.

D380. Streeter, A.J., Bjorge, S.M., Axworthy, D.B., Nelson, S.D., Baillie, T.A., *Drug Metab. Dispos.* (1984) **12**, 565–576.

D382. Cone, E.J., Gorodetzky, C.W., Yousefnejad, D., Buchwald, W.F., Johnson, R.E., *Drug Metab. Dispos.* (1984) **12**, 577–581.

D383. Chism, J.P., Turner, M.J., Jr., Rickert, D.E., *Drug Metab. Dispos.* (1984) **12**, 596–602.

D384. Assandri, A., Perazzi, A., Fontanella, L., Ferrari, P., Ripamonti, A., Tarzia, G., Tuan, G., Martinelli, E., *Drug Metab. Dispos.* (1984) **12**, 635–640.

D385. Bekersky, I., Popick, I., Colburn, W.A., *Drug Metab. Dispos.* (1984) **12**, 607–613.

D386. Marten, T.R., Bourne, G.R., Miles, G.S., Shuker, B., Rankine, H.D., Dutka, V.N., *Drug Metab. Dispos.* (1984) **12**, 652–660.

D387. Yagen, B., Foureman, G.L., Ben-Zvi, Z., Ryan, A.J., Henandez, O., Cox, R.H., Bend, J.R., *Drug Metab. Dispos.* (1984) **12**, 389–395.

D388. McQuinn, R.L., Quarforth, G.J., Johnson, J.D., Banitt, E.H., Pathre, S.V., Chang, S.F., Ober, R.E., Conard, G.J., *Drug Metab. Dispos.* (1984) **12**, 414–420.

D389. Yang, S.K., Chou, M.W., Evans, F.E., Fu, P.P., *Drug Metab. Dispos.* (1984) **12**, 403–413.

D390. Chen, C., Lefers, R.C., Brough, E.L., Gurka, D.P., *Drug Metab. Dispos.* (1984) **12**, 421–426.

D391. Monks, T.J., Lau, S.S., Highet, R.J., *Drug Metab. Dispos.* (1984) **12**, 432–437.

D392. Hilbert, J., Pramanik, B., Symchowicz, S., Zampaglione, N., *Drug Metab. Dispos.* (1984) **12**, 452–459.

D393. Borondy, P.E., Michniewicz, B.M., *Drug Metab. Dispos.* (1984) **12**, 444–451.

D401. Anderson, M.M., Mays, J.B., Mitchum, R.K., Hinson, J.A., *Drug Metab. Dispos.* (1984) **12**, 179–184.

D402. Monks, T.J., Lau, S.S., Pohl, L.R., Gillette, J.R., *Drug Metab. Dispos.* (1984) **12**, 193–198.

D410. Waring, R.H., Mitchell, S.C., *Xenobiotica* (1985) **15**, 459–468.

D411. Mårdh, G., Luehr, C.A., Vallee, B.L., *Proc. Natl. Acad. Sci. U.S.A.* (1985) **82**, 4979–4982.

D412. Kobari, T., Namekawa, H., Kato, Y., Yamada, S., *Xenobiotica* (1985) **15**, 469–476.

D413. Paterson, P., Fry, J.R., *Xenobiotica* (1985) **15**, 493–502.

D414. Walle, T., Christ, D.D., Walle, U.K., Wilson, M.J., *J. Chromatogr.* (1985) **341**, 213–216.

D415. Kulkarni, A.P., Strohm, B.H., Houser, W.H., *Xenobiotica* (1985) **15**, 513–519.

D416. Imhof, D.A., Logan, C.J., Dodds, P.F., *Biochem. Pharmacol.* (1985) **34**, 3009–3010.

D417. Isobe, N., Matsuo, M., Miyamoto, J., *J. Pesticide Sci.* (1984) **9**, 105–115.

D418. Doherty, M.D'A., Cohen, G.M., Gant, T.W., Naish, S., Riley, P.A., *Biochem. Pharmacol.* (1985) **34**, 3167–3172.

D419. Nagamatsu, K., Terao, T., Toki, S., *Biochem. Pharmacol.* (1985) **34**, 3143–3146.

D420. Moss, E.J., Neal, G.E., *Biochem. Pharmacol.* (1985) **34**, 3193–3197.

D421. Burke, M.D., Thompson, S., Elcombe, C.R., Halpert, J., Haaparanta, T., Mayer, R.T., *Biochem. Pharmacol.* (1985) **34**, 3337–3345.

D422. Bhaird, N.N.A., Fowler, C.J., Thorberg, O., Tipton, K.F., *Biochem. Pharmacol.* (1985) **34**, 3599–3601.

D423. Koster, A.S., Frankhuijzen-Sierevogel, A.C., Noordhoek, J., *Biochem. Pharmacol.* (1985) **34**, 3527–3532.

D424. Shelton, R.C., Karoum, F., Chuang, L.-W., Wyatt, R.J., *Biochem. Pharmacol.* (1985) **34**, 3623–3626.

D425. Poussen, H.K., Lerche, A., Skovgaard, L.T., *Biochem. Pharmacol.* (1985) **34**, 3729–3733.

D426. Servin, A.L., Jacquot, C., Lussiana, J.P., Viel, C., *Xenobiotica* (1985) **15**, 553–566.

D427. Rickert, D.E., de Bethizy, J.D., Glover, M.R., Kedderis, G.L., *Biochem. Pharmacol.* (1985) **34**, 4163–4168.

D428. Le Fur, J.M., Labaune, J.P., *Xenobiotica* (1985) **15**, 567–577.

D429. Dostert, P., Strolin Benedetti, M., Rovei, V., Smith, R.L., *Xenobiotica* (1985) **15**, 799–803.

D430. Wright, D.J., Robinson, H.K., Holder, G.M., Ryan, A.J., Bonin, A.M., Baker, R.S.U., *Xenobiotica* (1985) **15**, 825–834.

D431. Trennery, P.N., Waring, R.H., *Xenobiotica* (1985) **15**, 813–823.

D432. Davis, P.J., Guenthner, L.E., *Xenobiotica* (1985) **15**, 845–857.

D433. Subrahmanyam, V.V., O'Brien, P.J., *Xenobiotica* (1985) **15**, 873–885.

D434. Cerniglia, C.E., Freeman, J.P., Evans, F.E., *Arch. Microbiol.* (1984) **138**, 283–286.

D435. Mansuy, D., Dansette, P.M., Foures, C., Jaouen, M., *Biochem. Pharmacol.* (1984) **33**, 1429–1435.

D436. Baker, G.B., Coutts, R.T., Nazarali, A.J., Danielson, T.J., Rubens, M., *Proc. West. Pharmacol. Soc.* (1984) **27**, 523–525.

D437. Kloss, M.W., Rosen, G.M., Rauckman, E.J., *Psychopharmacology* (*Berlin*) (1984) **84**, 221–224.

D438. Narimatsu, S., Watanabe, K., Shimonishi, T., Matsubara, K., Yamamoto, I., Yoshimura, H., *J. Pharmacobio.-Dyn.* (1984) **7**, 671–676.

D439. Hege, H.G., Lietz, H., Weymann, J., *Arzneim.-Forsch.* (1984) **34**, 972–979.

D440. Sasaki, K., Suzuki, T., Takeda, M., Uchiyama, M., *Bull. Econ. Environ. Contam. Toxicol.* (1984) **33**, 281–288.

D441. Toeda, K., Miyamoto, T., Kato, S., Yamamoto, I., *Nippon Noyaku Gakkaishi* (1984) **9**, 305–315.

D442. Christou, M., Wilson, N.M., Jefcoate, C.R., *Carcinogenesis* (1984) **5**, 1239–1247.

D443. Kaneko, H., Izumi, T., Matsuo, M., Miyamoto, J., *Nippon Noyaku Gakkaishi* (1984) **9**, 269–274.

D444. Kari, F.W., Kauffman, F.C., Thurman, R.G., *Cancer Res.* (1984) **44**, 5073–5078.

D445. Köster, J., Bussmann, R., Barz, W., *Arch. Biochem. Biophys.* (1984) **234**, 513–521.

D446. Sabourin, P.J., Hodgson, E., *Chem.-Biol. Interact.* (1984) **51**, 125–139.

D447. Allaben, W.T., King, C.M., *J. Biol. Chem.* (1984) **259**, 12128–12134.

D448. Boyd, J.A., Eling, T.E., *J. Biol. Chem.* (1984) **259**, 13885–13896.

D449. Bild, G.S., Morris, J.C., *Arch. Biochem. Biophys.* (1984) **235**, 41–47.

D450. McClanahan, R.H., Robertson, L.W., *J. Nat. Prod.* (1984) **47**, 828–834.

D451. Brahme, N.M., Gonzalez, J.E., Rolls, J.P., Hessler, E.J., Mizsak, S., Hurley, L.H., *J. Am. Chem. Soc.* (1984) **106**, 7873–7878.

D452. Rontani, J.F., Mattei, G., Bertrand, J.C., Giusti, G., *Biotechnol. Lett.* (1984) **6**, 441–444.

D453. Eriksson, K.E., Gupta, J.K., Nishida, A., Rao, M., *J. Gen. Microbiol.* (1984) **130**, 2457–2464.

D454. Roitner, M., Schalkhammer, T., Pittner, F., *Appl. Biochem. Biotechnol.* (1984) **9**, 483–488.

D455. Stafford, H.A., Lester, H.H., *Plant Physiol.* (1984) **76**, 184–186.

D456. Halldin, M.M., Widman, M., Agurell, S., Hollister, L.E., Kanter, S.L., *Cannabinoids: Chem. Pharmacol. Ther. Aspects* (1982) 211–218.

D457. Weenen, H., Van Maanen, J.M.S., De Planque, M.M., McVie, J.G., Pinedo, H.M., *Eur. J. Cancer Clin. Oncol.* (1984) **20**, 919–926.

D458. Clements, J.A., Critchley, J.A.J.H., Prescott, L.F., *Br. J. Clin. Pharmacol.* (1984) **18**, 481–485.

D459. Hassell, T.M., Maguire, J.H., Cooper, C.G., Johnson, P.T., *Epilepsia (N.Y.)* (1984) **25**, 556–563.

D460. Mizuma, T., Araya, H., Hayashi, M., Awazu, S., *J. Pharmacobio.-Dyn.* (1984) **7**, 784–789.

D461. Kahl, S., Bitman, J., Rumsey, T.S., *Domest. Anim. Endocrinol.* (1984) **1**, 279–290.

D462. Divincenzo, G.D., Hamilton, M.L., Reynolds, R.C., Ziegler, D.A., *Toxicology* (1984) **33**, 9–18.

D463. Hafsah, Z., Tahara, S., Mizutani, J., *Nippon Noyaku Gakkaishi* (1984) **9**, 117–123.

D464. Foerster, R.E., Würzler, F.E., *Arch. Toxicol.* (1984) **56**, 12–17.

D465. Swain, L., Melius, P., *Comp. Biochem. Physiol. C: Comp. Pharmacol. Toxicol.* (1984) **79C**, 151–158.

D466. Feutry, A., Letouze, R., *Phytochemistry* (1984) **23**, 1557–1559.

D467. Gold, M.H., Kuwahara, M., Chiu, A.A., Glenn, J.K., *Arch. Biochem. Biophys.* (1984) **234**, 353–362.

D468. Hagemann, M., Grisebach, H., *FEBS Lett.* (1984) **175**, 199–202.

D469. McIntire, W., Hopper, D.J., Craig, J.C., Everhart, E.T., Webster, R.V., Causer, M.J., Singer, T.P., *Biochem. J.* (1984) **224**, 617–621.

D470. Beecher, C.W.W., Kelleher, W.J., *Tetrahedron Lett.* (1984) **25**, 4595–4598.

D471. Stotz, G., Spribille, R., Forkmann, G., *J. Plant Physiol.* (1984) **116**, 173–183.

D472. Kumari, A.a., Mahadevan, A., *Z. Pflanzenkrankh. Pflanzenschutz* (1984) **91**, 383–390.

D473. Müller, R., Thiele, J., Klages, U., Lingens, F., *Biochem. Biophys. Res. Commun.* (1984) **124**, 178–182.

D474. Schliefer, D., Komossa, D., Barz, W., *Z. Naturforsch. C: Biosci.* (1984) **39C**, 882–887.

D475. Durham, D.R., *J. Bacteriol.* (1984) **160**, 778–780.

D476. Della Corte, L., Giovannini, M.G., Sgaragli, G.P., *Arch. Toxicol., Suppl.* (1984) **7**, 307–310.

D477. Hammond, A.C., Slyter, L.L., Carlson, J.R., Wong, L.P., Breeze, R.G., *Am. J. Vet. Res.* (1984) **45**, 2247–2250.

D478. Matsubara, K., Kagawa, M., Fukui, Y., *Forensic Sci. Int.* (1984) **26**, 169–180.

D479. Dromgoole, S.H., Furst, D.E., Paulus, H.E., *J. Pharm. Sci.* (1984) **73**, 1657–1659.

D480. Fluck, D.S., Rappaport, S.M., Eastmond, D.A., Smith, M.T., *Arch. Biochem. Biophys.* (1984) **235**, 351–358.

D481. El-Bayoumy, K., Hecht, S.S., Sackl, T., Stoner, G.D., *Carcinogenesis* (1984) **5**, 1449–1452.

D482. Engström, K., Riihinmaaki, V., Laine, A., *Int. Arch. Occup. Environ. Health* (1984) **54**, 355–363.

D483. Yang, S.K., Fu, P.P., *Biochem. J.* (1984) **223**, 775–782.

D484. Raabe, T., Appel, M., Lingens, F., *FEMS Microbiol. Lett.* (1984) **25**, 61–64.

D485. Pons, J.L., Rimbault, A., Darbord, J.C., Leluan, G., *Ann. Microbiol. (Paris)* (1984) **135B**, 219–222.

D486. Tsang, C.P.W., Grunder, A.A., *Steroids* (1984) **43**, 71–84.

D487. Vakeria, D., Vivian, A., Fewson, C.A., *J. Gen. Microbiol.* (1984) **130**, 2893–2903.

D488. Para, G., Rifai, S., Baratti, J., *Biotechnol. Lett.* (1984) **6**, 703–708.

D489. Faed, E.M., Dobbs, B.R., Lee, D., *Arch. Int. Pharmacodyn. Ther.* (1984) **272**, 4–16.

D490. Ross, D., Albano, E., Nilsson, U., Moldeus, P., *Biochem. Biophys. Res. Commun.* (1984) **125**, 109–115.

D491. Baumann, J., v. Bruchhausen, F., Wurm, G., *Arzneim.-Forsch.* (1984) **34**, 1278–1282.

D492. Hilbert, J.M., Gural, R.P., Symchowicz, S., Zampaglione, N., J., *Clin. Pharmacol.* (1984) **24**, 457–462.

D493. Wefers, H., Komai, T., Sies, H., *Biochem. Pharmacol.* (1984) **33**, 4081–4085.

D495. Letelier, M.E., Sanchez, E., Del Villar, E., *Gen. Pharmacol.* (1984) **15**, 403–406.

D496. Quistad, G.B., Mulholland, K.M., Skiles, G., Jamieson, G.C., *J. Agric. Food Chem.* (1985) **33**, 95–98.

D497. Chiu, P.-L., Mushtaq, M., Weems, H.B., Yang, S.K., *Biochem. Biophys. Res. Commun.* (1984) **124**, 114–120.

D498. Subramanian, V., Vaidyanathan, C.S., *J. Bacteriol.* (1985) **160**, 651–655.

D499. Belvedere, G., Elovaara, E., Vainio, H., *Toxicol. Lett.* (1984) **23**, 157–162.

D500. Sugiyama, Y., Stolz, A., Sugimoto, M., Kuhlenkamp, J., Yamada, T., Kaplowitz, N., *Biochem. J.* (1984) **224**, 947–953.

D501. Kaschke, V., Haufler, U., Zöllner, R., *Hoppe-Seyler's Z. Physiol. Chem.* (1984) **365**, 1435–1443.

D502. Milewich, L., Garcia, R.L., Gerrity, L.W., *J. Steroid Biochem.* (1984) **21**, 529–538.

D503. Andreoni, V., Galli, E., Galliani, G., *Syst. Appl. Microbiol.* (1984) **5**, 299–304.

D504. Baker, J.K., Little, T.L., *J. Med. Chem.* (1985) **28**, 46–50.

D506. Yamada, J., Horie, S., Watanabe, T., Suga, T., *Biochem. Biophys. Res. Commun.* (1984) **125**, 123–128.

D507. Grosa, G., Viola, F., Caputo, O., Biglino, G., *Pharm. Res.* (1984) 274–276.

D508. Young, I.M., Leone, R.M., Francis, P., Stovell, P., Silman, R.E., *J. Clin. Endocrinol. Metab.* (1985) **60**, 114–119.

D509. Johannessen, J.N., Chiueh, C.C., Burns, R.S., Markey, S.P., *Life Sci.* (1985) **36**, 219–224.

D510. Ohmori, S., Motohashi, K., Kitada, M., Kanakubo Y., Igarashi, T., Ueno, K., Kitagawa, H., *Biochem. Biophys. Res. Commun.* (1984) **125**, 1089–1095.

D511. Chou, M.W., Fu, P.P., *J. Toxicol. Environ. Health* (1984) **14**, 211–223.

D512. Prantl, F., Strasser, A., Aebi, M., Furter, R., Niederberger P., Kirschner, K., Huetter, R., *Eur. J. Biochem.* (1985) **146**, 95–100.

D513. Jezequel, S.G., Kaye, B., Higgins, I.J., *Biotech. Lett.* (1984) **6**, 567–570.

D514. Li, S.A., Klicka, J.K., Li, J.J., *Cancer Res.* (1985) **45**, 181–185.

D515. Banerji, A., Chintalwar, G.J., *Phytochemistry* (1984) **23**, 1605–1606.

D516. Schmidt, E., Knackmuss, H.J., *Appl. Microbiol. Biotechnol.* (1984) **20**, 351–355.

D517. Megalla, S.E., Mohran, M.A., *Mycopathologia* (1984) **88**, 27–29.

D518. Matsuzawa, Y., Nakase, T., *J. Pharmacobio.-Dyn.* (1984) **7**, 776–783.

D519. Castelli, M.G., Benfenati, E., Pastorelli, R., Salmona, M., Fanelli, R., *Food Chem. Toxicol.* (1984) **22**, 901–904.

D520. Mattioli, M., Gaiani, R., Bono, G., Chiesa, F., *Arch. Vet. Ital.* (1984) **35**, 22–26.

D521. Lo, M.W., Pond, S.M., Effeney, D.J., Silber, B.M., Riegelman, S., Tozer, T.M., *J. Pharmacokinet. Biopharm.* (1984) **12**, 401–412.

D523. Manis, M.O., Braselton, W.E., Jr., *Fundam. Appl. Toxicol.* (1984) **4**, 1000–1008.

D524. Ishimitsu, S., Fujimoto, S., Ohara, A., *Chem. Pharm. Bull.* (1984) **32**, 4645–4649.

D525. Shibata, D., Doi, E., *Plant Cell Physiol.* (1984) **25**, 1411–1419.

D526. Yamada, Y., Okada, N., *Phytochemistry* (1985) **24**, 63–65.

D527. Walle, T., Walle, U.K., Wilson, M.J., Fagan, T.C., Gaffney, T.E., *Br. J. Clin. Pharmacol.* (1984) **18**, 741–748.

D528. Shibasaki, J., Konishi, R., Takemura, M., Fukushima, E., Nakamura, J., Sasaki, H., *J. Pharmacobio.-Dyn.* (1984) **7**, 804–810.

D529. Churchill, F.C., Patchen, L.C., Campbell, C.C., Schwartz, I.K., Phuc Nguyen Dinh, Dickinson, C.M., *Life Sci.* (1985) **36**, 53–62.

D530. Jimenez, M., Garcia-Carmona, F., Garcia-Canovas, F., Iborra, J.L., Lozano, J.A., Martinez, F., *Arch. Biochem. Biophys.* (1984) **235**, 438–448.

D531. Visser, T.J., Otten, N.H., Mol, J.A., Docter, R., Hennemann, G., *Horm. Metab. Res., Suppl. Ser.* (1984) **14**, 35–41.

D532. McCarthy, D.J., Waud, W.R., Struck, R.F., Hill, D.L., *Cancer Res.* (1985) **45**, 174–180.

D534. Melancon, M.J., Lech, J.J., *Comp. Biochem. Physiol. C: Comp. Pharmacol. Toxicol.* (1984) **79C**, 331–336.

D535. Neilson, A.H., Allard, A.S., Reiland, S., Remberger, M., Taernholm, A., Viktor, T., Landner, L., *Can. J. Fish Aquat. Sci.* (1984) **41**, 1502–1512.

D536. Rueffer, M., Zenk, M.H., *Tetrahedron Lett.* (1985) **26**, 201–202.

D537. Appel, K.E., Ruehl, C.S., Spieghalder, B., Hildebrandt, A.G., *Toxicol. Lett.* (1984) **23**, 353–358.

D538. Pujar, B.G., Ribbons, D.W., *Arch. Environ. Microbiol.* (1985) **49**, 374–376.

D539. Nakazawa, H., Kumagai, H., Yamada, H., *Agric. Biol. Chem.* (1985) **49**, 159–165.

D540. Chen, W., Ohmiya, K., Shimizu, S., Kawakami, H., *Appl. Environ. Microbiol.* (1985) **49**, 211–216.

D541. Herbert, R.B., Mann, J., *J. Chem. Soc., Chem. Commun.* (1984) 1474–1475.

D542. Amin, A.R., Vyas, P., Modi, V.V., *Indian. J. Exp. Biol.* (1984) **22**, 220–221.

D543. Nonhebel, H.M., Bandurski, R.S., *Plant Physiol.* (1984) **76**, 979–983.

D544. Marks, T.S., Smith, A.R.W., Quirk, A.V., *Appl. Environ. Microbiol.* (1984) **48**, 1020–1025.

D545. Nonhebel, H.M., Hillman, J.R., Crozier, A., Wilkins, M.B., *J. Exp. Bot.* (1985) **36**, 99–109.

D546. Yeh, S.Y., *J. Pharm. Sci.* (1984) **73**, 1783–1787.

D547. Dieterle, W., Faigle, J.W., Kueng, W., Theobald, W., *Biopharm. Drug Res.* (1984) **5**, 377–386.

D548. Tomori, E., Elekes, I., Lang, T., Horvath, G., *Pol. J. Pharmacol. Pharm.* (1984) **36**, 423–430.

D549. Hattori, M., Sakamoto, T., Endo, Y., Kakiuchi, N., Kobashi, K., Mizano, T., Namba, T., *Chem. Pharm. Bull.* (1984) **32**, 5010–5017.

D550. Eichelbaum, M., Tomson, T., Tybring, G., Bertilsson, L., *Clin. Pharmacokinet.* (1985) **10**, 80–90.

D551. Caccia, S., Conti, I., Notarnicola, A., *J. Pharm. Pharmacol.* (1985) **37**, 75–77.

D552. van Ommen, B., van Bladeren, P.J., Temmink, J.H.M., Müller, F., *Biochem. Biophys. Res. Commun.* (1985) **126**, 25–32.

D553. Kubota, K., Uchimura, H., Mitsuhashi, T., Chiu, S.C., Kuzuya, N., Ito, K., Nagataki, S., *Life Sci.* (1985) **36**, 1033–1039.

D554. Fujii, T., Kaneda, T., *Eur. J. Biochem.* (1985) **147**, 97–104.

D555. Buki, K.G., Dang Quang Vinh, Horvath, I., *Acta Microb. Hung.* (1985) **32**, 65–73.

D557. Guillochon, D., Esclade, L., Cambou, B., Thomas, D., *Ann. N. Y. Acad. Sci.* (1984) **434**, 214–218.

D558. Negrel, J., Martin, C., *Phytochemistry* (1984) **23**, 2797–2801.

D559. Pelkonen, O., Sotaniemi, E.A., Ahokas, J.J., *Br. J. Clin. Pharmacol.* (1985) **19**, 59–66.

D560. Jung, B., Graf, H., Ullrich, U., *Biol. Chem. Hoppe-Seyler* (1985) **366**, 23–31.

D561. Nelson, W.L., Bartels, M.J., *J. Pharm. Sci.* (1985) **74**, 33–36.

D562. Nishiyama, M., Abe, M., Arisaka, T., Amaki, M., Ujiie, A., Naito, J., *Iyakuhin Kenkyu* (1985) **16**, 108–122.

D563. Dankovic, D., Billings, R.E., Seifert. W., Stillwell, W.G., *Mol. Pharmacol.* (1985) **27**, 287–295.

D564. Di Spirito, A.A., Taaffe, L.R., Lipscomb, J.D., Hooper, A.B., *Biochim. Biophys. Acta* (1985) **827**, 320–326.

D565. Mikami, N., Wakabayashi, N., Yamada, H., Miyamoto, J., *Pestic. Sci.* (1985) **16**, 46–58.

D566. Canonica, L., Nali, M., Rindone, B., Tollari, S., Marchesini, A., *Gazz. Chim. Ital.* (1985) **115**, 1–5.

D567. Eling, T.E., Mason, R.P., Sivarajah, K., *J. Biol. Chem.* (1985) **260**, 1601–1607.

D568. Williamson, J., Van Orden, D., Rosazza, J.P., *Appl. Environ. Microbiol.* (1985) **49**, 563–567.

D569. Hatakka, A.I., *Arch. Microbiol.* (1985) **141**, 22–28.

D570. van Hogezand, R.A., van Hees, P.A.M., Zwanenburg, B., van Rossum, J.M., van Tongeren, J.H.M., *Gastroenterology* (1985) **88**, 717–722.

D571. Howard, P.C., Flammang, T.J., Beland, F.A., *Carcinogenesis* (1985) **6**, 243–249.

D572. Brochmann-Hanssen, E., *Planta Med.* (1984) **50**, 343–345.

D573. Larsson, R., Ross, D., Nordenskjoeld, M., Lindeke, B., Olsson, L.I., Moldeus, P., *Chem.-Biol. Interact.* (1984) **52**, 1–14.

D574. Kawada, T., Iwai, K., *Agric. Biol. Chem.* (1985) **49**, 441–448.

D575. Heller, W., Britsch, L., Forkmann, G., Grisebach, H., *Planta* (1985) **163**, 191–196.

D577. Prost, O., Nicollier, M., Laurent, R., Adessi, G.L., *Arch. Dermatol. Res.* (1985) **277**, 195–200.

D578. Galliani, G., Rindone, B., Dagnino, G., Salmona, M., *Eur. J. Drug Metab. Pharmacokinet.* (1984) **9**, 289–293.

D579. Borchers, F., Achtert, G., Hausleiter, H.J., Zeugner, H., *Eur. J. Drug Metab. Pharmacokinet.* (1984) **9**, 325–346.

D580. Camara, B., D'Harlingue, A., *Plant Cell Rep.* (1985) **4**, 31–32.

D581. Mümmler, S., Rüffer, M., Nagakura, N., Zenk, M.H., *Plant Cell Rep.* (1985) **4**, 36–39.

D582. Münz, M., Von Deimling, O., *Electrophoresis* (1985) **6**, 175–178.

D583. Whittemore, R.M., Pearce, L.B., Roth, J.A., *Biochemistry* (1985) **24**, 2477–2482.

D584. De Luca, V., Ibrahim, R.K., *Arch. Biochem. Biophys.* (1985) **238**, 606–618.

D585. Oatis, J.E., Jr., Walle, T., Daniell, H.B., Gaffney, T.E., Knapp, D.R., *J. Med. Chem.* (1985) **28**, 822–824.

D586. Wong, K.P., Yeo, T., Rauff, A., *Biog. Amines* (1985) **2**, 73–81.

D587. Armstrong, K.E., Wattenberg, L.W., *Cancer Res.* (1985) **45**, 1507–1510.

D588. Hodgson, R.M., Seidel, A., Bochnitschek, W., Glatt, H.R., Oesch, F., Grover, P.L., *Carcinogenesis* (1985) **16**, 135–139.

D589. De Luca, V., Ibrahim, R.K., *Arch. Biochem. Biophys.* (1985) **238**, 596–605.

D590. Seed, J.R., Hall, J.E., Sechelski, J., *J. Protozool.* (1985) **32**, 20–25.

D591. Miyoshi-Koshio, T., Suzuki, K., *Kenkyu Kiyo-Tokyo Kasei Daigaku* (1983) **23**, 7–14.

D592. Lee, C.W., Lucas, S., Desmazeaud, M.J., *FEMS Microbiol. Lett.* (1985) **26**, 201–205.

D593. Sahasrabudhe, S.R., Modi, V.V., *Biochem. Int.* (1985) **10**, 525–529.

D594. de Herder, W.W., Hazenberg, M.P., Otten, M.H., Pennock-Schröder, A.M., Visser, T.J., *FEMS Microbiol. Lett.* (1985) **27**, 79–83.

D595. Aoki, K., Uemori, T., Shinke, R., Nishira, H., *Agric. Biol. Chem.* (1985) **49**, 1151–1158.
D596. Beerhues, L., Wiermann, R., *Z Naturforsch. C: Biosci.* (1985) **40C**, 160–165.
D597. Sandberg, G., Haellgren, J.E., *Plant Cell. Rep.* (1985) **4**, 100–103.
D598. Sato, M., Hiraoka, A., *Chem. Pharm. Bull.* (1985) **33**, 1289–1292.
D599. Verdeil, J.L., Bister-Miel, F., Agier, C., Plat, M., Guignard, J.L., Viel, C., *C. R. Acad. Sci. Ser. 3* (1985) **300**, 305–308.
D600. Wang, C.Y., Zukowski, K., Lee, M.-S., *Biochem. Pharmacol.* (1985) **34**, 837–841.
D601. Mallett, A.K., King, L.J., Walker, R., *Biochem. Pharmacol.* (1985) **34**, 337–342.
D602. Decolin, M., Walther, B., Villoutreix, J., Batt, A.M., Siest, G., *Biochem. Pharmacol.* (1985) **34**, 401–402.
D603. Mesnil, M., Testa, B., Jenner, P., *Biochem. Pharmacol.* (1985) **34**, 435–436.
D604. Boobis, A.R., Murray, S., Hampden, C.E., Davies, D.S., *Biochem. Pharmacol.* (1985) **34**, 65–71.
D605. Gollmer, L., Graf, H., Ullrich, V., *Biochem. Pharmacol.* (1984) **33**, 3597–3602.
D606. Yu, P.H., Bailey, B.A., Durden, D.A., Boulton, A.A., *Biochem. Pharmacol.* (1986) **35**, 1027–1036.
D607. Barańczyk-Kuźma, A., Szymczyk, T., *Biochem. Pharmacol.* (1986) **35**, 995–999.
D608. McCague, R., Seago, A., *Biochem. Pharmacol.* (1986) **35**, 827–834.
D609. Guenthner, T.M., *Biochem. Pharmacol.* (1986) **35**, 839–845.
D610. Okuda, H., Hiratsuka, A., Nojima, H., Watabe, T., *Biochem. Pharmacol.* (1986) **35**, 535–538.
D611. Jocelyn, P.C., Cronshaw, A.D., *Biochem. Pharmacol.* (1986) **35**, 631–635.
D612. Waschek, J.A., Fielding, R.M., Tozer, T.N., Effeney, D.J., Pond, S.M., *Biochem. Pharmacol.* (1986) **35**, 544–546.
D613. Bonham Carter S.M., Glover, V., Sandler, M., *Biochem. Pharmacol.* (1986) **35**, 1205–1206.
D614. Yinon, J., Hwang, D.-G., *Toxicol. Lett.* (1985) **26**, 205–209.
D615. Steffens, P., Nagakura, N., Zenk, M.H., *Tetrahedron Lett.* (1984) **25**, 951–952.
D616. Shibasaki, J., Inoue, Y., Kadosaki, K., Sasaki, H., Nakamura, J., *J. Pharmacobio.-Dyn.* (1985) **8**, 989–995.
D617. Stafford, H.A., Lester, H.H., *Plant Physiol.* (1985) **78**, 791–794.
D618. Reinecke, D.M., Bandurski, R.S., *Biochem. Biophys. Res. Commun.* (1981) **103**, 429–433.
D619. Nonhebel, H.M., Kruse, L.I., Bandurski, R.S., *J. Biol. Chem.* (1985) **260**, 12685–12689.
D620. Mcnamee, C.G., Durham, D.R., *Biochem. Biophys. Res. Commun.* (1985) **129**, 485–492.
D621. Marks, T.S., Wait, R., Smith, A.R.W., Quirk, A.V., *Biochem. Biophys. Res. Commun.* (1984) **124**, 669–674.
D623. Lee, C.-W., Desmazeaud, M.J., *FEMS Microbiol. Lett.* (1986) **33**, 95–98.
D624. Cerniglia, C.E., Freeman, J.P., White, G.L., Helfich, R.H., Miller, D.W., *Appl. Environ. Microbiol.* (1985) **50**, 649–655.
D625. Hattori, M., Endo, Y., Takebe, S., Kobashi, K., Fukasaku, N., Namba, T., *Chem. Pharm. Bull.* (1986) **34**, 158–167.
D626. Warrander, A., Allen, J.M., Andrews, R.S., *Xenobiotica* (1985) **15**, 891–897.
D627. Heitkamp, M.A., Freeman, J.P., McMillan, D.C., Cerniglia, C.E., *Appl. Environ. Microbiol.* (1985) **50**, 265–273.
D628. Elmhirst, T.R.D., Chipman, J.K., Ribeiro, O., Hirom, P.C., Millburn, P., *Xenobiotica* (1985) **15**, 899–906.
D629. Hackett, A.M., Griffiths, L.A., Wermeille, M., *Xenobiotica* (1985) **15**, 907–914.

D630. Veronese, M.E., McLean, S., D'Souza, C.A., Davies, N.W., *Xenobiotica* (1985) **15**, 929–940.

D631. Fischbach, T., Lenk, W., *Xenobiotica* (1985) **15**, 915–927.

D632. Bässmann, H., Böttcher, J., Schüppel, R., Wray, V., *Xenobiotica* (1985) **15**, 941–952.

D633. Rovei, V., Strolin-Benedetti, M., Daniel, C., Moulin, A., Donath, A., *Xenobiotica* (1985) **15**, 953–964.

D634. Meredith, P.A., Kelman, A.W., McSherry, D.R., Vincent, J., Reid, J.L., *Xenobiotica* (1985) **15**, 979–985.

D635. Midgley, I., Chasseaud, L.F., Taylor, I.W., Walmsley, L.M., Fowkes, A.G., Darragh, A., Lambe, R.F., Bonn, R., *Xenobiotica* (1985) **15**, 965–977.

D636. Vignier, V., Berthou, F., Dreano, Y., Floch, H.H., *Xenobiotica* (1985) **15**, 991–999.

D637. Gorrod, J.W., Gooderham, N.J., *Xenobiotica* (1985) **15**, 1021–1031.

D638. Maduagwu, E.N., *Xenobiotica* (1985) **15**, 1061–1068.

D639. Morino, A., Nakamura, A., Nakanishi, K., Tatewaki, N., Sugiyama, M., *Xenobiotica* (1985) **15**, 1033–1043.

D640. Cascia, S., Fossati, T., Mancinelli, A., *Xenobiotica* (1985) **15**, 1111–1119.

D641. Farina, P., Torti, L., Urso, R., Horton, J.K., Gescher, A., D'Incalci, M., *Biochem. Pharmacol.* (1986) **35**, 209–215.

D642. Eckert, K.-G., Eyer, P., *Biochem. Pharmacol.* (1986) **35**, 325–329.

D643. Patel, B.N., Mackness, M.I., Nwosu, V., Connock, M.J., *Biochem. Pharmacol.* (1986) **35**, 231–235.

D644. Fabre, G., Seither, R., Goldman, I.D., *Biochem. Pharmacol.* (1986) **35**, 1325–1330.

D645. Assandri, A., Galliani, G., Zerilli, L., Tuan, G., Tarzia, G., Barone, D., *Biochem. Pharmacol.* (1986) **35**, 1459–1467.

D646. Crooks, P.A., Godin, C.S., Nwosu, C.G., Ansher, S.S., Jakoby, W.B., *Biochem. Pharmacol.* (1986) **35**, 1600–1603.

D647. Wolf, C.R., Macpherson, J.S., Smyth, J.F., *Biochem. Pharmacol.* (1986) **35**, 1577–1581.

D648. Kuo, G.Y., Hwang, B.Y.-H., Staiger, D.B., *Biochem. Pharmacol.* (1986) **35**, 1613–1615.

D649. Matthews, H.B., Chopade, H.M., Smith, R.W., Burka, L.T., *Xenobiotica* (1986) **16**, 1–10.

D650. Skjevrak, I., Solheim, E., Scheline, R.R., *Xenobiotica* (1986) **16**, 35–45.

D651. Machinist, J.M., Bopp, B.A., Anderson, D.J., Granneman, G.R., Sonders, R.C., Tolman, K., Buchi, K., Rollins, D., *Xenobiotica* (1986) **16**, 11–20.

D652. Raxworthy, M.J., Youde, I.R., Gulliver, P.A., *Xenobiotica* (1986) **16**, 47–52.

D653. Hutson, D.H., Logan, C.J., *Xenobiotica* (1986) **16**, 87–93.

D654. Groenewoud, G., Hundt, H.K.L., *Xenobiotica* (1986) **16**, 99–107.

D655. Ranganathan, V., Miki, K., Gold, M.H., *Arch. Biochem. Biophys.* (1985) **241**, 304–314.

D656. Campbell, N.R.C., Sundaram, R.S., Werness, P.G., Van Loon, J., Weinshilboum, R.M., *Clin. Pharmacol. Ther.* (*St. Louis*) (1985) **37**, 308–315.

D657. Hümpel, M., Krause, W., Hoyer, G.-A., Wendt, H., Pommerenke, G., *Eur. J. Drug Metab. Pharmacokinet.* (1984) **9**, 347–357.

D658. Inoue, T., Yasuda, T., Suzuki, S., Kishi, T., Niwaguchi, T., *Xenobiotica* (1986) **16**, 109–121.

D659. Pacifici, G.M., Bencini, C., Rane, A., *Xenobiotica* (1986) **16**, 123–128.

D660. Yamada, H., Oguri, K., Yoshimura, H., *Xenobiotica* (1986) **16**, 137–141.

D661. Yoshida, T., Yamada, Y., Yamamoto, T., Kuroiwa, Y., *Xenobiotica* (1986) **16**, 129–136.

D662. Ferdinandi, E.S., Sehgal, S.N., Demerson, C.A., Dubuc, J., Zilber, J.,

Dvornik, D., Cayen, M.N., *Xenobiotica* (1986) **16**, 153–166.

D663. Rumyantseva, G.V., Sushkov, D.G., Weiner, L.M., *Xenobiotica* (1986) **16**, 167–175.

D664. Dieterle, W., Faigle, J.W., Küng, W., Theobald, W., *Xenobiotica* (1986) **16**, 181–191.

D665. Emudianughe, T.S., Oduleye, S.O., Ebadan, E.E., Eneji, S.D., *Xenobiotica* (1986) **16**, 177–179.

D666. Boobis, A.R., Whyte, C., Davis, D.S., *Xenobiotica* (1986) **16**, 233–238.

D667. Baker, M.H., Foster, A.B., Leclercq, F., Jarman, M., Rowlands, M.G., Turner, J.C., *Xenobiotica* (1986) **16**, 195–203.

D668. Hutt, A.J., Caldwell, J., Smith, R.L., *Xenobiotica* (1986) **16**, 239–249.

D669. Weli, A.M., Lindeke, B., *Xenobiotica* (1986) **16**, 281–288.

D670. Cantineau, R., Kremers, P., De Graeve, J., Gielen, J.E., Lambotte, R., *J. Steroid Biochem.* (1985) **22**, 195–201.

D671. Mikami, N., Yoshimura, J., Kaneko, H., Yamada, H., Miyamoto, J., *Pestic. Sci.* (1985) **16**, 33–45.

D672. Chen, Y.P., Glenn, A.R., Dilworth, M.J., *Arch. Microbiol.* (1985) **141**, 225–228.

D673. Whittaker, M., Sokolove, P.M., Thurman, R.G., Kauffman, F.C., *Cancer Lett.* (*Shannon, Irel.*) (1985) **26**, 145–152.

D674. Allison, N., O'Donnell, M.J., Hoey, M.E., Fewson, C.A., *Biochem. J.* (1985) **227**, 753–757.

D675. Hoff, T., Liu, S.-Y., Bollag, J.-M., *Appl. Environ. Microbiol.* (1985) **49**, 1040–1045.

D676. Pirl, J.N., Bell, C.L., Lu, M.C., Bauer, L., *Arzneim.-Forsch.* (1985) **35**, 578–584.

D677. Morgan, J.G., McGettigan, S., Donlon, J., *Biochem. Soc. Trans.* (1985) **13**, 464–465.

D678. Köppel, C., Tenczer, J., *Arzneim.-Forsch.* (1985) **35**, 634–635.

D679. Johnson, J.A., Benziger, D.P., *Antimicrob. Agents Chemother.* (1985) **27**, 774–781.

D680. Namkung, M.J., Porubek, D.J., Nelson, S.D., Juchau, M.R., *J. Steroid Biochem.* (1985) **22**, 563–5676.

D682. Weische, A., Leistner, E., *Tetrahedron Lett.* (1985) **26**, 1487–1490.

D683. Mukharji, I., Silverman, R.B., *Proc. Natl. Acad. Sci. U.S.A.* (1985) **82**, 2713–2717.

D684. Matern, U., Potts, J.R.M., Hahlbrock, K., *Arch. Biochem. Biophys.* (1982) **208**, 233–241.

D685. Yoshizako, F., Ueno, T., Morishima, I., Karakawa, T., Sugiura, S., Araki, T., *Agric. Biol. Chem.* (1985) **49**, 877–879.

D686. Karegoudar, T.B., Pujar, B.G., *Curr. Microbiol.* (1984) **11**, 321–324.

D687. Brown, S.A., *Can. J. Biochem. Cell Biol.* (1985) **63**, 292–295.

D688. Pascal, R.A., Jr., Oliver, M.A., Chen, Y.C.J., *Biochemistry* (1985) **24**, 3158–3165.

D689. Srivastava, M., Dua, R.D., *Indian J. Biochem. Biophys.* (1985) **22**, 13–17.

D690. Bollag, J.M., Liu, S.Y., *Pestic. Biochem. Physiol.* (1985) **23**, 261–272.

D691. Tanaka, H., Naito, S., *Yakurigaku Zasshi* (1985) **85**, 357–366.

D692. Rainsford, K.D., *J. Pharm. Pharmacol.* (1985) **37**, 341–345.

D694. Bakke, J., Struble, C., Gustafsson, J.A., Gustafsson, B., *Proc. Natl. Acad. Sci. U.S.A.* (1985) **82**, 668–671.

D695. Hinson, J.A., Freeman, J.P., Potter, D.W., Mitchum, R.K., Evans, F.E., *Mol. Pharmacol.* (1985) **27**, 574–577.

D696. McIntire, W., Hopper, D.J., Singer, T.P., *Biochem. J.* (1985) **228**, 325–335.

D697. Akhtar, M.H., Hamilton, R.M.G., Trenholm, H.L., *J. Agric. Food Chem.* (1985) **33**, 610–617.

D698. Hinderer, W., Seitz, H.U., *Arch. Biochem. Biophys.* (1985) **240**, 265–272.

D699. Camara, B., *Methods Enzymol.* (1985) **111**, 544–547.

D701. Ishimitsu, S., Fujimoto, S., Ohara, A., *Chem. Pharm. Bull.* (1984) **32**, 2439–2441.

D702. Stafford, H.A., Lester, H.H., Porter, L.J., *Phytochemistry* (1985) **24**, 333–338.

D703. Lehmann, W.D., Heinrich, H.C., *Clin. Chim. Acta* (1985) **147**, 261–266.

D704. Grohmann, M., Trendelenburg, U., *Naunyn-Schmiedeberg's Arch Pharmacol.* (1985) **329**, 264–270.

D705. Gottschall, D.W., Penney, D.A., Traiger, G.J., Hanzlik, R.P., *Toxicol. Appl. Pharmacol.* (1985) **78**, 332–341.

D706. Kay, C.J., Palmer, J.M., *Biochem. J.* (1985) **228**, 309–318.

D707. Lilienblum, W., Bock-Hennig, B.S., Bock, K.W., *Mol. Pharmacol.* (1985) **27**, 451–458.

D708. Pognat, J.F., Enreille, A., Chabard, J.L., Busch, N., Berger, J.A., *Drug Metab. Dispos.* (1986) **14**, 147–154.

D709. Haribabu, B., Vaidyanathan, C.S., *Curr. Sci.* (1985) **54**, 419–421.

D710. Kolkmann, R., Leistner, E., *Tetrahedron Lett.* (1985) **26**, 1703–1704.

D711. Yoshigako, F., Nishimura, A., Ueno, T., *Agric. Biol. Chem.* (1985) **49**, 1879–1880.

D712. Li, J.J., Purdy, R.H., Appelman, E.H., Klicka, J.K., Li, S.A., *Mol. Pharmacol.* (1985) **27**, 559–565.

D713. Lind, L., *Arch. Biochem. Biophys.* (1985) **240**, 226–235.

D715. Mizukami, H., Terao, T., Ohashi, H., *Planta Med.* (1985) **51**, 104–107.

D716. Heimark, L.D., Trager, W.F., *J. Med. Chem.* (1985) **28**, 503–506.

D718. Kono, K., Yoshida, Y., Yamagata, H., Watanabe, M., Takeda, Y., Murao, M., Doi, K., Takatsu, M., *Ind. Health* (1985) **23**, 37–45.

D719. Lehmann, W.D., Schiebel, H.M., Böttcher, J., Bässmann, H., Schüppel, R., *Biomed. Mass Spectrom.* (1985) **12**, 215–220.

D720. Smith, R.W., Matthews, H.B., Parker, C.E., Hass, J.R., *Biomed. Mass Spectrom.* (1985) **12**, 208–214.

D721. Vaught, J.B., Glowinski, I.B., King, C.M., *Carcinogenesis* (1985) **6**, 495–499.

D722. Coles, B., Meyer, D.J., Ketterer, B., Stanton, C.A., Garner, R.C., *Carcinogenesis* (1985) **6**, 693–697.

D723. Malejka-Giganti, D., Ryzewski, C.N., *Carcinogenesis* (1985) **6**, 687–692.

D724. Lee, C.W., Desmazeaud, M.J., *J. Gen. Microbiol.* (1985) **131**, 459–467.

D725. Harper, D.B., *Int. J. Biochem.* (1985) **17**, 677–683.

D726. Waite, J.H., *J. Mar. Biol. Assoc. U.K.* (1985) **65**, 359–371.

D728. Head, J., Irvine, R.J., Barone, S., Stitzel, R.E., de la Lande, I.S., *J. Pharmacol. Exp. Ther.* (1985) **234**, 184–189.

D729. Stock, M., Bernasch, A., Schuette, H.R., *Nahrung* (1985) **29**, 473–479.

D730. Lindquist, B., Warshawski, D., *Experientia* (1985) **41**, 767–769.

D731. Loeppky, R.N., Tomasik, W., Kovacs, D.A., Outram, J.R., Byington, K.H., *IARC Sci. Publ.* (1984) **57**, 429–436.

D732. Lindquist, B., Warshawski, D., *Biochem. Biophys. Res. Commun.* (1985) **130**, 71–75.

D733. Richtand, N.M., Inagami, T., Misono, K., Kuczenski, R., *J. Biol. Chem.* (1985) **260**, 8465–8473.

D734. Gill, J.A., Naharro, G., Villanueva, J.R., Martin, J.F., *J. Gen. Microbiol.* (1983) **131**, 1279–1287.

D735. Hummel, B.C.W., Walfish, P.G., *Biochim. Biophys. Acta* (1985) **841**, 173–185.

D736. Wong, K.P., Wong, K.F., *Biog. Amines* (1985) **2**, 93–100.

D737. Grbić-Galić, D., Young, L.Y., *Appl. Environ. Microbiol.* (1985) **50**, 292–297.

D738. Hammons, G.J., Guengerich, F.P., Weis, C.C., Beland, F.A., Kadlubar, F.F., *Cancer Res.* (1985) **45**, 3578–3583.

D739. Baumann, P., Gaillard, J.M., Jonzier-Perey, M., Gerber, C., Bouras, C., *Pharmacopsychiatry* (1985) **18**, 91–93.

D740. Pasanen, M., Pelkonen, O., *Gen. Pharmacol.* (1985) **16**, 361–366.

D741. Yu, P.H., Rozdilsky, B., Boulton, A.A., *J. Neurochem.* (1985) **45**, 836–843.

D742. Eckhardt, K., Schumann, G., Graefe, U., Ihn, W., Wagner, C., Fleck, W.F., Thrum, H., *J. Antibiot.* (1985) **38**, 1096–1097.

D743. Shibata, D., Doi, E., *Plant Cell Physiol.* (1984) **25**, 1421–1429.

D744. Kraft, B., Barz, W., *Appl. Environ. Microbiol.* (1985) **50**, 45–48.

D745. Dutton, M.F., Ehrlich, K., Bennett, J.W., *Appl. Environ. Microbiol.* (1985) **49**, 1392–1395.

D746. Amund, O.O., Higgins, I.J., *Antonie van Leeuwenhoek* (1985) **51**, 45–56.

D747. Sundberg, B., Sandberg, G., Jensen, E., *Physiol. Plant.* (1985) **64**, 438–444.

D748. Yu, P.H., Walz, W., *Neurochem. Res.* (1985) **10**, 983–992.

D750. Svardal, A.M., Scheline, R.R., *Mol. Cell. Biochem.* (1985) **67**, 171–179.

D751. Miller, E., Schreier, P., *Food Chem.* (1985) **17**, 143–154.

D752. Davi, H., Guyonnet, J., Sales, Y., Cautreels, W., *Arzneim.-Forsch.* (1985) **35**, 1061–1065.

D753. Caldwell, J., Anthony, A., Cotgreave, I.A., Sangster, S.A., Sutton, J.D., Bernard, B.K., Ford, R.A., *Food Chem. Toxicol.* (1985) **23**, 559–566.

D754. Suzuki, T., Hashimura, Y., Takeyama, S., *Chem. Pharm. Bull.* (1985) **33**, 2549–2552.

D755. Akiyama, H., Kido, S., Shimizu, T., *Arzneim.-Forsch.* (1985) **35**, 1133–1140.

D756. Lai, C.C., Miller, J.A., Miller, E.C., Liem, A., *Carcinogenesis* (1985) **6**, 1037–1045.

D757. Maguire, J.H., Wilson, J.C., *J. Chromatogr.* (1985) **342**, 323–332.

D758. Renganathan, V., Miki, K., Gold, M.H., *Arch. Biochem. Biophys.* (1985) **241**, 304–314.

D759. Holland, H.L., Popperl, H., Ninniss, R.W., Chenchuiah, P.C., *Can. J. Chem.* (1985) **63**, 1118–1120.

D760. Wheeler, M.H., Stipanovic, R.D., *Arch. Microbiol.* (1985) **142**, 234–241.

D761. Hopper, D.J., Jones, H.G., Elmorsi, E., Rhodes-Roberts, M.E., *J. Gen. Microbiol.* (1985) **131**, 1807–1814.

D762. Palfi-Ledniczky, M., Szinai, I., Ujszaszy, K., Holly, S., Weisz, I., Kemeny, V., Tudos, H., Horvath, M., Otvos, L., *Stud. Org. Chem. (Amsterdam)* (1984) **18**, 329–333.

D763. Tegyey, Z., Vereczkey, L., Tamas, J., Rohricht, J., Kisfaludy, L., Otvos, L., *Stud. Org. Chem. (Amsterdam)* (1984) **18**, 367–371.

D764. Kobayashi, T., Yoshida, K., Mitani, M., Torii, H., Tanayama, S., *J. Pharmacobio.-Dyn.* (1985) **8**, 448–456.

D765. Tatsumi, K., Yamada, H., Yoshimura, H., Nishizawa, Y., Sakai, M., Mizuo, H., Yamato, C., *Hiroshima J. Med. Sci.* (1984) **33**, 669–678.

D766. Himeno, A., Kunisada, K., Niwa, M., Ozaki, M., *Jpn. J. Pharmacol.* (1985) **39**, 91–98.

D767. Roy, S.D., Hawes, E.M., McKay, G., Hubbard, J.W., Midha, K.K., *Can. J. Physiol. Pharmacol.* (1985) **63**, 778–781.

D768. Huynh, V.-B., Crawford, R.L., *FEMS Microbiol. Lett.* (1985) **28**, 119–123.

D769. Watabe, T., Hiratsuka, A., Ogura, K., Endoh, K., *Biochem. Biophys. Res. Commun.* (1985) **131**, 694–699.

D770. Tsuji, H., Ogawa, T., Bando, N., Sasaoka, K., *Biochem. Biophys. Res. Commun.* (1985) **130**, 633–639.

D771. Starovoytov, I.I., Selifonov, S.A., Nefedova, M.Y., Adanin, V.M., *FEMS Microbiol. Lett.* (1985) **28**, 183–186.

D772. Frömel, S., De Vlaming, P., Stotz, G., Wiering, H., Forkmann, G., Schram, A.W., *Theor. Appl. Genet.* (1985) **70**, 561–568.

D774. Noren, B., Strömberg, S., Ericsson, O., Vangbo, B., Gruells. M., Widlund, L., Lindeke, B., *Acta Pharm. Suec.* (1985) **22**, 131–146.

D775. Muhi-Eldeen, Z., Kawahara, M., Dakkuri, A., Hussain, A., *Int. J. Pharm.* (1985) **26**, 15–23.

D776. Möller, L., Nilsson, L., Gustafsson, J.Å., Rafter, J., *Environ. Int.* (1985) **11**, 363–368.

D777. Yoshizawa, I., Watanabe, K., Kurosawa, S., Nakagawa, S., *J. Steroid Biochem.* (1985) **23**, 239–241.

D778. Thomas, R., Williams, D.J., *J. Chem. Soc., Chem. Commun.* (1985) 802–803.

D779. Malarczyk, E., *Acta Biochim. Pol.* (1984) **31**, 383–395.

D780. Emmons, G.T., Campbell, I.M., Bentley, R., *Biochem. Biophys. Res. Commun.* (1985) **131**, 956–960.

D782. Heller, W., Forkmann, G., Britsch, L., Grisebach, H., *Planta* (1985) **165**, 284–287.

D783. Beyer, K.H., Friese, J., *Arch. Pharm. (Weinheim, Ger.)* (1985) **318**, 855–863.

D784. Noren, B., Stromberg, S., Ericsson, O., Olsson, L.I., Moses, P., *Biomed. Mass Spectrom.* (1985) **12**, 367–279.

D785. Iguchi, H., Tone, H., Ishikura, T., taneuchi, T., Umezawa, H., *Cancer Chemother. Pharmacol.* (1985) **15**, 132–140.

D786. Nouws, J.F.M., Vree, T.B., Baakman, M., Driessens, F., Smulders, A., Holtkamp, J., *J. Vet. Pharmacol. Ther.* (1985) **8**, 303–311.

D787. Vollmer, K.O., Liedtke, B., Poisson, A., Von Hodenberg, A., Steinbrecher, W., *Eur. J. Drug Metab. Pharmacokinet.* (1985) **10**, 61–69.

D788. Vollmer, K.O., Poisson, A., *Eur. J. Drug Metab. Pharmacokinet.* (1985) **10**, 139–145.

D789. Vollmer, K.O., Poisson, A., *Eur. J. Drug Metab. Pharmacokinet.* (1985) **10**, 71–76.

D790. Groopman, J.D., Donahue, P.R., Zhu, J., Chen, J., Wogan, G.N., *Proc. Natl. Acad. Sci. U.S.A.* (1985) **82**, 6492–6496.

D792. van Bladeren, P.J., Sayer, J.M., Ryan, D.E., Thomas, P.E., Levin, W., Jerina, D.M., *J. Biol. Chem.* (1985) **260**, 10226–10235.

D793. Ichinose, H., Kojima, K., Togari, A., Kato, Y., Parvez, S., Parvez, H., Nagatsu, T., *Anal. Biochem.* (1985) **150**, 408–414.

D794. Robeson, D.J., Cook, D.R., *Z. Naturforsch. C: Biosci.* (1985) **40C**, 740–742.

D796. Jimenez, M., Garcia-Carmona, F., Garcia-Canovas, F., Iborra, J.L., Lozano, J.A., *Int. J. Biochem.* (1985) **17**, 885–890.

D797. Peter, M.G., Vaupel, W., *J. Chem. Soc., Chem. Commun.* (1985) 848–850.

D798. Fischer, V., West, P.R., Nelson, S.D., Harvison, P.J., Mason, R.P., *J. Biol. Chem.* (1985) **260**, 11446–11450.

D799. Rane, A., Gawronska-Szklarza, B., Svensson, J.O., *J. Pharmacol. Exp. Ther.* (1985) **234**, 761–765.

D800. Mori, M., Fujita, M., Kozuka, H., *J. Pharm. Pharmacol.* (1985) **37**, 819–820.

D801. Toon, S., Heimark, L.D., Trager, W.F., O'Reilly, R.A., *J. Pharm. Sci.* (1985) **74**, 1037–1040.

D802. Wei, C.I., Marshall, M.R., Hsieh, D.P.H., *Food Chem. Toxicol.* (1985) **23**, 809–819.

D803. Wei, C.I., Hsieh, D.P.H., *Food Chem. Toxicol.* (1985) **23**, 821–825.

D804. Sultatos, L.G., Minor, L.D., *Toxicology* (1985) **36**, 159–169.

D805. La Voie, E.J., Coleman, D.T., Rice, J.E., Geddie, N.G., Hoffmann, D., *Carcinogenesis* (1985) **6**, 1483–1488.

D806. Tanaka, A.K., Umetsu, N., Fukuto, T.R., *J. Agric. Food Chem.* (1985) **33**, 1049–1055.

D807. Baranczyk-Kuzma, A., Borchardt, R.T., Pinnick, C.L., *Acta Biochem. Pol.* (1985) 32, 35–45.

D808. Milewich, L., Garcia, R.L., *J. Clin. Endocrinol. Metab.* (1985) **61**, 812–816.

D809. Barrowman, M.M., Fewson, C.A., *Curr. Microbiol.* (1985) **12**, 235–240.

D810. Crabb, T.A., Soilleux, S.L., *J. Chem. Soc., Perkin Trans. 1* (1985) 1381–1385.

D811. Ascherl, M., Eyer, P., Kampffmeyer, H., *Biochem. Pharmacol.* (1985) **34**, 3755–3763.

D812. Poulsen, H.E., Lersche, A., Skovgaard, L.T., *Biochem. Pharmacol.* (1985) **34**, 3729–3733.

D813. Chang, S.-L., Levy, R.H., *Drug Metab. Dispos.* (1986) **14**, 281–286.

D814. Streeter, A.J., Baillie, T.A., *Biochem. Pharmacol.* (1985) **34**, 2871–2876.

D815. Singh, P.D.A., Jackson, J.R., James, S.P., *Biochem. Pharmacol.* (1985) **34**, 2207–2209.

D816. Mulder, G.J., Brouwer, S., Weitering, J.G., Scholtens, E., Pang, K.S., *Biochem. Pharmacol.* (1985) **34**, 1325–1329.

D817. Chen, Q.Y., Schwarz, L.R., *Biochem. Pharmacol.* (1985) **34**, 1363–1366.

D818. Billings, R.E., Fischer, L.J., *Drug Metab. Dispos.* (1985) **13**, 312–327.

D819. Koble, D.L., Koons, J.C., Fischer, L.J., *Drug Metab. Dispos.* (1985) **13**, 305–311.

D820. Lin, J.H., Yeh, K.C., Duggan, D.E., *Drug Metab. Dispos.* (1985) **13**, 321–326.

D821. Chiba, K., Peterson, L.A., Castagnoli, K.P., Trevor, A.J., Castagnoli, N., Jr., *Drug Metab. Dispos.* (1985) **13**, 342–347.

D822. Sweeny, D.J., Weiner, M., *Drug Metab. Dispos.* (1985) **13**, 377–379.

D823. Jung, D., Nanavaty, M., Prasad, P., *Drug Metab. Dispos.* (1985) **13**, 359–363.

D824. Christ, D.D., Walle, T., *Drug Metab. Dispos.* (1985) **13**, 380–381.

D825. Kannan, R., Miller, S., Singh, B.N., *Drug Metab. Dispos.* (1985) **13**, 646–650.

D826. Cossum, P.A., Rickert, D.E., *Drug Metab. Dispos.* (1985) **13**, 664–668.

D827. Chism, J.P., Rickert, D.E., *Drug Metab. Dispos.* (1985) **13**, 651–657.

D828. Noker, P.E., Hill, D.L., Kalin, J.R., Temple, C.G., Jr., Montgomery, J.A., *Drug Metab. Dispos.* (1985) **13**, 677–681.

D829. Nocerini, M.R., Yost, G.S., Carlson, J.R., Liberato, D.J., Breese, R.G., *Drug Metab. Dispos.* (1985) **13**, 690–694.

D830. Mishima, M., Tanimoto, Y., Oguri, K., Yoshimura, H., *Drug Metab. Dispos.* (1985) **13**, 716–721.

D831. Miller. M.J., Sipes, I.G., Perry, D.F., Carter, D.E., *Drug Metab. Dispos.* (1985) **13**, 527–531.

D832. Reinke, L.A., Moyer, M.J., *Drug Metab. Dispos.* (1985) **13**, 548–552.

D833. Melancon, M.J., Williams, D.E., Buhler, D.R., Lech, J.J., *Drug Metab. Dispos.* (1985) **13**, 542–547.

D834. Monks, T.J., Lau, S.S., Highet, R.J., Gillette, J.R., *Drug Metab. Dispos.* (1985) **13**, 553–559.

D835. Chan, R.L., Chaplin, M.D., *Drug Metab. Dispos.* (1985) **13**, 566–571.

D836. Lin, J.H., Yeh, K.C., Duggan, D.E., *Drug Metab. Dispos.* (1985) **13**, 602–607.

D837. Ruenitz, P.C., Bagley, J.R., *Drug Metab. Dispos.* (1985) **13**, 582–586.

D838. Yamaguchi, T., Nakamura, Y., *Drug Metab. Dispos.* (1985) **13**, 614–619.

D839. Furuuchi, S., Naito, K., Otsuka, M., Harigaya, S., *Drug Metab. Dispos.* (1985) **13**, 620–626.

D840. Yang, H.-Y.L., Majesky, M.W., namkung, M.J., Juchau, M.R., *Drug Metab. Dispos.* (1986) **14**, 293–298.

D841. Yang, H.-Y.L., namkung, M.J., Nelson, W.L., Juchau, M.R., *Drug Metab. Dispos.* (1986) **14**, 287–292.

D842. Dostal, L.A., Aitio, A., Harris, C., Bhatia, A.V., Hernandez, O., Bend, J.R., *Drug Metab. Dispos.* (1986) **14**, 303–309.

D843. Hilbert, J.M., Ning, J., Symchowicz, S., Zampaglione, N., *Drug Metab. Dispos.* (1986) **14**, 310–312.

D844. Hoffmann, K.-J., *Drug Metab. Dispos.* (1986) **14**, 341–348.

D845. Loh, A.C., Williams, T.H., Tilley, J.W., Sasso, G.J., Szuna, A.J., Carbone, J.J., Toome, V., Leinweber, F.-J., *Drug Metab. Dispos.* (1986) **14**, 325–330.

D846. Mitchell, A.D., Paulson, G.D., Zaylskie, R.G., *Drug Metab. Dispos.* (1986) **14**, 155–160.

D847. Manis, M.O., Braselton, W.E., Jr., *Drug Metab. Dispos.* (1986) **14**, 166–174.

D848. Hamilton, M., Kissinger, P.T., *Drug Metab. Dispos.* (1986) **14**, 5–12.

D849. Levine, W.G., *Drug Metab. Dispos.* (1986) **14**, 13–18.

D850. Roberts, S.A., Price, V.F., Jollow, D.J., *Drug Metab. Dispos.* (1986) **14**, 25–33.

D851. Raza, H., Levine, W.G., *Drug Metab. Dispos.* (1986) **14**, 19–24.

D852. Sisenwine, S.F., Tio, C.O., *Drug Metab. Dispos.* (1986) **14**, 41–45.

D853. Massarella, J.W., Loh, A.C., Williams, T.H., Szuna, A.J., Sandor, D., Bressler, R., Leinweber, F.-J., *Drug Metab. Dispos.* (1986) **14**, 59–64.

D854. Loewen, G.R., McKay, G., Verbeeck, R.K., *Drug Metab. Dispos.* (1986) **14**, 127–131.

D855. Bartels, M.J., Horning, E.C., Horning, M.G., *Drug Metab. Dispos.* (1986) **14**, 97–101.

D856. Darnerud, P.O., Brandt, I., Klasson-Wehler, E., Bergman, Å., D'Argy, R., Dencker, L., Sperber, G.O., *Xenobiotica* (1986) **16**, 295–306.

D857. Franck, G., Beyer, J., *Xenobiotica* (1986) **16**, 291–294.

D858. Sultatos, L.G., Minor, L.D., *Drug Metab. Dispos.* (1986) **14**, 214–220.

D859. Talaat, R.E., Nelson, W.L., *Drug Metab. Dispos.* (1986) **14**, 201–207.

D860. Sgaragli, G., Ninci, R., Della Corte, L., Valoti. M., Nardini, M., Andreoli, V., Moneti, G., *Drug Metab. Dispos.* (1986) **14**, 263–266.

D861. Kawai, R., Fujita, S., Suzuki, T., *Drug Metab. Dispos.* (1986) **14**, 277–279.

D862. Banerji, A., Chintalivar, G.J., *Phytochemistry* (1983) **22**, 1945–1947.

D863. Stafford, H.A., Lester, H.H., *Plant Physiol.* (1984) **76**, 184–186.

D864. Roquet, F., Montay, G., Goeffon, Y., *Proc. Int. Congr. Chemother., 13th* (1983) **5**, 110/10–110/13.

D865. Sugumaran, M., *Biochemistry* (1986) **25**, 4489–4492.

D866. Stafford, H.A., Lester, H.H., *Plant Physiol.* (1982) **70**, 695–698.

D867. Beerhues, L., Wiermann, R., *Z. Naturforsch. C: Biosci.* (1985) **40C**, 160–165.

D868. Kahana, Z.E., Lapidot, A., *Anal. Biochem.* (1985) **149**, 549–554.

D869. Nouws, J.F.M., Vree, T.B., Baakman, M., Driessens, F., Breukink, H.J., Mevius, D., *Am. J. Vet. Res.* (1986) **47**, 642–649.

D870. Allison, N., O'Donnell, M.J., Fewson, C.A., *Biochem. J.* (1985) **231**, 407–416.

D871. Koster, A.S., Frankhuijzen-Sierevogel, A.K., Mentrup, A., *Biochem. Pharmacol* (1986) **35**, 1981–1985.

D872. Frederick, C.B., Weis, C.C., Flammang, T.J., Martin, C.N., Kadlubar, F.F., *Carcinogenesis* (1985) **6**, 959–965.

D873. Nielsen, P., Friis, C., Gyrd-Hansen, N., Rasmussen, F., *Biochem. Pharmacol* (1986) **35**, 2509–2512.

D874. Gynther, J., Lapinjoki, S.P., Airaksinen, M.M., Peura, P., *Biochem. Pharmacol* (1986) **35**, 2671–2675.

D875. Boobis, A.R., Seddon, C.E., Davies, D.S., *Biochem. Pharmacol* (1986) **35**, 2961–2965.

D876. Hobkirk, R., Cardy, C.A., *Can. J. Biochem. Cell Biol.* (1985) **63**, 785–791.

D877. Ribeiro, O., Kirkby, C.A., Hirom, P.C., Millburn, P., *Carcinogenesis* (1985) **6**, 1507–1511.

D878. Steffens, P., Nagakura, N., Zenk, M.H., *Phytochemistry* (1985) **24**, 2577–2583.

D879. Vara, J., Perez-Gonzalez, J.A., Jimenez, A., *Biochemistry* (1985) **24**, 8074–8081.

D880. Keller, B., Keller, E., Lingens, F., *Biol. Chem. Hoppe-Seyler* (1985) **366**, 1063–1066.

D881. Dornberger, K., Hübener, R., Ihn, W., Thrum, H., Radics, L., *J. Antibiot.* (1985) **38**, 1219–1225.

D882. Kabuto, S., Kimata, H., Yonemitsu, M., Suzuki, J., *Xenobiotica* (1986) **16**, 307–315.

D883. Cummings, S.W., Ansari, G.A.S., Guengerich, F.P., Crouch, L.S., Prough, R.A., *Cancer Res.* (1985) **45**, 5671–5624.

D884. Lamparczyk, H.S., McKay, S., Farmer, P.B., Cory, P.D., Grover, P.L., Sims, P., *Xenobiotica* (1986) **16**, 325–333.

D885. Sisenwine, S.F., Liu, A.L., Tio, C., Kimmel, H., Freeland, G., *Xenobiotica* (1986) **16**, 335–340.

D886. Krell, H.W., Sandermann, H., Jr., *J. Agric. Food Chem.* (1986) **34**, 194–198.

D887. Macholz, R., Kujawa, M., Schulze, J., Lewerenz, H.J., Schnaak, W., *Arch. Toxicol., Suppl.* (1985) **8**, 373–376.

D888. Posner, J., Bye, A., Dean, K., Peck, A.W., Whiteman, P.D., *Eur. J. Clin. Pharmacol.* (1985) **29**, 97–103.

D889. Rush, W.R., Alexander, O., Hall, D.J., Cairncross, L., Dow, R.J., Graham, D.J.G., *Xenobiotica* (1986) **16**, 341–349.

D890. Thimmaiah, K.N., Sethi, V.S., *Cancer Res.* (1985) **45**, 5382–5385.

D891. Lin, J.H., Hooke, K.F., Yeh, K.C., Duggan, D.E., *J. Pharmacol. Exp. Ther.* (1985) **235**, 402–406.

D892. Degen, G.H., McLachlan, J.A., *Chem.-Biol. Interact.* (1985) **54**, 363–375.

D893. Rice, J.E., Coleman, D.J., Hosted, T.J., Jr., La Voie, E.J., McCaustland, D.J., Wiley, J.C., Jr., *Cancer Res.* (1985) **45**, 5421–5425.

D894. Nali, M., Rindone, B., Tollari, S., Andreoni, V., Treccani, V., *Experientia* (1985) **41**, 1351–1353.

D895. Müller-Enoch, D., Bügetten, E., Nonnenmacher, A., *Z. Naturforsch. C: Biosci.* (1985) **40C**, 682–684.

D896. Helmlinger, J., Rausch, T., Hilgenberg, W., *Phytochemistry* (1985) **24**, 2497–2502.

D897. Dodion, P., Egorin, M.J., Engisch, K.L., Bachur, N.R., *Cancer Res.* (1985) **45**, 5352–5357.

D899. Garrett, E.R., Derendorf, H., Mattha, A.G., *J. Pharm. Sci.* (1985) **74**, 1203–1214.

D900. Yoshida, M., Hara, I., *Ind. Health* (1985) **23**, 239–243.

D901. Juhl, U., Witte, I., Butte, W., *Bull. Environ. Contam. Toxicol.* (1985) **35**, 596–601.

D902. Marshall, P.S., Morris, S.R., Threlfall, D.R., *Phytochemistry* (1985) **24**, 1705–1711.

D903. De France, B.F., Carter, M.H., Josephy, P.D., Bryant, D.W., McCalla, D.R., *Mutat. Res.* (1985) **144**, 159–163.

D904. Jay, M., De Luca, V., Ibrahim, R.K., *Eur. J. Biochem.* (1985) **153**, 321–325.

D905. Akao, T., Akao, T., Mibu, K., Hattori, M., Namba, T., Kobashi, K., *J. Pharmacobio.-Dyn.* (1985) **8**, 800–807.

D906. Goromaru, T., Katashima, M., Matsuura, H., Yoshimura, N., *Chem. Pharm. Bull.* (1985) **33**, 3922–3928.

D907. Roy, S., Hawes, E.M., McKay, G., Korchinski, E.D., Midha, K.K., *Clin. Pharmacol. Ther.* (*St. Louis*) (1985) **38**, 128–133.

D908. Laizure, S.C., De Vane, C., Stewart, J.T., Dommisse, C.S., Lai, A.A., *Clin. Pharmacol. Ther.* (*St. Louis*) (1985) **38**, 586–589.

D909. Garcha, G., Imrie, P.R., Marley, E., Thomas, D.V., *Br. J. Pharmacol.* (1985) **86**, 877–888.

D910. Castro, M.I., Braverman, L.E., Alex, S., Wu, C.F., Emerson, C.H., *J. Clin. Invest.* (1985) **76**, 1921–1926.

D911. Massé, R., Badr, M., Ayotte, C., Sylvestre, M., *Toxicol. Environ. Chem.* (1985) **10**, 225–246.

D912. Awasthi, Y.C., Singh, S.V., Das, M., Mukhtar, H., *Biochem. Biophys. Res. Commun.* (1985) **133**, 863–867.

D913. Assandri, A., Bernareggi, A., Ripamonti, A., Perazzi, A., Toja, E., *Eur.*

J. Drug Metab. Pharmacokinet. (1985) **10**, 161–170.

D914. Awaluddin, A.B., McLean, S., *Aust. J. Zool.* (1985) **33**, 693–698.

D915. Honma, S., Akutsu, R., Iwamura, S., Kawabe, Y., Tsukamoto, K., *Oyo Yakuri* (1985) **30**, 555–563.

D916. Ingebrigtsen, K., Skaare, J.U., Nafstad, I., Grave, K., Foerde, M., *Gen. Pharmacol.* (1986) **17**, 19–21.

D917. D'Ari, L., Barker, H.A., *Arch. Microbiol.* (1985) **143**, 311–312.

D918. Wessendorf, J., Rettenmaier, H., Lingens, F., *Biol. Chem. Hoppe-Seyler.* (1985) **366**, 945–951.

D919. Asano, Y., Nakazawa, A., *Agric. Biol. Chem.* (1985) **49**, 3631–3632.

D920. Chen, W., Supanwong, K., Ohmiya, K., Shimizu, S., Kawakami, K., *Appl. Environ. Microbiol.* (1985) **50**, 1451–1456.

D921. Hyman, M.R., Sansome-Smith, A.W., Shears, J.H., Wood, P.M., *Arch. Microbiol.* (1985) **143**, 302–306.

D922. Seong, B.L., Son, H.J., Mheen, T.I., Park, Y.H., Han, M.H., *J. Ferment. Technol.* (1985) **63**, 515–522.

D923. Bennett, J.L., Updegraff, D.M., Pereira, W.E., Rostad, C.E., *Microbios Lett.* (1985) **29**, 147–154.

D924. Vigevani, A., Ballabio, M., Gioia, B., Marsili, L., Vioglio, S., Franceschi, G., *J. Antibiot.* (1985) **38**, 1799–1802.

D925. Lane, E.A., Levy, R.H., *J. Pharmacokinet. Biopharm.* (1985) **13**, 373–386.

D926. Ames, M.M., Selassie, C.D., Woodson, L.C., Van Loon, J.A., Hansch, C., Weishilboum, R.M., *J. Med. Chem.* (1986) **29**, 354–358.

D927. Krechniak, J., Klosowska, B., *Bromatol. Chem. Toksykol.* (1985) **18**, 186–190.

D928. Sakurai, I., Suzuki, H., Shimizu, S., Yamamoto, Y., *Chem. Pharm. Bull.* (1985) **33**, 5141–5143.

D929. Ötük, G., *J. Ferment. Technol.* (1985) **63**, 501–506.

D930. Cavalieri, R.R., Gavin, L.A., Cole, R., De Vellis, J., *Brain Res.* (1986) **364**, 382–385.

D932. Bakke, J.E., Larsen, G.L., *Chemosphere* (1985) **14**, 1749–1754.

D933. Matsui, M., Nagai, F., *J. Pharmacobio.-Dyn.* (1985) **8**, 1048–1053.

D935. Tsuda, Y., Kawai, K., Nakajima, S., *Chem. Pharm. Bull.* (1985) **33**, 4657–4661.

D936. Fedorak, P.M., Westlake, D.W.S., *Appl. Environ. Microbiol.* (1986) **51**, 435–437.

D937. Matsumoto, K., Miyazaki, H., Fujii, T., Yoshida, K., Amejima, H., Hashimoto, M., *Arzneim.-Forsch.* (1986) **36**, 40–46.

D938. Hidvegi, S., Ecobichon, D.J., *Can. J. Physiol. Pharmacol.* (1986) **64**, 72–76.

D940. Tomoda. A., Yamaguchi, J., Kojima, H., Amemiya, H., Yoneyama, Y., *FEBS Lett.* (1986) **196**, 44–48.

D941. Lautier, J., Chanal, J.L., Delhon, A., *Comp. Biochem. Physiol. C: Comp. Pharmacol. Toxicol.* (1986) **83C**, 143–147.

D942. Lee, P.W., Stearns, S.M., Powell, W.R., Stoutamire, D.W., Payne, G.B., Woodward, M.D., Burton, W.B., Silveira, E.J., Ehmann, A., *J. Agric. Food Chem.* (1986) **34**, 162–170.

D943. Kimmel, E.C., Casida, J.E., Ruzo, L.O., *J. Agric. Food Chem.* (1986) **34**, 157–161.

D945. Gräslund, A., Waern, F., Jernström, B., *Carcinogenesis* (1986) **7**, 167–170.

D946. Maier, W., Krauspe, R., Gröger, D., *J. Biotechnol.* (1985) **3**, 155–166.

D947. Larson, R.L., Bussard, J.B., *Plant Physiol.* (1986) **80**, 483–486.

D948. Anderson, J.A., *Phytochemistry* (1986) **25**, 103–106.

D950. Forkmann, G., De Vlaming, P., Spribille, R., Wiering, H., Schram, A.W., *Z. Naturforsch. C: Biosci.* (1986) **41**, 179–186.

D951. Ritchie, J.C., Crothers, M.J., Shah, R.R., Idle, J.R., Smith, R.L., *Xenobiotica* (1986) **16**, 503–509.

D952. Eadsforth, C.V., Coveney, P.C., Hutson, D.H., Logan, C.J., Samuel, A.J., *Xenobiotica* (1986) **16**, 555–566.

D953. Bowes, S.G., Renwick, A.G., *Xenobiotica* (1986) **16**, 543–553.

D954. Rosenbaum, S.E., Lindup, W.E., Orton, T.C., *Xenobiotica* (1986) **16**, 567–573.

D955. Gorrod, J.W., Christou, M., *Xenobiotica* (1986) **16**, 575–585.

D956. Achari, R., Drissel, D., Hulse, J.D., *Clin. Chem.* (*Washington, DC*) (1986) **32**, 374–376.

D957. de Vries, J.X., Zimmermann, R., Harenberg, J., *Eur. J. Clin. Pharmacol.* (1986) **29**, 591–594.

D958. Elsom, L.F., Biggs, S.R., Chasseaud, L.F., Hawkins, D.R., Pulsford, J., Darragh, A., *Eur. J. Drug Metab. Pharmacokinet.* (1985) **10**, 209–215.

D960. Sund, R.B., Lauterbach, F., *Acta Pharmacol. Toxicol.* (1986) **58**, 74–83.

D961. Achari, R., Drissel, D., Matier, W.L., Hulse, J.D., *J. Clin. Pharmacol.* (1986) **26**, 44–47.

D962. Gill, J.H., Bonin, A.M., Podobna, E., Baker, R.S.U., Duke, C.C., Rosario, C.A., Ryan, A.J., Holder, G.M., *Carcinogenesis* (1986) **7**, 23–31.

D963. Wise, R.W., Zenser, T.V., Rice, J.R., Davis, B.B., *Carcinogenesis* (1986) **7**, 111–115.

D964. Boniwell, J.M., Butt, V.S., *Z. Naturforsch. C: Biosci.* (1986) **41**, 56–60.

D965. Stegeman, J.J., James, M.O., *Mar. Environ. Res.* (1985) **17**, 122–124.

D966. Mays, D.C., Rogers, S.L., Guiler, R.C., Sharp, D.E., Hecht, S.G., Staubus, A.E., Gerber, N., *J. Pharmacol. Exp. Ther.* (1986) **236**, 364–373.

D967. Samejima, M., Saburi, Y., Yoshimoto, T., Fukuzumi, T., Nakagawa, T., *Mokuzai Gakkaishi* (1985) **31**, 956–958.

D968. Ebeid, M.M., Dimova, S., Kutacek, M., *Biol. Plant.* (1985) **27**, 413–416.

D969. Holland, H.L., Carter, I.M., Chenchaiah, P.C., Khan, S.K., Munoz, B., Ninniss, R.W., Richards, D., *Tetrahedron Lett.* (1985) **26**, 6409–6412.

D970. Lodola, E., Borgia, M., Longo, A., Pocchiari, F., Pataccini, R., Sher, D., *Arzneim.-Forsch.* (1986) **36**, 345–348.

D971. Masciadri, R., Angst, W., Arigoni, D., *J. Chem. Soc., Chem. Commun.* (1985) 1573–1574.

D972. Neal, G.E., Judah, D.J., Green, J.A., *Toxicol. Appl. Pharmacol.* (1986) **82**, 454–460.

D973. Gallop, M.A., Eggers, N.J., Rep.-N.Z., *Dep. Sci. Ind. Res., Chem. Div.* (1985) CD 2363.

D974. Helferich, W.G., Baldwin, R.L., Hsieh, D.P.H., *J. Anim. Sci.* (1986) **62**, 697–705.

D975. Bode, R., Lippoldt, A., Birnbaum, D., *Biochem. Physiol. Pflanz.* (1986) **181**, 189–198.

D976. Fujimori, K., Yaguchi, M., Mikami, A., Matsuura, T., Furukawa, N., Oae, S., Iyanagi, T., *Tetrahedron Lett.* (1986) **27**, 1179–1182.

D977. Baranczyk-Kuzma, A., *Biochem. Med. Metab. Biol.* (1986) **35**, 18–30.

D978. Asada, Y., Sawa, Y., Tanizawa, K., Soda, K., *J. Biochem.* (1986) **99**, 1101–1110.

D979. Zimmermann, L., Bergstrom, J., Joernvall, H., *Clin. Nephrol.* (1986) **25**, 94–100.

D980. Herbert, R.B., Knagg, E., *Tetrahedron Lett.* (1986) **27**, 1099–1102.

D981. Rosazza, J.P., Reeg, S., Yang, L.M., *Enzyme Microb. Technol.* (1986) **8**, 161–165.

D982. Mérillon, J.M., Doireau, P., Guillot, A., Chénieux, J.C., Rideau, M., *Plant Cell Rep.* (1986) **5**, 23–26.

D983. Hauffe, K.D., Hahlbrock, K., Scheel, D., *Z. Naturforsch. C: Biosci.* (1986) **41**, 228–239.

D984. Jimenez, M., Garcia-Canovas, F., Garcia-Carmona, F., Tudela, J., Iborra, J.L., *Int. J. Biochem.* (1986) **18**, 39–47.

D986. Potter, D.W., Miller, D.W., Hinson, J.A., *Mol. Pharmacol.* (1986) **29**, 155–162.
D987. Cashman, J.R., Ziegler, D.M., *Mol. Pharmacol.* (1986) **29**, 163–167.
D989. Iida, S., Inokuchi, T., Ogawa, K., Kobayashi, K., Komatsu, T., Iwamoto, M., Izawa, O., Ninomiya, K., Ohlsuka, Y., Kikumoto, R., *Oyo Yakuri* (1986) **31**, 155–168.
D990. Pocchiari, F., Pataccini, R., Castelnovo, P., Longo, A., Casagrande, C., *Arzneim.-Forsch.* (1986) **36**, 334–340.
E1. Rakhit, A., Hurley, M.E., Tipnis, V., Coleman, J., Rommel, A., Brunner, H.R., *J. Clin. Pharmacol.* (1986) **26**, 156–164.
E2. Kato, Y., Kogure, T., Sato, M., Murata, T., Kimura, Y., *Toxicol. Appl. Pharmacol.* (1986) **82**, 505–511.
E3. Glover, V., Gibb, C., Sandler, M., *Neurosci. Lett.* (1986) **64**, 216–220.
E4. Tatsumi, K., Kitamura, S., Narai, N., *Cancer Res.* (1986) **46**, 1089–1093.
E5. Ueji, M., Tomizawa, C., *Nippon Noyaku Gakkaishi* (1985) **10**, 691–696.
E6. Ansher, S.S., Jakoby, W.B., *J. Biol. Chem.* (1986) **261**, 3996–4001.
E7. Marriage, H.J., Harvey, D.J., *J. Chromatogr.* (1986) **354**, 383–392.
E8. Tahara, S., Nakahara, S., Ingham, J.L., Mizutani, J., *Nippon Nogei Kagaku Kaishi* (1985) **59**, 1039–1044.
E10. Poon, G., Chui, Y.C., Law, F.C.P., *Xenobiotica* (1986) **16**, 795–800.
E11. Ganes, D.A., Hindmarsh, K.W., Midha, K.K., *Xenobiotica* (1986) **16**, 781–794.
E12. Krechniak, J., Kłosowska, B., *Xenobiotica* (1986) **16**, 809–815.
E13. Samuelsen, O.B., Brenna, J., Solheim, E., Scheline, R.R., *Xenobiotica* (1986) **16**, 845–852.
E14. Edwards, V.T., Jones, B.C., Hutson, D.H., *Xenobiotica* (1986) **16**, 801–807.
E15. Sidjimov, A.K., Marekov, N.L., *Phytochemistry* (1986) **25**, 565–567.
E16. Matsunaga, Y., Nambu, K., Ohe, Y., Miyazaki, H., Hashimoto, M., *Arzneim.-Forsch.* (1986) **36**, 453–456.
E17. Miyazaki, H., Matsunaga, Y., Nambu, K., Ohe, Y., Hashimoto, M., *Arzneim.-Forsch.* (1986) **36**, 443–452.
E18. Hege, H.G., Lietz, H., Weymann, J., *Arzneim.-Forsch.* (1986) **36**, 467–474.
E19. Ivie, G.W., Christopher, R.J., Munger, C.E., Coppock, C.E., *J. Anim. Sci.* (1986) **62**, 681–690.
E20. Okuyama, A., Ohuchi, S., Tanaka, T., Aoyaki, T., *Biochem. Int.* (1986) **12**, 485–491.
E21. Karns, J.S., Mulbry, W.W., Nelson, J.O., Kearney, P.C., *Pestic. Biochem. Physiol.* (1986) **25**, 211–217.
E23. Kochs, G., Grisebach, H., *Eur. J. Biochem.* (1986) **155**, 311–318.
E24. Camoretti-Mercado, B., Frydman, R.B., *Eur. J. Biochem.* (1986) **156**, 317–325.
E25. Zimmerman, J.K., Brown, T.M., *J. Agric. Food Chem.* (1986) **34**, 516–520.
E26. Grothusen, J.R., Bryson, P.K., Zimmerman, J.K., *J. Agric. Food Chem.* (1986) **34**, 513–515.
E27. Hofmann, K.H., *J. Basic Microbiol.* (1986) **26**, 109–111.
E28. Knogge, W., Weissenböck, G., *Planta* (1986) **167**, 196–205.
E29. Stipanovic, R.D., Stoessl, A., Stothers, J.B., Altman, D.W., Bell, A.A., Heinstein, P., *J. Chem. Soc., Chem. Commun.* (1986) 100–102.
E30. Takikawa, O., Yoshida, R., Kido, R., Hayashi, O., *J. Biol. Chem.* (1986) **261**, 3648–3653.
E31. Masunaga, S., Urushigawa, Y., Yonezawa, Y., *Water Res.* (1986) **20**, 477–484.
E32. Schatz, F., Haberl, H., Battig, F., Jobstmann, D., Schulz, G., Nefzger, M., Czok, R., Nikiforov, A., *Arzneim.-Forsch.* (1986) **36**, 248–255.

E36. Rosetti, V., Lombard, A., Buffa, M., Cassone, M.C., *IRCS Med. Sci.* (1986) **14**, 256–257.

E37. Lindberg, R.L.P., Syvalahti, E.K.G., *Clin. Pharmacol. Ther. (St. Louis)* (1986) **39**, 378–383.

E40. Dannan, G.A., Porubek, D.J., Nelson, S.D., Waxman, D.J., Guengerich, F.P., *Endocrinology (Baltimore)* (1986) **118**, 1952–1960.

E41. Takase, I., Omori, T., Minoda, Y., *Agric. Biol. Chem.* (1986) **50**, 681–686.

E42. van den Tweel, W.J.J., Smits, J.P., de Bont, J.A., *Arch. Microbiol.* (1986) **144**, 169–174.

E43. Gessner, W., Brossi, A., Bembenek, M.E., Fritz, R.R., Abell, C.W., *FEBS Lett.* (1986) **199**, 100–102.

E44. Kovar, K.A., El-Bolkiny, M.N., *Arch. Pharm. (Weinheim, Ger.)* (1986) **319**, 280–282.

E45. Iwasaki, K., Shiraga, T., Tada, K., Noda, K., Noguchi, H., *Xenobiotica* (1986) **16**, 717–723.

E46. Holland, H.L., Khan, S.H., Richards, D., Riemland, E., *Xenobiotica* (1986) **16**, 733–741.

E47. Sudo, K., Okazaki, O., Tsumura, M., Tachizawa, H., *Xenobiotica* (1986) **16**, 725–732.

E48. Dieterle, W., Faigle, J.W., Kriemler, H.-P., Winkler, T., *Xenobiotica* (1986) **16**, 743–752.

E49. Schütz, H., Feldmann, K.F., Faigle, J.W., Kriemler, H.-P., Winkler, T., *Xenobiotica* (1986) **16**, 769–778.

E51. Iwasaki, K., Shiraga, T., Noda, K., Tada, K., Noguchi, H., *Xenobiotica* (1986) **16**, 651–659.

E52. Inoue, T., Suzuki, S., *Xenobiotica* (1986) **16**, 691–698.

E53. Kammerer, R.C., Schmitz, D.A., *Xenobiotica* (1986) **16**, 681–690.

E54. van Ommen, B., Adang, A.E.P., Brader, L., Posthumus, M.A., Müller, F., van Bladeren, P.J., *Biochem. Pharmacol.* (1986) **35**, 3233–3238.

E55. Sullivan, J.P., McDonnell, L., Hardiman, O.M., Farrell, M.A., Phillips, J.P., Tipton, K.F., *Biochem. Pharmacol.* (1986) **35**, 3255–3260.

E56. Guenthner, T.M., *Biochem. Pharmacol.* (1986) **35**, 3261–3266.

E57. Nakamura, A., Tatewaki, N., Morino, A., Sugiyama, M., *Xenobiotica* (1986) **16**, 1079–1089.

E58. Hall, B.E., James, S.P., *Xenobiotica* (1986) **16**, 609–614.

E60. Esclade, L., Guillochon, D., Thomas, D., *Xenobiotica* (1986) **16**, 615–624.

E61. Bauer, E., McDougall, J., Cameron, B.D., *Xenobiotica* (1986) **16**, 625–633.

E62. Nagamatsu, K., Ohno, Y., Ikebuchi, H., Takahashi, A., Terao, T., Takanaka, A., *Biochem. Pharmacol.* (1986) **35**, 3543–3548.

E63. Maurizis, J.C., Mandelmont, J.C., Parry, D., Dauzonne, D., Royer, R., Chabard, J.L., *Xenobiotica* (1986) **16**, 635–643.

E64. Radulovic, L.L., La Ferla, J.J., Kulkarni, A.P., *Biochem. Pharmacol.* (1986) **35**, 3473–3480.

E65. Alexander, M.S., Husney, R.M., Sagone, A.L., Jr., *Biochem. Pharmacol.* (1986) **35**, 3649–3651.

E66. Fournel, S., Caldwell, J., *Biochem. Pharmacol.* (1986) **35**, 4153–4159.

E67. Hara, A., Hasebe, K., Hayashibara, M., Matsuura, K., Nakayama, T., Sawada, H., *Biochem. Pharmacol.* (1986) **35**, 4005–4012.

E68. Okuda, H., Miwa, K., Nojima, H., Watabe, T., *Biochem. Pharmacol.* (1986) **35**, 4573–4576.

E69. Ishimitsu, S., Fujimoto, S., Ohara, A., *Chem. Pharm. Bull.* (1986) **34**, 768–774.

E70. Ribeiro, O., Kirkby, C.A., Hirom, P.C., Millburn, P., *Biochem. Soc. Trans.* (1986) **14**, 305.

E72. Britsch, L., Grisebach, H., *Eur. J. Biochem.* (1986) **156**, 569–577.

E73. Shimao, M., Ninomiya, K., Kuno, O., Kato, N., Sakazawa, C., *Appl. Environ. Microbiol.* (1986) **51**, 268–275.

E74. Haim, N., Roman, J., Nemec, J., Sinha, B.K., *Biochem. Biophys. Res. Commun.* (1986) **135**, 215–220.
E75. Anderson, J.A., *Phytochemistry* (1986) **25**, 1115–1117.
E76. Dougan, D.H.F., Duffield, A.M., Duffield, P.H., Wade, D.N., *Br. J. Pharmacol.* (1986) **88**, 285–290.
E77. Brenneisen, R., Geisshüsler, S., Schorno, X., *J. Pharm. Pharmacol.* (1986) **38**, 298–300.
E78. Hammond, J.M., Hersey, R.M., Walega, M.A., Weisz, J., *Endocrinology (Baltimore)* (1986) **118**, 2292–2299.
E79. Patterson, L.H., Hall, G., Nijjar, B.S., Khatra, P.K., Cowan, D.A., *J. Pharm. Pharmacol.* (1986) **38**, 326.
E80. Stock, B.H., Bend, J.R., Eling, T.E., *J. Biol. Chem.* (1986) **261**, 5959–5964.
E81. Fischer, V., Mason, R.P., *Chem.-Biol. Interact.* (1986) **57**, 129–142.
E82. Thakker, D.R., Levin, W., Yagi, H., Yeh, H.J.C., Ryan, D.E., Thomas, P.E., Conney, A.H., Jerina, D.M., *J. Biol. Chem.* (1986) **261**, 5404–5413.
E83. Flescher, J.W., Myers, S.R., Bergo, C.H., Blake, J.W., *Chem.-Biol. Interact.* (1986) **57**, 223–233.
E84. Koop, D.R., *Mol. Pharmacol.* (1986) **29**, 399–404.
E85. Omori, T., Sugimura, K., Ishigooka, H., Minoda, Y., *Agric. Biol. Chem.* (1986) **50**, 931–937.
E86. Haemmerli, S.D., Leisola, M.S.A., Sanglard, D., Fiechter, A., *J. Biol. Chem.* (1986) **261**, 6900–6903.
E87. Bonner, W.D., Jr., Clarke, S.D., Rich, P.R., *Plant Physiol.* (1986) **80**, 838–842.
E88. Tsurumi, S., Wada, S., *Plant Cell Physiol.* (1986) **27**, 559–562.
E89. Ballet, F., Robert, J., Bonma, M.E., Vrignaud, P., Infante, R., *Pharmacol. Res. Commun.* (1986) **18**, 343–347.
E90. Frydman, A.M., Le Roux, Y., Lefebvre, M.A., Djebbar, F., Fourtillan, J.B., Gaillot, J., *J. Antimicrob. Chemother.* (1986) **17** (Suppl. B), 65–79.
E91. Bechtel, W.D., Mierau, J., Brandt, K., Förster, H.J., Pook, K.H., *Arzneim.-Forsch.* (1986) **36**, 578–586.
E92. Slikker, W., Jr., Holder, C.L., Lipe, G.W., Korfmacher, W.A., Thompson, H.C., Jr., Bailey, J.R., *J. Anal. Toxicol.* (1986) **10**, 87–92.
E93. Toftgard, R., Haaparanta, T., Halpert, J., *Toxicology* (1986) **39**, 225–231.
E94. Iyayi, C.B., *Microbios* (1986) **45**, 153–159.
E95. Nakahara, S., Tahara, S., Mizutani, J., Ingham, J.L., *Agric. Biol. Chem.* (1986) **50**, 863–873.
E96. Drotar, A., Fall, R., *Pestic. Biochem. Physiol.* (1986) **25**, 396–406.
E97. Lippoldt, A., Bode, R., Birnbaum, D., *J. Basic Microbiol.* (1986) **26**, 145–154.
E99. Kawahara, K., Awaji, T., Uda, K., Sakai, Y., Hashimoto, Y., *Eur. J. Drug Metab. Pharmacokinet.* (1986) **11**, 29–32.
E100. Buu, N.T., Duhaime, J., Kuchel, O., *Am. J. Physiol* (1986) **250**, F975-F977.
E101. Ahmed, S.A., Martin, B., Miles, E.W., *Biochemistry* (1986) **25**, 4233–4240.
E102. Ohe, T., Watanabe, Y., *Agric. Biol. Chem.* (1986) **50**, 1419–1426.
E103. Dua, R.D., Rao, V., *Indian J. Biochem. Biophys.* (1986) **23**, 110–113.
E104. Tanaka, K., Murakami, M., Greer, M.A., *Biochem. Biophys. Res. Commun.* (1986) **137**, 863–868.
E105. Iida, S., Izawa, O., Kitazawa, K., Iwamoto, M., Kobayashi, K., Komatsu, T., Hosokawa, K., Inokuchi, T., Taniguchi, M., et al. *Oyo Yakuri* (1986) **31**, 767–784.
E106. Amundi, I.M., Kauffman, F.C., El-Mouelhi, M., Thurman, R.G., *Mol. Pharmacol.* (1986) **29**, 599–605.
E107. Nieto, M.J., Hernandez Hernandez, J.M., Lanao, J.M., Fernandez, C., Dominguez-Gil, A., *Eur. J. Drug Metab. Pharmacokinet.* (1986) **11**, 23–27.

E109. Pacifici, G.M., Giuliani, L., Calcaprina, R., *Pharmacology* (1986) **33**, 103–109.

E110. Saito, K., Shinohara, A., Kamataki, T., Kato, R., *J. Biochem.* (1986) **99**, 1689–1697.

E112. Khatibi, A., Hassan, M.M.A., *Arab. Gulf J. Sci. Res.* (1986) **4**, 269–276.

E113. Niimura, T., Otani, C., Yamaha, T., *Shokuhin Eiseigaku Zasshi* (1986) **27**, 163–170.

E114. Foster, A.C., White, R.J., Schwarcz, R., *J. Neurochem.* (1986) **47**, 23–30.

E115. Nazareth, S., Mavinkurve, S., *Can. J. Microbiol.* (1986) **32**, 494–497.

E116. Schreiber, J., Mottley, C., Sinha, B.K., Kalyanaraman, B., Mason, R.P., *J. Am. Chem. Soc.* (1987) **109**, 348–351.

E117. Kutney, J.P., Choi, L.S.L., Hewitt, G.M., Salisbury, P.J., Singh, M., *Appl. Environ. Microbiol.* (1985) **49**, 96–100.

E118. Mottley, C., Mason, R.P., *J. Biol. Chem.* (1986) **261**, 16860–16864.

E119. Torii, H., Yoshida, K., Kobayashi, T., Tsukamoto, T., Tanayama, S., *J. Pharmacobio.-Dyn.* (1985) **8**, 457–467.

E120. Gerats, A.G.M., Bussard, J., Coe, E.H., Jr., Larson, R., *Biochem. Genet.* (1984) **22**, 1161–1169.

E121. Macholz, R.M., Knoll, R., Lewerenz, H.-J., Plass, R., *Arch. Toxicol.* (1982) **50**, 85–88.

E122. Tsumuri, S., Wada, S., *Plant Cell Physiol.* (1986) **27**, 1513–1522.

E124. Watkins, G.M., Whelpton, R., Buckley, D.G., Curry, S.H., *J. Pharm. Pharmacol.* (1986) **38**, 506–509.

E125. Larson, R., Bussard, J.B., Coe, E.H., Jr., *Biochem. Genet.* (1986) **24**, 615–624.

E126. Vicente, C., Gonzalez, A., Estrella Legaz, M., *Biochem. Syst. Ecol.* (1986) **14**, 283–286.

E127. Krowech, G., Caldera-Munoz, P.S., Straub, K., Castagnoli, N., Jr., Correia, M.A., *Chem.-Biol. Interact.* (1986) **58**, 29–40.

E129. Jedrychowski, M., Nilsson, E., Bieck, P.R., *Psychopharmacology (Berlin)* (1986) **88**, 529–530.

E130. Köppel, C., Tenczer, J., Wagemann, A., *Arzneim.-Forsch.* (1986) **36**, 1116–1118.

E131. Kao, L.R., Birnbaum, L.S., *J. Toxicol. Environ. Health* (1986) **18**, 441–458.

E132. Milewich, L., MacDonald, P.C., Guerami, A., Midgett, W.T., Lassiter, W.L., Carr, B.R., *J. Clin. Endocrinol. Metab.* (1986) **63**, 180–191.

E133. Grbic-Galic, D., *FEMS Microbiol. Ecol.* (1986) **38**, 161–169.

E134. Matsumura, E., Yamamoto, E., Numata, A., Kawano, T., Shin, T., Murao, S., *Agric. Biol. Chem.* (1986) **50**, 1355–1357.

E135. Fujiyoshi, T., Ikeda, K., Yamaura, T., Saito, M., Maeda, E., Iida, H., Hosono, M., Uematsu, T., *Yakuri to Chiryo* (1986) **14**, 2225–2233.

E136. Kurimoto, T., Ichida, S., Sato, S., Hirayama, H., *Showa Igakkai Zasshi* (1985) **45**, 739–749.

E137. Takashina, H., Horiuchi, M., Fujimura, K., Morishita, M., Kuwano, M., Uemura, O., Iso, T., *Iyakuhin Kenkyu* (1986) **17**, 527–533.

E138. Larkin, M.J., Day, M.J., *J. Appl. Bacteriol.* (1986) **60**, 233–242.

E139. Sahasrabudhe, S.R., Lala, D., Modi, V.V., *Can. J. Microbiol.* (1986) **32**, 535–538.

E140. Ohta, T., Miyata, W., Hirobe, M., *Chem. Pharm. Bull.* (1986) **34**, 2494–2500.

E141. Yu, D.K., Dimmitt, D.C., Lamman, R.C., Giesing, D.H., *J. Pharm. Sci.* (1986) **75**, 582–585.

E143. Coutts, R.T., Nazarali, A.J., Baker, G.B., Pasutto, F.M., *Can. J. Physiol. Pharmacol.* (1986) **64**, 724–728.

E144. Mondschein, J.S., Hersey, R.M., Weisz, J., *Endocrinology (Baltimore)* (1986) **119**, 1105–1112.

E145. Miyagawa, E., Harada, T., Motoke, Y., *Agric. Biol. Chem.* (1986) **50**, 1527–1531.

E146. Rubio, M.A., Engesser, K.H., Knackmuss, H.J., *Arch. Microbiol.* (1986) **145**, 116–122.

E147. Youdim, M.B.H., Feldman, S.C., Pappas, G.D., Pollard, H.B., *Brain Res.* (1986) **381**, 300–304.

E148. Baker, G.B., Hampson, D.R., Coutts, R.T., Micetich, R.G., Hall, T.W., Rao, T.S., *J. Neural Transm.* (1986) **65**, 233–243.

E149. Shetty, H.U., Nelson, W.L., *J. Med. Chem.* (1986) **29**, 2004–2008.

E150. Baker, J.K., McChesney, J.D., Jorge, L.F., *Pharm. Res.* (1986) **3**, 132–141.

E151. Sun, Y., Li, Q., Chen, Q., *Nanjing Yaoxueyuan Xuebao* (1986) **17**, 132–135.

E152. Kuo, B.S., Ritschel, W.A., *Pharm. Res.* (1986) **3**. 173–177.

E153. Chiccarelli, F.S., Morrison, J.A., Cosulich, D.B., Perkinson, N.A., Ridge, D.N., Sum, F.W., Murdock, K.C., Woodward, D.L., Arnold, E.T., *Cancer Res.* (1986) **46**, 4858–4861.

E155. Kirsten, R., Nelson, K., Neff, J., Haerlin, R., Steinijans, V., Radtke, H.W., *Eur. J. Clin. Pharmacol.* (1986) **30**, 549–552.

E156. Omori, T., Ishigooka, H., Minoda, Y., *Agric Biol. Chem.* (1986) **50**, 1513–1518.

E158. Vajda, M., Tegyey, Z., Kozma, M., Vereczkey, L., *J. Pharm. Biomed. Anal.* (1986) **4**, 497–503.

E159. Sugiyama, M., Ando, T., Okuyama, Y., Sugimoto, T., Choukai, S., *Arzneim.-Forsch.* (1986) **36**, 1229–1236.

E160. Broggini, M., Sommacampagna, B., Paolini, A., Dolfini, E., Donnelli, M.G., *Cancer Treat. Rep.* (1986) **70**, 697–702.

E161. Iijima, H., Ebizuka, Y., Sankawa, U., *Chem. Pharm. Bull.* (1986) **34**, 3534–3537.

E162. Tsuja, H., Ogawa, T., Bando, N., Sasaoka, K., *J. Biol. Chem.* (1986) **261**, 13203–13209.

E163. Markus, A., Krekel, D., Lingens, F., *J. Biol. Chem.* (1986) **261**, 12883–12888.

E164. McClure, N.C., Venables, W.A., *J. Gen. Microbiol.* (1986) **132**, 2209–2218.

E166. Jacob, J., Schmoldt, A., Grimmer, G., *Cancer Lett.* (*Shannon, Irel.*) (1986) **32**, 107–116.

E167. Fujihara, M., Nakamura, T., Takahashi, M., Hirakoso, K., Furuuchi, S., Harigaya, S., *Oyo Yakuri* (1986) **32**, 57–69.

E168. Mikesell, M.D., Boyd, S.A., *Appl. Environ. Microbiol.* (1986), **52**, 861–865.

E169. Gorrod, J.W., Bayraktar, N.N., *Anticancer Res.* (1986) **6**, 729–731.

E171. Beck, O., Faull, K.F., Repke, D.M., *Naunyn-Schmiedeberg's Arch. Pharmacol.* (1986) **333**, 307–312.

E173. Hopper, D.J., *Biochem. J.* (1986) **239**, 469–472.

E174. Katayama, T., Nakatsubo, F., Higuchi, T., *Mokuzai Gakkaishi* (1986) **32**, 535–544.

E175. El-Mansi, E.M.T., *J. Appl. Bacteriol.* (1986) **60**, 545–555.

E176. Bossert, I.D., Young, L.Y., *Appl. Environ. Microbiol.* (1986) **52**, 1117–1122.

E177. Demassieux, S., Romain, Y., Bordeleau, L., Gravel, D., Carriere, S., *Biochem. Cell Biol.* (1986) **64**, 1044–1048.

E178. Ansher, S.S., Cadet, J.L., Jakoby, W.B., Baker, J.K., *Biochem. Pharmacol.* (1986) **35**, 3359–3363.

E179. Müller, S., Schmidt, A., *Z. Naturforsch. C: Biosci.* (1986) **41**, 820–824.

E180. Hrazdina, G., Lifson, E., Weeden, N.F., *Arch. Biochem. Biophys.* (1986) **247**, 414–419.

E181. Cerniglia, C.E., Kelly, D.W., Freeman, J.P., Miller, D.W., *Chem.-Biol. Interact.* (1986) **57**, 203–216.

E182. Krumholz, L.R., Crawford, R.L., Hemling, M.E., Bryant, M.P., *Prog. Clin. Biol. Res.* (1986) **213**, 211–214.

E184. Kobayashi, S., Gao, Y., Ong, R.L., Pittman, C.S., *Life Sci.* (1986) **38**, 2231–2238.

E185. Kobayashi, S., Gao, Y., Pittman, C.S., *Endocrinol. Jpn.* (1985) **32**, 781–792.

E186. Toth, L.A., Scott, M.C., Elchisak, M.A., *Life Sci.* (1986) **39**, 519–526.

E187. Johnson, P.R.L., Zoegler, D.M., *J. Biochem. Toxicol.* (1986) **1**, 15–27.

E188. Rudas, P., Pethes, G., *Acta Vet. Hung.* (1985) **33**, 185–188.

E189. Samain, E., Albagnac, G., Dubourguier, H.C., *Arch. Microbiol.* (1986) **144**, 242–244.

E190. Kachi, H., Sassa, T., *Agric. Biol. Chem.* (1986) **50**, 1669–1671.

E191. Kim, D.-H., Konishi, L., Kobashi, K., *Biochim. Biophys. Acta* (1986) **872**, 33–41.

E192. Santoro, P., Parisi, G., *J. Exp. Zool.* (1986) **239**, 169–173.

E193. Boye, N., *Acta Endocrinol.* (*Copenhagen*) (1986) **112**, 536–540.

E194. Anderson, R.J., Yoon, J.K., Sinsheimer, E.G., Jackson, B.L., *Neuroendocrinology* (1986) **44**, 117–124.

E195. Romain, Y., Demassieux, S., D'Angelo, G., Gyger, M., Carriere, S., *Can. J. Physiol. Pharmacol.* (1986) **64**, 1197–1201.

E196. McNabb, F.M.A., Lyons, L.J., Hughes, T.E., *Comp. Biochem. Physiol., A: Comp. Physiol.* (1986) **85A**, 249–255.

E197. Vargas, F., My Dam Trung Toung, Schwartz, J.C., *J. Enzyme Inhib.* (1986) **1**, 105–112.

E198. Whittemore, R.M., Pearce, L.B., Roth, J.A., (1986) *Arch. Biochem. Biophys.* **249**, 464–467.

E200. Nakamura, J., Inoue, Y., Sasaki, H., Shibasaki, J., *Chem. Pharm. Bull.* (1986) **34**, 2624–2627.

E201. Jindal, S.P., Lutz, T., *J. Anal. Toxicol.* (1986) **10**, 150–155.

E202. Koruna, I., Ryska, M., Kleinerová, E., *Zfi.-Mitt.* (1986) **115**, 149–150.

E203. Fischer, V., Harman, L.S., West, P.R., Mason, R.P., *Chem.-Biol. Interact.* (1986) **60**, 115–127.

E204. Krug, M., Straube, G., *J. Basic Microbiol.* (1986) **26**, 271–281.

E205. Stock, B.H., Schreiber, J., Guenat, C., Mason, R.P., Bend, J.R., Eling, T.E., *J. Biol. Chem.* (1986) **261**, 15915–15922.

E206. Helander, A., Tottmar, O., *Biochem. Pharmacol.* (1987) **36**, 1077–1082.

E207. Huang, C.-Y., Cridland, J.S., Ivanetich, K.M., *Biochem. Pharmacol.* (1987) **36**, 689–697.

E208. Nakamura, J., Nakamura, T., Podder, S.K., Sasaki, J., Shibasaki, J., *Biochem. Pharmacol.* (1987) **36**, 1171–1174.

E209. Ankley, G.T., Reinert, R.E., Mayer, R.T., Burke, M.D., Agosin, M., *Biochem. Pharmacol.* (1987) **36**, 1379–1381.

E210. Sitaram, B.R., Talomisin, R., Blackman, G.L., McLeod, W.R., *Biochem. Pharmacol.* (1987) **36**, 1503–1508.

E211. Campbell, N.R.C., Van Loon, J.A., Weinshilboum, R.M., *Biochem. Pharmacol.* (1987) **36**, 1435–1446.

E212. Sitaram, B.R., Lockett, L., Talomsin, R., Blackman, G.L., McLeod, W.R., *Biochem. Pharmacol.* (1987) **36**,1509–1512.

E213. Parr, I.B., McCague, R., Leclercq, G., Stoessel, S., *Biochem. Pharmacol.* (1987) **36**, 1513–1519.

E214. Powis, G., see, K.L., Santone, K.S., Melder, D.C., Hodnett, E.M., *Biochem. Pharmacol.* (1987) **36**, 2473–2479.

E215. Dull, B.J., Salata, K., Van Langenhove, A., Goldman, P., *Biochem. Pharmacol.* (1987) **36**, 2467–2472.

E216. Cumming, P., Boyes, B.E., Martin, W.R.W., Adam, M., Ruth, T.J., McGeer, E.C., *Biochem. Pharmacol.* (1987) **36**, 2527–2531.

E217. Meunier, G., Bernadou, J., Meunier, B., *Biochem. Pharmacol.* (1987) **36**, 2599–2604.

E218. La Cagnin, L.B., Colby, H.D., Dalal, N.S., O'Donnell, J.P., *Biochem. Pharmacol.* (1987) **36**, 2667–2672.

E219. Meerman, J.H.N., Nijland, C., Mulder, G.J., *Biochem. Pharmacol.* (1987) **36**, 2605–2608.
E220. Nakajima, T., Okino, T., Sato, A., *Biochem. Pharmacol.* (1987) **36**, 2799–2804.
E221. Heubel, F., Gerstner, E., Grotjahn, H., *Biochem. Pharmacol.* (1987) **36**, 2867–2869.
E222. Johnson, C., Beedham, C., Stell, J.G.P., *Xenobiotica* (1987) **17**, 17–24.
E223. van der Straat, R., de Vries, J., de Boer, H.J.R., Vromans, R.M., Vermeulen, N.P.E., *Xenobiotica* (1987) **17**, 1–9.
E224. Mulder, G.J., Groen, K., Tijdens, R.B., *Xenobiotica* (1987) **17**, 85–92.
E225. Radulovic, L.L., Kulkarni, A.P., Dauterman, W.C., *Xenobiotica* (1987) **17**, 105–114.
E226. Sagami, I., Ohmachi, T., Fujii, H., Watanabe, M., *Xenobiotica* (1987) **17**, 189–198.
E227. Gorrod, J.W., Gooderham, N.J., *Xenobiotica* (1987) **17**, 165–177.
E229. Welch, R.M., Lai, A.A., Schroeder, D.H., *Xenobiotica* (1987) **17**, 287–298.
E230. Gau, W., Kurz, J., Petersen, U., Ploschke, H.J., Wuensche, C., *Arzneim.-Forsch.* (1986) **36**, 1545–1549.
E231. Caputo, O., Viola, F., Grosa, G., Rocco, F., Biglino, G., *Eur. J. Drug Metab. Pharmacokinet.* (1986) **11**, 91–99.
E232. Tasker, R.A.R., Nakatsu, K., *Can. J. Physiol. Pharmacol.* (1986) **64**, 1160–1163.
E233. Bodd, E., Solberg Christophersen, A., Fongen, U., *Acta Pharmacol. Toxicol.* (1986) **59**, 252–253.
E234. Schmauder, H.-P., Gröger, D., *Planta Med.* (1986) 395–397.
E235. Boberg, E.W., Miller, E.C., Miller, J.A., *Chem.-Biol. Interact.* (1986) **59**, 73–97.
E236. Kamil, W.M., Dewick, P.M., *Phytochemistry* (1986) **25**, 2089–2092.
E237. Kamil, W.M., Dewick, P.M., *Phytochemistry* (1986) **25**, 2093–2102.
E238. Clement, B., Zimmermann, M., *Xenobiotica* (1987) **17**, 659–667.
E239. Kamimura, H., Enjoji, Y., Sasaki, H., Kawai, R., Kaniwa, H., Niigata, K., Kageyama, S., *Xenobiotica* (1987) **17**, 645–658.
E240. Raza, H., Levine, W.G., Roy Chowdhury, N., Roy Chowdhury, J., *Xenobiotica* (1987) **17**, 669–677.
E241. Strong, H.A., Renwick, A.G., George, C.F., Liu, Y.F., Hill, M.J., *Xenobiotica* (1987) **17**, 685–696.
E242. Fry, J.R., *Xenobiotica* (1987) **17**, 751–758.
E243. Paulson, G.D., Feil, V.J., Macgregor, J.T., *Xenobiotica* (1987) **17**, 697–707.
E244. Tonda, K., Kamata, S., Hirata, M., *Xenobiotica* (1987) **17**, 759–768.
E245. Gott, D.M., Griffiths, L.A., *Xenobiotica* (1987) **17**, 423–434.
E246. Klasson-Wehler, E., Bergman, Å., Kowalski, B., Brandt, I., *Xenobiotica* (1987) **17**, 477–486.
E247. Sullivan, H.R., Hanasono, G.K., Miller, W.M., Wood, P.G., *Xenobiotica* (1987) **17**, 459–468.
E248. Jezo, I., Zemek, J., *Chem. Pap.* (1986) **40**, 279–281.
E249. Lenk, W., Sterzl, H., *Xenobiotica* (1987) **17**, 499–512.
E250. Bougerolle, A.M., Chabard, J.L., Dordain, G., Frydman, A.M., Gaillot, J., Piron, J.J., Petit, J., Berger, J.A., *Eur. J. Drug Metab. Pharmacokinet.* (1986) **11**, 113–118.
E251. Siefert, H.M., Maruhn, D., Maul, W., Förster, D., Ritter, W., *Arzneim.-Forsch.* (1986) **36**, 1496–1502.
E252. Raha, C., Hart-Anstey, M., Cheung, M.S., Bresnick, E., *J. Toxicol. Environ. Health* (1986) **19**, 55–64.
E253. Nguyen-Legros, J., Martin-Martinelli, E., Simon, A., Denoroy, L., Vigny, A., *Exp. Eye Res.* (1986) **43**, 575–584.
E254. Dock, L., Waern, F., Martinez, M., Grover, P.L., Jernstrom, B., *Chem.-Biol. Interact.* (1986) **58**, 301–318.

E255. Babason, J.R., Russo-Rodriguez, S.E., Wattley, R.U., Bergstein, P.L., Rastetter, W.H., Liber, H.L., Andon, B.M., Thilly, W.G., Wogan, N., *Toxicol. Appl. Pharmacol.* (1986) **85**, 355–366.

E256. Rueffer, M., Amann, M., Zenk, M.H., *Plant Cell Rep.* (1986) **3**, 182–185.

E257. Tottmar, O., *Anal. Biochem.* (1986) **158**, 6–11.

E258. Toshimitsu, N., Hamada, H., Kojima, M., *J. Ferment. Technol.* (1986) **64**, 459–462.

E259. Ogasawara, Y., Kusama, M., *Kenkyu Kiyo-Tokyo Kasei Daigaku* (1985) **25**, 167–172.

E260. de Bont, J.A.M., Vorage, M.J.A.W., Hartmans, S., van den Tweel, W.J.J., *Appl. Environ. Microbiol.* (1986) **52**, 677–680.

E261. Mangold, J.B., Mangold, B.L.K., Spina, A., *Biochim Biophys. Acta* (1986) **874**, 37–43.

E262. Gupta, J.K., Jebsen, C., Kneifel, H., *J. Gen. Microbiol.* (1986) **132**, 2793–2799.

E263. Heresztyn, T., *Arch. Microbiol.* (1986) **146**, 96–98.

E264. Olsen, M., Mirocha, C.J., Abbas, H.K., Johansson, B., *Poult. Sci.* (1986) **65**, 1905–1910.

E265. Steiert, J.G., Crawford, R.l., *Biochem. Biophys. Res. Commun.* (1986) **141**, 825–830.

E266. Granneman, G.R., Snyder, K.M., Shu, V.S., *Antimicrob. Agents Chemother.* (1986) **30**, 689–693.

E267. Reid, J.J., Stitzel, R.E., Head, R.J., *Naunyn-Schmiedeberg's Arch. Pharmacol.* (1986) **334**, 17–28.

E268. Durand, M., *Neurosci. Lett.* (1986) **72**, 174–178.

E269. Gonmori, K., Rao, K.S.P., Mehendale, H.M., *Exp. Lung Res.* (1986) **11**, 295–305.

E270. Powlowski, J.B., Dagley, S., Massey, V., Ballou, D.P., *J. Biol. Chem.* (1987) **262**, 69–74.

E271. Yamazaki, Y., Maeda, H., *Agric. Biol. Chem.* (1986) **50**, 2621–2631.

E272. Kyle, M.E., Kocsis, J.J., *Chem.-Biol. Interact.* (1986) **59**, 325–335.

E273. Sun, Y., Chen, Q., *Yaoxue Xuebao* (1986) **21**, 748–752.

E274. Latriano, L., Goldstein, B.D., Witz, G., *Proc. Natl. Acad. Sci. U.S.A.* (1986) **83**, 8356–8360.

E275. Weems, H.B., Fu, P.P., Yang, S.K., *Carcinogenesis* (1986) **7**, 1221–1230.

E276. Hasmann, M.J., Seidl, P.H., Engelhardt, G., Schliefer, K.H., *Arch. Microbiol.* (1986) **146**, 275–279.

E277. Cichowicz, D.J., Shane, B., *Biochemistry* (1986) **26**, 504–512.

E278. Miles, E.W., *Biochemistry* (1986) **26**, 597–603.

E279. Kergomard, A., Renard, M.F., *Agric. Biol. Chem.* (1986) **50**, 2913–2914.

E280. Petersen, M., Seitz, H.U., *J. Plant Physiol.* (1986) **125**, 383–390.

E281. Shirai, K., *Agric. Biol. Chem.* (1986) **50**, 2875–2880.

E282. Abdel-Maksoud, H.M., Neumeyer, J.L., Trainor, T.M., Davis, P.J., Vouros, P., *Stud. Org. Chem. (Amsterdam)* (1986) **26**, 357–370.

E283. Dayer, P., Leemann, T., Küpfer, A., Kronbach, T., Meyer, U.A., *Eur. J. Clin. Pharmacol.* (1986) **31**, 313–318.

E284. Tsuda, M., Oikawa, S., Okamura, Y., Kimura, K., Urabe, T., Nakajima, M., *Chem. Pharm. Bull.* (1986) **34**, 4457–4466.

E285. Podder, S.K., Nakashima, M., Nakamura, T., Sasaki, H., Nakamura, J., Shibasaki, J., *J. Pharmacobio.-Dyn.* (1986) **9**, 917–922.

E286. Muntaz, M., Moizuddin, M., Menzer, R.E., *J. Agric. Food Chem.* (1986) **34**, 929–936.

E287. Wang, L.-H., Hamzah, R.Y., Yu, Y., Tu, S.-C., *Biochemistry* (1987) **26**, 1099–1104.

E288. Suzuki, K., Itoh, M., *Plant Cell Physiol.* (1986) **27**, 1451–1460.

E289. Yu, Y., Wang, L.-H., Tu, S.-C., *Biochemistry* (1987) **26**, 1105–1110.

E290. van den Tweel, W.J.J., ter Burg, N., Kok, J.B., de Bont, J.A.M., *Appl. Microbiol. Biotechnol.* (1986) **25**, 289–294.

E291. Ward, L.A., Johnson, K.A., Robinson, I.M., Yokoyama, M.T., *Appl. Environ. Microbiol.* (1987) **53**, 189–192.

E292. Sugeno, K., Norikura, R., Yoshimori, T., Koike, M., Futaguchi, S., Tanaka, H., Inazawa, K., Muzojiri, K., *Iyaguhin Kenkyu* (1986) **17**, 1053–1062.

E293. Berry, D.F., Madsen, E.L., Bollag, J.M., *Appl. Environ. Microbiol.* (1987) **53**, 180–182.

E294. Shinozaki, Y., Urakubo, G., *Iyakuhin Kenkyu* (1986) **17**, 1070–1077.

E295. Mico, B.A., Swagzdis, J.E., Federowicz, D.A., Straub, K., *J. Pharm. Sci.* (1986) **75**, 929–933.

E296. Rietveld, E.C., Hendrickx, M.M.P., Seutter-Berlage, F., *Arch. Toxicol.* (1986) **59**, 228–234.

E297. Eastmond, D.A., Smith, M.T., Ruzo, L.O., Ross, D., *Mol. Pharmacol.* (1986) **30**, 674–679.

E298. Millner, G.C., Fu, P.P., Cerniglia, C.E., *J. Toxicol. Environ. Health* (1986) **19**, 519–530.

E299. Rueffer, M., Zenk, M.H., *Tetrahedron Lett.* (1986) **27**, 5603–5604.

E300. Fujii, I., Iijima, H., Tsukita, S., Ebizuka, Y., Sankawa, U., *J. Biochem.* (1987) **101**, 11–18.

E301. Stinson, E.E., Moreau, R.A., *Phytochemistry* (1986) **25**, 2721–2724.

E302. Oda, T., Sato, Y., *Chem. Pharm. Bull.* (1986) **34**, 4635–4640.

E303. Van den Tweel, W.J.J., Janssens, R.J.J., De Bont, J.A.M., *Antonie van Leeuwenhoek* (1986) **52**, 309–318.

E304. Crespi-Perellino, N., Guicciardo, A., Malyszko, G., Minghetti, A., *J. Nat. Prod.* (1986) **49**, 814–822.

E305. DeWeerd, K.A., Suflita, J.M., Linkfield, T., Tiedje, J.M., Pritchard, P.H., *FEMS Microbiol. Ecol.* (1986) **38**, 331–339.

E306. Wiese, G., Granbow, H.J., *Phytochemistry* (1986) **25**, 2451–2455.

E307. Lee, T.T., Starratt, A.N., *Phytochemistry* (1986) **25**, 2457–2461.

E308. Ortiz de Montellano, P.R., Grab, L.A., *Mol. Pharmacol.* (1986) **30**, 666–669.

E309. Chidgey, M.A.J., Kennedy, J.F., Caldwell, J., *Food Chem. Toxicol.* (1986) **24**, 1267–1272.

E310. Neidlein, R., Kleiser, M., *Arzneim.-Forsch.* (1987) **37**, 32–37.

E311. Susilo, R., Rommelspacher, H., *Naunyn-Schmiedeberg's Arch Pharmacol.* (1987) **335**, 70–76.

E312. Kvasničková, E., Nobilis, M., Šroler, A., Báčová, E., Hais, I.M., *J. Chromatogr.* (1987) **387**, 559–561.

E313. Wolf, M.E., Mosnaim, A.D., Callaghan, O.H., Chevesich, J., Caro, M., *Life Sci.* (1987) **40**, 489–494.

E314. Godin, C.S., Crooks, P.A., Damani, L.A., *Toxicol. Lett.* (1986) **34,** 217–222.

E315. Schraa, G., Boone, M.L., Jetten, M.S.M., van Neerven, A.R.W., Colberg, P.J., Zehnder, A.J.B., *Appl. Environ. Microbiol.* (1986) **52**, 1374–1381.

E316. Onishi, N., Yokozeki, K., Hirose, Y., Kubota, K., *Agric. Biol. Chem.* (1987) **51**, 291–292.

E317. Arjmand, M., Sandermann, H., Jr., *Pestic. Biochem. Physiol.* (1987) **27**, 173–181.

E318. Bellucci, G., Berti, G., Chiappe, C., Lippi, A., Marioni, F., *J. Med. Chem.* (1987) **30**, 768–773.

E319. Matthews, D.E., Weiner, E.J., Matthews, P.S., Van Etten, H.D., *Plant Physiol.* (1987) **83**, 365–370.

E320. Vachta, J., Valter, K., Gold-Aubert, P., *Eur. J. Drug Metab. Pharmacokinet.* (1986) **11**, 195–204.

E322. Pelsy, F., Leroux, P., Heslot, H., *Pestic. Biochem. Physiol.* (1987) **27**, 182–188.

E323. Ogushi, S., Koga, S., ito, K., Makino, Y., Ando, M., Tsuru, D., *Agric. Biol. Chem.* (1986) **50**, 3093–3100.

E324. Apajalahti, J.H.A., Salkinoja-Salonen, M.J., *J. Bacteriol.* (1987) **169**, 675–681.

E325. Grbic-Galic, D., *J. Appl. Bacteriol.* (1986) **61**, 491–497.

E326. Ogunbona, F.A., Fabayo, A.B., *Dev. Drugs Mod. Med.* (1986) 324–330.

E327. Parrott, S., Jones, S., Cooper, R.A., *J. gen. Microbiol.* (1987) **133**, 347–351.

E328. Misra, A.L., Pontani, R.B., Vadlamani, N.L., *J. Pharm. Pharmacol.* (1987) **39**, 225–227.

E329. Furukawa, K., Arimura, N., *J. Bacteriol.* (1987) **169**, 924–927.

E330. Duffel, M.W., Janss, M.N., *Adv. Exp. Med. Biol.* (1986) **197**, 415–422.

E331. Hongo, S., Ito, H., Takeda, M., Sato, T., *Enzyme* (1987) **36**, 232–238.

E332. Ueno, T., Morishima, I., Sugiura, S., Araki, T., Yoshizako, F., *Agric. Biol. Chem.* (1987) **51**, 947–948.

E333. Ellaiah, P., Krishna, K.T., *Indian Drugs* (1987) **24**, 192–195.

E334. Karoum, F., Chuang, L.W., *Biog. Amines* (1987) **4**, 1–7.

E335. van den Tweel, W.J.J., Kok, J.B., de Bont, J.A.M., *Appl. Environ. Microbiol.* (1987) **53**, 810–815.

E336. Agarwal, S.C., basu, S.K., Vora, V.C., Mason, J.R., Pirt, S.J., *Biotechnol. Bioeng.* (1987) **29**, 783–785.

E337. Numazawa, M., Osawa, Y., *J. Steroid Biochem.* (1987) **26**, 137–143.

E338. Mizukami, H., Hirano, A., Ohashi, H., *Plant Sci.* (Limerick, Irel.) (1987) **48**, 11–15.

E339. Nilsson, G.E., Tottmar, O., *J. Neurochem.* (1987) **48**, 1566–1572.

E340. Takimoto, Y., Ohshima, M., Miyamoto, J., *Ecotoxicol. Environ. Saf.* (1987) **13**, 126–134.

E341. Vargas, F., Schwartz, J.C., *FEBS Lett.* (1987) **211**, 234–238.

E342. Papadopoulos, A.S., Crammer, J.L., *Methodol. Surv. Biochem. Anal.* (1986) **16**, 207–210.

E343. Gurujeyalakshmi, G., mahadevan, A., *Zentralbl. Mikrobiol.* (1987) **142**, 187–192.

E344. Tonda, K., Hirata, M., *J. Pharmacobio.-Dyn.* (1987) **10**, 15–20.

E345. Foster, G.D., Crosby, D.G., *Environ. Toxicol. Chem.* (1986) **5**, 1059–1070.

E346. Spicer, L.J., Walega, M.A., Hammond, J.M., *Biol. Reprod.* (1987) **36**, 562–571.

E347. Carstam, R., Edner, C., Hansson, C., Lindbladh, C., Rorsman, H., Rosengren, E., *Acta Derm.-Venereol., Suppl.* (1986) **126**, 1–12.

E348. Karai, N., Kato, T., Katsuyama, M., nakamura, M., Katsube, J., *Life Sci.* (1987) **40**, 2261–2268.

E349. Yinon, J., Hwang, D.G., *J. Energ. Mater.* (1986) **4**, 305–313.

E350. Hashimoto, T., Yamada, Y., *Eur. J. Biochem.* (1987) **164**, 277–285.

E351. Khouri, H.E., Ishikura, N., Ibrahim, R.K., *Phytochemistry* (1986) **25**, 2475–2479.

E352. Helmlinger, J., Rausch, T., Hilgenberg, W., *Phytochemistry* (1987) **26**, 615–618.

E353. Effenberger, F., Ziegler, T., Förster, S., *Angew. Chem.* (1987) **99**, 491–492.

E354. Jenkins, R.O., Stephens, G.M., Dalton, H., *Biotechnol. Bioeng.* (1987) **29**, 873–883.

E355. Van den Tweel, W.J.J., De Bont, J.A.M., *J. Gen. Microbiol.* (1987) **133**, 745–754.

E356. Takahashi, Y., *Jpn. J. Vet. Sci.* (1986) **48**, 999–1002.

E357. Isaac, J.E., Robins, R.J., Rhodes, M.J.C., *Phytochemistry* (1987) **26**, 393–399.

E358. Tatsumi, K., Ishigai, M., *Arch. Biochem. Biophys.* (1987) **253**, 413–418.

E359. Thiele, J., Müller, R., Lingens, F., *FEMS Microbiol. Lett.* (1987) **41**, 115–119.

E360. Smithers, G.W., jahansouz, H., Kofron, J.L., Himes, R.H., Reed, G.H., *Biochemistry* (1987) **26**, 3943–3948.

E361. McIntire, W.S., Hopper, D.J., Singer, T.P., *Biochemistry* (1987) **26**, 4107–4117.
E362. Yeulet, S.E., Mantle, P.G., *FEMS Microbiol. Lett.* (1987) **41**, 207–210.
E363. Rooda, S.J.E., Van Loon, M.A.C., Visser, T.J., *J. Clin. Invest.* (1987) **79**, 1740–1748.
E364. Yang, S.K., Mushtaq, M., Chiu, P.L., Weems, H.B., *Adv. Exp. Med. Biol.* (1986) **197**, 809–818.
E365. Ghersi-Egea, J.F., Walther, B., Decolin, D., Minn, A., Siest, G., *Neuropharmacology* (1987) **26**, 367–372.
E366. Gold, B., Farber, J., Rogan, E., *Chem.-Biol. Interact.* (1987) **61**, 215–228.
E367. Ribeiro, O., Kirkby, C.A., Hirom, P.C., Millburn, P., *Adv. Exp. Med. Biol.* (1986) **197**, 833–842.
E368. Simpsin, H.D., Green, J., Dalton, H., *Biochem. J.* (1987) **244**, 585–590.
E369. Morohoshi, N., Wariishi, H., Muraiso, C., Nagai, T., Haraguchi, T., *Mokuzai Gakkaishi* (1987) **33**, 218–225.
E370. Kraft, B., Schwenen, L., Stöckl, D., Barz, W., *Arch. Microbiol.* (1987) **147**, 201–206.
E371. Naoi, M., Takahashi, T., Kuno, N., Nagatsu, T., *Biochem. Biophys. Res. Commun.* (1987) **143**, 482–488.
E373. Bhatnagar, D., McCormick, S.P., Lee, L.S., Hill, R.A., *Appl. Environ. Microbiol.* (1987) **53**, 1028–1033.
E374. Malarczyk, E., Korszen-Pilecka, I., Rogalski, J., Leonowicz, A., *Phytochemistry* (1987) **26**, 1321–1324.
E375. Chen, W., Ohmiya, K., Shimizu, S., Kawakami, H., *J. Ferment. Technol.* (1987) **65**, 221–224.
E376. Campbell, I.M., Gallo, M.A., Jones, C.A., La Sitis, P.R., Rosato, L.M., *Phytochemistry* (1987) **26**, 1413–1415.
E377. Schmidt, E., Vasić-Rački, D., Wandrey, C., *Appl. Microbiol. Biotechnol.* (1987) **26**, 42–48.
E380. Vakkuri, O., Tervo, J., Luttinen, R., Ruotsalainen, H., Rahkamaa, E., Leppaluoto, J., *Endocrinology (Baltimore)* (1987) **120**, 2453–2459.
E381. Yang, S.K., Mushtaq, M., Weems, H.B., *Arch. Biochem. Biophys.* (1987) **255**, 48–63.
E382. Wand, M.D., Thompson, J.A., *Adv. Exp. Med. Biol.* (1986) **197**, 337–346.
E383. Appel, K.E., Görsdorf, S., Scheper, T., Bauszus, M., Hildebrandt, A.G., *Arch. Toxicol.* (1987) **60**, 204–208.
E384. Rougraff, P.M., Paxton, R., *Comp. Biochem. Physiol. B: Comp. Bioochem.* (1987) **86B**, 601–606.
E385. Kanda, M., Hori, K., Kurotsu, T., Miura, S., Saito, Y., *J. Biochem.* (1987) **101**, 871–878.
E386. Boulton, A.A., Davis, B.A., *Biomed. Environ. Mass Spectrom.* (1987) **14**, 207–211.
E387. Santoro, P., Parisi, G., *Insect Biochem.* (1987) **17**, 635–638.
E388. Krumholz, L.R., Crawford, R.L., Hemling, M.E., Bryant, M.P., *J. Bacteriol.* (1987) **169**, 1886–1890.
E389. Wadleigh, R., Yu, S.J., *Insect Biochem.* (1987) **17**, 759–764.
E390. Brown, R.T., Polinsky, R.J., Kopin, I.J., *J. Lab. Clin. Med.* (1987) **110**, 119–124.
E391. Yokozeki, K., Nakamori, S., Yamanaka, S., Eguchi, C., Mitsugi, K., Yoshinaga, F., *Agric. Biol. Chem.* (1987) **51**, 715–719.
E392. Preiss, U., Engelhardt, G., Wallnoefer, P.R., Muecke, W., *Chemosphere* (1987) **16**, 963–968.
E393. Idaka, E., Ogawa, T., Horitsu, H., *Bull. Environ. Contam. Toxicol.* (1987) **39**, 100–107.
E394. Yang, S.K., Mushtaq, M., Weems, H.B., Miller, D.W., Fu, P.P., *Biochem. J.* (1987) **245**, 191–204.
E395. Idaka, E., Ogawa, T., Horitsu, H., *Bull. Environ. Contam. Toxicol.* (1987) **39**, 108–113.
E396. Reeve, C.D., Hopper, D.J., Carver, M.A., *Biochem. Soc. Trans.* (1987) **15**, 934–935.

E397. Wong, D.L., Yamasaki, L., Ciaranello, R.D., *Brain Res.* (1987) **410**, 32–45.

E398. Vaccaro, A.M., Salvioli, R., Muscillo, M., Renola, L., *Enzyme* (1987) **37**, 115–126.

E399. Andres, H.H., Vogel, R.S., Tarr, G.E., Johnson, L.L., Weber, W.W., *Mol. Pharmacol.* (1987) **31**, 446–456.

E400. Walsh, C.T., Erion, M.D., Walts, A.E., Delany, J.J. III, Berchtold, G.A., *Biochemistry* (1987) **26**, 4734–4745.

E401. Palcic, M.M., Shen, S.J., Schleicher, E., Kumagai, H., Sawada, S., Yamada, H., Floss, H.G., *Z. Naturforsch C: Biosci.* (1987) **42**, 307–318.

E402. Bourbonnais, R., Paice, M.G., *Appl. Microbiol. Biotechnol.* (1987) **26**, 164–169.

E403. Kobayashi, S., Nakano, M., Kimura, T., Schaap, AP., *Biochemistry* (1987) **26**, 5019–5022.

E404. Gurujeyalakshmi, G., Mahadevan, A., *Appl. Microbiol. Biotechnol.* (1987) **26**, 289–293.

E405. Ernstsen, A., Sandberg, G., Crozier, A., Wheeler, C.T., *Planta* (1987) **171**, 422–428.

E406. Chastain, J.L., McCormick, D.B., *J. Nutr.* (1987) **117**, 468–475.

E407. Barrett, F.M., *Can. J. Zool.* (1987) **65**, 1158–1166.

E408. Zürrer, D., Cook, A.M., Leisinger, T., *Appl. Environ. Microbiol.* (1987) **53**, 1459–1463.

E409. Orosz, D., Cohen, G., *Eur. J. Pharmacol.* (1987) **136**, 441–444.

E410. Setnikar, I., Arigoni, R., Chiste, R., Makovec, F., Revel, L., *Arzneim.-Forsch.* (1987) **37**, 698–702.

E411. Marriage, H.J., Harvey, D.J., *Res. Commun. Subst. Abuse* (1986) **7**, 89–97.

E412. Rogers, S.M., Back, D.J., Orme, M.L., *Br. J. Clin. Pharmacol.* (1987) **23**, 727–734.

E413. Schwartz, D.E., Eckert, G., Ekue, J.M.K., *Chemotherapy* (*Basel*) (1987) **33**, 305–308.

E414. Chou, M.W., Heflich, R.H., Fu, P.P., *Carcinogenesis* (1986) **7**, 1837–1844.

E415. Nagatsu, T., Togari, A., Naoi, M., Mino, N., Sawazaki, Y., Yoshida, E., Kanbayashi, N., Sakurai, E., Ohtawa, M., et al. *Biog. Amines* (1987) **4**, 107–117.

E416. Negishi, T., Nakano, M., Kobayashi, S., Kim, C.H., *Bull. Environ. Contam. Toxicol.* (1987) **39**, 294–298.

E417. Kamath, A.V., Dasgupta, D., Vaidyanathan, C.S., *Biochem. Biophys. Res. Commun.* (1987) **145**, 586–595.

E418. Amann, M., Wanner, G., Zenk, M.H., *Planta* (1986) **167**, 310–320.

E419. Cleveland, T.E., Lax, A.R., Lee, L.S., Bhatnagar, D., *Appl. Environ. Microbiol.* (1987) **53**, 1711–1713.

E420. Leete, E., *Can. J. Chem.* (1987) **65**, 226–228.

E421. Lewer, P., Bandurski, R.S., *Phytochemistry* (1987) **26**, 1247–1250.

E423. Asano, Y., Nakazawa, A., *Agric. Biol. Chem.* (1987) **51**, 2035–2036.

E424. Holder, C.L., Thompson, H.C., Jr., Gosnell, A.B., Siitonen, P.H., Korfmacher, W.A., Cerniglia, C.E., Miller, D.W., Casciano, D.A., Slikker, W., Jr., *J. Anal. Toxicol.* (1987) **11**, 113–121.

E425. Meuldermans, W., Hendrickx, J., Lauwers, W., Hurkmans, R., Swysen, E., Heykants, J., *Drug Dev. Res.* (1986) **8**, 37–51.

E426. Anundi, I., Kauffmann, F.C., El-Mouelhi, M., Thurman, R.G., *Arch. Toxicol.* (1987) **60**, 69–72.

E427. Mihailova, D., Velkov, M., Zhivkova, Z., *Eur. J. Drug Metab. Pharmacokinet.* (1987) **12**, 25–30.

E428. Zaidi, S.S.A., *Bull. Environ. Contam. Toxicol.* (1987) **39**, 327–333.

E429. Cossum, P.A., Rickert, D.E., *Toxicol. Lett.* (1987) **37**, 157–163.

E430. Weische, A., Johanni, M., Leistner, E., *Arch. Biochem. Biophys.* (1987) **256**, 212–222.

E431. Lee, F.T., Adams, J.B., *Biochem. Biophys. Res. Commun.* (1987) **144**, 569–575.
E432. Kobashi, K., Kim, D.-H., Morikawa, T., *J. Protein Chem.* (1987) **6**, 237–244.
E433. Hinderer, W., Flentje, U., Barz, W., *FEBS Lett.* (1987) **214**, 101–106.
E434. Keeran, W.S., Lee, R.F., *Arch. Biochem. Biophys.* (1987) **255**, 233–243.
E435. Fett, W.F., Osman, S.F., Dunn, M.F., *Appl. Environ. Microbiol.* (1987) **53**, 1839–1845.
E436. De-Eknamkul, W., Ellis, B.E., *Phytochemistry* (1987) **26**, 1941–1946.
E437. Tsurumi, S., *Shokubutsu no Kagaku Chosetsu* (1986) **21**, 126–133.
E438. Korfmacher, W.A., Holder, C.L., Betowski, L.D., Mitchum, R.K., *J. Anal. Toxicol.* (1987) **11**, 182–184.
E439. Katz, J., Finlay, T.H., Banerjee, S., Levitz, M., *J. Steroid Biochem.* (1987) **26**, 687–692.
E440. Lloyd, S.C., Foster, P.M.D., *Arch. Toxicol. Suppl.* (1986) **11**, 281–284.
E441. Leone, A.M., Francis, P.L., Silman, R.E., *J. Pineal Res.* (1987) **4**, 253–266.
E442. McCaffery, A.R., McDowell, P.G., *Pestic. Sci.* (1987) **19**, 185–196.
E443. Schramm, P., Rausch, T., Hilgenberg, W., *Plant Physiol.* (1987) **84**, 541–544.
E444. Taylor, D.C., Wightman, F., *Phytochemistry* (1987) **26**, 1279–1288.
E445. Asano, Y., Nakazawa, A., Endo, K., *J. Biol. Chem.* (1987) **262**, 10346–10354.
E447. Sedlmaier, H., Müller, F., Keller, P.J., Bacher, A., *Z. Naturforsch. C: Biosci.* (1987) **42**, 425–429.
E448. Yu, P.H., Sloley, B.D., *Comp. Biochem. Physiol. C, Comp. Pharmacol. Toxicol.* (1987) **87C**, 315–319.
E449. Van Eldere, J.R., De Pauw, G., Eyssen, H.J., *Appl. Environ. Microbiol.* (1987) **53**, 1655–1660.
E450. Yokozeki, K., Kubota, K., *Agric. Biol. Chem.* (1987) **51**, 721–728.
E451. Hummel, W., Schütte, H., Schmidt, E., Wandrey, C., Kula, M.R., *Appl. Microbiol. Biotechnol.* (1987) **26**, 409–416.
E452. Iwamura, S., Shibata, K., kawabe, Y., tsukamoto, K., Honma, S., *J. Pharmacobio.-Dyn.* (1987) **10**, 229–235.
E453. Rottero, A.E., Kissinger, P., *Biomed. Chromatogr.* (1987) **2**, 24–29.
E454. Ueji, M., Tomizawa, C., *Nippon Noyaku Gakkaishi* (1987) **12**, 269–271.
E455. Mohapatra, N., MacNair, P., Bryant, B.J., Ellis, S., Rudo, K., Sangaiah, R., Gold, A., Nesnow, S., *Mutat. Res.* (1987) **188**, 323–324.
E456. Ramaswamy, S.G., Jakoby, W.B., *J. Biol. Chem.* (1987) **262**, 10039–10043.
E457. Weische, A., Garvert, W., Leistner, E., *Arch. Biochem. Biophys.* (1987) **256**, 223–231.
E458. Gunata, Y.Z., Sapis, J.-C., Moutounet, M., *Phytochemistry* (1987) **26**, 1573–1575.
E459. van Bladeren, P.J., Balani, S.K., Sayer, J.M., Thakker, D.R., Boyd, D.R., Ryan, D.E., Thomas, P.E., Levin, W., Jerina, D.M., *Biochem. Biophys. Res. Commun.* (1987) **145**, 160–167.
E460. Garcia-Carmona, F., Cabanes, J., Garcia-Canovas, F., *Biochem. Int.* (1987) **14**, 1003–1013.
E461. Komives, E.A., Ortiz de Montellano, P.R., *J. Biol. Chem.* (1987) **262**, 9793–9802.
E462. Sariaslani, F.S., McGee, L.R., Ovenall, D.W., *Appl. Environ. Microbiol.* (1987) **53**, 1780–1784.
E463. Baggi, G., Barbieri, P., Galli, E., Tollari, S., *Appl. Environ. Microbiol.* (1987) **53**, 2129–2132.
E464. Fernstad, R., Pousette, A., Carlstrøm, K., *Gastroenterol.* (1987) **22**, 644–648.
E465. Watanabe, K., Yoshizawa, I., *J. Pharmacobio.-Dyn.* (1987) **10**, 302–308.
E466. Ueji, M., Tomizawa, C., *Nippon Noyaku Gakkaishi* (1987) **12**, 245–251.
E467. Raha, C.R., Cheung, M.S., Bresnick, E., *Toxicol. Lett.* (1987) **37**, 229–233.

E471. Lilienblum, W., Platt, K.L., Schirmer, G., Oesch, F., Bock, K.W., *Mol. Pharmacol.* (1987) **32**, 173–177.
E472. Okuno, E., Köhler, C., Schwarcz, R., *J. Neurochem.* (1987) **49**, 771–780.
E473. MacKintosh, R.W., Fewson, C.A., *Biochem. Soc. Trans.* (1987) **15**, 1048–1049.
E474. Treston, A.M., Plilippides, A., Jacobsen, N.W., Eadie, M.J., Hooper, W.D., *J. Pharm. Sci.* (1987) **76**, 496–501.
E475. Kasuya, F., Igarashi, K., Fukui, M., *J. Pharm. Sci.* (1987) **76**, 303–305.
E476. Ruenitz, P.C., Arrendale, R.F., George, G.D., Thompson, C.B., Mokler, C.M., Nanavati, N.T., *Cancer Res.* (1987) **47**, 4015–4019.
E477. Shibata, K., Igusa, R., Inoue, K., Makouyama, H., Nakajima, J., Fujino, A., Sekiya, T., Uchide, M., *Oyo Yakuri* (1987) **33**, 765–774.
E478. Richieri, P.R., Buckpitt, A.R., *J. Pharmacol. Exp.Ther.* (1987) **242**, 485–492.
E479. Van Maanen, J.M.S., De Vries, J., Pappie, D., Van den Akker, E., Vincent, M., Lafleur, M., Retel, J., van der Greef, J., Pinedo, H.M., *Cancer Res.* (1987) **47**, 4658–4662.
E480. Plakunov, I., Smolarck, T.A., Fischer, D.L., Wiley, J.C., Jr., Baird, W.M., *Carcinogenesis* (1987) **8**, 59–66.
E481. Möller, L., Rafter, J., Gustafsson, J.-Å., *Carcinogenesis* (1987) **8**, 637–645.
E482. Kobayashi, M., Frenzel, T., Lee, J.P., Zenk, M.H., Floss, H.G., *J. Am. Chem. Soc.* (1987) **109**, 6184–6185.
E483. Mushtaq, M., Yang, S.K., *Carcinogenesis* (1987) **8**, 705–709.
E484. Dorby, J.M., Taylor, D.G., Hopper, D.J., *J. Gen. Microbiol.* (1987) **133**, 2137–2146.
E485. Shimamura, M., Hayashi, T., Naruse, H., *Biomed. Res.* (1987) **8**, 247–251.
E486. Okuda, T., Itoh, S., Yamazaki, M., Nakahama, H., Fukuhara, Y., Orita, Y., *Chem. Pharm. Bull.* (1987) **35**, 3516–3518.
E487. Dobashi, G., Ota, H., *Jpn. Kokai Tokkyo Koho JP* 61,271,992 (2–12–1986).
E488. Tanaka, T., Kaneo, Y., Goromaru, T., Iguchi, S., *Chem. Pharm. Bull.* (1987) **35**, 3519–3522.
E489. Wagner, J.G., Rogge, M.C., Natale, R.B., Albert, K.S., Szpuznar, G.J., *Biopharm. Drug Dispos.* (1987) **8**, 405–425.
E490. Hsin, C.Y., Coats, J.R., *Nippon Noyaku Gakkaishi* (1987) **12**, 405–413.
E491. Proksch, P., Isman, M.B., Witte, L., Hartmann, T., *Phytochemistry* (1987) **26**, 2227–2230.
E492. Wong, K.P., *Biog. Amines* (1987) **4**, 189–200.
E493. Rüttimann, C., Seelenfreund, D., Vicuña, R., *Enzyme Microb. Technol.* (1987) **9**, 526–530.
E494. Yourtee, D.M., Bean, T.A., Kirk-Yourtee, C.L., *Toxicol. Lett.* (1987) **38**, 213–224.
E495. Köppel, C., Tenczer, J., Arndt, I., Ibe, K., *Arzneim.-Forsch.* (1987) **37**, 1062–1064.
E496. Gower, J.D., Wills, E.D., *Chem.-Biol. Interact.* (1987) **63**, 63–74.
E497. Namboordiri, M.A.A., Brownstein, M.J., Weller, J., Voisin, P., Klein, D.C., *J. Pineal Res.* (1987) **4**, 235–246.
E500. Taylor, D.C., Wightman, T., *Phytochemistry* (1987) **26**, 2167–2179.
E502. Hasegawa, Y., Higuchi, K., Obata, H., Yoshizako, F., Nishimura, A., Chubachi, M., Tokuyama, T., *Nippon Nogei Kagaku Kaishi* (1987) **61**, 107–112.
E503. Hasegawa, H., Yanagisawa, M., Inoue, F., Yanaihara, N., Ichiyama, A., *Biochem. J.* (1987) **248**, 501–509.
E504. Walton, M.I., Workman, P., *Biochem. Pharmacol.* (1987) **36**, 887–896.
E505. Susilo, R., Damm, H., Rommelspacher, H., *Neurosci. Lett.* (1987) **81**, 325–330.
E506. Sitaram, B.R., Lockett, L., Blackman, G.L., McLeod, W.R., *Biochem. Pharmacol.* (1987) **36**, 2235–2237.

E507. Krumholz, L.R., Bryant, M.P., *J. Bacteriol.* (1988) **170**, 2472–2479.
E508. Shapiro, B., Gross, M.D., *Med. Pediat. Oncol.* (1987) **15**, 170–177.
E509. Apajalahti, J.H., Salkinoja-Salonen, M.S., *J. Bacteriol.* (1987) **169**, 5125–5130.
E510. Grech-Belanger, O., Turgeon, J., Lalande, M., *Res. Commun. Chem. Pathol. Pharmacol.* (1987) **58**, 53–62.
E512. Cleveland, T.E., Bhatnagar, D., *Can. J. Microbiol.* (1987) **33**, 1108–1112.
E513. Laham, S., Potvin, M., *Drug Chem. Toxicol.* (1987) **10**, 209–225.
E514. McMillan, D.C., Fu, P.C., Cerniglia, C.E., *Appl. Environ. Microbiol.* (1987) **53**, 2560–2566.
E515. Bokkenheuser, V.D., Shackleton, C.H.L., Winter, J., *Biochem. J.* (1987) **248**, 953–956.
E516. Seymour, M.A., Millburn, P., Tait, G.H., *Biochem. Soc. Trans.* (1987) **15**, 1108–1109.
E517. Sharma, V., *Indian J. Chem., Sect. B* (1987) **26B**, 589–591.
E518. Susilo, R., Höfle, G., Rommelspacher, H., *Biochem. Biophys. Res. Commun.* (1987) **148**, 1045–1052.
E519. Emons, G., Merriam, G.R., Pfeiffer, D., Loriaux, D.L., Ball, P., Knuppen, R., *J. Steroid Biochem.* (1987) **28**, 499–504.
E520. Ahokas, J.T., Davies, C., Jacobsen, N., Karki, N.T., Philippides, A., Treston, A.M., *Pharmacol. Toxicol. (Copenhagen)* (1987) **61**, 184–190.
E522. Yang, S.K., Bao, Z.P., *Mol. Pharmacol.* (1987) **32**, 73–80.
E523. McKay, S., Grover, P.L., *Cancer Lett. (Shannon, Irel.)* (1987) **37**, 71–79.
E525. Sugita, O., Miyairi, S., Sassa, S., Kappas, A., *Biochem. Biophys. Res. Commun.* (1987) **147**, 1245–1250.
E526. Tahara, S., Ingham, J.L., *Z. Naturforsch.C: Biosci.* (1987) **42**, 1050–1054.
E528. Levett, P.N., *Lett. Appl. Microbiol.* (1987) **5**, 71–73.
E529. Hrazdina, G., *Bull. Liason-Groupe Polyphenols* (1986) **13**, 27–32.
E530. Gouyette, A., Voisin, E., Auclair, C., Paoletti, C., *Anticancer Res.* (1987) **7**, 823–827.
E531. Negrel, J., Jeandet, P., *Phytochemistry* (1987) **26**, 2185–2190.
E532. Kim, Y.C., Matthews, H.B., *Fundam. Appl. Toxicol.* (1987) **9**, 409–414.
E533. Yu, K.O., Olson, C.T., Ferry, M.J., Serve, M.P., *Biomed. Environ. Mass Spectrom.* (1987) **14**, 649–651.
E534. Kelley, M., Vessey, D.A., *Toxicol. Appl. Pharmacol.* (1987) **91**, 295–298.
E535. Schwartz, H., Chu, I., Villeneuve, D.C., Benoit, F.M., *J. Toxicol. Environ. Health* (1987) **22**, 341–350.
E536. Kühnl, T., Koch, U., Heller, W., Wellmann, E., *Arch. Biochem. Biophys.* (1987) **258**, 226–232.
E538. Sugita, O., Sassa, S., Miyairi, S., Fishman, J., Kubota, I., Noguchi, T., Kappas, A., *Biochemistry* (1988) **27**, 678–686.
E539. Lundh, T.J.O., Pettersson, H., Kiessling, K.H., *J. Agric. Food Chem.* (1988) **36**, 22–25.
E540. Thakker, D.R., Boehlert, C., Mirsadeghi, S., Levin, W., Ryan, D.E., Thomas, P.E., Yagi, H., Pannell, L.K., Sayer, J.M., Jerina, D.M., *J. Biol. Chem.* (1988) **263**, 98–105.
E541. Boyd, D.R., McMordie, R.A.S., Porter, H.P., Dalton, H., Jenkins, R.O., Howarth, O.W., *J. Chem. Soc., Chem. Commun.* (1987) 1722–1724.
E542. Jones, G.H., *J. Bacteriol.* (1987) **169**, 5575–5578.
E543. Rossiter, J.T., Williams, S.R., Cass, A.E.G., Ribbons, D.W., *Tetrahedron Lett.* (1987) **28**. 5173–5174.
E544. Cleveland, T.E., Bhatnagar, D., Foell, C.J., McCormick, S.P., *Appl. Environ. Microbiol.* (1987) **53**, 2804–2807.
E545. Reddy, A.R., Britsch, L., Salamini, F., Saedler, H., Rohde, W., *Plant Sci. (Limerick, Irel.)* (1987) **52**, 7–13.

E546. Coutts, R.T., Torok-Both, G.A., Chu, L.V., Tam, Y.K., Pasutto, F.M., *J. Chromatogr.* (1987) **421**, 267–280.

E547. Kawai, S., Umezawa, T., Shimada, M., Higuchi, T., Koide, K., Nishida, T., Morohoshi, N., Haraguchi, T., *Mokuzai Gakkaishi* (1987) **33**, 792–797.

E548. Maessen, P.A., Mross, K.B., Pinedo, H.M., Van der Vijgh, W.J.F., *Cancer Chemother. Pharmacol.* (1987) **20**, 85–87.

E549. Magyar, K., *Pol. J. Pharmacol. Pharm.* (1987) **39**, 107–112.

E550. Fonyo, M., Guttmann, A., Magyar, K., *Pol. J. Pharmacol. Pharm.* (1987) **39**, 129–132.

E552. Vree, T.B., Vree, M.L., Nouws, J.F.M., *Vet. Q.* (1987) **9**, 378–381.

E553. Vree, T.B., Vree, M.L., Nouws, J.F.M., *Vet. Q.* (1987) **9**, 381–384.

E555. Dahl-Puustinen, M.L., Bertilsson, L., *Pharmacol. Toxicol.* (*Copenhagen*) (1987) **61**, 342–346.

E556. Platt, K.L., Reischmann, I., *Mol. Pharmacol.* (1987) **32**, 710–722.

E557. Forkmann, G., Ruhnau, B., *Z. Naturforsch. C: Biosci.* (1987) **42**, 1146–1148.

E558. Lewis, N.G., Dubelsten, P., Eberhardt, T.L., Yamamoto, E., Towers, G.H.N., *Phytochemistry* (1987) **26**, 2729–2734.

E559. Hamerski, D., Matern, U., *Eur. J. Biochem.* (1988) **171**, 369–375.

E560. Christinaki, H., Bister-Miel, F., Hammoumi, A., Bury, M., Guignard, J.L., Viel, C., *Phytochemistry* (1987) **26**, 2991–2994.

E561. Friedrich, C.A., Morizot, D.C., Siciliano, M.J., Ferrell, R.E., *Biochem. Genet.* (1987) **25**, 657–669.

E562. Weitz, C.J., Faull, K.F., Goldstein, A., *Nature* (1987) **330**, 674–677.

E564. Dumont, E., Von Bahr, C., Perry, T.L., Jr., Bertilsson, L., *Pharmacol. Toxicol.* (*Copenhagen*) (1987) **61**, 335–341.

E565. Straube, G., *J. Basic Microbiol.* (1987) **27**, 229–232.

E567. Wackett, L.P., Kwart, L.D., Gibson, D.T., *Biochemistry* (1988) **27**, 1360–1367.

E569. Finocchiaro, L., Callebert, J., Launay, J.M., Jallon, J.M., *J. Neurochem.* (1988) **50**, 382–387.

E570. Stich, K., Forkmann, G., *Z. Naturforsch. C: Biosci.* (1987) **42**, 1193–1199.

E571. Jacob, J., Grimmer, G., Raab, G., Emura, M., Riebe, M., Mohr, U., Jr., *Cancer Lett.* (*Shannon, Irel.*) (1987) **38**, 171–180.

E572. Wariishi, H., Morohoshi, N., Haraguchi, T., *Mokuzai Gakkaishi* (1987) **33**, 892–898.

E573. Shu, Y.Z., Hattori, M., Akao, T., Kobashi, K., Namba, T., *Wakan Iyaku Gakkaishi* (1987) **4**, 82–87.

E574. Ishimitsu, S., Fujimoto, S., Ohara, A., *Chem. Pharm. Bull.* (1988) **36**, 279–285.

E575. Campbell, N.R.C., Van Loon, J.A., Sundaram, R.S., Ames, M.M., Hansch, C., Weinshilboum, R., *Mol. Pharmacol.* (1987) **32**, 813–819.

E576. Hall, M., Grover, P.L., *Cancer Lett.* (*Shannon, Irel.*) (1987) **38**, 57–64.

E577. Rueffer, M., Zenk, H., *Tetrahedron Lett.* (1987) **28**, 5307–5310.

E578. El-Sayed, M.A., Valadon, L.R.G., El-Shanshoury, A. el R., *Microbios Lett.* (1987) **36**, 85–95

E579. Welle, R., Grisebach, H., *Z. Naturforsch. C: Biosci.* (1987) **42**, 1200–1206.

E580. De Luca, V., Cutter, A.J., *Plant Physiol.* (1987) **85**, 1099–1102.

E581. Bhakuni, D.S., Jain, S., Chaturvedi, R., *Tetrahedron* (1987) **43**, 3975–3982.

E582. Burke, A.B., Millburn, P., Huckle, K.R., *Biochem. Soc. Trans.* (1988) **16**, 24–25.

E583. Smithgall, T.E., Harvey, R.G., Penning, T.M., *J. Biol. Chem.* (1988) **263**, 1814–1820.

E586. Yoshimoto, T., Tamesa, Y., Gushi, K., Murayama, N., Tsuru, D., *Agric. Biol. Chem.* (1988) **52**, 217–225.

E587. Hummel, W., Schuette, H., Schmidt, E., Kula, M.R., *Appl. Microbiol. Biotechnol.* (1987) **27**, 283–291.

E588. Hitchcock, M.J.M., Katz, E., *Arch. Biochem. Biophys.* (1988) **261**, 148–160.

E589. Johnston, J.B., Ranganathan, V., *Enzyme Microb. Technol.* (1987) **9**, 706–708.

E590. Gouyette, A., *Biomed. Environ. Mass Spectrom.* (1988) **15**, 243–247.

E593. Eaton, D.L., Monroe, D.H., Bellamy, G., Kalman, D.A., *Chem. Res. Toxicol.* (1988) **1**, 108–114.

E594. Kolkmann, R., Leistner, E., *Z. Naturforsch. C: Biosci.* (1987) **42**, 1207–1214.

E595. Kinouchi, T., Nishifuji, K., Ohnishi, Y., *Microbiol. Immunol.* (1987) **31**, 1145–1159.

E596. MacKintosh, R.W., Fewson, C.A., *Biochem. J.* (1988) **250**, 743–751.

E597. Itoh, N., Hasan, A.K.M., Izumi, Y., Yamada, H., *Eur. J. Biochem.* (1988) **172**, 477–484.

E598. Harvey, D.J., *Biomed. Environ. Mass Spectrom.* (1988) **15**, 117–122.

E599. Isobe, N., Ohe, A., Ito, S., Saito, K., Nakatsuka, I., Yoshitake, A., Miyamoto, J., *Nippon Noyaku Gakkaishi* (1987) **12**, 721–728.

E600. Kung, M.P., Spaulding, S.W., Roth, A., *Endocrinology (Baltimore)* (1988) **122**, 1195–1200.

E601. Kaplan, M.M., Pan, C., Gordon, P.R., Lee, J.K., Gilchrist, B.A., *J. Clin. Endocrinol. Metab.* (1988) **66**, 815–822.

E602. Olsen, M., Visconti, A., *Toxicol. In Vitro* (1988) **2**, 27–29.

E603. Sadler, A., Subrahmanyam, V.V., Ross, D., *Toxicol. Appl. Pharmacol.* (1988) **93**, 62–71.

E604. Schumacher, H.M., Zenk, M.H., *Plant Cell Rep.* (1988) **7**, 43–46.

E605. Geissler, J.F., Harwood, C.S., Gibson, J., *J. Bacteriol.* (1988) **170**, 1709–1714.

E606. de Boer, L., Harder, W., Dijkhuizen, L., *Arch. Microbiol.* (1988) **149**, 459–465.

E607. Zeyer, J., Kocher, H.P., *J. Bacteriol.* (1988) **170**, 1789–1794.

E608. Kajbaf, M., Gorrod, J.W., *Eur. J. Drug Metab. Pharmacokinet.* (1987) **12**, 285–290.

E609. Pierce, D.M., Warrington, S.J., Franklin, R.A., *Biopharm. Drug Dispos.* (1988) **9**, 147–157.

E610. Smolarek, T.A., Morgan, S.L., Moynihan, C.G., Lee, H., Harvey, R.G., Baird, W.M., *Carcinogenesis* (1987) **8**, 1501–1509.

E611. Djuric, Z., Coles, B., Fifer, E.K., Ketterer, B., Beland, F.A., *Carcinogenesis* (1987) **8**, 1781–1786.

E612. Myers, S.R., Blake, J.W., Flesher, J.W., *Biochem. Biophys. Res. Commun.* (1988) **151**, 1441–1445.

E613. Kandaswami, C., Kumar, S., Dubey, S.K., Sikka, H.C., *Carcinogenesis* (1987) **8**, 1861–1866.

E614. Kerscher, F., Franz, G., *J. Plant Physiol.* (1988) **132**, 110–115.

E615. Raju, S.G., Vaidyanathan, C.S., *J. Indian Inst. Sci.* (1986) **66**, 511–520.

E616. Omori, T., Ishigooka, H., Minoda, Y., *Agric, Biol. Chem.* (1988) **52**, 503–509.

E617. Zedan, H.H., Rasmy, S.A., *Egypt. J. Microbiol. (Special Issue)* (1986) 1–8.

E618. Henderberg, A., Bennett, J.W., Lee, L.S., *J. Gen. Microbiol.* (1988) **134**, 661–667.

E619. Ayabe, S., Udagawa, A., Furuya, T., *Arch. Biochem. Biophys.* (1988) **261**, 458–462.

E620. Lauwers, W., Le Jeune, L., Meuldermans, W., *Biomed. Environ. Mass Spectrom.* (1988) **15**, 323–328.

E621. Khoo, B.Y., Sit, K.H., Wong, K.P., *Life Sci.* (1988) **42**, 1165–1171.

E622. Hande, K., Anthony, L., Hamilton, R., Bennett, R., Sweetman, B., Branch, R., *Cancer Res.* (1988) **48**, 1829–1834.

E623. Brown, N.K., Harvey, D.J., *Biomed. Environ. Mass Spectrom.* (1988) **15**, 389–398.

E624. Watkins, P.J., Gorrod, J.W., *Eur. J. Drug Metab. Pharmacokinet.* (1987) **12**, 245–251.

E625. Brown, N.K., Harvey, D.J., *Biomed. Environ. Mass Spectrom.* (1988) **15**, 403–410.

E626. Beck, O., Jernström, B., Martinez, M., Repke, D.B., *Chem.-Biol. Interact.* (1988) **65**, 97–106.

E627. Hamilton, M., *J. Chromatogr.* (1988) **431**, 103–111.

E628. Göber, B., Dressler, K., Franke, P., *Pharmazie* (1988) **43**, 96–98.

E629. Hübler, G., Haustein, K.O., Scheithauer, S., *Pharmazie* (1988) **43**, 91–98.

E630. Binder, T.P., Duffel, M.W., *Mol. Pharmacol.* (1988) **33**, 477–479.

E631. Silberzahn, P., Gaillard, J.L., Quincey, D., Dintinger, T., Al-Timimi, I., *J. Steroid Biochem.* (1988) **29**, 119–125.

E634. Tahara, S., Misumi, E., Mizutani, J., Ingham, J.L., *Z. Naturforsch.* (1987) **42C**, 1055–1062.

E635. Schwartz, H., Chu, I., Villeneuve, D.C., Viau, A., Benoit, F.M., *J. Toxicol. Environ. Health* (1985) **15**, 603–607.

E636. Hrazdina, G., *Biochim. Biophys. Acta* (1988) **955**, 301–309.

E637. Short, C.R., Flory, W., Hsieh, L.C., Barker, S.A., *J. Vet. Pharmacol. Ther.* (1988) **11**, 50–55.

E638. Takimoto, Y., Ohshima, M., Miyamoto, J., *Ecotoxicol. Environ. Saf.* (1987) **13**, 104–117.

E639. Takimoto, Y., Ohshima, M., Miyamoto, J., *Ecotoxicol. Environ. Saf.* (1987) **13**, 118–125.

E640. Taylor, D.C., Wightman, F., *Phytochemistry* (1987) **26**, 2155–2166.

E641. Haigler, B.E., Nishino, S.F., Spain, J.C., *Appl. Environ. Microbiol.* (1988) **54**, 294–301.

E642. Hummel, W., Schütte, H., Kula, M.-R., *Anal. Biochem.* (1988) **170**, 397–401.

E643. Karoum, F., *Br. J. Pharmacol.* (1987) **90**, 335–345.

E644. Oda, T., Sato, Y., *Chem. Pharm. Bull.* (1985) **33**, 1077–1082.

E645. Weli, A.M., Höök, B.B., Lindeke, B., *Acta Pharm. Suec.* (1985) **22**, 249–264.

E646. Blaich, G., Pfaff, E., Metzler, M., *Biochem. Pharmacol.* (1987) **36**, 3135–3140.

E647. Beretta, C., Fedini, L., Stracciari, J.M., Montesissa, C., *Biochem. Pharmacol.* (1987) **36**, 3107–3114.

E648. Dull, B.J., Salata, K., Goldman, P., *Biochem. Pharmacol.* (1987) **36**, 3772–3774.

E649. Kammerer, R.C., Lampe, M.A., *Biochem. Pharmacol.* (1987) **36**, 3445–3452.

E650. Lennard, M.S., Tucker, G.T., Woods, H.F., Iyun, A.O., Eichelbaum, M., *Biochem. Pharmacol.* (1988) **37**, 97–98.

E652. Hande, K., Bennett, R., Hamilton, R., Grote, T., Branch, R., *Cancer Res.* (1988) **48**, 5692–5695.

E653. Rüttiman, C., Seelenfreund, D., Vicuña, R., *Enzym. Microb. Technol.* (1987) **9**, 526–530.

E654. Dubé, L.M., Mousseau, N., McGilvery, I.J., *J. Chromatogr.* (1988) **430**, 103–111.

E656. Formosa, P.J., Bray, T.M., Kubow, S., *Can. J. Physiol. Pharmacol.* (1988) **66**, 1524–1530.

E657. Chan, R.L., Henzl, M.R., Le Page, M.E., La Fargue, J., Nerenberg, C.A., Anik, S., Chaplin, M.D., *Clin. Pharmacol. Ther.* (*St. Louis*) (1988) **44**, 275–282.

E658. Delclos, K.B., El-Bayoumy, K., Hecht, S.S., Walker, R.P., Kadlubar, F.F., *Carcinogenesis* (1988) **9**, 1875–1884.

E659. Ishaaya, I., Degheele, D., *Pestic. Biochem. Physiol.* (1988) **32**, 180–187.

E660. Petersen, M., Alfermann, A.W., *Z. Naturforsch. C: Biosci.* (1988) **43**, 501–504.

E661. Welle, R., Grisebach, H., *FEBS Lett.* (1988) **236**, 221–225.

E662. Peters, A., Schneider-Poetsch, H.A.W., Schwartz, H., Weissenböck, G., *J. Plant Physiol.* (1988) **133**, 178–182.
E663. Mackintosh, R.W., Fewson, C.A., *Biochem. J.* (1988) **255**, 653–661.
E664. Nielson, A.H., Allard, A.S., Hynning, P.Å., Remberger, M., *Appl. Environ. Microbiol.* (1988) **54**, 2226–2236.
E665. Stevenson, D.E., Wright, J.N., Akhtar, M., *J. Chem. Soc., Perkin Trans 1* (1988) 2043–2052.
E666. Fuska, J., Proksa, B., Fuskova, A., Khandlova, A., *Acta Biotechnol.* (1988) **8**, 291–293.
E667. Bernardi, R., Bravo, P., Cardillo, R., Ghiringelli, D., Resnati, G., *J. Chem. Soc., Perkin Trans. 1* (1988) 2831–2834.
E668. Dealler, S.F., Hawkey, P.M., Millar, M.R., *J. Clin. Microbiol.* (1988) **26**, 2152–2156.
E669. van den Tweel, W.J.J., Widjojoatmodjo, M.N., de Bont, J.A.M., *Arch Microbiol.* (1988) **150**, 471–476.
E670. Meksuriyen, D., Cordell, G.A., *J. Nat. Prod.* (1988) **51**, 893–899.
E671. Raman, A., Christou, M., Gorrod, J.W., *Eur. J. Drug Metab. Pharmacokinet.* (1987) **12**, 279–283.
E672. Young, W.F., Jr., Gorman, C.A., Weinshilboum, R.M., *Endocrinology* (*Baltimore*) (1988) **122**, 1816–1824.
E673. Ryrfeldt, Å., Nilsson, E., Tunek, A., Svensson, L.-Å., *Pharm. Res.* (1988) **5**, 151–155.
E674. Laham, S., Potvin, M., Robinet, M., *Chemosphere* (1988) **17**, 517–524.
E675. Möller, L., Törnquist, S., Beije, B., Rafter, J., Toftgård, R., Gustafsson, J.Å., *Carcinogenesis* (1987) **8**, 1847–1852.
E676. Fawaz, F., Jones, G.H., *J. Biol. Chem.* (1988) **263**, 4602–4606.
E677. Mozier, N.M., McConnell, K.P., Hoffmann, J.L., *J. Biol. Chem.* (1988) **263**, 4527–4531.
E678. Okada, N., Shinmyo, A., Okada, H., Yamada, Y., *Phytochemistry* (1988) **27**, 979–982.
E679. van den Tweel, W.J.J., Smits, J.P., de Bont, J.A.M., *Arch. Microbiol.* (1988) **149**, 207–213.
E680. Agier, C., Bury, M., El-Ech, P., Cousson-Fonfraid, S., Bister-Miel, F., Guignard, J.L., Viel, C., *C. R. Acad. Sci., Ser. 3* (1988) **306**, 265–269.
E681. Tabata, M., Umetani, Y., Oya, M., Tanaka, S., *Phytochemistry* (1988) **27**, 809–813.
E682. Uetrecht, J., Zahid, N., *Chem. Res. Toxicol.* (1988) **1**, 148–151.
E683. Howell, S.R., Kotkoskie, L.A., Dills, R.L., Klaassen, C.D., *J. Pharm. Sci.* (1988) **77**, 309–313.
E684. Wu, E., Shinka, T., Caldera-Munoz, P., Yoshizumi, H., Trevor, A., Castagnoli, N., Jr., *Chem. Res. Toxicol.* (1988) **1**, 186–194.
E685. Uetrecht, J., Zahid, N., Shear, N.H., Biggar, W.D., *J. Pharmacol. Exp. Ther.* (1988) **245**, 274–279.
E686. Bosma, T.N.P., van der Meer, J.R., Schraa, G., Tros, M.E., Zehnder, A.J.B., *FEMS Microbiol. Ecol.* (1988) **53**, 223–229.
E687. Ziegler, D.M., Ansher, S.S., Nagata, T., Kadlubar, F.F., Jakoby, W.B., *Proc. Natl. Acad. Sci. U.S.A.* (1988) **85**, 2514–2517.
E688. Moeller, L., Corrie, M., Midtvedt, T., Rafter, J., Gustafsson, J.A., *Carcinogenesis* (1988) **9**, 823–830.
E689. Roy, A.K., El-Bayoumy, K., Hecht, S.S., *Carcinogenesis* (1988) **9**, 255–258.
E691. Mashaly, R.I., Habib, S.L., El-Deeb, S.A., Salem, M.H., Safwat, M.M., *Z. Lebensm.-Unters. Forsch.* (1988) **186**, 118–124.
E693. Nissinen, E., Tuominen, R., Perhoniemi, V., Kaakkola, S., *Life Sci* (1988) **42**, 2609–2614.
E694. Shukla, O.P., *Biol. Mem.* (1987) **13**, 115–129.
E695. Robinson-White, A., Costa, J.L., Launay, J.-M., Fay, D.P., *Microvasc. Res.* (1988) **35**, 363–367.

E696. Mikov, M., Caldwell, J., Dolphin, C.T., Smith, R.L., *Biochem. Pharmacol.* (1988) **37**, 1445–1449.
E697. Köppel, C., Tenczer, J., Ibe, K., *J. Chromatogr.* (1988) **427**, 144–150.
E698. Engesser, K.H., Cain, R.B., Knackmuss, H.J., *Arch. Microbiol.* (1988) **149**, 188–197.
E699. Hollingsworth, S., Pass, M.A., *Comp. Biochem. Physiol. B: Comp. Biochem.* (1988) **90B**, 99–101.
E700. de Weerd, K.A., Saxena, A., Nagle, D.P., Jr., Suflita, J.M., *Appl. Environ. Microbiol.* (1988) **54**, 1237–1242.
E701. Kutney, J.P., Botta, B., Boulet, C.A., Buschi, C.A., Choi, L.S., Golinski, J., Gumulka, M., Hewitt, G.M., Lee, G., McHugh, M., Nakano, J., Nikaido, T., Onodera, J-i., Perez, I., Salisbury, P., Singh, M., Suen, R., Tsukamoto, H., *Heterocycles* (1988) **27**, 629–637.
E702. Stich, K., Forkmann, G., *Z. Naturforsch. C: Biosci.* (1988) **43**, 311–314.
E703. Stich, K., Forkmann, G., *Phytochemistry* (1988) **27**, 785–789.
E704. Short, C.R., Barker, S.A., Hsieh, L.C., Ou, S.P., McDowell, T., *Res. Vet. Sci.* (1988) **44**, 215–219.
E705. Baker, J.K., McChesney, J.D., *J. Pharm. Sci.* (1988) **77**, 380–382.
E706. Morris, M.E., Yuen, V., Tang, B.K., Pang, K.S., *J. Pharmacol. Exp. Ther.* (1988) **245**, 614–624.
E707. Matsuki, Y., Dan, J., Fukuhara, K., Ito, T., Nambara, T., *Chem. Pharm. Bull.* (1988) **36**, 1431–1436.
E708. Barron, M.G., James, M.O., *Mar. Environ. Res.* (1988) **24**, 85–88.
E709. Fuller, R.W., Snoddy, H.D., Perry, K.W., *J. Pharm. Pharmacol.* (1988) **40**, 439–441.
E710. Kennish, J.M., Gillis, D., Hotaling, K., *Mar. Environ. Res.* (1988) **24**, 69–71.
E711. Arora, P.K., Riachi, N.J., Harik, S.I., Sayre, L.M., *Biochem. Biophys. Res. Commun.* (1988) **152**, 1339–1347.
E713. Knezovich, J.P., Lawton, M.P., Harrison, F.L., *Mar. Environ. Res.* (1988) **24**, 89–91.
E714. Reynolds, L.J., Garcia, G.A., Kozarich, J.W., Kenyon, G.L., *Biochemistry* (1988) **27**, 5530–5538.
E715. Murata, K., Iba, Y., Inoue, Y., Kimura, A., *Biochem. Biophys. Res. Commun.* (1988) **153**, 767–772.
E716. Caputo, O., Grosa, G., Balliano, G., Rocco, F., Biglino, G., *Eur. J. Drug Metab. Pharmacokinet.* (1988) **13**, 47–51.
E717. Salonen, J.S., Suolinna, E.M., *Eur. J. Drug Metab. Pharmacokinet.* (1988) **13**, 53–58.
E718. Pompa, G., Mintesissa, C., Dilauro, F.M., *J. Vet. Pharmacol. Ther.* (1988) **11**, 197–203.
E719. Salonen, J.S., Vuorilehto, L., Eloranta, M., Karjalainen, A.J., *Eur. J. Drug Metab. Pharmacokinet.* (1988) **13**, 59–68.
E720. Yang, S.C., Johannessen, J.N., Markey, S.P., *Chem. Res. Toxicol.* (1988) **1**, 228–233.
E721. El-Bayoumy, K., Shiue, G.H., Hecht, S.S., *Chem. Res. Toxicol.* (1988) **1**, 243–247.
E722. Goodbody, A.E., Endo, T., Vukovic, J., Kutney, J.P., Choi, L.S.L., Misawa, M., *Planta Med.* (1988) **54**, 136–140.
E723. Goswami, A., Macdonald, T.L., Hubbard, C., Duffel, M.W., Rosazza, J.P.N., *Chem. Res. Toxicol.* (1988) **1**, 238–242.
E724. Schlieper, D., Barz, W., *Phytochemistry* (1987) **26**, 2495–2498.
E725. Fawaz, F., Jones, G.H., *J. Biol. Chem.* (1988) **263**, 4602–4606.
E726. Bruhn, C., Bayly, R.C., Knackmuss, H.J., *Arch. Microbiol.* (1988) **150**, 171–177.
E727. Pieper, D.H., Reineke, W., Engesser, K.H., Knackmuss, H.J., *Arch. Microbiol.* (1988) **150**, 95–102.

E728. Meuldermans, W., Hendrickx, J., Woestenborghs, R., Van Peer, A., Lauwers, W., De Cree, J., Heykants, J., *Arzneim.-Forsch.* (1988) **38**, 789–794.

E729. Monteiro, A.M., Crozier, A., Sandberg, G., *Planta* (1988) **174**, 561–568.

E731. Spain, J.C., Gibson, D.T., *Appl. Environ. Microbiol.* (1988) **54**, 1399–1404.

E732. Galneder, E., Rueffer, M., Wanner, G., Tabata, M., Zenk, M.H., *Plant Cell Rep.* (1988) **7**, 1–4.

E733. Katayama, Y., Nishikawa, S., Murayama, A., Yamasaki, M., Morohoshi, N., Haraguchi, T., *FEBS Lett.* (1988) **233**, 129–133.

E734. Nouws, J.F.M., Mevius, D., Vree, T.B., Baakman, M., Degen, M., *Am. J. Vet. Res.* (1988) **49**, 1059–1065.

E735. Shirota, K., Fujisawa, H., *J. Neurochem.* (1988) **51**, 426–434.

E736. Dorisse, P., Gleye, J., Loiseau, P., Puig, P., Edy, A.M., Henry, M., *J. Nat. Prod.* (1988) **51**, 532–536.

E737. Gonzalez, B., Olave, I., Calderon, I., Vicuna, R., *Arch. Microbiol.* (1988) **149**, 389–394.

E738. Bennett, G.J., Lee, H.H., *J. Chem. Soc., Chem. Commun.* (1988) 619–620.

E739. Susilo, R., Rommelspacher, H., *Naunyn-Schmiedeberg's Arch. Pharmacol.* (1988) **337**, 566–571.

E740. Smithgall, T.E., Harvey, R.G., Penning, T.M., *Cancer Res.* (1988) **48**, 1227–1232.

E741. Loewen, G.R., Herman, R.J., Ross, S.G., Verbeeck, R.K., *Br. J. Clin. Pharmacol.* (1988) **26**, 31–39.

E742. Gajendiran, N., Mahadevan, A., *Plant Soil* (1988) **108**, 263–266.

E743. Kato, N., Konishi, H., Uda, K., Shimao, M., Sakazawa, C., *Agric. Biol. Chem.* (1988) **52**, 1885–1886.

E744. Townsend, C.A., Christensen, S.B., Davis, S.G., *J. Chem. Soc., Perkin Trans. 1* (1988) 839–861.

E745. Endo, T., Goodbody, A., Vukovic, J., Misawa, M., *Phytochemistry* (1988) **27**, 2147–2149.

E746. Yoshizako, F., Ueno, T., Morishima, I., Karakawa, T., Sugiura, S., Araki, T., *Agric. Biol. Chem.* (1985) **49**, 877–879.

E747. Olsen, M., Visconti, A., *Toxicol. in Vitro* (1988) **2**, 27–29.

E748. van Bladeren, P.J., Balani, S.K., Sayer, J.M., Thakker, D.R., Boyd, D.R., Ryan, D.E, Thomas, P.E., Levin, W., Jerina, D.M., *Biochem. Biophys. Res. Commun.* (1987) **145**, 160–167.

E749. Yang, S.K., Mushtaq, M., Weems, H.B., Miller, D.W., Fu, P.P., *Biochem. J.* (1987) **245**, 191–204.

E750. Ikeya, Y., Taguchi, H., Mitsuhashi, H., Sasaki, H., Matsuzaki, T., Aburada, M., Hosoya, E., *Chem. Pharm. Bull.* (1988) **36**, 2061–2069.

E751. Norpoth, K., Struecker, W., Krewet, E., Mueller, G., *Int. Arch. Occup. Environ. Health* (1988) **60**, 163–168.

E752. Müller, R., Oltmanns, R.H., Lingens, F., *Biol. Chem. Hoppe-Seyler* (1988) **369**, 567–571.

E753. Howard, P.C., Reed, K.A., Koop, D.R., *Cancer Res.* (1988) **48**, 4261–4265.

E754. Sekine, M., Ichikawa, T., Kuga, N., Kobayashi, M., Sakurai, A., Syono, K., *Plant Cell Physiol.* (1988) **29**, 867–874.

E756. Khouri, H.E., Tahara, S., Ibrahim, R.K., *Arch. Biochem. Biophys.* (1988) **262**, 592–598.

E757. Platt, K.L., Petrovic, P., Seidel, A., Beermann, D., Oesch, F., *Chem.-Biol. Interact.* (1988) **66**, 157–175.

E758. Ishikura, N., Murakami, H., Fujii, Y., *Plant Cell Physiol.* (1988) **29**, 795–799.

E759. Schladt, L., Hartmann, R., Wörner, W., Thomas, H., Oesch, F., *Eur. J. Biochem.* (1988) **176**, 31–37.

E761. Häggblom, M.M., Apajalahti, J.H.A., Salkinoja-Salonen, M.S., *Appl. Environ. Microbiol.* (1988) **54**, 1818–1824.

E762. Agarwal, M., Shukla, O.P., *J. Microb. Biotechnol.* (1988) **3**, 18–24.

E763. Bhatnagar, D., Cleveland, T.E., *Biochimie* (1988) **70**, 743–747.

E764. Khouri, H.E., De Luca, V., Ibrahim, R.K., *Arch. Biochem. Biophys.* (1988) **265**, 1–7.
E765. Rideau, M., Morard, P., Gansser, C., Chenieux, J.C., Viel, C., *Pharmazie* (1988) **43**, 332–334.
E766. Sakurai, I., Miyajima, H., Akiyama, K., Shimizu, S., Yamamoto, Y., *Chem. Pharm. Bull.* (1988) **36**, 2003–2011.
E767. Yabe, K., Ando, Y., Hamasaki, T., *Appl. Environ. Microbiol.* (1988) **54**, 2101–2106.
E768. Isobe, M., Sone, T., Takabatake, E., *Chem. Pharm. Bull.* (1988) **36**, 2267–2269.
E769. Pras, N., Wichers, H.J., Bruins, A.P., Malingre, T.M., *Plant Cell, Tissue Organ Cult.* (1988) **13**, 15–26.
E770. Ueda, M., Chang, C.C., Ohno, M., *Toxicon* (1988) **26**, 695–706.
E771. van den Tweel, W.J.J., Smits, J.P., Ogg, R.H.L.P., de Bont, J.A.M., *Appl. Microbiol. Biotechnol.* (1988) **29**, 224–230.
E772. Pearce, C.J., Doyle, T.W., Forenza, S., Lam, K.S., Schroeder, D.R., *J. Nat. Prod.* (1988) **51**, 937–940.
E773. Sharma, V., *Indian J. Chem., Sect. B* (1988) **27B**, 742–743.
E774. Stothers, J.B., Stoessl, A., *Can. J. Chem.* (1988) **66**, 2816–2818.
E775. Feigel, B.J., Knackmuss, H.J., *FEMS Microbiol. Lett.* (1988) **55**, 113–117.
E776. Takeshita, T., Oosumi, K., *Jpn. Kokai Tokkyo Koho* 63207391 *JP* (26–8-1988).
E777. Bhakuni, D.S., Jain, S., *J. Chem. Soc., Perkin Trans. 1* (1988) 1929–1931.
E778. Brewster, K., Harrison, J.M., Leadbeater, L., Newman, J., Upshall, D.G., *Xenobiotica* (1987) **17**, 911–924.
E779. Yoshimura, H., Yonemoto, Y., Yamada, H., Koga, N., Oguri, K., Saeki, S., *Xenobiotica* (1987) **17**, 897–910.
E780. Kinoshita, H., Tohira, Y., Sugiyama, H., Sawada, M., *Xenobiotica* (1987) **17**, 925–939.
E781. Yoshida, T., Oguro, T., Kuroiwa, Y., *Xenobiotica* (1987) **17**, 957–963.
E782. Laws, P.G.M., Rodrigues, L., Penny, R.H.C., *Xenobiotica* (1987) **17**, 973–979.
E783. Inoue, T., Suzuki, S., *Xenobiotica* (1987) **17**, 965–971.
E784. Krumbiegel, G., Hallensleben, J., Mennicke, W.H., Rittmann, N., Roth, H.H., *Xenobiotica* (1987) **17**, 981–991.
E785. Ganes, D.A., Midha, K.K., *Xenobiotica* (1987) **17**, 993–999.
E786. Woolf, T.F., Adams, J.D., *Xenobiotica* (1987) **17**, 839–847.
E787. Emudianughe, T.S., Caldwell, J., Smith, R.L., *Xenobiotica* (1987) **17**, 815–821.
E788. Rogiers, V., Paeme, G., Sonck, W., Vercruysse, A., *Xenobiotica* (1987) **17**, 849–857.
E789. Wood, S.G., John, B.A., Chasseaud, L.F., Bonn, R., Grote, H., Sandrock, K., Darragh, A., Lambe, R.F., *Xenobiotica* (1987) **17**, 881–896.
E790. Aschbacher, P.W., Struble, C.B., *Xenobiotica* (1987) **17**, 1047–1055.
E791. Baba, T., Yamada, H., Oguri, K., Yoshimura, H., *Xenobiotica* (1987) **17**, 1029–1038.
E800. Barańczyk-Kuźma, A., Szymczyk, T., *Biochem. Pharmacol.* (1987) **36**, 3141–3146.
E801. Blaich, G., Pfaff, E., Metzler, M., *Biochem. Pharmacol.* (1987) **36**, 3135–3140.
E802. Štefek, M., Beneš, L., Jergelová, M., Ščasnár, V., Turi-Nagy, L., Kočiš, P., *Xenobiotica* (1987) **17**, 1067–1073.
E803. Scott, J.G., Palmer, C.J., Casida, J.E., *Xenobiotica* (1987) **17**, 1085–1093.
E804. Oka, M., Ishimura, Y., Tsunematsu, T., Minakuchi, K., Ohuchi, T., Matsumoto, K., *Biochem. Pharmacol.* (1987) **36**, 3205–3208.
E805. Kammerer, C., Lampe, M.A., *Biochem. Pharmacol.* (1987) **36**, 3445–3452.

E806. Baba, T., Yamada, H., Oguri, K., Yoshimura, H., *Biochem. Pharmacol.* (1987) **36**, 4171–4173.

E807. Dull, B.J., Salata, K., Goldman, P., *Biochem. Pharmacol.* (1987) **36**, 3772–3774.

E808. Müller, A., Gabriel, H., Sies, H., Terlimden, R., Fischer, H., Römer. A., *Biochem. Pharmacol.* (1988) **37**, 1103–1109.

E809. Benuck, M., Reith, M.E.A., Lajtha, A., *Biochem. Pharmacol.* (1988) **37**, 1169–1172.

E810. Cumming, P., Häuser, M., Martin, W.R.W., Grierson, J., Adam, M.J., Ruth, T.J., McGeer, E.G., *Biochem. Pharmacol.* (1988) **37**, 247–250.

E811. Okuda, H., Nojima, H., Watanabe, N., Miwa, K., Watabe, T., *Biochem. Pharmacol.* (1988) **37**, 970–973.

E812. Sangster, S.A., Caldwell, J., Hutt, A.J., Anthony, A., Smith, R.L., *Xenobiotica* (1987) **17**, 1223–1232.

E813. Kasokat, T., Nagel, R., Urich, K., *Xenobiotica* (1987) **17**, 1215–1221.

E814. Klebovich, I., Vereczkey, L., Tóth, E., Tamás, J., Mák, M., Jalsovszky, G., Holly, S., *Xenobiotica* (1987) **17**, 1247–1258.

E815. Tse, F.L.S., Orwig, B.A., Jaffe, J.M., Dain, J.G., *Xenobiotica* (1987) **17**, 1259–1267.

E816. Foster, G.D., Crosby, D.G., *Xenobiotica* (1987) **17**, 1393–1404.

E817. Servin, A.L., Wicek, D., Oryszczyn, M.P., Jacquot, C., Lussiana, J.P., Christinaki, H., Viel, C., *Xenobiotica* (1987) **17**, 1381–1391.

E818. Terashita, S., Tokuma, Y., Fujiwara, T., Shiokawa, Y., Okumura, K., Noguchi, H., *Xenobiotica* (1987) **17**, 1415–1425.

E819. Grootveld, M., Halliwell, B., *Biochem. Pharmacol.* (1988) **37**, 271–280.

E820. Mikov, M., Caldwell, J., Dolphin, C.T., Smith, R.L., *Biochem. Pharmacol.* (1988) **37**, 1445–1449.

E821. Sawada, H., Hara, A., Nakayama, T., Nakagawa, M., Inoue, Y., Hasebe, K., Zhang, Y., *Biochem. Pharmacol.* (1988) **37**, 453–458.

E822. Beckett, A.H., Navas, G.E., Hutt, A.J., *Xenobiotica* (1988) **18**, 61–74.

E823. Kaneko, H., Takamatsu, Y., Okuno, Y., Abiko, J., Yoshitake, A., Miyamoto, J., *Xenobiotica* (1988) **18**, 11–19.

E824. Crooks, P.A., Godin, C.S., Damani, L.A., Ansher, S.S., Jakoby, W.B., *Biochem. Pharmacol.* (1988) **37**, 1673–1677.

E825. DeLuca, J.G., Dysart, G.R., Rasnick, D., Bradley, M.O., *Biochem. Pharmacol.* (1988) **37**, 1731–1739.

E826. Waechter, F., Bentley, P., Bieri, F., Muakkassah-Kelly, S., Stäubli, W., Villermain, M., *Biochem. Pharmacol.* (1988) **37**, 3897–3903.

E827. Tan, T.M.T., Wong, K.P., *Biochem. Pharmacol.* (1988) **37**, 2859–2862.

E828. Pandey, R.N., Armstrong, A.P., Hollenberg, P.F., *Biochem. Pharmacol.* (1989) **38**, 2181–2185.

E829. Hashizume, K., Yamatodani, A., Yamamoto, T., Wada, H., Ogihara, T., *Biochem. Pharmacol.* (1989) **38**, 1891–1895.

E830. Clement, B., Zimmermann, M., *Biochem. Pharmacol.* (1989) **38**, 4747–4752.

E831. Kolodziejczyk, P., Reszka, K., Lown, J.W., *Biochem. Pharmacol.* (1989) **38**, 803–809.

E832. Masai, E., Katayama, Y., Nishikawa, S., Yamasaki, M., Morohoshi, N., Haraguchi, T., *FEBS Lett.* (1989) **249**, 348–352.

E833. Delclos, K.B., El-Bayoumy, K., Casciano, D.A., Walker, K.P., Kadlubar, F.F., Hecht, S.S., Shivapurkar, N., Mandal, S., Stoner, G.D., *Cancer Res.* (1989) **49**, 2909–2913.

E834. Dalrymple, P.D., Nicholls, P.J., *Xenobiotica* (1988) **18**, 75–81.

E835. Guillouzo, A., Bégué, J.-M., Maurer, G., Koch, P., *Xenobiotica* (1988) **18**, 131–139.

E836. Beresford, A.P., Macrae, P.V., Stopher, D.A., *Xenobiotica* (1988) **18**, 169–182.

E837. Kamimura, H., Kawai, R., Kudo, H., *Xenobiotica* (1988) **18**, 141–149.

E838. Lavrijsen, K., Van Houdt, J., Meuldermans, W., Knaeps, F., Hendrickx, J., Lauwers, W., Hurkmans, R., Heykants, J., *Xenobiotica* (1988) **18**, 183–197.

E839. Lindeke, B., Ericsson, Ö, Jönsson, Å., Noren, B., Strömberg, S., Vangbo, B., *Xenobiotica* (1987) **17**, 1269–1278.

E840. Roy, A.B., Curtis, C.G., Powell, G.M., *Xenobiotica* (1987) **17**, 1299–1313.

E841. Görge, G., Beyer, J., Urich, K., *Xenobiotica* (1987) **17**, 1293–1298.

E842. Nakaoka, M., Hakusui, H., *Xenobiotica* (1987) **17**, 1329–1339.

E843. Souhaili El Amri, H., Fargetton, X., Delatour, P., Batt, A.M., *Xenobiotica* (1987) **17**, 1159–1168.

E844. Edwards, R., Millburn, P., Hutson, D.H., *Xenobiotica* (1987) **17**, 1175–1193.

E845. Brown, M.A., Casida, J.E., *Xenobiotica* (1987) **17**, 1169–1174.

E846. Sutfin, T.A., Gabrielsson, M., Regårdh, C.G., *Xenobiotica* (1987) **17**, 1203–1214.

E847. Niwa, T., Tokuma, Y., Noguchi, H., *Xenobiotica* (1988) **18**, 217–224.

E848. Beresford, A.P., McGibney, D., Humphrey, M.J., Macrae, P.V., Stopher, D.A., *Xenobiotica* (1988) **18**, 245–252.

E849. Gjertsen, F.B., Solheim, E., Scheline, R.R., *Xenobiotica* (1988) **18**, 225–234.

E850. Koike, M., Norikura, R., Iwatani, K., Sugeno, K., Takahashi, S., Nakagawa, Y., *Xenobiotica* (1988) **18**, 257–268.

E851. Cerniglia, C.E., Hansen, E.B., Jr., Lambert, K.J., Korfmacher, W.A., Miller, D.W., *Xenobiotica* (1988) **18**, 301–312.

E852. Moreland, T.A., Astoin, J., Lepage, F., Tombret, F., Levy, R.H., Baillie, T.A., *Drug Metab. Dispos.* (1986) **14**, 654–662.

E853. Larsen, G.L., Bakke, J.E., Feil, V.J., Huwe, J.K., *Xenobiotica* (1988) **18**, 313–322.

E854. Suckow, R.F., Smith, T.M., Perumal, A.S., Cooper, T.B., *Drug Metab. Dispos.* (1986) **14**, 692–697.

E855. Canfell, P.C., Castagnoli, N., Fahey, M.R., Hennis, P.J., Miller, R.D., *Drug Metab. Dispos.* (1986) **14**, 703–708.

E856. Shamsuddin, Z.A., Rahimtula, A.D., *Drug Metab. Dispos.* (1986) **14**, 724–732.

E857. Hultin, T.A., May, C.M., Moon, R.C., *Drug Metab. Dispos.* (1986) **14**, 714–717.

E858. Taylor, W.G., *Drug Metab. Dispos.* (1986) **14**, 532–539.

E859. Gombar, C.T., Straub, K., Levandoski, P., Gutzait, L., Swagzdis, J., Garvie, C., Joseph, G., Potts, B.D., Mico, B.A., *Drug Metab. Dispos.* (1986) **14**, 540–548.

E860. Harrison, L.I., Schuppan, D., Rohlfing, S.R., Hansen, A.R., Gerster, J.F., Hansen, C.S., Funk, M.L., Ober, R.E., *Drug Metab. Dispos.* (1986) **14**, 555–558.

E861. La Cagnin, L.B., Colby, H.D., O'Donnell, J.P., *Drug Metab. Dispos.* (1986) **14**, 549–554.

E862. Bäärnhielm, C., Backman, Å., Hoffmann, K.J., Weidolf, L., *Drug Metab. Dispos.* (1986) **14**, 613–618.

E863. Schillings, R.T., Sisenwine, S.F., *Drug Metab. Dispos.* (1986) **14**, 405–412.

E864. Gerardy, B.M., Kapusta, D., Dumont, P., Poupaert, J.H., *Drug Metab. Dispos.* (1986) **14**, 477–481.

E865. Young, R.A., Mehendale, H.M., *Drug Metab. Dispos.* (1986) **14**, 423–429.

E866. Peffer, R.C., Abraham, D.J., Zemaitis, M.A., Wong, L.K., Alvin, J.D., *Drug Metab. Dispos.* (1987) **15**, 305–311.

E867. Ziegler, M.G., Kennedy, B., Elayan, H., *Experientia* (1989) **45**, 718–720.

E868. Mays, D.C., Hecht, S.G., Unger, S.E., Pacula, C.M., Climie, J.M., Sharp, D.E., Gerber, N., *Drug Metab. Dispos.* (1987) **15**, 318–328.

E869. Sasame, H.A., Liberato, D.J., Gillette, J.R., *Drug Metab. Dispos.* (1987) **15**, 349–355.
E870. Zoetemelk, C.E.M., van Hove, W., van der Laan, W.L.J., Meeteren-Wälchli, A., van der Gen, A., Breimer, D.D., *Drug Metab. Dispos.* (1987) **15**, 418–425.
E871. Ioannou, Y.M., Burka, L.T., Sanders, J.M., Moorman, M.P., Matthews, H.B., *Drug Metab. Dispos.* (1987) **15**, 367–373.
E871a. Koike, M., Norikura, R., Futaguchi, S., Yamaguchi, T., Sugeno, K., Iwatani, K., Ikenishi, Y., Nakagawa, Y., *Drug Metab. Dispos.* (1987) **15**, 426–428.
E871b. Newton, J.F., Straub, K.M., Dewey, R.H., Perchonock, C.D., Leonard, T.B., McCarthy, M.E., Gleason, J.G., Eckhardt, R.D., *Drug Metab. Dispos.* (1987) **15**, 161–167.
E872. Samuels, A.R., Theilmann, L., Stollman, Y.R., Wolkoff, A.W., Bhargava, M.M., *Drug Metab. Dispos.* (1987) **15**, 184–188.
E873. Newton, J.F., Straub, K.M., Kuo, G.Y., Perchonock, C.D., McCarthy, M.E., Gleason, J.G., Lynn, R.K., *Drug Metab. Dispos.* (1987) **15**, 168–176.
E874. Dulik, D.M., Fenselau, C., *Drug Metab. Dispos.* (1987) **15**, 195–199.
E875. Mehvar, R., Jamali, F., Watson, M.B.W., Skelton, D., *Drug Metab. Dispos.* (1987) **15**, 250–255.
E876. Sinsheimer, J.E., Freeman, C.J., *Drug Metab. Dispos.* (1987) **15**, 267–269.
E877. Kobayashi, T., Ando, H., Sugihara, J., Harigaya, S., *Drug Metab. Dispos.* (1987) **15**, 262–266.
E878. Lertratanangkoon, K., Horning, M.G., *Drug Metab. Dispos.* (1987) **15**, 1–11.
E879. Budinsky, R.A., Roberts, S.M., Coats, E.A., Adams, L., Hess, E.V., *Drug Metab. Dispos.* (1987) **15**, 37–43.
E880. Tweedie, D.J., Burke, M.D., *Drug Metab. Dispos.* (1987) **15**, 74–81.
E881. Chui, Y.C., Addison, R.F., Law, F.C.P., *Drug Metab. Dispos.* (1987) **15**, 44–50.
E882. Hansen, E.B., Jr., Cerniglia, C.E., Korfmacher, W.A., Miller, D.W., Heflich, R.H., *Drug Metab. Dispos.* (1987) **15**, 97–106.
E883. Le Boeuf, E., Grech-Bélanger, O., *Drug Metab. Dispos.* (1987) **15**, 122–126.
E884. Malejka-Giganti, D., Magat, W.J., Adelmann, A.M., Decker, R.W., *Drug Metab. Dispos.* (1987) **15**, 760–766.
E885. Chow, S.A., Fischer, L.J., *Drug Metab. Dispos.* (1987) **15**, 740–748.
E886. El Mouelli, M., Ruelius, H.W., Fenselau, C., Dulik, D.M., *Drug Metab. Dispos.* (1987) **15**, 767–772.
E887. Felder, T.B., McLean, M.A., Vestal, M.L., Lu, K., Farquhar, D., Legha, S.S., Shah, R., Newman, R.A., *Drug Metab. Dispos.* (1987) **15**, 773–778.
E888. Nystrom, D.D., Rickert, D.E., *Drug Metab. Dispos.* (1987) **15**, 821–825.
E889. Lau, S.S., Monks, T.J., *Drug Metab. Dispos.* (1987) **15**, 801–807.
E890. Thompson, J.A., Malkinson, A.M., Wand, M.D., Mastovich, S.L., Mead, E.W., Schullek, K.M., Laudenschlager, W.G., *Drug Metab. Dispos.* (1987) **15**, 833–840.
E891. Lertratanangkoon, K., Horning, E.C., Horning, M.G., *Drug Metab. Dispos.* (1987) **15**, 857–867.
E892. Breyer-Pfaff, U., Prox, A., Wachsmuth, H., Yao, P., *Drug Metab. Dispos.* (1987) **15**, 882–889.
E893. Kobayashi, T., Sugihara, J., Harigaya, S., *Drug Metab Dispos.* (1987) **15**, 877–881.
E894. Prox, A., Breyer-Pfaff, U., *Drug Metab. Dispos.* (1987) **15**, 890–896.
E895. Meuldermans, W., Hendrickx, J., Lauwers, W., Hurkmans, R., Swysen, E., Thijssen, J., Timmerman, P., Woestenborghs, R., Heykants, J., *Drug Metab. Dispos.* (1987) **15**, 905–913.
E896. Ferdinandi, E.S., Cochran, D., Gedamke, R., *Drug Metab. Dispos.* (1987) **15**, 921–924.
E897. Harvey, D.J., Marriage, H.J., *Drug Metab. Dispos.* (1987) **15**, 914–920.
E898. Corpet, D.E., Bories, G.F., *Drug Metab. Dispos.* (1987) **15**, 925–927.

E899. Fullerton, F.R., Greenman, D.L., Young, J.F., *Drug Metab. Dispos.* (1987) **15**, 602–607.

E900. James, M.O., Pritchard, J.B., *Drug Metab. Dispos.* (1987) **15**, 665–670.

E901. Mroszczak, E.J., Lee, F.W., Combs, D., Sarnquist, F.H., Huang, B.-L., Wu, A.T., Tokes, L.G., Maddox, M.L., Clo, D.K., *Drug Metab. Dispos.* (1987) **15**, 618–626.

E902. Paulson, G.D., Feil, V.J., *Drug Metab. Dispos.* (1987) **15**, 671–675.

E903. McKay, S., Farmer, P.B., Cary, P.D., Grover, P.L., *Drug Metab. Dispos.* (1987) **15**, 682–694.

E904. Honma, S., Iwamura, S., Kobayashi, R., Kawabe, Y., Shibata, K., *Drug Metab. Dispos.* (1987) **15**, 551–559.

E905. Kelly, D.W., Slikker, W., Jr., *Drug Metab. Dispos.* (1987) **15**, 460–465.

E906. Buckpitt, A.R., Castagnoli, N., Jr., Nelson, S.D., Jones, A.D., Bahnson, L.S., *Drug Metab. Dispos.* (1987) **15**, 491–498.

E907. Yeh, S.Y., *Drug Metab. Dispos.* (1987) **15**, 466–472.

E908. Graham, D.J.M., Hama, K.M., Smith, S.A., Kurz, L., Chaplin, M.D., Hall, D.J., *Drug Metab. Dispos.* (1987) **15**, 565–570.

E909. Sisenwine, S.F., Tio, C.O., Liu, A.L., Politowski, J.F., *Drug Metab. Dispos.* (1987) **15**, 579–580.

E910. Ohta, S., Masumoto, H., Takeuchi, K., Hirobe, M., *Drug Metab. Dispos.* (1987) **15**, 583–584.

E911. Burke, A., Millburn, P., Huckle, K.R., Hutson, D.H., *Drug Metab. Dispos.* (1987) **15**, 581–582.

E912. Nelson, W.L., Olsen, L.D., Beitner, D.B., Pallow, R.J., Jr., *Drug Metab. Dispos.* (1988) **16**, 184–188.

E913. Shou, M., Yang, S.K., *Drug Metab. Dispos.* (1988) **16**, 173–183.

E914. Hall, I.H., Reynolds, D.J., Wyrick, S.D., Maguire, J.H., Shrewsbury, R.P., *Drug Metab. Dispos.* (1988) **16**, 196–200.

E915. Talaat, R.E., Nelson, W.L., *Drug Metab. Dispos.* (1988) **16**, 207–211.

E916. Gustavson, L.M., Nelson, W.L., *Drug Metab. Dispos.* (1988) **16**, 217–221.

E917. Talaat, R.E., Nelson, W.L., *Drug Metab. Dispos.* (1988) **16**, 212–216.

E918. Dain, J.G., Jaffe, J.M., *Drug Metab. Dispos.* (1988) **16**, 238–242.

E919. Di Monte, D., Shinka, T., Sandy, M.S., Castagnoli, N., Jr., Smith, M.T., *Drug Metab. Dispos.* (1988) **16**, 250–255.

E920. Weil, A., Caldwell, J., Strolin-Benedetti, M., *Drug Metab. Dispos.* (1988) **16**, 302–309.

E921. Pfeifer, A., Neumann, H.-G., *Drug Metab. Dispos.* (1988) **16**, 276–284.

E922. Martin, R.J., Downer, R.G.H., *Can. J. Zool.* (1989) **67**, 1495–1499.

E923. Borgulya, J., Bruderer, H., Bernauer, K., Zürcher, G., Da Prada, M., *Helv. Chim. Acta* (1989) **72**, 952–968.

E924. Straub, K., Davis, M., Hwang, B., *Drug Metab. Dispos.* (1988) **16**, 359–366.

E925. Gole, D.J., Pirat, J.-L., Kamenka, J.-M., Domino, E.F., *Drug Metab. Dispos.* (1988) **16**, 386–391.

E926. Meuldermans, W., Hendrickx, J., Lauwers, W., Hurkmans, R., Mostmans, E., Swysen, E., Bracke, J., Knaeps, A., Heykants, J., *Drug Metab. Dispos.* (1988) **16**, 410–419.

E927. Meuldermans, W., Van Peer, A., Hendrickx, J., Lauwers, W., Swysen, E., Bockx, M., Woestenborghs, R., Heykants, J., *Drug Metab. Dispos.* (1988) **16**, 403–409.

E928. Dostal, L.A., Guthenberg, C., Mannervik, B., Bend, J.R., *Drug Metab. Dispos.* (1988) **16**, 420–424.

E929. Quon, C.K., Mai, K., Patil, G., Stampfli, H.F., *Drug Metab. Dispos.* (1988) **16**, 425–428.

E930. Yeh, S.Y., Hsu, F.-L., *Drug Metab. Dispos.* (1988) **16**, 499–502.

E931. Baier-Weber, B., Prox, A., Wachsmuth, H., Breyer-Pfaff, U., *Drug Metab. Dispos.* (1988) **16**, 490–496.

E932. Chaudhary, A.K., Hubbard, J.W., McKay, G., Midha, K.K., *Drug Metab. Dispos.* (1988) **16**, 506–508.

E933. Nusser, E., Nill, K., Breyer-Pfaff, U., *Drug Metab. Dispos.* (1988) **16**, 509–511.

E934. Yeung, J.M., Taylor, W.G., *Drug Metab. Dispos.* (1988) **16**, 600–604.

E935. Egger, H., Itterly, W., John, V., Rodebaugh, R., Shimanskas, C., Stancato, F., Kapoor, A., *Drug Metab. Dispos.* (1988) **16**, 568–575.

E936. Cashman, J.R., Proudfoot, J., Pate, D.W., Högberg, T., *Drug Metab. Dispos.* (1988) **16**, 616–622.

E937. Fournel-Gigleux, S., Hamar-Hansen, C., Motassim, N., Antoine, B., Mothe, O., Decolin, D., Caldwell, J., Siest, G., *Drug Metab. Dispos.* (1988) **16**, 627–634.

E938. Volp, R.F., *Drug Metab. Dispos.* (1988) **16**, 650–652.

E939. Wal, J.-M., Peleran, J.-C., Perdu, E., Rao, D., Bories, G., *Drug Metab. Dispos.* (1988) **16**, 635–639.

E940. Kuo, G.Y., Hwang, B.Y.-K., Lynn, R.K., *Drug Metab. Dispos.* (1988) **16**, 793–798.

E941. Quistad, G.B., Saunders, A.L., Skinner, W.S., Collier, K.D., Sakai, D.H., Reuter, C.C., *Drug Metab. Dispos.* (1988) **16**, 818–922.

E942. Otton, S.V., Brinn, R.U., Gram, L.F., *Drug Metab. Dispos.* (1988) **16**, 15–17.

E943. Nelson, W.L., Olsen, L.D., *Drug Metab. Dispos.* (1988) **16**, 834–841.

E944. Wu, W.N., Hills, J.F., Chang, S.Y., Ng, K.T., *Drug Metab. Dispos.* (1988) **16**, 69–77.

E945. Narimatsu, S., Matsubara, K., Shimonishi, T., Watanabe, K., Yamamoto, I., Yoshimura, H., *Drug Metab. Dispos.* (1988) **16**, 156–161.

E946. Greenspan, M.D., Yudkovitz, J.B., Alberts, A.W., Argenbright, L.S., Arison, B.H., Smith, J.L., *Drug Metab. Dispos.* (1988) **16**, 678–682.

E947. Kerger, B.D., Roberts, S.M., James, R.C., *Drug Metab. Dispos.* (1988) **16**, 672–677.

E948. Tocco, D.J., deLuna, F.A., Duncan, A.E.W., Ramjit, H.G., Pitzenberger, S.M., Hsieh, J.Y., *Drug Metab. Dispos.* (1988) **16**, 690–696.

E949. Tocco, D.J., Duncan, A.E.W., deLuna, F.A., Wells, G., *Drug Metab. Dispos.* (1988) **16**, 697–700.

E950. Thompson, K.L., Chang, M.N., Chabala, J.C., Chiu, S.-H.L., Eline, D., Hucker, H.B., Sweeney, B.M., White, S.D., Arison, B.H., Smith, J.L., *Drug Metab. Dispos.* (1988) **16**, 737–743.

E951. Parli, C.J., Evenson, E., Potts, B.D., Beedle, E., Lawson, R., Robertson, D.W., Leander, J.D., *Drug Metab. Dispos.* (1988) **16**, 707–711.

E952. Yoshikawa, M., Endo, H., Otsuka, M., Yamaguchi, I., Harigaya, S., *Drug Metab. Dispos.* (1988) **16**, 754–758.

E953. Tee, L.B.G., Minchin, R.F., Ilett, K.F., *Carcinogenesis* (1988) **9**, 1869–1874.

E956. Lukič, V., Walterová, D., Husek, A., Gašič, O., Šimánek, V., *Acta Univ. Palacki. Olomuc. Fac. Med.* (1988) **120**, 429–434.

E957. Omori, T., Hatakeyama, K., Kodama, T., *Appl. Microbiol. Biotechnol.* (1988) **29**, 497–500.

E958. Kozawa, T., Maruyama, Y., *J. Bacteriol.* (1988) **170**, 5778–5784.

E959. Para, G., Baratti, J., *Biocatalysis* (1988) **2**, 39–50.

E960. Kohler, H.P.E., Kohler-Staub, D., Focht, D.D., *Appl. Environ. Microbiol.* (1988) **54**, 2683–2688.

E962. Singer, S.S., Palmert, M.R., Redman, M.D., Leahy, S., Feeser, T.C., Lucarelli, M.J., Volkwein, L.S., Bruns, M., *Hepatology* (*Baltimore*) (1988) **8**, 1511–1520.

E963. Martinéz Ortiz, F., Tudela Serrano, J., Rodríguez López, J.N., Varón Castellanos, R., Lozano Teruel, J.A.,

García-Cánovas, F., *Biochim. Biophys. Acta* (1988) **957**, 158–163.
E964. Bhatnagar, D., Ullah, A.H.J., Cleveland, T.E., *Prep. Biochem.* (1988) **18**, 321–349.
E965. Gal, J., Zirrolli, J.A., Lichtenstein, P.S., *Res. Commun. Chem. Pathol. Pharmacol.* (1988) **62**, 3–17.
E966. Timmers, R.J.M., Granneman, J.C.M., Lambert, J.D., Van Oordt, P.G.W.J., *Gen. Comp. Endocrinol.* (1988) **72**, 190–203.
E967. Smart, D., *Int. J. Parasitol.* (1988) **18**, 747–752.
E968. Martyn, P., Smith, D.L., Adams, J.B., *Mol. Cell. Endocrinol.* (1988) **60**, 7–13.
E969. Schladt, L., Thomas, H., Hartmann, R., Oesch, F., *Eur. J. Biochem.* (1988) **176**, 715–723.
E970. Lau, Y.-S., Crampton, J.M., Wilson, J.A., *Life Sci.* (1988) **43**, 1459–1464.
E971. Raha, C., Williamson, D., Thakran, P., Casida, D., Iversen, P., *J. Chromatogr.* (1988) **450**, 325–336.
E972. Baker, M.A., Zeman, E.M., Hirst, V.K., Brown, J.M., *Cancer Res.* (1988) **48**, 5947–5952.
E973. Cowan, D.A., Ozkirimli, S., Beckett, A.H., *Acta Pharm. Suec.* (1988) **25**, 141–150.
E974. Mack, J.P.G., Mulvena, D.P., Slaytor, M., *Plant Physiol.* (1988) **88**, 315–320.
E975. Saul, S., Sugumaran, M., *FEBS Lett.* (1988) **237**, 155–158.
E977. Bokkenheuser, V.D., Winter, J., *Prog. Clin. Biol. Res.* (1988) **280**, 143–145.
E978. Frenzel, T., Beale, J.M., Kobayashi, M., Zenk, M.H., Floss, H.G., *J. Am. Chem. Soc.* (1988) **110**, 7878–7880.
E979. Chiba, K., Kubota, E., Miyakawa, T., Kato, Y., Ishizaki, T., *J. Pharmacol. Exp. Ther.* (1988) **246**, 1108–1115.
E980. Ohe, T., Watanabe, Y., *Agric. Biol. Chem.* (1988) **52**, 2409–2414.
E982. Frankenberger, W.T., Jr., Poth, M., *Soil Biol. Biochem.* (1988) **20**, 299–304.
E983. Chen, R.-Y., Chang, T.-C., Liu, M.-S., *Agric. Biol. Chem.* (1988) **52**, 2137–2142.
E985. Griggs, D.J., Wise, R., Kirkpatrick, B., Ashby, J.P., *J. Antimicrob. Chemother.* (1988) **22**, 191–194.
E986. Hauptmann, J., Paintz, M., Kaiser, B., Richter, M., *Pharmazie* (1988) **43**, 559–560.
E987. Nakamura, J., Shiota, H., Sasaki, H., Shibasaki, J., *J. Pharmacobio.-Dyn.* (1988) **11**, 625–629.
E988. Lim, H.K., Foltz, R.L., *Chem. Res. Toxicol.* (1988) **1**, 370–378.
F1. Fuska, J., Proksa, B., Khandlova, A., Sturdikova, M., *Acta Biotechnol.* (1988) **8**, 485–489.
F2. Frandsen, J.C., Bone, L.W., *Comp. Biochem. Physiol. C: Comp. Pharmacol. Toxicol.* (1988) **91C**, 385–387.
F3. Asai, H., Onozaki, H., Imaseki, H., *Agric. Biol. Chem.* (1988) **52**, 2741–2746.
F4. Tanaka, H., Ohdo, T., Urushihora, M., Uchida, K., *Asahi Garasu Kenkyu Hokoku* (1988) **38**, 123–135.
F5. Neidlein, R., Pichler, B., Hege, G., Weymann, J., *Arch. Pharm. (Weinheim, Ger.)* (1988) **321**, 773–780.
F6. Ahmad, B., Powell, J.W., *J. Drug Metab. Pharmacokinet.* (1988) **13**, 177–183.
F7. Watanabe, K., Tanaka, T., Yamamoto, I., Yoshimura, H., *Biochem. Biophys. Res. Comun.* (1988) **157**, 75–80.
F8. Honkakoski, P., Kojo, A., Raunio, H., Pasanen, M., Juvonen, R., Lang, M.A., *Arch. Biochem. Biophys.* (1988) **267**, 558–567.
F9. Sasse, F., Witte, L., Berlin, J., *Planta Med.* (1987) **53**, 354–359.
F10. Mozayani, A., Coutts, R.T., Danielson, T.J., Baker, G.B., *Res. Commun. Chem. Pathol. Pharmacol.* (1988) **62**, 397–406.
F11. Kobayashi, H., Okumura, S., Kosaka, Y., Kobayashi, S., Inoue, A., Oka, T., Nakamizo, N., *Arzneim.-Forsch.* (1988) **38**, 1753–1756.
F14. Kiang, C.-H., Lee, C., Kushinsky, S., *Drug Metab. Dispos.* (1989) **17**, 43–48.
F15. Madsen, E.L., Bollag, J.-M., *Arch. Microbiol.* (1989) **151**, 71–76.

F16. Tremaine, L.M., Stroh, J.G., Ronfeld, R.A., *Drug Metab. Dispos.* (1989) **17**, 58–63.

F17. Niwa, T., Tokuma, Y., Nakagawa, K., Noguchi, H., *Drug Metab. Dispos.* (1989) **17**, 64–68.

F18. Ishida, T., Kumagai, Y., Ikeda, Y., Ito, K., Yano, M., Toki, S., Mihashi, K., Fujioka, T., Iwase, Y., Hachiyama, S., *Drug Metab. Dispos.* (1989) **17**, 77–81.

F19. Renberg, L., Simonsson, R., Hoffmann, K.-J., *Drug Metab. Dispos.* (1989) **17**, 69–76.

F20. Newton, J.F., Yodis, L.P., Keohane, D., Eckardt, R., Dewey, R., Dent, J., Mico, B., *Drug Metab. Dispos.* (1989) **17**, 174–179.

F21. Jeffcoat, A.R., Perez-Reyes, M., Hill, J.M., Sadler, B.M., Cook, C.E., *Drug Metab. Dispos.* (1989) **17**, 153–159.

F22. Lim, W.H., Hooper, W.D., *Drug Metab. Dispos.* (1989) **17**, 212–217.

F23. Rashed, M.S., Streeter, A.J., Nelson, S.D., *Drug Metab. Dispos.* (1989) **17**, 355–359.

F24. Iwakawa, S., Suganuma, T., Lee, S.-F., Spahn, H., Benet, L.Z., Lin, E.T., *Drug Metab. Dispos.* (1989) **17**, 414–419.

F25. Chen, T.H., Kuslikis, B.I., Braselton, W.E., Jr., *Drug Metab. Dispos.* (1989) **17**, 406–413.

F26. Coan, M., Simmons, R., *Drug Metab. Dispos.* (1989) **17**, 420–425.

F27. Dahl-Puustinen, M.-L., Dumont, E., Bertilsson, L., *Drug Metab. Dispos.* (1989) **17**, 433–436.

F28. Miura, Y., Chisima, S., Takeyama, S., *Drug Metab. Dispos.* (1989) **17**, 455–462.

F29. Klassen Wehler, E., Bergman, Å., Brandt, I., Darnerud, P.O., Wachtmeister, C.A., *Drug Metab. Dispos.* (1989) **17**, 441–448.

F30. Maurer, H., Kleff, I., *Arzneim.-Forsch.* (1988) **38**, 1843–1845.

F31. Rashed, M.S., Nelson, S.D., *Chem. Res. Toxicol.* (1989) **2**, 41–45.

F32. Setnikar, I., Makovec, F., Christe, R., Giachetti, C., Zanolo, G., *Arzneim.-Forsch.* (1988) **38**, 1815–1819.

F33. Norén, B., Strömberg, S., Ericsson, O., Lindeke, B., *Acta Pharm. Suec.* (1988) **25**, 281–292.

F34. Gofschtein-Gandman, L.V., Keynan, A., Milner, Y., *J. Bacteriol.* (1988) **170**, 5895–5900.

F35. Hirose, M., Hagiwara, A., Inoue, K., Ito, N., Kaneko, H., Saito, K., Matsunaga, H., Isobe, N., Yoshitake, A., Miyamoto, J., *Toxicology* (1988) **53**, 33–43.

F36. Hafsah, Z., Tahara, S., Mizutani, J., *Nippon Noyaku Gakkaishi* (1987) **12**, 617–625.

F37. Hakamatsuka, T., Noguchi, H., Ebizuka, Y., Sankawa, U., *Chem. Pharm. Bull.* (1988) **36**, 4225–4228.

F38. Phillips, R.S., Ravichandran, K., Von Tersch, R.L., *Enzyme Microb. Technol.* (1989) **11**, 80–83.

F39. Cahill, G.M., Besharse, J.C., *Proc. Natl. Acad. Sci. U.S.A.* (1989) **86**, 1098–1102.

F40. Rossi, L., Mcgirr, L.G., Silva, J., O'Brien, P.J., *Mol. Pharmacol.* (1988) **34**, 674–681.

F41. Stopher, D.A., Beresford, A.P., Macrae, P.V., Humphrey, M.J., *J. Cardiovasc. Pharmacol.* (1988) **12**, S55-S59.

F42. Bui, Q.D., Weisz, J., *Endocrinology (Baltimore)* (1989) **124**, 1085–1087.

F43. Yamamoto, I., Gohda, H., Narimatsu, S., Yoshimura, H., *J. Pharmacobio.-Dyn.* (1988) **11**, 833–838.

F44. Hattori, M., Kanda, T., Shu, Y.-Z., Akao, T., Kobashi, K., Namba, T., *Chem. Pharm. Bull.* (1988) **36**, 4462–4466.

F45. Kojima, M., Wakako, T., *J. Biochem.* (1989) **105**, 265–270.

F46. Müller, U., Lingens, F., *Biol. Chem. Hoppe-Seyler* (1988) **369**, 1031–1043.

F47. Gurujeyalakshmi, G., Oriel, P., *Appl. Environ. Microbiol.* (1989) **55**, 500–502.

F48. Merkel, S.M., Eberhard, A.E., Gibson, J., Harwood, C.S., *J. Bacteriol.* (1989) **171**, 1–7.
F49. Lee, W.K.,, Lee, J.Y., Kang, K.Y., Cho, M.J., *Han'guk Saenghwa Hakhoechi* (1988) **21**, 519–524.
F50. Boutagy, J., Rumble, F., Danagan, F., *J. Chromatogr.* (1989) **487**, 483–488.
F51. Vasudevan, N., Mahadevan, A., *FEMS Microbiol. Lett.* (1988) **56**, 349–353.
F52. Van Pelt, F.N.A.M., Mennes, W.C., Hassing, I.G.A.M., Blaauboer, B.J., *J. Chromatogr.* (1989) **487**, 489–495.
F54. Pascoe, G.A., Calleman, C.J., Baille, T.A., *Chem.-Biol. Interact.* (1988) **68**, 85–98.
F55. Maurer, H., *Arzneim.-Forsch.* (1989) **39**, 101–103.
F56. Cunningham, M.L., Burka, L.T., Matthews, H.B., *Drug Metab. Dispos.* (1989) **17**, 612–617.
F57. Potts, B.D., Gabriel, S., Parli, C.J., *Drug Metab. Dispos.* (1989) **17**, 656–661.
F58. Thongnopnua, P., Massarella, J.W., Zimmermann, C.L., *Drug Metab. Dispos.* (1989) **17**, 473–480.
F59. Woolf, T.F., Black, A., Hicks, J.L., Lee, H., Huang, C.C., Chang, T., *Drug Metab. Dispos.* (1989) **17**, 662–668.
F60. Bai, S.A., Easterling, D.E., Wilson, M.J., Walle, U.K., Haney, C.A., Peet, N.P., Pruett, J.K., Walle, T., *Drug Metab. Dispos.* (1989) **17**, 495–505.
F61. Feng, P.C.C., Patanella, J.E., *Pestic. Biochem. Physiol.* (1989) **33**, 16–25.
F62. Sterjiades, R., Pelmont, J., *Appl. Environ. Microbiol.* (1989) **55**, 340–347.
F63. Hampp, N., Zenk, M.H., *Phytochemistry* (1988) **27**, 3811–3815.
F65. Ziegler, K., Buder, R., Winter, J., Fuchs, G., *Arch. Microbiol.* (1989) **151**, 171–176.
F66. Kneusel, R.E., Matern, U., Nicolay, K., *Arch. Biochem. Biophys.* (1989) **269**, 455–462.
F67. Haigler, B.E., Spain, J.C., *Appl. Environ. Microbiol.* (1989) **55**, 372–379.
F68. Sugumaran, M., Saul, S.J., Semensi, V., *Arch. Insect Biochem. Physiol.* (1988) **9**, 269–281.
F69. Miyauchi, S., Sugiyama, Y., Iga, T., Hanano, M., *Chem. Pharm. Bull.* (1989) **37**, 475–480.
F70. Grant, D.M., Lottspeich, F., Meyer, U.A., *FEBS Lett.* (1989) **244**, 203–207.
F71. Nesnow, S., Ross, J., Mohapatra, N., Gold, A., Sangaiah, R., Gupta, R., *Mutat. Res.* (1989) **222**, 223–235.
F72. Fleurence, J., Negrel, J., *Phytochemistry* (1989) **28**, 733–736.
F74. Yeoh, H.H., Tan, T.K., Chua, S.L., Lim, G., *MIRCEN J. Appl. Microbiol. Biotechnol.* (1988) **4**, 425–430.
F75. Haddock, J.D., Ferry, J.G., *J. Biol. Chem.* (1989) **264**, 4423–4427.
F76. Renganathan, V., *Appl. Environ. Microbiol.* (1989) **55**, 330–334.
F77. Habu, N., Samejima, M., Yoshimoto, T., *Mokuzai Gakkaishi* (1988) **34**, 1026–1034.
F78. Beresford, A.P., Macrae, P.V., Alker, D., Kobylecki, R.J., *Arzneim.-Forsch.* (1989) **39**, 201–209.
F79. Bhatnagar, D., Cleveland, T.E., Lax, A.R., *Arch. Environ. Contam. Toxicol.* (1989) **18**, 434–438.
F80. Hendrickson, H.P., Scott, D.O., Lunte, C.E., *J. Chromatogr.* (1989) **477**, 401–408.
F81. Kruse, E., Hartley-Asp, B., *In Vivo* (1988) **2**, 371–374.
F82. Tremaine, L.M., Welch, W.M., Ronfeld, R.A., *Drug Metab. Dispos.* (1989) **17**, 542–550.
F83. Dock, L., Martinez, M., Jernström, B., *Carcinogenesis* (1989) **10**, 245–249.
F84. Xu, X., Hirayama, H., Pang, K.S., *Drug Metab. Dispos.* (1989) **17**, 556–563.
F85. Nakano, M., Mizojiri, K., *Drug Metab. Dispos.* (1989) **17**, 564–566.

F86. De Vries, M.H., Hofman, G.A., Koster, A.S., Noordhoek, J., *Drug Metab. Dispos.* (1989) **17**, 573–578.

F87. Kassam, J.P., Tang, B.K., Kadar, D., Kalow, W., *Drug Metab. Dispos.* (1989) **17**, 567–572.

F88. Reeve, C.D., Moreman, A.D., Hopper, D.J., *FEMS Microbiol. Lett.* (1988) **52**, 251–254.

F89. Bossert, I.D., Whited, G., Gibson, D.T., Young, L.Y., *J. Bacteriol.* (1989) **171**, 2956–2962.

F90. Saul, S.J., Sugumaran, M., *FEBS Lett.* (1989) **251**, 69–73.

F92. Bhatnagar, R.K., Hendry, A.T., Shanmugam, K.T., Jensen, R.A., *J. Gen. Microbiol.* (1989) **135**, 353–360.

F93. Adams, D., Ribbons, D.W., *Appl. Biochem. Biotechnol.* (1988) **17**, 231–244.

F94. Kühnl, T., Koch, U., Heller, W., Wellman, E., *Plant Sci. (Limerick, Irel.)* (1989) **60**, 21–25.

F96. Cohen, B.M., Lipinski, J.F., Waternaux, C., *Psychopharmacology (Berlin)* (1989) **97**, 481–488.

F97. Setnikar, I., Makovec, F., Chistè, R., Giachetti, C., Zanolo, G., *Arzneim.-Forsch.* (1989) **39**, 328–334.

F98. Woodward, M.D., Stearns, S.M., Lee, P.W., *J. Agric. Food Chem.* (1989) **37**, 787–791.

F99. Anderson, M.G., Monypenny, D., Rickards, R.W., Rothschild, J.M., *J. Chem. Soc., Chem. Commun.* (1989) 311–313.

F100. Negishi, M., Lindberg, R., Burkhart, B., Ichikawa, T., Honkakoski, P., Lang, M., *Biochemistry* (1989) **28**, 4169–4172.

F101. Vasudevan, N., Mahadevan, A., *J. Biotechnol.* (1989) **9**, 107–116.

F102. Barron, D., Ibrahim, R.K., *Z. Naturforsch. C: Biosci.* (1988) **43**, 625–630.

F103. Ballard, D.G.H., Courtis, A., Shirley, I.M., Taylor, S.C., *Air Force Off. Sci. Res.* (1987) AFOSR-TR-87–1884.

F104. Bremnes, R.M., Slørdal, L., Wist, E., Aarbakke, J., *Cancer Res.* (1989) **49**, 2460–2464.

F105. Sone, T., Isobe, M., Takabatake, E., Ozawa, N., Watabe, T., *J. Pharmacobio-Dyn.* (1989) **12**, 149–158.

F106. Adriaens, P., Kohler, H.P.E., Kohler-Staub, D., Focht, D.D., *Appl. Environ. Microbiol.* (1989) **55**, 887–892.

F107. Higson, F.K., Focht, D.D., *Appl. Environ. Microbiol.* (1989) **55**, 946–952.

F108. Katsumata, T., Shige, H., Ejiri, S., *Phytochemistry* (1989) **28**, 359–362.

F109. Tatsumi, K., Kitamura, S., Amano, S., Ueda, K., *Cancer Res.* (1989) **49**, 2059–2064.

F110. Asano, Y., Endo, K., *Appl. Microbiol. Biotechnol.* (1988) **29**, 523–527.

F111. Johanni, M., Hofmann, P., Leistner, E., *Arch. Biochem. Biophys.* (1989) **271**, 495–501.

F112. Pras, N., Hesselink, P.G.M., Guikema, W.M., Malingre, T.M., *Biotechnol. Bioeng.* (1989) **33**, 1461–1468.

F114. Trinidad, A., Kirlin, W.G., Ogolla, F., Andrews, A.F., Yerokun, T., Ferguson, R.J., Brady, P.K., Hein, D.W., *Drug Metab. Dispos.* (1989) **17**, 238–247.

F115. Yerokun, T., Kirlin, W.G., Trinidad, A., Ferguson, R.J., Ogolla, F., Andrews, A.F., Brady, P.K., Hein, D.W., *Drug Metab. Dispos.* (1989) **17**, 231–237.

F116. Sundaram, R.S., Szumlanski, C., Otterness, D., Van Loon, J.A., Weinshilboum, R.M., *Drug Metab. Dispos.* (1989) **17**, 255–264.

F117. Rettie, A.E., Eddy, A.C., Heimark, L.D., Gibaldi, M., Trager, W.F., *Drug Metab. Dispos.* (1989) **17**, 265–270.

F118. Vane, F.M., Buggé, C.J.L., Rodriguez, L.C., *Drug Metab. Dispos.* (1989) **17**, 275–279 and 280–285.

F119. Pohorecki, R., Rayburn, W., Coon, W.W., Domino, E.F., *Drug Metab. Dispos.* (1989) **17**, 271–274.

F120. Bartels, M.J., Smith, F.A., *Drug Metab. Dispos.* (1989) **17**, 286–291.

F121. Lindberg, C., Roos, C., Tunek, A., Svensson, L.Å., *Drug Metab. Dispos.* (1989) **17**, 311–317.

F122. Inaba, T., Kovacs, J., *Drug Metab. Dispos.* (1989) **17**, 330–333.

F123. Nakano, M., Nakajima, Y., Iwatani, K., Ikenishi, Y., Nakagawa, Y., Sugeno, K., *Drug Metab. Dispos.* (1989) **17**, 323–329.

F124. Craig, J.C., Gruenke, L.D., North-Root, H., Corbin, N.C., *Drug Metab. Dispos.* (1989) **17**, 345–347.

F125. Subrahmanyam, Y., Sadler, A., Suba, E., Ross, D., *Drug Metab. Dispos.* (1989) **17**, 348–349.

F126. Sobrero, A.F., McGuire, J.J., Bertino, J.R., *Biochem. Pharmacol.* (1988) **37**, 997–1001.

F127. Müller, A., Gabriel, H., Sies, H., Terlinden, R., Fischer, H., Römer, A., *Biochem. Pharmacol.* (1988) **37**, 1103–1109.

F128. Tunek, A., Levin, E., Svensson, L.-Å., *Biochem. Pharmacol.* (1988) **37**, 3867–3876.

F129. Robbie, M.A., Palmer, B.D., Denny, W.A., Wilson, W.R., *Biochem. Pharmacol.* (1990) **39**, 1411–1421.

F130. Neau, E., Dansette, P.M., Adronik, V., Mansuy, D., *Biochem. Pharmacol.* (1990) **39**, 1101–1107.

F131. McCague, R., Parr, I.B., Leclerq, G., Leung, O.-T., Jarman, M., *Biochem. Pharmacol.* (1990) **39**, 1459–1465.

F132. Dansette, P.M., Amar, C., Smith, C., Pons, C., Mansuy, D., *Biochem. Pharmacol.* (1990) **39**, 911–918.

F133. Hiratsuka, A., Yokoi, A., Iwata, H., Watabe, T., Satoh, K., Hatayama, I., Sato, K., *Biochem. Pharmacol.* (1989) **38**, 4405–4413.

F134. Davis, W.B., Mohammed, B.S., Mays, D.C., She, Z.-W., Mohammed, J.R., Husney, R.M., Sagone, A.L., *Biochem. Pharmacol.* (1989) **38**, 4013–4019.

F135. Oda, Y., Imaoka, S., Nakahira, Y., Asada, A., Fujimori, M., Fujita, S., Funae, Y., *Biochem. Pharmacol.* (1989) **38**, 4439–4444.

F136. Myers, S.R., Flesher, J.W., *Biochem. Pharmacol.* (1989) **38**, 3199–3206.

F137. Kelder, P.P., de Mol, N.J., Janssen, L.H.M., *Biochem. Pharmacol.* (1989) **38**, 3593–3599.

F138. Eckert, K.-G., Elbers, F.R., Eyer, P., *Biochem. Pharmacol.* (1989) **38**, 3253–3259.

F139. Asaoka, K., Takahashi, K., *Biochem. Pharmacol.* (1989) **38**, 2977–2983.

F140. Lambert, C., Parks, B.K., Kitteringham, N.R., *Biochem. Pharmacol.* (1989) **38**, 2853–2858.

F141. Okuda, H., Nojima, H., Watanabe, N., Watabe, T., *Biochem. Pharmacol.* (1989) **38**, 3003–3009.

F142. Hiratsuka, A., Yokoi, A., Sebata, N., Watabe, T., Satoh, K., Hatayama, I., Sato, K., *Biochem. Pharmacol.* (1989) **38**, 2609–2613.

F143. Villegas, R.J.A., Kojima, M., *J. Biol. Chem.* (1986) **261**, 8729–8733.

F144. Rizzo, J.D., Davis, P.J., *J. Pharm. Sci.* (1989) **78**, 183–189.

F145. Moriguchi, T., Villegas, R.J.A., Kondo, T., Kojima, M., *Plant Cell Physiol.* (1988) **29**, 1221–1226.

F148. Koop, D.R., Laethem, C.L., Schnier, G.G., *Toxicol. Appl. Pharmacol.* (1989) **98**, 278–288.

F150. Melikian, A.A., Bagheri, K., Hecht, S.S., Hoffmann, D., *Chem.-Biol. Interact.* (1989) **69**, 245–257.

F151. Sachde, A.G., Al-Bakir, A.Y., Abdul-Raheem, J.A.K., *J. Food Biochem.* (1989) **12**, 241–251.

F152. Ghosh, D.K., Ghoshal, K., Mandal, A., Chaudhuri, J., *Med. Sci. Res.* (1989) **17**, 183–184.

F153. Sawada, H., Hara, A., Nakagawa, M., Tsukada, F., Ohmura, M., Matsuura, K., *Int. J. Biochem.* (1989) **21**, 367–375.

F154. Polter, C., Ludwig, E., Uhlemann, E., *Zentralbl. Mikrobiol.* (1988) **143**, 561–567.

F156. Kondo, K., Ogura, M., Midorikawa, Y., Kozawa, M., Tsujibo, H., Baba, K., Inamori, Y., *Agric. Biol. Chem.* (1989) **53**, 777–782.

F157. Kondo, K., Ogura, M., Midorikawa, Y., Kozawa, M., Tsujibo, H., Inamori, Y., *Agric. Biol. Chem.* (1989) **53**, 1171–1172.

F158. Susilo, R., Rommelspacher, H., Höfle, G., *J. Neurochem.* (1989) **52**, 1793–1800.

F159. Leung, L.Y., Baillie, T., *Biomed. Environ. Mass Spectrom.* (1989) **18**, 401–404.

F160. Pakusch, A.-E., Kneusel, R.E., Matern, U., *Arch. Biochem. Biophys.* (1989) **271**, 488–494.

F161. Smith, M.R., Ratledge, C., *Appl. Microbiol. Biotechnol.* (1989) **30**, 395–401.

F162. Beerhues, L., Forkmann, G., Schoepker, H., Stotz, F., Wiermann, R., *J. Plant Physiol.* (1989) **133**, 743–746.

F163. Shields, M.S., Montgomery, S.O., Chapman, P.J., Cuskey, S.M., Pritchard, P.H., *Appl. Environ. Microbiol.* (1989) **55**, 1624–1629.

F164. Halkier, B.A., Scheller, H.V., Moeller, B.L., *Ciba Found. Symp.* (1988) **140**, 49–66.

F165. Schnellmann, R.G., Monks, T.J., Mandel, L.J., Lau, S.S., *Toxicol. Appl. Pharmacol.* (1989) **99**, 19–27.

F167. Delclos, K.B., El-Bayoumy, K., Casciano, D.A., Walker, R.P., Kadlubar, F.F., Hecht, S.S., Shivapurkar, N., Mandal, S., Stoner, G.D., *Cancer Res.* (1989) **49**, 2909–2913.

F168. Fortnagel, P., Harms, H., Wittich, R.M., Francke, W., Krohn, S., Meyer, H., *Naturwissenschaften* (1989) **76**, 222–223.

F169. Takeuchi, F., Tsubouchi, R., Izuta, S., Shibata, Y., *J. Nutr. Sci. Vitaminol.* (1989) **35**, 111–122.

F170. Ling, K.-H., Hanzlik, R.P., *Biochem. Biophys. Res. Commun.* (1989) **160**, 844–849.

F171. Van Steveninck, J., Koster, J.F., Dubbelman, T.M.A.R., *Biochem. J.* (1989) **259**, 633–637.

F172. Kirlin, W.G., Trinidad, A., Yerokun, T., Ogolla, F., Ferguson, R.J., Andrews, A.F., Brady, P.K., Hein, D.W., *Cancer Res.* (1989) **49**, 2448–2454.

F173. Larson, R.L., *Biochem. Physiol. Pflanz.* (1989) **184**, 453–460.

F174. Mattano, S.S., Land, S., King, C.M., Weber, W.W., *Mol. Pharmacol.* (1989) **35**, 599–609.

F175. Ishikawa, T., Okuno, E., Kawai, J., Kido, R., *Comp. Biochem. Physiol. B: Comp. Biochem.* (1989) **93B**, 107–111.

F176. Cashman, J.R., Proudfoot, J., Ho, Y.K., Chin, M.S., Olsen, L.D., *J. Am. Chem. Soc.* (1989) **111**, 4844–4852.

F177. Knoll, G., Winter, J., *Appl. Microbiol. Biotechnol.* (1989) **30**, 318–324.

F178. Kojima, M., Uritani, I., *Plant Cell Physiol.* (1972) **13**, 311–319.

F179. Villegas, R.J., Shimokawa, T., Okuyama, H., Kojima, M., *Phytochemistry* (1987) **26**, 1577–1581.

F180. Kojima, M., Kondo, T., *Agric. Biol. Chem.* (1985) **49**, 2467–2469.

F181. Vree, T.B., Vree, J.B., Kolmer, E.B., Hekster, Y.A., Nouws, J.F.M., Shomoda, M., *Vet. Q.* (1989) **11**, 125–128.

F182. Khouri, H.E., Tahara, S., Ibrahim, R.K., *Arch. Biochem. Biophys.* (1988) **262**, 592–598.

F184. Khouri, H.E., De Luca, V., Ibrahim, R.K., *Arch. Biochem. Biophys.* (1988) **265**, 1–7.

F185. Ohe, T., Ohmoto, T., Kobayashi, Y., Sato, A., Watanabe, Y., *Agric. Biol. Chem.* (1985) **54**, 669–675.

F186. Faleev, N.G., Ruvinov, S.B., Demidkina, T.V., Myagkikh, I.V., Gololobov, M.Y., Bakhmutov, V.I., Belikov, V.M., *Eur. J. Biochem.* (1988) **177**, 395–401.

F187. Lim, H.K., Foltz, R.L., *Chem. Res. Toxicol.* (1989) **2**, 142–143.

F188. Locher, H.H., Leisinger, T., Cook, A.M., *J. Gen. Microbiol.* (1989) **135**, 1969–1978.

F189. Prabhakaran, K., Harris, E.B., *Med. Sci. Res.* (1989) **17**, 507.

F190. Gould, S.J., Shen, B., Whittle, Y.G., *J. Am. Chem. Soc.* (1989) **111**, 7932–7938.

F191. Agarwal, M., Roy, U., Shukla, O.P., *Biol. Mem.* (1988) **14**, 187–198.

F192. Ellaiah, P., Devi, M.R., *Indian Drugs* (1989) **26**, 434–437.

F193. Smith, T.J., Powell, J.R., Hollander, A.L., Fuchs, M.S., *Comp. Biochem. Physiol. B: Comp. Biochem.* (1989) **93B**, 721–725.

F194. Yamamoto, I., Gohda, H., Narimatsu, S., Yoshimura, H., *J. Pharmacobio-Dyn.* (1989) **12**, 488–494.

F195. Wallin, H., Holme, J.A., Becher, G., Alexander, J., *Carcinogenesis* (1989) **10**, 1277–1283.

F196. Baranczyk-Kuzma, A., Audus, K.L., Borchardt, R.T., *Neurochem. Res.* (1989) **14**, 689–691.

F197. Børresen, T., Klausen, N.K., Larsen, L.M., Sørensen, H., *Biochim. Biophys. Acta* (1989) **993**, 108–115.

F198. Cerniglia, C.E., Campbell, W.L., Freeman, J.P., Evans, F.E., *Appl. Environ. Microbiol.* (1989) **55**, 2275–2279.

F199. Rahouti, M., Seigle-Murandi, F., Steiman, R., Eriksson, K.E., *Appl. Environ. Microbiol.* (1989) **55**, 2391–2398.

F200. Oltmanns, R.H., Müller, R., Otto, M.K., Lingens, F., *Appl. Environ. Microbiol.* (1989) **55**, 2499–2504.

F201. Takeshita, M., Yoshida, S., Kiya, R., Higuchi, N., Kobayashi, Y., *Chem. Pharm. Bull.* (1989) **37**, 615–617.

F202. Sylvestre, M., Mailhiot, K., Ahmad, D., Masse, R., *Can. J. Microbiol.* (1989) **35**, 439–443.

F203. Winter, J., Moore, L.H., Dowell, V.R., Jr., Bokkenheuser, V.D., *Appl. Environ. Microbiol.* (1989) **55**, 1203–1208.

F205. Sloley, B.D., *Neurol. Neurobiol.* (1988) **42A**, 313–318.

F206. Norström, A., Andersson, B., Levin, J.O., Naeslund, P.H., Wallen, M., Loef, A., *Chemosphere* (1989) **18**, 1513–1523.

F207. Dahl-Puustinen, M.-L., Perry, T.L., Jr., Dumont, E., Von Bahr, C., Nordin, C., Bertilsson, L., *Chem. Pharmacol. Ther. (St. Louis)* (1989) **45**, 650–656.

F208. Chung, S.J., *Han'guk Yongyang Siklyong Hakhoechi* (1988) **17**, 348–357.

F209. Barry, C.E. III, Nayar, P.G., Begley, T.P., *Biochemistry* (1989) **28**, 6323–6333.

F210. Varin, L., *Bull. Liason-Groupe Polyphenols* (1988) **14**, 248–257.

F212. Ungell, A.L., *Fish Physiol. Biochem.* (1989) **6**, 157–165.

F213. Bassoli, A., Di Gregorio, G., Rindone, B., Tollari, S., Chioccara, F., Salmona, M., *Gazz. Chim. Ital.* (1988) **118**, 763–768.

F214. Achtert, G., Borchers, F., Jacquot, C., Christen, M.O., *Eur. J. Drug Metab. Pharmacokinet.* (1989) **14**, 29–34.

F215. Cummings, J., Doble, J.A., Bibby, M.C., Farmer, P., Evans, S., Kerr, D.J., Kaye, S.B., Smyth, J.F., *Cancer Res.* (1989) **49**, 3587–3593.

F216. Williams, F.M., Mutch, E.M., Nicholson, E., Wynne, H., Wright, P., Lambert, D., Rawlins, M.D., *J. Pharm. Pharmacol.* (1989) **41**, 407–409.

F217. Barwell, C.J., Basma, A.N., Lafi, M.A.K., Leake, L.D., *J. Pharm. Pharmacol.* (1989) **41**, 421–423.

F218. Rutgers, M., Heusdens, F.A., Bonthuis, F., Visser, T.J., *Endocrinology (Baltimore)* (1989) **125**, 433–443.

F219. Watanabe, K., Wada, Y., Takanashi, K., Yoshizawa, I., *Yakugaku Zasshi* (1989) **109**, 329–334.

F220. Scriba, G.K.E., Borchardt, R.T., *J. Neurochem.* (1989) **53**, 610–615.

F221. González, B., Vicuña, R., *J. Bacteriol.* (1989) **171**, 2401–2405.

F222. Speth, A.R., Hund, H.K., Lingens, F., *Biol. Chem. Hoppe-Seyler* (1989) **370**, 591–599.

F223. El-Mansi, E.M.T., Hopper, D.J., *Biochem. Soc. Trans.* (1989) **17**, 898–899.

F224. Chen, Y.P., Dilworth, M.J., Glenn, A.R., *Arch. Microbiol.* (1989) **151**, 520–523.

F225. Deobald, L.A., Crawford, D.L., *Appl. Biochem. Biotechnol.* (1988) **20–21**, 153–163.

F226. Ueki, A., Willoughby, J., Glover, V., Sandler, M., *Biochem. Pharmacol.* (1989) **38**, 2383–2385.

F227. Berhow, M.A., Vandercook, C.E., *Phytochemistry* (1989) **28**, 1627–1630.

F228. Naoi, M., Matsuura, S., Takahashi, T., Nagatsu, T., *Biochem. Biophys. Res. Commun.* (1989) **161**, 1213–1219.

F229. Marekov, N., Sidzhimov, A., *Izv. Khim.* (1988) **21**, 246–255.

F230. Clement, B., Zimmermann, M., Schmitt, S., *Arch. Pharm. (Weinheim, Ger.)* (1989) **322**, 431–435.

F231. Elmarakby, S.A., Duffel, M.W., Rosazza, J.P.N., *J. Med. Chem.* (1989) **32**, 2158–2162.

F232. Sjödin, P., Wallin, H., Alexander, J., Jägerstad, M., *Carcinogenesis* (1989) **10**, 1269–1275.

F234. Kang, E.S., Deaton, P.R., Epstein, D., Ingram, L.A., Mirvis, D.M., *Gen. Pharmacol.* (1989) **20**, 223–227.

F235. Terlinden, R., Fischer, H., Roemer, A., *Proc. Eur. Congr. Biopharm. Pharmacokinet., 3rd* (1987) **2**, 296–301.

F236. Köppel, C., Tenczer, J., Wagemann, A., *J. Chromatogr.* (1989) **491**, 432–438.

F237. Dahl-Puustinen, M.-L., Aaberg-Wistedt, A., Bertilsson, L., *Pharmacol. Toxicol.* (1989) **65**, 37–39.

F238. Inoue, O., Seiji, K., Ikeda, M., *Bull. Environ. Contam. Toxicol.* (1989) **43**, 220–224.

F239. Ohshima, N., Kotaki, H., Saitoh, Y., Nakagawa, F., Tamura, Z., *J. Pharmacobio-Dyn.* (1989) **12**, 371–377.

F240. Short, C.R., Hardy, M.L., Barker, S.A., *Toxicology* (1989) **57**, 45–58.

F241. Hara, A., Mouri, K., Nakagawa, M., Nakamura, M., Nakayama, T., Matsuura, K., Sawada, H., *J. Biochem.* (1989) **106**, 126–132.

F242. Kochs, G., Grisebach, H., *Arch. Biochem. Biophys.* (1989) **273**, 543–553.

F243. Varin, L., Ibrahim, R.K., *Plant Physiol.* (1989) **90**, 977–981.

F244. Habu, N., Komatsu, T., Samejima, M., Yoshimoto, Y., *Mokuzai Gakkaishi* (1989) **35**, 348–355.

F245. Carter, G.T., Goodman, J.J., Torrey, M.J., Border, D.B., Gould, S.J., *J. Org. Chem.* (1989) **54**, 4321–4323.

F246. Tahara, S., Ingham, J.L., Mizutani, J., *Phytochemistry* (1989) **28**, 2079–2084.

F247. Engesser, K.H., Schulte, P., *FEMS Microbiol. Lett.* (1989) **60**, 143–147.

F248. Foerlin, L., Andersson, T., Wachtmeister, C.A., *Comp. Biochem. Physiol. B: Comp. Biochem.* (1989) **93B**, 653–656.

F249. Markantonis, S.L., Kyroudis, A., Beckett, A.H., *Biochem. Med. Metab. Biol.* (1989) **42**, 1–8.

F251. Manniti, F.S., Dilberto, E.J., Jr., *J. Neurochem.* (1989) **53**, 890–897.

F252. Mori, M., Kawajiri, T., Sayama, M., Miyahara, T., Kazuka, H., *Chem. Pharm. Bull.* (1989) **37**, 1904–1908.

F253. Ottoboni, S., Caldera, P., Trevor, A., Castagnoli, N., Jr., *J. Biol. Chem.* (1989) **264**, 13684–13688.

F254. Vree, T.B., Vree, M.L., Beneken Kolmer, E.W.J., Hekster, Y.A., Shimoda, M., Nouws, J.F.M., Yoshioka, T., Hoji, K., *Jpn. J. Vet. Sci.* (1989) **51**, 364–368.

F255. Ushiyama, M., Asada, Y., Yoshikawa, T., Furuya, T., *Phytochemistry* (1989) **28**, 1859–1869.

F256. Rao, S.S., Vijayaraghavan, R., Suryanarayana, M.V.S., Ramachandran, P.K., *J. Chromatogr.* (1989) **493**, 210–216.

F257. Wariishi, H., Valli, K., Renganathan, V., Gold, M.H., *J. Biol. Chem.* (1989) **264**, 14185–14191.

F258. Ihn, W., Janke, D., Tresselt, D., *J. Basic Microbiol.* (1989) **29**, 291–297.

F259. Eckardt, K., Schumann, G., Wagner, C., Stengel, C., Ihn, W., Novak, J., Tresselt, D., *J. Basic Microbiol.* (1989) **29**, 277–279.

F260. McGuire, S.M., Brobst, S.W., Graybill, T.L., Pal, K., Townsend, C.A., *J. Am. Chem. Soc.* (1989) **111**, 8308–8309.

F261. Chakraborty, R.S., Hubbard, J.W., Hawes, E.M., McKay, G., Cooper, J.K., Gurnsey, T., Korchinski, E.D., Midha, K.K., *Eur. J. Clin. Pharmacol.* (1989) **37**, 45–48.

F262. Weil, A., Caldwell, J., Guichard, J.P., Picot, G., *Chirality* (1989) **1**, 197–201.

F263. Foetsch, G., Pfeifer, S., Bartoszek, M., Franke, P., Hiller, K., *Pharmazie* (1989) **44**, 555–558.

F265. Kuroki, G.W., Conn, E.E., *Proc. Natl. Acad. Sci. U.S.A.* (1989) **86**, 6978–6981.

F266. Malarczyk, E., *Acta Microbiol. Pol.* (1989) **38**, 45–53.

F268. Kittell, B.L., Helinski, D.R., Ditta, G.S., *J. Bacteriol.* (1989) **171**, 5458–5466.

F271. Elliott, J., Callingham, B.A., Sharman, D.F., *Br. J. Pharmacol.* (1989) **98**, 507–514.

F272. Wu, J.T., Williams, K.I.H., *Steroids* (1989) **54**, 401–419.

F273. Rao, S.S., Swamy, R.V., Ramachandran, P.K., *Toxicology* (1989) **58**, 81–89.

F274. Moore, N.L., Mariam, D.H., Williams, A.L., Dashek, W.V., *J. Ind. Microbiol.* (1989) **4**, 349–363.

F275. Kandaswami, C., Rutkowski, J.P., Kumar, S., Sikka, H.C., *Comp. Biochem. Physiol. C: Comp. Pharmacol. Toxicol.* (1989) **93C**, 269–274.

F276. Knoch, M., Van Pée, K.H., Vining, L.C., Lingens, F., *J. Gen. Microbiol.* (1989) **135**, 2493–2502.

F277. Hund, H.-K., Bär, G., Lingens, F., *Z. Naturforsch. C: Biosci.* (1989) **44**, 797–801.

F278. Green, J., Dalton, H., *J. Biol. Chem.* (1989) **264**, 17698–17703.

F279. Annan, N., Paris, R., Jordan, F., *J. Am. Chem. Soc.* (1989) **111**, 8895–8901.

F280. Laws, I., Mantle, P.G., *J. Gen. Microbiol.* (1989) **135**, 2679–2692.

F281. Yabe, K., Ando, Y., Hashimoto, J., Hamasaki, T., *Appl. Environ. Microbiol.* (1989) **55**, 2172–2177.

F282. Timmers, R.J.M., Lambert, J.G.D., *Fish Physiol. Biochem.* (1989) **7**, 201–210.

F283. Renganathan, V., Johnston, J.B., *Appl. Microbiol. Biotechnol.* (1989) **31**, 419–424.

F284. Lewis, C.J., Havler, M.E., Humphrey, M.J., Lloyd-Jones, J.G., McCleavy, M.A., Muir, N.C., Waltham, K., *Xenobiotica* (1988) **18**, 519–532.

F285. Groen, K., Warrander, A., Miles, G.S., Booth, B.S., Mulder, G.J., *Xenobiotica* (1988) **18**, 511–518.

F286. Maniara, W., Levin, S.S., Schleyer, H., Cooper, D.Y., *Xenobiotica* (1988) **18**, 545–550.

F287. Spinelli, R., Fracasso, C., Guiso, G., Garattini, S., Caccia, S., *Xenobiotica* (1988) **18**, 573–584.

F289. Wallace, G., Casabé, N., Wood, E., Zerba, E., *Xenobiotica* (1988) **18**, 351–355.

F290. Teel, R.W., Martin, R.M., *Xenobiotica* (1988) **18**, 397–405.

F291. Brown, N.K., Harvey, D.J., *Xenobiotica* (1988) **18**, 417–427.

F292. Mennicke, W.H., Görler, K., Krumbiegel, G., Lorenz, D., Rittmann, G., *Xenobiotica* (1988) **18**, 441–447.

F293. Simons, P.J., Cockshott, I.D., Douglas, E.J., Gordon, E.A., Hopkins, K., Rowland, M., *Xenobiotica* (1988) **18**, 429–440.

F294. Bakke, J.E., Davison, K.L., *Xenobiotica* (1988) **18**, 1057–1062.

F295. Kurchbayashi, H., Tanaka, A., Hamaha, T., Tatahashi, A., *Xenobiotica* (1988) **18**, 1039–1048.

F296. Bakke, J.E., Larsen, G.L., Struble, C., Feil, V.J., Brandt, I., Brittebo, E.B., *Xenobiotica* (1988) **18**, 1063–1075.

F297. Kammerer, R.C., Schmitz, D.A., *Xenobiotica* (1988) **18**, 1085–1096.

F298. Lang, B., Iba, M.M., *Xenobiotica* (1988) **18**, 893–904.

F299. Mesnil, M., Testa, B., Jenner, P., *Xenobiotica* (1988) **18**, 1097–1106.

F300. Cummings, A.J., Copas, D.K., Kaelin, A.C., Rosankiewicz, J.R., Landon, M.J., *Xenobiotica* (1988) **18**, 905–917.

F301. Duchene, P., Bernouillet, C., Bromet-Petit, M., Mosser, J., Fenion, C., Gaudin, D., Virelizier, H., *Xenobiotica* (1988) **18**, 919–928.

F302. Davison, K.L., Bakke, J.E., *Xenobiotica* (1988) **18**, 941–948.

F303. Yoshimura, H., Yoshimitsu, T., Yamada, H., Koga, N., Oguri, K., *Xenobiotica* (1988) **18**, 929–940.

F304. Gristwood, W., Wilson, K., *Xenobiotica* (1988) **18**, 949–954.

F305. Hesk, D., Hutson, D.H., Logan, C.J., *Xenobiotica* (1988) **18**, 955–966.

F306. Berry, D., Parke, D.V., *Xenobiotica* (1988) **18**, 857–868.

F307. Bakke, J.E., Larsen, G.L., Feil, V.J., Brittebo, E.B., Brandt, I., *Xenobiotica* (1988) **18**, 817–829.

F308. Harrelson, J.G., Wong, Y.J.N., *Xenobiotica* (1988) **18**, 1239–1247.

F309. Tephly, T., Green, M., Puig, J., Irshaid, Y., *Xenobiotica* (1988) **18**, 1201–1210.

F310. Bernstein, J.R., Franklin, R.B., *Xenobiotica* (1988) **18**, 1335–1345.

F311. Fischer, H., Terlinden, R., Löhr, J.P., Römer, A., *Xenobiotica* (1988) **18**, 1347–1359.

F312. McCormick, D.J., Fitzgerald, T.C., McKillop, D., *Xenobiotica* (1988) **18**, 1401–1412.

F313. Ross, D.A., Osborne, P.M., Pue, M.A., Blake, T.J., Chenery, R.J., Metcalf, R., *Xenobiotica* (1988) **18**, 1373–1387.

F314. Walls, C.M., Gray, A., Vose, C.W., Robinson, Y., Lopez, N., Brownsill, R.D., Steiner, J.A., *Xenobiotica* (1988) **18**, 1413–1423.

F315. Rizzo, J.D., Davis, P.J., *Xenobiotica* (1988) **18**, 1425–1437.

F316. Degen, P.H., Dieterle, W., Schneider, W., Theobald, W., Sinterhauf, U., *Xenobiotica* (1988) **18**, 1449–1455.

F317. Jaworski, T.J., Hawes, E.M., McKay, G., Midha, K.K., *Xenobiotica* (1988) **18**, 1439–1447.

F318. Miyamoto, G., Sasabe, H., Tominaga, N., Uegaki, N., Tominaga, M., Shimizu, T., *Xenobiotica* (1988) **18**, 1143–1155.

F319. Kamali, F., Gescher, A., Slack, J.A., *Xenobiotica* (1988) **18**, 1157–1164.

F320. Faigle, J.W., Böttcher, I., Godbillon, J., Kriemler, H.P., Schlompf, E., Schneider, W., Schweizer, A., Stierlin, H., Winkler, T., *Xenobiotica* (1988) **18**, 1191–1197.

F321. Abu-El-Haj, S., Allahyari, R., Chavez, E., Fraser, I.M., Strother, A., *Xenobiotica* (1988) **18**, 1165–1178.

F322. Huckle, K.R., Warburton, P.A., Forbes, S., Logan, C.J., *Xenobiotica* (1989) **19**, 51–62.

F323. van Cantfort, J., de Graeve, J., Puozzo, C., *Xenobiotica* (1989) **19**, 11–23.

F324. Sayama, M., Mori, M.-A., Ishida, M., Okumura, K., Kozuka, H., *Xenobiotica* (1989) **19**, 83–92.

F325. Montessisa, C., Stracciari, J.M., Fadini, L., Beretta, C., *Xenobiotica* (1989) **19**, 97–100.

F326. Štefek, M., Beneš, L., Zelník, V., *Xenobiotica* (1989) **19**, 143–150.

F327. Behrendt, H., Kampffmeyer, H.G., *Xenobiotica* (1989) **19**, 131–141.

F328. Huo, J.-Z., Aldous, S., Campbell, K., Davies, N., *Xenobiotica* (1989) **19**, 161–170.

F329. Oguri, K., Tanimoto, Y., Mishina, M., Yoshimura, H., *Xenobiotica* (1989) **19**, 171–178.

F330. Drummond, L., Caldwell, J., Wilson, H.K., *Xenobiotica* (1989) **19**, 199–207.

F331. Muralidharan, G., Midha, K.K., McKay, G., Hawes, E.M., Inaba, T., *Xenobiotica* (1989) **19**, 189–197.

F332. Castegnaro, M., Bartsch, H., Bereziat, J.C., Arvella, P., Michelon, J., Broussolle, L., *Xenobiotica* (1989) **19**, 225–230.

F333. Bellucci, G., Chiappe, C., Marioni, F., Simonelli, C., *Xenobiotica* (1989) **19**, 279–285.

F334. Fujimaki, M., Ishigaki, N., Hakusui, H., *Xenobiotica* (1989) **19**, 609–625.

F335. Mori, M.-A., Uy, N., Sakai, K., Inoue, M., Miyahara, T., Kozuka, H., *Xenobiotica* (1989) **19**, 287–299.

F336. Sherratt, A.J., Damani, L.A., *Xenobiotica* (1989) **19**, 379–388.

F337. Hümpel, M., Sostarek, D., Gieschen, H., Labitzky, C., *Xenobiotica* (1989) **19**, 361–377.

F338. Mitchell, M.C., Hamilton, R., Wacker, L., Branch, R.A., *Xenobiotica* (1989) **19**, 389–400.

F339. Ozaki, O., Kurata, T., Tachizawa, H., *Xenobiotica* (1989) **19**, 419–429.

F340. Foster, B.C., Wilson, D.L., McGilveray, I.J., *Xenobiotica* (1989) **19**, 445–452.

F341. Malejka-Giganti, D., Magat, W.J., Decker, R.W., *Xenobiotica* (1989) **19**, 431–444.

F342. Lenk, W., Riedl, M., *Xenobiotica* (1989) **19**, 453–475.

F343. Nakagawa, Y., Tayama, S., *Xenobiotica* (1989) **19**, 499–507.

F344. Vree, T.B., Beneken Kolmer, E.W.J., Martea, M., Bosch, R., Hekster, Y.A., Shimoda, M., *Pharm. Weekbl. Sci. Ed.* (1990) **12**, 51–59.

F345. Foster, B.C., Coutts, R.T., Pasutto, F.M., *Xenobiotica* (1989) **19**, 531–538.

F346. Foster, B.C., Buttar, H.S., Qureshi, S.A., McGilveray, I.J., *Xenobiotica* (1989) **19**, 539–546.

F347. Richards, R.P., Gordon, B.H., Ings, R.M.J., Campbell, D.B., King, L.J., *Xenobiotica* (1989) **19**, 547–553.

F348. Štefek, M., Beneš, L., *Xenobiotica* (1989) **19**, 627–634.

F349. Linhart, I., Hanuš, V., Novák, J., Šmejkal, J., Pech, P., *Xenobiotica* (1989) **19**, 645–653.

F350. Murray, G.R., Whiffen, G.M., Franklin, R.A., Henry, J.A., *Xenobiotica* (1989) **19**, 669–675.

F351. Muirhead, M.R., Somogyi, A.A., Lloyd, C., Bochner, F., *Xenobiotica* (1989) **19**, 655–660.

F352. Bolton, J.L., Sevestre, H., Ibe, B.O., Thompson, J.A., *Chem. Res. Toxicol.* (1990) **3**, 65–70.

F353. Kojima, M., Poulton, J.E., Thayer, S.S., Conn, E.E., *Plant Physiol.* (1979) **63**, 1022–1028.

F354. Vree, T.B., Vree, M.L., Beneken Kolmer, E.W.J., Hekster, Y.A., Shimoda, M., Nouws, J.F.M., Yoshioka, T., Hoji, K., *Jpn. J. Vet. Sci.* (1989) **51**, 364–368.

F355. Cashman, J.R., Williams, D.E., *Mol. Pharmacol.* (1990) **37**, 333–339.

F356. Vree, T.B., Vree, M.L., Beneken Kolmer, E.W.J., Hekster, Y.A., Shimoda, M., Nouws, J.F.M., Yoshioka, T., Hoji, K., *Jpn. J. Vet. Sci.* (1989) **51**, 369–372.

F357. Eckardt, K., Schumann, G., Wagner, C., Stengel, C., Ihn, W., Novák, J., Tresselt, D.K., *J. Basic Microbiol.* (1989) **5**, 277–279.

F358. Hanzlik, R.P., Ling, K.-H.J., *J. Org. Chem.* (1990) **55**, 3992–3997.

F360. Platt, K.L., Petrovic, P., Seidel, A., Beermann, D., Oesch, F., *Chem.-Biol. Interact.* (1988) **66**, 157–175.

F361. Cummings, J., Matheson, L.M., Smyth, J.F., *J. Chromatogr.* (1990) **528**, 43–53.

F362. Topall, G., Laborit, H., *J. Pharm. Pharmacol.* (1989) **41**, 789–791.

F363. Fischer, V., Baldeck, J.P., Wiebel, F.J., *Mol. Toxicol.* (1987) **1**, 341–350.

F364. Coughtrie, M.W.H., Ask, B., Rane, A., Burchell, B., Hume, R., *Biochem. Pharmacol.* (1989) **38**, 3273–3280.

F365. Eelkman Rooda, S.J., Otten, M.H., van Loon, M.A.C., Kaptein, E., Visser, T.J., *Endocrinology (Baltimore)* (1989) **125**, 2187–2197.

F366. Chalmers, R.M., Fewson, C.A., Biochem. J., (1989) **263**, 913–919.

F367. De Carolis, E., Ibrahim, R.K., *Biochem. Cell Biol.* (1989) **67**, 763–769.

F369. Heide, L., Nishioka, N., Fukui, H., Tabata, M., *Phytochemistry* (1989) **28**, 1873–1877.

F370. Shiotsu, Y., Samejima, M., Habu, N., Yoshimoto, T., *Mokuzai Gakkaishi* (1989) **35**, 826–831.

F371. Gay, G., Rouillon, R., Bernillon, J., Favre-Bonvin, J., *Can. J. Bot.* (1989) **67**, 2235–2239.

F372. Ludwig-Müller, J., Hilgenberg, W., *Phytochemistry* (1989) **28**, 2571–2575.

F373. Brzezińska-Ślebodzińska, E., Ślebodziński, A.B., Krysin, E., *J. Dev. Physiol.* (1989) **11**, 351–353.

F374. Christou, M., Jovanovich, M.C., Jefcoate, C.R., *Carcinogenesis* (1989) **10**, 1883–1890.

F375. Laham, S., Potvin, M., *Toxicol. Environ. Chem.* (1989) **24**, 57–69.

F376. Kawabata, T.T., White, K.L., Jr., *Cancer Res.* (1989) **49**, 5816–5822.

F377. Nakagawa, M., Tsukada, F., Nakagama, T., Matsuura, K., Hara, A., Sawada, H., *J. Biochem.* (1989) **106**, 633–638.

F378. Svenson, A., Kjeller, L.O., Rappe, C., *Chemosphere* (1988) **19**, 585–587.

F379. Singh, R.L., *Biochem. Int.* (1989) **19**, 1109–1116.

F380. Coleman, A.M., Hindsgaul, O., Palcic, M.M., *J. Biol. Chem.* (1989) **264**, 19500–19505.

F381. Reeve, C.D., Carver, M.A., Hopper, D.J., *Biochem. J.* (1989) **263**, 431–437.

F382. Myers, S.R., Blake, J.W., Flesher, J.W., *Chem.-Biol. Interact.* (1989) **71**, 393–401.

F383. Häggblom, M.M., Janke, D., Salkinoja-Salonen, M.S., *Microb. Ecol.* (1989) **18**, 147–159.

F384. Roth, K., Hildebrand, M., Beyer, K.H., *Eur. J. Drug Metab. Pharmacokinet.* (1989) **14**, 133–138.

F385. Benfenati, E., Farina, P., Colombo, T., De Bellis, G., Capodiferro, M.V., D'Incalci, M., *Cancer Chemother. Pharmacol.* (1989) **24**, 354–358.

F386. Saul, S.J., Sugumaran, M., *FEBS Lett.* (1989) **255**, 340–344.

F387. Dumasia, M.C., Houghton, E., *Biomed. Environ. Mass Spectrom.* (1987) **18**, 1030–1033.

F388. Nakagawa, M., Harada, T., Hara, A., Nakayama, T., Sawada, H., *Chem. Pharm. Bull.* (1989) **37**, 2852–2854.

F389. Hara, A., Harada, T., Nakagawa, M., Matsuura, K., Nakayama, T., Sawada, H., *Biochem. J.* (1989) **264**, 403–407.

F390. Arfmann, H.A., Abraham, W.R., *Z. Naturforsch. C: Biosci.* (1989) **44**, 765–770.

F391. Nakagawa, M., Matsuura, K., Hara, A., Sawada, H., Bunai, Y., Ohya, I., *J. Biochem.* (1989) **106**, 1104–1109.

F392. Vora, K.A., Modi, V.V., *Indian J. Exp. Biol.* (1989) **27**, 967–971.

F393. Martin, R.J., Downer, R.G.H., *Arch. Insect Biochem. Physiol.* (1989) **11**, 29–45.

F394. de Boer, L., van Rijssel, M., Euverink, G.J., Dijkhuizen, L., *Arch. Microbiol.* (1989) **153**, 12–18.

F395. Youngster, S.K., McKeown, K.A., Jin, Y.Z., Ramsay, R.R., Heikkila, R.E., Singer, T.P., *J. Neurochem.* (1989) **53**, 1837–1842.

F396. Preisig, C.L., Matthews, D.E., Van Etten, H.D., *Plant Physiol.* (1989) **91**, 559–566.

F397. Elayan, H., Kennedy, B., Ziegler, M.G., *Gastroenterology* (1990) **98**, 152–155.

F398. Yabe, K., Ando, Y., Terakado, N., Hashimoto, J., Nakajima, H., Hamasaki, T., *Bioact. Mol.* (1989) **10**, 21–28.

F399. Aislabie, J., Rothenburger, S., Atlas, R.M., *Appl. Environ. Microbiol.* (1989) **55**, 3247–3249.
F400. Engesser, K.H., Strubel, V., Christoglou, K., Fischer, P., Rast, H.G., *FEMS Microbiol. Lett.* (1989) **65**, 205–209.
F401. Fetzner, S., Müller, R., Lingens, F., *Biol. Chem. Hoppe-Seyler* (1989) **370**, 1173–1182.
F402. Hsu, T., Daniel, S.L., Lux, M.F., Drake, H.L., *J. Bacteriol.* (1990) **172**, 212–217.
F403. Sivamurthy, K., Pujar, B.G., *J. Ferment. Bioeng.* (1989) **68**, 375–377.
F404. Mountfort, D.O., White, D., Asher, R.A., *Appl. Environ. Microbiol.* (1990) **56**, 245–249.
F405. McQueen-Mason, S.J., Hamilton, R.H., *Plant Cell Physiol.* (1989) **30**, 999–1005.
F406. Swahn, C.G., Persson, A., Pauli, S., *Hum. Psychopharmacol.* (1989) **4**, 297–301.
F407. Ho, D.H., Covington, W.P., Lin, J., Brown, N.S., Newman, R.A., *Res. Commun. Chem. Pathol. Pharmacol.* (1989) **66**, 255–267.
F408. Christ, D.D., Walle, T., *J. Pharmacol. Exp. Ther.* (1989) **251**, 949–955.
F409. Lu, X.L., Yang, S.K., *Mol. Pharmacol.* (1989) **36**, 932–938.
F410. Lemoine, A., Johann, M., Cresteil, T., *Arch. Biochem. Biophys.* (1990) **276**, 336–342.
F411. Sugumaran, M., Kundzicz, H., Bedell-Hogan, D., Schinkmann, K., *Arch. Insect Biochem. Physiol.* (1989) **11**, 109–125.
F412. Norén, B., Strömberg, S., Ericsson, O., Lindeke, B., *Chem.-Biol. Interact.* (1989) **71**, 325–337.
F413. Neugebauer, G., Wittenbrink-Dix, A.M., Woelke-Seidl, E., Kaufmann, B., Ponton, T., Dahmen, W., Mosberg, H., Nieder, N., Besenfelder, E., *Arzneim.-Forsch.* (1989) **39**, 1336–1339.
F414. Ireland, A., Priddle, J.D., Jewell, D.P., *Clin. Sci.* (1990) **78**, 105–111.
F415. Dickinson, R.G., King, A.R., Verbeeck, R.A., *Clin. Exp. Pharmacol. Physiol.* (1989) **16**, 913–924.
F416. Schlosser, M.J., Kalf, G.E., *Chem.-Biol. Interact.* (1989) **72**, 191–207.
F417. Kamath, A.V., Vaidyanathan, C.S., *Appl. Environ. Microbiol.* (1990) **56**, 275–280.
F419. Lewinsohn, E., Britsch, L., Mazur, Y., Gressel, J., *Plant Physiol.* (1989) **91**, 1323–1328.
F432. Watanabe, K., Narimatsu, S., Yamamoto, I., Yoshimura, H., *Biochem. Biophys. Res. Commun.* (1990) **166**, 1308–1312.
F433. Rouchaud, J., Gustin, F., Van de Steene, F., Pelerents, C., Gillet, J., De Proft, M., Seutin, E., Benoit, F., Ceustermans, N., Vanparijs, L., *Meded. Fac. Landbouwwet, Rijksuniv. Gent* (1989) **54**, 263–267.
F434. Brune, A., Schink, B., *J. Bacteriol.* (1990) **172**, 1070–1076.
F435. Kurihara, N., Ikemoto, Y., Okutani, S., Clark, A.G., *J. Labelled Compd. Radiopharm.* (1989) **27**, 945–953.
F436. Chakraborty, B.S., Hawes, E.M., McKay, G., Hubbard, J.W., Korchinski, E.D., Midha, K.K., Choc, M.G., Robinson, W.T., *Drug Metab. Drug Interact.* (1988) **6**, 425–437.
F437. Baker, J.K., Yarber, J.D., Nanayakkara, N.P.D., McChesney, J.D., Homo, F., Landau, I., *Pharm. Res.* (1990) **7**, 91–95.
F438. Coimbra-Farges, R., Puget, A., Monsarrat, B., Moisand, C., Meunier, J.C., *Life Sci.* (1990) **46**, 663–669.
F439. McCoy, E.C., Rosenkranz, H.S., Howard, P.C., *Mutat. Res.* (1990) **243**, 141–144.
F440. Hartmans, S., Smits, J.P., van der Werf, M.J., Volkering, F., de Bont, J.A.M., *Appl. Environ. Microbiol.* (1989) **55**, 2850–2855.
F443. Engesser, K.H., Auling, G., Busse, J., Knackmuss, H.J., *Arch. Microbiol.* (1990) **153**, 193–199.

F444. Robertson, I.G.C., Kestell, P., Dormer, R.A., Paxton, J.W., *Drug Metab. Drug Interact.* (1988) **6**, 371–381.

F445. Ushiyama, M., Kumagai, S., Furuya, T., *Phytochemistry* (1989) **28**, 3335–3339.

F446. Pue, M.A., Ransley, J.A., Writer, D.J., Dean, A.J., Franklin, E.R., Beattie, I.G., Ross, D.A., *Eur. J. Drug Metab. Pharmacokinet.* (1989) **14**, 209–219.

F447. Neidlein, R., Volland, C., Strein, K., *Arzneim.-Forsch.* (1990) **40**, 130–136.

F448. Göber, B., Almardini, M.A., Franke, P., *Pharmazie* (1989) **44**, 847–851.

F449. Jacob, J., Grimmer, G., Raab, G., Emura, M., Riebe, M., Mohr, U., *Int. J. Environ. Anal. Chem.* (1990) **38**, 221–230.

F450. Mori, M., Kawajiri, T., Sayama, M., Taniuchi, Y., Miyahara, T., Kozuka, H., *Xenobiotica* (1989) **19**,731–741.

F451. Birkholz, D.A., Coutts, R.T., Hrudey, S.E., *Xenobiotica* (1989) **19**, 695–710.

F452. Beloborodov, V.L., Rodionov, A.P., Tyukavkina, N.A., Klimov, A.V., Kaverina, N.V., Kolesnik, Y.A., Gritsenko, A.N., Lesina, V.P., *Xenobiotica* (1989) **19**, 755–767.

F453. Cocciara, G., Strolin-Benedetti, M., Vicario, G.P., Ballabio, M., Gioia, B., Vioglio, S., Vigevani, A., *Xenobiotica* (1989) **19**, 769–780.

F454. Shukla, O.P., *Microbios* (1989) **59**, 47–63.

F455. Elovaara, E., Engström, K., Häyri, L., Hase, T., Aitio, A., *Xenobiotica* (1989) **19**, 945–960.

F456. Tjørnelund, J., Hansen, S.H., Cornett, C., *Xenobiotica* (1989) **19**, 891–899.

F457. Ishida, T., Toyota, M., Asakawa, Y., *Xenobiotica* (1989) **19**, 843–855.

F458. Scriba, G.K.E., Borchardt, R.T., *Brain Res.* (1989) **501**, 175–178.

F459. Katayama, Y., Nishida, T., Morohoshi, N., Kuroda, K., *Microbiol. Lett.* (1989) **61**, 307–313.

F461. Hsieh, D.P.H., Wan, C.C., Billington, J.A., *Mycopathologia* (1989) **107**, 121–126.

F462. Ludwig-Müller, J., Hilgenberg, W., *Physiol. Plant.* (1989) **77**, 613–619.

F463. Buder, R., Fuchs, G., *Eur. J. Biochem.* (1989) **185**, 629–635.

F464. Cappiello, M., Giuliani, L., Pacifici, G.M., *Pharmacology* (1990) **40**, 69–76.

F465. Buder, R., Ziegler, K., Fuchs, G., Langkau, B., Ghisla, S., *Eur. J. Biochem.* (1989) **185**, 637–643.

F467. Vlahov, V., Badian, M., Verho, M., Bacracheva, N., *Eur. J. Pharmacol.* (1990) **38**, 61–65.

F468. Tschech, A., Fuchs, G., *Arch. Microbiol.* (1989) **152**, 594–599.

F469. Järvenpää, P., *J. Steroid Biochem.* (1990) **35**, 289–292.

F470. Jordan, P.M., Spencer, J.B., *J. Chem. Soc., Chem. Commun.* (1990) 238–242.

F471. Chen, C.-S., Chen, T., Shieh, W.R., *Biochim. Biophys. Acta* (1990) **1033**, 1–6.

F472. Tucker, M.A., Bisesi, M.S., Smith, T.J., *J. Biochem. Toxicol.* (1989) **4**, 267–268.

F473. Xie, F., Wang, H., Li, J., Shu, H., Jiang, J., Chang, J., Hsieh, Y., *Biomed. Chromatogr.* (1990) **4**, 43–46.

F474. Mutlib, A.E., Nelson, W.L., *J. Pharmacol. Exp. Ther.* (1990) **252**, 593–599.

F475. Tanimoto, Y., Ohkuma, T., Oguri, K., Yoshimura, H., *J. Pharmacobio-Dyn.* (1990) **13**, 136–141.

F476. Maggs, J.L., Morgan, P., Hussain, F., Page, P.C.B., Park, B.K., *Xenobiotica* (1990) **20**, 45–54.

F477. Shipley, L.A., Coleman, M.D., Brewer, T.G., Ashmore, R.W., Theoharides, A.D., *Xenobiotica* (1990) **20**, 31–44.

F478. Roy, S.D., *Xenobiotica* (1990) **20**, 55–70.

F479. White, D.A. Heffron, F., Miciak, A., Middleton, B., Knights, S., Knight, K., *Xenobiotica* (1990) **20**, 71–79.

F480. Jaworski, T.J., Hawes, E.M., McKay, G., Midha, K.K., *Xenobiotica* (1990) **20**, 107–115.

F481. Inoue, T., Suzuki, S., *Xenobiotica* (1990) **20**, 99–106.

F482. Steventon, G.B., Heafield, M.T.E., Waring, R.H., Williams, A.C., Sturman, S., Green, M., *Xenobiotica* (1990) **20**, 117–122.

F483. Rush, W.R., Alexander, O.F., Hall, D.J., Dow, R.J., Tokes, L., Kurz, L., Graham, D.J.M., *Xenobiotica* (1990) **20**, 123–132.

F484. Terashita, S., Tokuma, Y., Sekiguchi, M., Noguchi, H., *Xenobiotica* (1989) **19**, 1221–1229.

F485. Dahl, A.R., Waruszewski, B.A., *Xenobiotica* (1989) **19**, 1201–1205.

F486. Kotake, A.N., Kuwahara, S.K., Burton, E., McCoy, C.E., Goldberg, L.I., *Xenobiotica* (1989) **19**, 1247–1254.

F487. Koch, P., Hirst, D.R., von Wartburg, B.R., *Xenobiotica* (1989) **19**, 1255–1265.

F488. Koga, N., Beppu, M., Ishida, C., Yoshimura, H., *Xenobiotica* (1989) **19**, 1307–1318.

F489. Rietjens, I.M.C.M., Vervoort, J., *Xenobiotica* (1989) **19**, 1297–1305.

F490. Takeuchi, M., Nakano, M., Mizojiri, K., Iwatani, K., Nakagawa, Y., Kikuchi, J., Terui, Y., *Xenobiotica* (1989) **19**, 1327–1336.

F491. Woolf, T.F., Black, A., Chang, T., *Xenobiotica* (1989) **19**, 1369–1377.

F492. Harvey, D.J., *Xenobiotica* (1989) **19**, 1437–1447.

F493. Takayama, F., Iwasawa, Y., Saito, K., Shiratori, K., Ohtawa, M., *Xenobiotica* (1989) **19**, 1407–1420.

F494. Mishima, M., Kobayashi, S., Abe, S., Yamato, C., *Xenobiotica* (1990) **20**, 135–146.

F495. De Vries, J.X., Walter-Sack, I., Ittensohn, A., Weber, E., *Xenobiotica* (1989) **19**, 1461–1470.

F496. Drummond, L., Caldwell, J., Wilson, H.K., *Xenobiotica* (1990) **20**, 159–168.

F497. Samara, E., Bialer, M., Harvey, D.J., *Xenobiotica* (1990) **20**, 177–183.

F498. Mayo, B.C., Chaseaud, L.F., Hawkins, D.R., Taylor, I.W., Legeai, J., *Xenobiotica* (1990) **20**, 233–246.

F499. Fischer, I.U., von Unruh, G.E., Dengler, H.J., *Xenobiotica* (1990) **20**, 209–222.

F500. Sayo, H., Saito, M., *Xenobiotica* (1990) **20**, 247–253.

F501. Inoue, T., Tanaka, K., *Xenobiotica* (1990) **20**, 265–271.

F502. Walle, T., Walle, U.K., Cowart, T.D., Conradi, E.C., Gaffney, T.E., *Xenobiotica* (1990) **20**, 321–331.

F503. Harvey, D.J., Mechoulam, R., *Xenobiotica* (1990) **20**, 303–320.

F504. Ibrahim, A.-R.S., Abul-Hajj, Y.J., *Xenobiotica* (1990) **20**, 363–373.

F505. Davison, K.L., Bakke, J.E., Larsen, G.L., *Xenobiotica* (1990) **20**, 375–383.

F506. Paulson, G.D., Feil, V.J., McDonald, J.J., *Xenobiotica* (1990) **20**, 395–400.

F507. Nakano, M., Takeuchi, M., Kawahara, S., Mizojiri, K., *Xenobiotica* (1990) **20**, 385–393.

F508. Christensen, A., Kissmeyer, A.-M., Eilertsen, E., Rastrup-Andersen, N., Ahnfelt-Rønne, I., *Xenobiotica* (1990) **20**, 417–434.

F509. Tucker, M.A., Smith, T.J., *J. Appl. Toxicol.* (1990) **10**, 73–74.

F510. Belisario, M.A., Pecce, R., Della Morte, R., Arena, A.R., Cecinato, A., Ciccioli, P., Staiano, N., *Carcinogenesis* (1990) **11**, 213–218.

F511. Chakraborty, C., Davis, D.L., Dey, S.K., *J. Steroid Biochem.* (1990) **35**, 209–218.

F512. Křen, V., Flieger, M., Sajdl, P., *Appl. Microbiol. Biotechnol.* (1990) **32**, 645–650.

F513. Koizumi, M., Shimizu, M., Kobashi, K., *Chem. Pharm. Bull.* (1990) **38**, 794–796.

F514. Sloley, B.D., Downer, R.G.H., *Insect Biochem.* (1990) **20**, 165–171.

F515. Zhang, X., Morgan, T.V., Wiegel, J., *FEMS Microbiol. Lett.* (1990) **67**, 63–66.

F516. Kawabata, T.T., White, K.L., Jr., *Cancer Res.* (1989) **49**, 5816–5822.

F517. Williams, F.M., Moore, U., Seymour, R.A., Mutch, E.M., Nicholson, E., Wright, P., Wynne, H., Plain, P.G.,

Rawlins, M.D., *Br. J. Clin. Pharmacol.* (1989) **28**, 703–708.

F518. Vervoort, J., De Jager, P.A., Steenbergen, J., Rietjens, J.M.C.M., *Xenobiotica* (1990) **20**, 657–670.

F519. Mori, M.-A., Sakai, U.-Y.K., Inoue, M., Miyahara, T., Kozuka, H., *Xenobiotica* (1990) **20**, 653–656.

F520. Tanaka, K., Inoue, T., Kadota, S., Kikuchi, T., *Xenobiotica* (1990) **20**, 671–681.

F521. Oguri, K., Hanioka, N., Yoshimura, H., *Xenobiotica* (1990) **20**, 683–688.

F522. Hanna, P.E., El-Ghandour, A.M., McCormack, M.E., *Xenobiotica* (1990) **20**, 739–751.

F523. Breyer-Pfaff, U., Becher, B., Nusser, E., Nill, K., Baier-Weber, B., Zaunbrecher, D., Wachsmuth, H., Prox, A., *Xenobiotica* (1990) **20**, 727–738.

F524. Mutlib, A.E., Nelson, W.L., *Drug Metab. Dispos.* (1990) **18**, 309–314.

F525. Miller, N.E., Thomas, D., Billings, R.E., *Drug Metab. Dispos.* (1990) **18**, 304–308.

F526. Harvey, D.J., *Drug Metab. Dispos.* (1990) **18**, 350–355.

F527. Mutlib, A.E., Nelson, W.L., *Drug Metab. Dispos.* (1990) **18**, 331–337.

F528. Feng, P.C.C., Wilson, A.G.E., McClanahan, R.H., Patanella, J.E., Wratten, S.J., *Drug Metab. Dispos.* (1990) **18**, 373–377.

F529. Broly, F., Libersa, C., Lhermitte, M., *Drug Metab. Dispos.* (1990) **18**, 362–368.

F530. Rickert, D.E., Held, S.D., *Drug Metab. Dispos.* (1990) **18**, 5–9.

F531. Christ, D.D., Walle, U.K., Oatis, J.E., Jr., Walle, T., *Drug Metab. Dispos.* (1990) **18**, 1–4.

F532. Jajoo, H.K., Mayol, R.F., LaBudde, J.A., Blair, I.A., *Drug Metab. Dispos.* (1990) **18**, 28–35.

F533. Brodfuehrer, J.I., Chapman, D.E., Wilk, T.J., Powis, G., *Drug Metab. Dispos.* (1990) **18**, 20–27.

F534. Flescher, J.W., Myers, S.R., *Drug Metab. Dispos.* (1990) **18**, 163–167.

F535. Weil, A., Caldwell, J., Strolin-Benedetti, M., *Drug Metab. Dispos.* (1990) **18**, 115–120.

F536. Buonarati, M., Jones, A.D., Buckpitt, A., *Drug Metab. Dispos.* (1990) **18**, 183–189.

F537. Hultin, T.A., Filla, M.S., McCormick, D.L., *Drug Metab. Dispos.* (1990) **18**, 175–179.

F538. Morrison, R.A., Singhvi, S.M., Peterson, A.E., Pocetti, D.A., Migdalof, B.H., *Drug Metab. Dispos.* (1990) **18**, 253–257.

F539. Yoshikawa, M., Endo, H., Komatsu, K., Fujihara, M., Takaiti, O., Kagoshima, T., Umehara, M., Ishikawa, H., *Drug Metab. Dispos.* (1990) **18**, 212–217.

F540. Funaki, T., Soons, P.A., Guengerich, F.P., Breimer, D.D., *Biochem. Pharmacol.* (1989) **38**, 4213–4216.

F541. Vree, T.B., Vree, J.B., Beneken Kolmer, N., Hekster, Y.A., Shimoda, M., Nouws, J.F.M., Yoshioka, T., Hoji, K., *Vet. Q.* (1989) **11**, 138–143.

F543. Nakamura, J., Shiota, H., Sasaki, H., Shibasaki, J., *Chem. Pharm. Bull.* (1989) **37**, 2537–2538.

F544. Nakamura, J., Shiota, H., Haraguchi, Y., Sasaki, H., Shibasaki, J., *J. Pharmacobio-Dyn.* (1988) **11**, 53–57.

F545. Nakamura, J., Shiota, H., Sasaki, H., Shibasaki, J., *J. Pharmacobio-Dyn.* (1988) **11**, 625–629.

F546. Nakamura, J., Haraguchi, Y., Sasaki, H., Shibasaki, J., *J. Pharmacobio-Dyn.* (1989) **12**, 602–607.

F547. Halkier, B.A., Olsen, C.E., Møller, B.L., *J. Biol. Chem.* (1989) **264**, 19487–19494.

F548. Janke, D., Ihn, W., Tresselt, D., *J. Basic Microbiol.* (1989) **29**, 305–314.

F549. Uetrecht, J., Zahid, N., Rubin, R., *Chem. Res. Toxicol.* (1988) **1**, 74–78.

F550. Neidlein, R., Volland, C., Kramer, W., *Arch. Pharm. (Weinheim, Ger.)* (1990) **323**, 259–265.

F551. Vree, T.B., Vree, J.B., Beneken Kolmer, N., Hekster, Y.A., Shimoda, M., Nouws,

J.F.M., Yoshioka, T., Hoji, K., *Vet. Q.* (1989) **11**, 186–190.

F552. Vree, T.B., Vree, J.B., De Jonge, P.A.M., Plaum, M.J.M., Verwey, C.P.W.G.M., Hekster, Y.A., Shimoda, M., Nouws, J.F.M., *J. Vet. Pharmacol. Ther.* (1989) **12**, 459–462.

F553. Vree, T.B., Klimowicz, A., Vree, J.B., Shimoda, M., *Vet. Q.* (1990) **12**, 60–63.

F554. Plakas, S.M., James, M.O., *Drug Metab. Dispos.* (1990) **18**, 552–556.

F555. Nagata, K., Martin, B.M., Gillette, J.R., Sasame, H.A., *Drug Metab. Dispos.* (1990) **18**, 557–564.

F556. Hirayama, H., Pang, K.S., *Drug Metab. Dispos.* (1990) **18**, 580–587.

F557. Samara, E., Bialer, M., Harvey, D.J., *Drug Metab. Dispos.* (1990) **18**, 571–579.

F558. Hirayama, H., Morgado, J., Gasinska, I., Pang, K.S., *Drug Metab. Dispos.* (1990) **18**, 588–594.

F559. Howard, M.O., Newton, J.F., Keohane, D.J., Yodis, L.P., Saverino, C.M., Qualls, C.W., Jr., Schwartz, L.W., *Drug Metab. Dispos.* (1990) **18**, 607–612.

F560. Ruenitz, P.C., Nanavati, N.T., *Drug Metab. Dispos.* (1990) **18**, 645–648.

F561. Walker, D.G., Hutchison, M., Shepard, T.A., Osborne, P.M., Allenby, S.M., Webb, A.J., Viney, N.J., Pue, M.A., Chenery, R.J., Wood, P.A., Ross, D.A., *Drug Metab. Dispos.* (1990) **18**, 613–620.

F562. Myers, S.R., Flesher, J.W., *Drug Metab. Dispos.* (1990) **18**, 664–669.

F563. Ogolla, F., Ferguson, R.J., Kirlin, W.G., Trinidad, A., Andrews, A.F., Mpezo, M., Hein, D.W., *Drug Metab. Dispos.* (1990) **18**, 680–685.

F564. Magdalou, J., Chajes, V., Lafaurie, C., Siest, G., *Drug Metab. Dispos.* (1990) **18**, 692–697.

F565. Cho, A.K., Hiramatsu, M., Distefano, E.W., Chang, A.S., Jenden, D.J., *Drug Metab. Dispos.* (1990) **18**, 686–691.

F566. Tanaka, M., Ono, K., Hakusui, H., Takegoshi, T., Watanabe, Y., Kanao, M., *Drug Metab. Dispos.* (1990) **18**, 698–703.

F567. Lavrijsen, K., Van Houdt, J., Van Dyck, D., Hendrickx, J., Lauwers, W., Hurkmans, R., Bockx, M., Janssen, C., Meuldermans, W., Heykants, J., *Drug Metab. Dispos.* (1990) **18**, 704–710.

F568. Schrenk, D., Bock, K.W., *Drug Metab. Dispos.* (1990) **18**, 720–725.

F569. Pichard, L., Gillet, G., Fabre, I., Dalet-Beluche, I., Bonfils, C., Thenot, J.-P., Maurel, P., *Drug Metab. Dispos.* (1990) **18**, 711–719.

F570. Shu, Y.-Z., Hubbard, J.W., Cooper, J.K., McKay, G., Kochinski, E.D., Kumar, R., Midha, K.K., *Drug Metab. Dispos.* (1990) **18**, 735–741.

F571. Scatina, J.A., Wells, D.S., Kimmel, H.B., Kemper, C.J., Sisenwine, S.F., *Drug Metab. Dispos.* (1990) **18**, 746–752.

F572. Rashed, M.S., Myers, T.G. Nelson, S.D., *Drug Metab. Dispos.* (1990) **18**, 765–770.

F573. Caputo, O., Grosa, G., Ceruti, M., Rocco, F., Biglino, G., *Drug Metab. Dispos.* (1990) **18**, 771–778.

F574. Zhang, K., Lepage, F., Cavier, G., Astoin, J., Rashed, M.S., Baillie, T.A., *Drug Metab. Dispos.* (1990) **18**, 794–803.

F575. Hsu, R.S., Shutske, G.M., Dileo, E.M., Chesson, S.M., Linville, A.R., Allen, R.C., *Drug Metab. Dispos.* (1990) **18**, 779–783.

F576. Grislain, L., Gele, P., Bertrand, M., Luijten, W., Bromet, N., Salvadori, C., Kamoun, A., *Drug Metab. Dispos.* (1990) **18**, 804–808.

F577. Yeh, S.Y., *Drug Metab. Dispos.* (1990) **18**, 453–461.

F578. Brayshear, W.T., Kuhnert, B.R., Wei, R., *Drug Metab. Dispos.* (1990) **18**, 488–493.

F579. Kumagai, Y., Todaka, T., Yamano, S., Toki, S., *Drug Metab. Dispos.* (1990) **18**, 462–466.

F580. McMahon, T.F., Diliberto, J.J., Birnbaum, L.S., *Drug Metab. Dispos.* (1990) **18**, 494–503.

F581. Sanins, S.M., Adams, W.J., Kaiser, D.G., Halstead, G.W., Baillie, T.A., *Drug Metab. Dispos.* (1990) **18**, 527–533.

F582. Slatter, J.G., Abbott, F.S., Burton, R., *Xenobiotica* (1990) **20**, 999–1024.

F583. Samara, E., Brown, N.K., Harvey, D.J., *Drug Metab. Dispos.* (1990) **18**, 548–549.

F584. Fujimaki, M., Hakusui, H., *Xenobiotica* (1990) **20**, 1025–1034.

F585. Cook, C.S., Campion, J.G., Hriber, J.D., Karim, A., *Xenobiotica* (1990) **20**, 1065–1080.

F586. Bakke, J.E., Davison, K.L., Larsen, G.L., *Xenobiotica* (1990) **20**, 801–807.

F587. Fujimaki, Y., Hashimoto, K., Sudo, K., Tachizawa, H., *Xenobiotica* (1990) **20**, 1081–1094.

F588. Patel, D.K., Notarianni, L.J., Bennett, P.N., *Xenobiotica* (1990) **20**, 847–854.

F589. Cheymol, G., Cheymol, A., Jozefczak, C., Lecoq, V., Lecoq, B., Jaillon, P., *Xenobiotica* (1990) **20**, 855–860.

F590. Reddy, C.S.G., Acosta, D., Davis, P.J., *Xenobiotica* (1990) **20**, 1281–1297.

F591. Brien, J.F., Jimmo, S., Brennan, F.J., Armstrong, P.W., Abdollah, H., *Drug Metab. Dispos.* (1990) **18**, 846–851.

F592. Yoshihara, S., Tatsumi, K., *Drug Metab. Dispos.* (1990) **18**, 876–881.

F593. Vree, T.B., Beneken Kolmer, W.J., Hekster, Y.A., Shimoda, M., Ono, M., Miura, T., *Drug Metab. Dispos.* (1990) **18**, 852–858.

F594. Stiff, D.D., Zemaitis, M.A., *Drug Metab. Dispos.* (1990) **18**, 888–894.

F595. Monsarrat, B., Mariel, E., Cros, S., Garès, M., Guénard, D., Guéritte-Voegelein, F., Wright, M., *Drug Metab. Dispos.* (1990) **18**, 895–901

F596. Gerding, T.K., Drenth, B.F.H., De Zeeuw, R.A., Tepper, P.G., Horn, A.S., *Drug Metab. Dispos.* (1990) **18**, 923–928.

F597. Elfarra, A.A., Hwang, I.Y., *Drug Metab. Dispos.* (1990) **18**, 917–922.

F598. Chapman, D.E., Moore, T.J., Michener, S.R., Powis, G., *Drug Metab. Dispos.* (1990) **18**, 929–936.

F599. Narimatsu, S., Watanabe, K., Matsunaga, T., Yamamoto, I., Imaoka, S., Funae, Y., Yoshimura, H., *Drug Metab. Dispos.* (1990) **18**, 943–948.

F600. Nerland, D.E., Pierce, W.M., Jr., *Drug Metab. Dispos.* (1990) **18**, 958–961.

F601. Franz, P.M., Anliker, L., Callaghan, J.T., De Sante, K.A., Dhahir, P.H., Nelson, R.L., Rubin, A., *Drug Metab. Dispos.* (1990) **18**, 968–973.

F602. Vyas, K.P., Kari, P.H., Ramjit, H.G., Pitzenberger, S.M., Hichens, M., *Drug Metab. Dispos.* (1990) **18**, 1025–1030.

F603. Wong, B.K., Woolf, T.F., Chang, T., Whitfield, L.R., *Drug Metab. Dispos.* (1990) **18**, 980–986.

F604. Mutlib, A.E., Talaat, R.E., Slatter, J.G., Abbott, F.S., *Drug Metab. Dispos.* (1990) **18**, 1038–1045.

F605. Thomas, B.F., Martin, B.R., *Drug Metab. Dispos.* (1990) **18**, 1046–1054.

F606. Lapka, R., Šmolík, S., Rejholec, V., *Drug Metab. Dispos.* (1990) **18**, 1060–1064.

F607. Yeung, P.K., Mosher, S.J., Quilliam, M.A., Montague, T.J., *Drug Metab. Dispos.* (1990) **18**, 1055–1059.

F608. Brown, N.K., Harvey, D.J., *Drug Metab. Dispos.* (1990) **18**, 1065–1070.

F609. Shu, Y.-Z., Hubbard, J.W., McKay, G., Midha, K.K., *Drug Metab. Dispos.* (1990) **18**, 1096–1099.

F610. Muheim, A., Leisola, M.S.A., Schoemaker, H.E., *J. Biotechnol.* (1990) **13**, 159–167.

F611. Wong, F.A., Flor, S.C., *Drug Metab. Dispos.* (1990) **18**, 1103–1104.

F612. Gygax, D., Ghisalba, O., Treichler, H., Nuesch, J., *J. Antibiot.* (1990) **43**, 324–326.

F613. van Afferden, M., Schacht, S., Klein, J., Trüper, H.G., *Arch. Microbiol.* (1990) **153**, 324–328.

F615. Grund, E., Knorr, C., Eichenlaub, R., *Appl. Environ. Microbiol.* (1990) **56**, 1459–1464.

F616. John, N.J., Terlinden, R., Fischer, H., Sies, H., Evers, M., *Chem. Res. Toxicol.* (1990) **3**, 199–203.

F617. Servé, M.P., Ferry, M.J., Yu, K.O., Olson, C.T., Hobson, D.W., *J. Toxicol. Environ. Health* (1990) **29**, 409–416.

F618. Lafont, O., Cave, C., Ménager, S., Miocque, M., *Eur. J. Med. Chem.* (1990) **25**, 61–66.

F619. Maurer, H., Wollenberg, P., *Arzneim.-Forsch.* (1990) **40**, 460–462.

F620. Caputo, O., Grosa, G., Ceruti, M., Viola, F., Rocco, F., *Eur. J. Drug Metab. Pharmacokinet.* (1989) **14**, 263–268.

F621. Thomas, B.H., Whitehouse, L.W., Solomonraj, G., Paul, C.J., *J. Pharm. Sci.* (1990) **79**, 321–325.

F622. Elayan, H., Kennedy, B., Ziegler, M.G., *Arch. Dermatol. Res.* (1990) **282**, 194–197.

F623. Fukami, M.H., Haavik, J., Flatmark, T., *Biochem. J.* (1990) **268**, 525–528.

F624. El-Mansi, E.M.T., Hopper, D.J., *FEMS Microbiol. Lett.* (1990) **66**, 147–152.

F625. Abou-Donia, M.B., Suwita, E., Nomeir, A.A., *Toxicology* (1990) **61**, 13–25.

F626. Shimizu, E., Ichise, H., Yorifuji, T., *Agric. Biol. Chem.* (1990) **54**, 851–853.

F627. Ball, S.E., Forrester, L.M., Wolf, C.R., Back, D.J., *Biochem. J.* (1990) **267**, 221–226.

F628. Horvath, M., Ditzelmüller, G., Loidl, M., Streichsbier, F., *Appl. Microbiol. Biotechnol.* (1990) **33**, 213–216.

F629. Boyd, D.R., Dorrity, M.R.J., Malone, J.F., McMordie, R., Austin, S., Sharma, N.D., Dalton, H., Williams, P., *J. Chem. Soc., Perkin Trans. 1* (1990) 489–494.

F630. Ghosal, S., Unnikrishnan, S., Singh, S.K., *Phytochemistry* (1989) **28**, 2535–2537.

F631. Nakamura, J., Mizuma, T., Hayashi, M., Awazu, S., *J. Pharm. Pharmacol.* (1990) **42**, 207–208.

F632. Akhtar, M.H., *J. Agric. Food Chem.* (1990) **38**, 1417–1422.

F633. Cannon, J.G., Walker, K.A., Montanari, A., Long, J.P., Flynn, J.R., *J. Med. Chem.* (1990) **33**, 2000–2006.

F634. Van der Vijgh, W.J.F., Maessen, P.A., Pinedo, H.M., *Cancer Chemother. Pharmacol.* (1990) **26**, 9–12.

F635. Jacquot, C., Payan-Lepain, E., Christen, M.O., Achtert, G., Hausleiter, H.J., Borchers, F., *Eur. J. Drug Metab. Pharmacokinet.* (1989) **14**, 249–256.

F636. Real, M.D., Ferre, J., *J. Biol. Chem.* (1990) **265**, 7407–7412.

F637. Barzaghi, N., Leone, L., Monteleone, M., Tomasini, G., Perucca, E., *Eur. J. Drug Metab. Pharmacokinet.* (1989) **14**, 293–298.

F638. Corbett, M.D., Corbett, B.R., Quintana, S.J., Hannothiaux, M.H., Wei, C.I., *Chem. Res. Toxicol.* (1990) **3**, 296–300.

F639. Patel, T.R., Hameed, N., Martin, A.M., *Arch. Microbiol.* (1990) **153**, 438–443.

F640. Wittich, R.M., Schmidt, S., Fortnagel, P., *FEMS Microbiol. Lett.* (1990) **67**, 157–160.

F641. Deluca, M.E., Hudlicky, T., *Tetrahedron Lett.* (1990) **31**, 13–16.

F642. Ishiwata, K., Fukuhara, N., Shimada, M., Makiguchi, N., Soda, K., *Biotechnol. Appl. Biochem.* (1990) **12**, 141–149.

F643. Lu, X.L., Yang, S.K., *Chirality* (1990) **2**, 1–9.

F646. Whiteman, P.A., Abraham, E.P., Balarin, J.E., Fleming, M.D., Schofield, C.J., Sutherland, J.D., Willis, A.C., *FEBS Lett.* (1990) **262**, 342–344.

F647. Chen, Y.P., Lovell, C.R., *Appl. Environ. Microbiol.* (1990) **56**, 1971–1973.

F648. Viani, A., Temellini, A., Tusini, G., Pacifici, G.M., *Hum. Exp. Toxicol.* (1990) **9**, 65–69.

F649. Harpel, M.R., Lipscomb, J.D., *J. Biol. Chem.* (1990) **265**, 6301–6311.

F650. Fortnagel, P., Harms, H., Wittich, R.M., Krohn, S., Meyer, H., Sinnwell, V., Wilkes, H., Francke, W., *Appl. Environ. Microbiol.* (1990) **56**, 1148–1156.

F651. Kato, N., Konishi, H., Masuda, M., Joung, E., Shimao, M., Sakazawa, C., *J. Ferment. Bioeng.* (1990) **69**, 220–223.

F652. Masento, M.S., Taylor, G.W., Watson, D., Seidel, A., Bochnitschek, W., Oesch, F., Grover, P.L., *Chem.-Biol. Interact.* (1990) **74**, 163–178.
F653. Barthelmebs, M., Caillette, A., Ehrhardt, J.P., Velly, J., Imbs, J.L., *Kidney Int.* (1990) **37**, 1414–1422.
F655. Sembiring, T., Winter, J., *Appl. Microbiol. Biotechnol.* (1990) **33**, 233–238.
F657. Grootwassink, J.W.D., Balsevich, J.J., Kolenovsky, A.D., *Plant Sci. (Limerick, Irel.)* (1990) **66**, 11–20.
F658. You, I.S., Murray, R.I., Jollie, D., Gunsalus, I.C., *Biochem. Biophys. Res. Commun.* (1990) **169**, 1049–1054.
F659. Ludwig-Müller, J., Rausch, T., Lang, S., Hilgenberg, W., *Phytochemistry* (1990) **29**, 1397–1400.
F660. v. d. Wildenberg, H.M., Delbressine, L.P.C., Kaspersen, F.M., Wagenaars, G.N., Jacobs, P.L., *Arzneim.-Forsch.* (1990) **40**, 540–544.
F661. Rahman, M.M.A., Dewick, P.M., Jackson, D.E., Lucas, J.A., *Phytochemistry* (1990) **29**, 1841–1846.
F662. Thimmaiah, K.N., Sethi, V.S., *Microchem. J.* (1990) **41**, 313–319.
F663. Cummings, J., Matheson, L.M., Maurice, L., Smyth, J.F., *J. Pharm. Pharmacol.* (1990) **42**, 242–246.
F664. Leoni, O., Palmieri, S., Lattanzio, V., Van Sumere, C.F., *Food Chem.* (1990) **38**, 27–39.
F666. Zellner, G., Kneifel, H., Winter, J., *Appl. Environ. Microbiol.* (1990) **56**, 2228–2233.
F667. Siebertz, R., Proksch, P., Witte, L., *Phytochemistry* (1990) **29**, 2135–2138.
F668. Thimmaiah, K.N., Sethi, V.S., *Microchem. J.* (1990) **41**, 320–326.
F669. Kadir, H.A.,Knowles, C.O., *Comp. Biochem. Physiol. C: Comp. Pharmacol. Toxicol.* (1989) **94C**, 465–468.
F670. Kunimoto, S., Komuro, K., Nosaka, C., Tsuchiya, T., Fukatsu, S., Takeuchi, T., *J. Antibiot.* (1990) **43**, 556–565.
F671. Xu, X., Tang, B.K., Pang, K.S., *J. Pharmacol. Exp. Ther.* (1990) **253**, 965–973.
F672. Tsurusaki, K., Watanabe, S., Sakurai, N., Kuraishi, S., *Physiol. Plant.* (1990) **79**, 221–225.
F673. Ibrahim, A.R., Abul-Hajj, Y.J., *J. Nat. Prod.* (1990) **53**, 644–656.
F674. Duran, M., Van Vossen, R., Bruinvis, L., Ketting, D., Dorland, L., De Klerk, J.B.C., *Prog. Clin. Biol. Res.* (1988) **321**, 419–426.
F675. Seymour, M.A., Teale, P., Horner, M.W., *Biomed. Environ. Mass Spectrom.* (1990) **19**, 447–449.
F676. Baran, H., Schwarcz, R., *J. Neurochem.* (1990) **55**, 738–744.
F677. Hallén, B., Gralls, M., Brötell, H., Strömberg, S., *Pharmacol. Toxicol. (Copenhagen)* (1990) **66**, 373–381.
F678. Ottoboni, S., Carlson, T.J., Trager, W.F., Castagnoli, K., Castagnoli, N., Jr., *Chem. Res. Toxicol.* (1990) **3**, 423–427.
F679. Andries, M.J., Lucier, G.W., Goldstein, J., Thompson, C.L., *Mol. Pharmacol.* (1990) **37**, 990–995.
F680. Effenberger, F., Hörsch, B., Förster, S., Ziegler, T., *Tetrahedron Lett.* (1990) **31**, 1249–1252.
F682. Maurer, H.H., Fritz, C.F., *Arzneim.-Forsch.* (1990) **40**, 564–566.
F683. Ishiguka, T., Komiya, I., Hiratsuka, A., Watanabe, T., *Mol. Pharmacol.* (1990) **37**, 983–989.
F684. Holder, C.L., Siitonen, P.H., Slikker, W., Jr., Branscomb, C.J., Korfmacher, W.A., Thompson, H.C., Jr., Cerniglia, C.E., Cosnell, A.B., Lay, J.O., Jr., *J. Anal. Toxicol.* (1990) **14**, 247–251.
F685. Takahashi, K., Haginaka, J., Tamagawa, S., Nishihata, T., Yasuda, H., Katagi, T., *J. Pharm. Pharmacol.* (1990) **42**, 356–357.
F686. Karaszkiewicz, J.W., Kalf, G.F., *Biochim. Biophys. Acta* (1990) **1035**, 223–229.

F687. Turbek, C.S., Li, D., Choi, G.H., Schardl, C.L., Smith, D.A., *Phytochemistry* (1990) **29**, 2841–2846.

F688. Pritchard, M.P., Hawksworth, G.M., *Biochem. Soc. Trans.* (1990) **18**, 1215–1216.

F689. Saimmaime, I., Coulomb, C., Coulomb, P.J., Roggero, J.P., *Plant Physiol. Biochem.* (Paris) (1990) **28**, 323–331.

F691. Monks, T.J., Highet, R.J., Lau, S.S., *Mol. Pharmacol.* (1990) **38**, 121–127.

F692. Falany, C.N., Vazquez, M.E., Heroux, J.A., Roth, J.A., *Arch. Biochem. Biophys.* (1990) **278**, 312–318.

F693. Nanbo, T., Fukasaku, N., Yamaguchi, M., Takagi, H., Ueda, T., Yokoshima, T., *Chem. Pharm. Bull.* (1990) **38**, 1753–1754.

F694. Ohtomi, M., Ichikawa, T., Deguchi, T., *Biomed. Res.* (1990) **11**, 263–267.

F695. Reeve, C.D., Carver, M.A., Hopper, D.J., *Biochem. J.* (1990) **269**, 815–819.

F696. Topp, E., Akhtar, M.H., *Can. J. Microbiol.* (1990) **36**, 495–499.

F697. Engesser, K.H., Fietz, W., Schulte, P., Knackmuss, H.J., *FEMS Microbiol. Lett.* (1990) **69**, 317–321.

F698. Lundell, T., Leonowicz, A., Rogalski, J., Hatakka, A., *Appl. Environ. Microbiol.* (1990) **56**, 2623–2629.

F699. Tanahashi, T., Zenk, M.H., *Phytochemistry* (1990) **29**, 1113–1122.

F700. Locniskar, A., Greenblatt, D.J., *Biopharm. Drug Dispos.* (1990) **11**, 499–506.

F701. Cheynier, V., Rigaud, J., Moutounet, M., *Phytochemistry* (1990) **29**, 1751–1753.

F703. Tsuda, T., Aoki, S., Kojima, M., Harada, H., *Comp. Biochem. Physiol. C: Comp. Pharmacol. Toxicol.* (1990) **95C**, 145–150.

F704. Harvey, D.J., Brown, N.K., *Biomed. Environ. Mass Spectrom.* (1990) **19**, 545–553.

F706. Berger, B.J., Lombardy, R.J., Marbury, G.D., Bell, C.A., Dykstra, C.C., Hall, J.E., Tidwell, R.R., *Antimicrob. Agents Chemother.* (1990) **34**, 1678–1684.

F707. So, K.O., Wong, P.K., Chan, K.Y., *Toxic. Assess.* (1990) **5**, 221–235.

F708. Harvey, D.J., Brown, N.K., *Biomed. Environ. Mass Spectrom.* (1990) **19**, 559–567.

F709. Harvey, D.J., Brown, N.K., *Res. Commun. Subst. Abuse* (1990) **11**, 27–37.

F710. Kinouchi, T., Nishifuji, K., Ohnishi, Y., *Carcinogenesis* (1990) **11**, 1381–1387.

F711. Ferrero, M.A., Reglero, A., Martin-Villacorta, J., Fernandez-Canon, J.M., Luengo, J.M., *J. Antibiot.* (1990) **43**, 684–691.

F712. Robinson, S.P., Langan-Fahey, S.M., Johnson, D.A., Jordan, V.C., *Drug Metab. Dispos.* (1991) **19**, 36–43.

F713. Niemeijer, N.R., Gerding, T.K., De Zeeuw, R., *Drug Metab. Dispos.* (1991) **19**, 20–23.

F714. Holsztynska, E.J., Weber, W.W., Domino, E.F., *Drug Metab. Dispos.* (1991) **19**, 48–53.

F715. Maurizis, J.C., Nicolas, C., Verny, M., Ollier, M., Faurie, M., Payard, M., Veyre, A., *Drug Metab. Dispos.* (1991) **19**, 94–99.

F716. Bories, G.F., Perdu-Durand, E.F., Sutra, J.F., Tulliez, J.E., *Drug Metab. Dispos.* (1991) **19**, 140–143.

F717. Gan, K.N., Smolen, A., Eckerson, H.W., LaDu, B.N., *Drug Metab. Dispos.* (1991) **19**, 100–106.

F718. Sinz, M.W., Remmel, R.P., *Drug Metab. Dispos.* (1991) **19**, 149–153.

F719. Shu, Y.-Z., Hubbard, J.W., McKay, G., Hawes, E.M., Midha, K.K., *Drug Metab. Dispos.* (1991) **19**, 154–152.

F720. Jackson, C.-J.C., Hubbard J.W., McKay, G., Cooper, J.K., Hawes, E.M., Midha, K.K., *Drug Metab. Dispos.* (1991) **19**, 188–193.

F721. Taskinen, J., Wikberg, T., Ottoila, P., Kanner, L., Lotta, T., Pippuri, A., Bäckström, R., *Drug Metab. Dispos.* (1991) **19**, 178–183.

F722. Cheng, H., Stuhler, J.D., Dorrbecker, B.R., Gordon, W.P., *Drug Metab. Dispos.* (1991) **19**, 212–217.

F723. Watanabe, K., Hirahashi, N., Narimatsu, S., Yamamoto, I., Yoshimura, H., *Drug Metab. Dispos.* (1991) **19**, 218–221.

F724. Supko, J.G., Malspeis, L., *Drug Metab. Dispos.* (1991) **19**, 263–273.

F725. Kikuchi, K., Nagatsu, Y., Makino, Y., Mashino, T., Ohta, S., Hirobe, M., *Drug Metab. Dispos.* (1991) **19**, 257–262.

F726. El Mouelhi, M., Bock, K.W., *Drug Metab. Dispos.* (1991) **19**, 304–308.

F727. Knadler, M.P., Hall, S.D., *Drug Metab. Dispos.* (1991) **19**, 280–282.

F728. Goon, D., Klaassen, C.D., *Drug Metab. Dispos.* (1991) **19**, 340–347.

F729. Ehlhardt, W.J., Howbert, J.J., *Drug Metab. Dispos.* (1991) **19**, 366–369.

F730. Okazani, O., Kurata, T., Hakusui, H., Tachizawa, H., *Drug Metab. Dispos.* (1991) **19**, 376–380.

F731. Ehlhardt, W.J., *Drug Metab. Dispos.* (1991) **19**, 370–375.

F732. Mayol, R.F., Jajoo, H.K., Klunk, L.J., Blair, I.A., *Drug Metab. Dispos.* (1991) **19**, 394–399.

F733. Sanins, S.M., Adams, W.J., Kaiser, D.G., Halstead, G.W., Hosley, J., Barnes, H., Baillie, T.A., *Drug Metab. Dispos.* (1991) **19**, 405–410.

F734. Grech-Bélanger, O., Turgeon, J., Lalande, M., Bélanger, P.M., *Drug Metab. Dispos.* (1991) **19**, 458–461.

F735. Walle, T., Walle, U.K., *Drug Metab. Dispos.* (1991) **19**, 448–453.

F736. Clarke, J.T.R., Bier, D.M., *Metabolism* (1982) **31**, 999–1005.

F737. Matsumura, M., Osada, K., Aiba, S., *Biochim. Biophys. Acta* (1984) **786**, 9–17.

F738. Bolton, J.L., Thompson, J.A., *Drug Metab. Dispos.* (1991) **19**, 467–472.

F739. Zhang, H.X., Sultatos, L.G., *Drug Metab. Dispos.* (1991) **19**, 473–477.

F740. Ling, K.H., Chiou, C.M., Tseng, Y.L., *Drug Metab. Dispos.* (1991) **19**, 587–595.

F741. Patel, D.K., Woolley, J.L., Jr., Shockcor, J.P., Johnson, R.L., Taylor, L.C., Sigel, C.W., *Drug Metab. Dispos.* (1991) **19**, 491–497.

F742. Myers, S.R., Flesher, J.W., *Biochem. Pharmacol.* (1991) **41**, 1683–1689.

F743. Sjöberg, P., Egestad, B., Klasson-Wehler, E., Gustafsson, J., *Biochem. Pharmacol.* (1991) **41**, 1493–1496.

F744. Mennes, W.C., van Holsteijn, C.W.M., Timmerman, A., Noordhoek, J., Blaauboer, B.J., *Biochem. Pharmacol.* (1991) **41**, 1203–1208.

F745. Moorhouse, K.G., Dodds, P.F., Hutson, D.H., *Biochem. Pharmacol.* (1991) **41**, 1179–1185.

F746. Mikus, G., Somogyi, A.A., Bochner, F., Eichelbaum, M., *Biochem. Pharmacol.* (1991) **41**, 757–762.

F747. Walton, M.I., Workman, P., *Biochem. Pharmacol.* (1990) **39**, 1735–1742.

F748. Melega, W.P., Luxen, A., Perlmutter, M.M., Nissenson, C.H.K., Phelps, M.E., Barrio, J.R., *Biochem. Pharmacol.* (1990) **39**, 1853–1860.

F749. Moorhouse, K.G., Logan, C.J., Hutson, D.H., Dodds, P.F., *Biochem. Pharmacol.* (1990) **39**, 1529–1536.

F750. McCague, R., Parr, I.B., Haynes, B.P., *Biochem. Pharmacol.* (1990) **40**, 2277–2283.

F751. Khoo, B.Y., Sit, K.H., Wong, K.P., *Biochem. Pharmacol.* (1990) **40**, 1021–1026.

F752. Lin, W.-H., Roth, J.A., *Biochem. Pharmacol.* (1990) **40**, 629–635.

F753. Kim, D.-H., Yang, Y.-S., Jakoby, W.B., *Biochem. Pharmacol.* (1990) **40**, 481–487.

F754. Humfrey, C.D.N., Steventon, G.B., Sturman, S.G., Waring, R.H., Griffiths, B., Williams, A.C., *Biochem. Pharmacol.* (1990) **40**, 2562–2564.

F755. Ghabriel, H., Prakash, C., Tacke, U.G., Blair, I.A., Wilkinson, G.R., *Drug Metab. Dispos.* (1991) **19**, 596–599.

F756. Liu, K., Guengerich, F.P., Yang, S.K., *Drug Metab. Dispos.* (1991) **19**, 609–613.

F757. Woolley, J.L., De Angelis, D.V., Crouch, R.C., Shockcor, J.P., Sigel, C.W., *Drug Metab. Dispos.* (1991) **19**, 600–608.

F758. Obata, T., Egashira, T., Yamanaka, Y., *Biochem. Pharmacol.* (1990) **40**, 1689–1693.

F759. Kamal, A., *Biochem. Pharmacol.* (1990) **40**, 1669–1671.

F760. Coll, R.J., Cesar, D., Hynes, J.B., Shane, B., *Biochem. Pharmacol.* (1991) **42**, 833–838.

F761. Kumagai, Y., Wickham, K.A., Schmitz, D.A., Cho, A.K., *Biochem. Pharmacol.* (1991) **42**, 1061–1067.

F763. Remmel, R.P., Sinz, M.W., *Drug Metab. Dispos.* (1991) **19**, 630–636.

F764. McEuen, S.F., Miller, M.G., *Drug Metab. Dispos.* (1991) **19**, 661–666.

F765. Lu, X.L., Guengerich, F.P., Yang, S.K., *Drug Metab. Dispos.* (1991) **19**, 637–642.

F766. Woolf, T.F., Bjorge, S.M., Black, A.E., Holmes, A., Chang, T., *Drug Metab. Dispos.* (1991) **19**, 696–702.

F767. Harvey, D.J., Brown, N.K., *Drug Metab. Dispos.* (1991) **19**, 714–716.

F768. Clement, B., Zimmermann, M., *Biochem. Pharmacol.* (1988) **37**, 4747–4752.

F769. Luo, H., Hawes, E.M., McKay, G., Korchinski, E.D., Midha, K.K., *Drug Metab. Dispos.* (1991) **19**, 722–724.

F770. Barańczyk-Kuźma, A., Ciszewska-Pilczyńska, A., *Biochem. Pharmacol.* (1988) **38**, 2927–2929.

F771. Fischer, V., Wiebel, F.J., *Biochem. Pharmacol.* (1990) **39**, 1327–1333.

F772. McCague, R., Parr, I.B., Leclerq, G., Leung, O.T., Jarman, M., *Biochem. Pharmacol.* (1990) **39**, 1459–1465.

F774. Thompson, D., Constantin-Teodosiu, D., Egestad, B., Mickos, H., Moldéus, P., *Biochem. Pharmacol.* (1990) **39**, 1587–1595.

F778. Li, P.P., Warsh, J.J., Godse, D.D., Guttman, M., *Neurochem. Int.* (1987) **10**, 537–545.

F779. Miners, J., Lillywhite, K.J., Yoovathaworn, K., Pongmarutai, M., Birkett, D.J., *Biochem. Pharmacol.* (1990) **40**, 595–600.

F780. van den Tweel, W.J.J., Smits, J.P., de Bont, J.A.M., *Arch. Microbiol.* (1988) **149**, 207–213.

F781. Madsen, E.L., Francis, A.J., Bollag, J.-M., *Appl. Environ. Microbiol.* (1988) **54**, 74–78.

F782. Bhatnagar, D., Cleveland, T.E., *Biochimie* (1988) **70**, 743–747.

F783. Lee, M., Nohta, H., Ohkura, Y., *J. Chromatogr.* (1986) **378**, 329–336.

F784. Pearce, C.J., Doyle, T.W., Forenza, S., Lam, K.S., Schroeder, D.R., *J. Nat. Prod.* (1988) **51**, 937–940.

F785. Stothers, J.B., Stoessl, A., *Can. J. Chem.* (1988) **66**, 2816–2818.

F786. Berger, R.G., Drawert, F., *Z. Naturforsch.* (1988) **43C**, 485–490.

F787. El-Bayoumy, K., Shiue, G.-H., Hecht, S.S., *Chem. Res. Toxicol.* (1988) **1**, 243–247.

F788. Sekine, M., Ichikawa, T., Kuga, N., Kobayashi, M., Sakurai, A., Syono, K., *Plant Cell Physiol.* (1988) **29**, 867–874.

F789. Welle, R., Grisebach, H., *Arch. Biochem. Biophys.* (1989) **272**, 97–102.

F790. Inoue, O., Seiji, K., Nakatsuka, H., Watanabe, T., Ikeda, M., *Bull. Environ. Contam. Toxicol.* (1989) **43**, 74–79.

F791. Elchisak, M.A., Hausner, E.A., *Life Sci.* (1984) **35**, 2561–2569.

F792. Saul, S.J., Sugumaran, M., *FEBS Lett.* (1989) **249**, 155–158.

F793. Saul, S.J., Sugumaran, M., *J. Biol. Chem.* (1990) **265**, 16992–16999.

F794. Murthy, S.A., Shetty, H.U., Nelson, W.L., Jackson, P.R., Lennard, M.S., *Biochem. Pharmacol.* (1990) **40**, 1637–1644.

F795. Tilgmann, C., Kalkkinen, N., *Biochem. Biophys. Res. Commun.* (1991) **174**, 995–1002.

F796. Van Uden, W., Pras, N., Vossebeld, E.M., Mol. J.N.M., Malingre, T.M., *Plant Cell, Tissue Organ Cult.* (1990) **20**, 81–87.

F797. Tilgmann, C., Kalkkinen, N., *FEBS Lett.* (1990) **264**, 95–99.

F798. Beffa, R., Martin, H.V., Pilet, P.-E., *Plant Physiol.* (1990) **94**, 485–491.

F799. Morris, M.E., Hansel, S.B., *J. Chromatogr.* (1990) **532**, 285–293.

F800. Okuno, E., Du, F., Ishikawa, T., Tsujimoto, M., Nakamura, M., Schwarcz, R., Kido, R., *Brain Res.* (1990) **534**, 37–44.

F801. Johansen, M., Rasmussen, K.E., Christophersen, A.S., *J. Chromatogr.* (1990) **532**, 277–284.

F802. Sakane, S., Chopra, I.J., *Endocrinology (Baltimore)* (1990) **127**, 2709–2715.

F803. Ahmad, M., Alam, M., Mahmood, A., *Proc. Pak. Acad. Sci.* (1990) **27**, 1–10.

F804. Ishii, K., Fuse, M., Mizuyoshi, H., Yamamuro, W., *Nippon Shokakibyo Gakkai Zasshi* (1990) **87**, 1423–1429.

F805. Hiramatsu, M., Kumagai, Y., Unger, S., Cho, A.K., *J. Pharmacol. Exp. Ther.* (1990) **254**, 521–527.

F806. Ohkubo, T., Takahashi, A., Nambara, T., *J. Steroid Biochem.* (1990) **36**, 501–503.

F807. Rutgers, M., Heusdens, F.A., Bouthuis, F., Rooda, S.-J.E., Visser, T.J., *Endocrinology (Baltimore)* (1990) **127**, 1617–1624.

F808. Sutherland, J.B., Freeman, J.P., Selby, A.L., Fu, P.P., Miller, D.W., Cerniglia, C.E., *Arch. Microbiol.* (1990) **154**, 260–266.

F809. Howard, P.C., Aoyama, T., Bauer, S.L., Gelboin, H.V., Gonzalez, F.J., *Carcinogenesis* (1990) **11**, 1539–1542.

F810. Ferrer, J.C., Calzada, V., Bonet, J.J., *Steroids* (1990) **55**, 390–394.

F811. Connors, N.C., Bartel, P.L., Strohl, W.R., *J. Gen. Microbiol.* (1990) **136**, 1887–1894.

F812. Andreoni, V., Baggi, G., Bernasconi, S., Foglieni, C., Pelizzoni, F., *Appl. Microbiol. Biotechnol.* (1990) **33**, 633–636.

F813. Nakamichi, K., Shibatani, T., Yamamoto, Y., Sato, T., *Appl. Microbiol. Biotechnol.* (1990) **33**, 637–640.

F814. Yoshioka, H., Nagasawa, T., Hasegawa, R., Yamada, H., *Biotechnol. Lett.* (1990) **12**, 679–684.

F815. Bruce, N.C., Wilmot, C.J., Jordan, K.N., Trebilcock, A.E., Stephens, L.D.G., Lowe, C.R., *Arch. Microbiol.* (1990) **154**, 465–470.

F816. Efange, S.M.N., Michelson, R.H., Remmel, R.P., Boudreau, R.J., Dutta, A.K., Freshler, A., *J. Med. Chem.* (1990) **33**, 3133–3138.

F817. De-Eknamkul, W., Zenk, M.H., *Tetrahedron Lett.* (1990) **31**, 4855–4858.

F818. Ghauri, F.Y.D., Wilson, I.D., Nicholson, J.K., *Methodol. Surv. Biochem. Anal.* (1990) **20**, 321–324.

F819. Narimatsu, S., Akustu, Y., Matsunaga, H., Watanabe, K., Yamamoto, I., Yoshimura, H., *Biochem. Biophys. Res. Commun.* (1990) **172**, 607–613.

F821. Goswami, A., Rosenberg, I.N., *Biochem. Biophys. Res. Commun.* (1990) **173**, 6–12.

F822. Jones, D.C.N., Cooper, R.A., *Arch. Microbiol.* (1990) **154**, 489–495.

F823. Holland, H.L., Brown, F.M., Conn, M., *J. Chem. Soc., Perkin Trans. 2* (1990) 1651–1655.

F824. Funk, C., Brodelius, P.E., *Plant Physiol.* (1990) **94**, 102–108.

F825. Igarashi, K., Akai, M., *Agric. Biol. Chem.* (1990) **54**, 2143–2144.

F826. Petrushka, L., Rebrin, I., Grimm, U., Herrmann, F.H., *Clin. Chim. Acta* (1990) **193**, 65–77.

F828. Nakao, M., Goto, Y., Matsuki, Y., Hiratsuka, A., Watabe, T., *Chem. Pharm. Bull.* (1990) **38**, 2561–2566.

F829. Alam, M., Ahmad, M., Rasheed, A., Saleem, M., Javaid, M.K., Ikram, S., *Indian J. Exp. Biol.* (1990) **28**, 940–942.

F830. Moustafa, M.A., El-Eman, A.A., Subbagh, H.I.E., Sharaf El-Din, M.K., *Arzneim.-Forsch.* (1990) **40**, 1076–1078.

F831. Schoerlin, M.P., Da Prada, M., *Acta Psychiatr. Scand., Suppl.* (1990) **360**, 108–110.

F833. Pfeifer, S., Gedan, A., Franke, P., Klemm, W., Morgenroth, U., *Pharmazie* (1990) **45**, 603–607.

F835. Nakamura, S., Ito, Y., Fukushima, T., Sugawara, Y., Ohashi, M., *J. Pharmacobio-Dyn.* (1990) **13**, 612–621.

F837. Bui, Q.D., Weisz, J., Wrighton, S.A., *J. Steroid Biochem. Mol. Biol.* (1990) **37**, 285–293.

F838. Edwards, R., Millburn, P., Hutson, D.H., *Pestic. Sci.* (1990) **30**, 159–181.

F839. Burstyn, J.N., Iskandar, M., Brady, J.F., Jon, M., Cho, A.K., *Chem. Res. Toxicol.* (1991) **4**, 70–76.

F840. Henneböle, J., Parlar, H., *Chemosphere* (1990) **21**, 919–922.

F841. King, L.C., Kohan, M.J., George, S.E., Lewtas, J., Claxton, L.D., *J. Toxicol. Environ. Health* (1990) **31**, 179–192.

F842. Tanake, M., Kojima, M., *Arch. Biochem. Biophys.* (1991) **284**, 151–157.

F843. De Carolis, E., Chan, F., Balsevich, J., De Luca, V., *Plant Physiol.* (1990) **94**, 1323–1329.

F844. Kluge, C., Tschech, A., Fuchs, G., *Arch. Microbiol.* (1990) **155**, 68–74.

F845. Hashim, M.F., Hakamatsuka, T., Ebizuka, Y., Sankawa, U., *FEBS Lett.* (1990) **271**, 219–222.

F846. Ludwig-Müller, J., Schramm, P., Hilgenberg, W., *Physiol. Plant.* (1990) **80**, 472–478.

F847. Sánchez-Bravo, J., Ortuño, A., Botía, J.M., Del Río, J.A., Caballero, M., Acosta, M., Sabater, F., *Plant Growth Regul.* (1990) **9**, 315–327.

F848. Imaoka, S., Enomoto, K., Oda, Y., Asada, A., Fujimori, M., Shimada, T., Fujita, S., Guengerich, F.P., Funae, Y., *J. Pharmacol. Exp. Ther.* (1990) **255**, 1385–1391.

F849. Garay, R.P., Labaune, J.-P., Mesangeau, D., Nazaret, C., Imbert, T., Moinet, G., *J. Pharmacol. Exp. Ther.* (1990) **255**, 415–422.

F850. Roy, D., *Biochem. Int.* (1990) **22**, 849–857.

F851. Pogány, E., Wallnöfer, P.R., Ziegler, W., Mücke, W., *Chemosphere* (1990) **21**, 557–562.

F852. Klein, J., Post, K., Thomas, H., Wörner, W., Setiabudi, F., Frank, H., Oesch, F., Platt, K.L., *Chem.-Biol. Interact.* (1990) **76**, 211–226.

F853. Flescher, J.W., Myers, S.R., *Life Sci.* (1991) **48**, 843–850.

F854. Frenzel, T., Zenk, M.H., *Phytochemistry* (1990) **29**, 3505–3511.

F855. Loeffler, S., Zenk, M.H., *Phytochemistry* (1990) **29**, 3499–3503.

F856. Bauder, R., Tshisuaka, B., Lingens, F., *Biol. Chem. Hoppe-Seyler* (1990) **371**, 1137–1144.

F857. Pettigrew, C.A., Haigler, B.E., Spain, J.C., *Appl. Environ. Microbiol.* (1991) **57**, 157–162.

F859. Vasudevan, N., Mahadevan, A., *Ital. J. Biochem.* (1990) **39**, 285–293.

F860. Preisig, C.L., Bell, J.N., Sun, Y., Hrazdina, G., Matthews, D.E., VanEtten, H.D., *Plant Physiol.* (1990) **94**, 1444–1448.

F862. Köppel, C., Tenczer, J., *J. Chromatogr.* (1991) **562**, 207–211.

F863. Brier, M.E., Bowsher, R.R., Mayer, P.R., Henry, D.P., *Life Sci.* (1991) **48**, 901–907.

F864. Burka, L.T., McGown, S.R., Tomer, K.B., *Chemosphere* (1990) **21**, 1231–1242.

F865. Borneman, W.S., Hartley, R.D., Himmelsbach, D.S., Ljungdahl, L.G., *Anal. Biochem.* (1990) **190**, 129–133.

F866. Muheim, A., Waldner, R., Sanglard, D., Reiser, J., Schoemaker, H.E., Leisola, M.S.A., *Eur. J. Biochem.* (1991) **195**, 369–375.

F867. Sannia, G., Limongi, P., Cocca, E., Buonocore, F., Nitti, G., Giardina, P., *Biochim. Biophys. Acta* (1991) **1073**, 114–119.

F868. Hopper, D.J., Bossert, I.D., Rhodes-Roberts, M.E., *J. Bacteriol.* (1991) **173**, 1298–1301.

F869. Wohlfarth, G., Geerligs, G., Diekert, G., *J. Bacteriol.* (1991) **173**, 1414–1419.

F870. Utkin, I.B., Yakimov, M.M., Matveeva, L.N., Kozlyak, E.I., Rogozhin, I.S., Solomon, Z.G., Bezborodov, A.M., *FEMS Microbiol. Lett.* (1991) **77**, 237–241.

F871. Liu, S.Y., Freyer, A.J., Bollag, J.-M., *J. Agric. Food Chem.* (1991) **39**, 631–636.

F872. Baumert, A., Creche, J., Rideau, M., Chenieux, J.C., Groeger, D., *Plant Physiol. Biochem.* (Paris) (1990) **28**, 587–592.

F873. Hogan, M.E., Manners, G.D., *J. Chem. Ecol.* (1991) **17**, 167–174.

F874. Gaudet, S., Palkovits, M., Namboodiri, M.A.A., *Brain Res.* (1991) **539**, 355–357.

F875. Sugumaran, M., Saul, S.J., Dali, H., *Arch. Insect Biochem. Physiol.* (1990) **15**, 255–269.

F876. Kumagai, Y., Todaka, T., Toki, S., *J. Pharmacol. Exp. Ther.* (1990) **255**, 504–510.

F877. Aschbacher, P.W., Feil, V.J., *J. Agric. Food Chem.* (1991) **39**, 146–149.

F879. Forrester, L.M., Neal, G.E., Judah, D.J., Glancey, M.J., Wolf, C.R., *Proc. Natl. Acad. Sci. U.S.A.* (1990) **87**, 8306–8310.

F880. Bott, G., Schmidt, M., Rommel, T.O., Lingens, F., *Biol. Chem. Hoppe-Seyler* (1990) **371**, 999–1003.

F881. Linhart, I., Novak, J., *J. Chromatogr.* (1990) **530**, 283–294.

F882. Behne, D., Kyriakopoulos, A., Meinhold, H., Köhrle, J., *Biochem. Biophys. Res. Commun.* (1990) **173**, 1143–1149.

F883. Allard, A.S., Hynning, P.A., Lindgren, C., Remberger, M., Neilson, R.H., *Appl. Environ. Microbiol.* (1991) **57**, 77–84.

F884. Robinson, H.K., Duke, C.C., Holder, G.M., Ryan, A.J., *Xenobiotica* (1990) **20**, 457–470.

F885. Samara, E., Bialer, M., Bar-On, H., Harvey, D.J., *Xenobiotica* (1990) **20**, 447–455.

F886. Salonen, J.S., Eloranta, M., *Xenobiotica* (1990) **20**, 471–480.

F887. Frohlich, D.R., Boeker, E.A., Brindley, W.A., *Xenobiotica* (1990) **20**, 481–487.

F888. Gerding, T.K., Drenth, B.F.H., de Zeeuw, R.A., Tepper, P.G., Horn, A.S., *Xenobiotica* (1990) **20**, 525–536.

F889. Gerding, T.K., Drenth, B.F.H., Roosenstein, H.J., de Zeeuw, R.A., Tepper, P.G., Horn, A.S., *Xenobiotica* (1990) **20**, 515–524.

F890. Schlenk, D., Buhler, D.R., *Xenobiotica* (1990) **20**, 563–572.

F891. Renner, G., Hopper, C., *Xenobiotica* (1990) **20**, 573–582.

F892. Stefek, M., Ransom, R.W., DiStefano, E.W., Cho, A.K., *Xenobiotica* (1990) **20**, 591–600.

F893. Foster, B.C., Nantais, L.M., Wilson, D.L., By, A.W., Zamecnik, J., Lodge, B.A., *Xenobiotica* (1990) **20**, 583–590.

F894. Bakke, J.E., Feil, V.J., Mulford, D.J., *Xenobiotica* (1990) **20**, 601–605.

F895. Balajee, S., Mahadevan, A., *Xenobiotica* (1990) **20**, 607–617.

F896. Falkner, F.C., Twomey, T.M., Borger, A.P., Garg, D., Weidler, D., Gerber, N., Browder, I.W., *Xenobiotica* (1990) **20**, 645–652.

F897. Mori, A., Ishiyama, I., Akita, H., Suzuki, K., Mitsuoka, T., Oishi, T., *Xenobiotica* (1990) **20**, 629–634.

F898. Higaki, J., Kikuchi, J., Iyenishi, Y., Hirata, M., *Xenobiotica* (1991) **21**, 47–51.

F899. Liira, J., Elovaara, E., Raunio, H., Riihimäki, V., Engström, K., *Xenobiotica* (1991) **21**, 53–64.

F900. Struble, C.B., *Xenobiotica* (1991) **21**, 85–95.

F901. Eriksson, U.G., Hoffmann, K.-J., Simonsson, R., Regårdh, C.G., *Xenobiotica* (1991) **21**, 75–84.

F902. Kasuya, F., Igarashi, K., Fukui, M., *Xenobiotica* (1991) **21**, 97–109.

F903. Temellini, A., Franchi, M., Giuliani, L., Pacifici, G.M., *Xenobiotica* (1991) **21**, 171–177.

F904. Waldmeier, F., Kaiser, G., Ackermann, R., Faigle, J.W., Wagner, J., Barner, A., Lasseter, K.C., *Xenobiotica* (1991) **21**, 251–261.

F905. Delatour, P., Benoit, E., Besse, S., Boukraa, A., *Xenobiotica* (1991) **21**, 217–221.

F906. Lacarelle, B., Marre, F., Durand, A., Davi, H., Rahmani, R., *Xenobiotica* (1991) **21**, 317–329.

F907. Hermans, J.J.R., Thijssen, H.H.W., *Xenobiotica* (1991) **21**, 295–307.

F908. Richardson, K.A., Edwards, V.T., Jones, B.C., Hutson, D.H., *Xenobiotica* (1991) **21**, 371–382.

F909. Jackson, C.-J.C., Hubbard, J.W., Midha, K.K., *Xenobiotica* (1991) **21**, 383–393.

F910. Watt, J.A., King, A.R., Dickinson, R.G., *Xenobiotica* (1991) **21**, 403–415.

F911. Tanimoto, Y., Ohkuma, T., Oguri, K., Yoshimura, H., *Xenobiotica* (1991) **21**, 395–402.

F912. Karreth, S., Lenk, W., *Xenobiotica* (1991) **21**, 417–428.

F913. Bolaji, O.O., Babalola, L.P., Dixon, P.A.F., *Xenobiotica* (1991) **21**, 447–450.

F914. Peters, M.M.C.G., Walters, D.G., van Ommen, B., van Bladeren, P.J., Lake, B.G., *Xenobiotica* (1991) **21**, 499–514.

F915. Strolin Benedetti, M., Pianezzola, E., Fraier, D., Castelli, M.G., Dostert, P., *Xenobiotica* (1991) **21**, 473–480.

F916. Meinertz, J.R., Gingerich, W.H., Allen, J.L., *Xenobiotica* (1991) **21**, 525–533.

F917. Zheng, J., Hanzlik, R.P., *Xenobiotica* (1991) **21**, 535–546.

F918. Nouws, J.F.M., Mevius, D., Vree, T.B., Degen, M., *Vet. Q.* (1989) **11**, 78–86.

F919. Schellens, J.H.M., van Haelst, I.M.M., Houston, J.B., Breimer, D.D., *Xenobiotica* (1991) **21**, 547–555.

F920. Nwankwo, J.O., *Xenobiotica* (1991) **21**, 569–574.

F921. Takayama, F., Saito, K., Ishii, Y., Shiratori, K., Ohtawa, M., *Xenobiotica* (1991) **21**, 557–568.

F922. Saito, K., Kaneko, H., Sato, K., Yoshitake, A., Yamada, H., *Xenobiotica* (1991) **21**, 575–582.

F923. Mulford, D.J., Bakke, J.E., Feil, V.J., *Xenobiotica* (1991) **21**, 597–603.

F924. Tjørnelund, J., Hansen, S.H., Cornett, C., *Xenobiotica* (1991) **21**, 605–612.

F925. Thomson, K.L., Chang, M.N., Bugianesi, R.L., Ponpipom, M.M., Arison, B.H., Hucker, H.B., Sweeney, B.M., White, S.D., Chabala, J.C., *Xenobiotica* (1991) **21**, 613–625.

F926. Dickinson, R.G., King, A.R., Hansen-Møller, J., *Xenobiotica* (1991) **21**, 635–640.

F927. Yoshimura, T., Tanaka, S., Sakurai, H., Nagai, Y., Yamada, K., Katayama, S., Abe, S., Yamatsu, I., *Xenobiotica* (1991) **21**, 627–634.

F928. Nouws, J.F.M., Firth, E.D., Vree, T.B., Baakman, M., *Am. J. Vet. Res.* (1987) **48**, 392–402.

F929. Nouws, J.F.M., Vree, T.B., Mevius, D., Degen, M., *J. Vet. Pharmacol. Ther.* (1989) **12**, 50–57.

F930. Elovaara, E., Engström, K., Nakajima, T., Park, S.S, Gelboin, H.V., Vainio, H., *Xenobiotica* (1991) **21**, 651–661.

F931. Ludwig-Müller, J., Hilgenberg, W., *Physiol. Plant.* (1990) **80**, 541–548.

F933. Yang, S.K., Bao, Z.-P., *Mol. Pharmacol.* (1987) **32**, 73–80.

F934. Fawaz, F., Jones, G.H., *J. Biol. Chem.* (1988) **263**, 4602–4606.

F935. Yang, S.K., Mushtaq, M., Weems, H.B., Miller, D.W., Fu, P.P., *Biochem. J.* (1987) **245**, 191–204.

F936. van Bladeren, P.J., Balani, S.K., Sayer, J.M., Thakker, D.R., Boyd, D.R., Ryan, D.E., Thomas, P.E., Levin, W., Jerina, D.M., *Biochem. Biophys. Res. Commun.* (1987) **145**, 160–167.

F937. Mikov, M., Caldwell, J., Dolphin, C.T., Smith, R.L., *Biochem. Pharmacol.* (1988) **37**, 1445–1449.

F938. Yokozeki, K., Hirose, Y., Kubota, K., *Agric. Biol. Chem.* (1987) **51**, 737–746.

F940. Yokozeki, K., Sano, K., Eguchi, C., Iwagami, H., Mitsugi, K., *Agric. Biol. Chem.* (1987) **51**,729–736.

F941. Yokozeki, K., Sano, K., Eguchi, C., Yamada, K., Mitsugi, K., *Agric. Biol. Chem.* (1987) **51**, 363–369.

F942. Appel, K.E., Görsdorf, S., Scheper, T., Ruf, H.H., Rühl, C.S., Hildebrandt, A.G., *J. Cancer Res. Clin. Oncol.* (1987) **113**, 131–136.

F943. Yokozeki, K., Nakamori, S., Eguchi, C., Yamada, K., Mitsugi, K., *Agric. Biol. Chem.* (1987) **51**, 355–362.

F944. Barrett, F.M., *Arch. Insect Biochem. Physiol.* (1987) **5**, 99–118.

F945. Mushtaq, M., Yang, S.K., *Carcinogenesis* (1987) **8**, 705–709.

F946. Nagatsu, T., Togari, A., Naoi, M., Mino, N., Sawazaki, Y., Yoshida, E., Kanbayashi, N., Sakurai, E., Ohtawa, M., Takeuchi, T., Umezawa, H., *Biog. Amines* (1987) **4**, 107–117.

F947. Lewer, P., *J. Chem. Soc., Perkin Trans. 1* (1987) 753–757.

F948. Tsuda, Y., *Chem. Pharm. Bull.* (1987) **35**, 2554–2557.

F950. Halldin, C., Långström, B., *J. Labelled Compd.* (1986) **28**, 715–722.

F951. Constantinidis, J., Geissbuhler, F., Gaillard, J.M., Aubert, C., Hovaguimian, T., Tissot, R., *Psychopharmacologia* (1975) **41**, 201–209.

F953. Möller, L., Rafter, J., Gustafsson, J.-Å., *Carcinogenesis* (1987) **8**, 637–645.

F954. Vree, T.B., Hekster, Y.A., Nouws, J.F.M., Baakman, M., *Ther. Drug Monit.* (1986) **8**, 434–439.

F955. Nouws, J.F.M., Mevius, D.J., Vree, T.B., Baars, A.M., Laurensen, J., *Vet. Q.* (1988) **10**, 156–163.

F956. Köppel, C., Tenczer, J., Ibe, K., *Arzneim.-Forsch.* (1987) **37**, 1304–1306.

F957. Bruins, A.P., Pras, N., *Anal. Chim. Acta* (1984) **163**, 91–100.

F958. Pras, N., Hesselink, P.G.M., ten Tusscher, J., Malingré, T.M., *Biotechnol. Bioeng.* (1989) **34**, 214–222.

F959. Josephy, P.D., Eling, T.E., Mason, R.P., *Mol. Pharmacol.* (1983) **23**, 766–770.

F960. Harrelson, W.G., Jr., Mason, R.P., *Mol. Pharmacol.* (1982) **22**, 239–242.

F961. Sabater, F., Sánchez-Bravo, J., Acosta, M., *Rev. Esp. Fisicol.* (1983) **39**, 169–174.

F962. Tasayco J., M.L., Prestwich, G.D., *Insect Biochem.* (1990) **20**, 691–700.

G1. Karreth, S., Lenk, W., *Xenobiotica* (1991) **21**, 709–724.

G2. Krause, W., Duka, T., Matthes, H., *Xenobiotica* (1991) **21**, 763–774.

G3. Shu, Y.-Z., Kingston, D.G.I., Van Tassell, R.L., Wilkins, T.D., *Xenobiotica* (1991) **21**,737–750.

G4. Eyanagi, R., Hisanari, Y., Shigematsu, H., *Xenobiotica* (1991) **21**, 793–803.

G5. Heilmair, R., Karreth, S., Lenk, W., *Xenobiotica* (1991) **21**, 805–815.

G6. Fouda, H.G., Lukaszewicz, J., Clark, D.A., Hulin, B., *Xenobiotica* (1991) **21**, 925–934.

G7. Galtier, P., Alvinerie, M., Plusquellec, Y., Tufenkji, A.E., Houin, G., *Xenobiotica* (1991) **21**, 917–924.

G8. Nicholls, D., Gescher, A., Griffin, R.J., *Xenobiotica* (1991) **21**, 935–943.

G9. Srinivasan, K., Levine, W.G., Bhargava, M.M., *Xenobiotica* (1991) **21**, 961–969.

G10. Pacifici, G.M., Vannucci, L., Bencini, C., Tusini, G., Mosca, F., *Xenobiotica* (1991) **21**, 1113–1118.

G11. Tanaka, A., Deguchi, Y., Ishii, I., Terai, T., *Xenobiotica* (1991) **21**, 1119–1125.

G12. Atsumi, R., Suzuki, W., Hakusui, H., *Xenobiotica* (1991) **21**, 1159–1169.

G13. Hall, L.R., Hanzlik, R.P., *Xenobiotica* (1991) **21**, 1127–1138.

G14. Abul-Hajj, Y.J., Ghaffari, M.A., Mehrotra, S., *Xenobiotica* (1991) **21**, 1171–1177.

G15. Wilson, K., Chissick, H., Fowler, A.M., Frearson, F.J., Gittins, M., Swinbourne, F.J., *Xenobiotica* (1991) **21**, 1179–1183.

G16. Simons, P.J., Cockshott, I.D., Douglas, E.J., Gordon, E.A., Knott, S., Ruane, R.J., *Xenobiotica* (1991) **21**, 1243–1256.

G17. Henrikus, B.M., Breuer, W., Kampffmeyer, H.G., *Xenobiotica* (1991) **21**, 1229–1241.

G18. Luo, H., Hawes, E.M., McKay, G., Korchinski, E.D., Midha, K.K., *Xenobiotica* (1991) **21**, 1281–1288.

G19. Wada, S., Yoshimitsu, T., Koga, N., Yamada, H., Oguri, K., Yoshimura, H., *Xenobiotica* (1991) **21**, 1289–1300.

G20. Foster, B.C., Litster, D.L., Lodge, B.A., *Xenobiotica* (1991) **21**, 1337–1346.

G21. Mori, M.-A., Kobayashi, M., Yumoto, Y., Nakafuku, K., Miyahara, T., Kozuka, H., *Xenobiotica* (1991) **21**, 1301–1309.

G23. Coutts, R.T., Hussain, M.S., Baker, G.B., Daneshtalab, M., Micetich, R.G., *Xenobiotica* (1991) **21**, 1393–1405.

G24. Gottmans, H., Kroker, R., Ungemach, F.R., *Xenobiotica* (1991) **21**, 1431–1439.

G25. Coutts, R.T., Jamali, F., Malek, F., Peliowski, A., Finer, N.N., *Xenobiotica* (1991) **21**, 1407–1418.

G26. Needham, D., Challis, I.R., *Xenobiotica* (1991) **21**, 1473–1482.

G27. Jaworski, T.J., Hawes, E.M., Hubbard, J.W., McKay, G., Midha, K.K., *Xenobiotica* (1991) **21**, 1451–1459.

G28. Macdonald, J.I., Dickinson, R.G., Reid, R.S., Edom, R.W., King, A.R., Verbeeck, R.K., *Xenobiotica* (1991) **21**, 1521–1533.

G29. Herber, R., Perrin, R., Ziegler, J.-M., *Xenobiotica* (1991) **21**, 1493–1499.

G30. Dickinson, R.G., Verbeeck, R.K., King, A.R., *Xenobiotica* (1991) **21**, 1535–1546.

G31. Scatina, J.A., Lockhead, S.R., Cayen, M.N., *Xenobiotica* (1991) **21**, 1591–1604.

G32. Dady, J.M., Bradbury, S.P., Hoffman, A.D., Voit, M.M., Olson, D.L., *Xenobiotica* (1991) **21**, 1605–1620.

G33. Fujimaki, M., Shintani, S., Hakusui, H., *Drug Metab. Dispos.* (1991) **19**, 749–753.

G34. Yu, P.H., Davis, B.A., Durden, D.A., *Drug Metab. Dispos.* (1991) **19**, 830–834.

G35. Brammer, K.W., Coakley, A.J., Jezequel, S.G., Tarbit, M.H., *Drug Metab. Dispos.* (1991) **19**, 764–767.

G36. Walle, U.K., Walle, T., *Drug Metab. Dispos.* (1991) **19**, 838–840.

G37. El Mouehli, M., Schwenk, M., *Drug Metab. Dispos.* (1991) **19**, 844–845.

G38. Roberts, S.M., Harbison, R.D., James, R.C., *Drug Metab. Dispos.* (1991) **19**, 1046–1051.

G39. Tajima, K., Hashizaki, M., Yamamoto, K., Mizutani, T., *Drug Metab. Dispos.* (1991) **19**, 1028–1033.

G40. Hernández, J.S., Powers, S.P., Weinshilboum, R.M., *Drug Metab. Dispos.* (1991) **19**, 1071–1079.

G41. Murthy, S.S., Nelson, W.L., Shen, D.D., Power, J.M., Cahill, C.M., McLean, A.J., *Drug Metab. Dispos.* (1991) **19**, 1093–1100.

G42. Adusumalli, V.E., Wong, K.K., Kucharczyk, N., Sofia, R.D., *Drug Metab. Dispos.* (1991) **19**, 1135–1138.

G43. Yang, J.T., Adusumalli, V.E., Wong, K.K., Kucharczyk, N., Sofia, R.D., *Drug Metab. Dispos.* (1991) **19**, 1126–1134.

G44. Woolley, J.L., De Angelis, D.V., Grace, M.E., Liao, S.H.T., Crouch, R.C., Sigel, C.W., *Drug Metab. Dispos.* (1991) **19**, 1139–1146.

G45. Tedford, C.E., Ruperto, V.B., Barnett, A., *Drug Metab. Dispos.* (1991) **19**, 1152–1159.

G46. Iwakawa, S., Spahn, H., Benet, L.Z., Lin, E.T., *Drug Metab. Dispos.* (1991) **19**, 853–857.

G47. Stearns, R.A., Doss, G.A., Miller, R.R., Chiu, S.-H.L., *Drug Metab. Dispos.* (1991) **19**, 1160–1162.

G48. Blanz, J., Mewes, K., Ehninger, G., Proksch, B., Waidelich, D., Greger, B., Zeller, K.-P., *Drug Metab. Dispos.* (1991) **19**, 871–880.

G49. Eriksson, U.G., Lundahl, J., Bäärnhielm, C., Regårdh, C.G., *Drug Metab. Dispos.* (1991) **19**, 889–894.

G50. Kuhnz, W., Hoyer, G.-A., Backhus, S., Jakobs, U., *Drug Metab. Dispos.* (1991) **19**, 920–925.

G51. Ishida, T., Yano, M., Toki, S., *Drug Metab. Dispos.* (1991) **19**, 895–899.

G52. Yoshikawa, M., Nishiyama, S., Endo, H., Togo, Y., Takaiti, O., *Drug Metab. Dispos.* (1991) **19**, 960–965.

G53. Edwards, D.M.F., Marrari, P., Efthymiopoulos, C., Basileo, G., Fraier, D., Pianezzola, E., Strolin-Benedetti, M., *Drug Metab. Dispos.* (1991) **19**, 938–945.

G55. Garrigou-Gadenne, D., Durand, A., Thenot, J.P., Morselli, P.L., *Drug Metab. Dispos.* (1991) **19**, 574–579.

G58. Tanaka, Y., Kadoh, Y., Ishikawa, H., *Xenobiotica* (1992) **22**, 51–55.

G59. Stiff, D.D., Robicheau, J.T., Zemaitis, M.A., *Xenobiotica* (1992) **22**, 1–11.

G60. Hryk, R., Beyer, J., Urich, K., *Xenobiotica* (1992) **22**, 77–81.

G61. Gray, P.A., Park, G.R., Cockshott, I.D., Douglas, E.J., Shuker, B., Simons, P.J., *Xenobiotica* (1992) **22**, 105–114.

G62. Wu, W.N., Pritchard, J.F., Ng, K.T., Hills, J.F., Uetz, J.A., Yorgey, K.A., McKown, L.A., O'Neill, P.J., *Xenobiotica* (1992) **22**, 153–169.

G63. Rücker, G., Neugebauer, M., Zhong, D., *Xenobiotica* (1992) **22**, 143–152.

G64. Bakke, J.E., Huwe, J.K., Mulford, D.J., Bergman, Å., *Xenobiotica* (1992) **22**, 199–210.

G65. Davi, H., Bonnet, J.M., Berger, Y., *Xenobiotica* (1992) **22**, 171–184.

G66. Delbressine, L.P.C., Moonen, M.E.G., Kaspersen, F.M., Jacobs, P.L., Wagenaars, G.L., *Xenobiotica* (1992) **22**, 227–236.

G67. Rush, W.R., Hall, D.J., Graham, D.J.M., Selby, I.A., *Xenobiotica* (1992) **22**, 237–246.

G68. Kariya, K., Okumura, H., Lee, E., Hirata, M., *Xenobiotica* (1992) **22**, 319–323.

G69. Lin, G., Hawes, E.M., McKay, G., Midha, K.K., *Xenobiotica* (1992) **22**, 303–317.

G70. Linhart, I., Mitera, J., Vosmanská, M., Šmejkal, J., Pech, P., *Xenobiotica* (1992) **22**, 335–343.

G71. Dulik, D.M., Huwe, J.K., Bakke, J.E., Connors, M.S., Fenselau, C., *Xenobiotica* (1992) **22**, 325–334.

G72. Nordholm, L., Wassmann, O., Nielsen, P.G., Thøgersen, H., Grønvald, F., *Xenobiotica* (1992) **22**, 345–356.

G73. Fentem, J.H., Fry, J.R., *Xenobiotica* (1992) **22**, 357–367.

G74. Treston, A.M., Hooper, W.D., *Xenobiotica* (1992) **22**, 385–394.

G75. Oh, E.S., Hong, S.K., Kang, G.I., *Xenobiotica* (1992) **22**, 377–384.

G76. Kumagai, Y., Schmitz, D.A., Cho, A.K., *Xenobiotica* (1992) **22**, 395–403.

G77. Lanusse, C.F., Nare, B., Gascon, L.H., Prichard, R.K., *Xenobiotica* (1992) **22**, 419–426.

G78. Cashman, J.R., Celestial, J.R., Leach, A.R., *Xenobiotica* (1992) **22**, 459–469.

G79. Mori, M.-A., Kobayashi, M., Sakai, K., Nakafuku, K., Mori, Y., Kozuka, H., *Xenobiotica* (1992) **22**, 451–457.

G80. Davison, K.L., Bakke, J.E., Larsen, G.L., *Xenobiotica* (1992) **22**, 479–485.

G81. Higaki, K., Nakano, M., *Drug Metab. Dispos.* (1992) **20**, 350–355.

G82. Suzuki, T., Narimatsu, S., Fujita, S., Masubuchi, Y., Umeda, S., Imaoka, S., Funae, Y., *Drug Metab. Dispos.* (1992) **20**, 367–373.

G83. Akhtar, M., Mahadevan, S., *Drug Metab. Dispos.* (1992) **20**, 356–359.

G84. Nicoll-Griffith, D., Yergey, J., Trimble, L., Williams, H., Rasori, R., Zamboni, R., *Drug Metab. Dispos.* (1992) **20**, 383–389.

G85. Chen, I.-W., Dorley, J.M., Ramjit, H.G., Pitzenberger, S.M., Lin, J.H., *Drug Metab. Dispos.* (1992) **20**, 390–395.

G86. Luo, G., Qato, M.K., Guenthner, T.M., *Drug Metab. Dispos.* (1992) **20**, 440–445.

G87. Hernández, J.S., Watson, R.W.G., Wood, T.C., Weinshilboum, R.M., *Drug Metab. Dispos.* (1992) **20**, 413–422.

G88. Mutlib, A.E., Abbott, F.S., *Drug Metab. Dispos.* (1992) **20**, 451–460.

G89. Wong, J.M.Y., Mahon, W.A., Kalow, W., Inaba, T., *Drug Metab. Dispos.* (1992) **20**, 465–466.

G90. Yang, W., Davis, P.J., *Drug Metab. Dispos.* (1992) **20**, 38–46.

G91. Egestad, B., Sjöberg, P., *Drug Metab. Dispos.* (1992) **20**, 470–472.

G92. Ball, S.E., Maurer, G., Zollinger, M., Ladona, M., Vickers, A.E.M., *Drug Metab. Dispos.* (1992) **20**, 56–63.

G93. Stehly, G.R., Plakas, S.M., *Drug Metab. Dispos.* (1992) **20**, 70–73.

G94. Yang, J.T., Morris, M., Wong, K.K., Kucharczyk, N., Sofia, R.D., *Drug Metab. Dispos.* (1992) **20**, 84–88.

G95. Narimatsu, S., Watanabe, K., Matsunaga, T., Yamamoto, I., Imaoka, S., Funae, Y., Yoshimura, H., *Drug Metab. Dispos.* (1992) **20**, 79–83.

G96. Swanson, S.P., Catlow, J., *Drug Metab. Dispos.* (1992) **20**, 102–107.

G97. Uetrecht, J., Sokoluk, B., *Drug Metab. Dispos.* (1992) **20**, 120–123.

G98. Zhoo, N.-S., Sladek, N.E., *Drug Metab. Dispos.* (1992) **20**, 134–136.

G99. Schaefer, W.H., *Drug Metab. Dispos.* (1992) **20**, 130–133.

G101. Byczkowski, J.Z., Kulkarni, A.P., *Xenobiotica* (1992) **22**, 609–618.

G102. Ibrahim, A.-R., Abul-Hajj, Y.J., *J. Nat. Prod.* (1990) **53**, 1471–1478.

G103. Martin, R.J., Downer, R.G.J., *Arch. Insect Biochem. Physiol.* (1989) **11**, 29–45.

G104. Mappouras, D.G., Fragoulis, E.G., *Comp. Biochem. Physiol. B: Comp. Biochem.* (1990) **97B**, 301–306.

G105. Docampo, R., Moreno, S.N.J., Mason, R.P., *J. Biol. Chem.* (1983) **258**, 14920–14925.

G107. Higson, F.K., Focht, D.D., *Appl. Environ. Microbiol.* (1990) **56**, 3678–3685.

G108. Agarwal, R., Boyd, D.R., Sharma, N.D., Dalton, H., Gray, D.J., *J. Chem. Soc., Chem. Commun* (1990) 1711–1713.

G110. Kaneo, Y., Morichika, T., Tanaka, T., Ozawa, K., Kojima, S., Suguro, K., Abiko, K., *Iyakuhin Kenkyu* (1990) **21**, 694–702.

G111. May, D.G., Porter, J.A., Uetrecht, J.P., Wilkinson, G.R., Branch, R.A., *Clin. Pharmacol. Ther. (St. Louis)* (1990) **48**, 619–627.

G113. Harvey, D.J., Samara, E., Mechoulam, R., *J. Chromatogr.* (1991) **562**, 299–322.

G114. Kimura, T., Yamamoto, I., Watanabe, K., Tateoka, Y., Ho, I.K., *Res. Commun. Chem. Pathol. Pharmacol.* (1991) **71**, 27–48.

G115. Kawahara, F., Ooie, T., Nagatsu, Y., Uchida, H., *Chemotherapy (Tokyo)* (1990) **38** (Suppl. 2), 122–134.

G116. Scheper, T., Appel, K.E., Schunack, W., Somogyi, A., Hildebrandt, A.G., *Chem.-Biol. Interact.* (1991) **77**, 81–96.

G117. Kanekal, S., Plopper, C., Morin, D., Buckpitt, A., *J. Pharmacol. Exp. Ther.* (1991) **256**, 391–401.

G118. Myers, S.R., Flescher, J.W., *Chem.-Biol. Interact.* (1991) **77**, 203–221.

G119. Girod, P.A., Zryd, J.P., *Phytochemistry* (1991) **30**, 169–274.

G120. Anderson, J.A., Chung, C.H., Cho, S.H., *Mycopathologia* (1990) **111**, 39–45.

G121. Safran, M., Leonard, J.L., *J. Biol. Chem.* (1991) **266**, 3233–3238.

G122. Kelley, I., Freeman, J.P., Evans, F.E., Cerniglia, C.E., *Appl. Environ. Microbiol.* (1991) **57**, 636–641.

G123. Saul, S.J., Dali, H., Sugumaran, M., *Arch. Insect Biochem. Physiol.* (1991) **16**, 123–138.

G124. Kokubo, T., Ambe, Y., Nakamura, M., Yamakawa, T., Noguchi, H., Kodama, T., *Agric. Biol. Chem.* (1991) **55**, 613–614.

G125. Ozawa, S., Abu-Zeid, M., Kawakubo, Y., Toyama, S., Yamazoe, Y., Kato, R., *Carcinogenesis* (1990) **11**, 2137–2144.

G126. Brownsill, R., Wallace, D., Taylor, A., Campbell, B., *J. Chromatogr.* (1991) **562**, 267–277.

G127. Arima, N., Kato, Y., *J. Pharmacobio-Dyn.* (1990) **13**, 739–744.

G128. Hirano, S., Tohyama, C., Mitsumori, F., Ito, H., Suzuki, K.T., *Arch. Environ. Contam. Toxicol.* (1991) **20**, 67–72.

G129. Abou-Donia, M.B., Nomeir, A.A., Bower, J.H., Makkawy, H.A., *Toxicology* (1990) **65**, 61–74.

G130. Subrahmanyam, V.V., Kolachana, P., Smith, M.T., *Arch. Biochem. Biophys.* (1991) **286**, 76–84.

G131. Arinc, E., Aydogmus, A., *Comp. Biochem. Physiol. B: Comp. Biochem.* (1990) **97B**, 455–460.

G132. Pedras, M.S.C., Taylor, J.L., *J. Org. Chem.* (1991) **56**, 2619–2621.

G133. Zbaida, S., Levine, W.G., *Biochem. Pharmacol.* (1990) **40**, 2415–2423.

G134. Ray, L., Majundar, S.K., *Indian J. Exp. Biol.* (1991) **29**, 282–283.

G135. Hashimoto, T., Hayashi, A., Amano, Y., Kohno, J., Iwanari, H., Usuda, S., Yamada, Y., *J. Biol. Chem.* (1991) **266**, 4648–4653.

G136. Bauer, W., Zenk, M.H., *Tetrahedron Lett.* (1991) **32**, 487–490.

G137. Kodama, T., Ishida, H., Kokubo, T., Yamakawa, T., Noguchi, H., *Agric. Biol. Chem.* (1990) **54**, 3283–3288.

G138. Gehlert, R., Kindl, H., *Phytochemistry* (1991) **30**, 457–460.

G139. Ghini, A.A., Burton, G., Gros, E.G., *Phytochemistry* (1991) **30**, 779–784.

G140. Kato, M., Morikawa, T., Miyoshi, T., *Jpn. Kokai Tokkyo Koho* JP 02207793 (17–8–1990).

G141. Okuno, E., Nakamura, M., Schwarcz, R., *Brain Res.* (1991) **542**, 307–312.

G142. Brown, N.D., Phillips, L.R., Leader, H., Chiang, P.K., *J. Chromatogr.* (1991) **563**, 466–471.

G143. Brown, N.K., Harvey, D.J., *Eur. J. Drug Metab. Pharmacokinet.* (1990) **15**, 253–258.

G144. Mross, K., Mayer, U., Hamm, K., Burk, K., Hossfeld, D.K., *Eur. J. Clin. Pharmacol.* (1990) **39**, 507–513.

G145. Yousif, M.Y., Fitzgerald, R.L., Narasimhachari, N., Rosecrans, J.A., Blanke, R.V., Glennon, R.A., *Drug Alcohol Depend.* (1990) **26**, 127–135.

G146. Belanger, P.M., St-Hilaire, S., *Can. J. Physiol. Pharmacol.* (1991) **69**, 400–405.

G147. Zaleski, J., Steward, A.R., Sikka, H.C., *Carcinogenesis* (1991) **12**, 167–174.

G148. Hummel, W., *Appl. Microbiol. Biotechnol.* (1990) **34**, 15–19.

G149. Werner, P., Cohen, G., *FEBS Lett.* (1991) **280**, 44–46.

G150. Sigmund, H.U., Kaufman, S., *J. Biol. Chem.* (1991) **266**, 2903–2910.

G151. Blaschke, M., Kretzer, A., Schaefer, C., Nagel, M., Andreesen, J.R., *Arch. Mikrobiol.* (1991) **155**, 164–169.
G152. Dolfing, J., Tiedje, J.M., *Appl. Environ. Microbiol.* (1991) **57**, 820–824.
G153. Spain, J.C., Gibson, D.T., *Appl. Environ. Microbiol.* (1991) **57**, 812–819.
G154. Bhatnagar, D., Cleveland, T.E., Kingston, D.G., *Biochemistry* (1991) **30**, 4343–4350.
G155. Schnell, S., Schink, B., *Arch. Microbiol.* (1991) **155**, 183–190.
G156. Sahasrabudhe, A.V., Modi, V.V., *Indian J. Exp. Biol.* (1991) **29**, 252–255.
G159. Seyfried, B., Tschech, A., Fuchs, G., *Arch. Microbiol.* (1991) **155**, 249–255.
G161. Buchert, J., Niemela, K., *J. Biotechnol.* (1991) **18**, 1–12.
G163. Delatour, P., Benoit, E., Garnier, F., Besse, S., *J. Vet. Pharmacol. Ther.* (1990) **13**, 361–366.
G164. Yamamoto, K., Tajima, K., Takemura, M., Mizutani, T., *Chem. Pharm. Bull.* (1991) **39**, 512–514.
G165. Sweatman, T.W., Seshadri, R., *Cancer Chemother. Pharmacol.* (1990) **27**, 205–210.
G167. Genthner, B.R.S., Townsend, G.T., Chapman, P.J., *FEMS Microbiol. Lett.* (1991) **78**, 265–269.
G168. Asano, Y., Yamada, A., Kato, Y., Yamaguchi, K., Hibino, Y., Hirai, K., Kondo, K., *J. Org. Chem.* (1990) **55**, 67–71.
G169. Kawaguchi, M., Kobayashi, M., Sakurai, A., Syono, K., *Plant Cell Physiol.* (1991) **32**, 143–149.
G170. Rieman, J.D., McNabb, F.M.A., *Gen. Comp. Endocrinol.* (1991) **82**, 53–59.
G171. Bhaird, N.N.A., Tipton, K.E., *Biochem. Soc. Trans.* (1991) **19**, 20S.
G172. Kakeya, H., Sakai, N., Sugai, T., Ohta, H., *Tetrahedron Lett.* (1991) **32**, 1343–1346.
G173. Oh, K.H., Tuovinen, O.H., *J. Ind. Microbiol.* (1990) **6**, 275–278.
G174. Klempier, N., de Raadt, A., Faber, K., Griengl, H., *Tetrahedron Lett.* (1991) **32**, 341–344.
G175. Kiang, C.H., Lee, C., *J. Chromatogr.* (1991) **565**, 339–347.
G176. Arima, N., *J. Pharmacobio-Dyn.* (1990) **13**, 724–732.
G178. Arima, N., *J. Pharmacobio-Dyn.* (1990) **13**, 733–738.
G180. Whited, G.M., Gibson, D.T., *J. Bacteriol.* (1991) **173**, 3017–3020.
G181. Bruce, N.C., Wilmot, C.J., Jordan, K.N., Stephens, L.D.G., Lowe, C.R., *Biochem. J.* (1991) **274**, 875–880.
G182. Weissenfels, W.D., Beyer, M., Klein, J., Rehm, H.J., *Appl. Microbiol. Biotechnol.* (1992) **34**, 528–535.
G183. Oh, K.H., Tuovinen, O.H., *FEMS Microbiol. Lett.* (1991) **79**, 141–146.
G184. Saul, S.J., Sugumaran, M., *Arch. Insect Biochem. Physiol.* (1990) **15**, 237–254.
G185. Rao, J.R., Sharma, N.D., Hamilton, J.T.G., Boyd, D.R., Cooper, J.E., *Appl. Environ. Microbiol.* (1991) **57**, 1563–1565.
G186. Bechtel, W.D., *Eur. J. Drug Metab. Pharmacokinet.* (1990) **15**, 287–293.
G187. Herchuelz, A., Gangji, D., Derenne, F., Jeanniot, J.P., Douchamps, J., *Br. J. Clin. Pharmacol.* (1991) **31**, 73–76.
G188. Hutt, V., Theodor, R., Pabst, G., Dilger, C., Mosberg, H., Fischer, W., Jaeger, H., *Arzneim.-Forsch.* (1991) **41**, 367–372.
G189. Scherling, D., Bühner, K., Krause, H.P., Karl, W., Wünsche, C., *Arzneim.-Forsch.* (1991) **41**, 392–398.
G190. Haycock, J.W., *J. Neurochem.* (1991) **56**, 2139–2142.
G191. Tanimoto, Y., Kaneko, H., Ohkuma, T., Oguri, K., Yoshimura, H., *J. Pharmacobio-Dyn.* (1991) **14**, 161–169.
G192. Barthelmebs, M., Vailly, B., Velly, J., Ehrhardt, J.D., Grima, M., Imbs, J.L., *Am. J. Hypertens.* (1990) **3**, 18S-21S.

G193. Inoue, M., Maejima, T., Sanai, S., Okamoto, R., Hashimoto, H., *J. Antibiot.* (1991) **44**, 435–440.

G194. Glatt, H., Henschler, R., Phillips, D.H., Blake, J.W., Steinberg, P., Seidel, A., Oesch, F., *Environ. Health Perspect.* (1990) **88**, 43–48.

G195. Thomas, H., Schladt, L., Doehmer, J., Knehr, M., Oesch, F., *Environ. Health Perspect.* (1990) **88**, 49–55.

G196. Taylor, M.G., Massey, V., *J. Biol. Chem.* (1991) **266**, 8291–8301.

G197. de la Fuente, G., Perestelo, F., Rodriguez-Pérez, A., Falcón, M.A., *Appl. Environ. Microbiol.* (1991) **57**, 1275–1276.

G198. Schnell, S., Brune, A., Schink, B., *Arch. Microbiol.* (1991) **155**, 511–516.

G199. Suarez, M., Gibello, A., Allende, J.L., Martin, M., Ferrer, E., Garrido-Pertierra, A., *Appl. Microbiol. Biotechnol.* (1991) **34**, 677–682.

G201. Broomhead, A.J., Rahman, M.M.A., Dewick, P.M., Jackson, D.E., Lucas, J.A., *Phytochemistry* (1991) **30**, 1489–1492.

G202. Grislain, L., Gele, P., Bromet, N., Luijten, W., Volland, J.P., Mocaer, E., Kamoun, A., *Eur. J. Drug Metab. Pharmacokinet.* (1990) **15**, 339–345.

G203. Shen, J., Wanwimolruk, S., Roberts, M.S., *J. Chromatogr.* (1991) **565**, 309–320.

G204. Wong, Y.W.J., Davis, P.J., *J. Pharm. Sci.* (1991) **80**, 305–308.

G205. Löffler, F., Müller, R., Lingens, F., *Biochem. Biophys. Res. Commun.* (1991) **176**, 1106–1111.

G206. Topp, E., Akhtar, M.H., *Appl. Environ. Microbiol.* (1991) **57**, 1294–1300.

G207. Kato, N., Joung, E.H., Yang, H.C., Masuda, M., Shimao, M., Yanase, H., *Agric. Biol. Chem.* (1991) **55**, 757–762.

G208. McNally, K.J., Harper, D.B., *J. Gen. Microbiol.* (1991) **137**, 1029–1032.

G209. Zhao, Z., Leung, L.Y., Trevor, A., Castagnoli, N., Jr., *Chem. Res. Toxicol.* (1991) **4**, 426–429.

G210. Brown, N.K., Harvey, D.J., *Biol. Mass Spectrom.* (1991) **20**, 324–328.

G211. Shimamura, H., Nakai, S., Yamamoto, K., Hitomi, N., Yumioka, E., *Shoyakugaku Zasshi* (1990) **44**, 265–268.

G212. Whited, G.M., Gibson, D.T., *J. Bacteriol.* (1991) **173**, 3010–3016.

G213. Tenhaken, R., Salmen, H.C., Barz, W., *Arch. Microbiol.* (1991) **155**, 353–359.

G214. Martin, M., Gibello, A., Fernandez, J., Ferrer, E., Garrido-Pertierra, A., *J. Gen. Microbiol.* (1991) **137**, 621–628.

G215. Heilemann, J., Strack, D., *Phytochemistry* (1991) **30**, 1773–1776.

G216. Endo, F., Katoh, H., Matsuda, I., *J. Inherited Metab. Dis.* (1990) **13**, 780–782.

G217. Bhavnani, B.R., Gerulath, A.H., *J. Steroid Biochem. Mol. Biol.* (1991) **38**, 433–439.

G218. Ishida, C., Koga, N., Hanioka, N., Sakei, H.K., Yoshimura, H., *J. Pharmacobio-Dyn.* (1991) **14**, 276–284.

G219. Rietjens, I.M.C.M., Vervoort, J., *Chem.-Biol. Interact.* (1991) **77**, 263–281.

G220. Gallagher, E.P., Kedderis, G.L., Di Giulio, R.T., *Biochem. Pharmacol.* (1991) **42**, 139–145.

G221. Radjendirano, V., Bhat, M.A., Vaidyanathan, C.S., *Arch. Biochem. Biophys.* (1991) **288**, 169–176.

G222. Ohshima, T., Takada, H., Yoshimura, T., Esaki, N., Soda, K., *J. Bacteriol.* (1991) **173**, 3943–3948.

G223. Tobias, R.B., Larson, R.L., *Biochem. Physiol. Pflanz.* (1991) **187**, 243–250.

G224. Katase, T., Bollag, J.M., *Soil Sci.* (1991) **151**, 291–296.

G225. Dangel, W., Brackmann, R., Lack, A., Mohamed, M., Koch, J., Oswald, B., Seyfried, B., Tschech, A., Fuchs, G., *Arch. Microbiol.* (1991) **155**, 256–262.

G226. Che, Q.M., Akao, T., Hattori, M., Kobashi, K., Namba, T., *Chem. Pharm. Bull.* (1991) **39**, 704–708.

G227. Sander, P., Wittich, RM., Fortnagel, P., Wilkes, H., Francke, W., *Appl. Environ. Microbiol.* (1991) **57**, 1430–1440.

G228. Halkier, M.A., Moller, B.L., *Plant Physiol.* (1991) **96**, 10–17.

G229. Yazaki, K., Heide, L., Tabata, M., *Phytochemistry* (1991) **30**, 2233–2236.

G230. Frischer, H., Mellovitz, R.L., Ahmad, T., Nora, M.V., *J. Lab. Clin. Med.* (1991) **117**, 468–476.

G231. Blanz, J., Mewes, K., Ehninger, G., Proksch, B., Greger, B., Waidelich, D., Zeller, K.P., *Cancer Res.* (1991) **51**, 3427–3433.

G233. Salonen, J.S., *Pharmacol. Toxicol.* (1991) **69**, 1–4.

G234. Kajbaf, M., Jahanshahi, M., Lamb, J.H., Gorrod, J.W., Naylor, S., *Xenobiotica* (1992) **22**, 641–655.

G235. Sayama, M., Inoue, M., Mori, M.-A., Maruyama, Y., Kozuka, H., *Xenobiotica* (1992) **22**, 633–640.

G236. Robertson, I.G.C., Palmer, B.D., Paxton, J.W., Shaw, G.T., *Xenobiotica* (1992) **22**, 657–669.

G237. Tjeerdema, R.S., Crosby, D.G., *Xenobiotica* (1992) **22**, 681–690.

G238. Mori, M.-A., Kobayashi, M., Uemura, H., Miyahara, T., Kozuka, H., *Xenobiotica* (1992) **22**, 701–708.

G239. Tonda, K., Hirata, M., *Xenobiotica* (1992) **22**, 691–699.

G240. Smith, D.A., Rasmussen, H.S., Stopher, D.A., Walker, D.K., *Xenobiotica* (1992) **22**, 709–719.

G241. Pfandl, B., Mörike, K., Winne, D., Schareck, W., Breyer-Pfaff, U., *Xenobiotica* (1992) **22**, 721–730.

G242. Thakker, D.R., Boelert, C., Levin, W., Ryan, D.E., Thomas, P.E., Yagi, H., Jerina, D.M., *Arch. Biochem. Biophys.* (1991) **288**, 54–63.

G243. Shou, M., Yang, S.K., *Carcinogenesis* (1990) **11**, 933–940.

G244. Locher, H.H., Leisinger, T., Cook, A.M., *Biochem. J.* (1991) **274**, 833–842.

G245. Bailey, B.A., Larson, R.L., *Plant Physiol.* (1991) **95**, 792–796

G246. Yamanaka, K., Iino, H., Oikawa, T., *Agric. Biol. Chem.* (1991) **55**, 989–996.

G247. Sandermann, H., Jr., Schmitt, R., Eckey, H., Bauknecht, T., *Arch. Biochem. Biophys.* (1991) **287**, 341–350.

G248. Latchinian-Sadek, L., Ibrahim, R.K., *Phytochemistry* (1991) **30**, 1767–1771.

G249. Yu, P.H., Davis, B.A., Durden, D.A., *Prog. Neuro-Psychopharmacol. Biol. Psychiatry* (1991) **15**, 307–312.

G250. Stearns, R.H., Miller, R.R., Doss, G.A., Chakravarty, P.K., Rosegay, A., Gatto, G.J., Chiu, S.-H.L., *Drug Metab Dispos.* (1992) **20**, 281–287.

G251. Mackness, M.I., Arrol, S., Darrington, P.N., *Biochem. Soc. Trans.* (1991) **19**, 304S.

G252. Ahmad, D., Sylvestre, M., Sondossi, M., Masse, R., *J. Gen. Microbiol.* (1991) **137**, 1375–1385.

G253. Harms, H., Wilkes, H., Sinnwell, V., Wittich, R.M., Figge, K., Francke, W., Fortnagel, P., *FEMS Microbiol. Lett.* (1991) **81**, 25–29.

G255. Bokern, M., Wray, V., Strack, D., *Planta* (1991) **184**, 261–270.

G257. Pabsch, K., Petersen, M., Rao, N.N., Alfermann, A.W., Wandrey, C., *Recl. Trav. Chim. Pays-Bas* (1991) **110**, 199–205.

G258. Koh, M.H., Gorrod, J.W., *Drug Metab. Drug Interact.* (1989) **7**, 253–272.

G259. Koh, M.H., Gorrod, J.W., *Drug Metab. Drug Interact.* (1989) **7**, 273–285.

G260. Fujitani, T., Yoneyama, M., Ogata, A., Ueta, T., Mori, K., Ichikawa, H., *Food Chem. Toxicol.* (1991) **29**, 265–274.

G261. Zill, G., Engelhardt, G., Wohner, B., Wallnoefer, P.R., *Mycotoxin Res.* (1990) **6**, 31–40.

G262. Latchinian-Sadek, L., Ibrahim, R.K., *Arch. Biochem. Biophys.* (1991) **289**, 230–236.

G263. Smitskamp-Wilms, E., Brussee, J., Van der Gen, A., Van Scharrenburg, G.J.M., Sloothaak, J.B., *Recl. Trav. Chim. Pays-Bas* (1991) **110**, 209–215.

G264. Tahara, S., Ingham, J.L., Mitzutani, J., *Z. Naturforsch. C: Biosci.* (1991) **46**, 341–348.

G265. Fujimoto, S., Ishimitsu, S., Hirayama, S., Kawakami, N., Ohara, A., *Chem. Pharm. Bull.* (1991) **39**, 1598–1600.

G266. Moore, R., Millington, D.S. Norwood, D., Kodo, N., Robinson, P., Glasgow, J.F.T., *J. Inherited Metab. Dis.* (1990) **13**, 325–329.

G267. Kelley, I., Freeman, J.P., Cerniglia, C.E., *Biodegradation* (1990) **1**, 283–290.

G268. Fukuto, J.M., Kumagai, Y., Cho, A.K., *J. Med. Chem.* (1991) **34**, 2871–2876.

G269. Hussain, M.S., Coutts, R.T., Baker, G.B., Micetich, R.G., Daneshtalab, M., *Prog. Neuro-Psychopharmacol. Biol. Psychiatry* (1991) **15**, 285–289.

G270. Okumura, T., Morooka, N., Yamamoto, H., Ohno, S., *Iyakuhin Kenkyu* (1991) **22**, 225–236.

G271. Okumura, T., Asano, T., Adachi, T., Hirano, K., *Iyakuhin Kenkyu* (1991) **22**, 270–273.

G272. Xun, L., Orser, C.S., *J. Bacteriol.* (1991) **173**, 4447–4453.

G273. Edwards, R., Dixon, R.A., *Arch. Biochem. Biophys.* (1991) **287**, 372–379.

G274. Hund, H.K., De Beyer, A., Lingens, F., *Biol. Chem. Hoppe-Seyler* (1990) **371**, 1005–1008.

G275. Vasudevan, N., Mahadevan, A., *J. Appl. Bacteriol.* (1991) **70**, 169–176.

G276. Hermans, J.T.R., Thijssen, H.H.W., *Drug Metab. Dispos.* (1992) **20**, 268–274.

G277. Shukla, O.P., *Microbios Lett.* (1990) **45**, 89–96.

G278. Bisaillon, J.G., Lepine, F., Beaudet, R., Sylvestre, M., *Appl. Environ. Microbiol.* (1991) **57**, 2131–2134.

G280. Ohmoto, T., Sakai, K., Hamada, N., Ohe, T., *Agric. Biol. Chem.* (1991) **55**, 1733–1737.

G281. Bisaillon, J.G., Lepine, F., Beaudet, R., *Can. J. Microbiol.* (1991) **37**, 573–576.

G282. El-Bayoumy, K., Upadhyaya, P., Date, V., Sohn, O.S., Fiala, E.S., Reddy, B., *Chem. Res. Toxicol.* (1991) **4**, 560–565.

G283. Howes, A.W., Dutton, M.F., Chuturgoon, A.A., *Mycopathologia* (1991) **113**, 25–29.

G284. Mansuy, D., Valadon, P., Erdelmeier, I., Lopez-Garcia, P., Amar, C., Girault, J.P., Dansette, P.M., *J. Am. Chem. Soc.* (1991) **113**, 7825–7826.

G285. Delatour, P., Jaussard, P., Courtot, D., Fau, D., *J. Vet. Pharmacol. Ther.* (1991) **14**, 209–212.

G286. Huwer, T., Altmann, H.J., Grunow, W., Lenhardt, S., Przybylski, M., *Chem. Res. Toxicol.* (1991) **4**, 586–590.

G287. Raynaud, F., Pevet, P., *Neurosci. Lett.* (1991) **123**, 172–174.

G289. Häusler, E., Petersen, M., Alfermann, A.W., *Z. Naturforsch. C: Biosci.* (1991) **46**, 371–376.

G290. Ribeiro, P., Pigeon, D., Kaufman, S., *J. Biol. Chem.* (1991) **266**, 16207–16211.

G290a. Dix, T.A., Kuhn, D.M., Benkovic, S.J., *Biochemistry* (1987) **26**, 3354–3361.

G291. Lack, A., Tommasi, I., Aresta, M., Fuchs, G., *Eur. J. Biochem.* (1991) **197**, 473–479.

G292. Hopper, W., Mahadevan, A., *Appl. Microbiol. Biotechnol.* (1991) **35**, 411–415.

G293. Dutton, P.J., Hughes, L.A., Foster, D.O., Burton, G.W., Ingold, K.U., *Free Radical Biol. Med.* (1990) **9**, 435–439.

G294. Kimura, R., Ohishi, N., Kato, Y., Yamada, S., Sato, M., *Drug Metab. Dispos.* (1992) **20**, 161–171.

G295. Schrenk, D., Ingelman-Sundberg, M., Bock, K.W., *Drug Metab. Dispos.* (1992) **20**, 137–141.

G296. Hansen, K.T., Thomsen, K.F., Bundgaard, H., *Drug Metab. Dispos.* (1992) **20**, 172–178.

G297. Tatsumi, K., Kitamura, S., Kato, M., Hiraoka, K., *Drug Metab. Dispos.* (1992) **20**, 226–233.

G298. Bornheim, L.M., Lasker, J.M., Rauchy, J.L., *Drug Metab. Dispos.* (1992) **20**, 241–246.

G299. Gregus, Z., Fekete, T., Varga, F., Klaassen, C.D., *Drug Metab. Dispos.* (1992) **20**, 234–240.

G300. Weidolf, L., Karlsson, K.-E., Nilsson, I., *Drug Metab. Dispos.* (1992) **20**, 262–267.

G302. Marre, F., Fabre, G., Lacarelle, B., Bourrie, M., Catalin, J., Berger, Y., Rahmani, R., Cano, J.-P., *Drug Metab. Dispos.* (1992) **20**, 316–321.

G303. Kamimura, H., Ogata, H., Takahara, H., *Drug Metab. Dispos.* (1992) **20**, 309–315.

G304. Tracy, T.S., Hall, S.D., *Drug Metab. Dispos.* (1992) **20**, 322–327.

G305. Sweeny, D.J., Bouska, J., Machinist, J., Bell, R., Carter, G., Cepa, S., Nellans, H.N., *Drug Metab. Dispos.* (1992) **20**, 328–329.

G306. Duffel, M.W., Modi, R.B., King, R.S., *Drug Metab. Dispos.* (1992) **20**, 339–341.

G307. Walle, T., Walle, U.K., *Drug Metab. Dispos.* (1992) **20**, 333–336.

G308. Tomlinson, P.W., Ramji, J.V., Filer, C.W., *Xenobiotica* (1992) **22**, 799–814.

G309. Gilbert, P.J., Hartley, T.E., Troke, J.A., Turcan, R.G., Vose, C.W., Watson, K.V., *Xenobiotica* (1992) **22**, 775–787.

G310. Lavrijsen, K., Van Houdt, J., Van Dyck, D., Hendrickx, J., Bockx, M., Hurkmans, R., Meuldermans, W., Le Jeune, L., Lauwers, W., Heykants, J., *Xenobiotica* (1992) **22**, 815–836.

G311. Wiltshire, H.R., Harris, S.R., Prior, K.J., Kozlowski, U.M., Worth, E., *Xenobiotica* (1992) **22**, 837–857.

G313. Hughes, K.M., Lang, J.C.T., Lazare, R., Gordon, D., Stanton, S.L., Malone-Lee, J., Geraint, M., *Xenobiotica* (1992) **22**, 859–869.

G314. Yeung, P.K.F., Mosher, S.J., Pollak, P.T., *Eur. J. Drug Metab. Pharmacokinet.* (1991) **16**, 69–74.

G317. Gong, D.W., Ozawa, S., Yamazoe, Y., Kato, R., *J. Biochem.* (1991) **110**, 226–231.

G318. Zhao, Z., Quan, W., *J. Environ. Sci.* (1991) **3**, 127–132.

G319. Winter, J., Popoff, M.R., Grimont, P., Bokkenheuser, V.D., *Int. J. Syst. Bacteriol.* (1991) **41**, 355–357.

G320. Shin, M., Shinguu, T., Sano, K., Umezawa, C., *Chem. Pharm. Bull.* (1991) **39**, 1792–1795.

G321. Fuchs, K., Schreiner, A., Lingens, F., *J. Gen. Microbiol.* (1991) **137**, 2033–2039.

G322. Edwards, R., Dixon, R.A., *Phytochemistry* (1991) **30**, 2597–2606.

G324. Heim, M.H., Blum, M., Beer, M., Meyer, U.A., *J. Neurochem.* (1991) **57**, 1095–1099.

G325. Zandbergen, P., van der Linden, J., Brussee, J., van der Gen, A., *Synth Commun.* (1991) **21**, 1387–1391.

G326. Di Monte, D.A., Wu, E.Y., Irwin, I., Delanney, L.E., Langston, J.W., *J. Pharmacol. Exp. Ther.* (1991) **258**, 594–600.

G327. Borneman, W.S., Ljungdahl, L.G., Hartley, R.D., Akin, D.E., *Appl. Environ. Microbiol.* (1991) **57**, 2337–2344.

G328. Pieper, D.H., Kuhm, A.E., Stadler-Fritzsche, K., Fischer, P., Knackmuss, H.J., *Arch. Microbiol.* (1991) **156**, 218–222.

G329. Heitkamp, M.A., Freeman, J.P., Miller, D.W., Cerniglia, C.E., *Arch. Microbiol.* (1991) **156**, 223–230.

G330. Gu, J.D., Berry, D.F., *Appl. Environ. Microbiol.* (1991) **57**, 2622–2627.

G331. Madsen, T., Aamand, J., *Appl. Environ. Microbiol.* (1991) **57**, 2453–2458.

G332. Poon, H.C., Schmidt, B.M., *Clin. Biochem.* (1991) **24**, 149–152.

G333. Dörner, C., Schink, B., *Arch. Microbiol.* (1991) **156**, 302–306.

G334. Matsunaga, Y., Miyazaki, H., Ohe, Y., Nambu, K., Furukawa, H., Yoshida, K., Hashimoto, M., *Arzneim.-Forsch.* (1991) **41**, 747–759.

G335. Harvey, D.J., Brown, N.K., *Biol. Mass Spectrom.* (1991) **20**, 275–285.

G336. Dean, R.A., Christian, C.D., Sample, R.H.B., Bosron, W.F., *FASEB J.* (1991) **5**, 2735–2739.

G337. Rutgers, M., Heusdens, F.A., Visser, T.J., *Endocrinology (Baltimore)* (1991) **129**, 1375–1381.

G338. Fentem, J.H., Fry, J.R., Whiting, D.A., *Biochem. Biophys. Res. Commun.* (1991) **179**, 197–203.

G339. Sugawara, K., Tahara, S., Mizutani, J., *Agric. Biol. Chem.* (1991) **55**, 1799–1804.

G340. Morrison, V.M., Burnett, A.K., Forrester, L.M., Wolf, C.R., Craft, J.A., *Chem.-Biol. Interact.* (1991) **79**, 179–196.

G341. Bechtold, A., Berger, U., Heide, L., *Arch. Biochem. Biophys.* (1991) **288**, 39–47.

G342. Altenschmidt, U., Oswald, B., Fuchs, G., *J. Bacteriol.* (1991) **173**, 5494–5501.

G343. Yoshioka, H., Nagasawa, T., Yamada, H., *Eur. J. Biochem.* (1991) **199**, 17–24.

G344. Yabe, K., Ando, Y., Hamasaki, T., *Agric. Biol. Chem.* (1991) **55**, 1907–1911.

G345. Preisig, C.L., Van Etten, H.D., Moreau, R.A., *Arch. Biochem. Biophys.* (1991) **290**, 468–473.

G346. Miessner, M., Crescenzi, O., Napolitano, A., Prota, G., Andersen, S.O., Peter, M.G., *Helv. Chim. Acta* (1991) **74**, 1205–1212.

G348. Volland, C., Benet, L.Z., *Pharmacology* (1991) **43**, 53–60.

G349. Liu, H.E., Vincent-Viry, M., Galteau, M.M., Guenguen, R., Magdalou, J., Nicolas, A., Leroy, P., Siest, G., *Eur. J. Clin. Pharmacol.* (1991) **41**, 153–159.

G350. Koopman-Kimenai, P.M., Vree, T.B., Hasenbos, M.A.W.M., Weber, E.W.G., Verweij-Van Wissen, C.P.W.G.M., Booij, L.H.D.J., *Pharm. Weekbl., Sci. Ed.* (1991) **13**, 142–147.

G351. Lloyd, R.V., Duling, D.R., Rumyantseva, G.V., Mason, R.P., Bridson, P.K., *Mol. Pharmacol.* (1991) **40**, 440–445.

G352. Hinchman, C.A., Matsumoto, H., Simmons, T.W., Ballatori, N., *J. Biol. Chem.* (1991) **266**, 22179–22185.

G353. Kawashima, T., Satomi, O., Awata, N., *J. Pharmacobio-Dyn.* (1991) **14**, 449–459.

G354. Lim, H.K., Foltz, R.L., *Chem. Res. Toxicol.* (1991) **4**, 626–632.

G355. Nakanishi, Y., Murakami, S., Shinke, R., Aoki, K., *Agric. Biol. Chem.* (1991) **55**, 1281–1289.

G356. Schmidt, V., Wittich, R.-M., Fortnagel, P., *Arch. Microbiol.* (1991) **156**, 213–217.

G357. Tétaz, F., Barrelle, M., Béguin, C.G., Pelmont, J., *New J. Chem.* (1991) **15**, 587–591.

G358. Dolfing, J., Tiedge, J.M., *Arch. Microbiol.* (1991) **156**, 356–361.

G359. Pavel, S., Holden, J.L., Riley, P.A., *Pigm. Cell Res.* (1989) **2**, 421–426.

G360. Agarwal, R., Medrano, E.E., Khan, I.U., Nordlund, J.J., Mukhtar, H., *Carcinogenesis* (1991) **12**, 1963–1966.

G361. Bauer, S.L., Howard, P.C., *Carcinogenesis* (1991) **12**, 1545–1549.

G363. Faulds, C.B., Williamson, G., *J. Gen. Microbiol.* (1991) **137**, 2339–2345.

G364. Löffler, F., Müller, R., *FEBS Lett.* (1991) **290**, 224–226.

G365. Clement, B., Lustig, K.L., Ziegler, D.M., *Prog. Pharmacol. Clin. Pharmacol.* (1991) **8**, 127–136.

G366. Draths, K.M., Frost, J.W., *J. Am. Chem. Soc.* (1991) **113**, 9361–9363.

G367. Daniel, S.L., Keith, E.S., Yang, H., Lin, Y.S., Drake, H.L., *Biochem. Biophys. Res. Commun.* (1991) **180**, 416–422.

G369. Oren, A., Gurevich, P., Henis, Y., *Appl. Environ. Microbiol.* (1991) **57**, 3367–3370.

G370. Huang, J., Dunford, H.B., *Arch. Biochem. Biophys.* (1991) **287**, 257–262 and (1991) **289**, 405.

G371. Grossi, P., Pellegatti, M., Vigelli, G., Ayrton, J., Duncan, B.A., Evans, G.L., Maltas, J., *Xenobiotica* (1992) **22**, 899–916.

G372. Tanaka, Y., Fujiwara, T., Esumi, Y., *Xenobiotica* (1992) **22**, 941–947.

G373. Paulson, G.D., Feil, V.J., Giddings, J.M., Lamoureux, C.H., *Xenobiotica* (1992) **22**, 925–939.

G374. Nishimoto, M., Yanagida, G.K., Stein, J.E., Baird, W.M., Varanasi, U., *Xenobiotica* (1992) **22**, 949–961.

G375. Dring, L.G., Bye, A., Constable, D.A., Lang, J.C.D., *Xenobiotica* (1992) **22**, 963–971.

G376. Chou, R.C., Wyss, R., Huselton, C.A., Wiegand, U.-W., *Xenobiotica* (1992) **22**, 993–1002.

G377. Woollen, B.H., Marsh, J.R., Laird, W.D.J., Lesser, J.E., *Xenobiotica* (1992) **22**, 983–991.

G378. Chapman, D.E., Michener, S.R., Powis, G., *Xenobiotica* (1992) **22**, 1015–1028.

G379. Beyer, C., Spiteller, G., *Xenobiotica* (1992) **22**, 1029–1035.

G380. Levine, W.G., Stoddart, A., Zbaida, S., *Xenobiotica* (1992) **22**, 1111–1120.

G381. Kato, R., Yamazoe, Y., Yasumori, T., *Xenobiotica* (1992) **22**, 1083–1092.

G382. McNulty, M.J., Deal, D.L., Nelson, F.R., Weller, S., Chandrasurin, P., Shockcor, J., Findlay, J.W.A., *Drug Metab. Dispos.* (1992) **20**, 679–687.

G384. Newton, J.F., Straub, K.M., Dewey, R.H., Perchonock, C.D., McCarthy, M.E., Gleason, J.G., Lynn, R.K., *Drug Metab. Dispos.* (1992) **20**, 479–484.

G385. Turgeon, J., Paré, J.R.J., Lalande, M., Grech-Bélanger, O., Bélanger, P.M., *Drug Metab. Dispos.* (1992) **20**, 762–769.

G386. Gordon, W.P., Cheng, H., Larsen, D.L., Ragner, J.A., Landmesser, N.G., *Drug Metab. Dispos.* (1992) **20**, 596–602.

G387. Yang, J.T., Wong, K.K., Kucharczyk, N., Sofia, R.D., *Drug Metab. Dispos.* (1992) **20**, 536–546.

G388. Fischer, V., Baldeck, J.-P., Tse, F.L.S., *Drug Metab. Dispos.* (1992) **20**, 603–607.

G389. Kajbaf, M., Gorrod, J.W., *Prog. Pharmacol. Clin. Pharmacol.* (1991) **8**, 13–21.

G390. Akingbule, O.I., Damani, L.A., *Prog. Pharmacol. Clin. Pharmacol.* (1991) **8**, 59–64.

G392. Cashman, J.R., *Prog. Pharmacol. Clin. Pharmacol.* (1991) **8**, 117–126.

G393. Finocchiaro, L.M.E., Nahmod, V.E., Launay, J.M., *Biochem. J.* (1991) **280**, 727–731.

G395. Gaillard, J.L., *Comp. Biochem. Physiol. B: Comp. Biochem.* (1991) **100B**, 107–115.

G396. Oberhänsli, T., Défago, G., Haas, D., *J. Gen. Microbiol.* (1991) **187**, 2273–2279.

G398. Scherling, D., Karl, W., Ahr, H.J., Kern, A., Siefert, H.M., *Arzneim.-Forsch.* (1991) **41**, 1009–1021.

G400. Niebch, G., Locher, M., Peter, G., Borbe, H.O., *Arzneim.-Forsch.* (1991) **41**, 1027–1032.

G401. Bironaite, D., Cenas, N., Kulys, J., *Biochim. Biophys. Acta* (1991) **1060**, 203–209.

G402. Coutts, R.T., Baker, G.B., Malek, F., Hussain, M.S., *Res. Commun. Chem. Pathol. Pharmacol.* (1991) **74**, 15–24.

G403. Sanchez-Torres, J.M., Pedraz, J.L., Gascon, A.R., Begona, C.M., Hernandez, R.M., Dominguez-Gil, A., *Boll. Chim. Farm.* (1991) **130**, 172–177.

G404. Van Berkel, W.J.H., Van den Tweel, W.J.J., *Eur. J. Biochem.* (1991) **201**, 585–592.

G405. Locher, H.H., Malli, C., Hooper, S.W., Vorherr, T., Leisinger, T., Cook, A.M., *J. Gen. Microbiol.* (1991) **137**, 2201–2208.

G406. Sutherland, J.B., Selby, A.L., Freeman, J.P., Evans, F.E., Cerniglia, C.E.,

Appl. Environ. Microbiol. (1991) **57**, 3310–3316.

G407. Spanggord, R.J., Spain, J.C., Nishino, S.F., Mortelmans, K.E., *Appl. Environ. Microbiol.* (1991) **57**, 3200–3205.

G408. Bailey, A.D., Van Staden, J., *Plant Physiol. Biochem.* (1991) **29**, 403–408.

G409. Yeh, S.Y., *J. Pharm. Sci.* (1991) **80**, 815–819.

G410. On, N.T., Rance, D.J., Damani, L.A., *Prog. Pharmacol. Clin. Pharmacol.* (1991) **8**, 65–71.

G411. Pattichis, K., Kajbaf, M., Gorrod, J.W., *Prog. Pharmacol. Clin. Pharmacol.* (1991) **8**, 31–40.

G412. Stavrianakis, I.G., Soper, C.J., *Pharmazie* (1991) **46**, 340–341.

G414. Turesky, R.J., Markovic, J., Bracco-Hammer, I., Fay, L.B., *Carcinogenesis* (1991) **12**, 1847–1855.

G415. Caputo, O., Grosa, G., Ceruti, M., Rocco, F., Biglino, G., *Eur. J. Drug Metab. Pharmacokinet.* (1991) **16**, 113–118.

G416. White, H., Feicht, R., Huber, C., Lottspeich, F., Simon, H., *Biol. Chem. Hoppe-Seyler* (1991) **372**, 999–1005.

G417. Zaki, K., Nasser, H., Farid, T., *Egypt. J. Biochem.* (1991) **9**, 233–247.

G418. Hammel, K.E., Green, B., Gai, W.Z., *Proc. Natl. Acad. Sci. U.S.A.* (1991) **88**, 10605–10608.

G419. Borlakoglu, J.T., Haegele, K.D., Reich, H.J., Dils, R.R., Wilkins, J.P.G., *Int. J. Biochem.* (1991) **23**, 1427–1437.

G420. Peterson, M.S., *Phytochemistry* (1991) **30**, 2877–2881.

G422. Noma, Y., Ojamita, Y., Takahashi, Y., Asakawa, Y., *Phytochemistry* (1991) **30**, 2969–2972.

G423. Ishikawa, T., Okuno, E., Tsujimoto, M., Nakamura, M., Kido, R., *Adv. Exp. Med. Biol.* (1991) **294**, 567–572.

G424. Gallert, C., Knoll, G., Winter, J., *Appl. Microbiol. Biotechnol.* (1991) **36**, 124–129.

G425. Iwasa, K., Kamigauchi, M., Takao, N., *Phytochemistry* (1991) **30**, 2973–2975.

G426. Yamamoto, I., Gohda, H., Narimatsu, S., Watanabe, K., Yoshimura, H., *Pharmacol. Biochem. Behav.* (1991) **40**, 541–546.

G427. Holland, H.L., Rand, C.G., Viski, P., Brown, F.M., *Can. J. Chem.* (1991) **69**, 1989–1993.

G428. Stanley, J.S., York, J.L., Benson, A.M., *Cancer Res.* (1992) **52**, 58–63.

G429. Jacobs, M.H.J., Van den Wijngaard, A.J., Pentenga, M., Janssen, D.B., *Eur. J. Biochem.* (1991) **202**, 1217–1222.

G430. Nawaz, M.S., Heinze, T.M., Cerniglia, C.E., *Appl. Environ. Microbiol.* (1992) **58**, 27–31.

G431. Heising, S., Brune, A., Schink, B., *FEMS Microbiol. Lett.* (1991) **84**, 267–272.

G432. Angeles de le Torre, M., Gomez-Alarcon, G., *Microbios* (1991) **68**, 119–127.

G433. Enreille, A., Pognat, J.F., Galmier, M.J., Lartigue-Mattei, C., Chabard, J.L., Busch, N., Berger, J.A., *Eur. J. Drug Metab. Pharmacokinet.* (1991) **16**, 161–172.

G434. Fetzner, S., Müller, R., Lingens, F., *J. Bacteriol.* (1992) **174**, 279–290.

G435. Gandini, R., Montalto, C., Castoldi, D., Monzani, V., Nava, M.L., Scaricabarozza, I., Vargiu, G., Bartosek, I., *Farmaco* (1991) **46**, 1071–1079.

G436. Valli, K., Brock, B.J., Joshi, D.K., Gold, M.H., *Appl. Environ. Microbiol.* (1992) **58**, 221–228.

G437. Tripathi, C.R.M., Basu, S.K., Vora, V.C., Mason, J.R., Pirt, S.J., *Res. Ind.* (1991) **36**, 159–160.

G438. Faison, B.D., Clark, T.M., Lewis, S.N., Ma, C.Y., Sharkey, D.M., Woodward, C.A., *Appl. Biochem. Biotechnol.* (1991) **28–29**, 237–251.

G439. van Tellingen, O., Beijnen, J.H., Baurain, R., ten Bokkel Huinink, W.W., van der

Woude, H.R., Nooyen, W.J., *J. Chromatogr.* (1992) **574**, 293–298.

G440. Klein, J., Seidel, A., Heinz, F., Oesch, F., Platt, K.L., *Chem.-Biol. Interact.* (1991) **79**, 287–303.

G442. Lecoq, S., Chalvet, O., Strapelias, H., Grover, P.L., Phillips, D.H., Duquesne, M., *Chem.-Biol. Interact.* (1991) **80**, 261–279.

G443. Watkins, B.E., Suzuki, M., Wallin, H., Wakabayashi, K., Alexander, J., Vanderlaan, M., Sugimura, T., Esumi, H., *Carcinogenesis* (1991) **12**, 2291–2295.

G444. Ishikura, N., Yang, Z.Q., *Z. Naturforsch. C: Biosci.* (1991) **46**, 1003–1010.

G445. Wan, L., Xue, W., Schneider, J., Reilman, R., Radike, M., Warshawsky, D., *Chem.-Biol. Interact.* (1992) **81**, 131–147.

G446. Kutacek, M., Terziivanova-Dimova, S., *Biol. Plant.* (1991) **33**, 408–416.

G447. Trincone, A., Nicolaus, B., Lama, L., Gambacorta, A., J. Chem. Soc., *Perkin Trans. 1* (1991), 2841–2844.

G448. Higson, F.H., Focht, D.D., *Appl. Environ. Microbiol.* (1992) **58**, 194–200.

G449. Miranda, M., Bonfigli, A., Zarivi, O., Ragnelli, A.M., Pacioni, G., Botti, D., *Plant Sci.* (1992) **81**, 175–182.

G450. Oi, Y., Kawada, T., Watanabe, T., Iwai, K., *J. Agric. Food Chem.* (1992) **40**, 467–470.

G451. Fang, J., Gorrod, J.W., *Toxicol. Lett.* (1991) **59**, 117–123.

G452. Nasrin, S., Ichinose, H., Nagatsu, T., *Biochim. Biophys. Acta* (1992) **1118**, 318–322.

G453. Suzuki, T., Fujita, S., Narimatsu, S., Masubuchi, Y., Tachibana, M., Ohta, S., Hirobe, M., *FASEB J.* (1992) **6**, 771–776.

G454. Groenewegen, P.E.J., van den Tweel, W.J.J., de Bont, J.A.M., *Appl. Microbiol. Biotechnol.* (1992) **36**, 541–547.

G455. Vree, T.B., Vree, J.B., Beneken Kolmer, E.W.J., Hekster, Y.A., *Vet. Q.* (1991) **13**, 218–224.

G456. Pangilinan, N.C., Ivie, G.W., Clement, B.A., Beier, R.C., Uwayjan, M., *J. Chem. Ecol.* (1992) **18**, 253–270.

G457. Wiese, C., Cogoli-Greuter, M., Weinreich, R., Winterhalter, K.H., *J. Neurochem.* (1992) **58**, 219–226.

G458. Halpin, C., Knight, M.E., Grima-Pettenati, J., Goffner, D., Boudet, A., Schuch, W., *Plant Physiol.* (1992) **98**, 12–16.

G459. Noma, Y., Akehi, E., Miki, N., Asakawa, Y., *Phytochemistry* (1992) **31**, 515–517.

G460. Harvey, D.J., Samara, E., Mechoulam, R., *Pharmacol. Biochem. Behav.* (1991) **40**, 523–532.

G461. Munder, M., Vogt, J.R.A., Vogler, B., Renz, P., *Eur. J. Biochem.* (1992) **204**, 679–683.

G462. Cocchiara, G., Battaglia, R., Pevarello, P., Strolin Benedetti, M., *Eur. J. Drug Metab. Pharmacokinet.* (1991) **16**, 231–239.

G463. Hays, M.T., Cavalieri, R.R., *Metab. Clin. Exp.* (1992) **41**, 494–497.

G464. Wittich, R.M., Wilkes, H., Sinnwell, V., Francke, W., Fortnagel, P., *Appl. Environ. Microbiol.* (1992) **58**, 1005–1010.

G466. Ohnishi, K., Tanabe, H., Hayashi, S., Suemitsu, R., *Biosci. Biotechnol. Biochem.* (1992) **56**, 42–43.

G467. Yamada, Y., Kikuzaki, H., Nakatani, N., *Biosci. Biotechnol. Biochem.* (1992) **56**, 153–154.

G470. Iwasa, K., Kamigauchi, M., Takao, N., *J. Nat. Prod.* (1992) **55**, 491–495.

G471. Mohn, W.W., Kennedy, K.J., *Appl. Environ. Microbiol.* (1992) **58**, 1367–1370.

G472. Becher, B., Fischer, W., Taneri, Z., Scholz, E., Mueller, W.E., Breyer-Pfaff, U., *Psychopharmacology* (1992) **106**, 303–310.

G473. Hiltunen, M., Soederhaell, K., *Exp. Mycol.* (1992) **16**, 44–51.

G474. Narro, M.L., Cerniglia, C.E., Van Baalen, C., Gibson, D.T., *Appl. Environ. Microbiol.* (1992) **58**, 1351–1359.

G475. Koizumi, M., Akao, T., Kadota, S., Kikuchi, T., Okuda, T., Kobashi, K., *Chem. Pharm. Bull.* (1991) **39**, 2638–2643.

G476. Narro, M.L., Cerniglia, C.E., Van Baalen, C., Gibson, D.T., *Appl. Environ. Microbiol.* (1992) **58**, 1360–1363.

G477. Ohta, A., Shimada, M., *Mokuzai Gakkaishi* (1991) **37**, 748–752.

G478. Kuhm, A.E., Stolz, A., Knackmuss, H.J., *Biodegradation* (1991) **2**, 115–120.

G479. Fujii, I., Chen, Z.G., Ebizuka, Y., Sankawa, U., *Biochem. Int.* (1991) **25**, 1043–1049.

G480. Stolz, A., Nörtemann, B., Knackmuss, H.J., *Biochem. J.* (1992) **282**, 675–680.

G481. Koch, J., Fuchs, G., *Eur. J. Biochem.* (1992) **205**, 195–202.

G482. Stich, K., Eidenberger, T., Wirst, F., Forkmann, G., *Planta* (1992) **187**, 103–108.

G483. Kutacek, M., Terziivanova-Dimova, S., *Biol. Plant.* (1991) **33**, 395–407.

G484. O'Malley, D.M., Porter, S., Sederoff, R.R., *Plant Physiol.* (1992) **98**, 1364–1371.

G485. Onodera, R., Ueda, H., Nagasawa, T., Okuuchi, K., Chaen, S., Mieno, M., Kudo, H., *Anim. Sci. Technol.* (1992) **63**, 23–31.

G486. Suzuki, T., Narimatsu, S., Fujita, S., Masubuchi, Y., Umeda, S., *Biochem. Pharmacol.* (1991) **42**, 2241–2244.

G487. Lee, T.T., Starratt, A.N., *Physiol. Plant.* (1992) **84**, 209–216.

G488. Morgan, M., Maggs, J.L., Bulman-Page, P.C., Hussain, F., Park, B.K., *Biochem. Pharmacol.* (1992) **43**, 985–994.

G489. Michnovicz, J.J., Rosenberg, D.W., *Biochem. Pharmacol.* (1992) **43**, 1847–1852.

G490. Patel, D.K., Shockcor, J.P., Chang, S.Y., Sigel, C.W., Huber, B.E., *Biochem. Pharmacol.* (1991) **42**, 337–346.

G491. Myers, S.R., Flesher, J.W., *Biochem. Pharmacol.* (1991) **41**, 1683–1689.

G492. Coughtrie, M.W.H., Sharp, S., *Biochem. Pharmacol.* (1991) **42**, 1497–1501.

G493. Foster, B.C., Litster, D.L., Wilson, D.L., Ormsby, E., Dawson, D.A., *Xenobiotica* (1992) **22**, 1221–1228.

G494. Marchant, N.C., Breen, M.A., Wallace, D., Bass, S., Taylor, A.R., Ings, R.M.J., Campbell, D.B., Williams, J., *Xenobiotica* (1992) **22**, 1251–1266.

G495. Cotler, S., Chang, D., Henderson, L., Garland, W., Town, C., *Xenobiotica* (1992) **22**, 1229–1237.

G496. Simons, P.J., Cockshott, I.D., Glen, J.B., Gordon, E.A., Knott, S., Ruane, R.J., *Xenobiotica* (1992) **22**, 1267–1273.

G497. Ariyoshi, N., Koga, N., Oguri, K., Yoshimura, H., *Xenobiotica* (1992) **22**, 1275–1290.

G498. Sugiyanto, Sharping, C.E., McManus, M.E., Birkett, D.J., Holder, G.M., Ryan, A.J., *Xenobiotica* (1992) **22**, 1299–1307.

G499. Ishida, T., Matsumoto, T., *Xenobiotica* (1992) **22**, 1291–1298.

G500. Ohi, H., Takahara, E., Ohta, S., Hirobe, M., *Xenobiotica* (1992) **22**, 1329–1337.

G501. John, B.A., Chasseaud, L.F., Wood, S.G., Forlot, P., *Xenobiotica* (1992) **22**, 1339–1351.

G502. Sadeque, A.J.M., Eddy, A.C., Meier, G.P., Rettie, A.E., *Drug Metab. Dispos.* (1992) **20**, 832–839.

G503. Tajima, K., Hashizaki, M., Yamamoto, K., Mizutani, T., *Drug Metab. Dispos.* (1992) **20**, 816–820.

G504. Mutlib, A.E., Chui, Y.C., Young, L.M., Abbott, F.S., *Drug Metab. Dispos.* (1992) **20**, 840–848.

G505. Balani, S.K., Pitzenberger, S.M., Kauffman, L.R., Arison, B.H., Ramjit, H.G., Goldman, M.E., O'Brien, J.A., King, J.D., Hoffman, J.M., Rooney, C.S., Theoharides, A.D., *Drug Metab. Dispos.* (1992) **20**, 869–876.

G506. Hezari, M., Davis, P.J., *Drug Metab. Dispos.* (1992) **20**, 882–888.

G507. Benoit, E., Cresteil, T., Riviere, J.-L., Delatour, P., *Drug Metab. Dispos.* (1992) **20**, 877–881.

G508. Beedham, C., Al-Tayib, Y., Smith, J.A., *Drug Metab. Dispos.* (1992) **20**, 889–895.

G509. Styczynski, P.B., Green, M.D., Coffman, B., Tephly, T.R., *Drug Metab. Dispos.* (1992) **20**, 896–901.

G510. Deal, D.L., Chandrasurin, P., Shockcor, J., Minick, D.J., Findlay, J.W.A., McNulty, M.J., *Drug Metab. Dispos.* (1992) **20**, 920–927.

G511. Zbaida, S., Brewer, C.F., Levine, W.G., *Drug Metab. Dispos.* (1992) **20**, 902–908.

G512. McNulty, M.J., Deal, D.L., Page, T.L., Chandrasurin, P., Findlay, J.W.A., *Drug Metab. Dispos.* (1992) **20**, 928–935.

G513. Kiss, A.D., Haggard, K., Veltman, J.C., Ziegler, D.M., *Drug Metab. Dispos.* (1992) **20**, 948–953.

G514. Ehlhardt, W.J., Zimmermann, J.L., Ray, J.E., Toth, J.E., *Drug Metab. Dispos.* (1992) **20**, 958–961.

G515. Town, C., Chang, D., Henderson, L., Garland, W.A., *Drug Metab. Dispos.* (1992) **20**, 954–957.

G516. Foster, B.C., McLeish, J., Wilson, D.L., Whitehouse, L.W., Zamecnik, J., Lodge, B.A., *Xenobiotica* (1992) **22**, 1383–1394.

G517. Henrikus, B.M., Kampffmeyer, H.G., *Xenobiotica* (1992) **22**, 1357–1366.

G518. Boeren, S., Tyrakowska, B., Vervoort, J., De Hoffman, E., Teunis, K., Van Veldhuizen, A., Rietjens, I.M.C.M., *Xenobiotica* (1992) **22**, 1403–1423.

G519. Yoshida, T., Tabuchi, T., Andoh, K., *Xenobiotica* (1992) **22**, 1459–1470.

G520. Kutacek, M., Terzuvanova-Dimova, S., *Biol. Plant.* (1991) **33**, 408–416.

G521. Déchelotte, P., Sabouraud, A., Sandouk, P., Hackbarth, I., Schwenk, M., *Drug Metab. Dispos.* (1993) **21**, 13–17.

G522. Patel, D.K., Lewis, R.P., Hinton, M.L, Schroeder, D.H., Shockcor, J.P., Johnson, R.L., Sigel, C.W., *Drug Metab. Dispos.* (1993) **21**, 62–70.

G523. Walle, T., Walle, U.K., Thornburg, K.R., Schey, K.L., *Drug Metab. Dispos.* (1993) **21**, 76–80.

G524. Ebner, T., Burchell, B., *Drug Metab. Dispos.* (1993) **21**, 50–55.

G525. Clement, B., Lustig, K.L., Ziegler, D.M., *Drug Metab. Dispos.* (1993) **21**, 24–29.

G526. Sylvia, L.A., Gerig, J.T., *Drug Metab. Dispos.* (1993) **21**, 105–113.

G527. Wikberg, T., Vuorela, A., Ottoila, P., Taskinen, J., *Drug Metab. Dispos.* (1993) **21**, 81–92.

G528. Tracy, T.S., Wirthwein, D.P., Hall, S.D., *Drug Metab. Dispos.* (1993) **21**, 114–120.

G529. Cho, A.K., Hiramatsu, M., Schmitz, D.A., Landaw, E.M., Chang, A.S., Ramamurthy, S., Jenden, D.J., *Drug Metab. Dispos.* (1993) **21**, 125–132.

G530. Thomassin, J., Battaglia, R., Allievi, C., Castelli, M.G., Strolin Benedetti, M., *Drug Metab. Dispos.* (1993) **21**, 151–155.

G531. Taylor, W.G., Hall, T.W., Vedres, D.D., *Drug Metab. Dispos.* (1993) **21**, 133–140.

G532. Edwall, B., Arvidsson, A., Lake-Bakaar, D., Lanbeck-Vallén, K., Yisak, W., *Drug Metab. Dispos.* (1993) **21**, 171–177.

G533. Alton, K.B., Chan, T.-M., Pramanik, B., *Drug Metab. Dispos.* (1993) **21**, 198–199.

G534. Nerita, M., Nagai, E., Hagiwara, H., Aburada, M., Yokoi, T., Kamataki, T., *Xenobiotica* (1993) **23**, 5–10.

G535. Gerlach, M., Schwelle, N., Lerbs, W., Luckner, M., *Phytochemistry* (1985) **24**, 1935–1939.

G536. Mori, M.-A., Kobayashi, M., Uemura, H., Mori, Y., Miyahara, T., Kozuka, H., *Xenobiotica* (1993) **23**, 11–18.

G537. de Jongh, J., Buser, H.R., Poiger, H., *Xenobiotica* (1993) **23**, 19–26.

G538. Lin, G., Hawes, E.M., McKay, G., Korchinski, E.D., Midha, K.K., *Xenobiotica* (1993) **23**, 37–52.

G539. Lin, G., Hawes, E.M., McKay, G., Cooper, S.F., Korchinski, E.D., Midha, K.K., *Xenobiotica* (1993) **23**, 27–36.

G540. Fujimaki, Y., Sudo, K., Hakusui, H., *Xenobiotica* (1993) **23**, 61–70.

G541. Thomas, B.H., Whitehouse, L.W., Solomonraj, G., Paul, C.J., *Xenobiotica* (1993) **23**, 99–105.

G542. Sayama, M., Mori, M.-A., Maruyama, Y., Inoue, M., Kozuka, H., *Xenobiotica* (1993) **23**, 123–131.

G543. Strolin Benedetti, M., Fraier, D., Pianezzola, E., Castelli, M.G., Dostert, P., Gianni, L., *Xenobiotica* (1993) **23**, 115–121.

G544. Da Silva Morais, M., Augusto, O., *Xenobiotica* (1993) **23**, 133–139.

G545. Gregus, Z., Fekete, I., Varga, F., Klaassen, C.D., *Xenobiotica* (1993) **23**, 141–153.

G546. Palette, C., Dubor, F., Rovei, V., Advenier, C., *Xenobiotica* (1993) **23**, 181–191.

G547. Clement, B., Kunze, T., *Xenobiotica* (1993) **23**, 155–167.

G548. Lenk, W., Rosenbauer-Thilman, R., *Xenobiotica* (1993) **23**, 241–257.

G549. Canivenc-Lavier, M.-C., Brunold, C., Siess, M.-H., Suschetet, M., *Xenobiotica* (1993) **23**, 259–266.

G550. Lanusse, C.E., Nare, B., Poichard, R.K., *Xenobiotica* (1993) **23**, 285–295.

G551. Spoonamore, J.E., Frohlich, D.R., Wells, M.A., *Xenobiotica* (1993) **23**, 279–284.

G552. Davison, K.L., Larsen, G.L., *Xenobiotica* (1993) **23**, 297–305.

G553. Krause, W., Düsterberg, B., Jacobs, U., Hoyer, G.-A., *Drug Metab. Dispos.* (1993) **21**, 203–208.

G554. El Dareer, S.M., Tillery, K.F., Rose, L.M., Posey, C.F., Struck, R.F., Stiller, S.W., Hill, D.L., *Drug Metab. Dispos.* (1993) **21**, 231–235.

G555. Boersma, M.G., Cnubben, N.H.P., Van Berkel, W.J.H., Blom, M., Vervoort, J., Rietjens, I.M.C.M., *Drug Metab. Dispos.* (1993) **21**, 218–230.

G556. Franz, P.M., Mattiuz, E.L., Hatcher, B.L., Desante, K.A., Breau, A.P., Occolowictz, J.L., Dorman, D.E., Schmid, C.R., Goldberg, M.J., Rubin, A., *Drug Metab. Dispos.* (1993) **21**, 249–254.

G557. Hezari, M., Davis, P.J., *Drug Metab. Dispos.* (1993) **21**, 259–267.

G558. Kim, M., Shen, D.D., Eddy, A.C., Nelson, W.L., *Drug Metab. Dispos.* (1993) **21**, 309–317.

G559. Yeleswaram, K., Rurak, D.W., Kwam, E., Hall, C., Doroudian, A., Abbott, F.S., Axelson, J.E., *Drug Metab. Dispos.* (1993) **21**, 284–292.

G560. Wikberg, T., Taskinen, J., *Drug Metab. Dispos.* (1993) **21**, 325–333.

G561. Sandberg, J.A., Murphey, L.J., Olsen, G.D., *Drug Metab. Dispos.* (1993) **21**, 390–395.

G562. Smith, P.C., Liu, J.H., *Xenobiotica* (1993) **23**, 337–348.

G564. Howell, S.R., Husbands, G.E.M., Scatina, J.A., Sisenwine, S.F., *Xenobiotica* (1993) **23**, 349–359.

G565. Schlemper, B., Siegers, D.J., Paxton, J.W., Robertson, I.G.C., *Xenobiotica* (1993) **23**, 361–371.

G566. Witkamp, R.F., Nijmeijer, S.M., Mennes, W.C., Rozema, A.W., Noordhoek, J., Van Miert, A.S.J.P.A.M., *Xenobiotica* (1993) **23**, 401–410.

G567. Kenyon, E.M., Calabrese, E.J., *Xenobiotica* (1993) **23**, 373–381.

G568. Yoshino, H., Matsunaga, H., Kaneko, H., Yoshitake, A., Nakatsuka, I., Yamada, H., *Xenobiotica* (1993) **23**, 609–619.

G569. Gorrod, J.W., Fang, J., *Xenobiotica* (1993) **23**, 495–508.

G570. Mori, M.-A., Uemura, H., Kobayashi, M., Miyahara, T., Kozuka, H., *Xenobiotica* (1993) **23**, 709–716.

G571. Jodynis-Liebert, J., *Xenobiotica* (1993) **23**, 693–701.

G572. Borel, A.G., Abbott, F.S., *Drug Metab. Dispos.* (1993) **21**, 415–427.

G573. Yun, C.-H., Okerholm, R.A., Guengerich, F.P., *Drug Metab. Dispos.* (1993) **21**, 403–409.

G574. Lennernäs, H., Renberg, L., Hoffmann, K.-J., Regårdh, C.-G., *Drug Metab. Dispos.* (1993) **21**, 435–440.

G575. Witkamp, R.F., Nijmeijer, S.M., Yun, H., Noordhoek, J., Van Miert, A.S.J.P.A.M., *Drug Metab. Dispos.* (1993) **21**, 441–446.

G576. Vickers, A.E.M., Connors, S., Zollinger, M., Biggi, W.A., Larrauri, A., Vogelaar, J.P.W., Brendel, K., *Drug Metab. Dispos.* (1993) **21**, 454–459.

G577. Vickers, S., Duncan, C.A., Kari, P.H., Homnick, C.F., Elliott, J.M., Pitzenberger, S.M., Hichens, M., Vyas, K.P., *Drug Metab. Dispos.* (1993) **21**, 467–473.

G578. Shukla, U.A., Marathe, P.H., Pittman, K.A., Barbhaiya, R.H., *Drug Metab. Dispos.* (1993) **21**, 502–507.

G579. Cashman, J.R., Yang, Z., Yang, L., Wrighton, S.A., *Drug Metab. Dispos.* (1993) **21**, 492–501.

G580. Tyrakowska, B., Boeren, S., Geurtsen, B., Rietjens, I.M.C.M., *Drug Metab. Dispos.* (1993) **21**, 508–519.

G581. Uchida, T., Usui, T., Teramura, T., Watanabe, T., Higuchi, S., *Drug Metab. Dispos.* (1993) **21**, 524–529.

G582. Roberts, S.M., Harbison, R.D., James, R.C., *Drug Metab. Dispos.* (1993) **21**, 537–541.

G583. Robertson, I.G.C., Palmer, B.D., Paxton, J.W., Bland, T.J., *Drug Metab. Dispos.* (1993) **21**, 530–536.

G584. Balani, S.K., Goldman, M.E., Kauffman, L.R., Varga, S.L., O'Brien, J.A., Smith, S.J., Olah, T.V., Ramjit, H.G., Schorn, T.W., Pitzenberger, S.M., Williams, T.M., Rooney, C.S., Theoharides, A.D., *Drug Metab. Dispos.* (1993) **21**, 598–604.

G585. Dain, J.G., Fu, E., Gorski, J., Nicoletti, J., Scallen, T.J., *Drug Metab. Dispos.* (1993) **21**, 567–572.

G586. Smith, D.J., Feil, V.J., Huwe, J.K., Paulson, G.D., *Drug Metab. Dispos.* (1993) **21**, 624–633.

G587. Mani, C., Gelboin, H.V., Park, S.S., Pearce, R., Parkinson, A., Kupfer, D., *Drug Metab. Dispos.* (1993) **21**, 645–657.

G588. Stearns, R.A., Miller, R.R., Arison, B.H., Rosegay, A., Smith, J.L., Mantlo, N.B., Chiu, S.H.L., *Drug Metab. Dispos.* (1993) **21**, 670–676.

G589. Mani, C., Hodgson, E., Kupfer, D., *Drug Metab. Dispos.* (1993) **21**, 657–661.

G590. Krause, W., Kühne, G., Jakobs, U., Hoyer, G.-A., *Drug Metab. Dispos.* (1993) **21**, 682–689.

G591. Adusumalli, V.E., Choi, Y.M., Romanyshyn, L.A., Sparadoski, R.E., Jr., Wickmann, J.K., Wong, K.K., Kucharczyk, N., Sofia, R.D., *Drug Metab. Dispos.* (1993) **21**, 710–716.

G592. Boyle, G.W., McKillop, D., Phillips, P.J., Harding, J.R., Pickford, R., McCormick, A.D., *Xenobiotica* (1993) **23**, 781–798.

G593. Ogawa, N., Mizuno, K., Fukushima, K., Suwa, T., Satoh, T., *Xenobiotica* (1993) **23**, 747–759.

G594. Shimada, K., Kimura, E., Yasui, Y., Tanaka, H., Matsushita, S., Hagihara, H., Nagakura, M., Kawahisa, M., *Appl. Environ. Microbiol.* (1992) **58**, 1577–1582.

G595. Nakagawa, Y., Tayama, S., Moore, G.A., Moldeus, P., *Biochem. Pharmacol.* (1992) **43**, 1431–1437.

G596. Apanovitch, D., Kitareewan, S., Walz, F.G., Jr., *Biochem. Biophys. Res. Commun.* (1992) **184**, 338–346.

G597. Maggs, J.L., Hussain, F., Page, P.C.B., Park, B.K., *J. Steroid Biochem. Mol. Biol.* (1992) **42**, 77–85.

G598. Leelayoova, S., Marbury, D., Rainey, P.M., MacKenzie, N.E., Hall, J.E., *J. Protozool.* (1992) **39**, 350–358.

G599. Czekala, N.M., Roocroft, A., Bates, M., Allen, J., Lasley, B.L., *Zoo Biol.* (1992) **11**, 75–80.

G600. Anhalt, S., Weissenböck, G., *Planta* (1992) **187**, 83–88.

G601. Bevan, D.R., Sadler, V.M., *Carcinogenesis* (1992) **13**, 403–407.
G602. Konishi-Imamura, L., Kim, D.H., Kobashi, K., *Chem. Pharm. Bull.* (1991) **39**, 2994–2998.
G603. Suzuki, S., Hirahara, T., Shim, J.K., Horinouchi, S., Beppu, T., *Biosci. Biotechnol. Biochem.* (1992) **56**, 84–89.
G604. Lake, B.G., Gaudin, H., Price, R.J., Walters, D.G., *Food Chem. Toxicol.* (1992) **30**, 105–115.
G605. Lake, B.G., Osborne, D.J., Walters, D.G., Price, R.J., *Food Chem. Toxicol.* (1992) **30**, 99–104.
G606. Cribb, A.E., Spielberg, S.P., *Clin. Pharmacol. Ther.* (1992) **51**, 522–526.
G607. Wilcocks, R., Ward, O.P., *Biotechnol. Bioeng.* (1992) **39**, 1058–1063.
G608. Samara, E., Bialer, M., Harvey, D.J., *Eur. J. Drug. Metab. Pharmacokinet.* (1991) **16**, 305–313.
G609. Vokounova, M., Vacek, O., Kunc, F., *Folia Microbiol.* (1992) **37**, 122–127.
G610. Maxwell, C.A., Edwards, R., Dixon, R.A., *Arch. Biochem. Biophys.* (1992) **293**, 158–166.
G611. Nair, R.V., Nettikumara, A.N., Cortez, C., Harvey, R.G., Di Giovanni, J., *Chem. Res. Toxicol.* (1992) **5**, 532–540.
G612. Ishiko, H., Shigeoka, S., Nakano, Y., Mitsunaga, T., *Phytochemistry* (1992) **31**, 1499–1500.
G613. Goetghebeur, M., Nicolas, M., Brun, S., Galzy, P., *Biosci. Biotechnol. Biochem.* (1992) **56**, 298–303.
G614. Santini, F., Chopra, I.J., Wu, S.Y., Solomon, D.H., Chua Teco, G.N., Guadalupe, N., *Pediat. Res.* (1992) **31**, 541–544.
G615. Psenak, M., *Biologia* (1991) **46**, 1109–1112.
G616. Valoti, M., Tipton, K.F., *Biochem. Pharmacol.* (1992) **43**, 945–951.
G617. Silveira, P.F., Schiripa, L.N., Picarelli, Z.P., *Comp. Biochem. Physiol. B: Comp. Biochem.* (1992) **102B**, 119–122.
G620. Sugumaran, M., Giglio, L., Kundzicz, H., Saul, S., Semensi, V., *Arch. Insect Biochem. Physiol.* (1992) **19**, 271–283.
G621. Brune, A., Schnell, S., Schink, B., *Appl. Environ. Microbiol.* (1992) **58**, 1861–1868.
G622. Hammel, K.E., Gai, W.Z., Green, B., Moen, M.A., *Appl. Environ. Microbiol.* (1992) **58**, 1832–1838.
G623. Lobos, J.H., Leib, T.K., Su, T.M., *Appl. Environ. Microbiol.* (1992) **58**, 1823–1831.
G624. Bugos, R.C., Chiang, V.L.C., Campbell, W.H., *Phytochemistry* (1992) **31**, 1495–1498.
G625. Haaf, H., Metzler, M., Li, J.J., *J. Steroid Biochem. Mol. Biol.* (1992) **42**, 389–397.
G626. Zheng, J., Hanzlik, R.P., *Chem. Res. Toxicol.* (1992) **5**, 561–567.
G627. Bestetti, G., Galli, E., Leoni, B., Pelizzoni, F., Sello, G., *Appl. Microbiol. Biotechnol.* (1992) **37**, 260–263.
G628. Savidge, R.A., Randeniya, P.V., *Biochem. Soc. Trans.* (1992) **20**, 229S.
G629. Augustijns, P., Verbeke, N., *Arzneim.-Forsch.* (1992) **42**, 825–828.
G630. Flowers-Geary, L., Harvey, R.G., Penning, T.M., *Chem. Res. Toxicol.* (1992) **5**, 576–583.
G631. Li, N.Y., Gorrod, J.W., *J. Biopharm. Sci.* (1991) **2**, 231–241.
G632. Gallert, C., Winter, J., *Appl. Microbiol. Biotechnol.* (1992) **37**, 119–124.
G633. Chen, Z.-G., Fujii, I., Ebizuka, Y., Sankawa, U., *Arch. Microbiol.* (1992) **158**, 29–34.
G634. Salonen, J.S., Vuorilehto, L., Gilbert, M., Maylin, G.A., *Eur. J. Drug Metab. Pharmacokinet.* (1992) **17**, 13–20.
G635. Wiese, C., Cogoli-Greuter, M., Argentini, M., Mäder, T., Weinreich, R., Winterhalter, K.H., *Biochem. Pharmacol.* (1992) **44**, 99–105.
G636. Noorström, A., Löf, A., Aringer, L., Samuelson, R., Andersson, B., Levin, J.O., Näslund, P., *Chemosphere* (1992) **24**, 1553–1561.

G637. Vasudevan, N., Mahadevan, A., *Res. Microbiol.* (1992) **143**, 333–339.
G639. Sambourin, P.J., Muggenburg, B.A., Couch, R.C., Lefler, D., Lucier, G., Birnbaum, L.S., Henderson, R.F., *Toxicol. Appl. Pharmacol.* (1992) **114**, 277–284.
G640. Wilcocks, R., Ward, O.P., Collins, S., Dewdney, N.J., Hong, Y., Prosen, E., *Appl. Environ. Microbiol.* (1992) **58**, 1699–1704.
G641. Xun, L., Topp, E., Orser, C.S., *J. Bacteriol.* (1992) **174**, 2898–2902.
G642. Lochmeyer, C., Koch, J., Fuchs, G., *J. Bacteriol.* (1992) **174**, 3621–3628.
G643. Gobbetti, M., Rossi, J., *J. Dairy Sci.* (1992) **75**, 924–934.
G644. Grund, E., Denecke, B., Eichenlaub, R., *Appl. Environ. Microbiol.* (1992) **58**, 1874–1877.
G645. Griffiths, D.A., Brown, D.E., Jezequel, S.G., *Appl. Microbiol. Biotechnol.* (1992) **37**, 169–175.
G646. Miller, K.W., *Appl. Environ. Microbiol.* (1992) **58**, 2176–2179.
G647. Hinteregger, C., Leitner, R., Loidl, M., Ferschl, A., Streichsbier, F., *Appl. Microbiol. Biotechnol.* (1992) **37**, 252–259.
G648. Sekine, Y., Oto, A., Otake, A., Suzuki, Y., *Sagame Joshi Daigaku Kiyo, Shizenkai* (1991) **553**, 51–57.
G649. Klempier, N., De Raadt, A., Griengl, H., Heinisch, G., *J. Heterocycl. Chem.* (1992) **29**, 93–95.
G650. Castanares, A., Wood, T.M., *Biochem. Soc. Trans.* (1992) **20**, 275S.
G651. Woolf, T.F., Black, A., Sedman, A., Chang, T., *Eur. J. Drug Metab. Pharmacokinet.* (1992) **17**, 21–27.
G652. Altenschmidt, U., Fuchs, G., *J. Bacteriol.* (1992) **174**, 4860–4862.
G653. Robinson, G.K., Stephens, G.M., Dalton, H., *Biocatalysis* (1992) **6**, 81–100.
G654. Rhyu, M.R., Nishimura, T., Kato, Y., Okitani, A., Kato, H., *Eur. J. Biochem.* (1992) **208**, 53–59.
G655. Klein, J., Thomas, H., Post, K., Wörner, W., Oesch, F., *Eur. J. Biochem.* (1992) **205**, 1155–1162.
G656. Agematsu, H., Shibamoto, N., Nishida, H., Okamoto, R., Shin, T., Murao, S., *Biosci. Biotechnol. Biochem.* (1992) **56**, 1176–1177.
G657. Higson, F.K., Focht, D.D., *Appl. Environ. Microbiol.* (1992) **58**, 2501–2504.
G658. Nic A'Bhaird, N., Tipton, K.F., *J. Neural Transm. Suppl.* (1990) **32**, 359–368.
G659. Kunapuli, S.P., Vaidyanathan, C.S., *J. Indian Inst. Sci.* (1991) **71**, 503–513.
G660. Goffner, D., Joffroy, I., Grima-Pettenati, J., Halpin, C., Knight, M.E., Schuch, W., Brudet, A.M., *Planta* (1992) **188**, 48–53.
G661. Robertson, J.B., Spain, J.C., Haddock, J.D., Gibson, D.T., *Appl. Environ. Microbiol.* (1992) **58**, 2643–2648.
G662. De Jong, E., Van Berkel, W.J.H., Van Der Zwan, R.P., De Bont, J.A.M., *Eur. J. Biochem.* (1992) **208**, 651–657.
G663. To-Figueras, J., Gomez-Catalan, J., Rodamilans, M., Corbella, J., *Hum. Exp. Toxicol* (1992) **11**, 271–273.
G664. Olson, P.E., Qi, B., Que, L., Jr., Wackett, L.P., *Appl. Environ. Microbiol.* (1992) **58**, 2820–2826.
G665. Oda, S., Kikuchi, Y., Nanishi, Y., *Biosci. Biotechnol. Biochem.* (1992) **56**, 1216–1220.
G666. Bergman, B., Dohrmann, J.K., Kahl, R., *Toxicology* (1992) **74**, 127–133.
G667. Madsen, T., Licht, D., *Appl. Environ. Microbiol.* (1992) **58**, 2874–2878.
G668. Guillen, F., Martinez, A.T., Martinez, M.J., *Eur. J. Biochem.* (1992) **209**, 603–611.
G669. Grifoli, M., Casellas, M., Bayona, J.M., Solanas, A.M., *Appl. Environ. Microbiol.* (1992) **58**, 2910–2917.

G670. Klos, C., Koob, M., Kramer, C., Dekant, W., *Toxicol. Appl. Pharmacol.* (1992) **115**, 98–106.

G671. Tedford, C.E., Ruperto, V.B., Coffin, M., Libonati, M., Barnell, A., *Drug Dev. Res.* (1992) **26**, 389–403.

G672. Doy, B.W., Sahali, Y., Hutchins, D.A., Wildschütte, M., Pastorelli, R., Nguyen, T.T., Naylor, S., Skipper, P.L., Wishnok, J.S., Tannenbaum, S.R., *Chem. Res. Toxicol.* (1992) **5**, 779–786.

G673. Snyderwine, E.G., Welti, D.H., Fay, L.B., Würzner, H.P., Turesky, R.J., *Chem. Res. Toxicol.* (1992) **5**, 843–851.

G674. Katayama, A., Ukai, T., Nomura, K., Kuwatsuka, S., *Biosci. Biotechnol. Biochem.* (1992) **56**, 1520–1521.

G675. Gilissen, R.A.H.J., Meerman, J.H.N., *Life Sci.* (1992) **51**, 1255–1260.

G676. Uotila, J.S., Kitunen, V.H., Saastamoinen, T., Coote, T., Häggblom, M.M., Salkinoja-Salonen, M.S., *J. Bacteriol.* (1992) **174**, 5669–5675.

G678. Swanson, P.E., *Appl. Environ. Microbiol.* (1992) **58**, 3404–3406.

G679. Maloney, S.E., Maule, A., Smith, R.W., *Arch. Microbiol.* (1992) **158**, 282–286.

G680. Liu, D., Maguire, R.J., Pacepavicius, G.J., Nagy, E., *Environ. Toxicol. Water Qual.* (1992) **7**, 355–372.

G681. Hayashi, T., Kanetoshi, A., Nakamura, M., Tamura, M., Shirahama, H., *Biochem. Pharmacol.* (1992) **44**, 489–493.

G682. Berger, B.J., Naiman, N.A., Hall, J.E., Peggins, J., Brewer, T.G., Tidwell, R.R., *Antimicrob. Agents Chemother.* (1992) **36**, 1825–1831.

G683. Constantino, L., Rosa, E., Iley, J., *Biochem. Pharmacol.* (1992) **44**, 651–658.

G684. Khanna, P., Kaur, S., Sanwal, G.G., Ali, B., *J. Pharmacol. Exp. Ther.* (1992) **262**,12 25–1231.

G685. Santini, F., Hurd, R.E., Chopra, I.J., *Endocrinology* (1992) **131**, 1689–1694.

G687. Wu, S.Y., Polk, D., Wong, S., Reviczky, A., Vu, R., Fisher, D., *Endocrinology* (1992) **131**, 1751–1756.

G688. Bakke, J.E., Mulford, D.J., Huwe, J.K., Feil, V.J., Davison, K.L., *Xenobiotica* (1993) **23**, 949–960.

G689. Fitzpatrick, P.J., Sheehan, D., *Xenobiotica* (1993) **23**, 851–861.

G690. Nakasa, H., Komiya, M., Ohmori S., Rikihisa, T., Kitada, M., *Drug Metab. Dispos.* (1993) **21**, 777–781.

G691. Dixon, C.M., Saynor, D.A., Andrew, P.D., Oxford, J., Bradbury, A., Tarbit, M.H., *Drug Metab. Dispos.* (1993) **21**, 761–769.

G692. Okereke, C.S., Kadry, A.M., Abdel-Rahman, M.S., Davis, R.A., Friedman, M.A., *Drug Metab. Dispos.* (1993) **21**, 788–791.

G693. Huskey, S.-E.W., Miller, R.R., Chiu, S.-H.L., *Drug Metab. Dispos.* (1993) **21**, 792–799.

G694. Uetrecht, J.P., Shear, N.H., Zahid, N., *Drug Metab. Dispos.* (1993) **21**, 830–834.

G695. Hussein, Z., Mulford, D.J., Bopp, B.A., Granneman, G.R., *Drug Metab. Dispos.* (1993) **21**, 805–810.

G696. Nicoll-Griffith, D.A., Chauret, N., Yergey, J.A., Trimble, L.A., Favreau, L., Zamboni, R., Grossman, S.J., Drey, J., Herold, E., *Drug Metab. Dispos.* (1993) **21**, 861–867.

G697. Woolf, T.F., Pool, W.F., Bjorge, S.M., Chang, T., Goel, O.P., Purchase, C.F. III, Schroeder, M.C., Kunze, K.L., Trager, W.F., *Drug Metab. Dispos.* (1993) **21**, 874–882.

G698. Blanz, J., Renner, U., Schmeer, K., Ehninger, G., Zeller, K.-P., *Drug Metab. Dispos.* (1993) **21**, 955–961.

G699. Jayyosi, Z., Villoutreix, J., Ziegler, J.M., Batt, A.M., De Maack, F., Siest, G., Thomas, P.E., *Drug Metab. Dispos.* (1993) **21**, 939–945.

G700. Ferrarese, A., De Palma, F., Celeste, S., Marini, P.P., Caccia, S., *Xenobiotica* (1993) **23**, 1007–1016.

G701. Yang, W., Jiang, T., Acosta, D., Davis, P.J., *Xenobiotica* (1993) **23**, 973–982.

G702. Smith, D.J., Skiles, G.L., Appleton, M.L., Carlson, J.R., Yost, G.S., *Xenobiotica* (1993) **23**, 1025–1044.

G703. De Wolf, W., Seinen, W., Hermens, J.L.M., *Xenobiotica* (1993) **23**, 1045–1056.

G704. Yoshino, H., Kaneko, H., Nakatsuka, I., Yamada, H., *Xenobiotica* (1993) **23**, 1075–1084.

G705. Lin, G., Hawes, E.M., McKay, G., Korchinski, E.D., Midha, K.K., *Xenobiotica* (1993) **23**, 1059–1074.

G706. Griffiths, D.A., Brown, D.E., Jezequel, S.G., *Xenobiotica* (1993) **23**, 1085–1100.

G707. Vallès, B., Coassolo, P., De Sousa, G., Aubert, C., Rahman, R., *Xenobiotica* (1993) **23**, 1101–1111.

G708. McKillop, D., Boyl, G.W., Cockshott, I.D., Jones, D.C., Plillips, P.J., Yates, R.A., *Xenobiotica* (1993) **23**, 1241–1253.

G709. Anfossi, P., Roncada, P., Stracciari, G.L., Montana, M., Pasqualucci, C., Montesissa, C., *Xenobiotica* (1993) **23**, 1113–1123.

G710. Krause, W., Kühne, G., *Xenobiotica* (1993) **23**, 1277–1288.

G711. Su, P., Coutts, R.T., Baker, G.B., Daneshtalab, M., *Xenobiotica* (1993) **23**, 1289–1298.

G712. Nakamura, J., Mizuma, T., Hayashi, M., Awazu, S., *Chem. Pharm. Bull.* (1992) **40**, 1964–1965.

G713. Levêque, D., Jehl, F., Quoix, E., Monteil, H., *Xenobiotica* (1993) **23**, 1325–1333.

G714. Bellino, F.L., *Steroids* (1992) **57**, 507–510.

G715. Bock, K.W., Gschaidmeier, H., Seidel, A., Baird, S., Burchell, B., *Mol. Pharmacol.* (1992) **42**, 613–618.

G716. Miyamoto, K., Ohta, H., *Eur. J. Biochem.* (1992) **210**, 475–481.

G717. Laignelet, L., Riviere, J.L., Lhuguenot, J.C., *Food Chem. Toxicol.* (1992) **30**, 575–583.

G718. Goetz, F.E., Harmuth, L.J., *FEMS Microbiol. Lett.* (1992) **97**, 45–49.

G719. Boopathy, R., Kulpa, C.F., *Curr. Microbiol.* (1992) **25**, 235–241.

G720. Kariya, S., Isozaki, S., Narimatsu, S., Suzuki, T., *Biochem. Pharmacol.* (1992) **44**, 1471–1474.

G721. Westerhausen, D., Herrmann, S., Hummel, W., Steckham, E., *Angew. Chem.* (1992) **104**, 1496–1498.

G722. Bennett, M.J., Sherwood, W.G., Bhala, A., Hale, D.E., *Clin. Chim. Acta* (1992) **210**, 55–62.

G723. Arakawa, H., Clark, W.G., Psenak, M., Coscia, C.J., *Arch. Biochem. Biophys.* (1992) **299**, 1–7.

G724. Layh, N., Stolz, A., Förster, S., Effenberger, F., Knackmuss, H.J., *Arch. Microbiol.* (1992) **158**, 405–411.

G725. Porthuluri, J.V., Freeman, J.P., Evans, F.E., Cerniglia, C.E., *Appl. Environ. Microbiol.* (1992) **58**, 3654–3659.

G726. Lutterbach, R., Stöckigt, J., *Helv. Chim. Acta* (1992) **75**, 2009–2011.

G729. Loeffler, F., Mueller, R., Lingens, F., *Biol. Chem. Hoppe-Seyler* (1992) **373**, 1001–1007.

G731. Nakamura, J., Asai, K., Nishida, K., Sasaki, H., *Chem. Pharm. Bull.* (1992) **40**, 2164–2168.

G732. Yoshizumi, M., Church, T., Masuda, Y., Katoh, I., Oka, M., *Biochem. Pharmacol.* (1992) **44**, 2263–2265.

G733. Kerlan, V., Dreano, Y., Bercovici, J.P., Beaune, P.H., Floch, H.H., Berthou, F., *Biochem. Pharmacol.* (1992) **44**, 1745–1756.

G734. den Besten, C., Ellenbroek, M., van der Ree, M.A.E., Rietjens, I.M.C.M., van Bladeren, P.J., *Chem.-Biol. Interact.* (1992) **84**, 259–275.

G735. Wang, C.Y., Zukowski, K., Lee, M.S., Sone, T., *Carcinogenesis* (1992) **13**, 2017–2020.

G736. Stadler, R., Zenk, M.H., *J. Biol. Chem.* (1993) **268**, 823–831.

G737. Chu, I., Ng, K.M.E., Benoit, F.M., Moir, D., *J. Environ. Sci. Health, Pt. B* (1992) **B27**, 729–749.

G738. Fleming, P.E., Mocek, U., Floss, H.G., *J. Am. Chem. Soc.* (1993) **115**, 805–807.

G740. Filipovic, D., Paulsen, M.D., Loida, P.J., Sligar, S.G., Ornstein, R.L., *Biochem. Biophys. Res. Commun.* (1992) **189**, 488–495.

G741. Visser, T.J., Kaptein, E., van Raaij, J.A.G.M., Joe, C.T.T., Ebner, T., Burchell, B., *FEBS Lett.* (1993) **315**, 65–68.

G742. Yoshida, T., *Chem. Express* (1993) **8**, 33–36.

G743. Orzechowski, A., Schrenk, D., Bock, K.W., *Carcinogenesis* (1992) **13**, 2227–2232.

G744. Sone, T., Yamaguchi, T., Isobe, M., Takabatake, E., Adachi, T., Hirano, K., Wang, C.Y., *Chem. Pharm. Bull.* (1992) **40**, 2857–2859.

G745. Savidge, R., Udagama-Randeniya, P., *Phytochemistry* (1992) **31**, 2959–2966.

G746. Xun, L., Topp, E., Orser, C.S., *J. Bacteriol.* (1992) **174**, 5745–5747.

G747. Petersen, M., Häusler, E., Karwatzki, B., Meinhard, J., *Planta* (1993) **189**, 10–14.

G748. Alary, J., Cravedi, J.P., Baradat, M., Carrera, G., *Biochem. Pharmacol.* (1992) **44**, 2139–2147.

G749. Müller, M.J., Zenk, M.H., *Planta Med.* (1992) **58**, 524–527.

G750. Mizoguchi, T., Nanjo, H., Umemura, T., Nishinaka, T., Iwata, C., Imanishi, T., Tanaka, T., Terada, T., Nishihara, T., *J. Biochem.* (1992) **112**, 523–529.

G751. Spencer, J.B., Jordan, P.M., *Biochem. J.* (1992) **288**, 839–846.

G752. Long, M.T., Hopper, D.J., Trudgill, P.W., *FEMS Microbiol. Lett.* (1993) **106**, 111–116.

G753. Monna, L., Omori, T., Kodama, T., *Appl. Environ. Microbiol.* (1993) **59**, 285–289.

G754. Shigeoka, S., Ishiko, H., Nakano, Y., Mitsunaga, T., *Biochim. Biophys. Acta* (1992) **1128**, 220–226.

G755. Pothuluri, J.V., Freeman, J.P., Evans, F.E., Moorman, T.B., Cerniglia, C.E., *J. Agric. Food Chem.* (1993) **41**, 483–488.

G756. Kim, D.H., Kim, H.S., Kobashi, K., *J. Biochem.* (1992) **112**, 456–460.

G757. Hein, D.W., Doll, M.A., Gray, K., Rustan, T.D., Ferguson, R.J., *Cancer Res.* (1993) **53**, 509–514.

G758. Ludwig-Müller, J., Epstein, E., *Curr. Plant Sci. Biotechnol. Agric.* (1992) **13**, 188–193.

G759. Borneman, W.S., Ljungdahl, L.G., Hartley, R.D., Akin, D.E., *Appl. Environ. Microbiol.* (1992) **58**, 3762–3766.

G760. Kawaguchi, M., Fujioka, S., Sakurai, A., Yamaki, Y.T., Syono, K., *Plant Cell Physiol.* (1993) **34**, 121–128.

G761. Stich, K., Eidenberger, T., Wurst, F., Forkmann, G., *Z. Naturforsch. C: Biosci.* (1992) **47**, 553–560.

G762. Ademola, J.I., Wester, R.C., Maibach, H.I., *J. Pharm. Sci.* (1993) **82**, 150–154.

G763. Rosetti, V., Lombard, A., Buffa, M., Miglietta, A., Oliviero, A., Gadoni, E., *Eur. J. Drug Metab. Pharmacokinet.* (1992) **17**, 183–186.

G764. Larsson, P., Tjaale, H., *Carcinogenesis* (1993) **14**, 1–6.

G765. Mendrala, A.L., Langvardt, P.W., Nitschke, K.D., Quast, J.F., Nolan, R.J., *Arch. Toxicol.* (1993) **67**, 18–27.

G766. Ninnekar, H.Z., *Biochem. Int.* (1992) **28**, 97–103.

G767. Ghosh, D.K., Mandal, A., Chaudhuri, J., *FEMS Microbiol. Lett.* (1922) **98**, 229–233.

G768. Yalpani, N., Schulz, M., Davis, M.P., Balke, N.E., *Plant Physiol.* (1922) **100**, 457–463.

G769. Varin, L., Ibrahim, R.K., *J. Biol. Chem.* (1992) **267**, 1858–1863.

G770. Yamazaki, Y., Kobayashi, H., *Chem. Express* (1993) **8**, 97–100.

G771. Eltis, L.D., Hofmann, B., Hecht, H.-J., Lünsdorf, H., Timmis, K.N., *J. Biol. Chem.* (1993) **268**, 2727–2732.
G772. Haddock, J.D., Ferry, J.G., *J. Bacteriol.* (1993) **175**, 669–673.
G773. Middelhoven, W.J., Coenen, A., Kraakman, B., Sollewijn Gelpke, M.D., *Antonie van Leeuwenhoek.* (1992) **62**, 181–187.
G774. Smekal, O., Yasin, M., Fewson, C.A., Reid, G.A., Chapman, S.K., *Biochem. J.* (1993) **290**, 103–107.
G775. Steward, A.R., Maslanka, R., Kumar, S., Sikka, H.C., *Chemosphere.* (1992) **25**, 1215–1220.
G776. Bar, T., Okon, Y., *Can. J. Microbiol.* (1993) **39**, 81–86.
G777. Kajbaf, M., Sepai, O., Lamb, J.H., Naylor, S., *J. Chromatogr., Biomed. Appl.* (1992) **583**, 63–76.
G778. Bauer, G., Lingens, F., *Biol. Chem. Hoppe-Seyler* (1992) **373**, 699–705.
G779. Cruz, S.M., Scott, M.N., Merritt, A.K., *J. Agric. Food Chem.* (1993) **41**, 662–668.
G780. Kempton, L.R.C., Isaac, R.E., Pillmoor, J.B., Willis, R.J., *Insect Biochem. Mol. Biol.* (1992) **22**, 777–783.
G781. Galliano, H., Heller, W., Sandermann, H., Jr., *Phytochemistry* (1993) **32**, 557–563.
G782. Coulter, C., Kennedy, J.T., McRoberts, W.C., Harper, D.B., *Appl. Environ. Microbiol.* (1993) **59**, 706–711.
G783. Haddock, J.D., Nadim, L.M., Gibson, D.T., *J. Bacteriol.* (1993) **175**, 395–400.
G784. Patel, T.L., Hameed, N., Armstrong, S., *J. Basic Microbiol.* (1992) **32**, 233–240.
G785. Feigel, B.J., Knackmuss, H.J., *Arch. Microbiol.* (1993) **159**, 124–130.
G786. Haber, M.T., Nashed, N.T., Jerina, D.M., *Bioorg. Med. Chem. Lett.* (1992) **2**, 1729–1734.
G787. Wood, A.M., Truscott, R.J.W., *Exp. Eye Res.* (1993) **56**, 317–325.
G789. Hartley, R., Quinn, M., Green, M., Levene, M.I., *Br. J. Clin. Pharmacol.* (1993) **35**, 314–317.
G790. Cnubben, N.H.P., Vervoort, J., Veeger, C., Rietjens, I.M.C.M., *Chem.-Biol. Interact.* (1992) **85**, 151–172.
G791. Jones, G.H., *J. Biol. Chem.* (1993) **268**, 6831–6834.
G793. Sterjiades, R., *Biotechnol. Appl. Biochem.* (1993) **17**, 77–90.
G794. Hinton, R.J., Ramsay, J.R., Atkinson, T., Bown, K.J., Hammond, P.M., *Ann. N.Y. Acad. Sci.* (1992) **672**, 224–229.
G795. Brune, A., Schink, B., *Arch. Microbiol.* (1992) **157**, 417–424.
G797. Hashidoko, Y., Urashima, M., Yoshida, T., Mizutani, J., *Biosci. Biotechnol. Biochem.* (1993) **57**, 215–219.
G798. Clement, B., Jung, F., Pfunder, H., *Mol. Pharmacol.* (1993) **43**, 335–342.
G799. Haritos, V.S., French, J.R.J., Ahokas, J.T., *Chemosphere* (1993) **26**, 1291–1299.
G800. Tasaki, Y., Makino, Y., Ohta, S., Hirobe, M., *Adv. Neurol.* (1993) **60**, 231–233.
G802. Jin, H.J., Park, A.K., Lee, S.S., *Arch. Pharmacol. Res.* (1992) **15**, 215–219.
G803. De Beyer, A., Lingens, F., *Biol. Chem. Hoppe-Seyler* (1993) **374**, 101–109.
G804. Large, A.T., Connock, M.J., *Comp. Biochem. Physiol. B: Comp. Biochem.* (1993) **104B**, 489–491.
G805. Castanares, A., McCrae, S.I., Wood, T.M., *Enzyme Microb. Technol.* (1992) **14**, 875–884.
G806. Boopathy, R., Kulpa, C.F., Wilson, M., *Appl. Microbiol. Biotechnol.* (1993) **39**, 270–275.
G807. Cocciara, G., Battaglia, R., Fontana, E., Strolin Benedetti, M., *Drug Metab. Drug Interact.* (1992) **10**, 185–197.
G809. Radulovic, L., Woolf, T., Bjorge, S., Taylor, C., Reily, M., Bockbrader, H., Chang, T., *Chem. Res. Toxicol.* (1993) **6**, 341–344.
G810. Soares-da-Silva, P., *Gen. Pharmacol.* (1993) **24**, 75–81.

G811. Malejka-Giganti, D., Ritter, C.L., Willmott, D., *Carcinogenesis* (1993) **14**, 341–346.

G812. Block, D.W., Lingens, F., *Biol. Chem. Hoppe-Seyler* (1992) **373**, 249–254.

G813. Bärenwald, G., Schneider, B., Schütte, H.R., *Phytochemistry* (1993) **32**, 523–526.

G814. Sambandam, T., Mahadevan, A., *World J. Microbiol. Biotechnol.* (1993) **9**, 37–44.

G815. Hibino, T., Shibata, D., Umezawa, T., Higuchi, T., *Phytochemistry* (1993) **32**, 565567.

G816. Ahmad, S.A., Hopkins, T.L., *Comp. Biochem. Physiol. B: Comp. Biochem.* (1993) **104B**, 515–519.

G817. Prieto, M.A., Perez-Aranda, A., Garcia, J.L., *J. Bacteriol.* (1993) **175**, 2162–2167.

G818. Kelley, I., Freeman, J.P., Evans, F.E., Cerniglia, C.E., *Appl. Environ. Microbiol.* (1993) **59**, 800–806.

G819. Knöss, W., Glombitza, K.W., *Phytochemistry* (1993) **32**, 1119–1123.

G820. Armstrong, S., Patel, D.R., *Can. J. Microbiol.* (1993) **39**, 175–179.

G821. Arfmann, H.A., Abraham, W.R., *Z. Naturforsch. C: Biosci.* (1993) **48**, 52–57.

G822. Stockinger, J., Hinteregger, C., Loidl, M., Ferschl, A., Streichsbier, F., *Appl. Microbiol. Biotechnol.* (1992) **38**, 421–428.

G823. Bevinakatti, B.G., Ninnekar, H.Z., *Appl. Microbiol. Biotechnol.* (1992) **32**, 273–275.

G824. Withyachumnarkul, B., Pongsa-Asawapaiboon, A., Poolsanguan, B., *Comp. Biochem. Physiol. B: Comp. Biochem.* (1993) **104B**, 449–454.

G825. Rojas, M.G., Stipanovic, R.D., Williams, H.J., Vinson, S.B., *Environ. Entomol.* (1992) **21**, 518–526.

G826. Yamazaki, Y., Kobayashi, H., *Tetrahedron: Asymmetry* (1993) **4**, 1287–1294.

G827. Boopathy, R., Kulpa, C.F., *Can. J. Microbiol.* (1993) **39**, 430–433.

G828. Yamazaki, Y., Maruyama, A., Hosono, K., Higashihara, T., Kobayashi, H., *Z. Naturforsch.* (1993) **48C**, 451–456.

G829. Goodwin, B.L., Ruthven, C.R.J., Sandler, M., *Biochem. Pharmacol.* (1993) **46**, 2109–2110.

G830. Perego, R., Beccaglia, P., Angelini, M., Villa, P., Cova, D., *Xenobiotica* (1993) **23**, 1345–1352.

G831. Maurizis, J.C., Rapp, M., Madelmont, J.C., Gillardin, J.M., Lepage, F., Labarre, P., Dupuy, J.M., Veyre, A., *Xenobiotica* (1993) **23**, 1335–1343.

G832. Battaglia, R., Strolin Benedetti, M., Mantegani, S., Castelli, M.G., Cocchiara, G., Dostert, P., *Xenobiotica* (1993) **23**, 1377–1389.

G833. Yabuki, M., Mine, T., Iba, K., Nakatsuka, I., Yoshitake, A., *Xenobiotica* (1993) **23**, 1367–1375.

G834. Temellini, A., Mogavero, S., Giulianotti, P.C., Pietrabissa, A., Mosca, F., Pacifici, G.M., *Xenobiotica* (1993) **23**, 1427–1433.

G835. De Vries, J.X., Walter-Sack, I., Voss, A., Forster, W., Ilisistegui Pons, P., Stoetzer, F., Spraul, M., Ackermann, M., Moyna, G., *Xenobiotica* (1993) **23**, 1435–1450.

G836. Masubuchi, Y., Kagimoto, N., Narimatsu, S., Fujita, S., Suzuki, T., *Drug Metab. Dispos.* (1993) **21**, 1012–1016.

G837. Fabre, G., Julian, B., Saint-Aubert, B., Joyeux, H., Berger, Y., *Drug Metab. Dispos.* (1993) **21**, 978–985.

G838. Manchee, G.R., Barrow, A., Kulkarni, S., Palmer, E., Oxford, J., Colthup, P.V., Maconochie, J.G., Tarbit, M.N., *Drug Metab. Dispos.* (1993) **21**, 1022–1028.

G839. Grossman, S.J., Herold, E.G., Drey, J.M., Alberts, D.W., Umbenhauser, D.R., Patrick, D.H., Nicoll-Griffith, D., Chauret, N., Yergey, J.A., *Drug Metab. Dispos.* (1993) **21**, 1029–1036.

G840. Turcan, R.G., Hillbeck, D., Hartley, T.E., Gilbert, P.J., Coe, R.A.J., Troke,

J.A., Vose, C.W., *Drug Metab. Dispos.* (1993) **21**, 1037–1047.

G841. Chung, J.G., Levy, G.N., Weber, W.W., *Drug Metab. Dispos.* (1993) **21**, 1057–1063.

G842. Zheng, W., Winter, S.M., Mayersohn, M., Bishop, J.B., Sipes, I.G., *Drug Metab. Dispos.* (1993) **21**, 1091–1097.

G843. Zhao, Y., Snel, C.A.W., Mulder, G.J., Pang, K.S., *Drug Metab. Dispos.* (1993) **21**, 1070–1078.

G844. Poon, G.K., Chui, Y.C., McCague, R., Lønning, P.E., Feng, R., Rowlands, M.G., Jarman, M., *Drug Metab. Dispos.* (1993) **21**, 1119–1124.

G845. Mannens, G., Huang, M.-L., Meuldermans, W., Hendrickx, J., Woestenborghs, R., Heykants, J., *Drug Metab. Dispos.* (1993) **21**, 1134–1141.

G846. Milne, R.W., Sloan, P.A., McLean, C.F., Mather, L.E., Natiol, R.L., Runciman, W.B., Rutten, A.J., Somogyi, A.A., *Drug Metab. Dispos.* (1993) **21**, 1151–1156.

G847. Yoshida, T., Tabuchi, T., Andoh, K., *Drug Metab. Dispos.* (1993) **21**, 1142–1146.

G848. Boutin, J.A., Meunier, F., Lambert, P.-H., Hennig, P., Bertin, D., Serkiz, B., Volland, J.-P., *Drug Metab. Dispos.* (1993) **21**, 1157–1166.

G849. Yabuki, M., Shono, F., Nakatsuka, I., Yoshitake, A., *Drug Metab. Dispos.* (1993) **21**, 1167–1169.

G850. Lee, K., Vandenberghe, Y., Herin, M., Cavalier, R., Beck, D., Li, A., Verbeke, Lesne, M., Roba, J., *Xenobiotica* (1994) **24**, 25–36.

G851. Riley, P., Hanzlik, R.P., *Xenobiotica* (1994) **24**, 1–16.

G852. Critchley, D.P.J., Rance, D.J., Beedham, C., *Xenobiotica* (1994) **24**, 37–47.

G853. Wang, X.-Q., Seinen, W., De Wolf, W., *Xenobiotica* (1994) **24**, 59–69.

G854. Midgley, I., Hood, A.J., Proctor, P., Chasseaud, L.F., Irons, S.R., Cheng, K.N., Brindley, C.J., Bonn, R., *Xenobiotica* (1994) **24**, 79–92.

G855. Hildebrandt, R., Wagner, B., Preiss-Nowzohour, K., Gundert-Remy, U., *Xenobiotica* (1994) **24**, 71–77.

G856. Eisenhardt, E.U., Bickel, M.H., *Drug Metab. Dispos.* (1994) **22**, 26–30.

G857. Ohtani, M., Kotaki, H., Uchino, K., Sawada, Y., Iga, T., *Drug Metab. Dispos.* (1994) **22**, 2–7.

G858. Eisenhardt, E.U., Bickel, M.H., *Drug Metab. Dispos.* (1994) **22**, 31–35.

G859. Hussain, M.D., Tam, Y.K., Gray, M.R., Coutts, R.T., *Drug Metab. Dispos.* (1994) **22**, 36–42.

G860. Meuldermans, W., Hendrickx, J., Mannens, G., Lavrijsen, K., Janssen C., Bracke, J., Le Jeune, L., Lauwers, W., Heykants, J., *Drug Metab. Dispos.* (1994) **22**, 129–138.

G861. Pollack, G.M., Spencer, A.P., Horton, T.L., Brouwer, K.L.R., *Drug Metab. Dispos.* (1994) **22**, 120–123.

G862. Mayol, R.F., Cole, C.A., Colson, K.E., Kerns, E.H., *Drug Metab. Dispos.* (1994) **22**, 171–174.

G863. Kumar, G.N., Oatis, J.E., Jr., Thornberg, K.R., Heldrich, F.J., Hazard, E.S. III, Walle, T., *Drug Metab. Dispos.* (1994) **22**, 177–179.

G864. Sandker, G.W., Vos, R.M.E., Delbressine, L.P.C., Sloof, M.J.H., Meijer, D.K.F., Groothuis, G.M.M., *Xenobiotica* (1994) **24**, 143–155.

G865. Goodwin, B.L., Ruthven, C.R.J., Sandler, M., *Xenobiotica* (1994) **24**, 129–141.

G866. Colletti, A.E., Krieter, P.A., *Drug Metab. Dispos.* (1994) **22**, 183–188.

G867. Balani, S.K., Kauffman, L.R., Arison, B.H., Olah, T.V., Goldman, M.E., Varga, S.L., O'Brien, J.A., Ramjit, H.G., Rooney, C.S., Hoffman, J.M., Pitzenberger, S.M., *Drug Metab. Dispos.* (1994) **22**, 200–205.

G868. Marathe, P.H., Shen, D.D., Nelson, W.L., *Drug Metab. Dispos.* (1994) **22**, 237–247.

G869. Rubio, F., Jensen, B.K., Henderson, L., Garland, W.A., Szuna, A., Town, C., *Drug Metab. Dispos.* (1994) **22**, 211–215.

G870. Fischer, V., Vickers, A.E.M., Heitz, F., Mahadevan, S., Baldeck, J.-P., Minery, P., Tynes, R., *Drug Metab. Dispos.* (1994) **22**, 269–274.

G871. Yoshida, T., *Drug Metab. Dispos.* (1994) **22**, 275–280.

G872. Mayol, R.F., Cole, C.A., Luke, G.M., Colson, K.L., Kerns, E.H., *Drug Metab. Dispos.* (1994) **22**, 304–311.

G873. Yabuki, M., Mine, T., Iba, K., Nakatsuka, I., Yoshitake, A., *Drug Metab. Dispos.* (1994) **22**, 294–297.

G874. Law, F.C.P., Meng, J.X., He, Y.T., Chui, Y.C., *Xenobiotica* (1994) **24**, 221–229.

G875. Mewes, K., Blanz, J., Freund, S., Ehninger, G., Zeller, K.-P., *Xenobiotica* (1994) **24**, 199–213.

G876. Barber, E.D., Fox, J.A., Giordano, G.J., *Xenobiotica* (1994) **24**, 441–450.

G877. Kashiyama, E., Todaka, T., Odomi, M., Tanokura, Y., Johnson, D.B., Yokoi, T., Kamataki, T., Shimizu, T., *Xenobiotica* (1994) **24**, 369–377.

G879. Jennison, T.A., Wozniak, E., Nelson, G., Urry, F.M., *J. Anal. Toxicol.* (1993) **17**, 208–210.

G880. Hamada, H., Fuchikami, Y., Jansing, R.L., Kaminsky, L.S., *Phytochemistry* (1993) **33**, 599–600.

G881. Morishita, N., Tomono, Y., Hasegawa, J., Yuzuriha, T., Giese, U., *Drug Invest.* (1993) **5**, 296–301.

G882. Veronese, M.E., Miners, J.O., Rees, D.L.P., Birkett, D.J., *Pharmacogenetics* (1993) **3**, 86–93.

G883. von Moltke, L.L., Greenblatt, D.J., Harmatz, J.S., Shader, R.I., *Pharmacology* (1993) **47**, 268–276.

G884. Lertratangkoon, K., Horning, E.C., Horning, M.G., *Res. Commun. Chem. Pathol. Pharmacol.* (1993) **80**, 259–282.

G885. Borlakoglu, J.T., Wilkins, J.P.G., *Comp. Biochem. Physiol. C: Comp. Pharmacol. Toxicol.* (1993) **105C**, 119–125.

G886. Ademola, J.I., Wester, R.C., Maibach, H.I., *Toxicol. Appl. Pharmacol.* (1993) **121**, 78–86.

G887. Mitchell, C.E., Bechtold, W.E., Belinsky, S.A., *Carcinogenesis* (1993) **14**, 1161–1166.

G888. Koshiba, T., Mito, N., Miyakado, M., *J. Plant Res.* (1993) **106**, 25–29.

G889. Schmitz, A.R., Squibb, K.S., *Toxicol. Appl. Pharmacol.* (1993) **121**, 58–70.

G890. Mohamed, M.-E.S., Fuchs, G., *Arch. Microbiol.* (1993) **159**, 554–562.

G891. Maloney, S.E., Maule, A., Smith, A.R.W., *Appl. Environ. Microbiol.* (1993) **59**, 2007–2013.

G892. Tanaka-Kawai, H., Yomoda, S., *Clin. Chim. Acta* (1993) **215**, 127–138.

G893. Armstrong, S.M., Patel, T.R., *J. Basic Microbiol.* (1993) **33**, 75–81.

G894. Montemartini, M., Santomé, J.A., Cazzulo, J.J., Nowicki, C., *Biochem. J.* (1993) **292**, 901–906.

G895. Huang, Z., Dostal, L., Rosazza, J.P.N., *Appl. Environ. Microbiol.* (1993) **59**, 2244–2250.

G896. Crooks, G.P., Copley, S.D., *J. Am. Chem. Soc.* (1993) **115**, 6422–6423.

G897. Ahmad, S.A., Hopkins, T.L., *Insect Biochem. Mol. Biol.* (1993) **23**, 581–589.

G898. Ehlhardt, W.J., Sullivan, H.R., Wood, P.G., Woodland, J.M., Hamilton, M., Hamilton, C., Cornpropst, D., Grindey, G.B., Worzalla, J.F., Bewley, J.R., Todd, G.C., Howbert, J.J., *J. Pharm. Sci.* (1993) **82**, 683–688.

G899. Kuo, B.S., Poole, J.C., Hwang, K.K., Cheng, H., *J. Pharm. Sci.* (1993) **82**, 694–698.

G901. Eltom, S.E., Guard, C.L., Schwark, W.S., *J. Vet. Pharmacol. Ther.* (1993) **16**, 141–151.

G902. Soares-da-Silva, P., Fernandez, M.H., Pestana, M., *Acta Physiol. Scand.* (1993) **148**, 347–351.

G903. Takeno, S., Hirano, Y., Kitamura, A., Sakai, T., *Toxicol. Appl. Pharmacol.* (1993) **121**, 233–238.

G904. Yasin, M., Fewson, C.A., *Biochem. J.* (1993) **293**, 455–460.

G905. Stocker, A., Rüttimann, A., Woggon, W.-D., *Helv. Chim. Acta* (1993) **76**, 1729–1738.

G906. Berlin, J., Rügenhagen, C., Greidziak, N., Kuzovkina, I.N., Witte, L., Wray, V., *Phytochemistry* (1993) **33**, 593–597.

G907. Hailes, A.M., Bruce, N.C., *Appl. Environ. Microbiol.* (1993) **59**, 2166–2170.

G908. Heinig, R., Delbressine, L.P.C., Kaspersen, F.M., Blaschke, G., *Arzneim.-Forsch.* (1993) **43**, 709–715.

G909. Xu, B.Q., Bjørneboe, Å., Ripel, A., Aasmundstad, T.A., Christophersen, A.S., Mørland, J., *Pharmacol. Toxicol.* (1993) **73**, 35–40.

G910. Jurková, M., Wurst, M., *FEMS Microbiol. Lett.* (1993) **111**, 245–250.

G911. Huang, Z., Dostal, L., Rosazza, J.P.N., *J. Biol. Chem.* (1993) **268**, 23954–23958.

G912. Abe, K., Nakada, Y., Suzuki, A., Yumioka, E., *Shoyakugaku Zasshi* (1993) **47**, 43–46.

G913. Nawaz, R., Ilahi, A., Iqbal, Z., *Pak. Vet. J.* (1993) **13**, 19–22.

G914. Whitaker, D.P., Goosey, M.W., *Biochem. J.* (1993) **295**, 149–154.

G915. Faulds, C.B., Williamson, G., *Biotechnol. Appl. Biochem.* (1993) **17**, 349–359.

G916. Pirrung, M.C., Chen, J., Rowley, E.G., McPhail, A.T., *J. Am. Chem. Soc.* (1993) **115**, 7103–7110.

G917. Guy, J., Lepine, F., Beaudet, R., Sylvestre, M., *Can. J. Microbiol.* (1993) **39**, 642–648.

G918. Nakazawa, H., Sano, K., Matsuda, K., Mitsuga, K., *Biosci. Biotechnol. Biochem.* (1993) **57**, 1210–1211.

G919. Renz, P., Endres, B., Kurz, B., Marquardt, J., *Eur. J. Biochem.* (1993) **217**, 1117–1121.

G920. Liang, P.H., Yang, G., Dunaway-Mariano, D., *Biochemistry* (1993) **32**, 12245–12250.

G921. Gurevich, P., Oren, A., Sarig, S., Henis, Y., *Water Sci. Technol.* (1993) **27**, 89–96.

G922. Pieper, D.H., Knackmuss, H.J., Timmis, K.N., *Appl. Microbiol. Biotechnol.* (1993) **39**, 563–567.

G923. Gallert, C., Winter, J., *Appl. Microbiol. Biotechnol.* (1993) **39**, 627–631.

G924. Dugelay, I., Gunata, Z. Sapis, J.C., Baumes, R., Bayonove, C., *J. Agric. Food Chem.* (1993) **41**, 2092–2096.

G925. Jacqz-Agrain, E., Cresteil, T., *Dev. Pharmacol. Ther.* (1992) **18**, 161–168.

G926. Fernandes, M.H., Soares-da-Silva, P., *Adv. Biosci.* (1993) **88**, 21–30.

G927. El-Mekkawy, S., Meselhy, M.R., Kawata, Y., Kadota, S., Hattori, M., Namba, T., *Planta Med.* (1993) **59**, 347–350.

G928. Hlavica, P., Künzel-Mulas, U., *Biochim. Biophys. Acta* (1993) **1158**, 83–90.

G929. Mohamed, M.E.S., Seyfried, B., Tschech, A., Fuchs, G., *Arch. Microbiol.* (1993) **159**, 563–573.

G930. Kometani, T., Tanimoto, H., Nishimura, T., Okada, S., *Biosci. Biotechnol. Biochem.* (1993) **57**, 1290–1293.

G931. Stupperich, E., Konle, R., *Appl. Environ. Microbiol.* (1993) **59**, 3110–3116.

G932. Goto, H., Yoshida, M., Nakazono, N., *J. Toxicol. Sci.* (1993) **18**, 103–109.

G933. Chung, B.C., Kim, D.H., Jung, B.H., Eom, K., Slikker, W., Jr., Park, J., *Xenobiotica* (1994) **24**, 451–459.

G934. Bochner, F., Somogyi, A.A., Chen, Z.R., *Xenobiotica* (1994) **24**, 543–552.

G935. Konishi, M., Shiono, Y., Ohno, M., Takahashi, H., Aoki, T., *Xenobiotica* (1994) **24**, 591–601.

G936. Kestell, P., Rewcastle, G.W., Baguley, B.C., *Xenobiotica* (1994) **24**, 635–647.

G937. Bioque, G., Abián, J., Bulbena, O., Roselló-Catafau, J., Gelpí, E., *Xenobiotica* (1994) **24**, 613–621.

G938. Wu, W.N., Hills, J.F., Egan, R.S., *Xenobiotica* (1994) **24**, 649–662.

G939. Clement, B., Schultze-Mosgau, M.H., Richter, P.H., Besch, A., *Xenobiotica* (1994) **24**, 671–688.

G940. Zhang, K., Tang, C., Rashed, M., Cui, D., Tombret, F., Botte, H., Lepage, F., Levy, R.H., Baillie, T.A., *Drug Metab. Dispos.* (1994) **22**, 544–553.
G941. Rivera, M.I., Hinojosa, L.M., Hill, B.A., Lau, S.S., Monks, T.J., *Drug Metab. Dispos.* (1994) **22**, 503–510.
G942. Tang, C., Zhang, K., Lepage, F., Levy, R.H., Baillie, T.A., *Drug Metab. Dispos.* (1994) **22**, 554–560.
G943. Vage, C., Svensson, C.K., *Drug Metab. Dispos.* (1994) **22**, 572–577.
G944. Chikhale, P.J., Borchardt, R.T., *Drug Metab. Dispos.* (1994) **22**, 592–600.
G945. Cravedi, J.P., Baradat, M., Debrauwer, L., Alary, J., Tulliez, J., Bories, G., *Drug Metab. Dispos.* (1994) **22**, 578–583.
G946. Krieter, P.A., Colletti, A.E., Doss, G.A., Miller, R.R., *Drug Metab. Dispos.* (1994) **22**, 625–630.
G947. Cussac, C., Mounetou, E., Rapp, M., Madelmont, J.C., Maurizis, J.C., Labarre, P., Chollet, P., Chabard, J.L., Godeneche, D., Baudry, J.P., Veyre, A., *Drug Metab. Dispos.* (1994) **22**, 637–642.
G948. Strandgården, K., Gunnarsson, P.O., *Xenobiotica* (1994) **24**, 703–711.
G949. Huskey, S.-E.W., Doss, G.A., Miller, R.A., Schoen, W.R., Chiu, S.-H.L., *Drug Metab. Dispos.* (1994) **22**, 651–658.
G950. Tjeerdema, R.S., Luckrich, K.L., Stevens, E.M., *Xenobiotica* (1994) **24**, 749–757.
G951. Ulgen, M., Gorrod, J.W., Barlow, D., *Xenobiotica* (1994) **24**, 735–748.
G952. Koga, N., Nishimura, N., Kuroki, H., Masuda, Y., Yoshimura, H., *Xenobiotica* (1994) **24**, 775–783.
G953. Soffers, A.E.M.F., Veeger, C., Rietjens, I.M.C.M., *Xenobiotica* (1994) **24**, 759–774.
G954. Van Iersel, M.L.P.S., Henderson, C.J., Walters, D.G., Price, R.J., Wolf, C.R., Lake, B.G., *Xenobiotica* (1994) **24**, 795–803.
G955. Naitoh, T., Horie, T., Nagato, S., Kagaya, T., Kubota, A., Akasaka, K., *Xenobiotica* (1994) **24**, 819–826.
G956. Steensma, A., Beamand, J.A., Walters, D.G., Price, R.J., Lake, B.G., *Xenobiotica* (1994) **24**, 893–907.
G957. Wada, S., Shimizudani, T., Yamada, H., Oguri, K., Yoshimura, H., *Xenobiotica* (1994) **24**, 853–861.
G958. Fujimaki, M., *Drug Metab. Dispos.* (1994) **22**, 700–708.
G959. Chen, I.-W., Lin, J.H., *Drug Metab. Dispos.* (1994) **22**, 693–699.
G960. Tephly, T.R., Coffman, B., Styczynski, P., Rios, G., Charkowski, D.M., Vanrollins, M., McQuade, R.D., Tedford, C.E., *Drug Metab. Dispos.* (1994) **22**, 713–718.
G961. Hooper, W.D., Pool, W.F., Woolf, T.F., Gal, J., *Drug Metab. Dispos.* (1994) **22**, 719–724.
G962. Dow, J., Piriou, F., Wolf, E., Dulery, B.D., Haegele, K.D., *Drug Metab. Dispos.* (1994) **22**, 738–749.
G963. Luo, G., Guenther, T.M., *Drug Metab. Dispos.* (1994) **22**, 731–737.
G964. Mizutani, T., Yoshida, K., Kawazoe, S., *Drug Metab. Dispos.* (1994) **22**, 750–755.
G965. Coutts, R.T., Bolaji, O.O., Su, P., Baker, G.B., *Drug Metab. Dispos.* (1994) **22**, 756–760.
G966. Pochopin, N.L., Charman, W.N., Stella, V.J., *Drug Metab. Dispos.* (1994) **22**, 770–775.
G967. Takahara, E., Nagata, O., Kato, H., Ohta, S., Hirobe, M., *Drug Metab. Dispos.* (1994) **22**, 761–769.
G968. Green, M.D., Oturu, E.M., Tephly, T.R., *Drug Metab. Dispos.* (1994) **22**, 799–805.
G969. Hargreaves, M.B., Jones, B.C., Smith, D.A., Gescher, A., *Drug Metab. Dispos.* (1994) **22**, 806–810.
G970. Klos, C., Dekant, W., *Xenobiotica* (1994) **24**, 965–976.
G971. Tajima, K., Hashizaki, M., Yamamoto, K., Mizutani, T., *Drug Metab. Dispos.* (1994) **22**, 815–818.

G972. den Besten, P.J., Lemaire, P., Livingstone, D.R., *Xenobiotica* (1994) **24**, 989–1001.
G973. Davison, K.L., Larsen, G.L., Feil, V.J., *Xenobiotica* (1994) **24**, 1003–1012.
G974. Vree, T.B., van den Biggelaar-Martesa, M., Verwey-van Wissen, C.P.W.G.M., Vree, J.B., Guelen, P.J.M., *Biopharm. Drug Dispos.* (1993) **14**, 491–502.
G975. Sneyd, J.R., Simons, P.J., Wright, B., *Xenobiotica* (1994) **24**, 1021–1028.
G976. Abolfathi, Z., Fiset, C., Gilbert, M., Moericke, K., Bélanger, P.M., Turgeon, J., *J. Pharmacol. Exp. Ther.* (1993) **266**, 1196–1201.
G977. Kroemer, H.K., Gauter, J.C., Beaune, P., Henderson, C., Wolf, C.R., Eichelbaum, M., *Naunyn-Schmiedeberg's Arch. Pharmacol.* (1993) **348**, 332–337.
G978. Hayes, J.D., Judah, D.J., *Cancer Res.* (1993) **53**, 3887–3894.
G979. Akhtar, M.H., Mahadevan, S., Russell, F., *J. Environ. Sci. Health Pt. B* (1993) **B28**, 527–543.
G980. Haigler, B.E., Spain, J.C., *Appl. Environ. Microbiol.* (1993) **59**, 2239–2243.
G981. Londry, K.L., Fedorak, P.M., *Appl. Environ. Microbiol.* (1993) **59**, 2229–2238.
G982. Ishikura, N., Nakamura, S., Mato, M., Yamamoto, K., *Bot. Mag. Tokyo* (1992) **105**, 83–94.
G983. Altenschmidt, U., Oswald, B., Steiner, E., Herrmann, H., *J. Bacteriol.* (1993) **175**, 4851–4858.
G984. Matsumoto, S., Tagawa, M., Hatoyama, T., Fujii, T., Miyazaki, H., Sekine, Y., *Arzneim.-Forsch.* (1993) **43**, 1103–1108.
G985. Weaver, R.J., Dickins, M., Burke, M.D., *Biochem. Pharmacol.* (1993) **46**, 1183–1197.
G986. Darbyshire, J.F., Caldwell, J., *Chirality* (1993) **5**, 191–198.
G987. Robertson, I.G.C., Palmer, B.D., Shaw, G.J., *Biol. Mass Spectrom.* (1993) **22**, 661–665.
G988. Suzuki, E., Saegusa, K., Anjo, T., Matsuki, Y., Nambara, T., *Steroids* (1993) **58**, 540–546.
G989. Kawaji, A., Ohara, K., Takabatake, E., *Anal. Biochem.* (1993) **214**, 409–412.
G990. Clement, B., Immel, M., Schmitt, S., Steinmann, U., *Arch. Pharm. (Weinheim, Ger.)* (1993) **326**, 807–812.
G991. Bolaji, O.O., Coutts, T., Baker, G.B., *Res. Commun. Pathol. Pharmacol.* (1993) **82**, 111–120.
G992. Walle, T., Kumar, G.N., McMillan, J.M., Thornburg, K.R., Walle, U.K., *Biochem. Pharmacol.* (1993) **46**, 1661–1664.
G993. Bhavnani, B.R., Cecutti, A., *J. Clin. Endocrinol. Metab.* (1993) **77**, 1269–1274.
G994. Banoo, R., Matheison, D.W., *Indian J. Biochem. Biophys.* (1993) **30**, 229–233.
G995. Benson, A.M., *Biochem. Pharmacol.* (1993) **46**, 1217–1221.
G996. Ino, I., Yamaguchi, M., *Phytochemistry* (1993) **33**, 1415–1417.
G997. Muimo, R., Isaac, R.E., *Comp. Biochem. Physiol. B: Comp. Biochem.* (1993) **106B**, 969–976.
G998. Chu, A., Dinkova, A., Davin, L.B., Bedgar, D.L., Lewis, N.G., *J. Biol. Chem.* (1993) **268**, 27026–27033.
G999. Faulds, C.B., Williamson, G., *Carbohydr. Polym.* (1993) **21**, 153–155.
G1000. Rhys-Williams, W., Taylor, S.C., Williams, P.A., *J. Gen. Microbiol.* (1993) **139**, 1967–1972.
H1. Ashburn, S.P., Han, X., Liehr, J.G., *Mol. Pharmacol.* (1993) **43**, 534–541.
H2. Yoshida, T., *Chem. Express* (1993) **8**, 225–228.
H3. Bartels, M.J., Timchalk, C., Smith, F.A., *Biol. Mass Spectrom.* (1993) **22**,194–200.
H4. Clemens, S., Hinderer, W., Wittkampf, U., Barz, W., *Phytochemistry* (1993) **32**, 653–657.
H5. De Carolis, E., De Luca, *J. Biol. Chem.* (1993) **268**, 5504–5511.

H6. Armstrong, S., Patel, T.R., *J. Basic Microbiol.* (1992) **32**, 363–367.

H7. Kozikowski, A.P., Chen, C., Wu, J.P., Shibuya, M., Kim, C.G., Foloss, H., *J. Am. Chem. Soc.* (1993) **115**, 2482–2488.

H8. Mohan, K., Khanna, S., *FEMS Microbiol. Lett.* (1993) **108**, 361–366.

H9. Albro, P.W., Corbett, J.T., Schroeder, J.L., *Comp. Biochem. Physiol. C: Comp. Pharmacol. Toxicol.* (1993) **104C**, 335–344.

H10. Roberts-Thomson, S.J., McManus, M.E., Tukey, R.H., Gonzalez, F.F., Holder, G.M., *Biochem. Biophys. Res. Commun.* (1993) **192**, 1373–1379.

H11. Kren, V., Sedmera, P., Havlicek, V., Fiserova, A., *Tetrahedron Lett.* (1992) **33**, 7233–7236.

H12. Itoh, N., Morinaga, N., Kouzai, T., *Biochem. Mol. Biol. Int.* (1993) **29**, 785–791.

H13. Gorny, N., Wahl, G., Brune, A., Schink, B., *Arch. Microbiol.* (1922) **158**, 48–53.

H14. Judah, D.J., Hayes, J.D., Yang, J.C., Lian, L.Y., Roberts, G.C.K., Farmer, P.B., Lamb, J.H., Neal, G.E., *Biochem. J.* (1993) **292**, 13–18.

H16. Katsura, H., Kitamura, S., Tatsumi, K., *Arch. Biochem. Biophys.* (1993) **302**, 356–361.

H17. Petersen, M., Metzger, J.W., *Phytochem. Anal.* (1993) **4**, 131–134.

H18. Stolz, A., Knackmuss, H.J., *FEMS Microbiol. Lett.* (1993) **108**, 219–224.

H19. Koch, J., Eisenreich, W., Bacher, A., Fuchs, G., *Eur. J. Biochem.* (1993) **211**, 649–661.

H20. Jones, K.H., Trudgill, P.W., Hopper, D.J., *Appl. Environ. Microbiol.* (1993) **59**, 1125–1130.

H21. Ouchiyama, N., Zhang, Y., Omori, T., Kodama, T., *Biosci. Biotechnol. Biochem.* (1993) **57**, 455–460.

H22. Kohler, H.-P.E., Schmid, A., van der Maarel, M., *J. Bacteriol.* (1993) **175**, 1621–1628.

H23. Mobley, D.P., Finkbeiner, H.L., Lockwood, S.H., Spivack, J., *Tetrahedron* (1993) **49**, 3273–3280.

H24. Gates, V.L., Tjeerdema, R.S., *Pestic. Biochem. Physiol.* (1993) **46**, 161–170.

H26. Adachi, J., Naito, T., Ueno, Y., Ogawa, Y., Ninomiya, I., Tatsuno, Y., *Arch. Toxicol.* (1993) **67**, 284–289.

H27. Shofer, S.L., Tjeedema, R.S., *Pestic. Biochem. Physiol.* (1993) **46**, 85–95.

H28. Komatsu, T., Omori, T., Kodama, T., *Biosci. Biotechnol. Biochem.* (1993) **57**, 864–865.

H29. Gubatz, S., Wiermann, R., *Bot. Acta* (1992) **105**, 407–413.

H30. Corey, P.F., *Bioorg. Med. Chem. Lett.* (1993) **3**, 175–178.

H31. Schulz, M., Schnabl, H., Manthe, B., Schweinhofen, B., Casser, I., *Phytochemistry* (1993) **33**, 291–294.

H32. Walton, M.I., Workman, P., *J. Pharmacol. Exp. Ther.* (1993) **265**, 938–947.

H33. Lemoine, A., Gautier, J.C., Azoulay, D., Kiffel, L., Belloc, C., Guengerich, F.P., Maurel, P., Beaune, P., Leroux, J.P., *Mol. Pharmacol.* (1993) **43**, 827–832.

H34. De Caro, L., Ghizzi, A., Zunino, M.T., *Arzneim.-Forsch.* (1993) **43**, 651–655.

H35. Narimatsu, S., Kariya, S., Isozaki, S., Ohmori, S., Kitada, M., Hosokawa, S., Masubuchi, Y., Suzuki, T., *Biochem. Biophys. Res. Commun.* (1993) **193**, 1262–1268.

H36. Zhu, B.T., Liehr, J.G., *Arch. Biochem. Biophys.* (1991) **304**, 248–256.

H37. Selifonov, S.A., Griffoll, M., Gurst, J.E., Chapman, P.J., *Biochem. Biophys. Res. Commun.* (1993) **193**, 67–76.

H38. Babu, S.R., Wongsurawat, V.J., Zenser, T.V., Davis, B.B., *Carcinogenesis* (1993) **14**, 893–897.

H39. Knaak, J.B., Al-Bayati, M.A., Raabe, O.G., Blancato, J.N., *Toxicol. Appl. Pharmacol.* (1993) **120**, 106–113.

H40. Muimo, R., Isaac, R.E., *Biochem. Soc. Trans.* (1993) **21**, 295S.

H41. Sund, B., Lauterbach, F., *Pharmacol. Toxicol.* (1993) **72**, 84–89.

H42. Taylor, B.F., Amador, J.A., Levinson, H.S., *FEMS Microbiol. Lett.* (1993) **110**, 213–216.

H43. Saito, K., Suzuki, H., Takamatsu, S., Murakoshi, I., *Phytochemistry* (1992) **32**, 87–91.

H45. Li, C.L.J., James, M.O., *Aquat. Toxicol.* (1993) **26**, 57–71.

H46. Epstein, E., Sagee, O., Zeker, A., *Plant Growth Regul.* (1993) **13**, 31–40.

H47. Pothuluri, J.V., Freeman, J.P., Evans, F.E., Cerniglia, C.E., *Appl. Environ. Microbiol.* (1993) **59**, 1977–1980.

H48. Nose, M., Fujimoto, T., Nishibe, S., Ogihara, Y., *Planta Med.* (1993) **59**, 131–134.

H49. Karigar, C.S., Pujar, B.G., *FEMS Microbiol. Lett.* (1993) **110**, 59–64.

H50. Boyd, D.R., Sharma, N.D., Dorrity, M.R.J., Hand, M.V., McMordie, R.A.S., Malone, J.F., Porter, H.P., Dalton, H., Chima, J., Sheldrake, G.N., *J. Chem. Soc., Perkin Trans. 1* (1993) 1065–1071.

H51. Biehl, M.L., Prelusky, D.B., Koritz, G.D., Hartin, K.E., Buck, W.B., Trenholm, H.L., *Toxicol. Appl. Pharmacol.* (1993) **121**, 152–159.

H52. Cavin, J.F., Andioc, V., Etievant, P.X., Divies, C., *Am. J. Enol. Vitic.* (1993) **44**, 76–80.

H53. Jurima-Romet, M., Crawford, K., Cyr, T., Inaba, T., *Drug Metab. Dispos.* (1994) **22**, 849–857.

H54. Masubuchi, Y., Hosokawa, S., Horie, T., Suzuki, T., Ohmori, S., Kitada, M., Narimatsu, S., *Drug Metab. Dispos.* (1994) **22**, 909–915.

H55. Alton, K.B., Chan, T.-M., Pramanik, B.N., *Drug Metab. Dispos.* (1994) **22**, 866–872.

H56. Babu, S.R., Lakshmi, V.M., Zenser, T.V., Davis, B.B., *Drug Metab. Dispos.* (1994) **22**, 922–927.

H57. Dehal, S.S., Kupfer, D., *Drug Metab. Dispos.* (1994) **22**, 937–946.

H58. Lee, H.S., Jin, C.B., Chong, H.S., Yun, C.H., Park, J.S., Kim, D.H., *Xenobiotica* (1994) **24**, 1053–1062.

H59. Heyn, H., Bankmann, Y.G., Anders, M.W., *Drug Metab. Dispos.* (1994) **22**, 973–974.

H60. Schlenk, D., Bevers, R.J., Vertino, A.M., Cerniglia, C.E., *Xenobiotica* (1994) **24**, 1077–1083.

H61. Kukan, M., Bezek, Š., Pool, W.F., Woolf, T.F., *Xenobiotica* (1994) **24**, 1107–1117.

H62. Wu, W.-N., Hills, J.F., Chang, S.Y., *Xenobiotica* (1994) **24**, 1133–1148.

H63. Clarke, S.E., Austin, N.E., Bloomer, J.C., Haddock, R.E., Higham, F.C., Hollis, F.J., Nash, M., Shardlow, P.C., Tasker, T.C., Woods, F.R., Allen, G.D., *Xenobiotica* (1994) **24**, 1119–1131.

H64. Blinovsky, A.M., Markaryan, A.N., *Enzyme Microb. Technol.* (1993) **15**, 965–973.

H65. Folsom, B.R., Stierli, R., Schwarzenbach, R.P., Zeyer, J., *Environ. Sci. Technol.* (1994) **28**, 306–311.

H66. Vree, T.B., Van Ewijk-Benken Kolmer, E.W.J., Wuis, E.W., Hekster, Y.A., Broekman, M.M.M., *Pharm. World Sci.* (1993) **15**, 197–202.

H67. Watanabe, K., Narimatsu, S., Matsunaga, T., Yamamoto, I., Yoshimura, H., *Biochem. Pharmacol.* (1993) **46**, 405–411.

H68. Hattori, M., Akao, T., Kobashi, K., Namba, T., *Pharmacology* (1993) **47** (S1), 125–133.

H69. Wang, M.Y., Liehr, J.G., *J. Biol. Chem.* (1994) **269**, 284–291.

H70. Iwasa, K., Kamigauchi, M., Takao, N., Cashman, M., *J. Nat. Prod.* (1993) **56**, 2053–2067.

H71. Mackenbrock, U., Gunia, W., Barz, W., *J. Plant Physiol.* (1993) **142**, 385–391.

H72. Tanaka, H., Morimoto, S., Shoyama, Y., *J. Nat. Prod.* (1993) **56**, 2068–2072.

H73. Goodwin, B.L., Ruthven, C.R.J., Sandler, M., *Biochem. Pharmacol.* (1994) **47**, 2294–2297.

H74. Rozman, E., Galceran, M.T., Anglada, L., Albet, C., *J. Pharm. Sci.* (1994) **83**, 252–254.

H76. Nakamura, S., Sugawara, S., Fukushima, T., Ito, Y., Ohashi, M., Tahaiti, O., *Biol. Pharm. Bull.* (1993) **16**, 635–646.

H77. Clement, B., Schultze-Mosgau, M.-H., Wohlers, H., *Biochem. Pharmacol.* (1993) **46**, 2249–2267.

H78. Tai, H.L., McReynolds, J.H., Goldstein, J.A., Eugster, H.P., Sengstag, C., Alworth, W.L., Olson, J.R., *Toxicol. Appl. Pharmacol.* (1993) **123**, 34–42.

H79. Benoit, E., Buronfosse, T., Moroni, P., Delatour, P., Riviere, J.-L., *Biochem. Pharmacol.* (1993) **46**, 2337–2341.

H80. Winer, L., Riov, J., Goren, R., *Physiol. Plant.* (1993) **89**, 220–226.

H81. Bünz, P.V., Cook, A.M., *J. Bacteriol.* (1993) **175**, 6467–6475.

H82. Hofrichter, M., Günther, T., Fritsche, W., *Biodegradation* (1993) **3**, 415–421.

H83. Resnick, S.M., Torok, D.S., Gibson, D.T., *FEMS Microbiol. Lett.* (1993) **113**, 297–302.

H84. Nowak, J.Z., Szymanska, B., Zawilska, J.B., Bialek, B., *J. Pineal Res.* (1993) **15**, 35–42.

H85. Pacifici, G.M., Eligi, M., Giuliani, L., *Eur. J. Clin. Pharmacol.* (1993) **45**, 483–487.

H86. Klasson Wehler, E., Lindberg, L., Jönsson, C.-J., Bergman, Å., *Chemosphere* (1993) **27**, 2397–2412.

H87. Cui, Y., Wang, M., *Eur. J. Metab. Pharmacokinet.* (1993) **18**, 155–160.

H88. Snyderwine, E.G., Buonarati, M.H., Felton, J.S., Turteltaub, K.W., *Carcinogenesis* (1993) **14**, 2517–2522.

H89. Suemori, A., Kurane, R., Tomizuka, N., *Biosci. Biotechnol. Biochem.* (1993) **57**, 1482–1486.

H90. Smekal, O., Reid, G.A., Chapman, S.K., *Biochem. J.* (1994) **297**, 647–652.

H91. Packdibamrung, K., Misono, H., Harada, M., Nagata, S., Nagasaki, S., *J. Biochem.* (1993) **114**, 930–935.

H92. Jin, C.J., Miners, J.O., Burchell, B., Mackenzie, P.I., *Carcinogenesis* (1993) **14**, 2637–2639.

H93. Benner, D.B., Tjeerdema, R.S., *J. Biochem. Toxicol.* (1993) **8**, 111–117.

H94. Hisatsuka, K., Sato, M., *Biosci. Biotechnol. Biochem.* (1994) **58**, 213–214.

H96. Ritter, C.L., Malejka-Giganti, D., *Carcinogenesis* (1994) **15**, 163–170.

H97. Ryu, J.C., Song, Y.S., Kim, M., Cho, J.-H., Yun-Choi, H.S., *Arch. Pharmacol. Res.* (1993) **16**, 213–218.

H98. Nikolova, P., Ward, O.P., *Biotechnol. Lett.* (1994) **16**, 7–10.

H99. Grotz, V.L., Ji, S., Kline, S.A., Goldstein, B.D., Wirtz, G., *Toxicol. Lett.* (1994) **70**, 281–290.

H100. Holder, C.L., Slikker, W., Thompson, H.C., Jr., *Comp. Biochem. Physiol. C: Pharmacol. Toxicol. Endocrinol.* (1994) **107C**, 159–164.

H101. Reinecke, T., Kindl, H., *Phytochemistry* (1994) **35**, 63–66.

H102. Gössner, A., Daniel, S.L., Drake, H.L., *Arch. Microbiol.* (1994) **161**, 126–131.

H103. Raza, H., Montague, W., *Comp. Biochem. Physiol. C: Pharmacol. Toxicol. Endocrinol.* (1994) **107C**, 379–386.

H104. Gibson, S.A., Suflita, J.M., *Biodegradation* (1993) **4**, 51–57.

H105. Akao, T., Che, Q.M., Kobashi, K., Yang, L., Hattori, M., Namba, T., *Appl. Environ. Microbiol.* (1994) **60**, 1041–1043.

H106. Boteju, L.W., Hanna, P.E., *Carcinogenesis* (1993) **14**, 1651–1657.

H108. Karigar, C.S., Ganji, S.H., Pujar, B.G., *Curr. Microbiol.* (1993) **27**, 177–180.

H110. Gorny, N., Schink, B., *Arch. Microbiol.* (1994) **161**, 25–32.

H111. Ganji, S.H., Karigar, C.S., Pujar, B.G., *World J. Microbiol. Biotechnol.* (1993) **9**, 597–598.

H112. Trenz, S.P., Engesser, K.H., Fischer, P., Knackmuss, H.J., *J. Bacteriol.* (1994) **176**, 789–795.
H113. Ueng, T.H., Ueng, Y.F., Chou, M.W., *Toxicol. Lett.* (1994) **70**, 89–99.
H114. Fischer, R.S., Bonner, C.A., Boone, D.R., Jensen, R.A., *Arch. Microbiol.* (1993) **160**, 440–446.
H115. Eisenhofer, G., Finberg, J.P.N., *J. Pharmacol. Exp. Ther.* (1994) **268**, 1242–1251.
H116. Andreotti, G., Cubellis, M.V., Nitti, G., Sannia, G., Mai, X., Marino, G., Adams, M.W.W., *Eur. J. Biochem.* (1994) **220**, 543–549.
H117. Schoenfeld, A., Schulz, M., Petrizka, M., Gassmann, B., *Nahrung* (1993) **37**, 498–500.
H118. Schmidt, B., Thiede, B., Rivero, C., *Pestic. Sci.* (1994) **40**, 231–238.
H119. Spivack, J., Leib, T.K., Lobos, J.H., *J. Biol. Chem.* (1994) **269**, 7323–7329.
H121. Armstrong, S.M., Patel, R., Monacha, S., *Microbios* (1993) **76**, 93–103.
H122. Langner, A., Kemper, S., Nerlich, C., Franke, P., Pfeifer, S., *Pharmazie* (1994) **49**, 169–175.
H123. Tanaka, K., Oda, Y., Asada, A., Fujimori, M., Funae, Y., *Biochem. Pharmacol.* (1994) **47**, 1061–1066.
H124. Kumagai, H., Amemiya, M., Naganawa, H., Sawa, T., Ishizuka, M., Takeuchi, T., *J. Antibiot.* (1994) **47**, 440–446.
H126. Rekoslavskaya, N.I., Bandurski, R.S., *Phytochemistry* (1994) **35**, 905–909.
H127. Chen, Y.P., Lovell, C.R., *Arch. Microbiol.* (1994) **161**, 191–195.
H128. Hashimoto, S., Mizobuchi, M., Kuroda, T., Okabe, H., Mizojiri, K., Takahashi, S., Kikuchi, J., Terui, Y., *Xenobiotica* (1994) **24**, 1177–1193.
H129. Chichester, C.H., Buckpitt, A.R., Chan, A., Plopper, C.G., *Mol. Pharmacol.* (1994) **45**, 664–672.
H130. Uchida, T., Nakamura, E., Usui, T., Imasaki, H., Kawakami, R., Watanabe, T., Higuchi, S., *Xenobiotica* (1994) **24**, 1223–1236.
H131. Tanaka, K., Inoue, T., Kanai, M., Kikuchi, T., *Xenobiotica* (1994) **24**, 1237–1243.
H132. Mason, J.P., Caldwell, J., Dring, L.G., *Xenobiotica* (1995) **25**, 71–80.
H133. Madden, S., Spaldin, V., Hayes, R.N., Woolf, T.F., Pool, W.F., Park, B.K., *Xenobiotica* (1995) **25**, 103–116.
H135. Idowu, O.R., Peggins, J.O., Brewer, T.G., Kelley, C., *Drug Metab. Dispos.* (1995) **23**, 1–17.
H136. Idowu, O.R., Peggins, J.O., Brewer, T.G., *Drug Metab. Dispos.* (1995) **23**, 18–27.
H137. Chauret, N., Li, C., Ducharme, Y., Trimble, L.A., Yergey, J.A., Ramachandran, C., Nicoll-Griffith, D.A., *Drug Metab. Dispos.* (1995) **23**, 65–71.
H138. Nyarko, A.K., Harvison, P.J., *Drug Metab. Dispos.* (1995) **23**, 107–112.
H139. Renner, U., Blanz, J., Freund, S., Waidelich, D., Ehninger, G., Zeller, K.-P., *Drug Metab. Dispos.* (1995) **23**, 94–101.
H140. Tsuruda, L.S., Lamé, M.W., Jones, A.D., *Drug Metab. Dispos.* (1995) **23**, 129–136.
H141. Shore, L.J., Fenselau, C., King, A.R., Dickinson, R.G., *Drug Metab. Dispos.* (1995) **23**, 119–123.
H142. Sweeny, D.J., Nellans, H.N., *Drug Metab. Dispos.* (1995) **23**, 149–153.
H143. Moroni, P., Buronfosse, T., Longin-Sauvageon, C., Delatour, P., Benoit, E., *Drug Metab. Dispos.* (1995) **23**, 160–165.
H144. Ring, J.A., Ghabrial, H., Ching, M.S., Shulkes, A., Smallwood, R.A., Morgan, D.J., *Drug Metab. Dispos.* (1995) **23**, 190–196.
H145. Balani, S.K., Pitzenberger, S.M., Schwartz, M.S., Ramjit, H.G., Thompson, W.J., *Drug Metab. Dispos.* (1995) **23**, 185–189.
H146. Stearns, R.A., Chakravarty, P.K., Chen, R., Chiu, S.-H.L., *Drug Metab. Dispos.* (1995) **23**, 207–215.

H147. Van Beijsterveldt, L.E.C., Geerts, R.J.F., Wuyts, K., Woestenborghs, R.J.H., Meuldermans, W.E.G., Heykants, J.J.P., *Drug Metab. Dispos.* (1995) **23**, 216–222.

H148. Poon, K., Pang, K.S., *Drug Metab. Dispos.* (1995) **23**, 255–260.

H149. Griffin, R.J., Burka, L.T., *Drug Metab. Dispos.* (1995) **23**, 232–239.

H150. Watanabe, K., Matsunaga, T., Yamamoto, I., Yashimura, H., *Drug Metab. Dispos.* (1995) **23**, 261–265.

H151. Balani, S.K., Arison, B.H., Mathai, L., Kauffman, L.R., Miller, R.R., Stearns, R.A., Chen, I.-W., Lin, J.H., *Drug Metab. Dispos.* (1995) **23**, 266–270.

H152. Ware, J.A., Divakaruni, P., Svensson, C.K., *Drug Metab. Dispos.* (1995) **23**, 295–297.

H153. Yun, C.-H., Lee, H.S., Lee, H., Rho, J.K., Jeong, H.G., Guengerich, F.P., *Drug Metab. Dispos.* (1995) **23**, 285–289.

H154. Lake, B.G., Sauer, M.J., Esclangon, F., Beamand, J.A., Price, R.J., Walters, D.G., *Xenobiotica* (1995) **25**, 133–141.

H155. Peters, L.D., Livingstone, D.R., Shenin-Johnson, S., Hines, R.N., Schlenk, D., *Xenobiotica* (1995) **25**, 121–131.

H156. Fowler, S.M., England, P.A., Westlake, A.C.G., Rouch, D.R., Nickerson, D.P., Blunt, C., Braybrook, D., West, S., Wong, L.-L., Flitsch, S.L., *J. Chem. Soc., Chem. Commun.* (1994) 2761–2762.

H157. Hölscher, D., Schneider, B., *J. Chem. Soc., Chem. Commun.* (1995) 525–526.

H158. Rizk, P.N., Hanzlik, R.P., *Xenobiotica* (1995) **25**, 143–150.

H159. Jewell, H., Maggs, J.L., Harrison, A.C., O'Neill, P.M., Ruscoe, J.E., Park, B.K., *Xenobiotica* (1995) **25**, 199–217.

H160. Morse, D.C., Van Bladeren, P.J., Klasson Wehler, E., Brouwer, A., *Xenobiotica* (1995) **25**, 245–260.

H162. Luo, H., Hawes, E.M., McKay, G., Korchinski, E.D., Midha, K.K., *Xenobiotica* (1995) **25**, 291–301.

H163. Ogawa, N., Hirose, T., Fukushima, K., Suwa, T., Satoh, T., *Xenobiotica* (1995) **25**, 283–290.

H164. Chesters, N.C.J.E., O'Hagan, D., Robins, R.J., Kastelle, A., Floss, H.G., *J. Chem. Soc., Chem. Commun.* (1995) 129–130.

H165. Allen, C.C.R., Boyd, D.R., Dalton, H., Sharma, N.D., Haughey, S.A., McMordie, R.A.S., McMurray, B.T., Sheldrake, G.N., Sproule, K., *J. Chem. Soc., Chem. Commun.* (1995) 119–120.

H166. Blake, B.L., Rose, R.L., Mailman, R.B., Levi, P.E., Hodgson, E., *Xenobiotica* (1995) **25**, 377–393.

H167. Agúndez, J.A.G., Martínez, C., Benítez, J., *Xenobiotica* (1995) **25**, 417–427.

H168. Clement, B., Jung, F., *Xenobiotica* (1995) **25**, 443–455.

H169. Villaverde, C., Alvarez, A.I., Redondo, P., Voces, J., Del Estal, J.L., Prieto, J.G., *Xenobiotica* (1995) **25**, 433–441.

H170. Fujimaki, Y., Hosokami, T., Ono, K., *Xenobiotica* (1995) **25**, 501–510.

H171. Smith, D.J., Giddings, J.M., Feil, V.J., Paulson, G.D., *Xenobiotica* (1995) **25**, 511–520.

H172. Lemke, L.E., McQueen, C.A., *Drug Metab. Dispos.* (1995) **23**, 559–565.

H173. Neighbors, S.M., Soine, W.H., *Drug Metab. Dispos.* (1995) **23**, 548–552.

H174. Small, D.S., McNamara, P.J., *Drug Metab. Dispos.* (1995) **23**, 465–472.

H175. Richards, L.E., Sun, J.-H., *Drug Metab. Dispos.* (1995) **23**, 600–602.

H176. Labroo, R.B., Thummel, K.E., Kunze, K.L., Podoll, T., Trager, W.F., Kharasch, E.D., *Drug Metab. Dispos.* (1995) **23**, 490–496.

H177. Walle, T., Walle, U.K., Kumar, G.N., Bhalla, K.N., *Drug Metab. Dispos.* (1995) **23**, 506–512.

H178. Milne, R.W., McLean, C.F., Mather, L.E., Nation, R.L., Runciman, W.B., Rutten, A.J., Somogyi, A.A., *Drug Metab. Dispos.* (1995) **23**, 334–342.

H179. Lin, D.-X., Lang, N.P., Kadlubar, F.F., *Drug Metab. Dispos.* (1995) **23**, 518–524.

H180. Hamdoune, M., Mounie, J., Magdalou, J., Masmoudi, T., Goudonnet, H., Escousse, A., *Drug Metab. Dispos.* (1995) **23**, 343–348.

H181. Lavrijsen, K., van Dyck, D., van Houdt, J., Hendrickx, J., Monbaliu, J., Woestenborghs, R., Meuldermans, W., Heykants, J., *Drug Metab. Dispos.* (1995) **23**, 354–362.

H182. Poon, G.K., Walter, B., Lønning, P.E., Horton, M.N., McCague, R., *Drug Metab. Dispos.* (1995) **23**, 377–382.

H183. Webster, L.K., Ellis, A.G., Kestell, P., Rewcastle, G.W., *Drug Metab. Dispos.* (1995) **23**, 363–368.

H184. Cribb, A.E., Spielberg, S.P., Griffin, G.P., *Drug Metab. Dispos.* (1995) **23**, 406–414.

H193. Yun, C.-L., Lee, H.S., Lee, H., Rho, J.K., Jeong, H.G., Guengerich, F.P., *Drug Metab. Dispos.* (1995) **23**, 285–289.

H194. Mugford, C.A., Tarloff, J.B., *Drug Metab. Dispos.* (1995) **23**, 290–294.

H196. Ling, K.-H.J., Leeson, G.A., Burmaster, S.D., Hook, R.H., Reith, M.K., Cheng, L.K., *Drug Metab. Dispos.* (1995) **23**, 631–636.

H197. Lin, L.Y., Kumagai, Y., Hiratsuka, A., Narimatsu, S., Suzuki, T., Funae, Y., DiStefano, E.W., Cho, A.K., *Drug Metab. Dispos.* (1995) **23**, 610–614.

H198. Hansen, K.T., Stentoft, K., *Xenobiotica* (1995) **25**, 611–622.

H199. Seaton, C.L., Tjeerdema, R.S., *Xenobiotica* (1995) **25**, 553–562.

H200. McKillop, D., Simons, P.J., Cockshott, I.D., Hill, S.J., Harding. J.R., Cooper, K.J., Jones, D.C., *Xenobiotica* (1995) **25**, 623–624.

H201. Prueksaritanont, T., Balani, S.K., Dwyer, L.M., Ellis, J.D., Kauffman, L.R., Varga, S.L., Pitzenberger, S.M., Theoharides, A.D., *Drug Metab. Dispos.* (1995) **23**, 688–695.

H202. Chan, E., McLachlan, A.J., Rowland, M., *Eur. J. Pharm. Sci.* (1994) **1**, 189–193.

H203. Cannell, R.J.P., Knaggs, A.R., Dawson, M.J., Manchee, G.R., Eddershaw, P.J., Waterhouse, I., Sutherland, D.R., Bowers, G.D., Sidebottom, P.J., *Drug Metab. Dispos.* (1995) **23**, 724–729.

H204. Derewlany, L.O., Knie, B., Koren, G., *J. Pharmacol. Exp. Ther.* (1994) **269**, 761–765.

H205. den Besten, C., Bennik, M.M.H., van Iersel, M., Peters, M.A.W., Teunis, C., van Bladeren, P.J., *Chem.-Biol. Interact.* (1994) **90**, 121–137.

H206. Arunachalam, U., Massey, V., *J. Biol. Chem.* (1994) **269**, 11795–11801.

H207. Akhtar, M.H., Mahadevan, S., Paquet, A., *J. Environ. Sci. Health, Pt. B* (1994) **B29**, 369–394.

H208. Kiljunen, E., Kanerva, L.T., *Tetrahedron: Asymmetry* (1994) **5**, 311–314.

H209. Ruoff, W.L., Dziuk, P.J., *Domest. Anim. Endocrinol.* (1994) **11**, 197–208.

H210. Hasspieler, B.M., Di Giulio, R.T., *Toxicol. Appl. Pharmacol.* (1994) **125**, 184–191.

H211. Bhavnani, B.R., Cecutti, A., Wallace, D., *Steroids* (1994) **59**, 389–394.

H212. Romanov, V., Hausinger, R.P., *J. Bacteriol.* (1994) **176**, 3368–3374.

H213. Hou, C.T., Jackson, M.A., Bagley, M.O., Becker, L.A., *Appl. Microbiol. Biotechnol.* (1994) **41**, 178–182.

H214. White, K.N., Vale, V.L., Hope, D.B., *Biochem. Soc. Trans.* (1994) **22**, 220S.

H215. Boopathy, R., Kulpa, C.F., *Can. J. Microbiol.* (1994) **40**, 273–278.

H216. Yang, G., Liang, P.-H., Dunaway-Mariano, D., *Biochemistry* (1994) **33**, 8527–8531.

H217. Bünz, P.V., Falchetto, R., Cook, A.M., *Biodegradation* (1993) **4**, 171–178.

H218. Youn, W.G., Hyun, S.H., Kim, D.H., Kim, N.J., Hong, N.D., *Yakhak Hoechi* (1993) **37**, 262–269.

H219. Keener, W.K., Arp, D.J., *Appl. Environ. Microbiol.* (1994) **60**, 1914–1920.

H220. Sugimoto, Y., Yoshida, A., Uchida, S., Inanaga, S., Yamada, Y., *Phytochemistry* (1994) **36**, 679–683.

H221. Stringfellow, W.T., Aitken, M.D., *Can. J. Microbiol.* (1994) **40**, 432–438.

H222. van Iersel, M., Walters, D.G., Price, R.J., Lovell, D.P., Lake, B.G., *Food Chem. Toxicol.* (1994) **32**, 387–390.

H223. Hooper, N.M., Hesp, R.J., Tieku, S., Boesten, W.H.J., Toniolo, C., Kamphuis, J., *J. Agric. Food Chem.* (1994) **42**, 1397–1401.

H224. Ravis, W.R., Valaer, A.K., *Life Sci.* (1994) **54**, PL519-PL524.

H225. Kamimura, M., *Sei Marianna Ika Daigaku Zasshi* (1993) **21**, 77–86.

H226. Brownsill, R., Combal, J.P., Taylor, A., Bertrand, M., Luitjen, W., Walther, B., *Rapid Commun. Mass Spectrom.* (1994) **8**, 361–365.

H227. Morita, K., Mizuochi, M., Yamaji, A., Yokoyama, T., *Biol. Pharm. Bull.* (1994) **17**, 531–534.

H228. McKown, L.A., Wu, W.N., O'Neill, P.J., *J. Pharm. Biomed. Anal.* (1994) **12**, 771–775.

H229. Wu, W.N., Masucci, J.A., *J. Pharm. Biomed. Anal.* (1994) **12**, 569–571.

H230. Maser, E., Bannenberg, G., *Biochem. Pharmacol.* (1994) **47**, 1805–1812.

H231. Maruyama, W., Nakahara, D., Naoi, M., *J. Neural Transm.: Parkinson's Dis. Dementia Sect.* (1994) **7**, 21–33.

H232. Gibello, A., Ferrer, E., Martin, M., Garrito-Pertierra, A., *Biochem. J.* (1994) **301**, 145–150.

H234. Funayama, M., Arakawa, H., Yamamoto, R., Nishino, T., Shin, T., Murao, S., *Biosci. Biotechnol. Biochem.* (1994) **58**, 817–821.

H235. Jiménez-Cervantes, C., Solano, F., Kobayashi, T., Urabe, K., Hearing, V.J., Lozano, J.A., Garcia-Borrón, J.C., *J. Biol. Chem.* (1994) **269**, 17993–18001.

H236. Jones, K.H., Trudgill, P.W., Hopper, D.J., *Appl. Environ. Microbiol.* (1994) **60**, 1978–1983.

H237. Fiedler, F., Hinz, H., *Eur. J. Biochem.* (1994) **222**, 75–81.

H238. Haigler, B.E., Nishino, S.F., Spain, J.C., *J. Bacteriol.* (1994) **176**, 3433–3437.

H239. Junker, F., Field, J.A., Bangerter, F., Ramsteiner, K., Kohler, H.-P., Joannou, C.L., Mason, J.R., Leisinger, T., Cook, A.M., *Biochem. J.* (1994) **300**, 429–436.

H240. Inoue, M., Inoue, T., Okami, M., Sayama, M., Hirai, Y., *Chem. Lett.* (1994) 1315–1316.

H241. Guillen, F., Martinez, A.T., Martinez, M.J., Evans, C.S., *Appl. Microbiol. Biotechnol.* (1994) **41**, 465–470.

H242. Hara, M., Tanaka, S., Tabata, M., *Phytochemistry* (1994) **36**, 327–332.

H243. De Lombaert, S., Singh, K., Blanchard, L., Soliman, V.F., *Bioorg. Med. Chem. Lett.* (1994) **4**, 899–902.

H244. Chiba, K., Saitoh, A., Koyama, E., Tani, M., Hayashi, M., Ishizaki, T., *Br. J. Clin. Pharmacol.* (1994) **37**, 237–242.

H245. Dixon, C.M., Park, G.R., Tarbit, M.H., *Biochem. Pharmacol.* (1994) **47**, 1253–1257.

H246. Henrikus, B., Kampffmeyer, H.G., *Skin Pharmacol.* (1994) **6**, 246–252.

H247. Sone, T., Zukowski, K., Land, S.J., Kin, C.M., Martin, B.M., Pohl, L.R., Wang, C.Y., *Carcinogenesis* (1994) **15**, 595–599.

H248. Genet, R., Denoyelle, C., Menez, A., *J. Biol. Chem.* (1994) **169**, 18177–18184.

H250. Swahn, C.-G., Halldin, C., Farde, L., Sedvall, G., *Hum. Psychopharmacol.* (1994) **9**, 25–31.

H251. Resnick, S.M., Gibson, D.T., *Biodegradation* (1993) **4**, 195–203.

H252. Mahajan, M.C., Phale, P.S., Vaidyanathan, C.S., *Arch. Microbiol.* (1994) **161**, 425–433.

H253. Warhurst, A.M., Fewson, C.A., *Appl. Environ. Microbiol.* (1994) **60**, 1137–1145.

H254. Grifoll, M., Selifonov, S.A., Chapman, P.J., *Appl. Environ. Microbiol.* (1994) **60**, 2438–2449.

H255. Zenk, M.H., *Tetrahedron Lett.* (1994) **35**, 3897–3900.

H256. Monegier, B., Gaillard, C., Sablé, S., Vuilhorgne, M., *Tetrahedron Lett.* (1994) **35**, 3715–3718.

H257. Edwards, R., *J. Plant Physiol.* (1994) **143**, 609–614.

H258. Guan, X., Kaiser, D.G., Halstead, G.W., Baillie, T.A., *J. Pharmacol. Exp. Ther.* (1994) **269**, 1166–1175.

H259. Paulikova, I., Helia, O., *Pharmazie* (1994) **49**, 531–532.

H260. Harris, J.W., Rahman, A., Kim, B.-R., Guengerich, F.P., Collins, J.M., *Cancer Res.* (1994) **54**, 4026–4035.

H261. Singer, S.S., *Chem.-Biol. Interact.* (1994) **92**, 33–45.

H262. Gorski, J.C., Jones, D.R., Wrighton, S.A., Hall, S.D., *Biochem. Pharmacol.* (1994) **48**, 173–182.

H263. Low, C.-M., Ulgen, M., Gorrod, J.W., *J. Pharm. Pharmacol.* (1994) **46**, 585–590.

H264. Koss, G., Weider, T., Seubert, A., *Food Chem. Toxicol.* (1994) **32**, 605–610.

H265. Arensdorf, J.J., Focht, D.D., *Appl. Environ. Microbiol.* (1994) **60**, 2884–2889.

H266. Ibe, B.O., Raj, J.U., *Exp. Lung Res.* (1994) **20**, 207–222.

H267. Nair, A.J., Rao, J.M., Sudhakaran, P.R., *Plant Sci.* (1994) **96**, 217–221.

H268. Sutherland, J.B., Evans, F.E., Freeman, J.P., Williams, A.J., Deck, J., Cerniglia, C.E., *Mycologia* (1994) **86**, 117–120.

H269. Stindl, A., Keller, U., *Biochemistry* (1994) **33**, 9358–9364.

H270. Jain, R.K., Dreisbach, J.H., Spain, J.C., *Appl. Environ. Microbiol.* (1994) **60**, 3030–3032.

H271. Wood, A.M., Truscott, R.J.W., *Vision Res.* (1994) **34**, 1369–1374.

H272. Hirose, J., Kimura, N., Suyama, A., Kobayashi, A., Hayashida, S., Furukama, K., *FEMS Microbiol. Lett.* (1994) **118**, 273–278.

H273. Cnubben, N.P.H., Peelen, S., Borst, J.-W., Vervoort, J., Veeger, C., Rietjens, I.M.C.M., *Chem. Res. Toxicol.* (1994) **7**, 590–598.

H274. Okuuchi, K., Tomita, Y., Onodera, R., *Anim. Sci. Technol.* (1993) **64**, 1182–1186.

H275. Maladkar, N.K., *Amino Acids* (1992) **3**, 235–241.

H276. Pacifici, G.M., Temellini, A., Castiglioni, M., D'Alessandro, C., Ducci, A., Giuliani, L., *Chem.-Biol. Interact.* (1994) **92**, 219–231.

H277. van Berkel, W.J.H., Michel, E.H.M., Middelhoven, W.J., Vervoort, J., Rietjens, I.M.C.M., *FEMS Microbiol. Lett.* (1994) **121**, 207–216.

H278. Derewlany, L.O., Koren, G., *J. Lab. Clin. Med.* (1994) **124**, 134–141.

H279. Crooks, G.P., Copley, S.D., *Biochemistry* (1994) **33**, 11645–11649.

H280. Rao, J.R., Cooper, J.E., *J. Bacteriol.* (1994) **176**, 5409–5413.

H282. Hadley, M.R., Švajdlenka, E., Damani, L.A., Oldham, H.G., Tribe, J., Camilleri, P., Hutt, A.J., *Chirality* (1994) **6**, 91–97.

H283. Wikberg, T., Ottoila, P., Taskinen, J., *Eur. J. Drug Metab. Pharmacokinet.* (1993) **18**, 359–367.

H284. Rieble, S., Joshi, D.K., Gold, M.H., *J. Bacteriol.* (1994) **176**, 4838–4844.

H285. Sahali, Y., Kidd, L.R., Skipper, P.L., Tannenbaum, S.R., *Cancer Lett.* (1994) **83**, 299–303.

H286. Pare, P.W., Wang, H.-B., Davin, L.B., Lewis, N.G., *Tetrahedron Lett.* (1994) **35**, 4731–4734.

H287. Lázníček, M., Lázníčková, A., *Eur. J. Drug Metab. Pharmacokinet.* (1994) **19**, 21–26.

H288. Boopathy, R., Manning, J., Montemagno, C., Rimkus, K., *Can. J. Microbiol.* (1994) **40**, 787–790.

H289. Menzel, S., Waibel, R., Brune, K., Geisslinger, G., *Biochem. Pharmacol.* (1994) **48**, 1056–1058.

H290. Richter, P., Kauffman, F.C., Zaleski, J., *Toxicol. Lett.* (1994) **74**, 79–90.

H291. Hart, R.W., Xiao, Y., Von Tungeln, L.S., Chou, W.M., Zhan, D.-J., Fu, P.P., *Polycyclic Aromat. Compd.* (1994) **7**, 19–26.
H292. Van Dongen, R.T.M., Crul, B.J.P., Koopman-Kimenai, P.M., Vree, T.B., *Br. J. Clin. Pharmacol.* (1994) **38**, 271–273.
H293. Gorny, N., Schink, B., *Arch. Microbiol.* (1994) **162**, 131–135.
H295. Babu, S.R., Lakshmi, V.M., Owens, I.S., Zenser, T.V., Davis, B.B., *Carcinogenesis* (1994) **15**, 2003–2007.
H296. Nishimura, T., Kometani, T., Takii, H., Terada, Y., Okada, S., *J. Ferment. Bioeng.* (1994) **78**, 31–36.
H297. Junker, F., Leisinger, T., Cook, A.M., *Microbiology (Reading U.K.)* (1994) **140**, 1713–1722.
H298. Matsuzaki, R., Fukui, T., Sato, H., Ozaki, Y., Tanizawa, K., *FEBS Lett.* (1994) **351**, 360–364.
H299. Russ, R., Müller, C., Knackmuss, H.-J., Stolz, A., *FEMS Microbiol. Lett.* (1994) **122**, 137–144.
H300. Duskova, J., Jahodar, L., Dusek, J., *Pharmazie* (1994) **49**, 624.
H301. Rueffer, M., Zenk, M.H., *Phytochemistry* (1994) **36**, 1219–1223.
H302. Beutling, D., *Arch. Lebensmittelhyg.* (1993) **44**, 83–84 and 86–87.
H303. Tellitu, I., Badia, D., Dominguez, E., Garcia, F.J., *Tetrahedron: Asymmetry* (1994) **5**, 1567–1578.
H304. Glissen, R.A.H.J., Hume, R., Meerman, J.H.N., Coughtrie, M.W.H., *Biochem. Pharmacol.* (1994) **48**, 837–840.
H305. Markus, B., Kwon, C.-H., *J. Pharm. Sci.* (1994) **83**, 1729–1734.
H306. Mai, X., Adams, M.W.W., *J. Biol. Chem.* (1994) **269**, 16726–16732.
H307. Bouvrette, P., Luong, J.H.T., *Appl. Biochem. Biotechnol.* (1994) **48**, 61–74.
H308. Leinhos, V., Udagama-Randeniya, P.V., Savidge, R.A., *Phytochemistry* (1994) **37**, 311–315.
H309. Falconnier, B., Lapierre, C., Lesage-Meessen, L., Yonnet, G., Brunerie, P., Colonna-Ceccaldi, B., Corrieu, G., Asther, M., *J. Biotechnol.* (1994) **37**, 123–132.
H310. Menting, J.G.T., Scopes, R.K., Stevenson, T.W., *Plant Physiol.* (1994) **106**, 633–642.
H311. Pahernik, S.A., Schmid, J., Sauter, T., Schildberg, F.W., Koebe, H.-G., *Xenobiotica* (1995) **25**, 811–823.
H312. He, H., McKay, G., Midha, K.K., *Xenobiotica* (1995) **25**, 857–872.
H313. Nakaoka, M., Katoh, K., Hakusui, H., *Xenobiotica* (1995) **25**, 847–855.
H314. Hughes, M.F., Hall, L.L., *Xenobiotica* (1995) **25**, 873–883.
H315. Reith, M.K., Sproles, G.D., Cheng, L.K., *Drug Metab. Dispos.* (1995) **23**, 806–812.
H316. Dain, J.G., Nicoletti, J., *Xenobiotica* (1995) **25**, 951–961.
H317. Hadley, M.R., Oldham, H.G., Damani, L.A., Hutt, A.J., *Chirality* (1994) **6**, 98–104.
H318. Haritos, V.S., Brumley, C.M., Holdway, D.A., Ahokas, J.T., *Xenobiotica* (1995) **25**, 963–971.
H321. Cook, C.S., Ames, G.B., Hribar, J.D., Liu, N.W.K., Sweeney, D.L., Schoenhard, G.L., Karim, A., *Xenobiotica* (1995) **25**, 981–991.
H322. Kokkinakis, D.M., Moschel, R.C., Pegg, A.E., Dolan, M.E., Schold, S.C., Jr., *Chem. Res. Toxicol.* (1994) **7**, 762–769.
H323. Barrow, A., Camp, S.J., Dayal, S., Jenner, W.N., Lashmar, D., Oxford, J.M., Palmer, E., Scully, N.L., Curtis, G.C., Hughes, H.M., Park, G.W., *Xenobiotica* (1995) **25**, 993–1007.
H324. Surapaneni, S., Khahil, S.K.W., *Chirality* (1994) **6**, 479–483.
H325. Monsarrat, B., Alvinerie, P., Gares, M., Wright, M., Dubois, J., Gueritte-Voegelein, F., Guenard, D., Donehower, R., Rowinsky, E., *Cell. Pharmacol.* (1993) **1** (Suppl. 1), S77-S81.
H328. Boopathy, R., *Arch. Microbiol.* (1994) 162–172.

H329. Cerniglia, C.E., Gibson, D.T., Dodge, R.N., *Appl. Environ. Microbiol.* (1994) **60**, 3931–3938.

H330. Goffner, D., Campbell, M.M., Campargue, C., Clastre, M., Borderies, G., Boudet, A., Boudet, A.M., *Plant Physiol.* (1994) **106**, 625–632.

H331. Sibbesen, O., Koch, B., Halkier, B.A., Moller, B.L., *Proc. Natl. Acad. Sci. U.S.A.* (1994) **91**, 9740–9744.

H332. Sutherland, J.B., Freeman, J.P., Williams, A.J., Cerniglia, C.E., *Exp. Mycol.* (1994) **18**, 271–274.

H333. Chen, C.L., Sangiah, S., Chen, H., Berlin, K.D., Garrison, G.L., Scherlag, B.J., Lazzara, R., *J. Liq. Chromatogr.* (1994) **17**, 3681–3694.

H334. Corkery, D.M., O'Connor, K.E., Buckley, C.M., Dobson, A.D.W., *FEMS Microbiol. Lett.* (1994) **124**, 23–28.

H336. Rafii, F., Selby, A.L., Newton, R.K., Cerniglia, C.E., *Appl. Environ. Microbiol.* (1994) **60**, 4263–4267.

H337. Reichlin, F., Kohler, H.-P.E., *Appl. Environ. Microbiol.* (1994) **60**, 4587–4591.

H338. Solano, F., Martinez-Liarte, J.H., Jimenez-Cervantes, C., Garcia-Barron, J.C., Lozano, J.A., *Biochem. Biophys. Res. Commun.* (1994) **204**, 1243–1250.

H339. Czich, A., Bartsch, I., Dogra, S., Hornhardt, S., Glatt, H.R., *Chem.-Biol. Interact.* (1994) **92**, 119–128.

H340. Schleissner, C., Olivera, E.R., Fernandez-Valverde, M., Luengro, J.M., *J. Bacteriol.* (1994) **176**, 7667–7676.

H341. Suemori, A., Nakajima, K., Kurane, R., Nakamura, Y., *FEMS Microbiol. Lett.* (1995) **125**, 31–36.

H342. Suchar, L.A., Chang, R.L., Rosen, R.T., Lech, J., Conney, A.H., *J. Pharmacol. Exp. Ther.* (1995) **272**, 197–206.

H343. Swain, E., Poulton, J.E., *Plant Physiol.* (1994) **106**, 1285–1291.

H344. An, D., Gibson, D.T., Spain, J.C., *J. Bacteriol.* (1994) **176**, 7462–7467.

H345. Alexander, J., Fossum, B.H., Holme, J.A., *Environ. Health Perspect. Suppl.* (1994) **102** (Suppl. 6), 109–114.

H346. Degrassi, G., De Laureto, P.P., Bruschi, C.V., *Appl. Environ, Microbiol.* (1995) **61**, 326–332.

H347. Zaborina, O., Latus, M., Eberspächer, J., Golovleva, L.A., Lingens, F., *J. Bacteriol.* (1995) **177**, 229–234.

H348. Koga, J., Syono, K., Ichikawa, T., Adachi, T., *Biochim. Biophys. Acta* (1994) **1209**, 241–247.

H349. Tuominen, H., Östin, A., Sandberg, G., Sundberg, B., *Plant Physiol.* (1994) **106**, 1511–1520.

H350. Wu, W.N., McKown, L.A., O'Neill, P.J., *J. Pharm. Sci.* (1995) **84**, 185–189.

H351. Todorova, M., Werner, C., Hesse, M., *Phytochemistry* (1994) **37**, 1251–1256.

H352. Nagata, T., Fukuda, Y., *J. Pharm. Pharmacol.* (1994) **46**, 1004–1012.

H353. Ponousis, C., Kettle, A.J., Phillips, D.R., *Biochem. Pharmacol.* (1994) **48**, 2223–2230.

H354. Konishi-Imamura, L., Sato, M., Kohi, K., Kadota, S., Namba, T., Kobashi, K., *Biol. Pharm. Bull.* (1994) **17**, 1018–1022.

H356. Edlin, D.A.N., Narbad, A., Dickinson, J.R., Lloyd, D., *FEMS Microbiol. Lett.* (1995) **125**, 311–316.

H357. Kozlovsky, S.A., Zaitsev, G.M., Kunc, F., *Folia Microbiol.* (*Prague*) (1994) **39**, 399–403.

H358. Whited, G.M., Downie, J.C., Hudlicky, T., Fearnley, S.P., Dudding, T.C., Olivo, H.F., Parker, D., *Bioorg. Med. Chem.* (1994) **2**, 727–734.

H360. Williams, M.L., Lennard, M.S., Martin, I.J., Tucker, G.T., *Carcinogenesis* (1994) **15**, 2733–2738.

H361. Wu, R.-F., Ichikawa, Y., *FEBS Lett.* (1995) **358**, 145–148.

H362. Stiborova, M., Asfaw, B., Frei, E., Schmeiser, H.H., Wiessler, M., *Collect. Czech. Chem. Commun.* (1994) **59**, 2727–2733.

H363. Goodwin, B.L., Ruthven, C.R.J., Sandler, M., *Biochem. Pharmacol.* (1994) **47**, 2294–2297.

H364. Huang, Z., Dostal, L., Rosazza, J.P.N., *J. Bacteriol.* (1994) **176**, 5912–5918.

H365. Amann, T., Roos, P.H., Huh, H., Zenk, M.H., *Heterocycles* (1995) **40**, 425–440.

H366. Kobayashi, M., Suzuki, T., Fujita, T., Masuda, M., Shimizu, S., *Proc. Natl. Acad. Sci. U.S.A.* (1995) **92**, 714–718.

H367. Hyun, Y.-L., Davidson, V.L., *Biochemistry* (1995) **34**, 816–823.

H368. Bauer, I., de Beyer, A., Tshisuaka, B., Fetzner, S., Lingens, F., *FEMS Microbiol. Lett.* (1994) **117**, 299–304.

H369. O'Connor, K., Buckley, C.M., Hartmans, S., Dobson, A.D.W., *Appl. Environ. Microbiol.* (1995) **61**, 544–548.

H370. Banoo, R., Mathieson, D.W., *J. Bangladesh Acad. Sci.* (1994) **18**, 57–61.

H371. Sakakibara, Y., Masahito, S., Nakajima, H., Liu, M.-C., *Biochem. J.* (1995) **305**, 993–998.

H372. Easton, C.J., Hutton, C.A., *J. Chem. Soc., Perkin Trans. 1* (1994) 3545–3548.

H374. Stull, T.L., Hyun, L., Sharetzky, C., Wooten, J., McCauley, J.P., Jr., Smith, A.B. III, *J. Biol. Chem.* (1995) **270**, 5–8.

H375. Resnick, S.M., Torok, D.S., Lee, K., Brand, J.M., Gibson, D.T., *Appl. Environ. Microbiol.* (1994) **60**, 3323–3328.

H376. Akerboom, T.P.M., Sies, H., Ziegler, D.M., *Arch. Biochem. Biophys.* (1995) **316**, 220–226.

H377. Tsui, B.C.H., Feng, J.D.Z., Buckley, S.J., Yeung, P.K.F., *Eur. J. Drug Metab. Pharmacokinet.* (1994) **19**, 369–373.

H378. Youn, H.-D., Kim, K.-J., Maeng, J.-S., Han, Y.-H., Jeong, I.-B., Jeong, G., Kang, S.-O., Hah, Y.C., *Microbiology (Reading U.K.)* (1995) **141**, 393–398.

H379. Roberts-Thomson, S.J., McManus, M.E., Tukey, R.H., Gonzalez, F.J., Holder, G.M., *Cancer Res.* (1995) **55**, 1052–1059.

H380. Nagy, P.L., Marolewski, A., Benkovic, S.J., Zalkin, H., *J. Bacteriol.* (1995) **177**, 1292–1298.

H381. Ralet, M.-C., Faulds, C.B., Williamson, G., Thibault, J.-F., *Carbohydr. Res.* (1994) **263**, 257–269.

H382. Forbes, B.J.R., Hamilton, G.A., *Bioorg. Chem.* (1994) **22**, 343–361.

H383. Lam, W.W.Y., Bugg, T.D.H., *J. Chem. Soc., Chem. Commun.* (1994) 1163–1164.

H384. Emanuele, J.J., Fitzpatrick, P.F., *Biochemistry* (1995) **34**, 3716–3723.

H385. Hamid, M.A., Iwaku, M., Hoshino, E., *Arch. Oral. Biol.* (1994) **39**, 967–972.

H386. Rani, N.L., Lalithakumari, D., *Can. J. Microbiol.* (1994) **40**, 1000–1006.

H387. Turner, J.E., Allison, N., *J. Appl. Bacteriol.* (1995) **78**, 125–133.

H388. Wunder, T., Kremer, S., Sterner, O., Anke, H., *Appl. Microbiol. Biotechnol.* (1994) **42**, 636–641.

H389. Van Berkel, W.J., Fraaije, M.W., De Jong, E., *PCT Int. Appl. WO* 9,502,062 A1(19–1–1995).

H390. Pirmahoned, M., Williams, D., Madden, S., Templeton, E., Park, B.K., *J. Pharmacol. Exp. Ther.* (1995) **272**, 984–990.

H391. Ulgen, M., Kajbaf, M., Lamb, J.H., Jahanshahi, M., Gorrod, J.W., *Eur. J. Drug Metab. Pharmacokinet.* (1994) **19**, 343–348.

H392. Loza, M.C., *J. Steroid Biochem. Mol. Biol.* (1995) **52**, 277–280.

H393. Viau, C., Vyskocil, A., *Sci. Total Environ.* (1995) **163**, 187–190.

H394. Nagai, T., Serrano, A.E., Jr., Nagayama, F., *Fish Sci.* (1994) **60**, 445–448.

H395. Funk, M., Gath, I., Seidel, A., Oesch, F., Platt, K.L., *Chem.-Biol. Interact.* (1995) **95**, 189–201.

H396. Schläfli, H.R., Weiss, M.A., Leisinger, T., Cook, A.M., *J. Bacteriol.* (1994) **176**, 6644–6652.

H397. Niemetz, R., Altenschmidt, U., Brucker, S., Fuchs, G., *Eur. J. Biochem.* (1995) **227**, 161–168.

H398. Kis, K., Volk, R., Bacher, A., *Biochemistry* (1995) **34**, 2883–2892.

H399. Suarez, M., Ferrer, E., Garrido-Pertierra, A., Martin, M., *FEMS Microbiol. Lett.* (1995) **126**, 283–290.

H400. Middelhoven, W.J., Sollewijn, G., Maarten, D., *Antonie van Leeuwenhoek* (1995) **67**, 217–219.

H401. Hara, M., Morio, H., Yazaki, K., Tanaka, S., Tabata, M., *Phytochemistry* (1995) **38**, 89–93.

H402. Wetzel, A., Sandermann, H.J., Jr., *Arch. Biochem. Biophys.* (1994) **314**, 323–328.

H403. Prosen, E., Ward, O.P., Collins, S., Dewdney, N.J., Hong, Y., Wilcocks, R., *Biocatalysis* (1993) **8**, 21–29.

H404. Sawchuk, R.J., Maloney, J.A., Cartier, L.L., Rackley, R.J., Chan, K.K.H., Lau, H.S.L., *Pharm. Res.* (1995) **12**, 756–762.

H405. Vree, T.B., van der Ven, A.J.A.M., Koopmans, P.P., Kolmer, E.W.J., van, E.-B., van Wissen, C.P.W.G.M.V., *Clin. Drug Invest.* (1995) **9**, 43–53.

H406. Yasuda, T., Kano, Y., Saito, K., Ohsawa, K., *Biol. Pharm. Bull.* (1994) **17**, 1369–1374.

H407. Soraci, A., Benoit, E., Jaussaud, P., Lees, P., Delatour, P., *Am. J. Vet. Res.* (1995) **56**, 358–361.

H408. Yasuda, T., Kano, Y., Saito, K., Ohsawa, K., *Biol. Pharm. Bull.* (1995) **18**, 300–303.

H409. Noda, A., Ono, Y., Inokuchi, K., Eto, S., Noda, H., Muro, H., Uchida, T., *Biol. Pharm. Bull.* (1994) **17**, 1433–1435.

H410. Surh, Y.-J., Ahn, S.H., Kim, K.-C., Park, J.-B., Sohn, Y.W., Lee, S.S., *Life Sci.* (1995) **56**, PL305-PL311.

H411. Gu, Y., Knaebel, D.B., Korus, R.A., Crawford, R.L., *Environ. Sci. Technol.* (1995) **29**, 1622–1627.

H412. Fujita, S., bin Saari, N., Maegawa, M., Tetsuka, T., Tono, T., *J. Agric. Food Chem.* (1995) **43**, 1138–1142.

H413. Singer, S.S., Hagedorn, J.E., Smith, D.M., Williams, J.L., *J. Pineal Res.* (1995) **18**, 49–55.

H414. Falkenhagen, H., Stoeckigt, J., *J. Naturforsch. C: Biosci.* (1995) **50**, 45–53.

H415. Phale, P.S., Mahajan, M.C., Vaidyanathan, C.S., *Arch. Microbiol.* (1995) **163**, 42–47.

H416. Takeshita, M., Miura, M., Unuma, Y., Iwaim, S., Sato, I., Arai, T., Kosaka, K., *Res. Commun. Mol. Pathol. Pharmacol.* (1995) **88**, 123–126.

H417. Connors, M.S., Malfatti, M.A., Felton, J.S., *Chem.-Biol. Interact.* (1995) **96**, 185–202.

H418. Matsunaga, T., Iwawaki, Y., Watanabe, K., Yamamoto, I., Kageyama, T., Yoshimura, H., *Life Sci.* (1995) **56**, 2089–2095.

H419. Romberg, R.W., Lee, L., *J. Anal. Toxicol.* (1995) **19**, 157–162.

H420. Peelen, S., Rietjens, I.M.C.M., Boersma, M.G., Vervoort, J., *Eur. J. Biochem.* (1995) **227**, 284–291.

H421. Edwards, S.L., Davidson, V.L., Hyun, Y.-L., Wingfield, P.T., *J. Biol. Chem.* (1995) **270**, 4293–4298.

H422. Hanlon, S.P., Carpenter, K., Hassan, A., Cooper, R.A., *Biochem. J.* (1995) **306**, 627–630.

H423. Sibbesen, O., Koch, B., Halkier, B.A., Moeller, B.L., *J. Biol. Chem.* (1995) **270**, 3506–3511.

H424. Ganji, S.H., Karigar, C.S., Pujar, B.G., *Biodegradation* (1995) **6**, 61–66.

H425. Nikolova, P., Ward, O.P., *Biocatalysis* (1994) **9**, 329–341.

H426. Andreoni, V., Bernasconi, S., Bestetti, G., *Appl. Microbiol. Biotechnol.* (1995) **42**, 830–835.

H427. Stano, J., Nemec, P., Weissova, K., Kovacs, P., Kakoniova, D., Liskova, D., *Phytochemistry* (1995) **38**, 859–860.

H428. Brown, R.T., Carter, N.E., Scheinmann, F., Turner, N.J., *tetrahedron Lett.* (1995) **36**, 1117–1120.

H429. Aso, Y., Yamamoto, K., Yoshinaga, T., Yamamoto, H., Yamagami, T., *Biosci. Biotechnol. Biochem.* (1995) **59**, 277–281.

H430. Caputo, O., Rocco, F., Grosa, G., *Eur. J. Drug Metab. Pharmacokinet.* (1994) **19**, 303–310.

H431. Pichard, L., Curi-Pedrosa, R., Bonfils, C., Jacqz-Aigrain, E., Domergue, J., Joyeux, H., Cosme, J., Guengerich, F.P., Maurel, P., *Mol. Pharmacol.* (1995) **47**, 410–418.

H432. Tsunematsu, H., Horie, H., Futatsuka, Y., Yamamoto, M., *Drug Delivery Syst.* (1994) **9**, 331–338.

H433. Gilbert, R.B., Peng, P.I., Wong, D., *J. Anal. Toxicol.* (1995) **19**, 84–86.

H434. Chou, H.-C., Lang, N.P., Kadlubar, F.F., *Carcinogenesis* (1995) **16**, 413–417.

H435. Ciskanik, L.M., Wilczek, J.M., Fallon, R.D., *Appl. Environ. Microbiol.* (1995) **61**, 998–1003.

H436. Schmidt, S.R., Müller, C.R., Kress, W., *Eur. J. Biochem.* (1995) **228**, 425–430.

H437. Andreotti, G., Cubellis, M.V., Nitti, G., Sannia, G., Mai, X., Adams, M.W.W., Marino, G., *Biochim. Biophys. Acta* (1995) **1247**, 90–96.

H438. Suemori, A., Kurane, R., Tomizuka, N., *Biosci. Biotech. Biochem.* (1993) **57**, 1487–1491.

H439. Kim, D.-H., Kim, B., Kim, H.-S., Sohng, I.-S., Kobashi, K., *Biol. Pharm. Bull.* (1994) **17**, 1326–1328.

H441. Suemori, A., Kurane, R., Tomizuka, N., *Biosci. Biotech. Biochem.* (1993) **57**, 1781–1783.

H443. Woodland, J.M., Elhardt, W.J., *Drug Metab. Dispos.* (1995) **23**, 910–915.

H445. Spaldin, V., Madden, S., Adams, D.A., Edwards, R.J., Davies, D.S., Park, K., *Drug Metab. Dispos.* (1995) **23**, 929–934.

H446. Stresser, D.M., Williams, D.E., Griffin, D.A., Bailey, G.S., *Drug Metab. Dispos.* (1995) **23**, 965–975.

H447. Mutlib, A.E., Strupczewski, J.T., Chesson, S.M., *Drug Metab. Dispos.* (1995) **23**, 951–964.

H448. Rozman, E., Galceran, M.T., Anglada, L., Albet, C., *Drug Metab. Dispos.* (1995) **23**, 976–981.

H449. Adams, P.C., Rickert, D.E., *Drug Metab. Dispos.* (1995) **23**, 982–987.

H450. Ruenitz, P.C., Bai, X., *Drug Metab. Dispos.* (1995) **23**, 993–998.

H451. Vistelle, R., Lamiable, D., Morin, E., Trenque, T., Kaltenbach, M., *Drug Metab. Dispos.* (1995) **23**, 988–992.

H453. Orzechowski, A., Schwarz, L.R., Schwegler, U., Bock, K.W., Snyder, R., Schrenk, D., *Xenobiotica* (1995) **25**, 1093–1102.

H454. Braun, L., Fulceri, R., Benedetti, A., Mandl, J., Bánhegyi, G., *Xenobiotica* (1995) **25**, 1073–1079.

H455. Shiraga, T., Iwasaki, K., Takeshita, K., Matsuda, H., Niwa, T., Tozuka, Z., Hata, T., Guengerich, F.P., *Xenobiotica* (1995) **25**, 1063–1071.

H456. Hansen, E.B., Jr., Cho, B.P., Korfmacher, W.A., Cerniglia, C.E., *Xenobiotica* (1995) **25**, 1081–1092.

H459. Caccia, S., Bergami, A., Fracasso, C., Garattini, S., Campbell, B., *Xenobiotica* (1995) **25**, 1143–1150.

H460. Hanioka, H., Hamamura, M., Kakino, K., Ogata, H., Jinno, H., Takahashi, A., Nishimura, T., Ando, M., *Xenobiotica* (1995) **25**, 1207–1217.

H461. Tomigahara, Y., Onogi, M., Saito, K., Shiba, K., Kaneko, H., Nakatsuka, I., Yamada, H., *Xenobiotica* (1995) **25**, 1195–1206.

H462. Schmid, J., Busch, U., Trummlitz, G., Prox, A., Kaschke, S., Wachsmuth, H., *Xenobiotica* (1995) **25**, 1219–1236.

H463. Tanaka, M., Ishikawa, F., Hakusui, H., *Xenobiotica* (1995) **25**, 1247–1257.

H464. Nicoll-Griffith, D.A., Gupta, N., Twa, S.P., Williams, H., Trimble, L.A., Yergey, J.A., *Drug Metab. Dispos.* (1995) **23**, 1085–1093.

H465. Hosokawa, M., Endo, T., Fujisawa, M., Hara, S., Iwata, N., Sato, Y., Satoh, T., *Drug Metab. Dispos.* (1995) **23**, 1022–1027.

H466. Iwaki, M., Bischer, A., Nguyen, A.C., McDonagh, A.F., Benet, L.Z., *Drug Metab. Dispos.* (1995) **23**, 1099–1103.

H467. Lau, S.S., Kleiner, H.E., Monks, T.J., *Drug Metab. Dispos.* (1995) **23**, 1136–1142.

H468. Holmes, E., Sweatman, B.C., Bollard, M.E., Blackledge, C.A., Beddell, C.R., Wilson, I.D., Lindon, J.C., Nicholson, J.K., *Xenobiotica* (1995) **25**, 1269–1281.

H469. Machinist, J.M., Mayer, M.D., Shet, M.S., Ferrero, J.L., Rodrigues, A.D., *Drug Metab. Dispos.* (1995) **23**, 1163–1174.

H470. Breyer-Pfaff, U., Nill, K., *Xenobiotica* (1995) **25**, 1311–1325.

H471. Sakamoto, K., Nakamura, Y., Aikoh, S., Baba, T., Perregaard, J., Pedersen, H., Moltzen, E.K., Mulford, D.J., Yamaguchi, T., *Xenobiotica* (1995) **25**, 1327–1343.

H472. Mason, J.P., Dring, L.G., Caldwell, J., *Xenobiotica* (1995) **25**, 1371–1380.

H473. Rieutord, A., Stupans, I., Shenfield, G.M., Gross, A.S., *Xenobiotica* (1995) **25**, 1345–1354.

H474. Yamaguchi, Y., Baba, T., Touchi, A., Matsubara, T., *Drug Metab. Dispos.* (1995) **23**, 1195–1201.

H475. Taylor, W.G., Hall, T.W., Vedres, D.D., *Drug Metab. Dispos.* (1995) **23**, 1188–1194.

H476. Dixon, C.M., Colthup, P.V., Serabjit-Singh, C.J., Kerr, B.M., Boehlert, C.C., Park, G.R., Tarbit, M.H., *Drug Metab. Dispos.* (1995) **23**, 1225–1230.

H477. Schmid, J., Busch, U., Heinzel, G., Bozler, G., Kaschke, S., Kummer, M., *Drug Metab. Dispos.* (1995) **23**, 1206–1213.

H478. Pichard, L., Gillet, G., Bonfils, C., Domergue, J., Thénot, J.-P., Maurel, P., *Drug Metab. Dispos.* (1995) **23**, 1253–1262.

H479. Anderson, C.D., Wang, J., Kumar, G.N., McMillan, J.M., Walle, U.K., Walle, T., *Drug Metab. Dispos.* (1995) **23**, 1286–1290.

H480. Herber, R.H., Hercelin, B., van Cantfort, J., de Graeve, J., Fournel-Gigleux, S., Taguchi, T., Magdalou, J., *Drug Metab. Dispos.* (1995) **23**, 1305–1314.

H481. Zhang, Z., Fasco, M.J., Huang, Z., Guengerich, F.P., Kaminsky, L.S., *Drug Metab. Dispos.* (1995) **23**, 1339–1345.

H482. Chen, G.-P., Poulsen, L.L., Ziegler, D.M., *Drug Metab. Dispos.* (1995) **23**, 1390–1393.

H483. Chen, G., Baron, J., Duffel, M.W., *Drug Metab. Dispos.* (1995) **23**, 1346–1353.

H484. Fisher, M.B., Lawton, M.P., Atta-Asafo-Adjei, E., Philpot, R.M., Rettie, A.E., *Drug Metab. Dispos.* (1995) **23**, 1431–1433.

H485. Zhang, D., Evans, F.E., Freeman, J.P., Duhart, B., Jr., Cerniglia, C.E., *Drug Metab. Dispos.* (1995) **23**, 1417–1425.

H486. Klekotka, P.A., Halpert, J.R., *Drug Metab. Dispos.* (1995) **23**, 1434–1435.

H487. Singh, Y., Cooke, J.B., Hinton, D.E., Miller, M.G., *Drug Metab. Dispos.* (1996) **24**, 7–14.

H488. Taylor, A.R., Brownsill, R.D., Grandon, H., Lefoulon, F., Petit, A., Luijten, W., Kopelman, P.G., Walther, B., *Drug Metab. Dispos.* (1996) **24**, 55–64.

H489. Los, L.E., Welsh, D.A., Herold, E.G., Bagdon, W.J., Zacchei, A.G., *Drug Metab. Dispos.* (1996) **24**, 28–33.

H490. Barbhaiya, R.H., Dandekar, K.A., Greene, D.S., *Drug Metab. Dispos.* (1996) **24**, 91–95.

H491. Slatter, J.G., Feenstra, K.L., Hauer, M.J., Kloosterman, D.A., Parton, A.H., Saunders, P.E., Scott, G., Speed, W., *Drug Metab. Dispos.* (1996) **24**, 65–73.

H492. Smith, D.J., Appleton, M.L., Carlson, J.R., Yost, G.S., *Drug Metab. Dispos.* (1996) **24**, 119–125.

H493. Rodrigues, A.D., Kukulka, M.J., Roberts, E.M., Ouellet, D., Rodgers, T.R., *Drug Metab. Dispos.* (1996) **24**, 126–136.

H494. Koopman-Kimenai, P.M., Booij, L.H.D.J., Dirksen, R., Nijhuis, G.M.M., *Eur. J. Clin. Pharmacol.* (1991) **41**, 375–378.

H495. Sutherland, J.B., Selby, A.L., Freeman, J.P., Evans, F.E., Cerniglia, C.E., *Appl. Environ. Microbiol.* (1991) **57**, 3310–3316.

H496. Johansen, M., Rasmussen, K.E., Christophersen, A.S., Skuterud, B., *Acta Pharm. Nord.* (1991) **3**, 91–94.

H497. Johansen, M., Tonnesen, F., Rasmussen, K.E., *J. Chromatogr.* (1992) **573**, 283–288.

H499. Delatour, P., Garnier, F., Benoit, E., Caude, I., *Res. Vet. Sci.* (1991) **50**, 134–138.

H500. Auburger, G., Winter, J., *Appl. Microbiol. Biotechnol.* (1992) **37**, 789–795.

H501. Pavel, S., Holden, J.L., Riley, P.A., *Pigm. Cell Res.* (1989) **2**, 421–426.

H502. Vree, T.B., van der Ven, A.J.A.M., Koopmans, P.P., van Ewijk-Beneken Kolmer, E.W.J., Verwey-van Wissen, C.P.W.G.M., *Clin. Drug Invest.* (1995) **9**, 43–53.

H503. Cresteil, T., Monsarrat, B., Alvinerie, P., Tréluyer, J.M., Vieira, I., Wright, M., *Cancer Res.* (1994) **54**, 386–392.

H504. Monsarrat, B., Alvinerie, P., Wright, M., Dubois, J., Guéritte-Voegelein, F., Guénard, D., Donehower, R.C., Rowinsky, E.K., *J. Natl. Cancer Inst. Monographs* (1993) **15**, 39–46.

H505. Royer, I., Alvinerie, P., Armand, J.P., Ho, L.K., Wright, M., Monsarrat, B., *Rapid Commun. Mass Spectrom.* (1995) **9**, 495–502.

H506. Schindowski, A., Wittich, R.-M., Fortnagel, P., *FEMS Microbiol. Lett.* (1991) **84**, 63–70.

H508. Mohn, W.W., Tiedje, J.M., *Arch. Microbiol.* (1991) **157**, 1–6.

H509. Mohn, W.W., Kennedy, K.J., *Appl. Environ. Microbiol.* (1992) **58**, 2131–2136.

H510. Pillonel, C., Hunziker, P., Binder, A., *J. Exp. Bot.* (1992) **43**, 299–305.

H511. Wong, K.O., Wong, K.P., *Xenobiotica* (1996) **26**, 17–26.

H512. Lappin, G.J., Pritchard, D., Moore, R.B., Laird, W.J.D., *Xenobiotica* (1996) **26**, 65–77.

H513. Shockcor, J.P., Silver, I.S., Wurm, R.M., Sanderson, P.N., Farrant, R.D., Sweatman, B.C., Lindon, J.C., *Xenobiotica* (1996) **26**, 41–48.

H514. Hissink, A.M., Van Ommen, B., Van Bladeren, P.J., *Xenobiotica* (1996) **26**, 89–105.

H515. Mori, M.-A., Shoji, M., Dohrin, M., Kawagoshi, T., Honda, T., Kozuka, H., *Xenobiotica* (1996) **26**, 79–88.

H516. van der Aar, E.M., Buikema, D., Commandeur, J.N.M., te Koppele, J.M., van Ommen, B., van Bladeren, P.J., Vermeulen, N.P.E., *Xenobiotica* (1996) **26**, 143–155.

H517. Brunelle, F.M., Verbeeck, R.K., *Xenobiotica* (1996) **26**, 123–131.

H518. Teramura, T., Tokunaga, T., Matsumoto, H., Watanabe, T., Higuchi, S., *Xenobiotica* (1996) **26**, 177–187.

H519. Cupid, B.C., Beddell, C.R., Lindon, J.C., Wilson, I.D., Nicholson, J.K., *Xenobiotica* (1996) **26**, 157–176.

H520. Damani, L.A., Mitterhauser, M., Lin, G., Ho, Y.P., Zolle, I., *Xenobiotica* (1996) **26**, 211–219.

H521. Jackson, P.J., Brownsill, R.D., Taylor, A.R., Resplandy, G., Walther, B., Schwietert, H.R., *Xenobiotica* (1996) **26**, 221–228.

H522. Bollard, M.E., Holmes, E., Blackledge, C.A., Lindon, J.C., Wilson, I.D., Nicholson, J.K., *Xenobiotica* (1996) **26**, 255–273.

H523. Nakabeppu, H., Asada, M., Oda, T., Shinozaki, Y., Yajima, T., *Xenobiotica* (1996) **26**, 229–239.

H524. Jauch, R., Schmid, P., Fischli, W., Meister, W., Maurer, R., Wendt, G., *Xenobiotica* (1996) **26**,285–295.
H525. Sánchez, S.F., Alvarez, K.I., Lanusse, C.E., *Xenobiotica* (1996) **26**, 307–320.
H526. Tanaka, K., Inoue, T., Tezuka, Y., Kikuchi, T., *Xenobiotica* (1996) **26**, 347–354.
H527. Hozoe, K., Mae, T., Yamashita, K., Fujii, K., Yamane, T., Hidaka, T., Ohashi, T., *Xenobiotica* (1996) **26**,321–332.
H528. Soeishi, Y., Matsushima, H., Teraya, Y., Watanabe, T., Higuchi, S., Kaniwa, H., *Xenobiotica* (1996) **26**, 355–365.
H529. Ho, C.C., Lin, T.H., Lai, Y.S., Chung, J.G., Levy, G.N., Weber, W.W., *Drug Metab. Dispos.* (1996) **24**, 137–141.
H530. Schoenig, G.P., Hartnagel, R.E., Osimitz, T.G., Llanso, S., *Drug Metab. Dispos.* (1996) **24**, 156–163.
H531. Paliwal, J.K., Gupta, R.C., *Drug Metab. Dispos.* (1996) **24**, 148–155.
H532. Sugihara, K., Kitamura, S., Tatsumi, K., *Drug Metab. Dispos.* (1996) **24**, 199–202.
H533. Sun, E.L., Feenstra, K.L., Bell, F.P., Sanders, P.E., Slatter, J.G., Ulrich, R.G., *Drug Metab. Dispos.* (1996) **24**, 221–231.
H534. Fujita, S., Masuda, M., Shimamoto, Y., Hoshi, H., Kariya, S., Kazusaka, A., Suzuki, T., *Drug Metab. Dispos.* (1996) **24**, 254–259.
H535. Pillai, U.A., Ziegler, T.L., Wang, D.X., Kattnig, M.J., McClure, T., Liebler, D.C., Mayersohn, M., Sipes, I.G., *Drug Metab. Dispos.* (1996) **24**, 238–244.
H536. Appanna, G., Tang, B.K., Müller, R., Kalow, W., *Drug Metab. Dispos.* (1996) **24**, 303–306.
H537. Gygi, S.P., Colón, F., Raftogianis, R.B., Galinsky, R.E., Wilkins, D.G., Rollins, D.E., *Drug Metab. Dispos.* (1996) **24**, 282–287.
H538. Chiba, M., Hensleigh, M., Nishime, J.A., Balani, S.K., Lin, J.H., *Drug Metab. Dispos.* (1996) **24**, 307–314.
H539. Coffman, B.L., Rios, G.R., Tephly, T.R., *Drug Metab. Dispos.* (1996) **24**, 329–333.
H541. Green, M.D., Tephly, T.R., *Drug Metab. Dispos.* (1996) **24**, 356–363.
H542. Shimada, T., Yamazaki, H., Guengerich, F.P., *Xenobiotica* (1996) **26**, 395–403.
H543. Griffin, R.J., Rutt, D.B., Henesy, C.M., Harvison, P.J., *Xenobiotica* (1996) **26**, 369–380.
H544. Bogan, D.P., Deasy, B., O'Kennedy, R., Smyth, M.R., *Xenobiotica* (1996) **26**, 437–445.
H545. Matsusue, K., Ariyoshi, N., Oguri, K., Koga, N., Yoshimura, Y., *Xenobiotica* (1996) **26**, 405–414.
H546. Street, J.C., Lee, J.S., Jarema, M.A.C., *Xenobiotica* (1996) **26**, 447–457.
H547. Beaumont, K.C., Causey, A.G., Coates, P.E., Smith, D.A., *Xenobiotica* (1996) **26**, 459–471.
H548. Oda, Y., Furuichi, K., Tanaka, K., Hiroi, T., Imaoka, S., Asada, A., Fujimori, M., Funae, Y., *Anesthesiology* (1995) **82**, 214–220.
H549. Lave, T., Efthymiopoulos, C., Christmann, D., Jung, L., Koffel, J.-C., *Clin. Drug Invest.* (1995) **9**, 150–155.
H550. Piscitelli, S.C., Thibault, A., Figg, W.D., Tompkins, A., Headee, D., Lieberman, R., Samid, D., Myers, C.E., *J. Clin. Pharmacol.* (1995) **35**, 368–373.
H551. Madsen, H., Nielsen, K.K., Broesen, K., *Br. J. Clin.Pharmacol.* (1995) **39**, 433–439.
H552. Yamamoto, Y., Tsuda, M., Uda, K., Matsushita, H., Shindo, T., Kawaguchi, Y., *Yakubutsu Dotai* (1995) **10**, 205–210.
H553. Nicholls, A.W., Caddick, S., Wilson, I.D., Farrant, R.D., Lindon, J.C., Nicholson, J.C., *Biochem. Pharmacol.* (1995) **49**, 1155–1164.
H554. Zhu, J-s., Scott, G.K., *Biochem. Mol. Biol. Int.* (1995) **35**, 423–432.
H555. Takac, S., Akay, B., Oezdamar, T.H., *Enzyme Microb. Technol.* (1995) **17**, 445–452.

H556. Okamoto, T., Yazaki, K., Tabata, M., *Phytochemistry* (1995) **38**, 83–88.

H557. Facchini, P.J., De Luca, V., *Phytochemistry* (1995) **38**, 1119–1126.

H558. Vic, G., Biton, J., Le Beller, D., Michel, J.-M., Thomas, D., *Biotechnol. Bioeng.* (1995) **46**, 109–116.

H560. Lee, S.C., Renwick, A.G., *Biochem. Pharmacol.* (1995) **49**, 1567–1576.

H561. Lee, S.C., Renwick, A.G., *Biochem. Pharmacol.* (1995) **49**, 1557–1565.

H562. Deleu, D., Sarre, S., Ebinger, G., Michotte, Y., *J. Pharmacol. Exp. Ther.* (1995) **273**, 1323–1331.

H563. Stampfli, H.F., Quon, C.Y., *Res. Comun. Mol. Pathol. Pharmacol.* (1995) **88**, 87–97.

H565. Cravedi, J.P., Perdu-Durand, E., Baradat, M., Alary, J., Debrauwer, L., Bories, G., *Chem. Res. Toxicol.* (1995) **8**, 642–648.

H566. Land, S.J., King, C.M., *Environ. Health Perspect. Suppl.* (1994) **102** (Suppl. 6), 91–93.

H567. Oktay, M., Kufrevioglu, I., Kocacaliskan, I., Sakiroglu, H., *J. Food Sci.* (1995) **60**, 494–496.

H568. Bellucci, G., Chiappe, C., Ingrosso, G., *Chirality* (1994) **6**, 577–582.

H569. Misono, H., Okuda, H., Shin, K., Nagata, S., Nagasaki, S., *Biosci. Biotechnol. Biochem.* (1995) **59**, 339–340.

H570. Jones, K.H., Trudgill, P.W., Hopper, D.J., *Arch. Microbiol.* (1995) **163**, 176–181.

H571. Dagnino, D., Schripsema, J., Verpoorte, R., *Phytochemistry* (1995) **39**, 341–349.

H572. Jarman, M., Poon, G.K., Rowlands, M.G., Grimshaw, R.M., Horton, M.N., Potter, G.A., McCague, R., *Carcinogenesis* (1995) **16**, 683–688.

H573. Trincone, A., Pagnotta, E., *Biotechnol. Lett.* (1995) **17**, 45–48.

H574. Kenyon, E.M., Seely, M.E., Janszen, D., Medinsky, M.A., *J. Toxicol. Environ. Health* (1995) **44**, 219–233.

H575. Gschaidmeier, H., Seidel, A., Burchell, B., Bock, K.W., *Biochem. Pharmacol.* (1995) **49**, 1601–1606.

H576. He, Z., Wiegel, J., *Eur. J. Biochem.* (1995) **229**, 77–82.

H577. Xie, H.-G., Huang, S.-L., Zhou, H.-H., *J. Chromatogr. B: Biomed. Appl.* (1995) **668**, 125–131.

H578. Hara, M., Yazaki, K., Tanaka, S., Tabata, M., *Phytochemistry* (1995) **38**, 1131–1135.

H580. Gibson, D.T., Resnick, S.M., Lee, K., Brand, J.M., Torok, D.S., Wackett, L.P., Schocken, M.J., Haigler, B.E., *J. Bacteriol.* (1995) **177**, 2615–2621.

H581. Nicotra, A., Mammi, C., Senatori, O., *Biog. Amines* (1995) **11**, 261–268.

H582. Preisig-Müller, R., Gnau, P., Kindl, H., *Arch. Biochem. Biophys.* (1995) **317**, 201–207.

H583. Koga, N., Kikuichi-Nishimura, N., Hara, T., Harada, N., Ishii, Y., Yamada, H., Oguri, K., Yohimura, H., *Arch. Biochem. Biophys.* (1995) **317**, 464–470.

H584. Firkusny, L., Kroemer, H.K., Eichelbaum, M., *Biochem. Pharmacol.* (1995) **49**, 1777–1784.

H585. Iwasa, K., Kondoh, Y., Kamigauchi, K., *J. Nat. Prod.* (1995) **58**, 379–391.

H586. Ono, S., Tsutsui, M., Gonzalez, F.J., Satoh, T., Masubuchi, Y., Horie, T., Suzuki, T., Narimatsu, S., *Pharmacogenetics* (1995) **5**, 97–102.

H587. Molinero, P., Osuna, C., Guerrero, J.M., *J. Endocrinol.* (1995) **146**, 105–111.

H588. Yamamoto, S., Ueki, T., Tomita, K., Ohnuma, T., Sawada, Y., Fukagawa, Y., Oki, T., *Biosci. Biotechnol. Biochem.* (1995) **59**, 1057–1061.

H589. Faulds, C.B., Williamson, G., *Biochem. Soc. Trans.* (1995) **23**, 253S.

H591. Benavente-Garcia, O., Castillo, J., Sabater, F., Del Rio, J.A., *Plant Physiol. Biochem.* (1995) **33**, 263–271.

H592. Dingemanse, J., Jorga, K.M., Schmitt, M., Gieschke, R., Fotteler, B., Zuercher, G., Da Prada, M., van Brummelen, P.,

Clin. Pharmacol. Ther. (1995) **57**, 508–517.

H593. Silverman, P., Seskar, M., Kanter, D., Schweizer, P., Metraux, J.-P., Raskin, I., *Plant Physiol.* (1995) **108**, 633–639.

H594. Zhou, H.-H., Wood, A.J.J., *Clin. Pharmacol. Ther.* (1995) **57**, 518–524.

H595. Aasmundstad, T.A., Xu, B.Q., Johansson, I., Ripel, A., Bjoerneboe, A., Christophersen, A.S., Bodd, E., Moerland, J., *Br. J. Clin. Pharmacol.* (1995) **39**, 611–620.

H596. Balian, J.D., Sukhova, N., Harris, J.W., Hewett, J., Pickle, L., Goldstein, J.A., Woosley, R.L., Flockhart, D.A., *Clin. Pharmacol. Ther.* (1995) **57**, 662–669.

H597. Soraci, A., Benoit, E., Delatour, P., *J. Vet. Pharmacol. Ther.* (1995) **18**, 167–171.

H598. Hals, P.-A., Dahl, S.G., *Eur. J. Drug. Metab. Pharmacokinet.* (1995) **20**, 61–71.

H599. Little, B.B., Roe, D.A., Stettler, R.W., Bohman, V.R., Westfall, K.L., Sobhi, S., *Am. J. Obstet. Gynecol.* (1995) **172**, 1441–1445.

H600. Lakshmi, V.M., Bell, D.A., Watson, M.A., Zenser, T.V., Davis, B.B., *Carcinogenesis* (1995) **16**, 1565–1571.

H601. Ariyoshi, N., Oguri, K., Koga, N., Yoshimura, H., Funae, Y., *Biochem. Biophys. Res. Commun.* (1995) **212**, 455–460.

H602. Latus, M., Seitz, H.-J., Eberspächer, J., Lingens, F., *Appl. Environ. Microbiol.* (1995) **61**, 2453–2460.

H603. Ueki, T., Sawada, Y., Fukagawa, Y., Oki, T., *Biosci. Biotechnol. Biochem.* (1995) **59**, 1062–1068.

H605. Langkau, B., Vock, P., Massey, V., Fuchs, G., Ghisla, S., *Eur. J. Biochem.* (1995) **230**, 676–685.

H606. Sanvoisin, J., Langley, G.J., Bugg, T.D.H., *J. Am. Chem. Soc.* (1995) **117**, 7836–7837.

H607. Raiber, S., Schröder, G., Schröder, J., *FEBS Lett.* (1995) **361**, 299–302.

H608. Malherbe, P., Alberati-Giani, D., Köhler, C., Cesura, A.M., *FEBS Lett.* (1995) **367**, 141–144.

H609. Holt, D.E., Hurley, R., Harvey, D., *Biol. Neonate* (1995) **67**, 230–239.

H610. Biegert, T., Altenschmidt, U., Ekershorn, C., Fuchs, G., *Arch. Microbiol.* (1995) **163**, 418–423.

H611. Brock, B.J., Rieble, S., Gold, M.H., *Appl. Environ. Microbiol.* (1995) **61**, 3076–3081.

H612. Shin, K.-S., *Mycoscience* (1995) **36**, 31–35.

H613. Asada, Y., Watanabe, A., Ohtsu, Y., Kuwahara, M., *Biosci. Biotechnol. Biochem.* (1995) **59**, 1339–1341.

H614. Hamachi, I., Fujita, A., Kunitake, T., *Chem. Lett.* (1995) 657–658.

H615. Monica, M., *Med. Biol. Environ.* (1993) **21**, 563–569.

H616. Biegert, T., Fuchs, G., *Arch. Microbiol.* (1995) **163**, 407–417.

H617. Faleev, N.G., Gogoleva, O.I., Dementieva, I.S., Zakomirdina, L.N., Belikov, V.M., *Biochem. Mol. Biol. Int.* (1995) **35**, 1037–1040.

H618. Omori, T., Saiki, Y., Kasuga, K., Kodama, T., *Biosci. Biotechnol. Biochem.* (1995) **59**, 1195–1198.

H619. Kirk, C.D., Chen, L., Imeson, H.C., Cossins, E.A., *Phytochemistry* (1995) **39**, 1309–1317.

H620. Back, N., Ahonen, M., Soinila, S., Kivilaakso, E., Kiviluoto, T., *Histochem. Cell. Biol.* (1995) **104**, 63–67.

H621. Betry, P., Fliniaux, M.A., Mackova, M., Gillet, F., Macek, T., Jacquin-Dubreuil, A., *J. Plant Physiol.* (1995) **146**, 503–507.

H622. Jamis-Dow, C.A., Klecker, R.W., Katki, A.G., Collins, J.M., *Cancer Chemother. Pharmacol.* (1995) **36**, 107–114.

H623. Stettler, R.W., Bohman, V.R., Little, B.B., Westfall, K.L., Sobhi, S., *J. Matern.-Fetal Med.* (1995) **4**, 135–138.

H624. Ali-Sadat, S., Mohan, K.S., Walia, S.K., *FEMS Microbiol. Ecol.* (1995) **17**, 169–176.

H625. Nakabeppu, H., Nakajima, A., Kamikawaji, Y., Shinozaki, Y., *Arzneim.-Forsch.* (1995) **45**, 766–770.

H627. Ueki, T., Sawada, Y., Fukagawa, Y., Oki, T., *Biosci. Biotechnol. Biochem.* (1995) **59**, 1069–1075.

H628. Baranczyk-Kuzma, A., Barszczewska, I., Borchardt, R.T., Audus, K.L., *Biog. Amines* (1995) **11**, 281–293.

H629. Loeffler, S., Deus-Neumann, B., Zenk, M.H., *Phytochemistry* (1995) **38**, 1387–1395.

H630. Dehmel, U., Engesser, K.-H., Timmis, K.N., Dwyer, D.E., *Arch. Microbiol.* (1995) **163**, 35–41.

H631. Vannelli, T., Hooper, A.B., *Biochemistry* (1995) **34**, 11743–11749.

H632. Fujita, Y., Yaegashi, T., Sawada, S., Oyama, H., Yoshimoto, T., Tsuru, D., *Biol. Pharm. Bull.* (1995) **18**, 648–652.

H633. Liu, Q., Bonness, M.S., Liu, M., Seradge, E., Dixon, R.A., Mabry, T.J., *Arch. Biochem. Biophys.* (1995) **321**, 397–404.

H634. Watanabe, K., Kayano, Y., Matsunaga, T., Yamamoto, I., Yoshimura, H., *Biol. Pharm. Bull.* (1995) **18**, 696–699.

H635. Mays, D.C., Pawluk, L.J., Apseloff, G., Davis, W.B., She, Z.-W., Sagone, A.L., Gerber, N., *Biochem. Pharmacol.* (1995) **50**, 367–380.

H636. Yoshikawa, M., Nishiyama, S., Takaiti, O., *Hypertens. Res.* (1995) **18**, S211-S213.

H637. Vree, T.B., Schoondermark-van de ven, E., Verwey-van Wissen, C.P.W.G.M., Baars, A.M., Swolfs, A., van Galen, P.M., Amatdjais-Groenen, H., *J. Chromatogr. B: Biomed. Appl.* (1995) **670**, 111–123.

H638. Fujimaki, Y., Sudo, K., Hakusui, H., *Arzneim.-Forsch.* (1995) **45**, 748–752.

H639. Koopman-Kimenai, P.M., Vree, T.B., Booij, L.H.D.J., Hasenbos, M.A.W.M., *Biopharm. Drug Dispos.* (1995) **16**, 507–520.

H640. Taylor, L.C.E., Singh, R., Chang, S.Y., Johnson, R.L., Schwartz, J., *Rapid Commun. Mass. Spectrom.* (1995) **9**, 902–910.

H641. Atsumi, R., Okazaki, O., Hakusui, H., *Biol. Pharm. Bull.* (1995) **18**, 1024–1026.

H642. Ni, S., Fredrickson, J.K., Xun, L., *J. Bacteriol.* (1995) **177**, 5135–5139.

H643. Elmarakby, S.A., Shappell, N.W., Kumar, S., Sikka, H.C., *Aquat. Toxicol.* (1995) **33**, 1–15.

H644. Xun, L., Wagnon, K.B., *Appl. Environ. Microbiol.* (1995) **61**, 3499–3502.

H645. Loeffler, F., Lingens, F., Mueller, R., *Biodegradation* (1995) **6**, 203–212.

H646. Pelletier J.N., MacKenzie, R.E., *Biochemistry* (1995) **34**, 12673–12680.

H647. Retey, J., *Proc. Natl. Acad. Sci. U.S.A.* (1995) **92**, 8433–8437.

H648. Li, R., Kenyon, G.L., *Anal. Biochem.* (1995) **230**, 37–40.

H649. Selifonov, S.A., Gurst, J.E., Wackett, L.P., *Biochem. Biophys. Res. Commun.* (1995) **213**, 759–767.

H650. Schmidt, B., Rivero, C., Thiede, B., *Phytochemistry* (1995) **39**, 81–84.

H651. Fujita, T., Hatamoto, H., Iwasaki, T., Takafuji, S., *Phytochemistry* (1995) **39**, 1085–1089.

H652. Bugge, A., Aasmudstad, T., Aasen, A.J., Christophersen, A.S., Morgenlie, S., Mørland, J., *Acta Chem Scand.* (1995) **49**, 380–384.

H653. Kumagai, S., Matsunaga, I., *Int. Arch. Occup. Environ. Health* (1995) **67**, 211–217.

H654. Sparreboom, A., Huizing, M.T., Boesen, J.J.B., Nooijen, W.J., van Tellingen, O., Beijnen, J.H., *Cancer Chemother. Pharmacol.* (1995) **36**, 299–304.

H655. Wu, W.-N., Mutter, M.S., *J. Pharm. Biomed. Anal.* (1995) **13**, 857–867.

H656. Shimoda, K., Noguchi, T., Ozeki, Y., Morita, S., Shibasaki, M., Someya, T., Takahashi, S.,

Neuropsychopharmacology (1995) **12**, 323–333.

H657. Huang, T., Schirch, V., *J. Biol. Chem.* (1995) **270**, 22296–22300.

H658. Vogt, T., Taylor, L.P., *Plant Physiol.* (1995) **108**, 903–911.

H659. Nothe, C., Hartmans, S., *Biocatalysis* (1994) **10**, 219–225.

H660. Knops, M., Schuphan, I., Schmidt, B., *Plant Sci.* (*Shannon, Irel.*) (1995) **109**, 215–224.

H662. Suarez, M., Martin, M., Ferrer, E., Garrido-Pertierra, A., *Arch. Microbiol.* (1995) **164**, 70–77.

H663. Barber, J.T., Sharma, H.A., Ensley, H.E., Polito, M.A., Thomas, D.A., *Chemosphere* (1995) **31**, 3567–3574.

H664. Genet, R., Benetti, P.-H., Hammadi, A., Menek, A., *J. Biol. Chem.* (1995) **270**, 23540–23545.

H665. Bonting, C.F.C., Schneider, S., Schmidtberg, G., Fuchs, G., *Arch. Microbiol.* (1995) **164**, 63–69.

H666. Gilbert, J.A., Bates, L.A., Ames, M.M., *Biochem. Pharmacol.* (1995) **50**, 845–850.

H667. Reichling, J., Kemmerer, B., Sauer-Guerth, H., *Pharm. World Sci.* (1995) **17**, 113–119.

H668. Takahashi, E., Nakamichi, K., Furui, M., *J. Ferment. Bioeng.* (1995) **80**, 247–250.

H669. Hearn, W.L., Pablo, J., Hime, G.W., Mash, D.C., *J. Anal. Toxicol.* (1995) **19**, 427–434.

H670. Aberg, G., Ciofalo, V.B., Pendleton, R.G., Ray, G., Weddle, D., *Chirality* (1995) **7**, 383–387.

H671. Hegadorn, K.M., Baker, G.B., Coutts, R.T., *J. Pharmacol. Toxicol. Methods* (1995) **34**, 117–123.

H672. Grifoll, M., Selifonov, S.A., Chapman, P.J., *Appl. Environ. Microbiol.* (1995) **61**, 3490–3493.

H673. Yamamoto, I., Nagai, K., Watanabe, K., Matsunaga, T., Yoshimura, H., *J. Pharm. Pharmacol.* (1995) **47**, 683–686.

H674. van der Krol, D., Schuphan, I., Thiede, B., Schmidt, B., *Pestic. Sci.* (1995) **45**, 143–152.

H675. McKeown, D.S.J., McNicholas, C., Simpson, T.J., Willet, N.J., *J. Chem. Soc., Chem. Commun.* (1996) 301–302.

H676. Schmidt, D., Stöckigt, J., *Planta Med.* (1995) **61**, 254–258.

H677. Bennett, R.N., Hick, A.J., Dawson, G.W., Wallsgrove, R.M., *Plant Physiol.* (1995) **109**, 299–305.

H678. Torok, D.S., Resnick, S.M., Brand, J.M., Cruden, D.L., Gibson, D.T., *J. Bacteriol.* (1995) **177**, 5799–5805.

H679. Jensen, J.B., Egsgaard, H., Van Onckelen, H., Jochimsen, B.U., *J. Bacteriol.* (1995) **177**, 5762–5766.

H680. Kutchan, T.M., Dittrich, H., *J. Biol. Chem.* (1995) **270**, 24475–24481.

H681. Leon, J., Shulaev, V., Yalpani, N., Lawton, M.A., Raskin, I., *Proc. Natl. Acad. Sci. U.S.A.* (1995) **92**, 10413–10417.

H682. Hyun, Y.-L., Davidson, V.L., *Biochim. Biophys. Acta* (1995) **1251**, 198–200.

H683. Nagai, T., Hamada, M., Kai, N., Tanoue, Y., Nagayama, F., *Comp. Biochem. Physiol. B: Biochem. Mol. Biol.* (1995) **112B**, 265–270.

H684. González, B., Brezny, R., Herrera, M., Joyce, T.W., *World J. Microbiol. Biotechnol.* (1995) **11**, 536–540.

H685. Hong, G.-X., Pang, S.F., *Comp. Biochem. Physiol. B: Biochem. Mol. Biol.* (1995) **112B**, 251–255.

H686. Yamanaka, M., Shimomura, K., Sasaki, K., Yoshihira, K., Ishimaru, K., *Phytochemistry* (1995) **40**, 1149–1150.

H687. Benedito, E., Jimenez-Cervantes, C., Cubillana, J.D., Solano, F., Lozano, J.A., Garcia-Barron, J.C., *Biochim. Biophys. Acta* (1995) **1252**, 217–224.

H688. Yeleswaran, K., Volk, K.J., Rourick, R., Catt, J.D., Mattson, R.J., McGoff, K.B., Knipe, J.O., *Res. Commun. Mol. Pathol. Pharmacol.* (1995) **89**, 27–44.

H689. Takeshita, M., Miura, M., Unuma, Y., Iwai, S., Sato, I., Arai, T., Kosaka, K., *Res. Commun. Mol. Pathol. Pharmacol.* (1995) **89**, 351–356.

H690. Spiess, E., Sommer, C., Görisch, H., *Appl. Environ. Microbiol.* (1995) **61**, 3884–3888.

H691. Sgaragli, G.P., Valoti, M., Palmi, M., Frosini, M., Giovannini, M.G., Bianchi, L., Della Corte, L., *J. Pharm. Pharmacol.* (1995) **47**, 782–790.

H692. Armstrong, S.M., Sankey, B.M., Voordouw, G., *Biotechnol. Lett.* (1995) **17**, 1133–1136.

H693. Seeger, M., Timmis, K.N., Hofer, B., *FEMS Microbiol Lett.* (1995) **133**, 259–264.

H694. Ludwig-Müller, J., Hilgenberg, W., *Physiol. Plant.* (1995) **94**, 651–660.

H695. Zheng, L., Poulton, J.E., *Plant Physiol.* (1995) **109**, 31–39.

H696. Carney, J.R., Rinehart, K.L., *J. Nat. Prod.* (1995) **58**, 971–985.

H697. Ball, S.E., Thiel, V.E., Tio, C.O., Lim, H.-K., Kao, J., Sisenwine, S.F., Scatina, J., *Drug Metab. Dispos.* (1996) **24**, 383–389.

H697a. Kunze, K.L., Wienkers, L.C., Thummel, K.E., Trager, W.F., *Drug Metab. Dispos.* (1996) **24**, 414–421.

H697b. Raza, H., King, R.S., Squires, R.B., Guengerich, F.P., Miller, D.W., Freeman, J.P., Lang, N.P., Kadlubar, F.F., *Drug Metab. Dispos.* (1996) **24**, 395–400.

H698. Black, D.J., Kunze, K.L., Wienkers, L.C., Gidal, B.E., Seaton, T.L., McDonnell, N.D., Evans, J.S., Bauwens, J.E., Trager, W.F., *Drug Metab. Dispos.* (1996) **24**, 422–428.

H699. Walker, D.K., Alabaster, C.T., Congrave, G.S., Hargreaves, M.B., Hyland, R., Jones, B.C., Reed, L.J., Smith, D.A., *Drug Metab. Dispos.* (1996) **24**, 447–455.

H700. Sharer, J.E., Wrighton, S.A., *Drug Metab. Dispos.* (1996) **24**, 487–494.

H701. Madden, S., Maggs, J.L., Park, B.K., *Drug Metab. Dispos.* (1996) **24**, 469–479.

H702. Zhang, G.-Q., Mckay, G., Hubbard, J.W., Midha, K.K., *Xenobiotica* (1996) **26**, 541–550.

H703. Nakasa, H., Ohmori, S., Kitada, M., *Xenobiotica* (1996) **26**, 495–501.

H704. Robertson, I.G.C., Bland, T.J., Palmer, B.D., *Xenobiotica* (1996) **26**, 559–569.

H705. Manchee, G.R., Eddershaw, P.J., Ranshaw, L.E., Herriott, D., Park, G.R., Bayliss, M.K., Tarbitt, M.H., *Drug Metab. Dispos.* (1996) **24**, 555–559.

H706. Butterworth, M., Lau, S.S., Monks, T.J., *Drug Metab. Dispos.* (1996) **24**, 588–594.

H707. Mathews, J.M., Black, S.L., Matthews, H.B., *Drug Metab. Dispos.* (1996) **24**, 579–587.

H708. Shou, M., Yang, S.K., *Drug Metab. Dispos.* (1996) **24**, 595–601.

H709. Sanwald, P., David, M., Dow, J., *Drug Metab. Dispos.* (1996) **24**, 602–609.

H710. Roy-de Vos, M., Mayer, J.M., Etter, J.-C., Testa, B., *Xenobiotica* (1996) **26**, 571–582.

H711. Wienkers, L.C., Wurden, C.J., Storch, E., Kunze, K.L., Rettie, A.E., Trager, W.F., *Drug Metab. Dispos.* (1996) **24**, 610–614.

H712. Burka, L.T., Sanders, J.M., Matthews, H.B., *Xenobiotica* (1996) **26**, 597–611.

H713. Yamaguchi, Y., Norikura, R., Nakanishi, M., Touchi, A., Yoshimori, T., Murakami, T., Baba, T., Mizojiri, K., Matsubara, T., *Xenobiotica* (1996) **26**, 613–626.

H714. Soeishi, Y., Matsushima, H., Watanabe, T., Higuchi, S., Cornelissen, K., Ward, J., *Xenobiotica* (1996) **26**, 637–645.

H715. Bolton, G.C., Keogh, J.P., East, P.D., Hollis, F.J., Shore, A.D., *Xenobiotica* (1996) **26**, 627–636.

H716. Bessems, J.G.M., Te Koppele, J.M., Van Dijk, P.A., Van Stee, L.L.P., Commandeur, J.N.M., Vermeulen, N.P.E., *Xenobiotica* (1996) **26**, 647–666.

H717. Rey-Grobellet, X., Eeckhoutte, C., Sutra, J.F., Alvinerie, M., Galtier, P., *Xenobiotica* (1996) **26**, 765–778.

H718. McNally, W.P., Pool, W.F., Sinz, M.W., Dehart, P., Ortwine, D.F., Huang, C.C., Chang, T., Woolf, T.F., *Drug Metab. Dispos.* (1996) **24**, 629–633.
H719. Murthy, S.S., Mathur, C., Kvalo, L.T., Lessor, R.A., Wilhelm, J.A., *Xenobiotica* (1996) **26**, 779–792.
H720. Nyberg, S.L. Mann, H.J., Hu, M.Y., Payne, W.D., Hu, W.-S., Cerra, F.B., Remmel, R.P., *Drug Metab. Dispos.* (1996) **24**, 643–648.
H721. Sparreboom, A., Van Tellingen, O., Scherrenburg, E.J., Boesen, J.J.B., Huizing, M.T., Nooijen, W.J., Versluis, C., Beijnen, J.H., *Drug Metab. Dispos.* (1996) **24**, 655–663.
H722. Gregus, Z., Fekete, T., Halászi, É., Klaassen, C.D., *Drug Metab. Dispos.* (1996) **24**, 682–688.
H723. Lê, K.-T., Maurice, H., du Souich, P., *Drug Metab. Dispos.* (1996) **24**, 711–716.
H724. Leung, K.H., Cohn, D.A., Miller, R.R., Doss, G.A., Stearns, R.A., Simpson, R.E., Feeney, W.P., Chiu, S.-H.L., *Drug Metab. Dispos.* (1996) **24**, 753–760.
H725. Bounds, S.V.J., Caldwell, J., *Drug Metab. Dispos.* (1996) **24**, 717–724.
H726. Caraco, Y., Tateishi, T., Guengerich, F.P., Wood, A.J.J., *Drug Metab. Dispos.* (1996) **24**, 761–764.
H727. Wang, R.W., Liu, L., Cheng, H., *Drug Metab. Dispos.* (1996) **24**, 786–791.
H728. Salmon, F., Kohl, W., *Xenobiotica* (1996) **26**, 803–811.
H729. Mae, T., Hosoe, K., Fujii, K., Yamashita, K., Yamane, T., Hidaka, T., Ohashi, T., *Xenobiotica* (1996) **26**, 793–802.
H730. Fujimaki, Y., Hakusui, H., Yamazoe, Y., *Xenobiotica* (1996) **26**, 821–830.
H731. Morioka, Y., Ohmizo, M., Harada, M., Goto, K., Naito, S., Tsutsumi, K., *Xenobiotica* (1996) **26**, 853–861.
H732. Worboys, P.D., Brennan, B., Bradbury, A., Houston, J.B., *Xenobiotica* (1996) **26**, 897–907.
H733. Paulson, G.D., Feil, V.J., *Xenobiotica* (1996) **26**, 863–875.
H734. Adams, P.E., Feil, V.J., Paulson, G.D., *Xenobiotica* (1996) **26**, 921–933.
H735. Jezequel, S.G., Uden, S., Wastall, P., *Xenobiotica* (1996) **26**, 963–975.
H736. Ensslin, H.K., Maurer, H.H., Gouzoulis, E., Hermle, L., Kovar, K.-A., *Drug Metab. Dispos.* (1996) **24**, 813–820.
H737. Nakaoka, M., Tsumura, M., Ichikawa, E., Suzuki, W., Hakusui, H., Nakazawa, T., *Xenobiotica* (1996) **26**, 983–994.
H738. Tirona, R.G., Pang, K.S., *Drug Metab. Dispos.* (1996) **24**, 821–833.
H739. Bort, R., Ponsoda, X., Carrasco, E., Gómez-Lechón, M.J., Castell, J.V., *Drug Metab. Dispos.* (1996) **24**, 834–841.
H740. Kasuya, F., Igarashi, K., Fukui, M., Nokihara, K., *Drug Metab. Dispos.* (1996) **24**, 879–883.
H741. Thompson, D.C., Perera, K., London, R., *Drug Metab. Dispos.* (1996) **24**, 866–871.
H742. Walker, D.K., Beaumont, K.C., Stopher, D.A., Smith, D.A., *Xenobiotica* (1996) **26**, 1101–1111.
H743. Acheampong, A.A., Chien, D.-S., Lam, S., Vekich, S., Breau, A., Usansky, J., Harcourt, D., Munk, S.A., Nguyen, H., Garst, M., Tang-Liu, D., *Xenobiotica* (1996) **26**, 1035–1055.
H744. Avgerinos, A., Axarlis, S., Dragatsis, J., Karidas, T., Malamataris, S., J. Chromatogr. B., *Biomed. Appl.* (1995) **673**, 142–146.
H745. Edlund, P.O., *J. Mass Spectrom.* (1995) **30**, 1380–1392.
H746. Martel, F., Azevdo, I., *J. Auton. Pharmacol.* (1995) **15**, 309–319.
H747. Ikegami, F., Matsunae, K., Hisamitsu, M., Kurihara, T., Yamamoto, T., Murakoshi, I., *Biol. Pharm. Bull.* (1995) **18**, 1531–1534.
H748. Gibello, A., Ferrer, E., Sang, J., Martin, M., *Appl. Environ. Microbiol.* (1995) **61**, 4167–4171.

H749. Nakamura, K., Inoue, Y., Matsuda, T., Ohno, A., *Tetrahedron Lett.* (1995) **36**, 6263–6266.

H750. Scigelova, M., Macek, T., Minghetti, A., Mackova, M., Sedmera, P., Prikrylova, V., Kren, V., *Biotechnol. Lett.* (1995) **17**, 1213–1218.

H751. Faulds, C.B., Kroon, P.A., Saulnier, L., Thibault, J.-F., Williamson, G., *Carbohydr. Polym.* (1995) **27**, 187–190.

H752. Wright, J.L., Downing, D.M., Feng, M.R., Hayes, R.N., Heffner, T.G., MacKenzie, R.G., Meltzer, L.T., Pugsley, T.A., Wise, L.D., *J. Med. Chem.* (1995) **38**, 5007–5014.

H753. Basavaiaha, D., Raju, S.B., *Synth. Commun.* (1995) **25**, 3293–3306.

H754. Sams, R.A., Gerken, D.F., Detra, R.L., Stanley, S.D., Wood, W.E., Tobin, T., Yang, J.-M., Tai, H.-H., Jeganathan, A., Watt, D.S., *J. Pharm. Sci.* (1996) **85**, 79–84.

H755. McLean, M.R., Bauer, U., Amaro, A.R., Robertson, L.W., *Chem. Res. Toxicol.* (1996) **9**, 158–164.

H756. Lee, N.-S., Kim, B.-T., Kim, D.-H., Kobashi, K., *J. Biochem.* (1995) **118**, 796–801.

H757. Jung, K.-H., Lee, J.-Y., Kim, H.-S., *Biotechnol. Bioeng.* (1995) **48**, 625–630.

H758. Hintermann, E., Jenoe, P., Meyer, U.A., *FEBS Lett.* (1995) **375**, 148–150.

H759. Ellis, E.M., Hayes, D.J., *Biochem. J.* (1995) **312**, 535–541.

H760. Allison, N., Turner, J.E., Wait, R., *FEMS Microbiol. Lett.* (1995) **134**, 213–219.

H761. Martini, R., Murray, M., *Chem. Res. Toxicol.* (1996) **9**, 268–276.

H762. Guidetti, P., Eastman, C.L., Schwarcz, R., *J. Neurochem.* (1995) **65**, 2621–2632.

H763. Binder, W.H., Kählig, H., Schmid, W., *Tetrahedron: Asymmetry* (1995) **6**, 1703–1710.

H764. Murakami, K., Tanimura, H., Ishimoto, K., Uchiyama, K., Yukawa, H., Onishi, H., Iwakura, K., *Drugs* (1995) **49** (Suppl. 2), 331–332.

H765. Gieg, L.M., Otter, A., Fedorak, P.M., *Environ. Sci. Technol.* (1996) **30**, 575–585.

H766. Koopman-Kimenai, P.M., Vree, T.B., Booij, L.H.D.J., Dirksen, R., *Int. J. Clin. Pharmacol. Ther.* (1995) **33**, 442–448.

H767. Fukuoka, M., Satoh, M., Tanaka, A., *Arch. Toxicol.* (1995) **70**, 1–9.

H768. Hudecova, S., Strakova, Z., Krizanova, O., *Int. J. Biochem. Cell Biol.* (1995) **27**, 1357–1363.

H769. Rietjens, I.M.C.M., Steensma, A., Den Besten, C., van Tintelen, G., Haas, J., van Ommen, B., van Bladeren, P., *Eur. J. Pharmacol., Environ. Toxicol. Pharmacol. Sect.* (1995) **293**, 293–299.

H770. Nellaiah, H., Morisseau, C., Archelas, A., Furstoss, R., Baratti, J.C., *Biotechnol. Bioeng.* (1996) **49**, 70–77.

H771. Shin, H.S., Rogers, P.L., *Biotechnol. Bioeng.* (1996) **49**, 52–62.

H772. Blake, T.L., Reasor, M.J., *Toxicol. Lett.* (1995) **81**, 223–228.

H773. Griffin, R.J., Burka, L.T., Demby, K.B., *Hum. Exp. Toxicol.* (1995) **14**, 779–786.

H774. Asami, M., Takasaki, W., Iwabuchi, H., Haruyama, H., Wachi, K., Terada, A., Tanaka, Y., *Biol. Pharm. Bull.* (1995) **18**, 1584–1589.

H775. Prueksaritanont, T., Dwyer, L.M., Cribb, A.E., *Biochem. Pharmacol.* (1995) **50**, 1521–1525.

H776. Ahmadi, M., Nicholls, P.J., Smith, H.J., Spencer, P.S.J., Preet-Ryatt, M.S., Spragg, B.P., *J. Pharm. Pharmacol.* (1995) **47**, 876–878.

H777. Suh, D.-Y., Yang, H.O., Kim, Y.C., Han, B.H., *Arzneim.-Forsch.* (1995) **45**, 1071–1074.

H778. Weinz, C., Blaschke, G., *J. Chromatogr. B: Biomed. Appl.* (1995) **674**, 287–292.

H779. Schneider, J., Grosser, R., Jayasimhulu, K., Xue, W., Warshawsky, D., *Appl. Environ. Microbiol.* (1996) **62**, 13–19.

H780. Stabile, M.R., Hudlicky, T., Meisels, M.L., Butora, G., Gum, A.G., Fearnley, S.P., Thorpe, A.J., Ellis, M.R., *Chirality* (1995) **7**, 556–559.

H781. Wajant, H., Förster, S., Selmar, D., Effenberger, F., Pfitzenmaier, K., *Plant Physiol.* (1995) **109**, 1231–1238.

H782. Radler, S., Wermeille, M., Blaschke, G., *Arzneim.-Forsch.* (1995) **45**, 1086–1092.

H783. Kern, C., Mautz, D.S., Bernards, C.M., *Anesthesiology* (1995) **83**, 1078–1081.

H784. Sindhu, R.K., Kikkawa, Y., *Toxicol. Lett.* (1995) **81**, 5–13.

H785. French, C.E., Bruce, N.C., *Biochem. J.* (1995) **312**, 671–678.

H786. Oguri, K., Kurogi, A., Yamabe, K.-i., Tanaka, M., Yoshisue, K., Ishii, Y., Yoshimura, H., *Arch. Biochem. Biophys.* (1996) **325**, 159–166.

H787. Rose, A., Glässen, W.G., Hopp, W., Seitz, H.U., *Planta* (1996) **198**, 397–403.

H788. Mautz, D.S., Shen, D.D., Nelson, W.L., *Pharm. Res.* (1995) **12**, 2053–2056.

H789. Suemori, A., Nakajima, K., Kurane, R., Nakamura, Y., *Appl. Microbiol. Biotechnol.* (1995) **43**, 470–472.

H790. Floreani, M., Carpenedo, F., *Gen. Pharmacol.* (1995) **26**, 1757–1764.

H791. Tijmes, M., Pedraza, R., Valladares, L., *Steroids* (1996) **61**, 65–68.

H792. Koehl, W., Amin, S., Staretz, M.E., *Cancer Res.* (1996) **56**, 316–324.

H793. McCord, A., Boyle, S.P., Knowler, J.T., Burnett, A.K., Craft, J.A., *Chem.-Biol. Interact.* (1996) **99**, 29–40.

H794. West, M.G., Horne, D.W., Appling, D.R., *Biochemistry* (1996) **35**, 3122–3132.

H795. Dodds, P.F., Chou, S.C., Ranasinghe, A., Coleman, R.A., *J. Lipid Res.* (1995) **36**, 2493–2503.

H796. Anzenbacher, P., Niwa, T., Tolbert, L.M., Sirimanne, S.R., Guengerich, F.P., *Biochemistry* (1996) **35**, 2512–2520.

H798. Xu, C.L., Sim, M.K., *Biochem. Pharmacol.* (1995) **50**, 1333–1337.

H801. Nicholls, A.W., Lindon, J.C., Caddick, S., Farrant, R.D., Wilson, I.D., Nicholson, J.K., *Xenobiotica* (1997) **27**, 1175–1186.

H802. Clement, B., Demesmaeker, M., *Drug Metab. Dispos.* (1997) **25**, 1266–1271.

H803. Horsmans, Y., Saliez, A., van den Berge, V., Desager, J.-P., Geubel, A.P., Pauwels, S., Lambotte, L., *Drug Metab. Dispos.* (1997) **25**, 1257–1259.

H804. Balani, S.K., Xu, X., Pratha, V., Koss, M.A., Amin, R.D., Dufresne, C., Miller, R.R., Arison, B.H., Doss, G.A., Chiba, M., Freeman, A., Holland, S.D., Schwartz, J.I., Lasseter, K.C., Gertz, B.J., Isenberg, J.I., Rogers, J.D., Lin, J.H., Baillie, T.A., *Drug Metab. Dispos.* (1997) **25**, 1282–1287.

H805. Shet, M.S., McPhaul, M., Fisher, C.W., Stallings, N.R., Estabrook, R.W., *Drug Metab. Dispos.* (1997) **25**, 1298–1303.

H807. Born, S.L., Rodriguez, P.A., Eddy, C,L., Lehman-McKeeman, L.D., *Drug Metab. Dispos.* (1997) **25**, 1318–1324.

H808. Dangman, E., Stolz, A., Kuhm, A.E., Hammer, A., Feigel, B., Noisommit-Rizzi, N., Reuss, M., Knackmuss, H.-J., *Biodegradation* (1996) **7**, 223–229.

H809. Schafer, A., Harms, H., Zehnder, A.J.B., *Biodegradation* (1996) **7**, 249–255.

H810. Lewis, T.A., Goszynski, S., Crawford, R.L., Korus, R.A., Admassu, W., *Appl. Environ. Microbiol.* (1996) **62**, 4669–4674.

H811. Hosokawa, R., Nohara, R., Fujibayashi, Y., Okuda, K., Ogino, M., Hata, T., Fujita, M., Tamaki, N., Konishi, J., Sasayama, S., *J. Nucl. Med.* (1996) **37**, 1836–1840.

H812. McCord, A., Dunlop, N., Stagg, R.M., Craft, J.A., *Fish Physiol. Biochem.* (1996) **15**, 421–430.

H813. Morawski, B., Eaton, R.W., Rossiter, J.T., Guoping, S., Griengl, H., Ribbons, D.W., *J. Bacteriol.* (1997) **179**, 115–121.

H814. Hissink, A.M., Oudshoorn, M.J., Van Ommen, B., Haenen, G.R.M.M., Van

Bladeren, P.J., *Chem. Res. Toxicol.* (1996) **9**, 1249–1256.

H815. Adams, P.C., Rickert, D.E., *Biopharm. Drug Dispos.* (1996) **17**, 675–698.

H816. Kobayashi, K., Hatayama-Hirayama, K., Tobita, S., *J. Gen. Appl. Microbiol.* (1996) **42**, 401–411.

H817. Liu, L., Kumarathasan, P., Guenette, J., Vincent, R., *Am. J. Physiol.* (1996) **271**, L995-L1003.

H818. Schrenk, D., Orzechowski, A., Schwarz, L.R., Snyder, R., Burchell, B., Ingelman-Sundberg, M., Bock, K.W., *Environ. Health Perspect. Suppl.* (1996) **104**, 1183–1188.

H819. Stiborova, M., Hansikova, H., Frei, E., *Cancer Lett.* (1996) **110**, 11–17.

H820. Olesen, M.R., Jochimsen, B.U., *Plant Soil.* (1996) **186**, 143–149.

H821. Hissink, A.M., Dunnewijk, van Ommen, B., van Bladeren, P.J., *Chem.-Biol. Interact.* (1997) **103**, 17–33.

H822. VandenHeuvel, W.J.A., Wislocki, P.G., Hirsch, M.P., Porter, N.K., Ambrose, R.T., Robillard, K.A., *J. Agric. Food Chem.* (1997) **45**, 985–989.

H823. Meulenberg, R., Pepi, M., De Bont, J.A.M., *Biodegradation* (1996) **7**, 303–311.

H824. Semba, H., Mukouyama, M., Sakano, K., *Appl. Microbiol. Biotechnol.* (1996) **46**, 432–437.

H825. Linhart, I., Weidenhoffer, Z., Mraz, J., Smejkal, J., Mladkova, I., *Toxicol. Lett.* (1997) **90**, 199–205.

H826. Simpson, R.M., Nonhebel, H.M., Christie, D.L., *Planta* (1997) **201**, 71–77.

H827. Gibson, J., Dispensa, M., Harwood, C.S., *J. Bacteriol.* (1997) **179**, 634–642.

H828. Misra, S., Sharma, C.B., *Biomed. Chromatogr.* (1997) **11**, 50–53.

H829. Long, M.T., Bartholomew, B.A., Smith, M.J., Trudgill, P.W., Hopper, D.J., *J. Bacteriol.* (1997) **179**, 1044–1050.

H830. Gregory, A.C.E., Edwards, R., *Acta Hortic.* (1994) **381**, 291–297.

H831. Curir, P., Pasini, C., Dolci, M., *Acta Hortic.* (1994) **381**, 605–610.

H832. Moore, B.S., Walker, K., Tornus, I., Handa, S., Poralla, K., Floss, H.G., *J. Org. Chem.* (1997) **62**, 2173–2185.

H833. Edwards, R., *Acta Hortic.* (1994) **381**, 383–389.

H834. D'Annibale, A., Celletti, D., Felici, M., Di Mattia, E., Giavannozzi-Sermanni, G., *Acta Biotechnol.* (1996) **16**, 257–270.

H835. Ichihara, N., Okada, M., Nakagawa, H., Takeda, M., *Insect Biochem. Mol. Biol.* (1997) **27**, 241–246.

H836. Woodcroft, K.J., Webb, C.D., Yao, M., Weedon, A.C., Bend, J.R., *Chem. Res. Toxicol.* (1997) **10**, 589–599.

H837. Morimoto, K., Tatsumi, K., *Chemosphere* (1997) **34**, 1277–1283.

H838. Phelps, C.D., Young, L.Y., *Microb. Ecol.* (1997) **33**, 206–215.

H839. Driskell, W.J., Hill, R.H., Jr., *Bull. Environ. Contam. Toxicol.* (1997) **58**, 929–933.

H841. Rekoslavskaya, N.I., Yur'eva, O.V., Shibanova, L.A., Salyaev, R.K., *Russ. J. Plant Physiol.* (*or Fiziol. Rast.*) (1997) **44**, 196–203.

H842. Antoine, P., Tallieu, X., Thonart, P., *Biochem. Biotechnol.* (1997) **63–65**, 707–717.

H844. Dean-Ross, D., Cerniglia, C.E., *Appl. Microbiol. Biotechnol.* (1996) **46**, 307–312.

H845. Itoh, M.T., Hattori, A., Sumi, Y., *J. Chromatogr. B: Biomed. Sci. Appl.* (1997) **692**, 217–221.

H846. Zhou, L.X., Pihlstrom, B., Hardwick, J.L., *Clin. Pharmacol. Ther.* (1996) **60**, 191–198.

H847. Pothuluri, J.V., Evans, F.E., Doerge, D.R., Churchwell, M.I., Cerniglia, C.E., *Arch. Environ. Contam. Toxicol.* (1997) **32**, 117–125.

H848. Hadley, M.R., Oldham, H.G., Camilleri, P., Damani, L.A., Hutt, A.J., *Chirality* (1996) **8**, 430–440.

H850. Narimatsu, S., Gotoh, M., Masubichi, Y., Horie, T., Ohmori, S., Kitada, M., Kageyama, T., Asaoka, K., Yamamoto, I., Suzuki, T., *Biol. Pharm. Bull.* (1996) **19**, 1429–1433.

H852. Kariya, S., Isozaki, S., Uchino, K., Suzuki, T., Narimatsu, S., *Biol. Pharm. Bull.* (1996) **19**, 1511–1524.

H853. Sidelmann, U.G., Braumann, U., Hoffmann, M., Spraul, M., Lindon, J.C., Nicholson, J.K., Hansen, S.H., *Anal. Chem.* (1997) **69**, 607–612.

H854. Mori, M.-A., Sayama, M., Shoji, M., Inoue, M., Kawagoshi, T., Maeda, M., Honda, T., *Xenobiotica* (1997) **27**, 1225–1236.

H855. Zalko, D., Debrauwer, L., Bories, G., Tulliez, J., *Chem. Res. Toxicol.* (1997) **10**, 197–204.

H856. Yan, J.-H., Hubbard, J.W., McKay, G., Midha, K.K., *Xenobiotica* (1997) **27**, 1245–1257.

H857. Leng, G., Leng, A., Kühn, K.-H., Lewalter, J., Pauluhn, J., *Xenobiotica* (1997) **27**, 1273–1283.

H858. Monostory, K., Jemnitz, K., Vereczkey, L., Czira, G., *Drug Metab. Dispos.* (1997) **25**, 1370–1378.

H859. Moody, D.E., Alburges, M.E., Parker, R.J., Collins, J.M., Strong, J.M., *Drug Metab. Dispos.* (1997) **25**, 1347–1353.

H860. Surapaneni, S.S., Clay, M.P., Spangle, L.A., Paschal, J.W., Lindstrom, T.D., *Drug Metab. Dispos.* (1997) **25**, 1383–1388.

H861. Green, M.D., Bélanger, G., Hum, D.W., Bélanger, A., Tephly, T.R., *Drug Metab. Dispos.* (1997) **25**, 1389–1394.

H862. Court, M.H., Von Moltke, L.L., Shader, R.I., Greenblatt, D.J., *Biopharm. Drug Dispos.* (1997) **18**, 213–226.

H863. Mizojiri, K., Okabe, H., Sugeno, K., Esumi, Y., Takaichi, M., Miyake, T., Seki, H., Inaba, A., *Arzneim.-Forsch.* (1997) **47**, 59–69.

H864. Schmider, J., Greenblatt, D.J., Fogelman, S.M., Von Moltke, L.L., Shader, R.I., *Biopharm. Drug Dispos.* (1997) **18**, 227–240.

H865. Irshaid, Y.M., Al-Hadidi, H.F., Latif, A., Awwadi, F., Al-Zoubi, M., Rawashdeh, N.M., *Eur. J. Drug Metab. Pharmacokinet.* (1996) **21**, 301–307.

H866. Beaumont, K.C., Cussans, N.J., Nichols, D.J., Smith, D.A., *Xenobiotica* (1998) **28**, 63–75.

H867. Chesné, C., Guyomard, C., Guillouzo, A., Schmid, J., Ludwig, E., Sauter, T., *Xenobiotica* (1998) **28**, 1–13.

H868. Kakiki, M., Ishihara, H., *Xenobiotica* (1998) **28**, 77–91.

H869. Mitchell, S.C., Waring, R.H., *Xenobiotica* (1998) **28**, 93–101.

H870. Stroomberg, G.J., Reuther, C., Kozin, I., van Brummelen, T.C., van Gestel, C.A.M., Gooijer, C., Cofino, W.P., *Chemosphere* (1996) **33**, 1905–1914.

H872. Bjoerk, P., Aaman, P., Hindemith, A., Odh, G., Jacobsson, L., Rosengren, E., Rorsman, H., *Eur. J. Haematol.* (1996) **57**, 254–256.

H873. Paje, M.L., Couperwhite, I., *World J. Microbiol. Biotechnol.* (1996) **12**, 653–654.

H874. Barkovskii, A.L., Adriaens, P., *Appl. Environ. Microbiol.* (1996) **62**, 4556–4562.

H875. He, X.-Z., Dixon, R.A., *Arch. Biochem. Biophys.* (1996) **336**, 121–129.

H876. Hückelhoven R., Schuphan, I., Thiede, B., Schmidt, B., *J. Agric. Food Chem.* (1997) **45**, 263–269.

H877. Kitamura, S., Tatsumi, K., *Biol. Pharm. Bull.* (1996) **19**, 1524–1526.

H878. Fujita, S., bin Saari, N., Maegawa, M., Samura, N., Hayashi, N., Tono, T., *J. Agric. Food Chem.* (1997) **45**, 59–63.

H879. Lange, B., Kremer, S., Sterner, O., Anke, H., *Can. J. Microbiol.* (1996) **42**, 1179–1183.

H880. Eaton, S.L., Resnick, S.M., Gibson, D.T., *Appl. Environ. Microbiol.* (1996) **62**, 4388–4394.

H881. Itoh, N., Yoshida, K., Okada, K., *Biosci. Biotechnol. Biochem.* (1996) **60**, 1826–1830.

H882. Toyoda, N., Kaptein, E., Berry, M.J., Harney, J.W., Larsen, P.R., Visser, T.J., *Endocrinology* (1997) **138**, 213–219.

H883. Dinkova-Kostova, A.T., Gang, D.R., Davin, L.B., Bedgar, L., Chu, A., Lewis, N.G., *J. Biol. Chem.* (1996) **271**, 29473–29482.

H884. Wieser, M., Wagner, B., Eberspächer, J., Lingens, F., *J. Bacteriol.* (1997) **179**, 202–208.

H885. Allen, C.C.R., Boyd, D.R., Larkin, M.J., Reid, K.A., Sharma, N.D., Wilson, K., *Appl. Environ. Microbiol.* (1997) **63**, 151–155.

H886. Bailly, C., Cormier, F., Bao Do, C., *Plant Sci.* (1997) **122**, 81–89.

H887. Masubuchi, Y., Iwasa, T., Rujita, S., Suzuki, T., Horie, T., Narimatsu, S., *J. Pharm. Pharmacol.* (1996) **48**, 925–929.

H888. Sakakibara, Y., Katafuchi, J., Takami, Y., Nakayama, T., Suiko, M., Nakajima, H., Liu, M.-C., *Biochim. Biophys. Acta* (1997) **1355**, 102–106.

H889. Clement, B., Mladek, C., *Pharmazie* (1996) **51**, 882–885.

H890. Kamendulis, L.M., Brzezinski, M.R., Pindel, E.V., Bosron, W.F., Dean, R.A., *J. Pharmacol. Exp. Ther.* (1996) **279**, 713–717.

H891. Almahidi, J., Moslemi, S., Drosdowsky, M.A., Seralini, G.-E., *J. Steroid Biochem. Mol. Biol.* (1996) **59**, 55–61.

H892. French, C.E., Hailes, A.M., Rathbone, D.A., Bruce, N.C., *Ann. N. Y. Acad. Sci.* (1996) **799**, 97–101.

H893. Kiemer, P., Tsisuaka, B., Fetzner, S., Lingens, F., *Biol. Fertil. Soils* (1996) **23**, 307–313.

H894. Baik, W., Kim, D.I., Koo, S., Rhee, J.U., Shin, S.H., Kim, B.H., *Tetrahedron Lett.* (1997) **38**, 845–848.

H895. Tamogami, S., Rakwal, R., Kodama, O., *FEBS Lett.* (1997) **401**, 239–242.

H897. Cavin, J.-F., Barthelmebs, L., Guzzo, J., Van Beeumen, J., Samyn, B., Travers, J.-F., Diviès, C., *FEMS Microbiol. Lett.* (1997) **147**, 291–295.

H898. Sassanella, T.M., Fukumori, F., Bagdasarian, M., Hausinger, R.P., *Appl. Environ. Microbiol.* (1997) **63**, 1189–1191.

H899. Ronfeld, R.A., Tremaine, L.M., Wilner, K.D., *Clin. Pharmacokinet.* (1997) **32** (Suppl. 1), 22–30.

H900. Fernández, Cañón, J.M., Peñalva, M.A., *Anal. Biochem.* (1997) **245**, 218–221.

H901. Casellas, M., Griffoll, M., Bayona, J.M., Solanas, A.M., *Appl. Environ. Microbiol.* (1997) **63**, 819–826.

H902. Basse, C.W., Lottspeich, F., Steglich, W., Kahmann, R., *Eur. J. Biochem.* (1996) **242**, 648–656.

H903. Ibrahim, A.-R.S., *Saudi Pharm. J.* (1997) **5**, 52–55.

H904. Maybruck, W.M., Markaverich, B.M., *Steriods* (1997) **62**, 321–330.

H905. Byard, J., Marshall, D.E., Houghton, E., Barker, P.J., Gower, D.B., *Biochem. Soc. Trans.* (1977) **25**, 66S.

H906. Mühlenbeck, U., Barz, W., *Phytochemistry* (1997) **44**,865–867.

H907. Riley, P.A., Cooksey, C.J., Johnson, C.I., Land, E.J., Latter, A.M., Ramsden, C.A., *Eur. J. Cancer* (1997) **33**, 135–143.

H908. Ichimura, M., Furuno, T., Takahashi, T., Dixon, R.A., Ayabe, S., *Phytochemistry* (1997) **44**, 991–995.

H909. Kalman, S., Metcalf, K., Eintrei, C., *Reg. Anesth.* (1997) **22**, 131–136.

H910. Nagai, T., Hamada, M., Kai, N., Tanoue, Y., Nagayama, F., *Comp. Biochem. Physiol. B: Biochem. Mol. Biol.* (1997) **116B**, 161–165.

H911. Takada, S., Matsueda, T., Kondo, R., Sakai, K., *Organohalogen Compd.* (1996) **28**, 64–67.

H912. Shimizu, E., Ohta, K., Takayama, S., Kitagai, Y., Tanizawa, K., Yorifuji, T., *Biosci. Biotechnol. Biochem.* (1997) **61**, 501–505.

H914. Beck, H.C., *FEMS Microbiol. Lett.* (1997) **149**, 233–238.

H915. Aoki, K., Takenaka, S., Murakami, S., Shinke, R., *Microbiol. Res.* (1997) **152**, 33–38.

H917. Johansen, S.S., Licht, D., Arvin, E., Mosbaek, H., Hansen, A.B., *Appl. Microbiol. Biotechnol.* (1997) **47**, 292–300.

H920. Yamano, S., Ito, K.-i., Ogata, S., Toki, S., *Arch. Biochem. Biophys.* (1997) **341**, 81–88.

H921. Benedito, M.A.C., *J. Pharm. Pharmacol.* (1997) **49**, 273–276.

H922. Morisseau, C., Nellaiah, H., Archelas, A., Furstoss, R., Baratti, J.C., *Enzyme Microb. Technol.* (1997) **20**, 446–452.

H923. Mol, K.A., Van der Geyten, S., Darras, V.M., Visser, T.J., Kühn, E.R., *Endocrinology* (1997) **138**, 1787–1793.

H924. O'Donnell, J.P., Khosla, N.B., Dalvie, D.K., *Xenobiotica* (1998) **28**, 153–166.

H925. Yamazaki, H., Inoue, K., Shimada, T., *Xenobiotica* (1998) **28**, 103–115.

H926. Kajbaf, M., Rossato, P., Barnaby, J.R., Pellegatti, M., *Xenobiotica* (1998) **28**, 167–178.

H927. Kobayashi, H., Kobayashi, S., *Xenobiotica* (1998) **28**, 179–197.

H928. Holladay, J.W., Dewey, M.J., Yoo, S.D., *Drug Metab. Dispos.* (1998) **26**, 20–24.

H929. Örn, U., Klasson-Wehler, E., *Xenobiotica* (1998) **28**, 199–211.

H930. Zalko, D., Perdu-Durand, E., Debrauwer, L., Bec-Ferte, M.-P., Tulliez, J., *Drug Metab. Dispos.* (1998) **26**, 28–35.

H931. Ehlhardt, W.J., Woodland, J.M., Baughman, T.M., Vandenbranden, M., Wrighton, S.A., Kroin, J.S., Norman, B.H., Maple, S.R., *Drug Metab. Dispos.* (1998) **26**, 42–51.

H932. van Breeman, R.B., Nikolic, D., Bolton, J.L., *Drug Metab. Dispos.* (1998) **26**, 85–90.

H933. Coffman, B.L., King, C.D., Rios, G.R., Tephly, T.R., *Drug Metab. Dispos.* (1998) **26**, 73–77.

H934. Wang, M., Dickinson, R.G., *Drug Metab. Dispos.* (1998) **26**, 98–104.

H935. Mayr, C.A., Sami, S.M., Remers, W.A., Dorr, R.T., *Drug Metab. Dispos.* (1998) **26**, 105–109.

H936. Maurizis, J.-C., Rapp, M., Azim, E.M., Gaudreault, R.C., Veyre, A., Madelmont, J.-C., *Drug Metab. Dispos.* (1998) **26**, 146–151.

H937. Wu, W.N., Masucci, J.A., Caldwell, G.W., Carson, J.R., *Drug Metab. Dispos.* (1998) **26**, 115–125.

H938. Priymenko, N., Garnier, F., Ferre, J.-P., Delatour, P., Toutain, P.-L., *Drug Metab. Dispos.* (1998) **26**, 170–176.

H939. Naitoh, T., Kawaguchi, S., Kakiki, M., Ohe, H., Kajiwara, A., Horie, T., *Xenobiotica* (1998) **28**, 269–280.

H940. Huwe, J.K., Fell, V.J., Larsen, G.L., *Organohalogen Compd.* (1996) **29**, 462–467.

H941. Yamaguchi, K., Yasuzawa, T., Sakai, T., Kobayashi, S., *Xenobiotica* (1998) **28**, 281–291.

H942. Koga, N., Kikuichi, N., Kanamaru, T., Kuroki, H., Matsusue, K., Ishida, C., Ariyoshi, N., Oguri, K., Yoshimura, H., *Organohalogen Compd.* (1996) **29**, 472–475.

H943. Kuroki, H., Ishimura, N., Fuzino, K., Masada, Y., Koga, N., Yoshimura, H., *Organohalogen Compd.* (1996) **29**, 481–484.

H944. Effenberger, F., Schwaemmle, A., *Biocatal. Biotransform.* (1997) **14**, 167–179.

H945. Johnson, W.W., Yamazaki, H., Shimada, T., Ueng, Y.-F., Guengerich, F.P., *Chem. Res. Toxicol.* (1997) **10**, 672–676.

H946. Alva, K.S., Kumar, J., Marx, K.A., Tripathy, S.K., *Macromolecules* (1997) **30**, 4024–4029.

H947. Bennett, R.N., Kiddle, G., Wallsgrove, R.M., *Phytochemistry* (1997) **45**, 59–66.

H948. Kader, F., Rovel, B., Girardin, M., Metche, M., *J. Sci. Food Agric.* (1997) **73**, 513–516.

H949. Gearhart, D.A., Neafsey, E.J., Collins, M.A., *Neurochem. Res.* (1997) **22**, 113–121.

H950. Iribarne, C., Picart, D., Dreano, Y., Bail, J.-P., Berthou, F., *Life Sci.* (1997) **60**, 1953–1964.

H951. Gonzalez, D., Schapiro, V., Seoane, G., Hudlicky, T., *Tetrahedron: Asymmetry* (1997) **8**, 975–977.

H952. Meth-Cohn, O., Wang, M.-X., *J. Chem. Soc., Perkin Trans. 1* (1997) 1099–1104.

H953. Lehning, A., Fock, U., Wittich, R.-M., Timmis, K.N., Pieper, D.H., *Appl. Environ. Microbiol.* (1997) **63**, 1974–1979.

H954. Krix, G., Bommarius, A.S., Drauz, K., Kottenhahn, M., Schwarm, M., Kula, M.-R., *J. Biotechnol.* (1997) **53**, 29–39.

H955. Li, T., Rosazza, J.P.N., *J. Bacteriol.* (1997) **179**, 3482–3487.

H956. Guitton, J., Burronfosse, T., Sanchez, M., Desage, M., *Anal. Lett.* (1997) **30**, 1369–1378.

H957. Fiorella, P.D., Spain, J.C., *Appl. Environ. Microbiol.* (1997) **63**, 2007–2015.

H958. Amin, M.R., Onodera, R., *J. Gen. Appl. Microbiol.* (1997) **43**, 1–7.

H959. Tamura, H., Miyawaki, A., Inoh, N., Harada, Y., Mikoshiba, K., Matsui, M., *Chem.-Biol. Interact.* (1997) **104**, 1–9.

H960. Amin, M.R., Onodera, R., *J. Gen. Appl. Microbiol.* (1997) **43**, 9–15.

J1. Lenz R., Zenk, M. H., *J. Biol. Chem.* (1995) **270**, 31091–31096.

J2. Chang, Y.-C., Nair, M. G., *J. Nat. Prod.* (1995) **58**, 1892–1896.

J4. Daufman, R., Sloley, D., *Insect Biochem. Mol. Biol.* (1996) **26**, 101–106.

J5. Kitao, S., Matsudo, T., Saitoh, M., Horiuchi, T., Sekine, H., *Biosci. Biotechnol. Biochem.* (1995) **59**, 2167–2169.

J6. Wanmimolruk, S., Wong, S.-M., Zhang, H., Coville, P.F., Walker, R.J., *J. Pharm. Pharmacol.* (1995) **47**, 957–963.

J7. Wanmimolruk, S., Wong, S.-M., Zhang, H., Coville, P.F., *J. Liquid Chromatogr. Relat. Technol.* (1996) **19**, 293–305.

J8. Takeshita, M., Miura, M., Unuma, Y., Iwai, S., Sato, I., Hongo, T., Arai, T., Kosaka, K., *Heterocycles* (1996) **42**, 831–836.

J9. Fowler, A.M., Chissick, H.H., Frearson, M.J., Wilson, K., *Biochem. Soc. Trans.* (1995) **23**, 604S.

J10. Kimura, N., Kato, H., Nishi, A., Furukawa, K., *Biosci. Biotechnol. Biochem.* (1996) **60**, 220–223.

J11. Anderson, A.H., Kuttab, S., Castagnoli, N., Jr, *Biochemistry* (1996) **35**, 3335–3340.

J12. Lepine, F., Bisaillon, J.-G., Milot, S., Khalid, T.H., Beaudet, R., Villemur, R., *Appl. Environ. Microbiol.* (1996) **62**, 809–814.

J13. Kurosaki, F., *Phytochemistry* (1996) **41**, 1023–1027.

J14. Shou, M., Krausz, K.W., Gonzalez, F.J., Gelboin, H.V., *Arch. Biochem. Biophys.* (1996) **328**, 201–207.

J15. Gaisser, S., Heide, L., *Phytochemistry* (1996) **41**, 1065–1072.

J16. Fraije, M.W., Veeger, C., Van Berkel, W.J.H., *Eur. J. Biochem.* (1995) **234**, 271–277.

J17. Bellucci, G., Chiappe, C., Cordoni, A., *Tetrahedron: Asymmetry* (1996) **7**, 197–202.

J18. Lefebvre, M., Homsey, W., Caille, G., du Suich, P., *Pharm. Res.* (1996) **13**, 124–128.

J19. Gill, H.J., Tingle, M.D., Park, B.K., *Br. J. Clin. Pharmacol.* (1995) 40, 531–538.

J20. Moltke, L.L., Greenblatt, D.J., Harmatz, J.S., Duan, S.X., Harrel, L.M., Cotreau-Bibbo, M.M., Pritchard, G,A., Wright, C.E., Shader, R.I., *J. Pharmacol. Exp. Ther.* (1996) **276**, 370–379.

J21. Ring, B.J., Catlow, J., Lindsay, T.J., Gillespie, T., Roskos, L.K., Cerimele, B.J., Swanson, S.P., Hamman, M.A., Wreighton, S.A., *J. Pharmacol. Exp. Ther.* (1996) **276**, 658–666.

J22. Soto-Urzua, L., Xochinua-Corona, Y.G., Flores-Encanacion, M., Baca, B.E., *Can. J. Microbiol.* (1996) **42**, 294–298.

J23. Suemori, A., Nakajima, K., Kurane, R., Nakamura, Y., *J. Ferment. Bioeng.* (1996) **81**, 133–137.

J24. Klyosov, A.A., *Biochemistry* (1996) **35**, 4457–4467.

J26. Borejsza-Wysocki, W., Hrazdina, G., *Plant Physiol.* (1996) **110**, 791–799.

J27. Rani, M., Prakash, Sobti, R.C., Jain, R.K., *Biochem. Biophys. Res. Commun.* (1996) **220**, 377–381.

J28. St-Pierre, M.V., Pang, K.S., *J. Pharmacokinet. Biopharm.* (1995) **23**, 243–266.

J29. Zhang, K.E., Naue, J.A., Arison, B., Vyas, K.P., *Chem. Res. Toxicol.* (1996) **9**, 547–554.

J30. Shepherd, J.A., Poon, W.W., Myles, D.C., Clarke C.F., *Tetrahedron Lett.* (1996) **37**, 2395–2398.

J31. Street, C.A., Di, W.L., Peniston-Bird, J.F., Patel, S., Sadler, P., Silman, R.E., *J. Pineal Res.* (1996) **20**, 98–114.

J32. Bubeck, B., Tshisuaka, B., Fetzner, S., Lingens, F., *Biochim. Biophys. Acta* (1996) **1293**, 39–44.

J33. Stephan, I., Tshisuaka, B., Fetzner, S., Lingens, F., *Eur. J. Biochem.* (1996) **236**, 155–162.

J34. D'Cunha, G.B., Satyanarayan, V., Nair, P.M., *Phytochemistry* (1996) **42**, 17–20.

J35. Beerhues, L., *FEBS Lett.* (1996) **383**, 264–266.

J36. Resnick, S.M., Gibson, D.T., *Appl. Environ. Microbiol.* (1996) **62**, 1364–1368.

J37. Mashetty, S.B., Manohar, S., Karegoudar, T.B., *Indian J. Biochem. Biophys.* (1996) **33**, 145–148.

J38. Finnison, K.W., Eales, J.G., *Gen. Comp. Endocrinol.* (1996) **101**, 145.

J39. Lam, T.B.T., Iiyama, K., Stone, B.A., *Phytochemistry* (1996) **41**, 1507–1510.

J40. Ahmad, S.A., Hopkins. T.L., Kramer, K.J., *Insect Biochem. Mol. Biol.* (1996) **26**, 49–57.

J41. Feierman, D.E., Lasker, J.M., *Drug Metab. Dispos.* (1996) **24**, 932–939.

J42. Heyn, H., White, R.B., Stevens, J.C., *Drug Metab. Dispos.* (1996) **24**, 948–954.

J43. Ghosal, A., Satoh, H., Thomas, P.E., Bush, E., Moore, D., *Drug Metab. Dispos.* (1996) **24**, 940–947.

J44. Ekström, G., Gunnarsson, U-B., *Drug Metab. Dispos.* (1996) **24**, 955–961.

J45. Halldin, M.M., Bredberg, E., Angelin, B., Arvidsson, T., Askemark, Y., Elofsson, S., Widman, M., *Drug Metab. Dispos.* (1996) **24**, 962–981.

J46. Sallustio, B.C., Fairchild, B.A., Shanahan, K., Evans, A.M., Nation, R.L., *Drug Metab. Dispos.* (1996) **24**, 984–989.

J47. Bort, R., Ponsoda, X., Carrasco, E., Gómez-Lechón, M.J., Castell, J.V., *Drug Metab. Dispos.* (1996) **24**, 969–975.

J48. Darbyshire, J.F., Iyer, K.R., Grogan, J., Korzekwa, K.R., Trager, W.F., *Drug Metab. Dispos.* (1996) **24**, 1038–1045.

J49. Torchin, C.D., McNeilly, P.J., Kapetanovic, I.M., Strong, J.M., Kupferberg, H.J., *Drug Metab. Dispos.* (1996) **24**, 1002–1008.

J50. Lin, J.H., Chiba, M., Balani, S.K., Chen, I.-W. Kwei, G.Y.-S., Vastag, K.J., Nishime, J.A., *Drug Metab. Dispos.* (1996) **24**, 1111–1120.

J51. Karam, W.G., Goldstein, J.A., Lasker, J.M., Ghanayem, B.I., *Drug Metab. Dispos.* (1996) **24**, 1081–1087.

J52. Yuan, J.H., Birkmeier, J., Yang, D.C., Hribar, J.D., Liu, N., Bible, R., Hajdu, E., Rock. M., Schoenhard, G., *Drug Metab. Dispos.* (1996) **24**, 1124–1133.

J53. Mutlib, A.E., Jureak, J., Hrib, N., *Drug Metab. Dispos.* (1996) **24**, 1139–1150.

J54. Chapman, D.T., Crout, D.H.G., Mahmoudian, M., Scopes, D.I.C., Smith, P.W., *J. Chem. Soc., Chem. Commun.* (1996) 2415–2416.

J55. Jones, N.E., England, P.A., Rouch, D.A., Wong, L.-L., *J. Chem. Soc., Chem. Commun.* (1996) 2413–2414.

J56. Parker, R.J., Collins, J.M., Strong, J.M., *Drug Metab. Dispos.* (1996) **24**,1167–1173.

J57. Meldahl, A.C., Nithipatikom, K., Lech, J.J., *Xenobiotica* (1996) **26**, 1167–1180.

J58. Balani, S.K., Woolf, E.J., Hoagland, V.L., Sturgill, M.G., Deutsch, P.J., Yeh, K.C., Lin, J.H., *Drug Metab. Dispos.* (1996) **24**, 1389–1394.

J58. Bajpal, M., Roskos, L.K., Shen, D.D., Levy, R.H., *Drug Metab. Dispos.* (1996) **24**, 1401–1403.

J60. Linhart, I., Weidenhoffer, Z., Sedmera, P., Polášek, M., Šmejkal, J., *Xenobiotica* (1996) **26**,1263–1272.

J61. Rekka, E.A., Soldan, M., Belai, I., Netter, K.J., Maser, E., *Xenobiotica* (1996) **26**, 1221–1229.

J62. Ayres, D.W., Barrow, A., Scully, N.L., Curtis, G.C., Hughes, H.M., *Xenobiotica* (1996) **26**, 1273–1282.

J63. Juchau, M.R., Chapman, D.E., Yang, H.-Y.L., Lee, Q.P., Namkung, M.J., Hinds, T.R., *Drug Metab. Dispos.* (1996) **24**, 1362–1368.

J64. Vincent, S.H., Painter, S.K., Krieter, P.A., Colletti, A.E., Lyszack, E., Rosegay, A., Dean, D., Luffer-Atlas, D., Miller, R.R., Cioffe, C., Chiu, S.-H.L., *Drug Metab. Dispos.* (1996) **24**, 1369–1377.

J65. Ring, J.A., Ghabrial, H., Ching, M.S., Shulkes, A., Smallwood, R.A., Morgan, D.J., *Drug Metab. Dispos.* (1996) **24**, 1378–1384.

J66. Kusano, K., Seko, T., Tanaka, S., Shikata, Y., Ando, T., Ida, S., Hosokawa, M., Satoh, T., Yuzuriha, T., Horie, T., *Drug Metab. Dispos.* (1996) **24**, 1186–1191.

J68. Roy, S.K., Moschel, R.C., Dolan, M.E., *Drug Metab. Dispos.* (1996) **24**, 1205–1211.

J69. Dalvie, D.K., Khosla, N.B., Navetta, K.A., Brighty, K.E., *Drug Metab. Dispos.* (1996) **24**, 1231–1240.

J70. Coutts, R.T., Bach, M.V., Baker, G.B., *Xenobiotica* (1997) **27**, 33–47.

J71. Prueksaritanont, T., Gorham, L.M., Ellis, J.D., Fernandez-Metzler, C., Deluna, P., Gehret, J.R., Strong, K.L., Hochman, J.H., Askew, B.C., Duggan, M.E., Gilbert, J.D., Lin, J.H., Vyas, K.P., *Drug Metab. Dispos.* (1996) **24**, 1263–1271.

J72. Waldmeier, F., Flesch, G., Müller, P., Winkler, T., Kriemler, H.-P., Bühlmayer, P., De Gasparo, M., *Xenobiotica* (1997) **27**, 59–71.

J73. Redembino, N., Dessalles, M.-C., Trouvin, J.-H., Loiseau, P.M., Gayral, P., Mahuzier, G., Rapp, M., Labarre, P., Godeneche, D., Madelmont, J.-C., Maurizis, J.-C., Veyre, A., Chabard, J.-L., *Xenobiotica* (1997) **27**, 73–85.

J74. Coffman, B.L., Rios, G.R., King, C.D., Tephly, T.R., *Drug Metab. Dispos.* (1997) **25**, 1–4.

J75. Gilham, D.E., Cairns, W., Paine, M.J.I., Modi, S., Poulsom, R., Roberts, G.C.K., Wolf, C.R., *Xenobiotica* (1997) **27**, 111–125.

J76. Maurizis, J.C., Madelmont, J.C., Rapp, M., Marijinen, M.C., Cerf, M.C., Gillardin, J.M., Lepage, F., Veyre, A., *Drug Metab. Dispos.* (1997) **25**, 33–39.

J77. Yoon, E.J., Shim, H.J., Lee, J.J., Lee, S.D., Kim, W.B., Yang, J., Lee, M.G., *Drug Metab. Dispos.* (1997) **25**, 66–74.

J78. Kassahun K., Mattiuz, E., Nyhart, E., Jr., Obermeyer, B., Gillespie, T., Murphy, A., Goodwin, R.M., Tupper, D., Callaghan, J.T., Lemberger, L., *Drug Metab. Dispos.* (1997) **25**, 81–93.

J79. Yamaguchi, Y., Kirita, S., Baba, T., Aoyama, J., Touchi, A., Tukey, R.H.,

Guengerich, F.P., Matsubara, T., *Drug Metab. Dispos.* (1997) **25**, 75–80.

J80. Kumar, G.N., Dubberke, E., Rodrigues, A.D., Roberts, E., Dennisen, J.F., *Drug Metab. Dispos.* (1997) **25**, 110–115.

J81. Ohe, T., Mashino, T., Hirobe, M., *Xenobiotica* (1997) **27**, 116–122.

J82. Cannell, R.J.P., Rashid, T., Ismail, I.M., Sidebottom, P.J., Knaggs, A.R., Marshall, P.S., *Xenobiotica* (1997) **27**, 147–157.

J83. Sinz, M.W., Black, A.E., Bjorge, S.M., Holmes, A., Trivedi, B.K., Woolf, T.F., *Drug Metab. Dispos.* (1997) **25**, 123–130.

J84. Taramura, T., Watanabe, T., Higuchi, S., Hashimoto, K., *Xenobiotica* (1997) **27**, 203–216.

J85. Richards, L.E., Pieniaszek H.J., Jr., Shatzmiller, S., Page, G.O., Blom, K.F., Read, J.M., Davidson, A.F., Confalone, P.N., *Xenobiotica* (1997) **27**, 217–229.

J86. Damani, L.A., Mitterhauser, M., Zolle, I., Lin, G., Oehler, E., Ho, Y.P., *Nucl. Med. Biol.* (1995) **22**, 1067–1074.

J87. Roy, S.K., Kulkarni, A.P., *Xenobiotica* (1997) **27**, 231–241.

J88. Nicotra, A., Senatori, O., Santangelo, G., *Comp. Biochem, Physiol.* (1993) **106C**, 479–482.

J89. Gorski, J.C., Jones, D.R., Wrighton, S.A., Hall, S.D., *Xenobiotica* (1997) **27**, 243–256.

J90. Zhang, D., Hansen, E.B., Jr, Deck, J., Heinze, T.M., Henderson, A., Korfmacher, W.A., Cerniglia, C.E., *Xenobiotica* (1997) **27**, 301–315.

J91. Pacifici, G.M., Guilianetti, B., Quelici, M.C., Spisni, R., Nervi, M., Giuliani, L., Gomeni, R., *Xenobiotica* (1997) **27**, 279–286.

J92. Fouda, H.G., Avery, M.J., Dalvie, D., Falkner, F.C., Melvin, L.S., Ronfeld, R.A., *Drug Metab. Dispos.* (1997) **25**, 140–148.

J93. Jung, F., Richardson, T.H., Raucy, J.L., Johnson, E.F., *Drug Metab. Dispos.* (1997) **25**, 133–139.

J94. Staretz, M.E., Murphy, S.E., Patten, C.J., Nunes, M.G., Koehl, W., Amin, S., Koenig, L.A., Guengerich, F.P., Hecht, S.S., *Drug Metab. Dispos.* (1997) **25**, 154–162.

J95. Datta, K., Sherblom, P.M., Kulkarni, A.P., *Drug Metab. Dispos.* (1997) **25**, 196–205.

J96. Chang, M., Sood, V.K., Wilson, G.J., Kloosterman, D.A., Sanders, P.E., Hauer, M.J., Fagerness, P.E., *Drug Metab. Dispos.* (1997) **25**, 228–242.

J97. Prakash, C., Kamel, A., Anderson, W., Howard, H., *Drug Metab. Dispos.* (1997) **25**, 206–218.

97a. King, C.D., Rios, G.R., Green, M.D., MacKenzie, P.I., Tephly, T.R., *Drug Metab. Dispos.* (1997) **25**, 251–255.

97b. Fitzsimmons, M.E., Collins, J.M., *Drug Metab. Dispos.* (1997) **25**, 256–266.

J98. Shiiyama, S., Soejima-Ohkuma, T., Honda, S., Kumagai, Y., Cho, A.K., Yamada, H., Oguri, K., Yoshimura, H., *Xenobiotica* (1997) **27**, 379–387.

J99. Oddy, EA., Manchee, G.R., Coughtrie, M.W.H., *Xenobiotica* (1997) **27**, 369–377.

J100. Gorrod, J.W., Sai, Y., *Xenobiotica* (1997) **27**, 389–399.

J101. Maggs, J.L., Pirmohamed, M., Kitteringham, N.R., Park, B.K., *Drug Metab. Dispos.* (1997) **25**, 275–280.

J102. Bolton, R.M., Ahokas, J.T., *Xenobiotica* (1997) **27**, 573–586.

J103. Wiltshire, H.R., Sutton, B.M. Heeps, G., Betty, A.M., Angus, D.W., Madigan, M.J., Sharp, S.R., *Xenobiotica* (1997) **27**, 539–5565.

J104. Li, C.-L.J., James, M.O., *Xenobiotica* (1997) **27**, 609–626.

J105. Kobayashi, H., Kobayashi, S., Dalrymple, P.D., Wood, S.G., Chasseaud, L.F., *Xenobiotica* (1997) **27**, 579–608.

J106. Lee, H.S., Jeong, S., Kim, K., Kim, J.H., Lee, S.K., Kang, B.H., Roh, J.K., *Xenobiotica* (1997) **27**, 423–429.

J107. Vargas, M., Franklin, M.R., *Xenobiotica* (1997) **27**, 413–421.

J108. McNeilly, P.J., Torchin, C.D., Anderson, L.W., Kapetanovic, I.M., Kupferberg, J., Strong, J.M., *Xenobiotica* (1997) **27**, 431–444.

J130. Ueda, K., Yamaoka, K., Rodriguez, M.E., Shibukawa, A., Nakagawa, T., *Drug Metab. Dispos.* (1997) **25**, 281–286.

J131. Nakamoto, T., Oda, Y., Imaoka S., Funae, Y., Fujimori, M., *Drug Metab. Dispos.* (1997) **25**, 296–300.

J132. Moussa, C., Houziaux, P., Danree, B., Azerad, R., *Drug Metab. Dispos.* (1997) **25**, 311–316.

J133. Moussa, C., Houziaux, P., Danree, B., Azerad, R., *Drug Metab. Dispos.* (1997) **25**, 301–310.

J134. Iosifidou, E.G., Haagsma, N., Olling, M., Boon, J.H., Tanck, M.W.T., *Drug Metab. Dispos.* (1997) **25**, 317–320.

J135. Boberg, Angerbauer, R., Fey, P., Kanhai, W.K., Karl, W., Kern, A., Ploschke, J., Radtke, M., *Drug Metab. Dispos.* (1997) **25**, 321–331.

J136. Shaw, J.P., Louie, M.S., Krishnamurty, V.V., Arimilli, M.N., Jones, R.J., Bidgood, A.M., Lee, W.A., Cundy, K.C., *Drug Metab. Dispos.* (1997) **25**, 362–366.

J137. Breyer-Pfaff, U., Fischer, D., Winne, D., *Drug Metab. Dispos.* (1997) **25**, 340–345.

J138. Franklin, T.J., Jacobs, V.N., Jones, G., Pile, P., *Drug Metab. Dispos.* (1997) **25**, 367–370.

J139. Lehman-McKeeman, L.D., Johnson, D.R., Caudill, D., Stuard, S.B., *Drug Metab. Dispos.* (1997) **25**, 384–389.

J140. Dalvie, D.K., Khosla, N., Vincent, J., *Drug Metab. Dispos.* (1997) **25**, 423–427.

J141. Salvi, A., Carrupt, P.-A., Mayer, J.M., Testa, B., *Drug Metab. Dispos.* (1997) **25**, 395–398.

J142. Yu, Z., Hayton, W.L., Chan, K.K., *Drug Metab. Dispos.* (1997) **25**, 431–436.

J143. Dahms, M., Lotz, R., Lang, W., Renner, U., Bayer, E., Spahn-Langguth, H., *Drug Metab. Dispos.* (1997) **25**, 442–452.

J144. Lakshmi, V.M., Zenser, T.V., Davis, B.B., *Drug Metab. Dispos.* (1997) **25**, 481–488.

J145. Sipe, H.J., Jr, Corbett, J.T., Mason, R.P., *Drug Metab. Dispos.* (1997) **25**, 468–480.

J146. Denissen, J.F., Grabowski, B.A., Johnson, M.K., Buko, A.M., Kempf, D.J., Thomas, S.B., Surber, B.W., *Drug Metab. Dispos.* (1997) **25**, 489–501.

J147. Tugnait, M., Hawes, E.M., McKay, G., Rettie, A.E., Haining, R.L., Midha, K.K., *Drug Metab. Dispos.* (1997) **25**, 542–547.

J148. White, R.D., Shea, D., Stegeman, J.J., *Drug Metab. Dispos.* (1997) **25**, 564–572.

J149. Laurenzana, E.M., Owens, S.M., *Drug Metab. Dispos.* (1997) **25**, 557–563.

J150. Matting, E., Franklin, R., Gillespie, T., Murphy, A., Bernstein, J., Chiu, A., Hotten, T., Kassahun, K., *Drug Metab. Dispos.* (1997) **25**,573–583.

J151. Pool, W.F., Reily, M.D., Bjorge, S.M., Woolf, T.F., *Drug Metab. Dispos.* (1997) **25**, 590–597.

J152. Shin, H.-S., *Drug Metab. Dispos.* (1997) **25**, 657–662.

J153. Dain, J.G., Nicoletti, J., Ballard, F., *Drug Metab. Dispos.* (1997) **25**, 603–609.

J154. Kitani, M., Miyamoto, G., Nagasawa, M., Yamada, T., Matsubara, J., Uchida, M., Odomi, M., *Drug Metab. Dispos.* (1997) **25**, 663–674.

J155. Freitag, D.G., Foster, R.T., Coutts, R.T., Pickard, M.A., Pasutto, F.M., *Drug Metab. Dispos.* (1997) **25**, 685–692.

J156. Ehlhardt, W.J., Woodland, J.M., Toth, J.E., Ray, J.E., *Drug Metab. Dispos.* (1997) **25**, 701–708.

J157. Woodland, J.M., Barnett, C.J., Dorman, D.E., Gruber, J.M., Shih, C., Spangle, L.A., Wilson, T.M., Elhardt, W.J., *Drug Metab. Dispos.* (1997) **25**, 693–700.

J158. Rao, G.P., Davis, P.J., *Drug Metab. Dispos.* (1997) **25**, 709–715.

J159. Kashiyama, E., Yokoi, T., Odomi, M., Funae, Y., Inoue, K., Kamataki, T., *Drug Metab. Dispos.* (1997) **25**, 716–724.

J160. Sauer, J.-M., Smith, R.L., Bao, J., Kattnig, M.J., Kuester, R.K., McClure, T.D., Mayersohn, M., Sipes, I.G., *Drug Metab. Dispos.* (1997) **25**, 732–739.

J161. Sidelman, U.G., Christiansen, E., Krogh, L., Cornett, C., Tjørnelund, J., Hansen, S.H., *Drug Metab. Dispos.* (1997) **25**, 725–731.

J162. Olesen, O.V., Linnet, K., *Drug Metab. Dispos.* (1997) **25**, 740–744.

J163. Rodriguez, R.J., Acosta, D., Jr., *Drug Metab. Dispos.* (1997) **25**, 772–777.

J164. Kim, H., Wang, R.S., Elovaara, E., Raonio, H., Pelkonen, O., Aoyama, T., Vainio, H., Nakajima, T., *Xenobiotica* (1997) **27**, 657–665.

J165. Yamano, S., Takahashi, A., Todaka, T., Toki, S., *Xenobiotica* (1997) **27**, 645–656.

J166. Nordelmann, L., Tjørnelund, J., Hansen, S.H., Cornett, C., Sidelmann, U.G., Braumann, U., Christensen, F., Christensen, S.B., *Xenobiotica* (1997) **27**, 667–680.

J167. Kakiki, M., Tadano, K., *Xenobiotica* (1997) **27**, 723–732.

J168. Yoshida, H., Kohno, Y., Endo, H., Hasegawa, M., Yamaguchi, J., Tomisawa, K., Suwa, T., *Biol. Pharm. Bull.* (1996) **19**, 424–429.

J169. Yasuda, T., Mizunuma, S., Kano, Y., Saito, K., Ohsawa, K., *Biol. Pharm. Bull.* (1996) **19**, 413–417.

J170. Clement, B., Demesmaeker, M., Linne, S., *Chem. Res. Toxicol.* (1996) **9**, 682–688.

J171. Butterworth, M., Lau, S., Monks, T., *Chem. Res. Toxicol.* (1996) **9**, 793–799.

J172. Coleman, T., Ellis, S.W., Martin, I.J., Lennard, M.S., Tucker, G.T., *J. Pharmacol. Exp. Ther.* (1996) **277**, 685–690.

J173. Fava, F., Baldoni, F., Marchetti, L., *Lett. Appl. Microbiol.* (1996) **22**, 275–279.

J174. Bhagwat, S.V., Bhamre, S., Boyd, M.R., Ravindranath, V., *Biochem. Pharmacol.* (1996) **51**, 1469–1475

J175. Boll, M., Fuchs, G., *Eur. J. Biochem.* (1995) **234**, 921–933.

J176. Sugumaran, M., Tan, S., Sun, H.-L., *Arch. Biochem. Biophys.* (1996) **329**, 175–180.

J177. Xun, L., *J. Bacteriol.,* (1996) **178**, 2645–2649.

J178. Romanov, V., Hausinger, R.P., *J. Bacteriol.* (1996) **178**, 2656–2661.

J179. Rasmussen, S., Wolff, C., Rudolph, H., *Phytochemistry* (1996) **42**, 81–87.

J180. Chakraborty, A., Chowdhury, B.K., *Indian J. Chem., Sect. B: Org. Chem. Incl. Med. Chem.* (1996) **35B**, 617–620.

J181. Haidour, A., Ramos, J.L., *Environ. Sci. Technol.* (1996) **30,** 2365–2370.

J182. Pearce, R.E., Rodrigues, A.D., Goldstein, J.A., Parkinson, A., *J. Pharmacol. Exp. Ther.* (1996) **277**, 805–816.

J183. Brunelle, F.M., Raoof, A.A., De Ville De Goyet, J., Verbeeck, R.K., *Biopharm. Drug Dispos.* (1996) **17**, 311–317.

J184. He, Z., Wiegel, J., *J. Bacteriol.,* (1996) **178**, 3539–3543.

J185. Heuer, S., Vogt, T., Boehm, H., Strack, D., *Planta* (1996) **199**, 244–250.

J186. Spiess, E., Goerisch, H., *Arch. Microbiol.* (1996) **165**, 201–205.

J187. Billaud, C., Lecorni, D., *J. Agric. Food Chem.* (1996) **44**, 1668–1675.

J188. Kemmerer, B., Reichling, J., *Phytochemistry* (1996) **42**, 397–403.

J189. Minamisawa, K., Ogawa, K.-I., Fukuhara, H., *Plant Cell Physiol.* (1996) **37,** 449–453.

J190. Rakwal, R., Hasegawa, M., Kodama, O., *Biochem. Biophys. Res. Commun.* (1996) **222**, 732–735.

J191. Biegert, T., Fuchs, G., Heider, J., *Eur. J. Biochem.* (1996) **238**, 661–668.

J192. Shin, H.S., Rogers, P.L., *Appl. Microbiol. Biotechnol.* (1995) **44**, 7–14.

J193. Zezalei, L., Hadar, Y., Fu, P.P., Freeman, J.P., Cerniglia, C.E., *Appl. Envron. Microbiol.* (1996) **62**, 2547–2553.

J194. Kiljunen, E., Kanerva, L.T., *Tetrahedron: Asymmetry* (1996) **7**, 1105–1116.

J195. Pool, W.F., Woolf, T.F., Reily, M.D., Caprathe, B.W., Emmerling, M.R., Jaen, J.C., *J. Med. Chem.* (1996) **39**, 3014–3018.

J196. Noguera, D., Freedman, D.L., *Appl. Envron. Microbiol.* (1996) **62**, 2257–2263.

J197. Ramauge, M., Pallad, S., Esfandiari, A., Gavaret, J.-M., Lennon, A.-M., Pierre, M., Courtin, F., *Endocrinology* (1996) **137**, 3021–3025.

J198. Pond, A.L., Coyne, C.P., Chambers, H.W., Chambers, J.E., *Biochem. Pharmacol.* (1996) **52**, 363–369.

J199. Stupperich, E., Konle, R., Eckerskorn, C., *Biochem. Biophys. Res. Commun.* (1996) **223**, 770–777.

J200. Nielsen, K.K., Flinois, J.P., Beaune, P., Brosen, K., *J. Pharmacol. Exp. Ther.* (1996) **277**, 1659–1664.

J201. Urano, Y., Higuchi, T., Hirobe, M., *J. Chem. Soc., Perkin Trans. 2* (1996) 1169–1173.

J202. Gich, I., Bayes, M., Barbanoj, M.J., Artigas, R., Tost, D. Mauleon, D., *Clin. Drug Invest.* (1996) **11**, 347–353.

J203. Kraeling, M.E.K., Lipicky, R.J., Bronaugh, R.L., *Skin Pharmacol.* (1996) **9**, 221–230.

J204. Lembert, N., *Biochem. J.,* (1996) **317**, 273–277.

J205. Taura, F., Morimoto, S., Shoyama, Y., *J. Biol. Chem.* (1996) **271**, 17411–17416.

J206. Santha, R., Rao, N.A., Vaidyanathan, C.S., *Biochim. Biophys. Acta* (1996) **1293**, 191–200.

J207. Sugaya, S., Sakai, S., *Physiol. Plant.* (1996) **97**, 433–439.

J208. Lin, G., Chu, K.-W., Damani, L.A., Hawes, E.M., *J. Pharm. Biomed. Anal.* (1996) **14**, 727–738.

J210. Manninen, A., Auriola, S., Vartianinen, M., Liesivuori, J., Turunen, T., Pasanen, M., *Arch. Toxicol.* (1996) **70**, 579–584.

J211. Suemori, A., Nakajima, K., Kurane, R., Nakamura, Y., *Lett. Appl. Microbiol.* (1996) **23**, 1–4.

J212. Kato, Y., Yamada, H., Asano, Y., *J. Mol. Catal. B: Enzym.* (1996) **1**, 151–160.

J213. Bonting, C.F.C., Fuchs, G., *Arch. Microbiol.* (1996) **165**, 402–408.

J214. Vickers, A.E.M., Fischer, V., Connors, M.S., Biggi, W.A., Heitz, F., Baldeck, J.P., Brendel, K., *Eur. J. Drug Metab. Pharmacokinet.* (1996) **21**, 43–50.

J215. Koch, A., *J. Pharm. Biomed. Anal.* (1996) **14**, 1335–1338.

J216. Bosetti, A., Bianchi, D., Andriollo, N., Cidaria, D., Cesti, P., Sello, G., Di Gennaro, P., *J. Chem. Technol. Biotechnol.* (1996) **66**, 375–381.

J218. Hammer, A., Stolz, A., Knackmuss, H.J., *Arch. Microbiol.* (1996) **166**, 92–100.

J219. Lee, K., Gibson, D.T., *Appl. Environ. Microbiol.* (1996) **62**, 3101–3106.

J220. Kondo, T., Yoshida, K., Yoshimura, Y., Motohashi, M., Tanayama, S., *J. Mass Spectrom.* (1996) **31**, 873–878.

J221. Hauschild, J.E., Masai, E., Sugiyama, K., Hatta, T., Kimbara, K., Fukuda, M., Yano, K., *Appl. Environ. Microbiol.* (1996) **62**, 2940–2946.

J222. Ahmed, S., Imai, T., Otagiri, M., *Pharm. Sci.* (1996) **1**, 429–432.

J223. Rivory, L.P., Riou, J.-F., Haaz, M.-C., Sable, S., Viulhorgne, M., Commercon, A., Pond, S.M., Robert, S.M., *Cancer Res.* (1996) **56**, 689–694.

J224. Apanovitch, D., Walz, F.G., Jr., *Biochim. Biophys. Acta* (1996) **1291**, 16–26.

J225. Paine, M.F., Shen, D.D., Kunze, K.L., Perkins, J.D., Marsh, C.L., McVicar, J.P., Barr, D.M., Gillies, B.S., Thummel, K.E., *Clin. Pharmacol. Ther.* (1996) **60**, 14–24.

J226. Hazen, S.L., Hsu, F.F., Mueller, D.M., Crowley, J.R., Heinecke, J.W., *J. Clin. Invest.* (1996) **98**, 1283–1289.

J227. Lin, G., Mc Kay, G., Midha, K.K., *J. Pharm. Biomed. Anal.* (1996) **14**, 1561–1577.

J228. Steventon, G.B., Mitchell, S.C., Waring, R.H., *Drug Metab. Drug Interact.* (1996) **13**, 111–117.

J229. Giurnazi, A.M., Garle, M.J., Lal, K., Fry, J.R., *J. Pharm. Pharmacol.* (1996) **48**, 729–733.

J230. Spanggord, R.J., Gordon, G.R., Schocken, M.J., Starr, R.I., *Environ. Toxicol. Chem.* (1996) **15**, 1655–1663.

J231. Kim, S.H., Kronsatad, J.W., Ellis, B.E., *Phytochemistry* (1996) **43**, 351–357.

J232. Löffler, F.E., Stanford, R.A., Tiedje, J.J., *Appl. Environ. Microbiol.* (1996) **62**, 3809–3813.

J233. Kamei, A., *Biol. Pharm. Bull.* (1996) **19**, 1223–1226.

J234. Brugger, R., Alia, B.G., Reichel, C., Waibel, R., Menzel, S., Brune, K., Geisslinger, G., *Biochem. Pharmacol.* (1996) **52**, 1007–1013.

J235. Lee, M.H., Zhang, Z.-H., MacKinnon, C.H., Baldwin, J.E., Crouch, N.P., *FEBS Lett.* (1996) **393**, 269–272.

J237. Harper, D.B., McRoberts, W.C., Kennedy, J.T., *Appl. Environ. Microbiol.* (1996) **62**, 3366–3370.

J239. Hannah, P., Malizia, A., Glover, V., Bridges, P., Sandler, M., *Drug Metab. Drug Interact.* (1996) **32**, 119–128.

J240. Rivory, L.P., Bowles, M.R., Robert, J., Pond, S.M., *Biochem. Pharmacol.* (1996) **52** 1103–1111.

J241. Lajtha, A., Sershen, H., Cooper, T., Hashim, A., Gaal, J., *Neurochem. Res.* (1996) **21**, 1155–1160.

J243. Hao, Z., Charles, D.J., Yu, L., Simon, J.E., *Phytochemistry* (1996) **43**, 735–739.

J244. Tseng, C.-Y., Wang, S.-L., Lai, M.-L., Huang, J., *Clin. Pharmacol. Ther.* (1996) **60**, 177–182.

J245. Petushkova, N.A., Bachmanova, G.I., Archakov, A.I., *In Vitro Toxicol.* (1996) **9**, 291–295.

J246. Suarez, M., Ferrer, E., Martin, M., *FEMS Microbiol. Lett.* (1996) **143**, 89–95.

J247. Barta, I.C., Boeger, P., *Pestic. Sci.* (1996) **48**, 109–116.

J248. Terashima, K., Taka, S., Usami, Y., Adachi, T., Sugiyama, T., Katagiri, Y., Hirano, K., *Pharm. Res.* (1996) **13**, 1327–1330.

J249. Haigler, B.E., Suen, W.-C., Spain, J.C., *J. Bacteriol.* (1996) **178**, 6019–6024.

J250. Shimada, A., Shishido, H., Nakamura, I., *Amino Acids* (1996) **11**, 83–89.

J251. Daubaras, D., Saido, K., Chakrabarty, A.M., *Appl. Environ. Microbiol.* (1996) **62**, 4276–4279.

J252. Mosca, L., Foppoli, C., Coccia, R., Rosei, M.A., *Pigm. Cell Res.* (1996) **9**, 117–125.

J253. Briggs, B.S., Kreuzman, A.J., Whitesitt, C., W.-W., Zmijewski, M., *J. Mol. Catal. B: Enzym.* (1996) **2**, 53–69.

J254. Chou, J.-C., Kuleck. G.A., Cohen, J.D., Mulbry, W.W., *Plant Physiol.* (1996) **112**, 1281–1287.

J255. Lendenmann, U., Spain J.C., *J. Bacteriol.* (1996) **178**, 6227–6232.

J256. Turner, J.E., Allison, N., Fewson, C.A., *Arch. Microbiol.* (1996) **166**, 252–259.

J257. Teramura, T., Fukunaga, Y., van Hoogdalem, E.J., Watanabe, T., Higuchi, S., *Xenobiotica* (1997) **27**, 885–900.

J258. Ariyoshi, N., Koga, N., Yoshimura, H., Oguri, K., *Xenobiotica* (1997) **27**, 973–983.

J259. Sensa, C., Kishimoto, W., Sakai, K., Nagakura, A., Igarashi, T., *Xenobiotica* (1997) **27**, 913–922.

J260. Koerts, J., Soffers, A.E.M.F., de Kraker, J.-W., Cnubben, N.P.H., Rietjens, I.M.C.M., *Xenobiotica* (1997) **27**, 801–807.

J261. Garras, A., Elholm, M., Slebolda, J., Frøyland, L., Osmundsen, H., Berge, R.K., *Xenobiotica* (1997) **27**, 781–799.

J262. King, J.N., Maurer, M., Morrison, C.A., Mauron, C., Kaiser, G., *Xenobiotica* (1997) **27**, 819–829.

J263. Harahuchi, K., Kato, Y., Masuda, Y., Kimura, R., *Xenobiotica* (1997) **27**, 831–842.

J264. Lin, L.Y., Di Stefano, E.W., Schmitz, D.A., Hsu, L., Ellis, S.W., Lennard, M.S., Tucker, G.T., Cho, A.K., *Drug Metab. Dispos.* (1997) **25**, 1059–1064.

J265. Chiba, M., Xu, X., Nishime, J.A., Balani, S.K., Lin, J.H., *Drug Metab. Dispos.* (1977) **25**,1022–1031.

J266. Labroo, R.B., Paine, M.F., Thummel, K.E., Kharasch, E.D., *Drug Metab. Dispos.* (1997) **25**, 1072–1080.

J267. Springate, J., Zamlauski-Tucker, M.J., Lu, H., Chan, K.K., *Drug Metab. Dispos.* (1997) **25**, 1081–1082.

J268. Vincent, S.H., Painter, S.K., Luffer-Atlas, D., Karanam, B.V., McGowan, E., Cioffe, C., Doss, G., Chiu, S.-H.L., *Drug Metab. Dispos.* (1997) **25**, 932–939.

J269. Rodrigues, A.D., Surber, B.W., Yao, Y., Wong, S.L., Roberts, E.M., *Drug Metab. Dispos.* (1997) **25**, 1097–1100.

J270. Luffer-Atlas, D., Vincent, S.H., Painter, S.K., Arison, B.H., Stearns, R.A., Chiu, S.-H.L., *Drug Metab. Dispos.* (1997) **25**, 940–952.

J271. Amore, B.M., Kalhorn, T.F., Skiles, G.L., Hunter, A.P., Bennett, G.D., Finnell, R.H., Nelson, S.D., Slattery, J.T., *Drug Metab. Dispos.* (1997) **25**, 953–962.

J272. Oldham, H.G., Clarke, S.E., *Drug Metab. Dispos.* (1997) **25**, 970–977.

J273. Utkin, I., Koudriakova, T., Thompson, T., Cottrell, C., Iatsimirskaia, E., Barry, J., Vouros, P., Gerber, N., *Drug Metab. Dispos.* (1997) **25**, 963–969.

J274. Prueksaritanont, T., Gorham, L.M., Breslin, M.J., Hutchinson, J.H., Hartman, G.D., Vyas, K.P., Baillie, T.A., *Drug Metab. Dispos.* (1997) **25**, 978–984.

J275. Chang, M., Sood, V.K., Kloosterman, D.A., Hauer, M.J., Fagerness, P.E., Sanders, P.E., Vrbanac, J.J., *Drug Metab. Dispos.* (1997) **25**, 814–827.

J276. Bloomer, J.C., Clarke, S.E., Chenery, R.J., *Drug Metab. Dispos.* (1997) **25**, 840–844.

J277. Chang, M., Sood, V.K., Wilson, G.J., Kloosterman, D.A., Sanders, P.E. Hauer, M.J., Zhang, W., Branstetter, D.G., *Drug Metab. Dispos.* (1997) **25**, 828–839.

J278. Haraguchi, K., Kato, Y., Kimura, R., Masuda, Y., *Drug Metab. Dispos.* (1997) **25**, 845–852.

J279. Prakash, C., Kamel, A., Gummerus, J., Wilner, K., *Drug Metab. Dispos.* (1997) **25**, 863–872.

J280. Nakamura, A., Hirota, T., Morino, A., Shimada, T., Uematsu, T., *Xenobiotica* (1997) **27**, 995–1003.

J281. Prakash, C., Kamel, A., Cui, D., *Drug Metab. Dispos.* (1997) **25**, 891–901.

J282. Jamis-Dow, C.A., Katki, A.G., Collins, J.M., Klecker, R.W., *Xenobiotica* (1997) **27**, 1015–1024.

J283. Jones, B.C., Tyman, C.A., Smith, D.A., *Xenobiotica* (1997) **27**, 1025–1037.

J284. Naitoh, T., Kakiki, M., Kozaki, T., Mishima, M., Yuzurika, T., Horie, T., *Xenobiotica* (1997) **27**, 1071–1089.

J285. Slatter, J.G., Feenstra, K.L., Hauer, M.J., Sanders, P.E., Vrabanac, J.J., Scott, G., Speed, W., *Xenobiotica* (1997) **27**, 1039–1051.

J286. Kaye, B., Brearley, C.J., Cussans, N.J., Herron, M., Humphrey, M.J., Mollatt, A.R., *Xenobiotica* (1997) **27**, 1091–1102.

J287. Goodwin, B.L., Ruthven, C.R.J., Sandler, M., *Gen. Pharmacol.* (1997) **28**, 535–543.

J288. Jang, G.R., Benet, L.Z., *Drug Metab. Dispos.* (1997) **25**, 1119–1122.

J289. Resnick, S.M., Gibson, D.T., *Appl. Environ. Microbiol.* (1996) **62**, 3355–3359.

J290. Shirley, M.A., Wheelan, P., Howell, S.R., Murphy, R.C., *Drug Metab. Dispos.* (1997) **25**, 1144–1149.

J291. Slatter, J.G., Su, P., Sams, J.P., Schaaf, L.J., Wienkers, L.C., *Drug Metab. Dispos.* (1977) **25**, 1157–1164.

J292. Oddy, E.A., Manchee, G.R., Freeman, N.M., Ward, M.A., Coughtrie, M.W.H., *Drug Metab. Dispos.* (1977) **25**, 1205–1210.
J293. Sauer, J.-M., Bao, J., Smith, R.L., Kuester, R.K., Mayersohn, M., Sipes, I.G., *Drug Metab. Dispos.* (1997) **25**, 1184–1190.
J294. Taft, D.R., Iyer, G.R., Behair, L., DiGregori, R.V., *Drug Metab. Dispos.* (1997) **25**, 1225–1218.
J295. Muth, E.A., Moyer, J.A., Haskins, J.T., Andree, T.H., Husbands, G.E.M, *Drug Dev. Res.* (1991) **23**, 191–199.
J296. Umehara, K., Kudo, S., Odomi, M., *Xenobiotica* (1997) **27**, 1121–1129.
J297. Chiba, M., Nshime, J.A., Linn, J.H., *Drug Metab. Dispos.* (1997) **25**, 1219–1222.
J298. Wilson, A.A., Wang, J., Koch, P., Walle, T., *Xenobiotica* (1997) **27**, 1147–1154.
J299. Renwick, A.B., Mistry, H., Ball, S.E., Walters, D.G., Kao, J., Lake, B.G., *Xenobiotica* (1998) **28**, 337–348.
J300. Nielsen, S.E., Breinholt, V., Justesen, U., Cornett, C., Dragsted, L.O., *Xenobiotica* (1998) **28**, 389–401.
J301. Scarfe, G.B., Wright, B., Clayton, E., Taylor, S., Wilson, I.D., Lindon, J.C., Nicholson, J.K., *Xenobiotica* (1998) **28**, 373–388.
J302. Kiljunen, E., Kanerva, L.T., *Tetrahedron: Asymmetry* (1997) **8**, 1225–1234.
J303. Preuss, S., Wittneben, D., Lorber, K.E., *Toxicol. Environ. Chem.* (1997) **58**, 179–195.
J304. Li, L., Popko, J.L., Zhang, X.-H., Osakabe, K., Tsai, C.-J., Joshi, C.P., Chang, V.L., *Proc. Natl. Acad. Sci. USA* (1997) **94**, 5461–5466.
J305. Hamid, A., Uematsu, H., Sato, N., Kota, K., Iwaku, M., Hoshino, E., *J. Antimicrob. Chemother.* (1997) **39**, 129–134.
J306. Barillas, W., Beerhues, L., *Planta* (1997) **202**, 112–116.
J307. Lombard, A., Rossetti, V., Cassone, M.C., *Chirality* (1997) **9**, 317–319.
J308. Sohn, D.-R., Kwon, J.-T., Kim, H.-K., Ishizaki, T., *Clin. Pharmacol. Ther.* (1997) **61**, 574–582.
J309. Tatsimirskaia, E., Tulebaev, S., Storozhuk, E., Utkin, I., Smith, D., Gerber, N., Koudriakova, *Clin. Pharmacol. Ther.* (1997) **61**, 554–562.
J310. Sarabia, S.F., Zhu, B.T., Kurosawa, T., Tohma, M., Liehr, J.C., *Chem. Res. Toxicol.* (1997) **10**, 767–771.
J311. Ferraris, M., Dosanjh, M.K., Marafante, E., *Polycyclic. Aromat. Compd.* (1996) **11**, 91–98.
J312. Siddiqui, M.A., Fujiwara, S., Imanaka, T., *Mol. Gen. Genet.* (1997) **254**, 433–439.
J313. Jeon, B.-J., Lee, K.-H., Ryu, H.-S., You, B.-J., *J. Food Sci. Nutr.* (1996) **1**, 111–116.
J314. Doukyu, N., Aono, R., *Extremophiles* (1997) **1**, 100–105.
J315. Rehmann, K., Steimberg, C., Kettrup, A., *Polycyclic Aromat. Compd.* (1996) **11**, 125–130.
J317. Koch, M., Noll, H., Steinberg, C., Kettrup, A., *Polycyclic Aromat. Compd.* (1996) **11**, 131–135.
J318. Laloue, H., Weber-Lotfi, F., Lucau-Danila, A., Guillemaut, P., *Plant Physiol.* (1997) **35**, 341–346.
J319. Finnson, K.W., Eales, J.G., *Comp. Biochem. Physiol. C, Pharmacol. Toxicol. Endocrinol.* (1997) **117C**, 193–199.
J320. Stich, K., Halbwirth, H., Wurst, F., Forkmann, G., *Z. Naturforsch. C: Biosci.* (1997) **52**, 153–158.
J321. Bratton, S.B., Lau, S.S., Monks, T.J., *Chem. Res. Toxicol.* (1997) **10**, 859–865.
J322. Chung, J.G., Lee, J.H., Ho, C.C., Chou, J.M., Teng, H.H., Hung, C.F., Huang, S.C., *J. Food Biochem.* (1997) **20**, 481–490.
J323. Osman, S., Luthra, S.K., Brady, F., Hume, S.P., Brown, G., Harte, R.J.A., Mathews, J.C., Denny, W.A., Baguley,

B.C., Jones, T., Price, P.M., *Cancer Res.* (1997) **57**, 2172–2180.

J324. Pacifici, G.M., D'alessandro, C., Gucci, A., Giuliani, L., *Arch. Toxicol.* (1997) **71**, 477–481.

J325. Haritos, V.S., Ching, M.S., Ghabrial, H., Ahokas, J.T., *J. Chromatogr., B: Biomed. Sci. Appl.* (1997) **693**, 327–337.

J326. Brunella, A., Graf, M., Kittelman, M., Laumen, K., Ghisalba, O., *Appl. Microbiol. Biotechnol.* (1997) **47**, 515–520.

J327. Michon, T., Chenu, M., Kellershon, N., Desmadril, M., Gueguen, J., *Biochemistry,* (1997) **36**, 8504–8513.

J328. Dellus, V., Heller, W., Sandermann, H. Jr., Scalbert, A., *Phytochemistry* (1997) **45**, 1415–1418.

J329. Takenaka, S., Murakami, S., Shinke, R., Hatakeyama, K., Yukawa, H., Aoki, K., *J. Biol. Chem.* (1997) **272**, 14727–14732.

J330. Stevens, G.J., Hitchcock, K., Wang, Y.K., Coppola, G.M., Versace, R.W., Chin, J.A., Shapiro, M., Suwanrumpha, S., Mangold, B.L.K., *Chem. Res. Toxicol,* (1997) **10**, 733–741.

J331. Madhu, C., Duff, S., Baumgarten, V., Rix, P., Small, D., Tang-Liu, D., *J. Pharm. Sci.* (1997) **86**, 972–974.

J332. Bianchi, D., Bosetti, A., Cidaria, D., Bernardi, A., Gagliardi, I., D'amico, P., *Appl. Microbiol. Biotechnol.* (1997) **47**, 596–599.

J333. Murdter, T.E., Sperker, B., Kivisto, M., Fritz, P., Friedel, G., Linder, A., Bosslet, K., Toomes, H., Dierkesmann, R., Kroemer, H.K., *Cancer Res.* (1997) **57**, 2440–2445.

J334. Kiljunen, E., Kanerva, L.Y., *Tetrahedron: Asymmetry* (1997) **8**, 1551–1557.

J335. Hammond, D.K., Zhu, B.T., Wang, M.Y., Ricci, M.J., Liehr, J.G., *Toxicol. Appl. Pharmacol.* (1997) **145**, 54–60.

J336. Lin, J., Cashman, J.R., *Chem. Res. Toxicol.* (1997) **10**, 842–852.

J337. Kumagai, S., Matsunaga, I., *Ind. Health.* (1997) **35**, 399–403.

J338. Morawski, B., Casy, G., Illaszewicz, C., Griengl, H., Ribbond, D.W., *J. Bacteriol.* (1997) **179**, 4023–4029.

J339. Jeffers, M.R., McRoberts, W.C., Harper, D.B., *Microbiology* (1997) **143**, 1975–1981.

J340. Reinscheid, U.M., Baure, M.P., Mueller, R., *Biodegradation* (1996–1997) **7**, 455–461.

J341. Maloney, S.E., Marks, T.S., Sharp, R.J., *Lett. Appl. Microbiol.* (1997) **24**, 441–444.

J342. Roig-Sagues, A.X., Hernandez-Herrero, M., Rodriguez-Jerez, J.J., Lopez-Sabater, E.I., Mora-Ventura, M.T., *J. Food Saf.* (1997) **17**, 13–22.

J343. Yabuki, M., Shimakura, J., Ito, M., Kanamoru, H., Iba, K., Nakatsuka, I., *Eur. J. Drug Metab. Pharmacokinet.* (1997) **22**, 25–33.

J344. Mars, A., Kasberg, T., Kaschabek, S.R., van Agteren, M.H., Janssen, D.B., Reineke, W., *J. Bacteriol.* (1997) **179**, 4530–4537.

J345. Soby S., Bates, R., Van Etten, H., *Phytochemistry* (1997) **45**, 925–929.

J346. Abelskov, A.K., Smith, A.T., Rasmussen, C.B., Dunford, H.B., Welinder, K.G., *Biochemistry,* (1997) **36**, 9453–9463.

J347. Bezalel, L., Hadar, Y., Cerniglia, C.E., *Appl. Environ. Microbiol.* (1997) **63**, 2495–2501.

J348. Guray, T., Guvenc, T., *Biochem. Arch.* (1997) **13**, 107–111.

J349. Petersen, M., *Phytochemistry* (1997) **45**, 1165–1172.

J351. Lindstrom, A.B., Yeowell-O'connell, K., Waidyanatha, S., Golding, B.T., Tornero-Velez, R., Rappaport, S.M., *Carcinogenesis* (1997) **18**, 1637–1641.

J352. Doerge, D.R., Divi, R.L., Churchwell, M.I., *Anal. Biochem.* (1997) **250**, 10–17.

J353. Wiyakrutta, S., Meevootison, V., *J. Biotechnol.* (1997) **55**, 193–203.

J354. Fraaije, M.W., van Berkel, W.J.H., *J. Biol. Chem.* (1997) **272**, 18111–18116.

J356. Cho, H.-Y., Soda, K., *Biosci. Biotechnol. Biochem.* (1997) **61**, 1216–1218.

J357. Di Leece, C., Accarino, M., Bolognese, F., Galli, E., Barbieri, P., *Appl. Environ. Microbiol.* (1997) **63**, 3279–3281.

J358. Yamada, K., Tojo, K., *Eur. J. Pharm. Biopharm.* (1997) **43**, 253.

J359. Jankowski. K., Delaforge, M., Jaouen, M., Virelizier, H., *Spectroscopy* (1996–1997) **13**, 163–174.

J360. Yamakawa, T., Ohta, S., *Biochem. Biophys. Res. Commun.* (1997) **236**, 676–681.

J361. Kaufmann, F., Wohlfarth, G., Diekert, G., *Arch. Microbiol.* (1997) **168**, 136–142.

J362. Miyaki, Y., Yamamoto, K., Osawa, T., *J. Agric. Food Chem.* (1997) **45**, 3738–3742.

J363. Hotzl, B.K., Thomas, H., *Chirality* (1997) **9**, 367–372.

J364. Siegers, C.P., Siegers, J.P., Pentz, R., Bodinet, C., Freudenstein, J., *Pharm. Pharmacol. Lett.* (1997) **7**, 90–92.

J365. Li, S.-M., Wang, Z.-X., Heide, L., *Phytochemistry* (1997) **46**, 27–32.

J366. Zissi, U., Lyberatos, G., Pavlou, S., *J. Ind. Microbiol. Biotechnol.* (1997) **19**, 49–55.

J367. Johansen, S.S., Arvin, E., Mosbaek, H., Hansen, A.B., *Environ. Toxicol. Chem.* (1997) **16**, 1821–1828.

J368. Finkel'stein, Z.I., Baskunov, B.P., Vavilova, L.N., Golovleva, L.A., *Microbiology (Moscow)* (1997) **66**, 402–407.

J369. Stillwell, W.G., Kidd, L.C.R., Wishnok, J.S., Tannenbaum, S.R., Sinha, R., *Cancer Res.* (1997) **57**, 3457–3464.

J370. Di Gennaro, P., Bestetti, G., Galli, E., Orsini, F., Pelizzoni, F., Sello, G., *Tetrahedron Lett.* (1997) **38**, 6267–6270.

J371. Olesen, O.V., Linnet, K., *J. Chromatogr, B: Biomed. Sci. Appl.* (1997) **698**, 209–216.

J372. Guo, Z.-w., Salamonczyk, G.M., Han, K., Machiya, K., Sih, C.J., *J. Org. Chem.* (1997) **62**, 6700–6701.

J373. Niedan, V., Scholer, H.F., *Chemosphere* (1997) **35**, 1233–1241.

J374. Brady, T.C., Kind, A.J., Hyde, W.H., Favrow, M., Hill, D.W., *Drug Metab. Dispos.* (1998) **26**, 294 -298.

J375. Postlind, H., Danielson, Å., Lindgren, A., Andersson, S.H.G., *Drug Metab. Dispos.* (1998) **26**, 289–293.

J376. Draper, A.J., Madan, A., Latham, J., Parkinson, A., *Drug Metab. Dispos.* (1998) **26**, 305–312.

J377. Mohri, K., Okada, K., Benet, L.Z., *Drug Metab. Dispos.* (1998) **26**, 332–337.

J378. Tougou, K., Nakamura, A., Watanabe, S., Okuyama, Y., Morino, A., *Drug Metab. Dispos.* (1998) **26**, 355–359.

J379. Coldham, N.G., Sivapathasundaram, S., Dave, M., Ashfield, L.A., Pottinger, T.G., Goodall, C., Sauer, M.J., (1997) 347–354.

J380. Kuso, S., Umehara, K., Morita, S., Uchida, M. Miyamoto, G., Odomi, M., *Xenobiotica* (1998) **28**, 507–514.

J381. Lampen, A., Bader, A., Bestmann, T., Winkler, M., Witte, L., Borlak, J.T., *Xenobiotica* (1998) **28**, 429–441.

J382. Arfmann, H.-A., Timmis, K.N., Wittich, R.-M., *Appl. Environ. Microbiol.* (1997) **63**, 3458–3462.

J383. Klecker, R.W., Collins, J.M., *J. Cardiovasc. Pharmacol.* (1997) **30**, 69–74.

J384. Rasmussen, S., Rudolph, H., *Phytochemistry* (1997) **46**, 449–453.

J385. Ahn, S.H., Duffel, M.W., Rosazza, J.P.N., *J. Nat. Prod.* (1997) **60**, 1125–1129.

J386. Hedli, C.C., Hoffmann, M.J., Ji, S., Thomas, P.E., Snyder, R., *Toxicol. Appl. Pharmacol.* (1997) **146**, 60–68.

J387. Jung, B.H., Park, Y.H., Park, J., Chung, B.C., *Arch. Pharmacol. Res.* (1997) **20**, 358–362.

J388. Rafii, F., Sunderland, J.B., Hansen, E.B., Jr., Cerniglia, C.E., *Clin. Infect. Dis.* (1997) **25** (Suppl. 2), S121-S122.
J389. Yoshioka, T., Ohno, H., Uematsu, T., *Eur. J. Biochem.* (1997) **248**, 58–62.
J390. Briganti, F., Pessione, E., Giunta, C., Scozzafava, A., *FEBS Lett.* (1997) **416**, 61–64.
J391. Sanderink, G.-J., Bournique, B., Stevens, J., Petry, M., Martinet, M., *J. Pharmacol. Exp. Ther.* (1997) **282**, 1465–1472.
J392. Suske, W.A., Held, M., Schmid, A., Fleischmann, T., Wubbolts, M.G., Kohler, H.-P.E., *J. Biol. Chem.* (1997) **272**, 24257–24265.
J393. Chen, H., Mc Cormick, D.B., *Biochim. Biophys. Acta* (1997) **1342**, 116–118.
J394. Yasuda, T., Ohsawa, K., *Annu. Rep. Tohoku Coll. Pharm.* (1996) **43**, 159–164.
J395. Monserrate, E., Haggblom, M.M., *Appl. Environ. Microbiol.* (1997) **63**, 3911–3915.
J396. Oldfield, C., Pogrebinsky, O., Simmonds, J., Olson, E.S., Kulpa, C.F., *Microbiology* (1997) **143**, 2961–2973.
J397. Schneider, S., Mohamed, M.E.-S., Fuchs, G., *Arch. Microbiol.* (1997) **168**, 310–320.
J398. Bianchi, D., Bernardi, A., Bosetti, A., Bortolo, R., Cidaria, D., Crespi, E., Gagliardi, I., *Appl. Microbiol. Biotechnol.* (1997) **48**, 363–366.
J399. Yasukochi, T., Fukase, K., Suda, Y., Takagaki, K., Endo, M., Kusumoto, S., *Bull. Chem. Soc. Jpn.* (1997) **70**, 2719–2725.
J400. Stain-Texier, F., Sandouk, P., Scherrmann, J.-M., *Drug Metab. Dispos.* (1998) **26**, 383–387.
J401. Stewart, K., Spedding, P.L., Otterburn, M.S., Lewis, D.M., *J. Appl. Polym. Sci.* (1997) **66**, 2359–2363.
J402. Vickers, S., Duncan, C.A., Slaughter, D.E., Arison, B.H., Greber, T., Olah, T.V., Vyas, K.P., *Drug Metab. Dispos.* (1998) **26**, 388–395.
J403. Chando, T.J., Everett, D.W., Kahle, A.D., Starrett, A.M., Vachharajani, N., Shyu, W.C., Kripalani, K.J., Barbhaiya, R.H., *Drug Metab. Dispos.* (1998) **26**, 408–417.
J404. Bartels, M.J., Mc Nett, D.A., Timchalke, C., Mendrala, A.L., Christenson, W.R., Sangha, G.K., Brzak, K.A., Shabranz, S.N., *Xenobiotica* (1998) **28**, 579–594.
J405. Estra-Rodgers, L., Levy, G.N., Weber, W.W., *Drug Metab. Dispos.* (1998) **26**, 502–505.
J406. Maire-Gauthier, R., Buronfosse, T., Magdalou, J., Herber, R., Besse, S., Delatour, P., Benoit, E., *Xenobiotica* (1998) **28**, 595–604.
J407. Plakas, S.M., El Said, K.R., Bencsath, F.A., Musser, S.M., Hayton, W.L., *Xenobiotica* (1998) **28**, 605–616.
J408. Jatoe, S.D., Gorrod, J.W., *Arch. Toxicol.* (1997) **60**, 65–68.
J409. Coles, B., Srai, S.K.S., Ketterer, B., Waynforth, B., Kadlubar, F.F., *Chem.-Biol. Interact.* (1983) **43**, 123–129.
J410. Berger, B.J., Reddy, V.V., Le, S.T., Lombardy, R.J., Hall, J.E., Tidwell, R.R., *J. Pharmacol. Exp. Ther.* (1997) **256**, 883–889.
J411. Goodwin, B.L., Ruthven, C.R.J., Sandler, M., *Gen. Pharmacol.* (1998) **31**, 437–440.
J412. Thibault, R., Debrauwer, L., Rao, D., Cravedi, J.-P., *Xenobiotica* (1998) **28**, 595–604.
J413. Mitchell, S.C., Gillham, J., Jackson, W.F.M., Preston, S.L., Porter, E.R., Zhang, A.Q., *Xenobiotica* (1998) **28**, 715–722.
J414. Kajbaf, M., Marchioro, C., Barnaby, J.R., Pellegatti, M., *Xenobiotica* (1998) **28**, 785–794.
J415. Bessems, J.G.M., de Groot, M.J., Baede, E.J., te Koppele, J.M., Vermeulen, N.P.E., *Xenobiotica* (1998) **28**, 855–875.
J416. Green, M.D., King, C.D., Mojarrabi, B., Mackenzie, P.I., Tephly, T.R., *Drug Metab. Dispos.* (1998) **26**, 507–512.

J417. Ohkawa, T., Goto, S., Miki, S., Sato, A., Kuroda, T., Iwatani, K., Takeuchi, M., Nakano, M., *Xenobiotica* (1998) **28**, 877–886.

J418. Prueksaritanont, T., Deluna, P., Gorham, L.M., Ma, B., Cohn, D., Pang, J., Xu, X., Leung, K., Lin, J., *Drug Metab. Dispos.* (1998) **26**, 520–527.

J419. Andersson, S.H.G., Lindgren, A., Postlind, H., *Drug Metab. Dispos.* (1998) **26**, 528–535.

J420. Koudriaova, T., Iatsimirskaia, E., Utkin, I., Gangl, E., Vouros, P., Storozhuk, E., Orza, D., Marinina, J., Gerber, N., *Drug Metab. Dispos.* (1998) **26**, 552–561.

J421. Mueller, S.O., Stopper, H., Dekant, W., *Drug Metab. Dispos.* (1998) **26**, 540–546.

J422. Hashizume, T., Mise, M., Terauchi, Y, O, L., Fujii, T., Miyazaki, H., Inaba, T., *Drug Metab. Dispos.* (1998) **26**, 566–571.

J423. Ritzinger, S., Fang, J., Baker, G.B., *Drug Metab. Dispos.* (1998) **26**, 572–575.

J424. Voorman, R.L., Maio, S.M., Hauer, M.J., Sanders, P.E., Payne, N.A., Ackland, M.J., *Drug Metab. Dispos.* (1998) **26**, 631–639.

J425. Busch, U., Schmid, J., Heinzel, G., Schmaus, H., Baierl, J., Huber, C., Roth, W., *Drug Metab. Dispos.* (1998) **26**, 576–584.

J426. Boberg, M., Angerbauer, R., Kanhai, W.K., Karl, W., Kern, A., Radtke, M., Steinke, W., *Drug Metab. Dispos.* (1998) **26**, 640–652.

J427. Ju, C., Uetrecht, J.P., *Drug Metab. Dispos.* (1998) **26**, 676–680.

J428. Irkman, S.K., Zhang, M.-Y., Horwatt, P.M., Scatina, J., *Drug Metab. Dispos.* (1998) **26**, 720–723.

J429. Yoshioka, T., Uematsu, T., *Drug Metab. Dispos.* (1998) **26**, 705–710.

J430. Reith, K., Keung, A., Toren, P.C., Cheng, L., Eller, M.G., Weir, S.J., *Drug Metab. Dispos.* (1998) **26**, 732–738.

J431. Emary, W.B., Toren, P.C., Mathews, B., Huk, K., *Drug Metab. Dispos.* (1998) **26**, 725–731.

J432. Sjöquist, B., Basu, S., Byding, P., Bergh, K., Stjernschantz, J., *Drug Metab. Dispos.* (1998) **26**, 745–754.

J433. Black, A.E., Sinz, M.W., Hayes, R.N., Woolf, T.F., *Drug Metab. Dispos.* (1998) **26**, 755–763.

J434. Haaz, M.-C., Riché, C., Rivory, L.P., Robert, J., *Drug Metab. Dispos.* (1998) **26**, 769–775.

J435. Obach, R.S., Pablo, J., Mash, D.C., *Drug Metab. Dispos.* (1998) **26**, 764–768.

J436. Ko, J.-W., Desta, Z., Flockhart, D.A., *Drug Metab. Dispos.* (1998) **26**, 775–778.

J437. Kobayashi, K., Yamamoto, T., Chiba, K., Tani, M., Shimada, N., Ishizaki, T., Kuroiwa, Y., *Drug Metab. Dispos.* (1998) **26**, 818–821.

J438. Aeschlimann, C., Küpfer, A., Schefer, H., Cerny, T., *Drug Metab. Dispos.* (1998) **26**, 883–890.

J439. Stresser, D.M., Kupfer, D., *Drug Metab. Dispos.* (1998) **26**, 868–874.

J440. Zalko, D., Debrauwer, L., Bories, G., Tulliez, J., *Drug Metab. Dispos.* (1998) **26**, 891–899.

J441. Griffin, R.J., Harvisen, P.J., *Drug Metab. Dispos.* (1998) **26**, 907–913.

J442. Kamimura, H., Oishi, S., Matsushima, H., Watanabe, T., Higuchi, S., Hall, M., Wood, S.G., Chasseaud, L.F., *Xenobiotica* (1998) **28**, 909–922.

J443. Schaefer, W.H., Politowski, J., Hwang, B., Dixon, F., Jr., Goalwin, A., Gutzait, L., Anderson, K., Debrosse, C., Bean, M., Rhodes, G.R., *Drug Metab. Dispos.* (1998) **26**, 958–969.

J444. Wu, W.N., Masucci, J.A., Caldwell, G.W., *Xenobiotica* (1998) **28**, 957–972.

J445. Larsen, G.L., Huwe, J.K., Bakke, J.E., *Xenobiotica* (1998) **28**, 973–984.

J446. Machinist, J.M., Mayer, M.D., Roberts, E.M., Surber, B.W., Rodrigues, A.D., *Drug Metab. Dispos.* (1998) **26**, 970–976.

J447. Hong, S.K., Rankin, G.O., *Xenobiotica* (1998) **28**, 985–994.

J448. Feng, M.R., Loo, J., Wright, J., *Drug Metab. Dispos.* (1998) **26**, 982–988.

J449. Chang, M., Sood, V.K., Wilson, G.J., Kloosterman, D.A., Sanders, P.E., Schuette, M.R., Judy, R.W., Voorman, R.L., Maio, S.M., Slatter, J.G., *Drug Metab. Dispos.* (1998) **26**,1008–1018.

J450. Wynalda, M.A., Hauer, M.J., Wienkers, L.C., *Drug Metab. Dispos.* (1998) **26**, 1048–1051.

J451. Matsunaga, T., Kishi, N., Tanaka, H., Watanabe, K., Yoshimura, H., Yamamoto, I., *Drug Metab. Dispos.* (1998) **26**, 1041–1047.

J452. Montesissa, C., Anfossi, P., Biancotto, G., Angeletti, R., *Xenobiotica* (1998) **28**, 1049–1060.

J453. Hillenweck, A., Corpet, D.E., Killeen, J.C. Jr, Bliss, M. Jr., Cravedi, J.P., *Xenobiotica* (1998) **28**, 1017–1028.

J454. Kapetanovic, I.M., Torchin, C.D., Thompson, C.D., Miller, T.A., Mcneilly, P.J., Macdonald, T.L., Kupferberg, H.K., Perhach, J.L., Sofia, R.D., Strong, J.M., *Drug Metab. Dispos.* (1998) **26**, 1089–1095.

J455. Savidge, R.D., Bui, K.H., Birmingham, B.K., Morse, J.L., Spreen, R.C., *Drug Metab. Dispos.* (1998) **26**, 1069–1076.

J456. Eiermann, B., Edlund, P.O., Tjenberg, A., Dalén, P., Dahl, M.-L., Bertilsson, L., *Drug Metab. Dispos.* (1998) **26**, 1096–1101.

J457. Schneider, R.P., Fouda, H.G., Inskeep, P.B., *Drug Metab. Dispos.* (1998) **26**, 1149–1159.

J458. Maier, U.H., Zenk, M.H., *Tetrahedron Lett.* **38**, 7357–7360.

J459. Schneider, R.P., Davenport, C.J., Hoffmaster, K.A., Inskeep, P.B., *Drug Metab. Dispos.* (1998) **26**, 1160–1166.

J460. Cooksey, C.J., Garratt, P.J., Land, E.J., Pavel, S., Ramsden, C.A., Riley, P.A., Smit, N.P.M., *J. Biol. Chem.* (1997) **272**, 26226–26235.

J461. Eppinck, M.H.M., Boeren, S.A., Vervoort, J., Van Berkel, W.J.W., *J. Bacteriol.* (1997) **179**, 6680–6687.

J462. Hammadi, A., Menez, A., Genet, R., *Tetrahedron* (1997) **53**, 16115–16122.

J463. Soarez-Da-Silva, P., Vieira-Coelho, M.A., Serrao, M.P., *Biochem. Pharmacol.* (1997) **54**, 1037–1046.

J464. Eisenhofer, G., Aneman, A., Friberg, P., Hooper, D., Fandriks, L., Lonroth, H., Hunyady, B., Mezey, E., *J. Clin. Endocrinol. Metab.* (1997) **82**, 3864–3871.

J466. Makdessi, K., Lechner, U., *FEMS Microbiol. Lett.* (1997) **157**, 95–101.

J467. Fernandez-Medarde, A., Luengo, J.M., *FEMS Microbiol. Lett.* (1997) **157**, 47–53.

J468. Hollender, J., Hopp, J., Doth, W., *Appl. Envron. Microbiol.* (1997) **63**, 4567–4572.

J469. Ostermann, A., Gallus, C., Schinck, B., *Curr. Microbiol.* (1997) **35**, 270–273.

J470. Guidetti, P., Okuno, E., Schwarcz, R., *J. Neurosci. Res.* (1997) **50**, 457–465.

J471. Ünsalan, S., Ülgen, M., *FABAD Farm. Bilimber Derg.* (1997) **22**, 103–109.

J472. Naoi, M., Maruyama, W., Matsubara, K., Hashizume, Y., *Neurosci. Lett.* (1997) **235**, 81–84.

J473. Varin, L., Chamberland, H., Lafontaine, J.G., Martine, R., *Plant J.* (1997) **12**, 831–837.

J474. Drzyzga, O., Blotevogel, K.-H., *Curr. Microbiol.* (1997) **35**, 343–347.

J475. Rosengren, E., Aman, P., Thelin, S., Hansson, C., Ahlfors, S., Bjork, P., Jacobsson, L., Rorsman, H., *FEBS Lett.* (1997) **417**, 85–88.

J476. Razo-Flores, E., Luijten, M., Donlon, B., Lettinga, G., Field, J., *Water Sci. Technol.* (1997) **36**, (6–7 Anaerobic Digestion VIII), 65–72.

J477. Gemechu-Hatewu, M., Platt, K.-L., Oesch, F., Hacker, H.-J., Bannasch, P., Steinberg, P., *Int. J. Cancer* (1997) **73**, 587–591.

J479. O'connor, K.E., Dobson, A.D.W., Hartmans, S., *Appl. Environ. Microbiol.* (1997) **63**, 4287–4291.

J580. Palmer, L., Andersson, L., Andersson, T., Stenberg, U., *J. Pharm. Biomed. Anal.* (1997) **16**, 155–165.

J481. Hauck, C., *Rapid Commun. Mass Spectrom.* (1997) **11**, 1824.

J482. Swarts, H.J., Master, T., Verhagen, F.J.M., Field, J.A., Wijnberg, J.B.P.A., *Phytochemistry,* (1997) **46**, 1011–1013.

J483. Sepic, E., Briceli, M., Leskovsek., H., *J. Appl. Microbiol.* (1997) **83**, 561–568.

J484. Hopper, W., Mahadevan, A., *Biodegradation* (1997) **8**, 159–165.

J485. Fukuda, K., Koito, M., Osabe, M., Sakaguchi, A., Fukuhara, N., *Jpn. Kokai Tokkyo Koho JP* (9–2–1997) 09,313,195.

J486. Kitayama, T., *Tetrahedron: Asymmetry* (1997) **8**, 3765–3774.

J487. Sasayama, H., Yamazaki, T., Fukuzawa, S., Fujii, T., Matsuura, A., *Jpn. Kokai Tokkyo Koho JP* (2–12–1997) 09,308,493.

J488. Nagahori, H., Matsui, M., Tomigahara, Y., Matsunaga, H., Kaneko, H., Nakatsuka, I., *J. Pestic. Sci.* (1997) **22**, 293–298.

J489. Ross, F.C., Botting, N.P., Leeson, P.D., *Tetrahedron* (1997) **53**, 15761–15770.

J490. Ciotti, M., Marrone, A., Potter, C., Owens, I.S., *Pharmacogenetics* (1997) **7**, 485–495.

J491. Genther, B.R.S., Townsend, G.T., Blattmann, B.O., *Appl. Environ. Microbiol.* (1997) **63**, 4698–4703.

J492. Jewell, C.S.E., Mayeaux, M.H., Winston, G.W., *Comp. Biochem. Physiol.: C, Pharmacol., Toxicol. Endocrinol.* (1997) **118C**, 369–374.

J493. Neau, B., Roux, J., Marty, J.P., *Arzneim.-Forsch.* (1998) **48**, 34–38.

J494. Kim-Kang, H., Robinson, R.A., Wu, J., *J. Agric. Food. Chem.* (1998) **46**, 707–717.

J495. Kaptein, E., van Haasteren, G.A.C., Linkels, E., de Greef, W.J., Visser, T.J., *Endocrinology* (1997) **138**, 5136–5143.

J496. Iwasaka, K., Kim, C.W., *Phytochemistry* (1997) **46**, 1359–1363.

J497. Angerer, J., Schildbach, M., Kramer, A., *Arch. Toxicol.* (1998) **72**,119–123.

J498. Stalford, A.C., Maggs, J.L., Gilchrist, T.L., Park, B.K., *Steroids*, (1997) **62**, 750–761.

J499. Donaghy, J., McKay, A.M., *J. Appl. Microbiol.* (1997) **83**, 718–726.

J500. Kalil, M.S., Stephens, G.M., *Biotechnol. Lett.* (1997) **19**, 1165–1168.

J501. Shah, M.M., Campbell, J.A., *Biochem. Biophys. Res. Commun.* (1997) **241**, 794–796.

J502. Sagner, S., Shen, Z.-W., Dews-Neumann, B., Zenk, M.H., *Phytochemistry* (1998) **47**, 375–387.

J503. Winston, G.W., Mayeaux, M.H., Hefferman, L.M., *Mar. Environ. Res.* (1998) **45**, 89–100.

J504. Kahn, R.A., Back, S., Svendsen I., Halkier, B.A., Moller, B.L., *Plant Physiol.* (1997) **115**, 1661–1670.

J505. Kawakubo, Y., Ohkido, M., *J. Dermatol. Sci.* (1998) **16**, 99–103.

J506. Roldan, M.D., Blasco, R., Caballero, F.J., Castillo, F., *Arch. Microbiol.* (1998) **169**, 36–42.

J507. Smelt, V.A., Mardon, H.J., Redman, C.W.G., Sim, E., *Eur. J. Drug Metab. Pharmacokinet.* (1997) **22**, 403–408.

J508. Asano, Y., Kato, Y., *FEMS Microbiol. Lett.* (1998) **158**, 185–190.

J509. Schmidt, W., Beerhues, L., *FEBS Lett.* (1997) **420**, 143–146.

J510. Li, J, Li, G., *Insect Biochem. Mol. Biol.* (1997) **27**, 859–867.

J511. Yabe, K., Matsushima, K.-I., Koyama, T., Hamasaki, T., *Appl. Environ. Microbiol.* (1998) **64**, 166–171.

J512. Park, H.-S., Lee, J.Y., Kim, H.S., *Biotechnol. Bioeng.* (1998) **58**, 339–343.

J513. Lim, H.-W., Park, S.-S., Lim, C.-J., *Mol. Cells* (1997) **7**, 715–710.

J514. Rosengren, E., Thelin, A., Aman, P., Hansson, C., Jacobsson, L., Rorsman, H., *Melanoma Res.* (1997) **7**, 517–518.

J515. Kaschabek, S.R., Kasberg, T., Müller, D., Mars, A.E., Janssen, D.B., Reineke, W., *J. Bacteriol.* (1998) **180**, 296–302.

J516. Kato, M., Chu, A., Davin, L.B., Lewis, N.G., *Phytochemistry* (1997) **47**, 583–591.

J517. Itoh, N., Hayashi, K., Okada, K., Ito, T., Mizuguchi, N., *Biosci. Biotechnol. Biochem.* (1997) **61**, 2058–2062.

J518. Saari, R.E., Hausinger, R.P., *Biochemistry* (1998) **37**, 3035–3042.

J519. Chen, R., Chaen, H., Hosokawa, K., *Res. Commun. Biochem. Cell Mol. Biol.* (1997) **1**, 304–322.

J520. Chang, F.C., Chung, J.G., *Curr. Microbiol.* (1998) **36**, 125–130.

J521. Dudareva, N., Raguso, R.A., Wang, J., Ross, J.R., Pichersky, E., *Plant Physiol.* (1998) **116**, 599–604.

J522. Thompson, J.E., Basarab, G.S., Pierce, J., Hodge, C.N., Jordan, D.B., *Anal. Biochem.* (1998) **256**, 1–6.

J523. Küçüküzel, I., Ülgen, M., Gorrod, J.W., *Eur. J. Drug Metab. Pharmacokinet.* (1997) **22**, 351–358.

J524. Parales, J.V., Parales, R.E., Resnick, S.M., Gibson, D.T., *J. Bacteriol.* (1998) **160**, 1194–1199.

J525. Kim, D.-H., Jung, E.-A., Sohng, I.-S., Han, J.-A., Kim, T.-H., Han, M.J., *Arch. Pharmacol.* (1998) **21**, 17–23.

J526. Yang, X., Kulkarni, A.P., *J. Biochem. Mol. Toxicol.* (1998) **12**, 175–183.

J527. Boll, M., Fuchs, G., *Eur, J. Biochem.* (1998) **251**, 946–954.

J528. Brimer, L., Cicalini, A.R., Federici, F., Petruccioli, M., *Arch. Microbiol.* (1998) **169**, 106–112.

J529. Baciocchi, E., Belvedere, S., Bietti, M., Lanzalunga, O., *Eur. J. Org. Chem.* (1998) 299–302.

J530. England, P.A., Harford-Cross, C.F., Stevenson, J.-A., Rouch, D.A., Wong, L.-L., *FEBS Lett.* (1998) **424**, 271–274.

J531. Venkatakrishnan, K., Greenblatt, D.J., Von Moltke, L.L., Schmider, J., Harmatz, J.S., Shader, R.I., *J. Clin. Pharmacol.* (1998) **38**, 112–121.

J532. Ramaprasad, P., Kasinathan, C., *Gen. Pharmacol.* (1998) **30**, 555–559.

J533. Ito, Miyazaki, T., Makino, K., Kurose, M., *Jpn. Kokai Tokkyo Koho JP* (14–4–1998) 10 94,399.

J534. Asano, Y., Tanetani, M., *Arch. Microbiol.* (1998) **169**, 220–224.

J535. Anderson, R.J., Kudlacek, P.E., Clemens, D.L., *Chem.-Biol. Interact.* (1998) **109**, 53–67.

J536. Hudlicky, Gonzalez, D., Stabile, M., Endoma, M.A.A., Deluca, M., Parker, D., Gibson, D.T.,Resnick, S.M., Whited, G.M., *J. Fluorine Chem.* (1998) **89**, 23–30.

J537. Boyd, D.R., Sharma, N.D., Carroll, J.G., Malone, J.F., Macherracher, D.G., Allen, C.C.R., *Chem. Commun.* (1998) 683–684.

J538. Tan, W., Li, Q., Mckay, G., Semple, H.A., *J. Pharm. Biomed. Anal.* (1998) **16**, 991–1003.

J539. Leuschner, R.G., Heidel, M., Hammes, W.P., *Int. J. Food Microbiol.* (1998) **39**, 1–10.

J540. Bohmer, S., Messner, K., Srebotnik, E., *Biochem. Biophys. Res. Commun.* (1998) **244**, 233–238.

J541. Chen, W., Koenigs, L.L., Thompson, S.J., Peter, R.M., Rettie, A.E., Trager, W.F., Nelson, S.D., *Chem. Res. Toxicol.* (1998) **11**, 295–301.

J542. Coulet, M., Eeckhoutte, C., Larrieu, G., Sutra, J.-F., Hoogenboom, L.A.P., Huveneers-Oorsprong, M.B.M., Kuiper, H.A., Castell, J.V., Alvenerie, M., Galtier, P., *J. Agric. Food Chem.* (1998) **46**, 742–748.

J543. Lee, H.S., Jeong, T.C., Kim, J.H., *J. Chromatogr., B: Biomed. Sci. Appl.* (1998) **705**, 367–372.

J544. Kraemer, T., Vernaleken, I., Maurer, H.H., *J. Chromatogr., B: Biomed. Sci. Appl.* (1998) **702**, 93–102.

J545. Hovinen, J., Silvennoinen, R., Vilpo, J., *Chem. Res. Toxicol.* (1998) **11**, 91–93.

J546. Lautala, P., Kivimaa, M., Salomies, H., Elovaara, E., Taskinen, J., *Pharm. Res.* (1997) **14**, 1444–1448.

J547. Tunc, N., Yilmaz, F., Ulgen, M., Gorrod, J.W., *Eur. J. Drug Metab. Pharmacokinet.* (1997) **22**, 345–349.

J548. Clement, B., Demesmaeker, M., *Arch. Pharm.* (1997) **330**, 303–306.

J549. Cupid, B.C., Holmes, E., Wilson, I.D., Lindon, J.C., Nicholson, J.K., *Xenobiotica* (1999) **29**, 27–42.

J550. Meunier, C.J., Verbeek, R.K., *Drug Metab. Dispos.* (1999) **27**, 26–31.

J551. Scarfe, G.B., Wright, B., Clayton, E., Taylor, S., Wilson, I.D., Lindon, J.C., Nicholson, J.K., *Xenobiotica* (1999) **29**, 77–91.

J552. Sugiyama, K., Wang, T.-C.L., Simpson, J.T., Rodriguez, L., Kador, P.F., Sato, S., *Drug Metab. Dispos.* (1999) **27**, 60–67.

J553. Sogorb, M.A., Sánchez, I., López-Rivadulla, M., Céspides, V., Vilanova, E., *Drug Metab. Dispos.* (1999) **27**, 53–59.

J554. Nakamura, A., Hirota, T., Morino, A., Imaoka, S., Funae, Y., Yamamoto, Y., Tasaki, T., Masuda, M., Kazusaka, A., Fujita, S., *Xenobiotica* (1999) **29**, 243–252.

J555. Shiraga, T., Kaneko, H., Iwasaki, K., Tozuka, Z., Suzuki, A., Hata, T., *Xenobiotica* (1999) **29**, 217–229.

J556. Bayliss, M.K., Bell, J.A., Jenner, W.N., Park, G.R., Wilson, K., *Xenobiotica* (1999) **29**, 253–268.

J557. Walker, D.K., Ackland, M.J., James, G.C., Muirhead, G.J., Rance, D.J., Wastall, P., Wright, P.A., *Xenobiotica* (1999) **29**, 297–310.

J558. Scarfe, G.B., Tugnait, M., Wilson, I.D., Nicholson, J.K., *Xenobiotica* (1999) **29**, 205–216.

J559. Niwa, T., Sato, R., Yabusaki, Y., Ishibashi, F., Katagiri, M., *Xenobiotica* (1999) **29**, 187–193.

J560. Madyastha, K.M., Gaikwad, N.W., *Drug Metab. Dispos.* (1999) **27**, 74–80.

J561. Ramji, J.V., Keogh, J.P., Blake, T.J., Broom, C., Chenery, R.J., Citerone, D.R., Lewis, V.A., Taylor, A.C., Yeulet, S.E., *Xenobiotica* (1999) **29**, 311–325.

J562. Kumar, G.N., Venkata, J., Lee, R.D., Whittern, D.N., Uchic, J., Thomas, S., Johnson, P., Grabowski, B., Sham, H., Betebenner, D., Kempf, D.J., Dennissen, J.F., *Drug Metab. Dispos.* (1999) **27**, 86–91.

J563. Kitamura, S., Sugihara, K., Tatsumi, K., *Drug Metab. Dispos.* (1999) **27**, 92–97.

J564. Thomas, B.F., Burgess, J.P., Coleman, D.P., Scheffler, N.M., Jeffcoat, A.R., Dix, K.J., *Drug Metab. Dispos.* (1999) **27**, 147–157.

J565. Dix, K.J., Coleman, D.P., Jeffcoat, A.R., *Drug Metab. Dispos.* (1999) **27**, 138–146.

J566. Konishi, T., Nishikawa, H., Kitamura, S., Tatsumi, K., *Drug Metab. Dispos.* (1999) **27**, 158–160.

J567. Dupont, I., Berthou, F., Bodenez, P., Bardou, L., Guirriec, C., Stephan, N., Dreano, Y., Lucas, D., *Drug Metab. Dispos.* (1999) **27**, 322–326.

J568. Nicoll-Griffith, D.A., Falgueyret, J.-P., Silva, J.M., Morin, P.-E., Trimble, L., Chan, C.-C., Clas, S., Leger, S., Wang, Z., Yergey, J.A., Riendeau, D., *Drug Metab. Dispos.* (1999) **27**, 403–409.

J569. Tang, W., Stearns, R.A., Bandiera, S.M., Zhang, Y., Raab, C., Braun, M.P., Dean, D.C., Pang, J., Leung, K.H., Doss, G.A., Strauss, J.R., Kwei, G.Y., Rushmore, T.H., Chiu, S.-H.L., Baillie, T.A., *Drug Metab. Dispos.* (1999) **27**, 365–372.

J570. Kawashima, K., Hosoi, K., Naruke, T., Shiba, T., Kitamura, M., Watabe, T., *Drug Metab. Dispos.* (1999) **27**, 422–428.

J571. Shiraga, T., Hata, T., Yamazoe, Y., Ohno, Y., Iwasaki, K., *Xenobiotica* (1999) **29**, 341–347.

J572. Constantino, L., Iley, J., *Xenobiotica* (1999) **29**, 409–416.

J573. Tettey, J.N.A., Skellern, G.G., Midgley, J.M., Grant, M.H., Wilkinson, R., Pitt, A.R., *Xenobiotica* (1999) **29**, 349–360.

J574. Kumar, S., Riggs, K.W., Rurak, D.W., *Drug Metab. Dispos.* (1999) **27**, 463–470.

J575. Deisinger, P.J., English, J.C., *Drug Metab. Dispos.* (1999) **27**, 442–448.

J576. Zhou, S.Y., Fleischer, D., Pao, L.H., Li, C., Winward, B., Zimmerman, E.M., *Drug Metab. Dispos.* (1999) **27**, 479–485.

J577. Pan, M.-H., Huang, T.-M., Lin, J.-K., *Drug Metab. Dispos.* (1999) **27**, 486–494.

J578. Baek, M., Chung, H.S., Kim, Y., Kim, D.-H., *Drug Metab. Dispos.* (1999) **27**, 510–516.

J579. Tagawa, Y., Miwa, K., Tsukuda, R., Yoshimura, Y., Tanayama, S., Tanigawara, Y., *Drug Metab. Dispos.* (1999) **27**, 495–501.

J580. Sauer, M.J., Dave, M., Lake, B.G., Manchee, G.R., Howells, L.C., Coldham, N.G., Xenobiotica, (1999) **29**, 483–497.

J581. Wu, W.N., McKown, L.A., Takucs, A.R., Jones, W.J., Reitz, A.B., *Xenobiotica* (1999) **29**, 453–466.

J582. Cravedi, J.P., Lafuente, A., Baradat., M., Hillenweck. A., Perdu-Durand, E., (1998) 499–509.

J583. Habig, W.H., Pabst, M.J., Jakoby, W.B., *J. Biol. Chem.* (1974) **249**, 7130–7139.

J584. Pless, D., Gouze, J.N., Senay, C., Herber, R., Leroy, P., Barberousse, V., Fournel-Gigleux, S., Magdalou, J., *Drug Metab. Dispos.* (1999) **27**, 588–595.

J585. Trubetskoy, O.V., Shaw, P.M., *Drug Metab. Dispos.* (1999) **27**, 555–557.

J586. Hiller, A., Nguyen, N., Strassburg, C.P., Li, Q., Jainta, H., Pechstein, B., Ruus, P., Engel, J., Tukey, R.H., Kronbach, T., *Drug Metab. Dispos.* (1999) **27**, 605–612.

J587. Henesy, C.M., Kellner-Weibel, G.L., Tarliff, J.B., Harvison, P.J., *Drug Metab. Dispos.* (1999) **27**, 674–680.

J588. Redondo, P.A., Álvarez, A.I., Garcia, J.L., Larrodé, O.M., Marino, G., Prieto, J.G., *Drug Metab. Dispos.* (1999) **27**, 736–740.

J589. Dehal, S.S., Kupfer, D., *Drug Metab. Dispos.* (1999) **27**, 681–688.

J590. Raune, B., Furlan, V., Taburnet, A.M., Farinotti, R., *Xenobiotica*, (1999) **29**, 595–601.

J591. Zalko, D., Bories, G., Tulliez, J., *J. Agric. Food Chem.* (1998) **46**, 1935–1943.

J592. Schmitt, M.M., Schuler, E., Braun, M., Haring, D., Schreier, P., *Tetrahedron Lett.* (1998) **39**, 2945–2946.

J593. Sridhar, M., Vadivel, S.K, Bhalerao, U.T., *Synth. Commun.* (1998) **28**, 1499–1502.

J594. Bertoldi, M., Dominici, P., Moore, P.S., Maras, B., Voltattorni, C.B., *Biochemistry* (1998) **37**, 6552–6561.

J595. Paek, N.S., Kang, D.J., Lee, H.S., Lee, J.J., Choi, Y.J., Kim, T.H., Kim, K.W., *Biosci. Biotechnol. Biochem.* (1998) **62**, 588–589.

J596. Selifonov, S.A., Chapman, P.J., Akkerman, S.B., Gurst, J.E., Bortiatynski, J.M., Nanny, M.A., Hatcher, P.G., *Appl. Environ. Microbiol.* (1998) **64**, 1447–1453.

J597. Jager, W., Zembsch, B., Wolschann, P., Pittenauer, E., Sanderowicz, A.M., Sausville, E.A., Sedlacek, H.H., Graf, J., Thalhammer, T., *Life Sci.* (1998) **62**, 1861–1873.

J598. Iwabuchi, T., Harayama, S., *J. Biol. Chem.* (1998) **273**, 8332–8336.

J599. Wang, K., Ramji, S., Bhathena, A., Lee, C., Riddick D.S., *Xenobiotica* (1999) **29**, 155–170.

J600. Takenaga, N., Ishii, M., Nakajima, S., Hasegawa, T., Iwasa, R., Ishizaki, H., Kamei, T., *Drug Metab. Dispos.* (1999) **27**, 205–212.

J602. Takenaga, N., Hasegawa, T., Ishii, M., Ishizaki, H., Hata, H., Kamei, T., *Drug Metab. Dispos.* (1999) **27**, 213–220.

J603. Scarfe, G.B., Tugnait, M., Wilson, I.D., Nicholson, J.K., *Xenobiotica* (1999) **29**, 205–216.

J604. Calabrese, C.R., Loadman, P.M., Lim, L.S.E., Bibby, M.C., Double, J.A., Brown, J.E., Lamb, J.H., *Drug Metab. Dispos.* (1999) **27**, 240–245.

J605. Bourrié, M., Meunier, V., Berger, Y., Fabre, G., *Drug Metab. Dispos.* (1999) **27**, 288–296.

J606. Kudo, S., Okumura, H., Miyamoto, G., Ishizaki, T., *Drug Metab. Dispos.* (1999) **27**, 303–308.

J607. Kumar, S., Riggs, F.W., Rurak, D.W., *Drug Metab. Dispos.* (1999) **27**, 297–302.

J608. Okushio, K., Suzuki, M., Matsumoto, N., Nanjo, F., Hara, Y., *Drug Metab. Dispos.* (1999) **27**, 309–316.

J609. Kudo, S., Okumura, H., Miyamoto, G., Ishizaki, T., *Drug Metab. Dispos.* (1999) **27**, 303–308.

J610. Carlile, D.J., Stevens, A.J., Ashforth, E.I., Waghela, D., Houston, J.B., *Drug Metab. Dispos.* (1999) **27**, 216–221.

J611. Bourland, J.A., Martin, D.K., Mayersohn, M., *Drug Metab. Dispos.* (1999) **27**, 203–206.

J612. Monsarrat, B., Chatelut, E., Royer, I., Alvinierie, P., Dubois, J., Dezeuse, A., Roche, H., Cros, S., Wright, M., Canal, P., *Drug Metab. Dispos.* (1999) **27**, 229–233.

J613. Qiu, Y., Burlingame, A.L., Benet, L.Z., *Drug Metab. Dispos.* (1998) **26**, 246–256.

J614. Ouyang, Y.-B., Lane, W.S., Moore, K.L., *Proc. Natl. Acad. Sci. U.S.A.* (1998) **95**, 2896–1901.

J615. Nakajima, M., Inoue, T., Shimada, N., Tokudome, S., Yamamoto, T., Kuroiwa, Y., *Drug Metab. Dispos.* (1998) **26**, 261–266.

J616. Kishimoto, W., Hiroi, T., Sakai, K., Funae, Y., Igarashi, T., *Res. Commun. Mol. Pathol. Pharmacol.* (1997) **98**, 273–292.

J617. McGurk, K.A., Brierley, C.H., Burchell, B., *Biochem. Pharmacol.* (1998) **55**, 1005–1012.

J618. Tegtmeier, M., Legrum, W., *Arch. Pharm.* (1998) **331**, 143–148.

J619. Ludwig, E., Wolfinger, H., Ebner, T., *J. Chromatogr., B, Biomed. Sci. Appl.* (1998) **707**, 347–350.

J620. Lukkari, E., Taavitsainen, P., Juhakoski, A., Pelkonen, O., *Pharmacol. Toxicol.* (1998) **82**, 161–166.

J621. Gillooly, D.J., Robertson, A.G.S., Fewson, C.A., *Biochem. J.* (1998) **330**, 1375–1381.

J622. Peres, C.M., Naveau, H., Agathos, S.N., *Appl. Microbiol. Biotechnol.* (1998) **49**, 343–349.

J623. Cavin, J.-F., Dartois, V., Divies, C., *Appl. Environ. Microbiol.* (1998) **64**, 1466–1471.

J624. Sheng, D., Joshi, D.K., Gold, M.H., *Arch. Biochem. Biophys.* (1998) **352**, 121–129.

J626. Wimmer, G., Halbwirth, H., Wurst, F., Forkmann, G., Stich, K., *Phytochemistry* (1998) **47**, 1013–1016.

J627. Klaus, K., Barz, W., *Phytochemistry* (1998) **47**, 1045–1048.

J628. Carpendo, R., Mannaioni, G., Moroni, F., *J. Neurochem.* (1998) **70**, 1998–2003.

J629. Kirkwood, L.C., Nation, R.L., Somogyi, A.A., *Clin. Exp. Pharmacol. Physiol.* (1998) **25**, 266–270.

J630. Fraiz, F.J., Pinto, R.M., Costas, M.J., Avalos, M., Cannales, J., Cabezas, A., Camerselle, J.C., *Biochem. J.* (1998) **330**, 881–888.

J631. Itoh, K., Kitade, Y., Yatome, C., *Bull. Environ. Contam. Toxicol.* (1998) **60**, 786–790.

J632. Sakamoto, T., Ichihara, N., Takeda, M., *Appl. Entom. Zool.* (1998) **33**, 97–104.

J633. Asano, H., Takeda, M., *Appl. Entom. Zool.* (1998) **33**, 127–132.

J634. Ware, J.A., Reilly, T.P., Svensson, C.K., *Biochem. Pharmacol.* (1998) **55**, 1475–1479.

J635. Craig, D., Moody, P.C.E., Bruce, N.C., Scrutton, N.S., *Biochemistry* (1998) **37**, 7598–7607.

J636. Van der Plas, L.H.W., Hagendoorn. M.J.M., Jamar, D.C.L., *J. Plant Physiol.* (1998) **152**, 235–241.

J637. Schrenk, C., Steinberg, C.E.W., *Environ. Sci. Pollut. Res. Int.* (1998) **5**, 83–88.

J638. Nishizawa, C., Ohta, T., Egashira, Y., Sanada, H., *Food Sci. Technol. Int., Tokyo* (1998) **4**, 94–97.

J639. Guner, S., Colak, A., Ozen, T., Senturk, H.B., *Indian J. Biochem. Biophys.* (1998) **35**, 11–15.

J641. Salama, A.K.M., *J. Environ. Sci. Health, Part B* (1998) **B33**, 253–266.

J642. Neusser, D., Schmidt, H., Spizek, J., Novotna, J., Peschke, U., Kaschabek, S., Tichy, P., Piepersberg, W., *Arch. Microbiol.* (1998) **169**, 322–332.

J643. Kim, C.-G., Yu, T.-W., Frygle, C.B., Handa. S., Floss, H.G., *J. Biol. Chem.* (1998) **273**, 6030–6040.

J644. Sakakibara, Y., Takami, Y., Nakayama, T., Suiko, M., Liu, M.-C., *J. Biol. Chem.* (1998) **273**, 6242–6247.

J645. Xu, J.F., Feng, P.S., *J. Biotechnol.* (1998) **61**, 69–73.

J646. Rehmann, K., Noll, H.P., Steinberg, C.E.W., Kettrup, A.A., *Chemosphere* (1998) **36**, 2977–2992.

J647. Huang, Z., Guengerich, F.P., Kaminsky, L.S., *Carcinogenesis*, (1998) **19**, 867–872.

J648. Sutherland, J.B., Freeman, J.P., Williams, A.J., *Appl. Microbiol. Biotechnol.* (1998) **49**, 445–449.

J649. Murakami, S., Nakanishi, Y., Kodama, N., Takenaka, S., Shinke, R., Aoki, K., *Biosci. Biotechnol. Biochem.* (1998) **62**, 747–752.

J650. Narbad, A., Gasson, M.J., *Microbiology* (1998) **144**, 1397–1405.

J651. Leuthner, B., Leutwein, C., Schulz, H., Horth, P., Haehnel, W., Schiltz, E., Schagger, H., Heider, J., *Mol. Microbiol.* (1998) **28**, 615–628.

J652. Hsieh, S.-E., Lo, H.-H., Chung, J.-G., *Curr. Microbiol.* (1998) **36**, 353–360.

J653. Kang, B.-S., Ha, J.Y., Lim, J.-C., Lee, J., Kim, C.K., Min, K.R., Kim, Y., *Biochem. Biophys. Res. Commun.* (1998) **245**, 791–796.

J654. Riegert, U., Heiss, G., Fischer, P., Stolz, A., *J. Bacteriol.,* (1998) **180**, 2849–2853.

J655. Piskula, M.K., Terao, J., *J. Nutr.* (1998) **128**, 1172–1178.

J656. Boyd, D.R., Sharma, N.D., Byrne, B., Hand, M.V., Malone, J.F., Sheldrake, G.N., Blacker, J., Dalton, H., *J. Chem. Soc., Perkin Trans. 1* (1998) 1935–1944.

J657. Balashov, S.V., Urinbasarov, M.U., Adanin, V.M., Il'yasov, P.K., Boronin, A., *Prikl. Biokhim. Mikrobiol.* (1998) **3**, 142–147.

J659. Carrera, G., Alary, J., Melgar, M.J., Lamboeuf, Y., Pipy, B., *Arch. Environ. Contam. Toxicol.* (1998) **35**, 89–96.

J660. Lui, E., Chirakal, R., Firnau, G., *J. Labelled Compd. Radiopharm.* (1998) **41**, 503–521.

J661. Mironowicz, A., *Phytochemistry* (1998) **47**, 1531–1534.

J662. Ireland, L.S., Harison, D.J., Neal, G.E., Hayes, J.D., *Biochem. J.* (1998) **332**, 21–34.

J663. Fox, S.D., Roman, J.M., Issaq, H.J., Nims, R.W., *Arch. Environ. Contam. Toxicol.* (1998) **35**, 104–108.

J664. Drijfhout, F.P., Fraaije, M.W., Jongejan, H., Van Berkel, W.J.H., Franssen, M.C.R., *Biotechnol. Bioeng.* (1998) **59**, 171–177.

J665. Schneider, S., Fuchs, G., *Arch. Microbiol.* (1998) **169**, 509–516.

J666. Mitrochkine, A.A., Eydoux, F., Gil, G., Reglier, M., *Eur. J. Org. Chem.* (1998) 1171–1176.

J667. Kaufmann, F., Wohlfarth, G., Diekert, G., *Eur., J. Biochem.* (1998) **253**, 706–711.

J668. Li, J., Li, G., *J. Liq. Chromatogr. Relat. Technol.* (1998) **21**, 1511–1525.

J669. Hammann, R., Kutzner, H.J., *J. Basic Microbiol.* (1998) **38**, 207–220.

J670. Brauman, A., Muller, J.A., Garcia, J.-L., Brune, A., Schink, B., *Int. J. Syst. Bacteriol.* (1998) **48**, 215–221.

J671. Griffin, R.J., Godfrey, V.B., Burka, L.T., *Toxicol. Sci.* (1998) **42**, 73–81.

J672. Carrillo, J.A., Ramos, S.I., Agundez, J.A.G., Martinez, C., Benitez, J., *Ther. Drug Monit.* (1998) **20**, 319–324.

J673. Belpaire, F.M., Wijnant, P., Temmermann, A., Rasmussen, B.B., Brosen, K., *Eur. J. Clin. Pharmacol.* (1998) **54**, 261–264.

J674. Iwabuchi, T., Inomata-Yamauchi, Y., Katsuta, A., Harayama, S., *J. Mar. Biotechnol.* (1998) **6**, 86–90.

J675. Wright, A.W.E., Nocente, M.-L., Smith, M.T., *Life Sci.* (1998) **63**, 401–411.

J676. Kuilman, M.E.M., Maas, R.F.M., Judah, B.J., Fink-Gremmels, J., *J. Agric. Food Chem.* (1998) **46**, 2707–2713.

J677. Chung, J.G., *Curr. Microbiol.* (1998) **37**, 70–73.

J678. Espin, J.C., Garcia-Ruiz, P.A., Tudela, J., Varon, R., Garcia-Canovas, F., *J. Agric. Food Chem.* (1998) **46**, 2968–2975.

J679. Kadiyala, V., Spain, J.C., *Appl. Environ. Microbiol.* (1998) **64**, 2479–2484.

J680. Fraaije, M.W., Van Den Heuvel, R.H.H., Roelofs, J.C.A.A., van Berkel, W.J.H., *Eur. J. Biochem.* (1998) **253**, 712–719.

J681. Sam, E., Sarre, S., Michotte, Y., Verbeke, N., *Eur. J. Drug Metab. Pharmacokinet.* (1998) **23**, 137–142.

J682. Cleij, M., Archelas, A., Furstoff, R., *Tetrahedron: Asymmetry* (1998) **9**, 1839–1842.

J683. Chou, H.-C., Ozawa, S., Fu, P.P., Lang, N.P., Kadlubar, F.F., *Carcinogenesis* (1998) **19**, 1071–1076.

J684. Nucaro, E., Jodra, M., Russell, E., Anderson, L., Dennison, P., Dufton, M., *Toxicon* (1998) **36**, 1173–1187.

J685. Zbaida, S., Du, Y., Shannon, D., Laudicinia, D., Thonoor, C.M., Nga, K., Blumenkrantz, N., Patrick, J.E., Cayen, M.N., Friary, R., Seidl, V., Chan T.-M., Pramanik, B., Spangler, M., Mc Phail, A.T., *Biopharm. Drug Dispos.* (1998) **19**, 315–332.

J686. Nakazawa, T., Ohsawa, K., *J. Nat. Prod.* (1998) **61**, 993–996.

J687. Oelschlaeger, H., Klinger, W., Rothley, D., Seeling, A., Bockhard, H., Hofmann, B., Machts, H., Riederer, H., Rackur, H., *Pharmazie* (1998) **53**, 620–624.

J688. Johnsen N.M., Nyholm, S.H., Hang, K., Scholz, T., Holme, J.A., *Chem.-Biol. Interact.* (1998) **113**, 217–237.

J689. Du, L., Halkier, B.A., *Phytochemistry* (1998) **48**, 1145–1150.

J690. Morand, C., Crespy, V., Manach, C., Besson, C., Demigne, C., Remesy, C., *Am. J. Physiol.* (1998) **275**, R212-R219.

J691. Coquoz, J.-L., Buchala, A., Metraux, J.-P., *Plant Physiol.* (1998) **117**, 1095–1101.

J692. Cody, J.T., Valtier, S., *J. Anal. Toxicol.* (1998) **22**, 299–309.

J693. Wu, W.-N., *J. Pharm. Biomed. Anal.* (1998) **17**, 689–697.

J694. Lee, H.-I., Raskin, I., *Phytopathology* (1998) **88**, 692–697.

J695. Duroux, L., Delmotte, F.M., Lancelin, J.-M., Kervais, G., Jay-Allemand, C., *Biochem. J.* (1998) **333**, 275–283.

J696. Deschler, C., Duran, R., Junqua, M., Landou, C., Salvado, J.-C., Goulas, P., *J. Mol. Catal. B: Enzym.* (1998) **5**, 423–428.

J697. Lee, I.-Y., Volm, T.G., Rosazza, J.P.N., *Enzym. Microb. Technol.* (1998) **23**, 261–266.

J698. Murata, T., Itoh, T., Usui, T., *Glycoconjugate J.* (1998) **15**, 575–582.

J699. Takenaka, S., Murakami, S., Shinke, R., Aoki, K., *Arch. Microbiol.* (1998) **170**, 132–137.

J700. Van Eekert, M.H.A., Van Ras, N.J.P., Mentink, G.H., Rijnaarts, H.H.M., Stams, A.J.M., Field, J.A., Schraa, G., *Environ. Sci. Technol.* (1998) **32**, 3299–3304.

J701. St-Pierre, B., Laflamme, P., Alarco, A.-M., De Luca, V., *Plant J.* (1998) **14**, 703–713.

J702. Potrawfke, T., Timmis, K.N., Wittich, R.-M., *Appl. Environ. Microbiol.* (1998) **64**, 3798–3806.

J703. Strachan, P.D., Freer, P.D., Fewson, C.A., *Biochem. J.* (1998) **333**, 741–747.

J704. Platzer, P., Schaden, S., Thalhammer, T., Hamilton, G., Rosenberg, B., Silgoner, I., Jager, W., *Anticancer Res.* (1998) **18**, 2695–2700.

J705. Zaborina, O., Daubaras, D.L., Zago, A., Xun, L., Saido, K., Klem, T., Nikolic, D., Chakrabarty, A.M. *J. Bacteriol.* (1998) **180**, 4667–4675.

J707. Achterholt, S., Priefert, H., Steinbuchel, A., *J. Bacteriol.* (1998) **180**, 4387–4391.

J708. Verhagen F.J.M., Swarts, H.J., Wijnberg, J.P.B.A., Field, J.A., *Appl. Environ. Microbiol.* (1998) **64**, 3225–3231.

J709. Malcherek, K., Breuer, J., Schuphan, I., Schmidt, B., *J. Plant Physiol.* (1998) **153**, 192–199.

J710. Seah, S.Y.K., Terracina, G., Bolin, J.T., Riebel, P., Snieckus, V., Eltis, L.D., *J. Biol. Chem.* (1998) **273**, 22943–22949.

J711. Ohmoto, T., Kinoshita, T., Moriyoshi, K., Sakai, K., Hamada, N., Ohe, T., *J. Biochem.* (*Tokyo*) (1998) **124**, 591–597.

J712. Wang, S.X., Ellis, B.E., *Phytochemistry* (1998) **49**, 307–318.

J713. Ostin, A., Kowalyczk, M., Bhalerao, R.P., Sandberg, G., *Plant Physiol.* (1998) **118**, 285–296.

J714. Peng, H.W., Cheng, F.C., Huang, Y.T., Chen, C.F., Tsai, T.H., *J. Chromatogr., B: Biomed. Sci. Appl.* (1998) **714**, 369–374.

J715. Muranaka, T., Miyata, M., Ito, K., Tachibana, S., *Phytochemistry* (1998) **49**, 491–496.

J716. Mazerska, Z., Gorlewska, K., Kraciuk, A., Jerzy, K., *Chem.-Biol. Interact.* (1998) **115**, 1–22.

J717. Yasuda, T., Ohsawa, K., *Biol. Pharm. Bull.* (1998) **21**, 953–957.

J718. Coldham, N.G., Dave, M., Sauer, M.T., *J. Mass Spectrom.* (1998) **33**, 803–810.

J719. Kawashiro, T., Yamashita, K., Zhao, X.-J., Koyama, E., Tani, M., Chiba, K., Ishizaki, T., *J. Pharmacol. Exp. Ther.* (1998) **286**, 1294–1300.

J720. Sütfeld, R., *Phytochemistry* (1998) **49**, 451–459.

J721. Leszczak, J.-P., Tran-Minh, C., *Biotechnol. Bioeng.* (1998) **60**, 356–361.

J722. Ogata, J., Teramoto, S., Yoshitama, K., *J. Plant Res.* (1998) **111**, 213–216.

J723. Li, K., Frost, J.W., *J. Am. Chem. Soc.* (1998) **120**, 10545–10546.

J724. Beisswanger, R., Corbeil, D., Vannier, C., Thiele, C., Dohrmann, U., Kellner, R., Ashman, K., Niehrs, C., Huttner, W.B., *Proc. Natl. Acad. Sci. USA* (1998) **95**, 11134–11139.

J725. Grace, J.M., Skanchy, D.J., Aguilar, A.J., *Xenobiotica* (1999) **29**, 703–717.

J726. Duhart, B.T., Jr., Zhang, D., Deck, J., Freeman, J.P., Cerniglia, C.E., *Xenobiotica* (1999) **29**, 733–746.

J727. Bach, M.V., Coutts, R.T., Baker, G.B., *Xenobiotica* (1999) **29**, 719–732.

J728. Azuma, R., Komuro, M., Korsch, B.H., Andre, J.C., Onnagawa, O., Black, S.R., Mathews, J.M., *Xenobiotica* (1999) **29**, 747–762.

J729. Kobayashi, K., Ishizuka, T., Shimada, N., Yoshimura, Y., Kamijima, K., Chiba, K., *Drug Metab. Dispos.* (1999) **27**, 763–766.

J730. Lanza, D.L., Code, E., Crespi, C.L., Gonzalez, F.J., Yost, G.S., *Drug Metab. Dispos.* (1999) **27**, 798–803.

J731. Kitamura, S., Okamoto, Y., Takeshita, M., Ohta, S., *Drug Metab. Dispos.* (1999) **27**, 767–769.

J732. Weber, C., Gasser, R., Hopfgartner, G., *Drug Metab. Dispos.* (1999) **27**, 810–815.

J733. Kobayashi, K., Nakajima, M., Oshima, K., Shimada, N., Yokoi, T., Chiba, K., *Drug Metab. Dispos.* (1999) **27**, 860–865.

J734. Kumar, G.N., Dykstra, J., Roberts, E.M., Jayanti, V.K., Hickman, D., Uchic, J., Yao, Y., Surber, B., Thomas, S.,

Granneman, G.R., *Drug Metab. Dispos.* (1999) **27**, 902–908.

J735. Rodriguez, R.J., Proteau, P.J., Marquez, B.L., Hetherington, C.L., Buckholz, C.J., O'Connell, K.L., *Drug Metab. Dispos.* (1999) **27**, 880–886.

J736. Black, A.E., Hayes, R.N., Roth, B.D., Woo, P., Woolf, T.F., *Drug Metab. Dispos.* (1999) **27**, 916–923.

J737. Omai, T., Taketani, M., Suzu, T., Kusube, K., Otagiri, M., *Drug Metab. Dispos.* (1999) **27**, 942–946.

J738. Kashiyama, E., Yokoi, T., Odomi, M., Kamataki, T., *Xenobiotica* (1999) **29**, 815–826.

J739. Sugimoto, T., Furuta, S., Suzuki, M., Sano, H., *Xenobiotica* (1999) **29**, 803–814.

J740. Watanabe, M., Tateishi, T., Tanaka, M., Kumai, T., Kobayashi, S., *Xenobiotica* (1999) **29**, 839–846.

J741. Wild, M.J., Mc Killop, D., Butters, C.J., *Xenobiotica* (1999) **29**, 849–857.

J742. Nakjajima, M., Kobayashi, K., Oshima, K., Shimada, N., Tokudome, S., Chiba, K., Yokoi, T., *Xenobiotica* (1999) **29**, 885–898.

J743. Martínez, L., Serafini, M.T., Pretel, M.J., Garcia-Soret, A., Moragon, T., Puig, S., *Xenobiotica* (1999) **29**, 859–871.

J744. Yamano, S., Nakamoto, N., Toki, S., *Xenobiotica* (1999) **29**, 917–930.

J745. Taguchi, K., Konishi, T., Nishikawa, H., Kitamura, S., *Xenobiotica* (1999) **29**, 899–907.

J746. Gorski, J.C., Jones, D.R., Hamman, M.A., Wrighton, S.A., Hall, S.D., *Xenobiotica* (1999) **29**, 931–944.

J747. Pieniaszek, H.J., Jr., Davidson, A.F., Chaney, J.E., Shum, L., Robinson, C.A., Mayersohn, M., *Xenobiotica* (1999) **29**, 945–955.

J748. Mutlib, A.E., Klein, J.T., *J. Pharmacol. Exp. Ther.* (1998) **286**, 1285–1293.

J749. Ismail, I.M., Dear, G.J., Mutch, P.J., Davies, L.H., Plumb, R.S., Sweatman, B.C., *Xenobiotica* (1999) **29**, 957–967.

J750. Mae, T., Konishi, E., Hosoe, K., Hidaka, T., *Xenobiotica* (1999) **29**, 1073–1087.

J751. Matsui, K., Taniguchi, S., Yoshimura, T., *Xenobiotica* (1999) **29**, 1059–1072.

J752. Wu, W.N., McKown, L.A., Moyer, M.D., Johannsen, T.B., Takacs, A.R., *Xenobiotica* (1999) **29**, 1089–1100.

J753. Raleigh, S.M., Wanogho, E., Burke, M.D., Patterson, L.H., *Xenobiotica* (1999) **29**, 1115–1122.

J754. Strolin-Benedetti, M., Brogin, G., Bani, M., Oesch, F., Hengstler, J.G., *Xenobiotica* (1999) **29**, 1171–1180.

J755. Dalvie, D.K., O'Donnell, J.P., *Xenobiotica* (1999) **29**, 1123–1140.

J756. Sevoz, C., Rousselle, C., Benoît, E., *Xenobiotica* (1999) **29**, 1007–1016.

J757. Coller, J.K., Somogyi, A.A., Bochner, F., *Xenobiotica* (1999) **29**, 973–986.

J758. Dalgaard, L., Larsen, C., *Xenobiotica* (1999) **29**, 1033–1041.

J759. Stevens, G.J., Payton, M., Sim, E., McQueen, C.A., *Drug Metab. Dispos.* (1999) **27**, 966–971.

J760. Zhao, Z., Koeplinger, K.A., Peterson, T., Conradi, R.A., Burton, P.S., Suarato, A., Heinrikson, R.L., Tomasselli, A.G., *Drug Metab. Dispos.* (1999) **27**, 992–998.

J761. Koeplinger, K.A., Zhao, Z., Peterson, T., Leone, J.W., Schwende, F.S., Heinrikson, R.L., Tomasselli, A.G., *Drug Metab. Dispos.* (1999) **27**, 986–991.

J762. Mankowski, D.C., *Drug Metab. Dispos.* (1999) **27**, 1024–1028.

J763. Vickers, A.E.M., Sinclair, J.R., Zollinger, M., Heitz, F., Glänzel, U., Johanson, L., Fischer, V., *Drug Metab. Dispos.* (1999) **27**, 1029–1038.

J764. Falany, C.N., Falany, J.L., Wang, J., Hedström, J., Chelpin, H.v.E., Swedmark, S., *Drug Metab. Dispos.* (1999) **27**, 1057–1063.

J765. Mutlib, A.E., Chen, H., Nemeth, G., Gan, L.-S., Christ, D.D., *Drug Metab. Dispos.* (1999) **27**, 1045–1056.

J766. Zenser, T.V., Lakshmi, V.M., Davis, B.B., *Drug Metab. Dispos.* (1999) **27**, 1064–1067.

J767. Becquemont, L., Mouajjah, S., Escaffre, O., Beaune, P., Funck-Brentano, C., Jaillon, P., *Drug Metab. Dispos.* (1999) **27**, 1068–1073.

J768. Suzuki, A., Iida, I., Tanaka, F., Akimoto, M., Fukushima, K., Tani, M., Ishizaki, T., Chiba, K., *Drug Metab. Dispos.* (1999) **27**, 1254–1259.

J769. Akira, K., Negishi, E., Sakuma, C., Hasimoto, T., *Drug Metab. Dispos.* (1999) **27**, 1248–1253.

J770. Yamazaki, H., Shibata, A., Suzuki, M., Nakajima, M., Shimada, N., Guengerich., F.P., Yokoi, T., *Drug Metab. Dispos.* (1999) **27**, 1260–1266.

J771. Mey, U., Wachsmuth, H., Breyer-Pfaff, U., *Drug Metab. Dispos.* (1999) **27**, 1281–1292.

J772. Gelderblom, H., Verweij, J., Brouwer, E., Pillay, M., De Bruijn, P., Nooter, K., Stoter, G., Sparreboom, *Drug Metab. Dispos.* (1999) **27**, 1300–1305.

J773. Court, M.H., Hay-Kraus, B.L., Hill, D.W., Kind, A.J., Greenblatt, D.J., *Drug Metab. Dispos.* (1999) **27**, 1293–1299.

J774. Chen, H., Brzezinski, M.R., Fantel, A.G., Juchau, M.R., *Drug Metab. Dispos.* (1999) **27**, 1306–1308.

J775. Mutlib, A.E., Chen, H., Nemeth, G.A., Markwalder, J.A., Seitz, S.P., Gan, L.S., Christ, D.D., *Drug Metab. Dispos.* (1999) **27**, 1319–1333.

J776. Vickers, S., Theoharides, A.D., Arison, B., Balani, S.K., Cui, D., Duncan, C.A., Ellis, J.D., Gorham, L.M., Polsky, S.L., Prueksaritanont, T., Ramjit, H.G., Slaughter, D.E., Vyas, K.P., *Drug Metab. Dispos.* (1999) **27**, 1360–1366.

J777. Wienkers, L.C., Allievi, C., Hauer, M.J., Wynalda, M.A., *Drug Metab. Dispos.* (1999) **27**, 1334–1340.

J778. Huskey, S.-E.W., Luffer-Atlas, D., Dean, B.J., McGowan, E.M., Feeny, W.P., Chiu, S.-H.L., *Drug Metab. Dispos.* (1999) **27**, 1367–1373.

J779. Rosenborg, J., Larsson, P., Tegnér, K., Hallström, G., *Drug Metab. Dispos.* (1999) **27**, 1104–1116.

J780. Diaz, G.J., Skordos, K.W., Yost, G.S., Squires, E.J., *Drug Metab. Dispos.* (1999) **27**, 1150–1156.

J781. Paulson, S.K., Engel, L., Reitz, B., Bolton, S., Burton, E.G., Maziasz, T.J., Yan, B., Schoenhard, G.L., *Drug Metab. Dispos.* (1999) **27**, 1133–1142.

J782. Moody, J.D., Freeman, J.P., Cerniglia, C.E., *Drug Metab. Dispos.* (1999) **27**, 1157–1164.

J783. Cheng, Z., Radominska-Pandya, A., Tephly, T.R., *Drug Metab. Dispos.* (1999) **27**, 1165–1170.

J784. Chow, T., Hiroi, T., Imaoka, S., Chiba, K., Funae, Y., *Drug Metab. Dispos.* (1999) **27**, 1200–1204.

J785. Scarfe, G.B., Lindon, J.C., Nicholson, J.K., Wright, B., Clayton, E., Wilson, I.D., *Drug Metab. Dispos.* (1999) **27**, 1171–1178.

J786. Jacobsen, P.B., Larsen, Å., Konarboland, R., Skotland, T., *Drug Metab. Dispos.* (1999) **27**, 1205–1213.

J787. Nakajima, M., Nakamura, S., Tokudome, S., Shimada, N., Yamazaki, H., Yokoi, T., *Drug Metab. Dispos.* (1999) **27**, 1381–1391.

J788. Dong, J.Q., Etheridge, A.S., Smith, P.C., *Drug Metab. Dispos.* (1999) **27**, 1423–1428.

J789. Matsui, K., Mishima, M., Nagai, Y., Yuzuriha, T., Yoshimura, T., *Drug Metab. Dispos.* (1999) **27**, 1406–1414.

J790. Kobayashi, K., Abe, S., Nakajima, M., Shimada, N., Tani, M., Chiba, K., Yamamoto, T., *Drug Metab. Dispos.* (1999) **27**, 1429–1433.

J791. Halladay, J.S., Sauer, J.-M., Sipes, I.G., *Drug Metab. Dispos.* (1999) **27**, 1456–1465.

J792. Prueksaritanont, T., Lu, P., Gorham, L., Sternfeld, F., Vyas, K.P., *Xenobiotica* (2000) **30**, 47–59.

J793. Payan, J.-P., Beydon, D., Cossec, B., Ensminger, A., Fabry, J.-P., Ferrari, E., *Drug Metab. Dispos.* (1999) **27**, 1470–1478.

J794. Ishiguro, N., Senda, C., Kishimoto, W., Sakai, K., Funae, Y., Igarashi, T., *Xenobiotica* (2000) **30**, 71–80.

J795. Mathews, J.M., De Costa, K.S., *Drug Metab. Dispos.* (1999) **27**, 1499–1504.

J796. Juvonen, R.O., Gynther, J., Pasanen, M., Alhava, E., Poso, A., *Xenobiotica* (2000) **30**, 81–92.

J797. Naito, S., Nishimura, M., *Xenobiotica* (2000) **30**, 103–109.

J798. Jemnitz, K., Verez, Z., Monostory, K., Vereczkey, L., *Drug Metab. Dispos.* (2000) **28**, 34–37.

J799. Black, S.R., Mathews, J.M., *Drug Metab. Dispos.* (2000) **28**, 1–4.

J800. Äbelö, A., Andersson, T.B., Bredberg, U., Skånberg, I.,Weidolf, L., *Drug Metab. Dispos.* (2000) **28**, 58–64.

J801. Kuriya, S.-I., Ohmori, S., Hino, M., Ishii, I., Nakamura, H., Senda, C., Igarashi, T., Kiuchi, M., Kitada, M., *Drug Metab. Dispos.* (2000) **28**, 73–78.

J802. Vyas, K.P., Halpin, R.A., Geer, L.A., Ellis, J.D., Liu, L., Cheng, H., Chavez-Eng, C., Matuszewski, B.K., Varga, S.L., Guiblin, A.R., Rogers, J.D., *Drug Metab. Dispos.* (2000) **28**, 89–95.

J803. Davi, H., Tronquet, C., Miscoria, G., Perrier, L., DuPont, P., Caix, J., Simiand, J., Berger, Y., *Drug Metab. Dispos.* (2000) **28**, 79–88.

J804. Hegstad, S., Lundanes, E., Holme, J.A., Alexander, J., *Xenobiotica* (1999) **29**, 1257–1272.

J805. Galijatovic, A., Otake, Y., Walle, U.K., Walle, T., *Xenobiotica* (1999) **29**, 1241–1246.

J806. Fujimaki, Y., Arai, N., Inaba, T., *Xenobiotica* (1999) **29**, 1273–1282.

J807. Molden, E., Christensen, H., Sund, R.B., *Drug Metab. Dispos.* (2000) **28**, 107–109.

J809. Born, S.L., Hu, J.K., Lehman-McKeeman, L.D., *Drug Metab. Dispos.* (2000) **28**, 218–223.

J811. El-Bayoumy, K., Upadhuana, P., Sohn, O.-S., Rosa, J.G., Fiala, E.S., *Carcinogenesis* (1998) **19**, 1603–1607.

J812. Laempe, D., Eisenreich, W., Bacher, A., Fuchs, G., *Eur.J. Biochem.* (1998) **255**, 618–627.

J813. Spatzenegger, M., Ecker, G., Jaeger, W., *Sci. Pharm.* (1998) **66**, 159–168.

J814. Zaks, A., Dodds, D.R., *Appl. Biochem. Biotechnol.* (1998) **73**, 205–214.

J816. Tsou, M.F., Chung, J.G., Wu, L.T., Cheng, K.S., Hung, C.F., *Microbios* (1998) **94**, 133–143.

J817. Beller, H.R., Spormann, A.M., *J. Bacteriol.* (1998) **180**, 5454–5457.

J818. Cabello-Hurtado, F., Durst, F., Jorrin, J.V., Werck-Reichhart, D., *Phytochemistry* (1998) **49**, 1029–1036.

J819. Ribnicky, D.M., Shulaev, V., Raskin, I., *Plant Physiol.* (1998) **118**, 565–572.

J820. Mc Dougall, *J. Plant Physiol.* (1998) **153**, 539–544.

J821. Day, A.J., Du Pont, M.S., Ridley, S., Rhodes, M., Rhodes, M.J.C., Morgan, M.R.A., Williamson, G., *FEBS Lett.* (1998) **436**, 71–75.

J822. Nozue, M., Souri, M., Arakawa, D., Kojima, M., *J. Plant Physiol.* (1998) **153**, 552–557.

J823. Pezet, R., *FEMS Microbiol. Lett.* (1998) **167**, 203–208.

J824. Lange, J., Hammer, E., Specht, M., Franke, W., Schauer, F., *Appl. Microbiol. Biotechnol.* (1998) **50**, 364–368.

J825. Hoyle, A.J., Bunch, A.W., Knowles, C.T., *Enzyme Microb. Technol.* (1998) **23**, 475–482.

J826. Bannon, P., Yu, P., Cook, J.M., Roy, L., Villeneuve, J.-P., *J. Chromatogr. B: Biomed. Sci. Appl.* (1998) **715**, 387–393.

J827. Cristofol, C., Navarro, M., Franquelo, C., *Vet. Parasitol.* (1998) **78**, 223–31.

J828. Lee, P.C., Marquardt, M., Lech, J.J., *Toxicol. Lett.* (1998) **99**, 117–126.
J829. Valoti, M., Frosini, M., Palmi, M., De Matteis, F., Sgaragli, G., *J. Pharm. Pharmacol.* (1998) **50**, 1005–1011.
J830. Fujigaki, S., Saito, K., Takemura, M., Fujii, H., Wada, H., Noma, A., Seishima, M., *Arch Biochem. Biophys.* (1998) **358**, 329–335.
J831. Koseki, T., Furuse, S., Iwano, K., Matsuzawa, H., *Biosci. Biotechnol. Biochem.* (1998) **62**, 2032–2034.
J832. Huwe, J.K., Feil, V.J., Larsen, G.L., Wiener, C., *Chemosphere* (1998) **37**, 1885–1893.
J833. Buffoni, F., Dowling, T.G., Cambi, S., *Int. J. Mol. Med.* (1998) **2**, 187–195.
J834. Ali, S., Fernandez, Lafuente, R., Cowan, D.A., *Enzyme Microb. Technol.* (1998) **23**, 462–468.
J835. Kim, J.H., Stanbury, K.H., Walker, N.J., Trush, M.A., Strickland, P.T., Sutter, T.R., *Carcinogenesis* (1998) **19**, 1847–1853.
J836. Van den Heuvel, R.H.H., Fraaije, M.W., Laane, C., Van Berkel, W.J.H., *J. Bacteriol.* (1998) **180**, 5646–5651.
J837. Suiko, M., Sakakibara, Y., Awan-Khan, R., Sakaida, H., Yoshikawa, H., Ranasinghe, J.G.S., Liu, M.-C., *J. Biochem.* (1998) **124**, 707–711.
J838. Yokoyama, M., Yamaguchi, Y., *Koho JP* (2–12–1998) 10 313,856.
J839. Yamaguchi, T., Nakajima, Y., Mizobuchi, M., Inazawa, K., Kanazu, T., Kadono, K., Ohkawa, T., Iwanati, K., *Arzneim.-Forsch.* (1998) **48**, 995–1006.
J841. Larsen, G.L., Hakk, H., Klasson-Wehler, E., Oern, U., Bergman, A., *Organohalogen Compd.* (1998) **37**, 413–416.
J842. Cherqui, A., Duvic, B., Reibel, C., Brehelin, M., *Insect Biochem. Mol. Biol.* (1998) **28**, 839–848.
J843. Wunderwald, U., Hofrichter, M., Kreisel, G., Fritsche, W., *Biodegradation* (1997–8) **8**, 379–385.
J844. Rosing, H., Van Zomeren, D.M., Doyle, E., Bult, A., Beijnen J.H., *Anti-Cancer Drugs* (1998) **9**, 587–592.
J845. Benedito, M.A.C., *Braz. J. Med. Biol. Res.* (1998) **31**, 1113–1118.
J846. Shimodara, K., Shibasaki, M., Inaba, T., Cheung, S.W., Someya, T., Takahashi, S., *Pharmacol. Toxicol.* (1998) **83**, 164–168.
J847. Rabus, R., Heider, J., *Arch. Microbiol.* (1998) **170**, 377–384.
J848. Bouchard, M., Krishnan, K., Viau, C., *Arch. Toxicol.* (1998) **72**, 475–482.
J849. Peng, H.-W., Huang, Y.-T., Chen, C.-F., Tsai, T.H., *Planta Med.* (1998) **64**, 779.
J850. Khalil, E.M., De Angelis, J., Cole, P.A., *J. Biol. Chem.* (1998) **273**, 30321–30327.
J851. Zhang, T., Roberge, C., Reddy, J., Connors, N., Chartrain, M., Buckland, B., Greasham, R., *Enzyme Microb. Technol.* (1999) **24**, 86–95.
J853. Hashidoko, Y., Tahara, S., *Arch. Biochem. Biophys.* (1998) **359**, 225–230.
J854. Bast, G.E., Kampffmeyer, H.G., *Skin Pharmacol. Appl. Skin Physiol.* (1998) **11**, 250–257.
J855. Ertl, R.P., Alworth, W.L., Winston, G.W., *J. Biochem. Mol. Toxicol.* (1999) **13**, 17–27.
J856. Sepic, E., Bricelj, M., Leskovsek, H., *J. Appl. Microbiol.* (1998) **85**, 746–754.
J857. Kobayashi, M., Goda, M., Shimizu, S., *FEBS Lett.* (1998) **439**, 325–328.
J858. Jimenez, M., Garcia-Carmona, F., *J. Agric. Fd. Chem.* (1999) **47**, 56–60.
J859. Zeida, M., Wieser, M., Yoshida, T., Sugio, T., Nagasawa, T., *Appl. Environ, Microbiol.* (1998) **64**, 4743–4747.
J860. Barriault, D., Durand, J., Maaroufi, H., Eltis, L.D., Sylvestre, M., *Appl. Environ, Microbiol.* (1998) **64**, 4637–4642.
J861. Morimoyo, S., Komatsu, K., Taura, F., Shoyama, Y., *Phytochemistry* (1998) **49**, 1525–1529.

J862. Joyce, K.B., Jones, A.E., Scott, R.J., Biddlecombe, R.A., Pleasance, S., *Rapid Commun. Mass Spectrom.* (1998) **12**, 1899–1910.

J863. Panday, V.R.N., Rosing, H., Huizing, M.T., Koopman, F.J., Van Warmerdem, L.J.C., Ten Bokkel Huinink, W.W., Beijnen, J.H., *J. Oncol. Pharm. Pract.* (1998) **4**, 159–164.

J864. Frank, C., Schwarz, U., Matthies, C., Drake, H.L., *Arch. Microbiol.* (1998) **170**, 427–434.

J865. Peterson, T.G., Ji, G.-P., Kirk, M., Coward, L., Falany, C.N., Barnes, S., *Am. J. Clin. Nutr.* (1998) **68**, 1505S-1511S.

J866. Mountfield, R.J., Panduro, A.M., Wassman, O., Thompson, M., John, B., Van Der Merbel. N., *Xenobiotica* (2000) **30**, 1–17.

J867. Chiba, M., Nishime, J.A., Neway, W., Lin, Y., Lin, J.H., *Xenobiotica* (2000) **30**, 117–129.

J868. Bowers, G.D., Bayliss, M.J., Donnelly, M.C., Fellows, I., Ismail, I.M., Mookhherge, C.R., *J. Pharm. Biomed. Anal.* (1998) **18**, 461–470.

J869. Ohashi, N., Furunushi, S., Yoshikawa, M., *J. Pharm. Biomed. Anal.* (1998) **18**, 325–334.

J870. Haritos, V.S., Ching, M.S., Ghabrial, H., Gross, A.S., Taavitsainen, P., Pelkonen, O., Battaglia, S.E., Smallwood, R.A., Ahokas, J.T., *Pharmacogenetics* (1998) **8**, 423–432.

J871. Yaich, M., Popon, M., Medard, Y., Aigrain, E.J., *Pharmacogenetics* (1998) **8**, 449–451.

J872. Schliemann, W., Steiner, U., Strack, D., *Phytochemistry* (1998) **49**,1593–1598.

J873. Tiseo, P.J., Perdomo, C.A., Friedhoff, L.T., *Br. J. Clin. Pharmacol.* (1998) **46** (Suppl. 1), 19–24.

J874. Saez, L.P., Castillo, F., Caballero, F.J., *Curr. Microbiol.* (1999) **38**, 51–56.

J875. Haselden, J.N., Hutson, D.H., Dodds, P.F., *Biochem. Pharmacol.* (1998) **56**, 1591–1598.

J876. Kitamura, S., Terada, A., Inoue, N., Kamio, H., Ohta, S., Tatsumi, K., *Biol. Pharm. Bull.* (1998) **21**, 1344–1347.

J877. Haselden, J.N., Dodds, P.F., Hutson, D.H., *Biochem. Pharmacol.* (1998) **56**, 1599–1606.

J878. Yoshigae, Y., Imai, T., Taketani, M., Otagiri, M., *Chirality* (1999) **11**, 10–13.

J879. Lanchote, V.L., Cesarino, E.J., Santos, V.J., Mere, Y., Jr., Santos, S.R.C.J., *Chirality* (1999) **11**, 29–32.

J880. Jimenez, M., Escribo-Celrian, J., Garcia-Carmona, F., *Biochim. Biophys. Acta* (1998) **1425**, 534–542.

J881. Fawcett, J.P., Kennedy, J., Kumar, A., Ledger, R., Zacharias, M., *Chirality* (1999) **11**, 50–55.

J882. Schmidt, S., Fortnagel, P., *Microbiol. Res.* (1998) **153**, 331–335.

J883. Lim, Y.-L., Yoshimura, T., Soda, K., Eskari, N., *J. Ferment. Bioeng.* (1998) **86**, 400–402.

J884. Thompson, M.A., Weinshilboum, R.M., *J. Biol. Chem.* (1998) **273**, 34502–34510.

J885. Snell, D., Colby, J., *Enzyme Microb. Technol.* (1999) **24**, 160–163.

J886. Luukkanen, L., Kilpelaeinen, I., Kangas, H., Ottoila, P., Elovaara, E., Taskinen, J., *Bioconjugate Chem.* (1999) **10**, 150–154.

J887. Zuurbier, K.W.M., Leser, J., Berger, T., Hofte, A.J.P., Schroder, G., Verpoorte, R., Schroder, J., *Phytochemistry* (1998) **49**, 1945–1951.

J888. Rotzinger, S., Fang, J., Coutts, R.T., Baker, G.B., *Biol. Psychiatry* (1998) **44**, 1185–1191.

J889. Arlander, E., Ekström, G., Alm, C., Carillo, J.A., Bielenstein, M., Bottiger, Y., Bertilsson, L., Gustafsson, L.L., *Clin. Pharmacol. Ther.* (1998) **64**, 484–491.

J890. Furukawa, H., Wieser, M., Morito, H., Sugio, T., Nagasawa, T., *Arch. Microbiol.* (1998) **171**, 37–43.

J891. Parmar, D. Dhawan, A., Seth, P.K., *Mol. Cell Biol.* (1998) **189**, 201–205.
J892. Pasta, P., Carrea, G., Monzani, E., Gaggero, N., Colonna, S., *Biotechnol. Bioeng.* (1999) **62**, 489–493.
J893. Itoh, N., Mizuguchi, N., Mabuchi, M., *J. Mol. Catal. B: Enzym.* (1999) **6**, 41–50.
J895. Nadeau, L.J., Sayler, G.S., Spain, J.C., *Arch. Microbiol.* (1998) **171**, 44–49.
J896. Chartrain, M., Jackey, B., Taylor, C., Sandford, V., Gbewonyo, K., Lister, L., Dimichele, L., Hirsch, C., Heimbuch, B., Maxwell, C., Pascoe, D., Buckland, B., Greasham, R., *J. Ferment. Bioeng.* (1998) **86**, 550–558.
J897. Zhao, Z., Mabic, S., Kuttab, S., Franot, C., Castagnoli, K., Castagnoli, N., Jr., *Bioorg. Med. Chem.* (1998) **6**, 2531–2539.
J898. Matsumoto, S., Takahashi, M., Kitadai, N., Katae, H., *J. Vet. Med. Sci.* (1998) **60**, 1259–1261.
J899. Bahnmaier, A.H., Woesle, B., Thomas, H., *Chirality* (1999) **11**, 160–165.
J900. Boocock, D.J., Maggs, J.L., White, I.N.H., Park, B.K., *Carcinogenesis* (1999) **20**, 153–160.
J901. Kolanczyk, R., Schmieder, P., Bradbury, S., Spizzo, T., *Aquat. Toxicol.* (1999) **45**, 47–61.
J902. Friederich, S., Maier, U.H., Deus-Neumann, B., Asakawa, Y., Zenk, M.H., *Phytochemistry* (1999) **50**, 589–598.
J903. Haigler, B.E., Johnson, G.R., Suen, W.-C., Spain, J.C., *J. Bacteriol.* (1999) **181**, 965–972.
J904. Iba, M., Fung, J., *In Vitro Mol. Toxicol.* (1998) **11**, 285–297.
J905. Ruggiero, C.E., Dooley, D.M., *Biochemistry* (1999) **38**, 2892–2898.
J906. Sabordo, L., Sallustio, B.C., Evans, A.M., Nation, R.L., *J. Pharmacol. Exp. Ther.* (1999) **288**, 414–420.
J907. Rama, A., Mougin, C., Boyer, F.-D., Kollmann, A., Malosse, C., Sigoillot, J.-C., *Biotechnol. Lett.* (1998) **20**, 1101–1104.
J908. Geng, W., Pang, K.S., *J. Pharmacol. Exp. Ther.* (1999) **288**, 597–606.
J909. Mesdjian, E., Seree, E., Charvet, B., Mirrione, A., Bourgarel-Rey, V., Desobry, A., Barra, Y., *Life Sci.* (1999) **64**, 827–835.
J910. Colrat, S., Latche, A., Guis, M., Pech, J.-C., Bouzayen, M., Fallot, J., Roustan, J.-P., *Plant Physiol.* (1999) **119**, 621–626.
J911. Saari, R.E., Hogan, D.A., Hausinger, R.P., *J. Mol. Catal. B: Enzym.* (1999) **6**, 421–428.
J912. Ikemoto, M., Sasaki, M., Haradahira, T., Tada, T., Omura, H., Furuya, Y., Watanabe, Y., Suzuki, K., *Appl. Radiat. Isot.* (1999) **50**, 715–721.
J913. Zhuo, X., Gu, J., Zhang, Q.-Y., Spink, D.C., Kaminsky, L.S., Ding, X., *J. Pharmacol. Exp. Ther.* (1999) **288**, 463–471.
J914. Shiraga, T., Iwasaki, K., Hata, T., Yoshinari, K., Nagata, K., Yamazoe, Y., Ohno, Y., *Arch. Biochem. Biophys.* (1999) **362**, 265–274.
J915. Van Bekkum, Y.M., van den Broek, P.H.H., Scheepers, P.T.J., Noordhoek, J., Bos, R.P., *Chem.-Biol. Interact.* (1999) **117**, 15–33.
J916. Zhang, Z., Hillas, P.J., Ortiz de Montellano, P.R., *Arch. Biochem. Biophys.* (1999) **363**,19–26.
J917. Groenewald, E.G., Van Der Westhuizen, A.J., *Biol. Plant.* (1988) **41**, 475.
J918. Chiffoleau-Giraud, V., Spangenberg, P., Dion, M., Rabiller, C., *Eur. J. Org. Chem.* (1999), 757–763.
K1. Fang, J., Bourin, M., Baker, G.B., *Naunyn-Schmiedeberg's Arch. Pharmacol.* (1999) **359**, 147–151.
K2. Park, H.-S., Lim, S.-J., Chang, Y.K., Livingston, A.G., Kim, H.-S., *Appl. Environ. Microbiol.* (1999) **65**, 1083–1091.
K3. Clay, M.P., Hanssen, B.R., Surapaneni, S.S., Lindstrom, T.D., *Chirality* (1999) **11**, 233–240.

K4. Feng, Y., Khoo, H.E., Poh, C.L., *Appl. Environ. Microbiol.* (1999) **65**, 946–950.
K5. Mars, A.E., Kingma, J., Kaschabek, S.R., Reineke, W., Janssen, D.B., *J. Bacteriol.* (1999) **181**, 1309–1318.
K6. Lin, G., Han, S., Li, Z., *Tetrahedron* (1999) **55**, 3531–3540.
K7. Chae, J.-C., Ahn, K.-J., Kim, C.-K., *J. Microbiol. Biotechnol.* (1998) **8**, 692–695.
K8. Kojima, J., Fujino, H., Abe, H., Yosimura, M., Kanda, H., Kimata, H., *Biol. Pharm. Bull.* (1999) **22**, 142–150.
K10. Okushio, K., Suzuki, M., Matsumoto, N., Nanjo, F., Hara, Y., *Biosci. Biotechnol. Biochem.* (1999) **63**, 430–432.
K11. Fu, W., Oriel, P., *Extremophiles* (1999) **3**, 45–53.
K12. Watt, K.C., Buckpitt, A.R., *Drug Metab. Dispos.* (2000) **28**, 376–378.
K13. Minegishi, K., Nambaru, S., Fukuoka, M., Tanaka, A., Nishimaki-Mogami, T., *Arch. Toxicol.* (1991) **65**, 273–282.
K14. Gantenbein, M., Attolini, L., Bruguerolle, B., Villard, P.-H., Puyoou, F., Durand, A., Lacarelle, B., Hardwigsen, J., Le-Treut, Y.-P., *Drug Metab. Dispos.* (2000) **28**, 383–385.
K15. Sevoz, C., Benoit, E., Buronfosse, T., *Drug Metab. Dispos.* (2000) **28**, 398–402.
K16. Bayse, G.S., Jinadu, L.A., Shaw, K.L., Wiley, K.L., *Drug Metab. Dispos.* (2000) **28**, 487–492.
K17. Slatter, J.G., Schaaf, J., Sams, J.P., Feenstra, K.L., Johnson, M.G., Bombardt, P.A., Cathcart, K.S., Verburg, M.T., Pearson, L.K., Compton, L.D., Miller, L.L., Baker, D.S., Pesheck, C.V., Lord, R.S., III, *Drug Metab. Dispos.* (2000) **28**, 423–433.
K18. Doerge, D.R., Chang, H.C., Churchwell, M.I., Holder, C.L., *Drug Metab. Dispos.* (2000) **28**, 298–307.
K19. Paulson, S.K., Hribar, J.D., Liu, N.W.K., Hajdu, E., Bible, R.H., Jr., Piergies, A., Karim, A., *Drug Metab. Dispos.* (2000) **28**, 308–314.
K20. Law, M.Y.L., Slawson, M.H., Moody, D.E., *Drug Metab. Dispos.* (2000) **28**, 348–353.
K21. Kraemer, T., Bickeboeller-Friedrich, J., Maurer, H.H., *Drug Metab. Dispos.* (2000) **28**, 339–347.
K22. Westler, M.R., Lasker, J.M., Johnson, E.F., Raucy, J.L., *Drug Metab. Dispos.* (2000) **28**, 354–359.
K23. Carmona, G.N., Jufer, R.A., Goldberg, S.R., Gorelick, D.A., Greig, N.H., Yu, Q.-S., Cone, E.J., Schindler, C.W., *Drug Metab. Dispos.* (2000) **28**, 367–371.
K24. Tuvesson, H., Strandgården, K., Nordle, Ö., Gunnarsson, P.O., Seidegård, J., Persson, R., *Xenobiotica* (2000) **30**, 371–380.
K25. Kitamura, S., Mita, M., Matsuda, K., Ohta, S., Tatsumi, K., *Xenobiotica* (2000) **30**, 359–369.
K26. Virkel, G., Lifschitz, A., Soraci, A., Sansinanea, A., Lanusse, C., *Xenobiotica* (2000) **30**, 381–393.
K27. Tomigahara, Y., Onogi, M., Saito, K., Isobe, N., Kaneko, H., Nakatsuka, I., *Xenobiotica* (2000) **30**, 395–406.
K28. Smith, D.J., Feil, V.J., Paulson, D.G., *Xenobiotica* (2000) **30**, 427–440.
K29. Dear, G.J., Ismail, I.M., Mutch, P.J., Plumb, R.S., Davies, L.H., Sweatman, B.C., *Xenobiotica* (2000) **30**, 407–426.
K30. Paulson, S.K., Zhang, J.Y., Breau, A.P., Hribar, J.D., Liu, N.W.K., Jessen, S.M., Lawal, Y.M., Cogburn, J.N., Gresk, C.J., Markos, C.S., Maziasz, T.J., Schoenhard, G.L., Burton, E.G., *Drug Metab. Dispos.* (2000) **28**, 514–521.
K31. Mountfield, R.J., Kiehr, B., John, B.A., *Drug Metab. Dispos.* (2000) **28**, 503–513.
K32. Yoshihara, S., Harada, K., Ohta, S., *Drug Metab. Dispos.* (2000) **28**, 538–543.
K33. Fisher, M.B., Campanale, K., Ackermann, B.L., Vandenbranden, M., Wrighton, S.A., *Drug Metab. Dispos.* (2000) **28**, 560–566.

K34. Lewis, A.J., Otake, Y., Walle, U.K., Walle, T., *Xenobiotica* (2000) **30**, 253–261.

K35. Lightfoot, T., Ellis, S.W., Mahling, J., Ackland, M.J., Blaney, F.E., Bijloo, G.J., De Groot, M.J., Vermeulen, N.P.E., Blackburn, G.M., Lennard, M.S., Tucker, G.T., *Xenobiotica* (2000) **30**, 219–233.

K36. Sanders, J.M., Chen, L.-J., Burka, L.T., Matthews, H.B., *Xenobiotica* (2000) **30**, 263–272.

K37. Bach, M.V., Coutts, R.T., Baker, G.B., *Xenobiotica* (2000) **30**, 297–306.

K38. Markowitz, J.S., Devane, C.L., Boulton, D.W., Nahas, Z., Risch, S.C., Diamond, F., Patrick, K.S., *Drug Metab. Dispos.* (2000) **28**, 620–624.

K39. Xu, J., Purcell, W.M., Ray, D.E., *Xenobiotica* (2000) **30**, 307–315.

K40. Zhang, K.E., Kari, P.H., Davis, M.B., Doss, G., Baillie, T.A., Vyas, K.P., *Drug Metab. Dispos.* (2000) **28**, 633–642.

K41. King, C., Finley, B., Franklin, R., *Drug Metab. Dispos.* (2000) **28**, 661–663.

K42. Vashishtha, S.C., Hawes, E.M., McKay, G., McCann, D.J., *Drug Metab. Dispos.* (2000) **28**,1009–1013.

K43. Sohlenius-Sternbeck, A.-K., Von Euler Chelpin H., Orzechowski, A., Halldin, M.M., *Drug Metab. Dispos.* (2000) **28**, 695–700.

K44. Wynalda, M.A., Hauer, M.J., Wienkers, L.C., *Drug Metab. Dispos.* (2000) **28**, 1014–1017.

K45. Zhu, B.T., Patel, U.M., Cai, M.X., Conney, A.H., *Drug Metab. Dispos.* (2000) **28**, 1024–1030.

K46. Frame, L.T., Ozawa, S., Nowell, S.A., Chou, L.-C., Delongchamp, R.R., Doerge, D.R., Lang, N.P., Kadlubar, F.F., *Drug Metab. Dispos.* (2000) **28**, 1063–1068.

K47. Ngui, J.S., Tang, W., Stearns, R.A., Shou, M., Miller, R.R., Zhang, Y., Lin, J.H., Baillie, T.A., *Drug Metab. Dispos.* (2000) **28**, 1043–1050.

K48. Obach, R.S., *Drug Metab. Dispos.* (2000) **28**, 1069–1076.

K49. Rodriguez, R.J., Miranda, C.L., *Drug Metab. Dispos.* (2000) **28**, 1083–1086.

K50. Yeung, C.K., Lang, D.H., Thummel, K.E., Rettie, A.E., *Drug Metab. Dispos.* (2000) **28**, 1107–1111.

K51. Coleman, T., Spellman, E.F., Rostami-Hodjegan, R., Lennard, M.S., Tucker, G.T., *Drug Metab. Dispos.* (2000) **28**, 1094–1099.

K52. Niwa, T., Shiraga, T., Mitani, Y., Terakawa, M., Tokuma, Y., Kagayama, A., *Drug Metab. Dispos.* (2000) **28**, 1128–1134.

K53. Liu, Z.C., Uetrecht, J.P., *Drug Metab. Dispos.* (2000) **28**, 726–730.

K54. Abernethy, D.R., Wainer, I.W., Anacleto, A.I., *Drug Metab. Dispos.* (2000) **28**, 760–765.

K55. Cox, P.J., Ryan, D.A., Hollis, F.J., Harris, A.-M., Miller, A.K., Vousden, M., Cowley, H., *Drug Metab. Dispos.* (2000) **28**, 772–780.

K56. Dieckhaus, C.M., Miller, T.A., Sofia, R.D., Macdonald, T.L., *Drug Metab. Dispos.* (2000) **28**, 814–822.

K57. Desta, Z., Soukhova, N., Mahal, S.K., Flockhart, D.A., *Drug Metab. Dispos.* (2000) **28**, 789–800.

K58. Breyer-Pfaff, U., Mey, U., Green, M.D., Tephly, T.R., *Drug Metab. Dispos.* (2000) **28**, 869–872.

K59. Winter, H.R., Wang, Y., Unadkat, J.T., *Drug Metab. Dispos.* (2000) **28**, 865–868.

K60. Yan, Z., Nikelly, J.G., Killmer, L., Jr., Tarloff, J.B., *Drug Metab. Dispos.* (2000) **28**, 880–886.

K61. Umehara, K., Kudo, S., Hirao, Y., Morita, S., Uchida, M., Odomi, M., Miyamoto, G., *Drug Metab. Dispos.* (2000) **28**, 887–894.

K62. Frankmoelle, W.P., Medina, J.C., Shan, B., Narbut, M.R., Beckmann, H., *Drug Metab. Dispos.* (2000) **28**, 951–958.

K63. Cuttle, L., Munns, A.J., Hogg, N.A., Scott, J.R., Hooper, W.D., Dickinson,

R.G., Gillam, E.M., *Drug Metab. Dispos.* (2000) **28**, 945–950.

K64. Wang, J.-S., Backman, J.T., Taavitsainen, P., Neuvonen, P.J., Kivistö, K.T., *Drug Metab. Dispos.* (2000) **28**, 959–965.

K65. Äbelö, A., Andersson, T.B., Antonsson, M., Naudot, A.K., Skånberg, I., Weidolf, L., *Drug Metab. Dispos.* (2000) **28**, 966–972.

K66. Oldfield, M.F., Bennett, R.N., Kiddle, G., Wallsgrove, R.M., Botting, N.P., *Plant Physiol. Biochem.* (1999) **37**, 99–108.

K67. Dockens, R.C., Santone, K.S., Mitroka, J.G., Morrison, R.A., Jemal, M., Greene, D.S., Barbhaiya, R.H., *Drug Metab. Dispos.* (2000) **28**, 973–980.

K68. Palmer, D.R.J., Garrett, J.B., Sharma, V., Meganathan, R., Babbitt, P.C., Gerlt, J.A., *Biochemistry* (1999) **38**, 4252–4258.

K69. Galton, V.A., Martinez, E., Hernandez, A., St. Germain, E.A., Bates, J.M., St. Germain, D.L., *J. Clin. Invest.* (1999) **103**, 979–987.

K70. Donaghy, J.A., Kelly, P.F., McKay, A., *J. Sci. Food Agric.* (1999) **79**, 453–456.

K71. Zong, L., Inoue, M., Nose, M., Kojima, K., Sakaguchi, N., Isuzugawa, K., Takeda, T., Ogihara, Y., *Biol. Pharm. Bull.* (1999) **22**, 326–329.

K72. Lesage-Meessen, L., Stentelaire, C., Lomascolo, A., Sigoillot, J.-C., Asther, M., *J. Sci. Food Agric.* (1999) **79**, 487–490.

K73. Tin Leung, K., Campbell, S., Gan, Y., White, D.C., Lee, H., Trevors, J.T., *FEMS Microbiol. Lett.* (1999) **173**, 247–253.

K74. Carrillo, J.R., Monroy, O., *Av. Ing. Quim.* (1997) **7**, 201–206.

K75. Chikaoka, Y., Tamura, M., *J. Biochem.* (1999) **125**, 634–640.

K76. Bishop, L.P.D., Maggs, J.L., O'Neill, P.M., Park, B.K., *J. Pharmacol. Exp. Ther.* (1999) **289**, 511–520.

K77. De Voogt, P., Bleeker, E.A.J., van Vlaardingen, P.L.A., Fernández, A., Slobodnik, J., Wever, H., Kraak, M.H.S., *J. Chromatogr. B, Biomed. Sci. Appl.* (1999) **724**, 265–274.

K79. Chae, Y.-H., Thomas, T., Guengerich, F.P., Fu, P.P., El-Bayoumy, K., *Cancer Res.* (1999) **59**, 1473–1480.

K80. Katsileva, E., Wray, V., Pieper, D.H., Wittich, R.-M., *Appl. Environ. Microbiol.* (1999) **65**, 1405–1412.

K81. Liu, W.-H., Li, R.-M., Cheng, T.-L., Tang, T.-Y., *Food Sci. Agric. Chem.* (1999) **1**, 16–21.

K82. Duchene, P., Tran, G., Ladure, P., Olliver, R., Buzas, A., Merour, J.-Y., Houin, G., *Arzneim.-Forsch.* (1999) **49**, 504–508.

K83. Sadzuka, Y., Hirotsu, S., Hirota, S., *Jpn. J. Cancer Res.* (1999) **90**, 226–232.

K84. Odou, P., Barthélémy, C., Chatelier, D., Luyckx, M., Brunet, C., Dine, T., Gressier, B., Cazin, M., Cazin, J.C., Robert, H., *Eur. J. Drug Metab. Pharmacokinet.* (1999) **24**, 1–7.

K85. Esteban, A., Pérez-Mateo, M., *Eur. J. Drug Metab. Pharmacokinet.* (1999) **24**, 9–13.

K86. Virkel, G., Lifschitz, A., Pis, A., Lanusse, C., *Eur. J. Drug Metab. Pharmacokinet.* (1999) **24**, 31–37.

K87. Degterev, I.A., Nogueira, P.C.L., Marsaioli, A.J., Verces, A.E., *Eur. J. Drug Metab. Pharmacokinet.* (1999) **24**, 15–22.

K88. Augustijns, P., Peeters, L., Croes, K., Verbeeck, R.K., *Eur. J. Drug Metab. Pharmacokinet.* (1999) **24**, 105–108.

K89. Rios, M., Habecker, B., Sasaoka, T., Eisenhofer, G., Tian, H., Landis, S., Chikaraishi, D., Roffler-Tarlov, S., *J. Neurosci.* (1999) **19**, 3519–3526.

K90. Kester, M.H.A., Kaptein, E., Roest, T.J., Van Dijk, C.H., Tibboel, D., Meinl, W., Glatt, H., Coughtrie, M.W.H., Visser, T.J., *J. Clin. Endocrinol. Metab.* (1999) **84**, 1357–1364.

K91. Menhard, B., Zenk, M.H., *Phytochemistry* (1999) **50**, 763–774.

K92. Mitra, A., Kitamura, Y., Gasson, M.J., Narbad, A., Parr, A.J., Payne, J., Rhodes, M.J.C., Sewter, C., Walton, N.J., *Arch. Biochem. Biophys.* (1999) **365**, 10–15.

K93. Steiner, U., Schliemann, W., Böhm, H., Strack, D., *Planta* (1999) **208**, 114–124.

K94. Chao, Y.-P., Lai, Z.J., Chen, P., Chern, J.-T., *Biotechnol. Prog.* (1999) **15**, 453–458.

K95. Lehoux, I.E., Mitra, B., *Biochemistry* (1999) **38**, 5836–5848.

K96. Jiang, Y.-M., *Food Chem.* (1999) **66**, 75–79.

K97. Boyd, D.R., Sharma, N.D., Hempenstall, F., Kennedy, M.A., Malone, J.F., Allen, C.C.R., Resnick, S.M., Gibson, D.T., *J. Org. Chem.* (1999) **64**, 4005–4011.

K98. Vitali, A., Monache, G.D., Zappia, G., Misiti, D., Gacs-Baitz, E., Botta, B., *Heterocycles* (1999) **50**, 721–730.

K99. Kaneko, S., Ishiwata, K., Hatano, K., Omura, H., Ito, K., Senda, M., *Appl. Radiat. Isot.* (1999) **50**, 1025–1032.

K101. Zhang, Z., Flurkey, W.H., *J. Food Biochem.* (1999) **23**, 95–108.

K102. Myllynen, P., Pienimaki, P., Raunio, H., Vahakangas, K., *Hum. Exp. Toxicol.* (1998) **17**, 668–676.

K103. Fogelman, S.M., Schmider, J., Venkatakrishnan, K., Von Moltke, L.L., Harmatz, J.S., Shader, R.I., Greenblatt, D.J., *Neuropsychopharmacology* (1999) **20**, 480–490.

K104. Walker, K., Ketchum, R.E.B., Hezari, M., Gatfield, D., Goleniowski, M., Barthol, A., Croteau, R., *Arch. Biochem. Biophys.* (1999) **364**, 273–279.

K105. Baur, A., Köhrle, J., *Eur. J. Endocrinol.* (1999) **140**, 367–370.

K106. Motoda, S., *J. Biosci. Bioeng.* (1999) **87**, 137–143.

K107. Martin, G., Dijols, S., Capeillere-Blandia, C., Artaud, I., *Eur. J. Biochem.* (1999) **261**, 533–538.

K108. King, C.D., Rios, G.R., Assouline, J.A., Tephly, T.R., *Arch. Biochem. Biophys.* (1999) **365**, 156–162.

K109. Moran, G.R., Entsch, B., Palfrey, B.A., Ballou, D.P., *Biochemistry* (1999) **38**, 6292–6299.

K110. Homann, M.J., Morgan, W.B., *U.S.* US 5,919,672 (6–7–1999).

K111. Sugihara, K., Kitamura, S., Ohta, S., Tatsumi, K., *Xenobiotica* (2000) **30**, 457–467.

K112. Boogaard, P.J., De Kloe, K.P., Bierau, J., Kuiken, G., Borkulo, P.E.D., Van Sittert, N.J., *Xenobiotica* (2000) **30**, 485–502.

K113. Boogaard, P.J., Denneman, M.A., Van Sittert, N.J., *Xenobiotica* (2000) **30**, 469–483.

K114. Noguchi, K., Mera, A., Watanabe, T., Higuchi, S., Chiba, K., *Xenobiotica* (2000) **30**, 503–513.

K115. Cook, C.S., Lonien, M., Hribar, J., *Xenobiotica* (2000) **30**, 523–533.

K116. Hay Kraus, B.L., Greenblatt, D.J., Venkatakrishnan, K., Court, M.H., *Xenobiotica* (2000) **30**, 575–588.

K117. Mae, T., Inaba, T., Konishi, E., Hosoe, K., Hidaka, T., *Xenobiotica* (2000) **30**, 565–574.

K118. De Santi, C., Pietrabissa, A., Spisni, R., Mosca, F., Pacifici, G.M., *Xenobiotica* (2000) **30**, 609–617.

K119. Beaumont, K., Harper, A., Smith, D.A., Abel, S., *Xenobiotica* (2000) **30**, 627–642.

K120. Kawai, K., Hirota, T., Muramatsu, S., Tsuruta, F., Ikeda, T., Kobashi, K., Nakamura, K.-I., *Xenobiotica* (2000) **30**, 707–715.

K121. Kimura, T., Miki, M., Watanabe, K., Kondo, S., Ho. I.K., Yamamoto, I., *Xenobiotica* (2000) **30**, 643–653.

K122. Scarfe, G.B., Lindon, J.C., Nicholson, J.K., Martin, P., Wright, B., Taylor, S., Lenz, E., Wilson, I.D., *Xenobiotica* (2000) **30**, 717–729.

K123. Paulson, S.K., Zhang, J.Y., Jessen, S.M., Lawal, Y., Liu, N.W.K., Dudkowski,

C.M., Wang, Y.-F., Chang, M., Yang, D., Findlay, J.W.A., Berge, M.A., Markos, C.S., Breau, A.P., Hribar, J.D., Yuan, J., *Xenobiotica* (2000) **30**, 731–744.

K124. De Santi, C., Pietrabissa, A., Spisni, R., Mosca, F., Pacifici, G.M., *Xenobiotica* (2000) **30**, 857–866.

K125. Gilissen, R.A.H.J., Ferrari, L., Barnaby, R.J., Kajbaf, M., *Xenobiotica* (2000) **30**, 843–856.

K126. Pedersen, R.T., Hill, E.M., *Xenobiotica* (2000) **30**, 867–879.

K127. Hakk, H., Larsen, G., Bergman, Å, Örn, U., *Xenobiotica* (2000) **30**, 881–890.

K128. Tuvesson, H., Wienkers, L.C., Gunnarsson, P.O., Seidegård, J., Persson, R., *Xenobiotica* (2000) **30**, 905–914.

K129. Sohlenius-Sternbeck, A.-K., Floby, E., Svedling, M., Orzechowski, A., *Xenobiotica* (2000) **30**, 891–903.

K130. Störmer, E., Von Moltke, L.L., Shader, R.I., Greenblatt, D.J., *Drug Metab. Dispos.* (2000) **28**, 1168–1175.

K131. Ramu, K., Lam, G.N., Hughes, H., *Drug Metab. Dispos.* (2000) **28**, 1153–1161.

K132. Hesse, L.M., Venkatakrishnan, K., Court, M.H., Von Moltke, L.L., Duan, S.X., Shader, R.I., Greenblatt, D.J., *Drug Metab. Dispos.* (2000) **28**, 1176–1183.

K133. Margolis, J.M., O'Donnell, J.P., Mankowski, D.C., Ekins, S., Obach, R.S., *Drug Metab. Dispos.* (2000) **28**, 1187–1191.

K134. Czernik, P.J., Little, J.M., Barone, G.W., Raufman, J.-P., Radominska-Pandya, A., *Drug Metab. Dispos.* (2000) **28**, 1210–1216.

K135. Kostrubsky, V.E., Sinclair, J.F., Ramachandran, V., Venkataramanan, R., Wen, Y.H., Kindt, E., Galchev, V., Rose, K., Sinz, M., Strom, S.C., *Drug Metab. Dispos.* (2000) **28**, 1192–1197.

K136. Faucette, S.R., Hawke, R.L., Lecluyse, E.L., Shord, S.S., Yan, B., Laethem, R.M., *Drug Metab. Dispos.* (2000) **28**, 1222–1230.

K137. Mushiroda, T., Douya, R., Takahara, E., Nagata, O., *Drug Metab. Dispos.* (2000) **28**, 1231–1237.

K138. Reddy, G.V.B., Gold, M.H., *Biochem. Biophys. Res. Commun.* (1999) **257**, 901–905 and (1999) **259**, 232.

K139. Halpin, R.A., Geer, L.A., Zhang, K.E., Marks, T.M., Dean, D.C., Jones, A.N., Melillo, D., Doss, G., Vyas, K.P., *Drug Metab. Dispos.* (2000) **28**, 1244–1254.

K140. Kahn, V., Ben-Shalom, N., Zakin, V., *J. Food Biochem.* (1999) **23**, 75–94.

K141. Kobayashi, K., Katayama-Hirayama, K., Tobita, S., *Kankyo Kagaku* (1999) **9**, 23–28.

K142. Schofield, P.C., Robertson, I.G.C., Paxton, J.W., McCrystal, M.R., Evans, B.D., Kestell, P., Baguley, B.C., *Cancer Chemother. Pharmacol.* (1999) **44**, 51–58.

K143. Burd, V.N., Van Pee, K.-H., *Bioorg. Khim.* (1998) **24**, 462–464.

K144. Anderson, J.J., Shalaby, L.M., Berg, D.S., *J. Agric. Food Chem.* (1999) **47**, 2439–2446.

K146. Ajay Kumar, R., Gunasekaran, P., Lakshmanan, M., *J. Basic Microbiol.* (1999) **39**, 161–168.

K147. Breuil, A.-C., Jeandet, P., Adrian, M., Chopin, F., Pirio, N., Meunier, P., Bessis, R., *Phytopathology* (1999) **89**, 298–302.

K148. Benedito, M.A.C., *Braz. J. Med. Biol. Res.* (1999) **32**, 275–278.

K149. Oikawa, A., Ebisui, K., Sue, M., Ishihara, A., Iwamura, H., *Z. Naturforsch. C: J. Biosci.* (1999) **54**, 181–185.

K150. Qiu, F., Chen, Y., Kano, Y., Yao, X., *Yaoxue Xuebao* (1999) **34**, 222–226.

K152. Tomigahara, Y., Onogi, M., Kaneko, H., Nakatsuka, I., Yamane, S., *J. Agric. Food Chem.* (1999) **47**, 2429–2438.

K153. Flodin, C., *Phytochemistry* (1999) **51**, 249–255.

K154. Dhawan, S., Hensley, P., Osborne, J.C., Jr., Fleming, P.J., *J. Biol. Chem.* (1986) **261**, 7680–7684.

K155. Esaki, H., Watanabe, R., Onozaki, H., Kawakishi, S., Osawa, T., *Biosci. Biotechnol. Biochem.* (1999) **63**, 851–858.

K156. Klinman, J.P., Krueger, M., Brenner, M., Edmondson, D.E., *J. Biol. Chem.* (1984) **259**, 3399–3402.

K157. Saxena, A., Hensley, P., Osborne, J.C., Jr., Fleming, P.J., *J. Biol. Chem.* (1985) **260**, 3386–3392.

K158. Faulds, C.B., Williamson, G., *J. Sci. Food Agric.* (1999) **79**, 450–452.

K160. Ziegler, D.M., Mitchell, C.H., *Arch. Biochem. Biophys.* (1972) **150**, 116–125.

K161. Emoto, C., Yamazaki, H., Yamasaki, S., Shimada, N., Nakajima, M., Yokoi, T., *Xenobiotica* (2000) **30**, 943–953.

K162. Shimada, T., Gillam, E.M.J., Oda, Y., Tsumura, F., Sutter, T.R., Guengerich, F.P., Inoue, K., *Chem. Res. Toxicol.* (1999) **12**, 623–629.

K163. Jeppesen, U., Eap, C.B., Powell, K., Baumann, P., Brosen, K., *Pharmacogenetics* (1999) **9**, 55–60.

K164. Furukawa, H., Wieser, M., Morita, H., Sugio, T., Nagasawa, T., *J. Biosci. Bioeng.* (1999) **87**, 285–290.

K165. Minoletti, C., Dijols, S., Dansette, P.M., Mansuy, D., *Biochemistry* (1999) **38**, 7828–7836.

K166. Katayama, Y., Nishikawa, S., Hotta, Y., *Jpn. Kokai Tokkyo Koho* JP 11 225,787 (24–8–1999).

K167. DiGennaro, P., Colmegna, A., Galli, E., Sello, G., Pelizzoni, F., Bestetti, G., *Appl. Environ. Microbiol.* (1999) **65**, 2794–2797.

K168. Narimatsu, S., Yamamoto, S., Kato, R., Masubuchi, Y., Horie, T., *Biol. Pharm. Bull.* (1999) **22**, 567–571.

K169. DePietro, F.R., Fernstrom, J.D., *Brain Res.* (1999) **831**, 72–84.

K170. Tong, J., Chambers, J.E., *J. Biochem. Mol. Toxicol.* (1999) **13**, 261–268.

K171. Hillenweck, A., Corpet, D.E., Killeen, J.C., Jr., Bliss, M., J., Cravedi, J.-P., *J. Agric. Food Chem.* (1999) **47**, 2898–2903.

K172. Hoffmann, M.J., Ji, S., Hedli, C.C., Snyder, R., *Toxicol. Sci.* (1999) **49**, 40–47.

K173. Rhee, S.-K., Fuchs, G., *Eur. J. Biochem.* (1999) **262**, 507–515.

K174. Chung, J.G., *Microbios* (1999) **98**, 15–25.

K175. Paniego, N.B., Zuurbier, K.W.M., Fung, S.-Y., van der Heiden, R., Scheffer, J.J.C., Verpoorte, R., *Eur. J. Biochem.* (1999) **262**, 612–616.

K176. Hook, I.L., Ryan, S., Sheridan, H., *Phytichemistry* (1999) **51**, 621–627.

K177. Wieser, M, Furukawa, H., Morita, H., Yoshida, T., Nagasawa, T., *Tetrahedron: Asymmetry* (1999) **10**, 1627–1630.

K178. Li, H., Luo, S., Zhou, T., *Phytother. Res.* (1999) **13**, 236–238.

K179. Zhang, J., Burnell, J.C., Dumaual, N., Bosron, W.F., *J. Pharmacol. Exp. Ther.* (1999) **290**, 314–318.

K180. Ciotti, M., Basu, N., Brangi, M., Owens, I.S., *Biochem. Biophys. Res. Commun.* (1999) **260**, 199–202.

K181. Alkatheeri, N.A., Wasfi, I.A., Lambert, M., *J. Vet. Pharmacol. Ther.* (1999) **22**, 127–135.

K182. Duchêne, P., Laneury, J.P., Tran, G., Ladure, P., Olivier, R., Buzas, A., Merour, J.-Y., Houin, G., *Arzneim.-Forsch.* (1999) **49**, 699–704.

K183. Wu, W.-N., McKown, L.A., Yorgey, K.A., Pritchard, J.F., *J. Pharm. Biomed. Anal.* (1999) **20**, 687–695.

K184. Dhillon, J.K., Shivaraman, N., *Can. J. Microbiol.* (1999) **45**, 201–208.

K185. Castaneda-Acosta, J., Bounds, P.L., Winston, G.W., *J. Biochem. Mol. Toxicol.* (1999) **13**, 279–286.

K187. Barriault, D., Sylvestre, M., *Appl. Microbiol. Biotechnol.* (1999) **51**, 592–597.

K189. Palatini, P., *Biochem. Int.* (1985) **11**, 565–572.

K190. Emtiazi, G., Habibi, M.H., Mazaherian, M., *Proc. World Pet. Congr. 15th* (1998) 939–941.

K191. Chang, K.-H., Liang, P.-H., Beck, W., Scholten, J.D., Dunaway-Mariano, D., *Biochemistry* (1992) **31**, 5605–5610.

K192. Qiu, F., Tan. Q.S., Chen, Y.J., Kano, Y., Yao, X.S., *Chin. Chem. Lett.* (1998) **9**, 549–552.

K193. Fischer-Colbie, R., Schachinger, M., Zangerle, R., Winkler, H., *J. Neurochem.* (1982) **38**, 725–732.

K194. Rodrigo, L., Gil, F., Hernandez, A.F., Pla, A., *Chem.-Biol. Interact.* (1999) **119–120**, 263–275.

K195. Hasegawa, M., Matsubara, K., Fukushima, S., Maseda, C., Uezono, T., Kimura, K., *Forensic Sci. Int.* (1999) **101**, 95–106.

K196. Barriault, D., Vedadi, M., Powlowski, J., Sylvestre, M., *Biochem. Biophys. Res. Commun.* (1999) **260**, 181–187.

K197. Reddy, J., Lee, C., Neeper, M., Greasham, R., Zhang, J., *Appl. Microbiol. Biotechnol.* (1999) **51**, 614–620.

K198. Yao, Q., Wang, M., *Yaoxue Xuebao* (1999) **34**, 303–307.

K199. Albani, D., Parisi, B., Carbonera, D., Cella, R., *Plant Mol. Biol.* (1985) **5**, 363–372.

K200. Haraguchi, K., Hirose, Y., Masuda, Y., Kato, Y., Kimura, R., *Fukuoka Igaku Zasshi* (1999) **90**, 210–219.

K201. Kamali, F., Ball, D.E., McLauchlin, W.S., Seymour, R.A., *J. Periodontal Res.* (1999) **34**, 145–153.

K202. Hawari, J., Halasz, A., Beaudet, S., Paquet, L., Ampleman, G., Thiboutot, S., *Appl. Environ. Microbiol.* (1999) **65**, 2977–2986.

K203. Miller, E.S., Jr., Peretti, S.W., *Green Chem.* (1999) **1**, 143–152.

K204. Ha, Y.M., Jung, Y.H., Kwon, D.Y., Kim, Y.-C., Kim, Y., Kim, C.-K., Min, K.H., *J. Microbiol. Biotechnol.* (1999) **9**, 249–254.

K205. Kim, D.-H., Kim, S.-Y., Park, S.-Y., Han, M.J., *Biol. Pharm. Bull.* (1999) **22**, 749–751.

K206. Reich, T., Schmidt, S., Fortnagel, P., *FEMS Microbiol. Lett.* (1999) **176**, 477–482.

K207. Bowers, N.I., Boyd, D.R., Sharma, N.D., Goodrich, P.A., Groocock, M.R., Blacker, A.J., Goode, P., Dalton, H., *J. Chem. Soc., Perkin Trans. 1* (1999) 1453–1462.

K208. Chen, G.-W., Hung, C.-F., Chang, S.-H., Lin, J.-G., Chung, J.-G., *Microbios* (1999) **98**, 159–174.

K209. Yasuda, T., Kon, R., Nakazawa, T., Ohsawa, K., *J. Nat. Prod.* (1999) **62**, 1142–1144.

K210. Londry, K.L., Suflita, J.M., Tanner, R.S., *Can. J. Microbiol.* (1999) **45**, 458–463.

K211. Kvasnickova, E., Szotakova, B., Wsol, V., Trejtnar, F., Skalova, L., Hais, I.M., Kuchar, M., Poppova, M., *Exp. Toxicol. Pathol.* (1999) **51**, 352–356.

K212. Bort, R., Macé, K., Boobis, A., Gómez-Lechón, M.-J., Pfeifer, A., Castell, J., *Biochem. Pharmacol.* (1999) **58**, 787–796.

K213. Qiu F., Chen, Y.J., Kano, Y., Yao, X.S., *Chin. Chem. Lett.* (1998) **9**, 393–395.

K214. Hund, H.-K., Breuer, J., Lingens, F., Hüttermann, J., Kappl, R., Fetzner, S., *Eur. J. Biochem.* (1999) **263**, 871–878.

K215. Hartmann, S., Hultschig, C., Eisenreich, W., Fuchs, G., Bacher, A., Ghisla, S., *Proc. Natl. Acad. Sci. U.S.A.* (1999) **96**, 7831–7836.

K216. Nielsen, J.S., Moller, B.L., *Arch. Biochem. Biophys.* (1999) **368**, 121–130.

K217. Osakabe, K., Tsao, C.C., Li, L., Popko, J.L., Umezawa, T., Carraway, D.T., Smeltzer, R.H., Joshi, C.P., Chiang, V.L., *Proc. Natl. Acad. Sci. U.S.A.* (1999) **96**, 8955–8960.

K219. Boyle, A.W., Higgblom, M.M., Young, L.Y., *FEMS Microbiol. Ecol.* (1999) **29**, 379–387.

K220. Vessey, D.A., Kelley, M., Warren, R.S., *Biochim. Biophys. Acta* (1999) **1428**, 455–462.

K221. Riegert, U., Heiss, G., Kuhm, A.E., Muller, C., Contzen, M., Knackmuss, H.-J., Stolz, A., *J. Bacteriol.* (1999) **181**, 4812–4817.

K222. Van den Ban, E.C.D., Willemen, H.M., Wassink, H., Laane, C., Haaker, H., *Enzyme Microb. Technol.* (1999) **25**, 251–257.

K224. Murphy, A., Taiz, L., *Plant Physiol. Biochem. (Paris)* (1999) **37**, 431–443.

K225. Schmitt, B., Schneider, B., *Phytochemistry* (1999) **52**, 45–53.

K226. Qiu, F., Gu, M.Z., Tan, Q.S., Chen, Y.J., Kano, Y., Yao, X.S., *Chin. Chem. Lett.* (1998) **9**, 957–959.

K227. Schlitt, S.C., Lester, G., Russell, P.J., *J. Bacteriol.* (1971) **117**, 1117–1120.

K228. Milo, R.E., Duffner, F.M., Muller, R., *Extremophiles* (1999) **3**, 185–190.

K229. Kim, K., Suk, H., *J. Biochem. Mol. Biol.* (1999) **32**, 321–325.

K230. Xu, Y., Mitra, B., *Biochemistry* (1999) **38**, 12367–12376.

K231. Costes, D., Wehtje, E., Adlercreutz, P., *Enzyme Microbiol. Technol.* (1999) **25**, 384–391.

K232. Fukuyama, Y., Matoishi, K., Iwasaki, M., Takizawa, E., Miyazaki, M., Mamoru, M., Ohta, H., Hanzawa, S., Kakidani, H., Sugai, T., *Biosci. Biotechnol. Biochem.* (1999) **63**, 1664–1666.

K233. Cyr, L.V., Newton, M.G., Phillips, R.S., *Bioorg. Med. Chem.* (1999) **7**, 1497–1503.

K234. Baden, K.L., Valtier, S., Cody, J.T., *J. Anal. Toxicol.* (1999) **23**, 511–517.

K235. Lai, W.G., Zahid, N., Uetrecht, J.P., *J. Pharmacol. Exp. Ther.* (1999) **291**, 292–299.

K236. Thibaut, R., Debrauwer, L., Rao, D., Cravedi, J.P., *Sci. Total Environ.* (1999) **233**, 193–200.

K237. Yabe, K., Nakamura, M., Hamasaki, T., *Appl. Environ. Microbiol.* (1999) **65**, 3867–3872.

K238. Beller, H.R., Spormann, A.M., *FEMS Microbiol. Lett.* (1999) **178**, 147–153.

K239. Ross, J.R., Nam, K.H., D'Auria, J.C., Pichersky, E., *Arch. Biochem. Biophys.* (1999) **367**, 9–16.

K240. Nam, K.L., Dudareva, N., Pichersky, E., *Plant Cell Physiol.* (1999) **40**, 916–923.

K241. Johnson, H.A., Spormann, A.M., *J. Bacteriol.* (1999) **181**, 5662–5668.

K242. Finnson, K.W., McLeese, J.M., Eales, J.G., *Gen. Comp. Endocrinol.* (1999) **115**, 387–397.

K243. Karlsson, A., Ejlertsson, J., Nezirevic, D., Svensson, B.H., *Anaerobe* (1999) **5**, 25–35.

K244. Takemoto, M., Achiwa, K., *Tetrahedron Lett.* (1999) **40**, 6595–6598.

K246. Larkin, M.J., Allen, C.C.R., Kulakov, L.A., Lipscomb, D.A., *J. Bacteriol.* (1999) **181**, 6200–6204.

K247. Van de Pas, B.A., Smidt, H., Hagen, W.R., Van der Oost, J., Schraa, G., Stams, A.J.M., De Vos, W.M., *J. Biol. Chem.* (1999) **274**, 20287–20292.

K248. Wang, J., Liu, P., Yi, Q., *Process Biochem.* (1999) **34**, 745–749.

K249. Gengan, R.M., Chuturgoon, A.A., Mulholland, D.A., Dutton, M.F., *Mycopathologia* (1999) **144**, 15–22.

K250. Nakamura, Y., Teramoto, S., Yoshitama, K., *Z. Naturforsch C: J. Biosci.* (1999) **54**, 501–507.

K251. Tanabe, J., Sue, M., Ishihara, A., Iwamura, H., Biosci., Biotechnol., *Biochem.* (1999) **63**, 1614–1617.

K252. Roos, J., Effenberger, F., *Tetrahedron: Asymmetry* (1999) **10**, 2817–2828.

K253. Zhang, X., Young, L.Y., *Appl. Environ. Microbiol.* (1997) **63**, 4759–4764; *Appl. Environ. Microbiol.* (1999) **65**, 2279.

K254. Wang, J.-S., Wen, X., Backman, J.T., Taavitsainen, P., Neuvonen, P.J., Kivistö, K.T., *Pharmacol. Toxicol. (Copenhagen)* (1999) **85**, 157–161.

K255. Bun, H., Disdier, B., Aubert, C., Catalin, J., *Fundam. Clin. Pharmacol.* (1999) **13**, 577–581.

K256. Anderson, C., Brumhard, B., Ditgens, K., Reiner, H., *Pflanzenschutz-Nachr. Bayer* (1999) **52**, 233–258.

K257. Ozawa, S., Shimizu, M., Katoh, T., Miyajima, A., Ohno, Y., Matsumoto, Y., Fukuoka, M., Tang, Y.-M., Lang, N.P., Kadlubar, F.F., *J. Biochem.* (1999) **126**, 271–277.

K258. Lovell, D.P., Van Iersel, M., Walters, D.G., Price, R.G., Lake, B.G., *Pharmacogenetics* (1999) **9**, 239–250.

K259. Gunmalla, S., Broadbent, J.R., *J. Dairy Sci.* (1999) **82**, 2070–2077.

K260. Stroomberg, G.J., De Knecht, J.A., Ariese, F., Van Gestel, C.A.M., Velthorst, N.H., *Environ. Toxicol. Chem.* (1999) **18**, 2217–2224.

K261. Nativelle, C., Picard, K., Valentin, I., Lhuguenot, J.C., Chagnon, M.C., *Food Chem. Toxicol.* (1999) **37**, 905–917.

K263. Sapota, A., Szymańska, J.A., Czerski, B., Piotrowski, J.K., *Chemosphere* (1999) **39**, 2229–2238.

K264. LeCorre, P., Dollo, G., Chevanne, F., LeVerge, R., *Eur. J. Pharm. Sci.* (1999) **9**, 79–84.

K265. Koga, N., Kanamaru, T., Oishi, N., Kuroki, H., Haraguchi, K., Masuda, Y., Yoshimura, H., *Organohalogen Compd.* (1999) **42**, 181–184.

K266. Liu, B.-H., Bhatnagar, D., Chiu, F.S., *Nat. Toxins* (1999) **7**, 63–69.

K267. Adams, J.D., Jr., Sayer, J.M., Chadha, A., Shirai, N., Lehr, R.E., Kumar, S., Jerina, D.M., *Chem.-Biol. Interact.* (1999) **122**, 117–135.

K268. Molinari, F., Gandolfi, F., Aragozzini, F., Leon, R., Prazeres, D.M.F., *Enzyme Microb. Technol.* (1999) **25**, 729–735.

K269. Genthner, B.R.S., *Biodegradation* (1999) **10**, 27–33.

K270. Pal, A., Seidel, A., Xia, H., Hu, X., Srivastava, K., Oesch, F., Singh, S.V., *Carcinogenesis* (1999) **20**, 1997–2001.

K271. Krieger, C.J., Beller, H.R., Reihard, M., Spormann, A.M., *J. Bacteriol.* (1999) **181**, 6403–6410.

K272. Soucek, P., *J. Chromatogr, B: Biomed. Sci. Appl.* (1999) **734**, 23–29.

K273. Schaller, M., Schaffhauser, M., Sans, N., Wermuth, B., *Eur. J. Biochem* (1999) **265**, 1056–1060.

K274. Johnson, W.H., Jr., Czerwinski, R.M., Stamps, S.L., Whitman, C.P., *Biochemistry* (1999) **38**, 16024–16033.

K275. Ro, H.-S., Miles, E.W., *J. Biol. Chem.* (1999) **274**, 31189–31194.

K276. Guitton, C., Kinowski, J.-M., Abbar, M., Chabrand, P., Bressolle, F., *J. Clin. Pharmacol.* (1999) **39**, 721–728.

K277. Moreno-Arribas, V., Lonvaud-Funel, A., *FEMS. Microbiol. Lett.* (1999) **180**, 55–60.

K278. Manohar, S., Kim, C.-K., Karegoudar, T.B., *J. Microbiol. (Seoul)* (1999) **37**, 73–79.

K279. Andhare, J., Shastri, N.V., *Res. Bull. Panjab Univ., Sci.* (1997) **47**, 123–134.

K281. Centeno, M.L., Fernandez, B., Feito, I., Rodriguez, A., *J. Plant Growth Regul.* (1999) **18**, 81–88.

K282. Kleerebezem, R., Pol, L.W.H., Lettinga, G., *Biodegradation* (1999) **10**, 63–73.

K283. Fujimoto, H., Matsuo, I., Isomura, M., Miyasoto, M., Ajisaka, K., *J. Appl. Glycosci.* (1999) **46**, 135–141.

K284. Balashova, N.V., Kosheleva, I.A., Golovchenko, N.P., Boronin, A.M., *Process Biochem.* (1999) **35**, 291–296.

K285. Shao, L., Kumar, G., Lenhart, J.L., Smith, P.J., Payne, G.F., *Enzyme Microb. Technol.* (1999) **25**, 660–668.

K286. Srivastava, S.K., Hu, X., Xia, H., Awasthi, S., Amin, S., Singh, S.V., *Arch. Biochem. Biophys.* (1999) **371**, 340–344.

K287. Gribaldo, L., Catalini, P., Erminio, M., *Drug Metab. Drug Interact.* (1999) **15**, 279–291.

K288. Platzek T., Lang, C., Grohmann, G., Gi, U.S., Baltes, W., *Hum. Exp. Toxicol.* (1999) **18**, 552–559.

K289. Engst, S., Kuusk, V., Efimov, I., Cronin, C.N., McIntire, W.S., *Biochemistry* (1999) **38**, 16620–16628.
K290. Akashi, T., Fukuchi-Mizutani, M., Aoki, T., Ueyama, Y., Yonekura-Sakakibara, K., Tanaka, Y., Kusumi, T., Ayabe, S.-I., *Plant Cell Physiol.* (1999) **40**, 1182–1186.
K291. Van Laere, K.M.J., Hartemink, R., Beldman, G., Pitson, S., Dijkema, C., Schols, H.A., Voragen, A.G.J., *Appl. Microbiol. Biotechnol.* (1999) **52**, 681–688.
K292. Muller, J.A., Galushko, A.S., Kappler, A., Schink, B., *Arch. Microbiol.* (1999) **172**, 287–294.
K293. Mampel, J., Ruff, J., Junker, F., Cook, A.M., *Microbiology* (1999) **145**, 3255–3264.
K294. Kleingeist, B., Bocker, R., Geisslinger, G., Brugger, R., *J. Pharm. Pharm. Sci.* (1998) **1**, 38–46.
K295. Rao, S.R., Ravishankar, G.A., *Process Biochem. (Oxford)* (1999) **35**, 341–348.
K296. Kucukguzel, I, Ulgen, M., *J. Fac. Pharm. Istanbul Univ.* (1998) **32**, 69–86.
K297. Gulerman, N., Rollas, S., Ulgen, M., *Boll. Chim. Farm.* (1998) **137**, 140–143.
K298. Jorga, K., Foteler, B., Heizmann, P., Gasser, R., *Br. J. Clin. Pharmacol.* (1999) **48**, 513–520.
K299. Janus, K., Antoszek, J., *J. Vet. Pharmacol. Ther.* (1999) **22**, 163–169.
K300. Wanwimolruk, S., Zhang, H., Coville, P.F., Saville, D.J., Davis, L.S., *Comp. Biochem. Physiol., Part C: Pharmacol. Toxicol. Endocrinol.* (1999) **124**, 301–307.
K301. Olesen, O.V., Linnet, K., *Pharmacology* (1999) **59**, 298–309.
K302. Leutwein, C., Heider, J., *Microbiology* (1999) **145**, 3265–3271.
K303. Durren, R.L., McIntosh, C.A., *Phytochemistry* (1999) **52**, 793–798.
K304. Lu, J., Nakajima-Kambe, T., Shigeno, T., Ohbo, A., Nomura, N., Nakahara, T., *J. Biosci. Bioeng.* (1999) **88**, 293–299.
K305. Buckland, B.C., Drew, S.W., Connors, N.C., Chartrain, M.M., Lee, C., Salmon, P.M., Gbewonyo, K., Zhou, W., Gailliot, P., Singhvi, R., Olewinski, R.C., Jr., Sun, W.-J., Reddy, J., Zhang, J., Jackey, B.A., Taylor, C., Goklen, K.E., Junker, B., Greasham, R.L., *Metab. Eng.* (1999) **1**, 63–74.
K307. Yamamoto, I., *Yakubutsu Dotai* (1999) **14**, 111–119.
K308. Akashi, T., Aoki, T., Ayabe, S.-I., *Plant Physiol.* (1999) **121**, 821–828.
K309. Lambert, N., Kroon, P.A., Faulds, C.B., Plumb, G.W., McLauchlan, W.R., Day, A.J., Williamson, G., *Biochim. Biophys. Acta* (1999) **1435**, 110–116.
K310. Strauss, U.T., Faber, K., *Tetrahedron: Asymmetry* (1999) **10**, 4079–4081.
K311. Lister, D.L., Kanugo, G., Rathbone, D.A., Bruce, N.C., *FEMS Microbiol. Lett.* (1999) **181**, 137–144.
K312. Šepič, E., Leskovšek, H., *Analyst (Cambridge, UK)* (1999) **124**, 1765–1769.
K313. Jones, P.R., Moller, B.L., Hoj, P.B., *J. Biol. Chem.* (1999) **274**, 35483–35491.
K314. Mazumder, R., Logan, J.R., Mikell, A.T., Jr., Hooper, S.W., *J. Ind. Microb. Biotechnol.* (1999) **23**, 476–483.
K315. Ohshiro, T., Kojima, T., Torii, K., Kawasoe, H., Izumi, Y., *J. Biosci. Bioeng.* (1999) **88**, 610–616.
K316. Krupenko, S.A., Wagner, C., *J. Biol. Chem.* (1999) **274**, 35777–35784.
K317. Chosungnoen, N., Wiyakrutta, S., Intararuangsorn, S., Fujiyama, K., Seki, T., Meevootisom, V., *Biotechnol. Sustainable Util. Biol. Resour. Trop.* (1998) **12**, 300–306.
K318. Allende, J.L., Gibello, A., Fortun, A., Mengs, G., Ferrer, E., Martin, M., *Curr. Microbiol.* (2000) **40**, 34–39.
K319. Maxwell, S., Yu, J., *Process. Biochem.* (2000) **35**, 551–556.
K320. Roda, G., Riva, S., Daniels, B., *Tetrahedron: Asymmetry* (1999) **10**, 3239–3249.

K321. Balani, S.K., Xu, X., Arison, B.H., Silva, M.V., Gries, A., DeLuna, F.A., Cui, D., Kari, P.H., Ly, T., Hop, C.E.C., Singh, R., Wallace, M.A., Dean, D.C., Lin, J.H., Pearson, P.G., Baillie, T.A., *Drug Metab. Dispos.* (2000) **28**, 1274–1278.

K322. Matsunaga, T., Kishi, N., Higuchi, S., Watanabe, K., Ohshima, T., Yamamoto, I., *Drug Metab. Dispos.* (2000) **28**, 1291–1296.

K323. Thijssen, H.H., Flinois, J.-P., Beaune, P.H., *Drug Metab. Dispos.* (2000) **28**, 1284–1290.

K324. Ohyama, K., Nakajima, M., Nakamura, S., Shimada, N., Yamazaki, H., Yokoi, T., *Drug Metab. Dispos.* (2000) **28**, 1303–1310.

K325. Donato, M.T., Vitala, P., Rodriguez-Antonia, C., Lindfors, A., Castell, J.V., Raunio, H., Gómez-Lechón, M.J., Pelkonen, O., *Drug Metab. Dispos.* (2000) **28**, 1321–1326.

K326. Jacobsen, W., Christians, U., Benet, L.Z., *Drug Metab. Dispos.* (2000) **28**, 1343–1351.

K327. Billeke, S., Draganov, D., Counsell, R., Stetson, P., Watson, C., Hsu, C., LaDu, B.N., *Drug Metab. Dispos.* (2000) **28**, 1335–1342.

K328. Komatsu, T., Yamazaki, H., Asahi, S., Gillam, E.M., Guengerich, F.P., Nakajima, M., Yokoi, T., *Drug Metab. Dispos.* (2000) **28**, 1361–1368.

K329. Jacobsen, W., Kuhn, B., Soldner, A., Kirchner, G., Sewing, K.-F., Kollman P.A., Benet, L.Z., Christians, U., *Drug Metab. Dispos.* (2000) **28**, 1369–1378.

K330. Dong, H., Haining, R.L., Thummel, K.E., Rettie, A.E., Nelson, S.D., *Drug Metab. Dispos.* (2000) **28**, 1397–1400.

K331. Lautala, P., Ethell, B.T., Taskinen, J., Burchell, B., *Drug Metab. Dispos.* (2000) **28**, 1385–1389.

K332. Reist, M., Roy-de Vos, M., Montseny, J.-P., Mayer, J.M., Carrupt, P.-A., Berger, Y., Testa, B., *Drug Metab. Dispos.* (2000) **28**, 1405–1410.

K333. Umehara, K., Kudo, S., Hirao, Y., Morita, S., Ohtani, T., Uchida, M., Miyamoto, G., *Drug Metab. Dispos.* (2000) **28**, 1417–1424.

K334. Yamada, S., Okura, T., Kimura, R., Deguchi, Y., Suzuki, Y., Kobayashi, T., Aisaka, K., *Drug Metab. Dispos.* (2000) **28**, 1487–1492.

K335. Zhou, S., Paxton, J.W., Tingle, M.D., Kestell, P., *Drug Metab. Dispos.* (2000) **28**, 1449–1456.

K336. Antonian, L., Zhang, H., Yang, C., Wagner, G., Shawver, L.K., Shet, M., Ogilvie, B., Madan, A., Parkinson, A., *Drug Metab. Dispos.* (2000) **28**, 1505–1512.

K337. Křen, V., Ulrichová, J., Kosina, P., Stevenson, D., Sedmera, P., Přikrylová, V., Halada, P., Šimánek, V., *Drug Metab. Dispos.* (2000) **28**, 1513–1517.

K338. Trepanier, L.A., Miller, J.L., *Xenobiotica* (2000) **30**, 1111–1121.

K339. Venhorst, J., Onderwater, R.C.A., Meerman, J.H.N., Commandeur, J.N.M., Vermeulen, N.P.E., *Drug Metab. Dispos.* (2000) **28**, 1524–1532.

K340. Larsen, G.L., Feil, V.J., Bakke, J.E., *Xenobiotica* (2000) **30**, 1153–1157.

K341. Breyer-Pfaff, U., Nill, K., *Biochem. Pharmacol.* (2000) **59**, 249–260.

K342. Brzakovic, B., Pokrajac, M., Dzoljic, E., Levic, Z., Varagic, V.M., *Eur. J. Drug Metab. Pharmacokinet.* (1999) **24**, 233–236.

K343. Inselmann, G., Volkmann, A., Heidelmann, H.T., *Antimicrob. Agents Chemother.* (2000) **44**, 131–133.

K344. Suiko, M., Sakakibara, Y., Liu, M.-C., *Biochem. Biophys. Res. Commun.* (2000) **267**, 80–84.

K345. Powley, M.W., Carlson, G.P., *Toxicology* (1999) **139**, 207–217.

K346. Arend, J., Warzecha, H., Stöckigt, J., *Phytochemistry* (2000) **53**, 187–193.

K347. Marcinek, H., Weyler, W., Deus-Neumann, B., Zenk, M.H., *Phytochemistry* (2000) **53**, 201–207.

K348. Raschke, H., Fleischmann, T., Van Der Meer, J.R., Kohler, H.-P.E., *Appl. Environ. Microbiol.* (1999) **65**, 5242–5246.

K349. Boeriu, C.G., Kaycsa, A.D., Peter, F., Preda, G.M., Eisler, S.R., Deretey, E., Oana, D.F.R., *Rom.* RO 108252 (31–3–1994).

K350. Renwick, A.B., Surry, D., Price, R.J., Lake, B.G., Evans, D.C., *Xenobiotica* (2000) **30**, 955–969.

K351. Kobayashi, S., Uyama, H., Oguchi, T., *Jpn. Kokai Tokkyo Koho* JP 2000 41690 (15–2–2000).

K352. Bollard, M., Caldwell, J., Taylor, G.W., Strolin-Benedetti, M., Semeraro, C., *Xenobiotica* (2000) **30**, 983–991.

K353. Ozawa, S., Ohta, K., Miyajima, A., Kurebayashi, H., Sunouchi, M., Shimizu, M., Murayama, N., Matsumoto, Y., Fukuoka, M., Ohno, Y., *Xenobiotica* (2000) **30**, 1005–1017.

K354. De Santi, C., Pietrabissa, A., Mosca, F., Pacifici, G.M., *Xenobiotica* (2000) **30**, 1047–1054.

K355. Althaus, M., Retzow, A., Castell, J.V., Gómez-Lechón, M.-J., Amalou, Z., Rose, T., Appel, K., *Xenobiotica* (2000) **30**, 1033–1045.

K356. Mutlib, A.E., Diamond, S., Shockcor, J., Way, R., Nemeth, G., Gan, L., Christ, D.D., *Xenobiotica* (2000) **30**, 1091–1110.

K357. Thompson, C.D., Barthen, M.T., Hopper, D.U., Miller, T.A., Quigg, M., Hudspeth, C., Montouris, G., Marsh, L., Perhach, J.L., Sofia, R.D., Macdonald, T.L., *Epilepsia* (1999) **40**, 769–776.

K358. Breyer-Pfaff, U., Nill, K., *Adv. Exp. Med. Biol.* (1999) **463**, 473–480.

K359. Hu, J., Sajan, M., Kulkarni, A.P., *Int. J. Toxicol.* (1999) **18**, 251–257.

K360. Matsumoto, S., Nakai, M., Yoshida, M., Katae, H., *J. Vet. Pharmacol. Ther.* (1999) **22**, 286–289.

K361. Xu, Z.-H., Wang, W., Zhao, X.-J., Huang, S.-L., Zhu, B., He, N., Shu, Y., Liu, Z.-Q., Zhou, H.-H., *Br. J. Clin. Pharmacol.* (1999) **48**, 416–423.

K362. Teramoto, M., Futamata, H., Harayama, S., Watanabe, K., *Mol. Gen. Genet.* (1999) **262**, 552–558.

K363. Coldham, N.G., Howells, L.C., Santi, A., Montesissa, C., Langlais, C., King, L.J., Macpherson, D.D., Sauer, M.J., *J. Steroid Biochem. Mol. Biol.* (1999) **70**, 169–184.

K364. Wayment, D.G., Bhadra, R., Lauritzen J., Hughes, J.B., Shanks, J.V., *Int. J. Phytorem.* (1999) **1**, 227–239.

K365. Pfaar, U., Gygax, D., Gertsch, W., Winkler, T., Ghisalba, O., *Chimia* (1999) **53**, 590–593.

K366. Schubert, T., Kula, M.-R., Muller, M., *Synthesis* (1999) 2045–2048.

K368. Linhart, I., Gut, I., Smejkal, J., Novak, J., *Chem. Res. Toxicol* (2000) **13**, 36–44.

K369. Ohe, T., Sakai, K., Ohmoto, T., *Kagaku to Kogyo (Osaka)* (1999) **73**, 433–437.

K370. Eilers, A., Rüngeling, E., Stündl, U.M., Gottschalk G., *Appl. Microbiol. Biotechnol.* (1999) **53**, 75–80.

K372. Samanta, S.K., Chakraborti, A.K., Jain,.K., *Appl. Microbiol. Biotechnol.* (1999) **53**, 98–107.

K373. Varela, E., Böckle, B., Romero, A., Martínez, A.T., Martínez, M.J., *Biochim. Biophys. Acta* (2000) **1476**, 129–138.

K374. Jiménez, M., García-Carmona, F., *Arch. Biochem. Biophys.* (2000) **373**, 255–260.

K375. Minami, Y., Nishimura, O., Hara-Nishimura, I., Nishimura, M., Matsubara, H., *Plant Cell Physiol.* (2000) **41**, 218–225.

K376. Briganti, F., Randazzo, D., Scozzafava, A., Berti, D., Baglioni, P., Di Gennaro, P., Galli, E., Bestetti, G., *J. Mol. Catal. B: Enzym.* (1999) **7**, 263–272.

K377. Song, D.-J., Ho, Y., Hsu, K.-Y., *Biopharm. Drug Dispos.* (1999) **20**, 263–270.

K378. Zorko, M., Gottlieb, H.E., Zakelj-Mavric, M., *Steroids* (2000) **65**, 46–53.

K379. Bostrom, M., Becedas, L., DePierre, J.W., *Chem.-Biol. Interact.* (2000) **124**, 103–118.
K380. Swoboda, K.J., Hyland, K., Goldstein, D.S., Kuban, K.C.K., Arnold, L.A., Holmes, C.S., Levy, H.L., *Neurology* (1999) **53**, 1205–1211.
K381. Kitamura, S., Sugihara, K., Nakatani, K., Ohta, S., Oh-hara, T., Ninomiya, S., Green, C.E., Tyson, C.E., *IUBMB Life* (1999) **48**, 607–611.
K382. Mohammed, N., Onodera, R., Khan, R.I., *J. Gen. Appl. Microbiol.* (1999) **45**, 143–147.
K383. Hamana, K., Niitsu, M., *J. Gen. Appl. Microbiol.* (1999) **45**, 149–153.
K385. Hirabayashi, H., Takizawa, S., Fukuyama, N., Nakazawa, H., Shinohara, Y., *Brain Res.* (2000) **852**, 319–325.
K386. Kulys, J., Vidziunaite, R., *Biologija* (1998) 37–43.
K387. Svoboda, Z., Nobilis, M., Kvetina, J., Lemr, K., *Acta Med. (Hradec Kralove, Czech. Rep.)* (1999) **42**, 73–78.
K388. Abourashed, E.A., El-Feraly, F.S., Khan, I.A., Hufford, C.D., *Chem. Pharm. Bull.* (2000) **48**, 45–47.
K389. Perloff, M.D., Von Moltke, L.L., Court, M.H., Kotegawa, T., Shader, R.I., Greenblatt, D.J., *J. Pharmacol. Exp. Ther.* (2000) **292**, 618–628.
K390. Khan, R.I., Onodera, R., Amin, M.R., Mohammed, N., *Amino Acids* (1999) **17**, 335–346.
K391. Karegoudar, T.B., Chae, J.-C., Kim, C.-K., *J. Microbiol. (Seoul)* (1999) **37**, 123–127.
K392. Soledale, M., Pedras, C., Khan, A.Q., *Phytochemistry* (2000) **53**, 59–69.
K393. Taguchi, G., Fujikawa, S., Yazawa, T., Kodaira, R., Hayashida, N., Shimosaka, M., Okazaki, M., *Plant Sci. (Shannon, Irel)* (2000) **151**, 153–161.
K394. Zocher, F., Enzelberger, M.M., Bornscheuer, U.T., Hauer, B., Wohlleben B., Schmid, R.D., *J. Biotechnol.* (2000) **77**, 287–292.
K395. Fenton, B., Valverde-R, C., *Gen. Comp. Endocrinol.* (2000) **117**, 77–88.
K396. Hwang, P., Chow, T., Adrian, N.R., *Environ. Toxicol. Chem.* (2000) **19**, 836–841.
K397. Matsunaga, T., Shibayama, K., Higuchi, S., Tanaka, H., Watanabe, K., Yamamoto, I., *Biol. Pharm. Bull.* (2000) **23**, 43–46.
K399. Scordo, M.G., Spina, E., Facciola, G., Avenoso, A., Johansson, I., Dahl, M.-L., *Psychopharmacology (Berlin)* (1999) **147**, 300–305.
K400. Raj, H.R.G., Parmar, V.S., Jain, S.C., Goel, S., Tyagi, Y.K., Sharma, S.K., Olsen, C.E., Wengel, J., *Bioorg. Med. Chem.* (2000) **8**, 233–237.
K401. Valtier, S., Cody, J.T., *J. Forensic Sci.* (1999) **44**, 17–22.
K402. Nagashima, S., Hirotani, M., Yoshikawa, T., *Phytochemistry* (2000) **53**, 533–538.
K404. Çelik, A., Cullis, P.M., Lloyd Raven, E., *Arch. Biochem. Biophys.* (2000) **373**, 175–181.
K405. Harris, R.M., Waring, R.H., Kirk, C.J., Hughes, P.J., *J. Biol. Chem.* (2000) **275**, 159–166.
K406. Li, T., Rosazza, J.P.N., *Appl. Environ. Microbiol.* (2000) **66**, 684–687.
K407. Chauhan, A., Jain, R.K., *Biochem. Biophys. Res. Commun.* (2000) **267**, 236–244.
K408. Moller, N., Meek, S., Bigelow, M., Andrews, J., Nair, K.S., *Proc. Natl. Acad. Sci. U.S.A.* (2000) **97**, 1242–1246.
K409. Maghsoudi, S., Kheirolomoom, A., Vossoughi, M., Tanaka, E., Katoh, S., *Biochem. Eng. J.* (2000) **5**, 11–16.
K410. Cunane, L.M., Chen, Z.-W., Shamala, N., Mathews, F.S., Cronin, C.N., McIntire, W.S., *J. Mol. Biol.* (2000) **295**, 357–374.

K411. Kato, Y., Nakamura, K., Sakiyama, H., Mayhew, S.G., Asano, Y., *Biochemistry* (2000) **39**, 800–809.

K412. Lakshmi, V.M., Hsu, F.F., Davis, B.B., Zenser, T.V., *Chem. Res. Toxicol.* (2000) **13**, 96–102.

K413. Shin, S.-K., Oh, Y.-S., Kim, S.-J., *J. Microbiol (Seoul)* (1999) **37**, 185–192.

K414. Itoh, S., Onishi, S., *Early Hum. Dev.* (2000) **57**, 15–23.

K415. Voirol, P., Jonzier-Perez, M., Porchet, F., Reymond, M.J., Janzer, R.C., Bouras, C., Strobel, H.W., Kosel, M., Eap, C.B., Baumann, P., *Brain Res.* (2000) **855**, 235–243.

K416. Bhadra, R., Spanggord, R.J., Wayment, D.G., Hughes, J.B., Shanks, J.V., *Wetlands Rem., Int. Conf.* (1999) 127–132.

K417. Ijiri, Y., Kobayashi, T., Suzuki, K., Ohi, K., Itoh, H., Wakamiya, E., Suzuki, S., Tamai, H., *J. Appl. Ther. Res.* (1999) **2**, 281–289.

K419. Hover, C.G., Kulkarni, A.P., *Chem.-Biol. Interact.* (2000) **124**, 191–203.

K420. Roh, J.H., Wouters, J., Depiereux, E., Yukawa, H., Inui, M., Minami, H., Suzuki, H., Kumagai, H., *Biochem. Biophys. Res. Commun.* (2000) **268**, 293–297.

K421. Annweiler, E., Richnow, H.H., Antranikian, G., Hebenbrock, S., Garms, C., Franke, S., Franke, W., Michaelis, W., *Appl. Environ. Microbiol.* (2000) **66**, 518–523.

K422. Rhofir, C., Hawari, J., *J. Chromatogr. A* (2000) **873**, 53–61.

K423. Tanaka, T., Kondou, K., Kouno, I., *Phytochemistry* (2000) **53**, 311–316.

K424. Demir, A.S., Dünnwald, T., Iding, H., Pohl, M., Müller, M., *Tetrahedron: Asymmetry* (1999) **10**, 4769–4774.

K425. Besse, P., Baziard-Mouysset, G., Boubekeur, K., Palvadeau, P., Veschambre, H., Payard, M., Mousset, G., *Tetrahedron: Asymmetry* (1999) **10**, 4745–4754.

K426. Naito, S., Nishimura, M., Jin, Y., *Eur. J. Metab. Pharmacokinet.* (1999) **24**, 279–286.

K427. Vorwieger, G., Bergmann, R., Syhre, R., Fuchtner, F., Steinbach, J., Brust, P., Johansen, B., *Wiss.-Tech. Ber.-Forschungszent. Rossendorf* (1999) 17–20.

K428. Brandt, V., Geerlings, A., Tits, M., Delaude, C., Van der Heijden, R., Verpoort, R., Angenot, L., *Plant Physiol. Biochem. (Paris)* (2000) **38**, 187–192.

K429. Kataoka, M., Honda, K., Shimizu, S., *Eur. J. Biochem.* (2000) **267**, 3–10.

K430. Delserone, L.M., McCluskey, K., Matthews, D.E., Vanetten, H.E., *Physiol. Mol. Plant Pathol.* (1999) **55**, 317–326.

K431. Tuynman, A., Spelberg, J.L., Kooter, I.M., Shoemaker, H.E., Wever, R., *J. Biol. Chem.* (2000) **275**, 3025–3030.

K432. Hong, H.B., Chang, Y.S., Choi, S.D., Park, Y.H., *Water Res.* (2000) **34**, 2404–2407.

K433. Rodriguez, M.J., Lebrero, J.L.A., Alvarez, E., *Biocatal. Biotransform.* (1999) **17**, 45–60.

K434. Diaz, G.J., Squires, E.J., *J. Agric. Food Chem.* (2000) **48**, 833–837.

K435. Ulgen, M., Ercak, K., Kucukguzel, I., Gorrod, J.W., *Boll. Chim. Farm.* (1997) **136**, 600–604.

K436. Leguen, I., Carlsson, C., Perdu-Durand, E., Prunet, P., Part, P., Cravedi, J.P., *Aquat. Toxicol.* (2000) **48**, 165–176.

K437. Hakamata, W., Nishio, T., Oku, T., *Carbohydr. Res.* (2000) **324**, 107–115.

K438. Lessard, E., Hamelin, B.A., Labbe, L., O'Hara, G., Belanger, P.M., Turgeon, J., *Pharmacogenetics* (1999) **9**, 683–696.

K439. Munir, I.Z., Dordick, J.S., *Enzyme Microb. Technol.* (2000) **26**, 337–341.

K440. Dalen, P., Dahl, M.-L., Eichelbaum, M., Bertilsson, L., Wilkinson, G.R., *Pharmacogenetics* (1999) **9**, 679–706.

K441. Warrington, J.S., Poku, J.W., Von Moltke, L.L., Shader, R.I., Harmatz,

J.S., Greenblatt, D.J., *J. Pharmacol. Exp. Ther.* (2000) **292**, 1024–1031.

K442. Schneider, J., Grosser, R.J., Jayasimhulu, K., Xue, W., Kinkle, B., Warshawsky, D., *Can. J. Microbiol.* (2000) **46**, 269–277.

K443. Todaka, T., Yamano, S., Toki, S., *Arch. Biochem. Biophys.* (2000) **374**, 189–197.

K445. Johnson, G.R., Jain, R.K., Spain, J.C., *Arch. Microbiol.* (2000) **173**, 86–90.

K446. Philipp, B., Schink, B., *Arch. Microbiol.* (2000) **173**, 91–96.

K447. Brau, F., Berengo, J.-C., Min, K.-L., Steghens, J.-P., *Biochem. Biophys. Res. Commun.* (2000) **270**, 247–253.

K449. Iwasaki, A., Yamada, Y., Ikenaka, Y., Yasohara, Y., Hasegawa, J., *PCT Int. Appl.* WO 0026 351 (11–5–2000).

K450. Valoti, M., Fusi, F., Frosini, M., Pessina, F., Tipton, K.F., Sgaragli, G.P., *Eur. J. Pharmacol.* (2000) **391**, 199–206.

K452. Chen, C.-L., Potthast, A., Rosenau, T., Gratzl, J.S., Kirkman, A.G., Nagai, D., Miyakoshi, T., *J. Mol. Catal. B: Enzym.* (2000) **8**, 213–219.

K453. Andlauer, W., Kolb, J., Stehle, P., Furst, P., *J. Nutr.* (2000) **130**, 843–846.

K454. Barbieri, C., Caruso, E., D'Arrigo, P., Fantoni, G.P., Servi, S., *Tetrahedron: Asymmetry* (1999) **10**, 3931–3937.

K455. Cichowicz, D.J., Shane, B., *Biochemistry* (1987) **26**, 513–521.

K456. Tahara, S., Saitoh, F., Mizutani, J., *Z. Naturforsch. C: Biosci.* (1993) **48**, 16–21.

K457. Wu, Z., Daniel, S.L., Drake, H.L., *J. Bacteriol.* (1988) **170**, 5747–5750.

K458. Rancourt, S.L., Walker, V.K., *Biochim. Biophys. Acta* (1990) **1039**, 261–268.

K459. Demassieux, S., Bordeleau, L., Gravel, D., Carrière, S., *Life Sci.* (1987) **40**, 183–191.

K460. Sabban, E.L., Goldstein, M., Levin, B.E., *Ann. N.Y. Acad. Sci.* (1986) **463**, 134–136.

K461. Mohammad, A.M.S., Al-Chalabi, K., Abood, S.A., *J. Exp. Bot.* (1989) **40**, 693–699.

K462. Monard, C., Lamiable, D., Vistelle, R., Morin, E., Ratanasavanh, D., *Pharmacol. Res.* (2000) **42**, 87–92.

K463. Gow, P.J., Ghabrial, H., Treepongkaruna, S., Shulkes, A., Smallwood, R.A., Morgan, D.J., Ching, M.S., *J. Pharm. Sci.* (2000) **89**, 36–44.

K464. Ogiso, T., Uchiyama, K., Suzuki, H., Yoshimoto, M., Tanino, T., Iwaki, M., Uno, S., *Biol. Pharm. Bull.* (2000) **23**, 482–486.

K465. MacCallum, J., Cummings, J., Dixon, J.M., Miller, R.W., *Br. J. Cancer* (2000) **82**, 1629–1635.

K466. Orabi, K.Y., Walker, L.A., Clark, A.M., Hufford, C.D., *J. Nat. Prod.* (2000) **63**, 685–687.

K467. Oliveira, E.J., Watson, D.G., *FEBS Lett.* (2000) **471**, 1–6.

K468. Zhang, M., Fawcett, J.P., Kennedy, J.M., Shaw, J.P., *Br. J. Clin. Pharmacol.* (2000) **49**, 152–157.

K469. de Jager, A.D., Hundt, H.K.L., Hundt, A.F., Swart, K.J., Knight, M., Roberts, J., *J. Chromatogr., B: Biomed. Sci. Appl.* (2000) **740**, 247–251.

K470. Prakash, C., Kamel, A., Cui, D., Whalen, R.D., Miceli, J.J., Tweedie, D., *Br. J. Clin. Pharmacol.* (2000) **49** (Suppl 1), 35S-42S.

K471. Phillips, L.R., Bramhall, C., Buckley, J., Wolfe, T., House, T.L., Stinson, S.F., *J. Chromatogr., B: Biomed. Sci. Appl.* (2000) **741**, 205–211.

K472. Sweet, R.A., Pollock, B.G., Mulsant, B.H., Rosen, J., Sorisio, D. Kirshner, M., Henteleff, R., DeMichele, M.A.A., *Clin. Psychopharmacol.* (2000) **20**, 181–187.

K473. Von Moltke, L.L., Greenblatt, D.J., Granda, B.W., Grassi, J.M., Schmider, J., Harmatz, J.S., Shader, R.I., *Psychopharmacology* (1999) **145**, 113–122.

K475. Kreth, K.-P., Kovar, K.-A., Schwab, M., Zanger, U.M., *Biochem. Pharmacol.* (2000) **59**, 1563–1571.

K476. Jurima-Romet, M., Casley, W.L., Leblanc, C.A., Nowakowska, M., *Toxicol. In Vitro* (2000) **14**, 253–263.

K477. Naito, S., Nishmura, M., Nogawa, H., Yoshitsugu, H., Tamao, Y., Asano, S., Nomura, Y., *Pharm. Pharmacol. Commun.* (2000) **6**, 179–186.

K478. Gu, J.-K., Zhong, D.-F., Chen, X.-Y., Zhou, H., Shen, J.-C., Lin, Y., *Gaodeng Xuexiao Huaxue Xuebao* (2000) **21**, 690–693.

K479. Bohets, H., Lavrijsen, K., Hendrickx, J., van Houdt, J., van Genechten, V., Verboven, P., Meuldermans, W., Heykants, J., *Br. J. Pharmacol.* (2000) **129**, 1655–1657.

K480. Taavitsainen, P., Anttila, M., Nyman, L., Karnani, H., Salonen, J.S., Pelkonen, O., *Pharmacol. Toxicol.* (2000) **86**, 215–221.

K481. Momen, A.Z.M.R., Hoshino, T., *Biosci. Biotechnol. Biochem.* (2000) **64**, 539–549.

K482. Boyle, S.P., Craft, J.A., *Chem.-Biol. Interact.* (2000) **125**, 209–220.

K483. Reichenbecher, W., Philipp, B., Suter, M.J.-F., Schink, B., *Arch. Microbiol.* (2000) **173**, 206–212.

K484. Rao, K.N., *Indian J. Biochem. Biophys.* (2000) **37**, 121–129.

K485. Chung, J.G., Lee, J.H., Ho, H.C., Ho, C.C., Lai, J.M., Chang, S.H., Hung, C.F., *Toxicol. Environ. Chem.* (1999) **69**, 19–29.

K486. Takeshi, K., Kawahata, Y., *Jpn. Kokai Tokkyo Koho JP* 2000 166586 (20–6–2000).

K488. Tassaneeyakul, W., Tassaneeyakul, W., Vannaprasaht, S., Yamazoe, Y., *Br. J. Clin. Pharmacol.* (2000) **49**, 139–144.

K489. Carlson, G.P., Mantick, N.A., Powley, M.W., *J. Toxicol. Environ. Health, Part A* (2000) **59**, 591–595.

K490. Ferry, G., Loynel, A., Kucharczyk, N., Bertin, S., Rodriguez, M., Delagrange, P., Galizzi, J.-P., Jacoby, E., Volland, J.-P., Lesieur, D., Renard, P., Canet, E., Fauchère, J.-L., Boutin, J.A., *J. Biol. Chem.* (2000) **275**, 8794–8805.

K491. Bühler, B., Schmid, A., Hauer, B., Witholt, B., *J. Biol. Chem.* (2000) **275**, 10085–10092.

K492. Attieh, J., Kleppinger-Sparace, K.F., Nunes, C., Sparace, S.A., Saini, H.S., *Plant, Cell Environ.* (2000) **23**, 165–174.

K493. Muller, J.A., Schink, B., *Arch. Microbiol.* (2000) **173**, 288–295.

K494. Nishino, S.F., Paoli, G.C., Spain, J.C., *Appl. Environ. Microbiol.* (2000) **66**, 2139–2147.

K495. Kim, S.-J., Hwang, S., Kim, Y.-C., *J. Microbiol. (Seoul)* (1999) **37**, 224–228.

K496. Leblond, J.D., Applegate, B.M., Menn, F.-M., Schultz, T.W., Sayler, G.S., *Environ. Toxicol. Chem.* (2000) **19**, 1235–1246.

K497. Tong, Z., James, M.O., *Arch. Biochem. Biophys.* (2000) **376**, 409–419.

K498. Moss, T., Howes, D., Williams, F.M., *Food Chem. Toxicol.* (2000) **38**, 361–370.

K499. Choroba, O.W., Williams, D.H., Spencer, J.B., *J. Am. Chem. Soc.* (2000) **122**, 5389–5390.

K500. Nakagawa, Y., Tayama, S., *Arch. Toxicol.* (2000) **74**, 99–105.

K501. Sundin, A., Eriksson, B., Bergstrom, M., Bjurling, P., Lidner, K.-J., Oberg, K., Langstrom, B., *Nucl. Med. Biol.* (2000) **27**, 33–41.

K503. Sriubolmas, N., Khampha, W., Wiyakrutta, S., Panphut, W., Laowanapiban, P., Plainkum, P., Ruangrungsi, N., Meevootisom, V., *Biotechnol. Sustainable Util. Biol. Resour. Trop.* (2000) **14**, 165–182.

K504. Narimatsu, S., Yamamoto, S., *Yakubutsu Dotai* (2000) **15**, 39–45.

K505. Ader, P., Wessmann, A., Wolffram, S., *Free Radical Biol. Med.* (2000) **28**, 1056–1067.

K506. Petersen, M.B., Friis, C., *Am. J. Vet. Res.* (2000) **61**, 573–576.

K507. Blakledge, C.A., Partridge, E.A., Wilson, I.D., Nicholson, J.K., *J. Pharm. Biomed. Anal.* (2000) **22**, 1023–1028.

K508. Hamman, M.A., Haehner-Daniels, B.D., Wrighton, S.A., Rettie, A.E., Hall, S.D., *Biochem. Pharmacol.* (2000) **60**, 7–17.

K509. Bock, C., *Landbauforsch. Voelkenroede, Sonderh.* (1999) **207**, 1–121.

K510. Zhang, K., Cai, R., Chen, D., Mao, L., *Anal. Chim. Acta* (2000) **413**, 109–113.

K511. Renoux, A.Y., Sarrazin, M., Hawari, J., Sunahara, G.I., *Environ. Toxicol. Chem.* (2000) **19**, 1473–1480.

K512. Sue, M., Ishihara, A., Iwamura, H., *Plant Sci. (Shannon, Irel.)* (2000) **155**, 67–74.

K513. Yoo, S.D., Kang, E., Jun, H., Shin, B.S., Lee, K.C., Lee, K.H., *Chem. Pharm. Bull.* (2000) **48**, 798–801.

K514. Kato, Y., Ooi, R., Asano, Y., *Appl. Environ. Microbiol.* (2000) **66**, 2290–2296.

K515. Yoshii, K., Kobayashi, K., Tsumuji, M., Tani, M., Shimada, N., Chiba, K., *Life Sci.* (2000) **67,** 175–184.

K516. Vander Mijnsbrugge, K., Meyermans, H., Van Montegu, M., Bauw, G., Boerjan, W., *Planta* (2000) **210**, 589–598.

K517. Jenkins, G.N., Stachulski, A.V., Scheinmann, F., Turner, N.J., *Tetrahedron: Asymmetry* (2000) **11**, 413–416.

K518. Jiao, K., Suri, W., Zhang, S.-s., Sun, G., *Anal. Chim. Acta* (2000) **413**, 71–78.

K519. Saito, A., Iwabuchi, T., Harayama, S., *J. Bacteriol.* (2000) **182**, 2134–2141.

K520. Zhao, J.-S., Singh, A., Huang, X.-D., Ward, O.P., *Appl. Environ. Microbiol.* (2000) **66**, 2336–2342.

K521. Fischer, G.M., Nemeti, B., Farkas, V., Debreceni, B., Laszlo, A., Schaffer, Z., Somogyi, C., Sandor, A., *Biochim. Biophys. Acta* (2000) **1501**, 200–210.

K522. Seah, S.Y.K., Labbé, G., Nerdinger, S., Johnson, M.R., Snieckus, V., Eltis, L.D., *J. Biol. Chem.* (2000) **275**, 15701–15708.

K523. Rana, S.V.S., Gupta, S., *J. Environ. Biol.* (2000) **21**, 105–109.

K524. Dünnwald, T., Demir, A.S., Siegert, P., Pohl, M., Müller, M., *Eur. J. Org. Chem.* (2000) 2161–2170.

K525. Knights, K.M., Gasser, R., Klemisch, W., *Biochem. Pharmacol.* (2000) **60**, 507–516.

K526. Raeside, J.I., Christie, H.L., Forster, L., Renaud, R.L., *Biochem. Biophys. Res. Commun.* (2000) **273**, 505–508.

K527. Kim, D.-H., Park, E.-K., Bae, E.-A., Han, M.K., *Biol. Pharm. Bull.* (2000) **23**, 830–833.

K528. Doddamani, H.P., Ninnekar, H.Z., *Curr. Microbiol.* (2000) **41**, 11–14.

K529. Ronen, Z., Abeliovich, A., *Appl. Environ. Microbiol.* (2000) **66**, 2372–2377.

K530. Moise, P.A., Birmingham, M.C., Schentag, J.J., *Drugs Today* (2000) **36**, 229–244.

K531. Piskula, M.K., *J. Nutr.* (2000) **130**, 1766–1771.

K532. Santos, A., Zanetta, S., Cresteil, T., Deroussent, A., Pein, F., Raymond, E., Vernillet, L., Risse, M.-L., Boige, V., Gouyette, A., Vassal, G., *Clin. Cancer Res.* (2000) **6**, 2012–2020.

K533. Miyake, Y., Shimoi, K., Kumazawa, S., Yamamoto, K., Kinae, N., Osawa, T., *J. Agric. Food Chem.* (2000) **48**, 3217–3224.

K534. Dickert, S., Pierik, A.J., Linder, D., Buckel, W., *Eur. J. Biochem.* (2000) **267**, 3874–3884.

K535. Awad, H.M., Boersma, M.G., Vervoort, J., Rietjens, I.M.C.M., *Arch. Biochem. Biophys.* (2000) **378**, 224–233.

K536. Inoue, O., Kanno, E., Kakizaki, M., Watanabe, T., Higashikawa, K., Ikeda, M., *Ind. Health* (2000) **38**, 195–204.

K537. Eaton, R.W., *Appl. Environ. Microbiol.* (2000) **66**, 2668–2672.

K538. Tripathi, C.K.M., Bihari, V., Tyagi, R.D., *Process Biochem. (Oxford)* (2000) **35**, 1247–1251.

K539. Stiborová, M., Schmeiser, H.H., Frei, E., *Phytochemistry* (2000) **54**, 353–362.

K540. Sato, Y., Tsukamoto, T., Sato, M., *J. For. Res.* (1999) **4**, 275–280.

K541. Pedras, M., Soledale, C., Okanga, F.I., *Can. J. Chem.* (2000) **78**, 338–346.

K542. Hay, A.G., Focht, D.D., *FEMS Microbiol. Ecol.* (2000) **31**, 249–253.

K543. Van Aken, B., Stahl, J.D., Naveau, H., Agathos, S.N., Aust, S.D., *Biorem. J.* (2000) **4**, 135–145.

K544. Miyakoda, H., Tabata, M., Onodera, S., Takeda, K., *J. Health Sci.* (2000) **46**, 269–274.

K545. Iwahashi, K., Anemo, K., Nakamura, K., Yoshihara, E., Igarashi, K., *Clin. Chim. Acta* (2000) **298**, 193–195.

K546. Kelly, V.P., Ireland, L.S., Ellis, E.M., Hayes, J.D., *Biochen. J.* (2000) **348**, 389–400.

K547. Arnett, C.M., Parales, J.V., Haddock, J.D., *Appl. Environ. Microbiol.* (2000) **66**, 2928–2933.

K548. Schroder, P., Abele, C., Gohr, P., Stuhlfauth-Roisch, U., Grosse, W., *Adv. Exp. Med. Biol.* (1999) **467**, 637–644.

K549. Bui, V., Vidar Hansen, T., Stenstrom, Y., Ribbons, D.W., Hudlicky, T., *Perkin 1* (2000) 1669–1672.

K550. Murch, S.J., KrishnaRaj, J., Saxena, P., *Plant Cell Rep.* (2000) **19**, 698–704.

K551. Pruett, J.H., Oehler, D.D., Kammlah, D.M., Guerrero, F.D., *J. Econ. Entomol.* (2000) **93**, 920–924.

K552. Zhao, J.-S., Ward, O.P., *Can. J. Microbiol.* (2000) **46**, 643–652.

K553. Chauhan, A., Samanta, S.K., Jain, R.K., *J. Appl. Microbiol.* (2000) **88**, 764–772.

K554. Mutlib, A., Chen, H., Shockcor, J., Espina, R., Chen, S., Cao, K., Du, A., Nemeth, G., Prakash, S., Shimoga, P., Gan, L.-S., *Chem. Res. Toxicol.* (2000) **13**, 775–784.

K555. Gu, J., Zhong, D., Guo, J., Zhou, H., Shen, J., *Xaoxue Xuebao* (2000) **35**, 181–184.

K556. Taguchi, G., Imura, H., Maeda, Y., Kodaira, R., Hayashida, N., Shimosaka, M., Okazaki, M., *Plant Sci. (Shannon, Irel.)* (2000) **157**, 105–112.

K557. Cho, M.-H., Paik, Y.-S., Hahn, T.-R., *J. Agric. Food Chem.* (2000) **48**, 3917–3921.

K558. Brunhuber, N.M.W., Thoden, J.B., Blanchard, J.S., Vanhooke, J.L., *Biochemistry* (2000) **39**, 9174–9187.

K559. Stamps, S.L., Taylor, A.B., Wang, S.C., *Biochemistry* (2000) **39**, 9671–9678.

K560. Basak, A., Nag, A., Bhattacharya, G., Mandal, S., Nag, S., *Tetrahedron: Asymmetry* (2000) **11**, 2403–2407.

K561. Shalit, M., Katzir, N., Larkov, O., Burger, Y., Shalekhet, F., Laslochkin, E., Ravid, U., Amar, O., Edelstein, M., Lewinsohn, E., *Acta Hortic.* (2000) **510**, 455–461.

K562. Honey, S., O'Keefe, P., Drahushuk, A.T., Olson, J.R., Kumar, S., Sikka, H.C., *Comp. Biochem. Physiol., Part C: Toxicol. Pharmacol.* (2000) **126C**, 285–292.

K563. Furuya, Y., Sawada, H., Hirahara, T., Ito, K., Oshiro, T., Izumi, Y., *Biosci. Biotechnol. Biochem.* (2000) **64**, 1486–1493.

K564. Meckenstock, R.U., Annweiler, E., Michaelis, W., Richnow, H.H., Schink, B., *Appl. Environ. Microbiol.* (2000) **66**, 2743–2747.

K565. Iwahara, K., Honda, Y., Watanabe, T., Kuwahara, M., *Appl. Microbiol. Biotechnol.* (2000) **54**, 104–111.

K566. Boll, M., Laempe, D., Eisenreich, W., Bacher, A., Mittelberger, T., Heinze, J., Fuchs, G., *J. Biol. Chem.* (2000) **275**, 21889–21895.

K567. Barthelmebs, L., Divies, C., Cavin, J.-F., *Appl. Environ. Microbiol.* (2000) **66**, 3368–3375.

K568. Keiser, J., Ericsson, Ö., Burri, C., *Clin. Pharmacol. Ther.* (2000) **67**, 478–488.

K569. Kobana, K., Watanabe, K., Kimura, T., Matsunaga, T., Kondo, S., Yamamoto, I., *Biol. Pharm. Bull.* (2000) **23**, 917–921.

K570. White, R.D., Shea, D., Schlezinger, J.J., Hahn, M.E., Stegeman, J.J., *Comp. Biochem. Physiol., Part C: Toxicol. Pharmacol.* (2000) **126C**, 267–284.

K572. Morishige, T., Tsujita, T., Yamada, Y., Sato, F., *J. Biol. Chem.* (2000) **275**, 23398–23405.

K573. Li, T., Juteau, P., Beaudet, R., Lepine, F., Villemur, R., Bisaillon, J.-G., *Can. J. Microbiol.* (2000) **46**, 856–859.

K574. Li, Y., Meselhy, M.R., Wang, L.-Q., Ma, C.-M., Nakamura, N., Hattori, M., *Chem. Pharm. Bull.* (2000) **48**, 1239–1241.

K575. Yasuda, T., Inaba, A., Ohmori, M., Endo, T., Kubo, S., Ohsawa, K., *J. Nat. Prod.* (2000) **63**, 1444–1446.

K576. Matoishi, K., Kakidani, H., Suzuki, M., Sugai, T., Ohta, H., Hanzawa, S., *Chem. Commun.* (2000) 1519–1520.

K577. Diem, S., Bergmann, J., Herderich, M., *J. Agric. Food Chem.* (2000) **48**, 4913–4917.

K578. Johnson, J.A., Herring, V.L., Wolfe, M.S., Relling, M.V., *J. Pharmacol. Exp. Ther.* (2000) **294**, 1099–1105.

K579. Cho, M.-H., Hahn, T.-R., *Arch. Biochem. Biophys.* (2000) **382**, 238–244.

K580. Khanna, R., Morton, C.L., Danks, M.K., Potter, P.M., *Cancer Res.* (2000) **60**, 4725–4728.

K581. Szökõ, É, Kalász, H., Magyar, K., *Eur. J. Drug Metab. Pharmacokinet.* (1999) **24**, 315–319.

K582. Nielsen, S.E., Breinholt, V., Cornett, C., Dragsted, L.O., *Food Chem. Toxicol.* (2000) **38**, 739–746.

K583. He, Z., Spain, F., *J. Ind. Microb. Biotechnol.* (2000) **25**, 25–28.

K584. Berestovskaya, Y.Y., Ignatov, V.V., Markina, L.N., Kamenev, A.A., Makarov, O.E., *Microbiology (Moscow)* (2000) **69**, 397–400.

K585. Obel, N., Scheller, H.V., *Anal. Biochem.* (2000) **286**, 38–44.

K587. Moularak, A.S., Rosenkrans, C.F., Jr., *Biochem. Biophys. Res. Commun.* (2000) **274**, 746–749.

K588. Harris, R.M., Picton, R., Singh, S., Waring, R.H., *Life Sci.* (2000) **67**, 2051–2057.

K589. Bressler, D.C., Fedorak, P.M., Pickard, M.A., *Biotechnol. Lett.* (2000) **22**, 1119–1125.

K590. Niku-Paavola, M.-L., Viikari, L., *J. Mol. Catal. B: Enzym.* (2000) **10**, 435–444.

K591. Kambourakis, S., Draths, K.M., Frost, J.W., *J. Am. Chem. Soc.* (2000) **122**, 9042–9043.

K592. Carlson, G.P., *J. Toxicol. Environ. Health, Part A* (2000) **61**, 709–717.

K593. Simpson, T.J., Bandumathie Weerasooriya, M.K., *Perkin 1* (2000) 2771–2775.

K594. Attieh, J., Sparace, S.A., Saini, H.S., *Arch. Biochem. Biophys.* (2000) **380**, 257–266.

K595. El-Bachá, R.S., Leclerc, S., Netter, P., Magdalou, J., Minn, A., *Life Sci.* (2000) **67**, 1735–1745.

K596. van den Heuvel, R.H.H., Fraiije, M.W., Ferrer, M., Mattevi, A., van Berkel, W.J.H., *Proc. Natl. Acad. Sci. U.S.A.* (2000) **97**, 9455–9460.

K597. Veeresham, C., Rao, M.A., Chitimalla, P.B., Mamatha, R., Kokate, C.K., *Indian Drugs* (2000) **37**, 86–89.

K598. Datta, J., Maiti, A.K., Modak, D.P., Chakrabartty, P.K., Bhattacharyya, P., Ray, P.K., *J. Gen. Appl. Microbiol.* (2000) **46**, 59–67.

K599. Abu-Quare, A.W., Abou-Donia, M.B., *Toxicology* (2000) **150**, 119–127.

K600. Chua, M.-S., Kashiyama, E., Bradshaw, T.D., Stinson, S.F., Brantley, E., Sausville, E.A. Stevens, M.F.G., *Cancer Res.* (2000) **60**, 5196–5203.

K601. Kawai, K., Tsuruta, F., Iwabuchi, H., Ishikawa, M., Haruyama, H. Ikenaga,

H., Ikeda, T., Nakamura, K., *Yakubutsu Dotai* (2000) **15**, 89–100.

K602. Dettling, M., Sachse, C., Brockmoller, J., Schley, J., Muller-Oerlinghausen, B., Pickersgill, I., Rolfs, A., Schaub, R.T., Schmider, J., *Psychopharmacology* (2000) **152**, 80–86.

K603. Mitamura, K., Yatera, M., Shimada, K., *J. Chromatogr., B: Biomed. Sci. Appl.* (2000) **748**, 89–96.

K604. Pope, S.A.S., Clayton, P.T., Muller, D.P.R., *Arch. Biochem. Biophys.* (2000) **381**, 8–15.

K605. Müller, A., Weiler, E.W., *Biol. Chem.* (2000) **381**, 679–686.

K606. Elsworth, C., Gill, M., Saubern, S., *Phytochemistry* (2000) **55**, 23–27.

K607. Bhushan, B., Samanta, S.K., Jain, R.K., *Lett. Appl. Microbiol* (2000) **31**, 5–9.

K608. Clement, B., Behrens, D., Moeller, W., Cashman, J.R., *Chem. Res. Toxicol.* (2000) **13**, 1037–1045.

K609. Weimer, T.L., Reddy, A.P., Harttig, U.D., Stamm, S.C., Miller, M.R., Baird, W., Hendricks, J., Bailey, G., *Toxicol. Sci.* (2000) **57**, 217–228.

K610. Pinyakong, O., Habe, H., Supaka, N., Pinpanichkarn, P., Juntongjin, K., Yoshida, T., Furihata, K., Nojiri, H., Yamane, H., Omori, T., *FEMS Microbiol. Lett.* (2000) **191**, 115–121.

K611. Neuser, F., Zorn, H., Berger, R.G., *J. Naturforsch. C: J. Biosci.* (2000) **55**, 560–568.

K612. Johanson, G., Ernstgård, L., Gullstrand, E., Löf, A., Osterman-Golkar, S., Williams, C.C., Sumner, S.C.J., *Toxicol. Appl. Pharmacol.* (2000) **168**, 36–49.

K613. Parales, R.E., Resnick, S.M., Yu, C.-L., Boyd, D.R., Sharma, N.D., Gibson, D.T., *J. Bacteriol.* (2000) **182**, 5495–5504.

K614. Bacher, D., Specht, M., Hammer, E., Francke, W., Schauer, F., *Appl. Environ. Microbiol.* (2000) **66**, 4528–4531.

K615. Bock, M., Kneifel, H., Schoberth, S.M., Sahm, H., *Acta Biotechnol.* (2000) **20**, 189–201.

K617. Rotcenkovs, G., Kanerva, L.T., *J. Mol. Catal. B: Enzym.* (2000) **11**, 37–43.

K618. Daniel, W.A., Syrek, M., Haduch, A., Wojcikowski, J., *Br. J. Pharmacol.* (2000) **131**, 287–295.

K619. Klette, K.L., Anderson, C.J., Poch, G.K., Nimrod, A.C., ElSohly, M.A., *J. Anal. Toxicol.* (2000) **24**, 550–556.

K620. Li, X., Clemens, D.L., Anderson, R.J., *Biochem. Pharmacol.* (2000) **60**, 1713–1716.

K621. Spink, B.C., Katz, B.H., Hussain, M.M., Pang, S., Connor, S.P., Aldous, K.M., Gierthy, J.F., Spink, D.C., *Carcinogenesis* (2000) **21**, 1947–1957.

K622. Pradeepkumar, S., Karegoudar, T.B., *J. Microbiol. Biotechnol.* (2000) **10**, 518–521.

K623. Gillam, E.M.J., Notley, L.M., Cai, H., De Voss, J.J., Guengerich, F.P., *Biochemistry* (2000) **39**, 13817–13824.

K625. Zhong, R., Morrison, W.H. III, Himmelsbach, D.S., Poole, F.L. II, Ye, Z.-H., *Plant Physiol.* (2000) **124**, 563–577.

K626. Gatfield, I.-L., Hilmer, J.-M., *Eur. Pat. Appl. EP* 1,061,132 (20–12–2000).

K627. Hur, H.-G., Rafii, F., *FEMS Microbiol. Lett.* (2000) **192**, 21–25.

K628. Sojo, M.M., Nunez-Delicado, E., Sanchez-Ferrer, A., Garcia-Carmona, F., *J. Agric. Food Chem.* (2000) **48**, 5543–5547.

K630. Akashi, T., Sawada, Y., Aoki, T., Ayabe, S.-I., *Biosci., Biotechnol., Biochem.* (2000) **64**, 2276–2279.

K631. Gearhart, D.A., Collins, M.A., Lee, J.M., Neafsey, E.J., *Neurobiol. Dis.* (2000) **7**, 201–211.

K632. Tan, N.C.G., Borger, A., Slenders, P., Svitelskaya, A., Lettinga, G., Field, J.A., *Water Sci. Technol.* (2000) **42**, 5–6.

K633. Nuesslein, B., Kurzmann, M., Bauer, R., Kreis, W., *J. Nat. Prod.* (2000) **63**, 1615–1618.

K634. Ceccato, A., Vanderbist, F., Pabst, J.-Y., Streel, B., *J. Chromatogr., B: Biomed. Sci. Appl.* (2000) **748**, 65–76.

K635. Tolokán, A., Horváth, V., Horvai, G., Egresi, A., Balogh, N.K., Klebovich, I., *J. Chromatogr., A* (2000) **896**, 279–290.

K636. Hofmann, U., Pecia, M., Heinkele, G., Dilger, K., Kroemer, H.K., Eichelbaum, M., *J. Chromatogr., B: Biomed. Sci. Appl.* (2000) **748**, 113–123.

K637. Strickmann, D.B., Blaschke, G., *J. Chromatogr., B: Biomed. Sci. Appl.* (2000) **748**, 213–219.

K638. Meier, H., Blaschke, G., *J. Chromatogr., B: Biomed. Sci. Appl.* (2000) **748**, 221–231.

K639. Sabatino, L., Iervasi, G., Ferrazzi, P., Francesconi, D., Chopra, I.J., *Life Sci.* (2000) **68**, 191–202.

K640. Sakuno, E., Yabe, K., Hamasaki, T., Nakajima, H., *J. Nat. Prod.* (2000) **63**, 1677–1678.

K641. Sanada, K., Kawaguchi, A., Furuya, T., Hamada, H., *Plant Biotechnol. (Tokyo)* (2000) **17**, 321–323.

K642. Parshina, S.N., Kleerebezem, R., van Kempen, E., Nozhevnikova, A.N., Lettinga, G., Stams, A.J.M., *Process Biochem. (Oxford, U.K.)* (2000) **36**, 423–429.

K643. Larsen, B.R., Amado, R., *Lebensm.-Wiss. Technol.* (2000) **33**, 424–430.

K644. Sekiya, K., Tezuka, Y., Tanaka, K., Prasain, J.K., Namba, T., Katayama, K., Koizumi, T., Maeda, M., Kondo, T., Kadota, S., *J. Ethnopharmacol.* (2000) **71**, 401–409.

K645. Kuhnle, G., Spencer, J.P.E., Schroeter, H., Shenoy, B., Debnam, E.S., Srai, S.K.S., Rice-Evans, C., Hahn, U., *Biochem. Biophys. Res. Commun.* (2000) **277**, 507–512.

K647. Shu, Y., Wang, L.-S., Xu, Z.-H., He, N., Xiao, W.-M., Wang, W., Huang, S.-L., Zhou, H.-H., *J. Pharmacol. Exp. Ther.* (2000) **295**, 844–851.

K648. Savi, P., Pereillo, J.M., Uzabiaga, M.F., Combalbert, J., Picard, C., Maffrand, J.P., Pascal, M., Herbert, J.M., *Thromb. Haemostasis* (2000) **84**, 891–896.

K649. Wu, S.-Y., Huang, W.-S., Fisher, D.A., Florsheim, W.H., Germain, D.L., Galton, V.A., *Pediatr. Res.* (2000) **48**, 847–851.

K650. Murfitt, L.M., Kolosova, N., Mann, C.J., Dudareva, N., *Arch. Biochem. Biophys.* (2000) **382**, 145–151.

K651. Snyder, R.W., Maness, S.C., Gaido, K.W., Welsch, F., Sumner, S.C.J., Fennell, T.R., *Toxicol. Appl. Pharmacol.* (2000) **168**, 225–234.

K652. Dhar, K., Rosazza, J.P.N., *Appl. Environ. Microbiol.* (2000) **66**, 4877–4882.

K653. Eppink, M.H.M., Cammaart, E., Van Wassenaar, D., Middelhoven, W.J., Van Berkel, W.J.H., *Eur. J. Biochem.* (2000) **267**, 6832–6840.

K654. Chartrain, M., Ikemoto, N., Taylor, C., Stahl, S., Sandford, V., Gbewonyo, K., Chirdo, C., Maxwell, C., Osoria, J., Buckland, B., Greasham R., *J. Biosci. Bioeng.* (2000) **90**, 321–327.

K655. Tanaka, M., Hirokane, Y., *J. Biosci. Bioeng.* (2000) **90**, 341–343.

K656. Gilissen, R.A.H.J., Barnaby, R.J., Kajbaf, M., *Drug Metab. Drug Interact.* (2000) **16**, 173–189.

K657. Dünnwald, T., Müller, M., *J. Org. Chem.* (2000) **65**, 8608–12.

K658. Anzellotti, D., Ibrahim, R.K., *Arch. Biochem. Biophys.* (2000) **382**, 161–172.

K659. Nakayama, T., Yonekura-Sakakibara, K., Sato, T., Kikuchi, S., Fukui, Y., Fukuchi-Mizutani, M., Ueda, T., Nakao, M., Tanaka, Y., Kusumi, T., Nishino, T., *Science* (2000) **290**, 1163–1166.

K660. DeBlassio, J.L., Delong, M.A., Glufke, U., Kulathila, R., Merkler, K.A., Vederas, J.C., Merkler, D.J., *Arch. Biochem. Biophys.* (2000) **383**, 46–55.

K661. Kobayashi, K., Tagawa, S., Mogi, T., *Biochemistry* (2000) **39**, 15620–15625.
K662. Wenker, M.A.M., Kežić, S., Monster, A.C., de Wolff, F.A., *Toxicol. Appl. Pharmacol.* (2000) **169**, 52–58.
K663. Thompson, J.E., Fahnestock, S., Farrall, L., Liao, D.-I., Valent, B., Jordan, D.B., *J. Biol. Chem.* (2000) **275**, 34867–34872.
K664. Kawai, H., Kotake, Y., Ohta, S., *Chem. Res. Toxicol.* (2000) **13**, 1294–1301.
K665. Tomoeda, K., Awata, H., Matsuura, T., Matsuda, I., Ploechl, E., Milovac, T., Boneh, A., Scott, C.R., Danks, D.M., Endo, F., *Mol. Genet. Metab.* (2000) **71**, 506–510.
K666. Keating, T.A., Marshall, C.G., Walsh, C.T., *Biochemistry* (2000) **39**, 15513–15521.
K667. Van den Hurk, P., James, M.O., *Mar. Environ. Res.* (2000) **50**, 11–15.
K668. Dudai, N., Larkov, O., Putievsky, E., Lerner, H.R., Ravid, U., Lewinsohn, E., Mayer, A.M., *Phytochemistry* (2000) **55**, 375–382.
K669. Zurhelle, G., Petz, M., Mueller-Seitz, E., Siewert, E., *J. Agric. Food Chem.* (2000) **48**, 6392–6396.
K670. Maruyama, W., Strolin-Benedetti, M., Naoi, M., *Neurobiology (Budapest)* (2000) **8**, 55–68.
K672. Bonnin, E., Lesage-Meessen, L., Asther, M., Thibault, J.-F., *Afinidad* (2000) **57**, 357–364.
K673. Law, M.Y.L., Moody, D.E., *Toxicol. Lett.* (2000) **117**, 139–144.
K674. Zhang, J.Y., Wang, Y., Dudkowski, C., Yang, D.-C., Chang, M., Yuan, J., Paulson, S.K., Breau, A.P., *J. Mass Spectrom.* (2000) **35**, 1259–1270.
K676. Nakagawa, Y., Suzuki, T., Tayama, S., *Toxicology* (2000) **156**, 27–36.
K677. Shin, K.-S., Lee, Y.-J., *Arch. Biochem. Biophys.* (2000) **384**, 109–115.
K678. de Raadt, A., Fetz, B., Griengl, H., Klingler, M.F., Kopper, I., Krenn, B., Munzer, D.F., Ott, R.G., Plachota, P., Weber, H.J., Braunegg, G., Mosler, W., Saf, R., *Eur. J. Org. Chem.* (2000) 3835–3847.
K679. Jarvis, A.P., Schaaf, O., Oldham, N.J., *Planta* (2000) **212**, 119–126.
K680. Kucukguzel, S.G., Ulgen, M., *Farmaco* (2000) **55**, 624–630.
K681. Arukwe, A., Goksoyr, A., Thibaut, R., Cravedi, J.P., *Mar. Environ. Res.* (2000) **50**, 141–145.
K682. Sietmann, R., Hammer, E., Moody, J., Cerniglia, C.E., Schauer, F., *Arch. Microbiol.* (2000) **174**, 353–361.
K683. Bonifacio, M.J., Viera-Coelho, M.A., Borges, N., Sares-da-Silva, P., *Arch. Biochem. Biophys.* (2000) **384**, 361–367.
K684. Ouyang, J., Shao, X., Li, J., *Plant J.* (2000) **24**, 327–333.
K686. Ladona, M.G., Gonzalez, M.L., Rane, A., Peter, R.M., de la Torre, R., *Life Sci.* (2000) **68**, 431–443.
K687. Ambekar, C.S., Cheung, B., Lee, J., Chan, L.C., Kumana, C.R., *Eur. J. Clin. Pharmacol.* (2000) **56**, 405–409.
K688. Hubbard, B.K., Thomas, M.G., Walsh, C.T., *Chem. Biol.* (2000) **7**, 931–942.
K689. Jonas, U., Hammer, E., Haupt, E.T.K., Schauer, F., *Arch. Microbiol.* (2000) **174**, 393–398.
K690. Lee, S.-E., Campbell, B.C., *Arch. Insect Biochem. Physiol.* (2000) **45**, 166–174.
K691. Clews, J., Cooksey, C.J., Garratt, P.J., Land, E.J., Ramsden, C.A., Riley, P.A., *Perkin 1* (2000) 4306–4315.
K692. Lewinsohn, E., Ziv-Raz, I., Dudai, N, Tadmor, Y., Lastochkin, E., Larkov, O., Chaimovitsh, D., Ravid, U., Putievsky, E., Pichersky, E., Shoham, Y., *Plant Sci. (Shannon, Irel.)* (2000) **160**, 27–35.
K693. Gulerman, N.N., Oruc, E.E., Kartal, F., Rollas, S., *Eur. J. Drug Metab. Pharmacokinet.* (2000) **25**, 103–108.
K694. Rezende, R.P., Dias, J.C.T., Ferraz, V., Linardi, V.P., *J. Basic Microbiol.* (2000) **40**, 389–392.
K695. Fang, J., *Eur. J. Drug Metab. Pharmacokinet.* (2000) **25**, 109–114.

K696. Gil, M., Haidour, A., Ramos, J.L., *Biodegradation* (2000) **11**, 49–53.

K697. Yasuda, T., Endo, T., Kubo, S., Ohsawa, K., *J. Tohoku Pharm. Univ.* (1999) **46**, 127–130.

K698. Tanner, A., Hopper, D.J., *J. Bacteriol.* (2000) **182**, 6565–6569.

K699. Bui, V.P., Hansen, T.V., Stenstrøm, Y., Hudlicky, T., *Green Chem.* (2000) **2**, 263–265.

K700. Pfitzner, A., Stöckigt, J., *Phytochemistry* (1982) **21**, 1585–1588.

K701. Fujii, I., Ebizuka, Y., Sankawa, U., *J. Biochem.* (1988) **103**, 878–883.

K702. Fischer, D., Ebenau-Jehle, C., Grisebach, H., *Arch. Biochem. Biophys.* (1990) **276**, 390–395.

K703. Wheeler, M.H., Greenblatt, G.A., *Exp. Mycol.* (1988) **12**, 151–160.

K704. Thompson, J.E., Basarab, G.S., Andersson, A., Lindqvist, Y., Jordan, D.B., *Biochemistry* (1997) **36**, 1852–1860.

K705. Dittrich, H., Kutchan, T.M., *Proc. Natl. Acad. Sci. U.S.A.* (1991) **88**, 9969–9973.

K706. Saito, K., Komamine, A., *Eur. J. Biochem.* (1976) **68**, 237–243.

K707. Stadler, R., Kutchan, T.M., Zenk, M.H., *Phytochemistry* (1989) **28**, 1083–1086.

K708. Saito, K., Komamine, A., *Eur. J. Biochem.* (1978) **82**, 385–392.

K709. Cuskey, S.M., Peccoraro, V., Olsen, R.H., *J. Bacteriol.* (1987) **169**, 2398–2404.

K710. Dyckerhoff, H., Armbraster, R., *Hoppe-Seyler's Z. Physiol. Chem.* (1933) **219**, 38–56.

K711. Halkier, B.A., Møller, B.L., *J. Biol. Chem.* (1990) **265**, 21114–21121.

K712. Hopkinson, D.A., Mestriner, M.A., Cortner, J., Harris, H., *Ann. Hum. Genet.* (1973) **37**, 119–137.

K713. Soffer, R.L., Hechtman, P., Savage, M., *J. Biol. Chem.* (1973) **248**, 1224–1230.

K714. Harada, I., Noguchi, T., Kido, R., *Hoppe-Seyler's Z. Physiol. Chem.* (1978) **359**, 481–488.

K715. Mavrides, C., Orr, W., *J. Biol. Chem.* (1975) **250**, 4128–4233.

K716. van den Tweel, W.J.J., Smits, J.P., Ogg, R.L.H.P., de Bont, J.A.M., *Appl. Microbiol. Biotechnol.* (1988) **29**, 224–230.

K717. Harada, I., *Wakayama Igaku* (1980) **31**, 61–68.

K718. Cooper, A.J.L., Meister, A., *J. Biol. Chem.* (1974) **249**, 2554–2561.

K719. Cooper, A.J.L., *Anal. Biochem.* (1978) **89**, 451–460.

K720. Gráwe, W., Strack, D., *Z. Naturforsch. C: Biosci.* (1986) **41**, 28–33.

K721. Strack, D., Gross, W., *Plant Physiol.* (1990) **92**, 41–47.

K722. Strack, D., Keller, H., Weissenböck, G., *J. Plant Physiol.* (1987) **131**, 61–73.

K723. Strack, D., Gross, W., Wray, V., Grotjahn, L., *Plant Physiol.* (1987) **83**, 475–478.

K724. Dahlbender, B., Strack, D., *Phytochemistry* (1986) **25**, 1043–1046.

K725. Matern, U., Feser, C., Hammer, D., *Arch. Biochem. Biophys.* (1983) **226**, 206–217.

K726. Cammann, J., Denzel, K., Schilling, G., Gross, G.G., *Arch. Biochem. Biophys.* (1989) **273**, 58–63.

K727. Matern, U., Feser, C., Heller, W., *Arch. Biochem. Biophys.* (1984) **235**, 218–227.

K728. Ebel, J., Schaller-Hekeler, B., Knobloch, K.-H., Wellman, E., Grisebach, H., Hahlbrock, K., *Biochim. Biophys. Acta* (1974) **362**, 417–424.

K729. Fahn, W., Laussermair, E., Deus-Neumann, G., Stoeckigt, J., *Plant Cell Rep.* (1985) **4**, 337–340.

K730. Kurosaki, F., Nishi, A., *FEBS Lett.* (1988) **227**, 183–186.

K731. Kühnl, T., Koch, U., Heller, W., Wellmann, E., *Plant Sci.* (1989) **60**, 21–25.

K732. Bhatnagar, D., McCormick, S.P., Lee, L.S., Hill, R.A., *Appl. Environ. Microbiol.* (1987) **53**, 1028–1033.

K733. Eilert, U., Wolters, B., *Plant Cell, Tissue Organ Cult.* (1989) **18**, 1–18.

K734. Corcuera, L.J., Bandurski, R.S., *Plant Physiol.* (1982) **70**, 1664–1666.

K735. Duffel, M., Jakoby, W.B., *J. Biol. Chem.* (1981) **256**, 11123–11127.

K736. Gross, G.G., *FEBS Lett.* (1982) **148**, 67–70.

K737. Gross, G.G., *Phytochemistry* (1982) **22**, 2179–2182.

K738. Köster, J., Barz, W., *Arch. Biochem. Biophys.* (1981) **212**, 98–104.

K739. Sutter, A., Ortmann, R., Grisebach, H., *Biochim. Biophys. Acta* (1972) **258**, 71–87.

K741. Hino, F., Okazaki, M., Miura, Y., *Plant Physiol.* (1982) **69**, 810–813.

K742. Altenschmidt, U., Ekerskorn, C., Fuchs, G., *Eur. J. Biochem.* (1990) **194**, 647–653.

K743. Martinez-Blanco, H., Reglro, A., Rodriguez-Aparicio, L.B., Luengo, J.M., *J. Biol. Chem.* (1990) **265**, 7084–7090.

K744. Meganathan, R., Bentley, R., *J. Bacteriol.* (1979) **140**, 92–98.

K745. Drake, J.C., Baram, J., Allegra, C.J., *Biochem. Pharmacol.* (1990) **39**, 615–618.

K746. Michalczuk, L., Bandurski, R.S., *Biochem. J.* (1982) **207**, 273–281.

K747. Bartsch, H., Dworkin, C., Miller, E.C., Miller, J.A., *Biochim. Biophys. Acta* (1973) **304**, 42–55.

K748. Daumy, G.O., McColl, A.S., *J. Bacteriol.* (1982) **149**, 384–385.

K749. Fargetton, X., Galtier, P., Delatour, P., *Vet. Res. Commun.* (1986) **10**, 317–324.

K750. Schenk, T., Müller, R., Mörsberger, F., Otto, M.K., Lingens, F., *J. Bacteriol.* (1989) **171**, 5487–5491.

K751. Hinderer, W., Flentje, U., Barz, W., *FEBS Lett.* (1987) **214**, 101–106.

K752. Frenzel, T., Zhou, P., Floss, H.G., *Arch. Biochem. Biophys.* (1990) **278**, 35–40.

K753. Eaton, R.W., Ribbons, D.W., *J. Bacteriol.* (1982) **151**, 48–57.

K754. Kamoda, S., Habu, N., Samejima, S., Yoshimoto, T., *Agric. Biol. Chem.* (1989) **53**, 2757–2761.

K755. Solano, F., Jiménez-Cervantes, C., Martínez-Liarte, J.H., García-Borrón, J.C., Jara, J.R., Lozano, J.A., *Biochem. J.* (1996) **313**, 447–453.

K756. Lenz, R., Zenk, M.H., *Eur. J. Biochem.* (1995) **233**, 132–139.

K757. Lenz, R., Zenk, M.H., *Tetrahedron Lett.* (1995) **36**, 2449–2452.

K758. Gerardy, R., Zenk, M.H., *Phytochemistry* (1993) **34**, 125–132.

K759. Gerardy, R., Zenk, M.H., *Phytochemistry* (1993) **32**, 79–86.

K760. Hirsch, W., Schrägger, H., Fuchs, G., *Eur. J. Biochem.* (1998) **251**, 907–915.

K762. Saller, E., Laue, H., Schläfli Oppenberg, H.R., Cook, A.M., *FEMS Microbiol. Lett.* (1996) **130**, 97–102.

K763. Neidle, E.L., Hartnett, C., Ornston, N., Bairoch, A., Rekik, M., Harayama, S., *Eur. J. Biochem.* (1992) **204**, 113–120.

K764. Lehmann, M., Tshisuaka, B., Fetzner, S., Lingens, F., *J. Biol. Chem.* (1995) **270**, 14420–14429.

K765. Lehmann, M., Tshisuaka, B., Fetzner, S., Röger, P., Lingens, F., *J. Biol. Chem.* (1994) **269**, 11254–11260.

K766. Tshisuaka, B., Kappl, R, Hüttermann, J., Lingens, F., *Biochemistry* (1993) **32**, 12928–12934.

K767. Peschke, B., Lingens, F., *Biol. Chem. Hoppe-Seyler* (1991) **372**, 1081–1088.

K768. Sauter, M., Tshisuaka, B., Fetzner, S., Lingens, F., *Biol. Chem. Hoppe-Seyler* (1993) **374**, 1037–1046.

K769. Schach, S.B., Tshisuaka, B., Fetzner, S., Lingens, F., *Eur. J. Biochem.* (1995) **232**, 536–544.

K770. Fetzner, S., Lingens, F., *Biol. Chem. Hoppe-Seyler* (1993) **374**, 363–376.

K771. De-Eknamkul, W., Zenk, M.H., *Phytochemistry* (1992) **31**, 813–821.

K772. Dross, F., Geisler, V., Lenger, R., Theis, F., Krafft, T., Fahrenholz, F., Kojoro, E., Duchêne, A., Tripier, D., Juvenal, K., Kröger, A., *Eur. J. Biochem.* (1992) **206**, 93–102.

K773. Rao, P.V., Krishna, C.M., Zigler, J.S., Jr., *J. Biol. Chem.* (1992) **267**, 96–102.
K774. Bauer, I., de Beyer, A., Tshisuaka, B., Fetzner, S., Lingens, F., *FEMS Microbiol. Lett.* (1995) **117**, 299–304.
K775. Junker, F., Leisinger, T., Cook, A.M., *Microbiology* (1994) **140**, 1713–1722.
K776. Kammerer, L., De-Eknamkul, W., Zenk, M.H., *Phytochemistry* (1994) **36**, 1409–1416.
K777. De-Eknamkul, W., Tanahashi, T., Zenk, M.H., *Phytochemistry* (1992) **31**, 2713–2717.
K778. Rosche, B., Tshisuaka, B., Fetzner, S., Lingens, F., *J. Biol. Chem.* (1995) **270**, 17836–17842.
K779. van Berkel, W.H.J., van den Tweel, W.J.J., *Eur. J. Biochem.* (1991) **201**, 585–592.
K780. Frenzel, T., Zenk, M.H., *Phytochemistry* (1990) **29**, 3491–3497.
K781. Clarke, C.F., Williams, W., Teruya, J.H., *J. Biol. Chem.* (1991) **266**, 16636–16644.
K782. Rueffer, M., Zumstein, G., Zenk, M.H., *Phytochemistry* (1990) **29**, 3727–3733.
K784. Lenz, R., Zenk, M.H., *Tetrahedron Lett.* (1994) **35**, 3897–3900.
K785. Sato, F., Tsujita, T., Katagiri, Y., Yoshida, S., Yamada, Y., *Eur. J. Biochem.* (1994) **225**, 125–131.
K786. Lenz, R., Zenk, M.H., *J. Biol. Chem.* (1995) **270**, 31091–31096.
K787. Mock, H.-P., Strack, D., *Phytochemistry* (1993) **332**, 575–579.
K788. Fujiwara, H., Tanaka, Y., Fukui, Y., Nakao, M., Ashikari, T., Kusumi, T., *Eur. J. Biochem.* (1997) **249**, 45–51.
K789. Latza, S., Gansser, D., Berger, R.G., *Phytochemistry* (1996) **43**, 481–485.
K790. Shimizu, S., Ogawa, J., Chung, M.C.-M., Yamada, H., *Eur. J. Biochem.* (1992) **209**, 375–382.
K792. Koga, J., *Biochim. Biophys. Acta* (1995) **1249**, 1–13.
K793. Costello, C.A., Kreuzman, A.J., Zmijewski, M.J., *Tetrahedron Lett.* (1996) **37**, 7469–7472.
K794. Miyamoto, K., Ohta, H., *Appl. Microbiol. Biotechnol.* (1992) **38**, 234–238.
K795. Selmar, D., Lieberei, R., Biehl, B., Conn, E.E., *Physiol. Plant.* (1989) **75**, 97–101.
K797. Turbek, C.S., Smith, D.A., Schardl, C.L., *FEMS Microbiol. Lett.* (1992) **94**, 187–190.
K798. Dunaway-Mariano, D., Babbitt, P.C., *Biodegradation* (1994) **5**, 259–276.
K800. Twilfer, H., Bernhardt, F.-H., Gersonde, K., *Eur. J. Biochem.* (1981) **119**, 595–602.
K801. Bernhardt, F.-H., Nastainczyk,W., Seydewitz, V., *Eur. J. Biochem.* (1977) **72**, 107–115.
K802. Burton, E.G., Sakami, W., *Biochem. Biophys. Res. Commun.* (1969) **36**, 228–234.
K803. Cormier, M.J., Crane, J.N., Jr., Nakano, Y., *Biochem. Biophys. Res. Commun.* (1967) **29**,747–752.
K804. Taylor, R.T., *Biochim. Biophys. Acta* (1971) **242**, 355–364.
K805. Ebel, J., Hahlbrock, K., Grisebach, H., *Biochim. Biophys. Acta* (1972) **268**, 313–326.
K807. Guest, J.R., Friedman, S., Foster, M.A., Tejerina, G., Woods, D.D., *Biochem. J.* (1964) **92**, 497–504.
K808. Kreuzaler, F., Ragg, H., Heller, W., Tesch, R., Witt, I., Hammer, D., Hahlbrock, K., *Eur. J. Biochem.* (1979) **99**, 89–96.
K809. Ibrahim, R.K., Grisebach, H., *Arch. Biochem. Biophys.* (1976) **176**, 700–708.
K810. McIntosh, C.A., Latchinian, L., Mansell, R.L., *Arch. Biochem. Biophys.* (1990) **282**, 50–57.
K811. Corcuera, L.J., Michalczuk, L., Bandurski, R.S., *Biochem. J.* (1982) **207**, 283–290.
K812. Cormier, M.J., Hori, K., Anderson, J.M., *Biochim. Biophys. Acta* (1974) **346**, 137–164.
K813. Hori, K., Anderson, J.M., Ward, W.W., Cormier, M.J., *Biochemistry* (1975) **14**, 2371–2376.

K814. Beckmann, J.D., Frerman, F.E., *Biochemistry* (1985) **24**, 3913–3921.

K815. Tsuji, F.I., Lynch, R.V.III, Stevens C.L., *Biochemistry* (1974) **13**, 5204–5209.

K816. Rieske, J.S., *Biochim. Biophys. Acta* (1976) **456**, 195–247.

K817. Hopkins, R.P., Drummond, E.C., Callaghan, P., *Biochem. Soc. Trans.* (1973) **1**, 989–991.

K819. Dagley S., Chapman, P.J., Gibson, D.T., *Biochem. J.* (1965) **97**, 643–650.

K820. Oka, T., Simpson, F.J., *Biochem. Biophys. Res. Commun.* (1971) **43**, 1–5.

K821. Fujioka, M., Wada, H., *Biochim. Biophys. Acta* (1968) **158**, 70–78.

K822. Hori, K., Wampler, J.E., Matthews, J.C., Cormier, M.J., *Biochemistry* (1973) **12**, 4463–4472.

K823. Frydman, R.B., Tomaro, M., Frydman, B., *Biochim. Biophys. Acta* (1972) **284**, 80–89.

K824. Kurylo-Borowska, Z., Abramsky, T., *Biochim. Biophys. Acta* (1972) **264**, 1–10.

K825. Cowles, J.R., Key, J.L., *Biochim. Biophys. Acta* (1972) **281**, 33–44.

K826. Greenberg, D.M., Wynston, L.K., Nagabhushanam, A., *Biochemistry* (1965) **4**, 1872–1878.

K827. Ortmann, R., Sutter, A., Grisebach, H., *Biochim. Biophys. Acta* (1972) **289**, 293–302.

K828. Giri, K.V., Appaji Rao, N., Cama, H.R., Kumar, S.A., *Biochem. J.* (1960) **75**, 381–386.

K829. Takahashi, I., Ojima, N., Ogura, K., Seto, S., *Biochemistry* (1978) **17**, 2696–2702.

K830. Rusnak, F., Faraci, W.S., Walsh, C.T., *Biochemistry* (1989) **28**, 6827–6835.

K831. Cormier, M.J., Hori, K., Karkhanis, Y.D., *Biochemistry* (1970) **9**, 1184–1188.

K832. Hösel, W., Surholt, E., Borgmann, E., *Eur. J. Biochem.* (1978) **84**, 487–492.

K833. Marcinowski, S., Grisebach, H., *Eur. J. Biochem.* (1978) **87**, 37–44.

K834. Cooksey, C.J., Garratt, P.J., Land, E.J., Ramsden, C.A., Riley, P.A., *Biochem. J.* (1998) **333**, 685–691.

K835. Lerch, C., Blaschke, G., *J. Chromatogr., B: Biomed Sci. Appl.* (1998) **708**, 267–275.

K837. Delk, A.S., Nagle, D.P., Jr., Rabinowitz, J.C., *J. Biol. Chem.* (1980) **255**, 4387–4390.

K841. McIntire, W., Edmondson, D.E., Singer, T.P., *J. Biol. Chem.* (1980) **255**, 6553–6555.

K842. Whitfield, C.D., Steers, E.J., Jr., Weissbach, H., *J. Biol. Chem.* (1970) **245**, 390–401.

K844. Wagner, C., Lusty, S.M., Jr., Kung, H.-F., Rogers, N.L., *J. Biol. Chem.* (1967) **242**, 1287–1293.

K846. Rao, M.M., Rebello, P.F., Pogell, B.M., *J. Biol. Chem.* (1969) **244**, 112–118.

K847. Hornemann, U., Speedie, M.K., Hurley, L.H., Floss, H.G., *Biochem. Biophys. Res. Commun.* (1970) **39**, 594–599.

K849. Hopkins, T.A., Selinger, H.H., White, E.H., Cass, M.W., *J. Am. Chem. Soc.* (1967) **89**, 7148–7150.

K850. White, E.H., McCapra, F., Field, G.F., McElroy, W.D., *J. Am. Chem. Soc.* (1963) **83**, 2402–2403.

K851. Katzen, H.M., Buchanan, J.M., *J. Biol. Chem.* (1965) **240**, 825–835.

K853. Kuno, S., Tashiro, M., Taniuchi, H., Horibata, K., Hayaishi, O., Seno, S., Tokuyama, T., Sakan, T., *Fed. Proc.* (1961) **20**, 3.

K854. Preddie, E.C., *J. Biol. Chem.* (1969) **244**, 3958–3968.

K855. Brot, N., Goodwin, J., *J. Biol. Chem.* (1968) **243**, 510–513.

K856. Katagiri, H., Imai, K., *J. Biochem.* (1959) **46**, 1119–1126.

K857. Ueda, H., Yoshihara, Y., Fukushima, N., Shiomi, H., Nakamura, A., Takagi, H., *J. Biol. Chem.* (1987) **262**, 8165–8173.

K858. McCullogh, W.G., Piligian, J.T., Daniel, I.J., *J. Am. Chem. Soc.* (1957) **79**, 628–630.

K859. Schmidt, J.C., Zalkin, H., *Biochemistry* (1969) **8**, 175–181.

K860. Yamada, M., Kurahashi, K., *J. Biochem.* (1969) **66**, 529–540.
K861. Bosmann, H.B., *Biochim. Biophys. Acta* (1972) **276**, 180–191.
K862. Schultz, J., Mosbach, K., *Eur. J. Biochem.* (1971) **22**, 153–157.
K863. Mackness, M.I., Thompson, H.M., Hardy, A.R., Walker, C.H., *Biochem. J.* (1987) **245**, 293–296.
K864. Attias, J., Bonnet, J.L., *Biochim. Biophys. Acta* (1972) **268**, 422–430.
K865. Attias, J., Durand, H., *Biochim. Biophys. Acta* (1973) **321**, 561–565.
K866. Westlake, D.W.S., *Can. J. Microbiol.* (1963) **9**, 211–220.
K867. Foulkes, J.G., Howard, R.F., Ziemiecki, A., *FEBS Lett.* (1981) **130**, 197–200.
K868. Hemscheidt, T., Zenk, M.H., *FEBS Lett.* (1980) **110**, 187–191.
K869. Fujioka, M., Morino, Y., Wada, H., *Methods Enzymol.* (1970) **17A**, 585–596.
K870. Levy, C.C., Weinstein, G.D., *Biochemistry* (1964) **3**, 1944–1947.
K873. Koch, G.L.E., Shaw, D.C., Gibson, F., *Biochim. Biophys. Acta* (1970) **212**, 375–386.
K874. Taniuchi, H., Hayaishi, O., *J. Biol. Chem.* (1963) **238**, 283–293.
K875. Moyed, H.S., Williamson, V., *J. Biol. Chem.* (1967) **242**, 1075–1077.
K876. Young, I.G., Gibson, F., *Biochim. Biophys. Acta* (1969) **177**, 401–411.
K877. Kutzbach, C., Stokstad, E.L.R., *Biochem. Biophys. Res. Commun.* (1968) **30**, 111–117.
K878. Farmer, V.C., Henderson, M.E.K., Russell, J.D., *Biochem. J.* (1960) **74**, 257–262.
K879. Bless, W., Barz, W., *FEBS Lett.* (1988) **235**, 47–50.
K880. Nair, P.M., Vining, L.C., *Arch. Biochem. Biophys.* (1964) **106**, 422–427.
K882. Mahadevan, S., *Arch. Biochem. Biophys.* (1963) **100**, 557–558.
K883. Akita, H., Furuichi, A., Koshiji, H., Horikoshi, K., Oishi, T., *Chem. Pharm. Bull.* (1984) **32**, 1342–1348.
K884. Shimomura, O., Johnson, F.H., *Proc. Natl. Acad. Sci. U.S.A.* (1975) **72**, 1546–1549.
K885. Hashimoto, T., Yamada, Y., *Plant Physiol.* (1986) **81**, 619–625.
K886. Ayabe, S.-I., Yoshikawa, T., Kobayashi, M., Furuya, T., *Phytochemistry* (1980) **19**, 2331–2336.
K887. Hashimoto, T., Kohno, J., Yamada, Y., *Phytochemistry* (1989) **28**, 1077–1082.
K888. Denzel, K., Schilling, G., Gross, G.G., *Planta* (1988) **176**, 135–137.
K889. Khouri, H.E., Ibrahim, R.K., *Phytochemistry* (1987) **26**, 2531–2535.
K890. Schulz, M., Weissenböck, G., *Phytochemistry* (1988) **27**, 1261–1267.
K891. McIntosh, C.A., Mansell, R.L., *Phytochemistry* (1990) **29**, 1533–1538.
K892. Bailey, B.A., Larson, R.L., *Plant Physiol.* (1989) **90**, 1071–1076.
K893. Zenk, M.H., Schmitt, J., *Naturwissenschaften* (1964) **51**, 510–511.
K894. Gross, G.G., *Z. Naturforsch. C: Biosci.* (1983) **38**, 519–523.
K895. Zenk, M.H., Schmitt, J.H., *Biochem. Z.* (1965) **342**, 54–65.
K896. Paszczyński, A., Trojanowski, J., *Microbios* (1977) **18**, 111–122.
K897. Rettenmaier, H., Lingens, F., *Biol. Chem. Hoppe-Seyler* (1985) **366**, 637–646.
K898. Zeyer, J., Kocher, H.P., Timmis, N., *Appl. Environ. Microbiol,* (1990) **52**, 334–339.
K899. Inoue, S., Kakoi, H., Goto, T., *Tetrahedron Lett.* (1976) 2971–2974.
K900. Behal, V., *CRC Crit. Rev. Biotechnol.* (1987) **5**, 295–318.
K902. Beevers, H., French, R.C., *Arch. Biochem. Biophys.* (1954) **50**, 427–439.
K903. Toriyama, N., *Nichidai Igaku Zasshi* (1965) **24**, 423–432.
K904. Leube, J., Grisebach, H., *Z. Naturforsch. C: Biosci.* (1983) **38**, 730–735.
K905. Zähringer, U., Schaller, E., Grisebach, H., *Z. Naturforsch. C: Biosci.* (1981) **36**, 234–241.
K906. Lerbs, W., Luckner, M., *J. Basic Microbiol.* (1985) **25**, 387–391.

K907. Havada, I., Noguchi, T., Kido, R., *Hoppe-Seyler's Z. Physiol. Chem.* (1978) **359**, 481–488.

K908. Pettersson, G., *Acta Chem. Scand.* (1965) **19**, 2013–2021.

K909. Schöbel, B., Pollmann, W., *Z. Naturforsch. C: Biosci.* (1980) **35**, 209–212.

K910. Kuroki, G., Lizotte, P.A., Poulton, J.E., *Z. Naturforsch. C: Biosci.* (1984) **39**, 232–239.

K911. Schöbel, B., Pollmann, W., *Z. Naturforsch. C: Biosci.* (1980) **35**, 699–701.

K914. Bauer, W., Zenk, M.H., *Phytochemistry* (1991) **30**, 2953–2961.

K915. Creighton, T.E., Yanofsky, C., *Methods Enzymol.* (1970) **179**, 365–380.

K916. Potts, J.R.M., Weklych, R., Conn, E.E., *J. Biol. Chem.* (1974) **249**, 5019–5026.

K917. Jeffrey, A.M., Knight, M., Evans, W.C., *Biochem. J.* (1972) **130**, 373–381.

K918. Hunter, T., Sefton, B.M., *Proc. Natl. Acad. Sci. U.S.A.* (1980) **77**, 1311–1315.

K919. Gallis, B., Bornstein, P., Brautigan, D.L., *Proc. Natl. Acad. Sci. U.S.A.* (1981) **78**, 6689–6693.

K921. Glass, N.L., Kosuge, T., *J. Bacteriol.* (1986) **166**, 598–603.

K922. Batie, C.J., LaHaie, E., Ballou, D.P., *J. Biol. Chem.* (1987) **262**, 1510–1518.

K923. Ramaswamy, S.G., Jakoby, W.B., *J. Biol. Chem.* (1987) **262**, 10039–10043.

K924. Krishna, R.V., Krishnaswamy, P.R., Rajagopal Rao, D., *Biochem. J.* (1971) **124**, 905–913.

K925. Nakani, M., *J. Biol. Chem.* (1967) **242**, 73–81.

K926. Hori, K., Wampler, J.E., Matthews, J.C., Cormier, M.J., *Biochemistry* (1973) **12**, 4463–4468.

K928. Moore, M.R., O'Brien, W.E., Ljungdahl, L.G., *J. Biol. Chem.* (1974) **249**, 5250–5253.

K930. Marres, C.A.M., Slater, E.C., *Biochim. Biophys. Acta* (1977) **462**, 531–548.

K931. Fahn, W., Gundlach, H., Deus-Neumann, B., Stoeckigt, J., *Plant Cell Rep.* (1985) **4**, 333–336.

K932. Stadler, R., Kutchan, T.M., Zenk, M.H., *Phytochemistry* (1989) **28**, 1083–1086.

K933. Stadler, R., Zenk, M.H., *Liebigs Ann. Chem.* (1990) 555–562.

K934. Ramasastri, B.V., Blakley, R.L., *J. Biol. Chem.* (1962) **237**, 1982–1988.

K935. Yeh, Y.-C., Greenberg, D.M., *Biochim. Biophys. Acta* (1965) **105**, 279–291.

K937. Tabor, H., Wyngarden, L., *J. Biol. Chem.* (1959) **234**, 1830–1846.

K938. Vančurová, I., Volc, J., Flieger, M., Neužil, J., Novotná, J., Vlach, J., Běhal, V., *Biochem. J.* (1988) **253**, 263–267.

K939. Schübel,H., Stöckigt, J., Feicht, R., Simon, H., *Helv. Chim. Acta* (1986) **69**, 538–547.

K940. Akita, H., Furuichi, A., Hiroko, K., Horikoshi, K., Oishi, T., *Chem. Pharm. Bull.* (1983) **31**, 4376–4383.

K941. Doctor, B.P., Mudd, J.A., *J. Biol. Chem.* (1963) **238**, 3677–3681.

K942. Rabinowitz, J.C., Pricer, W.E., Jr., *J. Biol. Chem.* (1962) **237**, 2898–2902.

K943. Grant, D.J.W., *Antonie Van Leeuwenhoek J. Microbiol. Serol.* (1969) **35**, 325–343.

K944. Reiner, A.M., *J. Biol. Chem.* (1972) **247**, 4960–4965.

K945. Ribbons, D.W., *Biochem. J.* (1966) **99**, 30P.

K946. Forrester, P.I., Gaucher, G.M., *Biochemistry* (1972) **11**, 1108–1114.

K947. Beecher, C.W.W., Kelleher, W.J., *Tetrahedron Lett.* (1984) **25**, 4595–4598.

K948. Goodwin, B.L., Kite, G.C., *J. Neural Transmiss.* (1998) **105**, 1265–1269.

K949. Langkau, B., Ghisla, S., Buder, R., Ziegler, K., Fuchs, G., *Eur. J. Biochem.* (1990) **191**, 365–371.

K950. Gunsalus, C.F., Stanier, R.Y., Gunsalus, I.C., *J. Bacteriol.* (1953) **66**, 548–553.

Part 2 Index

Some entries in the index are duplicates, differing only in names that are in common use for the same enzyme

Handbook of Biotransformations of Aromatic Compounds on CD-ROM

This CD-ROM allows users to search the entire contents of the Handbook using intuitive search software. In addition, this CD-ROM also contains the full text of an earlier edition of Dr Goodwin's book, *Handbook of Intermediary Metabolism of Aromatic Compounds* (1976), providing users with a hundred years' worth of literature. Users can search across both volumes simultaneously or on a single volume at a time.

System Requirements

Hardware:

Pentium III Processor or higher
64 MB RAM
30MB Hard disk space
24X speed CD-ROM drive or faster
SVGA monitor at 800 × 600, 256 colors
Windows 2000/XP

Software:

Internet Explorer 6.0 or higher
Netscape Navigator 7.1 or higher
MSXML 4.0 (provided on CD-ROM)

Installation Instructions

Insert the CD into your CD-ROM drive.
Do one of the following:

1. If the CD launches automatically, follow the on-screen prompts.

or

2. If the CD does not launch automatically.
Click the **Start** button.
Choose **Run** from the menu.
Click **Browse**.
Choose the CD-ROM drive and select **setup.exe**.
Click **OK** and follow the on-screen prompts.
Note: After clicking the Start button and choosing Run, you may also type, **D**:\setup.exe (where **D** is the letter of your CD-ROM drive) in the text box and press **Enter** or click **OK**.

Technical Support

Please contact:

Technical Support Desk
Electronic Publishing Division
CRC Press
2000 NW Corporate Boulevard
Boca Raton FL 33431-9868 USA
Tel: 800 272 7737 ext 6066
Tel: 561 361 6066 (outside USA)
Fax: 561 361 6056
E-mail: techsupport@crcpress.com